2025 9th IEEE Electron Devices Technology & Manufacturing Conference (EDTM 2025)

Hong Kong
9-12 March 2025

Pages 1-648

IEEE Catalog Number: CFP25J58-POD
ISBN: 979-8-3315-0417-5

**Copyright © 2025 by the Institute of Electrical and Electronics Engineers, Inc.
All Rights Reserved**

Copyright and Reprint Permissions: Abstracting is permitted with credit to the source. Libraries are permitted to photocopy beyond the limit of U.S. copyright law for private use of patrons those articles in this volume that carry a code at the bottom of the first page, provided the per-copy fee indicated in the code is paid through Copyright Clearance Center, 222 Rosewood Drive, Danvers, MA 01923.

For other copying, reprint or republication permission, write to IEEE Copyrights Manager, IEEE Service Center, 445 Hoes Lane, Piscataway, NJ 08854. All rights reserved.

*** This is a print representation of what appears in the IEEE Digital Library. Some format issues inherent in the e-media version may also appear in this print version.

IEEE Catalog Number:	CFP25J58-POD
ISBN (Print-On-Demand):	979-8-3315-0417-5
ISBN (Online):	979-8-3315-0416-8

Additional Copies of This Publication Are Available From:

Curran Associates, Inc
57 Morehouse Lane
Red Hook, NY 12571 USA
Phone: (845) 758-0400
Fax: (845) 758-2633
E-mail: curran@proceedings.com
Web: www.proceedings.com

TABLE OF CONTENTS

Energy-Efficient Voltage-Induced Self-Regulated Precessional MRAM with Low Write Error Rate $<10^{-9}$ 1
Stanislav Sin, Saeroonter Oh

A 2-D Noise Model for CMOS Single Photon Avalanche Diodes 4
Xuanyu Qian, Wei Jiang, M. Jamal Deen

InP/GaAsSb DHBT Emitter Etching Process Optimization with a Simultaneous $fT/FMAX = 451/914$ GHz and 86% Device Yield 7
M. Ebrahimi, S. Hamzeloui, F. Ciabattini, A. M. Arabhavi, O. Ostinelli, C. R. Bolognesi

A Novel Mixed Collector Structure IGBT Operating at High Temperature 10
Chang-Hao Wang, Hang Xu, Jian-Bin Guo, Xin-Ru Chen, Yafen Yang, David Wei Zhang

Controllable Oxygen Vacancies in NbOx MOTT Memristor for Tunable Spiking Neurons 13
Guolei Liu, Dingwei Li, Fanfan Li, Bowen Zhu

Study of the Characteristics of GaN Substrate-Based Microleds with Different Epitaxial Structures 16
Shan Huang, Yibo Liu, Feng Feng, Jingyang Zhang, Zichun Li, Man Hoi Wong, Zhaojun Liu

Impact of Oxide Quality in Self-Aligned Block Region on Hot Carrier Degradation in N-Type CFP-LDMOS with 0.18 μm Bipolar-CMOS-DMOS Technology 19
Qiao Teng, Yixian Song, Junzhe Kang, Kai Xu, Dawei Gao

Investigation of High-Sensitivity Pressure Sensor Integrated with InSnZnO Thin-Film Transistors 22
Mei Yang, Ming Li, Wei Huang, Rongsheng Chen

Investigation of Reliability and Optimization of Reprogram Process in 3D NAND Flash Memory Based on Physical Model 25
Jooyoung Lee, Jinil Yoo, Hyungcheol Shin

Circuit Polymorphism Enabled by RFET Devices Processed on Industrial FDSOI 28
N. Bhattacharjee, G. Galderisi, Y. He, V. Sessi, M. Drescher, V. Havel, M. Zier, M. Simon, K. Ruttloff, K. Li, A. Zeun, A.-S. Seidel, C. Metze, M. Grothe, S. Jansen, M. Wijvliet, S. Rai, A. Kumar, S. Slesazeck, J. Hoentschel, T. Mikolajick, J. Trommer

Optimizing SiN Composition for Enhanced Charge-Trapping in Next-Generation 3D NAND Flash Memories 31
Tomoya Nagahashi, Hajime Karasawa, Ryota Horiike, Atsushi Oshiyama, Kenji Shiraishi

Effect of Top Al2O3 Interlayer Thickness on the Memory Window of Fefets with TiN/Al2O3/Hf0.5Zr0.5O2 /SiOx /Si (MIFIS) Gate Structure 34
Tao Hu, Runhao Han, Xinpei Jia, Jia Yang, Zeqi Chen, Xiaoqing Sun, Junshuai Chai, Hao Xu, Xiaolei Wang, Wenwu Wang, Tianchun Ye

Erbium-Doped Aluminium Oxide Based Distributed Feedback Waveguide Laser in Thin Film Lithium Niobate Platform 37
Zhekang Zhang, Ruqi Wang, Renfei Kuang, Xifa Liang, Shijie Liang, Qingming Chen

Comparison of Border Trap Density in MIFM Capacitors with Different Interlayers Using the C−ln(f) Method... 40

Min Liao, Hao Xu, Xiaoyu Ke, Yuanyuan Zhao, Junshuai Chai, Xiaoqing Sun, Jinjuan Xiang, Xiaolei Wang, Wenwu Wang

Boosting Mobility of Oxide TFTs Via PVD and ALD Hybrid Process 43

Yuhang Zhang, Xiao Li, Shengjie Yang, Huan Yang, Tengyan Huang, Xinwei Wang, Lei Lu, Shengdong Zhang

High-Performance Flexible Graphene Field-Effect Transistors and Multistage Inverter Chains for Signal Amplification 46

Zhongyang Ren, Pei Peng, Jiaojiao Tian, Chenhao Xia, Muchan Li, Liming Ren, Fei Liu, Yunyi Fu

Oxide-Based Optical Synapses with Low Consumption for In-Sensor Reservoir Computing..................... 49

Jingyi Zhou, Biyi Jiang, Huansong Tang, Zhen Wang

A SAW Pressure Sensor Based on AlN Thin Film with a Novel Differential Structure: Balancing Sensitivity Enhancement and Temperature Decoupling 52

Aobei Chen, Dapeng Li, Ge Gao, Chun Hu, Dezhi Zheng

A BSIM Compact Model of Two-Dimensional Semiconductor Field Effect Transistors................................ 55

Jen-Hao Chen, Ahtisham Pampori, Chien-Ting Tung, Sayeef Salahuddin, Chenming Hu

Improved Ohmic Contact Resistance and DC Performance of InAlN/GaN HEMTs with Si-Incorporated Contact Scheme.. 58

Yang Jiang, Fangzhou Du, Xinyi Tang, Ziyang Wang, Kangyao Wen, Zhongrui Wang, Qing Wang, Hongyu Yu

BSIM-NN: A Machine Learning Compact Model for Fast IC Simulation 61

C. T. Tung, S. Salahuddin, C. Hu

975V/4.3M $\Omega \cdot$cm2 Enhancement-Mode (001) β−Ga2O3 Vertical Multi-Fin Power Transistors....................... 64

Gaofu Guo, Xiaodong Zhang, Chunhong Zeng, Tiwei Chen, Zhili Zou, Zheyuan Hu, Zhucheng Li, Dengrui Zhao, Yuhua Sun, Xianqi Dai, Baoshun Zhang

An Atomic Layer Etching Technique for MOCVD in-Situ SiNx 67

Fangzhou Du, Yang Jiang, Ziyang Wang, Xinyi Tang, Qing Wang, Hongyu Yu

Reliability-Aware Device and Programming Scheme Optimization for PBS/NBS-Immune IGZO-Based 2T0C DRAM 70

Zhidong Tang, Yanbo Su, Jianshi Tang, Yijia Fan, Yiwei Du, Mingcheng Shi, Yibei Zhang, Dong Wu, Bin Gao, He Qian, Huaqiang Wu

A 2T2R TCAM Based on RRAM and Accurate Compact Model for a Kilobit Word.. 73

Zhen Wang, Pengtao Li, Zijian Wang, Xuemeng Fan, Shengpeng Xing, Yishu Zhang

Prediction of Single Particle Output Response in the CTFET Inverter Based on Deep Learning Algorithm 76

Chen Chong, Hongxia Liu, Zexi Wang

Structure Design and Characteristics of 3T Sense-Switch Pflash for Computing-Inmemory 79

Wei Zhao, Jinghe Wei, Guozhu Liu, Yidan Wei, Yingqiang Wei, Zhiyuan Sui, Meijie Liu, Qi Xi, Xinhe Wang, Zongguang Yu, Juyan Xu

High-Performance Cu2O/Ga2O3 Heterojunction Diodes for Power Electronics.................... 82
Xiaohui Wang, Mujun Li, Minghao He, Chun-Zhang Chen, Haozhe Yu, Long Chen, Qing Wang, Hongyu Yu

Comprehensive Analysis of Oxidant Effects During ALD Process of Hf0.5Zr0.5O2 Ferroelectric Thin Films 85
Xiaopeng Li, Yilin Hou, Xiaoyu Dou, Yaoyu He, Yuwei Qu, Pengpeng Sang, Xuepeng Zhan, Xiaolei Wang, Jixuan Wu, Jiezhi Chen

Interface Engineering on Inorganic-Organic Hybrid Phototransistors for High-Responsive Near-Infrared Sensing 88
Dingwei Li, Huihui Ren, Yingjie Tang, Yitong Chen, Yan Wang, Qi Huang, Bowen Zhu

The Comprehensive Study of Difference Between Cryogenic Planar MOSFET and FinFET Variation and Band Tail States Assessment 91
Chenyang Zhang, Yewei Zhang, Yongkang Xue, Shuying Wang, Sheng Yang, Pengpeng Ren, Zhigang Ji, Ru Huang

Contact Engineering of 1L-WSe2 p-FETs: Straightforward WOx Doping and Thickness Dependence 94
Wenyu Wang, Yanda Zhu, Ziyi Wang, Yun Li, Jason C. C. Hwang, Xi Lin, Jiaqi Su, Jing-Kai Huang, Sean Li

Monolithic 3D Integration of 2D Devices 97
Saptarshi Das

Broadband Vis-NIR Neuromorphic Photodetector Based on PdSe2/Bi2O2Se Heterotransistor for Motion Detection.................... 100
Yu Zhu, Xinrui Guo, Shuo Liu, Junling Liu, Ru Huang, Ming He

Study of Threshold Voltage Degradation Mechanism of Ferroelectric Field-EffectTransistors (FeFETs) with TiN/SiO2/Hf0.5Zr0.5O2/SiOx/Si (MIFIS) GateStacks.................... 103
Zeqi Chen, Tao Hu, Xinpei J Jia, Runhao Han, Jia Yang, Yajing Ding, Xiaoqing Sun, Junshuai Chai, Hao Xu, Xiaolei Wang, Wenwu Wang, Tianchun Ye

Modeling of the Switching Characteristics of Ag/HfO2-Based Volatile Memristors 106
Yikang Huo, Chunsheng Jiang, Qin Xie

A Submersible Soft Robot with Ultrasonic Echolocation Capabilities.................... 109
Zhongming Chen, Qilin Hua, Jiaqiang Xu, Guozhen Shen

Recent Progress on Electronics and Optoelectronics Based on 2D Tellurium 112
Jiajia Zha, Haoxin Huang, Chaoliang Tan

Nanodiamond Quantum Thermometers and Their Biological Applications.................... 115
Masazumi Fujiwara

Enhancing Transduction Efficiency in CMOS-MEMS CMUTs Through Atomic Layer Deposition: A Preliminary Study 118
Tzu-Yun Huang, Ming-Huang Li

A 3.98 Ghz Aluminum Nitride Overmoded Bulk Acoustic Wave Resonator for Temperature Sensing Applications 121
Zhi-Qiang Lee, Kuan-Ting Chen, Chin-Yu Chang, Cheng-Chien Lin, Yung-Hsiang Chen, Yelehanka Ramachandramurthy Pradeep, Rakesh Chand, Yenshih Ho, Ming-Huang Li

Effect of Gate Voltage Rise Time on Gate Charges for GaN HEMTs with Partially Depleted P-GaN Cap Layer 124

Ruize Sun, Renjie Wu, Yunfei Ma, Wenhan Yuan, Qinan Guo, Wanjun Chen, Bo Zhang

Self-Rectifying Memristors with High Rectification Ratio for Security Primitives with Ultra-Low Bit-Errors 127

Guobin Zhang, Zijian Wang, Xuemeng Fan, Dawei Gao, Yishu Zhang

Competing Ferroelectric Polarization and Defect Migration Induced Resistive Switching in Van Der Waals β-In2Se3 130

Yinfeng Long, Saiyu Bu, Han Chen, Kai Liu, Shiyu Zhang, Lin Wang

Improving the Reliability of 40nm RRAM Chip by Pre-Cycle Operation 133

Ruofei Hu, Yilong Huang, Chengxiang Ma, Qianze Zheng, Kaimeng Liu, Yuelin Jiang, Siyu Chen, Jianshi Tang, Dong Wu, Bin Gao, He Qian, Huaqiang Wu

Investigation of Passivation Layers for Self-Aligned Top-Gate Amorphous InGaZnO Thin-Film Transistors with Metal-Reacted Low-Resistance Source/Drain 136

Yuhan Zhang, Jiye Li, Hao Peng, Huan Yang, Lei Lu, Shengdong Zhang

Enabling Highly-Efficient, Low-Latency Analog CAM Operations with Optimized MoS2 Flash Memory Devices 139

Guoyun Gao, Bo Wen, Ni Yang, Zhiyuan Du, Mingrui Jiang, Ruibin Mao, Yingnan Cao, Hongxia Xue, Pak San Yip, Qihan Liu, Dong-Keun Ki, Jinyao Tang, Paddy K. L. Chan, Hao Jiang, Han Wang, Lain-Jong Li, Can Li

An Artificial Thermoresponsive Nociceptor Based on Amorphous Carbon Threshold Memristor 142

Qiaoling Tian, Xiaoning Zhao, Zhongqiang Wang, Haiyang Xu, Yichun Liu

Single-Crystal Diamond MEMS for Extreme Sensors 145

Meiyong Liao

A Variability-Aware Data Preprocessing Method for Data-Driven Memristive Device Inverse Modeling 148

Zhoujie Pan, Yanming Liu, Daixuan Wu, Zihan Zhang, He Tian

4.6-Bits-Per-Cell Resistive Probabilistic-Bit Computing for High-Efficient Evaluation of Bio-Genomic Evolution Achieving 6.25x Acceleration of Data-Operations 151

Kai Wen Cheng, Huan-Hsiang Su, Yu-Hsien Lin, Jerry Chung, Y.-S. Hsieh, T.-Y. Chen, E Ray Hsieh, Yu-Cheng Lin, Ching-Ju Lin, Jen-Chou Tseng, Chien-Fan Wang, Wen-Ting Chu, Yu-Der Chih, Jonathan Chang

Passivation Effect on Atomic-Layer-Deposited Indium-Gallium-Zinc-Oxide Transistors 154

Zhiyu Lin, Jinxiu Zhao, Chen Wang, Mengwei Si

Characteristic Length of Transition-Metal Dichalcogenides Based Complementary Field-Effect Transistors 157

Xianmao Cao, Yiting Wu, Panpan Zhang

Dual Gate-Enhanced Mechanical-Electrical Stability of Flexible InGaZnO TFTs 160

Jilin Li, Yuhan Zhang, Runxiao Shi, Fion Sze Yan Yeung, Man Hoi Wong, Hoi Sing Kwok, Shengdong Zhang, Lei Lu

Flexible Carbon Nanotube Optoelectronic Neuromorphic Devices and Irradiation-Resistance Logic Circuits with Low-Work-Function Gate Electrodes 163

Meng Deng, Kaixiang Kang, Nianzi Sui, Jiaqi Li, Zebin Wang, Jianwen Zhao

A Compact Model for GIDL-Assisted Erase Transients of 3D MONOS Charge-Trap NAND Flash Memories 166

    Changhyeok Im, Sungju Kim, Hyungcheol Shin

From Monitoring to Modulation: Intelligent Wearable Devices for Health Sensing and Drug Delivery 169

    Yuanzhe Li, Shirong Qiu, Zhou Jiang, Cunman Liang, Yixin Qi, Ni Zhao

Ultrafast Photoresponse of Vertical Diodes Utilizing WSe2/ITO Schottky Junctions 172

    Xixi Jiang, Jingli Wang, Shukui Zhang, Qingqing Sun, Jie Wei Chen, Yang Chai

TDA (Thermal Design Automation) for Multiscale Thermal Managements of GaN HEMTs 175

    Bingyang Cao

The Observation of 2D Electron Gas at Ga2O3/IGZO Interface 178

    Jinxiu Zhao, Zhiyu Lin, Chen Wang, Mengwei Si

Embracing SemiverseTM Solutions: Semiconductor Virtual Fabrication and Its Applications 181

    Qingpeng Wang, Yujia Zhong, Lifei Sun, Benjamin Vincent, Ivan Chakarov, Joseph Ervin

GeTe9 Ovonic Threshold Switch Memristor-Centered 1S1T1R Cell as Bio-Realistic Stochastic Synapse 184

    Xinyu Wen, Kuan Wang, Lun Wang, Qi Chen, Hao Tong, Xiang-Shui Miao, Yuhui He

Development of an Inkjet-Printed Sensor Using Ink Locked Food-Based Nano Conductive Paste for Electrochemical Sensing of Insulin: A First Attempt 187

    Ramya K, Sanket Goel

Chirality Reversal Related Spin-Orbit Torque Switching in Nanoscale Perpendicular Ferromagnet 190

    Xue Zhang, Zhengde Xu, Zhifeng Zhu

Portable NV (Nitrogen Vacancy) Vector Magnetometer with High Sensitivity and Wide Dynamic Range 193

    Yingjie Yang, Jianghao Fu, Xuanhui Ren, Doudou Zheng, Yang Li, Hui Wang, Zongmin Ma, Jun Liu

MoS2 Transistors Based 2TOC DRAM Optimized with Optical Modulation and Read Pulse Compensation 196

    Gongpeng Lan, Dongdong Sun, Haotian Fang, Wuqing Fang, Yuanyuan Shi

Solution-Processed Oxide Semiconductors-Based Enhancement-Mode Thin-Film Transistor Circuits for Artificial Spiking Neurons 199

    Jiayi Mao, Qi Huang, Bowen Zhu

CMOS BEOL Compatible Metal-Oxides Logics and Memories for New Paradigm Computing in the Post-Moore Era 202

    Yida Li, Feichi Zhou

Endurance Optimization of 40nm RRAM Towards 106 Cycles by Tuning the Stoichiometry of TiN Bottom Electrode 205

    Chengxiang Ma, Qianze Zheng, Ruofei Hu, Yilong Huang, Kaimeng Liu, Yuelin Jiang, Siyu Chen, Jianshi Tang, Dong Wu, Bin Gao, He Qian, Huaqiang Wu

Optimized Electric Field Distribution and Dynamic Performance in GaN HEMTs Using Segmented-Extended $\boldsymbol{p}-\mathbf{GaN}$ Gate Structures 208

    Xinyue Dai, Haiyang Li, Qimeng Jiang, Xiaoping Wang, Yuxi Wan

Piezoresistive Internal Stress Sensing and Gas Detection Via Polymer Swelling 211
    Masaya Toda

Understanding the Impact of Discrete Trap Positions and Mapping the Hysteresis Dynamics in MoS_{2} FETs by Advanced TCAD Modeling 214
    Y. Z. Lv, Y. J. Chai, Yu. Yu. Illarionov

Flexible Pulse Waveform Sensor Array for Cuffless PWV Measurement.. 217
    Jiang Zhou, Liang Cunman, Zhao Ni

Physical Unclonable Function in Spiking Neural Network Based on in-Te Ovonic Threshold Switching Devices 220
    Huan Wang, Qihang Zhu, Hengyi Hu, Yi Li, Xiangshui Miao, Ming Xu

Direct Backside Contact Impact on 3-Dimensional Stacked FET SRAM Beyond 1nm Node......................... 223
    Mingyu Kim, Jaehyun Park, Sungil Park, Kyunghwan Lee, Deuk Ho Yeon, Daewon Ha, Hyungcheol Shin

Highly Sensitive and Polarization Selective Au Serrate Nanogratings Fabricated by Nanoimprint Lithography 226
    Li Chunxia, Zhu Shuyan

High-Performance InSnO/ZnO Heterojunction Transistors and Inverters 229
    Dengqin Xu, Tingchen Yi, Junchen Dong, Lifeng Liu, Zheng Zhou, Dedong Han, Xing Zhang

A 1.8TB/S HBM Heterogeneously Integrated GPU Design Exploring 2.5D Packaging Technology.............. 233
    Shuang Wang, Weiliang Chen, Xueqing Li, Chen Jiang, Huazhong Yang

9-KV p-GaN Gate HEMT with Gate Termination Extension Demonstrated on Sapphire Substrate for Improved Breakdown Voltage 236
    Jingjing Yu, Junjie Yang, Jiawei Cui, Hao Chang, Sihang Liu, Yunhong Lao, Xuelin Yang, Xiaosen Liu, Maojun Wang, Bo Shen, Jin Wei

Rapid and Extensive Conductance Modulation in MoOx Based Electrochemical Random-Access Memory for Spiking Neuromorphic Systems... 239
    Xiaoci Liang, Dongyue Su, Younian Tang, Bin Xi, Chunzhen Yang, Huixin Xiu, Jialiang Wang, Chuan Liu, Mengye Wang, Yang Chai

TCAD and Mixed-Mode Simulation Supporting the Development of Contact-Controlled Thin-Film Transistors and Circuits 242
    Eva Bestelink, Radu A. Sporea

Dynamic Clocked Comparator Based on Flexible LTPO Technology 245
    Chunxiu Wang, Jiaqiao Liang, Min Zhang

VO2 Memristor-Based Adaptive Neurons for Electromyography Signal Processing 248
    Zhiyuan Li, Jiaping Yao, Beining Zhang, Wei Tang, Xiangshui Miao, Rui Yang

Fabrication and Research of Wide-Bandgap Semiconductor AlN-Based Unipolar Memristors 251
    Haiming Qin, Xinpeng Wang, Dayu Zhou, Liang Zeng, Yi Liu, Yi Tong

Reliability Optimization in Hafnium Oxide Based Ferroelectric Field-Effect Transistors (FeFETs).............. 254
    Kechao Tang, Yuejia Zhou, Zhongxin Liang, Ru Huang

High-Performance 2D Fets with Single-Crystal Anatase TiO2 High-κ Dielectric 257
    Ni Yang, Ji Zhang, Yu-Ming Chang, Fangyuan Zheng, Lingqi Li, Chenyang Li, Jian Liu, Dong-Keun Ki, Yi Wan, Sean Li, Kah-Wee Ang, Jing-Kai Huang, Lain-Jong Li

All-Optical Modulated Artificial Synapse Based on Quantum Dots/Oxide Heterojunction for Neuromorphic Visual Simulation 260
    Yan Wang, Yingjie Tang, Yitong Chen, Dingwei Li, Huihui Ren, Guolei Liu, Fanfan Li, Qi Huang, Botao Ji, Bowen Zhu

Volumetric Super-Resolution Imaging Based on Dual Bessel Beams Sted Microscopy 263
    Renlong Zhang, Haoxian Zhou, Chenguang Wang, Xiaoyu Weng, Liwei Liu, Peng Xi, Junle Qu

Direct Evidence of Oxygen Vacancy Generation in Whole Gate Stacks Through Multiple Electrical and Atomic-Scale Physical Methods as the Cause of Endurance Failure in FeFETs 267
    Xianzhou Shao, Hao Xu, Saifei Dai, Fengbin Tian, Xiaoyu Ke, Jiahui Duan, Min Liao, Xinpei Jia, Xiaoqing Sun, Junshuai Chai, Jun Luo, Wenwu Wang, Xiaolei Wang

Selective-Attention Neuromorphic Vision Classifications Based on α-In2Se3Bi2O2Se Ferroelectric-Semiconductor Heterotransistors 270
    Xinrui Guo, Yu Zhu, Shuo Liu, Junling Liu, Ru Huang, Ming He

Comprehensive Investigation of the Disturb and Retention Issues in Scaled FeNAND Arrays 273
    Yuejia Zhou, Ru Huang, Kechao Tang

Prediction of Endurance Characteristics in Si FeFET with Ferroelectric Hf0.5Zr0.5O2 276
    Xinpei Jia, Tao Hu, Mingkai Bai, Xiaoqing Sun, Junshuai Chai, Hao Xu, Xiaolei Wang, Wenwu Wang, Tianchun Ye

A Novel FeFET-Based Multibit Content Addressable Memory Through Thermometer Encoding for Manhattan Distance Metric with High Area- And Energy-Efficiency 279
    Weikai Xu, Zeyu Zhang, Qianqian Huang, Ru Huang

Novel High-Efficiency Large-Area Optical-To-Optical Conversion Integrated Device for Optical Signal Processing 282
    Yahui Su, Shanjing Liu, Peixuan Song, Peiran Du, Hui Wang, Juan Li

Van Der Waals Interfacial Engineering for High-Performance Macroscopic Assembled Graphene (MAG)-Silicon Schottky Photodiodes 285
    Srikrishna Chanakya Bodepudi, Muhammad Abid Anwar, Muhammad Malik, Xiaolei Ding, Yance Chen, Yue Dai, Zongwen Li, Zhi-Xiang Zhang, Yunfei Xie, Wenzhang Fang, Huan Hu, Bin Yu, Yang Xu

Spintronic Stochastic Neuron-Based Deep Belief Networks for Image Classification 288
    Aijaz H. Lone, Meng Tang, Daniel N. Rahimi, Divyanshu Divyanshu, Camelia Florica, Selma Amara, Hossein Fariborzi, Gianluca Setti

G-Universe: A Collection of In-House Technology Computer-Aided Design Simulators as a Platform for Developing New Simulation Capabilities 291
    S.-M. Hong, I. K. Kim, S.-W. Jung, M.-S. Jang, P.-H. Ahn, T. Oh, G.-T. Jang, K.-W. Lee

Impact of Load-Si Thickness and Punch-Through Stopper Doping on the Electrical Performance of SOI Gate-All-Around Field-Effect Transistors 294
    Longyu Sun, Jiayi Zhang, Yan Li, Haoyan Liu, Xi Zhang, Yongliang Li

20 nm Gate-Length Normally-Off AlGaN/GaN HEMTs Enabled by SiNx Stress-Engineered Technique 297

    Chenkai Deng, Peiran Wang, Qing Wang, Hongyu Yu

RESURF Ga2O3-On-SiC Field Effect Transistors for Enhanced Breakdown Voltage 300

    Junting Chen, Junlei Zhao, Xiaohan Zhang, Jin Wei, Mengyuan Hua

Polyoxometalate-Doped Memristor with Redox Dynamics for Reliable Single-Component Artificial Neuron 303

    Shirui Zhu, Yan-Bing Leng, Guohua Zhang, Yu-Qi Zhang, Pengfei Han, Hecheng Cai, Ziyu Lv, Yongbiao Zhai, Ye Zhou, Su-Ting Han

Solder Fatigue Life Prediction Method Considering Intermetallic Compound Growth 306

    Yikang Wang, Xiaopeng Wu, Jiahao Hou, Can Liu, Yintang Yang

A Model-Driven Design Technology Co-Optimization (DTCO) with Multi-Objective Bayesian Algorithm for Advanced Technology 309

    Baokang Peng, Guoyao Cheng, Runsheng Wang, Ru Huang, Mansun Chan, Lining Zhang

Reverse Blocking Gan on Sic Hemts with Ultralow Dynamic Ron for 1200 V Power Applications 312

    Yutong Fan, Mengqiang Yuan, Yuqing Hu, Weihang Zhang, Yachao Zhang, Zhihong Liu, Yue Hao, Jincheng Zhang

Demonstration of Reliable Magnetic Shift Register Reading Using 50 nm MTJs on CMOS IC Towards 3D Ultra-High Density Memory 315

    M. Quinsat, Y. Ueda, N. Shimomura, S. Hashimoto, N. Umetsu, Y. Ootera, J. Iwata, H. Tokuhira, S. Miyano, M. Yoshikawa, T. Kondo, M. Saitoh, M. Kado

Artificial Hodgkin–Huxley Neurons Based on Ferro-Ionic CuInP2S6 318

    Fan Yang, Lei Liang, Qirui Zhang, Xuemei Wang, Qing Liu, Xiao Luo, Fucai Liu

High Mobility and Improved Subthreshold Characteristics of Ultra-Thin Channel IGZTO TFTs Down to 3 nm 321

    Kai Chen, Yi Jiang, Zhaolong He, Rui Zhang, Junkang Li, Yunlong Li

Design and Application of a Multi-Mode Signal Codetection System for Brain Signals 324

    Jianbo Jiang, Xueying Wang, Huiran Yang, Ziyi Zhu, Dujuan Zou, Siyuan Ni, Zhengyu Liang, Guopei Zhou, Zhitao Zhou, Liuyang Sun, Tiger H. Tao, Xiaoling Wei

Demonstration of EOT-Scaled Lg~25 nm FinFET Based on Thickness-Proportion Controlled HfO2-ZrO2-HfO2 Superlattice Gate Stacks 327

    Kun Zhong, Zhaohao Zhang, Siyuan Liu, Haiyuan Lyu, Huaxiang Yin

Global Stress Analysis in Fin Patterned Si/SiGe Multilayer Nanosheets for Nanosheet-Based CMOS Device Technology 330

    Amit Kumar Singh Chauhan, Imtiyaz Ahmad Khan, Kunal, Harsh Raju, Sanjeev Kumar Manhas

Van Der Waals Dielectrics and Electrodes in 2D Transistors for 2T0C DRAM 333

    Jianmiao Guo, Ziyuan Lin, Cong Wang, Tianqing Wan, Jianmin Yan, Yang Chai

Analysis of High Performance IGZO Thin-Film Transistors Under High-Gate and Drain Bias Stress 336

    Yupeng Lu, Yanyu Yang, Peng Wang, Jie Luo, Yunjiao Bao, Gaobo Xu, Huaxiang Yin

Ferroelectric In2Se3 Transistors with Multi-Terminal Plasticity for Highly Efficient Hardware Implementation of Reinforcement Learning 339
Yasai Wang, Weiwei Xiong, Jianmin Yan, Yue Zhou, Chaoyi Zhu, Xiangshui Miao, Yuhui He, Yang Chai

Enabling Artificial Spiking Sensory Neurons with a Single Flexible VO2 Mott Memristor for Neuromorphic Sensing 342
Chuan Yu Han, Shujing Zhao, Shengli Fang, Shi Quan Fan, Weihua Liu, Xin Li, Li Geng

Elemental Sn Promotes the Formation of Single-Crystal ε - Ga2O3 Films in MOCVD 345
Long Wang, Yao Wang, Qian Feng, Yachao Zhang, Jincheng Zhang, Yue Hao

Two-Dimensional ReSe2 Based Optoelectronic Synaptic Transistor 348
Wei Zeng, Jiyu Zhao, Hang Li, Guanglong Ding, Ye Zhou, Su-Ting Han

Capacitive Length-Extension Mode Resonators with Stress-Induced Gap-Closing Electrodes for Motional Resistance Reduction 351
Hao Yu, Yechen Miao, Fang Wang, Ke Sun, Yi Sun, Tiger H. Tao, Heng Yang

Thermal Stability of TiO2 Channel FE-VNAND: From Fabrication to High-Temperature Operation 354
Xujin Song, Dijiang Sun, Xiaoyan Liu, Jinfeng Kang

Rapid Customizing of Flexible Oects Arrays for Low-Cost Biosensing and Biocomputing 357
Xinyu Tian, Jing Bai, Dingyao Liu, Shiming Zhang

Realization of Wafer-Scale AlScN Ferroelectric Films and the Investigation of AlScN/N-GaN Ferroelectric Memristors 360
Mingrui Liu, Hang Zang, Shunpeng Lv, Zhiming Shi, Yuping Jia, Ke Jiang, Jianwei Ben, Dan Li, Xiaojuan Sun, Dabing Li

A Multi-Ion Sensing System on a Chip with Edge Computing Capability 363
Haolin Zhao, Zhancheng Mai, Kai Zhuang, Kai Wang

Surrogate MTJ Model for Early-Stage MRAM Macro Reliability Analysis 366
Quanhai Zhu, Hao Cai

Self-Heating and Bias Temperature Instability in Recessed-Gate GaN MIS-HEMTs with AlN/SiNx Bilayer Dielectric 369
Xin Wang, Hongyue Wang, Chen Wang, Jiayin He, Ju Gao, Bin Zhang, Ziheng Liu, Hongjie Peng, Chengkang Ao, Jiahui Yuan, Jinyan Wang

Enhancing Heat Dissipation of GaN HEMTs Based on Finite Element Simulation 372
Hongda Chen, Hongzhen Chen, Xiaohan He, Shiming Li, Mei Wu, Xiaohua Ma, Yue Hao

Fabrication and Performance Assessment of High-Mobility Sige Channel P-Type SOI FinFET Transistor 375
Zijing Zhang, Yuchen Wu, Yan Li, Huaizhi Luo, Fanyu Liu, Yongliang Li

BEOL-Compatible Photosensors by Atomic-Layer-Deposited ZnO Semiconductor Transistors and Their Monolithic 3D Integration 378
Yuyan Fan, Ziheng Wang, Yulong Dong, Zhiyu Lin, Danyang Chen, Shiyi Zhang, Jingquan Liu, Mengwei Si, Xiuyan Li

Optimizing Etch Processes for Enhanced Yield and Performance in SOT-MRAM Devices on 300 mm Wafers 381
Zhenghui Ji, Wenlong Yang, Guoxiu Qiu, Dandan Yang, Kaiyuan Zhou, Qingxiu Li, Qijun Guo, Enlong Liu, Shikun He

High-Frequency and Wideband RF Filters for 6G and Wi-Fi 7 (Invited Paper).. 384
Chengjie Zuo, Zhongbin Dai

Dynamic Performance Analysis of Ultra-Fast Inverter Based on Tri-Gate AlGaN/GaN MIS-HEMTs............ 387
Yunsong Xu, Weisheng Wang, Dechang Quan, Haotian Ji, Shenlei Ding, Yunzhou Jiang,
Kain Lu Low, Jiangmin Gu, Wen Liu

Achieving High Endurance Ferroelectricity in Hf0.5Zr0.5O2 Thin Films on Ge Substrate Through
Helium Ion Doping Engineering 390
Peiyuan Du, Huan Liu, Dongya Li, Chengji Jin, Hongrui Zhang, Di Wang, Yian Ding,
Bing Chen, Ran Cheng, Mengnan Ke, Xiao Yu, Yan Liu, Yue Hao, Genquan Han

Wide Temperature Behavior Analyses of SiGe HBTs Based on Small-Signal Parameter Extraction
Method 393
Lu Zhao, Guofang Yu, Jie Cui, Yue Zhao, Jun Fu, Yanyan Liu

In2Se3 FET-Based Neural Network Circuit Design for Complete Associative Learning 396
Weiwei Xiong, Yasai Wang, Xiangshui Miao, Yang Chai, Yuhui He

Data Retention in Co-Doped Hzo Fecaps: Roles of Fe Thickness and Thermal Budget............................... 399
Justine Barbot, Markus Peller, Isaac Emanuel Robert, Kerstin Bernert, Hannes Mähne,
Steffen Thiem, David Lehninger, Ayse Sünbül, Konrad Seidel, Thomas Kämpfe

Technologies of GaN Power Integration and Modeling 402
Sheng Li, Yanfeng Ma, Ran Ye, Siyang Liu, Weifeng Sun

Indium-Tin-Oxide Transistor with Maximum Transconductance Over 1100µS/µm and Cut-Off
Frequency of 23 GHz 405
Yuxuan Wang, Jiawei Xie, Zijie Zheng, Yuye Kang, Xuanqi Chen, Gerui Zheng, Rui Shao,
Kaizhen Han, Xiao Gong

Optoelectronic Multiplication Emulator Based on FPGA 408
Jinxian Li, Runyu Hu, Jiabin Shen, Zengguang Cheng, Peng Zhou

Integrated MEMS Diamond Quantum Magnetometer with Active Laser Noise Suppression.........................411
Nan Wang, Xiao Peng, Yaochen Zhu, Qihui Liu, Jiachen Han, Xin Chen, Xin Luo,
Yongquan Su, Lihao Wang, Yichen Liu, Hao Chen, Jiangong Cheng, Zhenyu Wu

A Transistor-Free Analog Content Addressable Memory with High Bit Density.. 414
Renhao Xue, Quanyi Tu, Mansun Chan, Xiwen Liu

Effect of the Channel Thickness on the Pbs Reliability and 1/F Noise of the ALD Ultrathin ITO
Field-Effect Transistor.. 417
Jiaming Zhao, Peiyan Hong, Xuefei Li

A Physics-Based Compact Model for Electromigration Failure Prediction and Dynamic IR-Drop
Evaluation.. 420
Chenglin Ye, Yizhan Liu, Zheng Zhou, Xiaoyan Liu

Multi-Mode Resonance and Temperature Behavior of GaN/SiC SAW Resonators 423
Guofang Yu, Renrong Liang, Deng Luo, Jianjun Chen, Yaqing Chi, Hanhan Sun, Bin
Liang

Enhanced ESD Protection Techniques for 10V Neurostimulator Circuits in 65nm CMOS
Technology 426
Tanay Das, Naef Ahmad, Laxmeesha Somappa, Sandip Lashkare

A Novel Transistor Free Design of SOT-MRAM Written by Unipolar Current with Record Bit Cell Size (15F2) 429

Meiyin Yang, Lei Zhao, Bowen Yang, Bowen Shen, Yanru Li, Peiyue Yu, Jianfeng Gao, Ruipeng Shi, Zhuangzhuang Ye, Shuo Xu, Yan Cui, Xiaolei Yang, Ming Wang, Shikun He, Kaiming Cai, Jun Luo

A Physics-Based Compact Model for Ambipolar Schottky-Barrier CNTFETs.................... 432

Rui Zhan, Bin Zhou, Zilin Teng, Yiheng Xue, Panpan Zhang, Jianhua Jiang

Wavelength-Dependent Reconfigurable Photo Memory Enabled by Organic-Gated Transistor.................... 435

Xiaokun Guo, Yaoqiang Zhou, Jianbin Xu

Deeply Scaled Gate Field Plate to Suppress Drain-Induced Dynamic Threshold Voltage Instability in Schottky-Type p-GaN Gate HEMT.................... 438

Chen Wang, Xin Wang, Junjie Yang, Jiayin He, Ju Gao, Chengkang Ao, Ziheng Liu, Hongjie Peng, Wenbo Xia, Jin Wei, Jinyan Wang

Investigation of Cryogenic Ultra-Low V_{TH} Mosfets and Beol for Power Efficiency Enhancement 441

Yuanke Zhang, Yuefeng Chen, Hengxu Guo, Jun Xu, Guoping Guo, Chao Luo

Low-Temperature Trench Ohmic Contact Suitable for Self-Aligned P-GaN HEMT.................... 444

Zhiqiang Xue, Mao Jia, Qian Xiao, Bin Hou, Ling Yang, Xiaohua Ma, Yue Hao

Enhancing Heat Dissipation in GaN-On-Diamond HEMTs Through Device-First Transfer Bonding.................... 447

Shiming Li, Mei Wu, Xiaohan He, Haolun Sun, Ling Yang, Xiaohua Ma, Yue Hao

5-Bit High-Linearity UV-Stimulated Synaptic Device Based on MoS2/GaN Heterostructure 450

Zijia Su, Yong Yan, Haiding Sun, Chengjie Zuo

Low Thermal Budget Ultrathin Ti Silicide for Advanced Backside Contact of Backside Power Delivery Network (BSPDN).................... 453

Hongxu Liao, Xijun Zhou, Fangze Liu, Lanyi Xie, Haixia Li, Jieyin Zhang, Jianjun Zhang, Xiaoyan Xu, Xia An, Heng Wu, Ru Huang, Ming Li

HRS Retention of 28 nm BEOL Integrated ReRAM.................... 456

Stefan Wiefels, Nils Kopperberg, Stephan Menzel

On the Evaluation of Remnant Polarization in 3D Cylindrical Hafnia-Based Ferroelectric Capacitors.................... 459

Yishan Wu, Puyang Cai, Junwei Guo, Haobo Lin, Xuepei Wang, Jinhao Liu, Boyao Cui, Yichen Wen, Maokun Wu, Runsheng Wang, Sheng Ye, Haibao Chen, Pengpeng Ren, Zhigang Ji, Ru Huang

Compact Modeling of Kink Effect in BULK MOSFETS at Cryogenic Temperatures.................... 462

Nisha Manzoor, Wajid Manzoor, Debashish Nandi, Aloke K. Dutta, Yogesh Singh Chauhan

Perovskite Based Artificial Vision System for Geometric Shape Recognition.................... 465

Shivam Kumar, Swapnadeep Poddar, Zhenghao Long, Zhiyong Fan

Fully Tunable In-Memory Eligibility Traces Based on Ferroelectric Semiconductor Field-Effect Transistors 468

Junling Liu, Shuo Liu, Xinrui Guo, Yu Zhu, Ru Huang, Ming He

Low-Voltage Multi-Level Flash Memory Based on Intra-Float-Gate Charge Transfer.................... 471

Yifan Chen, Haixia Li, Qing Wang, Zongwei Shang, Mingmin Shi, Xijun Zhou, Xiaoyan Xu, Xia An, Ru Huang, Ming Li

First Demonstration of Tunnel Fet-Based Physical Unclonable Function with Independent Entropy Source Through Ambipolar Current Modulation.. 474

Kaifeng Wang, Yingxi Zhou, Rundong Jia, Hongyan Han, Weihai Bu, Qianqian Huang, Ru Huang

Reconfigurable and Nonvolatile Graphene Photodetector Integrated onto Photonic Crystal Waveguide 477

Ruijuan Tian, Yu Zhang, Zhipei Sun, Xuetao Gan

Back-End-Of-Line Integration of Organic Thin-Film Transistor Active-Matrix on III-V Micro-LED Array for Video-Rate High-Resolution Displays.. 479

R. Shi, S. Ogier, J. Li, W. Tang, L. Deng, S. Li, X. Guo

Physics-Informed Neural Network for Predicting Out-Of-Training-Range TCAD Solution with Minimized Domain Expertise... 482

Albert Lu, Yu Foon Chau, Hiu Yung Wong

Optimization of RRAM Read Performance and Area Efficiency: A Large-Scale and Low-Parasitic Array with Novel Interconnection Schemes... 485

Shengyu Bao, Yuhang Yang, Zongwei Wang, Linbo Shan, Qishen Wang, Yimao Cai, Ru Huang

Multimode Transistors Based on Ion-Dynamic Capacitance... 488

Xiaoci Liang, Yiyang Luo, Yanli Pei, Mengye Wang, Chuan Liu

Pixel-To-Pixel Variance in Graphene-Silicon Photodetector Arrays 491

Muhammad Abid Anwar, Muhammad Malik, Srikrishna Chanakya Bodepudi, Xiaolei Ding, Yance Chen, Zongwen Li, Zhi-Xiang Zhang, Wenzhang Fang, Huan Hu, Bin Yu, Yang Xu

50 nm Gan Hemts Technology with High Frequency, Low Noise, and High JFoM 494

Yun Zhang, Jiaheng He, Changxin Mi, Xuankun Wu, Shujie Xie, Zhe Cheng, Lian Zhang

Enhanced Threshold Switching Devices Based on Conductive Filaments in SiOx Through Vertically Aligned MoS2 Layers 497

Jimin Lee, Sofia Cruces, Dennis Braun, Lukas Völkel, Ke Ran, Joachim Mayer, Alwin Daus, Max C. Lemme

Vertical Channel Gate-All-Around(VCG) CMOS Transistors with MBE in-Situ Doping Channel and TiN/HfO2 Gate Stacks... 500

Ran Bi, Haoran Zhao, Mingmin Shi, Jianhuan Wang, Jianjun Zhang, Xiaoyan Xu, Xia An, Heng Wu, Ru Huang, Ming Li

Dependence of Dislocation Density Distribution on Radial Temperature Gradient in 300MM Si Wafer During IGBT High Thermal Budget Process... 503

Jiuyang Yuan, Bozhou Cai, Yoshiji Miyamura, Wataru Saito, Shin-Ichi Nishizawa

Yield and Reliability Optimization of Analog RRAM for In-Memory Computing on a 28nm CMOS Platform 506

Siyao Yang, Bin Gao

Development of an On-Chip Millimeter-Wave Antenna 510

Ziqi Mei, Chao Liang, Enze Zhou, Yan Zhang, Ji Li, Rongbo Xie, Bingbai Li, Xiayu Wang, Chi Zhang, Rui You, Xiaoguang Zhao

Investigation of Enhanced Robustness Against Floating-Substrate-Induced Dynamic RON Degradation in 900-V p-GaN Gate HEMT Using Virtual-Body Technology 513
Hao Chang, Junjie Yang, Jingjing Yu, Jiawei Cui, Youyi Yin, Han Yang, Xuelin Yang, Jinyan Wang, Maojun Wang, Bo Shen, Jin Wei

A Comparative Analysis of Direct Leakage Current Compensation and Positive-Up-Negative-Down in the Characterization of Leaky Ferroelectric Structures 516
Tiang Teck Tan, Tian-Li Wu, Hsien-Yang Liu, Chen-Yu Yu, Laurent Grenouillet, Paolo La Torraca, Andrea Padovani, Francesco Maria Puglisi, Nagarajan Raghavan, Kin Leong Pey

Efficiency Analysis of Large-Area $\boldsymbol{\beta}-\text{Ga}_{2} \mathrm{O}_{3}$ Schottky Barrier Diodes for Dc-Dc Converter 519
Yuru Lai, Chenxi Li, Shengliang Cheng, Huaxing Jiang, Leidang Zhou, Zimin Chen, Yanli Pei, Gang Wang, Xing Lu

A Universal Method to Regulate Contact Resistance in Thin Film Transistors 522
Yanzhuo Wei, Guohui Li, Yanxia Cui, Hongwei Hao, Chen Chen, Dongdong Li, Shan-Ting Zhang

Breakthroughs in Undoped HfO_{2} Ferroelectric Capacitors Achieved Through Enhanced Nanocrystallite Seeding in as-Deposited Films Via $\mathbf{O}_{\mathbf{2}}$-Plasma Ald 525
Zongwei Shang, Changqing Ye, Hao Li, Xing Wu, Runsheng Wang, Ming Li, Ru Huang

High-Performance Broadband (300-1600 Nm) Si-Based Photodetector Enabled by 2.5D Out-Of-Plane Architecture Metasurface Enhancement 528
Yunfei Xie, Jin He, Zongwen Li, Zhi-Xiang Zhang, Xiaochen Wang, Feng Tian, Qianqian Zhang, Srikrishna Chanakya Bodepudi, Yuan Ma, Zijian Pan, Muhammad Abid Anwar, Bin Yu, Liaoyong Wen, Yang Xu

A Bionic Eye with Color Vision, Environmental Adaptivity and Neuromorphic Signal Processing Functions 531
Zhenghao Long, Xiao Qiu, Chak Lam Jonathan Chan, Zhibo Sun, Zhengnan Yuan, Zhiyong Fan

Positive and Negative Photoresponses in MoS2 Flakes Photogated by PN Junction Diode 534
Yumeng Liu, Zhengfang Fan, Jianyong Wei, Yizhuo Wang, Zhijuan Su, Yaping Dan

The Ultra-Thin Al2O3 Oxide Layer Via ALD Repaired Si Channel Interface and Optimizied Subthreshold Characteristics and Reduced Leakage Current by 96.2% 537
Renjie Jiang, Peng Wang, Lianlian Li, Qinkun Li, Zhongrui Wang, Hang Zhang, Huaxiang Yin, Qingzhu Zhang

Investigation of Al2O3/SiO2 Interface Charge for the Feasibility Study of Charge Sheet Super Junction 540
Swadhin Kumar Jena, Chiranjibi Padhee, Akshay K, Parlapalli Venkata Satyam

RRAM-Based Isotropic CNNs with High Robustness and Resource Utilization Rate 543
Wenyong Zhou, Yuan Ren, Jiajun Zhou, Chenchen Ding, Zhengwu Liu, Ngai Wong

Laser-Induced Low Temperature Dopant Segregation Schottky Barrier Mosfet for Monolithic-3D 546
Feixiong Wang, Yadong Zhang, Jinbiao Liu, Yunjiao Bao, Zhiyao Wang, Shuang Liu, Mingzheng Ding, Zhaohao Zhang, Qingzhu Zhang, Huaxiang Yin

A Heterogeneous FTJ-Based Computing-In-Memory Architecture for Vision Transformer Acceleration Via Hardware and Algorithm Co-Design 549
Letian Wang, Pinfeng Jiang, Zichong Zhang, Yifan Yang, Yilong Fang, Yi Wang, Mingde Zhu, Xiangshui Miao, Xingsheng Wang

Boosting 2D Transistor Performance Via TiS2 Van Der Waals Contact Engineering 552
Jialei Miao, Heng Zhang, Zheng Bian, Tianjiao Zhang, Yuda Zhao

Memristor-Based Approximate Adders for Efficient In-Memory Computing in Image Processing 555
Zhouchao Gan, Mengjie Li, Fan Yang, Fang Cheng, Xiangshui Miao, Xingsheng Wang

Fluorescent Nanodiamond Encapsulated Liposomes for Quantum Sensing in Nematode Worms 558
Takaki Arakawa, F. Kamada, K. Kinjo, K. Oshimi, M. Sara, M. Maeki, M. Tokeshi, M. Fujiwara

Enhancing Cad Workflows: Heterogeneous Graph Attention Networks for Efficient Routing Congestion Prediction in Chip Design 561
Qingyuan Yang, Mingkun Xu, Hongyi Li, Yu Du, Dahu Feng, Rong Zhao

Enabling Floating Body Effect in Bulk-Si Transistor for Area and Energy-Efficient Spiking Neuron 564
Shubham Patil, Hemant Hajare, Abhishek Kadam, Jay Sonawane, Shreyas Deshmukh, Veeresh Deshpande, Udayan Ganguly

Carbon Nanotube-Based High-Performance Bioelectronics 567
Youfan Hu

Controlling the Clamping Voltage in Punch-Through Diodes Via N+ Well and Contact Design for Low Voltage System Level ESD Protection 570
Praful Likhitkar, Navin Maheshwari, Sandip Lashkare

Demonstration of a 3D-Folded 2DEG-Channel Structure with Regrown AlGaN/GaN Heterostructures 573
Fuqiang Guo, Sen Huang, Xingyu Fu, Xuelin Yang, Qimeng Jiang, Xinguo Gao, Shuaiyu Chen, Ning Tang, Bo Shen

Highly Stable Zn Metal Anodes Enabled by In2O3 Coating for High-Performance Flexible Aqueous Zinc-Ion Batteries 576
Zhongqi Liang, Xiaohong Tan, Fan Xiao, Yufeng Jin, Guoshen Yang, Hang Zhou

Hardware Virtualization Technology for Multicore Brain-Inspired Chip 579
Dahu Feng, Rong Zhao

Improved Linearity by Double-Channel GaN HEMTs with the Ultra-Thin AlN Barrier Layer 582
Long Zhang, Qian Yu, Ling Yang, Meng Zhang, Chunzhou Shi, Xu Zou, Wenze Gao, Bin Hou

Simulation Optimization of Embedded Microchannel Heat Sink for GaN HEMTs 585
Bowen Yang, Shiming Li, Mei Wu, Ling Yang, Hao Lu, Bin Hou, Meng Zhang, Xiaohua Ma, Yue Hao

High Performance Pixel Development for Thin-Film Based Image Sensors 588
Jiwon Lee, Minhyun Jin

Spatial Microwave Magnetic Field Distribution Mapped by the Permalloy/Ta Bilayer Sensor 591
Peiwen Luo, Guanjun Zhang, Bin Peng, Wenxu Zhang

Characterization of a 1T-Floating Body DRAM Cell in Bulk Silicon MOSFETS for Cryogenic Memory Applications 595

    Hengxu Guo, Yuanke Zhang, Yuefeng Chen, Haoyu Sheng, Chi Fang, Guoping Guo, Chao Luo

Double-Gate Cu-MIC Poly-Ge1-XSnx TFTs on Glass Substrates Via the Gate-Last Process 598

    Daiki Goshima, Akito Kurihara, Akito Hara

From Flip FET to Flip 3D Integration (F3D): Maximizing the Scaling Potential of Wafer Both Sides Beyond Conventional 3D Integration 601

    Heng Wu, Haoran Lu, Wanyue Peng, Ziqiao Xu, Yanbang Chu, Jiacheng Sun, Falong Zhou, Jack Wu, Lijie Zhang, Weihai Bu, Jin Kang, Ming Li, Yibo Lin, Runsheng Wang, Xin Zhang, Ru Huang

Optimization of Scaling-Down Performance in Sub-100 nm AlGaN/GaN HFETs Based on Electron Velocity Modulation 604

    Mingyan Wang, Yuanjie Lv, Heng Zhou, Chao Liu, Peng Cui, Zhaojun Lin, Sen Huang

Monolithic Integration of GaN μ LED and Normally-Off P-GaN HEMT by Flip Chip Bonding.................... 607

    Jinxia Jiang, Ang Li, Guohao Yu, Han Yue, Zhongming Zeng, Baoshun Zhang

High Density 3D Integration of 2D Transistors Via Van Der Waals Lamination.. 610

    Donglin Lu, Quanyao Tao, Yuan Liu

Optimizing ALD-Deposited IGZO TFT Thermal Stability Through Compositional Adjustments 613

    Jianting Wu, Huajian Zheng, Min Guo, Yi Huang, Xiaoci Liang, Qian Wu, Chuan Liu

Fully Printed Polymer Gas Sensors Based on Machine Learning for Calibration-Free Mobile Sensing 616

    Siying Li, Sujie Chen, Qiuqi Zhang, Yuying Si, Xiaojun Guo

Re-Examination of Uniaxial Stress Effects in Ultra-Scaled GAAFETs 619

    Yusi Zhao, Huawei Tang, Rongzheng Ding, Yudong Lv, Yanbo Tang, Shaofeng Yu

GIT-Based Bipolar p-FET with Enhanced Conduction Capability on E-Mode GaN-On-Si HEMT Platform 622

    Chengcai Wang, Jinjin Tang, Junting Chen, Mengyuan Hua

93.6 cm2 V−1·s−1Homostructure a-IGZO Thin-Film Transistor with High-K Gate Dielectric Fabricated at Room Temperature.. 625

    Heng Yue Gong, Jia Cheng Li, Yang Hui Xia, Ya Dong Zhou, Hui Xia Yang, Yuan Xiao Ma, Ye Liang Wang

Impact of Cross-Sectional Current Crowding on Electromigration in Interconnects 628

    Yichen Wen, Shuying Wang, Xiaoman Yang, Hai-Bao Chen, Maokun Wu, Runsheng Wang, Zhigang Ji, Ru Huang

Effective Mass Engineering in Ultra-Scaled GAAFETs.. 631

    Yusi Zhao, Huawei Tang, Rongzheng Ding, Yudong Lv, Yanbo Tang, Shaofeng Yu

Unveiling the Role of Oxygen Vacancy Inhomogeneity in Enhancing the Reliability of Ferroelectric HfO2 ZrO2 Superlattice Structures.. 634

    Boyao Cui, Maokun Wu, Sheng Ye, Xuepei Wang, Yuchun Li, Yishan Wu, Yichen Wen, Jinhao Liu, Zhigang Ji, Hongliang Lu, David Wei Zhang, Runsheng Wang, Ru Huang

BEOL Electro-Biological Interface for 1024-Channel TFT Neurostimulator with Cultured DRG Neurons 637
Haobin Zhou, Bowen Liu, Taoming Guo, Hanbin Ma, Chen Jiang

Dual Functionality of MoS2 in nFET and pFET Through Contact Metal Selection 640
Kwok-Ho Wong, Mansun Chan

An Efficient Simulation-Time Aware Data-Driven Automatic Design Method for Analog Circuit 643
Shun-Qi Dai, Yuan Lei, Bei-Ping Yan

CMOS-Compatible Si3N4 Optical Waveguides: An Electromagnetic Study for Fabrication Considerations of SiO2 Cladding and Silicon Wafer Choice 646
Wenli Zhou, Rui Ray Yao, Sang Lam

Microwave Coherent Storage Based on the Long Lifetime Cavity Electromechanical System 649
Tiefu Li

Performance Analysis of N-Type Stacked-Vertical FET for Enhanced in Advanced CMOS Applications 652
Yonghwan Ahn, Junjong Lee, Seunghwan Lee, Sanguk Lee, Kyeongrae Cho, Minchan Kim, Rock-Hyun Baek

Threshold Switching Memristor-Based Spiking Neuron Modeling and Simulation 655
Pengyu Liu, Lekai Song, Kong-Pang Pun, Guohua Hu

Enhanced Gate Stack Reliability Test Framework for Gan Hemt in High-Stress Environments Leveraging on Ramp and Constant Voltage Stress Protocols 658
Jerry Joseph James, Nagarajan Raghavan

Energy Efficient Ground Enhanced Scheme for Large-Scale SOT-MRAM Arrays 661
Wen Wang, Haoran Du, Shuyu Wang, Bo Liu, Hao Cai

Modeling and Simulation of Prepulse and Pulse Front for 4H-SiC Drift Step Recovery Diode 664
D. Y. Guo, Y. Zhou, J. K. Guo, X. Y. Tang, L. J. Sun, Y. M. Zhang, Q. W. Song

Van Der Waals Epitaxy h-BN/AlN Back Barrier with Controllable Boron-Diffusion for High-Erformance AlGaN/GaN HEMTs 667
Haidi Wu, Jing Ning, Jincheng Zhang, Yue Hao

Multi-Output Virtual Metrology for Physical Vapor Deposition Using Projective Selection Algorithm 670
Amina Mevic, Andreas Laber, Sandor Szedmak, Dženana Đonko, Senka Krivic

On the Understanding of Temperature Dependence of Flicker Noise in Advanced FinFET Technology 673
Sheng Yang, Shuying Wang, Chenyang Zhang, Yewei Zhang, Yongkang Xue, Pengpeng Ren, Zhigang Ji

Overlay-Aware Variation Study of Flip FET and Benchmark with CFET 676
Wanyue Peng, Haoran Lu, Jingru Jiang, Jiacheng Sun, Ming Li, Runsheng Wang, Heng Wu, Ru Huang

A 12-Bit Fully Differential SAR/SS ADC Architecture Based on Scale Reference for High-Speed CMOS Image Sensors 679
Changhui Yang, Bailin Zhang, Weizhe Zhang, Nanbo Chen, Jingyang Chen, Ziyang Hu, Gang Wang, Peng Feng, Jian Liu, Nanjian Wu, Liyuan Liu

A High-Throughput Parasitic TRNG in Self-Rectifying Memristor Based CIM for Edge Secure Computing 682

Ying-Jie Yu, Sheng-Guang Ren, Yu-Yang Fu, Jian-Cong Li, Pu-Yi Zhang, Yi Li, Xiang-Shui Miao

Low-Complexity Method for Shortest Path Optimization Problems Based on Nanowire Memristor Network 685

Yanming Liu, Shenghao Wu, Ming Jian, Daixuan Wu, Yanxi Long, He Tian

Enhancing 3D DRAM Double-Gate Device Performance by Leveraging the Floating Body Effect 688

Jing Liang, Yong Yu, Feng Shao, Menglong Zhou, Tiantian Wei, Chunyang Li, Jianpeng Jiang, Jing Cai, Bryan Kang, Mingxu Liu, Xiangsheng Wang, Guilei Wang, Chao Zhao, Shaojun Wei

Categorization of Stacking Faults and Their Effects on I-V Characteristics in 2NM P-Type GAA Nanosheet Transistors 691

Seungjoon Eom, Sanguk Lee, Rock-Hyun Baek

A Low-Programming-Variation and High-Yield 2-Bit Programmable 1-Kb Self-Rectifying Memristor Crossbar Array 694

Jia-Yi Sun, Sheng-Guang Ren, Heng-Feng Zhang, Zhi-Li Cui, Yi-Bai Xue, Yu Zhang, Wen-Bin Zuo, Yi Li, Xiang-Shui Miao

Based on Vertical Structures for N-Type Organic Electrochemical Transistors 697

Wenjing Zhang, Songjia Han, Chuan Liu

Recent Research on Ultrawide Bandgap Semiconductor Ga2O3 and AlN 700

Dao Hua Zhang, Xian-Hu Zha, Rong-Jun Zhang, Shuang Li, Yan Liu, Yu-Xi Wan

Scalable In-Memory Walsh-Hadamard Transform for Image Compression 703

Jing Tian, Huai-Zhi Pei, Jian-Cong Li, Xiao-Di Huang, Yi-Bai Xue, Yi Li, Xiang-Shui Miao

A Novel De-Mirroring Approach for Bias-Dependent Capacitance Extraction in Nanosheet Fet Using Conformal Mapping 706

Deven H Patil, Sandeep Kumar, Sunil Rathore, S. Dasgupta, Navjeet Bagga

4F2/Bit Memristive Multi-Bit Content Addressable Memory Enabled by Nonlinear Encoding for in-Memory Similarity Search 709

Tong Hu, Yibai Xue, Yingjie Yu, Wenbin Zuo, Jiancong Li, Yi Li, Xiangshui Miao

Monolithic and Heterogeneous CTFT on Glass Substrate Using Ni-MIC Poly-Si TFT and Cu-MIC DG Poly-Ge TFT 712

Y. Ito, D. Goshima, A. Kurihara, A. Hara

High Endurance Nanoscale TiN Bottom Heater for Phase Change Memory 715

Ziqi Wan, Wencheng Fang, Li Xie, Xi Li, Ruobing Wang, Zhitang Song, Xilin Zhou

A Device to Circuit BTI and HCD Aging Analysis Framework: (Invited Paper) 718

Payel Chatterjee, Karansingh Thakor, Souvik Mahapatra

Valley Transistors as Graded Neurons for Accurate Action Recognition 721

Jiewei Chen, Wenxiao Wang, Yue Zhou, Yang Chai

High-Sensitivity Flexible X-Ray Detectors Based on Drop Casting 2D/3D Perovskite Thick Film 724

Liwen Qiu, Bosen Zhang, Qiang Lou, Maojun Sun, Hang Zhou

Modeling Nanowire Single-Photon Avalanche Detectors Via Deep Neural Networks 727
Boyang Zhang, Zhe Li, Zhongju Wang, Daoyi Dong, Lan Fu

Multi-Scale Thermal Modeling of 3D-Heterogeneous Integrated Processing-Near-Memory Chip for
Edge Large Language Model Inference.. 729
*Awang Ma, Bin Gao, Yuyao Lu, Yudeng Lin, Yiming Zhou, Xing Mou, Zhuodong Kang,
Zhipeng Kuang, Peng Yao, Jianshi Tang, He Qian, Huaqiang Wu*

Polarization Regulation in Algan Solar-Blind Ultraviolet Photodetectors 732
Bingxiang Wang, Ke Jiang, Tong Fang, Xiaojuan Sun, Dabing Li

Comprehensive Performance Evaluation of 18 Perfluoroelastomers O-Ring Models for
Semiconductor Manufacturing 735
Zhanfeng Guo, He Tian, Tian-Ling Ren

A Bio-Inspired Ionic Retina.. 738
Hongjie Zhang, Kai Xiao

ScAlN-Based Bulk Acoustic Wave Technology for 5G Filtering Applications................................... 741
*Chen Liu, Xinghua Wang, You Qian, Ying Zhang, Minghua Li, Peng Liu, Huamao Lin,
Qingxin Zhang, Yao Zhu*

Dual Charge-Trapping Nanowire Flash Transistor for Low-Voltage High-Precision Biomimetic
Synaptic Device with 1024 States 744
Qing Wang, Haixia Li, Ran Bi, Hongxu Liao, Baotong Zhang, Ru Huang, Ming Li

Consideration of VFET for Ultimate Logic Scaling: A Design Perspective....................................... 747
*Yimeng Wang, Yanbang Chu, Ziqiao Xu, Yu Liu, Rui Guo, Jiacheng Sun, Wanyue Peng,
Haoran Lu, Ming Li, Runsheng Wang, Heng Wu, Ru Huang*

Robot Collision Detection Acceleration with In-Memory Search Based on the Monolithic 3D
Integration of 2D Transistors and Vertical RRAMs.. 750
Yijian Zhang, Jiahao Sun, Maosong Xie, Yueyang Jia, Rui Yang

Enhanced NBTI Characteristics in HfO2/TiAlC RMG Stacks Via Low-Temperature H* Remote
Plasma Treatment After Interfacial Layer Growth 753
*Songyi Jiang, Qianqian Liu, Hong Yang, Mingyang Sun, Junjie Li, Jianfeng Gao, Shuai
Yang, Runsheng Wang, Jun Luo, Wenwu Wang*

Modeling Multi-Junction Tandem Solar Cell with High Bifacial Gain... 756
Mohammad Ajmain Fatin, Arpan Saha, Md. Shahriar, Mainul Hossain

Recent Advances in on-Chip Learning with Organic Neuromorphic Circuits................................... 759
Yoeri Van De Burgt

Pixelated Germanium-On-Silicon Photodetector with High Responsivity for Short Wavelength
Infrared Imaging... 762
Zhou Zhou, Chao Gao, Xin Jin, Xiaolin Liu, Kai Wang

Towards Spectrum Spurious-Free and Wideband SAW Devices Based on LN/AT-Quartz Layered
Structure 765
*Peisen Liu, Sulei Fu, Boyuan Xiao, Xinchen Zhou, Qiufeng Xu, Jiajun Gao, Shuai Zhang,
Rui Wang, Cheng Song, Fei Zeng, Weibiao Wang, Feng Pan*

Optoelectronic Switching Memory Device Based on WO3 Semiconductor Film Enables High-
Efficiency In-Sensing Reservoir Computing for Speech Recognition 768
Xun Li, Jie Li, Ke Xiang, Wenyang Xu, Yu Xu, Denshun Gu, Guangdong Zhou

High Dielectric Constant of HfO2 Technology for Memory Applications 771

Min-Hung Lee, Zhao-Feng Luo, Chun-Yu Liao, Kuo-Yu Hsiang, Jia-Yang Lee, Fu-Shen Chang, Yii-Tay Chang, Cheng-Hong Liu, Che-Chi Cheng

Optoelectronic Artificial Synaptic Device Based on Graphene-AlGaN Van Der Waals Junction 774

Yang Chen, Yuanyuan Yue, Bingchen Lv, Jin Zhang, Xiaoyu Wei, Xiaojuan Sun, Dabing Li

Impact of Titanium Nitride (TiN) Thickness Uniformity on the Reliability of n-FinFETs: A Comparative Study of ALD and PVD TiN Techniques 777

Mingyang Sun, Yunfei Shi, Hong Yang, Qianqian Liu, Qingzhu Zhang, Tao Yang, Junfeng Li, Huaxiang Yin, Xiaolei Wang, Jun Luo, Wenwu Wang

Dual-Gate Thin-Film Transistor-Based Multi-Parameter Sensor for Comprehensive Water Quality Monitoring in Aquatic Environments 780

Qiang Chen, Qiyi Su, Haolin Zhao, Zhancheng Mai, Kai Zhuang, Yitong Xu, Xinghui Liu, Kai Wang

Impact of Sb2Se3 Annealing on the Photoresponse of TiO2/Sb2Se3/Si Back-To-Back Photodiodes 783

Bangsen Ouyang, Jialiang Wang, Yang Chai

Design and TCAD Simulation of a Surface Super-Junction Based GaN HEMT with Enhanced Breakdown Voltage 786

Jiayi An, Fuqiang Guo, Zhongchen Ji, Sen Huang, Qimeng Jiang, Xinhua Wang, Ke Wei, Xinyu Liu

Joule Heating Effect on Quality Factor and Frequency Tuning of 2D MoS2 NEMS Resonators 790

Shuai Yuan, Zuheng Liu, Pengcheng Zhang, Rui Yang

Improved Cmos-Based Noise-Immune Sigmoid Activation Function for Neural Networks 793

Harshita Singh, Anant Singhal, Harshit Agarwal

Self-Heating and Process Induced Performance Barrier on Complementary Field Effect Transistor: A Reliability Perspective 796

Sandeep Kumar, Deven H Patil, Khushi Jain, Ankit Dixit, Sunil Rathore, Mohd. Shakir, Naveen Kumar, Vihar Georgiev, S. Dasgupta, Navjeet Bagga

Compact Modeling of Silicon Carbide (SiC) Power Fets 799

Karunesh Kumar Tripathi, Yogendra Machhiwar, Danish Raja, Harshit Agarwal

Performance Analysis of Advanced Ferroelectric $\text{HfO}_{2}\text{-}\text{ZrO}_{2}$ Superlattice Gate Stack Transistor with Multi-Phase Ferroelectric Order 802

Danish Raja, Yogendra Machhiwar, Karunesh Kumar Tripathi, Girish Pahwa, Harshit Agarwal

Novel Ferroelectric Tunnel Fet-Based Computing-In-Memory with In-Situ XOR Cipher-Encrypted and-Type Multiply-Accumulate for Secure Edge AI 805

Jin Luo, Zhiyuan Fu, Qianqian Huang, Ru Huang

Photoresponse Improvement in the Near-Infrared Region of Organic Photodetectors by Introducing a Trap Layer 808

Fangchen Zhu, Yang Wang

CMOS Compatible Spike Vision Sensor 811

Xiaolin Liu, Chao Gao, Bowei Jiang, Xin Jin, Zhou Zhou, Yihong Qi, Kai Wang

Dynamic Adaptive Visuomorphic Electronics 814

Le Wang, Yiru Wang, He Shao, Wei Huang, Haifeng Ling

Enhanced Thermal Stability of Ru Interconnects Using h-BN as Barrier Layers 817
Kun Chen, Zhuming Wang, Chen Wang, David Wei Zhang

Leveraging Ferroelectric Negative-Capacitance Effect for Energy Efficient Electronics 820
Jie-Ni Dai, Pin Su

Thermal Crosstalk Analysis in Advanced CMOS Circuits: Insights from 3nm Gate-All-Around
Transistor Technology 823
Sihao Chen, Honglin Wu, Runsheng Wang, Ru Huang, Lining Zhang

Flexible and Skin-Compatible rGO/PVDF Composite Sensor for Multi-Functional on-Skin
Applications... 826
Zi-Jia Su, Lu-Qi Tao, Li-Wei Liang, Zhi-Fei Xie, Yi Yang, Tian-Ling Ren

Modelling and Design of Short Channel Ferroelectric FETs with a Metal Interlayer Easing the
Multilevel Operation 829
Chiara Rossi, Daniel Lizzit, David Esseni

Effect of Pulse Schemes on Multi-Level Switching and Short-Term Instability in 1T1r
Configuration... 832
*Xiaohua Liu, Felix Cüppers, Dennis Nielinger, Susanne Hoffmann-Eifert, Rainer Waser,
Stefan Wiefels*

Comprehensive Modeling of Ferroelectric Tunnel Junctions: Variability Analysis and Device
Design.. 835
Jiajun Qiu, Ning Ji, Hao Li, Ning Feng, Runsheng Wang, Ru Huang, Lining Zhang

Cryogenic sEKV Compact Model Applied to 22 nm FDSOI Enabling Low-Temperature Circuit
Simulation 838
Hung-Chi Han, Yating Zou, Batuhan Keskin, Edoardo Charbon, Christian Enz

An Rram-Based Multi-Mode and Pipelined Pooling Scheme in Computing-In-Memory for
Convolutional Neural Networks 841
Yi Gao, Zongwei Wang, Lin Bao, Yimao Cai, Ru Huang

Temperature Dependent Back-Hopping in Spin Transfer Torque Switching of Perpendicular
Magnetic Tunnel Junctions.. 844
Yapeng Zhao, Tiaoyang Li, Yanqing Wu

Selective-Ferroelectricity-Defining Method Utilizing Directional Oxygen Plasma Treatment for
Optimizations of Hf0.5Zr0.5O2-Based Memory Devices.. 847
*Yi Ding, Mingcheng Shi, Yuyan Wang, Bowen Shen, Wen Sun, Benjamin Yang, Yanbo Su,
Huanan Liu, Jian Yuan, Jianshi Tang, Bin Gao, He Qian, Huaqiang Wu*

Beol-Compatible Multi-Layer ITO-ZnO-ITO Channel FETs Achieving Enhanced Mobility,
Positive V_{TH} Shift, and Improved PBTI.. 850
Ying Xu, Yiyuan Sun, Zijie Zheng, Xiao Gong

Synergistic Optimization of Thermal and Electrical Performances in Hetero-Integrated β-Ga2 O3
SBDs 853
Yinfei Xie, Yang He, Zhengyue Li, Wenhui Xu, Tiangui You, Xin Ou, Huarui Sun

Selecting Device, In-Memory Search, and Monolithic 3D Integration of RRAMs and 2D Devices
Towards High Reliability and High Efficiency: (Invited Paper) 856
Rui Yang, Yueyang Jia, Maosong Xie, Minliang Shen, Yijian Zhang, Yuzhuo Liu

Adaptive Update Precision with Reduced Iterative Write Cycles for Efficient Training Neural Networks on ECRAM Arrays 859

    Peihong Li, Peng Chen, Peng Lin, Gang Pan

Twisted-Placed Multilayer Stack for Inherent Suppression of Transverse Modes on Layered SAW Devices 862

    Boyuan Xiao, Sulei Fu, Peisen Liu, Xinchen Zhou, Qiufeng Xu, Jiajun Gao, Shuai Zhang, Rui Wang, Cheng Song, Fei Zeng, Weibiao Wang, Feng Pan

A Novel Low Loss Superjunction MOSFET with Hybrid Conduction Modes 865

    Yun Xia, Gang Chen, Yuxi Wan, Wanjun Chen, Ruize Sun, Xiaoming Wang, Yan Wang, Xiaoping Wang

Graphene/Silicon Pixel Array with Integrated Imaging and Readout System 868

    Yuan Ma, Zongwen Li, Youshui He, Qianqian Zhang, Yunfei Xie, Zhi-Xiang Zhang, Feng Tian, Muhammad Abid Anwar, Muhammad Malik, Srikrishna Chanakya Bodepudi, Bin Yu, Yuda Zhao, Yang Xu

Effect of Fin Dimensions on the Performance of Gan-On-Si Fin-HEMTs 871

    Mengdi Li, Jiejie Zhu, Lingjie Qin, Bowen Zhang, Yuxi Zhou, Mingchen Zhang, Yichong Ding, Yuchen Qian, Xiaohua Ma, Yue Hao

Surface Potential-Based Compact Model for IGZO-DRAM Enables the TCAD-To-SPICE Framework for Reliability-Aware DTCO Flow 874

    Xufan Li, Wenfeng Jiang, Chen Gu, Yue Zhao, Di Geng, Guanhua Yang, Lingfei Wang, Ling Li

Optimization of the Read Transistor in Hybrid 2T0C DRAM by Co-Modeling the Drain Current of a-IGZO and Low Temperature Poly-Silicon TFTs 877

    Haolin Li, Zheng Zhou, Xiaoyan Liu

Enhanced Performance of P-FeFETs with TiN:2.5nm/Mo/TiN Gate Stacks for 3-Bit-Per-Cell Operation, 3.5V Read-After-Write, and High Endurance (109 Cycles) in Compute-In-Memory Applications 880

    C.-Y. Tsai, M.-H. Hsiung, Y.-C. Chen, Y.-T. Tsai, Y.-T. Tang

Dual-Functional Volatile and Nonvolatile Resistive Switching Characteristics of $\mathbf{C U} / \mathbf{G A}_{\mathbf{2}} \mathbf{O}_{\mathbf{3}} / \mathbf{P T}$ Memristor Deposited Via Electron Beam Evaporation 883

    Nan He, Zi Li, Shuai Chen, Hao Zhang, Xinpeng Wang, Lei Wang, Yi Tong

Enabling Broader Memory Windows by Double-Gate Nanosheet Ferroelectric FETS for Next-Generation Non-Volatile Memory Storage 886

    F. Wu, C.-Y. Chiu, T.-Y. Lin, C.-H. Wu, V. P.-H. Hu, P. Su, C.-J. Su

On-Chip Photon-Mediated Magnon-Superconducting Qubit System and Its Quantum Application 889

    Jiacheng Liu, Ferris Prima Nugraha, Qiming Shao

Reduced Process Induced Threshold Voltage Variability in Bulk Negative Capacitance Junctionless Transistors 892

    Ruma S R, Vita Pi-Ho Hu, Manish Gupta

Optimal Transfer Learning Strategies for Property Predictions in Materials Science 895

    Reshma Devi, Keith T. Butler, Sai Gautam Gopalakrishnan

Invertible Prediction Model for Si3N4 Wet Etching Using DHF 897

    Koki Shibata, Takashi Ota, Koji Ando, Naoko Misawa, Chihiro Matsui, Ken Takeuchi

The Demonstration of Scalable-HZO/ZrO2 FeFET with Large Memory Window of 2.3V for 3Bit-Per-Cell, Immediate Read After Write, High Endurance of 109 Cycles, and the High Accuracy of 92% for Machine Learning 900
S.-T. Huang, H.-M. Chen, Y.-T. Tsai, C.-S. Pai, Y.-T. Tang

A Fan-Out Wafer-Level Packaging-Based THz Communication Transceiver System for High-Speed Chip-To-Chip Wireless Interconnect Applications 903
Si Rui Liu, Ya Fei Wu, Zong Rui He, Yu Jian Cheng, Yang Chai

Temperature-Dependent Characteristics of Field-Effect Mobility in MOCVD-Grown β-Ga2O3MOSFETs on Sapphire Substrate... 906
Chunxiao Yu, Yibo Wang, Yiyang Wu, Jun Zheng, Chenyu Liu, Xiaole Jia, Bochang Li, Haodong Hu, Cize Fang, Yan Liu, Yue Hao, Genquan Han

Content Addressable Memory Hierarchies for Computing in Memory 909
Paul-Philipp Manea, Nathan Leroux, John Paul Strachan

6-Inch GaN-On-Si Gold-Free Fabrication Technolgies for Monolithic Microwave Integrated Circuits (MMICs) 912
Xiaojin Chen, Jin Zhou, Peiyu Mao, Tong Wang, Zhenyuan Li, Hu Wei, Hanghai Du, Weichuan Xing, Weihang Zhang, Xiangdong Li, Zhihong Liu, Jincheng Zhang, Yue Hao

Electrical and Thermal Characterization of Hetero-Integrated β-Ga2O3-On-Diamond SBDs by Transfer Printing Technology 916
Zhenyu Qu, Tiancheng Zhao, Wenhui Xu, Haodong Jiang, Yeliang Wang, Xinbo Zou, Min Zhou, Tiangui You, Xin Ou

Electrical Transport at n-Ga2O3/N-SiC Hetero-Interface Constructed by Hydrophilic and Surface Activated Bonding... 919
Zhenyu Qu, Wenhui Xu, Haodong Jiang, Tiancheng Zhao, Yeliang Wang, Haowen Guo, Xinbo Zou, Min Zhou, Tiangui You, Xin Ou

Experimental Investigation of Inserted HfO2 Impact on VFb and Interface Via ALD for La2O3 Dipole-First Multi-VT Techniques 922
Yanzhao Wei, Jiaxin Yao, Yu Wang, Qingzhu Zhang, Huaxiang Yin

Multi-Physics Modeling of Au/MoS2/Au Memristors Combining Molecular Dynamics and Electro-Thermal Simulations 925
Mohit Tewari, Ashutosh Krishna Amaram, Tarun Agarwal

High Performance 4H-SiC DSRD with 1.4kV Peak Voltage, 400ps Risetime and 1-MHz Continuous Repetition-Rate 928
Yu Zhou, Jingkai Guo, Dengyao Guo, Fengyu Du, Lejia Sun, Xiaoyan Tang, Qingwen Song, Yuming Zhang

Neural Network Assisted MOSFETs Gate Dielectric Traps Extraction.. 931
Xiaoyan Liu, Jinghan Xu, Zheng Zhou

A Novel Superlattice HfO2-ZrO2 Ferroelectric Tunnel FET for Overall Improvement in Memory Window, EOT and Disturb Immunity... 934
Shaodi Xu, Zhiyuan Fu, Shengjie Cao, Yue Yu, Hao Zheng, Qianqian Huang, Ru Huang

A 512×256 TFT-Based Image Array Sensor with High Sensitivity, High Frame Rate and Wide Dynamic Range for Industrial Soft X-Ray Detection... 937
Weikang Yan, Haorong Xie, Xianming Li, Chao Gao, Qi Liu, Song Kang, Jun Chen, Kai Wang

Investigation of Effects of Non-Volatile Memory-Based Computation-In-Memory Non-Idealities and Model Size on Performance and Robustness of Small Language Model During Inference Phase............ 940
Adil Padiyal, Tao Wang, Naoko Misawa, Chihiro Matsui, Ken Takeuchi

Effect of Cu Microstructures on Cu/SiO2 Hybrid Bonding for 3D IC Heterogeneous Integration.................. 943
Chih Chen, Huai-En Lin, Wei-Lan Chiu, Hsiang-Hung Chang

Efficient Implementation of 16 Reconfigurable Boolean Logics Based on Memristors and Their Application in Image Edge Detection... 947
Zhouchao Gan, Yifeng Xiong, Fan Yang, Xiangshui Miao, Xingsheng Wang

A CMOS-Compatible MoS2 Transistor on Silicon-Rich Silicon Nitride as Multifunctional Neuromorphic Device 950
Xiangwei Su, Hongzhao Wu, Caijing Liang, Xinlong Zeng, Yuda Zhao

A Sensing-Computing System Based on High Uniformity Photolithographic Organic Thin Film Transistor... 953
Yaojie Zheng, Taoming Guo, Haobin Zhou, Chen Jiang

Physics-Based Circuit-Compatible Model of Polycrystalline Hafnia-Based 3D Ferroelectric Capacitor for High-Density Memory Applications 956
Minyue Deng, Chang Su, Jiayan Zhu, Liang Chen, Shengjie Cao, Qianqian Huang, Ru Huang

Van Der Waals Infrared Photon Detectors for Standard Blackbody Characterization 959
Yang Wang, Weida Hu, Peng Zhou

Remote Effect of Extra Metal Layer on TiN Electrode on the Ferroelectric Properties of Hf0.5Zr0.5O2 Thin Films... 962
Zhipeng Xue, Danyang Chen, Zikang Yao, Tianning Cui, Yulong Dong, Jingquan Liu, Mengwei Si, Xiuyan Li

Electrical Tunability in Band-To-Band-Tunneling Based Neuron for Low Power Neuromorphic Computing 965
Shubham Patil, Jayatika Sakhuja, Anmol Biswas, Hemant Hajare, Abhishek Kadam, Shreyas Deshmukh, Ajay Kumar Singh, Sandip Lashkare, Nihar Ranjan Mohapatra, Udayan Ganguly

Direct 3D Force Mapping Enabled by Flexible Single-Crystal Piezoelectric Sensor Array............................ 968
Jiefei Zhu, Changjian Zhou, Min Zhang

Hydrogel-Based Bipolar Synaptic Device for Artificial Neural Networks.. 971
Mengjiao Pei, Yun Li, Changjin Wan

Challenges in Accelerating Power SiC Device Commercialization 974
Victor Veliadis

Materials for Thermal Dissipation Applications in Stacked Devices 977
W.-Y. Woon, J.-H. Jhang, S. Vaziri, M. Malakoutian, K.-K. Hu, I. Dayte, C.-C. Shih, J.-F. Hsu, J.-P. Lin, Y. Wu, A. Kasperovich, R. Soman, J. Kim, H.-K. Wei, X. Bao, M. Nomura, S. Chowbhury, S. Sandy Liao

Thermal Analysis of Multi-Chiplet Heterogeneous Integration Based on the Lidless Fan-Out Package... 980
Yongbo Wu, Jiexun Yu, Changming Song, Zheqi Xu, Lin Tan, Qian Wang, Jian Cai

Toward Low-Thermal-Budget Processing and Low-Voltage Operating in Ferroelectric Stack Films by Introducing ZrO2 Middle Layer 983
Yinchi Liu, Hao Zhang, Jining Yang, Xun Lu, Shiyu Li, Yeye Guo, Chunlei Wu, Handong Zhu, Xuning Zhang, Zhenxin Wang, Yaoyi Wang, Ruli Zeng, Yiwen Yu, Wenjun Liu

Broadband Miniaturized Spectrometers with Van Der Waals Junctions 986
Md Gius Uddin, Susobhan Das, Abde Mayeen Shafi, Lei Wang, Xiaoqi Cui, Fedor Nigmatulin, Faisal Ahmed, Andreas C. Liapis, Weiwei Cai, Zongyin Yang, Harri Lipsanen, Tawfique Hasan, Hoon Hahn Yoon, Zhipei Sun

2-D FET Modeling: Why Incorporating the Gate- And Drain-Dependent Source Tunneling Barriers Matter 989
Cristine Jin Estrada, Zichao Ma, Lining Zhang, Mansun Chan

Interface Engineering Induced Digital to Analog Switching Transition in Hafnium Oxide-Based Memory Device 992
Cong Han, Miaocheng Zhang, Yi Liu, Xinpeng Wang, Hao Zhang, Yi Tong

MEMS Integrated with Self-Assembled Electrets 995
Daisuke Yamane

"Quantum Nanophotonics with Hexagonal Boron Nitride" 998
Igor Aharonovich

Operation Principles and Applications of Ultra-Sensitive Optical Detectors at Nanoscale: Facts and Artifacts 1000
Taras Plakhotnik

Ag:SiOx-Based Volatile Memristors for Dendritic Computations 1003
Ruiqi Chen, Yulin Feng, Nan Tang, Yiyang Chen, Hao Ai, Haozhang Yang, Zheng Zhou, Lifeng Liu, Xiaoyan Liu, Jinfeng Kang, Peng Huang

Synergetic Effect of Doping and Oxygen Vacancies to Realize Higher Permittivity and Lower Leakage for DRAM Capacitors: A First-Principles Study 1006
Ting Zhang, Maokun Wu, Miaojia Yuan, Yichen Wen, Yilin Hu, Pengpeng Ren, Sheng Ye, Runsheng Wang, Zhigang Ji, Ru Huang

Wafer-Scale Fabrication of Janus-MXene Films and Its Based Flexible Artificial Synapse 1009
Xin Liu, Conghui Zhang, Yuhang Yang, Peisong Liu, Shuiren Liu, Lingxian Meng, Fei Hui

Improved Breakdown Voltage and Leakage Current in βGa2O3 Schottky Barrier Diode Realized by N Ion-Implantation Edge Termination 1012
Hao Zhang, Xinlong Zhou, Yinchi Liu, Jining Yang, Handong Zhu, Xun Lu, Shiyu Li, Yeye Guo, Chunlei Wu, Xuning Zhang, Zhenxin Wang, Yaoyi Wang, Ruli Zeng, Yiwen Yu, Wenjun Liu

Miniature Optical Fiber Fabry–Pérot Interferometric Acoustic Sensor with 3D Micro-Printed Ortho-Planar Springs 1015
Shangming Liu, Peng Wang, Taige Li, A. Ping Zhang

Diode Microheaters for Scalable Actuation in Micro-Transfer Printing 1018
Jiajun Zhang, Qinhua Guo, Xiwen Liu, Yunda Wang

Spatiotemporal Encoding Based on Mott Spiking Neurons for Sound Localization 1021
Zihan Guo, Linbo Shan, Zongwei Wang, Xing Zhang, Yimao Cai, Ru Huang

Energy-Efficient Temperature-Calibration Readout Circuits with Thermal-State Sensible Sampling and Zoom Window Switch Scheme for RRAM-Based Analog Computing-In-Memory 1024
Zhuoya Chen, Zongwei Wang, Haisu Zhang, Linbo Shan, Qishen Wang, Xiyuan Tang, Yunyi Fu, Yimao Cai, Ru Huang

S-Band Internally Matched GaN Power Amplifiers 1027
Li Zhang, Xuefeng Zheng, Zheng Chen, Changcheng Zhang, Zhida Wu, Zhipeng Ren, Pengbo Du, Hanbin Qu

The Impact of DC Stress on the Recoverable Tetragonal-To-Orthorhombic Phase Transition in Hafnium Zirconium Oxide Capacitors 1030
Zhenyu Chen, Dan Lv, Zhiyu Lin, Dongdong Li, Mengwei Si

A Particle Swarm Optimization Algorithm Based Parameters Extraction Technique to Model Organic Light-Emitting Diode for Flexible Displays 1033
Jianhui Wanghe, Jiachen Kang, Wei Tang, Gufeng He

Modeling Dynamics-Rich Devices with the Dynamic Time Evolution Method (Invited) 1036
Yu Li, Ning Feng, Runsheng Wang, Ru Huang, Lining Zhang

Heterogeneous and Monolithic 3D (HM3D) Integration of III-V and CMOS for Next-Generation Wireless Communications 1039
Jaeyong Jeong, Yoon-Je Suh, Nahyun Rheem, Chan Jik Lee, Seong Kwang Kim, Bong Ho Kim, Joon Pyo Kim, Joonsup Shim, Minsik Park, Jeong-Taek Lim, Minkyoung Seong, Jooseok Lee, Kihyun Kim, Dae-Myeong Geum, Jongmin Kim, Woo-Suk Sul, Won-Chul Lee, Choul-Young Kim, Jongwon Lee, Sanghyeon Kim

First Demonstration of 4-Inch $\mathbf{G a N}$ on $\mathbf{S I O}_{\mathbf{2}}$ / $\mathbf{S I}(\mathbf{1 0 0})$ Monolithic Integration Materials by Ion-Cutting Technique with Hydrophilic Wafer Bonding at Elevated Temperature 1042
Jiaxin Ding, Jialiang Sun, Tiangui You, Xin Ou

Novel Hybrid Gate Ferroelectric Transistor-Based Weight Device with High Linearity and Symmetry for On-Chip Learning 1045
Yuxin Lin, Jin Luo, Zerui Chen, Zhiyuan Fu, Qianqian Huang, Ru Huang

Oxygen Vacancy-Zr Content Synergy for Morphotropic Phase Boundary Towards High-Performance DRAM Applications 1048
Jinhao Liu, Xuepei Wang, Maokun Wu, Boyao Cui, Yichen Wen, Yishan Wu, Sheng Ye, Pengpeng Ren, Runsheng Wang, Zhigang Ji, Ru Huang

Recent Advances in Compact Modeling for Advanced Semiconductor Technology 1051
Runsheng Wang, Baokang Peng, Lining Zhang, Ru Huang

Multi-Channel Intelligent Electronic Nose for Rapid Identification of Complex Hazardous Gases 1054
Wenyuan Liu, Jiachuang Wang, Fangyu Zhao, Nan Qin, Tiger H. Tao

Opportunities for Wide and Ultrawide Bandgap Devices with Heterogenous Integration 1057
Vanjari Sai Charan, Aditya K. Bhat, A. Anjali, Zequan Chen, Matthew D. Smith, James Pomeroy, Martin Kuball

Optimization of Short-Channel Top-Gate MoS2 FETs Via a Non-Transfer Fabrication 1059
Haojie Chen, Xinliu He, Jinshu Zhang, Jiahao Wang, Sen Wang, Yuchen Tian, Saifei Gou, Xiangqi Dong, Mingrui Ao, Qicheng Sun, Zhejia Zhang, Yan Hu, Jieya Shang, Yufei Song, Yuxuan Zhu, Wenzhong Bao

A Controlled Metal Doping Method Based on MoS2 Top-Gate Transistor.. 1062
Zhejia Zhang, Jingjie Zhou, Saifei Gou, Yuxuan Zhu, Xiangqi Dong, Mingrui Ao, Qicheng
Sun, Yuchen Tian, Jinshu Zhang, Yan Hu, Xinliu He, Haojie Chen, Yufei Song, Jieya
Shang, Zhengjie Sun, Xiaojun Tan, Wenzhong Bao

A Controllable and CMOS Compatible Doping Process for 2D Integrated Circuits................................ 1065
Yuchen Tian, Yan Hu, Shicheng Zeng, Yuxuan Zhu, Saifei Gou, Xiangqi Dong, Zhejia
Zhang, Jinshu Zhang, Qicheng Sun, Mingrui Ao, Xinliu He, Haojie Chen, Yufei Song,
Jieya Shang, Zihan Xu, Chuming Sheng, Zhengzong Sun, Wenzhong Bao

Research on Oxidizer Engineering of ALD for Industrial Production of ZrO2 Capacitor in Dram 1068
Xinyi Tang, Songming Miao, Yuanbiao Li, Guangwei Xu, Di Lu, Shibing Long

In-Material Multimodal Physical Computing for Multisensory Integration.. 1071
Ming He, Shuo Liu, Junling Liu, Lei Xu, Ru Huang

Biofuel Cell-Inspired Chemical Sensors for Monitoring Glutamate in Mammalian Central Nervous
System 1074
Jinghua Li

Boosted Performance of Atomic-Layer-Deposited Dual-Gate Indium-Gallium-Zinc-Oxide
Transistors 1077
Anyu Tong, Qianlan Hu, Min Zeng, Yuzhe Zhu, Wenjie Zhao, Zhiyu Wang, Yanqing Wu

Compact Modeling of GaN Based RF Switches: Invited Paper 1080
Yogesh Singh Chauhan, Mir Mohammad Shayoub, Ahtisham Pampori, Mohammad Sajid
Nazir

A Fusion of Optical Microscopy and Functional Nanomaterials for Subcellular-Scale
Thermodynamic Control of Muscle Contraction... 1083
Madoka Suzuki

An Energy-Efficient Microwave Magnetic Field Generator for NV Center Quantum
Magnetometers Based on an Array of Four Injection-Locked VCOs.. 1086
Hadi Lotfi, Qing Yang, Tarek Elrifai, Michal Kern, Jens Anders

Theoretical Design of Silicon-Based Nanostructures for Spin Qubits 1089
Yang Liu, Shan Guan, Jun-Wei Luo

Integration of 2D Ultrafast Flash Memory: From Device to Chip 1092
Zhenyuan Cao, Chunsen Liu

Robust OS-FeFETs with Crystallized Anatase-TiO2 Channel-Hafnia Ferroelectric Layer Stack for
Integration of 3D Memory Applications... 1095
J. F. Kang, X. J. Song, C. X. Yu, D. J. Sun, J. J. Zhang, S. Z. Li, X. Y. Liu

2D Novel Antiferroelectric Materials for Neuromorphic Computing 1098
Dongliang Yang, Linfeng Sun

Dual Driven Approaches for General Purposed Brain Inspired Computing...................................... 1101
Luping Shi, Yuqing Cong, Wei Zhang

Precise Transmission Matrix Measurement of a Multimode Fiber and Its Applications 1104
Yuwen Xiong, Zihao Ma, Yi Xu, Yuwen Qin

Performance Evaluation of 6T-Sram in Sub-3 nm Complementary Fet... 1107
Anirban Kar, Mahdi Benkhelifa, Yogesh Singh Chauhan, Hussam Amrouch

Experiment Investigation on La2O 3 Dipole-Last Cap-Less VFB Tuning Technology Based on Nitrogen Atmosphere..................1111
    Yu Wang, Jiaxin Yao, Yanzhao Wei, Qingzhu Zhang, Huaxiang Yin

A Digital Twin for Advanced Manufacturing of Materials..................1114
    Dipayan Sanpui, Anirban Chandra, Sukriti Manna, Henry Chan, Subramanian K. R. S. Sankaranarayanan

Strategies for Reliable Emerging Memories and Their Applications 1117
    Shinhyun Choi

A Hybrid Design Method of Lamb Wave Mode Filter Based on Machine Learning and COM Model 1119
    Lihang Liao, Chen Ma, Zhiyu Wang, Xiangyu Zou, Zhiyuan Wang, Xi He, Feixuan Huang, Qinghua Ren, Fengyuan Yang, Yiming Ma, Jianlin Chen, Nan Wang

Engineered Substrates for 3D RF Front Ends..................1122
    Luis Andia

Calculation Optimization of Double-Free-Layer Magnetic Tunnel Junction 1125
    Zifeng Wang, Lang Zeng

Single Photon Devices Using Layered Materials 1128
    Mayank Chhaperwal, Nithin Abraham, Kausik Majumdar

Impact of Dielectrics on Hysteresis and Bias Stress Stability in Oxide Semiconductor and 2D-Material Field-Effect Transistors..................1131
    Alwin Daus, Sumaiya Wahid, Qu?nh Th? Phùng, Eric Pop

Heterogeneous 3D CFET with Hybrid Channel Configuration..................1134
    Sanghyeon Kim, Seongkwang Kim, Hyeongrak Lim, Jaeyong Jeong, Youngkeun Park, Jaejoong Jeong, Joonpyo Kim, Bongho Kim, Daemyeong Geum, Younghyun Kim, Byung Jin Cho

Benchmarking of the BSIM-BULK for Cryo-CMOS Design 1137
    Wajid Manzoor, Nisha Manzoor, Yawar Hayat Zarkob, Debashish Nandi, Aloke K. Dutta, Yogesh Singh Chauhan

Oxidation of TiN Interface and Improvement of AlN Intercalation of ZrO2 Capacitor in DRAM 1140
    Songming Miao, Xinyi Tang, Yuanbiao Li, Guangwei Xu, Di Lu, Shibing Long

Heterogeneously Integrated Intelligent System for Learning at the Edge 1143
    Wenju Huo, Peng Chen, Peng Lin, Gang Pan

Novel GaN Integrated Photonics for Advanced Optical Communication and Imaging 1146
    Muhammad Hunain Memon, Huabin Yu, Haochen Zhang, Haiding Sun

Leveraging Mature Chip Manufacturing Techniques for Innovative Technology Development..................1149
    Yunlong Li, Kai Xu, Dianyu Qi, Yishu Zhang, Ran Tao, Yongyu Wu, Dawei Gao

Ultra-Low Temperature Solution Processed Organic Thin-Film Transistor for Flexible Integration..................1152
    Zikang Mei, Xiaojun Guo

A Compact 256-Channel CMOS Brain Surface Recording and Stimulation Array with Soft Electrodes 1155
    Lachlan Fraser, Muhammad Saif Ul Islam, Peijun Qin, Nigel Lovell, David Tsai

Hafnia-Based XP-FeRAM: A Novel High-Speed and Low-Power Cross-Point Ferroelectric Memory for Data-Intensive Applications 1158

Qianqian Huang, Shengjie Cao, Zhiyuan Fu, Ru Huang

Decoupling Polarization and Trap Charges by Direct Vmid Measurement for Insights into Dynamic Mechanisms of MFMIS-FeFET 1161

Wenpu Luo, Runteng Zhu, Hanyong Shao, Yuejia Zhou, Ru Huang, Kechao Tang

Multi-Functional Flexible Intelligent Glove for Gesture Recognition and Combustible Detection.................1164

Jiachuang Wang, Fangyu Zhao, Wenyuan Liu, Nan Qin, Tiger H. Tao

Piezoelectric Micromachined Ultrasonic Transducers for Advanced Sensing Applications 1167

Bowen Sheng, Lei Zhao, Jinghan Gan, Yufeng Gao, Jiao Xia, Junhao Wang, Chenyuan Zhang, Yipeng Lu

Epitaxial Growth of Stacking Faults-Free Hexagonal Bilayer MoS2...1170

Cheol-Joo Kim

Exploring the Potential of Hafnium Oxide-Based Ferroelectric Memories for Next-Generation Storage Class Memories...1173

Sourav De

Simulation of Germanium-Tin-Based $\boldsymbol{n}^{+} / \boldsymbol{i}$-Well Dot Single-Photon Avalanche Diode for Fiber-Optic Telecommunication Networks...1176

Harshvardhan Kumar, Advaita Sinha, P. Susthitha Menon

High-Frequency Capacitance Measurement Techniques and Their Applications in Memory Technology Development...1180

Jiang Qian, Dan Lv, Lu Wang, Ruiqi Ma, Shuai Kong, Shanting Zhang, Dongdong Li, Liang Zhao

Characterizing Building Blocks for Optoelectronic Devices Based on Two-Dimensional Materials by Resonant Raman Spectroscopy 1183

Ping-Heng Tan

Nanorods-Based Memristors: Advancing Bio-Inspired System and Neuromorphic Computing 1186

Ji Eun Kim, Suk Yeop Chun, Keunho Soh, Jung Ho Yoon

Novel Three-Dimensional DRAM Cell Architectures with IGZO-Channel and Key Technologies Toward Sub-10nm and Beyond 1189

Wonsok Lee, Daewon Ha, Y. Lee, S. Yoo, M. H. Cho, K. Yoo, S. M. Lee, S. Lee, M. Terai, T. H Lee, J. H. Bae, K. J. Moon, C. Sung, M. Hong, D. G. Cho, K. Lee, S. W. Park, K. Park, B. J. Kuh, P. Yun, S. Hyun, S. J. Ahn, J. Song

Evaluation of Insulator Candidates for Nanoelectronics Based on 2D Materials...1192

Mina Bahrami, Theresia Knobloch, Pedram Khakbaz, Mohammad Rasool Davoudi, Alexander Karl, Seyed Mehdi Sattari-Esfahlan, Dominic Waldhoer, Tibor Grasser

Physics of Operation of GaN Power Devices: Modeling Device and Circuit Effects Using Mit Virtual Source GaNFET (MVSG) Model...1195

Ujwal Radhakrishna, Daiyao Xu, Kaiman Chan, Tim Merkin, Ryan Fang, Johan Alant, Lan Wei

Magnetic Resonance Based Soft Electronic Implant for Wireless Electrotherapy and Thermal Ablation...1198

Sicheng Xing, Wubin Bai

Characterization of Self-Heating Using the AC Conductance Method (Invited Paper) 1207
  A. J. Scholten, R. M. T. Pijper, T. V. Dinh

Reliability in Heterogeneous Integration: A Theoretical View (Invited) 1210
  Zhiping Xu

Fundamentals and Future Challenges of SiC Power Devices 1213
  T. Kimoto, R. Ishikawa, K. Tachiki, X. Chi, K. Mikami, M. Kaneko

How Semiconductor Industry New Challenges Will Foster Engineered Substrates? 1216
  C. Figuet

Advanced Electronic Devices Empowering Opto-Sensors for Imaging and Perception 1219
  Wen Pan, Lai Wang, Jinpu Tang, Zhibiao Hao, Changzheng Sun, Bing Xiong, Jian Wang,
  Yanjun Han, Hongtao Li, Lin Gan, Yi Luo

Neuromorphic Multisensory Numerosity Perception Enhanced by a Tactile Glove 1222
  Hongwei Tan, Syed Ashraf, Sebastian Hannula, Zhong-Peng Lv, Bo Peng, Sebastiaan Van
  Dijken

2D MoS2 Thin-Film Transistors for Large-Area, Flexible Electronics 1225
  Jong-Hyun Ahn

Spintronic Foundation Cells for Scalable Unconventional Computing 1228
  Zhihua Xiao, Qiming Shao

An Efficient Pipeline Programming Scheme Based on 40nm PCM Compute-In-Memory Chip for
CNNs 1231
  Xile Wang, Longhao Yan, Xi Li, Yaoyu Tao, Zhitang Song, Yuchao Yang

Investigation on the Effect of Self-Clocking in MEMS Gyroscope 1234
  Xuewen Liu, Zhiyuan Wang, Hongsheng Li

High Density and High Reliability (H2DR) RRAM for Advanced Memory Technology 1237
  Zongwei Wang, Zimeng Wu, Lin Bao, Qishen Wang, Yuhang Yang, Shengyu Bao, Jingwei
  Sun, Cuimei Wang, Yimao Cai, Ru Huang

4H-SiC Semiconductor for EV and Beyond 1240
  Hongchao Liu

The Atomic Layer Etching Technique with Low Damage for p-GaN/AlGaN/GaN Structure 1243
  Xinyi Tang, Honghao Lu, Chun Fu, Chuying Tang, Fangzhou Du, Qing Wang, Hongyu Yu

Device Considerations for GaN Power Switching Transistors from Application and Reliability
Perspectives (Invited) 1246
  Zhikai Tang, Maik Peter Kaufmann, Sandeep Bahl, Chang Soo Suh, Jungwoo Joh, Dong
  Seup Lee, Tim Merkin, Jeffrey Morroni

HfO2-Based Ferroelectric Field-Effect Transistors for Next-Generation Storage and In-Memory
Computing Applications 1249
  Genquan Han, Chengji Jin, Jiajia Chen, Xiao Yu, Jiuren Zhou, Siying Zheng, Haoji Qian,
  Ran Cheng, Bing Chen, Yan Liu

Silicon Doping in Amorphous Gallium Oxide Films by Plasma-Enhanced Atomic Layer Deposition
for Dielectric and Optoelectronic Applications 1252
  Yongjie He, Jingxuan Wei, Jining Yang, Rongxu Bai, Hao Zhu, Qingqing Sun

On the Scalability of Nanosheet Oxide Semiconductor Transistors 1255
    Masaharu Kobayashi, Kaito Hikake, Xingyu Huang, Sunghun Kim, Kota Sakai, Zhuo Li,
    Tomoko Mizutani, Takuya Saraya, Toshiro Hiramoto, Takanori Takahashi, Mutsunori
    Uenuma, Yukiharu Uraoka

Reconfigurable Magnonic Devices for Spin-Wave Manipulation on the Nanoscale 1258
    Huajun Qin

The IHP OpenPDK Initiative: The Status and Roadmap 1261
    Wladek Grabinski, Mustafa Alchalabi, Sergei Andreev, Luca Benini, Mike Brinson,
    Matthias Bucher, Anton Datsuk, Frank K. Gurkaynak, Norbert Herfurth, Krzysztof Herman,
    Sebastien Martinie, Vinayak Pachkawade, Harald Pretl, Christoph Sandner, Rene Scholz,
    Jan Taro Svejda, Frank Vater, Christian Wittke

Heterogeneous Integration of Compound Semiconductor Materials and Devices by Ion-Cutting
Technique 1264
    Tiangui You, Tian Liang, Jiaxin Ding, Jialiang Sun, Shangyu Yang, Xin Ou

Impact of Charge Trapping at Defects on the Robustness of Electronic Circuits 1267
    Michael Waltl, Bernhard Stampfer, Roberto Orio

Solution-Processed Reduced-Dimensional Cesium Lead Halide Perovskites 1270
    Gaukhar Nigmetova, Zhuldyz Yelzhanova, Hryhorii Parkhomenko, Meruyert Tilegen,
    Askhat N. Jumabekov, Tri T. Pham, Annie Ng

Knowledge Discovery from Microscopic Image Data Using an Explainable AI "Extended Free
Energy Model" 1273
    Masato Kotsugi

Ferroelectric 3D NAND Storage (Invited) 1276
    Prasanna Venkatesan, Asif Khan

2D Materials for Neuromorphic Computing Devices 1279
    Max C. Lemme, Lukas Völkel, Sofia Cruces, Jimin Lee, Yuan Fa

Insulators for Devices Based on 2D Materials 1282
    Tibor Grasser, Dominic Waldhoer, Theresia Knobloch

Heterointegrated Ga2O3-On-SiC RF MOSFETS 1285
    Xinxin Yu, Wenhui Xu, Rui Shen, Bing Qiao, Zhonghui Li, Xin Ou, Jiandong Ye

Benefits of Using High-Resistivity Substrates for RF ICs 1288
    Xunyu Li, Cesar Roda Neve, Ionut Radu, Albert Wang

2D Materials for Future Physical Computing 1291
    Feng Miao

CMOS BEOL-Compatible Three-Dimensional Heterogeneous Integration with Emerging Devices
for Advanced Information Processing Systems 1293
    Heyi Huang, Feixiong Wang, Shuang Liu, Yanqin Li, Huaxiang Yin, Xiaolei Wang, Jun
    Luo

Phase Change Memory: From Technological Challenges to Materials Science 1296
    Ruobing Wang, Yichen Song, Xilin Zhou, Zhitang Song

Advancing Emerging Device and Architecture Innovations in the AI Era 1299
    Ming Liu

Author Index

Energy-Efficient Voltage-Induced Self-Regulated Precessional MRAM with Low Write Error Rate $<10^{-9}$

Stanislav Sin[1] and Saeroonter Oh[2], email:sroonter@skku.edu

[1]Dept. of Electrical and Computer Engineering, Sungkyunkwan University, Korea
[2]Dept. of Semiconductor Convergence Engineering, Sungkyunkwan University, Korea

ABSTRACT

Voltage-controlled precessional MRAM offers a good power-performance-area balance, positioning it as a promising intermediate memory between SRAM, DRAM, and flash. However, its practical implementation is hindered by the need for highly precise timing control of the incoming voltage pulse, making it susceptible to process, temperature, and other variations, compromising its robustness for memory applications. This study proposes voltage-induced self-regulated precessional (VISP) memory to address this issue. By using angular dependence of MTJ's resistance, energy asymmetry is created so that the precession is self-terminated without external control. The robustness of switching is evaluated using Fokker-Planck simulations, and it was found that write-error rate $< 10^{-9}$ can be achieved with a pulse width of ~ 26 nanoseconds, making VISP well-suited for storage-class memory.

Keywords: MRAM, storage-class memory, VCMA

INTRODUCTION

An ever-increasing demand for cheap and low-power memory for IoT devices creates new challenges for the semiconductor industry. Current memory devices have a large performance-capacity gap between cache, main, and storage memories [1]. SRAM is the fastest conventional memory (delay ~ 100 ps), but due to the high cost its capacity is < 1 MB. DRAM (used as main memory) has a capacity of a few tens of GBs with a delay on the order of ten nanoseconds, whereas flash memory has $\sim$ TB capacity but milliseconds order delay, as visualized by the memory hierarchy pyramid in Fig. 1. Emerging non-volatile magnetoresistive memory is well-positioned to fill these empty spaces [1]–[3].

A voltage-controlled precessional (VCP) MRAM has a sub-nanosecond delay, few fJ energy/bit consumption, and a minimal cell size of $6F^2$ [2]. However, it requires a very precise control of voltage pulse width, typically the deviation should be $\Delta t_p < 100$ ps. Due to manufacturing variations, thermal fluctuations of the magnetization vector $\boldsymbol{m}$, and small Δt_p, VCP has a relatively high write-error rate (WER), preventing the practical implementation of such memory. Consequently, a large number of studies is devoted to WER minimization, including

injecting spin-transfer torque [4], utilization of advanced writing circuits [5], high damping constant α [6], and built-in stray field [7]. However, these solutions increase memory complexity. This study presents a novel voltage-induced self-regulated precessional (VISP) memory, which breaks the energy symmetry using resistance-angle asymmetry so that the precession is self-terminated without external control. In previous work [8], WER $< 10^{-9}$ was reported with a mean delay < 10 ns; however, the effect of stack parameters on the minimum delay was not analyzed. In this paper, the WER of VISP memory is studied using Fokker-Planck (FP) simulations, which require significantly less computation time compared to the Monte Carlo (MC) simulations. Furthermore, optimization studies were conducted to minimize the delay and energy/bit while keeping the target WER.

VOLTAGE-INDUCED SELF-REGULATED PRECESSION

The VISP stack composition is shown in Fig. 2(a). It comprises a double-barrier MTJ, where the bottom and top magnetic layers are pinned and aligned antiparallel, while the free magnetic layer with two oxide barriers are sandwiched between them. It is known that charge induced at the magnet/oxide interface is capable of lowering energy barrier due to the voltage-controlled magnetic anisotropy (VCMA) $\Delta E \propto \xi V/t_{ox}$ [9]. Here ξ is the VCMA constant, V is the voltage, and t_{ox} is the oxide thickness. Because of the tunneling magnetoresistance effect, a constant supply current through an terminals T1 and T3 creates energy asymmetry favoring parallel orientation of the free layer (FL) and pinned layer 1 (PL1): $IR_p < IR_{ap} \rightarrow E_p < E_{ap}$, i.e. the FL will be aligned in $+z$-direction. The simplified free energy density plot is given in Fig. 2(b). Similarly, the current through T2 and T3 will switch FL parallel to the PL2, in $-z$-direction. The MTJ1 is used to read the reference state of the cell.

FOKKER-PLANCK SIMULATION FRAMEWORK

WER of the proposed memory is evaluated using FP approach. It models the time evolution of the spatial distribution of the probability function [10]. FP simulations allow WER computation with a precise resolution in a single 2D FDM/FEM simulation. However, only a monodomain model is possible. A

979-8-3315-0417-5/25 $31.00 © 2025 IEEE

monodomain approximation is reasonable for nanoscale magnetic films with uniform structure. The probability density in FP is formulated as:

$$\dot{P}(\boldsymbol{m}) = -\boldsymbol{\nabla} P(\boldsymbol{m}) \cdot \boldsymbol{L}(\boldsymbol{m}) + D\boldsymbol{\nabla}^2 \cdot P(\boldsymbol{m}) \qquad (1)$$

$$\boldsymbol{L}(\boldsymbol{m}) = -\frac{\gamma_0}{1+\alpha^2}\left[\boldsymbol{m} \times \boldsymbol{H} - \alpha \boldsymbol{m} \times (\boldsymbol{m} \times \boldsymbol{H})\right] \qquad (2)$$

$$D = \alpha \gamma_0 k_B T / [(1+\alpha^2)\mu_0^2 V_{fl} M_s] \qquad (3)$$

where $P(\boldsymbol{m})$ is the probability density, $\boldsymbol{L}(\boldsymbol{m})$ is the force due to the effective magnetic field, D is the diffusion constant. $\boldsymbol{H}$ is the effective field, γ_0 is the electron gyromagnetic ratio, $k_B T$ is the thermal potential, V_{fl} is the volume of FL, and M_s is the saturation magnetization. The probability is integrated over the upper hemisphere ($m_z > 0$).

To demonstrate the deterministic switching of VISP memory, zero temperature LLG simulations using parameters given in Table I [8] were performed. A time evolution of z-component of $\boldsymbol{m}$ at different values of current is shown in Fig. 3. As can be seen, a current of 3.0 µA is not sufficient to start a precession; hence, is suitable for the read operation. At 3.2 µA the precession exhibits a self-regulating behavior. It is also true for 3.8 µA, however, m_z shows three zero-crossings before the final relaxation.

For the FP equation, an initial spatial distribution of $P(\boldsymbol{m})$ is required. It was obtained by a stochastic LLG simulation with t_{sim} = 1 µs. An FDM solver in a spherical coordinate system with periodic boundary conditions was applied, a grid spacing was 120 × 120 points. LLG equation was calculated in Cartesian coordinates and transformed to the spherical ones. The comparison of FP and MC simulations shows a close match between them (see Fig. 4). WER dependence on the pulse width and current is shown in Fig. 5.

In this study, the switching delay is defined as the minimal pulse width with the target WER $< 10^{-9}$. At 3.3 µA the delay is $\sim$ 170 ns. Corresponding energy consumption of the VISP device is roughly 894 fJ. Such energy is closer to an STT memory, and to improve the performance, an optimization study was performed. First, the effect of changing α was investigated as shown in Fig. 6. The delay can be reduced to 65 ns at α = 0.15. Note that this point is an optimal value and delay will increase above it. As can be seen in Fig. 7, delay monotonically decreases with TMR rise. An effect of resistance sensitivity to bias is studied by changing the half-resistance voltage V_h (see Fig. 8). The V_h is a fitting parameter and it reflects how fast the resistance lowers at high bias voltage. As can be seen, the delay abruptly decreases at $V_h > 0.5$ V, and at V_h = 1.25 V it was 48 ns. Although changing resistive properties of the device is promising, it requires complex layer optimization, which somewhat limits this approach. An effect of the aspect ratio on delay is given in Fig. 9. The optimum

value of AR was 3.0 with the delay of 45 ns, which corresponds to 195 fJ/bit energy consumption. Fig. 10 shows that delay and energy can be further improved to $\sim$ 26 ns and $\sim$ 106 fJ for a device with α = 0.15, TMR = 300%, V_h = 1.0 V, and AR = 3.0. Moreover, higher ξ will also improves performance, since the required voltage (and hence the current) will be lower, and the effect of bias dependence could be neglected.

Benchmarking of the proposed memory to the STT and VCP MRAM is shown in Table II. Since STT cell is switched by a current, oxide thickness and resistance-area product were reduced to 1.1 nm and 32.98 $\Omega \cdot \mu m^2$. For VCP, the external field of 20 kA/m was generated using current-carrying wire. Although VCP memory is much faster and efficient, it requires an external field, pre-read pulse, and precise pulse timing. The latter issue leads to a high WER, which limits the practical application of VCP. VISP achieves the target WER, it has almost the same delay as STT, but energy consumption is one order of magnitude lower. Therefore, VISP MRAM has a good balance between energy-efficient VCP and reliable STT memory.

CONCLUSION

In conclusion, the WER of VISP device was studied using FP simulations. It was found that the proposed VISP memory cell is capable of meeting the stringent requirements of WER $< 10^{-9}$, for practical memory applications. The device with basic parameters demands a pulse width $\geq$ 180 ns, but changing the AR, TMR, V_h, and α improves delay to 26 ns. The mean power consumption/bit was $\sim$ 106 fJ. The proposed VISP MRAM holds considerable promise as a future storage-class memory solution, offering low WER, fast switching time, and low energy consumption.

ACKNOWLEDGMENT

This work was supported in part by the National Research Foundation of Korea (NRF) funded by the Ministry of Science and ICT (MSIT) of the Korea Government under grants 2022M3F3A2A01073562 and 2022M3F3A2A01072215.

REFERENCES

[1] S. Yu, "Semiconductor Memory Devices and Circuits" CRC Press, ISBN: 1000567613 (2022).
[2] V. P. K. Miriyala et al., IEEE Trans. Electron Devices, 66, pp. 944-949 (2019).
[3] S. Sin and S. Oh, Sci. Rep., vol. 12, no. 1, p. 19762, 2022.
[4] S. Kanai et al., Appl. Phys. Lett., 104, p. 212406 (2014).
[5] S. Wang et al., IEEE Trans. Emerg. Top. Comput., 9, pp. 402-413 (2021).
[6] R. Matsumoto et al., Phys. Rev. Appl., 18, p. 054069 (2022).
[7] Y. C. Wu et al., IEEE Symp. VLSI Technol., (2020).
[8] S. Sin and S. Oh, Sci. Rep., 13, p. 16084 (2023).
[9] C. Grezes et al., Appl. Phys. Lett., 108, p. 012403 (2016).
[10] Y. Xie et al., IEEE Trans. Electron Devices, 64, pp. 319-324 (2017).

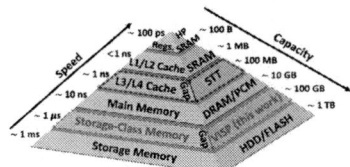

Fig. 1: Memory hierarchy pyramid sorted by speed and capacity.

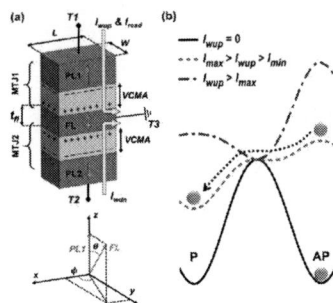

Fig. 2: (a) VISP stack composition and reference coordinate frame. (b) Simplified free energy plot at current through T1 and T3.

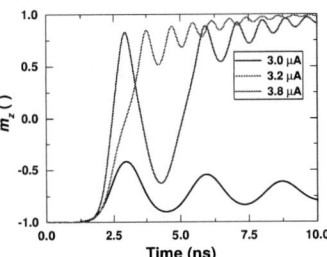

Fig. 3: Time evolution of m_z at different currents and $T = 0$ K.

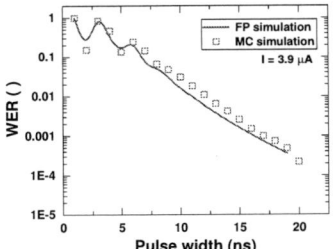

Fig. 4: WER calculated using FP and MC simulations.

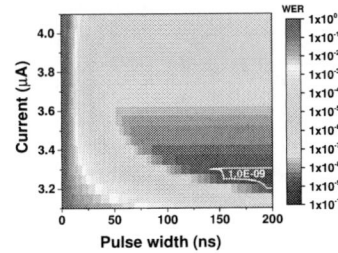

Fig. 5: Dependence of WER on pulse width and current magnitude.

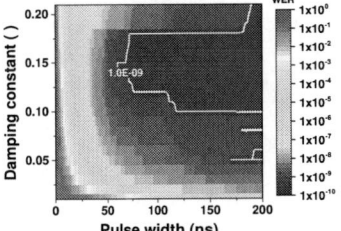

Fig. 6: WER as a function of pulse width and damping constant.

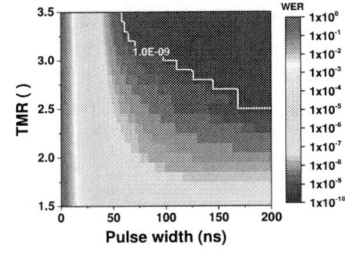

Fig. 7: WER as a function of pulse width and tunneling magnetoresistance.

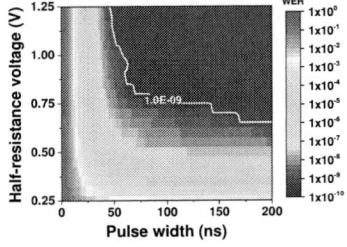

Fig. 8: WER as a function of pulse width and half-resistance voltage.

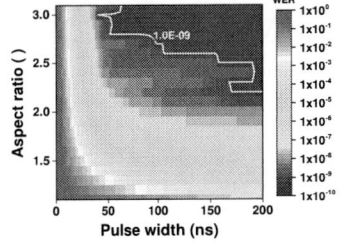

Fig. 9: WER as a function of pulse width and aspect ratio.

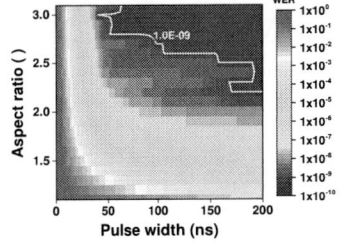

Fig. 10: WER for the optimized device with $\alpha = 0.15$, $TMR = 300\%$, $V_h = 1$ V, and $AR = 3$.

TABLE I: Simulation Parameters

Parameter	Value
M_s	6.25×10^6
$K_i (\Delta = 60)$	0.68 mJ/m^2
ξ	100 fJ/V·m
α	0.05
$L \times W$	50×110 nm^2
t_{fl}	3.0 nm
t_{ox}	1.65 nm
RA_p	758.7 $\Omega \cdot \mu$m^2
TMR	250%
V_h	0.65 V

TABLE II: MRAM Benchmarking

Figure of merit	STT	VCP	VISP (this work)
Delay (ns)	20.2	0.68	26
Energy (fJ)	5996.4	36.8	106
External field (kA/m)	No	20	No
Pre-read	No	Yes	No
Precise timing	No	Yes	No
WER	10^{-9}	10^{-6}	10^{-9}

A 2-D Noise Model for CMOS Single Photon Avalanche Diodes

Xuanyu Qian[1], Wei Jiang[1], and M. Jamal Deen[2,3]

[1]Microelectronics Thrust, Hong Kong University of Science and Technology (Guangzhou), China
[2]Department of Electrical and Computer Engineering, McMaster University, Canada
[3]Department of Biomedical Engineering, McMaster University, Canada

ABSTRACT

Recent advances have enabled single-photon avalanche diodes (SPADs) to be fully integrated with CMOS signal conditioning and processing circuits, paving the way for compact and efficient imaging applications. However, a significant challenge remains: the absence of accurate models for optimizing the SPAD design prior to fabrication. In this paper, we present a comprehensive noise model and compare simulation results with experimental measurements. Our findings highlight the effectiveness and the necessity of robust SPAD modeling for design optimization as a critical step before fabrication.

(Keywords: CMOS, SPAD, Models, and Simulation)

INTRODUCTION

Photon detection with single-photon capabilities holds great promise for enhancing fast imaging with high accuracy and efficiency. CMOS-based single-photon avalanche diodes (SPADs) in standard silicon technology benefit from high manufacturing volumes and competitive costs, making them suitable for integration into various imaging systems, including fluorescence lifetime imaging, positron emission tomography, Raman spectroscopy. While the integration of SPADs with circuits offers advantages in compact size and low power consumption over discrete optical sensing systems, accurately simulating their performance remains challenging due to the lack of CMOS library models.

To overcome this challenge, it is essential to conduct modeling of SPADs prior to the design of on-chip system. Improved modeling within standard CMOS technologies facilitates the optimization and design of high-performance imaging systems.

In this paper, we propose a modeling and simulation process aimed at optimizing the noise characteristics of SPADs, considering the influence of edge effects across different dimensions. While recent studies have laid the groundwork for modeling SPADs in CMOS technologies [1], this work specifically targets the prediction of key noise parameters. SPADs of varying sizes have been fabricated using standard 65 nm CMOS technology, and the experimental results are compared to statistical data derived from simulations of a 2D SPAD model, focusing particularly on the dark count rate (DCR).

MODELING PROCESS AND SIMULATION

A typical SPAD structure using TSMC's 65 nm standard CMOS technology was built using technology-computer-aided design (TCAD) tools, as illustrated in Fig. 1. The SPADs are reverse-biased above the breakdown voltage, with the active region encompassing the entire depletion region, where the electric field is significantly stronger than the field in the neutral region. To investigate the effects of dimensions on performance, the SPADs' diameters were varied from 6 μm to 20 μm.

Previously developed 1-D SPAD models effectively predict photon detection probability (PDP) and dark count rate (DCR) when edge effects are minimal and detection predominantly occurs in the central, uniform region. However, as the SPAD sizes decrease, the contribution from the edge region becomes more pronounced, leading to diminished accuracy in DCR predictions using 1-D models. To enhance simulation precision, this paper introduces a 2-D model.

The 2-D model is built upon the basic 1-D framework, but incorporates the variations in electric field direction, particularly near the edge [2]. To model the DCR of SPADs accurately, several key physical parameters, as detailed in Table I, are defined. While existing models can simulate dark carrier generation, no physical model currently describes the triggering probability - the likelihood of a carrier initiating an avalanche event. Therefore, we employ a statistical model for triggering, adapted to a 2-D framework, to improve the overall accuracy of our predictions.

Dark Count Rate (DCR) Model

The dark count rate (DCR) is a critical parameter that describes avalanches initiated by dark carriers and significantly influences the system's signal-to-noise ratio (SNR) [3], [4]. Therefore, accurately characterizing the DCR of a SPAD is essential. To model the DCR, we first examine the generation of carriers, which occurs through four primary mechanisms: direct thermal generation, trap-assisted tunneling, band-to-band tunneling, and Shockley-Read-Hall (SRH) recombination [5], [6]. In our simulations, we created 25 streamlines that follow the direction of the local electric field across half of the SPAD's structure, as illustrated in Fig. 2. The triggering probabilities for electrons and holes, denoted as P_n and P_p respectively, are calculated according

979-8-3315-0417-5/25 $31.00 © 2025 IEEE

to Eq. 1 and Eq. 2, using the respective ionization coefficients α_n and α_p.

The simulated triggering probability is then integrated with the carrier generation rates along each line to derive the DCR. Notably, the negligible probability observed along Line 25 indicates a significant reduction in electric field strength due to edge effects. The DCR values obtained for each line are averaged to provide an overall estimate of the SPAD's DCR level. This averaged value helps determine whether the 2-D DCR model is necessary for accurately assessing the SPAD's performance.

$$dP_n/dx = [1 - P_n] \times \alpha_n \times [P_n + P_p - P_n \times P_p] \quad (1)$$

$$dP_p/dx = [P_p - 1] \times \alpha_p \times [P_n + P_p - P_n \times P_p] \quad (2)$$

SPAD PIXEL AND MEASUREMENTS

To validate the 2-D noise model of the SPAD, five SPADs are designed and fabricated in standard 65 nm CMOS technology. We implemented a straightforward passive quench and reset circuit, as depicted in Fig. 3. A large resistor (50 $k\Omega$) is utilized to minimize the voltage drop across the SPAD after a significant avalanche current is triggered. When the bias voltage falls below the avalanche breakdown threshold, the avalanche event is quenched, allowing the voltage source to gradually recharge the SPADs' cathodes to its bias voltage in preparation for the next detection. In addition, a multistage buffer is employed to read out the output signal.

The layout of the five SPADs is illustrated in Fig. 3. The active region of each SPAD is designed in an octagonal shape to comply with technology design guidelines. The diameters of these five SPADs are 6 μm, 10 μm, 12 μm, 16 μm, and 20 μm.

A. Measurement Setup

The fabricated SPADs and associated circuits are packaged and bonded on a PCB board for testing. Before measuring the SPADs' DCR, the I-V curves of the SPADs are recorded and compared with the simulation results (see Fig. 4). The agreement between the simulated and measured I-V curves indicates that the model parameters used are reasonable. The complete DCR measurement setup is shown in Fig. 5. Two voltage sources are employed to bias the pins and SPADs. while the output waveform is captured using a LeCroy 625Zi oscilloscope. Based on the output pulses, the inter-arrival-time (IAT) is measured to calculate the DCR. All DCR measurements are conducted at a constant room temperature to eliminate any potential influence of temperature fluctuations on the results.

B. Characterization

The DCR was measured at various biasing voltages. A DCR measurement for a SPAD at an excess voltage of 0.7 V is presented in Fig. 6. Given that carriers are generated from random noise, the measured data should conform to a Poisson distribution [7]. By applying an exponential fit to the histogram of the IAT with a single time constant τ_1, the DCR was determined to be 13.386 kHz, which is consistent with our previous findings in the standard technology node [8]–[10]. As the triggering probability increases with higher biasing voltages, Fig. 7 compares the simulated DCR with the measured DCR across different bias levels, clearly demonstrating a strong correlation between the two.

CONCLUSION

In this paper, we proposed a modeling and simulation process of CMOS-based SPADs, and validated it with experimental measurements. The good agreement between the simulation results and the measured dark count rate (DCR) from the fabricated SPADs underscores the necessity and effectiveness of the 2-D model. The insights gained from these simulations enhance efficiency of optimizing SPAD integration with circuits. Future developments of the model could focus on achieving greater accuracy by incorporating precise process-related parameters. Additionally, other metrics, such as photon detection probability (PDP), could also benefit from the advancements in the 2-D modeling approach.

REFERENCES

[1] Y. Liu, R. Fan, Y. Zhao, J. Hu, R. Ma, and Z. Zhu, "Junction-Optimized SPAD With 50.6 % Peak PDP and 0.64 cps/μm^2 DCR at 2 V Excess Bias Voltage in 130 nm CMOS", IEEE Electron Device Letters, 45, pp. 308-311 (2024).

[2] C.-H. Liu, C.-A. Hsien, and S.-D. Lin, "2-D Photon-Detection-Probability Simulation and a Novel Guard-ring Design for Small CMOS Single-Photon Avalanche Diodes", IEEE Trans. Electron Devices, 69, pp. 2873-2878 (2021).

[3] W. Sun, Y. Wang, M. Liu, and Y. Yang, "A Back-Illuminated 4 μm P+/N-Well Single Photon Avalanche Diode Pixel Array With 0.36 Hz/μm^2 Dark Count Rate at 2.5 V Excess Bias Voltage", IEEE Electron Device Letters, 43, pp. 1519-1522 (2022).

[4] X. Qian, W. Jiang, and M. J. Deen, "Single Photon Detectors for Automotive LiDAR Applications: State-of-the-Art and Research Challenges", IEEE Journal of Selected Topics in Quantum Electronics, 30, pp. 1-20 (2024).

[5] A. Panglosse, P. Martin-Gonthier, O. Marcelot, C. Virmontois, O. Saint-Pé, and P. Magnan, "Dark Count Rate Modeling in Single-Photon Avalanche Diodes", IEEE Transactions on Circuits and Systems I: Regular Papers, 67, pp. 1507-1515 (2020).

[6] D. P. Palubiak and M. J. Deen, "CMOS SPADs: Design Issues and Research Challenges for Detectors, Circuits, and Arrays," IEEE Journal of Selected Topics in Quantum Electronics, 20, pp. 409–426 (2014)

[7] O. Marinov, M. J. Deen, and J. A. Jiménez-Tejada, "Low-frequency noise in downscaled silicon transistors: Trends, theory and practice," Physics Reports, 990, pp. 1–179 (2022).

[8] W. Jiang, R. Scott, and M. J. Deen, "High-Speed Active Quench and Reset Circuit for SPAD in a Standard 65 nm CMOS Technology", IEEE Photonics Technology Letters, 33, pp. 1431-1434 (2021)

[9] W. Jiang and M. J. Deen, "Random Telegraph Signal in n+/p-Well CMOS Single-Photon Avalanche Diodes", IEEE Trans. Electron Devices, 68, pp. 2764-2769 (2021)

[10] N. Faramarzpour, M. J. Deen, S. Shirani, and Q. Fang, "Fully Integrated Single Photon Avalanche Diode Detector in Standard CMOS 0.18- μm Technology," IEEE Transactions on Electron Devices, 55, pp. 760–767 (2008)

Table 1: Simulation Parameters of SPADs in TCAD.

Parameters	Value	Unit
N-well Doping	2.21×10^{17}	cm^{-3}
P+ Doping	1×10^{18}	cm^{-3}
Temperature	300	K
Ionization Model	Okuto	$N.A$
SRH Model	Default	$N.A$
BTBT Model	Schenk	$N.A$
Breakdown Limit	0.15	A
Electron Saturation Velocity	1.07×10^7	cm/s
Hole Saturation Velocity	8.37×10^6	cm/s
Voltage at Cathode	1-15	V

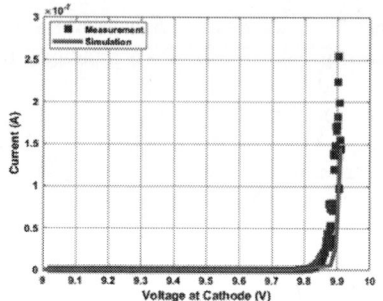

Fig. 4. Simulated (solid) and measured (squares) I-V of SPADs.

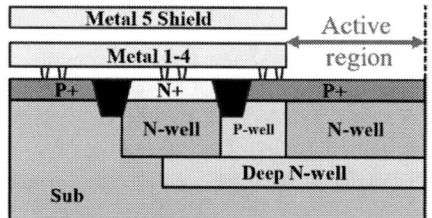

Fig. 1. Cross-sectional illustration of SPAD's half structure.

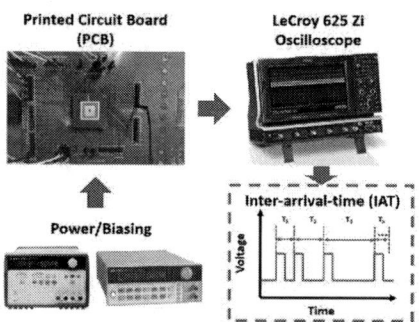

Fig. 5. Dark count rate (DCR) measurement setup.

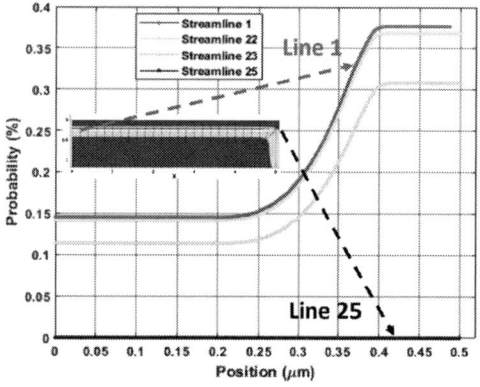

Fig. 2. Multiple streamlines created and the simulated triggering probability of each line.

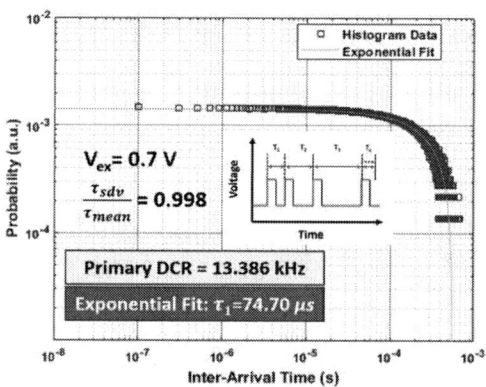

Fig. 6. Histogram of measured IAT and exponential fit to obtain DCR

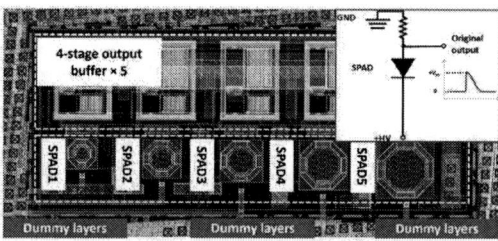

Fig. 3. Schematic (Top right) and layout of SPADs and output buffers.

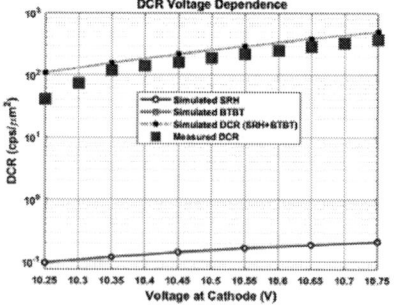

Fig. 7. Comparison of DCR between simulation and measurements.

979-8-3315-0417-5/25 $31.00 © 2025 IEEE

InP/GaAsSb DHBT Emitter Etching Process Optimization with a Simultaneous f_T/f_{MAX} = 451/914 GHz and 86% Device Yield

M. Ebrahimi[1], S. Hamzeloui[1], F. Ciabattini[1], A. M. Arabhavi[1], O. Ostinelli[1], and C. R. Bolognesi[*]

[1]Millimeter-Wave Electronics Group, ETH-Zurich, Switzerland, [*]*author email*: colombo@ieee.org

ABSTRACT

The present study contrasts three emitter etching process variants in InP/GaAsSb DHBT fabrication. Specifically, dry etching of the emitter mesa was studied for different inductively coupled plasma (ICP) powers with/without a prior Ar-sputtering electrode edge smoothing procedure facilitated in an electron-beam evaporator. Emitter etching significantly affects device yield and the scaling behavior of RF performance with emitter size. Excellent performance metrics are achieved with maximum cut-off frequencies f_T/f_{MAX} = 451/914 GHz and an 86% yield for a 300 W ICP power at 30 mTorr and 170°C (without Ar-pre-sputtering). The work highlights the importance of emitter etching on the scaling of performance metrics in the manufacturing of Type-II DHBTs.

Keywords: Double heterojunction bipolar transistors (DHBTs), InP/GaAsSb, Yield

INTRODUCTION

Transistor technologies with high cut-off frequencies and output power levels are necessary to exploit millimeter-wave frequencies. Type-II InP/GaAsSb Double Heterojunction Bipolar Transistors (DHBTs) show excellent properties in this frequency range due to their high current drive and, high current gain and maximum oscillation cut-off frequencies f_T/f_{MAX} [1], [2]. Recent reports include InP/GaAsSb DHBTs with f_T/f_{MAX} values up to 0.475/1.2 THz and high power-added efficiencies at 94 GHz [3]. It has been observed that f_{MAX} depends inversely on the square root of the effective time constant $(R_B C_{BC})_{eff}$ [4]. The reduction of capacitances and resistances is crucial to achieve a high for a given base-collector layer design. In this regard, emitter etching is a critical step in self-aligned DHBT manufacturing, because it affects the separation between the bottom of the emitter mesa and the base metal, i.e. the base access distance (and thus, the total base resistance).

Conventionally, InP chlorine-based ICP/RIE etching is often performed at a low chamber pressure of 4 mTorr [5], [6]. However, the low-pressure process can be problematic due to pressure gauge fluctuations in the chamber, which in turn lead to non-uniform emitter profiles and low device yields (Fig. 1). Non-uniformities in the emitter mesa profile lead to higher base access distance/resistance and consequently non-optimal RF performance metrics. Experiments were conducted on dummy emitter structures to optimize the emitter etching process by

- Sweeping ICP power;
- Changing the chamber pressure;
- Adjusting the ratio and flow of etchant gases (Cl_2 and N_2).

An ICP power lower than 300 W leads to a non-uniform emitter profile. On the other hand, increasing the ICP power to 400 W results in a rough surface morphology. However, when the ICP power is further increased to 500 W, the surface recovers to become smoother again. The chamber pressure was varied from 4 to 30 mTorr, and ultimately, 30 mTorr was chosen to maintain the gauge variation during etching below 10%. A higher Cl_2/N_2 ratio causes the surface to become extremely rough, while a ratio lower than 10% results in a non-uniform emitter profile. Furthermore, the RF power was varied to maintain a consistent etching rate. Based on these preliminary tests, two different ICP etching processes showing acceptable profile uniformity and base layer surface roughness were selected for device fabrication runs. Additionally, as discussed in [7], Ar-sputtering in an electron-beam evaporator prior to ICP etching to reduce edge roughness was also examined in this study. To summarize, three different emitter etching processes are contrasted below:

- Process A) 15 W RF power, 500 W ICP power without Ar-sputtering;
- Process B) 20 W RF power, 300 W ICP power without Ar-sputtering;
- Process C) 20 W RF power, 300 W ICP power with Ar-sputtering.

These processes were carried out at a pressure of 30 mTorr, a chamber temperature of 170°C, with a 10% $Cl_2/(Cl_2+N_2)$ ratio and a 50 sccm N_2+Cl_2 flow. This study monitored the emitter morphology, the DC and RF device performances, the process yield, and the device scalability resulting from different etching process conditions. The second emitter etching process (B) yielded the best results, with the highest f_T/f_{MAX} = 451/914 GHz for a (0.15×5) μm^2 emitter contact area. Out of 195 devices, 168 were functional, resulting in 86% yield. We believe that the high yield was achieved by protecting the base layer from ion bombardment (no Ar-pre-sputtering) and strong plasma. Furthermore, this process exhibited a superior base ideality factor among all the processes, with an average of 1.37 ± 0.09 for 121 devices that had $\beta > 20$ and $f_{MAX} > f_T$.

DEVICE FABRICATION

All the emitter dry etching processes compared in this work were performed on the same epilayers. The InP/GaAsSb

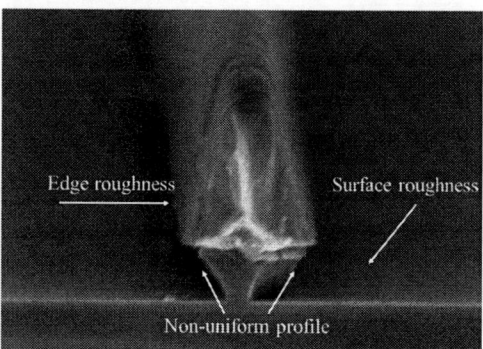

Fig. 1: SEM image of a dummy emitter structure with 4mTorr of ICP chamber pressure etching process.

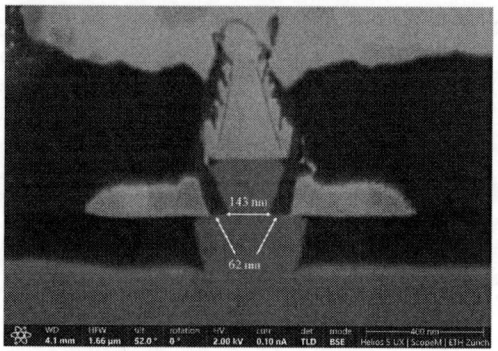

Fig. 2: FIB/SEM image of the completed device cross-section with an emitter width of 150 nm.

DHBT layers were grown at ETH Zurich by metal-organic chemical vapor deposition (MOCVD) on 2-inch semi-insulating InP substrates [2].

DHBTs were fabricated in a self-aligned triple mesa process with 0.15 to 0.3 emitter contacts and self-aligned 0.4 μm wide base electrodes defined by electron beam lithography (EBL). Ti/Pt/Au and Pt/Ta/Pt/Au metal stacks were electron-beam evaporated for the emitter and base electrodes, respectively. After emitter metal deposition, the emitter mesa is formed using a combination of dry and wet etching. Dry etching of the emitter mesa was performed using three different processes discussed in the previous section. All dry etching recipes were carried out at 170°C and 30 mTorr ICP chamber pressure. After dry etching of InGaAs/InP, the remaining InP was etched in an H_3PO_4/HCl-based solution to shape the emitter mesa. The emitter sidewalls and extrinsic base surface were later passivated by Al_2O_3 atomic layer deposition at 150°C. The Al_2O_3 is etched by ICP-RIE using the emitter metal as a shadow mask, allowing Al_2O_3 to remain in the base-emitter access region. The base layer is etched with a combination of dry (ICP-RIE) and wet etching. Two cycles of GaInAs/InP wet etching were performed to isolate devices from each other (mesa isolation), followed by a BCB-based etch-back planarization process. Finally, the probe pad metal was deposited for DC and RF characterization.

RESULTS AND DISCUSSION

The focused ion beam/scanning electron microscope (FIB/SEM) cross-section of a completed device with an emitter width of 150 nm is shown in Fig. 2. White arrows denote the precise measurements of the emitter width and the base access distance of the device. The measured base access distance in process B is 60 nm, while processes A and C have base access distances of 75 and 65 nm, respectively. The higher base access with processes A and C accounts for the poorer RF performance compared to process B [7]. Furthermore, process B shows a lower process bias and results in an emitter width that is closer to the targeted width compared to other processes. Moreover, out of 188, 195, and 179 transistors measured from

processes A, B, and C, 157, 168, and 97 transistors were found to be working and having the minmum gain of 20.

Fig. 3a) shows the typical Gummel and I_C-V_{CE} characteristics of (0.15 × 5) μm^2 InP/GaAsSb DHBTs for process B. The base and collector ideality factors are $\eta_B = 1.36$ and $\eta_C = 1.02$, respectively. With this geometry, the peak common-emitter gain is $\beta = 26$, whereas large-area emitter devices exhibit a maximum $\beta = 32$. According to Table I, all three chips have a similar peak DC current gain β of $\sim$30. However, a noticeable difference in the base ideality factor is observed. The average base ideality factor for processes A, B, and C are 1.43, 1.37, and 1.55, respectively.

The small-signal RF performance was measured using a PNA-X vector network analyzer from 0.2 to 50 GHz with an input power of -30 dBm. To set the reference planes to the probe tips, an off-wafer impedance standard substrate and line-reflect-reflect-match (LRRM) calibration were employed. The probe pad parasitics were de-embedded using on-chip OPEN and SHORT structures with the same layouts as the device pads. Mason's unilateral power gain U and the common-emitter short-circuit current gain $|h_{21}|^2$ were extracted from the measured S-parameters and are depicted in Fig. 3 for the same device (process B). Extrapolating from a single-pole fit using the entire frequency dataset, a peak $f_{MAX} = 914$ GHz was obtained with a simultaneous $f_T = 451$ GHz at $J_C = 10.1$ mA/μm^2 and $V_{CE} = 1.0$ V. The collector current dependency of f_T and f_{MAX} for $V_{CE} = 1.0$ V is shown in Fig. 3 b) inset for process B.

Table I summarizes the RF and DC performances for all the devices of various sizes that were measured and met the criteria of $\beta > 20$ and $f_{MAX} > f_T$. Although the f_T value for processes A and B is nearly identical, process C has the lowest value among the three processes which can be attributed to its higher collector to emitter area ratio (A_C/A_E) [5]. The average f_{MAX} of these devices are 652 ± 77, 750 ± 71, and 719 ± 61 GHz for processes A, B, and C, respectively.

Etching process B shows the highest device yield, and the best DC and RF performances among the processes considered here. Fig. 4 presents the average extracted f_{MAX} values as a function of emitter widths (W_E) for processes A, B, and C.

979-8-3315-0417-5/25 $31.00 © 2025 IEEE

(a)

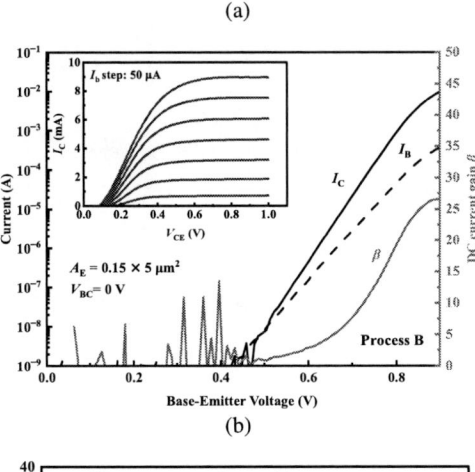

(b)

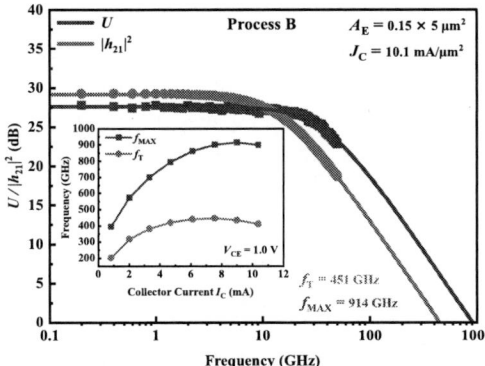

Fig. 3: a) Gummel characteristics of a $0.15 \times 5~\mu m^2$ DHBT from process B. The DC gain as a function of V_{BE} is also plotted. Inset: Corresponding I-V curves for I_B ranging from 0 to 400 μA. b) Measured small-signal current gain $|h_{21}|^2$ and Mason's unilateral gain U. Inset: Bias dependence of f_T/f_{MAX} versus current.

TABLE I: DC and RF performance comparison

Process	$f_{T, AVG}$ (GHz)	$f_{MAX, AVG}$ (GHz)	β_{AVG}	$\eta_{B, AVG}$
Process A[a]	475 ± 15	652 ± 77	28.4 ± 3.5	1.43 ± 0.06
Process B[b]	474 ± 12	750 ± 71	30.3 ± 2.6	1.37 ± 0.09
Process C[c]	458 ± 26	719 ± 61	29.6 ± 2.9	1.55 ± 0.09

[a]85, [b]121, and [c]82 devices measured

Process B results in near optimal scaling of f_{MAX} *vs.* W_E for emitter lengths L_E ranging from 3.5 to 10 μm according to the InP DHBT scaling analysis of Yu et al. [8]. Process B can therefore be applied to the fabrication of various emitter widths with little or no adjustment.

Conclusion

We demonstrated an optimized ICP emitter etching process for Type-II InP/GaAsSb DHBTs, resulting in a simultaneous $f_T/f_{MAX} = 451/914$ GHz with an 86% yield. We observed that increasing the ICP power adversely affects the RF performance, while it has no significant impact on the DC behavior and the yield. Additionally, we found that Ar-sputtering prior

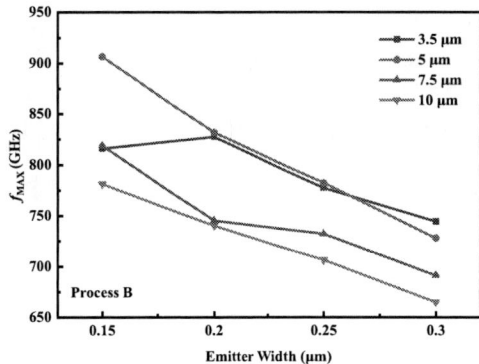

Fig. 4: f_{MAX} scaling trend of process B.

to the dry etching of the emitter in the ICP leads to reduced yield and base damage. The processes involving higher ICP power and Ar-sputtering were found to be detrimental to the base layer and DHBT scalability. Conversely, our newly developed emitter etching technique exhibits the most favorable scaling trend for DHBTs.

Acknowledgment

The authors would like to thank the staff of FIRST Lab at ETH Zurich, Zurich, Switzerland, for their support.

References

[1] S. Hamzeloui et al., "High Power InP/Ga(In)AsSb DHBTs for Millimeter-Wave PAs: 14.5 dBm Output Power and 10.4 mw/μm^2 Power Density at 94 GHz," IEEE Journal of Microwaves, vol. 2, no. 4, pp. 660-668, Oct. 2022.

[2] A. M. Arabhavi et al., "Scaling of InP/GaAsSb DHBTs: A simultaneous f_T/f_{MAX} = 463/829 GHz in a 10 μm long emitter," Proc. IEEE BCICTS, Oct. 2018, pp. 132–135.

[3] A. M. Arabhavi et al., "InP/GaAsSb Double Heterojunction Bipolar Transistor Emitter-Fin Technology With f_{MAX} = 1.2 THz," IEEE Transactions on Electron Devices, vol. 69, no. 4, pp. 2122-2129, April 2022.

[4] K. Kurishima, "An analytic expression of f_{MAX} for HBTs," IEEE Trans. Electron Devices, vol. 43, no. 12, pp. 2074–2079, Dec. 1996.

[5] R. Lovblom, "Development of Sub-Millimeter-Wave InP/GaAsSb Double Heterojunction Bipolar Transistors." PhD thesis, ETH Zurich, 2014.

[6] S. Topaloglu, "Process Technology for High-Speed InP Based Heterojunction Bipolar Transistors." PhD thesis, Universitat Duisburg-Essen, 2006.

[7] W. Quan et al., "Comparison of Ion-Milling and Ion-Sputtering to Remove Edge Roughness of EBL Defined Emitter Metallization in InP/GaAsSb DHBTs," Proc. Compound Manufacturing Tech. Conf. (CS MANTECH), May 2017, pp 20.11.

[8] D. Yu et al., "Design guideline for high-speed InP/InGaAs SHBT using a practical scaling law," Solid-State Electronics, vol. 50, no. 5, pp. 733-740, 2006.

A Novel Mixed Collector Structure IGBT Operating at High Temperature

Chang-Hao Wang, Hang Xu, Jian-bin Guo, Xin-Ru Chen, Yafen Yang and

David Wei Zhang

Abstract—This work proposes a novel insulated gate bipolar transistor (IGBT) with a mixed N-region collector (MNC-IGBT). The addition of a thin N-type doped region within the P-type collector serves to effectively reduce the amplification effect of the internal parasitic PNP transistor on the leakage current in the off-state. As a result, MNC-IGBT markedly reduces the leakage current and increases the operational temperature of the device. The simulation results demonstrate that the leakage current of the MNC-IGBT is reduced by 9% and 8%, respectively, at 175°C and 200°C in comparison to conventional CSTBT, while maintaining a higher breakdown voltage even at 215°C. Furthermore, the on-state voltage drop (V_{on}) and switching losses (E_{on}, E_{off})of MNC-IGBT are close to those of the CSTBT. The excellent performance of MNC-IGBT provides a reliable solution for increasing the operating temperature of IGBTs.

Index Terms—Trench gate IGBT, high operating temperature, leakage current, breakdown voltage

I. INTRODUCTION

Insulated gate bipolar transistors (IGBTs) are employed extensively in a multitude of fields[1]. However, there is a compelling need for IGBTs to operate at elevated temperatures. For every 25°C increase in the operating temperature of an IGBT, the rated power of the power conversion system can increase by 20-30% [2-4].

The primary internal factor impeding IGBT operation at ultra-high temperatures is the exponential increasing leakage current that can be attributed to two primary sources, as illustrated in Fig. 1[5]. The first is the generation current in the off-state, and the second is the amplification resulting from the parasitic PNP transistor formed by the base, drift, and collector regions [6].

To effectively reduce the current gain of the PNP transistor of IGBT, numerous studies have been conducted. For example, some researchers have employed the injection of deep-level impurities into the field stop layer (FS layer), leveraging the gradual ionization of these impurities with temperature increases [7-9]. Other scholars concentrate their efforts on identifying the optimal value of a specific process parameter [10-12]. Nevertheless, these methods also result in a reduction of the current gain of the PNP transistor in the on-state, which in turn leads to an increase in both Von and static power consumption.

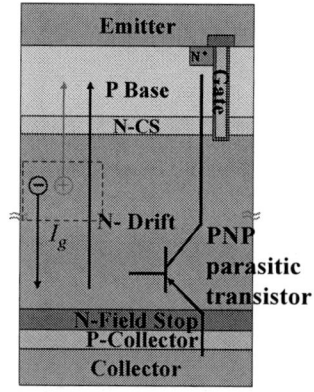

Fig 1. Source of leakage current of IGBT.

This paper proposes a novel IGBT with a mixed N-region collector (MNC-IGBT) and evaluates its electrical performance using Sentaurus TCAD. By leveraging the discrepancy in depletion region widths between the on-state and off-state of the N-type region in the collector, the MNC-IGBT effectively mitigates leakage current at elevated temperatures while sustaining V_{on}, thereby enhancing the maximum operational temperature of the device. The MNC-IGBT represents a reliable solution for the application of IGBTs in high-temperature environments.

II. DEVICE STRUCTURE AND MECHANISM

Fig. 2 provides a schematic illustration of the CSTBT and MNC-IGBT. In contrast to the conventional CSTBT, the MNC-IGBT incorporates an additional N-region within the P-doped collector to form a mixed collector structure. This configuration effectively mitigates the amplification effect of the parasitic PNP transistor on the generated current, thereby reducing the total leakage current in the device. The underlying mechanism is elucidated as follows:

When the CSTBT is in the off-state, the parasitic PNP transistor will amplify the leakage current (as illustrated in Fig. 3(a)). In contrast, in MNC-IGBT, a reverse-biased PN junction (referred to as J_{WELL}) is formed between the N-region and the P-doped collector under the FS layer. In the off-state condition, the depletion region of J_{WELL} penetrates into the similarly N-type FS layer. In this scenario, the amplification effect of the

PNP parasitic transistor on the current is eliminated (as illustrated in Fig. 3(b)).

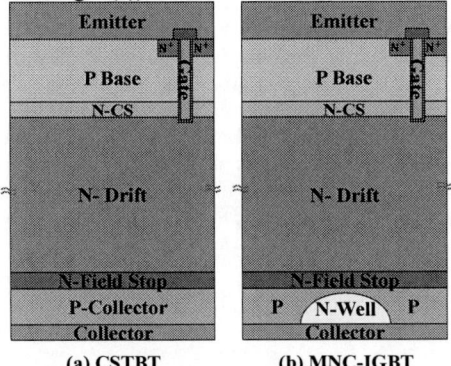

(a) CSTBT　　　　**(b) MNC-IGBT**

Fig 2. Schematic of (a) CSTBT and (b) MNC-IGBT.

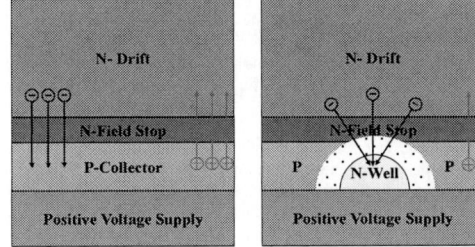

Fig 3. The different amplifications of generation current by parasitic transistors in CSTBT (a) and MNC-IGBT (b).

During the on-state operation, the application of a small positive voltage (i.e., the V_{on}) to the collector results in the depletion region of the N-region being too narrow to penetrate the FS layer. This guarantees that the MNC-IGBT will maintain a low V_{on} while effectively reducing the leakage current at elevated temperatures.

III. RESULT AND DISCUSSION

A Leakage Current Characteristics

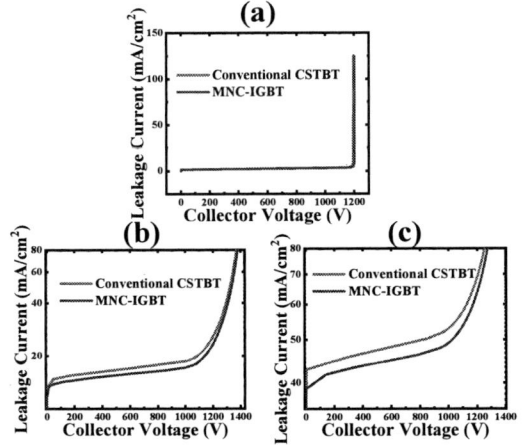

Fig 4. Leakage current density-collector voltage curves of MNC-IGBT and CSTBT at 25℃ (a), 175℃ (b), 200℃ (c).

Fig. 4 depicts the correlation between leakage current density and collector voltage of two devices across varying temperatures. At room temperature, both devices exhibit low leakage currents, and the difference is not readily discernible. As the temperature increases, Fig. 4(b) and (c) reveal a notable distinction, with the MNC-IGBT consistently demonstrating lower leakage currents across the entire collector voltage range.

Table I presents the variation in leakage current density with temperature at a fixed collector voltage (1000 V) for both devices. It is evident that MNC-IGBT displays a lower leakage current than CSTBT across the entire temperature range. At temperatures of 175°C and 200°C, the leakage current of the MNC-IGBT is reduced by 9% and 8% in comparison to the CSTBT. This evidence supports the assertion that the MNC-IGBT exhibits superior high-temperature leakage current characteristics.

TABLE I. LEAKAGE CURRENT DENSITY(MA/CM²) OF MNC-IGBT AND CSTBT AT DIFFERENT TEMPERATURES.

Temperature(°C)	MNC-IGBT	CSTBT
25	2.93674	3.22012
50	2.93649	3.16645
75	3.0666	3.27186
100	3.2423	3.511
125	3.8858	4.22161
150	6.82111	7.57258
175	17.12475	18.71188
200	49.14337	52.99512

The breakdown voltage (BV) temperature curves of MNC-IGBT and CSTBT are illustrated in Fig. 5. At an extreme temperature of 215°C, the BV of MNC-IGBT remains almost identical to its value at room temperature (1200V), indicating its stability. In contrast, the BV of CSTBT degrades by several tens of volts, which is sufficient to render the device inoperable.

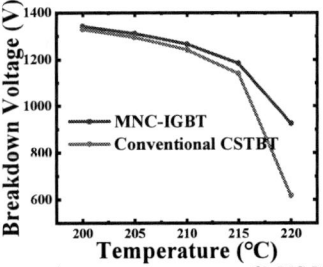

Fig 5. Breakdown voltage-temperature curves of MNC-IGBT and CSTBT.

B Output and Transfer Characteristics

MNC-IGBT exhibits comparable output and transfer characteristics to CSTBT. The measurement results indicate that the Von of MNC-IGBT and CSTBT are 1.65 V and 1.64 V, respectively, at a current density of 200 A/cm². Similarly, threshold voltages (V_{th}) are 4.61 V and 4.56 V for MNC-IGBT and CSTBT, respectively.

In summary, MNC-IGBT retains similar output and transfer characteristics to CSTBT, demonstrating a stability in V_{on} and V_{th}.

C Switching Characteristics

Device switching characteristics simulations are completed. The resulting turn-off curves are shown in Fig. 6. It can be seen that under high-temperature conditions, the collector voltage and current curves of the two devices overlap significantly. Moreover, Table II provides a comprehensive account of the switching loss of the two devices at varying temperatures.

The aforementioned discussions align with our expectations and offer compelling evidence for the stability of MNC-IGBT in practical applications, particularly with regard to its switching characteristics.

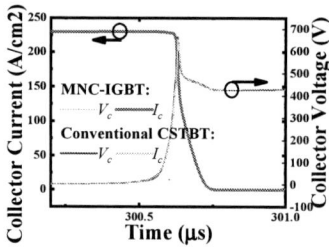

Fig 6. Collector current-voltage curve during the turn-off of MNC-IGBT and CSTBT.

TABLE II. SWITCHING LOSS OF MNC-IGBT AND CSTBT AT DIFFERENT TEMPERATURES.

Parameters	MNC-IGBT	CSTBT
E_{off} (mJ),25°C	16.69	16.69
E_{on} (mJ),25°C	56.07	56.07
E_{off} (mJ),200°C	18.90	18.90
E_{on} (mJ),200°C	41.09	41.09

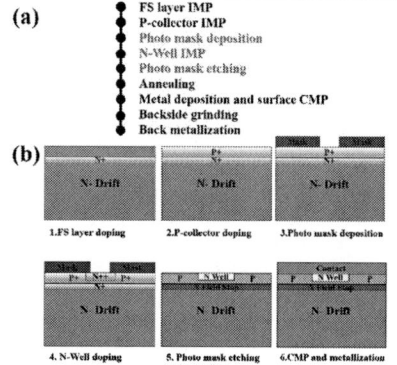

Fig. 7. (a) Device process flow of the proposed MNC-IGBT. The red and highlighted steps are the additional processes compared to the conventional CSTBT baseline. (b) Schematic process flow.

The primary fabrication steps and flowchart for manufacturing the collector-side structure of MNC-IGBT are illustrated in Fig.7. Notably, the process flow is fully compatible with existing CSTBT fabrication techniques.

IV. CONCLUSION

A novel IGBT with a mixed N-region collector (MNC-IGBT) has been proposed and electrically characterized through simulation. By employing the disparate depletion widths of the N-region during the on and off states, MNC-IGBT effectively curtails the leakage current and demonstrates a **9%** and **8%** reduction at 175°C and 200°C compared to CSTBT. Moreover, MNC-IGBT exhibits comparable V_{on} and switching losses to CSTBT, ensuring a stable and comprehensive performance. Finally, the proposed structure is fully compatible with existing CSTBT manufacturing processes, offering a potential solution for IGBT devices operating in ultra-high temperature environments.

V. REFERENCES

[1] Vinod Kumar Khanna, "IGBT Process Design and Fabrication Technology," in Insulated Gate Bipolar Transistor IGBT Theory and Design , IEEE, 2003, pp.411-463, doi: 10.1002/047172291X.ch8.

[2] R. W. Johnson, J. L. Evans, P. Jacobsen, J. R. Thompson and M. Christopher, "The changing automotive environment: high temperature electronics," IEEE Trans. Electronics Packaging Manufacturing, vol. 27, no. 3, pp. 164-176, July 2004.

[3] U. Schlapbach, M. Rahimo, C. von Arx, A. Mukhitdinov and S. Linder, "1200V IGBTs operating at 200°C? An investigation on the potentials and the design constraints," in Proc. 19th International Symposium on Power Semiconductor Devices and IC's (ISPSD), 2007, pp. 9-12.

[4] Th. Schütze, J. Biermann, R. Spanke and M. Pfaffenlehner, "High power IGBT modules with improved mechanical performance and advanced 3.3 kV IGBT3 chip technology," in Proc. PCIM Europe Conf., 2006.

[5] Z. Xu, M. Li, F. Wang and Z. Liang," Investigation of Si IGBT Operation at 200°C for Traction Applications," IEEE Trans. Power Electronics, vol. 28, no. 5, pp. 2604-2615, May 2013.

[6] B. J. Baliga, Fundamentals of Power Semiconductors Devices. US: Springer, 2010, pp. 762.

[7] Q. Zhang, K. Xiao, J. Wang, G. Deng and S. Liang, "A Novel IGBT with Double Buffer Layers for High Temperature Operation," *2022 IEEE 3rd China International Youth Conference on Electrical Engineering (CIYCEE)*, Wuhan, China, 2022, pp. 1-6, doi: 10.1109/CIYCEE55749.2022.9959005.

[8] S. Voss, H. . -J. Schulze and F. . -J. Niedernostheide, "Optimization of the temperature dependence of the anode-side current gain of IGBTs by field-stop design," *2010 22nd International Symposium on Power Semiconductor Devices & IC's (ISPSD)*, Hiroshima, Japan, 2010, pp. 141-144.

[9] A. P. -S. Hsieh *et al.*, "Field-stop layer optimization for 1200V FS IGBT operating at 200°C," *2014 IEEE 26th International Symposium on Power Semiconductor Devices & IC's (ISPSD)*, Waikoloa, HI, USA, 2014, pp. 115-118, doi: 10.1109/ISPSD.2014.6855989.

[10] H. . -J. Schulze, S. Voss, H. Huesken and F. . -J. Niedernostheide, "Reduction of the temperature dependence of leakage current of IGBTs by field-stop design," *2011 IEEE 23rd International Symposium on Power Semiconductor Devices and ICs*, San Diego, CA, USA, 2011, pp. 120-123, doi: 10.1109/ISPSD.2011.5890805.

[11] E. Buitrago *et al.*, "A Critical View of IGBT Buffer Designs for 200 °C Operation," *2019 31st International Symposium on Power Semiconductor Devices and ICs (ISPSD)*, Shanghai, China, 2019, pp. 47-50, doi: 10.1109/ISPSD.2019.8757638.

[12] E. Buitrago, A. Mesemanolis, C. Papadopoulos, C. Corvasce, J. Vobecky and M. Rahimo, "An advanced soft punch through buffer design for thin wafer IGBTs targeting lower losses and higher operating temperatures up to 200 °C," *2018 IEEE 30th International Symposium on Power Semiconductor Devices and ICs (ISPSD)*, Chicago, IL, USA, 2018, pp. 499-502, doi: 10.1109/ISPSD.2018.8393712.

979-8-3315-0417-5/25 $31.00 © 2025 IEEE

Controllable Oxygen Vacancies in NbO$_x$ Mott Memristor for Tunable Spiking Neurons

Guolei Liu[1], Dingwei Li[1], Fanfan Li[1], and Bowen Zhu[1,2,*]

[1] Key Laboratory of 3D Micro/Nano Fabrication and Characterization of Zhejiang Province, School of Engineering, Westlake University, Hangzhou, China. Email: zhubowen@westlake.edu.cn
[2] Westlake Institute for Optoelectronics, Westlake University, Hangzhou, China.

Abstract

Artificial spiking neurons based on NbO$_x$ memristors are widely used in neuromorphic systems, but their performance generally exhibits limited reconfigurability after the forming process. Herein, we present a method for post-forming the device through annealing to facilitate performance modulation. This post-annealing process facilitates the diffusion of oxygen vacancies into the NbO$_x$ layer, subsequently altering its performance characteristics. The proposed method enhances the tunability of artificial neurons and paves the way for adaptable neuromorphic devices.

Keywords: Mott memristor, artificial neuron, niobium oxide (NbO$_x$), oxygen vacancy, spiking neural network

Introduction

Neuromorphic computing has garnered considerable interest due to its inherent parallelism, ultra-low power consumption, and seamless integration of spike signals [1]. Central to this paradigm are artificial neurons, which are vital for information processing. Artificial neurons not only aggregate multiple inputs but also exhibit highly nonlinear responses. Additionally, the ability to reconfigure the relationship between input and output allows artificial neurons to adapt to environmental fluctuations or synaptic variations, thereby enhancing system accuracy [2].

While many artificial neurons have been implemented using analog circuits, challenges remain in achieving tunable and biologically relevant characteristics with complementary metal-oxide-semiconductor (CMOS) technology [3]. Therefore, developing artificial neuron devices that leverage unique physical mechanisms, such as VO$_2$, and NbO$_2$ memristors, is crucial for progressing neuromorphic computing and non-von Neumann architecture chips. NbO$_2$ memristor-based artificial neurons exhibit physical similarities to biological neurons, enhancing the potential to replicate biological network behavior in novel computing paradigms. However, the current lack of tunable performance in NbO$_2$ memristors presents a significant obstacle that needs to be addressed [4-7].

In this work, we introduce a Pt/Ti/NbO$_x$/Pt/Ti threshold-switching memristor featuring tunable threshold characteristics. This device exploits various mechanisms, such as the movement of oxygen vacancies. By annealing at 200 °C for 60 minutes, we can achieve varying activation thresholds; positive voltage sweeps reduce the threshold voltage (V$_{th}$), while negative voltage sweeps increase V$_{th}$. Our findings demonstrate the potential for engineering threshold-tunable memristors, which could enhance the configurability of artificial neurons. This adaptability may enable neuromorphic computing systems to better withstand environmental fluctuations and improve overall stability.

Methodology

The NbO$_x$ Mott memristors were fabricated on a Si substrate with a 100 nm SiO$_2$ layer. The bottom electrodes were defined using lithography, followed by the deposition of a Ti/Pt (5/35 nm) layer through electron-beam evaporation. The NbO$_x$ layer was sputtered at room temperature using an NbO$_{2+x}$ target in an Ar environment, and the top electrodes were subsequently deposited using the same process. Electrical characterizations were performed using an Agilent B1500A semiconductor parameter analyzer for DC tests, while pulse measurements were conducted with a Keysight pulse generator. The NbO$_x$ devices were subjected to an initial forming process, which induced a conductive path through Joule heating, creating a crystalline NbO$_2$ phase. And annealing at 200 °C help the diffusion of oxygen vacancies into the NbO$_x$ layer.

Results and discussion

A typical NbO$_x$-based locally active device used in this work has a simple metal/insulator/metal (MIM) structure, consisting of a NbO$_x$ layer sandwiched between two Ti/Pt electrodes. Fig. 1 illustrates the structural and elemental distributed characteristics of the NbO$_x$ memristor. The

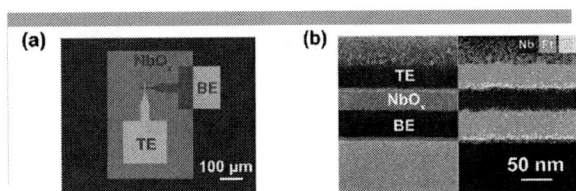

Fig. 1. (a) Optical microscope image of the Ti/Pt/NbO$_x$/Ti/Pt memristor (top electrode TE; bottom electrode BE), the effective area 5 μm × 5 μm. (b) Cross-sectional elemental mapping of Nb, Pt, and Ti in the device.

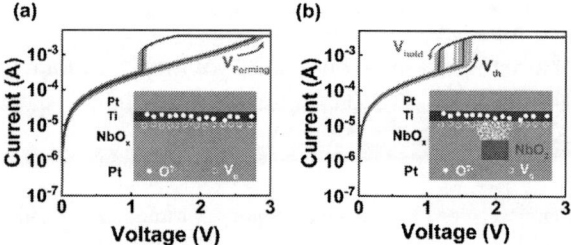

Fig. 2. (a) Device-to-device I-V characteristics of the NbO$_x$ memristor during the forming process. The inset depicts the initial state of the NbO$_x$ layer with oxygen vacancies (V$_o$) and oxygen ions (O^{2-}) distributed across the layer. (b) Device-to-device I-V characteristics during subsequent switching cycles. The inset illustrates the formation of a conductive NbO$_2$ filament within the NbO$_x$ layer, which drives the resistive switching behavior.

crossbar structure of the optical microscope is shown in Fig. 1(a), and elemental mapping confirms the presence of Nb, Pt, and Ti in the device is shown in Fig. 1(b).

All fresh devices require a positive voltage to activate the threshold-switching (TS) behavior. Fig. 2(a) displays the I-V characteristics of the NbO$_x$ memristor across 15 different operational devices. The device-to-device I-V characteristics during the forming process exhibit stability property. The inset highlight the initial state of the NbO$_x$ layer, which shows the distribution of oxygen vacancies (V$_o$) and oxygen ions (O^{2-}). This initial state is critical for the forming electrical behavior. Fig. 2(b) illustrates the I-V characteristics observed during subsequent switching cycles. The inset emphasizes the formation of a conductive for the Mott switching phenomenon. This transition is essential for the memristor's functionality, enabling effective modulation of resistance and thereby supporting the operation of artificial neurons.

Based on the above device, we implemented a leaky integrate-and-fire (LIF) neuron with a configurable activation function. Fig. 3 illustrates key performance as artificial neuron the based NbO$_x$ memristor. Fig. 3(a) presents the device I-V characteristics, including forming sweep, voltage sweep, and current sweep. The inset shows the circuit setup that includes a load resistor (R$_L$). Fig. 3(b) demonstrates the device's rapid response times during switching, achieving fast on/off switching times of 110 ns and 55 ns, which are critical for neuromorphic applications. Fig. 3 (c) captures the oscillation behavior of the memristor under varying input voltages (V$_{in}$), revealing its dynamic conductive properties. Finally, Fig. 3(d) depicts the relationship between oscillation frequency and applied voltage, indicating how voltage modulation can influence

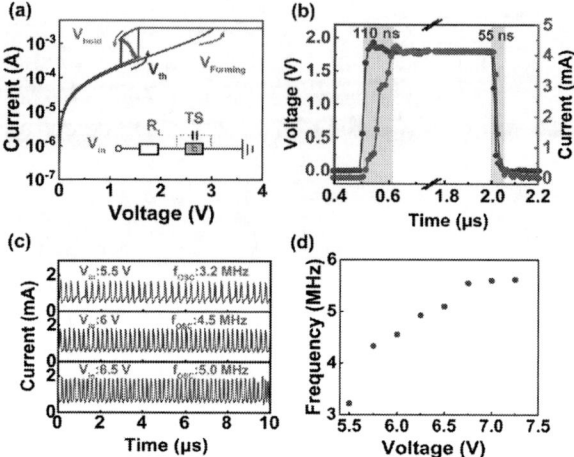

Fig. 3. (a) Device I-V characteristics of the NbO$_x$ memristor. The circuit diagram inset shows the setup with a load resistor (R$_L$). (b) Device response time during switching, demonstrating fast on/off switching times of 110 ns and 55 ns. (c) Oscillation behavior of the memristor for different input voltages. (d) The relationship between oscillation frequency and applied voltage.

the operational frequency of the memristor.

To further enhance the tunable characteristics of the switching threshold, Fig. 4 investigates the effects of annealing on the conductivity and electrical properties of the NbO$_x$ memristor with annealing. Fig. 4(a) illustrates the variation in conductivity after annealing at 200 °C for different durations, demonstrating how time affects the material property under applied voltage. Fig. 4(b) shows the

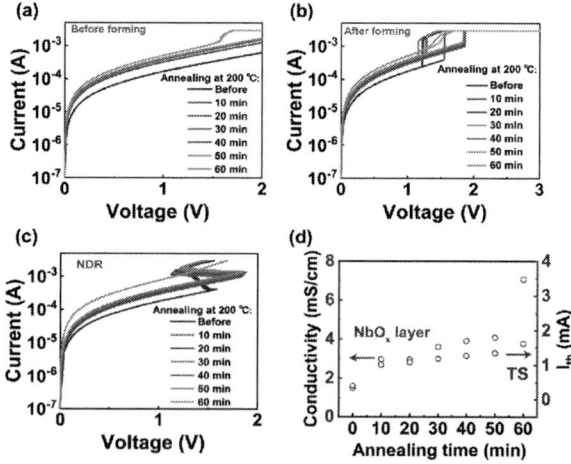

Fig. 4. (a) Conductivity variation after annealing at 200 °C for different duration times. (b) The change of I-V characteristics after the forming process. (c) Current sweep behavior following various annealing times. (d) The correlation between annealing time and both the conductivity of the NbO$_x$ layer and threshold current of the memristor.

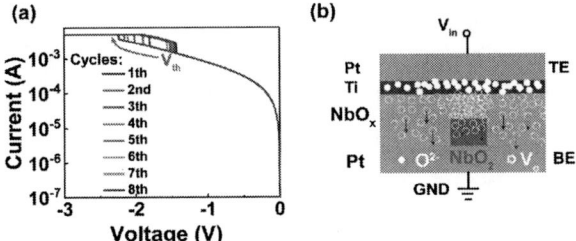

Fig. 5. (a) Negative voltage sweeps I-V characteristics of the device after 60 minutes of annealing. (b) Schematic illustration of the memristor's switching mechanism, depicting the migration of oxygen vacancies toward the BE under the applied voltage.

changes in I-V characteristics post-forming, indicating increase threshold current. Fig. 4(c) highlights the current sweep behavior for various annealing times. Finally, Fig. 4(d) illustrates the correlation between annealing time and both the conductivity of the NbO_x layer and the threshold current of the memristor, emphasizing the relationship between these parameters and the potential for tuning device performance through thermal treatment. influences the device's TS performance, thus enabling enhanced performance in artificial neuron functionalities.

Fig. 5 provides insights into the NbO_x memristor's behavior under specific conditions. Fig. 5(a) displays the I-V characteristics during negative voltage sweeps after 60 minutes of annealing, indicating the device's performance and stability in this operational regime. This analysis reveals how annealing can influence the electrical response of the memristor. Fig. 5(b) offers a schematic illustration of the switching mechanism and the critical role of oxygen vacancy dynamics in determining the device. This visual representation elucidates the fundamental processes detailing the migration of oxygen vacancies toward the BE under applied voltage.

Fig. 6(a) shows the I-V characteristics during positive voltage sweeps, illustrating the device's electrical response and the behavior of resistance during this operational phase.

This is essential for understanding the memristor's functionality and reliability under varying voltage applications. Fig. 6(b) provides a schematic illustration of the switching mechanism, depicting the migration of oxygen vacancies toward the TE under applied voltage. This illustration highlights the critical role of oxygen vacancy movement in enabling tunability resistive switching.

Conclusion

In conclusion, we propose a tunable TS memristor with bidirectionally adjustable threshold voltage under electrical stimulation. We demonstrate a viable method to modulate the V_{th} by annealing and pulse stimulation. Ultimately, this device facilitates the construction of configurable neurons suitable for neuromorphic computing. These results indicate that tunable threshold memristors are well-suited for developing configurable neurons, highlighting their significant potential in creating efficient and stable neuromorphic systems.

References

[1] W. Zhang *et al.*, "Neuro-inspired computing chips," *Nature Electronics*, vol. 3, no. 7, pp. 371-382, 2020, doi: 10.1038/s41928-020-0435-7.

[2] Y. Wang *et al.*, "A configurable artificial neuron based on a threshold-tunable TiN/NbOx/Pt memristor," *IEEE Electron Device Letters*, vol. 43, no. 4, pp. 631-634, 2022, doi: 10.1109/led.2022.3150034.

[3] X. Zhang *et al.*, "Fully memristive snns with temporal coding for fast and low-power edge computing," in *IEDM Tech. Dig.*, Dec. 2020, pp. 29.6.1-29.6.4.

[4] G. Liu *et al.*, "Experimental demonstration of coplanar NbOx Mott memristors for spiking neurons," *IEEE Electron Device Lett.*, vol. 45, no. 4, pp. 708-711, Feb. 2024, doi: 10.1109/led.2024.3362829.

[5] F. Li *et al.*, "A skin-inspired artificial mechanoreceptor for tactile enhancement and integration," *ACS Nano*, vol. 15, no. 10, pp. 16422-16431, Oct 26 2021, doi: 10.1021/acsnano.1c05836.

[6] M. Zhao *et al.*, "Silk protein based volatile threshold switching memristors for neuromorphic computing," *Advanced Electronic Materials*, vol. 8, no. 4, 2022, doi: 10.1002/aelm.202101139.

[7] Y. Li, Y. Ding, X. Zhang, S. Jia, W. Wang, Y. Li, M. Wang, H. Jiang, Q. Liu, N. Xu, and M. Liu, "Fatigue of NbOx-based locally active memristors—part II: mechanisms and modeling," *IEEE Transactions on Electron Devices*, pp. 1-7, 2023, doi: 10.1109/ted.2023.3322672..

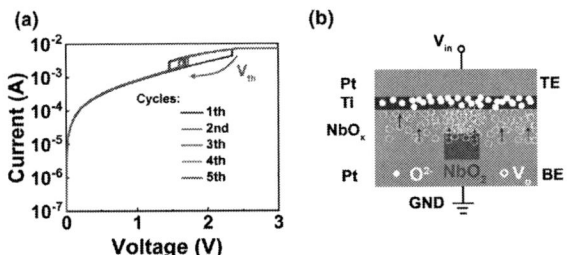

Fig. 6. (a) Positive voltage sweeps I-V characteristics of the device (b) Schematic illustration of the switching mechanism, depicting the migration of oxygen vacancies toward the TE under the applied voltage.

Study of the Characteristics of GaN Substrate-Based MicroLEDs with Different Epitaxial Structures

Shan Huang[1], Yibo Liu[1], Feng Feng[1], Jingyang Zhang[1], Zichun Li[1], Man Hoi Wong[1] and Zhaojun Liu[2, a]

1 State Key Laboratory of Advanced Displays and Optoelectronics Technologies, The Hong Kong University of Science and Technology, Hong Kong, China

2 Department of Electrical and Electronic Engineering, Southern University of Science and Technology, Shenzhen 518055, China

a) Author to whom correspondence should be addressed: liuzj@sustech.edu.cn

Abstract

This study provides an in-depth analysis of the performance of GaN substrate-based MicroLEDs with different epitaxial structures using Silvaco TCAD. The focus is on exploring how variations in Al content in the electron blocking layer, In content in the quantum well layers and doping concentration in p-GaN affect device performance. The results show that changes in these epitaxial structures impact key metrics such as threshold voltage, wall plug efficiency, internal quantum efficiency, and the ideality factor. These findings offer valuable insights for designing high-performance GaN substrate-based MicroLEDs.
Keywords: MicroLED, GaN

Introduction

Currently, Micro-LED, as the fourth-generation display technology, is garnering more and more attention [1,2,3]. Its benefits, including high brightness, high contrast, long lifespan, and low power consumption, have led to its widespread application in displays of various sizes [4,5,6].

Currently, sapphire substrates are widely used in GaN-based MicroLEDs. However, sapphire substrates still suffer from the effects of lattice mismatch and thermal stress between the GaN epitaxial layer and themselves. Considering these factors, GaN substrates have emerged as a promising alternative due to their high compatibility with GaN epitaxial layers and superior thermal conductivity [7,8].

For GaN substrate-based MicroLEDs, optimizing the structures of multi-quantum wells (MQWs), the electron blocking layer (EBL) and the p-GaN region is crucial during the design process. By adjusting the composition of the quantum wells, the design of the MQWs can be optimized to enhance light generation and capture, improving the luminous efficiency and color performance of the MicroLEDs [9]. An appropriate composition for the EBL can effectively prevent carrier leakage, ensuring that electrons and holes can recombine efficiently within the quantum wells, while also enhancing the stability of the light output of the MicroLEDs [10]. Optimizing p-GaN region can improve carrier injection efficiency and light emission efficiency [11].

In this study, device simulations were conducted using Silvaco TCAD to investigate the impact of the EBL, MQWs and p-GaN region on the performance of MicroLEDs with GaN substrate. The simulations were carried out by adjusting the Al content in $Al_xGa_{1-x}N$ (0<x<1) of the EBL, the In content in $In_yGa_{1-y}N$/GaN MQWs (0<y<1) and doping concentration in p-GaN. The results obtained provide theoretical guidance for the fabrication of high-performance MicroLEDs with GaN substrate.

Results and Discussion

In this work, Silvaco TCAD Atlas is used for the simulations of MicroLEDs. The schematic diagram of the MicroLED is shown in Fig. 1. The Si doping concentration of the n-GaN layer is 1×10^{18} cm^{-3}. The MQWs layers consist of two pairs of $In_yGa_{1-y}N$/GaN (7 nm)/GaN (2 nm) structures (0<y<1). The 200 nm thick p-GaN region and the 404 nm thick p-$Al_xGa_{1-x}N$ electron blocking layer (EBL) (0<x<1) are doped with Mg, and the doping concentration is 1×10^{18} cm^{-3}. The device is set to be 1×100 μm^2.

The simulation process is carried out by adjusting the epitaxial parameters, such that the Al content of EBL varies in steps from 0.10 to 0.35, the In content of MQWs varies in steps from 0.15 to 0.35 [12], and the doping concentration of p-GaN varies in steps from 1×10^{18} cm^{-3} to 9×10^{18} cm^{-3}.

When the anode current is 1 mA, the corresponding anode voltage is determined as the threshold voltage of the MicroLED. As shown in Fig. 2, with the increase of Al content in the EBL, the threshold voltage shows a slight increasing trend at different temperatures. As the temperature rises, the variation in the threshold voltage decreases, indicating that the influence of Al content on the concentration threshold voltage of MicroLEDs becomes

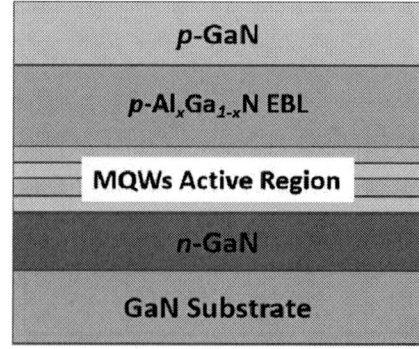

Fig. 1: Schematic diagram of the MicroLED with GaN substrate in the study.

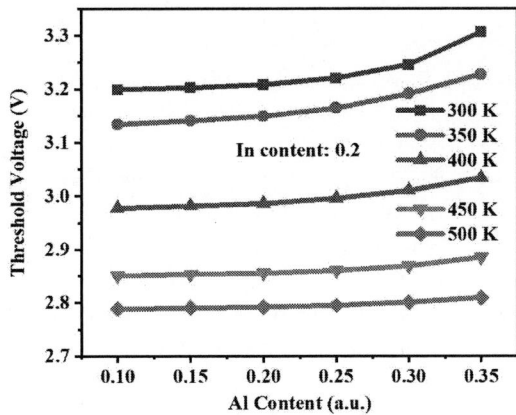

Fig. 2: The threshold voltage of MicroLEDs with different Al contents at various temperatures.

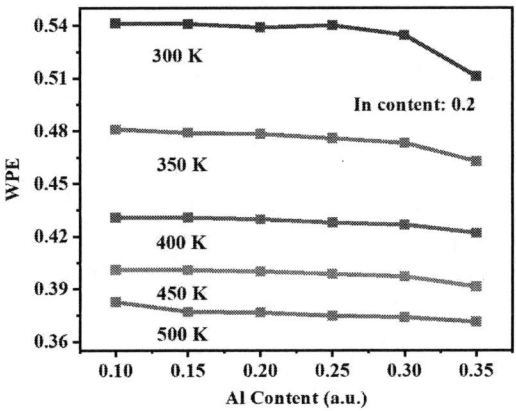

Fig. 3: WPE of MicroLEDs with different Al contents at various temperatures.

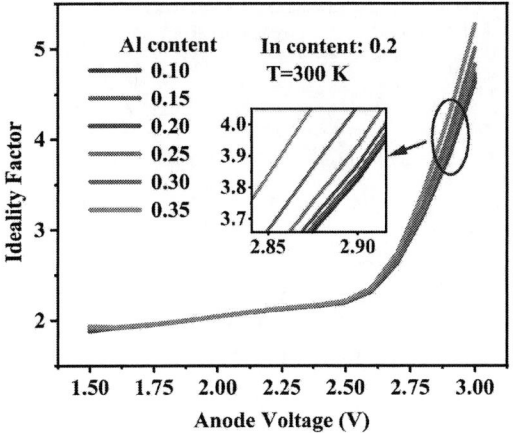

Fig. 4: Ideality factor of MicroLEDs with different Al contents.

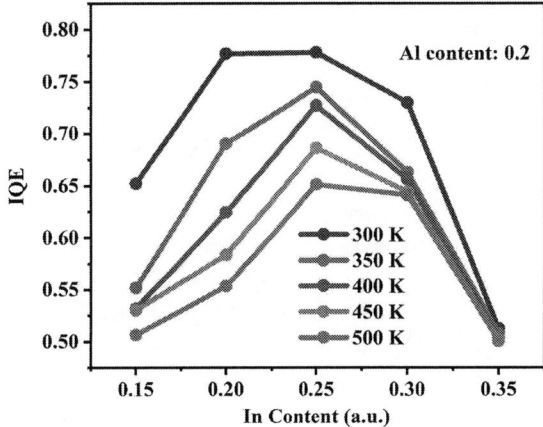

Fig. 5: IQE of MicroLEDs with different In contents at various temperatures.

smaller with increasing temperature.

Fig. 3 shows a decreasing trend of wall-plug efficiency (WPE) with increasing Al content, and this trend remains the same as the temperature gradually rises. Fig. 4 shows the variation of the ideality factor with voltage as the Al content increases from 0.1 to 0.35. It is found that at the same voltage, MicroLEDs with lower Al content have a lower ideality factor.

The aforementioned effects can be explained as follows: as the Al content in the EBL increases, the bandgap of the EBL widens, creating a higher barrier between the p-type region and the quantum wells. Consequently, a larger drive voltage is required to overcome this barrier, leading to an increase in the threshold voltage of the device. At the same time, the higher barrier also hinders electron injection into the quantum wells, resulting in a decrease in current injection efficiency, which in turn lowers the WPE and increases the ideality factor [10],[12].

As shown in Fig. 5, with the increase in In content, IQE initially increases and then decreases. It can be observed

that at different temperatures, IQE reaches its maximum value at $y = 0.25$. The reason for this phenomenon is that initially, the increase in In content reduces the bandgap of the quantum wells, enhancing the confinement of electrons and holes, and increasing the radiative recombination probability of carriers, thereby improving the IQE. However, as the In content continues to increase, lattice mismatch becomes more severe, leading to the formation of more defects, dislocations, and non-radiative recombination centers, which decrease the carrier recombination efficiency and result in a decline in IQE [13,14].

The performance of MicroLEDs with diferent p-type doping concentrations in p-GaN is also investigated. As shown in Fig. 6, when the doping concentration of p-GaN increases from 1×10^{18} cm^{-3} to 9×10^{18} cm^{-3}, the threshold voltage of the MicroLEDs slightly decreases, while IQE slightly increases. This can be explained that the increase in the doping concentration of p-GaN raises the hole concentration in the p-type layer, making it easier for holes

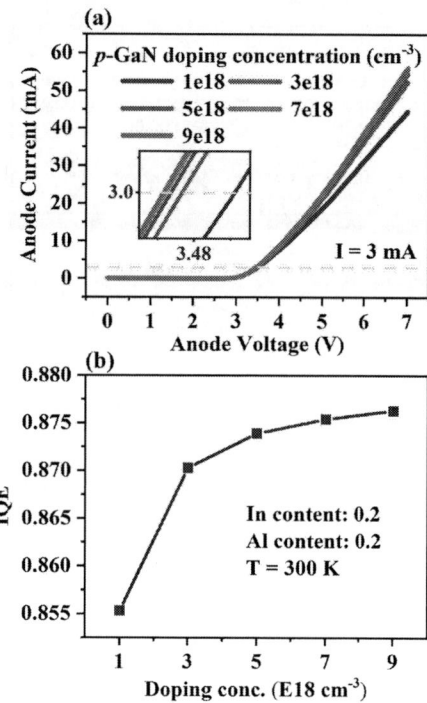

Fig. 6: (a) The current-voltage curves of MicroLEDs with various doping concentrations in p-GaN; (b) IQE of MicroLEDs with different doping concentrations in p-GaN.

to be injected into the quantum well region, thus increasing the chances of recombination with electrons. A higher hole concentration improves the injection balance between holes and electrons, reducing the driving voltage required to reach the threshold voltage, while also enhancing the probability of radiative recombination, thereby increasing IQE [11].

Conclusion

In this paper, simulations of GaN substrate-based MicroLEDs with different epitaxial parameters were conducted using Silvaco-TCAD. The results indicate that with the increase in Al content, the threshold voltage of the MicroLEDs increases, while WPE shows a slight decreasing trend, and the ideality factor gradually rises. As the In content increases, IQE first increases and then decreases, reaching a maximum at $y = 0.25$. Additionally, the increase in doping concentration of p-GaN leads to a slight decrease in threshold voltage and a slight increase in IQE. The above research provides theoretical guidance for the preparation and application of high-performance GaN substrate-based MicroLEDs.

Acknowledgments

This work was supported in part by the National Key R&D Program of China under Grant 2023YFB2806800, in part by the Fundamental and Applied Fundamental Research Fund of Guangdong Province under Grant 2021B1515130001, and in part by the Shenzhen Science and Technology Program under Grant JCYJ20220818100603007.

References

[1] Z. Liu, W. C. Chong, K. M. Wong, and K. M. Lau, "GaN-based LED micro-displays for wearable applications," *Microelectron. Eng.*, 148, pp. 98–103 (2015).

[2] Y. Liu, K. Zhang, B. -R. Hyun, H. S. Kwok and Z. Liu, "High-Brightness InGaN/GaN Micro-LEDs With Secondary Peak Effect for Displays," *IEEE Electron Device Letters*, 41, pp. 1380-1383 (2020).

[3] F. Feng et al., "AlGaN-Based Deep-UV Micro-LED Array for Quantum Dots Converted Display With Ultra-Wide Color Gamut," *IEEE Electron Device Letters*, 43, pp. 60-63 (2022).

[4] S. Huang et al., "Advances in Full-Color Microdisplays Based on MicroLED for AR and VR Applications," *IEEE Open Journal on Immersive Displays*, 1, pp. 127-134 (2024).

[5] Liu. Y, Wang. G, Feng. F et al., "Ultra-low-defect homoepitaxial micro-LEDs with enhanced efficiency and monochromaticity for high-PPI AR/MR displays," *PhotoniX*, 5, p. 23 (2024).

[6] Y. Liu et al., "Analysis of size dependence and the behavior under ultrahigh current density injection condition of GaN-based Micro-LEDs with pixel size down to 3 μm," *J. Phys. Appl. Phys.*, 55 (2022).

[7] X. Shan et al., "Comparison of Beyond 1 GHz C-Plane Freestanding and Sapphire-Substrate GaN-Based micro-LEDs for High-Speed Visible Light Communication," J. Lightwave Technol., 41, pp. 1480–1486 (2023).

[8] Li. Z, Liu. Y, Feng. F, Wong. M and Liu. Z, "P-154: Enhanced Thermal Stability and High Color Accuracy in GaN-on-GaN Homoepitaxy Micro-LEDs," *SID Symp. Dig. of Tech. Papers*, 55, pp. 1973-1976 (2024).

[9] A. Vaitkevičius, J. Mickevičius, D. Dobrovolskas et al., "Influence of quantum-confined Stark effect on optical properties within trench defects in InGaN quantum wells with different indium content," *Journal of Applied Physics*, 115, p. 213512 (2014).

[10] Sang-Heon Han et al., "Effect of electron blocking layer on efficiency droop in InGaN/GaN multiple quantum well light-emitting diodes," *Appl. Phys. Lett.*, 94, p. 231123 (2009).

[11] Lu, S., Li, J., Huang, K. et al., "Designs of InGaN Micro-LED Structure for Improving Quantum Efficiency at Low Current Density," *Nanoscale Res Lett* 16, 99 (2021).

[12] P. Ye et al., "Simulation Investigation On The Characteristics Of Gan-Based Multi-Quantum Wells Micro-Leds," *2023 China Semiconductor Technology International Conference (CSTIC)*, Shanghai, China, pp. 1-3 (2023).

[13] D. Cherns, S. J. Henley, F. A. Ponce, "Edge and screw dislocations as nonradiative centers in InGaN/GaN quantum well luminescence," *Appl. Phys. Lett.*, 78, pp. 2691–2693 (2001).

[14] J. J. Wierer, A. J. Fischer, D. D. Koleske, "The impact of piezoelectric polarization and nonradiative recombination on the performance of (0001) face GaN/InGaN photovoltaic devices," *Appl. Phys. Lett.*, 96, p. 051107 (2010).

Impact of Oxide Quality in Self-Aligned Block Region on Hot Carrier Degradation in n-Type CFP-LDMOS with 0.18 μm Bipolar-CMOS-DMOS Technology

Qiao Teng[1,#], Yixian Song[1,#], Junzhe Kang[3], Kai Xu[1,2,*], and Dawei Gao[1,*]

[1]College of integrated circuits, Zhejiang University, Hangzhou, China, 311200; [2]ZJU-Hangzhou Global Scientific and Technological Innovation Center, Zhejiang University, Hangzhou, China, 311215; [3]Department of Electrical and Computer Engineering, University of Illinois at Urbana-Champaign, Urbana, USA, 61801. *Email: dawei_gao@zju.edu.cn , xuk@zju.edu.cn.
#These authors contributed equally.

Abstract

In this work, the effect of oxide quality in the self-aligned block (SAB) region of n-LDMOS with contact field plate on hot carrier degradation (HCD) is investigated. Experiments and simulations confirm that the degradation of I_D and R_{sp} in on-state HCD is less than that in off-state HCD due to the capture competition mechanism between electrons and holes. This trapping mechanism is inhibited when the combination process is used to form high-quality oxides in the SAB region, improving HCD.

Keywords: LDMOS, Hot Carrier Degradation, Process.

Introduction

In recent years, the market size of portable electronic products has gradually increased. Power management integrated circuit (PMIC) chips used in high-voltage circuits for industrial automation systems, intelligent electronic devices, and automotive electronics have received extensive attention [1]. It acts as a power source for multiple modules to complete efficient energy conversion work. Bipolar-CMOS-DMOS (BCD) technology is considered a suitable technology for PMIC due to its high integration and ability to simultaneously meet different input voltage requirements [2]. Notably, high voltage lateral diffusion metal oxide semiconductor (LDMOS) is instrumental in delivering high power input and output due to its compatibility with traditional CMOS processes and low on-resistance (Ron). For high-voltage devices, on-state and off-state hot carrier degradation (HCD) is a severe reliability problem. In general, HCD is highly correlated with the length and concentration of the drift region. However, HCD in LDMOS with a contact field plate (CFP) is closely related to the oxide quality used as a blocking layer for the non-silicide region beneath the field plate. The interfacial state density (D_{it}) between the oxide in the self-aligned block (SAB) region and the silicon substrate significantly affects the HCD of LDMOS. At present, there are few studies focused on improving the HCD effect by optimizing the oxide quality in the SAB region.

In this work, the effect of oxide quality in the SAB region on the on-state and off-state HCD of a CFP-LDMOS is studied by combining experimental and simulation methods. The experimental results show that the combination of in-situ steam generation (ISSG) and plasma-enhanced chemical vapor deposition (PECVD) processes for the oxide growth in the SAB region can significantly improve HCD. It is also successfully proved that the improvement of HCD is caused by the reduction of the D_{it} in the SAB region through technology computer aided design (TCAD) simulation. In mass production, the results are conducive to improving the HCD of the CFP-LDMOS without adding additional special processes.

Experimental Details

The 30 V n-type CFP-LDMOS used in this study are fabricated with 0.18 μm BCD process technique. The gate length L_G is 0.2 μm, and the width W_G is 5 μm. The process flow of LDMOS is shown in Fig. 1. After the source/drain process is completed, a conventional PECVD process and the combination process of ISSG and PECVD are used to respectively form an oxide in the SAB region. Here, the oxide grown by ISSG in the SAB region is generated during the gate oxidation stage. The oxide is formed above the drift region as a dielectric layer of the field plate. The schematic of the studied LDMOS is shown in Fig. 2(a), with the corresponding transmission electron microscope (TEM) in Fig. 2(b), and the simulation result by Sentaurus TCAD is displayed in Fig. 2(c). The operation voltage of this device is V_D = 30 V and V_G = 5 V. For HCD measurement, the stress voltage applied to the drain is 35 V, with gate stress voltages of 0 V and 6.5 V, respectively. The device is stressed for 1000 s while interrupted in a log timescale to measure threshold voltage (V_T) at V_{DS} = 0.1 V by the constant current method and saturation current (I_D) at V_{DS} = 30 V.

Results and Discussion

As shown in Fig. 3, the I_D-V_D curves of the conventional process and the combination process are compared. It is found that the CFP-LDMOS fabricated with the combination process has a better drive current. The ISSG process in the combination method facilitates the formation of higher-quality oxides in the SAB region, reducing interface defects at the n-drift region, which increases carrier mobility when electrons flow through it and results in a higher I_D [3]. To consider the I_D uniformity of the combination process, 9 dies of CFP-LDMOS on the wafer are measured and compared with conventional processes. Fig. 4 shows the box plot of I_D, indicating that the I_D of the CFP-LDMOS treated with the combination process has an increase in overall uniformity. Furthermore, the combined process increased the I_D by 5% relative to the traditional process.

In addition to performance improvements, the HCD in both off-state and on-state is important for CFP-LDMOS. Fig. 5 and Fig. 6, respectively, show the degradation behavior of I_D and V_T at off-state HCD stress. It is found that the $\Delta I_D/I_D$ in the combination process is significantly weaker than that in the conventional process, while it has almost no effect on V_T. Furthermore, no significant shift is observed in V_T, but I_D degrades notably, indicating that only carrier injection occurred without the generation of additional traps [4]. Due to the injection of electrons, the specific on-resistance (R_{sp}) increases with stress time, as shown in Fig. 7. Under on-state HCD stress, with the increase of $V_{G,Stress}$, CFP-LDMOS fabricated by the conventional process show a greater I_D degradation than that in the combination process, as depicted in Fig. 8. It is worth noting that under high $V_{G,Stress}$ conditions, hot carriers are generated in the channel region, resulting in increasing interface defects and causing V_T shift. Since CFP-LDMOS has the same fabrication process in the channel region, there is no significant difference in V_T degradation, as shown in Fig. 9. Similar to I_D degradation, Fig. 10 shows R_{sp} degradation with stress time, indicating that high-quality interface characteristics in the SAB region contribute to reducing interface states and suppressing electron injection. Meanwhile, it is observed that the degradation of I_D and R_{sp} at on-state HCD stress is less significant than in the off-state HCD stress, suggesting the presence of a specific mechanism that accounts for this behavior.

To study this phenomenon, the TCAD simulation is performed to explain the results of the experiment. Fig. 11(a) and (b) show the impact ionization behavior of the device under high and low $V_{G,Stress}$, respectively. In the off-state HCD stress, impact ionization mainly occurs in the non-channel region, and a few non-hot electrons are injected into the oxide above the drift region to form interface-trapped charges, leading to an increase in R_{sp} and a reduction in I_D without interface degradation. In the on-state HCD stress, significant impact ionization occurs beneath the CFP and in the channel region, generating more electron-hole pairs, which leads to V_T shift and a decrease in I_D. Especially when $V_{G,Stress}$ increases, the maximum impact ionization takes place most intensely at the drift region/drain interface due to the kirk effect, and an abundance of holes originating from impact ionization are more susceptible to the effect of V_{GD} and trapped in the oxide of SAB region [5]. Due to the competition between hole and electron trapping, the degradations of I_D and R_{sp} at the on-state HCD are small. To further confirm this phenomenon, the simulated results of I_D–V_G characteristic with $V_D = 0.1$ V under different trapping situations. It can be seen from Fig. 12(a) that electron trapping in the SAB region results only in I_D degradation, whereas both I_D degradation and V_T shift occur simultaneously when electron trapping occurs in the channel region. Furthermore, hole trapping in the SAB region leads to an increase in I_D, as shown in Fig. 12(b). In summary, whether under off-state or on-state HCD stress conditions, the carrier trapping behavior of CFP-LDMOS has been suppressed when manufactured using the combination process due to the formation of high-quality oxide in the SAB region, which contributes to the improvement of HCD.

Conclusion

In summary, the performance and HCD are demonstrated to be effectively improved by using the combination process to form the oxide in the SAB region of CFP-LDMOS. This is because high-quality oxides with lower defect densities can more effectively inhibit the injection of electrons and holes. The simulation results further confirm that charge trapping at the SAB region affects only the I_D and R_{sp}, while charge trapping at the channel region impacts the V_T, as well as the I_D and R_{sp}. These results indicate essential guidelines for high-reliability CFP-LDMOS engineering.

Acknowledgments

The authors would also like to acknowledge the supports from the National Key R&D Program of China (2022YFF0605800), Zhejiang Jianbing Program (Grant No. 2024C01002, 2022C01063 and 2024SJCZX0030), the National Natural Science Foundation of China (Grant No. 62204217).

References

[1] D. Kim et al., "The Lowest On-Resistance and Robust 130nm BCDMOS Technology implementation utilizing HFP and DPN for mobile PMIC applications," in 2019 31st International Symposium on Power Semiconductor Devices and ICs (ISPSD), 2019, pp. 391-394.

[2] S. Yu et al., "Design and Simulation Optimization of an Ultra-Low Specific On-Resistance LDMOS Device," IEEE Journal of the Electron Devices Society, vol. 12, pp. 14-22, 2024.

[3] W.-C. Hung et al., "Defect Passivation and Reliability Enhancement by Low-Temperature-High-Pressure Hydrogenation in LDMOS With 0.13-μm Bipolar-CMOS-DMOS Technology," IEEE Electron Device Letters, vol. 44, no. 5, pp. 789-792, 2023.

[4] W.-C. Hung et al., "Abnormal Two-Stage Degradation Under Hot Carrier Injection With Lateral Double-Diffused MOS With 0.13-μm Bipolar-CMOS-DMOS Technology," IEEE Transactions on Electron Devices, vol. 70, no. 7, pp. 3419-3423, 2023.

[5] Y.-S. Lin, L.-H. Chen, T.-C. Chang, K.-J. Liu, C.-Y. Lin, and F.-M. Ciou, "The Relationship Between Resistive Protective Oxide (RPO) and Hot Carrier Stress (HCS) Degradation in n-Channel LD SOI MOSFET," IEEE Transactions on Electron Devices, vol. 68, no. 3, pp. 962-967, 2021.

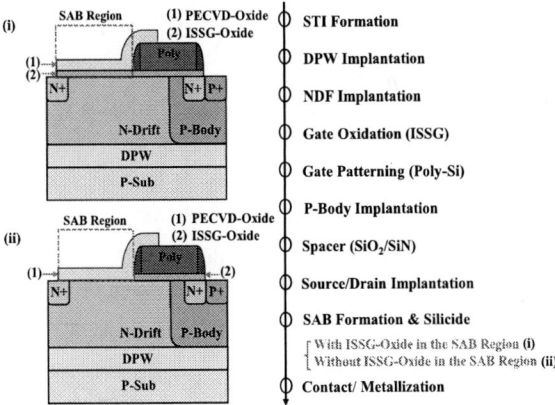

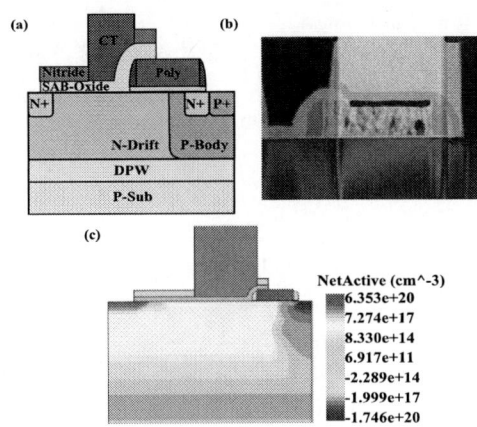

Fig. 1 Key process flow of CFP-LDMOS fabricated by 0.18 μm BCD process technique with or without ISSG-Oxide in the SAB region.

Fig. 2 (a) Schematic diagram, (b) TEM cross-sectional image, and (c) Simulated structural diagram of CFP-LDMOS.

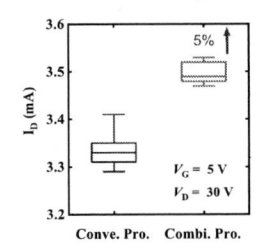

Fig. 3 The I_D-V_D characteristic of CFP-LDMOS in conventional process and combination process.

Fig. 4 The box plot of initial I_D in the CFP-LDMOS using 9 dies in two processes.

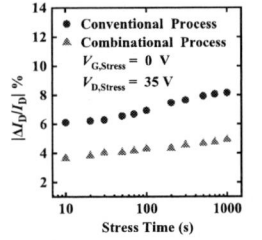

Fig. 5 I_D degradation with stress time in various processes at $V_{G,Stress}$ = 0 V, $V_{D,Stress}$ = 35 V.

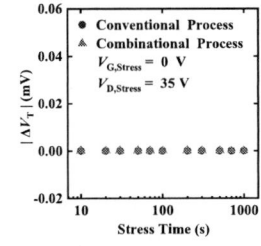

Fig. 6 V_T dose not degradation after HCD in different processes at $V_{G,Stress}$ = 0 V, $V_{D,Stress}$ = 35 V.

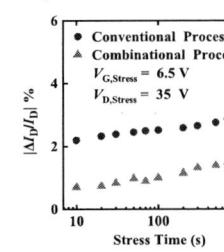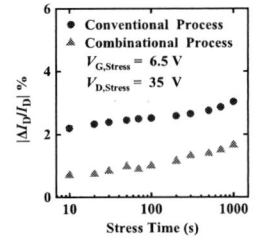

Fig.7 R_{sp} degradation as a function of stress time in two processes at $V_{G,Stress}$ = 0 V, $V_{D,Stress}$ = 35 V.

Fig. 8 Less I_D degradation under HCD stress in different processes at $V_{G,Stress}$ = 6.5 V, $V_{D,Stress}$ = 35 V.

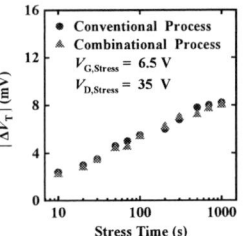

Fig. 9 V_T degradation increases with stress time in two processes at $V_{G,Stress}$ = 6.5 V, $V_{D,Stress}$ = 35 V.

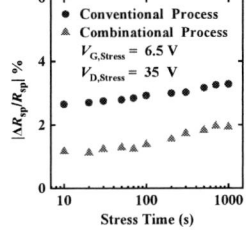

Fig. 10 Smaller R_{sp} degradation after HCD in two processes at $V_{G,Stress}$ = 6.5 V, $V_{D,Stress}$ = 35 V.

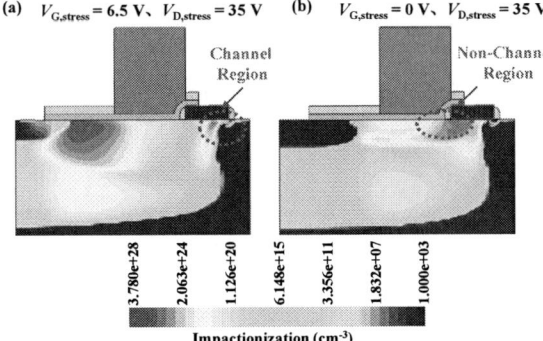

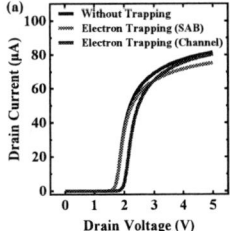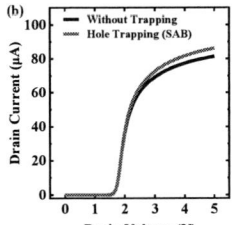

Fig. 12 Simulated I_D–V_G characteristic with V_D = 0.1 V under different trapping situations. (a) The electron trapping in the SAB region only reduces the I_D, whereas, in the channel region, it both increases the V_T and degrades the I_D. (b) The hole trapping in the SAB region only increases the I_D.

Fig. 11 (a) Strong impact ionization in the channel and drain at on-state stress. (b) Small impact ionization in the drift region at off-state stress.

979-8-3315-0417-5/25 $31.00 © 2025 IEEE

21

Investigation of High-Sensitivity Pressure Sensor Integrated with InSnZnO Thin-Film Transistors

Mei Yang[1], Ming Li[1], Wei Huang[1], Rongsheng Chen[1,2*]

[1]School of Microelectronics, South China University of Technology, Guangzhou, China.

[2]State Key Laboratory of Advanced Displays and Optoelectronics Technologies, Department of Electronic and Computer Engineering, The Hong Kong University of Science and Technology, Hong Kong, China

*Corresponding Author Email: chenrs@scut.edu.cn

Abstract

With the demand for flexible pressure sensors in wearable devices and medical applications, piezoelectric sensors stand out for their self-powering and fast response. This paper presents a PVDF/ZnO@MXene (PZM) sensor with high sensitivity (2.32 V N^{-1}), detection of different items, and fast response (46 ms). The PZM sensor integrated with ITZO TFT for signal amplification demonstrates stable performance, highlighting its potential for use in personalized identification, human-computer interaction, and soft robotics.
Keywords: Piezoelectric sensor, MXene, InSnZnO (ITZO), System integration research

Introduction

With the continuous advancement of electronic devices and smart terminals, flexible pressure sensors have garnered significant research interest in areas such as wearable healthcare devices, interactive displays, and humanoid robotics [1], [2]. Among various types, piezoelectric sensors exhibit self-powering properties, allowing them to produce voltage signals independently without external energy sources. Their mechanical robustness, simple design, minimal signal interference, and high sensitivity have driven notable progress in piezoelectric sensor technology in recent years [3].
Piezoelectric sensing materials are primarily categorized into organic and inorganic types, playing critical roles across various applications. While organic materials exhibit excellent flexibility, their piezoelectric performance is often suboptimal [4]. However, pure PVDF sensors typically yield low output voltage. A common strategy to enhance piezoelectric constants involves incorporating inorganic nanofillers into the organic matrix. Recent studies have demonstrated that strategic filler doping and structural design can significantly improve piezoelectric performance, paving the way for advanced wearable electronics using sustainable materials [5], [6]. The integration of InSnZnO thin-film transistors (ITZO TFTs) further amplifies weak signals generated by piezoelectric sensors, enhancing the overall efficiency of sensor systems.
In this work, we fabricated a flexible PVDF/ZnO@MXene (PZM) piezoelectric sensor using a hybrid approach that combines electrospinning, hydrothermal, and spin-coating techniques. Utilizing the inherent structural properties of PVDF nanofibers in synergy with ZnO nanoparticles and MXenes, the piezoelectric performance of the sensor can be significantly improved. To further enhance these features, ITZO TFTs with excellent electrical properties were integrated with the developed PZM piezoelectric sensor. The sensor system exhibits an output voltage of approximately 3.82 V and in-situ amplification of weak sensing signals by approximately 10 times, indicating its application potential in fields such as human-computer interaction and brain-computer interface.

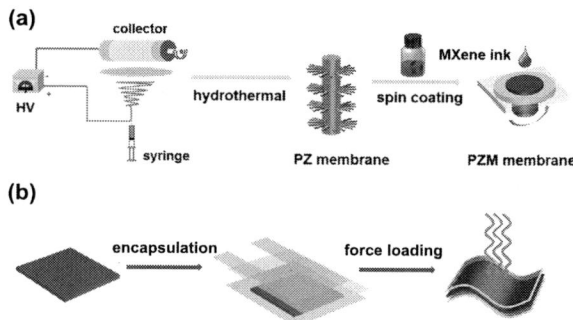

Fig. 1. (a) Schematic illustration of the PVDF/ZnO@MXene (PZM) composite membrane preparation, and (b) PZM sensor fabrication steps.

Experimental Section

A. Fabrication of PVDF/ZnO@MXene(PZM) Sensor

The schematic diagram in **Fig. 1(a)** presents the key fabrication processes of the PZM composite piezoelectric film. Initially, a flexible PVDF fiber membrane substrate is prepared using electrospinning technology. Following this, a hydrothermal reaction is performed in a high-pressure reactor at 90°C for 3, 5, and 7 hours to obtain a PVDF composite film, which facilitates the growth of ZnO nanorods, labeled as PZ. Lastly, the MAX phase precursor is etched following the method outlined in the literature to yield a multilayer MXene aqueous solution [7]. This solution is diluted to a specific mass fraction of 5% and subsequently spin-coated onto the PZ composite film, forming the PZM active layer for the piezoelectric sensor. As shown in **Fig. 1(b)**, the resulting PZM fiber membrane is cut into dimensions of 1.0 × 2.0 cm. Conductive copper foil is used as the top and bottom electrodes for the device, while the outermost layer is encapsulated with a polyurethane film for protection.

B. Characterization and Measurement

The surface morphology of the samples was analyzed using scanning electron microscopy (SEM). The presence of the β phase in the material was confirmed by Fourier transform infrared spectroscopy (FTIR), while X-ray diffraction (XRD) and X-ray photoelectron spectroscopy (XPS) were employed to investigate the material composition.

C. Sensor Testing and Integrated System Research

External force was applied using a universal testing machine (Mark-10, ESM 303). The PZM piezoelectric sensor is integrated with the gate of the ITZO TFT, forming a unified device structure. The TFT amplifies and controls the electrical signals, while the piezoelectric sensor detects mechanical stimuli and converts mechanical pressure into electrical signals. The electrical performance parameters of the device were measured using a semiconductor analyzer (Agilent B1500).

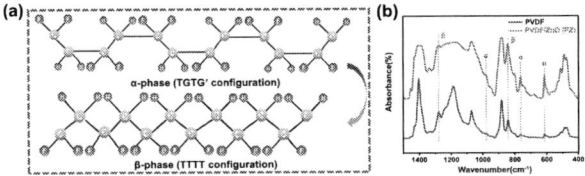

Fig. 2. (a) Schematic diagram illustrating the transition between the α and β conformations in PVDF. (b) FTIR spectra of pure PVDF and PVDF/ZnO (PZ) membrane.

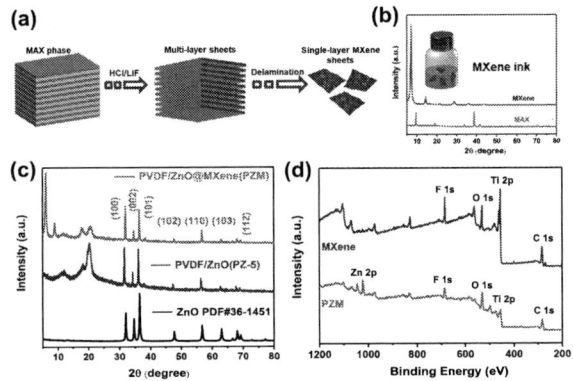

Fig. 3. (a) Schematic diagram of the preparation process of multilayer MXene aqueous solution by etching MAX phase raw materials. (b-c) XRD spectra of the MAX/MXene (b), the ZnO, PZ-5, and the PZM composite membrane (c). (d) XPS full spectrum of the PZM and MXene film.

Results and Discussion

A. Device Characterization

Fig. 2(a) presents a schematic depiction of its conformational structure. Semicrystalline PVDF exists in three conformations: the non-electroactive α phase, the semi-polar γ phase, and the highly electroactive β phase. These phases result from different arrangements of trans (T) and gauche (G) bond conformations, with the α phase adopting an alternating TGTG' structure and the β phase having an all-trans (TTTT) structure. In the α phase, antiparallel dipole arrangement renders it non-polar and electrically inactive, while the β phase has a higher dipole moment, leading to significant electrical activity [8]. The α-to-β phase transformation can be induced through treatments such as stretching and polarization. **Fig. 2(b)** presents the FTIR spectra of the PVDF fiber and PVDF/ZnO (PZ) composite membrane. The characteristic peaks of the PVDF α phase are observed at 613 cm⁻¹, 763 cm⁻¹, and 977 cm⁻¹, with the 763 cm⁻¹ peak corresponding to -CH₂- bending vibrations. In contrast, the β phase shows prominent peaks at 842 cm⁻¹ and 1278 cm⁻¹. A relatively mild etching method is used to prepare a multilayer MXene aqueous solution, as illustrated in **Fig. 3(a)**. **Fig. 3(b)** shows that the Al layer (2θ = 39.2°) in Ti₃AlC₂ (MAX) is completely etched away. In addition, MXene only left a strong characteristic peak (002) at 6.5° and a weak high-order diffraction peak (004) at 15.2°, indicating that the prepared MXene was fully etched with high purity based on the comparison of the peaks of MAX. **Fig. 3(c)** presents the XRD analysis of various materials, showing distinct ZnO peaks in the PZ-5 composite film. After introducing higher-crystallinity MXene, the PVDF peak height decreases significantly, but PZM retains a prominent ZnO diffraction peak, corresponding to the hexagonal wurtzite structure. XPS analysis, as shown in **Fig. 3(d)**, reveals a noticeable Zn 2p peak in the PZM composite, absent in MXene. In **Fig. 4** we delineate the narrow-scan Ti 2p and C 1s XPS spectrum of PZM. This spectrum delineates into Ti 2p₃/₂ (452.24 eV) and Ti 2p₁/₂ (457.91 eV) components. Furthermore, the narrow-scan C 1s XPS spectrum of PZM exhibits four distinctive components at 278.69, 281.58, 282.87, and 287.56 eV in **Fig. 4(b)**, attributed to C-Ti-Tₓ, C-C, C-O/CHₓ, and -COO functionalities, respectively.

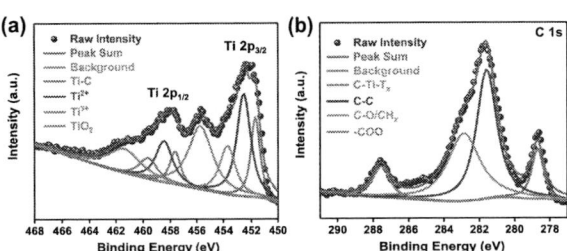

Fig. 4. (a) The Ti 2p XPS subdivision spectra of PZM. (b) The C 1s XPS subdivision spectra of PZM.

B. Piezoelectric Sensor Application

Fig. 5(a) illustrates the results of performance tests related to the piezoelectric sensor. Among them, PZM demonstrates the highest sensitivity, reaching 2.32 V N⁻¹. As shown in **Fig. 5(b)**, leaf swiping with a handheld leaf generates a signal of ~30 mV, demonstrating the sensor's ability to detect small signals with a response time of 46 ms. **Fig. 5(c)** shows the dynamic signals produced when objects are placed on or removed from the surface of the piezoelectric film, highlighting the distinct responses for common laboratory items (centrifuge tube, and marker). The PZM sensor exhibits clear piezoelectric output signals during these actions, demonstrating its capacity to detect external mechanical stimuli.

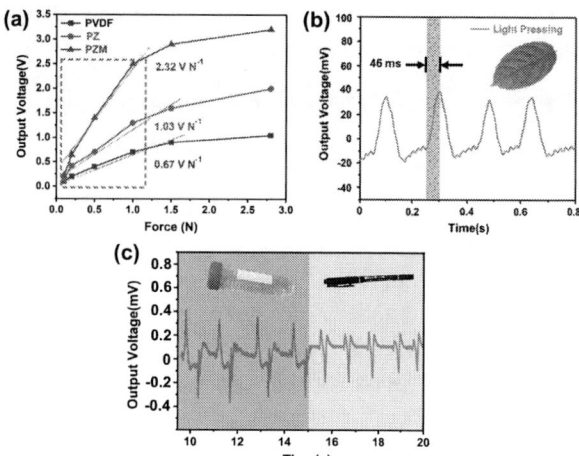

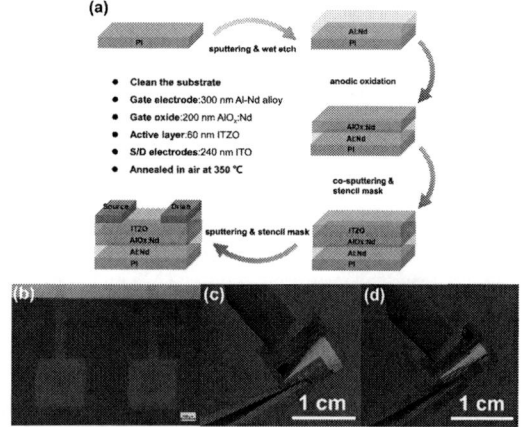

Fig. 5. (a) Sensitivity comparison of three types of fiber membrane-based sensors. (b) Display of output voltage under leaf swiping state. (c) Applications of PZM sensor in detecting objects.

C. Integrated ITZO TFTs Application

Fig. 6 shows a TFT preparation process flow chart and a physical picture of the obtained TFT. The preparation method of ITZO TFT and the circuit of sensor integration have been described in detail in the published work [9]. TFT exhibits the ability to endure varying degrees of bending, demonstrating excellent flexibility. **Fig. 7(a)** shows a circuit system diagram for sensor signal acquisition. The input pressure signal is finally collected by an oscilloscope. As shown in **Fig. 7(b)**, a light finger touch (F = 0.1N) generates an output voltage of 200 mV without the integrated TFT, whereas the signal is amplified to 1.8 V with the TFT integration. This amplification significantly enhances the sensor's sensitivity, allowing it to detect subtle changes more effectively and improving compatibility with subsequent electronic processing circuits.

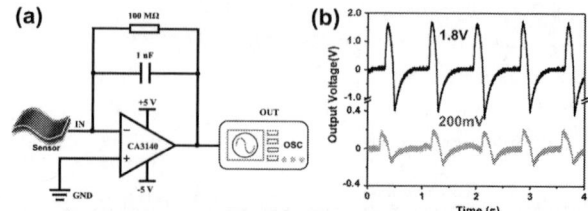

Fig. 7. (a) Sensor test circuit system diagram. (b) The amplification effect of TFT integration on small output signal (F = 0.1N).

Conclusion

In summary, a flexible piezoelectric sensor with a simple fabrication process was developed, featuring fast response (46 ms), high sensitivity (2.32 V N⁻¹), and low energy consumption, making it highly promising for wearable electronics. The sensor effectively identifies various items and amplifies weak signals by about 10 times when integrated with ITZO TFT. This enhancement supports future applications in human-computer interaction, soft robotics, and smart sensor networks, particularly for IoT integration.

Acknowledgments

This work was financially supported in part by the National Natural Science Foundation of China (62374060), and the Science and Technology Program of Guangzhou (2024A04J6314).

References

[1] P. Zhu et al., "Flexible 3D architectured piezo/thermoelectric bimodal tactile sensor array for E-skin application," *Adv. Energy Mater.*, vol. 10, no. 39, Aug. 2020, Art. no. 2001945.

[2] Y. Tang et al., "Flexible, Transparent, Active-Matrix Tactile Sensor Interface Enabled by Solution-Processed Oxide TFTs," IEDM, (2022).

[3] J. Yu et al., "Highly skin-conformal wearable tactile sensor based on piezoelectric-enhanced triboelectric nanogenerator," *Nano Energy*, vol. 64, Oct. 2019, Art. no. 103923.

[4] Y. Tan et al., "High-performance textile piezoelectric pressure sensor with novel structural hierarchy based on ZnO nanorods array for wearable application," *Nano Res.*, vol. 14, no. 11, pp. 3969–3976, Mar. 2021.

[5] K. I. Park et al., "Flexible nanocomposite generator made of BaTiO₃ nanoparticles and graphitic carbons," *Adv. Mater.*, vol. 24, no. 22, pp. 2999–3004, Jun. 2012.

[6] J. Liu et al., "Flexible and lead-free piezoelectric nanogenerator as self-powered sensor based on electrospinning BZT-BCT/P(VDF-TrFE) nanofibers," *Sens. Actuators A Phys.*, Mar. 2020.

[7] M. Naguib et al., "Two-dimensional nanocrystals produced by exfoliation of Ti₃AlC₂," *Adv. Mater.*, vol. 23, no. 37, pp. 4248–4253, Aug. 2011.

[8] X. Cai, T. Lei, D. Sun, and L. Lin, "A critical analysis of the α, β and γ phases in poly(vinylidene fluoride) using FTIR," *RSC Adv.*, vol. 7, no. 25, pp. 15382–15389, Mar. 2017.

[9] M. Yang et al., "A wearable piezoelectric sensor for static-dynamic signal detection and ITZO TFT integrated research", International Conference on Display Technology, (2024).

Fig. 6. (a) Preparation process of the ITZO TFT. (b) Partial image of the TFT under microscope. (c-d) Flexibility of the ITZO TFT.

Investigation of Reliability and Optimization of Reprogram Process in 3D NAND Flash Memory Based on Physical Model

Jooyoung Lee[1,2], Jinil Yoo[1,2], and Hyungcheol Shin[1,2]

[1]Seoul National University, Korea, [2]Integra Semiconductor, Ltd

ABSTRACT

In this study, we simulate a 3D Bandgap Engineered (BE-TOX) NAND Flash device using the Monte Carlo method. The simulation considers the Incremental Step Pulse Programming (ISPP) and Reprogram Scheme (Re-PGM), which are commonly used in the current industry. Through this research, we first investigate the formation process of the threshold voltage (V_{th}) distribution and analyze it from a physical perspective. Furthermore, we identify the trade-off relationship between reliability and performance. Finally, we propose strategies to optimize the program operation.

Keywords: 3D NAND Flash memories, Bandgap Engineering, Incremental Step Pulse Programming, Reprogram, Threshold Voltage Distribution

INTRODUCTION

The currently used 3D NAND Flash memory employs Multi-Level Cell (MLC) technology to increase bit density. However, storing multiple bits in a single cell introduces reliability issues [1]. Fig 1 shows the V_{th} distribution in a Quad-Level Cell (QLC) device, where narrower distribution widths of each Program Voltage (PV) level are more advantageous. [2]

To achieve this, ISPP and Re-PGM are applied in the industry. ISPP sequentially increases the voltage applied to the cell and verifies it, reflecting the program speed of each cell. Re-PGM divides the program operation into two stages to reduce cell-to-cell interference (Z-int). An initial rough intermediate distribution is created using Foggy PGM (which employs a large step pulse), followed by precise Fine PGM to generate the final distribution. In this study, we conduct simulations incorporating these methods into a physical compact model to analyze the V_{th} distribution.

PROCESS TECHNOLOGY

The Monte Carlo method was employed [3], incorporating cell variation to account for process deviations. Specifications of cell variation are shown in the appendix. We utilized a physical model of the 3D BE-TOX NAND Flash [4], and used Integra-NANDSim.C for simulation which was developed by Integra Semiconductor.

The simulation was conducted on 8,192 (2^{13}) memory cells [5], [6], based on a QLC device with one Erase state and fifteen Program states. ISPP noise was modeled using a Poisson distribution [7], and Random Telegraphic Noise (RTN) was incorporated based on the exponential fitting using a Laplacian random generator [8]. Z-int, the coupling effect between adjacent cells, where V_{th} distribution shift in the Victim Cell due to program of the Aggressor Cell was accounted for, was modeled using exponential function [9]. Equations are shown in Appendix, and flowchart is presented in Fig 2.

In the implementation of ISPP and Re-PGM, design variables include the number of levels in the foggy distribution, ISPP step voltage (V_{step}), foggy verify level, and fine verify level. The number of levels in the foggy distribution, which is closely related to program time and reliability, was set to 2, 4, 8, and 16 for comparison. For optimization, $V_{step,fine}$ and the foggy-fine level distance were expressed using the variable α, as shown in (1).

$$PV_{fine} - PV_{foggy} = \alpha V_{step,fine}. \quad (1)$$

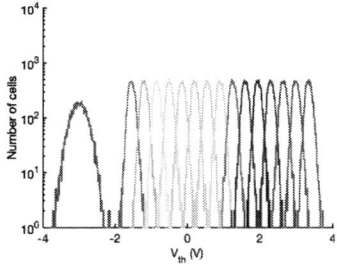

Fig. 1. V_{th} distribution of QLC device, 1 erase state and 15 program state.

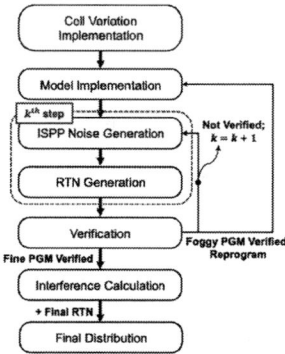

Fig. 2. Simulation flowchart, calculate ISPP noise and RTN for each step, and apply z-int for final distribution.

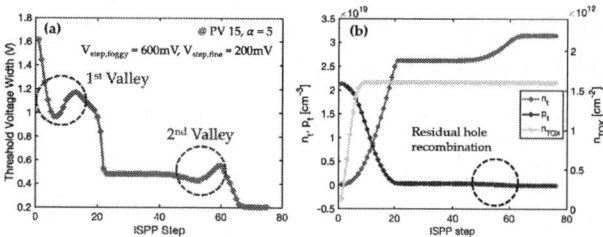

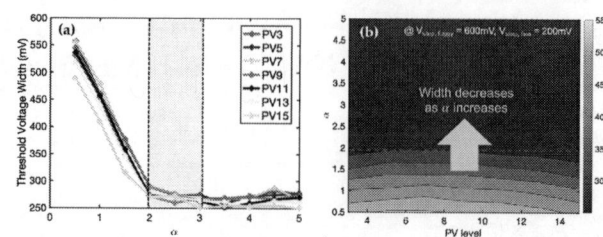

Fig. 3. (a) Distribution width and (b) Carrier concentration for ISPP step.

Fig. 5. (a) Distribution width for α, (b) Contour of distribution width, same trend for each PV level.

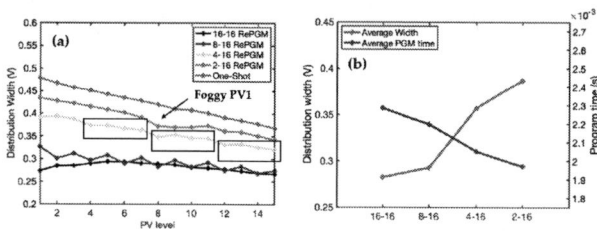

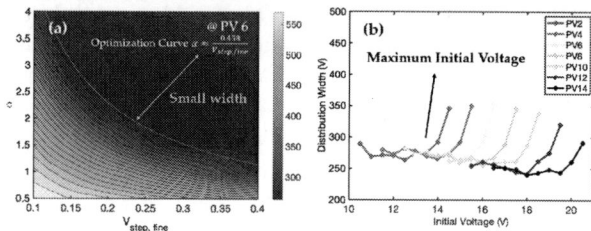

Fig. 4. (a) Distribution width for each foggy level setting, (b) Trade-off of reliability and performance.

Fig. 6. (a) Coutour plot and optimization curve, (b) Distribution width when increasing fine PGM initial voltage.

RESULTS

A. Process Analysis

In Fig 3 (a), the distribution width during the program process can be observed. Due to the two program stages, two valleys are shown. These valleys can be subdivided into four segments: (1) width decrease, (2) width increase, (3) width decrease again, and (4) abrupt width decrease. Segment (1) occurs due to the influence of residual holes, where cells biased toward the left with a higher concentration of holes in the initial distribution experience a faster program due to the hole-induced gate voltage supplementation effect [10]. Segment (2) arises from a decrease in recombination rate and an increase in electron trapping into unoccupied traps [11]. Segment (3) is attributed to feedback effects from already trapped electrons, while segment (4) corresponds to the Verify process. Fig 3 (b) illustrates the carrier concentration during the same process. The decrease in number of holes due to recombination and the injection of electrons are evident in both Foggy and Fine program stages. The trapped charges in the TOX reach a nearly saturated value after Foggy PGM.

B. Trade-off

Fig 4 (a) compares the distribution width based on the number of PV levels in Foggy PGM. As the number of PV levels in Foggy PGM decreases, the V_{th} shift increases, leading to a greater impact of Z-int and an increase in distribution width. Compared to the 16-16 Re-PGM, the 8-16, 4-16, and 2-16 Re-PGM setups exhibit patterns corresponding to 2, 4, and 8 levels, respectively. This results from the increase in the number of Fine levels (2, 4, 8) separated from a single Foggy level, with different foggy-fine level distances.

On the other hand, reducing the number of PV levels in Foggy PGM allows for larger $V_{step,foggy}$, which reduces the program time and improves device performance. This trade-off relationship is illustrated in Fig 4 (b).

C. Optimization

Fig 5 (a) shows the distribution width as a function of the parameter α (with $V_{step,fine}$ fixed). It can be observed that as α increases, the distribution width decreases until it reaches saturation. When α is small, the distance between the foggy distribution and the fine distribution is small, leading to an immediate reach to the fine verify level without reducing the distribution width. When the foggy-fine distance is sufficiently large, this issue is resolved, and the distribution width is no longer affected. However, if α is set unnecessarily high, the number of ISPP steps increases, resulting in a longer program time and reduced device speed, thus it is appropriate to use the minimum value at which saturation occurs. Fig 5 (b) displays a contour plot illustrating the trend with respect to the PV level. Each coordinate represents the distribution width. The same trend is observed across all PV levels.

Fig 6 (a) shows the contour plot for α and $V_{step,fine}$ at a fixed PV level. Each coordinate represents the distribution width. A local minimum region exists along the diagonal of the contour, and as the graph extends outward, the width increases again due to the influence of Z-int. Optimization curve was extracted from the innermost region for minimize program time. Since the curve takes the form of a rational function, the curve provides the optimal distance between the foggy verify level and the fine verify level as a constant, referring to equation 3.

At Fig 3 (a), a constant region can be observed at the early stage of Fine PGM. This leads to an increase in program time, which can be mitigated by raising the initial voltage of Fine PGM. However, an excessively high initial voltage may increase ISPP noise or cause the distribution to immediately reach the verify level. Therefore, adjustments should be made to avoid these issues. The curve in Fig 6 (b) shows the maximum initial voltage that does not affect the distribution width.

CONCLUSION

In this paper, we analyze the formation of the V_{th} distribution in 3D NAND Flash memory based on simulation and propose optimization methods. The study takes into account ISPP and Re-PGM, offering reference points and directions for improvement in memory device design. Furthermore, it provides simulation-based predictions that can guide the development of higher bit density technologies such as PLC and HLC.

APPENDIX

TABLE I: Cell Variation

Quantities	Mean [nm]	Standard Deviation [nm]
R_f (Filler Radius)	30	1.3
$t_{Channel}$ (Channel Thickness)	8	0.4
t_{TOX} (TOX Thickness)	4	0.23
t_{CTL} (CTL Thickness)	4	0.23
t_{BOX} (BOX Thickness)	8	0.23
L_{WL} (Wordline Length)	22	0.4

$$f_{\text{RTN}}(x|\mu, b) = \frac{1}{2b} \exp\left(-\frac{|x - \mu|}{b}\right) \quad (2)$$

where μ is center of distribution, and b is fitting parameter.

$$\Delta V_{T,Z-int.} = \left(a_1 e^{-a_2 V_{T,vic}^{before}} + a_3\right) \\ \times a_4 \left(V_{T,agg}^{after} e^{a_5 V_{T,agg}^{after}} - V_{T,agg}^{before} e^{a_5 V_{T,agg}^{before}}\right). \quad (3)$$

where a_k (k = 1 ∼ 5) represent the fitting parameters

REFERENCES

[1] C. Monzio Compagnoni, A. Goda, A. S. Spinelli, P. Feeley, A. L. Lacaita, and A. Visconti, "Reviewing the evolution of the nand flash technology," *Proceedings of the IEEE*, vol. 105, no. 9, pp. 1609–1633, 2017.

[2] A. S. Spinelli, C. M. Compagnoni, and A. L. Lacaita, "Reliability of nand flash memories: Planar cells and emerging issues in 3d devices," *Computers*, vol. 6, no. 2, p. 16, 2017. [Online]. Available: https://www.mdpi.com/2073-431X/6/2/16

[3] C.-C. Hsieh, H.-T. Lue, T.-H. Hsu, P.-Y. Du, K.-H. Chiang, and C.-Y. Lu, "A monte carlo simulation method to predict large-density nand product memory window from small-array test element group (teg) verified on a 3d nand flash test chip," in *2016 IEEE Symposium on VLSI Technology*, 2016, pp. 1–2.

[4] M. Kim, S. Kim, and H. Shin, "A compact model for ispp of 3-d charge-trap nand flash memories," *IEEE Transactions on Electron Devices*, vol. 67, no. 8, pp. 3095–3101, 2020.

[5] K. Wang, G. Du, Z. Lun, W. Chen, and X. Liu, "Modeling of program vth distribution for 3-d tlc nand flash memory," *Journal of Physics: Conference Series*, vol. 62, no. 1, p. 42401, 2019.

[6] W. Hou, L. Jin, X. Jia, Z. Wang, Q. Wang, Z. Luo, D. Li, F. Xu, and Z. Huo, "Investigation of program noise in charge trap based 3d nand flash memory," *IEEE Electron Device Letters*, vol. 41, no. 1, pp. 30–33, 2020.

[7] C. Monzio Compagnoni, A. S. Spinelli, R. Gusmeroli, S. Beltrami, A. Ghetti, and A. Visconti, "Ultimate accuracy for the nand flash program algorithm due to the electron injection statistics," *IEEE Transactions on Electron Devices*, vol. 55, no. 10, pp. 2695–2702, 2008.

[8] C. Monzio Compagnoni, R. Gusmeroli, A. S. Spinelli, A. L. Lacaita, M. Bonanomi, and A. Visconti, "Statistical model for random telegraph noise in flash memories," *IEEE Transactions on Electron Devices*, vol. 55, no. 1, pp. 388–395, 2008.

[9] H. Jo, S. Ahn, and H. Shin, "Investigation and modeling of z-interference in poly-si channel-based 3-d nand flash memories," *IEEE Transactions on Electron Devices*, vol. 69, no. 2, pp. 543–548, 2022.

[10] J. Yoo and H. Shin, "A compact model-based threshold voltage distribution simulation of 3d gate-all-around (gaa) nand flash memories," in *2024 8th IEEE Electron Devices Technology Manufacturing Conference (EDTM)*, 2024, pp. 1–3.

[11] C. Monzio Compagnoni, R. Gusmeroli, M. Ghidotti, A. S. Spinelli, and A. Visconti, "Investigation of the electron-injection spread in barrier-engineered nand flash memories," *IEEE Electron Device Letters*, vol. 30, no. 7, pp. 769–771, 2009.

979-8-3315-0417-5/25 $31.00 © 2025 IEEE

Circuit Polymorphism Enabled by RFET Devices Processed on Industrial FDSOI

N. Bhattacharjee[1], G. Galderisi[1], Y. He[1], V. Sessi[2], M. Drescher[2], V. Havel[1], M. Zier [2], M. Simon[1], K. Ruttloff[2], K. Li[2], A. Zeun[2], A.-S. Seidel[2], C. Metze[2], M. Grothe[2], S. Jansen[2], M. Wijvliet[3], S. Rai[3], A. Kumar[3], S. Slesazeck[1], J. Hoentschel[2], T. Mikolajick[1,4], and J. Trommer[1]

[1] NaMLab gGmbH, 01187, Dresden, Germany [2] GlobalFoundries Fab1 LCC & Co. KG, Dresden, Germany [3] Chair of Processor Design, TU Dresden, 01187, Germany [4] Chair of Nanoelectronics, TU Dresden, 01187, Germany

Abstract

We present three-independent-gate Reconfigurable Field Effect Transistors, processed on a 300 mm industrial platform. The devices, able to function as both n-type and p-type transistors, were built on a GlobalFoundries fully-depleted silicon-on-insulator technology, and show highest symmetry between the on-state currents of both polarity modes, as well as a clearly defined multi-V_T behavior. Based on them, we show electrical transient measurements demonstrating the functionality of a highly reconfigurable logic gate, the RGATE, able to yield up to eight different logic functions using only four transistors. Furthermore, we developed a Verilog-A table model of the presented transistors, that we used to build a 2-bit adder/2-bit half subtract reconfigurable circuit demonstrating the functionality of a highly tiled, security-oriented architecture employing only RGATEs.

Introduction

With the market of specialized mobile devices overtaking classical processor design, the functional demands to CMOS technology platforms have changed. It is expected that this functional diversification trend will rapidly expand in the next years, setting new demands for specialized electronic devices, e.g., for artificial intelligence (AI) or automotive applications at the edge. GlobalFoundries' fully-depleted silicon-on-insulator (FDSOI) technologies already support a variety of add-on elements ranging from logic and mm-wave/RF to embedded non-volatile memory [1]. Here, we demonstrate the potential of Reconfigurable Field Effect Transistors (RFETs) to serve as new candidates for future add-on functionality, enhancing security of edge systems [2]. RFETs are a CMOS-compatible, Schottky barrier-based type of beyond-Moore emerging devices, characterized by multiple independent gates [3]. They can deliver both n-type and p-type electrical characteristics within the same transistor, and they also provide a multi-V_T behavior, when considered in their three-independent-gate version (TIG-RFET) [4]. Benefits arising from the reconfigurability features of these transistors, are however not always evident in the individual device metrics, but reveal their impact at the circuit level. In fact, when RFETs are combined into circuits, it is possible to realize polymorphic logic gates: these are especially relevant in hardware security applications, because of their inherent layout obfuscation properties [5] as well as their resistance against both delay [6] and power side-channel [7] attacks.

TIG-RFETs on Industrial FDSOI

The TIG-RFETs, fabricated on an FDSOI platform from GlobalFoundries, share nearly all process modules with the CMOS baseline technology, including the back-contact formation, the gate-first high-k metal gate (HKMG) stack, a set of spacers, and the complete back-end-of-line (BEOL). Modified source/drain contacts have been developed to allow the formation of silicide to silicon junctions reaching into the doping-free channel regions positioned below the outer gates. A design solution avoiding the need for additional masks was developed to prevent the silicide formation between the gates of a single device. Each gate is 110 nm long with a gate pitch of 220 nm. The voltage applied to the drain gate (DG) of a TIG-RFET is always used to set its polarity: it blocks the injection in the channel of the undesired carriers. A positive bias is applied for n-type operation, a negative bias for p-type operation. When the device is turned on and off with an input signal applied to the central gate (CG), and the source gate (SG) is also biased to program the device polarity, the RFET operates in the low-V_T mode. Applying the switching input to the source gate, and the programming bias also to the central gate (CG) sets the device to the high-V_T mode. Measured transfer and output characteristics of an exemplary device are shown in Fig. 1. A back-bias of 90 mV yields symmetric on-currents of 2.6 (2.5) μA/μm for n-type (p-type) programmed devices, respectively, independent of the threshold mode. Threshold voltages are equal to 0.32 V and -0.74 V for n-type and p-type devices operated in the low-V_T mode, respectively, and 1.02 V and -1.17 V for the n-type and p-type programmed devices, in the case of the high-V_T mode, respectively. TIG-RFETs can be combined in the four-transistor cell design shown in Fig.2 to realize a polymorphic logic gate able to yield up to 8 different logic functions, the RGATE [8]. Here, we replicate its functionality on an industrial process.

979-8-3315-0417-5/25 $31.00 © 2025 IEEE

Electrical transient measurement of NOT, NAND, NOR, XOR, XNOR, MUX, and 3MIN logic operators, obtained reconfiguring the same logic gate, are reported in Fig.3.

Modelling for Circuit Extrapolation

To enable more complex demonstrations of the capabilities of this technology, we developed a Verilog-A table model based on TCAD process simulations of a TIG-RFET, whose dimensions obey all core-design rules of the 22 nm FDSOI technology like the minimal gate pitch (2F). To yield symmetric on-state currents for both n-type and p-type polarities, we used the following parameters: 4.56 eV and 4.7 eV for the gate metal and the source/drain contacts work functions, respectively, $0.12 \cdot m_0$ and $0.134 \cdot m_0$ effective masses for electrons and holes, respectively. Default parameters for all the materials were used. The Verilog-A table model was populated by varying all the voltages between each of the terminals from -0.8 V to 0.8 V with a bias point granularity of 0.1 V. Linear extrapolation was used to produce the not extracted biasing, providing good convergence and speed performances for the SPICE model. Finally, both the TCAD and the Verilog-A model were verified against each other, showing excellent correlation as shown in Fig.4. We exploited the developed model to verify the functionality of the RGATE. The polymorphic logic gate was redesigned obeying core-device and BEOL design rules of the technology, producing a layout with an area of $612 \cdot F^2$.

Circuit Polymorphism for Hardware Security

A hardware security scenario that can benefit from the polymorphic properties of the RGATE is dynamic circuit obfuscation. Here, a given circuit representation can yield multiple valid functions exploiting the reconfiguration of some circuit sub-blocks, while keeping a functionality-invariant appearance. Due to their symmetry features between pull-up and pull-down networks, RFET-based logic gates are well suited to represent self-dual functions such as XOR/XNOR. This property can be exploited at logic synthesis level, using XOR-Majority-Graph (XMG) transformations, to map any given circuit with a certain subset of functions implementable with RGATEs. When circuits are mapped in this manner, it is possible to find common shared paths and paths of the corresponding self-dual function, and their representations can be merged, realizing a super-circuit whose functionality can be set by the inputs used to reconfigure the self-dual common gates, i.e., the keys. We designed and tested several reconfigurable logic circuits by the tiled connection of the presented polymorphic logic gates, as shown in Fig. 5. Often, mapping higher order logic functions in a combination of lower order ones allows to add a higher number of keys in a trade-off for area. While some circuits shared a natural subset of functions (like adder

and subtractor), logic optimization was beneficial to yield results including multipliers, in particular transforming OR-AND-INV gates into their fragments. MUXs were added if needed to allow for different unequal number of inputs. We selected the specific case of a polymorphic 2-bit adder/half-subtract circuit and verified it by simulating the waveforms for both configurations in SPICE, using the developed Verilog-A model, as shown in Fig. 6. A nine-digit key was inserted to switch between adder and subtract functions while producing wrong output patterns in the other cases. This approach promises to hinder layout reverse engineering and oracle-less attacks since no information on the intended functionality of the circuit can leak with the key itself.

Conclusion

We presented the fabrication and the electrical characteristics of TIG-RFETs on an industrial FDSOI technology, and we demonstrated the electrical functionality of a polymorphic logic gate, the RGATE. Based on this concept, we simulated an obfuscated reconfigurable circuit yielding 2-bit adder and 2-bit half subtract functionalities, developing a scaled Verilog-A table model of the individual devices.

Acknowledgments

This work was partially financed by the EU under the grant agreement no. 101135316, by the German Federal Ministry of Education and Research (BMBF) under the framework of VE-CirroStrato, and from the tax revenues on the basis of the budget adopted by the Saxon State Parliament.

References

[1] R. Carter, et al. "22nm FDSOI technology for emerging mobile, Internet-of-Things, and RF applications." *International Electron Devices Meeting* (2016). doi:10.1109/IEDM.2016.7838029

[2] S. Rai, et al. "Security promises and vulnerabilities in emerging reconfigurable nanotechnology-based circuits." *IEEE TECT* (2020). doi:10.1109/TETC.2020.3039375

[3] T. Mikolajick, et al. "Reconfigurable field effect transistors: A technology enablers perspective." *SSE* (2022). doi:j.sse.2022.108381

[4] M. De Marchi, et al. "Polarity control in double-gate, gate-all-around vertically stacked silicon nanowire FETs." *International Electron Devices Meeting* (2012). doi:10.1109/IEDM.2012.6479004

[5] P. Wu, et al. "Two-dimensional transistors with reconfigurable polarities for secure circuits." *Nat. Electron.* (2021). doi:10.1038/s41928-020-00511-7

[6] G. Galderisi, et al. "Reconfigurable Field Effect Transistor design solutions for delay-invariant logic gates." *IEEE EDL* (2022). 10.1109/LES.2022.3144010

[7] N. Kavand, et al. "REDCAP: Reconfigurable RFET-Based Circuits Against Power Side-Channel Attacks." *Design Automation and Test in Europe* (2024). doi:10.23919/DATE58400.2024.10546825

[8] G. Galderisi, et al. "The RGATE: An 8-in-1 Polymorphic Logic Gate Built from Reconfigurable Field Effect Transistors." *IEEE EDL* (2023). doi:10.1109/LED.2023.3347397

979-8-3315-0417-5/25 $31.00 © 2025 IEEE

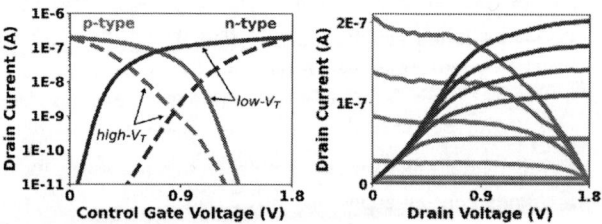
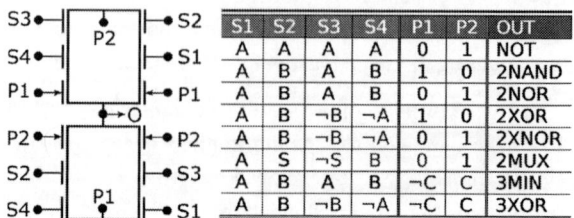

Fig. 1: Transfer (left) and output (right) characteristics of a single reconfigurable FET. N-type (p-) is shown in blue (red), low-V_T (high-V_T) in full (dashed) lines.

Fig. 2 : RGATE: symmetric four transistors polymorphic gate circuit schematic (left) and 8-in-1 logic functionalities look-up table (right).

S1	S2	S3	S4	P1	P2	OUT
A	A	A	A	0	1	NOT
A	B	A	B	1	0	2NAND
A	B	A	B	0	1	2NOR
A	B	¬B	¬A	1	0	2XOR
A	B	¬B	¬A	0	1	2XNOR
A	S	¬S	B	0	1	2MUX
A	B	A	B	¬C	C	3MIN
A	B	¬B	¬A	¬C	C	3XOR

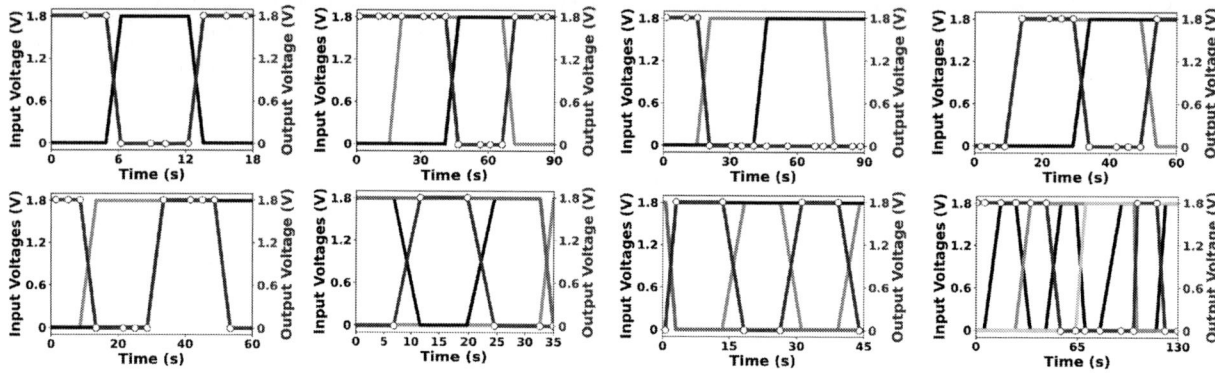

Fig. 3: Electrical measurements of NOT, NAND, NOR, XOR, XNOR, inverting MUX and 3MIN functionalities (left to right) embodied on the same logic gate build of 4 RFETs. The inputs are shown in grey-scale full lines, the output in a blue full line with hollow circles, while for the MUX the selecting input is a red full line. Time resolution is limited by the quasi-stationary measurement setup and the comparatively large pad parasitic capacitances of the test structures.

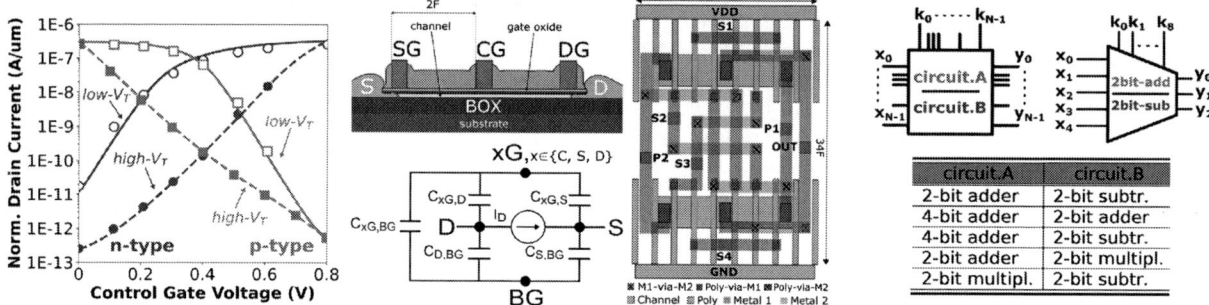

Fig. 4 TCAD (lines) and Verilog-A (symbols) resulting transfer curves (left). A constant back-gate voltage $V_{BG} = 0.56$ V is applied for adjusting symmetry. 2D TCAD process simulation of a RFET based on 22nm core-devices (top-center) and equivalent circuit representation of its Verilog-A table model (bottom-center). Layout of the polymorphic RGATE obeying the 22FDX® PDK design rules (right).

Fig. 5 : Merged circuits results: generic circuit representation (top-left). 2-bit adder/half-subtractor (top-right). Table of tested experimental designs (bottom).

circuit.A	circuit.B
2-bit adder	2-bit subtr.
4-bit adder	2-bit adder
4-bit adder	2-bit subtr.
2-bit adder	2-bit multipl.
2-bit multipl.	2-bit subtr.

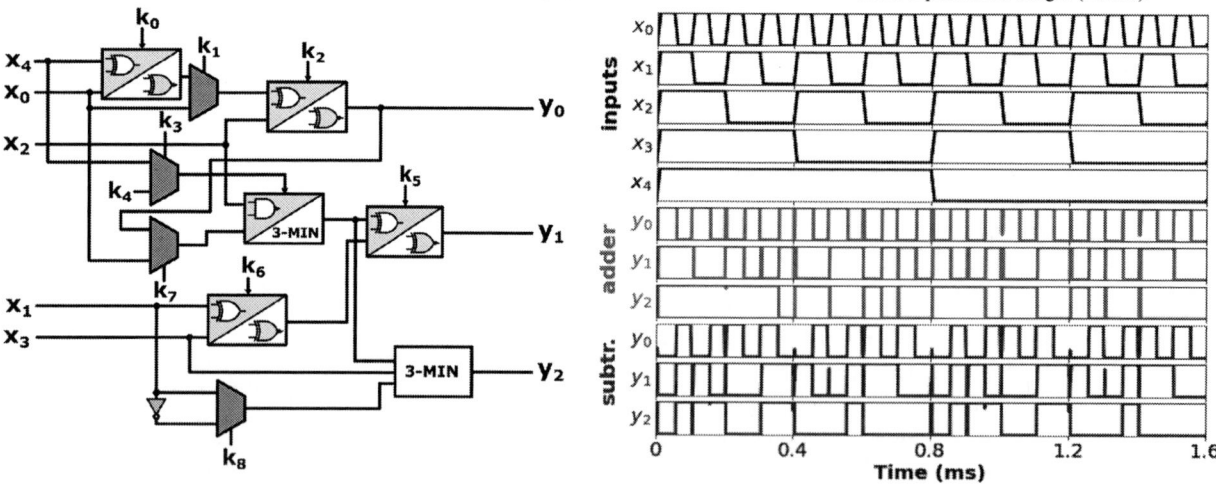

Fig. 6: Circuit diagram of 2-bit adder/half-subtractor built from 10 polymorphic gates (RGATEs), neglecting input inverters. Arrows indicate programming inputs (left). Simulated traces of 2-bit adder/half-subtractor, A = $(x1x0)$, B = $(x3x2)$, $C_{in} = x4$, all signals switch between 0 V and 0.8 V (right).

Optimizing SiN Composition for Enhanced Charge-Trapping in Next-Generation 3D NAND Flash Memories

Tomoya Nagahashi[1,2], Hajime Karasawa[2], Ryota Horiike[2], Atsushi Oshiyama[3], and Kenji Shiraishi[1,3]

[1]Graduate School of Engineering, Nagoya University, Aichi, Japan, nagahashi@imass.nagoya-u.ac.jp,
[2]KOKUSAI ELECTRIC CORPORATION, Toyama, Japan, [3]IMaSS, Nagoya University, Aichi, Japan

Abstract

This study investigates the optimization of charge-trap silicon nitride (CT-SiN) for 3D NAND flash memories. We employed density functional theory (DFT) calculations and experimental analysis to identify the material properties influencing electron-trap levels, trap density, and program/erase (P/E) endurance. Our results demonstrate that a near-stoichiometric N/Si ratio of 1.13 offers the deeper electron traps and higher trap density, leading to improved P/E endurance. This study paves the way for significantly improved performance of CT-SiN in future 3D NAND devices.

Keywords: 3D NAND, charge-trapping, and silicon nitride

Introduction

The current 3D NAND flash memories are predominantly of the charge-trap (CT) type, with the properties of the CT silicon nitride (CT-SiN) layer significantly influencing device performance. As device scaling progresses, the 3D NAND structure becomes increasingly stacked. To achieve further device-scaling, improving electrical properties of CT-SiN is essential, such as charge-retention performance, multi-leveling, and program/erase (P/E) endurance [1–3]. Charge-retention performance will depend on the electron-trap levels and the existence of shallow levels, as shown in Fig. 1. The potential for realizing multi-leveling is strongly related to the concentration of electron-trap structure, as shown in Fig. 2. Regarding P/E endurance, Yamaguchi et al. proposed that reversible structural changes through P/E operation are necessary, as shown in Fig. 3 [4,5]. Previous theoretical studies for CT-SiN microscopic structures were based on β-Si_3N_4 crystal, suggesting that electron-trap structures are associated with N vacancies and its atomic substitutions [4–7]. However, CT-SiN is Si-rich amorphous SiN (a-SiN), and its microscopic structures are expected to differ from β-Si_3N_4 crystal. Consequently, we evaluated the microscopic structures and electrical properties of a-SiN using density functional theory (DFT) calculations and experimental analyses. This study has the following two objectives; (1) to clarify the physical origin of electron-trapping of CT-SiN, (2) to clarify the impact of Si-richness on the CT performance.

Methods

Atomistic simulation

The DFT calculations were performed using VASP [8]. We used GGA approximation with Perdew, Burke, and Ernzerhof (PBE) functional for geometry optimizations. The hybrid functional by Hyde, Scuseria, and Ernzerhof (HSE) [9] is used to calculate the energy levels accurately. We prepared Si-rich a-SiN models with different N/Si ratio: $Si_{48}N_{63}H_3$ (N/Si = 1.31), $Si_{52}N_{59}H_3$ (N/Si = 1.13), and $Si_{56}N_{55}H_3$ (N/Si = 0.98). To create an atomistic structural model, we performed first-principles molecular dynamics (MD) calculation. The systems were first heated up to 5000 K and equilibrated for 10 ps. Subsequently, we quenched at 50 K/ps until 300 K and optimized the structures. An example of the $Si_{52}N_{59}H_3$ structural model is shown in Fig. 4. Table I lists the structural parameters of three different composition $Si_xN_yH_3$. The radial distribution functions (RDF) shown in Fig. 5 were derived based on these models.

Deposition and Analyses of a-SiN

We prepared a-SiN films deposited on Si substrates using LP-CVD by supplying Si_2Cl_6 and NH_3 gases alternately at 923 K. The N/Si rations were controlled by adjusting the exposure amount of these gases. The N/Si ratios and the volume densities were measured by X-ray photoelectron spectroscopy (XPS) and X-ray reflectivity (XRR), respectively. The H concentrations were determined by secondary ion mass spectrometry (SIMS). The bond lengths of Si–N, Si–Si, and Si–N–Si were determined based on the X-ray absorption fine structure (XAFS) of Si K-edge. Electron spin resonance (ESR) was performed to evaluate Si-dangling bond (Si-db) concentrations from the intensities of g-value of 2.003. The bandgaps were characterized using reflection electron energy loss spectroscopy (REELS). Mercury probe CV measurements were performed for 20 nm a-SiN on n-type substrate to derive the flat band voltage shift (ΔV_{FB}). Table II presents the characterization of a-SiN films, which have approximately the same N/Si ratio as the simulation models of $Si_{52}N_{59}H_3$ and $Si_{56}N_{55}H_3$. As presented in Table I and II, the volume densities and H concentrations of simulation models are almost consistent with experimental results. The RDF curves of the simulation models shown in Fig. 5 indicate peak positions corresponding to Si–N, Si–Si, and Si–N–Si bond lengths of 1.74 Å, 2.34 Å, and 3.06 Å, respectively, and each value is almost equivalent to experimental result listed in Table II. These considerations demonstrate that the simulation models are reasonable.

979-8-3315-0417-5/25 $31.00 © 2025 IEEE

Results and discussion

A. Physical Origin of Electron-trapping in CT-SiN

Si-dbs are considered an electron-trap structure, as shown in Fig. 6(a). However, Si-dbs are not a major origin of electron-trap because of the low concentration of Si-db. Figure 7 plots the experimental data of Si-db and H concentrations in a-SiN films with various N/Si ratios. CT-SiN needs to be able to trap over 10^{19} cm^{-3} electron to function as devices accurately [1], however, the Si-db concentrations in a-SiN films are insufficient for electron-trapping. The low concentration of Si-db is due to H terminations (Fig. 6(b)). The energy levels and Kohn–Sham (KS) orbitals of $Si_{48}N_{63}H_3$ model are illustrated in Fig. 8. Here, q denotes the charging state of a unit-cell. If the charging state is neutral ($q = 0$), there is no state in the bandgap. When an electron is captured ($q = -1$), an electron occupied state appears around the mid-gap, which originates from the conduction band minimum (CBM) at $q = 0$. The trapped electron is located around a tetracoordinated Si atom with a larger bond angle. This electron trap site is named the *localized floating state* (LFS) [10]. Si atoms are positively charged due to electronegativity difference from N atoms. Thus, where there is electron-less space around Si atom due to large bond angle, this space can capture an electron, as shown in Fig. 9.

B. Impact of Si-richness

Figure 10 plots the formation energies [11] of $q = +1 \sim -2$ and KS orbitals of electron-trap levels. The charge-trap levels based on formation energies are shown in Table III. The electron-trap levels of $Si_{48}N_{63}H_3$, $Si_{52}N_{59}H_3$, and $Si_{56}N_{55}H_3$ are 3.68 eV ($q = -2$), 3.56 eV ($q = -1$), and 3.36 eV ($q = -1$), respectively. The KS orbitals indicate the LFS captures electrons in each model. Upon investigating the local structures around the LFS, we observed certain differences between these models, such as the number of Si atoms. As the Si atom is slightly positively charged, electrons trapped at the LFS are more stable as the local Si density increases (Fig. 11). Therefore, a higher Si composition of a-SiN results in deeper electron-trap levels. Figure 12 illustrates the change of the bond network structure of $Si_{48}N_{63}H_3$ and $Si_{52}N_{59}H_3$ models during P/E operation. Bond network rearrangement on P/E operation is undesirable, because it causes irreversible structural change and degradation of P/E endurance, as shown in Fig.3. Table IV refers to the structural reversibility of each model during P/E operation. $Si_{48}N_{63}H_3$ charging state of $q = -1$ is prohibited, and the state of $q = -2$ is only allowed during program operation as referred in Table III. However, when they enter the state of $q = -2$, it is difficult to return to the initial state of $q = 0$ due to bond network rearrangement. Conversely, the $Si_{52}N_{59}H_3$ and $Si_{56}N_{55}H_3$ models can be in the $q = -1$ state during the program operation, allowing for reversible structural changes during P/E operation. However,

when they enter the $q = -2$ state, reversible structural changes become difficult. To enhance P/E endurance, it is necessary to prevent the $q = -2$ state. A higher electron-trap density, achieved by increasing Si-richness, is desirable. A higher Si composition increases the number of LFSs, allowing many electrons to be trapped in the $q = -1$ state, thereby improving P/E endurance. In conclusion of this section, more Si-rich a-SiN is expected to have deeper electron-trap levels, higher electron-trap density, and greater P/E endurance.

C. Suitable Si Composition of CT-SiN

Bandgaps were experimentally measured to confirm the existence of shallow levels. Figure 13 shows the relationship between the N/Si ratio and the bandgap of deposited a-SiN films. The bandgap decreases sharply under an N/Si ratio of approximately 1.13, suggesting that bandgap narrowing is caused by the nucleation of shallow levels due to excessive Si. Bandgap narrowing effectively makes the electron-trap level shallower. Figure 14 shows the ΔV_{FB} for samples with different N/Si ratios. Flat band voltage (V_{FB}) depends on the amount of trapped charge, and thus ΔV_{FB} is affected by electron-trap density. ΔV_{FB} peaks at an N/Si ratio of 1.12, indicating that this sample has the highest electron-trap capacity among the samples. These results suggest that the optimal N/Si ratio is approximately 1.13. We show the summary of this study in Fig. 15. It was established that more Si-rich a-SiN has deeper electron-trap levels, higher electron-trap density, and greater P/E endurance. However, there is a limit to Si-richness at an N/Si ratio of approximately 1.13 due to shallow levels.

Conclusion

The electron trapping in Si-rich a-SiN is facilitated by localized floating state (LFS), which originates around a tetracoordinated Si atom. As the Si composition increases, the a-SiN exhibits deeper trap levels, higher electron-trap density, and greater P/E endurance due to the properties of LFS. Taking into account the impact of shallow levels, the optimal N/Si ratio is approximately 1.13. By fine-tuning the N/Si ratio towards 1.13, the performance of CT-SiN can be maximized.

References

[1] W. Chen et al., IEDM, pp. 5.5.1–5.5.4 (2015).

[2] A. Khakifirooz et al., ISSCC, pp. 424–426 (2021).

[3] X. Jia, IEEE J. Electron Devices Soc. 8, pp. 62–66 (2020).

[4] K. Yamaguchi et al., IEDM, pp. 11.5.1–11.5.4 (2009).

[5] K. Yamaguchi et al., IEDM, pp. 5.7.1–5.7.4 (2010).

[6] E. Vianello et al., IEDM, pp. 4.5.1–4.5.4 (2009).

[7] J. Wu et al., IEDM, pp. 4.5.1–4.5.4 (2017).

[8] G. Kresse et al., Phys. Rev. B p. 54 (1996).

[9] J. Heyd et al., J. Chem. Phys. 118, pp. 8207–8215 (2003).

[10] F. Nanataki et al., Phys. Rev. B 106, p. 155201 (2022).

[11] S. Wei, Comput. Mater. Sci. 30, pp. 337–348 (2004).

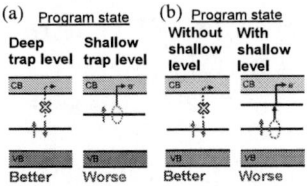

Fig. 1. Influence on retention performance (a) by depth of electron-trap level and (b) by shallow levels in CT-SiN.

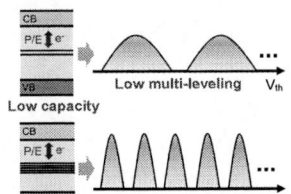

Fig. 2. Likelihood of realizing multi-leveling of NAND devices depending on electron-trap capacity of CT-SiN.

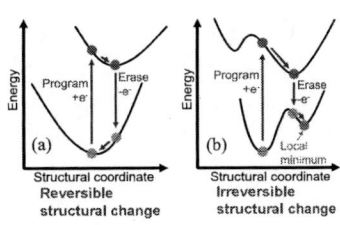

Fig.3. Schematic illustration of energy diagram on program/erase operation for (a) reversible structural change and (b) irreversible structural change. (Ref. [4,5])

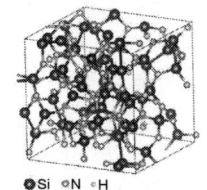

Fig. 4. Constructed simulation model of Si$_{52}$N$_{59}$H$_3$.

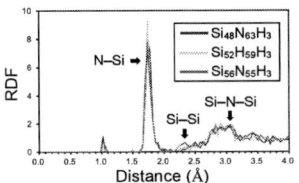

Fig. 5. Calculated radial distribution functions (RDF) of Si$_{48}$N$_{63}$H$_3$, Si$_{52}$N$_{59}$H$_3$, and Si$_{56}$N$_{55}$H$_3$ models. The peaks of 1.74 Å, 2.34 Å, and 3.06 Å correspond to N–Si, Si–Si, Si–N–Si bonds, respectively.

Table I. Structural parameters of the simulation models.

Simulation model	N/Si ratio	One side length of unit-cell (Å)	Density (g/cm³)	Hydrogen (/cm³)
Si$_{48}$N$_{63}$H$_3$	1.31	10.73	3.00	2.43×10²¹
Si$_{52}$N$_{59}$H$_3$	1.13	10.82	3.00	2.37×10²¹
Si$_{56}$N$_{55}$H$_3$	0.98	10.91	3.00	2.31×10²¹

Table II. Characterization of deposited films, which have approximately the same N/Si ratio as the simulation models of Si$_{52}$N$_{59}$H$_3$ and Si$_{56}$N$_{55}$H$_3$.

Sample	N/Si ratio	Density (g/cm³)	Hydrogen (/cm³)	Si–N (Å)	Si–Si (Å)	Si–N–Si (Å)
a-SiN (1)	1.13	2.96	8.1×10²¹	1.71		3.08
a-SiN (2)	0.97	2.94	3.0×10²¹	1.71	2.39	3.10

Experimental data

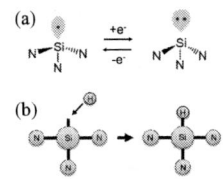
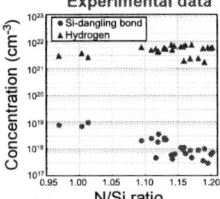

Fig. 6. (a) Electron-trap mechanism of Si-dangling bond (Si-db). (b) Cause of low Si-db concentration.

Fig. 7. Si-dangling bond and H concentrations of deposited a-SiN films with various N/Si ratios.

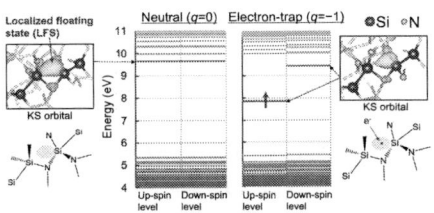

Fig. 8. Energy levels, Kohn–Sham (KS) orbitals, and their structures, at neutral state ($q = 0$) and electron-trap state ($q = -1$) of Si$_{48}$N$_{63}$H$_3$ model. KS orbitals are shown by yellow (+) and blue (−) blobs.

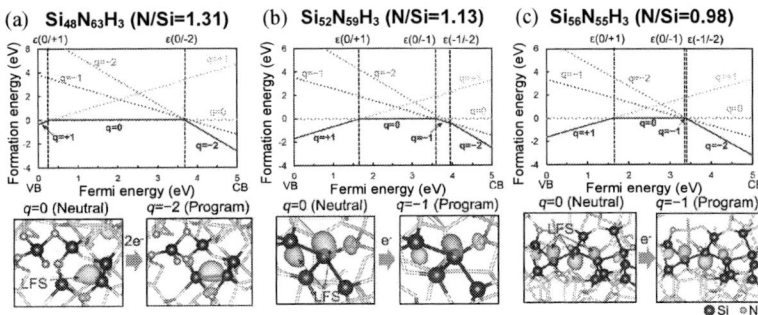

Fig. 10. Formation energies and KS orbitals of (a) Si$_{48}$N$_{63}$H$_3$, (b) Si$_{52}$N$_{59}$H$_3$, and (c) Si$_{56}$N$_{55}$H$_3$ models. The cross points (ε) of formation energies correspond to charge-trap levels. The obtained charge-trap levels are shown in Table III.

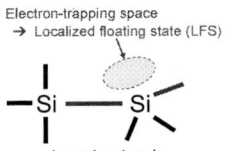

Fig. 9. Location where localized floating state (LFS) generates.

Table III. Charge-trap levels based on formation energies in Fig.10. The energies are calculated as the height from VBM of Si$_{48}$N$_{63}$H$_3$ model.

Simulation model	Hole-trap level ($q=+1$) (eV)	Electron-trap level ($q=-1$) (eV)	Electron-trap level ($q=-2$) (eV)
Si$_{48}$N$_{63}$H$_3$	0.28	3.80	3.68
Si$_{52}$N$_{59}$H$_3$	1.69	3.56	3.96
Si$_{56}$N$_{55}$H$_3$	1.61	3.36	3.41

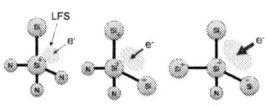

Fig. 11. Change in the depth of electron trap level with increasing local Si density.

Table IV. Structural reversibility on P/E operation based on the bond network rearrangement in Fig. 12.

Simulation model	Structural reversibility during P/E	
	$q=0 \rightleftarrows q=-1$	$q=0 \rightleftarrows q=-2$
Si$_{48}$N$_{63}$H$_3$	–	Irreversible
Si$_{52}$N$_{59}$H$_3$	Reversible	Irreversible
Si$_{56}$N$_{55}$H$_3$	Reversible	Irreversible

Fig. 12. Structural changes on program/erase operations of (a) Si$_{48}$N$_{63}$H$_3$ and (b) Si$_{52}$N$_{59}$H$_3$ models.

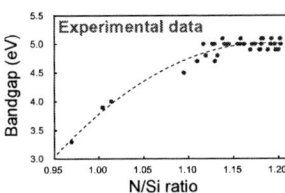

Fig. 13. Experimental bandgaps of deposited a-SiN films with various N/Si ratios.

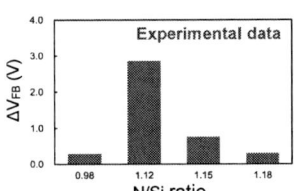

Fig. 14. Flat band voltage shift (ΔV_{FB}) for a-SiN films of N/Si = 0.98, 1.12, 1.15, and 1.18.

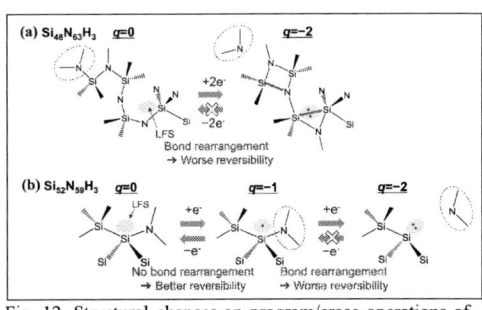

Simulation model	Amount of shallow levels	Depth of electron-trap level	Electron-trap density	Structural reversibility on P/E operation
N/Si=1.31 (Si$_{48}$N$_{63}$H$_3$)	Few	Shallow	Low	Irreversible
N/Si=1.13 (Si$_{52}$N$_{59}$H$_3$)	Few	Middle	Middle	$q=0 \rightleftarrows q=-1$: Reversible / $q=0 \rightleftarrows q=-2$: Irreversible
N/Si=0.98 (Si$_{56}$N$_{55}$H$_3$)	Many (Narrow bandgap)	Deep	High	$q=0 \rightleftarrows q=-1$: Reversible / $q=0 \rightleftarrows q=-2$: Irreversible

N/Si ratio	Retention ability	Multi-leveling	P/E endurance
N/Si=1.31	**Middle**	Worse	Worse
N/Si=1.13	Better	Better	**Middle**
N/Si=0.98	Worse	**Middle**	Better

Fig. 15. Summary of electron-trap properties, and benchmark of N/Si ratio for CT-SiN performance.

Effect of Top Al₂O₃ Interlayer Thickness on the Memory Window of FeFETs with TiN/Al₂O₃/Hf₀.₅Zr₀.₅O₂/SiOₓ/Si (MIFIS) Gate Structure

Tao Hu[1,2,3], Runhao Han[1,2,3], Xinpei Jia[1,2,3], Jia Yang[1,2,3], Zeqi Chen[1,2,3], Xiaoqing Sun[1,2,3], Junshuai Chai[1,2,3], Hao Xu[1,2,3*], Xiaolei Wang[1,2,3*], Wenwu Wang[1,2,3], and Tianchun Ye[1,2,3]

[1]Institute of Microelectronics, Chinese Academy of Sciences, Beijing 100029, China; [2]Key Laboratory of Fabrication Technologies for Integrated Circuits, Chinese Academy of Sciences, Beijing 100029, China; [3]School of Integrated Circuits, University of Chinese Academy of Sciences, Beijing 100049, China. (*E-mail: wangxiaolei@ime.ac.cn; xuhao@ime.ac.cn)

Abstract

In this work, we investigate the effect of top Al₂O₃ interlayer thickness on the memory window (MW) of Si channel ferroelectric field-effect transistors (Si-FeFETs) with TiN/Al₂O₃/Hf₀.₅Zr₀.₅O₂/SiOₓ/Si (MIFIS) gate structure. We find that the MW first increases and then remains almost constant with the increasing thickness of the top Al₂O₃ interlayer. This phenomenon is attributed to the smaller electric field of the ferroelectric Hf₀.₅Zr₀.₅O₂ in the MIFIS structure with a thicker top Al₂O₃ interlayer after a program operation, which makes the charges injected from the metal gate trapped at the top Al₂O₃/Hf₀.₅Zr₀.₅O₂ interface cannot remain.

Keywords: FeFETs, memory window, Hf₀.₅Zr₀.₅O₂, and MIFIS gate structure.

Introduction

Since the discovery of the ferroelectricity in doped-HfO₂ [1], Hafnia (HfO₂) based silicon channel ferroelectric field-effect transistors (HfO₂ Si-FeFETs) have been extensively studied due to its fast read/write speed, lower operating power consumption and excellent scaling capabilities [2, 3]. However, the large spontaneous polarization (P_s) of the ferroelectric doped-HfO₂ (~20-30 μC/cm²) results in the significant charge trapping and de-trapping phenomenon between the ferroelectric Hf₀.₅Zr₀.₅O₂/SiOₓ interface and the Si channel, which makes the MW decrease dramatically [4]. To suppress the charge injection from the Si channel to the ferroelectric Hf₀.₅Zr₀.₅O₂/SiOₓ interface, several studies have been conducted, such as reducing the spontaneous polarization of the ferroelectric [5], and applying high-κ interlayer [6]. Nevertheless, these methods do not significantly improve the MW, which remains limited to less than 2 V. This MW of 2 V is not satisfied for applications in multi-bit memory cells. Recently, inserting a dielectric interlayer (e.g., Al₂O₃ or SiO₂) between the metal gate and ferroelectric layer was found to be an effective method to significantly improve the MW [7]. The MW can achieve 4.1 V by inserting 3 nm Al₂O₃ and 6.4 V (or 8.3 V) by inserting 3.4 nm (or 4 nm) SiO₂ [8-10]. Moreover, the simulation results from [11] point out that using a thicker top interlayer is beneficial for the MW increase.

However, there are still no experimental studies on the effect of top Al₂O₃ interlayer thickness on the MW. Therefore, we experimentally report the effect of top Al₂O₃ interlayer

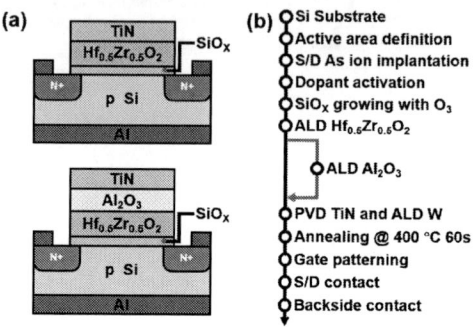

Fig. 1. (a) Schematic of the HfO₂ Si-FeFETs device structure and (b) fabrication process flow.

thickness on the MW of FeFETs with MIFIS gate stacks. We find that the MW first increases and then remains almost constant with the increasing thickness of the top Al₂O₃ interlayer. This phenomenon is attributed to the smaller electric field of the ferroelectric Hf₀.₅Zr₀.₅O₂ in the MIFIS with a thicker top Al₂O₃ interlayer after a program operation, which makes the charges injected from the metal gate cannot remain.

Experiments

Fig. 1(a-b) shows the structure of the gate stacks and the schematic of the fabrication process flow. There are two different gate stacks. One is TiN/Hf₀.₅Zr₀.₅O₂/SiOₓ/Si (MFIS) as the control sample. The other is TiN/Al₂O₃/Hf₀.₅Zr₀.₅O₂/SiOₓ/Si (MIFIS) with a 0.85, 1.7, 2.55, 4.5, 5.5, 8, or 13 nm top Al₂O₃ interlayer. Fig. 2(a-b) shows High-Resolution Transmission Electron Microscopy (HRTEM) images and Energy Dispersion Spectrometer (EDS) results for both MFIS and MIFIS structures. For the MIFIS structure, the presence of a peak concentration of Al at the TiN/Hf₀.₅Zr₀.₅O₂ interface confirms the presence of the top Al₂O₃ interlayer. Ferroelectric Hf₀.₅Zr₀.₅O₂ and top Al₂O₃ interlayer were deposited by atomic layer deposition (ALD) at 300 °C using TDMA-Hf, TDMA-Zr, TMA, and H₂O as precursors for Hf, Zr, Al, and O, respectively. The detailed fabrication procedure can be found in the previous report [10].

The gate length/width (L/W) of devices in this work is 5/150 μm. The electrical measures were performed by Keysight B1500A. The threshold voltage (V_{th}) is extracted by the constant current method at the drain current I_d = W/L × 10⁻⁷ A.

Results and Discussions

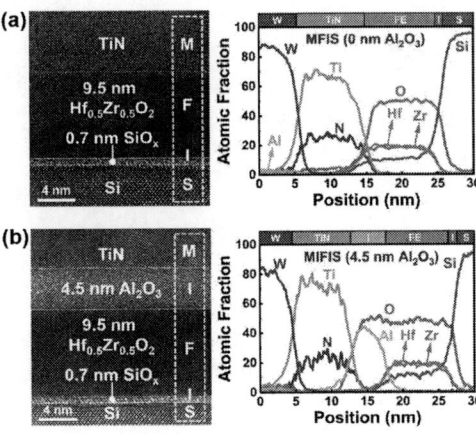

Fig. 2. HRTEM images and EDS of the (a) MFIS structure and (b) MIFIS structures with 4.5 nm top Al_2O_3 interlayer.

We investigate the dependence of the MW on the pulse amplitude. The pulse width is set as 100 μs. Fig. 3 shows the MW mapping results for the MFIS and MIFIS structure with 5.5 nm top Al_2O_3. For the MFIS or MIFIS structure, if the program pulse voltage goes beyond 5 V or 12 V, the devices break down. Thus, we find that the maximum MW is 8.4 V for the MIFIS structure with 5.5 nm top Al_2O_3, while the MW is 1.1 V for the MFIS structure. Thus, this indicates that the top Al_2O_3 interlayer between the metal gate TiN and ferroelectric $Hf_{0.5}Zr_{0.5}O_2$ can significantly improve the MW.

We investigate the effect of the top Al_2O_3 interlayer thickness on the maximum MW. For each sample, we repeated the above MW mapping measurement process. Fig. 4(a) and (b) show the dependence of the maximum MW and the corresponding threshold voltage (V_{th}) on top Al_2O_3 thickness. We can find that the maximum MW first increases and then remains almost constant with the increasing thickness of the top Al_2O_3 interlayer.

Firstly, we discuss the physical origin of stage I of the dependence of the maximum MW and corresponding V_{th} on top Al_2O_3 interlayer thickness. The MW enlargement of the MIFIS structure compared with the MFIS structure is attributed to the presence of the charges ($Q_{it'}$) injected from the metal gate trapped at the top $Al_2O_3/Hf_{0.5}Zr_{0.5}O_2$ interface, as described by [11]. The positive (or negative) charges trapped at the top $Al_2O_3/Hf_{0.5}Zr_{0.5}O_2$ interface after the program (or erase) operation cause the V_{th} to shift negatively (or positively). The shift of the V_{th} with respect to the MFIS sample is calculated by

$$\Delta V_{th} \approx -\frac{(Q_{it'} + Q_{it})}{\varepsilon_0 \varepsilon_{Al_2O_3}} d_{Al_2O_3} \quad (1)$$

Where $Q_{it'}$ is the charges injected from the metal gate trapped at the top $Al_2O_3/Hf_{0.5}Zr_{0.5}O_2$ interface, Q_{it} is the charges injected from the Si channel trapped at the $SiO_x/Hf_{0.5}Zr_{0.5}O_2$ interface, ε_0 is the vacuum dielectric

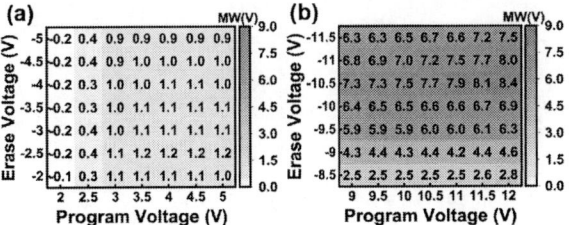

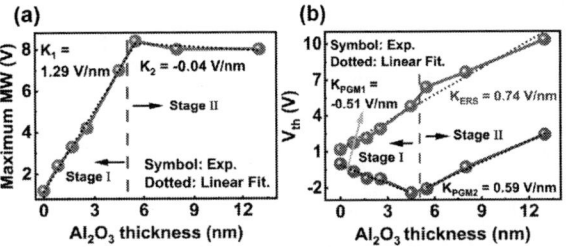

Fig. 3. The MW mapping of (a) MFIS and (b) MIFIS with 5.5 nm top Al_2O_3 as a function of the pulse amplitude.

Fig. 4. The dependence of (a) the maximum MW and (b) the V_{th} corresponding to maximum MW on the top Al_2O_3 interlayer thickness.

constant, $\varepsilon_{Al_2O_3}$ and $d_{Al_2O_3}$ are the relative dielectric constant and physical thickness of the top Al_2O_3. For MFIS structure and MIFIS structure, Q_{it} should be identical due to the same bottom SiO_x interlayer. The V_{th} after the program operation ($V_{th,PGM}$) is linearly dependent on top Al_2O_3 thickness at stage I, as shown in Fig. 4(b), which indicates that the $Q_{it'}$ after the program operation is identical in the MFIS structure and MIFIS structure. Similarly, the identical $Q_{it'}$ after the erase operation results in the linear dependence of the V_{th} after the erase operation ($V_{th,ERS}$) on top Al_2O_3 thickness at stage I. Finally, the above results lead to a linear dependence of the maximum MW vs. top Al_2O_3 thickness.

Secondly, we discuss the physical origin of stage II of the dependence of the maximum MW and corresponding V_{th} on top Al_2O_3 thickness. Likewise, as shown in stage I, the $V_{th,ERS}$ still increases linearly with the thickness of top Al_2O_3 at stage II, which means the $Q_{it'}$ after the erase operation is identical in the MIFIS structure with different top Al_2O_3 thicknesses. In addition, we find that the $V_{th,PGM}$ no longer decreases linearly with increasing thickness of top Al_2O_3 as in stage I, which results in the maximum MW remaining almost constant with increasing thickness of top Al_2O_3 interlayer.

Next, we discuss the physical origin of the dependence of the $V_{th,PGM}$ on top Al_2O_3 thickness at stage II. As the thickness of top Al_2O_3 increases, the Fermi energy level of the metal gate shifts to a lower energy position during the program operation [10], and hole trapping and/or electron de-trapping are more prone to occur. The dependence of the $V_{th,PGM}$ on top Al_2O_3 thickness at stage II is not caused by a reduction in the charge injected from the metal gate to the top $Al_2O_3/Hf_{0.5}Zr_{0.5}O_2$ interface during the program operation. Therefore, this dependence of the $V_{th,PGM}$ on top Al_2O_3

979-8-3315-0417-5/25 $31.00 © 2025 IEEE

35

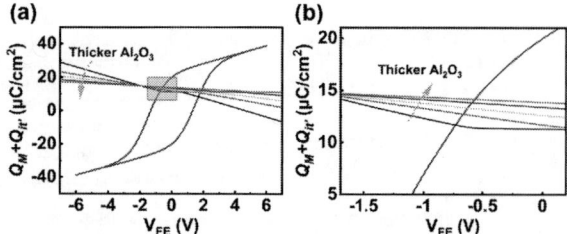

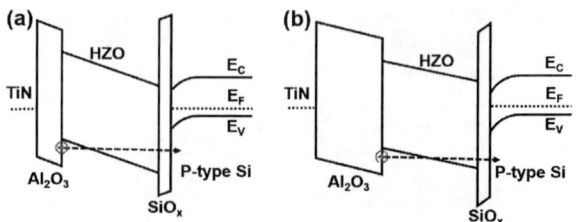

Fig. 5. (a) Hysteresis loop of $Q_M + Q_{it'} - V_{FE}$ and load-line at $V_g = 0$ V after the program operation for the different top Al_2O_3 thicknesses. (b) Enlarged view of the box section in (a).

Fig. 6. Energy band diagrams for the (a) thinner Al_2O_3 and (b) thicker Al_2O_3 cases at $V_g = 0$ V after the program operation.

thickness at stage II is attributed to the $Q_{it'}$ reduction when the gate voltage returns to 0 V.

We use the load-line and ferroelectric hysteresis loop to further explain the physical origin of $Q_{it'}$ reduction. The voltage distribution across the gate stacks is given as

$$V_g = \varphi_s + V_{FE} + V_{BIL} + V_{TIL} \qquad (2)$$

where φ_s, V_{FE}, V_{BIL} and V_{TIL} are the voltage drop across Si substrate, ferroelectric $Hf_{0.5}Zr_{0.5}O_2$, bottom SiO_x interlayer and top Al_2O_3 interlayer, respectively. According to the charge neutrality condition, we can obtain

$$Q_M + Q_{it'} + Q_{it} + Q_s = 0 \qquad (3)$$

where Q_M, Q_{it} and Q_s are the charge density of the metal gate, $SiO_x/Hf_{0.5}Zr_{0.5}O_2$ interface, and Si substrate, respectively. V_{TIL} is given as

$$V_{TIL} = \frac{Q_M}{C_{TIL}} \qquad (4)$$

where C_{TIL} is the capacitance of the top Al_2O_3 interlayer. V_{BIL} is given as

$$V_{BIL} = \frac{-Q_{Si}}{C_{BIL}} = \frac{Q_M + Q_{it'} + Q_{it}}{C_{BIL}} \qquad (5)$$

where C_{BIL} is the capacitance of the bottom SiO_x interlayer. Introducing (4) and (5) into (2), the expression of the load line for the MIFIS structure is given as

$$Q_M + Q_{it'} = \frac{C_{TIL}C_{BIL}}{C_{TIL} + C_{BIL}}(V_g - V_{FE} - \varphi_s + \frac{Q_{it'} + Q_{it}}{C_{TIL}}) - Q_{it} \qquad (6)$$

Based on the principle of the continuity of the electric displacement vector, $Q_M + Q_{it'}$ can be expressed as

$$Q_M + Q_{it'} = C_{FE}V_{FE} + P(V_{FE}) \qquad (7)$$

where C_{FE} is the background capacitance of the ferroelectric $Hf_{0.5}Zr_{0.5}O_2$. The ferroelectric hysteresis loop is obtained by adopting the Preisach model here [12].

Based on previous studies, the charge density trapped at the bottom $SiO_x/Hf_{0.5}Zr_{0.5}O_2$ interface during the program operation is $\sim 10^{14}$ cm^{-2} [13]. When the gate voltage returns to 0 V, about 70% of the trapped charges are still reserved [13]. Hence, the value Q_{it} is -11.4 $\mu C/cm^2$. According to Equation (1) and Fig. 4(b), the value $Q_{it'}$ is set as +14.8 $\mu C/cm^2$. Fig. 5 shows the hysteresis loop and load-line at $V_g = 0$ V after the program operation for the different top Al_2O_3

thicknesses. From Fig. 5(b), it can be found that the electric field of the ferroelectric $Hf_{0.5}Zr_{0.5}O_2$ gradually decreases with increasing thickness of the top Al_2O_3. This means that with increasing thickness of the top Al_2O_3, the charges trapped at the top $Al_2O_3/Hf_{0.5}Zr_{0.5}O_2$ interface face a smaller tunneling barrier when tunneling to the Si substrate or the $SiO_x/Hf_{0.5}Zr_{0.5}O_2$ interface, as shown in Fig. 6, which results in more charges tunneling to the Si substrate or the $SiO_x/Hf_{0.5}Zr_{0.5}O_2$ interface to recombine with electrons. The result leads to a decrease in $Q_{it'}$, which results in $Q_{it'} < |Q_{it}|$ for MIFIS structures with thicker top Al_2O_3 interlayer. Ultimately, the $V_{th,PGM}$ no longer decreases linearly as increasing thickness of the top Al_2O_3 at stage II.

Conclusions

We study the effect of the top Al_2O_3 thickness on the maximum MW of Si-FeFETs with MIFIS gate stacks. We find that the MW first increases and then remains almost constant with the increasing thickness of the top Al_2O_3. This phenomenon is attributed to the smaller electric field of the ferroelectric $Hf_{0.5}Zr_{0.5}O_2$ in the MIFIS structure with a thicker top Al_2O_3 after program operation, which makes the charges injected from the metal gate trapped at the top $Al_2O_3/Hf_{0.5}Zr_{0.5}O_2$ interface cannot remain.

Acknowledgments

This work was supported in part by the National Natural Science Foundation of China under Grant No. 92264104 and 52350195, and supported by the Postdoctoral Fellowship Program of CPSF under Grant No. GZC20232925.

References

[1] T. S. Böscke et al., Appl. Phys. Lett. 99, 102903 (2011).

[2] J. Müller et al., Appl. Phys. Lett. 99, 112901 (2011).

[3] M. Trentzsch et al., IEDM2016, p.294.

[4] E. Yurchuk et al., IEEE Trans. Electron Devices 63, 3501 (2016).

[5] S. Deng et al., IEEE Electron Device Lett. 41, 1348 (2020).

[6] A. J. Tan et al., IEEE Electron Device Lett. 42, 994 (2021).

[7] S. Yoon et al., VLSI2023, p.1.

[8] L. Fernandes et al., IEEE Electron Device Lett. 45, 1776 (2024).

[9] T. Hu et al., IEEE Electron Device Lett. 45, 825 (2024).

[10] T. Hu et al., IEEE Trans. Electron Devices Early Access, 1 (2024).

[11] S. Lim et al., IEDM2023, p.1.

[12] J. Bo et al., VLSI1997, p.141.

[13] K. Toprasertpong et al., IEDM2019, p.570.

Erbium-doped Aluminium Oxide based Distributed Feedback Waveguide Laser in Thin Film Lithium Niobate Platform

Zhekang Zhang[1], Ruqi Wang[1], Renfei Kuang[1], Xifa Liang[1], Shijie Liang[1], Qingming Chen[1]*

[1] School of Microelectronics Science and Technology, Sun Yat-Sen University, Zhuhai 519082, China

*Email: chenqm28@mail.sysu.edu.cn

Abstract

As the realization of a series of high-performance lithium niobate (LN) photonic integrated devices, LN has become one of the most promising platforms for photonic integrated circuits. While the highly compatible LN based on chip light sources are still lacking. It is an urgent need to develop C-band light source on LN chips. We theoretically studied an erbium-doped Al_2O_3 waveguide laser based on distributed feedback (DFB) cavity in the thin film lithium niobate (TFLN) platform. By utilizing chalcogenide glass (CHG) as the cladding material in a sandwich structure, a high confinement factor (52.5%) and laser power is attained. The Laser is pumped at 1480 nm and exhibit fundamental TE mode at 1533 nm, with lasers yield efficiencies up to 27.1% and lasering threshold as low as 4.4 mW in a 10-mm-long DFB cavity.

Keywords: Photonic Integrated Circuits; Erbium-doped Aluminium Oxide; Lithium Niobate; Distributed Feedback Waveguide Laser

Introduction

In the past decade, the thin-film lithium niobate (TFLN) has been extensively explored in integrated photonics due to its superior electro-optic effects and nonlinear properties [1]. For the integration of active photonic devices on the lithium niobate platform, it is feasible to introduce the erbium-doped aluminum oxide as gain medium, which facilitates high-efficiency, reliable, and stable on-chip amplifiers and lasers working at C-band [1-2]. Recently, different types of rare earth doped LN waveguide amplifiers and lasers have been proposed [3-5]. Compared with directly doped erbium ions in TFLN, Er^{3+}: Al_2O_3 is capable of attaining higher ion concentrations, hence achieving higher material gain and laser power [6]. Additionally, it can also avoid the influence of ion implantation on the electro-optical properties of TFLN, and facilitate the hybrid integration of active and passive device on lithium niobate platform, thereby achieving versatile photonic circuits [7].

In addition to the choice of gain medium, lithium niobate lasers require careful consideration of resonator cavity, including F-P cavities, mirroring, distributed Bragg reflectors (DBR), and distributed feedback (DFB) structures [8]. Among them, DFB resonators make it viable and effective to achieve narrow linewidth and stable single longitudinal-mode lasers without shortening the cavity or intra-cavity filters to obtain the cavity mode spacing that exceeds the material gain bandwidth [9].

In this work, we propose a single longitudinal-mode Er^{3+}: Al_2O_3 waveguide laser based on thin-film lithium niobate. By sandwiching the Er^{3+}: Al_2O_3 layer as the gain medium between the thin-film lithium niobate and chalcogenide glass, a slot waveguide in the vertical direction is formed for mode confinement. The overlap factor between the mode and the gain medium can be maximized to approximately 52.5%. For a 10 mm long DFB cavity pumped at the wavelength of 1480 nm, the laser yield efficiencies can reach up to 27.1% at the wavelength of 1533 nm with a pump threshold of 4.4 mW.

Structure Design

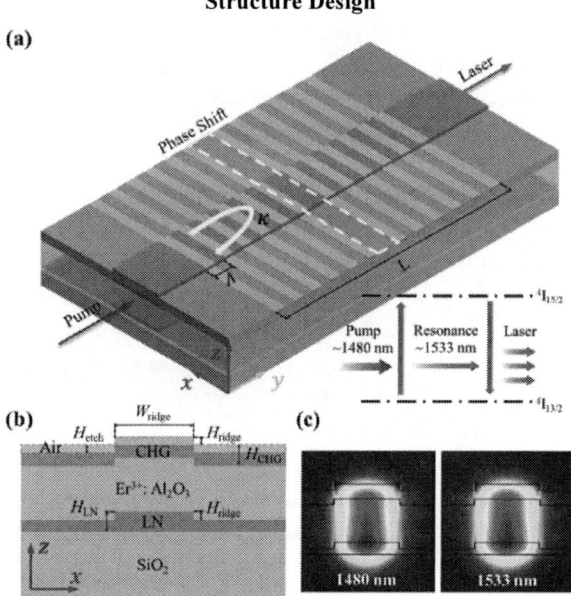

Fig. 1. (a) The schematic of the sandwich DFB structure and the working principle of the Er^{3+}: Al_2O_3 optical waveguide pumped at 1480 nm. (b) Cross-sectional view of the composite waveguide structure. (c) Simulated optical TE mode profiles for 1480 nm and 1533 nm.

The 3D schematic of the proposed waveguide laser is shown in Figure 1(a), where the DFB structure is composed of SiO_2 substrate, z-cut TFLN (n_e = 2.13, n_o = 2.21) as ridge waveguide, Er^{3+}: Al_2O_3 (n = 1.66) as gain medium and chalcogenide glass (n = 2.23) layer from bottom to top. Additionally, a surface corrugated Bragg grating is applied to CHG layer for realizing particular desired coupling coefficient and resonant wavelength. To achieve a single resonance exactly at the Bragg wavelength λ_B, the corrugated grating requires a phase shift of $\pi/2$ in the center, corresponding to one quarter of the Bragg wavelength. Figure 1(b) provides the structural parameters of the waveguide in cross section view: H_{LN} = 0.1 μm, H_{ridge} = 0.05 μm. Based on

979-8-3315-0417-5/25 $31.00 © 2025 IEEE

current fabrication conditions, the thickness of Er^{3+}: Al_2O_3 is set to 0.35 μm, and the doping concentration of Er^{3+} ions is 4.9×10^{20} cm^{-3} [6]. Due to the close refractive indices between LN and CHG, the mode field is mostly constrained in the Er^{3+}: Al_2O_3 layer with the overlap factor over 52%, as evidenced by the field distribution at pump and laser wavelength in Figure 1(c). In order to get higher laser power, the resonant frequency should be located at the maximum gain spectrum of erbium. Therefore, the period (Λ) of the grating is set to 465 nm with Bragg wavelength designed to 1533 nm.

Given the structural parameters of the LN ridge waveguide, the thickness of the cladding CHG determines the confinement factor $\Gamma_{s,p}$ of the optical field in the gain layer, which is defined by

$$\Gamma_{s,p} = \iint_A \phi_{s,p}(x,y)dxdy \qquad (1)$$

where $\Phi_{s,p}$ is the normalized transverse intensity distribution of fundamental mode with laser and pump wavelength, and A is the cross-sectional area of the active region. Note that the erbium ions are assumed to be uniformly distributed in alumina layer.

To characterize influence of the CHG layer thickness H_{CHG} on $\Gamma_{s,p}$, the Lumerical MODE based on eigenmode expansion method (EME) is utilized for numerical simulation of the mode field distribution in Figure 2(a). As the cladding thickness increases, the confinement factors for both the pump and laser light initially rise slowly, reaching a peak of 52.5% at a thickness of 100 nm, and then rapidly decrease. This phenomenon is due to the higher refractive index of CHG compared to alumina. As the thickness of CHG increases, the modal field, which is predominantly distributed in the lithium niobate layer, is drawn into the central gain layer, leading to an enhancement in the confinement factor. However, upon further increasing H_{CHG} to a point where it can support the fundamental mode propagation, the confinement factor is observed to decrease. This can also be explained by the increase in the effective refractive index with the increasing H_{CHG}.

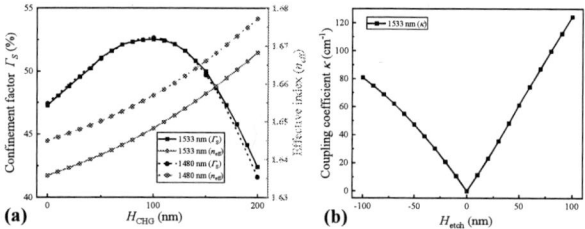

(a)   **(b)**

Fig. 2. (a) Confinement factor $\Gamma_{s,p}$ and effective index dependent on the cladding thickness H_{CHG}. Solid and dash line represent the laser and pump wavelength respectively. (b) κ dependent on the H_{etch} at 1533 nm.

A surface corrugated Bragg grating is used to construct the resonant cavity, where the lasing properties are determined

by parameter H_{etch}. The grating strength influences the intracavity laser power distribution and the performance of the laser, which is determined by the product of Bragg grating length (L) and the grating coupling coefficient (κ). The difference in refractive index, denoted as n_H and n_L, decides the coupling coefficient of the grating, which is defined by

$$\kappa = \frac{\Gamma_s (n_H^2 - n_L^2)}{\lambda_B n_{eff}} \sin(\pi D) \qquad (2)$$

where λ_B = 1533 nm is the designed Bragg wavelength and D = 0.5 is the duty cycle of the grating. Considering the maximized confinement factor, we take a cladding thickness of 100 nm as the baseline to calculate κ as shown in Figure 2(b). In this work, to achieve sufficient grating reflection while avoiding excessively high intracavity power, we set H_{etch} to 20 nm, corresponding to κ of 25 cm^{-1}.

Simulation and discussion

Figure 3 shows the passive response of Bragg grating (BRG) and DFB structure simulated by 3D finite-difference time-domain (FDTD) method and transfer matrix method (TMM) [10]. In a lossless Bragg grating, the peak power of the reflection at the Bragg wavelength is directly correlated with the dimensionless quantity κL, which is denoted as

$$R_{peak} = \tanh^2(\kappa L) \qquad (3)$$

where κL is approximately equal to 20, corresponding to a reflectivity of about 92% at 1533 nm.

Actually, the DFB structure introduces a phase-shift region equivalent to the length of grating period Λ at the center, distinct from a standard Bragg grating. Compared with the simulation in Lumerical FDTD according to the parameters given above, the Bragg wavelength and stopband fit with the result from TMM accurately.

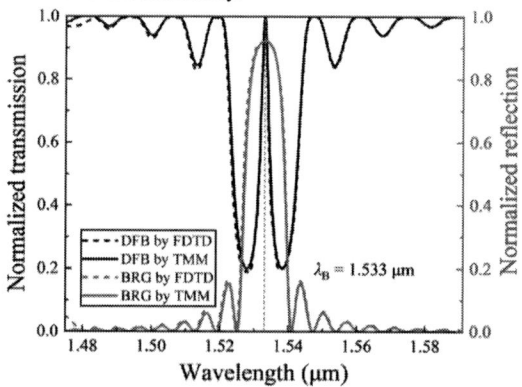

Fig. 3. The passive response for Bragg grating and DFB structure simulated by FDTD and TMM.

Figure 4 (a) presents the relationship between the output power of the hybrid sandwich DFB waveguide laser and pump power for different cavity length. Note that higher laser power can be realized in longer cavity length when pump power is unsaturated. But it results in a rapid increase for intracavity laser power and significant hole burning effects in the phase-shift region. In Figure 4 (b), the efficiency curve of

979-8-3315-0417-5/25 $31.00 © 2025 IEEE   38

the device is presented. The threshold pump power increase slightly from 4 mW to 4.7 mW when cavity lengthened from 2.5 mm to 12.5 mm. For a 10-mm-long cavity, the slope efficiency is 27.1% with lasing thresholds of 4.4 mW.

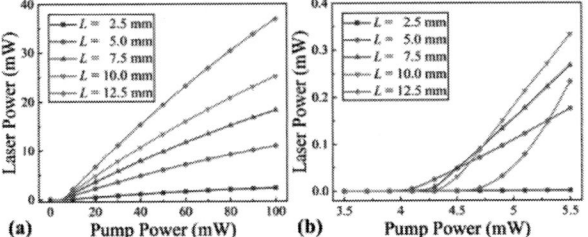

Fig. 4. Output power and threshold pump power of the laser with sandwich DFB cavity versus pump power at different cavity length.

Figure 5 shows the normalized population distribution in the cavity. Considering the Excited-state absorption (ESA) and cooperative up-conversion (CUC), the model is based on the energy transfer dynamics between four energy levels as $^4I_{15/2}$, $^4I_{13/2}$, $^4I_{11/2}$ and $^4I_{9/2}$, where the population density of Er^{3+} icon on 1st excited state (N_2) and ground state (N_1) decides the optical gain and pump absorption. When the N_2 exceeds N_1 and population inversion occurs, the optical gain escalates as the increase of the difference. Otherwise, the population inversion is suppressed with the depletion of pump power, and optical gain tends toward 0.

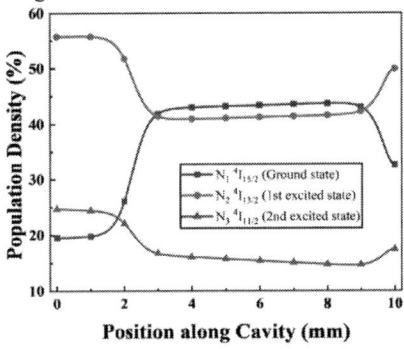

Fig. 5. The normalized population distribution for ground state, first excited state and second excited state in 10-mm-long DFB cavity when the pump power is 50 mW. The population density of $^4I_{9/2}$ (N_4) is less than 1% hence not represented in the figure.

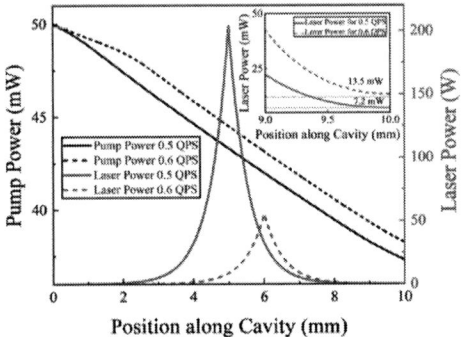

Fig. 6. The laser and pump power distribution curves in the DFB resonant cavity for phase shift at different position.

The intra-laser-cavity power along the length of the 10 mm cavity is illustrated in Figure 6 for pump powers of 50 mW. The laser power is almost confined in the phase shift region and decays exponentially around it. When the phase shift region is deviated by 1 mm from the center of cavity towards the output end, the peak laser inside the cavity can be significantly reduced. The offset of the phase shift region may reduce the consumption of pump power and increase the output laser power from 7.2 mW to 13.5 mW.

Conclusion

In this paper, we have designed a Er^{3+}: Al_2O_3 waveguide laser in TFLN platform based on DFB structure. It provides a single longitude laser at the wavelength of 1533 nm when pumped at 1480 nm. Take advantage of hybrid sandwich structure, we can realize high lasers yield efficiencies up to 27.1% and threshold power as low as 4.4 mW in the optical cavity length of 10 mm. In addition, the introduction of Er^{3+}: Al_2O_3 makes it possible to generate high-power lasers without doping the TFLN, maintaining its superior electro-optical properties for achieving high-speed tunable lasers on lithium niobate substrates in the future.

Acknowledgments

This work is supported by National Natural Science Foundation of China (NSFC) through grant number 62475296.

References

[1] G. Chen, N. Li, J. Ng, et al. "Advances in lithium niobate photonics: development status and perspectives." Advanced Photonics 4, 34003, 2022.

[2] Q. Luo, F. Bo, Y. Kong, G. Zhang, and J. Xu, "Advances in lithium niobate thin-film lasers and amplifiers: A review," Advanced Photonics 5, 2023.

[3] Y. Jia, J. Wu, X. Sun, et al., "Integrated Photonics Based on Rare-Earth Ion-Doped Thin-Film Lithium Niobate," Laser & Photonics Reviews 16, 2200059, 2022.

[4] Z. Xiao, K. Wu, M. Cai, et al., "Single-frequency integrated laser on erbium-doped lithium niobate on insulator," Opt. Lett. 46, 4128, 2021.

[5] Y. Liu, X. Yan, J. Wu, et al., "On-chip erbium-doped lithium niobate microcavity laser," Science China Physics, Mechanics & Astronomy 64, 234262, 2020.

[6] C. Wang, J. Song, Z. Ao, et al., "High-Gain Waveguide Amplifiers in $Ge_{25}Sb_{10}S_{65}$ Photonics Heterogeneous Integration with Erbium-Doped Al_2O_3 Thin Films," Laser & Photonics Reviews 18, 2300893, 2024.

[7] W. Hendriks, L. Chang, C. Emmerik, et al. "Rare-earth ion doped Al_2O_3 for active integrated photonics." Advances in Physics: X 6, 1833753, 2021.

[8] W. Bo, P. Zhou, and X. Wang. "Numerical Analyzation of Distributed Feedback Lasers with Different Phase Shift Structures." 2021 Asia Communications and Photonics Conference (ACP). IEEE, 2021.

[9] E. H. Bernhardi, H. Wolferen, L. Agazzi, et al. "Ultra-narrow-linewidth, single-frequency distributed feedback waveguide laser in Al_2O_3: Er^{3+} on silicon." Optics Letters 35, 2394, 2010.

[10] E. H. Bernhardi, Q. Lu, H. Wolferen, et al., "Monolithic distributed Bragg reflector cavities in Al_2O_3 with quality factors exceeding 10^6," Photonics and Nanostructures - Fundamentals and Applications 9, 225, 2011.

Comparison of border trap density in MIFM capacitors with different interlayers using the C-ln(f) method

Min Liao[1], Hao Xu[1], Xiaoyu Ke[1], Yuanyuan Zhao[1], Junshuai Chai[1], Xiaoqing Sun[1], Jinjuan Xiang[2], Xiaolei Wang[1], and Wenwu Wang[1]

[1]Institute of Microelectronics, Chinese Academy of Sciences, Beijing 100029, China

[2]Beijing Superstring Academy of Memory Technology, Beijing 100176, People's Republic of China

Abstract

The border trap density is compared in Metal/Interlayer/Ferroelectric/Metal (MIFM) capacitors with different interlayers by measuring Capacitance vs. logarithm of frequency, i.e. the C-ln(f) method. The C-ln(f) method is verified and modified based on the simulation results of the distributed bulk-oxide trap model. The MIFM capacitors were fabricated with different interlayers (HfO$_2$ or ZrO$_2$) between the HZO and top TiN electrode. It is found that the devices with HfO$_2$ or ZrO$_2$ interlayer have lower border trap density after wake-up cycling, demonstrating the potential for reliability improvement.

Keywords: Border trap, Ferroelectric, HZO, Capacitance–voltage, Capacitance–frequency, Reliability.

Introduction

Ferroelectric Hf$_{0.5}$Zr$_{0.5}$O$_2$ (HZO) is a key multi-functional material in current electronic device technology, offering advantages such as non-volatility, scalability, and compatibility with existing semiconductor processes [1-3]. However, reliability concerns arise from trap generation and charge trapping within HZO-based ferroelectric devices, which can significantly affect device performance and application.

The frequency dispersion of capacitance, measured through AC small-signal Capacitance-Voltage (C-V) demonstrates good linearity and has been effectively utilized to analyze border traps in Metal/Ferroelectric/Metal (MFM) structures [4, 5]. The validation of the Capacitance vs. logarithm of frequency (C-ln(f)) method for quantitatively extracting trap density remains unclear in the literature.

In this work, we verify the C-ln(f) method through simulations of the distributed bulk-oxide trap model [6]. Additionally, we propose a modified equation for a more accurate extraction of border trap density. We fabricate Metal/Interlayer/Ferroelectric/Metal (MIFM) capacitors with different interlayers (HfO$_2$ or ZrO$_2$) between HZO and the top electrode (TE). By employing the modified C-ln(f) method, we compare the border trap distributions and discuss the influence of interlayer materials on border trap generation after wake-up cycling.

Modified C-ln(f) Method

The equivalent circuit of the distributed bulk-oxide trap model is depicted in Fig. 1. To account for parasitic effects, we include the components of conductance G_p and Rp. The

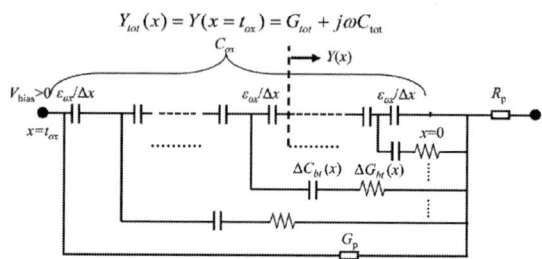

Fig. 1. The equivalent circuit of the distributed bulk-oxide trap model, refer to [6]. The parasitic effects could be considered by adding components of G_p and R_p.

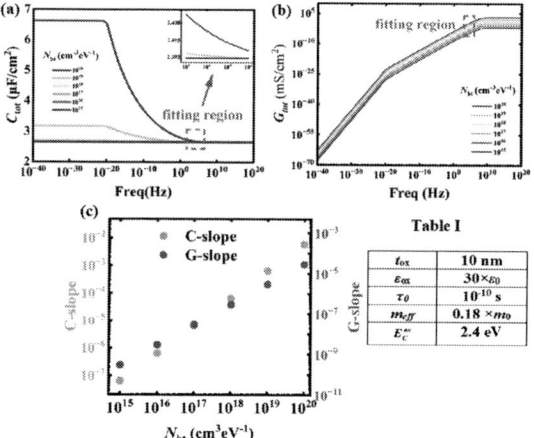

Fig. 2. (a) and (b) are the simulated C_{tot} and ln(G_{tot}) curves with different trap densities (N_{bt}) vs. ln(f), respectively. (c) The slope of C_{tot}-ln(f) and G_{tot}-f vs. N_{bt} are extracted from simulation results in the frequency range of 1 kHz to 1 MHz.

total admittance (Y_{tot}) of the distributed circuit, as experienced by the electrode under positive bias, is numerically calculated using a recursive procedure.

The typical simulation results of C_{tot} vs. ln(f) and ln(G_{tot}) vs. ln(f) with different border trap densities (N_{bt}) are given in Fig. 2(a) and 2(b), respectively. The default parameters used in this work are referenced to [8] and summarized in Table I, where N_{bt} is considered as a constant in cm^{-3} eV^{-1}.

The linear fitting slopes of C_{tot}-ln(f) and G_{tot}-f in the frequency range of 1 kHz to 1 MHz are plotted vs. N_{bt}, as shown in Fig. 2(c). This frequency range aligns with commonly used measurement configurations, ensuring the relevance of our findings. The observed linearity between the

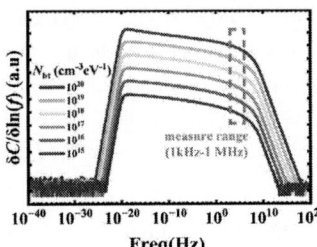

Fig. 3. $\partial C/\partial \ln(f)$ vs. $\ln(f)$. The simulation results show that as the frequency increases, the slope first increases and then decreases.

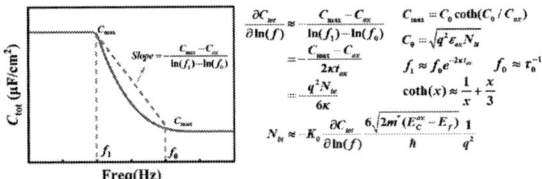

Fig. 4. The derivation for extraction of the N_{bt}. A three-term Taylor series allows N_{bt} to be determined from the slope of C_{tot} vs. $\ln(f)$. The average slope is approximately evaluated by the transition points of the C-$\ln(f)$ curve.

C_{tot}-$\ln(f)$ slope and N_{bt} supports the validation of the C-$\ln(f)$ method for trap density extraction. However, it is important to note that this linear relationship serves as an approximation within a narrow frequency range. According to Fig. 2(a), the $\partial C/\partial \ln(f)$ is calculated and shown in Fig. 3. It can be seen that as the frequency increases, the slope first increases and then decreases.

Next, the quantitative equation for the extraction of N_{bt} is derived. We mainly focus on the frequency dispersion of capacitance, due to the conductance being sensitive to parasitic effects. The derivation is shown in Fig. 4.

$$N_{bt} \approx -K_0 \frac{\partial C_{tot}}{\partial \ln(f)} \frac{6\sqrt{2m^*(E_C^{ox}-E_f)}}{\hbar} \frac{1}{q^2} \qquad (1)$$

In equation (1), N_{bt} is border trap density, K_0 is the additional coefficient, $\partial C/\partial \ln(f)$ is measured capacitance vs. $\ln(f)$, m^* is electron effective mass in the dielectric and E_C^{ox} is the energy of the top of the dielectric barrier. As the frequency increases, the first transition point happens when $f = f_1 \approx \tau_0^{-1}\exp(-2\kappa t_{ox})$, and $C_{max} \approx C_0\coth(C_0/C_{ox})$. The second transition point happens when $f = f \approx \tau_0^{-1}$, and $C_{min} \approx C_{ox}$ [6]. Meanwhile, considering the $\partial C/\partial \ln(f)$ in the range of 1 kHz to 1 MHz is smaller than the average slope (Fig. 3), an additional coefficient K_0 (>1) is added, and finally the equation (1) is obtained. Compared with previous work [7], equation (1) is expected for a more accurate N_{bt}.

Fig. 5 shows the effect of parameters on the simulation. Approximately, K_0 increases linearly with t_{ox}, while is nearly independent on ε_{ox} and N_{bt}. K_0 increases linearly with t_{ox}, while is nearly independent on ε_{ox} and N_{bt}. Thus, the expression of $K_0 = 2$ nm$^{-1}\cdot t_{ox}$ -10 is used for the trap analysis

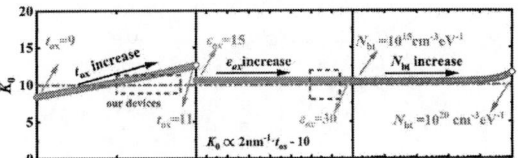

Fig. 5. The dependence of K_0 on t_{ox}, ε_{ox}, and N_{bt}.

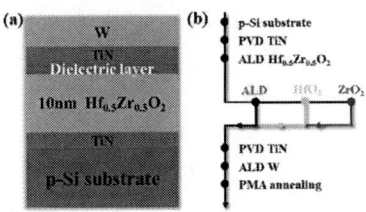

Fig. 6. (a) Schematic diagram of MFM and MIFM capacitors. (b) The process flow of MFM and MIFM capacitors. 1 nm dielectric layer (HfO$_2$ or ZrO$_2$) was deposited by ALD. The annealing temperature is 550°C.

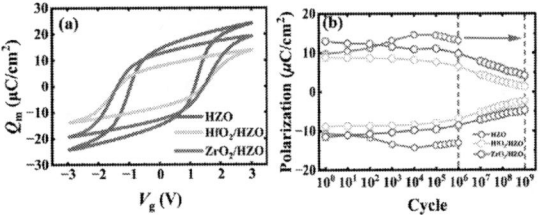

Fig. 7. (a) The Q_m-V_g curves of these different devices, a triangle wave of 3V at 1 kHz is used for the Q_m-V_g measurement. (b) A square wave of ±3 MV/cm amplitude at 100 kHz is used for the cycling.

in the experiment

Experiment

Fig. 6 shows the schematic diagram and process flow of MFM and MIFM capacitors. First, 20 nm TiN was deposited on (100)-oriented p-Si. Then 10 nm Hf$_{0.5}$Zr$_{0.5}$O$_2$ ferroelectric layer and 1 nm dielectric layer (including HfO$_2$, and ZrO$_2$) were deposited in turn by atomic layer deposition (ALD) at 300 °C. The oxygen source was H$_2$O. 20 nm TiN TE and 75 nm W were deposited by sputtering and ALD, respectively. Finally, post-metallization annealing (PMA) was performed. Here, the ferroelectric phase crystallization was achieved at the temperature of 550 °C by rapid-thermal-annealing (RTA) for 60 s in N$_2$ ambient to form an orthorhombic phase.

Results

The AC small signal C-V is measured by the B1500A semiconductor analyzer. Charge-voltage (Q-V) tests are performed utilizing TF3000. The Q_m-V_g and endurance are shown in Fig. 7. After wake-up cycling, compared with the HZO device, MIFM with interlayer shows reduced remanent polarization (P_r), increased coercive field (E_c), and significantly increased endurance.

The endurance tests are conducted using a rectangular

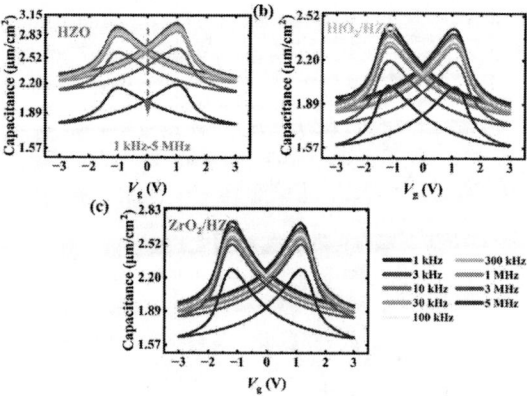

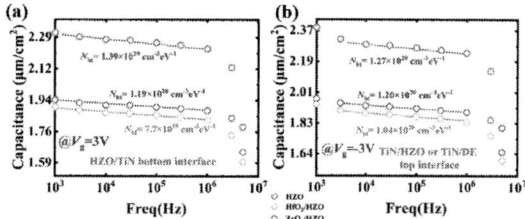

Fig. 8 (a)-(c) multi-frequency C-V curves of the three devices. When $f>1$ MHz, the rapid decrease in capacitance value is due to the parasitic effect caused by R_p.

Fig. 9 (a)-(b) show the C-ln(f) dispersion of capacitance of MFM and MIFM capacitors at ±3V and 1 kHz-5 MHz respectively. 1000 cycles for wake-up.

waveform with a frequency of 100 kHz. For the MFM capacitor, the breakdown occurred after 10^6 cycles. For the MIFM device with 1nm intercalation, no breakdown occurred after 10^9 cycles. The endurance has been improved by 3 orders of magnitude.

The measured multi-frequency C-V curves after wake-up cycling are shown in Fig. 8. The C-ln(f) relationship at ±3 V is shown in Fig. 9, which is used to extract N_{bt} from the near bottom electrode (BE) or TE, respectively. It is assumed that τ_0 is 10^{-10} s, then the trap detection depth is approximately 1.7 to 2.4 nm from the surface of each electrode. Thus, it is suggested that trap detection happens in all HZO rather than in the dielectric layer.

The extracted N_{bt} based on equation (1) is shown in Fig. 9. All three devices show good linear relationship in $\delta C/\delta$ ln(f) under the condition of 1 kHz-1 MHz. The N_{bt} of the bottom interface of TiN and HZO is extracted at $V_g =+3$V, and the N_{bt} of the upper interface of TiN and HZO or TiN and the dielectric layer is extracted under a $V_g=-3$V. On the one hand, both near TE and BE in devices with HfO$_2$ or ZrO$_2$ interlayer show reduced N_{bt}. On the other hand, in HZO devices,

$N_{bt,BE}>N_{bt,TE}$, while in devices with interlayers, the results are contrary. These results suggest that the interlayer is helpful in suppressing the trap generation in HZO, especially near BE, which may be critical for endurance improvement. It can be verified from the endurance characteristic Fig. 7(b).

Conclusion

In this study, we verified and modified the C-ln(f) method to enhance the accuracy of border trap density (N_{bt}) extraction. Through the fabrication of MFM and MIFM capacitors with different interlayers, we compared the border trap densities in the devices. Our findings indicate that devices with HfO$_2$ or ZrO$_2$ exhibit significantly reduced border trap densities. This reduction suggests a promising potential for improving the reliability of ferroelectric devices,

Acknowledgments

This work was supported in part by the National Natural Science Foundation of China under Grant Nos. 92264104 and 52350195, in part by National Key Research and Development Program of China under Grant No. 2022YFB4400300, in part by R&D Program of Beijing Municipal Education Commission under Grant No.KZ202210009014, in part by the Young Elite Scientists Sponsorship Program under Grant No. BYESS2023033, and Supported by the Postdoctoral Fellowship Program of CPSF under Grant No. GZC20232925.

References

[1] T. S. Böscke, J. Müller, D. Bräuhaus, U. Schröder, U. Böttger; "Ferroelectricity in hafnium oxide thin films", Appl. Phys. Lett., 99, 102903 (2011)

[2] J. Okuno et al., "SoC Compatible 1T1C FeRAM Memory Array Based on Ferroelectric Hf$_{0.5}$Zr$_{0.5}$O$_2$," IEEE Symp. on VLSI Tech., p. 1-2 (2020).

[3] M. Saitoh et al., "HfO$_2$-based FeFET and FTJ for Ferroelectric-Memory Centric 3D LSI towards Low-Power and High-Density Storage and AI Applications," IEDM., pp. 18.1.1-18.1.4 (2022).

[4] LYU, X., et al., "Ferroelectric and anti-ferroelectric hafnium zirconium oxide: Scaling limit, switching speed and record high polarization density." IEEE Symp. on VLSI Tech., p. T44-T45 (2019)

[5] Y. Yuan et al., "A Distributed Bulk-Oxide Trap Model for Al$_2$O$_3$ InGaAs MOS Devices," IEEE Trans. on Electron Devices., 55, pp. 2100-2106 (2012).

[6] M. A. Ebrish, D. A. Deen and S. J. Koester, "Border trap characterization in metal-oxide-graphene capacitors with HfO$_2$ dielectrics," 71st Device Research Conference (2013).

[7] L. Vandelli, A. Padovani, L. Larcher, R. G. Southwick, W. B. Knowlton and G. Bersuker, "A Physical Model of the Temperature Dependence of the Current Through SiO$_2$/HfO$_2$ Stacks," IEEE Trans. on Electron Devices. 58, pp. 2878-2887 (2011)

979-8-3315-0417-5/25 $31.00 © 2025 IEEE

Boosting Mobility of Oxide TFTs via PVD and ALD Hybrid Process

Yuhang Zhang[1], Xiao Li[1], Shengjie Yang[1], Huan Yang[1], Tengyan Huang[1], Xinwei Wang[2], Lei Lu[1], and Shengdong Zhang[1]

[1]School of Electronic and Computer Engineering, Peking University, Shenzhen, China
[2]School of Advanced Materials, Peking University, Shenzhen, China
E-mail: zhangsd@pku.edu.cn

Abstract

The physical vapor deposited (PVD) oxide semiconductor TFTs often suffer from the trade-off between positive threshold voltage (V_{th}) and high mobility. In this work, the hybrid active layer of sputtered InZnO (IZO) and atomic layer deposited (ALD) InGaO (IGO) is investigated. After ALD residuals are thermally removed, a high-mobility quantum well can be formed at the well-defined PVD-IZO/ALD-IGO heterojunction. Compared to the PVD-IZO TFT, the PVD-ALD one shows considerably elevated mobility, and maintains low off-current and positive V_{th}.

Keywords: heterojunction, InZnO/InGaO TFT, oxide semiconductor and PVD-ALD

Introduction

Oxide semiconductor (OS) thin-film transistors (TFTs) boast several advantages, including ultralow off-state current and low processing temperatures. The promising potentials in advanced displays [1], flexible electronics [2], and three-dimensional integrated circuits [3] demand mobility evolving beyond that of mainstream InGaZnO (IGZO) [4]. While the mobility of monolayer OS TFT can be increased by compositional tuning [5] and crystallization [6] often at the costs of deteriorating other key parameters, such as threshold voltage (V_{th}) and off-state current, the multilayer channel can effectively combine the merits of diversified OS compositions [7]-[15].

The band diagram of the heterojunction channel is of paramount importance to the enhancements of performance metrics. Thus, the bilayer channel demands precise control over the composition and thickness of layers and their interface [8] [9]. Physical vapor deposition (PVD) is prevalent for OSs due to rapid deposition rates and large-area scalability but lacks the controlling flexibilities on composition and thickness. As comparison, atom layer deposition (ALD) provides accurate thickness controllability and flexible composition tuning because of chemically self-limiting growth behavior. However, the long growth cycle limits the application of full ALD in the case of relative thick film preparation.

This paper introduces a novel bilayer channel by combining ALD OS with PVD OS. Utilizing the precision of ALD in controlling both composition and film thickness, an ultrathin indium gallium oxide (IGO) layer was fabricated by ALD to accurately match the band structure of PVD indium zinc

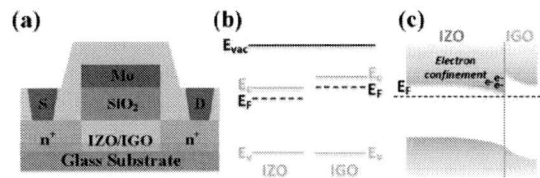

Fig. 1 (a) Schematic device structure of IZO/IGO bilayer self-aligned top-gate thin-film transistor. (b) Relative band structure of IZO, IGO. (c) Schematic band diagram at IZO/IGO hetero-interface.

oxide (IZO). This configuration facilitates the formation of a two-dimensional electron gas (2DEG) transport channel with enhanced mobility at the interface between the PVD and ALD films. Following the mitigation of residual impurities within the ALD IGO layer, the mobility was elevated from 24.09 cm^2/Vs in the single-layer IZO channel to 31.88 cm^2/Vs in the IZO/IGO heterojunction channel, maintaining a V_{th} around 0 V.

Experiments

Fig.1 (a) illustrates the schematic structure of the IZO/IGO bilayer self-aligned top-gate TFT. First, a 5-nm-thick IGO film was deposited by ALD on a glass substrate. Then, a 10-nm-thick IZO film was deposited by PVD to form the bilayer channel. Single layer samples composed solely of ALD IGO or PVD IZO were also prepared for comparison. After photolithography and wet etching of the active layers, a 200-nm-thick silicon dioxide (SiO$_2$) dielectric layer was deposited at 300 °C by plasma-enhanced chemical vapor deposition (PECVD). The samples were annealed for 90 minutes at 300 °C in an oxygen atmosphere. Following annealing, a 120-nm-thick molybdenum (Mo) was sputtered and patterned to form the top-gate electrodes. The gate dielectric layer was then etched using the top-gate electrode pattern for self-alignment, exposing parts of the active layer that were treated with argon plasma to form highly conductive source/drain (S/D) regions. Finally, a 200-nm-thick SiO$_2$ was deposited as a passivation layer by PECVD at 300 °C followed by the contact hole opening, and a 50-nm-thick Mo was sputtered and patterned as contact electrodes. The electrical characteristics of all TFTs were measured using a B1500 semiconductor parameter analyzer under ambient conditions of room temperature and darkness, with channel width and length set at 100 μm and 100 μm.

Here, the ALD IGO film was intentionally fabricated with a

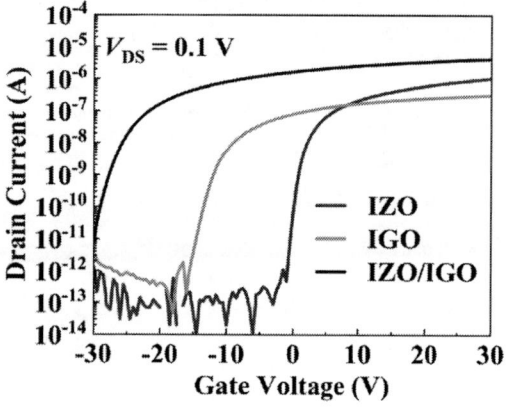

Fig. 2 The transfer curves of IZO single-layer, IGO single-layer and IZO/IGO bilayer TFTs at V_{DS} of 0.1 V.

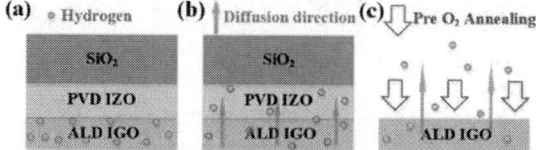

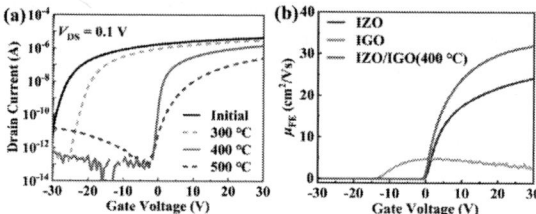

Fig. 3 (a) Schematic diagram of H residue in the ALD IGO layer and (b) diffusion throughout the channel during annealing. (c) Schematic diagram of oxygen annealing to remove the H residues in the ALD IGO before IZO deposition.

Fig. 4 (a) Transfer curves of IZO/IGO bilayer TFTs with ALD IGO pre-treated under various temperatures. (b) Field-effect mobility curves of IZO single layer, IGO single layer TFT and IZO/IGO(400 °C) bilayer TFTs.

high Ga and low In content to match the band structure of the PVD IZO. As shown in Fig.1 (b), this ALD IGO film has a higher conduction band (E_C) and wider band gap (E_g) than the PVD IZO, due to the Ga_2O_3-induced E_g widening effect [16]. It can be inferred that mobile electrons migrate from the E_C of ALD IGO to that of PVD IZO, inducing the accumulation of free carriers near PVD IZO. As illustrated in Fig.1(c), a high-mobility quantum well is expected to form at this PVD-IZO/ALD-IGO heterojunction interface.

Results and Discussion

Fig. 2 compares the transfer characteristic curves of PVD IZO single-layer, ALD IGO single-layer, and IZO/IGO bilayer TFTs at V_{DS} = 0.1 V. The sputtered IZO single-layer TFT exhibits a positive V_{th} of 0.67 V, an on-current of 1.06 µA and an off-current below 1 pA. In comparison, the ALD IGO single-layer TFT has a negative V_{th} of -12 V, a larger off-current and a much lower on-current of 0.32 µA. Compared with these two single-layer devices, the PVD IZO/ALD IGO bilayer TFT exhibits a significantly larger on-current of 4 µA but exhibits a very negative V_{th} of -27.5 V. Optimized sputtering oxygen partial pressure and post-annealing treatment enables the PVD IZO single-layer TFT a suitable V_{th} and a moderate on-current. The ALD IGO layer with high Ga content and low In content was designed specifically to achieve a wide band gap, and its lower mobility was anticipated. However, the IZO/IGO bilayer TFT exhibits a severe negative shift in V_{th}, more negative than either single layer. This differs from previous expectation on 2DEG heterojunction and necessitates further investigation on the unexpected introduction of a significant number of free electrons in the channel.

Previous study on ALD oxide as gate insulators of OS TFTs [17] has shown that a significant amount of H from unreacted ligands of precursor or unreacted OH groups can diffuse into PVD OS channel, acting as donor dopants and leading to conductive channel behavior in resulting transistor without modulation capability. Therefore, it is reasonable to speculate

that the bottom IGO film contains a significant amount of H residues from ALD process, which may diffuse into the above PVD IZO during post-annealing, as illustrated in Fig. 3 (a) and (b). Due to the heavy electron doping from these unintended H impurities, the IZO/IGO TFTs present the extremely negative V_{th}. To address the issue of H residues, the ALD IGO layer was pre-annealed in an oxygen atmosphere at 300 °C, 400 °C or 500 °C before depositing the subsequent PVD IZO. As demonstrated in Fig.3 (c), this heat treatment processing can drive the residual H out of the ALD IGO film into the ambient environment.

The transfer characteristics of the IZO/IGO bilayer TFTs with ALD IGO layers pre-annealed at different temperatures were investigated in Fig.4 (a). It is clear that the V_{th} of the bilayer devices shifts towards positive values as the temperature increases. When the pre-annealing temperature increases to 400 °C, the V_{th} improves to 0.33 V and the off-current keeps round 0.1 pA. Compared with the PVD IZO single-layer TFT, the optimized IZO/IGO(400 °C) bilayer TFT increases the on-current from 1.06 µA to 1.43 µA. The optimized temperature to remove H residues is 400 °C and an excessive temperature deteriorates the performance. Fig. 4 (b) compares the field-effect mobility curves of the IZO single-layer, IGO single-layer and IZO/IGO(400 °C) bilayer TFTs. The optimized bilayer TFT demonstrates a significantly higher mobility compared to both single-layer TFTs. Three key electrical characteristics of these TFTs are summarized in Table 1, including V_{th}, mobility, and sub-threshold swing (SS). The optimized IZO/IGO(400 °C) bilayer TFT achieves the highest mobility of 31.89 cm²/Vs, while maintaining a V_{th} of 0.33 V and a moderate SS of 454 mV/Dec. We believe that a high-mobility quantum well has been formed at the

979-8-3315-0417-5/25 $31.00 © 2025 IEEE

Table 1: Electrical parameters of IZO, IGO single-layer and IZO/IGO(400 °C) bilayer TFTs

Channel	V_{th}(V)	μ_{FE}(cm^2/Vs)	SS(mV/Dec)
IZO	0.67	24.09	415
IGO	-12.00	4.84	624
IZO/IGO (400 °C)	0.33	31.88	454

IZO/IGO(400 °C) heterojunction interface as expected in Fig.1 (c), resulting in the significant mobility improvement.

Conclusion

The performances of oxide TFTs have been significantly improved by forming a heterojunction active channel by combining the PVD IZO and ALD IGO. Removing the undesirable ALD residues through thermal treatment enables the formation of a high-mobility quantum well at the PVD-IZO/ALD-IGO interface. Compared to the PVD IZO single-layer TFTs, this well-defined IZO/IGO heterojunction elevates the mobility increasing from 24 cm^2/Vs to 32 cm^2/Vs, while maintaining a low off-state current below 0.1 pA and a positive V_{th} of 0.33 V. This advancement highlights the potential of the PVD-ALD combination for developing high-performance heterojunction TFTs.

Acknowledgments

This work was conducted in Guangdong Provincial Center for Oxide Semiconductor Devices and ICs, and supported financially by the National Key Research and Development Program of China under Grant No. 2022YFB3607200, and the Shenzhen Municipal Scientific Program under Grant No. JCYJ20220818100808019 and SGDX20211123145404006, and the Guangdong Basic and Applied Basic Research Foundation No. 2024B0101120003 and 2023TQ07A463.

References

[1] J.-S. Park, H. Kim, and I.-D. Kim, "Overview of electroceramic materials for oxide semiconductor thin film transistors," Journal of Electroceramics. 32, pp. 117–140, (2014).

[2] A. Panca, J. Panidi, H. Faber, S. Stathopoulos, T. D. An- thopoulos, and T. Prodromakis, "Flexible oxide thin film transistors, memristors, and their integration," Advanced Functional Materials, 33, p. 2213762, (2023).

[3] S. Datta, S. Dutta, B. Grisafe, J. Smith, S. Srinivasa, and H. Ye, "Back-end-of-line compatible transistors for monolithic 3-D integration," IEEE micro, 39, pp. 8–15, (2019).

[4] K. Nomura, H. Ohta, A. Takagi, T. Kamiya, M. Hirano, and H. Hosono, "Room-temperature fabrication of trans- parent flexible thin-film transistors using amorphous oxide semiconductors," nature, 432, pp. 488–492, (2004).

[5] J. K. Jeong, J. H. Jeong, H. W. Yang, J.-S. Park, Y.- G. Mo, and H. D. Kim, "High performance thin film transistors with cosputtered amorphous indium gallium zinc oxide channel," Applied Physics Letters, 91, p. 113505, (2007).

[6] Y. Shin, S. T. Kim, S. Kim, M. Y. Kim, S. Oh, and J. K. Jeong, "The mobility enhancement of indium gallium zinc oxide transistors via low-temperature crystallization using a tantalum catalytic layer," Scientific Reports, 7, p. 10885, (2017).

[7] S. I. Kim, C. J. Kim, J. C. Park, I. Song, S. W. Kim, H. Yin, E. Lee, J. C. Lee, and Y. Park, "High performance oxide thin film transistors with double active layers," in 2008 IEEE International Electron Devices Meeting. (IEDM), pp. 1–4, (2008).

[8] C.-Y. Park, S.-P. Jeon, J. B. Park, H.-B. Park, D.-H. Kim, S. H. Yang, G. Kim, J.-W. Jo, M. S. Oh, M. Kim et al., "High-performance ITO/a-IGZO heterostructure TFTs enabled by thickness-dependent carrier concentration and band alignment manipulation," Ceramics International, 49, pp. 5905–5914, (2023).

[9] H. J. Seul, J. H. Cho, J. S. Hur, M. H. Cho, M. H. Cho, M. T. Ryu, and J. K. Jeong, "Improvement in carrier mobility through band-gap engineering in atomic-layer- deposited In-Ga-Zn-O stacks," Journal of Alloys and Compounds, 903, p. 163876, (2022).

[10] W. Pan, G. Zhang, X. Liu, K. Song, L. Ning, S. Li, L. Chen, X. Zhang, T. Huang, H. Yang, X. Zhou, S. Zhang and L. Lu, "Achieving High Performance of ZnSnO Thin-Film Transistor via Homojunction Strategy," Micromachines, 14, p. 2144, (2023).

[11] P. Wang, H. Yang, J. Li, X. Zhang, L. Wang, J. Xiao, B. Zhao, S. Zhang, and L. Lu, "Synergistically Enhanced Performance and Reliability of Abrupt Metal-Oxide Heterojunction Transistor," Advanced Electronic Materials, 9, p. 2200807, (2023).

[12] H. Yang, X. Zhou, L. Lu and S. Zhang, "Investigation to the Carrier Transport Properties in Heterojunction-Channel Amorphous Oxides Thin-Film Transistors Using Dual-Gate Bias," IEEE Electron Device Letters, 44, pp. 68-71, (2023).

[13] H. Yang, Z. Ma, L. Lu and S. Zhang, "High-Performance Dual-Gate a-IGZO/a-IZO Thin-Film Transistors," SID Symposium Digest of Technical Papers, 55, pp. 698-700, (2024).

[14] B. Chang, X. Deng, J. Tao, H. Yang and S. Zhang, "Self-aligned top-gate amorphous oxide thin-film transistors with IZO/IGZO stacked active layer and Al reacted source/drain region," 2019 IEEE International Conference on Electron Devices and Solid-State Circuits (EDSSC), pp. 1-3, (2019).

[15] H. Yang, B. Wang, W. Dong, Z. Ma, W. Pan, L. Lu and S. Zhang, "Energy-Band-Dependent Mobility in Heterojunction Amorphous Oxide Semiconductor Thin-Film Transistors," SID Symposium Digest of Technical Papers, 54, pp. 461-463, (2023).

[16] J. Kim, J. Bang, N. Nakamura, and H. Hosono, "Ultra- wide bandgap amorphous oxide semiconductors for NBIS-free thin-film transistors," APL Materials, 7, p. 022501, (2019).

[17] C. H. Choi, T. Kim, S. Ueda, Y.-S. Shiah, H. Hosono, J. Kim, and J. K. Jeong, "High-performance indium gallium tin oxide transistors with an Al$_2$O$_3$ gate insulator deposited by atomic layer deposition at a low temperature of 150°C: roles of hydrogen and excess oxygen in the Al$_2$O$_3$ dielectric film," ACS Applied Materials & Interfaces, 13, pp. 28 451– 28 461, (2021).

High-Performance Flexible Graphene Field-Effect Transistors and Multistage Inverter Chains for Signal Amplification

Zhongyang Ren[1#], Pei Peng[1#], Jiaojiao Tian[1], Chenhao Xia[1], Muchan Li [1], Liming Ren[1], Fei Liu[1*], Yunyi Fu[1*]

[1] School of Integrated circuits, Peking University, Beijing 100871, China

[*] Author to whom any correspondence should be addressed. [#] Equal contribution

Email: yyfu@pku.edu.cn; feiliu@pku.edu.cn

Abstract

Graphene is a promising material in the realm of flexible electronics, offering exceptional properties for advanced device fabrication. In this study, we present an optimized fabrication process for flexible graphene field-effect transistors (GFETs), achieving an intrinsic voltage gain of 6 dB. Furthermore, we demonstrate the development of multistage graphene inverter chains, successfully realizing a three-stage graphene inverter cascade on a flexible substrate, integrating six flexible GFETs with a total gain of 15.6 dB. The potential of these graphene-based inverter chains for signal amplification is illustrated through the successful amplification of heart rate signals, showcasing their application in flexible bioelectronics.

Keywords: Graphene, Flexible Electronics, Field-Effect Transistors, Inverter Chain, Signal Amplification, Heart Rate Signal.

Introduction

Graphene's exceptional flexibility and electrical properties make it a highly promising material for use in flexible field-effect transistors (FETs) and circuits [1]. However, one of the main challenges is the absence of a bandgap in graphene, which leads to high output conductance and limits its switching capabilities [2]. On flexible substrates, thermal deformation further constrains the performance of graphene field-effect transistors (GFETs), reducing the allowable drain voltage and increasing the leakage current density, which can cause device failure [3]. As a result, GFETs on flexible substrates typically exhibit intrinsic voltage gains less than 1, which limits their application in integrated circuits, such as graphene inverter chains, ring oscillators, and voltage amplifiers that require higher voltage gains for proper operation [4]. The development of flexible graphene-based multi-stage inverters is of considerable significance for enhancing the application of graphene in flexible electronic circuits. The direct coupling between inverter stages enhances low-frequency performance, while the cascaded configuration increases overall voltage gain. This makes flexible graphene multi-stage inverters suitable for amplifying and buffering output signals in flexible low-frequency biosensors or wearable devices and circuits.

Additionally, the buffering function isolates the sensor from the load, preventing performance degradation due to low-resistance loads. To enable these applications, it is critical to investigate input/output matching techniques for graphene inverters and to develop multistage configurations that can operate effectively on flexible substrates. Such advances are vital for extending the use of graphene in flexible circuit applications.

This work focuses on optimizing the fabrication process of flexible GFETs, characterizing their electrical properties, and examining the impact of substrate bending on device performance. We establish a large-signal model of GFETs and explore the key factors influencing input/output matching in graphene inverters are studied. A method for achieving efficient input/output matching is proposed, and a three-stage graphene inverter chain, integrating six GFETs, is experimentally demonstrated. Furthermore, we investigate the potential application of the inverter chain for low-frequency signal amplification, showcasing its ability to buffer and amplify heart rate signals, which is essential for flexible biomedical devices.

Experimental

To achieve high intrinsic gain in GFETs, we investigated gate oxide preparation and flexible substrate pre-treatment. Previous studies show that increasing the gate oxide capacitance (C_{ox}) enhances GFETs' output saturation characteristics, thereby improving voltage gain [5]. A gate oxide layer with a thickness of approximately 5 nm and capacitance up to 1.48 $\mu F/cm^2$ was formed using metal oxide methods [6]. By evaporating and oxidizing 3 nm of Yttrium (Y), Y_2O_3 was formed as the gate oxide. To improve substrate thermal conductivity and resist thermal deformation, a 100 nm Al_2O_3 layer was deposited on a PEN substrate, forming an Al_2O_3/PEN composite.

Fabrication steps: 1. Graphene Patterning: Defined the active region using electron beam lithography and oxygen plasma etching. 2. Source/Drain Electrode Fabrication: Source/drain regions were patterned, followed by Pd/Au evaporation and lift-off to form contacts. 3. Gate Dielectric Preparation: 3 nm of Y was evaporated, followed by oxidation to form ~5 nm Y_2O_3. 4. Top Gate Electrode

Fabrication: Gate electrodes were formed by Ti/Au evaporation after electron beam lithography (**Figure 1a-d**). The resulting flexible GFETs structure is shown in **Figure 1e**, with optical and microscope images of the device array in **Figure 1(f-g)**, demonstrating the flexibility and transparency of the devices.

Results and Discussion

Performance and intrinsic voltage gain of flexible GFETs: To evaluate the impact of substrate bending on GFETs' electrical performance, GFETs on flexible PEN substrates were tested by attaching them to a semicylinder with a radius (r) of 12.5 mm (**Figure 2a**). Electrical measurements were conducted at room temperature and atmospheric pressure, with bending along the GFET channel (**Figure 2b**). The hysteresis of the transfer characteristics under a 12.5 mm bending radius is shown in **Figure 2c**, with minimal effect on hysteresis observed. Output characteristics (**Figure 2d**) show a maximum current density of ~0.3 mA/μm, likely due to the improved thermal conductivity of the Al_2O_3/PEN substrate (Al_2O_3: ~2 W/m°C, PEN: ~0.1-0.2 W/m°C). The transfer characteristics in **Figure 2e** show a peak transconductance of ~240 μS/μm. The intrinsic voltage gain, calculated as $A_v = g_m/g_d$ under specific bias conditions (V_d and V_g), is shown in **Figure 2f**, with a maximum intrinsic gain of approximately 2 (6 dB) when both gate and drain voltages are around 1V.

Graphene Inverter Chain: The circuit diagram of the GFET-based inverter is shown in **Figure 3a**, consisting of two GFETs (T1 and T2, **Figure 3b**). Input/output matching is crucial for cascading stages in inverter chains. Graphene inverters achieve maximum gain at the midpoint between the Dirac voltages of the two GFETs. To bring the initial Dirac point closer to zero, appropriate gate metal selection was made, leveraging the work function difference between the metal and graphene channel to adjust the Fermi level to the Dirac point. **Figure 3c** shows the Dirac point statistics for GFETs with Ti/Au (red points) and Pd/Au (blue points) gate metals. Ti/Au gates yielded Dirac points closer to zero. **Figure 3d** demonstrates that a mismatch deviation $|\Delta V|$ below 0.13V occurs when using Ti/Au gates, maintaining good input/output matching. The measured input-output characteristic and corresponding gain are shown in **Figure 3e**, with a voltage gain of approximately 1.9, close to the intrinsic voltage gain of a single GFET, and the gain peak occurring near the point where input and output voltages are equal, indicating good matching.

Three-stage graphene inverter chain: Increasing the number of cascaded stages further amplifies the circuit gain. A schematic of the three-stage cascaded graphene flexible inverter is shown in **Figure 4a**, with the circuit diagram in **Figure 4b**. The optical image of the fabricated three-stage inverter (comprising six GFETs) is shown in **Figure 4c**.

Voltage waveforms at three frequencies (0.5Hz, 100Hz, 10kHz) are presented in **Figures 4(d-f)**, demonstrating a gain of ~6 (15.6 dB) across all frequencies.

Heart Rate Signal Amplification: As shown in the test setup in **Figure 5a**, low-amplitude heart rate signals from a heart rate sensor were input into the three-stage graphene flexible inverter. **Figure 5b** shows the measured output signal, with a significant amplification relative to the original heart rate signal (input signal), achieving a voltage gain of 5.4 (12.5mV/2.3mV). The results demonstrate that the multi-stage graphene inverter can be used for amplifying low-frequency bio-signals. Furthermore, the inverter's high input impedance and low output impedance make it suitable for isolating sensors from loads, preventing low-resistance loads from affecting sensor performance.

Conclusions

This work successfully optimized the fabrication process for flexible graphene field-effect transistors (GFETs), achieving an intrinsic voltage gain of 6 dB. Input/output matching techniques for graphene-based inverters were investigated, and a three-stage graphene inverter chain (comprising six GFETs) was realized on a flexible substrate, with a total gain of 15.6 dB. The potential of the multi-stage inverter for signal buffering and amplification was demonstrated, including its successful application in amplifying heart rate signals. These results provide a solid foundation for advancing the integration of flexible GFET circuits, paving the way for future developments in flexible electronics.

Acknowledgments

The authors gratefully acknowledge the support from the National Natural Science Foundation of China (Grant No. T2293703 and Grant T2293700) and the Instrumental Analysis Fund of Peking University (No. KF-2305−07).

References

[1] Krishnan S K, Nataraj N, Meyyappan M, et al. Graphene-based field-effect transistors in biosensing and neural interfacing applications: recent advances and prospects[J]. Analytical Chemistry, 95(5): 2590-2622 (2023).

[2] Huang T J, Ankolekar A, et al. Investigating the Device Performance Variation of a Buried Locally Gated Al/Al2O3 Graphene Field-Effect Transistor Process[J]. Applied Sciences, 13(12): 7201 (2023).

[3] Ning J, Wang Y, Feng X, et al. Flexible field-effect transistors with a high on/off current ratio based on large-area single-crystal graphene[J]. Carbon, 163: 417-424 (2020).

[4] Bianchi M, Guerriero E, Fiocco M, et al. Scaling of graphene integrated circuits[J]. Nanoscale, 7(17): 8076-8083 (2015).

[5] Han S J, Reddy D, et al. Current saturation in submicrometer graphene transistors with thin gate dielectric: Experiment, simulation, and theory[J]. ACS nano, 6(6): 5220-5226 (2012).

[6] Cai Q, Ye J, Jahannia B, et al. Comprehensive study and design of graphene transistor[J]. Micromachines, 15(3): 406 (2024).

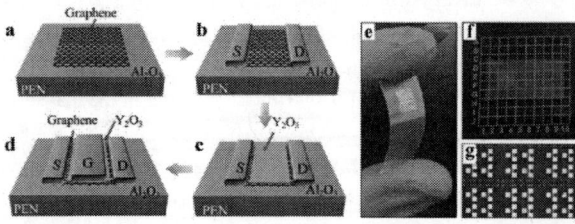

Figure 1: Flexible GFET fabrication process: (a-d) Steps for defining graphene, source/drain electrodes, gate dielectric, and gate metal. D, S, and G represent drain, source, and gate, respectively. (e) Optical image of GFET array on flexible substrate. (f-g) Optical microscope images showing device flexibility and transparency.

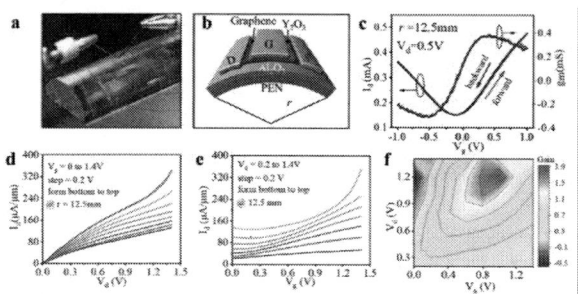

Figure 2: Performance and intrinsic voltage gain of flexible GFETs: (a) Test setup for bending the substrate. (b) Bending schematic. (c) Hysteresis under 12.5 mm bending. (d) Transfer characteristics. (e) Output characteristics. (f) Relationship between intrinsic voltage gain and bias (V_g, V_d).

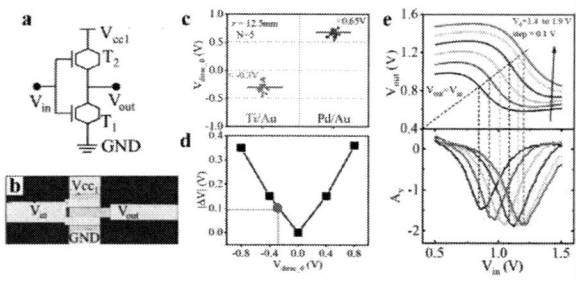

Figure 3: Two-stage graphene inverter chain and voltage amplification characteristics: (a) Circuit diagram. (b) Optical image of the fabricated graphene inverter. (c) Dirac point statistics for different gate metals. (d) Mismatch $|\Delta V|$ vs. initial Dirac voltage (V_{dirac_0}). (e) Input-output characteristic (top) and corresponding gain (bottom).

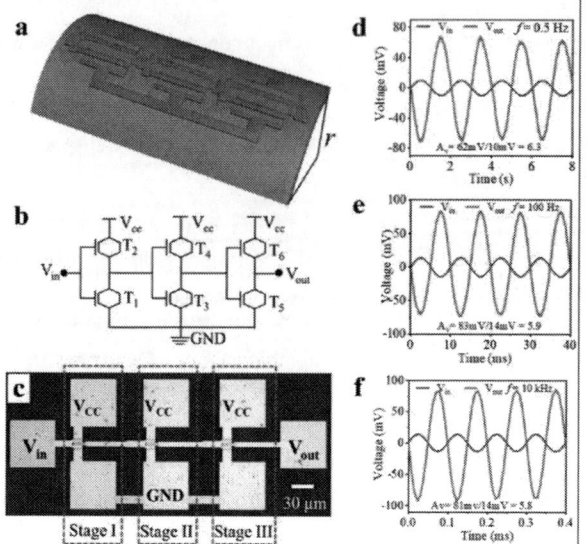

Figure 4: Three-stage graphene inverter chain and voltage amplification characteristics: (a) Circuit structure. (b) Circuit diagram. (c) Optical image of the three-stage inverter. (d-f) Voltage waveforms at input frequencies of 0.5 Hz, 100 Hz, and 10 kHz, respectively.

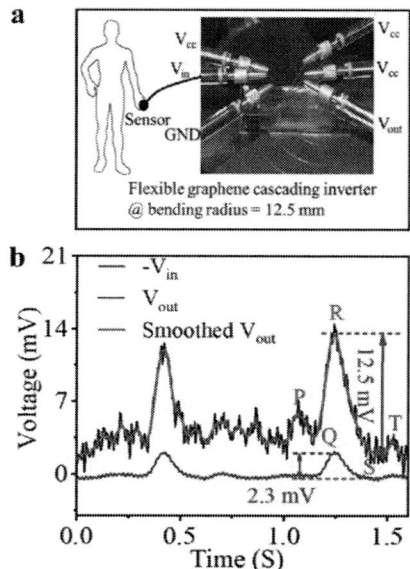

Figure 5: Demonstration of heart rate signal amplification using the graphene inverter chain: (a) Test setup. (b) Comparison of input and output waveforms.

979-8-3315-0417-5/25 $31.00 © 2025 IEEE

Oxide-based Optical Synapses with Low Consumption for In-Sensor Reservoir Computing

Jingyi Zhou[1], Biyi Jiang[1], Huansong Tang[1], Zhen Wang[1]*

[1]School of Microelectronics, Southern University of Science and Technology, Shenzhen, China
*E-mail: wangz8@sustech.edu.cn

Abstract

In this work, we propose a two-terminal ITO/NiO/HfO$_2$/Pd optical synapse that has tunable temporal dynamics under light stimuli. The optical synapse exhibits light intensity and frequency-dependent nonlinear synaptic short-term plasticity (STP) at a low read voltage of 0.05 V, indicating the potential for achieving in-sensor reservoir computing (RC) with reduced power consumption. The RC system, utilizing our optical synapses, can operate at an ultra-low energy consumption level of 15 fJ while achieving a remarkably high recognition accuracy of 95.95 %.

Keywords: Optical Synaptic Device, Reservoir Computing and Oxide-Based Memristor

Introduction

Reservoir Computing (RC), as a novel neuromorphic computing model, employs artificial synapses for the implementation of reservoir hardware, which provides high efficiency, simplicity, and ease of training [1-2], leading to significant advancements in processing spatiotemporal information [3-5]. The RC systems, however, primarily rely on conventional electrical synaptic devices that necessitate additional sensors and therefore have limitations such as computational latency, and energy consumption [6-8]. These issues can be effectively resolved by employing optoelectronic synaptic devices (*e.g.* organic, two-dimensional (2D) and oxide-based), which can integrate data acquisition, memory, and processing functions into a single device. Organic and 2D optical synaptic devices, despite significant progress in constructing RC hardware, still pose a major challenge in large-scale integration. On the contrary, oxide-based optical synaptic devices offer advantages such a simple structure, large scalability, high reliability and complementary compatibility with metal–oxide–semiconductor (CMOS), exhibiting considerable potential in future neuromorphic vision systems.

In this work, we construct a vertically stacked ITO/NiO/HfO$_2$/Pd optical synapse. At a voltage as low as 0.05 V, the optical synaptic device with NiO/HfO$_2$ double-layer structure exhibits exceptional photocurrent response, which can be effectively modulated by light intensity and frequency. Additionally, the optical synaptic device, when integrated into an RC system, demonstrates

high efficiency with an energy consumption of only 15 fJ per operation while maintaining a recognition accuracy of 95.95%.

Results and Discussion

A. Device Fabrication

The schematic diagram of a two-terminal ITO/NiO/HfO$_2$/Pd device is illustrated in Fig. 1a. The details of fabrication process, depicted in Fig. 1b, encompasses the following sequential steps. Initially, a 30 nm-thick Pd film was deposited using electron-beam evaporation onto a Si/SiO$_2$ substrate. Then, the dielectric layer HfO$_2$ with a thickness of 5 nm and the active layer NiO with a thickness of 40 nm were sequentially deposited via radio frequency (RF) sputtering. Finally, the ITO top electrode (40 nm thickness) was sputtered onto the stack, at a deposition rate of 0.3 Å s^{-1}, and patterned using hard masks with a dimension of 200 microns.

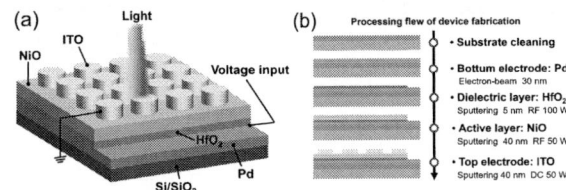

Fig. 1: (a) Schematic diagram and (b) the fabrication process of a two-terminal ITO/NiO/HfO$_2$/Pd optical synaptic device.

B. Tunable Photosynaptic Characteristics of The Optical Synapses

The representative *IV* characteristics of the NiO/HfO$_2$-based synaptic device are illustrated in Fig. 2a. By sweeping from 0 V to 1 V and back to 0 V, the device undergoes a gradual transition from a high resistance state to a low resistance state. The reset process is achieved by scanning from 0 V to 1 V and then back to 0 V. Notably, after 100 consecutive scans, the synaptic device exhibits low cycle-to-cycle volatility.

The photocurrent response at various pulse intensities (0.22, 0.42, 0.67, 0.90, 1.10, and 1.38 µW/cm²) with a pulse width of 500 ms and a read voltage of 0.05 V is shown in Fig. 2b. As the light intensity increased from 0.22 µW/cm² to 1.38 µW/cm², the device exhibits an increase in photocurrent

979-8-3315-0417-5/25 $31.00 © 2025 IEEE

from 34.6 pA to 85.1 pA, accompanied by a gradual non-linear relaxation process after the removal of light. To further investigate the synaptic short-term plasticity (STP) of optical synapses, various light pulse widths ranging from 500 to 1000 ms are applied under a light intensity of 1.38 $\mu W/cm^2$. The peak current output of the device increases from 43.5 pA to 89.9 pA as schematically depicted in Fig. 2c, when extending the pulse width from 50 ms to 1000 ms. By employing a simple exponential model for curve fitting, we obtain the characteristic time parameter t_0 in Fig. 2d, which is approximately equal to 1.06 s. The findings demonstrate that the response of the dynamic memristor is not solely dependent on the immediate current input, but also influenced by the historical profile of the input signal.

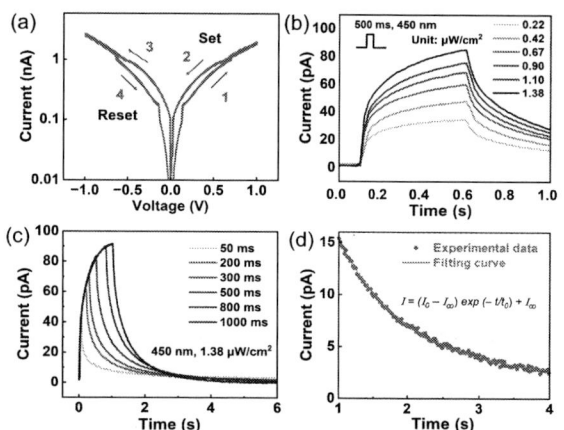

Fig. 2: (a) The *I-V* characteristic curves of the device was obtained after 100 repeated cycles. The arrows indicate the direction of the voltage scan. (b) The photocurrents were taken at various light intensities (0.22, 0.42, 0.67, 0.90, 1.10, and 1.38 $\mu W/cm^2$) and the wavelengths of 450 nm, each lasting for a width of 500 ms at a read voltage of 0.05 V. (c) The short-term potentiation effect stimulated by increased light pulse widths (0.05, 0.2, 0.3, 0.5, 0.8 and 1 s, intensity: 1.38 $\mu W/cm^2$) at 0.05 V read voltage. (d) The current decay over time follows an exponential relationship with a fitting characteristic time t_0 of 1.06 s.

The device in Fig. 3a is subjected to irradiation by a sequence of 10 light pulses with the intensities of 0.22, 0.56, 0.90, 1.23, and 1.38 $\mu W/cm^2$ respectively, as depicted. The application of these light pulses results in an increase in the device's photocurrent from 28.3 pA to 69.1 pA, demonstrating the light-intensity dependent nonlinear synaptic plasticity. Fig. 3b further explores the spiking-rate-dependent plasticity (SRDP) behaviour. After the 10 sequential light pulses within a frequency range of 5 Hz-13 Hz and an intensity of 1.38 $\mu W/cm^2$, the photocurrents gradually decay back to their initial values.

C. In-Sensor Reservoir Computing Based on Optical Synapses for Handwritten Digit Recognition

The RC model encompasses a reservoir layer and a

simulated output layer, as illustrated in Fig. 4a. The utilization of the optical synaptic devices as reservoirs enables the RC systems significantly reduces the network scale and computational resources needed for encoding input signals into a high-dimensional computing space. Fig. 4b illustrates the encoding process of the optical synaptic device, demonstrating the transformation of 16 optical input patterns (0000-1111) into 16 distinct current outputs.

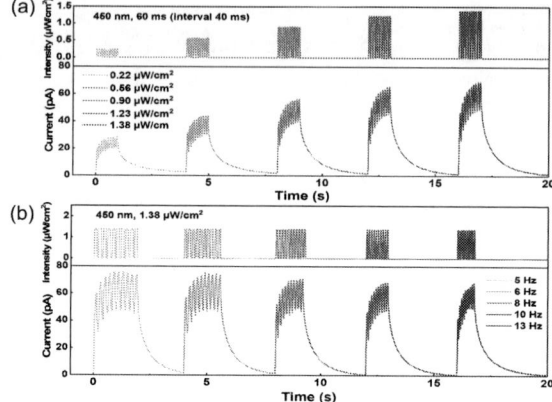

Fig. 3: (a) The photocurrent triggered by 10 light pulses, with varying intensities ranging from 0.22 $\mu W/cm^2$ to 1.38 $\mu W/cm^2$, a pulse width of 60 ms and an interval of 40 ms, respectively. (b) The photocurrent triggered by 10 light pulses with a pulse frequency range of 5 Hz to 13 Hz and an intensity of 1.38 $\mu W/cm^2$. All measurements were performed at 0.05 V read voltage.

The detailed encoding scheme of the optical synapse is illustrated in Fig. 4c, where '1' denotes the optical input (200 ms, 1.38 $\mu W/cm^2$) and '0' represents no light. By applying four sequential pulses, the optical synapse effectively encodes the 16 different input patterns into distinct current levels. As can be seen from Fig. 4d, the average difference between output current states of our optical synaptic device is 5.44 pA, thereby facilitating noise resilient and enhancing recognition accuracy.

This RC system utilizes the Mixed National Institute of Standards and Technology (MINST) database for a handwritten digit recognition task. The RC system leverages the device's optical encoding capability to effectively convert input data into a high-dimensional space, achieving a fourfold reduction in data size. Fig.4e displays the sample input images (28 × 28 pixels) and the corresponding encoded images (28 × 7 pixels) based on our optical synaptic device. The input sample image contains two distinct intensity levels: 0 (0 $\mu W/cm^2$) and 1(1.38 $\mu W/cm^2$). Four optical pulses, corresponding to four consecutive pixels in each image row, are input to one device. The device encodes these four optical pulses using 16 possible patterns (0000 to 1111) into 16 distinct conductance levels, thus achieving a fourfold reduction in data.

979-8-3315-0417-5/25 $31.00 © 2025 IEEE

The RC system demonstrates a remarkable recognition accuracy of 95.95 % in the handwritten digit recognition task, as illustrated in Fig.4f, thereby emphasizing the effectiveness of the optical synapse within this specific application.

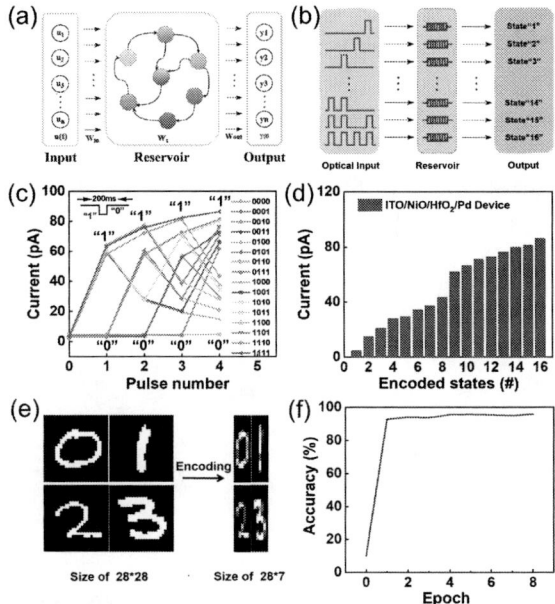

Fig. 4: (a) Schematic diagram of the RC architecture. (b) The optical encoding process of the optical synapse. (c) The current response of 16 encoded states in a 4-bit RC system, which is encoded by a 4-pulse sequence with a pulse width of 200 ms and an intensity of 1.38 μW/cm^2 at a read voltage of 0.05 V. (d) The output currents corresponding to 16 encoded states. (e) Four input sample images from the MNIST dataset and the encoded images based on optical synapse. (f) Recognition accuracy of the NiO/HfO$_2$-based RC system.

Conclusion

In this work, we demonstrate a two-terminal oxide-based optical synaptic device with light-intensity and frequency tunable STP. The optical synaptic device exhibits a significantly large photocurrent response, demonstrating exceptional capabilities in recognition and noise reduction within the RC model. Each optical encoding operation in the RC system consumes 15 fJ, highlighting their effectiveness and potential in the field of optoelectronic neuromorphic computing.

Acknowledgments

This work was supported by the National Key Research and Development Project of China (Grant No. 2023YFB2806300), National Natural Science Foundation of China (Grant No. 62104091, 52273246, 62174074), Guangdong Natural Science Foundation (Grant No. 2022A1515011064), Shenzhen Science and Technology Program (Grant No. JCYJ20220530115204009, JCYJ20220530115014032), Zhujiang Young Talent Program (Grant No. 2021QN02X362), and SUSTech SME-Pixelcore Neuromorphic In-sensor Computing Joint Lab.

References

[1] Y. Zhong, J. Tang, X. Li, B. Gao, H. Qian and H. Wu, "Dynamic memristor-based reservoir computing for high-efficiency temporal signal processing", Nature Communications, 12(1), 408 (2021).

[2] Du, F. Cai, M. Zidan, W. Ma, S. Lee, and W. Lu, "Reservoir computing using dynamic memristors for temporal information processing", Nature Communications, 8, 2204 (2017).

[3] Z. Zhang, X. Zhao, X. Zhang, X. Hou, X. Ma, S. Tang, Y. Zhang, G. Xu, Q. Liu, and S. Long, "In-sensor reservoir computing system for latent fingerprint recognition with deep ultraviolet photo-synapses and memristor array", Nature Communications, 13, 6590 (2022).

[4] Y. Zhong, J. Tang, X. Li, X. Liang, Z. Liu, Y. Li, Y. Xi, P. Yao, Z. Hao, B. Gao, H. Qian, and H. Wu, "A memristor-based analogue reservoir computing system for real-time and power-efficient signal processing", Nature Electronics, 5, 672–681 (2022).

[5] K. Liu, T. Zhang, B. Dang, L, Bao, L. Xu, C. Cheng, Z. Yang, R. Huang, and Y. Yang, "An optoelectronic synapse based on α-In$_2$Se$_3$ with controllable temporal dynamics for multimode", Nature Electronics, 5(11), 761–773 (2022).

[6] M. Pickett, G. Medeiros-Ribeiro, and R. Williams, "A scalable neuristor built with Mott memristors", Nature Materials, 12(2), 114–117 (2013).

[7] Q. Hua, H. Wu, B. Gao, M. Zhao, Y. Li, X. Li, X. Hou, M. Chang, P. Zhou, H. Qian, "A threshold switching selector based on highly ordered Ag nanodots for X-Point memory applications", Advanced Science, 2019, 6, 1900024 (2020).

[8] W. Yi, K. Tsang, S. Lam, X. Bai, J. Crowell, and E. Flores, "Biological plausibility and stochasticity in scalable VO$_2$ active memristor neurons", Nature Communications, 9, 4661 (2018).

A SAW Pressure Sensor based on AlN thin film with a Novel Differential Structure: Balancing Sensitivity Enhancement and Temperature Decoupling

Aobei Chen[1], Dapeng Li[1], Ge Gao[2], Chun Hu[2] and Dezhi Zheng[1,2,*]

[1]Beihang University, China, [2]Beijing Institute of Technology, China
*Correspondence: zhengdezhi@bit.edu.cn

Abstract

Two SAW devices were integrated on a pressure-sensitive diaphragm based on an AlN film substrate. The placement of the devices was carefully designed according to the frequency variation behavior of the SAW devices and the strain distribution across the diaphragm, forming an effective differential structure. Experimental results demonstrated the beneficial effects of this design: the pressure sensitivity increased from 2.2267 ppm/kPa to 24.8767 ppm/kPa, while the temperature sensitivity decreased from 21.2572 ppm/°C to 3.8763 ppm/°C.

Keywords: Temperature decoupling, SAW pressure sensor, AlN thin film

Introduction

Surface Acoustic Wave (SAW) pressure sensors are characterized by a wide operating temperature range, simple structure, and ease of fabrication, making them widely applicable in scenarios such as atmospheric detection in complex environments, tire pressure monitoring in vehicles, and chemical process control [1-2]. SAW pressure sensors utilizing AlN thin film composite structures as the piezoelectric substrate offer notable advantages, including a straightforward design and good compatibility with MEMS processes [3]. However, AlN films are often grown on Si or SOI substrates, and the thermal mismatch and temperature characteristics of these substrates can result in a high temperature coefficient (TC) for the sensor, limiting its application scenarios [4]. In existing research, some have attempted to reduce the TC by adding multiple layers of SiO_2 on the Si substrate and the interdigital transducer (IDT) [5]. However, this approach increases production costs and offers limited temperature compensation effectiveness. Additionally, some researchers have created differential structures in strain sensors by applying forces in different directions to two identical SAW structures [6]. Nevertheless, they merely adhered two sensors together without achieving integration and did not consider the coordination between the SAW sensor and the pressure-sensitive diaphragm.

To address the aforementioned issues, we designed and fabricated a differential SAW pressure sensor based on an AlN thin film composite structure. This differential structure cleverly exploits the relationship between SAW device frequency variations and the spatial distribution of diaphragm strain, significantly reducing temperature sensitivity while greatly enhancing pressure sensitivity. Experimental results demonstrate that this differential structure increased the pressure sensitivity (at 25°C) from 2.2267 ppm/kPa to 24.8767 ppm/kPa (increased by a factor of 11.17), while the temperature sensitivity (at 110kPa) was reduced from 21.2572 ppm/°C to 3.8673 ppm/°C (reduced to 1/5.50 of the original).

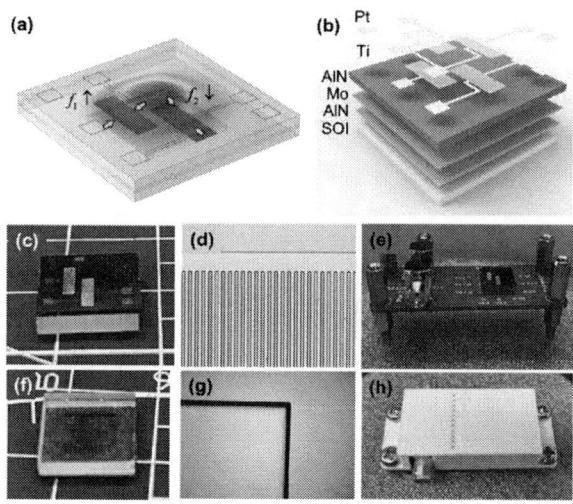

Fig. 1. Sensor design, fabrication, and packaging. (a) Diaphragm deformation under pressure induces frequency changes in the SAW; (b) Materials used in each layer of the sensor; (c) and (f) show the top and bottom views of the sensor mounted on the SiO_2 base, respectively; (d) and (g) present the optical microscope magnifications of the IDT and the diaphragm; (e) Sensor packaging method; (h) Protective casing for the sensor.

Design and Manufacturing

The variation of the SAW device's resonant frequency can be expressed as [7]:

$$\Delta f/f = (\gamma_1 + \gamma_2 \cos 2\theta)\varepsilon. \tag{1}$$

Where γ_1 and γ_2 are the strain coefficients of acoustic velocity in the substrate, θ is the angle between the direction of the IDT and the strain loading direction, and ε is the diaphragm strain. Therefore, placing the SAW devices at different locations on the diaphragm can produce varying frequency responses. The overall structure of the sensor is illustrated in Fig. 1 (a). The diaphragm is designed as a square with a side length of 5800 μm and a thickness of 200 μm.

979-8-3315-0417-5/25 $31.00 © 2025 IEEE

Two SAW devices are positioned at the edges of the diaphragm, with one aligned parallel and the other perpendicular to the IDT direction under strain, resulting in opposite frequency shifts (f_1 increases, f_2 decreases).

The two SAW devices have a period of $\lambda = 10.4\ \mu m$ and an acoustic aperture of 1150 μm, with 150 pairs of IDTs and 300 reflector electrodes. The layer structure of the sensor is shown in Fig. 1 (b). First, a 15 μm thick AlN seed layer is grown on a single-sided polished SOI (1.5μm Si/1μm SiO$_2$/500μm Si) substrate using atomic layer deposition, followed by the growth of 0.25μm Mo/1μm AlN using chemical vapor deposition. Deep reactive ion etching is then used to etch through the AlN layer, creating a pathway that connects to the Mo bottom electrode (which serves as ground, providing electromagnetic shielding and improving signal quality). A 30nm Ti/80nm Pt electrode layer is subsequently deposited on the AlN using reactive ion sputtering, and the electrodes are patterned using a lift-off process. Finally, the sensor chip is flipped, and the back is etched 300 μm deep via deep silicon etching to form the diaphragm.

We securely bonded the fabricated sensor chip to a 2mm-thick SiO$_2$ base using epoxy resin, forming a pressure reference cavity to enable absolute pressure measurement. Additionally, the SiO$_2$ base provides stress isolation and temperature compensation, which further enhances the stability of the chip. After completing the sensor fabrication, we used an optical microscope to inspect the IDT and diaphragm, as shown in Fig. 1 (d) and (g). The inspection reveals that both the IDT and diaphragm were well fabricated and met the required specifications. The sensor was affixed to a PCB using acrylic adhesive, and gold wire bonding was employed to extract the electrode signals. The assembled structure is shown in Fig. 1 (e) and (h).

Results and Discussion

A. Experimental Setup

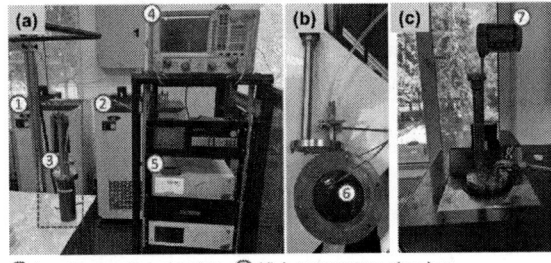

① Low-temperature chamber ② High-temperature chamber
③ Vacuum test chamber ④ Vector Network Analyzer ⑤ Pressure controller
⑥ SAW pressure sensor ⑦ Standard thermometer

Fig. 2. Sensor testing system. (a) Overview of the integrated temperature and pressure testing system; (b) SAW pressure sensor placed inside the vacuum test chamber; (c) Vacuum test chamber placed inside the temperature chamber.

To evaluate the pressure and temperature sensitivities of the sensor, we built an integrated temperature and pressure testing system, with the overall hardware configuration shown in Fig. 2 (a). The low-temperature chamber (AVIC CIMM-TH-80) and high-temperature chamber (AVIC CIMM-TH-0) are used to create precise and stable temperature environments, with a temperature range of $-45 \sim 80\ ^\circ C$. The vacuum test chamber, made of copper, facilitates rapid heat exchange and reduces the time required to reach thermal equilibrium. The sensor signals are extracted from the chamber through a vacuum flange and input into a vector network analyzer (Agilent N5247A) for resonant frequency readings. The pressure inside the chamber is precisely controlled by a pressure controller (Druck PACE6000), with a pressure variation range of 0~300 kPa. During testing, the sensor is placed inside the vacuum chamber (Fig. 2 (b)), which is then sealed and placed inside the temperature chamber (Fig. 2 (c)), enabling integrated temperature and pressure testing.

B. Pressure Sensitivity

During the testing process, the temperature inside the vacuum test chamber was stabilized at 25 °C, and the internal pressure was gradually increased from 5 kPa to 200 kPa in 50 kPa intervals. The S11 parameters of the left and right SAW devices were measured using the VNA to further analyze the changes in their resonant frequencies. As shown in Fig. 3, although the structural parameters of the two sets of devices were designed to be identical, they exhibited different Q-values. This discrepancy is attributed to factors such as fabrication errors, uneven AlN growth, and residual stress during the bonding process.

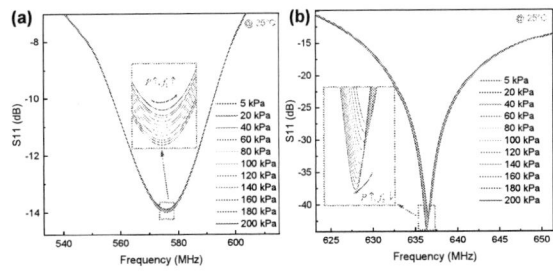

Fig. 3. S11 parameters of the sensor measured using the VNA. (a) Right SAW device; (b) Left SAW device.

Subsequently, we extracted the frequencies of the two SAW devices at different pressures from the S11 curves and fitted the data using the least squares method. The results are shown in Fig. 4. Both devices exhibit good linearity, with R^2 values of 0.9985 and 0.9972, respectively. The left SAW device, which is strained along the length of the IDT, showed a smaller frequency shift, with a pressure sensitivity of 0.1692 ppm/kPa. In contrast, the right SAW device had a pressure sensitivity of -2.2267 ppm/kPa. The sensor employs a differential design, where the frequencies of the two SAW devices are combined using a frequency synthesizer, with the final output being $f_2 - f_1$. After applying the differential frequency, the pressure sensitivity was enhanced to -24.8767 ppm/kPa. The performance improvement is due to two factors: first, the opposite frequency shift directions of the two devices under pressure, and second, the differential design significantly reduces the baseline frequency, resulting in a substantial increase in the relative frequency change.

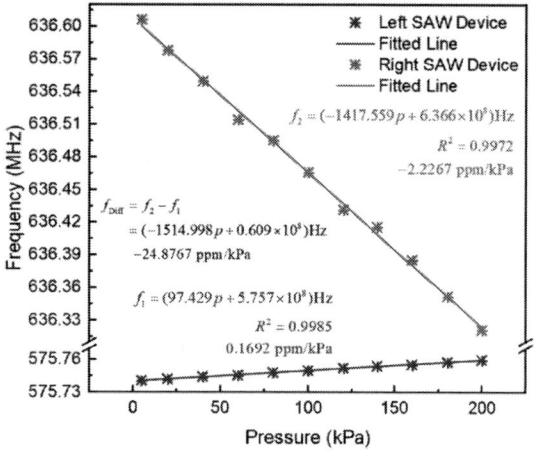

Fig. 4. Pressure-frequency data and fitting results of two SAW devices (at 25°C).

B. Temperature Sensitivity

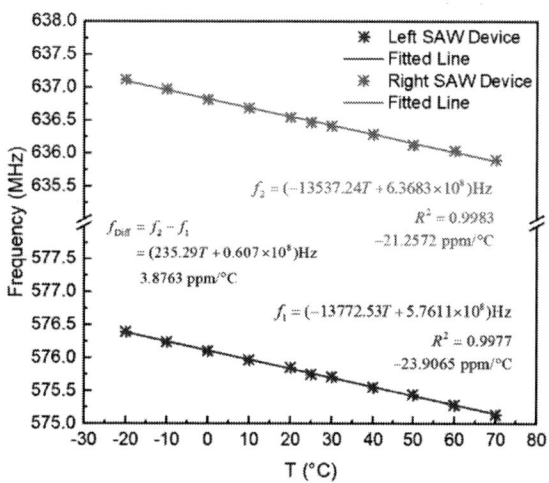

Fig. 5. Temperature-frequency data and fitting results of two SAW devices (at 100kPa).

Similar to the pressure sensitivity testing method, we stabilized the chamber pressure at 100 kPa and then adjusted the temperature chamber to vary the internal temperature from $-20 \sim 70$ °C in 10 °C intervals. The frequency response data

and fitting results for the two SAW devices at different temperatures are shown in Fig. 5. Both devices exhibit a frequency decrease with increasing temperature, and their TC are relatively close. Before applying the differential method, the TC values for the left and right devices were -23.9065 ppm/°C and -21.2572 ppm/°C, respectively. After applying the differential method, the sensor's TC was significantly reduced to 3.8763 ppm/°C.

Conclusion

In summary, this study integrated two SAW devices onto the diaphragm of a pressure sensor, forming a differential structure. Experimental results confirmed the beneficial effect of this design, which enhanced pressure sensitivity while reducing the influence of temperature. This work provides a novel approach for temperature decoupling in SAW sensors, and the differential structure can be further applied to other sensor types.

Acknowledgments

The authors gratefully acknowledge the support provided by the National Key Research and Development Program of China (2022YFB3207700) in this study.

References

[1] X. Liang, L. Zhang, Q. Tan, W. Cheng, D. Hu, S. Li, L. Jing, and J. Xiong, "Temperature, pressure, and humidity saw sensor based on co-planar integrated LGS," *Microsystems & Nanoengineering*, vol. 9, no. 1, p. 110, 2023.

[2] D. Mandal and S. Banerjee, "Surface acoustic wave (SAW) sensors: Physics, materials, and applications," *Sensors*, vol. 22, no. 3, p. 820, 2022.

[3] F. Wang, F. Xiao, D. Song, L. Qian, Y. Feng, B. Fu, C. Dong, C. Li, and K. Zhang, "Research of micro area piezoelectric properties of AlN films and fabrication of high frequency saw devices," *Microelectronic Engineering*, vol. 199, pp. 63–68, 2018.

[4] S. Pan, M. M. Memon, J. Wan, T. Wang, and W. Zhang, "The influence of temperature on the pressure sensitivity of surface acoustic wave pressure sensor," *Sensors and Actuators A: Physical*, vol. 332, p. 113183, 2021.

[5] T. Wang, X. Mu, A. B. Randles, Y. Gu, and C. Lee, "Diaphragm shape effect on the sensitivity of surface acoustic wave based pressure sensor for harsh environment," *Applied Physics Letters*, vol. 107, no. 12, 2015.

[6] Y. Yang, B. Peng, Z. Sun, F. Huang, and W. Zhang, "Study on the sensitivity of diaphragm-type saw pressure sensor," *IEEE Sensors Journal*, 2023.

[7] M. M. Memon, S. Pan, J. Wan, T. Wang, B. Peng, and W. Zhang, "Sensitivity enhancement of saw pressure sensor based on the crystalline direction," *IEEE Sensors Journal*, vol. 22, no. 10, pp. 9329–9335, 2022.

A BSIM Compact Model of Two-Dimensional Semiconductor Field Effect Transistors

Jen-Hao Chen[1], Ahtisham Pampori[1], Chien-Ting Tung[1], Sayeef Salahuddin[1] and Chenming Hu[1]

[1] University of California, Berkeley, CA, USA. (Email: jenhaochen@berkeley.edu)

Abstract

In this work, a compact model for two-dimensional semiconductor field effect transistors (2DFETs), based on the BSIM-CMG framework, is developed. We start with modeling the channel free charge density near the source side and the drain side, and achieve an explicit drain to source current expression. We incorporate this core model for 2DFETs with the sub models, including mobility degradation, velocity saturation, subthreshold swing degradation, short channel effect and self-heating in BSIM-CMG. This model is validated against measurements from state-of-the-art MoS$_2$-based FETs.

Keywords: 2D semiconductor, BSIM-CMG, Compact model.

Introduction

In this work, we develop a charge-based compact model for two-dimensional semiconductor field-effect transistors (2DFETs). The model starts by establishing an explicit relationship between gate-source voltage and free charge density. The drain current is then expressed in terms of free charge density near the source and drain. Leveraging the well-established BSIM-CMG framework, our 2DFET core model achieves a good fitting without incorporating trap charges. This approach reduces the complexity of the core model and avoids potential convergence issues related to model complexity, as compared to previous works [1-3].

Model Framework

A. Core Model for Channel Charge Density

Using the constant density of states in 2D material, the free charge density Q$_m$ can be obtained by integrating the product of the density of states and Fermi-Dirac distribution function:

$$Q_m = q \int_{E_C}^{\infty} D \cdot F(E) dE = q \cdot D \cdot kT \cdot exp\left(\frac{(-E_C + E_F)}{kT}\right) \quad (1)$$

where D is the density of states of 2D materials, F(E) is Fermi-Dirac distribution function, k is Boltzmann constant, T is the temperature, E$_c$ is the conduction band edge, and E$_F$ is the Fermi level. Note that the free charge density Q$_m$ is obtained using Boltzmann approximation to simplify this model. Replacing E_C with -qφ_s and E_F with -qV$_{ch}$, the free charge density can be further expressed as follows:

$$Q_m = q \cdot D \cdot kT \cdot exp\left(\frac{(\varphi_s - V_{ch})}{Vt}\right) \quad (2)$$

where φ_s is the semiconductor potential, V$_{ch}$ is the channel potential (i.e. V$_{ch,source}$ = 0 and V$_{ch,drain}$= V$_{DS}$), and Vt is the thermal voltage.

Applying Gauss' law and the boundary condition at the interface between the dielectric layer and the channel material, we obtain the following relationship:

$$C_{ox}(V_{gs} - V_{FB} - \varphi_s) = q \cdot D \cdot kT \cdot exp\left(\frac{(\varphi_s - V_{ch})}{Vt}\right) + q \cdot N_D \quad (3)$$

where C$_{ox}$ is oxide capacitance per unit area, V_{FB} is the flat band voltage, and N_D is the doping concentration per unit area. By substituting equation (2) into equation (3), we obtain a closed-form relationship between gate-source voltage and free charge density as follows:

$$C_{ox}\left(V_{gs} - V_{FB} - Vt \cdot ln\left(\frac{Q_m}{q \cdot D \cdot kT}\right) - V_{ch}\right) = Q_m + q \cdot N_D \quad (4)$$

Here, we use Halley's method [4] to solve for the free charge density Q$_m$, starting with an initial guess Q$_{m0}$ given by:

$$Q_{m0} = C_{ox} \cdot Vt \cdot ln\left(1 + exp\left(\frac{V_{gs} - V_{FB} - V_{ch}}{Vt}\right)\right) \quad (5)$$

By applying Halley's method twice, an explicit free charge density near the source end (Q$_{m,s}$) and the drain end (Q$_{m,d}$) can be obtained when V$_{ch}$ = 0 and V$_{ch}$ = V$_{DS}$. As shown in Fig. 1, the free charge density calculated by this core model agrees well with the numerical solution for equation (4).

B. Core Model for Drain-to-Source Current

In this model for drain current, it is assumed that the current is primarily governed by the drift-diffusion mechanism. The drain current of a 2DFET can be obtained using Pao–Sah's double integral [5]:

$$I_{ds} = \frac{\mu W}{L} \int_{Q_{m,s}}^{Q_{m,d}} Q_m \cdot \frac{dV_{ch}}{dQ_m} dQ_m \quad (6)$$

where W is the effective device width, L is the gate length, and the μ is the carrier mobility. By taking a derivative on both sides of equation (4) and substituting the derivative of V$_{ch}$ with respect to Q$_m$ into Pao-Sah's integral, the drain current can be expressed in terms of free charge density near the source and the drain:

$$I_{ds} = \frac{\mu W C_{ox}}{L}\left(\frac{1}{2} \cdot (q_{is}^2 - q_{id}^2) + Vt \cdot (q_{is} - q_{id})\right) \quad (7)$$

where q$_{is,(d)}$ = Q$_{m,s(d)}$/C$_{ox}$ are the normalized free charge density near the source side and the drain side.

C. Series Resistance Model for Schottky Contact

With an ultra-thin body, it is challenging to heavily dope two-dimensional materials to achieve Ohmic source/drain contacts. As a result, 2D semiconductor FETs often experience a gate voltage-dependent Schottky contact resistance in practice. [6] To address this issue, an additional term for Schottky contact resistance, as described in [7], was incorporated into the existing series resistance model in BSIM-CMG [8] as:

$$R_{schs} = R_{schs0} \cdot exp\left(\frac{-\left(V_{gseff} - V_{schref}\right)}{nsch \cdot Vt}\right) \tag{8}$$

$$R_{schd} = R_{schd0} \cdot exp\left(\frac{-\left(V_{gdeff} - V_{schref}\right)}{nsch \cdot Vt}\right) \tag{9}$$

in which R_{schs0} and R_{schd0} are two resistance parameters, V_{schref} is the barrier height at zero effective gate voltage, and n_{sch} represents the ideality factor of the Schottky contact.

Model Validation

In this section, the core model for charge density and drain-to-source current is integrated with BSIM-CMG sub-models. These sub-models include mobility degradation, velocity saturation, series resistance, subthreshold swing degradation, short-channel effects, and self-heating [8]. This model is validated against notable published data of MoS_2-based FETs. The parameter extraction in this section is performed using Keysight's ICCAP suite, with ADS as the underlying simulator.

In Fig. 2, this model is validated using data from a dual-gated MoS_2 FET. This MoS_2 FET, with an Lg = 10nm, has source and drain contacts made of yttrium, which greatly lowers the Schottky barrier height at the contacts. [9] It demonstrates a remarkable ON current level (~1mA/um) and good subthreshold behavior. The Coulomb scattering and the subthreshold swing degradation sub-models in BSIM-CMG allow this model to capture the effects of possible trap charges in the subthreshold region. Additionally, as shown in Fig. 3, this model accurately fits the data for another dual-gated MoS_2 FET with Lg = 20nm.

Fig. 4 shows a good validation with data from two back-gated MoS_2 FETs with different gate lengths (Lg = 55nm and 75nm). In these MoS_2 FETs, the 2D channel is equipped with an interfacial layer of hBN to reduce interface trap density and improve subthreshold swing behavior. [10] As seen in Fig. 4(a), this model provides a good I_D-V_G fitting with Lg = 55nm in both the subthreshold and the above threshold regions. Also, as seen in Fig. 4(b), this model accurately fits I_D-V_D with Lg = 75nm in both the linear and the saturation regions.

Figure 5 shows another model validation using data from a single nanosheet MoS_2 FET with a gate length of Lg = 60nm.[11] As illustrated in Fig. 5, our model reproduces the transfer and output characteristics. The model can fit the output characteristics in the linear region (low drain bias), which demonstrates our 2DFET core model, along with the mobility degradation model, effectively capture the 2DFET device characteristics.

Figure 6 shows a validation of the model using data from two back-gated MoS_2 FETs with a gate length of Lg = 100nm. These MoS_2 FETs, with Lg = 100nm, have contacts made of bismuth, making the source and drain contacts exhibit ohmic behavior. [12] Our 2DFET core model, with short channel effects incorporated through the BSIM-CMG framework, captures the DIBL effect, as shown in Fig. 6(a).

Due to the thin nature of the 2DFET channel and its Van der Waals properties, heat dissipation is challenging. [6] This makes 2DFETs prone to self-heating effects. Our 2DFET model, with the self-heating model from the BSIM-CMG framework, is also validated against experimental data from back-gated MoS_2 FETs experiencing self-heating effects. As seen in Figs. 7 and 8, our model, with the self-heating model and the Schottky contact resistance model, accurately matches the I_D-V_G and I_D-V_D characteristics of the measured 1L-MoS_2 device.

Conclusion

A compact model of 2DFETs, based on the BSIM-CMG framework, is proposed. It starts with developing a channel free charge density model and achieves an explicit drain current formula in terms of channel-free charge density near the source and the drain. This core model of 2DFETs is incorporated with BSIM-CMG sub-models, including mobility degradation, velocity saturation, series resistance, subthreshold swing degradation, short channel effect and self-heating. This model is extensively validated against measurements from state-of-the-art MoS_2-based FETs. With the BSIM-CMG framework, this simpler core model without trap charges can achieve an accurate fitting, significantly reducing the complexity of the core model.

Acknowledgments

This work is supported by the Berkeley Device Modeling Center.

References

[1] Cao Wei, Kang Jiahao, Liu Wei and Banerjee, Kaustav. "A Compact Current–Voltage Model for 2D Semiconductor Based Field-Effect Transistors Considering Interface Traps, Mobility Degradation, and Inefficient Doping Effect." in IEEE Transactions on Electron Devices, vol. 61. pp. 4282-4290. 2014.

[2] C. Yadav, A. Agarwal and Y. S. Chauhan, "Compact Modeling of Transition Metal Dichalcogenide based Thin body Transistors and Circuit Validation," in IEEE Transactions on Electron Devices, vol. 64, no. 3, pp. 1261-1268, March 2017

[3] S. V Suryavanshi et al., "S2DS: physics-based compact model for circuit simulation of two-dimensional semiconductor devices including non-idealities," J. Appl. Phys., vol. 110, no. 22, pp. 224503(1)-224503(10), Dec. 2016.

[4] Halley's Method. [Online]. Available: https://mathworld.wolfram.com/HalleysMethod.html

[5] H. Pao and C. Sah, "Effects of diffusion current on characteristics of metal–oxide (insulator)–semiconductor transistors," Solid State Electron., vol. 9, no. 10, pp. 927–937, Oct. 1966.

[6] Li, W., Gong, X., Yu, Z. et al. Approaching the quantum limit in two-dimensional semiconductor contacts. Nature 613, 274–279 (2023).

[7] G. Pahwa, S. Salahuddin and C. Hu, "An All-Region BSIM Thin-Film

Transistor Model for Display and BEOL 3-D Integration Applications," in IEEE Transactions on Electron Devices, vol. 71, no. 8, pp. 4701-4709, Aug. 2024.

[8] BSIM-CMG Technical Manual. [Online]. Available: https://bsim.berkeley.edu/models/bsimcmg/

[9] Jianfeng Jiang, et al. "Yttrium-induced phase-transition technology for forming perfect ohmic contact in two-dimensional MoS2 transistors", PREPRINT (Version 1) available at Research Square. 08 February 2023.

[10] H. -Y. Lan, J. Appenzeller and Z. Chen, "Dielectric Interface Engineering for High-Performance Monolayer MoS$_2$ Transistors via hBN Interfacial Layer and Ta Seeding," 2022 International Electron Devices Meeting (IEDM), San Francisco, CA, USA, 2022, pp. 7.7.1-7.7.4.

[11] Y. -Y. Chung et al., "Monolayer-MoS Stacked Nanosheet Channel with C-type Metal Contact," 2023 International Electron Devices Meeting (IEDM), San Francisco, CA, USA, 2023, pp. 1-4.

[12] Y. Lin et al., "Contact Engineering for High-Performance N-Type 2D Semiconductor Transistors," 2021 IEEE International Electron Devices Meeting (IEDM), San Francisco, CA, USA, 2021, pp. 37.2.1-37.2.4.

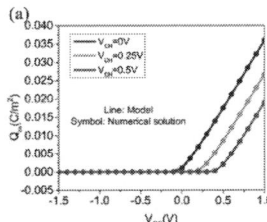 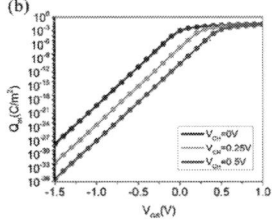

Fig. 1. Free charge density as a function of applied gate–source voltage with three different V_{ch}. (a) Linear scale. (b) Log scale. C_{ox}:0.03835F/m^2, V_{FB}=0, D=2.496*10^{37} (J*m2)$^{-1}$. N_D=10^{16} m^{-2}.

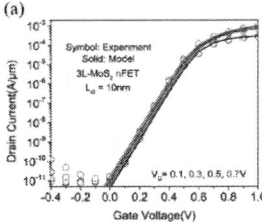 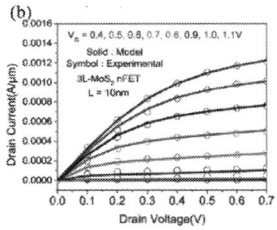

Fig. 2. Mode validation for a short channel double-gated 3-layers MoS$_2$ FET with L_G = 10nm. [9] (a) I_D-V_G characteristics with four different drain voltages. (b) I_D-V_D characteristics with eight different gate voltages.

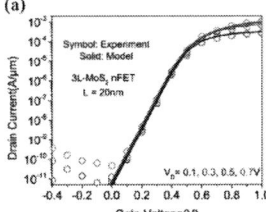 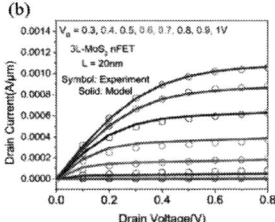

Fig. 3. Mode validation for a short channel double-gated 3-layers MoS$_2$ FET with L_G = 20nm. [9] (a) I_D-V_G characteristics with four different drain voltages. (b) I_D-V_D characteristics with eight different gate voltages.

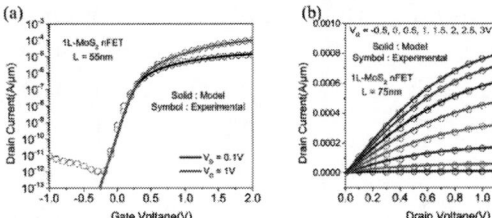

Fig. 4. Mode validation for two back-gated 1-layer MoS$_2$ FETs [10] (a) I_D-V_G characteristics of L_G=55nm. (b) I_D-V_D characteristics of L_G=75nm

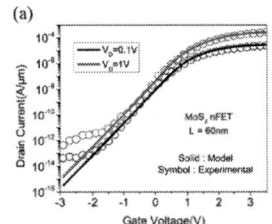 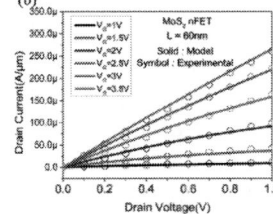

Fig. 5. Mode validation for a single nanosheet MoS$_2$ FET with L_G = 60nm. [11] (a) I_D-V_G characteristics with two different drain voltages. (b) I_D-V_D characteristics with six different gate voltages.

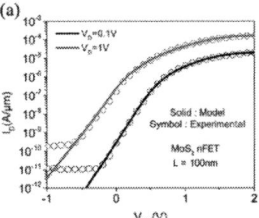 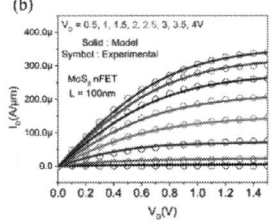

Fig. 6. Mode validation for back-gated 1-layer MoS$_2$ FETs with L_G = 100nm. [12] (a) I_D-V_G characteristics with two different drain voltages. (b) I_D-V_D characteristics with eight different gate voltages.

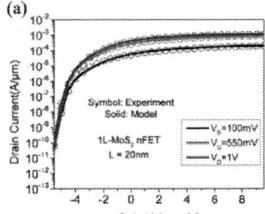 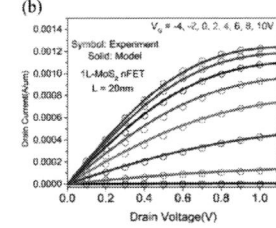

Fig. 7. Mode validation for a back-gated 1-layer MoS$_2$ FETs suffering from self- heating effect. [6] (a) I_D-V_G characteristics with three different drain voltages. (b) I_D-V_D characteristics with eight different gate voltages.

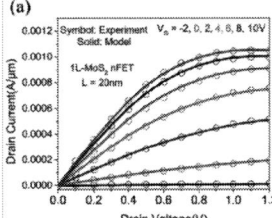 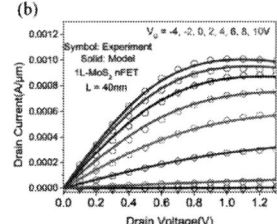

Fig. 8. Mode validation for two additional back-gated 1-layer MoS$_2$ FETs suffering from self- heating effect. [6] I_D-V_D characteristics with gate length of (a) 20nm and (b) 40nm.

Improved Ohmic Contact Resistance and DC Performance of InAlN/GaN HEMTs with Si-incorporated Contact Scheme

Yang Jiang[1,2], Fangzhou Du[1], Xinyi Tang[1], Ziyang Wang[1], Kangyao Wen[1], Zhongrui Wang[1],
Qing Wang[1*], Hongyu Yu[1*]

(*Email: wangq7@sustech.edu.cn; yuhy@sustech.edu.cn)

[1] School of Microelectronics, Southern University of Science and Technology, Shenzhen, 518055, China
[2] Department of Electrical and Electronic Engineering, The University of Hong Kong, Hong Kong

Abstract

In this work, we propose a Si-incorporated ohmic contact scheme. An ultra-low contact resistance of <0.1 Ω·mm is demonstrated, representing an 82.7% reduction compared to the conventional scheme. This excellent ohmic contact significantly enhances the DC performance of InAlN/GaN HEMTs. Specifically, the maximum drain current ($I_{DS,max}$) increases by over 30%, and the on-resistance (R_{ON}) decreases by over 20%. These results present a promising approach to promote GaN HEMTs in high-power and high-frequency switching applications.

Keywords: InAlN/GaN HEMTs, Si-incorporated Ohmic Contact, Contact Resistance, DC Performance

Introduction

InAlN/GaN high electron mobility transistors (HEMTs) have garnered significant attention due to their numerous advantages over conventional AlGaN/GaN HEMTs.[1] Firstly, the strong spontaneous polarization and larger conduction band discontinuity of the InAlN layer result in higher two-dimensional electron gas (2-DEG) density, which lead to lower on-resistance (R_{ON}) and higher maximum drain current ($I_{DS,max}$).[2] Additionally, a lattice-matched $In_{0.17}Al_{0.83}N$/GaN heterostructure effectively eliminates the piezoelectric effect, thereby enhancing device reliability.[3] Therefore, InAlN has emerged as an ideal alternative strain-free barrier layer.

Achieving optimal DC performance in InAlN HEMTs critically depends on establishing ohmic contacts with low resistance. Extensive investigations into utilizing regrowth, barrier recess, and ion implantation approaches show promise in overcoming this obstacle.[4-6] However, these techniques involve trade-offs between contact resistance (R_c) and fabrication cost, they also suffer from potential plasma or ion damage. Therefore, a robust and low-resistance ohmic contact using a metal stacking approach on the as-grown epilayer is urgently needed.[7]

In this work, a Si-incorporated ohmic contact is proposed to offer a promising damage-free and economical scheme. We fabricate conventional InAlN HEMTs and InAlN MIS-HEMTs with and without Si-incorporated ohmic contact. The contact resistance is significantly reduced with the Si-incorporated scheme, achieving an ultra-low contact resistance of 0.096 Ω·mm. Moreover, improvements in surface morphology are observed through Atomic Force Microscope (AFM) characterization. The DC performance, including transfer curves, output curves, and breakdown performance, is measured using a Keithely-4200-SCS parameter analyzer. The $I_{DS,max}$ increases by 33.9% and 35.4% in conventional InAlN HEMTs and InAlN MIS-HEMTs, respectively. Additionally, the R_{ON} decreases by 21.4% and 27.7%, respectively. The DC performance comparison of conventional InAlN HEMTs and InAlN MIS-HEMTs is also investigated.

Experiments

The lattice-matched $In_{0.17}Al_{0.83}N$/GaN epitaxy wafer was grown on Si. The epilayers comprise a 10 nm $In_{0.17}Al_{0.83}N$ barrier layer, a 0.8 nm AlN spacer layer, a 1000 nm GaN channel layer, and a 300 nm buffer layer. The Hall measurement showed a 2-DEG density of 1.34×10^{13} cm^{-2}, electron mobility of 1910 cm^2/V·s, and sheet resistance of 242 Ω/sq.

As shown in Fig. 1, the fabrication process started with mesa isolation using inductively coupled plasma (ICP) dry etching with a Cl_2/BCl_3/Ar gas mixture. Next, the native oxidation layer on the InAlN surface was removed in a diluted hydrochloric acid solution (deionized water: HCl=4:1). An ohmic contact stack of Ti/Al/Ti/Au (without Si-incorporated) and Si/Ti/Al/Ti/Au (with Si-incorporated) was deposited by e-beam evaporation (EBE), followed by rapid thermal annealing at 830 °C for 45 s under N_2 ambient. For conventional InAlN HEMTs, the N_2O surface treatment was performed employing plasma enhanced chemical vapor deposition (PECVD), which was expected to create a thin oxidation layer formation through N_2O plasma oxidation. For InAlN MIS-HEMTs, an Al_2O_3 gate dielectric layer was deposited using atomic layer deposition (ALD). A Ni/Au (20/130 nm) gate electrode metal was deposited by EBE. The pad window was opened by dry etching with a Cl_2/BCl_3 gas mixture, and the test pad metal (Ti/Au 20/130 nm) was deposited by EBE. Finally, the post device annealing (PDA) process was carried out in N_2 ambient at 350 °C for 10 min. The gate-source spacing, gate length, gate width, and gate-drain spacing were L_{GS}=4 μm, L_G=3 μm, W_G=150 μm and L_{SD}=11 μm, respectively.

Results and Discussion

Fig. 2(a) and 2(b) show the optical microscope images of Ti/Al/Ti/Au and Si/Ti/Al/Ti/Au ohmic contacts after annealing, respectively. The corresponding AFM measurement results in Fig. 2(c) and 2(d) indicate that a lower surface roughness and better surface morphology are achieved with the Si-incorporated scheme. The circular transmission line model (CTLM) is employed for the measurement of ohmic contact resistance (R_c) and sheet resistance (R_{sh}).[8] The CTLM measured I-V curves with spacing of 5 μm-30 μm for Ti/Al/Ti/Au and Si/Ti/Al/Ti/Au ohmic contacts are shown in Fig. 3(a) and 3(b). The statistical mean value of R_c reduces from 0.55 Ω·mm to 0.096 Ω·mm using the Si-incorporated scheme, which is 82.7% lower than the conventional scheme. Moreover, the R_{sh} reduces from 345.34 Ω/sq to 292.35 Ω/sq. These results indicate that the Si-incorporated ohmic contact improves the tunneling efficiency through deliberate doping of the barrier layer into n-type, facilitating better ohmic contact results.

The electrical characterization of conventional InAlN HEMTs without (w/o) and with (w) Si-incorporated ohmic contact is shown in Fig. 4. Similar transfer characteristics under a drain-source voltage (V_{DS}) of 10V are observed. Fig. 4(c) and 4(d) show the output characteristics (V_{GS} sweeps from -3V to 2V with 1V step) for the devices without and with Si-incorporated ohmic contact, respectively. A higher $I_{DS,max}$ and lower R_{ON} are observed with the Si-incorporated scheme. The extracted mean value of $I_{DS,max}$ is 204.56 mA/mm (w/o) and 274.03 mA/mm (w). The R_{ON} is found to decrease from 31.64 Ω·mm (w/o) to 24.85 Ω·mm (w). The improved DC performance is also observed in InAlN MIS-HEMTs as shown in Fig. 5. The extracted mean value of $I_{DS,max}$ is 505.53 mA/mm (w/o) and 373.41 mA/mm (w). The R_{ON} is found to decrease from 22.44 Ω·mm (w/o) to 16.22 Ω·mm (w). Using Si-incorporated ohmic contact, the $I_{DS,max}$ increases by over 30%, and the R_{ON} decreases by over 20% in both conventional InAlN HEMTs and InAlN MIS-HEMTs, which proves the proposed approach boosts the DC performance. The performance comparison of conventional InAlN HEMTs and InAlN MIS-HEMTs in Fig. 6 demonstrates that the Al_2O_3 layer serves as an effective passivation layer, reducing the surface defect states and further improving the DC performance and breakdown performance of InAlN HEMTs.[9-10]

Conclusion

In this work, we demonstrate that using Si-incorporated ohmic contact could effectively improve the surface morphology and lower the contact resistance. The improved DC performance, specifically $I_{DS,max}$ and R_{ON}, is proved in conventional InAlN HEMTs and InAlN MIS-HEMTs structures. By implementing an effective passivation layer, device performance can be further improved. The proposed ohmic contact scheme shows great potential for application in future device fabrication optimization.

Acknowledgments

This work was supported by National Natural Science Foundation of China (Grant No: 62274082). Research on mechanism of source/drain ohmic contact and the related GaN p-FET (Grant No: 2023A1515030034). Research on high-reliable GaN power device and the related industrial power system (Grant No: KZQB-KCZYZ-2021052). Study on the reliability of GaN power devices (Grant No: JCYJ20220818100605012). Research on novelty low-resistance Source/Drain ohmic contact for GaN p-FET (Grant No: JCYJ20220530115411025). Research on the key technology of 1200 V SiC MOSFETs (Grant No: JSGG20220831094404008). "5G Frontier" project Micro-Nano processing platform (Grant No: K2023390010). High level of special funds (G03034K004). The authors also acknowledge the assistance of SUSTech Core Research Facilities.

References

[1] A. Hambitzer, "InAlN/GaN-Based High Electron Mobility Transistors," ETH Zurich, 2019.

[2] J. Kuzmik, "Power electronics on InAlN/(In) GaN: Prospect for a record performance," IEEE Electron Device Letters, vol. 22, no. 11, pp. 510-512, 2001.

[3] N. Islam, M. F. P. Mohamed, M. F. A. J. Khan, S. Falina, H. Kawarada, and M. Syamsul, "Reliability, applications and challenges of GaN HEMT technology for modern power devices: A review," Crystals, vol. 12, no. 11, p. 1581, 2022.

[4] Y. Yue, Z. Hu, J. Guo, B. Sensale-Rodriguez, G. Li, R. Wang, F. Faria, T. Fang, B. Song, and X. Gao, "InAlN/AlN/GaN HEMTs with regrown ohmic contacts and f_T of 370 GHz," IEEE Electron Device Letters, vol. 33, no. 7, pp. 988-990, 2012.

[5] J. Bergsten, A. Malmros, M. Tordjman, P. Gamarra, C. Lacam, M. di Forte-Poisson, and N. Rorsman, "Low resistive Au-free, Ta-based, recessed ohmic contacts to InAlN/AlN/GaN heterostructures," *Semiconductor Science and Technology,* vol. 30, no. 10, pp. 105034, 2015.

[6] J. Bergsten, A. Malmros, M. Tordjman, P. Gamarra, C. Lacam, M. di Forte-Poisson, and N. Rorsman, "Low resistive Au-free, Ta-based, recessed ohmic contacts to InAlN/AlN/GaN heterostructures," Semiconductor Science and Technology, vol. 30, no. 10, pp. 105034, 2015.

[7] Y. Jiang, F. Du, J. He, Z. Qiao, C. Tang, X. Tang, Z. Wang, Q. Wang, and H. Yu, "Microscopic formation mechanism of Si/Ti5Al1/TiN ohmic contact on non-recessed i-InAlN/GaN heterostructures with ultra-low resistance," Applied Physics Letters, vol. 121, no. 21, 2022.

[8] Y. Jiang, Z. Qiao, F. Du, G. Yang, M. Fan, X. Tang, Q. Wang, and H. Yu, "Achieving A Low Contact Resistivity of 0.11 Ω·mm for Ti_5Al_1/TiN S/D Contact on $Al_{0.2}Ga_{0.8}$N/AlN/GaN Structure without Barrier Recess." in 2021 5th IEEE Electron Devices Technology & Manufacturing Conference (EDTM), 2021: IEEE, pp. 1-3.

[9] A. Malmros, P. Gamarra, M.-A. di Forte-Poisson, H. Hjelmgren, C. Lacam, M. Thorsell, M. Tordjman, R. Aubry, and N. Rorsman, "Evaluation of thermal versus plasma-assisted ALD Al_2O_3 as passivation for InAlN/AlN/GaN HEMTs," IEEE Electron Device Letters, vol. 36, no. 3, pp. 235-237, 2015.

[10] D. Xu, K. Chu, J. Diaz, M. Ashman, J. Komiak, L. M. Pleasant, C. Creamer, K. Nichols, K. Duh, and P. Smith, "0.1-μm Atomic Layer Deposition Al_2O_3 Passivated InAlN/GaN High Electron-Mobility Transistors for E-Band Power Amplifiers," *IEEE Electron Device Letters,* vol. 36, no. 5, pp. 442-444, 2015.

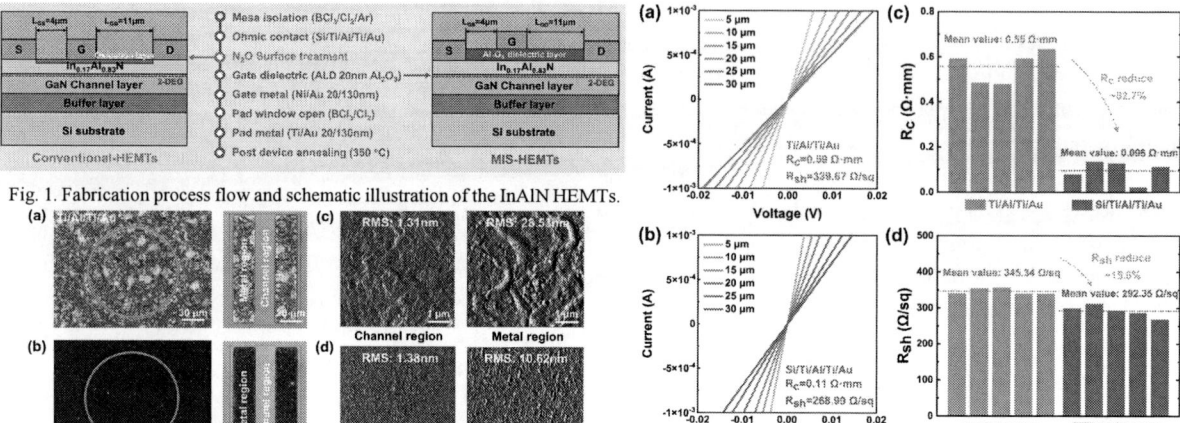

Fig. 1. Fabrication process flow and schematic illustration of the InAlN HEMTs.

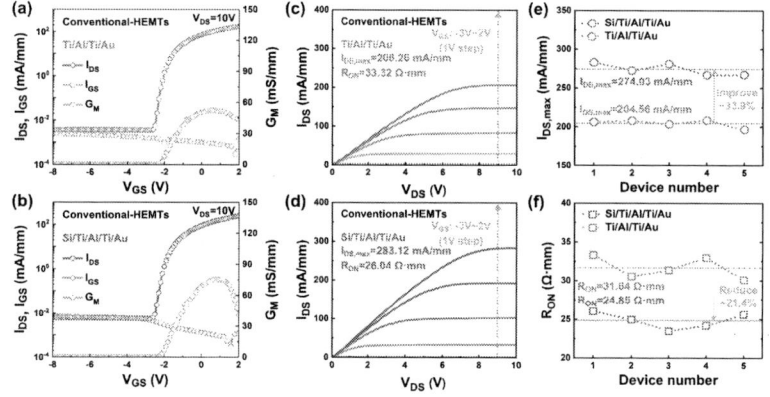

Fig. 2. (a) and (b) optical microscope images and (c) and (d) corresponding AFM measurement results of Ti/Al/Ti/Au and Si/Ti/Al/Ti/Au ohmic contact after annealing, respectively.

Fig. 3. The CTLM measured I-V curves with spacing of 5 μm-30 μm for (a) Ti/Al/Ti/Au and (b) Si/Ti/Al/Ti/Au ohmic contact. The statistical comparison of (c) R_c and (d) R_{sh}.

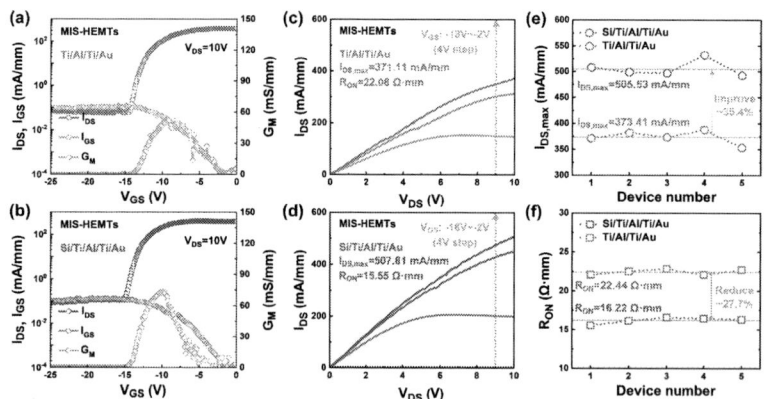

Fig. 4. The electrical characterization of conventional InAlN HEMTs with Ti/Al/Ti/Au and Si/Ti/Al/Ti/Au ohmic contact. (a) and (b) Transfer characteristics, (c) and (d) output characteristics, (e) and (f) The statistical $I_{DS,max}$ and R_{ON} results of device-to-device variation.

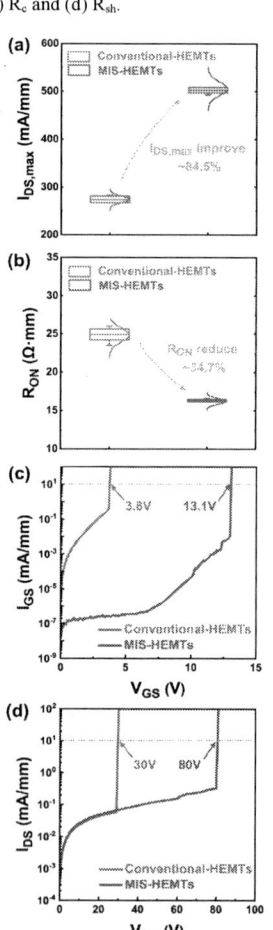

Fig. 5. The electrical characterization of InAlN MIS-HEMTs with Ti/Al/Ti/Au and Si/Ti/Al/Ti/Au ohmic contact. (a) and (b) Transfer characteristics, (c) and (d) output characteristics, (e) and (f) The statistical $I_{DS,max}$ and R_{ON} results of device-to-device variation.

Fig. 6. The performance comparison of conventional InAlN HEMTs and InAlN MIS-HEMTs. (a) $I_{DS,max}$, (b) R_{ON}, (c) gate breakdown, (d) off-state breakdown.

BSIM-NN: A Machine Learning Compact Model for Fast IC Simulation

C. T. Tung[1], S. Salahuddin[1] and C. Hu[1]

[1]University of California, Berkeley, CA. USA, email: cttung@berkeley.edu

Abstract

We present BSIM-NN, a comprehensive neural network (NN)-based compact model of FET IV and CV characteristics, including geometry and temperature dependence, self-heating (SH), non-quasi-static (NQS) effect, variability, parasitic capacitances, and noises. The model uses NNs to replace the analytical equations of the industry-standard BSIM-CMG FinFET/GAA model with proven accuracy. We demonstrate the model's accuracy in DC, AC, transient, RF, and noise simulations and speedup of digital and analog IC simulations.

Keywords: Compact model, machine learning, neural network

Introduction

The number of transistors in ICs keeps increasing and the physics of transistors becomes more complex. Faster compact models are needed, and neural network (NN)/machine learning (ML)-based compact models can model complex device physics (implicitly) by matrix multiplication with the potential to reduce circuit simulation time. Many studies have been conducted to build NN-based compact models [1-4]. However, they only model DC characteristics, and none addresses NQS, SH, and noises which are essential for logic, analog, and RF designs. We present a fast and comprehensive NN-based transistor compact model, BSIM-NN, that can perform DC, AC, transient, RF, and noise simulations. NN-compatible models for NQS, SH, thermal noise, and flicker noise are developed using the hybrid physics-ML method.

Core Model

A. IV Model

Our core quasi-static (QS) IV model uses transformed terminal currents (I_D, I_G) as NN outputs (y_1, y_2, y_3) with the transfer functions (1) and (2). We design the transfer functions to optimize the model accuracy. a in (1) is a parameter to tune the training accuracy. The inputs are V_{GS}, V_{DS}, L (gate length), W (total width), EOT, and T_0 (ambient temperature). Work function or threshold voltage difference is modeled by a voltage shift in V_G. An inverse square root unit (ISRU) function is chosen as the activation function tested to be more computationally efficient than the more popular sigmoid and tanh functions [5]. The loss function of the IV model includes errors in $y_{1,2,3}$, gm, gds, gm', and gds' [1].

$$I_D = \tanh(\alpha V_{DS})e^{y_1}, \quad y_1 = \ln(I_D / \tanh(\alpha V_{DS})) \quad (1)$$

$$y_{2,3} = \ln(\pm I_G / 2 + \sqrt{I_G^2 + \Delta^2} / 2 + I_0) \quad (2)$$

B. QV Model

Our quasi-static QV model uses terminal charges (Q_G, Q_D, Q_S) for the NN outputs to ensure charge conservation. We make the charge model trainable solely with capacitance data (C_{gg}, C_{gd}, C_{dg}, C_{dd}, C_{sg}, C_{sd}). Offset charges ($Q_{G,S,D0}$) are added to the loss function [6] to solve the charge-shifting issue reported in [3] when trained with only capacitance data. $Q_{G,D,S0}$ are the estimated $Q_{G,D,S}$ at $V_{GS}=V_{DS}=0$.

Non-Quasi-Static Effect, Self-Heating & Noise

A. Non-Quasi-Static

To model the NQS effect, we use a relaxation time-based charge deficit model [6]. The charge deficit model is simple and efficient for IC simulations. The charge deficit model uses the trained terminal charges (Q). Physics-based fringing (Q_f) and overlapping charges (Q_{ov}) are added to the extract Q_i so that $Q_i=Q-Q_f-Q_{ov}$. The deficit charge Q_{def} is calculated by (3) where $Q_{i,G}$ is the intrinsic gate charge and t is the transit time. The deficit current is $-Q_{def}/t$ at the gate and $-x_DQ_{def}/t$ at the drain. x_D is the partition factor that equals $Q_{i,D}/Q_{i,G}$.

$$dQ_{def} / dt = -dQ_{i,G} / dt - Q_{def} / \tau, \quad (3)$$

B. Self-Heating

SH effect causes the device temperature (T) to deviate from the ambient temperature (T_0) and needs to be taken seriously in modeling. People used to perform pulsed IV measurements to remove SH, separately extract SH parameters, or self-consistently extract SH parameters with other model parameters. We invent a simpler temperature relaxation (TR) model that directly uses the DC data $I(T_0)/Q(T_0)$ including SH to train the ML model [7]. The TR model maps the easy-to-measure DC data $I(T_0)/Q(T_0)$ from T_0 to T as in (4) to recover the SH-free IV(T) and QV(T) which can be found at T_0', while the dynamic SH simulation is performed by (5). R_{TH} is the thermal resistance of the trained/measured device. R'_{TH} is the thermal resistance of the device in IC simulation. C_{TH} is the thermal capacitance. $DR_{TH}= R_{TH}-R'_{TH}$. The SH-free characteristics extracted with the TR model can also be used with the conventional SH model for circuit simulations ($DT(t)+(R_{TH}C_{TH})dDT(t)/dt=R_{TH}I_DV_{DS}$).

$$T_0'(t) = T_0 - \delta T(t) \quad (4)$$

$$d(R_{TH}I_DV_{DS} - \delta T) / dt = (\delta T - \Delta R_{TH}I_DV_{DS}) / R'_{TH} C_{TH} \quad (5)$$

C. Noise

Noise is important in analog and RF circuit designs. Instead of training models for noise or using complex physics-based equations to calculate noises, we use simple empirical equations to keep the model efficient. The channel thermal

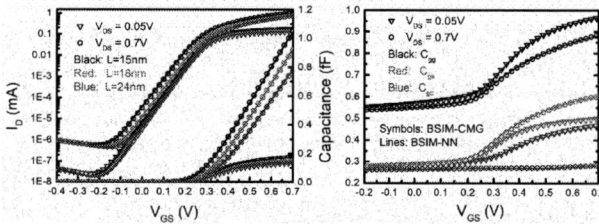

Fig. 1. IV verification at different gate length and CV verification for C_{gg}, C_{gs}, and C_{gd}.

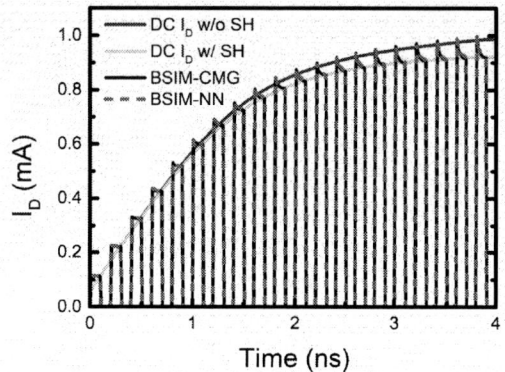

Fig. 2. Pulse simulation of BSIM-CMG and BSIM-NN considering the self-heating effect.

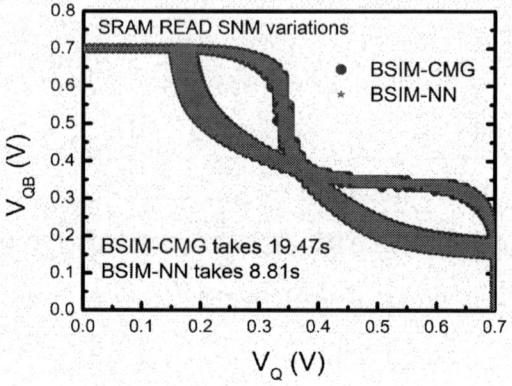

Fig. 3. SRAM READ SNM variation simulation of BSIM-CMG and BSIM-NN. BSIM-NN is more than twice as fast as BSIM-CMG.

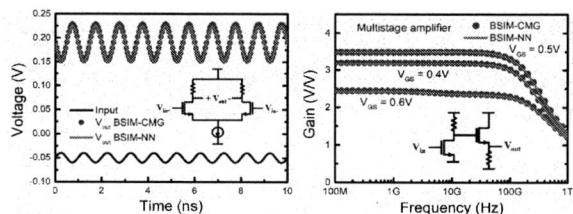

Fig. 4. A differential amplifier simulation and a multistage amplifier frequency response simulation of BSIM-CMG and BSIM-NN.

noise (S_{th}) is proportional to gds at the linear region, and gm at the saturation region [8]. We use an empirical equation (6) to sum the model's gm and gds to calculate S_{TH} where a_{1-7} are fitting parameters. Flicker noise (S_{fn}) or 1/f noise is empirically modeled by (7) where b_{1-7} are fitting parameters. The S_{fn} model in (7) can adjust the subthreshold and inversion region noise separately. The gate shot noise is modeled by $\overline{i_g^2} = 2qI_G$.

$$S_{th} = 4kT[a_1 gm + \frac{a_2 gm}{1 + e^{(V_{DS} - V_{DSAT})/a_3}} + a_4 gds(1 + a_5 e^{\frac{-|V_{DS}|}{a_6 V_{GS}^{a_7}}})] \quad (6)$$

$$S_{fn} = [(b_1 |I_D|^{b_2})^{-1} + (b_3 |I_D|^{b_4 (1 + b_5 e^{-|V_{DS}|/b_6})})^{-1}]^{-1} / f^{b_7} \quad (7)$$

Circuit Simulation

BSIM-NN applies to FinFET/GAA and other emerging FETs, e.g., NCFET [1, 2]. Here, we demonstrate the model trained and tested with a calibrated BSIM-CMG model of an advanced FinFET device [1]. BSIM-NN is coded into Verilog-A and simulated with Hspice. Fig. 1 shows the validation of the DC IV characteristics at L=15nm, 18nm, and 24nm with excellent accuracy and the accurate CV fitting for C_{gg}, C_{gs}, and C_{gd}. Fig. 2. shows the pulsed $I_D V_D$ simulation where the TR model captures the current degradation due to self-heating during the pulse.

To further demonstrate the model's capability in IC simulation, we show statistical SRAM simulations of READ signal noise margin (SNM) in Fig. 3. Variations of L, H_{FIN},

T_{FIN}, and EOT are simulated, and the model reproduces the same SNM as BSIM-CMG with faster simulation speed. In Fig. 4, we test the BSIM-NN with analog circuits: a differential amplifier and a multistage amplifier. The DC and AC gain of them match the BSIM-CMG results. BSIM-NN accurately simulates the AC responses of these circuits at various biases in the presence of NQS and SH effects. For digital circuit, a killer NOR gate simulation (Fig. 5) is performed to examine the NQS model and BSIM-NN correctly shows the transient response with no unphysical behavior.

More AC and RF simulations are examined. BSIM-NN can give the correct RF Pin-Pout and harmonic balance as shown in Fig. 6. The temperature dependence of the power gain and the slope for each harmonic component are predicted very well. Fig. 7 compares the H21 and f_T generated by the BSIM-CMG and the BSIM-NN. The results show that the NN model is accurate for high-frequency and RF simulations. To test the noise model, noise simulations of BSIM-NN and BSIM-CMG are shown in Fig. 8. The model accurately captures the frequency dependence of noise from flicker to thermal noise of a common-source amplifier. The noise spectrum of a 3-stage ring oscillator (RO) is also simulated and BSIM-NN produces the same results as BSIM-CMG.

In addition to the speed comparison in Fig. 3, we compare the simulation time of a 1001-stage NAND RO simulation of BSIM-NN and BSIM-CMG without the bypass in HSPICE. The simulation is 0.59hr for BSIM-NN and 2.85hr for BSIM-CMG. Thus, there is a 481% speed boost with a more complicated circuit. Depending on the simulator setting and the model optimization, the speed boost could be even more [5].

Conclusion

We demonstrate the first comprehensive ML transistor compact model, BSIM-NN, that includes IV, QV, NQS, SH, noises, and variability. The model can be used for digital, analog, and RF circuit designs. It produces accurate simulations with faster simulation speed compared to the standard compact model for FinFET and GAA, BSIM-CMG.

Acknowledgments

This work was supported by the Berkeley Device Modeling Center, University of California, Berkeley, CA, USA.

References

[1] C. -T. Tung and C. Hu, "Neural Network-Based BSIM Transistor Model Framework: Currents, Charges, Variability, and Circuit Simulation," in *IEEE Transactions on Electron Devices*, vol. 70, no. 4, pp. 2157-2160, April 2023

[2] C. -T. Tung, M. -Y. Kao and C. Hu, "Neural Network-Based Modeling With High Accuracy and Potential Model Speed," in *IEEE Transactions on Electron Devices*, vol. 69, no. 11, pp. 6476-6479, Nov. 2022.

[3] J. Wang, Y. -H. Kim, J. Ryu, C. Jeong, W. Choi and D. Kim, "Artificial Neural Network-Based Compact Modeling Methodology for Advanced Transistors," in *IEEE Transactions on Electron Devices*, vol. 68, no. 3, pp. 1318-1325, March 2021.

[4] Z. Yang, A. D. Gaidhane, K. Anderson, G. Workman and Y. Cao, "Graph-Based Compact Model (GCM) for Efficient Transistor Parameter Extraction: A Machine Learning Approach on 12 nm FinFETs," in *IEEE Transactions on Electron Devices*, vol. 71, no. 1, pp. 254-262, Jan. 2024.

[5] C. -T. Tung, S. Salahuddin and C. Hu, " A SPICE-compatible Neural Network Compact Model for Efficient IC Simulations," *2024 International Conference on Simulation of Semiconductor Processes and Devices (SISPAD)*, San Jose, CA, USA, 2024.

[6] C. -T. Tung, S. Salahuddin and C. Hu, "Non-Quasi-Static Modeling of Neural Network-Based Transistor Compact Model for Fast Transient, AC, and RF Simulations," in *IEEE Electron Device Letters*, vol. 45, no. 7, pp. 1277-1280, July 2024.

[7] C. -T. Tung, A. Pampori, C. Kumar Dabhi, S. Salahuddin and C. Hu, "A Novel Neural Network-Based Transistor Compact Model Including Self-Heating," in *IEEE Electron Device Letters*, vol. 45, no. 8, pp. 1512-1515, Aug. 2024.

[8] Bing Wang, J. R. Hellums and C. G. Sodini, "MOSFET thermal noise modeling for analog integrated circuits," in *IEEE Journal of Solid-State Circuits*, vol. 29, no. 7, pp. 833-835, July 1994.

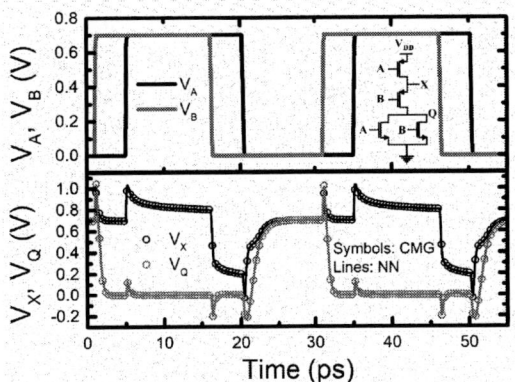

Fig. 5. A killer NOR gate simulation of BSIM-CMG and the NN model.

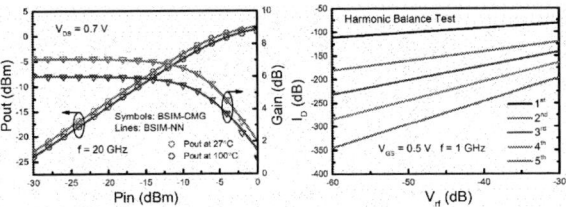

Fig. 6. An Rf Pout Pin simulation at 27°C and 100°C of BSIM-CMG and BSIM-NN and the harmonic balance test of BSIM-NN.

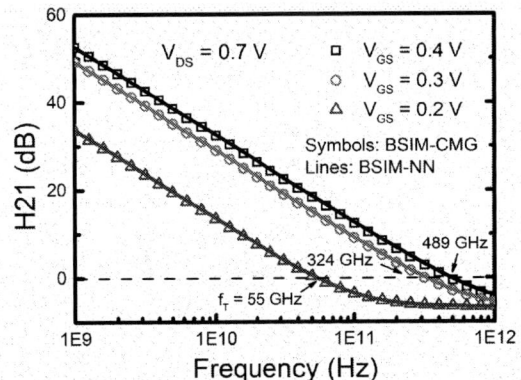

Fig. 7. H21 simulation of BSIM-CMG and BSIM-NN.

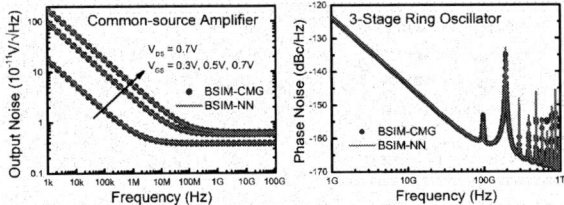

Fig. 8. Output noise simulation validation of a common-source amplifier and a 3-stage ring oscillator.

975V/4.3mΩ·cm² Enhancement-mode (001) β-Ga₂O₃ Vertical Multi-fin Power Transistors

Gaofu Guo[1,2], Xiaodong Zhang[1], Chunhong Zeng[1], Tiwei Chen[1], Zhili Zou[1], Zheyuan Hu[1],
Zhucheng Li[1], Dengrui Zhao[1], Yuhua Sun[1], Xianqi Dai[2], and Baoshun Zhang[1]

[1] Nanofabrication facility, Suzhou Institute of Nano-Tech and Nano-Bionics, Chinese Academy of Sciences (CAS), Suzhou, Jiangsu 215123, China

[2] School of Physics, Henan Normal University, Xinxiang, Henan 453007, China

Abstract

This work reports a high-performance enhancement-mode β-Ga₂O₃ multi-fin field-effect transistor (FinFET). Utilizing a non-metal mask etching and self-aligned process based on photoresist planarization. The β-Ga₂O₃ FinFET exhibited a maximum output current density of 361.5 A/cm², a peak transconductance of 127 S/cm², a specific on-resistance of 4.3 mΩ·cm², a breakdown voltage of 975 V, and a power figure of merit (PFOM) of 0.22 GW/cm². These results demonstrate the feasibility of β-Ga₂O₃ power transistors for large-area applications.

Keywords: enhancement-mode (E-mode), β-Ga₂O₃ vertical power transistors

Introduction

Gallium oxide (Ga₂O₃) has emerged in recent years as a promising high-power oxide semiconductor owing to its large-size, high-quality substrates that can be obtained via melt growth[1]. The monoclinic β-Ga₂O₃, being the most stable phase, offers an impressive wide bandgap of up to 4.9 eV, an expected high breakdown electric field of up to 8 MV/cm, and an intrinsic bulk electron mobility limit of 250 cm²/V·s [2], [3], [4]. Vertical devices have become dominant in high-current, high-voltage power device applications due to their ability to achieve higher power density in a smaller footprint[5]. Moreover, the vertical conductive channel allows the breakdown voltage to scale with the drift layer thickness, enabling higher breakdown voltages without increasing chip area[6], [7], [8]. However, the lack of p-type doping has significantly hindered the development of enhancement-mode power transistors[9], [10], [11].

To date, normally-off vertical Ga₂O₃ field-effect transistors (FET) can be categorized into current-blocking layer-based transistors and geometrically confined FinFETs[6], [12], [13]. The former includes Current Aperture Vertical Electron Transistors (CAVETs), Vertical Diffused Barrier FETs (VDBFET), and U-shaped trenchgate metal-oxide-semiconductor field-effect transistor (UMOSFET)[5], [5], [14]. Last year, we demonstrated a field-plated UMOSFET with an ion-implanted CBL layer, achieving a current density of up to 702.3 A/cm². However, the device exhibited constrained gate control due to the limited activation of dopants from high-energy ion implantation. In this work, we demonstrate an enhancement-mode (E-mode) FinFET with a breakdown voltage of up to 975 V and a specific on-resistance of 4.3 mΩ·cm². The submicron channels and double-gate control enable normally-off operation and excellent gate control without p-type doping.

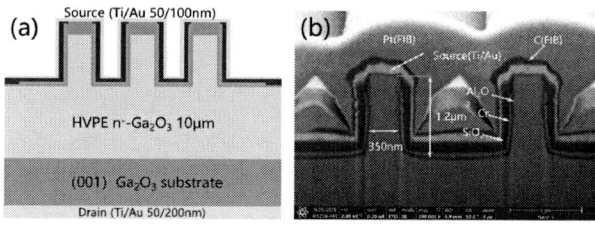

Fig. 1: (a) Schematic cross-section of Ga₂O₃ vertical fin transistors with multiple fins. (b) Scanning electron microscopy (SEM) cross-section image of a fin channel.

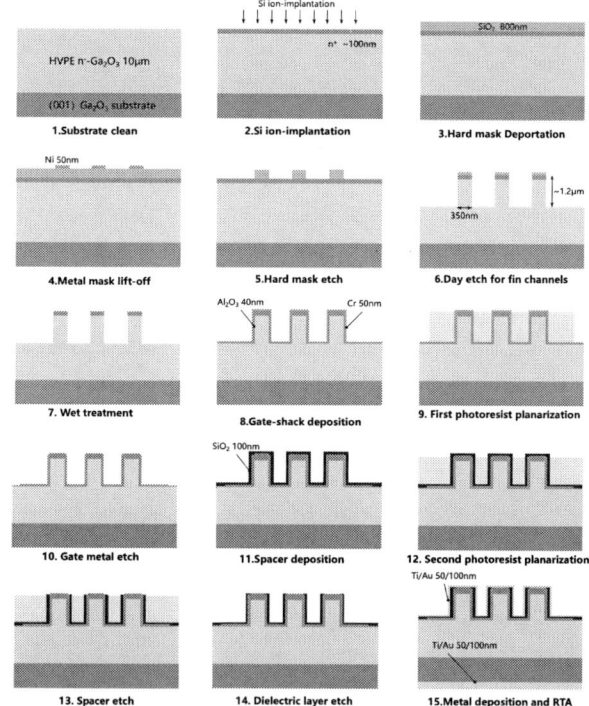

Fig. 2: Baseline process flow of Ga₂O₃ vertical FinFET. The main steps include ion implantation, fin etching, gate stacking, and two photoresist flattening and opening processes.

Device Fabrication

A β-Ga₂O₃ FinFET was fabricated on a (001) Sn-doped β-Ga₂O₃ substrate with a 10 μm n⁻ Ga₂O₃ drift layer grown via halide vapor phase epitaxy (HVPE). The net doping concentration in the HVPE drift layer was approximately 2×10^{16} cm^{-3}. The complete process flow for FinFET fabrication is shown in Fig. 2. Si ion implantation was performed on the wafer surface to form an n$^+$ layer, followed by annealing at 900°C for 15 minutes to activate the dopants for subsequent ohmic contact formation. Since Cl-based ICP etching in the Nanofabrication facility does not allow metal masks, electron-beam lithography (EBL) was used to define the fins and 50 nm of Ni was sputtered as a metal mask for hard mask etching. After completing the hard mask etching, the metal mask was removed. Fins were then formed using BCl$_3$/Ar dry etching, followed by a 20-minute BOE (7:1) treatment to remove plasma-induced damage. A 40 nm Al$_2$O$_3$ gate Dielectric was deposited via plasma-enhanced atomic layer deposition (PEALD), followed by the sputtering of 50 nm Cr as the gate metal. A self-aligned resist planarization process was used to remove the top gate metal and dielectric layers selectively. Subsequently, 100 nm SiO$_2$ was deposited by plasma-enhanced chemical vapor deposition (PECVD) as a spacer layer, and the n$^+$ Ga$_2$O$_3$ layer was exposed through another planarization process. Finally, Ti/Au source/drain contacts were deposited on the top and bottom regions. Fig. 1(a) shows a schematic cross-section of the device, and Fig. 1(b) shows a 52° cross-section image of the complete device taken in a focused ion beam (FIB) scanning electron microscope (SEM) system. The channel width was measured to be 350 nm, and the vertical gate length was 1.2 μm. Fig. 3 shows SEM images of the key steps in the device fabrication process, starting with fin etching. The photos include two critical photoresist planarization steps, followed by the top etching and contact opening processes after planarization.

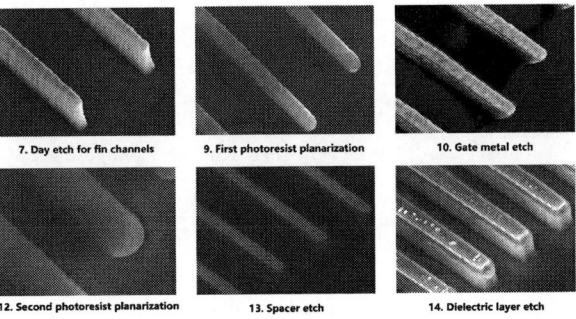

Fig. 3: SEM images of key manufacturing processes. Critical process monitoring for the Fin etching process and two photoresist flattening and hole opening processes

Results and Discussion

The electrical characteristics of the FinFET were measured at room temperature using a Keysight 1505A power device analyzer. Fig. 4(a) presents the I$_D$-V$_{GS}$ characteristics of the device, with a threshold voltage (V$_{th}$) of 0.87 V (V$_{DS}$=10 V). The peak transconductance is 124 S/cm². The output

characteristic of the FinFET is shown in Fig. 4(b), with V$_{DS}$ swept from 0 V to 10 V and V$_{GS}$ ranging from 0 V to 3 V in 0.125 V increments. The specific on-resistance (R$_{on, sp}$) is 4.3 mΩ·cm² at V$_{GS}$=3 V, and the maximum current density is 760 A/cm² (V$_{GS}$=3 V, V$_{DS}$=10 V). Due to drain-induced barrier lowering (DIBL), the output current does not exhibit saturation when the drain voltage increases. Figure 3(c) shows the transfer curve and the semi-log plot of gate leakage. The device displays an on/off ratio of 7×10^6 and a subthreshold swing of 100 mV/dec. However, the gate metal deposition and subsequent source/drain metal annealing resulted in noticeable leakage currents. To suppress DIBL effects, reverse breakdown characteristics were measured at V$_{GS}$ = -1 V, as shown in Figure 4(c). The results indicate that gate and drain leakage currents remain low until device breakdown occurs.

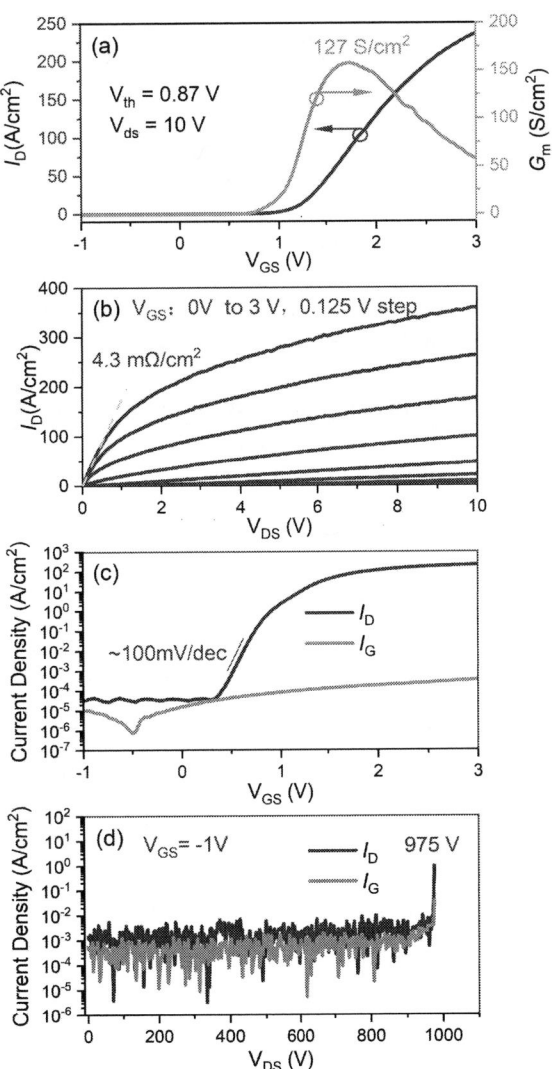

Fig. 5: Electrical Performance of the FinFET. (b)Transfer I$_D$-V$_{GS}$ characteristics (V$_{DS}$=10 V); (b) Output I$_D$-V$_{GS}$ characteristics; (c) Semi-logarithmic transfer characteristics; (d) Reverse breakdown characteristics (V$_{GS}$ = -1 V).

Fig. 5 presents a benchmark comparison of $R_{(on, sp)}$ and V_{br} for Ga$_2$O$_3$ vertical power transistors. The FinFET developed in this work achieves a power figure of merit (PFOM) of 0.22 GW/cm^2, outperforming other devices. It also exhibits an $R_{(on, sp)}$ of 4.3 mΩ·cm^2, the lowest reported among all Ga$_2$O$_3$ transistors.

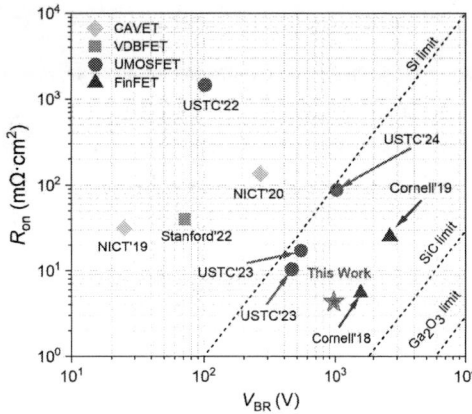

Fig. 5: Benchmark plot of Vbr and $R_{(on, sp)}$ for state-of-the-art vertical β-Ga$_2$O$_3$ MOSFETs.

Conclusion

This study demonstrates a high-performance normally-off Ga$_2$O$_3$ vertical power transistor fabricated through high-selectivity etching and two photoresist planarization steps. The FinFET exhibits a threshold voltage (Vth) of 0.86 V, a specific on-resistance (Ron) of 4.3 Ω, and a breakdown voltage (Vbr) of 975 V. However, a significant leakage current was observed, potentially caused by gate dielectric degradation during the 350°C SiO$_2$ deposition and high-temperature annealing of the source.

Acknowledgments

This work was supported in part by the National Key Research and Development Program of China under Grants 2021YFB3600200 and 2021YFC2203400; in part by the Key Laboratory Construction Project of Nanchang under Grant 2020-NCZDSY-008; in part by the Jiangxi Province Double Thousand Plan under Grant S2019CQKJ2638; in part by the National Natural Science Foundation of China under Grant dnos. 62074053, U23A20358, and 62234007. The authors thank the Nano Fabrication Facility and Vacuum Interconnected Nanotech Workstation (NANO-X) of Suzhou Institute of Nano-Tech and Nano-Bionics, Chinese Academy of Sciences, for their technical support.

References

[1] A. J. Green *et al.*, "β-Gallium oxide power electronics," *APL Materials*, vol. 10, no. 2, p. 029201, Feb. 2022, doi: 10.1063/5.0060327.

[2] Shivani, D. Kaur, A. Ghosh, and M. Kumar, "A strategic review on gallium oxide based power electronics: Recent progress and prospects," *Materials Today Communications*, vol. 33, p. 104244, Dec. 2022, doi: 10.1016/j.mtcomm.2022.104244.

[3] A. Waseem, Z. Ren, H.-C. Huang, K. Nguyen, X. Wu, and X. Li, "A Review of Recent Progress in β-Ga$_2$O$_3$ Epitaxial Growth: Effect of Substrate Orientation and Precursors in Metal–Organic Chemical Vapor Deposition," *Phys Status Solidi A*, p. 2200616, Dec. 2022, doi: 10.1002/pssa.202200616.

[4] S. J. Pearton *et al.*, "A review of Ga$_2$O$_3$ materials, processing, and devices," *Appl Phys Rev*, vol. 5, no. 1, p. 011301, Mar. 2018, doi: 10.1063/1.5006941.

[5] Y. Ma *et al.*, "702.3 A·cm^{-2}/10.4 mΩ·cm^2 β-Ga$_2$O$_3$ U-Shape Trench Gate MOSFET With N-Ion Implantation," *IEEE Electron Device Lett.*, vol. 44, no. 3, pp. 384–387, Mar. 2023, doi: 10.1109/LED.2023.3235777.

[6] Z. Hu *et al.*, "1.6 kV Vertical Ga$_2$O$_3$ FinFETs With Source-Connected Field Plates and Normally-off Operation," in *2019 31st International Symposium on Power Semiconductor Devices and ICs (ISPSD)*, Shanghai, China: IEEE, May 2019, pp. 483–486. doi: 10.1109/ISPSD.2019.8757633.

[8] Y. Qin *et al.*, "2 kV, 0.7 mΩ•cm^2 Vertical Ga$_2$O$_3$ Superjunction Schottky Rectifier with Dynamic Robustness," in *2023 International Electron Devices Meeting (IEDM)*, San Francisco, CA, USA: IEEE, Dec. 2023, pp.1–4. doi:10.1109/IEDM45741.2023.10413795.

[12] W. Li *et al.*, "2.44 kV Ga$_2$O$_3$ vertical trench Schottky barrier diodes with very low reverse leakage current," in *2018 IEEE International Electron Devices Meeting (IEDM)*, San Francisco, CA: IEEE, Dec. 2018, p. 8.5.1-8.5.4. doi: 10.1109/IEDM.2018.8614693.

[13] Z. Hu *et al.*, "Enhancement-Mode Ga$_2$O$_3$ Vertical Transistors With Breakdown Voltage >1 kV," *IEEE Electron Device Lett*, vol. 39, no. 6, pp. 869–872, Jun. 2018, doi: 10.1109/LED.2018.2830184.

[14] X. Zhou *et al.*, "Enhancement-mode β-Ga$_2$O$_3$ U-shaped gate trench vertical MOSFET realized by oxygen annealing," *Appl Phys Lett*, vol. 121, no. 22, p. 223501, Nov. 2022, doi: 10.1063/5.0130292.

979-8-3315-0417-5/25 $31.00 © 2025 IEEE

An Atomic Layer Etching Technique for MOCVD *in-situ* SiN$_x$

Fangzhou Du[1], Yang Jiang[1], Ziyang Wang[1], Xinyi Tang[1], Qing Wang[1*], Hongyu Yu[1*]

[1]School of Microelectronics, Southern University of Science and Technology, Shenzhen 518055, China
(wangq7@sustech.edu.cn; yuhy@sustech.edu.cn)

Abstract

This work systematically established the atomic layer etching (ALE) technique for MOCVD *in-situ* SiN$_x$. Achieving ultra-low etching damage, with ~110% and ~220% improvement in surface morphology (RMS roughness) compared to as-grown and continuous etching wafers. Ultra-high etching precision with etching per cycle (EPC) of 1.13 nm/cycle was obtained. The mechanism of ALE was validated. This technique optimized the patterning of *in-situ* SiN$_x$ passivation/dielectric on GaN, laying the foundation for realizing high-performance Si-based GaN HEMTs.

Keywords: Atomic Layer Etching, MOCVD *In-situ* SiN$_x$

Introduction

Compared to *ex-situ* SiN$_x$, *in-situ* SiN$_x$ grown by metal-organic chemical vapor deposition (MOCVD) can effectively prevent the as-grown GaN barrier from exposure to air. Additionally, due to the high temperature and low growth rate epitaxial process, *in-situ* SiN$_x$ exhibits superior film quality [1]. The *in-situ* SiN$_x$ layer offers an advantageous choice of passivation/dielectric layer for GaN HEMTs [1,2].

The etching technique is critical for patterning during device fabrication. Various conventional etching approaches have been proposed for SiN$_x$ film, such as fluorocarbon-based [3] and SF$_6$-based continuous etching [4]. Recently, atomic layer etching (ALE) technique has been introduced to address the issues of ion damage and poor etching precision associated with continuous etching [5]. A series of ALE approaches (CH$_x$F$_y$-based modification step + Ar plasma removal step + O$_2$ plasma ashing step) for *ex-situ* SiN$_x$ have been established [6,7]. However, there is still a lack of research on ALE processes for MOCVD *in-situ* SiN$_x$ materials.

This paper presented a three-step ALE technique for *in-situ* SiN$_x$ etching, featuring a recipe with extremely low etching damage and high precision. The process for optimizing the etching parameters was discussed systematically, and the mechanism of each ALE step was further validated. The development of this technique filled the R&D gap in the field of ALE and provided a viable *in-situ* SiN$_x$ passivation/dielectric etching solution for the preparation of high-performance Si-based GaN HEMTs.

Experiments

The 6" wafer was purchased and diced into 2 x 2 cm^2 samples for etching. It consisted of a 50 nm MOCVD *in-situ* SiN$_x$ grown on the Si substrate, as shown in Fig.1 (a). In Fig. 2, the X-ray photoelectron spectroscopy (XPS) measurements for the as-grown surface exhibited high-intensity Si-N bonds, indicating the presence of *in-situ* SiN$_x$. The minor intensity of Si-O bonds, introduced by water and oxygen in the air, further suggested the effectiveness of the *in-situ* SiN$_x$ layer in preventing the influence of air on the underlying layer.

Fig. 1 (b) to Fig. 1 (e) illustrated the three-step cycle of the ALE process. Given that O$_2$ plasma consumed photoresist (PR), a thick PR (2 μm) was employed as the etch mask. As the modification step, a CHF$_3$/Ar gas mixture generated an HFC polymer layer on the *in-situ* SiN$_x$ surface, creating a reactive layer (Si-F, C-N) at the polymer/SiN$_x$ interface. Subsequently, the Ar plasma removal step etched both the reactive layer and partial polymer layer. Finally, an O$_2$ plasma ashing step was performed to eliminate the residual C-rich polymer on the surface. The etching process was carried out in Corial 210 IL Inductively Coupled Plasma-Reactive Ion Etching (ICP-RIE) system. Etching depth and roughness were characterized using atomic force microscopy (AFM). XPS was employed to investigate the surface conditions following each step.

Results and Discussion

Initially, the modification step was systematically established using XPS measurements. Fig.3 demonstrated the Si2p spectra corresponding to different modification times. With the increase of time, the peak intensity of Si2p decreased progressively. The literature reported that the Si2p peak intensity ratios of the modified and as-grown samples could be calculated to the HFC polymer thickness [6]:

$$d_{HFC} = -\lambda_{Si2p} \cdot \ln\left(\frac{I_{Si2p}}{I_{Si2p}^{ref}}\right) \qquad (1)$$

Here, λ_{Si2p} was the mean free path of Si2p photoelectrons in the HFC polymer layer, which was typically set to 2.5 nm. I_{Si2p} and reference of I_{Si2p} corresponded to the Si2p peak intensities of the modified and as-grown samples, respectively. The calculated HFC thicknesses were displayed in Table 1, and the linear fitting of the data was shown in Fig.4. To avoid difficulties in subsequent Ar plasma removal due to an excessively thick HFC layer, a duration time of 5 s was selected for the modification step.

The removal step was also optimized to enhance the etching performance further. Table 2 listed the parameters setting of the three-step ALE, including gas component, gas flow rate, RF/ICP power, and reaction time. 20 s gas transition time was set between each step. The parameters of the modification and ashing steps had been established, while the removal step was set up with 15 sets of variables: three different RF powers corresponding to five removal times ranging from 20 s to 100 s (step = 20 s). After 20 cycles etching, AFM was employed to measure the average etching per cycle (EPC) and RMS roughness for each ALE sample, as illustrated in Fig.5. The optimal parameters for the removal step were ultimately determined as RF power = 25 W and removal time = 60 s.

To further investigate the mechanism of the three-step ALE process, XPS measurements of the C1s element were applied throughout the ALE. Fig.6 (a) showed the as-grown sample, with the C1s spectrum consisting of C-C (284.8 eV) and C-O-C (286.2 eV) peaks. The C-O-C peak likely arose from water and oxygen in the air. After the modification step, as shown in Fig. 6(b), C-F (287.4 eV), $C-F_2$ (288.7 eV), and $C-F_3$ (290.0 eV) peaks emerged. The C-O-C peak was replaced by the C-N (286.4 eV) peak (the positions of the two peaks nearly overlapped), indicating that HFC polymer had been deposited, and the polymer/SiN_x interface was modified. Following Fig. 6(c), after the removal step, the $C-F_2$ and $C-F_3$ peaks were erased, and the C-N and C-F peaks were significantly reduced, demonstrating that the *in-situ* SiN_x reactive interface and HFC layer were etched effectively. The intensity of the C-C peak was unchanged, which was caused by the residual C-rich polymer. In Fig. 6(d), after the ashing step, the C-O-C/C-N and C-F peaks were kept at a low level, while the C-C peak drastically decreased, suggesting that the O_2 removed the residual C-rich polymer successfully. The XPS results confirmed the effectiveness of each ALE step.

Fig. 7 illustrated the 3D surface morphology (5 x 5 μm^2) of the samples with and without etching. Fig. 7(a) displayed the as-grown sample with RMS roughness of 0.21 nm. Fig. 7(b) showed the sample after 20 cycles ALE and before PR mask cleaning, where the RMS roughness is approximately 0.1 nm, similar to the sample after mask cleaning (Fig. 7(c)). Fig. 7(d) presented the sample after SF_6-based continuous etching at RF = 25 W. At a similar etching depth as the 20 cycles ALE, the RMS roughness reached 0.32 nm. Compared to as-grown and continuous etching samples, the ALE technique improved surface morphology by ~110% and ~220%, respectively, demonstrating its advantage of ultra-low etching damage. The etching depth for 0-20 cycles (step = 5 cycles) was shown in Fig. 8. The high linearity fitting results (EPC = 1.13 nm/cycle) confirmed the high etching precision of the ALE technique.

Conclusion

In this work, a three-step ALE recipe for *in-situ* SiN_x was systematically established based on ICP-RIE. The mechanism of each ALE step was verified through XPS. Ultra-low etching damage was achieved, with surface morphology improved by ~110% and ~220% compared to as-grown and continuous etching wafers. Meanwhile, high etching precision was attained, with EPC of 1.13 nm/cycle based on high linearity fitting. This technique provided an excellent solution for patterning *in-situ* SiNx passivation/dielectric layers on high-performance Si-based GaN HEMTs.

Acknowledgments

This work was supported by Fabrication of Normally-Off GaN Devices based on In-situ SiN_x Passivation and Selective Area Growth Recessed-Gate Techniques and the Reliability Study (National Natural Science Foundation of China, Grant No: 62274082), Research on mechanism of Source/Drain ohmic contact and the related GaN p-FET (Grant No: 2023A1515030034), Research on high-reliable GaN power device and the related industrial power system (Grant No: HZQB-KCZYZ-2021052), Study on the reliability of GaN power devices (Grant No: JCYJ20220818100605012), Research on novelty low-resistance Source/Drain ohmic contact for GaN p-FET (Grant No: JCYJ20220530115411025), NSQKJJ under grant K2023390010, and High level of special funds (G03034K004).

References

[1] H. Jiang, C. Liu, Y. Chen, X. Lu, C. W. Tang, and K. M. Lau, "Investigation of *In Situ* SiN as Gate Dielectric and Surface Passivation for GaN MISHEMTs", IEEE Trans. Electron Devices 64(3): 832-839 (2017).

[2] A. Siddique, R. Ahmed, J. Anderson, and E. L. Piner, "Effect of reactant gas stoichiometry of *in-situ* SiN_x passivation on structural properties of MOCVD AlGaN/GaN HEMTs", Journal of Crystal Growth 517: 28-34 (2019).

[3] R. J. Gasvoda, Z. Zhang, S. Wang, E. A. Hudson, and S. Agarwal, "Etch selectivity during plasma-assisted etching of SiO_2 and SiN_x: Transitioning from reactive ion etching to atomic layer etching", J. Vac. Sci. Technol. A 38(5): 050803 (2020).

[4] C. Reyes-Betanzo, S. Moshkalyov, J. Swart, and A. Ramos, "Silicon nitride etching in high-and low-density plasmas using $SF_6/O_2/N_2$ mixtures", J. Vac. Sci. Technol. A 21(2): 461-469 (2003).

[5] K. J. Kanarik, T. Lill, E. A. Hudson, S. Sriraman, S. Tan, J. Marks, V. Vahedi, and R. A. Gottscho, "Overview of atomic layer etching in the semiconductor industry", J. Vac. Sci. Technol. A 33(2): 020802 (2015).

[6] A. Hirata, M. Fukasawa, K. Kugimiya, K. Nagaoka, K. Karahashi, S. Hamaguchi, and H. Iwamoto, "Mechanism of SiN etching rate fluctuation in atomic layer etching", J. Vac. Sci. Technol. A 38(6): 062601 (2020).

[7] A. Hirata, M. Fukasawa, K. Kugimiya, K. Nagaoka, K. Karahashi, S. Hamaguchi, and H. Iwamoto, "On-wafer monitoring and control of ion energy distribution for damage minimization in atomic layer etching processes", Jpn. J. Appl. Phys. 59(SJ): SJJC01 (2020).

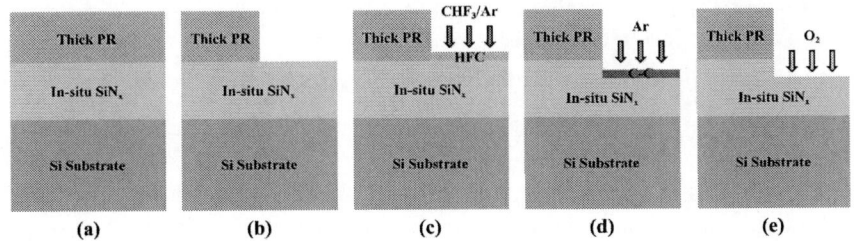

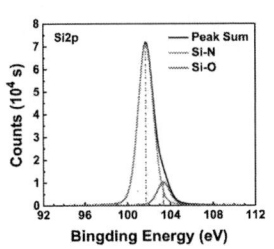

Fig. 1: Schematic of cross-section of *in-situ* SiN$_x$/Si epitaxy during ALE: (a) photoresist deposition as mask, (b) patterning definition, (c) CHF$_3$/Ar modification step, (d) Ar removal step, and (d) O$_2$ ashing step

Fig. 2: XPS results of Si element for as-grown *In-situ* SiN$_x$/Si wafer

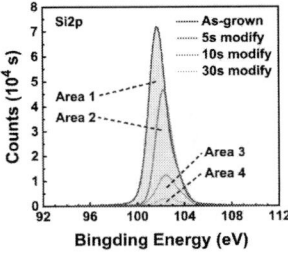

Fig. 3: XPS of Si element for samples with different modification time

Table 1: XPS peak intensity of Si element and calculated HFC ploymer thickness under different modification time

Area Number	Modification Time (s)	Si2p XPS peak Intensity	HFC Layer Thickness (nm)
Area 1	0	147768.84	0
Area 2	5	87562.43	1.31
Area 3	10	28397.16	4.12
Area 4	30	7093.84	7.59

Fig. 4: Linear fitting of the HFC thickness vs modification time

Table 2: Table of parameter settings for each step of ALE (including variables)

ALE/Etching Step	Gas Component	Gas Flow (sccm)	RF Power (W)	ICP Power (W)	Duration Time (s)
Modification	CHF$_3$/Ar	5:150	0	100	5
Removal	Ar	150	5, 15, 25	0	20, 40, 60, 80, 100
Ashing	O$_2$	50	50	100	10

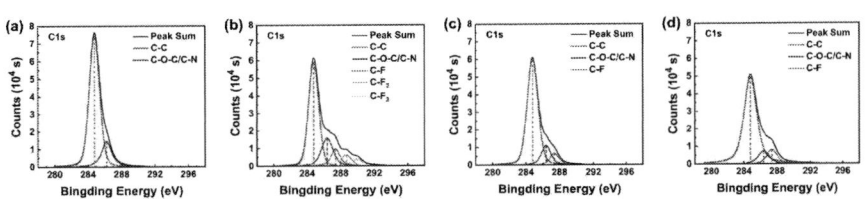

Fig. 6: XPS measurements of C element for (a) as-grown sample, (b) sample after modification step, (c) sample after removal step, and (d) sample after ashing step

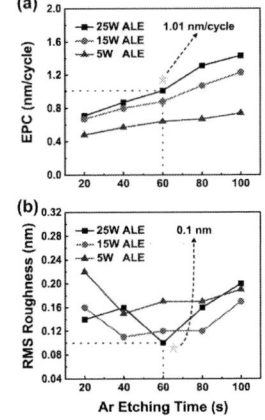

Fig. 5: (a) EPC and (b) RMS roughness measured by AFM for ALE recipes with different RF power and duration time

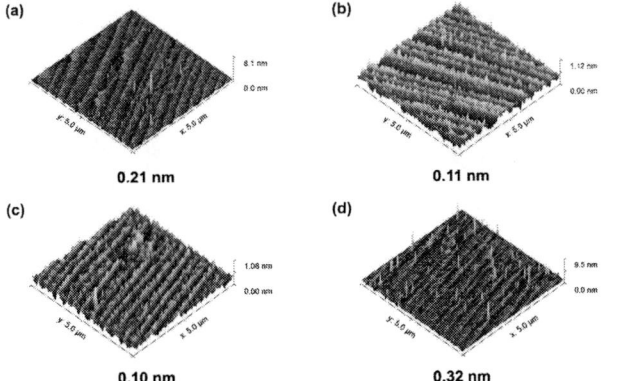

Fig. 7: AFM results of (a) as-grown sample, (b) sample after ALE before cleaning PR, (c) sample after ALE and cleaning PR, and (d) sample after continuous etching

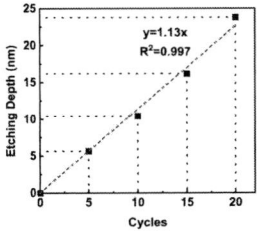

Fig. 8: Etching depth measured by AFM vs etching cycles

979-8-3315-0417-5/25 $31.00 © 2025 IEEE

Reliability-Aware Device and Programming Scheme Optimization for PBS/NBS-Immune IGZO-based 2T0C DRAM

Zhidong Tang†, Yanbo Su, Jianshi Tang*, Yijia Fan, Yiwei Du, Mingcheng Shi, Yibei Zhang, Dong Wu, Bin Gao, He Qian, and Huaqiang Wu

School of Integrated Circuits, Beijing Advanced Innovation Center for Integrated Circuits, BNRist, Tsinghua University
*Email: jtang@tsinghua.edu.cn;

Abstract

This work develops a reliability-aware device and programming scheme optimization methodology for IGZO-based 2T0C DRAM cells. The read and write transistors were fabricated by adopting different compositions and layer thicknesses of IGZO channel. Meanwhile, the reliability degradation map in the full $\{V_{GS,str}, V_{DS,str}\}$ space showed that the degradation of PBS can be mitigated by applying a $V_{DS,str}$, from which a PBS-mitigated programming scheme was developed. Finally, the PBS/NBS immunity of 2T0C DRAM cells were demonstrated.

Keywords: 2T0C DRAM, Oxide Semiconductor (OS), InGaZnOx (IGZO), reliability.

Introduction

IGZO-FETs are viewed as promising transistors for 3D stacked 2T0C DRAM cells due to their extremely low leakage, superior uniformity, and BEOL compatibility [1]. The 2T0C DRAM cells comprise a write transistor and a read transistor, storing charges in the storage node (SN) and determining the state of V_{SN} by measuring the current through the read transistor. An effective 2T0C DRAM structure necessitates a write transistor with positive V_{th} and a read transistor with substantial drive current. In 2T0C DRAM cell, the read and write transistors are also subjected to both positive and negative bias stress (PBS/NBS) during operational modes.

Typically, the IGZO-based 2T0C DRAM cell faces a trade-off issue between read and write transistors when single material recipe is utilized. This is primarily because IGZO-FETs face challenges in maintaining voltage bias stress reliability while concurrently aiming to achieve high drive current and positive V_{th} [2]. The high drive current of IGZO-FETs can be ascribed to abundant donor-like oxygen vacancies (V_O), which originate from weak metal to oxygen (M-O) bonds, contributing to high mobility and high I_{on}/I_{off}. However, excessive V_O also causes undesirable negative initial V_{th}, leading to high I_{off} at zero V_{GS}. Moreover, these V_O defects can be unstable and can interact with the defects of gate oxides under voltage bias stress, posing elusive reliability issues.

In this work, to break the trade-off issue of read and write transistors, we propose IGZO-based 2T0C DRAM cell with two different IGZO channel recipes by adopting different doping compositions and layer thicknesses. In addition to the optimized device recipes, we also introduced a programming scheme aimed at alleviating the aging process of 2T0C DRAM cells. As a result, the impacts of PBS/NBS on the IGZO-based 2T0C cell have been significantly mitigated.

Device Optimization Scheme and Sample Fabrication

For 2T0C DRAM cells, the read and write transistors have unique requirements. The read transistor demands high drive current and PBS stability, while the write transistor necessitates high V_{th} and NBS stability. Therefore, two distinct recipes have been devised, each utilizing different channel compositions and t_{ch} of the IGZO channel, as depicted in Fig. 1b.

Previous work [3] demonstrated that the threshold voltage of OS-FETs follows $V_{th} = (\varphi_M - \chi_s)/q - qN_{ch}t_{ch}/C_G$, where N_{ch} is positively correlated with the In content and inversely correlated with the Ga content for IGZO, and t_{ch} is the channel thickness. Thus, by adjusting the N_{ch} and t_{ch} of the IGZO-FETs, recipe #1 with higher drive current and recipe #2 with more positive V_{th} can be obtained.

Furthermore, N_{ch} and t_{ch} significantly influence the performance of PBS and NBS of IGZO-FETs. As described in [4], high N_{ch} in OS-FETs channel enhances PBS stability, which results in a more negative initial V_{th} and a reduced $\triangle V_{th}$ under PBS. Given that the read transistor requires a high drive current and primarily undergoes PBS condition, recipe #1, with its highly In% doped channel composition, fulfills the requirement. Additionally, the NBS immunity can be strengthened by increasing t_{ch}. Based on the subgap DOS model, IGZO-FETs with a thicker channel will have a diminished surface electric field, which in turn suppresses the formation of shallow donor states under NBS condition [5]. As the write transistor is predominantly subjected to NBS, the thickness of the IGZO channel should be increased, as indicated in recipe #2.

Fig. 1c shows the process flow of fabricating the IGZO-FETs. 1nm Ti/20nm Pd back gate was deposited by e-beam evaporation. Next, 10 nm HfO_2 was deposited by ALD as gate dielectric layer, followed by ICP etching. IGZO channel was deposited by ALD with different component ratios and number of cycles. Specifically, recipe #1, featuring a 10 nm In:Ga:Zn=2:2:1 IGZO channel, was used for the read transistor, while recipe #2, with a 15 nm In:Ga:Zn=2:4:1 IGZO channel, was utilized for the write transistor. Then, 1nm Ti/45nm Pd was evaporated as source/drain contacts.

979-8-3315-0417-5/25 $31.00 © 2025 IEEE

The IGZO channel was patterned by wet etching. Fig. 1d shows the cross-sectional transmission electron microscopy (TEM) images of device structure for the implementation of 2T0C DRAM cell. Connecting the gate of one IGZO-FET to the other IGZO-FET's drain completes the 2T0C DRAM cell.

Device Characterizations of 2T0C DRAM Cell

Both IGZO-FETs fabricated with the two different recipes were thoroughly characterized in the dark environment at room temperature. Fig. 2a shows the device transfer curves of recipe #1 and #2. It shows that recipe #1 achieved higher drive current of $I_{Dsat} \sim 10 \ \mu A/\mu m$, which is suitable for the read transistor of the 2T0C to achieve large V_{SN} storage window. In comparison, recipe #2 achieved lower leakage with more positive $V_{th} \sim 0.87$ V, which is favorable for long data retention time.

Fig. 3 displays the evolution of $\triangle V_{th}$ under PBS. It is seen that both devices exhibited positive V_{th} shifts when the stress time increased at room temperature. When the overdrive stress gate voltage increased, recipe #1 showed better PBS immunity than recipe #2 as displayed in Fig. 3a. It should also be noted that the $\triangle V_{th}$ curves did not follow the power law behavior with respect to overdrive voltage $V_{OV}=(V_{GS}-V_{th0})$ as typical MOSFET BTI, where $\triangle V_{th}=At_{str}^n(V_{GS}-V_{th0})^\gamma$. It indicated that the IGZO-FETs not only suffered from oxide charge trapping induced $\triangle V_{th}$, but also IGZO channel-dominated effects including V_O-related and H-related degradation. On the other hand, as displayed in Fig. 3b, the $\triangle V_{th}$ shifts followed t_{str}-dependent power law trend, where $\triangle V_{th}=A \exp(-E_a/kT)t_{str}^n$ for both recipes #1 (n~0.28) and #2 (n~0.43) for different V_{ov}, indicating better PBS immunity for recipe #1 as the read transistor. Moreover, the negative V_{WWL} (optional for longer data retention time) would impose NBS on the write transistor during the read and hold operation. Negative V_{th} shifts were observed for IGZO-FETs as the negative bias stress voltages increased at room temperature (Fig. 4a). As illustrated in Fig. 4b, the t_{str}-dependent power law trend $\triangle V_{th}=AV_{GS,str}^\gamma t_{str}^n$ was also observed in NBS for both recipe #1 (n~0.32) and #2 (n~0.18). With a lower time exponent, recipe #2 exhibited better NBS immunity as the write transistor. Based on the above analysis, recipes #1 and #2 are optimal for making the read and write transistors respectively.

Next, the impact of additional $V_{DS,str}$ stresses on the IGZO-FET's electrical characteristics was investigated under $V_{GS,str}$. Fig. 5a and 5b showed the V_{th} and I_{Dsat} degradation of IGZO-FETs in {$V_{GS,str}$, $V_{DS,str}$} stress space. Strikingly, the most significant V_{th} and I_{Dsat} degradation was observed under high $V_{GS,str}$ and zero $V_{DS,str}$ (pure PBS region). But applying an additional $V_{DS,str}$ can effectively mitigate the impact of PBS. Fig. 5c indicated that there was no remarkable degradation on SS and g_m in the full {$V_{GS,str}$, $V_{DS,str}$} stress space, suggesting a different mechanism from the hot-carrier degradation of traditional Si MOSFETs.

Programming Scheme & PBS/NBS immunity of 2T0C Cell

Based on the experimental observations from Fig. 5, we discovered that the approach of increasing V_{DS} bias can also be applied to the IGZO 2T0C cells. As illustrated in Fig. 6a, when the V_{SN} node holds a voltage, we can raise the V_{DS}/V_{RWL} of the read transistor to V_{DD}. This approach has several benefits. First, increasing V_{RWL} can increase the V_{SN} storage window. Fig. 6b showed that increasing V_{RWL} from 0.1 V to 2 V improved the V_{SN} storage window by nearly 10×. Second, we found that applying a large $V_{DS,str}$ to the read transistor also mitigated V_{SN} degradation when the read transistor was stressed under PBS. When the V_{SN} maintains data "1" for a long period of time, the stress voltage applied on the gate of the read transistor leads to the degradation of the I_{RWL} and hence a negative $\triangle V_{SN}$ as depicted in Fig. 7a. By imposing a higher V_{RWL}, the V_{SN} maintain time (i.e. the time when V_{SN} drops by 5% from the $V_{SN,ini}$ due to the degradation of I_{RWL}) increased by 173% and 301% respectively for $V_{SN,ini}$ =2 V and $V_{SN,ini}$ =4 V as illustrated in Fig. 7b, demonstrating outstanding PBS immunity.

Moreover, the exceptional NBS immunity of the write transistor (recipe #2) enables a long retention time over 200 seconds under 10 write-read cycles. Fig. 8b revealed that during the write-read operations (Fig. 8a), the V_{SN} remained stable when the write transistor was subjected to continuous NBS stress (V_{WWL}= -1V & -2V). The retention time (i.e. the duration for the V_{SN} drops by 0.1V from the initial V_{SN} due to leakage current) displayed in Fig. 8c indicated that this 2T0C cell has excellent retention performance and NBS immunity.

Conclusion

To summarize, we have developed a reliability-aware device and programming scheme optimization methodology, and demonstrated high reliable IGZO-based 2T0C DRAM cells by integrating two kinds of recipes according to the different requirements for the write and read transistors. By implementing an elevated V_{RWL} programming scheme, the V_{SN} storage window was enlarged by 10×, and the V_{SN} maintain time was increased by 173% and 301% under PBS. Finally, the optimized 2T0C cell achieved excellent NBS-immune retention property under endurance test.

Acknowledgments

This work was in part supported by the NSFC 92264201 and 62025111.

References

[1] A. Belmonte, *et al.*, *IEDM, 2020.* [2] Y. Shiah, *et al.*, *Nature Electronics,* 2021. [3] S. Liu, *et al.*, *IEDM, 2023.* [4] Q. Jiang, *et al.*, *VLSI, 2024.* [5] D. Kong, *et al.*, *IEEE EDL, 2011.* [6] S. Liu, *et al.*, *VLSI, 2024.*

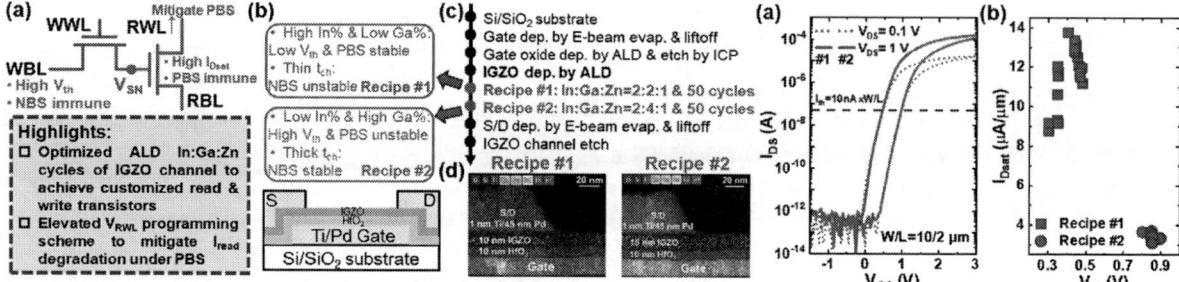

Fig. 1. (a) This work presents a device and programming scheme to break the trade-off of IGZO-based 2T0C DRAM cells for read and write transistors. (b) The doping level and layer thickness (t_{ch}) of the IGZO channel is made to achieve the different requirements for read and write transistors in IGZO 2T0C cells. (c) Fabrication process flow of IGZO-FETs. (d) Cross-sectional TEM images of the IGZO FET.

Fig. 2. (a) Transfer characteristics of IGZO-FETs from recipe #1 and #2. The V_{th} and I_{Dsat} are extracted at V_{GS} @ I_{DS}=10nA×W/L and I_{DS} @ V_{GS}=2V & V_{DS}=1V respectively. (b) Device statistics of recipe #1 and #2, clearly presenting the V_{th} and I_{Dsat} trade-off for IGZO-FETs.

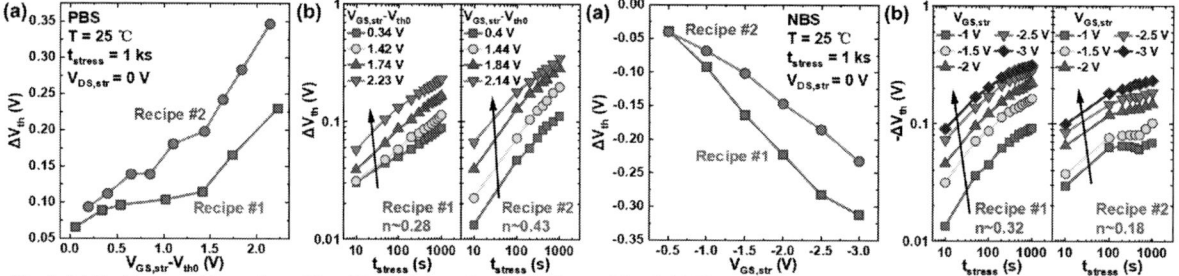

Fig. 3. (a) Positive overdrive-voltage ($V_{OV}=V_{GS}-V_{th}$) dependence of ΔV_{th} for recipes #1 and #2 after 1 ks stress time at room temperature. (b) The degradation of V_{th} under different PBS for recipes #1 and #2. The exponent n is fitted using the following equation $\Delta V_{th}=At_{str}^{n}(V_{GS}-V_{th0})^{\gamma}$.

Fig. 4. (a) Negative V_{GS} stress voltage dependence of ΔV_{th} for recipe #1 and #2 after 1 ks stress time at room temperature. (b) The degradation of V_{th} under different NBS for recipe #1 and #2. The exponent n is fitted using the following equation: $\Delta V_{th}=AV_{GS}^{\gamma}t_{str}^{n}$.

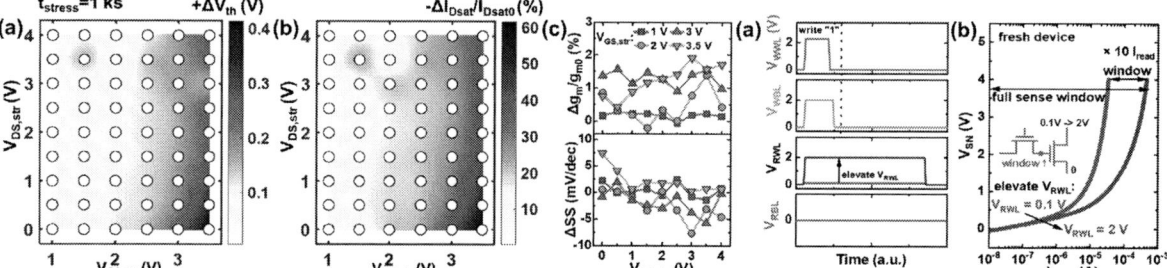

Fig. 5. The (a) V_{th} and (b) I_{Dsat} degradation at the cumulative stress time of 1 ks in different {$V_{GS,str}$, $V_{DS,str}$} stress space. The experimental data is derived from 54 measured devices of Recipe #2. Each dot represents an individual data point from the experiment. The values between experiments were interpolated using cubic function. (c) The g_m and SS remain almost unchanged at cumulative stress time of 1 ks in different {$V_{GS,str}$, $V_{DS,str}$} stress space.

Fig. 6. (a) The waveform of the proposed V_{RWL} elevated programming scheme. (b) The V_{SN} sense window can be enlarged by 10× with elevated V_{RWL}.

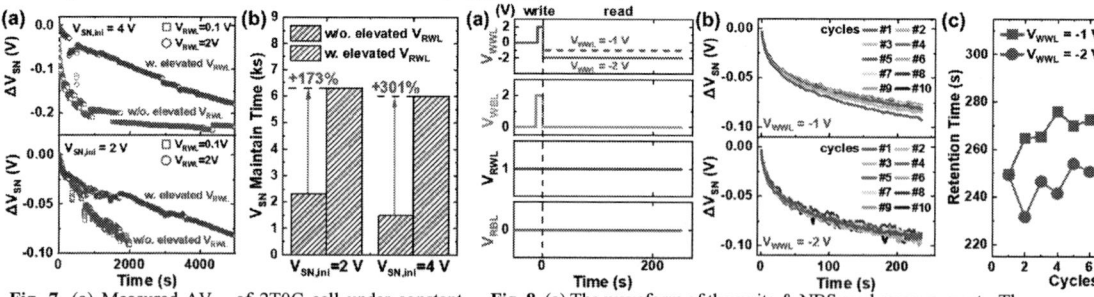

Fig. 7. (a) Measured ΔV_{SN} of 2T0C cell under constant voltage bias. With elevated V_{RWL} programming scheme, the PBS degradation of read transistor can be mitigated. (b) The V_{SN} maintain time is increased with elevated V_{RWL}.

Fig. 8. (a) The waveform of the write & NBS read measurements. The measurements are repeated for continuous 10 cycles. (b) ΔV_{SN} evolution for V_{WWL} = -1 & -2 V during the read operation. (c) The retention performance of the proposed 2T0C cell under 10 write & NBS read cycles.

979-8-3315-0417-5/25 $31.00 © 2025 IEEE

A 2T2R TCAM Based on RRAM and Accurate Compact Model for A Kilobit Word

Zhen Wang[1,2], Pengtao Li[1,2], Zijian Wang[1,2], Xuemeng Fan[1,2], Shengpeng Xing[1,2] and Yishu Zhang[1,2*]

[1] College of Integrated Circuits, Zhejiang University, China
[2] ZJU-Hangzhou Global Scientific and Technological Innovation Center, China

Abstract

Ternary content-addressable memory (TCAM) enables massively parallel search operations but faces challenges in achieving high storage density and area efficiency. This paper proposes a 2-transistor 2-resistor (2T2R) TCAM array architecture using RRAM cells. It can maintain sensing margins for word lengths exceeding 2048 bits with RRAM devices. To validate the concept, Ta/ZnO/TaO$_x$/Pt RRAM devices are experimentally characterized and a compact model is optimized to fit the measured I-V data with 99.32 % accuracy. Circuit simulations are also conducted to verify its functions.

Keywords: memristor, simulation and TCAM.

Introduction

The rapid growth of data volumes has necessitated novel computational architectures to address the limitations of von Neumann architectures [1]. TCAM based on static random-access memories (SRAMs) is introduced to support ternary logic states {0, 1, X}, providing flexibility for precise and fuzzy matching on large datasets in parallel [2]. Nevertheless, as array sizes increase to meet contemporary demands, it exhibits high area overhead but low energy efficiency. Therefore, computational memory architectures leveraging emerging non-volatile memories such as resistive RAM (RRAM), magnetic RAM (MRAM), phase-change memory (PCM) and ferroelectric memory (FeRAM) have been proposed to increase density and reduce power[3-5]. Although, FeRAM and MRAM show insufficient scalability and PCM cell needs more power for switching [6]. RRAM devices have appeared as a potential candidate for the forthcoming flexible non-volatile memory (NVM) device due to their distinctive features such as scalability, higher speed operation, CMOS compatibility, and low power consumption [7].

In this work, we present a TCAM array constructed from Ta/ZnO/TaO$_x$/Pt RRAM devices in a 2T2R cell design. Based on the experimental results, we optimize the accuracy of the RRAM compact model to around 99.32%. The functions of the array are validated based on accurate RRAM compact model. The TCAM can maintain sensing margins for word lengths exceeding 2048 bits through a threshold-adjustable invertor.

Experiment

A p-type (100) silicon wafer with a 300 nm thermally grown SiO$_2$ layer is used as the substrate. After cleaning and drying, a 60-nm-thick Platinum (Pt) bottom electrode (BE) is deposited by direct current (DC) magnetic sputtering and patterned. A 50 nm silicon nitride layer is then deposited via radio frequency (RF) sputtering to prevent crosstalk between devices. Photolithography defines top electrode (TE) shapes. 20 nm TaO$_x$ is deposited by RF sputtering as the first switching layer. 10 nm ZnO is deposited by RF sputtering to form the ZnO/TaO$_x$ bilayer switching medium. 50 nm Ta TE is deposited by DC sputtering. Depositions are done in situ using a DISCOVERY-635 sputtering system. Layers are patterned using photolithography. The completed RRAM stack and SEM image are shown in Fig.1. Electrical characterization of the fabricated cells will evaluate the switching and data retention performance of the ZnO/TaO$_x$ bilayer system.

Result and discussion

The electrical characteristics of the Ta/ZnO/TaO$_x$/Pt RRAM device are shown in Fig.2. The device demonstrates good uniformity across tested cells, stable resistance states and low switching voltages less than 1 V which ensures a sufficient distinction during search operations in the TCAM without incurring undesired state inversion. Fig.2(d) shows the set and reset operations occurring within 110 ns and 220 ns respectively under the successive pulses of ±2 V with 500 ns width, satisfying TCAM speed requirements.

Then we build an accurate compact model based on these analyses. In order to fit well in different regimes, we optimize it as below:

$$\frac{dw(t)}{dt} = \begin{cases} k_{on}\left(\dfrac{v(t) - v_{on}}{m_{on}}\right)^{\alpha_{on}} \cdot f(t), & 0 < v_{on} \leq v \\ 0, & v_{off} < v < v_{on} \\ k_{off}\left(\dfrac{v(t) - v_{off}}{m_{off}}\right)^{\alpha_{off}} \cdot f(t), & v \leq v_{off} < 0 \end{cases} \quad (1)$$

where k_{off}, k_{on}, m_{on}, m_{off}, α_{off}, and α_{on} are fitting parameters, v_{on} and v_{off} are threshold voltages, and $w(t)$ represents the length of CF between 0 and 30 nm. The function $f(t)$ is relative to the variation of $v(t)$. As shown in Fig.3(a), this revised approach improves the fitting rate of DC curve to

---

979-8-3315-0417-5/25 $31.00 © 2025 IEEE

99.32% and we also fit the pulse curve well in Fig.3(b).

Based on the model of the fabricated device and 65-nm TSMC PDK, a 2T2R TCAM architecture is proposed. As shown in Fig.4, the gates within the same column share a search line (SL) for comparison operations while rows connected to word lines (WL) are set to 0.8 V to avoid inadvertently switching RRAM states and provide enough voltage for comparison. The match line (ML) is tied to the common nodes of the two 1T1R cells and pulled to the ground through a transistor (N1) controlled by a 0.4 V clock signal. The voltage of ML will be tuned through a 'main' inverter (M-inv), of which the threshold voltage is adjusted by a 'sub' inverter (S-inv). The voltage of SA_{out} represents the searching results. The cell integrates RRAM devices into a 65 nm CMOS backend to create two 1T1R sub-cells, shown in Fig.5.

During search, N1 is pull down and the search data is applied simultaneously to all SLs to be compared against stored data, determining the ML voltage in a single clock cycle. If all the data match with the search data, then the ML becomes low. If there is at least one mismatch, the ML will be pulled up. Fig.6 verifies the functions for searching the three states {0, 1, X}. In short words, ML clearly distinguishes "match" and "mismatch" states. Through our design, voltage across RRAM is less than 0 (ML < WL in read operation), which avoids an undesirable change in the HRS. And in the back end of line (BEOL), the process flows of fabricating R1 and R2 can be unified (from bottom electrode to top electrode), which satisfies a united polarity for R1 and R2.

However, the resistance decreases for long parallel cells, reducing the sense margin. Through the S-inv transforming the ML to the signal marked as Sub connected to the substrate of the M-inv, the threshold voltage of the M-inv is tuned as shown in Fig.7 (a). When the ML rises, the Sub signal falls, making the threshold voltage of the M-inv moves right. As a result, for a worse ML, M-inv can still transform it to a high level. Fig.7 (b) and (c) shows the simulations of the trends of ML and SA_{out} versus the word length. Even when 2048 bits, SA_{out} has a discrimination of 500 mV, sufficient for sense margin. The comparison results with some other designs are shown in Fig.7 (d).

Conclusion

This work demonstrates a 2T2R TCAM design based on a $Ta/ZnO/TaO_x/Pt$ RRAM device. Leveraging an optimized accurate model, the 2T2R TCAM architecture is simulated, showing reliable sense margin and enhanced density. Compatibility with standard CMOS fabrication and high density positions this RRAM-based approach as

a promising solution for future TCAMs needed in networking applications.

References

[1] Z. Shen, C. Zhao, Y. Qi, W. Xu, Y. Liu, I.Z. Mitrovic, L. Yang, and C. Zhao, "Advances of RRAM Devices: Resistive Switching Mechanisms, Materials and Bionic Synaptic Application," Nanomaterials-basel, vol. 10, no. 8, Art. no. 8, Aug. 2020.

[2] D. R. B. Ly, J-P. Noel, B. Giraud, P. Royer, E. Esmanhotto, N. Castellani, T. Dalgaty, J-F. Nodin, C. Fenouillet-Beranger, E. Nowak and E. Vianello, "Novel 1T2R1T RRAM-based Ternary Content Addressable Memory for Large Scale Pattern Recognition," in 2019 IEEE International Electron Devices Meeting (IEDM), Dec. 2019, p. 35.5.1-35.5.4.

[3] J. Li, R. K. Montoye, M. Ishii, and L. Chang, "1 Mb 0.41 µm² 2T-2R Cell Nonvolatile TCAM With Two-Bit Encoding and Clocked Self-Referenced Sensing," IEEE Journal of Solid-State Circuits, vol. 49, no. 4, pp. 896–907, Apr. 2014.

[4] W. Xu, T. Zhang, and Y. Chen, "Design of Spin-Torque Transfer Magnetoresistive RAM and CAM/TCAM with High Sensing and Search Speed," IEEE Transactions on Very Large Scale Integration (VLSI) Systems, vol. 18, no. 1, pp. 66–74, Jan. 2010.

[5] Ava J. Tan , Korok Chatterjee, Jiuren Zhou, Daewoong Kwon, Yu-Hung Liao, Suraj Cheema, Chenming Hu and Sayeef Salahuddin, "Experimental Demonstration of a Ferroelectric HfO₂-Based Content Addressable Memory Cell," IEEE Electron Device Lett., vol. 41, no. 2, pp. 240–243, Feb. 2020.

[6] G. W. Burr, B. N. Kurdi, J. C. Scott, C. H. Lam, K. Gopalakrishnan, and R. S. Shenoy, "Overview of candidate device technologies for storage-class memory," Ibm J Res Dev, vol. 52, no. 4.5, pp. 449–464, Jul. 2008.

[7] Z. Wang, Y. Song, G. Zhang, Q. Luo, K. Xu, D. Gao, B. Yu, D. Loke, S. Zhong, and Y. Zhang, "Advances of embedded resistive random access memory in industrial manufacturing and its potential applications," Int J Extreme Manuf, vol. 6, no. 3, p. 032006, Mar. 2024.

[8] X. Wang, Y. Qu, F. Yang, L. Zhao, C. Lee, and Y. Zhao, "A Highly Compact Nonvolatile Ternary Content Addressable Memory (TCAM) With Ultralow Power and 200-ps Search Operation," IEEE Trans. Electron Devices, vol. 69, no. 8, pp. 4259–4264, Aug. 2022.

[9] M.F. Chang, C.C. Lin, A. Lee, Y.N. Chiang, C.C. Kuo, G.H. Yang, H.J. Tsai, T.F. Chen, and S.S. Sheu, "A 3T1R Nonvolatile TCAM Using MLC ReRAM for Frequent-Off Instant-On Filters in IoT and Big-Data Processing," IEEE Journal of Solid-State Circuits, vol. 52, no. 6, pp. 1664–1679, Jun. 2017.

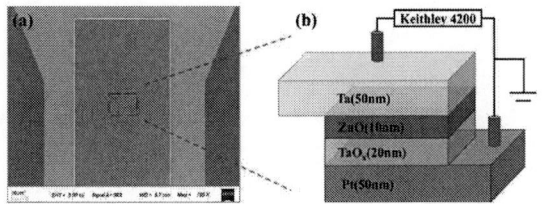

Fig. 1: (a) The high-resolution SEM image and (b) 3D structure of the device.

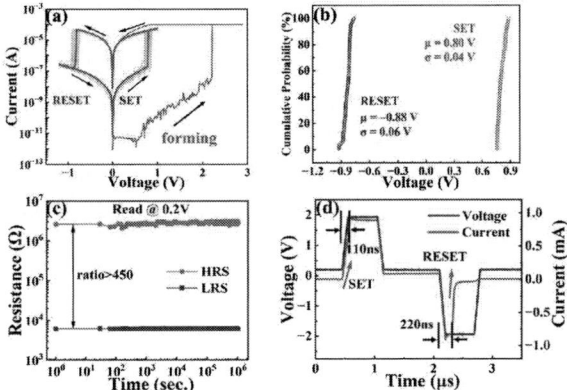

Fig. 2: (a) I-V characteristics during RS switching (SET/RESET) processes for 100 cycles (b) The cumulative probability of threshold voltage. (c) The retention measurement of the device. (d) Changes of current of the device under successive ±2 V pulses with a constant read voltage of 0.2 V to monitor the device status.

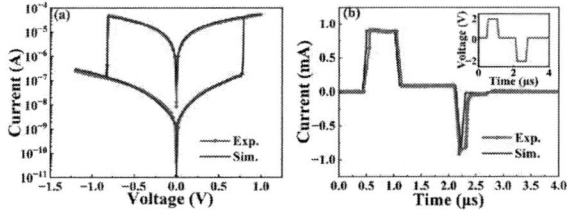

Fig. 3: (a) The compact model fits to experimental data for I-V curve. (b) The compact model fits to experimental data for pulse curve.

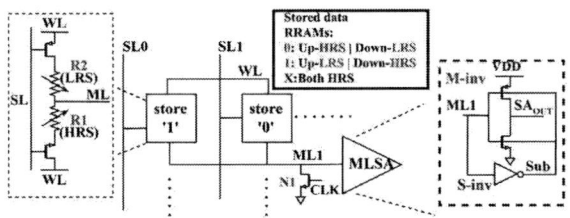

Fig. 4: Circuit diagram of the simulated TCAM array with the data encoding definitions shown at the top right corner.

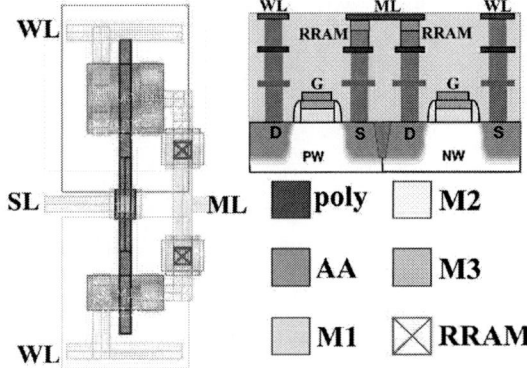

Fig. 5: Layout (left) and profile (right) of the proposed 2T2R bit-cell.

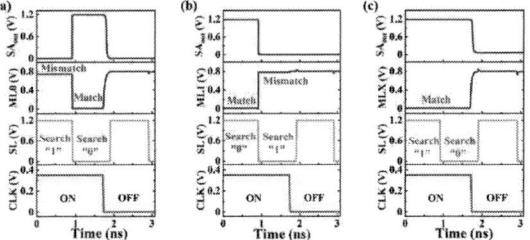

Fig. 6: (a) The ML voltage of match and mismatch for searching '0'. (b) The ML voltage of match and mismatch for searching '1'. (c) The ML voltage of match and mismatch for searching 'X'.

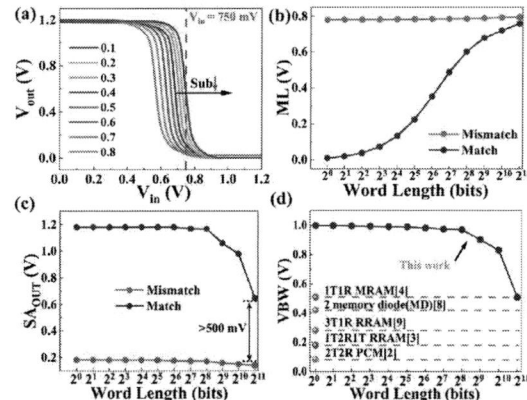

Fig. 7: (a) The change of threshold of the M-inv versus Sub. (b) Simulation of ML voltage with increasing word lengths. (c) Simulation of SA voltage with increasing word lengths. (d) Simulation of SA_out voltage difference between match and 1-bit-mismatch (VBW) of this work compared with other works.

979-8-3315-0417-5/25 $31.00 © 2025 IEEE

Prediction of single particle output response in the CTFET inverter based on deep learning algorithm

Chen Chong[1], Hongxia Liu[1], Zexi Wang[1]

[1]Xidian university, China,

Abstract

A deep learning algorithm network model is established to predict the single-particle output response curve of complementary tunnel field-effect transistor (CTFET) inverters. The prediction results show that when the time is less than 10^{-12}s, the relative error rate of the predicted value is less than 1%. In comparison experiments with four other traditional machine learning methods (decision tree, K-nearest neighbor algorithm, ridge regression, linear regression), the deep learning algorithm shows the smallest average error percentage.

Keywords: Deep Learning, Single particle effect, CTFET inverter

Introduction

Since the space irradiation environment will cause damage to semiconductor devices, which will result in the reversal of the logical state of digital integrated circuits, it is of great significance to predict the irradiation effect in digital integrated circuits in the aerospace field [1].

Tunnel field-effect transistor(TFET) devices can obtain sub-threshold swing less than 60mV/dec and achieve low power consumption [2], so they have been widely studied. There are many studies on irradiation based on CMOS inverter [3], but there are few studies on irradiation of CTFET inverter based on TFET device.

Compared with traditional prediction methods based on physical models, deep learning models have better generalization ability and adaptability, especially when facing complex irradiation conditions [4]. For example, in 2019, Wang Hai et al. proposed the deep learning potential energy model, which not only provides excellent performance in the prediction of balanced material properties, but also captures the physical phenomena when atoms are very close to each other [5]. In 2022, Dean B et al. irradiated a COTS microcontroller unit (MCU) with 10 keV X-rays and trained a convolutional neural network (CNN) model based on the noise signatures generated inside the clock module. Thus, MCUs are accurately divided into initial MCUs and irradiation MCUs, and the TID of MCUrs is predicted by regression model [6]. The above are some researches on the prediction of semiconductor irradiation effect using deep learning algorithm, but there are few researches on the prediction of single particle effect based

on CTFET inverter circuit. Therefore, this paper establishes a deep learning algorithm prediction model for the first time to predict the single particle output response in CTFET inverter, and the prediction curve fits well with the simulation curve.

Methodology

A. CTFET inverter

Fig.1(a) shows the schematic diagram of the N-type SSTGTFET device structure, which adopts the design of groove gate and stacked heterogeneous source[7]. In the N-type SSTGTFET device, the channel doping concentration is 5×10^{14}cm^{-3}, the source doping concentration is 1×10^{20}cm^{-3}, the drain doping concentration is 1×10^{18}cm^{-3}, and the thickness of the gate oxide layer is 2nm. The P-type SSTGTFET device only needs to change the source region to N+ doping and the drain region to P+ doping.

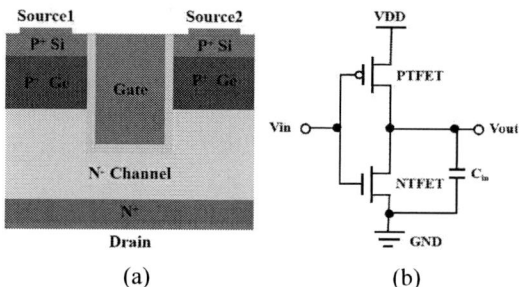

Fig.1. (a)Structure diagram of the N-type SSTGTFET device and (b)Circuit diagram of the CTFET inverter

Fig.1(b) shows the schematic diagram of the CTFET inverter. The CTFET inverter can make up for the shortcomings of PMOS inverter and NMOS inverter, and neither negative power supply nor a large area of load resistance.

B. Physical model of single particle effect

By taking the output of the device model as the input of the circuit simulation, mixed-mode simulation can take into account the influence of the device characteristics on the circuit behavior. This simulation method can simulate the TFET models of various structures, and the degradation of circuits subjected to single particle irradiation can be studied by adding heavy ion models. In this paper, the irradiation effect of single particle in the CTFET inverter is

979-8-3315-0417-5/25 $31.00 © 2025 IEEE

studied by using the HeavyIon model in Sentaurus software. The HeavyIon model are added to the SDEVICE module, and the related parameters are shown in Table 1.

Table 1. Device parameters used for the simulation

Parameter Name	Explanation	Unit
LET	Linear energy transfer	MeV·cm²/mg
W	Incident width of heavy ions in the device	μm
L	Incident length of heavy ions in the device	μm
angle	Incident angle of heavy ions in the device	°
Direction	Incident direction of heavy ions in the device	(0,0,1)
Location	Incident location of heavy ions in the device	-
Time	Incident time of heavy ions in the device	s
Gaussian	Gaussian distribution	-

C. Construction of Deep Learning algorithm predictive network model

According to the relevant parameters mentioned in the single particle simulation model of the CTFET inverter, linear transmission energy(LET), single particle incident width(W), incident length(L) and incident angle(angle) are selected as four characteristic variables. The values of characteristic variables are shown in Table 2.

Table 2. Values of characteristic variables

Characteristic variable	Value
LET(MeV·cm²/mg)	10, 20, 30, 40, 50, 60, 70, 80, 90, 100
W(μm)	0.01, 0.02, 0.03, 0.04, 0.05
L(μm)	0.02, 0.04, 0.06, 0.08, 0.1
angle(°)	30, 45, 60, 90

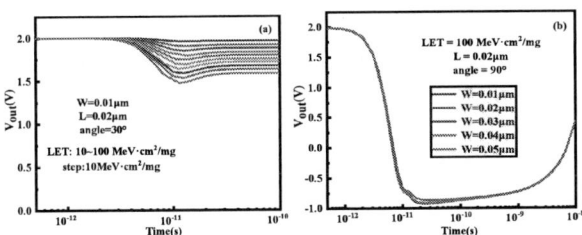

Fig. 2. Partial single particle output response curve data set

After the characteristic variables were designed, the CTFET inverter was simulated in mixed mode by Sentaurus and the data set was established. According to the simulation results, LET, W, L, angle are selected as the network inputs to predict the single particle output response of CTFET inverter. The network output is the single particle output response curve. A total of 1000 sets of data were obtained for the prediction of the single particle output response of

the CTFET inverter. Fig. 2 shows the partial curve data set of the simulated single-particle output response.

The loss function is used to measure the performance of the model. The loss function used in the training network in this paper is the absolute error loss function (L1Loss) :

$$L1Loss = \frac{1}{N}\sum_{i=1}^{N}|x_i - \hat{x}_i| \qquad (1)$$

Where N is the total number of samples, x_i is the simulation value of the i th sample, and $\hat{x}_i$ is the predicted value of the i th sample.

Pytorch framework is used to build the network model, the network structure is mainly full connection and convolution, and the optimizer uses Adam. The number of neurons in the input layer of the network is equal to the feature dimension of the input data. The output layer of the network is a 3×75 matrix.

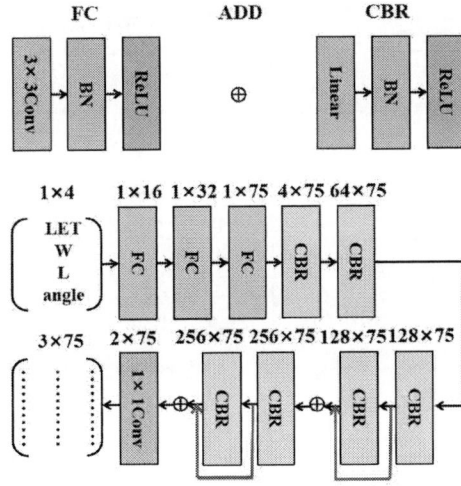

Fig. 3. Architecture diagram of neural network algorithm for predicting single particle output response of the CTFET inverter

As shown in Fig. 3, the four parameters LET, W, L, and angle are set as inputs, and the output is a 3×75 matrix. First, there are three FC modules, each consisting of a full connection layer, a BatchNormalization layer, and the ReLU activation function. The FC layer extends the 1×4 matrix to the 1×75 matrix. Next, six CBR modules are introduced, each consisting of a 3×3 1d convolution layer (which performs convolution operations on one-dimensional data), a BatchNormalization layer, and the ReLU activation function. The outputs of the third CBR layer and the fourth CBR layer are added, and the outputs of the fifth CBR layer and the sixth CBR layer are added. This helps solve the problem of disappearing gradients and exploding gradients, while also helping the model converge faster. Finally, a 1×1 1d convolution layer is used to reduce the dimension of the output of the previous layer, and the predicted value of the curve is obtained. The BatchNormalization layer and ReLU activation functions in the FC and CBR layers serve to speed up the convergence of the network, mitigate the problem of disappearing gradients.

Results

A. Deep learning algorithms predict results

After building a neural network model to predict the output response of a single particle, the prediction results of the network are analyzed. Six random test sets were selected for simulation prediction and compared with the actual simulation curve, where the blue area represents the error value of the output voltage of the two curves. It can be observed from Fig. 4 that the fitting effect of the two curves is better and the error value is small. Especially in the case of small time, the error of the output voltage is almost zero.

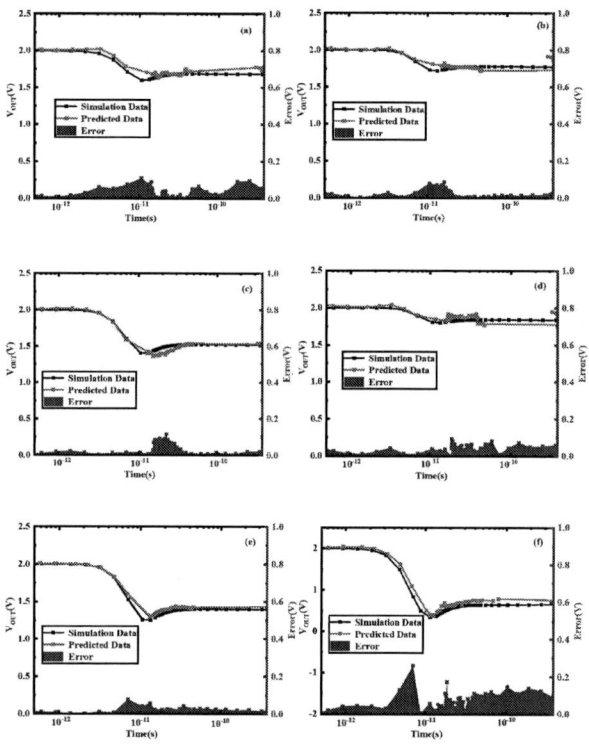

Fig. 4. Comparison of simulation curve and prediction curve of six sets of single particle output response

B. Deep Learning versus other machine learning methods

Four traditional machine learning methods, namely decision tree, K-nearest neighbor algorithm, ridge regression and linear regression, are selected. By comparing these methods with deep learning methods to make predictions on 100 sets of data in the test set, we get the average percentage of error shown in Fig. 5. It can be seen from the figure that both linear regression and decision tree have the worst performance in predicting the single-particle output response of inverter, and their average relative error percentage is higher than other algorithms. The average relative error percentage results of K-nearest neighbor

algorithm and ridge regression are similar. Deep learning algorithms show optimal performance with the smallest average error percentage.

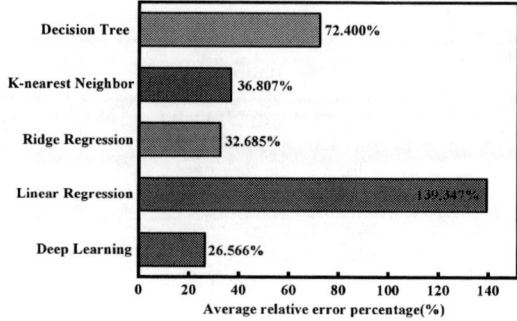

Fig. 5. Comparison of performance of predicting single particle output response under different machine learning methods

Conclusion

In summary, an algorithm model for predicting the output response curve of the CTFET inverter is built by studying the output response of the CTFET inverter. The deep learning network prediction results show the predicted curve and simulation curve fit well. Compared with four machine learning methods: decision tree, K-nearest neighbor algorithm, ridge regression and linear regression, it can be seen that the average relative error percentage of deep learning algorithm is 26.566%, which is the smallest. Therefore, the application of deep learning algorithms in predicting single-particle irradiation of circuits provides a new direction for radiation prediction methods.

References

[1] Velazco R, Franco F J., "Single event effects on digital integrated circuits: Origins and mitigation techniques" 2007 IEEE International Symposium on Industrial Electronics, pp. 3322-3327 (2007).

[2] Choi W Y , Park B G , Lee J D ,et al, "Tunneling Field-Effect Transistors (TFETs) With Subthreshold Swing (SS) Less Than 60 mV/dec", IEEE Electron Device Letters, pp. 28 743-745(2007).

[3] Gao T, et al, "Simulation of total ionizing dose effects technique for cmos inverter circuit", Micromachines, 14, pp. 1438 (2023).

[4] Etxegarai G, López A, et al, "An analysis of different deep learning neural networks for intra-hour solar irradiation forecasting to compute solar photovoltaic generators' energy production", Energy for Sustainable Development, 68, pp. 1-17 (2022).

[5] Wang H, Guo X, Zhang L, et al., "Deep learning inter-atomic potential model for accurate irradiation damage simulations", Applied Physics Letters, 114, (2019).

[6] Dean B, Peyton T, Carpenter J L, et al., "Machine Learning Approaches for Analysis of Total Ionizing Dose in Microelectronics", 2022 22nd European Conference on Radiation and Its Effects on Components and Systems(RADECS), pp. 1-7(2022).

[7] Chong C, Liu H, Du S, et al., "Study on the simulation of biosensors based on stacked source trench gate TFET", Nanomaterials, 13, p. 531(2023).

Structure design and characteristics of 3T sense-switch pFlash for computing-in-memory

Wei Zhao[1,2,3†*], Jinghe Wei [1,2,3†], Guozhu Liu[1,2,3*], Yidan Wei[1,2], Yingqiang Wei[1,2], Zhiyuan Sui[12], Meijie Liu[1,2], Qi Xi[1,2], Xinhe Wang[4], Zongguang Yu[1], and Juyan Xu[1]

[1]The 58th Research Institute of China Electronics Technology Group Corporation, Wuxi, China,
[2]National Key Laboratory of Integrated Circuits and Microsystems, Wuxi, China,
[3]Key Laboratory of Aerospace Integrated Circuits and Microsystem, Ministry of Industry and Information Technology, Nanjing, China,
[4]School of Integrated Circuit Science and Engineering, Beihang University, Beijing, China.
*Email: zhaow09@sina.cn; gzliucetc@163.com. †These authors contributed equally.

Abstract

In this paper, we propose a novel structure for a 3T sense-switch pFlash (SSpF) that is specially designed for computing-in-memory. By leveraging the effect of programming pulse width modulation, the 3T-SSpF can be adjusted to more than a hundred conductance states. Furthermore, the conductance level and conductance variation of a 256k 3T-SSpF array are evaluated, and a typical neural network is implemented and applied to the MNIST and CIFAR-10 datasets. The network shows reasonable performance, with a loss of accuracy less than 1%; this approach will therefore enable the realization of more accurate NVM-based neural networks for edge inference.

Keywords: NVM, Flash, Computing-in-memory

Introduction

There has been an exponential growth in demand for huge computing power as machine learning algorithms have undergone rapid development. However, as CMOS technology node has been scaled down to 5 nm/3 nm, Moore's law is now faced with several bottlenecks, such as high technical thresholds. In addition, the "memory wall" and "power wall" restrict the power efficiency of the von Neumann architecture. Ways of improving computing power through innovation in architecture have therefore attracted increasing amounts of attention in recent years.

Various types of NVM devices have been investigated for computing-in-memory, based on two technical routes. The first is known as nonvolatile emerging memory (NEM), and includes resistive random-access memory (RRAM), phase change RAM (PCRAM) etc. The second route involves conventional forms of NVM, such as Flash. Most of the NEM-based computing-in-memory accelerators are still at the laboratory stage, with immature processes and limited storage capacities. In contrast, conventional NVM-based accelerators adopt standard NVM cells provided by foundries, and have not fully exploited the potential benefit of computing-in-memory architecture.

In this paper, we propose a novel structure for a 3T sense-switch pFlash (SSpF) that is specially designed for computing-in-memory. The 3T-SSpF consists of a select MOS transistor and a sense-switch Flash transistor. Compared to traditional 2T Flash, 3T-SSpF exhibits a larger conductance window and higher uniformity in terms of weight storage. By leveraging the effect of program pulse width modulation, 3T-SSpF can be adjusted to more than 128 conductance states. A pair of 3T-SSpF cells is designed to store weights covering the range −128 to 127, equivalent to an 8-bit weight. A 256k 3T-SSpF array is fabricated with a 180 nm embedded Flash technology node. We also evaluate the performance of a CNN in which the 3T-SSpF array is adopted, on the MNIST and CIFAR-10 datasets.

Structure Design and Fabrication of 3T-sspF

A 2D schematic diagram of the structure of 3T-SSpF is shown in **Fig. 1(a)**. The 3T-SSpF cell consists of one pMOS transistor (T1) and two p-channel (P-ch) floating gate (FG) transistors in the same p-well, referred to here as a sense transistor (T2) and a switch transistor (T3), respectively. T1 and T2 are in series, while T2 and T3 share the same FG and the same control gate (CG), and are mainly used for program/erase and weight storage, respectively. **Figs. 1(b) and 1(c)** show the 3D schematic and the scanning electron microscope (SEM) image of the 3T-SSpF cell, respectively.

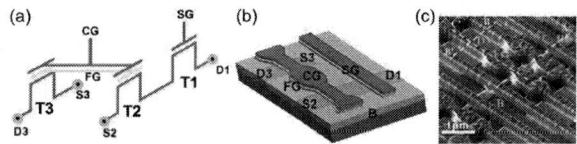

Fig. 1. (a) 2D schematic structure, (b) 3D schematic structure and (c) Scanning electron microscope (SEM) image of the 3T Sense-Switch pFlash (3T-SSpF) cell.

The weight stored in T3 can be adjusted by injecting/removing electrons into/from the FG, through band-to-band tunneling-induced hot electron (BBHE) program and Fowler–Nordheim (FN) tunneling erase operations, respectively. A schematic diagram of the program/erase mechanism is shown in **Fig. 2**. For erase operation, the bias voltages applied to the n-well, the source side of T2, the drain side of T1, and the select gate (SG) of T1 are positive; the bias voltage applied to the CG of T1 is

979-8-3315-0417-5/25 $31.00 © 2025 IEEE

negative. For program operation, the bias voltages applied to the drain side of T1, the SG of T1, and the source side of T2 are negative; the bias voltage applied to the CG of T1 is positive. The shared charge in the FG can effectively adjust the channel ability of T3. Thus, the program operation of T1 directly modulates the weight storage of T3 [3]. **Fig. 3** shows a transmission electron microscope (TEM) cross-sectional image of the fabricated 3T-SSpF cell.

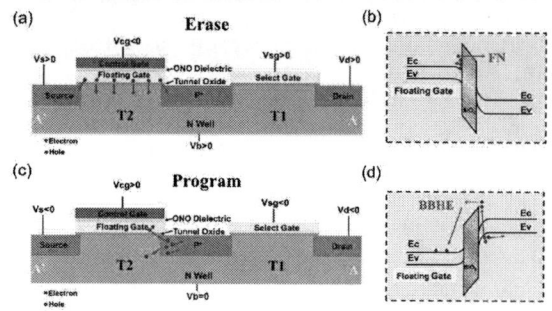

Fig. 2. (a) Cross-sectional view of the AA' side and (b) Band diagram of the 3T-SSpF for erase operation. (c) Cross-sectional view of the AA' side and (d) Band diagram of the 3T-SSpF for program operation.

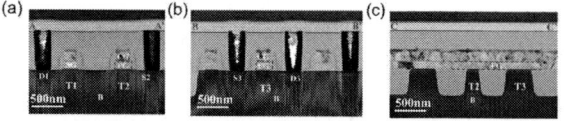

Fig. 3. Transmission electron microscope (TEM) image of (a) AA' and (b) BB' and (c) CC' cutline cross-sectional image of the 3T-SSpF cell.

Electrical Measurement Results

The transfer characteristic curves for 3T-SSpF in 151 intermediate conductance states are plotted in **Fig. 4(a)**. **Fig. 4(b)** illustrates the pulse width modulation effect of 3T-SSpF. After 650 program pulses of different widths (20, 22.5 and 25 ns), the 3T-SSpF indicates different conductance states. The large conductance window and high precision programming characteristics of 3T-SSpF enable more accurate computing-in-memory to be achieved. **Fig. 4(c)** shows the cumulative probability distribution for 22 3T-SSpF cells with respect to 64 independent conductance states, and it can be seen that all the curves are separated, with no overlap. The coefficient of variation for the 3T-SSpF cell is less than 0.5% as shown in **Fig. 4(d)**, thus guaranteeing 8-bit weight storage.

Fig. 5(a) shows the retention test of 3T-SSpF cell under different read voltages. The read current is stable over the test duration of 20,000 s, indicating excellent weight retention characteristic. **Fig. 5(b)** shows the evaluation result of 3T-SSpF cell under BL and WL crosstalk. The stable read current of the unselected 3T-SSpF cell illustrates the advantages of high reliability characteristic of 3T-SSpF array.

A 3T-SSpF array consisting of 128×2048 pairs of 3T-SSpF cells and an analog-to-digital converter (ADC) read-out array was fabricated with a 180 nm embedded Flash technology node. As depicted in **Fig. 6**, each 3T-SSpF cell in the array

was connected to seven lines: a select gate word line (SGWL), a control gate word line (CGWL), a select bit line (sBL), a read bit line (rBL), a select source line (sSL), a read source line (rSL), and a bulk terminal to the p-substrate. In program mode of multi-level T3, multi-level storage within T3 can be implemented. In other words, the threshold voltage of T3 can be programmed at multiple levels. A 256k 3T-SSpF array can store 256k weights for MAC operations, where each pair of adjacent 3T-SSpF cells represent a signed weight. The 3T-SSpF cells connected to rSL+ store the positive parts of the weights, whereas the 3T-SSpF cells connected to rSL− store the corresponding negative parts of the weights.

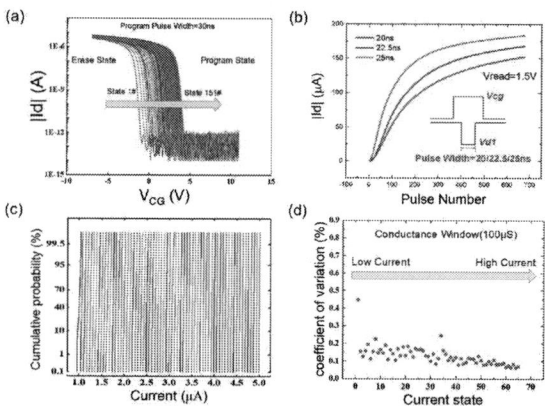

Fig. 4. (a) The transfer characteristic curves of the 3T-SSpF cells in the array under identical pulses during program process. (b) Conductance behavior of the 3T-SSpF cells in the array under different pulse widths. (c) Cumulative probability distribution and (d) Coefficient of variation of 22 3T-SSpF cells with respect to 64 independent conductance states.

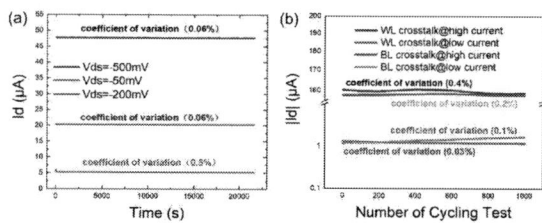

Fig. 5. (a) Retention test of 3T-SSpF cell under different read voltages. (b) Evaluation of 3T-SSpF cell under BL/WL crosstalk, respectively.

In computing mode, multiple 8-bit input data are converted to multiple analog voltages applied to the rBLs. In this way, multiple activated 3T-SSpFs cells can generate a sum current in each rSL+ and rSL− (**Fig. 6**). To obtain the final quantized digital MAC value, the two adjacent rSLs (i.e., rSL+ and rSL−) are sent to a differential current-type analog shift amplifier (AS-Amp) and an 8-bit ADC. Hence, the current differences between the positive partial sum current and the negative partial sum current are quantized to an 8-bit MAC value.

Fig. 7(a) shows the program behavior of the 3T-SSpF array. and illustrates the pulse width modulation effect of each 3T-SSpF cell in the array. The program speed of 3T-SSpF depends on the program clock frequency, corresponding to the program pulse width. Under a program pulse width of 10 ns, 2000 conductance states of 3T-SSpF are obtained as shown in **Fig. 7(b)**. **Fig. 8** illustrates the program behavior of the 3T-SSpF cells in 9 different regions of the array. All of the output values reach 255, the upper limit of 8-bit ADC, indicating good consistency of the 3T-SSpF array.

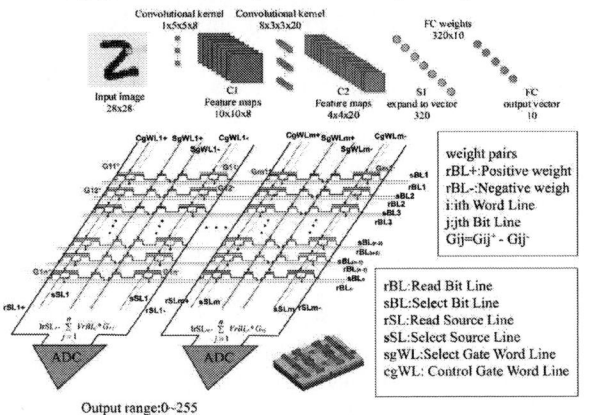

Fig. 6. Schematic structure of the 3T-SSpF array for computing-in-memory.

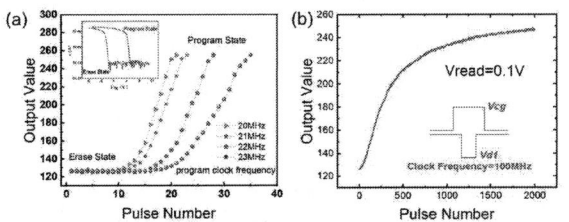

Fig. 7. (a) Program behavior of the 3T-SSpF cell storing positive weight in the array under different pulse widths. (b) Program behavior of the 3T-SSpF cell storing positive weight in the array under 2000 identical pulses.

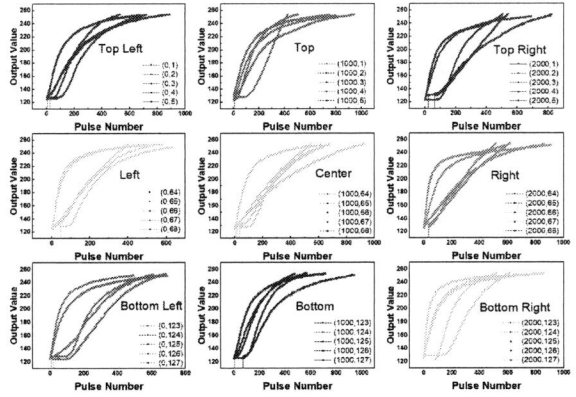

Fig. 8. Program behavior of the 3T-SSpF cells storing positive weight in 9 different regions of the array under identical pulses.

On-chip Implementations of AI Task

Fig. 9(a) shows the evaluation result of CNN in which a 256k 3T-SSpF array is adopted, when applied to the MNIST dataset. To guarantee the accuracy, a weight precision of 4-bit is the minimum requirement. We also evaluate the

influence of weight variation on the accuracy of CNN when applied to the CIFAR-10 dataset, as shown in **Fig. 9(b)**. Under the measured device conditions (for a coefficient of variation of 3T-SSpF cell less than 0.5%), the network maintains reasonable performance, with a loss of accuracy less than 1%.

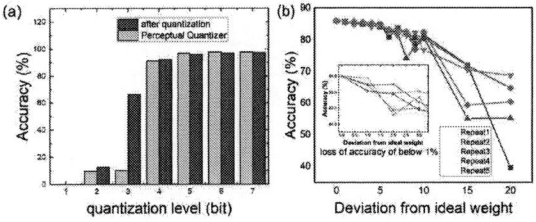

Fig. 9. Evaluation of CNN adopting 256kb 3T-SSpF array for (a) the MNIST and (b) CIFAR-10 datasets.

Conclusion

In this paper, we have proposed a novel structure of 3T-SSpF that is specially designed for computing-in-memory. By utilizing individual sense and switch transistors for program operation and weight storage, respectively, the 3T-SSpF has higher flexibility in terms of conductance state adjustment, and has an 8-bit weight storage capability with a very low coefficient of variation less than 0.5%. The excellent weight retention characteristic of 3T-SSpF guarantee long life span for edge AI tasks which usually implement fixed neural networks, thus requiring few program/erase operations. This paper has illustrated the tremendous potential of the novel 3T-SSpF in regard to the development of more accurate NVM-based neural networks for edge inference applications such as industrial control and medical diagnosis.

Table 1. Comparison with state-of-the-art NVM cell for computing-in-memory.

	Nature Electronics 2022[4]	IEDM 2023[5]	This work
Technology	180nm	500nm	180nm
Memory cell	Y-Flash	2T-Flash	3T-SSpF
Conductance states	1500	16	2000
Conductance window	10^3	10^5	10^6

Acknowledgments

This work was supported by the NSFC grants 62174150, 62204233, Key Technologies R&D Program of Jiangsu (Prospective and Key Technologies for Industry BE2023005), and Fundamental Research Funds for the Central Universities under Grant NJ2023020.

References

[1] Q. Liu, et al., *ISSCC*, 2020, pp 500-502.
[2] P. Yao, et al., *Nature*, 2020, pp 641-646.
[3] G. Liu, et al., *Microelectronics Reliability*, 2019, 13514.
[4] W. Wang, et al., *Nature Electronics*, 2022, pp 870-880.
[5] J. Kim, et al., *IEDM*, 2023, pp 1-4.

High-performance Cu_2O/Ga_2O_3 Heterojunction Diodes for Power Electronics

Xiaohui Wang[1,2], Mujun Li[1], Minghao He[1,3], Chun-Zhang Chen[2], Haozhe Yu[1], Long Chen[1], Qing Wang[1,4], Hongyu Yu[1,4]

[1]School of Microelectronics, Southern University of Science and Technology, Shenzhen 518055, China

[2]Peng Cheng Laboratory, Shenzhen 518000, P. R. China

[3]Department of Electrical and Computer Engineering, National University of Singapore, Singapore

[4]Engineering Research Center of Integrated Circuits for Next-Generation Communications, Ministry of Education, Southern University of Science and Technology, Shenzhen 518055, China

a) E-mail: wangq7@sustech.edu.cn ; b) E-mail: yuhy@sustech.edu.cn

Abstract

In this work, high-performance Cu_2O/Ga_2O_3 heterojunction diodes (HJDs) are demonstrated via the construction of a double-layer of Cu_2O with graded hole concentrations. Taking advantages of the single-layer HJDs featuring p^+ Cu_2O layer and p^- Cu_2O layer, respectively, the double-layer HJD achieves a high breakdown voltage of 2430 V, further yielding a power figure-of-merit of 0.91 GW/cm². This offers a promising pathway for advancing the application of Ga_2O_3-based HJDs in future power electronics.

Keywords: Ga_2O_3, Heterojunction diodes, Power electronics

Introduction

Gallium oxide (Ga_2O_3) has emerged as a promising candidate for applications in high-voltage power electronics, owing to its ultra-wide bandgap of 4.5-4.8 eV, high Baliga's figure-of-merit, and high critical breakdown field (8 MV/cm) superior to Si, SiC, and GaN materials [1]. Additionally, the availability of melt-grown single crystal Ga_2O_3 substrates enhances their potential for cost-effective and high-quality homoepitaxy [2]. However, the lack of p-type doping capability still displays a critical limitation for Ga_2O_3 bipolar power electronics. To address this challenge, the integrating n-type Ga_2O_3 with p-type semiconductors presents an effective strategy for fabricating p-n heterojunction diodes (HJDs), thereby paving the way for high-voltage power electronics applications.

Cuprous oxide (Cu_2O), as a p-type oxide semiconductor, is a promising candidate with a bandgap of 2.1 eV and high mobility. The first kilovolt-class p-Cu_2O/n-Ga_2O_3 HJD was previously reported by sputtering a Cu_2O thin layer onto Ga_2O_3 [3]. Ampere-class Cu_2O/Ga_2O_3 trench heterojunction barrier Schottky diode exhibited a maximum current of 3.5 A, and a breakdown voltage (BV) of 986 V, delivering high-performance Cu_2O/Ga_2O_3 power devices with great potential for the practical applications [4]. Cu_2O/Ga_2O_3 HJDs with a BV of 1015 V and a specific on-resistance ($R_{on,sp}$) of 8.32 mΩ.cm² has been reported, and its electrical properties have been thoroughly investigated [5]. These findings indicate that the quality of p-type layers significantly impacts the electrical properties of HJDs. Therefore, a careful design of the Cu_2O layers is critical for enhancing the performance of HJDs.

This study presents Cu_2O/Ga_2O_3 HJDs with a double-layer (DL) design of Cu_2O. By adjusting the hole concentration in Cu_2O layers, the peak electric field occurs in the bulk of the DL HJD, resulting in a BV of up to 2430 V. It also exhibits a $R_{on,sp}$ of 6.52 mΩ·cm², yielding a high power figure-of-merit ($PFOM=BV^2/R_{on,sp}$) compared to reported Cu_2O/Ga_2O_3 HJDs. These results indicate that Cu_2O/Ga_2O_3 HJDs has great potential for power devices.

Device Fabrication

The Ga_2O_3 epitaxial wafer, purchased from Novel Crystal Technology, Inc., Japan, consists of a 10-μm Si-doped n-type drift layer and a 650-μm Sn-doped (001) substrate. Figure 1 illustrates Cu_2O/Ga_2O_3 HJDs and reference Schottky barrier diode (SBD) fabricated on the same wafer. The fabrication flow started from organic and acid cleaning. Mesa isolation was achieved by inductively coupled plasma (ICP) etching, using BCl_3 as the etchant, followed by a 10-minute treatment with piranha solution ($H_2SO_4:H_2O_2 = 4:1$) to remove surface impurities. Ti/Au (20/100 nm) metal stack was deposited on the backside of the wafer by e-beam evaporation and annealed at 510°C for 1 min in N_2 ambient to form the backside ohmic contacts. Subsequently, a double-layer of Cu_2O was sputtered on the Ga_2O_3 drift layer by the radio frequency magnetron sputtering technique and lift-off process at room temperature, with high-purity (99.99%) Cu_2O as the target. The flux ratios of Ar/O_2 were 50:1 and 50:3 for p^- Cu_2O layer and p^+ Cu_2O layer, respectively. Finally, a Ni/Au metal stack (50/50 nm) was deposited onto the Cu_2O layer. The device features a p^+ Cu_2O layer and p^- Cu_2O layer, with their hole concentrations and mobility of 3.9×10^{19} cm⁻², 0.12 cm²/V·s and 3.7×10^{17} cm⁻², 0.94 cm²/V·s, respectively, as mentioned in previous work. Figure 2 shows the top-view of optical microscope image of the fabricated DL HJD. Additionally, all the Ga_2O_3 diodes were fabricated with the same anode metal (radius=50 μm) to guarantee the same forward current.

Results and Discussion

Figure 3 shows a STEM image of DL HJD, revealing that the Cu_2O layers are in a polycrystalline state. Figure 4 presents the comparison of the three-dimensional surface morphology measured by atomic force microscopy (AFM). Figure 4(a) shows the untreated Ga_2O_3 wafer, exhibiting a root mean square (RMS) surface roughness of 0.12 nm. The RMS in p^+ and p^- Cu_2O layers are 0.33 nm and 0.30 nm, respectively. These results indicate that adjusting O_2 flux effectively tunes the hole concentration in the Cu_2O layer without significantly affecting its surface morphology.

Figure 5(a) shows the linear-scale J–V characteristics and the extracted differential $R_{on,sp}$ of Ga_2O_3 SBD, p^+ HJD, p^- HJD, and DL HJD. The $R_{on,sp}$ of the SBD is 5.83 $m\Omega\cdot cm^2$. The $R_{on,sp}$ value of the p^+ HJD and p^- HJD are 6.43 $m\Omega\cdot cm^2$ and 6.58 $m\Omega\cdot cm^2$, respectively. The increase in $R_{on,sp}$ for p^- HJD is primarily attributed to the reduced conductivity of the Cu_2O layer, which is caused by its low hole concentration. The $R_{on,sp}$ value for DL HJD is 6.52 $m\Omega\cdot cm^2$, which is between p^+ and p^- HJDs. Figure 5(b) shows the semi-log scale J–V characteristics, showing that all Ga_2O_3 devices achieve high rectification ratios over 10^8. The turn-on voltage (V_{ON}) is determined to be 0.8 V for SBD, 1.7 V for p^+ HJD, 1.5 V for p^- HJD, and 1.8 V for DL HJD. Figure 5(c) depicts the reverse J–V characteristics. The BV of SBD is only of 420 V, while the p^+ HJD and p^- HJD exhibits higher BVs of 1910 V and 1130 V, respectively. This enhancement is attributed to the increased depletion depth in the PN junction. Notably, the DL HJD achieves a significantly higher BV of 2430 V. The improved performance is ascribed to the p^- Cu_2O layer, which smooths the electric field peak within the device, while the p^+ Cu_2O layer contributes to a reduction in $R_{on,sp}$ by lowering the Ni/Cu_2O contact resistance. The BV statistics for the four samples are plotted in Figure 5(d), demonstrating that the devices have stable repeatability.

TCAD simulation was performed to investigate the impact of single-layer and double-layer Cu_2O with varying hole concentrations on the electric field distribution in reverse-biased HJDs. Figure 6 compares the electric field distribution for all devices. The profiles along the cutline at Ga_2O_3 surface are extracted and shown in Figure 7. The peak electric field of the SBD is located at the anode edge with a value of 6.9 MV/cm. For the p^+ HJD, the peak electric field occurs at the p^+ Cu_2O edge and decreases to 3.8 MV/cm, while the p^- HJD exhibits a peak electric field at the anode edge, reduced to 1.7 MV/cm. The DL HJD shows a decrease in peak electric field to 2.2 MV/cm at the foot edge of the p^+ Cu_2O within the device bulk, which is consistent with the previously discussed BV results [6].

Figure 8 depicts the benchmark plot of $R_{on,sp}$ versus BV for state-of-the-art Ga_2O_3 diodes. Table 1 summarizes the $R_{on,sp}$ and BV value for the four samples analyzed in this work, and it also calculates their PFOM. The fabricated DL HJD achieves a PFOM of 0.91 GW/cm^2, indicating a relatively high value for Cu_2O/Ga_2O_3 HJDs. Furthermore, the performance of the DL HJD is comparable to that of extensively researched NiO/Ga_2O_3 HJDs.

Conclusion

This study presents an optimization strategy for the Cu_2O/Ga_2O_3 HJDs by adjusting the hole concentration in Cu_2O layers. The fabricated DL HJD exhibits a remarkable BV of 2430 V, a $R_{on,sp}$ of 6.52 $m\Omega\cdot cm^2$, and a V_{ON} of 1.8 V. Furthermore, it yields a high PFOM of 0.91 GW/cm^2. This work provides a practical and effective approach to enhancing the performance of Cu_2O/Ga_2O_3 HJDs.

Acknowledgments

This work was supported by National Natural Science Foundation of China (Grant No: 62274082). Research on mechanism of source/drain ohmic contact and the related GaN p-FET (Grant No: 2023A1515030034). Research on high-reliable GaN power device and the related industrial power system (Grant No: KZQB-KCZYZ-2021052). Study on the reliability of GaN power devices (Grant No: JCYJ20220818100605012). Research on novelty low-resistance Source/Drain ohmic contact for GaN p-FET (Grant No: JCYJ20220530115411025). Research on the key technology of 1200 V SiC MOSFETs (Grant No: JSGG20220831094404008). "5G Frontier" project Micro-Nano processing platform (Grant No: K2023390010). High level of special funds (G03034K004). The authors also acknowledge the assistance of SUSTech Core Research Facilities.

References

[1] S. H. Sun, C. L. Wang, S. Alghamdi, "Recent advanced ultra-wide bandgap β-Ga_2O_3 material and device technologies," *Adv. Electron. Mater.*, 2024 May 2024.

[2] E. G. Víllora, K. Shimamura, Y. Yoshikawa, "Large-size β-Ga_2O_3 single crystals and wafers," *J. Cryst. Growth,* vol. 270, no. 3-4, pp. 420-426, Oct 2004.

[3] T. Watahiki, Y. Yuda, A. Furukawa, "Heterojunction p-Cu_2O/n-Ga_2O_3 diode with high breakdown voltage," *Appl. Phys. Lett.,* vol. 111, no. 22, Nov 2017, Art no. 222104.

[4] H. M. A. Takatsuka, K. Sasaki and A. Kuramata, "Fabrication of ampere-class p-Cu_2O-n-β-Ga_2O_3 trench heterojunction barrier Schottky diodes and double-pulse evaluation," *35th Int. Symp. Power Semiconductor Devices and ICs (ISPSD),* vol. Hong Kong, pp. 342-345, June 2023.

[5] Y. Jia, S. Sato, A. Traoré, "Electrical properties of vertical Cu_2O/β-Ga_2O_3 (001) p-n diodes," *AIP Advances,* vol. 13, no. 10, Oct 2023, Art no. 105306.

[6] C. Liao, X. Lu, T. L. Xu, "Optimization of NiO/β-Ga_2O_3 heterojunction diodes for high-power application," *IEEE Trans. on Electron Devices,* vol. 69, no. 10, pp. 5722-5727, Oct 2022.

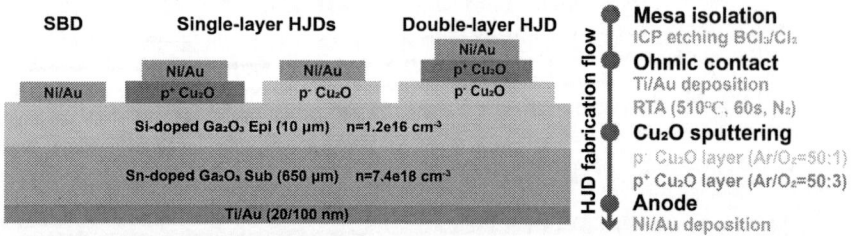
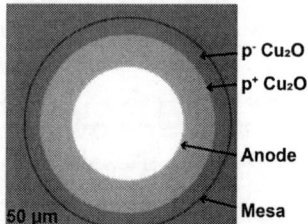

Fig. 1. Cross-section schematic of Cu$_2$O/Ga$_2$O$_3$ HJDs with various devices structures and their fabrication flow.

Fig. 2. Top-view of optical microscope image of the fabricated DL HJD.

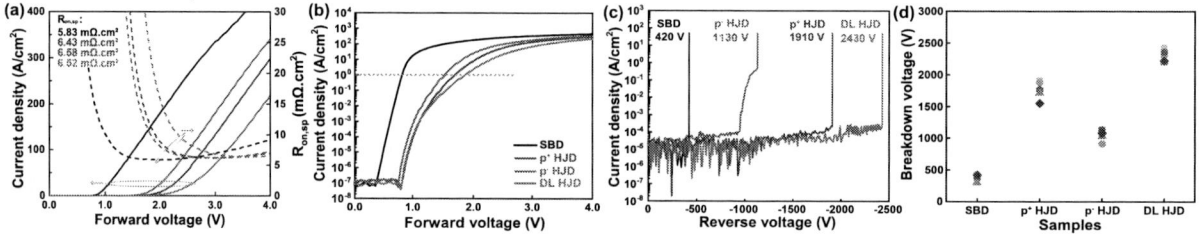

Fig. 3. STEM image of the fabricated DL HJD.

Fig. 4. AFM images of (a) untreated Ga$_2$O$_3$ wafer, (b) p$^+$ Cu$_2$O (50 sccm Ar + 3 sccm O$_2$), and (c) p$^-$ Cu$_2$O (50 sccm Ar + 1 sccm O$_2$).

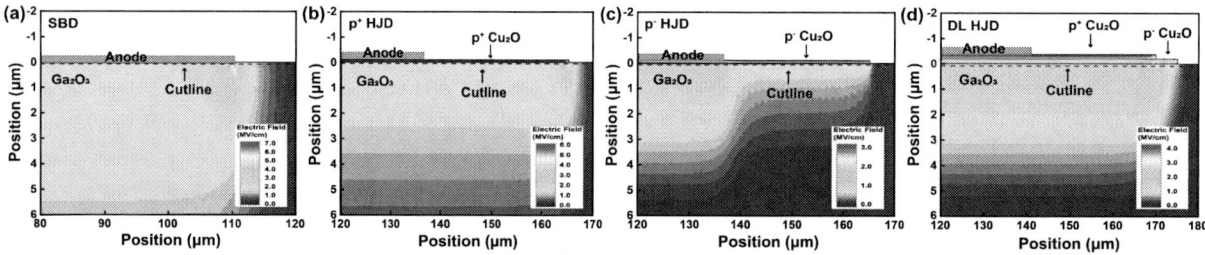

Fig. 5. (a) Linear plots of J–V characteristics and the extracted R$_{on,sp}$ vs forward voltage. (b) Semi-logarithmic plots of the J–V characteristics. (c) Reverse J–V characteristics of all devices. (d) Statistics of the BV for all devices across the five samples.

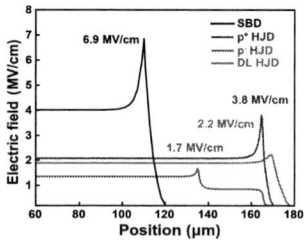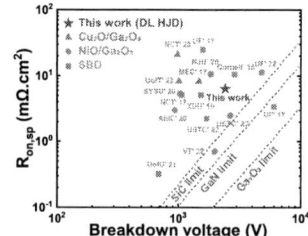

Fig. 6. Electric field distribution for (a) SBD, (b) p$^+$ HJD, (c) p$^-$ HJD, and (d) DL HJD.

Table 1: Summary of R$_{on,sp}$ and BV for the Ga$_2$O$_3$ diodes.

Samples	R$_{on,sp}$ (mΩ.cm^2)	BV (V)	PFOM (GW/cm^2)
SBD	5.83	420	0.03
p$^+$ HJD	6.43	1910	0.57
p$^-$ HJD	6.58	1130	0.19
DL HJD	6.52	2430	0.91

Fig. 7. Extracted electric field profiles along the cutline at the Ga$_2$O$_3$ surface.

Fig. 8. Benchmark plot for reported state-of-the-art Ga$_2$O$_3$ diodes.

979-8-3315-0417-5/25 $31.00 © 2025 IEEE   84

Comprehensive Analysis of Oxidant Effects during ALD Process of $Hf_{0.5}Zr_{0.5}O_2$ Ferroelectric Thin Films

Xiaopeng Li[1], Yilin Hou[1], Xiaoyu Dou[1], Yaoyu He[1], Yuwei Qu[1], Pengpeng Sang[1], Xuepeng Zhan[1], Xiaolei Wang[2], Jixuan Wu[1, *], and Jiezhi Chen[1, *]

[1]School of Information Science and Engineering, Shandong University, China;
[2]Institute of Microelectronics of Chinese Academy of Sciences, China.

*E-mail: jixuanwu@sdu.edu.cn; chen.jiezhi@sdu.edu.cn

ABSTRACT

To guide the process optimization of HfO_2-based ferroelectric (FE) devices, the effects of oxidants (O_3/H_2O) during atomic layer deposition (ALD) process of $Hf_{0.5}Zr_{0.5}O_2$ films are systematically investigated. The results indicate that using the stronger oxidant (O_3) leads to higher remanent polarization (Pr) but poorer endurance, with significant leakage deterioration after wakeup and a pronounced imprint effect during cycling. In contrast, the weaker oxidant (H_2O) results in lower Pr but improves polarization switching speed. The influence of oxidants on pre-existing defects and interfacial reactions may be potential key factors. Our findings provide valuable insights for process optimization tailored to specific application requirements.

Keywords: *ferroelectricity, $Hf_{0.5}Zr_{0.5}O_2$, ALD, and oxidant*

I. INTRODUCTION

HfO_2-based ferroelectric (FE) thin films have shown great promise for embedded non-volatile memory applications, benefitting from excellent scalability, high response speed, low power consumption, as well as compatibility with CMOS processing [1]. In recent years, significant advancements in device performance have been achieved through process optimizations, such as dopant type/concentration, capping effects, and deposition/rapid thermal annealing conditions [2-3]. However, research on atomic layer deposition (ALD), a critical technology for thin film fabrication, has primarily focused on the effects of deposition temperature, with oxidants remaining relatively understudied. Current reports often favor the use of strong oxidants to enhance FE performance, overlooking the trade-offs between different characteristics [4-5]. Additionally, specific application scenarios have distinct requirements, for instance, neuron computing emphasizes response speed, while multi-level storage prioritizes storage window [6]. Therefore, it is essential to systematically investigate the effects of oxidants on various properties to identify the most suitable ALD oxidant for specific application contexts.

In this work, the effects of two common oxidants (O_3 and H_2O) on the performance of $Hf_{0.5}Zr_{0.5}O_2$ (HZO) thin films have been systematically analyzed. The results show that the stronger oxidant (O_3) enhances remanent polarization (Pr) and long-term reliability, but exhibits poorer endurance, with rapid deterioration in imprint effect and leakage current

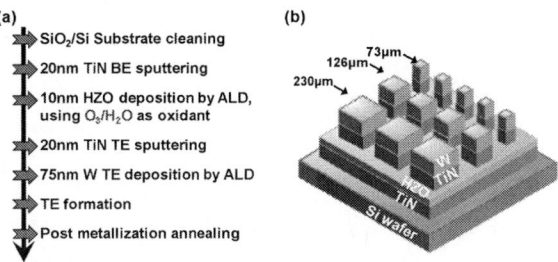

Fig. 1. *(a) Key process to fabricate the MFM HZO capacitor. (b) Schematic illustration of the HZO capacitor.*

during cycling. In contrast, the weaker oxidant (H_2O) improves anti-imprint performance and polarization switching speed. Potential physical mechanisms are discussed, providing valuable insights for the understanding and design of HfO_2-based devices.

II. DEVICE FABRICATION AND MEASUREMENTS

Fig. 1 shows the process steps and diagram for fabricating the HZO capacitor. First, a 20nm TiN is sputtered by physical vapor deposition (PVD) as bottom electrode (BE). Next, a 10nm HZO thin film with a Hf:Zr ratio of 1:1 is deposited by ALD at 300°C, using either O_3 or H_2O as the oxidant, with a pulse time of 1s. Then, a 20nm TiN is sputtered via PVD, followed by the deposition of a 75nm W via ALD as top electrode (TE). After patterning TE, the capacitor is annealed for 60s at 550°C in nitrogen atmosphere to crystallize HZO film. Additionally, all measurements are conducted on capacitors with voltage applied from TE, and Pr values are extracted using the positive-up-negative-down (PUND) technique to minimize leakage and displacement interference.

III. RESULTS AND DISCUSSION

A. *Ferroelectric Properties Analysis*

Fig. 2(a) shows P-V and I-V hysteresis loops of O_3- and H_2O-based capacitors. Wakeup effect can be observed in both capacitors. The stronger oxidant device shows higher Pr. After wakeup, 2Pr value for O_3-based capacitor reaches ~33.6μC/cm², which is higher than that for H_2O-based capacitor (~26.0μC/cm²). Endurance under rectangular pulse cycling (3V, 100μs) is shown in **Fig. 2(b)**. Compared to the ~10^9 cycles for H_2O-based capacitor, the shorter endurance observed for O_3-based capacitor is in contradiction with previous reports that enhanced cycling endurance using stronger oxidants [7-8].

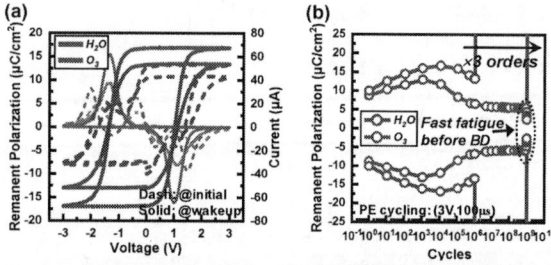

Fig. 2. *(a) P-V and I-V hysteresis loops of O_3- and H_2O-based capacitors at initial and wakeup stages. (b) Cycling endurance characteristics of O_3- and H_2O-based capacitors.*

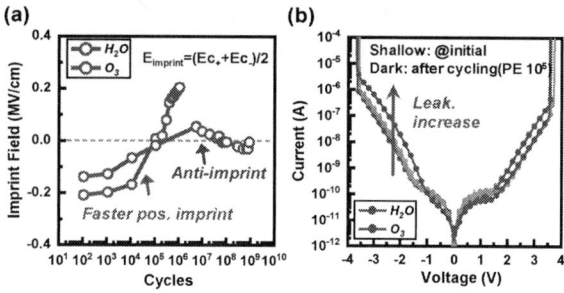

Fig. 3. *(a) Imprint field versus cycles. (b) Leakage current versus voltage at different cycling stages (initial and PE 10^5).*

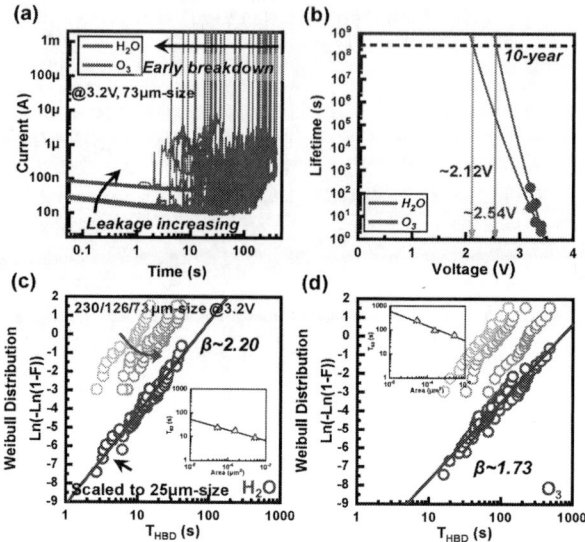

Fig. 4. *(a) I-t traces under CVS from multiple samples at 3.2V and 73μm-size. (b) Operation voltage extrapolation for ten-year lifetime at 63.2% failure. (c)&(d) Poisson area scaling with CVS: (c) H_2O-based capacitor, and (d) O_3-based capacitor, at 3.2V. Inset: T_{63} (time to breakdown at 63.2%) versus electrode area.*

Considering that endurance is related to defects, we further analyze the imprint effect and leakage current to explore the underlying causes. **Fig. 3(a)** presents the variation of imprint field ($E_{imprint}=(E_{C+}+E_{C-})/2$) during cycling. It is observed that the positive imprint field of O_3-based capacitor significantly increases with cycling, whereas this phenomenon is not observed in H_2O-based capacitor. Additionally, time-zero dielectric breakdown (TZDB) measurements indicate that the leakage current under negative voltage for O_3-based capacitor rapidly increases after cycling (PE 10^5 cycles), which is also not observed in H_2O-based capacitor (**Fig. 3(b)**). This may be related to the formation of TiO_x/TiO_xN_y interfacial layer at BE interface during film deposition with O_3 as oxidant, which leads to the formation and accumulation of oxygen vacancies ($V_O^{0/2+}$) at the bottom interface during subsequent cycling [9-10]. On one hand, this creates a positive built-in electric field, resulting in a positive imprint; on the other hand, the release of electrons on the BE side under negative bias exacerbates the leakage current [11-12].

B. *Long-Term Reliability Analysis*

Since time-dependent dielectric breakdown (TDDB) is caused by stress-induced defects, constant voltage stress (CVS) method is employed to reveal the evolution of leakage paths and analyze defect, as shown in **Fig. 4**. **Fig. 4(a)** shows the I-t curves under same voltage and area conditions (3.2V, 73μm-size). It can be observed that O_3-based capacitor exhibits lower leakage current and longer time-to-breakdown (T_{BD}) compared to H_2O-based capacitor, indicating fewer pre-existing defects. **Fig. 4(b)** shows the extrapolated ten-year lifetime prediction at 63.2% failure. The extract ten-year

breakdown voltage for O_3-based capacitor is approximately 2.54V, significantly higher than that of H_2O-based capacitor (2.12V). **Figs. 4(c)&(d)** display the statistics of T_{BD} with Weibull distribution for three electrode areas (230/126/73μm-size). To exclude additional random variables, which may be caused by random variations in thickness across samples, the Poisson area scaling method is applied, resulting in Weibull slope values of 1.73 and 2.20 for O_3- and H_2O-based capacitors, respectively. According to the percolation model, the Weibull slope is proportional to the defect generation rate, indicating that O_3-based capacitor has a slower defect generation rate under CVS [13]. Overall, O_3-based capacitor exhibits excellent long-term reliability, consistent with recent reports on the oxidizer engineering of ALD for ZrO_2-based DRAM [14].

C. *Polarization Switching Characteristics Analysis*

As an important indicator for FE device applications, the polarization switching kinetics of capacitors are investigated, as shown in **Fig. 5**. Since ALD-deposited fluorite-structured HZO ferroelectrics are nanoscale polycrystalline films, the time-dependent polarization switching characteristics is analyzed using nucleation limited switching (NLS) model [15]. It is described as:

$$\frac{\Delta P(t)}{2Pr} = \int \left[1 - exp\left\{-\left(t/\tau\right)^n\right\}\right] \cdot \frac{A}{\pi} \cdot \left[\frac{\omega}{(log\tau - log\tau_1)^2 + \omega^2}\right] \cdot d(log\tau)$$

where A is a normalized constant, $log\tau$ is a Lorentzian distribution of switching time, $log\tau_1$ is the median logarithmic of the distribution, n is 2 for thin film, and ω is the half-width at half-maximum of Lorentzian distribution.

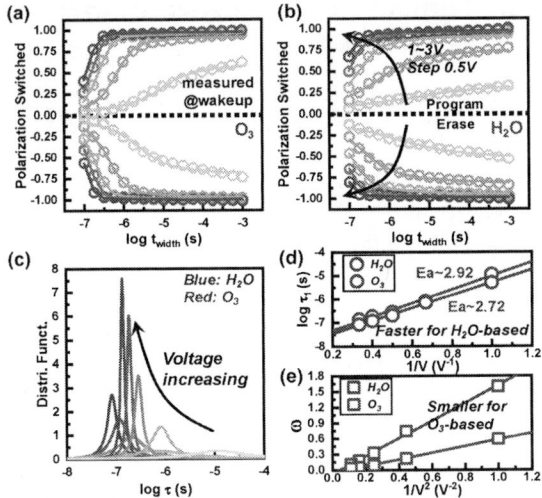

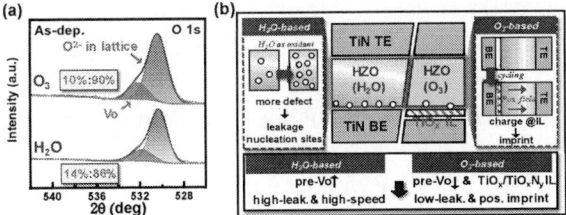

Fig. 5. *(a)&(b) Measured polarization switching kinetics of capacitors (a) O₃-based, and (b) H₂O-based. The symbols are experimental data, and the solid lines are the fitted results based on the NLS model. (c) The Lorentzian distribution extracted from (a)&(b). (d) logτ_1 versus 1/V. (e) ω versus 1/V². Note, (d)&(e) are the results of the program state.*

Figs. 5(a)&(b) show the measured results (symbols) and fitting curves (lines) at wakeup stage for O₃- and H₂O-based capacitors, demonstrating a good fit with NLS model. The extracted Lorentzian distributions at varying voltages are shown in **Fig. 5(c)**. **Figs. 5(d)&(e)** illustrate the switching time (logτ_1) as a function of the inverse of external voltage (1/V) and ω as a function of the inverse of the square of the external voltage (1/V²). At an external voltage of 3V, logτ_1 for O₃- and H₂O-based capacitors are -6.86 and -7.08, respectively, with corresponding ω values of 0.04 and 0.11. Furthermore, the differences between O₃- and H₂O-based capacitors become even more pronounced at lower external voltages. The smaller logτ_1 for H₂O-based capacitor is advantageous for rapid operations in memory applications. However, for O₃-based capacitor, the smaller ω promotes more uniform polarization switching conditions, combined with its higher Pr, which enhances its suitability for multi-level storage applications.

According to polarization switching theory, the time required for polarization switching is determined by the nucleation time. When the external voltage is sufficient to overcome the critical nucleation radius, the nucleation sites (usually oxygen vacancies) nucleate and propagate rapidly, completing the successful polarization switching. As a reasonable hypothesis, the impact of oxidants on polarization switching dynamics may relate to their influence on oxygen vacancies [8,16]. **Fig. 6(a)** shows the O 1s spectra for O₃- and H₂O-based HZO films, where the higher proportion of Vo in H₂O-based HZO strongly supports this hypothesis. This also explains the higher initial leakage and faster defect generation rate observed in H₂O-based capacitors.

Fig. 6. *(a) XPS O 1s spectra of O₃-based and H₂O-based HZO. (b) Schematic diagram of the performance impact of H₂O and O₃ oxidants.*

IV. Conclusions

In summary, the effects of oxidants during the ALD process on HZO capacitors' performance are systematically compared. The stronger oxidant (O₃) enhances Pr and long-term reliability under constant voltage stress but results in serious leakage and imprint during cycling, potentially reducing endurance. Conversely, the weaker oxidant (H₂O) improves switching speed and anti-imprint performance, though with relatively lower Pr. Pre-existing defects and interfacial reactions are potential underlying factors (**Fig. 6(b)**). These findings offer valuable insights for optimizing the process of HfO₂-based ferroelectric devices.

Acknowledgment

This work was supported by China Key Research and Development Program under Grant (2023YFB4402400), National Natural Science Foundation of China (Nos. 62034006, 92264201, U23B2040), Natural Science Foundation of Shandong Province (ZR2023LZH007, ZR2023QF054), China Postdoctoral Science Foundation (GZC20231435), TaiShan Scholars (TSQN202306059), and MIND project (MINDXZ202407).

References

[1] J. Okuno et al., A highly reliable 1.8 V 1 Mb Hf₀.₅Zr₀.₅O₂-based 1T1C FeRAM Array with 3-D Capacitors, IEDM, 2023; [2] M. Zeng et al., First Demonstration of Annealing-Free Top Gate La:HZO-IGZO FeFET with Record Memory Window and Endurance, IEDM, 2023; [3] F. Huang et al., First Observation of Ultra-high Polarization (~ 108 μC/cm²) in Nanometer Scaled High Performance Ferroelectric HZO Capacitors with Mo Electrodes, VLSI, 2023; [4] A. Hsain et al., Role of Oxygen Source on Buried Interfaces in Atomic-Layer-Deposited Ferroelectric Hafnia–Zirconia Thin Films, ACS Appl. Mater. Inter., 14(37), 2022; [5] S. J. Kim et al., Effect of hydrogen derived from oxygen source on low-temperature ferroelectric TiN/Hf₀.₅Zr₀.₅O₂/TiN capacitors, APL, 115(18), 2019; [6] K. Ni et al., A Novel Ferroelectric Superlattice Based Multi-Level Cell Non-Volatile Memory, IEDM, 2019; [7] Y. C. Jung et al., Robust low-temperature (350 °C) ferroelectric Hf₀.₅Zr₀.₅O₂ fabricated using anhydrous H₂O₂ as the ALD oxidant, APL, 121(22), 2022; [8] J. Kim et al., "Toward Low-Thermal-Budget Hafnia-Based Ferroelectrics via Atomic Layer Deposition," ACS Applied Electronic Materials, 5(9), 2023; [9] Y. Jeong et al., Oxygen Vacancy Control as a Strategy to Enhance Imprinting Effect in Hafnia Ferroelectric Devices, IEEE TED, 70(1), 2023; [10] T. Onaya et al., Role of Oxidant Gas for Atomic Layer Deposition of HfxZr₁₋xO₂ Thin Films on Ferroelectricity of Metal-Ferroelectric-Metal Capacitors, ECS Trans. 113(51), 2024; [11] X. Li et al., Imprint-Correlated Retention Loss in Hf₀.₅Zr₀.₅O₂ Ferroelectric Thin Film Through Wide-Temperature Characterizations, IEEE TED, 71(9), 2024; [12] X. Li et al., In-Depth Investigation of Seed Layer Engineering in Ferroelectric Hf₀.₅Zr₀.₅O₂ Film: Wakeup-Free Achievement and Reliability Mechanisms, IEEE TED, 71(2), 2024; [13] E. Y. Wu, Facts and Myths of Dielectric Breakdown Processes—Part I: Statistics, Experimental, and Physical Acceleration Models, IEEE TED, 66(1), 2019; [14] X. Tang et al., Oxidizer Engineering of ALD for Efficient Production of ZrO₂ Capacitors in DRAM, IEEE EDL, 2024; [15] W. Wei et al., In-depth Understanding of Polarization Switching Kinetics in Polycrystalline Hf₀.₅Zr₀.₅O₂ Ferroelectric Thin Film: A Transition From NLS to KAI, IEDM, 2021; [16] A. Kashir, S. Oh, and H. Hwang, Defect Engineering to Achieve Wake-up Free HfO₂Based Ferroelectrics, Adv. Eng. Mater., 23(1), 2021.

Interface Engineering on Inorganic–Organic Hybrid Phototransistors for High-responsive Near-infrared Sensing

Dingwei Li[1], Huihui Ren[1], Yingjie Tang[1], Yitong Chen[1], Yan Wang[1], Qi Huang[2], Bowen Zhu[1,2]

[1] Key Laboratory of 3D Micro/Nano Fabrication and Characterization of Zhejiang Province, School of Engineering, Westlake University, Hangzhou, China,
[2] Westlake Institute for Optoelectronics, Hangzhou, China.

Abstract

Flexible and high-responsive phototransistors are highly demanded for wearable near-infrared (NIR) image sensors. Hybrid phototransistors utilizing the photogating effect, featuring separate sensing and transport layers, show considerable potential. However, optimizing the photogating effect remains a significant challenge. Here, we proposed a facile interface reduction processing to oxide semiconductors, enabling high-performance In_2O_3/Y6 heterojunction phototransistors with enhanced photo-gating effect for NIR sensing.

Keywords: organic semiconductors, metal oxide, hybrid phototransistor

Introduction

Due to its deep penetration and high contrast, near-infrared (NIR) light is highly valuable for emerging applications such as wearable medical monitoring, non-invasive biomedical imaging, and biometric extraction [1-3]. To achieve high-performance NIR sensors, careful design of light-responsive materials and carrier transport structures is essential. This involves ensuring appropriate band alignment, optimizing trap states, and enabling efficient electron-hole pair dissociation [4].

Heterojunction phototransistors featuring hybrid photosensitive and carrier transport channels have been developed to enhance NIR response via the photogating effect [5, 6]. Metal oxide and organic semiconductor heterojunctions have shown significant promise for high-performance NIR sensors due to their favorable energy band alignment, high mobility, and strong light absorbance, as well as their solution processability and flexibility [7, 8]. However, engineering the interface to maximize the photogating effect, particularly by extending the lifetime of one carrier to increase photoconductive gain, remains a challenge [9, 10].

In this study, we employ a type-II heterojunction comprising In_2O_3 and Y6 to enhance NIR sensor performance. Furthermore, we preprocess the In_2O_3 film using $NaBH_4$ to introduce additional trap sites, which extend carrier lifetime and consequently increase the photoconductive gain. Moreover, we have demonstrated the flexible device on the polyimide substrate, demonstrating the potential in wearable electronics applications.

Results and discussion

Fig. 1(a) illustrates the fabrication process and structure of the hybrid phototransistor. The In_2O_3 thin-film transistor (TFT) was initially fabricated with a bottom-gate, top-contact configuration. The p^{++} Si substrate, with a 100 nm thick layer of SiO_2, served as the bottom gate and dielectric. The channel consisted of a solution-processed In_2O_3 film (~8 nm), and Al (50 nm) was deposited as the source/drain (S/D) contacts and the channel width/length is 1000 μm/100 μm. Following this, the In_2O_3 TFT was immersed in a 0.2 M $NaBH_4$ solution for 60 seconds to

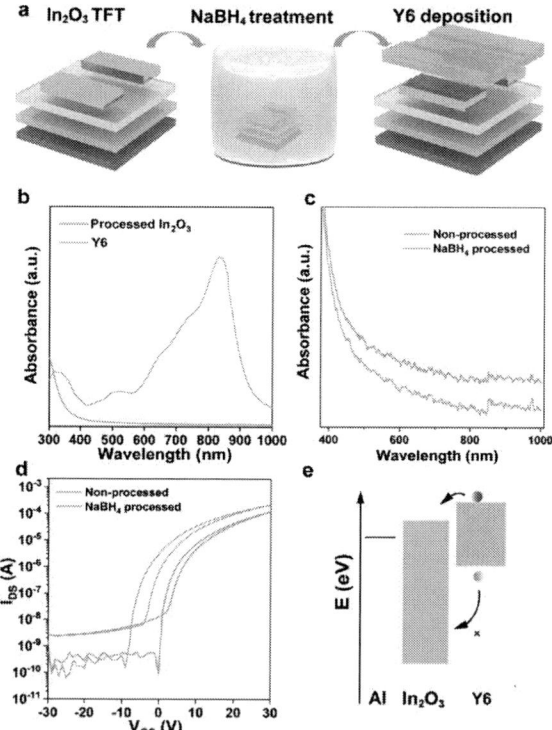

Fig. 1. (a) Fabrication flow of the hybrid phototransistor device. (b) Absorbance spectra of processed In_2O_3 and Y6 films. (c) Enlarged absorbance spectra comparing the processed and non-processed In_2O_3 films in the visible light range. (d) Transfer curves of non-processed and processed hybrid phototransistors. (e) Band alignment schematic of the heterojunction.

that the higher oxygen vacancy concentration contributes to

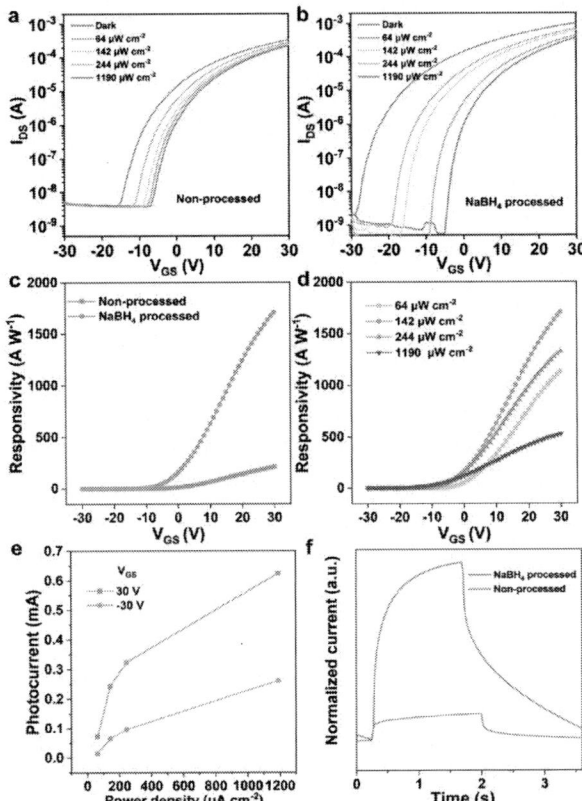

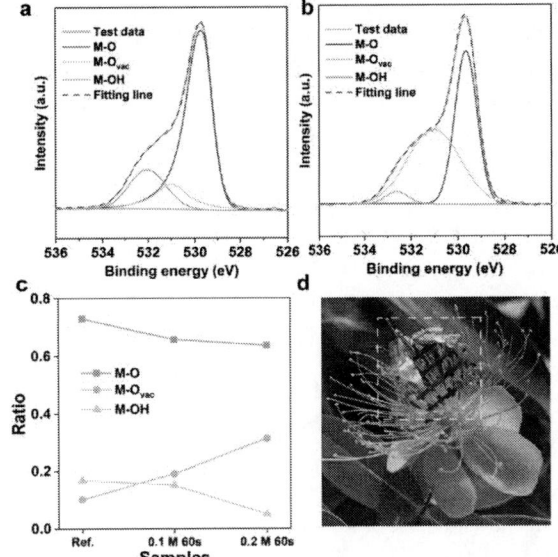

Fig. 3. (a, b) XPS results of non-processed and 0.2 M 60 s NaBH₄ processed oxide films. (c) Extracted XPS results of oxide films with varying NaBH₄ concentrations and dipping times. (d) Flexible device exfoliated from the polyimide substrate.

Fig. 2. (a, b) Transfer curves of non-processed and processed hybrid devices under various NIR power densities. (c) Extracted responsivity of non-processed and processed devices at different gate voltages. (d) Gate-voltage-dependent responsivity at different power densities. (e) The photocurrent at different gate voltage and power densities. (f) Dynamic photoresponse of non-processed and processed devices.

facilitate the reduction process, creating trap sites on the surface of the In_2O_3. Finally, the 8 mg mL⁻¹ solution of Y6 was deposited onto the In_2O_3 to form the hybrid phototransistor. Fig. 1(b) displays the UV-Vis spectra of the In_2O_3 and Y6 films. The In_2O_3 film exhibits high transparency in the visible and infrared regions, while Y6 shows an absorbance peak at 800 nm. Fig. 1(c) shows the visible region absorbance of the processed and non-processed devices. The processed hybrid phototransistors exhibit overall reduced absorbance, which can be attributed to the increased concentration of oxygen vacancies. These vacancies can be ionized, releasing free electrons that occupy sub-gap states.

Fig. 1(d) presents the transfer curves, where the processed device demonstrates a widened hysteresis window and a negatively shifted threshold voltage, indicating an increase in trapping sites and enhanced conductivity. This observation aligns with the absorbance spectra, suggesting

increased conductivity. Additionally, Fig. 1(e) illustrates the band alignment of the heterojunction, showing that the LUMO of Y6 is closely aligned with the conduction band of In_2O_3, allowing efficient electron transfer. However, the large energy gap between the valence band and the HOMO creates a significant barrier for hole transfer, causing photo-induced holes to be more likely blocked at the interface and captured by trap sites as the photogate.

Fig. 2(a, b) present the photoresponse of the reference and processed hybrid devices under varying power densities of 800 nm NIR irradiation. The processed device exhibits a significantly enhanced photoresponse compared to the reference device. Fig. 2(c) illustrates the responsivity (R) of both devices under ~142 μW/cm² NIR irradiation, showing gate-voltage-dependent responsivity due to the photogating effect. The NaBH₄-processed device achieves a maximum responsivity of approximately 1700 A W⁻¹, which is markedly higher than the ~209 A W⁻¹ observed in the non-processed device. The gate-dependent responsivity at various power densities is shown in Fig. 2(d). As the power density increases, the responsivity initially rises and then decreases at 30 V gate voltage, with the highest value observed at approximately 140 μW cm⁻² under NIR illumination. The subsequent decrease in responsivity at higher power densities can be attributed to the gradual saturation of trapping sites. Fig. 2(e) illustrates the photocurrent as a function of power density at different gate voltages. The photocurrent exhibits a sublinear relationship with power density, further indicating the dominant role of the photogating effect. Additionally, the response times of both processed and non-processed devices were assessed under identical NIR power densities, as shown in Fig. 2(f). The processed device exhibited a higher photocurrent, but with significantly prolonged rise time (1.41 s vs. 0.07 s) and fall time (1.73 s vs. 0.20 s) compared to the non-processed device. This extended photoresponse time further supports the conclusion that the increased carrier lifetime is the primary contributor to the improved responsivity.

Subsequently, X-ray photoelectron spectroscopy (XPS) was performed on In_2O_3 films subjected to different processing conditions (Fig. 3(a, b)). The extracted results indicate a continuous increase in oxygen vacancy as the concentration or processing time of $NaBH_4$ treatment increases (Fig. 3(c)). Films processed for 60 seconds in a 0.2 M $NaBH_4$ solution displayed nearly double the ratio of oxygen vacancies compared to untreated films. Oxygen vacancies in metal oxides are typically associated with donor-like trap states, suggesting that $NaBH_4$ processing enhances the presence of such trap sites in the oxide film. Given the low-temperature, low-cost solution fabrication process, we fabricated flexible devices to demonstrate their potential for future wearable electronics. These devices were fabricated on a polyimide substrate and mechanically exfoliated, as depicted in Fig. 3(d), highlighting the device's suitability for flexible applications.

Conclusion

In conclusion, we employed $NaBH_4$ solution to process the oxide film before organic layer deposition. This treatment increased the density of trapping sites by enhancing oxygen vacancy formation, thereby amplifying the photogating effect in the heterojunction hybrid devices compared to untreated counterparts. Furthermore, we demonstrated the fabrication of flexible devices, showcasing their potential for future wearable electronic applications.

References

[1] P. C. Y. Chow and T. Someya, "Organic Photodetectors for Next-generation Wearable Electronics," *Adv. Mater.*, 32, e1902045 (2020).

[2] F. Zhou and Y. Chai, "Near-sensor and in-sensor computing," *Nat. Electron.*, 3, 664 (2020).

[3] T. Yokota, K. Fukuda, and T. Someya, "Recent Progress of Flexible Image Sensors for Biomedical Applications," *Adv. Mater.*, 33, e2004416 (2021).

[4] D. Li *et al.*, "Flexible and Air-Stable Near-Infrared Sensors Based on Solution-Processed Inorganic–Organic Hybrid Phototransistors," *Adv. Funct. Mater.*, 31, 2105887 (2021).

[5] F. Wang *et al.*, "Recent Progress on Electrical and Optical Manipulations of Perovskite Photodetectors," *Adv. Sci.*, 8, e2100569 (2021).

[6] D. Shao *et al.*, "Organic-inorganic Heterointerfaces for Ultrasensitive Detection of Ultraviolet Light," *Nano Lett.*, 15, 3787-92 (2015).

[7] D. Li *et al.*, "Solution-Processed Organic–Inorganic Semiconductor Heterostructures for Advanced Hybrid Phototransistors," *ACS Appl. Electron. Mater.*, 5, 578 (2023).

[8] D. Li *et al.*, "An Active-Matrix Synaptic Phototransistor Array for In-Sensor Spectral Processing," *Adv. Sci.*, 2406401 (2024).

[9] H. Fang and W. Hu, "Photogating in Low Dimensional Photodetectors," *Adv. Sci.*, 4, 1700323 (2017).

[10] A. Morteza Najarian, M. Vafaie, B. Chen, F. P. García de Arquer, and E. H. Sargent, "Photophysical properties of materials for high-speed photodetection," *Nat. Rev. Phys.*, 6, 219-230 (2024).

The Comprehensive Study of Difference between Cryogenic Planar MOSFET and FinFET Variation and Band Tail States Assessment

Chenyang Zhang[1,2], Yewei Zhang[1,2], Yongkang Xue[1,2], Shuying Wang[1,2], Sheng Yang[1,2], Pengpeng Ren[*,1,2], *Member, IEEE*, Zhigang Ji[*,1,2,4], *Member, IEEE* and Ru Huang[3,4], *Fellow, IEEE*

[1]National Key Laboratory of Science and Technology on Micro/Nano Fabrication, SJTU, China and PKU, China, [2]Department of Micro/Nano Electronics, SJTU, China, [3]The Institute of Microelectronics, PKU, China, [4]Institute of Electronic Design Automation, PKU, Wuxi, China
E-mail: pengpengren@sjtu.edu.cn, zhigangji@sjtu.edu.cn

Abstract

In this work, we compare the variation between planar MOSFETs and FinFET and extract band tail states using a physical model for comparison. The results clearly show that the threshold voltage (V_{th}) and subthreshold swing (SS) variations of both planar MOSFETs and FinFET increase as temperature decreases and as device size shrinks. However, a comparison of device variability reveals that FinFET exhibits greater variability in both V_{th} and SS at cryogenic temperature. Furthermore, the extracted band tail states parameters indicate that FinFET exhibits more band tail states, this may be the reason for the larger variation. Our comparison demonstrates that FinFET experience higher variability than planar MOSFETs. This suggests that designing devices for advanced process nodes at cryogenic temperature demands greater attention to device variations to prevent device failure.
(Keywords: Cryogenic MOSFET, band tail states, variability, FinFET, planar MOSFET)

Introduction

As server-level data computing evolves, so does the need for powerful computing performance and low power consumption. However, traditional computational methods are experiencing bottlenecks due to increasing data volumes and lack of arithmetic power [1]. Quantum computers have the potential to solve these problems that are intractable by conventional computers. Quantum bits (qubits) are the basic unit of a quantum computer, and in order to control the proper functioning of the qubits, the quantum computing system must operate at cryogenic temperatures, and in addition, the quantum computing system must contain a large amount of memory for storing the computational procedures and information for quantum error correction. More than thousands of qubits should be used to address practical problems, thus requiring the adoption of an electronic control interface operating at cryogenic temperature [2]. Nanometer CMOS technology can control complex circuits and will be the first choice for implementing such cryogenic controllers.

For such circuits to operate properly at cryogenic temperatures, it is necessary to understand not only the typical characteristics of Cryo-CMOS devices, but also their variability. Recent work has focused on the evaluation of the performance and variation of planar MOSFETs at cryogenic temperatures, and has not reported on the variation of advanced process nodes such as FinFET. Current research suggests that the variation will increase in low temperature devices, usually thought to be related to the band tail states [3]. There are only a few reports on the use of physical models to extract band tail state parameters [4, 5], and it is necessary to extract band tail states parameters for different sizes and devices in order to understand the relationship between band-tailed state and device parameters, which can help cryogenic circuit design and fabrication.

In this article, by comparing the variations of planar MOSFETs and FinFET at cryogenic temperatures, it is shown that FinFET exhibits stronger variability. Additionally, it is demonstrated that the variation in both devices increases as device size decreases or the number of FINs is reduced. The IV curves at different temperatures are fitted using a physical model to extract the parameters of the band tail states. It is found that FinFET exhibits significantly more pronounced band tail states compared to planar MOSFETs, and these states follow a similar dimensional trend as device size and variation. This study highlights the relationship between band tail states and device size, providing valuable insights for designing circuits optimized for cryogenic temperatures.

Measurement Setup and Device Description

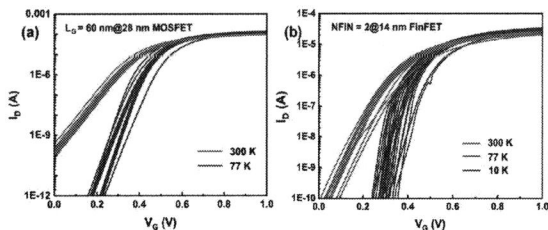

Fig. 1 The same devices of I_D-V_G characteristics of **(a)** 28 nm planar MOSFET (L_G = 60 nm) and **(b)** 14 nm FinFET (NFIN = 2) from different dies.

In this article, we have characterized industrial 28 nm planar n-MOSFETs (gate length L_G = 60, 100 and 600 nm, and width W = 1000 nm) and 14 nm n-FinFET (NFIN = 2, 4, 8, 12). The chip measurements are carried out with the help of

MicroXact CPS-200 cryogenic probe station. The transfer characteristics of the nMOS devices in the linear regime (V_{DS} = 0.05 V, V_{GS} = 0 – 1 V) are tested at 10, 77 and 300 K.

Fig. 1(a-b) shows the I_D -V_G (Drain current vs Gate Voltage) characteristics of the same planar MOSFET and FinFET from different dies characterized at cryogenic temperature. It can be clearly seen that the same device on different dies has different IV characteristic curves, where a decrease in temperature will lead to significant fluctuations in current characteristics. The results of other sizes and FIN numbers are not presented one by one in this article.

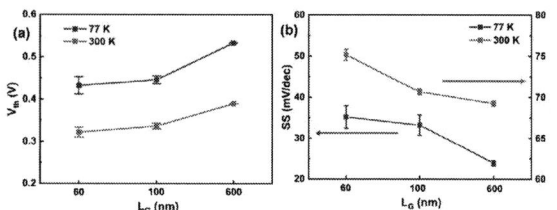

Fig. 2 (a) V_{th} and **(b)** SS of 28 nm planar MOSFETs vary in mean and variance with device size at different temperatures.

Results and Discussion

The extracted planar n-MOSFET parameters as a function of gate length are shown in **Fig. 2(a-b)**. As shown in **Fig. 2a**, as the temperature decreases, V_{th} at the same gate length will increase, and the variation will also increase, which is consistent with the trend reported in the [2]. In addition, as the device size decreases, the mean and variance of V_{th} will increase. However, SS will decrease as the temperature decreases (**Fig. 2b**), and the variation will increase as the device size decreases. The general trends in terms of gate length and temperature scaling are as expected.

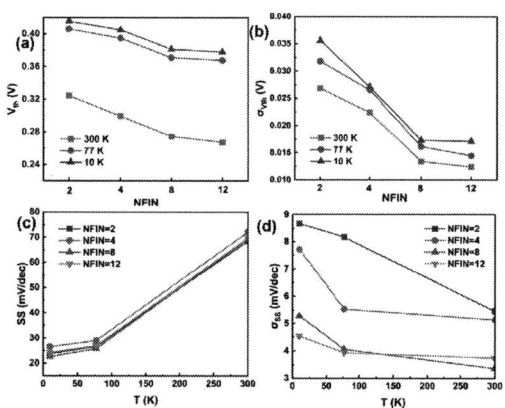

Fig. 3 The variation trend of **(a)** V_{th}, **(b)** σ_{Vth}, **(c)** SS and **(d)** σ_{SS} of 14 nm FinFET with the number of FINs at different temperatures.

In addition, to compare the differences in variation between planar MOSFET and FinFET at cryogenic temperature, as far as we know, the variation of FinFET was tested for the first time at cryogenic temperature. From a temperature

perspective, the V_{th} of any FinFET with NFIN will increase with decreasing temperature and show a saturation trend (**Fig. 3a**). Furthermore, it is obvious showed that as the NFIN decreases, V_{th} increases, this is consistent with the trend in the article [6], which may be related to IR drop. In **Fig. 3b**, we can observe similar trends to V_{th} in the variation of V_{th}, whether with temperature or NFIN. However, SS does decrease with decreasing temperature (**Fig. 3c**), but there is no clear trend in the change of NFIN, which may be due to the fact that the change of SS depends not only on the band tail states but also on the interface states. In **Fig. 3d**, the variation of SS will increase with the decrease of NFIN and temperature.

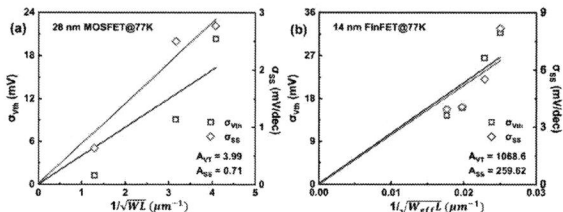

Fig. 4 Pelgrom plots of **(a)** 28 nm MOSFET and **(b)** 14 nm FinFET at 77 K for σ_{Vth} and σ_{SS} with A_{VT} and A_{SS}.

The well-known area dependence of the threshold-voltage and current-factor variability is described by the Pelgrom law, and also has been proven that can be used at cryogenic temperature [2, 7]. As shown in **Fig. 4(a-b)**, we can clearly see that the V_{th} and SS variations of FinFET are much higher than those of planar MOSFET, which means that they not only increase with the decrease of device size, but also change with the change of device configuration. In addition, the variation of the device is closely related to the device process, so the significant differences between planar MOSFET and FinFET processes may also lead to differences in variation, we will not study this in this article for the time being.

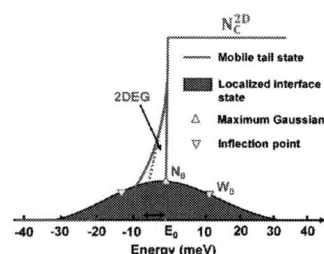

Fig. 5 Model employed for DOS, including band tail states and localized states.

To further understand the difference in band tail states between FinFET and MOSFET, we fit the obtained IV curve using the current band tail states model. As shown in **Fig. 5**, density of states (DOS) will be simplified to 2DEG, and due to the crystalline disorder, strain, residual impurities, DOS does not terminate at the band edge, but enters the bandgap in the form of exponential decay [8, 9], forming band tail states.

Furthermore, the Gaussian distribution of localized states is centered around the sharp band edge at the silicon/oxide interface. This model has been proven to explain many phenomena at cryogenic temperatures, such as SS saturation [8], inflection phenomenon [8], and so on. In addition, the model has been successfully used to extract band tail states in MOSFET [4], FinFET [10], TFT [5] and other devices, proving the correctness and rationality of our use of this model for extraction.

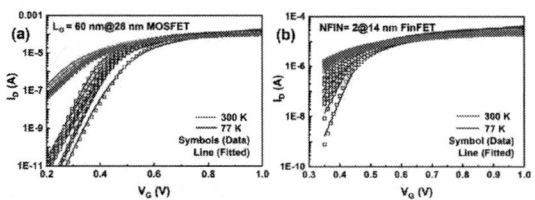

Fig. 6 Measured data (symbol) and fitted (line) data of **(a)** 28 nm MOSFET and **(b)** 14 nm FinFET, which can prove that the model and experiment can be well matched.

In order to ensure the correctness of the fitting, we will fit the IV curves at different temperatures together. This means that except for the different migration rates, the band tail state parameters and localized states parameters will be consistent, which ensures that our model can guarantee accuracy at different temperatures. As shown in **Fig. 6(a-b)**, we successfully fitted the IV curves of MOSFET and FinFET, and the good fit of data and experiment proved the accuracy of our model. The fitting results of other channel length and NFIN were not presented here one by one. The results we extracted are shown in **Fig. 7**. For MOSFETs, Wt increase with the decrease of device size (**Fig. 7a**).

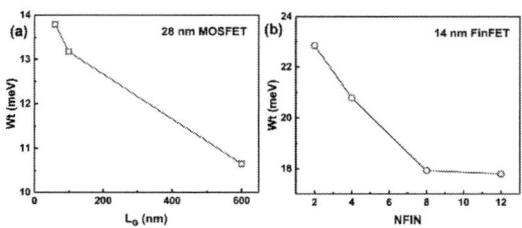

Fig. 7 The relationship between band tail states parameters (Wt) and **(a)** the channel length of MOSFET and **(b)** the number of FINs in FinFET.

Moreover, Wt in FinFET will decrease with the increase of NFINs (**Fig. 7b**). The above results are consistent with the variation trend of the device, proving the correctness of our extraction. Comparing the differences between the two devices, we can clearly see that the band tail states of FinFET is significantly higher than that of MOSFET, as we mentioned earlier. Bnad tail states are usually related to surface roughness and impurities, and planar MOSFET are usually considered to be caused by impurities. However, the 3D structure of FinFET introduces more surface roughness,

resulting in more band tail states, which exacerbates the variation at low temperatures. This work successfully extracted the band tail state parameters of different devices and discovered the relationship between device size and band tail states, providing guidance for the design of cryogenic temperature devices.

Conclusion

In this article, we compared the variations in threshold voltage (V_{th}) and subthreshold swing (SS) of MOSFETs with different channel lengths and FinFET with varying numbers of FINs. Our findings indicate that as the channel length or fin number decreases, the device variation increases, with FinFET showing greater variability compared to MOSFETs. The extracted band tail state parameters align with these variation trends, revealing that FinFET exhibits more pronounced band tail states than MOSFETs, leading to greater variability. This suggests that in circuit design, careful consideration must be given to the increased variability caused by device size reduction, as it may severely impact performance at cryogenic temperature.

Acknowledgments

This work was supported in part by the National Natural Science Foundation of China under Grant 92164205, 61927901, 62125401, 62027818, 61874034, and 11974320.

We would also like to thank Mr. Zhen Zhang and Mr. Jingze Yan from PRIMARIUS, for their contribution towards measurements.

References

[1] P. A. t. Hart *et al.*, in *ESSDERC 2019 - 49th European Solid-State Device Research Conference (ESSDERC)*, 23-26 Sept. 2019 2019, pp. 98-101

[2] P. A. T. Hart *et al.*, *IEEE Journal of the Electron Devices Society*, vol. 8, pp. 797-806, 2020

[3] P. S. Huang *et al.*, in *2024 IEEE Workshop on Microelectronics and Electron Devices (WMED)*, 29-29 March 2024 2024, pp. 1-4

[4] S. Takagi *et al.*, in *2023 IEEE 23rd International Conference on Nanotechnology (NANO)*, 2-5 July 2023 2023, pp. 10-14

[5] Y. C. Chen *et al.*, *IEEE Transactions on Electron Devices*, pp. 1-6, 2024

[6] Q. Zhang *et al.*, *IEEE Transactions on Electron Devices*, vol. 61, no. 2, pp. 643-646, 2014

[7] P. A. T. Hart *et al.*, *IEEE Journal of the Electron Devices Society*, vol. 8, pp. 263-273, 2020

[8] A. Beckers *et al.*, *IEEE Electron Device Letters*, vol. 41, no. 2, pp. 276-279, 2020

[9] A. Beckers *et al.*, *IEEE Transactions on Electron Devices*, vol. 67, no. 3, pp. 1357-1360, 2020

[10] X. Zhang *et al.*, *IEEE Transactions on Electron Devices*, pp. 1-6, 2023

Contact Engineering of 1L-WSe$_2$ p-FETs: Straightforward WO$_x$ Doping and Thickness Dependence

Wenyu Wang[1], Yanda Zhu[1], Ziyi Wang[1], Yun Li[2,4], Jason C. C. Hwang[2], Xi Lin[1], Jiaqi Su[1], Jing-Kai Huang[3*], Sean Li[1*]

[1]School of Materials Science and Engineering, UNSW Sydney, Australia, [2]Research and Prototype Foundry, USYD, Sydney, Australia, [3]Department of Systems Engineering, CityUHK, Hong Kong, [4]Present: School of Physics, USYD, Sydney, Australia

(Email: sean.li@unsw.edu.au, jkhuang@cityu.edu.hk, wenyu.wang@unsw.edu.au)

Abstract

This study presents a straightforward WO$_x$ doping strategy for the contact region of 1L-WSe$_2$ FETs, utilizing e-beam evaporated WO$_3$ bulk. We investigate the dopant thickness dependence, revealing that at an optimal thickness of 20 nm, the maximum on current $I_{on, max}$ increases by ~100 times, while maintaining an on-off ratio exceeding seven orders. This strategy paves the way for developing Ohmic contacts in 1L-WSe$_2$ FETs.

Keywords: Ohmic contact, p-FETs, doping.

Introduction

Transition metal dichalcogenides (TMDs) are proposed channel materials for field effect transistors (FETs) as their excellent electrical performance [1]. WSe$_2$ is particularly well-suited for p-type FETs, as its Fermi level (E_F) is positioned close to the midpoint of the bandgap, enabling p-type behaviour with high work function metals [2]. However, achieving Ohmic contact is a key challenge for high-performance p-FETs. Palladium (Pd) is favoured as a contact metal for WSe$_2$ [3]. But Pd-contacted WSe$_2$ p-FETs continue to exhibit high contact resistance.

To reduce Schottky barrier height (SBH) and facilitate the tunnelling of holes at the contacts, surface transfer doping is employed, utilizing high work function transition metal oxides as effective p-dopants. Nevertheless, previous studies have limitations. For example, MoO$_x$ [4] is a common choice, but its lattice constant differs slightly from that of WSe$_2$. In contrast, WO$_x$, the native oxide of WSe$_2$, has gained attention recently. However, WO$_x$ is typically obtained by oxidizing WSe$_2$ through UV-ozone [5] or O$_2$ plasma [6], making it complex to optimize treatment conditions, and the treatment can potentially damage the surface or introduce defects in the WO$_x$. In this study, we propose a straightforward method of depositing WO$_x$ via e-beam evaporated WO$_3$ on the source and drain of 1L-WSe$_2$ FETs to enhance the contact performance. We also investigate the effect of WO$_x$ thickness to optimize doping strategies.

WO$_x$ is a desirable contact p-dopant

As underlined in Fig. 1(a), the work function Φ_m of WO$_x$, up to 6.9 eV, higher than Φ_m of Pd (5.22 eV), exceeds the electron affinity (χ) of WSe$_2$ (3.5–4 eV), placing the conduction band edge (CBE) of WO$_x$ far below the valence band edge (VBE) of WSe$_2$, indicating that the spontaneous electrons transfer from WSe$_2$ to WO$_x$ is favourably advantageous, leading to p-doping of WSe$_2$. Fig. 1(b) and (c) illustrate the alteration of the energy-band diagram between WSe$_2$ and Pd caused by WO$_x$ doping. The interlayer WO$_x$ introduces the hole doping in WSe$_2$, resulting in a further upward band bending. Consequently, an effective p-type doping is induced.

To verify the conductivity of Pd/WO$_x$, we examined the current–voltage performance of Pd/WO$_x$ stacks consisting of 10, 30, 50 nm WO$_x$, which were deposited onto an Au substrate via e-beam evaporation, as the inset of Fig. 2(a) which displays the linear current–voltage characteristics, confirming the ohmic behaviour of Pd/WO$_x$. Fig. 2(b) presents the total resistance of these stacks as a function of WO$_x$ thickness, showing that the resistance of a single contact remains below 1.9 m$\Omega\cdot$cm^2. Such a high conductivity of WO$_x$ is ascribed to its large stoichiometry (with $x \approx 3$).

Device Fabrication and Electrical Performance

1L-WSe$_2$ FETs were fabricated using global back-gated structures, as illustrated in Fig. 3(a). The 1L-WSe$_2$ film was grown using chemical vapour deposition and then transferred onto ~50 nm Al$_2$O$_3$/Si substrates. Source/drain was defined via e-beam lithography, followed by Pd evaporated as the contact metal. Fig. 3(b) presents the top-view scanning electron microscopy image. For WO$_x$ doped 1L-WSe$_2$ FETs, WO$_x$ was deposited by e-beam evaporation beneath the Pd contact (Fig. 3(c)).

We prepared WO$_x$ doped 1L-WSe$_2$ FETs with varying WO$_x$ thicknesses and compared their maximum on current ($I_{on, max}$). The channel length $L_{ch} = 1\,\mu$m and $V_{ds} = 1$ V. The thickness-dependent $I_{on, max}$ is summarized in Fig. 4(a), showing that the $I_{on, max}$ of undoped 1L-WSe$_2$ FETs (0 nm WO$_x$) is ~ 0.01 μA/μm. It gradually increases over 1 μA/μm as the WO$_x$ thickness increases from 0 to 20 nm, before decreasing back to ~0.01 μA/μm at 30 nm. The highest $I_{on, max}$ is observed at 20 nm WO$_x$ doping, indicating that WO$_x$ enhances by ~ 100 times. This is because the effectiveness of p-doping improves with increasing WO$_x$ thickness up to 20 nm. However, beyond this thickness, the benefits of p-doping are outweighed by the increased serial WO$_x$ resistance, which is confirmed in Fig. 2(b), the thicker WO$_x$ thin film leads to higher Pd/WO$_x$ resistance. Fig. 4(b) shows the transfer curves of representative 20 nm WO$_x$ doped 1L-WSe$_2$ FET, achieving the highest $I_{on, max}$ of 4.37 μA/μm, and the on-off ratio remained over 7 orders. The output curves in Figs. 4(d)-(e) also indicates such an improvement. It demonstrates that WO$_x$ doping facilitates hole injection through the p-doping of WSe$_2$, and the optimal WO$_x$ dopant thickness is 20 nm.

Characterization Confirmation

We confirmed the surface transfer doping of WO$_x$ doped 1L-WSe$_2$ via Raman and photoluminescence (PL) spectroscopy. The Raman spectrum is shown in Fig. 6(a). The Raman shift around 250 cm^{-1} is attributed to mixed modes of E$^1_{2g}$ and A$_{1g}$, which blue-shifted by ~1cm^{-1} after doping, consistent with previous study [7], which suggests that reduction of the electron density in 1L-WSe$_2$. The effect of WO$_x$ doping on the PL spectrum is depicted in Fig. 6(b). We found PL intensity was increased by WO$_x$ doping. The PL peak of undoped WSe$_2$ can be deconvoluted into exciton (≈1.64 eV) and trion (≈1.59 eV) peaks, with the trion is considered negative [8]. The WO$_x$ doping increased the intensity of the exciton peak while reducing that of the trion peak, resulting in a blue shift. The enhanced exciton peak corresponds to a reduction in negative charges due to holes supplied from WO$_x$, which decreases the ratio of negative trions.

To further understand the changes in 1L-WSe$_2$ surface properties induced by WO$_x$ doping, X-ray photoelectron spectroscopy (XPS) measurements of W$_{4f}$ and Se$_{3d}$ were performed. In Fig. 7(a), XPS spectra of W$_{4f}$ show two W-containing phases: peaks at 33 eV and 35.2 eV are attributed to WSe$_2$, while peaks at 37 eV and 39.1 eV are associated with WO$_x$ [9]. The doublet peaks at 55.2 eV and 56 eV are related to the Se as shown in Fig. 7(b). XPS spectra of the undoped and WO$_x$ doped films reveal that both W$_{4f}$ and Se$_{3d}$ peaks shifted to lower energies ($\Delta E \geq 0.7$ eV) after doping, summarized in Table 1. Similar results were reported by Ji et al. [7], indicating that the E_F of WSe$_2$ moves toward the VBE, resulting in a p-doping effect.

Conclusion

In conclusion, we demonstrated the straightforward WO$_x$ surface transfer doping via e-beam evaporation. It created effective p-doped contacts for 1L-WSe$_2$ FETs, resulting in a hundred times improvement in $I_{on, max}$. As well, the optimal WO$_x$ thickness was identified as 20 nm. This simple strategy can be applied to other TMDs to further develop high-performance devices.

Acknowledgments

We thank the facilities as well as the scientific and technical assistance of the RPF at USYD, part of the ANFF, the ANFF and the units within the MWAC at UNSW Sydney; W. W. thanks the Australian Government RTP Scholarship; Y.Z. in UNSW Sydney for assistance in device fabrication.

Reference

[1] Radisavljevic, B., et al., Single-layer MoS$_2$ transistors. Nature nanotechnology, 2011. 6(3): p. 147-150.

[2] Das, S. and J. Appenzeller, WSe$_2$ field effect transistors with enhanced ambipolar characteristics. Applied physics letters, 2013. 103(10).

[3] Oberoi, A., et al., Toward high-performance p-type two-dimensional field effect transistors: contact engineering, scaling, and doping. ACS nano, 2023. 17(20): p. 19709-19723.

[4] Chen, Y.-H., et al., P-Type Ohmic Contact to Monolayer WSe2 Field-Effect Transistors Using High-Electron Affinity Amorphous MoO$_3$. ACS Applied Electronic Materials, 2022. 4(11): p. 5379-5386.

[5] Yang, S., G. Lee, and J. Kim, Selective p-doping of 2D WSe$_2$ via UV/ozone treatments and its application in field-effect transistors. ACS Applied Materials & Interfaces, 2020. 13(1): p. 955-961.

[6] Lan, H.-Y., et al. Wafer-scale CVD Monolayer WSe$_2$ p-FETs with Record-high 727 μA/μm I on and 490 μS/μm g max via Hybrid Charge Transfer and Molecular Doping. in 2023 International Electron Devices Meeting (IEDM). 2023. IEEE.

[7] Ji, H.G., et al., Chemically tuned p - and n - type WSe$_2$ monolayers with high carrier mobility for advanced electronics. Advanced Materials, 2019. 31(42): p. 1903613.

[8] Li, Z., et al., Revealing the biexciton and trion-exciton complexes in BN encapsulated WSe$_2$. Nature communications, 2018. 9(1): p. 3719.

[9] Boscher, N.D., C.J. Carmalt, and I.P. Parkin, Atmospheric pressure chemical vapor deposition of WSe$_2$ thin films on glass—highly hydrophobic sticky surfaces. Journal of Materials Chemistry, 2006. 16(1): p. 122-127.

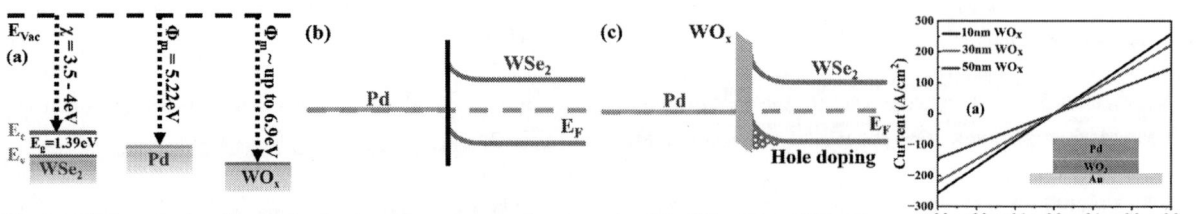

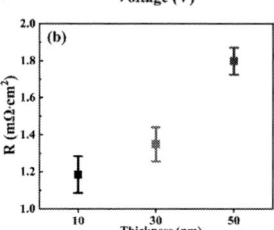

Fig. 1. (a) Valence band and conduction band positions relative to the vacuum level for WSe$_2$, Pd and WO$_x$. The electron affinity (χ) of WSe$_2$ is between 3.5 and 4 eV, the work function (Φ_m) of Pd is 5.22 eV, and the Φ_m of WO$_x$ can reach up to 6.9 eV. (b) Energy-band diagram of undoped WSe$_2$ in contact with Pd. (c) The introduction of hole doping by the WO$_x$ interlayer between WSe$_2$ and the Pd contact.

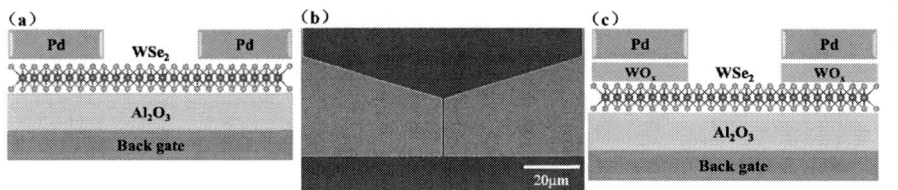

Fig. 3. (a) Cross-sectional schematic illustration of the undoped 1L-WSe$_2$ FET. (b) The scanning electron microscopy image of the undoped 1L-WSe$_2$ FET in top view. (c) Cross-sectional schematic illustration of the WO$_x$ doped 1L-WSe$_2$ FET.

Fig. 2. (a) Current−voltage curves for Pd/WO$_x$ stacks, with an inset showing a schematic of the stacks. (b) Resistance of the Pd/WO$_x$ stacks as a function of WO$_x$ thickness.

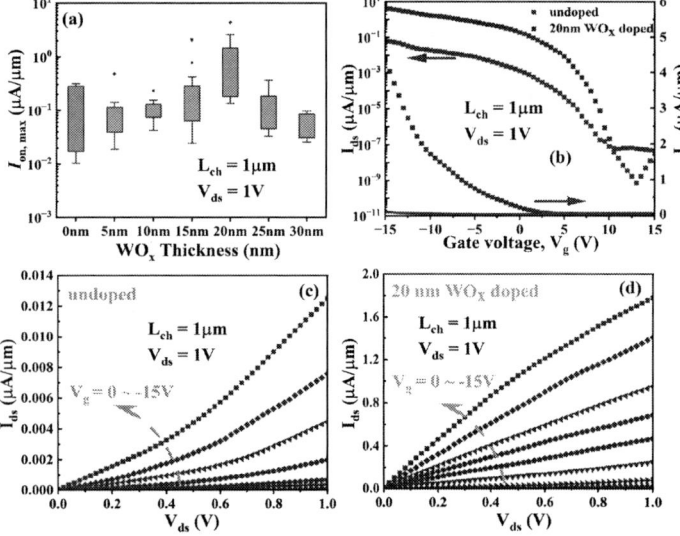

Fig. 4. (a) The thickness-dependent maximum on current ($I_{on,\,max}$) of WO$_x$ doped 1L-WSe$_2$ FETs, compared to the maximum on current $I_{on,\,max}$ of undoped 1L-WSe$_2$ FETs. (b) The transfer curves for representative undoped and 20 nm WO$_x$ doped 1L-WSe$_2$ FET displayed in both log and linear scales. The output curves for representative (c) undoped and (d) 20 nm WO$_x$ doped 1L-WSe$_2$ FET.

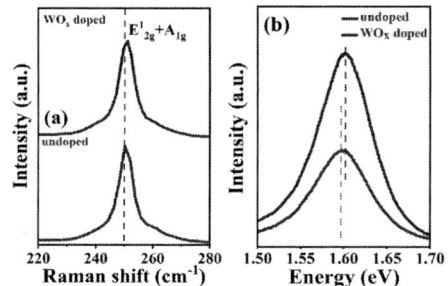

Fig. 5. (a) Raman spectra and (b) PL spectra of undoped and WO$_x$ doped 1L-WSe$_2$.

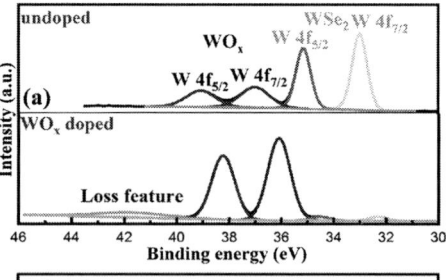

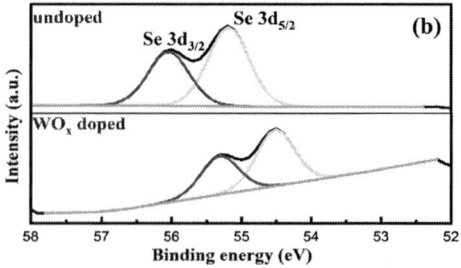

Fig 6. The XPS spectra of undoped and WO$_x$ doped 1L-WSe$_2$: (a) W$_{4f}$ and (b) Se$_{3d}$.

Table 1: binding energy details and shifts for W$_{4f}$ and Se$_{3d}$ between undoped and WO$_x$ doped 1L-WSe$_2$.

	WO$_x$ (eV)		WSe$_2$ (eV)			
	W 4f$_{5/2}$	W 4f$_{7/2}$	W 4f$_{5/2}$	W 4f$_{7/2}$	Se 3d$_{3/2}$	Se 3d$_{5/2}$
undoped	39.1	37	35.2	33	56	55.2
WO$_x$ doped	38.2	36	34.5	32.3	55.3	54.5
ΔE	0.9	1	0.7	0.7	0.7	0.7

Monolithic 3D Integration of 2D Devices

Saptarshi Das, PhD.

Ackley Professor of Engineering, Engineering Science and Mechanics, Electrical Engineering and Computer Science, Materials Science and Engineering, Materials Research Institute, Pennsylvania State University, University Park, PA, USA

Monolithic three-dimensional (3D) integration of 2D materials represents a major breakthrough in semiconductor technology, offering higher device density and multifunctionality [1]. Recently, we achieved large-scale, wafer-level 3D integration of MoS_2 and WSe_2 field-effect transistors (FETs) across multiple tiers [2]. This work includes 2-tier integration of MoS_2 FETs with over 10,000 devices per tier and 3-tier integration of both n-type MoS_2 and p-type WSe_2 FETs, with approximately 500 devices per tier (Fig. 1). The 3-tier architecture combines multifunctional capabilities, such as logic, memory storage, and sensing, into a single chip. Aggressively scaled MoS_2 FETs with channel lengths as short as 45 nm further demonstrate the scalability of this approach. The entire integration process is BEOL-compatible, using low-temperature techniques to prevent degradation of lower-tier devices, marking a significant step toward highly dense, multifunctional 3D integrated circuits based on 2D materials.

In another study, we focused on complementary metal-oxide-semiconductor (CMOS) logic circuits using monolithic 3D integration of tungsten diselenide (WSe_2) p-type FETs stacked on WSe_2 n-type FETs [3]. This work integrates 340 n-type FETs in tier 1 and 340 p-type FETs in tier 2, connected by 300 nm vias with a pitch below 1 μm (Fig. 2). The resulting CMOS logic gates, such as inverters and NAND/NOR gates, exhibit high performance and scalability, showing that 2D materials can complement existing silicon technologies while maintaining a low thermal budget of 200°C. These developments open the door to high-density, low-power 3D integrated circuits, addressing critical challenges in interconnect latency and power consumption.

More recently, we introduced a monolithic 3D integration platform incorporating graphene-based chemisensors with molybdenum disulfide (MoS_2) memtransistors in a two-tier architecture. The vertically stacked 2D materials, separated by less than 50 nm, enable near-sensor computing with an interconnect density of 62,500 I/O per mm². The manufacturing process is fully BEOL-compatible, employing a low thermal budget of 200°C. This architecture reduces latency in near-sensor computing, showcasing the potential for advanced, heterogeneous integration of 2D materials.

References

[1] D. Jayachandran, N. U. Sakib, and S. Das, "3D integration of 2D electronics," *Nature Reviews Electrical Engineering,* pp. 1-17, 2024.

[2] D. Jayachandran, R. Pendurthi, M. U. K. Sadaf, N. U. Sakib, A. Pannone, C. Chen, *et al.,* "Three-dimensional integration of two-dimensional field-effect transistors," *Nature,* vol. 625, pp. 276-281, 2024/01/01 2024.

[3] R. Pendurthi, N. U. Sakib, M. U. K. Sadaf, Z. Zhang, Y. Sun, C. Chen, *et al.,* "Monolithic three-dimensional integration of complementary two-dimensional field-effect transistors," *Nature Nanotechnology,* vol. 19, pp. 970-977, 2024/07/01 2024.

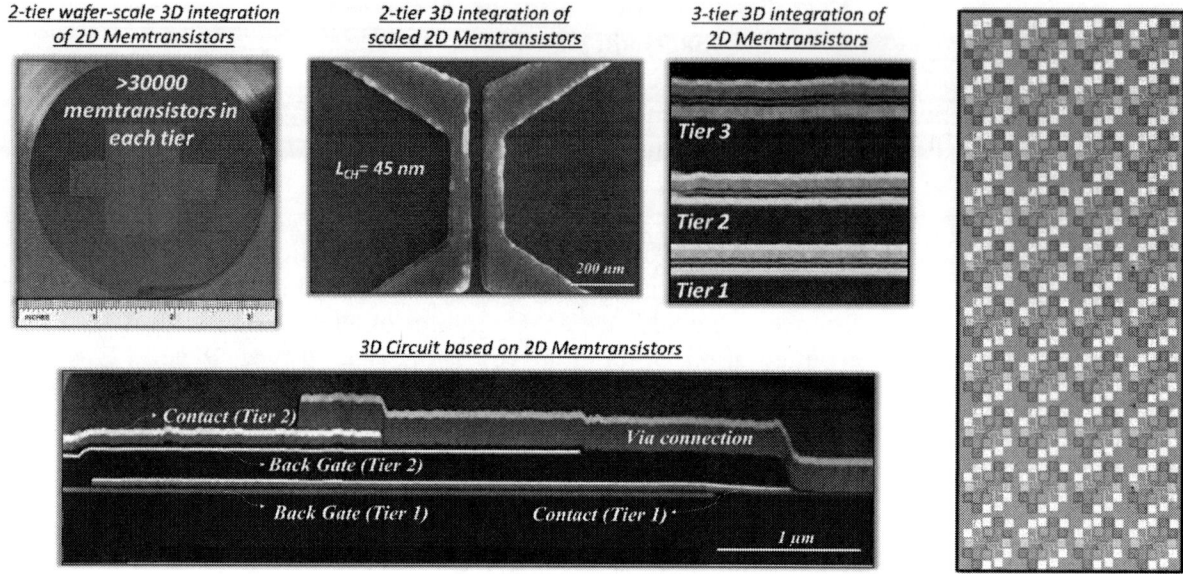

Figure 1. Monolithic three-dimensional (3D) integration of 2D materials. *Reproduced with permission from Nature.*

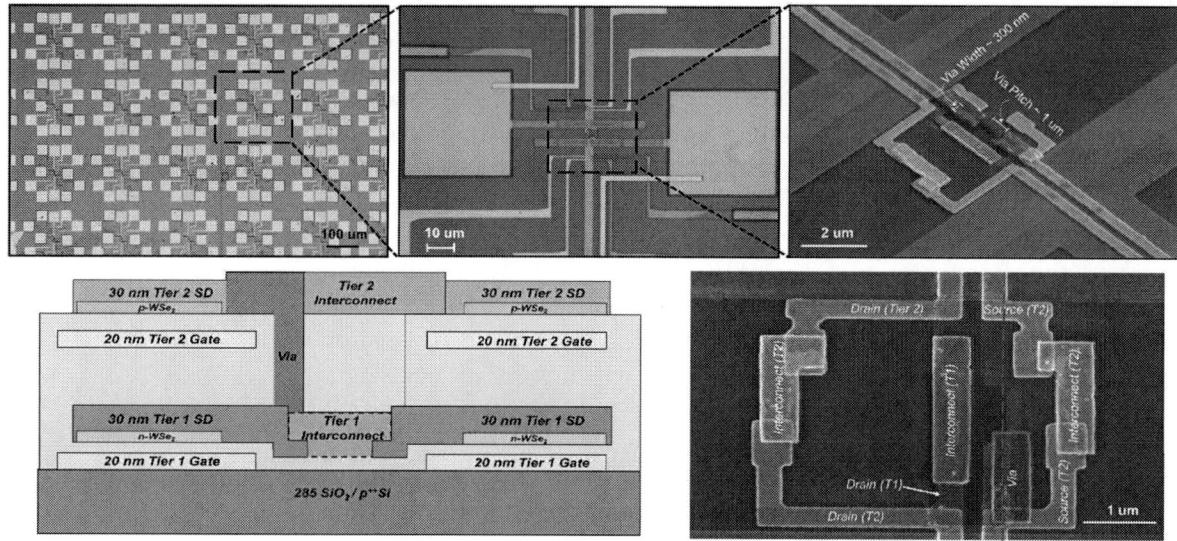

Figure 2. Monolithic three-dimensional (3D) integration of complementary 2D materials using dense vias. *Reproduced with permission from Nature Nanotechnology.*

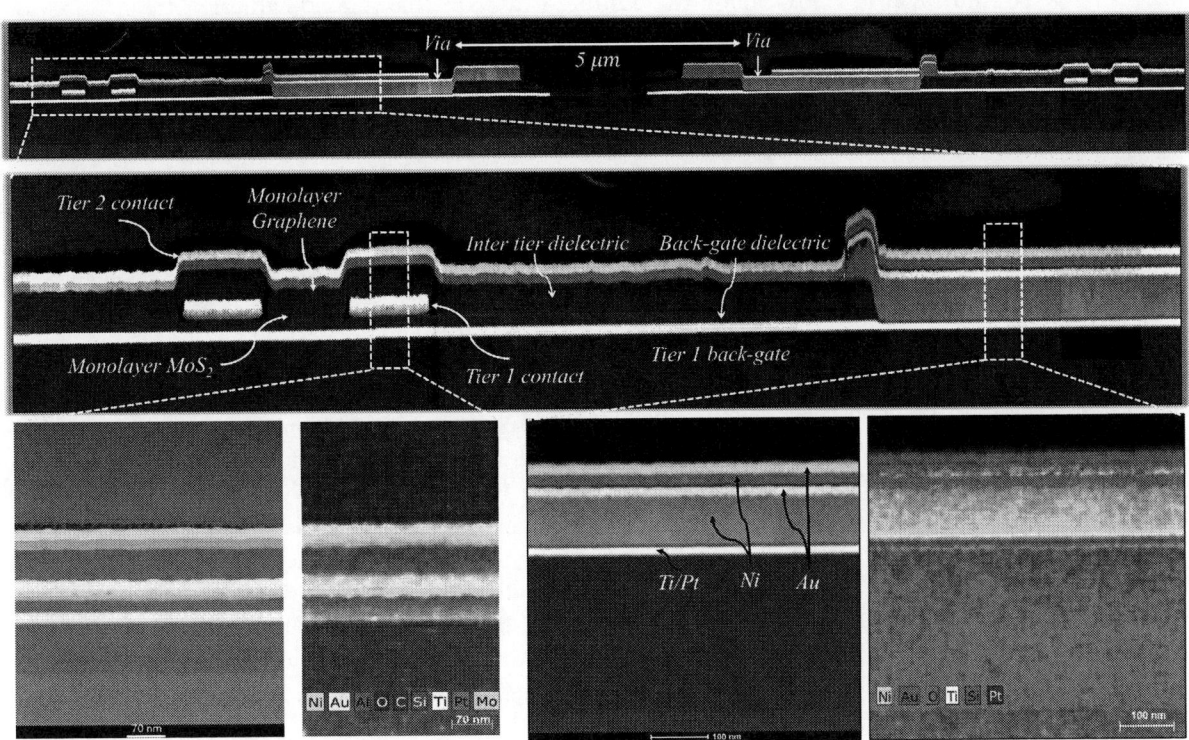

Figure 3. Monolithic and heterogeneous three-dimensional (3D) integration of graphene and MoS₂. *Reproduced with permission from Nature Electronics.*

Broadband Vis-NIR Neuromorphic Photodetector Based on PdSe$_2$/Bi$_2$O$_2$Se Heterotransistor for Motion Detection

Yu Zhu[1], Xinrui Guo[1], Shuo Liu[1], Junling Liu[1], Ru Huang[1], and Ming He[1,2,*]

[1]School of Integrated Circuits, Beijing Advanced Innovation Center for Integrated Circuits, Peking University, Beijing 100871, China; [2]Frontiers Science Center for Nano-optoelectronics, Peking University, Beijing 100871, China. *E-mail: minghe@pku.edu.cn

Abstract

Herein, we present a high-performance, broad-spectrum neuromorphic phototransistor utilizing PdSe$_2$/Bi$_2$O$_2$Se van der Waals (vdWs) heterostructures, which extends the spectral sensing range from 300 nm to 1550 nm, achieving an exceptional photoresponsivity on the order of 10^5 A/W. Remarkably, the PdSe$_2$/Bi$_2$O$_2$Se phototransistor exhibits multifunctional capabilities, enabling tunable memory window from 10 ms to 10^4 ms through gate voltage tuning. We successfully demonstrate its application in recognizing and classifying multi-target motion speeds ranging from 10 km/h to 200 km/h. These findings offer a promising pathway for the advancement of high-performance neuromorphic photodetectors aimed at tracking motion objectives.

(Keywords: Neuromorphic Photodetector, Broadband Vis-NIR, Heterotransistor, Motion Detection)

Introduction

Multifunctional neuromorphic photodetector that integrates sensing, memory, and processing capabilities presents a compelling approach for advancing motion sensing hardware [1,2]. Traditional motion detection relies on discrete units, necessitating extensive data transfers between front-end sensors and back-end processors. This architecture often leads to large computational overhead and delays due to the exchange of redundant data. Although near-sensing architectures have emerged, they continue to grapple with the dual challenge of achieving rapid response times alongside long-term memory retention [3]. Recent advancement in neuromorphic photodetectors, inspired by retinomorphic vision, have adopted the short-term memory (STM) for preprocessing functions [4,5]. However, these devices usually experience limitations in photoresponse performance attributable to material or structural constraints. The simultaneous realization of wide-spectrum photodetection and reconfigurable optical-memory functionalities marks a critical frontier in sensor technology, yet this domain needs be adequately explored.

In this work, we develop PdSe$_2$/Bi$_2$O$_2$Se vdWs heterostructure phototransistors that exhibit broad spectral coverage and exceptional in-sensor preprocessing performance. Specifically, the interplay between the internal electric field in the heterojunction and gate-regulated interface defect states facilitates modulation of both photoresponse speed and photo-memory timescales. This unique mechanism endows the device with rapid sensing and preprocessing capabilities. Furthermore, we demonstrate multi-target motion speed recognition using a 3×3 device array, displaying the potential of this architecture for advanced sensing applications.

Device Fabrication

The fabrication process begins with the transfer of Bi$_2$O$_2$Se nanosheets from mica substrates onto HfO$_2$/p-Si substrates via a polymethyl methacrylate assisted method (**Fig. 1a**). Following this, PdSe$_2$ nanosheets that obtained through mechanical exfoliation are dry-transferred onto Bi$_2$O$_2$Se. The resulting heterojunction undergoes thermal annealing at 150 °C for 60 minutes in a vacuum atmosphere. Finally, source and drain electrodes (Ti/Au=5/50 nm) are fabricated using standard electron beam lithography (EBL).

Results and Discussion

The neuromorphic photodetector was manufactured in the bottom-gated architecture of PdSe$_2$/Bi$_2$O$_2$Se vdWs heterostructure phototransistors. The cross-sectional transmission electron microscopy (TEM) image presented in **Fig. 1b** reveals a distinct interface between the layered Bi$_2$O$_2$Se and the PdSe$_2$ films. To assess the crystallinity of the individual films as well as the PdSe$_2$/Bi$_2$O$_2$Se vdWs heterojunction at the overlap region, Raman spectroscopy was employed (**Fig. 1c**). The analysis of the Raman spectra and sharp X-ray diffraction (XRD) data shown in **Fig. 1d** indicates that both the Bi$_2$O$_2$Se and PdSe$_2$ films exhibit high crystal quality.

Fig. 2a depicts the real-time photoresponses of PdSe$_2$/Bi$_2$O$_2$Se heterostructure phototransistors under 532 nm illumination. The type-II band alignment at the heterojunction establishes a built-in electric field that efficiently separates photogenerated carriers, facilitating rapid collection by the electrodes and yielding a response time of 10 ms that well-suited for real-time visual information processing [6]. In **Fig. 2b**, the relationship between photoresponsivity ($R=I_{ph}/(P·S)$) and incident light power density is illustrated, where I_{ph} is the photocurrent, P is the light power intensity, and S is the illuminated channel area, revealing an impressive photoresponse performance of up to 10^5 A/W [7]. **Fig. 2c** shows the real-time photoresponse of the heterojunction phototransistor across a wavelength range from 532 nm to 1550 nm, with

979-8-3315-0417-5/25 $31.00 © 2025 IEEE

continuous light pulses applied at the interval of 2 s. The photosensor exhibits stable and repeatable photoresponses over an extensive spectral range encompassing visible to near-infrared (Vis-NIR) bands. The corresponding photoresponsivity of the heterostructure phototransistor throughout this wavelength spectrum is depicted in **Fig. 2d**. The enhanced response performance in these wavebands is attributed to the narrow bandgap of the photosensitive material and the robust interlayer coupling within the vdWs heterojunction.

The interplay of defect-carrier capture and slow de-capture processes at the Bi_2O_2Se/Hf_2O interface within the heterojunction phototransistor under back-gate voltage bias, enables a photoresponse time that ranges from a swift 10 ms to an extended photomemory duration of 1.5×10^4 ms (**Fig. 3a**). Curve fitting of the photoresponse and memory relaxation time across the visible to near-infrared spectrum is shown in **Fig. 3b**, highlighting the reconfigurable photoresponse characteristics of the heterojunction phototransistor. The photoresponse data collected at varying power densities (**Fig. 3c**), alongside the photoresponse at different light spike widths with V_G=0.1 V (**Fig. 3d**), reveals that the heterostructure phototransistor exhibits stable short-term potentiation (STP) response characteristics akin to postsynaptic current output, thereby facilitating its function as a neuromorphic photodetector.

As shown in **Fig. 4a**, the secondary excitatory postsynaptic current (EPSC) peak elicited by a paired optical stimulus surpasses the initial peak, a phenomenon commonly associated with facilitation behavior. The paired pulse facilitation (PPF) ratio of the heterostructure synaptic phototransistor was further examined using a series of paired optical spikes with varying time intervals (Δt), as depicted in **Fig. 4b**. This PPF ratio is intricately linked to synaptic plasticity and plays a crucial role in visual information processing within biological neural systems. In **Fig. 4c**, the application of successive optical spike sequences at different frequencies with V_G=0.1 V results in a consistent increase in EPSC. This cumulative effect becomes increasingly pronounced with higher optical spike frequencies (**Fig. 4d**), leading to an enhanced EPSC response. This photo-induced dynamic behavior underscores the potential of the $PdSe_2/Bi_2O_2Se$ heterostructure phototransistor for advanced dynamic visual perception applications.

Utilizing the retinomorphic characteristics of light spikes for extensive dynamic modulation, the device is positioned as visual neural network hardware capable of in-sensor multi-target speed recognition. By introducing optical spikes of 10 ms, 20 ms, and 200 ms from left to right across each row of a 3×3 sensor array, three representative object velocities of 200 km/h, 100 km/h,

and 10 km/h are simulated, reflecting real traffic scenarios [4,5]. Figs. **5a-c** present the optical spike response of the sensor array under gate voltage biases of 0.1 V, 0.5 V, and 1 V, respectively. Notably, the memory of shorter optical spikes intensifies at higher gate voltage biases, allowing the sensor array to sensitively detect 10 ms spikes at V_G=1 V. Conversely, at lower V_G settings, the array exhibits superior differentiation of broader optical spikes. This adaptability underscores the potential of the device for effective dynamic modulation in real-time visual recognition applications.

The confusion matrix presented in **Fig. 5d** is generated through mean-square calculations of the photoresponse results across various gate voltage modes. This matrix facilitates the classification of three distinct moving speeds associated with humans, cars, and planes, thereby illustrating the potential for in-situ multi-target speed recognition within edge sensing hardware. **Table 1** further elucidates the comparative processing capabilities of the proposed neuromorphic photodetector against other sensor methodologies.

Conclusion

In conclusion, we have demonstrated reconfigurable neuromorphic photodetectors with broad optoelectronic performances of $PdSe_2/Bi_2O_2Se$ vdWs heterostructure. The ability to tune photomemory across various gate voltages affords a promising avenue for developing high-performance in-sensor motion recognition hardware.

Acknowledgments

This work was supported by the National Key R&D Program of China 2022YFB4400100, the Natural Science Foundation of China (92164205, 62074004, 61927901), and the 111 Project (B18001).

References

[1] Zhang, Z. et al. All-in-one two-dimensional retinomorphic hardware device for motion detection and recognition. Nat. Nanotechnol. 17, 27–32 (2022).

[2] Zhang, Z. et al. In-sensor reservoir computing system for latent fingerprint recognition with deep ultraviolet photo-synapses and memristor array. Nat. Commun. 13, 6590 (2022).

[3] Fu, Z. et al. Novel Energy-efficient Hafnia-based Ferroelectric Processing-in-Sensor with in-situ Motion Detection and Four-quarter Mutipilcation. 2022 International Electron Devices Meeting (IEDM), San Francisco, CA, USA, 24.5.1-24.5.4 (2022).

[4] Bhattacharjee. et al. Emulating synaptic response in n- and p-channel MoS2 transistors by utilizing charge trapping dynamics. Sci. Rep. 10, 12178 (2020).

[5] Zhou, Y. et al. Computational event-driven vision sensors for in-sensor spiking neural networks. Nat. Electron. 6, 870–878 (2023).

[6] He, M., et al., Sub-10mK-Resolution Thermal-Bolometric Integrated FET-Type Sensors Based on Layered Bi_2O_2Se Semiconductor Nanosheets. 2020 IEEE International Electron Devices Meeting (IEDM), San Francisco, CA, USA, 26.1.1-26.1.4 (2020).

[7] He, M. et al. Ultrasensitive Retinomorphic Dim-Light Vision with In-Sensor Convolutional Processing Based on Reconfigurable Perovskite-Bi_2O_2Se Heterotransistors. 2023 International Electron Devices Meeting (IEDM), San Francisco, CA, USA, 33.3.1-33.3.4 (2023).

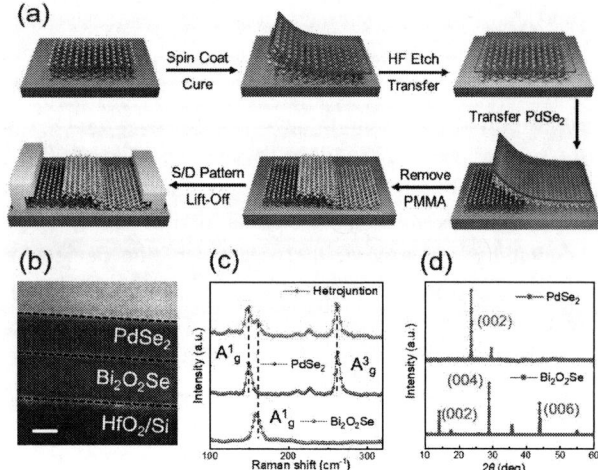

Figure 1. (a) Schematic illustration of manufacturing PdSe$_2$/Bi$_2$O$_2$Se vdWs heterostructure phototransistor; (b) transmission electron microscope (TEM) images; (c) Raman and (d) X-ray diffraction (XRD) patterns of PdSe$_2$/Bi$_2$O$_2$Se heterostructure.

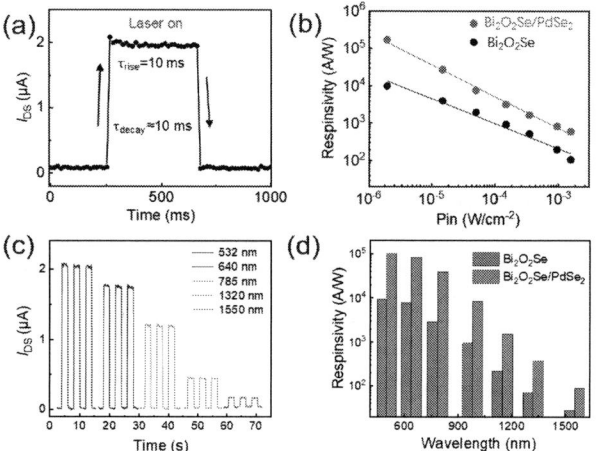

Figure 2. (a) Characteristics of photoresponse speed; (b) dependence of the photoresponsivity (R) on the incident light power density; (c) photoelectric response in the Vis-NIR range and (d) the corresponding photoresponsivity of phototransistor in the wavelength range of 532 nm to 1550 nm.

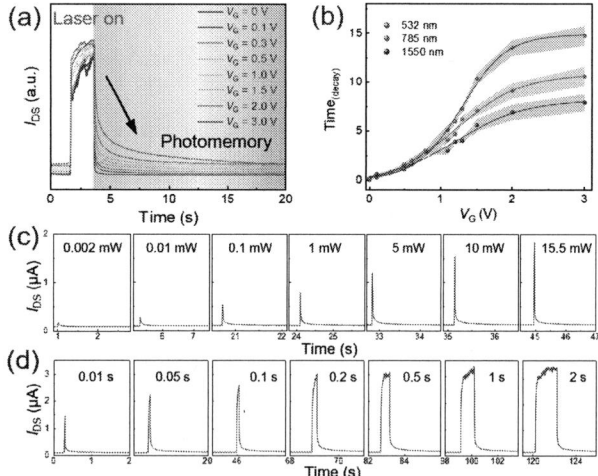

Figure 3. (a) Photoresponse memory relaxation in device regulated by gate voltage (V$_G$); (b) photoresponse relaxation time at different wavelengths; Light pulse response of (c) different power density and (d) width at V$_G$ = 0.1 V.

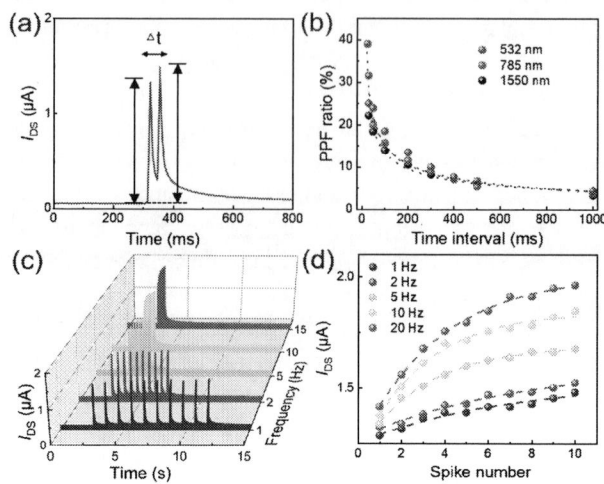

Figure 4. (a) Characteristics of paired pulse facilitation (PPF); (b) spike-interval ($\triangle t$)-dependent PPF ratio curves at different wavelengths; (c) excitatory postsynaptic current (EPSC) dynamics triggered by multi-pulse of different frequencies, and (d) spike-number-dependent EPSC variation.

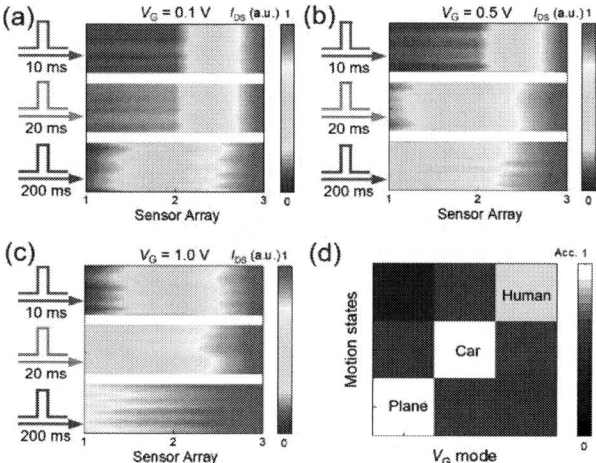

Figure 5. Input optical spikes of 10 ms, 20 ms, and 200 ms from left to right to simulate objects moving at different speeds, and the output signal distribution of the 3×3 device array collected at (a) V$_G$ = 0.1 V, (b) V$_G$ = 0.5 V, and (c) V$_G$ = 1 V, respectively; (d) confusion matrix of the speed recognition.

Table 1. Benchmark of the PdSe$_2$/Bi$_2$O$_2$Se neuromorphic photodetector with other photodetectors.

Device type	R (A/W)	Response waveband	Response time (ms)	Modulation ratio
WSe$_2$/h-BN/BP [1]	10^3	Visable	100 s	/
a-GaOx [2]	10^2	UV	10-5000	10^2
FE-HZO [3]	10^{-1}	NIR	10-100	10
Re-doped MoS$_2$ [4]	/	/	100-10000	10^2
WSe$_2$ with capacitors [5]	10^4	Visable	5-500	10^2
PdSe$_2$/Bi$_2$O$_2$Se (this work)	10^5	Vis-NIR	10-15000	1.5×10^3

Study of Threshold Voltage Degradation Mechanism of Ferroelectric Field-Effect Transistors (FeFETs) with TiN/SiO₂/Hf₀.₅Zr₀.₅O₂/SiOₓ/Si (MIFIS) Gate Stacks

Zeqi Chen[1,2,3#], Tao Hu[1,2,3#], Xinpei Jia[1,2,3], Runhao Han[1,2,3], Jia Yang[1,2,3], Yajing Ding[1,2,3], Xiaoqing Sun[1,2,3], Junshuai Chai[1,2,3], Hao Xu[1,2,3], Xiaolei Wang[1,2,3*], Wenwu Wang[1,2,3], and Tianchun Ye[1,2,3]

[1]Institute of Microelectronics, Chinese Academy of Sciences, Beijing 100029, China. [2]Key Laboratory of Fabrication Technologies for Integrated Circuits, Chinese Academy of Sciences, Beijing 100029, China. [3]The School of Integrated Circuits, University of Chinese Academy of Sciences, Beijing 100049, China.

(*E-mail: wangxiaolei@ime.ac.cn, chaijunshuai@ime.ac.cn. #These authors contributed equally.)

Abstract

In this work, we investigate the degradation of the threshold voltage after the erase operation in HfO₂-based silicon channel ferroelectric field-effect transistors (HfO₂ Si-FeFETs) with TiN/SiO₂/Hf₀.₅Zr₀.₅O₂/SiOₓ/Si (MIFIS) gate structure. We find that after the erase operation, the increased hole density trapped at the SiOx/Hf0.5Zr0.5O2 interface and decreased ferroelectric polarization, are key factors leading to the degradation of the threshold voltage after the erase operation (Vth_ERS) for the FeFETs with MIFIS structure. In addition, the low capacitance value of the top SiO₂ interlayer further exacerbates the impact of increased hole density on the V_{th_ERS}.

Keywords: FeFETs, charge trapping and de-trapping, endurance, MIFIS gate structure.

Introduction

Since the discovery of ferroelectricity in doped hafnium oxide materials [1], HfO₂-based Si channel ferroelectric field-effect transistors (HfO₂ Si-FeFETs) have been widely studied due to their fast read/write speed and low power consumption [2, 3]. These excellent properties make HfO₂ Si-FeFETs very promising for applications in 3D NAND. However, the memory window (MW) of HfO₂ Si-FeFETs is about 2 V, which cannot meet the requirements of multi-bit memory cells for applications in 3D NAND [4].

Recently, inserting dielectric layers between the metal gate and ferroelectric layers can significantly expand the MW of FeFETs [5, 6, 7]. The MW enlargement is attributed to the presence of the charges ($Q_{it'}$) injected from the metal gate trapped at the top interlayer/ferroelectric interface. Based on this MIFIS gate structure, Lim et al. have achieved a MW of 5.5 V, and Yoon et al. have achieved a MW of 10.54 V [5, 6]. Our team also achieved a MW of 6.3 V by inserting a 3.4 nm top SiO₂ interlayer [8]. However, the MW of FeFETs with MIFIS gate structure decreases significantly after about $10^3 \sim 10^4$ endurance cycles. Our research finds that the threshold voltage after program operation (V_{th_PGM}) remains almost constant, while the threshold voltage after the erase operation (V_{th_ERS}) decreases during endurance cycling significantly, which

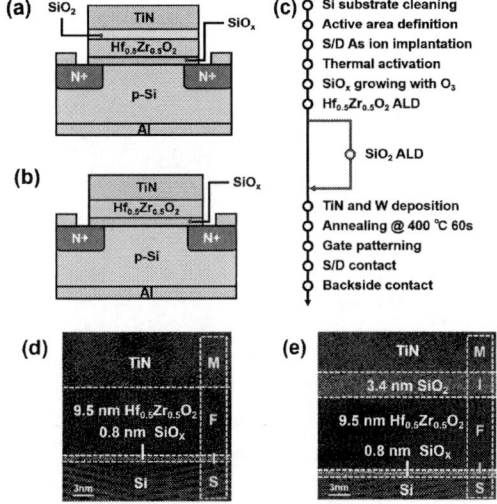

Fig. 1. The device structure of (a) MFIS structure and (b) MIFIS structure. (c) The fabrication process flow for both MIFIS structure and MFIS structure. HRTEM images of (d) the MFIS structure and (e) MIFIS structures.

leads to the rapid degradation of the MW of FeFETs with MIFIS gate stacks. Due to the strong coupling of ferroelectric polarization (P), the charges trapped at the top SiO₂/Hf₀.₅Zr₀.₅O₂ interface ($Q_{it'}$) and the charges trapped at the bottom SiOₓ/Hf₀.₅Zr₀.₅O₂ interface (Q_{it}), the degradation mechanism of the V_{th_ERS} during endurance cycling in MIFIS structure is still unclear.

In this work, we fabricated the Si-FeFETs with MFIS gate structure and MIFIS gate structure. We discuss the impact of three factors on the V_{th_ERS} during endurance cycling: (i) ferroelectric polarization P, (ii) the Q_{it} injected from the silicon substrate trapped at the bottom SiOₓ/Hf₀.₅Zr₀.₅O₂ interface, and (iii) the $Q_{it'}$ injected from the metal gate trapped at top SiO₂/Hf₀.₅Zr₀.₅O₂ interface. Finally, we find that during endurance cycling, the increased holes injected from the silicon substrate and the decreased ferroelectric polarization result in a significant decrease in V_{th_ERS}. Moreover, the low capacitance value of the top SiO₂ interlayer amplifies the impact of the Q_{it} on V_{th_ERS}.

Experiments

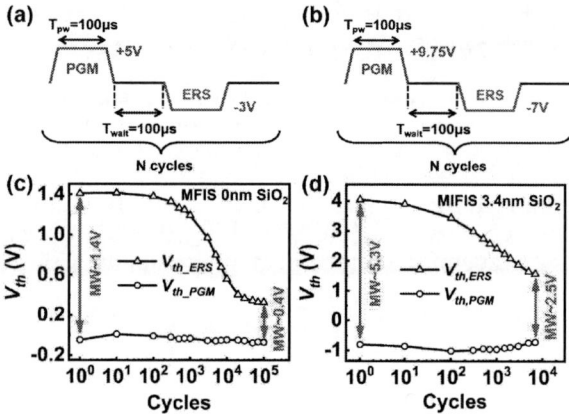

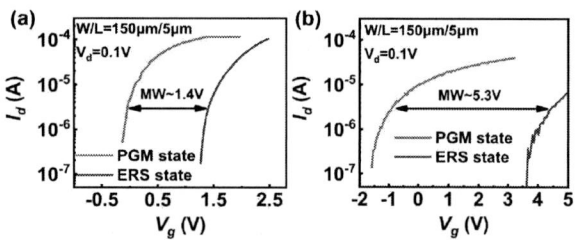

Fig. 2. Endurance measurement waveforms for the (a) MFIS structure and (b) MIFIS structure. Endurance characteristics of the (c) MFIS structure and (d) MIFIS structure.

Fig. 3. I_d–V_g curves of the maximum MW during endurance cycling for the (a) MFIS structure and (b) MIFIS structure.

There are two types of gate structures in this work, i.e., TiN/Hf$_{0.5}$Zr$_{0.5}$O$_2$/SiO$_x$/Si (MFIS) gate structure and TiN/SiO$_2$/Hf$_{0.5}$Zr$_{0.5}$O$_2$/SiO$_x$/Si (MIFIS) gate structure, as shown in Fig. 1(a-b). Both devices were prepared using the gate-last process in an 8-inch p-type silicon wafer. The fabrication process flow is shown in Fig. 1(c). For the MIFIS structure, 9.5 nm Hf$_{0.5}$Zr$_{0.5}$O$_2$ and 3.4 nm top SiO$_2$ interlayer are grown by atomic layer deposition (ALD) at 300 °C, while only 9.5 nm Hf$_{0.5}$Zr$_{0.5}$O$_2$ is grown by ALD for MFIS structure. After the growth of the metal gate TiN and W, these devices were annealed at 400 °C for 60 s in the N$_2$ atmosphere using rapid thermal annealing (RTA) to form an orthorhombic phase.

Fig. 1(d-e) shows high-resolution transmission electron microscopy (HRTEM) images of the MFIS structure and MIFIS structure. The gate length/width (L/W) of these devices in this work is 5/150 μm. The electrical measures were conducted by Keysight B1500A. The threshold voltage (V_{th}) is extracted by the constant current method.

Results and Discussions

We measured the endurance characteristics of both the MFIS structure and the MIFIS structure. The pulse width is set as 100 μs. The program voltage and erase voltage for the MFIS structure and MIFIS structure are +5/-3 V and +9.75/-7 V, respectively, as shown in Fig. 2(a-b). During

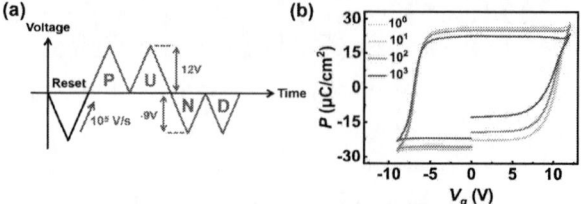

Fig. 4. (a) PUND measurement waveform. (b) PUND measurement result for the MIFIS structure during endurance cycling.

endurance cycling, the MIFIS structure shows a maximum MW of 5.3 V, and the MFIS structure shows a maximum MW of 1.4 V, as shown in Fig. 3. This indicates that the top SiO$_2$ interlayer can effectively increase the MW of FeFETs. Fig. 2(c-d) shows the measurement results of endurance characteristics under the above pulse amplitude. We find that after 10^4 cycles, the V_{th_ERS} of the MIFIS structure is rapidly reduced from the initial 4 V to 1.6 V, while V_{th_PGM} remains almost constant, which results in a sharp decrease in the MW. The V_{th_ERS} of the MIFIS structure can be given as

$$V_{th\_ERS} = \frac{P - Q_{it}}{C_{FE}} + \frac{Q_{it'} - Q_{it}}{C_{TIL}} + c_0 \qquad (1)$$

where c_0 represents the V_{th} of the MIFIS structure in the absence of ferroelectric polarization as well as the trapped charge, which is a constant value during endurance cycling. C_{TIL} and C_{FE} are the capacitance of the top SiO$_2$ interlayer and ferroelectric Hf$_{0.5}$Zr$_{0.5}$O$_2$, respectively. Three factors affect the V_{th_ERS} here: ferroelectric polarization, top interface trapped $Q_{it'}$ and bottom interface trapped Q_{it}, we will discuss their effect on the V_{th_ERS} during endurance cycling separately.

A. ferroelectric polarization

We use the PUND measurement to characterize the changes in ferroelectric polarization for the MIFIS structure during endurance cycling. Fig. 4(a) shows the waveform of PUND measurement. The asymmetric voltage amplitude is due to the use of a P-type silicon substrate. Fig. 4(b) shows the PUND measurement results. It can be seen that the ferroelectric polarization gradually decreases with the increasing number of cycles, which results in the V_{th_ERS} decreases during endurance cycling. However, the ferroelectric polarization increases due to the wake-up effect during 10^5 cycles for the MFM structure [9]. Therefore, we infer that because the fixed charge at the interlayer/Hf$_{0.5}$Zr$_{0.5}$O$_2$ increases with the number of electrical cycles, it causes pinning of ferroelectric domains, which leads to a decrease in ferroelectric polarization [10]. According to Equation (1), we find that the decrease in ferroelectric polarization is one of the reasons for the degradation of the V_{th_ERS}.

B. $Q_{it'}$ injected from the metal gate

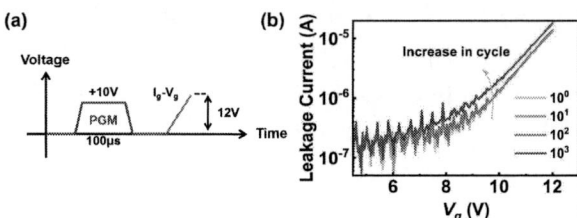

Fig. 5. (a) Measurement waveforms of leakage current. (b) Leakage current measurement result for the MIFIS structure.

We discuss the impact of the change in $Q_{it'}$ trapped at the top $SiO_2/Hf_{0.5}Zr_{0.5}O_2$ interface on V_{th_ERS} during endurance cycling. We conduct the leakage current measurement of the gate stacks during endurance cycling for the MIFIS structure. Fig. 5 shows the waveform and results of the leakage current measurement. The leakage current of the gate stacks gradually increases with the electric cycles. This indicates that applying electric cycles causes an increase in defects of the gate stacks, which leads to an increase in trap density at the top $SiO_2/Hf_{0.5}Zr_{0.5}O_2$ interface. In addition, the degradation problem of the tunnel oxide caused by repeated charge tunneling in NAND has been widely discussed. Some studies have shown that a large number of traps are generated in the oxide during endurance cycles [11]. Thus, for the MIFIS structure, the charges injected from the metal gate to the top $SiO_2/Hf_{0.5}Zr_{0.5}O_2$ interface are not reduced during endurance cycling. This results in an increase in V_{th_ERS}, which is contrary to the observed experimental phenomenon. Therefore, $Q_{it'}$ is not responsible for the rapid degradation of the V_{th_ERS}.

C. Q_{it} injected from the silicon substrate

The injected charge Q_{it} at the bottom $SiO_x/Hf_{0.5}Zr_{0.5}O_2$ interface can be considered using the MFIS structure. For the MFIS structure, the expression of V_{th_ERS} is given as

$$V_{th\_ERS} = \frac{P - Q_{it}}{C_{FE}} + c_1 \qquad (2)$$

where c_1 represents the V_{th} of the MFIS structure in the absence of ferroelectric polarization as well as the trapped charge, which is a constant value during endurance cycling.

From Fig. 2(c), we find that the degradation of V_{th_ERS} is much more significant than that of V_{th_PGM} during endurance cycling for the MFIS structure. Yang et al. studied the dependence of the ferroelectric polarization on the number of electric cycles and found the ferroelectric polarization almost remains constant for the MFIS structure [12]. Thus, the degradation of V_{th_ERS} is attributed to an increase in Q_{it} during endurance cycling. A similar phenomenon has also been reported. Shao et al. observed that the holes injected from the channel increase with the increasing number of electric cycles, which is the key factor contributing to

endurance fatigue for the MFIS structure [13]. The presence of large spontaneous polarization results in the bottom SiO_x experiencing almost the same electric field in both the MFIS and MIFIS structures, close to its breakdown electric field. Therefore, the increase in Q_{it} in the MIFIS structure also leads to the degradation of V_{th_ERS}.

Furthermore, based on Equations (1) and (2), we observe that in the MIFIS structure, the same change in Q_{it} results in a more significant degradation of V_{th_ERS} compared to the MFIS structure, since C_{TIL} is smaller than C_{FE}. Thus, the top interlayer not only facilitates the formation of a large MW utilizing the $Q_{it'}$ but also amplifies the adverse effects of the change in Q_{it} on the endurance characteristics of the MIFIS structure. To improve the endurance characteristics of HfO_2 Si-FeFETs with the MIFIS structure, we should focus on the reduction of the charge Q_{it} injected from the silicon channel to the bottom $SiO_x/Hf_{0.5}Zr_{0.5}O_2$ interface.

Conclusion

In this work, we investigate the degradation mechanism of the V_{th_ERS} in HfO_2 Si-FeFETs with MIFIS gate structure. We find that as the increasing number of electric cycles, the increased hole density after the erase operation trapped at the $SiO_x/Hf_{0.5}Zr_{0.5}O_2$ interface and decreased ferroelectric polarization of the ferroelectric $Hf_{0.5}Zr_{0.5}O_2$ are key factors leading to the degradation of the V_{th_ERS} for the FeFETs with MIFIS structure. In addition, the low capacitance value of the top SiO_2 interlayer further exacerbates the adverse impact of increased hole density on the V_{th_ERS} of the MIFIS structure.

Acknowledgments

This work was supported in part by the National Natural Science Foundation of China under Grant Nos. 92264104 and 52350195, and Supported by the Postdoctoral Fellowship Program of CPSF under Grant No. GZC20232925.

References

[1]   T. S. Böscke et al., Appl. Phys. Lett. 99, 102903 (2011).

[2]   J. Müller et al., Appl. Phys. Lett. 99, 112901 (2011).

[3]   M. Trentzsch et al., IEDM2016, p.294.

[4]   K. Florent et al., IEDM 2018, pp. 2.5.1-2.5.4.

[5]   S. Lim et al., IEDM2023, pp. 1-4.

[6]   S. Yoon et al., VLSI2023, pp. 1-2.

[7]   T. Hu et al., IEEE Electron Device Lett. 45, 825 (2024).

[8]   T. Hu et al., IEEE Trans. Electron Devices Early Access, 1 (2024).

[9]   Mittmanne et al., Microelectron. Eng.178, 48 (2017).

[10]   M. Pešić et al., Adv. Funct. Mater. 26 (2016).

[11]   H. Jo et al., IEEE Trans. Electron Devices. 71, 1852 (2024).

[12]   J. Yang et al., IEEE Trans. Electron Devices Early Access, 1 (2024).

[13]   X. Shao et al., IEEE Trans. Electron Devices. 70, 3043 (2023).

Modeling of the Switching Characteristics of Ag/HfO$_2$-based Volatile Memristors

Yikang Huo[1,2], Chunsheng Jiang[1,2], and Qin Xie[1,2]

[1] Guangxi Key Laboratory of Brain-inspired Computing and Intelligent Chips, Guangxi Normal University, Guilin, China, **email**:csjiang13@163.com

[2] Key Laboratory of Integrated Circuits and Microsystems (Guangxi Normal University), Education Department of Guangxi Zhuang Autonomous Region, Guilin, China

Abstract

The Ag/HfO$_2$-based volatile memristor shows great potential for use in low-power logic transistors and neuromorphic computing. In this paper, we present a compact current-voltage (I-V) model for Ag/HfO$_2$-based threshold switching memristors, based on the formation and rupture of conductive filaments. From this model, we derive analytical expressions for the threshold voltage (V_{th}) and hold voltage (V_h) that describe the switching characteristics for the first time. The calculated results align closely with experimental data. Finally, we systematically analyze the influence of various process parameters on V_{th} and V_h.

Keywords: Ag/HfO$_2$-based volatile memristor, Switching characteristics and Compact modeling

Introduction

The Ag/HfO$_2$-based threshold switching (TS) device offers several advantages, including ultra-low off-state leakage current (~0.1 pA), an exceptionally high on/off current ratio (>6), and a simple structure. These features make it well-suited for use in steep slope devices [1-2] and artificial neurons [3-4]. The TS device exhibits distinct steep threshold switching behavior, driven by the formation and rupture of conductive filaments[1]. A compact current-voltage (I-V) model for Ag/HfO$_2$-based memristors is crucial for understanding their working mechanism, optimizing device structures, and enabling circuit simulations. However, compact models for TS devices remain scarce to date. In this paper, an analytical I-V model for Ag/HfO$_2$-based volatile memristor is developed relying on the Frenkel-Poole emission and Flower-Nordheim tunneling mechanisms. This model is capable of describing the dynamic evolution process of conductive filaments.

Model Description

The formation of conductive filaments in TS devices occurs in three stages: (a) when an external voltage (V_{app}) is applied, active silver (Ag) atoms are oxidized into mobile metal cations (Ag$^+$): Ag → Ag$^+$; (b) these Ag$^+$ ions then migrate toward the inert bottom electrode; (c) as more Ag+ ions accumulate, they gradually form metal nanoclusters that bridge the top and bottom electrodes, resulting in the formation of a conductive filaments. This transition causes the device to switch from a high-resistance state (HRS) to a low-resistance state (LRS). Note that the formation of the conductive filaments is driven by the competition between ion drift, induced by the applied electric field (E), and the intrinsic diffusion of ions. If the electric field (E) is not sufficiently strong, the filaments will break spontaneously, causing Ag$^+$ ions to reduce to Ag atoms. Thus, the filament's formation is governed by ion migration under the electric field, while its rupture is driven by ion diffusion. This competition between ion drift and ion diffusion can be described the following equation [5]:

$$\frac{dt_{cf}}{dt} = \mu_i E - D_s \frac{t_{cf}}{t_{ox}^2} \tag{1}$$

Here, t_{cf} represents the total length of the metal cluster (i.e. conductive filament), and t_{ox} refers to the thickness of the oxide layer. Meanwhile, μ_i denotes the ion mobility, and D_s represents the diffusion coefficient. As noted in our previous work [1], the current (I_{fp}) in HRS of TS devices follows the Frenkel-Poole emission model.

$$I_{fp} = \alpha_{fp} \sinh\left(\gamma_{fp}\sqrt{E}\right), E = V_m/t_{ox,eq} \tag{2}$$

where both α_{FP} and γ_{fp} are fitting parameters extracted from the experimental data [1]. V_m is the voltage drop on the equivalent oxide layer (i.e. $t_{ox,eq} = t_{ox} - t_f$). E represents the electric field within the insulating oxide layer. On the other hand, the LRS current (I_{fnt}) of TS devices is described by Fowler-Nordheim tunneling mechanism:

$$I_{fnt} = \frac{E^2}{\beta_{fnt}} exp\left(-k\frac{1}{E}\right) \tag{3}$$

$$k = \frac{8\pi\varphi_B\sqrt{2qm^*\varphi_B}}{3h} \tag{4}$$

In the above equation, β_{fnt} is a fitting parameter, while φ_B represents the contact potential between the functional oxide layer and the active metal electrode, arising from their work function difference. Here, q is the elementary charge of an electron, m^* is the effective mass of the tunneling electrons, and h is Planck's constant.

Finally, the parasitic series resistance of the contact electrodes is denoted as R. Consequently, the current-voltage relationship can be expressed as follows:

$$V_a = I \cdot R + V_m, I = I_{fp} + I_{fnt} \tag{5}$$

In this expression, V_a represents the total externally applied voltage, and I is the conduction current through the TS device.

Results and Discussion

Fig. 1(a) illustrates the I-V curve for both simulated and

experimentally measured data [1] of a TS device. The default values of the model parameters are provided in Table 1. The simulation results align closely with the experimental data, validating our model. Notably, the device exhibits pronounced steep switching and hysteretic characteristics. As shown in Fig. 1(b), the steep I-V behavior arises from the rapid formation and rupture of conductive filaments, with both the threshold and hold voltages (V_{th1} and V_h) being consistent across t_{cf}. In HRS, the Frenkel-Poole emission current is the dominant mechanism, whereas in LRS, the Fowler-Nordheim tunneling current takes precedence, as illustrated in Fig. 2.

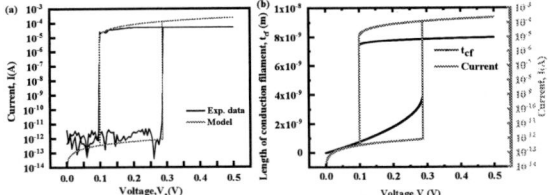

Fig.1 (a) Comparison of simulation data and experimental data of a TS device; (b) t_{cf} and I varying with the applied voltage of the same device.

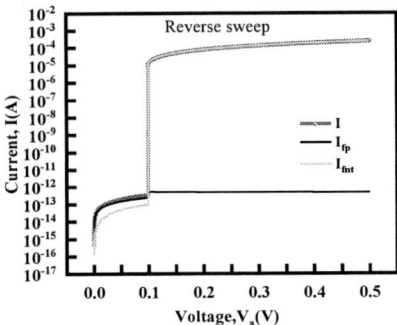

Fig. 2 The Frenkel-Poole emission current (I_{fp}), Fowler-Nordheim tunneling current (I_{fnt}) and the total current (I) with respect to V_a.

The threshold voltage (V_{th}) and hold voltage (V_h) are key parameters for describing the hysteretic switching behavior of Ag/HfO$_2$-based diffusive memristors. Analytical expressions for V_{th} and V_h can be derived from the developed I-V model, allowing for the exploration of how different processes affect these voltages. Firstly, the threshold voltage (V_{th}) can be derived as follows. In the high-resistance state (HRS), the total conduction current (I) is small, so the voltage drop across the contact resistance (R) can be neglected. As a result, the applied voltage (V_a) is approximately equal to V_m. Consequently, Eq. (1) simplifies to:

$$\frac{u_i V_a}{t_{ox} - t_{cf}} - \frac{D_s t_{cf}}{t_{ox}^2} = 0 \qquad (6)$$

It can be easily observed that when $t_{cf} = 0.5 t_{ox}$, the applied voltage V_a reaches the threshold voltage V_{th}. In other words,

$$V_{th} = \frac{1}{4}\frac{D_s}{\mu_i} \qquad (7)$$

Interestingly, a linear relationship between V_{th} and the term D_s/u_i is predicted in (7).

The hold voltage (V_h) can be determined as follows. As illustrated in Fig. 1, the hysteretic I-V characteristic arises due to the negative differential resistance (NDR) effect. Therefore, V_h can be derived using the following equation:

$$\frac{dV_a}{dI}\Big|_{V_a=V_h} = 0 \qquad (8)$$

As indicated in Fig. 2, I_{fnt} is the dominant component of the total conduction current in LRS. In other words,

$$I \approx I_{fnt} = \frac{1}{\beta_{fnt}} E_c^2 \exp\left(-k\frac{1}{E_c}\right) \qquad (9)$$

Here, E_c represents the critical electric field when $V_a = V_h$. For a static I-V sweep, Eq. (1) is reduced to:

$$\mu_i E = D_s \frac{t_{ox} - \frac{V_m}{E_c}}{t_{ox}^2} \qquad (10)$$

By solving equations (5) and (8)-(10), an implicit equation for E_c is obtained:

$$2\frac{\mu_i E_c t_{ox}^2}{D_s} - t_{ox} = (2E_c + k)\frac{R}{\beta_{fnt}}\exp\left(-k\frac{1}{E_c}\right) \qquad (11)$$

Fortunately, Eq. (11) can be solved analytically using r-Lambert function[6]:

$$\frac{1}{E_c} = \frac{2}{k} + \frac{1}{k}W_{r,-2}\left[-\frac{2t_{ox}\beta_{fnt}}{R}\cdot\left(\frac{1}{k}+\frac{u_i t_{ox}}{D_s}\right)\cdot e^{-2}\right] \qquad (12)$$

$$V_h = E_c t_{ox} - \frac{u_i t_{ox}^2}{D_s}\cdot E_c^2 + \frac{1}{\beta_{fnt}}E_c^2 \exp\left(-\frac{k}{E_c}\right)R \qquad (13)$$

It is important to note that the parabolic term is the dominant factor in Eq. (13).

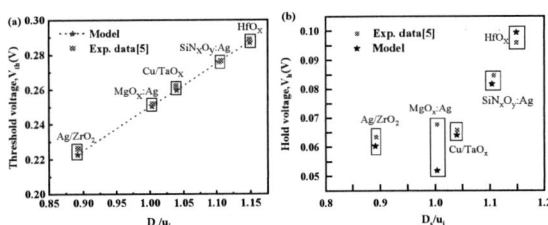

Fig. 3 (a) Comparison between V_{th} computed from (7) and experimental data. (b) Comparison between the V_h calculated from (13) and the experimental data.

To confirm the accuracy of equations (7) and (13), we compare the simulated results with experimental data from [5] for TS devices made from different materials. Notably, equation (7) precisely predicts the linear relationship between V_{th} and the term D_s/u_i, as observed in the experimental measurements and illustrated in Fig. 3(a). Figure 3(b) presents a comparison between the calculated V_h from equation (13) and the experimental data from [5], showing a strong agreement between the two. Since $W_{r,-2}(x)$ in equation (12) is a monotonically increasing function, E_c decreases as the ratio D_s/u_i increases. Consequently,

979-8-3315-0417-5/25 $31.00 © 2025 IEEE

according to equation (13), V_h increases as D_s/u_i becomes larger.

We now examine the effects of various processes on V_{th} and V_h using the developed model. Fig. 4(a) shows the hysteretic I-V characteristics of a TS device for different values of R. According to (7), R has no effect on V_{th}, as the voltage drop across R can be neglected when the device is in the off-state. However, for V_h, as R increases, the voltage drop across R also increases when the device works in the on-state, requiring a higher V_h to sustain the formation of conduction filaments. Fig. 4(b) shows the I-V curves of a TS device for different t_{ox} values. From equation (7), t_{ox} has no effect on V_{th}. However, as t_{ox} increases, V_h decreases significantly because the diffusion effect is weakened in this scenario, as described by (1).

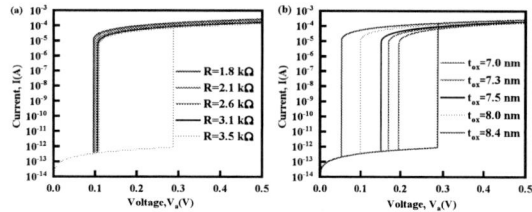

Fig. 4 (a) I-V characteristics for different R obtained from analytical model. (b) I-V characteristics for various different t_{ox}.

Fig. 5(a), (b), (c), and (d) illustrate the effects of α_{fp}, β_{fnt}, φ_B and m^* on the I-V characteristics of a TS device, respectively. As seen from equation (7), none of these parameters affect V_{th}. Additionally, α_{fp} has no influence on V_h, as the Frenkel-Poole emission current can be neglected in the low-resistance state (LRS). From equation (12), E_c increases as β_{fnt} rises, which in turn reduces V_h, as described by equation (13). Both ϕ_B and m^* increase the value of k in (4), and since E_c is approximately proportional to k, they both enhance E_c and consequently lower V_h, as indicated by (13).

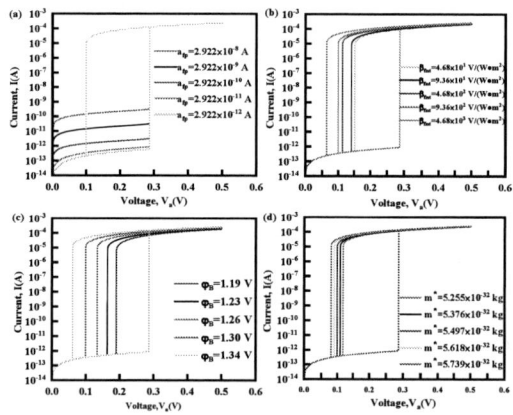

Fig. 5 (a) I-V characteristics for different α_{fp}. (b) I-V characteristics for various different β_{fnt}.(c) I-V characteristics for various different φ_B.(d) I-V characteristics for various different m^*.

Conclusion

In this work, a compact I-V model for Ag/HfO$_2$ -based volatile memristors is developed, with analytical expressions for the threshold and hold voltages derived for the first time. The proposed model shows strong agreement with experimental data. Additionally, the effects of various parameters on V_h and V_{th} are systematically analyzed using the developed I-V model and analytical expressions. The conclusions drawn from this study can be applied to optimize transistors and memory devices based on Ag/HfO$_2$ -based diffusive memristors.

Acknowledgments

This work is supported in part by Guangxi Science and Technology Base and Talent Special Project (No. AD22035213), Guilin Innovation Platform and Talent Special Plan (No. 20220125-1), the National Science Foundation of Guangxi province (No. 2023GXNSFBA026021), and Outstanding Youth Fund of Guangxi Normal University (No.2022TD005).

Tabel 1 Default model parameters

Parameter	Physical Meanings	Values
$\alpha_{fp}(A)$	The first fitting parameters in PF emission model	2.922×10^{-11}
$\gamma_{fp}(V^{-0.5})$	The second fitting parameters in PF emission model	1.3443×10^{-6}
$\mu_i(m^2(V \cdot s)^{-1})$	Ionic migration coefficient	7.94×10^{-11}
$D_S(m^2 s^{-1})$	Diffusion coefficient	9.122×10^{-11}
$t_{ox}(m)$	Thickness of oxide layer	8×10^{-9}
$\beta_{fnt}(V\Omega m^{-2})$	The fitting parameters in F-N tunneling model	9.36×10^2
$\varphi_B(V)$	Difference in work function	1.30
$m^*(kg)$	Effective mass of the tunneling electrons	5.497×10^{-32}
$q(C)$	Electron charge	1.6×10^{-19}
$h(J \cdot s)$	Planck constant	6.62×10^{-34}
$R(\Omega)$	Series resistance	2.1 k

References

[1] W. Cheng, R. Liang, and Q. Hua et al., "A novel steep slope hybrid InGaZnO TFT with negative DIBL improvement based on the Ag/HfO$_2$ threshold switching device", Appl. Phys. Express, 12, pp. 091002(2019).

[2] Q. Hua, G. Gao, and C. Jiang et al., "Atomic threshold-switching enabled MoS2 transistors towards ultralow-power electronics", Nat. Commun., 11, pp. 7207(2020).

[3] Zhang Z., Gao S., and Z. Li et al., "Artificial LIF neuron with bursting behavior based on threshold switching device", IEEE Trans. Electron Devices, 70, pp. 1374(2023).

[4] Q. Hua, C. Jiang, and W. Hu, "AG/HFO$_2$-based threshold switching memristor as an oscillatory neuron", 5th IEEE Electron Devices Technology & Manufacturing Conference (EDTM) (2021).

[5] Y. Dai, J. Zou, and Z. Feng et al., "Modeling of a diffusive memristor based on the DT-FNT mechanism transition", Semicond. Sci. Technol., 37, pp. 095001(2022).

[6] I. Mező, "On the structure of the solution set of a generalized Euler–Lambert equation", J. Math. Anal. Appl., 455, pp. 53(2017).

A Submersible Soft Robot with Ultrasonic Echolocation Capabilities

Zhongming Chen[1], Qilin Hua[1,*], Jiaqiang Xu[1], Guozhen Shen[1,*]

[1] School of Integrated Circuits and Electronics, Beijing Institute of Technology, China
Email: huaqilin@bit.edu.cn; gzshen@bit.edu.cn

Abstract

Underwater soft robots are being developed to address the limitations of traditional underwater robots, such as heavy weight, susceptibility to rust, and lack of flexibility. They can replace humans in tasks like underwater search and rescue, exploration and salvage, and pipeline inspection and maintenance. Recognizing the importance of echolocation, this paper proposes a submersible soft robot with flexible ultrasonic array for bathymetric measurement, underwater topography imaging, and underwater defect detection, offering a new solution for profound exploration under harsh conditions.

Keywords: Flexible electronics, Ultrasonic imaging and Soft robotics

Introduction

Submersible robots, designed for underwater operations, such as underwater search and rescue missions, resource exploration, urban pipeline inspections, and more. Advanced soft robotics [1], recognized for their enhanced flexibility, adaptability, eco-friendliness, and lightweight nature, have garnered significant attention as a promising avenue for underwater exploration. And the use of echo-based sensors [2] plays a pivotal role in locating and detecting objects underwater. Benefiting from the progress in flexible electronics [3-4], traditional rigid ultrasonic arrays are transitioning towards more flexible configurations for three-dimensional (3D) imaging of complex surfaces and objects [5]. Ultrasonic detection methods offer inherent benefits such as high sensitivity, deep penetration, and cost-effectiveness. The development and introduction of flexible ultrasonic arrays make it possible to organically integrate them with soft robots.

To address challenging scenarios like weld and inherent defects in underwater pipelines, we proposed a versatile programmable ultrasonic soft robot, which is inspired by the locomotion of worms and jellyfish. Driven by gas, this robot maneuvers effectively in aquatic environments, equipped with a flexible ultrasonic array for enhanced detection capabilities during its movements. Leveraging the unique material properties and robust encapsulation of the soft robot, it excels in underwater missions, and promising applications in underwater terrain imaging, bathymetry, and related tasks. This technology would present extensive opportunities across industrial inspections, disaster response efforts, and underwater surveillance applications.

Device Design and Fabrication

Fig. 1a illustrates the assembly process of the ultrasonic soft robot. The sensing module of the ultrasonic array employs 1-3 composite piezoelectric ceramics (1-3 composite) with prominent directionality, low acoustic impedance, and outstanding electromechanical coupling coefficients. These ceramics are adhered to Ni-tape patterned electrodes cut by a laser to establish a conductive path. And the upper and lower layers are then encapsulated by polydimethylsiloxane (PDMS) and polyimide (PI), respectively. The Ecoflex precursor liquid is injected into a 3D-printed mold for gas chamber formation and soft robot encapsulation. Directional bending of the pneumatic tentacles is achieved through secondary PDMS packaging at the bottom layer. Lastly, the ultrasonic array and soft robot are bonded using PDMS to finalize the ultrasonic soft robot assembly. In Fig. 1b, a closer view of the ultrasonic array reveals a departure from the common one-to-one addressing method (shared ground electrode). Instead, a more efficient row and column addressing approach is employed, reducing the signal output lines from N*N to N+N.

Characterization of ultrasonic soft robot

As shown in Fig. 1c, impedance spectrum tests were performed on a series of 1-3 composites, revealing an average resonant frequency of 5.18 MHz across the nine elements with significant consistency. Subsequent simulation of the ultrasonic beam from a single element (Fig. 1d) demonstrated enhanced directionality and penetration depth. The time-frequency domain echo analysis in Fig. 1e shows a device center frequency of 5.26 MHz with a frequency bandwidth (BW) at −6dB measuring 17% [6].

A series of characterizations of the soft robot is presented in Fig. 1f-h. Fig. 1f shows the low-cycle fatigue tests conducted on the inflated material Ecoflex under varying strains, highlighting its elastic deformation capabilities up to 150%, ensuring prolonged reusability of the soft robot. A comparison of Young's modulus between the inflated

979-8-3315-0417-5/25 $31.00 © 2025 IEEE

material Ecoflex and the bending material PDMS is presented in Fig. 1g, indicating that the former, with a lower Young's modulus, is more predisposed to inflation and expansion, causing the tentacles to bend towards the PDMS side [7]. To enhance the bond between the two materials, a microstructure design was implemented, as shown in Fig. 1h. The microstructure, unlike the smooth surfaces of the materials, increases the contact area, fortifying adhesion and reducing contact friction, thereby facilitating smoother movements of the soft robot.

Ultrasonic echolocation capabilities

We carried out a series of tests in water on the assembled ultrasonic soft robot, as illustrated in Fig. 2a, featuring a four-tentacles design that allows movement in eight directions for adept navigation through intricate terrains.

As shown in Fig. 2b and c, the ultrasonic soft robot was placed on a path with aluminum alloy defect test blocks, with signals collected from three elements (2, 5), (1, 5), and (2, 4) for analysis. Using the formula $x = vt/2$, where x represents the defect distance from the surface, v is the ultrasound propagation speed in the aluminum alloy (6300 m/s), and t is the ultrasound flight time, we determined that the (2, 5) element identified defects at 1.25 cm and 3.30 cm on test block 1 (red data graph). The (1, 5) and (2, 4) elements further delineated defect orientation, pinpointing a defect at 1.35 cm in the direction of (2, 5) to (2, 4), and another at 3.30 cm in the direction of (2, 5) to (1, 5). The actual location of the defect is 1.00 cm and 3.50 cm, and the accuracy rate is 80% and 94%, respectively. Similarly, defects in test block 2 (black data graph) at 2.33 cm and test block 3 (blue data graph) at 2.11 cm were detected with accuracies of 97% and 95%, respectively. Further analysis, the size of the defect can be obtained, which will be expanded in the future. Moreover, as shown in Fig. 2d, the ultrasonic soft robot demonstrated diving exploration capabilities, offering real-time updates on water bottom distances, and showcasing its potential for underwater detection tasks.

Conclusion

A submersible soft robot with ultrasonic echolocation capabilities is proposed and assembled, which organically combines the flexible ultrasonic array with the soft robot, realizes defect detection in narrow pipes, complex terrain and semi-enclosed space, and shows the ability of underwater detection. In the future, highly integrated underwater soft robots are expected to become more intelligent and capable of performing more complex and precise tasks.

Acknowledgments

The authors are thankful for the support from the National Key Research and Development Program of China (2023YFC3603500), the National Natural Science Foundation of China (62374018, and 61904012), Beijing Natural Science Foundation (L223006), the Fundamental Research Funds for the Central Universities, and Beijing Institute of Technology Research Fund Program for Young Scholars.

References

[1] R. F. Shepherd et al., "Multigait soft robot," Proceedings of the National Academy of Sciences, vol. 108, no. 51, pp. 20400-20403, (2011).

[2] B. Wang et al., "Body-Integrated Ultrasensitive All-Textile Pressure Sensors for Skin-Inspired Artificial Sensory Systems," Small Science, vol. 4, no. 9, p. 2400026, (2024).

[3] Q. Hua and G. Shen, "Low-dimensional nanostructures for monolithic 3D-integrated flexible and stretchable electronics," Chemical Society Reviews, 10.1039/D3CS00918A vol. 53, no. 3, pp. 1316-1353, (2024).

[4] X. Chen, J. A. Rogers, S. P. Lacour, W. Hu, and D.-H. Kim, "Materials chemistry in flexible electronics," Chemical Society Reviews, vol. 48, no. 6, pp. 1431-1433, (2019).

[5] Z. Zhang, Y. Wang, D. Mei, and J. Jin, "Highly sensitive and stretchable ultrasonic transducer array for object internal characteristics detection in robotics," IEEE Transactions on Instrumentation and Measurement, vol. 72, pp. 1-9, (2023).

[6] H. Hu et al., "Stretchable ultrasonic transducer arrays for three-dimensional imaging on complex surfaces," Science Advances, vol. 4, no. 3, p. eaar3979, (2018).

[7] Y. Yang, Y. Wang, M. Lin, M. Liu, and C. Huang, "Bio-inspired facile strategy for programmable osmosis-driven shape-morphing elastomer composite structures," Materials Horizons, vol. 11, no. 9, pp. 2180-2190, (2024).

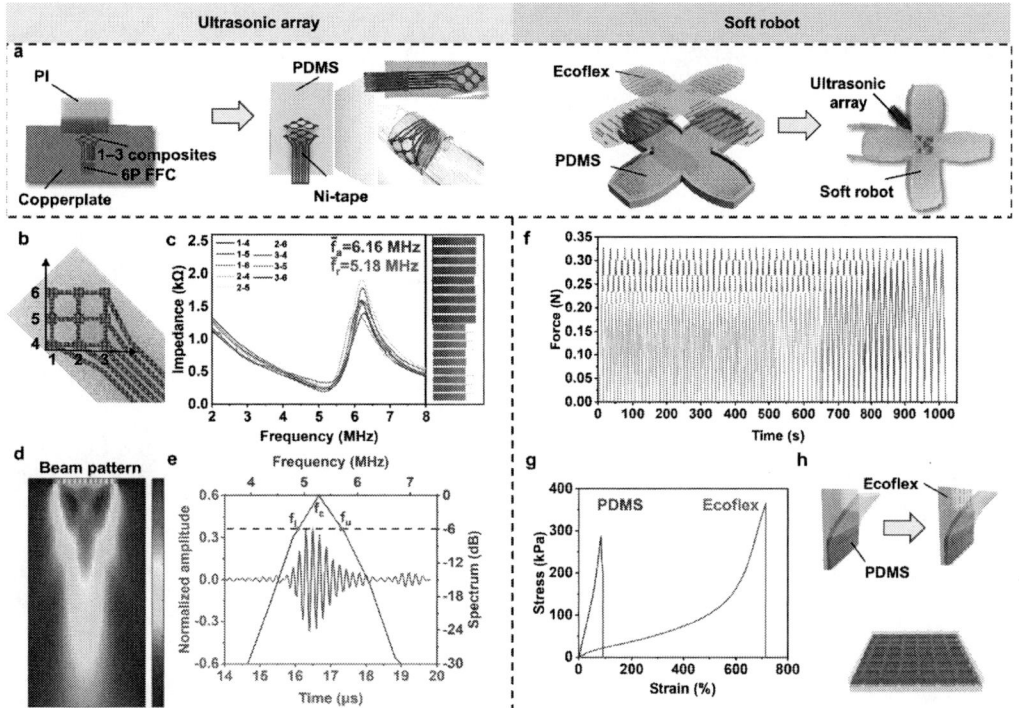

Fig. 1: (a) Diagram of assembling ultrasonic soft robot. (b) Diagram of the ultrasonic array and label method. (c) The impedance spectra, resonant frequency, and anti-resonant frequency of the 1-3 composites. (d) Beam characterization of single 1-3 composite. (e) Frequency spectra of the ultrasonic array. (f) Low cycle fatigue test of Ecoflex under different strains. (g) PDMS and Ecoflex tensile fracture test. (h) PDMS and Ecoflex bonding diagram.

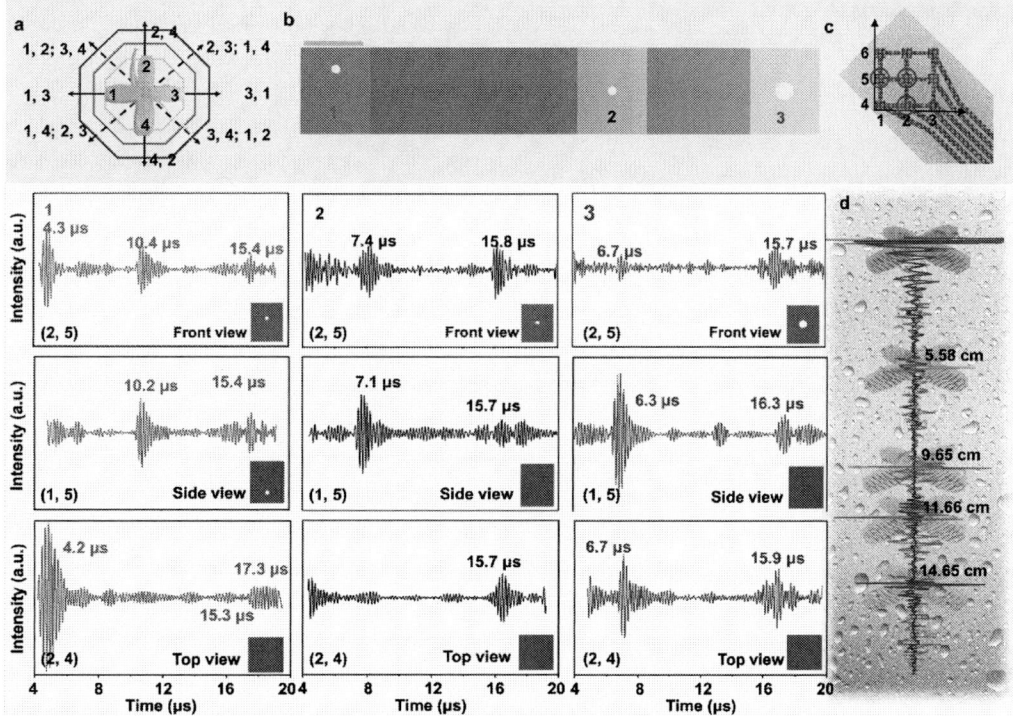

Fig. 2: (a) Diagram of multi-directional motion of soft robots. (b) Diagram of 3D printing of metal defect path. (c) The echo data of 1-3 composites at coordinates (2, 5), (1, 5), (2, 4). (d) Real-time echolocation of ultrasonic soft robot diving.

Recent Progress on Electronics and Optoelectronics Based on 2D Tellurium

Jiajia Zha, Haoxin Huang, Chaoliang Tan[*]

Department of Electrical Engineering, City University of Hong Kong, Hong Kong SAR 999077, China
*Email: chaoltan@cityu.edu.hk

Abstract

Tellurium (Te), an emerging *p*-type van der Waals (vdW) semiconductor, has attracted great attention since its rediscovery. Its unique quasi-one-dimensional (1D) crystal structure grants it high hole mobilities and tunable bandgaps, covering short wavelength infrared (SWIR) spectrum. These properties lay the foundation for its applications in advanced electronics and optoelectronics, including *p*-type field-effect transistors (FETs), infrared (IR) photodetectors, and memory devices. Here, we summarize the recent progress made by our group on Te-enabled devices.

Keywords: Tellurium, Electronics and Optoelectronics

Introduction

Although tellurium (Te), a chalcogenide, was discovered more than two centuries ago, it did not attract much attention until its successful preparation in two-dimensional (2D) morphology [1, 2]. 2D Te is composed of one-dimensional (1D) helical Te molecular chains, stacked in a hexagonal array via van der Waals (vdW) forces [3]. This unique crystal structure endows 2D Te with many intriguing properties, making it a promising candidate for applications in advanced devices. For example, unlike common *n*-type 2D semiconductors, such as those represented by transition metal dichalcogenides (TMDs) thin films, 2D Te exhibits *p*-type transport behaviors with high hole mobilities exceeding one thousand cm^{2}V^{-1}s^{-1}. Moreover, the bandgap of 2D Te increases from ~0.3 eV to ~1.0 eV as the thickness decreases to monolayer limit, covering the short wavelength infrared (SWIR) spectrum. Distinct from the well-known *p*-type black phosphorus (bP), the high air stability of Te enhances its compatibility with various device fabrication technologies. In addition, 2D Te presents considerable thermoelectric, piezoelectric, and ferroelectric properties, which could encourage it to find applications in energy harvesting. Leveraging these fascinating properties, our group, along with many other researchers, has reported various electronics and optoelectronics enabled by 2D Te. These developments significantly enrich the 2D device family and promote the step of 2D semiconductors into practical applications. In this paper, we will briefly summarize our recently published works on electronics and optoelectronics based on 2D Te.

Electronics Based on 2D Te

The small hole effective mass of 2D Te leads to ultrahigh field-effect hole mobility, highlighting its suitability for use in high-performance *p*-type field-effect transistors (FETs). Considering the organic residue on the surfaces of the solution-phase prepared Te nanoflakes may impede subsequent device performance, a modified chemical vapor deposition (CVD) method has been developed to prepare single-crystalline Te nanobelts [4].

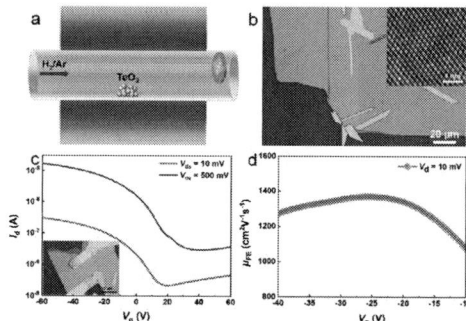

Fig. 1 CVD-grown Te nanobelts for *p*-type FET with hole mobility [4].

As shown in Fig. 1a, vdW hexagonal boron nitride (h-BN) nanoflakes with atomically flat surfaces have been introduced as the growth substrate and the single-crystalline nature of the prepared samples was identified by high-resolution transmission electron microscopy (HRTEM) images (Fig. 1b). Thanks to the high crystal quality of Te nanobelt channel and the significantly reduced scattering centers at the interface between channel and h-BN dielectric, the subsequent *p*-type FET based on the prepared sample (Fig. 1c) achieves a room-temperature field-effect hole mobility up to 1370 cm^{2}V^{-1}s^{-1} (Fig. 1d).

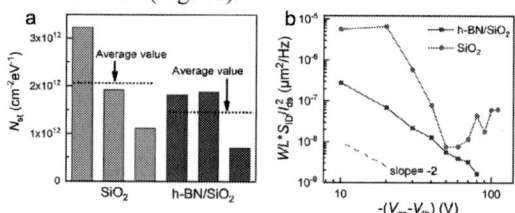

Fig. 2 Comparison of the surface trap density and normalized power spectral density (PSD) for Te FETs on SiO$_2$ and h-BN/SiO$_2$ substrates [5].

Further study on the mechanism of current fluctuation in Te FET reveals that the introduction of h-BN layer can not only reduce the surface trap density (N_{st}) (Fig. 2a), but also decrease the low frequency noise (LFN) in hole accumulation area (Fig. 2b) [5].

979-8-3315-0417-5/25 $31.00 © 2025 IEEE

Although Te FETs exhibit high carrier mobilities, the narrow bandgap of the Te channel significantly limits their current on/off ratios. Alloying Te with selenium (Se) could widen the bandgap of Te and therefore provide a promising solution.

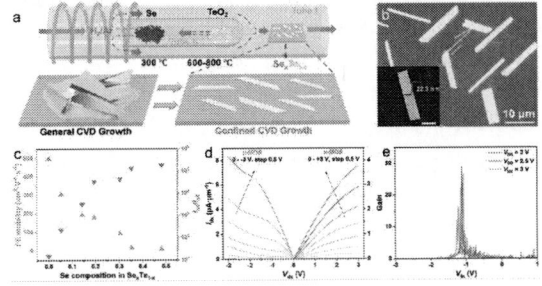

Fig. 3 Se_xTe_{1-x} alloy-based p-FET and 2D inverters [6].

As shown in Fig. 3a, single-crystalline Se_xTe_{1-x} alloy nanosheets with a tunable composition ranging from $x = 0.14$ to 0.49 can be prepared using a precursor-confined CVD system [6]. These products exhibit 2D morphology, with typical thicknesses in tens of nanometers and lateral sizes of tens of micrometers (Fig. 3b). The field-effect hole mobilities and on/off current ratios are highly related to Se composition (Fig. 3e) and $Se_{0.3}Te_{0.7}$ nanosheets exhibit high on/off current ratios up to 4×10^5 and hole mobilities up to 120 $cm^2V^{-1}s^{-1}$. Integrating p-type $Se_{0.3}Te_{0.7}$ and n-type MoS_2 nanosheets with comparable performance (Fig. 3d), a 2D inverter was demonstrated that achieves a gain of 30 at the operation voltage of $V_{dd} = 3$ V was demonstrated (Fig. 3e).

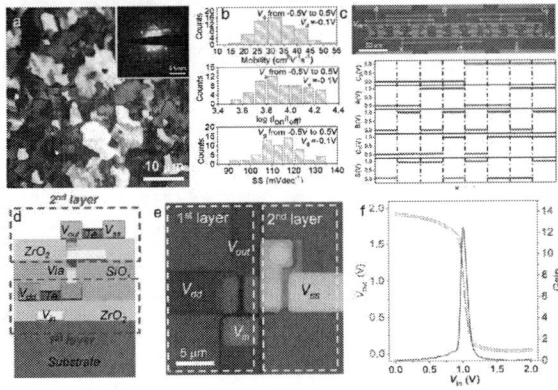

Fig. 4 Logic gates and 3D inverters based on Te thin films [7].

In addition to single p-FETs based on Te/Se_xTe_{1-x} alloy nanoflakes, logic gates and three-dimensional (3D) inverters can be constructed based on thermally evaporated Te thin films [7]. Fig. 4a provides the polarized light microscopy image of an evaporated 9-nm Te film and its polycrystalline nature was corroborated by HRTEM characterization. The Te FET arrays fabricated from Te thin film exhibit considerable effective hole mobilities of ~30 $cm^2V^{-1}s^{-1}$, on/off current ratios of ~10^4, and subthreshold swing (SS) values of ~110 $mVdec^{-1}$ (Fig. 4b). Based on 35 Te FETs, a full adder was successfully demonstrated, functioning properly with a maximum output voltage loss of only 6% as shown in Fig. 4c. The prepared Te thin films also facilitate the construction of

3D inverters, which achieve a gain of ~12 at an operating voltage of 2 V (Fig. 4d–f).

Optoelectronics Based on 2D Te

Leveraging its superior infrared (IR) response, 2D Te has also been employed as the building block for optoelectronic devices working in the IR region [8].

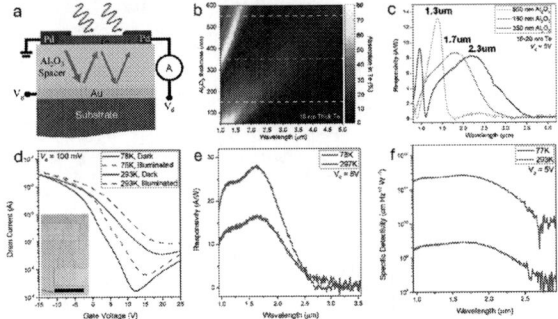

Fig. 5 SWIR photodetectors enabled by Te nanoflakes [3].

Fig. 5a illustrates the SWIR photodetector enabled by hydrothermally synthesized Te nanoflakes and the specifically designed aluminum oxide (Al_2O_3) optical spacer functions as the gate dielectric and optical spacer simultaneously [3]. After controlling the thickness of the optical spacer layer, the photoresponsivity of the detector could be enhanced and the peak position in photoresponse spectra could be tuned (Fig. 5b and 5c). The fabricated device exhibits obvious photocurrent under the illumination of 1.7-μm laser at both 78 K and 293 K (Fig. 5d). The low-temperature peak responsivity and specific detectivity (D^*) at $\lambda = 1.7$ μm reach 27 AW^{-1} (Fig. 5e) and 2.6×10^{11} Jones $(cmHz^{1/2}W^{-1})$ (Fig. 5f), respectively.

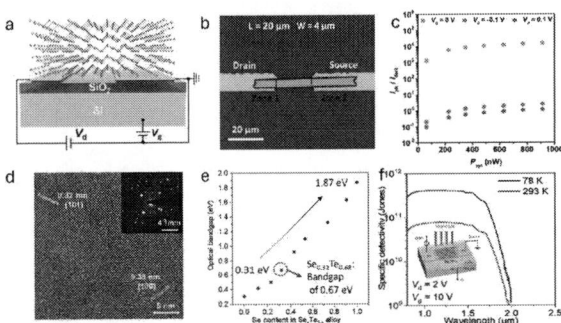

Fig. 6 Asymmetrically contacted Te SWIR photodetector [9] and Se_xTe_{1-x} SWIR photodetector [10].

The narrow bandgap of 2D Te makes it an ideal candidate for high-performance IR photodetectors; however, thermally excited carriers across this narrow bandgap at room temperature can lead to a large dark current, ultimately degrading the performance of Te-based IR photodetectors. The introduction of the asymmetrical contacts in Te SWIR, enabling the operation at zero bias, helps alleviate this issue. As shown in Fig. 6a and 6b, the significantly different contact areas generate a considerable photocurrent in Te photodetector at zero bias by raising the electric field intensity difference in Te channel near the drain and the source electrodes. The zero-bias photocurrent/dark current ratio reaches 1.57×10^4 (Fig. 6c) and the response speed is as fast as 720 μs.

Similarly, alloying Te with Se leads to a larger bandgap, thereby reducing the dark current as well. In Se_xTe_{1-x} SWIR photodetector, the polycrystalline Se_xTe_{1-x} thin films were also prepared via a thermal evaporation method (Fig. 6d) [10]. The optical bandgap of the products can be tuned from 0.31 eV to 1.87 eV by controlling the Se composition (Fig. 6e). As shown in Fig. 6f, the fabricated Se_xTe_{1-x} SWIR photodetector achieves a room-temperature specific detectivity of 6.5×10^{10} Jones at 1.55 μm (Fig. 6f).

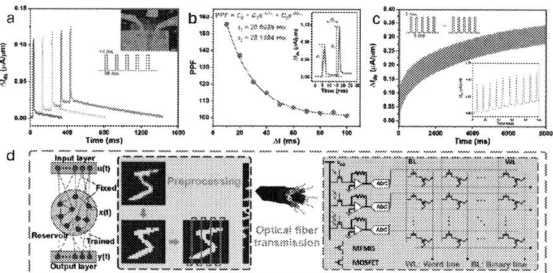

Fig. 7 Optoelectronic memory based on Te for an in-sensor RC [11].

Besides IR photodetectors, 2D Te can also be used to construct optoelectronic memory devices. After stacking Te with ferroelectric copper indium diphosphorus hexasulfide ($CuInP_2S_6$: CIPS) nanoflake, an electronic/optoelectronic memory device (Fig. 7a) [11]. This device exhibits long-term electronic memory behaviors triggered by electrical pulses and short-term optoelectronic memory behaviors (Fig. 7b and 7c). Leveraging the rich dynamics measured in this device, a fully memristive in-sensor reservoir computing (RC) system (Fig. 7d) was simulated for simultaneously sense, decode and learn messages transmitted through optical fibers.

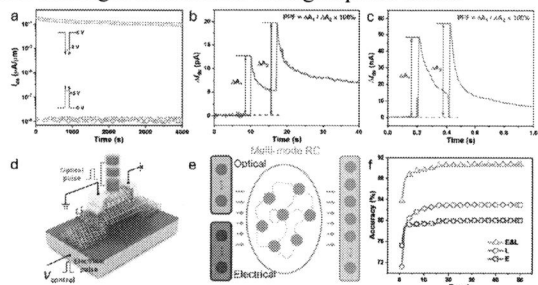

Fig. 8 A Te-based floating gate memory device for multimodal RC [12].

In addition to channel materials, 2D Te can also be applied as charge trapping to enable a multifunctional floating gate memory device for multimodal RC [12]. Under intense electrical stimulation, the proposed device behaves like a flash memory device with considerable endurability and retention (Fig. 8a). The nonvolatile memory states of the device can also be modulated by intense optical pulses. After weakening the intensity of the stimulation, the nonvolatile memory behaviors observed in this device degrade into volatile memory ones (Fig. 8b and 8c).

Leveraging the volatile electronic and optoelectronic memory behaviors, a multimodal RC (Fig. 8d and 8e) was simulated. Which achieves a high recognition accuracy of 90.77% for event-type multimodal handwritten digit (Fig. 8f).

Conclusion

In this paper, we make a summary of recent progress on electronics and optoelectronics based on 2D Te. As a *p*-type vdW semiconductor with many intriguing properties, 2D Te holds the great promise to advance 2D devices in the future.

Acknowledgments

C.T. thanks the funding support from the National Natural Science Foundation of China – Excellent Young Scientists Fund (Hong Kong and Macau) (52122002), ECS Scheme (21201821), and General Research Fund (11200122) from the Research Grant Council of Hong Kong.

References

[1] Y. Wang *et al.*, "Field-Effect Transistors Made from Solution-Grown Two-Dimensional Tellurene," Nat. Electron., 1, pp. 228-236 (2018).

[2] J. Zha *et al.*, "Electronics and Optoelectronics Based on Tellurium," Adv. Mater., p.2408969 (2024).

[3] M. Amani *et al.*, "Solution-Synthesized High-Mobility Tellurium Nanoflakes for Short-Wave Infrared Photodetectors," ACS Nano, 12, p7253-7263 (2018).

[4] P. Yang *et al.*, "Growth of Tellurium Nanobelts on h-BN for p-Type Transistors with Ultrahigh Hole Mobility," Nano-Micro Lett., 14, p. 109 (2022).

[5] P. Yang *et al.*, "Mechanisms of Current Fluctuation in High-Mobility p-Type Tellurium Field-Effect Transistors," IEEE T. Electron. Dev., 71, p.6417-6423 (2024).

[6] H. Huang *et al.*, "Precursor-Confined Chemical Vapor Deposition of 2D Single-Crystalline Se_xTe_{1-x} Nanosheets for p-Type Transistors and Inverters," ACS Nano, 18, p.17293-17303 (2024).

[7] C. Zhao *et al.*, "Evaporated Tellurium Thin Films for p-Type Field-Effect Transistors and Circuits," Nat. Nanotechnol., 15, p.53-58 (2020).

[8] J. Zha *et al.*, "A Perspective on Tellurium-Based Optoelectronics," Appl. Phys. Lett., 125, p.070504 2024.

[9] H. Wang *et al.*, "Asymmetrically Contacted Tellurium Short-Wave Infrared Photodetector with Low Dark Current and High Sensitivity at Room Temperature," Adv. Optical Mater., 11, p.2301508 (2023).

[10] C. Tan *et al.*, "Evaporated Se_xTe_{1-x} Thin Films with Tunable Bandgaps for Short-Wave Infrared Photodetectors," Adv. Mater., 32, p.2001329 (2020).

[11] J. Zha *et al.*, "Electronic/Optoelectronic Memory Device Enabled by Tellurium-Based 2D van der Waals Heterostructure for in-Sensor Reservoir Computing at the Optical Communication Band," Adv. Mater., 35, p.2211598 (2023).

[12] J. Zha *et al.*, "A 2D Heterostructure-Based Multifunctional Floating Gate Memory Device for Multimodal Reservoir Computing," Adv. Mater., 36, p.2308502 (2024).

Nanodiamond quantum thermometers and their biological applications

Masazumi Fujiwara

[1]Department of Chemistry, Okayama University, Japan
masazumi@okayama-u.ac.jp

Abstract

Fluorescent nanodiamonds (FNDs) are stable and nontoxic fluorescent probes, which are suited to chemical and biological analysis. Recently they have been intensively studied for quantum applications because their electron-spin-dependent fluorescence can provide us with a sensitive measurement tool for nanoscale information of magnetic field and temperature. Here, we report on our recent efforts for developing nanoscale thermometry using FNDs particularly for biological applications.

Keywords: Nanodiamond, Quantum Sensing, Biosensing

Introduction

Temperature, a fundamental quantity in thermodynamics, is inherently local due to the nature of contact probes used in traditional thermometers. Recent advancements in nanoscale thermometry have led to experimental observations that challenge the limitations of macroscopic thermodynamics [1, 2]. This has spurred the development of theories for nano- and microscale thermodynamics and ignited debates about temperature distribution in living cells [3, 4]. Consequently, there is a growing interest in exploring novel nanoscale thermometry methods. Diamond quantum sensors, based on quantum-enhanced sensing technology, offer a promising solution for high-precision nanoscale thermometry [5, 6]. These sensors utilize nitrogen-vacancy (NV) spin systems in diamonds, which are sensitive to environmental magnetic fields. Since the first demonstration of diamond quantum thermometry in 2013 [7], researchers have explored its applications in various fields, including microfluidics, joule heating, and biological systems.

In this talk, I will present our recent work on fluorescent nanodiamond (FND) quantum sensing, focusing on thermometry. I will discuss our experiments probing temperature in stem cells and nematode worms, as well as our efforts to integrate FND quantum sensing into chip-based bioassays. Finally, I will present the development of quantum-grade FNDs for enhanced performance in thermometry and other applications.

Temperature probing in biological samples

Our research focused on utilizing FNDs as nanoscale thermometers to investigate temperature dynamics in biological systems. By labeling stem cells and nematode worms with FNDs, we were able to monitor intracellular and organismal temperature variations in real-time.

For example, in our stem cell experiments, we precisely measured the temperatures inside cells using FNDs and observed that cell culture temperature influenced their differentiation ability, which was confirmed by several molecular biomarkers [8]. In nematode worms, we successfully integrated a particle tracking algorithm into FND thermometry [9] and demonstrated real-time temperature measurements in vivo [10] (Fig. 1). These findings highlight the importance of understanding temperature dynamics at the micro or nanoscale in biological

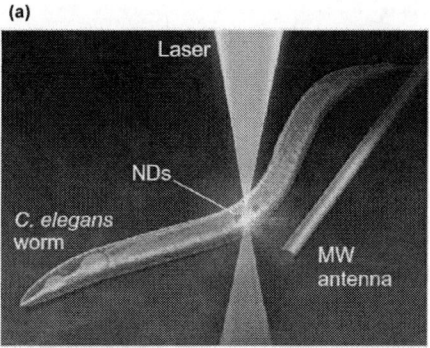

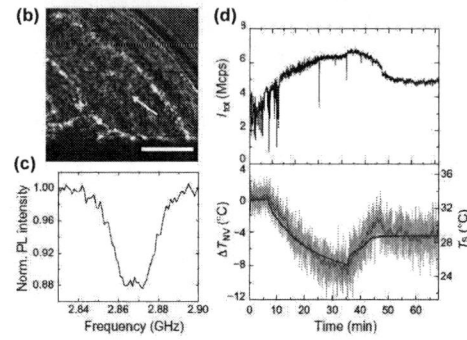

Fig. 1. (a) Schematic of FND quantum thermometers probing inside the worms. (b, c) FNDs and its ODMR spectrum in the worm body. (d) Time profiles of fluorescence intensity (top) and probed temperature.

systems. While there remain technical challenges such as quantification of temperature and particle-by-particle inhomogeneity of sensing performance, FND-based thermometry offers a powerful tool for studying these processes and could contribute to advances in fields such as regenerative medicine, neuroscience, and developmental biology.

On-chip integration of FND quantum sensing

The above thermometry demonstration needs to be more effectively optimized for modern bioassay tools and platforms. To achieve this, we focused on developing a microwave delivery architecture on glass-chip bioassay devices [11]. We designed a notch-shaped coplanar waveguide antenna on a glass plate, which provides a large detection area and broadband microwave excitation (Fig. 2). Unlike previous approaches, our antenna offers a scalable and versatile platform for integrating FND-based quantum sensing into various bioassay platforms. By integrating this antenna with different types of bioassay platforms such as glass-bottom dishes and multi-well plates, we demonstrated uniform ODMR detection for FNDs labeled in cultured cells, tissue, and nematode worms. This versatility highlights the potential of our technology for a wide range of applications, including diagnostics and cell screening. Furthermore, our approach offers several advantages over traditional methods. The large detection area of our antenna enables efficient detection of FNDs in biological samples for a wide range of cell numbers and sizes. The broadband microwave excitation ensures that we can capture a wide range of ODMR signals, providing information about the properties of the FNDs and their interactions with the surrounding environment. It is anticipated that this notch-shaped microwave architecture can be extended to various bioassay chip devices for further bio quantum sensing applications [12-14].

Quantum-grade FNDs

Improving the sensitivity of FND quantum sensing is crucial to the success of the technology because the sensitivity with FNDs has been a few orders of magnitude worse than that with NV-bulk diamonds [15, 16]. To address this, we engineered novel FNDs to possess bulk-like NV spin properties and bright fluorescence, which are essential for achieving high sensitivity and bioimaging compatibility [17] (Fig. 3). By controlling the spin impurities of ^{13}C and N, while increasing the NV concentration, we were able to significantly improve the spin relaxation times of the NVs in these FNDs. This resulted in a 5-fold increase in T_1 and an 11-fold increase in T_2 compared to conventional type-Ib NDs. These enhanced spin properties, combined with the bright fluorescence of our FNDs, make them ideal for a wide range of quantum biosensing applications. For example, they can be used for high-resolution imaging of biological samples, sensitive detection of magnetic fields, and precise measurements of temperature and pH. Furthermore, our FNDs exhibit narrow ODMR spectra, which require significantly less microwave excitation power to achieve comparable signal-to-noise ratios compared to conventional NDs. This reduction in power application is crucial for minimizing potential damage to biological samples and improving the overall sensitivity of quantum biosensing measurements.

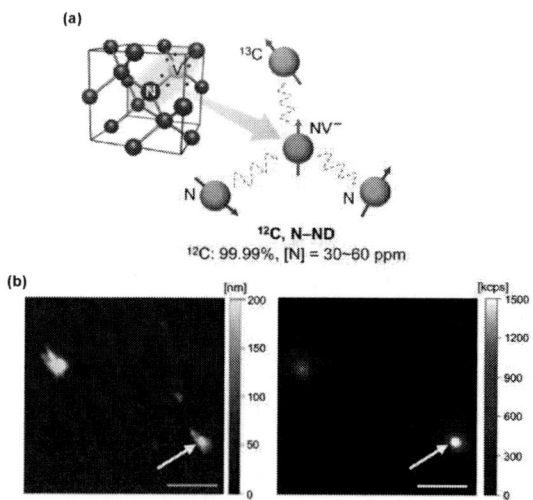

Fig. 3. (a) Illustrations of the NV crystal structure and the interaction of NV with the spin bath of N and ^{13}C. (b) AFM topography image of a single grid engraved on a coverslip and PL image. Scale bar: 10 μm.

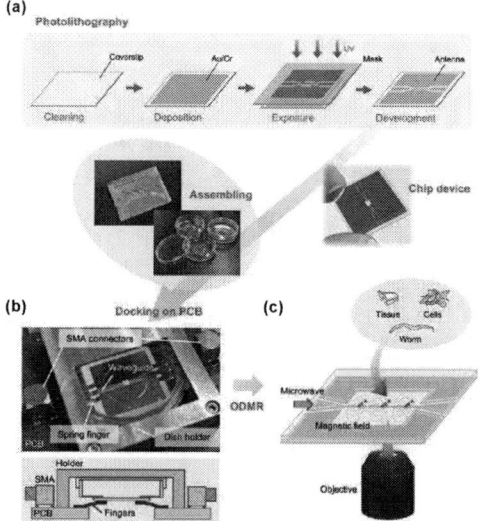

Fig. 2. (a) Chip fabrication scheme. (b) Photograph of a dish-style chip device docked to a PCB circuit board. (c) Schematic spatial arrangement of biological samples relative to the waveguide pattern and microscope objective.

Conclusion

In this talk, I have presented our recent advancements in FND-based quantum sensing, focusing on thermometry. We have demonstrated the successful application of FNDs for precise temperature measurements in biological systems, including stem cells and nematode worms. Additionally, we have developed a novel microwave antenna for on-chip integration of FND quantum sensing, enabling scalable and versatile bioassay platforms. Finally, we have engineered quantum-grade FNDs with enhanced spin properties and bright fluorescence, which significantly improve the sensitivity and performance of FND-based quantum sensors.

These advancements pave the way for a wide range of applications in fields such as biomedical research, diagnostics, and environmental monitoring. By continuing to develop and refine FND-based quantum sensing technologies, we can unlock new possibilities for exploring the complex and fascinating world of biological processes at the nanoscale.

Acknowledgments

The author gratefully acknowledges the support provided to my group and collaborators from the following sources:

- JSPS-KAKENHI (20H00335, 20KK0317, 21H05599)
- JST-ASPIRE (JPMJAP2339)
- NEDO (JPNP20004)
- AMED (JP23zf0127004)
- JST (JPMJMI21G1)
- RSK Sanyo Foundation
- Asahi Glass Foundation

References

[1] T. L. Hill, 2013, Thermodynamics of Small Systems, Parts I & II (New York: Dover)

[2] K. Sekimoto, 2010, Stochastic Energetics Lecture Notes in Physics (Berlin: Springer)

[3] M. Suzuki and T. Plakhotnik, Biophys. Rev. **12**, 593 (2020).

[4] S. Ghonge and D. C. Vural, J. Stat. Mech. Theory Exp. 073102 (2018).

[5] M. Fujiwara et al., Nanotech. **32**, 482002 (2021).

[6] G. Q. Liu et al., Acc. Chem. Res. **56**, 95 (2023).

[7] G. Kucsko et al., Nature **500**, 54 (2013).

[8] H. Yukawa et al., Nanoscale Adv. **2**, 1859 (2020).

[9] M. Fujiwara et al., Phys. Rev. Research **2**, 043415 (2020).

[10] M. Fujiwara et al., Sci. Adv. **6**, eaba9636 (2020).

[11] K. Oshimi et al., Lab Chip **22**, 2519 (2022).

[12] R. Allert et al., Lab Chip **22**, 4831 (2022).

[13] M. Fujiwara, Biomicrofluidics **17**, 54107 (2023).

[14] B. S. Miller et al., Nature **587**, 588 (2020).

[15] H. S. Knowles et al., Nat. Mater. **13**, 21 (2014).

[16] B. D. Wood, Phys. Rev. B **105**, 205401 (2022).

[17] Oshimi et al., arXiv:2312.17603.

Enhancing Transduction Efficiency in CMOS-MEMS CMUTs Through Atomic Layer Deposition: A Preliminary Study

Tzu-Yun Huang and Ming-Huang Li

Department of Power Mechanical Engineering, National Tsing Hua University, Hsinchu, Taiwan

Abstract

We investigate the enhancement of electrostatic transduction in CMOS-MEMS capacitive micromachined ultrasound transducers (CMUTs) by employing atomic layer deposition (ALD) to modify the transduction gaps between the bottom electrode and the membrane. In this preliminary study, the effectiveness of the ALD deposition was experimentally evaluated using high-frequency CMUT arrays centered at 5.2, 7.9, and 13 MHz that implemented using the TSMC 0.18 μm standard CMOS process. With a 50 nm layer of aluminum oxide (Al_2O_3) coating, a 7–15 dB improvement in peak admittance is observed from CMUT array measurements.

Keywords: Capacitance micromachined ultrasonic transducer (CMUT), CMOS-MEMS, atomic layer deposition (ALD).

Introduction

Recent advancements in capacitive micromachined ultrasonic transducer (CMUT) have been widely employed in many different fields including biomedical diagnostics and environmental sensing [1]. For the emerging applications such as intravascular ultrasound (IVUS) and intracardiac echocardiography (ICE), high-frequency CMUTs are desirable for high spatial resolution [2]. However, the design of high-frequency CMUTs often requires reducing the membrane dimensions while increasing its thickness, which can result in decreased sensitivity and compromised performance.

An intuitive yet effective method to enhance the performance of a general capacitive transducer is to reduce the gap between the fixed electrode and the vibrating membrane, thereby increasing the capacitive coupling (η_e), as expressed in (1) [3][4],

$$\eta_e = \gamma \frac{\varepsilon_o A}{d_{\text{eff}}^2} V_p, \tag{1}$$

where γ is the mode shape correction factor, ε_o is the permittivity in vacuum, A is the electrode area of the electromechanical structure, V_P is the dc-biasing, and d_{eff} is the effective gap. From (1), it is evident that η_e is inversely proportional to d_{eff}^2. Thus, when the CMUT operates in pulse-echo mode, its transduction efficiency becomes proportional to the d_{eff}^4. This implies that reducing d_{eff} to half of its original value can lead to a 16-fold increase in transduction efficiency. However, achieving such minimal

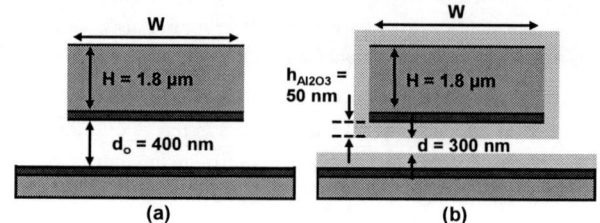

Fig. 1: (a) Conventional TiN-C CMOS-MEMS resonant transducer. (b) TiN-C CMOS-MEMS resonant transducer with ALD dielectric film coating.

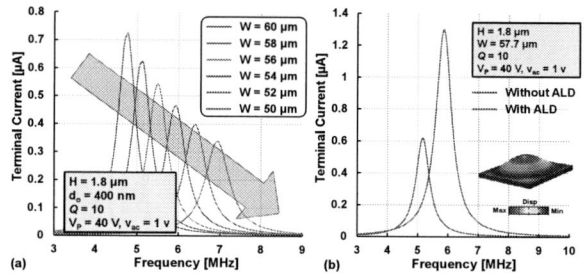

Fig. 2: FEM simulations of the CMOS-MEMS CMUT. (a) Motional current vs. resonant frequency. (b) Motional current of conventional and ALD-coated CMUTs.

gap dimensions through micromachining poses significant challenges, particularly in terms of manufacturing repeatability and yield.

To achieve improved CMUT performance, this study employs atomic layer deposition (ALD) to partially fill the transduction gaps of the prototyped CMOS-MEMS CMUT arrays with a dielectric material (Al_2O_3 in this case). While this technique has been previously explored in capacitive resonator studies [5][6], its application to resonant devices fabricated using CMOS-MEMS platforms remains novel. A previously developed titanium nitride composite (TiN-C) platform was selected as the fabrication process in this study to achieve a nominal gap size of 400 nm [7]. As a result, a reduced gap was successfully achieved using a 50 nm ALD layer of Al_2O_3, which was verified in three groups of CMUT arrays centered at around 5.2, 7.9, and 13 MHz.

CMOS-MEMS CMUT

The CMUTs are designed and fabricated using the previously reported TiN-C CMOS-MEMS process, where the transducer gap spacing of 400 nm is defined by the thickness of the back-end-of-line (BEOL) metallization layer [8]. Fig. 1(a) shows a conceptual cross-sectional view of the CMUT. To verify the effectiveness of the ALD dielectric coating, three CMUT arrays of different sizes were designed and characterized in this study. Each CMUT pixel consists of a 4×4 array and features unit-cell widths of 57.7 μm, 46.7 μm,

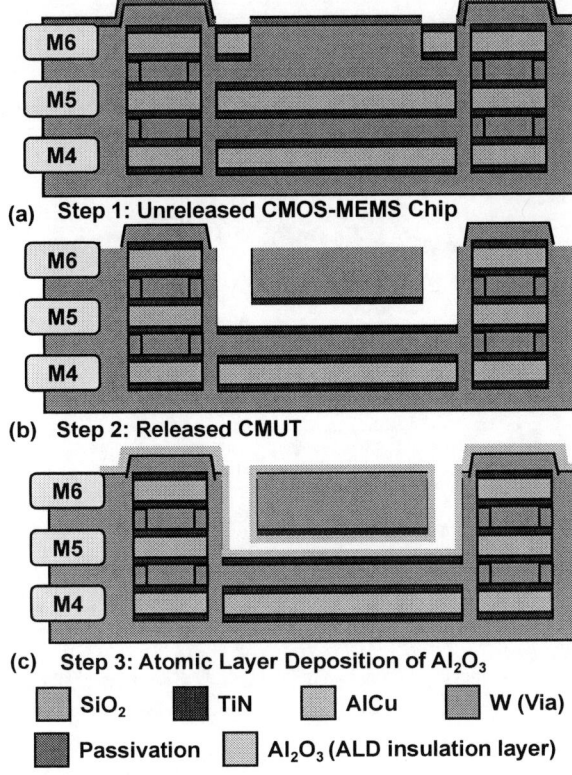

(a) Step 1: Unreleased CMOS-MEMS Chip

(b) Step 2: Released CMUT

(c) Step 3: Atomic Layer Deposition of Al₂O₃

	SiO₂		TiN		AlCu		W (Via)
	Passivation		Al₂O₃ (ALD insulation layer)				

Fig. 3: CMOS-MEMS CMUT fabrication process. (a) Unreleased CMOS chip from foundry. (b) Released CMUT before Al₂O₃ coating. (c) Released CMUT after Al₂O₃ coating.

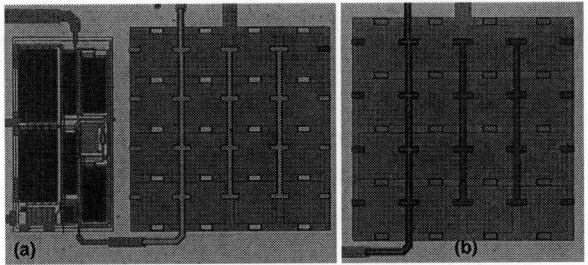

Fig. 4: Optical photos of the CMOS-MEMS CMUTs. (a) Before Al₂O₃ coating. (b) After Al₂O₃ coating.

and 35.6 µm, achieving center frequencies of approximately 5.2, 7.9, and 13 MHz, respectively. To achieve η_e enhancement, a 50 nm-thick of Al₂O₃ is deposited, as shown in Fig. 1(b). Due to the conformal coating characteristic of the ALD process, both sides of the transduction gap were covered by the Al₂O₃ thin film, resulting in an expected remaining physical gap of 300 nm.

The effective gap after partial gap-filling can be evaluated by using

$$d_{\text{eff}} = \frac{2h_{Al2O3}}{\varepsilon_{r,Al2O3}} + d, \qquad (2)$$

where h_{Al2O3} and $\varepsilon_{r,Al2O3}$ are the thickness and relative dielectric constant of the Al₂O₃ thin film, respectively, and d is the dimension of the remaining gap. The dielectric constant of Al₂O₃ varies with the deposition conditions, and a value of $\varepsilon_{r,Al2O3}$ of around 5 is used in our calculations and simulations [9]. This results in an effective gap of approximately 320 nm, which is very close to the remaining physical gap. As a result, the motional resistance of the CMUT is expected to be reduced by 2.5 times after the ALD process.

Fig. 2(a) and (b) show simulation results based on finite-element method (FEM). In Fig. 2(a), CMUTs with varying lateral dimensions (W) were simulated. The results show that the peak output current decreases progressively as the frequency increases under a constant driving voltage, which

is attributed to the smaller transduction area and the heightened mechanical stiffness at smaller W. In Fig. 2(b), a comparison of CMUT performance before and after ALD is presented. The enhancement in peak current by a factor of 2.1 is close to the theoretical predictions.

Fabrication & Post Process

The CMUT arrays were implemented using the TSMC 0.18 µm 1P6M standard CMOS platform, followed by a post-CMOS process. Fig. 3(a) shows a cross-sectional schematic of the chip prepared by the foundry, where the passivation layer (Si₃N₄) is partially removed to expose the top sacrificial metal layer (M6). The sacrificial metal was then removed using a heated piranha etchant (H₂SO₄:H₂O₂ = 3:1) at 105°C. A subsequent dry-etching step was employed to remove the underlying SiO₂, exposing the second sacrificial metal layer (M5). Finally, an aluminum etchant (Al etch-716H) with high selectivity was used to remove the sacrificial metal. The released device schematic is shown in Fig. 3(b). The remaining TiN layers from M5 serve as electrodes, with an initial gap spacing of approximately 400 nm.

To further reduce the gap, a Picosun R-200 ALD system was used to deposit Al₂O₃ over the released CMUT samples, using H₂O and trimethylaluminum (TMA) as the oxygen and aluminum sources. Under processing conditions of 10 hPa and 150°C, a 50 nm-thick Al₂O₃ film was deposited over 2.5 hours. Fig. 3(c) shows the CMUT schematic after the ALD process, where the remaining physical gap is expected to be around 300 nm.

Measurement result

Fig. 4(a) and (b) show the optical images of the CMOS-MEMS CMUT array before and after the ALD process. Standalone CMUT arrays, without CMOS amplifiers, were chosen for the subsequent measurements.

In this preliminary study, a total of six CMUT arrays from two distinct CMUT chips (one with ALD partial gap filling) were characterized in vacuum as resonators, using a vector network analyzer (Keysight E5071C) for S-parameter measurements and a Keithley 2612B for DC biasing (V_P). Fig. 5(a)-(c) present the measured transmission response (S₂₁) of the CMUTs under a V_P of 40V for performance comparison, where the frequency difference between the CMUTs with and without ALD Al₂O₃ is primarily attributed to variations in the

979-8-3315-0417-5/25 $31.00 © 2025 IEEE

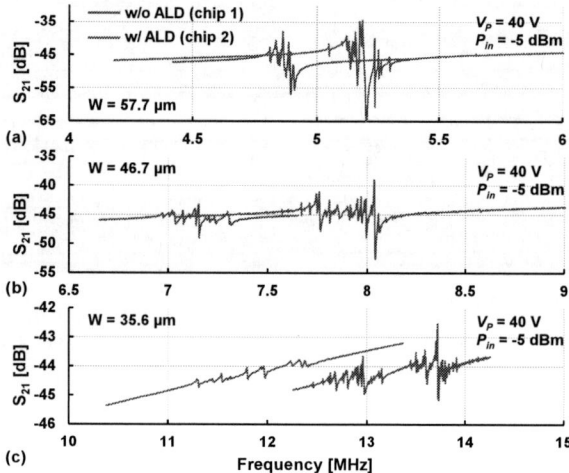 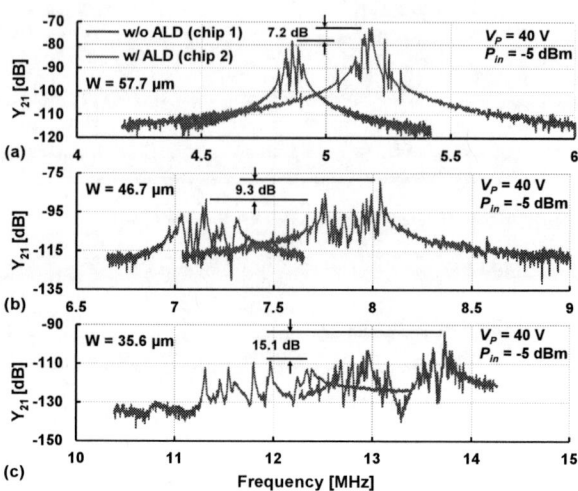

Fig. 5: Measured S_{21} of CMOS-MEMS CMUTs in vacuum, operating at (a) 4.7 – 5.3 MHz, (b) 7 – 8 MHz, and (c) 11.5 – 13.5 MHz

Fig. 6: Measured Y_{21} of CMOS-MEMS CMUTs in vacuum with numerical de-embedding, operating at (a) 4.7 – 5.3 MHz, (b) 7 – 8 MHz, and (c) 11.5 – 13.5 MHz. The differences in peak admittance are labeled.

post-CMOS process, as they are from different chips. It is apparent that the ALD Al_2O_3 effectively enhances the S_{21} of the CMUTs.

Furthermore, the background of the measurement spectrum, caused by the CMUT capacitance (C_o), was numerically removed for clearer motional signal analysis [10]. Fig. 6(a)-(c) depict the admittance of the devices after C_o removal. Although the multiple resonance modes of the CMUT array make parameter extraction challenging, it is evident that the peak admittances improved by 7 – 15 dB in the Al_2O_3-coated CMUT arrays, highlighting the effectiveness of this approach.

Conclusion

In this work, we investigate the use of the ALD technique to enhance CMUT performance in the CMOS-MEMS platform for the first time by reducing the equivalent gap distance, addressing the challenges of high-frequency CMUT elements with smaller sizes. This resulted in significant improvements in both the transmission and de-embedded admittance responses across three different CMUT arrays centered at 5.2, 7.9, and 13 MHz. Underwater measurements for the Al_2O_3-coated CMUTs are ongoing.

Acknowledgments

This research is supported by the NSTC of Taiwan under grants 113-2628-E-007-016-MY3. The CMOS chip fabrication was supported by Taiwan Semiconductor Research Institute (TSRI) and TSMC, Hsinchu, Taiwan. The post-CMOS process is supported by the CNMM of National Tsing Hua University, Hsinchu, Taiwan. The ALD process was supported by Wei-Kai Sung and Prof. Weileun Fang at National Tsing Hua University, Hsinchu, Taiwan.

References

[1] J. Joseph, B. Ma, and B. T. Khuri-Yakub, "Applications of capacitive micromachined ultrasonic transducers: a comprehensive review,"

IEEE Trans. Ultrason. Ferroelect. Freq. Contr., vol. 69, no. 2, pp. 456-467, Feb. 2022.

[2] G. Gurun *et al.*, "Single-chip CMUT-on-CMOS front-end system for real-time volumetric IVUS and ICE imaging," *IEEE Trans. Ultrason. Ferroelect. Freq. Contr.*, vol. 61, no. 2, pp. 239-250, Feb. 2014.

[3] H.-Y. Chen, S.-S. Li, and M.-H. Li, "A low impedance CMOS-MEMS capacitive resonator based on metal-insulator-metal (MIM) capacitor structure," *IEEE Electron Device Lett.*, vol. 42, no. 7, pp. 1045-1048, May. 2021.

[4] G. G. Yaralioglu *et al.*, "Calculation and measurement of electromechanical coupling coefficient of capacitive micromachined ultrasonic transducers," *IEEE Trans. Ultrason. Ferroelect. Freq. Contr.*, vol. 50, no. 4, pp. 449-456, Apr. 2003.

[5] T. J. Cheng and S. A. Bhave, "High-Q, low impedance polysilicon resonators with 10 nm air gaps," in *Proc. IEEE 23rd Int. Conf. Micro Electro Mechanical Systems (MEMS)*, Hong Kong, China, 2010, pp. 695-698.

[6] L.-W. Hung *et al.*, "Capacitive transducer strengthening via ALD-enabled partial-gap filling," in *Proc. Solid-State Sensors, Actuators, and Microsystems Workshop (Hilton Head)*, Hilton Head Island, South Carolina, June 1-5, 2008, pp. 208-211.

[7] C.-Y. Chen, M.-H. Li, A. A. Zope, and S.-S. Li, "A CMOS-integrated MEMS platform for frequency stable resonators - Part I: Fabrication, implementation and characterization," *J. Microelectromech. Syst.*, vol. 28, no. 5, pp. 744-754, Oct. 2019.

[8] T.-H. Hsu, A. A. Zope, M.-H. Li and S.-S. Li, "A compact monolithic CMUT receiver front-end in a TiN-C CMOS-MEMS platform," in *Proc. 2020 IEEE International Ultrasonics Symposium (IUS)*, Las Vegas, NV, USA, 2020, pp. 1-4.

[9] Y. Luo *et al.*, "Effect of deposition temperature on Al_2O_3 films deposited by atomic layers deposition," in *Proc. IEEE Int. Vac. Electron. Conf. (IVEC)*, Chengdu, China, 2023, pp. 1-2

[10] J. E.-Y. Lee, A. A. Seshia, "Direct parameter extraction in feedthrough-embedded capacitive MEMS resonators," *Sensor Actuat. A-Phys.*, vol. 167, no. 2, pp. 237–244, Jun. 2011.

979-8-3315-0417-5/25 $31.00 © 2025 IEEE

A 3.98 GHz Aluminum Nitride Overmoded Bulk Acoustic Wave Resonator for Temperature Sensing Applications

Zhi-Qiang Lee[1], Kuan-Ting Chen[1], Chin-Yu Chang[2], Cheng-Chien Lin[1], Yung-Hsiang Chen[3], Yelehanka Ramachandramurthy Pradeep[4], Rakesh Chand[4], Yenshih Ho[3], and Ming-Huang Li[1]

[1]Department of Power Mechanical Engineering, National Tsing Hua University, Hsinchu, Taiwan
[2]Institute of NanoEngineering and MicroSystems, National Tsing Hua University, Hsinchu, Taiwan
[3]Vanguard International Semiconductor Corporation, Hsinchu, Taiwan
[4]Vanguard International Semiconductor Corporation Singapore PTE. Ltd., Singapore

Abstract

In this work, we present a novel microelectromechanical systems (MEMS) temperature sensor based on an aluminum nitride (AlN) overmoded bulk acoustic wave resonator (OBAR). Utilizing the overmoded operation of a AlN bulk acoustic wave resonator with a thick metal electrode, a large temperature coefficient of frequency (TCF) of -105 ppm/°C is achieved, while maintaining a high quality factor (Q) of 300 and an electromechanical coupling coefficient (k_t^2) of 4.34% at 3.98 GHz. This TCF is approximately 4.5X higher than that of a regular AlN thin-film bulk acoustic resonator (FBAR). Furthermore, based on the dual-oscillator beat frequency sensing topology using OBAR and FBAR resonators characterized in this study, a large TCF$_{beat}$ of -1,088 ppm/°C is obtained through theoretical predictions.

Keywords: Overmoded bulk acoustic resonator (OBAR), temperature sensor, aluminum nitride (AlN), temperature coefficient of frequency (TCF).

Introduction

Temperature sensors, essential in various applications including industrial, consumer, military, and space domains [1], [2], [3], are in higher demand within electronic systems due to emerging trends of the Artificial Intelligence of Things (AIoT) [4]. Existing technologies, including resistive temperature detectors (RTDs) [5] and CMOS-based smart temperature sensors [6], have been widely deployed to target high-precision and low-cost markets, respectively. Alternatively, MEMS resonators have strong potential to provide excellent temperature sensing accuracy while maintaining low costs due to their large temperature coefficient of frequency (TCF), high quality factor (Q), and batch-fabrication capability [7], [8], [9].

In addition to commonly used flexural resonators operating in the kHz and MHz ranges, thin-film bulk acoustic wave resonators (FBARs) in the GHz range are also promising candidates for temperature sensor development [10]. The high Q and high k_t^2 characteristics of FBARs help reduce motional impedance, enabling the development of low-power and low-noise oscillators for self-sustained temperature sensing. However, the TCF of FBARs is primarily determined by the piezoelectric material, resulting

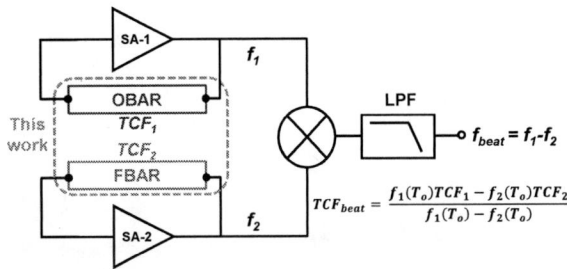

Fig. 1: Conceptual schematic of beat frequency temperature sensing based on dual resonators.

in a temperature-frequency responsivity of only around -30 ppm/°C for AlN [11] and -39 ppm/°C for Al$_{0.7}$Sc$_{0.3}$N [12].

In this work, a thick aluminum-copper (AlCu) film is deposited on a suspended AlN thin film as a novel overmoded bulk acoustic wave resonator (OBAR) structure for temperature sensing. The large temperature coefficient of elasticity (TCE) of AlCu [13], combined with the significant energy distribution in the AlCu film due to overmoded operation, increases the OBAR's TCF to -105 ppm/°C at 3.9 GHz. Additionally, the AlN MEMS platform used in this study is capable of fabricating multiple FBARs and OBARs on the same chip [14], [15], enabling a dual-oscillator sensing scheme to further boost responsivity, as shown in Fig. 1 [7], [16]. As a result, a large beat frequency responsivity (TCF$_{beat}$) of -1088 ppm/°C is theoretically estimated using a 3.6 GHz FBAR (TCF = -23 ppm/°C) and a 3.9 GHz OBAR (TCF = -105 ppm/°C), demonstrating the potential of the proposed approach.

Design and Analysis

To evaluate the TCF enhancement from the thick metal layer, Fig. 2 presents the cross-sectional view and mode shape of the devices designed in this platform. The regular FBAR device comprises a 0.6 μm AlN structural layer, sandwiched between two 140 nm molybdenum (Mo) electrodes, along with a top AlN passivation layer, while the proposed thick-metal resonator features an additional 1 μm AlCu thick layer applied directly onto the Mo top electrode.

A 2D finite element method (FEM) simulation using COMSOL Multiphysics was performed to investigate the temperature sensitivity provided by these devices based on the following TCE values from the literature: TCE$_{SiO2}$ = 180 ppm/°C [13], TCE$_{AlCu}$ = -620 ppm/°C [13], and TCE$_{Mo}$ = -127

979-8-3315-0417-5/25 $31.00 © 2025 IEEE

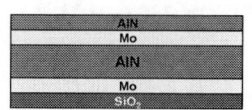

| (a) Regular resonator | (b) Thick-metal resonator |

Fig. 2: The cross-sectional view of the layer stacking (a) regular and (b) thick-metal resonator.

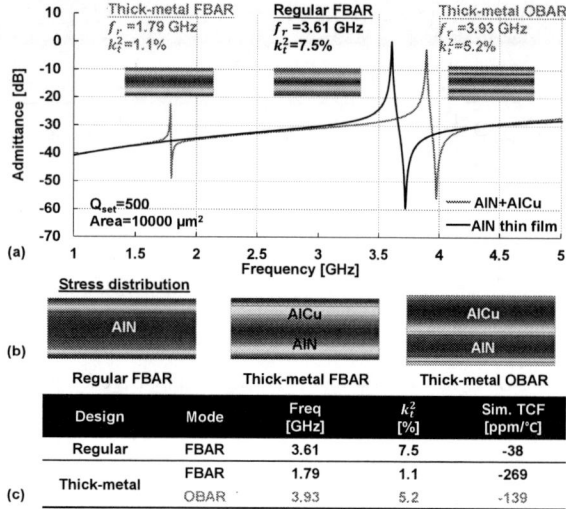

Fig. 3: (a) Simulated admittance of the BAW mode for both regular and thick-metal resonators. (b) Simulated stress distribution. (c) Extracted performance of the BAW modes.

Design	Mode	Freq [GHz]	k_t^2 [%]	Sim. TCF [ppm/°C]
Regular	FBAR	3.61	7.5	-38
Thick-metal	FBAR	1.79	1.1	-269
	OBAR	3.93	5.2	-139

ppm/°C [17]. Fig. 3(a) and (b) illustrate the simulated admittance and the stress distribution of the FBAR and OBAR modes from both resonators. For the regular FBAR resonator operated at 3.61 GHz, it exhibits a TCF of -38 ppm/°C, which is primarily governed by the intrinsic TCE of the AlN. On the other hand, for the proposed thick-metal resonator, the TCF of its FBAR and OBAR modes reaches -269 ppm/°C and -139 ppm/°C, respectively, which demonstrate a significant improvement in temperature sensitivity.

Although a large TCF is achieved from the FBAR mode operation of the proposed resonator, the thick and asymmetric structure results in a lower operation frequency of 1.79 GHz and a substantial degradation in k_t^2 due to ineffective stress distribution within the AlN layer. As a result, the overmoded operation of the proposed resonator appears to be a better choice, which demonstrates a high operating frequency of 3.93 GHz and a high k_t^2 of 5.2%. As the AlCu thickness (1 μm) is comparable to the half-wavelength of the targeted operation frequency ($\lambda/2 \sim 0.63$ μm), the OBAR mode allows a better stress distribution across the AlN layer, preserving the k_t^2 while still enhancing the TCF by 3.65X compared to a regular FBAR. The comparison of the regular and proposed devices is summarized in Fig. 3(c).

Measurement Results

Both the regular and thick-metal resonators were designed and fabricated on an 8-inch AlN piezoelectric MEMS

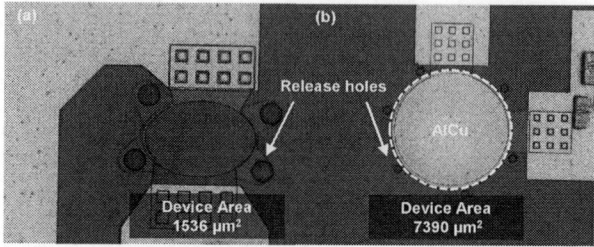

Fig. 4: Optical images of the fabricated (a) regular and (b) thick-metal resonator.

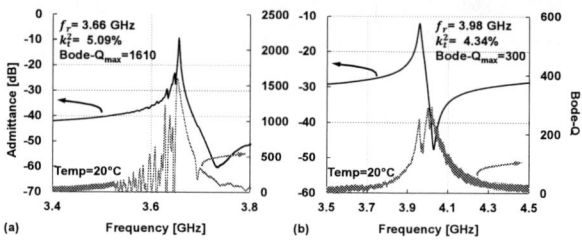

Fig. 5: Measured admittance and extracted Bode-Q of the (a) regular FBAR and (b) thick-metal OBAR at 20°C.

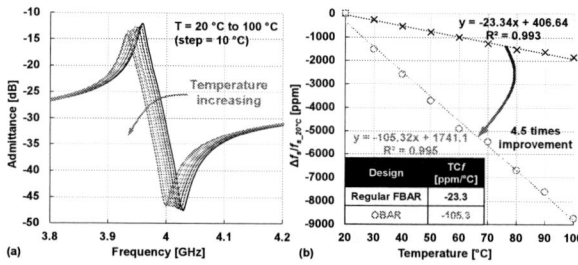

Fig. 6: (a) Measured admittance of the thick-metal OBAR under different temperature ranging from 20°C to 100°C. (b) Extracted TCF of the regular FBAR and thick-metal OBAR.

platform, provided by Vanguard International Semiconductor Corporation (VIS), using 8-inch wafers. Fig. 4 shows the optical images of the regular resonator [Fig. 4(a)] and the thick-metal resonator [Fig. 4(b)].

The electrical characteristics of the fabricated resonators were measured by a Keysight E5071C vector network analyzer (VNA) with standard RF probing. The admittance and extracted Bode-Q under 20°C ambient condition for the regular FBAR and the proposed OBAR are illustrated in Fig. 5(a) and (b), respectively. The regular FBAR shows a k_t^2 of 5.09% and the Bode-Q of 1,610 at 3.66 GHz while the OBAR device shows a k_t^2 of 4.34% and Bode-Q of 300 at 3.98 GHz. The degradation in the Bode-Q of the OBAR device may be attributed to the acoustic losses present in the top-deposited AlCu thick layer.

Next, the resonators were measured under varying ambient temperatures using a temperature-controlled probe station, ranging from 20°C to 100°C in increments of 10°C. The measured admittance curves of the OBAR device at different temperatures are illustrated in Fig. 6(a). The results indicate that increasing the ambient temperature leads to a significant reduction in the operating frequency, which can be primarily attributed to the material softening effect caused by the negative TCE of AlN and AlCu.

Additionally, the variations in resonance frequency of both

979-8-3315-0417-5/25 $31.00 © 2025 IEEE

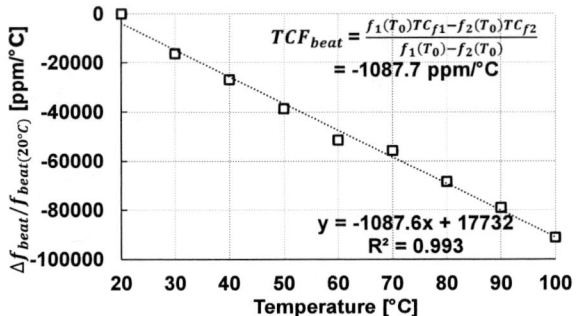

$$TCF_{beat} = \frac{f_1(T_0)TC_{f1} - f_2(T_0)TC_{f2}}{f_1(T_0) - f_2(T_0)}$$
$$= -1087.7 \text{ ppm/°C}$$

$$y = -1087.6x + 17732$$
$$R^2 = 0.993$$

Fig. 7: The extracted TCF_{beat} based on dual-oscillator beat frequency sensing concept using the fabricated thick-metal OBAR and regular FBAR devices.

devices were extracted and are illustrated in Fig. 6(b). The targeted OBAR device exhibits a TCF of -105.3 ppm/°C, which is 4.5 times higher than that of the regular FBAR (-23.3 ppm/°C), in line with the trend observed in the FEM simulation. The discrepancy between the simulated and measured TCF values for both devices is attributed to uncertainties in the material constants. Furthermore, the frequency variation with respect to temperature shows excellent linearity for both devices ($R^2 > 0.99$).

Moreover, as the two resonators shown in Fig. 6(c) are fabricated on the same chip, the temperature responsivity can be further enhanced based on the dual-oscillator sensing scheme. Using the conceptual dual-oscillator system illustrated in Fig. 1, the TCF of the beat frequency (TCF_{beat}) was theoretically estimated using open-loop measurement data, as shown in Fig. 7. A very high TCF_{beat} of -1,088 ppm/°C was theoretically obtained, highlighting the potential of implementing temperature sensors with a dual-oscillator system.

Conclusion

In this work, a 3.98 GHz AlN OBAR with a thick metal electrode is implemented in an 8-inch AlN MEMS platform as a GHz temperature sensor. Due to the large TCE of the AlCu loading layer, the OBAR demonstrates a high Q of 300 and a k_t^2 of 4.34%, with a large TCF of -105 ppm/°C, which is 4.5 times higher than that of a regular FBAR operating in a similar frequency range. A dual-oscillator sensing scheme is also proposed to further demonstrate the potential of GHz temperature sensors, with a theoretically estimated TCF_{beat} of -1088 ppm/°C.

Acknowledgments

This work was supported in part by Vanguard International Semiconductor (VIS) Corporation Joint Development Project, and in part by the NSTC of Taiwan (NSTC 112-2221-E-007-103-MY3).

References

[1] C. D. Matthus et al., "Feasibility of 4H-SiC p-i-n diode for sensitive temperature measurements between 20.5 K and 802 K," *IEEE Sensors J.*, vol. 19, no. 8, pp. 2871-2878, Apr. 2019.

[2] A. Bakker, "CMOS smart temperature sensors - an overview," in *Proc. IEEE SENSORS*, Orlando, FL, USA, 2002, pp. 1423-1427.

[3] J. Kohler et al., "Modular multifunctional silicon microsystems for spacecraft applications," in Proc., 14th Int. Conf. on Solid-State Sensors, Actuators and Microsystems Conference, Lyon, France, 2007, pp. 1673-1676.

[4] N. N. Misra et al., "IoT, big data, and artificial intelligence in agriculture and food Industry," *IEEE Internet Things J.*, vol. 9, no. 9, pp. 6305-6324, May 2022.

[5] Y. Huang et al., "Ultrasmall and ultrathin Ni-based resistance temperature detector integrated in micro thermoelectric devices for in situ temperature measurement," *IEEE Sensors J.*, vol. 23, no. 17, pp. 19409-19416, Sept. 2023.

[6] Z. Tang et al., "A sub-1 V capacitively biased BJT-based temperature sensor with an inaccuracy of ±0.15 °C (3σ) From -55 °C to 125 °C," *IEEE J. Solid-State Circuits*, vol. 58, no. 12, pp. 3433-3441, Dec. 2023.

[7] C. M. Jha et al., "High resolution microresonator-based digital temperature sensor," *Appl. Phys. Lett.*, vol. 91, Art. no. 074101, Aug. 2007.

[8] C. Tu et al., "Highly sensitive temperature sensor based on coupled-beam AlN-on-Si MEMS resonators operating in out-of-plane flexural vibration modes," *Research*, vol. 2022, Art. no. 9865926, Aug. 2022.

[9] Y. Yuan et al., "A piezoelectric MEMS resonant temperature sensor with 10-μk resolution and 0.06-pJK^2 resolution FOM," in *Proc., the 37th IEEE Micro Electro Mechanical Systems*, Austin, TX, USA, 2024, pp. 983-986.

[10] J.-H. Lin and Y.-H. Kao, "Wireless temperature sensing using a passive RFID tag with film bulk acoustic resonator," in *Proc. IEEE Int. Ultrason. Symp.*, Beijing, China, 2008, pp. 2209-2212.

[11] R. Ruby et al., "Positioning FBAR technology in the frequency and timing domain," *IEEE Trans. Ultrason. Ferroelect. Freq. Contr.*, vol. 59, no. 3, pp. 334-345, Mar. 2012.

[12] J. Wang, M. Park and A. Ansari, "Thermal characterization of ferroelectric aluminum scandium nitride acoustic resonators," in *Proc., the 34th IEEE Micro Electro Mechanical Systems*, Gainesville, FL, USA, 2021, pp. 214-217.

[13] M.-H. Li et al., "A monolithic CMOS-MEMS oscillator based on an ultra-low-power ovenized micromechanical resonator," *J. Microelectromech. Syst.*, vol. 24, no. 2, pp. 360-372, Apr. 2015.

[14] Y.-M. Huang et al., "S-band micromechanical resonant impedance transformers based on aluminum nitride FBARs," IEEE Trans. Microw. Theory Techn., vol. 71, no. 10, pp. 4193-4205, Oct. 2023.

[15] C.-Y. Chang et al., "A 3.6 GHz radio frequency circulator based on AlN FBAR filters," in *Proc. IEEE Ultrason., Ferroelectr., Freq. Control Joint Symp.*, Taipei, Taiwan, Sep. 22-26, 2024, pp. 1-4.

[16] M.-H. Li et al., "Design and characterization of a dual-mode CMOS-MEMS resonator for TCF manipulation," *J. Microelectromech. Syst.*, vol. 24, no. 2, pp. 446-457, Apr. 2015.

[17] J. M. Dickinson and P. E. Armstrong, "Temperature dependence of the elastic constants of molybdenum," *J. Appl. Phys.* vol.38, pp. 602–606, Feb. 1967.

Effect of Gate Voltage Rise Time on Gate Charges for GaN HEMTs with Partially Depleted p-GaN Cap Layer

Ruize Sun[1,2], Renjie Wu[1], Yunfei Ma[1], Wenhan Yuan[1], Qinan Guo[1], Wanjun Chen[1*], and Bo Zhang[1]

[1]State Key Laboratory of Electronic Thin Films and Integrated Devices, University of Electronic Science and Technology of China, Chengdu 611731, China
[2]Institute of Electronic and Information Engineering of UESTC, Dongguan 523808, China
*Corresponding author E-mail: wjchen@uestc.edu.cn

Abstract

This study explores the effect of gate voltage rise time (t_r) on the gate charge (Q_G) and performance of p-GaN HEMT devices under different temperature conditions. Experiments and TCAD simulations reveal a positive correlation between Q_G and t_r, with the impact weakening at higher temperatures. The incomplete ionization of magnesium acceptors in the p-GaN layer, along with the effect of t_r on conduction band energy (E_C) and two-dimensional electron gas (2DEG) density, are key factors. These findings provide important insights for loss analysis and the optimization of gate driving strategies for GaN transistors.

Keywords: GaN HEMT, Gate Charge, p-GaN, Ionization

Introduction

To drive GaN power transistors efficiently and reliably, the rise time (t_r) of the gate voltage (V_G) and the slew rate are commonly adjusted using gate resistors and current management circuits within the gate loop. However, these modifications can impact the device's performance, particularly the threshold voltage (V_{TH}). Relevant studies indicate that the V_{TH} of p-GaN HEMT devices varies under different static and dynamic stresses; specifically, V_{TH} decreases under significant static bias stress but increases under rapid dynamic stress. The instability of V_{TH} arises from the time-dependent storage and release mechanisms of charges (electrons and holes) in the p-GaN layer [1-3]. Additionally, TCAD simulations have demonstrated that the lag in dynamic V_{TH} is attributed to the hole charging and discharging processes in the p-GaN layer [4, 5]. Further research indicates that p-GaN HEMTs experience a negative shift in V_{TH} and exhibit high gate current under conditions of high dV_G/dt [6].

The performance variation of p-GaN HEMT devices at high conversion rates is always caused by the storage/release of charges (electrons and holes) in p-GaN under the gate in a short period of time. In order to further explore the reasons for the change of performance, we investigate the influence of t_r on gate charges Q_G for GaN HEMTs with partially depleted p-GaN cap layers at the nanosecond time scale. Experiments and TCAD simulation are implemented to investigate the mechanism for the dynamic increase in Q_G at higher slew rate. The t_r- and temperature-dependent

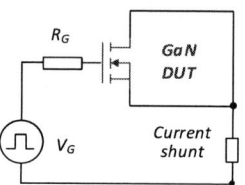

Fig. 1: Schematic of Q_G measurement circuit. The non-inductive co-axial current shunt SSDN-414-10 with resistance of 0.0969 Ω is used.

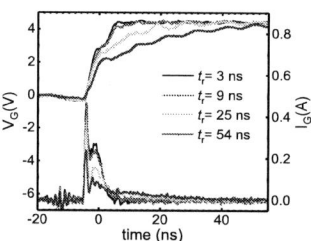

Fig. 2: Typical waveforms of V_G and I_G with rise time t_r of 3/9/25/54 ns at 298K.

incomplete ionization of acceptors in p-GaN is examined, offering valuable insights for loss analysis and guiding optimal gate driving strategies.

Experiment Setup

The device examined in this study is a p-GaN gate power transistor from the GaN System GS66504B series. A total of eight experimental groups were established, varying in gate bias (4.5 V and 5.5 V) and rise time t_r (3, 9, 25, and 54 ns). These groups were compared to one another. To characterize and quantitatively analyze the gate charge (Q_G) of the device, the circuit configuration shown in Fig. 1 was implemented. The gate bias, ranging from 0 to 4.5 V or 5.5 V, was applied using the ADuM4121 gate driver IC, while the gate resistor (R_G) was adjusted to achieve the desired rise times of 3, 9, 25, and 54 ns. At a room temperature of 278 K and a gate bias of 4.5 V, typical waveforms of the gate voltage (V_G) and gate current (I_G) at various rise times ($t_r = 3, 9, 25,$ and 54 ns) were obtained from measurements of ten devices, as illustrated in Fig. 2.

Result and Discussion

Under conditions of 298 K, varying gate bias, and different rise times (t_r), the gate charges of the device were measured, as shown in Fig. 3. By comparing the gate charge (Q_G) of the device under test (DUT) with the stored charges of multilayer ceramic capacitors (MLCC) at different rise times, a clear

979-8-3315-0417-5/25 $31.00 © 2025 IEEE

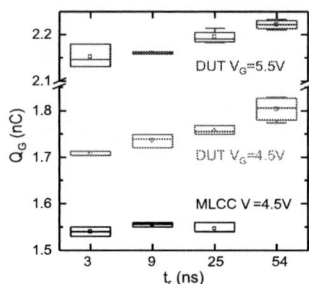

Fig. 3: Relationship between Q_G and t_r at 298 K for a GaN DUT with a V_G of 4.5/5.5 V and an MLCC capacitor with an applied voltage of 4.5 V. The Q_G of DUT has a positive dependence on t_r, indicating dynamic charge transportations rather than the capacitive behavior alone like the MLCC.

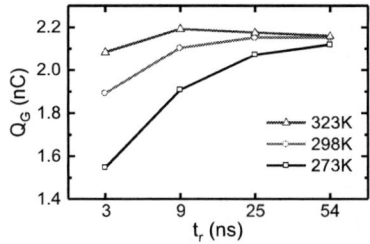

Fig. 4: Positive correlation between Q_G and t_r at 273 K, 298 K and 323K in TCAD simulation, which is consistent with the experimental results.

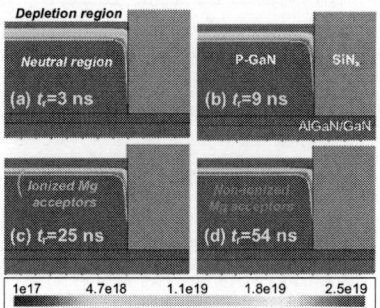

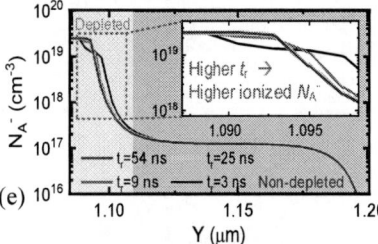

Fig.5: (a)-(d) Simulated profiles of ionized acceptors concentration (N_A^-) with t_r of 3/9/25/54 ns at 273K, showing t_r-dependent degree of incomplete ionization; (e) the profiles along the cut-line$|_y$ at the center of p-GaN layer, where inset shows higher ionized N_A^- for higher t_r in depleted p-GaN.

distinction emerged: the positive correlation between Q_G (=$\int I_G dt$). and t_r for the DUTs differs from the charge performance observed in MLCCs. This suggests a reliance on charge transport mechanisms rather than purely capacitive behavior. The relationship between Q_G and t_r at different temperatures (273, 298, and 323 K), as simulated by *Sentaurus* TCAD, is shown in Fig. 4. This illustrates the positive correlation between Q_G and t_r, consistent with Fig. 3. It was observed that dQ_G/dt_r decreases at higher temperatures, exhibiting a negative temperature coefficient, which suggests that the impact of t_r on Q_G diminishes at elevated temperatures (323 K). If the temperature continues to rise, the Q_G-t_r curve may even approach a horizontal level.

Considering the incomplete ionization of magnesium acceptors in p-GaN, simulations with varying rise times were conducted at a temperature of 273 K. Fig. 5(a) to (d) illustrate the simulated profiles of ionized acceptor concentration (N_A^-) at this temperature for rise times of 3, 9, 25, and 54 ns. Fig. 5(e) presents the profile along the cut-line$|_y$ at the center of the p-GaN layer, indicating that in the depleted p-GaN region, a higher concentration of ionized N_A^- is observed at larger t_r, while a lower concentration is noted at smaller t_r.

To further explore why the influence of t_r on Q_G diminishes with increasing temperature, and considering the characteristics of the crystal lattice, the correlation between simulated conduction band energy (E_C) and t_r at temperatures of 278, 298, and 323 K is illustrated in Fig. 6. At 273 K, E_C varies significantly with different t_r, decreasing as t_r increases. In contrast, at the higher temperature of 323 K, there is minimal difference in E_C across varying t_r. Lower temperatures accentuate the impact of t_r on E_C due to reduced thermal energy from the lattice.

Additionally, the incomplete ionization of acceptors in p-GaN also influences the density of the two-dimensional electron gas (2DEG) (n_e). The relationship between n_e in p-GaN HEMT devices and V_G at different rise times (t_r = 3, 9, 25, and 54 ns) is depicted in Fig. 7. As shown in Fig. 7, n_e decreases as t_r decreases. Fig. 3, 4, and 7 collectively indicate that as t_r increases, N_A^- increases, E_C decreases, and n_e subsequently increases. This relationship among the three parameters aligns with the underlying physical mechanisms. Magnesium dopants, with impurity levels deeper than the thermal energy (kT), contribute to the t_r and temperature-dependent behavior of N_A^- and Q_G. This phenomenon is further illustrated in Fig. 8(a). The incomplete ionization of Magnesium doping results in variations of gate charge (ΔQ_G) in the p-GaN/AlGaN/GaN gate stack under the influence of V_G, t_r, and temperature.

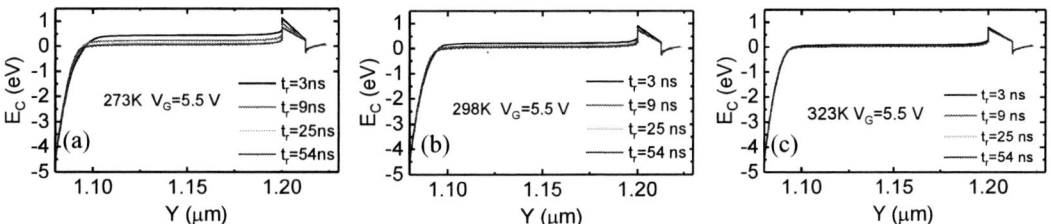

Fig.6: Relationship between simulated E_C and t_r at 278/298/323 K. At lower temperature, the effect of t_r on E_C is accentuated due to lower thermal energy.

979-8-3315-0417-5/25 $31.00 © 2025 IEEE

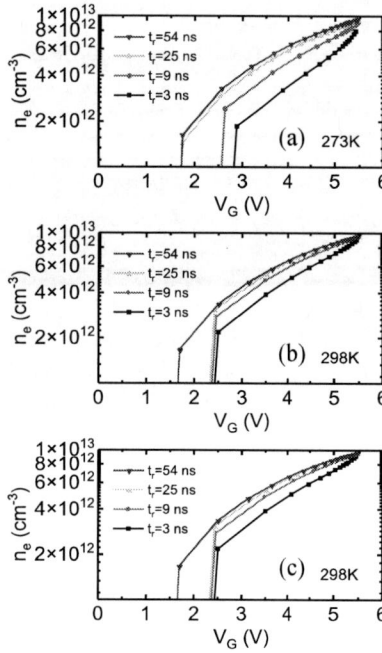

Fig.7: Relationship between simulated 2DEG density n_e and t_r at temperature of 278/298/323 K. The effect of t_r on n_e accentuated at lower temperature.

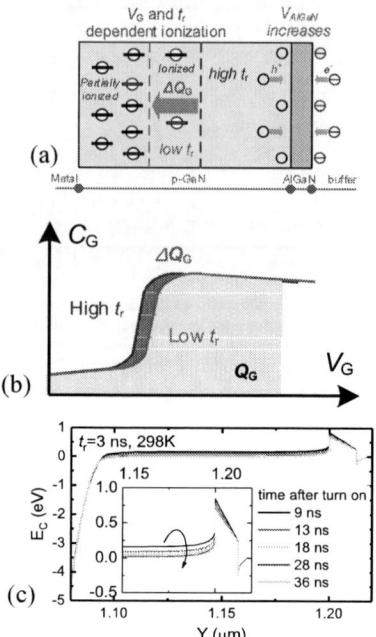

Fig.8: (a) V_G-, t_r- and temperature-dependent incomplete ionization of acceptors induced change of gate charge ΔQ_G in p-GaN/AlGaN/ GaN gate stacks; (b) t_r induced change of Q_G; (c) transient simulation with t_r of 3 ns at 298K shows more acceptor impurities will be ionized then the energy band will return to its equilibrium state.

Magnesium, as a deep-level acceptor, has an ionization energy significantly higher than the thermal energy provided by the lattice at room temperature. Consequently, only a portion of Magnesium atoms can become ionized under normal operating conditions. This incomplete ionization profoundly affects the dynamic behavior of the device, as it directly influences the charge density in the p-GaN layer, thereby impacting the Q_G. As the t_r increases, more Magnesium atoms become ionized, leading to an increase in the N_A^- and subsequently enhancing Q_G, as shown in Fig. 8(b). Experimental results demonstrate a positive correlation between Q_G and t_r. Furthermore, when t_r is larger, the depletion region within the p-GaN layer narrows, allowing for higher Q_G; conversely, a smaller t_r results in a wider depletion region, insufficient ionization, and limits the device's ability to maintain the V_G under high switching rates, as shown in Fig. 6. This reduction in voltage drop across the gate layer results in a decrease in both Q_G and the density of the 2DEG (Fig. 7), which directly affects the device's V_{TH}, as a reduced number of free electrons in the channel alters V_{TH}. Fig. 8(c) illustrates that even with a t_r of 3 ns, E_C gradually returns to a stable state under transient conditions (within the time range of 9 to 36 ns). This indicates that after the pulse is applied, the ionization of magnesium continues at 298 K, and this process does not stop abruptly but progresses gradually over time. As more magnesium atoms become ionized, E_C slowly returns to equilibrium. With the ongoing ionization, both Q_G and n_e steadily increase until they reach a stable state. According to TCAD simulation, the activation energy (E_A) of magnesium in p-GaN layer is approximately 0.23eV, with a cross-section of 1×10^{-14} cm², consistent with values reported in other studies [7,8]. These findings underscore the critical influence of rise time and temperature on the charge dynamics of p-GaN HEMT devices, revealing that optimizing t_r is a key strategy for enhancing performance at high switching frequencies. This is particularly important for the design of gate drivers and control circuits in power electronics applications, where reducing losses and improving switching performance are essential.

Conclusion

Through experiments and TCAD simulations, the influence of t_r and temperature on Gate HEMTs with Partially Depleted p-GaN is studied. Q_G increases with the increase of t_r, and the higher the temperature, the smaller the influence of t_r on Q_G. A lower t_r results in a wider depletion region, resulting in a lower voltage drop in the barrier layer, which results in a lower Q_G and lower 2DEG sheet density.

References

[1] J. He, G. Tang, et al. "VTH Instability of p-GaN Gate HEMTs Under Static and Dynamic Gate Stress." *IEEE EDL* 39(10), 2018.

[2] H. Xu et al., "Dynamic Interplays of Gate Junctions in Schottky-type p-GaN Gate Power HEMTs during Switching Operation," *International Symposium on Power Semiconductor Devices and ICs (ISPSD). IEEE*, 2022: 325-328.

[3] Q. Wu et al., "Charge Control in Schottky-Type p-GaN Gate HEMTs With Partially and Fully Depleted p-GaN Conditions," *IEEE TED* 69(5), 2022.

[4] Tallarico, A. N., et al. "TCAD modeling of the dynamic VTH hysteresis under fast sweeping characterization in p-GaN gate HEMTs." *IEEE TED* 69(2), 2021.

[5] L. Sayadi, G. Iannaccone, S. Sicre, O. Häberlen and G. Curatola, "Threshold Voltage Instability in p-GaN Gate AlGaN/GaN HFETs," in *IEEE TED* 65(6), 2018.

[6] Tang, Shun-Wei, et al. "Capacitance-Dependent VTH Instability Under a High dVg/dt Event in p-GaN Power HEMTs." *IEEE EDL* 43(10), 2022.

[7] Sakurai, Hideki, et al. "Highly effective activation of Mg-implanted p-type GaN by ultra-high-pressure annealing." *APL* 115.14 (2019).

[8] Ke, Wen-Cheng, et al. "Effects of growth conditions on the acceptor activation of Mg-doped p-GaN." *Materials Chemistry and Physics* 133.2-3 (2012).

Self-Rectifying Memristors with High Rectification Ratio for Security Primitives with Ultra-low bit-errors

Guobin Zhang[1,2,3], Zijian Wang[1,2,3], Xuemeng Fan[1,2,3], Dawei Gao[1,2,3], Yishu Zhang[1,2,3*]

[1]College of Integrated Circuits, Zhejiang University, Hangzhou 311200, China

[2]ZJU-Hangzhou Global Scientific and Technological Innovation Center, Hangzhou 310027, China

[3]Zhejiang ICsprout Semiconductor Co., Ltd, Hangzhou, Zhejiang 310027, China

(*Email: zhangyishu@zju.edu.cn)

Abstract

Self-rectifying memristors (SRMs) become vital for security primitives with high accuracy. This study introduces a Pt/HfO$_2$/WO$_{3-x}$/TiN SRM with a rectification ratio over 10^6, addressing sneak path currents. The SRM's conductive behavior is influenced by the WO$_{3-x}$ layer thickness. Traps in WO$_{3-x}$ suppress negative current and boost positive current. The proposed SRMs enable over 21 Gbit crossbar arrays and ultra-low bit-errors, showing great potential for ultra-large-scale integration in the field of security hardware.

(Keywords: Self-rectifying Memristor, Rectification Ratio and Physical unclonable function)

Introduction

In the era of the Internet of Things (IoT), the widespread adoption of intelligent devices and the ubiquity of internet connectivity have enabled the collection, transmission, and processing of data in real time. [1]. However, the broad implementation of IoT devices has brought heightened security risks to light. They are tasked with processing personal data and also have the responsibility of overseeing vital corporate information and essential infrastructure control[2]. Therefore, the security of IoT systems has become the focus of research and practice. To cope with these challenges, IoT system security technologies and strategies need to be updated and improved. As one of renowned security primitives, physical unclonable function (PUF) provides authentication through the intrinsic properties of the devices and circuits to prevent unauthorized devices from accessing the network, and thus, enhances network security[3, 4].

Memristors have been regarded as promising candidates for security hardware due to their unique properties[5]. However, the sneak path currents in the crossbar structures, causing crosstalk between devices, limit the large-scale integration of memristors[6]. Several mainstream approaches including the use of a transistor (1T1R)[7], a diode (1D1R)[8], and a selector (1S1R)[9] have been proposed to suppress sneak path currents. Consequently, SRMs have gained attention as a superior alternative to replace these components for much lower area consumption. Recently, several researches regarding SRMs have been conducted, demonstrating great and stable rectification ratios (RR)[10]. High RR enables large array scale with sufficient reading margin, ensuring the reading accuracy of the security hardware. Therefore, security primitives such as PUF based on SRMs needs to be deeply characterized for ultra-large-scale integration and ultra-low bit-errors.

In this study, we introduce a novel bilayer Pt/HfO$_2$/WO$_{3-x}$/TiN SRM with a high RR of ~10^6. The rectification mechanism is explained, and the SRMs are shown to maintain ultra-low bit-errors with a high RR, advantageous for security primitives with ultra-large scale and high accuracy.

Results and Discussion

To investigate the array sizes feasible by 3×3 μm^2 SRMs with the proposed Pt/HfO$_2$/WO$_{3-x}$/TiN structure, we initially fabricated 32×32 small-scale crossbar arrays (**Fig. 1(a)**). Firstly, the bottom electrode (BE) was fabricated on a four-inch p-type (100) silicon wafer with a 300 nm thermal oxide layer. The corresponding microstructure of each device is shown in **Fig. 1(b)** through high-resolution transmission electron microscopy (HR-TEM). Then we utilized a Keithley 4200A SCS semiconductor parameter analyzer to conduct the direct current (DC) measurements on the proposed SRMs for 100 sweeping cycles. As shown in **Fig. 1(c)**, the positive current exhibits a pronounced nonlinearity, surging to near μA level after an initial pA level, upon sweeping the input voltage from 0 V to 2 V in the direction indicated by the arrows. The operative voltage, 2 V, is almost the lowest among state-of-art SRMs[10], ensuring great endurance and low energy consumption. Conversely, when the input voltage is scanned from 0 V to -2 V, the negative current remains constrained below the pA level, indicating an overall RR of ~10^6 for the SRMs with the 1/3 operative voltage scheme[11]. Further, to elucidate the impact of thin-film WO$_{3-x}$ thickness on the electrical characteristics of SRMs, we modulated the WO$_{3-x}$ film thickness while fixing the HfO$_2$ layer at 10 nm (**Fig. 1(d)**). Notably, the negative currents of the SRMs are increasingly suppressed with increasing the WO$_{3-x}$ layer thickness, while the positive current is promoted more significantly, underscoring the importance of the WO$_{3-x}$ layer as a resistive layer in the proposed SRMs. As the thickness of the WO$_{3-x}$ layer exceeds 40 nm, the DC characteristics hardly change, so the optimal value of the WO$_{3-x}$ film thickness is at least about 40 nm. Finally, we evaluated the retention characteristics of the SRMs to verify the reliability of the fabricated array for data storage and in-memory computing applications, as **Fig. 1(e)** and **1(f)** demonstrated the great retention and endurance characteristics of the SRMs.

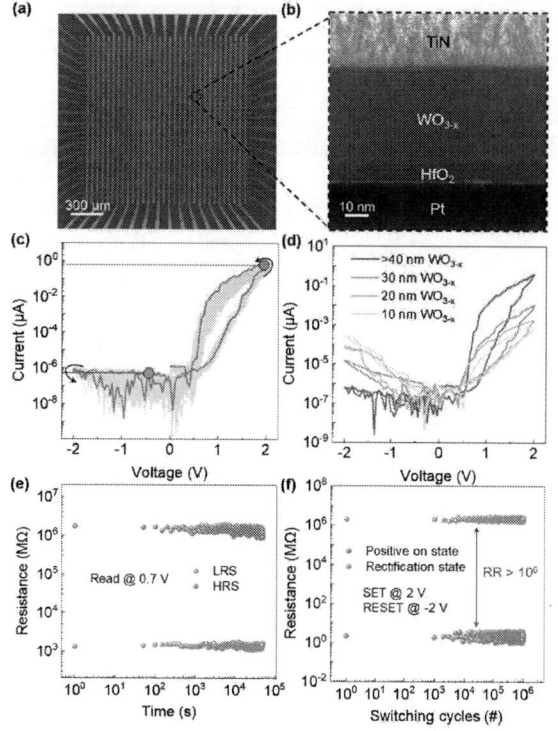

Fig. 1: (a) The SEM image of the 32 × 32 crossbar array comprised of SRMs. (b) The TEM image of the microstructure of SRMs. (c) The basic DC characteristics of the proposed memristors with the structure of Pt/HfO$_2$/ WO$_{3-x}$ /TiN for 100 cycles. (d) The DC characteristics of the SRMs with the different thickness of WO$_{3-x}$. (e) The retention characteristics of the SRMs. (f) The endurance characteristics of the SRMs over 10^6 switching cycles.

To elucidate the switching mechanisms underlying the three structures, we conducted a detailed analysis. When a negative bias is applied to the SRMs (**Fig. 2(a)**), the electrons from the Pt cathode partially traverse the HfO$_2$ barrier layer, reaching the HfO$_2$/WO$_{3-x}$ interface, where they are subsequently trapped by oxygen vacancies in the WO$_{3-x}$ resistive layer. At the same time, the large potential barrier of the Pt/HfO$_2$ interface blocks the electrons from undergoing Schottky emission, resulting in the suppression of negative current below 1 pA with negligible negative P-F emission. When a positive bias is applied to SRMs (**Fig. 2(b)**), the electrons from the TiN cathode are initially blocked by the TiN/WO$_{3-x}$ barrier and simultaneously injected into the WO$_{3-x}$ barrier layer, corresponding to the HRS region below 0.7 V in the I-V curve. As the positive bias increases, the accumulation of trapped electrons in the WO$_{3-x}$ resistive layer reaches the threshold of P-F emission, leading to a current leap phenomenon dominated by P-F emission. At this point, electrons are emitted from the WO$_{3-x}$ and pass through the HfO$_2$ by F-N tunneling mechanism to reach the Pt anode,

realizing the formation of a conductive path and an RR of ~10^6. Additionally, the mechanism of the WO$_{3-x}$ resistive layer thickness effect on the positive and negative current of SRMs as shown in Fig. 1(d), can be explained as follows: with increased thickness, the number of oxygen vacancy defects in the resistive layer rises, resulting in a higher probability of electron trapping after passing through the resistive layer. Under negative bias, fewer electrons can be emitted from the traps, leading to a gradual limitation of negative current below 1 pA. Conversely, under positive bias, a larger number of trapped electrons facilitates a stronger "forceful emission", analogous to an "electron slingshot", resulting in a larger positive current.

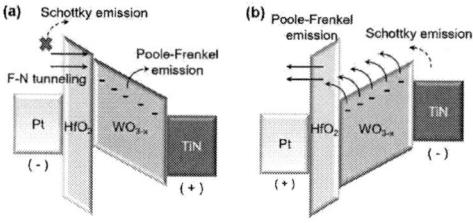

Fig. 2: (a) Pt/HfO$_2$/ WO$_{3-x}$ /TiN with negative bias and (b) Pt/HfO$_2$/ WO$_{3-x}$ /TiN with positive bias.

To verify the advantages of high RR of the proposed SRMs, we simulated their array-scale scalability. As shown in **Fig. 3(a)**, we used the 1/3V scheme to program and read the entire crossbar array. The current in the array flows through the selected path and the sneak path. For SRMs, the current only passes through the selected cells. On the contrary, for non-self-rectifying cells, the current is shunted by the sneak path in parallel with the selected cell, causing crosstalk. Wherein, two 1/3-selected cells and one unselected cell are connected in series with each other to form a sneak path in parallel with the selected cell. We calculated the read margin (RM) with the number of word lines and bit lines in the crossbar array by one bit-line pull-up strategy[12], presenting the scalability of SRMs. The calculation result is illustrated in the **Fig. 3(b)**. Before RM decreases below 10%, the array size could be regarded as reasonable metrics. Thus, the SRMs could achieve over 10^5 crossbar lines and the huge array size of ~ 21 Gbit (145146 × 145146), indicating the proposed SRMs with high RR have outstanding potential for high integration density, which is vital to in-memory computing.

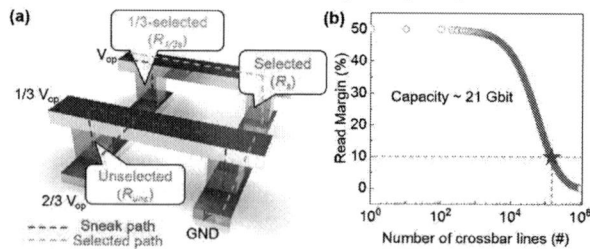

Fig. 3: (a) The sneak path and selected path in the crossbar array. (b) The calculation result of RM with array size based on the proposed SRMs.

In order to show the application potential of the proposed SRMs, we employed crossbar arrays of SRMs to lay out the PUF, a classical and important example of in-memory computation. The PUF's response should be consistent across repeated starts. A high bit error rate means that there are more erroneous bits in the response, which reduces the reliability of the PUF and affects its performance as a hardware security primitive. **Fig. 4(a)** demonstrates the PUF maps generated by crossbar arrays with and without SR. In this case, the PUF maps generated by the arrays without SR appear to have a large number of "bit flips"[13], which are further caused by read errors due to inter-cell crosstalk. On the contrary, crossbar arrays composed of SRMs can effectively reduce the problem of sneak path currents, thereby greatly alleviating or even completely eliminating crosstalk between cells. **Fig. 4(b)** and **4(c)** show the statistical distributions of read errors corresponding to the occurrence of PUF maps with and without SR, respectively. It is clear that SRMs are able to optimize read errors down to as low as 0.059% without the use of any post-processing, which is absolutely state-of-the-art in the field of PUFs, whether or not they are based on memristors[3, 7]. The proposed SRMs with a high RR draw a promising blueprint for further realization of in-memory computing for next-generation security hardware with ultra-low bit-errors and ultra-large-scale integration in the future.

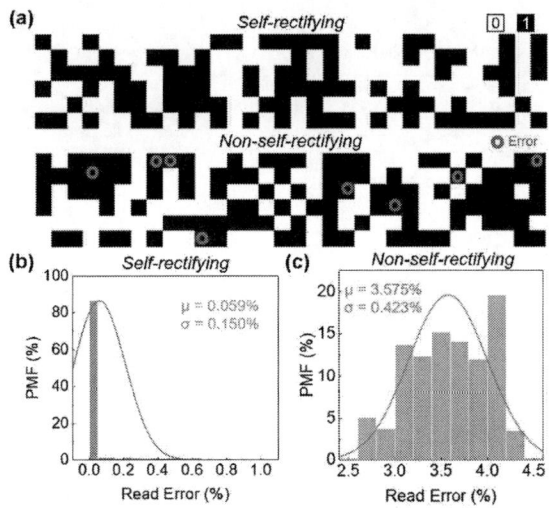

Fig. 4: (a) 32 × 6 partial PUF maps based on crossbar arrays with and without SR. (b) Distribution of read errors based on SR crossbar arrays. (c) Distribution of read errors based on non-SR crossbar arrays.

Conclusion

This study presents a proof-of-concept SRM based on the Pt/HfO$_2$/WO$_{3-x}$/TiN structure, with a blocking layer thickness of 10 nm and a resistive layer thickness of 40 nm. Notably, this SRM achieves a remarkable rectification ratio exceeding 10^6. We validate the proposed high RR using simulated scalability. According to the results of RM, the crossbar array based on the proposed SRMs can be scaled up to ~21 Gbit hyper-scale. These SRMs are capable of addressing the sneak path current issue with significantly mitigating crosstalk between cells, further greatly alleviating bit errors of PUFs. Our SRMs offer a promising blueprint for achieving advanced security primitives with ultra-low bit-errors and ultra-large-scale integration, which are vital for the future of secure in-memory computing.

Acknowledgments

The authors thank Jiabao Sun of ZJU Micro-Nano Fabrication Center for the support. This work is supported by the National Natural Science Foundation of China (Grants No.62204219) and Major Program of Natural Science Foundation of Zhejiang Province (Grants No. LDT23F0401).

References

[1] C. Wang *et al.*, "Parallel in-memory wireless computing," *Nature Electronics,* Article vol. 6, no. 5, pp. 381-389, 2023

[2] Y.-C. Chiu *et al.*, "A CMOS-integrated spintronic compute-in-memory macro for secure AI edge devices," *Nature Electronics,* vol. 6, no. 7, pp. 534-543, 2023

[3] D. Zhong *et al.*, "Twin physically unclonable functions based on aligned carbon nanotube arrays," *Nature Electronics,* vol. 5, no. 7, pp. 424-432, 2022

[4] Y. Gao, S. F. Al-Sarawi, and D. Abbott, "Physical unclonable functions," *Nature Electronics,* vol. 3, no. 2, pp. 81-91, 2020

[5] M. Lanza *et al.*, "Memristive technologies for data storage, computation, encryption, and radio-frequency communication," *Science,* Review vol. 376, no. 6597, pp. 1066-+, 2022

[6] S. Kannan, J. Rajendran, R. Karri, and O. Sinanoglu, "Sneak-Path Testing of Crossbar-Based Nonvolatile Random Access Memories," *Ieee Transactions on Nanotechnology,* Article vol. 12, no. 3, pp. 413-426, 2013

[7] B. Gao *et al.*, "Concealable physically unclonable function chip with a memristor array," *Science advances,* vol. 8, no. 24, p. 7753, 2022

[8] Z.-J. Liu, J.-Y. Gan, and T.-R. Yew, "ZnO-based one diode-one resistor device structure for crossbar memory applications," *Applied Physics Letters,* Article vol. 100, no. 15, p. 153503, 2012

[9] N. K. Upadhyay *et al.*, "A Memristor with Low Switching Current and Voltage for 1S1R Integration and Array Operation," *Advanced Electronic Materials,* Review vol. 6, no. 5, p. 1901411, 2020

[10] S.-G. Ren *et al.*, "Self-Rectifying Memristors for Three-Dimensional In-Memory Computing," *Advanced Materials,* Review vol. 36, no. 4, p. 2307218, 2024

[11] K. Jeon *et al.*, "Purely self-rectifying memristor-based passive crossbar array for artificial neural network accelerators," *Nature Communications,* Article vol. 15, no. 1, p. 129, 2024

[12] X. Zhao *et al.*, "Self-Rectifying Al2O3/TaOx Memristor With Gradual Operation at Low Current by Interfacial Layer," *Ieee Transactions on Electron Devices,* Article vol. 68, no. 12, pp. 6100-6105, 2021

[13] S. Diao *et al.*, "Dynamic Ag nanoclusters inside atomically thin SiOx enable stochastic memristors for physical unclonable functions," *Ceramics International,* Article vol. 49, no. 12, pp. 20901-20906, 2023

Competing ferroelectric polarization and defect migration induced resistive switching in van der Waals β'-In₂Se₃

Yinfeng Long[1], Saiyu Bu[2], Han Chen[1], Kai Liu[1], Shiyu Zhang[1], Lin Wang[1*]

[1]Shanghai Jiao Tong University, China
[2]Peking University, China

Abstract

This work systematically investigates the in-plane resistive switching behavior of representative van der Waals (vdW) ferroelectric β'-In₂Se₃. Besides resistive switching resulting from ferroelectric polarization reversal, the critical role of Se vacancy migration is unveiled in determining the overall electrical characteristics of β'-In₂Se₃ devices through time-dependent current evolution, in-situ electric force microscopy (EFM) and density functional theory (DFT) calculations. By considering the interplay between switchable bound charges and mobile defects, a comprehensive physical picture of the complex hysteretic behavior of β'-In₂Se₃ devices is established.

Keywords: β'-In₂Se₃, van der Waals ferroelectrics, ferroelectric semiconductor, ionic conduction, resistive switching

Introduction

Two-dimensional vdW ferroelectric materials offer unique advantages in addressing the miniaturization, integration, and multifunctional challenges of next-generation ferroelectric devices[1-3]. However, the complex resistive switching phenomena pose a formidable challenge in elucidating underlying physical mechanisms and hindering ferroelectric device design for optimized performance. Particularly, previous research on vdW ferroelectric devices primarily focused on ferroelectric polarization switching, which involves modulating interface or tunneling barriers to influence the threshold switching voltage and on-off ratio[4-6]. Crucially, defect dynamics are widely known to impact the electrical properties of both semiconductors and ferroelectric materials. In this work, diverse resistive switching behaviors with opposite polarities were observed in β'-In₂Se₃ devices. By systematically analyzing these seemingly contradictory phenomena, we identified ferroelectric switching and defect migration as the primary mechanisms influencing resistive switching. The critical role of defect migration in resistive switching was substantiated through a combination of characterization techniques. A phase diagram illustrating the competitive interplay between ferroelectric switching and defect migration as a function of scan rate and electric field was constructed. Our findings provide a comprehensive understanding of the coupling and competition between dipole order (ferroelectricity) and ionic disorder (ionic conduction) in vdW ferroelectric materials, offering new avenues for manipulating and optimizing β'-In₂Se₃ devices for practical applications.

Results and Discussion

In **Fig. 1a**, in-plane PFM images exhibit a 180° phase difference between domains, suggesting in-plane ferroelectricity of exfoliated β'-In₂Se₃ nanoflakes. **Fig. 1b** illustrates a schematic of the β'-In₂Se₃ device architecture. The thickness of exfoliated nanoflakes ranged from 20 to 40 nm, while fabricated device channel lengths typically varied between 1 and 2 μm. Ti/Au (10 nm/50 nm) were used as source and drain contacts. An atomic layer deposition (ALD)-grown Al₂O₃ capping layer (15 nm) was employed to enhance device stability.

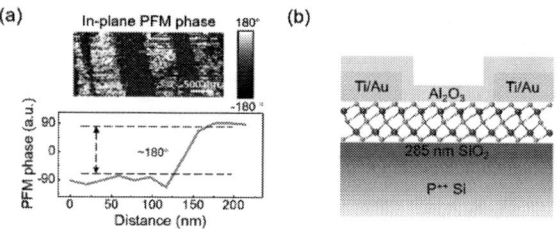

Fig. 1. (a) In-plane PFM phase image (above) of an exfoliated β'-In₂Se₃ nanoflake and the averaged line profile (below) extracted from the red rectangle in the PFM phase image. (b) Schematic of the β'-In₂Se₃ device configuration.

Fig. 2a presents representative type I *I-V* hysteresis loops displaying a "counter-eightwise" polarity for various voltage sweep ranges. In both positive and negative voltage regimes, the device resistance transitioned from a high-resistance state (HRS) to a low-resistance state (LRS). This resistive switching behavior can be explained by the classic "back-to-back" ferroelectric Schottky diode model[4, 5]. Regardless of voltage polarity, the current flow is dominated by the contact under reverse bias. The observed linear relationship between $\ln(I)$ and $V^{1/4}$ within a moderate voltage bias range in **Fig. 2b** supports the thermionic emission model incorporating image-force-induced barrier lowering. The energy band diagram corresponding to the resistive switching process is illustrated in **Fig. 2c-d**. When a small positive V_{ds} bias is applied and the ferroelectric polarization points from the source to the drain terminal, the negative bound charges elevate the barrier height of the reverse-biased source contact, impeding electron transport and leading to HRS state of the device. Upon increasing V_{ds}, ferroelectric polarization is reversed. The subsequent positive bound charges reduce the barrier height of the reverse-biased source contact, facilitating electron transport and transitioning the device to the LRS state. As V_{ds} enters the negative regime, ferroelectric polarization remains oriented from drain to source, but the drain contact becomes the reverse-biased diode. A similar analysis predicts a return to the HRS state, until the device transitions back to the LRS state upon ferroelectric polarization reversal.

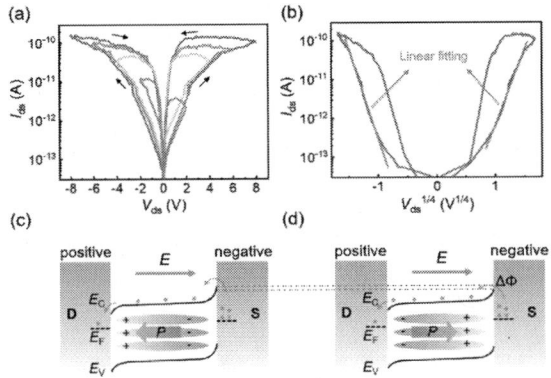

Fig. 2. (a) Type I I-V hysteresis loop curves for various voltage sweep ranges. Arrows indicate the switching directions. (b) Ln(I_{ds}) as a function of $V_{ds}^{1/4}$ from the data in (a). (c-d) Energy band diagram illustrating the ferroelectric modulated Schottky barrier at four different states.

Another type of hysteretic loop with an "eightwise" polarity (type II) was frequently observed, as depicted in **Fig. 3a**. Both positive and negative voltage sweeps induced transitions from the LRS to the HRS, with negative differential resistance (NDR) phenomena observed under specific conditions, which is modulated by scan rates. The distinct characteristics of type II hysteresis loops suggest the involvement of other factors influencing resistive switching. The scan rate-modulated characteristics of current, hysteresis window and NDR effect are reminiscent of the ionic memristors[7-9]. The linear relationship between ln(I) and V at significant voltage biases supports the model of device influenced by mobile defects dynamics. To further explore the device physics, DC voltage stressing experiments were conducted while monitoring the current (**Fig. 3b**). During both positive-bias and negative-bias stage, the current increased continuously before gradually decreasing, which aligns with the character of dynamic process of defects migration. Compared to the nanosecond timescales of electron conduction and ferroelectric inversion, ion diffusion is significantly slower. Consequently, ionic current initially increases gradually. The subsequent current decrease arises from defect accumulation near the contact interface. As

defect accumulation progresses, the increasing built-in electric field counteracts the external field, reducing the net electric field and ionic current. And a higher voltage accelerates the process of the device current to reach new equilibrium, as a result of a stronger electric field which lowers the diffusion barrier and enhances the defects migration. *In-situ* EFM (**Fig. 3c-d**) was performed to monitor the evolution of the channel surface potential upon bias switching. A tens-of-seconds relaxation time is recorded for the surface potential to establish equilibrium within the channel for devices exhibiting type-II hysteresis, which is closed associated with the defect accumulation and redistribution process within the device channel.

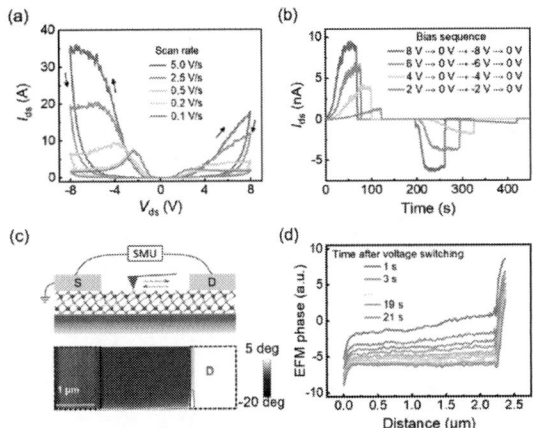

Fig. 3. (a) Type II I-V hysteresis loop curves with a full-cycle at various scan rates. Arrows indicate the switching directions. (b) Time-dependent current evolution during various DC bias sequences. (c) Schematic of the *in-situ* EFM measurement with an SMU for applying bias voltage to the device (above). A continuous *in-situ* EFM scan along a fixed line within the device channel when the bias is switched from 5 V to -5 V (below). (d) Line profiles of EFM phase extracted from the dashed red rectangle in (c).

The crucial role of Se vacancy migration in device hysteresis is displayed by DFT. Theoretical calculation reveals lower formation energy of Se vacancy in β'-In$_2$Se$_3$ grown under Se-rich conditions (**Fig. 4a-b**). A model combining Se vacancy migration and electron transport is proposed to elucidate the device characteristics. When a positive V_{ds} is applied, the total current comprises both electronic and ionic components. As the V_{ds} sweep progresses, the increasing built-in electric field induced by the migration and accumulation of Se vacancies weakens the ionic current (**Fig. 4c**). Concurrently, the trapping process resulting from charge accumulation induced trapping states at the interface impede the electron transport. Both two induce a resistive transition from LRS to HRS (**Fig. 4d**). Notably, negative V_{ds} biases can reconfigure the Se vacancy distribution, resetting the device from HRS to LRS, resulting in bipolar resistive switching.

Fig. 5a-c show the observation of transitions between these two conduction mechanisms in devices influenced by Se

979-8-3315-0417-5/25 $31.00 © 2025 IEEE

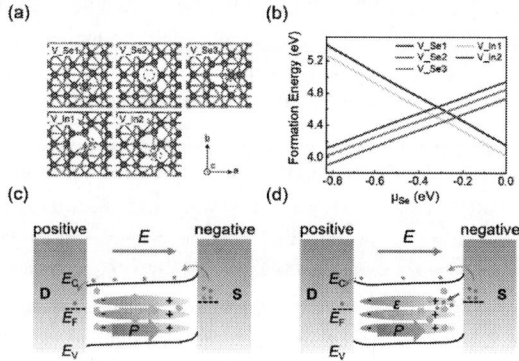

Fig. 4. Structure (a) and formation energies (b) of single-atom vacancy defects in β'-In$_2$Se$_3$. The red dashed box indicates the vacancy defect position. (c-d) Energy band diagrams elucidating the different states in the type II *I-V* hysteresis sweep process.

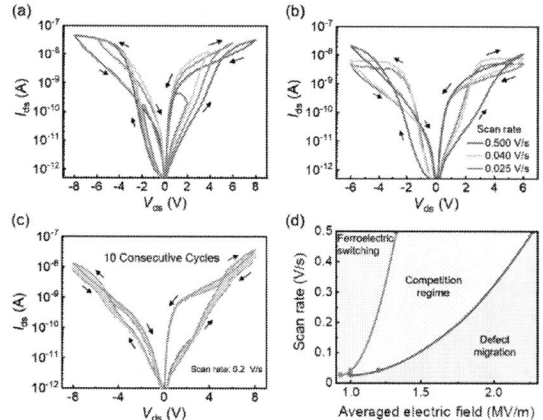

Fig. 5. (a) Typical *I-V* loop curves for various voltage sweep ranges. (b) *I-V* loop curves with a full-cycle at a series of scan rates. (c) Bipolar consecutive voltage sweeps of the devices. Arrows indicate the switching directions. (d) Phase diagram of the resistive switching mechanism as a function of scan rate and averaged electric field.

vacancies. The resistance transition sequence shifts to "HRS-LRS-HRS-LRS" in both positive and negative branches. The complex resistive switching behavior arises from the competitive interplay between ferroelectric switching and defect migration. A phase diagram in **Fig. 5d** depicts the resistive switching mechanism transition as a function of voltage bias and averaged electric field. At high scan rates and low electric field, the high diffusion barrier hinders Se vacancy migration, resulting in ferroelectric switching dominance. Conversely, at low scan rates and high electric field, defect migration is significantly enhanced. The intermediate region represents a competitive state where both mechanisms contribute to the electrical behavior. We note that these complicated hysteresis behaviors have been frequently observed in vdW ferroelectric devices, yet poorly elucidated. Our study deepens the understanding of underlying device physical process. The ability to

reconfigure device operation mechanisms suggests the potential for designing multifunctional ferro-ionic devices. For example, the opposite polarities of resistive switching might endow the devices with the capability to implement neuromorphic computing with different operation modes in response to similar stimuli.

Conclusion

We have comprehensively investigated the resistive switching behavior of β'-In$_2$Se$_3$ devices. Beyond Schottky barrier modulation associated with ferroelectric polarization reversal, theoretical and experimental analyses reveal the critical role of Se vacancy migration in inducing distinct resistive switching characteristics. The competition between these two mechanisms, influenced by electric field and scan rate, is elucidated. By establishing a phase diagram for resistive switching in β'-In$_2$Se$_3$, this work provides insights into the complex electrical behavior of vdW ferroelectric devices. The findings highlight the potential of defect engineering in β'-In$_2$Se$_3$ as well as other vdW ferroelectric materials for developing advanced non-volatile device applications.

Acknowledgments

This work is supported by National Natural Science Foundation of China (Grant No. 52302187).

References

[1] M. Wu *et al.*, "Two-Dimensional van der Waals Ferroelectrics: Scientific and Technological Opportunities," ACS Nano, vol. 15, no. 6, pp. 9229-9237 (2021).

[2] C. Wang, L. You, D. Cobden, and J. Wang, "Towards two-dimensional van der Waals ferroelectrics," Nat. Mater., vol. 22, no. 5, pp. 542-552 (2023).

[3] T. Jin *et al.*, "Ferroelectrics-Integrated Two-Dimensional Devices toward Next-Generation Electronics," ACS Nano, vol. 16, no. 9, pp. 13595-13611 (2022).

[4] F. Xue *et al.*, "Unraveling the origin of ferroelectric resistance switching through the interfacial engineering of layered ferroelectric-metal junctions," Nat. Commun., vol. 12, no. 1, p. 7291 (2021).

[5] L. Wang *et al.*, "Exploring Ferroelectric Switching in α-In$_2$Se$_3$ for Neuromorphic Computing," Adv. Funct. Mater., vol. 30, no. 45, p. 2004609 (2020).

[6] J. Wu *et al.*, "High tunnelling electroresistance in a ferroelectric van der Waals heterojunction via giant barrier height modulation," Nat. Electron., vol. 3, no. 8, pp. 466-472 (2020).

[7] X. Jiang *et al.*, "Manipulation of current rectification in van der Waals ferroionic CuInP$_2$S$_6$," Nat. Commun., vol. 13, no. 1, p. 574 (2022).

[8] D. Zhang *et al.*, "Anisotropic Ion Migration and Electronic Conduction in van der Waals Ferroelectric CuInP$_2$S$_6$," Nano Lett., vol. 21, no. 2, pp. 995-1002 (2021).

[9] Y. Du *et al.*, "Symmetrical Negative Differential Resistance Behavior of a Resistive Switching Device," *ACS Nano*, vol. 6, no. 3, pp. 2517-2523 (2012)

Improving the Reliability of 40nm RRAM Chip by Pre-cycle Operation

Ruofei Hu[1,†], Yilong Huang[1,†], Chengxiang Ma[1], Qianze Zheng[1], Kaimeng Liu[1], Yuelin Jiang[1], Siyu Chen[1], Jianshi Tang[1*], Dong Wu[1], Bin Gao[1], He Qian[1], Huaqiang Wu[1]

[1]School of Integrated Circuits, Beijing Advanced Innovation Center for Integrated Circuits, Tsinghua University, Beijing. [†]Equal Contributions, *Email: jtang@tsinghua.edu.cn

Abstract

This work proposes a practical method, pre-cycle, to enhance the comprehensive reliability of resistive random-access memory (RRAM) on a 40nm CMOS platform. Through 100 pre-cycles, RRAM can achieve endurance > 10k cycles, retention > 10 years at 85°C, and read disturbance > 10^9 cycles. The operational cost of pre-cycle is similar to forming, making it feasible for practical applications. This method provides new possibilities for the further integration of RRAM towards more advanced technology nodes.

Keywords: Resistive Random-Access Memory (RRAM), embedded non-volatile memory (eNVM), reliability

Introduction

RRAM has been widely regarded as a promising solution for the next generation of embedded non-volatile memory (eNVM), due to its advantages in low cost, excellent scaling capability, fast operation and low operating voltage [1]. Despite of extensive studies on RRAM-based eNVM [2, 3], reliability still remains a critical issue hindering the RRAM commercialization in advanced technology node.

In this work, we propose an effective method using pre-cycle to improve the comprehensive reliability of RRAM on 40nm CMOS platform. A 1kb one-transistor-one-resistor (1T1R) RRAM array is designed and used as the test vehicle. We carry out a systematic study on the pulse number and voltage amplitude of the pre-cycle operation, and evaluate its impact on the reliability of RRAM. By adopting the pre-cycle operation, it achieves a high endurance > 10k cycles, retention > 10 years at 85°C, and read disturbance > 10^9 cycles altogether.

Experimental Platform and Device Operations

Fig. 1 shows the resistive switching curve of the 1T1R RRAM under DC voltage sweep. The inset of Fig. 1 shows the schematic diagram of 1T1R structure and the optical image of the 1kb 1T1R array fabricated on the commercial 40nm Si CMOS platform. A thermal enhanced layer (TEL) is capped on the resistive switching layer (RSL) to improve the RRAM performance [4]. The RSL/TEL layer is sandwiched by the top electrode (TE) and bottom electrode (BE). The programming scheme of RRAM is shown in Fig. 2. Pulses with fixed voltages and a pulse width of 100ns are used for SET/RESET. In the following, "programming" refers to the SET/RESET operation with verify, where a SET/RESET pulse is followed by a read pulse to verify whether the RRAM current reaches the target value (Fig. 2(a)). SET/RESET pulses are repeatedly applied until the read current reaches the target or the pulse number reaches the upper limit. "Pre-cycle" refers to alternate SET/RESET pulse series applied to RRAM after typical forming operation. There is no verify pulses during the pre-cycles (Fig. 2(b)). One pre-cycle consists of one SET pulse and one RESET pulse.

Experimental Results and Discussions

A. Influence of Pre-cycle on Retention

In Fig. 3, the influence of pre-cycle number on retention is studied. After forming and pre-cycle operation, the RRAM devices are programmed to high-resistance state (HRS) and low-resistance state (LRS) respectively, and then baked at 175°C for 3.75h to accelerate resistance drift. It can be seen that RRAM devices without pre-cycle operation exhibit more pronounced resistance drift, especially for LRS. More than 10% of the LRS devices drifted and overlapped with HRS after baking, resulting in an unacceptably high bit error rate. By adopting the pre-cycle operation, the retention of RRAM is significantly improved. As the pre-cycle number increases to 100, a large memory window (MW) is obtained between the HRS and LRS of baked RRAM devices. Not only has the LRS retention been significantly improved, but also even the HRS drift has been suppressed. When the pre-cycle number increases to 1000, the improvement effect on retention tends to saturate. Therefore, 100 appears to be the optimal number for pre-cycles. Since no verify is needed in the pre-cycle, the cumulative pulse width of 100 pre-cycles is only 20μs, which is close to the cumulative pulse width of a typical forming pulse [5]. In other words, the pre-cycle following forming will not result in excessive time overhead, making it both applicable and effective in practice.

We further study the effect of voltage amplitude for pre-cycle. For comparison, the voltage condition for typical programming is used as baseline here. Fig. 4 shows the effect of different SET voltage conditions for pre-cycle. As the SET pulse becomes weaker (lower amplitude), the retention of both HRS and LRS becomes worse. The degradation of HRS retention is even larger than that of LRS retention. Stronger SET pulse, on the other hand, improves the LRS retention at the cost of HRS retention degradation. The MW under stronger SET pulse is still smaller than baseline. Similarly, Fig. 5 shows the effect of different RESET voltage conditions for pre-cycle. Weaker RESET pulse also results in drastic degradation of both HRS and LRS retention. The resistance drift under weak RESET 2 condition is even close to the RRAM devices without pre-cycle operation, indicating the improvement effects of pre-cycle has diminished. On the other hand, stronger RESET pulse has little impact on retention, but it could potentially deteriorate endurance [6], so it is not favorable in practice. The above results reveal that the voltage condition of pre-cycle has an important impact on the retention improvement by pre-cycle. SET/RESET pulse conditions close to the ones used for actual programming yield the best optimization results.

B. Comprehensive Evaluation of Reliability

Furthermore, the impact of pre-cycle on endurance is also studied. Fig. 6 shows the endurance of 1kb RRAM without pre-cycle operation. We can see a small proportion of RRAM devices exhibit temporary programming failures in the early programming cycles. Despite failures occurred in first few cycles, these devices can still be programmed to the target in the subsequent cycles. This indicates such early failures are not permanent. Fig. 7 plots the programming failure rate during the endurance test. For RRAM without pre-cycle, the early programming failures usually occur within the first 100 programming cycles while the permanent failures occur after 1k cycles. In comparison, RRAM devices with 100 pre-cycle operations have no early failures but only permanent failures. This suggests that, although there is no verify in pre-cycle, it can still program most devices and eliminate early failures. In addition, the pre-cycle has no negative impact on the permanent fail rate. As shown in Fig.8, RRAM subject to 100 pre-cycles can reach the programming target within 1k cycles, and still maintain a large MW after 10k cycles.

Following the endurance test, we carry out retention and read disturbance tests on the same batch of 1kb RRAM, to comprehensively evaluate the reliability performance. As shown in Fig. 9, a valid MW can still be maintained after baking at 175°C for 37.5h. According to our previous work [2], this result is equivalent to 10-year retention at 85°C. Then we applied 10^9 read pulses to the RRAM devices that have been cycled and baked. The read current distribution exhibits only minor changes after 10^9 read cycles, and the MW sustained read disturbance. This further proves the excellent stability of RRAM with pre-cycle operation.

Mechanism of Pre-cycle

The mechanism of pre-cycle is attributed to the robustness of the conductive filament (CF) morphology corresponding to certain RRAM resistance state. As shown in Fig. 11, the transition between HRS and LRS can be depicted by the Arrhenius plot. This process follows the statistical physics principles. During SET/RESET, the electric field drives the migration of oxygen ions and oxygen vacancies inside the RSL, resulting in the growth and rupture of CF. This microscopic phenomenon is reflected in the Arrhenius plot as

the electrical constraints indicated by the dashed line (Fig. 11(a) and (b)). The electrical constraints alter the energy barrier between the HRS and LRS states and drive resistance switching towards a certain resistance state. Because of the stochastic nature in the CF formation [7], a certain proportion of RRAM devices only form a weak CF after forming (Fig. 11(c)). For these devices, the low energy barrier leads to poor retention, especially for LRS. Through repeated SET/RESET operations during pre-cycle, the CF is gradually strengthened (Fig. 11(d)), where the higher energy barrier enhances the RRAM retention after pre-cycle. Because the SET/RESET conditions determine the programmed resistance value, there is a relationship between the HRS/LRS and the electrical constraints in the Arrhenius plot. Therefore, to obtain the robust CF corresponding to certain HRS/LRS, the conditions of pre-cycle need to be close to that of programming. The early programming failures in Fig. 6 can also be explained by the stochastics during repeated resistance state transitions. Robustness in resistance transition between HRS and LRS can also be enhanced through the alternating application of corresponding electrical constraints during pre-cycle.

Conclusion

In this work, pre-cycle is proposed and demonstrated as an efficient and effective method to improve the comprehensive reliability of RRAM on 40nm CMOS platform. With 100 pre-cycles, RRAM can achieve excellent reliability, including endurance > 10k cycles, retention > 10 years at 85°C and read disturbance > 10^9 cycles. The effects of pre-cycle number and pulse conditions on the RRAM reliability are systematically studied, and the underlying mechanism is also analyzed. This work facilitates RRAM integration towards even more advanced technology nodes.

Acknowledgments

This work was in part supported by the Semiconductor Technology Innovation Center (Beijing) Corporation under Grant QYJS-2021-0801-B and Grant QYJS-2021-0802-B.

References

[1] H. Wu et al., Proceedings of the IEEE, vol. 105, no. 9, pp. 1770-1789, 2017, doi: 10.1109/jproc.2017.2684830.

[2] Y. He et al., IEEE International Memory Workshop, pp. 1-4, 2024, doi: 10.1109/IMW59701.2024.10536978.

[3] Y. C. Huang et al., IEEE International Solid-State Circuits Conference, pp. 288-290, 2024, doi: 10.1109/ISSCC49657.2024.10454367.

[4] W. Wu et al., IEEE Electron Device Letters, vol. 38, no. 8, pp. 1019-1022, 2017, doi: 10.1109/led.2017.2719161.

[5] X. Li et al., IEEE Transactions on Electron Devices, vol. 70, no. 2, pp. 499-505, 2023, doi: 10.1109/ted.2022.3232313.

[6] M. Zhao et al., Applied Physics Reviews., vol. 7, no. 1, p. 011301, 2020, doi: 10.1063/1.5124915.

[7] B. Gao et al., IEEE International Electron Devices Meeting, pp. 4.4.1-4.4.4, 2017, doi: 10.1109/IEDM.2017.8268326.

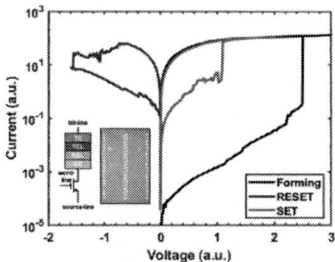

Fig. 1: DC I-V curve of 1T1R RRAM. The inset shows the schematic diagram of 1T1R structure and the optical image of the 1kb 1T1R array.

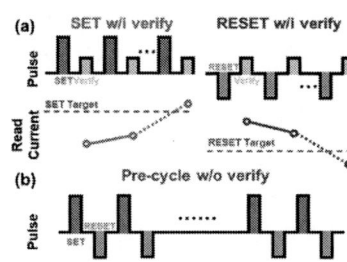

Fig. 2: Schematic diagram of (a) SET/RESET with verify and (b) pre-cycle operation without verify.

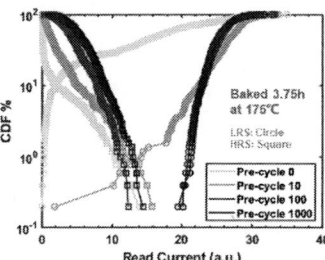

Fig. 3: Retention of 1kb RRAM with different pre-cycle numbers. The devices are baked at 175°C for 3.75h.

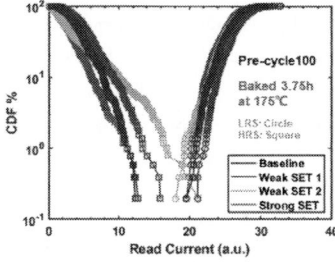

Fig. 4: Retention of 1kb RRAM with different SET conditions. The voltage condition of weak SET 2 is weaker than weak SET 1.

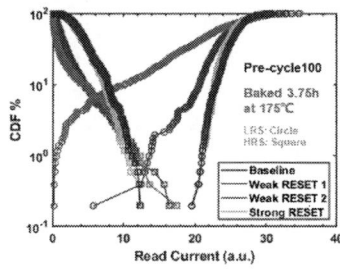

Fig. 5: Retention of 1kb RRAM with different RESET conditions. The voltage condition of weak RESET 2 is weaker than weak RESET 1.

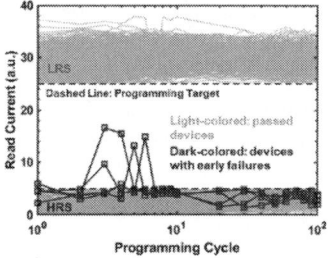

Fig. 6: Endurance of 1kb RRAM without pre-cycle operation. A fraction of devices show temporary programming failures in early cycles.

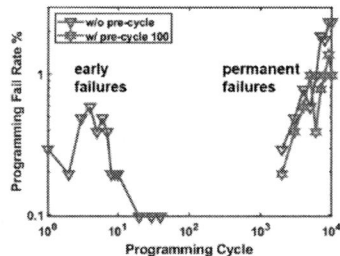

Fig. 7: The programming fail rate curves of 1kb RRAM. Pre-cycle operation eliminates failures in early programming cycle.

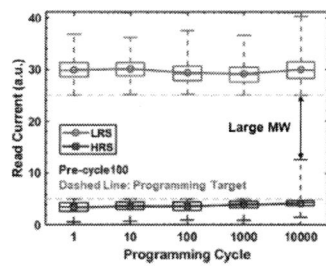

Fig. 8: Endurance of 1kb RRAM subjected to 100 pre-cycles. The dashed error bar shows the maximum and minimum of read current.

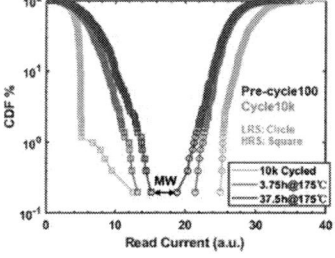

Fig. 9: Retention of 1kb RRAM subjected to 100 pre-cycles and 10k programming cycles. The devices are baked at 175°C for 37.5h.

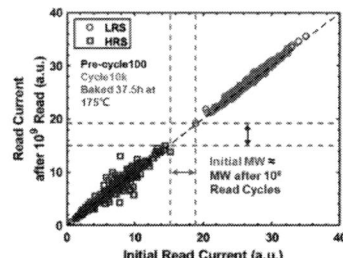

Fig. 10: Read disturbance test of 1kb RRAM. The read current after the endurance and retention test is consistent with the initial value.

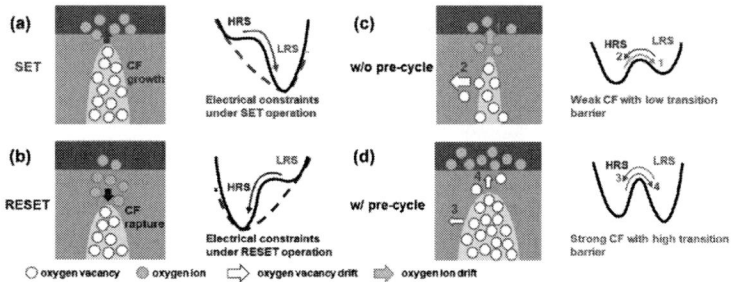

Fig. 11: CF morphology and Arrhenius plot of RRAM under (a) SET and (b) RESET operation, as well as RRAM (c) with and (d) without pre-cycle operation. Resistive switching is subject to corresponding electrical constraints. Pre-cycle operation facilitates robust CF and resistance state.

Investigation of Passivation Layers for Self-aligned Top-Gate Amorphous InGaZnO Thin-Film Transistors with Metal-Reacted Low-Resistance Source/Drain

Yuhan Zhang[1], Jiye Li[1], Hao Peng[1], Huan Yang[1], Lei Lu[1], Shengdong Zhang[1]

[1]School of Electronic and Computer Engineering, Peking University, Shenzhen, China

E-mail: zhangsd@pku.edu.cn

Abstract

The metal reaction approach has been successfully utilized to fabricate high-performance self-aligned top-gate (SATG) oxide TFTs with low-resistance source/drain (S/D) regions. However, the stability remains a concern, as the metal-reacted S/D regions are susceptible to degradation during thermal processing. This work systematically investigated the passivation effects of PECVD SiO_2, Si_3N_4, and sputtered Al_2O_3. Results indicate that the 150 °C PECVD Si_3N_4 passivation layer (PL) offers SATG TFTs superior stability under thermal, environmental and bias stress conditions, along with excellent scalability and well-maintained low S/D resistance, compared to other PLs.

Keywords: self-aligned top-gate, thin-film transistors, passivation layer, source/drain resistance, channel length shrinking, thermal and environmental stability.

Introduction

In recent years, amorphous oxide semiconductor (AOS) thin-film transistors (TFTs) have been extensively investigated and are considered as the cornerstone for the next-generation displays, owing to their relatively high mobility, extremely-low off current, good uniformity and low temperature fabrication [1]. Furthermore, other AOS-based electronic applications such as flexible electronics [2], sensors [3] and three-dimensional (3D) memories [4] have also attracted widespread attention.

Among the various TFT architectures, the self-aligned top-gate (SATG) TFT stands out as the preferred choice for advanced AOS applications, due to its smaller footprint, superior scalability and low parasitic capacitance [5]. However, the major challenge in SATG TFT fabrication is the formation of self-aligned highly conductive source/drain (S/D) regions. Commonly used S/D doping techniques include plasma treatment [6], hydrogen doping [7] and metal reaction [8], [9]. While the metal reaction method has been demonstrated to provide lower S/D resistance (R_{SD}) and minimal channel length shrinking ($2\Delta L$) [8], the low R_{SD} would not survive from subsequent thermal processes and harsh environments [10], [11]. Therefore, to achieve high-performance SATG AOS TFTs, it is essential to establish a reliable passivation process that guarantees the thermal and environmental stability of the metal-treated source/drain regions.

In this work, several PLs for SATG a-IGZO TFTs with magnesium (Mg) doped S/D regions were thoroughly

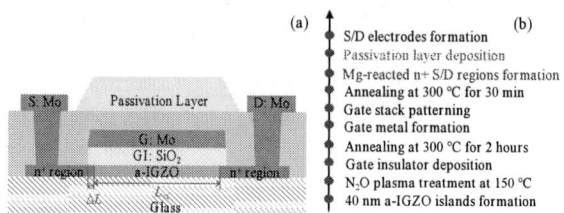

Fig. 1: (a) The schematic cross-section and (b) process flow of the SATG a-IGZO TFT with Mg-reacted n+ S/D regions.

investigated. The effects of these PLs on R_{SD} and $2\Delta L$ were analyzed and their thermal and environmental stabilities were assessed. The transistors with a 150 °C PECVD SiN_x PL demonstrate excellent stability against thermal and environmental stress. This is evidenced by a highly stable R_{SD} of 8.4 Ω·cm and $2\Delta L$ of 0.61 μm, along with good bias stress stability.

Experiment

The schematic cross-section and major process flow of the SATG a-IGZO TFTs with Mg-reacted S/D regions are presented in Fig. 1. First, a 40-nm a-IGZO active layer was deposited on the glass substrate at room temperature (RT) using direct-current (DC) magnetron sputtering with a ceramic target having a molecular ratio of In_2O_3: Ga_2O_3: ZnO = 1:1:2 mol%. Following the formation of the active islands, a nitrous oxide (N_2O) plasma treatment was conducted at 150 °C. Next, a 200-nm silicon dioxide (SiO_2) was deposited at 300 °C as the gate insulator (GI) using plasma-enhanced chemical vapor deposition (PECVD), following which a thermal annealing process was performed at 300 °C for two hours. Afterwards, a 100-nm molybdenum (Mo) was deposited as the gate electrode (G) and the Mo/SiO_2 gate stack was subsequently patterned. Following a second thermal annealing in an O_2 ambient for 30 minutes, a thin Mg layer was deposited by sputtering on a heated substrate to form Mg-reacted heavily doped S/D regions [8]. The wafers were then rinsed in hot water to remove any excess Mg. Various PLs were deposited, including a 200-nm SiO_2 deposited by PECVD at 150 °C, a 200-nm Si_3N_4 deposited by PECVD at either 150 °C or 300 °C and a 100-nm Al_2O_3 deposited by reactive sputtering at RT. Finally, the Mo S/D electrodes were formed by lift-off process.

The thermal stability test was conducted after an additional annealing process at 200 °C in air. The environmental stability was evaluated after exposing the wafers to 80 °C and 80% relative humidity (RH) for 24 hours. The electrical

979-8-3315-0417-5/25 $31.00 © 2025 IEEE

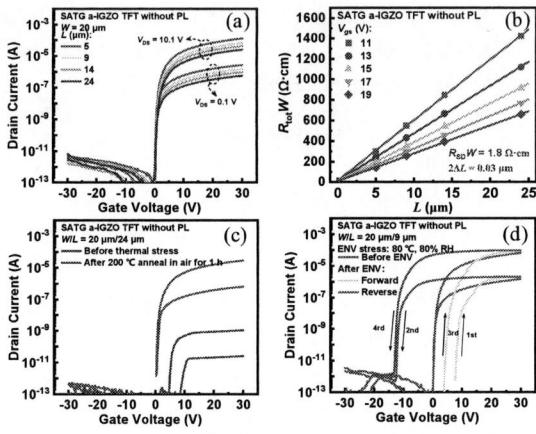

Fig. 2: (a) The L scalability and (b) the $R_{tot}W$ of the SATG TFTs without PL with L decreasing from 24 μm to 5 μm, and the transfer characteristics of the SATG TFTs without PL after (c) thermal stability test and (d) environmental stability test.

characteristics of the SATG a-IGZO TFTs were measured using an Agilent B1500 semiconductor parameter analyzer at RT in a dark environment.

Result and Discussion

Fig. 2(a) shows the channel length (L) scalability of the Mg-reacted SATG a-IGZO TFTs without PLs. The transfer curves exhibit a negligible negative threshold voltage (V_{th}) shift as L decreases from 24 μm to 5 μm. As shown in Fig. 2(b), the R_{SD} and $2\Delta L$ are calculated to be 1.8 Ω·cm and 0.03 μm, respectively, using the transmission line method (TLM). These characteristics indicate that the Mg-treated SATG TFTs have excellent downscaling capability and sufficiently low R_{SD}. Generally, metal-reacted S/D regions are more stable than those formed by other methods [8]. However, as illustrated in Fig. 2(c), the on-state current (I_{on}) of the device without PLs significantly decreases after the thermal process at 200 °C in air, accompanied by a positive V_{th} shift. In addition, as shown in Fig. 2(d), the transistors also experience severe deterioration after environmental stress test, including significant hysteresis and negative V_{th} shift. As mentioned, these degradation phenomena could be attributed to the diffusion of O_2 and H_2O into the S/D and channel regions under extreme conditions, highlighting the necessity of high-quality PLs to act as a barrier against ambient O_2 and H_2O.

Fig. 3 illustrates the transfer curves and R_{SD} of SATG devices with four different PLs. The performance parameters are summarized in Table 1. As shown in Fig. 3(b) and (f), the device with a 200-nm Si_3N_4 PL deposited at 300 °C exhibits a notable negative V_{th} shift when L is scaled down to 5 μm. The R_{SD} and $2\Delta L$ increase to 18.8 Ω·cm and 2.57 μm, respectively, indicating significant lateral diffusion of carriers from the S/D regions. As shown in Fig. 3(d), the transistor with Al_2O_3 deposited by sputtering shows a significant off current (I_{off}), likely due to the damage to the sidewalls of the GI during the sputtering process. In contrast, the transistor with a Si_3N_4 PL deposited at 150 °C shows a

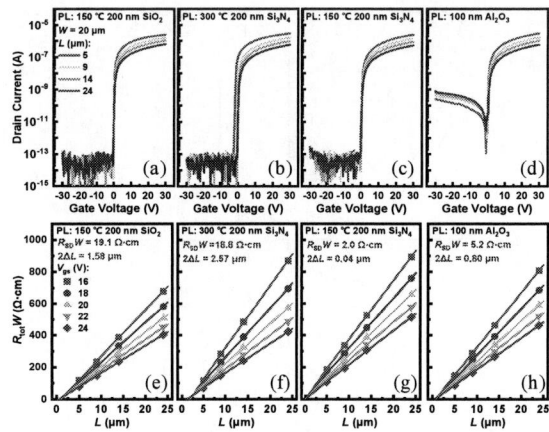

Fig. 3: The L scalability and the $R_{tot}W$ of the SATG TFTs with different PL, (a)(e) 150 °C 200 nm SiO_2, (b)(f) 300 °C 200 nm Si_3N_4, (c)(g) 150 °C 200 nm Si_3N_4 and (d)(h) RT 100 nm Al_2O_3.

Table 1 Performance parameters of the SATG TFTs.

SATG TFTs (20 μm/5 μm)	μ_{FE} (cm²/Vs)	SS (V/dec)	V_{th} (V)	I_{OFF} (μA)	$R_{SD}W$ (Ω·cm)	$2\Delta L$ (μm)
Without PL	14.6	0.19	0.50	<10⁻¹³	1.8	0.03
150 °C SiO_2	11.4	0.15	0.22	<10⁻¹³	19.1	1.58
300 °C Si_3N_4	16.1	0.23	-1.37	<10⁻¹³	18.8	2.57
150 °C Si_3N_4	15.3	0.18	0.36	<10⁻¹³	2.0	0.04
Al_2O_3	16.0	0.20	0.41	<10⁻⁹	5.2	0.80

well-maintained R_{SD} of 2.0 Ω·cm and $2\Delta L$ of 0.04 μm, which is comparable to the as-fabricated devices.

Fig. 4 shows the thermal stability of the fabricated SATG TFTs. As shown in Fig. 4(a) and (b), the SiO_2 and Al_2O_3-passivated transistors show severe degradation of I_{on} after undergoing annealing at 200 °C in air for 1 hour, indicating that the Mg-reacted heavily doped S/D regions have been deteriorated by the thermal process. Although the R_{SD} and $2\Delta L$ are slightly increased to 8.4 Ω·cm and 0.61 μm, respectively, the low-temperature Si_3N_4-passivated transistor maintains good device performance, as shown in Fig. 4(c) and (d). This could be attributed to the excellent protective capability of Si_3N_4 against H_2O and O_2.

Fig. 5 shows the bias stress stability of the fabricated transistors with different PLs before and after exposure to an environmental stress test at 80 °C and 80% RH for 24 hours. As plotted in Fig. 5(a), the devices without PLs demonstrate excellent stability under both negative bias stress (NBS) and positive bias stress (PBS). After the deposition of PLs, the SATG TFTs exhibit a slightly positive V_{th} shift (ΔV_{th}) under PBS.

To investigate the protection capability of the PLs against moisture, all the wafers were placed in an environmental chamber at 80 °C and 80% RH for 24 hours. As shown in Fig. 5(b), (c) and (d), after exposure to the humid environment, the SiO_2 and Al_2O_3-passivated transistors show abnormal negative V_{th} shift under PBS, even leading to device failure, while the Si_3N_4-passivated transistor maintains good device

979-8-3315-0417-5/25 $31.00 © 2025 IEEE

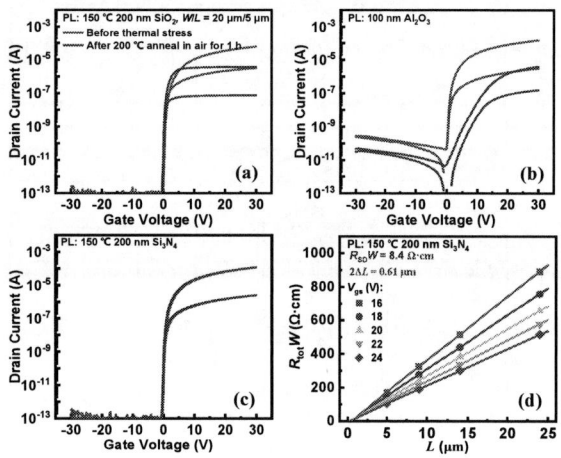

Fig. 4: The transfer characteristics of the SATG TFTs with different PL, (a) 150 °C 200 nm SiO_2, (b) 150 °C 200 nm Si_3N_4 and (c) RT 100 nm Al_2O_3.

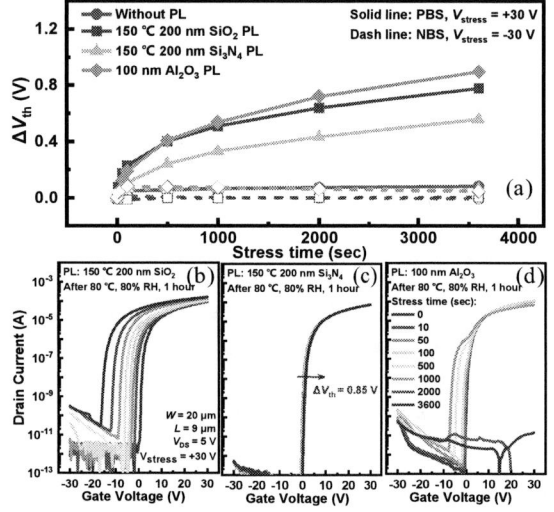

Fig. 5: (a) The ΔV_{th} versus time of the SATG TFTs with and without PLs under PBS and NBS. The transfer curves of the SATG TFTs with (b) 150 °C 200 nm SiO_2, (c) 150 °C 200 nm Si_3N_4 and (d) RT 100 nm Al_2O_3 under PBS after environmental stability test.

performance with $\Delta V_{th} < 0.85$ V under PBS. The abnormal negative ΔV_{th} observed in the SiO_2 and Al_2O_3-passivated transistors could be attributed to the insufficient barrier capability of the PLs against ambient moisture, resulting in the formation of mobile impurity ions or polar groups from H_2O molecules absorbed by the active layer [11]. The environmental stress test further validates the excellent quality of the low-temperature PECVD-deposited Si_3N_4 PL for SATG a-IGZO TFTs with metal-reacted heavily doped S/D regions.

Conclusion

The passivation layer for SATG a-IGZO TFTs with metal-reacted heavily doped S/D regions has been successfully optimized to provide adequate thermal and environmental stability. The low-temperature Si_3N_4-passivated transistor exhibits superior performance, including a stable R_{SD} of 8.4 $\Omega\cdot$cm and a minimal $2\Delta L$ of 0.61 μm after the thermal stress test, along with excellent stability under PBS ($\Delta V_{th} < 0.85$ V)

following the environmental stress test. These robust characteristics make SATG TFTs an attractive choice for emerging electronic applications.

Acknowledgments

This work was conducted in Guangdong Provincial Center for Oxide Semiconductor Devices and ICs, and supported financially by Ministry of Science and Technology Key Research and Development Program Grant 2022YFB3607200, Shenzhen Municipal Scientific Program Grant JCYJ20220818100808019.

References

[1] K. Ide, K. Nomura, H. Hosono, and T. Kamiya, "Electronic Defects in Amorphous Oxide Semiconductors: A Review," *physica status solidi (a),* vol. 216, no. 5, (2019).

[2] X. Xiao, L. Zhang, Y. Shao, X. Zhou, H. He, and S. Zhang, "Room-Temperature-Processed Flexible Amorphous InGaZnO Thin Film Transistor," *ACS Appl Mater Interfaces,* vol. 10, no. 31, pp. 25850-25857, (2018).

[3] S. P. S. Jeon, I. Song, J.-H. Hur, J. Park, S. Kim, S. Kim, H. Yin, E. Lee, S. Ahn, H. Kim, C. Kim, and U.-I. Chung, "180 nm gate length amorphous InGaZnO thin film transistor for high density image sensor applications," *IEDM Tech. Dig.,* pp. 21.3.1–21.3.4., (2010).

[4] Y. Zhang, J. Li, J. Li, T. Huang, Y. Guan, Y. Zhang, H. Yang, M. Chan, X. Wang, L. Lu, and S. Zhang, "3-Masks-Processed Sub-100 nm Amorphous InGaZnO Thin-Film Transistors for Monolithic 3D Capacitor-Less Dynamic Random Access Memories," *Adv. Electron. Mater.,* vol. 9, no. 8, (2023).

[5] Y. Zhang, J. Li, Y. Zhang, H. Yang, Y. Guan, M. Chan, L. Lu, and S. Zhang, "Deep Sub-Micron Self-Aligned Bottom-Gate Amorphous InGaZnO Thin-Film Transistors With Low-Resistance Source/Drain," *IEEE Electron Device Lett.,* vol. 44, pp. 1300-1303, (2023).

[6] J. Li, Y. Zhang, J. Wang, H. Yang, X. Zhou, M. Chan, X. Wang, L. Lu, and S. Zhang, "High-Performance Self-Aligned Top-Gate Amorphous InGaZnO TFTs With 4 nm-Thick Atomic-Layer-Deposited AlOx Insulator," *IEEE Electron Device Lett.,* vol. 43, pp. 729-732, (2022).

[7] H.-C. Chen, J.-J. Chen, K.-J. Zhou, G.-F. Chen, C.-W. Kuo, Y.-S. Shih, W.-C. Su, C.-C. Yang, H.-C. Huang, C.-C. Shih, W.-C. Lai, and T.-C. Chang, "Hydrogen Diffusion and Threshold Voltage Shifts in Top-Gate Amorphous InGaZnO Thin-Film Transistors," *IEEE Trans. Electron Devices,* vol. 67, no. 8, pp. 3123-3128, (2020).

[8] H. Yang, X. Zhou, H. Fu, B. Chang, Y. Min, H. Peng, L. Lu, and S. Zhang, "Metal Reaction-Induced Bulk-Doping Effect in Forming Conductive Source-Drain Regions of Self-Aligned Top-Gate Amorphous InGaZnO Thin-Film Transistors," *ACS Appl Mater Interfaces,* vol. 13, no. 9, pp. 11442-11448, Mar 10. (2021).

[9] H. Peng, B. Chang, H. Fu, H. Yang, Y. Zhang, X. Zhou, L. Lu, and S. Zhang, "Top-Gate Amorphous Indium-Gallium-Zinc-OxideThin-Film Transistors With Magnesium Metallized Source/Drain Regions," *IEEE Trans. Electron Devices,* vol. 67, pp. 1619-1624, (2020).

[10] S. Y. Hong, H. J. Kim, D. H. Kim, H. Y. Jeong, S. H. Song, I. T. Cho, J. Noh, P. S. Yun, S. W. Lee, K. S. Park, S. Yoon, I. B. Kang, and H. I. Kwon, "Study on the Lateral Carrier Diffusion and Source-Drain Series Resistance in Self-Aligned Top-Gate Coplanar InGaZnO Thin-Film Transistors," *Sci Rep,* vol. 9, pp. 6588, Apr 29. (2019).

[11] J.-S. Park, J. K. Jeong, H.-J. Chung, Y.-G. Mo, and H. D. Kim, "Electronic transport properties of amorphous indium-gallium-zinc oxide semiconductor upon exposure to water," *Appl. Phys. Lett.,* vol. 92, (2008).

Enabling Highly-Efficient, Low-Latency Analog CAM Operations with Optimized MoS₂ Flash Memory Devices

Guoyun Gao[1,#], Bo Wen [1,#], Ni Yang[3], Zhiyuan Du[1], Mingrui Jiang[1], Ruibin Mao [1], Yingnan Cao[4], Hongxia Xue[5], Pak San Yip[3], Qihan Liu[6], Dong-Keun Ki[5], Jinyao Tang[4], Paddy K. L. Chan[3], Hao Jiang[6], Han Wang[1,2], Lain-Jong Li[2] and Can Li[1,2,*]

[1]Dept. EEE, [2]Inst. Mind, [3]Dept. ME, [4]Dept. Chem, [5]Dept. Phys, The University of Hong Kong, Hong Kong SAR, China; [6]FICS, Fudan University, Shanghai, China; [#]Equal contribution; *email: canl@hku.hk

Abstract—Emerging non-volatile memory-based analog content addressable memories promise massively parallel data search and explainable edge processing. Still, their performance improvement has been bottlenecked by silicon-based transistor performance. In response, we optimize atomic thin MoS₂ flash memories with a record-high readout current (60 μA/μm) and a large current ON/OFF ratio (>10⁹) for low-latency, low column interference, and high energy efficiency of analog CAM cells, enabled by Antimony (Sb) semimetal contact for lower contact resistance and better gate controllability. We experimentally demonstrated the range search operation with our fabricated MoS₂ analog CAM array, promising the potential of atomically thin semiconductors for developing next-generation high-speed and low-power edge devices.

I. INTRODUCTION

In the era of Big Data, the data transfer bottleneck in data-intensive applications demands in-memory computing hardware, for efficient processes on edge devices. Often ignored, content addressable memories (CAM) are the earliest in-memory computing with commercial success because they offer extremely high throughput but at the expense of high energy consumption and large area overhead. Therefore, applications are limited to highly specialized scenarios like network routing. Analog CAM [1-2] has been proposed recently to increase memory density and reduce energy consumption, and the range search opens the possibility for efficient probabilistic computing, explainable tree-based model inference, and more [2]. However, silicon-based transistors are approaching their physical limit, have bottlenecked the performance, and prevented three-dimensional (3D) stack-ability. For example, the limited on-state current and current ON/OFF ratio leads to large latency, limited sense margin, low power efficiency, and small array size.

Two-dimensional (2D) transition metal dichalcogenides (TMDs), *e.g.*, MoS₂-based flash memory or floating-gate field-effect transistors (FGFETs), are promising to tackle the problems owing to their scalability, intrinsic immunity to short-channel effects, high theoretical mobility, large on-state current, large current ON/OFF ratio, 3D stack-ability, and small drain/gate capacitance [3-5]. However, the device performance based on those 2D materials is often limited in practice, because of the large contact resistance, owing to the severe Fermi level pinning effect or metal-induced gap state (MIGS) at the metal-semiconductor interface [6]. This limitation results in limited output current and high latency in computing applications.

In this work, we introduce the *first* analog CAM array built using 2D material MoS₂ flash memory devices and experimentally demonstrate the range search operations. To enable the low-latency and high-efficiency operation of the analog CAM, we optimized the MoS₂ flash memory for ultrahigh readout current by using semimetal Sb contact.

II. ANALOG CAM WITH OPTIMIZED MoS₂ FGFET

Analog CAM functions as a look-up table, which returns the address of the match data for a given search input. Each analog CAM cell searches for a range stored in memory in parallel, and at the exact location where the data is stored. **Fig. 1** shows the circuit schematic of an analog CAM array and one cell, which consists of a pair of FGFETs connected in parallel between the match line (ML) and ground line (GND). At the beginning, ML is pre-charged at a high-level voltage. When an input voltage is applied to the data line (DL) or the gate of the left transistor (T₁), an opposite value is applied to the gate of the right transistor (T₂) simultaneously via an analog inverter. If the input voltage on DL(V_{DL}) is smaller than the threshold voltage V_{th1} of T₁, the left current path is off (no discharge current), ML keeps being charged. Otherwise, ML will be discharged quickly. Thus, T₁ sets the upper bound of the match range. Similarly, when the opposite voltage -V_{DL} is larger than the threshold voltage -V_{th2} of T₂, the right current path is off (no discharge current), ML keeps being charged. Otherwise, ML will be discharged quickly. Thus, T₂ sets the lower bound of the match range. When the above conditions are met, i.e., $V_{DL} < V_{th1}$, $V_{DL} < V_{th2}$, the cell returns a match.

Despite its promise, achieving good performance with traditional silicon-based technologies is still challenging. This is partly due to the large off-state leakage current and limited current ON/OFF ratio. **Fig.2** illustrates that the off-state leakage current reduces the sense margin, limiting the size of each analog CAM array. Additionally, the limited on-state current, combined with large drain capacitance, causes high latency to maintain an acceptable sense margin.

This work uses optimized MoS₂ flash memory to address these challenges. **Fig. 3** shows the optical image of our fabricated analog CAM array with 256 MoS₂ flash memories. For those 2D flash memories, 5/10nm Ti/Au was used as the control gate followed by oxygen plasma treatment. 2 nm Al (or HfO₂) was used as the floating gate (or charge trapping layer). 10-15nm HfO₂ and 5nm Al₂O₃ were deposited by atomic layer deposition (ALD) as the blocking layer and tunneling layer, respectively. Large-scale monolayer MoS₂ film was grown by

979-8-3315-0417-5/25 $31.00 © 2025 IEEE

chemical vapor deposition (CVD), and transferred by wet method, then patterned by photolithography and reactive ion etching (RIE). 20/30nm Sb/Au were deposited as contact electrodes for low contact resistance through a thermal evaporator. The device showed an extremely small leakage current due to the enhanced electrostatic controllability with 2D channel, and we optimized the contact quality of those devices for high output current and low latency search.

For the contact electrodes, semimetal Sb was used to reduce the contact resistance because of the ideal contact interface with depressed Fermi level pinning effect and MIGS, which contributed to the Ohmic contact (**Fig.** 4). From I_D-V_G curves of 50 MoS_2 flash memories (**Fig.** 5), one sees that good device uniformity and excellent current-delivery capability with a high on-state current I_{ON} and large current ON/OFF ratio with an ideal Sb-MoS_2 Ohmic contact. We fabricated a series of short-channel devices with channel lengths (L_{CH}) ranging from 200nm to 500nm. All show a large current ON/OFF ratio ($>10^9$), and a small subthreshold swing (SS~100mV/dec). Inset is the statistics of readout current, showing a linearly increased trend of the average readout current by scaling down the L_{CH}, reaching a recorded high value of 60 μA/μm at V_G of 2V and V_D of 1V. A large readout current could be achieved by further scaling down L_{CH} to sub-100nm to achieve a higher readout current and shorter latency for analog CAM search operations.

Fig. 6 summarizes the benchmarked readout current and the current ON/OFF ratio of 2D flash memories, with all devices evaluated at a V_D of 1V. Our device with Sb contact excels in both metrics at the same time, beating those previously reported state-of-the-art 2D non-volatile memories, including MoS_2 FGFET or mem-transistor [3,18-20], MoS_2 ferroelectric FET (FeFET) [21], InSe FGFET [22], and MoS_2 1T1R with metal contact [16].

As a non-volatile memory, the device shows a large memory window (3V) with continuously tunable threshold voltage V_{th} (**Fig.** 7). Here we just programmed the device into eight states as a brief demonstration. More states are plausible as seen from the clear margin between states. The V_{th} of eight stored states remains stable over 2000 s (**Fig.** 8), which is long enough for the analog CAM search operation.

III. EXPERIMENTAL DEMONSTRATION OF ANALOG CAM OPERATIONS

We experimentally demonstrate the programming and search operation of the MoS_2 FGFET CAM array operation. **Fig.** 9 shows the picture of our home-built experimental setup consisting of a source measurement unit (SMU), a microcontroller (MCU), a test board, and a probe card that lands on an analog CAM array. Control signals are generated from SUM controlled by an MCU and applied to the analog CAM array via a probe card. To show the range search function of the analog CAM, we first programmed the analog CAMs to search for arbitrary ranges, including match ranges with tunable upper or lower bound and different locations (**Fig.** 10). The search experiment is conducted by searching an input that

returns only one match. **Fig.** 11 shows the simulated discharging behaviors for the search operation in an analog CAM array with all-match and different numbers of mismatched bits, increasing from 0 bit to 8 bis. From all MLs, where only one ML is kept high to indicate a match. The speed is limited by the RC in off-chip measurements, but the experimentally calibrated model suggests that Sb/Au contact MoS_2 flash memory shows a significantly reduced latency to picoseconds during a search operation. Lastly, we compared the benchmark of energy per search and latency for 1 bit mismatch (**Fig.** 12). Our MoS_2 FGFET analog CAM exhibits reduction both in energy and latency due to the optimized performance, compared with the previously reported RRAM-based analog CAM [1], ferroelectric CAM in digital/analog mode [15], and MoS_2 TCAM [16].

IV. CONCLUSIONS

We, for the first time, fabricated MoS_2 flash memories with Sb semimetal contact for performance optimizations of analog CAM. Our device shows high current-delivery capability, exhibiting a recorded high readout current and large current ON/OFF ratio, which is critical for constructing analog CAM with low latency and high efficiency. We demonstrated the programming and inference operation in the analog CAM array by both experiment and simulation. We expect large-scale 2D analog CAM with monolithic 3D integration to be a promising build block for in-memory and near-sensor processing with explainable machine learning models.

ACKNOWLEDGMENT: This work was supported in part by RGC (27210321, C1009-22GF, T45-701/22-R), NSFC (62122005); ACCESS – an InnoHK center by ITC, and Croucher Foundation. This fabrication was performed in part in the Central Fabrication Laboratory (CFL) at HKU and the Nanosystem Fabrication Facility of HKUST.

REFERENCE: [1] C. Li, *et al*, Nat Comm 11, 1638 (2020). [2] X. Yin, *et al.*, ArXiv abs/2110.02495 (2021). [3] G. M. Marega, *et al*, Nature 587, 72 (2020). [4] B. Radisavljevic, *et al*, Nat. Nanotechnol. 6, 147 (2011). [5] S. B. Desai *et al*, Science 354, 99 (2016). [6] Shen, *et al*. Nature 593, 211–217 (2021). [7] K. Huang, *et al*, IEEE Electron Device Lett,41, 10, 1600-1603, (2020). [8] Chou A. S., *et al*. IEDM, 2021: 7.2. 1. [9] A. Sanne, *et al*, IEEE ISCAS 2017: 1-4. [10] H.-Y. Chang, *et al*, Adv. Mater. 28, 1818-1823(2016). [11] R. Yang, *et al*, IEDM 2017: 19.5.1. [12] Z. Yu, *et al*, IEDM 2017: 23.6.1. [13] C. D. *et al*, IEDM 2016: 5.6.1 [14] Li, W., *et al*. Nature 613, 274–279 (2023).[15] X. Yin et al., IEEE Trans Electron Devices, 67, 7, 2785-2792 (2020) [16] Yang, R, et al. Nat Electron 2, 108–114 (2019).[17] W. Li, et al., IEDM 2021, 37.3.1.[18] Yu, J., *et al*. Nat Commun 14, 5662 (2023). [19] Vu, Q., *et al*. Nat Commun 7, 12725 (2016). [20] Dodda, A., *et al*. Nat Commun 13, 3587 (2022). [21] Ning, H., *et al*. Nat. Nanotechnol. 18, 493–500 (2023). [22] Wu, L., *et al*. Nat. Nanotechnol. 16, 882–887 (2021). [23] Liu, X., *et al*. Device 2, 100218 (2024).

I. Analog CAM Concept Calls for New Device Innovations

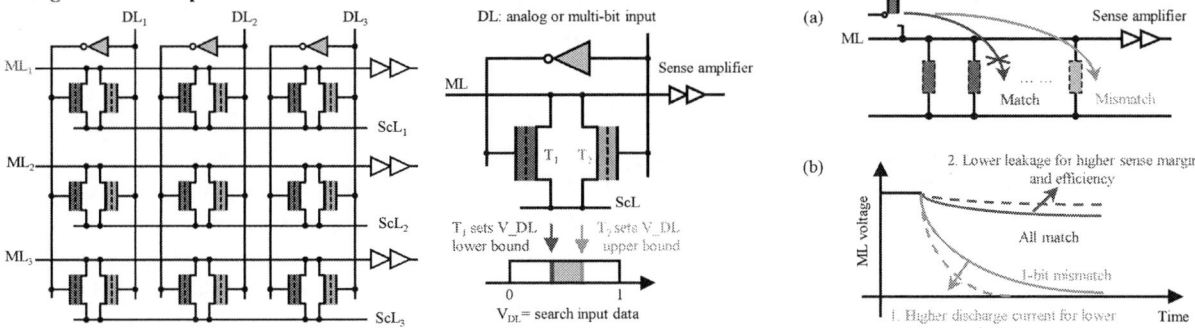

Fig.1. The circuit schematic of analog CAM cell and array. Columns are DLs that accept analog input, while rows are MLs that are pre-charged and return results after searching operation. ML keeps charged at a high voltage level to indicate a match. Otherwise, the ML discharges to a low voltage level to indicate a mismatch. A higher discharge current leads to lower latency, while a higher ON/OFF ratio results in a larger sense margin.

Fig. 2. (a)This illustration shows a single TCAM word. The figure highlights two distinct current paths: one for a mismatched cell, where maximizing the current is preferred, and another for a matched cell, where minimizing the leakage current through the matched cell is the goal for optimal performance.(b)This plot shows that a higher current through the ML pull-down transistor improves latency. At the same time, lower leakage, or a higher ON/OFF ratio, boosts sense margin and power efficiency.

II. Analog CAM Arrays with MoS$_2$ flash memory with Semimetal Contacts

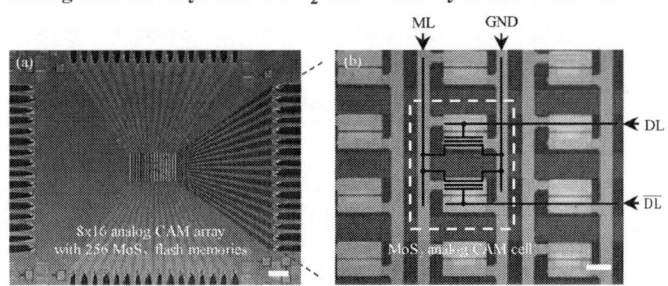

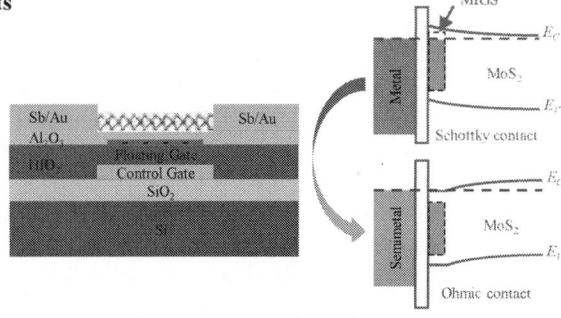

Fig. 3. (a) The photograph of the fabricated analog CAM 8×16 array with 256 MoS$_2$ flash memories. Scale bar: 200 μm . (b) Zoomed-in image showing how MoS$_2$ FGFETs (L_{CH}/W_{CH}=0.5/10μm) are connected to construct an analog CAM array. Scale bar: 10 μm .

Fig. 4 (a) Cross-sectional schematic of the MoS$_2$ back gate flash memory. (b) The band structure of the Sb-MoS$_2$ contact, illustrating the formation of an Ohmic contact and a low Schottky barrier with depressed MIGS.

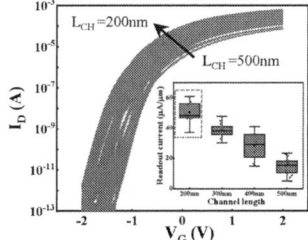

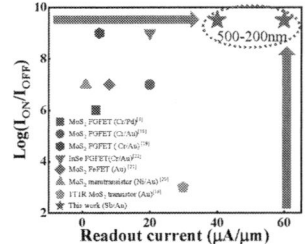

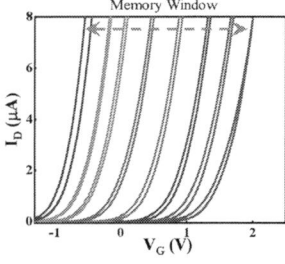

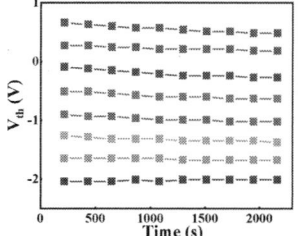

Fig. 5. Statistics of readout current I_{ON} at V_G of 2 V for 50 flash memories with L_{CH} of 200-500nm.

Fig. 6. Benchmark for our MoS$_2$ flash memories, all devices measured under the same V_D of 1V.

Fig. 7. (a) I_D-V_G curves for 8 programmed states, showing a large memory window.

Fig. 8. (b) The V_{th} of eight stored states measured over 2000s, showing stable multi-level states.

III. Analog CAM Array based on MoS$_2$ Flash Memories for High-Efficiency and Low-Latency

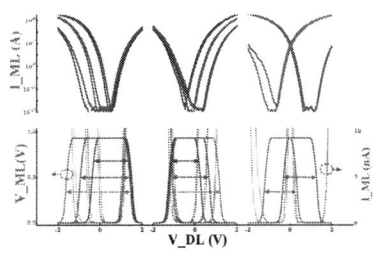

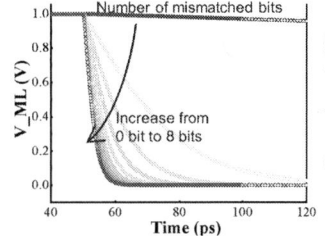

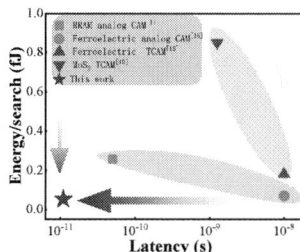

Fig. 9. Picture of the home-built experimental setup consisting of a microcontroller (MCU), test board, and probe card that lands on an analog CAM array.

Fig. 10. The experimental discharging current with different input (i.e., I_{ML}-V_{DL} plot) for one 2D FGFET analog CAM cell, and their corresponding sensed match line (ML) voltage for simulated range search.

Fig. 11. Simulated discharging behaviors for the search operation in an analog CAM array with all-match and different numbers of mismatched bits, increasing from 0 bit to 8 bis.

Fig. 12. Benchmark of search energy and latency. MoS$_2$ analog CAM exhibits reduction both in energy and latency, compared with previously reported similar works.

979-8-3315-0417-5/25 $31.00 © 2025 IEEE

An Artificial Thermoresponsive Nociceptor Based on Amorphous Carbon Threshold Memristor

Qiaoling Tian[1], Xiaoning Zhao[1], Zhongqiang Wang[1], Haiyang Xu[1], Yichun Liu[1]

[1] Northeast Normal University, China

Abstract

In this work, an artificial thermoresponsive nociceptor (TRN) based on amorphous carbon (a-C) memristor with temperature-modulated threshold switching (TS) behavior was reported. Both the response current and relaxation time increase with elevated temperatures. The device can emulate biological TRN features including threshold, relaxation, no adaption, and sensitization. As proof of concept, reflex arc system with thermal pain conduction is constructed by integrating this artificial TRN with robot arm.

Keywords: Thermoresponsive nociceptor, threshold switching and reflex arc system

Introduction

Thermoresponsive nociceptor (TRN) is a receptor in biological sensory nervous system that can respond to noxious thermal-stimulus and transmit signals to the brain [1]. It allows humans to consciously recognize pain feel and generate appropriate actions to avoid potential injuries [2, 3]. Therefore, the development of artificial TRN is of great significance for biorealistic applications such as humanoid robots. The artificial TRN should exhibit "threshold" and "relaxation" characteristics in the normal state, which is activated by temperature exceeding the threshold value (T_T) and requires a period of relaxation time to reset to its initial state [4]. In addition, "no adaptation" and "sensitization" are other important features of TRN in the abnormal state [5]. No adaptation means that the TRN maintains sensitivity to experienced noxious temperature. For sensitization feature, the injured TRN decreases the activation threshold value (allodynia) and increases the intensity of response (hyperalgesia). At present, various emerging devices have been developed for artificial TRN [6-8]. Among them, volatile memristors with threshold switching (TS) dynamics play the role of the threshold function of the nociceptor. However, the emulation of important sensitization feature of injured tissue under temperature stimulus was not reported. On the other hand, external thermoelectric modules are required as temperature sensor, which increases the complexity of auxiliary circuits. Hence, it is important to develop artificial TRN that can sense temperature and simultaneously emulate four features of biological TRN.

In this work, an amorphous carbon (a-C) memristor with temperature-modulated TS behavior was proposed to realize artificial TRN. The Ag/a-C/Pt memristors show TS behavior at low compliance current. The response current and relaxation time of the device increased with elevated temperature. Furthermore, artificial TRN with temperature sensing and signal processing functions were successfully realized. The Ag/a-C/Pt devices emulated the pain threshold, relaxation, no adaptation and sensitization features of the biological TRN under noxious temperature. As a proof of concept, the reflex arc system was constructed by integrating a robot arm to realize reflex action under thermal stimulus.

Device Fabrication

The preparation process of Ag/a-C/Pt devices is as follows: Firstly, the Pt bottom electrodes with 10 μm-wide were patterned on SiO$_2$/Si substrates by thermal evaporation and ultraviolet photolithography process. Secondly, the a-C film with thickness of 50 nm was deposited by radio frequency sputtering carbon target (purity: 99.99%) at room temperature. Finally, the Ag top electrodes (10 μm-wide) were patterned on a-C film to form Ag/a-C/Pt devices.

Results and Discussion

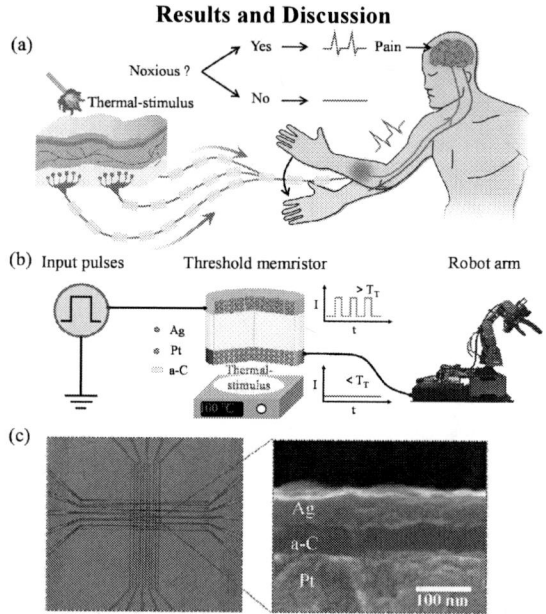

Fig. 1. Schematic illustration of (a) biological TRN and (b) artificial TRN. (c) The microscope image and cross-sectional SEM image of the Ag/a-C/Pt device.

As shown in Fig. 1(a), biological TRN convert noxious thermal-stimulus exceeding T_T into electrical signals, which are transmitted to the spinal cord and further initiate an action to avoid injury. The Ag/a-C/Pt memristor with temperature-modulated TS behavior was proposed to emulate biological TRN, and reflex arc system was constructed by connecting robot arm (Fig. 1(b)). Fig. 1(c)

979-8-3315-0417-5/25 $31.00 © 2025 IEEE

shows the microscope image of Ag/a-C/Pt crossbar array and cross-sectional scanning electron microscopy (SEM) image.

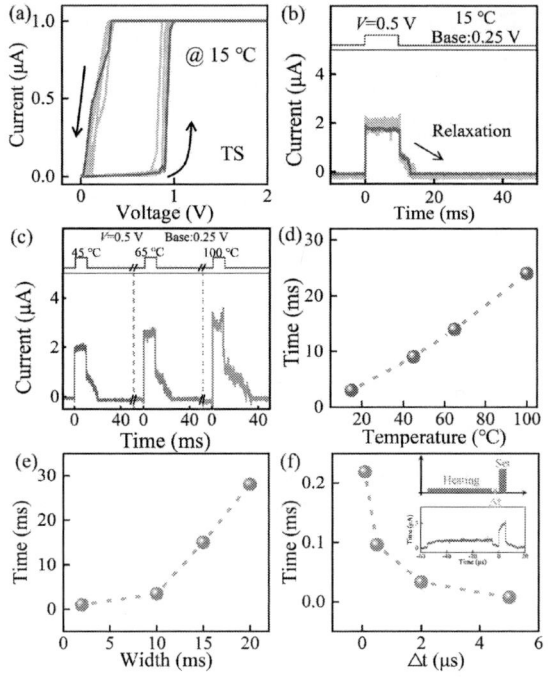

Fig. 2. (a)-(c) The I-V curves and relaxation characteristics of Ag/a-C/Pt device in DC and pulse modes at different temperatures. (d) and (e) Relaxation time distribution of voltage pulse with different temperatures and widths. (f) The relaxation modulation of heating and set pulses under different intervals.

As shown in Fig. 2(a), the device exhibits typical TS behavior at temperature of 15 °C. As the applied voltage gradually decreased, the device spontaneously returned from high conductance state (HCS) to low conductance state (LCS). Fig. 2(b)-(d) shows the response current of the device applying voltage pulses at different temperatures (15 °C, 45 °C, 65 °C and 100 °C). When voltage pulses were applied, the response currents suddenly increased. As the voltage pulse ended, a period of relaxation time was required to gradually reset to the initial state. The response current and relaxation time increased with elevated temperature, which successfully emulated the relaxation feature of biological TRN. The voltage pulses with longer width leaded to an increase of relaxation time, as shown in Fig. 2(e). Figure 2(f) shows that the relaxation time increases with the decrease of the non-overlapping pulse (heating and set) interval (Δt), which can be attributed to the accumulation of significant thermal effects caused by the heating pulse in the shorter Δt. Considering the electrochemical activity nature of Ag, the TS characteristics of the Ag/a-C/Pt memristor can be attributable to spontaneous rupture of small-sized Ag CFs due to their high surface energy, which have been extensively discussed in previous reports [9]. More importantly, high temperature

can promote the formation of large-sized CFs to increase the relaxation time [10]. These results suggest that temperature-modulated TS behavior of Ag/a-C/Pt device is beneficial for designing artificial TRN.

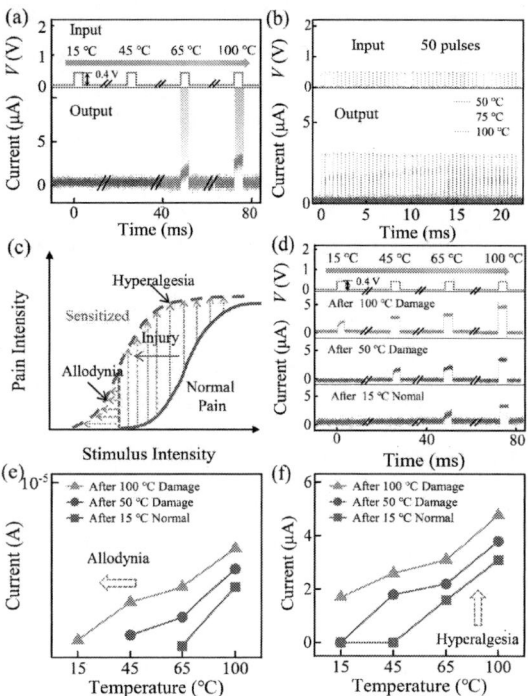

Fig. 3. (a) and (b) The threshold and no adaption feature of device at different temperatures. (c) Schematic diagram of allodynia and hyperalgesia features of biological TRN. (d) Response current and subthreshold characteristic of the device at different temperatures, and the artificial (e) allodynia and (f) hyperalgesia features.

The features of the artificial TRN were studied. Fig. 3(a) shows the threshold characteristic of device by applying voltage pulse at different temperatures (15 °C, 45 °C, 65 °C and 100 °C). The response current can be observed at the temperature exceeding 65 °C (T_T) and the current increases with elevated temperature, which is consistent with the threshold feature of biological TRN. As we all know, the innocuous perception temperature range of humans is 15-45 °C. Fig. 3(b) shows the response current of Ag/a-C/Pt devices at noxious temperatures (50 °C, 75 °C and 100 °C). The device maintained the levels of response currents under the same noxious temperature stimulus, emulating continuous pain feel due to no adaptation feature. Fig. 3(c) shows the schematic diagram of sensitization feature of biological TRN. Compared with normal state (solid blue line-allodynia feature), the activation threshold temperature decreased (defined as subthreshold value) and the response intensity increased for the injured state (red dotted line-hyperalgesia feature). After undergoing normal and damage processes, voltage pulses were applied to the device at different temperatures (15 °C, 45 °C, 65 °C and 100 °C) to evaluate the sensitivity of injured TRN (Fig.

979-8-3315-0417-5/25 $31.00 © 2025 IEEE   143

3(d)). Fig. 3(e) and (f) show the corresponding maximum response currents extracted at different temperatures at log and linear scales, respectively. Compared to the normal TRN, the subthreshold temperature shifted to lower value (allodynia) and stronger response (hyperalgesia) with the increase of injured level for TRN. This result indicates that Ag/a-C/Pt device has the ability to sense temperature and simultaneously emulate the four features of biological TRN.

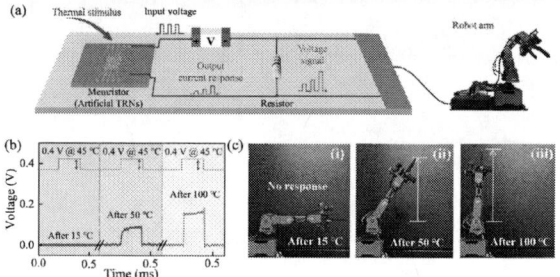

Fig. 4. (a) Schematic diagram of reflex arc with thermal pain conduction composed of Ag/a-C/Pt device, resistor, and robot arm. (b) The voltage pulse (V=0.4 V, W=0.2 ms) drops on the resistor and (c) digital picture of the different robot arm actions at innocuous temperature of 45 °C.

To demonstrate the potential of the artificial TRN, a reflex arc was constructed by integrating the Ag/a-C/Pt device, a resistor (R_{load}), and a robot arm (Fig. 4(a)). The voltages of R_{load} were used to control the robot arm by Python programming. After the device was subjected to different temperatures (15 °C, 50 °C and 100 °C), voltage pulses were subsequently applied to evaluate the injured level at innocuous temperature of 45 °C. Fig. 4(b) shows the voltage dropped on the R_{load}, which is used to robot arm to rise the set height. When the device was subjected to innoxious temperatures of 15 °C, the voltage dropped across the R_{load} can hardly be monitored and the robot arm stayed at the initial state (Fig. 4(c)-(i)). When the device was subjected to noxious temperatures (50 °C and 100 °C), the increased voltage dropped on the R_{load} can drive the robot arm to rise, emulating pain feeling and feedback movements (Fig. 4(c)-(ii) and (iii)). The results demonstrate that the Ag/a-C/Pt device has great significance in the application of humanoid robots with biological TRN sensing capability.

Conclusion

In summary, a Ag/a-C/Pt memristor with temperature-modulated TS behavior was proposed to emulate essential features of biological TRN. The response current and relaxation time of devices increased with elevated temperature. These features enable the memristor to successfully emulate the features of TRN, including threshold, relaxation, no adaption, and sensitization. A reflex arc with thermal pain conduction is constructed by integrating the artificial TRN with robot arm.

Acknowledgments

This work was supported financially in parts by the "Science and Technology Development Plan Project of Jilin Province, China" (Grant Nos. 20240101018JJ), the National Natural Science Foundation of China (Grant Nos. 52072065, 52272140, 52372137, and U23A20568), the fund from Jilin Province (Grant No. 20220502002GH), and the Fundamental Research Funds for the Central Universities (Grant No. 2412023YQ004).

References

[1] A. E. Dubin et al., "Nociceptors: the sensors of the pain pathway", J. Clin. Invest., 120, pp. 3760 (2010).

[2] J. Zhu et al., "An artificial spiking nociceptor integrating pressure sensors and memristors", IEEE Electron Device Lett., 43, pp. 962 (2022).

[3] S. Shrivastava et al., "Emulating synaptic and nociceptive behavior via negative photoconductivity of a memristor", IEEE Trans. Electron Devices, 70, pp. 3530 (2023).

[4] D. B. Tillman et al., "Response of C fibre nociceptors in the anaesthetized monkey to heat stimuli: correlation with pain threshold in humans", J. Physiol., 485, pp. 767 (1995).

[5] M. S. Gold et al., "Nociceptor sensitization in pain pathogenesis", Nat. Med., 16, pp. 1248 (2010).

[6] D. Dev et al., "Artificial nociceptor using 2D MoS$_2$ threshold switching memristor", IEEE Electron Device Lett., 41, pp. 1440 (2020).

[7] X. Chen et al., "An oxide-based bilayer ZrO$_2$/IGZO memristor for synaptic plasticity and artificial nociceptor", IEEE Trans. Electron Devices, 70, pp. 1001 (2023).

[8] Y. Shi et al., "Neuro-inspired thermoresponsive nociceptor for intelligent sensory systems", Nano Energy, 113, p. 108549 (2023).

[9] X. Zhang et al., "An artificial neuron based on a threshold switching memristor", IEEE Electron Device Lett., 39, pp. 308 (2017).

[10] S. Kim et al., "Compact Two-State-Variable Second-Order Memristor Model", Small, 12, pp. 3320 (2016).

Single-crystal Diamond MEMS for Extreme Sensors

Meiyong Liao

National Institute for Materials Science, Japan

Abstract

Diamond has been a potential rival of Si for MEMS in terms of the outstanding mechanical, electrical, thermal and chemical properties. These extreme properties not only enable developing highly reliable MEMS devices but innovating the sensitivity, precision, and stability. Here, we show the advanced research on single-crystal diamond (SCD) MEMS achieved in our lab. These include the invention of the smart-cut technology for fabricating SCD MEMS structures, energy dissipation mechanisms, and sensing/switch applications.

Keywords: Diamond, MEMS, Sensors

Introduction

The semiconductor electronics technology has witnessed the expansion from conventional silicon material to wide bandgap semiconductors such as GaN and SiC, and then to ultra-wide bandgap semiconductor such as AlN, BN, Ga_2O_3 and diamond. Similarly, MEMS has been experiencing the same development trend from silicon to diamond. In both fields, silicon semiconductor is still the mainstream. While silicon MEMS has been growing explosively thanks to the maturity in CMOS technology. MEMS based on the emerging wide-bandgap semiconductors is still at the initial stage.

Diamond is traditionally known for the highest mechanical hardness and Young's modulus for mechanical tools. It has also been realized that diamond exhibits outstanding semiconducting properties super to the existing semiconductor materials due to the ultra-wide bandgap energy, high carriers mobilities, high blocking voltages and the highest thermal conductivity[1]. In addition to these well documented mechanical and electronic properties, the known dopants in diamond have deep energy levels, greatly reducing the energy loss from electron-phonon interaction[2]. The non-existence of native oxides further mitigates the surface energy loss. Therefore, high resonance frequency and high quality (Q) factor diamond MEMS resonators can be achieved, enabling the development of MEMS sensors breaking the limits of silicon MEMS.

Nevertheless, the mechanical hardness and chemical inertness make micromachining diamond difficult. In this work, we present the single-crystal diamond (SCD) MEMS starting from the material micromachining and MEMS physis to sensing applications developed in our lab.

Micromachining of single-crystal diamond

At this stage, it is still difficult to deposit high-quality SCD epilayers on foreign substrates (Si etc). Therefore, Silicon based MEMS micromachining process is not applicable for SCD. Uniquely, the allotropes of carbon offer the route to transform sp3-hybridized crystal diamond to sp2-hybridized graphite-like carbon. This feature forms the basis for the fabrication of SCD MEMS structures, which has been called as the smart-cut method. The strategy here is to use high-energy ion implantation into a SCD substrate, which leads to the formation of buried graphite-like carbon within the SCD substrate[3]. By using lithography and etching process, the SCD MEMS structures can be released. This method enables the mass production of SCD MEMS with well controlled dimensions and internal stress. The fabrication process is briefly illustrated in Fig.1(a). Fig.1(b) shows the optical images of SCD MEMS cantilevers produced by the smart-cut method.

Energy dissipation and quality factors of SCD MEMS

The energy dissipation determines the quality (Q) factor of a MEMS resonator, which, in turn, influences the ultimate device performance, such as the sensitivity, precision, frequency stability, and the noise level of the devices. The energy dissipations in diamond MEMS resonators include both intrinsic and extrinsic loss. The overall Q factor can be generally written as

$$\frac{1}{Q} = \frac{1}{Q_{air}} + \frac{1}{Q_{clamp}} + \frac{1}{Q_{TED}} + \frac{1}{Q_{MD}} + \frac{1}{Q_{surface}} \quad (1)$$

where Q_{air}, Q_{clamp}, Q_{TED}, Q_{MD}, and $Q_{surface}$ represents the dissipation sources from air damping, clamping loss, thermoelastic damping (TED), mechanical defects (MD), and surface loss, respectively. The extrinsic loss Q_{air} and Q_{clamp}, can be reduced by changing the measurement environments, such as in vacuum to avoid air damping, optimizing the device geometry to reduce the clamping loss, etc. The material properties and crystal quality define the intrinsic loss. The materials properties such as the thermal conductivity, thermal expansion coefficient affect the Q_{TED}. Reducing the crystal defects such as the grain boundaries, dislocations, and impurities are important to improve the Q factor. In SCD MEMS structures fabricated by the smart-cut method, ion-implantation induced point defects degrade the Q factor. Therefore, the Q factor in such a case is at the order of $1,000$. The strategies to reduce the effect of these defects include increasing the SCD epilayer thickness, high-temperature annealing, and removing the damaged layer in the MEMS structures. Increasing the epilayer thickness is the simplest way, which can increase the Q factor by over 10 times[4]. However, increasing the thickness also increases the damping loss, limiting the Q factors. By annealing the SCD MEMS structure in an ultra-high vacuum at high temperatures over 900°C, the Q factor was improved by twice[5]. These two methods can not totally remove the damaged layer, the Q factor is still at the order of 10,000. The most efficient way is the combination of the growth of a high quality SCD epilayer and the removal the damaged layer. Eventually, the Q factor over 1 million was achieved at room temperature (Fig. 2) for the smart-cut method fabricated SCD cantilevers[6], which is the same level as that of the SCD cantilever by thinning a SCD plate bonded with a foreign substrate.

Extreme SCD MEMS sensors

The SCD MEMS showed excellent thermal stability up to 1000 K and low thermal coefficient of resonance frequency (TCF) ~ 5ppm, much less than those of Si and III-nitrides. The high thermal stability enables the development high-temperature and high-reliability MEMS sensors and switches.

We developed all-electrical SCD MEMS magnetic sensors able to work up to 773 K[7]. In this magnetic sensor, a high-Curie temperature magnetostrictive FeGa thin film (~ 80 nm) was deposited on the SCD cantilevers beams (Fig.3). To enhance the adhesion between diamond and FeGa, a thin Ti layer was deposited. When applying an external magnetic field on the SCD cantilever beam, the force induced by the magnetostrictive FeGa film is transferred to the beam, inducing the shift of the resonance frequency of the cantilever (Fig.3b and Fig.3d). This is something equal to the change of the Young's modulus (E). Therefore, this physical principle is also called ΔE effect. The FeGa/Ti/SCD MEMS structure allows the device sensing magnetic field up to 773K at least. The noise level of the device remained as low as 10 nT/Hz$^{0.5}$ even at 773K.

Due to the high sensitivity of MEMS for mass sensing. The surface states information of a semiconductor can be revealed in an alternative way from the conventional ways such as X-ray photoelectron spectroscopy. Oxygen-terminated diamond surface is commonly utilized and vital for semiconductor diamond electronic devices such as Schottky diodes, photodiodes, and metal-oxide-field effect transistor (MOSFETs). By annealing the SCD cantilevers in an ultra-high vacuum, we measured the resonance frequency shift, based on which the mass loss (~pg) and thickness (~nm) of the surface absorbents of the surface terminations were disclosed[8].

By using the robust mechanical hardness and wear resistance, diamond nanomechanical system (NEMS) switch was developed, as shown in Fig. 4. The structure of the NEMS switch resembles a transistor, having source, drain and gate electrodes. Since the off-current is nearly zero, the subthreshold swing is nearly zero, overcoming the thermodynamic limit (60mV/dec) of a field-effect transistor at room temperature. Therefore, the standby power consumption of the SCD NEMS switch is nearly zero. The diamond NEMS can also operate at high temperatures [3].

Conclusion

We developed the smart-cut method for the mass fabrication of SCD MEMS structures. This process is highly controllable and controllable. The Q factor the resulting SCD MEMS cantilever beam was over 1 million at room temperature. The SCD MEMS was used for various applications such as high-reliability high-temperature MEMS magnetic sensors, surface state analysis of a semiconductor and NEMS switch. Diamond MEMS paves the way for innovating the conventional MEMS in either performance and sensitivity.

Acknowledgments

The author gratefully acknowledges the contributions of Z. Zhang, H. Sun, H. Wu, X. Shen, G. Chen, and K. Gu to this work. This work was supported by JSPS KAKENHI (Grant Number 24H00287, 22K18957) and Bilateral joint research between JSPS/CAS.

References

[1] C. E. Nebel, "CVD diamond: a review on options and reality," *Functional Diamond,* vol. 3, no. 1, p. 2201592, 2023/12/31 2023, doi: 10.1080/26941112.2023.2201592.

[2] H. Sun *et al.*, "Effect of deep-defects excitation on mechanical energy dissipation of single-crystal diamond," *Physical review letters,* vol. 125, no. 20, p. 206802, 2020.

[3] M. Liao *et al*, "Suspended single-crystal diamond nanowires for high-performance nanoelectromechanical switches," *Advanced Materials,* vol. 22, no. 47, pp. 5393-5397, 2010.

[4] M. Liao *et al* "Improvement of the quality factor of single crystal diamond mechanical resonators," *Japanese Journal of Applied Physics,* vol. 56, no. 2, p. 024101, 2017.

[5] G. Chen, M. Liao *et al*, "Disclosing the annihilation effect of ion-implantation induced defects in single-crystal diamond by resonant MEMS," *Diamond and Related Materials,* vol. 138, p. 110240, 2023/10/01 2023, doi: https://doi.org/10.1016/j.diamond.2023.110240.

[6] H. Wu, M. Liao *et al.*, "Reducing intrinsic energy dissipation in diamond-on-diamond mechanical resonators toward one million quality factor," *Physical Review Materials,* vol. 2, no. 9, p. 090601, 2018.

[7] Z. Zhang, M. Liao *et al.*, "Enhancing Delta E Effect at High Temperatures of Galfenol/Ti/Single-Crystal Diamond Resonators for Magnetic Sensing," *ACS Applied Materials & Interfaces,* vol. 12, no. 20, pp. 23155-23164, 2020/05/20 2020, doi: 10.1021/acsami.0c06593.

[8] K. Gu M. Liao *et al*, "Oxygen-termination effect on the surface energy dissipation in diamond MEMS," *Carbon,* vol. 225, p. 119159, 2024/05/01 2024, doi: https://doi.org/10.1016/j.carbon.2024.119159.

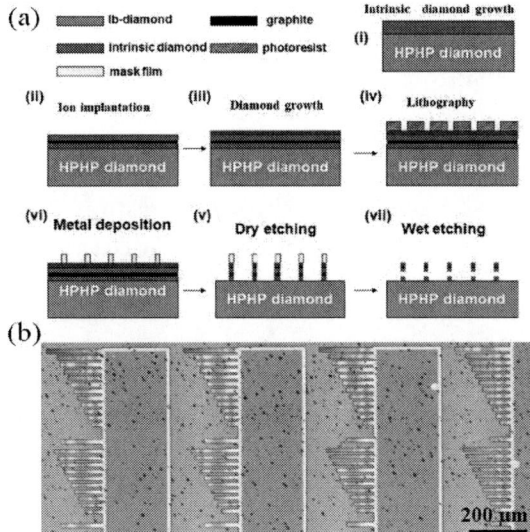

Fig. 1:(a) Fabrication of SCD MEM strucures (i) diamond growth, (ii) ion implantation, (iii) diamond growth, (iv)-(vii) photothography and structrures release. (b) Opticla image of SCD cantiever beams.

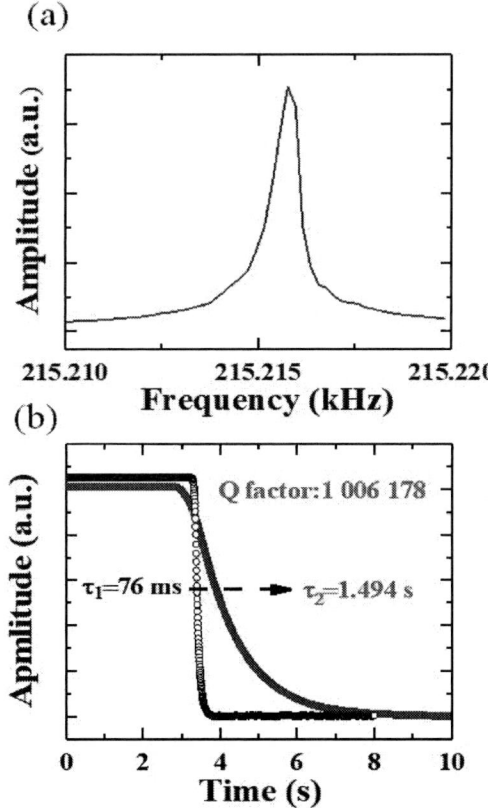

Fig. 2: (a) resonance frequecy spectra of a 140 μm long SCD cantilever (b) ring-down measurents of the cantilever before and after removing the damaged layer.

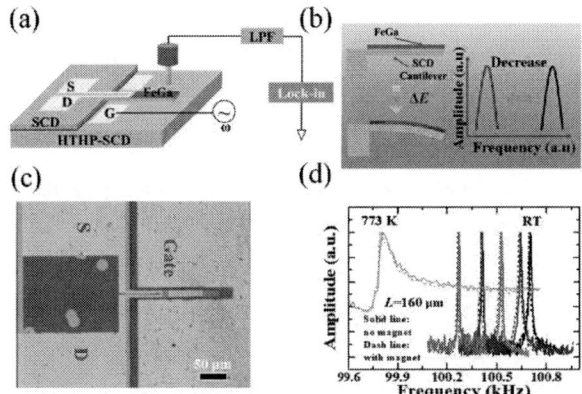

Fig 3: On-chip diamond MEMS magnetic sensor. (a) all-electrical actuation and readout scheme, (b) physics principle for magnetic sensing, (c) optical image of the FeGa/Ti/diamond on-chip MEMS magnetic sensor, (d) magnetic sensing up to 773 K.

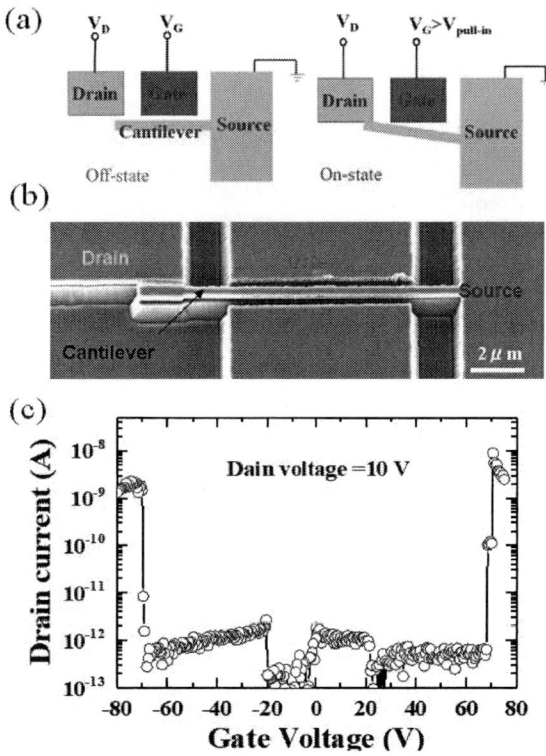

Fig.4: Diamond NEMS switch (a) switching principle: the gate voltage inducing the deflection of the beam to connect the source and drain, (b) optical image of a SCD NEMS switch (c) Electrical switching proeprties.

A variability-aware data preprocessing method for data-driven memristive device inverse modeling

Zhoujie Pan[#1], Yanming Liu[#1], Daixuan Wu[#1], Zihan Zhang[2], He Tian*[1], IEEE Senior Member

[1] School of Integrated Circuits, Tsinghua University, China; [2] XingJian College, Tsinghua University, China (Email: tianhe88@tsinghua.edu.cn)

Abstract

Data-driven learning methods have great potential in optimizing memristive devices, yet its application in memristor inverse modeling is still facing big challenge due to the scarcity of experimental data and device randomness. Here, we proposes a novel data preprocessing method including data generation, data augmentation and statistical preprocessing. With the aid of above techniques, we manage to achieve less than 5% relative errors in the test regression problem just need a few original I-V curves. Comparing with relevant studies, this work just needs a few of original experimental or simulated I-V curves which is achievable under laboratory conditions.

Keywords: memristor modeling, statistical preprocessing

Introduction

Memristor is one of the most active research directions in the field of microelectronic devices in recent years for the property of in-memory computing and high-density integration. Various algorithms such as convolutional neural networks[1], cellular automata[2], and p-bits[3] have been embedded into memristor arrays for in-memory computing operations[4].

However, in the case of memristors and memristors-based devices, although there are some attempts about traditional device modeling based on neural network [5~6], the feasibility of establishing an accurate, inverse device model still needs to be figured out. One of the major factors limiting the application of inverse model in memristor field is the high cost of acquiring high quality data, The property of memristors and memristors-based devices also causes obstacles for the application of machine learning in this field. Because of the general existing of device-to-device variation and cycle-to-cycle variation in each device[7], it is harder to use data to train the learning model directly. The uncertainty phenomenon blurs the decision boundaries, making it harder to train an efficient model.

In this work, we propose a solution to these challenges through an integrated approach involving data collection, augmentation, preprocessing, and network selection, thus enabling ML algorithms to effectively contribute to memristor design. Our proposed strategy has been validated both theoretically and practically by reversing the prediction of gate thickness based on the I-V curve of a gate-controlled memristor (GCBRAM)[8]. We present a comprehensive comparison of prediction accuracy using our preprocessing approach against traditional methods, and examine the performance of various learning models to provide practical guidance.

Method Development

The framework of this study is concluded in **Fig.1**. Experiment results of memristor is shown in **Fig.1(a~b)**, showing cycle-to-cycle variation. **Fig.1(d)** shows the framework of our methods. Since the simulation algorithm of memristors is difficult to produce a large number of auxiliary data like TCAD, which is due to two main reasons. First, the simulation algorithm in the field of memristor generally simulating the evolution of particles, which consumes a large amount of computing resources. At the same time, due to the cycle-to-cycle variation of memristors, a process parameter often needs to run more than once. Comparing with other strategies, our work executes data augmentation only by original dataset itself to solve the problem. We also put forward a corresponding data preprocessing method to deal with the cycle-to-cycle variation of memristor.

The effect of noise on the performance of learning algorithm are shown in **Fig.2**. According to **Fig.2(a)**, we regarded I-V curve as a high-dimensional variable determined by three main parts: the controllable part (also known as the learning target), the uncontrollable parts introduced during fabrication process, and the gaussian noise part introduced by measuring error and device intrinsic property. The second part is hard to eliminate by mathmatical methods, and can be controlled by more careful controlling fabrication conditions. So in this work, we mainly discussed about the gaussian noise part, which may confuse the learning model, making it harder to train.

A theoretical discussion is shown in **Fig.2(b)~(d)**. **Fig.2(b)** shows the effect of unrepresentative sampling on population estimates. We can find that on the magnitude of 20 sampling points, extreme values can significantly affect the population estimates. **Fig.2(c)~(d)** shows when sampling from two different normal population, some extreme values will go into the other decision regions, we call them "misleading points", which existence may confuse the classification boundary. Thus, if we can find a way to eliminate the "misleading points", while reserve the "representative points", it would be more easier for a learning algorithm to learn information from those samples.

Based on this perspective, we propose a preprocessing method based on the idea of statistical test as shown in **Fig.2(e)**. Since both the population variance and the mean are unknown, standard statistical tests cannot be applied. The outline of our preprocessing algorithm is shown in **Fig.3**.

Since the variance calculation of the sample is susceptible to extreme values, resulting in too loose estimates, we put forward the concept of "self-consistent" to guide the pretreatment. It is characterized by the unbiased estimation of the population mean based on the sampled population mean, and then the data are statistically tested in order of the distance from the sample mean from high to low. For each statistical test, all the data in the sample except this data are used to estimate the population distribution. If this data deviates from the preset allowable range of significance, the data will be discarded. And so on until the end of the cycle.

Method Validation

In order to demonstrate the effectiveness of our preprocessing algorithm, we build a regression problem shown in **Fig.4 (a)**. The result is shown in **Fig.4 (b)**. We can find that by applying appropriate statistical preprocessing methods, the performance of learning algorithm improves. To learn more from I-V curve, we also apply a autoencoder to extract the useful information, which structure is shown in **Fig.4(c)**.

In this study, we conducted supervised learning for a regression problem that predicted gate thickness from device I-V curves when applying the same gate voltage (shown in **Fig.5**). Four different learning algorithms were trained on the dataset. With preprocessing methods mentioned above, we also compare the model behavior whether preprocessing methods is adopted. In general, we compared the training effect of the same model in three parts: training directly with I-V curve, training after data augmentation (step **I**), training further after experiencing statistical processing and feature extraction (step **II**). In order to simulate the magnitude of the data in the actual scenario, we used only a few hundred I-V curves as the initial data.

The original I-V curve dataset is shown in **Table.1**. The dataset is split into three parts: A: training dataset, B: test dataset 1, demonstrating the model's interpolation capabilities, C: test dataset 2, demonstrating the model's extrapolation capabilities.

The result is visualized in **Fig.6,** showing the prediction results of different models. We can conclude that CNN model need feature engineering to refine the information, MLP model have a fair enough behavior as long as with enough training data. Nevertheless, MLP model is a fundamental model and it's very hard to further improve model performance by simply adding the hidden layer or neurons. GBDT model as a tree based model, it can have a accurate performance on B testset, but when it comes to extrapolation capabilities, its performance would decrease and hard to improve by data augmentation methods. Thus, we using attention mechanism to combine the advantage of feature engineering and conserving original data with the structure of attention unet. Its performance is shown in **Fig.6(e)**, which shows good performance both on B testset

and C testset.

Conclusion

This study develops a complete preprocessing pipeline consists of data generation, augmentation and statistical preprocessing to help establishing a robust inverse model of memristor device. Comparing to relevant studies, our work do not need plenty of original data and have a rigorous mathematical background. To verify the effectiveness of our work, we construct a regression problem between gate thickness and devices I-V curves of gate controlled conductive bridge memristors. The result shows that our method can significantly raise the prediction accuracy of learning models on the magnitude of several hundred initial data points.

Acknowledgments

This work was supported in part by STI 2030—Major Projects under Grant 2022ZD0209200, in part by Beijing Natural Science Foundation-Xiaomi Innovation Joint Fund (L233009), in part by National Natural Science Foundation of China under Grant No. 62374099, in part by Independent Research Program of School of Integrated Circuits, Tsinghua University.

References

[1] Yao, P., Wu, H., Gao, B. et al. Fully hardware-implemented memristor convolutional neural network. Nature 577, 641–646 (2020). https://doi.org/10.1038/s41586-020-1942-4

[2] Liu, Y., Tian, H., Wu, F. et al. Cellular automata imbedded memristor-based recirculated logic in-memory computing. Nat Commun 14, 2695 (2023). https://doi.org/10.1038/s41467-023-38299-7

[3] Woo, K.S., Kim, J., Han, J. et al. Probabilistic computing using Cu0.1Te0.9/HfO2/Pt diffusive memristors. Nat Commun 13, 5762 (2022) . https://doi.org/10.1038/s41467-022-33455-x

[4] Xu N , Park T , Yoon K J , et al. In-memory Stateful Logic Computing using Memristors: Gate, Calculation, and Application[J]. physica status solidi (RRL) - Rapid Research Letters, 2021.

[5] J. Hutchins et al., "A Generalized Workflow for Creating Machine Learning-Powered Compact Models for Multi-State Devices," in IEEE Access, vol. 10, pp. 115513-115519, 2022, doi: 10.1109/ACCESS.2022.3218333.

[6] Y. Zhang, G. He, K. -T. Tang, Y. Li and G. Wang, "GEM. A Generalized Memristor Device Modeling Framework Based on Neural Network for Transient Circuit Simulation," in IEEE Transactions on Computer-Aided Design of Integrated Circuits and Systems, vol. 42, no. 3, pp. 834-846, March 2023, doi: 10.1109/TCAD.2022.3188961

[7] Liao Z , Fu J , Wang J . Level Scaling and Pulse Regulating Methods to Mitigate Cycle-to-cycle Variation in Memristor Based Learning System[J]. 2020.

[8] Y. Liu, Z. Pan and H. Tian, "Reconfigurable Logic Operation Scheme Based on Gate-Tunable CBRAM," in IEEE Transactions on Electron Devices, vol. 70, no. 1, pp. 88-92, Jan. 2023, doi: 10.1109/TED.2022.3225134.

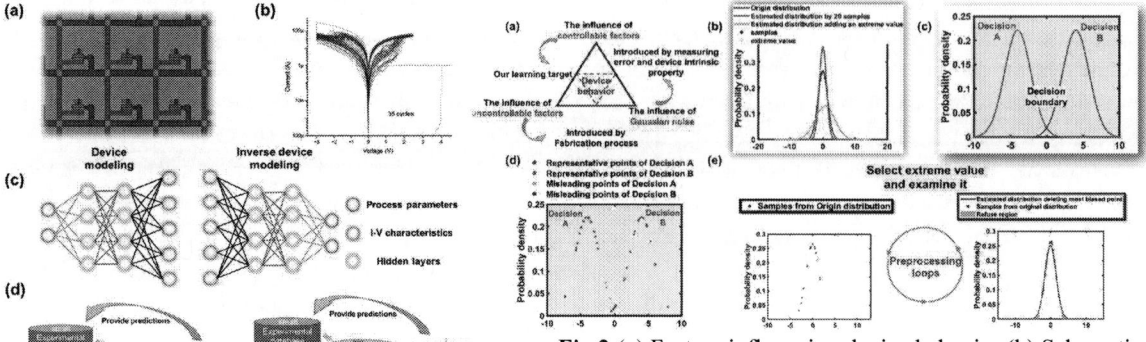

Fig.1 (a) Image of MoS₂ 1T1R device. (b) Experimental I–V curves of 1T1R under DC mode, showing cycle-cycle variation. (c) The difference between device modeling and device inverse modeling. (d) A framework of traditional strategy (left) and our strategy.

Fig.2 (a) Factors influencing device behavior (b) Schematic of sampling from normal distribution, the orange line shows the estimated distribution when existing extreme values(c) Problem of distinguish two diverse distribution (d) Sampling from two diverse distribution, showing misleading points and representative points (e) Schematic of statistical preprocessing methods.

```
Algorithm 1 Algorithm for data preprocessing
Input: Input sample set D = {x₁, x₂, ..., xₙ} Input sample sets
Output: Output sample set {o₁, o₂, ..., oₙ}
1: Calculating sample mean μ₀ and sample variance S₀
2: Sort input element by its distance from sample mean and build a priority queue Q(biggest first
   dequeue)
3: while Q! = ∅ do
4:    item = Q.dequeue() // fetch the farthest element still in queue
5:    Calculating sample mean μ₁ and sample variance S₁ of D \ {item}
6:    Introducing the Cumulative Distribution Function(CDF) of t-distribution
7:    if CDF₍ₙ₋₁₎{item−μ₁/√S₁} < α or CDF₍ₙ₋₁₎{item−μ₁/√S₁} > 1 − α then
8:       Delete item from D
9:       Sort D \ {item} again and refresh Q
10:   else
11:      Break
12:   end if
13: end while
14: o = DataAugmentation(D)
15: return o
```

Fig.3 Pseudocode for preprocessing algorithm

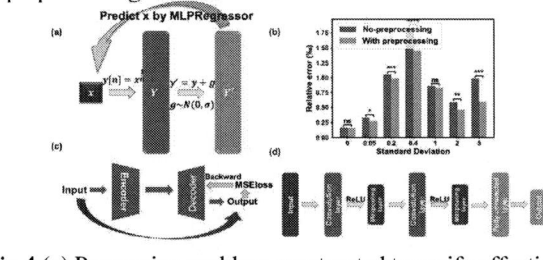

Fig.4 (a) Regression problem constructed to verify effectiveness of statistical preprocessing (b) Comparison of prediction accuracy whether statistical preprocessed or not. (c) Schematic of autoencoder (d) Schematic of Convolutional neural network designed for regression problem

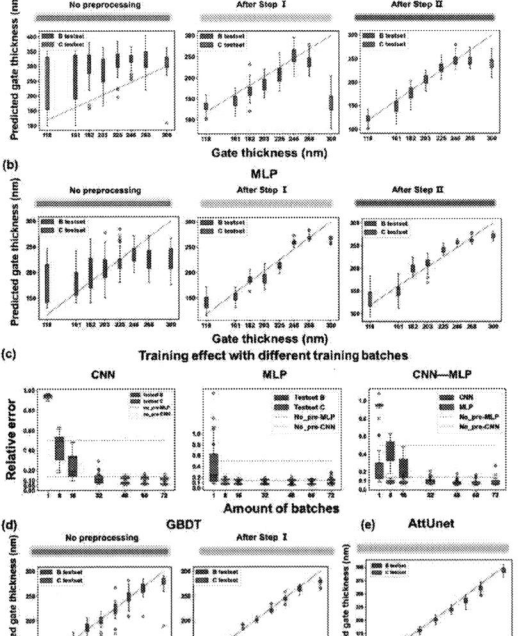

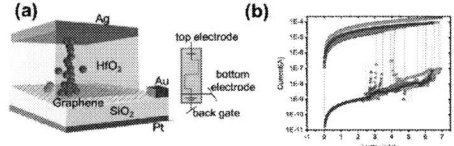

Fig.5 (a) Schematic of gate-controlled memristors (b) Repeated simulated I-V curves of gate-controlled memristors

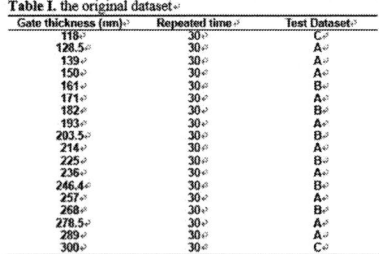

Table I. the original dataset

Gate thickness (nm)	Repeated time	Test Dataset
118	30	C
128.5	30	A
139	30	A
150	30	A
161	30	B
171	30	B
182	30	B
193	30	A
203.5	30	A
214	30	A
225	30	A
236	30	B
246.4	30	B
257	30	A
268	30	B
278.5	30	A
289	30	A
300	30	C

Fig.6 (a) The box plots of prediction results using CNN model as regressor. The blue box shows the prediction result of testset B, while the red box shows the prediction result of testset C, the line represents perfect fit. (b) The box plots of prediction results using MLP. (c) The relationship between data magnitude after data augmentation and prediction errors of MLP and CNN (d) The box plots of prediction results using GBDT model. (e) The box plots of prediction results using attention Unet model as regressor.

4.6-bits-per-cell Resistive Probabilistic-bit Computing for High-efficient Evaluation of Bio-Genomic Evolution Achieving 6.25x Acceleration of Data-operations

Kai Wen Cheng[1], Huan-Hsiang Su[2], Yu-Hsien Lin[2], Jerry Chung[3], Y.-S. Hsieh[4], T.-Y. Chen[3], E Ray Hsieh[2,*], Yu-Cheng Lin[5], Ching-Ju Lin[5], Jen-Chou Tseng[5], Chien-Fan Wang[5], Wen-Ting Chu[5], Yu-Der Chih[5], Jonathan Chang[5]

[1] Coll. of Semiconductor Research, National Tsing Hua University, Hsinchu, Taiwan, [2]Dept. of Electrical Engineering, National Central University, Taoyuan, Taiwan, [3]Department of Computer Science and Information Engineering, National Central University, Taoyuan, Taiwan, [4]Dept. of Department of Computer Science and Information Engineering, National Cheng Kung University, Tainan, Taiwan, [5]TSMC, Hsinchu, Taiwan, *email: erayhsieh@cc.ncu.edu.tw

Abstract

The multiple-resistive-probabilistic-bit-per-cell has been firstly proposed to accelerate the computing of bio-genomic evolution, especially on the super virus (SARS-Cov-2) genome. With the assistance of the gradual-SET/RESET operations, the 4.6-bits-per-cell (25-states) 1T1R array has been to demonstrate the effectiveness of this concept. The genomic evolution includes two mechanisms: the genomic mutation and searching. Therefore, a mutation engine of the SARS-Cov-2 has been constructed on this platform and conducts the phylogenetic tree of the virus. Meanwhile, for the genomic searching, "4.6-bits-GenePara Accelerator (4.6-GPA)" has been developed to accelerate 6.25x of the data operations. As a result, the probabilistic-bits contain great potential to predict the occurrence of the super virus to prevent the next pandemic.

Introduction

$\mathscr{P}$robabilistic-bit (p-bit) [1] computing has rapidly become a novel research topic since it considers bits not only as exactly digits but flipping bits between bit-"1" and "0" with a certain probability. The p-bits can promptly generate high-quality real-random-numbers without sophisticated software algorithm calculation, i.e., the true-random-number-generator (TRNG) [2], which enables the p-bit to compute Probability and Statistic in a more energy-and-cost-efficient way. Therefore, this unique property of the p-bit has been successfully applied to the cryptography [3]; Bayesian DNN [4], and the quantum-commuting emulator. [5]

It is of great interest for us to propose a novel multiple-bit-per-cell by a 1-Mb 1T1R RRAM array for the genome evolution of super viruses because how to timely and effectively predict and discover the virus's phylogeny really matters after the SARS-Cov-2 (Covid-19) has wreaked havoc during pandemic.

First of all, in this work, we realize 2048 memory states in the 1T1R cell with the resistance-linearly-tuning property by performing the gradual SET/RESET scheme. Then, we accomplish 25 resistance states stored in each cell of the total 100k cells in the array, and guarantee 100k cycles for each resistance state, followed by the multiple-probabilistic-bit operations in the 4 low-resistance-states, which can be used as the mutation emulators in the virus's genome. Furthermore, we provide the result of a phylogenetic tree for the SARS Cov-2 virus lineages based on the multiple-p-bit mutation emulator engine. Finally, the data operation efficiency of this engine has been evaluated, achieving 6.25x of improvement, w.r.t. results of the software approaches.

Design details of the 1Mb 1T1R Macro

Fig. 1 (a) shows 2 unit-cells. Each cell comprises 1-nMOSFET (T) and 1-RRAM. The RRAM top-electrode links to the BL. Gate electrodes and source-terminals of 2 nMOSFETs in 2 neighbor unit-cells share the WL and SL, respectively. The BL, WL, and SL go to the BL, WL, and SL decoders/drivers, respectively. The other end of the BL connects to a current-reference sense amplifier. **Fig. 1 (b)** shows an aerial view of the test-chip photo-diagram. The array comprises 1 million unit-cells, fabricated by a low-power 40-nm CMOS technology.

Resistance Gradual-tuning Operations

Fig.2 shows the CDFs after a constant and a gradual FORMing (V_{BL}= 3V to 3.6V), respectively. The latter shows higher 1st LRS values after cells are just FORMed in comparison to the former. Then, in **Fig. 3**, we applied the gradual SET and RESET to those cells after the gradual FORMing was just performed. V_{BL}=1.2 V to 2 V for the gradual SET; V_{SL}= 1.2 V to 2 V for the gradual RESET. The results indicate that the LRS values can be gradually SET from 20 kΩ to 4 kΩ; the HRS values can be gradually RESET from 40 kΩ to 1MΩ. Moreover, the resistance gradual-tuning-range is expanded from 13.3x to 225x. Most importantly, we demonstrated 2048 memory-states **Fig. 4** in a 1T1R cell. And

979-8-3315-0417-5/25 $31.00 © 2025 IEEE

Fig. 5 shows the near-ideal linearity of the current gradual-tuning characteristics.

4.6-bit-per-cell 1T1R Array

Accurately controlled continually tuned resistance values have been characterized with 200x of the HRS/LRS ratio. **Fig. 7** shows statistical gradual-tuning resistance values. **Fig. 8** shows the CDFs of 25 levels, in terms of 4.6-bit-per-cell, collected by 100k cells, where 10 resistance-levels are distributed in the LRS branch from 30 kΩ to 3kΩ; the other 15 levels are in the HRS branch from 200 kΩ to 1 MΩ. **Fig. 9** further shows each level has been endured to 10^6 times. **Fig. 10 (a)** and **(b)** evaluates the characteristics of p-bit in the LRS and HRS, respectively. The LRS shows apparent electrical-variations, and each state is also coupled and overlapped together. Thus, the LRS will be used as p-bits for the mutation emulator of the virus. On the other hand, for the HRS, the electrical variations are much smaller and the gaps are clearly created between each state. That the HRS will be used for the memory bits for permanent storage result.

Multiple-p-bit-per-cell for Genome Mutation Emulation

To emulate the genome mutation events, the error table of the resistance states is constructed, **Table 1 (a)** and **(b)**. The RNA-virus, SARS-Cov-2, has been tested, **Fig. 11**. The engine mutation **Fig. 12** comprises two 1T1R-cells. The BLs of 2 cells go to the inputs of a current-reference SA. E.g., 1 cell stores the LRS1 (8kΩ), and the other stores the LRS2 (13kΩ). Since the LRS1< the LRS2, in most of cases, the S.A. always outputs the bit-"0". However, when the case that the LRS2 is smaller than the LRS1 due to electrical variations, the S.A. will suddenly output the bit-"1". We define the distance of the bit-string between two bits-"1" of the S.A. output signals as the "decoupled distance", **Fig. 13**, which is a set of true-random-numbers. Since the bits-"1" are rare events in the S.A. outputs, which are emulated as mutation-events for the genome evolution **Fig. 14**. Moreover, the good quality of the cryptographic parameters, Hamming-weight (HW) **Fig. 15**, and both inter and intra Hamming-distance (HD) **Fig. 16** have been confirmed. Finally, we can calculate the phylogenetics algorithm [6] and provide the phylogenetic tree of the SARS Cov-2 virus with different lineages **Fig. 17**.

High-speed of RNA Searching for the 4.6-p-bit-per Cell

we investigate the efficacy of a 4.6-bit-pe-cell for enhancing genomic searching, known as the "4.6-bit GenePara Accelerator (4.6-GPA)" **Fig. 18**. Unlike the 1-bit encoding, we activate 4.6-p-bit-per-cell diagonally to perform searches. The 4.6-GPA allows for an arrangement within a 4x4 RRAM-based Hamming Distance Unit (RHU), enabling 2 characters to be aligned in a single BL. This configuration represents 3 outcomes (0, 1, and 2) for Hamming distance calculations and results in a 4x throughput, compared to a 1-bit cell array. Furthermore, the 4.6-GPA model is applied to RNA pairwise approximate search **Fig. 19**, where RNA sequences are diagonally aligned. A low comparison score signifies a high match, and a high score indicates a low match. Comparative analyses **Fig. 20** reveal that the 4.6-GPA outperforms elder genome methods [7, 8]; 6.25x improvement of operation efficiency.

Conclusion

Consequently, we have sucussfully demonstrated the 4.6-bits-per-cell resistive probabilistic-bit computing for the bio-genome evolution on a 1-Mb 1T1R array using the genome of the SARS Cov-2 virus. The gradual FORMing, SET, and RESET schemes have been applied to tune the memory states to 2048 states in one cell, achieving 4.6-p-bit-per-cell in 100k cells. Then, the LRS values have been utilized to the p-bits for the mutation events of the genome, and the phylogenetic tree of the SARS Cov-2 virus has been constructed. Finally, we have shown that our genomic searching model (4.6-GPA) can achieve 6.25x improvement of the operation efficiency.

Acknowledgements:

This work was supported and sponsored by the joint-development-project of the TSMC and National Central University and by NSTC (National Science and Technology Council) 110-2636-E-008-007 and NSTC 110-2636-E-008-008, Taiwan.

References

[1] H. V. Ravish Aradhya, J. Fadnavis and S. G. Gojanur, "Memory Design and Verification of SRAM-based Energy Efficient Ternary Content Addressable Memory," International Conference on Information Systems and Computer Networks (ISCON), Mathura, India, 2021, pp. 1-7.

[2] E. R. Hsieh, P. H. Huang, M. L. Miu, S. Y. Huang, S. M. Lu and R. Y. Lyu, "Emulator of Gene Mutation: Interfering True Random Number Generators Using Resistive-Gate Non-Volatile-Memories," in IEEE Electron Device Letters, vol. 43, no. 11, pp. 1870-1873, Nov. 2022.

[3] Bhat, R., Sunitha, N.R. & Iyengar, S.S. A probabilistic public key encryption switching scheme for secure cloud storage. Int. j. inf. tecnol. 15, 675–690 (2023).

[4] M. Lin, Y., Zhang, Q., Gao, B. et al. Uncertainty quantification via a memristor Bayesian deep neural network for risk-sensitive reinforcement learning. Nat Mach Intell 5, 714–723 (2023).

[5] Chowdhury, S., Camsari, K.Y. & Datta, S. Accelerated quantum Monte Carlo with probabilistic computers. Commun Phys 6, 85 (2023).

[6] Felsenstein, Joseph. "Numerical Methods for Inferring Evolutionary Trees." The Quarterly Review of Biology 57, no. 4 (1982): 379–404.

[7] F. Zokaee, M. Zhang and L. Jiang, "FindeR: Accelerating FM-Index-Based Exact Pattern Matching in Genomic Sequences through ReRAM Technology," International Conference on Parallel Architectures and Compilation Techniques (PACT), Seattle, WA, USA, 2019, pp. 284-295.

[8] Ting Wu, Chin-Fu Nien, Kuang-Chao Chou, Hsiang-Yun Cheng. RePAIR: A ReRAM-based Processing-in-Memory Accelerator for Indel Realignment. 2022 Design, Automation & Test in Europe Conference & Exhibition, Antwerp, Belgium, March 14-23, 2022. Pages 400-405

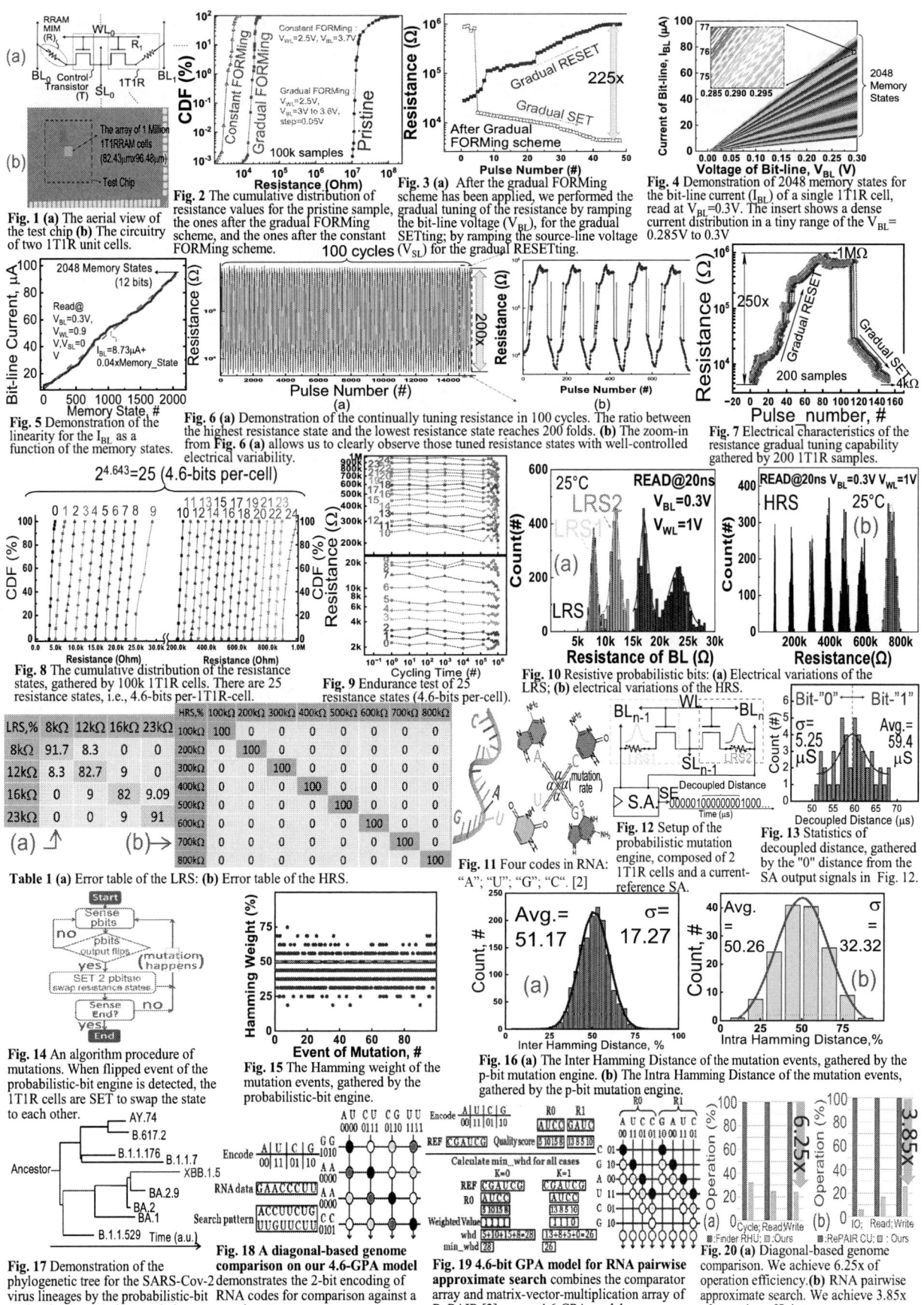

Fig. 1 (a) The aerial view of the test chip **(b)** The circuitry of two 1T1R unit cells.

Fig. 2 The cumulative distribution of resistance values for the pristine sample, the ones after the gradual FORMing scheme, and the ones after the constant FORMing scheme.

Fig. 3 (a) After the gradual FORMing scheme has been applied, we performed the gradual tuning of the resistance by ramping the bit-line voltage (V_{BL}), for the gradual SETting; by ramping the source-line voltage (V_{SL}) for the gradual RESETting.

Fig. 4 Demonstration of 2048 memory states for the bit-line current (I_{BL}) of a single 1T1R cell, read at $V_{BL}=0.3V$. The insert shows a dense current distribution in a tiny range of the $V_{BL}=0.285V$ to $0.3V$

Fig. 5 Demonstration of the linearity for the I_{BL} as a function of the memory states.

Fig. 6 (a) Demonstration of the continually tuning resistance in 100 cycles. The ratio between the highest resistance state and the lowest resistance state reaches 200 folds. **(b)** The zoom-in from **Fig. 6 (a)** allows us to clearly observe those tuned resistance states with well-controlled electrical variability.

Fig. 7 Electrical characteristics of the resistance gradual tuning capability gathered by 200 1T1R samples.

Fig. 8 The cumulative distribution of the resistance states, gathered by 100k 1T1R cells. There are 25 resistance states, i.e., 4.6-bits per-1T1R-cell.

Fig. 9 Endurance test of 25 resistance states (4.6-bits per-cell).

Fig. 10 Resistive probabilistic bits: **(a)** Electrical variations of the LRS; **(b)** electrical variations of the HRS.

LRS,%	8kΩ	12kΩ	16kΩ	23kΩ
8kΩ	91.7	8.3	0	0
12kΩ	8.3	82.7	9	0
16kΩ	0	9	82	9.09
23kΩ	0	0	9	91

HRS,%	100kΩ	200kΩ	300kΩ	400kΩ	500kΩ	600kΩ	700kΩ	800kΩ
100kΩ	100	0	0	0	0	0	0	0
200kΩ	0	100	0	0	0	0	0	0
300kΩ	0	0	100	0	0	0	0	0
400kΩ	0	0	0	100	0	0	0	0
500kΩ	0	0	0	0	100	0	0	0
600kΩ	0	0	0	0	0	100	0	0
700kΩ	0	0	0	0	0	0	100	0
800kΩ	0	0	0	0	0	0	0	100

Table 1 (a) Error table of the LRS: **(b)** Error table of the HRS.

Fig. 11 Four codes in RNA: "A"; "U"; "G"; "C". [2]

Fig. 12 Setup of the probabilistic mutation engine, composed of 2 1T1R cells and a current-reference SA.

Fig. 13 Statistics of decoupled distance, gathered by the "0" distance from the SA output signals in Fig. 12.

Fig. 14 An algorithm procedure of mutations. When flipped event of the probabilistic-bit engine is detected, the 1T1R cells are SET to swap the state to each other.

Fig. 15 The Hamming weight of the mutation events, gathered by the probabilistic-bit engine.

Fig. 16 (a) The Inter Hamming Distance of the mutation events, gathered by the p-bit mutation engine. **(b)** The Intra Hamming Distance of the mutation events, gathered by the p-bit mutation engine.

Fig. 17 Demonstration of the phylogenetic tree for the SARS-Cov-2 virus lineages by the probabilistic-bit engine for the mutation emulation.

Fig. 18 A diagonal-based genome comparison on our 4.6-GPA model demonstrates the 2-bit encoding of RNA codes for comparison against a search pattern query.

Fig. 19 4.6-bit GPA model for RNA pairwise approximate search combines the comparator array and matrix-vector-multiplication array of RePAIR [2] on our 4.6-GPA model.

Fig. 20 (a) Diagonal-based genome comparison. We achieve 6.25x of operation efficiency.**(b)** RNA pairwise approximate search. We achieve 3.85x of operation efficiency.

979-8-3315-0417-5/25 $31.00 © 2025 IEEE

Passivation effect on atomic-layer-deposited indium-gallium-zinc-oxide transistors

Zhiyu Lin, Jinxiu Zhao, Chen Wang, Mengwei Si*

Department of Electronic Engineering and State Key Laboratory of Radio Frequency Heterogeneous Integration, Shanghai Jiao Tong University, Shanghai, China.

Email: mengwei.si@sjtu.edu.cn

Abstract

In this work, passivation effect on ALD IGZO transistors was studied by characterizing the X-ray photoelectron spectroscopy (XPS) and positive bias temperature instability (PBTI). IGZO devices without passivation, with H_2O-growth Ga_2O_3 passivation and O_3-growth SiO_2 passivation were fabricated and characterized to compare the passivation effect. It is found that H_2O-growth Ga_2O_3 introduces a large amount of oxygen vacancies (V_O) and degrades IGZO device performance dramatically. While O_3-growth SiO_2 passivation has much less damage on IGZO and shows decent performance. However, devices with O_3-growth SiO_2 passivation still show degraded PBTI performance compared with devices without passivation, suggesting fast trap generation still exists due to the growth of SiO_2 passivation.
Keywords: IGZO, trap, passivation, reliability

Introduction

Indium-gallium-zinc oxide (IGZO) transistor has low leakage current, decent mobility, high reliability, low cost, and large-area fabrication capability[1-9]. Furthermore, the low thermal budget makes it attractive in monolithic 3-dimensional integration applications[10-13]. However, to be utilized as back-end-of-line (BEOL) compatible transistor, passivation layer is necessary. For bottom gate transistor, passivation layer directly grows on oxide semiconductor and affects the channel. In particular, atomic-layer-deposited (ALD) passivation layer introduces defects during metal-organic precursor reaction[14], [15]. In this work, we fabricated IGZO devices with different passivation layers and studied the impact of passivation growth on IGZO devices.

Experiments

Fig. 1(a) shows the schematic cross-sectional diagram of a bottom gate ALD IGZO transistor. The fabrication process flow is summarised in Fig. 1(b). ALD IGZO Devices were fabricated on the Si wafer with thermally oxidized SiO_2 substrates. TiN gate was deposited by PVD method and patterned by photolithography and lift-off process. 7 nm Al_2O_3 and 10 nm IGZO were grown by ALD successively, as gate insulator and channel layer, respectively. After patterning of IGZO, Ni was deposited by PVD as source/drain. To study the passivation effect on bottom-gate IGZO transistors, different passivation layers were grown on the devices. ALD growth SiO_2 with O_3 as oxygen precursor was utilized as passivation layer for stack 1 devices. ALD growth Ga_2O_3 with H_2O as oxygen precursor was utilized as passivation layer for stack 2 devices. While stack 0 devices are unpassivated for comparison. Electrical measurements were conducted with Keysight B1500 semiconductor parameter analyser and probe station under vacuum to avoid the impact of oxygen and water vapour.

Results and discussion

Fig. 2 compares the I_D-V_{GS} characteristics of devices with different passivation stacks and channel length (L) of 2 μm. Device without passivation layer has a decent field-effect mobility (μ_{FE}) of 15.82 cm^2/V·s and a positive threshold voltage of 0.67 V. After Ga_2O_3 passivation layer growth, a sudden increase in I_D is observed and device shows poor on/off ratio, which may be result in Ga precursor reacts with IGZO and generates oxygen vacancy (V_O) during the ALD Ga_2O_3 passivation layer growth. Furthermore, H_2O as oxygen precursor might introduce H which act as donor site in IGZO thin film. In the case of SiO_2 passivation layer the device shows slight decrease in μ_{FE} and positive shift in V_{TH}. This phenomenon could be explained by the higher oxidizability of O_3 than H_2O.

Fig. 3 demonstrates the scaling metrics of V_{TH}, μ_{FE}, and SS of device before and after SiO_2 passivation layer growth. V_{TH} has slight change after SiO_2 passivation layer growth, indicating that SiO_2 growth has low damage on IGZO channel. μ_{FE} decreases after SiO_2 passivation layer growth, which might be caused by O_3 oxidization effect and passivation of V_O. SS has slight change after SiO_2 passivation layer growth, indicating that SiO_2 passivation layer growth has almost no impact on the IGZO/gate insulator interface.

To further study the passivation effect of different passivation stacks, X-ray photoelectron spectroscopy was performed to investigate the valence state of different elements at the passivation layer/IGZO interface. In the case of IGZO thin film without passivation layer as shown in Fig. 4(a), In 3d peak shows no shoulder but only one peak at about 445 eV corresponding to In^{3+} state. In the case of H_2O growth Ga_2O_3/IGZO interface as shown in Fig. 4(b), In 3d peak shows significant shoulder at about 443.6 eV, which is related with metallic In^0 state. The shoulder proves that during the ALD Ga_2O_3 growth process, In-O bonds break and V_O generate. However, the Ga and Zn peak has no obvious shoulder, which may be originated from higher Ga-O and Zn-O bond energy.

979-8-3315-0417-5/25 $31.00 © 2025 IEEE

Although the SiO$_2$ passivation has minor impact on the DC performance of IGZO device, during long-time operation some traps might be involved. Fig. 5(a)-(b) compares the positive bias temperature instability (PBTI) performance measured at room temperature and stress voltage of 1 V, 2 V, and 3 V for devices without passivation and devices with SiO$_2$ passivation, respectively. Both device without passivation and device with SiO$_2$ passivation show good stability under PBTI stress with ΔV_{TH} lower than 0.2 V after 1000 s stress. However, the time-dependent ΔV_{TH} degradation trend is significantly different between device without passivation and device with SiO$_2$ passivation. During the PBTI stress in device with SiO$_2$ passivation a sudden increase of ΔV_{TH} at short stress time occurred under V_{stress} from 1 V to 3 V, which phenomenon was invisible in device without passivation. The PBTI performance indicates that SiO$_2$ passivation layer introduced some fast traps which have little impact on device DC performance but degrade significantly under PBTI stress.

Fig. 6 illustrates the possible mechanism of fast trap reaction during PBTI stress of device with SiO$_2$ passivation layer. O$_3$ growth SiO$_2$ might introduce a large amount of peroxide state (O_2^{2-}). During the PBTI stress, the energy barrier of the transition from O_2^{2-} to O^{2-} is lowered and this transition captures two electrons simultaneously[16], [17]. O^{2-} act as acceptor-like state, which result in a fast positive shift of ΔV_{TH}.

Conclusion

In this work, ALD IGZO devices with different passivation layers were fabricated and characterized. The impact of defects generation during ALD passivation layer growth on IGZO channel was studied by combining XPS and PBTI measurement. By analysing the XPS In 3d peak, we concluded H$_2$O growth Ga$_2$O$_3$ passivation layer introduced high density V$_O$ and leads to a normally-on operation. Device with O$_3$ growth SiO$_2$ passivation layer demonstrates decent mobility with an enhancement-mode operation. But O_2^{2-} generation would cause a fast ΔV_{TH} shift during PBTI stress.

Acknowledgments

This work was supported by National Key R&D Program of China (No. 2022YFB3606900), National Natural Science Foundation of China (No. 62274107, 92264204) and Shanghai Pilot Program for Basic Research-Shanghai Jiao Tong University under Grant 21TQ1400212.

References

[1] K. Nomura et al., "Room-temperature fabrication of transparent flexible thin-film transistors using amorphous oxide semiconductors," Nature, 432, pp. 488 (2004).

[2] J. Sheng et al., "Amorphous IGZO TFT with High Mobility of 70 cm^2/(V s) via Vertical Dimension Control Using PEALD," ACS Appl Mater Interfaces, 11, pp. 40300 (2019).

[3] T. Ono et al., "Hydrogen absorption method using HfO$_x$ in crystalline In-Ga-Zn Oxide FETs for NVM applications," IEDM Tech. Dig., pp. 877 (2020).

[4] N. On, B. K. Kim, S. Lee, E. H. Kim, J. H. Lim, and J. K. Jeong, "Hot Carrier Effect in Self-Aligned In-Ga-Zn-O Thin-Film Transistors with Short Channel Length," IEEE Trans. Electron Devices, 67, pp. 5544 (2020).

[5] N. Lv et al., "Suppression of the Short-Channel Effect in Dehydrogenated Elevated-Metal Metal- Oxide (EMMO) Thin-Film Transistors," IEEE Trans. Electron Devices, 67, pp. 3001 (2020).

[6] Y. S. Shiah et al., "Mobility–stability trade-off in oxide thin-film transistors," Nature Electronics, 4, pp. 800 (2021).

[7] A. Chasin et al., "Understanding and modelling the PBTI reliability of thin-film IGZO transistors," IEDM Tech. Dig., pp. 657 (2021).

[8] M. Si, Z. Lin, A. Charnas, and P. D. Ye, "Scaled Atomic-Layer-Deposited Indium Oxide Nanometer Transistors with Maximum Drain Current Exceeding 2 A/mm at Drain Voltage of 0.7 v," IEEE Electron Device Lett., 42, pp. 184 (2021).

[9] Q. Kong et al., "New Insights into the Impact of Hydrogen Evolution on the Reliability of IGZO FETs: Experiment and Modeling," IEDM Tech. Dig., pp. 699 (2021).

[10] C. Wang et al., "Extremely Scaled Bottom Gate a-IGZO Transistors Using a Novel Patterning Technique Achieving Record High G$_m$ of 479.5 S/m (V$_{DS}$ of 1 V) and f$_T$ of 18.3 GHz (V$_{DS}$ of 3 V)," Proc. Symp. VLSI Technol. pp. 294 (2022).

[11] J. Zhang, D. Zheng, Z. Zhang, A. Charnas, Z. Lin, and P. D. Ye, "Ultrathin InGaO Thin Film Transistors by Atomic Layer Deposition," IEEE Electron Device Lett., pp. 273 (2022).

[12] J. Li et al., "High-Performance Self-Aligned Top-Gate Amorphous InGaZnO TFTs With 4 nm-Thick Atomic-Layer-Deposited AlO$_x$ Insulator," IEEE Electron Device Lett., 43, pp. 729 (2022).

[13] W. Pan et al., "Multiple effects of hydrogen on InGaZnO thin-film transistor and the hydrogenation-resistibility enhancement," J. Alloys Compd., 947, pp. 169509 (2023).

[14] S. Y. Lee, J. Kim, A. Park, J. Park, and H. Seo, "Creation of a Short-Range Ordered Two-Dimensional Electron Gas Channel in Al$_2$O$_3$/In$_2$O$_3$ Interfaces," ACS Nano, 11, pp. 6040 (2017).

[15] M. Si, Z. Lin, Z. Chen, and P. D. Ye, "High-Performance Atomic-Layer-Deposited Indium Oxide 3-D Transistors and Integrated Circuits for Monolithic 3-D Integration," IEEE Trans. Electron Devices, 68, pp. 6605 (2021).

[16] S. Choi et al., "Excessive Oxygen Peroxide Model-Based Analysis of Positive-Bias-Stress and Negative-Bias-Illumination-Stress Instabilities in Self-Aligned Top-Gate Coplanar In–Ga–Zn–O Thin-Film Transistors," Adv. Electron. Mater., 8, pp. 2101062 (2022).

[17] C. K. Chen et al., "Negative-U Defect Passivation in Oxide-Semiconductor by Channel Defect Self-Compensation Effect to Achieve Low Bias Stress V$_{TH}$ Instability of Low-Thermal Budget IGZO TFT and FeFETs," IEDM Tech. Dig., pp. 41-2 (2023).

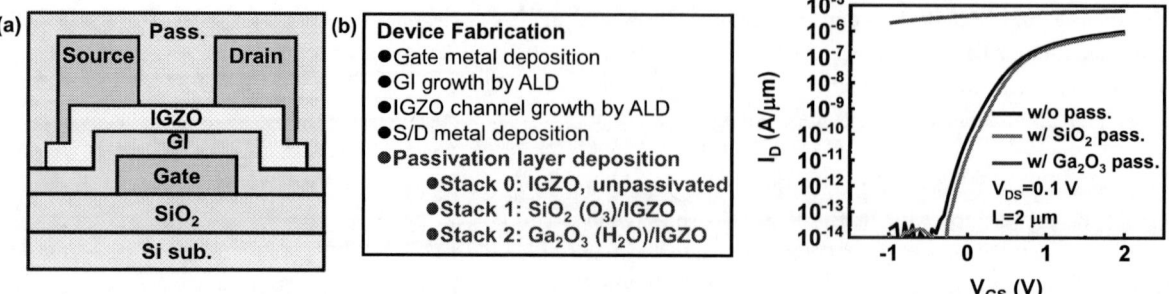

Fig. 1: (a) Schematic diagram of an ALD IGZO transistor with passivation layer. (b) Device fabrication process flow of ALD IGZO transistors with different passivation stacks.

Fig. 2: I_D-V_{GS} characteristics of IGZO transistors with different passivation stacks.

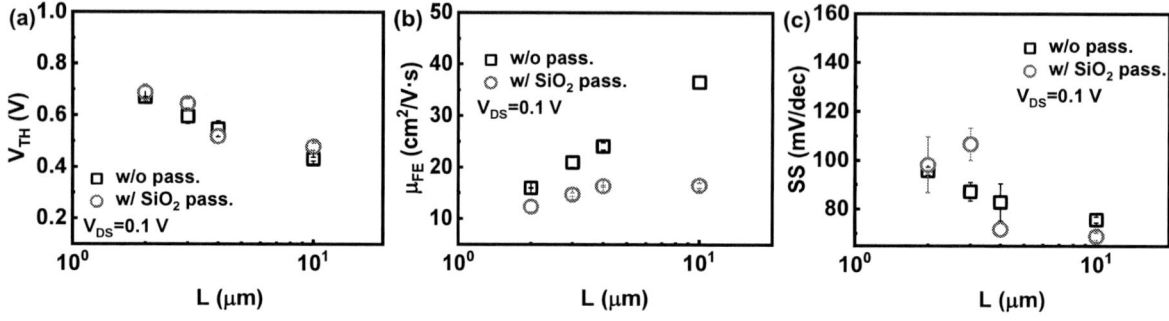

Fig. 3: Scaling metrics of (a) V_{TH}, (b) μ_{FE}, (c) SS of ALD IGZO transistors without and with ALD growth SiO_2 passivation layer. Each points includes at least 4 devices.

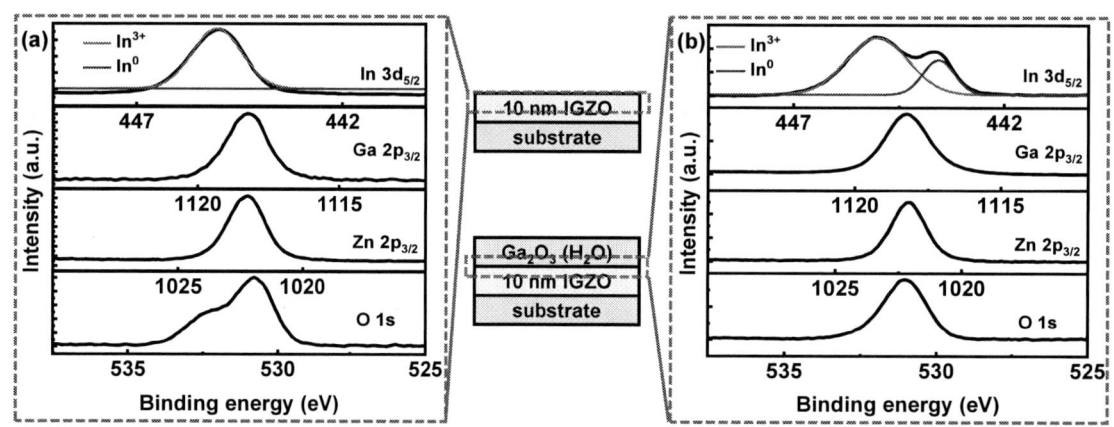

Fig. 4: XPS results of In 3d, Ga 2p, Zn 2p, Si 2p, O 1s spectrum of (a) passivation stack 0 and (b) stack 2. In 3d spectra of sample with H_2O growth Ga_2O_3 passivation exhibits obvious shoulder at about 443.6 eV, while the one without passivation exhibits only one peak at about 445.0 eV.

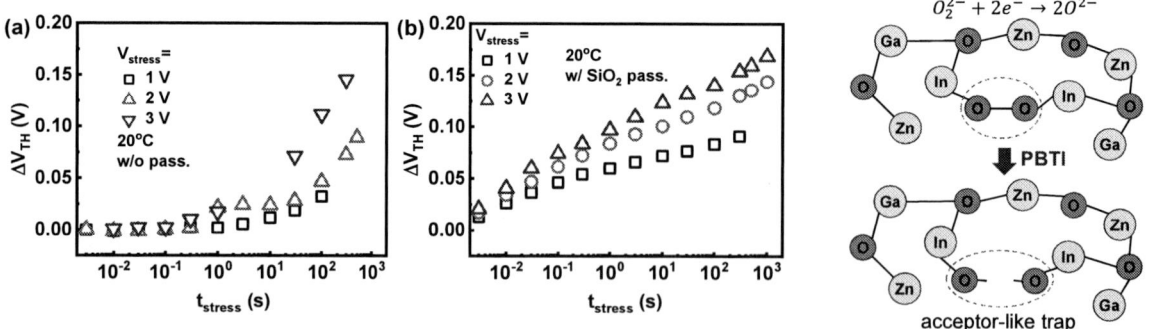

Fig. 5: ΔV_{TH} versus t_{stress} at different V_{stress} of IGZO devices (a) without passivation layer and (b) with O_3-growth SiO_2 passivation layer. More fast traps contribute to the PBTI of devices with passivation layer.

Fig. 6: Mechanism of trap generation in PBTI of devices with SiO_2 passivation layer.

Characteristic Length of Transition-metal Dichalcogenides based Complementary Field-effect Transistors

Xianmao Cao[1], Yiting Wu[2], Panpan Zhang[1,*]

[1]School of Integrated Circuits, Beijing University of Posts and Telecommunications，
Beijing 100876, China.

[2]School of Mechanical Engineering, Shanghai Jiao Tong University,
Shanghai 200240, China

（*email：tanji_ic@bupt.edu.cn）

Abstract

Complementary FETs (CFETs) have been demonstrated to deliver higher integration density than FinFETs for sub-2 nm technology nodes. Herein, analytical models for the characteristic length of transition-metal dichalcogenide (TMD) based CFETs are proposed to uncover the origin of asymmetric immunity of their top and bottom transistor to short channel effects (SCEs). The substrate imposes penalty to the electrostatic integrity of the bottom transistor. This work offers guidelines on TMD-based CFET design principles in terms of material screening and their associated stacking order.

Keywords: Complementary Field-effect Transistor (CFET), Transition-metal dichalcogenides (TMDs), electrostatic integrity (EI), characteristic length, scaling

Introduction

Transition-metal dichalcogenides (TMDs) as channel materials in conjunction with emerging device architectures are being extensively explored to continue the progression of Moore's Law. The complementary field-effect transistor (CFET) has been proposed as an innovative architecture for the beyond 2 nm technology node to overcome the scaling limitations of traditional planar devices [1] , [2]. Vertically stacking n- and p-type transistors above each other to eliminate the n-to-p separation bottleneck reduces the overall cell footprint [3], [4]. What's more, the inherent layered structure of TMDs allows precise control of the thickness at the atomic level, which makes them promising candidates for constructing CFETs in terms of the electrostatic control. However, the bottom transistor usually appears with deteriorated subthreshold performance while scaling down, characterized by an inferior subthreshold swing (SS) and more pronounced drain-induced-barrier lowering (DIBL) effect compared to its top counterpart. Besides the damage from the top transistor process, physical insight is still absent into the vulnerability of the bottom one to short-channel effects (SCEs). This work proposes an analytical model for the characteristic length of CFET (λ), confirmed by Technology Computer-Aided Design (TCAD) simulations (Sentaurus is used here), providing guidance on the CFET design principles regarding material screening and their associated stacking order.

Model Setup

Fig. 1 illustrates the CFET structure with MoS_2 and $MoTe_2$ serving as the n- and p-type channel material respectively (borrowed from Refs [5] and [6]), while HfO_2 and gold act as the high-κ dielectric and common gate of the nFET and pFET respectively. A TCAD model for the proposed TMD-based CFET is first developed and calibrated to generate the electrical characteristics as the baseline, which agrees well with the experimental results as shown in Fig. 2. The band-to-band tunneling mechanism is turned on to investigate its impact on the subthreshold performance. Note that interface states are eliminated here to unveil the intrinsic subthreshold behavior. Relevant parameters for material properties and device geometry are summarized in Table.1 [7], [8].

Results and Discussion

With the calibrated TCAD model as a starting point, the channel length of CFETs were scaled down to 100, 50 and 30 nm respectively and transfer characteristics were simulated as shown in Fig. 3 . It shows that subthreshold characteristics of pFETs deteriorate dramatically while scaling down, exhibiting inferior SS and more significant DIBL than nFETs. To quantify the gap, the values of SS and DIBL are extracted and plotted in Figs. 4 (a) and (b) respectively with L_{ch} varying from 20 to 300 nm, indicating the poor scalability of pFETs for $L_{ch} < 100$ nm regime. To investigate asymmetrical immunities for SCEs within the CFET architecture, the characteristic length models for both nFET and pFET are derived based on the two-dimensional Poisson equation, which characterizes the distance over which the nearby mobile charges can attenuate any local perturbation of the potential.

In addition to the effects of device scaling, we investigate the impact of the dielectric environment associated with the vertical structure on the horizontal leakage in the CFET. Fig. 5 shows the cross-sectional view of a CFET. By solving the two-dimensional Poisson equation, the general form of the characteristic length for CFETs can be derived as Eq. 1,

979-8-3315-0417-5/25 $31.00 © 2025 IEEE

$$\lambda_{n/p} = \sqrt{\frac{t_{ch,n/p}^2 \left(1 + \frac{2C_{ch,n/p}}{C_{ox,bn/p}}\right)}{2\left(1 + \frac{C_{ox,fn/p}}{C_{ch,n/p}} + \frac{C_{ox,fn/p}}{C_{ox,bn/p}}\right)}} \qquad (1)$$

where t denotes the material thickness, C denotes the capacitance, and n/p denotes the channel type, while the letter "ox", "ch", "f" and "b" in the subscript denote gate oxide, channel, front-side and back-side interface of the channel, respectively. Note that these definitions hold throughout the manuscript unless otherwise stated. Although Eq. 1 applies to both transistors of CFETs, for nFETs, the back-channel oxide capacitance $C_{ox,bn}$ is much smaller than both the front-channel oxide capacitance $C_{ox,fn}$ and n-channel capacitance $C_{ch,n}$, which can be omitted. Therefore, the expression of the characteristic length for nFETs can be reduced down to Eq. 2 as follows:

$$\lambda_n = \sqrt{\frac{2\varepsilon_{ch,n} t_{ch,n} t_{ox,fn}}{\varepsilon_{ox,fn}}} \qquad (2)$$

Here, ε represents the dielectric constant, while the counterpart for pFETs still holds as the general form:

$$\lambda_p = \sqrt{\frac{t_{ch,p}^2 \left(1 + \frac{2C_{ch,p}}{C_{ox,bp}}\right)}{2\left(1 + \frac{C_{ox,fp}}{C_{ch,p}} + \frac{C_{ox,fp}}{C_{ox,bp}}\right)}} \qquad (3)$$

The divergence mainly originates from the asymmetric dielectric environments that each transistor experiences. Eq. 3 implies that the substrate imposes an electrostatic penalty to the bottom transistor with the presence of $C_{ox,bp}$, which cannot be omitted due to the weak screening effects of the corresponding channel. Substituting the derived expression for the characteristic length and the parabolic function into the 2-D Poisson equation yields the surface potential for the CFET that shows good agreement with the TCAD simulated results, as shown in the Fig.6, confirming the validity of the proposed analytical model.

Substituting the relevant geometry and dielectric parameters into Eqs. 2 and 3, characteristic lengths of the devices selected for simulating the static behaviors of nFET and pFET in this study are derived to be 5.3 and 11.2 nm, respectively. With the characteristic length serving as the scaling factor, we benchmark the scalability of nFETs and pFETs in terms of subthreshold characteristics, i.e. SS and DIBL as shown in Fig. 7. As highlighted in the blue shaded region, the deterioration of subthreshold characteristics of pFETs is more sensitive to channel length scaling. For instance, in the sub-100 nm range, Figs. 7(a) and (b) indicate that the SS of nFETs gradually degrades over the $L_{ch}/2\lambda$ range of 2 to 10, down to 170 mV/dec at $L_{ch} = 20$ nm, while the SS of pFETs drops sharply to 386 mV/dec over the $L_{ch}/2\lambda$ range of $1 \sim 4.5$. A similar situation emerges for the DIBL

comparison illustrated in Figs.7(c) and (d).

In general, different characteristic lengths enforce inconsistent lower bounds on the scaling of n- and p-type transistors in CFET configuration. Especially, the bottom transistor (i.e. pFET here) exhibits poorer scaling potential due to the inevitable capacitance induced by the substrate oxide, which participates in the determination of the channel surface potential and degrades the EI. It is illustrated in Fig. 8 that low-dimensional channel materials usually have low density of states (DOS), and the resulted low carrier density within the channel cannot effectively prevent the electric field from penetrating the substrate, leading to asymmetric dielectric environments experienced by top and bottom transistor of CFETs respectively. Therefore, we note that relatively lower channel capacitance and higher top-gate capacitance for the bottom transistor are preferred to enable matched scaling capability of CFETs.

Conclusion

The boundary condition is crucial in determining the characteristic length. Analytical characteristic length models are proposed to shed light on the unmatched scalability of the n- and p-type transistor in CFET configuration incorporating MoS_2 and $MoTe_2$ as channel materials, by taking into account the specific dielectric environment of each transistor. In particular, we show that the presence of substrate capacitance imposes a penalty on the EI of CFETs, making the choice and stacking order of channel materials matter. Characteristic length models for the bottom and top transistors have universal application scenarios and are not limited to TMD materials. Overall, a physics-rich but concise picture is extracted for unveiling the role of vertical structures on affecting the horizontal leakage and allowing the design space exploration towards ultrascaled, highly robust and heterogeneous integration of CFETs in future microelectronics chip.

Acknowledgments

This work is supported by the Start-up Funding of Beijing University of Posts and Telecommunications.

References

[1] Ryckaert. J, *et al.* , *VLSI*, p. 141-142, (2018) .

[2] Brunet. L, *et al.*, *VLSI*, p. 1-2, (2016) .

[3] L. Tong, *et al.*, *Nature Electronics*, vol. 6, no. 1, pp. 37–44 (2023).

[4] X. Zhu, *et al.*, *IEEE Transactions on Computer Aided Design of Integrated Circuits and Systems*, (2023).

[5] Y. Xia, *et al.*, *Small*, vol. 18, no. 20, p. 2107650 (2022).

[6] X. Jia, *et al.*, *Small*, vol. 19, no. 19, p. 2207927(2023).

[7] A. Laturia, *et al.*, *npj 2D Materials and Applications*, vol. 2,no. 1, p. 6(2018).

[8] H. M. Hill, *et al.*, *Nano letters*, vol. 16, no. 8, pp. 4831–4837(2016).

TABLE I. Device parameters of CFET used in Sentaurus.

Geometrical parameters			Material parameters		
Parameters	Description	Values	Parameters	Description	Values
L_{ch}	Channel length	20 nm-300 nm	ε_{MoS_2}	Dielectric constant of MoS_2	6.9
W_{ch}	Channel width	1 μm	ε_{MoTe_2}	Dielectric constant of $MoTe_2$	10.4
t_{chp}	Thickness of $MoTe_2$	9 nm	ε_{HfO_2}	Dielectric constant of HfO_2	20
t_{chn}	Thickness of MoS_2	3 nm	ε_{SiO_2}	Dielectric constant of SiO_2	3.4
t_{oxfn}	Thickness of n-channel gate oxide	30 nm	E_{g,MoS_2}	Bandgap of MoS_2	2.1 eV
t_{oxfp}	Thickness of p-channel gate oxide	30 nm	$E_{g,MoTe_2}$	Bandgap of $MoTe_2$	1.1 eV
t_{sub}	Thickness of SiO_2 substrate	285 nm	χ_{MoS_2}	Electron affinity of MoS_2	4.25 eV
			χ_{MoTe_2}	Electron affinity of $MoTe_2$	4.2 eV
			$m^*_{MoS_2}$	Effective mass(e/h) of MoS_2	0.44/0.54 (m^*)
			$m^*_{MoTe_2}$	Effective mass(e/h) of $MoTe_2$	0.64/0.78 (m^*)

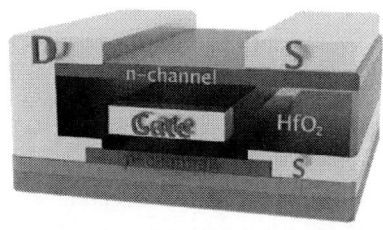

Fig. 1: Device structure diagram of the CFET with MoS_2 and $MoTe_2$ being incorporated as the n- and p-type channel material, respectively.

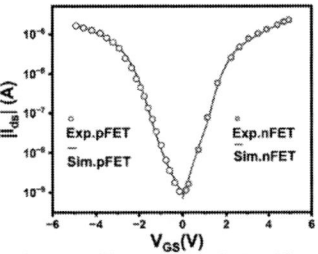

Fig. 2: Calibrated TCAD I_{ds}–V_{gs} characteristic curves versus Ref.6.

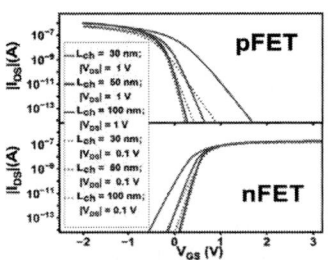

Fig. 3: The simulated transfer curves for the nFET and pFET, respectively, with L_{ch} = 30, 50, and 100 nm.

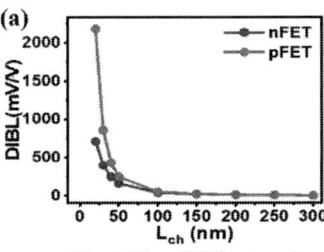
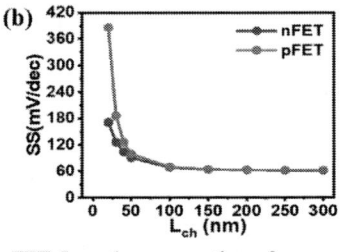

Fig. 4: The nFET outperforms pFET from the perspective of subthreshold behaviors.(a) DIBL versus L_{ch} within the range of 20 to 300 nm. (b) SS versus L_{ch} within the range of 20 to 300 nm.

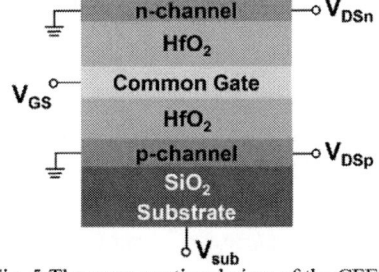

Fig. 5: The cross-sectional view of the CFET.

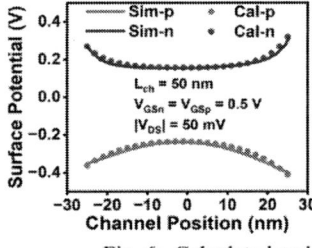
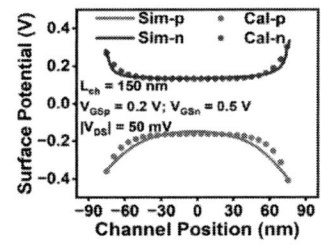
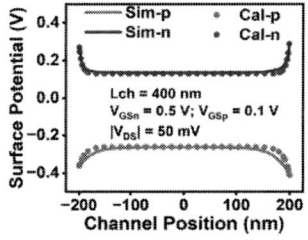

Fig. 6: Calculated and Sentaurus simulated surface potentials for different L_{ch} of 50, 150 and 400 nm.

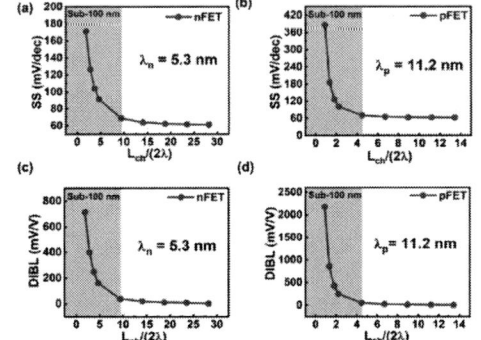

Fig. 7: Scalability of nFETs and pFETs with characteristic length as the scaling factor. (a) SS of the nFET versus $L_{ch}/(2\lambda)$. (b) SS of pFET versus $L_{ch}/(2\lambda)$. (c) DIBL of nFET versus $L_{ch}/(2\lambda)$. (d) DIBL of pFET versus $L_{ch}/(2\lambda)$.

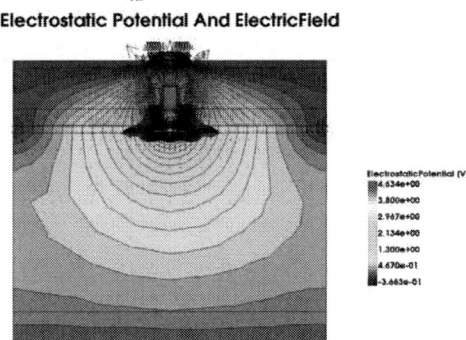

Fig.8: Electrostatic potential and electric field (vectors) distributions of CFET

979-8-3315-0417-5/25 $31.00 © 2025 IEEE

Dual Gate-Enhanced Mechanical-Electrical Stability of Flexible InGaZnO TFTs

Jilin Li[1], Yuhan Zhang[1], Runxiao Shi[2], Fion Sze Yan Yeung[2], Man Hoi Wong[2], Hoi Sing Kwok[2], Shengdong Zhang[1], Lei Lu[1,#]

[1] School of Electronic and Computer Engineering, Shenzhen Graduate School, Peking University, Shenzhen, China.

[2] The State Key Laboratory of Advanced Displays and Optoelectronics Technologies, The Hong Kong University of Science and Technology (HKUST), Hong Kong, China.

(Email: lulei@pku.edu.cn)

Abstract

The electrical characteristics and mechanical bias stress stabilities of dual-gate (DG) amorphous InGaZnO (a-IGZO) thin-film transistors (TFTs) were investigated. The same device exhibits diversified degradation and recovery behaviors under BG, TG and DG modes. The coupling electric filed of DG demonstrates superior stabilities against bias and mechanical stresses. Moreover, compared to the partially recovery of large degradations under BG and TG modes, the V_{th} shift (ΔV_{th}) is only 0.92 V under bending stress and can be mostly recovered, suggesting nonuniform degradations along the channel thickness direction.

Keywords: Dual-gate thin-film transistors; amorphous InGaZnO; flexible; degradation recovery; mechanical stability

Introduction

In recent years, flexible electronic devices, known for their lightweight and durable properties, have garnered significant research interest. As a fundamental component of flexible electronics, a-IGZO thin-film transistors (TFTs) offer several advantages, including high carrier mobility, excellent large-area uniformity, and low cost [1-3]. These features make them widely utilized in flexible displays [4] and circuit applications [5]. During working, flexible electronics are subjected to prolonged mechanical stress, such as bending stress in dynamic [8] or static [6,9] conditions. Consequently, the reliability of a-IGZO TFTs under mechanical loads is a critical concern. Additionally, the dual-gate (DG) structure is considered more appealed due to its enhanced gate control capabilities and improved stability compared to single-gate designs [7]. Typically, the characteristics of TFTs degrade progressively during stress application. Once the stress is removed, the performance of the degraded TFTs generally recovers to some extent. This phenomenon, commonly referred to as "recovery," is frequently observed under various electrical bias stress conditions in reliability studies of devices, including a-IGZO TFTs subjected to mechanical stress [9]. While some research has explored the recovery behavior of TFTs through thermal annealing after mechanical stress [10], there is a lack of reports on the recovery behavior of a-IGZO TFTs under electrical bias stress experiencing static mechanical load. Furthermore, the degradation of devices, taking recovery into account, has been less thoroughly investigated. To develop DG TFTs with high stability against bending stress, it is essential to comprehensively examine the effects of static stress on DG a-IGZO TFTs to uncover the underlying mechanisms.

The aim of our research is to explore the effects of static tensile stress on DG a-IGZO TFTs. The varied operating modes of DG TFTs position the carrier transport path at different locations within the AOS layer, revealing both recoverable and unrecoverable components of the degradations induced by tensile stress, which will further inform effective coping strategies.

Experimental Details

The schematic structure of the dual-gate a-IGZO TFT is shown in Figure. 1, and the processing step is as follows: First, deposit 200 nm SiN_x on PI glass by PECVD at 300°C, followed by the deposition of 300 nm SiO_2 as the buffer layer. The reaction gases for SiN_x are SiH_4, N_2, and NH_3, while the reaction gases for SiO_2 are SiH_4 and N_2O. Next, 100 nm molybdenum (Mo) on the buffer layer deposited by DC magnetron sputtering and patterned by wet etching to form the bottom gate electrode. Then, deposit 200 nm SiO_2 as the bottom gate insulator (BGI) by PECVD at 300°C. TFTs are then annealed at 300°C for 1.5 hours in O_2 atmosphere. The active layer is 40 nm a-IGZO deposited by DC magnetron sputtering, followed by annealing at 350°C for 1 hour in O_2 atmosphere. After annealing, CF_4 can be used for F plasma treatment, and the active layer was patterned by wet etching. A ceramic target, which consists of $In_2O_3:ZnO_2:Ga_2O_3$ at a ratio of 1:1:1, was used. Subsequently, deposit 200 nm SiO_2 as the top gate insulator (TGI) after N_2O plasma treatment at 150°C, followed by a 3-hour annealing at 300°C. Then, deposit 100 nm Mo and 30 nm ITO through DC magnetron sputtering and pattern them by wet etching to form the top gate electrode. A ceramic target, which consists of $In_2O_3:SnO_2$ at a ratio of 9:1, was used. Next, remove TGI by dry etching and make S/D conductive by Ar bombardment. Afterward, grow 200 nm SiO_2 as a passivation layer (PL) by PECVD at 150°C. Form contact holes for S/D and gate electrodes by dry etching, and then deposit 100 nm Mo as the contact metal through DC magnetron sputtering and pattern it using wet etching. Finally, peel the device off with the PI film from the glass substrate by Laser Lift-off (LLO).

The dual gate TFT has three operation modes with different gate biases, including BG, TG, and DG modes, as shown in Figure. 2. The B1500 (Agilent Technologies) semiconductor parameter analyzer was used to test the electrical characteristics of TFTs indifferent operating modes.

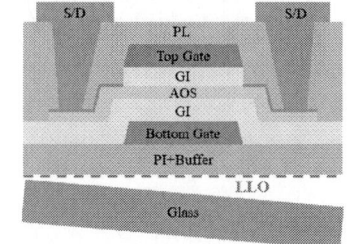

Fig. 1: The schematic structure of the dual-gate TFTs

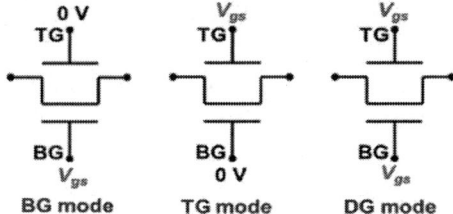

Fig. 2: Three different operating modes of dual-gate TFTs

Results and Discussion

The electrical performance of TFTs subjected to bending-recovery cycle show degradation-recovery phenomenon of TG, BG and DG mode, respectively corresponding to Figure. 3 (a), (b) and (c). The bending axis is parallel to the channel with a 10 mm-radius of curvature shown in Figure. 3 (d). The bending state is to be bent on the above for 1 h and the flat recovery is to be flat for 24 h. The electrical performance of each mode underwent degradation including positive ΔV_{th}, increased SS, decreased I_{on}, but clockwise hysteresis only not in DG mode, all of which may result from the increased electron traps and reduced shallow donor V_O induced by tensile mechanical stress. The hysteresis indicates that there must be something different between DG mode and the other modes. Under flat recovery, the transfer characteristic of DG mode nearly achieved full recovery, while the other modes did not do as DG mode with the degraded SS as shown in Figure. 3 (a), (b) and (c), which points out the difference of DG mode. However it could be observed form the parameters detailed in Table 1 that there is still SS deteriorations unrecovered of DG mode with the μ_{lin} and V_{th} restored, which is not apparent in Figure. 3 (c). But under bending-recovery cycle, the ΔSS and ΔV_{th} of DG mode is always smaller than the sum divided by 2 of those of BG and TG modes, which proves that the DG mode of dual-gate TFT is the most robust under bending stress.

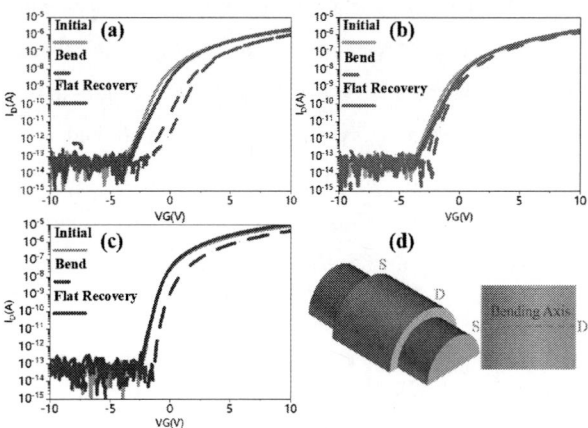

Fig. 3: I_D-V_G curve at V_D=10.1 V of each operating mode of dual-gate TFTs with W/L=10/20 μm after bending-recovery test and the bending direction: a) TG mode; b) BG mode; c) DG mode; d) the schematic of bending axis parallel to the direction of source to drain with a 10 mm-radius

Table 1: List of key parameters of each operating mode under bending-recovery test

		μ_{lin}(cm²/Vs)	SS(mv/dec)	V_{th}(V)
BG	Ini.	10.2	625	-1.23
	Bend	10.5	748	-0.61
	Rec.	10.7	710	-0.8
TG	Ini.	12.39	573	-1.2
	Bend	11.73	782	0.95
	Rec.	12.4	760	-0.71
DG	Ini.	12.6	308	-1.12
	Bend	11.4	392	-0.2
	Rec.	12.7	370	-1.15

To further investigate the degradation caused by mechanical stress, positive bias stress (PBS) was also tested during bending-recovery cycle, as shown in Figure. 5. The $|\Delta Vth|$ under PBS showed the same degradation-recovery phenomenon in each operating mode. Under bending condition, the $|\Delta V_{th}|$ under PBS of BG, TG, and DG modes increased by 4.6, 4.3, and 3.9 V, respectively, namely that the tensile stress made the transfer characteristics negatively drift more under PBS, which suggests that mechanical stress can causes deteriorations not shown in transfer characteristics. The larger $|\Delta V_{th}|$ may be possible due to the formation of acceptor-like defects within the AOS resulted from the tensile stress, such as O_i with the reaction of $O_i + 2e^- \rightarrow O_i^{2-}$ under PBS. In the recovery, the $|\Delta V_{th}|$ under PBS of BG, TG and DG increased by 2.6, 2.5, and 1.7 V, respectively. Although the recovery under PBS of DG Mode is not as perfect as that in

979-8-3315-0417-5/25 $31.00 © 2025 IEEE

the transfer characteristics, it still suffers the least impact from mechanical stress. The difference of mechanical stress stability between the DG mode and the other modes may be explained by the stress concentration area located at the AOS/GI interface extending into the bulk and the bulk transport mechanism under DG mode, which minimizes defects induced by mechanical stress within the bulk as shown in Figure 5.

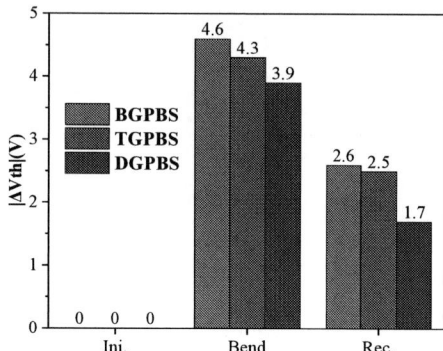

Fig. 4: Contrast of PBS induced $|\Delta V_{th}|$ of each operating mode of dual-gate TFTs under bending-recovery test with the gate bias stress at V_G=20 V

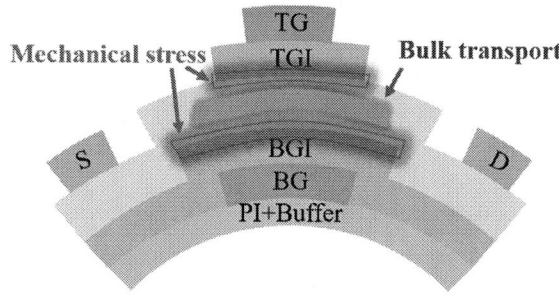

Fig. 5: Mechanical stress concentration area and the bulk transport under DG Mode

Conclusion

For flexible DG a-IGZO TFTs, device degradation under mechanical stress is a critical concern. This study analyzed the electrical performance and PBS stability under tensile stress with 10 mm-bending radius. Unrecoverable deteriorations were observed in BG- and TG-mode AOS TFTs under tensile stress, while the purely mechanical stress-induced degradation is fully recoverable in DG mode. As a contrast, the DG degradation still exhibited the strongest resilience to the combination of PBS and tensile bending but can only be mostly recovered. The corresponding degradation and recovery mechanisms were investigated. The superior stabilities against mechanical bias stresses of DG-mode AOS TFT inherently benefit from the bulk transport of carriers away from the AOS/GI interface, which is more vulnerable to both electrical and mechanical stresses.

Acknowledgments

This work was supported financially by Guangdong S&T Program (2024B0101120003), Shenzhen Municipal Scientific Program (KJZD20230923114111021) and Shenzhen Science and Technology Program (KJZD20230923115005009). This work was conducted in Shenzhen POC center of Flexible Electronics.

References

[1] K. Nomura, H. Ohta, A. Takagi, T. Kamiya, M. Hirano, and H. Hosono, "Room-temperature fabrication of transparent flexible thin-film transistors using amorphous oxide semiconductors," Nature, vol. 432, no. 7016, pp. 488-92, Nov 25 2004, doi: 10.1038/nature03090.

[2] T. Kamiya, K. Nomura, and H. Hosono, "Present status of amorphous In-Ga-Zn-O thin-film transistors," Sci Technol Adv Mater, vol. 11, no. 4, p. 044305, Aug 2010, doi: 10.1088/1468-6996/11/4/044305.

[3] J. Troughton and D. Atkinson, "Amorphous InGaZnO and metal oxide semiconductor devices: an overview and current status," Journal of Materials Chemistry C, vol. 7, no. 40, pp. 12388-12414, 2019, doi: 10.1039/C9TC03933C.

[4] A. Srivastava, D. Dubey, M. Goswami, and K. Kandpal, " Mechanical strain and bias-stress compensated, 6T-1C pixel circuit for flexible AMOLED displays, " Microelectronics Journal, vol. 117, p. 105266, 2021, doi:10.1016/j.mejo.2021.105266.

[5] H. Çeliker, F. D. Roose, M. Willegems, S. Smout, W. Dehaene and K. Myny, "Analysis and Comparison of Logic Architectures for Digital Circuits in a-IGZO Thin-Film Transistor Technologies," IEEE Journal of Solid-State Circuits, pp. 1-13, Dec 06 2023, doi: 10.1109/JSSC.2023.3334328.

[6] Han, C., H. Kim, D. Kim, J. Shin, Y. Park, C. Byun, and B. Choi. C, "Electrical Reliability of Flexible Low-Temperature Polycrystalline Oxide Thin-Film Transistors Under Mechanical Stress," IEEE Transactions on Electron Devices, vol. 70, no. 2, pp. 527-531, Feb. 2023, doi: 10.1109/TED.2022.3229292.

[7] K. Nomura, T. Kamiya, and H. Hosono, "Highly stable amorphous In-Ga-Zn-O thin-film transistors produced by eliminating deep subgap defects," Applied Physics Letters, vol. 99, no. 5, p. 053505, 2011/08/01 2011, doi: 10.1063/1.3622121.

[8] W. Jiang, M. X. Wang, H. S. Wang, and D. L. Zhang. "Spontaneous Degradation of Flexible Poly-Si TFTs Subject to Dynamic Bending Stress," IEEE Transactions on Electron Devices, vol. 66, no. 5, pp. 2214-2218, May 2019, doi: 10.1109/ted.2019.2907042.

[9] J. Yang, T. C. Chang, B. W. Chen, P. Y. Liao, H. C. Chiang and Q. Zhang, "Effects of Mechanical Stress on Flexible Dual-Gate a-InGaZnO Thin-Film Transistors, " Phys. Status Solidi A, vol. 215, no. 1, p. 1700426, 2018, doi:10.1002/pssa.201700426.

[10] M. M. Hasan, M. M. Billah, M. N. Naik, J. G. Um and J. Jang, "Bending Stress Induced Performance Change in Plastic Oxide Thin-Film Transistor and Recovery by Annealing at 300 °C," IEEE Electron Device Letters, vol. 38, no. 8, pp. 1035-1038, Aug. 2017, doi: 10.1109/LED.2017.2718565.

Flexible carbon nanotube optoelectronic neuromorphic devices and irradiation-resistance logic circuits with low-work-function gate electrodes

Meng Deng[1,2,3], Kaixiang Kang[2,3], Nianzi Sui[2,3], Jiaqi Li [2,3], Zebin Wang[1,2,3] and Jianwen Zhao [2,3]*

[1]Institute of Nano Science and Technology, University of Science and Technology of China, No. 166 RenAi Road, Suzhou Industrial Park, Suzhou, Jiangsu Province 215123, PR China

[2]Key Laboratory of Semiconductor Display Materials and Chips, Division of Nanodevices and Related Nanomaterials, Suzhou Institute of Nano-Tech and Nano-Bionics, Chinese Academy of Sciences, No. 398 Ruoshui Road, Suzhou Industrial Park, Suzhou, Jiangsu Province, 215123, PR China

[3]School of Nano-Tech and Nano-Bionics, University of Science and Technology of China, No. 398 Ruoshui Road, Suzhou Industrial Park, Suzhou, Jiangsu Province, 215123, PR China

Email: jwzhao2011@sinano.ac.cn

Abstract

Semiconducting SWCNTs (sc-SWCNTs) are one of promising materials for flexible optoelectronic neuromorphic devices and irradiation-resistance logic circuits because of their excellent electrical properties and strong C-C structure, flexibility, high stability, and low process temperatures. Here, we will report some new progress of high-performance flexible carbon nanotube thin film transistors (TFTs) and circuits in our group, including optoelectronic neuromorphic devices and logic circuits with the recorded power consumption, excellent irradiation resistance and strong self-repair ability after optimizing the designed device structures, and using high-capacitance dielectric materials and low-work-function gate electrodes (Yttrium~3.23 eV and Aluminium~4.2 eV).

Keywords: Carbon nanotube, Thin film transistors (TFTs), Optoelectronic neuromorphic devices, Flexible electronics, Irradiation resistance

Introduction

Semiconducting single-walled carbon nanotubes (sc-SWCNTs) are considered outstanding candidates for fabricating flexible three-terminal artificial synapse devices and circuits with low power consumption and irradiation resistance, given their exceptionally high charge carrier mobility, saturation rates, efficient gate control, and excellent mechanical properties[1, 2] . However, the fabrication of neuromorphic devices and thin film transistors (TFTs) using solution-processed sc-SWCNTs often exhibits the depletion mode, undoubtedly exacerbating power consumption in future large scale neural computing applications. Presently, extensive research has been conducted to modulate the device threshold voltage. For instance, utilizing aluminium (Al) and titanium (Ti) as gate electrodes, chemical doping, and self-assembly processes have been explored to achieve low-power-consumption enhancement-mode transistor devices. However, these methods often need the complicated fabrication process, and as-prepared transistor devices exhibit small threshold voltage modulation window and low fault tolerance, hindering their large-scale integration in integrated circuits and neuromorphic chip. Additionally, due to the adsorption behavior of carbon nanotubes, exposure to light and so on, SWCNT transistor devices may still exhibit depletion mode, as the threshold voltage tuning window is limited[3] . Therefore, it is necessary to develop synaptic transistor devices with a broader threshold voltage tuning window and high fault tolerance in diverse manufacturing scenarios, enabling ultra-low power consumption artificial vision system.

Furthermore, with the continuous development of aerospace technology, the performance requirements and application demands for transistor devices and integrated circuits are gradually increasing. However, various types of ionizing radiation present in these extreme environments can cause severe radiation damage to the semiconductor channel layer, gate oxide, and surrounding insulators of transistors, leading to performance degradation or even failure of integrated circuits[4, 5] . Therefore, it is necessary to develop new methods and technologies (such as altering device structure, materials, and packaging) to further enhance the radiation resistance of devices and circuits.

Here, high-performance optoelectronic neuromorphic devices (the single-pulse power consumption of artificial synaptic devices at wavelengths of 620 nm is only 33.9 aJ) and irradiation-resistance logic circuits (The total ionizing dose (TID) can reach 5.5 Mrad at the irradiation dose rate of 360 rad/s based on flexible SWCNT TFTs have been demonstrated by tuning the work function of gate electrodes, optimizing the device structures and so on.

Flexible enhanced optoelectronic synaptic devices and arrays

The conventional method for preparing artificial optoelectronic synaptic transistors involves spin-coating photosensitive materials onto the device channel area to form the hetero-structure. However, this often leads to a decrease in conductivity in the device channel area, causing severe degradation in electrical performance such as on-state current, hysteresis, threshold voltage, and subthreshold swing. Here, we successfully constructed an Ag_2S optoelectronic synaptic transistor by incorporating Ag_2S quantum dots (QDs) into the ionic liquids cross-linking-poly(4-vinylphenol) (ILs-c-PVP) solid-state electrolytes to form a hybrid photosensitive medium materials. Using metallic Y as the gate electrode reduces the transistor's threshold voltage, constituting an enhanced SWCNT TFT arrays. The device structure and transfer curves of SWCNT TFTs with different gate electrodes are shown in Figures 1(a) and 1(b). The Y-gate SWCNT TFTs feature high switch ratio ($>10^5$), negative threshold voltage (-1 V) and low operating voltage (-2 V~0 V).

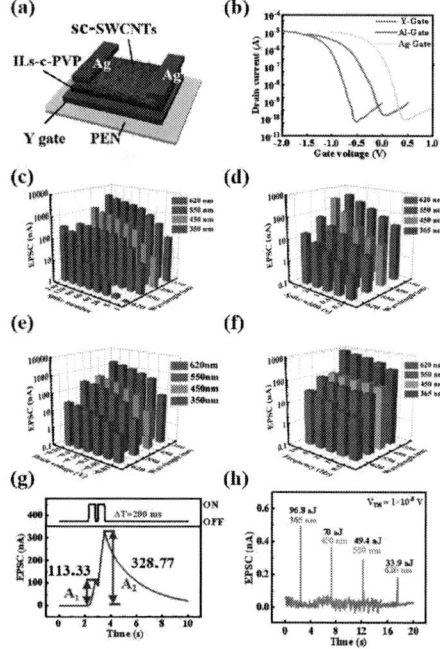

Figure 1 (a) Schematic of a single device structure, (b) the transfer curves of SWCNT TFTs with different gate electrodes, (c) EPSC variations with different pulse numbers (1~240 pulses) under 365 nm to 620 nm light pulses (3 mW/cm²), (d) EPSC variations with different pulse widths (0.2s to 10 s), (e) EPSC variations with different V_{DS} (-0.01 V to -0.25 V), (f) EPSC variations with different pulse frequencies (0.5 Hz to 10 Hz), (g) PPF index under 365 nm light pulses, (h) Single pulse power consumption with different pulse lights ($V_{DS}=1\times10^{-5}$ V, and $\Delta T=20$ ms).

We tested the synaptic characteristics under four different light pulses ranging from 365 nm to 620 nm. Taking UV light pulses (365 nm, 3 mW/cm²) as an example, as shown in Figure 1(c), under V_{DS} = -0.1 V, continuous light stimulation was applied to the device

ranging from a single pulse to 240 pulses. This resulted in continuous excitatory postsynaptic current (EPSC) that increased with each light stimulus. As the number of pulses gradually increased, the device's EPSC also continuously increased from 108.9 nA for a single pulse to 3915.6 nA for 240 multiple pulses, maintaining excellent synaptic plasticity. This indicates that under different light pulse stimuli, the yttrium gate carbon nanotube optoelectronic synaptic devices based on Ag_2S QDs can achieve a transition from short-term potentiation (STP) to long-term potentiation (LTP). We verified the synaptic plasticity models under different numbers of light pulses, pulse widths, V_{DS} values, and pulse frequencies, as shown in Figures 1(c), 1(d), 1(e), and 1(f), respectively. It can be observed that pulse frequencies, widths, and V_{DS} values across four different wavelengths, the EPSC values significantly increase with increasing numbers of light pulses, demonstrating good synaptic plasticity.

Additionally, taking 365 nm light pulses as an example, we applied two consecutive light pulses (pulse width of 200 ms) and tested the PPF index. As shown in Figure 1(g), our device achieved an ultra-high PPF index of 290% under 365 nm light stimulation. Furthermore, at a bias voltage of 1×10^{-5} V and a pulse width of 20 ms, the single-pulse power consumption of artificial synaptic devices at wavelengths of 365 nm, 450 nm, 550 nm, and 620 nm were 96.8 aJ, 70 aJ, 49.4 aJ, and 33.9 aJ, respectively, as shown in Figure 1(h). This demonstrates the potential application of our devices in ultra-low power consumption synaptic devices.

Flexible Radiation-Hardened Carbon Nanotube TFTs and Inverters

We developed radiation-hardened flexible single-wall carbon nanotube (SWCNT) thin-film transistors (TFTs) and circuits, utilizing an atomic layer deposition (ALD)-grown 20 nm thick alumina (Al_2O_3) film, ionic liquid-c-PVP solid-state electrolyte as dual dielectric layers, and low-work-function aluminum (Al) as the gate electrode. The source (S), drain (D), and gate (G) of the transistor are connected via photolithography/lift-off on a polyimide substrate. The sources of two TFTs are combined to form the output voltage (V_{OUT}) of the inverter, while their common gate serves as the input voltage (V_{IN}). The drains act as the ground and power supply for the inverter, respectively. Figures 2(c) and (d) show that the TFTs exhibit near 0 V flat-band voltage, high switch ratio ($>10^5$), and low operating voltage.

Additionally, we studied the radiation resistance of

979-8-3315-0417-5/25 $31.00 © 2025 IEEE

flexible SWCNT TFTs and CMOS-like inverters in an atmospheric environment, with the irradiation dose rate maintained at 360 rad/s. We measured the electrical properties of SWCNT TFTs and logic circuits after irradiation and post-irradiation heat recovery, with results shown in Figures 2(c), 2(d), and 2(e). After irradiation, the transistor's threshold voltage deviated by 0.5 V, and the inverter's starting inversion voltage shifted by 0.3 V. The on/off current ratio remained relatively stable; this indicates that external strong radiation energy did not cause the generation of dielectric trapped charges or interface trapped charges within the device, and the carbon nanotube transistors could still operate normally and recover after heating. This suggests that our prepared flexible carbon nanotube logic circuits can withstand over 5.5 Mrad of total ionizing dose (TID) irradiation in a real space environment.

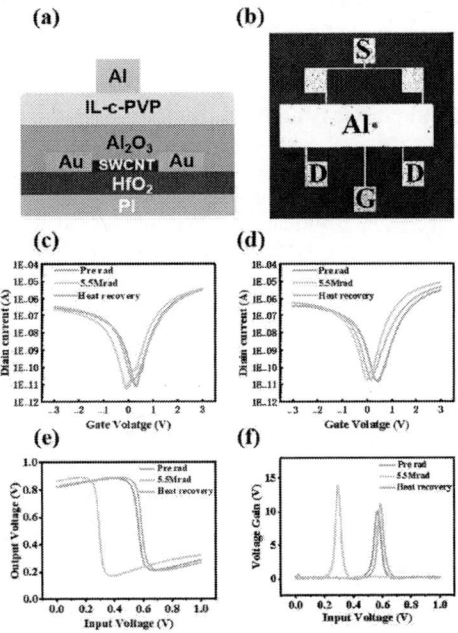

Figure 2 (a) Transistor structure, (b) optical image of a flexible inverter, (c) flexible TFT transfer characteristic curves, (d) the flexible inverter output characteristic curves, and (e) the flexible inverter gain curves before and after irradiation with 5.5 M rad radiation at a dose rate of 360 rad/s, and after thermal recovery post-irradiation.

Conclusion

Based on the use of low-work-function metal Y and Al as the gate electrodes, we developed enhancement-mode broadband photoelectric synaptic SWCNT TFTs by incorporating photosensitive Ag_2S QDs into ILs-c-PVP to form a sensitive dielectric layer and irradiation-hardened logic circuits. The Y-gate TFTs not only operate effectively

under illumination but also feature high switch ratios ($>10^5$) and low operating voltage (-2 V~0 V). Additionally, the device demonstrated excellent synaptic characteristics. The Y-gate carbon nanotube TFT based on Ag_2S QDs exhibited broadband photocurrent response (365 nm ~ 620 nm), ultra-high PPF index (290%), ultra-low single light pulse power consumption (33.9 aJ), and STP to LTP transition. Furthermore, irradiation-hardened flexible top-gate SWCNT TFTs and logic circuits are obtained using low-work-function Al as the gate electrodes and the TIDs can be up to 5.5 Mrad at a dose rate of 360 rad/s. This work suggests that SWCNT TFTs are candidates for ultra-low-power visual bionics chips with irradiation resistance and have significant potential for extreme applications.

Acknowledgments

This work was supported by Natural Science Foundation of China (62274174), the National Key Research and Development Program of China (2020YFA0714700), Key Research and Development Program of Jiangsu Province (BK20232009), a fellowship from the China Postdoctoral Science Foundation (NO: 2023M742559). The authors are grateful for the technical support for Nano-X from Suzhou Institute of Nano-Tech and Nano-Bionics, Chinese Academy of Sciences.

References

[1] J. Si, P. Zhang, C. Zhao, D. Lin, L. Xu "A Carbon-nanotube-based Tensor Processing Unit", Nature Electronics, 7, pp.684-693, 2024.

[2] X. Wang, M. Zhu, X. Li, Z. Qin, G. Lu "Ultralow-Power and Radiation-Tolerant Complementary Metal-Oxide-Semiconductor Electronics Utilizing Enhancement-Mode Carbon Nanotube Transistors on Paper Substrates", Advanced Materials, 34, pp.2204066, 2022.

[3] Z. Wang, M. Li, H. Yang, S. Shao, J. Li "Enhancement-Mode Carbon Nanotube Optoelectronic Synaptic Transistors with Large and Controllable Threshold Voltage Modulation Window for Broadband Flexible Vision Systems", ACS Nano, 18, pp.14298-14311, 2024.

[4] N. Zhang, J. Li, N. Sui, K. Kang, M. Deng "Flexible Solid-Electrolyte-Gated-Dielectric Carbon Nanotube Thin Film Transistors and Integrated Circuits with the Recorded Radiation Tolerance and Reparability", Nano Letters, 24, pp.7688-7697, 2024.

[5] M. Zhu, J. Zhou, P. Sun, L.-M. Peng and Z. Zhang "Analyzing Gamma-Ray Irradiation Effects on Carbon Nanotube Top-Gated Field-Effect Transistors", ACS Applied Materials & Interfaces, 13, pp.47756-47763, 2021.

A Compact Model for GIDL-assisted Erase Transients of 3D MONOS Charge-Trap NAND Flash Memories

Changhyeok Im[1], Sungju Kim[1], and Hyungcheol Shin[1,2]

[1]Seoul National University, Korea, [2]Integra Semiconductor, Ltd.

ABSTRACT

This paper presents an enhanced compact model for describing the time dynamics of GIDL-assisted erase in MONOS 3D NAND flash strings. Compare to previous models, our approach accounts for both electron back-tunneling from metal gate and electron emission from CTN under high electric fields. While increasing the thickness of BOX reduces electron back tunneling, it also slows the erase transient speed. An increase in the gate work function (WF) can suppress back tunneling and delay saturation, thereby improving the overall erase characteristics.

Keywords: GIDL-assisted erase, compact model, MONOS charge-trap NAND flash memories

INTRODUCTION

As 3D NAND flash technology advances to accommodate higher bit density, the necessity for accurate compact models describing NAND memory behavior grows. While recent studies have focused on modeling the program and retention mechanism, relatively little attention has been paid to the erase operation that employ gate-induced drain leakage (GIDL). Despite the development in compact models to explain GIDL-assisted erase processes [1], [2], these models often overlook the interaction between the channel and the cell during erase operation. This work introduces a novel model for the erase mechanism in metal-oxide-nitride-oxide-silicon (MONOS) memory devices, which builds on prior research to address these gaps. The model incorporates electron back-tunneling from the gate and electron emission through tunneling, enhancing previous approaches by analyzing the device's electrostatics through the Poisson equation. Furthermore, it investigates the impact of blocking oxide (BOX) thickness and gate metal WF on device performance, offering insights on optimizing these parameters to improve efficiency.

MODEL DESCRIPTION

The proposed model builds on prior works [1], [2], especially adding electron back-tunneling and modifying electron emission process. Fig. 1 shows architecture of conventional MONOS that model assumed. Each parameter used in this work shown in Table 1. Fig. 2 presents a schematic of the erase mechanism depicted in the proposed model. Holes generated via GIDL tunnel

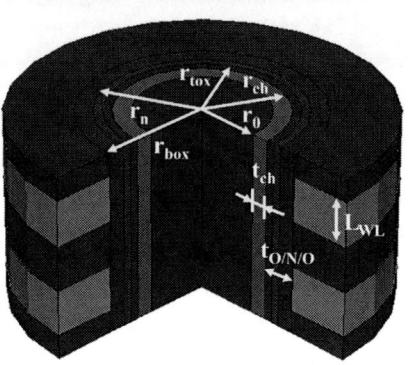

Fig. 1. Conventioanl MONOS 3D NAND flash memory cell architecture.

TABLE I: Nominal value of the NAND flash cell parameters assumed in this compact model.

Parameters	Values	Parameters	Values
r_0	17.5 nm	t_{box}	4.5 nm
t_{ch}	10 nm	L_{WL}	50 nm
t_{tox}	4 nm	$\Phi_{tungsten}$	4.6 eV
t_n	4 nm	V_{BL}	16 V

through the tunneling oxide (TOX) under high electric field, injecting into the CTN and generating a hole current density, derived from [1], [3]. Similarly, electron back-tunneling follows the hole injection mechanism. The electrostatics of the cell are determined by solving the 1-D Poisson equation in cylindrical coordinates, as expressed in [3]. In the erase process, a bias voltage (V_{BL}) is applied to the channel while the gate is grounded. By applying the boundary conditions across each layer, we can obtain the electrostatic potential and electric field distribution as

$$
\begin{cases}
V_{box}(r) = C_1 \ln\left(\frac{r}{r_{box}}\right) \\
V_n(r) = C_1 \ln\left(\frac{r_n}{r_{box}}\right) + C_2 \ln\left(\frac{r}{r_n}\right) \\
\quad + \frac{qn_t}{4\varepsilon_n}\left(r^2 - r_n^2\right) \\
V_{tox}(r) = V_{BL} - C_3 \ln\left(\frac{r_{ch}}{r}\right)
\end{cases}
\tag{1}
$$

$$
\begin{cases}
E_{box}(r) = -C_1 \frac{1}{r} \\
E_n(r) = -C_2 \frac{1}{r} - \frac{qn_t}{2\varepsilon_n} r \\
E_{tox}(r) = -C_3 \frac{1}{r}
\end{cases}
\tag{2}
$$

where $V_{box}(r)$, $V_n(r)$, and $V_{tox}(r)$ represent the electro-

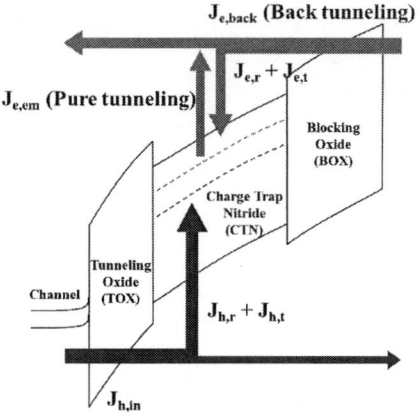

Fig. 2. Concept of the proposed erase model. Blue-line is hole mechanism used in previous model and red-line is electron mechanism newly proposed in this model.

static potential of the BOX, CTN, and TOX. Similarly, $E_{\mathrm{box}}(r)$, $E_{\mathrm{n}}(r)$, and $E_{\mathrm{tox}}(r)$ denote the electric field of each layer. Explicit expressions for constants are provided in the appendix. From Eq. 2, the maximum electric field, F_i, occurs at the metal gate and BOX interface [3]. The electron tunneling process from the metal gate to the BOX can be described by Fowler-Nordheim tunneling written as [3], [4]

$$J_{e,back} = AF_{eq}^2 \exp\left[-\frac{B}{F_{eq}}\right] \quad (3)$$

A and B were calculated as $A = 4.22 \times 10^{-7} \mathrm{A \cdot V^{-2}}$ and $B = 297 \mathrm{MV \cdot cm^{-1}}$, respectively, assuming an effective electron mass of $0.39 m_0$, where m_0 is the free electron mass. In order to simplify the calculations, the equivalent electric field (F_{eq}) derived through the WKB approximation is expressed in terms of F_i [3]. The tunneling occurs opposite to the increasing radius, resulting in an inverse relationship between F_{eq} and F_i. Fig. 3 shows that as r_{box} decreases, F_{eq} shifts proportionally with F_i, which can be fitted to the equation of

$$F_{eq} = -F_i + \frac{V_1}{r_{\mathrm{box}}} \quad (4)$$

where V_1 is a fitting parameter set to 1.05V. Electron emission from CTN contributes to the electron current density [1] as $J_{e,em} = qn_{t,e}e_n$, where $n_{t,e}$ is trapped electron concentration and e_n is electron emission rate. In low electric fields, Poole-Frenkel (PF) emission is the predominant mechanism, whereas in high fields, pure tunneling is the primary process. Given the high electric field present during the erase transient, pure tunneling is the dominant mechanism [5]. Assuming pure tunneling, e_n is calculated according to the following equation of

$$e_n = \nu_0 \cdot \exp\left[-\frac{4\sqrt{2m_e^*}E_T^{\frac{3}{2}}}{3\hbar q F_N}\right] \quad (5)$$

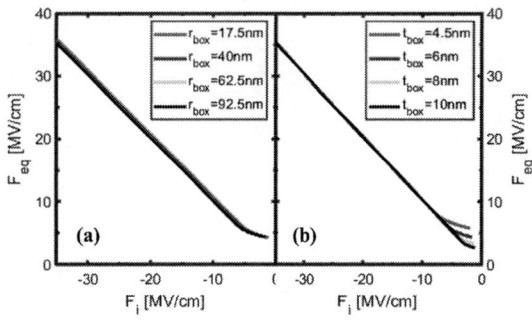

Fig. 3. F_{eq}-versus-F_i relationship for different (a) r_{BOX}'s and (b) t_{BOX}'s.

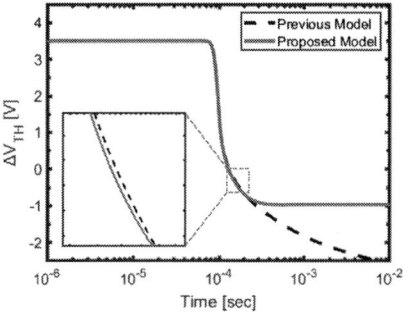

Fig. 4. Comparison of previous model and proposed model. Proposed model has adopted electron back tunneling pure tunneling.

where ν_0 is attempt-to-escape frequency, m_e^* is effective mass of electron, E_T is trap level from conduction band, and F_N is average electric field in nitride. The fraction of hole current captured by the nitride was determined under the assumption that holes can either neutralize traps that are filled with electrons or be captured in vacant traps, thereby becoming positively charged [1]. Similarly, Injected electron current in the CTN either recombines with holes in positively charged traps or are trapped in the neutral traps, resulting in a negative charge, expressed by the following equation of

$$\frac{dn_{t,e}}{dt} = \frac{I_e\sigma_{e,t}\left(N_{t,e} - n_{t,e}\right) - I_e\sigma_{e,r}n_{t,h}}{2\pi q\left(r_{\mathrm{tox}} + \frac{t_\mathrm{n}}{2}\right)L_{\mathrm{WL}}} - \frac{I_{e,em}}{q} \quad (6)$$

$$\frac{dn_{t,h}}{dt} = \frac{I_h\sigma_{h,t}\left(N_{t,h} - n_{t,h}\right) - I_h\sigma_{h,r}n_{t,e}}{2\pi q\left(r_{\mathrm{tox}} + \frac{t_\mathrm{n}}{2}\right)L_{\mathrm{WL}}} \quad (7)$$

where $n_{t,e}$ and $n_{t,h}$ represent the number of trapped electrons and holes under the CTN, while $N_{t,e}$ and $N_{t,h}$ denote the total number of available traps for each. $\sigma_{h,r}$ and $\sigma_{h,t}$ are the hole capture cross sections (CCS) for traps filled with electrons and empty traps and $\sigma_{e,r}$ and $\sigma_{e,t}$ are the electron CCS for traps filled with holes and unoccupied traps. I_e, I_h, and I_{em} are the electron back-tunneling current, hole injection current, and electron emission current, respectively, and L_{WL} is word line length. Note that when negative values occur,

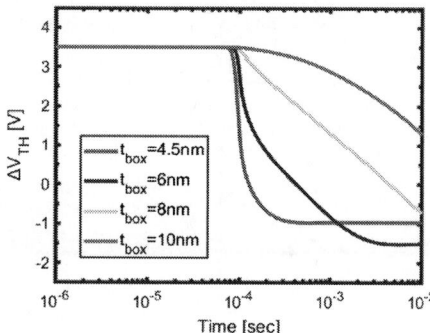

Fig. 5. Simulation results of erase transients for ΔV_{TH} with different values of t_{box}.

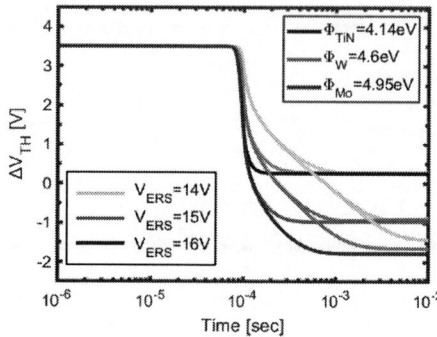

Fig. 6. Simulation results of erase transients for ΔV_{TH} with different gate metal work functions and erase bias condition.

$n_{t,e}$ and $n_{t,h}$ are set to zero, since emitted electrons and holes cannot exceed the injected amount. The threshold voltage shift is then calculated as in [1].

SIMULATION RESULTS

Fig. 4 has compared erase simulations from the proposed model and the previous model [2], based on threshold voltage plots. The proposed model has predicted earlier saturation of the erase transient due to electron back-tunneling [4], impacting the final stages of the erase process without affecting overall erase speed. In contrast, pure tunneling has influenced the erase speed and has delayed saturation induced by back-tunneling, as shown in the insert. A comparative analysis of the PF emission versus the combination of pure tunneling and back-tunneling models has demonstrated these behaviors. Fig. 5 has shown that back-tunneling is primarily influenced by BOX thickness, with a thicker BOX reducing back-tunneling and improving the erase process. However, increasing the BOX thickness has decelerated the erase speed, limiting its effectiveness. Adjusting the gate metal WF has been shown to suppress back-tunneling [4]. Fig. 6 has investigated the impact of WF using various metals such as tungsten, titanium nitride, and molybdenum. The model has illustrated that higher WF has suppressed back-tunneling and has delayed reaching saturation, thereby enhancing the erase characteristics. However, changes in erase voltage have become less efficient when back-tunneling is present, risking incomplete cell erase [5]. The WF primarily has affected back-tunneling, thus delaying the attainment of saturation states and enabling the device to reach fully erased states without reducing erase speed. In contrast to low WF metals, high WF metals such as molybdenum have not reached saturation at the same ΔV_{th}, exhibiting a behavior similar to that of the reference [5].

CONCLUSION

The proposed compact model explains the erase mechanism of 3D MONOS NAND flash memory, focusing on electron back-tunneling from the metal gate and emission in the CTN layer. By solving the Poisson equation, it predicts early erase saturation due to back-tunneling and investigates the impact of BOX thickness and gate WF control in order to overcome the adverse effects of back-tunneling. This model thus provides an accurate characteristics of MONOS memory and offers insights for enhancing performance.

APPENDIX

The explicit constants of electrostatic potential and electric field are described as

$$
\begin{aligned}
C_1 &= \frac{V_{BL}}{\alpha} + \frac{qn_t}{2\varepsilon_n\alpha}\left[r_n^2 \ln\left(\frac{r_{tox}}{r_n}\right) \right. \\
&\quad \left. - \frac{1}{2}\left(r_{tox}^2 - r_n^2\right)\left(1 + 2\frac{\varepsilon_n}{\varepsilon_{tox}}\ln\left(\frac{r_{ch}}{r_{tox}}\right)\right) \right] \\
C_2 &= C_1\frac{\varepsilon_{box}}{\varepsilon_n} - \frac{qn_t}{2\varepsilon_n}r_n^2 \\
C_3 &= C_1\frac{\varepsilon_{box}}{\varepsilon_n} + \frac{qn_t}{2\varepsilon_{tox}}\left(r_{tox}^2 - r_n^2\right) \\
\alpha &= \frac{\varepsilon_{box}}{\varepsilon_n}\ln\left(\frac{r_{tox}}{r_n}\right) + \ln\left(\frac{r_n}{r_{box}}\right) + \frac{\varepsilon_{box}}{\varepsilon_n}\ln\left(\frac{r_{ch}}{r_{tox}}\right)
\end{aligned}
$$

REFERENCES

[1] G. Malavena, A. Mannara, A. L. Lacaita, A. Sottocornola Spinelli, and C. Monzio Compagnoni, "Compact modeling of GIDL-assisted erase in 3-D NAND Flash strings," J. Comput. Electron., 18, pp. 561–568 (2019).

[2] K. Lee and H. Shin, "Distinguishing capture cross-section parameter between GIDL erase compact model and TCAD," Jpn. J. Appl. Phys., 60, pp. 124002 (2021).

[3] S. M. Amoroso, C. M. Compagnoni, A. Mauri, A. Maconi, A. S. Spinelli, and A. L. Lacaita, "Semi-Analytical Model for the Transient Operation of Gate-All-Around Charge-Trap Memories," IEEE Trans. Electron Devices, 58, pp. 3116–3123 (2011).

[4] S. H. Jeon, J. H. Han, J. H. Lee, S. M. Choi, H. S. Hwang, and C. W. Kim, "Impact of metal work function on memory properties of charge-trap flash memory devices using fowler-nordheim P/E mode," IEEE Electron Device Lett., 27, pp. 486–488 (2006).

[5] M. Kim, S. Kim, and H. Shin, "A Compact Model for ISPP of 3-D Charge-Trap NAND Flash Memories," IEEE Trans. Electron Devices, 67, pp. 3095–3101 (2020).

From Monitoring to Modulation: Intelligent Wearable Devices for Health Sensing and Drug Delivery

Yuanzhe Li[1], Shirong Qiu[1], Zhou Jiang[1,2], Cunman Liang[1,2], Yixin Qi[1], Ni Zhao[1,2]

[1] Department of Electronic Engineering, The Chinese University of Hong Kong, Hong Kong SAR
[2] Hong Kong Centre for Cerebro-cardiovascular Health Engineering (COCHE), Hong Kong SAR

Abstract

In this talk, I present advancements in intelligent wearable devices (IWDs) that integrate sensing and drug delivery functions to address critical challenges in calibration, treatment efficacy, and device footprint. Three approaches are introduced: a dual-modality piezoelectric microsystem for long-term blood pressure monitoring, a polymeric electrode for ultrasound-enhanced iontophoresis, and a novel organic electrochemical transceiver for seamless sensing and therapeutic agent release. These devices advance IWDs toward effective, adaptive physiological monitoring and modulation.

Keywords: Sensors, Wearable medical devices

Introduction

Recent developments in wearable medical devices have shifted focus from simple data acquisition to active intervention, including drug delivery, thereby creating intelligent wearable devices (IWDs) with real-time sensing and responsive therapeutic functionalities. Challenges remain in the areas of calibration for personalized measurements, effective drug release, and streamlined device architecture. Addressing these limitations, I present three solutions: a demographic-adaptive blood pressure (BP) monitoring system, a polymeric electrode for efficient transdermal drug delivery, and an integrated organic electrochemical transceiver (OECTc) that consolidates sensing and release functions into one compact device. These technologies collectively promote advanced and practical applications of IWDs in continuous health monitoring and adaptive therapy.

Results and Discussion

The first approach, a conformal and stretchable dual-modality piezoelectric microsystem (DMPM), enables long-term BP monitoring without recalibration. Incorporating pulse waveform pressure sensors and an ultrasound transducer, DMPM tracks local pulse wave velocity and vessel diameter, thereby adapting to individual vascular profiles. This device is accompanied by an adaptive BP extraction algorithm that incorporates demographic feature learning and time decay compensation to adjust for dynamic conditions and avoid signal distortion. Compared to traditional cuffless monitors, DMPM excels in continuous BP measurement with Grade A accuracy, contributing to its suitability for everyday, wearable applications.

The second approach introduces a biocompatible polymeric electrode optimized for iontophoresis, essential for drug administration via the skin. By doping the PEDOT polymer with dopamine, this device achieves prolonged stability under high current densities. Integrated with a flexible piezoelectric transducer, the electrode facilitates ultrasound-enhanced iontophoresis, ensuring precise and efficient drug delivery while maintaining skin compatibility. This advancement supports transdermal drug delivery applications requiring high performance and reduced discomfort.

Finally, I discuss the OECTc—a monolithically fabricated IWD that performs both sensing and drug release within a single device structure. Leveraging the electronic wettability switch (EWS) in poly(3-hexylthiophene), the OECTc modulates therapeutic molecule release rates based on applied voltages, while concurrently maintaining biosensing capabilities. This dual functionality enables sense-and-release operations in real time, highlighting the potential for applications in managing diabetes and neurotransmitter modulation. The OECTc's compact, all-in-one design exemplifies an efficient and scalable approach to IWDs, minimizing energy consumption and fabrication complexity.

Conclusion

The three novel approaches outlined advance IWD technology by enhancing continuous monitoring, transdermal drug delivery, and integrated functionality. Together, these devices enable personalized, precise, and adaptive healthcare interventions, providing a pathway toward a new generation of wearable systems that not only monitor physiological conditions but also actively modulate them. This work sets the foundation for future IWDs that are adaptable to various clinical and physiological needs, significantly impacting fields such as chronic disease management and neurotherapy.

Acknowledgments

The studies presented here were supported by the C-type Project from Science, Technology and Innovation Commission of Shenzhen Municipality (Reference No. 202205303000149), the Hong Kong Innovation and Technology Fund (ITS/218/22), and the Hong Kong Centre for Cerebro-cardiovascular Health Engineering under the InnoHK Scheme of Hong Kong SAR.

979-8-3315-0417-5/25 $31.00 © 2025 IEEE

979-8-3315-0417-5/25 $31.00 © 2025 IEEE

Ultrafast Photoresponse of Vertical Diodes Utilizing WSe₂/ITO Schottky Junctions

Xixi Jiang[1,2], Jingli Wang[4], Shukui Zhang[1,2], Qingqing Sun[3], Jie Wei Chen[1,2], Yang Chai[1,2]

[1]Department of Applied Physics, The Hong Kong Polytechnic University, Hong Kong, China. ychai@polyu.edu.hk. [2]Joint Research Center of Microelectronics, The Hong Kong Polytechnic University, Hong Kong, China. [3] School of Microelectronics, Fudan University, Shanghai, 200433, China. [4] Frontier Institute of Chip and System, Fudan University, Shanghai, 200433, China

Abstract

Schottky diodes are widely utilized in optoelectronics. In this study, we fabricated a vertical photodiode using a WSe₂/ITO Schottky junction. The device exhibited an impressively low dark current of 0.7 pA and an ultrafast light response time of 13 ns, with a broadband photo-response across the entire visible spectrum. These exceptional results offer a promising approach for developing high-performance photodiodes in the future.

Keywords: Schottky diode; tungsten diselenide (WSe₂); ultrafast photoresponse

Introduction

Schottky diodes have been widely used in high-frequency switching applications, rectifiers, demodulators, photodetectors, sensors.[1] At present, the commercialized Schottky diodes are mostly made of bulk materials like Si, 4H-SiC, ZnO and GaN. To form a good ohmic contact, a thin semiconductor epitaxial layer usually deposited on a highly doped substrate in the device fabrication. However, the thick substrate makes further miniaturize difficult. The scheme of applying 2D semiconductor materials to Schottky diodes is feasible because 2D materials can be easily stacked on the bottom metal electrode through van der Waals interaction, ohmic contact or Schottky contact can be formed by changing the height of Schottky barrier between different metals and semiconductors.

In this work, we developed a vertical Schottky diode with ITO/p-WSe₂/Pt device structure. Due to the outstanding optical properties of multilateral p-WSe₂, the photodetector achieves an impressively low dark current of just 0.7 pA, a rapid light response of approximately 13 ns, and a broadband photo-response spanning the entire visible spectrum. These results provide a new way to realize optoelectronic devices with excellent performance in the future.

Device Fabrication

The Pt/ p-WSe₂ /ITO Schottky diodes were fabricated as below. Firstly, a trench with a depth of 20 nm was fabricated by UV-lithography and reactive ion etching, the Ti/ Pt metal electrodes were deposited by sputtering to filling and leveling the trench. Then, the p-WSe₂ flakes were prepared by mechanical exfoliation and placed on the bottom electrode by transferring. Finally, the top electrode ITO (~90 nm) were formed by electron beam lithography and sputtering.

Results and Discussion

Figure 1 shows a schematic diagram of the ITO/p-WSe₂/Pt vertical Schottky diode. The device was fabricated by stacking a p-WSe₂ flake onto the Pt bottom electrode through van der Waals interaction, and the ITO top electrode is prepared on the area where the p-WSe₂ flake overlaps with Pt electrode. The energy band diagram of ITO/p-WSe₂/Pt Schottky junction is shown in Figure 2. According to the measurement results of ultraviolet electron spectroscopy (UPS), Figure 3 shows the Fermi level (E_F) of the charge carrier extraction layer which is obtained by subtracting the binding energies of the secondary electron cutoffs (in the range of 15-18 eV) from the excitation energy (21.2 eV) of HeI UPS spectra. And Figure 4 demonstrates the energy difference between the valance band maximum (E_{VB}) and E_F which was derived from the low binding energy tails (in the range below 4 eV). Conduction band (E_{CB}) edge values were obtained by adding the optical band gaps (E_g) to E_{VB}. The valence band maximum (E_{VB}) and Fermi energy (E_F) of the bulk p-WSe₂ crystal is -5.8 eV and -5.5 eV, respectively. The bandgap (E_g) of the p-WSe₂ is about 1.3 eV according to a published work.[2] So the conduction band minimum (E_c) of the p-WSe₂ is about -4.5V. The calculation method of band edge position has been discussed in our previous work.[3] When p-WSe₂ contacts the bottom Pt electrode, the Schottky barrier height is as low as 0.15 eV, which provides an approximate ohmic contact. On the other hand, the Schottky barrier height of ITO/p-WSe₂ junction is about 1.1 eV, which shows a typical Schottky contact. Although the energy band diagram will be slightly different due to the existence of Fermi pinning at the interface between the electrodes and 2D material, this supplies a solid foundation for achieving a vertical Schottky diode with high performance.

The STEM image and the corresponding energy-dispersive X-ray spectrometry (EDS) elemental map of the ITO/p-WSe₂/Pt heterojunction are shown in Figure 5. This demonstrates the uniform distribution of In, O, Sn, W, Se, Pt and Ti with no chemical reaction at the interface observed. The bottom Pt electrode and the top ITO electrode have formed good contact with p-WSe₂ flake. The thickness of the p-WSe₂ flake was found to be 48 nm. The thicknesses of ITO electrode and Pt electrode are 90 nm and 20 nm, respectively. Besides, we can see that the interface of the device is very clean and there is no residual photoresist, which is very important for obtaining high-performance optoelectronic devices.

Figure 6 shows I-V curves of the device under dark and illumination of 520 nm laser with incident power of 1 mW (the spot diameter is 2~3 um). We could find that a distinct photocurrent was generated under illumination, and the dark current was as low as 0.7 pA, which was benefited from the high Schottky barrier at the ITO/p-WSe$_2$ interface. At a light intensity of 1 mW, the photoresponsivity of the device is approximately 1mA/W. Figure 7 shows the photocurrent of the device under different light wavelengths at zero bias voltage, with a constant light intensity of 30 µW for the light source. Due to the limitation of the grating in wavelength adjustable laser, only the photoelectric response in the range of 500-1000 nm can be measured. According to previous work reports [4], the device in the spectral range of 400-500 nm should also have good photoelectric response. These results show that the device can exhibit high sensitivity across the entire visible spectrum.

Figure 8 shows the time response measurement of the Pt/p-WSe$_2$/ITO photodiode under illumination with 520 nm wavelength (100 mW/cm^2 power density) at $V_{bias} = 0$ V. We can find that the device will respond quickly as the light is turned on or off. The time-resolved photocurrent is shown in Figure 9, the rise time of 12 ns and the fall time of 13 ns are achieved, if it is not for the limitation of the emission frequency of pulse lasers (pulse width of 39 ns) and the sampling frequency of oscilloscopes, it may be possible to obtain a more faster response speed. As shown in Table 1, these results are very excellent compare to most of the photodetectors that have been reported.

The spatially resolved photocurrent mapping of the Pt/p-WSe$_2$/ITO photodiode under illumination with a 520 nm laser beam at $V_{bias} = 0$ V is shown in Figure 10. These results demonstrate that photocurrent is mainly generated in the region where Pt, p-WSe$_2$ and ITO are stacked, and there is almost no significant photocurrent generation outside the overlapping region.

Conclusion

In summary, we have developed a Schottky diode using the ITO/p-WSe$_2$/Pt vertical structure. The energy band alignment of the device structure has been analyzed, which shows a typical Schottky contact at the ITO/p-WSe$_2$ junction and an approximate ohmic contact at the p-WSe$_2$/Pt junction. Thanks to the excellent optical characteristics of multilateral p-WSe$_2$, the device with a broadband photo-response in the visible light region. The photo-detector achieves a dark current as low as 0.7 pA and a very fast light response of 13 ns. These findings suggest that our device concept and structure have paved the way for high-performance optoelectronic devices.

Acknowledgments

The authors gratefully acknowledge financial support from the China Postdoctoral Science Foundation (NO. 2020TQ0065), MOST National Key Technologies R&D Programme (SQ2022YFA1200118-04), Research Grant Council of Hong Kong (CRS_PolyU502/22) and The Hong Kong Polytechnic University (1-ZE1T, YXBA and WZ4X) , National Science Foundation of China (NSFC) for Distinguished Young Scholar (62425405).

References

[1] Georgiadou, D. G.; Semple, J.; Sagade, A. A.; Forstén, H.; Rantakari, P.; Lin, Y.-H.; Alkhalil, F.; Seitkhan, A.; Loganathan, K.; Faber, H.; Anthopoulos, T. D., "100 GHz zinc oxide Schottky diodes processed from solution on a wafer scale". Nature Electronics, 3 (11), pp. 718-725 (2020).

[2] Yang, S. J.; Park, K. T.; Im, J.; Hong, S.; Lee, Y.; Min, B. W.; Kim, K.; Im, S., "Ultrafast 27 GHz cutoff frequency in vertical WSe$_2$ Schottky diodes with extremely low contact resistance". Nature Communnication, 11 (1574), pp. 1-9 (2020).

[3] Jiang, X.; Shi, X.; Zhang, M.; Wang, Y.; Gu, Z.; Chen, L.; Zhu, H.; Zhang, K.; Sun, Q.; Zhang, D. W., "A Symmetric Tunnel Field-Effect Transistor Based on MoS2/Black Phosphorus/MoS$_2$ Nanolayered Heterostructures". ACS Applied Nano Materials, 2 (9), pp. 5674-5680 (2019).

[4] Luo, M.; Wu, F.; Long, M.; Chen, X., "WSe$_2$/Au vertical Schottky junction photodetector with low dark current and fast photoresponse". Nanotechnology, 29 (44), pp. 444001 (2018).

[5] Li, W.; Liu, L.; Tao, Q.; Chen, Y.; Lu, Z.; Kong, L.; Dang, W.; Zhang, W.; Li, Z.; Li, Q.; Tang, J.; Ren, L.; Song, W.; Duan, X.; Ma, C.; Xiang, Y.; Liao, L.; Liu, Y., "Realization of Ultra-Scaled MoS$_2$ Vertical Diodes via Double-Side Electrodes Lamination". Nano Lett, 22 (11), pp. 4429-4436 (2022).

[6] Gong, F.; Fang, H.; Wang, P.; Su, M.; Li, Q.; Ho, J. C.; Chen, X.; Lu, W.; Liao, L.; Wang, J.; Hu, W., "Visible to near-infrared photodetectors based on MoS$_2$ vertical Schottky junctions". Nanotechnology, 28 (48), pp. 484002 (2017).

[7] Wei, X.; Yan, F.; Lv, Q.; Shen, C.; Wang, K., "Fast gate-tunable photodetection in the graphene sandwiched WSe$_2$/GaSe heterojunctions". Nanoscale, 9 (24), pp. 8388-8392 (2017).

[8] Tang, Y.; Wang, Z.; Wang, P.; Wu, F.; Wang, Y.; Chen, Y.; Wang, H.; Peng, M.; Shan, C.; Zhu, Z.; Qin, S.; Hu, W., "WSe$_2$ Photovoltaic Device Based on Intramolecular p-n Junction". Small 2019, 15 (12),

Table 1: Performance comparisons of the photodiodes using 2D Material

Device Structure	I_{dark} (pA)	Response Time (µs)	Ref.
ITO/WSe$_2$/Au	1	50	[4]
Pt/MoS$_2$/Ag	100	0.02	[5]
ITO/MoS$_2$/Au	1	64	[6]
WSe$_2$/GaSe	100	30	[7]
WSe$_2$ p-n junction	1000	200	[8]
This work	**0.7**	**0.013**	

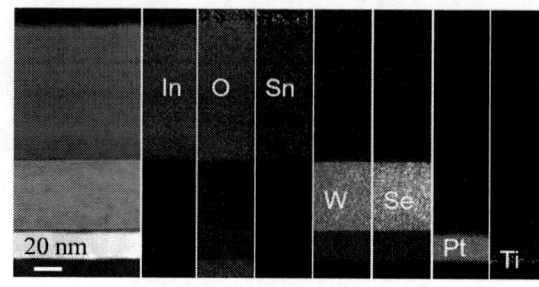

Figure 5. STEM image and the corresponding EDS elemental map of the ITO/p-WSe$_2$/Pt heterojunction

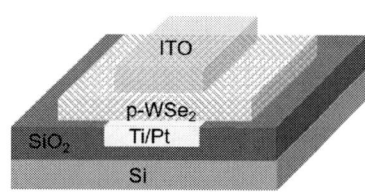

Figure 1. Schematic diagram of an ITO/p-WSe$_2$/Pt vertical Schottky diode.

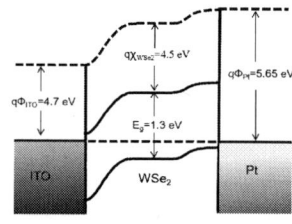

Figure 2. Band diagram of ITO/p-WSe$_2$/Pt Schottky junction.

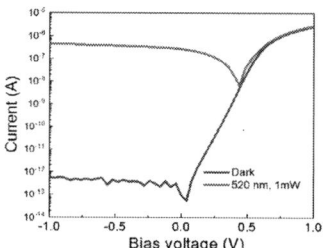

Figure 6. I-V curves of the device at dark condition and under illumination.

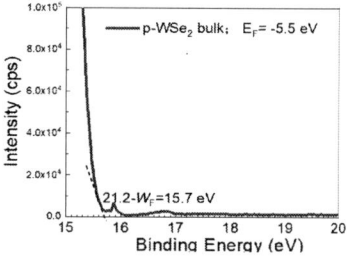

Figure 3. UPS spectrum used for calculating E_F of p-WSe$_2$.

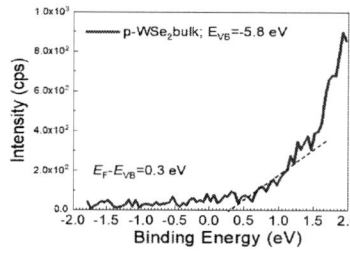

Figure 4. UPS spectrum used for calculating E_{VB} of p-WSe$_2$.

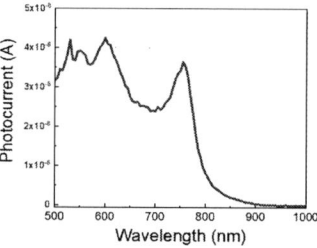

Figure 7. Photocurrent versus wavelength with 500 nm to 1000 nm.

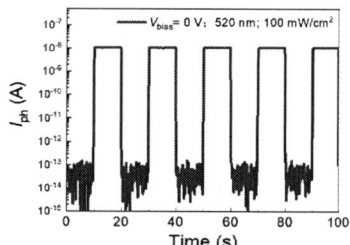

Figure 8. Photoswitching response of the devise at Vbias = 0 V.

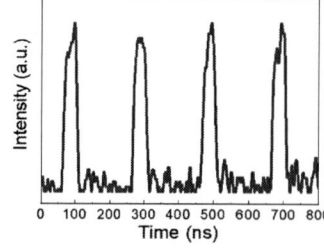

Figure 9. Time resolved photo-response of the device at V_{bias} = 0 V.

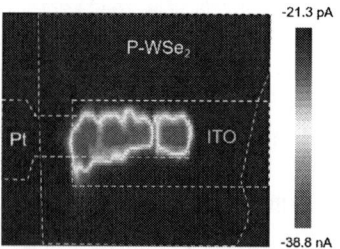

Figure 10. Spatial photocurrent mapping of the Pt/p-WSe$_2$/ITO photodiode

979-8-3315-0417-5/25 $31.00 © 2025 IEEE

TDA (Thermal Design Automation) for Multiscale Thermal Managements of GaN HEMTs

Bingyang Cao[1]

[1]The Key Laboratory of Thermal Science and Power Engineering of Education of Ministry, Department of Engineering Mechanics, Tsinghua University, Beijing 100084, China

Abstract

A TDA (Thermal Design Automation) system is proposed for multiscale thermal simulation and design of Gallium Nitride (GaN) High Electron Mobility Transistors (HEMTs). By integrating first-principle calculations, machine learning potential-driven lattice dynamics, phonon Monte Carlo, and finite element methods, etc., TDA enables precise device-level electro-thermal simulations and designs. This approach has the potential to significantly enhance the accuracy and efficiency of thermal management for GaN HEMTs and other electronic devices.

Keywords: TDA (Thermal Design Automation), GaN HEMTs, multiscale simulation, thermal management

Introduction

Gallium Nitride (GaN) high-electron-mobility transistors (HEMTs) are essential for high-frequency and high-power applications due to their excellent electronic properties[1]. However, significant overheating has emerged as a major limitation for GaN HEMTs, affecting both performance and reliability[2]. Effective thermal management, aimed at dissipating heat from hotspots and reducing thermal resistance, is therefore crucial for optimizing device design.

Current thermal management strategies focus primarily on external approaches, such as air cooling, microchannel cooling, and jet impingement cooling[3]. These methods typically act as heat sinks installed externally to the package of pre-designed devices. However, as device power densities increase and external cooling technologies mature, the main thermal bottleneck has shifted to internal thermal resistance, particularly in the heat conduction path from the transistor junction to the substrate. Thus, reducing internal thermal resistance through near-junction thermal management is crucial for overcoming the thermal challenges of GaN HEMTs.

Device-level thermal simulation and design of GaN HEMTs present several challenges[4]. Firstly, heat generation primarily arises from Joule heating due to electron-phonon interactions. Accurate heat profile prediction and understanding thermal effects require electro-thermal co-simulations. After the heat is generated, it spreads from the small hotspot to the entire device, leading to significant thermal spreading resistance. Since the layer size and thermal source dimensions are comparable to phonon mean-free-paths (MFPs), Fourier's law of heat conduction fails, necessitating consideration of phonon ballistic transport. Furthermore, modeling the phonon transport in the whole device is computationally demanding, thus necessitating multiscale approaches. Moreover, GaN HEMTs involve multiple thin-film epitaxial layers, including the substrate, buffer, and barrier layers, etc. Thermal boundary resistance (TBR) between different layers is a critical factor in heat dissipation paths, influenced significantly by phonon scattering at interfaces, which is difficult to predict due to variations in interfacial layers and manufacturing processes.

Consequently, the accurate thermal simulation and design of GaN HEMTs require electro-thermal simulation, precise modeling of phonon thermal spreading, TBR prediction, and a multiscale methodology integrating various simulation techniques. However, current mainstream approaches, which rely solely on macroscopic method-based commercial software, fall short in effective thermal design. To address these challenges, we propose a TDA (Thermal Design Automation) system for multiscale thermal management of GaN HEMTs, integrating various methods to tackle these issues. The remainder of this paper provides an overview of the TDA system, including its background and applications.

Results and Discussion

A. Self-heating effects

The self-heating effect in GaN HEMTs arises from the scattering of high-energy carriers with the lattice, which intensifies in regions of high electric fields. This loss of carrier kinetic energy can create localized hotspots, which in GaN HEMTs typically occur in the drain-side region beneath the gate, as shown in Fig. 1(a). Figure 1(b) illustrates the heat generation pattern beneath the heterojunction interface, as highlighted by the green dashed box in Fig. 1(a), with a lateral characteristic size of approximately 100 nm. Furthermore, heat generation in HEMT devices is bias-dependent, and in the unsaturated state, uniform heat distribution across the channel should not be overlooked[5].

To reflect the self-heating effect, electro-thermal simulations are essential. The most widely used electro-thermal simulation method relies on the drift-diffusion model, which involves solving the Poisson equation and the heat conduction equation to determine the electric field and temperature distribution. Alternatively, the heat conduction equation can be replaced with phonon transport simulations, providing a more accurate representation of phonon dynamics. This approach better captures self-heating effects and enables precise temperature predictions.

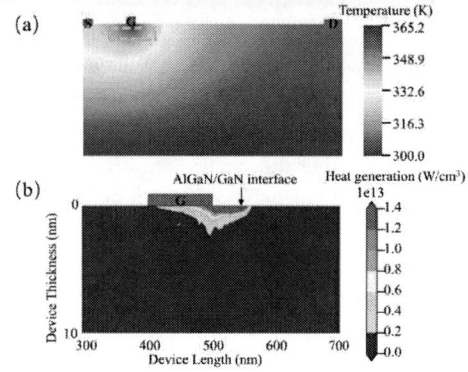

Fig. 1: (a) Temperature and (b) heat generation distribution in GaN HEMTs.

B. Non-Fourier heat conduction and heat spreading

Thermal spreading resistance occurs when heat is transferred from a small heat source region to a larger area. As shown in Fig. 1, the heat generation is concentrated at the drain-side gate edge with a narrow width, which is much smaller than the device dimensions. Consequently, significant thermal spreading resistance dominates the heat transfer in GaN HEMTs. Most studies on thermal spreading resistance are based on Fourier's law of heat conduction:

$$\mathbf{q} = -\kappa \nabla T , \qquad (1)$$

where $\mathbf{q}$ is the heat flux, κ is the thermal conductivity, and T is the temperature. Fourier's law applies only to macroscopic systems where the characteristic size is much larger than the phonon MFPs. In GaN HEMTs, the heat source size is comparable to phonon MFPs, resulting in significant ballistic phonon transport, which can largely elevate hotspot temperatures[6]. The GaN HEMTs consist of multiple epitaxial layers with thicknesses ranging from tens of nanometers to micrometers, also comparable to phonon MFPs. The frequent phonon-boundary scattering reduces the effective thermal conductivities of the nanofilms, which must be considered in thermal simulations of GaN HEMTs.

To account for ballistic phonon transport, the phonon Boltzmann transport equation (BTE) should be solved. Under the single-mode relaxation time approximation, the phonon BTE is given by[7]:

$$\frac{\partial f}{\partial t} + \mathbf{v} \cdot \nabla f = \frac{f^{eq} - f}{\tau} , \qquad (2)$$

where f is the phonon distribution function, f^{eq} is the Bose-Einstein distribution function, $\mathbf{v}$ is the phonon group velocity, and τ is the phonon relaxation time. Phonon Monte Carlo (MC) is an efficient approach to solving the phonon BTE, which can adopt ab initio properties obtained from first-principles calculations as input for accurate full-band phonon transport predictions.

C. Thermal boundary resistance

TBR takes up a large portion of the total thermal resistance of GaN HEMTs. Thermal boundary conductance (the inverse of TBR) can be determined by the Landauer formula:

$$G = \frac{1}{V} \sum_{\lambda} \hbar \quad \mathrm{T} \, \frac{\partial f^{eq}}{\partial T} , \qquad (3)$$

in which V is the volume of the primitive cell, λ represents the phonon mode, ω_λ is the phonon angular frequency, $v_{\lambda z}$ is the normal component of the phonon group velocity, T_λ is the phonon transmittance at the interface. Accurately predicting mode-resolved phonon transmittance is challenging due to the influence of interfacial atomic structures.

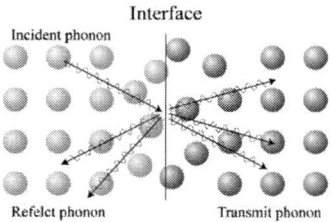

Fig. 2. Illustration of the lattice dynamics method.

The lattice dynamics (LD) method computes the mode-resolved transmittance. Yang et al. improved the stability by introducing energy-conservation constraints and enhancing the algorithm[8]. As shown in Fig. 2, the inputs include atomic-scale modeling of the interfacial structure and interatomic forces with quantum mechanical precision, enabled by machine learning (ML) potentials. This atomic-scale modeling allows LD to consider the subtle effects of interfacial structures on TBR, such as amorphous layers. The outputs include the mode-resolved phonon scattering spectrum, phonon transmittance, and TBR, which bridges the gap between full-band phonon MC simulations and ab initio material properties when involving multiple materials.

D. TDA and applications

As illustrated before, accurate thermal simulation and design of GaN HEMTs require electro-thermal simulation, phonon BTE-based modeling of near-junction thermal spreading, and accurate TBR prediction. The proposed TDA system integrates different simulation techniques for multiscale thermal management, as shown in Fig. 3. Heat generation is predicted by solving fundamental semiconductor equations. In regions near the heat source, where ballistic effects are

979-8-3315-0417-5/25 $31.00 © 2025 IEEE

significant, full-band phonon MC simulations are conducted. This accurately predicts temperatures of the hotspot caused by near-junction phonon thermal spreading, and the effective thermal conductivities of epitaxial layers along with the TBC between layers. Leveraging ab initio phonon properties and mode-resolved transmittance from first-principle calculations and ML-driven LD, respectively, the full-band MC simulation has the capability to account for the effects of doping, isotopes, defects, and interfacial atomic structures, etc. The MC methods are then coupled with electrical simulations for electrothermal simulations and co-designs. For device-level multiscale simulations, FEM can be coupled with MC through the boundary conditions[9].

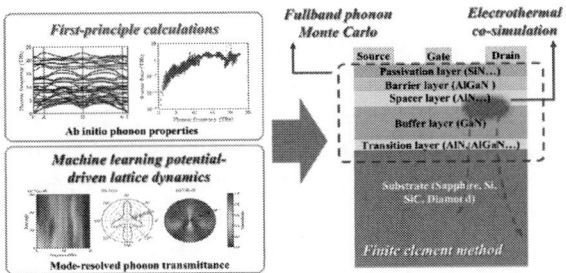

Fig. 3: TDA framework for multiscale thermal simulations of GaN HEMTs

TDA can be employed for multiscale thermal simulation and design of GaN HEMTs, enabling the prediction of effective thermal properties, investigation of electrothermal transport mechanisms, tuning of interfaces, and improvement of device designs. We have applied the TDA system in several scenarios. First, we compared the effective thermal conductivities and TBCs in GaN/AlN/SiC systems predicted by TDA, demonstrating good agreement with experimental results obtained using the multi-pulse thermoreflectance imaging method[10]. Then, we conducted electrothermal simulations using TDA and found that an asymmetric field plate structure has significant potential for mitigating self-heating effects in HEMT devices. With a slant angle of 6° and an FP length of 1200 nm, the maximum heat generation density was reduced by 50%. MC simulation results indicated that the slanted field plate reduced hotspot temperature rise by over 33%, which is double the optimization predicted by Fourier's law (16%). Additionally, experimental characterization combined with MACE ML potential-driven LD modeling and multiscale thermal simulations revealed that regulating the amorphous transition layer reduced the total thermal resistance of gallium oxide devices by more than 20%. These examples highlight just a few applications of TDA, which has many other potential uses in thermal management and device optimizations.

Conclusion

We propose a TDA system for the multiscale thermal management of GaN HEMTs. This approach integrates first-principle calculations, ML potential-driven LD, full-band phonon MC simulations, and FEM, enabling electrothermal co-design, accurate predictions of effective thermal properties, precise modeling of self-heating and near-junction phonon thermal spreading processes, TBR tuning and multiscale device-level thermal designs for GaN HEMTs and other wide-bandgap semiconductor power devices. The proposed TDA framework holds large potential to improve the accuracy and efficiency of thermal management for these devices.

Acknowledgments

This study was financially supported by the National Natural Science Foundation of China (Nos. 52425601, 52327809, 52250273), National Key Research and Development Program of China (No. 2023YFB4404104), and Beijing Natural Science Foundation (No. L233022).

References

[1] M. Meneghini, C. De Santi, I. Abid, M. Buffolo, M. Cioni, R. A. Khadar, L. Nela, N. Zagni, A. Chini, F. Medjdoub, and G. Meneghesso, "GaN-based power devices: Physics, reliability, and perspectives", J. Appl. Phys., 130(18) (2021).

[2] J. Cho, Z. Li, M. Asheghi, and K. E. Goodson, "Near-junction thermal management: Thermal conduction in gallium nitride composite substrates", Annual Review of Heat Transfer, 18 (2015).

[3] Z. Zhang, X. Wang, and Y. Yan, "A review of the state-of-the-art in electronic cooling", e-Prime - Advances in Electrical Engineering, Electronics and Energy, 1, 100009 (2021).

[4] Y. C. Hua, Y. Shen, Z. L. Tang, D. S. Tang, X. Ran, and B. Y. Cao, "Near-junction thermal managements of electronics", Advances in Heat Transfer, 56, 355-434 (2023).

[5] Y. Shen, X.S. Chen, Y.C. Hua, H.L. Li, L. Wei, and B.Y. Cao, "Bias dependence of non-Fourier heat spreading in GaN HEMTs", IEEE Trans. Electron Devices, 70(2): 409-417 (2023).

[6] Y.C. Hua, H.L. Li, and B.Y. Cao, "Thermal spreading resistance in ballistic-diffusive regime in GaN HEMTs", IEEE Trans. Electron Devices, 66(8): 3296-3301 (2019).

[7] H. Bao, J. Chen, X.K. Gu, and B.Y. Cao, "A review of simulation methods in micro/nanoscale heat conduction", ES Energy & Environment, 1: 16-55 (2018).

[8] H.A. Yang, B.Y. Cao, "Mode-resolved phonon transmittance using lattice dynamics: Robust algorithm and statistical characteristics", J. Appl. Phys., 134: 155302 (2023).

[9] H.L. Li, Y. Shen, Y.C. Hua, S.L. Sobolev, and B.Y. Cao, "Hybrid Monte Carlo-diffusion studies of modeling self-heating in ballistic-diffusive regime for GaN HEMTs", J. Electron. Packag., 145: 011203 (2023).

[10] Z.K. Liu, G. Yang, and B.Y. Cao, "Pulsed thermoreflectance imaging for thermophysical properties measurement of GaN epitaxial heterostructures", Rev. Sci. Instrum., 94(9): 094902 (2023).

The Observation of 2D Electron Gas at Ga_2O_3/IGZO Interface

Jinxiu Zhao, Zhiyu Lin, Chen Wang, Mengwei Si*

Department of Electronic Engineering and State Key Laboratory of Radio Frequency Heterogeneous Integration, Shanghai Jiao Tong University, Shanghai, China.

Email: mengwei.si@sjtu.edu.cn

Abstract

In this work, we demonstrate a novel oxide semiconductor thin-film transistor (TFT) with Ga_2O_3/IGZO heterojunction structure by atomic layer deposition (ALD). A high-density two-dimensional electron gas (2DEG) over $6 \times 10^{13}/cm^2$ is achieved, by forming a Ga_2O_3/IGZO interface, the density of which is beyond the gate control limitation, leading to a low channel resistance. The proposed Ga_2O_3/IGZO heterojunction TFT enables a new approach to enhance the drive current of oxide TFTs beyond conventional material engineering for high mobility.

Keywords: IGZO, heterostructure, TFT

Introduction

Since the first indium gallium zinc oxide (IGZO) thin-film transistors (TFTs) in 2004 [1], IGZO has gained prominence due to its exceptional electronic properties, including low leakage, large area electrical uniformity, and good transparency [2-4], which attribute IGZO widely used in displays. However, new driving current requirements have recently arisen due to the demands of next-generation displays, MicroLEDs typically use higher current densities ($1-1000$ A cm^{-2}) than OLED and LCD (<1 A cm^{-2}) counterparts that push the boundaries of TFT technology [5]. Post deposition annealing, increasing indium content and inducing crystallization are common ways to optimize film quality to boost the current of IGZO TFTs [6-9]. However, these methods still face inherent limitations that can restrict performance, such as negative threshold voltage, poor stability, and poor uniformity.

To address these limitations, we propose a novel approach by employing a Ga_2O_3/IGZO heterojunction structure, fabricated using atomic layer deposition (ALD). This innovative structure allows for the formation of a high-density two-dimensional electron gas (2DEG) at the Ga_2O_3/IGZO interface, achieving an electron density exceeding 6×10^{13} cm^{-2} which significantly reduces channel resistance and markedly enhances the drive current performance of the TFTs. This study shows that IGZO TFTs could be highly promising for high current driving.

Experiments

The structure of the devices is shown in Fig. 1. The devices were fabricated on Si wafer with thermally oxidized SiO_2. Gate metal was grown and patterned by photolithography and dry etching. Then HfO_2 and IGZO were fabricated by ALD function as insulator and channel, respectively. The Ga_2O_3 was grown after IGZO channel patterning. The patterning of Ga_2O_3 was achieved simultaneously with nickel (Ni) source/drain lithography, where the 2.38% TMAH developer could etch Ga_2O_3 under the S/D region while but IGZO was retained. Thermal-deposited 40 nm Ni source and drain electrodes were patterned formed by thermal evaporation and a lift-off method. After device fabrication, TFTs were subjected to air annealing for 10 min at 250 °C. The electrical properties of the devices were measured using a B1500A semiconductor parameter analyzer at room temperature. Hall mobility (μ_H) and sheet carrier density (n_{2D}) at channel region were analysed by hall measurement carried out in physical property measurement system (PPMS) (DynaCool-14T).

Results and discussion

Fig. 3 shows transfer characteristics and field-effect mobility (μ_{FE}) extracted at low V_{DS} of Ga_2O_3/IGZO heterojunction TFTs which a channel length (L_{ch} of 40 μm). The output characteristics curve is presented in Fig. 4. As can be seen, a high μ_{FE} of 95.50 $cm^2/(V \cdot s)$ was achieved with a positive $V_{TH}=0.73$ V, in contrast to the mobility of conventional IGZO (In/Ga/Zn = 1:1:1) TFT, which is only $10-15$ $cm^2/(V \cdot s)$ [10]. The high I_D and mobility imply that there are high concentration carriers in the channel region. To further characterize the carrier density at channel region, the gated Hall measurement is performed at 300 K to evaluate the μ_H and n_{2D} at different V_{GS} under ± 1 T magnetic field. Fig. 6 shows the extracted μ_H under ± 1 T to be ~ 17 $cm^2/(V \cdot s)$ at high V_{GS}, which is consistent with the μ_{FE} of IGZO TFTs without Ga_2O_3. The extracted n_{2D} is shown in Figure 5, which increases linearly at high V_{GS}, exhibiting a maximum n_{2D} of 6×10^{13} $/cm^2$. Ga_2O_3 has a larger band gap than IGZO, a small conduction band offset is likely formed between the Ga_2O_3 and IGZO region, so a high-density 2DEG are expected to concentrate in the Ga_2O_3 and IGZO interface shown in Fig. 2. The maximum carrier density that can be modulated by V_{GS} is determined by the equation $C_{ox} \cdot (V_{GS}-V_{TH})$, which is green underlined in Fig. 5. Carriers that exceed the gate control limit between 0-1 V, indicate the existence of another mechanism regulating carriers. This phenomenon could potentially be attributed to the presence of a Schottky barrier at drain region, when the gate voltage decreases, the barrier height increases, effectively switching the device off. In contrast, as the gate voltage increases, the barrier height decreases, allowing carriers to easily traverse the reduced barrier corresponding the linear increase in n_{2D} as a function of V_{GS}. We noticed that carrier density also slightly exceeds the gate control limit between 1-2 V, but this may be resulted from the experimental noises and needs further confirmation. The formation mechanism of Schottky barrier may involve either pinning or de-pinning of the Fermi level at Ni/IGZO interface [11, 12].

Conclusion

In this work, we demonstrate an attractive result in IGZO TFTs through the development of a Ga_2O_3/IGZO heterojunction structure, fabricated by ALD. The formation of a high-density 2DEG exceeding 6×10^{13}/cm² at the Ga_2O_3/IGZO interface surpassing conventional gate control limitations. This innovation leads to a marked reduction in channel resistance and facilitates enhanced drive currents, thereby establishing the Ga_2O_3/IGZO heterojunction TFT as a promising candidate for applications that require high mobility.

Acknowledgments

This work was supported by by STI 2030-Major Projects under Grant 2022ZD0210600, National Natural Science Foundation of China (No. 62274107, 92264204) and Shanghai Pilot Program for Basic Research-Shanghai Jiao Tong University under Grant 21TQ1400212.

References

[1] K. Nomura, H. Ohta, A. Takagi, T. Kamiya, M. Hirano, and H. Hosono, "Room-temperature fabrication of transparent flexible thin-film transistors using amorphous oxide semiconductors," Nature, vol. 432, no. 7016, pp. 488-492, 2004.

[2] T. Kamiya, K. Nomura, and H. Hosono, "Present status of amorphous In–Ga–Zn–O thin-film transistors," Science and Technology of Advanced Materials, vol. 11, no. 4, pp. 044305, 2010.

[3] E. Fortunato, P. Barquinha, and R. Martins, "Oxide semiconductor thin-film transistors: a review of recent advances," Adv Mater, vol. 24, no. 22, pp. 2945-86, 2012.

[4] M. Coll, J. Fontcuberta, M. Althammer, M. Bibes, H. Boschker, A. Calleja, G. Cheng, M. Cuoco, R. Dittmann,

and B. Dkhil, "Towards oxide electronics: a roadmap," Applied surface science, vol. 482, pp. 1-93, 2019.

[5] K. Behrman, and I. Kymissis, "Micro light-emitting diodes," Nature Electronics, vol. 5, no. 9, pp. 564-573, 2022.

[6] C. Peng, S. Yang, C. Pan, X. Li, and J. Zhang, "Effect of two-step annealing on high stability of a-IGZO thin-film transistor," IEEE Transactions on Electron Devices, vol. 67, no. 10, pp. 4262-4268, 2020.

[7] T. Mudgal, N. Walsh, R. G. Manley, and K. D. Hirschman, "Impact of annealing on contact formation and stability of IGZO TFTs," ECS Transactions, vol. 61, no. 4, pp. 405, 2014.

[8] H. J. Yang, H. J. Seul, M. J. Kim, Y. Kim, H. C. Cho, M. H. Cho, Y. H. Song, H. Yang, and J. K. Jeong, "High-performance thin-film transistors with an atomic-layer-deposited indium gallium oxide channel: A cation combinatorial approach," ACS applied materials & interfaces, vol. 12, no. 47, pp. 52937-52951, 2020.

[9] H. Han, S. Jang, D. Kim, T. Kim, H. Cho, H. Shin, and C. Choi, "Memory characteristics of thin film transistor with catalytic metal layer induced crystallized indium-gallium-zinc-oxide (IGZO) channel," Electronics, vol. 11, no. 1, pp. 53, 2021.

[10] A. Illiberi, B. Cobb, A. Sharma, T. Grehl, H. Brongersma, F. Roozeboom, G. Gelinck, and P. Poodt, "Spatial atmospheric atomic layer deposition of $In_xGa_yZn_zO$ for thin film transistors," ACS applied materials & interfaces, vol. 7, no. 6, pp. 3671-3675, 2015.

[11] H. Qian, C. Wu, H. Lu, W. Xu, D. Zhou, F. Ren, D. Chen, R. Zhang, and Y. Zheng, "Bias stress instability involving subgap state transitions in a-IGZO Schottky barrier diodes," Journal of Physics D: Applied Physics, vol. 49, no. 39, pp. 395104, 2016.

[12] A. Chasin, S. Steudel, F. Vanaverbeke, K. Myny, M. Nag, T.H. Ke, S. Schols, G. Gielen, J. Genoe, and P. Heremans, "UHF IGZO Schottky diode," in IEDM, 2012, pp. 12.4. 1-12.4. 4.

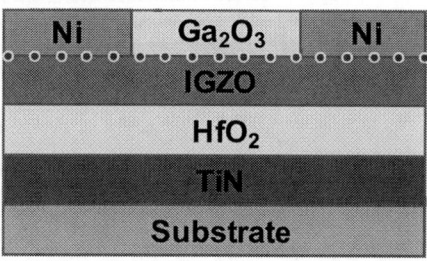

Fig. 1. Schematic diagram of a Ga$_2$O$_3$/IGZO heterojunction transistor.

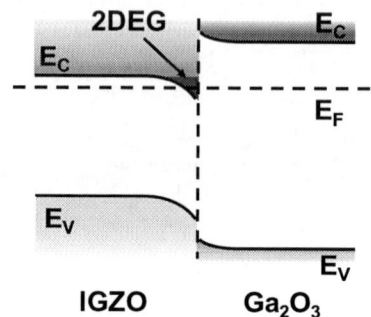

Fig. 2. Band diagram at Ga$_2$O$_3$/IGZO interface.

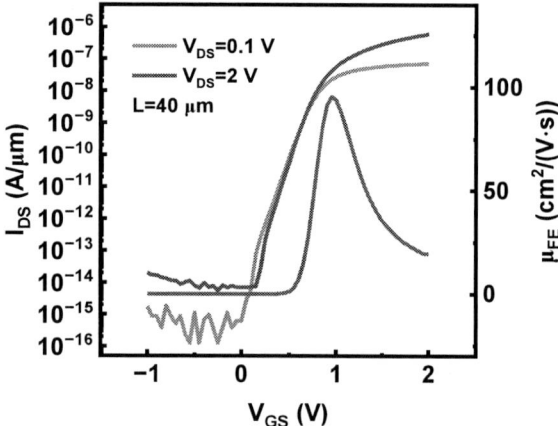

Fig. 3. Measured I$_D$ and μ$_{FE}$ as a function of V$_{GS}$ of the Ga$_2$O$_3$/IGZO heterojunction transistor.

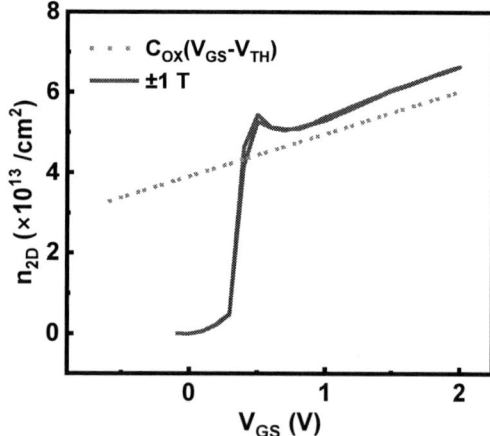

Fig. 4. I$_D$-V$_{DS}$ curves of the Ga$_2$O$_3$/IGZO heterojunction transistor.

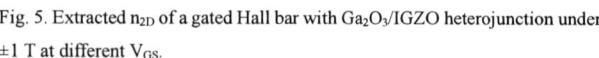

Fig. 5. Extracted n$_{2D}$ of a gated Hall bar with Ga$_2$O$_3$/IGZO heterojunction under ±1 T at different V$_{GS}$.

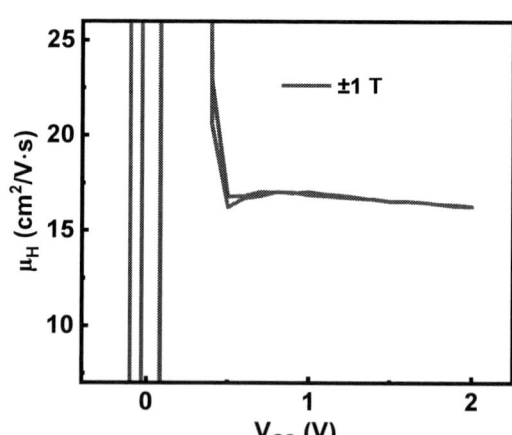

Fig. 6. Extracted μ$_H$ of a gated Hall bar with Ga$_2$O$_3$/IGZO heterojunction under ±1 T at different V$_{GS}$.

Embracing Semiverse™ Solutions: Semiconductor Virtual Fabrication and Its Applications

Qingpeng Wang*, Yujia Zhong, Lifei Sun, Benjamin Vincent, Ivan Chakarov and Joseph Ervin

Lam Research, Shanghai, China
*E-mail:Qingpeng.Wang@lamresearch.com

Abstract

In this paper, we share several applications of semiconductor virtual fabrication. Topics include prototyping/visualization, unit process optimization, process targeting and window checks, and stress evolution. We will demonstrate how virtual fabrication will be used to shape the future of semiconductor manufacturing, while saving time and money during semiconductor development.

Keywords: Virtual Fabrication, prototyping, process window, stress evolution, process modeling

Introduction

Artificial Intelligence (AI) and Machine Learning (ML) have evolved quickly in recent years, driven by advancements in the semiconductor industry [1]. However, semiconductor manufacturing itself is still a capital-intensive, technology-intensive, and workforce-intensive industry. As semiconductor processes advance, integration schemes become more and more complex, and each process step needs tighter control. Potential variability sources have increased, and the necessary silicon experimentation has exploded exponentially. This has not only slowed down the R&D process, but also raised the financial burden of developing technology and the R&D engineering workload [2].

Semiconductor technology development should be people-oriented and eco-friendly. Semiconductor manufacturing has been crucial for building the information society and fostering AI and ML growth. Now, it's time for AI and ML to give back to the semiconductor manufacturing industry. In this article, we will explore how virtual fabrication (part of Lam's Semiverse™ Solutions) and AI/ML can potentially reduce engineering workloads, lower R&D costs, and decrease process development time.

Virtual Fabrication and Typical Applications

SEMulator3D® is a software platform for semiconductor process modeling and virtual fabrication [3]. It can create "virtual twins" of actual semiconductor devices, replicates an actual fab and its operations. Some SEMulator3D virtual fabrication applications that have been published and studied in recent years by our group are shown below.

A. Prototyping and Visualization

As the demand for greater semiconductor device density has increased, structures for both logic and memory devices have moved from 2D to 3D. Understanding and prototyping a device structure or process flow in 3D has become essential for semiconductor process development. Virtual fabrication can help with this task. Figure 1 displays the complex 3D structures of (a) advance logic GAA with self-aligned backside contact, (b) 3D NAND with hybrid bonding and (c) 3D DRAM integration flow created by virtual fabrication [4]. Virtual fabrication is a technique to more easily and vividly visualize these complicated structures and flows.

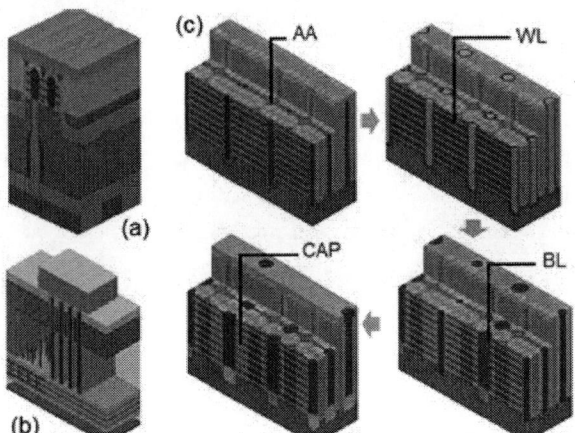

Fig. 1: (a) GAA, (b) 3D NAN and (c) 3D DRAM process.

B. Unit Process Development

As mentioned earlier, unit process control has become more difficult at advanced nodes. Many tuning controls are available on advanced semiconductor manufacturing equipment, and these controls can be used to vary process recipes and meet precise manufacturing requirements. However, finding the optimal instrumentation settings in an exponentially large tuning space is difficult for R&D engineers, and requires an increasing amount of trial-and-error silicon experimentation. Virtual fabrication can accelerate process optimization and decrease wafer-based experimentation. Figure 2 provides an example of how virtual fabrication can be used in a bowed via to optimize a W dep-etch-dep (DED) fill process and obtain a void-free structure with dep/etch amount split (DOE1), DEDED split (DOE2) and incoming bowing profile split (DOE3). Figure 3 displays a study on how to achieve a better fin profile with

979-8-3315-0417-5/25 $31.00 © 2025 IEEE 181

less bowing by hard mask height, sidewall angle and space CD split [5]. Using a virtual DOE, the best process paths can be explored prior to performing DOEs in silicon. Si wafers and process development time can be reduced, accelerating time to solution.

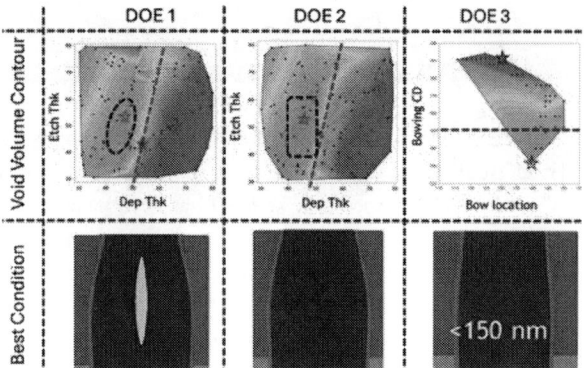

Fig. 2: W DED fill optimization DOEs to pursue a void free structure

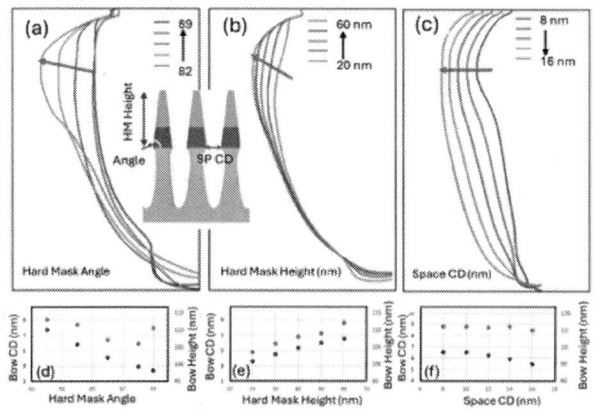

Fig. 3: Fin profile tuning with less bowing.

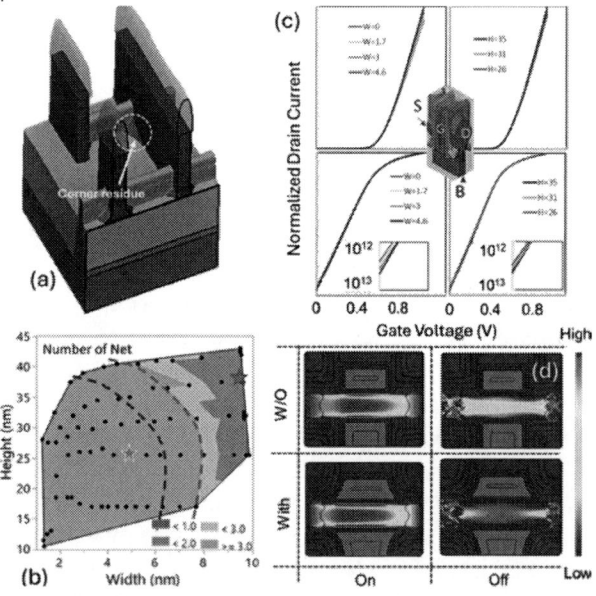

Fig. 4: Poly residue process window and device performance.

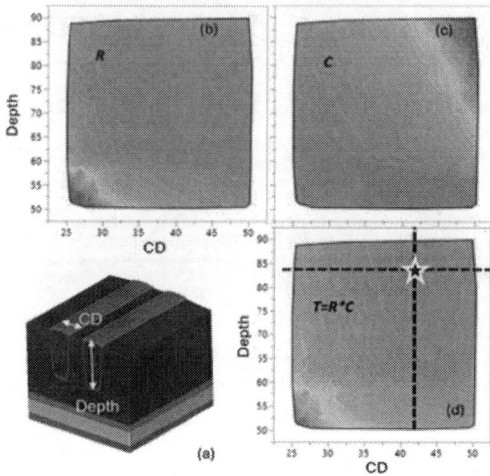

Fig. 5: BEOL metal profile optimization for better RC.

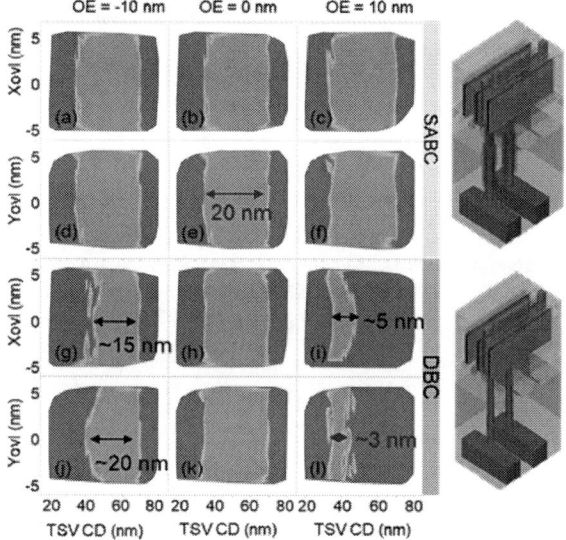

Fig. 6: Process window comparison between DBC and SABC of GAA structures.

C. Process Targeting and Window Enhancement

The number of process steps required to build advanced node technologies has increased as device density has increased. Properly targeting process specifications and identifying process windows for multiple steps are important since variability at each process step must be strictly limited or controlled. Again, virtual fabrication can be used to identify specifications for process steps and process windows to minimize variability and improve yield. Figure 4 displays a virtual experiment where poly residue effects were virtually reproduced in a poly etch process, to explore allowable process windows and device performance [6]. In an additional study (Figure 5), a metal profile (cd/depth) optimization study was undertaken to obtain the optimal trade-off between metal resistance and capacitance (similar results were reported in [7]). Figure 6 highlights a process window comparison between a direct backside contact (DBC) scheme and a self-aligned backside contact (SABC) scheme

979-8-3315-0417-5/25 $31.00 © 2025 IEEE   182

used in a GAA structure for different TSV over etch (OE), TSV CD and overlay conditions [8]. With the help of virtual fabrication, the 2 alternatives were compared by quantifying their performance and process window settings to choose the optimal scheme.

D. Stress Evolution at Advance Nodes

Our most recent study found that stress-induced deformation of GAA structures at advanced nodes can create potential performance and yield issues. Using virtual fabrication techniques, we investigated the impact of stress evolution and stress induced deformation in a GAA structure undergoing multiple process steps (Figure 7). We were able to quantify the expected deformation.

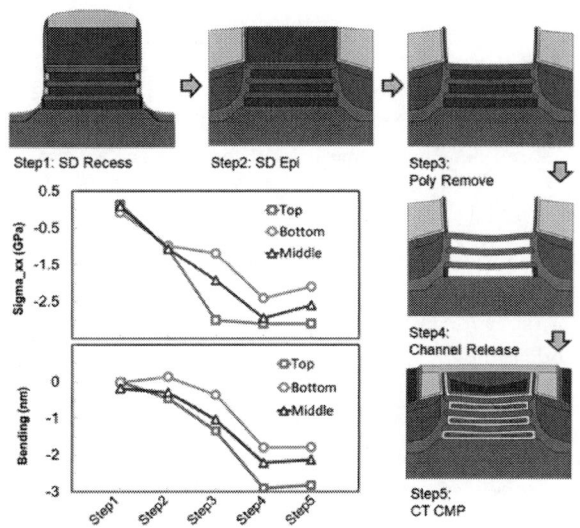

Fig. 7: Stress Evolution and nanosheet bending in a GAA flow.

In each of these examples, we have demonstrated how virtual fabrication can improve semiconductor engineering at multiple stages of technology development. This technique can be used to visualize the step-by-step process of fabricating complicated 3D structures, explore unit process windows, define optimal process conditions and obtain in-spec conditions using virtual DOEs, and perform yield enhancement by process windows optimization. Virtual fabrication can mimic actual fabrication-based testing, but obtain data faster and less expensively. This data can also be fed into AI or ML models to accelerate semiconductor development.

Conclusion

In this paper we highlighted multiple applications where virtual fabrication can assist semiconductor development, from visualization, to prototyping, to pathfinding, along with unit process tuning, process targeting, process window optimization and stress evolution. The capabilities of virtual fabrication and virtual twins during semiconductor development are limited only by the imagination. It is time to embrace a seamless physical-virtual semiconductor ecosystem.

References

[1] K. F. Lee and Q. C., AI 2041: Ten visions for our future. Crown Currency, 14 Sept. 2021.

[2] A. Mallik et al., "Economics of semiconductor scaling: a cost analysis for advanced technology node," in Proc. Symp. VLSI Technol., pp. T202-T203, IEEE, 2019.

[3] http://www.coventor.com/products/semulator3d.

[4] K. S. Choi et al., "A Three Dimensional DRAM (3D DRAM) Technology for the Next Decades," in Proc. IEEE Symp. VLSI Technol. Circuits (VLSI), pp. 1-2, IEEE, 2024.

[5] L. Sun, P. Lyu, Q. Wang, Q. Zhong, K. Wang, and Y. Chi, "Investigation of FIN bowing formation mechanism during STI etching by virtual fabrication," in Proc. China Semicond. Technol. Int. Conf. (CSTIC), pp. 01-03, IEEE, 2022.

[6] Q. Wang, D. Yu Chen, C. Li, R. Bao, J. Huang, and J. Ervin, "The Effects of Poly Corner Etch Residue on Advanced FinFET Device Performance," in Proc. China Semicond. Technol. Int. Conf. (CSTIC), pp. 1-3, IEEE, 2021.

[7] P. Lyu, Q. Wang, and L. Sun, "Optimization of Metal Line Thickness & CD and Effect on RC Delay," in Proc. China Semicond. Technol. Int. Conf. (CSTIC), pp. 1-3, IEEE, 2022.

[8] Q. Wang, S. Sumant Sarkar, and T. Oh, "A Novel Self-Aligned Backside Contact Architecture for Advanced Logic Nodes," in Int. Conf. Solid State Devices Mater. (SSDM2024), PS-01-01, 20 Sept. 2024.

GeTe$_9$ Ovonic Threshold Switch Memristor-centered 1S1T1R cell as Bio-realistic Stochastic Synapse

Xinyu Wen[1], Kuan Wang[2], Lun Wang[1], Qi Chen[1], Hao Tong[1]*, Xiang-Shui Miao[1] and Yuhui He[1]*

1School of Integrated Circuit, Huazhong University of Science and Technology, Wuhan, China

2Department of Integrated Power Systems and Device Technology, Hubei Jiufengshan Laboratory, Wuhan, China

*heyuhui@hust.edu.cn, tonghao@hust.edu.cn

Abstract

Biological synapses conserve significant energy within the brain's complex networks due to their stochastic switching behavior, which is crucial for neuromorphic probabilistic computing. To replicate this, GeTe$_9$ OTS devices were fabricated, exhibiting switching mechanisms consistent with biological synapses, thereby providing the necessary randomness for stochastic synapses. Furthermore, we implemented a bio-realistic stochastic synapse using the proposed "1S1T1R" structure. The random switching characteristics were successfully validated through Cadence simulations.

(Keywords: neuromorphic probabilistic computing, stochastic synapse, 1S1T1R and Ovonic threshold switch)

Introduction

As Moore's law approaches its limits and Dennard scaling comes to an end, neuromorphic computers present a new energy-efficient architecture. [1] To emulate biological neural networks and develop ultra-low-power, brain-inspired systems, efforts have focused on various dimensions, including fundamental devices, hardware circuits, computing architectures, and software algorithms. [2] Electronic synapses, as a core hardware unit in neuromorphic computing, play a critical role in building hardware systems for Neural Networks such as DNNs and SNNs.

However, biological synaptic release is inherently probabilistic, which conserves a substantial amount of energy in the brain's complex synaptic networks. [3] As a result, stochastic synapses with probabilistic switching more closely resemble biological synapses. In terms of algorithmic architecture, stochastic synapses are also a key component in current brain-inspired probabilistic computing systems. Numerous studies have demonstrated the implementation of brain-inspired probabilistic computing tasks that quantify uncertainty using various forms of stochastic synapses. [4, 7] In this paper, we investigated the intrinsic randomness of GeTe$_9$ Ovonic Threshold Switch (OTS) devices. The stable probabilistic switching characteristics of these devices were validated through pulse amplitude and pulse width modulation tests. Building on this, we proposed a "1S1T1R" stochastic synapse architecture, with its practical applicability confirmed through Cadence simulations.

Results and Discussion

A. GeTe$_9$ Ovonic Threshold Switch implementation

Fig1 illustrates the alignment between the switching mechanism of OTS devices and biological stochastic synapses. At the biological level, when an action potential reaches the synapse, there is a probabilistic chance that vesicles carrying information will be generated and transmitted to the postsynaptic membrane. [3] Similarly, at the device level, when a signal pulse arrives, initially localized defects become delocalized to quenched defects, creating a probability that the device will activate. [5] Thus, OTS devices are inherently well-suited to serve as a source of randomness for stochastic synapses.

As shown in Fig 2a, 574 GeTe$_9$ OTS devices with varying hole sizes (diameters of 250, 500, 750, and 1250 nm) were fabricated on a 1 cm × 1 cm silicon wafer with a via-hole structure. Fig. 2b presents an enlarged unit featuring a 500 nm pore and the 150 × 150 μm² pattern of the Tellurium Electrode (TE) layer. Fig. 2c displays the pore details under a scanning electron microscope, while additional details under a transmission electron microscope are shown in Fig. 2d. Meanwhile, Fig. 2e provides a clear schematic illustration. The thickness of the GeTe$_9$ functional layer in this batch is approximately 40 nm.

B. Basic electrical characterization tests

The basic electrical characteristics of the W/GeTe9/W device were measured using the Agilent B1500A semiconductor analyzer. Fig 3 illustrates the DC I-V test of the devices, with a compliance current (I_{cc}) of 100μA set to prevent device damage. In this series of tests, a voltage sweep from 0 to 3V was applied to the device, with the switching process and the forming process represented by the red and orange lines in Fig 3, respectively. It is worth noting that subsequent tests on the device's low-resistance state indicate an on/off ratio of approximately 1×10^4.

During the endurance test, a 2kΩ resistor was connected in series with the device for protection. As shown in Fig 4, the GeTe$_9$ device demonstrated endurance up to 2×10^{10} cycles. The switching-on process, depicted in Fig 5, shows that the device could be activated within 6.2 ns. In Fig 6, the on-state resistance of different OTS devices was measured over five distinct cycles, exhibiting good consistency. In conclusion, the OTS device exhibits excellent endurance and exceptionally fast activation speed, making it a reliable choice for stochastic synapses that demand rapid information transmission and frequent switching.

C. Intrinsic stochasticity tests of the device

The opening and closing of OTS devices are governed by the localization and delocalization of defect states. [5] Specifically, the variability of the threshold voltage across different cycles reflects the intrinsic randomness of the device. As shown in the inset of Fig 7, a $GeTe_9$ OTS device was connected in series with a 200Ω resistor to measure the threshold voltage. By applying triangular waves with a 2V amplitude and 20μs width, the threshold voltage exhibited variability. To further analyze the device's randomness, the threshold voltage distribution was measured over 300 consecutive cycles using the same input method. The segmented statistical results are presented in Fig 8, where the distribution of the threshold voltage closely fits a normal distribution. This stable distribution confirms the feasibility of utilizing the device as a reliable source of randomness.

To maximize the utilization of the device's inherent randomness, we conducted tests on pulse amplitude and pulse probability modulation. The primary testing method is illustrated in Fig 9a, where the circuit current is measured during the last 1μs of each pulse. Fig 9b shows the results of the pulse amplitude modulation tests, where the amplitude of the applied pulses was incremented from 1.25V by 0.05V up to 1.50V, ultimately achieving a regulated probabilistic switching range from 10% to 100%. Notably, the maximum probability deviation at each amplitude was only around 10%. The pulse width modulation test, shown in Fig 9c, was performed with a fixed amplitude of 1.4V. Increasing the pulse width from 400ns to 50μs raised the device's switching probability from 0.28 to 0.62. As demonstrated in Fig 9d, a nearly 1000-fold increase in pulse width was required to elevate the switching probability from 0% to 100%. Therefore, voltage amplitude modulation is more effective and better suited for controlling the overall switching probability of stochastic synapses.

D. Stochastic synapse implementation

The most widely used hardware synapse structure in Deep Neural Networks (DNNs) is the "1T1R" configuration, where 'T' represents a transistor and 'R' denotes non-volatile memory devices such as RRAM. Fig10a presents MC-Dropconnect, an algorithm commonly used in Bayesian brain-inspired computing systems, which has been validated for efficient probabilistic computation. [6] Building on this, we propose the "1S1T1R" structure, as shown in Fig10b. In this design, randomness is introduced to the 1T1R synapse via the OTS device, and the switching probability of the stochastic synapse is controlled by adjusting the input pulse amplitude through V_{pr}.

The simulation validation was performed on the Cadence platform, where the OTS device was modeled using Verilog-A based on previous test characteristics. The synaptic memristor was replaced with a standard resistor. Specifically, R_{syn} was set to 3kΩ and R_L to 200Ω. As shown in Fig11, a 2V, 10μs triangular wave was applied through V_{pr} to ensure the transistor gate was open, while a constant voltage of 1V was applied at V_d. The switching characteristics of the stochastic synapse functioned as expected, and the on-off ratio improved to 10^6 due to the transistor. When V_{pr} was set to 1.4V with a continuous 10μs pulse, the stochastic synapse demonstrated a 50% switching probability, consistent with the test results in Fig. 9b. In conclusion, the probabilistically adjustable stochastic synapse has been fully implemented in hardware. As shown in Table 1, our work is compared with other implementations of stochastic synapses.

Conclusion

We propose a hardware implementation of stochastic synapses based on $GeTe_9$ OTS devices. Initially, a series of tests were conducted to investigate the randomness characteristics of OTS devices. Furthermore, the stochastic synapse based on the proposed "1S1T1R" structure was validated as effective, suggesting its potential for future probabilistic neuromorphic systems.

Acknowledgments

The work is supported by the National Key Research and Development Program of China (No.2023YFB4502200), Natural Science Foundation of China (No. 92164204 & 62374063)

References

[1] Schuman C D, Kulkarni S R, Parsa M, et al. Opportunities for neuromorphic computing algorithms and applications[J]. Nature Computational Science, 2022, 2(1): 10-19.

[2] Wang Z, Wu H, Burr G W, et al. Resistive switching materials for information processing[J]. Nature Reviews Materials, 2020, 5(3): 173-195.

[3] Borst J G G. The low synaptic release probability in vivo[J]. Trends in neurosciences, 2010, 33(6): 259-266.

[4] Lin Y, Zhang Q, Gao B, et al. Uncertainty quantification via a memristor Bayesian deep neural network for risk-sensitive reinforcement learning[J]. Nature Machine Intelligence, 2023, 5(7): 714-723.

[5] Chai Z, Shao W, Zhang W, et al. GeSe-based ovonic threshold switching volatile true random number generator[J]. IEEE Electron Device Letters, 2019, 41(2): 228-231.

[6] Gal Y. Uncertainty in deep learning[J]. 2016.

[7] Dutta S, Detorakis G, Khanna A, et al. Neural sampling machine with stochastic synapse allows brain-like learning and inference[J]. Nature Communications, 2022, 13(1): 2571.

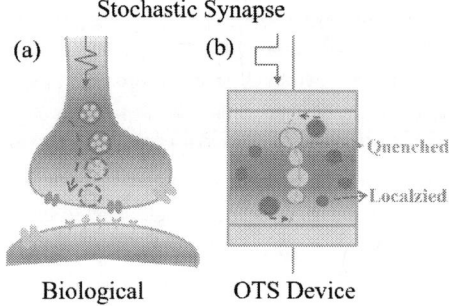

Stochastic Synapse

(a)   (b)

Biological   OTS Device

Fig. 1: Schematic illustration of stochastic synapses at (a) biological and (b) device levels.

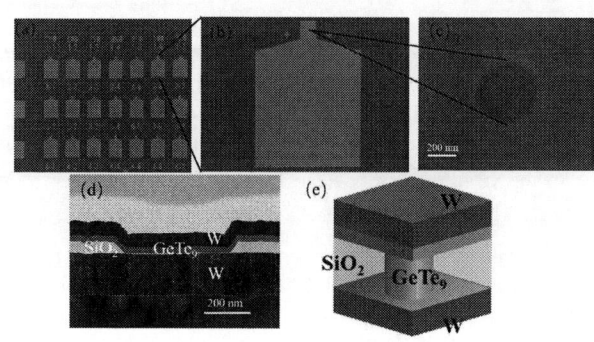

Fig. 2: Scanning electron micrograph of (a) GeTe₉ OTS devices on silicon wafer and enlarged images of (b) a single unit and (c) a 500nm via-hole. (d)Transmission Electron Microscopy image and (e)schematic diagram of the via-hole structure.

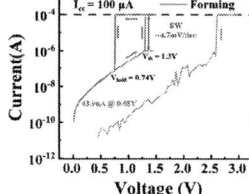

Fig.3: Basic I-V of OTS devices. The current is limited to 100μA.

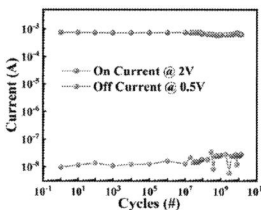

Fig.4: Endurance of OTS devices.

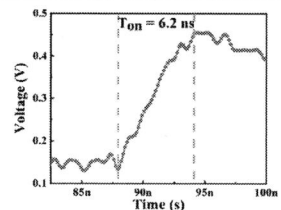

Fig.5: Enlarged switch-on time of OTS devices within 10 ns.

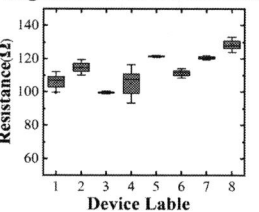

Fig.6: The d2d variation and c2c variation of R_{on} of OTS devices.

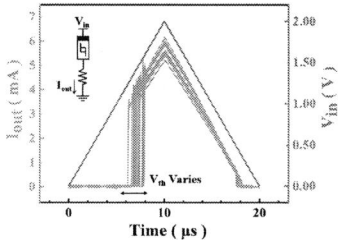

Fig.7: Under the input of a 2V triangular pulse with 20μs width, the current characteristics of the device across different cycles are illustrated.

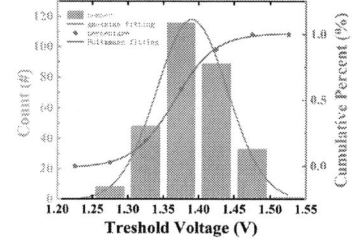

Fig.8: The threshold voltage (Vth) statistical distribution obtained of the device under 300 consecutive triangular wave tests as fig7. The threshold voltage overall exhibits a normal distribution.

	This Work	*Ref 7*
Structure	1S1T1R	1S1FeFET
Selector Materials	GeTe₉	Ag/HfO₂
Weight Realization	\	FeFET
Endurance	2×10^{10}	10^8
Switch-on Time	6.2ns	28ns
Probability Tuned	√	\

Table.1: Comparison of the implemented stochastic synapses.

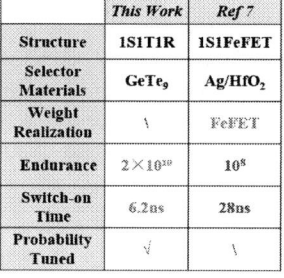

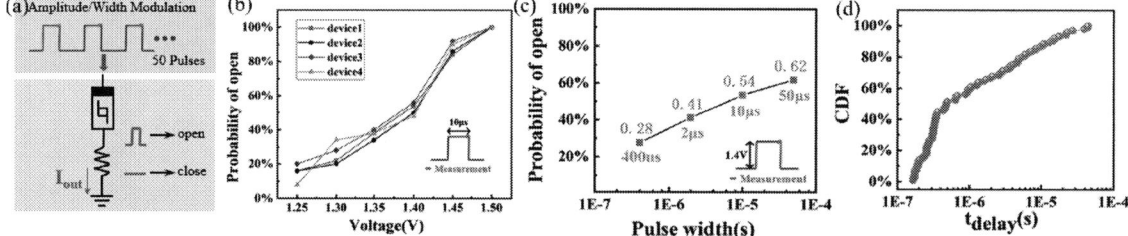

Fig.9: The measurement of the probability of OTS open. (a) Schematic diagram of the measurement. (b) Under pulse amplitude modulation, the open probability of 4 OTS devices. (c) Under pulse amplitude modulation, the open probability of devices. (d) The cumulative distribution function of t_{delay} obtained from the test of (c).

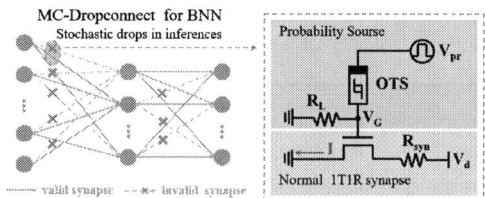

Fig.10: The stochastic synapses in Bayesian neuromorphic system. The structure within the dashed box represents the 1S1T1R hardware unit of the stochastic synapse.

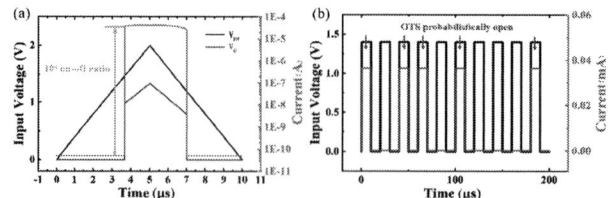

Fig.11: The simulation results in Cadence based on the probabilistic synapse circuit diagram shown in Fig.10. (a) The synapse's current response under a 2V,10 μs triangular wave input. (b) The current response under 10 consecutive 1.4V pulses.

Development of an Inkjet-Printed Sensor using Ink locked Food-Based Nano Conductive paste for Electrochemical sensing of Insulin: A First attempt

Ramya K[1,2]

[1]MEMS, Microfluidics and Nanoelectronics (MMNE) Lab, Birla Institute of Technology and Science (BITS) Pilani, Hyderabad Campus, Hyderabad 500078

[2]Department of Electrical and Electronics Engineering, Birla Institute of Technology and Science (BITS) Pilani, Hyderabad Campus,500078

Hyderabad,India

p20210416@hyderabad.bits-pilani.ac.in

Sanket Goel[1,2]

[1]MEMS, Microfluidics and Nanoelectronics (MMNE) Lab, Birla Institute of Technology and Science (BITS) Pilani, Hyderabad Campus, Hyderabad 500078

[2]Department of Electrical and Electronics Engineering, Birla Institute of Technology and Science (BITS) Pilani, Hyderabad Campus,500078

Hyderabad,India

sgoel@hyderabad.bits-pilani.ac.in

Abstract—This work introduces the first inkjet-printed electrochemical sensor utilizing a food-based nano conductive paste (FN-CoP) for insulin detection. FN-CoP, composed of activated charcoal, vegetable gelatin, and oral rehydration salts, is developed in-house as an eco-friendly and biocompatible alternative to traditional carbon pastes. Inkjet printing allows for precise electrode patterning while minimizing material waste. FN-CoP demonstrates high conductivity (1484 S/m) and when validated for electrochemical insulin sensing, it shows excellent sensitivity, with a limit of detection (LoD) of 1.8 pM. Its strong adhesion and compatibility across various substrates (glass, polyamide, cloth) indicate its potential for broader applications in biosensing and energy-related technologies.

Keywords—Inkjet-printing, Food based conductive paste, Miniaturized Electrochemical sensor, Insulin detection, Biosensing, Energy applications.

I. INTRODUCTION

Biosensors have become vital in healthcare diagnostics, environmental monitoring, and energy applications due to their ability to detect biological analytes with high sensitivity and specificity. The increasing demand for these devices has prompted the development of sustainable and biocompatible materials that reduce environmental impact while maintaining performance. This shift addresses the limitations of conventional fabrication processes, which often involve toxic chemicals and significant material waste. Electrochemical sensors, widely used for detecting analytes like glucose and insulin, are traditionally fabricated using screen-printing techniques with carbon-based conductive pastes. However, this method faces challenges in terms of electrode resolution, material waste, and biocompatibility [1].

Screen-printing is simple and cost-effective, but its limitations hinder advanced biosensing applications. It struggles to achieve high-resolution electrode patterns due to constraints of the screen mesh size and the viscosity of conductive pastes, leading to poor electrode geometry and uneven material distribution. This negatively impacts sensor performance, particularly in precision applications like insulin detection. Moreover, screen-printing requires excess material

to ensure adequate deposition, which increases material costs and creates unnecessary wastage. Furthermore, traditional carbon-based conductive pastes used in screen-printing often contain toxic chemicals, posing environmental hazards and health risks, thus limiting their viability for biosensing [2][3].

To overcome these limitations, researchers have explored alternative fabrication methods offering finer control over sensor architecture while reducing reliance on harmful materials. Inkjet printing has gained attention as a versatile method that allows precise material deposition on various substrates. It offers advantages over screen-printing, including high-resolution patterns, reduced material waste, and the inhouse prepared FN-CoP offers elimination of toxic chemicals. However, the effectiveness of inkjet printing in fabricating electrochemical sensors has been limited by the lack of suitable conductive pastes that can meet the demands of conductivity, consistency, and biocompatibility. While functional in screen-printing, traditional carbon-based pastes often underperform in inkjet printing due to material deposition challenges and pattern consistency [4].

This work introduces a novel approach to electrochemical sensor fabrication by developing a food-based nano conductive paste (FN-CoP) formulated specifically for inkjet printing. FN-CoP represents a significant advancement over traditional carbon-based pastes by using eco-friendly, biocompatible materials that are both sustainable and effective for biosensing applications. The paste comprises three key components: activated charcoal (AC), vegetable gelatin, and oral rehydration salts (ORS). Activated charcoal, with its high surface area and conductivity, serves as the primary conductive material, making it ideal for electrochemical applications. Its nanoscale size ensures uniform deposition during inkjet printing, enabling precise patterning and high electrode resolution. Vegetable gelatin, a natural polymer, acts as a binder that provides structural integrity while maintaining biocompatibility and biodegradability. ORS, typically composed of electrolytes like sodium chloride and potassium chloride, enhances the pastes ionic conductivity, further improving its performance in electrochemical sensing.

This research represents the first successful attempt to fabricate electrodes using inkjet printing with an in-house made conductive ink from food-based materials. The development of this food-based paste offers a sustainable alternative to traditional carbon-based materials, providing the necessary biocompatibility for biosensing while maintaining high conductivity and sensitivity. The inkjet printing process is optimized for FN-CoP, enabling precise control over electrode patterning on various substrates, including glass, polyethylene terephthalate (PET), Cloth and polyamide sheets. This process not only allows for high-resolution electrode fabrication but also minimizes material waste, meeting the demand for eco-friendly fabrication methods. The inkjet printing process is optimized by adjusting the viscosity and particle size of the FN-CoP formulation. The paste viscosity is controlled in a range of 1,600 to 4,000 mPa.s, ensuring smooth inkjet flow and preventing nozzle clogging (Table 1). Additionally, the scanning electron microscope (SEM) revealed averaging around 64 nm (Figure 3(a)), of uniform material distribution across substrates, crucial for producing high-performance electrochemical sensors. The resulting electrodes exhibited excellent conductivity, with a measured value of 1484 S/m, suitable for biosensing applications. This level of conductivity is particularly important for biosensing applications like insulin detection, where sensitivity and rapid response times are crucial.

The electrochemical performance of the inkjet-printed FN-CoP sensors is validated through insulin detection experiments. The electrodes are modified with titanium dioxide (TiO_2) nanofibers to enhance sensitivity, and cyclic voltammetry (CV) is used to measure oxidation potential of insulin. An oxidation peak at 0.7 V confirmed high sensitivity of the sensor for insulin detection. Chronoamperometry studies further demonstrated the selectivity of the paste, with no significant interference from other bioanalytes. The sensor achieved a limit of detection (LoD) of 1.8 pM and a limit of quantification (LOQ) of 6.11 pM.

The successful application of FN-CoP in inkjet-printed electrochemical sensors opens new possibilities for sustainable biosensor development. The biocompatibility of the paste, combined with the precision of inkjet printing, makes it a promising candidate for various applications, including water monitoring and energy devices. This work represents a significant advancement in biosensing and lays the groundwork for future research into sustainable materials in sensor fabrication.

II. RESULTS AND FINDINGS

A. Ink Formulation and Characterization

After evaluating various formulations, the mixture containing 16.12% activated charcoal (AC), 6.45% vegetable gelatin, and 77.43% oral rehydration salts (ORS) is identified as the optimal composition as shown in Table 1. This specific formulation provided the ideal balance between viscosity, which is in the range of 1,600-4,000 mPa.s using Digital Rotational Viscometer (LMDV-60, India), and electrical conductivity, measured as 1484 S/m using Liquid Conductivity meter (Elico CM 180, Hyderabad, India) Both these parameters are critical for ensuring the FN-CoP is compatible with inkjet printing processes. Scanning Electron Microscopy (SEM) analysis of the paste revealed an average particle size of 64 nm, which contributed to the uniform deposition of materials on various substrates, ensuring high-quality electrode patterning. The nanoscale dimensions of the particles also played a role in enhancing the surface area, a key factor in boosting the sensors sensitivity.

B. Electrode Fabrication via Inkjet Printing

The inkjet printing process is systematically optimized to produce electrodes on three different substrates: glass, polyethylene terephthalate (PET), and polyamide sheets using benchtop Voltera V-One (Voltera, USA) PCB inkjet printer. The optimized FN-CoP is loaded in into the nozzle of the printer, where droplets are accurately deposited onto the substrates in pre-defined patterns as shown in Figure 1. This demonstrates excellent compatibility with all these substrates, forming uniform, high-resolution electrode patterns. A crucial advantage of this method is the ability to fabricate electrodes with minimal material waste while maintaining high reproducibility, offering a significant improvement over traditional screen-printing technique.

TABLE 1: OPTIMIZED FN-CoP COMPOSITION AND PROPERTIES

FN-CoP Composition				
Activated charcoal (AC)	*Veg Gelatin*	*Oral rehydration Solution (ORS)*	*Viscosity (mPa.s)*	*Conductivity (S/m)*
20.41 %	8.16 %	71.43 %	5,000-10,000	1787
18.97 %	7.59 %	73.43 %	5,000-9,000	1709
17.54 %	7.02 %	75.43 %	3,000-5,000	1622
16.12 %	**6.45 %**	**77.43 %**	**1,600-4,000**	**1484**

The FN-CoP formulation also displayed excellent ink-lock properties, contributing to robust adhesion and durability across all substrates. These features of FN-CoP not only reduce the environmental impact of the fabrication process but also ensure long-term stability of the sensors.

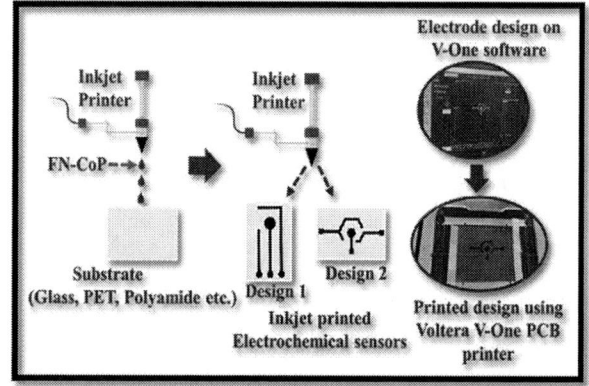

Figure 1: Fabrication of Electrodes Using Inkjet Printing with Developed Food-Based Nano Conductive Paste (FN-CoP).

C. Electrochemical Performance and Insulin Sensing

The electrochemical performance of the inkjet-printed FN-CoP sensors is validated through insulin detection experiments using Potentiostat (BioLogic , SP-150, France). Figure 2(a) shows the successful modification of electrode with TiO_2 in $K_3[Fe(CN)_6]$. Cyclic voltammetry (CV) experiments revealed that the insulin oxidation potential occurred at 0.7 V for various concentration, highlighting the sensitivity of the FN-CoP electrodes as shown in Figure 2(b,c) . To assess the selectivity of the sensors, chronoamperometry studies are conducted using a variety of bioanalytes (Mix 1-9). The results confirmed that FN-CoP-modified electrodes displayed minimal interference from other bioanalytes, reinforcing the high selectivity of the paste for insulin detection as shown in Figure 2(d). The FN-CoP electrodes achieved a limit of detection (LoD) of 1.8 pM and a limit of quantification (LOQ) of 6.11 pM, indicating that the sensors are highly sensitive and suitable for detecting insulin at very low concentrations.This level of sensitivity positions FN-CoP as a viable material for biosensing applications where precise detection of analytes is crucial, such as in glucose or hormone monitoring.

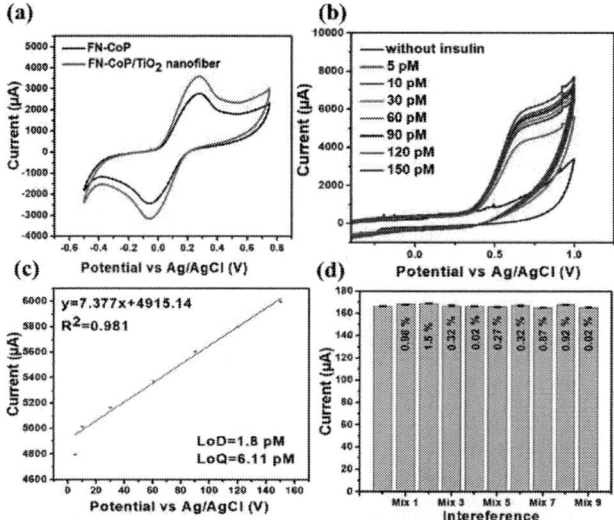

Figure 2 (a) CV response of FN-CoP and Modified FN-CoP at 20 mVs⁻¹ in K₃[Fe(CN)₆]. (b,c) Calibration curve for various concentration of Insulin (5 pM, 10 pM, 30 pM, 60 pM, 90 pM, 120 pM and 150 pM) at 20 mVs⁻¹ in PBS on FN-CoP/TiO₂. (d) Error bar showing Interference studies of various bio analytes by Chronoamperometry technique at E_{ox}=0.7 V in PBS (Insulin, **Mix 1**=Insulin+Ascorbic acid(AA), **Mix 2**-Mix 1+Dopamine(DA), **Mix 3**-Mix 2+Uric acid (UA), **Mix 4**- Mix 3+Fructose(F), **Mix 5**- Mix 4+Nitrate(NO₃⁻), **Mix 6**- Mix 5+Nitrite (NO₂⁻), **Mix 7**- Mix 6+Creatinine, **Mix 8**- Mix 7+Lactic acid, **Mix 9**- Mix 8+L-Tyrosine.

D. Substrate Adhesion and, Stability

The stability of the inkjet-printed FN-CoP electrodes is rigorously tested to evaluate their potential for long-term use. Electrodes (FN-CoP/TiO₂) are printed on glass substrate, and subjected to multiple washing cycles to simulate environmental conditions and assess the robustness of the adhesion. After four wash cycles, the FN-CoP/TiO₂

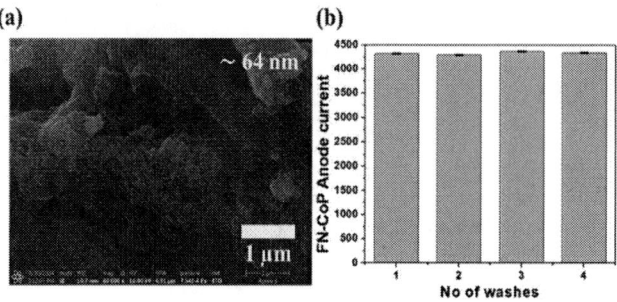

Figure 3 (a) SEM micrograph of FN-CoP (b) Error bar from CV response of inkjet printed FN-CoP/TiO2 electrodes at a scan rate of 20 mV/s in PBS

solution for 5pM of insulin, demonstrating stability over multiple washing cycle

electrodes showed no significant degradation, demonstrating excellent stability and strong adhesion to the substrates tested (Figure 3). These results indicate that FN-CoP is a highly reliable material for the fabrication of durable electrochemical sensors. The robust performance across various substrates, coupled with the eco-friendly nature of the paste, underscores the potential for using FN-CoP in a range of biosensing and energy-related applications.

CONCLUSION

This work demonstrates the successful development of inkjet-printed electrochemical sensors using non-toxic inhouse prepared food-based nano conductive paste (FN-CoP), marking a advancement in sustainable sensor fabrication. Leveraging the precision of inkjet printing, this study achieves high-resolution, uniform electrode patterns with minimal material waste on substrates such as glass, PET, cloth and polyamide. The optimized FN-CoP formulation, comprising 16.12% activated charcoal, 6.45% vegetable gelatin, and 77.43% oral rehydration salts (ORS), not only ensures biocompatibility but also provides excellent adhesion and durability, maintaining strong performance until 4 wash cycles. With a conductivity of 1484 S/m and a limit of detection (LoD) of 1.8 pM for insulin, this method presents a scalable, eco-friendly alternative to conventional screen-printing techniques, offering enhanced sensitivity and selectivity for biosensing and for future potential applications in energy harvesting.

ACKNOWLEDGMENTS

The authors are grateful to the Central Analytical Laboratory of the BITS Pilani Hyderabad campus for their contribution as well as DST-SERB Power Grant Program SPG/2021/001087 and ministry of science and technology, Indo-Austria (WTZ), DST/IC/Austria/P-3/2021) for providing financial help.

REFERENCES

[1] E. S. Hosseini, S. Dervin, P. Ganguly, and R. Dahiya, "Biodegradable Materials for Sustainable Health Monitoring Devices," 2021. doi: 10.1021/acsabm.0c01139.

[2] J. Muñoz and M. Baeza, "Customized Bio-functionalization of Nanocomposite Carbon Paste Electrodes for Electrochemical Sensing: A Mini Review," 2017. doi: 10.1002/elan.201700087.

[3] R. R. Suresh et al., "Fabrication of screen-printed electrodes: opportunities and challenges," 2021. doi: 10.1007/s10853-020-05499-1.

[4] P. Rewatkar, P. K. Enaganti, M. Rishi, S. Mukhopadhyay, and S. Goel, "Single-step inkjet-printed paper-origami arrayed air-breathing microfluidic microbial fuel cell and its validation," Int J Hydrogen Energy, vol. 46, no. 71, 2021, doi: 10.1016/j.ijhydene.2021.08.102.

Chirality Reversal Related Spin-Orbit Torque Switching in Nanoscale Perpendicular Ferromagnet

Xue Zhang, Zhengde Xu, and Zhifeng Zhu[†]

School of Information Science and Technology, ShanghaiTech University, Shanghai 201210, China,

[†]E-mail: zhuzhf@shanghaitech.edu.cn

Abstract

The miniaturization of magnetic memory to nanometer scale requires a theoretical model to accurately predict the switching current. Previous models show large discrepancy with experimental results in studying the spin-orbit torque switching of perpendicular magnets. In this study, we find that magnetization follows a smooth transition during the switching, which allows for a reduced switching current. Guided by this refined physical picture, we identify the reversal of precession chirality as a pivotal factor and develop an analytical model which shows good alignment with the experimental data. Our work provides a more comprehensive understanding of the current-induced magnetization switching mechanism, which will benefit the developments in spin-orbit torque magnetic random-access memory.

Keywords: nanoscale ferromagnet, spin-orbit torque

Introduction

In recent years, spin-orbit torque (SOT) devices have gained attention for their lower switching current and faster operation speed [1-3]. Typically, there are three types of SOT devices categorized by the direction of the easy axis, i.e., type x, y, and z. Type x and y require an elliptical sample to stabilize the magnetization within the thin film plane, which presents challenges in the device fabrication when a small sample to sample variation is required. In contrast, the demagnetizing field assists the switching in type z, and the device can have circular shape since the in-plane components along the x and y axes are on an equal footing. However, since the spin polarization (σ) and the equilibrium magnetization are noncollinear in type z, the deterministic switching requires both the SOT and the assisted magnetic field along the current direction, and thus one cannot simply balance the SOT and damping-like torque to obtain the threshold current (J_{th}). The previous study [4] concluded that $J_{th} = 2eM_st_F \times (H^{eff}_K/2 - H_{ext}/\sqrt{2})/\hbar\theta_{SH}$, where $H^{eff}_K = H_K - H_d$ is the effective field, and t_F is the thickness of the FM layer. Based on the same principle, modified equations have been proposed to include the effect of field-like torque (FLT), thermal fluctuation and second-order perpendicular magnetic anisotropy [5-7]. However, the predicted J_{th} remains significantly higher than experimental values, highlighting the need for a more accurate analytical model.

In this work, we employ the macrospin model and present numerical results for SOT induced switching of perpendicular magnetization. The trajectory of magnetization reveals a smooth transition, indicating a smaller switching current in both the numerical and analytical models. Based on this observation, we conduct a theoretical analysis of the switching process and identify the alteration of precession chirality as a key factor in achieving deterministic switching. Our results demonstrate excellent agreement between numerical, analytical, and experimental results for nanoscale devices.

Methodology

Fig. 1 illustrates the device studied in this work. The sample size is 50 nm × 50 nm × 1.2 nm, in which the ferromagnetic (FM) layer has perpendicular magnetization that is stabilized by the crystalline anisotropy field (H_K) along the z direction and the demagnetizing field (H_d) which is calculated based on the sample geometry. The charge current (J_c) is applied along the x direction through the heavy-metal (HM) layer, which produces a spin current (J_s) that flows into the FM layer with a resultant SOT. The direction of σ can be determined by the following equation: $J_c = \theta_{SH}\sigma \times J_c$, i.e., $\sigma = -y$ when J_c is along the x direction. The spin-Hall angle $\theta_{SH} = 0.03$ is used in this work [8].

The magnetization dynamics is described by the Landau-Lifshitz-Gilbert-Slonczewski (LLGS) equation:
$$\partial\mathbf{m}/\partial t = -\gamma\mathbf{m} \times (\mathbf{H}_{eff} + \mathbf{H}_{ext}) + \alpha\mathbf{m} \times \partial\mathbf{m}/\partial t - \gamma B_{D,SOT}\mathbf{m} \times (\mathbf{m} \times \boldsymbol{\sigma}). \quad (1)$$
$\gamma = 1.76 \times 10^{11}$ s^{-1}T^{-1} is the gyromagnetic ratio, and $H_{eff} = H_K - H_d = 0.5$ T is the effective magnetic field. The damping constant $\alpha = 0.02$ [8]. $B_{D,SOT} = \hbar\theta_{SH}J_c/2eM_st_F$ represents the strength of the damping-like SOT. The saturation magnetization $M_s = 1.3$ T [8].

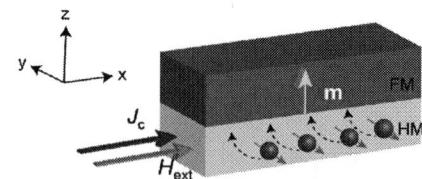

Fig. 1. Illustration of the device structure.

Results and Discussion

SOT switching of perpendicular magnetization has a key feature: both SOT and $\mathbf{H}_{ext}$ are indispensable. We confirm this by applying $\mathbf{J}_c$ and $\mathbf{H}_{ext}$ separately. As shown in Fig. 2(a), when only $\mathbf{H}_{ext}$ is applied, the magnetization remains at the initial state. In Fig. 2(b), in the absence of $\mathbf{H}_{ext}$, the magnetization remains the same under small $\mathbf{J}_c$, and the SOT pulls it to $\boldsymbol{\sigma}$ when $\mathbf{J}_c$ is large at 4.5×10^{12} A/m^2. In contrast, when a small $\mathbf{H}_{ext} = -20$ mT is applied, the magnetization can be easily switched under a small $\mathbf{J}_c = 2.604\times10^{12}$ A/m^2 [see Fig. 2(c)]. In addition, the switching polarity reverses when $\mathbf{H}_{ext}$ is reversed. These results show that our model correctly captures the key features of the SOT induced switching.

It is worth noting that the experimental J_{th} is 2.76×10^{12} A/m^2 [8], which is almost the same as our result. In contrast, the previous model predicts $J_{th} = 76\times10^{12}$ A/m^2 [4], in which $H_{eff}^K = 0.5$ T is used. This value is 28 times larger than the experimental result. Since samples with diameter $D = 50$ nm were used in the experiment, one can assume that the switching is uniform. Therefore, our numerical results agree excellently with the experiment, indicating that the macrospin model is indeed suitable to describe the magnetization switching of nanoscale magnet. The remaining challenge is to resolve the disagreement between the analytical model and the experimental results.

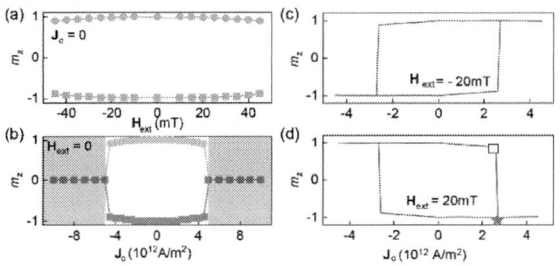

Fig. 2. (a) m_z as a function of $\mathbf{H}_{ext}$ when $\mathbf{J}_c = 0$. The dots and squares represent the sample with the initial state $m_z = 1$ and $m_z = -1$, respectively. (b) m_z as a function of $\mathbf{J}_c$ when $\mathbf{H}_{ext} = 0$. m_z-$\mathbf{J}_c$ loop with (c) negative and (d) positive $\mathbf{H}_{ext}$. The yellow square and the star mark points before and after the switching.

Fig. 3(a) shows the direct current (DC)-induced switching in our work, where the switching trajectory exhibits a smooth magnetization transition from up to down. In contrast to the analytical model proposed in Ref. 4, which suggests that the magnetization first aligns with $\boldsymbol{\sigma}$ before switching to the opposite state. Our findings in Fig. 2(b) show that this alignment requires a substantial J_c, specifically 4.5×10^{12} A/m^2. This accounts for the overlarge J_{th} predicted by the previous analytical model.

We also plot the time evolution for different pulse widths and J_c values. As shown in Figs. 3(b, c), both the 5 ns pulse (as used in Ref. 4) and the 10 ns pulse (as used in experiment Ref. 8) yield the same deterministic switching results, confirming that our results are independent of the pulse duration. As shown in Fig. 3(d), when a very large J_c is applied, the magnetization is pulled to the $\boldsymbol{\sigma}$ direction, consistent with observation in Ref. 4. Therefore, we conclude

that Ref. 4 is suitable for high J_c scenarios.

In contrast to the previous model, our numerical simulation demonstrates a smooth magnetization transition from up to down state. Comparing these two switching pictures, it becomes evident that the key to resolve the disagreement between the experiment and the analytical model lies in developing a model based on the smooth switching picture. To identify the critical switching condition, we analyze the magnetization trajectory before and after J_{th} [marked as a square and star in Fig. 2(d)]. As shown in Fig. 4(a), when J_c is close to J_{th}, the magnetization precesses without triggering the switching. It is noticed that the precession direction remains in the counterclockwise direction. When $J_c = J_{th}$, the switching happens, and the trajectories are shown in Figs. 3(a) and 4(b). Before reaching this point, the magnetization precesses in the counterclockwise direction, similar to the previous non-switching case. However, after crossing this point, the precession direction reverses. By varying both the polarity and magnitude of $\mathbf{J}_c$ and $\mathbf{H}_{ext}$, we consistently observe that successful switching is accompanied by this reversal of precession chirality. Thus, we identify it as a critical condition for deterministic switching. Additionally, we find that near to the critical point, m_x, m_y and m_z are stable, and m_y is almost zero (i.e., the azimuthal angle $\varphi = 0$). We use this as the other critical condition.

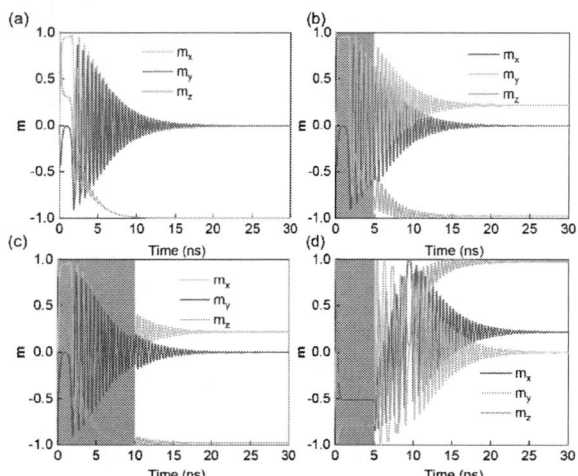

Fig. 3. Time evolution of $\mathbf{m}$ when $J_c = 2.604\times10^{12}$ A/m^2 with (a) DC current, (b) 5 ns pulse, and (c) 10 ns pulse. (d) The magnetization dynamics under $J_c = 7.5\times10^{12}$ A/m^2 with 5 ns pulse.

To derive the analytical expression of J_{th} based on these observations, we convert the LLG equation into the Landau-Lifshitz (LL) form:

$$\partial\mathbf{m}/\partial t\,(1+\alpha^2) = -\gamma\mathbf{m}\times\mathbf{H}_{ext} - \gamma\alpha\mathbf{m}\times(\mathbf{m}\times\mathbf{H}_{ext}) - \gamma\mathbf{m}\times\mathbf{H}_{eff} - \gamma\alpha\mathbf{m}\times(\mathbf{m}\times\mathbf{H}_{eff}) + \gamma\alpha B_{D,SOT}\mathbf{m}\times\boldsymbol{\sigma} - \gamma B_{D,SOT}\mathbf{m}\times(\mathbf{m}\times\boldsymbol{\sigma}).$$

As a result, there are six torques in this system, which we label as torques 1 through 6 in order. We start in the state with $\mathbf{m}$ along the $+\mathbf{z}$-direction, where torque 3 ($-\gamma\mathbf{m}\times\mathbf{H}_{eff}$)

and torque 4 ($-\gamma\alpha\mathbf{m}\times(\mathbf{m}\times\mathbf{H}_{\text{eff}})$) are zero. The rest of the torques pull $\mathbf{m}$ into the vector space with $\mathbf{x} > 0$, $\mathbf{y} < 0$, and $\mathbf{z} > 0$. Subsequently, $\mathbf{m}$ gradually moves closer to the $\mathbf{x}$-$\mathbf{z}$ plane until the chirality reverses. In this space, the rotation direction is determined by torques 1, 3, and 6 while torques 2, 4, and 5 compete to switch $\mathbf{m}$ across the $\mathbf{x}$-$\mathbf{y}$ plane. We then express $\mathbf{m} = [\sin\theta\cos\varphi, \sin\theta\sin\varphi, \cos\theta]$ where θ is the polar angle. $\mathbf{H}_{\text{eff}} = [-H_{\text{d},x}m_x, -H_{\text{d},y}m_y, (H_k - H_{\text{d},z})m_z]$ where H_d is the strength of the demagnetizing field. To fulfill the condition that the precession chirality is reversed, one needs the total torque from 1, 3, and 6 to be zero which is shown in Fig. 4(b). With the additional requirement that $\varphi = 0$, we arrive at an analytical expression of J_{th} as:

$$J_{\text{th}} = \cos\theta(\sin\theta(H_k - H_{\text{d},z} + H_{\text{d},x}) - H_{\text{ext}})2t_{\text{FM}}M_s e/\theta_{\text{SH}}\hbar.$$

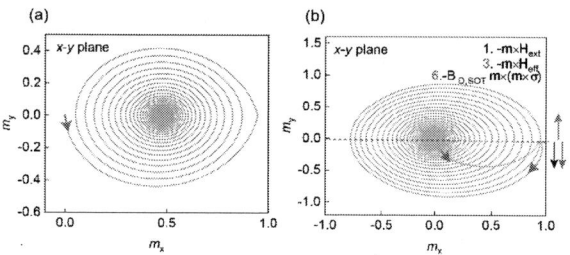

Fig. 4. The projection of the magnetization dynamics on the $\mathbf{x}$-$\mathbf{y}$ plane with (a) $\mathbf{J}_c = 2.603\times10^{12}$ A/m^2 and (b) $\mathbf{J}_c = 2.604\times10^{12}$ A/m^2.

We then compare our results with experimental data for devices of various sizes. As shown in Fig. 5, J_{th} consistently decreases as the device size increases. Notably, both our analytical model and macrospin simulation results show excellent agreement with the experimental data [8].

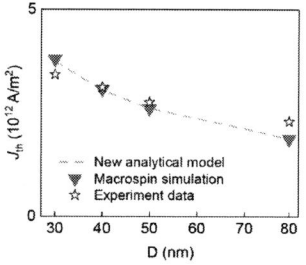

Fig. 5. J_{th} as a function of device size.

Conclusion

In this work, we revisit the spin-orbit torque switching of nanoscale perpendicular magnet. We observed that the spin-orbit torque and external magnetic field act cooperatively, producing a smooth magnetization transition. The refined physical picture guarantees a smaller switching current which is comparable to the experimental results. We further identify the chirality reversal of precession as a key condition for the deterministic switching, based on which we derived a new analytical expression for the switching current that shows remarkable agreement with experiments.

Acknowledgments

We acknowledge the support from the National Key R&D Program of China (Grant No. 2022YFB4401700), National Natural Science Foundation of China (Grants Nos. 12104301). The simulation conducted in this work is supported by SIST Computing Platform at ShanghaiTech University.

References

[1] I. M. Miron, K. Garello, G. Gaudin, P.-J. Zermatten, M. V. Costache, S. Auffret, S. Bandiera, B. Rodmacq, A. Schuhl, P. Gambardella, Perpendicular switching of a single ferromagnetic layer induced by in-plane current injection, Nature. **476**, 189 (2011).

[2] L. Liu, C.-F. Pai, Y. Li, H. Tseng, D. Ralph, R. Buhrman, Spin-torque switching with the giant spin Hall effect of tantalum, Science. **336**, 555 (2012).

[3] S. Fukami, T. Anekawa, C. Zhang, H. Ohno, A spin–orbit torque switching scheme with collinear magnetic easy axis and current configuration, Nat. Nanotechnol. **11**, 621 (2016).

[4] K.-S. Lee, S.-W. Lee, B.-C. Min, K.-J. Lee, Threshold current for switching of a perpendicular magnetic layer induced by spin Hall effect, Appl. Phys. Lett. **102**, 112410 (2013).

[5] T. Taniguchi, S. Mitani, M. Hayashi, Critical current destabilizing perpendicular magnetization by the spin Hall effect, Phys. Rev. B. **92**, 024428 (2015).

[6] S. J. Yun, K.-J. Lee, S. H. Lim, Critical switching current density induced by spin Hall effect in magnetic structures with first-and second-order perpendicular magnetic anisotropy, Sci. Rep. **7**, 15314 (2017).

[7] K.-S. Lee, S.-W. Lee, B.-C. Min, K.-J. Lee, Thermally activated switching of perpendicular magnet by spin-orbit spin torque, Appl. Phys. Lett. **104**, 072413 (2014).

[8] C. Zhang, S. Fukami, H. Sato, F. Matsukura, H. Ohno, Spin-orbit torque induced magnetization switching in nano-scale Ta/CoFeB/MgO, Appl. Phys. Lett. **107**, 012401 (2015)

Portable NV (Nitrogen Vacancy) Vector Magnetometer with High Sensitivity and Wide Dynamic Range

Yingjie Yang[123], Jianghao Fu[123], Xuanhui Ren[123], Doudou Zheng[123],
Yang Li[123], Hui Wang[123], Zongmin Ma[123]*, Jun Liu[123]*

[1]North University of China Institute of Instrument and Electronics, China
[2]Key Laboratory of Quantum Sensing and Precision Measurement, China
[3]The State Key Laboratory of Dynamic Measurement Technology, China

Abstract

In this study, we propose a compact, integrated vector magnetometer that can simultaneously measure the dynamic magnetic field at the nitrogen vacancy (NV) centers in diamond. The device is capable of converting the detected magnetic field into its respective components within a spatial coordinate system. The system employs a multi-channel phase-lock technique to retrieve magnetic field information along each NV axis from fluorescence signals captured by a single photodetector, enabling real-time vector measurements. The stability and dynamic range of the system is improved by adjusting the microwave frequency in real time.

Keywords: nitrogen-vacancy (NV) centers, integrated vector magnetometer, simultaneous vector measurement

Introduction

Portable，rugged，and highly sensitive magnetometers are extensively utilized in industrial and geophysical applications, including magnetic navigation[1],[2], [3],[4], battery scanning, surveying[5], and biomagnetism field detection [6],[7],[8],[9]. NV centers are strong contenders for vector magnetic field measurement applications [10],[11]. In NV magnetometers, the diamond can be aligned along any of the four crystallographic orientations in the tetrahedral diamond lattice, which itself has a vector function. The rigidity of the diamond lattice ensures that these axes are precisely defined, minimizing issues related to non-linearity and non-orthogonality [12],[13],[14],[15]. Utilizing feedback control can greatly expand the dynamic range [16],[17],[18]. These characteristics render NV magnetometers particularly well-suited for deployment on mobile platforms.

Several integrated or partially integrated NV magnetometers have demonstrated sensitivities at the nanotesla level and below, including some configurations utilizing flux controllers in molly centering machines [18],[19]. However, not all of these systems possess vector and feedback control capabilities. Platform testing has achieved vector sensitivities as high as 50 pT/$\sqrt{\text{Hz}}$. Nevertheless, due to inherent platform limitations, these systems are not suitable for mobile or out-of-field measurements. While some integrated vector magnetometers have achieved portable integration with vector measurement capabilities, they typically rely on the traditional frequency-hopping method, wherein the microwave (MW) source sequentially and continuously resonates along different NV axes. This approach is limited to sensing static or slowly varying magnetic; when the magnetic varies more rapidly than the frequency-hopping interval, inaccuracies in vector calculation arise. Additionally, the sensitivity of such vector magnetometers is inherently unsatisfactory because the sensor's sensing of the external magnetic field is temporarily lost during each frequency-hopping address, resulting in a certain amount of dead time for vector field measurements, including the time it takes to change the MW frequency or to drive away from the resonance, which reduces both the speed and bandwidth of the measurement.

In this study, we propose a portable NV vector magnetometer with dimensions of 162.5mm × 55mm × 63mm, capable of conducting magnetic field measurements both indoors and outdoors while mounted on a mobile platform. The device utilizes feedback control and simultaneous readings along four NV directions to capture all Cartesian components of the magnetic in real time, enabling the reconstruction of the complete magnetic field vector. Our measurement scheme achieves high sensitivity，and a wide dynamic range. The dynamic range of the AC magnetic field is ± 148.8uT in the X, ± 151.2uT in the Y, and ± 152.5uT in the Z. Additionally, the three-axis sensitivity is 0.93nT/in the X, 0.76nT/Hz in the Y, and 0.54nT/Hz in the Z. This approach circumvents the inefficiencies associated with conventional frequency hopping measurement techniques and provides a robust foundation for outdoor vectorial experiments.

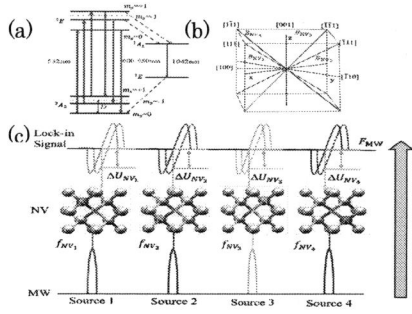

Fig. 1. (a) Energy level diagram of the NV centers in diamond. (b) Diagram showing the NV symmetry axes and the orientations of the laboratory coordinate system relative to the diamond lattice vectors X, Y, and Z. (c) Schematic diagram of the simultaneous vector measurement process. Four MW signals probe the resonance frequencies of the four NV axes in real time, and the resonance offset frequency is derived from the positively correlated offset voltage difference.

displacements; the lower panel shows the corresponding rms values with a sketch illustrating the orientation of the NV centers corresponding to each measured frequency offset.

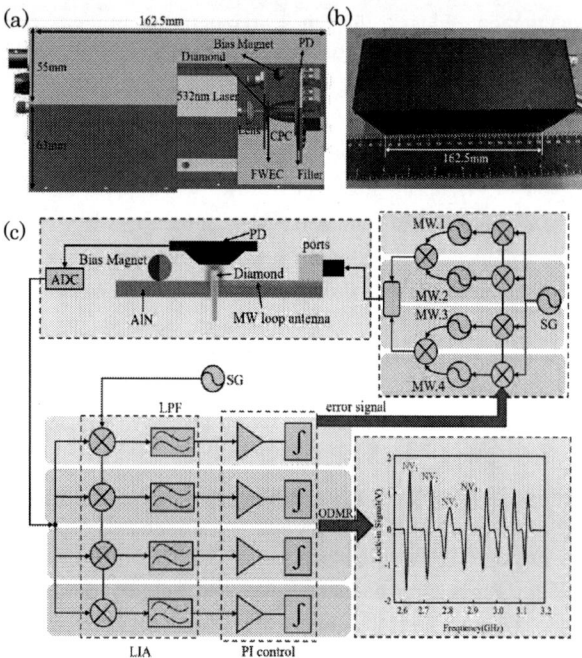

Fig. 2 (a) Three-dimensional sectional view of the portable integrated probe; (b) physical representation of the portable integrated probe; (c) schematic diagram of the experimental setup.

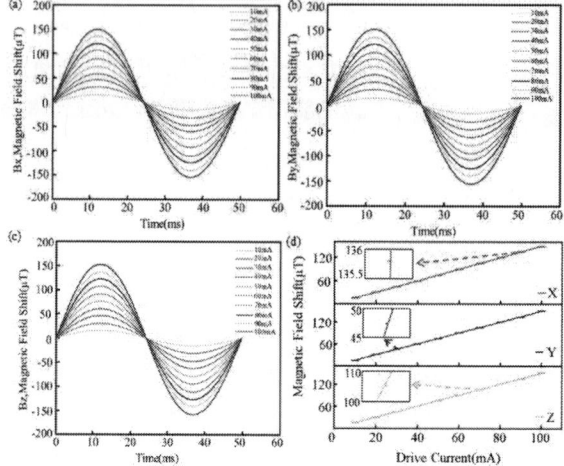

Fig.4. Dynamic range and linearity of AC magnetic field measurements. (a) Magnetic field detection range following the X of the magnetometer. (b) Magnetic field detection range following the Y of the magnetometer. (c) Magnetic field detection range following the Z of the magnetometer. (d) Linearity of the magnetometer measurements following the X, Y, and Z

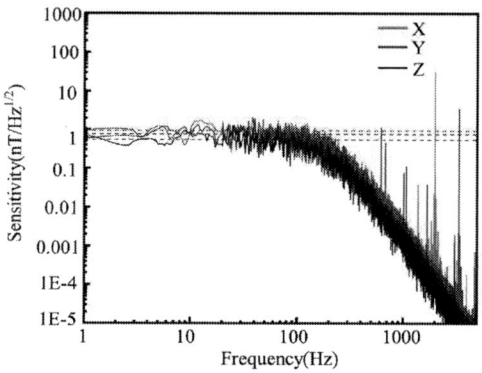

Fig. 5 Triaxial sensitivity of the vector magnetometer.

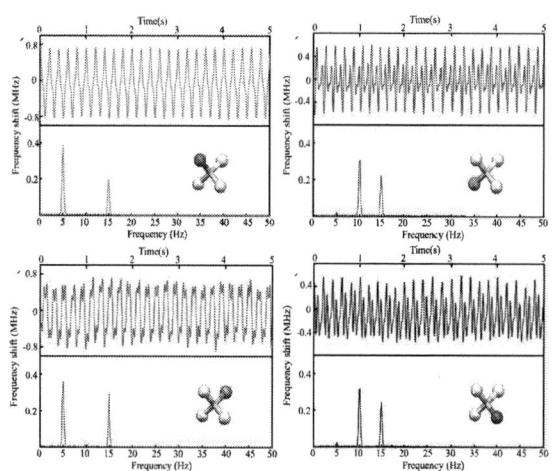

Fig. 3 (a)-(d) show the frequency shifts associated with the magnetic field. The above graphs show the time domain signals for 5 seconds of LIA acquisition. These show the frequency offsets associated with magnetic field

Conclusion

These measurements demonstrate that our portable NV magnetometer effectively measures rapidly changing AC magnetic fields and performs vector decomposition within a spatial coordinate system. The vector measurements are in excellent agreement with fluxgate measurements, exhibiting errors in the range of 2% to 3%. Furthermore, by incorporating a feedback mechanism and synchronizing vector measurements, the magnetometer achieves pT sensitivities with values of 0.93, 0.76, and 0.54 nT/$\sqrt{\text{Hz}}$ following the X, Y, and Z axes, separately. The AC magnetic field detection ranges are $\pm148.8\mu T$ following the X, $\pm151.2\mu T$ following the Y, and $\pm152.5\mu T$ following the Z. In future work, achieving higher sub-Hz vector sensitivities will be essential for detecting magnetic anomaly fields, particularly when the magnetometer is deployed on vehicles operating at significant altitudes above the Earth's surface . As the robustness of our feedback control system improves, it is expected to become a key technique for this purpose. Additional sensitivity gains can be obtained by enhancing the green-red photon conversion efficiency, . Furthermore, employing dual resonance magnetometry, which involves addressing and tracking all eight ODMR peaks simultaneously, can provide increased stability and sensitivity.

Acknowledgments

The authors gratefully acknowledge the contributions of J. E. Lilienfeld, J. Bardeen, W. Brattain, and W. Shockley to the semiconductor industry.

References

[1] W. L. Webb, "Aircraft navigation instruments," Electr. Eng., vol. 70, no. 5, pp. 384–389, May 1951, doi: 10.1109/EE.1951.6432396

[2] J. E. Lenz, "A review of magnetic sensors," Proc. IEEE, vol. 78, no. 6, pp. 973–989, Jun. 1990, doi: 10.1109/5.56910.

[3] J. Lenz and S. Edelstein, "Magnetic sensors and their applications," IEEE Sensors J., vol. 6, no. 3, pp. 631–649, Jun. 2006, doi: 10.1109/JSEN.2006.874493.

[4] C. J. Cochrane, J. Blacksberg, M. A. Anders, and P. M. Lenahan, "Vectorized magnetometer for space applications using electrical readout of atomic scale defects in silicon carbide," Sci Rep, vol. 6, no. 1, p. 37077, Nov. 2016, doi: 10.1038/srep37077.

[5] D. T. Germain-Jones, "Post-war developments in geophysical instrumentation for oil prospecting," J. Sci. Instrum., vol. 34, no. 1, pp. 1–8, Jan. 1957, doi: 10.1088/0950-7671/34/1/302.

[6] D. R. Glenn et al., "Single-cell magnetic imaging using a quantum diamond microscope," Nat Methods, vol. 12, no. 8, pp. 736–738, Aug. 2015, doi: 10.1038/nmeth.3449.

[7] J. F. Barry et al., "Optical magnetic detection of single-neuron action potentials using quantum defects in diamond," Proc. Natl. Acad. Sci. U.S.A., vol. 113, no. 49, pp. 14133–14138, Dec. 2016, doi: 10.1073/pnas.1601513113.

[8] T. Wang, Y. Zhou, C. Lei, J. Luo, S. Xie, and H. Pu, "Magnetic impedance biosensor: A review," Biosensors and Bioelectronics, vol. 90, pp. 418–435, Apr. 2017, doi: 10.1016/j.bios.2016.10.031.

[9] H. C. Davis et al., "Mapping the microscale origins of magnetic resonance image contrast with subcellular diamond magnetometry," Nat Commun, vol. 9, no. 1, p. 131, Jan. 2018, doi: 10.1038/s41467-017-02471-7.

[10] J. M. Taylor et al., "High-sensitivity diamond magnetometer with nanoscale resolution," Nature Phys, vol. 4, no. 10, pp. 810–816, Oct. 2008, doi: 10.1038/nphys1075.

[11] V. M. Acosta et al., "Diamonds with a high density of nitrogen-vacancy centers for magnetometry applications," Phys. Rev. B, vol. 80, no. 11, p. 115202, Sep. 2009, doi: 10.1103/PhysRevB.80.115202.

[12] L. M. Pham et al., "Magnetic field imaging with nitrogen-vacancy ensembles," New J. Phys., vol. 13, no. 4, p. 045021, Apr. 2011, doi: 10.1088/1367-2630/13/4/045021.

[13] H. Clevenson, L. M. Pham, C. Teale, K. Johnson, D. Englund, and D. Braje, "Robust high-dynamic-range vector magnetometry with nitrogen-vacancy centers in diamond," Applied Physics Letters, vol. 112, no. 25, p. 252406, Jun. 2018, doi: 10.1063/1.5034216.

[14] J. M. Schloss, J. F. Barry, M. J. Turner, and R. L. Walsworth, "Simultaneous Broadband Vector Magnetometry Using Solid-State Spins," Phys. Rev. Applied, vol. 10, no. 3, p. 034044, Sep. 2018, doi: 10.1103/PhysRevApplied.10.034044.

[15] C. Zhang et al., "Vector magnetometer based on synchronous manipulation of nitrogen-vacancy centers in all crystal directions," J. Phys. D: Appl. Phys., vol. 51, no. 15, p. 155102, Apr. 2018, doi: 10.1088/1361-6463/aab2d0.

[16] C. Wang et al., "Realization of high-dynamic-range broadband magnetic-field sensing with ensemble nitrogen-vacancy centers in diamond," Review of Scientific Instruments, vol. 94, no. 1, p. 015109, Jan. 2023, doi: 10.1063/5.0089908.

[17] K.-M. C. Fu, G. Z. Iwata, A. Wickenbrock, and D. Budker, "Sensitive magnetometry in challenging environments," AVS Quantum Science, vol. 2, no. 4, p. 044702, Dec. 2020, doi: 10.1116/5.0025186.

Yingjie Yang received a M.S. degree from the North University of China, in 2021. He is currently pursuing a Ph.D. degree at the North University of China. His research interests include quantum sensing.e-mail:857385922@qq.com

Zongmin Ma received a Ph.D. degree from Osaka University in 2013. Currently, he works as a professor at the State Key Laboratory of Dynamic Testing Technology of the North University of China. His research interests include quantum sensing and ultra-precision measuring technology. e-mail:mzmncit@163.com

Jun Liu received a Ph.D. degree from the Beijing Institute of Technology in 2005. He works as a professor at the State Key Laboratory of Dynamic Testing Technology of the North University of China. His research interests are in the fields of quantum sensing,inertial devices and navigation, and dynamic testing technology.e-mail: liuj@nuc.edu.cn

MoS₂ transistors based 2T0C DRAM Optimized with Optical Modulation and Read Pulse Compensation

Gongpeng Lan[1,#], Dongdong Sun[1,#], Haotian Fang[1], Wuqing Fang[1], Yuanyuan Shi[1,*]

[1]School of Microelectronics, University of Science and Technology of China, Hefei, 230026, China
[#]Equal contribution, [*]Corresponding email: yuanyuanshi@ustc.edu.cn

Abstract

The capacitor-less two-transistor configuration (2T0C) brings further possibility for dynamic random access memory (DRAM) scaling down. Monolayer molybdenum disulfide (ML MoS₂) has a relatively wide band gap and low leakage current, which can be applied to 2T0C cell. In this work, 1-2L MoS₂ transistors based 2T0C DRAM cells are fabricated. The threshold voltages (V_{th}) of MoS₂ transistors are modulated via a simple illumination process to make an operation balance between write transistors and read transistors. Moreover, the capacitive coupling effect at the storage-node (SN) could be optimized by a read pulse compensation. Finally, we demonstrate 2T0C DRAM cell based on 1-2L MoS₂ with a retention time (t_{ret}) of 31.1s and low leakage current below 10^{-15}A/μm, at the hold voltage (V_{hold}) of 0V.

Keywords: MoS₂, 2D semiconductors, MOSFET, 2T0C, DRAM

Introduction

In modern computing systems, DRAM is the main memory cell, but the scalability is limited by the necessary high-aspect-ratio capacitors. Conventional 1T1C DRAM also faces serious power consumption problem due to the need for frequent data refresh and write-back operations [1]. This problem results from the high off-state leakage current of the Si-based transistors and becomes more serious as the scaling down [2], making it difficult for Si-based DRAM to maintain the original refresh time of 64ms under advanced technology nodes.

The off-state leakage current can be effectively reduced by using wide bandgap semiconductors. With extremely low leakage current, the storage capacitance can be reduced to the gate oxide capacitance of the transistor, enabling a new design of capacitorless 2T0C DRAM structure [3][5]. This structure has been proven to achieve a long retention time, and brings in the unique advantage of non-destructive reading operation. Coupled with the absence of capacitors, 2T0C provides an efficient solution for achieving high-density 3D DRAM [6].

Compared to Si (E_g=1.12eV), ML MoS₂ has a wider band gap (E_g~1.8eV) [7], which expected a lower off-state leakage current. In addition to its monolayer thickness, short channel effects (SCEs) and gate-induced drain leakage (GIDL) are expected to be effectively suppressed [8]. These reasons make MoS₂ suitable for low leakage DRAM cells. In the previous work [9], mechanical exfoliated multilayer MoS₂ based 2T0C DRAM cell is fabricated, but it shows a short retention time (~1.3s) even with applying a large negative hold voltage (-1.8V), which consumes additional power to maintain the charge retention. Besides, the mechanical exfoliated materials are difficult for large-scale integration.

In this study, we demonstrate the feasibility of large-scale integrated MoS₂-based 2T0C DRAM cells by transferring 1-2L MoS₂ grown by CVD. 1-2L MoS₂ is selected to ensure a relatively large bandgap since the bandgap of MoS₂ is thickness dependent [7]. The V_{th} of MoS₂ transistor can be effectively modulated by an illumination process, and finally it is controlled at a slightly positive position. This enables the transistors can be turned off at 0V. In addition, we apply an optimized pulse scheme on the read transistor to offset a sudden drop of the voltage at SN (V_{SN}), which is caused by the capacitive coupling effect. Finally, the t_{ret} at the V_{hold} of 0V is improved to 31.1s, with a low off-state current (I_{off}) below 10^{-15}A/μm (measurement limit).

Experiments

Fig. 1 describes the fabrication process flow and the structure schematic of 2T0C DRAM cell based on MoS₂. The local back gate is patterned by lithography and 40nm TiN is deposited by sputtering. Then 21nm HfO₂ is grown by ALD at 250°C as the gate dielectric. Monolayer MoS₂ film is wet transferred as channel and patterned through RIE. Next, the via connecting the source electrode of write transistor and the gate electrode of read transistor (*i.e.* SN), is fabricated by etching away HfO₂ through ICP-RIE. Finally, 20nm Ni/40nm Au is deposited through e-beam evaporation to form the source, drain electrodes (S/D) and the SN interconnection. All measurements are performed with the source measurement units (SMUs) of the Keysight B1500A semiconductor analyzer.

Results and Discussion

Fig. 2 shows the optical microscope images of the fabricated 2T0C DRAM arrays and a zoom-in cell. The transfer characteristic (black curve in Fig. 3a) of single MoS₂ transistor is firstly measured in dark environment. However, the V_{th} is highly positive of ~5.2V extracted at 0.1pA*W/L, which means that the turning on voltage and the writing data voltage need to be high. This is not conducive to reducing operating power consumption, so the V_{th} is need to be adjusted to an appropriate position.

For write transistor, we expect the V_{th} to be slightly positive of write transistors so that the device can be turned off at 0V. At V_{hold}= 0V, the actual non-volatile retention characteristic without additional power consumption can be investigated. It is worth to note that most t_{ret} reported so far can only be achieved at negative V_{hold}. For read transistor, the slightly negative V_{th} is expected to ensure the read margin under reasonably low supply voltage.

The V_{th} of MoS$_2$ transistor shifts negatively due to photo-generated carriers under a light illumination. At fixed light illumination condition (LED illumination of the probe station's optical microscope system), the V_{th} can be regulated via tuning the illumination duration (Fig. 3a). Fig. 3b summarizes the V_{th} dependence of MoS$_2$ transistors with light illumination duration. Controlling the illumination duration from 0s (dark) to 30s modulates the V_{th} from 5.2V to 0.1V. Balancing the requirements between read and write transistors, the illumination duration of 10s is finally chosen for all the transistors with different device dimensions (Fig. 4), which also shows the excellent stability of this regulation.

After modulating the V_{th} via illumination, we characterize the 2T0C memory performance with the pulse parameters shown in Fig. 5a-b. For write operation, write bit line (WBL) pulse inputs 1V slightly earlier than write word line (WWL) pulse, and it also turns off slightly later than WWL pulse. Furthermore, WWL raises to 1.5V to turn the write transistor on and write data '1' to the gate of read transistor. After that, V_{hold} of WWL is set to 0V to switch off the write transistor and 1V (read bit line, RBL) is used to read current data. Fig. 5c-d shows the read current (I_{read}) collected at RBL and the corresponding changes of V_{SN} after WBL pulse, respectively. The t_{ret} of data '1' is around 1.81s, extracted by the time required when the V_{SN} drops 0.1V from the initial value.

In fact, the initial V_{SN} is less than 1V due to the coupling effect caused by the parasitic capacitance from the overlap of gate and S/D. Fig. 6 shows the parasitic capacitance in 2T0C cell. A few electrons (negative charge) are injected into the SN through C_{WWL-SN} during the WWL pulse falling. So, there is a sudden drop on V_{SN} at the end of write operation, which has a serious impact on data storage. In addition, RBL pulse could also cause voltage sudden change on V_{SN} [7], as show in Eq. (1):

$$\Delta V_{SN} = \frac{\Delta V_{WWL} C_{WWL-SN} + \Delta V_{RWL} C_{RWL-SN}}{C_\Sigma} + \frac{\alpha \Delta V_{RWL} C_{ox}}{C_\Sigma} \quad (1)$$

where C_Σ=C_{WWL-SN}+C_{RWL-SN}+C_{RBL-SN}+C_{ox}, C_{ox} is the gate oxide capacitance of read transistor, α equals to 0 and 0.5 when the data of the 2T0C DRAM is data '0' and data '1', respectively. We can utilize this property to reduce the impact of capacitive coupling effect. As shown in Fig. 7a-b, there is

a pulse with opposite polarity applied to the RBL during the write operation (WWL pulse). Thanks to the opposite polarity of the pulses, the RBL pulse rises when the WWL pulse falls, thus cancelling out the impact of capacitive coupling. Fig. 7c-d show the comparison of I_{read} and V_{SN} with and without the RBL pulses. With the RBL pulses, initial I_{read} increases from 3nA to 7.5nA, and the corresponding V_{SN} increases from 0.8V to 1V, which proves the successful optimization.

However, t_{ret} is reduced to 1.09s under the criterion of ΔV_{SN}= 0.1V since the I_{off} increases with the higher initial V_{SN}. It is not suitable to use this criterion to evaluate t_{ret} for different initial V_{SN}. Since the data written is 1V, it can be considered valid until the V_{SN} is reduced to 0.5V. So, here we take 0.5V as the failure voltage value (V_F), and extract the t_{ret} according to the time when the V_{SN} drops to V_F= 0.5V. Under this criterion, t_{ret} is improved from 18.2s (without RBL pulse) to 31.1s (with RBL pulse). The result shows that the RBL pulse compensation can effectively improve memory performance. Table 1 compares the performance of all the reported MoS$_2$-based 2T0C DRAM cells, where we demonstrate the long t_{ret} with V_{hold}= 0V, rather than applying an extra voltage to hold the charge.

Conclusion

In summary, we demonstrate and optimize the 1-2L MoS$_2$-based 2T0C DRAM cell with light illumination and an improved pulse strategy. By controlling the light illumination duration, the V_{th} of MoS$_2$ transistors with different L_{ch} can be effectively regulated to ~0.2V-0.35V, making the transistors could be turned off at V_{hold}= 0V. An additional RBL pulse is utilized when the MoS$_2$ transistors based 2T0C DRAM cell is measured, which could offset the V_{SN} sudden drop due to the capacitive coupling effect. This helps a successful setting of the initial V_{SN} to the write voltage. Finally, the t_{ret} at V_{hold}= 0V is improved from 18.2s to 31.1s, with a low leakage current below 10^{-15}A/μm. In addition, the role of optical modulation in the 2T0C structure may be further explored, which could pave a new way to multi-functional in-sensor computing architectures.

Acknowledgments

This work is supported by National Natural Science Foundation of China (Grant Nos. 62374155 and 62411560158). The authors acknowledge the support of the Instruments Center for Physical Science, University of Science and Technology of China (USTC) and USTC Center for Micro- and Nanoscale Research and Fabrication.

References

[1] I. Bhati et al., IEEE Trans. Comput., 65, 108-121 (2016).
[2] D.-i. Bae et al., IEDM, pp. 28.1.1-28.1.4 (2016).
[3] A. Belmonte et al., IEDM, pp. 28.2.1-28.2.4 (2020).
[4] Q. Hu et al., Adv. Mater., 35, 2210554 (2023).
[5] L. Zheng et al., IEEE Electron Device Lett., 44, 1284-1287 (2023).
[6] X. Duan et al., IEEE Trans. Electron Devices, 69, 2196-2202 (2022).
[7] K. F. Mak et al., Phys. Rev. Lett., 105, 136805 (2010).
[8] C. Kshirsagar et al., Device Research Conference, 187-188 (2014).
[9] C. U. Kshirsagar et al., ACS Nano, 10, 8457-8464 (2016).

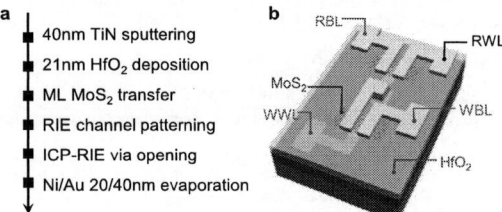

Fig. 1. Device process flow of 1-2L MoS₂ transistors based 2T0C DRAM. (a) Fabrication process flow of the MoS₂-based 2T0C DRAM cell. (b) Structure schematic of the 2T0C DRAM cell.

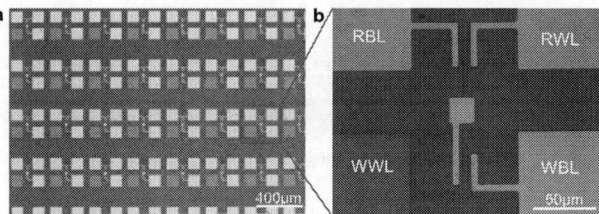

Fig. 2. 1-2L MoS₂ transistor based 2T0C DRAM arrays. (a) Optical microscope images of MoS₂-based 2T0C DRAM arrays and (b) a zoom-in cell.

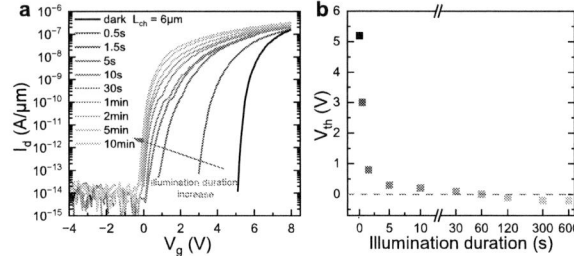

Fig. 3. V_{th} modulation of 1-2L MoS₂ transistor via light illumination duration. (a) Transfer curves of the transistor under dark and light illumination conditions. (b) V_{th} dependence on the light illumination duration, extracted from (a).

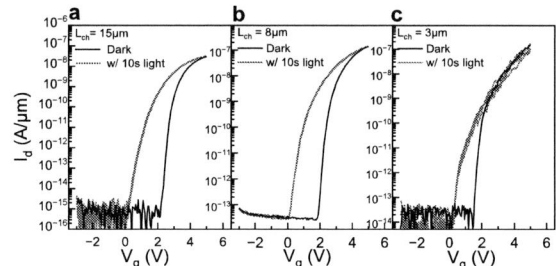

Fig. 4. Reproducible V_{th} modulation process of 1-2L MoS₂ transistors with different L_{ch}. Transfer curves of the transistors with L_{ch} of (a) 15μm, (b) 8μm and (c) 3μm under dark and 10s light illumination conditions.

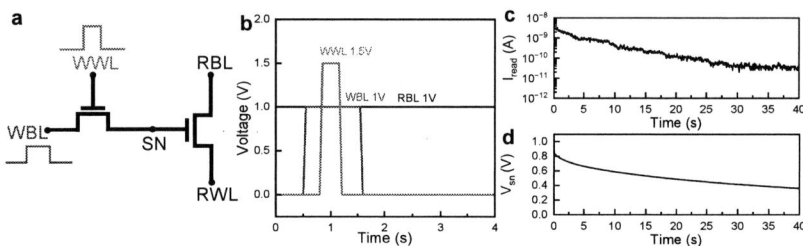

Fig. 5. 1-2L MoS₂ transistors based 2T0C DRAM operation. (a) Circuit and pulse schematics of the 2T0C DRAM cell. (b) Measurement pulse parameters for 2T0C DRAM with V_{hold}= 0V. (c) I_{read} and (d) V_{SN} versus time characteristics collected during read operation.

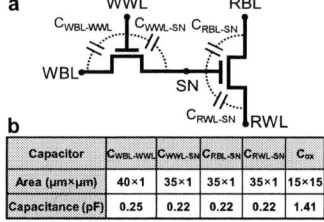

Fig. 6. Parasitic capacitance in 2T0C DRAM cell. (a) Schematic of the overlap nodes with parasitic capacitance. (b) The calculated parasitic at each overlap node.

Capacitor	$C_{WBL-WWL}$	C_{WWL-SN}	C_{RBL-SN}	C_{RWL-SN}	C_{ox}
Area (μm×μm)	40×1	35×1	35×1	35×1	15×15
Capacitance (pF)	0.25	0.22	0.22	0.22	1.41

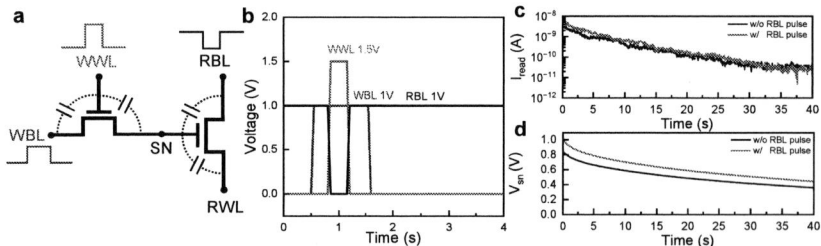

Fig. 7. 1-2L MoS₂ transistors based 2T0C DRAM operation with a read pulse compensation. (a) Circuit and optimized pulse schematics of the 2T0C DRAM cell. (b) Optimized measurement pulse with an additional RBL pulse. (c) I_{read} and (d) V_{SN} versus time characteristics with (red) and without (black) RBL pulse.

Table 1. Performance benchmark of our 1-2L MoS₂ transistors based 2T0C DRAM cell.

Ref.	Large scale	MoS₂ thickness	L_{ch} (μm)	V_{hold} (V)	Initial V_{SN} (V)	t_{on} (s) @V_r=0.5V	t_{on} (s) @I_{hold}=50%
[9]	NO	5-7nm (7-10L)	0.5	-1.8	N/A	N/A	1.3
this work	YES	0.7-1.4nm (1-2L)	15	w/o RWL pulse	0	0.8	18.2	2.71
					-1	0.8	14.4	2.54
				w/ RWL pulse	0	1.0	21.1	1.01
					-1	0.9	19.7	1.75

979-8-3315-0417-5/25 $31.00 © 2025 IEEE

Solution-processed oxide semiconductors-based enhancement-mode thin-film transistor circuits for artificial spiking neurons

Jiayi Mao[1], Qi Huang[2], and Bowen Zhu[1,2*]

[1] Key Laboratory of 3D Micro/Nano Fabrication and Characterization of Zhejiang Province, School of Engineering, Westlake University, Hangzhou, China. Email: zhubowen@westlake.edu.cn
[2] Westlake Institute for Optoelectronics, Westlake University, Hangzhou, China.

Abstract

Artificial spiking neurons have ascended as a distinguished solution for sensory information processing. However, the complexity and inherent rigidity of CMOS neuron circuits constrain their energy efficiency and adaptability for flexible integration. In this work, we successfully achieved an artificial spiking neuron circuit using enhancement-mode solution-processed indium oxide (In_2O_3) thin-film transistors (TFTs). This circuit encodes sensory information into spike signals, utilizing integration and fire along with frequency modulation, thereby enabling more energy-efficient processing.

Keywords: In_2O_3, TFT, artificial neuron, neuromorphic circuit

Introduction

Neurons possess the ability to integrate and transmit external sensory information, encoding it into electrical pulse signals characterized by event-driven and parallel processing [1, 2]. This allows individuals to adeptly process sensory data while conserving energy. Currently, sensory information is sampled through analog-to-digital converters (ADCs) [3], leading to delays and redundant data. To mitigate this, artificial neurons mimicking biological sensing mechanisms have been developed to encode sensory information into spike pulses and thus enhance the efficiency of sensory data processing.

Various spike-based artificial neuron circuits have been proposed, ranging from Hodgkin–Huxley models to adaptive integrate-and-fire models [4-6]. However, these circuits predominantly rely on CMOS technology, which poses compatibility challenges with biological systems and flexible sensors due to its complexity and rigidity. In contrast, oxide semiconductors emerge as promising candidates for constructing TFT neuron circuits, offering superior electrical performance, excellent uniformity, and compatibility with back-end-of-line (BEOL) processes for monolithic integration. Especially, solution-based methods present significant potential for flexible electronics, given their low cost and processing flexibility. However, owing to the pronounced n-doping of the semiconductor films, solution-processed metal oxide semiconductors generally function as depletion-mode devices, thereby hindering the development of artificial spiking neurons.

In this paper, we developed enhancement-mode oxide TFT devices by optimizing the annealing temperature in the solution process and successfully constructed an oxide TFT circuit based artificial spiking neuron device. This circuit enables effective frequency modulation of spike pulses in response to external stimuli and is anticipated to integrate with flexible sensors, forming a unified sensing-computing architecture for efficient, energy-conserving sensory information processing.

Methodology

The fabrication process of In_2O_3 TFTs is illustrated in Fig. 1. TFTs were fabricated on a silicon substrate with a 100 nm silicon oxide layer. Following the patterning of the bottom gate electrode, a sequential deposition of 5 nm chromium (Cr) and 50 nm gold (Au) was performed via electron beam evaporation. A 30 nm aluminum oxide layer was then deposited as the gate dielectric using an atomic layer deposition system. In the subsequent phase, indium nitrate was dissolved in 2-methoxyethanol (2-ME) to produce an In_2O_3 precursor solution. To promote an exothermic combustion reaction, equimolar amounts of acetylacetone (AcAc) and ammonium hydroxide ($NH_3 \cdot H_2O$) were incorporated as additives. The precursor solution was spin-coated onto the aluminum oxide layer and patterned to define the channel region. Subsequently, a wet etching technique was employed to remove the material outside this channel region. Finally, the devices were annealed in air at 300 °C and 280 °C for 1 hour each. The source and drain electrodes were also deposited via electron beam evaporation, utilizing 5 nm Cr and 50 nm Au as electrode materials.

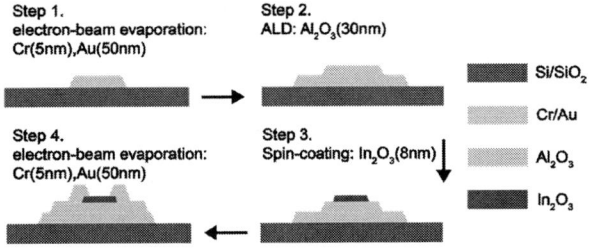

Fig. 1. The fabrication process of In_2O_3 TFT.

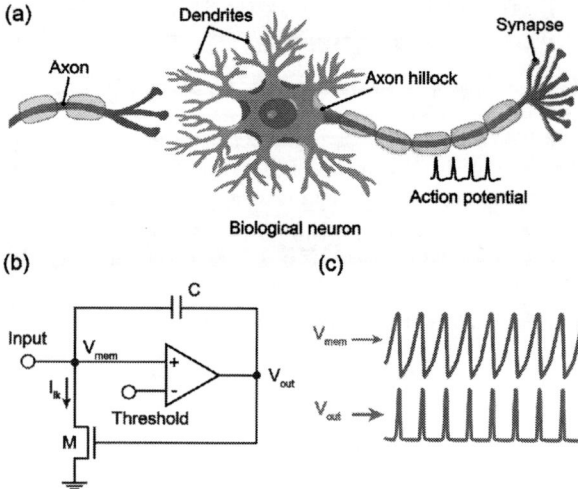

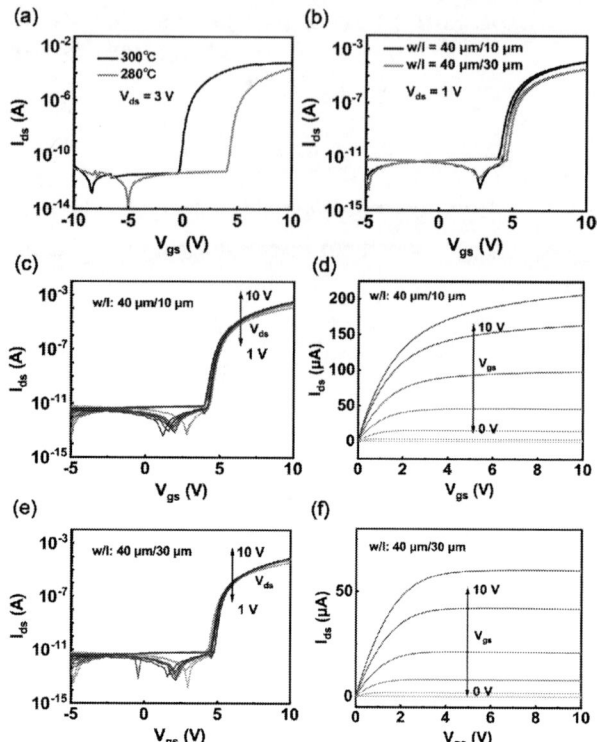

Fig. 2. (a) Schematic diagram of biological neuron. (b) Schematic diagram of artificial neuron circuit. (c) The simulation results of artificial neuron circuit in LTSPICE when a DC current is applied.

Fig. 3. (a) Characteristic transfer curves of TFTs at two different annealing temperatures. (b) Transfer characteristic curve of two aspect ratios. (c) Transfer characteristic curve and (d) output characteristic curve of 400 μm/10 μm In_2O_3 TFT. (e) Transfer characteristic curve and (f) output characteristic curve of 400 μm/30 μm In_2O_3 TFT.

Results and discussion

Biological neurons orchestrate the swift propagation of electrochemical signals along the axon through action potentials (see Fig. 2(a)). Interconnected via synapses, these neurons employ neurotransmitters to facilitate intricate communication and modulation of information. In analogy with the behavior of a biological neuron, we developed an axon-hillock circuit architecture. Illustrated in Fig. 2(b), this configuration comprises an amplifier, a capacitor, and a transistor. To evaluate its feasibility, we executed simulations utilizing LTSPICE. The resultant data are depicted in Fig. 2(c), where the first row (red curve) elucidates the waveform of the membrane potential (V_{mem}), while the second row (green curve) conveys the output voltage waveform (V_{out}). These findings substantiate that the circuit effectively implements spike frequency modulation of the input signal.

To realize artificial spiking neuron circuits, we fabricated In_2O_3 TFT with varying aspect ratios and threshold voltages, as depicted in Fig. 3. Through annealing at distinct temperatures, the In_2O_3 TFTs could achieve a diverse range of threshold voltages. Fig. 3(a) showcases the In_2O_3 TFTs subjected to annealing at 300 °C and 280 °C. Notably, as the annealing temperature diminishes, the threshold voltage of the transistors escalates from -0.4 V to an impressive 4 V, significantly contributing to the realization of artificial spiking neurons. Fig. 3(b) demonstrates the transfer characteristics of TFTs with differing aspect ratios, where the black and green curves represent TFTs with aspect ratios of 400 μm/10 μm and 400 μm/30 μm, respectively, under an applied V_{ds} of 1 V. Furthermore, we displayed the transfer curves of these two aspect ratios across varying V_{ds} (Fig.

3(c-e)), with V_{ds} incrementally increased from 1 to 10 V in 1 V steps. We also presented the output curves at different V_{gs}, applying V_{gs} from 0 to 10 V in 1 V increments (Fig. 3(d-f)). The experimental results indicate that the threshold voltages of TFTs with varying aspect ratios exhibit remarkable consistency under identical conditions, while fluctuations in V_{ds} exert a negligible influence on the threshold voltage of the same device.

The neuron's circuit's pivotal element is an amplifier composed of inverters. To simplify the circuit design, we utilized a diode-connected common-source amplifier, as depicted in Fig. 4(a). In this configuration, M1 serves as the load transistor with an aspect ratio of 400 μm/30 μm, while M2 functions as the driver transistor with an aspect ratio of 400 μm /10 μm. Under a supply voltage of 3 V, the voltage transfer curve (VTC) is presented in Fig. 4(b), demonstrating a gain of 2.5.

Building upon the inverter, we constructed an artificial spiking neuron circuit (5T1C) using In_2O_3 TFTs, as illustrated in Fig. 5(a). In this circuit, the aspect ratio of M1 and M3 is 400 μm / 30 μm, while the aspect ratio of M2, M4, and M5 is 400 μm/ 10 μm. M1, M2, M3, and M4 serve as non-inverting amplifier, formed by cascading two

979-8-3315-0417-5/25 $31.00 © 2025 IEEE   200

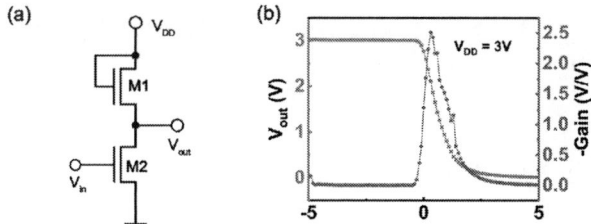

Fig. 4. (a) Common-source amplifier circuit with diode-connect load. (b) Measured output voltage (V_{out}) as function of input voltage (V_{in}) while using 3V supply (V_{DD}) (green curve); The gain of amplifier (blue curve).

first-stage inverting amplifiers. Capacitor C provides positive feedback for the circuit, while M5 serves as the negative feedback component. When the input current to the artificial neuron increases, the output voltage of the non-inverting amplifier demonstrates a slight upward trend. This output voltage is positively fed back to the V_{mem} node via capacitor C, and when the accumulated voltage at the V_{mem} node reaches the amplifier's threshold, the leakage transistor M5 activates, causing the output voltage to drop, thereby generating spike pulses.

Fig. 5(b) illustrates the output pulse waveforms corresponding to various input currents, indicating that as the input increases, the output pulse frequency rises, while the output pulse amplitude remains nearly constant, ranging from 3.6 to 4.3 V. We extracted the relationship between output pulse frequency and input current, as shown in Fig. 5(c), demonstrating that the output pulse frequency is directly proportional to the input current.

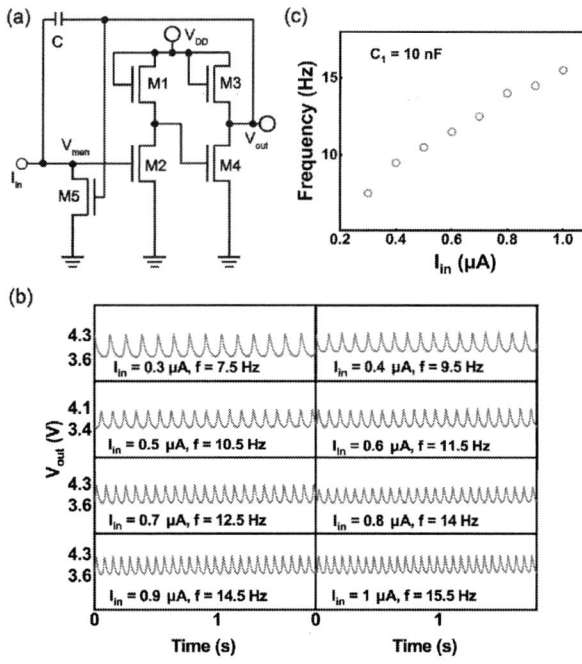

Fig. 5. (a) Artificial neuron (5T1C) circuit. (b) Measured output waveforms with different input currents, ranging from 0.3 to 1μA. (c) Extracted output frequency as function of input current (I_{in}).

Conclusion

In summary, we achieved an artificial spiking neuron circuit by utilizing solution-processed enhancement-mode In_2O_3 TFTs, capable of encoding external stimuli into electrical pulse signals while facilitating spike frequency modulation. Furthermore, this circuit exhibits significant potential for integration with flexible sensors, thus achieving low-energy and efficient sensory computing and processing.

References

[1] F. Liu *et al.*, "Neuro-Inspired Electronic Skin for Robots", Science robotics, 7, pp. eabl7344 (2022)

[2] H. Markram *et al.*, "A History of Spike-Timing-Dependent Plasticity", Front. Synaptic Neurosci., 3, pp. 4 (2011)

[3] Y. He *et al.*, "An Implantable Neuromorphic Sensing System Featuring near-Sensor Computation and Send-on-Delta Transmission for Wireless Neural Sensing of Peripheral Nerves", IEEE J. Solid-State Circuits, 57, pp. 3058-3070 (2022)

[4] F. Danneville *et al.*, "A Sub-35 Pw Axon-Hillock Artificial Neuron Circuit", Solid-State Electronics, 153, pp. 88-92 (2019)

[5] Z. Gao. "Study on the Modeling of a Neuromorphic Transistor Circuit System." Paper presented at the 2023 3rd International Conference on Electronic Information Engineering and Computer Communication (EIECC), 2023.

[6] G. Indiveri *et al.*, "Neuromorphic Silicon Neuron Circuits", Front. Neurosci., 5, pp. 73 (2011)

CMOS BEOL Compatible Metal-Oxides Logics and Memories for New Paradigm Computing in the Post-Moore Era

Yida Li and Feichi Zhou

School of Microelectronics, Southern University of Science and Technology, Shenzhen, China

Email: liyd3@sustech.edu.cn

Abstract

Low temperature processable electronic materials have been garnering increasing research attention due to its compatibility in deposition on a large variety of substrates, and ability for monolithic 3D (M3D) integration for high device density. This makes it suitable for a plethora of applications ranging from conventional applications in high performance computing, to unique applications such as in bioelectronics. Among them, metal-oxide have emerged as pivotal materials with it able to function as memory and logic devices. Here, we present briefly the development of metal-oxides technology in logic devices and related circuits, as well as emerging memory devices for new computing paradigm. The challenges and future outlook for the use of metal-oxides in future electronics are then concluded.

Keywords: Metal-oxides, Low temperature processing

Introduction

The demand for data-driven artificial intelligence (AI) such as the next-generation deep-learning neural network (DNN) accelerators and neuromorphic computing has grown dramatically over the past decades, for conventional applications in high performance computing, to unique applications such as in bioelectronics. However, issues arise due to traditional von-Neumann computing architecture with disjoint memory and processing units suffers from huge memory latency and limited data bandwidth. The challenges are increasingly amplified with the emergence of next generation AI utilizing large language model such as the powerful ChatGPT. In order to surpass this bottleneck, three-dimensional (3D) computing architecture that encompasses embedded logic and memory devices is touted as promising solution (Fig 1a), but requires new electronic materials able to be processed at a low temperature (Fig 1b) [1]. Metal-oxides are potential candidates due to its multi-functionality in logic and memory device (Fig. 1c). Most importantly, its low processing temperature (<400 °C) capability satisfy its integration in CMOS backend of line (BEOL), as well as able to be deposited over a wide variety of substrates required by different applications [2].

Despite its advantages, the performance of metal-oxides has still not reached a level required for commercial implementation. Areas requiring impending development include that such as material selection, growth optimization, device engineering, device integration approaches and related circuit design etc. Here, we present briefly on the above-mentioned efforts in metal-oxides technology development. Firstly, we will touch upon the development of defects engineered n- and p- type metal-oxides semiconductors thin-film transistors (TFT), as well as its related circuits design and demonstration. Secondly, we will move on to discuss the progress of metal-oxide memories, with focus on resistive random-access memory (RRAM) and

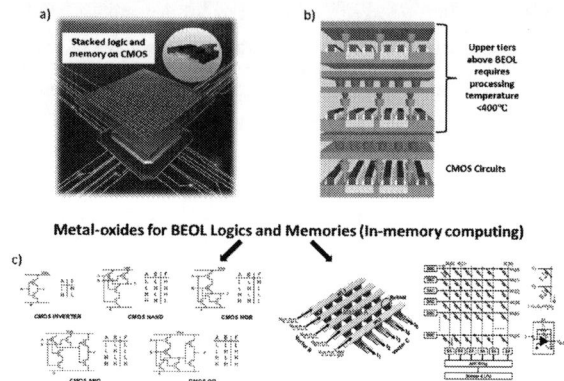

Fig. 1: a) Illustration showing M3D integration of logics and memories above CMOS BEOL, b) Materials integrated above CMOS BEOL must be strictly processed below 400 °C, and c) Metal oxides can be implemented in logic circuits and memory array for storage or in-memory in CMOS BEOL.

ferroelectric (FE) random-access memory (RAM). While brief, this paper aims to provide a holistic view in metal-oxide device technology development in recent years and its future outlook, to provide insights in the design of metal-oxide-based devices and circuits beyond Si technology for emerging applications in the post Moore-era.

Metal-oxide Semiconductors

Metal-oxide semiconductors (M-OS) typically possess smaller energy bandgap of 2 to 4 eV as compared to its insulator counterpart, and allowing decent TFT conduction while in the accumulation mode. Unlike conventional silicon (Si) where a doping process transforms it into n- or p- type polarity, polarity of M-OS is determined by the material system, driven primarily by conduction in either the conduction or valance band. N-type M-OS has been much more reported than p-type M-OS due to its significantly better electrical performance. Besides its use as a select transistor for high-density memory in CMOS BEOL, researchers also toyed with the idea of implementing them in BEOL logic circuit. While n- type only pseudo-CMOS circuit can realize the same function as its true CMOS circuit counterpart, they suffer greatly in the PPAC metrics, consisting of power consumption, performance, area and cost. Hence, development in p- type M-OS is equally as critical, and both of their progress presented as follows.

A. n- type Metal-oxide Semiconductors

N-type M-OS includes candidates such as WO_3, In_2O_3, ZnO, and indium gallium zinc oxide (IGZO) etc, with desirable electrical characteristics including high carrier mobility (μ), and excellent electrostatic control for ultra-low leakage current. Of the n-type M-OS, Zn-based M-OS are more popular choice of study due to its superior electrical performance as well as its stability over others. A significant breakthrough for ZnO TFT was reported in 2017 where an $\mu_{electron}$ of 345 cm²/V·s is achieved on a flexible substrate, owing to a novel HfLaO passivation that shields the ZnO layer effectively from ambient moisture that accelerates its degradation [3]. This achievement is significant because it now appears that there is a way to build logic devices from alternative electronic materials at significantly lower temperature, but with performance that comes close to that of Si transistor. It becomes apparent that as long as the ZnO layer can be reasonably sealed hermetically, performance and reliability can be ensured. Shortly, IGZO started to gain increasing research attention an alternative channel layer to ZnO. Despite its slightly worse-off $\mu_{electron}$, its significantly hardy nature to ensure performance integrity even without a proper passivation layer makes it an extremely attractive material for large-scale implementation in circuits. Thereafter, multitudes of reports investigating the performance enhancement of these Zn-based M-OS TFTs started to emerge, reporting on optimization in stoichiometry and defects modulation for improved $\mu_{electron}$ through material growth approaches [3,4] and post annealing conditions [5], to device structure engineering - nanowires [6], metal-semiconductor contact resistance reduction [7], ultra-scaled channel length [8], passivation approaches [9], and to novel integration process demonstrating their potentials in M3D stacked architecture [10].

B. p- type Metal-oxide Semiconductors
While p- type M-OS TFT is critical for oxide-based CMOS circuits, related reports are far lesser. This is due to the intrinsic issues of p- type M-OS, where they possess strong localized valence band maxima (VBM) in anisotropic O 2p orbitals, that typically leads to inferior electrical performance. Till date, only limited p-type oxide semiconductors have been identified, such as SnO, Cu_2O, and NiO. Among them, SnO with an indirect bandgap of 2.7 eV, is considered to be the most promising due to its low formation energy of Sn vacancies (V_{Sn}) which can induce hole carriers. Furthermore, the hybridization of Sn 5s and O 2p orbitals results in a comparatively low hole effective mass in the VBM for decent hole transport properties. First principles simulation of SnO phase puts its μ_{hole} upper limit to 60 cm²/V·s, allowing easier device matching with n- type M-OS TFT for circuit design. However, despite the theoretical evidence in the commendable performance limits, current research efforts in contacts engineering [11], passivation approaches [12], and band structure engineering via interlayer integration [13] still are not able to achieve significant progress, with best reported μ_{hole} of SnO TFTs still falls in the low 10's of cm²/V·s. To further enhance the device performance, understanding of the metastable SnO phase formation in the channel layer for the desired phase composition is needed.

C. Metal-oxide Semiconductors based Circuits
The escalation to utilizing M-OS to the circuit level is natural for practical use. With significantly better n- type TFT, most reports are on unipolar circuit design, but requires a pseudo-CMOS design approach incorporating both pseudo enhancement and pseudo depletion designs for proper circuit functionality. In a recent reported work, an atomic layer deposition (ALD) optimized ZnO TFTs achieving record-

high $\mu_{electron}$ was experimentally implemented in novel pseudo-CMOS inverter and ring oscillator designs with excellent performance [2]. However, while intended functions can be properly realized, it is to be realized that such an approach incurs additional area and power penalty, and thus not suitable for very large-scale integration (VLSI) circuits or applications requiring ultra-low power budget. While a true CMOS circuit design utilizing both n- and p-type transistors is still preferred and has seen some research efforts, the demonstrated circuit (inverter) is still plagued with issues such as highly size unmatched n- and p- type TFTs, high operating voltages, and high static power consumption etc. [14]. Hence, it is clear that the underwhelming electrical performance of p- type TFT wanes its commercialization interests, and requires timely attention for a truly high-performance M-OS based CMOS circuits.

Metal-oxide Memories
Metal-oxide memories belong to an emerging class of memories with new operating mechanisms. Having a much shorter history then conventional memories, metal-oxide memories suffer from the lack of testament in their long-term reliability. However, owing to their operating mechanism and low temperature process, they are particularly suitable for use in new computing paradigm in the form of in-memory computing and high-density memory integration. Of the emerging memories, metal-oxides with insulative properties are used as RRAM with defects modulated filament for switching, while suitable material system processed at the right condition can exhibit FE polarization for use as FERAM. Progress of both memories are as presented.

A. Resistive Random-access Memory
Being a simple two-terminal device, metal-oxides based RRAMs have an extremely small footprint for high-density integration in cross-point form in an array. In most investigation of RRAM devices, mono-oxides are a popular choice due to its simple binary compound composition and easier understanding of the operating mechanism. However, the room for further performance metrics improvements such as operating voltage, speed and endurance is also limited. To break this limit, advanced process techniques such as doping [15], layering [16], and post deposition treatment process [17] are promising approaches, and have seen considerable success. One important area in which RRAM finds its position is its use as weights in artificial neural network (ANN). With appropriate input signals, RRAM array can be used in direct vector matrix multiplication (VMM) using concept of in-memory computing, boasting a power efficiency that is of orders of magnitude higher than conventional von-Neumann computing architecture [18]. Although current studies of RRAM performance improvements are still on-going, small-scale offering of this type of memory are now available in commercial foundries, thus shedding light in its potential for ramped-up production in the near future.

B. Ferroelectric Random-access Memory
FE memory demonstrates unique electrical properties, holding tremendous potential in the field of non-volatile storage and in-memory computing [19]. Traditional FE memory using PZT have met with both cost and scaling issues that restricts its growth in the memory market [20].

979-8-3315-0417-5/25 $31.00 © 2025 IEEE   203

The recent demonstration of improved performance of $Hf_xZr_{1-x}O_2$ (HZO) FE thin-film, whose versatility to be integrated in various device structures such as capacitors and transistors, and CMOS compatibility have generated increasing research interests in this material system [21]. HZO FE capacitors are also typically referred to as FERAM, are potentially more promising candidates than RRAM for implementation in in-memory computing system due to its much larger reliability, ultra-low static power consumption and IR-drop free behavior [22]. HZO FE films are usually fabricated via ALD, followed by a post-annealing step to transform the initial non-ferroelectric phase into the orthorhombic ferroelectric phase [23]. The annealing process is a key tuning knob for modulating the performance of HZO thin films, with parameters such as temperature, time, and pressure affecting their behavior [24]. In short, FERAM studies using HZO FE film has a much shorter history than RRAM, but given its promises in cost, performance and reliability advantages, the future for large-scale implementation of FERAM appears to be bright.

Conclusion

In summary, metal-oxides, with its unique material properties, relatively matured deposition process, and ability to be processed at low temperature opens up tremendous opportunities in various applications requiring different types of substrates, as well as M3D stacking for added functionalities and high-density device integration. We provide a quick overview of the status of metal-oxides based logic devices (n- and p- type TFTs), circuits, and memory devices (RRAM and FERAM), its current research direction, as well as the challenges and near future outlook. With systematic investigation of metal-oxides encompassing new material system and devices, novel architecture, device modelling and simulation, and co-design of the devices at a system level, the potential for metal-oxides to be implemented at a commercial level looks to be immense.

Acknowledgments

This work was supported by the National Natural Science Foundation of China (62174074, 62274081, 52273246), Shenzhen Fundamental Research Program (JCYJ20220530115014032, JCYJ20220530115204009), Zhujiang Young Talent Program (2021QN02X362), Guangdong Provincial Department of Education Innovation Team Program (2021KCXTD012), SUSTech SME-Pixelcore Neuromorphic In-sensor Computing and SME-CIMCube Joint Lab. We would also like to acknowledge the Core Research Facilities (CRF) at SUSTech for the facilities used, and the technical support provided by the staff and engineers at the CRF.

References

[1] M. M. Sabry Aly et. al., "Energy-Efficient Abundant-Data Computing: The N3XT 1,000x", Computer, 48, pp. 24-33 (2015).

[2] W. Wang et. al., "CMOS Backend-of-Line Compatible Memory Array and Logic Circuitries Enabled by High Performance Atomic Layer Deposited ZnO Thin-Film Transistor", Nat. Commun., 14 (2023).

[3] S. C. Wei, and A. Chin, "Remarkably High Mobility Thin-Film Transistor on Flexible Substrate by Novel Passivation Material", Sci. Rep., 7 (2017).

[4] X. Chen et. al., "Transparent and Flexible Thin-Film Transistors with High Performance Prepared at Ultralow Temperatures by Atomic Layer Deposition", Adv. Electron. Mater., 5 (2019).

[5] S. Subhranu et. al., "Low Subthreshold Swing and High Mobility Amorphous Indium–Gallium–Zinc-Oxide Thin-Film Transistor With Thin HfO_2 Gate Dielectric and Excellent Uniformity", IEEE Electron Dev. Lett., 41, pp. 856-859 (2020).

[6] K. Han et. al., "Indium-Gallium-Zinc-Oxide (IGZO) Nanowire Transistors", IEEE Trans. on Electron Dev., 68, pp. 6610-6616, (2021).

[7] J. Lu et. al., "Contact Resistance Reduction of Low Temperature Atomic Layer Deposition ZnO Thin Film Transistor Using Ar Plasma Surface Treatment", IEEE Electron Dev. Lett., 43, pp. 890-893, (2022).

[8] U. Chand et. al., "Sub-10nm Ultra-thin ZnO Channel FET with Record-High 561 $\mu A/\mu m$ ION at VDS 1V, High μ-84 cm^2/V-s and 1T-1RRAM Memory Cell Demonstration Memory Implications for Energy-Efficient Deep-Learning Computing", IEEE Symposium on VLSI Technology and Circuits (VLSI Technology and Circuits), pp. 326-327, (2022).

[9] W. Wang et. al., "Air Stable High Mobility ALD ZnO TFT with HfO_2 Passivation Layer Suitable for CMOS-BEOL Integration", China Semiconductor Technology International Conference (CSTIC), pp. 1-4, (2022).

[10] U. Chand et. al., "2-kbit Array of 3-D Monolithically-stacked IGZO FETs with Low SS-64mV/dec, Ultra-low-leakage, Competitive μ-57 cm^2/V-s Performance and Novel nMOS-Only Circuit Demonstration", Symposium on VLSI Technology, pp. 1-2, (2021).

[11] S. -M. Hsu et. al., "Mobility Enhancement in P-Type SnO Thin-Film Transistors via Ni Incorporation by Co-Sputtering", IEEE Electron Device Lett., 43, pp. 228-231, (2022).

[12] T. Kim et. al., "High Mobility P-Channel Tin Monoxide Thin-Film Transistors with Hysteresis-Free Like Behavior", Appl. Phys. Lett., 121 (2022).

[13] T. Zhang et. al., "Multi-Operating Mode Field-Effect Transistors Based on SnO/SnS Heterostructures and CMOS-Like Inverter Applications", Adv. Electron. Mater., 9 (2023).

[14] J. H. Lee et. al., "Cu_2O p-type Thin-Film Transistors with Enhanced Switching Characteristics for CMOS Logic Circuit by Controlling Deposition Condition and Annealing in the N_2 Atmosphere", ACS Appl. Electron. Mater., 5 (2023).

[15] J. Lan et. al., "Improved Performance of Hf_xZn_yO-Based RRAM and its Switching Characteristics down to 4 K Temperature", Adv. Electron. Mater., 9 (2023).

[16] Y. Fang et. al., "Improvement of HfO_x-Based RRAM Device Variation by Inserting ALD TiN Buffer Layer", IEEE Electron Dev. Lett., 39, pp. 819-822 (2018).

[17] H. Chen et. al., "Performance Optimization of Atomic Layer Deposited HfO_x Memristor by Annealing With Back-End-of-Line Compatibility", IEEE Electron Dev. Lett., 43, pp. 1141-1144 (2022).

[18] H. Veluri et. al., "High-Throughput, Area-Efficient, and Variation-Tolerant 3-D In-Memory Compute System for Deep Convolutional Neural Networks", IEEE Internet of Things Journal, 8, pp. 9219-9232 (2021).

[19] R. Yang, "In-Memory Computing with Ferroelectrics", Nature Electronics, 3, pp. 237–238 (2020).

[20] R. Moazzami et. al., "Electrical Characteristics of Ferroelectric PZT Thin Films for DRAM Applications", IEEE Trans. Electron Dev., 39, pp. 2044-2049 (1992).

[21] T. Mikolajick et. al., "The Past, the Present, and the Future of Ferroelectric Memories", IEEE Trans. Electron Dev., 67, pp. 1434-1443 (2020).

[22] Y.-C. Luo et. al., "Experimental Demonstration of Non-volatile Capacitive Crossbar Array for In-memory Computing", IEEE International Electron Devices Meeting (IEDM), pp. 1-4 (2021).

[23] M. H. Park et. al., "Review and perspective on ferroelectric HfO2-based thin films for memory applications", MRS Commun., 8, pp. 795–808 (2018).

[24] Y. H. Yeh et. al., "Impact of deposition temperature on electrical properties of HZO-based FeRAM", J. of Appl. Phys., 135 (2024)

Endurance Optimization of 40nm RRAM towards 10^6 cycles by Tuning the Stoichiometry of TiN Bottom Electrode

Chengxiang Ma[1,†], Qianze Zheng[1,†], Ruofei Hu[1], Yilong Huang[1], Kaimeng Liu[1], Yuelin Jiang[1], Siyu Chen[1], Jianshi Tang[1*], Dong Wu[1], Bin Gao[1], He Qian[1], Huaqiang Wu[1]

[1]School of Integrated Circuits, Beijing Advanced Innovation Center for Integrated Circuits, Tsinghua University, Beijing. [†]Equal Contribution, *Email: jtang@tsinghua.edu.cn

Abstract

This work proposes tuning the stoichiometry of TiN bottom electrode (BE) as an effective approach to improve the endurance of 40nm resistive random-access memory (RRAM). By adopting an inert TiN BE, the optimized RRAM achieves a high endurance close to 10^6 cycles, matching the performance of embedded Flash (eFlash). The mechanism for the impact of TiN stoichiometry on the endurance performance is clarified through electrical tests and material characterizations.

Keywords: RRAM, Endurance, TiN Bottom electrode, Stoichiometry Engineering

Introduction

RRAM has attracted enormous interests in both academia and industry as a promising candidate to replace eFlash for next generation of eNVM owing to its high density, excellent scalability and CMOS compatibility, low operating voltage, etc [1]. Despite these advantages, the endurance of RRAM still lags behind mature eFlash technology, which is typically much less than 10^6 cycles [2]. Although a variety of work has been carried out to improve RRAM endurance [3-5], few of them are applicable for mass integration in advanced technology nodes.

In this work, the endurance of 40nm RRAM array is significantly improved by carefully tuning the stoichiometry of TiN BE. For direct comparison, two 1kb RRAM arrays with different stoichiometry of TiN BE are fabricated and measured. In particular, the impact of the TiN stoichiometry on the array-level endurance performance is investigated, and the underlying mechanism is further analyzed. By using an inert TiN BE, an outstanding endurance of 10^6 cycles is achieved at the array level, catching up with the performance of eFlash.

Experimental Platform and Test Vehicles

Fig. 1 shows the schematic of the fabricated 1kb one-transistor-one-resistor (1T1R) RRAM array. The access transistors and the peripheral circuits are fabricated using a commercial 40nm Si CMOS process. The RRAM has a thermal enhanced layer (TEL) on top of the resistive switching layer (RSL) for better performance [6]. Both the top electrode (TE) and BE of RRAM are made of TiN. Because the TE is not in direct contact with the RSL and has

little impact on the resistive switching [6, 7], this work only tunes the TiN stoichiometry of the BE and studies its impact on the RRAM characteristics. As shown in Fig. 2, we apply a gate-last incremental step pulse programming (ISPP) scheme [8]. The pulse width for SET/RESET operations is 100ns. Besides electrical measurements, we also use electron energy-loss spectroscopy (EELS) to characterize the TiN BE. Analysis of the EELS spectrum is performed using the Gatan Microscopy Suite (GMS) 3 software.

Experimental Results and Discussion

In this work, we fabricate RRAM devices with two types of BE by tuning the condition of TiN reactive sputtering. Fig. 3 shows the EELS spectrum of these two types of TiN. The N-K and Ti-L edges are extracted to obtain the stoichiometry of TiN. According to the elemental analysis, one type of TiN has a Ti:N ratio of 1:0.6 (Fig. 3(a)), indicating an active TiN BE, while the other has a Ti:N ratio of 1:1.1 (Fig. 3(b)), indicating an inert TiN BE. Except for the BE, the other functional layers remain the same. Both the active and inert TiN BE have good electrical contact to the underlying metal layers. The RRAM devices with two types of TiN BE exhibit the same switching on/off ratio, thus the same programming targets of high-resistance state (HRS) and low-resistance state (LRS) are used in the following experiments.

Fig. 4 compares the endurance between RRAM with active and inert TiN BE, where the thickness of the RRAM film stack remains the same. The RRAM devices with active TiN BE show poor endurance (Fig. 4(a)), as ~5% of RRAM devices exhibit overlap between the HRS and LRS, resulting in bit errors. On the contrary, tuning the TiN BE to be inert can significantly improve the endurance of RRAM (Fig. 4(b)). For RRAM with inert TiN BE, a large memory window (MW) can be maintained even after 5×10^5 programming cycles. After 10^6 cycles, there is still a small MW without current overlap in the range of ±3σ at the 1kb array level. In addition, these two sets of RRAM devices have the same type of cycling failure, that is, the failed devices cannot be RESET to HRS. This indicates the endurance performance of both types of RRAM is closely related to the rupture process of the conductive filaments (CF) inside the RSL. The detailed mechanism analysis is elaborated in the following.

In order to further explore the impact of TiN BE on RRAM, we measure the initial read current of 1kb RRAM array, as

979-8-3315-0417-5/25 $31.00 © 2025 IEEE

shown in Fig. 5. The initial read current refers to the read current of RRAM devices measured before the forming operation. The average initial read current of RRAM with active TiN BE is one order higher than that of RRAM with inert TiN BE. Furthermore, ~3% of RRAM devices with active TiN BE have initial read current even higher than the forming target. These devices exhibit forming-free behavior but are easier to fail in the early programming cycles. Then, we perform DC I-V sweep on the 1T1R device to study the conduction mechanism. The forming curve of RRAM is shown in Fig. 6. The I-V curves for both types of RRAM fits well with the classical Poole-Frenkel emission mechanism. This indicates the stoichiometry of TiN BE has a minor impact on the conduction mechanism of RSL, and the difference in the initial read current may be attributed to the difference in the effective RSL thickness. In Fig. 7, We fabricate RRAM arrays with different RSL thickness and extract the forming voltage for each batch of RRAM. It can be seen that the forming voltage exhibits a linear correlation with the RSL thickness. This is because the onset of the forming process is determined by the critical electric field inside the RSL [9]. Moreover, the two fitting lines for the RRAM devices with active and inert TiN BE have the same slope of ~10MV/cm, which is equivalent to the critical electric field intensity for forming. The two fitting lines only have an offset in the intercept of ~0.3V. This suggests that the difference between the active and inert TiN BE results in an effective RSL thickness reduction of ~3Å for pristine (without forming) RRAM devices.

Besides the effective RSL thickness for pristine RRAM, the stoichiometry of TiN BE also has an influence on the resistive switching characteristics during the programming cycles. As shown in Fig. 8, for RRAM with active TiN BE, the endurance performance improves by increasing the thickness of the RSL. This result can be explained by the difference in the amount of oxygen ions stored at the TEL [3]. However, the endurance of RRAM with active TiN BE is still inferior to RRAM with inert TiN BE, even when the effective thickness of RSL is higher than the latter (Active TiN-3). This proves that the active TiN BE contributes to the degradation of the endurance performance other than reducing the effective RSL thickness.

Mechanism Analysis

Based on the above experimental results, the mechanism of the endurance performance difference between RRAM with active and inert TiN BE is summarized in Fig. 9. Compared to the inert TiN BE with high chemical stability, the active TiN BE is more prone to scavenge oxygen ions from the RSL and undergoes oxidation reaction [10]. This process generates extra oxygen vacancies and induces an oxygen-deficient interface between the BE and the RSL. This interface has Ohmic conduction characteristic and causes the thinning of the effective RSL for pristine RRAM devices (Fig. 9(a)). This explains why the RRAM with active TiN BE has a higher initial read current and lower forming voltage. In addition, fewer oxygen ions are driven by electric field and stored at the TEL during the forming process for such RRAM (Fig. 9(b)). On top of that, the active TiN BE also continuously absorbs oxygen ions from the RSL as repeated programming cycles are applied. Because of this, RRAM with active TiN BE tends to run out of oxygen ions stored at the TEL more quickly and eventually fail to rupture the CF of RRAM (Fig. 9(c)). This results in the degraded endurance. Conversely, tuning the BE TiN to be inert can slow down the consumption of stored oxygen ions. This brings significant improvement in endurance at the cost of slight increase of forming voltage.

Conclusion

In this work, we achieve a significant improvement in endurance of 40nm RRAM by tuning the stoichiometry of BE TiN. The RRAM with inert TiN BE can maintain a relatively large MW after 5×10^5 programming cycles, and a small MW still exists without read current overlap after 10^6 cycles. We carry out comprehensive investigations on the impact of different types BE TiN, and clarify the mechanism of tuning TiN stoichiometry. This work can further facilitate the integration of embedded RRAM in more advanced technology nodes beyond 40nm.

Acknowledgments

This work was in part supported by the Semiconductor Technology Innovation Center (Beijing) Corporation under Grant QYJS-2021-0801-B and Grant QYJS-2021-0802-B.

References

[1] H. Wu et al., Proceedings of the IEEE, vol. 105, no. 9, pp. 1770-1789, 2017, doi: 10.1109/jproc.2017.2684830.

[2] H. Hidaka, Springer Cham, ISBN: 978-3-319-55305-4, 2018, doi: 10.1007/978-3-319-55306-1.

[3] R. Hu et al., IEEE Electron Device Letters, vol. 44, no. 4, pp. 618-621, 2023, doi: 10.1109/LED.2023.3250449.

[4] T. Kempen et al., IEEE International Memory Workshop (IMW), pp. 1-4, 2021, doi: 10.1109/imw51353.2021.9439591.

[5] L. Wu et al., IEEE Silicon Nanoelectronics Workshop (SNW), pp. 77-78, 2024, doi: 10.1109/SNW63608.2024.10639217.

[6] W. Wu et al., IEEE Electron Device Letters, vol. 38, no. 8, pp. 1019-1022, 2017, doi: 10.1109/led.2017.2719161.

[7] W. Wu et al., IEEE Symposium on VLSI Technology, pp. 103-104, 2018, doi: 10.1109/VLSIT.2018.8510690.

[8] M. Zhao et al., IEEE International Electron Devices Meeting (IEDM), pp. 20.2.1-20.2.4, 2018, doi: 10.1109/IEDM.2018.8614664.

[9] A. Beck et al., Applied Physics Letters, vol. 77, no. 1, pp. 139-141, 2000, doi: 10.1063/1.126902.

[10] I. Montero et al., Surface Science, vol. 251-252, pp. 1038-1043, 1991, doi: 10.1016/0039-6028(91)91147-P.

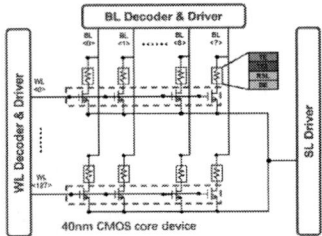

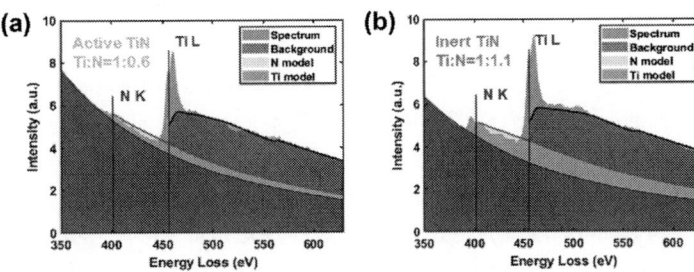

Fig. 1: The schematic of the 1kb 1T1R RRAM array. The transistors and peripheral circuits are fabricated on the 40nm CMOS platform.

Fig. 3: The EELS spectrum of (a) the active TiN and (b) the inert TiN. The intensity of N-K and Ti-L edges is extracted using the GMS 3 software. The elemental analysis shows that the Ti:N ratio of active TiN and inert TiN is 1:0.6 and 1:1.1 respectively.

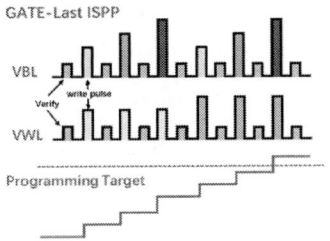

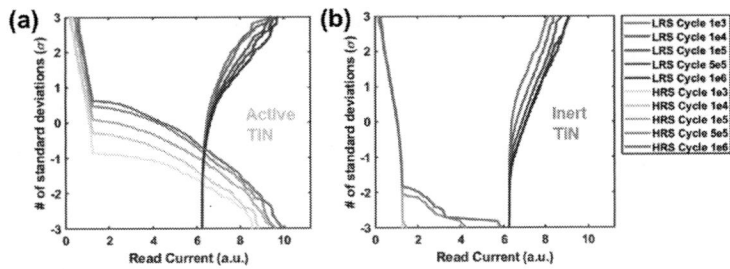

Fig. 2: Schematic diagram of the gate-last ISPP programming scheme. Here the SET operation is illustrated as an example.

Fig. 4: Endurance of 1kb RRAM array with (a) active TiN BE and (b) inert TiN BE. Except for the BE material, the other functional layers of these two sets of RRAM remain the same. The TiN BE stoichiometry has a significant impact on the endurance performance of RRAM.

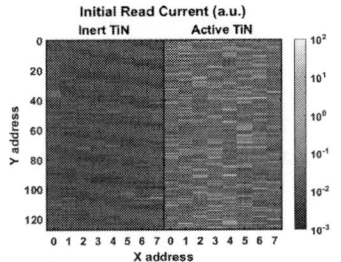

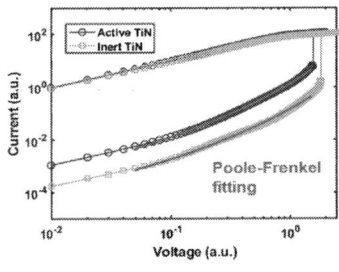

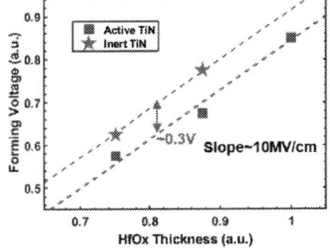

Fig. 5: Initial read current, i.e. the read current before forming, of the 1kb RRAM with active and inert TiN BE.

Fig. 6: DC I-V curve during forming. Both RRAM with active and inert TiN BE follow the Poole-Frenkel conduction mechanism.

Fig. 7: The relationship between the forming voltage and the RSL thickness of 1kb RRAM array with active and inert TiN BE.

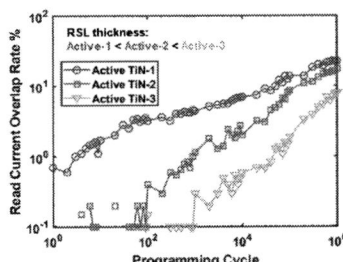

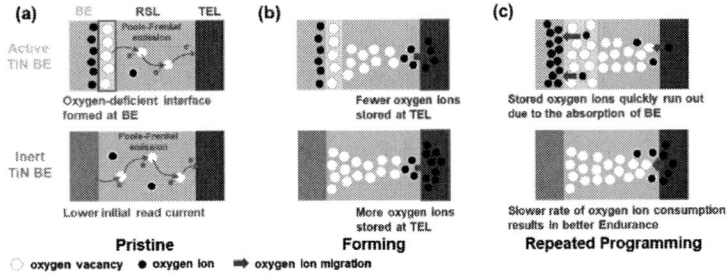

Fig. 8: Endurance of 1kb RRAM array with different RSL thicknesses. The order of RSL thickness for different groups has been noted.

Fig. 9: Mechanism comparison of RRAM with active and inert TiN BE. High chemical reactivity of active TiN generates an oxygen-deficient interface between the BE and the RSL, resulting in (a) higher initial read current, (b) lower forming voltage, and (c) degraded endurance.

Optimized Electric Field Distribution and Dynamic Performance in GaN HEMTs using Segmented-Extended *p*-GaN Gate Structures

Xinyue Dai[1], Haiyang Li[1], Qimeng Jiang[2], Xiaoping Wang[1], Yuxi Wan[1]

[1] Shenzhen Pinghu Laboratory, Shenzhen, China

[2] Institute of Microelectronics of Chinese Academy of Sciences, Beijing, China

daixinyue@phlab.com.cn

Abstract

A segmented-extended *p*-GaN gate (SEP) structure for AlGaN/GaN HEMTs was developed to address electric field challenges and enhance dynamic R_{ON}. The SEP structure, combining traditional and extended *p*-GaN gate designs, demonstrated improvements in OFF-state electric field distribution and dynamic R_{ON} compared to traditional designs. 3D TCAD simulations revealed that reduced segmented dimensions significantly benefit OFF-state electric field modulation, providing insights into optimizing both reliability and dynamic performance in GaN power devices.

Keywords: *p*-GaN HEMT, electric field, TCAD simulation.

Introduction

AlGaN/GaN HEMTs, with high electron mobility and breakdown voltage, offer strong potential for next-generation power applications. While 2-dimensional electron gas (2DEG) formation at the AlGaN/GaN interface allows low R_{ON} and fast switching, its sensitivity to OFF-state field distribution introduces reliability challenges [1]. Improper field management can lead to high-voltage leakage and dynamic R_{ON} degradation [2]. Traditional methods like source or gate field plates (FP) mitigate peak electric fields (*E*-field) but introduce parasitic charges and demand high-quality passivation [3-4]. To address these limitations, we introduce lightly-doped drain (LDD) technology [5-6] into *p*-GaN HEMTs by extending a thin *p*-GaN layer toward the drain to regulate 2DEG concentration. The proposed *p*-GaN extension maintains conduction current density and enhances OFF-state field distribution. However, the extended *p*-GaN (EP) layer also adds parasitic capacitance and resistance, potentially degrading dynamic R_{ON} at high frequencies [7]. As a solution, a segmented-extended *p*-GaN (SEP) structure, combining traditional (BSL) and EP structures was proposed. This design combines traditional *p*-GaN gate elements with the extended *p*-GaN approach, enabling to adjust parameters such as the duty cycle of the segments for optimal performance. Experiments show that SEP devices exhibit OFF-state performance between EP and BSL devices but have superior dynamic on-resistance, indicating effective mitigation of high-field effects.

In this paper, the benefit of SEP structure was verified further through TCAD simulations based on the experimental results.

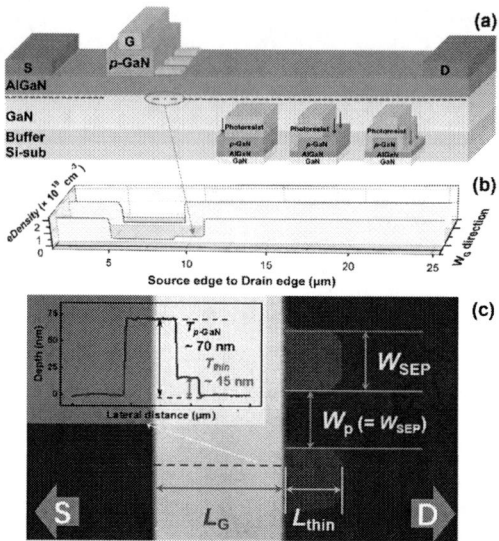

Fig. 1: (a) 3D schematic of Segmented-extended *p*-GaN gate (SEP) HEMT, and inset shows the three-step process to form the (S)EP structure. (b) Simulated electron density along the AlGaN/GaN heterojunctions at $V_{GS} = V_{DS} = 0$ V with or without extended *p*-GaN gate (EP). (c) AFM image of the fabricated SEP structure. The length of *p*-GaN gate (L_G) is 4μm, and the length of extended *p*-GaN (L_{thin}), width of segmented *p*-GaN (W_{SEP}), and width of interval (W_p) is all 2 μm. The inset figure is the actual AFM thickness data, with the thickness of *p*-GaN (T_{p-GaN}) and extended *p*-GaN layer (T_{thin}) are 70 nm and 15 nm, respectively.

A detailed analysis of the SEP structure using TCAD simulations was conducted, focusing on its modulation effects along the gate width direction based on the experimental findings [8].

Device Structure and Process

The epi-structure consist of a 70 nm *p*-GaN (with approximately 3×10^{19} cm^{-3} Mg doping), a 15 nm Al$_{0.22}$Ga$_{0.78}$N, and 200 nm unintentional-GaN, grown on a commercial 6-inch GaN-on-Si wafer. The different gate structures are fabricated by three-step lithography and ICP-etching processes to create the traditional *p*-GaN gate (BSL), extended-*p*-GaN gate (EP), and segmented-extended *p*-GaN gate (SEP) structures simultaneously. A ~15-nm *p*-GaN extension layer (T_{thin}) is confirmed by AFM measurements (inset in Fig. 1(c)). After forming the gate structures,

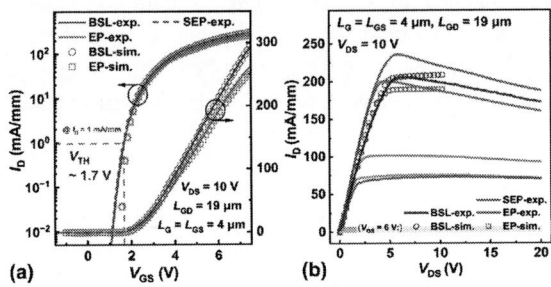

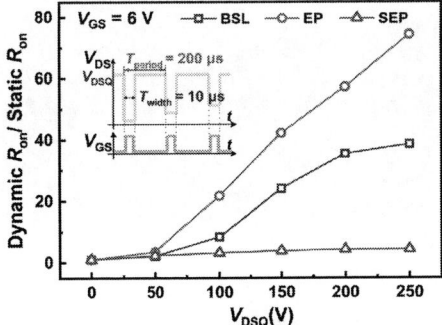

Fig. 2: Measured (line) and simulated (dot) (a) transfer characteristic and (b) output characteristics of the BSL-HEMT (black), EP-HEMT (red), and SEP-HEMT (blue) with $L_{GS} = L_G = 4$ μm, $L_{GD} = 19$ μm, the L_{thin} of EP and SEP is 2 μm and $W_{SEP} = W_p = 2$ μm.

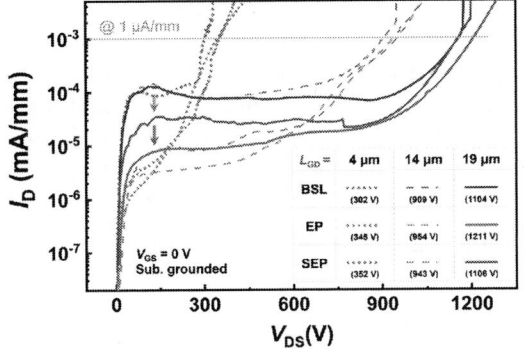

Fig. 3. The breakdown characteristics of the BSL-HEMT, EP-HEMT, and SEP-HEMT with different L_{GD} and gate structures at $V_{GS} = 0$ V. All the results are substrate grounded.

source/drain electrodes are defined by evaporating a Ti/Al/Ni/Au (20/150/55/45 nm) metal stack, followed by rapid thermal annealing in an N$_2$ environment at 850°C for 50 seconds. The contact resistance (R_c) was measured at 0.55 Ω·mm using the linear transmission line method. Next, a 3-nm ALD-Al$_2$O$_3$ layer and a 200-nm PECVD-SiO$_2$ layer are deposited for passivation, followed by multi-energy ion implantation for isolation. The passivation layer is then opened using fluorine-based ICP plasma etching and wet chemical treatment. Finally, the Ni/Au gate metal is deposited to complete the process.

Result and Discussions

A. Device Performance

Fig. 2 plots the transfer and output characteristics of the three target devices. For all devices, the gate-source distance L_{GS}, channel length L_G, gate-drain distance L_{GD}, and p-GaN extension L_{thin} (for EP and SEP only) are set to 4/4/19/2 μm, respectively, achieving normally-OFF operation with a 1.7 V threshold voltage. To maintain comparability among SEP, BSL, and EP structures, the segment ratio ($W_{SEP} : W_p$) in the SEP structure is fixed at 2:2 μm for this section. Fig. 3 illustrates the breakdown characteristics for each device under various L_{GD}. The (S)EP structure exhibits significantly

Fig. 4: Ratio of dynamic R_{ON}/ static R_{ON} of the BSL-HEMT, EP-HEMT and SEP-HEMT switched from different OFF-state base voltages (V_{DSQ}) at $T_{period} = 200$ μs, $T_{width} = 10$ μs. The R_{ON} was extracted at $V_{DS} = 1$ V and $V_{GS} = 6$ V.

leakage current reduced, especially at lower drain voltages, compared to the traditional design.

To investigate the dynamic performance of three devices, a double-pulsed I-V measurement with a pulse width of 10 μs and a pulse period of 200 μs was conducted, as shown in Fig. 4. Results show that the SEP-HEMT achieves the lowest dynamic R_{ON} among the EP and BSL structures. Although the EP-HEMT provides excellent leakage suppression, the Mg-doped extended p-GaN introduces parasitic effects that degrade device performance. In contrast, the SEP structure significantly enhances dynamic performance, with optimization effects on dynamic R_{ON} falling between those of the BSL and EP structures, yet outperforming both. This improvement is attributed to the reduced p-GaN underlying area, which minimizes parasitic effects and optimizes the OFF-state E-field distribution.

B. TCAD modeling and 3D simulation

To further study the optimization effect of SEP device, 2D structure of fabricated BSL and EP devices were modeled and calibrated using *Sentaurus TCAD*. It is seen in Fig. 2, that a good agreement is achieved between simulation results (dots) and experiment data (lines). Subsequently, 3D OFF-state simulations were conducted for SEP devices with varying $W_{SEP}(=W_p)$ dimensions. As shown in Fig. 5, the E-field distribution of different SEP structure in the OFF-state ($V_{GS} = 0$ V) and high drain voltage ($V_{DS} = 1200$ V), and is extracted at the surface of AlGaN barrier layer. The simulation results indicate that as W_{SEP} decreases, the lateral modulation (along the gate width direction) of the OFF-state peak E-field distribution in the SEP devices becomes increasingly pronounced. This suggests that fabricating SEP devices with smaller W_{SEP} could lead to further optimization of the OFF-state electrical characteristics. Additionally, the improvement in dynamic R_{ON} is expected to be maintained despite the reduced W_{SEP}. However, this assumption requires further experimental validation to confirm the correlation between smaller W_{SEP}

979-8-3315-0417-5/25 $31.00 © 2025 IEEE   209

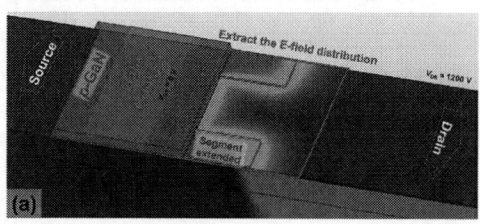

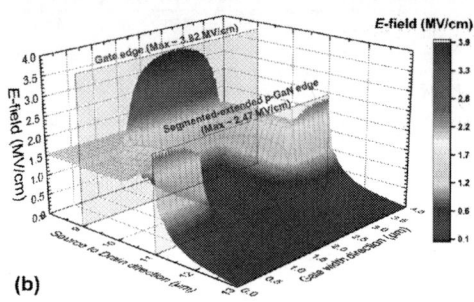

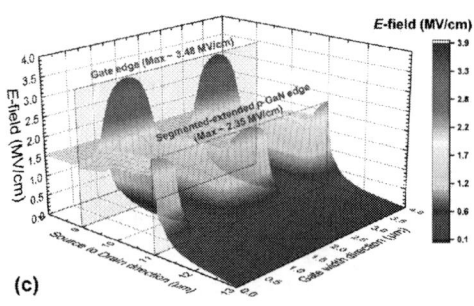

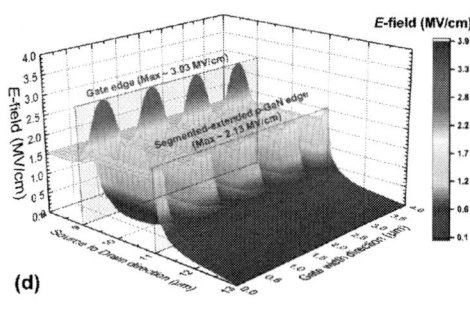

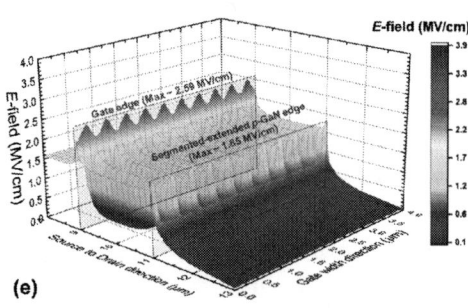

Fig. 5: (a) Extract simulated E-field distributions at the p-GaN gate and extended p-GaN structure of SEP-HEMTs with W_{SEP} of (b) 2 μm, (c) 1 μm, (d) 0.5 μm and (e) 0.2 μm in OFF-state (V_{GS} = 0 V, V_{DS} = 1200 V).

and enhanced device performance in both OFF-state and dynamic characteristics. These findings highlight the potential of fine-tuning W_{SEP} as a key factor in balancing peak E-field management and dynamic performance in SEP-based GaN power devices.

Conclusion

The SEP structure in AlGaN/GaN HEMTs demonstrated substantial performance advantages over baseline and extended p-GaN configurations. Experimental data indicate improved dynamic R_{ON}, with simulations confirming that smaller W_{SEP} values enhance OFF-state E-field distribution, potentially advancing device reliability. This suggests that tuning W_{SEP} could be a key strategy for balancing field management and dynamic characteristics. Future studies with refined W_{SEP} values may yield further optimization in E-field handling and dynamic R_{ON}. These findings establish SEP as a robust solution for high-performance GaN power devices, addressing both static and dynamic performance requirements.

References

[1] M. Meneghini, C. D. Santi, I. Abid, M. Buffolo, M. Cioni, R. A. Khadar, L. Nela, "GaN-based power devices: Physics, reliability, and perspectives," Journal of Applied Physics, 130, p. 181101, (2021).

[2] J. Si, J. Wei, W. Chen, and B. Zhang, "Electric Field Distribution Around Drain-Side Gate Edge in AlGaN/GaN HEMTs: Analytical Approach," IEEE Transactions on Electron Devices, 60, pp. 3223–3229, (2013).

[3] I. Rossetto, M. Meneghini, S.Pandey, M. Gajda, G.A.M. Hurkx, J.A. Croon, J. Sonsky, G. Meneghesso and E. Zanoni, "Field-Related Failure of GaN-on-Si HEMTs: Dependence on Device Geometry and Passivation," IEEE Trans. Electron Devices, vol. 64, no. 1, pp. 73–77, (2017).

[4] W. Saito, T. Nitta, Y. Kakiuchi, Y. Saito, K. Tsuda, I. Omura and M. Yamaguchi, "Suppression of Dynamic On-Resistance Increase and Gate Charge Measurements in High-Voltage GaN-HEMTs With Optimized Field-Plate Structure," IEEE Trans. Electron Devices, 54, pp. 1825–1830, (2007).

[5] S. Ogura, P. J. Tsang, W. W. Walker, D. L. Critchlow, and J. F. Shepard, "Design and Characteristics of the Lightly Doped Drain-Source (LDD) Insulated Gate Field-Effect Transistor," IEEE Journal of Solid-State Circuits, 15, pp. 424–432, (1980).

[6] D. Song, J. Liu, Z. Cheng, W. C. W. Tang, K. M. Lau, and K. J. Chen, "Normally Off AlGaN/GaN Low-Density Drain HEMT (LDD-HEMT) With Enhanced Breakdown Voltage and Reduced Current Collapse," IEEE Electron Device Lett., 28, pp. 189–191 (2007).

[7] X. Dai et al., "Suppression of Reverse Leakage in Enhancement-Mode GaN High-Electron-Mobility Transistor by Extended PGaN," Physica Status Solidi (a), 220, p. 2200692 (2023).

[8] X. Wei, X. Zhang, C. Sun, et al. "Improvement of breakdown voltage and ON-resistance in normally-OFF AlGaN/GaN HEMTs using etching-free p-GaN stripe array gate". IEEE Trans. Electron Devices, 65, pp. 5041-5047 (2021)

979-8-3315-0417-5/25 $31.00 © 2025 IEEE

Piezoresistive Internal Stress Sensing and Gas Detection via Polymer Swelling

Masaya Toda

Graduate School of Engineering, Tohoku University, Japan
6-6-01 Aramaki-aza-aoba, Aoba-ku, Sendai 980-8579.
Phone: +81-22-795-5810, E-mail: toda@tohoku.ac.jp

Abstract

This study presents a silicon-based nanomechanical piezoresistive gas sensor, enhanced by a suspended structure with functionalized polymers as gas absorbers with Si slits. The suspended design, exposing both sides to gas, increases sensitivity and expands the detection range by translating gas-induced stress into electrical signals. The stress, concentrated at the piezoresistor via cantilever lateral deflection, enhances strain, yielding approximately 20-40 times higher sensitivity than conventional sensors. Each sensor, coated with different polymers, shows high responsiveness to a variety of volatile compounds and gases. Keywords: Internal stress, polymer swelling, and gas sensing

Introduction

The recent years have seen a growing interest in odor sensors capable of analyzing complex gas components. Challenging issues still remain in the application of conventional gas sensors for the quantitative identification of complex gases. Piezoresistive gas sensors, which use a polymer as the sensitive part and measure the expansion of the sensitive part caused by gas adsorption, have been the target of much interest as a method for measuring gases. Baller et al. have been studying mechanical gas sensors using polymers since the 1990s [1], but the sensitivity is not sufficient for piezoresistive measurement, and optical measurement of cantilever deflection has been used. However, the sensitivity of piezoresistive measurement was not sufficient, and optical measurement of cantilever deflection was used, resulting in a large measurement system. Recently, a thin-film surface stress sensor (MSS) and membrane-type internal-stress sensor (MIS) using piezoresistivity proposed have brought nanomechanical gas sensors much closer to practical application [2-4]. Although it is difficult to perform complex analysis with a single polymer, MSS has the advantage of relatively high sensitivity while the sensor size is very small, ranging from several hundred μm to less than 1 mm. Although the selectivity of each polymer is not high, the small size of the sensor itself is expected to improve the discrimination performance of gas species through integration. This is an analytical method, which enables more precise gas component detection hand in hand with database-based AI technology as well as PCA analysis.

An array of sensors with different polymers on top is used to study the response of each polymer to a single sample of gas. By learning the obtained signal waveforms, it is possible to analyze gas components without identifying features. Since the response of the nanomechanical gas sensor is not based on a specific chemical reaction but on a wide range of gases, this multivariate analysis method can identify the composition and concentration of gases even if they have complex components. On the other hand, the practical sensitivity is still in the ppm order when limited to the polymer response, and the sensitivity of the sensor needs to be improved for use in food control and medical fields.

In this paper, as a structure of a highly sensitive nanomechanical gas sensor that concentrates stress on the piezoresistive part, we investigate a sensor structure in which the sensitive material is embedded in the substrate and its expansion is transmitted to the piezoresistive part in the in-plane direction of the sensor, and demonstrate the advantages of nanomechanical gas sensors, such as a size that can be integrated, high response performance, and wide range of target gas species. We aim to fabricate a high-sensitivity molecular sensor that has the advantages of a nanomechanical gas sensor, such as an integrated size, high response performance, and a wide range of target gases, while also enabling the realization of a novel gas sensor.

Design of Sensors

In order to increase the sensitivity of a piezoresistive gas sensor, it is important to find a structure that effectively transmits the expansion of the polymer to the piezoresistive part. We have previously reported a piezoresistive gas sensor with a slit structure in which a polymer is embedded inside the slit and the volumetric expansion force is used to measure gas concentration [5] (Figure 1).

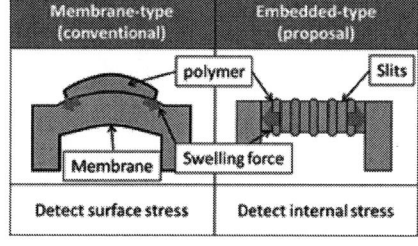

Fig. 1 Conceptual diagram of thin-film and embedded piezoresistive gas sensors

The piezoresistive effect, in which electrical resistance changes when stress is applied, is a phenomenon found in many materials, but silicon has long been used as a device to signal strain electrically due to its excellent mechanical properties. The change in resistance can be calculated as a function of stress, and in p-type silicon, the piezoresistance coefficient takes the largest value. p-type silicon piezoresistance is more important than the other two coefficients and can be approximated as follows

$$\frac{\Delta R}{R} \cong \frac{\pi_{44}}{2}(\sigma_l - \sigma_t) \qquad (1)$$

where σ_l and σ_t are the longitudinal and transverse stresses, and π_{44} is one of the stress coefficient in the piezoresistivity tensor of silicon. For p-type Si with Boron concentration of 7.8×10^{15} atoms/cm³, it is reported that $\pi_{44} = 138 \times 10^{-11}$ Pa⁻¹. p-type piezoresistors are generally the most commonly used.

In this paper, in comparing sensor structures, the sensitivity S of a structure is defined as a parameter like the gage factor in a strain gage, using the following equation.

$$S = \frac{\Delta R}{R} \Big/ \varepsilon_{\text{polymer}}, \qquad (2)$$

where $\varepsilon_{\text{polymer}}$ is the expansion coefficient of polymer.

Polymer expansion due to gas adsorption

Polymers are macromolecules formed by the polymerization of molecules. Small molecules dissolve into polymers by diffusion. This phenomenon is called sorption, and the molar concentration x_{gas} of a particular gas sorbed into a polymer can be approximated by Henry's law.

$$x_{gas} = \frac{p_{gas}}{H_{gas}} \qquad (3)$$

where p_{gas} is the partial pressure of the gas in air and H_{gas} is the Henry constant of the gas relative to the polymer. the Henry constant is obtained using the density of the polymer ρ_{polymer}, the molecular weight of the gas Mgas, the excess chemical potential $\mu_{\text{gas}}^{\text{ex}}$, Boltzmann's constant $\beta = 1/k_{\text{b}}T$ and the reciprocal of the absolute temperature T as follows.

$$H_{\text{gas}} \cong \frac{\rho_{\text{polymer}}RT}{M_{\text{gas}}} \lim_{x \to 0} (\exp(+\beta\mu_{\text{gas}}^{\text{ex}})) \qquad (4)$$

The mass fraction w_{gas} is derived from the molecular weight of the polymer and the gas as following.

$$w_{gas} = x_{gas} \times \frac{M_{gas}}{M_{polymer}} \qquad (5)$$

Assuming that the increase in mass upon sorption of gas molecules has the same specific gravity as that of the polymer, the rate of volume change that occurs is approximated.

$$\frac{\Delta V}{V} \cong w_{\text{gas}} \times \frac{1}{\rho_{\text{polymer}}} \qquad (6)$$

Since the expansion of the polymer caused by gas sorption is generally small, the coefficient of linear expansion when the partial pressure of the gas is Therefore, the coefficient of linear expansion is estimated as bellow.

$$\varepsilon_{\text{polymer}} \cong \frac{1}{3}\frac{\Delta V}{V} \qquad (7)$$

Comparison of sensor structures

The proposed sensor structure of MIS with in-plane displacement and internal stress sensing was compared with that of MSS with surface stress sensing. For each structure, the typical length of the sensitive material part is 300 μm. The names and values of each part of the device are shown in Fig.2. The piezoresistivity coefficient generated when the polymer expands with a linear expansion coefficient of 1×10^{-5} was examined. Both structures were made such that the typical length of the polymer part is and the dimensions of the support including the piezoresistive part are $5 \times 5 \times 5$ μm³. As a result, S, which represents the sensitivity of the proposed MIS structure, was nearly 20 times higher than that of MSS.

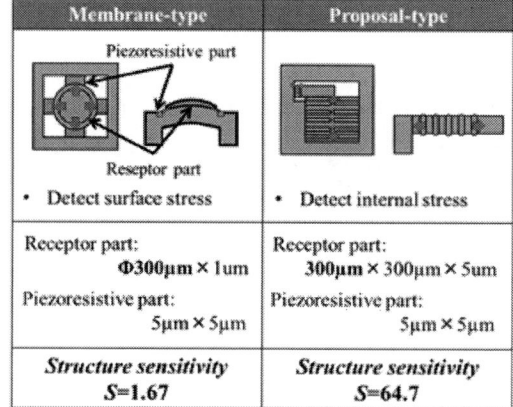

Fig. 2 Comparison results of sensitivity between thin-film and embedded types over the same area

Fabrication process

The fabrication process is shown in Fig.3. First, the surface of an SOI wafer (p-type, 5/0.5/300μm) is doped with boron by spin coating of spin-on dopant (SOD) and rapid thermal annealing. This was done to reduce the contact resistance between the Al electrode and Si. Then, Al sputtered on the substrate was patterning by wet etching and sintering to form the electrode. Next, silicon was patterned by photolithography and DeepRIE etching on both sides of the wafer. The sacrificial layer was removed by Vapor HF etching, and a protective film of Al₂O₃ was deposited by Atomic Layer Deposition (ALD). Finally, the device wiring and polymer coating were applied. The polymer was an acrylic polymer, applied. Figure 4 shows an image of the completed sensor. The polymer film was formed on the comb teeth.

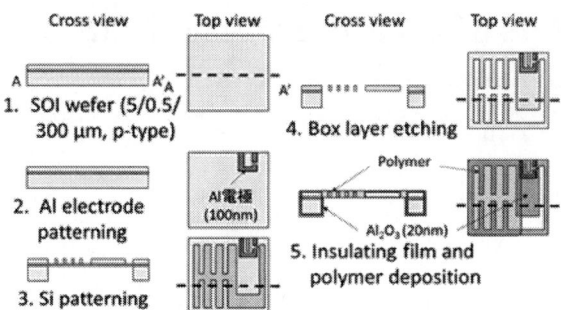

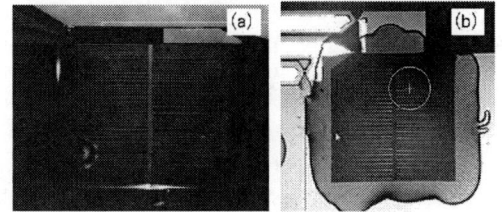

Fig.3 Comparison results of sensitivity between thin-film and embedded types over the same area

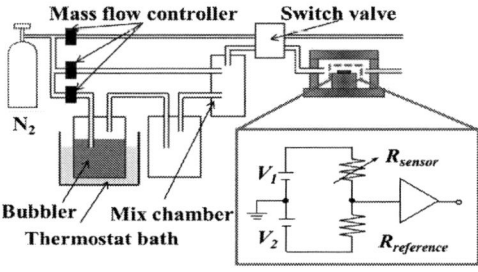

Fig.4   Fabricated sensors (a) dip-coated polymer on Si-slits, (b) ink-jet coated polymer on Si-slits.

Experimental

The gas concentration measurement of the fabricated sensor was evaluated. Fig.5 shows the gas measurement system. The sample gas is generated by mixing saturated vapor and dry N_2 gas using a bubbler, and the concentration of the gas flowing in the chamber is controlled at each flow rate by a flow controller. A reference resistor in the same substrate is used to offset the effects of noise in the piezoresistive element due to temperature changes and other factors. The sensor and reference resistor are incorporated in the bridge circuit shown in Figure 5, and voltage of different polarity is applied to each to adjust the output potential to be near 0 V. The voltage is amplified by a 5000-fold amplifier to detect strain.

Since the change in resistance caused by the piezo-resistance effect is generally very small, the relationship between the output voltage and the rate of change of piezo-resistance can be approximated by the following equation using the voltage applied to the sensor.

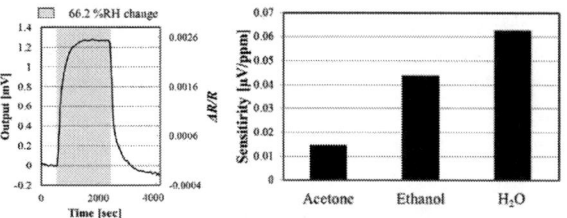

Fig.5 Schematic diagram of the experimental system.

Results and Discussion

The basic response performance of the fabricated gas sensor

was evaluated from the response to a gas containing water vapor. The sensitivity of the resistance change rate was 1.25×10^{-7} per ppm. Next, the response of the sensor to acetone and ethanol vapors was evaluated. The sensitivity to acetone and ethanol was determined to be 1.47×10^{-2} μV/ppm and 0.044 μV/ppm, respectively, and the resolution was 24.8ppm and 8.31ppm. Comparing the sensitivity of the sensor to each gas (Fig. 6), water, ethanol, and acetone were the most sensitive gases, in that order, indicating that the sensitivity increases with the polarity of the molecules.

Fig.6 (left) Response to humidity changes, (right) difference in sensitivity due to gas type.

Conclusion

In this paper, we report the design of a high-sensitivity piezoresistive nanomechanical gas sensor with a cantilever that displaces in the in-plane direction and concentrates stress on the piezoresistive part. The sensitivity of the sensor structure is S = 133, which is higher than that of existing thin-film sensors, and a piezo-resistance change of $\Delta R/R = 2.4 \times 10^{-3}$ per %RH can be generated for a 1% change in degree. The fabricated sensor showed different responses to water molecules, acetone, and ethanol. Further improvement of sensitivity using variety of polymer responses will be possible in the future.

References

[1] M. K. Baller, H. P. Lang, J. Fritz, C. Gerber, J. K. Gimzewski, U. Drechsler, *et al.*, "A cantilever array-based artificial nose," *Ultramicroscopy*, vol. 82, pp. 1-9, 2000.

[2] G. Imamura, K. Shiba, and G. Yoshikawa, "Smell identification of spices using nanomechanical membrane-type surface stress sensors," *Japanese Journal of Applied Physics*, vol. 55, p. 1102B3, 2016.

[3] F. Loizeau, H. P. Lang, T. Akiyama, S. Gautsch, P. Vettiger, A. Tonin, *et al.*, "Piezoresistive membrane-type surface stress sensor arranged in arrays for cancer diagnosis through breath analysis," pp. 621-624, 2013.

[4] Z. Wang, T. Hokama, M. Toda, M. Yamazaki, K. Moorthi, and T. Ono, "Highly Sensitive Structure of Nanomechanical Gas Sensor Based on Stress Concentration Generated by Cantilever Lateral Deflection," pp. 2107-2109, 2019.

[5] M. M. Hossain, M. Toda, T. Hokama, M. Yamazaki, K. Moorthi, and T. Ono, "Piezoresistive Nanomechanical Humidity Sensors Using Internal Stress In-Plane of Si-Polymer Composite Membranes," *IEEE Sensors Letters*, vol. 3, p. 2500404, 2019.

Understanding the impact of discrete trap positions and mapping the hysteresis dynamics in MoS_2 FETs by advanced TCAD modeling

Y.Z. Lv, Y.J. Chai, and Yu.Yu. Illarionov*

L2DON, MSE Department, Southern University of Science and Technology, Shenzhen, China

*illarionov@sustech.edu.cn

ABSTRACT

Locally increased defect densities in gate insulators, for instance due to not properly adjusted processing, could present a serious reliability problem for 2D FETs. By doing advanced TCAD modeling, here we for the first time demonstrate that hysteresis dynamics in MoS_2 FETs may depend on the trap position in oxide. Furthermore, we show that in some positions defects may cause a localized hysteresis of both signs and introduce the universal hysteresis mapping method to properly benchmark this behavior. Finally, we revisit previous experimental results for FAB MoS_2 FETs and show that our TCAD-inspired method can add more understanding to the observed hysteresis dynamics.

Keywords: MoS_2 FETs, defects, hysteresis, TCAD

INTRODUCTION

Transistors with 2D channels are now considered as a promising alternative or supplement [1] to traditional Si devices which are already close to their scaling limits. Recent research achievements on 2D FET technologies [2] have made it possible for the industry to start with their integration into the FAB process flows [3]–[5]. However, reliability limitations due to charge trapping by defects of various origins still present a serious obstacle for these new technologies [6]. In the first available FAB prototypes of 2D FETs the problem of defects is especially sound, since all these devices use traditional high-k insulators like HfO_2 which have fundamental defect bands [7]. In addition, they can suffer from process-induced defects which can be introduced, for instance, during etching [8]. Furthermore, in nanoscale 2D FETs the electrostatic impact of defects is larger [9]. Thus, proper understanding of defect dynamics and possible contributions of defected regions of the device is of key importance to enable reliable scaling in future.

While some previous studies have shed light on how to minimize the charge trapping in 2D FETs by selecting favourable channel/oxide combinations [6], [10], they typically assumed homogeneous distributions of defects in a gate insulator. At the same time, possible impact of locally increased defect densities or localized discrete traps on the charge trapping dynamics is still barely understood, though being highly relevant for scaled FAB 2D devices.

Here by doing TCAD modeling for nanoscale MoS_2 FETs we demonstrate that hysteresis of the I_D-V_G characteristics can be very sensitive to the defect localization in a gate insulator. In particular, oxide defects situated in some positions can simultaneously cause clockwise and counterclockwise hysteresis, depending on the sweep time used. Thus, we also introduce a more comprehensive method to benchmark abnormal hysteresis dynamics and show that it can be successfully used to get more information from our previous experimental results for FAB MoS_2 FETs [8].

TCAD MODELING APPROACH

We use the advanced TCAD simulator Minimos-NT [11] which describes carrier transport in the channel with the drift-diffusion model and employs the four-state non-radiative multiphonon (NMP) model [12] for the charge trapping by insulator defects. We consider Schottky-barrier top-gated MoS_2 FET with $L = 10\,nm$ and 5.13 nm thick HfO_2 gate insulator (Fig.1a) which corresponds to the equivalent oxide thickness (EOT) of 1 nm. A discrete donor trap (i.e. the charge either +1 or 0) is placed 0.065 eV below the conduction band edge of MoS_2 to make the trap active, as schematically shown in the band diagram provided in Fig.1b. The electrostatic impact of this trap is magnified by 500, which is equivalent to placing either the same number of identical traps or a single but strong defect into a certain position. In order to map the hysteresis dynamics versus the trap position, we simulate up and down I_D-V_G characteristics assuming sweep times varied from 10^{-7} to 10^8, while localizing the traps along both the channel length and oxide thickness. A typical simulated I_D-V_G curve with hysteresis is shown in Fig.1c, and the parameters of selected trap and NMP model are provided in Fig.1d.

UNIVERSAL HYSTERESIS MAPPING METHOD

In most previous studies the hysteresis width ΔV_H is typically extracted using a constant current method near the threshold voltage V_{th} and plotted against the reciprocal sweep time $1/t_{sw}$ [8], [13]. However, our TCAD modeling results suggest that in some cases the hysteresis can be localized within a narrow gate voltage range, which goes in line with previous experimental observations [8]. For instance, in Fig.2a either a clockwise hysteresis below V_{th} or a counterclockwise hysteresis above V_{th} can be present depending on t_{sw}. This would make the value of ΔV_H extracted at a randomly selected I_D ill-defined and can result in missing valuable information about charge trapping. Therefore, here we suggest a universal hysteresis mapping method which consists in extracting ΔV_H by scanning constant I_D from the value slightly above I_{OFF} to $0.99*I_{ON}$, as schematically shown in Fig.2b. Then it is convenient to operate with the universal hysteresis functions (UHFs) constructed as a piecewise maximum (upper UHF) and minimum (lower UHF) from the obtained family of $\Delta V_H(1/t_{sw})$ curves.

In Fig.2c,d,e we show the results obtained for 3 different trap positions using 3000 constant I_D values. In Fig.2c, which corresponds to the I_D-V_G curves of Fig.2a, we clearly see that the lower UHF has a maximum of the counterclockwise hysteresis (negative ΔV_H) at slow sweeps and the upper UHF has a maximum of clockwise hysteresis (positive ΔV_H) at moderate sweeps. Fig.2d corresponds to the case where only a clockwise hysteresis is present, which gives a maximum only in the upper UHF. Finally, Fig.2e depicts the situation when the hysteresis is dissolved into many random noise events present in I_D-V_G curves. While each particular noise event appears as a maximum of $\Delta V_H(1/t_{sw})$ with a random position, both UHFs are barely dependent on t_{sw} and simply express the maximum/minimum noise levels.

HYSTERESIS DYNAMICS VS. TRAP POSITION

In Fig.3 we perform a systematic analysis of the hysteresis dynamics for ten trap positions (Fig.3a) while scanning the trends along both the channel length (X axis) and oxide thickness (Y axis). If the trap is situated roughly less than 0.5 nm (Fig.2e) or more than 3.2 nm from the channel/oxide interface (Fig.3b), only discrete noise events can be present for all sweep times. This is likely because the trap is too fast and can exchange charges with the channel or top gate many times during the sweep. As a result, strongly localized hysteresis-like window with $\Delta V_H < 20$ mV may appear randomly at different I_D (Fig.3b, inset). In Fig.3c we show the upper and lower UHFs extracted for the traps situated in the middle of the channel with Y varied from 3.065 to 0.565 nm. The counterclockwise hysteresis disappears when the trap is moved closer to the channel/oxide interface, which suggests that this behavior originates from charge exchange with the top gate. Remarkably, for Y = 3.065 nm the counterclockwise hysteresis at slow t_{sw} is even larger if the trap is situated near source and drain electrodes (Fig.3d). At the same time, the clockwise hysteresis caused by charge trapping from the MoS_2 channel becomes faster in the lower part

of oxide. At Y = 0.565 nm it strongly localizes below V_{th} (Fig.3e) and finally disappears, being splitted into random noise (e.g. Fig.2e). Based on Fig.3d,e we can also note that donor traps situated in the middle of the channel result in a more negative V_{th} and cause larger clockwise hysteresis as compared to source/drain traps.

RELEVANCE TO EXPERIMENTAL OBSERVATIONS

Additional modeling results in Fig.4 confirm that both clockwise and counterclockwise hysteresis caused by the trap at Y = 3.065 nm are thermally activated, i.e. for $T = 450$ K upper and lower UHFs shift to faster sweep times, being consistent with experiments [13].

Finally, in Fig.5 we revisit experimental dataset for imec FAB MoS_2 FETs [8], in which clockwise and counterclockwise hysteresis are present simultaneously at slow t_{sw} (Fig.5a), with the latter being due to top gate edge defects. A single point extraction of ΔV_H previously allowed us to reveal this interplay only at high temperatures when it becomes more pronounced [8]. In contrast, here we can see a sizable counterclockwise contribution in the lower UHF already at 25°C (Fig.5b), which confirms the high sensitivity of our TCAD-inspired mapping method even to those hysteresis features which are barely visible.

CONCLUSION

In summary, by doing advanced TCAD modeling we demonstrated that the hysteresis in nanoscale MoS_2 FETs can be very sensitive to the position of discrete traps in a gate insulator. Considering that abnormal hysteresis dynamics can be observed for some trap positions, we suggested a universal hysteresis mapping method which is extremely sensitive to all charge trapping features, such as strongly localized hysteresis and even noise. As we have also verified that this TCAD-inspired method is applicable to benchmark experimentally observed hysteresis dynamics in FAB MoS_2 FETs, our findings could support future reliability studies of these new technologies.

ACKNOWLEDGMENT

The authors acknowledge the financial support from National Natural Science Foundation of China (NSFC) grant number W2432040, Guangdong Basic and Applied Basic Research Foundation 2024A1515010179 and Shenzhen Science and Technology Program 20231115150611001.

REFERENCES

[1] M. Lemme et al, *Nat. Commun.*, 13(1), 1392 (2022).
[2] S. Das et al, *Nat. Electron.*, 4(11), pp. 786–799 (2021).
[3] C.-C. Cheng et al, *IEEE Symp. on VLSI Tech.*, pp. T244–T245 (2019).
[4] I. Asselberghs et al, *IEDM Tech. Dig.*, pp. 40–2 (2020).
[5] K. P. O'Brien et al, *IEDM Tech. Dig.*, pp. 7–1 (2021).
[6] Yu.Yu. Illarionov et al, *Nat. Commun.*, 11, 3385 (2020).
[7] G. Rzepa et al, *IEEE Symp. on VLSI Tech.*, pp. 208–209 (2016).
[8] Yu.Yu. Illarionov et al, *npj 2D Mater. and Appl.*, 8(1), 8 (2024).
[9] B. Stampfer et al, *ACS Nano*, 12(6), 5368 (2018).
[10] T. Knobloch et al, *Nat. Electron.*, 4(2), 98 (2022).
[11] *GTS Minimos-NT*, Global TCAD Solutions, Vienna, Austria, 2022.
[12] A. Alkauskas et al, *Phys. Rev. B*, 90(7), 075202 (2014).
[13] Yu.Yu. Illarionov et al, *2D Mater.*, 3, 035004 (2016).

979-8-3315-0417-5/25 $31.00 © 2025 IEEE

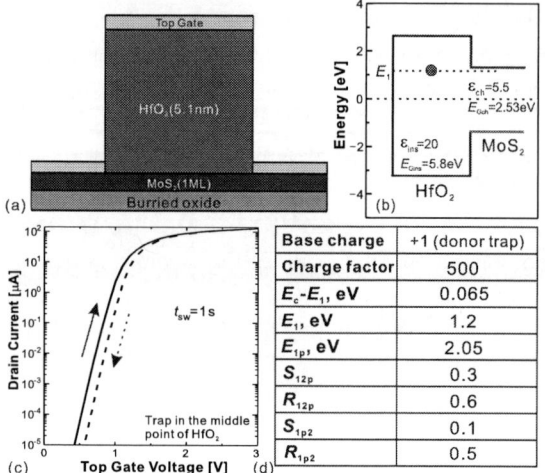

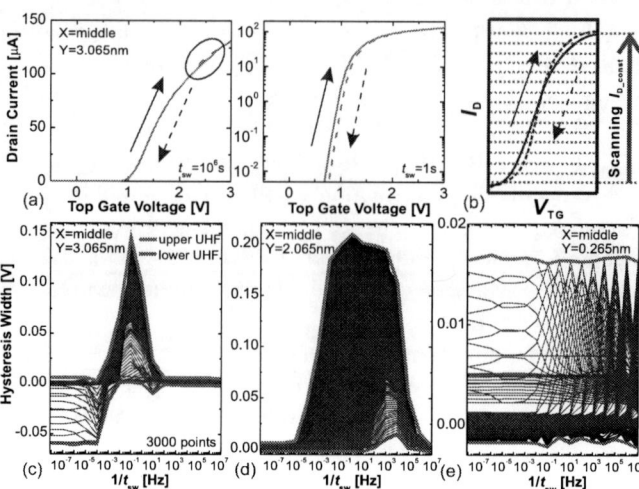

Fig. 1. (a) Schematic layout of MoS₂ FETs used in our TCAD simulations. (b) The corresponding band diagram in flat band condition with a donor trap placed slightly below the conduction band of MoS₂. Zero energy corresponds to the middle of the MoS₂ bandgap. (c) Typical simulated I_D-V_G curve with hysteresis. (d) Defect and NMP model parameters used in our TCAD setup to describe the charge trapping.

Fig. 2. (a) Simulated I_D-V_G curves which show either a localized counterclockwise (left) or a clockwise (right) hysteresis, depending on t_sw. (b) Schematic illustration of our universal hysteresis mapping for the I_D-V_G curve with localized hysteresis branches of different signs. (c,d,e) Typical series of $\Delta V_\mathrm{H}(1/t_\mathrm{sw})$ dependences concluded in between upper and lower UHFs extracted from TCAD simulated I_D-V_G curves for: (c) hysteresis of both signs from dataset of (b); (d) only clockwise hysteresis; (e) only noise.

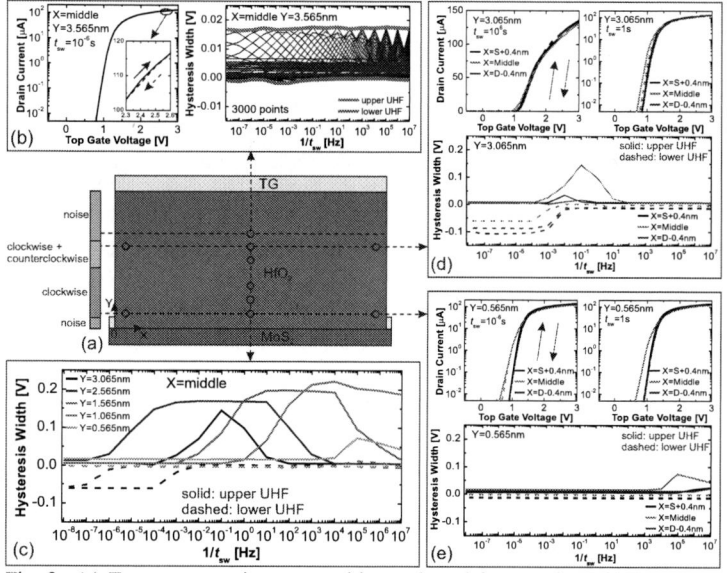

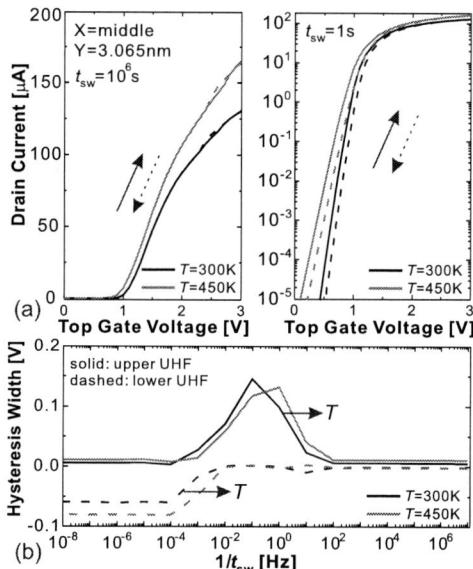

Fig. 3. (a) Ten representative trap positions selected for mapping of the hysteresis dynamics. Typical features are marked with colorbar on the left. (b) Only noise is present above Y = 3.2 nm and below 0.5 nm (Fig.2e). (c) Upper and lower UHFs extracted from Y = 3.065 nm to 0.565 nm for mid-channel traps. While at the top gate side a slow counterclockwise hysteresis is present, closer to the channel only a fast clockwise one can be observed. The hysteresis dynamics are mostly similar along X coordinate in both upper (d) and lower (e) parts of the oxide.

Fig. 4. (a) The I_D-V_G characteristics of the case shown in Fig.2c simulated considering $T = 300\,\mathrm{K}$ and 450 K. (b) The extracted upper and lower UHFs show that the maxima of both clockwise and counterclockwise hysteresis are shifted towards faster sweeps at higher T. This goes in line with thermal activation of charge trapping, i.e. the time constants become smaller.

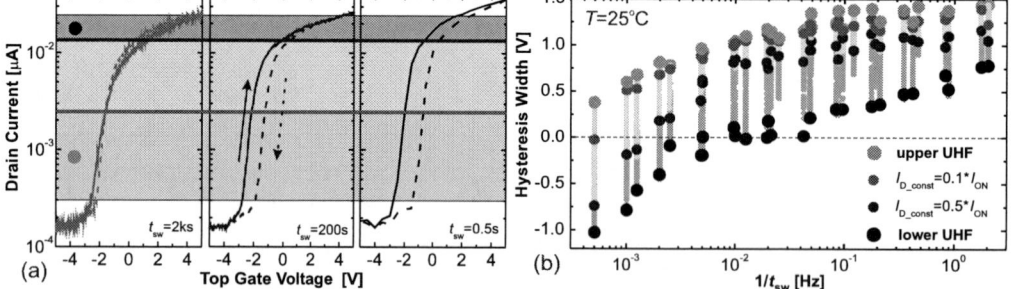

Fig. 5. (a) The I_D-V_G characteristics measured using different t_sw for FAB MoS₂ FETs [8]. (b) The hysteresis curves obtained using our universal mapping method. A clear counterclockwise feature is present in the lower UHF, which reflects slow sweep hysteresis above V_th.

Flexible pulse waveform sensor array
for cuffless PWV measurement

Jiang Zhou[1,2,3], Liang Cunman[4], Zhao Ni[2,3]

[1]Department of Biomedical Engineering, City University of Hong Kong, Hong Kong SAR China
[2]Department of Electronic Engineering, Chinese University of Hong Kong, Hong Kong SAR China
[3]Hong Kong Centre for Cerebro-cardiovascular Health Engineering (COCHE), Hong Kong SAR China
[4]Tianjin University, Key Laboratory of Mechanism Theory and Equipment Design
of Ministry of Education, Tianjin 300354, Peoples Republic of China

Abstract

Wearable devices are increasingly used for non-invasive pulse wave velocity (PWV) monitoring, with pressure sensors being the primary tools for tracking pulses. However, most existing pressure sensors require preloading through cuffs or tapes, thereby introducing variability due to tightness differences across users. Here we present a self-powered, cuffless device for PWV monitoring that overcomes these limitations by eliminating preloading requirements, thereby offering a more reliable and user-friendly solution for continuous PWV measurement.

Keywords: Pulse wave velocity, Pressure sensor, Cuffless

Introduction

Cardiovascular disease (CVD) remains the leading cause of morbidity and mortality worldwide, with its prevalence continuing to rise [1]. Arterial stiffness has emerged as a critical indicator for evaluating cardiovascular risk [2]. Pulse Wave Velocity (PWV) measures the speed at which a pulse wave travels through the arteries, is a crucial parameter used in assessing arterial stiffness and cardiovascular health. High PWV values are associated with increased risk of cardiovascular events and can serve as an early indicator of arterial aging and disease progression [3]. While the carotid to femoral PWV remains the gold-standard estimate of arterial stiffness, research developments supported by advances in technology have generated considerable interest in the measurement of local PWV. Unlike the average estimates provided by regional PWV, the local PWV provides an estimate close to the true propagation speed of blood pulse waves through a specific artery [4]. Recent research has shown a precise correlation between local PWV and vascular properties, biomechanical data, transmural blood pressure levels, and the pathophysiology of the examined artery [5].

With rapid advancements in cardiovascular physiology and medical measurement technologies, pressure sensors have become a powerful tool in health monitoring. Currently, flexible wearable pressure sensors are mainly divided into capacitive, piezoresistive, and piezoelectric [6]. Among these, piezoresistive and piezocapacitive are widely researched because of their static/dynamic dual sensing ability[7][8].

However, the requirement for an external power supply may increase the complexity and dependency of the equipment [9], thus limiting its mobility and clinical unity. Piezoelectric pressure sensors exhibit extraordinary properties that empower them to efficiently convert mechanical stress into electrical signals with exceptional precision and efficacy and without external power supply [10]. Nevertheless, in practice, such devices typically rely on preloading pressure with cuff or tape to ensure pressure signal acquisition, the variations in cuff or tape tightness among different individuals may lead to inaccuracies in measurements [11].

Here we present a self-powered, cuffless device for continuous local PWV monitoring. The device comprises two sets of pulse waveform (PWF) pressure sensors made from soft polymer piezoelectric material, enhanced by a cantilever structure to increase force sensitivity. A serpentine electrode design ensures the stretchability and flexibility of the device, enabling better adaptability to skin movement. The light weight and preload-free features of the sensor facilitate its application in long-term PWV monitoring.

Device design and fabrication process

For pulse detection, the sensitivity of the pressure sensors plays a crucial role in obtaining effective signals. A cantilever-based piezoelectric sensor offers a promising structure for enhancing pulse signal detection, providing highly sensitive output, as it can apply greater strain to the piezoelectric element under vibrational conditions [12]. Here, we compare two types of flexible PWF pressure sensor for non-invasive measurement of pulse wave: press mode and self-adhesion mode. The PWF sensors are made from polyvinylidene difluoride (PVDF) due to its excellent piezoelectric properties and flexibility. In the self-adhesive mode, the sensor exerts minimal pressure on the skin and blood vessels, enhancing wearer comfort. However, without a supporting layer like a belt or cuff, the self-adhesive mode reduces the force transmitted from the pulse to the sensor. Therefore, a higher force sensitivity is required to compensate for this decrease in force. To enhance the sensitivity of the pressure sensor, we implemented a modification in the sensor arrangement, transitioning from a

979-8-3315-0417-5/25 $31.00 © 2025 IEEE

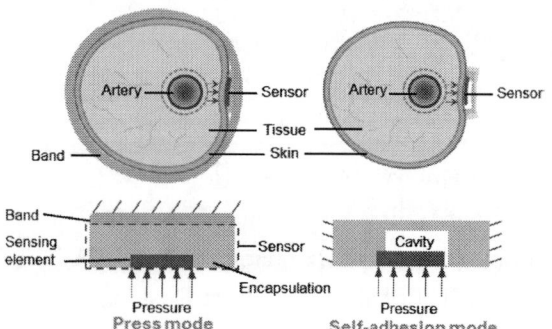

Figure 1 Two types of flexible PWF sensor: Press mode (left) with a band or tape and self-adhesion mode (right)

stacked configuration to a cantilever arrangement (Fig. 1). This alteration was aimed at optimizing the force sensitivity of the sensor.

Finite element analysis (FEA) using Comsol was conducted on the pressure sensor model in both the press and self-adhesion modes. The simulation involved fixing the lower surface of the device and applying a constant displacement load at the center of the piezoelectric beam's end. The results reveal that when a force is applied, the maximum force sensitivity of the stacked-type sensor is 0.15V/N, while the cantilever-type sensor showed a significantly higher sensitivity of 15.16V/N (Table 1). The force sensitivity influenced by the supporting layer is further analyzed. As the result shows, the sensor with 30 um exhibits the best force sensitivity (Table 2).

Table 1. FEA result of PWF sensor with/without cavity

w./w. o cavity	Force (N)	Output voltage (V)
With cavity	1×10^{-3}	0.0000152
Without cavity	1×10^{-3}	0.0015156

Table 2. FEA result of PWF sensor with different thickness of supporting layer

Thickness (μm)	Force (N)	Output voltage (V)
0	1×10^{-3}	0.0011682
30	1×10^{-3}	0.0015156
50	1×10^{-3}	0.0012592
100	1×10^{-3}	0.0009687

We then designed and fabricated the PWV monitoring device, as shown in Fig. 2a. The device consists of two sets of PWF sensors. The sensing part of the PVDF film is crafted into a rectangular shape (1mm*4mm) to function as a deformable beam while the other part is shaped into a serpentine structure to endure bending and tensile deformations. Positioned on a silicon rubber thin film (Ecoflex 00-30, Young's modulus: ~60 kPa), the PWF sensor feature a cavity (width: 3mm) made by high stiffness silicon rubber (PDMS, Young's modulus: ~1MPa) on the rectangular PVDF beam, forming a

piezo cantilever to enhance force sensitivity. The distance between two PWF pressure sensos is 12mm, which allows for easy differentiation of two pulse signals. By analyzing the measured pulse waveform signals, the PWV can be calculated by dividing the distance between the sensor sets by the time differential of the pulse wave signals. The entire device is encapsulated in low stiffness silicon rubber (Ecoflex 00-30) with dimensions of 40mm in length, 35mm in width and 0.425mm in thickness, and a total weight of less than 1g to prevent pressure on skin and blood vessels, reducing measurement errors from deformation. The entire device is stretchable and flexible, enabling easy attachment to human skin for long-term PWV measurement. The PWV monitoring device is fabricated by multi-step layered method with glass slides as temporary substates (Fig.2b). First spin coat PDMS on glass slide and laminate a PVDF film onto a PDMS-coated glass slide. The PWF sensor pattern is then defined by a laser cutter. After removing the excess PVDF material, the PWF sensors pattern are transferred on Ecoflex substrate spin-coated onto a dextran-coated glass by water-soluble tape. Bond the PDMS cavity to the pressure sensor using uncured Ecoflex as an adhesive to form the PWF sensor

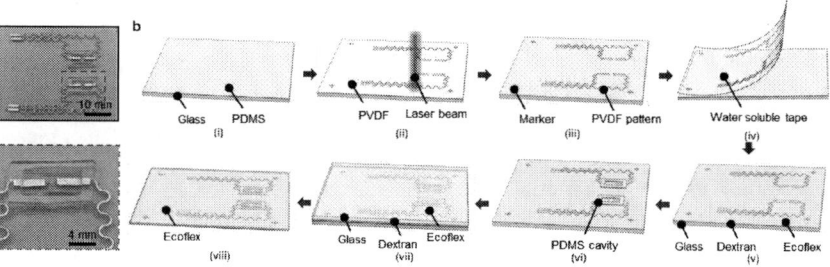

Figure 2 (a) Photo of PWV monitoring device; (b) Fabrication process of PWV monitoring device

array. Place the top surface of the cavity onto an uncured Ecoflex, pour Ecoflex and bake at 70 °C for 1 h, the glass slides are removed in warm water after the Ecoflex cures.

Results

To validate the FEA results, experimental characterization was conducted by applying square wave displacement loads of varying amplitudes using a commercial force sensor (Kistler Instruments 9119AA1), while recording the resulting voltage signals from the pressure sensors. The results shows that the stacked-type force sensor exhibits a force sensitivity of 0.20V/N, whereas the cantilever-type force sensor demonstrates a significantly higher force sensitivity of 32.43V/N (Fig. 3a-b). The measurement results for different supporting layer thicknesses shown in Fig. 3c suggest that the 30um-thick PWF pressure sensor exhibits the best force sensitivity. All these results align with the FEA prediction.

Next we perform local PWV measurements for the radial artery. Utilizing data captured by two sets of pressure sensors (Fig. 5a), we calculated the local PWV using a two-point method based on the time-shifted blood pulse. To mitigate the influence of reflected arterial pressure waves on the peak point of the pulse waveform, we utilized the first derivative maximum point as a robust indicator for assessing local PWV, thereby reducing the impact of arterial wave reflections [5]. The first derivatives of the two recorded pulse waveforms are depicted in Fig. 4b with a time difference of 2.97 ms between the two peak points illustrated in Fig. 4c. By continuously recording pulse waveforms for over 9 seconds (Fig. 4d), we were able to derive real-time local PWVs, with the mean local PWV calculated to be 4.03 m/s (Fig. 4e).

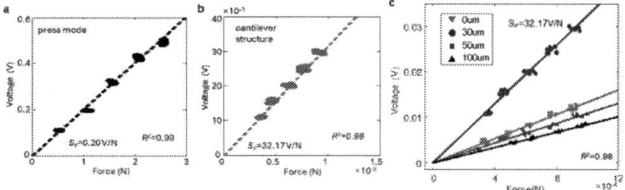

Figure 3 Experiment calibration of PWF pressure sensor; (a) pressure sensor with press mode; (b) Pressure sensor with cantilever structure; (c) PWF pressure sensor with different thickness of supporting layer, which aligns with FEA result.

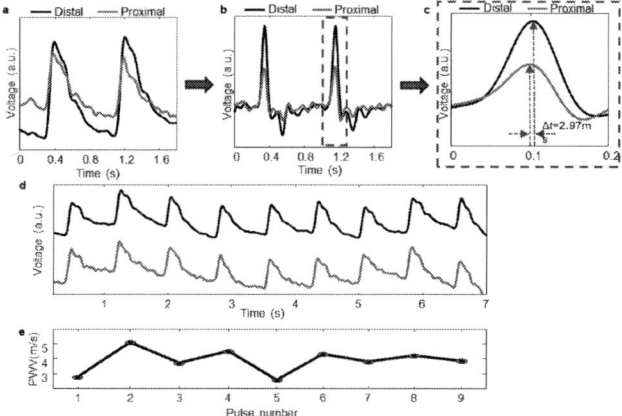

Figure 4 PWV measurement result (a) Pulse waveforms recorded by two sets of pressure sensors; (b) First derivatives of the pulse waveforms; (c) The time difference between first derivatives of the pulse waveforms measure from two sets of PWF sensors; (d) Continuous pulse waveforms recorded by two sets of pressure sensors; (e) PWV results calculated from the continuous pulse waveforms.

Conclusion

In this work, we have developed a self-powered cuffless local PWV monitoring device. The entire device is flexible, stretchable and light weight, allowing for seamless contact with human skin. Also, the device is easy to fabricate, facilitating its integration with more biomedical sensors for measuring additional artery parameters.

Acknowledgments

This project is supported by the C-type Project from Science, Technology and Innovation Commission of Shenzhen Municipality (Reference No. 202205303000149), Hong Kong Center for Cerebro-cardiovascular Health Engineering (COCHE) and National Natural Science Foundation of China (Grant no. 52305611).

References

[1] Roth, G. A. et al. Global burden of cardiovascular diseases and risk factors, 1990-2019 update from the GBD 2019 study. J. Am. Coll. Cardiol. 76, 2982–3021 (2020).

[2] An, D. W. et al. Derivation of an outcome-driven threshold for aortic pulse wave velocity: An individual-participant meta-analysis. Hypertension 80, 1949–1959 (2023).

[3] Ben-Shlomo, Y. et al. Aortic pulse wave velocity improves cardiovascular event prediction: an individual participant meta-analysis of prospective observational data from 17,635 subjects. J. Am. Coll. Cardiol. 63, 636–646 (2014).

[4] Marshall AG, Neikirk K, Afolabi J, Mwesigwa N, Shao B, Kirabo A, Reddy AK, Hinton A Jr. Update on the Use of Pulse Wave Velocity to Measure Age-Related Vascular Changes. Curr Hypertens Rep. 2024 Mar;26(3):131-140.

[5] P. M. Nabeel, V. R. Kiran, J. Joseph, V. V. Abhidev and M. Sivaprakasam, "Local Pulse Wave Velocity: Theory, Methods, Advancements, and Clinical Applications," in IEEE Reviews in Biomedical Engineering, vol. 13, pp. 74-112, 2020.

[6] Fangfang Gao, Xuan Zhao, Li Wang, A stretching-insensitive, self-powered and wearable pressure sensor, Nano Energy,Volume 91,2022,106695,ISSN 2211-2855.

[7] J. Wang et al. Sustainably powering wearable electronics solely by biomechanical energy, Nat. Commun (2016).

[8] L.Y. Chen et al. Continuous wireless pressure monitoring and mapping with ultra-small passive sensors for health monitoring and critical care. Nat. Commun (2014).

[9] Zhao, L., Liang, C., Huang, Y. et al. Emerging sensing and modeling technologies for wearable and cuffless blood pressure monitoring. npj Digit. Med. 6, 93 (2023).

[10] G. Fiori, F. Fuiano, A. Scorza, S. Conforto and S. A. Sciuto, "Non-Invasive Methods for PWV Measurement in Blood Vessel Stiffness Assessment," in IEEE Reviews in Biomedical Engineering, vol. 15, pp. 169-183, 2022.

[11] Li, J., Jia, H., Zhou, J. et al. Thin, soft, wearable system for continuous wireless monitoring of artery blood pressure. Nat Commun 14, 5009 (2023).

[12] Kang M-G, Jung W-S, Kang C-Y, Yoon S-J. Recent Progress on PZT Based Piezoelectric Energy Harvesting Technologies. Actuators. 2016; 5(1):5.

Physical Unclonable Function in Spiking Neural Network Based on In-Te Ovonic Threshold Switching Devices

Huan Wang[1], Qihang Zhu[1], Hengyi Hu[1], Yi Li[1,2], Xiangshui Miao[1,2], Ming Xu[1,2,*]

[1] Wuhan National Laboratory for Optoelectronics, School of Integrated Circuits, Huazhong University of Science and Technology, China. [2]Hubei Yangtze Memory Laboratories, Wuhan, China.

*Corresponding author e-mail: mxu@hust.edu.cn

Abstract

This work presents a threshold-adjustable leaky integrate-and-fire (LIF) neuron based on In_3Te_7 ovonic threshold switching (OTS) devices with high endurance and low leakage current. A two-layer spiking neural network (SNN) built with the neurons achieves 85% accuracy in handwritten digit recognition. Additionally, a physical unclonable function (PUF) based on the LIF neurons demonstrates high uniqueness and reliability. This hardware solution integrated with neuromorphic computing system offers a new approach for data security, which is crucial for AIoT.

Keywords: Ovonic threshold switching, physical unclonable function, neural network, InTe.

Introduction

With the exponential growth of data streams from internet of things (IoT) [1], there is an urgent need for an effective way to process massive data. Artificial intelligence of things (AIoT) combines machine intelligence and IoT technology, offering powerful capabilities for processing large volumes of data especially from edge devices, which has attracted widespread attention from industry and academia [2]. In AIoT, information security is a critical challenge, particularly in the data interactions between edge computing devices and cloud servers [3]. As a hardware security solution based on the inherent randomness of devices, physical unclonable functions (PUF) have a wide range of applications in countering security threats [4]. The variations of devices in PUFs define the mapping from input (challenge) to output (response), serving as a unique fingerprint for the hardware devices. Although various PUF schemes based on volatile and non-volatile memristors have been reported [5, 6], integrating them with neuromorphic computing systems often requires additional device arrays to serve as PUF units, which increases the chip area overhead. Therefore, developing new PUF entropy sources within the existing neuromorphic computing hardware framework is crucial for edge devices, as it significantly reduces the static power consumption and increases the integration density.

In this work, we developed a high-performance ovonic threshold switching (OTS) device based on In_3Te_7 materials, featuring a high on/off ratio of up to 10^5, leakage current (I_{off}) as low as 10^{-8} A, and endurance exceeding 10^6 cycles. Based on the In_3Te_7 devices, threshold-adjustable leaky integrate-and-fire (LIF) neurons were constructed, and a handwritten digit recognition task was completed by a two-layer fully connected spiking neural network (SNN). Finally, PUFs based on pulse firing frequency encoding were demonstrated, with repeated testing confirming the uniqueness and reliability.

Device Fabrication and Characterization

We fabricated OTS devices with a via-hole structure based on In_3Te_7, as shown in Fig. 1. The device features a via-hole with a diameter of 400 nm, patterned by electron beam lithography (EBL). Both the top and bottom electrodes are 100 nm tungsten (W), with a 20 nm In_3Te_7 functional layer in between, all fabricated through magnetron sputtering.

Fig. 2(a) shows the DC I-V characteristics of the OTS devices with a compliance current of 1 mA. Before the threshold switching, a forming voltage of 4 V was applied to initialize the devices. The OTS devices were reliably switched on at a threshold voltage (V_{th}) of 2.35 V, with an I_{off} of only 7 nA read at $1/2V_{th}$. To measure the dynamic response of the OTS devices, 3.5 V triangular pulses with a 10 μs width were applied, with a 330 Ω resistor connected in series to prevent surge currents. The pulsed I-V characteristics of the OTS devices were measured over 50 cycles, with an on-state current reaching 8 mA and the hold voltage around 0.8 V (Fig. 2(b)). The sampled output current with an oscilloscope reveals that the switching speed of the OTS device is below 100 ns (Fig. 3).

In Fig. 4, we recorded the threshold voltage distribution of 60 devices, with an average V_{th} of 2.32 V and significant device-to-device variations. The pure electronic theory of OTS devices propose that threshold switching mainly arises from electrons in defect states [7]. The variations in our devices are primarily attributed to different defect state paths formed in individual devices post-forming, laying the foundation for future PUF applications. We further evaluated the endurance of the OTS devices by applying over 10^6 consecutive pulses, as shown in Fig. 5(a). The V_{th} distribution of a single unit under 100 consecutive pulses demonstrated low cycle-to-cycle variations, which reduces the probability of error in the hardware system (Fig. 5(b)).

Leaky Integrate-and-fire Neurons

Neurons and synapses are the fundamental components of the biological nervous system and play a crucial role in learning and perception. The In_3Te_7 OTS device can serve as a threshold switch in LIF neuron circuits, due to its nonlinear threshold switching characteristics and excellent electrical properties. Fig. 6(a) shows the designed adjustable LIF neuron circuit in this study, which includes a resistive random access memory (RRAM) as an adjustable input resistance load and a capacitor (C_m) that mimics the membrane capacitance of biological neurons. An OTS device and an output resistance (R_{out}) serve as the discharge circuit for pulse emission, while a transistor in parallel with C_m acts as the control for the RRAM and provides an additional membrane potential leakage pathway. Under continuous input pulses with a period of 5 μs and an amplitude of 5 V, the circuit was observed to successfully perform LIF behavior (Fig. 6(b)).

Additionally, the output spike frequency of the LIF neuron is modulated by the input pulses and the input load resistance (Fig. 7). Fig. 8(a) illustrates a two-layer fully connected SNN based on the neurons, demonstrating handwritten digit recognition tasks in simulations. Fig. 8(b) shows the relationship between the number of training iterations and recognition accuracy. The network converges after 60 training iterations, achieving a final overall recognition accuracy of 85%.

Physical Unclonable Function

Due to device-to-device variations in parameters such as V_{th}, neuron units built with In_3Te_7 OTS devices can produce different responses under the same challenge, giving them the potential to serve as a 1-bit entropy source in PUFs (Fig. 9(a)). In practical applications of the above LIF neuron, the load RRAM setting is configured to a medium level. However, for PUF applications, the load resistance will be adjusted to a high level.

The output pulses from 130 neurons within 2 ms are shown in Fig. 9(b), and the distribution of intervals and the number of pulses approximates an exponential distribution. The interval is the reciprocal of the pulse frequency within 2 ms; if there are no output pulses within 2 ms, it records the time of the first pulse that occurs after this period. Thus, the presence of an output pulse within the 1 ms is used as the criterion to determine whether the output of the neuron unit is "0" or "1". Ten neurons in the output layer of the SNN are selected to provide outputs, with switches controlling whether they connect to the SNN for inference tasks or receive challenges from a cloud server. The 10 output-layer neurons collectively provide 10 bits of information, forming a weak PUF with 10-bit challenge-response pairs (CRPs).

An important criterion for evaluating the feasibility of a PUF scheme is the uniqueness and reliability. Uniqueness indicates that under the same challenge, different PUFs generate distinct responses. A total of 13 PUF groups were tested, as shown in Fig. 10(a). By calculating the normalized inter-chip Hamming distance (inter-HD) distribution among different PUF units (Fig.10(b)), we observed an approximately Gaussian distribution with an average value of

0.503, primarily ranging between 0.3 and 0.8. This significant difference between different PUFs confirms the excellent uniqueness of the design. Reliability emphasizes that under the repeated challenge, the ideal change in output response should be zero. The fourth PUF group was repeatedly tested under the same challenge (Fig. 11). The normalized inter-HD over 50 cycles indicates that, while some bit flips occur due to cycle-to-cycle variations in certain OTS devices, the overall distribution remains within the acceptable error range for the PUFs.

Conclusion

In conclusion, we demonstrated a novel threshold-adjustable LIF neuron based on In_3Te_7 OTS devices, which show both excellent electrical characteristics and reliable switching dynamics. Based on the LIF neurons, we implemented a SNN that successfully performed handwritten digit recognition, achieving an accuracy of 85% after 60 training iterations. Additionally, we introduced PUFs based on pulse firing frequency encoding from the output neurons in SNN, demonstrated high uniqueness and reliability. This work underscores the potential of integrating neuromorphic hardware with secure, lightweight PUF solutions to address both data processing and security challenges in AIoT applications.

Acknowledgments

This work was supported by National Science and Technology Major Project of China (Grant No. 2022ZD0117600); M.X. acknowledges the National Natural Science Foundation of China (Grant No. 62174060), Hubei Key Laboratory of Advanced Memories, and the Fundamental Research Funds for the Central Universities, HUST (No. 2021GCRC051).

References

[1] W. Zuo et al, "Volatile threshold switching memristor: An emerging enabler in the AIoT era," Journal of Semiconductors, 44, pp. 053102 (2023).

[2] Z. Zhang et al, "Artificial Intelligence-Enabled Sensing Technologies in the 5G/Internet of Things Era: From Virtual Reality/Augmented Reality to the Digital Twin," Advanced Intelligent Systems, 4, pp. 2100228 (2022).

[3] R. Carboni et al, "Stochastic Memory Devices for Security and Computing," Advanced Electronic Materials, 5, pp. 1900198 (2019).

[4] R. Pappu et al,"Physical one-way functions," Science, 297, pp. 2026-2030 (2002).

[5] R. Zhang et al, "Nanoscale diffusive memristor crossbars as physical unclonable functions," Nanoscale, 10, pp. 2721-2726 (2018).

[6] B. Gao et al, "Concealable physically unclonable function chip with a memristor array," Science advances, 8, pp. eabn7753 (2022).

[7] D. Ielmini et al, "Analytical model for subthreshold conduction and threshold switching in chalcogenide-based memory devices," J. Appl. Phys., 102, pp. 054517 (2007).

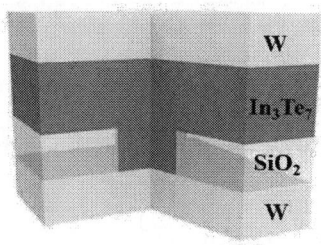

Fig. 1. Diagram of the In₃Te₇OTS device with a via-hole structure.

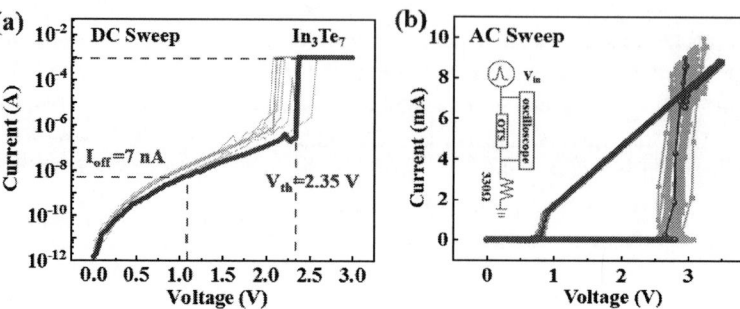

Fig. 2. (a) DC I-V sweeps of OTS devices with a 1 mA compliance current. The I_{off} is 7 nA and on/off ratio is ~10^5. (b) The AC I-V curves under 50 consecutive triangular pulses.

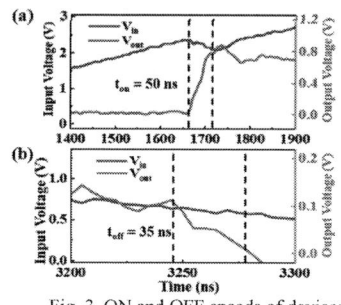

Fig. 3. ON and OFF speeds of devices, i.e., 50 ns and 35 ns, respectively.

Fig. 4. Device-to-device variations of 60 devices with an average V_{th} of 2.32 V.

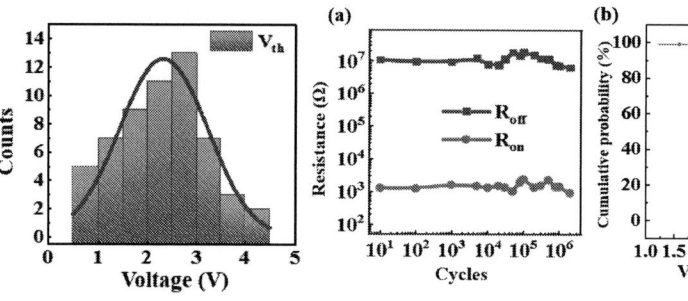 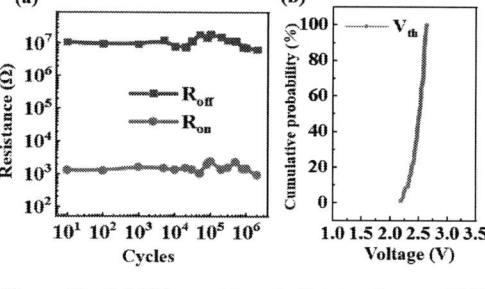

Fig. 5. (a) The on-state and off-state resistance of OTS devices during cycles. (b) Cycle-to-cycle variations.

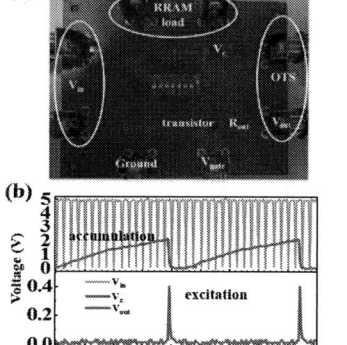

Fig. 6. The adjustable LIF neuron circuit and response.

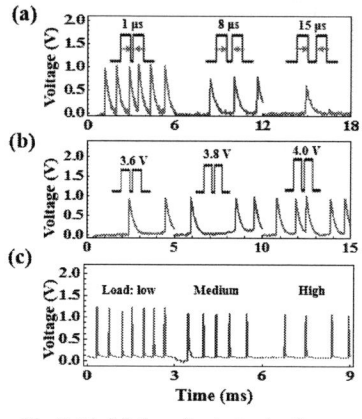

Fig. 7. Modulation of output pulse frequency by input pulses and load resistance.

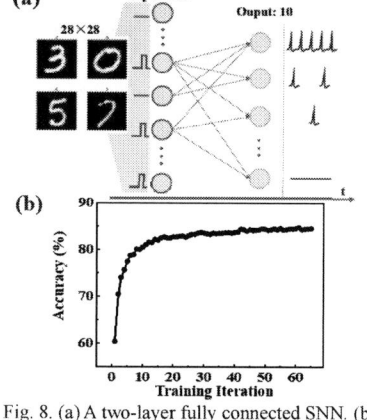

Fig. 8. (a) A two-layer fully connected SNN. (b)The accuracy of 85% after 60 training iterations.

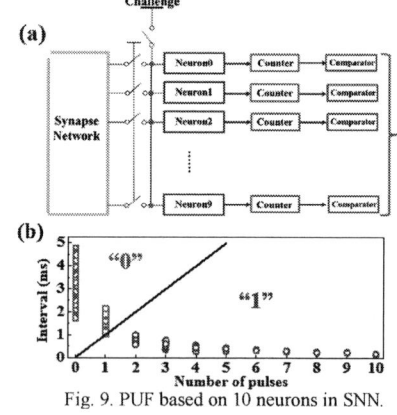

Fig. 9. PUF based on 10 neurons in SNN.

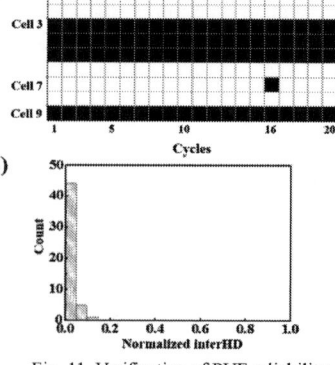

Fig. 10. Verification of PUF uniqueness.

Fig. 11. Verification of PUF reliability.

979-8-3315-0417-5/25 $31.00 © 2025 IEEE

Direct Backside Contact Impact on 3-dimensional Stacked FET SRAM Beyond 1nm Node

Mingyu Kim[1,2], Jaehyun Park[2], Sungil Park[2], Kyunghwan Lee[2], Deuk Ho Yeon[2], Daewon Ha[2], and Hyungcheol Shin[1,3]

[1]Seoul National University, Korea, [2]Samsung Electronics, Co. Ltd., [3]Integra Semiconductor, Ltd.

ABSTRACT

Direct Backside Contact (DBC) is crucial for reducing the logic standard cell and SRAM bitcell area in 3-Dimensional Stacked FET (3DSFET) beyond 1nm node. Among various Back Side Interconnection (BSI) methods, we propose using DBC for the PMOS device and front-side interconnection for the NMOS device. This approach achieves a 28% improvement in SRAM bitcell area scalability, a reduction in external resistance, and an increase in PD and PG transistor on-current by 7.2%, IREAD by 4.7%, and Gamma by 7.2%, demonstrating DBC's effectiveness for advanced CMOS scaling in 3DSFET architecture.

Keywords: 3-dimensional stacked FET (3DSFET), direct backside contact (DBC), back side interconnection (BSI)

INTRODUCTION

The transition from FinFET to MBCFET has been a key driver in advancing logic transistor scaling, with reduced fin depopulation and tighter metal pitch being central factors [1]. However, further scaling is limited by physical constraints, such as channel width, metal resistance, and short channel effects [2]. To address these issues, 3D Stacked Field Effect Transistor (3DSFET) has emerged as a promising solution, offering scalability beyond 1nm node [3]–[10]. The use of wafer Back Side Interconnections (BSI) for power and/or signal lines improves scaling for both logic standard cell and SRAM bitcells, which are critical components in CMOS integrated circuits [11]. 3DSFET, with its vertically stacked architecture, challenges conventional scaling trends and offers significant benefits for both logic standard cell and high-density SRAM bitcell area reduction [12]. The implementation of Direct Backside Contact (DBC) within this structure is essential for further development. This study will present potential DBC schemes, discussing their influence on SRAM bitcell area and the resulting changes in SRAM bitcell characteristic behavior for the next-generation technology based on the chosen direct backside contact scheme.

DIRECT BACKSIDE CONTACT STRUCTURE

This section focuses on backside interconnection structures and the most promising direct backside contact options, as presented by Dr. Ha [13]. These options include power tab cells, buried power rails, and direct

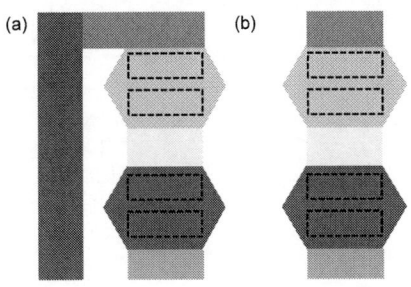

Fig. 1. Direct backside contact candidates (a) opt. a (b) opt. b

backside contact. Transitioning from power tab cells to direct backside contact minimizes the logic and SRAM bitcell areas and reduces IR drop. However, challenges with interface resistance remain. As shown in Fig. 1, two potential connections for the top NMOS device in the direct backside contact structure are presented: one connects it to the backside interconnection like the bottom PMOS, and the other to the front-side interconnection. Currently, there is limited research on how these changes in SRAM bitcell area and resistance affect the bitcell characteristics, which are essential for the 3DSFET architecture. This study addresses the need for further exploration of these connections and their impact on SRAM bitcell performance.

RESULT AND DISCUSSION

The following sections will discuss the changes in high-density SRAM bitcell area, resistance modeling and SRAM bitcell characteristics based on direct backside contact structure options, which are considered the most promising candidates within the 3DSFET architecture.

A. SRAM bitcell Layout

High-density SRAM layout was compared by direct backside contact scheme as shown in Fig. 2 (a) and (b). First of all, when explaining the options for each BSC scheme, in opt. a, the contact resistance and SRAM bitcell area are determined by the width of the top bottom connection contact VL layer. Reducing the width of the top and bottom connection contacts offers an advantage in terms of SRAM bitcell area, but increases contact resistance, requiring careful adjustment. In case of the opt. b scheme, the most significant difference is that eliminating the VL layer, which connects the upper and lower device, allows for a substantial reduction in the

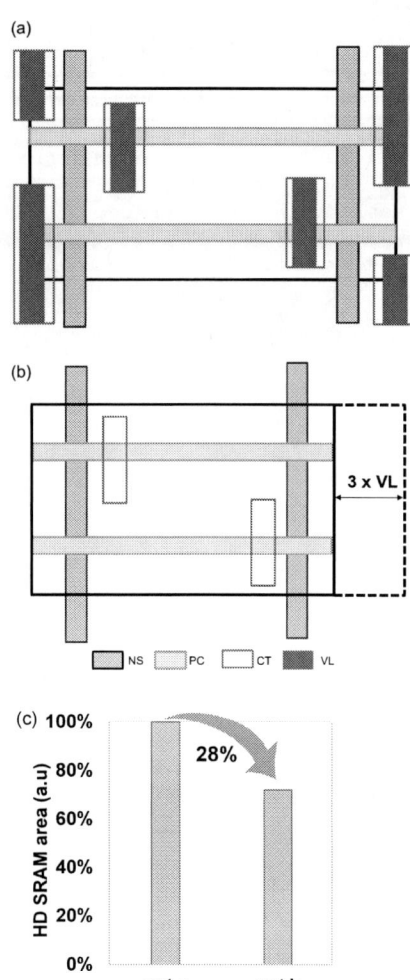

(a)

(b)

3 x VL

NS PC CT VL

(c) 100%

28%

Fig. 2. (a) SRAM bitcell layout using opt. a (b) SRAM bitcell layout using opt. b (c) SRAM area comparison

SRAM bitcell area. So by utilizing front-side contact for the NMOS device and direct backside contact for the PMOS device scheme, the SRAM bitcell area can be significantly reduced. In terms of Front-End-Of-Line (FEOL) SRAM layout design, we can eliminate the three-times VL Critical Dimension (CD) in one SRAM bitcell height. Consequently, as shown in Fig. 2 (c), we propose using opt. b direct backside contact option achieving a 28% higher scalability for the SRAM bitcell which highly innovative structure capable of achieving area scalability beyond a single node.

B. Back Side Contact Resistance Modeling

Fig. 3 shows the total resistance components for each contact scheme. In general, total resistance is composed of channel resistance and external resistance including

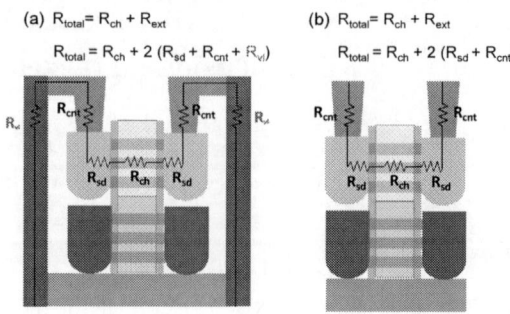

Fig. 3. (a) Opt. a total resistance modeling (b) Opt. b total resistance modeling

two source/drain (SD) and two contact resistances in opt. b direct backside contact as shown in Fig. 3 (b).

$$R_{\text{total}} = R_{\text{ch}} + R_{\text{ext}} = R_{\text{ch}} + 2\left(R_{\text{sd}} + R_{\text{cnt}}\right) \quad (1)$$

However, in the opt. a contact scheme, VL contact structure connecting the upper device is included, which introduces additional two VL resistances with contact resistance as shown in Fig. 3 (a).

$$R_{\text{total}} = R_{\text{ch}} + R_{\text{ext}} = R_{\text{ch}} + 2\left(R_{\text{sd}} + R_{\text{cnt}} + R_{\text{vl}}\right) \quad (2)$$

Thus, minimizing the VL resistance and additional CA resistance components is crucial for improving device performance. There are two main factors that can reduce VL resistance: one is the VL critical dimension (CD), and the other is the type of VL metal used. As the CD increases, the resistance component improves, but this comes at the cost of area efficiency. To address these problems, the CD was divided, and the contact metal resistance component was analyzed using metal A, the most commonly used metal material, and metal B, a promising future metal material due to its low resistivity.

$$R\left(\text{metal A or B}\right) = \frac{1}{A}\int_0^L \rho\left(x\right)dx \quad (3)$$

It can be observed that the additional resistance of VL and contact shows approximately $360\Omega\mu m$ for metal A at minimum VL CD, while for metal B, it shows approximately $100\Omega\mu m$ at minimum VL CD for reducing the SRAM bitcell area. This means that two Rvl resistances and additional contact resistance are added to the Rext resistance, increasing the overall external resistance. In the opt. a direct backside contact scheme, a transistor design that accounts for the additional VL external resistance is essential. Furthermore, it's important to analyze how the presence or absence of the external resistance component impacts the characteristics of both the transistor and the SRAM bitcell between opt. a and b direct backside contact.

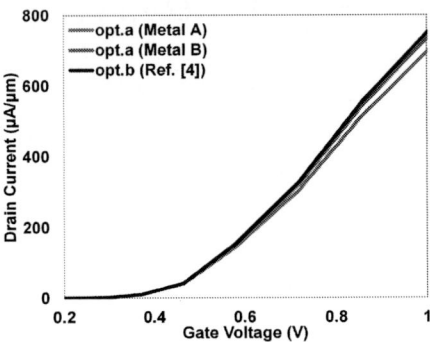

Fig. 4. PD/PG transistor Drain Current-Gate Voltage characteristics

C. SRAM bitcell Characteristics

In this section, we will evaluate the SRAM bitcell characteristics resulting from the different direct backside contact options and metal materials. First of all, to evaluate the external resistance impact on the PD/PG transistor level, the high-density SRAM transistor simulation is presented as shown in Fig. 4. The VL resistance (Rvl) reflects the presence of external resistance, resulting in a decrease in PD/PG transistor on-current of 2.1% for the metal A and 7.2% for the metal B. This indicates that a clear degradation in driving capability occurs when external resistors such as VL and contact resistance are added to the device [14]. Furthermore, to evaluate the direct backside contact external resistance and degradation in PD/PG transistor driving capability impact on the SRAM bitcell performance, the high-density SRAM bitcell simulation comparison is presented in Fig. 5 and Fig. 6. The results clearly indicate that opt. b contact scheme shows both 4.7% higher IREAD and 7.2% Gamma (PG/PU Ion ratio) improvement compared to the opt. a scheme. Degradation in PD/PG driving capability due to the additional external resistance leads to a decline in SRAM bitcell IREAD speed and write ability, especially Gamma (PG/PU Ion ratio). As a result, front-side interconnection for NMOS and direct backside contact for PMOS provides us not only the most competitive SRAM in terms of bitcell performance but also outstanding high-density SRAM bitcell area scalability.

CONCLUSION

This study proposes an optimal direct backside contact scheme to achieve the smallest SRAM bitcell area based on the 3DSFET architecture. Consequently, this structure also enhances the PD/PG transistor on-current and SRAM bitcell characteristics both IREAD speed and Gamma (PG/PU Ion ratio) for write ability by removing the additional VL and contact external resistance. Thus, this work paves the way for new possibilities in monolithic 3DSFET with DBC integration.

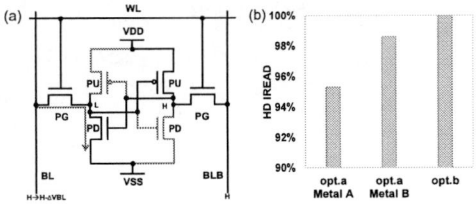

Fig. 5. (a) IREAD operation (b) Comparison of IREAD with different option

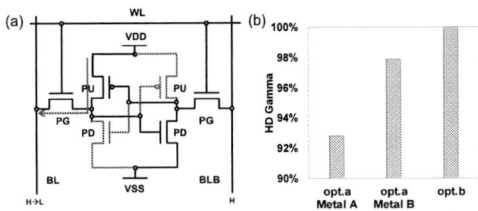

Fig. 6. (a) WRITE operation (b) Comparison of Gamma with different option

REFERENCES

[1] D. Ha and H-S. Kim, "Prospective Innovation of DRAM, Flash and Logic Technologies for DX Era", IEEE Symposium on VLSI Technology and Circuits (2022).

[2] J. Park et al., "First demonstration of 3-dimensional stacked FET with top/bottom source-drain isolation and stacked n/p metal gate", International Electron Device Meeting(IEDM) (2023).

[3] K. Kim, " The Smallest Engine Transforming Humanity: The Past, Present, and Future", International Electron Device Meeting(IEDM) (2021).

[4] J. Park et al., "Highly manufacturable Self-Aligned Direct Backside Contact (SA-DBC) and Backside Gate Contact (BGC) for 3-dimensional Stacked FET at 48nm gate pitch ", IEEE Symposium on VLSI Technology and Circuits (2024).

[5] P. Schuddinck et al., "PPAC of sheet-based CFET configurations for 4 track design with 16nm metal pitch". IEEE Symposium on VLSI Technology and Circuits (2021).

[6] M. Radosavljević et al., "Opportunities in 3-D stacked CMOS transistors", International Electron Device Meeting(IEDM) (2021).

[7] C. -Y. Huang et al., "3-D Self-aligned Stacked NMOS-on-PMOS Nanoribbon Transistors for Continued Moore's Law Scaling", International Electron Device Meeting(IEDM) (2020)

[8] S. Subramanian et al., "First Monolithic Integration of 3D Complementary FET (CFET) on 300mm Wafers", IEEE Symposium on VLSI Technology and Circuits (2021).

[9] S.-W Chang et al., "First Demonstration of CMOS Inverter and 6T-SRAM Based on GAA CFETs Structure for 3D-IC Applications", International Electron Device Meeting(IEDM) (2019).

[10] H. Mertens et al., "Nanosheet-based Complementary Field-Effect Transistors (CFETs) at 48nm Gate Pitch, and Middle Dielectric Isolation to enable CFET Inner Spacer Formation and Multi-Vt Patterning", IEEE Symposium on VLSI Technology and Circuits (2021).

[11] R. Chen et al., " Design and Optimization of SRAM Macro and Logic Using Backside Interconnects at 2nm node", International Electron Device Meeting(IEDM) (2021).

[12] M. Kim et al., " First demonstration of SRAM transistor based on 3-dimensional stacked FET with back side interconnection structure beyond 1nm node", IEEE Silicon Nano electronics Workshop (2024).

[13] D. Ha, International Electron Device Meeting(IEDM) Short course (2022).

[14] D-W. Lin "A Constant-Mobility Method to Enable MOSFET Series-Resistance Extraction", IEEE Elctron Device Letters (2007).

979-8-3315-0417-5/25 $31.00 © 2025 IEEE

Highly Sensitive and Polarization Selective Au Serrate Nanogratings Fabricated by Nanoimprint Lithography

Li Chunxia[1,2], Zhu Shuyan[2*]

[1]Shenzhen Institution of Information Technology, China,
[2]School of Microelectronics Science and Technology Sun Yat-sen University, China,
Email: zhushy27@mail.sysu.edu.cn

Abstract

In this work, we propose a low cost and high throughput novel asymmetric nanogratings fabricated by nanoimprint lithography. The proposed structure is 160 nm linewidth, 600 nm pitch and 1 cm^2 area. The proposed Au serrate nanogratings could induce surface charges under both TE and TM polarized light due to the asymmetric serrates. The measured sensitivity of serrate nanogratings is 561 nm/RIU. In addition, the resonance peak could be tunable by fabricating different nanostructures with the same nanoimprint mold.

Keywords: serrate nanograting, nanoimprint, nanoplasmonic

Introduction

Surface plasmon resonance (SPR) is the resonant oscillation of conductive electrons at the interface between noble metals and dielectric [1]. The SPR can be excited when the wave vector of surface plasmon is equal to incident light by prism coupling. The excited surface plasmons are sensitive to the changing of refractive index of materials around noble metals, making SPR as useful sensing technique for clinical, food safety, and environmental detection [2]. However, the prism coupling setup is complex and propagation distance of surface plasmon polarization at metal and dielectric interface is too large to be used for high spatial resolution multiplexing detection [2-3]. To overcome the above limitations, localized surface plasmon resonance (LSPR) has been proposed to excite the surface plasmons by directly incident light and allowing the plasmons enhanced in nanoscale region [4].

Stripe shaped nanogratings have been investigated over decades and widely used as optical filter and polarizer because their surface plasmons can only be excited by transverse magnetic (TM) polarized light [4]. But in the area of optical switch and multiplexed imaging, polarization selective multiplexed spectra are highly desired. A T-shaped nanoslits array was simulated to support polarization selective tunable multiplexed spectra by tailoring symmetric breaking [5]. Gold (Au) nanocross structure was also demonstrated to show multiplexed spectra by breaking the symmetry of nanocross [6]. Polarization selectivity was observed by a chiral shaped nanogratings without mirror symmetry [7]. In order to fabricate asymmetrical nanograting structures, nanofabrication methods such as electron beam lithography (EBL) and focus ion beam (FIB) were employed to allow custom design with asymmetric structures and nanoscale resolution. However, EBL and FIB are low throughput and high cost methods to fabricated nanostructures over large area. Typically, they will take many days to complete direct write nanopatterns.

Here, we use nanoimprint lithography (NIL) to fabricate a new asymmetric nanogratings named Au serrate nanogratings. NIL is a low cost and high throughput technique to fabricate nanostructures with high uniform over large area. The limitation of NIL is the mold which is very expensive and the resonance wavelength of plasmonic nanostructure fabricated by the mother mold is not tunable. In this study, we combine the NIL and angle evaporation to fabricated Au serrate nanogratings from a nanopillar mother mold. Unlike the conventional stripe nanogratings, the surface plasmon polaritons can't be excited by incident light with transverse electric (TE) polarization because the electric vector of TE polarized light was parallel to the stripe gratings. Both TE and TM polarized light can induce surface charges on Au serrate nanogratings because both of them can provide electric vector normal to some parts of serrate nanogratings.

Methods

Figure 1 shows the fabrication process of Au serrate nanogratings. PMMA and TU-7 resist were coated on glass and then intermediate polymer stamp (IPS) with 280 nm wide, 535 nm pitch, and 500 nm deep nanosquares was used to imprint TU-7 nanosquares. RIE plasma with 20 sccm O$_2$, 20 mTorr and 100 W rf power was used to etch TU-7 residual layer and PMMA film, followed by angle evaporation 2/20 nm Cr/Au with proper orientation of nanosquares. Part of the bottom layer which was shadowed by nanoquares would not be deposited Cr and Au films and then asymmetrical Au nanoholes were fabricated after lift-off process. The smallest linewidth of asymmetrical Au nanoholes was less than 50 nm and could be easily peel off by sonication. After peeling off connection tips of asymmetrical Au serrate nanogratings was fabricated. The extinction spectra were measured using an ultra-violet-visible-near infrared spectrophotometer (PE Lambda 750, PerkinElmer, USA). A normal incident light source under a wavelength range of 400 to 2400 nm was used to illuminate the Au serrate nanogratings.

979-8-3315-0417-5/25 $31.00 © 2025 IEEE

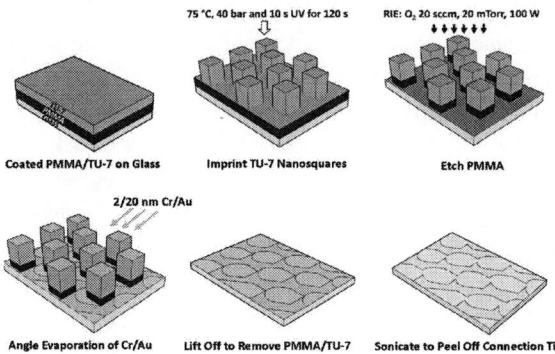

Fig.1: Fabrication process of Au serrate nanogratings by nanoimprint lithography.

The refractive index sensitivity (RIS) of the plasmonic sensors were measured by immersing the plasmonic layer in certified RI liquids (Cargille Laboratories, NJ, USA) with RI of 1.305, 1.404, and 1.604. Numerical solutions with finite difference time domain (FDTD) method (Vancouver, Canada) was employed to simulate the normalized extinction spectra and EM field distribution of Au serrate nanogratings. The normal incident light was a plane wave with wavelength of 400 to 2400 nm and polarization with perpendicular direction. The simulation region was 9 μm^2 in the x-y plane and periodic boundary conditions were adopted in both the x- and y-directions. Perfect match layer was applied in both the top and bottom along the z-direction to minimize the reflection errors. The transmission spectra was monitored by a frequency domain power monitor which was placed behind the simulated structure. A mesh size of 5 nm was used in most of the region except a smaller mesh size of 2 nm was used around the Au plasmonic nanostructures. The size of the Au serrate nanogratings were 160 nm linewidth and 600 nm pitch. The angle of Au serrate tips was 120 degree and the largest linewidth of Au serrate tips was 280 nm. The dielectric constant of Au was obtained from CRC handbook.

Results and Discussion

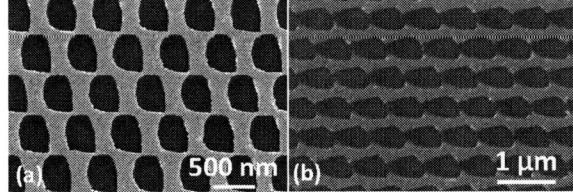

Fig.2: Micrographs of (a) Au asymmetrical nanoholes, (b) Au serrate nanogratings.

Micrographs of Au serrate nanogratings were shown in Fig.2. The normalized extinction spectra and simulated electromagnetic field (EM) intensity of Au serrate nanogratings were shown in Fig. 3.   As shown in Fig. 3(a),

Two resonance peaks at 782 and 1628 nm were observed with direction of electrical field perpendicular with nanogratings. Once the direction of electrical field parallel with nanogratings, Au serrate nanogratings only show one resonance peak at 760 nm. It was also observed that the EM field intensity of Au serrate nanogratings with perpendicular direction was higher than that with parallel direction. Fig.3(b) depicted that most of the surface plamons were accumulated on the tips of serrate nanogratings.

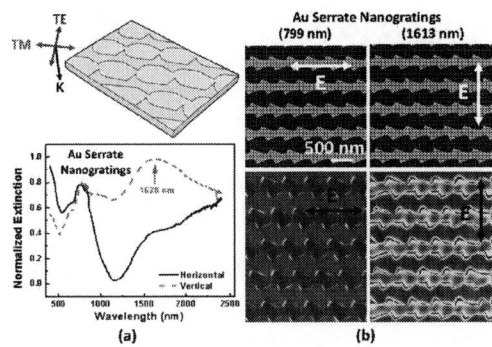

Fig.3: Extinction spectra of Au serrate nanogratings at different polarization.(b) Simulated electromagnetic field intensity of Au serrate nanogratings at different polarization.

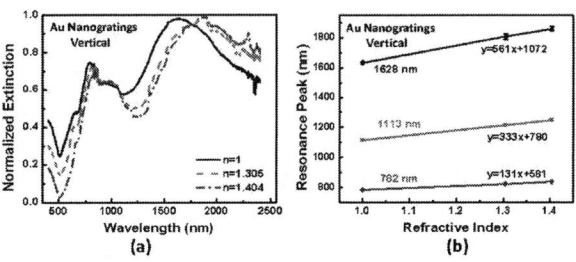

Fig.4: (a) Extinction spectra of Au nanogratings vertical. (b) Resonance peak position as a function of surrounding certified RI medium for Au nanogratings vertical.

Fig. 4(a) shows that the highest sensitivity of Au serrate nanogratings was 1628 nm and all the resonance peaks are red-shifted with RI increased from 1 to 1.404. The measured refractive index sensitivity is 561 nm/refractive index unit (RIU) and figure of merit (FOM) was 1.4. From Fig.4(b), there is a good linearity of the resonance peak position to the refractive index from 1 to 1.404. When the wavelength is increased from 782 nm to 1628 nm, the simulated coefficient is increased form 131 to 561.

Fig.5 shows the fabrication process, micrographs and normalized extinction of 2/20 nm Cr/Au nanoholes with 300 nm wide, 535 nm pitch. As shown in Fig. 5, IPS mold with 280 nm wide, 535 nm pitch and 500 nm deep nanopillars were used to pattern TU-7 photoresist, followed by etching residual TU-7 and PMMA sacrificial layer by RIE O_2 plasma. Cr and Au films were evaporated on glass and TU-7 nanopillars using an electron beam evaporator (JunSun EBS-500). Au nanoholes were formed in lift off process and the PMMA sacrificial layer was dissolved in Remover PG (MicroChem Corp., USA) solution at 60 °C for a few seconds. As shown in Fig.5, Au nanoholes with 300 nm wide, 535 nm pitch had one main resonance peak at 820 nm and the resonance peak was red-shifted linearly with increasing RI medium from 1.305 to 1.604. The measured RIS of Au nanoholes was 224 nm/RIU and FOM was 1.7.

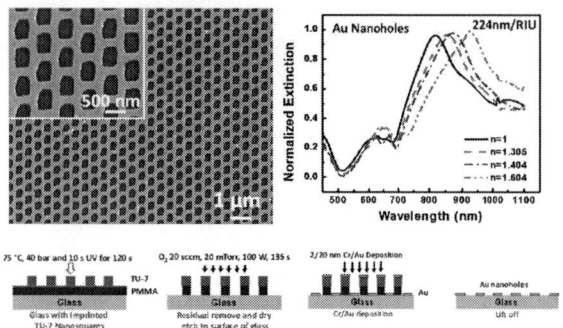

Fig. 5: Micrograph, sensitivity and fabrication process of 2/20 nm Cr/Au nanoholes with 300 nm wide, 535 nm pitch.

As shown in Fig.6, Au nanodots with 350 nm diameter and 535 nm pitch have one main resonance peak at 1221 nm and it was red-shifted linearly with increasing RI from 1 to 1.604. The measured RIS of Au nanodots was 436 nm/RIU and FOM was 0.8. Compared with Au nanodots, Au nanoholes could blue shift the resonance peak from 1221 nm to 820 nm and Au nanogratings fabricated by the same nanoimprint mold could blue shift with the resonance peak from 1628 nm to 1221 nm. The above results show that the resonance peak of nanoplamonic sensor could be tuned from 820 nm to 1628 nm by nanoimprint lithography with the same mother mold.

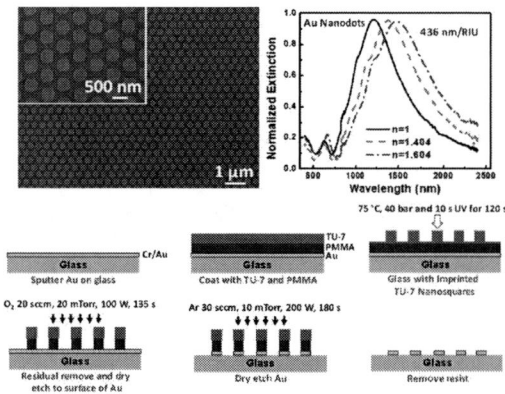

Fig. 6: Micrograph, sensitivity and fabrication process of 2/20 nm Cr/Au nanodots with 350 nm diameter, 535 nm pitch.

Conclusion

In summary, we propose a low cost and high throughput novel asymmetric Au serrate nanogratings fabricated by nanoimprint lithography. The proposed Au serrate nanogratings show highly sensitivity of 561 nm/RIU and polarization selective property. In addition, the Au nanoholes and nanodots using the same mother mold were also fabricated by namoimprint method. From the experiment results, Au nanogratings, nanodots and nanoholes demonstrated different resonance peaks of 1628/1221/820 nm and high sensitivity of 561/436/224 nm/RIU, respectively, suggesting that the nanoimprint lithograph with the same mother mold could fabricate different plasmonic nanostructures with the tuning ability.

Acknowledgments

This work was supported by the Guangdong Basic and Applied Basic Research Foundation (2023A1515011082), Guangdong Innovation and Entrepreneurship Team Project (2021ZT09X070), and the Fundamental Research Funds for the Central Universities Sun Yat-sen University (24qnpy155).

References

[1] J. Homola, S. S. Yee, and G. Gauglitz, "Surface plasmon resonance sensors: Review", Sens.Actuators B, 54, pp.3-15 (1999).

[2] D. Murugan, M. Tintelott, X.-T. Vu, and et. Al, "Recent Advances in Grating Coupled Surface Plasmon Resonance Technology," Adv. Optical Mater., 2401862, pp.1-29 (2024).

[3] X. Li, Q. Zhao, Y. N. Zhang, Y. Zhao and X. Zhou, "Porous Gold Nanocubes Particle-Sensitized U-Shaped Fiber-Optic Surface Plasmon Resonance Sensor for Ultra-Sensitive Detection of Virus RNA," IEEE Trans. Instrum. Meas., 73, pp.1-8 (2024).

[4] W. Ning et al., "An ultrasensitive J-shaped optical fiber LSPR aptasensor for the detection of Helicobacter pylori," Analytica Chim. Acta, 1278, pp.1-10 (2023).

[5] X. Zhang, Z. Li, J. J. Chen, S. Yue, and Q. H. Gong, "A dichroic surface-plasmon-polariton splitter based on an asymmetric T-shape nanoslit," Opt. Express 21, pp.14548-14554 (2013).

[6] G. Z. Li, H.J. Hu and L.J. Wu, "Tailoring Fano lineshapes using plasmonic nanobars for highly sensitive sensing and directional emission," Phys. Chem. Chem. Phys.,21, pp.252-259(2019).

[7] K. Konishi, T S, B. Bai, Y. Svirko, and M. Kuwata-Gonokami, "Effect of surface plasmon resonance on the optical activity of chiral metal nanogratings," Optics express, 15, pp.9575-9583 (2007).

High-performance InSnO/ZnO heterojunction transistors and inverters

Dengqin Xu[1], Tingchen Yi[1], Junchen Dong[2], Lifeng Liu[1*], Zheng Zhou[1], Dedong Han[1,3*], Xing Zhang[1,4*]

[1] School of Integrated Circuits, Beijing Advanced Innovation Center for Integrated Circuits, Peking University, Beijing 100871, China

[2] School of Information & Communication Engineering, Beijing Information Science and Technology University, Beijing 100101, China

[3] Beijing Superstring Academy of Memory Technology, Beijing 100176, China

[4] Peking University Shenzhen Graduate School, Shenzhen 518055, China

*Email: lfliu@pku.edu.cn; zhx@pku.edu.cn

Abstract

We demonstrate high-performance InSnO/ZnO (ITO/ZnO) heterojunction transistors and inverters. The core properties of the ITO/ZnO transistors include a high field-effect mobility of 74.92 cm^2/Vs, a small subthreshold swing of 123.96 mV/decade, and a large on/off current ratio of more than 10^7. The ITO/ZnO transistors also exhibit excellent stability, with a threshold voltage shift of -0.03 V and -0.18 V under negative bias stress (NBS) and negative bias illumination stress (NBIS), respectively. In addition, the inverters based on ITO/ZnO transistors show a gain of 15.25 V/V. Our findings manifest the huge potential of the heterojunction transistors in the field of integrated circuits.
Keywords: heterojunction transistor, inverters, stability

Introduction

Since the first demonstration of InGaZnO (IGZO) Thin film transistors (TFTs) in 2004, oxide transistor has been widely investigated for its high mobility, high transparency and low cost[1]. At present, IGZO display backplane has already been commercialized. Recently, researchers have noticed the potential applications of oxide transistors in integrated circuits (ICs), such as advanced memory and Monolithic 3D integration, for their low leakage current and low fabrication temperature[2],[3]. However, low mobility and poor stability of IGZO limit its application in ICs. To fulfil the demands of emerging applications, fabricating high mobility and high stability oxide transistor is of great importance.

InSnO (ITO) is an emerging channel material. Compared with IGZO, ITO has a higher electron density which leads to a higher mobility. This advantage makes ITO a candidate material for applications in ICs [4],[5]. Besides channel materials, heterojunction channel engineering is another effective approach to enhance electrical performance and stability of devices [6],[7].

In this study, we demonstrate high performance ITO/ZnO hetero-channel transistor with 300°C O$_2$ annealing treatment. The device shows excellent performance, including a high mobility of 74.92 cm^2/Vs, SS of 123.96 mV/decade and a low leakage current below 0.1pA/μm. Furthermore, stability of ITO/ZnO transistor including negative bias stress (NBS), negative bias illumination stress (NBIS) is investigated. Lastly, an inverter cell was fabricated, proving the potential application of ITO/ZnO transistors in ICs.

Experimental procedure

Scanning electron microscope (SEM) image in Fig.1(a) shows complete device structure of a 10μm channel length transistor. Schematic diagram and process flow of ITO/ZnO heterojunction transistor are shown in the Fig.1(b) and(c). On a substrate of Si/SiO$_2$, 100nm ITO film was deposited by RF sputtering as the gate electrode. And a 10nm HfO$_2$ dielectric was deposited by atomic layer deposition at 90°C. Then, 7nm ITO and 10nm ZnO hetero-channel layer was deposited by RF sputtering and ALD respectively. Lastly, stacked Ti/Au source drain electrodes were deposited by e-beam evaporation. After the device structure formation, it was annealed in an O$_2$ atmosphere at 300°C for 30mins. Performance of devices was tested using B1500A.

Result and discussion

To understand the mechanism of ITO/ZnO heterojunction channel, Ultraviolet photoelectron spectroscopy (UPS) analysis was performed in Fig. 2. Work function of ITO is 4.16eV and work function of ZnO is 4.3eV, which indicates that the ITO film has a higher electron density. Optical characteristics were measured to extract the band gap (Eg) of the two films. The Eg of ITO and ZnO are 3.18eV and 3.28eV respectively. Then the energy band diagrams exhibited in Fig.3(a), (b) were obtained. Due to the different fermi levels of the two materials, a conduction band offset is formed at the interface of the two materials, thereby forming a potential well to accumulate electrons. And these electrons in the well can be transported from the source to the drain, thus forming an extra current path which greatly enhances the mobility of ITO/ZnO transistor.

Transfer curves of transistors with different channel lengths at V_D of 0.1 V are shown in Fig. 4(a). As the channel length shrinks, the I_{on} of transistor increases as expected in Fig. 4(b). The field effect mobility of 10μm channel length ITO/ZnO transistor was extracted as 74.92 cm^2/Vs, the subthreshold swing (SS) as 123.96 mV/decade. And the transistor maintains a low leakage current below 0.1pA/μm as well as high on/off ratio over 10^7. The extracted parameters compared with similar works are shown in Table 1. The ITO/ZnO heterojunction transistor demonstrate higher performance in terms of mobility.

Contact resistance of ITO/ZnO transistor was extracted by transmission line method (TLM) shown in Equation (1). And the total resistance was extracted from output curves.

$$R_{total} = R_{channel} + 2R_{contact} \tag{1}$$

The fitted curve is shown in Fig. 5 and the contact resistance is represented by the intercept of the fitted curve as 7560(Ω·μm).

In addition to good electrical performance, the stability of transistor is also worth studying. In this work, NBS test was carried out to investigate the bias stability of the transistor. The test electric field is -10^6 V/cm and the test time ranges from 0 to 1000 seconds with a step of 200s. V_{th} demonstrates a positive shift of 0.09 V at first 200 seconds, and then shifts negatively to -0.03 V after 1000 seconds, as shown in Fig. 6(a). The good NBS stability is attributed to the passivation of the defects on the surface of ITO/ZnO by O_2 annealing.

Considering that the influence of light can aggravate the generation of photogenerated charge carriers, thus leading to instability problems, the NBIS performance of transistor was investigated. The device was tested under an electric field of -10^6 V/cm in combination with light irradiation of 5000 Lux. Fig. 6(b) illustrates the transfer curves of the transistor after negative bias and illumination. As expected, a more negative Vth shift of -0.18 V was observed under NBIS, compared to -0.03 V under NBS. Fig.6(c) summarizes the variation of Vth over time. And the negative shift of ITO/ZnO is less than that of IGZO transistors[13]. In a summary, ITO/ZnO transistor shows excellent bias and light stability.

At last, an inverter cell based on ITO/ZnO transistor was presented. The schematic of the inverter is shown in Fig. 7(a), and it consists of n-type depletion transistors. SEM image of the inverter structure is shown in Fig. 7(b), with the channel lengths of the two transistors being 5μm and 10μm, respectively. And the test result is shown in Fig. 7(c). The inverter operated within a voltage range of 0.5 V to 3 V. As V_{DD} increases, the gain of the inverter rises, reaching a maximum gain of 15.25 V/V as shown in Fig. 7(d). The result shows the potential of ITO/ZnO transistors for logic application.

Conclusion

In this work, we demonstrated high-performance O_2 annealed ITO/ZnO heterojunction transistors with field effect mobility of 74.92 cm^2/Vs, SS of 123.96 mV/decade, on/off ratio more than 10^7, and low leakage current less than 0.1(pA/μm). Besides, the transistor also performed well in stability test, with a V_{th} shift of -0.03 V in NBS and -0.18 V

in NBIS. We fabricated inverter cell and realized logic application of ITO/ZnO transistor. The results shows that high performance and high stability ITO/ZnO transistor have great application potential in ICs.

Acknowledgments

This work was supported by the R&D Program of Beijing Municipal Education Commission (KM202311232011), by the BJSAMT Project (SAMT-2022-PM02-14), and by the Shenzhen Science and Technology Innovation Committee (KQTD 2020082011310-5004).

References

[1] K. Nomura et al., "Amorphous Oxide Semiconductors for High-Performance Flexible Thin-Film Transistors," Japanese Journal of Applied Physics, 45, (2006).

[2] A. Belmonte et al., "Capacitor-less, Long-Retention (>400s) DRAM Cell Paving the Way towards Low-Power and High-Density Monolithic 3D DRAM," IEEE IEDM, (2020).

[3] X. Duan et al., "Novel Vertical Channel-All-Around (CAA) In-Ga-Zn-O FET for 2T0C-DRAM With High Density Beyond 4F2 by Monolithic Stacking," IEEE Transactions on Electron Devices, 69, (2022).

[4] X. Xu et al., "Amorphous Indium Tin Oxide Thin-Film Transistors Fabricated by Cosputtering Technique," IEEE Transactions on Electron Devices, 63, (2016).

[5] S. Li et al., "Nanometre-thin indium tin oxide for advanced high-performance electronics," Nat Mater, 18, (2019).

[6] J. Lee et al., "Heterojunction oxide thin film transistors: a review of recent advances," Journal of Materials Chemistry C, 11, (2023).

[7] X. Huang et al., "Enhancing the Carrier Mobility and Bias Stability in Metal–Oxide Thin Film Transistors with Bilayer InSnO/a-InGaZnO Heterojunction Structure," Micromachines, 15, (2024).

[8] Y. S. Kim et al., "Remarkable Stability Improvement with a High-Performance PEALD-IZO/IGZO Top-Gate Thin-Film Transistor via Modulating Dual-Channel Effects," Advanced Materials Interfaces, 9, (2022).

[9] M. Furuta et al., "Heterojunction channel engineering to enhance performance and reliability of amorphous In–Ga–Zn–O thin-film transistors," Japanese Journal of Applied Physics, 58, (2019).

[10] P. Wang et al., "Synergistically Enhanced Performance and Reliability of Abrupt Metal-Oxide Heterojunction Transistor," Advanced Electronic Materials, 9, (2022).

[11] H. J. Seul et al., "Atomic Layer Deposition Process-Enabled Carrier Mobility Boosting in Field-Effect Transistors through a Nanoscale ZnO/IGO Heterojunction," Applied Materials Interfaces, 12, (2020).

[12] J. -H. Yang et al., "Highly Stable AlInZnSnO and InZnO Double-Layer Oxide Thin-Film Transistors With Mobility Over 50 cm^2/Vs for High-Speed Operation," IEEE Electron Device Letters, 39, (2018).

[13] C. -C. Pan et al., "Improvement in Bias Stability of IGZO TFT With Etching Stop Structure by UV Irradiation Treatment of Active Layer Island," IEEE Journal of the Electron Devices Society, 8, (2020).

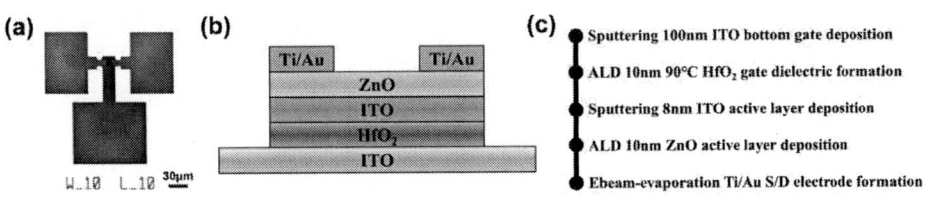

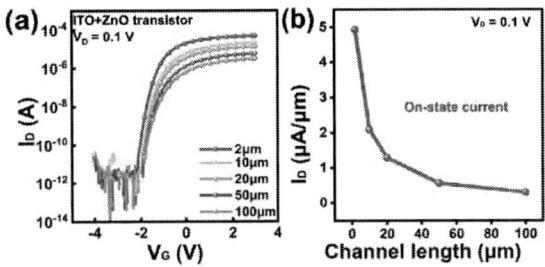

Fig.1. (a) SEM image of ITO/ZnO transistor with channel length of 10μm; **(b)** Schematic diagram; **(c)**fabrication process of ITO/ZnO heterojunction transistors

Fig.2. UPS analysis of ITO and ZnO film.

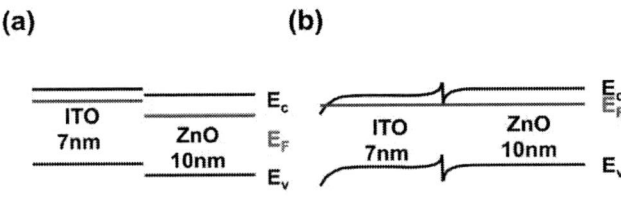

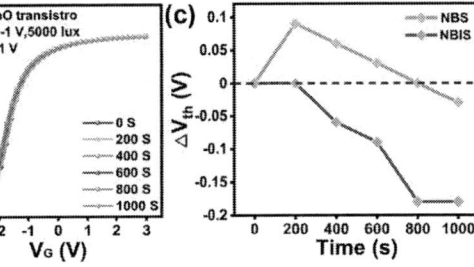

Fig.3. (a) Energy band diagram of single ITO and single ZnO film; **(b)** Energy band diagram of ITO/ZnO heterojunction films.

Fig.4. (a) Transfer curves of different channel length transistor at V_D of 0.1 V; **(b)** On-state current of ITO/ZnO transistor.

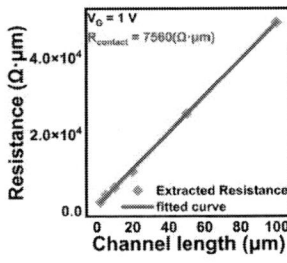

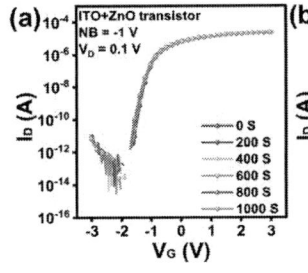

Fig.5. Contact resistance extracted by TLM method.

Fig.6. (a) NBS analysis of ITO/ZnO transistor; **(b)** NBIS analysis of ITO/ZnO transistor; **(c)** V_{th} shift of ITO/ZnO transistor

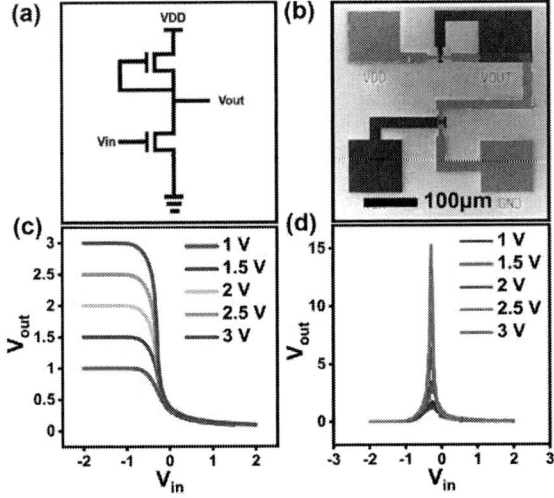

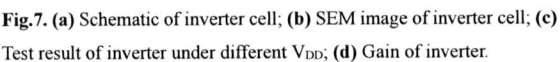

Fig.7. (a) Schematic of inverter cell; **(b)** SEM image of inverter cell; **(c)** Test result of inverter under different V_{DD}; **(d)** Gain of inverter.

Table 1: Benchmarking of ITO/ZnO transistor.

Sample	SS (mv/decade)	V_{on} (V)	μ (cm²/Vs)
This work	123.96	-2.1675	74.92
IGZO/IZO [8]	190	-1.33	40
HIIGZO/IGZO [9]	100	-0.9	24.7
IZO/IGZO [10]	170	0	21.6
IGO/ZnO [11]	260	-0.84	63.2
IZO/AITZO [12]	150	0.5	53.2

979-8-3315-0417-5/25 $31.00 © 2025 IEEE

979-8-3315-0417-5/25 $31.00 © 2025 IEEE

A 1.8TB/s HBM Heterogeneously Integrated GPU Design Exploring 2.5D Packaging Technology

Shuang Wang, Weiliang Chen, Xueqing Li, Chen Jiang, Huazhong Yang (Fellow IEEE)

Tsinghua University, China

Abstract

2.5D packaging supports high-speed data transmission, meeting performance requirements of GPU. However, signal integrity (SI) is a key challenge to packaging. This paper demonstrates a GPU with HBM and adopts 2.5D packaging technology. A series of design schemes are proposed, including 4-marker die-interposer alignment, 5-metal-layer connection with HBM by TSV. Moreover, Cup bump and octagon under bump metallurgy technologies are used in interposer to improve signal transmission quality. The measurement results illustrate these technologies can improve SI effectively, achieve 1.8 TB/s bandwidth in 2.5D packaging.

Keywords: 2.5D Packaging, GPU, HBM, Signal Integrity

Introduction

2.5D packaging improves chip integration and supports high-speed data transmission. It is a multi-die packaging technology based on high density interposer [1]. Interposer has been added between dies and a substrate to achieve multi-die integration. Through circuits on the interposer, the data transmission path between dies is significantly shortened, reducing latency and increasing bandwidth. This is crucial for applications that require extremely high bandwidth, such as GPU and CPU for artificial intelligence [2].

High bandwidth memory (HBM) is a high bandwidth storage chip. HBM2E is the latest version of HBM, with a maximum storage capacity of 16 gigabyte per HBM2E. It is composed of 16 gigabit single dies stacked in 8 layers, which can achieve a packaging capacity of 16 GB and ensure a stable data transmission speed of 3.2 Gbps. Each stack of HBM has a data transmission speed of 461 GB/s, and the transmit bandwidth of a HBM can reach 1.84 TB/s theoretically.

This paper demonstrates a GPU by the means of 2.5D packaging technology (Die on Wafer on Substrate with Interposer) is adopted. For high-performance GPU, the signal must transmit from HBM to PCB through interposer and substrate [3]. This is key challenge to signal integrity (SI) of 2.5D packaging in GPU.

Packaging Design

Figure 1 shows the top view and side view of 2.5D packaging. It can be seen that the packaging is divided into three layers, including substrate, interposer and dies. There are many bumps on the substrate, connected to the second layer of interposer. There are many connections inside the interposer, which are connected to the multiple dies on third layer. The dies on the third layer are laid flat and HBM dies

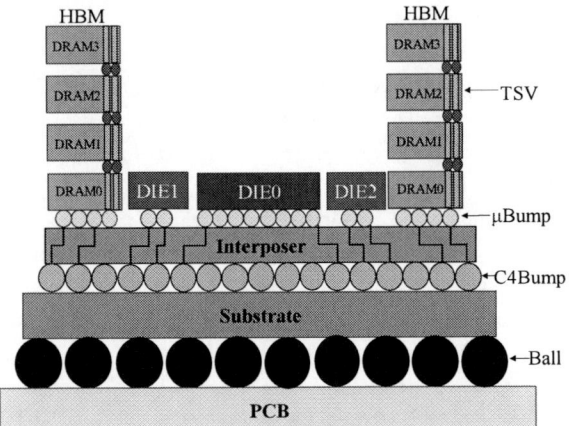

Fig. 1: 2.5D packaging with HBM in GPU design

placed in the corners. Each HBM die is stacked with 4 layers of DRAM. Each layer of DRAM is interconnected through TSV.

Figure 1 also shows the 3+1 levels (die-interposer-package-PCB) circuit design for SI. This work uses 30,000 C4bump and 230,000 µbump. Apart from die, package and PCB levels, it is necessary to place resistors and capacitors on the interposer. These circuits control the transmission quality of the signal by adjusting the parameters of resistors and capacitors.

Results and Discussion

A. 4-marker die-interposer alignment scheme

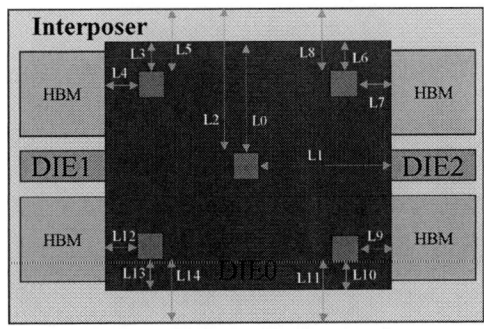

Fig. 2: 4-marker die-interposer alignment scheme diagram

Figure 2 shows the 4-marker alignment scheme in die and interposer. To keep SI through 3+1 levels circuit, location alignment between the die and interposer must be ensured. we propose a 4-mark areas alignment scheme on die and interposer. It is necessary to place the graphic database system (GDS) samples in the 4 alignment point areas of the die and interposer.

First of all, interposer offset on package was set to (0, 0). It means the center of interposer is match the center of substrate and the center of die is in the center of interposer directly. The alignment points were setting as L0, L1, and L2.

The coordinates of 4 alignment points include:

1) At the horizontal X coordinate of die 0, such as L4, L7, L9, and L12. The average distance of X locations was set to 4 mm, which is the 10% of interposer length.

2) In the vertical Y coordinate of die 0, such as L3, L6, L10, and L13. The average distance of Y locations was set to 3 mm, which is the 10% of interposer width.

3) At the vertical Y coordinate of interposer, such as L5, L8, L11, and L14. The average distance of Y locations was set to 6 mm, which is the 20% of interposer width.

After integrating dies into the interposer, we checked whether the sample patterns in the alignment point area of the dies and interposer are consistent. If consistent, it indicates that the dies and interposer are aligned [4].

B. Interposer 5-metal-layer connection with HBM by TSV

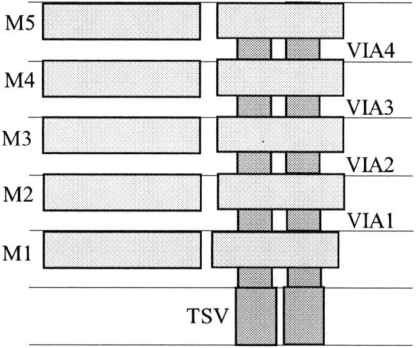

Fig. 3: Interposer 5-metal-layer connection design with HBM by TSV

Figure 3 shows interposer 5-metal-layer connection design with HBM by TSV. The metal layers inside the interposer are divided into 5 layers, each responsible for specific connection purposes, listing as follows.

1. Metal 1 and 2 (M1 and M2): these lower metal layers are used for local signal or short-distance signal transmission, with dense and fine connections.

2. Metal 3 and 4 (M3 and M4): the intermediate layer is used for long-distance signal transmission and some local power distribution.

3. Metal 5 (M5): the top layer metal is commonly used for power and ground distribution across the entire chip, as well as long-distance signal transmission.

In this chip, the HBM has 68 main interconnect signals. There are 48 signals output from the metal layers M5. Another 20 signals are output from the metal layers M3. Since M3 has exhausted all interconnect signals, the number of all HBM interconnect signals through VIA2 and VIA1 (see Figure 3) is 0.

Figure 4 shows interposer M5 layer signal eye diagram. The SI require eye width must be greater than 160 ps. Based on SI simulation and measurement results, M5 signal eye diagram is 189ps and can be used for long distance signal transmission.

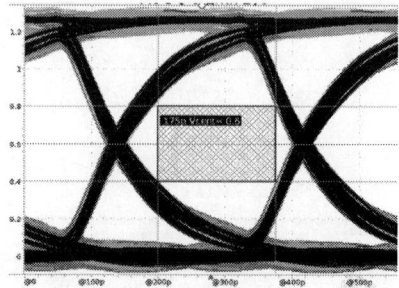

Fig. 4: Interposer M5 layer signal eye diagram

C. Interposer Cup bump and octagon UBM design

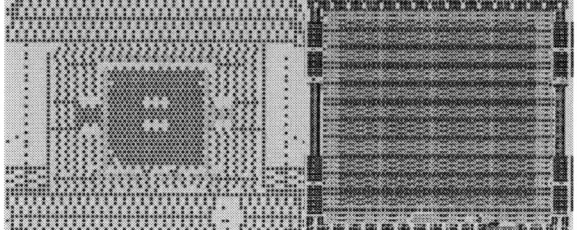

Fig. 5: Interposer Cup bump and octagon UBM design

Figure 5 shows interposer Cu-pillar (Cup) bump and octagon UBM design. In this work, silicon interposer is used, which has some key design technologies. Bump is responsible for achieving reliable electrical connections between dies and substrates. Compared to conventional solder bump, this work adopts Cup (Cu pillar) technology to improve signal transmission quality [5]. Cup bump is a new generation of chip interconnect technology used for connecting dies and substrates in integrated packaging processes and suitable for high-speed signal transmission. At the same time, the use of Cup bump in substrate design can reduce the number of substrate layers and achieve a reduction in overall packaging costs. Compared with other type bump, its overall packaging cost can be saved by about 20%. In order to place more bumps, the minimum bump pitch was set to 45 μm, which is reduced by 25% compared to the conventional pitch.

On the periphery of the entire chip, it is generally required to place seal ring. Its fundamental and main function is to prevent the chip from being damaged by mechanical stress during cutting. If the seal ring is grounded, it can also shield the chip from external interference. In this work, the seal ring was set to 20 mm × 20 mm. In order to keep DIE size, bounding box is used in interposer. The bounding box by interposer chip boundary was set to 910 μm × 910 μm. The gap between HBM and ASIC bounding box was set to 70 μm × 70 μm.

Under bump metallurgy (UBM) is an important method in the process of flip chip connection, which is part of the technology of making bumps on the chip surface [6]. The purpose of UBM is to improve the electrical and mechanical connections between chips and substrates, enhancing the reliability and performance of the connections. For the top of UBM, UBM size was set to 25.02 μm. UBM shape is octagon.

979-8-3315-0417-5/25 $31.00 © 2025 IEEE

However, the UBM bottom dimension is 82 μm × 105 μm, and its shape is Oval. The octagon UBM reduces the losses and interference during signal transmission by optimizing the structure and materials of the metal layer, improving the speed and stability of signal transmission. This design is particularly suitable for high-speed and high-frequency application scenarios.

D. GPU SI measurement in 2.5D packaging

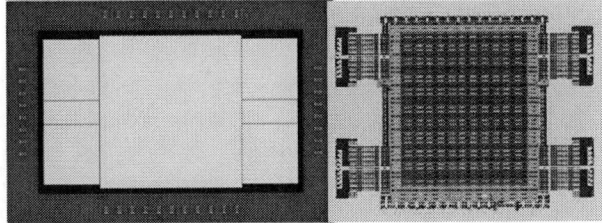

Fig. 6: GPU chip photo with 2.5D packaging technology

To verify the function of the proposed scheme, a GPU chip is designed with 2.5D packaging. As shown in Fig. 6, 2.5D packaging technology is adopted in this work. The dies on the third layer are laid flat and the HBM dies are placed at the corners.

TABLE I. Basic Specifications of GPU Chip with 2.5D Packaging Technology

Die size	700 mm²
Interposer size	1500 mm²
Interposer material	silicon
HBM size	64 GB
HBM bandwidth	1.8 TB/S
Package size	3210 mm²

The specifications have five indicators including die size, interposer size, material and package size. The interposer size is 1500 mm², which is 1.74× reticle size. In this work, the HBM bandwidth reaches 1.8 TB/s, and signal integrity of high-speed data transmission is the key challenge.

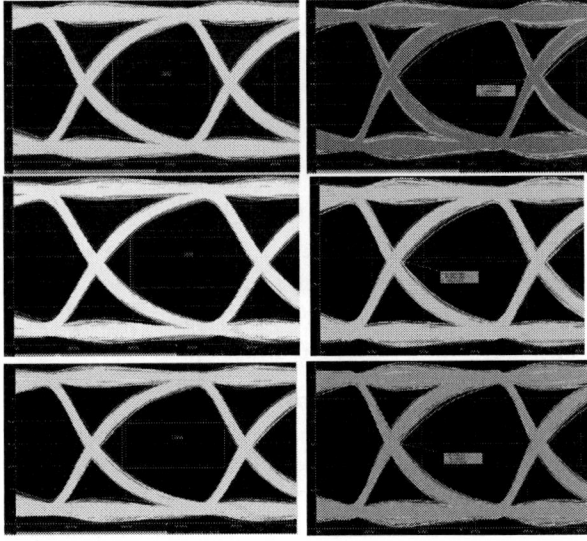

Fig. 7: Measurement result of GPU chip SI in FF, TT and SS corner

Figure 7 shows measurement results of chip SI. The signal was sent out from GPU, passing through the interposer, substrate, PCB board, and arriving the HBM finally [7]. The SI require eye width must be greater than 160 ps. The 1st group results were measured in FF corner with eye widths of 173 ps and 187 ps. The 2nd group results were measured in TT corner with eye widths of 172 ps and 185 ps. The 3rd group results were measured in SS corner with eye width is 166 ps and 180 ps. The average eye diagram is approximately 175 ps. The signal transmission is good and meets SI requirements.

Conclusion

This paper adopts 2.5D packaging technology in GPU design and HBM is integrated with TSV. To solve SI problem, a new scheme with 4-marker die-interposer alignment, 5-metal-layer connection with HBM by TSV design are proposed. Moreover, Cup bump and octagon UBM are used in interposer to improve signal transmission quality. The results illustrate these technologies can improve SI effectively and achieve 1.8 TB/s transmission in 2.5D packaging.

References

[1] Tummala, R.R., 2019, February. Moore's law for packaging to replace Moore's law for ICS. In 2019 Pan Pacific Microelectronics Symposium (Pan Pacific) (pp. 1-6).

[2] Dylan Stow, Yuan Xie, Taniya Siddiqua, Gabriel H. Loh. "Cost-Effective Design of Scalable High-Performance Systems Using Active and Passive Interposers." ICCAD 2017.

[3] Janak Sharda, "Thermal Modeling of 2.5D Integrated Package of CMOS Image Sensor and FPGA for Autonomous Driving", IEEE EDTM,69, pp. 380-387 (2023).

[4] Shuang Wang, Weiliang Chen, Xueqing Li, Leibo Liu, Huazhong Yang, "A 10TFLOPS Datacenter-Oriented GPU with 4-Corner Stacked 64GB Memory by The Means of 2.5D Packaging Technology." In ASSCC 2023.

[5] M. A. Smith, et al. "A low-temperature nickel silicide process for wafer bonding and high-density interconnects," IEEE Transactions on Components, Packaging and Manufacturing Technology, vol. 10, no. 5, pp. 908-916, (2020).

[6] A. Kumar and R. Dhiman, "Modeling and analysis of Cu-CNT composite through glass vias in 3D ICs," Proc. IEEE Electrical Design of Advanced Packaging and Systems (EDAPS), Urbana, IL, USA, 2021.

[7] Seongguk Kim, et al. "Signal Integrity and Computing Performance Analysis of a Processing-In-Memory of High Bandwidth Memory (PIM-HBM) Scheme," IEEE Transactions on Components, Packaging and Manufacturing Technology ,Vol.11, (2021)

[8] M. A. Smith, et al. "A low-temperature nickel silicide process for wafer bonding and high-density interconnects," IEEE Transactions on Components, Packaging and Manufacturing Technology, vol. 10, no. 5, pp. 908-916, (2020).

9-kV p-GaN Gate HEMT with Gate Termination Extension Demonstrated on Sapphire Substrate for Improved Breakdown Voltage

Jingjing Yu[1#], Junjie Yang[1#], Jiawei Cui[1], Hao Chang[1], Sihang Liu[1], Yunhong Lao[1], Xuelin Yang[2], Xiaosen Liu[3], Maojun Wang[1], Bo Shen[2] and Jin Wei[1*]

[1]School of Integrated Circuit, Peking University, China. [2]School of Physics, Peking University, China.
[3]School of Integrated Circuits, Tsinghua University, China. [#]Contributed equally to this work.
*Corresponding author: jin.wei@pku.edu.cn

Abstract

In this work, a 9-kV E-mode p-GaN gate HEMT with gate termination extension (GTE-HEMT) is demonstrated on sapphire substrate. The GTE-HEMT with L_{GD} = 77 μm exhibits a low R_{ON} of 39.5 Ω·mm and a specific R_{ON} of 34.76 mΩ·cm². Moreover, the gate termination extension effectively reduces the electric filed peak at the gate edge, leading to a high breakdown voltage of 9582 V and an excellent figure of merit (FOM) of 2.64 GW/cm².

(Keywords: p-GaN gate HEMT, kilovolt-level, gate termination extension, dynamic R_{ON})

Introduction

GaN power HEMTs have demonstrated their potential for high-voltage applications with low on-resistance (R_{ON}), due to the high critical electric field and high electron mobility in channel [1-2]. The breakdown voltage of GaN HEMTs increases linearly with the gate-to-drain distance, facilitating scaling to kilovolt-level (kV-level) applications [3-4]. Recently, GaN-on-sapphire technology has gained more attention for kV-level applications, benefiting from the elimination of vertical breakdown by the insulated sapphire substrate and the simplified buffer design due to enhanced mechanical strength of sapphire substrate. Several types of kV-level GaN power HEMTs have been demonstrated on sapphire substrate [4-7].

However, electric field crowding at the gate edge limits the breakdown voltage and degrades the performance of GaN power HEMTs [8-9]. Field-plate technologies are currently used to optimize the electric field distribution and suppress dynamic R_{ON} degradation [9-10]. However, for higher voltage applications, the number of field plates increases with the breakdown voltage, posing challenges in fabrication process and cost management.

In this paper, we proposed a straightforward method to reduce the electric field peak and then enhance the breakdown voltage utilizing gate extension termination. The gate extension termination is a thinned p-GaN layer connected to the p-GaN gate. In the OFF-state, the gate extension termination is supposed to be depleted to alleviate electric field crowding at the gate edge. Thus, with an L_{GD} = 77 μm, the proposed p-GaN gate HEMT with gate termination extension (GTE-HEMT) achieves a higher

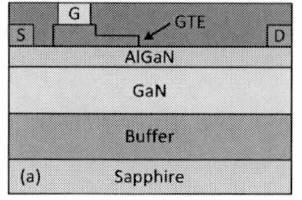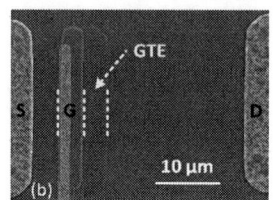

Fig. 1. (a) Schematic structure of the proposed GTE-HEMT. (b) SEM image of a fabricated GTE-HMET with L_{GD} = 27 μm and L_{GTE} = 4 μm.

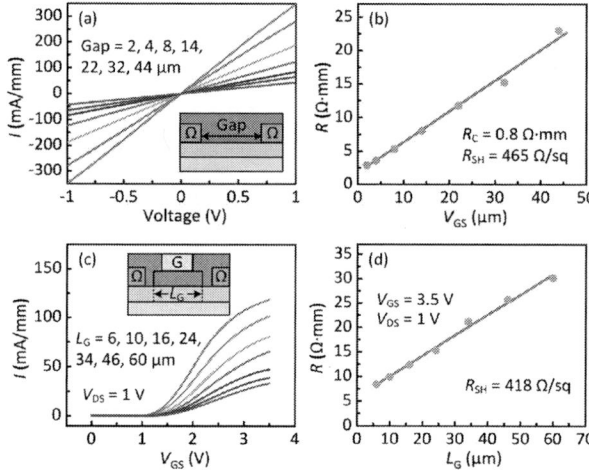

Fig. 2. (a)-(b) The measured sheet resistance (R_{SH}) under dielectric passivation and n-type contact resistance (R_C) using TLM. (c) Transfer characteristics of the devices with various L_G. (d) The calculated R_{SH} under p-GaN gate at V_{GS} = 3.5 V and V_{DS} = 1 V.

breakdown voltage (*BV*) of 9582 V compared to the conventional p-GaN gate HEMT (Conv-HEMT) with a *BV* of 7063 V. Furthermore, the GTE-HEMT demonstrates improved dynamic performance, with a low dynamic R_{ON}/static R_{ON} of 1.63 achieved after a 2000-V V_{DS-OFF} stress, measured by the B1505A/N1267A testing module.

Device Structure and Fabrication

Fig. 1 displays the structure and SEM image of a proposed GTE-HEMT with $L_{GS}/L_G/L_{GD}$ = 3/5/27 μm. GTE-HEMTs with various L_{GD} of 17, 27, 42, 62 and 77 μm were fabricated and investigated in this work. The gate extension termination is a 4-μm thinned p-GaN layer at the gate-to-drain side. In the OFF-state, the gate extension termination is expected to be

979-8-3315-0417-5/25 $31.00 © 2025 IEEE   236

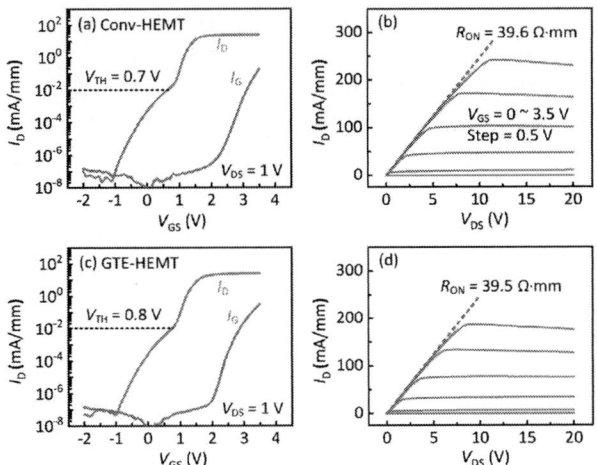

Fig. 3. (a)-(b) The transfer and output curves of a Conv-HEMT with $L_{GS}/L_G/L_{GD}$ = 3/5/77 μm. (c)-(d) The transfer and output curves of a GTE-HEMT with $L_{GS}/L_G/L_{GD}$ = 3/5/77 μm.

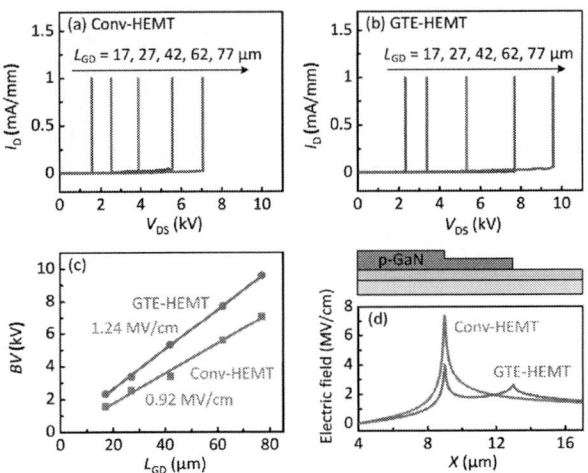

Fig. 4. The OFF-state I-V characteristics of (a) the Conv-HEMTs and (b) the GTE-HEMTs with various L_{GD}. (c) The summary of BVs in Conv-HEMTs and GTE-HEMTs. (d) The simulated electric field distribution in Conv-HEMT and GTE-HEMT with L_{GD} = 77 μm under 3000-V V_{DS} stress.

depleted to alleviate electric field crowding at the gate edge.

The devices are fabricated on a 2-inch GaN-on-sapphire wafer, consisting of a 10-nm p^{++}-GaN top layer for reducing the p-type contact resistance, an 80-nm p-GaN layer, a 15-nm Al$_{0.2}$Ga$_{0.8}$N barrier layer, a 200-nm u-GaN layer, and a buffer layer. Device fabrication started with selective p-GaN etching to define the gate region, followed by a second p-GaN etching to delineate the gate terminal extension. Subsequently, the fabrication process continued with passivation deposition, source/drain contacts formation, planar isolation, gate contacts formation and the deposition of probing pads.

Device Performance

Fig. 2(a)-(b) present the sheet resistance (R_{SH}) of the 2DEG under the dielectric passivation, as measured by transfer length measurement (TLM). The current demonstrates a linear increase with voltage, indicating good n-type ohmic contact. The source/drain contact resistance is 0.8 Ω·mm, and the R_{SH} at the access region is 465 Ω/sq. Additionally, the 2DEG sheet resistance under the p-GaN gate is extracted using transfer curve measurements with varying gate length (L_G) in Fig. 2(c)-(d). At V_{GS} = 3.5 V and V_{DS} = 1 V, the R_{SH} under the p-GaN gate is 418 Ω/sq, similar to that under the dielectric passivation.

Fig. 3 depicts the static I-V characteristics of a conventional p-GaN gate HEMT (Conv-HEMT) and a GTE-HEMT with $L_{GS}/L_G/L_{GD}$ = 3/5/77 μm. The E-mode GTE-HEMT has a threshold voltage (V_{TH}) of 0.9 V defined at I_D = 10 μA/mm and an R_{ON} of 39.5 Ω·mm, comparable to the Conv-HEMT. The GTE-HEMTs with L_{GD} of 17, 27, 42, 62 and 77 μm show R_{ON} of 13.2, 18.4, 26.0, 33.1 and 39.5 Ω·mm, respectively, corresponding to specific R_{ON} (R_{SP}) of 3.69, 6.99, 13.78, 24.16 and 34.76 mΩ·cm^2 (1.5-μm transfer length is considered for each ohmic contact).

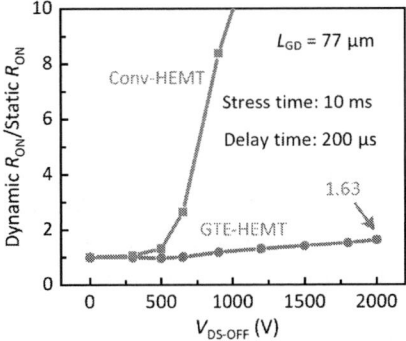

Fig. 5. The dynamic R_{ON} performance of a Conv-HEMT and a GTE-HEMT with L_{GD} = 77 μm. The GTE-HEMT presents slight dynamic R_{ON} degradation after 2000-V $V_{DS\text{-}OFF}$ stress.

Fig. 4 presents the breakdown voltage (BV) measurements of Conv-HEMTs and GTE-HEMTs with L_{GD} of 17, 27, 42, 62 and 77 μm using the B1505A/N1268A voltage expander. The maximum testing voltage of the B1505A/N1268A module is 10 kV. During the test, the devices are immersed in fluoride liquid to avoid surface flashing. For Conv-HEMTs, breakdown voltages are 1568, 2523, 3883, 5574 and 7063 V. Notably, the GTE-HEMTs achieve higher breakdown voltages of 2334, 3378, 5323, 7701 and 9582 V, resulting in a high BV/L_{GD} of approximately 1.24 MV/cm, demonstrating significant potential for kV-level applications.

In Fig. 4(d), the simulated electric field distribution for devices with L_{GD} = 77 μm under 3000-V V_{DS} stress is depicted. In the Conv-HEMT, a high electric field peak is observed at the gate edge, affecting the breakdown characteristics negatively. In contrast, the GTE-HEMT displays two electric field peaks at the edge of gate and gate extension termination. Depletion of the gate extension termination aids in reducing the electric field peak, leading to a higher breakdown voltage.

979-8-3315-0417-5/25 $31.00 © 2025 IEEE

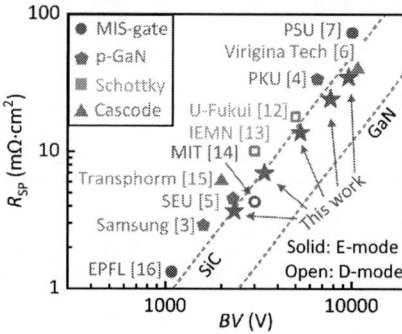

Fig. 6. Benchmark for R_{SP}-BV relationship of the proposed GTE-HEMTs with the state-of-the-art high-voltage GaN power transistors.

Additionally, the development of high-voltage GaN power HEMTs is hindered by dynamic R_{ON} degradation caused by surface and buffer trapping. Fig. 5 shows the dynamic R_{ON} performance of the Conv-HEMT and GTE-HEMT with $L_{GD} = 77$ μm. Dynamic R_{ON} is extracted 200 μs after 10-ms V_{DS-OFF} stress using the B1505/N1267A fast switching module [11]. In the Conv-HEMT, there is notable dynamic R_{ON} degradation with a dynamic R_{ON}/static R_{ON} ratio exceeding 10 times, primarily related to the electric filed peak at the gate edge [9]. However, in the GTE-HEMT, dynamic R_{ON} degradation is suppressed by reducing the electric field peak through gate termination extension. Consequently, the GTE-HEMT demonstrates a low dynamic R_{ON}/static R_{ON} of 1.63 after 2000-V V_{DS-OFF} stress.

In Fig. 6, the breakdown voltage (BV) versus specific ON-resistance (R_{SP}) of the proposed GTE-HEMTs is compared with the reported kV-level lateral GaN power HEMTs. The GTE-HEMTs showcase outstanding BV and R_{SP} characteristics. With $L_{GD} = 77$ μm, the GTE-HEMT demonstrates an impressive figure of merit (FOM = BV^2/R_{SP}) of 2.64 GW/cm², outperforming the SiC competitors. This highlights the potential of the GTE-HEMT as a promising solution for kV-level power devices, compatible with the established E-mode p-GaN gate HEMT technology.

Conclusion

In this work, we demonstrate an E-mode GTE-HEMT with breakdown voltage up to 9582 V and a FOM of 2.64 GW/cm². By adopting the gate termination extension to optimize electric field distribution, the GTE-HEMT showcases an improved average breakdown filed (BV/L_{GD}) of 1.24 MV/cm, surpassing the 0.91 MV/cm in conventional p-GaN gate HEMT. Superior dynamic performance is also obtained in the GTE-HEMT, with a low dynamic R_{ON}/static R_{ON} of 1.63 after the V_{DS-OFF} stress up to 2000 V. Moreover, the GTE-HEMT technology could be easily implemented using the widely-adopted p-GaN gate HEMT technology. Therefore, these results highlight the GTE-HEMT as an appealing approach for high-voltage applications with enhanced dynamic stability.

Acknowledgments

This work was supported by the National Natural Science Foundation of China under Grant 62174003.

References

[1] Y. Zhang et al., "Multidimensional device architectures for efficient power electronics," *Nat. Electron.*, vol. 5, no. 11, pp. 723-734 (2022).

[2] J. Wei et al., "GaN Power Integration Technology and Its Future Prospects," *IEEE Trans. Electron Devices*, vol. 71, no. 3, pp. 1365-1382 (2024).

[3] I. Hwang et al., "1.6 kV, 2.9 mΩ·cm² Normally-off p-GaN HEMT Device," in *Proc. ISPSD*, Bruges, Belgium, pp. 41-44 (2012).

[4] J. Cui et al., "6500-V E-mode Active-Passivation p-GaN Gate HEMT with Ultralow Dynamic R_{ON}," in *IEDM Tech. Dig.*, San Francisco, CA, USA, Sec. 26-1 (2023).

[5] S. Li et al., "1200V E-mode GaN Monolithic Integration Platform on Sapphire with Ultra-thin Buffer Technology," in *IEDM Tech. Dig.*, San Francisco, CA, USA, Sec. 9-5 (2023).

[6] M. Xiao et al., "Multi-Channel Monolithic-Cascode HEMT (MC2-HEMT): A New GaN Power Switch up to 10 kV," in *IEDM Tech. Dig.*, San Francisco, CA, USA, pp. 114-117 (2021).

[7] J. T. Kemmerling et al., "GaN Super-Heterojunction FETs with 10-kV Blocking and 3-kV Dynamic Switching," *IEEE Trans. Electron Devices*, vol. 71, no. 2, pp. 1153-1159 (2024).

[8] J. Cui et al., "High-Voltage E-Mode p-GaN Gate HEMT on Sapphire with Gate Termination Extension," *IEEE Trans. Electron Devices*, vol. 71, no. 3, pp. 1592-1597 (2024).

[9] T. Katsuno et al., "Direct observation of trapped charges under field-plate in p-GaN gate AlGaN/GaN high electron mobility transistors by electric field-induced optical second-harmonic generation," *Appl. Phys. Lett.*, vol. 110, p. 092101 (2017).

[10] X. Li et al., "1700 V High-Performance GaN HEMTs on 6-inch Sapphire With 1.5 μm Thin Buffer," *IEEE Electron Device Lett.*, vol. 45, no. 1, pp. 84-87 (2024).

[11] J. Yang et al., "Virtual-Body p-GaN Gate HEMT With Enhanced Ruggedness Against Hot-Electron-Induced Degradation," *IEEE Electron Device Lett.*, vol. 45, no. 5, pp. 770-773 (2024).

[12] J. H. Ng et al., "AlGaN/GaN HEMTs on free-standing GaN substrates with breakdown voltage of 5 kV and effective lateral critical field of 1 MV/cm," in *CS ManTech*, Miami, Florida, USA, pp. 215-218 (2016).

[13] N. Herbecq et al., "Above 2000V breakdown voltage at 600K GaN-on-silicon high electron mobility transistors," *Physica status solidi. A*, vol. 213, no. 4, pp. 873-877 (2016).

[14] H. Lee et al., "3000-V 4.3-mΩ·cm² InAlN/GaN MOSHEMTs With AlGaN Back Barrier," *IEEE Electron Device Lett.*, vol. 33, no. 7, pp. 982-984 (2012).

[15] G. Gupta et al., "1200V GaN Switches on Sapphire: A low-cost, high-performance platform for EV and industrial applications," in *IEDM Tech. Dig.*, San Francisco, CA, USA, pp. 835-838 (2022).

[16] L. Nela et al., "High-Performance Nanowire-Based E-Mode Power GaN MOSHEMTs With Large Work-Function Gate Metal," *IEEE Electron Device Lett.*, vol. 40, no. 3, pp. 439-442 (2019).

979-8-3315-0417-5/25 $31.00 © 2025 IEEE

Rapid and Extensive Conductance Modulation in MoO$_x$ based Electrochemical Random-Access Memory for Spiking Neuromorphic Systems

Xiaoci Liang[1], Dongyue Su[1], Younian Tang[2], Bin Xi[2], Chunzhen Yang[3], Huixin Xiu[4], Jialiang Wang[5], Chuan Liu[1]*, Mengye Wang[3]*, Yang Chai[5]*

[1]School of Electronics and Information technology, Sun Yat-Sen University, Guangzhou, China; [2]School of Materials Science and Engineering, Sun Yat-sen University, Guangzhou, China; [3]School of Materials, Sun Yat-Sen University, Shenzhen, China; [4]School of Materials Science and Engineering, University of Shanghai for Science and Technology, Shanghai, China; [5]Department of Applied Physics, The Hong Kong Polytechnic University, Hung Hom, Kowloon, Hong Kong, China. *Email: ychai@polyu.edu.hk, liuchuan5@mail.sysu.edu.cn, wangmengye@mail.sysu.edu.cn.

Abstract

We develop a MoO$_x$ based electrochemical random-access memory (ECRAM) for spiking neuromorphic computing. By controlling the electric-driven proton intercalation into MoO$_x$, the adsorbed proton leads to increased carrier concentration and conductance. The device exhibits an on/off ratio of 10^5 and retention over 10^3 seconds with a rapid response in a few microseconds. By constructing a two-stage ECRAMs hardware with linear synaptic and accumulative neuronal functionalities, an all-ECRAM spiking neural network is demonstrated for image recognition.

Keywords: MoO$_x$, ECRAM, Spiking neural network.

Introduction

Ion intercalation, facilitated by ion migration and redox reactions within semiconductors, can significantly alter microscopic properties such as atomic valence states and charge concentrations[1]. These transitions can profoundly impact the macroscopic conductance, which are essential for various electronic applications, such as ECRAM[2]. However, a significant challenge is achieving a large concentration range and stable ion intercalation, which is crucial for precise manipulation of device conductance, particularly in terms of achieving a wide conductance range and long retention times. [3]. Molybdenum oxide (MoO$_x$) is a metal oxide with variable valence. The valence of the metal can be modified during redox reactions, such as through annealing in a controlled atmosphere or via electrochemical ion intercalation and deintercalation, which in turn affects the conductance[4]. Due to the small atomic mass of proton, it can serve as mobile ions in electrolytes for rapid programming of three-terminal ECRAMs. Driven by external signals, protons can move into or out of the Van der Waals layered structure, allowing the conductance to be programmed to a target value and maintained for a long period. The ability to modulate conductance in MoO$_x$ through proton intercalation opens up new possibilities for neuromorphic computing. Spiking neural networks (SNNs), which mimic the temporal encoding of biological neurons, are considered promising for their potential low energy consumption and high computational efficiency. However, the implementation of temporal encoding and the achievement of high-precision tasks in these networks require materials that can support dynamic and stable conductance changes. By leveraging the conductance modulation capabilities of MoO$_x$, SNNs can potentially achieve the necessary precision and energy efficiency for advanced computational tasks.

In this study, we have successfully modulated the conductance of MoO$_x$ by controlling the inject ion current through electric field. The spectral evolution of MoO$_x$ was investigated before and after proton intercalation, revealing the reduction of Mo. The proton intercalation process enables stable proton adsorption on the MoO$_x$ surface, which in turn induces a significant increase in carrier density. This results in a nonvolatile increase in conductivity that begins within microseconds and can reach a change of five orders of magnitude, demonstrating the potential for rapid and substantial conductance modulation. Drawing inspiration from the manner in which biological neurons integrate signals from multiple synapses, an all-ECRAM array was constructed for the processing of sparse signals with memory-efficient temporal coding and exhibited high accuracy in image recognition tasks.

Performance of MoO$_x$ ECRAM

The MoO$_x$ was deposited by atomic layer deposition ALD and in situ annealed to obtain crystalized orthorhombic MoO$_3$ phase. The XPS measurement was carried out to analyze the chemical constituents and the valence band spectrum. The stoichiometric ratio of molybdenum oxide is calculated according to the proportion of molybdenum with different valence states[2]. Accordingly, the MoO$_x$ was marked as MoO$_{2.96}$ and H$_{0.22}$MoO$_{2.96}$ pre and post proton intercalation. The subpeak positions and atomic percentages of Mo3d (**Fig. 1,** right) indicate that 92% of Mo exists in the +6 oxidation state in MoO$_{2.96}$, whereas proton intercalation results in a 30% reduction of Mo to the +5 oxidation state in H$_{0.22}$MoO$_{2.96}$. The O1s peaks (**Fig. 1,** right) detect metal-oxygen bonds (M−O), oxygen vacancy (O$_v$) and hydroxyl groups (M-O-H). A

979-8-3315-0417-5/25 $31.00 © 2025 IEEE

pronounced M-O-H peak in $H_{0.22}MoO_{2.96}$ confirms that proton intercalation mainly involves the formation of O-H bonds. The first-principles calculations were performed in the Vienna Ab initio Simulation Package. **Fig. 2** reveal the energy bands and density of states for regular MoO_3 and proton-intercalated HMo_4O_{12}. In HMo_4O_{12}, a transition from low to high carrier concentration, signified with a narrowed band-gap and Fermi-energy within conduction band, and the emergence of a delocalized state at the conduction band minimum are observed, indicating intrinsically tunable conductivity through proton intercalation.

A three-terminal ECRAM device was fabricated (**Fig. 3a**). The conductance modulation in ECRAM was observed with inject ion currents ranging from 1 μA to 5 μA, utilizing a solid-state protonic electrolyte Nafion. Protons migrate from the Nafion to the MoO_x interface, where they undergo a charge transfer reaction and diffuse within the MoO_x. The electrochemical reaction is represented as: $nH^+ + ne^- + MoO_x \rightleftharpoons H_nMoO_x$. By adjusting the gate current, we could precisely control the proton flux density and the degree of proton intercalation in MoO_x. Notably, a high gate current of 5 μA results in a high H:Mo ratio of 0.22 and a conductance modulation ratio G_{max}/G_{min} exceeding 10^5 (**Fig. 3b**). The retention time for distinct storage states surpasses 1,000 seconds, showing good nonvolatility with an ultra-low coefficient of variation c_v ranging from 0.004 to 0.057 (**Fig. 4**). The c_v is defined as $c_v = \sqrt{\sum_{i=1}^{l}(G_i-\overline{G})^2/(l-1)}/\overline{G}$, where G_i is the conductance at different times and $\overline{G}$ is the average conductance. The response time of the proton intercalation was investigated by measuring the conductance under various electric stimulation durations. **Fig. 5a** displays the enhancement in current following a 10 μs V_g-pulse at 10 V. As the pulse width is increased from 10 μs to 100 ms, the average conductance change (ΔG) per pulse increases significantly, from 2.1 nS to 150 nS (**Fig. 5b**). Considering the minimum G of $MoO_{2.96}$ at approximately 1 nS, the device's response threshold is estimated to be around 2 μs for the shortest pulse.

SNN Demonstration

To verify the neuromorphic computing ability in the all-ECRAM devices, we fabricated a two-stage ECRAM device for SNN on the same substrate (**Fig. 6**). The two-stage device configuration mimics the biological process where processed action potentials from multiple synapses are integrated by neurons. The 1st-stage ECRAMs function as synapses, linearly modulating the incoming sparse pulse signals with the synaptic weights emulated by the channel conductance. The 2nd-stage ECRAMs, acting as neurons, integrate weighted signals arriving at different times from the synapses and generate the appropriate enhancement or attenuation of the channel conductance as the neuronal output.

During hardware experiments, to realize temporal coding algorithm, the conductance (G) of the 1st-stage ECRAMs, representing network weights, are adjusted through inject ion current and manipulate the input spatiotemporal pulse signals, and the conductance change (ΔG) in the 2nd-stage ECRAMs after pulse integration. We performed classification of the characters 3×3-pixels "o," "J," and "Y" with random grayscale noise. The noise results in increased or decreased grayscale. Unlike conventional ECRAM arrays that perform vector-matrix multiplication for inputs and weights, the pulse signals from noisy images are converted into spiking timings through temporal coding, where higher grayscale pixels lead to delayed pulse arrivals. The network contains 9 inputs and 3 outputs. The image data was introduced 20% random noise in gray scale and containing 999 images. According to the noisy gray scale, the pulse arrives at the first period when pixels with 0%-10% noise and arrives at the second period when pixels with 10%-20% noise. With the pulse input, the conductance of the ECRAMs were measured simultaneously. Through employing this coding strategy, nine pulses, acting as the input layer, are fed into the drain electrodes of the 1st-stage ECRAMs. The accumulated pulses are then injected to the gate electrodes of the 2nd-stage ECRAMs, inducing changes in conductance from G_0 to G_j, which function as the output layer for comparative analysis to classify the input images. The mean G_j/G_0 ratio, based on the summed pulse amplitude, linearly increases with amplitude (**Fig. 7**), ensuring the accuracy needed to classify noisy images.

The weights of the 1st-stage ECRAM were trained in software with learning rules adapted from error backpropagation to account for spike latencies[5]. The weights are normalized, evenly divided into four values, and then scaled to be compatible with hardware execution (**Fig. 8a**). The hardware implementation results are shown as a heatmap in **Fig. 8b**, which displays the G_j/G_0 values from the three channels as the output for various input "o", "J", "Y" images. The system achieves a 98.3% rate for noisy grayscale images, showing robustness in handling complex pattern recognition tasks.

Conclusion

Our research has revealed the role of proton intercalation in conduction modulation of MoO_x and its potential application in neuromorphic computing. This process, facilitated by the reduction in Mo, enables conductance changes within microseconds and up to five orders of magnitude. The integration of MoO_x ECRAMs has demonstrated units that mimic brain-like responses to sparse signals and enabled the realization of an all-ECRAM neuromorphic network, capable of processing pusle signals through temporal coding to achieve high-precision image classification, even under noisy conditions. This study inspires the MoO_x's potential for next-generation neuromorphic systems.

Acknowledgments

This project was supported by the National Key Research and Development Program of China (2021YFB3600701), the National Natural Science Foundation of China (U23B20166) and Guangdong Basic and Applied Basic Research Foundation (2024A1515030039).

References

[1]   F. Zhang *et al.*, "Nanoscale multistate resistive switching in WO3 through scanning probe induced proton evolution," *Nature Communications,* vol. 14, no. 1, p. 3950, (2023)

[2]   X. Yao *et al.*, "Protonic solid-state electrochemical synapse for physical neural networks," *Nat Commun,* vol. 11, no. 1, p. 3134, Jun 19 (2020)

[3]   I. A. de Castro *et al.*, "Molybdenum Oxides – From Fundamentals to Functionality," *Advanced Materials,* vol. 29, no. 40, p. 1701619, (2017)

[4]   Z. Wang *et al.*, "Toward a generalized Bienenstock-Cooper-Munro rule for spatiotemporal learning via triplet-STDP in memristive devices," *Nat Commun,* vol. 11, no. 1, p. 1510, Mar 20 (2020)

[5]   S. R. Kheradpisheh and T. Masquelier, "Temporal Backpropagation for Spiking Neural Networks with One Spike per Neuron," *International Journal of Neural Systems,* vol. 30, no. 06, p. 2050027, (2020)

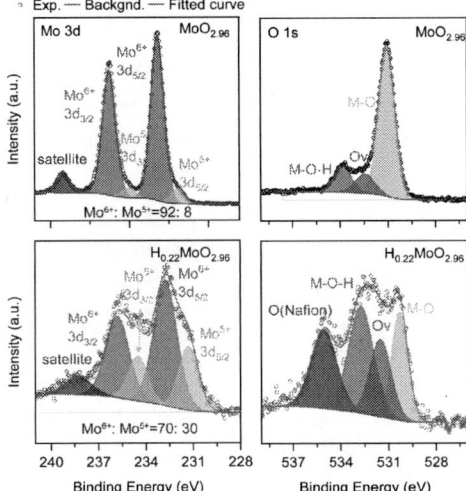

Fig. 1: XPS core-level spectra of Mo3d (left) and O1s (right) for $MoO_{2.96}$ (top), and $H_{0.22}MoO_{2.96}$ (bottom).

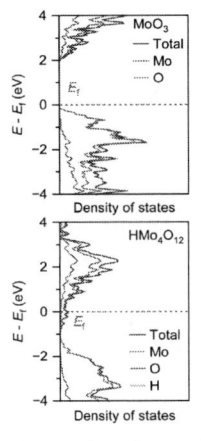

Fig. 2: Density of states for MoO_3 (top) and HMo_4O_{12} (bottom) calculated by density functional theory.

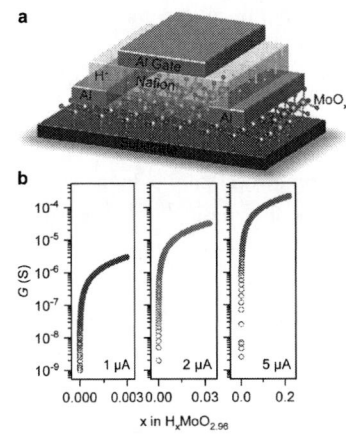

Fig. 3: **a**, Schematic of the ECRAM structure. **b**, Conductance modulation under long-term gate current bias.

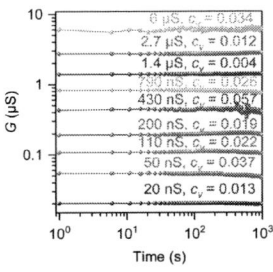

Fig. 4: Retention of 9 programmed analogue states and the corresponding variation coefficient c_v.

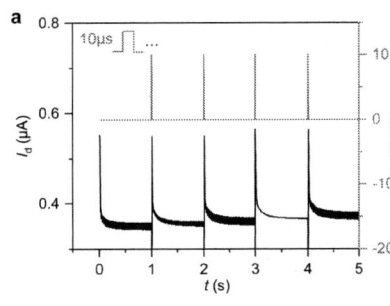

Fig. 5: **a**, Continuous readout current (I_d) update after programming by gate voltage pulses (10 V, 10 µs). **b**, Average change in conductance (ΔG) per pulse as a function of pulse width at an amplitude of 10 V.

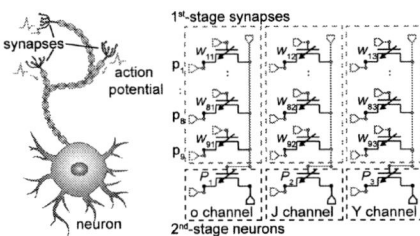

Fig. 6: Left: signals from the synapses are integrated and processed by the neurons; Right: circuit diagram of a two-stage SNN unit based on MoO_x ECRAM.

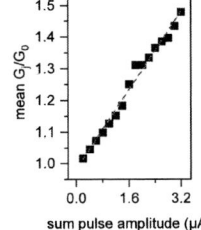

Fig. 7: Average G_j/G_0 plotted against total pulse amplitude.

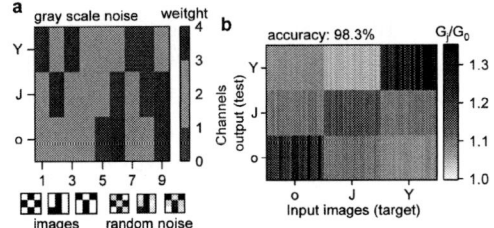

Fig. 8: **a**, Normalized weight distribution for gray scale noise conditions. **b**, G_j/G_0 as the output for the input various images with gray scale noise.

TCAD and mixed-mode simulation supporting the development of contact-controlled thin-film transistors and circuits

Eva Bestelink and Radu A. Sporea

Advanced Technology Institute, University of Surrey, Guildford, UK

Abstract

This paper presents recent advances in contact-controlled transistors supported by numerical simulation using Silvaco Atlas. In such devices, the source injection area, rather than the channel, governs drain current magnitude. Their correct design and application relies on deep understanding of internal operation; therefore technology computer-aided design (TCAD) is a useful optimization tool for established source-gated transistors (SGT) and emerging multimodal transistors (MMT) alike. In the absence of compact models, mixed-mode circuit simulations provide a valuable route toward application design.

Keywords: source-gated transistors, multimodal transistors, thin-film transistors.

Introduction

Contact-controlled transistors, such as the source-gated transistor (SGT) [1] are making inroads as pixel drivers [2], [3], supported by their long academic development heritage [4]–[6]. Minimal adaptation of existing processes should allow SGT fabrication alongside conventional TFT [5], brining an additional dimension to circuit design due to their reduced saturation voltage, high intrinsic gain, tolerance to geometrical variability and bias stress stability. Importantly, the overall design rules for such devices hold, irrespective of material system considered [1], [6], making them applicable to a wide range of processes.

Within the research community, efforts are largely focused on introducing new, high-performance materials, as well as on improvements to the quality of processing (interface quality, lateral resolution of patterning), with incremental and laborious progress.

A solution for accelerated development is the implementation of device structures with superior and/or additional functionality. The multimodal transistor (MMT, Fig. 1, left) [7] evolves the SGT concept, maintaining the benefits described above while enabling added operational versatility and leading to simpler, more robust circuits [8]–[10] suitable for both emerging and existing materials systems. In some cases, the structure is entirely compatible with some industrially deployed bottom-gate processes [11], [12]. The functionality afforded by the device structure may simplify the overall system design and reduce economic and environmental impacts related to manufacturing.

Multimodal transistor design and optimization using TCAD

MMTs rely on the separation of the control mechanisms for charge injection and for lateral transport [7], specifically comprising a rectifying source contact (as in SGTs) and gates which independently control the two processes (CG1 and CG2, respectively) (Fig, 1, left). Implementations in microcrystalline silicon (Fig. 1, right) confirm functionality [7].

TCAD simulations support the optimization of device operation by understanding the behavior of charge carriers in each region of the device under various biasing conditions. We have shown that, by tailoring layer thicknesses and properties, the transconductance achieved in saturation can be tailored from highly supralinear to linear [7], giving access to a wide range of current driving and signal conditioning applications, for both small- and large-signal inputs. It is, however, essential to understand how second-order generation, recombination, and coupling effects influence the practical device operation. Ref. [13] shows how biasing the channel at a suitable potential, depending on the application regime considered, significantly reduces overall hot carrier generation (Fig. 2) and thus improves intrinsic gain. In floating gate configuration, coupling between drain/channel potentials and the injection area can be practically eliminated using sacrificial channel control gates, which again leads to improved intrinsic gain and analog memory performance [7] (Fig. 3).

Multimodal transistor-based circuit design enabled by mixed-mode simulations

Mixed-mode (MM) numerical simulations, which physically simulate devices using TCAD and seamlessly integrate them as custom components in SPICE simulations, offer exceptional value when emerging device architectures do not yet have reliable compact models. Below are several examples in which MM studies using Silvaco Atlas have demonstrated the value of MMTs in simple, yet highly functional, circuit blocks.

A. Sensor signal conditioning and conversion - beyond digital information processing

Unique to MMT operation, the output current can be made proportional not only to the device width, but also to the input voltage applied to CG1 [7], [8]. Therefore, linear combinations of analog inputs can be weighed by the gate widths (Fig. 4, left). This is a useful property for analog computing applications when presenting analog inputs (e.g. multiply and accumulate – MAC), but also for compact, local sensor signal processing. specifically low-resolution multi-value logic digital to analog conversion (Fig. 4, right) [8]. Further applications which rely on the distinct operation enabled by the two types of input (analog for CG1-type gates, and digital or enable function for CG2) recommend the device for compact and robust implementations of useful functions such as ReLU [14] and XNOR [9].

B. High performance active-matrix display screens

Display pixel design is challenging when a.c. models are not available for the essential devices. MM allows the implementation of the MMT drive transistor as an Atlas element which can be simulated fully, including parasitic capacitance and second-order effects such as impact ionization, giving a credible illustration of circuit operation. Efforts so far have concentrated on demonstrating the MMT's ability to control the current amplitude and state of conduction independently. This has resulted in a compact design for a micro-LED pixel (Fig. 5) driven by pulse width modulation (PWM, Fig. 6, left) with the ability to simultaneously use amplitude modulation (PAM, Fig. 6, right) for tuning, in a unified 6T1M2C (six TFT, one MMT, two capacitors) implementation [10]. Functional characteristics have also been demonstrated in AMOLED pixel circuit designs in which the early saturation and flat saturated characteristics.

Economics and industrialization

The MMT structure is apparently more complex, due to the additional gating requirements. However, work in progress shows that the structure can be simplified without great detriment to overall performance.

Fig. 7 (left) shows a comparison between MMTs and conventional TFTs in several performance metrics. The reduced on-current, transconductance, and bandwidth may not be critical when applied to slow-moving signals. A benefit from increased functional abstraction is the potential reduction in overall circuit complexity, allowing for more compact designs which may require reduced processing in the back end of line, i.e. for metal interconnects, increasing both the functional density and the economic viability. It is expected that the contact-controlled nature of the device will bring superior yield as a result of improved tolerance to variability. Simultaneously, the functional abstraction should produce circuits with smaller footprint [10] for superior additional economic benefits (Fig. 7, right)

Conclusion and perspective

TCAD is instrumental in designing and optimizing advanced thin-film transistor structures with important functional and manufacturability advantages, the implementation of which will significantly expand the domain of viable applications for both existing and emerging thin-film technologies.

Acknowledgments

The author acknowledges funding from EPSRC through grants EP/R028559/1, EP/V002759/1, EP/Y000196/1.

References

[1] G. Wang *et al.*, "New opportunities for high-performance source-gated transistors using unconventional materials," *Adv. Sci.*, p. 2101473, 2021, doi: 10.1002/advs.202101473.

[2] X. Xu, R. A. Sporea, and X. Guo, "Source-gated transistors for power- and area-efficient AMOLED pixel circuits," in *IEEE/OSA Journal of Display Technology*, 2014, vol. 10, no. 11, pp. 928–933, doi: 10.1109/JDT.2013.2293181.

[3] S. Huang, J. Jin, J. Kim, W. Wu, A. Song, and J. Zhang, "IGZO Source-Gated Transistor for AMOLED Pixel Circuit," *IEEE Trans. Electron Devices*, vol. 70, no. 7, pp. 3637–3642, 2023, doi: 10.1109/TED.2023.3274501.

[4] J. M. Shannon, "Stable transistors in hydrogenated amorphous silicon," *Appl. Phys. Lett.*, vol. 85, no. 2, pp. 326–328, 2004, doi: 10.1063/1.1772518.

[5] R. A. Sporea, M. J. Trainor, N. D. Young, J. M. Shannon, and S. R. P. Silva, "Source-gated transistors for order-of-magnitude performance improvements in thin-film digital circuits," *Sci. Rep.*, vol. 4, p. 4295, 2014, doi: 10.1038/srep04295.

[6] E. Bestelink, U. Zschieschang, I. Bandara R M, H. Klauk, and R. A. Sporea, "The secret ingredient for exceptional contact-controlled transistors," *Adv. Electron. Mater.*, vol. 8, p. 2101101, 2021, doi: 10.1002/aelm.202101101.

[7] E. Bestelink, O. de Sagazan, L. Motte, B. Schultes, S. R. P. Silva, and R. A. Sporea, "Versatile thin-film transistor with independent control of charge injection and transport for mixed signal and analog computation," *Adv. Intell. Syst.*, vol. 3, p. 2000199, 2020, doi: 10.1002/aisy.202000199.

[8] E. Bestelink, O. de Sagazan, and R. A. Sporea, "P-18: Ultra-Compact Multi-Level Digital-to-Analog Converter based on Linear Multimodal Thin-Film Transistors," *SID Symp. Dig. Tech. Pap.*, vol. 51, no. 1, pp. 1375–1378, 2020, doi: 10.1002/sdtp.14141.

[9] E. Bestelink, O. de Sagazan, I. S. Pesch, and R. A. Sporea, "Compact Unipolar xnor/xor Circuit Using Multimodal Thin-Film Transistors," *IEEE Trans. Electron Devices*, vol. 68, no. 10, pp. 4951–4955, 2021, doi: 10.1109/TED.2021.3103491.

[10] E. Bestelink and R. A. Sporea, "Multimodal Transistor-Based 7T2C LTPS Pixel Circuit for Simultaneous PAM and PWM Control in µLED Display," *Dig. Tech. Pap. - SID Int. Symp.*, vol. 55, no. 1, pp. 493–496, 2024, doi: 10.1002/sdtp.17567.

[11] J. M. Shannon, C. Glasse, and S. D. Brotherton, "US7569435B2: Transistor manufacture," US7569435B2, 2004.

[12] C. Lius, K.-F. Lee, and N.-F. Hsu, "Display Device," US10784327B2, 2020.

[13] E. Bestelink, O. De Sagazan, L. Motte, and R. A. Sporea, "Suppression of hot-carrier effects facilitated by the multimodal thin-film transistor architecture," *Adv. Electron. Mater.*, vol. 7, p. 2100533, 2021, doi: 10.1002/aelm.202100533.

[14] I. Surekcigil Pesch, E. Bestelink, O. de Sagazan, A. Mehonic, and R. A. Sporea, "Multimodal transistors as ReLU activation functions in physical neural network classifiers," *Sci. Rep.*, vol. 12, no. 670, pp. 1–7, 2022, doi: 10.1038/s41598-021-04614-9.

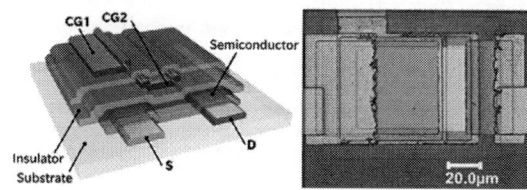

Fig. 1. Left: Schematic representation of a multimodal transistor (MMT) in which the source comprises a rectifying contact. Right: Optical image of a microcrystalline Si MMT. Reproduced from Ref. [7] under CC-BY 4.0.

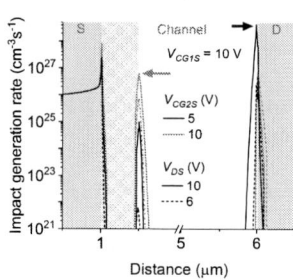

Fig. 2. Hot-carrier generation is mitigated through channel gate bias. Reproduced from Ref. [13] under CC-BY 4.0.

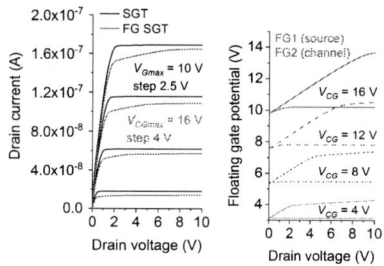

Fig. 3. SGTs are prone to deleterious capacitive coupling which degrades gain, whereas MMTs are immune because the channel potential couples to its respective control gate, which shields the source floating gate and retains high gain. Reproduced from Ref. [7] under CC-BY 4.0.

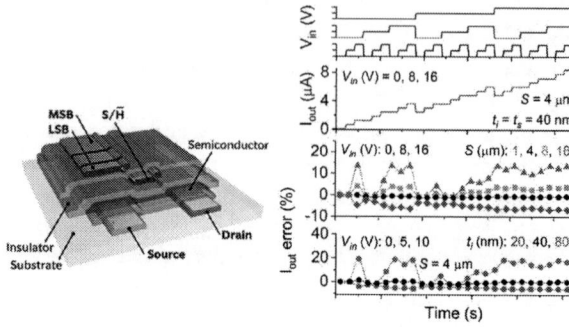

Fig. 4. Left: MMT with CG1 designed with binary weighted widths for digital-to-analog converter (DAC) operation. Right: Multiplying DAC operation. Reproduced from Ref. [7] under CC-BY 4.0.

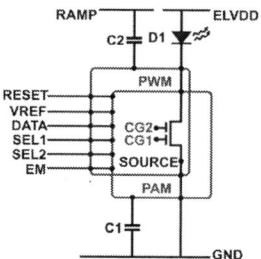

Fig. 5. MMT as pixel driver enables simultaneous PAM and PWM functionality. Reproduced from Ref. [10] with permission of John Wiley and Sons.

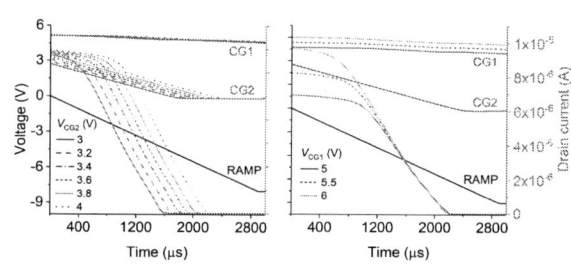

Fig. 6. Left: PWM operation controlled by CG2. Right: PAM controlled by CG1. Reproduced from Ref. [10] with permission of John Wiley and Sons.

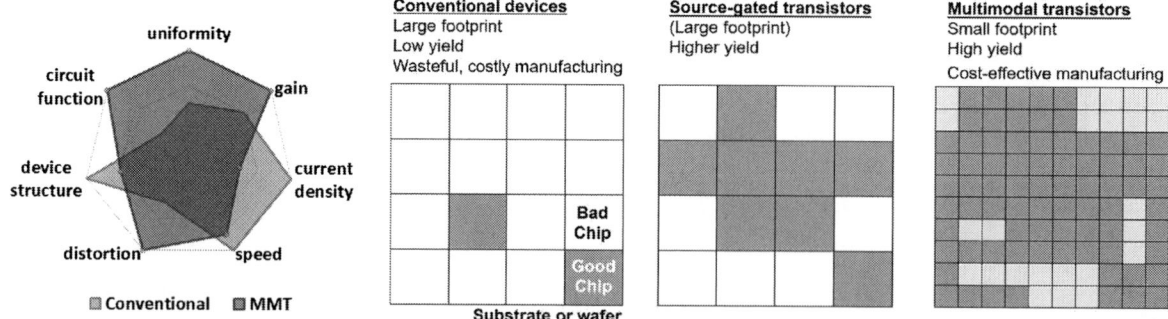

Fig. 7. Left: Comparison of conventional TFT vs MMT. Right: MMT improves yield due to compact functional circuits advantages in manufacturability.

979-8-3315-0417-5/25 $31.00 © 2025 IEEE   244

Dynamic Clocked Comparator based on Flexible LTPO Technology

Chunxiu Wang[1,†], Jiaqiao Liang[1,†], Min Zhang[2*]

[1]School of Electronic and Computer Engineering, Peking University, Shenzhen 518055, China

[2]School of Science and Engineering, The Chinese University of Hong Kong, Shenzhen 518172, China

[†]Equal contribution

*Email: mzhang@cuhk.edu.cn

Abstract

This paper presents a novel Low-Temperature Poly Oxide (LTPO) StrongARM comparator. The design and simulation of this work utilize 6 μm LTPO technology. The comparator achieves a operation frequency of 2 MHz, with a rail-to-rail output swing and an input common mode voltage (V_{inCM}) range of -4 V to 2 V input common mode voltage. The power consumption during each flip is 36 μW, with a delay of 33 ns under a 5 V power supply. Overall, this novel circuit design demonstrates its potential for low power, high performance flexible electronic systems.

Keywords: Flexible electronics, LTPO-TFT, High speed compactor, Flexible sensing systems.

Introduction

With the evolution of 5G networks, Artificial Intelligent (AI), and wearable electronics, the era of the Internet of Things (IoT) has arrived. Flexible electronics, as a novel technology, demonstrate significant potential in human-machine interaction, computation, display, energy generation and storage, and electrical textiles [1]. Thin-Film Transistor (TFT) is the key component for integrating flexible electronics.

Among various TFT technologies, the most common semiconductor materials for the channel include organic materials, amorphous silicon, Low-Temperature Poly Silicon (LTPS), and Amorphous Oxide Semiconductors (AOS). LTPS TFTs outperform traditional amorphous-silicon TFTs due to their high carrier mobility, making them suitable for high-speed devices. On the other hand, AOS such as amorphous Indium Gallium Zinc Oxide (IGZO) TFTs, can be fabricated at relatively lower temperature ($< 350°C$), making them more applicable for large-scale fabrication at lower cost. Low Temperature Poly Oxide (LTPO) technology integrates P-type LTPS TFT and N-Type IGZO TFT on the same substrate, which benefits from ultra-low leakage current with low-temperate processing of IGZO TFTs, and high driving capability of LTPS TFTs. This complementary structure enables more efficient and more complex TFT circuit integration.

In analog circuits, comparator serves as the fundamental quantization element of an Analog-to-Digital Converter (ADC) and is necessary for various ADC designs. Numerous studies have been conducted on comparators based on TFT technologies. Therefore, the novel LTPO technology brings up new possibilities for flexible electronic systems. In this work, we propose a comparator based on coplanar LTPO TFT technology, achieving a frequency of 2 MHz and a power consumption of 36 μW under 5 V power.

Section II

A. Device Design and Fabrication

The structure of the devices used in this study is shown in Fig. 1. The devices and circuits were fabricated on flexible polyimide (PI) substrate adhered to a glass substrate. The process of fabrication is listed as follows. A silicon oxide buffer layer is deposited prior to TFT fabrication to ensure device uniformity. The self-aligned technique is employed for the fabrication of LTPS and IGZO TFTs, which includes the following steps. First, deposing and patterning the active layer (LTPS/IGZO) for both channel and source drain regions, depositing the gate dielectrics, depositing and patterning the metal gate (Mo), and using the gate as a blocking layer to apply ion injection to the source and drain of the TFT. After the TFTs are fabricated, two layers of metal interconnects are added. The transfer characteristic is shown by Fig. 2.

B. Proposed Comparator

The schematic of the proposed comparator is illustrate in Fig. 3. The latched comparator forms the basis for the relative speed of the feedforward comparator. In this work, we choose to use LTPS TFTs as the tail current source and the differential input pair due to their high transconductance ($g_{m,LTPS}$) and large on-current, as the charge-discharge speed is highly dependent on the input current. The comparator consists of the components including a clock control transistor M1, a differential input pair (M2, M3), and two pairs of cross-coupled transistors (M4-M7), which effectively form a latch based on a pair of CMOS inverters and two pairs of reset charge switches (M8-M11). The output signals of the comparator are connected to a pair of inverters (M12-M15) for signal rectification. In case of small differences between the input terminals, the response time of the comparator increases. Furthermore, as the difference between input voltage decreases, the delay of the comparator increases. Thus, in the proposed circuits, a pre-amplifier (M16-M20) is added for voltage gain, enhancing the response

979-8-3315-0417-5/25 $31.00 © 2025 IEEE

speed of the comparator.

The operation phases of the comparator and the corresponding timing diagram are shown in Fig. 4. Details are listed below:

1) Reset Phase $\varphi1$: When the CLK is set to high, the input pair (M2, M3) is turned off, the reset switches (M8~M11) are activated, and the nodes X, Y, P, Q are discharge to V_{SS}.

2) Amplification Phase $\varphi2$: When the CLK is set to low, the circuit enters the linear amplification phase. The reset switches (M8-M11) are turned off, and the tail current source (M1) is activated. In the meantime, the input signals charge nodes P, Q via terminals INP and INQ. The duration of $\varphi2$ and its voltage gain are giving by:

$$T_{\varphi2} = \frac{c_{P,Q}V_{TH4,5}}{I_{CM}} \tag{1}$$

$$A_v = \frac{g_{ma}T_{\varphi2}}{c_{X,Y}} \tag{2}$$

where I_{CM} is the common mode current extracted from the capacitance of node P and Q, g_{ma} is the effective transconductance of M2 or M3, and V_{TH} is the threshold voltage of M4 or M5.

3) Regeneration Phase $\varphi3$: When the voltage of node P and Q reach to V_{TH4} and V_{TH5}, M4 or M5 will turn on first, allowing some of the current to flow from nodes P or Q toward the drain of M2 or M3, thus entering the regeneration phase. The cross-coupled P-type transistors initiate positive feedback, causing the output voltage to respond exponentially. The time constant of the regeneration phase is giving by:

$$\tau_{re} = \frac{c_{X,Y}}{g_{mp}(1-\frac{c_{X,Y}}{c_{P,Q}})} \tag{3}$$

where g_{mp} is the effective transconductance of M5 or M5.

When the nodes X or Y reach V_{TH6} or V_{TH7}, the cross-coupled N-Type TFTs M6 or M7 turn on, initiating the second phase of positive feedback. At this point, the inverter with the higher output voltage is rapidly pulled up to V_{DD}, while the inverter with the lower output voltage is quickly pulled down to V_{SS}.

C. Result and Discussion

To achieve low power consumption and minimal readout delay, the proposed comparator is optimized in size base on the 6 μm LTPO technology. Fig. 5 illustrates the delay characteristics of the latched comparator under different supply voltage and differential input voltages. With a supply voltage V_{DD} of 4 V and a differential voltage input ΔV_{in} of 100 mV, the comparator exhibits a readout delay of 51 ns. As V_{DD} increases from 4 V to 8 V, the delay decreases to 27 ns. Additionally, for a given supply voltage V_{DD}, the readout delay gradually decreases with increasing differential input

voltage. When the differential input voltage ΔV_{in} is 200 mV and the supply voltage V_{DD} is 8 V, the readout delay is 18.8 ns.

The common mode input voltage (V_{CM}) has significant impact on the performance of the comparator. Fig. 6 shows the delay characteristics of the comparator for different common mode and differential input voltages. The data indicate that the comparator achieves the best performance at the V_{CM} of -2 V. The performance of the comparator is also affected by the charge-discharge speed. When all transistors operate in the saturation zone, the circuit performs optimally; however, performance degrades when any transistor enters the linear region. In other words, as V_{CM} increases, the voltages at nodes P and Q also rise, forcing M4 and M5 entering their linear region. Consequently, the current flowing through the latch decreases, leading to a corresponding decline in performance.

In spite of this, the variation in power consumption with different differential input voltages (ΔV_{in}) is negligible and can be considered insignificant. Reducing the supply voltage can further decrease power consumption. Fig. 7 shows the power consumption trends of the comparator under various common mode and differential input conditions. Fig. 8 shows the layout of the proposed design.

Table1 summarizes the performance comparison of the proposed comparator compared with other reported studies on TFT based comparators. The results demonstrate a significant improvement in delay, power consumption, and clock frequency over the other TFT-based comparators.

Conclusion

The work has proposed a novel dynamic clocked comparator based on flexible LTPO technology. The performance of the design illustrates its potential for integration into flexible systems, achieving high performance and keeping low power at the same time.

Acknowledgments

This work was supported by National Key Research and Development (2022YFB3603600) and the National Natural Science Foundation of China (62074008).

References

[1] A. Yan et. al., Advanced. Functional. Materials., 34, 2304409 (2024).

[2] J. S. Park et. al., Thin Solid Films, 520, pp. 1679-1693 (2012).

[3] Y. Zhang et. al., IEEE Open Journal on Immersive Displays, 1, pp. 187-203, 2024.

[4] D. Raiteri et. al., Proceedings of the ESSCIRC (ESSCIRC), Bordeaux, France, pp. 141-144, 2012.

[5] K. Kim et. al. Society for Information Display Symposium Digest of Technical Papers, 45: pp. 1164-1167, 2014.

[6] D. Geng et. al., IEEE Electron Device Lett., 38, pp. 391-394 (2017).

[7] S. Shrivastava et. al., IEEE International Symposium on Circuits and Systems (ISCAS), pp. 1-5, 2023.

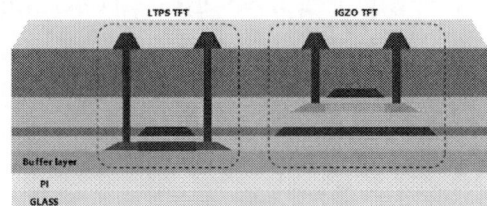

Fig. 1. Cross-section of the 6 μm LTPO technology device structure.

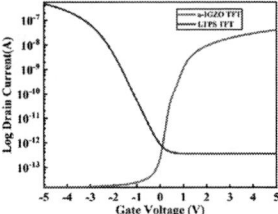

Fig. 2. Device characteristic of LTPS TFT and a-IGZO TFT (W/L = 6μm/6μm).

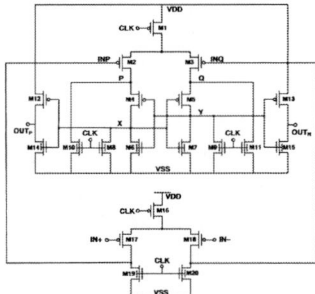

Fig. 3. Schematic of the proposed LTPO comparator.

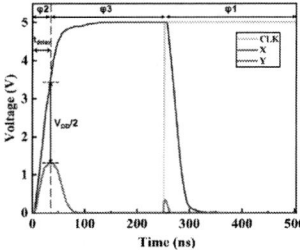

Fig. 4. Transient performance of proposed comparator circuit (V_{DD} = 5 V, V_{CM} = 0 V, V_{in} = 10 mV) by simulation.

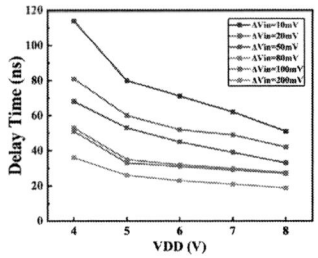

Fig. 5. Delay of the proposed comparator as a function of supply voltage (V_{cm} = -2 V) by simulation.

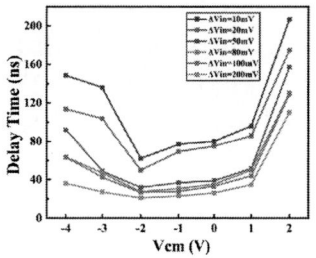

Fig. 6. Delay of the proposed comparator versus input common mode voltage (V_{cm}) at different input differential voltage(ΔV_{in}) by simulation.

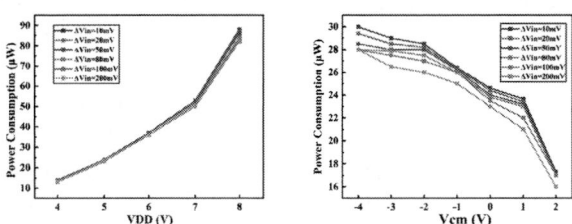

Fig. 7. Power consumption of the proposed comparator by simulation.

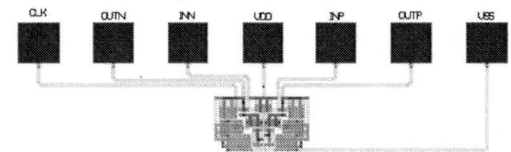

Fig. 8. Layout of the proposed comparator.

Type of TFT	Supply Voltage (V)	Clock frequency (kHz)	Power (μW)	Ref.
IGZO	20	0.1	>100	[4]
IGZO	50	0.1	-	[5]
IGZO	15	3.1	-	[6]
IGZO	4	5000	80	[7]
LTPO	5	2000	36	This work (simulate)

Table 1. Performance comparison of the proposed comparator.

979-8-3315-0417-5/25 $31.00 © 2025 IEEE   247

VO₂ memristor-based adaptive neurons for electromyography signal processing

Zhiyuan Li[1], Jiaping Yao[1], Beining Zhang[1], Wei Tang[1], Xiangshui Miao[1], and Rui Yang[1*]

[1]School of Integrated Circuits, Huazhong University of Science and Technology, Wuhan, China

*E-mail: yangrui@hust.edu.cn

Abstract

Hardware implementations of bio-plausible neurons are highly valuable for developing sparse and efficient neuromorphic computing systems. However, traditional CMOS-based neurons require complex auxiliary circuits and bulky capacitors. Here, we experimentally demonstrate a VO₂ memristor-based adaptive leaky integrate-and-fire (ALIF) neuron that enables advanced spike-frequency adaptation with ultra-low hardware cost (2T2R1C). These ALIF neurons serve as neural encoders, efficiently extracting temporal signal features with a sparse-spiking yet high-fidelity method. The spiking neural network (SNN) with ALIF achieves 97.1% accuracy in electromyography classification, showing great potential in human-computer interaction.

Keywords: Spike frequency adaptation, VO₂ memristor, leaky integrate-and-fire neuron, spiking neural network

Introduction

Neuromorphic computing particularly through spiking neural network (SNN), offers a transformative, energy-efficient solution for complex temporal tasks, closely emulating the architecture and operation of the human brain [1]. Artificial neurons, the core components of neuromorphic systems, emulating the key functions of biological neurons, such as potential accumulation and firing. While the leaky integrate-and-fire (LIF) model is the most fundamental, more advanced neuron dynamics, like spike frequency adaptation, have been shown to significantly improve neural information processing [2]. For the hardware implementation of neurons, hardware overhead are the critical evaluation criteria [3,4]. However, current neuron hardware implementation face challenges in meeting this metric, particularly in mimicking advanced biomimetic functions [5-9].

In this work, we demonstrate an adaptive LIF (ALIF) neuron with 2T2R1C configuration for SNN computing, capable of processing electromyography (EMG) signals using a sparse-spiking yet high-fidelity approach. The dynamic spike frequency adaption stem from the interaction between the VO₂ memristor-based 1T1R LIF unit and the adaptive unit. The spike adaptation speed can be tuned by adjusting the circuit parameters of the adaptive unit. Based on ALIF neurons, we constructed an ALIF-based SNN for electromyography (EMG) signal recognition. Encoding EMG signals with ALIF neurons leads to 4× reduction in spike production compared to LIF neurons, while improving classification accuracy on the EMG dataset from 94.6% to

97.1%. This work showing the great potential of memristors for constructing highly integrated, biologically inspired neurons and advancing efficient neuromorphic computing.

Device Fabrication and Experimental Set-up

The memristor in this work is fabricated with an Au/VO₂/Au planar structure, as shown in **Fig. 1(a)**. An 80 nm-thick VO₂ film was deposited on Si/SiO₂ substrate via magnetron sputtering, followed by a post rapid annealing process. Then, planar electrodes with a 600 nm gap and a 1 μm width were defined using EBL. Ti (10 nm)/Au (50 nm) electrodes were deposited using EBE. The dynamic characteristics of the device were measured using an Agilent B1500A and an oscilloscope. The neuronal circuit was established by mounting the packaged VO₂ memristors onto a custom-designed PCB.

Results and Discussion

Fig. 1(b) shows the typical volatile threshold switching behavior of the VO₂ memristor. Under consecutive 200 cycles of DC measurement, the device exhibits excellent uniformity in metallic-On and insulating-Off states. The threshold voltage (V_{TH}) is about 2.5 V, and the hold voltage (V_H) is about 0.43 V. The VO₂ memristor was then connected in series with the drain of a MOSFET to implement a LIF neuron model with a 1T1R structure (**Fig. 2(a)**). In this configuration, the inherent parasitic capacitance of the VO₂ memristor is utilized for charge integration, eliminating the need for external capacitors. The MOSFET operates in its variable resistance region, with its channel resistance (R_{c1}) serving as the integration resistor, which can be adjusted by varying the gate voltage (V_G). When V_G exceeds 1.6 V, regular neuronal oscillatory behavior is observed (**Fig. 2(b)**). The reconfigurable R_c under different V_G induces different $\tau_{integration} = (R_{c1}//R_{VO2})C_p$, resulting in different spiking frequencies of the neurons (**Fig. 3(a)**). The oscillatory frequency rises, as the input voltage (V_{IN}) increase. **Fig. 3(b)** shows the measured oscillatory frequency under different V_G and V_{IN}, demonstrating excellent neuronal encoding capabilities. Notably, increasing R_{c1} depresses the excitability of the neuron.

Based on the 1T1R structure of the LIF neuron, we added an adaptive control circuit to construct an efficient ALIF neuron with a 2T2R1C configuration (**Fig. 4(a)**). **Fig. 4(b)** illustrates the working flow of the ALIF neuron. Initially, the memristor is in high-resistance state, with the initial voltage of capacitor C_1 and C_2 set to 0 and V_{DD}, respectively. To

979-8-3315-0417-5/25 $31.00 © 2025 IEEE   248

achieve spike frequency adaptation, the output spike (V_{OUT}) from the 1T1R part drives M_2, which discharge C_2 and decrease V_G. As a result, the spiking frequency of the ALIF neuron is decreases during firing process. After a period of the adaptive time, $\tau_{adaptation} = (R_{c2}//R_L)C_p$, the charge and discharge capacity of C_2 reaches a dynamic balance, and V_G oscillates slightly around a fixed voltage value. Eventually, the neuron's spiking behavior stabilizes, achieving the adaptive functionality seen in biological neurons. **Fig. 5** demonstrate experimental results of the ALIF neuron. By applying two different constant voltage pulses to the adaptive and 1T1R parts, we measured the dynamic adaptation behavior of the neuron circuit. The adaptive waveforms of V_G and V_{OUT} are shown in **Fig. 5(a)**. As V_G decrease with each firing event, the spiking rate remains high initially, then progressively decreases at an increasing rate. Eventually, the firing frequency saturates at its lowest level. The number of spike and the inter-spike interval (ISI) are used to quantify the adaptive process. **Fig. 5(b)** shows the modulation of adaptive properties by circuit parameter V_{DD}. V_{DD}, the main excitation source of the adaptive feedback loop, directly determines the initial and steady-state values of the V_G of M_1. A higher V_{DD} induces a larger step increase in V_G, reducing the R_{C1} of M_1, which in turn lowers the ISI and increases the number of spikes. Moreover, the adaptive properties of the ALIF neuron can be further tuned by other circuit parameters, such as R_L, C_2, and M_2.

Using the above ALIF neuron as a neural encoder, we constructed a sparse and efficient SNN for EMG signal processing. **Fig. 6** shows the network structure. The 3-layer SNN ($8 \times 100 \times 4$) consists of an input spiking layer, a hidden layer, and an output classification layer. EMG signals are collected through 8-channel sensors and processed into spikes via the neural encoders. To demonstrate the coding advantage of the VO$_2$ memristor-based ALIF neurons, we compared different neuron models for encoding EMG signals. **Fig. 7(a)** shows a comparison of the total number of spikes in the encoding layers. When encoding EMG signals with ALIF neurons, the number of spikes generated is reduced by four times compared to LIF neurons, showing the sparse coding capability for extracting temporal signal features. Furthermore, the classification accuracy of ALIF-based SNN reach a maximum of 97.1%, which is superior to the LIF-based SNN (**Fig. 7(b)**).

Table I presents a comparison between our ALIF neuron and previously reported ALIF neurons. Compared to prior work, we experimentally demonstrate a memristor-based ALIF neuron for spike frequency adaptation at ultra-low hardware cost. Moreover, the ALIF neuron was employed as a neural encoder to process complex temporal physiological information, highlighting its potential applications in human-computer interaction.

Conclusion

In this work, we propose and experimentally demonstrate a VO$_2$ memristor-based ALIF neuron for spike frequency adaptation, for the first time. The ALIF neuron is designed by integrating a 1T1R neuron unit with an adaptive unit. Using the ALIF neurons, we developed a three-layer ALIF-based SNN to classify EMG datasets. The ALIF-based SNN achieved a classification accuracy of 97.1%, outperforming the LIF-based SNN (94.6%) while significantly reducing the number of spikes in the encoding layer. In addition, our ALIF neuron with its 2T2R1C configuration features an ultra-low hardware cost compared to previous works. These results show that compact VO$_2$ memristor-based ALIF neurons are a promising candidate for highly integrated, energy-efficient neuromorphic hardware.

Acknowledgments

This work was supported by the National Natural Science Foundation of China (Grant No. 62474073), the Natural Science Foundation for Distinguished Young Scholars of Hubei Province of China (Grant No. 2023AFA065) and Hubei Province Key Scientific and Technological Project (Grant No. 2022AEA001).

References

[1] P. A. Merolla et al., "A million spiking-neuron integrated circuit with a scalable communication network and interface," Science, 345, 6197 (2014).

[2] K. Boahen et al., "Neuromorphic silicon neuron circuits," Frontiers in Neuroscience, 5, 73 (2011).

[3] X. Zhang et al., "Experimental demonstration of conversion-based SNNs with 1T1R mott neurons for neuromorphic Inference," IEEE IEDM, pp. 6.7.1-6.7.4 (2019).

[4] Z. Li, et al., "Crossmodal sensory neurons based on high-performance flexible memristors for human-machine in-sensor computing system.," Nat. Commun., 15, 7275 (2024)

[5] G. Indiveri, F. Stefanini, and E. Chicca, "Spike-based learning with a generalized integrate and fire silicon neuron," IEEE International Symposium on Circuits and Systems (ISCAS), pp. 1951-1954 (2010).

[6] P. M. Ferreira, et al., "Energy efficient fJ/spike LTS e-Neuron using 55-nm node," in 2019 32nd Symposium on Integrated Circuits and Systems Design (SBCCI), pp. 1-6 (2019).

[7] C. Chen et al., "Bio-Inspired neurons based on novel Leaky-FeFET with ultra-low hardware cost and advanced functionality for all-ferroelectric neural network," in 2019 Symposium on VLSI Technology, pp. T136-T137 (2019).

[8] Q. Wei et al., "Artificial neuron with spike frequency adaptation based on mott memristor," in 5th IEEE Electron Devices Technology & Manufacturing Conference (EDTM), pp. 1-3 (2021).

[9] R. Yuan et al., "A neuromorphic physiological signal processing system based on VO$_2$ memristor for next-generation human-machine interface," Nat. Commun., 14, 3695 (2023).

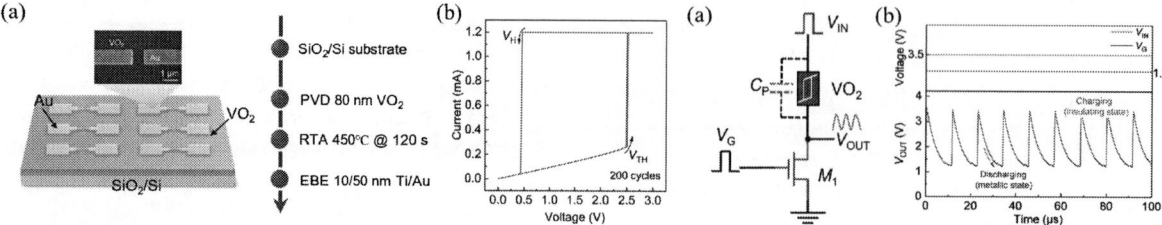

Fig. 1. Characteristics of the VO$_2$ memristor. (a) Schematic structure and fabrication process of the VO$_2$ planar memristor. Magnified SEM image shows a length of 800 nm. (b) Measured *I-V* curves of the VO$_2$ memristor repeated for 200 cycles. The V$_{TH}$ is about 2.5 V, and the V$_H$ is about 0.43 V.

Fig. 2. The VO$_2$ memristor-based LIF neuron. (a) Diagram of the VO$_2$ memristor-based 1T1R structure based on the LIF neuron model. (b) Measured LIF neuron response under a 3.5 V input voltage and a 1.6 V gate voltage.

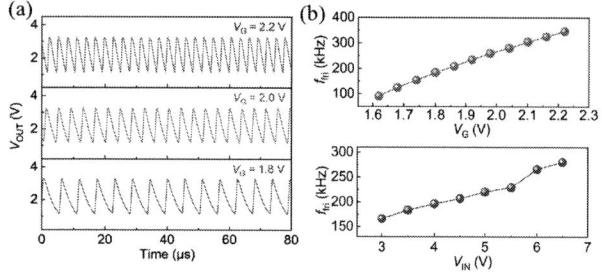

Fig. 3. Experimental results of the VO$_2$ memristor-based ITIR LIF neuron. (a) Measured LIF neuron response under different gate voltage V$_G$, where a higher V$_G$ leads to faster neuron firing. (b) The spike frequency varies with V$_G$ and V$_{IN}$. The f$_{fir}$ increases linearly with increasing V$_G$ and V$_{IN}$.

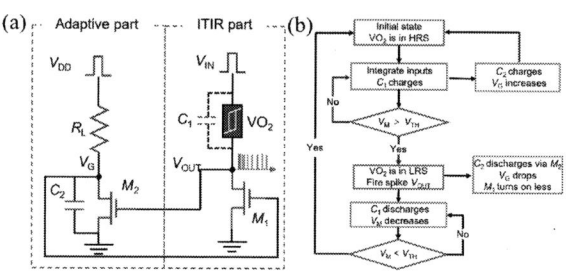

Fig. 4. Design of the VO$_2$ memristor-based ALIF neuron. (a) Circuit diagram of the VO$_2$ memristor-based ALIF neuron with 2T2R1C configuration. The ALIF neuron circuit consists of two parts (adaptive part, 1T1R part). (b) Working flowchart of the operation of the ALIF neuron.

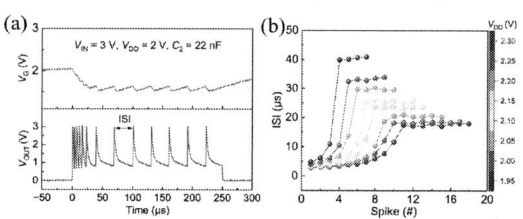

Fig. 5. Experimental results of the VO$_2$ memristor-based ALIF neuron. (a) Measured adaptive response of V$_G$ and V$_{OUT}$ when V$_{IN}$ (3 V) and V$_{DD}$ (2 V) are applied. The firing frequency dynamically adapts to the stimulus. (b) The evolution of ISI in terms of spike number under different V$_{DD}$.

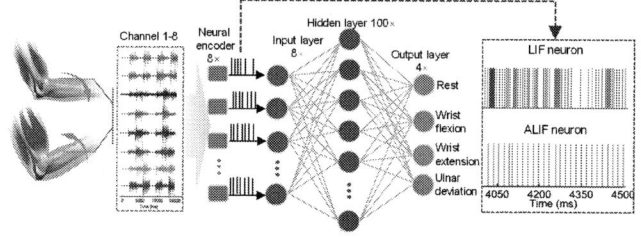

Fig. 6. The architecture of the neural encoder-based SNN for EMG signal recognition. In this neural network, LIF neurons and ALIF neurons are used as sensory neuron encoder to process EMG signal into spike, respectively. The spike number of ALIF neuron is much less than that of LIF neurons.

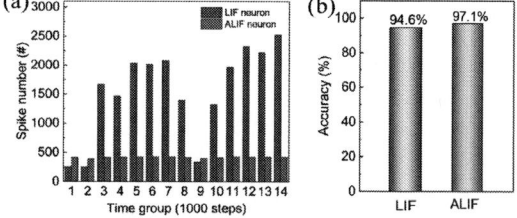

Fig. 7. Simulation results of the VO$_2$ encoder-based SNN recognizing EMG signal. (a) Spike number of the encoded EMG signal with different neuron model. (b) Comparison of the network accuracy of the LIF-based SNN and the ALIF-based SNN.

Table I. Comparison of our proposed 2T2R1C ALIF neuron and other ALIF neuron.

Type	Hardware cost	Circuit demonstration	ALIF application
CMOS [5]	2C+22T	Expt.	/
CMOS [6]	3C+10T+1R	Expt.	/
FeFET [7]	1C+6T	Expt.	/
NbO$_x$ [8]	1C+3T+1R	Sim.	MNIST dataset
VO$_2$ [9]	2C+3T+4R	Sim.	EEG & ECG
VO$_2$ [This work]	1C+2T+2R	Expt.	EMG

979-8-3315-0417-5/25 $31.00 © 2025 IEEE

Fabrication and research of wide-bandgap semiconductor AlN-based unipolar memristors

Haiming Qin[1,2,3], Xinpeng Wang[2,3], Dayu Zhou[4], Liang Zeng[5], Yi Liu[1], Yi Tong[2,3]*

[1]College of Integrated Circuit Science and Engineering, Nanjing University of Posts and Telecommunications, China, [2]Gusu Laboratory of Materials, China, [3]Suzhou Laboratory, China, [4]School of Materials Science and Engineering, Dalian University of Technology, China, [5]The Institute of Semiconductors, Chinese Academy of Sciences, China

*Email: tongyi2020@gusulab.ac.cn

Abstract

Wide-bandgap semiconductor aluminum nitride (AlN) has a wide range of applications. Here we designed memristors of Al (100 nm)/AlN (10 nm)/Al (100 nm) structure using plasma-enhanced atomic layer deposition (PEALD) and sputtering, which is compatible with CMOS technology and has the R_{OFF}/R_{ON} ratio of 3600 times. The special electrode design allows to exhibit interesting unipolarity, enabling SET and RESET operations through only unidirectional voltage. This research reports a new structure of AlN-based memristors, which also inspires further enriching the applications of wide-bandgap semiconductor AlN.

Keywords: Wide-bandgap, AlN and memristors

Introduction

The third-generation wide-bandgap semiconductor AlN has many excellent characteristics, such as a direct bandgap of up to 6.2 eV, high resistivity and breakdown field strength, and extremely high melting point and hardness. Simultaneously integrating thermoelectric, piezoelectric, and even ferroelectric (after doping) effects, it can be applied to many aspects such as light-emitting devices, high-frequency and high-power devices, sensors, and filters [1]. These advantages also cleverly meet the development needs of memristors. As the fourth basic circuit element, the resistance of memristors depends on the amount of charge flowing through them, so it has a memory function. Even after power failure, the state can still be maintained. This brain-like working method is of great significance in inspiring the next generation of computer architecture [2]. However, memristors face many challenges in their development, such as unstable performance under high-temperature conditions, large leakage current, low switching speed, and CMOS incompatibility, making them difficult to enter mass production [3].

In our previous research, we have proved that the third-generation wide-bandgap semiconductor SiC can be used as the functional layer of memristors, bringing better cycle endurance, lower operating voltage, and other characteristics, which are expected to be applied to artificial synapses [4]. Therefore, considering the compatibility of processes and

materials, we designed Al/AlN/Al structure, aiming to explore the potential application of wide-bandgap semiconductor AlN in memristors.

Experiment

The fabrication processes are shown in Fig. 1. 100 nm Al bottom electrode (BE) is deposited by direct current (DC) magnetron sputtering at room temperature (RT), with the power of 300 W, background vacuum of 8×10^{-6} Torr, Ar flow rate of 20 sccm, and sputtering rate of 2 Å/s. The pre-pasted metal hard mask can directly pattern BE. AlN was deposited as functional layer using PEALD, with liquid source trimethylaluminum (TMA) providing Al and gaseous source ammonia (NH_3) providing N. At chamber temperature of 300 °C, only one layer of atoms was grown at a time.

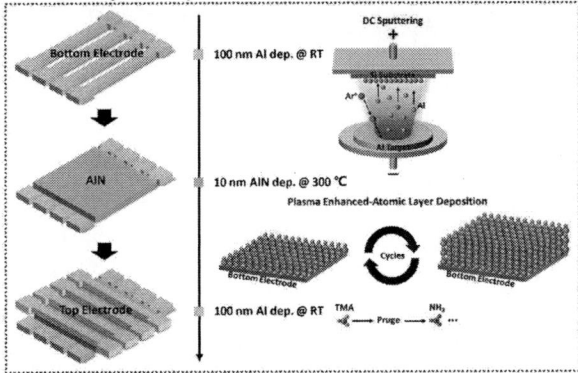

Fig. 1: Schematic diagram of structure and fabrication of AlN-based memristors.

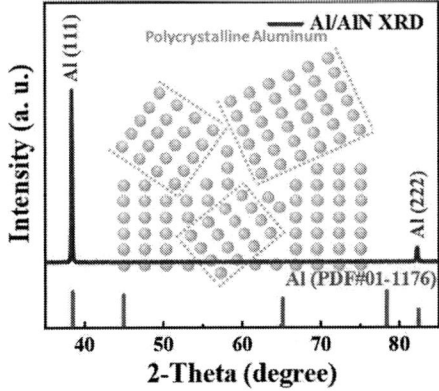

Fig. 2: XRD analysis of AlN and schematic diagram of polycrystalline Al.

After multiple cycles, thickness of 10 nm was finally achieved. 100 nm Al top electrode (TE) was deposited in the same way as BE, and 4×4 memristors crossbar arrays were finally formed through the pre-pasted metal hard mask.

Results and discussion

To investigate the crystalline quality of AlN, we conducted analysis of X-ray diffraction (XRD) before depositing TE, as shown in Fig. 2. Although strong peaks were observed, compared with the standard card PDF#01-1176, it was found that both signals matched Al of BE and had a polycrystalline structure [5]. We believe that there are two possible reasons for the absence of AlN peaks: (1) Insufficient thickness of AlN leads to very weak or even obscured diffraction signals, which are difficult to detect. (2) Under conditions of rough polycrystalline of Al and lower chamber temperatures, the crystallinity of AlN is low and exhibits amorphous state.

Next, we conducted electrical analysis using Keithley 4200A-SCS. Probes of the two channels directly contacted the pads of BE and TE to select the 100 μm unit in the array. A small bias voltage of 0.01 V was applied to read the initial state. We found that the resistance was very large after fabrication, which is consistent with the characteristics of the wide-bandgap semiconductor AlN itself. Then, we set small compliance current (CC) of 10 μA to prevent the thin AlN from being mistakenly broken down and attempted to apply dual voltage liner sweep from 0 V to 5 V. The I−V curve showed an interesting clockwise direction as shown in Fig. 3(a) Based on the results of XRD, we speculate that this is due to the presence of traps in amorphous AlN. During the

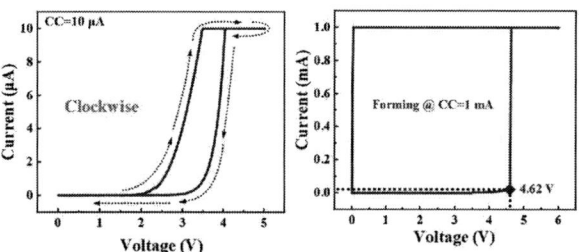

Fig. 3: (a) Clockwise I−V curve at CC=10 μA. (b) Forming process of AlN-based memristors at CC=1mA.

sweep from 0 V to 5 V with small CC, the trapped carriers obtain energy from the electric field and heat to overcome the binding barrier of the traps, thereby contributing to the current and causing it to rise. However, when sweeping from 5 V to 0 V, due to the decrease in electric field and dissipation of heat, carriers will be more quickly recaptured by traps and no longer contribute to the current [6], resulting in such clockwise nonlinear I−V curve.

Further increasing CC to 1 mA and setting dual voltage liner sweep from 0 V to 6 V, we obtained the expected counterclockwise hysteresis curve as shown in Fig. 3(b). The current rapidly increases to 1 mA near 4.62 V, indicating the

occurrence of switching behavior. The device switches from high-resistance state (HRS) to low-resistance state (LRS) and still maintains it when around 0 V. By comparing with the results mentioned later, we believe that this is the forming process, as it is the first switching behavior and the switching voltage of 4.62 V is significantly higher.

Then, we switched the device to HRS. Cyclic results after the forming process are shown in Fig. 4. The blue curve is the SET process, which still maintains CC of 1 mA with

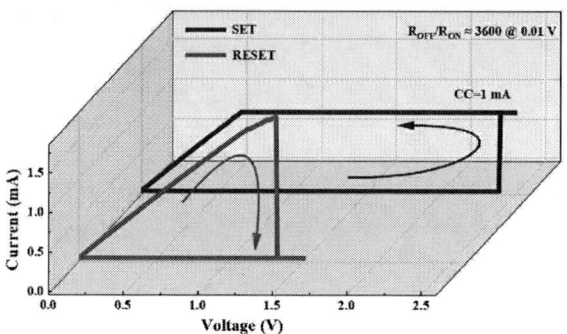

Fig. 4: I−V cycle curve after forming process. R_{OFF}/R_{ON}=3600 at 0.01 V.

dual voltage liner sweep from 0 V to 2.5 V. It can be seen that the switching behavior occurs near 2.4 V, showing a counterclockwise hysteresis curve. To demonstrate that the structure of Al/AlN/Al can switch from LRS to HRS, we started dual voltage liner sweep from 0 V to 1.5 V without specific CC until it reached the current limitation (0.1 A) of our machine. The reason why negative sweep is not performed is that our AlN-based memristors have a symmetrical electrode structure, and both are active metals. Based on existing research and experience, our devices may belong to unipolar memristor [7]. The conduction mechanism is determined by the formation and melting of Al conductive filament (CF), which we will analyze later in detail. It can be seen that without CC, the device exhibits switching behavior again at 1.32 V. Compared with the resistance at 0.01 V, the R_{OFF}/R_{ON} ratio is 3600 times. Only unidirectional operating voltage can achieve it, which is of great significance for simplifying the peripheral circuit design.

Based on the above results, we can find that such switching behavior is not determined by the polarity of the voltage, but the magnitude. The relationship of voltage is FORM>SET>RESET. As shown in Fig. 5, when we first started a sweep from 0 V to 6 V on the fresh AlN-based memristor, it was equivalent to a controlled breakdown. The local conductive path was formed inside, and the memristive effect was activated, laying the foundation for subsequent more unified switching characteristics. Once the local conduction path already exists, only the SET process from 0 V to 2.5 V is required, and CC of 1 mA is set to promote the migration of active metal Al from TE to BE along the local conduction path, forming stable CF and switching from HRS to LRS. Switching from LRS to HRS is necessary to ensure larger CC (0.1 A) and lower voltage (0 V to 1.5 V) to prevent thermal breakdown of the device. In this way, the originally stable CF will melt under the large Joule heat, and switch to HRS. The relationship of CC is RESET>SET [8].

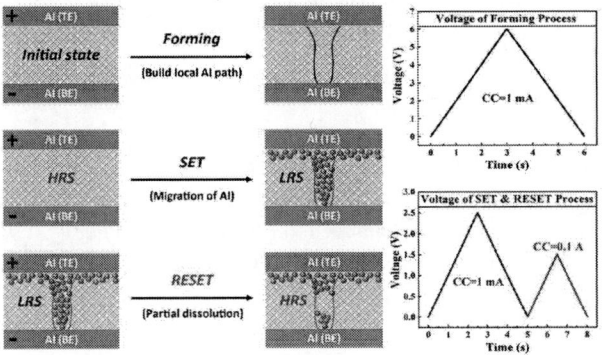

Fig. 5: Mechanism and dual voltage liner sweep diagram of Al/AlN/Al structure memristors.

Conclusion

In conclusion, we designed a wide-bandgap semiconductor AlN-based unipolar memristor with Al/AlN/Al structure. The low deposition temperature and mature material system are compatible with CMOS processes and have the R_{OFF}/R_{ON} ratio of 3600 times. The characteristic of resistive switching can be achieved by operating voltage in only one direction, which is of great significance for the simplification of peripheral circuit design when integrated with other devices in the future. At the same time, AlN-based memristors also show great scaling potential. Our research also further enriches the application of wide-bandgap semiconductor AlN.

Acknowledgments

This work was supported by the research fund of the Suzhou laboratory (No. SK-1202-2024-012), the 2030 Major Project of the Chinese Ministry of Science and Technology (Grant No. 2021ZD0201200).

References

[1] R. Yu, G. Liu, G. Wang, C. Chen, M. Xu, H. Zhou, T. Wang, J. Yu, G. Zhao, and L. Zhang, "Ultrawide-bandgap semiconductor AlN crystals: growth and applications", Journal of Materials Chemistry C, 9, pp. 1852-1873 (2021).

[2] Y. Sun, "The fourth fundamental circuit element: principle and applications", Journal of Physics D: Applied Physics, 55, pp. 253001 (2022).

[3] Y. Zhang, Z. Wang, J. Zhu, Y. Yang, M. Rao, W. Song, Y. Zhuo, X. Zhang, M. Cui, L. Shen, R. Huang, and J. Joshua Yang, "Brain-inspired computing with memristors: Challenges in devices, circuits, and systems", Applied Physics Reviews, 7, pp. 011308 (2020).

[4] H. Qin, S. Sun, N. He, P. Zhang, S. Chen, C. Han, R. Hu, J. Wu, W. Shao, M. Saadi, H. Zhang, Y. Hu, X. Wang, Y. Liu, L. Zeng, and Y. Tong, "Wide-bandgap semiconductor SiC-based memristors fabricated entirely by electron beam evaporation for artificial synapses", Applied Physics Letters, 125, pp. 143502 (2024).

[5] P. L. Zhang, P. Huang, and Z. Y. Jin, "Study of the Preferred Crystallographic Orientation of Polycrystalline Aluminum on Silicon Dioxide and Silicon", in 2007 International Workshop on Electron Devices and Semiconductor Technology (EDST), pp. 110-113, 2007.

[6] A. Kumar, M. Das, V. Garg, B. S. Sengar, M. T. Htay, S. Kumar, A. Kranti, and S. Mukherjee, "Forming-free high-endurance Al/ZnO/Al memristor fabricated by dual ion beam sputtering", Applied Physics Letters, 110, pp. 253509 (2017).

[7] H. Abunahla, M. A. Jaoude, C. J. O'Kelly, Y. Halawani, M. Al-Qutayri, S. F. Al-Sarawi, and B. Mohammad, "Switching characteristics of microscale unipolar Pd/Hf/HfO$_2$/Pd memristors", Microelectronic Engineering, 185, pp. 35-42 (2018).

[8] J. Molina-Reyes and L. Hernandez-Martinez, "Understanding the Resistive Switching Phenomena of Stacked Al/Al$_2$O3/Al Thin Films from the Dynamics of Conductive Filaments", 2017, pp. 8263904 (2017).

Reliability Optimization in Hafnium Oxide Based Ferroelectric Field-effect Transistors (FeFETs)

Kechao Tang[1,2], Yuejia Zhou[1], Zhongxin Liang[1], Ru Huang[1,2]

[1]School of Integrated Circuits, Peking University, Beijing, China. [2]Beijing Advanced Innovation Center for Integrated Circuits, Beijing, China (Email: tkch@pku.edu.cn)

Abstract

Hafnium oxide based ferroelectric field effect transistors (FeFETs) are promising candidates for next-generation memory, but are limited by reliability issues including endurance, variation and write disturb. This paper discusses the main causes of these reliability issues and our corresponding optimization strategies. A high endurance $> 5 \times 10^9$ is achieved via co-optimization of the ferroelectric and interlayer. Variation of V_{th} is reduced by 36% through ferroelectric grain size engineering. Nearly disturb-free operation is demonstrated with our proposed self-compensated write scheme.

Keywords: FeFETs, Memory, Reliability

Introduction

The discovery of HfO_2 based ferroelectrics (FE) re-invigorated the research in FE field-effect transistors (FeFETs). Featured by the good scalability, high speed and lower power consumption, HfO_2 based FeFETs are promising for advanced memory and compute-in-memory (CIM) applications [1, 2]. However, FeFETs are currently plagued by a series of reliability issues, including endurance, variation, and write disturb [3, 4]. These issues represent the reliability of FeFETs with respect to repeated cycling, device-to-device (D2D) selection, and array programming, respectively. The reliability concerns have received intensive research efforts, which led to basic understanding of their physical origin. The limited endurance (typically $< 10^6$) is mainly caused by the charge trapping and trap generation at the interface between FE and channel interlayer (IL). The variation of threshold voltage (V_{th}) from D2D is dominated by the limited domain numbers in scaled devices. Regarding the write disturb issue, it mainly arises from the accumulated bias stress applied on unselected cells.

In this paper, we report corresponding optimization strategies to comprehensively address the reliability issues in FeFETs. For the endurance, we introduced anti-ferroelectric (AFE) materials [5], as well as a FE-IL co-optimization approach to reduce the interlayer electric field (E_{IL}) [6]. For the variation, the HZO (ZrO_2 doped HfO_2) gate stack is engineered to prevent the formation of large grains, thus limiting the domain sizes [7]. For the write disturb, a self-compensated writing scheme is developed to reduce the accumulated bias stress on unselected cells [8].

Optimization of Endurance

A. E_{IL} Reduction by Anti-ferroelectric Materials

The limited endurance of FeFET is believed to correlate with the high E_{IL}. Compared to perovskite based FeFET, this problem is more severe for HZO FeFET, due to the much larger coercive field (E_c). Our corresponding strategy is to introduce AFE materials, which can be obtained by over-doping HZO with ZrO_2. As shown in Fig.3, one branch of AFE hysteresis can be used for memory. An effective E_c that is halved compared to HZO is obtained, thus reducing the E_{IL} for high endurance.

However, the ZrO_2 AFE cannot be directly integrated on Si channel FeFET, and no stable memory window (MW) can be measured. This is probably because the AFE is vulnerable to the large depolarization field, due the inevitable IL in Si-channel FeFET. The solution for this problem is to combine AFE with oxide semiconductor channel, like IGZO. With the stable oxide interface, an IL-free structure is available, as shown in Fig.4. The IL-free interface reduces the depolarization field and enables a MW window with > 10 year retention. With the low E_c AFE integrated in the IL-free device, a high endurance of 10^9 and a low operation voltage of 2V are achieved (Fig.5), which are highly beneficial for FeFETs operated with high speed and low power consumption.

B. Ferroelectric and Interlayer Co-optimization

High endurance FeFETs with Si channel are strongly desired, due to the scalable and mature process in FEOL. However, an IL is evitable, so an alternative approach other than above is required. We propose a novel holistic perspective to optimize the endurance in Si-based FeFETs. Fig.6 summarizes the target properties for the FE and IL. Besides their individual properties, combined effect including the interface quality and process compatibility also need to be considered.

Following this co-optimization strategy, we propose our materials of choice, which are HAO (Al_2O_3 doped HfO_2) for FE and Al_2O_3 for IL, respectively. HAO has lower E_c compared to HZO, consistent with first-principle calculations, while Al_2O_3 is a high-k oxide with good thermal stability. In addition, systematic trapping measurement indicates a low density of traps at FE-IL interface. The optimized device demonstrates a high endurance of 5×10^9 (Fig.7), which is around 3 orders of magnitude higher than typical Si-based FeFETs. Deep insight into the trapping dynamics is also provided via frequency-dependent electric measurements (Fig.8).

Optimization of Variation

The significant D2D variation is believed to arise from the limited domain numbers in scaled FeFETs. Since the grain boundaries in FE typically inhibit domain growth, reducing the size of FE crystal grains is one promising approach to mitigate the variation. We propose a method to engineer the grain size via insertion of a thin Al_2O_3 layer in the middle of HZO FE. This engineered FeFET, named HZO+HZO device, shows stronger polarization in FE and results a larger MW. More importantly, the D2D variation, as characterized by the V_{th} distribution sampled from 32 devices, is reduced by 36% compared to conventional HZO devices (Fig.9).

The improved variation is attributed to the reduction of grain sizes in FE (Fig.10), probably due to the prevention of crystalline nucleation by the Al_2O_3 border. The change of average grain size is evidenced by GIXRD, as shown in Fig.11. The average diameter of grain is inversely proportional to the width of the corresponding crystalline peak. A quantitative analysis demonstrates reduction of grain diameter from 13.6 nm to 9.9 nm via the inserted Al_2O_3 middle layer.

Optimization of Write Disturb

The proposed writing scheme for 1T MLC FeFET array is described in Fig.12. In the conventional writing approach, the entire array is first erased to 00 state and then programmed with an incremental pulse sequence row by row [13]. In our proposed scheme, the array is first set to middle state, followed by an alternative sequence of positive and negative pulses for writing. This setup results in alternating disturb bias stresses on unselected cells with opposite signs and similar magnitude. The impact of disturb stresses is thus self-compensating rather than accumulating, which effectively mitigates write-disturb.

The effect of self-compensated write scheme is tested on a 6×6 FeFET array. With $V_w/2$ applied on unselected bit lines and word lines as a general disturb inhibit setup, the variation of V_{th} under the self-compensated write scheme is compared with the conventional scheme. For devices in the first row, the V_{th} of the four states shows little variation after completely writing the entire array. This is in contrast to the devices with the conventional write scheme, demonstrating severe V_{th} shift, especially for the 00 states due to the disturb by accumulating positive pulses.

Conclusion and Prospect

In conclusion, we have discussed the key reliability issues limiting the application of HfO_2 based FeFETs. The endurance, variation and write disturb can be effectively improved following our optimization strategies. It is worth noting that the proposed methods are mutually compatible, and can be integrated for a final solution to address all the reliability issues. For example, the high-k channel IL can be combined with the Al_2O_3 middle interlayer, resulting a high reliability FeFET with 10^9 endurance and small D2D variations [7]. Error correction and endurance recovery operation can be merged with the self-compensated write flow, leading to a compact write scheme that addresses endurance, variation and write-disturb simultaneously (Fig.12) [8]. With the demand of reliability aligned with the application scenario, collective efforts in materials, devices, and circuit design are required to bring FeFETs into practical implementation.

Acknowledgments

This work was supported by National Key R&D Program of China (2022YFB4400300), National Natural Science Foundation of China (62274003, 61927901, and 92164203), 111 Project (B18001), and in part by the STIC under Grant QYJS-2022-1501-B. Part of the experiments were done at National Micro/Nano Fabrication Laboratory of Peking University.

References

[1] S. Dünkel et al., "A FeFET based super-low-power ultra-fast embedded NVM technology for 22nm FDSOI and beyond," in 2017 IEEE International Electron Devices Meeting (IEDM), 2017, pp. 19.7. 1-19.7. 4.

[2] Y. Zhou et al., "Hybrid-FE-Layer FeFET with High Linearity and Endurance Toward On-Chip CIM by Array Demonstration," IEEE Electron Device Letters, vol. 45, no. 2, pp. 276-279, 2024, doi: 10.1109/LED.2023.3346030.

[3] N. Zagni et al. "Reliability of HfO_2-based ferroelectric FETs: a critical review of current and future challenges." Proceedings of the IEEE 111.2 (2023): 158-184.

[4] K. Ni et al., "Write Disturb in Ferroelectric FETs and Its Implication for 1T-FeFET AND Memory Arrays," IEEE Electron Device Letters, vol. 39, no. 11, pp. 1656-1659, 2018, doi: 10.1109/LED.2018.2872347.

[5] Z. Liang et al., "A novel high-endurance FeFET memory device based on ZrO_2 anti-ferroelectric and IGZO channel." in 2021 IEEE International Electron Devices Meeting (IEDM), 2021, pp. 17.3.1-17.3.4.

[6] Y. Zhou et al., "Ferroelectric and Interlayer Co-optimization with In-depth Analysis for High Endurance FeFET," in 2022 International Electron Devices Meeting (IEDM), 2022, pp. 6.2.1-6.2.4.

[7] Y. Zhou et al., "A Reliable 2 bit MLC FeFET with High Uniformity and 10^9 Endurance by Gate Stack and Write Pulse Co-optimization." in 50th IEEE European Solid-State Electronics Research Conference (ESSERC), 2024.

[8] Y. Zhou et al., "A Compact Writing Scheme for the Reliability Challenges in 1T Multi-level FeFET Array: Variation, Endurance and Write Disturb," IEEE Electron Device Letters, Early Access, 2024, doi: 10.1109/LED.2024.3485803

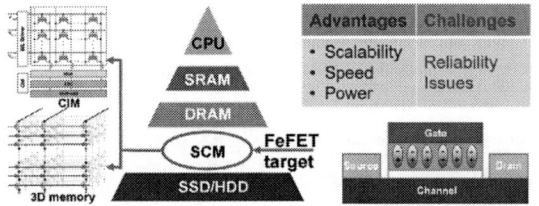

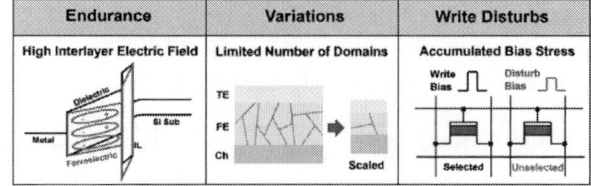

The Reliability Issues of FeFETs

Fig. 1. FeFET as a candidate SCM technology for memory and CIM applications. FeFET has advantages in scalability, speed and power, but is limited by reliability issues.

Fig. 2. The reliability issues in FeFETs and their main cause. Endurance: Charge trapping due to high interlayer electric field; Variation: Domain fluctuations in scaled devices; Write disturb: Accumulated bias stress on unselected cells in the array

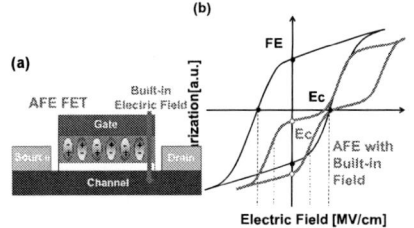

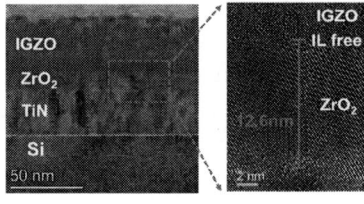

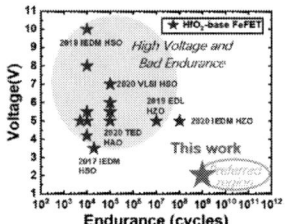

Fig. 3. Schematics of FeFET memory devices based on AFE gate oxide materials. (a) layout of the device structure. (b) P-V features of non-volatile storage achieved by AFE with built-in bias.

Fig. 4. HRTEM image of the ZrO_2-IGZO FET gate stack. No interlayer is detected at the ZrO_2-IGZO interface.

Fig. 5. The ZrO_2-IGZO AFE FET benchmarked with other work, featuring high endurance and low operation voltage.

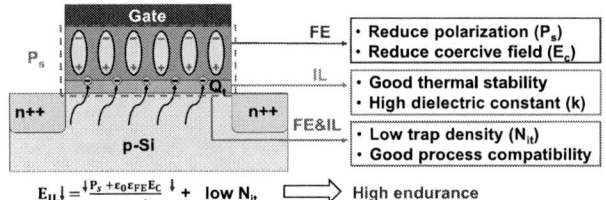

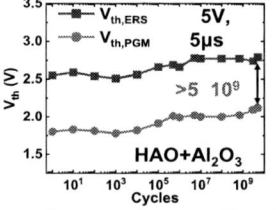

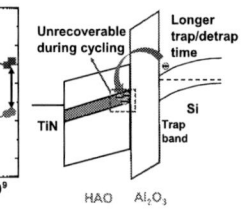

$$E_{IL}\downarrow = \frac{\downarrow P_s + \varepsilon_0\varepsilon_{FE}E_C \downarrow}{\varepsilon_0\varepsilon_{IL}\uparrow} + \text{low } N_{it} \Rightarrow \text{High endurance}$$

Fig. 6. Conceptual schematic of FE and IL layer co-optimization for high endurance FeFET. The individual properties of FE and IL layer, as well as their combined interface effect, needs to be considered from a holistic perspective.

Fig. 7. The proposed FeFET with HAO FE layer and Al_2O_3 IL, showing a high endurance > 5×10^9 (limited by test time).

Fig. 8. Schematic of the likely charge trapping mechanism in the high-endurance FeFET.

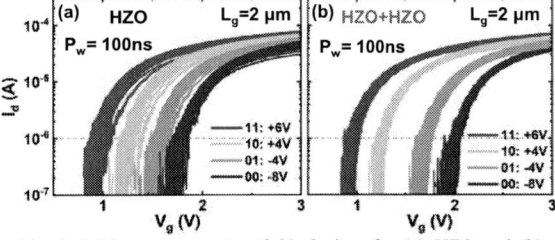

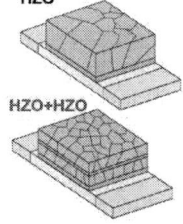

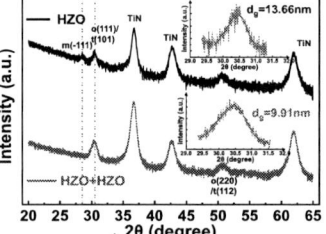

Fig. 9. I_d-V_g measurements of 32 devices for (a) HZO and (b) HZO+HZO FeFET. The HZO+HZO FeFET shows a larger MW and smaller I_d distribution.

Fig. 10. Schematic of the reduced variation due to grain size reduction

Fig. 11. Evidence of grain size reduction in HZO+HZO FeFET, characterized by the peak width in GIXRD.

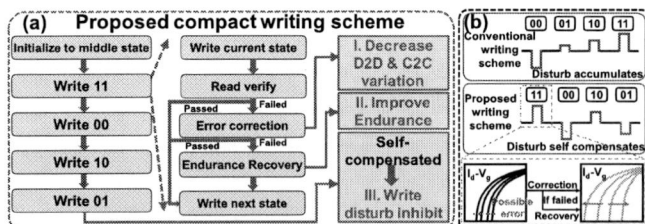

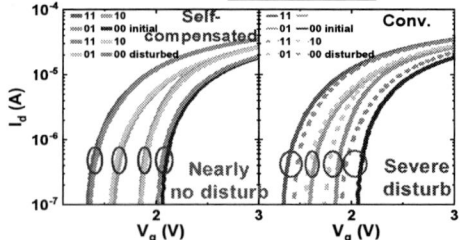

Fig. 12. (a) Proposed writing scheme for 1T MLC FeFET array, simultaneously addressing three key reliability issues. (b) Comparison of traditional writing scheme and the proposed writing scheme.

Fig. 23. Measured I_d-V_g curves of the FeFETs before and after array disturb. Nearly no disturb of V_{th} is achieved with the proposed writing scheme.

979-8-3315-0417-5/25 $31.00 © 2025 IEEE

High-Performance 2D FETs with Single-Crystal Anatase TiO$_2$ High-κ Dielectric

Ni Yang[1*], Ji Zhang[2], Yu-Ming Chang[1], Fangyuan Zheng[1], Lingqi Li[4], Chenyang Li[1], Jian Liu[5], Dong-Keun Ki[3], Yi Wan[1], Sean Li[2], Kah-Wee Ang[4], Jing-Kai Huang[5], Lain-Jong Li[1]

[1]Department of Mechanical Engineering, University of Hong Kong, Hong Kong, China
[2]School of Materials Science and Engineering, University of New South Wales, Sydney, Australia
[3]Department of Physics and HK Institute of Quantum Science and Technology, University of Hong Kong, Hong Kong, China
[4]Department of Electrical and Computer Engineering, National University of Singapore, Singapore,
[5]Department of Systems Engineering, City University of Hong Kong, Hong Kong, China
*Email: niyang96@connect.hku.hk

Abstract

Several bottlenecks must be resolved before two-dimensional (2D) materials-based transistors can be adopted for advanced electronics, particularly the challenge of coupling high-κ dielectrics with 2D semiconductors without adverse effects and continuing to scale their capacitance equivalent thickness (CET). In this work, we have developed a novel single-crystal dielectric and a corresponding integration approach to achieve 2D MoS$_2$ field-effect transistors (FETs) with outstanding performance, including steep SS (~ 68 mV/dec) and high ON/OFF ratio (> 10^7).

Keywords: Single-crystal oxide, High-k dielectric, 2D technology.

Introduction

2D layered semiconductors, characterized by their atomically thin bodies, have emerged as promising candidates for future channel materials in field-effect transistors. Despite the successful fabrication of high ON-current 2D FETs with scaled channel dimensions, significant challenges persist in scaling the gate insulator while maintaining a high-quality dielectric/channel interface. Two major strategies have been developed to address these issues: (1) incorporating interfacial buffer layers using gentle integration methods and (2) employing crystalline high-κ dielectrics such as hBN [1], CaF$_2$ [2], and SrTiO$_3$ [3]. However, state-of-the-art 2D FETs still suffer from considerable hysteresis, threshold voltage instability, and large variations in subthreshold swing, raising concerns about their reliability as indicated by bias temperature instability measurements. These issues are generally attributed to charge trapping at both the interface and gate dielectric body.

In this work, we demonstrate that large-area single-crystal anatase titanium dioxide (A-TiO$_2$), with a well-constructed and atomically flat surface, can serve as an ideal candidate for interfacing with 2D semiconductors. The developed approach effectively minimizes trapping states at the interface and within the dielectric body, resulting in a record-low hysteresis value compared to current silicon/high-k and other 2D dielectric technologies. Additionally, the high permittivity of A-TiO$_2$ facilitates the shrinking of the CET to the sub-1 nm scale. Our 2D FETs exhibit an excellent performance that ranks among the best-published results for similar device geometries, particularly with minor performance variations for batch-fabricated devices, which is critical for achieving stable and reliable operation of 2D-based logic. The A-TiO$_2$ has great potential to overcome the challenges currently faced by 2D FETs. The work also indicates that van der Waals (vdW) interface and high-quality dielectrics are critical for transistor devices based on 2D layer semiconductors.

Device Fabrication

The device structure and fabrication process flow are demonstrated in Fig. 1. Initially, the A-TiO$_2$/Sr$_3$Al2O6/SrTiO$_3$ epitaxial heterostructure is fabricated using pulsed-laser deposition. The heterolayer stack is then coated with a PMMA supporting layer and released by dissolving the water-soluble Sr3Al2O6 sacrificial layer from the SrTiO$_3$ substrate. The released A-TiO$_2$ membrane, supported by the PMMA layer, is then transferred onto the SiO$_2$/Si substrate with pre-defined buried Pt-gate electrodes. Subsequently, the PMMA supporting layer is removed by soaking it in a hot acetone solution for minutes. To further eliminate surface polymer residue, the transferred dielectric oxide is treated with mild oxygen plasma. Atomic force microscopy (AFM) is adapted to measure the thickness of A-TiO$_2$ dielectric layer in which the A-TiO$_2$ dielectric surface is atomically smooth with an ultra-low surface roughness (Ra ~ 0.2 nm), ensuring excellent interfacing with 2D semiconductor channels (Fig. 2). The single-crystal structure of the dielectric is verified by high-resolution transmission electron microscopy (HRTEM) as shown in Fig. 3. Next, a monolayer MoS2 film, grown by chemical vapor deposition (CVD), is then transferred onto the A-TiO$_2$ dielectric layer. The FET channel is defined using lithography and O$_2$ plasma etching. Source and drain contact areas are defined by e-beam lithography, followed by the deposition of 20 nm Sb and 40 nm Au using a thermal evaporator. The photo of the fabricated FET array is shown in Fig. 4.

Result and Discussion
Atomically sharp vdW interfaces

The cross-sectional HRTEM images of the device reveal two distinct vdW gaps at the Sb/MoS$_2$ and MoS$_2$/A-TiO$_2$ interfaces (Fig. 5). Parallel elemental mapping from energy-dispersive X-ray spectroscopy (EDS) aligns well with the device structure shown in Fig. 4. At the MoS_2/A-TiO_2 interface, the single-crystalline and well-defined dielectric surface provides a clean interface contact to MoS2, minimizing charged-impurity scattering and doping. The vdW interface between the 2D semiconductor channel and the dielectric effectively suppresses fringing-induced barrier lowering (FIBL) caused using high-κ dielectric, which in turn helps to mitigate drain-induced barrier lowering (DIBL) in scaled devices. Additionally, a pristine and intimate contact between the crystalline Sb and MoS_2, free from metal-induced defects, is observed, ensuring excellent charge injection from the contact metal.

Evaluation of single-crystal dielectric property

Fig. 6 shows the extracted CET values using a metal-insulator-metal (MIM) capacitor configuration, where the red dashed line represents the fitting curve of the dead-layer model, commonly observed in high-κ nanocapacitors. The estimated CET for 18 nm A-TiO_2 is around 1 nm (κ ≈ 65). Although the typical bandgap of A-TiO_2 is below 4 eV [4], which is a limitation for gate insulators, the vdW MoS_2/A-TiO_2 interface and the single-crystalline nature of our A-TiO_2 film with limited oxygen vacancies are expected to inhibit gate leakage pathways efficiently. Fig. 7 (a) demonstrates the current leakage characteristics of A-TiO_2 layers with varying thicknesses until the device breaks down. The 18 nm thick A-TiO_2, predominantly used in this study, exhibits a low leakage current density (J_{leak}) of $< 10^{-5}$ A/cm² when the applied voltage is below 2.5 V. The corresponding Jleak versus CET benchmark is presented in Fig. 7 (b), in which our result is clearly superior to the reported Si-based [5] and 2D-based data [2, 3, 6-10].

Performance of 2D vdW A-TiO_2/MoS_2 FETs

Fig. 8 (a) shows the transfer curves of 100 FETs with a channel length (L_{ch}) of 2 μm. Most devices (98%) exhibit over 7 orders of magnitude in ON/OFF switching ratio, with low gate leakage and minimal variation in threshold voltage (V_{th}). The average and minimum subthreshold swing (SS) values are 75 mV/dec and 68 mV/dec, respectively. Fig. 10b summarizes the SS versus CET for various 2D technologies, in which the shaded red corner is the latest IRDS projection for the 2028 HP specification (SS < 70 mV/dec, CET < 0.9 nm), highlighting that our results are among the best-reported values, meeting the desired performance metrics.

Conclusion

We successfully integrated large-area single-crystal A-TiO_2 as a gate dielectric for 2D MoS2 FETs, achieving exceptional performance metrics. The atomically flat surface of A-TiO_2 and high permittivity of A-TiO_2 the CET to sub-1 nm, enhancing device performance. Our A-TiO_2/MoS_2 FETs exhibit excellent electrical characteristics, including steep subthreshold swings, high ON/OFF ratios, and low leakage currents. The confirmed electrical stability indicates their potential for reliable long-term operation, highlighting single-crystal dielectrics as a key material for future 2D semiconductor technologies and scalable fabrication.

Acknowledgment

The authors acknowledge the support from University of Hong Kong, UNSW Sydney, National University of Singapore, and City University of Hong Kong.

References

[1] Chen, Tse-An, et al. "Wafer-scale single-crystal hexagonal boron nitride monolayers on Cu (111)." Nature 579.7798 (2020): 219-223.
[2] Illarionov, Yury Yu, et al. "Ultrathin calcium fluoride insulators for two-dimensional field-effect transistors." Nature Electronics 2.6 (2019): 230-235.
[3] Huang, Jing-Kai, et al. "High-κ perovskite membranes as insulators for two-dimensional transistors." Nature 605.7909 (2022): 262-267.
[4] Dette, Christian, et al. "TiO2 anatase with a bandgap in the visible region." Nano letters 14.11 (2014): 6533-6538.
[5] Robertson, John. "High dielectric constant oxides." The European Physical Journal-Applied Physics 28.3 (2004): 265-291.
[6] Li, Weisheng, et al. "Uniform and ultrathin high-κ gate dielectrics for two-dimensional electronic devices." Nature Electronics 2.12 (2019): 563-571.
[7] Dahal, Arjun, et al. "Seeding atomic layer deposition of alumina on graphene with yttria." ACS applied materials & interfaces 7.3 (2015): 2082-2087.
[8] Jeong, Seong-Jun, et al. "Thickness scaling of atomic-layer-deposited HfO2 films and their application to wafer-scale graphene tunnelling transistors." Scientific Reports 6.1 (2016): 20907.
[9] Liu, Kailang, et al. "A wafer-scale van der Waals dielectric made from an inorganic molecular crystal film." Nature Electronics 4.12 (2021): 906-913.
[10] Lu, Zheyi, et al. "Wafer-scale high-κ dielectrics for two-dimensional circuits via van der Waals integration." Nature communications 14.1 (2023): 2340.

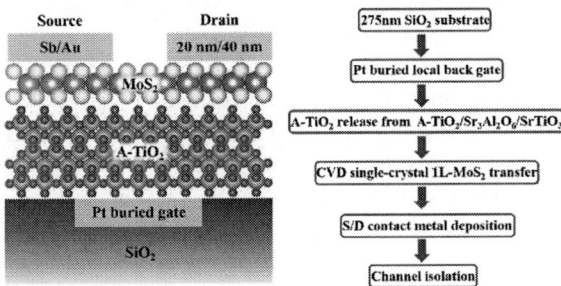

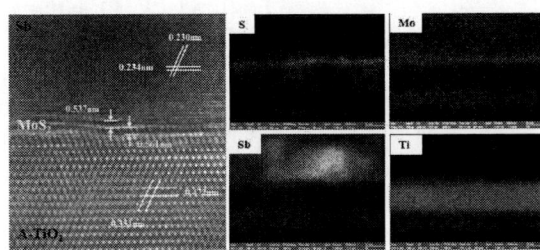

Fig. 1. The fabrication process flow of the device, and cross-section of Pt buried-gated MoS_2 FETs with a single-crystalline A-TiO_2 gate dielectric.

Fig. 5. Cross-sectional HRTEM image and corresponding EDS mapping of fabricated FET at contact region.

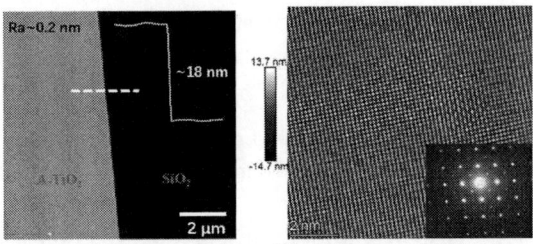

Fig. 2. AFM image of an 18-nm thick single crystalline A-TiO_2 surface with atomical flatness.

Fig. 3. Top-view HRTEM image and its diffraction pattern (inset) present the high crystallinity of A-TiO_2.

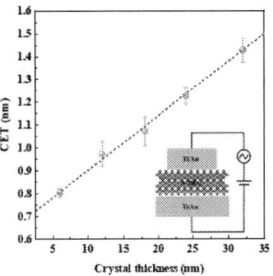

Fig. 6. CET as a function of different A-TiO_2 thicknesses extracted from MIM capacitors. (b) C-f curves of various A-TiO_2 thickness.

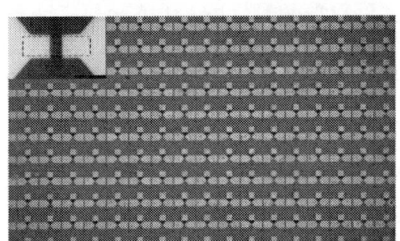

Fig. 4. Optical micrograph of batch-fabricated FET arrays. Inset: a magnified image of a single device. Scale bars, 500 μm and 10 μm (inset). The dashed square in the inset is the patterned MoS_2 channel.

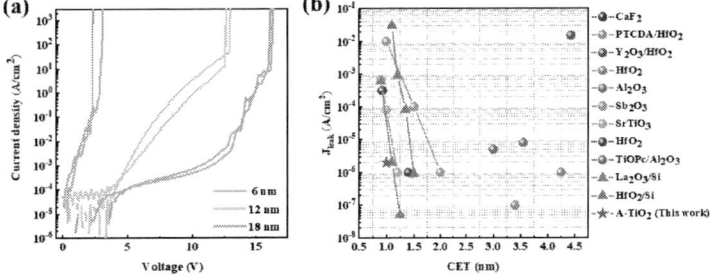

Fig. 7 (a) The leakage current characteristics of A-TiO_2 layers with varying thicknesses until device breakdown. (b) Benchmark of J_{leak} versus CET for various high-κ dielectric layers in 2D-based and Si-based FETs.

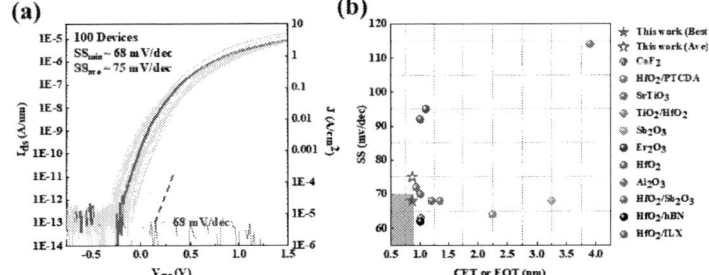

Fig. 8. (a) I_{ds}-V_{gs} characteristics of 100 1L-MoS_2 nFET at V_d= 1 V. (b) Comparison of SS values achieved by state-of-the-art CVD-prepared MoS_2 FETs with various CET or EOT.

All-Optical Modulated Artificial Synapse Based on Quantum dots/Oxide Heterojunction for Neuromorphic Visual Simulation

Yan Wang[1,2], Yingjie Tang[1,2], Yitong Chen[1,2], Dingwei Li[1,2], Huihui Ren[1,2], Guolei Liu[1,2], Fanfan Li[1,2], Qi Huang[3], Botao Ji,[2,3] and Bowen Zhu[2,3]

[1]Zhejiang University, Hangzhou 310024, China.
[2]Key Laboratory of 3D Micro/Nano Fabrication and Characterization of Zhejiang Province, School of Engineering, Westlake University, Hangzhou, China.
[3]Westlake Institute for Optoelectronics, Westlake University, Hangzhou, China.

Abstract

Artificial synapses capable of optical sensing and synaptic functions are essential for developing neuromorphic visual systems. We utilized quantum dots (QDs)/oxide heterostructure to achieve bidirectional synaptic weight modulation under ultraviolet and infrared light by the change of depletion layer width of the p-n junction. Such an all-optically controlled artificial synapse shows a promising prospect for future neuromorphic visual systems.

Keywords: all-optical, artificial synapse, heterojunction, phototransistor

Introduction

The physical distinction between memory and central processing units in traditional von Neumann architecture systems has encountered a bottleneck regarding energy efficiency and power consumption [1]. Inspired by the biological visual system, the neuromorphic visual system holds significant potential in artificial intelligence due to their ability to integrate learning, perception, processing, and memorization functionalities based on parallel operation [2-4].

As the basic unit of the nervous system, artificial synapse based on electronic devices with the capability of integrating optical sensing and synaptic functions have been confirmed crucial for neuromorphic visual systems [5]. However, reversible synaptic modulation, such as depression and potentiation in these devices, typically relies on a combination of optical and electrical stimuli. Using both signals together increases the complexity and burden on peripheral circuits, while also limiting the system's processing speed. Thus, optoelectronic synapses driven solely by light stimuli mimicking both inhibitory and excitatory offer significant potential for ultrafast computing, providing advantages such as low crosstalk, high bandwidth, and low power consumption.

Herein, we constructed optoelectronic synaptic devices by virtue of the PbS quantum dots (QDs)/IGZTO heterojunction to achieve bidirectional optoelectronic synaptic behaviors. The device can respond to ultra-violet (UV) and short-wavelength infrared (SWIR) light,

meanwhile the excitatory postsynaptic current (EPSC) and inhibitory postsynaptic current (IPSC) can be observed, respectively. Besides, neuromorphic characteristics of paired pulse facilitation/depression (PPF/PPD), short/long-term memory (STM/LTM), STM-to-LTM transition were mimicked successfully. These results underscore the potential of such p-n heterostructure for applications in fully optical-controlled neuromorphic systems.

Results and discussion

A schematic of the phototransistor based on the PbS QDs/IGZTO heterostructure is illustrated in Fig. 1(a). The IGZTO TFTs were fabricated with bottom-gate, top-contact architecture onto p-type Si/SiO₂ substrates, which were used as gate electrode and insulator, respectively. The IGZTO film (~10 nm) was deposited as a channel layer by co-sputtered IGZO and ITO [6]. Then an ITO layer (~80 nm) was deposited by sputtering as source/drain (S/D) electrodes with channel length (L) and width (W) of 50 μm and 500 μm, respectively. For PbS/IGZTO composite film to obtain heterostructure, the PbS QDs dispersed in toluene was spin-coated onto IGZTO TFTs and solidified at 90 °C for 1 min. To enhance conductivity of QDs film, the solid-state ligand exchange procedure was performed by covering the film with 0.15 mol L⁻¹ NH₄SCN in methanol for 1 min and spinning at 3000 rpm for 40 s. Then, the film was washed twice with pure methanol.

Fig. 1(b) shows the TEM cross-section image of the phototransistor. Fig. 1(c) exhibits the absorption spectra of the IGZTO and PbS/IGZTO thin films. The IGZTO film demonstrated high transparency across the visible to infrared spectrum, whereas the PbS/IGZTO heterostructure thin film showed a notable absorbance peak centered around 1550 nm in the SWIR range. Fig. 1(d) shows the photoluminescence (PL) properties of the thin films. A significant quenching in the PL intensity is observed in the PbS/IGZTO composite film compared to the single-layer PbS QDs, indicating an effective charge transfer between IGZTO and PbS QDs under illumination.

In biological neural networks, excitatory or inhibitory neurotransmitters will be released from presynapse to

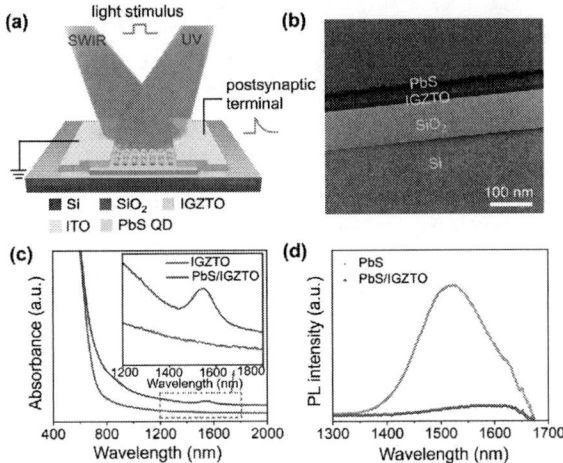

Fig. 1. (a) Schematic illustrating the phototransistor based on the PbS QDs/IGZTO heterostructure. (b) TEM image of the phototransistor. Absorption (c) and PL (d) spectra of single-layer thin film and PbS/IGZTO composite film, respectively.

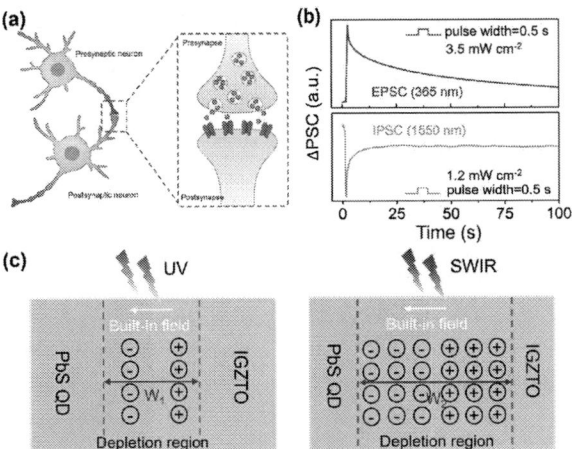

Fig. 2. (a) Schematic illustration of a biological neural network. (b) EPSC and IPSC of the device triggered by optical pulse with wavelengths of 365 nm and 1550 nm. (c) Schematic depletion region and carrier transfer dynamic diagram of the PbS/IGZTO heterojunction under UV (left) and SWIR (right) light illumination.

postsynapse under stimulation (Fig. 2(a)). Then the neurotransmitters can bind to the receptors, leading to an excitatory/inhibitory postsynaptic current (EPSC/IPSC) [7]. For our PbS/IGZTO heterojunction based artificial synaptic phototransistor, a bidirectional photoresponse along with an extended recovery time is observed, suggesting synaptic behaviors. As shown in Fig. 2(b), EPSC and IPSC are triggered by applying light pulses with 365 nm (3.5 mW cm^{-2}) and 1550 nm (1.2 mW cm^{-2}), respectively. In Fig. 2(c), we proposed the possible mechanism as depletion-region width modulation. In the PbS/IGZTO composite film, a built-in electric field is formed at their interface. Under UV light illumination, photogenerated electron—mainly derived from IGZTO, will transfer to the PbS QDs side，reaching equilibrium and creating a depletion region with a width of W1. This ongoing increase in carrier concentration leads to a positive response. While under infrared light illumination, photogenerated holes primarily originate from PbS QDs, disrupting the electric field equilibrium. Consequently, holes in QDs and electrons in IGZTO layer are further diffused towards each other，expanding the width of the depletion region (W2 > W1). This process significantly reduces the effective charge carriers, resulting in a negative photoresponse.

Paired pulse facilitation (PPF) is key type of short-term synaptic plasticity characterized by a transient increase in the probability of vesicular release in response to consecutive stimuli. The PPF behaviors of the PbS QDs/IGZTO-based synaptic device have been demonstrated. As illustrated in Fig. 3(a), the ΔPSC triggered by the second spike is noticeably enhanced than that induced by the first spike. A PPF index is introduced to quantify EPSC and IPSC, defined as the ratio of the amplitude of the second postsynaptic current (A2) to that of the first postsynaptic current (A1)—A2/A1. As shown in Fig. 3(b), the PPF index gradually decreases with the increase of the time interval (Δt) between two stimuli under both the stimulation of UV (365 nm) and SWIR (1550 nm) light. The decay of the PPF index with Δt can be fitted by a double-exponential decay equation [8], where τ1 and τ2 are the rapid and slow relaxation times. Here the fitted τ1 and τ2 values for the EPSC/IPSC are 0.38 s/0.70 s (τ1)

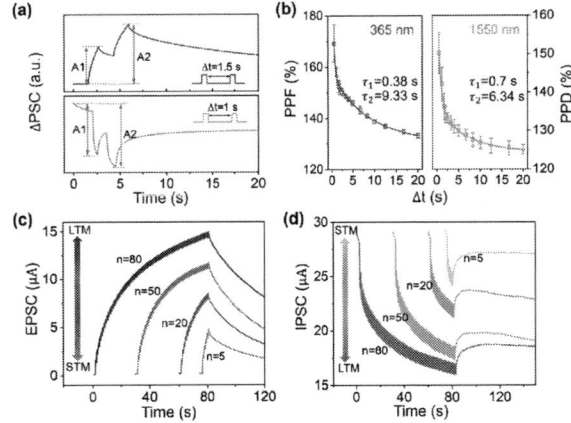

Fig. 3. (a) Changes in EPSC and IPSC induced by a pair of presynaptic light pulses with a pulse width of 1.5 s (365 nm) and 1 s (1550 nm). (b) The variation of PPF index with the interval of light pulse pairs. The STM-to-LTM transition induced by increasing the number of pulsed light stimuli under 365 nm (c) and 1550 nm (d) wavelength illumination.

979-8-3315-0417-5/25 $31.00 © 2025 IEEE

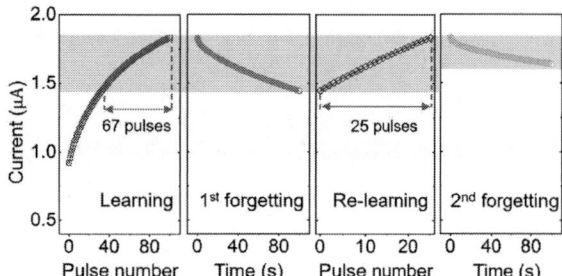

Fig. 4. Learning-experience behavior including learning, forgetting, and relearning processes with 0.5 s pulse duration and 0.5 s pulse interval. Light intensity is 1.56 mW cm⁻².

and 9.33 s/6.34 s (τ 2), respectively. In addition, applying pulsed light stimuli with varied number (n), various synaptic plasticity including short/long-term memory (STM/LTM), as well as STM-to-LTM transition have been well demonstrated in the PbS QDs/IGZTO heterojunction artificial synapse, as shown in Fig. 3(c), (d).

Meanwhile, the typical learning-experience behavior of human brains mimicked by the heterostructure-based synaptic device, which means that relearning the forgotten information always needs less time or training than the case of the first learning. As shown in Fig. 4, the first learning process was stimulated by 100 consecutive light pulses. Upon removal of light stimulation, the EPSC decayed spontaneously corresponding to the initial forgetting process, then the light pulses were reapplied to recover the potentiation. Interestingly, while the initial learning process required 67 pulses, only 25 pulses were needed to achieve the same level of potentiation during relearning. In contrast, the second forgetting process took longer than the first, suggesting an enhancement in potentiation during the relearning phase.

Conclusion

In summary, we developed an all-optical modulated artificial synapse by virtue of the PbS QDs/IGZTO heterojunction-based phototransistors. The heterostructure exhibited bidirectional synaptic weight modulation under UV and infrared light by depletion-region width modulation. Meanwhile, various functions of synapse such as PPF, STM, LTM, STM-to-LTM transition, and learning-experience behavior were simulated successfully. Such an all-optically controlled artificial synapse holds great potential for advancing future neuromorphic vision applications.

References

[1] P. A. Merolla et al., "A Million Spiking-neuron Integratedcircuit with a Scalable Communicationnetwork and Interface," *Science*, 345, 668, (2014).

[2] Y. He et al., "Indium−gallium−zinc−oxide Schottky Synaptic Transistors for Silent Synapse Conversion Emulation," *IEEE Electron Device Lett.*, 40, 139-142, (2018).

[3] S. M. Kwon et al., "Environment-Adaptable Artificial Visual Perception Behaviors Using a Light-Adjustable Optoelectronic Neuromorphic Device Array," *Adv. Mater.* 31, 1906433, (2019).

[4] F. Zhou et al., "Optoelectronic Resistive Random Access Memory for Neuromorphic Vision Sensors," *Nat. Nanotechnol.* 14, 776, (2019).

[5] G. A. Kerchner and R. A. Nicoll, "Silent Synapses and the Emergence of a Postsynaptic Mechanism for LTP," *Nature Rev. Neurosci.*, 9, 813–825, (2008).

[6] Y. Wang et al., "Room-temperature Fabrication of Flexible Oxide TFTs by Co-Sputtering of IGZO and ITO," *Flexible and Printed Electronics* 8, 035005, (2023).

[7] D. Li et al., "Schottky-Contact Hybrid Phototransistors with Bidirectional Photoresponses for Ultraviolet and Infrared Light Differentiating," *IEEE Electron Device Lett.* 43, 1515-1518, (2022).

[8] H. Tan et al., "Broadband Optoelectronic Synaptic Devices based on Silicon Nanocrystals for Neuromorphic Computing," *Nano Energy*, 52, 422-430, (2018).

Volumetric Super-Resolution Imaging based on Dual Bessel Beams STED Microscopy

Renlong Zhang[1], Haoxian Zhou[1], Chenguang Wang[2], Xiaoyu Weng[1], Liwei Liu[1], Peng Xi[3,*], Junle Qu[1,*]

[1] State Key Laboratory of Radio Frequency Heterogeneous Integration & Key Laboratory of Optoelectronic Devices and Systems, College of Physics and Optoelectronic Engineering, Shenzhen University, Shenzhen 518060, China, [2] State Key Laboratory of Integrated Optoelectronics, Key Laboratory of Advanced Gas Sensors of Jilin Province, College of Electronic Science and Engineering, Jilin University, Changchun, 130012, China, [3] Department of Biomedical Engineering, College of Future Technology, Peking University, Beijing 100871, China

* Correspondence should be addressed to P.X. (xipeng@pku.edu.cn) and J. Q(jlqu@szu.edu.cn).

Abstract

We developed a dual-Bessel beam STED (DB-STED) microscopy system for volumetric super-resolution imaging that surpasses optical diffraction limits and enhances high-throughput data acquisition. By precisely aligning a 0th order excitation Bessel beam with a 1st order depletion Bessel beam in both spatial and temporal dimensions, we achieved 69 nm lateral resolution across a 10 μm depth of field. This technique enables nanoscale visualization, significantly improving both volumetric imaging speed and resolution for advanced biomedical research.

Keywords: super-resolution; Volumetric imaging

Introduction

To gain deeper insights into dynamic nanoscale processes in organisms, optical microscopy with high spatial resolution and imaging speed in three dimensions (3D) is essential. Achieving both super-resolution and high-speed 3D imaging, however, remains a significant challenge.

Confocal and conventional multi-photon fluorescence microscopy are commonly used due to their ability to produce high-resolution, high-contrast 2D images. However, their 3D imaging capabilities are limited by the need for sequential section scanning, resulting in low volumetric imaging rates [1]. Recently, multi-photon microscopy with extended foci of Bessel beams has been developed to facilitate volumetric imaging, as demonstrated in zebrafish and mice models [2-4]. This advancement opens new opportunities for high-throughput imaging at greater speeds.

Despite this progress, these methods are inadequate for resolving fine cellular details, necessitating techniques that exceed the diffraction limit. Stimulated emission depletion (STED) microscopy is well-known for its ability to achieve nanometer-scale resolution[5, 6]. The integration of Bessel beam-based volumetric imaging with STED offers great potential for further improving both resolution and throughput.

In this study, we present a dual-Bessel beam STED (DB-STED) microscopy system that significantly enhances lateral resolution in volumetric imaging by integrating a 1st order Bessel beam for depletion with a 0th order excitation beam. Precise spatial alignment of the beams and a temporal delay of several hundred picoseconds between the hollow and solid Bessel beams are key to this enhancement [6, 7]. Our system enables high-throughput super-resolution volumetric imaging, demonstrating its ability to capture rapid nanoscale dynamics in 3D biological structures.

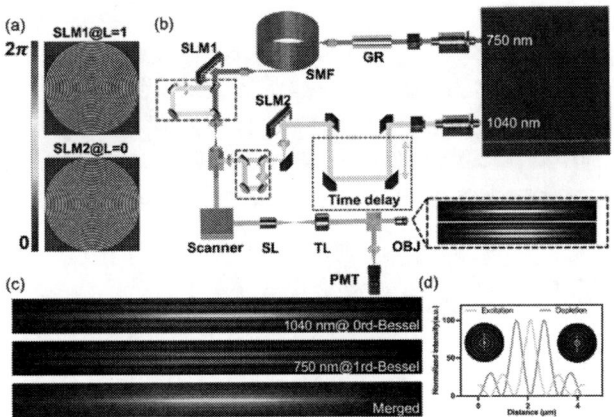

Figure 1. Schematic diagram of Dual Bessel Beam Volumetric Super-Resolution Microscopy .(a) the 1st axicon phase displayed by SLM1, the 0th axicon phase displayed by SLM2.(b) Simplified schematic of optical setup. SLM1 and SLM2 are programmed to generate 1st and 0th Bessel Beam, respectively, as shown in (a). The excitation and depletion light paths are enclosed in a dotted red box, where two flip mirrors are used to switch between Bessel beam and Gaussian beam. (c) Simulated convergence of the two Bessel beams in the x-z plane at the focal plane. The top section of panel (c) shows the solid zero-order Bessel beam from the 1040 nm TPEF light source, the middle section shows the hollow Bessel beam at 750 nm for depletion light, and the bottom section represents the schematic after their combination. (d) The lateral intensity profile of the two Bessel beams.

Volumetric super-resolution two-photon excited fluorescence (TPEF) microscopy based on dual Bessel beams

The DB-STED setup (Fig. 1(b)) is built on a custom two-photon STED microscope, integrating two spatial light modulators (SLMs) and a synchronized dual-output

979-8-3315-0417-5/25 $31.00 © 2025 IEEE

femtosecond laser. The depletion beam (750 nm) undergoes chirped dispersion through a glass rod and single-mode polarization-maintaining fiber, followed by modulation by SLM1 (red path, Fig. 1(b)), which displayed the axicon and 1st vortex phase combination pattern (Fig. 1(a)). This modulated light forms a sharp annular light through a Fourier lens. Simultaneously, SLM2 displayed the 0th order axicon phase pattern (green path, Fig. 1(b)) to modulate a 1040 nm femtosecond laser, also forming an annular light after passing through a lens. Both beams are collinearly aligned and directed into the microscope.

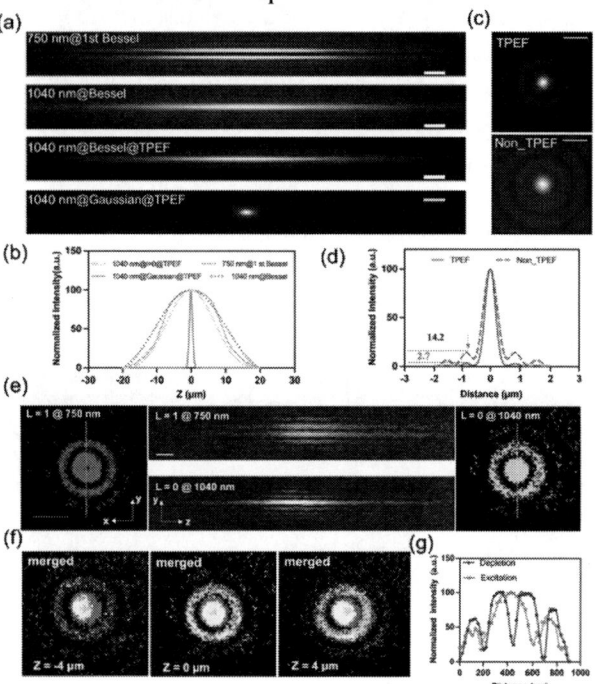

Figure 2. Simulated results and side-lobe energy analysis of Bessel foci. (a) Axial point spread function (PSF) of the 1st order depletion Bessel beam, single-photon and two-photon 0th order Bessel beams, and two-photon Gaussian beam, with axial full-width at half-maxima (FWHMs) of 18 μm, 20 μm, 13 μm, and 1.0 μm, respectively; scale bar = 2 μm. (b) Intensity profiles of axial PSFs from panel (a), showing the 750 nm depletion light covers the 1040 nm two-photon excitation range. (c) Lateral PSFs of 1040 nm non-TPEF and TPEF excitation Bessel beams. (d) Intensity profiles of lateral PSFs from panel (c), showing side-lobe intensities of 14.2% for non-TPEF and 2.7% for TPEF, compared to the main lobe; scale bar = 1 μm. (e) Lateral and axial scattering images of depletion and excitation beams using golden particles; scale bar = 500 nm and 2 μm in x-y and y-z views, respectively. (f) Merged results of excitation and depletion beams at z = -4 μm, 0 μm, and 4 μm. (g) Intensity profile along the dashed white line in panel (e).

Precise spatial and temporal alignment of the two Bessel beams is critical for optimal resolution. Spatial alignment

(both lateral and axial) is achieved by fine-tuning the reflective mirrors before the scanner (Fig. 1(c),(d)). Temporal synchronization is handled by a time overlap module with two reflective mirrors (Fig. 1(b)), mounted on a uniaxial linear displacement stage. Fine adjustments of the motorized stage, with a range of 20 cm, enable precise synchronization between the excitation (1040 nm) and depletion (750 nm) pulses, maximizing efficiency.

Principle of phase optimization DB-STED

To validate the DB-STED method, we constructed a tightly focused simulation model based on Richard & Wolf diffraction theory [8]. This model addresses the overlap between the depletion and excitation beams within the depth of field (DOF). By adjusting the axicon phase pattern parameters on both SLMs, we controlled the ring radius conjugated to the objective pupil, regulating the Bessel beam's DOF. As shown in Figure 2(a), the depletion beam's DOF slightly exceeds that of the two-photon excitation beam, ensuring efficient STED depletion within the axial excitation range. The Bessel beam's excitation length (13 μm) is approximately ten times greater than that of the Gaussian beam (Figure 2(b)), offering significantly higher throughput for volumetric super-resolution imaging. These findings were corroborated by gold particle scattering imaging, as shown in Figure 2(e), which confirmed precise spatial alignment of the laser beams.

To address the side lobe effects of Bessel beams on the signal-to-noise ratio, we compared side-lobe energy under TPEF and non-TPEF excitation. As shown in Figure 2(c), TPEF excitation significantly reduced side-lobe energy, with its intensity decreasing to 2.7% of the main lobe peak, compared to 14.2% for non-TPEF (Figures 2(f) and 2(g)). This demonstrates the efficiency of TPEF in suppressing side-lobe interference.

The performance analysis of DB-STED

To evaluate the performance of volumetric super-resolution imaging, we acquired 3D images (10 μm × 10 μm × 10 μm) of 50 nm fluorescent beads uniformly dispersed in agarose using DB-STED, Bessel TPEF (B-TPEF), and Gaussian foci TPEF microscopy (Fig. 3). The Gaussian foci TPEF image slices were acquired at 300 nm intervals, less than half the system's axial resolution of 860 nm, adhering to the Nyquist sampling principle. The slice stack spanned a range comparable to the axial length of the Bessel beam. Figure 3(a) illustrates the imaging process using Gaussian foci TPEF and DB-STED, with distinct pseudo-colors marking 2 μm intervals to highlight depth information (Fig. 3b).

Resolution measurements (Fig. 3c) yielded full width at half maximum (FWHM) values of 69 nm for DB-STED, 239 nm for B-TPEF, and 276 nm for Gaussian foci TPEF imaging. The annular illumination in Bessel beams sharpens the central peak, offering a modest resolution enhancement over Gaussian beams[9]. The extended DOF in Bessel beams also enhances information throughput, allowing DB-STED to capture the equivalent of 20 Gaussian foci TPEF frames in a single 10 μm × 10 μm × 10 μm volume, while maintaining super-resolution across the entire volume. Decorrelation analysis confirmed these results[10], showing resolutions of 74 nm for DB-STED, 263 nm for B-TPEF, and 314 nm for TPEF (Fig. 3d).

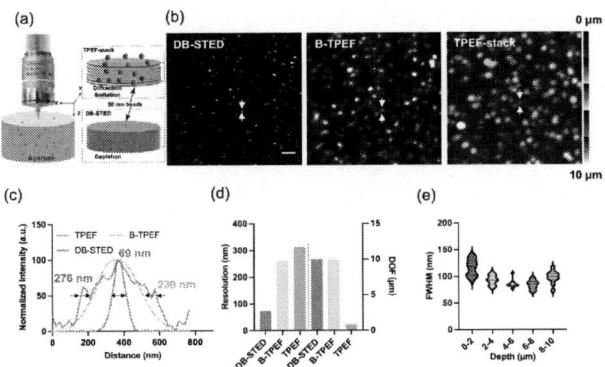

Figure 3. Volumetric super-resolution imaging of fluorescent beads encapsulated in agarose. (a) Scanning with Gaussian foci in the x-y plane captures a diffraction-limited optical section (red shaded region), while 2D scanning with Bessel foci in the x-y plane covers a 3D volume (blue shaded region), enabling super-resolution within the volume. (b) From left to right, the image sequence shows results from DB-STED, Bessel TPEF (B-TPEF), and Gaussian foci TPEF stacks; scale bar = 1 μm. Gaussian foci TPEF images are a composite of 20 slices with pseudocolors assigned at 2 μm intervals to indicate depth. All images were processed using Huygens deconvolution. (c) FWHM measurements of fluorescent beads marked in (a) show values of 69 nm for DB-STED, 239 nm for B-TPEF, and 276 nm for Gaussian foci TPEF. (d) The left section compares imaging resolutions using Decorrelation analysis. 74 nm for DB-STED, 263 nm for B-TPEF, and 314 nm for Gaussian foci TPEF. The right section compares the depth of field (DOF): 10 μm for DB-STED and B-TPEF, and 0.86 μm for Gaussian foci TPEF. (e) Statistical analysis of FWHM at different depths using DB-STED shows resolution improvements at depths between 2-8 μm, with comparatively lower resolution at 0-2 μm and 8-10 μm (n = 10).

We also analyzed the super-resolution performance at various axial positions (Fig. 3e), with the central portion of the Bessel beam achieving the greatest resolution improvement, averaging 83.9 nm. However, resolution varied slightly along the axial direction due to the uneven intensity distribution in the depletion Bessel beam, where reduced intensity at the edges diminished depletion efficiency, slightly compromising resolution at those positions.

Conclusion

In this study, we present a volumetric microscopy technique integrated with STED, enabling rapid acquisition of super-resolution structural details within a 10 μm axial range. This system, built on a custom two-photon STED microscope with dual Bessel beam generation, enhances imaging speed, reduces data size, and surpasses the diffraction limit, achieving a resolution of 69 nm. Compared to Gaussian beam foci TPEF, our approach accelerates volumetric imaging while minimizing photobleaching, particularly with the implementation of resonant-galvo scanning [11, 12]. With its high throughput and super-resolution capabilities, DB-STED is well-suited for fast imaging of biological processes, such as neuronal signal transmission and lipid droplets dynamics [13, 14], offering new insights into subcellular structures and dynamics. This technique holds great potential for advancing our understanding of complex biological phenomena.

Acknowledgements

This work has been partially supported by the National Key R&D Program of China (2021YFF0502900), National Natural Science Foundation of China (62127819/ 62375183/T2421003/62435011), Shenzhen Key Laboratory of Photonics and Biophotonics (ZDSYS20210623092006020), Shenzhen Science and Technology Program (JCYJ20220818100202005).

References

[1] Ji N., Freeman J., and Smith S. L. "Technologies for imaging neural activity in large volumes". Nature Neuroscience. 19 (9), 1154-1164. (2016)

[2] Chen W., Natan R. G., Yang Y., et al. "In vivo volumetric imaging of calcium and glutamate activity at synapses with high spatiotemporal resolution". Nature Communications. 12 (1), 6630. (2021)

[3] Chen W., Ge X., Zhang Q., et al. "High-throughput volumetric mapping of synaptic transmission". Nature Methods. 10.1038/s41592-024-02309-3. (2024)

[4] Lu R., Liang Y., Meng G., et al. "Rapid mesoscale volumetric imaging of neural activity with synaptic resolution". Nature Methods. 17 (3), 291-294. (2020)

[5] Dyba M., Jakobs S., and Hell S. W. "Immunofluorescence stimulated emission depletion microscopy". Nat Biotechnol. 21 (11), 1303-1304. (2003)

[6] Vicidomini G., Bianchini P., and Diaspro A. "STED super-resolved microscopy". Nature Methods. 15 (3), 173-182. (2018)

[7] Hell S. W. and Wichmann J. "Breaking the diffraction resolution limit by stimulated emission: stimulated-emission-depletion fluorescence microscopy". Opt. Lett. 19 (11), 780-782. (1994)

[8] Weng X., Song Q., Li X., et al. "Free-space creation of ultralong anti-diffracting beam with multiple energy oscillations adjusted using optical pen". Nat Commun. 9 (1), 5035. (2018)

[9] Welford W. T. "Use of Annular Apertures to Increase Focal Depth". J. Opt. Soc. Am. 50 (8), 749-753. (1960)

[10] Descloux A., Grußmayer K. S., and Radenovic A. "Parameter-free image resolution estimation based on decorrelation analysis". Nature Methods. 16 (9), 918-924. (2019)

[11] Velasco M. G. M., Zhang M., Antonello J., et al. "3D super-resolution deep-tissue imaging in living mice". Optica. 8 (4), 442-450. (2021)

[12] Ching-Roa V. D., Olson E. M., Ibrahim S. F., et al. "Ultrahigh-speed point scanning two-photon microscopy using high dynamic range silicon photomultipliers". Scientific Reports. 11 (1), 5248. (2021)

[13] Svoboda K., Denk W., Kleinfeld D., et al. "In vivo dendritic calcium dynamics in neocortical pyramidal neurons". Nature. 385 (6612), 161-165. (1997)

[14] Olzmann J. A. and Carvalho P. "Dynamics and functions of lipid droplets". Nature Reviews Molecular Cell Biology. 20 (3), 137-155. (2019)

Direct Evidence of Oxygen Vacancy Generation in Whole Gate Stacks Through Multiple Electrical and Atomic-Scale Physical Methods as the Cause of Endurance Failure in FeFETs

Xianzhou Shao[1,2,3], Hao Xu[1,2], Saifei Dai[1,2,3], Fengbin Tian[1,2,3], Xiaoyu Ke[1,2], Jiahui Duan[1,2,3], Min Liao[1,2,3], Xinpei Jia[1,2,3], Xiaoqing Sun[1,2], Junshuai Chai[1,2*], Jun Luo[1,2], Wenwu Wang[1,2], Xiaolei Wang[1,2*]

[1]Key Laboratory of Fabrication Technologies for Integrated Circuits, Chinese Academy of Sciences, Beijing, China
[2]Institute of Microelectronics, Chinese Academy of Sciences, Beijing, China
[3]School of Integrated Circuits, University of Chinese Academy of Sciences, Beijing, China

Abstract

We have experimentally confirmed that the oxygen vacancy (V_O) generation within whole gate stacks is the origin of the endurance failure in Si FeFET by multiple electrical methods and atomic-scale physical characterization techniques. Through gate leakage, low-frequency noise, and split I-V measurements, we found that defect generation occurs in whole gate stacks and electron-hole recombination during bipolar cycling plays a dominant role. Furthermore, we determined that the type of the generated defects is V_O based on scanning transmission electron microscopy (STEM) and electron energy-loss spectroscopy (EELS) characterizations.

Keywords: FeFET, Oxygen Vacancy and Endurance Failure

Introduction

Hafnium-based FeFET has emerged as a promising candidate for non-volatile memories (NVMs) due to the advantages of scalability, low power consumption, fast writing speed, and CMOS process compatibility [1]. However, limited endurance is still a bottleneck. The defect generation during endurance is considered the main cause [2]. However, the defect behaviors lack a comprehensive investigation including the defect type, the location of the defect, and the mechanism of defect generation as shown in Fig. 1.
In this study, we directly confirmed that V_O generation in whole gate stacks is the cause of endurance failure by multiple electrical methods and atomic-scale physical characterization techniques.

Device Fabrication and Measurements

Fig. 2 shows the structure and the fabrication process of Si FeFET and the detailed fabrication process can be found in other works [3]. Fig. 3 shows the waveforms and corresponding results. A wake-up process was applied before all measurements. The MW keeps stable before 10^3 cycles and decreases rapidly with cycling until it breaks down as shown in Fig. 3(c). The degradation of subthreshold swing (SS) during endurance in Fig. 3(b) and the significant increase of gate leakage at PGM and ERS states in Fig. 3(d) indicate the defect generation within gate stacks during endurance [2, 4].

Electrical characterization of defects

To investigate the defect evolution during endurance, we construct a gate leakage simulation framework [5] in Fig. 4. The results in Fig. 4(c) indicate that the defect generation occurs in SiO_x during endurance. In addition, we extract the defect density changes during endurance by the low-frequency noise method [6] as shown in Fig. 5(b). This shows the defects generated in the $Hf_{0.5}Zr_{0.5}O_2$ and SiO_x layers with cycling. To explore the causes of defect generation, we conduct the split I-V measurement as shown in Fig. 6. The current of the source/drain (S/D) reflects the electrons flowing by the Si substrate conduction band, while the current at body (B) reflects the holes flowing via the valance band [7]. The obvious open loop of charges in S/D (Q_{sd}) and B (Q_b) implies that electron-hole recombination occurs within gate stacks. Moreover, we measured the gate leakage changes at -4 V during endurance with unipolar (without recombination) and bipolar (with recombination) cycling pulse in Fig. 6(e). This indicates that the recombination process is the cause of the defect generation. In summary, we have concluded that the defect generation occurs in the whole gate stacks during endurance through multiple electrical methods and the electron-hole recombination induces the defect generation as shown in Fig. 7.

Atomic-scale physical characterization of defects

To further understand the defects' properties during endurance, we introduced STEM and EELS techniques for 10^0 cycle, 10^3 cycles, 10^4 cycles, and 10^5 cycles samples to atomic-scale characterization. Fig. 8 shows the results for the 10^0 cycle state sample near the $TiN/Hf_{0.5}Zr_{0.5}O_2$ interface. Fig. 8(c)-(e) shows the electron energy-loss near-edge structure (ELNES) of the Ti, N, and O at different positions (position 1 to position 13). We consider that the position where the O K-ELNES disappears is the bottom position of the TiN film (position 1), while the position where the Ti L-ELNES and N K-ELNES disappear is the top position of the $Hf_{0.5}Zr_{0.5}O_2$ film (position 13). Therefore, a 2 nm TiO_xN_y film appears. Similarly, we analyzed the properties between the $Hf_{0.5}Zr_{0.5}O_2$/p-Si interface as shown in Fig. 9(a)-(e). Thus, a 1.8 nm interlayer was found. To further confirm the chemical composition of this interlayer, we further analyze the O K-ELNES detailedly as shown in Fig. 9(e). For positions 26 and 30, there is no peak in the O K-ELNES of the SiO_x layer [8], while two split peaks appear in the O K-ELNES of the films containing Hf and Zr atoms at position 25. Thus, a 0.8 nm $Hf_xZr_ySi_zO_2$ layer and a 1 nm SiO_x layer are distinguished. Then, we investigate the variation of the interfacial layers during endurance as shown in Fig. 10(a)-(d). The thickness of the TiO_xN_y, $Hf_{0.5}Zr_{0.5}O_2$, and SiO_x layers remains almost unchanged. However, the thickness of the $Hf_xZr_ySi_zO_2$ layer increases with cycling. Moreover, the thickness variation of the SiO_x layer determined by STEM-HAADF images in Fig. 10(e) is consistent with EELS results in Fig. 10(d). For the information on the defects, we first study the defect distribution along in-plane and out-plane directions at 10^0 cycle state samples by the low energy loss spectra as shown in Fig. 11. The increase in peak B relative to peak A can reflect the increase in V_O concentration [9]. The results in Fig. 11(b) indicate that the V_O concentration decreases from the top region to the bottom region, while the results in Fig. 11(d) imply that the distribution of the V_O concentration is uniform along the in-plane direction. The V_O distribution of the $Hf_{0.5}Zr_{0.5}O_2$ layer is summarized in Fig. 11(e).
To investigate the V_O concentration changes of the $Hf_{0.5}Zr_{0.5}O_2$ layer during endurance, we study the low energy loss spectra of three positions (H1, H2, and H3) as shown in Fig. 12(a). The intensity of peak B increases with cycling shown in Fig. 12(b)-(d) indicating that the V_O generation occurs in the whole $Hf_{0.5}Zr_{0.5}O_2$ layer. We also study the V_O concentration in the $Hf_xZr_ySi_zO_2$ layer during endurance. The intensity of peak B significantly increases during endurance failure in Fig. 12(d). Although the intensity of peak B is also affected by Si atom concentration [10], the addition of Si atoms into $Hf_{0.5}Zr_{0.5}O_2$ favors the formation of Si-O chemical bonds due to the larger electronegativity of Si relative to both Hf and Zr [11], which can induce V_O generation. Thus, the V_O generation occurs in the $Hf_xZr_ySi_zO_2$ layer during endurance. For the defects in the SiO_x layer, the results in Fig. 4(c) have demonstrated that the defects in it increase with cycling. Moreover, our previous works have identified the type of this defect as V_O.
In summary, the V_O generation occurs throughout the entire gate stacks including $Hf_{0.5}Zr_{0.5}O_2$, $Hf_xZr_ySi_zO_2$, and SiO_x layers during endurance.

Conclusion

In this work, we directly determine that the V_O generation within the whole gate stacks is the origin of the endurance failure in Si FeFET combining multiple electrical methods with STEM and EELS techniques. A comprehensive endurance failure mechanism for Si FeFET is demonstrated in Figure 13.

References

[1] S. Dünkel et al., IEDM, 2017, pp.19.7.1-19.7.4. [2] N. B. Gong et al., EDL, 2018, pp. 15-18. [3] X. Shao et al., TED, 2023, pp. 3043-3050. [4] E. Yurchuk et al., IRPS, 2014, pp. 2E.5.1-2E.5.5. [5] F. Tian et al., TED, 2023, pp. 1040-1047. [6] J. Duan et al., TED, 2022, pp. 6547-6551. [7] X. Jia et al., TED, 2024, pp. 1845-1851. [8] D. A. Muller et al., Nature, pp. 758-761. [9] J. H. Jang et al., JAP, 2011, vol. 109, no. 2. [10] N. Ikarashi et al., JAP, 2003, pp. 480-486, 2003. [11] R. Shivaraman et al., Microscopy and Microanalysis, 2008, pp. 14-1.

Motivation: defect behaviors lack a comprehensive investigation

MFIS-FeFET fabrication

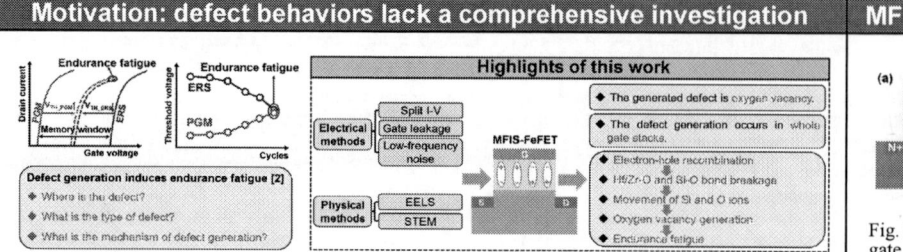

Fig. 1. The motivation of this work. We provide a comprehensive investigation of defect behaviors.

Fig. 2. (a) Schematic diagram of the gate stack structure and (b) fabrication process of the Si FeFET.

MW and gate leakage measurements: defect generation occurs during endurance

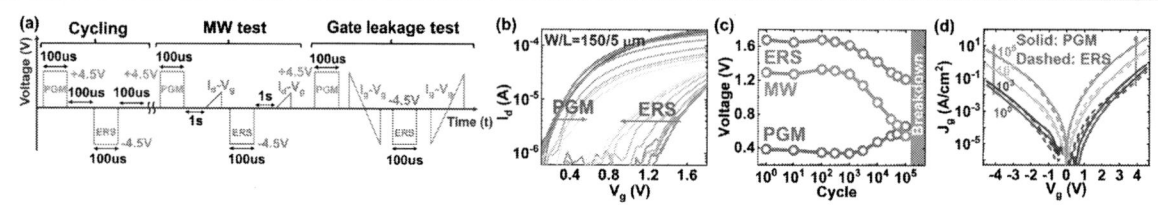

Fig. 3. (a) The waveforms used for cycling, MW test, and gate leakage test during endurance. (b) The I_d-V_g curves at PGM/ERS state during endurance. (c) The V_{th} and MW during endurance extracted by the constant current method (10 μA). (d) The gate leakage at PGM/ERS state during endurance.

Defect generation location and mechanism characterized by multiple electrical methods

Method 1: gate leakage measurement

Method 2: low-frequency noise measurement

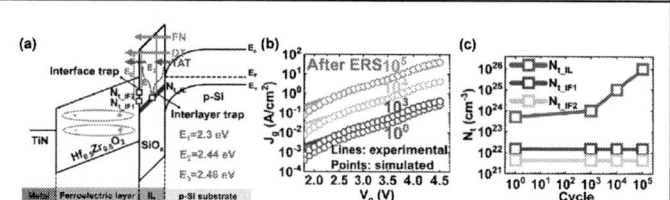

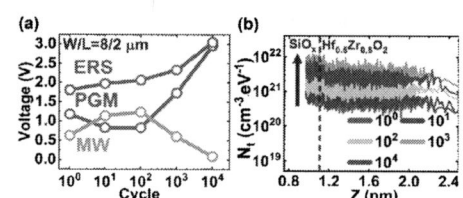

Fig. 4. (a) The band diagram for Si FeFET under positive gate bias. (b) The experimental and simulated gate leakage during endurance. (c) The extracted defect density changes at the FE/IL interface (N_{t_IF1} and N_{t_IF2}) and IL internal (N_{t_IL}) during endurance.

Fig. 5. (a) The V_{th} and MW changes during endurance. A +5 V/-5 V rectangular pulse with 50 μs width was used for endurance cycling. (b) The defect density changes within the $Hf_{0.5}Zr_{0.5}O_2$ and SiO_x layers during endurance.

Method 3: split I-V measurement

Conclusions summary

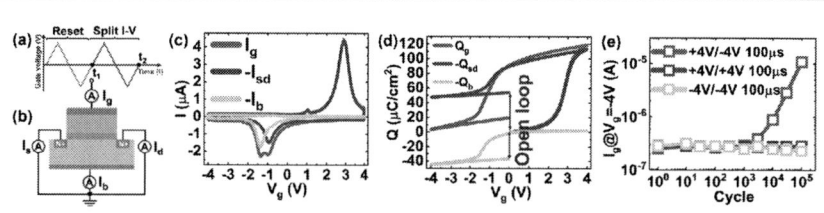

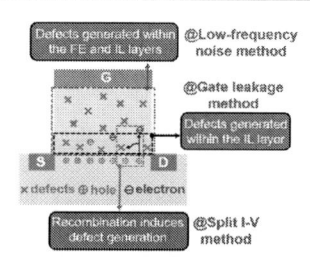

Fig. 6. (a) The waveforms of the split I-V measurement. A +4V/-4V triangular wave with a ramp rate of 10^3 V/s was used to reset the device at the ERS state. (b) The test configuration of the device terminals. (c) The current of four terminals during measurement. (d) The charges during measurement. (e) The gate leakage at V_g = -4 V during endurance with unipolar/bipolar cycling pulse.

Fig. 7. The summary of the electrical characterization methods results.

Defects information characterized by STEM-HAADF and EELS atomic-scale techniques

A TiN$_x$O$_y$ layer exists between the TiN and Hf$_{0.5}$Zr$_{0.5}$O$_2$ layers

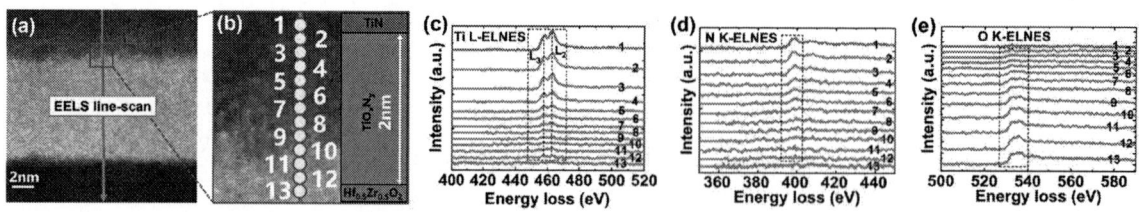

Fig. 8. (a) The STEM-HAADF image and EELS line-scan path. The enlarged STEM-HAADF image (b) the Ti L-ELNES (c), the N K-NLNES (d), and the O K-ELNES (e) near TiN/ Hf$_{0.5}$Zr$_{0.5}$O$_2$ interface.

A Hf$_x$Zr$_y$Si$_z$O$_2$ layer exists between the Hf$_{0.5}$Zr$_{0.5}$O$_2$ and SiO$_x$ layers

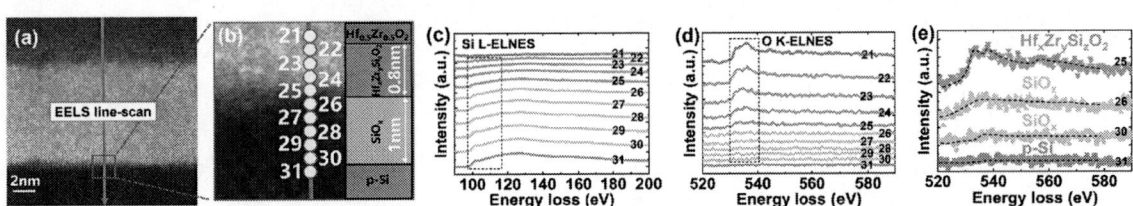

Fig. 9. (a) The STEM-HAADF image and EELS line-scan path. The enlarged STEM-HAADF image (b) the Si L-ELNES (c), and the O K-NLNES (d) near Hf$_{0.5}$Zr$_{0.5}$O$_2$/SiO$_x$ interface. (e) The O K-NLNES at positions 25, 26, 30, and 31. The black dashed line is a guideline for sight in (e).

The TiO$_x$N$_y$, Hf$_{0.5}$Zr$_{0.5}$O$_2$, and SiO$_x$ thicknesses remain unchanged, while the Hf$_x$Zr$_y$Si$_z$O$_2$ thickness increases with cycling

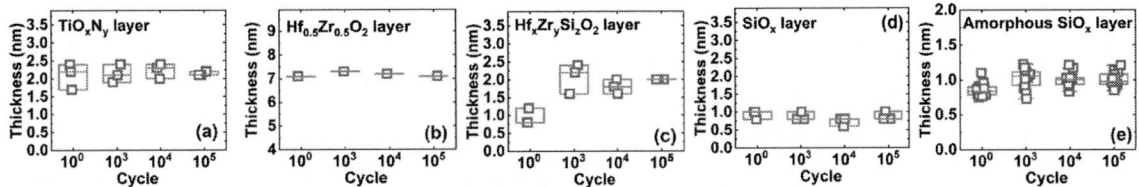

Fig. 10. The changes in the thickness of the (a) TiO$_x$N$_y$ layer, (b) Hf$_{0.5}$Zr$_{0.5}$O$_2$ layer, (c) Hf$_x$Zr$_y$Si$_z$O$_2$ layer, and (d) SiO$_x$ layer during endurance obtained by EELS line-scan measurement. (e) The amorphous SiO$_x$ thickness distribution during endurance obtained by STEM-HAADF images.

Confirmation of Hf$_{0.5}$Zr$_{0.5}$O$_2$ layer defect as Vo and the distribution of the V$_O$ concentration of the 10^0 cycle sample

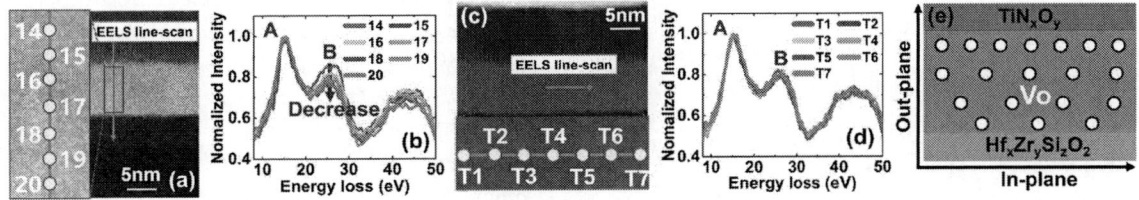

Fig. 11. (a) The STEM-HAADF images and the EELS line-scan path along the (a) out-plane and (c) in-plane directions of the Hf$_{0.5}$Zr$_{0.5}$O$_2$ layer. 7 positions are selected at intervals of 1 nm. The low energy loss spectra of 7 positions along the (b) out-plane and (d) in-plane directions. (e) Schematic diagram of the V$_O$ distribution inside the Hf$_{0.5}$Zr$_{0.5}$O$_2$ layer.

The V$_O$ generation within Hf$_{0.5}$Zr$_{0.5}$O$_2$ and Hf$_x$Zr$_y$Si$_z$O$_2$ layers during endurance

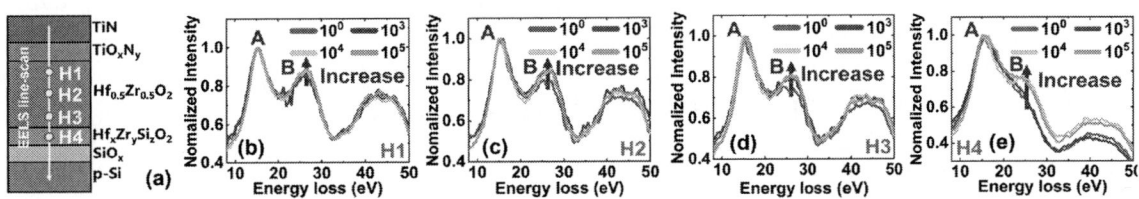

Fig. 12. (a) Schematic diagram of the gate stack and the EELS line-scan path along the out-plane direction. The EELS of the Hf$_{0.5}$Zr$_{0.5}$O$_2$ layer of the position (a) H1 near the top interface, (b) H2 at the middle, and (d) H3 near the bottom interface of the Hf$_{0.5}$Zr$_{0.5}$O$_2$ layer for four state samples. (e) The low energy loss spectra of four samples at the Hf$_x$Zr$_y$Si$_z$O$_2$ layer (H4).

Endurance failure mechanism of Si FeFET

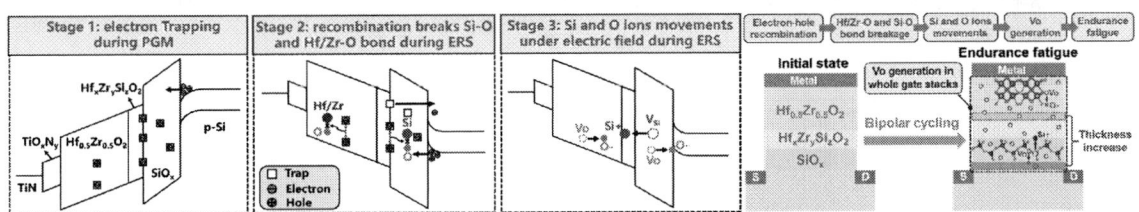

Fig. 13. Schematic diagram of the Vo generation process within whole gate stacks. Electron trapping occurs during PGM operation. The energy released by the electron-hole recombination process breaks the Si-O and Hf/Zr-O bonds during ERS operation. The Hf/Zr and O ions move towards the SiO$_x$ layer and Si substrate. The above processes induce the V$_O$ generated in whole gate stacks of Si FeFET. Finally, the endurance fatigue occurs.

979-8-3315-0417-5/25 $31.00 © 2025 IEEE

Selective-Attention Neuromorphic Vision Classifications Based on α-In₂Se₃/Bi₂O₂Se Ferroelectric-Semiconductor Heterotransistors

Xinrui Guo[1], Yu Zhu[1], Shuo Liu[1], Junling Liu[1], Ru Huang[1], and Ming He[1,2,*]

[1]School of Integrated Circuits, Beijing Advanced Innovation Center for Integrated Circuits, Peking University, Beijing 100871, China; [2]Frontiers Science Center for Nano-optoelectronics, Peking University, Beijing 100871, China. *E-mail: minghe@pku.edu.cn

Abstract

Herein, we present multifunctional neuromorphic vision device based on the α-In₂Se₃/Bi₂O₂Se ferroelectric-semiconductor heterojunction. Central to this innovation is the in-plane voltage that drives out-of-plane polarization coupling within the ferroelectric heterojunction, which facilitates non-volatile conductance alterations through modulating the interface charge configuration with the robust stability demonstrated over 10^4 cycles. Furthermore, the device exploits a photo-induced ferroelectric polarization reversal mechanism, resulting in a non-volatile light response that persists for over 10^3 s, accompanied with an impressive photoresponsivity of up to 10^5 A/W in the broad-spectrum range of 400-980 nm. We demonstrate the in-sensor pattern classification by the optoelectronic-ferroelectric driven spectrum-selective attention.

(Keywords: ferroelectric heterojunction; selective attention; pattern classification)

Introduction

Selective attention enables the efficient extraction of noteworthy information in biological systems by enhancing prominent features while suppressing non-salient ones, thereby improving the parallel processing of sensory data [1,2]. Conventional selective attention hardware often separates perception and processing units, resulting in constraints on computational power and efficiency [3,4]. Neuromorphic optoelectronic devices, inspired by biological mechanisms for integrated visual recognition, generally require continuous gate voltage regulation and demonstrate restricted memory dynamic ranges for accommodating varied visual inputs [5]. However, the critical function of selective attention for optical information remains formidable challenges for advancing visual perception systems.

In this work, we report multifunctional neuromorphic visual devices that incorporates selective attention using the α-In₂Se₃/Bi₂O₂Se ferroelectric-semiconductor heterostructure phototransistor, which enables photodetection across a broad spectral range of 400 – 980 nm with remarkable photoresponsivity of 1.5×10^5 A/W. More importantly, the phototransistor exploits the in-plane and out-of-plane ferroelectric polarization coupling, whereby the application of an in-plane voltage across the

vertical heterojunction modulates the interface band alignment, leading to change the non-volatile conductance with an endurance exceeding 10^4 cycles and a photo-memory retention surpassing 10^3 s. By employing the wavelength-selective photoresponse under positive ferroelectric polarization, we demonstrate in-sensor full-spectrum image classification capabilities.

Device Fabrication and Characterization

Fig. 1 outlines the preparation of the ferroelectric heterotransistor. CVD-grown Bi₂O₂Se nanosheet was transferred using a polymethyl methacrylate-assisted wet transfer technique. Mechanically exfoliated α-In₂Se₃ was transferred and aligned with the Bi₂O₂Se nanosheet. Electron beam lithography was then utilized to delineate the channel region, where Ti/Au electrodes (5/45 nm) were evaporated, followed by a standard lift-off procedure [5,6]. Selected area electron diffraction confirms the single-crystal nature and tetragonal structure of the synthesized Bi₂O₂Se nanosheet (**Fig. 2a**). X-ray diffractions for both Bi₂O₂Se and α-In₂Se₃ reveal sharp and intense crystallization peaks (**Fig. 2b**). Complementary Raman spectroscopy displays the A_{1g} mode at 159 cm⁻¹ for Bi₂O₂Se and the E_g mode at 89 cm⁻¹ alongside the A(LO+TO), A(TO), and A(LO) phonon modes at 104 cm⁻¹, 182 cm⁻¹, and 193 cm⁻¹, respectively, confirming the high-quality α-In₂Se₃/Bi₂O₂Se heterojunction (**Fig. 2c**).

Results and Discussion

Fig. 3a presents the output curves of the α-In₂Se₃/Bi₂O₂Se ferroelectric-semiconductor heterotransistor in its initial state, prior to ferroelectric polarization, as well as under positive and negative polarization conditions. The polarization state of α-In₂Se₃ can be modulated through various source-drain voltage (V_{ds}) pulse forms, allowing for precise control over the conductance of the heterojunction. **Fig. 3b** illustrates the non-volatility in conductance reduction of the heterotransistor as simulated by negative V_{ds} pulses of varying widths. It is evident that an increased pulse width leads to a greater reversal of ferroelectric domains, thereby producing more significant variations in current (**Fig. 3c**). Furthermore, **Fig. 3d** confirms that the heterotransistor maintains a stable and extensive dynamic range of conductivity states over 10^3 cycles of alternating $\pm V_{ds}$ pulses, thereby demonstrating its

979-8-3315-0417-5/25 $31.00 © 2025 IEEE

endurance. Notably, the ferroelectric operating voltage of the heterotransistor is 3 V, which facilitates the development of low-power devices. **Figs. 4a-c** illustrate the source-drain current (I_{ds}) modulated by blue light pulses (i.e., 405 nm) varying with the power density, the pulse width, and frequency, respectively. The photo-induced ferroelectric polarization reversal facilitates the transition of conductance from the short-term memory (STP) to the long-term memory (LTP). **Fig. 4d** depicts the dependence of photoresponsivity and specific detectivity on the light density at 405 nm with a V_{ds} of 0.5 V. The heterotransistor achieves the highest photoresponsivity of 1.5×10^5 A/W, largely surpassing the maximum photoresponsivity of 68.8 A/W observed in the α-In$_2$Se$_3$ device.

Fig. 5a compares the photoresponsivity of α-In$_2$Se$_3$ phototransistor and α-In$_2$Se$_3$/Bi$_2$O$_2$Se heterotransistor as a function of the light wavelength. The heterojunction band alignment significantly enhances the photoresponsivity across the broad spectrum of 400-980 nm by over three orders of magnitude. **Fig. 5b** depicts the time-dependent behavior of I_{ds} in the heterotransistor across three different states under the multi-pulse illumination at 405 nm. In the positive ferroelectric polarization state, substantial photo-induced polarization reversal occurs, leading to the enhanced non-volatile optical memory performance. **Fig. 5c** illustrates the response of the heterotransistor under the positive polarization across varied wavelengths of 405, 532, and 640 nm, respectively. It is noteworthy that the heterotransistor maintains a non-volatile optical memory lasting up to 10^3 s under the blue light of 405 nm. This notable duration is attributed to the high-energy photons generating a larger number of photogenic carriers, which facilitate the rearrangement of fixed polarized charge within the ferroelectric semiconductors, resulting in the robust non-volatile photoresponse. The current gain relationship with various V_{ds} and light pulses is depicted in **Fig. 5d**, further elucidating the dynamic capabilities of the ferroelectric-semiconductor heterotransistor.

We demonstrate an artificial visual perception array capable of image attention processing through the selective memory of α-In$_2$Se$_3$/Bi$_2$O$_2$Se ferroelectric-semiconductor heterotransistors (**Fig. 6a**). The device response to varying light wavelengths allows for both volatile and non-volatile behavior, facilitating the in-situ extraction of short-wavelength signals from complex environments. As depicted in **Fig. 6b**, a composite image featuring a blue bird nestled among green leaves is illuminated by three monochromatic lights (i.e., 405 nm, 532 nm, and 640 nm), resulting in an amplified signal for the blue bird while effectively suppressing the surrounding foliage. After a brief exposure, the features of the green leaves are entirely obscured, enabling the clear extraction of the "blue bird" (**Fig. 6c**).

Moreover, the α-In$_2$Se$_3$/Bi$_2$O$_2$Se heterotransistor is compatible with a perceptron network for image classifications by utilizing a dataset comprised of four distinct character patterns ("P," "K," "U," and "I"). Each image incorporates signals across various wavelengths (i.e., 405 nm, 532 nm, and 640 nm). Under positive polarizations, the conductance distribution reveals a marked responsiveness to short-wavelength wavelengths (**Fig. 6d**). The input patterns are encoded into conductance values, guiding the neural network to adjust V_{ds} accordingly. Upon completion of training, the output current distributions for different integrated patterns within the neural network reveal that each character corresponds to a distinct range of current distributions (**Fig. 6e**). With minimal overlap among the patterns, accurate identification based on output current is achieved, resulting in the classification accuracy of 96%, implying the selective-attention processing capability of α-In$_2$Se$_3$/Bi$_2$O$_2$Se heterotransistors for effective in-situ pattern recognition. **Table 1** provides a benchmark of characteristics of varied devices, emphasizing the promise of selective attention of α-In$_2$Se$_3$/Bi$_2$O$_2$Se heterotransistors in visual applications.

Conclusion

In conclusion, the α-In$_2$Se$_3$/Bi$_2$O$_2$Se ferroelectric-semiconductor heterotransistor demonstrates stable non-volatile features coupled with the robust photoresponse across broad spectrum. The photo-induced ferroelectric polarization inversion within the heterotransistor enables wavelength-selective optical memory responses. This integration offers a comprehensive device for perception, memory, and computing, specifically tailored for accurate pattern recognitions through the selective attention.

Acknowledgments

This work was supported by the National Key R&D Program of China 2022YFB4400100, the Natural Science Foundation of China (92164205, 62074004, 61927901), and 111 Project (B18001).

References

[1] Serences, J.T., et al., "Selective Visual Attention and Perceptual Coherence", Trends. Cogn. Sci., 10, 38-45 (2006).
[2] Chen, Y., et al. "All Two-Dimensional Integration-Type Optoelectronic Synapse Mimicking Visual Attention Mechanism for Multi-Target Recognition", Adv. Func. Mater., 33, 2209781 (2023).
[3] Bian, J., et al., "Neuromorphic Computing: Devices, Hardware, and System Application Facilitated by Two-Dimensional Materials", Appl. Phys. Rev., 8, 041313 (2021).
[4] Zhang, T., et al. "Improving the Efficiency of CMOS Image Sensors Through In-sensor Selective Attention", 2023 IEEE International Symposium on Circuits and Systems (ISCAS), Monterey, 1-4, (2023).
[5] He, M., et al., "Sub-10mK-Resolution Thermal-Bolometric Integrated FET-Type Sensors Based on Layered Bi$_2$O$_2$Se Semiconductor Nanosheets", 2020 International Electron Devices Meeting (IEDM), 26.1.1-26.1.4 (2020).
[6] He, M. et al. "Ultrasensitive Retinomorphic Dim-Light Vision with In-Sensor Convolutional Processing Based on Reconfigurable Perovskite-Bi$_2$O$_2$Se Heterotransistors", 2023 International Electron Devices Meeting (IEDM), 33.3.1-33.3.4 (2023).

Figure 1. Schematic illustration of manufacturing the α-In₂Se₃/Bi₂O₂Se ferroelectric-semiconductor heterotransistor.

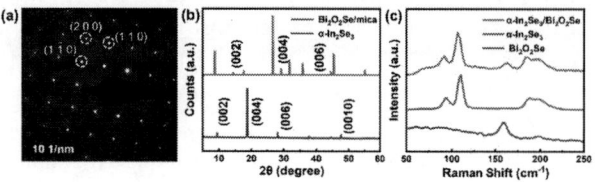

Figure 2. (a) Selected area electron diffraction of Bi₂O₂Se; (b) X-ray diffractions of Bi₂O₂Se and α-In₂Se₃; (c) Raman spectrums of Bi₂O₂Se, α-In₂Se₃, and the α-In₂Se₃/Bi₂O₂Se heterojunction.

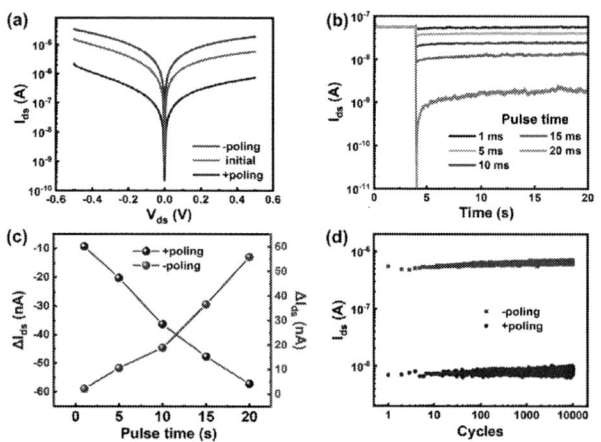

Figure 3. (a) I_{ds}-V_{ds} curves of α-In₂Se₃/Bi₂O₂Se heterotransistor under varied poling states; (b) I_{ds} of the heterotransistor regulated by electrical pulse signals at V_{ds} =-3 V; (c) effect of the pulse width on the height of I_{ds} under $\pm V_{ds}$ pulses; (d) cycling endurance test of α-In₂Se₃/Bi₂O₂Se heterotransistor.

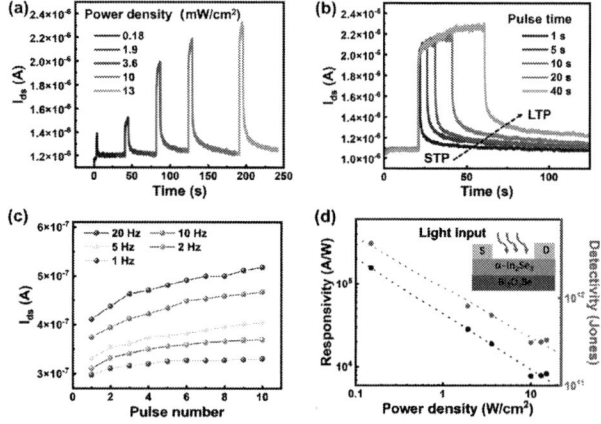

Figure 4. (a-c) I_{ds} of α-In₂Se₃/Bi₂O₂Se heterotransistor regulated by optical pulses under different (a) laser power density, (b) pulse width, and (c) pulse frequency; (d) laser power-density dependence of the photoresponsivity and the specific detectivity under the light illumination of 405 nm at V_{ds}=0.5 V.

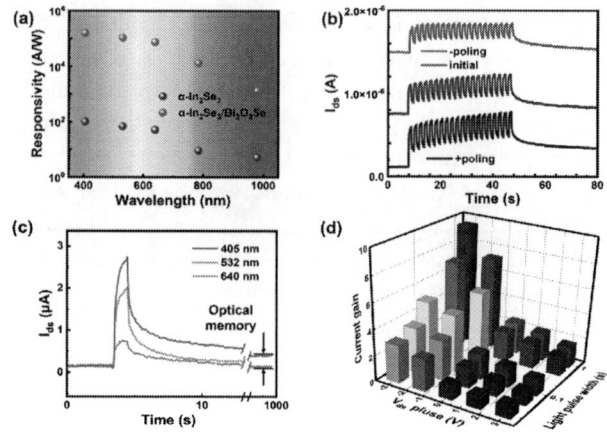

Figure 5. (a) Photoresponsivity of α-In₂Se₃ and α-In₂Se₃/Bi₂O₂Se heterotransistor as a function of the incident light wavelength; (b) I_{ds} of α-In₂Se₃/Bi₂O₂Se heterotransistor as a function of time under varied poling states with multi-pulse illuminations; (c) photoresponse of the heterotransistor to 405, 532, 640 nm under the positive poling state; (d) current gain after time period as function of 405 nm light pulse width and polarized voltage.

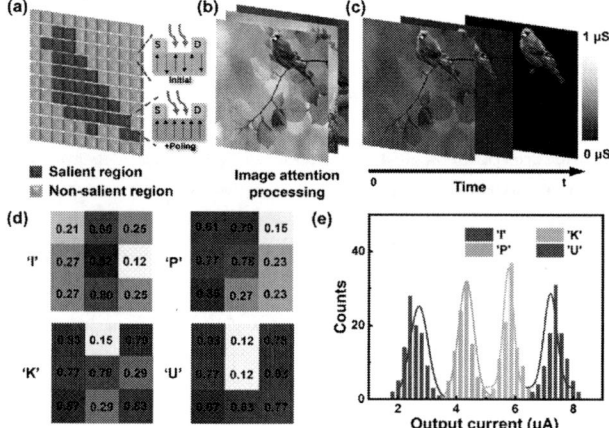

Figure 6. (a) Schematic illustration of α-In₂Se₃/Bi₂O₂Se based selective-attention image processing; (b) processing images with varied wavelengths of 405 nm, 532 nm, and 640 nm; (c) attention processing with the selective optical memory; (d) detected patterns with varied wavelengths and encoding conductance of 'I', 'P', 'K', 'U' patterns; (e) distribution of the output current in the 'I', 'P', 'K', 'U' patterns.

Table 1. Benchmark of the α-In₂Se₃/Bi₂O₂Se ferroelectric-semiconductor heterotransistor with other devices for selective-attention processing.

Materials	R (A/W)	Program voltage (V)	Optical memory (s)	Wavelength selectivity (nm)	Reference
α-In₂Se₃	3.43	10	>250	/	Adv. Funct. Mater., 2023, 34, 2306486
α-In₂Se₃/ BP	0.16	10	<1	/	Adv. Sci., 2023, 10, 2205813
MoS₂/CIPS	10	15	>10²	/	Adv. Mater., 2024, 36, 2403785
PₓOₓ/BP	3000	/	>10⁴	280-365	Adv. Mater., 2021, 33, 2004207
In₂S₃	473.6	/	>10³	359-671	Adv. Funct. Mater., 2024, 2407746
α-In₂Se₃/ Bi₂O₂Se	1.5× 10⁶	3	>10³	400-980	This work

979-8-3315-0417-5/25 $31.00 © 2025 IEEE   272

Comprehensive Investigation of the Disturb and Retention Issues in Scaled FeNAND Arrays

Yuejia Zhou[1], Ru Huang[1,2], Kechao Tang[1,2]*

[1] School of Integrated Circuits, Peking University, Beijing 100871, China. [2]Beijing Advanced Innovation Center for Integrated Circuits, Beijing 100871, China (*Email: tkch@pku.edu.cn).

Abstract

HfO_2-based ferroelectric field-effect transistor (FeFET) is a viable solution for next-generation 3D NAND technology. This work comprehensively investigates the array-level reliability of scaled FeNAND arrays, featuring the first exploration of the significant impact of adjacent cell interactions on both disturb and retention issues. Array-level disturb-free operation in 3 bits/cell storage is achieved via bias optimization. New insights into the mechanisms underlying cell-to-cell disturb and retention degradation are provided, highlighting the critical role of electric field and charge coupling between adjacent cells.

Keywords: FeNAND array, disturb issues, retention degradation.

Introduction

The prevailing charge-trap-flash (CTF)-based 3D NAND encounters significant obstacles in vertical pitch scaling, primarily constrained by cell-to-cell interference and the physical thickness of spacer oxides between adjacent cells [1]. To this end, the ferroelectric field-effect transistor (FeFET) based NAND (FeNAND) has emerged as a promising candidate for next-generation high-density memory technology [2]. By minimizing charge migration and operating voltage, the vertical pitch of 3D FeNAND can be further scaled, thereby enhancing memory density beyond the limit of NAND Flash [3]. In addition, recent advances in FeFETs with metal-insulator-ferroelectric-insulator-semiconductor (MIFIS) gate stack expands the MW to more than 15 V [4], primarily due to trapped charges in the gate interlayer (IL). Previous studies have predominantly focused on optimizing the memory performance and reliability of individual device. However, a comprehensive investigation into the reliability of FeNAND from array-level perspective, particularly regarding the disturb immunity and retention characteristics, is so far lacking. Notably, as FeNAND arrays are highly scaled, the increasing interaction between adjacent cells may lead to additional reliability issues. This FeFET-cell coupling involves new physical mechanisms and may be a bottleneck for FeNAND scaling, thus requiring in-depth investigation.

In this work, a scaled FeNAND array is fabricated based on the MIFIS FeFET, achieving stable 3 bits/cell (TLC) storage. The primary disturb issues, including V_{pass}, program and cell-to-cell (C2C) disturb, were experimentally examined. Margins for pass voltage (V_{pass}) and inhibit voltage ($V_{inhibit}$) are determined, enabling disturb-free operation. Device simulation provides physical insight into C2C disturb. Through systematic and in-depth characterization of the retention properties of the FeNAND array, a new understanding of the mechanisms behind retention degradation is obtained.

Fabrication and Property of the Scaled FeNAND Array

8×8 FeNAND arrays based on the MIFIS gate stack FeFET were experimentally fabricated, as shown in the SEM image of Fig.1 (a). Arrays with various gate length (L_g) and iWL down to 120 nm are prepared to evaluate the array-level reliability (Fig.1 (b)). Fig. 2 (a) and (b) depict the schematic and process flow of the MIFIS gate stack FeFET. The FeNAND arrays are fabricated on SOI substrate with junction-less structure, to fully reproduce the key features of 3D FeNAND technology. In addition to the two Al_2O_3 layers serving as gate and channel ILs, another thin Al_2O_3 layer is inserted within $Hf_{0.5}Zr_{0.5}O_2$ (HZO) FE layer to minimize the device-to-device (D2D) variation, as described in our previous work [5]. After sputtering TiN electrode, E-beam lithography is employed for gate patterning to shrink the L_g and iWL down to around 100 nm. The high-resolution TEM (HRTEM) image in Fig.2 (c) reveals excellent crystallization of the FE layer and well-defined interfaces in the gate stack, with thickness as designed. Fig. (3) presents the read I_d-V_g curves of the devices in the FeNAND array across various storage states, with $I_{on}/I_{off} > 10^6$. By extracting the threshold voltage (V_{th}) at a fixed current, a MW of about 4 V and stable 3 bits/cell storage are achieved.

Investigation of Disturb Issues in the Scaled FeNAND Array

The programming scheme of the FeNAND array, including three types of disturb issues affecting unselected cells, is illustrated in Fig. 4. Unlike full-block erasing, programming a specific cell requires the application of program voltage (V_{PGM}) to the selected word line (WL) and and V_{pass} to unselected WLs to access the target cell. Meanwhile, the selected bit line (BL) is grounded and $V_{inhibit}$ is applied to the unselected BLs to mitigate program disturb.

The V_{pass} disturb on cells along unselected WLs was first examined. The program efficiency (defined as the MW normalized to fully programmed state) of the target cell and the V_{th} shift of the pass cells as a function of V_{pass} is presented in Fig. 5 (a). A V_{pass} margin of -4V to -2.5V is determined, within which target cell can be efficiently programmed without disturbing pass cells. The cumulative effect of V_{pass} stress is investigated in Fig. 5 (b), indicating little V_{th} disturb after 10^6 cycles at V_{pass} = -3 V. For disturbed cells sharing the same WL as target cell, Fig. 6 shows the measured V_{th} as a

979-8-3315-0417-5/25 $31.00 © 2025 IEEE

function of $V_{inhibit}$ on unselected BLs. The program disturb is successfully avoided with $V_{inhibit} < -4.5V$ even under the worst case with a V_{PGM} of -12 V.

Subsequently, C2C disturb were systematically analyzed across four distinct data modes (PPP, EPE, EEE and PEP), with results shown in Fig. 7. Programming a target cell induces slight V_{th} shift in the neighboring cells, which is more pronounced for smaller iWL and higher V_{write}. In the worst case, V_{th} shift remains < 0.15 V (Fig. 7(a)). A COMSOL simulation of the FeNAND provides an in-depth physical perspective on C2C disturb, as shown in Fig.7 (b). The applied bias between adjacent cells induces coupled electric field, which is larger for smaller iWL. Based on the simulation, the schematic in Fig. 7 (c) illustrates the mechanism of C2C disturb. The coupled E-field includes a vertical component relative to the channel, causing partial reverse polarization flipping in PPP (or EEE) mode, while resulting in only slight disturb in PEP (or EPE) mode. This C2C disturb grows with decreasing iWL due to stronger coupled E-field.

After considering and addressing the disturbs, state map verification using the writing scheme with optimal V_{pass} and $V_{inhibit}$ is presented in Fig. 8. Negligible bit errors are detected in the high reliability TLC FeNAND array.

Investigation of Retention Degradation in the Scaled FeNAND Array

The retention properties of individual FeFET device are first measured at 85°C for up to 3000 s, with results shown in Fig. 9 (a). The FeFET exhibits distinct retention degradation behaviors across different storage states, where lower storage states (with lower V_{th}) demonstrate more pronounced retention instability. For quantitative comparison, the V_{th} shifts after 3000 s compared to the pristine state for 8 storage states are extracted in Fig. 9 (b). The V_{th} shift increases as lowering storage states, with the lowest P0 state exhibiting the largest shift of 0.32 V. This observation aligns with previous study [6], where the polarization in erased state is unstable and degrades more severely over time.

Subsequently, the retention degradation behaviors of the FeNAND arrays across various data modes are comprehensively investigated in Fig. 10. For the target cell in storage state P7, adjacent cells were set to different storage states corresponding to various data modes (PPP, EPE, and NPN), as characterized in Fig. 10(a). The retention behavior in PPP mode is more stable compared to EPE and NPN mode. Additionally, the inset plots the V_{th} shifts of target cells after 3000 s, indicating that V_{th} shift increases as the storage state of adjacent cells is lowered. When the target cell is in storage state P0, the V_{th} shift increases as the storage state of adjacent cells is raised (Fig.10 (b)). Further analysis of V_{th} shifts with respect to iWL across different data modes is provided in Fig. 11. The V_{th} shift increases with reduced iWL, particularly in EPE and PEP modes, where the V_{th} shifts reach 0.32 V and

0.59 V at iWL of 120 nm, respectively, suggesting more severe retention degradation.

The schematic in Fig. 12 summarizes the underlying physical mechanism for retention degradation in FeNAND array. Compared to the programmed state, the erased state features a more unstable polarization due to fewer positive trapped charges in the gate IL [7]. This instability and the vertical migration of the trapped charges lead to more severe retention degradation. For various data patterns, the different storage states between adjacent cells introduce a lateral electric field, causing the lateral migration of the trapped charge within gate IL of the cell out into the WL spacing, resulting in the V_{th} shift. This lateral migration is more pronounced as iWL shrinks, further degrading the retention.

Conclusion

This study presents a comprehensive investigation into the disturb and retention issues of experimentally prepared FeNAND arrays. Array-level disturb-free operation is successfully achieved in the TLC storage. Systematic characterization and simulation of scaled FeNAND arrays elucidate the underlying mechanisms, indicating the electric field and charge coupling between adjacent cells strongly affects disturb and retention. This work provides a new insight for further enhancing the reliability of FeNAND arrays.

Acknowledgments

This work was supported by National Key R&D Program of China (2022YFB4400300), NSFC (62274003, 61927901 and 92164203),111 Project (B18001) and National Micro/Nano Fabrication Laboratory of Peking University.

References

[1] A. Goda, "3-D NAND Technology Achievements and Future Scaling Perspectives," IEEE Transactions on Electron Devices, vol. 67, no. 4, pp. 1373-1381, 2020, doi: 10.1109/TED.2020.2968079.

[2] K. Florent et al., "Vertical Ferroelectric HfO2 FET based on 3-D NAND Architecture: Towards Dense Low-Power Memory," in 2018 IEEE International Electron Devices Meeting (IEDM), 1-5 Dec. 2018 2018, pp. 2.5.1-2.5.4, doi: 10.1109/IEDM.2018.8614710.

[3] S. Yoon et al., "Highly Stackable 3D Ferroelectric NAND Devices: Beyond the Charge Trap Based Memory," in 2022 IEEE International Memory Workshop (IMW), 15-18 May 2022 2022, pp. 1-4, doi: 10.1109/IMW52921.2022.9779278.

[4] S. Yoo et al., "Highly Enhanced Memory Window of 17.8V in Ferroelectric FET with IGZO Channel via Introduction of Intermediate Oxygen-Deficient Channel and Gate Interlayer," in 2024 IEEE Symposium on VLSI Technology and Circuits (VLSI Technology and Circuits), 16-20 June 2024 2024, pp. 1-2, doi: 10.1109/VLSITechnologyandCir46783.2024.10631534.

[5] Y. Zhou et al., "Hybrid-FE-Layer FeFET With High Linearity and Endurance Toward On-Chip CIM by Array Demonstration," IEEE Electron Device Letters, vol. 45, no. 2, pp. 276-279, 2024, doi: 10.1109/LED.2023.3346010.

[6] S. H. Kuk et al., "Unstable Retention Behavior in MIFIS FEFET Accurate Analysis of the Origin by Absolute Polarization Measurement," arxiv preprint 2024, doi: arxiv:2406.19618.

[7] G. Kim et al., "In-depth Analysis of the Hafnia Ferroelectrics as a Key Enabler for Low Voltage & QLC 3D VNAND Beyond 1K Layers: Experimental Demonstration and Modeling," in 2024 IEEE Symposium on VLSI Technology and Circuits (VLSI Technology and Circuits), 16-20 June 2024 2024, pp. 1-2, doi: 10.1109/VLSITechnologyandCir46783.2024.10631559.

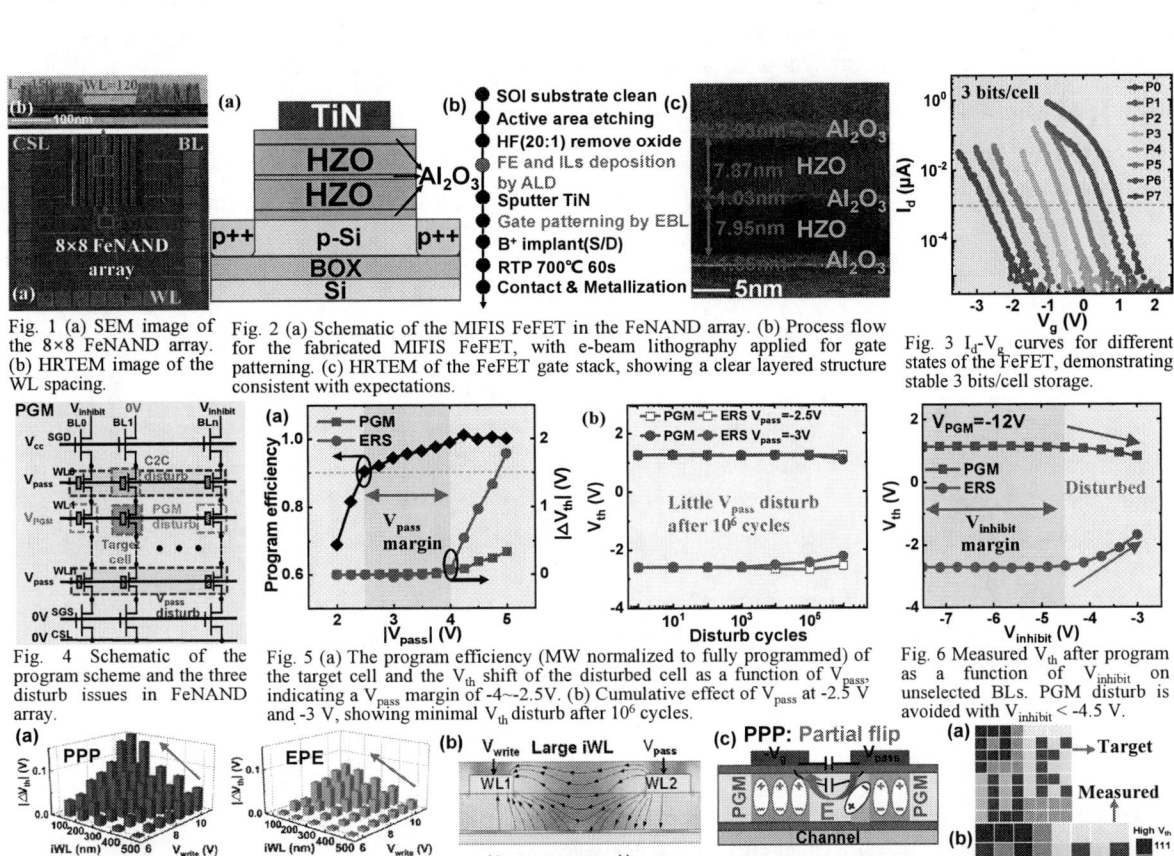

Fig. 1 (a) SEM image of the 8×8 FeNAND array. (b) HRTEM image of the WL spacing.

Fig. 2 (a) Schematic of the MIFIS FeFET in the FeNAND array. (b) Process flow for the fabricated MIFIS FeFET, with e-beam lithography applied for gate patterning. (c) HRTEM of the FeFET gate stack, showing a clear layered structure consistent with expectations.

Fig. 3 I_d-V_g curves for different states of the FeFET, demonstrating stable 3 bits/cell storage.

Fig. 4 Schematic of the program scheme and the three disturb issues in FeNAND array.

Fig. 5 (a) The program efficiency (MW normalized to fully programmed) of the target cell and the V_{th} shift of the disturbed cell as a function of V_{pass}, indicating a V_{pass} margin of -4~-2.5 V. (b) Cumulative effect of V_{pass} at -2.5 V and -3 V, showing minimal V_{th} disturb after 10^6 cycles.

Fig. 6 Measured V_{th} after program as a function of $V_{inhibit}$ on unselected BLs. PGM disturb is avoided with $V_{inhibit}$ < -4.5 V.

Fig. 7 C2C disturb. (a) Measured V_{th} shift of disturbed cell vs. V_{write} and iWL for different cases, which is <0.15 V in the worst case. (b) Simulated E-field is larger in smaller iWL. (c) Schematic of the C2C disturb, which is more severe in the PPP (or EEE) case, where the polarization is partially flipped by the coupled E-field. PPP and PEP denote cases where the programmed state is the same as or different from the state of the adjacent cells, respectively.

Fig. 8 State map verification of the 3 bits/cell FeNAND array, with nearly no error detected.

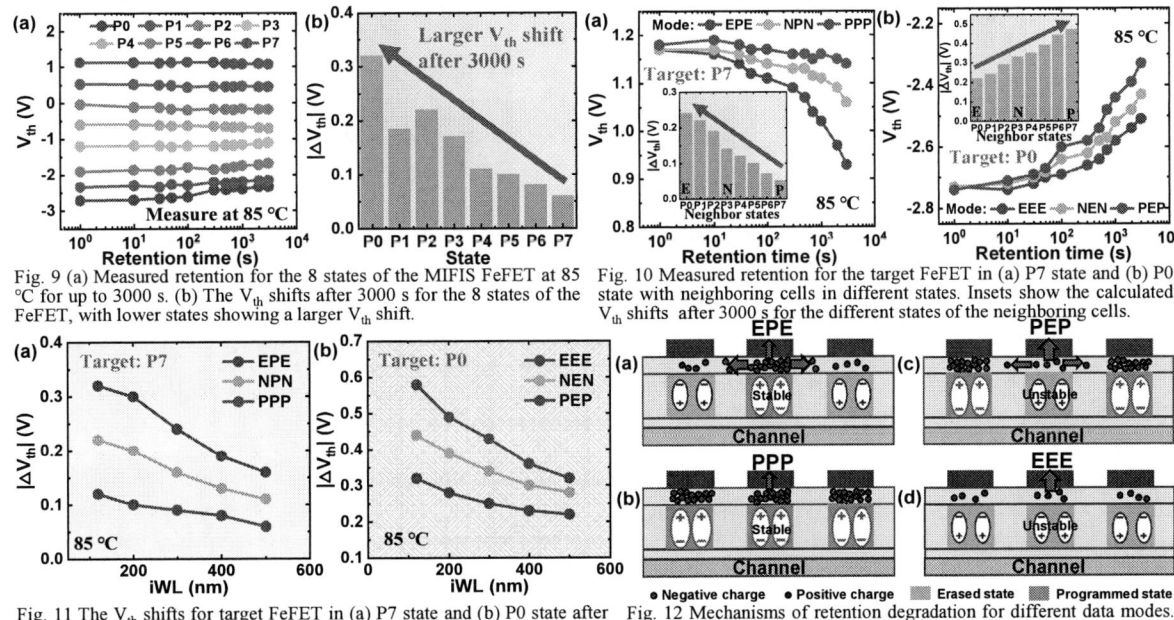

Fig. 9 (a) Measured retention for the 8 states of the MIFIS FeFET at 85 °C for up to 3000 s. (b) The V_{th} shifts after 3000 s for the 8 states of the FeFET, with lower states showing a larger V_{th} shift.

Fig. 10 Measured retention for the target FeFET in (a) P7 state and (b) P0 state with neighboring cells in different states. Insets show the calculated V_{th} shifts after 3000 s for the different states of the neighboring cells.

Fig. 11 The V_{th} shifts for target FeFET in (a) P7 state and (b) P0 state after 3000 s of retention at 85 °C as a function of iWL. The V_{th} shifts slightly increase as iWL decrease.

Fig. 12 Mechanisms of retention degradation for different data modes. The strong impact of neighbor states on retention behavior is attributed to the field-driven lateral charge migration.

979-8-3315-0417-5/25 $31.00 © 2025 IEEE

Prediction of Endurance Characteristics in Si FeFET with Ferroelectric Hf$_{0.5}$Zr$_{0.5}$O$_2$

Xinpei Jia[1,2,3], Tao Hu[1,2,3], Mingkai Bai[1,2,3], Xiaoqing Sun[1,2,3], Junshuai Chai[1,2,3], Hao Xu[1,2,3*], Xiaolei Wang[1,2,3*], Wenwu Wang[1,2,3], and Tianchun Ye[1,2,3]

[1]Institute of Microelectronics, Chinese Academy of Sciences, Beijing 100029, China. [2]Key Laboratory of Fabrication Technologies for Integrated Circuits, Chinese Academy of Sciences, Beijing 100029, China. [3]School of Integrated Circuits, University of Chinese Academy of Sciences, Beijing 100049, China.

(*E-mail: wangxiaolei@ime.ac.cn; xuhao@ime.ac.cn)

Abstract

In this work, we investigate the recombination charge during endurance cycles in Si ferroelectric field-effect transistor (FeFET) featuring Hf$_{0.5}$Zr$_{0.5}$O$_2$/SiO$_2$ gate stacks, utilizing split I-V measurements. Our findings indicate that when the memory window (MW) is reduced to 25% of its maximum value, the recombination charge during the fatigue process remains below approximately 10^6 µC/cm^2 across various pulse conditions. By analyzing the recombination charge per cycle, this work provides a prediction method for the endurance characteristics of the FeFET.

Keywords: Charge trapping, endurance fatigue, ferroelectric transistor, recombination charge.

Introduction

The discovery of ferroelectricity in doped hafnium oxides has renewed interest in ferroelectric field-effect transistor (FeFET), positioning hafnia-based FeFET as one of the promising emerging memory technologies due to their low power consumption, high write/read speed, CMOS compatibility, and scalability [1], [2]. However, their limited endurance, typically ranging from 10^4–10^6 cycles [3], [4], remains a major challenge for industry application [5], [6]. While recent studies have demonstrated improved endurance up to 10^{10} cycles [7], [8]. However, it is still much lower than the 10^{15} cycles required for RAM applications [6], [9]. Numerous studies have investigated the physical mechanisms behind endurance fatigue to improve endurance characteristics. Charge trapping, trap generation, and interfacial layer degradation are commonly identified as primary causes [10], [11]. Our recent work reveals that electron-hole recombination within the gate stacks induces trap generation, leading to the degradation of endurance characteristics [12].

In this work, we investigate the recombination charge during endurance cycles of the FeFET using split I-V measurements [12]. We analyze the endurance characteristics and the recombination charge during the endurance cycles under varying pulse amplitudes (V_m) and pulse widths (T_{pw}). We denote the recombination charge per cycle as Q_{re}, and the total recombination charge during the fatigue process as Q_{RE}. We find that with the increase of T_{pw}, the Q_{RE} is almost a constant value and with the increase of V_m, the Q_{RE} first

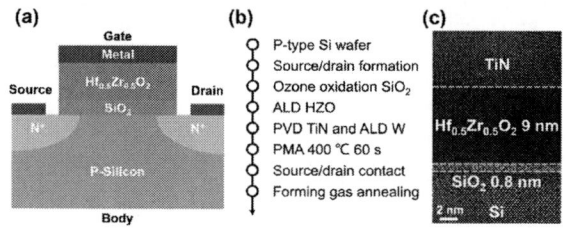

Fig. 1. (a) The schematic of the n-type FeFET. (b) The fabrication process flow. (c) The HRTEM image for the gate structure.

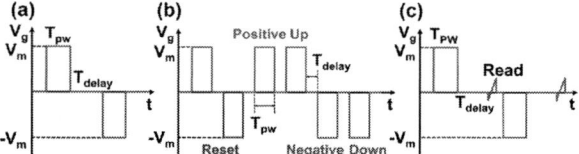

Fig. 2. The measurement waveforms of (a) cycling, (b) split I-V measurement, and (c) MW measurement.

increases and then decreases. The Q_{RE} is basically around 10^6 µC/cm^2 and below. By analyzing the Q_{re}, the endurance characteristics of the FeFET can be predicted.

Experimental

In this work, we studied the n-type FeFET with W/TiN/Hf$_{0.5}$Zr$_{0.5}$O$_2$/SiO$_2$/Si (MFIS) gate structure. The schematic of the FeFET and the fabrication process flow are shown in Fig. 1(a) and (b), respectively. The FeFETs were fabricated with the gate-last process. After diluted-HF clean, the gate-stack of W/TiN/Hf$_{0.5}$Zr$_{0.5}$O$_2$/SiO$_2$/Si was formed. The SiO$_2$ layer was grown via ozone oxidation. The Hf$_{0.5}$Zr$_{0.5}$O$_2$ layer was deposited via atomic layer deposition (ALD) at 300 °C using TEMA-Hf as Hf precursor, TEMA-Zr as Zr precursor, and H$_2$O as O precursor. After the deposition of TiN and W, these devices were annealed at 400 °C in N$_2$ for 60 s to form the orthorhombic phase. The gate length and width are 5 µm and 150 µm, respectively. The physical thickness of the Hf$_{0.5}$Zr$_{0.5}$O$_2$ and SiO$_2$ is 9 nm and 0.8 nm, respectively, as shown in Fig. 1(c).

Fig. 2 shows the waveforms for three parts: a) cycling, b) split I-V measurement, and c) memory window (MW) measurement. The split I-V measurement is performed with a Positive-Up Negative-Down (PUND) waveform to examine the current at individual terminals of the FeFET, then the Q_{re} for a program/erase (PGM/ERS) cycle can be obtained. To

979-8-3315-0417-5/25 $31.00 © 2025 IEEE

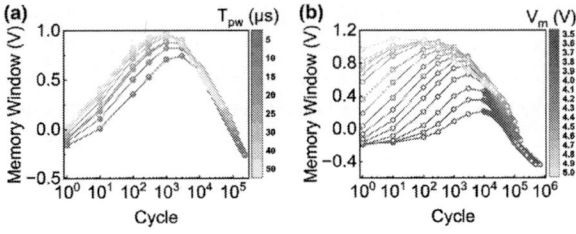

Fig. 3. The MW versus PGM/ERS cycles for (a) different T_{pw} where the V_m is 4 V and (b) different V_m where the T_{pw} is 10 µs.

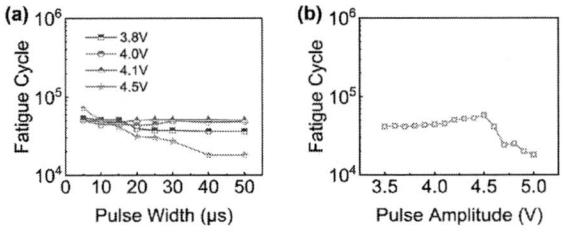

Fig. 4. (a) The FC for different T_{pw} where the V_m is 3.8 V, 4.0 V, 4.1 V, and 4.5 V. (b) The FC for different V_m where the T_{pw} is 10 µs.

meet the requirements of practical applications and ensure accurate measurement, the T_{pw} ranges from 5 µs to 50 µs, and the V_m ranges from 3.5 V to 5 V. The delay time (T_{delay}) between the pulses for the cycling and PUND waveforms is 10 µs, while the T_{delay} before I_d-V_g read for the MW measurement is 1 ms to obtain a stable threshold voltage (V_{th}).

Result and Discussion

We investigate the endurance characteristics under varying T_{pw} and V_m. Fig. 3 illustrates the MW vs. the number of cycles for different T_{pw} and V_m. As T_{pw} increases, the maximum MW increases slightly. In contrast, a higher V_m results in a more pronounced increase in the maximum MW. To compare the endurance characteristics under different pulse conditions, we define the fatigue cycle (FC) as the cycle where MW drops to 25% of its maximum. However, as the T_{pw} or V_m increases largely, the FeFET is more prone to breakdown, and thus MW may not decrease to 25% of its maximum value. In such cases, the FC is defined by breakdown cycle. To accurately compare the FC under different pulse conditions, we measure the MW and Q_{re} after every 1k PGM/ERS cycles, once the cycle exceeds 10k. This means that the resolution of PGM/ERS cycles is set as 1k.

Fig. 4(a) presents the FC for different T_{pw}. The FC remains nearly constant with increasing T_{pw} when V_m is 3.8 V, 4 V, and 4.1 V. The MW is primarily determined by the difference between the spontaneous polarization (P_s) and the trapped charge (Q_t). Therefore, although a larger T_{pw} accelerates defect generation, the decay rate of the MW is not necessarily faster. Thus, the variation of the FC with T_{pw} is reasonable. For V_m at 4.5 V, when the T_{pw} exceeds 5 µs, breakdown occurs before MW decreases to 25% of its maximum. As T_{pw} increases, the breakdown susceptibility rises, leading to a decrease in FC. Fig. 4(b) illustrates the FC for different V_m

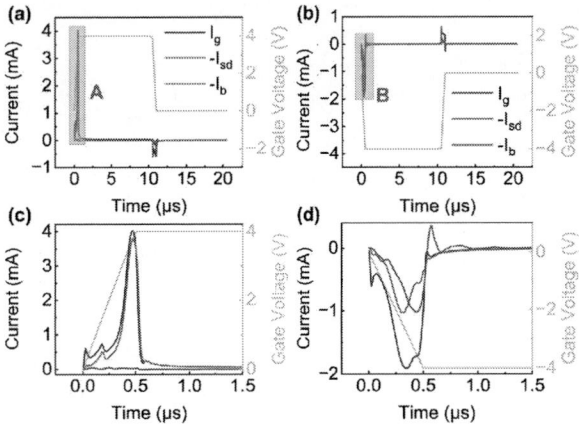

Fig. 5. The current characteristics of each terminal as a function of time, along with the variation in gate voltage over time for (a) Positive pulse and (b) Negative pulse. (c) and (d) are the enlarged views of the box section A and B, respectively.

where the T_{pw} is 10 µs. Initially, the FC increases with increasing V_m. Although an increase in V_m accelerates defect generation and device degradation, the variation in the MW does not solely depend on defect generation; rather, it is influenced by the interplay between Q_t and P_s [13]. Therefore, the observation that the FC increases with V_m when V_m is below 4.5 V is not surprising. However, as V_m continues to increase, the device begins to experience premature breakdown, leading to a decrease in the FC. In this case, electric field stress becomes the dominant factor affecting FC.

To analyze the factors determining the FC under different T_{pw} and V_m, we used the split I-V measurement to measure the transient currents of each terminal [12]. Fig. 5(a) and (b) show the current characteristics of each terminal and the corresponding gate voltage variations. Fig. 5(c) and (d) are the enlarged views of the box section A and B, respectively. The results are consistent with those obtained from triangular PUND waveform [12]. During the rising edge of the Positive pulse and the subsequent holding phase, the peak of the gate current (I_g) caused by ferroelectric polarization switching primarily corresponds to the peak of the source/drain current (I_{sd}), while the bulk current (I_b) is nearly zero, indicating all electrons flow in from the source/drain (S/D). During the falling edge of the Negative pulse and the subsequent holding phase, the peak of the I_g caused by ferroelectric polarization switching partially corresponds to the peak of the I_{sd} and partially to the peak of the I_b, signifying that some electrons flow out through the S/D while others recombine with trapped holes from the bulk. Therefore, the Q_{re} can be derived by analyzing the I_{sd} and the I_b. The falling-edge current in Fig. 5(a) and the rising-edge current in Fig. 5(b) exhibit some spikes and oscillations, but they do not affect the calculation of Q_{re}.

Next, we will discuss in detail the method for obtaining Q_{re}. The change in polarization charge due to ferroelectric

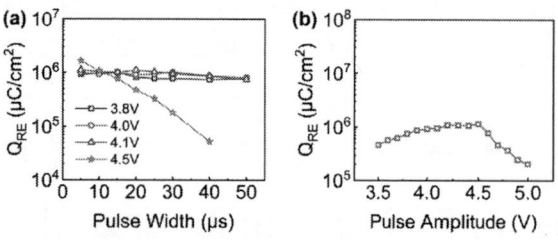

Fig. 6. The variation of Q_{RE} as a function of (a) T_{pw} where V_m is 3.8 V, 4.0 V, 4.1 V, and 4.5 V and (b) V_m where the T_{pw} is 10 µs.

switching can result in charge trapping/de-trapping in the gate stacks and alter the channel carrier concentration. However, the change in the channel carrier concentration is small [14], and thus the change in polarization charge is primarily reflected as the Q_t. Therefore, the integration of the pulse current reflects the Q_t. Due to the significant oscillation errors in the measured I_b, this work utilizes I_{sd} for the calculation of Q_{re}. The integrated charges of the I_{sd} for Positive, Up, Negative, and Down pulses are denoted as Q_{sdP}, Q_{sdU}, Q_{sdN}, and Q_{sdD}, respectively. The Up and Down pulses are used to eliminate the gate leakage charge and the back-switched charge. Thus, the charge trapped from the S/D into the gate during the Positive pulse, Q_{tsdP}, can be obtained as

$$Q_{sdP} = Q_{sdP} - Q_{sdU} \quad (1)$$

While the charge de-trapped from the gate to the S/D during the Negative pulse, Q_{tsdN}, can be obtained as

$$Q_{sdN} = Q_{sdN} - Q_{sdD} \quad (2)$$

Finally, the Q_{re} can be obtained as

$$Q_{re} = Q_{sdP} - Q_{sdN} \quad (3)$$

Fig. 6(a) presents the Q_{RE} as a function of T_{pw} for V_m is 3.8 V, 4 V, 4.1 V, and 4.5 V. For V_m is 3.8 V, 4 V, and 4.1V, as the T_{pw} increases, the Q_{RE} remains nearly constant. The energy released from electron-hole recombination and electric filed stress are known to accelerate defect generation [15], [16]. Prolonged electric field application further promotes defect generation, which accounts for the near-constant behavior of Q_{RE} as T_{pw} increases. For V_m = 4.5 V, as T_{pw} increases, Q_{RE} decreases due to breakdown occurring before MW reaches 25% of its maximum value. At this stage, the process is primarily governed by the electric field. With longer duration of electric field stress, breakdown is accelerated, resulting in a reduction of Q_{RE}. Fig. 6(b) shows the variation of Q_{RE} with increasing V_m. As V_m increases, Q_{RE} initially increases and then decreases. In the range of 3.5 V to 4.5 V, as the V_m increases, a larger number of defects must be generated. The increase in electric field alone is insufficient to produce such a large number of defects, resulting in a continuous increase in Q_{RE}. When V_m exceeds 4.5 V, the device enters premature breakdown. As the electric field continues to increase, the breakdown accelerates, with the electric field dominating the breakdown process, resulting in a decrease in Q_{RE}.

We next examine the practical applications of this study. At low V_m or T_{pw}, defect generation is primarily driven by

electron-hole recombination, leading to degradation in device endurance and affecting the FC. Consequently, as T_{pw} or V_m increases, Q_{RE} remains stable or increases. Under higher V_m or T_{pw}, device breakdown becomes the dominant factor influencing FC, causing Q_{RE} to decrease with further increases in T_{pw} or V_m. Thus, Q_{RE} reaches an upper limit of approximately 10^6 µC/cm² [Fig. 6]. The Q_{re} is approximately in the range of 10^0 to 10^1 µC/cm² [17], [18]. Using the maximum Q_{RE} of 10^6 µC/cm² and the minimum Q_{re} of 1 µC/cm², the estimated FC is around 10^6, consistent with the results in [17], [18]. Thus, suppressing charge trapping and de-trapping may not be critical for improving endurance characteristics. Instead, efforts should focus on preventing electron-hole recombination within the gate stacks. For RAM applications, achieving FC of 10^{15} requires reducing Q_{re} to 10^{-9} µC/cm², which poses a considerable challenge.

Conclusion

we investigate the recombination charge under varying pulse widths and amplitudes using the split I-V measurement. We find that the recombination charge during endurance cycles remains below approximately 10^6 µC/cm². These findings suggest that suppressing trapped charge recombination is crucial for improving endurance characteristics. Furthermore, by examining the recombination charge per cycle, the fatigue cycle of the FeFET can be estimated.

Acknowledgments

This work was supported in part by the National Natural Science Foundation of China under Grant No. 92264104 and 52350195, and supported by the Postdoctoral Fellowship Program of CPSF under Grant No. GZC20232925.

References

[1] T. S. Boscke et al., IEDM 2011, p. 24.5.1.

[2] T. Mikolajick et al., IEEE Trans. Electron Devices, 67, 1434 (2020).

[3] H.-K. Peng et al., Appl. Phys. Lett., 118, 103503 (2021).

[4] A. J. Tan et al., VLSI 2020, p. 1.

[5] N. Zagni et al., Proc. IEEE, 111, 158 (2023).

[6] W. Yang et al., Journal of Semiconductors, 44, 053101 (2023).

[7] Y. Zhou et al., IEDM 2022, p. 6.2.1.

[8] C.-Y. Liao et al., VLSI 2022, p. 1.

[9] J. P. B. Silva et al., APL Materials, 11, 089201, (2023).

[10] E. Yurchuk et al., IRPS 2014, p. 2E.5.1.

[11] E. Yurchuk et al., IEEE Trans. Electron Devices, 63, 3501 (2016).

[12] X. Jia et al., IEEE Trans. Electron Devices, 71, 1845 (2024).

[13] G. Kim et al., IEEE Trans. Electron Devices, 71, 6627 (2024).

[14] K. Toprasertpong et al., IEDM 2019, p. 23.7.1.

[15] J. W. McPherson et al., IEEE Trans. Electron Devices, 50, 1771 (2003).

[16] I. C. Chen et al., Journal of Applied Physics, 61, 4544 (1987).

[17] S.-H. Kuk et al., IEDM 2021, p. 33.6.1.

[18] N. Tasneem et al., IEDM 2021, p. 6.1.1.

A Novel FeFET-based Multibit Content Addressable Memory through Thermometer Encoding for Manhattan Distance Metric with High Area- and Energy-Efficiency

Weikai Xu[1#], Zeyu Zhang[1#], Qianqian Huang[1,2*] and Ru Huang[1,2*]

[1]School of Integrated Circuits, Peking University, Beijing 100871, China. [2]Beijing Advanced Innovation Center for Integrated Circuits, Beijing 100871, China. [#]Equal contribution.
(*Email: ruhuang@pku.edu.cn; hqq@pku.edu.cn)

Abstract

In this work, a novel ferroelectric FET (FeFET)-based multibit content addressable memory (MCAM) through modified thermometer encoding (TE), for Manhattan distance (MD) metric is proposed for the first time. By utilizing the multilevel storage of FeFET and the modified TE scheme, the arbitrary bit-length MD metric can be implemented with ultra-low hardware cost and power consumption. Moreover, an enhanced FeFET MCAM cell, which is more suitable for the proposed modified TE scheme, is further introduced to improve the MD metric precision and reliability when expanding the bit length of MD. Based on the proposed design, the MD metric is demonstrated with $4.5\times/34.8\times$ area-efficiency and $66.7\times/7.9\times$ energy-efficiency improvements compared with conventional CMOS-based digital/CAM design, showing its great potential for MD-based applications.

Keywords: Manhattan distance, multibit content addressable memory, thermometer encoding, ferroelectric FET

Introduction

Manhattan distance (MD), also known as L_1 norm, which is a distance metric used to quantify the absolute distance between two high-dimensional vectors, has been widely used in various applications, such as route planning, text processing and pattern recognition [1-3] (Fig. 1a). Generally, the accuracy of MD-based machine learning tasks can be enhanced as the bit length of MD increases, due to the finer granularity [4] (Fig. 1b). For hardware implementations of MD metric, the conventional digital-based design requires SRAM for storage and additional circuits for subtraction and absolute value calculation, suffering from high hardware cost and energy consumption [5] (Fig. 2a). On the other hand, the SRAM-based content addressable memory (CAM) can implement highly parallel MD metric within the CAM array by employing the thermometer encoding (TE) scheme, eliminating the need of extra calculation circuits [6] (Fig. 2b). However, the advantage of SRAM-based CAM for MD metric is limited to 4 bits, due to the exponentially increasing number of CAM cells and the high hardware cost of each SRAM-based CAM cell [6] (Fig. 2c).

In this work, a novel ferroelectric FET (FeFET)-based thermometer-encoded multibit CAM (TEMCAM) is proposed for the first time, which supports the MD metric with ultra-low hardware cost for arbitrary bit lengths.

Moreover, an enhanced TEMCAM cell is further proposed, enabling arbitrary configurable bit-length MD metric with improved metric quality. Based on the proposed design, the MD metric is demonstrated with significantly improved area- and energy-efficiency, indicating its great potential for distance-based machine learning.

Design of FeFET-based TEMCAM

Fig. 3a shows the proposed FeFET-based TEMCAM architecture, where multiple features of one high-dimensional vector (entry) share a single match line (ML), and stored entries share the search lines (SLs) of input vector (query) for simultaneous MD metric between the query and all entries in the TEMCAM array. In the TEMCAM architecture, each feature is represented by a group of MCAM cells with the proposed modified TE scheme. Different from the conventional TE-based CAM, which can only extend states by adding more bits (Fig. 2b), the modified TE scheme can further extend states by employing the MCAM cell, where each MCAM cell with m states can support m-1 incremental states, making it more efficient than the conventional TE scheme (Fig. 3c).

For hardware implementation, the MCAM cell is composed of two FeFETs with complementary programmed threshold voltages (V_{TH}, $V_{\overline{TH}}$) for one stored entry state, and corresponding complementary search voltages (V_{SL}, $V_{\overline{SL}}$) for one input query state (Fig. 3b). By carefully setting the values of V_{TH} and V_{SL} which will be discussed in detail in the next section, the FeFET-based MCAM cell enables the difference measure between multiple query and entry states by detecting the ML current (I_{ML}). Furthermore, by encoding the FeFET-based MCAM cells with the modified TE scheme for representing the whole feature, the total I_{ML} across all MCAM cells is approximately proportional to the MD between the query and entry (Fig. 3d), indicating the great capability for MD metric.

Characteristics and optimization of proposed FeFET-based TEMCAM

A. FeFET device used for TEMCAM

Fig.4a shows the structure of FeFET device, which is a promising non-volatile memory with high I_{ON}/I_{OFF} ratio, low write energy and good CMOS compatibility [7]. In this work, the calibrated FeFET model composed of multi-domain Preisach model of ferroelectric (FE) layer and BSIM MOSFET model [8] (Fig. 4b) is used for circuit simulation,

979-8-3315-0417-5/25 $31.00 © 2025 IEEE

to demonstrate and evaluate the proposed FeFET-based TEMCAM. The gradual polarization switching characteristics of FE capacitor is shown in Fig. 4c, which will lead to continuously modulated V_{TH} of FeFET for multi-level entry storage (Fig. 4d).

B. Characteristics of FeFET-based TEMCAM for MD

As mentioned above, to support the linear MD metric, the operating voltages (V_{SL}) of the FeFET-based MCAM cell should be carefully selected to achieve an incrementally linear I_{ML} in relation to MD. In this work, a parameter α according to the $I_{ML}(V_{SL})$ of FeFET is introduced for assessing the linearity of the MCAM cell as follows:

$$\alpha = \frac{I(V_0)}{I(V_0+\delta V)} + \sum_i \left(\left| \frac{I(V_0+i\cdot\delta V)}{I(V_0+\delta V)} - i \right| \right) \quad (1)$$

where δV represents the difference between two adjacent V_{SL}. The first term of Eq. (1) reflects whether the matched current (i.e., MD is 0) is small enough relative to mismatched currents of other MD values, and the second term indicates whether the mismatched current is proportional to the MD value. A configuration algorithm of MCAM cell is proposed to obtain the minimum α, and thus find the optimal V_{SL} for linear MD metric (Fig. 5a). Based on the configuring method and I_{ML}-V_{SL} curves of FeFET which is programmed to the far left state (Fig. 5b), the optimal V_{SL} values for 2-bit MCAM cell can be obtained (Fig. 5c). Benefiting from the proposed configuration algorithm, the FeFET-based MCAM cells can achieve approximately linear MD metric with different V_{ML}, and a better linearity is obtained with a higher V_{ML} due to the wider I_{ML}-V_{SL} linear region (Fig. 5de).

Moreover, by encoding MCAM cells with the proposed modified TE scheme, the expanded MD metric with more states of TEMCAM is further demonstrated (Fig. 6a). Fig. 6b shows the results of two MCAM cells encoded by modified TE, and the I_{ML} is approximately linear with the MD, along with only a slight sense margin (SM) degradation due to the different situations of the same MD value (Fig. 6c). Moreover, as the bit length of MD increases, the SM will not be further degraded (Fig. 6d), benefiting from the proposed modified TE scheme with a maximum of two different combinations of various MD values with 2-bit MCAM cell (Fig. 6e), indicating the capability of TEMCAM for MD metric with a greater bit length.

C. Optimized FeFET-based MCAM cell for TEMCAM

In practical MD metric, the I_{ML} of a matched CAM cell is not absolutely zero and will accumulate with the state expansion, resulting in the non-uniform distribution for various MD bit lengths. To further support configurable MD metric with arbitrary bit lengths, an enhanced FeFET-based MCAM cell is further proposed to address the above challenge in TEMCAM, that is to change the match conditions of the minimum and maximum states for each MCAM cell (Fig. 7ab). Taking 2-bit MCAM cell as example, by setting the lower V_0 and the higher V_{TH3}, the I_{ML} of the 0-0 and 3-3 match can be reduced to a negligible level, and thus obtaining a uniform distribution for various bit-length MD metric with improved precision. (Fig. 7cd). Moreover, due to the increased variety of combinations with the same MD value when further implementing MD metric of multiple features, the SM will diminish to an undetectable level, resulting in overlap (Fig. 8ab). Benefiting from the enhanced TEMCAM cell with uniform distribution, this overlap issue can also be effectively solved, indicating the improved reliability of the proposed TEMCAM (Fig. 8c).

Evaluation of proposed FeFET-based TEMCAM

Fig.9 shows the MD metric when expanding the state number per MCAM cell from 5-state to 8-state, both showing relatively good linearity due to the proposed configuration algorithm. Compared with other reports, the proposed FeFET-based TEMCAM achieves significant reduction in hardware overhead for MD metric with arbitrary bit lengths, and the improvement is more significant as the number of states per MCAM cell increases (Fig. 10a). Moreover, the proposed FeFET-based TEMCAM with 2-bit/3-bit MCAM cell can achieve $33.3\times/66.7\times$ and $3.9\times/7.9\times$ energy-efficiency improvement compared with conventional digital circuit and SRAM-based CAM for MD metric (Fig. 10b).

Conclusion

This work reports a novel FeFET-based TEMCAM for MD metric. Based on the proposed TEMCAM, the MD metric with arbitrary bit lengths can be implemented with ultra-low hardware cost and energy consumption, along with improved linearity and eliminated overlap. Furthermore, the area- and energy-efficiency can be further improved with the increased number of states per MCAM cell, showing its great potential for MD-based applications at the edge.

Acknowledgments

This work was supported by NSFC (61927901, 62374009) and 111 Project (B18001).

References

[1] J. Clempner et al., "Using the Manhattan distance for computing the multiobjective Markov chains problem," *Int. J. Comput. Math.*, 2018.

[2] S. Salihu et al. "Performance evaluation of manhattan and euclidean distance measures for clustering based automatic text summarization," *J. Eng. Technol.*, 2019.

[3] W. Zhang et al., "FNNWV: farthest-nearest neighbor-based weighted voting for class-imbalanced crowdsourcing," *Sci. China Inf. Sci.*, 2024.

[4] R. Suwanda et al., "Analysis of Euclidean Distance and Manhattan Distance in the K-Means Algorithm for Variations Number of Centroid K," *J. Phys. Conf. Ser.*, 2020.

[5] H. Mattausch et al., "An Architecture for Compact Associative Memories with Deca-ns Nearest-Match Capability up to large Distances," in *IEEE ISSCC*, 2001.

[6] H. Mattausch et al., "Fully-parallel pattern-matching engine with dynamic adaptability to Hamming or Manhattan distance," in *Symposium on VLSI*, 2002.

[7] W. Xu et al., "A Novel Ferroelectric FET Based Universal Content Addressable Memory With Reconfigurability for Area- and Energy-Efficient In-Memory-Searching System," *IEEE EDL*, 2024

[8] Z. Fu et al., "Device Modeling and Application Simulation of Ferroelectric-FETS with Dynamic Multi-Domain Behavior," in *CSTIC*, 2020.

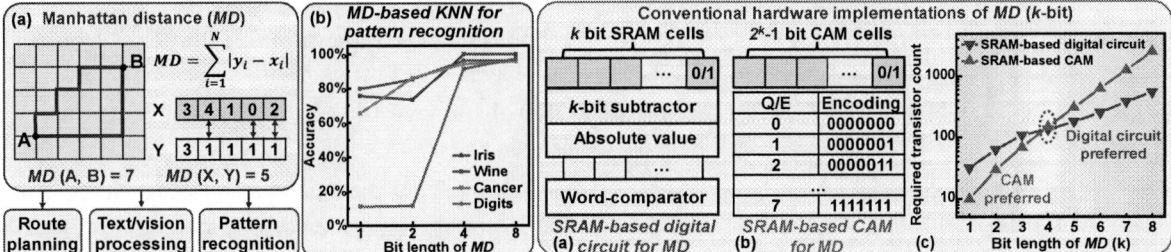

Fig. 1. (a) The definition of Manhattan distance (MD) in two-dimensional and multi-dimensional spaces, which is widely used for various applications. (b) The enhanced classification accuracy of MD-based K-Nearest Neighbors (KNN) algorithm for pattern recognition as the bit length of MD increases.

Fig. 2. (a) Architecture of conventional SRAM-based digital circuit for MD metric. (b) Architecture of conventional SRAM-based content addressable memory (CAM) for MD metric. (c) The number of transistors required in SRAM-based CAM exceeds that in SRAM-based digital circuit for MD metric when the bit length of MD is larger than 4.

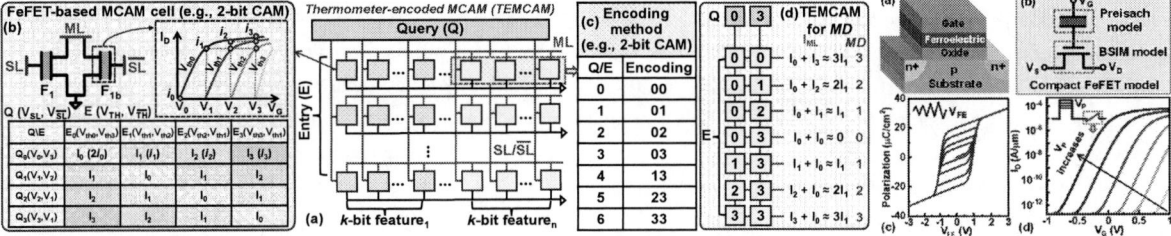

Fig. 3. (a) Design of proposed ferroelectric FET (FeFET)-based thermometer-encoded MCAM (TEMCAM) architecture. (b) FeFET-based MCAM cell used for TEMCAM and the truth table, taking 2-bit MCAM cell as an example. (c) Proposed modified thermometer encoding (TE) method for TEMCAM, taking 2-bit MCAM cell as an example. (d) The principle of MD metric based on the proposed TEMCAM.

Fig. 4. (a) FeFET device. (b) Compact FeFET model. (c) P-V loops with multilevel FE polarizations. (d) Modulated transfer curves of FeFET.

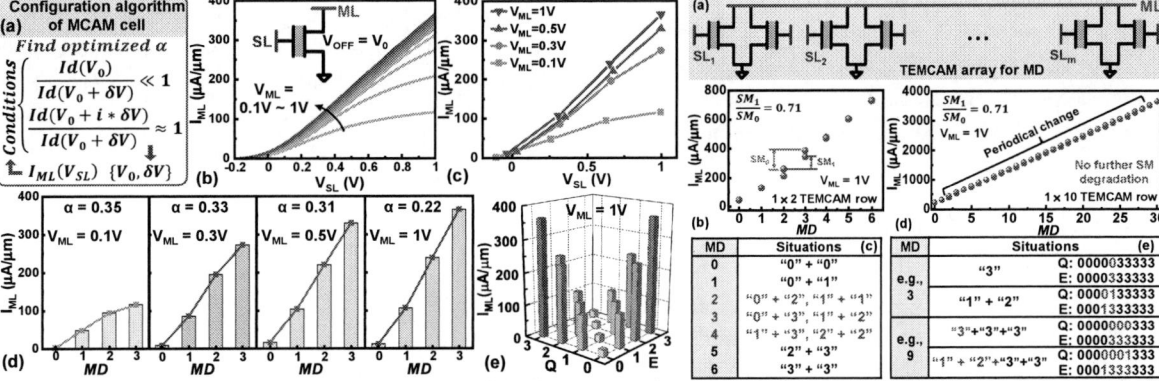

Fig. 5. (a) The proposed configuration algorithm of MCAM cell for linear MD metric. (b) The transfer curves of FeFET programmed to the far left state with ML voltage (V_{ML}) ranging from 0.1V to 1V. (c) Best configurations of MCAM cell for MD with multilevel V_{ML}. (d) ML currents (I_{ML}) corresponding to different MD with multilevel V_{ML}. (e) I_{ML} corresponding to different entry and query with V_{ML} of 1V.

Fig. 6. (a) The proposed TEMCAM array for MD metric. (b) The I_{ML} of two FeFET-based MCAM cells for MD metric. (c) The possible combinations of various MD values. (d) The I_{ML} ten FeFET-based MCAM cells for MD metric. (e) There are maximum of two possible combinations of various MD values.

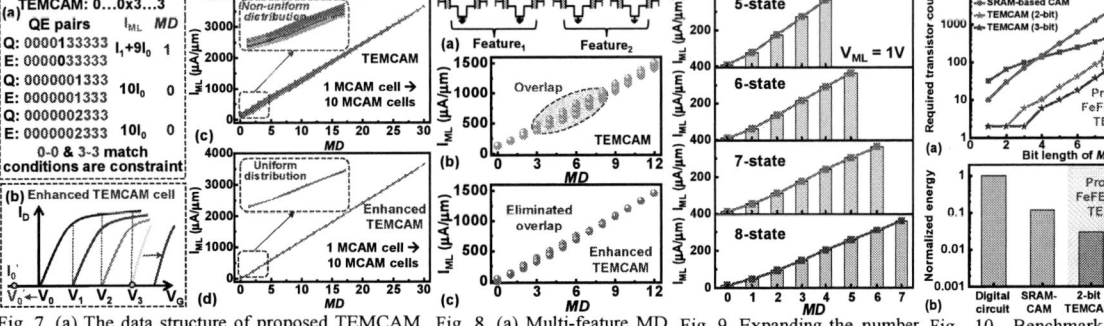

Fig. 7. (a) The data structure of proposed TEMCAM, with many 0-0 and 3-3 pairs for 2-bit MCAM cell. (b) The proposed enhanced TEMCAM cell. The I_{ML} from 1 MCAM cell to 10 MCAM cells, showing (c) overlap based on basic MCAM and (d) eliminated overlap of enhanced TEMCAM.

Fig. 8. (a) Multi-feature MD metric based on TEMCAM. The I_{ML} shows (b) overlap based on basic MCAM and (c) eliminated overlap of enhanced TEMCAM.

Fig. 9. Expanding the number of state per MCAM cell from 5 to 8 with V_{ML} of 1V, showing the good linearity for further improving the area-efficiency of TEMCAM.

Fig. 10. Benchmark of (a) required transistor count and (b) energy consumption of proposed FeFET-based TEMCAM with other reported works for MD metric.

979-8-3315-0417-5/25 $31.00 © 2025 IEEE

Novel High-Efficiency Large-Area Optical-to-Optical Conversion Integrated Device for Optical Signal Processing

Yahui Su[1], Shanjing Liu[1], Peixuan Song[1], Peiran Du[1], Hui Wang[2], Juan Li[1,*]

[1]Institute of Photoelectronic Thin Film Devices and Technology, Key Laboratory of Photoelectronic Thin Film Devices and Technology of Tianjin, Tianjin, China,

[2]China Computer Room Equipment Engineering Co., Ltd, Tianjin, China

lj1018@nankai.edu.cn

Abstract

Utilizing optoelectronic hybrid integration technology (OHIT), a large-area optical-to-optical conversion integrated device (OTO-CID) is developed for high-efficiency optical signal conversion. It includes a PV component for absorbing complex optical signals and an OLED for emitting refined signals. The OTO-CID achieved a maximum external quantum efficiency (EQE*, the parameter to characterize the optical-to-optical conversion efficiency) of 10.43% at 20% AM1.5, showcasing a promising direction for future optoelectronic integrated techniques.

Keywords: Optical-to-optical conversion integrated device, Perovskite solar cell modules, EQE*

Introduction

The escalating demand for efficient signal processing has underscored the significance of OHIT[1] in domains such as optical communication[2, 3] and consumer electronics[4]. The OTO-CID is capable of transforming complex light into specific wavelength emissions. It integrates perovskite solar cell modules (PSMs) and organic light-emitting diodes (OLEDs) to enhance the optical conversion efficiency. With challenges in scaling up due to crystallization uniformity and solvent residuals persisting[5], an anti-solvent spraying technique and a mechanical scribing process for series-connected PSMs are conducted to prepare large-area, high-efficiency PSMs. The OLEDs, with an integrated charge generation layer (CGL) with an ultra-thin Ag barrier, achieve high brightness and efficiency. The custom-built testing system confirms the high EQE* of 10.55% under LED lighting conditions and 10.43% under 20%AM1.5 conditions of OTO-CID, with corresponding luminance values of 1251.85 cd/m² and 2032.33 cd/m², respectively, showcasing its potential in optical signal processing and image sensing for next-generation portable devices.

Structure design and preparation of the OTO-CID

The OTO-CID is realized through the multipixel integration of PSMs and OLEDs. The device is constructed as a 4×4-pixel array, comprising 16 individual subunits on a 132 mm $\times$ 132 mm (174.24 cm²) substrate. Each subunits comprises a PSM, an OLED, and an attached OLED. Figure 1(a) depicts the structural diagrams of the OTO-CID and a subunit. The PSMs, with six series-connected sub-cells based on a n-i-p PSC architecture: ITO/SnO$_2$/FAPbI$_3$/Spiro-OMeTAD/Au, are prepared by the spin-coating method in the ambient environment. The red tandem top-emission OLEDs (TTE-OLEDs) with the layer stack: Ag/ITO/HI-V2:HI-D-J1(50,4%)/HTL-J1(150)/EBL(50)/HOST-R: RD-E1 (300, 5%)/ETL-N5:LiQ(150,1:0.7)/ETL-T1:EI-D3(150,3%)/Ag /HI-V2:HI-D-J1 (1050,4%)/HTL-J1(150) /EBL(50) / HOST-R: RD-E1(300,5%) / ETL-N5:LiQ (300,1:0.7)/Mg: Ag (200, 1:9)/CPL(700), is fabricated using the evaporation method. Specifically, the locations of the PSMs and OLEDs are determined by the patterning ITO through magnetron sputtering, which is predicated on establishing connections between the PSMs and OLEDs and reducing the transport distance for electrons generated within the PSMs. Following this, Ag/ITO (150 nm/10 nm) is sputtered and patterned in the OLED region to function as the cathodes, while the exposed ITO acts as the negative electrode for the PSMs. The PSM and OLED devices are prepared sequentially. The patterned ITO, Au electrode, and mechanical scribing technique process in concert to enable series connection of the PSMs. The positive electrodes of the PSMs are linked to the anodes of the OLEDs, while the negative electrodes are connected to the cathodes through the patterned ITO. The fabrication processes for the subunits and OTO-CID are aligned, as detailed above.

Enhancement of Crystallization Uniformity in Large-Area Perovskite Thin-Film Arrays

During the spin-coating process, the anti-solvents serve as extractants for perovskite precursor solvents crucial for the formation of dense and uniform perovskite films[6]. To address the challenge of crystallization uniformity in large-area perovskite films, we develop an innovative anti-solvent spraying technique equipped with a valve and multiple channels, as illustrated in Figure 1(b). The technique enables the rapid spreading of ethyl acetate (EA) across the entire perovskite wet film, significantly enhancing solvent extraction and promoting uniform perovskite crystallization. A comparison between the anti-solvent spraying technique (Target) and the conventional pipette method (Control) in terms of PSM performance is presented in Figure 1(e, f). The results demonstrate that the Target method exhibits superior stability and efficiency, with a power conversion efficiency (PCE) variance of 0.69, lower than the Control method variance of 2.16, facilitating the excellent crystallization uniformity of the PSMs and the functioning of the OLED.

Structural Design Optimization in Large-Area Perovskite Solar Cell Modules

The PV device with the integrated device enhances its capacity to drive OLED through the employment of a series-connected module of sub-cells[7]. In this study, the series-connected structure of PSMs is defined by mechanical scribing, a process critical to minimizing the dead zone[8], a significant factor that adversely affects light absorption in PSMs. As depicted in Figure 1(c, d), W_d, which represents the width of the dead zone, encompasses regions P1, P2, and P3.

Two primary scribing configurations are considered. Configuration 1(Figures 1(c)) results in a dead zone area of approximately $6.8 \times 10^{-6} cm^2$.Configuration 2 (Figure 1(d)) yields a reduced dead zone area of approximately $4.0 \times 10^{-6} cm^2$. Configuration 2 more effectively diminishes the dead zone area and series contact resistance. Furthermore, the J-V curves of the PSMs, as displayed in Figure 1(g), exhibit a high open-circuit voltage (V_{oc}) of 5.52V.

Highly-efficient Tandem Top-Emission OLED

The red tandem top-emission OLEDs aimed at cross-side light absorption and luminescence are applied to the OTO-CID, as depicted in Figure 1(h). Light-emitting units in series within tandem devices are integrated using a CGL, comprising an N-type doping layer/electron receiving layer/hole transport layer. To mitigate interface diffusion[9] in N-type doping layers, an ultra-thin Ag barrier layer is incorporated into the CGL. TTE-OLEDs achieve a maximum brightness of exceeding 140,000 cd/m² and the current efficiency reaches 93.3 cd/A at 1000 cd/m². Furthermore, the EQE of the TTE-OLED achieves 68.3%.

Performance of the Optical-to-Optical Conversion Integrated Device

The operation photograph of the OTO-CID is illustrated in Figure 2(a). A custom-built testing system facilitates the assessment of the OTO-CID, as depicted in Figure 2(b). The optical-to-optical conversion efficiency is characterized by measuring the EQE* and luminance under both LED and AM1.5 solar simulator illumination. The OTO-CID achieves a peak EQE* of 10.55% at a luminance of 1251.85 cd/m² under LED illumination, and 10.43% at a luminance of 2032.33 cd/m² under a 20% AM1.5 solar simulator. These findings demonstrate that the OTO-CID can operate effectively under low-light conditions, suggesting that the devices can provide high-efficiency optical-to-optical conversion in practical scenarios.

Figure 2(c, d) illustrates the correlation between luminance, real-time driving voltage of the PSMs, total current, and EQE* concerning the power of the LED illumination. It highlights that the PSMs sustain a robust driving voltage for OLED emission even under low-light conditions, with an EQE* stability of up to 10.51%. Furthermore, we investigate the relationship between EQE*, luminance, device power, and light transmittance under the AM1.5 solar simulator, as illustrated in Figure 2(e, f). An increase in light transmittance corresponds to higher brightness, due to the enhanced photon absorption of PSMs for OLED emission, leading to increased device power. The device exhibits its highest EQE* of 10.47% at a 20% light transmittance.

Conclusion

This study introduces a OTO-CID for photon-based signal conversion technologies. The OTO-CID, covering 174.2 cm² with 16 subunits, integrates the PSMs for obtaining complicated optical signals with series-connected structures and TTE-OLEDs for emitting polished signals. Enhancement of crystallization uniformity and structural design in large-area PSMs is enabled by a novel anti-solvent spraying technique and mechanical scribing respectively. The OTO-CID achieves 10.43% EQE* and 2032.33 Cd/m² under 20% AM1.5 conditions, demonstrating high-efficiency operation under ultra-low light intensities. This study enhances efficient optical signal processing and transmission, advancing perovskite-based optoelectronic hybrid integration technology.

References

[1] J. Vukusic, "Optoelectronics: An Introduction," *Optica Acta: International Journal of Optics,* vol. 31, no. 1, pp. 5-6, 1984.

[2] H. Yu *et al.*, "Dual-Functional Triangular-Shape Micro-Size Light-Emitting and Detecting Diode for On-Chip Optical Communication in the Deep Ultraviolet Band," *Laser & Photonics Reviews,* p. 2300789, 2024.

[3] N. Patle, A. B. Raj, C. Joseph, and N. Sharma, "Review of fibreless optical communication technology: History, evolution, and emerging trends," *Journal of Optical Communications,* vol. 45, no. 3, pp. 679-702, 2024.

[4] A. Higuchi and Y. Nakayama, "Cipher Modulation for Optical Camera Communication with Digital Signage," in *2024 IEEE 21st Consumer Communications & Networking Conference (CCNC),* 2024: IEEE, pp. 466-471.

[5] H. Zhou *et al.*, "Efficient and stable perovskite mini-module via high-quality homogeneous perovskite crystallization and improved interconnect," *Nature Communications,* vol. 15, no. 1, p. 6679, 2024.

[6] G. Shi *et al.*, "Manipulating solvent fluidic dynamics for large-area perovskite film-formation and white light-emitting diodes," *Nature Communications,* vol. 15, no. 1, p. 1066, 2024.

[7] F. Li, F. R. Lin, and A. K. Y. Jen, "Current state and future perspectives of printable organic and perovskite solar cells," *Advanced Materials,* vol. 36, no. 17, p. 2307161, 2024.

[8] Y. Jeong *et al.*, "Laser Scribing for Perovskite Solar Modules of Long-Term Stability," *Solar RRL,* vol. 8, no. 8, p. 2301040, 2024.

[9] J.-M. Wang *et al.*, "Toward ultra-high efficiency tandem OLEDs: Benzothiazole-based bipolar hosts and specific device architectures," *Chemical Engineering Journal,* vol. 472, p. 145023, 2023.

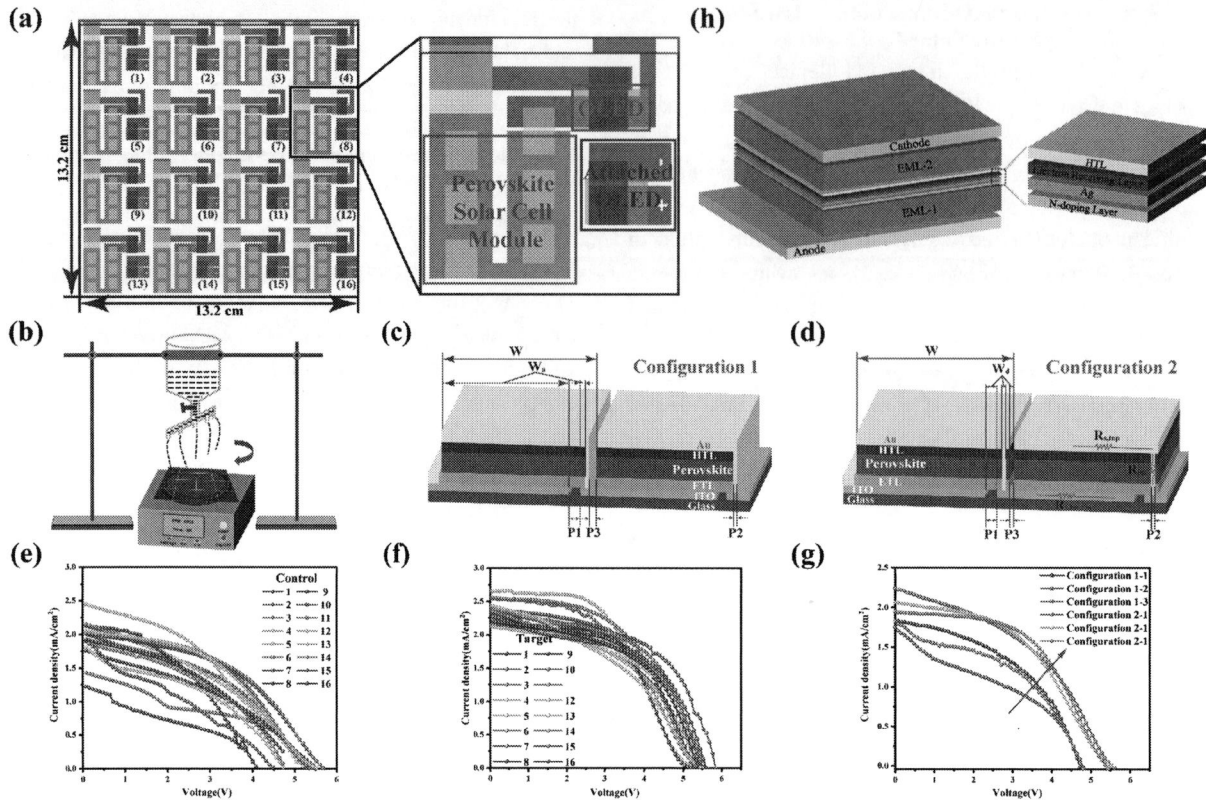

Fig. 1: (a) The structural diagrams of the OTO-CID and the subunit. (b) Schematic of the custom-built anti-solvent spraying technique. (c, d) The structural diagrams of scribing configurations 1 and configurations 2. (e, f) J-V curves of the control and target large-area PSMs. (g) J-V curves of the large-area PSMs prepared by configurations 1 and configurations 2. (h) The structural diagrams of the TTE-OLEDs.

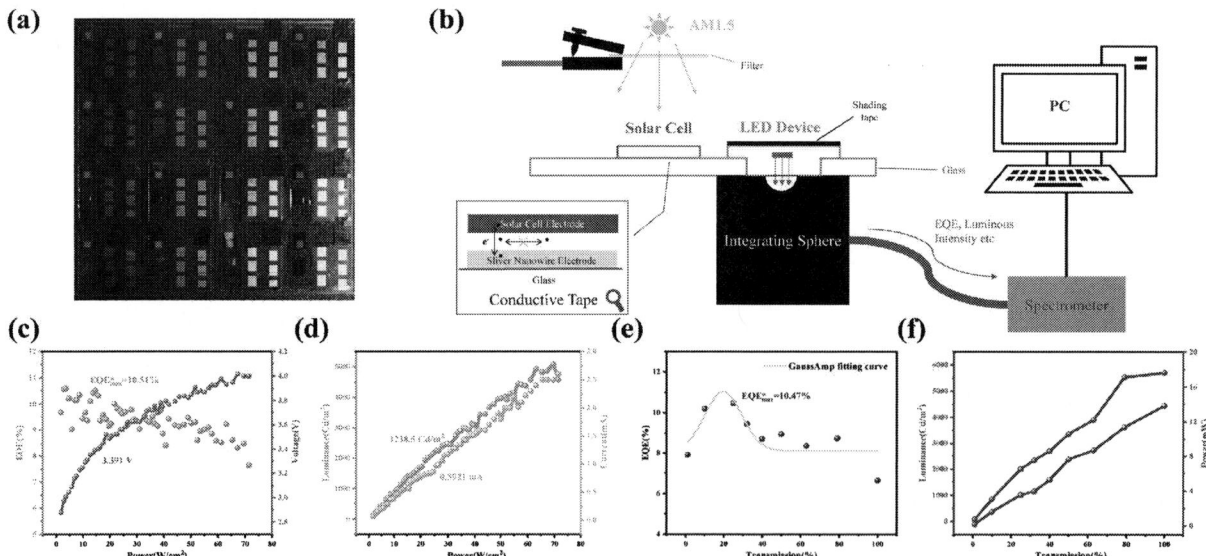

Fig. 2: (a) The operation photograph of the OTO-CID. (b) Schematic of the performance characterization and testing system. The relationship between the EQE*, voltage (c), and the luminance, current (d) concerning the different light intensities under the LED. The relationship among the EQE* (e) and the luminance, power (f) concerning the different transmissions under the AM1.5 solar simulator.

Van der Waals Interfacial Engineering for High-Performance Macroscopic Assembled Graphene (MAG)-Silicon Schottky Photodiodes

Srikrishna Chanakya Bodepudi[1,3]*, Muhammad Abid Anwar[1,3], Muhammad Malik[1], Xiaolei Ding[2], Yance Chen[1], Yue Dai[1], Zongwen Li[1], Zhi-Xiang Zhang[1], Yunfei Xie[1], Wenzhang Fang[1], Huan Hu[1,2], Bin Yu[1], Yang Xu[1,2]*

[1]College of Integrated Circuits, State Key Laboratory of Silicon and Advanced Semiconductor Materials, Zhejiang University, ZJU-HIC, Hangzhou, 310027, China; *Email: bodepudi@zju.edu.cn, yangxu-isee@zju.edu.cn

[2]ZJU-UIUC Institute, International Campus, Zhejiang University, Haining, 314400, China; [3]Authors contributed equally

Abstract

In 2D layered systems, converting excessive photoexcited energy into an electrical response before losing it to the substrate is crucial for energy-efficient photodetection. These hot electron energy losses at photoemission junctions limit the carrier transfer, sensitivity, and modulation frequencies. Our study showcases improved photoresponse in macroscopic assembled graphene (MAG)/Si Schottky junctions by inserting a graphene (Gr) layer that electronically decouples MAG from Si. This van der Waals (vdW) engineering interface effectively suppresses dark current while maintaining hot carrier re-thermalization in the MAG film, leading to enhanced photo thermionic emission (PTI). This vdW interfacial engineering technique has broader implications in energy-efficient electronics, optoelectronics, and memory devices.

Keywords: Graphene Schottky diode, van der Waals interface, hot carriers, photo-thermionic emission (PTI)

Introduction

Energy dissipation at various interfaces is one of the limiting factors in the performance of energy-efficient optoelectronics and memory devices, where interfacial barrier height, contact resistance, momentum mismatch, and electron-phonon coupling play vital roles [1,2,3]. Developing hybrid 2D-3D junctions can significantly suppress the dark current while exploiting the benefits of narrow bandgap atomically thin 2D layered systems such as graphene, black phosphorus, tellurium, palladium diselenide (PdSe2), etc. In these systems, charge carrier energy transfer to lattice heating at interfaces due to carrier concentration from high excitations compromises charge transfer efficiency, photoresponse, switching speeds, and detection bandwidths [2,3].

Interfacial engineering can suppress surface-defect or trap-related dark current by coating with high-k dielectrics, adding buffer layers, and band structure tuning with chemical treatment, or intercalation [4]. On the other hand, despite being free of dangling bonds and fermi-level pinning with a vdW gap, electronic energy dissipation into lattice heating at 2D-3D hybrid systems still leads to high dark currents and reduced detectivities [1,2]. This work addresses this issue by inserting decoupled interfacial graphene in hybrid 2D-3D Schottky diodes based on macroscopic assembled graphene (MAG) and silicon. This approach suppresses dark current, improves photoresponse, sensitivity, and bandwidth, and offers a simple vdW engineering method to enhance charge and photocarrier transfer in 2D layered integrated systems, with bulk (3D) semiconductors for logic, memory, and sensing applications.

High-quality macroscopic assembled graphene films

Figure 1(a-d) demonstrates the preparation steps of MAG. Our previous work extensively discusses MAG film's distinctive synthesis and transfer process, highlighting their seamless transfer without buckling or detachment [5, 6].

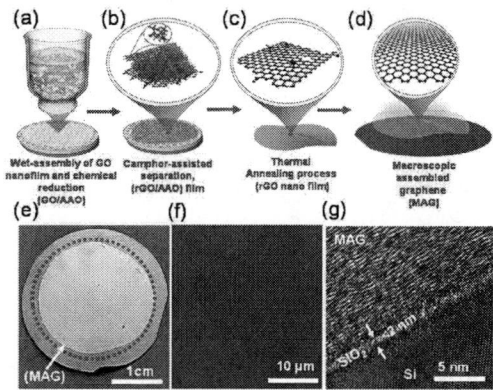

Fig. 1: (a-d) Preparation methods of MAG. (e) Optical image depicting 45nm macroscopic assembled graphene (MAG) film conformally transferred on a 2-inch Si wafer. (f, g) SEM and TEM image of the MAG film/Gr-Si interface.

The free-standing 45nm thick MAG nanofilms were transferred conformally onto 2-inch wafers, and no large-scale buckling or detachments were observed, as illustrated in Figure 1(e). The high-quality interface of the MAG/Gr-Si is also evident in the Scanning Electron Microscopy (SEM) and transmission electron microscopy (TEM) cross-sectional image in Figure 1 (f, g). A thin native SiO_2 (<2nm) can be observed. Since electrically decoupled layers often do not exhibit any additional physical distance, the decoupled Gr layer cannot be distinguished from MAG film at the interface in the cross-sectional image.

We have fabricated two sets of hybrid 2D-3D Schottky diodes for this study: (1) MAG nanofilm/Si and (2) MAG nanofilm/Gr-Si. A lightly doped n-type silicon with a 1–10 Ω.cm resistivity is used.

In addition to the standard steps for metallic contacts, fabrication involves photolithography to pattern silicon windows, and then SiO_2 is selectively etched with the buffer oxide etchant. Device-1 is prepared by transferring, MAG nanofilm onto the exposed silicon window to form a Schottky junction as shown in Figure 2 (a) with the transfer process described in our previous study [6]. For device-2, a single-layer CVD graphene is first transferred onto the exposed silicon window and patterned by a lithography and O_2 plasma etching followed by MAG nanofilm transfer with the same process used for device-1 as depicted in Figure 2 (b).

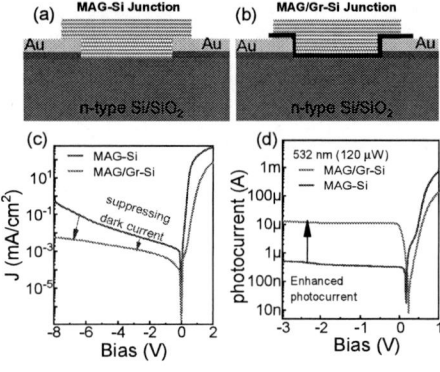

Fig. 2: (a) Device-1 MAG-Si (b) Device-2, MAG/Gr-Si Schottky diode. (c) The dark current of MAG-Si and MAG/Gr-Si device. The dark current in the MAG-Si junction has been suppressed by three orders of magnitude. (d) The enhanced photocurrent of ~1900% in Gr interfaced MAG/Si junction under the illumination of 532nm.

Suppressed Leakage Current and Photoresponse
As graphene (Gr) is separately transferred before MAG film transfer over Si, the orientation mismatch between interfacial layers leads to electronic decoupling between Gr and MAG nanofilm. This gives a decoupled interface between the MAG with graphite-like band structure and single-layer graphene with Dirac-like energy dispersion [9], leading to the momentum mismatch of carriers from MAG and Gr, enabling a tunneling transport through Gr to the Schottky interface [10]. This is evident with suppressed device current with a significant drop in the dark current

(~590% at -0.5V) in Gr-interfaced devices (Figure 2(c)). The decoupled graphene interface also significantly enhanced the photocurrent by 19×10^2% for 532nm at 120μW as shown in Figure 2(d). Photocurrent for varying incident power values has been measured for both types of devices (Figure. 3 (a, b)) and decoupled Gr-interfaced samples consistently show higher $\Delta I = I_{ph} - I_{Dark}$ than that of MAG-Si. We also observed three orders of improvement in the on/off ratio ($I_{ph}/I_{Dark} \sim 1000$) for the lowest measured power at 9μW. The significant increase in the photocurrent and responsivity compared to the MAG/Si device is presented in (Figure 3 (c d)). Higher photocurrent and responsivity indicate hot electron-induced photothermionic emission (PTI) while suppressing the energy losses at the MAG-Si interface [5, 6].

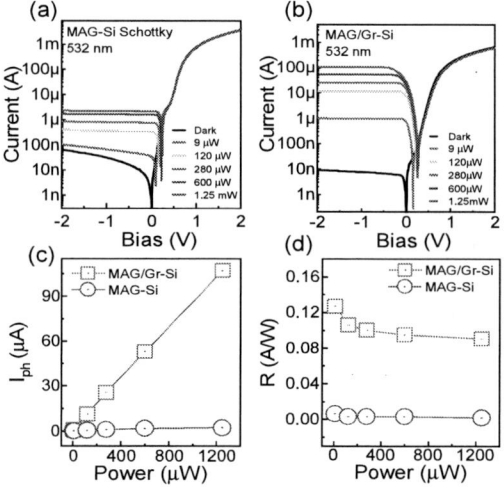

Fig. 3: (a, b) Photocurrent of MAG-Si and MAG/Gr-Si device at various incident power. (c) Photocurrent of MAG-Si and MAG/Gr-Si at different incident powers @ -2 V. The significant increase in the photocurrent in compared to MAG/Si indicates higher sensitivity. (d) Responsivity of MAG-Si and MAG/Gr-Si devices.

Decoupled Graphene Assisted Re-thermalization
We performed a transient photoresponse test with 200 femtoseconds (fs) pulsed laser (50kHz repetition rate) at 633nm wavelength, presented in Figure 4. We noticed that the Gr-interfaced MAG-Si junction shows a broader width of the photocurrent compared to that of MAG-Si, which enhanced hot carrier density in photothermionic emission (PTI), indicating that the PTI effect appears to dominate the photoemission in the device (Figure 4 (b)). In addition, a decrease in relaxation time (fall time, $\tau_f < 1 \mu s$)) in the transient photoresponse of Gr-interfaced MAG-Si junction compared to that of MAG-Si (~10 μs) under the same

979-8-3315-0417-5/25 $31.00 © 2025 IEEE

measurement conditions is observed (Figure. 4 (c)). Slower decay in MAG-Si results from device cooling, particularly interfacial heating via electron-substrate phonon coupling.

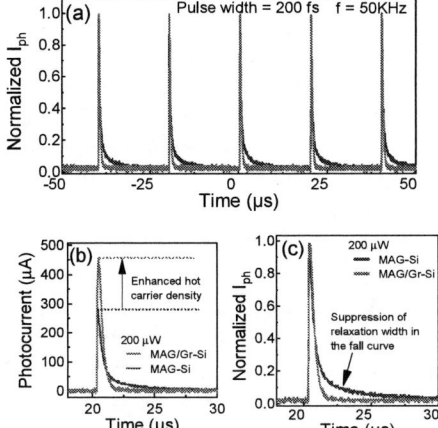

Fig. 4: (a) Transient photoresponse (200 fs pulse width, 50 kHz repetition rate under 633 nm) of MAG-Si and MAG/Gr-Si Device (b) The Photocurrent peak of MAG-Si and MAG/Gr-Si devices indicate that photothermionic emission (PTI) turns to be a dominating emission mechanism. (c) Transient photoresponse relaxation of MAG-Si and MAG/Gr-Si devices.

Adding graphene with a varying layer orientation than the MAG film at the interface decouples the MAG surface from Si and introduces an additional tunnel barrier for out-of-plane charge transport due to a momentum mismatch between photocarriers from MAG to Gr with linear energy dispersion [11]. This is evident from the significantly suppressed dark current observed in the presented case.

Based on the present work, the Gr-interfaced MAG-Si device provides two key features: (1) electronically decouples MAG from the Si interface and allows hot carriers with energies higher than (ϕ_B) into Si via tunneling through the Gr, and (2) effectively blocks the non-radiative energy transfer from hot electrons with energies lower than (ϕ_B) to the Si interface and preserves enhanced hot carrier density in MAG. Our preliminary studies suggest that in graphene-interfaced MAG-Si devices, the decoupled graphene effectively blocks the cooling of hot electrons via SPP [8] and preserves the hot electron energy within MAG [7]. This originates from the optical phonon cooling bottleneck effect, where excessive hot optical phonons re-thermalize electrons instead of cooling through acoustic phonons [7, 8].

This re-thermalization substantially increases the hot electron density and, thus, enhances the PTI in silicon.

To further confirm this, transient absorption spectroscopy studies are required. It is worth noting that Gr-interfaced samples show relatively better performance in key parameters like low dark current and short relaxation times compared to similar device schemes as displayed in Table 1.

Device Structure	J_{dark} (mA/cm²)	τ_{fall} (µs)
MLG(45nm)/Al₂O₃(2nm)/Ge [12]	0.4	1
MLG(45nm)/Si [6]	0.2	2
FLG(5nm)/Al₂O₃ (3nm)/Si [13]	33	10⁶
Gr/hBN(5nm)/Si [14]	0.0001	10³
GNR film(65nm)/Al₂O₃ (10nm)/Si [15]	0.1	310
MoS₂ (2nm)/Gr/Si [16]	0.096	10⁵
MLG(45nm)/Gr-Si [This work]	0.004	0.85

Table 1 Comparison of MAG/Gr-Si devices with some state-of-the-art hybrid 2D-3D photodiodes.

Conclusion

A decoupled interfacial graphene suppresses dark current and enhances photoresponse in MAG-Si Schottky junctions. This vdW engineering approach can potentially reduce energy losses in hybrid 2D-3D interfaces, benefiting energy-efficient logic, memory, and sensing.

Acknowledgments

We thank Chao Gao for providing MAG samples and Dawei Di for facilitating transient photocurrent and absorption studies. This work is funded by the start-up funding of ZJU-HIC Centre (04010000-K02013007, 02170000-K3F113024) and the National Key Research and Development Program of China (2022YFA1204900).

References

[1]   K. K. Paul, et al., Nat Rev Phys 2021, vol. 3, pp. 178–192.

[2]   P. V. Pham et al., Chem Rev 2022, vol. 122, pp. 6514-6613.

[3]   X. Zhang et al., Laser Photon Rev 2024, vol. 18, p. 2300936.

[4]   D. Wu et al., ACS Nano 2021, vol. 15, pp. 10119–10129.

[5]   W. Liu et al., Nat Electronics 2022, vol. 5, pp. 281–288.

[6]   Peng L, Liu L, Du S, et al., InfoMat 2022, vol. 4, p. e12309.

[7]   E. A. A. Pogna et al., ACS Nano 2021, vol. 15, pp. 11285–11295.

[8]   Tony. L, et al., Phys Rev B vol. 86, p. 045413.

[9]   Ferrari, A., Basko, D. Nature Nanotech 2013, vol. 8, pp. 235–246.

[10] Guohong Li, et al., Phys. Rev. Lett 2009, vol. 102, p. 176804.

[11] Xiaodan Zhu, et al., Nano Lett 2018, vol. 18, pp. 682–688.

[12] L. Liu et al., IEDM 2021, pp. 9.2.1-9.2.4.

[13] M. A. Rehman, et al., Carbon 2018, vol. 132, pp. 157-164.

[14] Won, U.Y., et al., Nano Res 2021, vol. 14, pp.1967-1972.

[15] Mingyang Wang, et al., Nano Lett 2024, vol. 24, pp. 165-171.

[16] Wondeok Seo, et al., App. Surf. Science 2022, vol. 604, p. 15448

Spintronic Stochastic Neuron -based Deep Belief Networks for Image Classification

Aijaz H. Lone[1], Meng Tang[1], Daniel N. Rahimi[1], Divyanshu Divyanshu[1],
Camelia Florica[2], Selma Amara[1], Hossein Fariborzi[1], Gianluca Setti[1]

[1]Department of Computer, Electrical and Mathematical Sciences and Engineering (CEMSE),

[2]Nanofabrication Core Lab,

King Abdullah University of Science and Technology (KAUST), Thuwal, 23955, Saudi Arabia

{aijaz.lone,meng.tang,divyanshu.divyanshu,camelia.florica,selma.amara,gianluca.setti}@kaust.edu.sa

Abstract

Spintronic devices hold promise for AI, particularly in Boltzmann computing architectures, due to their tunable stochastic switching. We demonstrate a SOT-tuned spintronic neuron integrated into a deep belief network for pattern classification. Its switching probability, driven by current, mirrors biological neuron activation through a sigmoid function. Integrated with RBM and logistic regression, the network achieves 97% classification accuracy, stable across temperature variations, underscoring spintronics' potential in robust AI applications.

Keywords: Spintronics, Boltzmann neuron and Neuromorphic computing

Introduction

The tremendous data generation in modern computing systems is accompanied by substantial energy dissipation, largely attributed to the Von Neumann bottleneck [1] and the inherent limitations of CMOS technology. As an alternative, neuromorphic or brain-inspired computing has gained traction over the past decade, showing promising advancements in addressing these challenges [2], [3]. Spintronic devices, on the other hand, have emerged as promising beyond CMOS technology [4]. Mainly due to the tunable energy barrier and memristive properties in spintronic devices, [4], novel computing architectures, such as neuromorphic computing [5], reservoir computing, and stochastic computing, based on these devices are being explored [6], [7]. The simulated and experimental demonstrations of spintronic synapse and neuron elements for neuromorphic computing are on emergence [4]–[7]. Although deterministic computing architectures based on CMOS or those beyond CMOS technologies have seen great success, the energy crises associated with this computing paradigm must be addressed. In deterministic computing architectures, noise-related stochasticity has been considered undesired. Adopting a deterministic computing approach may only be necessary for some applications. In many cases, leveraging stochastic behavior and integrating noise can provide significant advantages, enhancing performance and adaptability. Consequently, in recent years, there has been a growing interest in inherently probabilistic computing methods as an alternative to traditional deterministic computing [6], [8]. Some works have proposed [9] or demonstrated [10] the use of p-bit spintronic device-based RBMs for small pattern classification. However, p-bits are unstable against temperature and other process variations. Moreover, as the data becomes more complex, more sophisticated models like Restricted Boltzmann Machines (RBMs) and Deep Belief Networks (DBNs) become necessary to capture deeper patterns within the data. This paper demonstrates the experimental realization and development of a SOT-driven stochastic spintronic neuron device. The neuron switching probability follows the Boltzmann neuron-like characteristics, with input current as the driving variable. The measured device characteristics are modeled as the sigmoid neuron. The model is integrated within an RBM architecture as a Gibbs sampling node and tested with various classifiers as nonlinearity activation functions. The trained and tested classifiers include logistic regression, Multi-Layer Perceptron (MLP), and DBN. The neuron characteristics are measured for multiple devices and at variable temperatures to evaluate device-to-device variability and temperature's effect on classification accuracy. When trained and tested on MNIST dataset the RBM and DBN combination achieved the classification accuracy of around 97%. Thus, showing the applicability of these spintronic stochastic neurons in hardware DBNs.

Spintronic Stochastic Neuron

The stochastic Boltzmann neuron model introduced in this work is motivated by the probabilistic switching behavior observed in Ta/CoFeB/MgO/Ta spintronic device structure shown in Fig. 1(a). These devices consist of a Ta(5 nm) as heavy metal, in which a charge current is passed laterally in the x-axis. The charge current gets converted to the spin current via the spin-orbit coupling effect, which generates the spin-orbit torque on the magnetization vector of the CoFeB (0.85 nm) ferromagnetic layer deposited above the Ta. MgO (2 nm) acts as the barrier and symmetry-breaking layer, whereas the top Ta(2 nm) is the capping layer. These devices are fabricated using magnetron sputtering of the material stack at room temperature, followed by photolithography and ion-milling etch steps. The fabricated device structure

979-8-3315-0417-5/25 $31.00 © 2025 IEEE

optical image is shown in Fig. 1(b). The dimensions of the device nanotrack are $540\,\text{nm} \times 20\,\mu\text{m}$. We used the write-read scheme shown in Fig. 1(c) to program the neuron device in a stochastic region. First, the devices are reset into the state (0) by a fixed $I_N = -1.45\,\text{mA}$ current, and device Hall resistance is measured. After that, we apply the write current pulses, followed by the reading. The write current amplitude is increased in steps from $I_P = (1.3\,\text{mA to } 1.5\,\text{mA})$ for the device labeled (B1). Similarly, the same measurement scheme is followed in the rest of the devices. The measured devices show deterministic switching if the writing current amplitude is well above the critical current but when operated near the critical current region. These devices switched stochastically, as shown in Fig. 1(d); the switching probability was measured as a function of the writing current, which follows the sigmoid function characteristics. These characteristics form the foundation for our neuron model, where we modeled the probability of a neuron firing or switching as a custom Boltzmann sigmoid activation function.

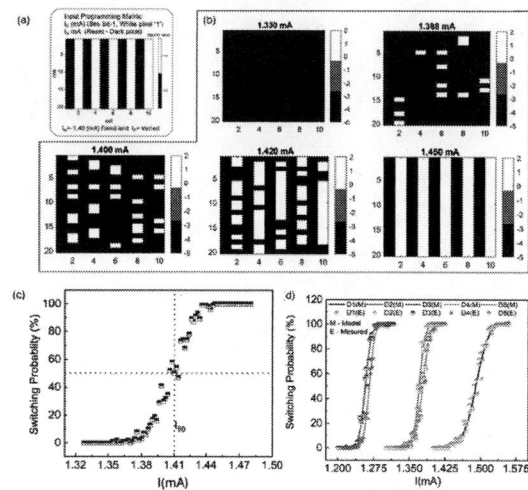

Fig. 2. (a) Writing programming matrix. (b) Device output matrix (response to the input programming) measured at different writing current densities. (c) SOT-driven stochastic switching and determination of I50. (d) Device switching probability as a function of the current for multiple devices (measured and modeled).

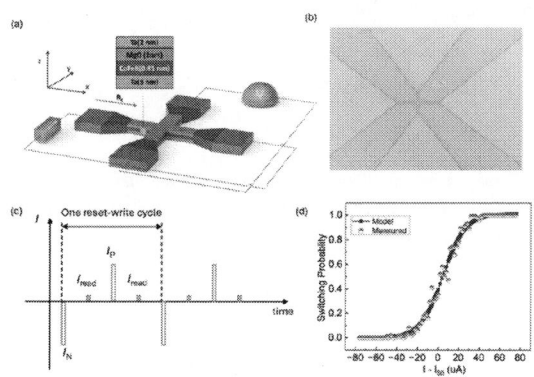

Fig. 1. (a) SOT controlled stochastic neuron device structure. (b) Optical microscope image of the device. (c) Read-write programming scheme. (d) The current controlled switching probability of the device is modeled as a Boltzmann sigmoid activation function.

A. Switching Probability and Sigmoid Function

Fig. 2(a) shows the device programming (writing) scheme implied to evaluate the switching probability as the function of time. As mentioned earlier, the I_P was increased gradually and device resistance was measured. Fig. 2(b) shows the device states bright (1) and dark (0) at increasing writing current amplitude. The resistance switching from 0 to 1 is increasing with the increasing current. The switching probability is defined as a ratio of 1s and total writing operations. The multiple measured devices (D1 to D5) exhibit the switching probability governed by the Boltzmann sigmoid function. After normalizing the probability to 1 and shifting the origin to

0. The characteristics are modeled using the following equation:

$$P(I) = \frac{1}{1 + e^{-kI}} \quad (1)$$

Where:
- I, and $k = 0.091\ A^{-1}$, represent the writing current, and slope of the probability function.

In our model, the input I represents the normalized and scaled pixel intensity of the MNIST image in terms of current, and the resulting $P(I)$ corresponds to the probability of the neuron switching to an active state for current I.

DEEP BELIEF NEURAL NETWORK (DBN) FOR MNIST IMAGE CLASSIFICATION

We integrated the stochastic sigmoid neuron model into a DBN image classification architecture. The DBN consisted of an input layer with 784 nodes, followed by 3 Restricted Boltzmann Machine (RBM) layers of size (500, 500, and 2000) and a classifier of 10 nodes. The RBM is a generative, unsupervised learning model that learns to reconstruct the input by minimizing the system's energy. The energy function of the RBM is defined as:

$$E(v, h) = -\sum_i a_i v_i - \sum_j b_j h_j - \sum_{i,j} v_i W_{ij} h_j \quad (2)$$

Where:
- v_i and h_j are the visible and hidden units, respectively.
- W_{ij} is the weight matrix connecting visible and hidden units.
- a_i and b_j are the biases for the visible and hidden units.

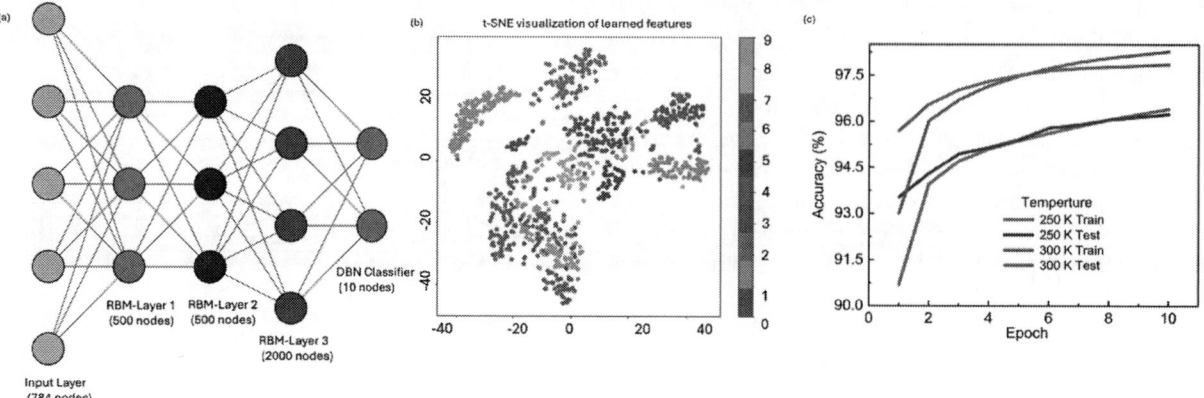

Fig. 3. (a) DBN Architecture. (b) t-SNE visualization of feature representations learned by the semi-supervised model, showing clear clustering of test samples by class. (c) Classification accuracy on MNIST dataset at 250 K and 300 K temperatures.

B. Stochastic Neuron in RBM

The stochastic neuron model is integrated into the RBM to compute the activation probabilities of the hidden units given the visible layer. In this context, the spintronic neuron's stochastic nature allows the RBM to explore a broader range of possible configurations, helping minimize the system's overall energy. During the Gibbs sampling process, the hidden states h_j are sampled from the computed probabilities:

$$h_j = P(h_j = 1|v_i) = \sigma(v_i W_{ij} + b_j) \qquad (3)$$

Where σ represents the custom stochastic sigmoid model of the device.

The RBM combined with logistic regression achieved an accuracy 85% on the MNIST dataset. This served as the baseline performance for further exploration. While the accuracy is solid, the model was limited in capturing non-linear relationships within the data. To improve the accuracy, we implemented a Deep Belief Network (DBN) by stacking multiple RBMs layer-wise, as shown in Fig. 3(a). Each RBM was trained unsupervised, and once all layers were pre-trained, the network was fine-tuned using backpropagation. The activations of hidden units represent extracted features, which the RBM uses to reconstruct the image. These extracted features shown in Fig. 3(b) enhance the classification performance by providing a concise and meaningful representation of the original images. After finetuning the classifier layer with learning rate of 0.001 for 60 epochs and batch size of 32, the DBN achieved the best performance, with an accuracy of 97% as shown in Fig. 3(c). This highlights the strength of deeper architectures and the effectiveness of the stochastic neuron model in feature extraction.

CONCLUSION AND FUTURE WORK

This paper demonstrates the integration of a stochastic spintronic sigmoid neuron within a DBN architecture for handwritten digit classification. The neuron introduces probabilistic activation into the learning process, enhancing the RBM's ability to extract meaningful features. When a classifier layer based on the neuron added to the RBM achieves the classification accuracy of around 97%, these results provide a clear roadmap for the circuit-level demonstration of the spintronic device-based DBNs or hardware DBNs.

REFERENCES

[1] Schuman *et al.*, "Opportunities for neuromorphic computing algorithms and applications," *Nature Computational Science*, vol. 2, pp. 10–19, 2022.

[2] B. Dieny *et al.*, "Opportunities and challenges for spintronics in the microelectronics industry," *Nat Electron*, vol. 3, no. 8, pp. 446–459, Aug. 2020.

[3] G. Finocchio *et al.*, "The promise of spintronics for unconventional computing," *Journal of Magnetism and Magnetic Materials*, vol. 521, p. 167506, Mar. 2021.

[4] A. Sengupta *et al.*, "Spin-orbit torque induced spike-timing dependent plasticity," *Applied Physics Letters*, vol. 106, no. 9, p. 093704, Mar. 2015.

[5] A. H. Lone *et al.*, "Magnetic tunnel junction based implementation of spike time-dependent plasticity learning for pattern recognition," *Neuromorph. Comput. Eng.*, vol. 2, no. 2, p. 024003, Jun. 2022.

[6] S. Chowdhury *et al.*, "A Full-Stack View of Probabilistic Computing With p-Bits: Devices, Architectures, and Algorithms," *IEEE J. Explor. Solid-State Comput. Devices Circuits*, vol. 9, no. 1, pp. 1–11, Jun. 2023.

[7] K. M. Song *et al.*, "Skyrmion-based artificial synapses for neuromorphic computing," *Nat Electron*, vol. 3, no. 3, pp. 148–155, Mar. 2020.

[8] B. Sutton *et al.*, "Intrinsic optimization using stochastic nanomagnets," *Sci Rep*, vol. 7, no. 1, p. 44370, Mar. 2017.

[9] J. Deng *et al.*, "Voltage-Controlled Spintronic Stochastic Neuron for Restricted Boltzmann Machine With Weight Sparsity," *IEEE Electron Device Lett.*, vol. 41, no. 7, pp. 1102–1105, Jul. 2020.

[10] N. S. Singh *et al.*, "Cmos plus stochastic nanomagnets enabling heterogeneous computers for probabilistic inference and learning," *Nature Communications*, vol. 15, p. 2685, 2024.

G-Universe: A Collection of In-House Technology Computer-Aided Design Simulators as a Platform for Developing New Simulation Capabilities

S.-M. Hong, I. K. Kim, S.-W. Jung, M.-S. Jang, P.-H. Ahn, T. Oh, G.-T. Jang, and K.-W. Lee

School of Electrical Engineering and Computer Science, Gwangju Institute of Science and Technology

ABSTRACT

In this work, G-Universe, a collection of in-house Technology Computer-Aided Design (TCAD) simulators for new simulation capabilities, is briefly introduced. After discussing the motivation of in-house TCAD development, its two major building blocks, G-Process and G-Device, are introduced with examples. Future development directions are discussed.
Keywords: TCAD, process emulation, device simulation

INTRODUCTION

G-Universe is a collection of TCAD simulators developed by the Semiconductor Device Simulation Laboratory (SDSL) at the Gwangju Institute of Science and Technology. Initially, it was developed as an internal device simulator for non-conventional transport simulations, including the convective derivative term [13]. Over time, it has evolved into a general TCAD framework. In this work, we provide a comprehensive overview of G-Universe.

The main research theme of the SDSL is to investigate and develop new TCAD simulation capabilities for the next-generation semiconductor device technology. Although there are several existing TCAD frameworks [2]–[6], which provide excellent simulation capabilities, those tools cannot be modified by an external user. Therefore, G-Universe has been built as an academic platform to develop new simulation capabilities.

The general structure of G-Universe, including its components, is shown in Fig. 1. Like other TCAD frameworks [2]–[6], there are two major building blocks, the process simulator(/emulator) and the device simulator. These tools are introduced with examples in subsequent sections.

G-PROCESS

For simple cases, a manual structure generation code can be adopted to generate a device structure. However, practical device structures adopted in industry are much more complicated. In order to yield meaningful analysis results, we need to consider realistic structures. It is clear that process simulation(/emulation) is required.

G-Process [7]–[9] is a process emulator based upon the three-dimensional (3D) multi-level-set method [10], as shown in Fig. 2. The sparse field level-set [10] is employed to compute the boundary movement efficiently. When the user requests, closed boundaries are extracted by using the marching cubes method [11].

In Fig. 3, the process emulation results for a CFET inverter, fabricated using the backside network with wrap-around contacts and buried power rails, are shown. Quite recently, a stress calculation routine has been implemented in G-Process. At time instances specified by the user, a 3D mesh for the process-emulated structure is constructed by using an external mesh generator [12]. Then, the finite-element method is applied to obtain the stress and strain profiles of the given structure. As shown in Fig. 4, the stress profile in the source/drain and channel regions of a PMOSFET can be simulated. After the process emulation is finished, the final 3D mesh is built and used in the device simulation.

G-DEVICE

G-Device [13], [13]–[19] is a device simulator and plays the central role in the G-Universe framework. It is a drift-diffusion simulator with additional simulation capabilities such as the density-gradient model. Examples of its applications are shown below.

The drift-diffusion device simulator provides a good initial, approximate solution for advanced transport solvers. As shown in Fig. 5, advanced transport solvers such as the NEGF (Non-Equilibrium Green Function) solver can be used in the stand-alone mode, where solutions at previous bias points are used to initiate the Newton-Raphson loop at the present bias point. Nevertheless, it is much more efficient to directly start the loop with the solution of G-Device at the same bias point [20]. Moreover, when the entire device structure including the source/drain regions is simulated, the conventional device simulator can be used to calculate carrier densities over the entire simulation domain.

Some materials newly introduced in the semiconductor industry may have several material parameters to be calibrated. In some cases, even new physical models should be implemented. In such cases, an in-house device simulator can be conveniently employed. As an example, Fig. 6 shows an IGZO (Indium Gallium Zinc Oxide) transistor [21] and its IV characteristics obtained from the G-Device simulation.

In Fig. 7, another advantage of in-house TCAD development is clearly shown. In [22], it has been demonstrated that multiple meshes having different types (for example, a fully 3D mesh and a quasi-two-dimensional (quasi-2D) mesh with cylindrical symmetry) can be combined in a single simulation. It can be used to investigate the impact of local geometrical deformation on the performance.

Our final example is related with the acceleration of TCAD device simulation. It has been shown that a quality approximate solution can be used to skip the time-consuming bias ramping procedure [14], [17], [18]. While there can be multiple ways to prepare such an accurate approximate solution (for example, pre-trained neural networks [14], [18]), any new acceleration algorithm can be tested only when we have an in-house device simulator. In Fig. 8, it is shown that the TCAD device simulation can be significantly accelerated with help of the quasi-1D model [17].

DISCUSSIONS

In addition to G-Process and G-Device, we need a program for visualizing the simulation results. Instead of developing yet another visualization program, an open-source visualization engine, ParaView [23], is adopted. The CGNS (CFD General Notation System) file format [24] is used to store the simulation results.

Since G-Universe has been developed by a small number of developers in the SDSL, there are still several missing simulation capabilities. Progresses in the code development will be presented elsewhere.

Let us discuss the future development directions. The growing demand for System-Technology Co-Optimization (STCO) [25] is clear and pressing. At the same time, from a TCAD developer's perspective, it is equally evident that the evolution path from traditional TCAD simulations to system-level performance assessments remains a bit unclear. Developing a well-established workflow that extends beyond Design-Technology Co-Optimization (DTCO) could be an exciting challenge ahead. G-Universe will be further developed to provide new simulation capabilities for addressing this challenge.

CONCLUSION

In this work, a comprehensive overview of G-Universe has been made. Representative applications of G-Universe have been demonstrated. It is expected that G-Universe can used as an academic platform for further development of new simulation capabilities.

ACKNOWLEDGMENT

This research was supported by the National Research Foundation of Korea (NRF) grant funded by the Korea government (NRF-2023R1A2C2007417).

REFERENCES

[1] S.-M. Hong and J.-H. Jang, "Numerical simulation of plasma oscillation in 2-D electron gas using a periodic steady-state solver", IEEE Trans. Electron Devices, 62, pp. 4192-4198 (2015).

[2] V. Moroz et al., "DTCO launches Moore's law over the feature scaling wall", IEDM Tech. Dig. (2020).

[3] P. Blaise et al., "Ab initio simulation of advanced materials and devices: Current challenges", IEDM Tech. Dig. (2019).

[4] G. Rzepa et al., "Performance and variability-aware SRAM design for gate-all-around nanosheets and benchmark with FinFETs at 3nm technology node", IEDM Tech. Dig. (2022).

[5] M. A. Stettler et al., "Industrial TCAD: Modeling atoms to chips", IEEE Trans. Electron Devices, 68, pp. 5350-5357 (2021).

[6] W. Choi et al., "Hierarchical simulation of monolithic CFETs using atomistic and continuum models", International Conference on Simulation of Semiconductor Processes and Devices (SISPAD) (2024).

[7] I. K. Kim, S. Cha, and S.-M. Hong, "Optimization of nitrogen ion implantation condition for β-Ga$_2$O$_3$ vertical MOSFETs via process and device simulation", IEEE Trans. Electron Devices, 69, pp. 6948-6955 (2022).

[8] I. K. Kim, S.-C. Han, G. Park, G.-T. Jang, and S.-M. Hong, "Effect of Si separator in Forksheet FETs on device characteristics investigated by using in-house TCAD process emulator and device simulator", International Conference on Simulation of Semiconductor Processes and Devices (SISPAD) (2023).

[9] S.-W. Jung, I. K. Kim, K.-W. Lee, and S.-M. Hong, "Simulation of monolithic CFET with split-gate structure", International Conference on Simulation of Semiconductor Processes and Devices (SISPAD) (2024).

[10] O. Ertl et al, "A fast level set framework for large three dimensional topography simulations", Comput. Phys. Commun., 180, pp. 1242-1250 (2009).

[11] T. S. Newman et al., "A survey of the marching cube algorithm", Computers & Graphics. (2006).

[12] H. Si, "TetGen, a Delaunay-based quality tetrahedron mesh generator", ACM Trans. Math. Softw., 41, pp. 11:1-11:36 (2015).

[13] S.-M. Hong and J.-H. Jang, "Numerical simulation of plasma oscillation in 2-D electron gate using a periodic steady-state solver", IEEE TED, 62, pp. 4192-4198 (2015).

[14] S.-C. Han, J. Choi, and S.-M. Hong, "Acceleration of semiconductor device simulation with approximate solutions predicted by trained neural networks", IEEE Trans. Electron Devices, 68, pp. 5483-5489 (2021).

[15] P.-H. Ahn and S.-M. Hong, "Quantum transport simulation with the first-order perturbation: Intrinsic AC performance of extremely scaled nanosheet MOSFETs in THz frequencies", IEDM Tech. Dig. (2021).

[16] K.-W. Lee and S.-M. Hong, "Derivation of a universal charge model for multigate MOS Structures with arbitrary cross sections", IEEE Trans. Electron Devices, 69, pp. 3014-3021 (2022).

[17] K.-W. Lee and S.-M. Hong, "Acceleration of semiconductor device simulation using compact charge model", International Conference on Simulation of Semiconductor Processes and Devices (SISPAD) (2022).

[18] S.-C. Han, I. K. Kim, and S.-M. Hong, "Accelerated simulation and performance optimization of 3D multigate logic transistors by using neural networks", IEEE Electron Devices Technology and Manufacturing Conference (EDTM) (2023).

[19] P.-H. Ahn and S.-M. Hong, "Geometric scattering describing mode-coupling effects in non-uniform cross-sections for non-equilibrium Green's function and multi subband Boltzmann transport equation solvers", IEDM Tech. Dig. (2023).

[20] M. Luisier, A. Schenk, and W. Fichtner, "Quantum transport in two-and three-dimensional nanoscale transistors: Coupled mode effects in the nonequilibrium Green's function formalism", Journal of Applied Physics, vol. 100, no. 4, (2006).

[21] T.-C. Fung et al., "Two-dimensional numerical simulation of radio frequency sputter amorphous In-Ga-Zn-O thin-film transistors", Journal of Applied Physics, vol. 106, no. 8, (2009).

[22] G.-T. Jang and S.-M. Hong, "Hybrid 2D/3D mesh for efficient device simulation of a locally tilted vertical NAND string", International Conference on Simulation of Semiconductor Processes and Devices (SISPAD) (2022).

[23] https://www.paraview.org/

[24] https://cgns.github.io/

[25] V. Moroz et al., "3DIC system-technology co-optimization with a focus on the interplay of thermal, power, timing, and stress effects", IEEE Symposium on VLSI Technology and Circuits. (2024).

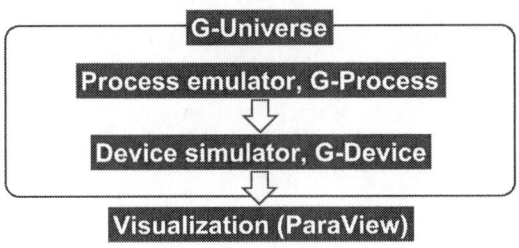

Fig. 1. General structure of G-Universe, a collection of in-house TCAD simulators. Its two major building blocks are G-Process and G-Device. Visualization is done by an external program.

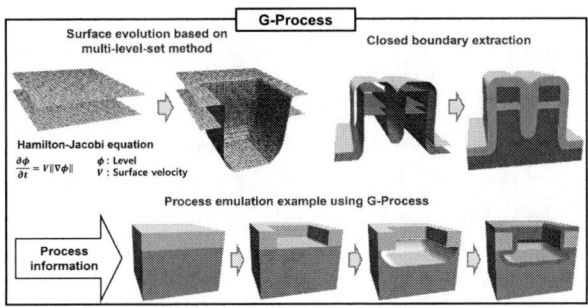

Fig. 2. Summary of the multi-level-set used in G-Process. It focuses on the surface evolution and closed boundaries are extracted based on the user request.

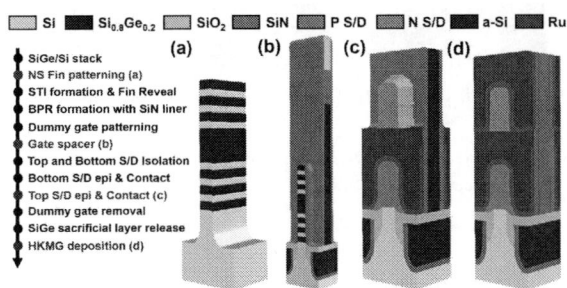

Fig. 3. Process emulation results for a CFET inverter using the backside network with wrap-around contacts and buried power rails.

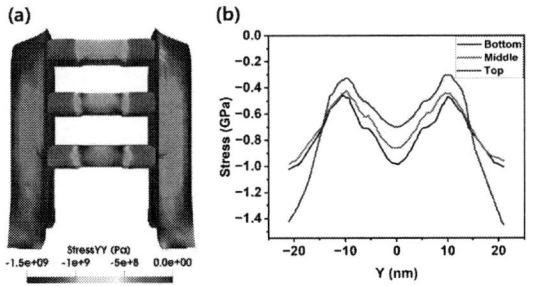

Fig. 4. Stress profile of a PMOS device. (a) 3D stress profile. (b) 1D stress profiles along the channel direction.

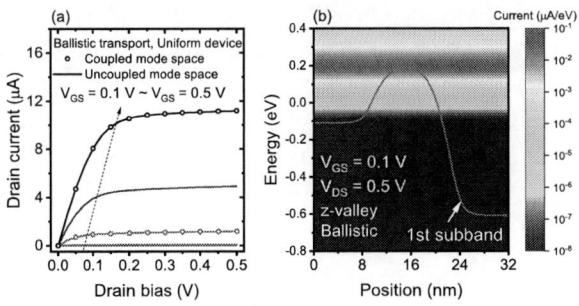

Fig. 5. (a) Output curve obtained from NEGF methods when the gate bias varies 0.1 V to 0.5 V. (b) Energy spectrum of current at the z-valley when the gate bias is 0.1 V and drain bias is 0.5 V.

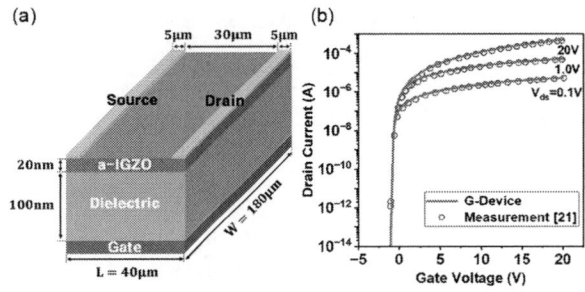

Fig. 6. (a) Structure of a-IGZO TFT. (b) Transfer characteristics at various drain voltages, 0.1V, 1.0V, and 20V.

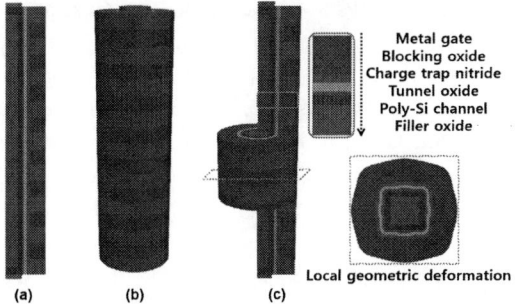

Fig. 7. Mesh types used in the simulation of 3D NAND flash memories. (a) Quasi-2D mesh structure, (b) full 3D mesh structure, and (c) hybrid mesh structure with local geometric deformation.

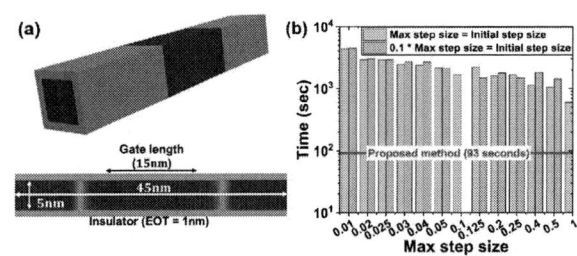

Fig. 8. (a) Structure of a square GAA MOSFET. (b) Computing time to obtain a solution at $V_G = V_D = 0.7$ V.

979-8-3315-0417-5/25 $31.00 © 2025 IEEE

Impact of Load-Si Thickness and Punch-Through Stopper Doping on the Electrical Performance of SOI Gate-All-Around Field-Effect Transistors

Longyu Sun[1,2], Jiayi Zhang[1,2], Yan Li[3], Haoyan Liu*[1,2,3], Xi Zhang[1,2,3], Yongliang Li*[1,2,3]

[1]Key Laboratory of Fabrication Technologies for Integrated Circuits, Chinese Academy of Sciences, China, [2]Institute of Microelectronics, Chinese Academy of Sciences, China, [3]School of Applied Science, Beijing Information Science and Technology University, China

*Email: liuhaoyan@ime.ac.cn, liyongliang@ime.ac.cn

Abstract

In this article, we systematically investigate the effects of load-Si thickness and punch-through stopper (PTS) on the electrical characteristics of SOI GAAFET through 3D technology computer-aided design (TCAD). When N_{PTS} increases to 5×10^{18} cm^{-3}, the leakage of SOI GAAFET with a thick load-Si has significantly suppressed. When T_{Si} is 5 nm or below, it is not necessary to apply the PTS scheme to achieve excellent electrical properties, thanks to stronger gate control and quantum effects.

Keywords: TCAD, Load-Si, PTS, SOI, GAAFET

Introduction

Gate-All-Around (GAA) field-effect transistors (FETs) have emerged as a promising candidate for continued device scaling in advanced technology nodes [1]. However, GAAFETs still face significant challenges, particularly regarding sub-fin leakage issues [2]. So far, several innovative solutions have been proposed, including punch-through stopper (PTS) [3], buried dielectric insulation (BDI) [4], and Silicon-on-Insulator (SOI) technologies [5]. Among these approaches, the SOI substrate, which incorporates an oxide layer (BOX) within the Si substrate, has demonstrated remarkable effectiveness in mitigating sub-fin leakage. However, the existing Si layer above the BOX (Load-Si) may lead to the formation of parasitic channel, because it should maintain a sufficient thickness for the following the stacked sacrificial layer/channel layer epitaxy of GAAFET. PTS strategy can be adopted in load-Si to effectively suppress leakage current of SOI GAAFET. However, during the high temperature load-Si oxidation and trimming process, undesirable doping impurity redistribution may happen, which will seriously affect the leakage suppression. Moreover, it is very difficult to precisely control impurity doping when PTS implantation is employed after thinning the load-Si to several tens of nanometers. Therefore, the electrical performance of SOI GAAFET needs to be further optimized in terms of load-Si thickness and PTS doping.

In this work, the effects of load-Si thickness and PTS on the electrical characteristics of SOI GAAFET using TCAD are systematically investigated in detail.

Simulation Structure and Methodology

Fig. 1(a) depicts the 3D schematic of SOI GAAFET, while Fig. 1(b-c) illustrate its cross-sectional views along X-X' and Y-Y' direction, respectively. In this article, we focus on P-SOI GAAFET, whose main parameters are referred from IRDS 2023, as shown in Table 1. The gate length (L_G) is set to 14 nm, the channel width (W_{CH}) is set to 25 nm, the channel thickness (T_{CH}) is set to 6 nm, and the inner spacer length (L_{IS}) is set to 6 nm. The channel doping concentration (N_{CH}) is set to 1×10^{16} cm^{-3}. The thickness of load-Si (T_{Si}) is set to 5-20 nm. For GAAFET with PTS scheme, the PTS doping concentration (N_{PTS}) is set to a series of values from 5×10^{17} cm^{-3} to 1×10^{19} cm^{-3}. Fig. 1(d) depicts the fabricated process flow of SOI GAAFET, compatible with bulk-Si GAAFET. The load-Si treatment comprises two parts: the thinning process and the PTS doping process. The remaining processes are consistent with those of the bulk-Si GAAFET.

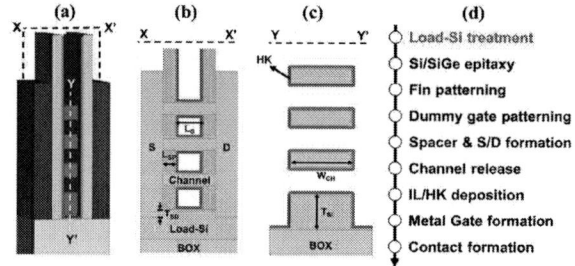

Fig. 1: (a) 3D schematic, (b) X-X' cross-section view, (c) Y-Y' cross-section view, and (d) process flow of SOI GAAFET.

Table 1: Geometric Parameters of SOI GAAFET

Parameters	Values
Gate Length (L_G)	14 nm
Channel Thickness (T_{CH})	6 nm
Channel Width (W_{CH})	12 nm
Inner Spacer Length (L_{IS})	6 nm
Channel Doping (N_{CH})	1×10^{15} cm^{-3}
S/D Doping (N_{SD})	3×10^{20} cm^{-3}
Equivalent Oxide Thickness (T_{OX}/T_{HfO2})	0.5 nm/1.28 nm
Load-Si Thickness (T_{Si})	5~20 nm
PTS Doping (N_{PTS})	5×10^{17}~1×10^{19} cm^{-3}

The electrical characteristics of SOI GAAFET were simulated by Sentaurus TCAD. The drift-diffusion and density-gradient models were considered for electrical simulation. The inversion and accumulation layer mobility models were used to simulate mobility degradation. The low-field ballistic and high-field saturation were also introduced for reliable simulation. Meanwhile, Slotboom bandgap narrowing model, Hurkx band-to-band tunneling (BTBT) model and recombination model were incorporated. The parameters of models were calibrated with the experimental result, as shown in Fig. 2.

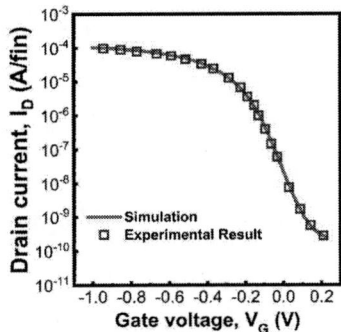

Fig. 2: The calibrated transfer characteristics of SOI GAAFET against experimental data from .

Results and Discussion

Fig. 3(a) demonstrates the I_D-V_G transfer characteristics of SOI GAAFET without PTS under various T_{Si}. The results clearly indicate that increasing T_{Si} leads to significant deterioration in both off-state current (I_{OFF}) and subthreshold swing (SS). This degradation can be primarily attributed to the presence of more leakage paths in thicker load-Si regions, which adversely affects the device's ability to suppress short-channel effects (SCE). Fig. 3(b) shows the I_D-V_G curves at T_{Si}=15nm for different N_{PTS}. The results indicate that when N_{PTS} increases to 5×10^{18} cm^{-3}, the I_{OFF} and SS have significantly improved. However, as N_{PTS} continues to increase, the I_{OFF} exhibits a trend of increase. Therefore, both T_{Si} and N_{PTS} have significant impacts on the DC electrical characteristics.

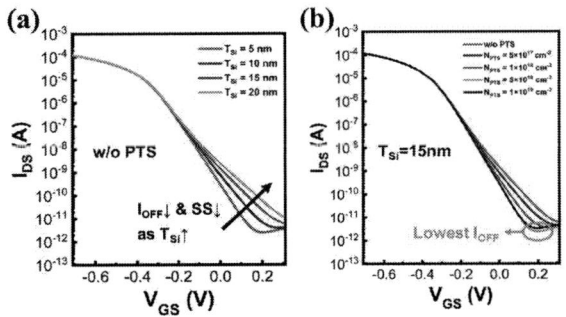

Fig. 3: (a) I_D-V_G curves of SOI GAAFET without PTS under different T_{Si}. (b) I_D-V_G curves of SOI GAAFET without PTS under different T_{Si}.

Fig. 4(a-b) show the effects of different N_{PTS} on I_{OFF} and SS at T_{Si} of 5 nm and 15 nm. Compared with T_{Si} of 15 nm, the I_{OFF} and SS are nearly immunity to N_{PTS} as T_{Si} is reduced to 5 nm, which may be related to its stronger gate control.

Therefore, when T_{Si} is thinned down to 5 nm or below, it is not necessary to apply the PTS scheme.

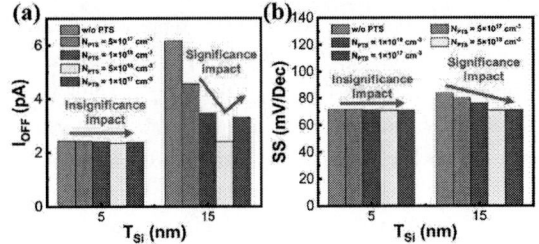

Fig. 4: (a) I_{OFF} and (b) SS of SOI GAAFET under different T_{Si} and N_{PTS}

Fig. 5 presents the hole current density distribution under the off-state state at T_{Si}=15nm. Without PTS, there is a significant PT current in the load-Si, which constitutes the main portion of I_{OFF}. When N_{PTS} reaches 5×10^{18} cm^{-3}, the PT current in the load-Si is effectively suppressed, leading to a substantial reduction in I_{OFF}. As shown in Fig. 6, the current in the load-Si only accounts for a small fraction of I_{OFF} under T_{Si} of 5nm, which is much smaller than the channel leakage. Therefore, although PTS can also reduce the leakage in the load-Si, it has no effect on the channel leakage and I_{OFF}.

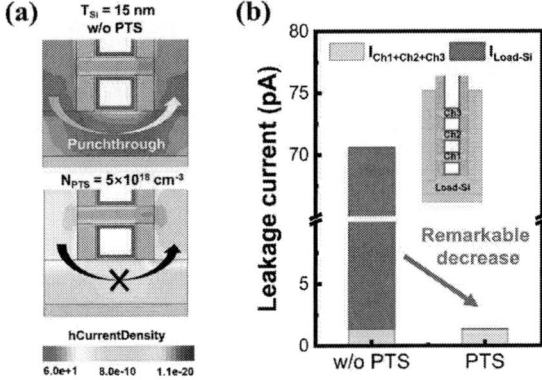

Fig. 5: (a) Off-state hole current density distribution at T_{Si} = 15nm with and without PTS. (b) $I_{Ch1+Ch2+Ch3}$ and $I_{Load-Si}$ values with and without PTS.

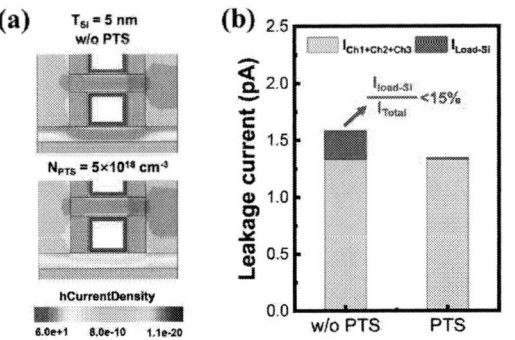

Fig. 6: (a) Off-state hole current density distribution at T_{Si} = 5nm with and without PTS. (b) $I_{Ch1+Ch2+Ch3}$ and $I_{Load-Si}$ values with and without PTS.

Fig. 7(a) illustrates the hole density distribution under T_{Si}=5nm. It can be observed that, compared to GAAFET without PTS, the incorporation of PTS does not significantly change the distribution of holes. Fig. 7(b) presents the energy band with and without PTS at T_{Si} of 5 nm and 15 nm, along the channel direction. It is evident that, compared to T_{Si}=15nm, the bandgap is wider at T_{Si} of 5 nm due to quantum effects. Meanwhile, when T_{Si} is 5nm, the energy bands are not influenced by PTS, further confirming that when T_{Si} is reduced to 5 nm or below, PTS doping is unnecessary to achieve perfect electrical properties.

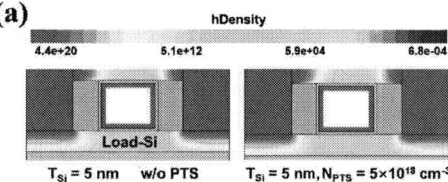

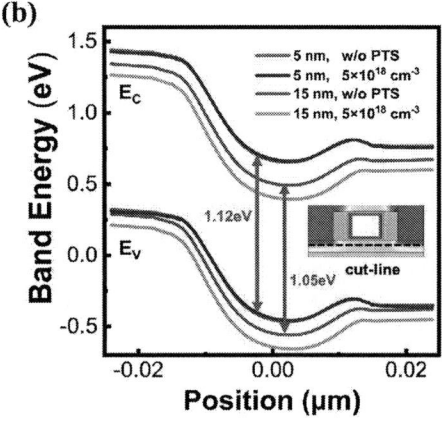

Fig. 7: (a) Off-state hole density distribution at T_{Si} = 5nm with and without PTS. (b) Band energies in the load-Si at T_{Si} = 5 and 15nm with and without PTS.

The impacts of T_{Si} and PTS on I_{ON} and C_{GG} were also investigated, as depicted in Fig. 8. I_{ON} is extracted at $|V_G|$ = $|V_{TH}|$ + 0.5V. Notably, when T_{Si}=5nm, I_{ON} exhibits a slight decreasing trend with the increase in N_{PTS}. It can be attributed to the higher doping concentration in the load-Si, which leads to stronger impurity scattering and consequently reduces mobility. Fig. 9 presents the current density distribution within the load-Si under the on-state state, both with and without PTS. It can be observed that increasing N_{PTS} results in a significant reduction in the current within the load-Si. Moreover, as illustrated in Fig. 8(b), C_{GG} gradually increases with the increase in T_{Si}. It indicates that a thinner load-Si not only obviates the need for PTS doping process but also decreases the device's capacitance.

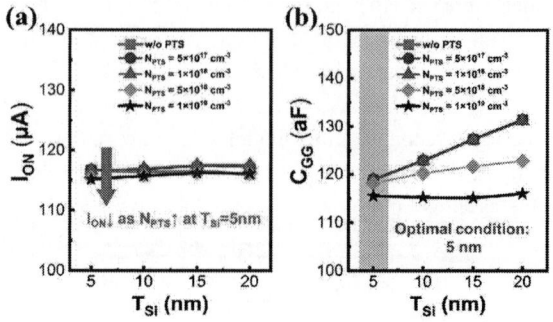

Fig. 8: (a) I_{ON} and (b) C_{GG} values according to T_{Si} under different N_{PTS}.

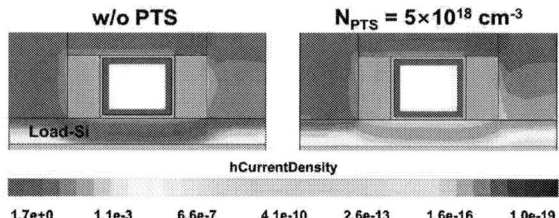

Fig. 9: On-state hole current density distribution at T_{Si} = 5nm with and without PTS.

Conclusion

In this work, we have demonstrated that reducing the thickness of the load-Si to 5 nm or below can not only achieve excellent electrical characteristics, but also eliminate the need for PTS doping process in SOI GAAFET, potentially simplifying the fabrication process.

Acknowledgments

This work is supported in part by the Strategic Priority Research Program of Chinese Academy of Sciences (Grant no. XDA0330304), and in part by the Joint Development Program (Grant no. QYJS-2023-3100-B).

References

[1] N. Loubet *et al.*, "Stacked nanosheet gate-all-around transistor to enable scaling beyond FinFET," in *2017 symposium on VLSI technology*, 2017: IEEE, pp. T230-T231.

[2] L. Liebmann, J. Zeng, X. Zhu, L. Yuan, G. Bouche, and J. Kye, "Overcoming scaling barriers through design technology cooptimization," in *2016 IEEE Symposium on VLSI Technology*, 2016: IEEE, pp. 1-2.

[3] H. Mertens *et al.*, "Gate-all-around MOSFETs based on vertically stacked horizontal Si nanowires in a replacement metal gate process on bulk Si substrates," in *2016 IEEE symposium on VLSI technology*, 2016: IEEE, pp. 1-2.

[4] J. Zhang *et al.*, "Full bottom dielectric isolation to enable stacked nanosheet transistor for low power and high performance applications," in *2019 IEEE International Electron Devices Meeting (IEDM)*, 2019: IEEE, pp. 11.6. 1-11.6. 4.

[5] Y.-G. Liaw, C.-W. Chen, W.-S. Liao, M.-C. Wang, and X. Zou, "Effects of ultra-thin Si-fin body widths upon SOI PMOS FinFETs," *Modern Physics Letters B*, vol. 32, no. 15, p. 1850157, 2018.

20 nm Gate-Length Normally-Off AlGaN/GaN HEMTs Enabled by SiN$_x$ Stress-Engineered Technique

Chenkai Deng[1,2#], Peiran Wang[2#], Qing Wang[2*], Hongyu Yu[2*]

(*Email: wangq7@sustech.edu.cn; yuhy@sustech.edu.cn)

[1] School of Electronic Information and Engineering, Harbin Institute of Technology, Harbin 150001, China

[2] School of Microelectronics, Southern University of Science and Technology, Shenzhen 518055, China

Abstract

In this work, we proposed a SiN$_x$ stress engineered technique to achieve normally off GaN HEMTs with recess-free. By introducing compressive stress SiN$_x$ passivation, a positive threshold voltage shift of 2.91 V was achieved in devices with a gate length of 20 nm compared to devices with stress-free passivation. Additionally, this approach effectively suppressed short-channel effects, ultimately resulting in a normally-off GaN devices with a threshold voltage of 1.13 V.
Keywords: AlGaN/GaN HEMTs, Normally-off, SCEs, SiN$_x$ stress-engineered and TCAD simulation

Introduction

Gallium Nitride (GaN), a third-generation wide-bandgap semiconductor, exhibits remarkable properties such as high saturated electron velocity, high critical breakdown field, and high-temperature stability [1]. Due to the strong spontaneous and piezoelectric polarization, GaN high-electron-mobility transistors (HEMTs) typically operate in a normally-on mode. However, using normally-on devices in switching circuits adds to design complexity and increases costs. In contrast, normally-off operation with a positive threshold voltage (V_{th}) enables a single-polarity power supply, eliminating the need for the dual-polarity supply required by normally-on devices. This approach not only saves space but also enhances safety and simplifies the overall circuit design [2].

To date, various techniques have been suggested to achieve normally off GaN devices, such recessed gates [3], fluorine ion implantation [4], P-type (e.g. p-GaN) gates [5], and tri-gate structure [6]. Fluorine ion implantation involves localized plasma processing beneath the gate, which can damage the surface barrier layer. At elevated temperatures, F ions diffuse easily, often leading to V_{th} instability. Recessed gates, p-type gates (e.g., p-GaN), and tri-gate structures require precise, low-damage etching processes. Inadequate etching, however, can lead to reliability issues, including increased leakage and V_{th} instability. In p-GaN gate HEMTs, the thick p-GaN cap layer increases the separation between the gate and channel, reducing peak transconductance and weakening gate control over the channel, potentially impacting overall device performance.

In this work, recess-free 20 nm gate length (L_g) normally-off GaN HEMTs with a threshold voltage of 1.13 V is realized by SiN$_x$ stress-engineered technique. TCAD Sentaurus simulations show that as gate dimensions decrease, stress levels increase, resulting in a more pronounced positive shift in the V_{th}. Additionally, the incorporation of SiN$_x$ stress effectively mitigates short-channel effects, counteracting the negative threshold voltage shift associated with gate scaling. These results underscore the potential of stress-engineering techniques in advancing normally off GaN HEMTs with reduced gate lengths.

Experiments

Fig. 1 shows the structure of normally-off AlGaN/GaN HEMTs utilizing a SiN$_x$ stress-engineered technique, as modeled in TCAD Sentaurus simulations. These simulations were calibrated according to the procedure in reference [7] and validated with experimental data from previous studies [8]. The device structure, shown in Fig. 1, consists of a 1.05 μm buffer layer, a 1 μm Al$_{0.07}$GaN back barrier layer, a 100 nm intrinsic GaN channel layer, a 1 nm AlN spacer, a 10 nm Al$_{0.25}$Ga$_{0.75}$N barrier layer, and 200 nm compressively stressed (-2 GPa) SiN$_x$ passivation layer. The reported devices have a L_g range from 20 nm to 500 nm, a gate width (W_g) of 100 μm, and equal gate-drain (L_{gd}) and gate-source (L_{gs}) spacings of 2 μm each.

Results and Discussion

Fig. 2 shows the stress distribution and 2DEG distribution of these devices. In devices with a gate length of 100 nm, the introduction of a -2 GPa (compressive) SiN$_x$ passivation layer effectively depletes the 2DEG beneath the gate. Fig 3(a) illustrates the polarization in an AlGaN/GaN heterostructure. The GaN layer, typically considered relaxed, exhibits spontaneous polarization (P_{SP}), while the AlGaN layer contains both spontaneous polarization (P_{SP}) and piezoelectric polarization (P_{PE}). Introducing a -2 GPa SiN$_x$ stress passivation layer effectively mitigates the polarization effects beneath the gate. As a result, this elevates the conduction band of the AlGaN/GaN heterojunction above the Fermi level, resulting in the depletion of the 2DEG beneath the gate and enabling a positive shift in V_{th}, as shown in Fig. 3(b-d). As predicted by the edge-force model [9], the stress field peaks at each gate edge. For the 20 nm gate dimension, the proximity of the gate edges causes the compression at the center of the gate to accumulate, resulting in a stronger compression compared to the 100 nm gate, as shown in Fig. 1(e). Figure 4a illustrates the stress distribution at the AlGaN/GaN heterojunction for -2 GPa SiN$_x$ devices with

979-8-3315-0417-5/25 $31.00 © 2025 IEEE

gate lengths ranging from 20 to 500 nm. As the gate length decreases, the stress becomes increasingly concentrated, rising from an initial -3 GPa to -9.5 GPa. Consequently, the 2DEG density beneath the heterojunction in the gate region decreases from approximately 3.7×10^{17} cm^{-3} to 2.4×10^3 cm^{-3}, as shown in Fig. 4(b). Typically, as the dimensions of GaN HEMTs are scaled down (with a decrease in L_g), short-channel effects (SCEs) become increasingly pronounced, leading to a negative shift in V_{th}. Consequently, achieving normally-off devices becomes more challenging when the gate length is reduced below 100 nm. In Fig. 5(a), for devices with stress-free SiN$_x$ passivation, the V_{th} drops from -0.88 V to -1.78 V as the gate length decreases from 500 nm to 20 nm. In contrast, for devices with -2 GPa SiN$_x$ passivation, the V_{th} increases as the gate length decreases from 500 nm to 20 nm, shifting from -0.19 V to 1.13 V, as shown in Fig. 5(b). This is primarily due to the enhanced effect of stress as the device dimensions are scaled down. The threshold voltages for different gate lengths were extracted using a linear fitting method. After the introduction of -2 GPa SiN$_x$ passivation, effective modulation of the threshold voltage was achieved for gate lengths below 500 nm. Specifically, at a gate length of 20 nm, the V_{th} exhibited a positive shift of 2.91 V. Figure 6 depicts the transfer characteristic curves of a device with a gate length of 20 nm under V_{ds} ranging from 1 V to 10 V. As V_{ds} increases, the V_{th} of the devices with stress-free SiN$_x$ passivation shifts negatively by over 4 V. In contrast, the devices with -2 GPa SiN$_x$ passivation experiences only a negative shift of about 1 V, demonstrating significant suppression of SCEs. In summary, the SiN$_x$ stress-engineering technique effectively neutralizes polarization effects, achieving a positive V_{th} shift, while also suppressing short-channel effects. This approach ultimately enables the realization of enhancement-mode GaN HEMTs with a V_{th} of 1.13 V at a gate length of 20 nm.

Conclusion

This study presents an innovative SiN$_x$ stress engineering technique for fabricating enhancement-mode GaN HEMTs without recess structures. By employing compressive stress SiN$_x$ passivation, a substantial positive V_{th} shift of 2.91 V was achieved in devices with a gate length of 20 nm compared to the devices with stress-free SiN$_x$ passivation, while effectively mitigating SCEs. As a result, a normally-off GaN device with a threshold voltage of 1.13 V was achieved. TCAD Sentaurus simulations reveal that as gate dimensions shrink, the induced stress increases, further stabilizing the threshold voltage. These results highlight the significant potential of SiN$_x$ stress engineered technique for high-performance normally-off GaN RF applications, simplifying device design while enhancing reliability and performance.

Acknowledgments

This work was supported by National Natural Science Foundation of China (Grant No: 62274082), Research on mechanism of Source/Drain ohmic contact and the related GaN p-FET (Grant No: 2023A1515030034), Research on high-reliable GaN power device and the related industrial power system (Grant No: HZQB-KCZYZ-2021052), Study on the reliability of GaN power devices (Grant No: JCYJ20220818100605012), Research on the key technology of 1200V SiC MOSFETs (Grant No: JSGG20220831094404008), Research on novelty low-resistance Source/Drain ohmic contact for GaN p-FET (Grant No: JCYJ20220530115411025), 5G Frontier" Project (Phase III) - Micro-Nano Processing Platform (Grant No: K2023390010), High level of special funds (Grant No:G03034K004).

References

[1] M. Asif Khan, A. Bhattarai, J. Kuznia, and D. Olson, "High electron mobility transistor based on a GaN/Al$_x$Ga$_{1-x}$N heterojunction," Applied Physics Letters, vol. 63, no. 9, pp. 1214-1215, 1993.

[2] Y. Cheng, Y. H. Ng, Z. Zheng, and K. J. Chen, "RF enhancement-mode p-GaN gate HEMT on 200 mm-Si substrates," IEEE Electron Device Letters, vol. 44, no. 1, pp. 29-31, 2022.

[3] W. Choi, O. Seok, H. Ryu, H.-Y. Cha, and K.-S. Seo, "High-Voltage and Low-Leakage-Current Gate Recessed Normally-Off GaN MIS-HEMTs With Dual Gate Insulator Employing PEALD-SiN$_x$/RF-Sputtered-HfO$_2$," IEEE Electron Device Letters, vol. 35, no. 2, pp. 175-177, 2013.

[4] Z. Feng, R. Zhou, S. Xie, J. Yin, J. Fang, B. Liu, W. Zhou, K. J. Chen, and S. Cai, "18-GHz 3.65-W/mm enhancement-mode AlGaN/GaN HFET using fluorine plasma ion implantation," IEEE electron device letters, vol. 31, no. 12, pp. 1386-1388, 2010.

[5] Y. Uemoto, M. Hikita, H. Ueno, H. Matsuo, H. Ishida, M. Yanagihara, T. Ueda, T. Tanaka, and D. Ueda, "A Normally-off AlGaN/GaN Transistor with R on A= 2.6 mΩ cm^2 and BV$_{ds}$= 640V using conductivity modulation," International Electron Devices Meeting. IEEE, pp. 1-4, 2006.

[6] M. Dwidar, A. Ofiare, E. Wasige, and A. Al-Khalidi, "Normally-off AlN/GaN HEMTs with a DIBL of 1.15 mV/V for RF Applications," 2023 18th European Microwave Integrated Circuits Conference (EuMIC), pp. 54-57, 2023.

[7] N. kumar Subramani, "Physics-based TCAD device simulations and measurements of GaN HEMT technology for RF power amplifier applications," Université de Limoges, 2017.

[8] C. Deng, W.-C. Cheng, X. Chen, K. Wen, M. He, C. Tang, P. Wang, Q. Wang, and H. Yu, "Current collapse suppression in AlGaN/GaN HEMTs using dual-layer SiNx stressor passivation," Applied Physics Letters, vol. 122, no. 23, 2023.

[9] C. E. Murray, "Mechanics of edge effects in anisotropic thin film/ substrate systems," Journal of Applied Physics, vol. 100, no. 10, 2006.

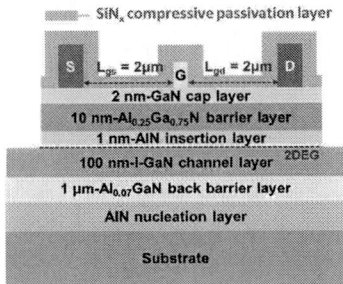

Fig 1. Schematic diagram of AlGaN/GaN HEMTs.

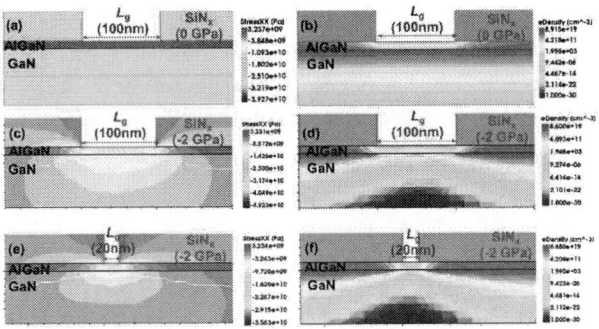

Fig 2. Stress distribution and 2DEG distribution of (a) and (b) stress-free SiN$_x$ devices with 100nm L_g, (c) and (d) -2GPa SiN$_x$ devices with 100nm L_g, (e) and (f) -2GPa SiN$_x$ devices with 20nm L_g.

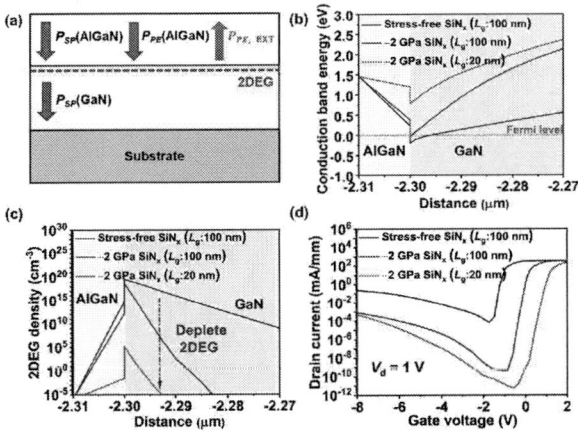

Fig 3. (a) Schematic diagram of polarization in an AlGaN/GaN heterojunction. (b) Conduction band energy, (c) 2DEG density at the heterojunction and (d) transfer characteristic curves under V_{ds} = 1 V of these devices.

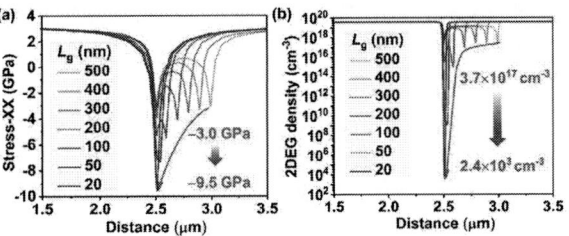

Fig 4. (a) Stress distribution along the XX direction and (b) 2DEG density at the AlGaN/GaN heterojunction beneath the gate for the -2 GPa SiN$_x$ devices with gate lengths ranging from 20 nm to 500 nm.

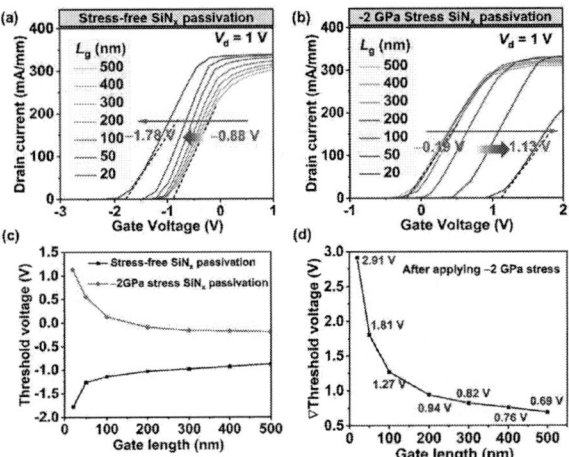

Fig 5. Transfer characteristic curves under V_{ds} = 1 V of (a) stress-free SiN$_x$ devices and (b) -2 GPa SiN$_x$ devices with gate lengths ranging from 20 nm to 500 nm. (c) Comparison of threshold voltages (extracted via linear fitting) for various device structures. (d) Threshold voltage variation after the introduction of -2 GPa SiN$_x$ for different L_g.

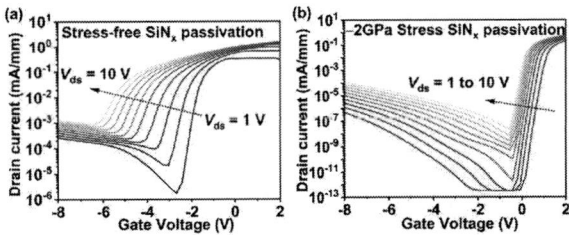

Fig 6. Transfer characteristic curves under V_{ds} = 1 V to 10 V of (a) stress-free SiN$_x$ devices and (b) -2 GPa SiN$_x$ devices with 20 nm L_g.

RESURF Ga₂O₃-on-SiC Field Effect Transistors for Enhanced Breakdown Voltage

Junting Chen[1], Junlei Zhao[1], Xiaohan Zhang[1], Jin Wei[2], and Mengyuan Hua[1,*]

[1]Department of EEE, Southern University of Science and Technology, Shenzhen, China. [2]School of Integrated Circuits, Peking University, Beijing, China. *E-mail: huamy@sustech.edu.cn

Abstract

Heterosubstrates have been extensively studied as a method to improve the heat dissipation of Ga_2O_3 devices. In this simulation work, we propose a novel role for p-type available heterosubstrates (e.g, SiC), as a component of a reduced surface field (RESURF) structure in Ga_2O_3 lateral FETs. The RESURF structure can eliminate the E-field crowding and contribute to higher breakdown voltage. The designing strategy of the p-type region, and the influence of interface conditions are systematically studied using TCAD modeling.

Keywords: Ga_2O_3, heterosubstrate, SiC, RESURF, field effect transistors

Introduction

Ga_2O_3 has attracted intensive research interests in power electronics in recent years [1, 2]. the Ga_2O_3-based power electronics have much higher Baliga's figure of merit compared to Si and other wide-bandgap semiconductors [1]. However, the intrinsically low thermal conductivity of Ga_2O_3 (0.1-0.3 $W·m^{-1}·K^{-1}$ [3]) becomes a major obstacle in the design of power electronics.

As a remedy for this thermal management challenge, heat dissipation of Ga_2O_3-based devices can be improved by utilizing heterogeneous semiconductor substrates with higher thermal conductivity, such as SiC (490 $W·m^{-1}·K^{-1}$). As a result of the effectively promoted heat dissipation, Ga_2O_3 FETs based on such heterosubstrates deliver improved thermal stability and suppressed self-heating effects [4, 5].

However, the investigations are mainly focused on the thermal properties and heat dissipation, whereas the electrical properties of the heterosubstrates and their potential benefits for the power electronics have not been extensively explored. In this work, we demonstrate the utilization of p-type doping of SiC heterosubstrate as a component of a RESURF structure to suppress E-field crowding in lateral Ga_2O_3 FETs.

RESURF Structure and Fabrication Feasibility

As shown in Fig. 1, the device structure (Fig. 1 (a)) and the E-field distribution (Fig. 1 (b)) of the conventional heterosubstrate Ga_2O_3-on-SiC FET is compared with that of a proposed selective-area p-SiC RESURF FET (Fig. 1 (c) and (d)). In the conventional FET, the E-field crowds at the drain-side gate corner, and the breakdown voltage (BV) of the device is 253 V. By introducing a selective-area p-SiC with optimal parameters (Fig. 1 (d)), the E-field is distributed over

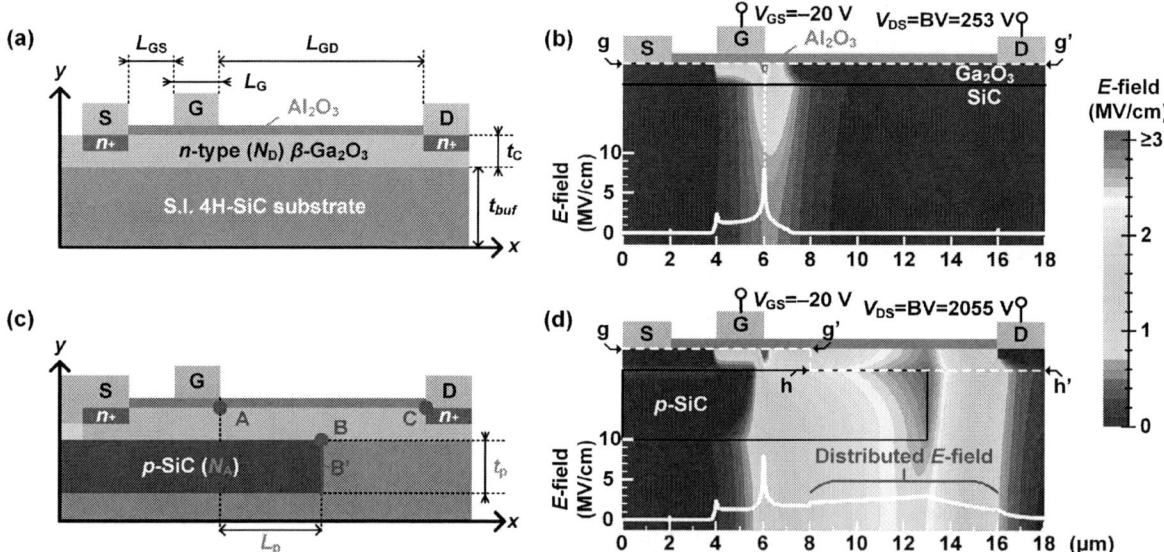

Figure 1. Schematic structures and E-field strengths at breakdown voltages of (a,b) conventional Ga_2O_3/SiC FET; and (c,d) FET with the n-Ga_2O_3/p-SiC RESURF structures. In (b), the E-field along cutline g-g' is shown in the inset. In (d), the E-field along cutline g-g' and h-h' is shown in the inset. The red dots (A, B, and C) and red line (B') in (c) are potential breakdown positions, and are critical in the following discussion. The values of key simulation parameters in (a): L_{GS}, L_G, L_{GD}, t_c, t_{buf}, N_D are 2 μm, 2 μm, 10 μm, 0.2 μm 1.5 μm, and 3×10^{17} cm^{-3}, respectively. One of the optimal values of N_A, L_p, t_p that used in (d) is 0.6×10^{17} cm^{-3}, 7 μm, 0.65 μm, respectively. The breakdown is defined as the E-field in the Ga_2O_3 reaches 8 $MV·cm^{-1}$, or the E-field in the SiC reaches 3 $MV·cm^{-1}$.

979-8-3315-0417-5/25 $31.00 © 2025 IEEE

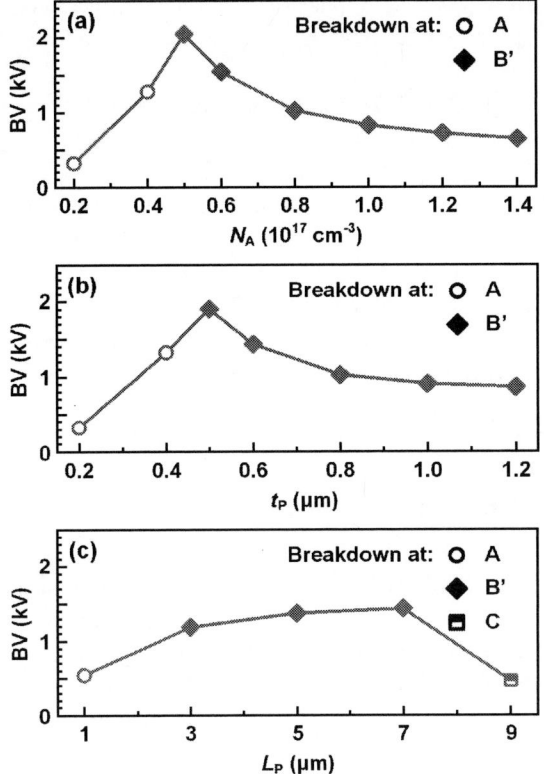

Figure 2. Dependence of BVs on the (a) N_A, (b) t_p, and (c) L_p.

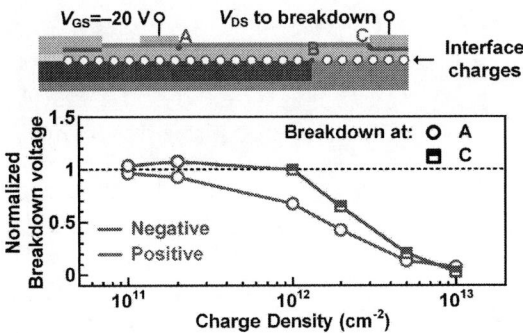

Figure 3. Effects of charge at the Ga_2O_3/SiC interfaces on the BVs. Fixed N_A=0.6×10^{17} cm^{-3}, t_p=0.6 μm, and L_p=7 μm.

a larger area instead of crowding at the gate corner, benefiting to a higher BV of 2055 V.

The fabrication of the proposed RESURF Ga_2O_3-on-SiC FETs is feasible based on existing techniques. The conventional Ga_2O_3-on-SiC FETs have been successfully fabricated [6-8]. Based on the existing devices, the fabrication of the proposed RESURF FETs requires two additional steps: (i) the selective area p-doping of the SiC substrate, and (ii) the Ohmic contact of the p-SiC. The selective-area p-doping technique is well-established and commonly used in SiC devices [9]. Moreover, the doping of the SiC substrate can be done before integrated to the Ga_2O_3 layer, so the engineering of p-SiC region is rather independent from the following fabrication of Ga_2O_3 devices. As for the Ohmic contact, forming Ohmic contact to the p-SiC needs high temperature (600-1100 °C) annealing process [10], whereas forming Ohmic contact to the n-Ga_2O_3 needs low temperature (400-600 °C) annealing process [11]. As a result, the Ohmic contact to the p-SiC should be formed before the formation of the Ohmic contact to the n-Ga_2O_3. Overall, utilizing the electrical properties of the heterosubstrate by the proposed RESURF structures would not add too much complexity to the device fabrication process.

Breakdown Regulation in RESURF Structures

The impacts of the three key parameters, including the N_A,

the thickness t_p, and the L_p of the p-SiC region (illustrated Fig. 1 (c)) are systemically studied.

Fig. 2 (a) shows the relationship between the BV and N_A. For the N_A in the range of 0.2×10^{17} to 0.5×10^{17} cm^{-3}, the BV is dominated by point A, and the BV increases with N_A increasing. Starting from the N_A of 0.5×10^{17} cm^{-3}, the breakdown at line B' dominates the breakdown of the devices, and the BV decreases with N_A increasing. The trend is explained as following. Without the RESURF structures, the E-field tends to crowd at point A, leading to low-voltage device breakdown. The p-doping can contribute negative space charge in the p-SiC to terminate a part of the E-field that originally directs to point A, and results in a fast reduction of the E-field at point A. However, as a compensation, the E-field at line B' increases. Since line B' (p-SiC) has a lower critical E-field (3 MV·cm^{-1}) compared to point A (Ga_2O_3, 8 MV·cm^{-1}), the breakdown point can transfer from point A to line B' when the p-SiC starts to terminate the E-field. Overall, the N_A of 0.5×10^{17} cm^{-3} is the nearly balanced value for the E-field at points A and B to simultaneously reach their critical E-fields.

Fig. 2 (b) shows the relationship between the BV and t_p. Similar to the trend in Fig. 2 (a), the BV increases with t_p increasing from 0.2 μm to 0.5 μm, owing to the reduction of E-field peak at point A. Then, the BV decreases with the t_p further increasing from 0.5 μm to 1.2 μm owing to the increase of the E-field peak at line B'.

Fig. 2(c) shows relationship between the BV and L_p. The p-SiC region with too short L_p cannot sufficiently block the E-field from crowding at point A, so the device breakdowns at point A at a low voltage when the L_p=1 μm. On the other hand, if the p-SiC is too close to the drain terminal (L_p=9 μm), the E-field crowds at point C, causing a low breakdown voltage. When the L_p is in a proper range (3-7 μm in this simulation), the E-field at point A and point B decreases monotonically with L_p increase, because a longer p-SiC benefits to a longer depletion region in the Ga_2O_3 channel, and thus a smaller E-field at a same V_{DS} bias.

Based on the analyses above, with the preset dimensions

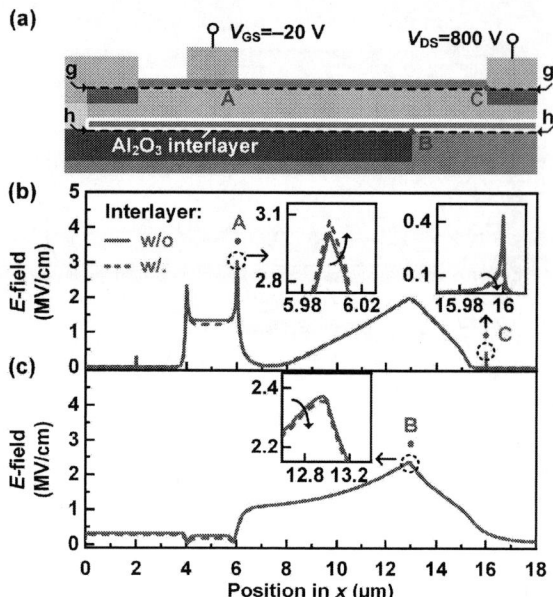

Figure 4. (a) Schematic illustration of Al_2O_3 interlayer. E-field strength along the cutlines (b) g-g' and (c) h-h' of the device with and without the 20-nm Al_2O_3 interlayer. Fixed N_A=0.8×10^{17} cm^{-3}, t_p=0.8 μm, and L_p=7 μm. V_{DS}=800 V. The E-field at point B is used to represent that along line B'. Because the E-field at point B is very close or equal (97-100% at V_{DS}=800 V) to the peak value of the E-field along line B'.

and doping concentration of the n-Ga_2O_3 channel epilayers, the optimal BV can be obtained when the point A in Ga_2O_3 and the line B' in SiC reach their critical E-field simultaneously. In the example shown in Fig. 1 (d), by carefully engineering the N_A and the dimensions of p-SiC, at V_{DS}=BV=2055 V, the peak E-field at point A is 7.91 MV·cm^{-1}, and that at line B' is 2.99 MV·cm^{-1}, indicating that point A and point B breakdown almost at the same time.

To mimic the realistic condition of device, the impact of net charges at the Ga_2O_3/SiC interface is studied. In Fig. 3, the BV with respect to the interface charge density is plotted. The BV is normalized to that of the device without interface charge. A small amount of negative charge ($2×10^{11}$ cm^{-2}) can act as field plates, which help blocking the point A from high E-field, thus slightly improve the breakdown voltage. As for the case where there is a large amount of negative interface charge, it is similar to the case where the L_p is very long (9 μm in Fig. 2(c)). Excessive negative interface charge will cause the E-field to crowd at point C, lowering the breakdown voltage. Positive interface charge has monotonic impact on the breakdown voltage. The positive interface charge compensates the negative space charge in the p-SiC region, and thus eliminates the improvement brought by the RESURF structures, causing the E-field to crowd at point A again. Moreover, the excessive positive interface charge can emit E-field to point A, further enhancing the E-field at point A. As a result, the BV decreases monotonically with positive

interface charge. In practical applications, the interface charge at the Ga_2O_3/SiC interface should be carefully controlled.

An additional Al_2O_3 interlayer at the Ga_2O_3/SiC interface (Fig. 4(a)) is often adopted to improve the interface binding quality and hence the thermal dissipation [7, 8]. The E-field distribution with and without the Al_2O_3 interlayer is compared in Fig. 4(b) (cutline g-g') and Fig. 6 (c) (cutline h-h'). The interlayer slightly weakens the impact of the RESURF structure, leading to a 1.7% increase of peak E-field at point A, a 0.4% decrease of peak E-field at point B, and a 67% decrease of peak E-field at point C. As points A and B are the dominate breakdown points in proper designed devices, and the influence of the interlayer at points A and B is quite small, the interlayer in the practical devices will not limit the adoption of the proposed RESURF structures.

Conclusion

In this work, the electrical properties of heat-dissipating Ga_2O_3-on-SiC heterosubstrate have been utilized to construct RESURF structures in Ga_2O_3 FETs. The proposed RESURF structures can evenly distribute the E-field to achieve higher BVs. With careful design of the N_A and dimensions of p-SiC, as well as eliminating the interface charges, the BV can be improved from 253 V to 2055 V with the RESURF structures. The Al_2O_3 interlayer in the existing heterosubstrate devices has minimal influence on the RESURF structures. This study provides a demonstration of unlocking the full potential of heat-dissipating heterosubstrates by leveraging their electrical properties.

Acknowledgments

This work was financially supported in part by the Guang Dong Basic and Applied Basic Research Foundation under Grant 2024A1515030224, Shenzhen Fundamental Research Program under Grant No. JCYJ20220530114615035 and No. 2023112115707001

References

[1] A. J. Green et al., APL Materials, vol. 10, no. 2, 2022.

[2] J. Zhang et al., Nature Communications, vol. 13, no. 1, p. 3900, 2022.

[3] M. D. Santia et al., Applied Physics Letters, vol. 107, no. 4, 2015.

[4] B. K. Mahajan et al., Applied Physics Letters, vol. 115, no. 17, 2019.

[5] S. A. O. Russell et al., IEEE Journal of the Electron Devices Society, vol. 5, no. 4, pp. 256-261, 2017.

[6] Y. Song et al., ACS Applied Materials & Interfaces, vol. 15, no. 5, pp. 7137-7147, 2023.

[7] Y. Wang et al., IEEE Transactions on Electron Devices, vol. 68, no. 3, pp. 1185-1189, 2021.

[8] W. Xu et al., in 2019 IEEE International Electron Devices Meeting (IEDM), pp. 12.5.1-12.5.4.

[9] P. Godignon et al., in Advancing Silicon Carbide Electronics Technology II, 2020, ch. 3, pp. 107–174.

[10] L. Huang et al., Journal of Crystal Growth, vol. 531, p. 125353, 2020.

[11] M.-H. Lee et al., Journal of Materials Research, vol. 36, no. 23, pp. 4771-4789, 2021.

Polyoxometalate-doped Memristor with Redox Dynamics for Reliable Single-component Artificial Neuron

Shirui Zhu[1†], Yan-Bing Leng[1†], Guohua Zhang[2], Yu-Qi Zhang[3], Pengfei Han[4], Hecheng Cai[4], Ziyu Lv[4], Yongbiao Zhai[4], Ye Zhou[5], and Su-Ting Han[1*]

[1]Department of Applied Biology and Chemical Technology, The Hong Kong Polytechnic University, Kowloon, Hong Kong, China, [2]Key Laboratory of Physics and Technology for Advanced Batteries (Ministry of Education), College of Physics, Jilin University, 130012, Changchun, China, [3]Institute of Microscale Optoelectronics, Shenzhen University, Shenzhen 518060, China, [4]College of Electronics and Information Engineering, Shenzhen University, Shenzhen 518060, China, [5]Institute for Advanced Study, Shenzhen University, Shenzhen 518060, China

[*]Email: suting.han@polyu.edu.hk

Abstract

This work reported a Polyoxometalate (POM)-doped $Ag/WSe_2/Ag$ memristor (PDM) with reversible redox dynamics to implement steady leaky integrate-and-fire (LIF) neuronal behaviors. Based on POM-assisted targeted reduction of Ag^+, the device demonstrates highly stable $I\text{-}V$ characteristics and rapid self-reset performance under various voltage sweeping ranges. Besides, pulse amplitude-dependent LIF characteristics can be simulated with a pulse width of 1 ms by a single PDM, facilitating simplifying the hardware implementations of spiking neural network.

Keywords: Memristor, Polyoxometalate-doping, Artificial neuron

Introduction

In view of the brain-inspired architecture and event-driven communication patterns, spiking neural network (SNN) provides a more realistic and energy-efficient alternative to develop high-performance neuromorphic computing [1]. Artificial neurons serve as the foundational elements in SNN implementation, emulating the accumulation and firing of biological membrane potential [2]. Utilizing conventional complementary metal-oxide-semiconductor (CMOS) hardware to construct artificial neurons requires extra reset circuits, inevitably leading to equipment redundancy and high-power consumption [3]. Threshold switching (TS) devices, most representative of metal conductive filament (CF) type memristors, characterized by inherent volatile, large ON/OFF ratio, and ultrafast switching properties, have emerged as promising candidates to replace the CMOS-based artificial neurons [4]. However, CF-type memristors display certain undesirable performances that hinder their suitability for simulating neuron behaviors within a single component [5]. Firstly, nano-scaled CFs are inadequately matched to the micrometer-scaled active region, contributing to the random growth of CFs and exhibiting a wide-range drift of threshold voltage (V_{TH}) and hold voltage (V_{hold}). Secondly, ultra-low V_{hold} may prevent the discharged device from returning to its initial state before the next stimuli. Therefore, resistance-capacitance circuits are employed to counteract the degradation of CF-type memristors in artificial neurons, thereby increasing the complexity of the hardware. Extensive research has been conducted on layered two-dimensional (2D) materials as the dielectric of memristors, which perform high ON/OFF ratio, lower V_{TH} and repeatable resistance switching behavior, owing to their capability of forming atomically thin tunneling barriers and exhibiting tunable electrical characteristics [6]. Furthermore, the smooth and dangling-bond-free surface of 2D materials is advantageous for constructing planar memristors, facilitating efficient ion charge transport and comprehensive characterization research [7]. POM is an inorganic cluster, coordinated by anionic metals and oxides with stable molecule structure and components [8]. These clusters exhibit reversible redox properties, enabling delocalization of large number of electrons to carry other cations and affect the performance of electrical applications [9]. Moreover, typically nanometer-sized POM clusters provide possibility for device chemical doping.

In this study, we proposed an $Ag/WSe_2/Ag$ memristor doped with $H_4PMo_{11}VO_{40}$ (POM) to simulate LIF neuronal behaviors without extra elements. Ag^+ can be target reduced within resistance dielectric by POM clusters, enabling highly stable $I\text{-}V$ characteristics and self-reset properties during different voltage ranges. Based on these performances, pulse amplitude-dependent neuronal spiking with a firing frequency of 6.85 Hz-18.75 Hz can be achieved steadily under the pulse width of 1 ms.

Experiments and Characterization

A. Device fabrication

The structure diagram of PDM is demonstrated in Fig. 1 (a). For the device fabrication (Fig. 1 (b)), the WSe_2 flakes were mechanically exfoliated from bulk materials and transferred on the pre-cleaning substrate (Si/300 nm SiO_2), serving as a resistive dielectric layer. The Ag electrodes (50 nm) were patterned by standard UV lithography process and thermal deposition sequentially on the WSe_2 flakes. After the lift-off process, completed Ag/WSe$_2$/Ag structure was dropped casting with prepared POM solution (solved in methyl alcohol, 0.05 mg/ml, 100 µl once) and experienced an annealing procedure under 80 ℃ (15 minutes) to accomplish the PDM.

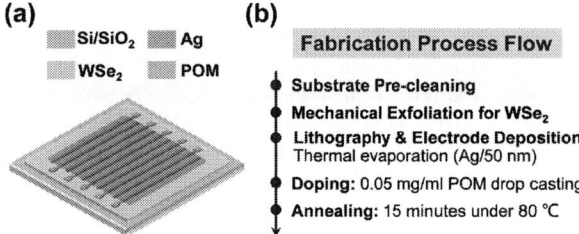

Fig.1 (a) The schematic diagram of POM-doped Ag/WSe$_2$/Ag device with planar structure. (b) The preparation process of devices.

B. Characterization

To demonstrate the device structure and doping distribution directly, the atomic force microscope (AFM) imaging was carried out before electrical programming. Fig. 2 (left) exhibits a 2 µm electrode spacing of the device. Focusing on the channel region, abundant POM clusters with the size of several tens of nanometers can be observed that disperse uniformly above the surface of WSe$_2$ channel, serving as dopants to assist in regulating the Ag$^+$ migration and redox dynamics (Fig. 2 right). Subsequently, the electrical measurements of memristor were implemented by Keithley 4200A-SCS (voltage-sweep mode) and Keysight B2902A (voltage-pulse mode) respectively.

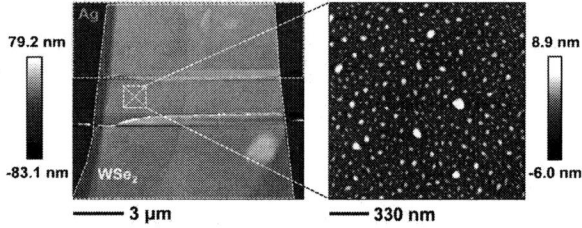

Fig. 2 AFM image of a PDM with an electrode spacing of 2 µm (left, scale bar: 3 µm) and partial channel (right, scale bar: 330 nm).

Results and Discussion

A. Resistance switching performance

According to electrical measurement, Fig. 3 (a) displays the voltage-sweep result of Ag/WSe$_2$/Ag device, which performs typical TS behavior. With the sweep range of 0 V-1 V, the device was switched abruptly from the initial high resistance state (HRS) to the low resistance state (LRS) at V_{TH}, consequently restricted by the compliance current (10 µA). After reversing the operating voltage to 0 V, Ag/WSe$_2$/Ag memristor spontaneously relaxed to initial HRS as the applied voltage is lower than V_{hold} (average value is 0.02 V). The distribution of V_{TH} within 30 cycles was recorded (Fig. 3 (a) insert), exhibiting a dispersive voltage range of 0.2 V-0.5 V, resulting from the stochastic growth of Ag CFs. In

contrast to the original Ag/WSe$_2$/Ag device, the PDM performs consecutive conductance changing under the applied voltage with the range of 1.0 V-2.0 V. Fig. 3 (b) illustrates 10 cycles of PDM under six voltage conditions respectively, achieving a remarkable restriction in the cycle-to-cycle deviation. Furthermore, during the voltage sweep from positive to 0 V, the current can be observed to return to 0.17 nA at 0.25 V, which is higher than the V_{hold} of Ag/WSe$_2$/Ag device, facilitating the self-reset performance of PDM.

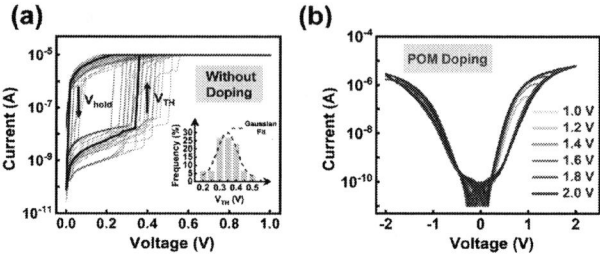

Fig. 3 (a) The resistance switching characteristic of Ag/WSe$_2$/Ag memristor with 30 cycles. Insert: The statistical distribution of V_{TH} within 30 cycles. (b) The resistance switching characteristic of PDM from the range of 1.0 V to 2.0 V with 10 cycles under each voltage sweep condition.

B. Single-component artificial neuron performance

Considering the reliable I-V characteristic, self-reset ability and reversible Ag$^+$ redox dynamics, PDM can be modulated to perform typical LIF neuronal behavior with appropriate pulse input without any extra capacitor and resistor. As depicted in Fig. 4, when a voltage pulse sequence with a width of 1 ms is applied, the current of device maintains a resting state initially, serving as the integration process. The PDM-artificial neuron cannot be transformed to the LRS until reaching the switching threshold point, exhibiting an abrupt current spike (threshold current: 10 nA), defined as neuron firing. After that, the output current leaks rapidly and reverses to the resting state, waiting for stimuli integration for the next firing. Moreover, As the amplitude of the input voltage pulse increases from 0.7 to 1.2 V, the average firing frequency of the artificial neuron gradually rises from 6.25 Hz to 18.75 Hz, demonstrating pulse amplitude-dependent LIF characteristics.

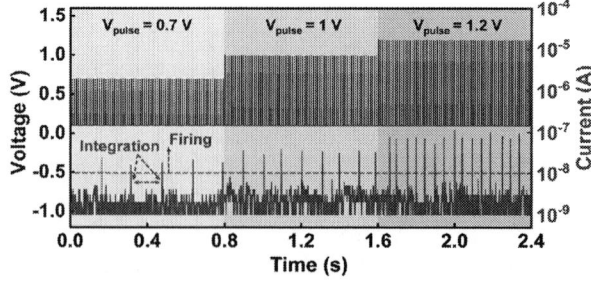

Fig. 4 Performance of PDM artificial neuron with the corresponding integration and refractory period under appropriate voltage input condition

(pulse width: 1 ms).

However, as the input voltage pulse width increases from 1 ms to 100 ms, the LIF neuronal behavior of PDM disappears. As shown in Fig. 5, the response of the device surges over the threshold current under each corresponding input voltage pulse with the same range of 0.7 V-1.2 V, failing to accomplish the charge accumulation of artificial neuron under the resting condition. Measurements indicate that properly setting the pulse width and duty cycle can effectively control the periodic growth and rupture of Ag CFs, thereby enabling the LIF neuronal behavior.

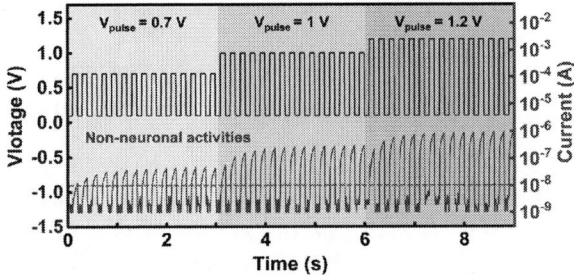

Fig. 5 Performance of PDM with non-neuronal activities under other voltage input condition (pulse width: 100 ms).

C. Mechanism of the device operation

In terms of the Ag/WSe$_2$/Ag device, the resistance switching mechanism is recognized for the rapid formation and rupture of Ag CFs. Under an operating voltage, the Ag electrode oxidizes to produce Ag$^+$, which migrates to the counter electrode driven by the electric field and is reduced back to Ag, resulting in device switching to the LRS (Fig. 6 (a)). The dangling-bond-free surface of layered WSe$_2$ is beneficial to the migration of Ag$^+$, enabling memristors to exhibit steep switching slopes and V$_{TH}$ lower than 0.5V [7]. Upon removal of voltage, the CFs spontaneously rupture, triggering the device to revert to HRS. In the case of PDM, e$^-$ is transformed from POM to Ag$^+$, facilitating the possible reduction to form nanoclusters within the dielectric layer (Fig. 6 (b)). The interfacial charge-transfer process further controls and enlarges the targeted reduction of Ag$^+$, enabling stable *I-V* characteristics of memristor. Moreover, after removing the operating voltage, PDM returns to its initial state rapidly because of the reversible multi-electron redox-active property of POM, enabling better self-reset ability of PDM to fulfill the requirement in neuronal leaky process.

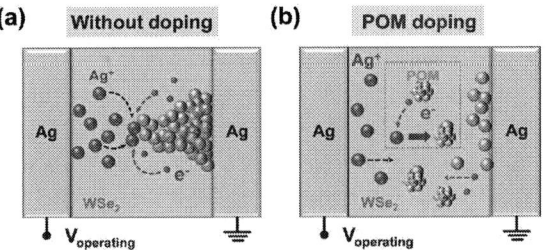

Fig. 6 Schematic diagram of the resistance switching mechanism within Ag/WSe$_2$/Ag memristor (a) and PDM (b).

Conclusion

In summary, this study presented a PDM to simulate LIF neuronal behaviors without any capacitor or resistance. POM-induced reversible redox activity allows Ag$^+$ to be reduced targeted within the dielectric layer rather than the cathode, accomplishing the stable *I-V* performance under different voltage sweeping ranges. The relaxation of conductance occurred at 0.25 V, exhibiting better self-reset property. These characteristics assist device in accomplish steady LIF neuronal firing with the average frequency of 6.85 Hz-18.75 Hz stimulated by voltage pulse from 0.7 V to 1.2 V, which offers a promising potential to reduce the hardware complexity associated with implementing spiking neural networks.

Acknowledgments

S. Zhu and Y. -B. Leng contributed equally to this work. This research was supported by the NSFC Program (Grant nos. 62122055, 62074104); Guangdong Provincial Department of Science and Technology (Grant No 2024B1515040002).

References

[1] Q. Xia and J. J. Yang, "Memristive Crossbar Arrays for Brain-inspired Computing", Nat. Mater., 18, pp. 309-323 (2019).

[2] J. Chen et al., "Optoelectronic Graded Neurons for Bioinspired In-sensor Motion Perception", Nat. Nanotechnol., 18, pp. 882-888 (2024).

[3] X. Zhang et al., "An Artificial Neuron Based on a Threshold Switching Memristor", IEEE Electron Device Lett., 39, pp. 308-311 (2018).

[4] Z. Lv et al., "Development of Bio-Voltage Operated Humidity-Sensory Neurons Comprising Self-Assembled Peptide Memristors", Adv. Mater., 15, 2405145. (2024).

[5] J. Bian et al., "A Stacked Memristive Device Enabling Both Analog and Threshold Switching Behaviors for Artificial Leaky Integrate and Fire Neuron," IEEE Electron Device Lett., 43, pp. 1436-1439 (2022).

[6] M. Wang et al., "Robust Memristors Based on Layered Two-dimensional materials", Nat. Electron., 1, pp. 130-136 (2018).

[7] X. Yang et al., "Highly Reproducible Van der Waals Integration of Two-dimensional Electronics on The Wafer Scale", Nat. Nanotechnol., 18, pp. 471-478 (2023).

[8] G. Zhang et al., "Polyoxometalate Accelerated Cationic Migration for Reservoir Computing", Adv. Funct. Mater., 32, 2204721, (2022).

[9] J. Kang et al., "Cluster-type Analogue Memristor by Engineering redox Dynamics for High-performance Neuromorphic Computing", Nat. Commun., 13, 4040 (2022).

Solder Fatigue Life Prediction Method Considering Intermetallic Compound Growth

Yikang Wang, Xiaopeng Wu*, Jiahao Hou, Can Liu, and Yintang Yang

Xidian University, Xi'an, China. (*Email: xpwu@mail.xidian.edu.cn)

ABSTRACT

This paper presents a method for predicting solder joint life by considering the growth of the intermetallic compound (IMC). The relationship between single-cycle damage and the number of cycles is established by combining fatigue life models with IMC growth equations. The cumulative damage theory is employed to integrate damage over time, enabling a quantitative evaluation of the IMC's impact on solder joint life. Keywords: Intermetallic compound, Cumulative damage theory and Thermal fatigue life

INTRODUCTION

Solder joints act as vital connections for electrical, mechanical, and thermal pathways between components and substrates, with reliability playing a key role in the lifespan of electronic products. During soldering, metallurgical reactions between the solder and under-bump metallization (UBM) produce an intermetallic compound (IMC), forming a strong bond. As shown in Fig. 1, typical solder joints vary in size [1]: the diameter of a BGA solder ball is 760 μm, a flip-chip C4 solder ball measures 100 μm, and a TSV micro-bump ranges from 10 to 20 μm. As the size of solder joints decreases, the proportion of IMC increases. Under high-temperature conditions, IMC grows significantly, requiring temperature-related reliability assessments. While experimental studies indicate that IMC growth reduces solder joint reliability [2], quantitative methods for life prediction remain lacking. This paper uses Ansys Workbench based on finite element method and MATLAB to simulate IMC growth and explore its effect on solder solder joint reliability. This approach enables a more accurate prediction of fatigue life, addressing design and process challenges for small solder joints.

PREDICTION METHOD

This section introduces a life prediction method that combines traditional fatigue life models with IMC growth equations through finite element simulations. The effect of IMC growth on single-cycle damage is evaluated, and the cumulative damage theory is applied to determine the total damage. In this paper, Engelmaier's modified Coffin-Manson model [3] is used for traditional life model, which is not described in detail here.

A. IMC Growth Formula

This section investigates the growth of Cu_6Sn_5 at the Cu/Sn interface under thermal loading. As shown in Eq. 1, the IMC's growth thickness under a single thermal stress is directly proportional to the square root of the temperature loading time [4].

$$d = \sqrt{2D_{Cu/Cu_6Sn_5}\frac{C_{Cu} - C_{Cu/Sn}}{C_{Cu/Cu_6Sn_5}}t + d_0^2} \quad (1)$$

Here, C_{Cu}, $C_{Cu/Sn}$, and C_{Cu/Cu_6Sn_5} represent the Cu atom concentrations in the Cu pad, solder, and IMC, respectively. D_{Cu/Cu_6Sn_5} denotes the thermal diffusion coefficient of Cu atoms in the IMC, and d_0 is the initial thickness of the IMC.

B. Cumulative damage theory

Palmgren-Miner's linear cumulative damage theory [5], shown in Eq. 2, assumes that under fixed loading conditions, single-cycle damage is the reciprocal of total life. For varying loads, cumulative damage is calculated as the sum of individual stress-induced damage,

$$D = \sum_{i=1}^{m} \frac{n_i}{N_i} \quad (2)$$

where n_i is the number of cycles under stress S_i, N_i is the fatigue life for the same stress, and m represents different load levels. When the load varies continuously over time, the cumulative damage can be approximated using the integral in Eq. 3.

$$D_{total} = \int_0^t D(N)\,dN \quad (3)$$

Here, $D(N)$ represents the damage at cycle N. When D_{total} equals 1, the solder joint is considered to have reached the end of its life.

C. Lifetime prediction method considering IMC growth

Models with different IMC thicknesses are established based on the chip geometry. The relationship between IMC thickness and temperature loading time is modeled using Eq. 1, and this relationship is converted into a function of temperature cycles. The Coffin-Manson model calculates traditional fatigue life for different IMC thicknesses. The reciprocal of this life gives the single-cycle damage. By combining these results, the model computes cumulative damage using the extended damage theory. The total number of cycles at which the cumulative damage equals 1 represents the predicted solder joint life.

VERIFICATION

This section verifies the proposed life prediction method by applying it to the C4 bump structure of the FCBGA196 chip.

D. Simulation model construction

Fig. 2 shows the finite element model of the chip, focusing on the C4 bumps while neglecting pin modeling for simplicity. The substrate connects to the silicon die through a pad and C4 bump, encased in a plastic binder. The model includes a 14 × 14 array of C4 bumps with a 450 μm pitch. The total height of the solder and IMC is 53 μm. Models with IMC thicknesses ranging from 2.5 to 15 μm were simulated, with a pure solder model as a reference.

The materials and dimensions of each component are summarized in Table 1 [6]. SAC305 solder follows the Anand viscoplastic model to account for plastic deformation during temperature cycling. Temperature cycles are applied following JESD22-A104, with a range of -55°C to 125°C, a 10°C/min ramp rate, and a 12-minute dwell at both peak and low temperatures.

E. Fatigue life calculation

The simulation results indicate that the highest stress and strain occur at the corners of the chip, farthest from the center. Fig. 3 shows stress distributions for models with various IMC thickness, revealing that increased IMC thickness concentrates stress on the solder layer, raising stress levels. Fig. 4 presents the relationship between IMC thickness and shear-plastic strain amplitude, fitted using an exponential model. Eq. 4 describes the fitted curve, with an average error of 4.09%, within the acceptable range.

$$\Delta\gamma = 0.002295 \times e^{0.09473d} + 0.0049 \qquad (4)$$

Here, d is the IMC thickness in μm and $\Delta\gamma$ is the shear-plastic strain amplitude. As calculated from Eq. 1, the relationship between the IMC thickness and time at 125°C is presented in Eq. 5.

$$d = \sqrt{0.065432t + 6.25} \qquad (5)$$

Since each cycle lasts one hour, t equals the number of cycles. Incorporating the temperature cycling conditions into the Coffin-Manson model yields the fatigue life, as presented in Eq. 6.

$$N_f = \frac{1}{2}\left(\frac{0.65}{\Delta\gamma}\right)^{2.4570534} \qquad (6)$$

By connecting the Eq. 4 to the Eq. 6, t is converted into cycle number N, and life N_f is converted into single-cycle damage by reciprocal. Fig. 5 illustrates the relationship between damage and the number of cycles, with solid and dashed lines representing models with and without IMC growth, respectively.

Based on the continuous cumulative damage formula shown in Eq. 3, when the damage accumulation reaches 1, it is regarded as failure, and the corresponding integration time is the final solder life. The shadow areas of red and blue in Fig. 5 are both 1. The red shadow represents the cumulative damage of the model considering IMC, corresponding to a lifetime of 5502 cycles. The blue shadow represents the cumulative damage of the model without IMC, corresponding to a lifetime of 7801 cycles. It was quantitatively characterized that the solder life would be reduced by about 29.5% when IMC growth was considered. Therefore, when evaluating the thermal fatigue life of small solder joints, the life prediction method considering IMC propsed in this paper is of great importance for reliability assessment.

CONCLUSION

This paper proposes a solder joint life prediction method that incorporates IMC growth, based on IMC growth model, Coffin Manson life calculation model and Miner linear cumulative damage theory. The method takes into account the effect of IMC growth on the fatigue life of solder joints in temperature cycle, and improves the calculation accuracy of fatigue life.

In this paper, the feasibility of the method is verified by simulation, and the quantitative analysis shows that IMC can reduce the thermal fatigue life of solder by about 29.5%, which verifies the necessity of the method for the prediction of thermal fatigue life of small solder joints.

REFERENCES

[1] Tu, King-Ning. "Reliability challenges in 3D IC packaging technology." Microelectronics Reliability 51.3 (2011).

[2] Yang, L. M., and Z. F. Zhang. "Influences of intermetallic compounds morphologies on fracture behaviors of Sn-3Ag-0.5 Cu/Cu solder joint." 2013 14th International Conference on Electronic Packaging Technology. IEEE, 2013.

[3] Chauhan, Preeti, et al. "Critical review of the Engelmaier model for solder joint creep fatigue reliability." IEEE Transactions on Components and Packaging Technologies 32.3 (2009).

[4] Liu, C. Y., et al. "Study of electromigration-induced Cu consumption in the flip-chip Sn/ Cu solder bumps." Journal of applied physics 100.8 (2006).

[5] Tepfers, Ralejs, Claes Fridén, and Leif Georgsson. "A study of the applicability to the fatigue of concrete of the Palmgren-Miner partial damage hypothesis." Magazine of Concrete Research 29.100 (1977).

[6] Ting-biao, Jiang, Du Chao, and Xu Long-hui. "Finite element analysis and fatigue life prediction of BGA mixed solder joints." 2007 International Symposium on High Density packaging and Microsystem Integration. IEEE, 2007.

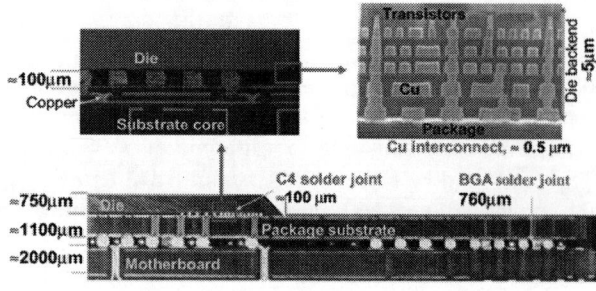

Fig. 1. Solder bumps at different locations

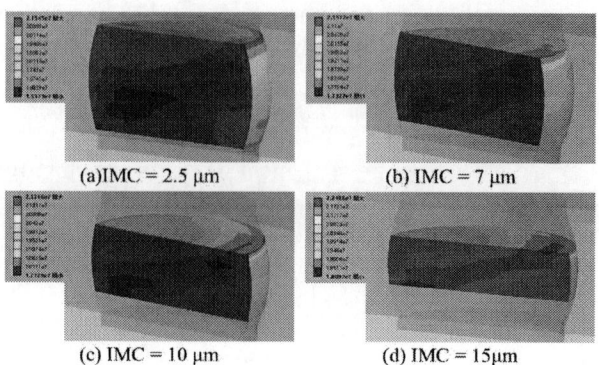

(a)IMC = 2.5 μm   (b) IMC = 7 μm

(c) IMC = 10 μm   (d) IMC = 15μm

Fig. 3. Stress contour plots of C4 bumps with different IMC thicknesses

TABLE I: The materials and dimensions of each component

Structure	Material	Shape: Dimension
Die	Si	Cube: Height 0.45mm, length 6.5mm, width 6.5mm
Substrate	Ceramic	Cube: Height 0.65mm, length 13mm, width 13mm
Pad	Cu	Cylinder: Height 6μm, diameter 79μm
C4 Bump	SAC305	Sphere: height 53μm, diameter 90μm
IMC	Cu_6Sn_5	
Mold Compound, Filling Material	EMC	Cube: Height 1.018mm, length 13mm, width 13mm

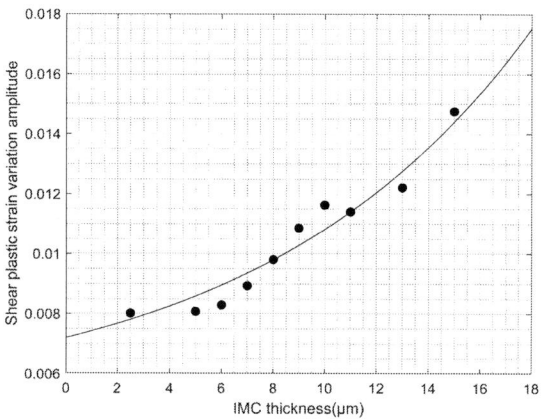

Fig. 4. Relationship between IMC thickness and shear-plastic strain

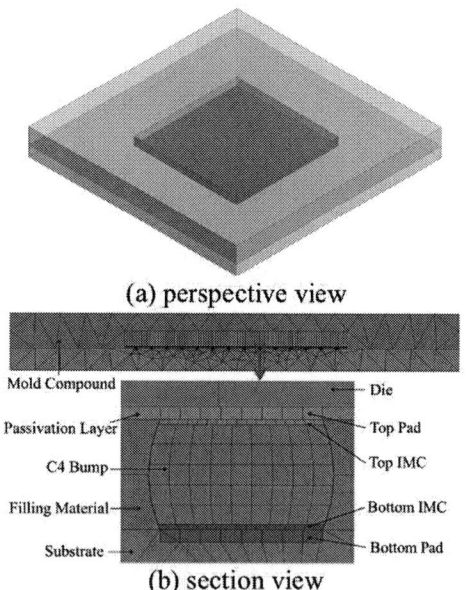

(a) perspective view

(b) section view

Fig. 2. Finite element model of chip with C4 bumps

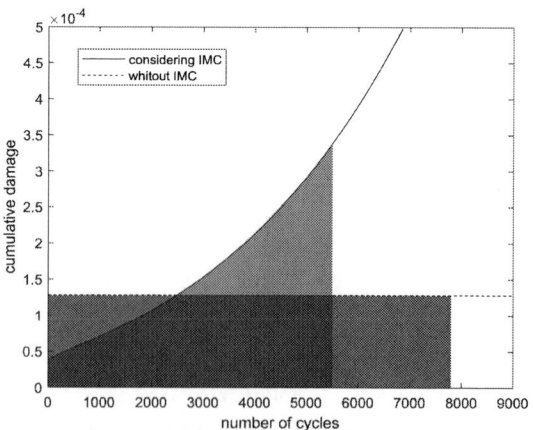

Fig. 5. Relationship between damage and cycle number

A Model-driven Design Technology Co-optimization (DTCO) with Multi-Objective Bayesian Algorithm for Advanced Technology

Baokang Peng[1], Guoyao Cheng[1], Runsheng Wang[2*], Ru Huang[2], Mansun Chan[3] and Lining Zhang[1*]

[1]School of Electronic and Computer Engineering, Peking University, China, *Email: eelnzhang@pku.edu.cn
[2]School of Integrated Circuits, Peking University, China, *Email: r.wang@pku.edu.cn
[3]Dept. of ECE, Hong Kong University of Science and Technology, Hong Kong SAR, China

Abstract

This work presents a multi-objective Bayesian (MOB) optimization technique for co-optimizing device parameters and digital standard cell libraries (SDC) to deeply explore the technology design space. In contrast to the traditional design based on intrinsic device delay and power, the developed framework unifies the SDC characterizations and the MOB optimization algorithm. With 7nm FinFET as an example, the proposed framework achieves a 20.4% improvement in the power-delay product (PDP) and a 74% increase in hypervolume compared to traditional methods. The basic SDC including NAND, NOR, and XOR are considered with the ASAP7 PDK, and the full library could be covered. This work provides a framework for automating the design-technology co-optimizations in advanced process nodes.

Keywords: Standard cell library, Bayesian optimization, design-technology co-optimization

Introduction

In recent decades, novel semiconductor architectures like FinFET, gate-all-around FET, and Complementary FET have emerged, offering enhanced performance and reduced power consumption [1-3]. To fully leverage these advances, design-technology co-optimization (DTCO) has been introduced, enabling fine-tuning of device parameters through optimized processes to maximize large-scale circuit performance [4]. The Process Design Kit (PDK) is a crucial link between manufacturing and circuit design, with standard cell libraries (SDC) as its core components. In large-scale circuits, synthesis tools translate design descriptions into logic netlists of SDC, which are then placed, routed, and optimized using automated tools. As a result, the quality of high-level digital designs heavily depends on SDC performance, making device parameter optimization essential for enhancing SDC effectiveness.

Previous studies have focused on optimizing intrinsic device performance by adjusting parameters like gate length (L_g), and oxide thickness (T_{ox}) to reduce the device delay and power [5-7]. Many of these efforts rely on black-box algorithms, such as NSGA-II and Bayesian optimization [6-7]. However, these objectives may not directly align with circuit-level Figures of Merit (FoMs). It is highly desirable to extend the target to the SDC FoMs, bridging the gap between device and circuit-level performance.

This work proposes a multi-objective Bayesian optimization approach for co-optimizing SDC and device parameters. Using the ASAP 7 nm FinFET PDK [8] as an example, our approach shows significant improvements in hypervolume and power-delay product (PDP) over conventional methods. This framework provides an efficient and scalable solution for DTCO.

Proposal of the DTCO Framework

A. Simulation Platform

A 7 nm predictive process design kit (PDK) called the ASAP7 PDK is used for public availability. Cadence Liberate [9] was used to characterize the timing, power, and signal integrity of the cell library by taking the extracted cell netlists, SPICE model, and template as inputs. For ease of simulation and analysis, three SDCs including the NAND, NOR, and XOR gates are selected as specific examples, with their average delay, power consumption, and leakage extracted. The library provided by the ASAP7 PDK served as the reference for comparison.

The device models employed featured LVT threshold voltages and TT process corners. The range and default values for the adjustable device parameters are listed in Table 1. The parameters shown in the table have minimal impact on the extracted cell netlists. Therefore, the effects of adjusting these parameters can be evaluated solely by modifying the SPICE model.

B. Proposed Framework

Fig. 1 provides a visual comparison of the traditional and our proposed workflow. It is clear from Fig. 1 that the traditional device optimization process considers only intrinsic delay $C_{gg}V_{dd}/I_{on}$ and dynamic power $C_{gg}V_{dd}^2f/2$, which often diverges significantly from actual circuit simulation results. In contrast, the proposed workflow directly extracts FoMs from the characterized SDC, resulting in higher accuracy. The specific optimization results are shown in Table 1, where the proposed workflow achieves a 20.4% improvement in PDP compared to the traditional method. Cadence Liberate is utilized to characterize the performance of SDC using device parameters. In each optimization iteration, Liberate provides performance metrics for the candidate device design, while the Bayesian optimizer determines the candidate design parameter set for the next iteration.

The workflow, as illustrated in Fig. 1(b), consists of multiple steps. Initially, a small batch of data points is

979-8-3315-0417-5/25 $31.00 © 2025 IEEE

generated within the parameter ranges specified in Table 1 using Sobol Sampling, an effective method for covering the design space with a limited number of samples. Subsequently, the performance of the SDC corresponding to these device parameter combinations is characterized and compared with the reference SDC from ASAP7 to extract the objective function, as shown in the following equation:

$$Obj_{1,2,3} = \sum_{i=1}^{S} \frac{\langle X_i \rangle}{\langle X_i^{ref} \rangle} / S \qquad (1)$$

Here, $Obj_{1,2,3}$ represent the three objective function values corresponding to delay, dynamic power consumption, and leakage associated with X, respectively. S denotes the total number of SDC to be characterized, $\langle X_i \rangle$ denotes the average delay, dynamic power, and leakage power for each SDC, while $\langle X_i^{ref} \rangle$ represents the corresponding metrics from the ASAP7 reference SDC.

Subsequently, the dataset D is used to train the surrogate model, for which Gaussian Process Regression (GPR) is employed. The acquisition function then determines the next device design candidate. In this framework, the Expected Hypervolume Improvement (EHVI) is used as the acquisition function, and for optimization with constraints, the Expected Hypervolume Improvement with Constraints (EHVIC) [10] is applied. During the iterative process, each iteration adds only one new device design. After its characterization, the corresponding dataset D is updated accordingly.

This iterative optimization process terminates once the maximum number of iterations is reached, returning the Pareto front objectives and the corresponding device parameter designs.

Results and Discussions

The device is optimized using the traditional workflow in Fig. 1(a), with specific optimization results after 200 iterations shown in Fig. 2(a). Device parameter sets are characterized using Liberate, yielding actual objective functions as defined in Eq. 1, with results in Fig. 2(b). It is clear that using intrinsic delay and power as objectives does not establish a well-defined Pareto front, nor do these metrics correlate directly with actual delay and power in the SDC. Consequently, optimizing for intrinsic delay and power can mislead the optimization direction, as seen in Fig. 2(b), where most samples show delay and power characteristics inferior to the reference SDC.

Fig. 3(a) shows the results after 200 iterations using the proposed workflow in Fig. 1(b), revealing a well-defined Pareto front with significant improvement over the reference SDC. Fig. 3(b) illustrates hypervolume progression, converging around the 60th iteration. As shown in Fig. 4, compared to traditional intrinsic delay and power optimization, the proposed workflow achieves a higher-quality Pareto front, with an 11.9% power reduction at the same delay and a 1.9% delay reduction at the same power.

For high-performance designs, delay improves by 6.9%, and for low-power designs, power consumption decreases by 12%. The hypervolume of the proposed workflow (0.022) surpasses that of the traditional method (0.0126) by 74%, highlighting its superior optimization efficiency. High-performance, optimal PDP, and low-power device parameters from the proposed workflow are shown in Fig. 5. The figure illustrates that the low-power design achieves reduced power consumption by increasing the threshold voltage via PHIG adjustments and decreasing HFIN and TFIN to lower drive strength, aligning well with fundamental device physics predictions. Additionally, we add a constraint that leakage power must not exceed 10% above the reference SDC, with the acquisition function adjusted to EHVIC. The constraint is defined as follows

$$Constraint = Obj_3 - 1.1 < 0 \qquad (2)$$

Fig. 6(a) shows optimization results under the constraint, where average dynamic power is significantly reduced. In Fig. 6(b), most points from the traditional method fail to meet the constraint, whereas the proposed workflow yields mostly feasible points, underscoring its effectiveness. Table 2 presents objective values and device parameters for both traditional and proposed workflows, demonstrating that the optimal solutions from the proposed workflow dominate those of the traditional workflow.

Conclusion

This paper presents a novel workflow for co-optimizing digital standard cell libraries (SDC) and device parameters using multi-objective Bayesian (MOB) optimization. To efficiently evaluate SDC performance, we characterized NAND, NOR, and XOR gates with Liberate, using average delay and power as Figures of Merit. The workflow achieved a 74% improvement in hypervolume and a 20.4% improvement in Power-Delay Product (PDP) over traditional methods based on intrinsic delay and power metrics. This scalable methodology also supports design-technology co-optimization for emerging device architectures, aiding the development of advanced process nodes.

Acknowledgments

This work is supported in part by the National Key Research and Development Program grant 2023YFB4402204, the NSFC grant 62474009, and in part by the Guangdong Basic and Applied Basic Research Foundation grant 2024B1515020064.

References

[1] M. G. Bardon et al., IEDM, 2016, p. 28.2.1-28.2.4. [2] S.-W. Chang et al., IEDM, 2019, p. 11.7.1-11.7.4. [3] B. Peng et al., TED, vol. 71, no. 1, pp. 461–467, 2024. [4] M. Liu, ISSCC, 2021, pp. 9–16. [5] H. Jeong et al., TED, pp. 1–7, 2024. [6] T. Wu and J. Guo, TED, vol. 68, no. 11, pp. 5476–5482. 2021. [7] A. Zhao et al., TCAD, pp. 1–1, 2024. [8] X. Xu, N. Shah, A. Evans, S. Sinha, B. Cline, and G. Yeric, ICCAD, 2017, pp. 999–1004. [9] Virtuoso Liberate Reference Manual, 2019. [10] M. Abdolshah, A. Shilton, S. Rana, S. Gupta, and S. Venkatesh, ICPR, 2018, pp. 3238–3243

Table. 1: Design parameter space, along with parameter set yielding the optimal PDP value with tradition and proposed workflow

Parameters	Default values	Value ranges	Traditional workflow	Proposed workflow
PHIG (NMOS)	4.307	4.302~4.312	4.311	4.312
PHIG (PMOS)	4.8681	4.8631~4.8731	4.8632	4.8631
HFIN [nm]	32	28~36	28	36
TFIN [nm]	6.5	5.8~7.2	5.8	5.8
EOT (NMOS) [nm]	1	0.9~1.1	0.9	0.9
EOT (PMOS) [nm]	1	0.9~1.1	1	0.9
Minimum PDP(Normalized)	-	-	0.7689	0.6385

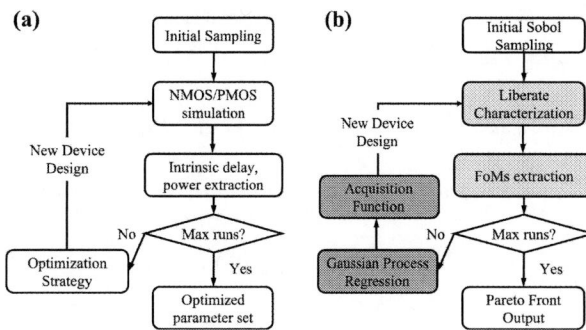

Fig. 1: (a) Traditional device optimization workflow, (b) Proposed device-circuit co-optimization workflow.

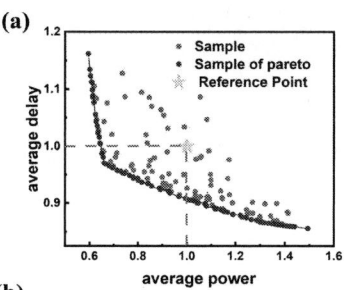

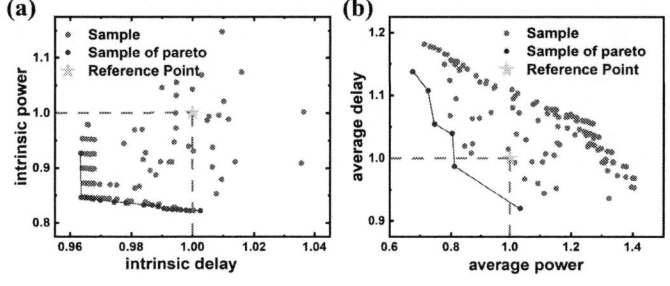

Fig. 2: Results using traditional intrinsic delay and power as optimization objectives: (a) Sampling points during the optimization process based on intrinsic delay and power, (b) Actual delay and power results of sample points in the standard cell library.

Fig. 3: Optimization results using the proposed workflow with average delay and power of SDC as objectives: (a) Sampling points during the optimization process, (b) Relationship between hypervolume and iteration.

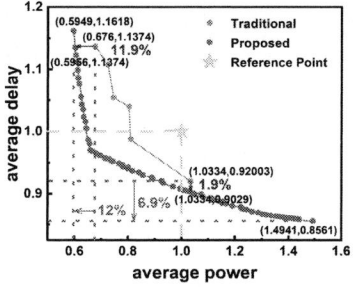

Fig. 4: Comparison of pareto front between the proposed workflow and the traditional approach.

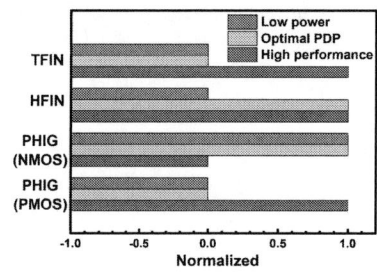

Fig. 5: Comparison of high-performance and low-power parameter sets optimized through the proposed workflow.

Fig. 6: (a) Sampling points for the traditional method, proposed workflow, and proposed workflow with leakage constraint, (b) 3D scatter plot comparison between the traditional method and the proposed workflow with leakage constraint.

Table. 2: Comparison of the optimal structural parameters, average delay, power, and constraints between the traditional and proposed workflows.

Configs	Optimal Structural Parameters						Avg. Power (Nor.)	Avg. Delay (Nor.)	Constraint (Nor.)
	PHIG (NMOS)	PHIG (PMOS)	HFIN [nm]	TFIN [nm]	EOT (NMOS) [nm]	EOT (PMOS) [nm]			
Traditional optimal value1	4.3106	4.8632	28	5.9	0.9	1	0.676	1.1374	-0.724
Proposed optimal value1	4.312	4.8631	28	5.8	0.9	0.9	0.6017	1.1347	-0.7836
Traditional optimal value2	4.311	4.8632	28	5.8	0.9	1	0.676	1.13743	-0.724
Proposed optimal value2	4.312	4.8631	36	5.8	0.9	0.9	0.6583	0.97	-0.6716
Traditional optimal value3	4.3044	4.8699	36	6.3	1	0.9	1.0344	0.92	-0.1879
Proposed optimal value3	4.3075	4.8631	36	6.6	0.9	0.9	0.9994	0.9088	-0.222

979-8-3315-0417-5/25 $31.00 © 2025 IEEE

Reverse Blocking GaN on SiC HEMTs with Ultralow Dynamic R_{ON} for 1200 V Power Applications

Yutong Fan[1,2], Mengqiang Yuan[1,2], Yuqing Hu[1,2], Weihang Zhang[1,2,*], Yachao Zhang[1,2], Zhihong Liu[1,2], Yue Hao[1,2], and Jincheng Zhang[1,2,*]

[1]State Key Discipline Lab of Wide Band Gap Semiconductor Technology, Xidian University, Xi'an 710071, China

[2]Guangzhong Wide Bandgap Semiconductor Innovation Center, Guangzhou institute of technology, Xidian University, Guangzhou 510555, China

*Corresponding authors: whzhang@xidian.edu.cn; jchzhang@xidian.edu.cn;

Abstract

This work demonstrates reverse-blocking (RB) GaN on SiC HEMTs and GaN-Si(100) Cascode Switches on a SiC substrate. The RB GaN HEMTs showed a turn-on voltage (V_{ON}) of 0.3 V and extremely low dynamic R_{ON} degradation at 2200 V, which were the best among all existing GaN power devices with reverse-blocking capability. Furthermore, GaN-Si(100) Cascode Switches were fabricated and showed great promise for future power IC (1200 - 1900 V) applications.

Keywords: Reverse Blocking, GaN HEMTs, dynamic R_{ON}, turn-on voltage, reverse leakage current.

Introduction

With the development of GaN power device manufacturing technology, more and more organizations pay much attention to kV-level GaN power devices to meet the demands of high-voltage applications such as EV drivers, EV charging, and PV inverters [1] - [8]. The low-cost Si substrate emerges as a highly enticing option for GaN power devices. However, the difficulty in growing high-quality GaN thick films on Si substrates hinders the development of 1200 V GaN-on-Si devices. Furthermore, high vertical leakage currents and limited vertical breakdown voltages are two major challenges in GaN-on-Si devices [9], [10]. However, SiC offers a higher electrical resistivity and thermal conductivity than Si [11]. GaN buffers with carbon doping or iron doping reduce lateral drain leakage, but the deep-level traps also lead to current collapse and increasing dynamic on-resistance [9], [10], [12]. AlN buffer is one of the effective ways to solve the above problems for 1200 V-1900 V applications. Furthermore, reverse-blocking devices are pivotal in numerous power applications, including AC-AC matrix converters, bidirectional switches, Class-S switch-mode amplifiers, multi-level inverters, and resonant converters [13] - [22].

In this work, the reverse-blocking D-mode GaN HEMTs on an AlN/SiC substrate (RB GaN on SiC HEMTs) were prepared. The fabricated devices exhibit an excellent trade-off between the reverse leakage current (I_R) and turn-on voltage (V_{ON}). Furthermore, the dynamic on-resistance (R_{ON}) degeneration sets a record value among all reported reverse-blocking D-Mode GaN power devices.

Material Structure and Device Fabrication

The $Al_{0.3}GaN/GaN/AlN$ epitaxial wafer used in this study was grown onto a 4-inch SiC substrate via metal-organic chemical vapor deposition (MOCVD). In proper growth order, the epitaxial layer consisted of a 440-nm AlN buffer layer, a 350-nm GaN channel layer, a 1-nm AlN interlayer, a 30-nm $Al_{0.3}Ga_{0.7}N$ barrier layer, and a 2 nm GaN capping layer. The schematic cross-sectional view of the fabricated device is presented in Fig. 1(a). The device fabrication steps consisted of the isolating mesa, Ohmic contact in the source area, drain-recessed etch, gate and drain metal contact, Al_2O_3/SiN layer passivation, via opening, and power metal deposition. A drain-recessed structure was fabricated via a BCl_3-based ICP etching process. Subsequently, the etching damage was repaired by annealing at 400°C in an N_2 atmosphere for 5mins. The recess depth was approximately 18 nm measured by atomic force microscopy (AFM) as shown in Fig. 1(b). Fig. 1(c) illustrates the surface morphology in the drain-recessed region. Moreover, a smooth etching morphology with a root mean square roughness value of 0.143 nm in a 2×2 μm² scan area was achieved by a low etching rate of 1.0 nm/min. An excellent drain morphology is vital for the V_{ON}. The gate metal and Schottky contact in the drain were fabricated through a one-step lithography process and electron beam evaporation (EBE) following a lift-off process. Subsequently, two types of GaN on SiC HEMTs with different drain contacts (a conventional ohmic drain and a Schottky drain) were fabricated on the same wafer. Schottky contact in the gate region and Schottky area in the drain region were formed using Ni/Au (50/200 nm) multilayer metals deposited by EBE. Fig.1(d) shows the scanning electron microscope (SEM) image of the smooth drain region of the fabricated devices. It could effectively reduce the peak electric field to improve the breakdown voltage. Then, the wafer was passivated with the 3 nm Al_2O_3 and 800 nm SiN layers. Finally, contact vias were opened via ion-beam etching, and the power metal Ni/Au (50/500 nm) used for interconnection was deposited using

979-8-3315-0417-5/25 $31.00 © 2025 IEEE

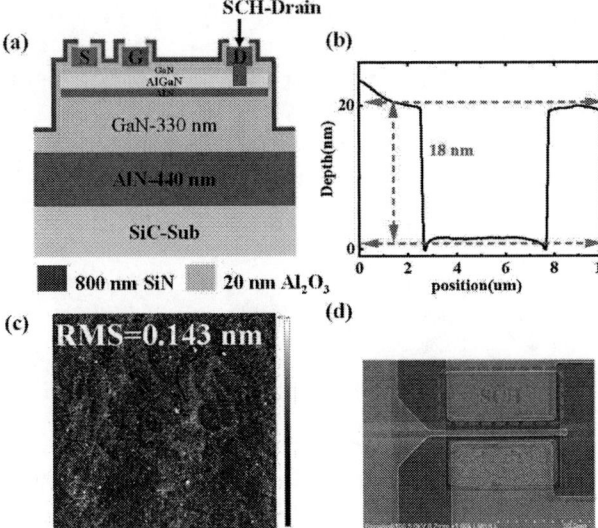

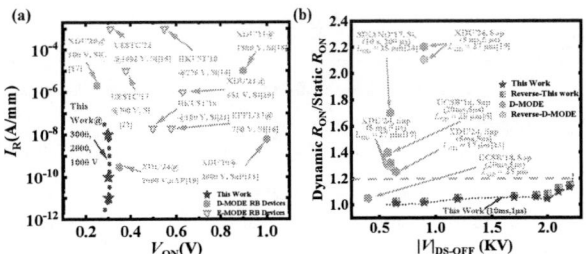

Fig. 1(a) The schematic cross-sectional view of the RB GaN on SiC HEMTs. (b) trench profile along the drain-recessed window. (c) Surface morphology in the drain-recessed region measured via AFM in a 2×2 μm² scan area, (d) SEM image of the RB GaN on SiC HEMTs.

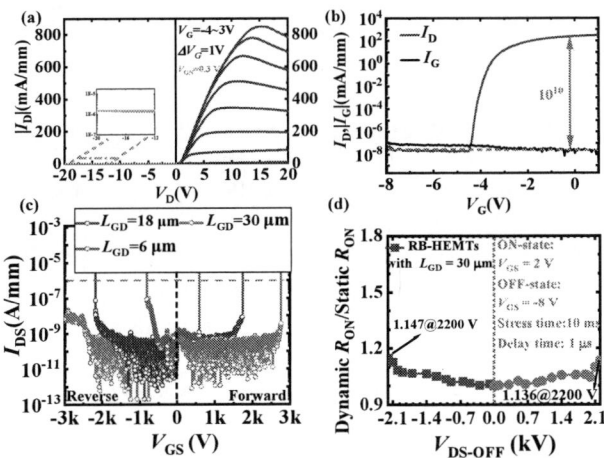

Fig. 2 (a) Output and (b) transfer characteristics of the RB GaN on SiC HEMTs. (c) Forward and reverse breakdown characteristics of the RBMHIC-switches with L_{GD} = 6, 18, and 30 μm at V_{GS} = -8 V. (b) The ratio of dynamic-R_{ON}/static-R_{ON} of the RB GaN on SiC HEMTs.

EBE. The fabricated devices have a gate length (L_G) of 2 μm, a gate width (W_G) of 50 μm, a gate-source distance (L_{GS}) of 2 μm, and gate-drain distances (L_{GD}) of 6/18/30 μm.

Results and Discussion

Fig 2(a) and (b) show the output and transfer characteristics of the RB GaN on SiC HEMTs with L_{GD} of 18 μm. The fabricated devices show a saturation drain current of 848 mA/mm, an I_{on}/I_{off} of 10^{10}, and a reverse drain current of 1×10^{-6} mA/mm. The V_{ON} of RB GaN on SiC HEMTs is 0.3 V at I_{DS} of 1 mA/mm. Fig.2 (c) shows the forward breakdown voltage (V_{FBR}) values of the RBMHIC-switches with L_{GD} of 6 μm, 18 μm, and 30 μm are 606 V, 1718 V, and 2794, respectively. Meanwhile, the reverse breakdown voltage (V_{RBR}) values of the RB GaN on SiC HEMTs with L_{GD} of 6

Fig. 3 (a) I_R and V_{ON} of the fabricated devices and the reverse blocking D-Mode GaN power devices reported in the literature. (b) Comparison of the dynamic characteristics of the RB GaN on SiC HEMTs to the state-of-the-art E/D-Mode GaN power devices.

μm, 18 μm, and 30 μm are -793 V, -2231 V, and -3000, respectively. The reverse leakage current (I_R) at V_{DS} of -3000 V is observed to be as low as 10^{-5} mA/mm. Furthermore, Fig.2(d) shows the dynamic characteristics of the RB GaN on SiC HEMTs, the forward and reverse ratio between dynamic-R_{ON} and static-R_{ON} was 1.136 and 1.147 after a 10 ms 1900 V drain stress voltage (V_{DS-OFF}) and a 10 ms -2200 V V_{DS-OFF} (delay time = 1 μs) for the RB GaN on SiC HEMTs, respectively, which set a record among all existing D-Mode GaN power devices with reverse blocking capability.

The I_R and V_{ON} of the RB GaN on SiC HEMTs are compared with E/D-Mode reverse blocking GaN power devices as presented in Fig. 3(a). The fabricated devices exhibit an excellent trade-off between I_R and V_{ON}. The I_R is observed to be as low as 10^{-5} mA/mm, 10^{-7}mA/mm, and 10^{-8} mA/mm at 3000 V, 2000 V, and 1000 V, respectively. Certainly, the devices presented in this work exhibit a small V_{ON} along with small I_R at V_{DS} = -3000 V. Besides, Fig.3 (b) presents the comparison of the dynamic characteristics of the RB GaN on SiC HEMTs to the state-of-the-art reverse-blocking D-Mode GaN power devices. The fabricated RB GaN on SiC HEMTs shows outstanding dynamic characteristics compared to the results reported in the literature.

Conclusion

In summary, The RB GaN On SiC HEMTs shows a low V_{ON} of 0.3 V, a high I_{ON}/I_{OFF} ratio of 10^{10}, a high V_{FBR} of 2794 V, and a high V_{RBR} of -3000 V. In addition, the devices exhibit the smallest I_R of $1 \times^{-5}$ mA/mm at V_{DS} = -3000 V with an ultralow forward and reverse dynamic R_{ON}/static R_{ON}. These excellent results indicate the massive potential of the D-Mode GaN devices for high-power applications.

Acknowledgments

This work was supported in part by the National Key Research and Development Program of China (Grant No. 2021YFB3601900), in part by the National Natural Science Foundation of China (Grants No. 62474137 and 62234002), and in part by the Guangdong Basic and Applied Basic Research Foundation (Grant No. 2024A1515013123).

References

[1] J. Ma, C. Erine, M. Zhu, N. Luca, P. Xiang et al., "1200 V Multi-Channel Power Devices with 2.8 Ω·mm ON-Resistance," IEDM Tech. Dig., p.4.1.1 (2019).

[2] M. W. Rahman, N. K. Kalarickal, H. LEE, T. RAZZAK, and S.RAJAN. "Integration of high permittivity $BaTiO_3$ with AlGaN/GaN for near-theoretical breakdown field kV-class transistors." Applied Physics Letters., pp. 119(2021).

[3] H. Wang, J.Wang, M. Li, Q. Cao, M.Yu et al., "823-mA/mm Drain Current Density and 945-MW/cm^2 Baliga's Figure-of-Meri Enhancement-Mode GaN MISFET With a Novel PEALD-AlN/LPCVD-Si_3N Dual-Gate Dielectric." IEEE Electron Device Lett., 39, pp. 1888-1891 (2018).

[4] B. Hult, J. Wang, M. Li, Q. Cao, M.Yu et al., "High Voltage and Low Leakage GaN-on-SiC MISHEMTs on a "Buffer-Free" Heterostructure." IEEE Electron Device Lett., 43, pp. 781-784 (2022).

[5] O S. Koksaldi, M. Thorsell, J. T. Chen and N. Rorsman. "N-Polar GaN HEMTs Exhibiting Record Breakdown Voltage Over 2000 V and Low Dynamic On-Resistance." IEEE Electron Device Lett., 39, pp. 1014-1017 (2018).

[6] Y. Wang, G. Lyu, J. Wei, Z. Zheng, J. He et al., "Characterization of Static and Dynamic Behavior of 1200 V Normally OFF GaN/SiC Cascode Devices." IEEE T IND ELECTRON, 67, pp. 10284-10294(2019).

[7] J. Kemmerling, R. Guan, M. Sadek, Y. Xiong, J. Song et al., "GaN Super-Heterojunction FETs With 10-k Blocking and 3-kV Dynamic Switching." IEEE Trans. Electron Devices, 71, pp. 1153-1159(2024).

[8] R. Hao, W. Li, K. Fu, G. Yu, L. Song et al., "Breakdown Enhancement and Current Collapse Suppression by High-Resistivity GaN Cap Laye in Normally-Off AlGaN/GaN HEMTs." IEEE Electron Device Lett., 38, pp. 1567-1570(2017).

[9] E. Bahat-Treidel, F. Brunner, O. Hilt, E. Cho, J. Wurflet al., "AlGaN/GaN/GaN: C back-barrier HFETs with breakdown voltage of over 1 kV and low $R_{ON} \times A$," IEEE Trans. Electron Devices, 57, pp. 3050–3058(2010).

[10] N. Remesh, N. Mohan, S. Raghavan, R. Muralidharan, and D. N. Nath, "Optimum carbon concentration in GaN-on-silicon for breakdown enhancement in AlGaN/GaN HEMTs," IEEE Trans. Electron Devices, 67, pp. 2311–2317(2020).

[11] D.-Y. Chen, A. Malmros, M. Thorsell, H. Hjelmgren, O. Kordinaeet al., "Microwave performance of 'buffer-free GaN-on-SiC high electron mobility transistors," IEEE Electron Device Lett., 41,pp. 828–831(2020).

[12] Y. Zhang, A. Dadgar, and T. Palacios, "Gallium nitride vertical power devices on foreign substrates: A review and outlook," J. Phys. D, Appl. Phys., 51(2018).

[13] Y. Wu, J. Zhang, S. Zhao, W. Zhang, Y. Zhan et al., "More Than 3000 V Reverse Blocking Schottky-Drain AlGaN-Channel HEMTs With >230 MW/cm2 Power Figure-of-Merit," IEEE Electron Device Lett., 40, pp. 1724-1727(2019).

[14] C. Zhou, W. Chen, E. L. Piner and K. J. Chen, "Schottky-Ohmic Drain AlGaN/GaN Normally Off HEMT With Reverse Drain Blocking Capability," IEEE Electron Device Lett., 31, pp. 668-670(2010).

[15] M. Xia, Y. Ma, V. Pathirana, K. Cheng, A. Xie, et al., "Multi-Channel Monolithic-Cascode HEMT (MC2-HEMT): A New GaN Power Switch up to 10 kV," in IEDM Tech. Dig., p. 5.5.1-5.5.4(2019).

[16] J. Ma, M. Zhu and E. Matioli, "900 V Reverse-Blocking GaN-on-Si MOSHEMTs With a Hybrid Tri-Anode Schottky Drain," IEEE Electron Device Lett., 38, pp. 1704-1707(2017).

[17] H. Wang, W. Mao, S. Zhao, M. Du, Y. Zhang et al., "1.3 kV Reverse-Blocking AlGaN/GaN MISHEMT With Ultralow Turn-On Voltage 0.25 V," IEEE J. Electron Devi, 9, pp. 125-129(2021).

[18] Y. Wu, W. Zhang, J. Zhang, S. Zhao, J. Luo et al., "Au-Free $Al_{0.4}Ga_{0.6}N/Al_{0.1}Ga_{0.9}N$ HEMTs on Silicon Substrate With High Reverse Blocking Voltage of 2 kV," IEEE Electron Device Lett., 68, pp. 4543-4549(2021).

[19] X. Liu, Y. Fan, R. Huang, Y. Wen, W. Zhang et al., "Low Turn-On Voltage and High-Power Figure-of-Merit GaN HEMTs With Reverse Blocking Capability," IEEE Trans. Electron Devices., 71, pp. 911-915(2024).

[20] H. Wang, W. Mao, J. Luo, C. Yang, J. Chen et al., "Experimental Demonstration of Monolithic Bidirectional Switch With Anti-Paralleled Reverse Blocking p-GaN HEMTs," IEEE Electron Device Lett., 42, pp. 1264-1267(2021).

[21] J. Le, J. Wei, G. Tang, Z. Zhang, Q. Qian et al., "Reverse-Blocking Normally-OFF GaN Double-Channel MOS-HEMT With Low Reverse Leakage Current and Low ON-State Resistance," IEEE Electron Device Lett., 39, pp. 1003-1006(2018).

[22] Shi Y, Chen W, Liu C, G. Hu, J. Liu et al. "A high-performance GaN E-mode reverse blocking MISHEMT with MIS field effect drain for bidirectional switch." in Proc. IEEE 29th Int. Symp. Power Semiconductor Devices IC's (ISPSD), p. 207(2017).

[23] Y. Fan, X. Liu, R. Huang, Y. Wen, W. Zhang et al., "High-breakdown-voltage (>3000 V) and low-power-dissipation $Al_{0.3}Ga_{0.7}N/GaN/Al_{0.1}Ga_{0.9}N$ double-heterostructure HEMTs with Ohmic/Schottky hybrid drains and Al_2O_3/SiO_2 passivation," Sci. China Inf. Sci., 6, pp. 781-784(2023).

[24] Z. Zhang, G. Yu, X. Zhang, X. Deng, S.Li, Y. Fan et al., "Studies on High-Voltage GaN-on-Si MIS-HEMTs Using LPCVD Si3N4 as Gate Dielectric and Passivation Layer," IEEE Trans. Electron Devices, 63, pp.731-738(2016).

Demonstration of Reliable Magnetic Shift Register Reading Using 50 nm MTJs on CMOS IC towards 3D Ultra-High Density Memory

M. Quinsat, Y. Ueda, N. Shimomura, S. Hashimoto, N. Umetsu, Y. Ootera, J. Iwata, H. Tokuhira, S. Miyano, M. Yoshikawa, T. Kondo, M. Saitoh, M. Kado

Kioxia Corporation, Yokohama, Japan

Abstract

We propose an ultra-high density memory based on a 3D magnetic shift register (3D-SR) using nanotubes compatible with existing process technologies. Through simulations, we confirm the remote reading operation of the 3D-SR when placing an in-plane magnetic tunnel junction sensor above the nanotube. We experimentally demonstrate the potential of this read method in 2D-SR by achieving external magnetic field-free Write/Shift/Read operations using CMOS sensing amplifiers and driving circuitry.

Keywords: 3D memory, magnetic shift register and CMOS

Introduction

The 2D magnetic shift register (SR) memory [1-2] is the solid state equivalent to HDD storage system where information is stored in magnetic nanowires (NW) in the form of up and down magnetic domain. 3D-SR based on high aspect ratio U-shaped ribbon-like NW [2] was proposed to achieve ultrahigh density, but the process for patterning such high aspect ratio NWs is still unknown. Also, field-free operation of 2D-SR devices has not been reported on CMOS IC making the prospect of such memories unlikely. In this paper, we report on a successful integration of SR devices on CMOS circuitry, and show for the first time the field-free fast I/O digital operation, thanks to remote in-plane magnetic tunnel junction (MTJ) readers and embedded metal writing lines (WLs), in Fig.1. We show that the demonstrated reading devices can be used "as-is" in a breakthrough ultra-high density 3D-SR concept using magnetic nanotubes (NT) inside memory holes with read/write elements at both ends, in Fig.2. The NT

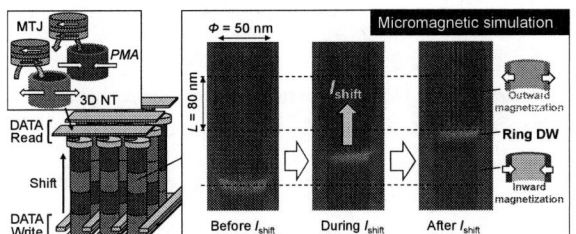

Fig. 2. Proposed 3D-SR with remote MTJ reading the magnetic NT information. Magnetic simulation results show that ring DWs are shifted in the NT. MTJ and NT are connected in series with a thick metallic spacer.

exhibits perpendicular magnetic anisotropy (PMA), which is ideal for high bit density. Simulation results show that ring-shaped domain walls (DWs) in 50 nm diameter NTs can be reliably shifted by spin orbit torque (SOT), assuming typical PMA material parameters. Highly packed micrometer deep memory holes are commonly achieved and, unlike existing 3D memories, our proposal does not require the formation of space-consuming word-line staircase structure to operate. Recently, the electrical DW shift motion in PMA Co/Pt nanowires deposited by ALD has been reported [3], which is a first step to the realization of 3D-SRs.

Remote MTJ Principle and its Demonstration in 2D-SR

A. Remote MTJ

In early 2D-SR proposals [1,4-6], the NW is an extended free-layer (FL) from a MTJ structure (Fig. 3a). Although such MTJs may potentially work both as a reader and writer, the etching of the reference layer (RL)/MgO /FL must be stopped within the ~1 nm thick MgO. Therefore, MTJ shorts and NW damages are hardly traded-off on very little choice of NW shapes and materials. Attempts to add an intermediate FL between the MgO and NW to relax process constraints have

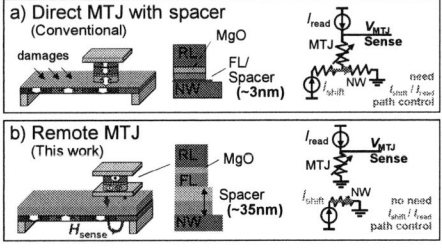

Fig.3. a) Existing 2D-SR demonstrations with direct MTJ stacking requiring trade-off between MTJ etching and NW damages. b) Proposed remote MTJ sensing where MTJ and NW are formed in different step. Proposed structure also simplifies the driving circuit for shift/read operation due to thick spacer.

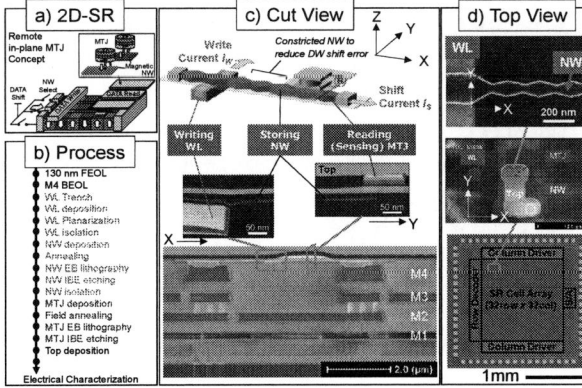

Fig. 1. a) 2D-SR concept with remote MTJs. 1024 2D-SR cells array chip b) process flow, c) cut and d) top views. WL, NW, and MTJ are formed using electron-beam (EB) lithography and ion beam etching (IBE) techniques.

979-8-3315-0417-5/25 $31.00 © 2025 IEEE

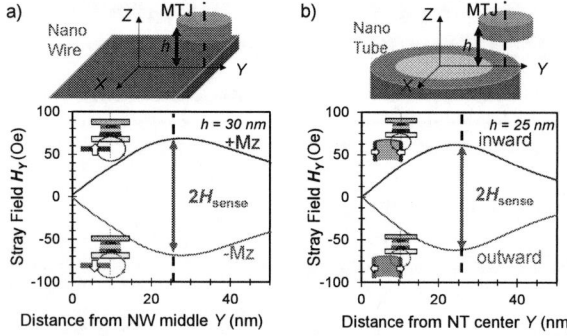

Fig.4. Stray field calculated as a function of offset position Y at a height a) h = 30 nm above the nanowire and b) at h = 25 nm above the nanotube.

not resulted in field-free operation with high MTJ output [4-6], and it is believed that yield issues would be a future challenging issue. In addition, control circuit must be optimized so as to produce different current paths in the 3 terminal device. Our proposed structure using MTJ to remotely sense the SR information allows both the independent optimization of NWs and MTJs structures and the separation of the read and shift current paths, in Fig.3(b). We simulated a NW that elongates in the X-direction, and the MTJ's position is varied in the NW width Y-direction at a height h = 30 nm in the Z direction (Fig. 4(a)). NW width in the Y-direction is 50 nm. Depending on the NW magnetic orientation $\pm M_Z$, the in-plane component of the magnetic stray field H_Y reaches a peak value of ± 65 Oe when the MTJ center is positioned above the edge of the NW in Y = 25 nm. Similar results are obtained for the 3D-SR (Fig. 4(b)), where the MTJ is placed at a height of h = 25 nm above the termination's edge of a 50 nm diameter NT. According to the magnetization pointing inward or outward to the NT center, peaks in-plane magnetic field of ± 60 Oe are obtained, which is large enough to be read by in-plane MTJs.

B. 2D-SR Fabrication on CMOS IC and MTJ remote sensing
The main fabrication process steps of 1024 2D-SR chips made on the Metal 4 of a 130 nm CMOS are shown in Fig.1(b) with detailed cut and top views. From bottom to top, the writing line (WL), is first formed and insulated from the NW patterned in straight and constricted wires to reduce the DW shift error mechanism [4]. The NW is encapsulated by insulating materials and the MTJ is formed on top of the edge of the NW. Detailed stack structures are shown in Fig.5(a). The WL/NW and NW/MTJ alignment and respective height were verified during and after process by SEM and TEM, Figs.5(b-c). We first fabricated the MTJ alone in the same chip architecture than the 2D-SR devices, omitting WL and NW fabrication steps, and measured the MTJ resistance R_{MTJ} as function of in-plane Y field H_Y sweep in Fig. 6(a). The tunnel magnetoresistance ratio (TMR) between parallel and antiparallel state of the MTJ is $TMR \sim 150\%$ and the resistance area product $RA \sim 100$ $\Omega\mu m^2$. To obtain suitable 0

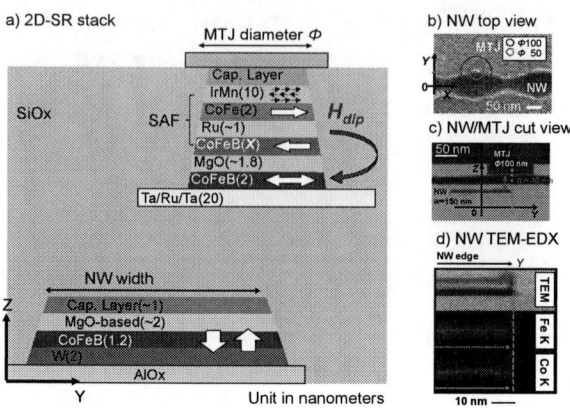

Fig.5. a) Detailed stack structure of the NW and MTJ. b) Top SEM image of the constricted NW. c) NW/MTJ offset cut view, $h \sim 40$ nm. d) TEM-EDX shows that the NW magnetic materials are present near the edge.

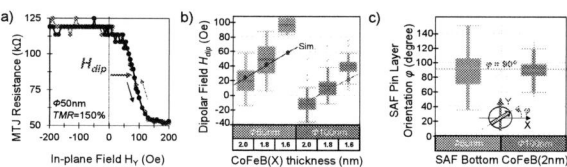

Fig.6. a) R_{MTJ} vs H_Y loop curve for MTJ device without underneath WL or NW. b) Stray field distributions with changing bottom CoFeB layer thickness and MTJ shape. c) SAF pin direction distribution is centered on Y-direction.

field sensitivity on the NW stray field, we choose the thickness of the bottom layer of the Synthetic Anti-Ferromagnetic (SAF) layer to 2 nm in order to exhibit the smallest dipolar field H_{dip}, for both the MTJ diameters Φ of 50 and 100 nm, as shown in Fig.6(b). We also verified in Fig. 6(c) that the SAF is correctly pinned in the Y-direction after process. The Fig.7(a) shows that overall shape of the R_{MTJ}-H_Y curve for a 2D-SR device having underneath WL and constricted NW are close to the reference MTJs without WL or NW (with similar TMR and RA), demonstrating the good integration of the SR device. The switching field between parallel and anti-parallel MTJ state is different depending on the NW magnetization direction $\pm M_Z$. This difference gives exactly $2H_{sense}$. The Fig, 7(b-c) shows $2H_{sense}$ as a function of the varying NW/MTJ Y-offset. The strength dependence of $2H_{sense}$ agrees well with the simulation with a maximum when MTJ is positioned near the NW's edge. Small diameter MTJs allow the pinpoint measure of the stray field with a fast varying amplitude near the NW's edges.

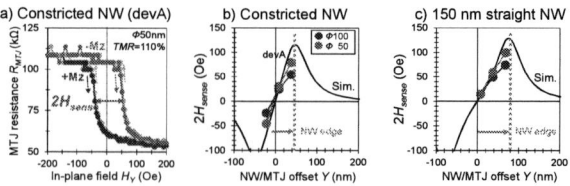

Fig.7. a) R_{MTJ} vs H_Y loop curves for an MTJ device with NW set in $+M_z$ (blue) or $-M_z$ (red) state. $2H_{sense}$ as a function of NW/MTJ offset Y for 50 and 100 nm MTJ with b) constricted and c) 150 nm wide straight NW shape.

979-8-3315-0417-5/25 $31.00 © 2025 IEEE   316

Field-Free Write/Shift/Read Operation of SR

Using the control IC schematically shown in Fig.8(a), we confirmed the DW writing in the NW by the Oersted field H_{WL} induced by the writing current flowing in the WL, in Fig. 8(b). We first initialized the R_{MTJ} to high value by applying a field $H_Z = -800$ Oe. At zero magnetic field we applied a 100 ns long pulse to the WL and we verified the switching field of the MTJ, i.e. the NW's magnetization switching underneath the MTJ, while increasing H_Z. We repeated the experiment for several writing voltage V_W. The switching field is reduced from initial coercive field $H_C \sim 350$ Oe to the DW moving field $H_{DW} \sim 100$ Oe, indicating the writting of DW in the NW for V_W above 1.9 V. Fig.8(c) shows the voltage of the sensing amplifier (S/A) circuit input V_{SAIN} from a Φ100 nm MTJ depending on the magnetization state $\pm M_z$. Reliable reading of the NW magnetization was achieved in less than 400 ns in the device with a V_{MTJ} bias of 0.3 V. Fig. 9 shows the demonstration of field-free operation of the shift register by sequentially writing 0 (W0) and 1 (W1), in between a

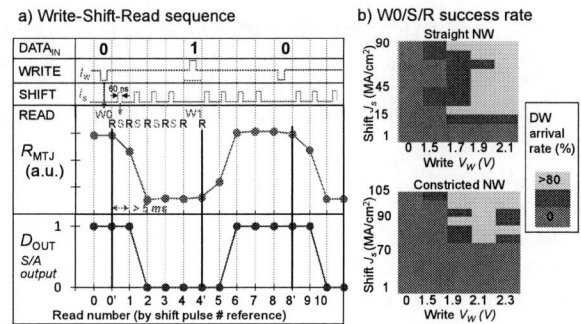

Fig.9. a) Read results during W(S/R)$_4$ cycles where R_{MTJ} is confirmed at each step by multi-meter (read time is 5 ms) and by sense amplifier output D_{OUT} (read time is 1 μs). b) DW arrival success rate in a W(SR)$_{10}$ sequence as a function of writing voltage V_W and J_S for a straight and a constricted device.

series of four 60 ns long shift pulse current (current density $J_s \sim 80$ MA/cm^2), while monitoring both R_{MTJ} and S/A output. Write-"W0"/Shift/Read sequences are abbreviated by their operation main letter, e.g. W0/S/R. During the W0/R/(S/R)$_4$ and W1/R/(S/R)$_4$ sequences, the MTJ changes its output each time after the second 2nd shift pulse following the input data writing. Finally, we show the success rate of the W0/(S/R)$_{10}$ sequence performed under 100 μs at zero field as a function of V_W and J_S for 150 nm straight and constricted NW's shape devices. We observed successful DW arrival for W/S/R sequence lasting about 700 ns which is limited by the read operation time from the S/A in our experiment (see Fig.8(d)).

	[1] (2011)	[4] (2017)	[5-6] (2021)	This Work
External Field Required	Yes	No	Yes	No
MTJ structure	Direct stacking	Direct stacking with spacer	Remote MTJ	
SR Magnetic Material	IMA	PMA	PMA	PMA
TMR ratio (%)	55	<0.1	80	110
MTJ diameter size (nm)	>140	N.D.	80	50
SR width size (nm)	185	N.D.	180	50-150
Read Sense Amplifier	-	-	-	Yes
3D type SR	U-Shape [2]	-	-	Tube

Table I Comparison of reported 2D-SR devices results with a focus on the MTJ read element. The demonstrated reading using S/A is also suitable for 3D-SR with magnetic nanotubes. (IMA: in-plane magnetic anisotropy)

Conclusion

Table I summarizes the improvements demonstrated in this work with respect to the previous 2D-SR reports with MTJs. We successfully demonstrated field-free W/S/R operation in 2D-SR devices fabricated on CMOS IC. The reading scheme is applicable to the proposed 3D-SR with magnetic nanotubes, which is a strong candidate for ultrahigh-density and highly-reliable solid state memory.

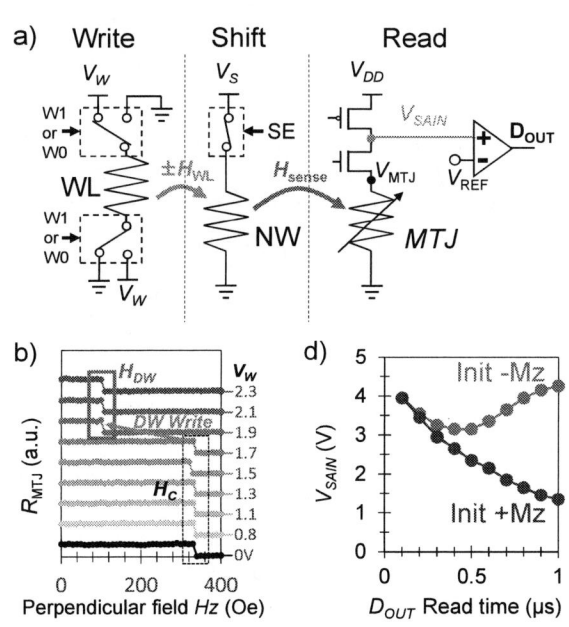

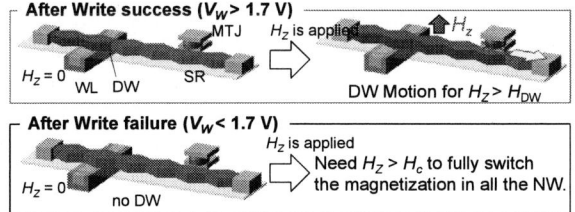

Fig.8. a) Write, shift, and read operation control circuit schematics. b) MTJ resistance R_{MTJ} vs H_z after a write pulse of intensity V_W is applied at zero field. c) After a write success, the device switching field is reduced from H_c to the DW motion field H_{DW}. d) S/A input voltage estimated during V_{REF} sweep while measuring D_{OUT} transitioning. The NW magnetization state is converted in digital output D_{OUT} in less than 1 μs (TMR ~ 110%).

References

[1] A. J. Annunziata et al., IEDM, pp. 539-542 (2011).

[2] S. S. P. Parkin et al., Science, vol. 320, no. 5873, pp. 190-194 (2008).

[3] M. Kado et al., IEEE Trans. on Mag., vol. 59, no. 11, 2101605 (2023).

[4] T. Kondo et al., VLSI-TSA, pp. 119-120 (2017).

[5] E. Raymenants et al., IEDM, pp. 685-688 (2021).

[6] E. Raymenants et al., Nature Electronics, vol. 4, pp. 392–398 (2021).

Artificial Hodgkin–Huxley Neurons based on Ferro-ionic CuInP$_2$P$_6$

Fan Yang[1], Lei Liang[1], Qirui Zhang[1], Xuemei Wang[1], Qing Liu[1], Xiao Luo[1], Fucai Liu[1]*

1 University of Electronic Science and Technology of China, China

Abstract

CuInP$_2$P$_6$ exhibits both ferroelectricity and ionic conductivity, which is an excellent candidate for mimicking artificial Hodgkin–Huxley (HH) neurons. We proposed an optoelectronic device composed of a Graphene/CuInP$_2$P$_6$/Graphene/h-BN. The unique combination of ionic conductivity and ferroelectricity in CuInP$_2$S$_6$ emulate the competition and collaboration of Na$^+$ and K$^+$ channels and replicate the action potential in neurons without any circuit elements. It showcases their promising capabilities in advancing artificial intelligence and neuromorphic computing systems.

Keywords: CuInP$_2$P$_6$, Ferro-ionic, HH neuron

Introduction

Neurons communicate and process information via action potential across their membranes. Developing realistic electrical devices that mimic the biological neuron is a key challenge in advancing artificial intelligence. Some models have been proposed to demonstrated how a neuron works, including the Hodgkin–Huxley (HH) model, the integrate-and-fire (IF) model, and the leaky integrate-and-fire (LIF) model. The HH model most closely resembles a biological neuron and is also the most complex, focusing on the variation of the conductance of the ionic channels and bioinspired spike generation [1]. Copper indium thiophosphate (CuInP$_2$P$_6$), which exhibits both ferroelectricity and ionic conductivity and dynamically responds to diverse external stimulation [2], is an excellent candidate for mimicking artificial neurons. These unique properties offer promise for replicating dynamic biological activities through intricate mechanism, especially the action potential. In this study, a CuInP$_2$P$_6$-based optoelectronic device is designed as an artificial HH neuron, and the ionic conductivity and ferroelectricity in CuInP$_2$P$_6$ emulate the competition and collaboration of Na$^+$ and K$^+$ channels in neurons. It showcases their promising capabilities in advancing artificial intelligence and neuromorphic computing systems.

Results and Discussion

The optoelectronic device, depicted in Fig. 1a, is composed of a Graphene/CuInP$_2$P$_6$/Graphene/h-BN structure. Layered CuInP$_2$P$_6$ exhibits both ferroelectricity and ionic conductivity at room temperature, whose resistance can be manipulated by external factors such as electric fields, temperature, and light [3]. An external electric field causes Cu$^+$ ions to migrate across the vdW gap, resulting in a deficiency of Cu$^+$ ions at the anode and an accumulation beneath the cathode. These Cu$^+$ ions affect the surface potential and modify the height of the interface Schottky barrier [4], causing a reversible rectifying resistive switching behavior (Fig. 1b). Notably, rectifying behavior is observed following an incentive voltage of $\pm$ 2 V applied for 30 s, as illustrated in Fig 1c. A persistent out-of-plane electric field enhances the concentration gradient of Cu$^+$ ions and establishes an internal electric field (E_{Ion}). It has been reported that photocarriers are separated by E_{Ion} and collected by electrodes (I_{Ion}) under light, giving rise to the photovoltaic phenomenon [5]. In addition, the dynamic response of this device is investigated by applying a series of pulses sequentially as shown in Fig. 1d. During the potentiating process, the current gradually increases with repeated stimulations, and larger amplitudes lead to faster potentiation. However, the current decays without external inputs due to the spontaneous relaxation of the ion concentration gradient.

Fig. 2a present the out-of-plane piezoresponse force microscopy (PFM) phase image of CuInP$_2$P$_6$. The evident reversal of phase contrast in the square pattern serves as a clear indicator of the polarization switching occurring under an electric field [6]. Fig.2b shows a butterfly-like amplitude loop and a 180° phase change in response to varying applied voltage. Additionally, the off-center shift of Cu$^+$ ions resulting from ferroelectricity induces a built-in electric field (E_{Ferro}) along the out-of-plane direction. In the presence of light, photogenerated electron-hole pairs are separated by E_{Ferro}, creating a ferroelectric photocurrent (I_{Ferro}). Fig. 2c presents the photovoltaic behavior under various polarization directions [7]. These data are measured when the photocurrent is maintained in a stable state, thus excluding the influence of I_{Ion}. However, there is an unusual phenomenon that some struggle to stabilize at a specific state without adequate incentive. As illustrated in Fig. 2d, this optoelectronic device exhibits a small positive photocurrent at zero-bias of pristine state (0th I-V cycle), which is attributed to an asymmetric barrier resulting from interface variations during device fabrication [8]. By applying a negative pulse (-4 V, 50 ms), the photocurrent transitions from positive to negative, indicating a reversal of polarization. Following illumination with light for 20 cycles of I-V curves (each I-V curve lasts 12 s), the photocurrent gradually returns to its initial value. Furthermore, a similar phenomenon is observed after applying the same stimulus and waiting for 120/300 s in the dark (Fig. 2e). This suggests that the relaxation process of reversal polarization readily occurs under light. Light-induced non-equilibrium carriers, generated through the transition between energy bands, result in ferroelectric polarization instability or even reversal [9]. In CuInP$_2$P$_6$, some Cu ions are trapped at non-equilibrium sites, such as vacancies, interstitials, and anti-site defects, requiring high energy to switch. The resulting defect dipoles create an imprint field that hinders polarization reversal, leading to the quick relaxation of switched polarization [10].

Upon the sequential application of an electric pulse and light stimuli, the photocurrent (I_{PV}) is promptly established, rapidly decays to negative direction, and ultimately stabilizes at a positive value (Fig. 3a). The unique combination of ferroelectricity and ionic conductivity emphasizes the potential of the device to replicate the action potential in neurons. When applying a voltage on device, some Cu ions overcome the energy barrier and migrate within the $CuInP_2P_6$. A sharp opposite ionic peak is observed at zero-bias following the voltage turn-off, representing the relaxation of ionic current due to an asymmetrical distribution [3]. Fig. 3c-d shows that the ionic peak current steadily increases with the rise of pulse width or amplitude, maintaining an opposite direction to the applied voltage. Upon exposure to the light, an immediate positive peak emerges. Simultaneously, as E_{Ion} gradually decreases, it is reflected in a sharp decline in I_{Ion}. The photocurrent peak also presents a similar trend with ionic current, which suggests that the photocurrent peak is directly influenced by the concentration gradient of Cu^+ ions induced by the electric pulse. Moreover, this relaxation process occurs in the dark without external stimuli, and the interval between the electric pulse and light stimuli directly influences the photocurrent peak. As the interval increases, the photocurrent peak gradually diminishes and nearly disappears at an interval of 150 s (Fig. 3b), corresponding to complete ion relaxation in the dark. Subsequently, I_{Ferro} resulted by reversed polarization becomes the predominant component of I_{PV}, with a direction opposite to I_{Ion}. Eventually, the polarization relaxes back to its initial state, and I_{PV} returns to a positive value and is stabilized.

The dynamic process of simulating action potential in a neuron is presented in Fig. 3d. In the resting state (Stage I), $CuInP_2P_6$ exhibits initial polarization states. The applied electric pulse results in an asymmetrical distribution of Cu ions and a reversal of polarization. Upon releasing the electric field and illuminating with light, the reversed polarization generates negative I_{Ferro} (Fig. 3e). Meanwhile, an asymmetrical distribution generates a larger positive I_{Ion}. The growth in coupled I_{PV} mirrors the rising phase of an action potential, akin to the opening of Na^+ channels (Stage II). Subsequently, I_{Ion} decreases as Cu ions spontaneously relax, leading to a sharp decline in I_{PV}. This process is similar to the falling phase (Stage III). Following this rapid relaxation, I_{Ferro} becomes the predominant component. While reversed polarization proves unstable under light, it reverses and stabilizes at initial polarization. Eventually, I_{PV} transitions

from a negative direction to a positive direction, corresponding to hyperpolarization (Stage IV). After an adequate relaxation period, the device can initiate firing with a new stimulus, illustrating its intrinsic resemblance to neuronal behavior.

Acknowledgments

This work was supported by the National Key Research & Development Program (2020YFA0309200), the National Natural Science Foundation of China (12161141015, 62074025, 62374043), Sichuan Science and Technology Program (2024YFHZ0264, 2024NSFSC1002), Sichuan Province Key Laboratory of Display Science and Technology, and Shanghai Oriental Talent Program-Youth Project (2022)

Conclusion

In summary, a novel optoelectronic device based on $CuInP_2P_6$ has been successfully adopted as artificial HH neurons. The unique combination of ionic conductivity and ferroelectricity within $CuInP_2P_6$ emulate the competition and collaboration of Na^+ and K^+ channels in neurons. We successfully demonstrate the coupling between these mechanisms and mimic the action potential in neurons without any circuit elements. It showcases their promising capabilities in advancing artificial intelligence and neuromorphic computing systems.

References

[1] H. Huang et al. Quasi‐Hodgkin‐Huxley Neurons with Leaky Integrate‐and‐Fire Functions Physically Realized with Memristive Devices. Advanced Materials, 2018 31(3): 1803849.

[2] Zhu H, et al. Highly Tunable Lateral Homojunction Formed in Two-Dimensional Layered $CuInP_2S_6$ via In-Plane Ionic Migration. ACS Nano, 2023, 17(2): 1239–1246.

[3] Neumayer S M, et al. Ionic Control over Ferroelectricity in 2D Layered van der Waals Capacitors. ACS Applied Materials & Interfaces, 2022, 14(2): 3018–3026.

[4] Li Y, et al. Enhanced bulk photovoltaic effect in two-dimensional ferroelectric $CuInP_2S_6$ Nature Communications, 2021, 12(1): 5896.

[5] Jiang X et al. Manipulation of current rectification in van der Waals ferroionic $CuInP_2S_6$. Nature Communications, 2022, 13(1): 574.

[6] Liu F, et al. Room-temperature ferroelectricity in $CuInP_2S_6$ ultrathin flakes. Nature Communications, 2016, 7(1): 12357.

[7] Wang Y, et al. A Three-Dimensional Neuromorphic Photosensor Array for Nonvolatile In-Sensor Computing. Nano Letters, 2023, 23(10): 4524–4532.

[8] Zhou S, et al. Anomalous polarization switching and permanent retention in a ferroelectric ionic conductor. Materials Horizons, 2020, 7(1): 263–274.

[9] Li T, et al. Optical control of polarization in ferroelectric heterostructures, Nature Communications, 2018, 9(1): 3344.

[10] Lipatov A, et al. Optoelectrical Molybdenum Disulfide (MoS_2)—Ferroelectric Memories. ACS Nano, 2015, 9(8): 8089–8098.

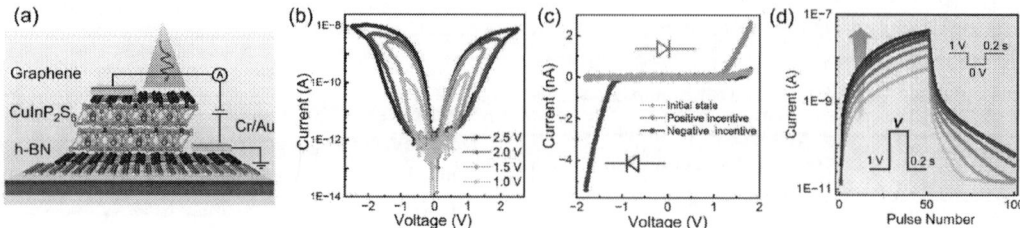

Fig. 1: Structure and ion migration characteristic. a: Illustration structure of Graphene/CuInP$_2$P$_6$/Graphene/h-BN. **b:** *I-V* curves are measured with a varying sweep rang. **c:** *I-V* curves ($\pm$ 1.8 V) in different rectification states. **d:** The variation of the current with pulse numbers. The pulse amplitude during the potentiating process ranges from +2 V to +3 V with an increase step of 0.2 V, and is 0 V in the relaxation process.

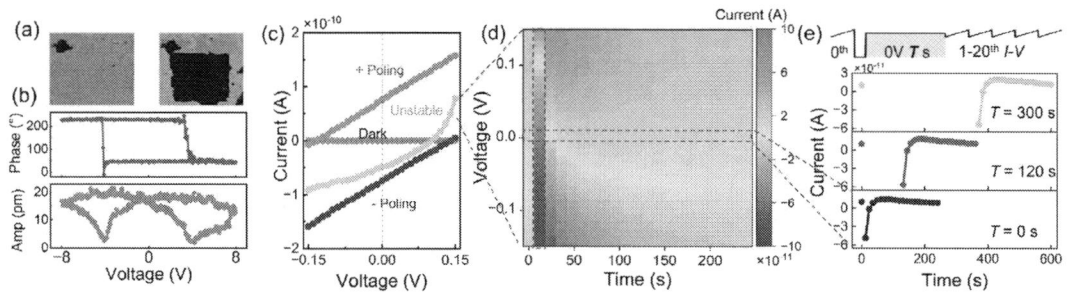

Fig. 2: Ferroelectric characteristic. a: PFM phase-maps of CuInP$_2$P$_6$. The application of a bias voltage to the microscope tip results in the polarization switching. **b:** PFM amplitude and phase hysteresis loop during the switching process for the CuInP$_2$P$_6$ under an electric field ($\pm$ 8 V). **c:** The ferroelectric photovoltaic behavior under electric polarization states within a reading bias of $\pm$ 0.15 V. **d:** The unstable photovoltaic behavior at initial state (0th *I-V* cycle) and relaxation process after an incentive of - 4 V with 50 ms (1 - 20th *I-V* cycle). **e:** The photocurrent at 0 V obtained from 0-20th *I-V* cycles after applying the same stimuli and waiting for 0/120/300 s in the dark.

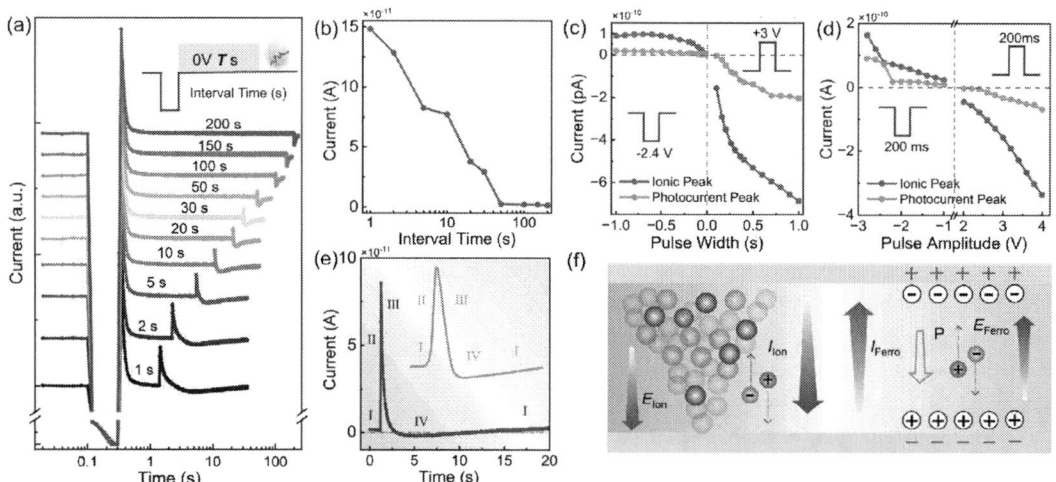

Fig. 3: Action potential simulation. a: The response of electric pulse (-4 V, 200 ms) followed by light stimuli with different interval time between the electric pulse and light stimuli. **b:** The variation of photocurrent peak with interval time. **c** and **d:** The variation of ionic current peak and photocurrent peak with different pulse width and amplitude, respectively. **e:** The optoelectronic device (blue curve) simulates the dynamic process of an action potential in a neuron (orange curve). I: Resting state; II: Hyperpolarization; III: Depolarization; IV: Undershoot. **f:** The mechanism of ferroelectricity and ionic conductivity in CuInP$_2$P$_6$.

979-8-3315-0417-5/25 $31.00 © 2025 IEEE

High Mobility and Improved Subthreshold Characteristics of Ultra-Thin Channel IGZTO TFTs Down to 3 nm

Kai Chen[1,3], Yi Jiang[1], Zhaolong He[2,3], Rui Zhang[1], Junkang Li[1*], Yunlong Li[1,3*]

[1]Zhejiang University, China, [2]University of Science and Technology of China, China

[3]Zhejiang ICsprout Semiconductor Co., Ltd, China

*Corresponding Email: lijunkang@zju.edu.cn; bta@zju.edu.cn

Abstract

In this work, the electric characteristics of IGZTO thin-film transistors (TFTs) with ultra-thin channel thicknesses (T_{CH}) down to 3 nm were systematically investigated. As T_{CH} decreases, a positive shift in threshold voltage (V_{th}) is observed, with the 3 nm device showing a V_{th} around 0.9 V and a field-effect mobility of 30.3 cm²(V·s)⁻¹. The 5 nm device, featuring an appropriate V_{th} of 0.36 V, achieves an optimal balance with high mobility, highlighting the potential of IGZTO TFTs in sub-10 nm process technology for low-power applications.

Keywords: IGZTO, TFT, Ultra-thin channel

Introduction

Amorphous oxide semiconductors (AOS) have emerged as promising materials for thin-film transistors (TFTs), offering high carrier mobility, good transparency, and compatibility with flexible substrates [1]. Among these, indium-gallium-zinc oxide (IGZO) has demonstrated excellent performance, making it a popular choice in display technologies and emerging electronic devices [2, 3]. However, as the demand for further miniaturization and improved efficiency increases, the limitations of IGZO, such as stability under long-term operation and further enhancement of mobility, become more apparent [4]. To address these challenges, researchers have turned to tin-doped IGZO (IGZTO), which offers improved mobility and stability, making it a promising candidate for next-generation TFTs [4].

IGZTO incorporates tin (Sn) into the IGZO matrix, potentially enhancing carrier mobility and offering better stability against bias stress and environmental influences [5]. Despite these advantages, the impact of T_{CH} on IGZTO TFT performance remains insufficiently explored, particularly as dimensions shrink to the ultra-thin regime. Understanding how T_{CH} influences key parameters like V_{th}, mobility, and subthreshold characteristics is crucial for optimizing IGZTO-based TFTs for next-generation applications [6].

In this study, IGZTO TFTs with T_{CH} ranging from 10 nm to as thin as 3 nm are comprehensively analyzed to understand the effects of ultra-thin scaling. A significant positive shift in V_{th} is observed as T_{CH} decreases, with the 3 nm device exhibiting a V_{th} around 0.9 V and a high mobility of 30.3 cm²(V·s)⁻¹. The 5 nm devices achieve a balance between high mobility (29-36 cm²(V·s)⁻¹) and low operating voltage (V_{th} around 0.3-0.4 V), positioning it as a potential candidate for scaled, low-power applications. These findings provide critical insights into the design and optimization of IGZTO TFTs, highlighting their applicability in ultra-thin, energy-efficient electronics.

Device Fabrication

Bottom gate IGZTO TFTs with L_{CH} of 1 μm and varying T_{CH} were fabricated as illustrated in **Fig. 1**. The process began with a heavily doped p-type silicon wafer as the gate electrode, onto which a 20 nm SiO₂ layer was thermally grown to serve as the gate insulator. IGZTO deposition was performed via RF magnetron sputtering at room temperature with an argon flow rate of 30 sccm and an RF power of 100 W, allowing for film thicknesses ranging from 10 nm to 3 nm by adjusting the sputtering time. After that, 50 nm of nickel (Ni) was thermally evaporated to form the source and drain electrodes, using the lift-off technique for precise patterning. Wet etching with 3% dilute hydrochloric acid (HCl) was employed to define the IGZTO channel and isolate individual devices. Finally, all devices were annealed in an oxygen atmosphere at 400°C for 30 minutes.

Fig. 2 presents a cross-sectional transmission electron microscopy (TEM) image of a fabricated IGZTO TFT with T_{CH} of 3 nm. The corresponding energy-dispersive X-ray spectroscopy (EDX) mapping verifies the distribution and presence of Ni, IGZTO, and SiO₂ layers within the device structure.

Results and Discussion

The transfer characteristics of IGZTO TFTs with a fixed channel length of 1 μm and varying T_{CH} were measured using the Keysight B1500 system at room temperature (300 K) in a dark environment, as presented in **Fig. 3**. The V_{th} values were extracted using the constant current method. As the T_{CH} decreases from 10 nm to 3 nm, the transfer curves show a clear positive shift in V_{th}. This shift could be attributed to a decrease in free carrier concentration in thinner channels, which may enhance the electrostatic control of the gate over the channel. Throughout the range of thicknesses, the gate leakage current remains consistently low, indicating strong gate control with minimal leakage paths.

The relationship between subthreshold swing (SS) and drain current density (I_D) for the various T_{CH} is illustrated in **Fig. 4(a)**. Across the range of current densities tested, the IGZTO TFTs show consistently low SS values, typically remaining below 100 mV dec⁻¹. Such low SS values suggest a high-quality interface between the IGZTO channel and the gate

979-8-3315-0417-5/25 $31.00 © 2025 IEEE

dielectric. **Fig. 4(b)** presents the extracted minimum SS values as a function of T_{CH}. Interestingly, a decrease in SS is observed as T_{CH} is reduced, which contrasts with common expectations that thinner channels may exhibit higher SS due to increased surface scattering and interface trap effects. This trend could be influenced by improved gate control over ultra-thin channels, which might mitigate the impact of interface traps and enhance the control over channel potential. Additionally, the deposition and annealing processes likely play a role in reducing interface defects, contributing to the observed subthreshold behavior at reduced thicknesses. However, further investigations are necessary to directly confirm these mechanisms.

To evaluate the field-effect mobility (μ_{FE}) of the IGZTO TFTs, the capacitance-voltage (C-V) characteristic of the 20 nm SiO_2 gate dielectric (configured in a 50 nm Ni/20 nm SiO_2/p+ Si structure) was measured, as shown in **Fig. 5(a)**. The measured capacitance values ranged between 175 and 150 nF/cm^2, likely influenced by parasitic capacitance effects resulting from the large area of the p+ silicon substrate. The extensive surface area of the heavily doped silicon substrate can enhance fringing fields and edge effects, thereby contributing to the observed capacitance values.

Using the measured capacitance data, the μ_{FE} was calculated for different T_{CH} using the linear mobility model. **Fig. 5(b)** illustrates the dependence of μ_{FE} on gate voltage (V_{GS}) for each T_{CH}. The results indicate that as the T_{CH} increases, the mobility improves, rising from 30.3 $cm^2(V \cdot s)^{-1}$ at T_{CH} of 3 nm to 49.6 $cm^2(V \cdot s)^{-1}$ at 10 nm. This increase in mobility could be attributed to the reduction in scattering effects within thicker channels, as thinner channels typically experience increased surface roughness and interface scattering, leading to reduced carrier transport efficiency.

Fig. 6 presents the output characteristics of IGZTO TFTs with varying T_{CH}, displaying I_D as a function of drain voltage (V_{DS}) for different V_{GS}. It can be observed that I_D increases with increasing T_{CH}. Among the different thicknesses tested, the device with a 10 nm T_{CH} exhibited the highest on-current (I_{on}), reaching up to 71 $\mu A \, \mu m^{-1}$. This increase in I_{on} with thicker channels can be primarily attributed to the negative shift in V_{th} as T_{CH} increases. A thicker channel results in a higher carrier concentration due to increased bulk conduction paths, which effectively lowers the energy barrier for carrier movement. This results in a negative shift in V_{th}, allowing the device to turn on at lower gate voltages and subsequently enhancing the current flow under the same V_{GS} conditions. Additionally, thicker channels also reduce scattering effects from surface roughness and interface traps, further improving carrier mobility and contributing to the increased current output.

In **Fig. 7**, the I_D-V_{GS} characteristics of 22 randomly selected devices with a T_{CH} of 5 nm were measured. The results demonstrate consistent performance across these devices,

with minimal variation observed in the transfer characteristics. Subsequent extraction of key electrical parameters in **Fig. 8** revealed that the V_{th} predominantly ranged from 0.3 V to 0.4 V, with an average value of 0.36 V. The SS values were found to be between 80 mV dec^{-1} and 95 mV dec^{-1}, yielding an average of 85.8 mV dec^{-1}. Additionally, the mobility ranged from 29.5 $cm^2(V \cdot s)^{-1}$ to 36.6 $cm^2(V \cdot s)^{-1}$, with an average of 31.8 $cm^2(V \cdot s)^{-1}$. These results demonstrate the good yield and manufacturability of the 5 nm devices. **Fig. 9** benchmarks the mobility and T_{CH} against a range of recently published oxide TFTs. This work achieved superior mobility values at T_{CH} of both 3 nm and 5 nm.

Conclusion

This study investigated the electric performance of IGZTO TFTs with ultra-thin T_{CH} ranging from 3 nm to 10 nm. It is found that reducing T_{CH} leads to a positive shift in V_{th} and improved gate control, with the 3 nm device achieving a V_{th} of 0.9 V and mobility of 30.3 $cm^2(V \cdot s)^{-1}$. The 5 nm devices balance high mobility of 31.8 $cm^2(V \cdot s)^{-1}$ with low V_{th} (~0.36 V), suitable for low-power applications. These results will provide valuable insights into optimizing T_{CH} for IGZTO TFTs, enhancing their potential in high-performance, energy-efficient electronics.

Acknowledgments

This work is supported by National Key R&D Program of China (2022YFF0605803), Zhejiang Key R&D project (2022C01063, 2023C01017). The authors also express their gratitude to the Micro-Nano Fabrication Center and the Independent Research Project of the State Key Laboratory of Silicon and Advanced Semiconductor Materials at Zhejiang University for their support.

References

[1] D. Geng et al., *Nature Electronics*, vol. 6, no. 12, pp. 963-972, 2023.

[2] K. Nomura et al., *nature*, vol. 432, no. 7016, pp. 488-492, 2004.

[3] A. Yan et al., *Advanced Functional Materials*, vol. 34, no. 3, p. 2304409, 2024.

[4] Y.-S. Shiah et al., *Nature Electronics*, vol. 4, no. 11, pp. 800-807, 2021.

[5] J. Wu et al., *2021 Symposium on VLSI Technology*, 2021: IEEE, pp. 1-2.

[6] M. J. Kim et al., *IEEE Transactions on Electron Devices*, vol. 69, no. 5, pp. 2409-2416, 2022.

[7] B. J. Park et al., *IEEE Electron Device Letters*, vol. 44, no. 11, pp. 1857-1860, 2023.

[8] S.-J. Park et al., *IEEE Electron Device Letters*, vol. 44, no. 4, pp. 642-645, 2023.

[9] Z. Lin et al., *IEEE Electron Device Letters*, vol. 44, no. 7, pp. 1136-1139, 2023.

[10] J. Zhang et al., *IEEE Transactions on Electron Devices*, vol. 70, no. 12, pp. 6651-6657, 2023.

[11] Z. Lin et al., *IEEE Transactions on Electron Devices*, vol. 71, no. 5, pp. 3002-3008, 2024.

[12] J.-Y. Lin et al., *IEEE Electron Device Letters*, vol. 45, no. 10, pp. 1851-1854, 2024.

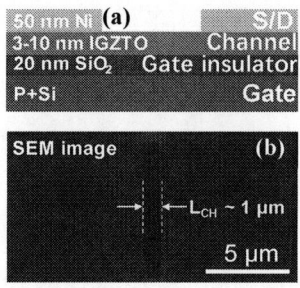

Fig. 1. (a) Schematic of the bottom gate IGZTO TFTs in this study. (b) SEM image of an IGZTO TFT with L_{CH} 1 μm.

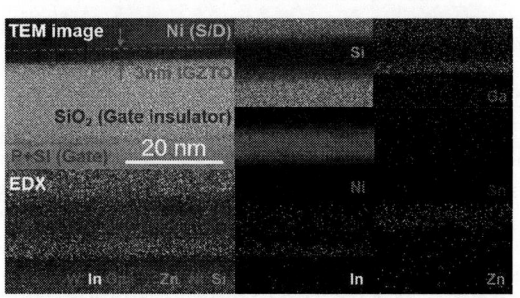

Fig. 2. HRTEM cross-section image of an IGZTO TFT with T_{CH} of 3 nm, SiO_2 thickness of 20 nm and EDX elemental mapping of W, Si, Ni, In, Ga, Sn, Zn element.

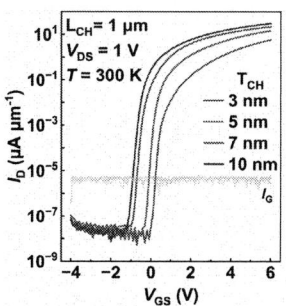

Fig. 3. I_D-V_{GS} characteristics of devices at V_{DS} of 1 V with different T_{CH} measured at room temperature (T=300K).

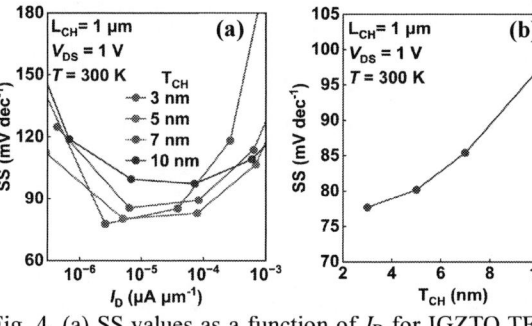

Fig. 4. (a) SS values as a function of I_D for IGZTO TFTs with varying T_{CH}. (b) Extracted minimum SS for different Tch. Thinner T_{CH} results in a lower minimum SS.

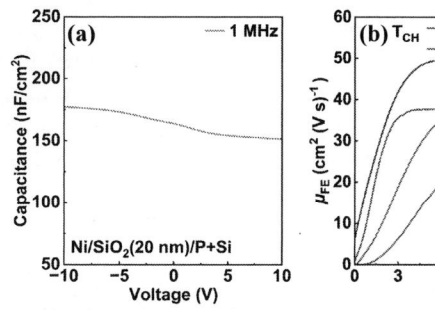

Fig. 5. (a) Capacitance-Voltage characteristic of 20 nm SiO_2 gate insulator (b) Variations in electron mobilities of the IGZTO TFTs with different T_{CH} as a function of V_{GS}.

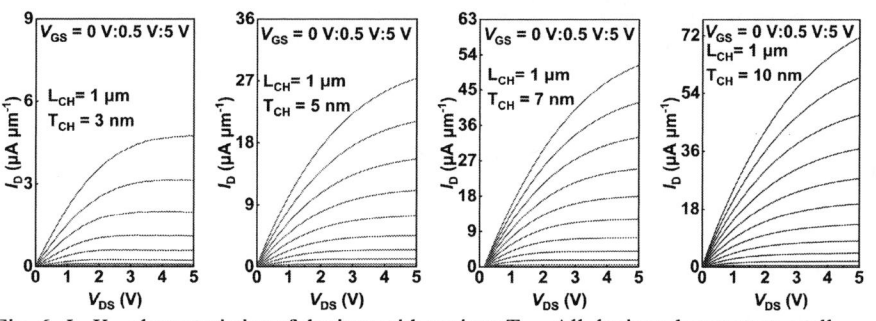

Fig. 6. I_D-V_{DS} characteristics of devices with various T_{CH}. All devices demonstrate well-defined saturation and pinch-off behavior. A noticeable reduction in I_{ON} is observed as the channel thickness decreases from 10 nm to 3 nm.

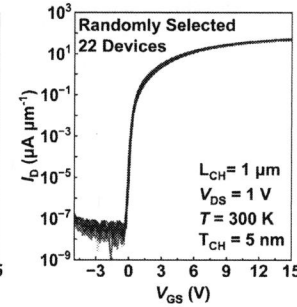

Fig. 7. Transfer curves of 22 randomly measured IGZTO TFTs with 5 nm T_{CH}.

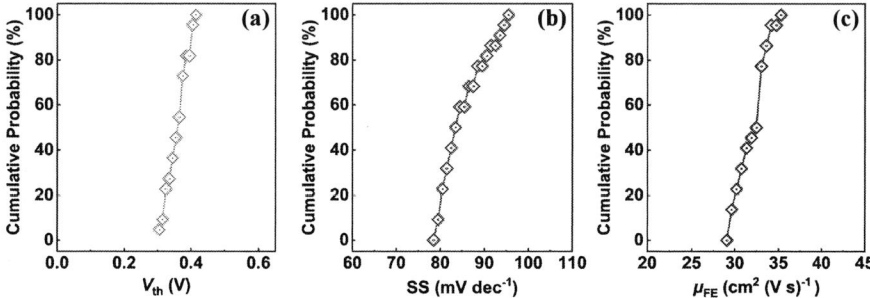

Fig. 8. Cumulative probability distributions of V_{th} (a), SS (b) and μ_{FE} (c) extracted from Fig. 7. Analysis of 22 devices shows V_{th} from 0.3 V to 0.4 V, SS from 80 mV dec^{-1} to 95 mV dec^{-1}, and μ_{FE} from 29 cm²(V·s)$^{-1}$ to 36 cm²(V·s)$^{-1}$, demonstrating good uniformity.

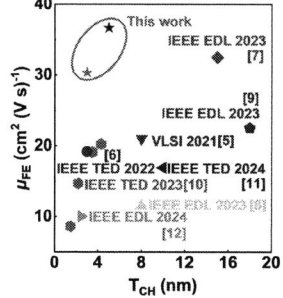

Fig. 9. Benchmark of μ_{FE} vs T_{CH}, showing improved μ_{FE} values achieved in this work.

979-8-3315-0417-5/25 $31.00 © 2025 IEEE

DESIGN AND APPLICATION OF A MULTI-MODE SIGNAL CO-DETECTION SYSTEM FOR BRAIN SIGNALS

Jianbo Jiang[1,2], Xueying Wang[1,2], Huiran Yang[1], Ziyi Zhu[1,2], Dujuan Zou[2,4], Siyuan Ni[1,2], Zhengyu Liang[1,2], Guopei Zhou[1,5], Zhitao Zhou[1,2], Liuyang Sun[2,4], Tiger H. Tao[1,2,3,4,6,7,8,9,10*] and Xiaoling Wei[1,2*]

[1]State Key Laboratory of Transducer Technology, Shanghai Institute of Microsystem and Information Technology, Chinese Academy of Sciences, Shanghai, China
[2]University of Chinese Academy of Sciences, Beijing, China
[3]School of Physical Science and Technology, Shanghai Tech University, Shanghai, China
[4]2020 X-Lab, Shanghai Institute of Microsystem and Information Technology, Chinese Academy of Sciences, Shanghai, China
[5]Wuhan Research Institute of Posts and Telecommunications
[6]Center of Materials Science and Optoelectronics Engineering, University of Chinese Academy of Sciences, Beijing, China
[7]Center for Excellence in Brain Science and Intelligence Technology, Chinese Academy of Sciences, Shanghai, China
[8]Neuroxess Co., Ltd. (Jiangxi), Nanchang, Jiangxi, China
[9]Guangdong Institute of Intelligence Science and Technology, Hengqin, Zhuhai, Guangdong, China
[10]Tianqiao and Chrissy Chen Institute for Translational Research, Shanghai, China

ABSTRACT

We present a multi-mode signal co-detection system integrating electrophysiology and electrochemical techniques, enabling simultaneous recording of neural and dopamine signals in real time from awake animals. The system features an integrated headstage with significantly reduced noise—only 32.8% of that found in commercial equipment (PalmSens4). The electrode, modified with PEDOT:PSS/IrOx, boasts a high charge storage capacity (CSC) of 36.3 mC/cm² and excellent dopamine sensitivity at 21.86 pA/µM, enhancing electroactive neurotransmitter detection.

Keywords: Multi-mode signal co-detection system, PEDOT:PSS/IrOx, High charge storage capacity (CSC)

INTRODUCTION

The integration of information technology and life sciences has accelerated the development of brain-computer interface technology, a crucial tool in neuroscience. Invasive neural electrodes, known for their deep brain penetration, offer precise single-neuron signal collection and are increasingly pivotal for studying brain and neural functions. Among these, electrodes made from flexible polymers or biomaterials provide superior mechanical adaptation to brain tissue, reducing immune responses and glial scar formation, thus enhancing accuracy and long-term stability. However, the limited number of channels in current neural electrodes restricts their ability to map and analyze complex brain circuits and functions. Most focus solely on detecting electrical signals, lacking the capability to monitor neurotransmitter changes in the brain's dynamic biochemical environment—changes linked to neurodegenerative diseases and mood disorders like depression. Therefore, developing multi-mode electrodes [1] that enable high-throughput co-detection of electrical signals and neurotransmitters is critically important.

Simultaneous detection of electrical and electrochemical signals is vital in neuroscience and neuroengineering. Recent advancements in electrode technology have facilitated widespread synchronous recordings across various animals. However, most experiments still depend on separate commercial instruments, like neural recording equipment and electrochemical workstations, complicating timely data analysis [2]. Thus, there is an urgent need for an integrated system to detect neuronal firing and neurotransmitter changes simultaneously. This study introduces a novel system for concurrent monitoring of neuronal electrical and dopamine activity in the mouse brain. It comprises an electrical-electrochemical synchronous acquisition headstage (E-EC headstage) and back-end signal acquisition equipment (Field Programmable Gate Array, FPGA), ensuring low-noise, high-quality capture of delicate brain signals.

METHODS

A. System design

The E-EC headstage described in this paper uses a discrete design stack package. Electrophysiological signal acquisition relies on the RHD2164 chip from Intan Technologies, known for its low input reference noise and integrated amplification and digitization capabilities. This enables low-noise, high-quality acquisition of weak EEG signals. Electrochemical signal acquisition is managed through a back-end FPGA, which utilizes the I2C protocol to control a digital-to-analog conversion (DAC) module. This module outputs the necessary voltage for various electrochemical methods, applied to the biochemical sensor's reference electrode (RE) pin. The target substance undergoes a redox reaction on the sensor's surface, transferring electrons and generating a current between the counter electrode (CE) and the working electrode (WE). This current is extracted through the WE pole. The weak current from the biochemical reaction is converted into a proportional voltage signal via a transimpedance amplifier (TIA). This signal is then buffered, amplified, and adjusted to a positive voltage offset. Finally, it undergoes digital conversion by an ADC, enabling the quantitative determination of the substance's concentration. Figure 1a shows the system test connection diagram, while Figure 1b illustrates the working principle. Figure 1c displays the core circuit of the potentiostat generating the reference electrode potential in the electrochemical system and the core circuit for pA-level weak current detection.

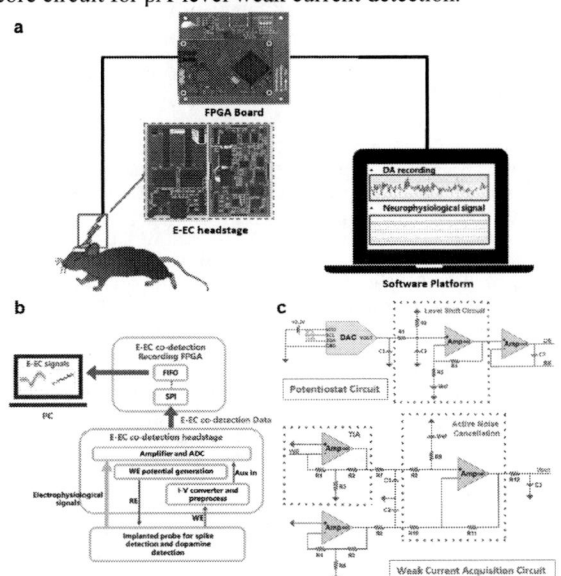

Figure 1: Multi-mode signal co-detection system. (a) Schematic illustration of the electrophysiological-electrochemical signal co-detection system. (b) Schematic representation of the working principle of the electrophysiological-electrochemical signal co-detection system. (c) Core circuit of the electrochemical signal acquisition module. It includes a potentiostat generating circuit for reference electrode (RE) and a neurotransmitter oxidation current detected by the working electrode (WE) amplifier circuit.

B. Modification of the flexible electrode

The electrochemical surface area (ESA) to geometric surface area (GSA) ratio of the multi-channel flexible electrode [3] was enhanced through PEDOT:PSS/IrOx modification. This significantly increases CSC and improves charge transfer efficiency, aiding in the detection of electrically active neurotransmitters [4, 5, 6]. Dopamine, which contains groups that can directly undergo redox reactions, facilitates electron gain, loss, and transfer under applied voltage. Nanomaterial modification on the electrode surface significantly increases the electrode surface area, enhancing the oxidation sites for electroactive substances and boosting response sensitivity. As shown in Figure 2a~d, optical and SEM images of different electrode modifications reveal that PEDOT:PSS/IrOx modification greatly increases the surface roughness compared to the commonly used PEDOT:PSS, thereby expanding the specific surface area of the electrode. Through the study and fitting of the EIS performance of the electrode, shown in Figure 2e, f, the modified electrode exhibited lower

electrochemical reaction impedance at low frequencies compared to both the unmodified and PEDOT:PSS electrodes. The CSC was further tested, revealing that electrodes with higher CSC can store and transfer more charge. The CSC of the electrode modified with PEDOT:PSS/IrOx reached an impressive 36.3 mC/cm² (Figure 2g, h). These results demonstrate that material modification can significantly increase the CSC of the microelectrode. Thus, the PEDOT:PSS/IrOx modification method notably enhances the electrochemical reaction capability of the electrode, highlighting its potential for improving electrode performance in dopamine detection.

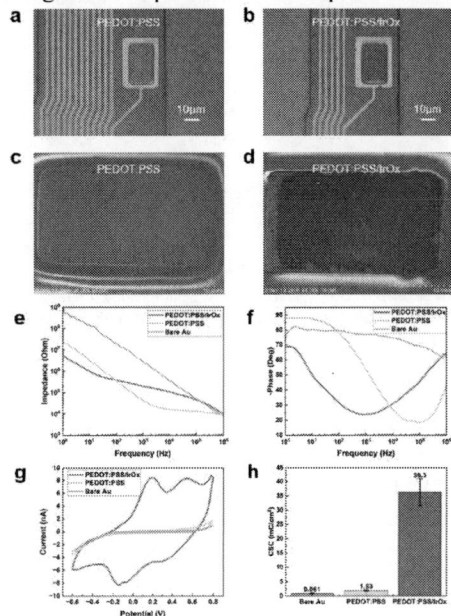

Figure 2: Electrochemical characterization of electrode channels. (a, b) Microscopic images of channels modified with PEDOT:PSS and PEDOT:PSS/IrOx. (c, d) SEM images of channels modified with PEDOT:PSS and PEDOT:PSS/IrOx. (e, f) Impedance and phase curves of electrodes modified by different methods. (g, h) Cyclic voltammetry (CV) curves and comparison of the CSC of electrodes modified by different methods.

EXPERIMENTAL RESULTS

A. Characterization in vitro

To compare the response of the PEDOT:PSS/IrOx modified electrode to dopamine (DA) with the PEDOT:PSS modified electrode, cyclic voltammetry (CV) responses were tested in PBS solution with the same DA concentration. The oxidation peaks in the CV curves were compared. CV tests were conducted on PEDOT:PSS/IrOx modified, PEDOT:PSS modified, and unmodified electrodes in PBS solution with a DA concentration of 20 µM [7] (Figure 3b). The PEDOT:PSS/IrOx modified electrode showed enhanced oxidation potential at +0.6 V, significantly improving its electrochemical reaction capability for DA detection.

A high-concentration DA (Sigma-Aldrich) stock solution was prepared and added drop by drop to 1×PBS solution to create a step concentration increase. During the test, changes in the oxidation current were continuously monitored (Figure 3c). Finally, the concentration and

979-8-3315-0417-5/25 $31.00 © 2025 IEEE

oxidation current were linearly fitted to determine the electrode's sensitivity. The calculated sensitivity is 21.86 pA/μM (Figure 3d), exceeding that achieved by other modification methods [8].

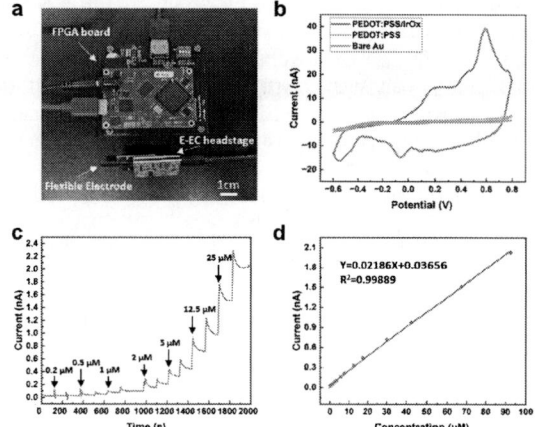

Figure 3: In vitro electrochemical signal acquisition. (a) Photograph of the system setup. (b) CV curve modified by different methods in a 20 μM dopamine solution. The PEDOT:PSS/IrOx modified channel shows an oxidation potential of +0.6 V. (c) Oxidation current responses to increasing concentrations of dopamine(DA). (d) Calibration curve of dopamine concentration vs. corresponding oxidation current. Electrode sensitivity is 21.86 pA/μM.

B. Experiment in vivo

The integrated multi-mode signal co-detection system enables simultaneous in vivo measurement of electrophysiology and dopamine levels without requiring complete anesthesia, preserving the animal's natural physiological state. This allows for high-quality, simultaneous recording of electrophysiological and dopamine electrochemical signals in awake or moving animals, enhancing the detection of dopamine concentration changes (Figure 4b). The electrode channel impedance, measured with a 1 kHz sine wave, is comparable to that of commercial headstages (Figure 4c). Additionally, noise from cable disturbances during recording is significantly reduced, achieving a noise level just 32.8% of that observed with traditional electrochemical workstations (Figure 4d).

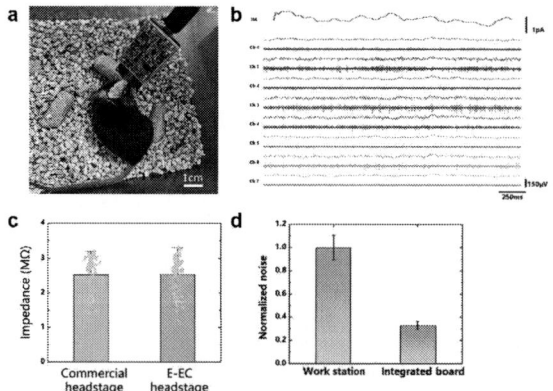

Figure 4: In vivo measurement results. (a) Photographs of the surgical procedure. (b) Waveform analysis using the integrated recording system for synchronous capture of neuronal electrical activity and DA concentration changes. (c) Comparison of electrode impedance at 1kHz detected by E-EC headstage and commercial headstage. (d) Noise level

comparison between commercial electrochemical workstation and E-EC headstage. the noise level was reduced to 32.8% at the commercial headstage.

CONCLUSION

In this paper, the developed multi-mode signal co-detection system successfully integrates electrophysiological and electrochemical techniques, offering real-time, low-noise recording capabilities. Enhanced electrode performance with PEDOT:PSS/IrOx modification provides high sensitivity and charge storage capacity, making it an excellent tool for studying neural and neurotransmitter activities in dynamic environments. This system is poised to significantly advance neuroscience research by providing robust and precise data acquisition, facilitating deeper insights into brain function and neurotransmitter dynamics.

ACKNOWLEDGEMENTS

This work was partially supported by National Science and Technology Major Project from the Minister of Science and Technology of China (Grant No. 2018AAA0103100), National Natural Science Foundation of China (Grant No. 61974154，62474182), Key Research Program of Frontier Sciences, CAS (Grant No. ZDBS-LY-JSC024), Shanghai Pilot Program for Basic Research—Chinese Academy of Science, Shanghai Branch (Grant No. JCYJ-SHFY-202201), Shanghai Municipal Science and Technology Major Project (Grant No. 2021SHZDZX), CAS Pioneer Hundred Talents Program, Shanghai Rising-Star Program (Grant No. 22QA1410900), the Innovative Research Team of High-level Local Universities in Shanghai, Youth Innovation Promotion Association, CAS (No. 2021231), Youth Innovation Promotion Association for Excellent Member, CAS (No. Y2023070), Shanghai Talent Program-Youth Project (No. 202312).

REFERENCES

[1] Xiao G, Song Y, Zhang Y, et al. ACS sensors, 2019, 4(8): 1992-2000.
[2] Xie J, Dai Y, Xing Y, et al. ACS sensors, 2023, 8(4): 1810-1818.
[3] Zhou Y, Yang H, Wang X, et al. Microsystems & Nanoengineering, 2023, 9(1): 88.
[4] Wang X, Jiang W, Yang H, et al. Micromachines, 2024, 15(4): 447.
[5] Fuentes-Rodriguez L, Abad L, Simonelli L, et al. The Journal of Physical Chemistry C, 2021, 125(30): 16629-16642.
[6] Xu S, Deng Y, Luo J, et al. Biosensors, 2022, 12(7): 546.
[7] Ni J, Wei H, Ji W, et al. ACS sensors, 2024, 9(5): 2447-2454.
[8] He E, Zhou Y, Luo J, et al. Biosensors and Bioelectronics, 2022, 209: 114263.

CONTACT

*Tiger H. Tao, tel: +86-21-62511070;
tiger@mail.sim.ac.cn
*Xiaoling Wei, tel: +86-21-62511070;
xlwei-jerry@mail.sim.ac.cn

Demonstration of EOT-Scaled L_g~25 nm FinFET based on Thickness-Proportion Controlled HfO$_2$-ZrO$_2$-HfO$_2$ Superlattice Gate Stacks

Kun Zhong[1,2], Zhaohao Zhang[1,2,*], Siyuan Liu[1,2], Haiyuan Lyu[1,2], Huaxiang Yin[1,2,*]

[1]Integrated Circuit Advanced Process R&D Center and State Key Laboratory of Fabrication Technologies for Integrated Circuits, Institute of Microelectronics of the Chinese Academy of Sciences

[2] University of Chinese Academy of Sciences, Beijing, China, 10049

*E-mail: zhangzhaohao@ime.ac.cn; yinhuaxiang@ime.ac.cn

Abstract

We demonstrate the equivalent oxide thickness (EOT)-scaled FinFETs with a physical gate length (L_g) of ~25 nm based on a thickness-proportional controlled HfO$_2$-ZrO$_2$-HfO$_2$ (TPC-HZH) superlattice gate stacks. The integrated gate oxides show an effective oxide thickness (EOT) of 7 Å after 450 °C annealing, indicating high compatibility with advanced HKMG FinFETs processes. Superior performance from the low EOT sustains on the L_g~25 nm TPC-HZH FinFET, showing a 38% higher driving current and a 54% higher transconductance compared with that of traditional HfO$_2$-based FinFET. Besides, enhanced SCE immunity, including SS and DIBL, of the TPC-HZH, indicates the great potential of TPC-HZH capability for future advanced node CMOS technology.

Keywords: Hafnium zirconium oxide, Gate stack, Equivalent oxide thickness, FinFET.

Introduction

The thinning of a gate dielectric's equivalent oxide thickness (EOT) represents a promising avenue for directly enhancing electrostatics and driving current in transistors while simultaneously reducing power consumption [1]. However, conventional EOT scaling technologies, such as interfacial layer (IL) scavenging or permittivity enhancement in high-κ films, have reached their limits and may potentially lead to mobility degradation and reliability issues.

Recently, the HfO$_2$-ZrO$_2$-HfO$_2$ (HZH) superlattice gate stack provides a new route for EOT scaling without compromising mobility and reliability, as there is no thinning of the IL thickness [2-3]. For advanced back-end-of-line (BEOL) process compatibility, our previous reports have demonstrated a thickness-proportion controlled HfO$_2$-ZrO$_2$-HfO$_2$ (TPC-HZH) superlattice gate stack with improved thermal stability [4]. Furthermore, the TPC-HZH has been successfully employed in the fabrication of advanced HKMG FinFETs with a gate length (L_g) of 500 nm and a drive current enhancement of 55%. However, since the underlying mechanism of the capacitance enhancement in the superlattice may come from the ferroelectric and anti-ferroelectric phase ratio [3-5], the ratio may change as the L_g scaling and thus the performance enhancement may destroy in advanced node devices.

In this work, the TPC-HZH FinFET with 25 nm L_g is fabricated with advanced HKMG processes. Driving current and transconductance enhancements, besides superior SCE immunity, have been confirmed on the devices of TPC-HZH FinFET, suggesting good potential for applications in advanced energy-efficient transistors.

Experimental Procedures

The structure and fabrication process of MOSCAPs are shown in Fig. 1(a) and (b), respectively. After the 0.8 nm SiO$_2$ interfacial layer was grown on Si substrates (8-12 Ω·cm) with ozone treatment, atomic layer deposition (ALD) was used to deposit TPC-HZH and conventional HfO$_2$. The deposition rates of the HfO$_2$ and ZrO$_2$ films were 0.75 and 0.64 Å/cycle, respectively. Subsequently, a 250°C post-deposition annealing (PDA) was performed to solidify the SiO$_2$-oxide interface. TiN and W were deposited by physical vapor deposition (PVD) as top electrodes. Finally, the HZH films were subjected to post-metallization annealing (PMA) under a range of thermal conditions (250, 350, and 450 °C,60 s, N$_2$) to explore the properties of the HZH gate structure under different thermal budgets.

The process flow and schematic of the TPC-HZH-based FinFETs are shown in Fig. 3(a) and (b), respectively. The flow is fundamentally compatible with the process used to fabricate FinFET devices [5-6], except for the ALD processes, which are the same as those for the MOSCAP fabrication. The L_g was defined by the direct-write electron beam. Capacitance-voltage (C-V) and leakage (J_g) characteristics were measured using a Keithley 4200 semiconductor parameter analyzer. As a control, a conventional 2.4-nm HfO$_2$-based MOSFET was also fabricated according to the above process.

Results and Discussion

Fig. 2(a) shows the C-V curves for traditional 2.4-nm HfO$_2$ and TPC-HZH super-lattice films at 450 °C annealing. It can be observed that TPC-HZH films exhibited clear capacitance enhancement. The EOTs, extracted from the devices using a capacitance-voltage simulator based on quantum mechanical effects (University of California, Berkeley) under various annealing temperatures are summarized and shown in Fig. 2(b). For annealing at 450 °C, the EOT of the conventional 2.4-nm HfO$_2$-based MOSCAP was 1.01 nm, the EOT of the 2.4-nm TPC-HZH-based MOSCAP was 0.71 nm, indicating a 29.7% decrease of the EOT.

Thanks to the superior EOT and excellent thermal stability, the 2.4-nm TPC-HZH was integrated into p-type FinFETs

979-8-3315-0417-5/25 $31.00 © 2025 IEEE

using advanced HKMG processing to investigate the scalability of the TPC-HZH technology. Fig. 3(c) and (d) present a cross-sectional transmission electron microscopy (TEM) view cut along the YY' and XX' direction, which characterizes the TPC-HZH-FinFET with a L_g of ~25 nm. Energy Dispersive X-ray spectrometer (EDX) mapping (Hf, Zr) of the layered HKMG structure indicates that the multilayer HKMG was distributed over the whole of the 3D fins and highly conformal and uniform. Notely, this fin has a sharper profile than previously reported owing to the use of a thinner hard mask for the fin etching.

Fig. 4(a) and 4(b) show the I_{DS}-V_{GS} and I_{DS}-V_{DS} characteristics of 25nm L_g FinFET with HfO_2 and TPC-HZH as the gate-stack, respectively. The value of I_{ON} was high for the TPC-HZH FinFET due to higher gate capacitance. The extracted transconductance (G_m) is shown in Fig. 5(a), which demonstrates that TPC-HZH-based FinFET exhibits a G_m enhancement of approximately 54%, showing better gate-control in TPC-HZH Superlattice FinFET. Fig. 5(b) shows the statistics I_{ON}-I_{OFF} of the TPC-HZH and HfO_2-FinFETs. A current enhancement of approximately 38% was obtained at the same level of the OFF-current. Besides the scaled EOT, the I_{ON} enhancement can also be attributed to the high mobility as shown in Fig. 6(a). The transistor's field-effect mobility was extracted from the G_m and is calculated as the equation, as shown in Eq. (1). ~15 cm²v⁻¹s⁻¹ maximum mobility can be obtained in both devices, indicating there is no mobility degradation in the TPC-HZH-based FinFET.

$$\mu = \frac{Lg_m}{WC_i}\frac{1}{V_{DS}} \qquad (1)$$

Furthermore, as the D_{it} of the TPC-HZH films was investigated on MOSCAP [7], the threshold voltage (V_{TH}) of the TPC-HZH and HfO_2-FinFETs, which were extracted from ~50 devices and shown in Fig. 6(b). The difference in V_{TH} between the two devices is ~50 mv, indicating that there is no significant drift in the V_{TH} and good interfacial charge density of the TPC-HZH FinFET.

Subsequently, the SCE characteristics of the devices were also investigated. The subthreshold swing (SS) and drain induced barrier low (DIBL) distributions of the TPC-HZH FinFETs and HfO_2-FinFETs are shown in Fig. 7(a) and (b), respectively. Specifically, the medians of the SS and DIBL were found to be respectively ~68.9 mV/dec and 17.5 mV/V for the TPC-HZH FinFETs, and ~73.2 mV/dec and 38.3 mV/V for the HfO_2-FinFETs with a L_g of ~25 nm. Although both devices exhibited distributions in SS and DIBL owing to variations in the fabrication process, statistical analysis revealed that the TPC-HZH FinFET exhibited enhanced SCE immunity and superior electrostatic control compared with the control device.

Conclusion

This work demonstrates EOT-scaled FinFETs with small L_g (~25nm) based on TPC-HZH superlattice gate stacks. A 38% I_{ON} boost and 54% G_m improvement were obtained in the TPC-HZH-based FinFET. In addition, TPC-HZH FinFET also enhances the SCE immunity including SS and DIBL, further validating the great potential of TPC-HZH to beyond conventional high dielectric constant materials based on HfO_2.

Acknowledgments

The authors acknowledge support from the National Natural Science Foundation of China (Grant No. 92064003, 91964202), and the Youth Innovation Promotion Association, Chinese Academy of Sciences under Grant 2023130.

References

[1] W. Cao, H. M. Bu, M. Vinet, M. Cao. S. Takagi, S. Hwang, T. Ghani, and K. Banerjee, "The future transistors," *Nature*, vol. 620, no. 7974, pp. 501–515, Aug. 2023, doi: 10.1038/s41586-023-06145-x.

[2] W. Li, L. C. Wang, S. S. Cheema, N. Shanker, C. Hu, and S. Salahuddin, "Enhancement in Capacitance and Transconductance in 90 nm nFETs with HfO2-ZrO2 Superlattice Gate Stack for Energy-efficient Cryo-CMOS," in *2022 International Electron Devices Meeting (IEDM)*, San Francisco, CA, USA, 2022, pp. 22.3.1-22.3.4, doi: 10.1109/IEDM45625.2022.10019496.

[3] N. Shanker, M. Cook, S.S. Cheema, W. Li, R. Rastogi, D. Pipitone, C. Chen, M. Smith, S. Meninger, F. Bauer, G. Pinelli, J. Hunt, S. Salahuddin, and M. Mohamed, "CMOS Demonstration of Negative Capacitance HfO2-ZrO2 Superlattice Gate Stack in a Self-Aligned, Replacement Gate Process," in *2022 International Electron Devices Meeting (IEDM)*, San Francisco, CA, USA, 2022, pp. 34.3.1-34.3.4, doi: 10.1109/IEDM45625.2022.10019472.

[4] K. Zhong, F. Zhang, Z. Zhang, Q. Zhang, Z. Hou, C. Zhao, G. Han, G. Xu, J. Li, Y. Liu, J. Xu, H. Yin, "Demonstration of EOT-Scaled FinFET Based on Thickness-Proportion Controlled HZH Superlattice Gate Stacks With Improved Thermal Stability (≥ 450 °C)," *IEEE Electron Device Letters*, vol. 620, no. 7, pp. 1193-1196, July 2024, doi: 10.1109/LED.2024.3401411

[5] Z. Zhang, G. Xu, Q. Zhang, Z. Hou, J. Li, Z. Kong, Y. Zhang, J. Xiang, Q. Xu, Z. Wu, H. Zhu, H. Yin, W. Wang, and T. Ye, "FinFET With Improved Subthreshold Swing and Drain Current Using 3-nm Ferroelectric Hf0.5r0.5O2," *IEEE Electron Device Letters*, vol. 40, no. 3, pp. 367-370, Mar. 2019, doi: 10.1109/LED.2019.2912413.

[6] Z. Zhang, Y. Luo, Y. Cui, H. Yang, Z. Wu, J. Xiang, Q. Liu, H. Yin, S. Mao, X. Wang, J. Li, Y. Zhang, Q. Luo, J. Gao, W. Xiong, J. Liu, Y. Li, J. Li, J. Luo, and W. Wang, "A Polarization-Switching, Charge-Trapping, Modulated Arithmetic Logic Unit for In-Memory Computing Based on Ferroelectric Fin Field-Effect Transistors," *ACS Appl. Mater. Interfaces*, vol. 14, no. 5, pp. 6967-6976, Jan. 2022, doi:10.1021/acsami.1c20189.

[7] K. Zhong, Z. Zhang, Y. Liu, Y. Zhang, H. Yin, "Interfacial Trap Density Investigation of HfO2-ZrO2 Superlattice Gate Stacks with Ultra-low Equivalent Oxide Thickness," *Solid State Devices and Materials (SSDM)*, Himeji, Japan, 2024

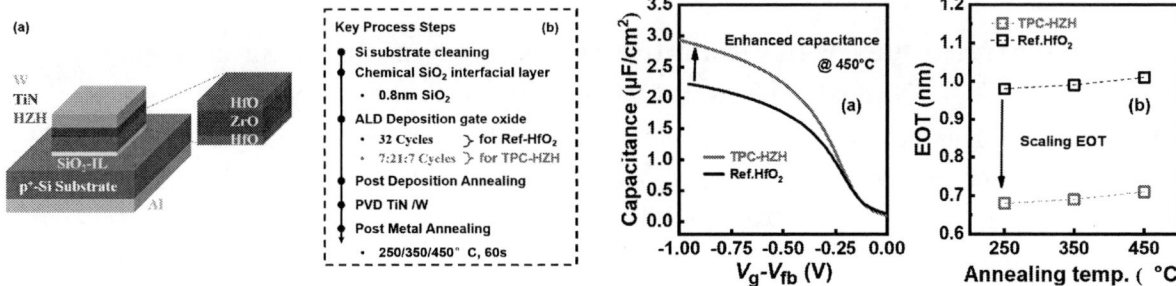

Fig. 1. (a) The schematic device structure and (b) key Process flow of HfO_2 /ZrO_2 /HfO_2 superlattice films-based metal oxide semiconductor capacitors.

Fig. 2. (a) Capacitance versus V_g-V_{fb} curves of MOS capacitors. (b) Extracted EOT from C-V curves based capacitors with various annealing temperatures.

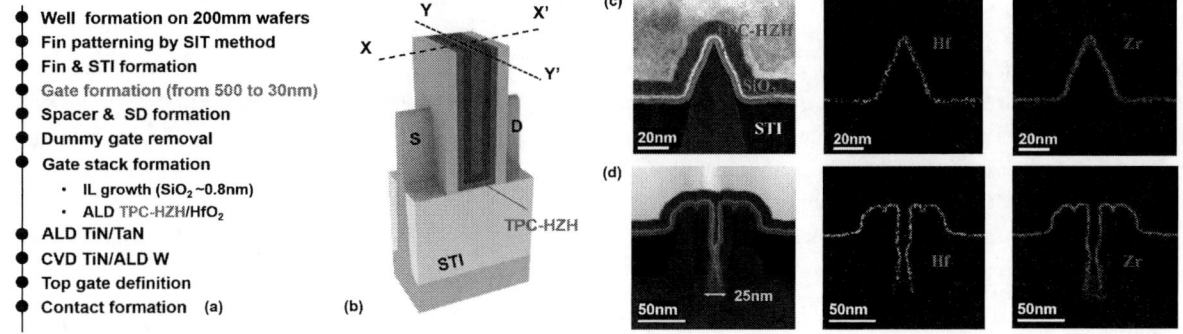

Fig. 3. (a) Process flow and (b) schematic diagram of TPC-HZH-based FinFET device structure. (c) Cross-sectional TEM image along the YY′ direction and EDX mapping (Hf, Zr) of the ~2.4 nm TPC-HZH-based FinFET. (d) TEM image along the XX′ direction and EDX mapping (Hf, Zr).

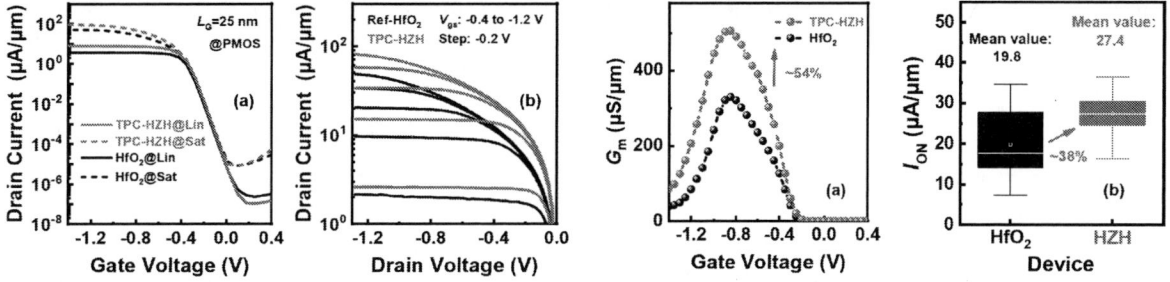

Fig. 4. (a) I_{DS}-V_{GS} and I_{DS}-V_{DS} characteristics of p-type FinFET with 25-nm L_g using TPC-HZH (red) and traditional HfO_2 (black) gate stack layers.

Fig. 5. (a) Extracted G_m and (b) I_{ON} from I_{DS}-V_{GS} curves of TPC-HZH and HfO_2 FinFET with 25-nm L_g.

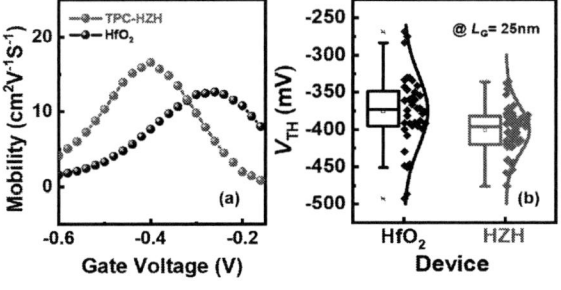

Fig. 6. (a) Field-effect mobility of TPC-HZH and HfO_2.(b) V_{TH} values of TPC-HZH and HfO_2-based FinFET.

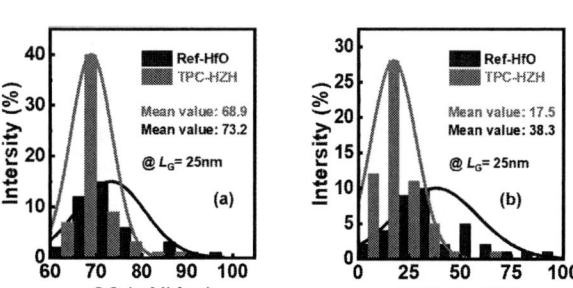

Fig. 7. (a) SS values and (b) DIBL values extracted for TPC-HZH and HfO_2 gate layer-based FinFET.

979-8-3315-0417-5/25 $31.00 © 2025 IEEE

Global Stress Analysis in Fin Patterned Si/SiGe Multilayer Nanosheets for Nanosheet-based CMOS Device Technology

Amit Kumar Singh Chauhan, Imtiyaz Ahmad Khan, Kunal, Harsh Raju, Sanjeev Kumar Manhas*

Indian Institute of Technology Roorkee, Roorkee, Uttarakhand, India

*sanjeev.manhas@ece.iitr.ac.in

Abstract

Si/SiGe nanosheet-based CMOS technology is essential for 3nm semiconductor nodes and below. This study investigates stress in Si/SiGe superlattice nanosheets after fin patterning on silicon substrates. SiGe nanosheets (Ge mole fraction 0.35) exhibit 2.17 GPa compressive stress, while Si nanosheets are initially stress-free. After patterning, tensile stress induced in Si nanosheets stress due to relaxation in SiGe. These findings provide insights into stress engineering, crucial for improving performance and reliability of nanosheet-based CMOS and heterostructure devices.

Keywords: Global Stress, Finite Element Method (FEM), Stacked Gate All Around (SGAA), Nanosheet (NS), SiGe

Introduction

The next-generation logic family needs the transition from FinFET to nanosheet-based CMOS transistors—such as stacked gate-all-around (SGAA) FETs, forksheet FETs, and nanosheet-based complementary FETs (CFETs)—at the 3nm technology node and beyond [1-4]. This transition is driven by the advantages of increased driving current, enhanced effective width, improved short-channel control, electrostatic properties, and 3D stacking. The fabrication process of these devices begins with the precise sequential stacking of silicon (Si) and silicon-germanium (SiGe) layers using epitaxy [2]. The next fabrication step of nanosheet FETs involves several intricate steps: fin formation, dummy gate formation, source/drain recess, channel release, epitaxy source/drain formation, etc [2]. Each step is vital in shaping the device's structure and ensuring optimal functionality. The lattice mismatch between these materials (Si/SiGe) generates global stress, which subsequently has a profound effect on carrier mobility. In this multilayer structure, either SiGe or Si nanosheets act as sacrificial layers, depending on the channel material used during the channel release process [3].

Fin formation is also crucial, as it leads to the relaxation and redistribution of the global strain/stress induced during Si/SiGe epitaxy NSs deposition. Investigating strain and stress redistribution at small scales is experimentally challenging due to its complexity. Simulation offers a more effective approach, providing detailed insights into strain and stress behaviors within channels and fins.

In this work, using 3D TCAD, the finite element method (FEM) is utilized to investigate the in-plane stress evolution

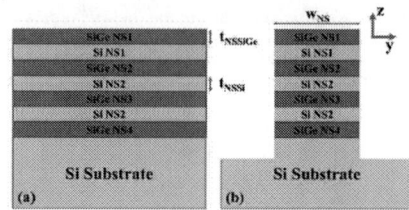

Fig. 1: (a) Two-dimensional schematic structure of blanket multilayers Si/SiGe NSs on Si substrate (b) Fin structure after patterning the blanket multilayer structure.

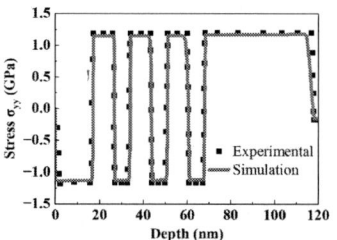

Fig. 2: Simulated and experimental results of the vertical profile of in-plane stress (σ_{yy}) from top to bottom in the center of the fin structure [6].

in Si nanosheets and the stress relaxation in silicon-germanium (SiGe) nanosheets throughout the fin patterning process. The study also examines the impact of fin dimensions, nanosheet length/width (w_{NS}), and nanosheet thickness (t_{NSSi} and t_{NSSiGe}), which are essential factors in strain engineering. As GAA nanosheet transistors are expected to be used in both logic transistors and memory access devices for future technology nodes, this study of stress profiles becomes essential. It plays a critical role in advancing stress engineering, which is key to optimizing device performance and reliability in next-generation technologies.

Structure and Simulation Method

We created a multilayer Si/SiGe NS structure by alternatingly depositing Si and SiGe layers on a Si substrate [Fig. 1(a)] and then patterned them into fin using Sentaurus Process Simulation as shown in Fig. 1(b)[5]. Stress simulation was performed using FEM and validated against experimental results from sample S2 in [6], which consists of a relaxed $Si_{0.8}Ge_{0.2}$ substrate, a 50 nm Si layer on the substrate, and a 10 nm Si/7 nm $Si_{0.65}Ge_{0.35}$ multilayer with 17 nm $Si_{0.65}Ge_{0.35}$ on top. The simulated in-plane stress (σ_{yy}) matched well with the experimental results, as shown in [Fig. 2] which ensured good accuracy and predictability of results in this work [6].

979-8-3315-0417-5/25 $31.00 © 2025 IEEE

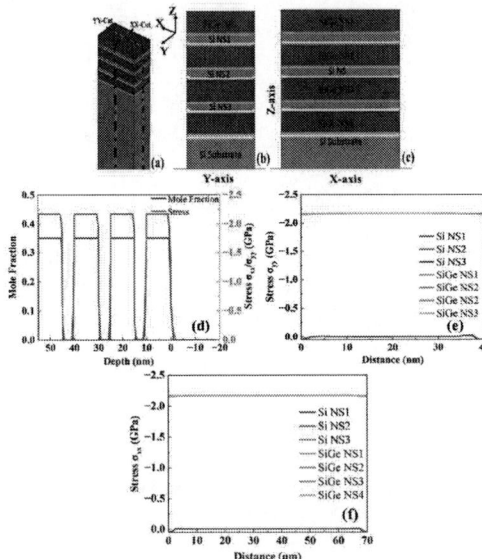

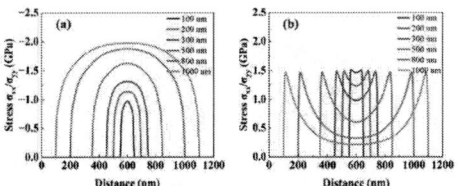

Fig. 4: Horizontal profile of in-plane stress (σ_{yy}/σ_{xx}) along the fin length/width for 100, 200, 300, 500, 800, and 1000nm fin length/width in the mid of the NSs (a) SiGe NS (SiGe NS3) (b) Si NS (Si NS2).

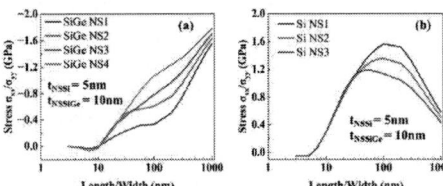

Fig. 5: Average in-plane stress (σ_{yy}/σ_{xx}) along the fin length/width in the mid of the NSs as a function of fin length/width (a) SiGe NSs (b) Si NSs.

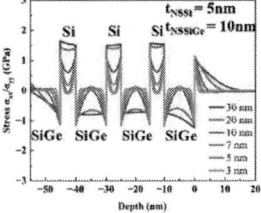

Fig. 6: Vertical profile of in-plane stress (σ_{xx}/σ_{yy}) from top to bottom in the center of the fin structure after pattering having length/width of fin (w_{NS}) 30nm, 20nm, 10nm, 7nm, 5nm and 3 nm for SiGe NSs thickness (t_{NSSiGe} = 10nm) and Si NSs thickness (t_{NSSi} = 5nm).

Fig. 3: (a) Three-dimensional schematic structure of blanket multilayers Si/SiGe NSs on Si substrate (b) 2D cross-section along the Y-axis (c) 2D cross-section along the X-axis (d) Vertical profile of Ge mole fraction and in-plane stress (σ_{xx}/σ_{yy}) at the mid of the 2D cross section from top to bottom (e) In-plane stress profile in the mid of Si/SiGe NSs along the Y-axis (f) In-plane stress profile in the mid of Si/SiGe NSs along the X-axis.

Results and Discussion

The three-dimensional (3D) schematic of the blanket multilayer Si/SiGe NSs structure (70nm x 40nm) is shown in Fig. 3(a), with Si and SiGe NSs having 5nm and 10nm thicknesses (t_{NSSi} and t_{NSSiGe}), respectively. The SiGe NSs have a Ge mole fraction of 0.35, as depicted in Fig. 3(d). Two-dimensional (2D) cross-sections of the 3D structure along the y-axis and x-axis are illustrated in Fig. 3(b) and Fig. 3(c), respectively. The vertical profile of Ge mole fraction and in-plane stress (σ_{xx}/σ_{yy}) due to lattice mismatch between Si and SiGe NSs is shown in Fig. 3(d). The horizontal profile of in-plane stress (σ_{yy}) along the width in Si and SiGe NSs (blanket) is shown in Fig. 3(e) and in-plane stress (σ_{xx}) along the length in Si and SiGe NSs (blanket) is shown in Fig. 3(f). The SiGe NSs exhibit a compressive stress of 2.17 GPa, while the Si NSs are stress-free. Approx. same stress values for SiGe and Si NSs were also calculated analytically using elasticity models [7-8].

The stress in blanket NSs depends solely on the Ge mole fraction in SiGe NSs for a fixed Si substrate, which is unaffected by NS dimensions or thickness. However, during the fin patterning of the Si/SiGe multilayer structure, the elastic relaxation of compressive stress in SiGe NSs transfers energy to Si NSs and induces tensile stress in the Si NSs. This elastic relaxation takes place due to edge relaxation and expansion of compressive SiGe NSs [5]. The horizontal profile of in-plane stress (σ_{xx}/σ_{yy}) along the fin length/width in mid of the Si NS (Si NS2) and SiGe NS (SiGe NS3) is shown in Fig.4. As shown in Fig. 4(a), the compressive stress in the SiGe NSs is relaxed fully at the edges of NSs and

accordingly the tensile stress in Si NSs induced strongly at the edges of Si NSs in comparison to the middle of the NSs as shown in Fig. 4(b). The relaxation of compressive stress in SiGe NSs and induced tensile stress in Si NSs depend not only on the length/width (w_{NS}) but also on the thickness of the NSs (t_{NSSi} and t_{NSSiGe}) and position of the NSs. The average in-plane stress (σ_{xx}/σ_{yy}) along the fin length/width in the middle of the NSs as a function of fin length/width (w_{NS}) and position is shown in Fig. 5. As the fin length/width decreases, the more relaxation takes place in SiGe NSs as demonstrated in Fig. 5(a). The bottom SiGe NS (SiGe NS4) relaxes less than the top SiGe NS (SiGe NS1). The stress difference between the SiGe NS1 and SiGe NS4 increases from 0.22 GPa at 1000nm to 0.734 GPa at 100nm fin length/width after that the difference decreases as the fin length/width decreases.

The tensile stress induced in Si NSs increases from 0.58 GPa in Si NS1 at 1000nm to 1.56 GPa at 100nm fin length/width and then decreases with decreasing the fin width/length as shown in Fig. 5(b). The tensile stress induced more in top Si NS (Si NS1) than in bottom Si NS (Si NS3). The stress difference between Si NS1 and Si NS3 increases from 0.1 GPa at 1000nm to 0.43 GPa at 100nm fin length/width and then the difference decreases to approx. zero at 30nm fin width/length, as demonstrated in Fig. 5(b).

The vertical profile of in-plane stress (σ_{xx}/σ_{yy}) from top to bottom in the center of the fin structure (5nm Si NSs and 10

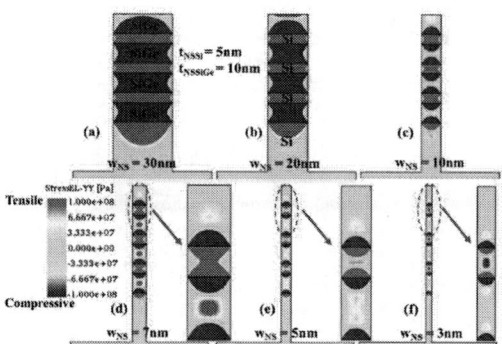

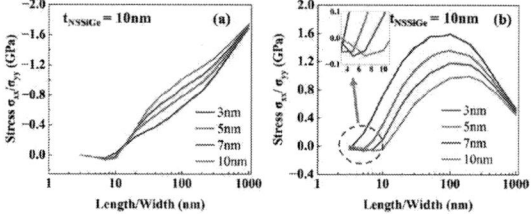

Fig. 7: 2D distribution of the in-plane stress (σ_{xx}) distribution in the patterned fin structure along the length/width having length/width (a) 30 nm (b) 20 nm (c) 10 nm (d) 7 nm (e) 5nm and (f) 3nm.

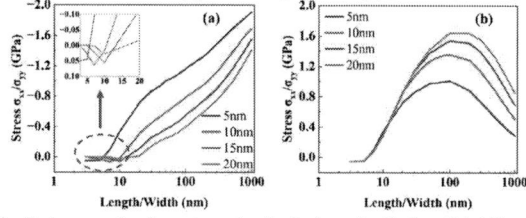

Fig. 8: Average in-plane stress (σ_{yy}/σ_{xx}) along the fin length/width as a function of fin length/width and thickness of Si NSs (t_{NSSi} = 3, 5, 7 and 10 nm) in the mid of the (a) SiGe NS (SiGe NS3) (b) Si NS (Si NS2).

Fig. 9: Average in-plane stress (σ_{yy}/σ_{xx}) along the fin length/width as a function of fin length/width and thickness of SiGe NSs (t_{NSSiGe}= 5, 10, 15, and 20nm) in the mid of the (a) SiGe NS (SiGe NS3) (b) Si NS (Si NS2).

nm SiGe NSs thickness) after patterning for 30nm, 20nm, 10nm, 7nm, 5nm, and 3nm fin length/width is shown in Fig. 6 and 2D distribution of in-plane stress in the fin structure along the length/width is shown in Fig.7. As the fin length/width decreases less than 20nm, the stress in SiGe NSs and Si NSs relaxes very much at the mid of the NSs and at the interface of Si/SiGe NSs there are very less changes in the in-plane stress. As the NSs thickness and length/width ratio (t_{NSSi}/w_{NSSi} = 1) become one, the opposite stress generated in the mid of Si NSs (from tensile stress to compressive stress) due to over-relaxation as demonstrated in Fig. 7(e)&(f). Similarly, for SiGe NSs, when the t_{NSSiGe}/w_{NS} becomes equal to one, the opposite stress generated in the mid of SiGe NSs (from compressive stress to tensile stress) without changing at the Si/SiGe interfaces as demonstrated in Fig. 7 (c)-(f).

Fig. 8 demonstrates the effect of Si NSs thickness on the in-plane stress profile along the length/width as a function of fin length/width. The effect of thickness is very small at 1000nm fin length/width but increases as the fin length/width decreases. The effect of Si NSs thickness is more dominant on the stress profile of Si NSs in comparison to SiGe NSs. The compressive stress of SiGe NSs relaxes more for thinner

Si NSs, as shown in Fig. 8(a). Accordingly, thinner Si NSs gain more tensile stress and retain it for lesser fin length/width, as shown in Fig.8(b).

Fig. 9 demonstrates the effect of SiGe NSs thickness on the in-plane stress profile along the length/width as a function of fin length/width. The compressive stress of thinner SiGe NSs relaxes less in comparison to thicker ones and retains the stress for smaller length/width, as shown in Fig. 9(a). Accordingly, Si NSs gain more tensile stress for thicker SiGe NSs, as shown in Fig.9(b). The effect of SiGe NSs thickness on the in-plane stress profile for Si NSs vanished for smaller fin length/width (less than 20 nm).

Conclusion

Fin patterning is the first crucial step in the fabrication of SGAA FETs, Forksheet FETs, and CFETs. During fin patterning, the relaxation of compressive stress in the SiGe nanosheets effectively transfers energy to the Si nanosheets, inducing tensile stress. This induced tensile stress, influenced by fin dimensions, the thickness of the Si/SiGe nanosheets, and Ge mole fraction in SiGe, leads to enhanced electron mobility in the Si layers. The thinner NSs retain the stress for the small fin length/width. This study offers an important insight into the stress evolution in superlattice nanostructures. These results are very crucial in the stress engineering at other fabrication steps of nanosheet-based CMOS transistors and heterostructure devices to enhance their performance and reliability.

References

[1] K. K. Bhuwalka *et al.*, "Optimization and Benchmarking FinFETs and GAA Nanosheet Architectures at 3-nm Technology Node: Impact of Unique Boosters," in *IEEE Transactions on Electron Devices*, vol. 69, no. 8, pp. 4088-4094, Aug. 2022

[2] N. Loubet et al., "Stacked nanosheet gate-all-around transistor to enable scaling beyond finFET," in Proc. Symp. VLSI Technol., pp. T17-5(2017).

[3] H. Mertens et al., "Forksheet FETs for advanced CMOS scaling: Forksheet-nanosheet co-integration and dual work function metal gates at 17nm N-P space," in Proc. Symp. VLSI Technol., pp. T2-1(2021).

[4] H. Mertens et al., "Nanosheet-based complementary field-effect transistors (CFETs) at 48nm gate pitch, and middle dielectric isolation to enable CFET inner spacer formation and multi-V_t patterning," in Proc. Symp. VLSI Technol., pp. T1-3(2023).

[5] SentaurusTM process user guide, version S-2021.06, Mountain View, California: Synopsys, Inc., 2021.

[6] S. Reboh et al., "Strain, stress, and mechanical relaxation in fin-patterned Si/SiGe multilayers for sub-7 nm nanosheet gate-all-around device technology," Appl. Phys. Lett., vol.112, no. 5, pp. 051901-5(2018).

[7] Y. Sun, Scott E. Thompson, *T.* Nishida "*Strain Effect in Semiconductors: Theory and Device Applications*" Edition 1, Springer New York Dordrecht Heidelberg London pp. 16-17 (2009).

[8] J. J. Wortman and R. A. Evans, "Young's Modulus, Shear Modulus, and Poisson's Ratio in Silicon and Germanium," *J. Appl. Phys.*, Vol. 36, pp. 153-156 (1965).

979-8-3315-0417-5/25 $31.00 © 2025 IEEE

Van der Waals Dielectrics and Electrodes in 2D Transistors for 2T0C DRAM

Jianmiao Guo[1,2], Ziyuan Lin[1,2], Cong Wang[1,2], Tianqing Wan[1,2], Jianmin Yan[1,2], and Yang Chai[1,2,*]

[1]Department of Applied Physics, The Hong Kong Polytechnic University, Kowloon, Hong Kong, China.
[2]Joint Research Centre of Microelectronics, The Hong Kong Polytechnic University, Kowloon, Hong Kong, China. Email: ychai@polyu.edu.hk

Abstract

Two-dimensional (2D) materials, with their atomic-scale thickness, hold promise for next-gen transistors and memory devices. However, integrating high-quality dielectrics on 2D semiconductors is challenging. We introduce a hexagonal boron nitride (h-BN)-assisted process for transferring van der Waals dielectrics and electrodes in 2D transistors. This method maintains clean interfaces, with aluminum oxide (Al_2O_3) showing exceptional flatness. The MoS_2 transistor achieves low interface trap density and leakage current, enabling efficient charge storage and facilitating high-density integration in DRAM applications.

Keywords: Van der Waals dielectric, MoS_2 transistor and 2T0C DRAM

Introduction

The rapid advancements in semiconductor technology have increased the demand for more efficient and miniaturized electronic devices.[1] Two-dimensional (2D) materials have emerged as promising candidates for future transistors and memory devices due to their unique properties.[2] However, a key challenge in fabricating these transistors is the integration of high-quality dielectric materials onto the surface of 2D semiconductors, which lack the dangling bonds required for traditional dielectric deposition methods.[3-5]

Various strategies have been employed to address this issue, such as modifying the surface of 2D materials through plasma treatments or by creating defect sites.[6-9] However, these approaches tend to damage the delicate 2D lattices and introduce trap states, which degrade their intrinsic properties. Another method involves using buffer layers to enable the deposition of high-κ dielectrics while protecting the 2D material. Unfortunately, these buffer layers often have low stability and insufficient dielectric constants, limiting their effectiveness. Moreover, this technique does not entirely prevent changes to the 2D material's properties during the dielectric deposition process, posing a continued challenge for producing damage-free top-gate transistors.

In this study, we present an h-BN-assisted van der Waals (vdW) integration strategy to fabricate top-gate 2D transistors with pristine, damage-free interfaces. By utilizing a monolayer of h-BN as a buffer layer, we pre-deposit the aluminum oxide (Al_2O_3) dielectric and electrodes, which are then transferred onto the target material through a one-step process. This approach ensures that the 2D material remains undamaged by high-energy deposition processes, resulting in clean dielectric/channel and contacts/channel interfaces. Using this method, we fabricated a MoS_2 transistor that exhibited an ON/OFF ratio of 10^8, a low leakage current ($\sim 10^{-7}$ A/cm), a small subthreshold swing (~ 70 mV dec^{-1}), and an extremely low interface trap density ($\sim 7 \times 10^{11}$ cm^{-2} eV^{-1}). Additionally, we demonstrated the fabrication of a 2T0C DRAM cell using two top-gate MoS_2 transistors, showing the potential for improved DRAM performance.

Experimental

Fig. 1a shows the fabrication process flow of the damage-free MoS_2 transistor. First, photolithography was employed to define the gate dielectric region, followed by depositing a 20 nm aluminum (Al) layer on monolayer h-BN using an electron-beam evaporator. After a lift-off process, the Al was annealed in air at 400 °C for 3 hours, forming a uniform Al_2O_3 dielectric layer. Next, gold (Au) source, gate, and drain electrodes were patterned via photolithography and deposited using an electron-beam evaporator (Fig. 1b). To fabricate the device, a PVA film was used to peel off the dielectric/electrode stack, which was then attached to a PDMS stamp on a glass slide (Fig. 1c). By using a 2D materials transfer platform, the device stack and MoS_2 flake were precisely aligned and adhered. The PVA film was dissolved with deionized water. Finally, the devices were annealed in a vacuum to improve contact quality (Fig. 1d). The same method was applied to fabricate an Au/MoS_2/h-BN/Al_2O_3/Au device on a quartz substrate for capacitance-voltage (C-V) testing. Electrical characterization was conducted in a Lakeshore TTPX probe station using a Keithley 4200A SCS parameter analyzer under atmospheric conditions.

Results and Discussion

The one-step transfer methodology offers an innovative approach to laminate the Al_2O_3 dielectric onto exfoliated MoS_2 using weak vdW forces. This method preserves the intrinsic properties of MoS_2, resulting in a top-gate MoS_2 transistor with an Al_2O_3 dielectric and electrode stack that exhibits excellent gate tunability. As shown in Fig. 2a, the Al_2O_3-gated MoS_2 transistor achieves an impressive ON/OFF ratio of 10^7 at a V_{DS} of 1 V. Additionally, the transistor exhibits a small SS of about 70 mV/decade (Fig. 2b), reflecting the low interface defect density. This favorable SS value is due to the superior interface quality maintained by the transfer process, which minimizes the introduction of

979-8-3315-0417-5/25 $31.00 © 2025 IEEE

impurities. Without back-gate voltage, the transistor shows an ON current of 0.4 µA µm^{-1} (Fig. 2c). Although this relatively low ON current is attributed to ungated regions between the gate and contact electrodes, increasing the back-gate voltage improves both the ON current and the ON/OFF ratio, confirming the dual-gate controllability of the device (Fig. 2d).

To analyze the interface trap density (D_{it}), high-low frequency C−V measurements were performed (Fig. 3a). The extracted D_{it} reveals an interface trap density of approximately 7×10^{11} eV^{-1}cm^{-2} in the accumulation region, confirming the pristine nature of the interfaces achieved through this damage-free vdW fabrication process. Furthermore, the gate leakage current of the top-gate MoS$_2$ transistor was measured, revealing an ultra-low gate leakage current density of less than 10^{-7} A/cm^2 (Fig. 3b). This low leakage current is significantly below the threshold for low-power complementary metal-oxide-semiconductor (CMOS) devices, greatly reducing the static power consumption of the transistors.

This h-BN-assisted vdW dielectric and electrode transfer method enables the creation of transistors with ultra-low leakage currents, a critical factor in extending dynamic random-access memory (DRAM) retention time. The transfer curves of the two transistors within a 2T0C DRAM cell show remarkably low leakage currents (Fig. 4a), which contribute to improved data retention capabilities. To optimize performance, the MoS$_2$ channel of the write transistor is designed to be thicker, providing a higher ON-state current for enhanced write speed. In contrast, the read transistor has a thinner MoS$_2$ channel, ensuring that it operates in the subthreshold region at a gate voltage of 0 V, which improves the sense margin.

As illustrated by the circuit diagram (the insert of Fig. 4b), the write and read operations of the 2T0C DRAM cell are separated, allowing the state of the DRAM to be read without disturbing the charge stored in the gate capacitor. A write speed test revealed that after a 10 ms write pulse, the I_{RBL} response clearly distinguishes between data "0" and data "1", demonstrating rapid write capability due to the high ON-current of the write transistor (Fig. 4b). However, the write speed is limited by the measurement equipment. Stability tests over 10 write/erase cycles showed that the current levels for data "1" remained consistent, highlighting the exceptional stability of the device (Fig. 4c).

By applying a longer 60 ms write pulse, the DRAM achieved a high current level of around 75 nA, allowing for multilevel storage capabilities. With a 60 ms pulse and varying write voltages, the device produced five distinct memory states, as indicated by stable output currents over a 45-second measurement period (Fig. 4d). These results showcase the potential of this technology for advanced multilevel DRAM applications.

Conclusion

In conclusion, the h-BN-assisted vdW integration strategy marks a significant advancement in fabricating damage-free top-gate 2D transistors with pristine interfaces. Using a monolayer h-BN as a buffer effectively shields 2D materials from high-energy deposition, preserving their intrinsic properties. The resulting MoS$_2$ transistors exhibit outstanding electrical performance, while the successful creation of a 2T0C DRAM cell highlights this approach's potential to enhance DRAM data retention and write speed. These findings underscore the promise of vdW integration techniques in advancing next-generation electronic devices, particularly those based on 2D materials.

Acknowledgments

This work is supported by MOST National Key Technologies R&D Programme (2022YFA1203804), National Natural Science Foundation of China (NSFC) for Distinguished Young Scholars (62425405), Research Grant Council of Hong Kong (15301023 and CRS_PolyU502/22), and the Hong Kong Polytechnic University (WZ4X and ZE1T).

References

[1] D. Akinwande et al., "Graphene and two-dimensional materials for silicon technology," Nature, 573, pp. 507-518 (2019).

[2] Y. Liu, X. Duan, H.-J. Shin, S. Park, Y. Huang, and X. Duan, "Promises and prospects of two-dimensional transistors," Nature, 591, pp. 43-53 (2021).

[3] Z. Lu et al., "Wafer-scale high-κ dielectrics for two-dimensional circuits via van der Waals integration," Nat. Commun., 14, pp. 2340 (2023).

[4] Y. Xu et al., "Scalable integration of hybrid high-κ dielectric materials on two-dimensional semiconductors," Nat. Mater., 22, pp. 1078-1084 (2023).

[5] L. Wang et al., "A general one-step plug-and-probe approach to top-gated transistors for rapidly probing delicate electronic materials," Nat. Nanotechnol., 17, pp. 1206-1213 (2022).

[6] J. Wang et al., "Integration of High-k Oxide on MoS2 by Using Ozone Pretreatment for High-Performance MoS2 Top-Gated Transistor with Thickness-Dependent Carrier Scattering Investigation," Small, 11, pp. 5932-5938 (2015).

[7] Y. Hu, H. Jiang, K. M. Lau, and Q. Li, "Chemical vapor deposited monolayer MoS2 top-gate MOSFET with atomic-layer-deposited ZrO2 as gate dielectric," Semicond. Sci. Technol., 33, pp. 045004 (2018).

[8] Y. Sheng et al., "Gate Stack Engineering in MoS2 Field-Effect Transistor for Reduced Channel Doping and Hysteresis Effect," Adv. Electron. Mater., 7, pp. 2000395 (2021).

[9] S. Yang, K. Liu, Y. Xu, L. Liu, H. Li, and T. Zhai, "Gate Dielectrics Integration for 2D Electronics: Challenges, Advances, and Outlook," Adv. Mater., 35, pp. 2207901 (2023)

[10] H. Wang et al., "Integrated Circuits Based on Bilayer MoS2 Transistors," Nano Lett., 12, pp. 4674-4680 (2012).

979-8-3315-0417-5/25 $31.00 © 2025 IEEE

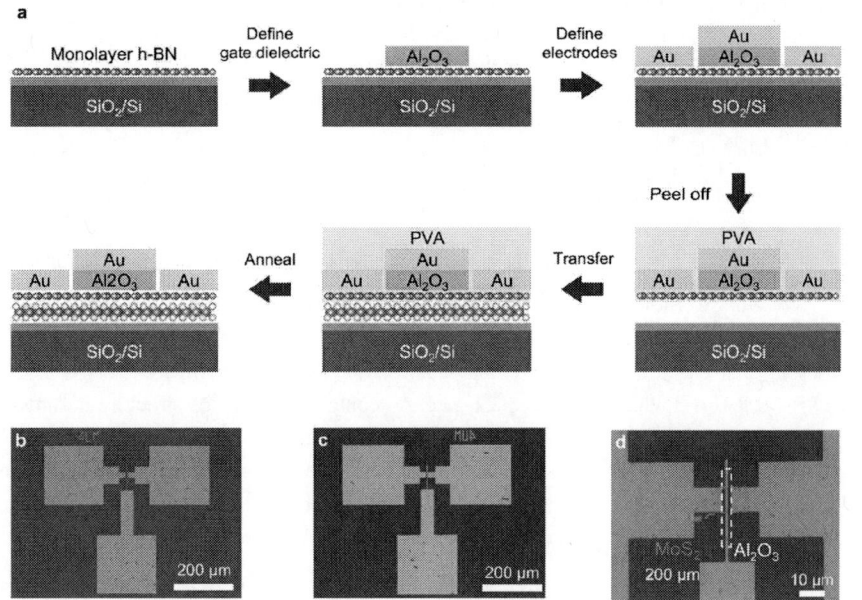

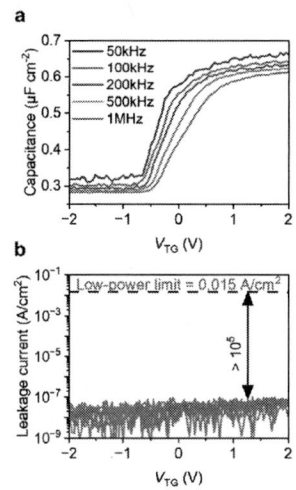

Fig. 3. (a) C-V curves of the MoS₂/h-BN/Al₂O₃ structure on a quartz substrate. (b) Gate leakage current density of the MoS₂ transistor. The red line represents the average leakage current density of nine devices (gray lines). The black dashed line represents the low-power limit of 0.015 A/cm².

Fig. 1. (a) Fabrication process flow of the damage-free MoS₂ transistor. (b) Optical image of source/gate/drain electrodes and gate dielectric stack on monolayer h-BN on sacrificial substrate. (c) Optical image of the peeled-off device stack on PVA. (d) Optical image of the damage-free MoS₂ transistor.

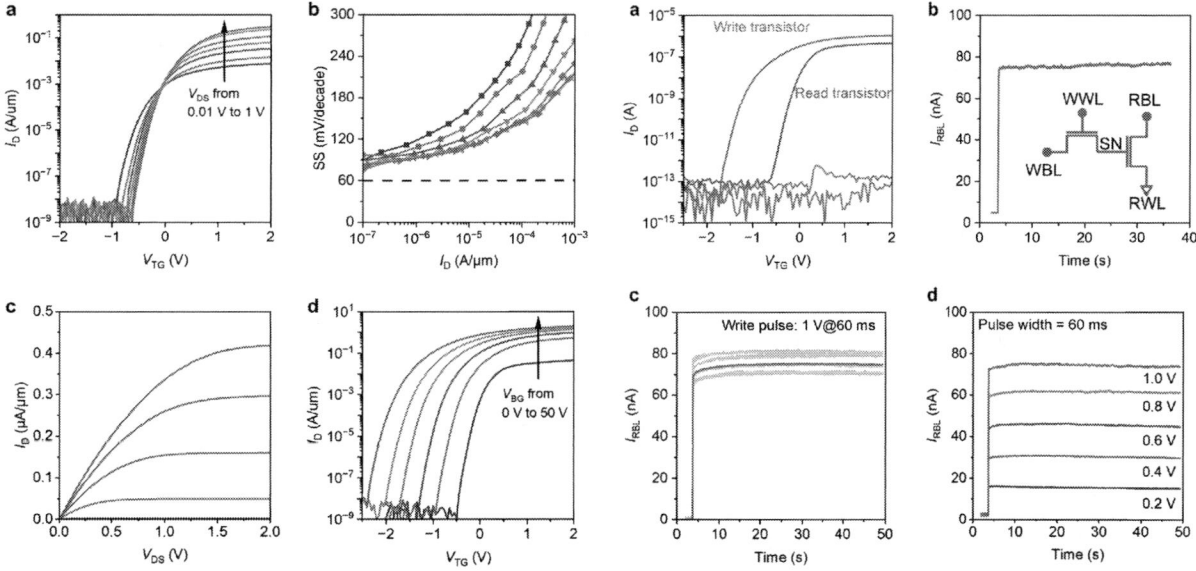

Fig. 2. (a) Transfer curves of the vdW integrated damage-free top-gate transistor with different drain voltages. (b) Extracted SS at different channel currents of the top-gate MoS₂ transistor. The black dashed line corresponds to the lower limit of 60 mV/decade at room temperature. (c) Output curves of the MoS₂ transistor. (d) Transfer curves of the vdW integrated damage-free top-gate transistor under various back-gate voltages.

Fig. 4. (a) Transfer curves of the write and read MoS₂ transistors. (b) I_{RBL} as a function of read time following the writing of data "1", with the write pulse width of 10 ms. The amplitude of the write pulse is fixed to be 1 V. The insert is the circuit diagram of a 2T0C DRAM. (c) 10 times write and erase operations of data "1". Each conducted with a pulse width of 60 ms and pulse amplitude of 1 V. (d) Data retention of the DRAM with V_{WBL} ranging from 0.2 V to 1.0 V.

979-8-3315-0417-5/25 $31.00 © 2025 IEEE   335

Analysis of High Performance IGZO Thin-Film Transistors Under High-gate and Drain Bias Stress

Yupeng Lu [1,2,3], Yanyu Yang[1,2,3], Peng Wang [1,2,3], Jie Luo[1,2,3]
Yunjiao Bao[1,2,3], Gaobo Xu[*2,3], Huaxiang Yin[*1,2,3]

[1] School of Integrated Circuits, University of Chinese Academy of Sciences, Beijing 100049, China;
[2] State Key Lab of Fabrication Technologies for Integrated Circuits, Institute of Microelectronics, Chinese Academy of Sciences, Beijing 100029, China;
[3] Integrated Circuit Advanced Process R&D Center, Institute of Microelectronics, Chinese Academy of Sciences Beijing 100029, China;
*Corresponding Author's Email: yinhuaxiang@ime.ac.cn

Abstract

This study investigated the electrical behavior of Indium Gallium Zinc Oxide thin-film transistors (a-IGZO TFTs) under high gate bias stress and high drain bias stress. In this research, we performed electrical behavior tests on devices with different gate widths, both before and after the application of stress. Post-stress application, the electrical performance of the devices exhibited a significant positive shift. As the stress applied to the gate and drain increases, the amount of charge trapped in the gate dielectric also increases, resulting in more pronounced electrical performance shifts. It is inferred that these phenomena are due to charge trapping in the gate dielectric during prolonged stress application.

Keywords—IGZO, reliability analysis, energy-band structure, electrical performance, thin-film transistor

Introduction

Indium Gallium Zinc Oxide thin-film transistors (a-IGZO TFTs) are emerging as the most promising devices for next-generation DRAMs due to their high mobility, extremely low off-state current, and exceptionally high on/off ratios[1][2]. Despite the excellent performance of IGZO TFTs, like other amorphous oxide TFTs, they exhibit stability degradation under electrical stress. Generally, when negative bias stress is applied, the characteristics of the devices do not show significant changes[3]. However, under positive bias stress, the transfer characteristics of the devices exhibit a positive shift[4].

With reference to previous research, the shift in transfer characteristics of devices under positive bias stress (PBS) is attributed to electron trapping either within the bulk gate insulator and/or at the interface[5]-[7], or within the channel layer[8]. Distinct from traditional PBS testing, which applies stress solely to the gate, this work applies stress simultaneously to both the gate and drain. Under these conditions, the transistor consistently operates in a saturation mode, closely resembling its long-term operational environment. In this work, we investigated the transfer characteristic shifts of devices with different gate widths under high gate and drain stress. Furthermore, we explained this phenomenon using band theory and carrier transport

mechanisms, providing insights for further research aimed at optimizing the lifespan of IGZO TFTs.

Experiment

The device designed for this experiment is an IGZO TFT with back-gate top contact. The device structure is shown in Figure 1 . The device is mainly divided into seven layers, including Si substrate layer, substrate isolation layer, back-gate metal layer, gate dielectric isolation layer, channel layer, source-drain metal layer, and passivation layer.

IGZO TFTs are fabricated on a p-type Si substrate covered by a 200 nm SiO2 overlay. The specific process flow is illustrated in Figure 2. Initially, a 200 nm SiO_2 isolation layer is deposited on the silicon substrate using a chemical vapor deposition (CVD) process. This is followed by the deposition of a 60 nm Mo layer as the gate metal layer using a physical vapor deposition (PVD) process. After photolithography using I-line, the back-gate Mo is patterned via dry etching, followed by the deposition of the gate dielectric layer. In this study, a 15 nm SiO_2 material is employed as the gate dielectric, deposited using the plasma-enhanced CVD (PECVD). The channel layer comprises 25 nm of IGZO deposited via PVD. After photolithography, wet etching with diluted HNO_3 is performed to pattern the active region. Subsequently, the growth and patterning of the source-drain metal layers are conducted, with 60 nm PVD Mo utilized as the source-drain electrodes. This is followed by the growth and through-hole etching of the passivation layer, employing a 100 nm CVD SiO_2 layer at 200°C. Finally, a 60 nm PVD Mo layer is deposited to fill the through-hole and serve as the pad metallic layer for electrode extraction, with photolithography and etching used to define the pad pattern.

Results and Discussion

The electrical performance of the fabricated devices was measured using a semiconductor parameter analyzer (Agilent 4156) under room temperature and dark conditions. Prior to testing, the devices underwent an annealing process in an air environment at 200°C for 70 minutes. The transfer characteristics of the prepared devices are shown in Figure 3, where the transfer curves for devices with channel widths of 10 μm, 50 μm, and 480 μm under different stress conditions are presented. For the stress testing, we measured the transfer

979-8-3315-0417-5/25 $31.00 © 2025 IEEE

characteristics under no stress, and then under the application of gate and drain stresses of 8V, 9V, and 10V for 4000s, respectively. As illustrated in the figure, all device sizes exhibited a positive shift in the transfer characteristic curves following the application of high gate and drain stresses.

Due to the prolonged application of high gate and drain stress, the transistors remained in a high-load operational state throughout the testing process. To monitor the testing process, we also recorded the variation of the output current (ID) with respect to the stress application duration, as shown in Figure 4. It can be observed that as the stress application time increases, the output current (ID) degrades correspondingly. This degradation is consistent with the results observed in the transfer characteristic curves after stress application.

In addition to the tests, we also evaluated the output characteristic curves of the fabricated devices under no stress conditions, as shown in Figure 5. During the testing, the range of V_{DS} was set from 0 to 10V, the range of V_{GS} was set from 0 to 10V, and the step size of V_{GS} was 2V. As depicted in the figure, the fabricated devices exhibited excellent performance, with no breakdown observed even under the extensive dynamic range of V_{DS} and V_{GS}.

Based on previous literature, researchers attribute the positive shift in the transfer characteristic curves under positive bias stress (PBS) to charge trapping in the gate dielectric[9][10]. Similarly, the positive shift in the transfer characteristic curves observed in our devices under prolonged high gate and drain stress can be explained by the same mechanism. As illustrated, when the transistors are subjected to prolonged high-load operation, mobile charges in the channel accumulate above the gate dielectric layer. During the fabrication process, defects are inevitably introduced at the interface between the gate dielectric layer and the channel layer. The process is shown in Figure 7 below. Consequently, the originally mobile free carriers become trapped at these defect sites, turning into fixed charges within the gate dielectric layer. This results in a negative potential within the channel, requiring a higher gate voltage to achieve inversion and compensate for the potential created by the fixed charges in the gate dielectric.

To better explain the reason for the drift of the transfer characteristic curve, we also explored the energy band structure of the IGZO TFT. The bias stability of IGZO TFTs can be explained by the following three electron trapping instability mechanisms[11][12], as shown in Figure 8, which are consistent with the instability and hysteresis trends observed in these SiO2/IGZO TFTs: a) Electron trapping within the IGZO channel layer. b) Electron injection and trapping within the gate insulator. c) Presence of deep-level states. Under a larger positive bias stress, process b) predominates, with a significant number of electrons entering the gate insulator and becoming defects within the insulator layer. Therefore, under high-gate and drain Bias Stress tests, a significant positive threshold voltage shift phenomenon is observed.

Conclusion

This paper provides a detailed account of the fabrication process, technical specifics, and reliability testing of IGZO TFTs. The devices we fabricated demonstrated exceptionally high dynamic operating range and performance advantages. Unlike traditional positive bias stress testing (PBS), we applied ultra-high voltages (10V) and extended duration (4000s) to both the gate and drain. The tests revealed that under extremely high load operating conditions, the performance of IGZO TFTs exhibited some degree of degradation. To address this phenomenon, we analyzed the carrier transport mechanisms and band structure within the channel region. This work offers reliable guidance for the design and fabrication of high-performance IGZO devices.

Acknowledgments

This work was supported by BJSAMT Project (SAMT-ZK-KT-22030102).

References

[1] Belmonte et al., "Capacitor-less, Long-Retention (>400s) DRAM Cell Paving the Way towards Low-Power and High-Density Monolithic 3D DRAM," 2020 IEEE International Electron Devices Meeting (IEDM)

[2] Belmonte et al., "Tailoring IGZO-TFT architecture for capacitorless DRAM, demonstrating > 103s retention, >1011 cycles endurance and Lg scalability down to 14nm," 2021 IEEE International Electron Devices Meeting (IEDM)

[3] G. Yan et al., "Mechanism Analysis of Ultralow Leakage and Abnormal Instability in InGaZnO Thin-Film Transistor Toward DRAM," in IEEE Transactions on Electron Devices

[4] K. Nomura, T. Kamiya, and H. Hosono, "Highly stable amorphous In–Ga–Zn–O thin-film transistors produced by eliminating deep subgap defects," Appl. Phys. Lett.

[5] D. A. Mourey, D. A. Zhao, J. Sun, and T. N. Jackson, "Fast PEALD ZnO thin-film transistor circuits," IEEE Trans. Electron Devices

[6] K. Hosino, D. Hong, H. Q. Chiang, and J. F. Wager, "Constant-voltage-bias stress testing of a-IGZO thin-film transistors," IEEE Trans. Electron Devices

[7] S. Kim, Y. W. Jeon, Y. Kim, D. Kong, H. K. Jung, M.-K. Bae, J.-H. Lee,B. D. Ahn, S. Y. Park, J.-H. Park, J. Park, H.-I. Kwon, D. M. Kim,and D. H. Kim, "Impact of oxygen flow rate on the instability under positive bias stresses in dc-sputtered amorphous InGaZnO thin-film transistors," IEEE Electron Device Lett.

[8] M. D. H. Chowdhury, P. Migliorato, and J. Jang, "Time-temperature dependence of positive gate bias stress and recoveryin amorphous indium-gallium-zinc-oxide thin-film-transistors," Appl.Phys. Lett.

[9] Y. Vygraneko, K. Wang, and A. Nathan, "Stable indium oxide thin-film transistors with fast threshold voltage recovery," Appl.Phys. Lett.

[10] R. B. M. Cross and M. M. De Souza, "Investigating the stability of ZnO thin film transistors," Appl. Phys. Lett.

[11] D. Zhao, D. A. Mourney, and T. N. Jackson, "Fast flexible plastic substrate ZnO circuits," IEEE Electron Device Lett.

[12] M. D. H. Chowdhury, P. Migliorato, and J. Jang, "Time-temperature dependence of positive gate bias stress and recoveryin amorphous indium-gallium-zinc-oxide thin-film-transistors," Appl.Phys. Lett

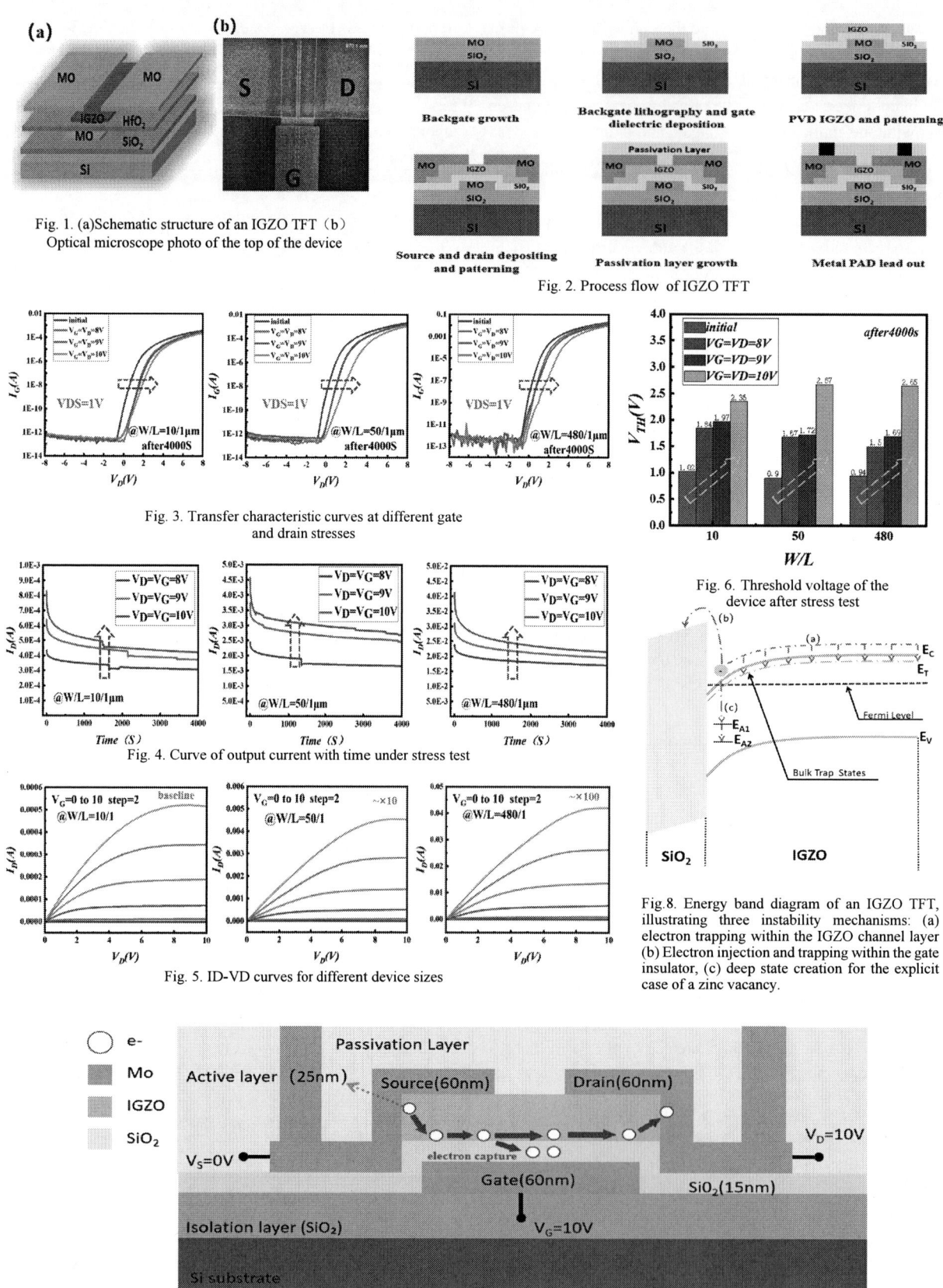

Fig. 1. (a)Schematic structure of an IGZO TFT （b）Optical microscope photo of the top of the device

Fig. 2. Process flow of IGZO TFT

Fig. 3. Transfer characteristic curves at different gate and drain stresses

Fig. 4. Curve of output current with time under stress test

Fig. 5. ID-VD curves for different device sizes

Fig. 6. Threshold voltage of the device after stress test

Fig.8. Energy band diagram of an IGZO TFT, illustrating three instability mechanisms: (a) electron trapping within the IGZO channel layer (b) Electron injection and trapping within the gate insulator, (c) deep state creation for the explicit case of a zinc vacancy.

Fig. 7. Schematic diagram of the gate dielectric charge trapping model

Ferroelectric In₂Se₃ transistors with multi-terminal plasticity for highly efficient hardware implementation of reinforcement learning

Yasai Wang[1,2], Weiwei Xiong[1], Jianmin Yan[2], Yue Zhou[2], Chaoyi Zhu[2], Xiangshui Miao[1], Yuhui He[1*] and Yang Chai[2*]

[1]School of Integrated Circuits, Huazhong University of Science and Technology, Wuhan, China.
[2]Department of Applied Physics, The Hong Kong Polytechnic University, Hong Kong, China.
*Email: heyuhui@hust.edu.cn, ychai@polyu.edu.hk

Abstract

Hardware implementation of reinforcement learning (RL) algorithms with silicon-based circuits requires complex designs with dozens of components. To simplify the conventional design, we demonstrate a synaptic cell with only a single transistor for in-situ RL weight updates. By exploiting the unique in-plane (IP) and out-of-plane (OOP) ferroelectric polarization coupling of In₂Se₃, we demonstrate multi-terminal synaptic plasticity in RL algorithm with In₂Se₃ FET, which can realize multi-terminal channel conductance tuning of V_{DS} (> 5 bits, G_{max}/G_{min}>6) and V_{GS} (>5 bits, G_{max}/G_{min}>25), respectively. Additionally, the ferroelectric relaxation naturally realizes the eligibility trace decay characteristic in the RL algorithm, which enhanced the performance of the algorithm. We experimentally demonstrate the In₂Se₃ FET arrays to solve the classical RL benchmark cart-pole task, exhibiting a way for realizing low-power (86 pJ per programming with eligibility trace) and highly area-efficient (10 μm²) hardware chip for reinforcement learning.

Introduction

As a major learning paradigm distinct from supervised and unsupervised learning, reinforcement learning (RL) obtains feedback through the interaction between the agent and the environment to update the strategy [1]. Reward-modulated spike-timing-dependent plasticity (R-STDP) is a brain-inspired RL rule (**Fig. 1a**) [2]. Compared with ANN-based algorithms that use error backpropagation and gradient descent to update weights, R-STDP is more biologically plausible by combining brain-inspired STDP with eligibility trace and reward signals to update weights.

Due to the complexity of the R-STDP update formula (**Fig. 1a**), the R-STDP hardware implementation based on silicon CMOS technology requires 18 transistors; and the eligibility trace hardware implementation based on emerging nonvolatile memory (NVMs) requires 3 Phase Change Memory (PCM) and additional synaptic unit (**Fig. 2a**) [3], which limits the development of RL hardware accelerators. In this work, we use a single In₂Se₃ transistor to perform the update calculation of R-STDP without resorting to additional computing resources (**Fig. 2b**). By utilizing the unique in-plane and out-of-plane ferroelectric polarization coupling characteristics of In₂Se₃, the STDP and reward parts of the algorithm can be achieved by modulation of IP and OOP polarization, respectively. Furthermore, the algorithm's eligibility trace decay function is mapped to the ferroelectric polarization relaxation effect caused by depolarization field (**Fig. 2b**). The RL SNN with

In₂Se₃ FETs can successfully complete the classic cart-pole benchmark, demonstrating the advantage of ultra-low hardware cost.

Multi-terminal programmable conductance in In₂Se₃ FET by OOP and IP polarization

Fig. 1b illustrates the schematic of In₂Se₃ FET with local bottom-gate structure. **Fig. 3** shows an SEM image of the In₂Se₃ FET array. Unlike typical FeFET, the transfer curve (I_{DS}-V_G) of In₂Se₃ FET shows clockwise hysteresis (**Fig. 4b**), which is a result from the channel band bending caused by OOP polarization of ferroelectric semiconductors (**Fig. 4a**). The large hysteresis window (±5 V, on/off ratio>10⁵) indicates that the device has a considerable conductance modulation range, which facilitates the subsequent R-STDP synaptic weight update. **Fig. 5** shows the analog conductance modulation under V_{GS} pulses. The device has excellent conductance modulation range (G_{max}/G_{min}>25), multiple conductance states (>32), and low cycle-to-cycle variation. Due to the IP polarization of In₂Se₃, the source-drain voltage can also affect the channel conductance. The I_{DS}-V_D scanning curve is shown in **Fig. 7b**, which also produces an obvious hysteresis window (on/off ratio>10). The window of the I_{DS}-V_D curve comes from the change of the contact barrier caused by the IP polarization (**Fig. 7a**). Under the above premise, V_{DS} pulses can also produce analog conductance modulation on channel conductance (**Fig. 8**). It is worth noting that since the thickness (~30 nm) of In₂Se₃ is much smaller than the channel length (1 μm), the equivalent electric field of V_{GS} is much higher than that of V_{DS}, which leads to strong modulation of V_{GS}. This characteristic aligns well with the requirements of the R-STDP algorithm in biological systems, where STDP signal (V_{DS}) modulation is weaker than reward signal (V_{GS}) modulation [2]. This compatibility facilitates our subsequent application of In₂Se₃ FET for R-STDP synapses.

Fig. 6 and **9** show the excitatory postsynaptic current (EPSC) and inhibitory postsynaptic current (IPSC) under two consecutive V_{GS} or V_{DS} pulses, respectively. The In₂Se₃ FET exhibits a mixture of long-term plasticity (LTP) and short-term plasticity (STP), which arise from the ferroelectric retention and relaxation induced by the depolarization field. The relaxation of the conductance is consistent with the required eligibility trace decay in R-STDP. Eligibility trace is widely used in many RL algorithms, enabling the algorithm to consider both past and future reward expectations to achieve optimal results. In₂Se₃ FET naturally achieves the relaxation of eligibility trace without the need for external units.

979-8-3315-0417-5/25 $31.00 © 2025 IEEE

R-STDP synapse with eligibility trace by In₂Se₃ FET

The simplified R-STDP rule is as follows [2]:

$$\Delta w = STDP(t) \cdot E(t) \cdot R(t)$$

where w is the synaptic weight, $STDP(t)$ is generated by the interaction of pre- and post-synaptic pulses, $E(t)$ represents the decay effect of the eligibility trace, and $R(t)$ represents the delayed reward signal. The timing signal diagram of weight update is shown in **Fig. 11a**. The schematic diagram of using In₂Se₃ FET to implement R-STDP in-situ update is shown in **Fig. 11b**: (1) V_{DS} represents the effective pulse composed of pre- and post-synaptic pulses, indicating the STDP signal; (2) The relaxation effect of the conductance after the pulse represents the decay of the eligibility trace; (3) The delayed V_{GS} pulse represents the delayed reward signal. The combined effect of V_{DS} and V_{GS} results in the final conductance update.

It is worth noting that the R-STDP update formula requires that the weight does not update when only one signal is present. The weight updates only when both STDP and R signals are present simultaneously. This requirement can be met by setting the amplitude of a single pulse near the ferroelectric threshold. The threshold test results for V_{DS} and V_{GS} are shown in **Fig. 10**, with values of 5.1 V and 2.3 V. More importantly, in R-STDP, the later the delayed reward signal, the more the eligibility trace decays, and the smaller the final weight update value. In₂Se₃ FETs achieve this through the time interval between V_{DS} and V_{GS} pulses. The larger the time interval, the more pronounced the ferroelectric relaxation, resulting in a weaker modulation. **Fig. 12** and **13** show the modulation results for different time intervals, which align with the expectations. Statistical results (**Fig. 13**) indicate that the greater the delay of reward signal, the weaker the weight modulation. This demonstrates that In₂Se₃ FET can realize the functionality of R-STDP synapses.

Table 1 presents a comparison between this work and other works in R-STDP synapses, highlighting the advantages of In₂Se₃ FET in terms of hardware costs and other aspects.

R-STDP synapse based SNN

Fig. 14 shows the schematic of array-level illustration of the SNN based on In₂Se₃ FET synapse, the synaptic weights are obtained from device array (**Fig. 3**). The SNN is trained to solve the cart-pole problem, where the output neurons exert force to the left or right to maintain the balance of the pole (**Fig. 15**). **Fig. 17** shows the time evolution of the cart position x and pole angle θ. After training, the pole successfully keeps balanced in 450 time steps. Furthermore, the synaptic weights connections to the left and right output neurons are shown in **Fig. 16**. To verify the impact of eligibility trace on training performance, we recorded the average reward (**Fig. 18**). The results indicate eligibility trace can lead to faster convergence and ultimately better performance. Without eligibility traces, the algorithm fails to converge. Therefore, the tunable eligibility trace of In₂Se₃ FET will bring significant convenience to the application of the algorithm.

Conclusion

We experimentally demonstrate in-situ updates for reinforcement learning with eligibility trace in a single In₂Se₃ FET. By coupled tuning of IP and OOP polarization, both V_{DS} (>5 bits, G_{max}/G_{min}>6) and V_{GS} (>5 bits, G_{max}/G_{min}>25) can achieve weight updates. The ferroelectric relaxation of In₂Se₃ can achieve eligibility trace, significantly improving the training efficiency of the cart-pole task. The In₂Se₃-based synapses provide a low-power (86 pJ per programming with eligibility trace) and high area-efficiency (10 μm²) method for edge computing in reinforcement learning.

Acknowledgment

This work is financially supported by the National Key Research and Development Program of China (No. 2023YFB4502200), NSFC No.62374063 and 92164204, Research Grant Council of Hong Kong (CRS_PolyU502/22), and the Hong Kong Polytechnic University (WZ4X and SB6M).

References

[1] Ota, K., et al. IEDM pp. 6-2, 2019. [2] Frémaux, N., et al. Frontiers in neural circuits, 9, 85(2016). [3] Demirağ, Y., et al. ISCAS pp. 1-5, 2021. [4] Zhou, Y., et al. Advanced Materials, 34, 48(2021). [5] Wang, H., et al. IEEE Trans. Biomed. Circuits. Syst., 16, 4(2022) [6] Nair, H., et al. ISVLSI pp. 266-271, 2021. [7] Luo, J., et al. IEEE Electron Device Letters, 202

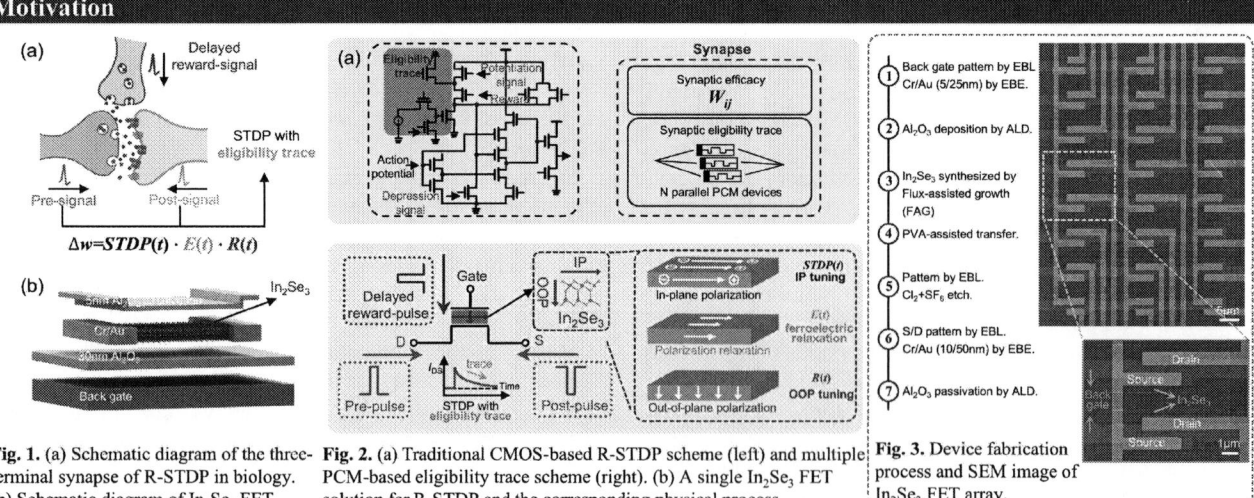

Motivation

Fig. 1. (a) Schematic diagram of the three-terminal synapse of R-STDP in biology. (b) Schematic diagram of In₂Se₃ FET.

Fig. 2. (a) Traditional CMOS-based R-STDP scheme (left) and multiple PCM-based eligibility trace scheme (right). (b) A single In₂Se₃ FET solution for R-STDP and the corresponding physical process.

Fig. 3. Device fabrication process and SEM image of In₂Se₃ FET array.

Multi-terminal programmable conductance in In$_2$Se$_3$ FET by OOP and IP

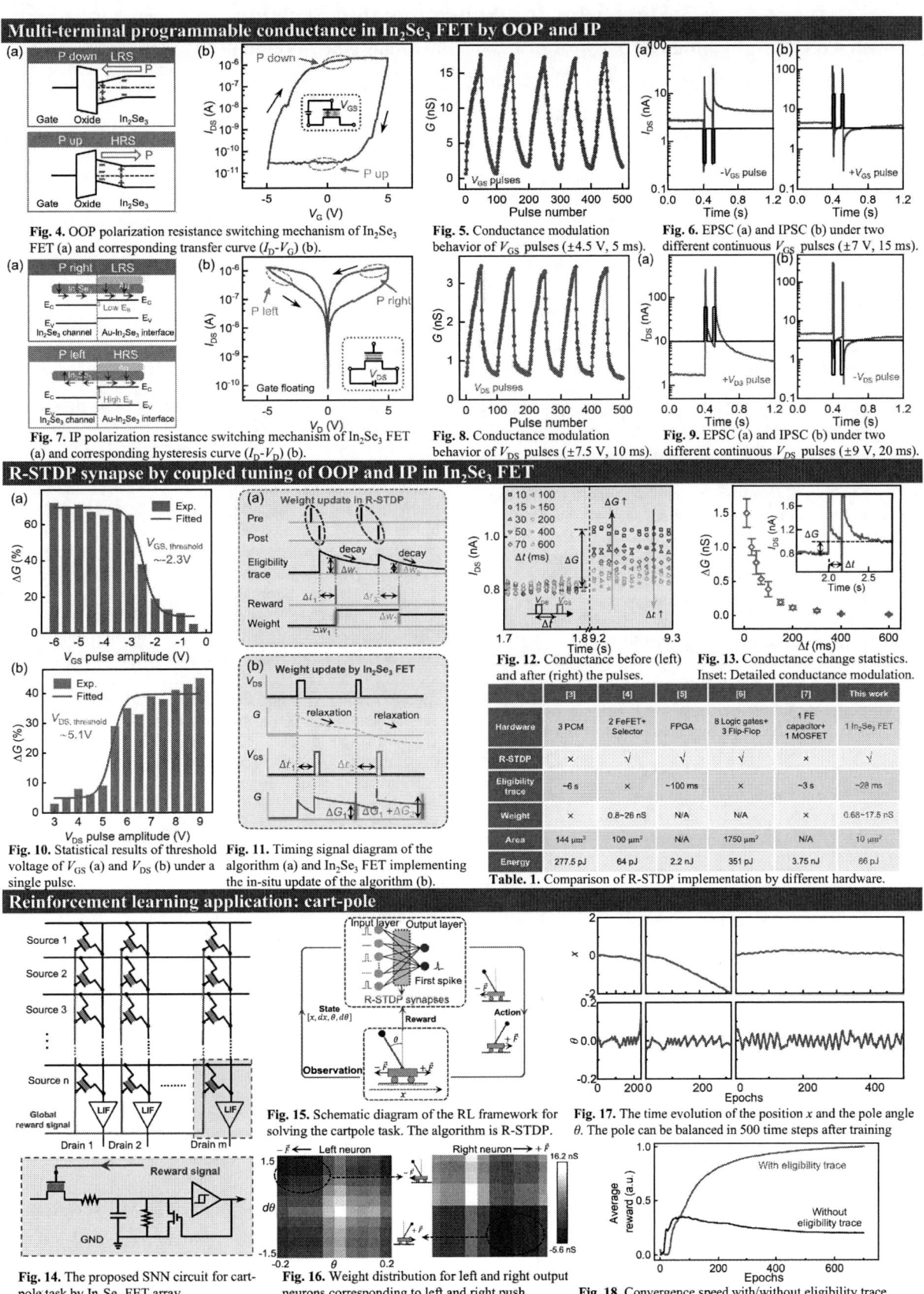

Fig. 4. OOP polarization resistance switching mechanism of In$_2$Se$_3$ FET (a) and corresponding transfer curve (I_D-V_G) (b).

Fig. 5. Conductance modulation behavior of V_{GS} pulses (±4.5 V, 5 ms).

Fig. 6. EPSC (a) and IPSC (b) under two different continuous V_{GS} pulses (±7 V, 15 ms).

Fig. 7. IP polarization resistance switching mechanism of In$_2$Se$_3$ FET (a) and corresponding hysteresis curve (I_D-V_D) (b).

Fig. 8. Conductance modulation behavior of V_{DS} pulses (±7.5 V, 10 ms).

Fig. 9. EPSC (a) and IPSC (b) under two different continuous V_{DS} pulses (±9 V, 20 ms).

R-STDP synapse by coupled tuning of OOP and IP in In$_2$Se$_3$ FET

Fig. 10. Statistical results of threshold voltage of V_{GS} (a) and V_{DS} (b) under a single pulse.

Fig. 11. Timing signal diagram of the algorithm (a) and In$_2$Se$_3$ FET implementing the in-situ update of the algorithm (b).

Fig. 12. Conductance before (left) and after (right) the pulses.

Fig. 13. Conductance change statistics. Inset: Detailed conductance modulation.

	[3]	[4]	[5]	[6]	[7]	This work
Hardware	3 PCM	2 FeFET+Selector	FPGA	8 Logic gates+3 Flip-Flop	1 FE capacitor+1 MOSFET	1 In$_2$Se$_3$ FET
R-STDP	×	√	√	√	×	√
Eligibility trace	~6 s	×	~100 ms	×	~3 s	~28 ms
Weight	×	0.8~26 nS	N/A	N/A	×	0.68~17.5 nS
Area	144 μm²	100 μm²	N/A	1750 μm²	N/A	10 μm²
Energy	277.5 pJ	64 pJ	2.2 nJ	351 pJ	3.75 nJ	86 pJ

Table. 1. Comparison of R-STDP implementation by different hardware.

Reinforcement learning application: cart-pole

Fig. 14. The proposed SNN circuit for cart-pole task by In$_2$Se$_3$ FET array.

Fig. 15. Schematic diagram of the RL framework for solving the cartpole task. The algorithm is R-STDP.

Fig. 16. Weight distribution for left and right output neurons corresponding to left and right push.

Fig. 17. The time evolution of the position x and the pole angle θ. The pole can be balanced in 500 time steps after training

Fig. 18. Convergence speed with/without eligibility trace.

Enabling Artificial Spiking Sensory Neurons with a Single Flexible VO₂ Mott Memristor for Neuromorphic Sensing

Chuan Yu Han[1*], Shujing Zhao[1], Shengli Fang[1], Shi Quan Fan[1], Weihua Liu[1], Xin Li[1], Li Geng[1].

[1] School of Microelectronics, Xi'an Jiaotong University, Xi'an 710049, P.R. China.

E-mail: hanchuanyu@xjtu.edu.cn

Abstract In this work, we present flexible VO₂ Mott memristors with excellent electrical properties, i.e. low operating current (<100 µA), good endurance (>10⁹) and good flexibility (bending radius ~5 mm) for constructing artificial spiking sensory neurons. As a result, an artificial spiking thermoreceptor (AST) and an artificial spiking photoreceptor (ASP) are both realized by using a single flexible VO₂ Mott memristor. With the stimulus temperature elevated from 20 °C to 40 °C, the spiking frequency of AST first increases, then falls off rather quickly to zero at the higher temperature of 45 °C, resembling the response characteristics of biological thermoreceptor. And the ASP has high spike-encoded photosensitivity and ultra-wide photosensing range (405~808 nm). Finally, the spike-based electronic retina is proposed by connecting an array of artificial spiking photoreceptors with multiple-input floating gate MOS transistors acting as optical nerves. Consequently, the color image is successfully segmented and the edges of objects in the image are clearly extracted through the electronic retina by simulation, thus providing a solid foundation for neuromorphic sensing system.

Keywords: Mott memristor, VO₂, artificial spiking thermoreceptor, artificial spiking photoreceptor

I. Introduction

Inspired by the powerful nervous system of human, spike-based machine intelligence using neuromorphic computing has aroused considerable research interest, aiming at realizing high artificial intelligence with high energy efficiency by emulating the functions of human's brain and peripheral nervous system[1]. Consequently, synatpic devices that emulate the the funcitons of biological synpases and artificial neurons that mimic the functions of neurons have been extensively investigated [2, 3]. Environmental stimuli are encoded into nerve spikes and transmitted to the brain via numerous afferent neurons and receptors. Thermoreceptors and photoreceptors, which convert environmental thermal and light stimuli into coded spikes, are essential components for interacting with the environment and realizing spike-based intelligent machines such as neurorobots and human-machine interfaces.

In this study, we experimentally demonstrate that flexible VO₂ Mott memristors can be utilized to construct artificial spiking thermoreceptors (ASTs) and artificial spiking photoreceptors (ASPs) with a single device. The flexible AST can adjust its output spike frequency in response to temperature changes, mimicking the response characteristics of biological warm receptors. The ASP exhibits high spike-encoded photosensitivity and an ultra-wide photosensing range (405~808 nm). When driven by a current source, the ASP can generate stable spikes that vary with the wavelength and intensity of the stimulus light. By integrating an array of artificial spiking photoreceptors with multiple-input floating gate MOS

transistors as optical nerves, an electronic retina based on spikes is achieved. This electronic retina can sense, encode, and preprocess information from color images in the form of spikes. As a result, color images are successfully segmented, and the edges of objects within the images are clearly extracted through the electronic retina, as demonstrated by simulation.

II. Result and discussion

A. Fabrication of the flexible VO₂ Mott memristors

Fig. 1 shows the fabrication process of the flexible VO₂ Mott memristors. First, the polyimide (PI) tape was adhered to a flexible polyethylene naphthalate (PEN) substrate, then both films were fixed on the SiO₂/Si carrier substrate by using polymethyl methacrylate (PMMA) during the fabrication process. Second, the Al₂O₃ (~3 nm) employed as buffer layer to increase the adhesion between the memristors and the substrate was deposited on the PI tape by using radio frequency (RF) magnetron sputterer. After that, the Ti/Pt (~ 5 nm/ 65 nm) as the bottom electrode (BE) and Al₂O₃ (~30 nm) as isolation layer were sputtered and patterned by using the lift-off process. Finally, the VO₂ (~ 200 nm) as switching layer and Pt (for the AST) or transparent ITO (for ASP) as top electrode (TE) were prepared by a RF magnetron sputterer using VO₂ and Pt or ITO targets in the Ar ambient at room temperature, and patterned by lift-off processes. **Fig. 2(a) and (b)** display the optical micrographs of the fabricated VO₂ Mott memristor. The memristors utilized in this study feature a via-hole Pt/VO₂/Pt or ITO cross-bar structure, with the via-hole in the isolation layer having a diameter of approximately 10 µm (as shown in the inset of **Fig. 2(b)**). The layer thicknesses were ascertained through cross-sectional transmission electron microscopy (TEM) imaging.

B. Characterization of the flexible VO₂ Mott memristors

After the forming process, the memristor shows stable bi-directional threshold switching characteristics (see **Fig. 3(a)** and **(b)**). **Fig. 3(c)** shows the current-voltage (I-V) curves under 1000 consecutive positive voltage sweeps, which nearly overlap with each other, indicating that the memristor has excellent stability. The stability of the memristors under the bending strain was investigated as shown in **Fig. 4**. The I-V curves of the memristor show no obvious change under small bending radius (~ 5 mm) and 100 bending cycles, indicating the good flexibility of fabricated device, thus enabling it suitable to the application of flexible AST and ASP.

C. Oscillator based on the flexible VO₂ Mott memristor

The oscillator based the flexible VO₂ Mott memristor (R_M) is shown in **Fig. 5(a)**. To make the oscillation easy to measure and characterize, a 10 nF capacitor (C) is connected in parallel with the R_M. Driven by the input current (I_in), stable oscillating spikes can be obtained at the output (V_{out}) and the spiking frequency is increased with the increase of I_{in} (**Fig. 5(b)**). The stability of oscillation is characterized as shown in **Fig. 6**. During the test, the device was switched between high-resistance state (HRS) and low-resistance state (LRS) over 10^9 times without obvious deterioration, indicating the good endurance and stability of fabricated device.

D. Artificial spiking thermoreceptor (AST)

In biology, the warm receptors fire action potentials according to the stimulus temperature in the range of ~20°C to ~40°C [4]. And their firing rates rise with the stimulus temperature elevated from ~20°C until it reaches a saturation value between 38 °C and 43 °C, and then falls off rather steeply at higher temperatures (**Fig. 7(a)**)[5]. Based on the fabricated VO$_2$ Mott memristor, a flexible artificial spiking warm receptor is proposed by using a single VO$_2$ memristor driven by the a current source as illustrated in the inset of **Fig. 7(b)**. The firing frequency of the flexible artificial warm receptor related to the stimulus temperature is shown in **Fig. 7(b)** With the stimulus temperature elevated from 20 °C to 40 °C, the spike frequency increases from 16.8 kHz to 60.2 kHz, then quickly decreases to zero at the higher temperature of 45 °C, resembling the response characteristics of biological warm receptors under temperature stimulus[6].

E. Artificial spiking photoreceptor (ASP)

According to the anatomy of human eye, there are three photoreceptors, i.e. sensitive to the color red, color green and color blue in the retina that are responsible for the perception of color, rendering us perceive the colorful world. The retina consists of photoreceptors and optical nerves. Each optical nerve is connected to a certain range of photoreceptors (receptive field)[7]. The photoreceptors sense the color information and then encode it into nerve spikes, while the optical nerves collect the spiking information, extract features of the visual world, encode them in frequency modulated spike trains, and then send them to visual brain centers to perceive the information.

To investigate the photoresponse characteristics of the ASP, the memristor was stimulated with blue (405 nm), green (520 nm), red (620 nm), and infrared (808 nm) lights of varying intensities. It is found that the I-V curves shift to the left as the light intensity increases, indicating that the proposed ASP is sensitive to the stimulus lights, for a broad photosensing range (405~808 nm). Under the same light intensity, the wavelength of the light also affects the I-V curve and the V_{th}/V_{hold}, as shown in **Fig. 8**. An ASP was constructed by connecting a single R_M and a C (=10 nF for easy measurement, can be the parasitic capacitance of the memristor) in parallel. **Fig.9(a)** shows the relationship between the frequency of output spikes and I_{in} for different wavelengths under the same stimulus intensity. The spiking frequency increases with increasing I_{in}, due to the decreasing charging time of the capacitor.

Fig.9(b) illustrates the output spikes generated by the VO$_2$ Mott memristor for different wavelengths of light under a constant light intensity of 8 mW, and an input current of 70 μA. These findings indicate that the creation of a spike-based electronic retina using a VO$_2$ Mott memristor and a transparent ITO top electrode is feasible[8].

F. Spike-based electronic retina

The spike-based electronic retina is proposed by connecting an array of artificial spiking photoreceptors with multiple-input floating gate MOS transistors that are called neuro-transistor as shown in inset of **Fig. 10(c)**. The image segmentation and edge extraction functions of spike-based electronic retina are demonstrated in **Fig. 10**. The original color image (see **Fig. 10(a)**) is sensed by the photoreceptors with different optical filters (red, green and blue) and then encoded into spikes, so the color image is segmented into three channels (**Fig. 10(b)**). In each channel, the spikes from several photoreceptors are collected by one neuro-transistor and again encoded into a train of spikes, so the edges are extracted by the electronic retina. As shown in **Fig. 10(d)**, after the processing, the edge information of the original color image is well preserved, proving that the proposed spike-based electronic retina can sense, encode, transmit and pre-process the visual information all in the spikes.

III. Conclusion

In this work, high-performance flexible VO$_2$ Mott memristors were fabricated. As a result, an artificial spiking thermoreceptor (AST) and an artificial spiking photoreceptor (ASP) are both realized by using a single flexible VO$_2$ Mott memristor. Finally, a spiking electronic retina was proposed by connecting an array of ASPs with multiple-input floating gate MOS transistors. Consequently, the edges of the color image are clearly extracted through the electronic retina by simulation, thus providing a solid foundation for spike-based neuromorphic sensing system.

Acknowledgment

This work was supported by the National Natural Science Foundation of China (Grant No. 62174130), the Strengthening Basic Disciplines Program (No. 2022-JCJQ-JJ-1108), the Key R&D plan of Shaanxi Province (Grant No. 2023-YBSF-407), and Xi'an Jiaotong University basic Scientific Research (Grant xzy012024148).

References

[1] K. Roy, A. Jaiswal, and P. Panda, "Towards spike-based machine intelligence with neuromorphic computing," *Nature*, vol. 575, no. 7784, pp. 607-617, 2019.

[2] Y. Kim, A. Chortos, W. Xu, Y. Liu, J. Y. Oh, D. Son, J. Kang, A. M. Foudeh, C. Zhu, Y. Lee, S. Niu, J. Liu, R. Pfattner, Z. Bao, and T. W. Lee, "A bioinspired flexible organic artificial afferent nerve," *Science*, vol. 360, no. 6392, pp. 998-1003, 2018.

[3] M. Rao, H. Tang, J. Wu, W. Song, M. Zhang, W. Yin, Y. Zhuo, F. Kiani, B. Chen, X. Jiang, H. Liu, H. Y. Chen, R. Midya, F. Ye, H. Jiang, Z. Wang, M. Wu, M. Hu, H. Wang, Q. Xia, N. Ge, J. Li, and J. J. Yang, "Thousands of conductance levels in memristors integrated on CMOS," *Nature*, vol. 615, no. 7954, pp. 823-829, 2023.

[4] Y. Zotterman, "Special senses: thermal receptors," *Annu. Rev. Physiol.*, vol. 15, pp. 357-72, 1953.

[5] E. Dodt and Y. Zotterman, "Mode of action of warm receptors," *Acta Physiol. Scand.*, vol. 26, no. 4, pp. 345-57, 1952.

[6] C. A. Y. Han, Z. R. Han, S. L. Fang, S. Q. Fan, J. Q. Yin, W. H. Liu, X. Li, S. Q. Yang, G. H. Zhang, X. L. Wang, and L. Geng, "Characterization and Modelling of Flexible VO$_x$ Mott Memristor for the Artificial Spiking Warm Receptor," *Adv. Mater. Interfaces*, vol. 9, no. 19, p. 2200394, 2022.

[7] J. K. Bowmaker, "Evolution of vertebrate visual pigments," *Vision Res*, vol. 48, no. 20, pp. 2022-2041, 2008.

[8] C. Y. Han, S. Zhao, S. L. Fang, W. Liu, W. M. Tang, P. T. Lai, C. Li, Y. X. Ma, J. Q. Song, X. Li, X. L. Wang, W. J. Ren, R. L. Wang, X. D. Huang, G. H. Zhang, and L. Geng, "A Flexible Artificial Spiking Photoreceptor Enabled by a Single VO$_2$ Mott Memristor for the Spike-Based Electronic Retina," *ACS Appl. Mater. Interfaces*, 2024.

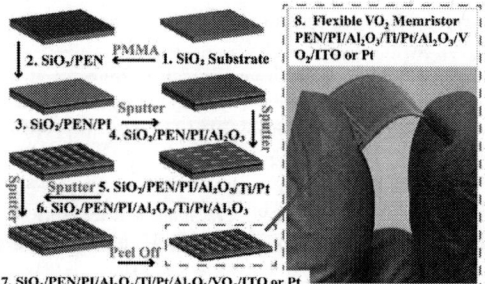

Fig. 1. Fabrication process of the flexible VO₂ Mott Memristors.

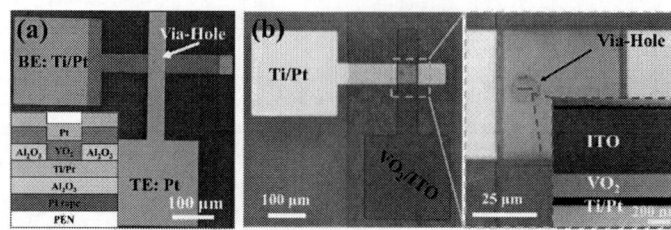

Fig. 2. Optical micrograph of the fabricated flexible VO₂ Mott memristor of AST (a) with Pt TE and ASP (b) with transparent ITO TE.

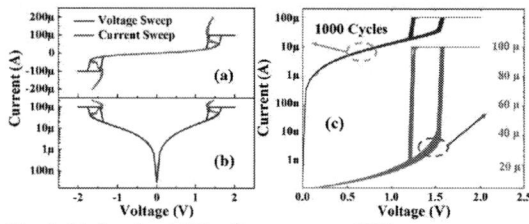

Fig. 3. Bi-directional DC voltage sweep and DC current sweeps of the VO₂ Mott memristor in linear scale (a) and in log scale (b). (c)1000 continuous DC positive voltage sweeps.

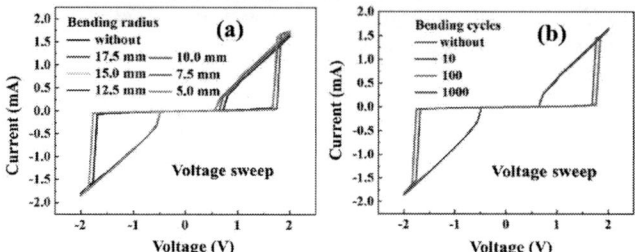

Fig. 4. (a) I–V characteristics of the flexible VO₂ Mott memristor after different bending radii and (b) bending cycles with the bending radius of 10.0 mm tested at room temperature. Inset of (a) shows the measurement platform.

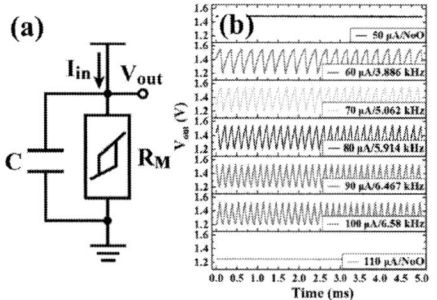

Fig. 5. (a) Oscillation circuit based on the VO₂ Mott memristor (R$_M$). (b) Output spiking rates of the oscillator under different I$_{in}$. C=10 nF is an external parallel capacitor.

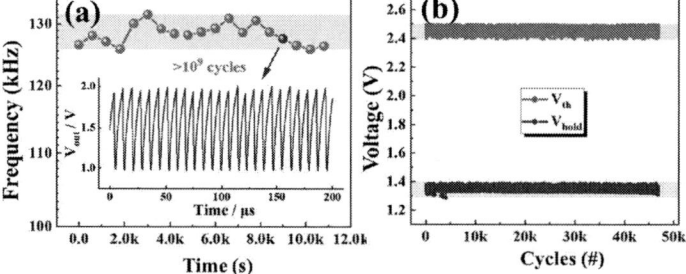

Fig. 6. Stability of oscillation spikes. And the device is switched between HRS and LRS ~×10⁹ times during the test (a). The distribution diagram of V$_{th}$ and V$_{hold}$ during the oscillation process (b).

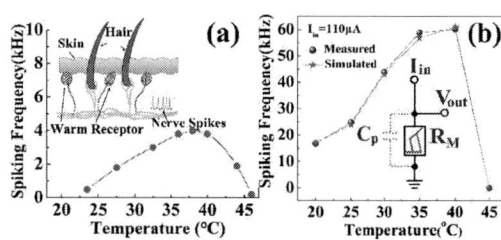

Fig.7. (a) Spike rate of warm receptor in response to the stimulus temperatures, (b) Relationship between the spiking frequency of the flexible AST and the stimulus temperatures.

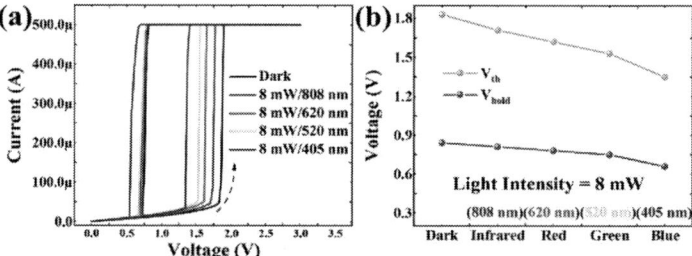

Fig. 8. (a) I-V curves of the ASP under the illumination of different wavelengths with the same light intensity. (b) Corresponding distributions of V$_{th}$ and V$_{hold}$.

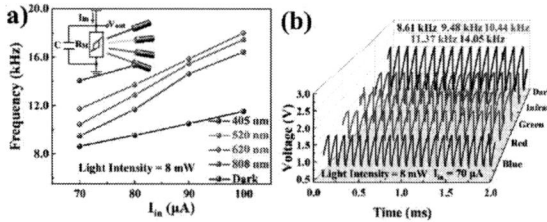

Fig. 9. (a) Relationship between the output spiking frequency of ASP and the input current (I$_{in}$) under light intensity of 8 mW with different wavelengths. (b) Output spikes of the artificial spiking photoreceptor under the intensity of 8 mW (with different wavelengths) and the input current of 70 μA.

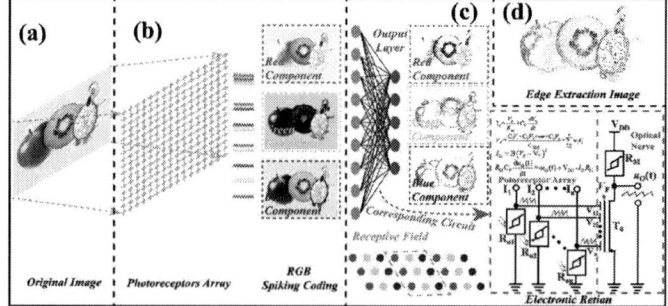

Fig. 10. Schematic illustration of color image segmentation and edge extraction. (a) Original color image. (b) Artificial spiking photoreceptors array for RGB spiking coding (segmentation). (c) Structure of the electronic retain for edge extraction. (d) Edge-extracted image after processing of electronic retain.

979-8-3315-0417-5/25 $31.00 © 2025 IEEE

Elemental Sn promotes the formation of single-crystal ε-Ga₂O₃ films in MOCVD

Long Wang[1,2,3], Yao Wang[1,2,3], Qian Feng[1,2,3], Yachao Zhang[1,2,3], Jincheng Zhang[1,2,3], Yue Hao[1,2,3]

[1]State Key Discipline Laboratory of Wide Band Gap Semiconductor Technology, School of Microelectronics, Xidian University, China, [2]Shaanxi Joint Key Laboratory of Graphene, School of Microelectronics, Xidian University, China, [3]State Key Laboratory of Wide Bandgap Semiconductor Devices and Integrated Technology, School of Microelectronics, Xidian University, China

Abstract

In MOCVD, there will be mixed phase in the preparation of ε-Ga₂O₃ using oxygen as oxygen source. The introduction of Sn element through a pulsed method can effectively resolve the issue of mixed phases in MOCVD with oxygen as the oxygen source. To obtain high-quality single-crystal ε-Ga₂O₃ thin films while increasing the growth rate.

Keywords: MOCVD, Pulsed Sn, ε-Ga₂O₃

Introduction

Ga₂O₃ demonstrates high potential for the development of high-temperature, high-power, radiation-resistant devices, as well as multifunctional optoelectronic devices such as solar-blind UV photodetectors and resonant filters [1]. Among the five isomers of Ga₂O₃ (α, β, γ, δ, ε), β-Ga₂O₃ is the most thermodynamically stable structure and has been most extensively studied, but ε-Ga₂O₃ has attracted considerable attention due to its unique properties. ε-Ga₂O₃ is the only crystal structure of Ga₂O₃ that exhibits spontaneous polarization, with a theoretical value of 23 μC/cm² [2]. This is approximately three times higher than the spontaneous polarization of AlN (8.1 μC/cm²) and eight times higher than that of GaN (2.9 μC/cm²)

To realize a two-dimensional electron gas, obtaining high-quality single-crystal ε-Ga₂O₃ epitaxial films is essential. In this regard, researchers have made efforts using techniques such as PLD, MBE, HVPE, and MOCVD. The advantage of MOCVD is that it can accurately control the growth rate and sample components, realize the preparation of high-purity samples and doping-controllable samples, easy realization of large-scale production, which is the main method to prepare ε-Ga₂O₃ epitaxial thin films at present. The choice of oxygen source is crucial for the preparation of single-crystal ε-Ga₂O₃ thin films using MOCVD. When researchers use oxygen or N₂O as oxygen source to prepare ε-Ga₂O₃, there is a challenge of ε and β phase coexistence, it impossible to obtain single-crystal ε-Ga₂O₃ thin films by merely adjusting the process parameters. Currently, high-quality single-crystal ε-Ga₂O₃ films can only be obtained when H₂O is used as the oxygen source [3]. High-purity oxygen is a simple, efficient and most commonly used oxygen source in the preparation of oxide semiconductors, so the goal of our research is to prepare a single-crystal ε-Ga₂O₃ thin film using high-purity oxygen as the oxygen source, aiming to solve the problem of mixing the two phases of β and ε.

In our experiments, we introduced the pulse Sn assisted growth ε-Ga₂O₃ in MOCVD, and the experimental results and corresponding analyses demonstrated that pulse Sn assisted growth successfully produce higher quality single-crystal ε-Ga₂O₃ thin films. It was also concluded that this approach not only promotes the formation of the ε-phase but also enhances the growth rate of ε-Ga₂O₃ thin films, and resulting in higher crystalline quality.

Experimentation

ε-Ga₂O₃ epitaxial films were prepared using independently developed MOCVD equipment, with sample A being without Sn assistance and sample B being a pulsed Sn-assisted sample. The Ga source used for sample preparation was Triethylgallium; The Sn source was TDMASn; N₂ (5N) was used as the carrier gas for both the Ga and Sn sources, while high-purity oxygen (5N) served as the oxygen source. The samples were epitaxially grown on 2-inch single-polished c-plane sapphire substrates, were sequentially ultrasonically cleaned in acetone, ethanol, and deionized water to remove surface impurities. Before sample preparation, the substrate were annealed in O₂ and N₂ atmospheres at 900 °C for 30 min. During sample preparation, the Ga source flow rate was maintained at 30 sccm, while the oxygen flow rate was controlled at 4000 sccm, the O/Ga ratio was kept consistent across all samples. The reaction temperature was set at 390 °C, with the pressure maintained at 40 Torr, and the growth time was 60 min. Figure 1 presents the different methods of Sn introduction for each of the two samples.

Results and discussion

The phase composition and crystallinity of the epitaxial films in different samples were analyzed using various modes of HRXRD, with the results shown in Figure 2. In figure2 (a), sample A shows a noticeable β and ε mixed phase, while sample B has a single-phase ε-Ga₂O₃, which match the standardized card PDF (04-027-0903) for ε-Ga₂O₃ [4]. Figure 2 (b) shows the HRXRD scanning results in ω mode, where a smaller FWHM value indicates higher crystalline quality of the film. The graph shows that the FWHM value for sample A is 0.87°, sample B is 0.53°. The results indicate that the addition of Sn improves the crystalline quality of the epitaxial films. The surface state of epitaxial film is an important factor to measure the quality of epitaxial films, and the surface flatness of the samples was observed by atomic force microscopy, the test range is 5 μm × 5 μm, as shown in figure 3. The test results were analyzed using NanoScope Analysis software, and the Root Mean Square (RMS) was calculated to be 2.76 nm, 1.77 nm for the two samples, respectively. Sample B exhibits the lowest RMS value. This suggests that the Sn-assisted preparation of ε-Ga$_2$O$_3$ effectively reduces the roughness of the epitaxial film and enhances its surface quality [5].

The role of Sn element in the formation of auxiliary ε-Ga$_2$O$_3$ epitaxial film can be clearly understood by analyzing the reaction involving Sn in the chamber and the associated changes in Gibbs free energy. The standard Gibbs free energy ΔG^0 [6] is calculated using the standard enthalpy and entropy of formation for the corresponding substance, with a negative value indicating a high probability of the reaction occurring.

$$Ga_2O + 2SnO_2 \leftrightarrow Ga_2O_3 + 2SnO$$

$$\Delta G^0 = -93.7 \ kJ/mol \qquad (1)$$

$$Ga_2O + 2SnO \leftrightarrow Ga_2O_3 + 2Sn$$

$$\Delta G^0 = -117.7 \ kJ/mol \qquad (2)$$

$$Ga_2O + SnO_2 \leftrightarrow Ga_2O_3 + Sn$$

$$\Delta G^0 = -105.7 \ kJ/mol \qquad (3)$$

The reaction process involving Sn in the chamber is shown in figure 4. During the formation of Ga$_2$O$_3$ thin films, the Ga source in the chamber bonds with the O source through Ga-O bonding. Simultaneously, partial Ga-O bond breakage occurs due to desorption, leading to the formation of the lower-valent oxide Ga$_2$O. When Sn atoms enter the chamber, they undergo oxidation, and the oxidized Sn can then oxidize the desorbed Ga$_2$O to Ga$_2$O$_3$, thereby promoting the growth of Ga$_2$O$_3$ thin films. After the reaction is completed the oxidized Sn is reduced to Sn atoms, which can be oxidized again to SnO and SnO$_2$, so the element Sn can be recycled to participate in the reaction. The introduction of Sn atoms inhibits the formation of desorbed Ga$_2$O, reduces the thermal decomposition process, and decreases the surface undulation, thereby improving the surface quality of the epitaxial film. This explains the reduced surface roughness observed in sample B during the AFM test.

To accurately analyze the phase composition in epitaxial films of different samples, the microstructures of sample A and sample B were examined using transmission electron microscopy. Figure 5 (a) shows a high resolution TEM image of sample A, revealing the presence of distinct mixed phases [7]. Figure 5 (a1) is an enlarged view of the atomic arrangement in I-region in figure 5 (a). The Fourier transform calculates the crystal plane spacing to be 0.464 nm, which corresponds to the (002) crystallographic orientation of ε-Ga$_2$O$_3$. Figure 5 (a2) is an enlarged view of the atomic

arrangement in II-region, which was calculated by Fourier transform to obtain a crystal plane spacing of 0.468 nm, corresponding to the (-201) crystallographic orientation of β-Ga$_2$O$_3$, which is marked with a green dashed line in figure 5 (a). Figure 5 (a3) shows the SAED image at the interface, where the positional relationship of the diffraction spots of the epitaxial film ε-Ga$_2$O$_3$ and α-Al$_2$O$_3$ substrate shows that the in-plane orientation relationship is [010] ε-Ga$_2$O$_3$∥[11-20] α-Al$_2$O$_3$, and the crystallographic orientations corresponding to the diffraction spots at different positions [8] have been labeled in the figure 5 (a3). Figure 5 (b) shows the high resolution TEM image of sample B, revealing a uniform atomic arrangement of the ε-Ga$_2$O$_3$ phase, with no evidence of other phases. Figure 5 (b1) and figure 5 (b2) are magnified images of regions I and II in figure 5 (b), respectively, both showing the regular atomic arrangement of ε-Ga$_2$O$_3$ phase. Figure 5 (b3) shows the SAED image of sample B, that the clear diffraction spots align well with the crystallographic orientation of ε-Ga$_2$O$_3$. This indicates that the formation of the β-phase can be effectively suppressed under pulsed Sn-assisted conditions. For this phenomenon is analyzed from the structural aspect of β-Ga$_2$O$_3$ and ε-Ga$_2$O$_3$. Figure 6 (a) and figure 6 (b) show the cell structure diagrams of β-Ga$_2$O$_3$ and ε-Ga$_2$O$_3$, respectively, with tetrahedral coordination for GaI (position of the blue ball in the Fig.6) and octahedral coordination for GaII (position of the blue ball in the Fig.6). In rutile SnO$_2$ crystals [9], both Sn and O atoms are bonded with octahedral coordination and the Sn-O bond length exceeds 2 Å. In gallium oxide, the Ga-O bond length in tetrahedral coordinated is less than 1.9 Å, while in octahedral coordinated, it exceeds 1.9 Å, so that Sn ions are more likely to occupy octahedral position.

Conclusion

In this paper, the ε-Ga$_2$O$_3$ film was prepared with sapphire as the substrate, and the Sn element was introduced by pulse, which resolved the issue of ε and β phases mixing in ε-Ga$_2$O$_3$ epitaxial film prepared by MOCVD with high-purity oxygen as the oxygen source, leading to the production of a high-quality single-crystal ε-Ga$_2$O$_3$ film. The role of Sn in the reaction is primarily to inhibit GaO desorption and promote the formation of ε-Ga$_2$O$_3$, while also being able to participate in the reaction cyclically. Sn ions mainly occupy the positions of Ga-O octahedral, which favors the formation of ε-Ga$_2$O$_3$ with more octahedral coordination numbers and contributes to the formation of high-quality single-crystal ε-Ga$_2$O$_3$ epitaxial films.

Acknowledgments

This work was supported by the National Natural Science Foundation of China (NSFC) under Grant No. U21A20503.

References

[1] CHEN Z M, LU X, TU Y J, et al. ε-Ga$_2$O$_3$: An Emerging Wide Bandgap

Piezoelectric Semiconductor for Application in Radio Frequency Resonators [J]. Advanced Science, 9(32):2203927 (2022).

[2] KANG H Y, YEOM M J, YANG J Y, et al. Epitaxial κ-Ga₂O₃/GaN heterostructure for high electron-mobility transistors [J]. Materials Today Physics, 2023, 31, 101002 (2022).

[3] DEMIR B, PETERSON R L. Improved heteroepitaxy of κ-Ga₂O₃ on c-plane sapphire by initial mist flow stabilization during mist chemical vapor deposition [J]. Thin Solid Films, 2024, 791, 140223 (2024).

[4] STEPANOV S I, NIKOLAEV V I, POLYAKOV A Y, et al. HVPE Growth and Characterization of Thick κ-Ga₂O₃ layers on GaN/Sapphire Templates [J]. Ecs Journal of Solid State Science and Technology, 12, 015002. (2023).

[5] MAZZOLINI P, VARLEY J B, PARISINI A, et al. Engineering shallow and deep level defects in κ-Ga₂O₃ thin films: comparing metal-organic vapour phase epitaxy to molecular beam epitaxy and the effect of annealing treatments [J]. Materials Today Physics, 45, 101463 (2024).

[6] KRACHT M, KARG A, SCHöRMANN J, et al. Tin-Assisted Synthesis of ε-Ga₂O₃ by Molecular Beam Epitaxy [J]. Physical Review Applied, 8, 054002 (2017).

[7] TANG W B, MA Y J, ZHANG X D, et al. High-quality (001) β-Ga₂O₃ homoepitaxial growth by metalorganic chemical vapor deposition enabled by in situ indium surfactant [J]. Applied Physics Letters, 120, 212103 (2022).

[8] ZHANG Y J, GONG Y Q, CHEN X H, et al. Unlocking the Single-Domain Heteroepitaxy of Orthorhombic κ-Ga₂O₃ via Phase Engineering [J]. Acs Applied Electronic Materials, 4: 461-468 (2022).

[9] CORA I, MEZZADRI F, BOSCHI F, et al. The real structure of ε-Ga₂O₃ and its relation to κ-phase [J]. Crystengcomm, 19, 1509-1516 (2017).

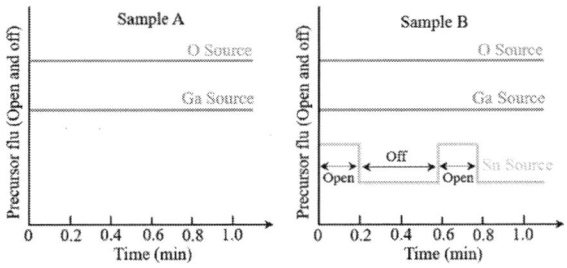

Fig. 1 Different introduction method of Sn elements in sample A, sample

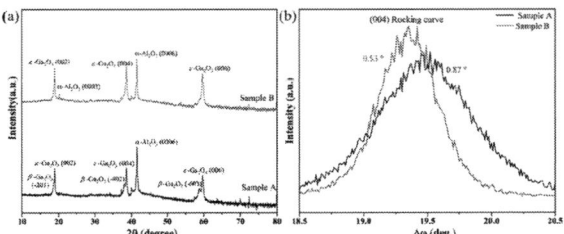

Fig. 2 HRXRD data in different scanning modes: (a) diffraction peaks of the samples in 2θ-ω scanning mode; (b) rocking curves of the (004) crystal planes for different samples.

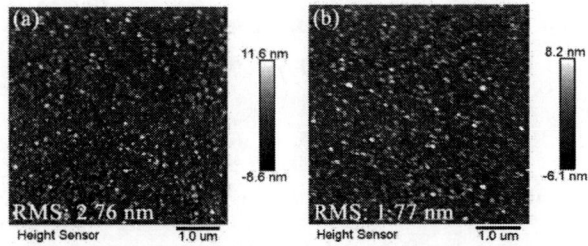

Fig. 3. AFM test results for different samples. (a) sample A; (b) sample B.

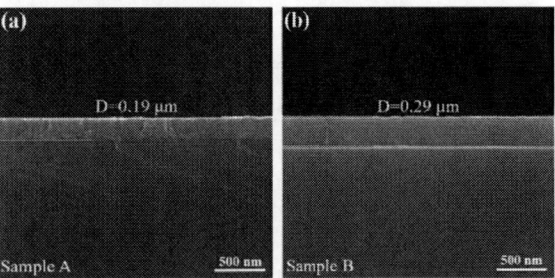

Fig. 4. Epitaxial film thicknesses of different samples grown for 60 min tested using SEM. (a) sample A; (b) sample B.

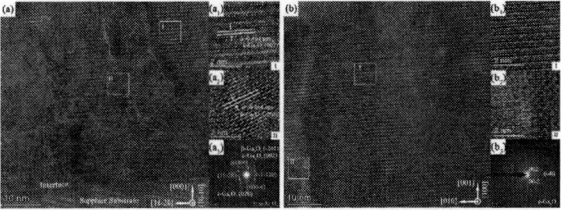

Fig. 5. (a) high-resolution TEM image of sample A; (a1) enlarged image of area I in sample A; (a2) enlarged image of area II in sample A; (a3) SAED image of sample A; (b) high-resolution TEM image of sample B; (b1) enlarged image of area I in sample B; (b2) enlarged image of area II in sample B; (b3) SAED image of sample B.

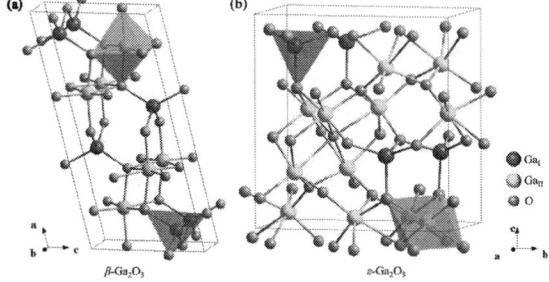

Fig. 6. (a) Schematic structure of the cell of β-Ga₂O₃; (b) Schematic structure of the cell of ε-Ga₂O₃.

Two-dimensional ReSe₂ based Optoelectronic Synaptic Transistor

Wei Zeng[1], JiYu Zhao[1], Hang Li[1], Guanglong Ding,[2*] Ye Zhou,[1,2*] Su-Ting Han[3*]

[1]Institute for Advanced Study, Shenzhen University, Shenzhen 518060, P. R. China.

[2]State Key Laboratory of Radio Frequency Heterogeneous Integration, Shenzhen University, Shenzhen 518060, PR China.

[3]Department of Applied Biology and Chemical Technology, The Hong Kong Polytechnic University, Hung Hom, Hong Kong SAR, P. R. China.

Abstract

The development of high-performance artificial synapses for neuromorphic computing offers a potential solution to overcome the von Neumann bottleneck. However, most artificial synapses rely on electrical manipulation, limiting their applications. In this study, we have created an optoelectronic synaptic transistor based on two-dimensional (2D) ReSe₂ that can simulate excitatory synaptic behaviors under light stimulations. Our proposed optoelectronic controlled synaptic device using 2D ReSe₂ provides a promising platform for versatile neuromorphic applications.

Keywords: Optoelectronic synaptic transistor, and 2D ReSe₂

Introduction

A crucial approach to achieving in-memory computing and overcoming von Neumann's bottleneck is by developing high-performance artificial synapses for neuromorphic computing systems. By considering the gate electrode and channel as pre- and post-synaptic neurons, respectively, transistor-type devices such as organic field-effect transistors (OFET) [1], electrolyte-gated transistors [2], floating gate transistors [3], and ferroelectric transistors [4] can be employed to emulate biological synaptic behaviors. Due to their distinctive structural features with separate control and read terminals, these transistors enable simultaneous programming and reading operations, thereby facilitating the realization of intricate neuromorphic functions. However, most of these transistors rely on electrical stimulation for synaptic emulation, resulting in lower bandwidth and significant energy consumption. The optoelectronic synaptic transistors, based on the photoelectric effect, can directly convert optical signals into electrical signals, attracting widespread attention due to their high efficiency, excellent sensory property, low power consumption, and large bandwidth. Yang et al. present a synapse transistor based on α-In₂Se₃, which exhibits controllable temporal dynamics under both electrical and optical stimuli, enabling multimodal and multiscale signal processing. [5]. Pan et al. proposed an ultra-low power consumption artificial photonic synapse based on a BP/CdS heterojunction structure. [6]. Therefore, the development of optoelectronic controlled artificial synapses holds significant implications for advancing neuromorphic computing and enhancing bionic intelligent perception.

The superior optical and electrical properties of two-dimensional (2D) transition metal sulfide (TMDC) materials with atomically thin have garnered significant attention in the field of optoelectronic devices [7]. Among them, ReSe₂ exhibits numerous exceptional properties, including layer-dependent electrical and optoelectronic responses, broad spectral sensitivity, and polarization-selective light reactions, which have found extensive applications in photocatalysis and polarization photodetector [8-10].

In this study, we present a synaptic transistor based on ReSe₂ that exhibits intrinsic bipolar characteristics. By harnessing the broad spectral response of ReSe₂, we successfully simulate artificial synapses modulated by optical means. The conductivity of the device can be reversibly modulated under irradiation with optical pulses at 445, 520, 637, and 808 nm. Furthermore, by adjusting the width of the light pulse, the device demonstrates synaptic behavior ranging from short-term potentiation (STP) to long-term potentiation (LTP). This work provides a viable reference for developing optoelectronic controlled synaptic transistors.

Experimental Details

Device fabrication: ReSe₂ crystal were mechanically exfoliated with blue thin film tape and transferred onto a 300 nm silicon dioxide substrate. The source and drain electrodes were prepared by patterned, metal thermal evaporation (thickness Au: 45 nm) and lift-off processes. The device has a channel distance of ~2 μm.

Electrical measurements: The electrical and synaptic properties of the prepared device were tested using Keysight B2902A semiconductor analyzer. 445, 520, 637 and 808 nm fiber lasers were used to test the photoelectric response of the device, and controlled by a signal generator (Gwenstel-AFG-2225).

Results and Discussion

Fig. 1(a) shows the structure of the synapse transistor. The device utilizes a bottom gate configuration, where the dielectric layer is a 300 nm SiO₂ layer and the gate is a highly p-type doped Si layer. We selected ReSe₂ as the semiconductor layer due to its excellent photoelectric properties and air stability. Meanwhile, gold was employed as the contact electrodes. Fig. 1(b) shows the atomic force microscopy (AFM) image of the device, where the thickness of the ReSe$_2$ and gold electrode is determined to be 3 and 45 nm. Few-layered 2D materials exhibit superior carrier transport properties owing to their atomically thin thickness. Fig. 1(c) and 1(d) shows transfer characteristic curves at different V$_{ds}$ ranging from 1 to 10 V and -1 to -10 V, indicating the bipolar feature with large on/off ratio of 10^4 of the 2D ReSe$_2$-based synaptic transistor. As the source drain bias changes, the threshold voltage values in the device transfer curves shift significantly, which may be attributed to the carrier mobility variations induced by depletion region narrows and the Joule heating effect at different source drain bias.

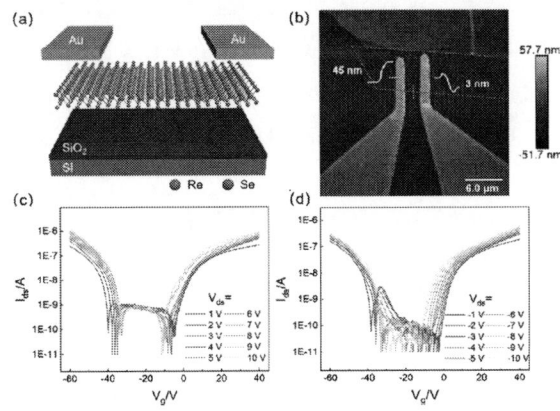

Fig. 1 (a) The device structure illustration of the 2D ReSe$_2$ based synaptic transistor. (b) AFM image of 2D ReSe$_2$ based synaptic transistor; the thickness of Au electrodes and ReSe$_2$ nanosheets are 45 and 3 nm, respectively. Scale bar: 6 μm. (c-d) Transfer curves of ReSe$_2$ based synaptic transistors under different V$_{ds}$.

The stimulation of the pre-synaptic neuron in the biological nervous system can elicit the release of neurotransmitters and induce an electro-responsive reaction in the post-synaptic neuron, thereby generating a nervous impulse. Due to the exceptional photoelectronic properties of 2D ReSe$_2$, the 2D ReSe$_2$ based synaptic transistor can mimic biological synapses by utilizing optical stimulation and device channel as pre- and post-synaptic neurons, respectively. Device channel conductance can be modulated by the optical pulses with different wavelengths and width. As shown in Fig. 2(c), under the stimulation of light pulse (V$_{ds}$ = 5 V, wavelength: 445 nm), the response current (i.e. post-synaptic current, PSC) demonstrates a gradual increase 33 to 45 nA as the illumination time extends from 50 to 1000 ms. Simultaneously, the similar excitatory PSC (EPSC) can be observed under the light with different wavelength from 445 to 808 nm [Fig. 2(b) to 2(d)]. Additionally, the prolonged exposure to light can induce significant photogenerated

carrier trapping at the interface between ReSe$_2$ and SiO$_2$, resulting in a transition from STP to LTP. The modulation effect of light on PSC demonstrates the potential for the application of 2D ReSe$_2$-based optoelectronic synaptic transistors in neuromorphic computing.

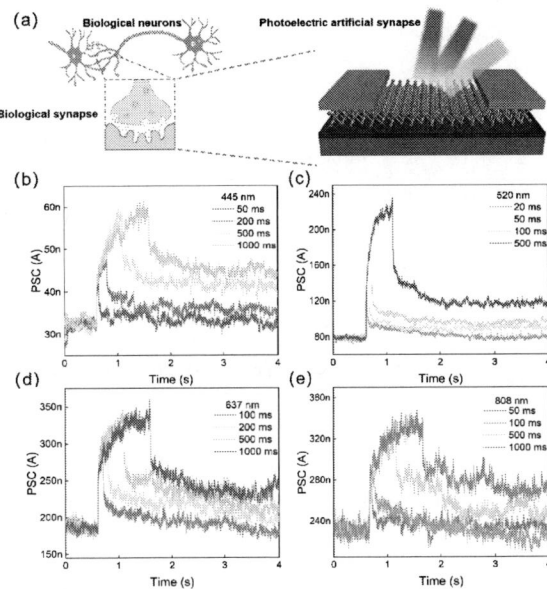

Fig. 2 (a) Schematic diagram of the signal transmission process in biological neurons and the optoelectronic synaptic transistor based on 2D ReSe$_2$. (b-e) The PSC responses of the ReSe$_2$ based synaptic transistor under the light stimulation with different pulse widths and wavelengths [(b) 445 nm, (c) 520 nm, (d) 637 nm and (e) 808 nm].

Conclusion

In summary, we have developed an optoelectronic synaptic transistor based on 2D ReSe$_2$ by leveraging its exceptional optoelectronic properties. The application of light illumination induces adjustable positive current responses and decay behavior in the ReSe$_2$-based synaptic transistor, enabling simulation of the EPSC function observed in biological synapses. By controlling the width of optical pulses, our fabricated synaptic transistor can transition from STP to LTP. Notably, due to the broad spectral response of ReSe$_2$, wide PSC responses can be achieved under optical pulses with wavelengths of 445 nm, 520 nm, 637 nm, and 808 nm. This work demonstrates promising prospects for developing artificial vision systems for future applications.

Acknowledgments

The authors acknowledge grants from the National Natural Science Foundation of China (Grant No. 62304137, 62122055, 62074104), Guangdong Basic and Applied Basic Research Foundation (Grant No. 2023A1515012479, 2024B1515040002), the Science and Technology Innovation Commission of Shenzhen (Grant No. JCYJ20220818100206013), the Project on Frontier and Interdisciplinary Research Assessment, Academic Divisions of the Chinese Academy of Sciences (Grant No. XK2023XXA002), and NTUT-SZU Joint Research Program.

References

[1] X. Chen, E. L. Li, X. H. Zhang, Q. Z. Chen, R. J. Yu, Y. Ye, H. P. Chen, T. L. Guo, "Printed Organic Synaptic Transistor Array for One-to-Many Neural Response", *IEEE Electron Device Lett.*, 43, pp. 394-397 (2022).

[2] X. X. Gao, J. Yin, J. Zhu, J. J. Chang, J. C. Zhang, Y. Hao, "Electrolyte-Gated Flexible MoS_2 Synaptic Transistors with Short-Term Plasticity", *IEEE Electron Device Lett.*, 45, pp. 605-608 (2024).

[3] G. C. Zhang, C. Ma, X. M. Wu, X. H. Zhang, C. S. Gao, H. P. Chen, T. L. Guo, "Transparent Organic Nonvolatile Memory and Volatile Synaptic Transistors Based on Floating Gate Structure", *IEEE Electron Device Lett.*, 43, pp. 733-736 (2022).

[4] T. Q. Lu, R. R. Liang, R. T. Zhao, Y. Yang, T. L. Ren, "Fabrication and Characterization of Ferroelectric HfZrO-based Synaptic Transistors with Multi-state Plasticity," in *2020 4th IEEE Electron Devices Technology & Manufacturing Conference* (EDTM) (2020).

[5] K. Q. Liu, T. Zhang, L. Bao, L. Y. Xu, C. D. Cheng, Z. Yuan, R. Huang, Y. C. Yang, "An optoelectronic synapse based on α-In_2Se_3 with controllable temporal dynamics for multimode and multiscale reservoir computing", *Nat. Electron*, 5, pp. 761-773 (2022).

[6] C. G. Zhu, H. W. Liu, W. Q. Wang, L. Xiang, J. Jiang, Q. Shuai, X. Yang, T. Zhang, B. Y. Zheng, H. Wang, D. Li, A. L. Lian, "Optical synaptic devices with ultra-low power consumption for neuromorphic computing", *Light-Sci. Appl.*, 11, pp. 3008-3017 (2022).

[7] S. F. Zeng, C. S. Liu, P. Zhou, "Transistor engineering based on 2D materials in the post-silicon era" *Nat. Rev. Electr. Eng.*, 1, pp. 335-348 (2024).

[8] M. Hafeez, L. Gan, H. Q. Li, Y. Ma, T. Y. Zhai, "Chemical Vapor Deposition Synthesis of Ultrathin Hexagonal $ReSe_2$ Flakes for Anisotropic Raman Property and Optoelectronic Application", *Adv. Mater.*, 28, pp. 8296-8301 (2016).

[9] C. Y. Liu, T. Zheng, K. X. Shu, S. Shu, Z. B. Lan, M. M. Yang, Z. Q. Zheng, N. J. Huo, W. Gao, J. B. Li, "Polarization-Sensitive Self-Powered Schottky Photodetector with High Photovoltaic Performance Induced by Geometry-Asymmetric Contacts" *ACS Appl. Mater. Interfaces*, 16, pp. 13914-13926 (2024).

[10] J. R. Ran, L. Chen, D. Y. Wang, A. Talebian-Kiakalaieh, Y. Jiao, M. A. Hamza, Y. Qu, L. Q Jing, K. Devay, S. Z. Qiu, "Atomic-Level Regulated 2D $ReSe_2$: A Universal Platform Boostin Photocatalysis", *Adv. Mater.*, 35, 2210164 (2023).

Capacitive Length-extension Mode Resonators with Stress-induced Gap-closing Electrodes for Motional Resistance Reduction

Hao Yu[1,2], Yechen Miao[1,2], Fang Wang[1,2], Ke Sun[1], Yi Sun[1], Tiger H. Tao[1,2], Heng Yang[1,2]

[1]Shanghai Institute of Microsystem and Information Technology, CAS, CHINA

[2]University of Chinese Academy of Sciences, Beijing, CHINA

Abstract

This work presents a capacitive length-extension mode resonator featuring novel gap-closing electrodes, which apply nonuniform stress to the structure of the movable electrodes to reduce the capacitive transduction gaps to sub-micrometer-level values after the device is released. This approach overcomes the limitations of deep etching. Sub-100 nm capacitive transduction gaps are obtained, and the equivalent impedance of the resonator is significantly reduced.

Keywords: capacitive silicon resonator; gap-closing electrode; motional resistance

Introduction

One significant challenge encountered by capacitive silicon resonators in oscillators is their large motional resistance R_m. This high R_m leads to elevated insertion losses, making it challenging for these resonators to satisfy oscillation requirements. Gap reduction considered as one of the best methods because the motional resistance is proportional to the fourth order of the gap width [1].

Numerous research studies have been conducted to reduce the width of the gap. The simplest approach to fabricating sub-micron capacitance gaps is through direct etching using deep reactive ion etching (DRIE) processes. However, because the aspect ratio of the deep-etching equipment is limited, the minimum capacitance gaps are determined by the thickness of the structures. Moreover, the process variation of DRIE is about $10 - 10^2$ nm level. The fabrication methods for a resonator with a small capacitive gap using a thin oxide film as a sacrificial layer have been presented in [2]. However, this approach is slightly complex and a high temperature process is required. Compared with the HARPSS process, alternative techniques based on novel gap-closing mechanisms offer a simpler approach [4, 5]. In this method, electrostatic forces are employed to pull the movable electrode toward the resonator in response to application of a bias voltage across the ends of the transduction gap to achieve a submicron gap. The advantage of this approach lies in the only requirement of additional bias voltage application to the electrodes without the need for complex fabrication steps. However, in this method, the movable electrode is likely to be pulled into unexpected positions under electrostatic forces, leading to structural failure.

In this paper, a novel gap-closing structure for a length-extensional mode resonator with a narrow gap using direct etching is proposed. By etching and refilling the high-stress polysilicon in a movable electrode, the electrode is driven toward the stopper electrode after the device is released because of the compressive stress of the polysilicon, reducing the capacitive transduction gaps to sub-micron level values. This approach does not require the application of a high bias voltage and any locking device. Resonators with 100-nm transduction gaps were fabricated using the proposed method. Furthermore, because the capacitance gap is determined by the difference between the initial and stopping gaps, the errors caused by process variations are compensated via a first-order correction.

Sensor Design and Simulation

A. Working principle

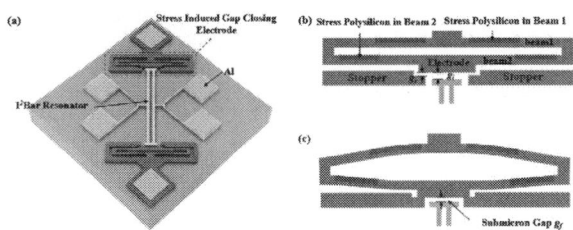

Fig. 1. Schematic and gap-closure mechanism principle of the resonator with stress-induced gap-closing electrodes. (a) Schematic of the device design. (b) Close-up schematic of one of the electrodes with insets illustrating the folded beam and stopper before HF etching. (c) Schematic illustrating stress inducing a contact point between the movable electrode and stopper after HF etching.

Fig. 1(a) shows a three-dimensional schematic of the device with stress-induced gap-closing electrodes. In this study, the I²BAR structure was selected as the vibrating unit of the resonator. The stress-induced gap-closing electrodes are symmetrically distributed at both ends of the resonator and are separated from the electrodes by a capacitive gap. Fig. 1(b) shows the non-closed state of the stress-induced gap-closing electrode, the stress-induced gap-closing electrode is supported via folded beams. Stopper structures are added and are initially separated for the electrode by a certain gap (g_s). The electrode and the resonator are separated by an initial capacitive gap (g_i). In the red region of the folded beam, polysilicon was selectively etched and refilled, generating a

979-8-3315-0417-5/25 $31.00 © 2025 IEEE

stress of approximately −270 MPa. Fig.1(c) shows the closed state of the stress-induced gap-closing electrode. Since the polysilicon layer is compressively stressed, the electrode deforms after release. When the deformation exceeds the gap (g_s) between the electrode and stopper, the electrode is in direct contact with the stoppers, and no locking device is required. In this case, the final capacitive transduction gap (g_f) between the electrode and the resonator is determined by the following equation:

$$g_f = g_i - g_s \qquad (1)$$

This final gap (g_f) is determined by the difference between the initial and stopping gaps, the errors caused by process variations are compensated via a first-order correction.

B. Simulation of the stress-induced gap-closing electrodes

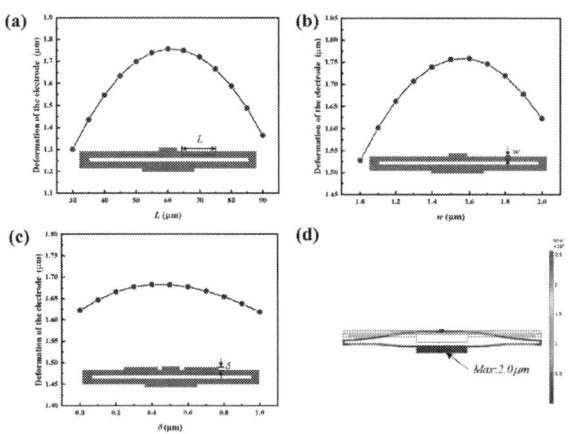

Fig. 2. Simulation results for design parameters when filling folded beam with polysilicon film. Simulation results for design parameters when filling folded beam with polysilicon film. (a) Relationship between electrode displacement and deformed polysilicon length. (b) Relationship between electrode deformation and deformed polysilicon width. (c) Relationship between the electrode deformation and misalignment.

Fig. 2 shows the results of simulating the influence of the length and width of the polysilicon film on the displacement of the electrode driven by compressive stress in the polysilicon. Fig. 2(a) shows that, when the length of the polysilicon film is approximately half the length of the beam, the electrode exhibits maximum deformation. Similarly, the deformation reaches its maximum when the width of the polysilicon is approximately half the width of the folded beam, as shown in fig. 2(b). The polysilicon film and beam were fabricated using a two-step photolithography process, which inevitably introduces misalignment errors. However, the simulation results, shown in fig. 2(c), show that the deformation is insensitive to the misalignment when it is in the range of [0, 1] μm and the widths of the stressed polysilicon and the beam are 2 μm and 3 μm, respectively. To

increase the deformation, we applied the same design to fill a short section of the folded beam with polysilicon. Fig. 2(d) indicates that the simulated deformation of the electrode is 2.0 μm with the parameters used in this paper are designed.

Sensor Fabrication and Experimental Results

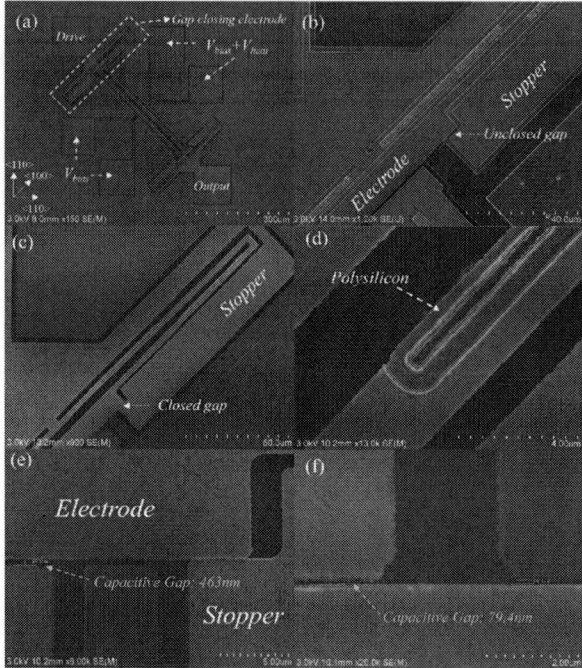

Fig. 3. SEM images of the resonator with stress-induced gap-closing electrodes. (a) Full-view image of the resonator. (b) Original unclosed state of the stress-induced gap-closing electrode before HF etching, and (c) closed state of the electrode after HF etching. (d) The high-stress polysilicon filled on the folding beam. (e) Final fabricated gap for the resonator with a 500-nm designed gap, and (f) final fabricated gap for the resonator with a 300-nm designed gap.

The resonators are fabricated on the silicon-on-insulator (SOI) wafers whose SOI layers were 6 μm thick. Fig. 3(a) presents an overview of the fabricated resonator by using scanning electron microscopy (SEM). Fig. 3(b) shows the device with the original unclosed gap before HF vapor etching. After the device is released, the electrode moves to come into contact with the stopper. It can be observed that the electrode tightly adheres to the stopper, reducing the capacitive gap from the original 1.5 μm to a final value of 463 nm, as shown in fig. 3(c) and (e). Notably, the device designed with a 300-nm gap resulted in an actual gap of approximately 80 nm, as shown in fig 3(f). This can be attributed to nonideal effects during the photolithography and etching processes. Fig. 3(d) illustrates the morphology of the filled polysilicon, showing that, owing to the inward growth starting from the sidewalls, the polysilicon leaves a partial depression in the middle.

The resonators were driven into resonance by electrostatic forces. An AC signal generated by a network analyzer was applied to the drive electrode, and a DC bias voltage (V_{bias}) was applied to one end of the resonator beam to generate a sufficiently large electrostatic force for resonator resonance. The output current signal was then passed through a transimpedance amplifier (*OPA656*) and connected to the input terminal of the network analyzer to measure the resonator output frequency signals.

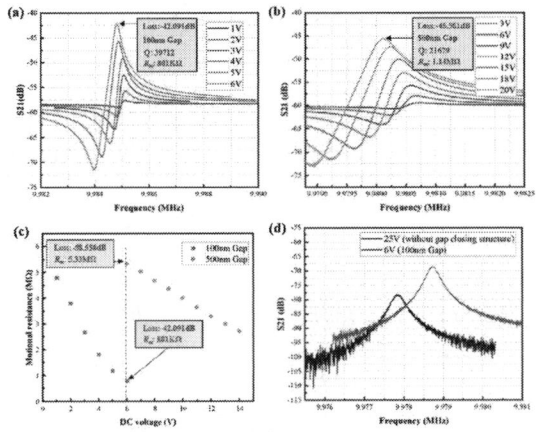

Fig. 4. Measurement of the resonators with 100nm capacitance gap and 500nm capacitance gap. (a) and (b) the resonance characteristics of the device with the 100-nm and 500-nm final capacitance gap, respectively. (c) Comparison of the motional resistance at 25°C for devices with two different capacitor gaps. (d) the resonance characteristics for a device without a stress-induced gap-closing structure.

Fig. 4 shows the resonators experimental results. The resonators with the stress-induced gap-closing electrodes were vacuum-sealed in ceramic tubes for testing. Fig. 4(a) illustrate the output characteristics of resonators with capacitor gaps of 100 nm at 25°C. As V_{bias} increased, the driving force and vibration amplitude of the resonant beam increased, leading to continuous improvement in the insertion loss. At a bias voltage of 6 V, the insertion loss was measured to be −42.091 dB, corresponding to a motional resistance of 801 KΩ. As shown in fig. 4(b), for the resonators with the 500 nm capacitor gap, a wider range of bias voltage can be applied. When the bias voltage is 20V at 25°C, the insertion loss is -45.561dB, corresponding to a motional resistance of 1.14 MΩ. Fig. 4(c) illustrates a comparison of the insertion loss at 25°C for devices with two different capacitor gaps (100 nm and 500 nm). At a 6V bias voltage, the motional resistance of the device with a 500-nm gap was 5.33 MΩ, which is six times higher than the motional resistance of the device with a 100-nm gap. For comparison, fig. 4(d) shows the test results for a device without a stress-induced gap-closing structure, which was fabricated on the same silicon wafer. Because of its high motional resistance, the capacitive detection method was found to be inapplicable.

Instead, the piezoresistive detection method was employed to measure the insertion loss of the device at a bias voltage of 25 V and heating voltage of 0.9 V, resulting in a value of −78.051 dB. Additionally, the resistive detection method was employed to measure the insertion loss of devices with a 100-nm capacitor gap at the same heating voltage, resulting in an insertion loss of −68.588 dB at a bias voltage of 6 V. We observed significant reductions in the motional resistance and bias voltage associated with the resonator with the stress-induced gap-closing electrode structure relative to the resonator without the gap-closing structure. Moreover, the quality factor (Q) of the two device types were similar, indicating that the gap-closing structure did not affect the Q value of the resonator.

Conclusion

This study presented a novel and efficient stress-induced gap-closing structure which does not require the application of a high bias voltage and any locking device for a capacitive silicon resonator to reduce the capacitive transduction gaps to sub-micrometer-level values. Two types of resonators with capacitor gaps of 100 nm and 500 nm were fabricated by using this mechanism. Upon application of a 20-V bias voltage, the resonator with a 500-nm gap exhibited an insertion loss of −45.561 dB and a motional resistance of 1.14 MΩ. The device with a 100-nm capacitor gap exhibited an insertion loss of −42 dB at a bias voltage of 6 V, which corresponds to a motional resistance of 801 KΩ, which is more than six times lower than that of the resonator with a 500-nm gap at the same voltage. The significant reductions in the resonator insertion loss, motional resistance, and operating voltage as well as no negative impact of the proposed mechanism on the achievable Q factor demonstrates the significant potential of the stress-induced gap-closing structure as a tool to reduce the motional resistance at low bias voltages.

Acknowledgments

This work was partially supported by National Science and Technology Major Project from the Minister of Science and Technology of China (grant nos. 2018AAA0103100)

References

[1] T. Mattila et al., 'A 12 MHz micromechanical bulk acoustic mode oscillator', Sensors and Actuators A: Physical, vol. 101, pp. 1–9, 2002.

[2] S. Pourkamali, A. Hashimura, R. Abdolvand, G. K. Ho, A. Erbil, and F. Ayazi, 'High-Q single crystal silicon HARPSS capacitive beam resonators with self-aligned sub-100-nm transduction gaps', J. Microelectromech. Syst., vol. 12, pp. 487–496, 2003.

[3] E. J. Ng et al., 'Stable pull-in electrodes for narrow gap actuation', in 2014 IEEE 27th International Conference on Micro Electro Mechanical Systems (MEMS), 2014, pp. 1281–1284.

[4] N. V. Toan, M. Toda, Y. Kawai, and T. Ono, 'A capacitive silicon resonator with a movable electrode structure for gap width reduction', J. Micromech. Microeng., vol. 24, p. 025006, 2014

Thermal Stability of TiO$_2$ Channel FE-VNAND: From Fabrication to High-Temperature Operation

Xujin Song[1,2], Dijiang Sun[1,2], Xiaoyan Liu[1,2] and Jinfeng Kang*[1,2]

[1]School of Integrated Circuits, Peking University, 100871, Beijing, China
[2]Beijing Advanced Innovation Center for Integrated Circuits, Beijing 100871, China

Abstract

In this work, we investigated the thermal stability of TiO$_2$ channel ferroelectric vertical NAND (FE-VNAND) devices during fabrication and high-temperature operation. Excellent thermally stable properties are demonstrated as follows: (1) Under various annealing temperature (550°C-650°C) and times, the fabricated FE-VNAND exhibit stable device properties with threshold voltage (V_{th}) variations less than 180mV and an on-off ratio over 10^6; (2) At high operation temperature up to 475K, the devices maintained excellent performance and reliability, with a 1.7V memory window (MW) and endurance over 10^8 cycles. The observed excellent thermal stability could be attributed to the crystalline nature of TiO$_2$ and its compatibility with HfO$_2$-based FE layers. (Keywords: NAND, TiO$_2$, Thermal stability)

Introduction

The continuous scaling of VNAND technology, involving more stacked layers and reduced vertical pitch, has driven higher bit density and lower costs [1]. As VNAND approaches its scaling limits [2], hafnium-based FE-VNAND has emerged as a promising candidate due to suitable polarization, coercive field, and highly scalable thickness [3]. Due to the interface layer (IL) induced reliability issues such as endurance degradation and read-after-write delay in Si channel FE-VNAND [4], IL-free oxide semiconductor (OS) channels have been introduced, significantly improving the reliability and performance of FE-VNAND devices [5], [6]. However, OS channels face significant thermal stability challenges during both fabrication and operation (Fig.1). High-temperature annealing required for the FE layer [7] and uneven temperature distribution within the stacked layers during annealing [8] can cause variations in device properties. During operation, pronounced self-heating effects can lead to elevated temperature [9], [10], potentially causing the OS channel degradation [11].

Recent studies have demonstrated that TiO$_2$ can withstand high-temperature annealing and is compatible with vertical channel integration [7], while the robustness to process condition during fabrication and thermal stability during operation are yet to be explored. In this work, we present a high-thermal-stability TiO$_2$ channel FE-VNAND device. We investigate the thermal stability during fabrication by varying the annealing temperature and number of cycles. Additionally, we evaluate the device's performance and reliability under high-temperature operating conditions through electrical measurements from room temperature up to 475 K.

Experiments

The fabrication process and schematic of the TiO$_2$ channel FE-VNAND are shown in Fig.2(a) and (b). HfLaO and TiO$_2$ layers were sequentially deposited using thermal atomic layer deposition (ALD). The first rapid thermal annealing step (RTA 1) was performed in vacuum at 550°C/600°C/650°C to simulate uneven temperature distribution during annealing. To further evaluate the fabrication thermal stability, the second and third annealing steps (RTA 2 and RTA 3) were conducted in N$_2$ atmosphere at 600°C on the devices previously annealed at 600°C in RTA 1. TEM images of the entire stack and a focused image of the vertical channel are shown in Fig.2(c). Electrical properties were measured from the bottom word line (WL), with the upper WL floated.

Results and Discussion

A. Fabrication thermal stability

The thermal stability of TiO$_2$ channel FE-VNAND is evaluated through their transfer characteristics and polarization properties. Fig.3(a) shows the I_D-V_G curves of devices under different annealing temperatures during RTA 1, exhibiting a distinct FE-induced MW, an on-off current ratio over 10^6, and stable V_{th} with variations less than 70 mV. For different RTA steps, as shown in Fig.3(b), the devices after RTA 2 and RTA 3 exhibit improved subthreshold swing (SS) due to reduced interface defects, while V_{th} remains relatively stable ($\Delta V_{th} < 180$mV). Polarization characterization using 3.5V, 10KHz triangle waves is shown in Fig.4. Correlated with the transfer curves, the polarization switching of devices under different temperature in RTA 1 remains stable. After RTA 3, a shift in the coercive voltage can also be attributed to the reduction of defects, where trapped charges may influence the switching dynamics of the HfLaO layer.

B. High-temperature operation stability

The operational thermal stability of the TiO$_2$ channel FE-VNAND was evaluated through device performance and reliability. Devices after RTA 3 were tested at various temperatures ranging from room temperature to 475 K (203°C), with the top WL fixed at 2V. The device exhibited stable on-off ratio, with an enlarged MW at high temperatures up to 475K (Fig.5). As the temperature increased, the MW expanded from 1.1V to 1.7V (Fig.6). With the bottom WL voltage V_G fixed at -0.5 V for non-destructive reading, the read current (I_{read}) showed an increased ratio between the program (PRG) and erase (ERS) state currents, reaching 10^4. The reliability of the TiO_2 channel FE-VNAND at high temperatures was also evaluated. Fig.8 shows the endurance test results under 3.5V, 1 MHz bipolar cycling. The device demonstrates endurance over 10^8 cycles at both 300K and 475K. Fig.9 illustrates the retention characteristics, showing that the device retains acceptable non-volatile characteristics after 1000 seconds. Despite faster degradation at high temperatures, the enlarged MW ensures that the device still exhibits better performance after endurance and retention tests at 475K.

To analyze the enhanced high-temperature performance, we extracted the field-effect mobility (μ_{FE}) and measured capacitance-voltage (CV) curves. As shown in Fig.10, the μ_{FE} exhibits minimal temperature dependence, indicating that the dominant transport mechanism in the TiO_2 channel is band-like transport rather than hopping, which typically shows strong exponential temperature dependence [12]. Fig.11 reveals that the capacitance in the accumulation region remains stable, suggesting that the carrier concentration in the TiO_2 channel is also stable. These results indicate that, despite the polycrystalline nature of the TiO_2 channel, the well-crystallized anatase phase, combined with a scaled gate length, supports band-like transport in the fabricated TiO_2 channel FE-VNAND. The combination of stable carrier concentration and trapped charge at the interface due to rising temperatures results in a positive shift in V_{th} [11], [12].

Table.1 benchmarks the TiO_2 channel FE-VNAND with previous ferroelectric devices in terms of thermal stability and basic device performance [3], [5], [6], [7], [11]. The device presented in this work not only exhibits the best thermal stability but also shows competitive performance.

Conclusion

In conclusion, we investigated the thermal stability of TiO_2 channel FE-VNAND devices in both fabrication and operational scenarios. During the fabrication process, the devices exhibited stable properties under various annealing temperatures and times. In high-temperature operation, the devices maintained excellent performance and reliability up to 475K. These results provide valuable insights for the robust fabrication and operation of FE-VNAND devices.

Acknowledgments

This work was supported by NSFC 92064001.

References

[1] J. Han, S. Kang, K. Kim, J. Jang, and J. Song, "Fundamental Issues in VNAND Integration Toward More Than 1K Layers," International Electron Devices Meeting (IEDM), pp. 1-5, (2023).

[2] S. Lim, T. Kim, I. Myeong, S. Park, S. Noh, S. M. Lee, J. Woo, H. Ko, Y. Noh, M. Choi, K. Lee, S. Han, J. Baek, K. Kim, J. Kim, D. Jung, K. Kim, S. Yoo, H. J. Lee, S. G. Nam, J. S. Kim, J. Park, C. Kim, S. Kim, H. Kim, J. Heo, K. Park, S. Jeon, W. Kim, D. Ha, Y. G. Shin, and J. Song, "Comprehensive Design Guidelines of Gate Stack for QLC and Highly Reliable Ferroelectric VNAND," International Electron Devices Meeting (IEDM), pp. 1-4 (2023).

[3] K. Florent, M. Pesic, A. Subirats, K. Banerjee, S. Lavizzari, A. Arreghini, L. D. Piazza, G. Potoms, F. Sebaai, S. R. C. McMitchell, M. Popovici, G. Groeseneken, and J. V. Houdt, "Vertical Ferroelectric HfO2 FET based on 3-D NAND Architecture: Towards Dense Low-Power Memory," International Electron Devices Meeting (IEDM), pp. 2.5.1-2.5.4, (2018).

[4] I. Myeong, S. Lim, T. Kim, S. Park, S. Noh, S. M. Lee, J. Woo, H. Ko, Y. Noh, M. Choi, K. Lee, S. Han, J. Baek, K. Kim, D. Jung, J. Kim, J. Park, S. Kim, H. Kim, I. Yoon, J. Kim, K. Kim, K. Park, B. J. Kuh, W. Kim, D. Ha, S. Ahn, J. Song, S. Yoo, H. J. Lee, D. H. Choe, S. G. Nam, and J. Heo, "A Comprehensive Study of Read-After-Write-Delay for Ferroelectric VNAND," International Reliability Physics Symposium (IRPS), pp. 9B.3-1-9B.3-6, (2024).

[5] Y. Feng, D. Zhang, C. Sun, Z. Zheng, Y. Chen, Q. Kong, G. Liu, Y. Kang, K. Han, Z. Zhou, G. Liang, K. Ni, J. Wu, J. Chen, and X. Gong, "First Demonstration of BEOL-Compatible 3D Vertical FeNOR," Symposium on VLSI Technology and Circuits (VLSI Technology and Circuits), pp. 1-2, (2024).

[6] M.-K. Kim, I.-J. Kim, and J.-S. Lee, "CMOS-compatible ferroelectric NAND flash memory for high-density, low-power, and high-speed three-dimensional memory," Science Advances, vol. 7, no. 3, p. eabe1341, (2021).

[7] S. Kabuyanagi, T. Hamai, M. Murase, T. Maeda, M. Saitoh, and S. Fujii, "A Vertical Channel-All-Around FeFET with Thermally Stable Oxide Semiconductor Achieving High ΔIon> 2μA/cell for 3D Stackable 4F2 High Speed Memory," Symposium on VLSI Technology and Circuits (VLSI Technology and Circuits), 16-20 June 2024 2024, pp. 1-2, (2024).

[8] D. Fan, Z. Xia, T. Yang, Y. Yang, L. Wu, K. Zhang, L. Liu, X. Chen, Y. Yan, W. Zhou, and Z. Huo, "An Emerging Local Annealing Method for Simultaneous Crystallization and Activation in Xtacking 3-D NAND Flash," IEEE Transactions on Semiconductor Manufacturing, vol. 36, no. 1, pp. 139-143, (2023).

[9] K. Wang, Z. Lun, W. Chen, X. Liu, and G. Du, "Impact of self-heating effect on the retention of 3-D NAND flash memory," International Symposium on the Physical and Failure Analysis of Integrated Circuits (IPFA), pp. 1-4, (2017).

[10] J. Y. Park, D. H. Yun, S. Y. Kim, and Y. K. Choi, "Suppression of Self-Heating Effects in 3-D V-NAND Flash Memory Using a Plugged Pillar-Shaped Heat Sink," IEEE Electron Device Letters, vol. 40, no. 2, pp. 212-215, (2019).

[11] C. Sun, Z. Zheng, K. Han, S. Samanta, J. Zhou, Q. Kong, J. Zhang, H. Xu, A. Kumar, C. Wang, and X. Gong, "Temperature-Dependent Operation of InGaZnO Ferroelectric Thin-Film Transistors With a Metal-Ferroelectric-Metal-Insulator- Semiconductor Structure," IEEE Electron Device Letters, vol. 42, no. 12, pp. 1786-1789, (2021).

[12] Y.-C. Liu, "Experimentally calibrated simulation of ZnO thin film transistors including traps," Ph.D., The Pennsylvania State University, United States -- Pennsylvania, 3583381, (2014).

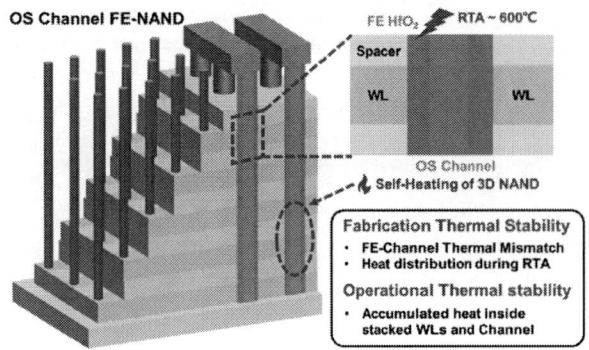

Fig.1 Thermal stability challenges encountered by the OS channel FE VNAND. (a) Fabrication issues include thermal mismatch between the FE layer and the channel, and uneven heat distribution during annealing. (b) Operational issues involve pronounced self-heating effects in VNAND.

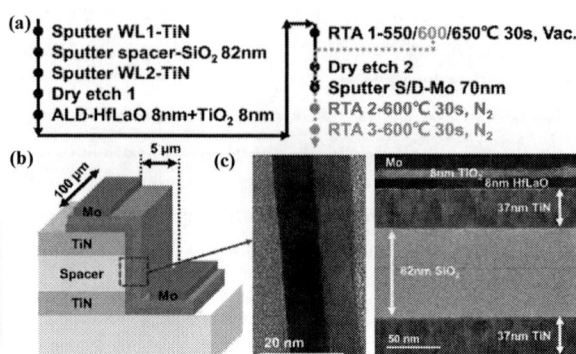

Fig.2 (a) Fabrication process flow of the FE VNAND. After RTA 1, the 600°C annealed devices underwent RTA 2 and RTA 3. (b) Schematic of the fabricated device. (c) TEM images of the channel region and the entire device, showing that both HfLaO and TiO_2 are crystallized after RTA.

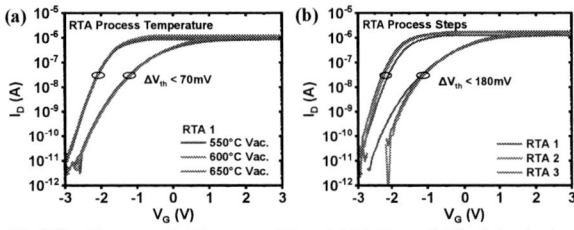

Fig.3 Double sweep I_D-V_G curves (V_D = 0.1 V, V_G = ±3 V) of the devices under (a) different annealing temperatures during RTA 1 and (b) after different annealing steps (RTA 1/2/3). The devices show V_{th} shift less than 70mV after annealing at 550-650°C and less than 180mV after different annealing steps. Additionally, RTA 2 and RTA 3 samples exhibit improved SS due to reduced defects after N_2 annealing.

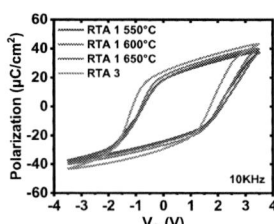

Fig.4 Polarization-voltage (PV) curves of the devices under various annealing conditions. The devices remain stable under different RTA temperatures, while a shift in coercive voltage is observed after N_2 annealing.

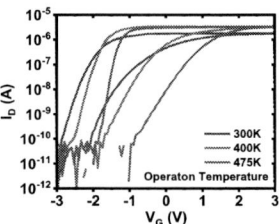

Fig.5 Double sweep I_D-V_G curves (V_D = 0.1 V, V_G = ±3 V) of the device operating at 300/400/475 K. As the operation temperature increases, the on-current is enhanced, and the MW is enlarged.

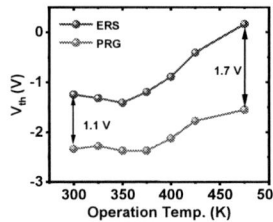

Fig.6 Extracted V_{th} of the PRG and ERS states at operation temperature from 300K to 475K. The MW increases from 1.1V to 1.7V.

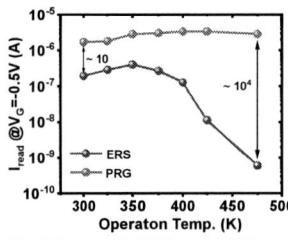

Fig.7 Extracted PRG and ERS state I_{read} with V_G=-0.5V and V_D=0.1V. The PRG/ERS current ratio increased from 10 to 10^4 as the temperature rises.

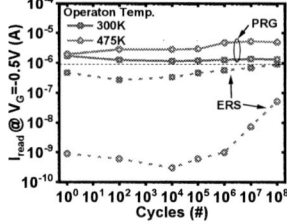

Fig.8 Endurance test at 300K and 475K, under 3.5V, 1MHz bipolar wave cycling. The device maintains endurance over 10^8 cycles at high temperature.

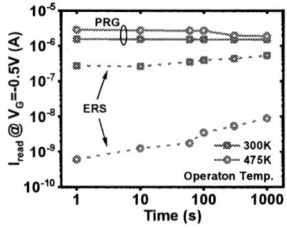

Fig.9 Retention test at 300 K and 475K. The device exhibits acceptable retention after 1000 seconds, with larger sense margin at 475K.

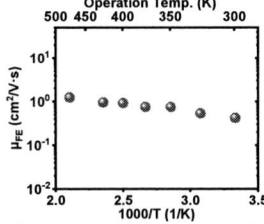

Fig.10 Arrhenius plot of the extracted μ_{FE} from the transfer curves with V_D=0.1 V. The minimal temperature dependence and low activation energy indicate band-like transport in the TiO_2 channel.

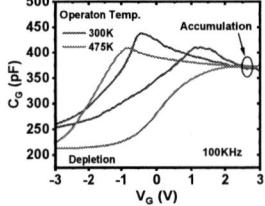

Fig.11 C-V curves measured from the gate under a 100 kHz AC signal. At the accumulation region (V_G=3V), the capacitance remains stable, indicating a stable carrier concentration from 300K to 475K.

Table.1 Benchmark of the TiO_2 FE-VNAND

	This work	[3]	[5]	[6]	[7]	[11]
Structure	**Vertical MFS**	Vertical MFIS	Vertical MFMIS	Vertical MFS	Vertical MFS	Planar MFMIS
Channel	**TiO_2**	Si	ZnO	InZnO	TiO_2	IGZO
Process Temp.	**600°C**	900°C	400°C	400°C	700°C	300°C
Operation Temp.	**203°C**	85°C	N.A.	N.A.	N.A.	85°C
MW @$V_{PRG/ERS}$	**1.7V @±3V**	2V @±10V	1.4V @±4V	1.6V @±4V	N.A.	3.5V @±5V
Endurance	**10^8**	10^4	10^7	10^8	10^6	10^5

979-8-3315-0417-5/25 $31.00 © 2025 IEEE

Rapid Customizing of Flexible OECTs Arrays for Low-Cost Biosensing and Biocomputing

Xinyu Tian[1], Jing Bai[1], Dingyao Liu[1], Shiming Zhang[1*]

[1] Department of Electrical and Electronic Engineering, The University of Hong Kong, Hong Kong SAR, China

e-mail: beszhang@hku.hk

Abstract

Organic electrochemical transistors (OECTs) are promising for next-generation biosensing and neuromorphic computing. With the growing demand for flexible OECTs in wearable bioelectronics, a low-cost, fast fabrication method is highly desired. Here, we propose a scalable printing-based fabrication method for rapid production of flexible all-solid-state OECTs. The proposed technology can produce devices with high yield, minimal variation, and high stability, which allows quick evaluation of OECT-relevant new materials, devices, and circuits.

Keywords: organic electrochemical transistors (OECT), PEDOT:PSS, wearable biocomputing and biosensing

Introduction

Organic electrochemical transistors (OECTs) are being pursued as promising candidates for emerging biosensing and bioelectronics applications due to their ability to amplify weak biosignals at extremely low power[1]. The typical OECT device structure consists of gate, source, and drain electrodes, an organic semiconductor channel, and an electrolyte. The electrolyte is in direct contact with the organic semiconductor channel, such as poly (3,4-ethylenedioxythiophene) poly (styrene-sulfonate) (PEDOT:PSS). When a positive/negative gate voltage (V_g) is applied, the cations/anions in the electrolyte are repulsed into the channel, and an electrochemical dedoping/doping process subsequently occurs. This process manipulates the conductivity of the channel. Besides, the ion-diffusion is a dynamic and non-linear process, which makes them potential hardware candidates to simulate the behavior of neurons[2, 3].

In recent years, flexible OECTs are being pursued to facilitate their use for emerging bioelectronic applications, such as medical wearables and implantables[4]. Despite the rapid growth, the manufacturing of flexible OECTs is time-consuming and suffers a low device yield, partially due to the complicated patterning of functional materials on plastic substrates[5]. Therefore, great efforts have been dedicated to simplifying the fabrication of flexible OECTs. Meantime, it is expected that fast turnaround time, low fabrication cost, and high device yield can be achieved at the same time.

Flexible printed circuit board (fPCB) manufacturing technologies have become successful in the past decades, allowing standardizable, low-cost, and large-scale production of flexible electronic circuits with short turnaround time, and are now widely used in the industry. However, the OECT community has not taken much advantage of this technology. A main challenge is that the fPCB uses copper (Cu) for electrodes, which is not favored in electrochemical (EC) systems due to the high redox activity in air and water. Besides, the fPCB process lacks a method of patterning semiconducting channel and electrolytes.

Here, we report a printing-based and scalable manufacturing method for rapidly prototyping flexible OECTs devices and circuits. The electrodes, interconnects, and insulators were first fabricated with the fPCB process, followed by patterning the channel and electrolyte by inkjet printing. The device features are ~100 μm. The device yield, device-to-device variation, and stability are comparable to the state-of-the-art methods, but the proposed process is more efficient in time and cost (< 10 USD/m²). As an example, we demonstrate this strategy for rapid prototyping of flexible and all-solid-state OECTs for neuromorphic computing applications.

Results and discussion

Fig.1 (a) shows the fabrication process of flexible OECTs, beginning with Cu electrodes patterning on the polyimide (PI) substrate, followed by the electroplating of a 20 nm gold layer for Cu protection, lastly encapsulating the electrodes with another layer of PI. The resultant electrode arrays are shown in Fig. 1 (b), containing 36 OECT units and 108 interconnects within a 5 cm * 8 cm area. Then, the organic semiconductor, PEDOT:PSS, was patterned between the source and drain electrodes as the channel with a customized inkjet printer, achieving a feature size ~100 μm. Verification tests indicate a device yield ~100%, without notable short, break, or delamination issues.

Despite the gold layer protection, Cu electrodes remain suffer from redox instability with the electrolyte, as aqueous electrolyte leakage leads to reactions with Cu. The reaction intensifies at positive gate voltages (> 0 V), limiting the device operation in water[6]. Nevertheless, we found that this phenomenon can be mitigated by using anhydrous ionic gel as electrolytes (Fig. 2 (a)). The gel was synthesized following the previous method[7], which utilizes ionic liquid as both mobile ions and solvents. As shown in the cyclic voltammetry (CV) curves in Fig. 2 (b), the results indicate the gel electrolytes can efficiently curb the corrosion of Cu, thus maintaining the high stability of the electrodes. The gel electrolyte can also be patterned by

979-8-3315-0417-5/25 $31.00 © 2025 IEEE

our customized inkjet printer, enabling rapid, scalable prototyping of all-solid-state flexible OECTs.

Fig. 3 summarizes the performance of the flexible all-solid-state OECTs, showing output, transfer, transient, and cyclic stability curves. The devices showed typical transistor characteristics, working in depletion mode (Fig.3 (a-b)). A high on/off ratio of ~1000 was extracted from the transfer curves (V_g from -0.2 V to 0.8 V, V_{ds} from -0.1 V to -0.6 V). Mobility of 1.1 cm^2 V^{-1} s^{-1} is calculated (at V_g = 0.3 V, V_{ds} = 0.2 V), indicating high device quality. The transfer curves showed minor hysteresis and good repeatability under cyclic scanning of V_g between -0.1 to 0.8 V, which is within the safe electrochemical window for Cu (Fig.3 (c)). Besides, the transient response remains unchanged after 100 gate pulse cycles, demonstrating high device stability thanks to the anhydrous gel electrolyte preventing Cu oxidation (Fig.3 (d)). The transient response also indicated a relatively fast doping/dedoping process of the channel due to the small geometrical size of the devices. To gain insight into the homogeneity of our assembled solid-state OECTs on fPCB, we measured the I_{ds} of an array consisting of 200 devices (Fig.4 (a)). The statistical results show a narrow Gaussian distribution, indicating good device homogeneity was achieved (>90% of samples showed currents between 0.4 and 0.5 mA), benchmarkable to those patterned with state-of-the-art perylene or orthogonal photoresists. Then, to evaluate the flexibility of the devices, bending tests were performed by laminating the devices on 3D-printed testbeds with varying bending curvatures (Fig.4 (b)). The transfer and transconductance curves of the devices showed negligible change upon increasing bending curvature, indicating the excellent conformability of the devices thanks to the thin thickness (200 μm) of the fPCB and the good adhesion between functional layers, e.g. PEDOT: PSS, and substrate (Fig.4 (c-d)).

The high stability of the fPCB-fabricated flexible electrode arrays in gel electrolytes permits their use in all-solid-state devices. A typical application of the device is to develop neuromorphic circuits to mimic the synaptic behaviors[3]. As shown in Fig.5 (a), we demonstrate the potential applications of those flexible OECT arrays with a specific neuromorphic computing framework, reservoir computing (RC). The Modified National Institute of Standards and Technology (MNIST) database was used as the test task with 16 flexible OECTs forming the reservoir layer. As shown in Fig.5 (b), the data from MNIST was firstly encoded to different binary gate input patterns, where '1' denotes applying a gate voltage (0.4 V) and '0' denotes no voltage is applied (0 V). Then the OECTs-based reservoir layer outputs 16 distinguishable I_{ds} values due to their excellent non-linear transient response (Fig. 5 (c)). A pre-trained single connected layer serves as the output, with performance evaluated by confusion analysis (Fig. 5 (d)). The results show a > 90% accuracy of flexible OECT-

based RC, permitting their deployment for practical applications such as wearable sensing and computing at the edge.

Conclusion

In conclusion, we presented a facile and scalable fabrication method that allows large-scale production of flexible all-solid-state OECTs. The electrode arrays and the encapsulation layers were fabricated using mature fPCB technology. The PEDOT:PSS channel and solid-state electrolyte were subsequently patterned on the fPCB with customized inkjet printing methods. Solid-state anhydrous gel was found that can avoid the undesired redox of the Cu electrodes of the fPCB. Statistical characterizations demonstrated a high device yield, high homogeneity, and high flexibility were obtained simultaneously. The short turnaround time, low cost, and scalability of the proposed fabrication methods pave the way for developing flexible solid-sate OECTs (and devices of similar kinds) for practical bioelectronic applications.

Acknowledgments

This work was supported by grants from the National Key R&D Program of China (2022YFE0202200), the Innovation and Technology Fund (Mainland-Hong Kong Joint Funding Scheme, MHP/053/21), and the Shenzhen-Hong Kong-Macau Technology Research Programme (SGDX20210823103537034). It is also partially funded by the Seed Fund for Basic Research, the Seed Funding for Strategic Interdisciplinary Research Scheme from the University of Hong Kong, and the RGC Germany/Hong Kong Joint Research Scheme (G-HKU707/22).

References

[1]   J. Rivnay, S. Inal, A. Salleo, R. M. Owens, M. Berggren, and G. G. Malliaras, "Organic electrochemical transistors," Nature Reviews Materials (2018).

[2]   F. Liao and Y. Chai, "In-sensor Computing Devices for Bio-inspired Vision Sensors", IEEE Electron Devices Technology and Manufacturing Conference (EDTM) (2022).

[3]   D. Liu et al., "A wearable in-sensor computing platform based on stretchable organic electrochemical transistors," Nature Electronics (2024).

[4]   J. Bai et al., "Coin-sized, fully integrated, and minimally invasive continuous glucose monitoring system based on organic electrochemical transistors," Science Advances (2024).

[5]   S. Wu, N. Phongphaew, and T. N. Ng, "Flexible Organic Electrochemical Sensors for Monitoring Marine Conditions," IEEE Electron Devices Technology and Manufacturing Conference (EDTM) (2022).

[6]   S. Zhang et al., "Water stability and orthogonal patterning of flexible micro-electrochemical transistors on plastic," Journal of Materials Chemistry C (2016).

[7]   Z. Lei and P. Wu, "A highly transparent and ultra-stretchable conductor with stable conductivity during large deformation," Nature Communications, (2019) .

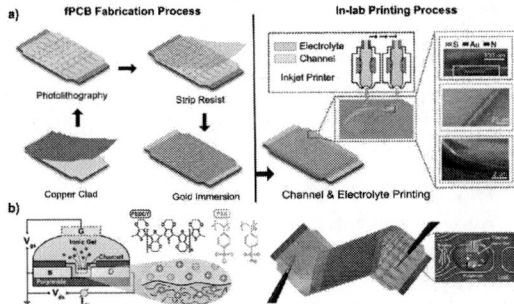

Fig.1: Fabrication of flexible OECT arrays with fPCB technology: a) shows the real images of the fPCB-fabricated electrodes. b) shows the schematic of an OECT device. c) shows the process flow of the fabrication. The electrodes were fabricated with fPCB technology, followed by inkjet printing the PEDOT:PSS channel and the gel-based electrolyte. The SEM image shows the side-view of the interfaces between PEDOT:PSS channel and the electrodes.

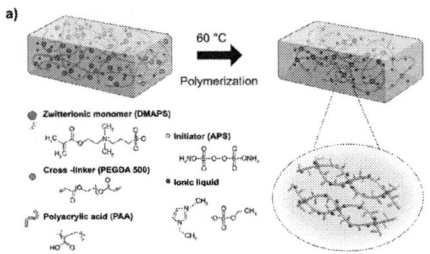

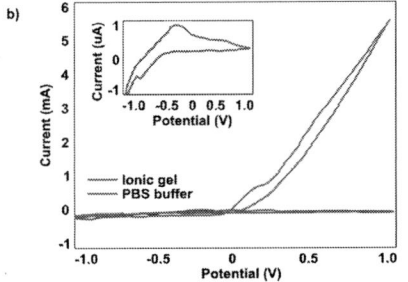

Fig.2: a) Schematic and chemical structure of relevant chemicals to synthesize DMAPS ionic gel. b) The comparison of CV curves of fPCB-fabricated electrodes (Au (200 nm)/Cu (30 μm)) with DMAPS and PBS as electrolytes.

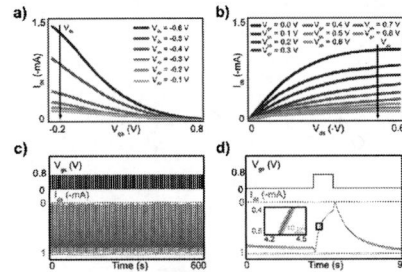

Fig.3: Electrical characterizations of OECTs fabricated with fPCB technology and ink-jet printing. a) The transfer curves at different. b) Output curves. c) Transient curves: a group of 60 pulses with 1.25 s width were applied at the gate electrode. d) The red solid line shows the average channel current response of the 60 pulses in c) and the red shadow shows the current variation range during the transient characterization (< 10 %).

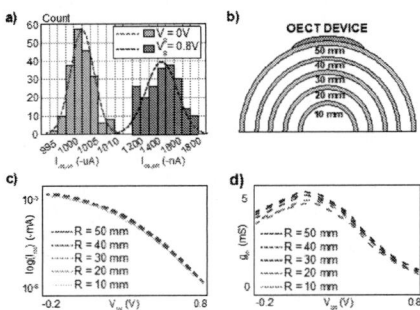

Fig.4: Homogeneity of flexibility characterizations of the OECTs. a) Comparison of channel currents at on and off status of 200 devices. b) Experimental setup of the flexibility test of OECTs. c-d) Transfer and transconductance curves under different bending radius (from 10 mm to 50 mm).

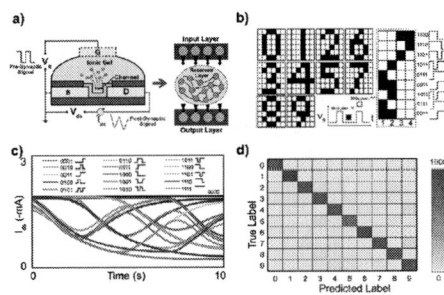

Fig.5: a) Evaluation of neuromorphic function (reservoir computing potential) of OECTs; b) Decoding of the ten digital numbers for RC validation. c) The I_d progression curves corresponding to different V_g patterns from 0000 to 1111. d) The comparison of confusion matrices of the MNIST test.

979-8-3315-0417-5/25 $31.00 © 2025 IEEE

Realization of Wafer-scale AlScN ferroelectric films and the Investigation of AlScN/n-GaN Ferroelectric Memristors

Mingrui Liu[1], Hang Zang[1], Shunpeng Lv[1], Zhiming Shi[1], Yuping Jia[1], Ke Jiang[1], Jianwei Ben[1], Dan Li[1], Xiaojuan Sun[1]*, and Dabing Li[1,2†]

[1]Key Laboratory of Luminescence Science and Technology, Chinese Academy of Sciences & State Key Laboratory of Luminescence and Applications, Changchun Institute of Optics, Fine Mechanics and Physics, Chinese Academy of Sciences, China, [2]Center of Materials Science and Optoelectronics Engineering, University of Chinese Academy of Sciences, Beijing, China

Abstract

The emergence of ferroelectricity in AlScN shows promise to expand the capability of next-generation III-N semiconductors, but it faces the challenge of large leakage current. Here, thermal annealing is proposed to reduce the leakage current by four orders and realize the 6-inch ferroelectric films. Furthermore, AlScN/n-GaN memristors were fabricated, which exhibit large ON/OFF ratios over 10^5 and reversible bipolar resistive switching. These results lay the foundation for future AlScN-based device applications.

Keywords: wafer-scale AlScN films, leakage regulation and memristors

Introduction

Recently, wurtzite-structured AlScN has attracted wide attention due to its combination of the excellent ferroelectric properties and advantages of wide-bandgap nitride materials, showing promise to expand the capability of next-generation III-N semiconductors, such as neuromorphic computing based on AlScN-based memristors. The P_r of AlScN is more than 2–6 times that of HfO_2 or perovskite ferroelectrics [1], which results in a stronger barrier regulation ability and a larger ON/OFF ratio than other ferroelectric memristors [2-3]. And the high Tc (> 1100 °C) ensures its stability in extreme environments.

However, AlScN confronts with the challenge of large leakage current, which restricts its further application[24]. Furthermore, AlScN-based memristors are still in their infancy. The resistive switching is mainly attributed to the strong barrier modulation caused by polarization flipping [4], which is not power-efficient due to its large coercive fields. In addition, the tunable ferroelectric depolarization field (E_{DP}) can efficiently modulate the transport of electron and nitrogen vacancies [5], which may affect memristor properties and should be further explored.

In this work, we grew wurtzite AlScN films on n-GaN substrates by sputtering and found that grain boundaries between abnormally oriented grains (AOGs) are the main leakage channel. Thermal annealing in pure N_2 can promote grain boundary migration and merge, reducing the leakage current of AlScN by four orders and causing some previously leaking regions to exhibit ferroelectric properties. Next, 6-inch AlScN ferroelectric films were achieved, which is of great significance to improve the ferroelectricity of AlScN and ensure the uniformity of devices. On the above basis, AlScN/n-GaN memristors based on the coexistence of energy band modulation and trap-assisted conduction mechanism were fabricated. Unlike typical ferroelectric memristors, the device presents obvious reproducible bipolar resistive switching characteristics without polarization flipping due to the extra trap-assisted conductive path, which effectively reduces energy consumption. Remarkably, multi-level ON/OFF ratios and reversible bipolar resistive switching characteristics can be achieved by regulating the magnitude and direction of ferroelectric polarization, indicating that different operating modes can be achieved on a single device, which shows promise for improving device integration density and information security.

Experimental Methods

170-nm $Al_{0.75}Sc_{0.25}N$ films were grown on commercial n-GaN/sapphire templates by the RF reactive magnetron sputtering from a metal alloy target in pure N_2 at 400 °C. The original samples were then transferred into a tubular furnace with a base vacuum pressure of 0.1 Pa and then annealed for one hour in pure N_2 with a pressure of 10^5 Pa. We characterized the crystal structure of the film based on XRD patterns. Microstructure characterizations and the chemical composition for the AlScN film were performed on a field-emission TEM (JEOL JEM-2100F, 200 kV) equipped with EDS. We used a Radiant Precision Premier II ferroelectric tester to study the ferroelectric properties and the leakage current density of the AlScN film before and after thermal annealing. Si-doped n-GaN contact layer with a carrier density higher than 10^{19} cm^{-3} can be used as the bottom electrode (BE), and Ti /Au (70/30 nm) circular top electrode (TE) with diameters of 100 μm were deposited by electron beam evaporation under a shadow mask.

Results and Discussion

A. Leakage Regulation in AlScN

To study the origin of leakage in AlScN films,

979-8-3315-0417-5/25 $31.00 © 2025 IEEE

high-resolution TEM and bright-field TEM cross-section measurements of the as-deposited AlScN film were performed. As shown in Fig.1a, the irregular brightness contrast indicates a large number of dislocations, which will lead to the distortion and twist of the grain, thus forming the grain boundary. As shown by the red dashed circles in Fig.1b, the columnar growth of the as-deposited AlScN grains with some AOGs can be observed. Molecular dynamics simulation and first-principles calculations suggest that grain boundaries will cause the defect states at Fermi level, lead to the large leakage current in the as-deposited AlScN films with AOGs (image-0), which can be effectively suppressed in the films with coincident-oriented grains (COGs, image-18), as shown in Fig.1c. Hence, the grain boundaries between AOGs might lead to the main leakage channel of AlScN films. Due to the migration and fusion of grain boundaries need to cross a certain energy barrier (Fig.1d), thermal annealing was proposed to promote the above process and reduce the leakage current. As shown by the TEM section inset Fig.1d, the density of AOGs decreased after annealing and the crystal structure becomes more ordered, which indicates a better c-axis orientation.

Next, we performed the polarization/displacement current density variations with an electric field (P-E/J-E) measurements to get further insights into the effect of post-deposition annealing on the ferroelectric properties of the AlScN film. For some regions of the as-deposited AlScN film with a large leakage current where the ferroelectricity is obscured, annealing at 400 °C in pure N_2 for one hour can reduce the leakage current by 4 orders of magnitude, as shown in Fig.2a. The leakage can be well fitted by hopping conduction (trap assisted tunneling) :

$$J = qanv \exp\left[\frac{qaE}{kT} - \frac{E_a}{KT}\right] \quad (1)$$

where a is the mean hopping distance (i.e., the mean spacing between trap sites), n is the electron concentration in the conduction, v is the frequency of thermal vibration of electrons at trap sites, E_a is the activation energy, T is the absolute temperature, and K is the Boltzmann constant. The trap spacing (a) was extracted by the slop of the linear part of log(J) versus E [6]. As shown in Fig.2b and c, The hopping distance for the as-deposited AlScN film with large leakage is 0.3 nm, which is reasonable because electron hopping is more likely to occur at such a small trap distance. With the increase of annealing temperature to 400 °C, the hopping distance increases to 0.7 nm due to the defect density decreasing, which ultimately leads to a decrease of leakage current. Furthermore, the ellipsoid-like P-E loop caused by leakage current transforms into a square shape, and the switching current peaks depicted in Fig. 2d by gray arrows also clearly support the ferroelectricity, which indicates that thermal annealing is of great significance to

improve the ferroelectricity of AlScN and ensure the uniformity of devices.

B. Wafer-scale AlScN films

Through the above regulation, 2 to 6-inch AlScN films were achieved. As shown in Fig.3a and b, the films are of uniform thickness with a standard deviation of less than 0.5. The average P_r ranges from 115 to 140 μC/cm², and the distribution difference of the E_c is less than 0.5 MV/cm (Fig. 3d-f). The Sc content at different positions is maintained at a uniform level of about 0.25 (Fig. 3c), and the characteristic diffraction peak for AlScN (002) with a 0.78° can be observed clearly at each region (Fig. 3g). The comparison of the (002) FWHM and Pr is shown in Fig. 3h -i, indicating that the wafer-scale AlScN film in this work preserves a high level of ferroelectric characteristics compared with previous reports.

C. AlScN/n-GaN Ferroelectric Memristors

On the basis of high-quality AlScN ferroelectric films, an Au/Ti/AlScN/n-GaN heterostructure memristor was fabricated (inset Fig. 4a). First, we applied a positive write voltage which is higher than the coercive voltage (Vc) to the TE to make the ferroelectric polarization of AlScN downward towards n-GaN (P_{down}). In Fig. 4a, without polarization flipping, the device presents bipolar resistive switching characteristics with the applied voltage being more than 65% [the "write" voltage is 20 V, and the Vc measured by the quasi-DC sweep is 58 V (not shown)] lower than the V_c. Furthermore, multi-level programming currents emerge with an increasing voltage range from 10 to 40 V, which is essential for high-density data storage and neuromorphic computing (Fig. 4b). Inset Fig. 4c shows the current-voltage (I-V) curves from 10 manually performed 40 V DC cycles, which indicate that the device is stable and repeatable. As shown in Fig. 4c, the retention properties were 10^3 s measured at a read voltage of 5 V after a write voltage of 40 V. Furthermore, the effect of ferroelectric polarization direction on resistance switching is also investigated. We applied a ferroelectric polarization of AlScN upward towards TE (P_{up}). Interestingly, the bipolar resistive switching characteristics were reversed as shown in Fig. 4d. Such controllable ON/OFF ratio and reversible bipolar resistive switching characteristics are attributed to the coexistence of heterointerface energy band modulation and trap-assisted conduction mechanisms, as shown in Fig. 4e. Fig. 4f shows the ON/OFF ratio extracted the I-V curves of P_{up} and P_{down}. The ON/OFF ratio read at 5.7 V is more than 10^5, which is three orders of magnitude larger than previous report [4].

Conclusion

In summary, thermal annealing was proposed to promote the migration and merge of grain boundaries in AlScN films, which inhibits the leakage channels and reduces the leakage current of by four orders, and realizes 6-inch AlScN ferroelectric materials. Furthermore, AlScN/n-GaN memristors based on the coexistence of energy band modulation and trap-assisted conduction mechanism were fabricated, which exhibit large ON/OFF ratios over 10^5 and reversible bipolar resistive switching. These results pave the way for obtaining high-quality AlScN and provide a perspective on AlScN-based memristors that help elucidate the resistive switching mechanism and the effect of ferroelectric polarization on device properties.

Acknowledgments

This work was supported by the National Key Research and Development Program of China (Grant No. 2022YFB3605600), the National Natural Science Foundation of China (Grant No. 62121005, 12204475), the Youth Growth Science and Technology Program of Jilin Province (Grant No. 20220508018RC).

References

[1] T. Mikolajick et al., "The past, the present, and the future of ferroelectric memories," IEEE T. Electron Dev., 67(4), pp. 1434-1443 (2020).

[2] T. Choi et al.,"Switchable Ferroelectric Diode and Photovoltaic Effect in BiFeO$_3$," Science, 324(5923), pp. 63-66(2009).

[3] Q. Luo et al., "A highly CMOS compatible hafnia-based ferroelectric diode," Nat. Commun. 11(1), pp. 1391-1398(2020).

[4] D. Wang et al., "An epitaxial ferroelectric ScAlN/GaN heterostructure memory," Adv. Electron. Mater., 8(9), pp. 2200005-1-2200005-8(2022).

[5] B. J. Choi et al., "High-speed and low-energy nitride memristors," Adv. Funct. Mater., 26(29), pp.5290-5296(2016).

[6] Liu, C. et al., Multiscale Modeling of Al$_{0.7}$Sc$_{0.3}$N-based FeRAM: The steep switching, leakage and selector-free array. In Proceedings of the 2021 IEEE International Electron Devices Meeting (IEDM), 2021.

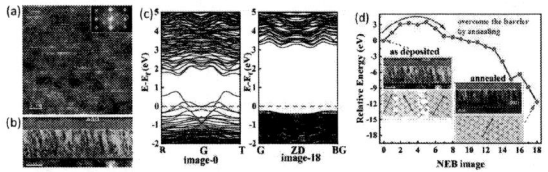

Fig. 1: (a)High-resolution TEM (top right corner: corresponding FFT) and (b) bright-field TEM cross-section of as-deposited AlScN film.(c)The band structure of AlScN with AOGs and COGs. (d)The relative energy versus the various microstructures.

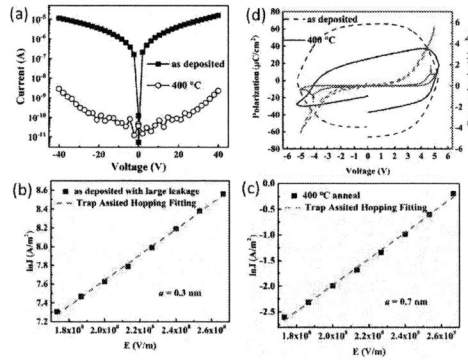

Fig.2 (a)Leakage current, and (b)-(c) its fitting, (d) P-E/J-E loops for AlScN film before and after annealing at 400 °C.

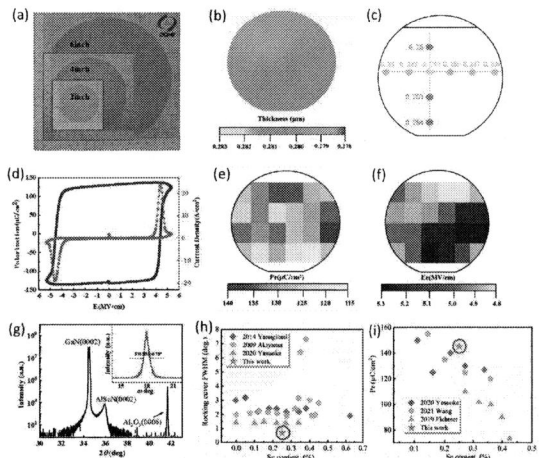

Fig.3 (a) 2 to 6-inch AlScN films; (b)thickness distribution mapping; (c) Sc content; (d)-(e) ferroelectricity; (f) XRD spectra for wafer-scale AlScN films. Comparison for (h) rocking curve FWHM; and (i) Pr of Al$_{1-x}$Sc$_x$N films.

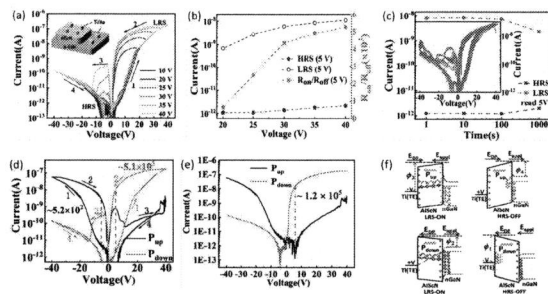

Fig.4 (a) Device structures and bipolar resistive switching characteristics. (b) Currents of high (low) resistance state and ON/OFF ratio versus the applied voltages. (c) Retention properties, inset: 10 cycles of I-V curves performed at 40 V. (d) Bipolar resistive switching characteristics for P$_{up}$ (black) and P$_{down}$ (red). (e) I-V characteristics for P$_{up}$ and P$_{down}$. (f) Schematic band structures of the high (low) resistance state.

A Multi-Ion Sensing System on a Chip with Edge Computing Capability

Haolin Zhao[1], Zhancheng Mai[1], Kai Zhuang[1], Kai Wang[1*]

[1] Guangdong Province Key Lab of Display Material and Technology, State Key Lab of Optoelectronic Materials and Technology, School of Electronics and Information Technology, Sun Yat-sen University, Guangzhou, China

*Corresponding author (wangkai23@mail.sysu.edu.cn)

Abstract

Ionic detection is very important in domestic water and industrial wastewater. In this work, we report on a multi-ion sensor chip that integrates Solid Contact Ion-Selective Electrodes, dual-gate Thin-Film Transistors, and an on-chip computing array for ANN network based on Multi-Layer Perceptron principles. The chip is designed to detect and compute the concentrations of four common ions—Ca^{2+}, Mg^{2+}, Na^+, K^+—with enhanced precision and reduced latency. Experimental results demonstrate an average ion detection error of 0.12 log(mol/L) after iterative training.

Keywords: Ion detection, TFT, edge computing

Introduction

Water is essential to life and indispensable in both daily activities and industrial processes [1]. The water typically contains not only H_2O but also various ions such as calcium (Ca^{2+}), magnesium (Mg^{2+}), and chloride (Cl^-), making water not only a vital resource but also a crucial medium for information. By analyzing the components in water, we can infer significant information related to environmental and industry activities. Calcium, magnesium, sodium and potassium are four common ions with relatively high concentrations in water and have substantial impacts on human health. Therefore, developing multi-ion sensors with high precision, low power consumption, and compactness is necessary. Traditional approaches often use multiple discrete sensors for various ions, resulting in high latency, cost, and poor portability. Additionally, due to the non-ideal nature of ion-selective membranes, signals in multi-ion scenarios are often coupled with interference from various ions, making it challenging to measure ion concentration. While Multi-Layer Perceptron (MLP) Artificial Neural Networks (ANNs) offers a suitable method for multi-ion concentration calculation, their computational cost is high because existing MLP implementations are typically connected to computers for calculations or uploaded to cloud servers, resulting in high latency, cost, and interference.

To address these issues, we propose a multi-ion sensor system on a chip (SoC) with edge computing capability. Compared to traditional ion-sensitive chips, the proposed chip features: (1) Ion-selective electrodes (ISEs) for detecting four common ions, offering good repeatability, stability, and response coefficients; (2) Signal amplification circuits using dual-gate thin-film transistors (TFTs) with an indium-gallium-zinc-oxide active layer, which reduce signal interference from the aqueous environment and electromagnetic interference; (3) A 3T2C matrix multiplication circuit based on the transfer characteristics of dual-gate TFTs, acting as neurons in ANNs, which can be expanded into an array to perform matrix multiplication for forward inference, thus reducing the computational complexity of digital. These parts are integrated onto a 20mm×20mm neuromorphic multi-ion sensor chip, allowing more accurate concentration calculation of multiple target ions and achieves sensing and computation on a single chip, thus avoiding errors introduced by signal transmission and transformation.

Device Model

Figure 1 compares this chip with traditional ion sensors, Figure 2 illustrates the architecture, which includes ISE electrodes, dual-gate TFT amplification circuits, differential amplification circuits, shift register circuits, and multiplication arrays. Figure 3 shows the signal-level flowchart.

Experimental

Figure 4 illustrates the detailed design and testing results for each component of the system. The first section focuses on the SC-ISE component, which is designed as follows: the ion sensing electrodes are circular, bare electrodes with a diameter of 2 μm. A rod-shaped reference electrode is positioned horizontally in the center of the sensing electrode array. The second section describes the common-source amplifier circuit constructed using dual-gate TFTs. The amplifier's amplification efficiency defined as Gm/Ids is limited by SS, which is analogous to the Landau limit, and is constrained by q/kT. The sub-threshold operation of dual-gate TFT transistor amplifier circuit is expected to surpass such a limit, achieving efficient exponential amplification of input signals. The principle involves adjusting the top-gate control coefficient γ in the dual-gate TFT to enhance amplification efficiency. The amplifier circuit employs a common-source configuration. This study also designs a biomimetic analog computation unit based on dual-gate TFTs to perform the core matrix multiplication. Based on the output and transfer characteristics of dual-gate TFTs, when both the top-gate and bottom-gate control the TFT's state simultaneously, the product of the input voltage and a gate voltage is linearly proportional to the multiple of the output current. Figure 4 presents the structure of the 3T2C multiplication unit. TFT1 serves as the main multiplication device, with TFT2 and TFT3 acting as switch transistors, controlled by scanning signals to charge the capacitors. The specific operational process is as follows: when Vscan inputs a high-level signal, TFT2 and TFT3 are activated to charge capacitors C1 and C2 with voltages VTgate and VBgate, respectively. After charging, Vscan is set to a low-level signal to turn off TFT2 and TFT3, locking the voltage in the capacitors. When a voltage Vin is applied, TFT1 outputs a current value controlled by the three-terminal input, which represents the unamplified multiplication result. With a capacitor capacity of 1 pF, the calculation rate of this unit can reach up to 20 MHz. To enhance the model's expressiveness and adaptability, each neuron in the neural network includes a bias term. The training results using multiplicative arrays are presented as shown in the inset of Figure 4.

To integrate a complete neural computation multi-ion sensing system on a single chip, this study incorporates an array of ion sensing units, signal acquisition circuits, shift drive units, and a 3T2C multiplication array on a 20mm × 20mm glass chip. The 3T2C multiplication circuit uses an 8-input dimension and a 16×9 array. Capacitors with a capacity of 1 pF are used, with switch transistors sized at W/L = 20μm/7μm, and multiplication transistors sized at W/L = 420μm/7μm. The NMMID-SoC employs TFT shift register circuits to replace the DAC-powered scanning voltage Vscan, providing pulse signals through two clock paths. The adjustable shift register signals supply the column scanning voltage. The hardware system equipped with the NMMID-SoC is shown in Figure 5(a). Prior to experimentation, standard mixed solutions of four ions {Ca^{2+}, Mg^{2+}, Na^+, K^+} with gradient concentrations are prepared. During training, the standard solution is applied to the electrodes, and the system is switched to ion concentration signal acquisition mode. The main control chip of the hardware system collects information from the sensing units and combines it with the standard solution ion concentration data from the host computer, storing it in a cache. The main control chip selects a specific coupling voltage-standard concentration combination from the cached training set and retrieves the required gate voltage from the multiplication table. The main control chip then controls the TFT shift register circuit to output the column scanning voltage Vscan, sequentially activating the gate voltages of the 3T2C column. Each weight and bias parameter is then locked onto the thin-film capacitors of the corresponding 3T2C units through the control circuit. The 3T2C multiplication array completes the forward inference process for the three hidden layers and the output layer of the neural network. Finally, the main control chip compares the results with the label values, using a custom backpropagation function to update the weights and bias parameters in the cache for the next test.

Testing results are shown in Figure 5(b). For concentration combinations {-2, -3, -4, -5}, the red dashed line represents the ideal concentration line. The figure illustrates that as the number of iterations increases, the predicted concentrations for the four ions approach the dashed line. At the 8000th iteration, the root mean square error (MSE) is 0.1341, indicating an average deviation of 0.1341 concentration log units per concentration, which is superior to the performance of digital MLP-ANN networks and is satisfactory for practical ion sensing. Furthermore, testing on unknown water samples yields results as presented in Figures 5(e) and 5(f), demonstrating that the NMMID-SoC performs well in detecting concentrations of calcium, magnesium, sodium, and potassium in various unknown water samples. Figure 5(g) provides an overview of the chip and compares it with other ion sensor chips. Figure 6 illustrates the complete integrated chip structure and parameters.

Conclusion

This paper presents a multi-ion sensor chip integrating SC-ISE, dual-gate Thin-Film Transistors, and a biomimetic analog computation unit based on MLP principles. The chip enables real-time detection of Ca^{2+}, Mg^{2+}, Na^+, and K^+ with an average detection error of 0.12 log(mol/L). This solution offers significant potential for real-time water quality monitoring and other environmental sensing applications.

Acknowledgments

This work was funded by Project Team of Foshan National Hi-tech Industrial Development Zone Industrialization Entrepreneurial Teams Program (2120197000110).

References

[1] Rylander R. Drinking water constituents and disease. J Nutr. 2008 Feb;138(2):423S-425S.

[2] S. Mizutani et al., "Development of amperometric ion sensor array for multi-ion detection," 2015 IEEE SENSORS, Busan, Korea (South), 2015, pp. 1-4.

[3] Y. Chen, Z. Tang, Y. Zhu, M. J. Castellano and L. Dong, "Miniature Multi-Ion Sensor Integrated With Artificial Neural Network," in IEEE Sensors Journal, vol. 21, no. 22, pp. 25606-25615, 15 Nov.15, 2021.

[4] W. Xu, L. Hong, J. Zheng, M. Li, Y. Hua and X. Zhao, "Wearable Smart Sensor System for Monitoring and Intelligent Prediction of Sodium Ions in Human Perspiration," in IEEE Internet of Things Journal, vol. 11, no. 5, pp. 8146-8155, 1 March1, 2024.

[5] M. Sophocleous, L. Contat-Rodrigo, E. García-Breijo and J. Georgiou, "Towards solid-state, thick-film K+ and Na+ ion sensors for soil quality assessment," 2020 IEEE SENSORS, Rotterdam, Netherlands, 2020, pp. 1-4.

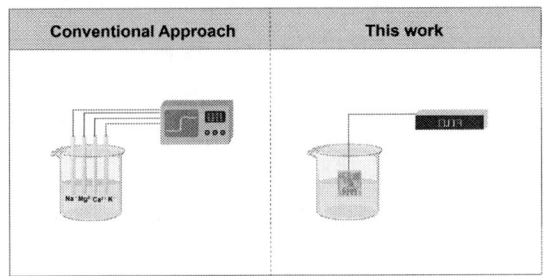

Fig. 1: Comparison of ion sensor chips

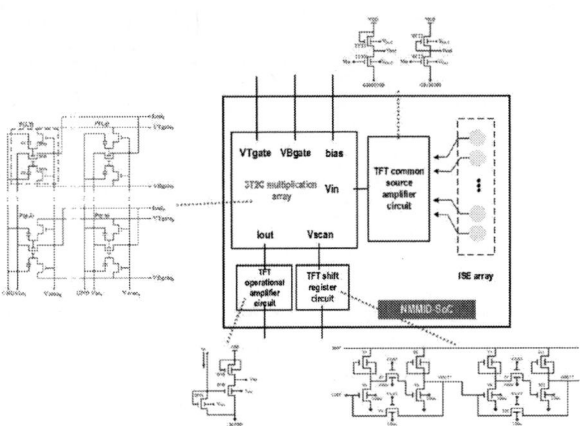

Fig. 2: System architecture and design

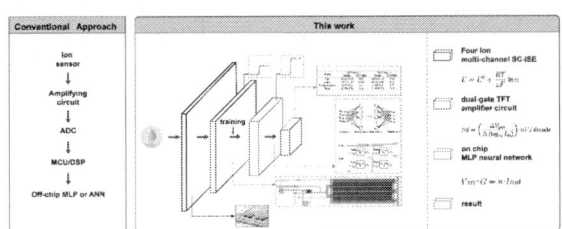

Fig. 3: Signal flow

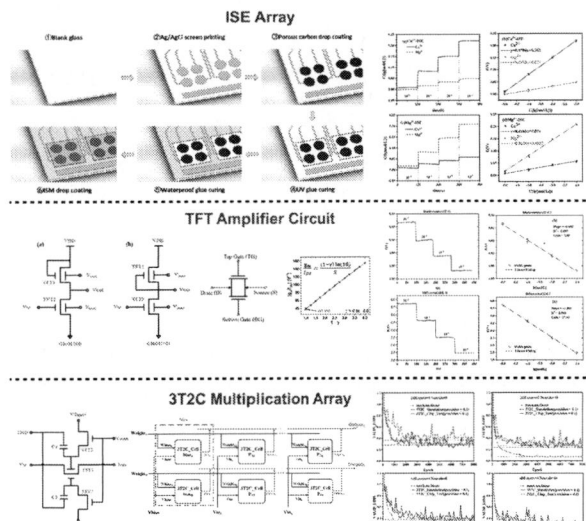

Fig. 4: Design and testing of each part

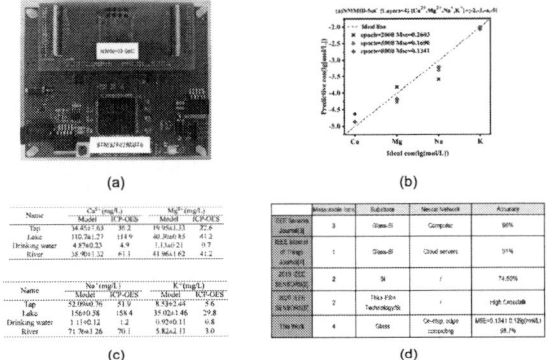

Fig. 5: Chip test (a)The hardware system of the integrated chip (b) The concentration combination is {-2,-3,-4,-5} test result(c)the concentration of calcium,magnesium,sodium and potassium ions in unknown water samples (d) Comparative analysis.

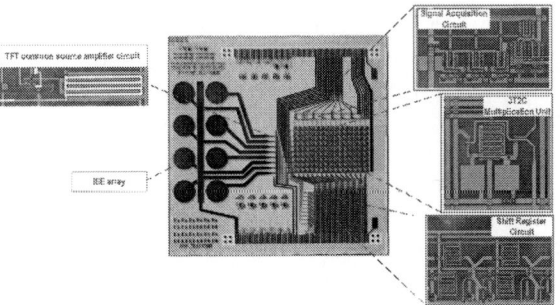

Chip Parameters	
Measurable Ions	4
Chip size	20mm×20mm
Average Ion Detection Error	0.12 lg(mol/L)
Multiplication Precision	0.1
Number of Hidden Layer Neurons	16
Multiplication Frequency	20MHz
Output Voltage Range	0~10V
Drive Voltage Range	-10V~10V

Fig. 6: Chip Structure and Parameter

Surrogate MTJ Model for Early-stage MRAM Macro Reliability Analysis

Quanhai Zhu and Hao Cai

School of Integrated Circuits, Southeast University, Nanjing, 210096, China

ABSTRACT

As emerging magnetic tunnel junction (MTJ) device has been gradually optimized, magnetic random-access memory (MRAM) macro performance requires early-stage estimation and implementation based on SPICE compatible model. This work analyzes the main physical characteristics of MTJ and proposes a novel reliable surrogate model for early-stage MRAM performance evaluation. The model adjusts write voltage and timing according to error correcting code (ECC) capability and facilitates the evaluation of write power consumption. The model is validated with a 55-nm CMOS process, with a critical dimension of 78-nm MTJ and a $0.216\,\mu m^2$ single bit-cell layout area. The surrogate model fits the MTJ magnetoresistance across a wide range of temperatures, voltages and prevents from endurance degradation. The simulated power consumption of write path can be reduced by 16.5%.

Keywords: Compact model, reliability evaluation, write error rate (WER) and error correcting code (ECC)

INTRODUCTION

Spin-transfer torque magnetic random-access memory (STT-MRAM) is a promising non-volatile storage, characterized by high speed, high density, and low power consumption. In the early-stage design of MRAM arrays, a SPICE-compatible compact model written in Verilog-A is necessary, reflecting some electrical characteristics of the magnetic tunnel junction (MTJ) [1], [2]. Previously, the MTJ parallel/anti-parallel state always alters under the critical switching current. Earlier works took into account factors such as thermal noise, the flipping process under irregular voltage excitation, endurance, and the breakdown process [3], [4]. Recent work has focused on model simulation speed, using a 1-D approximation instead of the traditional Landau-Lifshitz-Gilbert (LLG) equation, balancing physical principles with computational efficiency [5]. Until now, few models reflect the switching characteristics at different write error rates (WER) [6]. Fig. 1 shows the framework of our proposed surrogate model for early-stage MRAM reliability analysis. Compared with previous work, our model provides an in-depth analysis of the relationship between write error rate (WER), write pulse width, and voltage supply, offering references for circuit design.

MTJ COMPACT MODELING

A. Tunneling magnetoresistance

The MTJ structure includes a top free layer, a bottom reference layer, and an insulating barrier layer, typically made of CoFeB and MgO. The free layer's magnetic orientation can change with an external magnetic field or current, while the reference layer is fixed. In the parallel (P) state, low resistance indicates data "0," while in the anti-parallel (AP) state, high resistance represents data "1." The spin angle difference influences tunnel magneto-resistance (TMR).

$$R_P = \frac{\rho_P t_{ox} e^{\beta t_{ox}}}{A} \frac{\sin(\gamma_1 T)}{\gamma_1 T} \left[1 + \gamma_2 k_B T \ln\left(\frac{k_B T}{E_c}\right) \right]^{-1} \quad (1)$$

$$TMR_0 = (\alpha + 1) / \left(1 + 2Q \cdot \beta \cdot \ln\left(\frac{k_B \cdot T}{E_c}\right) \right) - 1 \quad (2)$$

$$TMR = \frac{TMR_0}{1 + \left(\frac{V_{MTJ}}{V_0}\right)^2} = \frac{R_{ap} - R_p}{R_p} \quad (3)$$

Rp is the resistance in the parallel state, ρ_P is the resistivity of the oxide layer when $V_{MTJ} = 0$. t_{ox} is the oxide layer thickness. α, β, γ_1, and γ_2 are fitting parameters, A is the area of the oxide layer, E_c is the barrier energy, T is the temperature, and k_B is the Boltzmann constant. Fig. 2(a) shows the electrical characteristics of our model based on the formulas above.

B. Switching Probability

We use the Landau-Lifshitz-Gilbert (LLG) equation to describe the reversal of the MTJ state. Notice that the differential equation is hard to solve during early-stage analysis; the following formula is typically used to describe the probability of switching failure [2]:

$$WER = \exp\left[-\frac{\tau_r}{\tau_{r0}} \cdot \exp\left(-\frac{\mu_0 M_s H_k V}{2 k_B T} \left(1 - \frac{I_g}{I_{c0}} \right) \right) \right] \quad (4)$$

where k_B is the Boltzmann constant, μ_0 is the permeability of free space, M_s is the saturation magnetization, H_k is the saturation magnetization, V is the volume of the free layer, τ_{r0} and I_{c0} are constants obtained through data fitting, τ_r is the width of the write pulse, and I_g is the current passing through the MTJ. The switching time τ_r is proportional to $-\ln(WER)$.

In the chip, there is a trade-off between write power consumption and write error rate (WER). The WER requirement for a single bit is related to the error correcting code (ECC) capability of the chip. By adjusting the write voltage according to different write circuits, it is possible to maximize the utilization of redundant ECC bits to reduce write power consumption without affecting the yield of the chip, thereby providing a reference for addressing the reliability issues of the chip.

Based on the physical formulas, our study categorizes the actual write distribution into three levels, with WER of approximately 2%, 0.2%, and 0.003% respectively. As shown in Fig. 3, they correspond to write error rates at 2, 3, and 4 standard deviations beyond the center of the normal distribution curve derived from the average of switching voltage. We calculated the τ_r corresponding to each WER. The model offers the write voltages and the corresponding necessary write times via a virtual pin. The relationship between the model and ECC capabilities will be elaborated in detail later.

The Variations in oxide layer thickness and critical dimensions lead to fluctuations in resistance. It is reflected in the process corner files of the model. In case of a critical dimension of 78 nm, the resistance values of this model under various temperatures, voltages, and states differ by less than 3% from the measured data.

MODEL VALIDATION

A. The write path

This Verilog-A model simulates write operations using an industrial 55 nm FDSOI process. As shown in Fig. 2(c), the MTJ bit-cell layout occupies $0.216\mu m^2$. The resistance and P/AP switching behavior depend on switching current, temperature, and WER. Fig. 4 illustrates the write path, including the low dropout regulator (LDO) for voltage supply.

When a memory row is selected, WEN and BEN are pulled high, activating the level shifter. The input value D_{in} is written to the bit-cell through transmission gates. The LDO supplies the voltage for the level shifter, regulated by V_{ref}. The model's virtual pin provides flip time data to adjust V_{ref}, subsequently affecting the output voltage of the level shifter.

Write power is calculated using the level shifter's V_{DDH}, write current, and virtual pin time. At 25°C, write power depends strongly on V_{DDH} (Fig. 5(a)). The model remains valid across process variations, aiding in optimal power point identification. Fig. 5(b) shows the write path's power distribution, with adjustments in V_{DDH} mainly affecting transistor and MTJ power, while power on IDO stays fixed.

B. Model choice through ECC (BCH code)

In actual circuit design, the write yield of the model can be adjusted based on the error correcting code (ECC) capability of the actual bank, as shown in Fig. 6(a). The ECC design is based on the Bose–Chaudhuri–Hocquenghem (BCH) codes. For example, in our previous bank design, we used a switching

voltage at WER $= 0.003\%$, with 64-bit ECC capability correcting 2-bit errors. It has a fixed switching time of 100 ns. However, using a model with WER $= 0.2\%$ is sufficient to realize a 99.99% accuracy rate for separate words after error correction.

With the adoption of the new model, the write circuit uses a model with WER $= 0.2\%$. By increasing the write voltage V_{DDH} from 1.57 V to 2.41 V and reducing the corresponding write time, the power consumption for writing a single bit is reduced from 32.4 pJ to 27.8 pJ, with a decrease of 16.5%, as shown in Fig. 5(a).

Fig. 6(b) shows the performance of the model's 3 tiers in terms of chip error rate across array storage sizes. For full-chip designs, the chip error rate tends to increase with the macro capacity. Our work designed a 1Mb macro using a model with WER $= 0.002$. It ensures that the chip error rate remains below 0.01.

C. model benefits for reliability

After a certain number of current pulses, the MgO insulating layer undergoes catastrophic failure, marked by a sudden and significant reduction in resistance. Due to the relatively low voltage during read operations, endurance degradation of MTJ is typically induced by write operations. The surrogate MTJ model significantly reduces the write voltage in MRAM arrays, enhancing the endurance of MTJs. By lowering the stress on the device during write operations, the model helps extend its lifespan. Furthermore, it enhances ECC effectiveness, contributing to the chip's long-term reliability.

CONCLUSION

This work presents a surrogate model for early-stage MRAM reliability analysis, using a 55-nm FDSOI transistor process with an MTJ of critical dimension 78 nm. The WER of the MTJ is deduced by physical principles and divided into three levels. The model maintains the time cost and power consumption in simulation, compared with the previous model. It achieves a 16.5% reduction in write power consumption while maintaining a 99.99% yield for an IO word width of 64 bits. By reducing write voltage, MTJ's endurance is extended, supporting the chip's long-term reliability.

ACKNOWLEDGMENTS

This work was supported by the Natural Science Foundation of Jiangsu Province (Grants No. BK20243042) and the National Natural Science Foundation of China (Grant No. 62274029).

REFERENCES

[1] C. -T. Tung et al., IEEE TED, 71(1), pp. 57-61(2024).
[2] Y. Wang et al., IEEE TED, 63(4), pp. 1762–1767 (2016).
[3] Gaul, Nishtha S., et al. IEEE IMW, pp. 1-4 (2022).
[4] Carboni, Roberto, et al. IEEE TED 65.6: 2470-2478 (2018).
[5] Tung, Chien-Ting, et al. IEEE TED (2023).
[6] L, Wu et al., IEEE TCAD, 41(11): 4991-5004 (2022).
[7] Müller, J., et al., IEEE EDTM (2024).

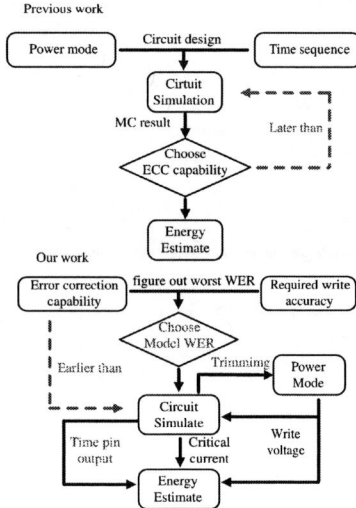

Fig. 1. Framework of early stage analysis and energy estimate in previous work and our work. Using the proposed model, we are able to choose model WER, trimming power mode and estimate power consumption.

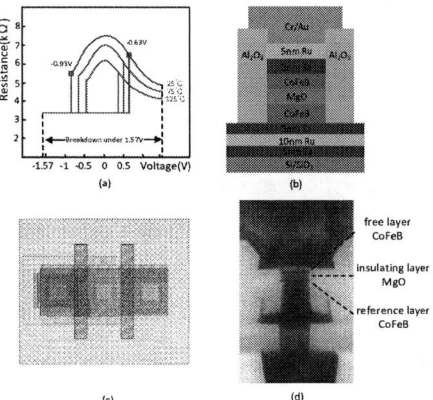

Fig. 2. (a) Electrical characteristics of MTJ in DC simulation. (b) Structure and layer materials of MTJ, with a critical dimension of 78 nm. (c) Layout of the MTJ bit-cell in $0.216\,\mu m^2$. (d) Electron microscope image of the MTJ.

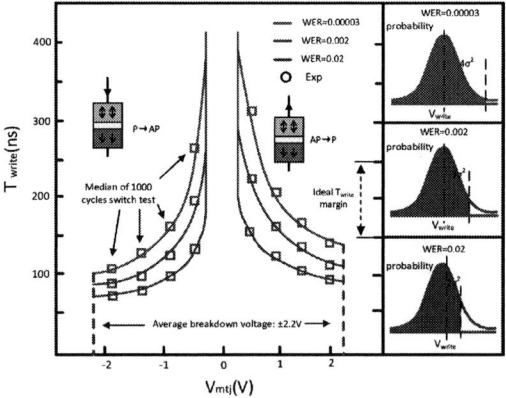

Fig. 3. Validation of the switching time model with the experiments. The lines represent the model results of different WER. The squares represent the experimental data. The WER correspond to different standard deviations in the distribution.

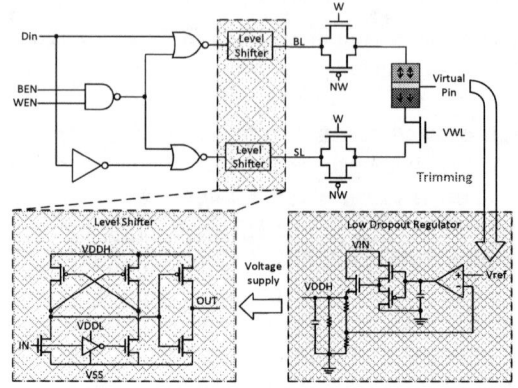

Fig. 4. The schematic of write circuit/path in MRAM memory bank. Adjust the VDDH of the MTJ write path by modifying the Vref of the IDO. It can be trimmed by the output of the virtual pin.

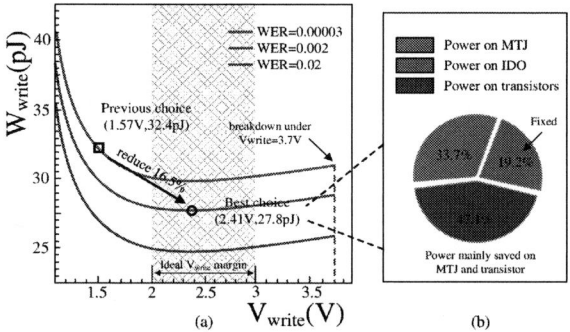

Fig. 5. (a) The estimate of power consumption with 64-bit ECC capability correcting 2-bit errors. The square represents the previous model with fixed switching time, and the circle represents the best choice in our model. (b) The energy consumption of the write circuit, where our model primarily reduces the power on the transistors and MTJ.

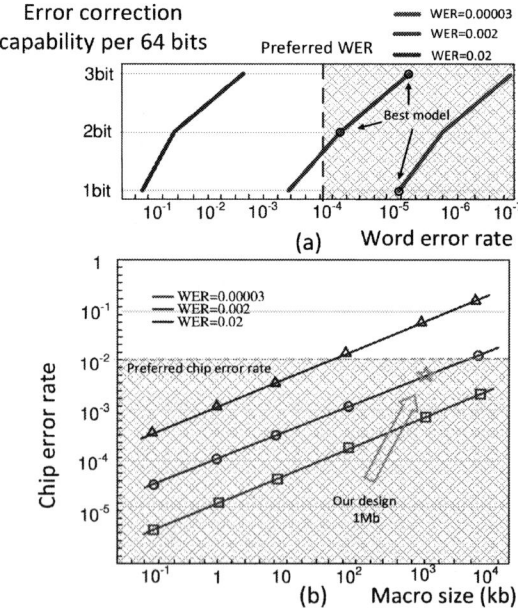

Fig. 6. (a) The data yield using 64-bit ECC. (b) The performance of the model's three tiers in terms of chip error rate across array storage sizes.

Self-Heating and Bias Temperature Instability in Recessed-Gate GaN MIS-HEMTs with AlN/SiNx Bilayer Dielectric.

Xin Wang[1], Hongyue Wang[2], Chen Wang[1], Jiayin He[1], Ju Gao[1], Bin Zhang[1], Ziheng Liu[1], Hongjie Peng[1], Chengkang Ao[1], Jiahui Yuan[2], *Jinyan Wang[1].

[1]School of Integrated Circuits, Peking University, Beijing 100871, China.

[2]China Electronic Product Reliability and Environmental Testing Research Institute, Guangzhou 510610, China.

Abstract

Self-heating and bias temperature instability (BTI) in E-mode recessed-gate GaN metal-insulator-semiconductor high electron mobility transistors (MIS-HEMT) were investigated. Thermoreflectance thermography was performed in the recessed-gate GaN MIS-HEMT, revealing a different heat distribution at the gate region and lower channel temperature compared with the D-mode GaN HEMT under the same DC power and drain bias. BTI measurements indicates that V_{TH} shift and R_{ON} degradation are more significant at elevated temperatures, due to the trapping/detrapping process of electrons at the AlN/SiN$_x$ dielectric layer interface.

Keywords: Recessed-gate GaN MIS-HEMT, Self-heating, Bias temperature instability.

Introduction

Recessed-gate GaN MIS-HEMT is an effective approach to realize normally-off operation [1]. With large gate swing, low gate leakage and high threshold voltage, this technology have garnered significant research attention. However, reliability issues, particularly V_{TH} shift [2] and R_{ON} degradation [3], hinder its further commercialization. Self-heating effect is a long-standing challenge for power semiconductor devices, which is closely related to device structures and aggravates these reliability issues. However, the self-heating effect and temperature dependent reliability issues in recessed-gate GaN MIS-HEMTs were seldom reported previously, requiring a more in-depth investigation.

In this work, we present a detailed study on the self-heating effect and BTI of recessed-gate GaN MIS-HEMTs. Channel temperature profiles were measured using thermoreflectance thermography under different DC bias conditions, with conventional Schottky-gate AlGaN/GaN HEMTs serving as a comparison. Positive and negative bias temperature instability (PBTI, NBTI) measurements at various temperatures were conducted to evaluate the impact of elevated temperature on V_{TH} shift and R_{ON} degradation.

Device Fabrication and Measurements

Fig.1 shows the structures of recessed and Schottky-gate devices, both were fabricated using commercial MOCVD-grown GaN-on-Si wafers. The hetero-structure comprises a 2.4nm GaN cap layer, a 24.5nm Al$_{0.25}$Ga$_{0.75}$N barrier layer, a 0.7nm AlN interlayer, a 420nm i-GaN channel layer and a 4.2μm GaN buffer layer. The self-terminating wet etching technique [5][6] was employed for recess etching in the fabrication of recessed-gate GaN MIS-HEMT. Subsequently, PEALD AlN and LPCVD SiN$_x$ were deposited. Ti/Al/Ni/Au ohmic contact and Ni/Au (50/100 nm) gate metal stacks were then formed successively. Schottky-gate HEMTs were fabricated with the same ohmic and gate metal structures but without passivation layers. The channel width of devices is 20μm. Transfer and output characteristics of the E-mode recessed-gate GaN MIS-HEMT are shown in Fig. 2. Thermoreflectance thermography was conducted using a Microsanj NT450 system to characterize the channel temperature. On-wafer channel temperature measurements were performed with the chip fixed on the calibration setup, and various DC-bias conditions were applied. Illumination at wavelengths of 365 nm and 530 nm was applied on the semiconductor and metal surfaces, respectively, to obtain high coefficient of thermal reflectance (C$_{th}$) [7][8].

PBTI and NBTI measurements were carried out by Keysight B1505A. As the gate stress was applied, there was a break periodically during which the transfer characteristics of the device under test (DUT) were measured with V_D=0.1V and V_G swept from 0V to 10V.

Results and Discussion

A. Self-heating Effect Characterization

Steady-state channel temperature characterization were performed under various DC-bias conditions, with V_D ranging from 5V to 45V and I_D controlled by V_G to maintain consistent power consumption for both device types. Detailed bias conditions are shown at the bottom of Fig. 5.

Fig. 3 shows the results of thermoreflectance thermography when $V_D = 30$V and $I_D = 8$ mA, where (a) and (d) present the results of AlGaN/GaN channel measured with 365nm laser, (b) and (e) show the results of gate metal surface measured with 530nm laser, and corresponding CCD images of device surface are shown in (c) and (f). In the HEMT, a hot region near the gate edge on the drain side could be observed. While no obvious hotspot appears hear in the recessed-gate MIS-HEMT, a higher ΔT is instead observed at the gate region.

Fig. 4 shows the detailed ΔT distribution along the channel under different DC-bias conditions. In the recessed-gate MIS-HEMT, as V_D and power increase, the ΔT increases significantly in the gate region, but it keeps low in the gate-drain access region. In contrast, an obvious hot region could be observed at the access region near the gate edge in the HEMT as V_D and power increases.

Fig. 5 shows the average and peak ΔT over the gate regions and access regions in both devices. For the MIS-HEMT, the hot region is mainly located in the gate region, while it is located at the access region for the HEMT. When the same V_D and power are adopted for both devices, the peak temperatures in these hot regions are comparable, but the average temperature in the gate region of MIS-HEMT is significantly lower than that of the gate-drain access region in the HEMT. The average ΔT in these two regions were adopted to evaluate thermal resistances (R_{th}), the R_{th} for the MIS-HEMT (2.15 K·mm/W) is only half that of the HEMT (4.06 K·mm/W).

B. PBTI and NBTI Measurements

As shown above, the self-heating of recessed-gate MIS-HEMT is located in the gate region. Trapping and detrapping of electrons are most likely to occur at the insulator interface here, which will significantly influence V_{TH} stability and R_{ON} degradation. To evaluate the impact of self-heating effect on this issue for recessed-gate GaN MIS-HEMT, PBTI and NBTI measurements were performed at elevated temperature. In PBTI measurements, $V_{G,Stress}$ of 20V was applied with the drain and source grounded, the results are shown in Fig. 6(a)(b) and Fig. 7(a)(b)(c). After 1800 seconds of positive gate stress, positive V_{TH} shifts of 0.38V and 0.62V were observed at 25°C and 100°C, which is due to electrons trapping under the gate. There is no obvious change in the subthreshold swing (SS), indicating that these electrons should be trapped at the AlN/SiN$_x$ interface, which is far from the channel and the SS won't be influenced significantly.

The maximum transconductance increases initially and then decreases while the on-resistance decreases first and then increases. The change of transconductance is mainly due to the resistance of gate-drain access region (R_D),

$$g_m = \frac{g_m'}{1+g_d' \cdot R_D} \tag{1}$$

where g_d' and g_m' are intrinsic conductance and transconductance that are determined by electron mobility. Therefore, the R_D should decrease at first due to detrapping of the original captured electrons at the AlN/SiN$_x$ interface traps, as shown in Fig. 9(a). This process dominates in the first period, so the R_{on} also increases in the first period. Then R_D increases as time in the second period, which might be due to electrons trapped by newly formed traps near the gate edge, and results in decreased g_m. In addition to the increased resistance at the channel under the gate caused by positive V_{TH} shift (ΔR_{ch}), the total on-resistance ($R_{on}=R_{ch}+R_D$) increases in the following period. At 100°C, the variation of g_m and R_{on} are more significant, indicating that these two processes are accelerated by the elevated temperature,

particularly the formation of new traps.

In PBTI measurements, $V_{G,Stress}$ of -10V was applied and the results of NBTI measurements are shown in Fig. 6 (c)(d) and Fig. 7 (d)(e)(f). A negative V_{TH} shift over time was observed without significant change in SS, due to the detrapping of electrons intrinsically located at the AlN/SiN$_x$ interface under the recessed gate. There is an increase in R_{on} while no significant change in g_m could be observed, indicating that the ΔR_{on} is mainly due to the positive ΔV_{TH}. As temperature increases, the detrapping process is accelerated, therefore more significant ΔV_{TH} and ΔR_{on} could be observed.

Conclusion

In this work, self-heating and BTI have been investigated in E-mode recessed-gate GaN MIS-HEMTs. Thermoreflectance thermography was performed for channel temperature characterization. The hot region is found to be located at the gate region instead of the gate-drain access region near the gate edge. Compared to the D-mode GaN-HEMTs, the recessed-gate MIS-HEMT exhibits lower channel temperature and thermal resistance under the same DC-bias conditions. Positive and negative ΔV_{TH} are observed in PBTI and NBTI measurements respectively, which is due to trapping and detrapping of electrons at AlN/SiN the interface. Detrapping of electrons and formation of new traps at the AlN/SiN$_x$ interface play critical roles in variation of R_{on} and g_m. At elevated temperature, these processes could be accelerated and result in more significant degradation.

References

[1] Saito, W. et al. Recessed-gate structure approach toward normally off high-Voltage AlGaN/GaN HEMT for power electronics applications. *IEEE Trans. Electron Devices*, 53(2), 356-362 (2006).

[2] Choi, W. et al. Improvement of V_{th} Instability in Normally-Off GaN MIS-HEMTs Employing PEALD-SiN$_x$ as an Interfacial Layer. *IEEE Electron Device Lett.*, 35(1), 30-32 (2014).

[3] Hu, Q. et al. Improved Current Collapse in Recessed AlGaN/GaN MOS-HEMTs by Interface and Structure Engineering. *IEEE Trans. Electron Devices*, 66(11), 4591-4596 (2019).

[4] He, J. et al. Performance and VTH Stability in E-Mode GaN Fully Recessed MIS-FETs and Partially Recessed MIS-HEMTs With LPCVD-SiNx/PECVD-SiNx Gate Dielectric Stack. *IEEE Trans. Electron Devices*, 65(8), 3185-3191 (2018).

[5] Xu, Z. et al. Fabrication of Normally Off AlGaN/GaN MOSFET Using a Self-Terminating Gate Recess Etching Technique. *IEEE Electron Device Lett.*, 34(7), 855-857 (2013).

[6] Xu, Z. et al. Demonstration of Normally-Off Recess-Gated AlGaN/GaN MOSFET Using GaN Cap Layer as Recess Mask. *IEEE Electron Device Lett.*, 35(12), 1197-1199 (2014).

[7] Tadjer, M. J. et al. GaN-On-Diamond HEMT Technology With T_{AVG}=176°C at $P_{DC,max}$=56W/mm Measured by Transient Thermoreflectance Imaging. *IEEE Electron Device Lett.*, 40(6), 881-884 (2019).

[8] Yuan, C., et al. A review of thermoreflectance techniques for characterizing wide bandgap semiconductors thermal properties and devices temperatures. *J. Appl. Phys.*, 132(22), Art no. 220701 (2022).

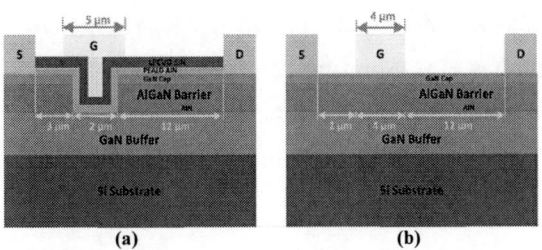

Fig. 1: **Illustration** of device structures. (a)**Recessed-gate** GaN MIS-HEMT; (b) Schottky-gate AlGaN/GaN HEMT.

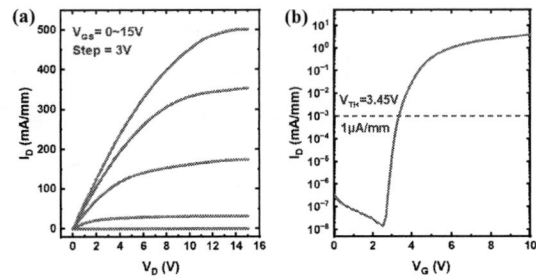

Fig. 2: (a) Output and (b) transfer characteristics of the recessed-gate GaN MIS-HEMT.

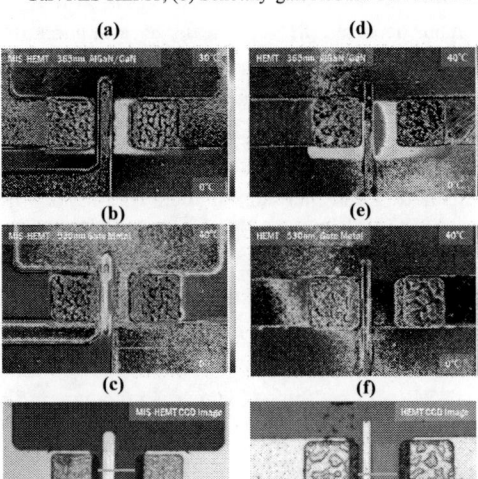

Fig. 3: Temperature rise profiles of thermoreflectance thermography measurements (V_D = 30V, I_D = 8 mA): (a) Access region of HEMT at 365 nm; (b) Gate of HEMT at 530 nm; (d) Access region of MIS-HEMT at 365 nm; (e) Gate of MIS-HEMT at 530 nm. CCD images of the device surface: (c) MIS-HEMT (f) HEMT.

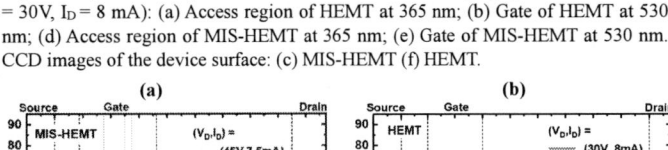

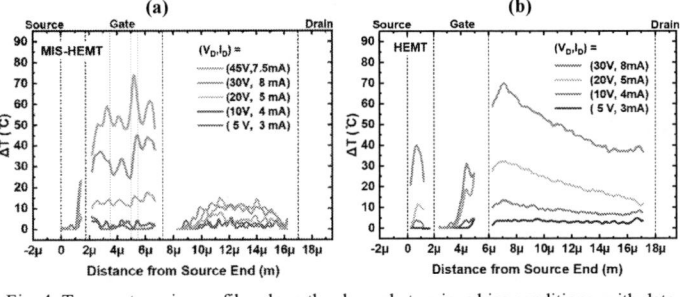

Fig. 4: Temperature rise profiles along the channel at various bias conditions, with data taken from regions shown by green lines in Fig. 3 (c) and (f). (a) Recessed-gate GaN MIS-HEMT (b) Schottky-gate AlGaN/GaN HEMT.

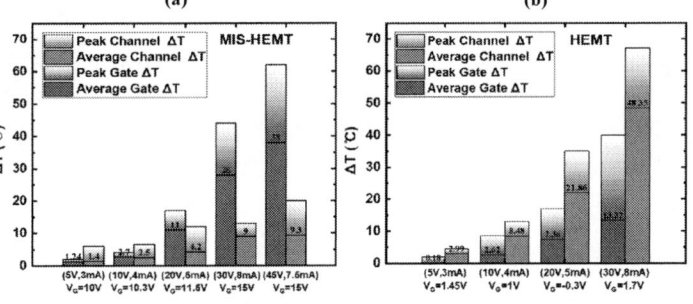

Fig. 5: Peak and averaged temperature rise for the entire gate-drain access region and gate region at various bias conditions, which are shown at the bottom. (a) Recessed-gate GaN HEMT (b) Schottky-gate AlGaN/GaN HEMT.

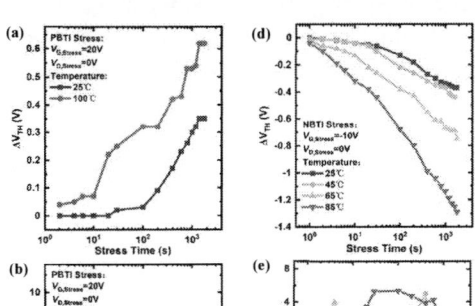

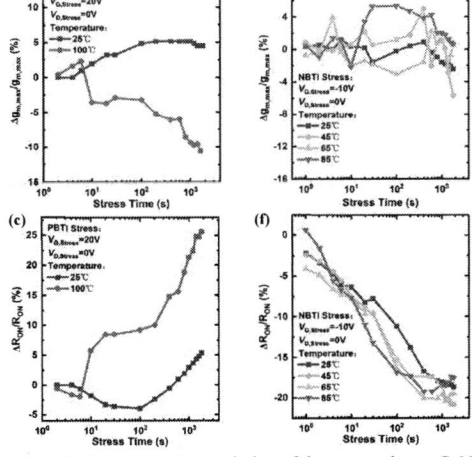

Fig. 7: Key parameters variation of the recessed-gate GaN MIS-HEMT device under test. (a) V_{TH} shift in PBTI; (b) Transconductance degradation in PBTI; (c) R_{ON} degradation in PBTI; (d) V_{TH} shift in NBTI; (e) Transconductance degradation in NBTI; (f) R_{ON} degradation in NBTI.

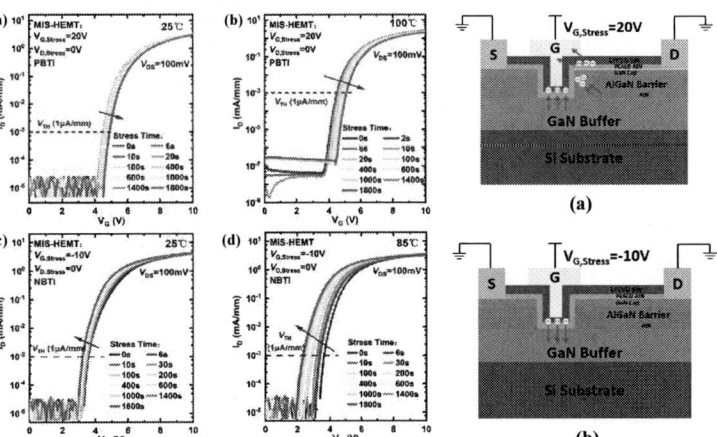

Fig. 6: Transfer characteristics in PBTI and NBTI measurements of recessed-gate GaN MIS-HEMT. (a) PBTI, 25°C; (b) PBTI, 100°C; (c) NBTI, 25°C; (d) NBTI, 100°C.

Fig. 9: Trapping and detrapping process under (a) PBTI and (b) NBTI stress.

979-8-3315-0417-5/25 $31.00 © 2025 IEEE

Enhancing Heat Dissipation of GaN HEMTs Based on Finite Element Simulation

Hongda Chen[1], Hongzhen Chen[1], Xiaohan He[1], Shiming Li[1],
Mei Wu[1], Xiaohua Ma[1], and Yue Hao[1]

[1]School of Microelectronics, Xidian University, Xi'an 710071, China

Abstract

In this paper, 3D thermal simulation model of GaN-on-diamond HEMT was established for studying its thermal management scheme. Through this simulation, it is concluded that considering electrical characteristics of the device, keeping the SiC layer at an appropriate thinning thickness is an effective scheme. In addition, the scheme for adding microchannel to the device is simulated and compared with the original device. The results show that the heat dissipation effect of this scheme is excellent.

Keywords: GaN HEMTs, heat dissipation, simulation, Diamond and microchannel

Introduction

GaN based high electron mobility transistor (HEMT) has a good application prospect in RF power electronics applications, but GaN HEMTs currently face serious heat accumulation problems. The high junction temperature of the device will lead to the decrease of electron mobility and the degradation of electrical properties [1, 2]. Therefore, in order to solve the heat dissipation problem of the device. There have been some successful heat dissipation schemes in the past, such as the active heat dissipation scheme: micro channel structure [3] and the passive scheme: replacing the original substrate with a diamond substrate with high thermal conductivity [4-6]. Besides, the passive scheme has an excellent cooling performance. Its structure diagram is shown in Figure 1.

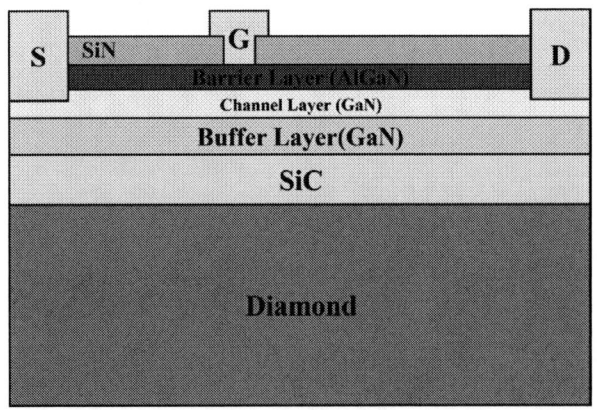

Figure 1: Structure of GaN HEMT on Diamond substrate

However, in the latter scheme, the thinning process of the original substrate (SiC) will have stress problems, resulting

in degradation of the electrical characteristics of the device [7]. When the SiC substrate is thinned to 200 μm, serious degradation has occurred [8]. It is worthy of noting that if the substrate is not completely removed, the heat dissipation potential of the diamond substrate cannot be fully utilized. Therefore, there is a trade-off between the thinning of the original substrate and the heat dissipation effect. And the excellent insulation and thermal conductivity of SiC which has a wider band gap can also improve the performance and reliability of the device [9].

Therefore, this paper studies the thickness that is beneficial to heat dissipation and does not significantly affect the electrical performance of the devices for the thinned SiC layer in GaN HEMTs by Finite Element simulation and the simulation of GaN HEMTs using the microchannel heat dissipation method at the corresponding thickness.

Simulation Details

The simulation principle is based on relevant research [10]. The simulation is done by COMSOL Multiphysics software. The thermal distribution of the GaN HEMT is simulated by the thermal conduction equation as follows:

$$\rho C_p \partial T/\partial t + \nabla \cdot (-k \nabla T) = Q \qquad (1)$$

Where ρ is the mass density, C_p is the heat capacity, k is the thermal conductivity, T is the temperature. Q is the joule heat generated by GaN devices. The time-dependent term $\partial T/\partial t$ was replaced with zero in the thermal steady-state simulation. The isothermal boundary condition of a constant temperature of 300 K was applied to the bottom of the substrate.

In order to study the influence of the thickness of SiC layer on the heat dissipation of the device, 3D thermal simulation analysis was carried out under different SiC layer thicknesses. In addition, GaN HEMTs devices on diamond substrate with microchannel structure were simulated and compared with the same devices without microchannel structure to study the improvement of microchannel structure on device heat dissipation. In the simulation, change the SiC layer thickness, and record the maximum temperature change of the entire device as the result. The structure in simulation is shown in Figure 2. It is noted that when studying the appropriate thickness of SiC layer, there is no microchannel structure in the simulation structure.

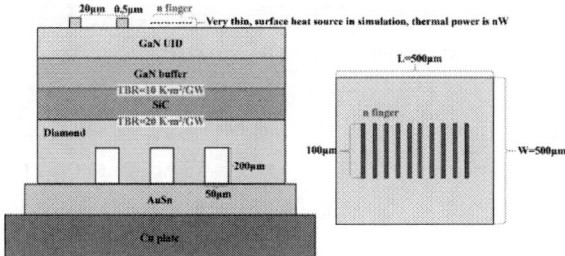

Figure 2: Structure in simulation

In the simulation structure, the interface thermal resistance of GaN/SiC is set to 10 m^{2}K/GW. The interface thermal resistance between SiC layer and diamond substrate is set to 20 m^{2}K/GW [10-11]. The material parameters used in the simulation are shown in Table 1 [10,12]. In addition, the thermal power of the heat source in the simulation is set to 10 W which has used 10 fingers, the cooling liquid used in the microchannel structure is pure water. The flow rate of water in the microchannel is 50 mL/Min, and the maximum pressure drop is 200 Kpa [3].

Table 1. Material parameters required for modeling

Material	Conductivity k(W/m^{2}K)	Mass density ρ(kg/m^3)	Heat capacity C_p(J/kg·K)	Thickness h(μm)
SiC	400[T]$^{-0.55}$	3210	690	0/1/5/10/15 /20/30/50/100 /200/350
Diamond	1500 [T]$^{-0.55}$	3510	509	350
GaN buffer	100 [T]$^{-1.42}$	6070	490	1
GaN UID	170 [T]$^{-1.42}$	6070	490	0.5
Au(Heat source)	315	19320	129	Very thin
AuSn	57	15500	129	50
Cu	387	8920	386	5000
Water	0.606	997	4180	N/A

Results and Discussion

When the SiC layer thickness is 5 μm, the simulation result is shown in Figure 3.

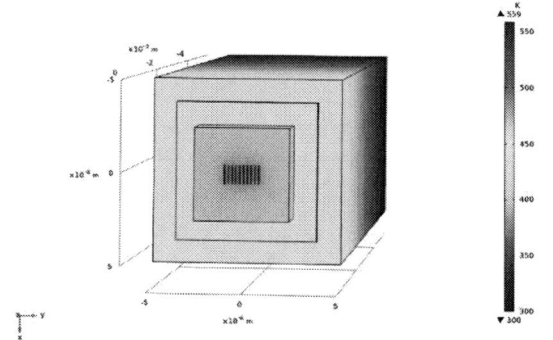

Figure 3: Simulation result for GaN HEMTs when SiC layer is 5 μm

The maximum temperature shown in Figure 3 can be used to evaluate the heat dissipation performance of the device at the corresponding SiC layer thickness. The relationship between the heat dissipation performance of GaN HEMTs device and the thickness of SiC layer can be observed by plotting all the recorded maximum temperatures as a linear plot.

A. SiC layer thickness influence

The maximum temperature of the device corresponding to different SiC layer thicknesses collected and the relationship between them is shown in Figure 4.

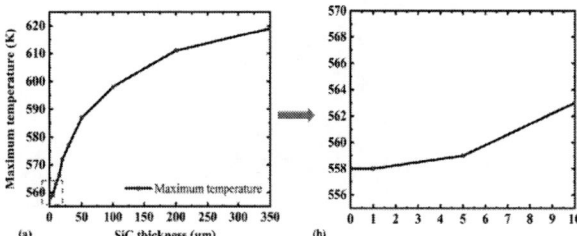

Figure 4: (a) Relationship between thickness of SiC layer and max temperature for GaN HEMTs. (b) The diagram of relationship when SiC layer thickness is at 0-10 μm

It can be concluded from the data in Figure 4(a) that there is a relationship between the heat dissipation performance of GaN HEMTs devices on diamond substrate and the thickness of SiC layer. When the thickness of SiC layer increases, the heat dissipation performance of the device decreases and the maximum temperature of the device increases. The cooling performance of the device decreases with the increase of SiC layer thickness. When the SiC layer thickness exceeds 50 μm, the maximum temperature gradient of the device decreases significantly. When the thickness of SiC layer is less than 10 μm, the variation trend of the maximum temperature of the device diminishes with the reduction in thickness. As shown in Figure 4(b), when the SiC layer thickness is less than 5 μm, the maximum temperature of the device barely changes any further.

From the view of heat dissipation, the thickness of SiC layer is smaller, the heat dissipation performance of device is better. However, the thickness of SiC layer not only affects the heat dissipation performance of the device, but also affects the electrical performance of the device. SiC layer can provide good electrical insulation performance, reduce leakage current and improve device reliability. But removing SiC layer completely will degrade its electrical insulation performance. Therefore, in order to support the structure of heterojunction and not completely remove the SiC layer, the appropriate thickness of the SiC layer should be determined to be about 5 μm for maximizing the heat dissipation potential of the diamond substrate.

B. Microchannel structure influence

After study of the passive scheme, this paper also investigates the improvement for the heat dissipation capacity of GaN HEMT devices by adding microchannel structure. The simulation model is shown in Figure 2. Under the same conditions, when the SiC layer thickness is 5 μm, the simulation result of GaN HEMT devices with microchannel structure is shown in Figure 5.

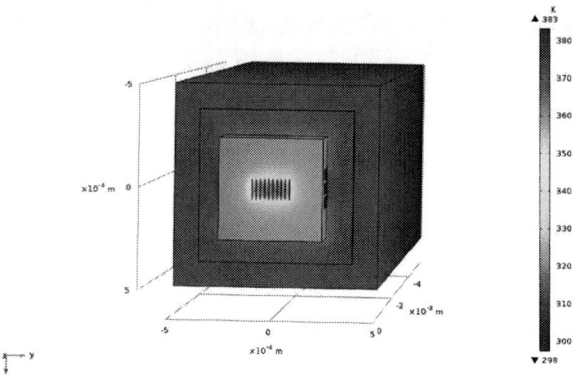

Figure 5: Simulation result for GaN HEMTs with microchannel when SiC layer is 5 μm

Compared with the simulation result without microchannel structure in Figure 2, the maximum device temperature of the former decreases by nearly 176 K. It shows that the microchannel structure greatly enhances the heat dissipation capacity of GaN HEMT devices. Moreover, according to the previous research, increasing the flow rate of the cooling liquid in the microchannel will also enhance the heat dissipation capacity of the device.

Although the microchannel enhances the heat dissipation performance of the device significantly, its manufacture is very difficult, and there are still connectivity difficulties and pump problems for its practical application. Therefore, in certain situations where reliability is critical, passive heat dissipation methods may still be preferred.

Conclusion

The 3D thermal analysis simulation of GaN-on-Diamond HEMTs power device was carried out. By changing the thickness of SiC layer in the structure, the simulation results show that the thickness of SiC layer is negatively correlated with the heat dissipation performance of the device. Reducing the thinning thickness of the SiC layer is a good cooling scheme. In addition, considering the electrical characteristics of the GaN HEMT device, the appropriate SiC thinning thickness should be about 5 μm.

Active heat dissipation schemes for GaN HEMT devices to enhance heat dissipation capacity by adding embedded microchannels are also investigated in this paper. By comparing the simulation results with those without microchannel, it can be concluded that the microchannel has

a great gain to the heat dissipation of the device. However, there are still many problems to be solved in practice.

Reference

[1] B.K. Schwitter, A.E. Parker, S.J. Mahon, A.P. Fattorini, M.C. Heimlich, Impact of bias and device structure on gate junction temperature in AlGaN/GaN-on-Si HEMTs, IEEE Trans. Electron Devices 61 (5) (May 2014) 1327–1334.

[2] A. Darwish, A.J. Bayba, H.A. Hung, Channel temperature analysis of GaN HEMTs with nonlinear thermal conductivity, IEEE Trans. Electron Devices 62 (3) (Mar. 2015) 840–846.

[3] X. Chen, F. N. Donmezer, S. Kumar and S. Graham, "A Numerical Study on Comparing the Active and Passive Cooling of AlGaN/GaN HEMTs," in IEEE Transactions on Electron Devices, vol. 61, no. 12, pp. 4056-4061, Dec. 2014.

[4] Qingzhi Wu et al., Performance Comparison of GaN HEMTs on Diamond and SiC Substrates Based on Surface Potential Model, 2017 ECS J. Solid State Sci. Technol. 6 Q171.

[5] T. Liu et al., "3-inch GaN-on-Diamond HEMTs With Device-First Transfer Technology," in IEEE Electron Device Letters, vol. 38, no. 10, pp. 1417-1420, Oct. 2017.

[6] A. E. Helou et al., "High-Resolution Thermoreflectance Imaging Investigation of Self-Heating in AlGaN/GaN HEMTs on Si, SiC, and Diamond Substrates," in IEEE Transactions on Electron Devices, vol. 67, no. 12, pp. 5415-5420, Dec. 2020.

[7] Piotr Caban, Wlodek Strupinski, Jan Szmidt, Marek Wojcik, Jaroslaw Gaca, Ozgur Kelekci, Deniz Caliskan, Ekmel Ozbay, Effect of growth pressure on coalescence thickness and crystal quality of GaN deposited on 4H–SiC, Journal of Crystal Growth, Volume 315, Issue 1, 2011, Pages 168-173, ISSN 0022-0248.

[8] B. Parvez et al., "Improvements From SiC Substrate Thinning in AlGaN/GaN HEMTs: Disparate Effects on Contacts, Access and Channel Regions," in IEEE Electron Device Letters, vol. 42, no. 5, pp. 684-687, May 2021.

[9] S. -I. Nishizawa, "SiC Materials and Devices for Future Green Society," 2024 8th IEEE Electron Devices Technology & Manufacturing Conference (EDTM), Bangalore, India, 2024, pp. 1-3.

[10] F. Guo, T. Li, H. Man, K. Liu and X. Wang, "Enhanced Heat Dissipation of GaN RF Devices Based on Double-diamond Structure," 2021 6th International Conference on Integrated Circuits and Microsystems (ICICM), Nanjing, China, 2021, pp. 55-60.

[11] J. Cho, E. Bozorg-Grayeli, D. H. Altman, M. Asheghi and K. E. Goodson, "Low Thermal Resistances at GaN–SiC Interfaces for HEMT Technology," in IEEE Electron Device Letters, vol. 33, no. 3, pp. 378-380, March 2012.

[12] Huaixin Guo, Yuechan Kong, Tangsheng Chen, Thermal simulation of high power GaN-on-diamond substrates for HEMT applications, Diamond and Related Materials, Volume 73, 2017, Pages 260-266.

Fabrication and Performance Assessment of High-Mobility SiGe Channel P-Type SOI FinFET Transistor

Zijing Zhang[1,2], Yuchen Wu[2,3], Yan Li[4], Huaizhi Luo[1,2,4], FanYu Liu[2,3], Yongliang Li*[1,2]

[1]Key Laboratory of Fabrication Technologies for Integrated Circuits, Chinese Academy of Sciences, Beijing100029. [2]Institute of Microelectronics, Chinese Academy of Sciences, Beijing100029, China. [3]Key Laboratory of Science and Technology on Silicon Devices, Chinese Academy of Sciences, Beijing100029, China. [4]School of Applied Science, Beijing Information Science and Technology University, China.

*E-mail: liyongliang@ime.ac.cn.

Abstract

In this work, we successfully fabricated P-type FinFET transistor featuring SiGe channel on a silicon-on-insulator (SOI) substrate by adopting the SiGe epitaxy and etching processes. Under the gate length of 40 nm, its subthreshold swing (SS) is 82 mV/dec and I_{on}/I_{off} ratio is ~10^6. Its excellent electrical and reliability characteristics reveal that this SiGe SOI FinFET is a potential candidate for the future advanced nodes.

Keywords: SiGe channel, SOI substrate, FinFET.

Introduction

With progresses of CMOS technology to the 3nm node and beyond, traditional Si-based devices face significant challenges, due to their performance improvements nearing the limit [1]. SiGe material has been demonstrated a promising candidate for p-MOSFET due to its higher effective hole mobility (μ_{eff}) [2]. However, compared to conventional Si material, its narrower bandgap makes the SiGe channel device more susceptible to tunneling effects, leading to increased leakage currents [3]. Silicon-on-insulator (SOI) substrate offers a viable solution to address these issues, providing additional benefits such as reduced source/drain parasitic capacitance, elimination of the latch-up effect and improved threshold voltage (V_{TH}) adjustability [4]. However, there are limited reports disclosing the fabrication of SiGe SOI FinFET with superior electrical and reliability.

In this work, SiGe SOI FinFET was successfully prepared by optimizing the epitaxy and etching processes. And its performance was assessed by multiple analysis methods.

Device Fabrication

The fabrication flow of SiGe SOI FinFET is schematically shown in Fig. 1. Firstly, the load-Si layer of SOI substrate was thinned down from 55nm to 20 nm by a thermal oxidation and wet etching of oxide removal. Then, a phosphorus ground-plane implantation was carried out in the load-Si layer to inhibit the leakage of the undesired load-Si parasitic channel. Subsequently, 60 nm SiGe with 5 nm Si-cap layer was deposited in a reduced pressure chemical vapor deposition system. The fin pattern was formed using the traditional self-aligned double patterning (SADP) technique under an optimal HBr/N_2/He plasma conditions. Following the dummy gate patterning and definition of spacers, source/drain (S/D) implantation were implemented. After the dummy gate removal, the stack of Al_2O_3/HfO_2/TiN/TaN/TiAlC was subsequently deposited by atomic layer deposition (ALD) to form the gate stack. Finally, the standard FinFET following processing was employed to complete the SiGe SOI FinFET device fabrication.

Results and Discussion

Figure 2 (a) and (b) show the analysis results of transmission electron microscopy (TEM) for the SiGe/Si-cap epitaxial layer. These results reveal that the SiGe/Si-cap layer has good crystal quality, without obvious mismatches or dislocations. Meanwhile, it also has a thin and clear interface. Moreover, the Secondary Ion Mass Spectrometry (SIMS) analysis was used to characterize the distribution and mole fraction of Si and Ge elements in the sample, as shown in Fig. 2 (c). It is found that the mole fraction of Ge is about 25%. Meanwhile, the gradient of the Ge at the interface is large, indicating that there is a thin and sharp interface, which is consistent with TEM analysis.

Figure 3 (a) and (b) show the etching profile of SiGe/Si-cap structure under present base-line and optimized etching condition. It can be found that the SiGe layer has obvious lateral etching and the etching rate of SiGe is significantly higher than that of Si under the present base-line etching condition, as shown in Fig. 3 (a). It can be attributed to the fact that the narrower band gap of SiGe leading to enhanced chemical reaction of SiGe etching [3], and present base-line etching condition can not provide sufficient passivation on the fin sidewall. Therefore, as shown in Figure 3(b), the optimized etching condition with HBr/N_2/He plasma was developed to attain a vertical profile of SiGe fin, thanks to N_2 was employed to realize a better sidewall passivation.

Figure 4 (a) presents the cross-sectional TEM image of the SiGe channel under the HK/MG stack at the end of fabrication processing. The total fin height is 85 nm, including a 5 nm Si-cap/60 nm SiGe/20nm load-Si and the

979-8-3315-0417-5/25 $31.00 © 2025 IEEE 375

fin width is 20 nm. Meanwhile, its EDS mapping results shown in Fig. 4 (b) reveal that the fin was uniformly encapsulated the Al_2O_3/HfO_2/TiN/TaN/TiAlC structure.

The transfer characteristics of SiGe and Si SOI FinFET are presented in Fig. 5. They were prepared under the same process, and the only difference is using the same thickness of Si epitaxy to replace SiGe film. It is found that SiGe SOI FinFET under the gate length of 40 nm achieved a higher I_{on} of 6.72×10^{-4} A/μm and a comparable SS of 82 mV/dec. Although its I_{off} was slightly worse due to narrower bandgap of SiGe, a higher I_{on}/I_{off} of $\sim 10^6$ was achieved, thanks to higher hole mobility and I_{on} of SiGe channel.

Figure 6 displays the transconductance (G_m) of SiGe and Si SOI FinFET. It is found that the G_{mmax} of SiGe device is significantly greater than that of Si device, due to the greater sensitivity of I_D to V_G in the saturation region, exhibiting a stronger control capability of gate.

As shown in Fig. 7, the mobility of SiGe and Si SOI FinFET was extracted by the split CV method, with CV testing conducted at 100 kHz and a hold time set to 0.05 s. The SiGe achieved a peak hole mobility of 368 cm^2/V·s, which is 60% higher than that of Si SOI FinFET. Moreover, Fig. 8 compares the previously reported mobility and SS of SiGe FinFET with this study. It is found that excellent mobility is achieved with reasonable SS, indicating the high-quality SiGe channel was attained using this fabrication process of SiGe SOI FinFET.

The hot carrier injection (HCI) testing was applied for SiGe SOI FinFET. The test was performed at the worst-case bias voltage with $V_G=V_D=$ -2V and a stress time of 1000 seconds, and electrical characterization was performed at specific intervals during the test. As shown in Fig. 9 (a), the HCI degradation in the linear region indicates that, with increasing stress time, the I_{on} shows insignificant degradation, while the V_{TH} exhibits a negative shift of 13mV. The change in V_{TH} is attributed to positive oxide defects generated due to HCI stress. The insignificant degradation in the I_{on} suggests that the mobility remains almost unaffected, indicating minimal increase in interface traps. This demonstrates that the interface has a low number of dangling bonds, confirming good interface quality of SiGe SOI FinFET. Moreover, as shown in Fig. 9 (b), in the saturation region, I_{DSAT} initially increases with stress time and then decreases, while the V_{TH} shifts from a negative to a positive drift. The initial negative V_{TH} drift is attributed to positive oxide traps caused by hole tunneling near the source. The subsequent positive V_{TH} shift is due to negative oxide traps injected by electron injection near the drain. These above results have further confirmed the long-term stability and high-quality channel of SiGe SOI FinFET.

Conclusion

SiGe SOI FinFET with excellent electrical performance

and reliability characteristics is successfully fabricated by adopting the SiGe epitaxy and etching processes, indicating that it is a potential candidate for the future advanced nodes.

Acknowledgments

This work is supported by the Strategic Priority Research Program of Chinese Academy of Sciences (Grant no. XDA0330304).

References

[1] T. Hiramoto, "Never-Ending CMOS Innovation: The Past, Present, and Future Perspectives," 2020 International Symposium on VLSI Technology, Systems and Applications (VLSI-TSA), Hsinchu, Taiwan, 2020, pp. 3-4.

[2] C. H. Lee et al., "Toward High Performance SiGe Channel CMOS: Design of High Electron Mobility in SiGe nFinFETs Outperforming Si," 2018 IEEE International Electron Devices Meeting (IEDM), San Francisco, CA, USA, 2018, pp. 35.1.1-35.1.4.

[3] K. Vanlalawmpuia, B. Bhowmick and M. Choudhury, "Optimization of electrical parameters in SiGe channel nMOSFET," 2017 Devices for Integrated Circuit (DevIC), Kalyani, India, 2017, pp. 231-235.

[4] Q. Zhang et al., "Novel GAA Si Nanowire p-MOSFETs With Excellent Short-Channel Effect Immunity via an Advanced Forming Process," in IEEE Electron Device Letters, vol. 39, no. 4, pp. 464-467.

[5] C. -H. Fu et al., "Enhanced Hole Mobility and Low TinvTinv for pMOSFET by a Novel Epitaxial Si/Ge Superlattice Channel," in IEEE Electron Device Letters, vol. 33, no. 2, pp. 188-190, Feb. 2012.

[6] D. Lei et al., "Enhanced Germanium-Tin P-Channel FinFET Performance using Post-Metal Anneal," 2018 EDTM, 2018, pp. 50-52.

[7] W. -S. Liao et al., "PMOS Hole Mobility Enhancement Through SiGe Conductive Channel and Highly Compressive ILD- SiNx Stressing Layer," in IEEE Electron Device Letters, vol. 29, no. 1, pp. 86-88, Jan. 2008.

[8] J. Hallstedt, M. von Haartman, P. . -E. Hellstrom, M. Ostling and H. H. Radamsson, "Hole mobility in ultrathin body SOI pMOSFETs with SiGe or SiGeC channels," in IEEE Electron Device Letters, vol. 27, no. 6, pp. 466-468, June 2006.

[9] K. Tachi et al., "Relationship between mobility and high-k interface properties in advanced Si and SiGe nanowires," 2009 IEEE International Electron Devices Meeting (IEDM), Baltimore, MD, USA, 2009, pp. 1-4.

979-8-3315-0417-5/25 $31.00 © 2025 IEEE

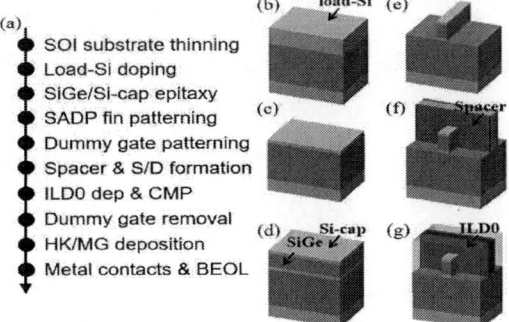

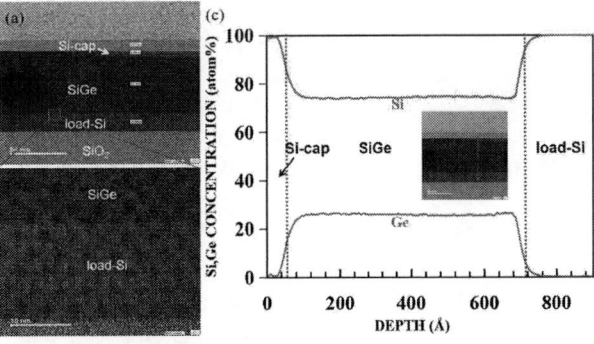

Fig. 1 (a) Key fabrication steps for realizing SiGe p-FinFETs. (b) SOI wafer preparing, (c) load-Si thinning, (d) SiGe/Si-cap epitaxy, (e) fin formation, (f) Spacer formation, (g) ILD0 deposition.

Fig. 2 (a) TEM image of SiGe/Si-cap epitaxial layer and (b) of the interface, showing the good crystal quality and a thin and clear interface. (c) SIMS analysis of SiGe/Si-cap epitaxial layer, and the mole fraction of Ge is about 25%.

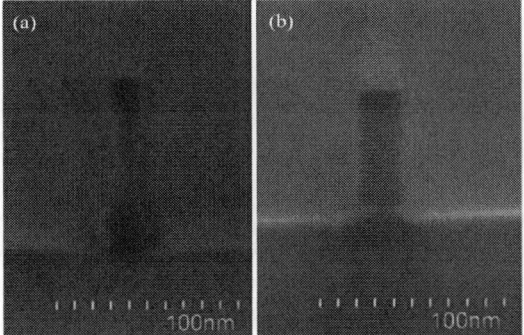

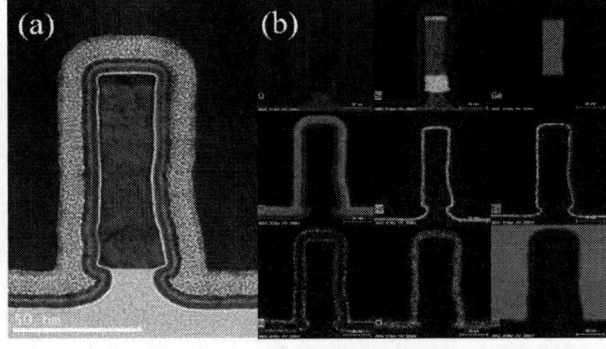

Fig. 3 Scanning electron microscope (SEM) images of fin profile under (a) $HBr/O_2/He$ plasma and (b) $HBr/N_2/He$ plasma.

Fig. 4 HAADF-STEM and EDS mapping results of the SiGe channel FinFET under the HK/MG stack at the end of processing.

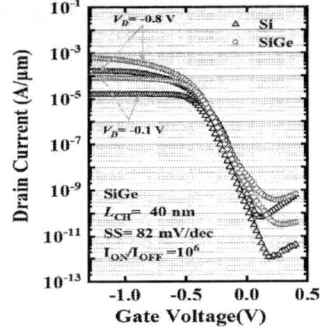

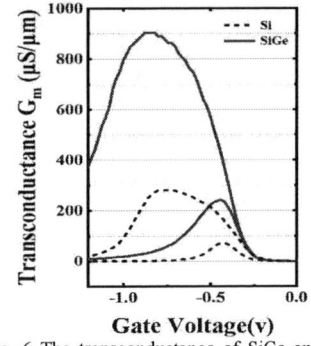

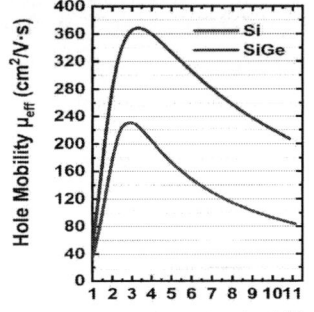

Fig. 5 The transfer characteristic of SiGe and Si SOI FinFET. SiGe SOI FinFET achieved a higher I_{on} and I_{on}/I_{off}.

Fig. 6 The transconductance of SiGe and Si SOI FinFET. It shows SiGe device has a stronger gate control capability than Si device.

Fig. 7 The mobility of SiGe and Si SOI FinFET extracted by Split CV. The peak mobility of SiGe achieves 368 cm²/V·s.

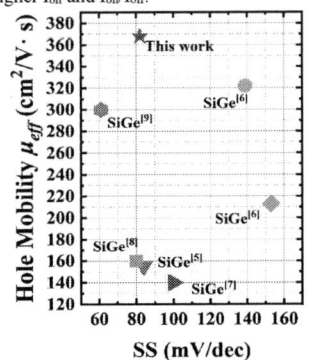

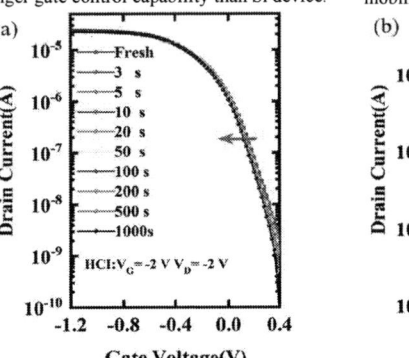

Fig. 8 Benchmarking of peak μ_{eff} and SS of this SiGe SOI FinFET with previously reported works.

Fig. 9 The hot carrier injection (HCI) testing applied for (a) the linear region and (b) the saturation region. linear region shows insignificant degradation. And in saturation region, I_D initially increases and then decreases, while the V_{TH} shifts from a negative to a positive drift.

979-8-3315-0417-5/25 $31.00 © 2025 IEEE   377

BEOL-Compatible Photosensors by Atomic-Layer-Deposited ZnO Semiconductor Transistors and Their Monolithic 3D Integration

Yuyan Fan[1,2,†], Ziheng Wang[3,†], Yulong Dong[1,2], Zhiyu Lin[3], Danyang Chen[1,2], Shiyi Zhang[2], Jingquan Liu[1], Mengwei Si[1,3*], and Xiuyan Li[1,2*]

[1] National Key Laboratory of Advanced Micro and Nano Manufacture Technology, Shanghai Jiao Tong University, Shanghai China,
[2] Department of Nano/Microelectronics, Shanghai Jiao Tong University, Shanghai China,
[3] Department of Electronic Engineering, Shanghai Jiao Tong University, Shanghai, China
[†]These authors contribute equally to this work, Email: mengwei.si@sjtu.edu.cn, xiuyanli@sjtu.edu.cn

Abstract

In this work, we demonstrate back-end-of-line (BEOL) compatible photosensors and their monolithic 3D integration with memory devices. By using the BEOL-compatible atomic layer deposition method, high-performance ZnO thin-film transistors with good photosensing properties are obtained. Furthermore, monolithic 3D integration of ZnO photosensor with low-temperature poly silicon transistor is demonstrated to achieve the capability of photosensing and storage in a 2T0C DRAM unit cell.

Keywords: oxide photosensor, BEOL-compatible, monolithic 3D integration

Introduction

Oxide semiconductors such as indium oxide (InO) [1-2], indium gallium zinc oxide (IGZO) [3-5], indium tungsten oxide (IWO) [6] have attracted intensive attention to be used as n-type channel materials. Particularly, atomic layer deposited (ALD) oxide semiconductors have attracted much interest due to high mobility, low off current (I_{OFF}), and good uniformity, with a low thermal budget under 250°C [6-8]. This is promising for achieving back-end-of-line (BEOL) compatible field-effect transistors (FETs) for monolithic 3D (M3D) integration of CMOS logic circuits, DRAM, and sensors. On the one hand, to integrate sensing, memory, and computation circuits through M3D integration can significantly improve the bandwidth of data transfer. On the other hand, such integration enables more functionality such as sensing and memory co-integration in a single cell, which is significant in fields such as motion detection and in-sensor computing [9].

In our previous work, M3D integration of oxide n-FETs stacked on low-temperature polysilicon (LTPS) FETs has been demonstrated [10], implementing M3D CMOS inverter and logic gates with improved area efficiency. In this work, we further demonstrate the high performance of BEOL-compatible photosensors with ALD-deposited ZnO transistors. Besides, M3D integration of photosensor with ZnO n-FET vertically stacked on LTPS n-FET is also demonstrated, showing a capability of photosensing and storage in a 2T0C DRAM unit cell. The proposed photosensing device and M3D integration process provide a potential candidate for future co-integration of sense, memory, and computing.

Device fabrication and characterization

ZnO FET was fabricated on Si/SiO$_2$ substrates to explore photosensing properties. Fig. 1(a) shows the schematic diagram of the transistors and the specific device fabrication process. 40 nm Ni was deposited as gate electrodes by thermal evaporation, patterned by photolithography. 13 nm Al$_2$O$_3$ as gate insulator was grown by ALD at 200 °C, followed by 12 nm ZnO channel deposition in situ. After via opening, source/drain metals were deposited and patterned, and the devices were annealed in air at 250°C. Fig. 1(b) schematically shows the measurement setup of photoelectric properties. Laser photo sources with four kinds of visible wavelengths, 405 (purple), 450 (blue), 520 (green) and 650 (red) nm, were used. The intensity of light was controlled by the input voltage of a signal generator and calibrated by a photometer. Fig. 1(c) illustrates the shift of threshold voltage (V_{th}) and mobility of ZnO FETs as a function of light intensity at different levels of 405 nm light, where the light energy exceeds the energy gap of ZnO. An obvious negative shift of V_{th} and mobility of 14.2 cm^2/V·s are observed, both indicating the high performance of ZnO FETs. Moreover, the I_D-V_G curves of ZnO FET at V_{DS}= 0.1V under different intensities of visible light with four wavelengths, respectively, are compared in Fig. 2(a)-(d). The negative shift of V_{th} is noticeable under each wavelength of light, although more obvious under blue and purple light with shorter wavelengths. And, the saturation in the V_{th} shift of I_D-V_G curves under large light intensities suggests that the photoelectric effect mainly originated from carrier activation by the illumination. These results demonstrate the high-resolution and wide-band light-sensing performance of ZnO FET.

M3D integration of ZnO FET for Photosensing and storage

In this section, we demonstrate the M3D integration of BEOL-compatible ZnO photosensors with LTPS FETs. Here, we propose to use a 2T0C DRAM cell (oxide n-FET + LTPS n-FET) to realize the sensing and memory co-integration. ZnO n-FET is used as a sensor and also as a write transistor of the 2T0C DRAM, whereas LTPS n-FET is used as a read transistor. The structure is schematically shown in Fig. 3. Fig. 4 shows the basic operation of the M3D 2T0C DRAM, where the charge is stored at the storage node (SN) after the write operation. For the operation of sensing and memory co-integration, as schematically shown in Fig. 5, state "0" is firstly written into the SN node (V_{WBL}=0), then a high voltage is applied to write bit-line (V_{WBL}). When the cell is exposed to light, the ZnO n-FET turns on and the V_{WBL} will be written into the SN node so that the SN node is in state "1". When the light is off, the SN node remains in state "1". Note that the LTPS n-FET is in the bottom layer so that does not have a photo response. Therefore, the above 2T0C cell can realize the sensing and memory functions in the same cell.

To experimentally demonstrate the function, blue and purple light are applied on ZnO FET after SN is set to state "0". The results in Fig. 6 clearly show the photosensing and storage function of the 2T0C DRAM cell in both cases, as expected. The slight drop of current on SN is likely caused by charge trapping effects, which does not affect the storage of detected photo information in state "1".

Conclusion

Photoelectric effects of BEOL-compatible ALD deposited ZnO n-FETs have been investigated and high performance of photosensing properties have been achieved in this work. In addition, the monolithic 3D integration of ZnO FET photosensor with LTPS n-FET as a 2T0C DRAM cell and the capability of photosensing and storage in the same cell have been demonstrated for the first time. Our work provides a promising way for the co-integration of sensing, memory, and computing.

Acknowledgments

This work is supported by STI 2030—Major Projects (Grant No. 2022ZD0210600), National Key R&D Program of China (Grant No. 2022YFB3606900), National Natural Science Foundation of China (62274107, 92264204), and the Multi-Glass Project (MPG) Platform from Tianma Microelectronics Co., Ltd.

References

[1] Si, Mengwei, et al., "Scaled indium oxide transistors fabricated using atomic layer deposition." , Nat. Electron., 5(3), pp. 164-170 (2022).

[2] Liu, Jialong, et al. "Low-power and scalable retention-enhanced IGZO TFT eDRAM-based charge-domain computing." 2021 IEEE International Electron Devices Meeting (IEDM), pp. 21-1, 2021.

[3] X. Duan, K. Huang, J. Feng, J. Niu, H. Qin, S. Yin, G. Jiao, D. Leonelli, X. Zhao, W. Jing, Z. Wang, Q. Chen, X. Chuai, C. Lu, W. Wang, G. Yang, D. Geng, L. Li, and M. Liu, "Novel Vertical Channel-All-Around (CAA) IGZO FETs for 2T0C DRAM with High Density beyond 4F² by Monolithic Stacking," in IEEE Int. Electron Devices Meet., pp. 222-225, 2021.

[4] Belmonte, Attilio, et al. "Lowest I OFF< 3×10^{-21} A/μm in capacitorless DRAM achieved by reactive ion etch of IGZO-TFT." 2023 IEEE Symposium on VLSI Technology and Circuits (VLSI Technology and Circuits), pp. 1-2, 2023.

[5] W. Chakraborty, B. Grisafe, H. Ye, I. Lightcap, K. Ni, and S. Datta, "BEOL compatible dual-gate ultra thin-body W-doped indium-oxide transistor with I_{on}= 370μA/μm, SS=73mV/dec and I_{on}/I_{off} Ratio> 4×10^9," in Proc. Symp. VLSI Technol., p. TH2.1., Jun. 2020.

[6] Si, Mengwei, et al., "Why In2O3 can make 0.7 nm atomic layer thin transistors.", Nano Lett., 21(1), pp. 500-506. (2020)

[7] M. Si, Z. Lin, Z. Chen, and P. D. Ye, "First Demonstration of Atomic-Layer-Deposited BEOL-Compatible In_2O_3 3D Fin Transistors and Integrated Circuits: High Mobility of 113 cm²/V•s, Maximum Drain Current of 2.5 mA/μm and Maximum Voltage Gain of 38 V/V in In_2O_3 Inverter," in Proc. Symp. VLSI Technol., p. T2-4, Jun. 2021.

[8] Zhang, Jie, et al. "First Demonstration of BEOL-Compatible Atomic-Layer-Deposited InGaZnO TFTs with 1.5 nm Channel Thickness and 60 nm Channel Length Achieving ON/OFF Ratio Exceeding 10 11, SS of 68 mV/dec, Normal-off Operation and High Positive Gate Bias Stability." 2023 IEEE Symposium on VLSI Technology and Circuits (VLSI Technology and Circuits), pp. 1-2, 2023.

[9] Zhang, Zhenhan, et al. "All-in-one two-dimensional retinomorphic hardware device for motion detection and recognition." Nat Nanotechnol., 17(1), pp.27-32. (2022).

[10] Tang, W., et al. "Monolithic 3D integration of vertically stacked CMOS devices and circuits with high-mobility atomic-layer-deposited In_2O_3 n-FET and polycrystalline Si p-FET: Achieving large noise margin and high voltage gain of 134 V/V.", International Electron Devices Meeting (IEDM), pp. 483-486, 2022.

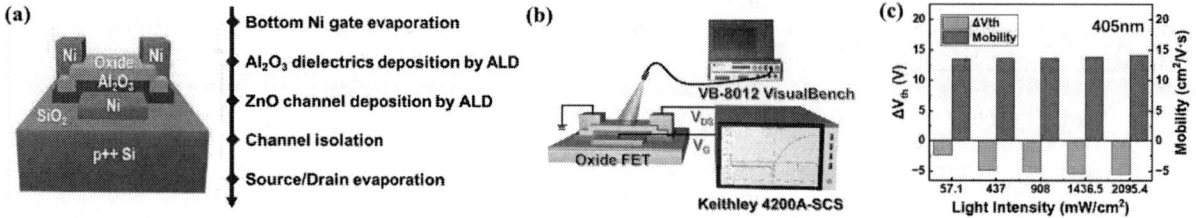

Fig. 1(a) The schematic diagram and fabrication process of ZnO FET. (b) The measurement setup of photoelectric properties of ZnO FET. (c) The shift of V_{th} and mobility of ZnO FET under different light intensities of 405nm light.

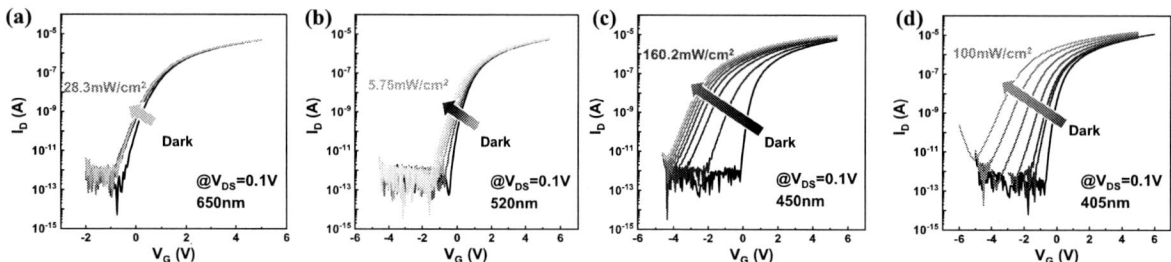

Fig. 2 The I_D-V_G curves of ZnO FET at V_{DS}= 0.1V and under different intensities of visible light with wavelength of (a) 650 nm (red), (b) 520 nm (green), (c) 450 nm (blue) and (d) 405nm(purple), respectively. A negative shift in V_T is noticeable with each light wavelength, and shorter wavelengths have a more significant impact. The V_T shift rises as light intensity increases initially and then saturates when it goes above certain values.

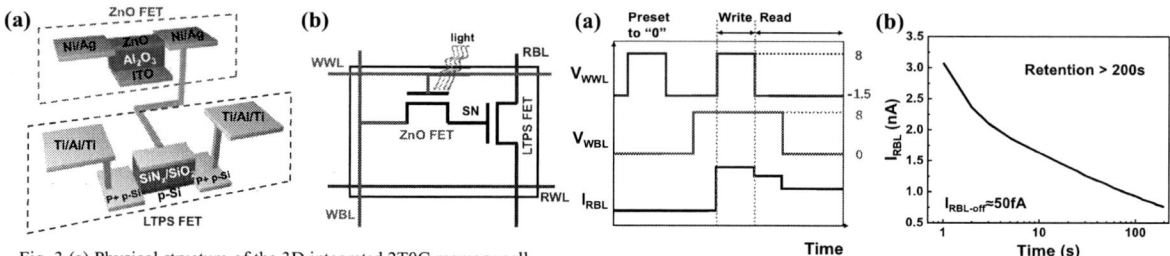

Fig. 3 (a) Physical structure of the 3D integrated 2T0C memory cell. (b) Equivalent circuit diagram of the 1-bit cell, where ZnO n-FET is used as sensor and also as write transistor of the 2T0C DRAM and LTPS n-FET is used as read transistor.

Fig. 4 (a) Timing diagram of write and read operation for the 2T0C cell, SN is first set to "0", and then a V_{WBL} of 8 V is written, after removing the V_{WBL}, the read current can be stored. (b) I_{RBL}-Time characteristic of the 3D integrated 2T0C cell, showcasing the remarkable storage capability of our integrated DRAM.

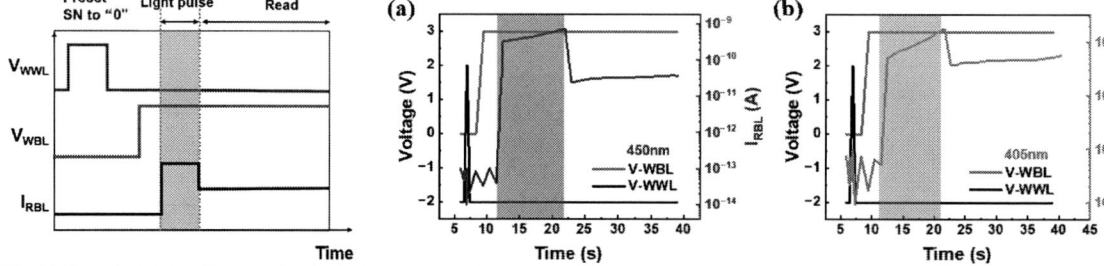

Fig. 5 Schematical timing diagram of optical write and electric read operation for the 2T0C cell. The sensing and memory functions of light pulse in a cell can be achieved.

Fig. 6 Experimental results of optical write and read operation for the 2T0C cell with light wavelength of (a) 450 nm and (b) 405 nm, respectively. The results clearly show the sensing and storage function of 2T0C DRAM cell in both cases.

979-8-3315-0417-5/25 $31.00 © 2025 IEEE

Optimizing Etch Processes for Enhanced Yield and Performance in SOT-MRAM Devices on 300 mm Wafers

Zhenghui Ji [1*], Wenlong Yang [1*], Guoxiu Qiu [1], Dandan Yang[1], Kaiyuan Zhou[1], Qingxiu Li[1], Qijun Guo[1], Enlong Liu[1#], Shikun He [1]

[1] Zhejiang Hikstor Technology Co. Ltd., Hangzhou 311305, China. jizhenghui@hikstor.com

[*]These authors contributed equally to this work. [#]Corresponding author

Abstract

To improve device performance and yield in magnetic tunnel junctions (MTJs) for spin-orbit-torque magnetic random access memory (SOT-MRAM), different ion beam etch (IBE) processes for MTJ device fabrication are developed and compared in detail. By introducing IBE in grazing incidence with silicon nitride (SiN) protection, footing reduction and metal redeposition removal can be achieved simultaneously, leading to more uniform device performance and high yield (>99.9%) on 300 mm wafer manufacturing platform. Integration of SOT-MTJs with the optimized IBE process provides a feasible way to streamline the mass production of SOT-MRAM.

Keywords: SOT-MRAM, Ion Beam Etch, High Yield, 300 mm wafer platform

Introduction

For its fast speed, high endurance, low-power dissipation and compatibility with CMOS processes, spin-transfer-torque magnetic random access memory (STT-MRAM) has attracted extensive consideration as one of the leading candidates of next generation non-volatile memory [1]. However, as a two-terminal device, high current density stresses the tunnel barrier (usually MgO) in the STT magnetic tunnel junctions (MTJs) during write operation and hence degrades its reliability. Thus, spin-orbit-torque MRAM (SOT-MRAM) with three-terminal device structure is proposed for its separate read and write paths, which can intrinsically solve the tunnel barrier aging issue, also for its sub-nanosecond switching and low power consumption.

However, the fabrication of SOT-MTJs still faces many problems due to its unique device structure. Firstly, the SOT channel is ultra-thin (less than 10 nm), which makes it challenging to stop MTJ etching exactly on the channel surface across 300 mm wafer during fabrication. Secondly, a steep sidewall with minor bottom footing is essential to address the degradation of exchange bias field ($\mu_0 H_{ex}$) in the commonly used top-pinned (TP) stack structure of SOT-MTJs [2]. Thirdly, the tunnel barrier in TP MTJs is next to the SOT channel and metal redeposition during etch process is more likely to happen, which will cause electrical short in single MTJ bit and induce yield loss.

In this paper, different ion beam etch (IBE) processes were optimized to solve the aforementioned issues. By splitting MTJ etch into steps with different incident angles and introducing intermediate silicon nitride (SiN) as protection layer, SOT-MTJs with excellent device performance and high yield have been fabricated, which showcases a promising outlook for large scale and high-yield SOT-MRAM fabrication.

Results and Discussion

The detailed integration of SOT-MTJs has been reported in our previous work [3]. In the following sections, the innovative optimization and enhancement of etch process will be focused on.

A. Etch process A and B

Process A consisted of two etch steps named main etch (ME) and trim, respectively. During this process, ME was aimed to pattern the devices following hard mask (HM) and remove MTJ film stack till the upper surface of SOT channel (4 nm tungsten in this work). Then trim process is applied to make a precise stop on the SOT channel across the whole wafer with the help of optical emission spectroscopy (OES) for its slow etching rate, as illustrated in Fig.1 (a). However, footing has emerged at the bottom of the MTJ pillar in process A, which is detrimental to $\mu_0 H_{ex}$ and the stability of MTJ switching .

To eliminate the issue of bottom footing, the ME process was divided into three distinct steps, utilizing various angles of $\theta_1/\theta_2/\theta_3$ in process B, as illustrated in Fig.1 (b). This approach was based on the difference in etching rate and selectivity at various etching angles. After ME reached near the SOT channel, the trim was subsequently applied as in process A. Due to the division of ME and typical trim, a steep MTJ profile with minor footing around 3.5 nm has been achieved in process B. The cross-section transmission electron microscopy (TEM) image of the SOT-MTJ with process A and B were compared in Fig.2 (a) and (b), showing dramatic reduction of footing. What's more, $\mu_0 H_{ex}$ extracted from the resistance-field hysteresis (R–H) loop has improved from 100 mT to 150 mT as shown in Fig.2 (c), which was conducive to device reliability. Nevertheless, a considerable number of short bits were obtained in MTJs from process B, as characterized by the scatter plot of TMR and MTJ resistance at parallel state (R_P) in Fig.2 (d). A rate of fail bits

979-8-3315-0417-5/25 $31.00 © 2025 IEEE

(categorized as TMR<80%) around 40% was calculated within 729 devices across the whole wafer.

B. Etch process C

To enhance device yield and maintain the MTJ profile, novel etch processes are essential. In process B, metal redeposition can occur during trim, albeit at a slow rate, as depicted in Fig.3 (a). Covering the SOT channel's upper surface with dielectric layers may mitigate this issue, like in Fig. 3 (b). However, this dielectric layer must not encapsulate the MgO barrier, in order to allow removal of metal residue from the ME process later. Thus, the new etch process should preserve the bottom dielectric layer and hence incorporate two key features: (1) larger thickness of the dielectric layer on the SOT channel than on the MTJ pillar sidewall, and (2) higher etching rate on the sidewall, which can be achieved through etching at a grazing angle.

Process C was proposed as explained below. Initially, the SOT-MTJ was patterned into pillars using process B, followed by a SiN encapsulation layer deposition. Due to step coverage, the SiN thickness on the bottom surface (25 nm) exceeded that on the MTJ pillar sidewall (15 nm), as shown in Fig.4 (a). However, the MTJ pillar will suffer from excessive sidewall damage if an etch at a grazing angle is applied immediately after SiN deposition. It is because the etch time to remove metal residue buried deeply in the bottom SiN encapsulation is too long, SiN on the sidewall will be completely consumed and MTJ pillars are exposed to ion beam. To address this, an intermediate grazing etch (grazing 1) with an incident angle of α_1 was introduced to equalize the SiN thickness to approximately 5 nm, as indicated by the EDS image in Fig.4 (b). Subsequently, a second grazing etch (grazing 2) with angle α_2 was used to quickly remove sidewall SiN while still preserving some on the bottom. This step also re-cleaned the MgO while effectively suppressed metal redeposition from the SOT channel. EDS result in Fig.4 (c) confirms no visible redeposition in the final MTJ pillars.

C. Device performance and yield

Finally, a high yield over 99.9% (728/729) was achieved with process C and not a single fail bit was found in the scatter plot of TMR-R_P in Fig.5 (a). The yield level was better than previous works [4]-[5]. As a comparison, different electrical properties of $\mu_0 H_{ex}$, TMR and R_{SOT} with process A, B and C were summarized in Fig.5 (b). Larger $\mu_0 H_{ex}$ in process B and C indicates a much steeper MTJ profile than process A, and larger TMR means further enhancement in device performance. However, the channel resistance (R_{SOT}) increases over 200 Ω in process C, which is induced by grazing etch and will be addressed in our follow-up work. In addition, electrical switching of a typical SOT-MTJ fabricated by process C was characterized with an external magnetic field ($\mu_0 H_{ext}$=20 mT) applied along the current flow direction. The critical switching current was tested with pulse width varying from 50 ns to 2 ns, as illustrated in Fig.5 (c). The switching current of I_{P-AP} (I_{AP-P}) from parallel to anti-parallel state (vice versa) presents gradually increases with decreasing pulse width and reaches 750 µA at 2 ns pulse.

Conclusion

In summary, three different fabrication methods of SOT-MTJs were introduced on 300 mm wafer. With a reasonable combination of ME and trim, a precise etching stop on SOT channel can be achieved across the wafer in process A. Through dividing ME into three steps with different angles, the footing issue has been addressed in process B. Finally, with intermediate SiN protection and etch at grazing angles, we have achieved a high device yield over 99.9% as well as high TMR (120%) and low critical switching current (750 µA at 2 ns) in process C. These approaches suggest a gradual improvement in SOT-MTJs performance, and our research provides valuable insights for further development in SOT-MRAM related manufacturing.

Acknowledgments

This work is supported by National Science and Technology Major Project (2020AAA0109003). We also acknowledge the support from Hikstor's pilot line.

References

[1] J. J. Kan, C. Park, C. Ching, J. Ahn, Y. Xie, M. Pakala and S. H. Kang, "A study on practically unlimited endurance of stt-mram," IEEE T Electron Dev, vol. 64, no. 9, pp. 3639–3646 (2017).

[2] X. G. Wang, D. Y. Xiao, J. P. Huang, T. H. Lee, Y. H. Zheng, K. Y. Cao, K. F. Dong, F. Jin, "High density hexagonal MTJ array with 72 nm pitch and 30nm CD by using advanced DRAM patterning solution and ion beam etch", AIP Adv, 12, 035152 (2022).

[3] W. Yang, Z. Ji, Y. Gao, K. Zhou, Q. Guo, D. Zeng, S. Wang, M. Wang, L. Shen, G. Chen, Y. Sun, E. Liu and S. He, "Achieving High Yield of Perpendicular SOT-MTJ Manufactured on 300 mm Wafers," IEEE Electron Device Lett, vol. 45, no. 11, pp. 2094-2097 (2024).

[4] N. Sato, G. A. Allen, W. P. Benson, B. Buford, A. Chakraborty, M. Christenson, T. A. Gosavi, P. E. Heil, N. A. Kabir, B. J. Krist, K. P. O'Brien, K. Oguz, R. R. Patil, J. Pellegren, A. K. Smith, E. S. Walker, P. J. Hentges, M. V. Metz, M. Seth, B. Turkot, C. J. Wiegand, H. J. Yoo and I. A. Young, "CMOS Compatible Process Integration of SOT-MRAM with Heavy-Metal Bi-Layer Bottom Electrode and 10ns Field-Free SOT Switching with STT Assist", 2020 Symposium on VLSI Technology Digest of Technical Papers, VLSI TMFS 3-271 (2020).

[5] M. Y. Song, C. M. Lee, S. Y. Yang, G. L. Chen, K. M. Chen, I. J. Wang, Y. C. Hsin, K. T. Chang, C. F. Hsu, S. H. Li, J. H. Wei, T. Y. Lee, M. F. Chang, X. Y. Bao, C. H. Diaz and S. J. Lin," High speed (1ns) and low voltage (1.5V) demonstration of 8Kb SOT-MRAM array", 2022 Symposium on VLSI Technology & Circuits Digest of Technical Papers, VLSI T11-3 (2022).

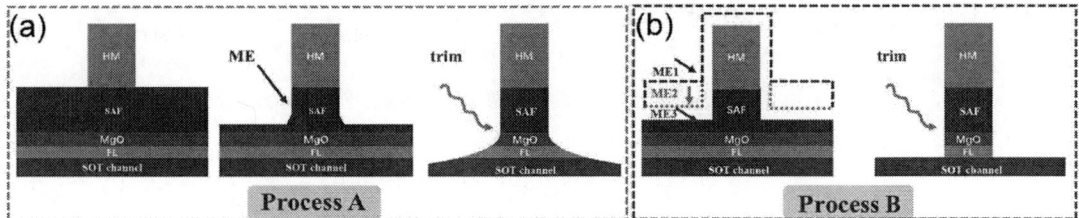

Fig. 1. (a) With a reasonable combination of ME and trim, a precise etching stop on SOT channel can be achieved across the wafer in process A. (b) Through dividing ME into three different angle steps, steep sidewall has been obtained in process B.

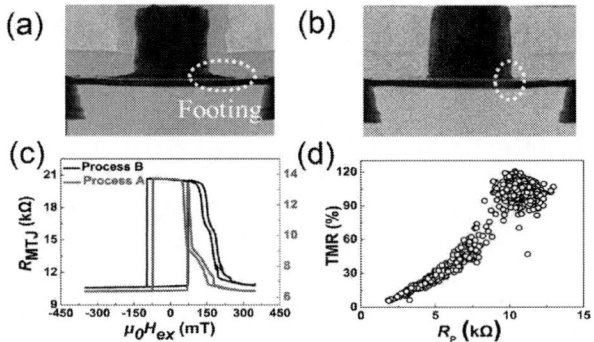

Fig. 2. TEM image of a MTJ pillar fabricated by (a) process A and (b) process B. (c) The R–H loop for a single device etched using process A and B. (d) Scatter plot of TMR-R_P for SOT-MTJ devices fabricated using process B.

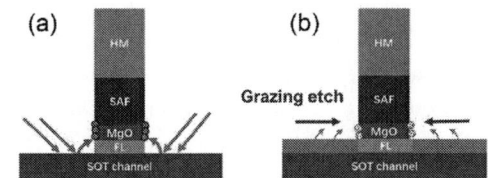

Fig. 3. (a) General short issue phenomenon with SOT metal sputtering. (b) Innovative etching process design to address short issues. With a dielectric layer covering on the SOT channel, metal redeposition from SOT channel can be forbidden and sidewall residues can be removed simultaneously through grazing etch process.

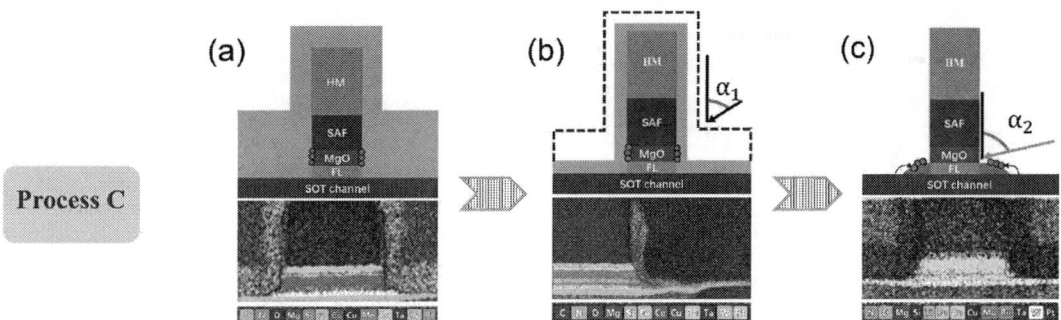

Fig. 4. Schematic representation of process C and the corresponding EDS images at each step. (a) SiN covering layer was deposited on patterned MTJ pillar with process B. (b) Grazing etch with angle α_1 was introduced to eliminate thickness difference between bottom and sidewall. (c) Grazing etch with angle α_2 was employed to facilitate secondary sidewall cleaning.

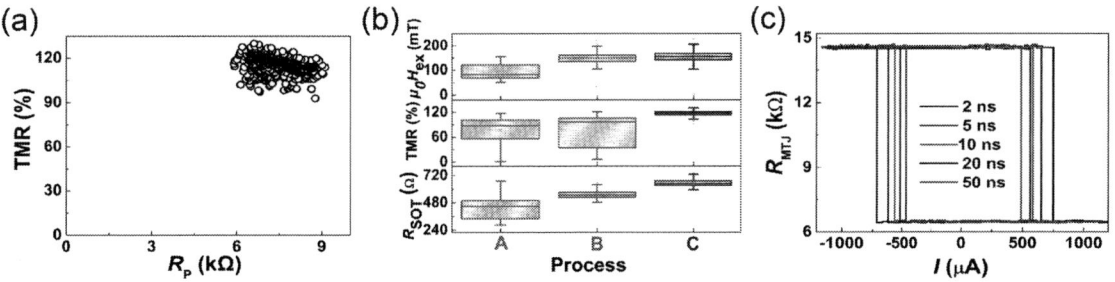

Fig. 5. (a) Scatter plot of TMR-R_P of SOT-MTJs fabricated by process C. (b) Comparison of μ_0H_{ex}, TMR and R_{SOT} between process A, B and C. (c) Electrical switching behavior of SOT-MTJs fabricated by process C with different pulse widths varying from 50 ns to 2 ns.

979-8-3315-0417-5/25 $31.00 © 2025 IEEE   383

High-Frequency and Wideband RF Filters for 6G and Wi-Fi 7

(Invited Paper)

Chengjie Zuo[1,2,3], *Senior Member, IEEE* and Zhongbin Dai[1]

[1] SRS Lab, School of Microelectronics, University of Science and Technology of China, Hefei, China

[2] YUNTA Technologies, Hefei, China

[3] ANUKI Technologies, Hefei, China

Abstract

This paper presents a brief review of our recent progress on high-frequency and wideband chip-level radio frequency (RF) filters, including Lamb-wave acoustic resonator performance advancements and new acoustic mode innovations. By focusing on three fundamental metrics of acoustic resonators, namely operating frequency (f), electromechanical coupling coefficient (k^2) and quality factor (Q), this paper describes our efforts in acoustic mode innovation, device structure optimization and fabrication process improvement for higher frequency and better performance. These breakthroughs open new paths to enable high-frequency Lamb-wave and SAW resonators with high k^2 for 6G (cmWave and mmWave) and Wi-Fi 7/8 wireless communications.

Keywords: Lamb wave resonators, coupled surface acoustic wave resonators, Q, k^2, 6G, Wi-Fi 7, cmWave, mmWave

Introduction

To address the challenges of global transition to 5G new radio (NR) Frequency Range 2 (FR2), 6G and Wi-Fi 7 wireless communication systems, radio frequency (RF) filters with high frequency, wide bandwidth and high rejection are ubiquitously needed to access the crowded electromagnetic spectrum and enhance the signal-to-noise ratio. All standard frequency bands for 5G, 6G and Wi-Fi 7 are above 3 GHz, and the bandwidths for these bands are usually above 500 MHz, or even 1 GHz [1-4]. In addition, the band allocations below 7 GHz have also become crowded, so co-existence problems emerge, e.g., the transition band between 5G n77 and n79 is only 200 MHz, and that between Wi-Fi 5 GHz and 6 GHz bands is as narrow as 110 MHz, which calls for new chip-level filter technologies that deliver high frequency, wideband and sharp roll-off at the same time.

Therefore, we proposed a Hybrid filter design based on the combination of high performance integrated passive device (IPD) and bulk acoustic wave (BAW) technologies to resolve the co-existence issue between 5G n77 and n79 bands [5]. Based on this Hybrid filter concept, we have also developed world-leading RF filter products for Wi-Fi 7 and achieved commercialization at several top-tier customers.

Nevertheless, as next-generation communication standards evolve towards higher frequencies, we still need to improve the Hybrid filter performances and the key is to innovate at the part of acoustic resonators. In this paper, we will introduce our recent work on Lamb wave resonators and coupled surface acoustic wave (SAW) resonators that achieved record-breaking high electromechanical coupling coefficient (k^2) at centimeter wave (cmWave) and millimeter wave (mmWave) frequencies, which pave the way for low-cost chip-level RF filter solution for future 5.5G, 6G, Wi-Fi 7, and Wi-Fi 8 wireless communications.

Lamb Wave Resonators

Building upon the success of suspended lithium niobate (LN) Lamb wave devices operating at sub-6 GHz, researchers are working to expand them to mmWave frequencies for 5G NR FR2 bands. However, the challenge for scaling LN acoustic resonators into mmWave range without sacrificing its k^2 lies in how to enable extremely thin (< 100 nm) suspended LN plate structure in terms of fabrication process and resonator design. We have proposed two techniques to address this challenge: fin-type anchor design and half-electrode reflector design.

A. Fin-Type Anchor Design

If using conventional etching holes and release methods, structure bending or cracking due to internal stress would directly hinder the fabrication of resonators based on ultra-thin LN thin films. Therefore, we proposed a new structure design with fin-type anchors to fix the two sides of the thin-film plate so as to overcome the above problems [6].

In Fig. 1, by thinning Z-cut LN thin film to below 100 nm (74 nm) by Ar+ plasma, first-order asymmetric (A1) mode Lamb wave resonator was implemented with high k^2 of 15.95%, whose operating frequency was around 25 GHz. It is also worth noting that spurious-free response has been achieved for both in-band (a large bandwidth of 2 GHz between f_s and f_p) and out-of-band (from 5 to 40 GHz) frequencies, as shown in Fig. 2.

This work demonstrated the largest k^2 and figure of merit ($FoM = k^2 \times Q$), when compared to other acoustic resonators in the 5G NR FR2 band at the time.

B. Half-Electrode Reflector Design

While fin-type anchor design resolved structure robustness issue and resulted in high k^2, acoustic resonators usually have a lower quality factor (Q) operating at higher frequencies. During the process when we were optimizing the fin-type anchor locations to minimize their impact on mode shape and quality factor, a piece of alternative thinking sparked and we proposed a new half-electrode design at both sides of the suspended plate [7]. In this way, while the plate is vibrating, the two side edges naturally have a zero-displacement boundary condition, because it is at the center of an interdigitated transducer electrode. With fin-type anchors to enforce zero-displacement condition, we can directly attach the anchors to the two sides, as shown in Fig. 3. As a demonstration, two resonators with and without the half-electrode reflector were designed and fabricated. The half-electrode design resulted in four times increase in quality factor and less spurious, as shown in Fig. 4.

Coupled SAW Resonators

Compared with bulk acoustic wave (BAW) and Lamb wave resonators, SAW resonators have the characteristics of simple process, mechanical stability and low cost, so they are more favored by the industry. The issue for SAW is that its operating frequency is limited by photolithography and it was widely believed SAW could only work well (with decent k^2 and Q) below 3 GHz. To break this ceiling, we proposed a novel coupled SAW technology to achieve record-breaking high k^2 of 42% at 5 GHz and 18% at 10 GHz, for the first time in the world.

A. Coupled Shear SAW Resonators

Different from conventional SAW resonators which utilize a single piezoelectric coefficient to transduce a 1-dimensional vibration mode, coupled shear SAW (CS-SAW) resonators were designed to combine two different shear piezoelectric coefficients (e_{16} and e_{34}) in one mechanical vibration mode, so that k^2 contributions from each piezoelectric coefficient add up [8]. The thickness to wavelength ratio (h_{LN}/λ) is the key design parameter here to make the electric field distributed both horizontally and vertically, and also the two shear piezoelectric coefficients couple together in one mode. Experimentally, as shown in Fig. 5, 300 nm X-cut single-crystal LN thin film on a SiC substrate was used to implement the CS-SAW resonators, and the interdigital transducer (IDT) wavelength was designed to be 600 nm for optimal coupling. Fig. 6 (a) and (b) plot the measured admittance and Bode-Q of the CS-SAW resonator when h_{LN}/λ is set to be 1/2. The k^2 was extracted to be 42.6%. This demonstration achieved the highest k^2 and high Q_{max} of 650 around 5 GHz.

B. Coupled Longitudinal and Shear SAW Resonators

As we learned from designing CS-SAW resonators, to further increase the operating frequency, we need to increase the elastic constant that participates in the mode shape. Therefore, we proposed a novel coupled longitudinal and shear SAW (LS-SAW) resonator [9], which introduces a longitudinal wave component (e_{11}) to increase the velocity of the coupled mode, as well as the operating frequency of the resonator. The corresponding longitudinal elastic constant (c_{11} of 221 GPa)

is four times that of the shear elastic constant (c_{55} of 56.8 GPa), which is the essence for the LS-SAW to operate above 10 GHz.

As shown in Fig. 7 and 8, the -52° LS-SAW resonator was fabricated on a 300 nm X-cut single-crystal LN thin film on a SiC substrate, while λ was fixed at 400 nm. The LS-SAW demonstrated the highest k^2 of 18%, Q_{max} of 218 and FoM of 39.2, when compared with other reported SAW resonators around 10 GHz.

Conclusion

This work introduces a variety of novel Lamb wave resonator structures and innovative coupled SAW resonator designs to achieve the highest k^2 and FoM in the cmWave and mmWave bands, which signify a significant step forward for the acoustic resonator field. The demonstrated breakthroughs pave the way for miniature-size chip-level RF filter solutions for 5.5G, 6G, Wi-Fi 7 and Wi-Fi 8 wireless communications.

Acknowledgments

This work was supported in part by the National Natural Science Foundation of China under Grant 62231023, in part by the University of Science and Technology of China (USTC) Center for Micro and Nanoscale Research and Fabrication, and in part by USTC Institute of Advanced Technology.

References

[1] K. Yang, C. He, J. Fang, X. Cui, H. Sun, Y. Yang, and C. Zuo, "Advanced RF filters for wireless communications," *Chip*, 2, pp. 100058 (2023).

[2] C. C. W. Ruppel, "Acoustic wave filter technology–a review," *IEEE Trans. Ultrason., Ferroelectr., Freq. Control*, 64, pp. 1390 (2017).

[3] W. Chen, X. Lin, J. Lee, A. Toskala, S. Sun, C. F. Chiasserini, and L. Liu, "5G-advanced toward 6G: Past, present, and future," *IEEE J. Sel. Areas Commun.*, 41, pp. 1592 (2023).

[4] C. Zuo, and K. Yang, "Hybrid filter for 5G and 6G applications," *IMWS-AMP 2022*.

[5] C. Zuo, C. He, W. Cheng, and Z. Wang, "Hybrid filter design for 5G using IPD and acoustic technologies," *IUS 2019*.

[6] K. Yang, F. Lin, J. Fang, J. Chen, H. Tao, H. Sun, and C. Zuo, "Nanosheet lithium niobate acoustic resonator for mmWave frequencies," *IEEE Electron Device Lett.*, 45, pp. 272 (2024).

[7] K. Yang, J. Fang, F. Lin, Y. Wang, J. Chen, M. Li, H. Tao, and C. Zuo, "12-GHz spurious-free fin-mounted Lamb wave resonator with half-electrode reflectors," *UFFC-JS 2024*.

[8] Z. Dai, X. Liu, H. Cheng, S. Xiao, H. Sun, and C. Zuo, "Ultra high Q lithium niobate resonator at 15-degree three-dimensional Euler angle," *IEEE Electron Device Lett.*, 43, pp. 1105 (2022).

[9] Z. Dai, H. Cheng, S. Xiao, H. Sun, and C. Zuo, "Coupled shear SAW resonator with high electromechanical coupling coefficient of 34% using X-cut LiNbO₃-on-SiC substrate," *IEEE Electron Device Lett.*, 45, pp. 720 (2024).

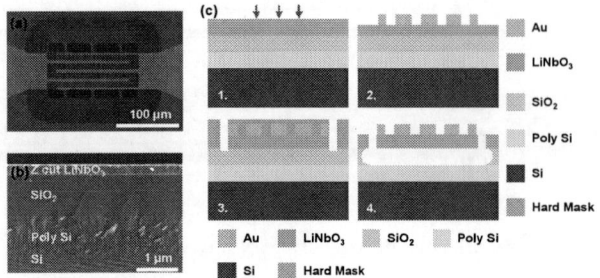

Fig. 1: Scanning electron microscopy (SEM) photos and fabrication process of the fin-mounted A1-mode resonator.

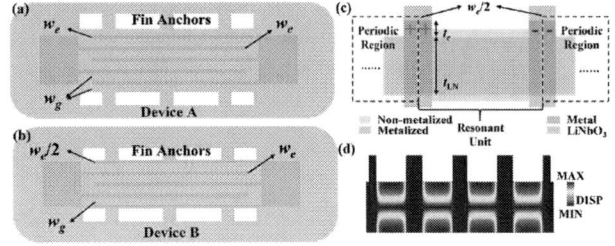

Fig. 3: (a), (b) Top view of the A1-mode fin-mounted Lamb wave resonator structures for traditional IDT design and half-electrode reflector design. (c), (d) Acoustic impedance equivalent model and vibration displacement diagram for a two-dimensional resonant unit of the Lamb wave resonator.

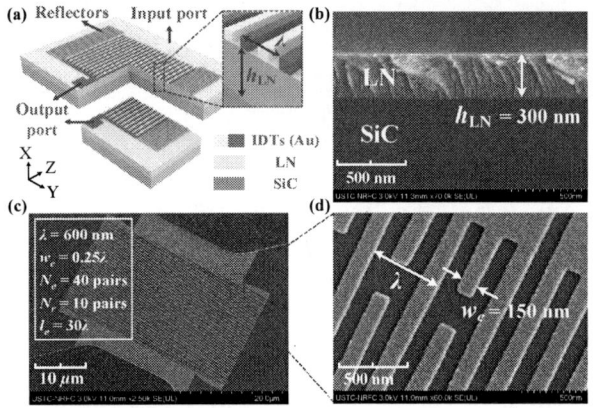

Fig. 5: Schematic diagram and scanning electron microscope (SEM) photos of the CS-SAW resonator design.

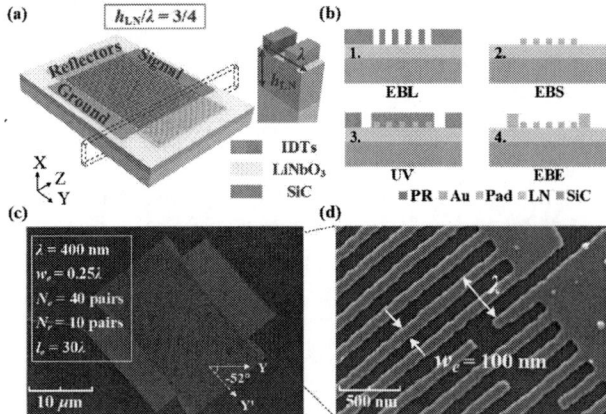

Fig. 7: Schematic diagram, fabrication process and scanning electron microscopy (SEM) photos of the LS-SAW design.

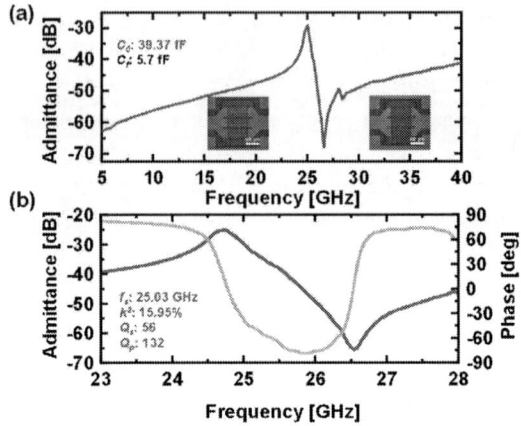

Fig. 2: (a) The A1-mode admittance curve in broadband (5-40 GHz). (b) Measured zoomed-in admittance response and phase of the DUT.

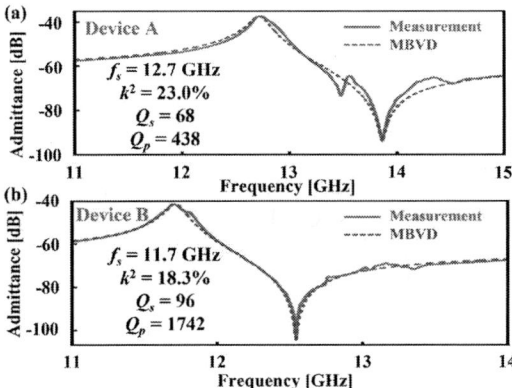

Fig. 4: Measurement results of the fabricated fin-mounted Lamb wave resonators: (a) resonator with traditional IDT; (b) resonator with half-electrode reflectors.

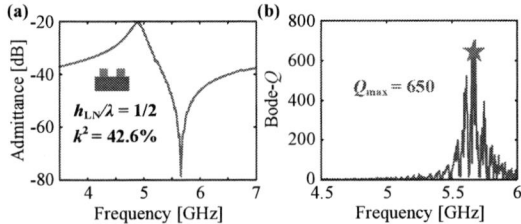

Fig. 6: (a), (b) Measured admittance and Bode-Q plots of the CS-SAW resonator with $h_{LN}/\lambda = 1/2$.

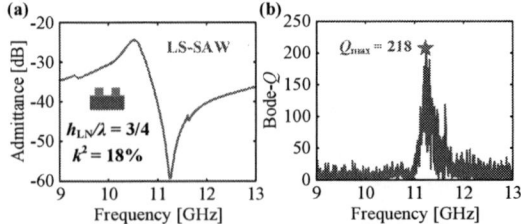

Fig. 8: (a), (b) Measured admittance and Bode-Q plots of the LS-SAW resonator with $h_{LN}/\lambda = 3/4$.

979-8-3315-0417-5/25 $31.00 © 2025 IEEE

Dynamic Performance Analysis of Ultra-fast Inverter based on Tri-gate AlGaN/GaN MIS-HEMTs

Yunsong Xu[1,2], Weisheng Wang[1,2], Dechang Quan[1], Haotian Ji[1], Shenlei Ding[1], Yunzhou Jiang[1],

Kain Lu Low[1,2], Jiangmin Gu[1,2*], Wen Liu[1,2]

[1] School of Advanced Technology, Xi'an Jiaotong-Liverpool University, Suzhou 215123, China,

[2] Department of Electrical Engineering and Electronics, University of Liverpool, Liverpool L69 3GJ, UK

*Email: Jiangmin.Gu@xjtlu.edu.cn

Abstract

This work demonstrates dynamic response characteristics of the tri-gate AlGaN/GaN high electron mobility transistor (HEMT) direct coupled FET logic (DCFL) inverter. The threshold voltage of the inverter is 2.74 V, the rise and fall time are 2.6 ns and 2.0 ns respectively with 1.8 ns and 0.2 ns propagation delay for each edge. The mechanisms for the excellent switching time performance are also revealed from the perspective of the device.

Keywords: GaN, Tri-gate MIS-HEMT, Monolithic integration and Logic circuit

Introduction

Emerging applications such as aerospace, hybrid vehicles, artificial intelligence servers place strong demands on the high-frequency and high-temperature performance of integrated circuits [1][2]. AlGaN/GaN HEMTs utilize the material properties of wide band-gap III-nitride semiconductors, have simple circuit design and highly symmetric switching characteristics, and have broad application potential [3][4].

Early experiments generally used DCFL to circumvent the lack of high-performance PMOS in the GaN platform [5]. It has been demonstrated that the enhancement and depletion-mode (E/D-mode) GaN-based inverter can operate at megahertz frequencies [6]. However, the gate recess process adopted in conventional MIS-HEMT introduces a larger on-resistance (R_{on}), which reduces the current density of the device and is not conducive to high-frequency applications. While the above problems can be improved by configuring additional bootstrap capacitors and diodes [7]. the relatively complex circuit structure is contrary to the concept of compact circuit design and high-power density.

The application of tri-gate structure in GaN power and RF has been reported [8][9], and its lower R_{on} makes it competitive with highly popular p-GaN HEMT platform. Tri-gate MIS-HEMT can achieve threshold modulation by changing the fin width (W_{fin}). In addition, since the gate length (L_g) of this structure is determined by the fin length (L_{fin}), the gate length can be reduced without losing gate control, thereby increasing the operating frequency [10]. While initial demonstrations are encouraging, comprehensive and systematic studies on the dynamic behavior of basic circuit units (e.g., inverters) using tri-gate MIS-HEMTs would be attracting.

In this work, a high-performance DCFL inverter is realized based on E-D-mode tri-gate MIS-HEMTs, which exhibits extremely short rise-fall times and is fully analyzed using equations based on physical device parameters.

Device Fabrication and Results

The E/D-mode AlGaN/GaN tri-gate MIS-HEMTs and inverters were fabricated on the same Si (111) substrate wafer, using the epitaxial structure and process flow described in [11]. Fig. 1 (a) shows the schematic diagram of the tri-gate MIS-HEMT structure. Fig. 2(a) shows the both DC output and transfer characteristics of E/D-mode HEMTs. The V_{th} of the E-mode HEMT is 1.1 V (W_{fin}=56 nm) and the maximum drain current density $I_{DS,max}$ is 426 mA/mm, while V_{th} is -1.8 V (W_{fin}=125 nm) and $I_{DS,max}$ is 555 mA/mm for the D-mode HEMT. Compared with the conventional GaN MIS-HEMTs, the proposed tri-gate MIS-HEMTs achieve higher $I_{DS,max}$ and lower R_{on}.

Characteristic and Discussion

In the inverter of this work, the E-mode tri-gate HEMTs as a driver and D-mode as an active load are set to W_g=100 μm and 10 μm, respectively, as depicted in Fig. 2(b). For the static voltage transfer characteristics, Fig. 3(a) shows the measurement of the GaN tri-gate E/D-mode inverter at a supply voltage V_{DD} of 10 V. The NM_L and NM_H noise margin of the inverter are 1.84 V and 7.01 V, respectively. The dynamic switching performance of the inverter based on the tri-gate MIS-HEMTs was characterized by connecting the input of the logic gates to a pulse generator, and the output to the high impedance port of an oscilloscope with a capacitive load (C_L) of approximately 50 pF. As shown in Fig. 3(b), the propagation delays are 1.8 ns for the rising edge and 0.2 ns for the falling edge, with a rise time of 2.6 ns and a fall time of 2.0 ns.

A. Gate Capacitance Density

The gate capacitance density plays an important role in the dynamic performance of the DCFL inverter. For the tri-gate structure, the main capacitance distribution of the gate and the equivalent circuit diagram are shown in Fig. 1 (b) and (c), respectively. Based on the relevant literature [12], the parameters related to capacitance density at the gate fin structure can be inferred as:

979-8-3315-0417-5/25 $31.00 © 2025 IEEE

$$C_{fin} = \frac{2\varepsilon_{GaN}\varepsilon_0}{\pi} ln\left(\frac{\pi\varepsilon_{Al2O3}}{4\varepsilon_{GaN}t_{Al2O3}}W_{fin} + 1\right) \quad (1)$$

$$\overline{C_{fin,total}} = \frac{1}{\left(\frac{t_{AlGaN}}{\varepsilon_{AlGaN}\varepsilon_0} + \frac{t_{Al2O3}}{\varepsilon_{Al2O3}\varepsilon_0}\right)} + \frac{2C_{fin}}{W_{fin}} \quad (2)$$

$$\overline{C_{space}} = \frac{\varepsilon_{Al2O3}\varepsilon_0}{t_{Al2O3}} \quad (3)$$

where C_{fin} is the capacitance density of the gate structure in the cross-sectional direction (A-A'). $\overline{C_{fin,total}}$ and $\overline{C_{space}}$ are the capacitance densities of the fin structure and the spacing groove, respectively. ε_X and ε_0 are the relative dielectric constant and absolute dielectric constant of the corresponding subscript material, respectively. t_X is the film thickness of the corresponding subscript material. Therefore, the overall gate capacitance density of the tri-gate structure can be expressed as:

$$\overline{C_G} = \alpha\overline{C_{fin,total}} + \beta\overline{C_{space}} \quad (4)$$

where α and β are the proportions of fin structure and spacer groove structures in the total gate, respectively.

B. Propagation Delay

Based on the physical structure analysis of tri-gate GaN MIS-HEMT and related literature [13]. The propagation delay is the maximum period the signal takes from the triggered input to the responded output. It is usually measured on the 50% of voltage range for the input and output signals. The propagation delay in the inverter design is dominantly occupied by the charging and discharging on the load capacitance. Consequently, they can be expressed as:

$$t_{PD,R} = \frac{C_L V_{DD}}{2I_{dch,D}} = \frac{C_L V_{DD}}{\mu_{eff,D}\overline{C_{G,D}}\frac{W_D}{L_D}\left(V_{DD} - |V_{TH,D}|\right)^2} \quad (5)$$

$$t_{PD,F} = \frac{C_L V_{DD}}{2I_{dch,E}} = \frac{C_L V_{DD}}{\mu_{eff,E}\overline{C_{G,E}}\frac{W_E}{L_E}\left(V_{DD} - V_{TH,E}\right)^2} \quad (6)$$

where $I_{dch,E}$ and $I_{dch,D}$ are the discharging current through the E/D-mode HEMTs, and $\overline{C_G}$ is the effective gate capacitance density of both HEMTs. C_L is the load capacitance.

C. Rise & Fall Time

Generally, the rise and fall time refers to the time interval for the input to switch between $V_{th,E}$ and $V_{DD}-|V_{th,D}|$. Since it is limited by the driving capability of the preceding logic gate, it can be estimated based on $t_{PD,R}$ and $t_{PD,F}$ and derived that:

$$t_{L\rightarrow H} = t_{PD,R} \frac{ln\left[\frac{V_{DD} - V_{TH,E}}{|V_{TH,D}|}\right]}{ln(2)} \quad (7)$$

$$t_{H\rightarrow L} = t_{PD,F} \frac{ln\left[\frac{V_{DD} - |V_{TH,D}|}{V_{TH,E}}\right]}{ln(2)} \quad (8)$$

Referring to (5)-(8), The $\overline{C_G}$ of E/D-mode HEMT and R_{on} of

D-mode as load are the key parameters affecting the dynamic switching performance of the inverter. The C_L of this work is mainly provided by the measurement instrument and is therefore not within the scope of discussion. Thanks to the larger $\overline{C_G}$ and lower R_{on} due to the tri-gate structure, the response time of the inverter is greatly shortened.

As shown in Table 1, the inverter of this study (including measured and derived results) is benchmarked with other inverters based on GaN platform. The derived results based on physical structure modeling analysis show a high degree of agreement with the measured data. The results reflect the great potential of inverters based on GaN tri-gate MIS-HEMTs for high-frequency robust operation.

Conclusion

The static and dynamic characteristics of the inverter using tri-gate AlGaN/GaN MIS-HEMT are characterized and analyzed. The results show that the inverter has extremely short rise-fall time and delay. In addition, the root cause of the switching time degradation using tri-gate devices is analyzed using equations based on physical device parameters. The high-performance DCFL inverter has shown great potential for deployment in high-frequency applications.

Acknowledgments

This work was supported by the National Natural Science Foundation of China (Grant No. 62374137), the Key Program Special Fund in Xi'an Jiaotong–Liverpool University under Grant KSF-T-07, Xi'an Jiaotong–Liverpool University Research Development Funds RDF-20-02-43 and RDF-22-01-110.

References

[1] S. R. Eisner *et al.*, *2021 IEEE Aerospace Conference (50100)*, IEEE, pp. 1–12, 2021.

[2] K. Hoo Teo *et al.*, *Journal of Applied Physics*, vol. 130, no. 16, 2021.

[3] F. Li *et al.*, *2023 35th International Symposium on Power Semiconductor Devices and ICs (ISPSD)*, IEEE, pp. 99–102, 2023.

[4] A. Li *et al.*, *IEEE Transactions on Electron Devices*, vol. 68, no. 6, pp. 2673–2679, 2022.

[5] T. Cosnier *et al.*, *2021 IEEE International Electron Devices Meeting (IEDM)*, 2021.

[6] R. Wang *et al.*, *IEEE Electron Device Letters*, vol. 44, no. 6, pp. 899–902, 2023.

[7] T.-W. Wang *et al.*, *IEEE Journal of Solid-State Circuits*, vol. 57, no. 12, pp. 3877–3888, 2022.

[8] T. Palacios *et al.*, *2021 33rd International Symposium on Power Semiconductor Devices and ICs (ISPSD)*, IEEE, pp. 6–10, 2021.

[9] Y. Zhang *et al.*, *Semiconductor Science and Technology*, vol. 36, no. 5, p. 054001, 2021.

[10] K.-S. Im, *et al.*, *IEEE Electron Device Letters*, vol. 41, no. 6, pp. 832–835, 2020.

[11] A. Li *et al.*, *2024 36th International Symposium on Power Semiconductor Devices and ICs (ISPSD)*, IEEE, pp. 1–3, 2024.

[12] Y. Zhao et al., *Physica Status Solidi (a)*, vol. 219, no. 3, 2022.

[13] Z. Zheng *et al.*, *Fundamental Research*, vol. 1, no. 6, pp. 661–671, 2021.

[14] B. Zhang *et al.*, *IEEE Electron Device Letters*, vol. 43, no. 7, pp. 1025–1028, 2022.

[15] M. Cui *et al.*, in *2019 International Conference on IC Design and Technology (ICICDT)*, IEEE, Jun. 2019, pp. 1–4, 2019.

[16] R. Wang *et al.*, *IEEE Electron Device Letters*, vol. 44, no. 6, pp. 899–902, 2023.

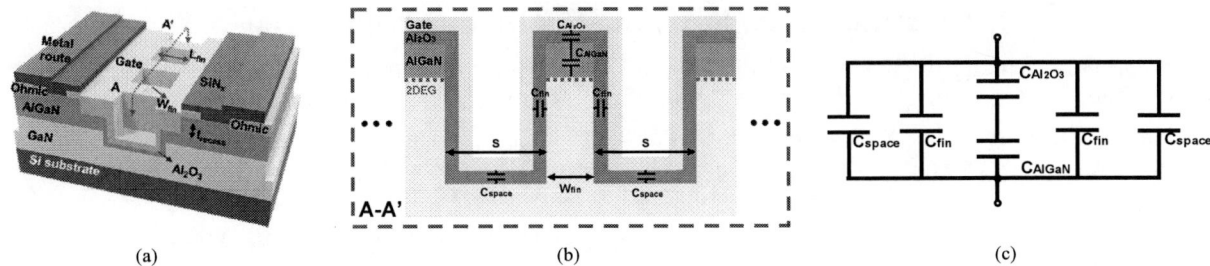

Fig. 1: The 3D (a) and A-A' cross-sectional (b) schematic of the tri-gate MIS-HEMTs used for DCFL inverter (SiN$_x$ on the gate metal is omitted for a clear demonstration of the fin structure). (c) Equivalent circuit diagram of gate structure capacitance distribution.

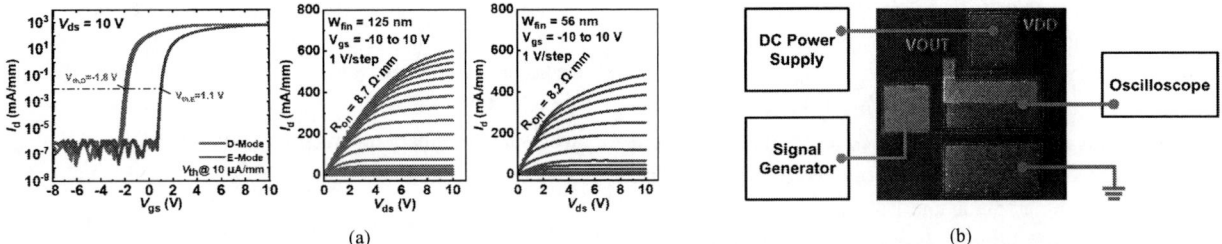

Fig. 2: (a) DC transfer and output characteristics of the D-mode and E-mode tri-gate MIS-HEMTs. (b) The false color SEM image of the DCFL inverter based on tri-gate MIS-HEMTs with dynamic test diagram.

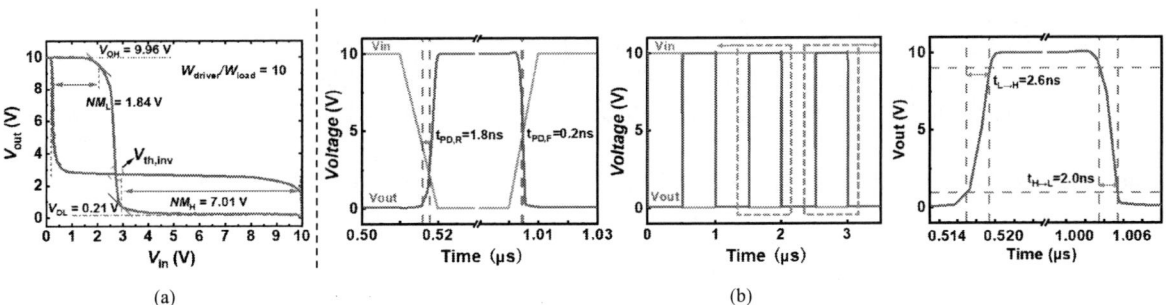

Fig. 3: Measured waveform of the inverter: (a) static voltage transfer characteristics, (b) dynamic switching characteristics with 1 MHz input signal.

Table 1

BENCHMARK OF STATE-OF-ART GAN DCFL INVERTER WITH DYNAMIC CHARACTERISTIC

	Ref.	Process	Vdd (V)	Vth,inv (V)	Propagation delay (ns) $\tau_{L\to H}$	Propagation delay (ns) $\tau_{H\to L}$	Rise Time t_{rise} (ns)	Fall Time t_{fall} (ns)	SC*
This	Measurement	GaN Tri-gate	10	2.74	1.8	0.2	2.6	2	16 67
Work	Derivation	MIS-HEMT			1.2	0.2	2.4	1.6	12.5
	[14]	GaN MIS-HEMT	15	5.1	178.15	12.5	353.1	96.6	0.06
	[15]	GaN MIS-HEMT	10	1.4	–	–	500	50	0.02
	[16]	GaN HEMT**	3	–	–	–	128	16	0.03

*SC (Symmetry Coefficient) is introduced to describe the symmetry of the rising and falling edges of the inverter (larger is better), which is defined as $Vdd/|t_{rise} - t_{fall}|$.

**p-GaN HEMT for E-mode and MIS-HEMT for D-mode.

979-8-3315-0417-5/25 $31.00 © 2025 IEEE   389

Achieving High Endurance Ferroelectricity in $Hf_{0.5}Zr_{0.5}O_2$ Thin Films on Ge Substrate through Helium Ion Doping Engineering

Peiyuan Du[1,2], Huan Liu[1,2,*], Dongya Li[1,2], Chengji Jin[1,2], Hongrui Zhang[2], Di Wang[2], Yian Ding[2], Bing Chen[1,2], Ran Cheng[3], Mengnan Ke[4], Xiao Yu[1,2,*], Yan Liu[1,*], Yue Hao[1], and Genquan Han[1,2]

[1]School of Microelectronics, Xidian University, Xi'an 710071, China; [2]Hangzhou Institute of Technology, Xidian University, Hangzhou 311231, China; [3]College of Integrated Circuits, Zhejiang University, Hangzhou 310000, China; [4]Faculty of Engineering, Chiba University, Chiba 263-8522, Japan

*E-mail: xdliuhuan@xidian.edu.cn; yuxiao@xidian.edu.cn; xdliuyan@xidian.edu.cn

Abstract

We have successfully demonstrated improved performance and reliability in ferroelectric (FE) HfZrO$_x$ (HZO) films through helium ion (He$^+$) doping. The doped HZO capacitors exhibit a reduced coercive field (E_c) and enhanced endurance with wake-up/fatigue-free characteristics. The He$^+$ doped metal-FE-semiconductor structure shows fatigue-recovery-free performance for over 10^{12} endurance cycles, extendable to over 10^{14} cycles, with stable remnant polarization and leakage. This method offers a promising approach for durable FE device applications.
Keywords: Ferroelectric, HZO and Endurance

Introduction

The discovery of HfZrO$_x$ (HZO)-based ferroelectric (FE) materials has attracted a lot of attention in devices such as ferroelectric random access memory (FeRAM) and ferroelectric field effect transistor (FeFET), owing to the high operation speed, low power consumption, excellent scalability, and CMOS process compatibility[1-2]. However, the primary concern for HZO-based devices is their reliability, specifically related to limited endurance and undesirable effects during cycling, including wake-up and fatigue[3-5]. Defects such as oxygen vacancies (V$_O$) play a crucial role in the ferroelectricity and reliability. Several methods have been proposed to effectively modulate these defects and enhance performance and reliability, including interface engineering, FE layer structure engineering and doping engineering[6-7]. While localized helium ion (He$^+$) bombardment has been reported to enhance ferroelectricity in HfO$_2$-based ferroelectric thin films[8], it is not compatible with CMOS processes. Besides, there is a lack of comprehensive evaluation and understanding of the mechanism and characteristics of He$^+$ doped devices.

On the other hand, Ge based metal-FE-semiconductor (MFS) structures have shown unique advantages. Unlike HfO$_2$/Si, where an (Hf)SiO$_x$ interfacial layer forms spontaneously, the interfaces between HZO/Ge are generally clean. This is due to the instability of (Hf)GeO$_x$ oxides, which readily dissociate at moderate annealing temperatures, resulting in oxide-free crystalline interfaces [9]. Additionally, as a low-bandgap semiconductor, Ge contains a high density of intrinsic free carriers that can effectively screen polarization charges, thereby stabilizing ferroelectric domains in MFS structures.

In this work, we first utilized a uniform He$^+$ doping technique to significantly enhance the endurance without wake-up and fatigue effect. Notably, the He$^+$ doped MFS structure demonstrated enhanced reliability, maintaining wake-up/fatigue-free performance over 10^{12} programming/erasing (P/E) cycles, which is extendable to over 10^{14} cycles, with stable remnant polarization (P_r) and leakage.

Experiments

Fig. 1 shows the structure and key fabrication process of the MFS capacitors used in this work. After the standard Ge-based cleaning procedure, 1 nm Al$_2$O$_3$, 1 nm TiO$_2$, and 10 nm HZO were sequentially deposited by using atomic layer deposition (ALD) at 280 °C. The Hf:Zr ratio is 1:1. The He$^+$ doping was conducted by ion implantation equipment AXCELIS NV-GSD HE. The doping energy was 15 keV, and the devices were implanted with doses of 10^{13} cm^{-2}. The undoped control sample was also prepared. After 100 nm TiN sputtering and patterning as top electrode (TE), the capacitors were annealed by rapid thermal annealing (RTA) for 30 s at 500 °C in nitrogen atmosphere.

The cross-sectional high-resolution transmission electron microscope (HRTEM) images of the He$^+$- implanted MFS structures, as shown in Fig. 2, illustrate that as the He$^+$ dose increases, the crystalline phase distribution of the polycrystalline HZO tends to become more uniform, and the grain size increases. The smooth and uniform interface can be observed between HZO film and Ge substrate, which would lead to improved reliability under higher P/E cycles.

Results and Discussion

He$^+$ doping engineering has been successfully demonstrated in HZO MFS structure. Fig. 3 (a) and (b) show the P-V loop of MFS devices with He$^+$ doping. The measurement frequency was set to 1 kHz, and the voltage was increased from 1.5 V to 5.5 V with a step of 0.5 V. A large $2P_r$ of approximately 40 μC/cm² was obtained in both device structures. Notably, after doping, no significant degradation in ferroelectricity was observed, and the coercive field (E_c) slightly decreased, as shown in Fig.3 (d), suggesting that He$^+$ can reduce the ferroelectric switching barrier. Additionally, while the undoped device exhibited significant leakage at 5.5 V, the 10^{13} cm^{-2} He$^+$ doped device maintained low leakage,

979-8-3315-0417-5/25 $31.00 © 2025 IEEE

indicating that the doping mitigated the conductive leakage path, allowing the device to remain stable under high voltage.

We examined the advantages of the doped device during high-voltage endurance cycling. Fig. 4 compares the endurance characteristics under different electric field conditions with P/E cycling at 2.5 MHz. The test voltages were set at 4 V, 4.5 V, and 5 V. The undoped devices broke down after 10^6 and 2×10^7 cycles, showing a noticeable wake-up effect. On the other hand, HZO MFS with He^+ doping showed a significant improvement for over 10^{11} cycles without any recovery operation, achieving an enhancement of approximately five orders of magnitude. At 4.5 V and 5 V, the devices have exhibited better P_r stability compared with the undoped ones. We proceeded with the endurance test at 4 V, compared to the previous test conditions, no significant reduction in P_r was observed. Moreover, without any wake-up or fatigue effects, the devices demonstrated high endurance up to 10^{12} cycles. Notably, the P_r and leakage current keep stable during cycling without the existence of wake-up and fatigue, suggesting that further endurance is achievable without recovery. This phenomenon suggests that the endurance of He^+-doped HZO can be projected to be over 10^{14} with optimized operation conditions.

The state of the device after endurance testing was further investigated in Fig. 5. Comparing the devices under different cycling conditions, it was found that the P_r value showed almost no degradation during electric field cycling. The most significant change occurred in the negative E_c (E_c^-) during the cycling process, where E_c^- shifted by about 0.5 V after 10^{12} cycles. In contrast, the positive E_c (E_c^+) showed almost no change. This can be attributed to the asymmetry in the MFS structure, where factors such as the interfacial layer thickness and defect content between HZO and the different electrodes may contribute to this phenomenon.

We also compared the leakage current of the undoped and 10^{13} cm^{-2} He^+ doped devices under different cycling conditions, as shown in Fig. 6. The undoped device showed significant leakage current after 10^7 cycles in the J-V graph. In contrast, our 10^{13} cm^{-2} He^+ doped device maintained the same leakage current state as the initial condition even after 10^{10} cycles. This once again demonstrates the optimization of the internal leakage paths due to doping, which is highly consistent with the trends observed in the P-V tests in Fig. 2.

Based on the above TEM characterization and electrical performance, we have illustrated the mechanism of He^+ doping, as illustrated in Fig. 7. Three aspects contribute to the enhancement of device performance after doping. First, the overall doping of the film leads to a more uniform distribution of defects within the device. After doping, the defects within the film are redistributed. According to [12],

this structure is beneficial for the charge trapping of internally charged defects. Another reason for endurance enhancement is grain boundary reduction. During the annealing process, the grain size increases and the number of grain boundaries decreases, which is related to the improved defect uniformity in the doped devices. The grain boundaries are important pathways for hard breakdown in HZO films, and the reduction corresponds to improved endurance performance, as shown in Fig. 4. Besides, the doping increases the thickness of the TiO_2 interfacial layer at the HZO and top electrode interface, as illustrated by the TEM images in Fig. 2. The increased thickness of the interfacial layer is considered as an indication of the enhanced oxidation state of the electrode. The formation of this interfacial layer helps reduce leakage current and enhance endurance. Moreover, TiO_2 contributes to reducing E_c and increasing the proportion of orthorhombic phase the in HZO layer [13]. We believe that these changes result in a comprehensive improvement of the HZO layer.

The benchmark in Table 1 of comprehensive ferroelectric properties illustrates that proposed HZO device with He^+ doping presenting the best overall endurance performance without wake-up and fatigue, achieving recovery-free endurance over 10^{12} and towards 10^{14} P/E cycling.

Conclusion

This study improves the performance and reliability of ferroelectric HZO films through He^+ doping. The He^+-doped HZO MFS structure exhibit lower E_c and prolonged endurance without wake-up and fatigue effects for over 10^{12} cycling without any recovery operations, which can potentially be extended towards 10^{14} P/E cycles, offering a promising avenue for enhancing the performance and reliability of ferroelectric applications.

Acknowledgments

The authors acknowledge support from the National Key Research and Development Project (Grant No. 2023YFB4402301), the National Natural Science Foundation of China (Grant No. 62374151, 62204226, 62025402, 62090033, 62174146, 91964202, 92064003 and 92264202), Major Program of Zhejiang Natural Science Foundation (Grant No. DT23F0402).

References

[1] E. R. Hsieh *et al.*, *VLSI*, pp. 1-2, 2023. [2] Z. Fu, *et al.*, *IEDM*, pp. 1-4, 2023. [3] S. Guo *et al.*, TED, pp. 3645-3650, 2024. [4] A. Senapati *et al.*, EDL, pp. 673-676, 2024. [5] T. Gong *et al.*, *VLSI*, 2021. [6] G. Kim *et al.*, *IEDM*, p. 5.4.1-5.4.4, 2022. [7] J. Chen *et al.*, *IEDM*, pp. 1-4, 2023. [8] S. Kang *et al.*, *Science*, pp. 731-738, 2022. [9] C. Zacharaki, *et al.*, APL, 2020. [10] J.H. Lee *et al.*, IEDM, pp. 1-4, 2022. [11] Yuanshen Qi *et al.*, APL, 2021. [12] C. Lo, *et al.*, EDL, pp. 224-227, 2022. [13] K. Tahara *et al.*, *VLSI*, 2021. [14] Y. Peng *et al.*, EDL., pp. 216–219, 2022.

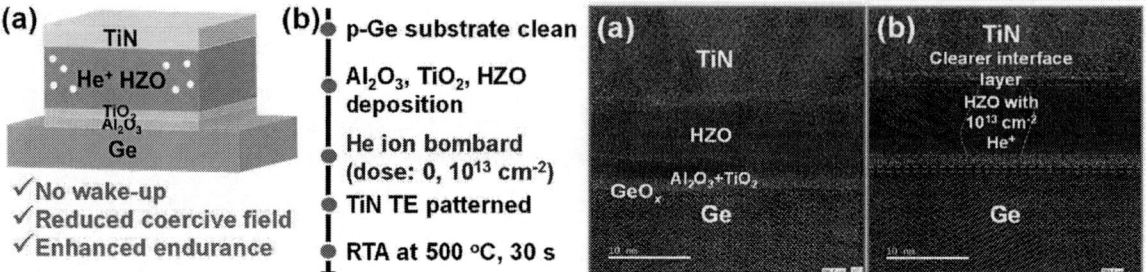

Fig. 1 (a) He$^+$ ion doping engineering in HZO based ferroelectric devices and (b) the key process flow of fabrication of HZO devices with He$^+$ doping.

Fig. 2 TEM image of (a) undoped and (b) 10^{13} cm^{-2} He$^+$ Ge MFS structure. The grain size increase and a distinct TiO$_2$ interface was formed between TiN and HZO after doping.

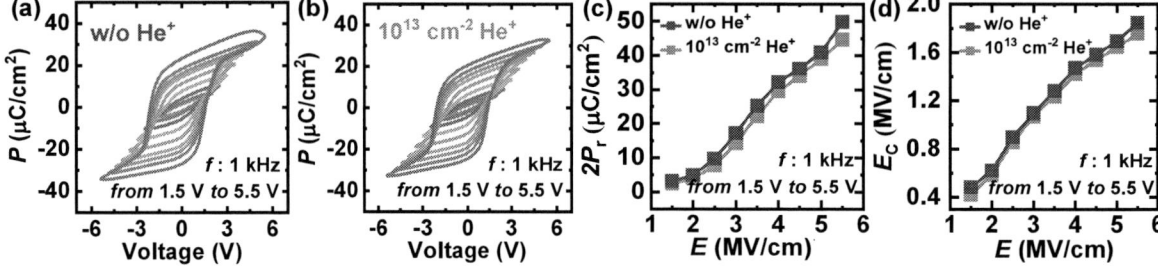

Fig. 3 P-V measurement of HZO devices with (a) undoped and (b) 10^{13} cm^{-2} doses, conducted at varying sweeping voltages. Fig (c) and (d) show the extracted $2P_r$ and E_c vs electric field, respectively. The value of $2P_r$ and E_c decreases slightly due to He$^+$ ion doped, which is related to the phase transition in the HZO layer after doping.

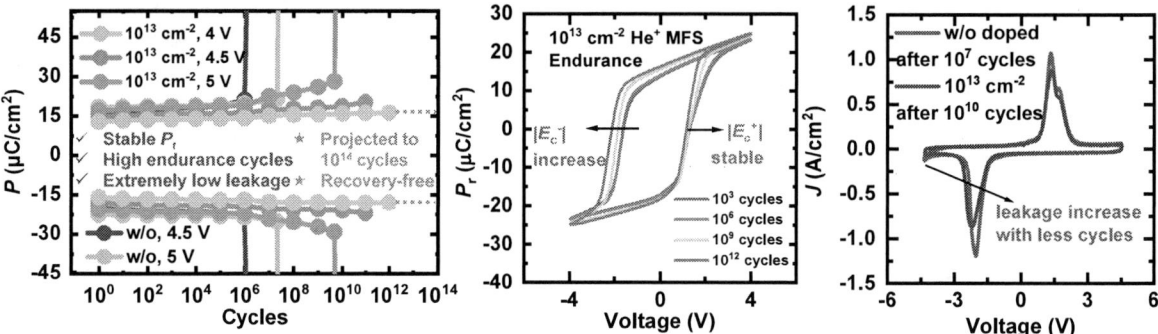

Fig. 4 Endurance characteristic of He-doped HZO MFS devices. The endurance of He$^+$-doped HZO can be projected to be more than 10^{14} cycles.

Fig. 5 Stable P-V hysteresis after 10^{12} cycles in the He-doped MFS structure. E_c^- increase clearly after endurance.

Fig. 6 Comparison of I-V after endurance measurement for devices with and without He$^+$ doping.

Fig. 7 Mechanistic explanation of endurance enhancement after doping. A uniform defect distribution and a thicker TiO$_2$ layer are important factors contributing to the improvement in endurance.

Table.1 benchmark of endurance characteristic in HZO-based FE devices.

Ref.	Initial $2P_r$ (μC/cm^2)	E_c @ E_{apply} (MV/cm)	Frequency (Hz)	Wake-up percentage	Cycles number	P_r loss	Recover Operation
[3]	7.5	1@2	1M	60%	10^{10}	60%	Yes
[4]	30	1.45@4	100k	5%	10^{11}	5%	No
[5]	14	0.7@1.5	500k	Free	10^8	42%	Yes
[12]	25	1.1@2.5	100k	5%	10^{10}	5%	No
[13]	28	1.2@4	100k	Free	10^{10}	26%	Yes
[14]	30	1.5@3	1M	N/A	10^{12}	40%	Yes
This work	32	1.25@4	2.5M	Free	> 10^{12}	~ 0	No

Wide Temperature Behavior Analyses of SiGe HBTs Based on Small-Signal Parameter Extraction Method

Lu Zhao[1,2], Guofang Yu[1], Jie Cui[1], Yue Zhao[1], Jun Fu[1], Yanyan Liu[2]

[1]School of Integrated Circuits, and Beijing National Research Center for Information Science and Technology (BNRist), Tsinghua University, Beijing 100084, China. (fujun@tinghua.edu.cn)
[2] Tianjin Key Laboratory of Efficient Utilization of Solar Energy, Nankai University, Tianjin, 300350, P.R. China

Abstract

This work analyzes the temperature characteristics of RF small-signal parameters for SiGe HBTs over a range of 80 K to 400 K. By extracting small-signal model parameters, we investigate how temperature variations affect device performance. The findings reveal significant enhancements in cut-off frequency from 189 GHz at 400 K to 406 GHz at 80 K, highlighting the suitability of SiGe HBTs for cryogenic applications and advanced RF circuit designs.
Keywords: SiGe HBTs, Wide temperatures, Parameter Extraction

Introduction

Silicon-Germanium (SiGe) heterojunction bipolar transistors (HBTs) have emerged as a formidable technology for radio-frequency (RF) and millimeter-wave applications, showing remarkable performance capabilities with a cutoff frequency (f_T) exceeding 300 GHz and a maximum oscillation frequency (f_{max}) surpassing 400 GHz at room temperature [1-4]. Advanced SiGe HBTs inherently exhibit exceptional suitability for low-temperature operation [5-6], as evidenced by the reported f_T of 710 GHz in [7] and f_{max} of 618 GHz in [8], both attained at a cryogenic temperature of 4K. These values signify a substantial enhancement of approximately 70% compared to their respective room temperature performances, highlighting the profound potential of SiGe HBTs over a wide temperature range. It is important to understand the temperature characteristics of SiGe HBTs over a wide temperature range for cryogenic applications.

As the device's operating temperature decreases, its static operating point shifts and its RF characteristics exhibit a pronounced temperature dependence. This requires a higher challenge for designing circuits over a wide temperature range. Consequently, gaining an in-depth understanding of the temperature's influence on the SiGe HBT's core characteristics is paramount.

This work delves into the DC and RF characteristics of SiGe HBTs from 80 K to 400 K. The temperature-dependent variation of the device's static operating point is analyzed. The parameters of a small signal model are extracted over a wide temperature range, and the relationship between model parameters and temperature is investigated. Finally, the S

parameters are simulated based on the extracted wide temperature range small signal model parameters.

Experimental Details and Small-Signal Model

A. Device fabrication and measurement configuration

The investigated NPN SiGe HBTs were manufactured using the GlobalFoundries 0.13 μm SiGe BiCMOS 8XP technology with a collector-base-emitter-base-collector (CBEBC) layout configuration. The emitter window had a width of 0.12 μm and a length of 5μm. Fig. 1 shows the layouts and optical images of the tested device (DUT) and dummy patterns of Open and Short. The Open and Short patterns do not include the active area, metal1 (M1) contacts, and M1 layer.

A Lakeshore Cryogenic probe station was utilized to achieve a broad temperature range from 80 K to 400 K. The DC characteristics were measured using a Keysight B1500A semiconductor parameter analyzer. The S-parameters were obtained through a Keysight N5247B vector network analyzer, with Short-Open Load-Through (SOLT) calibration being conducted prior to measurement.

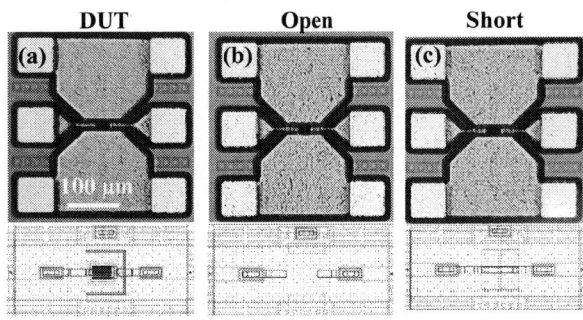

Fig.1: The optical images and layouts of (a) the tested SiGe HBT (DUT), (b) Open structure, and (c) Short structure.

B. Small-Signal Modeling of SiGe HBTs

Fig. 2 shows the schematic cross-sections of the SiGe HBT and the corresponding small-signal equivalent elements (The network of the substrate is not presented in the figure). These elements are divided into two parts, i.e., the extrinsic part of the circuit are deemed independent of bias, whereas the intrinsic part encompasses the bias-dependent parameters. The extraction of small-signal model parameters begins with the de-embedding of external

979-8-3315-0417-5/25 $31.00 © 2025 IEEE

parasitic capacitances and inductances (L_B, L_C, L_E) using Open and Short de-embedding structures. Subsequently, the "overdriven-I_B" method (high base current I_B) is utilized to compensate for over-calibration and to extract the parasitic resistances (R_{bx}, R_{cx}, R_{ex}). Following this step, the device is biased in the off-state to extract the overlap capacitances (C_{BEO} and C_{BCO}). After removing the extrinsic parasitic elements, the intrinsic part of the SiGe HBT is extracted using the T $\leftrightarrow$ Π transformations under the on-state conditions ($V_{CE} = 1.5V$, $V_{BE} = 0.8V \sim 1.2V$) [9].

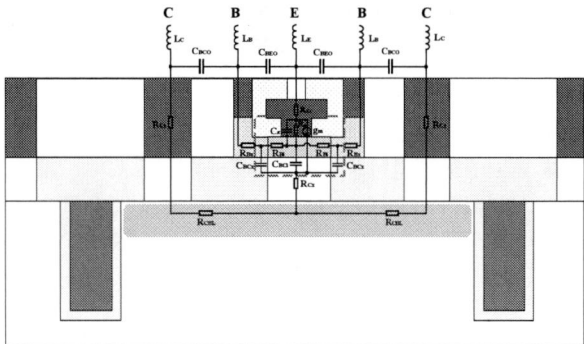

Fig.2: The schematic cross-sections of the SiGe HBT and the corresponding small-signal equivalent elements.

Results and Discussion

A. Gummel characteristics

Fig.3(a) presents the forward Gummel Characteristics of the fabricated SiGe HBT at $V_{CB} = 0$ V from 80 K to 400 K. The results show that the collector current (I_C) and base current both shift towards higher operating voltages, which indicates that the turn-on voltage of base-emitter voltage (V_{BE}) increases as the temperature decreases. A kink phenomenon can be observed in the $I_B - V_{BE}$ at cryogenic temperatures, which is attributed to the heterojunction barrier effect (HBE) under these conditions. The current gain ($\beta = I_C/I_B$) related to I_C is calculated and plotted in Fig.3(b). It can be clearly seen that as the temperature decreases, the peak value of the current gain gradually increases, but the profile of the current gain gradually narrows.

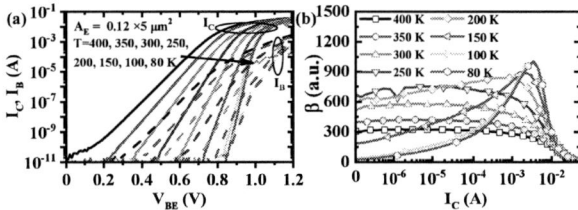

Fig.3: (a) Measured Gummel characteristics and (b) the extracted current gain (β) of SiGe HBTs from 80K to 400K.

B. The cut-off frequency (f_T)

The cutoff frequency f_T is extracted from the S parameter of the SiGe HBTs when it is operated under on-state

conditions with $V_{CE} = 1.5$ V. As depicted in Fig. 4, the value of f_T undergoes a notable increase as the temperature decreases, ultimately approaching saturation at temperatures below 150 K. The narrowing of the DC operating voltage range, in conjunction with the broadening of the operational current range for the f_T response, underscores a phenomenon where minute voltage variations can result in significant alterations in the static operating point, particularly under low-temperature conditions. The f_T demonstrates a robust correlation with transconductance (g_m) and capacitance (C_π). To investigate the underlying mechanism of temperature-dependent variations in f_T, we will further analyze the temperature relationship between g_m and C_π.

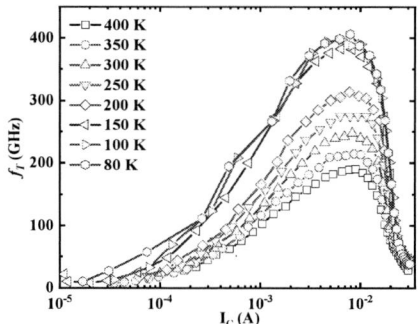

Fig.4: Extracted f_T as a function of I_C from 80 K to 400 K.

C. Temperature dependence of the g_m and C_π

The transconductance g_m is used to characterize the amplification ability of the device to signals. Fig.5(a) presents the g_m of the extracted SiGe HBT as a function of I_C from 80 K to 400 K. The results indicate that when the I_C is relatively small, the g_m gradually increases with the rise in I_C. However, the curve displays a prominent peak at approximately 18 mA, which is attributed to the kirk effect in the collector current [10]. Fig. 5(b) depicts that the maximum value of g_m is enhanced almost 2 times as the temperature reduces from 400 K to 80 K.

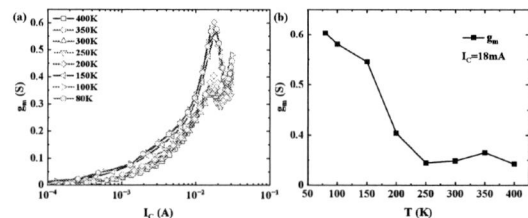

Fig.5: (a) Extracted g_m of SiGe HBTs as a function of I_C from 80 K to 400 K and (b) g_m maximum value as a function of the corresponding temperature

Fig.6(a) presents the extracted C_π as a function of I_C from 80 K to 400K. It can be clearly seen that C_π is proportional to the collector current. The values of C_π at I_C = 8 mA (where the peak value of f_T is obtained) are extracted and presented in Fig. 6(b). The results illustrate that C_π gradually decreases as the temperature lowers down, which indicates the increase of f_T.

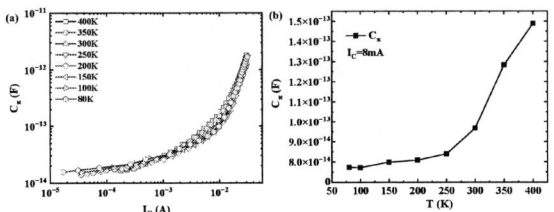

Fig.6: Extracted C_π of SiGe HBTs as a function of I_C from 80 K to 400 K and (b) the extracted C_π as a function of temperature at I_C = 8 mA

D. RF Simulation and analysis of SiGe HBT

Fig.7 presents the comparison of S-parameters between small-signal model simulation and measured data at 400 K, 300 K, 200 K and 80 K, respectively. It can be observed that the results show good agreement. The static operating point of the device increases from V_{BE} = 0.8 V to 1.05 V as the temperature decreases, accompanied by a gradual increase in S_{21} parameter. This observation holds significant importance for the design of circuits with a wide temperature range utilizing SiGe HBT technology.

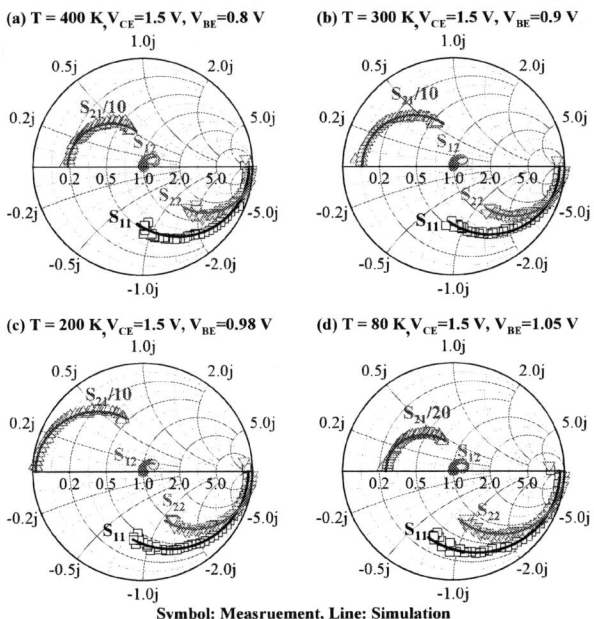

Fig.7: Comparison between model simulation and measured data of S-parameters at 400K,300k,200K and 80K

Conclusion

This work focuses on the temperature-dependent behavior of SiGe HBTs in the context of RF small-signal parameters. Conducted over a temperature range from 80 K to 400 K, the study employs a systematic extraction of small-signal model parameters to assess how temperature influences device performance. The results indicate notable improvements in f_T at lower temperatures, emphasizing the advantages of SiGe HBTs for cryogenic applications. The temperature dependence of the g_m and C_π are extracted to analyze the underlying influencing mechanism. Finally, the

comparison of S-parameters between small-signal model simulation and measured data from 80 K to 400 K are conducted.

Acknowledgments

This work was supported by Grant QYJS-2022-1700-B, and is partly supported by the National Natural Science Foundation of China (Grant number: 62474101 and 92064002).

References

[1] K. Ishimaru, "Future of Non-Volatile Memory From Storage to Computing," in IEDM Tech. Dig., 2019, pp.12.

[2] M. Khater, et al, "SiGe HBT technology with f_{max}/f_T =350/300 GHz and gate delay below 3.3 ps," in IEDM Tech. Dig., Dec. 2004, p. 247–250.

[3] M. D. Ganeriwala, et al, "A bottom-up scalable compact model for quantum confined nanosheet FETs, " IEEE Trans. Electron Devices, 69, pp. 380-387 (2022).

[4] P. Chevalier, et al., "A conventional double-polysilicon FSA-SEG Si/SiGe:C HBT reaching 400 GHz f_{MAX}," in Proc. IEEE Bipolar/BiCMOS Circuits Technol. Meeting, Oct. 2009, p. 1.1.

[5] H. Rücker, et al., "A 0.13 µm SiGe BiCMOS technology featuring f_T/f_{max} of 240/330 GHz and gate delays below 3 ps," in Proc. IEEE Bipolar/BiCMOS Circuits Technol. Meeting, Oct. 2009, p. 11.1.

[6] S. van Huylenbroeck, et al.," A 400 GHz f_{MAX} fully self-aligned SiGe:C HBT architecture," in Proc. IEEE Bipolar/BiCMOS Circuits Technol. Meeting, Oct. 2009, p. 1.2.

[7] A. J. Joseph, et al., " Operation of SiGe heterojunction bipolar transistors in the liquid-helium temperature regime," IEEE Electron Device Lett., vol. 16, no. 6, pp. 268–270, Jun. 1995.

[8] J. Yuan, et al., "On the performance limits of cryogenically-operated SiGe HBTs and its relation to scaling for terahertz speeds," IEEE Trans. Electron Devices, vol. 56, no. 5, pp. 1007–1019, May 2008.

[9] H. Y. Chen, et al., "Small-Signal Modeling of SiGe HBTs Using Direct Parameter-Extraction Method," IEEE Transactions on Electron Devices, vol. 53, no. 9, pp. 2287-2295, Sept. 2006.

[10] J. Yuan et al., ""An Investigation of Negative Differential Resistance and Novel Collector–Current Kink Effects in SiGe HBTs Operating at Cryogenic Temperatures," IEEE Transactions on Electron Devices, vol. 54, no. 3, pp. 504-516, March 2007.

979-8-3315-0417-5/25 $31.00 © 2025 IEEE

In₂Se₃ FET-Based Neural Network Circuit Design for Complete Associative Learning

Weiwei Xiong[1], Yasai Wang[1,2], Xiangshui Miao[1], Yang Chai[2], and Yuhui He[1*]

[1]School of Integrated Circuit, Huazhong University of Science and Technology, Wuhan, China
[2]Department of Applied Physics, The Hong Kong Polytechnic University, Hong Kong, China.
*heyuhui@hust.edu.cn

Abstract

Associative learning is a crucial mechanism in biological neural systems. In this work, a neural network circuit via In₂Se₃ FET for full-function associative learning was designed. A modulation dynamics model was developed by utilizing In₂Se₃ IP-OOP electrical characteristics. The synaptic plasticity of In₂Se₃ FET was simulated in HSPICE using Verilog-A implementation of the established model, and subsequently, the circuit-level simulations validated the complete Pavlov's dog associative learning functionality. This design presents a promising approach for future bio-inspired associative learning implementations.

Keywords: ferroelectric semiconductor transistor, associative learning, spice model.

Introduction

Due to the mismatch between the explosive growth in computing power demands in the intelligent era and the slowdown of Moore's Law scaling in the post-Moore era, traditional von Neumann architecture based on CMOS circuits have reached its limit [1-2]. Bio-inspired neuromorphic computing (NC) offers an effective solution to overcome the bottleneck known as "memory wall" through its characteristics of low power consumption and low latency. Associative learning, a crucial mechanism for biological adaptation to the environment [3], primarily relies on neural plasticity and is well suited to be implemented through NC architecture. Previous implementations of associative learning via neuromorphic devices typically encountered two challenges: (i) incomplete learning functionality, such as lacking extinction mechanism [4], and (ii) necessity for complex circuit designs to achieve full-function associative learning. As shown in Fig. 1, implementing complete associative learning functionality requires customized circuit modules and additional operational amplifiers, leading to significant power overhead.

In this work, we leverage the unique characteristics of In₂Se₃, which exhibits both in-plane (IP) and out-of-plane (OOP) ferroelectric polarization, to construct a three-terminal modulated ferroelectric semiconductor field effect transistor (FeSFET, as shown in Fig. 2). Its multi-terminal conductance modulation properties naturally facilitate associate learning with various function (Fig. 3). Based on the V_{DS} and V_G modulation characteristics, we established a dynamics model for its conductance evolution. We then described an In₂Se₃ FeSFET module in HSPICE using Verilog-A language.

Furthermore, we designed a compact, energy-efficient associative neural network circuit in HSPICE and demonstrated its excellent performance for fully-functional Pavlovs's dog associative learning.

Classical Conditioning Process

Classical conditioning is a prime example of associative learning and its complete process is shown in Fig. 4. Initially (Fig. 4a), individual unconditioned stimulus (US, e.g., bell) or conditioned stimulus (CS, e.g., food) only elicits unconditioned response (UR, e.g., audition) or conditioned response (CR, e.g., salivation), respectively. Repeated sequential presentation of US and CS (with an interstimulus interval defined as Δt) establishes a new association between US and CR (Fig. 4b). This associative learning process results in stable acquisition. After learning, US alone becomes sufficient to elicit CR. However, prolonged isolated US presentations result in backward conditioning (Fig. 4c), leading to progressive weakening and ultimate extinction of the US-CR association (Fig. 4d).

IP and OOP Electrical Modulation Behavior

In₂Se₃ is a unique ferroelectric semiconductor exhibiting both IP and OOP polarization. Its remnant polarization can be modulated by V_{DS} and V_G, which affects carrier distribution and channel conductance. As shown in Fig. 3, owing to three-terminal modulation, V_{DS} increases conductance during CS-US pairing, while feedback signals from salivation output generate V_{GS} to decrease conductance for backward conditioning.

Fig. 5 shows the IP electrical modulation of In₂Se₃ FET. The output characteristic exhibits memristive behavior (Fig. 5a). Individual V_{DS} pulse with various amplitudes induce conductance changes (ΔG), revealing an IP polarization threshold of 5.2V (Fig. 5b). The long-term potentiation (LTP) and depression (LTD) characteristics are shown in Fig. 5c and 5d, with experimental data (dots) and model fitting results (gray lines). The dynamics model is detailed in the next section.

Fig. 6 shows the OOP electrical modulation of In₂Se₃ FET. The transfer curves exhibit a clockwise hysteresis due to ferroelectric polarization (Fig. 6a). Gate-controlled LTP and LTD characteristics (Fig. 6c, d) show increased conductance under negative gate voltage.

Dynamics Model of In₂Se₃ FET

Based on the nonlinear conductance modulation measured in

Fig. 5c-d and 6c-d, we establish the following dynamics model, where α_p and α_n represent conductance modulation coefficients for SET and RESET process, respectively. V_{th_p} and V_{th_n} denote threshold voltages:

$$\frac{dG}{dt} = \begin{cases} \alpha_p\big(V(t)-V_{th\_p}\big)\big(G_{max}-G(t)\big), V(t) \leq V_{th\_p} \\ \alpha_n\big(V(t)-V_{th\_n}\big)\big(G(t)-G_{min}\big), V(t) \geq V_{th\_n} \\ 0, else \end{cases} \quad (1)$$

G_{max} scales linearly with voltage, with G_{min} fixed at around 1nS. The model fitting curves are shown as gray lines in Fig. 5c-d and 6c-d.

Associative Learning via In₂Se₃ FET

A. Device-Level Verification of Associative Functions
Fig. 7 demonstrates associative learning in the In₂Se₃ FET. (i) Starting from a high-resistance state (HRS), (ii) paired V_D and V_S pulses with an interstimulus interval of Δt (mimicking bell and food stimuli) increase conductance, indicating strengthened association. (iii,iv) Delayed feedback signal V_G (exceeds the V_{GS} threshold) enables extinction and backward conditioning. Spike-timing-dependent plasticity (STDP) is fundamental to associative learning. Using the dynamics model, STDP characteristics of In₂Se₃ FET are simulated in HSPICE, showing agreement with experimental data (Fig. 8).

B. Neural Network Circuit Design for Associative Learning
Utilizing the constructed In₂Se₃ FET module, we implemented a fully functional Pavlov's dog associative learning neural network circuit in HSPICE (Fig. 9). Subscripts *b*, *f*, *s*, and *a* represent "bell", "food", "salivation", and "audition", respectively. T_{bs} and T_{fa} denote synapses between bell-to-salivation and food-to- audition pathways. Diodes were incorporated to ensure unidirectional association.

C. Simulation Results and Comparison
The temporal simulation results of Pavlov's dog associative learning are shown in Fig. 10. The left panel displays the designed input waveforms. Amplitude of individual signal is below the modulation threshold. Initially (0-1s), both T_{bs} and T_{fa} are set to HRS. Individual bell or food signals trigger only corresponding neuronal responses, with device conductance remaining unchanged due to sub-threshold voltage between Node1 and Node2. During the associative learning process (1-6s), bell and food signals are input sequentially with interval Δt. When overlapped, voltage between Node1 and Node2 exceeds SET threshold. Due to diodes' reverse-blocking characteristic, voltage primarily drops on the transistor in the T_{bs} branch but on the diode in the T_{fa} branch, causing G_{bs} to increase gradually while G_{fa} remains constant. After learning (6-7 s), individual bell signal triggers salivation response, demonstrating acquisition function. Finally, continuous bell signals are applied again (7-9s).

Positive feedback V_G from activated N_s neuron reduces G_{bs} until bell signal no longer triggers N_s response, showing backward conditioning and extinction .
Table 1 illustrates the effects of different Δt, where N represents the number of salivation responses during the backward conditioning process. Table 2 compares key metrics between this work and previous associative learning circuits. Compared with prior works, our associative learning circuit, leveraging the unique coexistence of IP and OOP polarization in In₂Se₃, eliminates complex control modules. This design offers advantages including low power consumption and high scalability, facilitating expansion to larger-scale associative learning neural networks.

Conclusion

Based on the unique characteristic of coupled IP and OOP polarization in In₂Se₃, we established and validated a three-terminal modulation dynamics model. Subsequently, we designed a compact and concise associative learning neural network circuit, and simulated complete functionality of Pavlov's dog in HSPICE. This In₂Se₃ FET-based associative learning circuit demonstrates advantages of low power consumption and high efficiency, offering guidance for future neuromorphic associative learning designs.

Acknowledgments

The work is supported by the National Key Research and Development Program of China (No.2023YFB4502200), Natural Science Foundation of China (No. 92164204 & 62374063)

References

[1] M.-K. Song et al., "Recent Advances and Future Prospects for Memristive Materials, Devices, and Systems," ACS Nano, p. acsnano.3c03505, Jun. 2023, doi: 10.1021/acsnano.3c03505.

[2] C. C. Wanjura and F. Marquardt, "Fully nonlinear neuromorphic computing with linear wave scattering," Nat. Phys., vol. 20, no. 9, pp. 1434–1440, Sep. 2024, doi: 10.1038/s41567-024-02534-9.

[3] R. Hanssen et al., "Liraglutide restores impaired associative learning in individuals with obesity," Nat Metab, vol. 5, no. 8, pp. 1352–1363, Aug. 2023, doi: 10.1038/s42255-023-00859-y.

[4] X. Ji et al., "Mimicking associative learning using an ion-trapping non-volatile synaptic organic electrochemical transistor," Nat Commun, vol. 12, no. 1, p. 2480, Apr. 2021, doi: 10.1038/s41467-021-22680-5.

[5] J. Sun, G. Han, Z. Zeng, and Y. Wang, "Memristor-Based Neural Network Circuit of Full-Function Pavlov Associative Memory With Time Delay and Variable Learning Rate," IEEE Transactions on Cybernetics, vol. 50, no. 7, pp. 2935–2945, Jul. 2020, doi: 10.1109/TCYB.2019.2951520.

[6] Z. Xu, G. Chen, S. Chen, and H. Xu, "Mimicking Pain Conditioning Using an Electrolyte-Gated Organic Synaptic Transistor," Advanced Materials Technologies, vol. 9, no. 11, p. 2302047, 2024, doi: 10.1002/admt.202302047.

Background

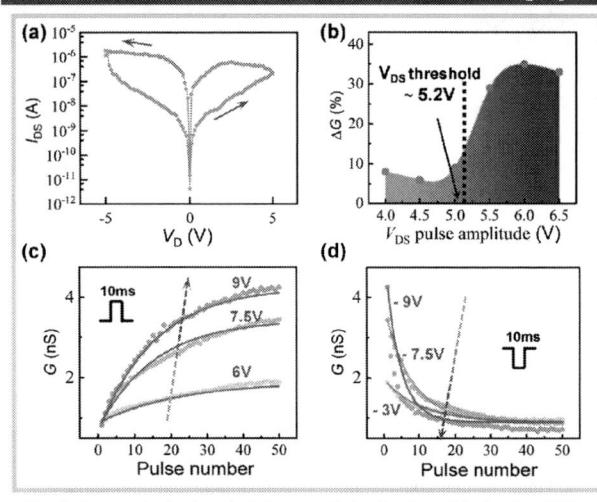

Fig. 1: Associative learning via memristor. " M " represents modules with specific functions.

Fig. 2: Schematic diagram of In₂Se₃ FeSFET

Fig. 3: Associative learning via In₂Se₃ FeSFET.

Fig. 4: Diagram of the complete process of associate learning: (a) initial, (b) associate learning, (c) backward conditioning, (d) extinction.

IP and OOP electrical modulation behavior of In₂Se₃ FET

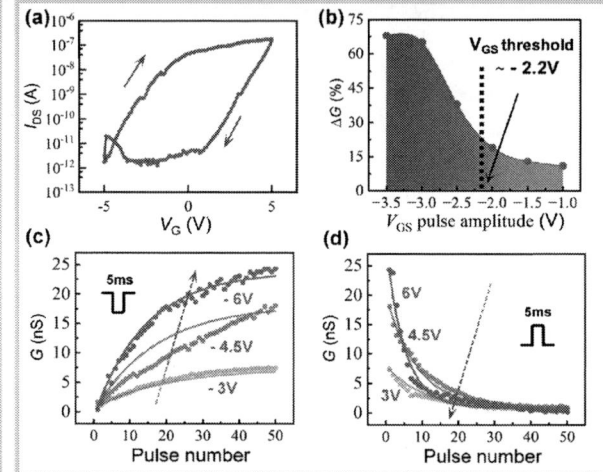

Fig. 5: Source-drain electrical characteristics dominated by in-plane (IP) polarization: (a) output characteristic curve (I_D-V_D), (b) ΔG under single V_{DS} pulse, dynamic model fitting curves (solid) and experiment data (ball) of (c) LTP characteristic and (d) LTD characteristic.

Fig. 6: Gate terminal electrical characteristics dominated out of plane polarization (OOP) polarization: (a) output characteristic curve (I_D-V_D), (b) ΔG under single V_{DS} pulse, dynamic model fitting curves (solid) and experiment data (ball) of (c) LTP characteristic and (d) LTD characteristic.

Associate learning via In₂Se₃ FET

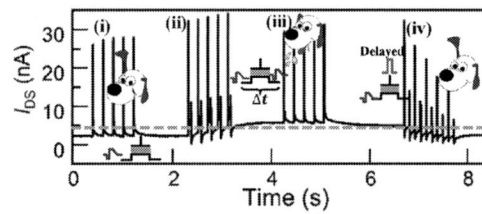

Fig. 7: The process of establishing associative learning (Single device).

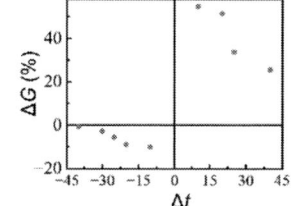

Fig. 8: STDP characteristics of In₂Se₃ FeSFET HSPICE model.

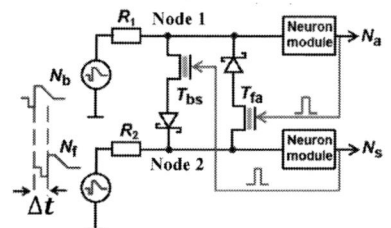

Fig. 9: Implementation of fully functional associative learning.

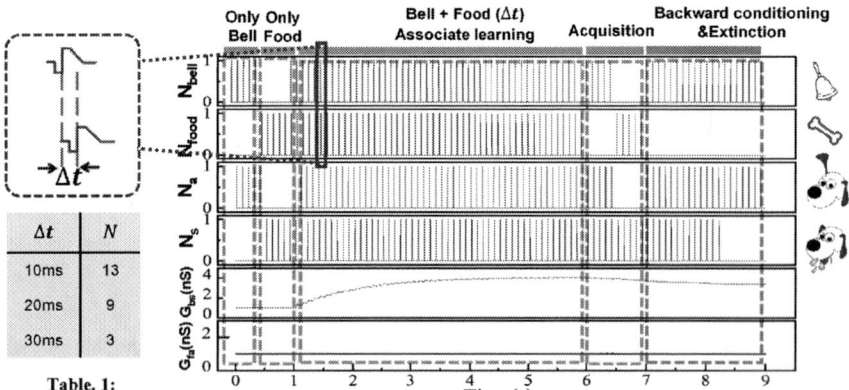

Δt	N
10ms	13
20ms	9
30ms	3

Table. 1: Influence of Δt

Fig. 10: Circuit simulation result of various functions in HSPICE.

	This work	Ref 5	Ref 6
Association	√	√	√
Extinction	√	√	×
Forgetting	√	√	√
Energy Consumption	√	×	√

Table. 2: Comparison of Different Circuits for Associative Learning

Data Retention in co-doped HZO FeCAPs: Roles of FE Thickness and Thermal Budget

Justine Barbot[1], Markus Peller[2], Isaac Emanuel Robert[2], Kerstin Bernert[2], Hannes Mähne[2], Steffen Thiem[2],
David Lehninger[3], Ayse Sünbül[3], Konrad Seidel[3], Thomas Kämpfe[3]
[1]X-FAB Global Services GmbH, Erfurt, Germany — justine.barbot@xfab.com — +49 361 427 8059,
[2]X-FAB Dresden GmbH Co. KG, Dresden, Germany,
[3]Fraunhofer Institute for Photonic Microsystems IPMS, Center Nanoelectronic Technologies CNT, Dresden, Germany

ABSTRACT

This study investigates data retention (DR) in co-doped Hafnium Zirconium Oxide (HZO) ferroelectric capacitors (FeCAPs) under high-temperature stress. Imprint, due to charge injection at an metal/ferroelectric (M/FE) interface, impacts DR, especially in opposite-state retention. The vulnerability to imprint is shown to be mitigated by high annealing temperature ($> 600°C$), operations under high program electric fields ($\simeq 4$ MV/cm), and thick FE layer ($t_{FE} \simeq 15$ nm). Two samples demonstrated DR exceeding 1000 hours at 125°C, with a model suggesting DR for over 10 years for the optimally annealed samples, supporting FeCAP in high-temperature applications.

Keywords: FeCAP, Hafnium Oxide, Data Retention

INTRODUCTION

HZO FeRAM has emerged as a promising candidate for non-volatile memory (NVM) technologies due to its scalability, low power consumption, and CMOS compatibility. However, ensuring long-term data retention, particularly under high-temperature stress, remains a significant obstacle, notably for automotive applications. A key factor affecting DR is imprint, a phenomenon where memory cells develop an unintended internal electric field over time, shifting the polarization state and reducing data processing. After describing the material, electrical characterization, and data analysis strategy, this article explains that imprint is mainly due to charge injection at the M/FE interface. This study then focuses on the impact of annealing temperature (T_{anneal}), FE thicknesses (t_{FE}), and operating electric fields on DR performance, aiming to provide insights into optimizing FeCAP reliability.

EXPERIMENTAL

Samples

Co-doped HZO MFM 432 μm-diameter FeCAPs were fabricated at Fraunhofer IPMS on Si-doped coupon substrates (see Fig. 1). The FE material consists of doped HfO_2 and ZrO_2 layers deposited by Atomic Layer Deposition, with thicknesses from 10 to 15 nm. The optimized TiN electrodes, deposited by physical vapor deposition, are topped with a thin Ti and Pt layer (Fig. 1). Additional deposition details are in [1]. FE layer crystallization was achieved by annealing from 400°C (1 h) to 800°C (5 mins).

Electrical Characterization

Fig. 2. shows the pulse sequences used for the measurements conducted using Keithley 4200 equipped with Pulse Measurement Units. For this study, 7 capacitors were used per samples and programming conditions (read and write voltage V_{WR}, 0 or 1 state).

At first, almost all capacitors were woken-up via a 10^4 triangular cycling at 5 kHz and at cycling voltage ($V_{cycling}$)

as shown on Table 1., except the sample S1 annealed at the highest temperature, which was found to be wake-up free. Then, at room temperature (RT), one half of the capacitors were poled in polarization +P (0 state), and the other half in -P (1 state) via respectively single positive and negative V_{WR}-high and 100 μs-wide triangular pulse. After bake at elevated temperature (125°C or 175°C), the states were measured at RT via read sequences mimicking the destructive FeRAM reading (see Fig.2.). When the state 1 (state 0) is read, the polarization is (not) switched and FE current noted I_{FE} (does not) adds to DE current noted I_{DE}.

In this study, for either 0 state or state 1, 3 retention states were evaluated to account for the rewrite after read needed in FeRAM: Same State (SS, state baked), New Same State (NSS, state baked restored after bake) and Opposite State (OS, state opposite to state baked). Therefore, as illustrated on Fig. 2. the read pulse sequence for state 0 (state 1) was made of 4 (5) pulses of $\pm V_{WR}$ 100 μs triangular pulses allowing to evaluate all conditions, and to then pole back the initial poled state. As illustrated in Fig. 3., after bake process, OS states are the most altered states. Indeed, half-hysteresis of OS state 0 (state 1) is more rapidly distorted by the bake process, underlying that polarization switching in the opposite state is hindered (increased), causing data loss.

DR Data Analysis Methodology

To quantify DR, Switched Polarization (P_{SW}) was determined by integrating the dynamic current (*i.e.* DE and FE currents) over each triangular read pulse. Fig.4.(a) gives an example of the evolution of P_{SW} for all states along bake time. OS states are less stable overtime, primarily due to imprint. Indeed, as illustrated in Fig.4.(b), the I-V curve after bake process suffers from a built-in voltage, called Voltage Offset (V_{off}), leading to a worst capture of the FE peaks and later to both write and read failures at constant V_{WR}.

Since imprint is causing failure, this study will focus on V_{off} of the capacitors rather than P_{SW} to investigate DR. The standard ways to estimate V_{off} is to extract the coercive voltages on I-V or P-V curves. However, after long bake time the built-in voltage is such that, this former method fails, because the FE peaks response is out of the read voltage window. To remedy to this, owing to the methodology presented by Tagantsev in [2], V_{off} was extrapolated from a linear relationship between P_{SW} and applied voltage determined through the sequencing of an I-V PUND and its integration (see an example on Fig.7. for S1).

V_{off} of OS 0 vs bake time curves are shown for different samples and measurement conditions on Fig.8. To interpret variations of these curves for different samples and measurements conditions, fits with the following equations from [2] were performed. V_{off} follows:

$$V_{off} = V_o \log_{10}(1 + t/\tau_o) \quad (1)$$

where V_o and τ_o are two fitting parameters defined as:

$$V_o \propto T\sqrt{\frac{P_o}{\epsilon*}}\ (2),\quad \epsilon* \propto \frac{t_{DE}}{t_{FE}}\ (3),\quad ln(\tau_o) \propto \frac{\Phi}{kT}\ (4).$$

P_o is the polarization of the first state written. Φ is an activation barrier reduced by an image force lowering, k the Boltzmann constant, T the temperature of bake, $\epsilon*$ relative permittivity of the stack. These equations assume that charge injection into an interfacial passive layer is the main root cause of imprint, omitting roles of crystallographic phase change and oxygen vacancy diffusion like in [3]. Such model was chosen as starting point for the discussion with regards to previous papers demonstrating the existence of passive layer at the TiN/HZO interface by material characterization [4], [5].

Hereafter, V_{off} vs bake time curves and their extrapolations allow to estimate the DR failure times of the FeCAPs. When V_{off} exceeds half of initial measured coercive voltage, the DR test could not anymore differentiate between a state 1 and state 0. This extrapolation methodology was validated by the high positive correlation found between the extrapolated and real failure recorded before 1000 h bake time, shown on Fig.9.

RESULTS AND DISCUSSION

The strategy for addressing DR in NVM involves performing a sequence of read, bake, renew read and renew bake time operations until the desired total data retention time is achieved. However, this cumulative bake strategy may seem not feasible in FeRAM due to its destructive read. In this study, each sample and voltage conditions, some capacitors were read at several baking time up to the final baking time, while others just once, aiming to conclude if mechanisms leading to DR failure are partially recoverable by read sequence. As reported in Fig.6., the difference between the P_{SW} of one time read capacitor and multi-time read capacitor for all states and conditions investigated in this study is for more than the third quartile of data below 1 $\mu C/cm^2$, which is below the resolution limit, and well below the variation of P_{SW} along the bake time. In that respect, it is confirmed that mechanisms causing failure -mainly imprint is not recoverable by a unique read sequence, and so that cumulative bake time procedure can be used to study FeCAP reliability.

At first glance, the validity of the charge injection model appears to be confirmed by the V_{off} vs bake time good fits in Fig.8.(a), which provides V_o and τ_o variations with baking temperature that align with Eq.2. and Eq.4. The higher the baking temperature, the more pronounced the imprint becomes, likely due to temperature-driven injection mechanisms such as Poole–Frenkel emission. Similarly, the charge injection model might explain the variation of V_o for different t_{FE} in Fig.8.(b). Thinner FE layers, have lower V_o, as expected with regards to the proportionality of V_o with $\frac{1}{\epsilon*}$ and P_o.

In Fig.8.(c), the effect of T_{anneal} on imprint is considered. It is evident that a low T_{anneal} exacerbates the imprint effect. Fits on Fig.8.(c) do not help to understand why so, especially if it is assumed that all samples have identical interfaces. Lower T_{anneal} reduces P_o and increases V_o, but Eq. 2. predicts that V_o should decrease. This discrepancy calls the assumption of identical interface for different T_{anneal} into

question. Additionally, a decrease of τ_o with the decrease of T_{anneal} suggest that annealing affects interface by modulating defect type and concentration. As suggested by [6], such modulation with annealing could be due to a change of the oxygen vacancy concentration at the interface.

In a similar manner, the effect of the electric field on the imprint can be explained via a consideration of interface change. It is known that high electric field cycling would diffuse oxygen vacancy through the FE layer, and thus reduce the chance of injection of charge at the interface, attenuating imprint [7]. Thus, it is clear that the injection model developed by Tagantsev is not a sufficient tool to understand physically the imprint behavior changes from sample-to-sample and electrical conditions since it does not describe with enough granularity the interfaces (defect type and concentration).

However, it was demonstrated that the model is a good tool to estimate failure times, then the failure time of S1 and S4 samples was predicted to be 10000 hours and 10 years at 125°C, a significant result compared to previously reported FeCAP retention studies as shown on table 2. These high DR for S1 and S4 underlines that thick FE layer or high non-BEOL thermal budget might be good strategy to enhance the DR. However both strategies require adaptations, the first one requires high voltage operations, while the second will require to develop high temperature treatment like Nanosecond Laser Anneal [8].

CONCLUSION

In this study, co-doped HZO FeCAP is investigated as a promising candidate for non-volatile memory applications, focusing on data retention (DR) under high-temperature stress. Imprint, caused by charge injection at the metal/ferroelectric interface, is identified as a potential contributor to DR degradation. By analyzing V_{off} versus bake time and fitting the data using the Tagantsev interface screening model, the study demonstrates that DR failure times can be predicted, with high annealed temperature samples showing retention beyond 10 years at 125°C. The results highlight potential strategies to enhance FeRAM reliability, though challenges remain in optimizing high-voltage operations and anneal treatments, especially since the most promising samples are unfortunately not BEoL compatible from a thermal budget perspective—suggesting that alternative approaches must be explored.

ACKNOWLEDGEMENT

This project has received funding from the ECSEL Joint Undertaking under grant agreement No 101007321. The JU receives support from the European Union's Horizon 2020 research, national innovation programme.

REFERENCES

[1] A. Sünbül, et al., Memories - Materials, Devices, Circuits and Systems, 8, (2024).
[2] A. K. Tagantsev, et al., J. Appl. Phys., 96, 11, pp6616-6623, (2004).
[3] F. P. G. Fengler, et al., J. Appl. Phys. 123, 204101 (2018)
[4] L. Baumgarten, et al., Appl. Phys. Lett, vol.118, 10, pp44-45, (2020).
[5] M. Müller, et al., ACS Appl. Electron. Mater., 2, 10, (2020).
[6] P. Yuan, et al., Nano Research, 15, 4, p3667-3674, (2022).
[7] J. Lee, et al., Nano Convergence, 10, 55, (2023).
[8] T. François, et al., IEEE VLSI, (2020).
[9] J. Laguerre, et al., IEEE IRPS, (2024).
[10] A. Sünbül, et al., pss, (2023).
[11] S. Müller, et al., TDMR, (2012).

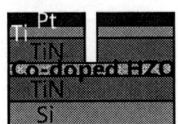

Fig. 1. Sketch of the co-doped HZO FeCAPs.

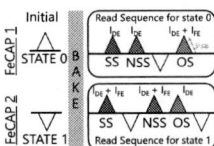

Fig. 2. V-t waveform applied to FeCAPs to evaluate retention.

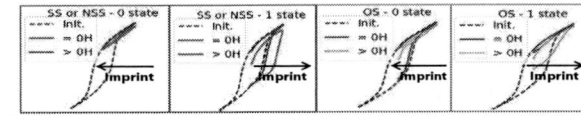

Fig. 3. P-V Hysteresis loops before setting the initial state *(black dashed lines)*. Half-hysteresis loops after bake measured via read of Fig. 2. *(= 0H bake time in purple solid lines, > 0H bake time in green or red solid lines)*. Shifts of the half-hysteresis along the x-axis indicate imprint during bake time at elevated temperatures. Identical half hysteresis curves were obtained for SS and OS. The greater distortion of the half-hysteresis loops for OS states underlines a greater vulnerability to imprint. *Axis are intentionally omitted in this graph, as it serves only as a hand-drawn illustration to represent the phenomenon.*

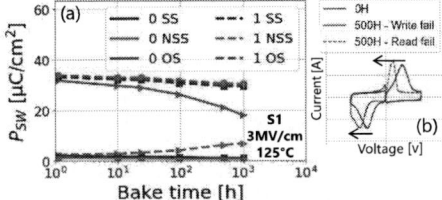

Fig. 4. (a) P_{SW} for SS, NSS and OS for 0 *(solid)* and 1 *(dashed)* states set at 3MV/cm along bake at 125°C. OS fails much quicker due to vulnerability of OS state to imprint. (b) Current vs Voltage of OS 1 for one capacitor on S1 3MV/cm bake @125°C showing both write and read fail.

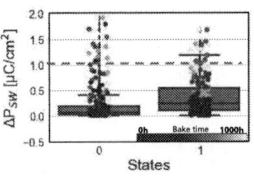

Fig. 6. Difference of Switched Polarization between a state read once and n times for all bake time ($\Delta P_{SW} = P_{SW}^{read\ n\ times} - P_{SW}^{read\ 1\ time}$) with the resolution limit of the measurements *(dotted)*. The different colours indicate the bake time.

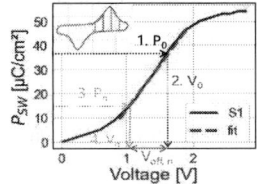

Fig. 7. Switching curve, derived from the stepped integration of PUND IV *(inset)*, is shown for the S1 sample, along with the resulting linear relationship *(red dashed line)*. The arrows illustrate how to determine V_{off} after n hours of bake, based on the switched polarization (P_n), and voltage (V_n).

Table 1. The table summarizes measured samples, test conditions (voltage, bake time, wake-up, anneal), fitted Taganstev model parameters (V_o, τ_o), goodness of fit (R^2), OS state failure time, and expected failure from V_{off}-bake time curves. Some columns are repeated to assist the reader with comparisons. Warm and cold colors indicate respectively max and min values for each case study.

Case study	Bake Temperature		Anneal Temperature				Ferroelectric Thickness			Electric Field	
Sample	1		1	5	2	2	2	3	4	2	
t_{FE} [nm]	10		10	10	10	10	10	12	15	10	
Anneal	800°C, 5mins		800°C, 5mins	650°C, 2mins	450°C, 1h	450°C, 1h	400°C, 1h			450°C, 1h	
2Pr [μC/cm] ∝ P_0	34		34	28	17	28	32		34	17	28
$V_{cycling}$, V_{WR} [V]	3		3			4	4.8	6		3	4
E_{WR} [MV/cm]	3		3			4				3	4
T_{bake} [°C]	125	175	125				125			125	
Experimental OS Fail	>1000h	@500h	>1000h	~ 200h	~ 2h	~ 500h	> 96h		>1000h	~ 2h	~ 500h
Extrapolated OS Fail	90000h	1000h	6000h	300h	2h	~ 500h	~ 700h		10000h	2h	~ 500h
V_o [V]	0.17	0.19	0.17	0.20	0.22	0.14	0.145		0.16	0.22	0.158
τ_o [s]	2.1	0.09	2.1	0.15	0.06	0.014	0.014		0.08	0.06	0.041
R^2	0.96	0.99	0.96	0.98	0.99	0.99	0.99		0.98	0.99	0.99
Figure	8 (a)		8 (b)				8 (c)			8 (d)	

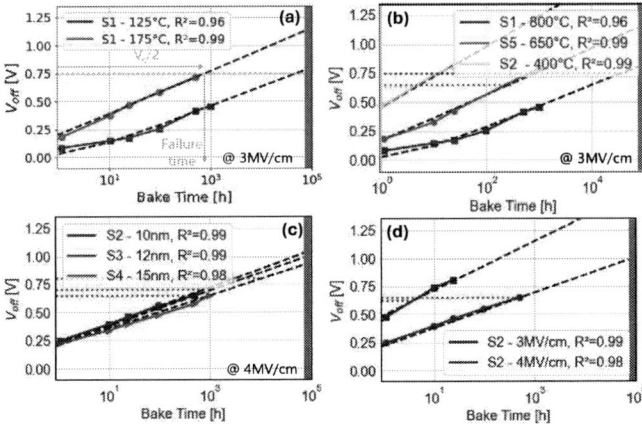

Table 2. Benchmark of FeCAPs retention.

Paper	FE	Anneal	WR	Wake-Up	OS test	Bake temperature	Bake time w/o failure	Extrapolated Retention
Ours	Co-doped HZO 10nm	800°C	3V→3MV/cm	No	Yes	125°C	1000h	90000h (10 years)
	Co-doped HZO 15nm	400°C	6V→4MV/cm	Yes	Yes	125°C	1000h	10000h
[9]	HZO 10nm	<500°C	4V→4MV/cm	Yes	Yes	125°C	27h	-
[10]	HZO 7nm	400°C	2.8V→4MV/cm	Yes	No	100°C	107h	-
[6]	HZO 10nm	800°C	4V→4MV/cm	No	No	150°C	166h	-
[11]	HSO 10nm	800°C	3V→3MV/cm	No	Yes	125°C	1000h	-

Fig. 8. Extrapolated Voltage offset (V_{off}) from measured P_{SW} of OS 1 along bake time at elevated temperature *(solid lines)* for different samples and measurement conditions (bake time and electric field). V_{off} vs bake time curves are fitted and extrapolated using Eq.1. *(black dashed lines)*. Goodness of the fits are given by R^2. Fitting parameters are detailed in Table 1. The failure time of OS states is estimated from the x-position of the crossing of the half coercive voltage (V_c, *orange dotted lines*) and the extrapolated curves. V_c values were taken from original P-V hysteresis loop measured before programming and bake. 10 years bake time is symbolized by the grey line. The graphs show how T_{bake} **(a)**, T_{anneal} **(b)**, t_{FE} **(c)**, and electric field **(d)** influence imprint.

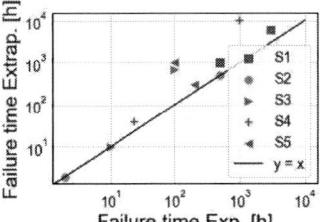

Fig. 9. Correlation plot of the experimental failure time vs the extrapolated failure time found using the methodology described in Fig.8.

Technologies of GaN Power Integration and Modeling

Sheng Li, Yanfeng Ma, Ran Ye, Siyang Liu, Weifeng Sun

Southeast University, China

Abstract

Gallium Nitride based high electron mobility transistors with p-type gate cap (p-GaN gate HEMTs) attract more and more attentions in the field of power electronics. To unleash the high frequency superiority of p-GaN gate HEMTs, monolithic power integration is developed, which faces the challenges of threshold voltage instability, device isolation and SPICE modeling. Novel hybrid gate technology, ultra-thin buffer technology and physics-based modeling methods are proposed to achieve stable threshold voltage, effective device isolation and precise circuit design.

Keywords: power GaN, power integration, SPICE model

Introduction

Thanks to the superiorities of Gallium Nitride (GaN) material, GaN based high electron mobility transistors (GaN HEMTs) exhibit high blocking voltage, low on-state resistance and high working frequency. Therefore, the power systems based on GaN HEMTs also have the advantages of high power density and high conversion efficiency. Nowadays, GaN devices are expected to have attractive applications in automotive electronics, data centers, industrial electronics, et al. The market about GaN devices also increases rapidly. However, during the fast switching transients, due to high dV/dt or di/dt condition, parasitic components cause overshoot and rings, therefore monolithic integration of GaN devices is expected to achieve safe high frequency switching [1][2]. To realize monolithic integration, stable threshold voltage (V_{th}), isolation between high-side and low-side area, and modeling of GaN power HEMTs are highly required.

Vth Instability

Stable V_{th} is important to integrated circuit design, especially for enhancement-mode GaN HEMT with p-type GaN gate cap (p-GaN gate HEMT) which works under high-speed switching condition. For traditional p-GaN gate HEMT, the contact between gate metal and p-GaN layer could be either ohmic- or Schottky-type. Compared to Ohmic-type p-GaN gate device, the Schottky type solution has expected desirable performance due to reduced gate leakage and enlarged operation voltage swing. Meanwhile, several comprehensive reliability studies on Schottky-type gate HEMT (Sch-HEMT) have also promoted related technology. Nevertheless, such a Schottky-type gate structure remains inherent risks [3]. The gate region of conventional Sch- HEMT is formed by two back-to-back diodes as seen in Fig. 1, which could trigger charge storage effect. The D1 is the metal/p-GaN diode, D2 is the p-GaN/AlGaN/GaN diode, C1 and C2 are the junction capacitances. Therefore, the depletion region could result in a "floating p-GaN layer" sandwiched between gate and channel. The charges induced during long time drain bias operation in the floating p-GaN layer are difficult to be released, resulting in obviously V_{th} shifts.

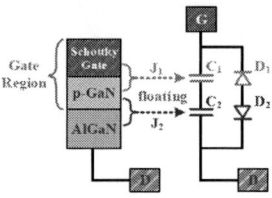

Fig. 1 Gate stack with p-type GaN gate cap.

In system applications, the increase of device V_{th} will lead to an increase of on-state resistance (R_{on}), which will directly affect the temperature rise of switches, ultimately reducing the conversion efficiency of the power system and even leading to the risk of bombing. Although Ohmic-type gate p-GaN HEMT (Ohm-HEMT) can obviate such effect, the relatively large gate leakage determinate a higher continuous gate current to maintain on-state operation voltage. Also, it will increase complexity of drive-circuit to achieve current drive mode using Ohm-HEMTs. Thereby, it is essential to provide a novel technology to realize devices with both enhanced Vth stability and reduced gate leakage current I_{gss}.

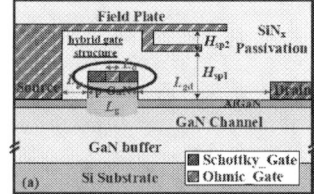

Fig. 2 Novel hybrid-gate p-GaN HEMT.

A novel hybrid gate p-GaN HEMT (Hyb-HEMT) technology based on the GaN-on-Si platform has been proposed [4]. In the proposed Hyb-HEMT, as seen in Fig. 2, the Ohmic-type gate region could provide a free-carrier "discharge path" to alleviate charge storage effect, effectively enhancing V_{th} stability. It is experimentally demonstrated that the Hyb-HEMT can achieve similar static performances and enhanced V_{th} stability. The results prove ΔV_{th} is reduced by about 89.5% under 7V V_{gs} stress, and is reduced by 91.4% under 200V V_{ds} stress.

Too larger Ohmic contact region would increase the gate leakage current. Generally, the leakage current along the Ohmic contact region can be controlled and pinched off by the adjacent Schottky contact regions. The results prove the

979-8-3315-0417-5/25 $31.00 © 2025 IEEE

gate leakage current I_{gss} is only increased by 18.3%, which is largely smaller than that in the pure-Ohmic gate contact device. Actually, according to our measurements, even at 125°C, the I_{gss} can be also controlled into the accepted range for practical application. Accelerated stress experiments and repetitive Unclamp-Inductive-Switching validation indicate improved gate reliability and de-trapping capability.

Half-bridge Integration

Monolithically integrating gate driver and half-bridge configuration is an effective approach to eliminate the parasitic inductive elements, hence becoming a trend of development. However, the state-of-art monolithic integration technology is within 650V level. Considering the rapid developments of power electronic systems that require power devices with a rated voltage above 1200V, it is meaningful to develop new technology to achieve 1200V GaN based monolithic half-bridge integration to improve the reliability of high voltage GaN power applications.

Due to conductive substrate caused vertical breakdown and crosstalk effects in conventional 650V GaN-on-Si platforms, it is hard to further improve the rated voltage level of monolithic integration. The GaN on ultra-thin buffer (GaN-on-UTB) technology with shallow trench isolation on Sapphire substrate to avoid the above problems, successfully achieving 1200V monolithic half-bridge integration [5].

The proposed monolithic half-bridge integration platform is presented in Figure 3. The 100 nm un-doped ultra-thin buffer is epitaxially grown on Sapphire substrate by MOCVD, followed by a 300 nm un-doped GaN channel layer and a 15 nm $Al_{0.23}GaN$ barrier layer. The top 100 nm p-GaN cap layer 700°C with N^2. The trench is formed by BCl_3/Cl_2 based dry etching, followed by implantation process. The trench is surrounded by implantation area to suppress the influences of sidewall traps along trench. Other fabrication processes are consistent with those of normal p-GaN HEMT.

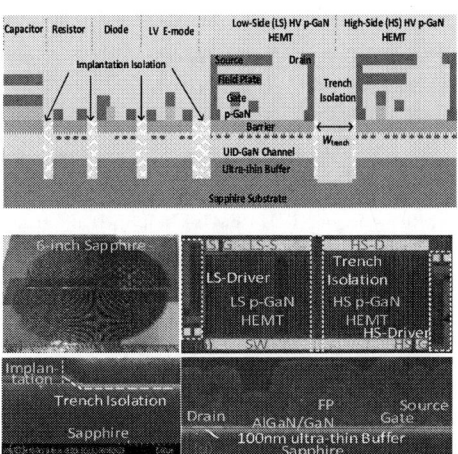

Fig. 3 1200V GaN-on-UTB power integration platform.

The breakdown voltage (BV) of trench isolation structures reaches over 3000V at 175°C with acceptable increase of leakage current, ensuring the safe isolation between high side (HS) and low side (LS) devices.

Substrate biasing effect is suppressed effectively by isolated Sapphire substrate, as shown in Fig. 4. To characterize crosstalk between HS and LS devices, a small V_{DS} is applied to LS p-GaN HEMT, then pulsed high voltage (HV) stress is applied to HS devices and substrate. Due to HV stress, the trapping effect occurs in LS p-GaN HEMT at stress stage. Accordingly, the crosstalk effect is monitored quickly after stress stage. By applying those setups, obvious crosstalk is observed for GaN-on-Si devices, while the crosstalk is eliminated effectively by shallow trench isolation for GaN-on-Sapphire devices, even at temperature of 175°C.

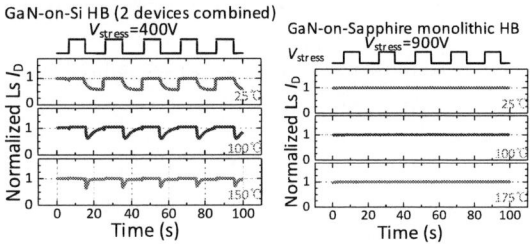

Fig. 4 Comparisons of cross talk effects between GaN-on-Si and GaN-on-UTB devices

The HV monolithic half-bridge circuit is evaluated. Hard switching setups are performed in Fig 5, the results indicate HV half-bridge circuit can work under 800V/1MHz conditions at 175°C.

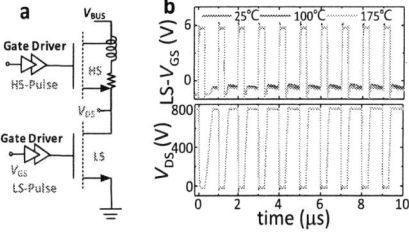

Fig. 5 Switching performances of GaN-on-UTB devices.

SPICE Model

To design a GaN based power circuit, SPICE models are also essential. Two physics based compact SPICE models for GaN HEMT, including ASM-HEMT and MVSG, have attracted growing attentions. ASM-HEMT is a kind of surface potential based model for GaN HEMT devices, which has good accuracy for GaN device simulation. However, the released version is generally suitable for D-mode HEMT, actually, a model to simulate the performances of E-mode p-GaN gate HEMT device is highly required. At this situation, a chare sharing E-mode GaN HEMT model (QSE-HEMT) for p-GaN gate HEMT device is proposed. In this model, Surface potential physics based model for p-GaN gate HEMT can be obtained by calculating the voltage drop on the p-GaN gate cap ($V_{\text{p-GaN}}$). Then, the channel cut-off voltage could be calculated. After putting these formula into the D-mode ASM-HEMT model, the entire enhanced model can be obtained, as presented in Fig. 6. Charge sharing theory is the fundaments of $V_{\text{p-gan}}$ calculation. the dynamic charges are the same between the two junction capacitors. As the charges in channel are changed.

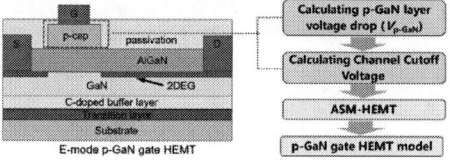

Fig. 6 QSE-HEMT modeling process.

As known, dynamic on-resistance ($R_{\text{on,dy}}$) phenomena frequently occur for GaN HEMT. Considering the influences of $R_{\text{on,dy}}$ on the GaN based power integration circuits, it would be very attractive to add $R_{\text{on,dy}}$ related reliability simulation function into the device model.

To describe the continuous variations of $R_{\text{on,dy}}$, a time-resolved electron mobility variation ($\Delta\mu_{\text{eff}}$) model is proposed [6], as seen in Fig. 7. Physical parameters including activation energy and voltage acceleration factor of traps in p-GaN HEMTs are extracted as the model parameters. Then, to achieve the goal of simulating $R_{\text{on,dy}}$, the proposed $\Delta\mu_{\text{eff}}$ model is incorporated into the surface potential based advanced SPICE model for GaN HEMT (ASM-HEMT). In order to make the model work, an iterative algorithm is build to integrate time-resolved μ_{eff} model into QSE-HEMT model core. Moreover, the above model is integrated into the commercial available EDA tool. The proposed SPICE model also performs good fitting results with measured dynamic R_{on} data, as shown in Fig. 7.

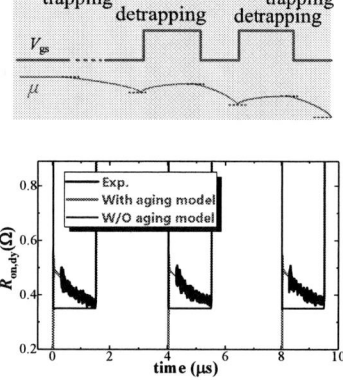

Fig. 7 SPICE modeling of dynamic Ron of GaN HEMT.

Conclusion

GaN based power integration is a popular trend of GaN power electronics. The challenges of GaN power integration developments include V_{th} instability, high voltage isolation between high-side and low-side area, and precise modeling of GaN power HEMTs. Among previous studies, a hybrid-gate technology was proposed to make the V_{th} of p-GaN gate HEMT more stable, a GaN-on-UTB technology with shallow trench was proposed to achieve effective half-bridge isolation for 1200V application, a time-resolved electron mobility variation ($\Delta\mu_{\text{eff}}$) model is proposed to achieve precise modeling of dynamic R_{on} of GaN HEMTs. Those works promoted the development of GaN based power integration.

References

[1] G. Lyu, J. Wei, W. Song, Z. Zheng, L. Zhang, J. Zhang, Y. Cheng, S. Feng, Y. N. Ng, T. Chen, J. Liu, R. Zeng, K. J. Chen, "A GaN Power Integration Platform Based on Engineered Bulk Si Substrate with Eliminated Crosstalk between High-Side and Low-Side HEMTs," 2021 IEEE International Electron Devices Meeting (IEDM), San Francisco, CA, USA, 2021, pp. 5.2.1-5.2.4

[2] J. Wei, Z. Zheng, G. Tang, H. Xu, G. Lyu, L. Zhang, J. Chen, M. Hua, S. Feng, T. Chen and K. J. Chen, "GaN Power Integration Technology and Its Future Prospects," in IEEE Transactions on Electron Devices, vol. 71, no. 3, pp. 1365-1382, March 2024.

[3] L. Sayadi, G. Iannaccone, S. Sicre, O. Häberlen and G. Curatola, "Threshold Voltage Instability in p-GaN Gate AlGaN/GaN HFETs," in IEEE Transactions on Electron Devices, vol. 65, no. 6, pp. 2454-2460, June 2018.

[4] Chi Zhang, Sheng Li, Siyang Liu, Weihao Lu, Yanfeng Ma, Jiaxing Wei, Long Zhang, Weifeng Sun, Denggui Wang, Jianjun Zhou, Song Bai, " Hybrid Gate p-GaN Power HEMTs Technology for Enhanced Vth Stability", 2022 International Electron Devices Meeting (IEDM), San Francisco, CA, USA, 2022, pp. 35.4.1-35.4.4.

[5] Sheng Li, Yanfeng Ma, Weihao Lu, Mingfei Li, Lixi Wang, Zikang Zhang, Tinggang Zhu, Yiheng Li, Jiaxing Wei, Long Zhang, Siyang Liu, Weifeng Sun, " 1200V E-mode GaN Monolithic Integration Platform on Sapphire with Ultra-thin Buffer Technology ", 2023 International Electron Devices Meeting (IEDM), San Francisco, CA, USA, 2023, pp. 1-4.

[6] Sheng Li , Yanfeng Ma, Chi Zhang, Weihao Lu, Mengli Liu, Mingfei Li, Lanlan Yang , Siyang Liu , Jiaxing Wei, Long Zhang, Weifeng Sun, and Jiaxin Sun. " Physics Based SPICE Modeling of Dynamic On-state Resistance of p-GaN HEMTs." in IEEE Transactions on Power Electronics, vol. 38, no. 7, pp. 7988-7992, July 2023.

Indium-Tin-Oxide Transistor with Maximum Transconductance over 1100 μS/μm and Cut-Off Frequency of 23 GHz

Yuxuan Wang[1#], Jiawei Xie[1#], Zijie Zheng[1], Yuye Kang[1], Xuanqi Chen[1], Gerui Zheng[1], Rui Shao[1], Kaizhen Han[1*], and Xiao Gong[1, 2*]

[1]Department of Electrical and Computer Engineering, National University of Singapore (NUS), 117583, Republic of Singapore

[2]Institute of Microelectronics (IME), Agency for Science, Technology and Research (A*STAR), 138634, Republic of Singapore

Email: elehank@nus.edu.sg, elegong@nus.edu.sg ([#]Two authors contribute equally to this work)

Abstract

In this work, we demonstrate high performance Indium-Tin-Oxide (ITO) field-effect transistors (FETs) featuring offset top-gate (OTG) design. A record-high maximum transconductance ($G_{m, max}$) of 1187 μS/μm among all reported ITO FETs and one of the best cut-off frequency (f_T) among amorphous oxide semiconductor (AOS) FETs reaching 23 GHz have been achieved by scaling the gate length (L_G) down to 100 nm. In addition, our investigation suggests that the substantial reduction in ITO sheet resistance (R_{sh}) after the atomic layer deposition (ALD) is essential to realize the high drive current observed in our devices.

Introduction

Advances in monolithic three-dimensional integration (M3DI) technology, which promises to deliver both high capacity and integration flexibility, are going to play a crucial role in driving hardware breakthroughs across various sectors, including artificial intelligence (AI), 5G/6G communication, and quantum computing. AOS materials, recognized for their decent mobility, cost-effective growth, and compatibility with back-end-of-line (BEOL) processes, are poised to be key technological enablers in advancing M3DI by facilitating a diverse range of device functionalities [1], [2]. Significant progress has been achieved in AOS-based applications, including driver circuits [3], memories [4-6], power management devices or circuits [7], and radio frequency (RF) devices [8-11]. However, the majority of these advancements have predominantly utilized a bottom-gate (BG) structure. In contrast, a top-gate (TG) transistor introduces a convenient offset design and unique gate shapes. A well-optimized TG structure holds promise for mitigating parasitic capacitance, thereby offering considerable advantages, particularly in on-chip RF applications. However, there still exists a gap between theoretical expectations and experimental validation of OTG AOSFETs, as well as the assessment of their RF characteristics.

In this work, we have demonstrated OTG ITO FETs with a record-high $G_{m, max}$ of 1187 μS/μm, surpassing all previously reported ITO FETs. Additionally, one of the best f_T of 23 GHz among all AOSFETs has been realized based on the excellent driving capability and minimized parasitic capacitance benefiting from the OTG design. Moreover, investigations based on refined transfer length model (RTLM) reveal a drastically decreased R_{sh} of ITO layer after ALD, which helps to enhance device performance by suppressing series resistance contributed by the gate offset (L_{os}) regions as shown in **Fig. 1 (a)**.

Device Fabrication

The 3D schematic of the ITO FET implemented in this study, along with its crucial fabrication process steps are summarized in **Fig. 1**. All devices measured in this work employed a symmetric gate offset design (the offset lengths from gate to source and gate to drain are designed to be the same), where L_{os} denotes the offset length. To eliminate substrate loss during high frequency measurements, the high-resistance Si substrate with 200 nm SiO_2 on top was used for device fabrication.

The process started from a 4 nm thick ITO channel deposition by RF sputtering. Smooth ITO surface having an RMS roughness as low as 0.2 nm was achieved and confirmed through AFM. Source/drain (S/D) regions along with the alignment markers were then defined by EBL and filled with a Ni/Pd dual stack by utilizing E-beam evaporator. After that, the active ITO area was patterned and wet etched using HCl. This was followed by a 10 nm HfO_2 ALD at 200 °C using TEMAHf and O_3 as the alternating precursors. Subsequently, the gate region was defined via EBL, followed by the deposition of a Pt/Ag/Au stack using E-beam evaporator and lifted off. Afterwards, the gate dielectric over source/drain metal was removed by DHF to facilitate the device measurement. Lastly, patterning and Ti/Ag/Au deposition were performed for the formation of GSG pads along with on-chip open and short structures. No additional annealing was introduced during the entire fabrication process since the 200 °C ALD served as an *in-situ* annealing. The TEM image depicting the device structure is shown in **Fig. 2 (a)**. The zoomed-in HRTEM image in **Fig. 2 (b)** confirms the precise dimensions of those key features, including the ~4 nm thick channel and the ~10 nm thick gate dielectric. The presence of In, Sn, O, Hf, Pd, Ni, Ag, and Pt elements can

979-8-3315-0417-5/25 $31.00 © 2025 IEEE

be observed in the EDX mapping [**Fig. 2 (c)**].

Results and Discussions

The DC performance characterization was conducted using a Keithley 4200 semiconductor analyzer at room temperature. Long-channel (LC) device without offset gate design was first measured to validate the TG fabrication process. **Fig. 3 (a)** shows the transfer characteristic curves of a TG ITO FET without offset gate design, which has both L_{SD} and L_G of 3 µm. The device possesses a decent subthreshold swing (SS) of 104 mV/decade and a high on/off ratio exceeding 10^8. To enhance device performance, both L_G scaling and offset gate design were implemented in short-channel (SC) devices. **Fig. 3 (b)** displays the drain current (I_D) and transconductance (G_m) of an OTG ITO FET with 200 nm L_{SD} and 100 nm L_G [L_{SD} and L_G are defined in **Fig. 1(a)**]. Notably, a high $G_{m, max}$ of 1187 µS/µm is observed. Output characteristics of one device exhibit an excellent on-state current (I_{ON}) of 924 µA/µm at a gate overdrive (V_{ov}) of 3 V and a drain to source voltage (V_{DS}) of 1 V [**Fig. 4**], which reaffirms the outstanding driving capability of the fabricated SC OTG devices.

Previously reported LC OTG AOSFETs suffer from lower I_{ON} and G_m, partly attributable to their extremely high source-drain series resistance (R_{SD}) [12], [13]. This elucidates why an overlapping area between gate and S/D is predominantly favored [8-10], [14], [15]. In this work, however, the dilemma between high R_{SD} and OTG design has been overcome and a record-high driving capability has been achieved among ITO FETs. To uncover the mechanism behind the phenomenon, RTLM structures with a gap width of 3 µm and gap lengths (L) varying from 100 nm to 1 µm were fabricated and measured [16]. Two batches of RTLM devices were fabricated: one with HfO_2 capping layer [**Fig. 5 (a)**], and the other without the capping but undergoing a 200 °C annealing having the same duration as the ALD process [**Fig. 5 (b)**].

According to the RTLM measurement results plotted in **Fig. 5 (c)**, the ITO film R_{sh} reduces by ~100 times from $(1.04 \pm 0.06) \times 10^6$ Ω/square to $(1.04 \pm 0.04) \times 10^4$ Ω/square after the ALD process, which is essential in achieving high performance OTG ITO FET obtained in this work. The optimization might be explained by O_3-enabled surface state passivation during ALD [17]. The surface states distributed over the ITO film could deplete the top $5-9$ Å of the film and degrade its conductivity [**Fig. 6**]. This impact magnifies when the film is extremely thin, such as the 4 nm thick ITO channel in this work.

The RF performance of OTG ITO TFT was further assessed using a Rohde & Schwarz ZVA 67 vector network analyzer. Standard TOSM off-wafer analyzer calibration and on-wafer de-embedding procedure with short and open structures were carried out. Leveraging the parasitic capacitance suppression enabled by the OTG design without compromising the high G_m, this work achieved a high f_T of 23 GHz (**Fig. 7**).

Table I summarizes those critical figures of merit for evaluating AOSFETs, including I_{ON}, $G_{m, max}$ and f_T, along with device dimensions. Our devices achieve the best $G_{m, max}$ without introducing an extremely scaled L_G, highlighting the substantial potential of further optimization. Moreover, this work demonstrates a f_T of 23 GHz, which is one of the highest among AOSFETs.

Conclusions

Our work presents the fabrication and the characterization of OTG ITO FETs. A record-high $G_{m, max}$ of 1187 µS/µm and one of the best f_T among AOSFETs reported to date reaching 23 GHz have been achieved. Our investigation also unveiled the underlying reasons behind the high Ion exhibited by the SC OTG ITO FETs, providing valuable insights guiding future device design.

Acknowledgments

This research is supported by Semiconductor Trailblazing Grant (A-8002620-00-00), Ministry of Education (MOE) Tier 2 (MOE-T2EP50221-0008), and MOE Tier 1 (A-8001168-00-00).

References

[1] K. Nomura *et al.*, *Nature*, 432, 488–492, 2004.

[2] E. Fortunato *et al.*, *Adv. Mater.*, 24, 2945–2986, 2012.

[3] P. G. Bahubalindruni *et al.*, *IEEE Trans. Circuits Syst. Regul. Pap.*, 64, 1118–1125, 2017.

[4] C. Sun *et al.*, *IEEE Trans. Electron Devices*, 69, 5262–5269, 2022.

[5] C. Sun *et al.*, in *VLSI Symp.*, pp. T7.4.1–T7.4.2, 2021.

[6] Z. Zheng *et al.*, in *VLSI Symp.*, pp. T12.2.1–T12.2.2, 2023.

[7] S. Deng *et al.*, in *IEDM*, pp. 1–4, 2023.

[8] S. Li *et al.*, in *IEDM*, pp. 40.5.1–40.5.4, 2020.

[9] C. Wang *et al.*, in *VLSI Symp.*, pp. 294–295, 2022.

[10] A. Charnas *et al.*, *IEEE Trans. Electron Devices*, 70, 532–536, 2023.

[11] D. Zheng *et al.*, in *VLSI Symp.*, pp. T11.1.1–T11.1.2, 2023.

[12] M. M. Billah *et al.*, *IEEE Electron Device Lett.*, 37, 1442–1445, 2016.

[13] S. Priyadarshi *et al.*, *IEEE Electron Device Lett.*, 43, 56–59, 2022.

[14] Q. Hu *et al*, in *IEDM*, pp. 1–4, 2023.

[15] Y. Kang *et al.*, in *VLSI Symp.*, pp. T11.2.1–T11.2.2, 2023.

[16] R. Dormaier and S. E. Mohney, *J. Vac. Sci. Technol. B*, 30, 031209, 2012.

[17] Y. Gassenbauer *et al.*, *Phys. Rev. B*, 73, 245312, 2006.

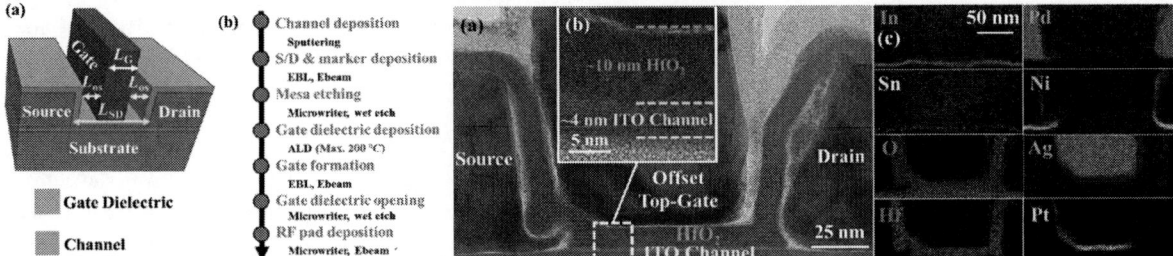

Fig. 1. (a) 3D structure of the OTG ITO FET fabricated in this work. (b) Fabrication process flow of the ITO FET.

Fig. 2. (a) TEM image showing an OTG ITO FET with 100 nm L_{SD} and 75 nm L_G. (b) HRTEM image of the channel region. (c) EDX mapping of main elements.

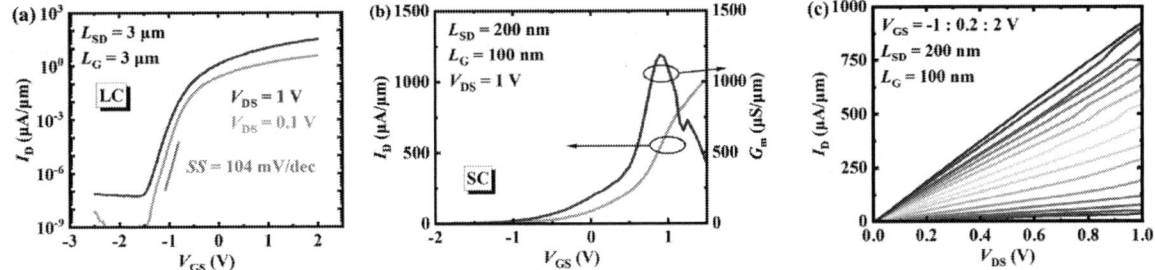

Fig. 3. (a) Transfer characteristic curves of the LC TG ITO FET with 3 μm L_{SD} and L_G, showing a decent SS of 104 mV/decade. (b) Transfer characteristic and $G_m - V_{GS}$ curves of the SC OTG ITO FET with 200 nm L_{SD} and 100 nm L_G.

Fig. 4. Output characteristic curves of the SC OTG ITO FET with 200 nm L_{SD} and 100 nm L_G.

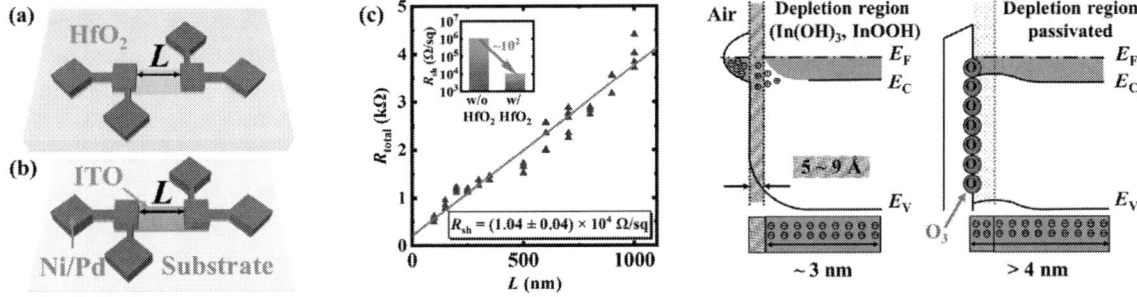

Fig. 5. Schematic of the RTLM structures utilized for R_{sh} extraction (a) with and (b) without HfO$_2$ capping layer. (c) Fitted R_{total} vs. L using RTLM structure covered with HfO$_2$.

Fig. 6. Band diagrams depicting the surface of ITO before and after HfO$_2$ passivation.

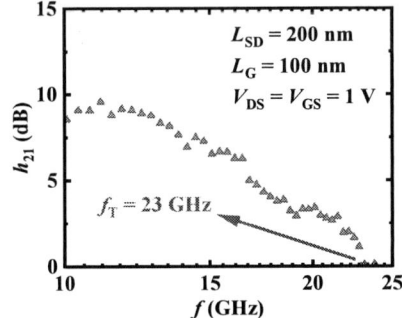

Fig. 7. $h_{21} - f$ curve of a SC OTG ITO FET with 200 nm L_{SD} and 100 nm L_G, showing a high f_T of 23 GHz.

Ref.	I_{ON} (μA/μm)	$G_{m, max}$ (μS/μm)	f_T (GHz)	Gate Structure	Gate Design	Channel Material	L_{SD} (nm)	L_G (nm)
*	924	1187	23	TG	Offset	ITO	200	100
[8]	1860	~1050	20	BG	Overlap	ITO	10	Large L_{OV}**
[9]	~1000	480	18.3	BG	Overlap	IGZO	12.3	Large L_{OV}
[10]	~1150	370	22.5	BG	Overlap	In$_2$O$_3$	150	350
[11]	~750	-	36	BG	Offset	In$_2$O$_3$	40	40
[14]	~1500	538	-	TG	Overlap	ITO	100	200
[15]	~1250	480	-	BG	Overlap	ITO	50	Large L_{OV}

* This work.
** L_{OV} refers to the overlap between gate and S/D.

Table I. Benchmark of AOSFETs fabricated on Si substrate with detailed gate structures and channel profiles.

Optoelectronic Multiplication Emulator
Based on FPGA

Jinxian Li[1,2], Runyu Hu[1], Jiabin Shen[2,3,*], Zengguang Cheng[1,2,*] and Peng Zhou[1,2,3,*]

[1]School of Microelectronics, Fudan University, Shanghai 200433, China; [2]Shaoxin Laboratory, Shaoxing 312000, China; [3]Institute of Optoelectronics, Fudan University, Shanghai 200433, China.
[*]E-mail: jiabin_shen@fudan.edu.cn, zgcheng@fudan.edu.cn, pengzhou@fudan.edu.cn

Abstract

Multiplication is a fundamental operation in artificial neural networks. We developed an optoelectronic multiplication emulator system, incorporating discrete optical components with an FPGA hardware. The system achieves high-speed, precise optical multiplications, validated by image convolution tests showing a mean error of 0.025. This emulator shows great potential for photonic neural network applications and establishes a strong foundation for advancing high-performance integrated photonic chips and optoelectronic computing systems.

Keywords: Optoelectronic Emulator, Photonic Multiplication, and Photonic computing

Introduction

The rapid advancement of technologies like artificial intelligence (AI) and deep learning is pushing the limits of traditional computing systems based on von Neumann architecture, leading to growing interest in neural processing unit (NPU) specifically optimized for AI tasks [1]. AI models, particularly Convolutional Neural Networks (CNNs) are highly effective at processing large datasets, with multiplication being the most computationally demanding operation [2]. Optimizing multiplication is therefore essential for improving system efficiency.

On the other hand, photonic computing offers significant advantages over conventional electronic computing, featuring low latency, large bandwidth and high energy efficiency [3]. Techniques like diffraction optics [4] and programmable on-chip photonic networks, such as Mach-Zehnder interferometers [5] and micro-ring resonators [6], have been developed for optical multiplications. However, these systems typically require continuous energy consumption and are limited to specific applications. Emerging non-volatile materials like phase-change materials (PCMs) [7, 8] and ferroelectric materials [9] offer promise for achieving near-zero energy consumption. Despite these advances, optoelectronic hybrid systems, with optical components handling linear processing and electronics managing the control and storage, are very promising for general purpose AI tasks [10, 11].

While Field-Programmable Gate Array (FPGA) based emulators have become crucial tools in integrated circuits design, specialized hardware devices or systems used for the simulation and verification of optoelectronic computing are still lacking [12]. In this work, we developed an optoelectronic multiplication emulator that integrates discrete optical components with a FPGA to perform image convolutions, paving the way for the design and verification of scalable, programmable optical computing systems.

Optoelectronic Emulator System Design

The optoelectronic hybrid emulator system, as shown in Fig. 1, consists of three main components: controller, processor, and converter. The FPGA acts as the system controller, managing operations and data storage. Input signal is loaded into the system via an electro-optic modulator (EOM), which converts electronic signals from the FPGA into analog optical signals, as depicted in Fig. 2(a). The EOM supports high-speed and high-precision data loading, achieving up to 7-bit precision at 20 kHz, with the potential to extend its bandwidth up to 100 GHz. Afterwards, the optical data is directed into a variable optical attenuator (VOA), with the attenuation level mapping the multiplication weight, as illustrated in Fig. 2(b). Each weight is held for 20 ms. The normal distribution of the weight error distribution is fitted, and the mean standard deviation is 0.0037. Optical multiplication is performed by propagating the EOM-modulated light signal through the VOA. Finally, a photodetector (PD) converts the optical multiplication result back into the electrical domain. Figure 2(c) demonstrates the multiplication result error, the mean error is -0.0459 and standard deviation is 0.0186, indication a high accuracy of the emulation system.

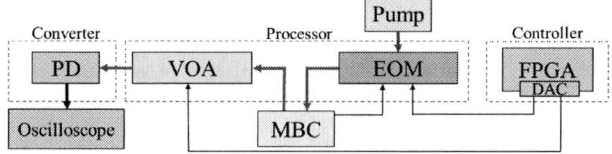

Fig. 1. Schematic diagram of optoelectronic multiplication emulator system. The optical signals serve as the main compute and transmission media in the processors (EOM and VOA), then compute results convert into electronic signals by a photodetector. The FPGA is responsible for the data storage and system control. Modulator bias controller (MBC) is employed to offset temporal deviations in the EOM, ensuring its stable operation over time.

979-8-3315-0417-5/25 $31.00 © 2025 IEEE   408

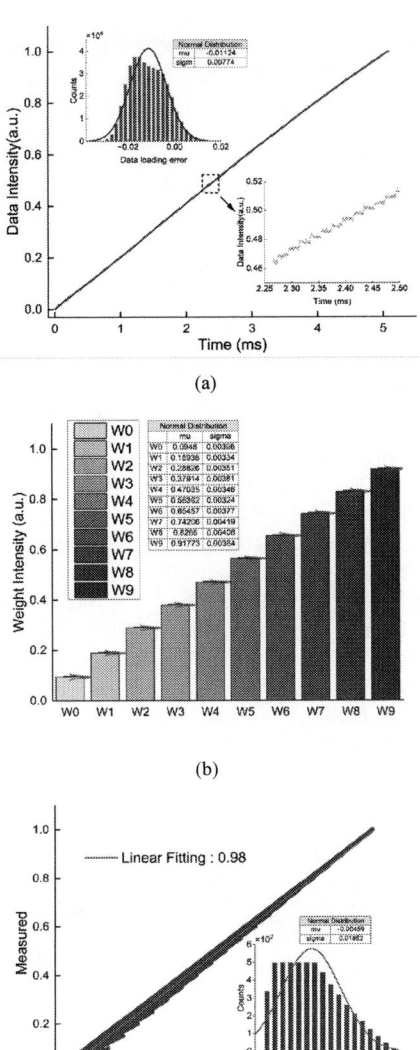

Fig. 2. Data distributions in data loading, weight loading and multiplication operations. (a) Loading data using an EOM, with normalized data intensity from 0 to 1. The top left inset displays the statistics of loading errors, and the bottom right inset provides an enlarged curve. (b) Configuration of various weights achieved by an VOA. Each weight performs a distribution analysis. (c) Comparison of expected and experimental multiplication results. The multiplication results are derived from scanning input data and weights, both ranging from 0 to 1. The inset shows the distribution of errors.

Experimental Results

Utilizing the emulator system, we performed the image filtering through photonic convolution. The image filtering scheme is shown in Fig. 3, where a 512×512 grayscale image was filtered by three kind of kernels (Blur, x-Edge and y-Edge). The pixel values of the image were converted into amplitudes of electrical signal and loaded into the EOM, while the convolution kernel values were adjusted on the VOA. The multiplication was performed optically, with the results detected by the PD. The final addition in the convolution was implemented in computer, generating the

filtered image. Fig. 4 shows the accuracy of the filtered image. The mean error between the experimental convolution result and the simulation by double-precision floating-point (64-bit) were -0.05541, -0.00808 and -0.0117, with mean standard deviation of 0.0263, 0.00691 and 0.01022. The average of mean error is -0.025, and standard deviation is 0.01448, indicating a high accuracy of the filtering by the emulator.

The experimental result confirms the suitability of the optoelectronic emulator for optical multiplication-based computations, such as image filtering and feature extraction. Additionally, it can be extended to perform operations in convolution layers and fully connected layers in neural networks. To further improve the emulation performance, enhancements such as optimizing the configurations of EOMs and VOAs for long-term stability, incorporating higher-precision components, and applying advanced signal processing algorithms could be explored.

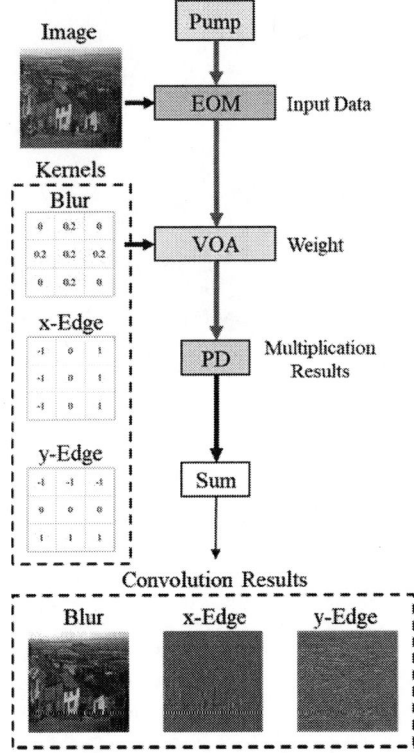

Fig. 3. The computational workflow for convolution processing by using the emulator system. Input data (pending images) are loaded into the system by an EOM, and weights (each value in convolution kernels) are deployed in the VOA. Multiplication results are converted into electronic signals from optics using a PD, and then they are summed to produce final results.

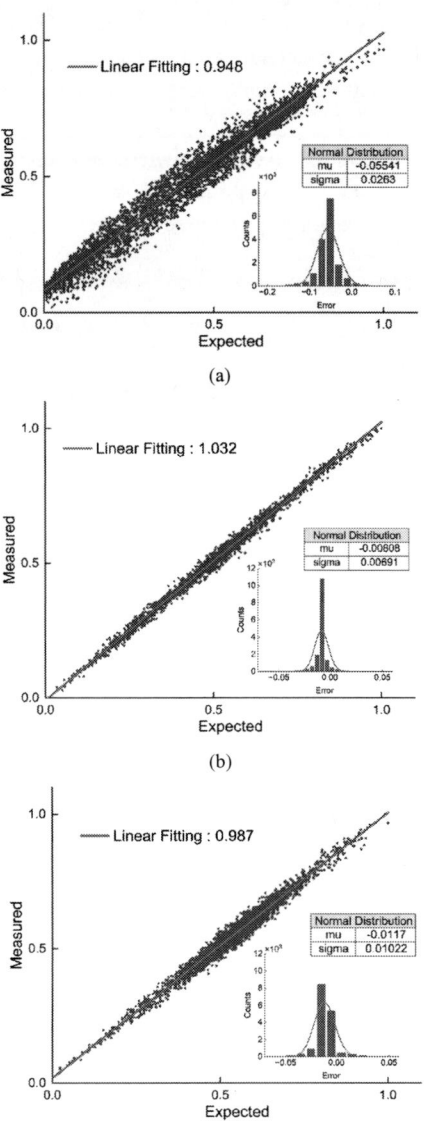

Fig. 4. Data distributions in experimental feature extraction tasks: (a) blur; (b) x-edge extraction; (c) y-edge extraction. The insets show corresponding error distributions.

Conclusion

In this work, we have demonstrated an optoelectronic hybrid emulator system capable of executing optical multiplications for tasks such as image filtering. The emulation system shows a high accuracy, as evidenced by the low error compared to simulations. This emulation system can be effectively applied to optical computations in neural networks, including convolutional and fully connected layers. Additionally, this development provides a crucial optoelectronic emulator for early-stage research and development of integrated photonic chips, laying a strong foundation for high-performance photonic chip advancement and paving the way for future innovations in integrated photonic systems.

Acknowledgments

We acknowledge the support of the Fundamental Research Funds for the Central Universities, National Natural Science Foundation of China (62074042, 62205066, 62204053) and the start-up funding from Fudan University.

References

[1] T. Chen, Z. Du, N. Sun, J. Wang, C. Wu, Y. Chen, and O. Temam, "DianNao," in Proceedings of the 19th international conference on Architectural support for programming languages and operating systems, pp. 269-284 (2014).

[2] K. Simonyan, and A. Zisserman, "Very Deep Convolutional Networks for Large-Scale Image Recognition," arXiv1409.1556 (2014).

[3] H. Zhou, J. Dong, J. Cheng, W. Dong, C. Huang, Y. Shen, Q. Zhang, M. Gu, C. Qian, H. Chen, Z. Ruan, and X. Zhang, "Photonic matrix multiplication lights up photonic accelerator and beyond," Light: Science & Applications, vol. 11, no. 1(2022).

[4] T. Fu, Y. Zang, Y. Huang, Z. Du, H. Huang, C. Hu, M. Chen, S. Yang, and H. Chen, "Photonic machine learning with on-chip diffractive optics," Nature Communications, vol. 14, no. 1(2023).

[5] Y. Shen, N. C. Harris, S. Skirlo, M. Prabhu, T. Baehr-Jones, M. Hochberg, X. Sun, S. Zhao, H. Larochelle, D. Englund, and M. Soljačić, "Deep learning with coherent nanophotonic circuits," Nature Photonics, vol. 11, no. 7, pp. 441-446 (2017).

[6] S. Xu, J. Wang, S. Yi, and W. Zou, "High-order tensor flow processing using integrated photonic circuits," Nature Communications, vol. 13, no. 1(2022).

[7] Z. Cheng, C. Ríos, W. H. P. Pernice, C. D. Wright, and H. Bhaskaran, "On-chip photonic synapse," Science Advances, vol. 3, no. 9(2017).

[8] J. Feldmann, N. Youngblood, M. Karpov, H. Gehring, X. Li, M. Stappers, M. Le Gallo, X. Fu, A. Lukashchuk, A. S. Raja, J. Liu, C. D. Wright, A. Sebastian, T. J. Kippenberg, W. H. P. Pernice, and H. Bhaskaran, "Parallel convolutional processing using an integrated photonic tensor core," Nature, vol. 589, no. 7840, pp. 52-58 (2021).

[9] J. Geler-Kremer, F. Eltes, P. Stark, D. Stark, D. Caimi, H. Siegwart, B. Jan Offrein, J. Fompeyrine, and S. Abel, "A ferroelectric multilevel non-volatile photonic phase shifter," Nature Photonics, vol. 16, no. 7, pp. 491-497 (2022).

[10] X. Xu, M. Tan, B. Corcoran, J. Wu, A. Boes, T. G. Nguyen, S. T. Chu, B. E. Little, D. G. Hicks, R. Morandotti, A. Mitchell, and D. J. Moss, "11 TOPS photonic convolutional accelerator for optical neural networks," Nature, vol. 589, no. 7840, pp. 44-51 (2021).

[11] Z. Xu, T. Zhou, M. Ma, C. Deng, Q. Dai, and L. Fang, "Large-scale photonic chiplet Taichi empowers 160-TOPS/W artificial general intelligence," Science, vol. 384, no. 6692, pp. 202-209 (2024).

[12] S. Ning, H. Zhu, C. Feng, J. Gu, Z. Jiang, Z. Ying, J. Midkiff, S. Jain, M. H. Hlaing, D. Z. Pan, and R. T. Chen, "Photonic-Electronic Integrated Circuits for High-Performance Computing and AI Accelerators," Journal of Lightwave Technology, pp. 1-26 (2024).

Integrated MEMS Diamond Quantum Magnetometer with Active Laser Noise Suppression

Nan Wang[1], Xiao Peng[1], Yaochen Zhu[1], Qihui Liu[1], Jiachen Han[2], Xin Chen[4], Xin Luo[2], Yongquan Su[1], Lihao Wang[1], Yichen Liu[1], Hao Chen[1,2*], Jiangong Cheng[1*] and Zhenyu Wu[1,2,3*]

[1] Shanghai Institute of Microsystem and Information Technology, China

[2] Shanghai University, China

[3] Shanghai Industrial μ Technology Research Institute, China

[4] University of Shanghai for Science and Technology, China.

Abstract

This study reports a miniaturized diamond sensor device fabricated with standard microfabrication for high-accuracy current sensing. The sensor incorporates a diamond quantum magnetometer and a piezo-driven variable optical attenuator (VOA) module for active laser noise suppression and thus magnetic detection signal-to-noise ratio (SNR) is improved by 22 times, achieving a sensitivity of $214\ pT\cdot Hz^{-1/2}$ with 200 mW laser power. By applying feedback control to actively tracking the fluorescence signal, the stability of detection sensitivity is enhanced by a factor of 9.16. On-site current measurements have been implemented with our diamond sensor, achieving a 0.2%FS accuracy in a range of 0A to 8 kA, which paves the way for current sensing in high-voltage power system.

Keywords: MEMS, Diamond magnetometer, Laser noise suppression, Outdoor electromagnetic sensing

Introduction

Quantum sensing based on nitrogen-vacancy (NV) centers in diamond has attracted widespread interest due to their high sensitivity, high stability in ambient conditions and large dynamic range. Recent studies have been focused on simultaneously improving the sensitivity and integration to accelerate the transition from laboratory setups to practical applications [1]. Optical detected magnetic resonance (ODMR) approach is commonly used for magnetic field measurement with diamond [2], but significant laser noise emerges accompanied by the transition from high laser power to fluorescence, which affects the sensitivity. It is essential to improve long-term stability while ensuring high detection accuracy of the sensor. As a result, common-mode-rejection (CMR) technique is employed to suppress laser noise for sensitivity optimization [3]. However, most of the research is based on discrete laboratory devices [4-5] or manually adjusted setups [6], which limits further integration and stability. To address this, we present a miniaturized module including a piezo-driven VOA integrated with diamond magnetometers for stable and high accuracy magnetic field sensing.

The working principle of the closed-loop diamond sensing system is shown in Fig. 1. Nitrogen-vacancy (NV) centers provide high-precision measurements of external magnetic fields by manipulating electron spin states and optical readout signals. The VOA was driven by a close-loop PID controller, whereby the fluorescence can be tracked in real-time to achieve high-precision CMR for laser noise suppression, compensating for laser fluctuation and fluorescence drift caused by external environmental interference. This device enhances the robustness and signal-to-noise ratio (SNR) for diamond magnetometers in practical engineering applications.

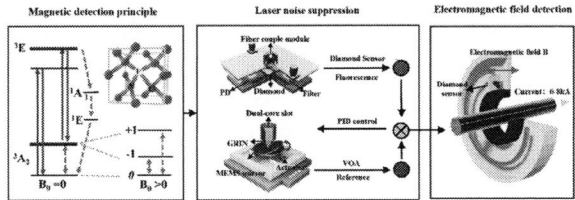

Fig. 1: Schematic of the close-loop diamond sensor.

Device fabrication

The key component of integrated diamond magnetometer is produced through the standard fabrication processes, involving a diamond chip and an integrated VOA.

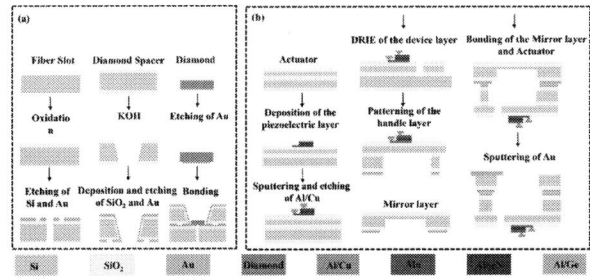

Fig. 2: Fabrication steps of (a) diamond chip and (b) VOA.

Fig.2 (a) demonstrates the integration of the diamond sensor head with a dimension of 0.1 cm³, including fabrication process of diamond-silicon heterogeneous integration and wafer-level bonding of functional components. The three substrates of the diamond sensor head are stacked as follows: a fiber slot for establishing the optical path between pump laser and the diamond, a spacer for enhancing fluorescence collection efficiency with a KOH etching, and the diamond chip with microwave transmission lines [7].

979-8-3315-0417-5/25 $31.00 © 2025 IEEE

The high-efficient laser noise suppression module is realized based on a piezoelectric micromirror, which exhibit high linearity and a large angular deflection. Fig. 2(b) illustrates the fabrication process for the micromirror. The fabrication of the cantilevers and the pillars connecting to the middle of the mirror is achieved through Deep Reactive Ion Etching (DRIE) to magnify the rotation angle of the mirror. The combination of the ultra-high precision patterning technology of the AlScN thin films with the non-Bosch DRIE process, ensures a high degree of consistency between the device and the design, thereby significantly reducing losses in optical and mechanical performance. Furthermore, this process utilizes a double stack structure to achieve a high device fill factor and greatly improves the preparation efficiency through wafer-level bonding [8].

Integrated system setup

Based on the devices fabricated as mentioned above, a board-level integration of a close-loop system based on all-fiber connections has been achieved. The immunity of the fiber to electromagnetic interference ensures the stability of the integrated device during testing. Additionally, the flexible multifunctional coupling based on fibers has optimized the system size, with a compact dimensions of $3.7 \times 5 \times 3.4$ cm³.

As shown in Fig.3, the pump laser is coupled into the diamond chip and VOA through a 9:1 fiber splitter. The differential signal, generated by a self-made differential amplifier circuit board, is then transmitted to a lock-in amplifier (LIA) for demodulation magnetic field demodulation.

The integrated device incorporates a closed-loop algorithm based on a proportional-integral-derivative (PID) controller, using the differential signal (U=0) as the feedback to adjust the driving voltage of the VOA, achieving a balance between fluorescence and reference. Through close-loop CMR, the laser noise in the demodulated signal is suppressed in real-time.

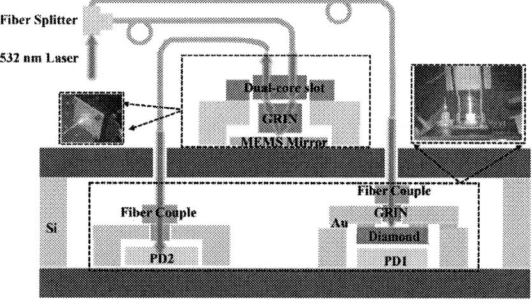

Fig. 3: Board level integrated system setup.

Active laser noise suppression device applied for diamond magnetometer

The performance of the MEMS mirror is characterized. As shown in Fig. 4, the mirror has a resonant frequency of ~830

Hz and a response time of approximately 0.6 ms, enabling effective tracking of high-speed fluorescence drift in rapidly changing AC fields. The optical angle range of the mirror is -30 to 30 mrad, which enables light intensity balance over a large dynamic range. Finally, a high resolution of ~29 μrad at 0.1 V, opens the way for high-precision CMR for the sensitivity optimization of the diamond magnetometer.

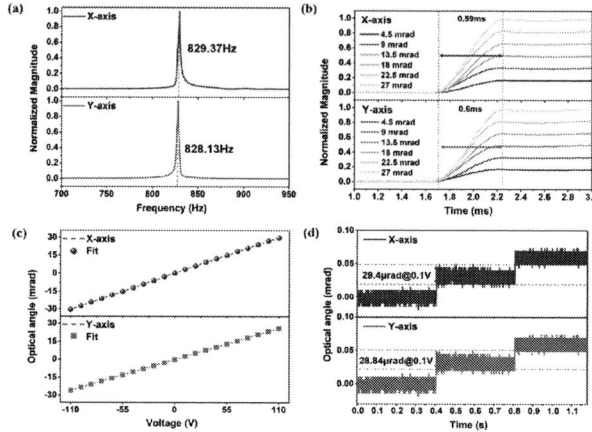

Fig. 4: Characterization of the MEMS mirror.

An optical path is established between the MEMS micromirror and the dual-core fiber via a gradient refractive index lens (GRIN), with multimode fibers served as the connecting medium to facilitate high optical transmission capability for the VOA. The tests presented in Fig. 5 indicate that an optical attenuation ratio of -3 dB to -20 dB is obtained, with a repeatability of 0.01 dB. This provides reliable support for achieving laser noise suppression in diamond magnetometers.

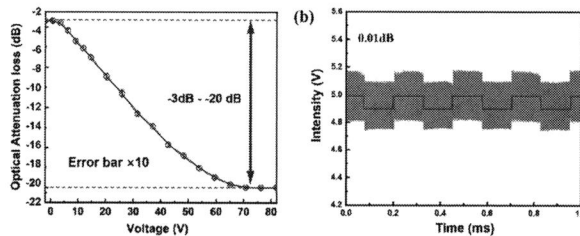

Fig. 5: Characterization of the VOA.

Due to the gain of the differential amplification circuit is 10^5 to 10^7, even small drifts in fluorescence or laser can result in significant laser noise coupling into the magnetic detection system, leading to the differential signal to deviate from zero and resulting in system differential mismatch. This random distribution of laser noise in the time domain affects the long-term sensitivity stability of the sensor.

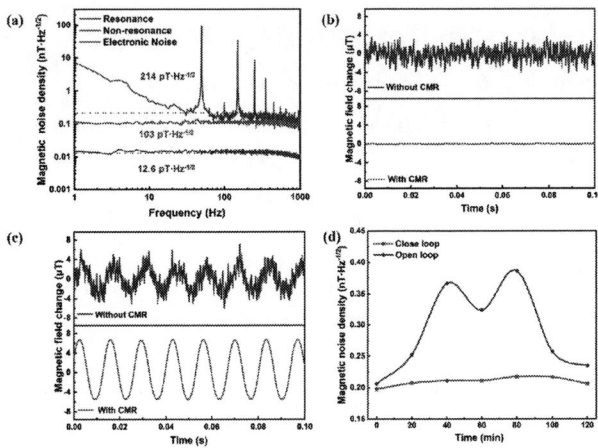

Fig. 6: Magnetic field detection of integrated MEMS diamond quantum Magnetometer with active laser noise suppression

The data recorded in Fig. 6(a) shows that, the sensitivity of the PID-based close-loop magnetic sensor is 214 pT·Hz$^{-1/2}$ at the resonance point with laser noise suppression. Fig. 6(b) and 6(c) indicate that, based on CMR, the signal-to-noise ratio (SNR) of both the DC magnetic field signal and the AC magnetic field signal has been improved by a factor of 22.

Figure 6(d) records a two-hour magnetic field detection, the sensitivity variation of the diamond sensor is compared with in open-loop and closed-loop mode. Mechanical vibrations in the environment and temperature fluctuations affect the stability of the optical signal for detection system. The close-loop PID controller can demodulate the signal from the laser fluctuation noise, ensuring that the detection intensity remains at a high-precision level over the long term. The experiments show that the stability of magnetic measurement under closed-loop conditions has improved by a factor of 9.16, effectively enhancing the robustness of the system.

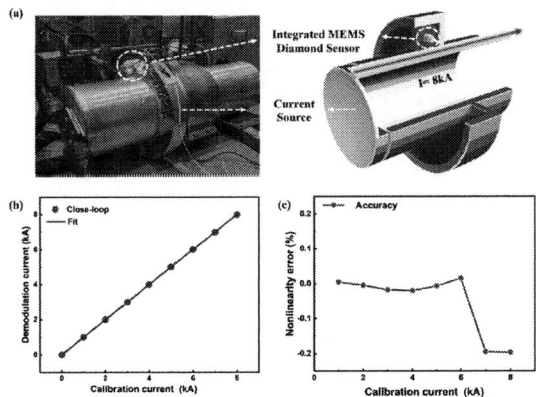

Fig. 7: Application of diamond magnetometers in high-voltage current measurement

The device is installed for high-voltage large current measurements, as shown in Fig. 7(a). The diamond sensor was secured to a tubular busbar with large current, and it calibrates the magnetic field-current conversion coefficient $k = 8.056$ μT/A through electromagnetic effect to achieve current sensing. The linearity of the current test is presented in Fig. 7(b). For the current test over a dynamic range of 0-8 kA, the linearity is essentially 100%, with the accuracy of 0.2% in Fig.7(c).

Conclusion

In this study, we propose a compact fiber-integrated MEMS diamond quantum magnetometers based on standard microfabrication for wide dynamic range current measurements. The diamond chip is integrated with a noise suppression module at the board level, forming a compact solid-state device that improves the detection SNR by 22 times and stability by 9 times. Current test over a dynamic range of 0-8 kA is achieved, with a detection accuracy of 0.2% FS. This research extends the application of diamond NV sensing from laboratory setups to outdoor detection, paving the way for future current applications, such as electric vehicle sensing systems.

Acknowledgments

The authors acknowledge the support from the National Key Research and Development Program of China (No. 2023YFB3209900).

References

[1] Xie, F, "A microfabricated fiber-integrated diamond magnetometer with ensemble nitrogen-vacancy centers". Applied. Physics. Letters., 120, 191104. (2022).

[2] Zhukov, I. V, "ODMR Spectroscopy of NV- Centers in Diamond Under High MW Power". Appl. Magn. Reson.,48, 1461-1469. (2017).

[3] Schloss, J. M, "Simultaneous Broadband Vector Magnetometry Using Solid-State Spins". Phys. Rev. A, 10, 034044. (2018).

[4] Hatano, Y, "Simultaneous thermometry and magnetometry using a fiber-coupled quantum diamond sensor". Appl. Phys. Lett,118, 034001. (2021).

[5] Graham, S. M, "Fiber-Coupled Diamond Magnetometry with an Unshielded Sensitivity of 30pT /√Hz". Phys. Rev. A,19, 044042. (2023).

[6] Stürner, F. M, "Integrated and Portable Magnetometer Based on Nitrogen-Vacancy Ensembles in Diamond". Adv. Quantum Technol.4, 2000111. (2021)

[7] Xie, F, "Miniaturized Diamond Quantum Magnetometer with Integrated Laser Source and All Electrical I/OS", IEEE MEMS., 577-580, (2024).

[8] Liu, Y. C, "AlScN Piezoelectric MEMS Mirrors with Large Field of View for LiDAR Application". Micromachines. 13, 1550. (2022).

A Transistor-Free Analog Content Addressable Memory with High Bit Density

Renhao Xue[1], Quanyi Tu[1], Mansun Chan[2], Xiwen Liu[1]*

[1] Thrust of Microelectronics, The Hong Kong University of Science and Technology (Guangzhou), Guangzhou, China.

[2] Dept of ECE, The Hong Kong University of Science and Technology, Kowloon, Hong Kong.

Email: xiwenliu@hkust-gz.edu.cn

Abstract

In this work, we design and demonstrate a nonvolatile analog content addressable memory (ACAM) from 2-complementary ferroelectric diodes (FeD) by experimentally calibrated TCAD simulations. Apart from the unique features of FeD, including high non-linearity, multi-conductive states and non-volatility, "data line" and "coding line" configurations are implemented, where data line takes care of both search line and match line functions over the traditional CAM structure, and coding line is used to enhance the expansion of match window. The designed ACAM achieves a high bit density (up to 5 bit/cell) parallel search with minimal overlap between match windows and the smallest footprint known (~ 0.12 μm^2/cell for 45 nm node).

Keywords: analog content addressable memory; ferroelectric diode; transistor-free; line programming

Introduction

In the dynamic field of artificial intelligence (AI), efficient processing of large volumes of data plays a pivotal role in large-scale data process and machine learning applications. Parallel search, in this regard, is widely acknowledged as a dependable and effective method of data processing [1]. Among the many designs used for search, content-addressable memory (CAM) is particularly suitable for performing operations that involve matching an input datum with a collection of data patterns stored in the CAM array [2]. When this matching operation is executed in parallel, it enables high-throughput comparisons with minimal latency, presenting a distinct advantage in applications such as network routing, associative computing, and database management systems [3]. From the point of view of the devices that make up the CAM architecture, it can be divided into two categories. One design is based on a complementary metal-oxide-semiconductor (CMOS) process. An example of such a design is the ternary CAM (TCAM) architecture, which is implemented using 16 transistors (i.e., configured with static random-access memory (SRAM). However, this architecture is characterized by larger cell area, higher power consumption, and lower memory density [4]. Another approach is based on non-volatile memory (NVM). NVM-based designs leverage the unique properties of non-volatility and multi-resistivity to enable energy-efficient, low-area, and high-speed parallel search. A representative example is the 2 FeFET-based ACAM architecture, which achieves different match windows (MW) by adjusting the threshold voltage [3]. However, there is still room for further optimization of the CAM, given the limitations of transistor components.

In this work, a transistor-free ACAM cell composed of 2 ferroelectric diode (FeD) is designed and verified through simulations. FeD's distinctive characteristics, including its high non-linearity, multi-state conductance [5], and non-volatility, make it an ideal candidate for achieving high parallel search capabilities [6]. Aside from device level, the search density (bit per cell) has been improved significantly by redesigning interconnected configuration of the memory array. By integrating the programming of FeD's resistive states (RS) with the coding line (CL), matching windows with an expanded range and minimal overlap have been achieved, surpassing the intrinsic limitations of the diode logics.

ACAM Cell Design and Mechanism

Fig. 1(a) illustrates the functioning of the TCAM architecture, where the efficiency of these systems is heavily reliant on the analog-to-digital conversion (ADC) process. This reliance, however, leads to a significant decrease in power efficiency and processing speed. The impact of this energy consumption is especially pronounced in edge computing environments, where power optimization is crucial. On the other hand, Fig. 1(b) and 1(c) demonstrate the concept and conventional ACAM cell design with transistors, where analog voltage values are supplied as the query to the ACAM and compared against the analog ranges encoded by analogically adjustable memristor's RS [1] or threshold voltage of FeFET devices [2].

The Schottky emission (SE) current, which is the dominant conduction mechanism in $Hf_{0.5}Zr_{0.5}O_2$ (HZO)-based FE diodes [6], becomes necessary for the formation of a match window because its value directly depends on the remnant polarization as well as different polarization directions give FeD rectification characteristics in different directions [5]. For ease of illustration, we refer to the polarization direction pointing towards the bottom electrode as quasi forward FeD (QF-FeD), and vice versa as quasi reverse FeD (QR-FeD). The proposed cell is made up of a QF-FeD and a QR-FeD as shown in Fig .2(a). While the conventional CAM structure includes distinct "search line (SL)" and "match line (ML)" components, in the proposed cell, as illustrated in Fig. 2(a), a component termed the "data line (DL)" is introduced by

---

979-8-3315-0417-5/25 $31.00 © 2025 IEEE

integrating the functionalities of both SL and ML into the DL. In this design, query data is input by pre-charging the DL to specific voltage levels, with each analog query data corresponding to a distinct voltage level. The matching result is determined by observing whether the potential of the DL decreases due to discharging: the DL remains high (indicating a match) only when two FeDs' channels are both cut off. The upper and lower limits of each matching window can be set by integrating the programming of FeD's states with CL's bias, as shown in Fig. 2(b), where the conductance of the FeD device can be programmed linearly by pulse train [5].

Fig. 3 presents the operation of a single proposed ACAM cell for "match", "mismatch" states with varying query data. In phase I, when the voltage of QR-FeD ($V_{DL}-V_{CL1}$) is below the threshold voltage, it is turned on, leading to a substantial discharge current between the DL and CL_1 predominantly across the QR-FeD. This results in a voltage drop in the DL, also producing a mismatch outcome. In phase II, both types of FeDs are in a cutoff state. Consequently, no discharging current will traverse through either of the FeDs between the DL and CL_1/CL_2, and the DL remains at a high level, denoting a match result. In phase III, when the voltage of QF-FeD ($V_{DL}-V_{CL2}$) is large enough to turn on this device, it leads to a substantial discharge current between the DL and CL_2, primarily through the QF-FeD, which in turn also causes a voltage drop in the DL that produces a mismatch outcome. Noted that the conducting of the FeD device is determined by the voltage drop between V_{DL} and V_{CL}, thus, the position of the MW can be programed through selection of CL's bias relative to V_{DL}.

Calibration and Result Discussion

TCAD tools are utilized to validate the proposed design, and we performed calibration of an individual FeD device using parameters and methodologies referenced in literature [6]. Fig. 4 (a) demonstrates the calibration of experimentally measured I-V characteristics with dual sweeping, successfully capturing key features such as ON current, and OFF current. Fig. 4(b) shows that $P–V$ curve for the calibrated device under quasi-static sweeping. More detailed material parameters for calibrated FeD are shown in Table I.

Fig. 5(a) demonstrates the schematic of $I_{FeD}-V_{DL}$ curve modulation by programming CL's bias, which is one of the important characteristics of proposed ACAM. Further, Fig. 5 (b) illustrates that $I_{FeD}-V_{DL}$ curve of different states of FeD, which is another important characteristic for proposed ACAM to be able to perform multi-bit search. Fig. 5 (c) and (d) then shows the modulation of CL's bias in the same state for FeD.

Fig. 6 (a) exhibits the $I_{DL}-V_{DL}$ curve of a single cell by modulating the FeD state only, without any CL's bias configuration ($V_{CL1}= V_{CL2}=0$ V). It can be seen that there is a large overlap between the different MWs, which is not beneficial to similarity search and will lead to a significant drop in search accuracy [3]. To address this issue, Fig. 6(b) and 6(c) present the matching outputs based on co-optimization through CL' bias programming and FeD's state configuration. The introduction of CL's bias configuration leads to a reduction in the MW's overlap, which subsequently enhances search accuracy [3]. Additionally, varying CL and DL configurations will position the MW differently (MWs $\in[-1V, 0V]$ at $V_{CL1}=0$ V & $V_{CL2}=-0.8$ V. MWs $\in[0V, 1V]$ at $V_{CL1}=0.8$ V & $V_{CL2}=0$ V). Thus, as shown in Fig. 6 (d) and (e), with the proper CL's configuration, higher bit densities (up to 5 bit/cell) can be achieved and are no longer dominated by the intrinsic number of states in the device. Fig. 6 (f) presents the simulated DL match/mismatch patterns, clearly demonstrating a significant difference. Such a small delay is owing to the transistor-free cell design, which reduces the effect of parasitic capacitance. Fig. 7 shows the benchmark between this work versus state-of-the-art CAM design [1], [3], [7]-[11]. Transistor-free cell and redesigning interconnection configuration ensure that this work achieves superior bit density and minimized area per cell.

Conclusion

In this work, we designed and simulated a compact and transistor-free ACAM architecture with high bit density, reduced latency, and minimized area overhead from only two FeDs. The transistor-free cell design, along with the redesigned interconnection configuration, enables this approach to achieve superior bit density and reduced area per cell.

Acknowledgements

X.L. acknowledges the support from the National Natural Science Foundation of China (No. 62404189), and the Start-up fund from the Hong Kong University of Science and Technology (Guangzhou). X.L. and M.C. also acknowledge the support from ACCESS – AI Chip Center for Emerging Smart Systems under the InnoHK funding of Hong Kong SAR.

References

[1] C. Li, et al. Nature communications, vol.11, no.1, pp. 1638 (2020).
[2] X. Yin, et al. Science Advances, vol.10, no.23, pp. eadk8471 (2024).
[3] X.W. Liu, et al. Cell Device, vol.2, no.2 (2024).
[4] Nii, K., et al. 2014 IEEE International Solid-State Circuits Conference Digest of Technical Papers (ISSCC), 2014: pp. 240–241
[5] X. W. Liu, et al. Nano letters, vol.22, no.18 pp. 7690-7698 (2022).
[6] Q. Luo, et al. Nature communications, vol.11, no.1, pp. 1391 (2020).
[7] B. Song, et al. IEEE Transactions on Circuits and Systems II: Express Briefs. vol.64, no. 6, pp. 700-704 (2016).
[8] C. Wang, et al. IEEE Transactions on Circuits and Systems I: Regular Papers. vol. 66, no. 4, pp.1454-1464 (2018).
[9] N. Kai, et al. Nature Electronics, vol. 2, no. 11 pp. 521-529 (2019).
[10] X. Yin, et al. IEEE Transactions on Electron Devices, vol. 67, no. 7 pp. 2785-2792, (2020).
[11] R. Ramin, et al. IEEE Transactions on Electron Devices, vol. 68, no. 1, pp.109-117, (2020).

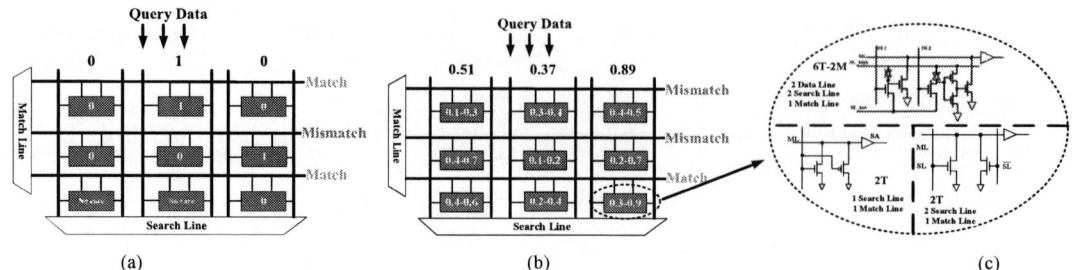

Fig. 1 Content addressable memory (CAM) architecture and its conceptual schematic. (a) Schematic diagram of ternary CAM (TCAM) architecture, which is currently the dominant CAM structure for parallel searching. It is used to compare digital queries and cell state in order to identify exact match or mismatch. However, it is worth noting that current TCAM designs face limitations in terms of both bit-density and functionality due to their operation primarily in the digital domain. (b) Schematic diagram of conventional analog CAM (ACAM) architecture. Instead of comparing for exact match or mismatch, ACAM compares analog query with analog cell state, which can significantly increase bit-density. Match line (ML) and search line (SL) are separate in conventional CAM architectures, and a representative cell design is shown in (c). The cell is basically composed of individual transistors (2T0M) or a combination of transistors and memristors (6T2M) [1]-[3].

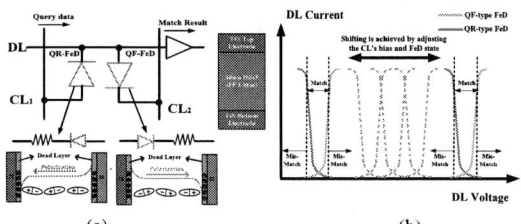

Fig. 2 Transistor-free cell design and the search matching principle. (a) 2-FeD ACAM cell for search operation. Unlike the conventional CAM structure in which "ML and "SL" are independent, the query data in this architecture is input from the left side of the DL, and the matching result is determined by whether the potential of the right side of the DL drops due to leakage. (b) The upper and lower limits of each matching window can be programed by both FeD state and coding line (CL) bias.

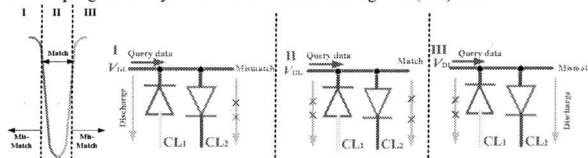

Fig. 3 Operation of a single ACAM cell comprising 2-FeDs for "match", "mismatch" states. Noted that the upper and lower limits of each matching window (i.e., the V_{DL} corresponding to when QF-FeD and QR-FeD are turned on) can be programmed by the FeD state and the bias of CL_1/CL_2.

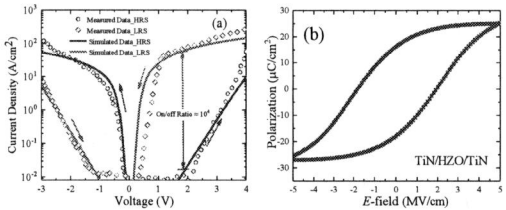

Fig 4 Calibration of TCAD models for TiN/HZO (10nm)/TiN FeD [6]. (a) Calibration for I-V curve with both polarization direction of FeD. (b) P–V curve for the calibrated device under quasi-static sweeping.

TABLE I
Key used for calibration

Parameter	FeD	Parameter	FeD
ε_{HZO}	30	P_S/ uC·cm^{-2}	22.8
P_S/ uC·cm^{-2}	28.5	E_c/ MV·cm^{-1}	1.76
φ_{TiN}/eV	4.5	χ_{HZO}/eV	2.7

Current transport model: Thermoelectric emission, direct tunneling, Fowler-Nordheim tunneling, trap-assisted tunneling

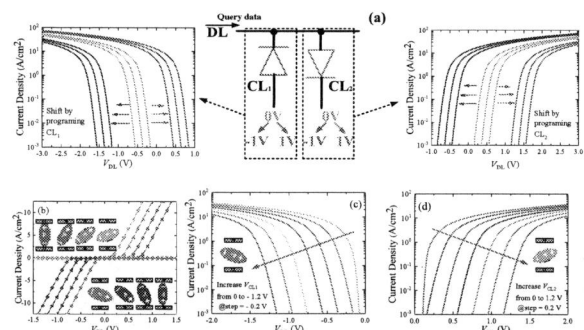

Fig. 5 The V_{DL} corresponding to when FeD are turned on can be programmed by the and CL's bias and FeD state. (a) Schematic of shifting I_{FeD}-V_{DL} curve by programming CL's bias. (b) I_{FeD}-V_{DL} curve of different FeD's state at $V_{CL1}= V_{CL2}=0$ V, where gradual switching can be achieved by modulating the pulses with stepwise voltages [5]. (c) I_{FeD}-V_{DL} curve of different V_{CL1} at the same QR-FeD state. (d) I_{FeD}-V_{DL} curve of different V_{CL2} at the same QF-FeD state.

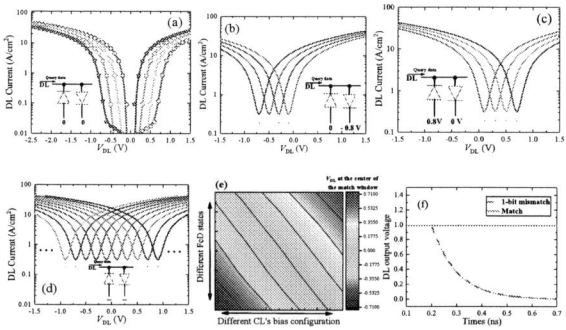

Fig. 6 Demonstration of 2 FeD ACAM performance. (a)-(c) Current through DL when sweeping voltage is applied on DL's input side at (a) $V_{CL1}= V_{CL2}=0$ V, (b) $V_{CL1}=0$ V/ $V_{CL2}=-0.8$ V, (c) $V_{CL1}=0.8$ V/ $V_{CL2}=0$ V. (d) By programming different CL's bias configurations, combined with different states of FeD, multiple matching windows (up to 32 MWs/cell) can be generated at different positions. (e) Window shift by appling different states and CL's bias. (f) Simulated DL discharge behavior during search operation, specifically for all-match and 1-bit-mismatch states.

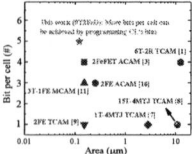

Fig. 7 Comparison of area and bit per cell for this work with state-of-the-arts. Area estimates for this work are based on 45 nm node process.

Effect of the Channel Thickness on the PBS Reliability and 1/f Noise of the ALD Ultrathin ITO Field-Effect Transistor

Jiaming Zhao[1,2], Peiyan Hong[1,2], Xuefei Li[1,2,*]

[1]Wuhan National High Magnetic Field Center and School of Integrated Circuits, Huazhong University of Science and Technology, Wuhan, China, [2]School of Integrated Circuits, Huazhong University of Science and Technology, Wuhan, China.

Email: xfli@hust.edu.cn

Abstract

In this work, we report ITO transistors by atomic layer deposition with channel thickness of 2.5 nm and 3.4 nm. The 3.4-nm-thick ITO FETs exhibit a high field-effect mobility μ (36 cm^2/V·s) and a negligible V_{th} shift of -30 mV under V_{th} + 2 V. Meanwhile, the 1/f noise behavior of 3.4-nm-thick ITO FETs can be described by the carrier number fluctuation model.

Keywords: Indium tin oxide (ITO), Atomic layer deposition (ALD), PBS reliability, Low-frequency noise.

Introduction

Sn-doped In$_2$O$_3$ (ITO) is widely used as a transparent electrode material in optoelectronics because it is a wide bandgap oxide semiconductor ranging from 3.5-4.3 V [1,2]. Owing to its degeneracy in the conduction band and unsatisfactory switching characteristics, its semiconductor behavior has rarely been reported. However, by reducing the thickness of the channel film, the conduction band can be modulated and the carrier density can be adjusted to achieve a high I_{on}/I_{off} ratio [3,4]. ITO FETs with ultrathin channel films can be considered promising candidates for BEOL-compatible transistors. Compared with other channel growth methods, such as physical vapor deposition (PVD), ALD can control the thickness precisely by changing the number of cycles [5]. Sn can be doped as a carrier suppressor because it has a stronger oxygen binder than In. The reliability of ITO FETs remains a concern when they are used as logic circuits. Determining whether ITO FETs can be applied commercially is crucial. Moreover, with the increasing necessary of device performance evaluation and circuit design, the 1/f noise mechanism becomes increasingly important [6].

In this work, we demonstrate ALD ultrathin ITO FET with 2.5 and 3.4 nm. As the thickness increases to 3.4 nm, ITO FETs achieve excellent subthreshold swing (SS) as low as 75 mV/dec, high mobility of up to 36 cm^2/V·s, and a high I_{on} of 467 µA/µm at a 500 nm channel length with a drain voltage (V_{ds}) of 3 V. The thickness-dependent V_{th} shift can be understood in terms of the conduction band shift and carrier number change. The high reliability ΔV_{th} was -30 mV after 1000 s with positive bias stress (PBS). The different V_{th} shift directions can be well explained by the E_{TNL} and trap generation. The low noise frequency (LNF) indicates that the thickness of the channel film decreases, the noise mechanism changes from the carrier number fluctuation to the mobility fluctuation.

Experiments

Fig. 1(a) shows the schematic device structure of an ITO transistor. The gate stack includes 8 nm HfO$_2$ as the gate dielectric, 2.5/3.4 nm ITO as the semiconducting channel, and 15/15 nm Ni/Au as the source/drain contact. Fig. 1(b) shows the process flow. The back gate and top contact structure were fabricated on p^{++} silicon substrates. HfO$_2$ (8 nm) was grown via ALD at 250 °C, and [(CH$_3$)$_2$N]$_4$Hf and H$_2$O were used as Hf and O precursors. 2.5-nm and 3.4-nm ITO films were deposited via ALD at 225 °C, using (CH$_3$)$_3$In, [(CH$_3$)$_2$N]$_4$Sn, H$_2$O and O$_3$ as In, Sn and O precursors. The film thickness was accurately controlled by adjusting the ALD cycles. Channel isolation was performed by wet etching after the region was defined by electron-beam lithography (EBL). Finally, 15/15 nm Ni/Au was deposited as source/drain contacts. The whole process was performed with a low thermal budget of 250 °C. The thickness of the ultrathin films was determined via atomic force microscopy. Fig. 1(c) shows that the thickness of the films were 2.5 nm and 3.4 nm. Fig. 1(d) shows atomically smooth and featureless surface morphologies. The surface root-mean-square (RMS) is 0.3 nm over a scan area of 2 × 2 µm^2.

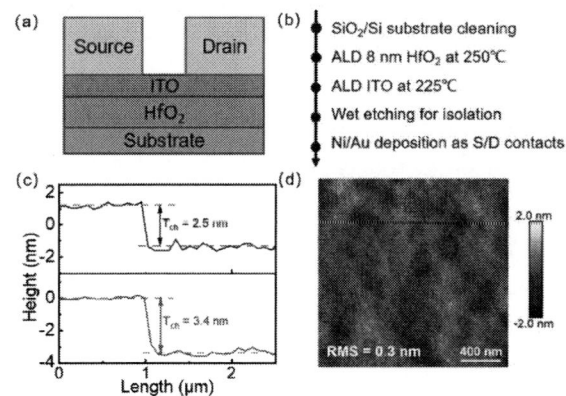

Fig. 1. (a) Schematic device structure of an ITO transistor. (b) Process flow of ITO TFTs. (c, d) AFM images of ITO films on SiO$_2$/Si substrates.

The electrical characteristics and PBS reliability of the ultrathin ITO FETs were measured via a Lakeshore vacuum probe station with an Agilent semiconductor parameter analyzer B1500A. The $1/f$ noise power spectral density (S_{id}) was measured at a low V_{ds} of 0.5 V for the 2.5-nm-thick ITO FETs and 0.2 V for the 3.4-nm-thick ITO FETs by an Agilent E4725A $1/f$ noise system with an Agilent E5052B.

Results and discussion

Fig. 2(a) shows the transfer characteristics and mobility of ITO FETs with 2.5 nm T_{ch}. The ultrathin 2.5-nm-thick ITO FETs exhibit a low mobility of 6.4 cm²/V·s, an SS of 85 mV/dec, a V_{th} of 1.7 V, and an I_{on}/I_{off} ratio > 10^7. The I_{on} can reach 13 µA/µm at a V_{gs} of 3 V and a V_{ds} of 3 V [Fig. 2(c)]. The channel thickness has a significant effect on the electrical characteristics compared with that of ITO FETs with a thickness of 3.4 nm. As the T_{ch} of the ITO FETs increases from 2.5 nm to 3.4 nm, the devices exhibit a negativity shifting V_{th} of -0.5 V, but five times higher μ values of 36 cm²/V·s, steeper SS of 75 mV/dec and I_{on}/I_{off} values also increase by one order of magnitude [Fig. 2(b)]. A higher I_{on} of 467 µA/µm at a V_{gs} of 3 V and a V_{ds} of 3 V can be achieved [Fig. 2(d)].

Owing to the notable V_{th} shifting positively with decreasing thickness of the channel film and Equation (1), both quantum confinement and the carrier number can be considered [7,8]. It is apparent that a thinner channel film results in a lower carrier number which is beneficial for obtaining an enhanced device. According to Equation (2), when ΔE_c increases rapidly, V_{th} largely shifts, whereas T_{ch} decreases. The coeffect of the two reasons lead to the over 2 V V_{th} shift.

$$\Delta V_{th} = \Delta V_{carrier\ number} + \Delta E_c \qquad (1)$$

$$\Delta E_c = \frac{\pi^2 \hbar^2}{2m^* T_{ch}^2} \qquad (2)$$

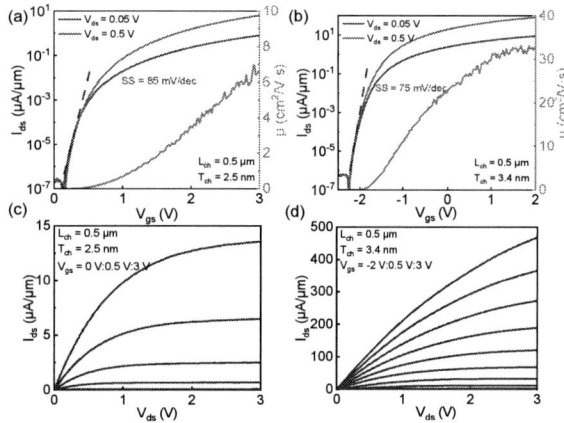

Fig. 2. I_{ds}-V_{gs}, mobility and I_{ds}-V_{ds} characteristics of ITO transistors with T_{ch} values of (a,c) 2.5 nm and (b,d) 3.4 nm and L_{ch} = 0.5 µm.

PBS measurement is the most important reability test for n-channel FETs. The n-type ITO channel was measured with

V_{bias} fixed at V_{th} +2 V for 1000 s. V_{th} degradation was evaluated via the linear extrapolation method. To avoid geometry dependence for different channel thinknesses, V_{ds}, channel lengths and widths are fixed at 0.05 V, 500 nm, and 1 µm, respectively. Fig. 3(a, c) shows the I_{ds}-V_{gs} characteristic curves, and a relatively large V_{th} shifts positively as T_{stress} increases with a 2.5-nm channel-thick ITO FETs. In contrast, 3.4-nm-thick ITO FETs show extraordinary reliability performance. Fig. 3(b, d) shows extremely small, almost negligible V_{th} degradation with 3.4-nm channel-thick ITO FETs. Notably, increasing the thickness of the ITO channel film improved the ΔV_{th} value from +0.28 V to -0.03 V. The negligible ΔV_{th} indicates the high interface quality between the ITO and HfO₂.

The phenomenon of different shift directions with 2.5-nm-thick and 3.4-nm-thick ITO FETs can be well explained by the physics model inducting the E_{TNL} [9]. E_{TNL} is a location where both acceptor-like and donor-like traps can compensate for each other. When the film is sufficiently thin, E_F is under E_{TNL}, which means that increasing T_{stress} induces acceptor-like traps. This will make V_{th} shift positively. As the thickness of the ITO channel film increases, the carrier concentration rapidly increases. This change makes it possible for the E_F to move up through the E_{TNL}. When the distance E_F exceeds E_{TNL} is short, the channel will generate a small donor-like trap and lead to a negligible negative V_{th} shift.

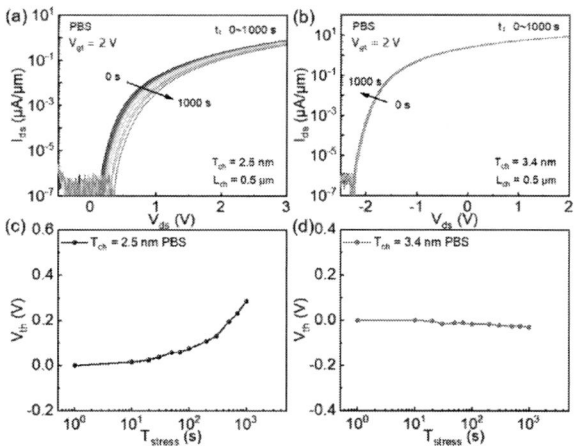

Fig. 3. Evolution of the transfer characteristics and V_{th} shift variation-dependent stress time in PBS for ITO TFTs with T_{ch} values of (a, c) 2.5 nm and (b, d) 3.4 nm. The stress condition is V_{ds} = 0.05 V.

Fig. 4(a, b) shows the normalized noise power spectrum density S_{id}/I_d^2 of the 2.5-nm-thick and 3.4-nm-thick ITO FETs as a function of frequency. Owing to the large changes in the values of I_{ds} and V_{th}, the measurement is performed at V_{ds} = 0.5 V for 2.5-nm-thick ITO FETs and V_{ds} = 0.2 V for 3.4-nm-thick ITO FETs. The LFN follows the $1/f$ noise behavior at different V_{gs}. ITO FETs with greater channel thicknesses exhibit better LFN performance at lower S_{id}/I_d^2 levels [8]. Fig. 4(c, d) show S_{id}/I_d^2 and $(g_m/I_d)^2$ vs. I_d. For 2.5-nm-thick ITO FETs, S_{id}/I_d^2 is proportional to $1/I_d$ indicating that the $1/f$ noise comes from the Hooge mobility fluctuation ($\Delta\mu$) model. For 3.4-nm-thick ITO FETs, S_{id}/I_d^2 is proportional to $(g_m/I_d)^2$ indicating that the $1/f$ noise comes from McWhorter's carrier number fluctuation (Δn) model. Another method to assess the $1/f$ noise mechanism involves finding the slope of the curve for S_{id}/I_d^2 vs. V_{gt} on a log scale. Fig. 4(e, f) shows that the S_{id}/I_d^2 values of our ITO TFTs with different thicknesses have two kinds of power law dependences on V_{gt}. The slope of the 2.5-nm-thick ITO TFTs is approximately -1, which is close to the prediction of the $\Delta\mu$ model. The opposite trend in which the slope is extracted to be -2 corresponds to the prediction of the Δn model for 3.4-nm thick ITO FETs.

Different mechanisms correspond to different sources of noise in which the device operates. The $\Delta\mu$ model reveals that dominant mechanism affecting $1/f$ noise is lattice scattering. This can in turn cause random mobility fluctuations. The Δn model reveals that dominant mechanism affecting $1/f$ noise is carrier trapping and detrapping by trap states [10].

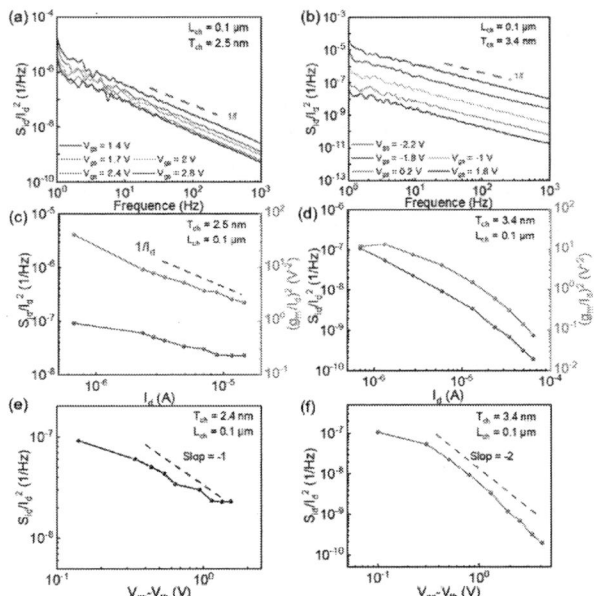

Fig. 4. Normalized noise power spectrum density S_{id}/I_d^2 vs. frequency for ITO TFTs with T_{ch} values of (a) 2.5 nm and (b) 3.4 nm. S_{id}/I_d^2 and $(g_m/I_d)^2$ vs. I_d with T_{ch} values of (c) 2.5 nm and (d) 3.4 nm. S_{id}/I_d^2 vs. V_{gt} with T_{ch} values of (e) 2.5 nm and (f) 3.4 nm. The S_{id}/I_d^2 values were extracted at 100 Hz.

Conclusion

In conclusion, ALD ultrathin ITO FETs with different film thicknesses and remarkable DC, reliability and LNF performance are demonstrated. This clearly shows that the channel thickness has a significant effect. The outstanding electric performance and PBS reliability are observed with 3.4-nm thick ITO FETs. For 3.4-nm thick ITO FETs, the noise model follows the Δn model, which means that carrier trapping and detrapping by trap states are the main mechanisms.

Acknowledgments

This work was financially supported by the National Key R&D Program of China (Grant No. 2020AAA0109005).

References

[1] S Li et al., "Nanometer-thin indium tin oxide for advanced high-performance electronics," *Nature Mater.*, vol. 18, no. 10, pp. 1091–1097, (2019).

[2] Z. Zhang et al., "Atomically Thin Indium-Tin-Oxide Transistors Enabled by Atomic Layer Deposition," *IEEE Transactions on Electron Devices*, vol. 69, no. 1, pp. 231-236, (2022).

[3] H Kim et al., "Effect of film thickness on the properties of indium tin oxide thin films." *Journal of Applied Physics.*, vol.88, no. 10, pp. 6021-6025.

[4] M. Si, Z. Lin, A. Charnas and P. D. Ye, "Scaled Atomic-Layer-Deposited Indium Oxide Nanometer Transistors with Maximum Drain Current Exceeding 2 A/mm at Drain Voltage of 0.7 V," *IEEE Electron Device Letters*, vol. 42, no. 2, pp. 184-187, (2021).

[5] Z. Zhang et al., "Atomically Thin Indium-Tin-Oxide Transistors Enabled by Atomic Layer Deposition," *IEEE Transactions on Electron Devices*, vol. 69, no. 1, pp. 231-236, (2022).

[6] H. He, X. Zheng and S. Zhang, "1/f Noise Expressions for Amorphous InGaZnO TFTs Considering Mobility Power-Law Parameter in Above-Threshold Regime," *IEEE Electron Device Letters*, vol. 36, no. 2, pp. 156-158, (2015).

[7] Y. Kang, K. Han, Y. Chen and X. Gong, "Thickness-Engineered Extremely thin Channel High Performance ITO TFTs with Raised S/D Architecture: Record-Low RSD, Highest Mobility (Sub 4 nm T_{CH} Regime), and High V_{TH} Tunability," *2023 IEEE Symposium on VLSI Technology and Circuits (VLSI Technology and Circuits)*, pp. 1-2, (2023).

[8] G. Liu et al., "Exploring the Impact of Channel Thickness Scaling on PBTI and Low-Frequency Noise in Ultrathin IGZO Transistors," *IEEE Transactions on Electron Devices*, vol. 71, no. 9, pp. 5407-5413, (2024).

[9] D. Zheng et al., "First Demonstration of BEOL-Compatible Ultrathin Atomic Layer-Deposited InZnO Transistors with GHz Operation and Record High Bias-Stress Stability," *2022 International Electron Devices Meeting (IEDM)*, pp. 4.3.1-4.3.4, (2022).

[10] Fung, Tze-Ching et al., Low frequency noise in long channel amorphous In–Ga–Zn–O thin film transistors. *Journal of Applied Physics.* (2010).

A physics-based compact model for electromigration failure prediction and dynamic IR-drop evaluation

Chenglin Ye[1], Yizhan Liu[2], Zheng Zhou[1]*, Xiaoyan Liu[1]

1.School of Integrated Circuits, Peking University, Beijing, 100871, China.
2.School of Software and Microelectronics, Peking University, Beijing, 100871, China.
Email: zhouzime@pku.edu.cn

Abstract

In this work, a novel physics-based compact model for electromigration (EM) failure statistical distribution prediction and dynamic IR-drop evaluation is proposed for multi-segment interconnects. The proposed model takes into account the atomic migration paths, void nucleation-formation-growth phase, as well as the Gaussian distribution of activation energy and the dimension of the void. It can accurately describe the statistical distributions of interconnect resistance degradation and EM time-to-failure (TTF) resulted by process varieties during nanofabrication as well as dynamic IR-drop along time. The results of the model well agreement to the experimental data.

Keywords: reliability, electromigration, time-to-failure, Gaussian distribution

Introduction

Electromigration has become a key concern for copper-based interconnects in recent years. Many EM analysis methods have been proposed to address the challenges. Exiting EM failure empirical prediction model Black's equation[1] and Blech's limit[2] would overestimate the TTF without the consideration of microscopic physical effects. Besides, they are only suitable to a single wire and are not functional in multi-segment interconnects. A finite element based method atomic flux divergence (AFD)[3] can work for multi-segment interconnects EM failure but can't predict the void location accurately, making the method inaccurate in complex multi-segment interconnects. Furthermore, the previous models are unable to reflect the statistical distribution patterns of EM-induced resistance degradation and TTF as a significant phenomenon that can be observed in practical measurement.

For predicting EM-induced failure statistical distribution and dynamic IR-drop in multi-segment interconnects accurately, we proposed a novel physics-based model. The model has features as follows: (1) Multi-physical effect as well as microscopic mechanisms of void formation and atomic migration are both considered in the model to simulate EM-induced resistance degradation accurately. (2) Rather than single R-t curve, the model can describe the statistical distribution of resistance degradation and the lognormal distribution of TTF, which exhibits well agreement with the experimental data. (3) We present the model for multi-segment interconnects EM-induced failure prediction. The developed model is capable of accurately predicting the locations of void formation, at the junctions or at the terminals, and the dynamic IR-drop in the layout, which provides valuable guidance for circuit design.

EM Modeling

The proposed methodology for EM-induced failure distribution prediction and dynamic IR-drop evaluation is shown in Fig.1. Multi-segment interconnects can be split into single wires and block resistance nodes. When the geometry, current and temperature are inputted, the model can provide the output including multi-segment R-t curve distribution, TTF distribution and dynamic IR-drop along time.

The dominant migration paths of Cu ions and void-induced resistance degradation are considered in the proposed model. Cu ions migrate along three main paths: the Cu/capping interfaces, the Cu/liner interfaces and the grain boundaries.[4] The effective drift velocity of Cu atoms can be written as shown in Fig.1, where H and W are the metal line height and width, respectively, the δ, D and Z donate the effective thickness, diffusivity and charge number. The subscript C, L and GB denote the Cu/Capping, Cu/liner interfaces and the grain boundaries, respectively. The void-induced resistance degradation includes three phases: the nucleation phase (Stage I), the formation phase (Stage II) and the growth phase (Stage III).[5] Nucleation and formation occur primarily in the nodes, also call as void form region (VFR). When the vacancy accumulation of the node reached a critical value, the void nucleates and begins to occupy the entire node, and the void begins to grow along the metal line region (MR). Electron wind, stress gradient, temperature, grain size effects are considered in the model. The vacancy accumulation rate in the VFR during Stage I, the void size rise rate during Stage II and the horizontal growth rate of the void during Stage III are written in the Fig.1(b), where f donates the oscillation frequency, E_A donates the activation energy, ΔE_{ew} and ΔE_{st} donate the electron wind correction and stress correction for E_A respectively, a donates the atom diameter, l_1 donates the critical size during the void formation phase and G is the modulation factor. The values of typical parameters are shown in Table 1.

As shown in Fig.2, the process variety and grain randomness lead to the normal distribution of activation energy and void formation volume and further lead to the lognormal distribution of the TTF, which can be observed in practical measurement.

With the proposed model, single & multi-segment EM failure can be predicted and dynamic IR-drop along time can be evaluated. Simulations and results for single wires and multi-segment interconnects are displayed as follows.

Experiment Results and Discussion

A. Single Straight Line

Fig.3(a) shows 50 simulation samples and average resistance shift for over time at $T = 350℃$. The length of the line is $150\mu m$ and the current density is $20mA/\mu m^2$. A very resistance abrupt increase can be observed in single sampling curve. Due to very high statistical sampling, the average curve shows a very smooth drift behavior. The TTF probability distribution function (PDF) is demonstrated in Fig.3(b). 200 samples are simulated and the results of the model well agreement to the experimental data. [6].

Fig.4 shows the PDF of 200 simulation sampling TTFs at $L = 150\mu m$ and $L = 50\mu m$ respectively. It can be observed that the MTTF of the longer line is smaller because the faster accumulation of vacancies results in the tension stress reaches the critical value faster at the cathode. And this result coincides to the experimental results. Fig.5 shows the PDF of TTF for different current densities. The simulation sampling number is 200. At a lower current density ($20mA/\mu m^2$), Less momentum is exchanged when an electron collides with atoms, causing the electromigration slower. And due to smaller electron wind force, the electron random motion is more obvious and the variance of TTF is larger.

The average R-increase trace and cumulative distribution function of TTFs at different temperatures are plotted in Fig.6. It can be observed that the electromigration failure process is accelerated when the simulation temperature increases. This is due to the increased diffusivity and probability of atomic activation at higher temperature.

B. Multi-segment Interconnects

Four different tested structures A, B, C and D are demonstrated in Fig.7 for multi-segment interconnects simulation, where A is a straight line at $L = 300\mu m$, B, C and D are bending lines divided by 2:1, 1:1 and 1:2 as the same length of A, respectively.

Fig.8 shows the single resistance samplings of A, B, C and D mentioned above. As shown in Fig.8(a), only one resistance jump can be observed in the structure A, B and C lines because there is only one void formation in node 1. Due to the bending nodes, the resistance of B and C lines will have an abrupt increase faster, but the abrupt change magnitudes of B and lines are less than that of the straight A. As shown in Fig.8(b), resistance of bending line D has two jumps when

voids are formed in the node 2 and 1, respectively. This is because L_{23} of D is longer and it will lead to faster vacancy accumulation in the bending line.

C. IR drop

Fig.9 shows the average R-t curve for the 4 structures. It is clear that the bending lines fail faster than the straight because the bending nodes have larger stress. Structure D fails fastest due to the growth of two voids. Fig 10 demonstrates the PDF of TTF for the 4 structures. Among these bending lines, the structure C divided 1:1 shows the strongest EM reliability. And the MTTFs of all bending lines are shorter than the straight line. In addition, the straight line has less TTF variety than the bending lines.

Dynamic IR-drop voltages of different PDNs using the four structures A-D under $VDD = 0.8V$ are demonstrated in Fig.11. The lowest IR-drop in the straight A can be found and the structure D bending line divided by 1:2 has higher IR-drop among these three bending lines B-D, which needs to be avoided in the PDN layout design. The simulation result show that for bending lines, we should make the two lengths of the two single segment as equal as possible.

Conclusion

In this work, a novel multi-physics based compact model for EM-induced multi-segment interconnection failure distribution prediction is proposed. The model has three features compared to previously mentioned models as follows. (a) Based on the void nucleation-formation-growth three phases, the microscopic metal ions migration and the multi-physical effect are investigated to account for the resistance degeneration. (b) The Gaussian distribution of the activation energy and void dimension are considered, and the probability distribution function and cumulative distribution function of TTFs can be demonstrated in the model. This distribution shows that TTF is usually lognormal distributed. The results of the model well agreement the experiment data. (c) Multi-segment interconnects EM-induced failure prediction is demonstrated, and the EM simulation for four different structures including a straight and three bending lines are taken for example. The dynamic IR-drop along the time for the 4 structures are demonstrated as well for the PND layout design guidance.

References

[1]   J. R. Black, TED, 1969.

[2]   I. A. Blech, J. Appl. Phys., 1976.

[3]   Dalleau D, et al., Microelectronics Reliability, 2001.

[4]   C. K. Hu., et al, Microelectronics Reliability, 2006.

[5]   L. Cai, et al., Science China Information Sciences, 2021.

[6]   J. Shuster-Passage et al., IRPS, 2024.

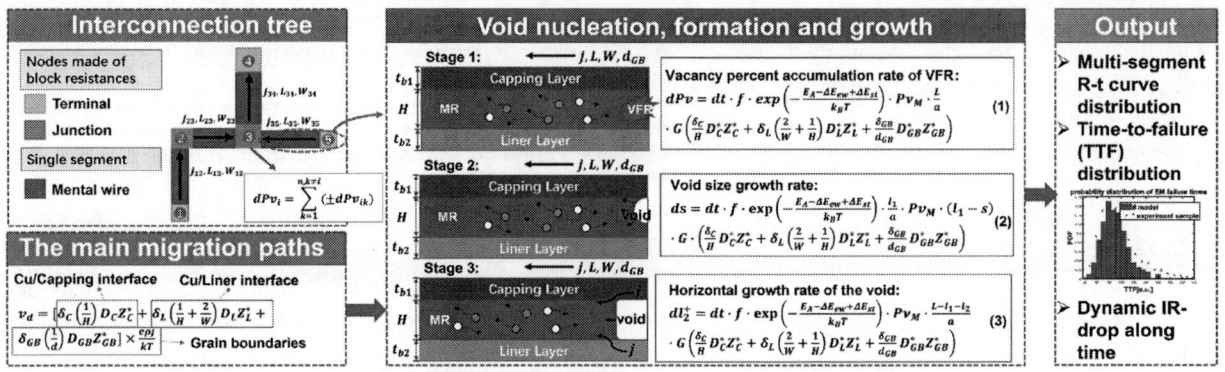

Fig.1 The physics theory and workflows of the method. The multi-segment interconnects can be split into nodes and single wires. The dominant migration paths of Cu ions and void-induced resistance degradation are considered. The accumulation of voids at nodes is related to adjacent lines. Outputs are provided including multi-segment R-t curve distribution, TTF distribution and dynamic IR-drop along time.

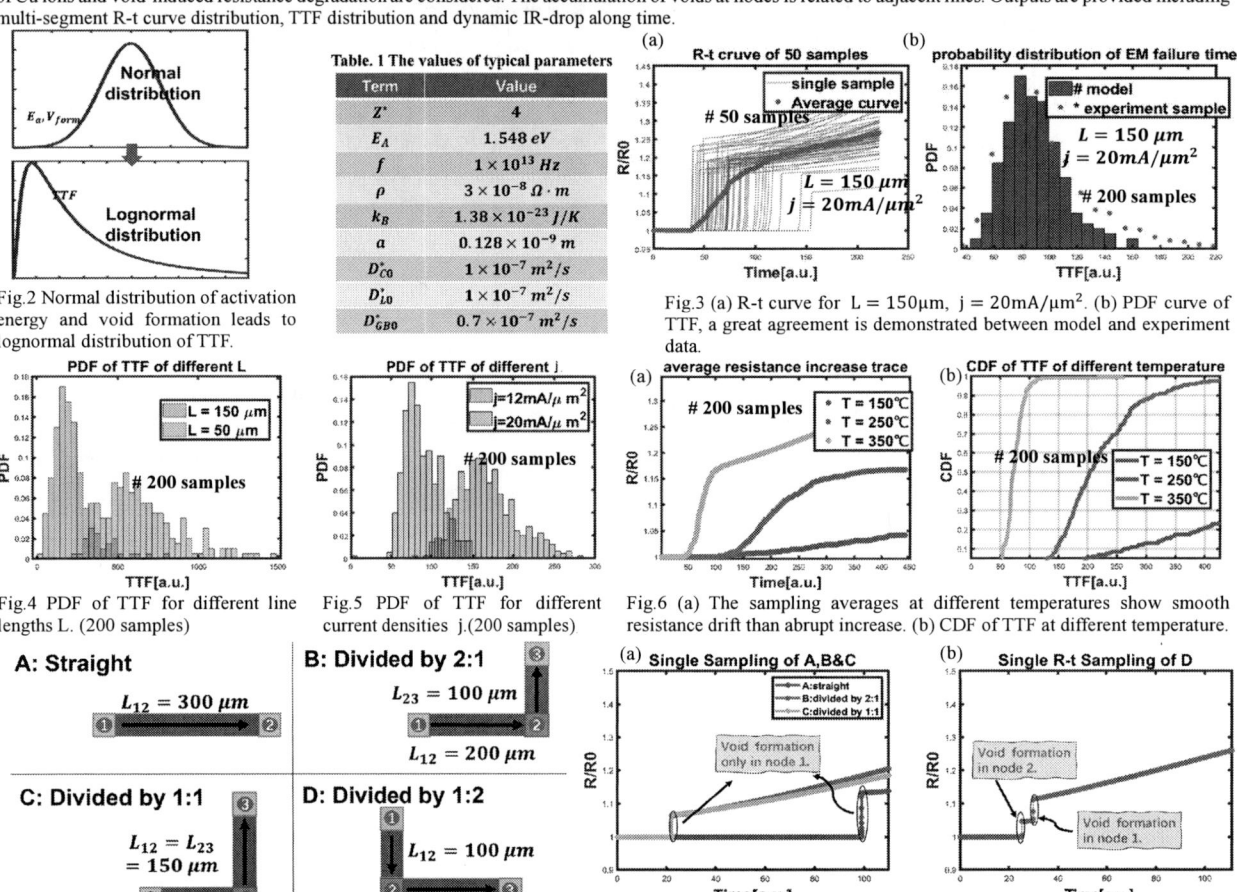

Fig.2 Normal distribution of activation energy and void formation leads to lognormal distribution of TTF.

Table. 1 The values of typical parameters

Term	Value
Z^*	4
E_A	$1.548\ eV$
f	$1 \times 10^{13}\ Hz$
ρ	$3 \times 10^{-8}\ \Omega \cdot m$
k_B	$1.38 \times 10^{-23}\ J/K$
a	$0.128 \times 10^{-9}\ m$
D^*_{C0}	$1 \times 10^{-7}\ m^2/s$
D^*_{L0}	$1 \times 10^{-7}\ m^2/s$
D^*_{GB0}	$0.7 \times 10^{-7}\ m^2/s$

Fig.3 (a) R-t curve for $L = 150\mu m$, $j = 20mA/\mu m^2$. (b) PDF curve of TTF, a great agreement is demonstrated between model and experiment data.

Fig.4 PDF of TTF for different line lengths L. (200 samples)

Fig.5 PDF of TTF for different current densities j.(200 samples)

Fig.6 (a) The sampling averages at different temperatures show smooth resistance drift than abrupt increase. (b) CDF of TTF at different temperature.

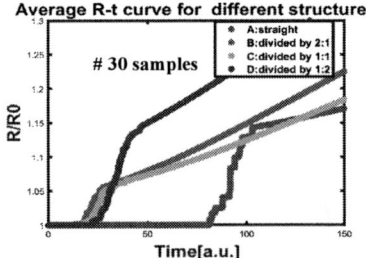

Fig.7 Four structures tested. A is a straight line, B, C and D are bending lines divided by 2:1, 1:1 and 1:2 respectively.

Fig.8 (a) Single R-t Sampling of A, B & C. Resistances have a jump due to the void formation only in node 1. (b) Single R-t sampling of D. Resistance has two jumps when voids are formed in node 2 & 1, respectively.

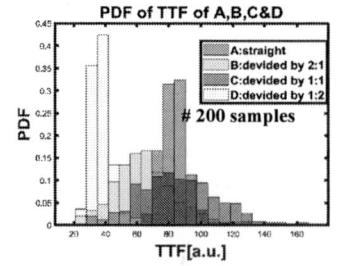

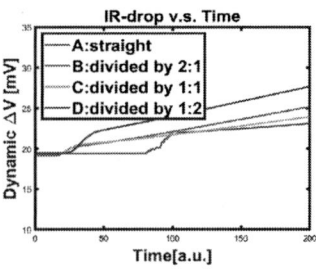

Fig.9 Average R-t curves for 4 structures in Fig.7. The bending lines fail faster than the straight because the bending nodes have larger stress. Structure D fails fastest due to the growth of two voids.

Fig.10 PDF of TTFs for 4 different structures A, B, C & D

Fig.11 Dynamic IR-drop for 4 different structures A, B, C and D.

Multi-Mode Resonance and Temperature Behavior of GaN/SiC SAW Resonators

Guofang Yu[1,2,3], Renrong Liang[3*], Deng Luo[1,2], Jianjun Chen[1,2], Yaqing Chi[1,2],
Hanhan Sun[1,2] and Bin Liang[1,2]

[1]College of Computer Science and Technology, National University of Defense Technology, Changsha 410073, China;

[2]Key Laboratory of Advanced Microprocessor Chips and Systems, College of Computer Science and Technology, National University of Defense Technology, Changsha 410073, China;

[3]School of Integrated Circuits, Tsinghua University, Beijing 100084, China. *Email: liangrr@mail.tsinghua.edu.cn

Abstract

This work delves into the fabrication and temperature characterization of multi-mode SAW resonators on GaN/SiC substrates. The Sezawa acoustic wave modes offers a pathway to enhance the product of resonance frequency and mechanical quality factor ($f_0 \times Q_m$). Through a multi-mode mBVD model, the temperature behavior of GaN/SiC SAW resonators is thoroughly investigated from 10 K to 500 K, demonstrating a high $f_0 \times Q_m$ of 6.21×10^{12} Hz and 9.71×10^{12} Hz at 10 K for Rayleigh and second Sezawa mode, respectively.

Keywords: GaN/SiC, surface acoustic wave, multi-mode, temperature behavior, mBVD

Introduction

Surface acoustic wave (SAW) devices have played an important role in the radio frequency (RF) front end of wireless communication applications due to their advantages of small chip size, high reliability, and low cost [1,2]. In addition, their sensitivity to various environmental factors, such as temperature, pressure, and humidity, makes SAW resonators highly valuable in sensing applications. Recently, gallium nitride (GaN) has been developed in the fabrication of gigahertz SAW resonators due to its excellent piezoelectric and mechanical properties along with thermal and chemical stability [3]. Sezawa (S) acoustic wave modes, emerging in layered structures of "slow on fast" such as GaN on silicon carbide (SiC), provide a promising solution to obtain a high product of resonance frequency and mechanical quality factor ($f_0 \times Q_m$) [4]–[7]. The S modes are guided modes that confine acoustic energy in the overlayer, resulting in the possibility of increasing the effective electromechanical coupling coefficient and quality factor of resonators without requiring substrate release [8].

For temperature-related SAW sensors, the relative sensitivity (SEN), defined as $SEN = df_0/dT$, is proportional to the resonance frequency. The SAW resonator with high Q_m exhibits great stability due to its enhanced capability to accurately select signals and suppress interference. GaN film on Si, SiC, and Sapphire can be incorporated into the fabrication of the SAW devices with high $f_0 \times Q_m$ [9]. Due to high acoustic velocity, and high thermal conductivity of SiC, the GaN/SiC SAW resonators provide a much-needed solution for temperature applications [3]. However, further research is still needed to understand how temperature affects the characteristics of GaN/SiC SAW resonators with multiple modes.

In this work, multi-mode SAW resonators with high $f_0 \times Q_m$ were demonstrated on GaN/SiC substrate. The resonance modes were identified by the finite element method (FEM) in COMSOL Multiphysics. The temperature behavior over a wide temperature range from 10 K to 500 K was characterized and investigated by a multi-mode modified Butterworth–Van-Dyke (mBVD) model, which indicates the reduction of losses with temperature decreasing.

Devices Fabrication and Characterization

As shown in Fig. 1a, a 100-nm AlN nucleation layer is initially deposited on a 4-inch semi-insulating 4H-SiC via metal-organic chemical vapor deposition. Then, a 2.6-μm GaN layer, comprising a 200-nm buffer layer, a 1.2-μm Fe-doped GaN layer, and a 1.2-μm unintentionally doped GaN (UID-GaN) layer, is grown. Fig. 1b presents a cross-sectional transmission electron micrograph (TEM) image of the grown GaN/SiC substrate, and Fig. 1c shows a high-resolution (HR) TEM image of the GaN film. Figs. 1d and 1e show the X-ray diffraction (XRD) analysis and a rocking curve with a full width at half maximum (FWHM) of 226.8 arcsec for the GaN (002). These results indicate good GaN growth quality.

SAW devices with wavelengths (λ) from 2 μm to 4 μm have been fabricated on the GaN/SiC substrate. Each device features 200 pairs of IDTs and 40 pairs of reflectors on each side, with an aperture of 30λ [10]. The electrodes are patterned by photolithography, followed by Ti (10 nm)/Al (90 nm) evaporation and lift-off process. Fig. 1f is a prepared one-port SAW resonator, and Figs. 1g and 1h present the scanning electron micrograph (SEM) images of the marked region. A Lakeshore Cryogenic probe station was used to achieve a wide temperature range from 10 K to 500 K. S-parameters were obtained via a vector network analyzer (Keysight N5247B) with Short-Open-Load-Through (SOLT) calibration performed before

979-8-3315-0417-5/25 $31.00 © 2025 IEEE

measurement.

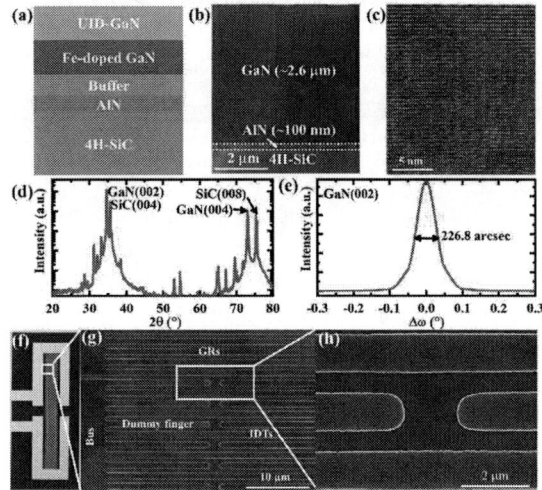

Fig. 1: (a) Layer structure of the developed GaN/SiC substrate, (b) cross-sectional TEM image, (c) HRTEM of GaN epilayer, (d) θ-2θ curve, (e) rocking curve of GaN(002), (f) optic image of a one-port SAW resonator, (g) and (h) enlarged SEM images of the marked region.

Results and Discussion

The S-parameters of one-port SAW resonators with the λ from 2 μm to 4 μm have been measured and are presented in Fig. 2. There are multiple resonant modes observed in these SAW resonators, resulting from the higher transverse velocity in the SiC substrate than that in the GaN epitaxy layer [8]. Resonance modes can be discerned based on the transverse velocity values in SiC and GaN (v_T^{GaN} and v_T^{SiC}) as well as the longitudinal velocity values in the GaN and SiC (v_L^{GaN} and v_L^{SiC}). The phase velocity of the surface acoustic waves (v_p) is obtained from the resonance frequency (f_0) and λ ($v_p = f_0\lambda$). The v_p depends on the normalized film-thickness-to-wavelength ratio ($hk = 2\pi h/\lambda$), where k represents the wave number ($2\pi/\lambda$), and h is the thickness of GaN film. When hk is small, the SAW extends deeply into the substrate, causing v_p to approach the higher velocity value of the SiC. Conversely, with a large hk, the SAW predominantly resides within the film region, leading to v_p being constrained by the lower velocity value of the GaN overlayer.

A 2D FEM model is performed on the scale of one acoustic wavelength with periodic boundary conditions in COMSOL Multiphysics. The material parameters for GaN and SiC are based on reference [11] and COMSOL material library, respectively. The admittance characteristics are simulated, and the propagation modes are analyzed. The corresponding total displacements for the 2 μm SAW resonator are depicted in Figs. 3a-d,

respectively. The total displacements of the first resonance at 1.93 GHz are almost inside the GaN layer, corresponding to the R mode. The S modes with the v_p larger than that of R mode are detected at 2.51 GHz (S_1), 3.17 GHz (S_2), and 3.66 GHz (S_3). Fig. 3(e) shows the excellent agreement between experimental and simulated values of the v_p vs. hk, with a relative error of less than 3.1% across all wavelengths. The S modes having v_p higher than the R mode can be obtained. Meanwhile, the Pseudo-Bulk (PB) mode appears in the resonator with the λ of 4 μm.

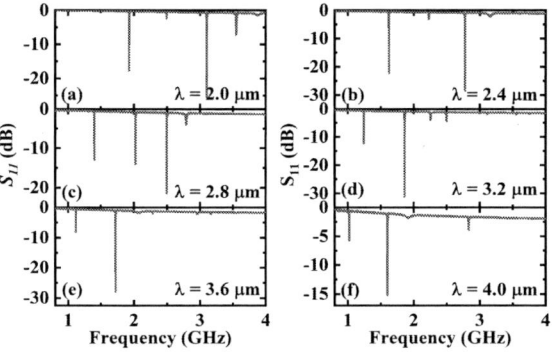

Fig. 2: Measured S_{11} for the one-port GaN/SiC SAW resonators with the λ from 2 μm to 4 μm at room temperature.

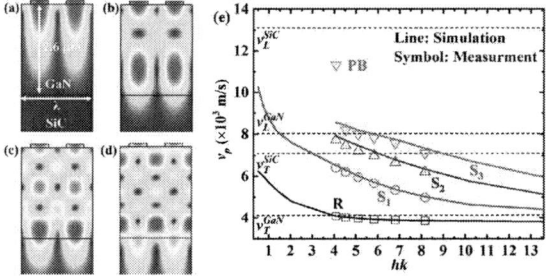

Fig. 3. FEM simulation results: total displacement of (a) R mode at 1.93 GHz, (b) S_1 mode at 2.51 GHz, (c) S_2 mode at 3.17 GHz, and (d) S_3 mode at 3.66 GHz, (e) Simulated and measured v_p vs. hk

The one-port SAW resonator with the λ of 2 μm has been characterized across a temperature range from 10 K to 500 K, presented in Fig. 4. The f_0 shifts with temperature (T) have been extracted, and the differences (Δf_0) related to the f_0 at 500 K have been calculated, as shown in Fig. 5a. Initially, the Δf_0 for all resonance modes increases linearly as the temperature decreases, then gradually exhibits nonlinearity and tends to saturate. The temperature coefficient of frequency (TCF) is defined as $1/f_0(T_0)\cdot\partial f_0/\partial T$. For the 300 K to 500 K temperature range, a linear fitting is performed to determine SEN and TCF. The sensitivity is higher for the S modes than the Rayleigh mode. However, due to the higher resonance frequency of S modes, the TCF is smaller than that of R mode. In addition, the admittance ratios (AR) for series and parallel resonance frequencies are shown in Fig. 5(b), indicating that a reduction in temperature can enhance the AR of various modes.

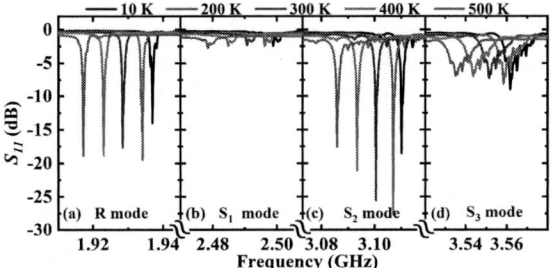

Fig. 4. The measured S_{11} of the one-port GaN/SiC SAW resonator with the λ of 2 μm from 10 K to 500 K.

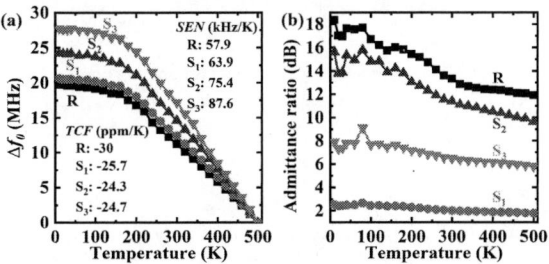

Fig. 5. The extracted (a) Δf_0 and (b) admittance ratio vs. temperature

A multi-mode mBVD model is developed to investigate the temperature influence on the GaN/SiC SAW resonator from 10 K to 500 K. The equivalent circuit is illustrated in Fig. 6(a). R_s is the series resistance of electrodes. The dielectric capacitance and resistance are represented by C_0 and R_0, respectively. The motional branches, comprising C_{mi}, L_{mi}, and R_{mi} (i, equal to 1 to 4, denotes R, S_1, S_2 and S_3, respectively), describe the mechanical resonances. A component L_s is introduced to consider the series inductance. Fig. 6(b) shows the admittance comparison between simulation and measurement results at 500 K, indicating the excellent accuracy of the multi-mode mBVD model.

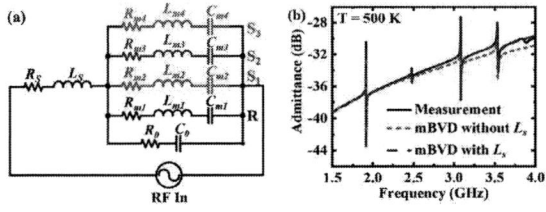

Fig. 6. (a) The equivalent circuit of the proposed multi-mode mBVD model and (b) the simulated admittance with and without L_s, and the measurement values are also plotted as a comparison.

The $Q_m = 1/R_{mi} \cdot \sqrt{L_{mi}/C_{mi}}$ is calculated using the extracted parameters. Fig. 7(a) demonstrates that the Q_m

value is significantly improved for all resonance modes as the temperature decreases. The results indicate high $f_0 \times Q_m$ values of 4.90×10^{12} Hz and 5.76×10^{12} Hz for R mode and S_2 mode, respectively, at 500 K. As the temperature decreases, the $f_0 \times Q_m$ values increase to $5. \times 10^{12}$ Hz and 5.79×10^{12} Hz at 300 K, and further to 6.21×10^{12} Hz and 9.71×10^{12} Hz at 10 K. Fig. 7(b) reveals that the prepared GaN/SiC SAW resonator exhibits excellent $f_0 \times Q_m$ values.

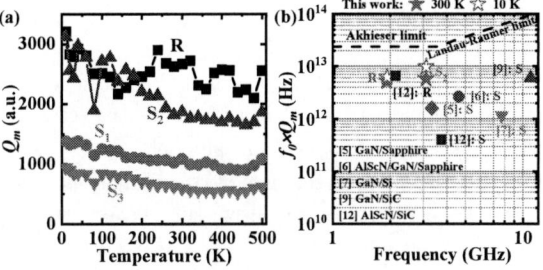

Fig. 7. (a) Q_m for various modes of the prepared GaN/SiC SAW resonator, and (b) comparison of $f_0 \times Q_m$ with recent works.

Conclusion

In this work, multi-mode SAW resonators were fabricated on a 4-inch GaN/SiC substrate, and an FEM model was constructed to identify the resonance modes. The results indicate that S mode on GaN/SiC provides a promising solution to improve $f_0 \times Q_m$. Furthermore, the behavior of the SAW resonator from 10 K to 500 K demonstrated enhanced performance with temperature decreasing. A multi-mode mBVD model was proposed to further analyze the temperature behavior.

Acknowledgments

This work is supported by the National Natural Science Foundation of China (Grant number: 62474101 and 92064002).

References

[1] R. Su, et al. IEEE Elec. Dev. Lett., 42(3): 438–441 (2021).

[2] A. Müller, et al. IEEE Elec. Dev. Lett., 31(12): 1398–1400 (2010).

[3] A. Qamar, et al. J. Micro. Syst., 29(5): 900–905 (2020).

[4] A. Ansari, et al. Dissertation, University of Michigan, 2016.

[5] M. M. Jaafar, et al. Appl. Phys. A, 125(11): 804 (2019).

[6] Y. Zheng, et al. 2022 IEEE MEMS, 1046–1049 (2022).

[7] A. Müller, et al. IEEE Elec. Dev. Lett. 36(12): 1299–1302 (2015).

[8] S. Valle, et al. Appl. Phys. Lett., 115(21): 212104 (2019).

[9] I. Ahmed, et al. IEEE TUFFC, 70(4): 291–301(2023).

[10] G. Yu, et al. IEEE EDTM, 1–3 (2023).

[11] M. Rais-Zadeh, et al. J. Micro. 23(6): 1252–1271(2014).

[12] V. J. Gokhale, et al. IEEE MEMS: 1262–1265(2020).

Enhanced ESD Protection Techniques for 10V Neurostimulator Circuits in 65nm CMOS Technology

Tanay Das[1], Naef Ahmad[2], Laxmeesha Somappa[2], Sandip Lashkare[1]

[1]Electrical Engineering, IIT Gandhinagar, 382355, India, [2]IIT Bombay, India

Abstract

A custom Electrostatic Discharge (ESD) protection circuit is essential for the reliability of 10-V neurostimulators implemented in a standard low-voltage CMOS process. The typical foundry-provided ESD diode cannot protect the ESD event without compromising on area. Here, we propose a custom ESD protection design in 65nm CMOS technology within the given area of 57 X 72 μm^2, limited to pad size. The dynamic resistance (R_{dyn}) of the proposed design is 3Ω which is $> 10X$ lower as compared to the foundry-provided design ($31.93\ \Omega$) providing substantially lower clamping voltage ($\sim 12V$) than the oxide breakdown limit ($16V$).

Keywords: Dynamic Resistance, Electrostatic Discharge

Introduction

In modern integrated circuits (ICs), the protection of electrostatic discharge (ESD) is crucial to preventing damage caused by high-voltage discharges, particularly in sensitive electronic components. This is especially crucial in high-voltage neurostimulator applications, where reliability, safety, and longevity are essential. Neurostimulators have high ESD sensitivity during operation and implantation since they interface directly with the human nervous system. ESD events can induce high transient currents that damage thin gate oxides and affect the junctions in CMOS devices damaging neurostimulator which are essentially the CMOS ICs [1], [2], [3]. Neurostimulators operating at high compliance voltages necessary for effective neural stimulation are at particular risk due to the high electric fields within the circuits [4]. Since these devices are implantable, robust ESD protection is required to ensure their long-term functionality and safety [5], [6], [7].

An overview of the neurostimulator with an ESD protection device is shown in Fig. 1 (a). Different devices, such as gate-grounded NMOS (GGNMOS) and diodes, can be used as ESD protection circuits. In the case of the neurostimulator, the operating voltage (V_{op}) is defined as $10V$. The failure voltage of the circuit ($V_{Failure}$), i.e., gate dielectric breakdown, is $16\ V$ for the 65nm CMOS technology. The ESD device must have the following properties to protect the core circuit from ESD events: (a) Breakdown voltage (V_{BD}) higher than V_{op} ($= 10V$), and (b) Clamping voltage (V_{Clamp}) less than the $V_{Failure}$ ($16V$). The ESD design window for the neurostimulator can be seen in Fig. 1 (b), where the red region indicates the IC failure region. During ESD events, if the ESD protection device has a higher clamping voltage ($>$ $16V$), then the gate oxide of the MOSFETs used for the

neurostimulator will be damaged. Hence, it is essential to limit the clamping voltage to $< 16V$. This is possible if the dynamic resistance of the ESD protection device is lower than 6Ω.

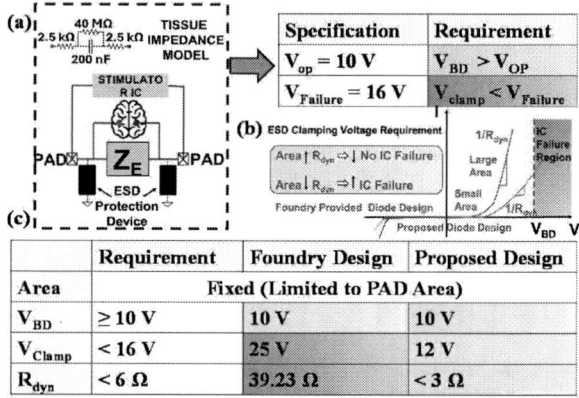

	Requirement	Foundry Design	Proposed Design
Area	**Fixed (Limited to PAD Area)**		
V_{BD}	$\geq 10\ V$	10 V	10 V
V_{Clamp}	$< 16\ V$	25 V	12 V
R_{dyn}	$< 6\ \Omega$	39.23 Ω	$< 3\ \Omega$

Figure 1: (a) This figure represents an ESD protection circuit for a brain-implanted neurostimulator. It includes a protection network for brain impedance (Z_E) positioned between two electrodes. The neurostimulator operates at 10V (V_{op}), with a failure voltage of 16V ($V_{Failure}$). The ESD device must have a breakdown voltage (V_{BD}) of 10V, and the clamp voltage (V_{Clamp}) must remain below the failure voltage. (b) Design window for ESD protection circuit (c) Table in the figure compares the foundry design with the proposed design specifications.

The conventional ESD protection techniques utilize an ESD detection circuit, control circuit, and clamp circuit, which is not area efficient (Fig. 2). Here, we propose an ESD protection circuit where two back-to-back diodes are used, which are area efficient without compromising ESD protection. The foundry-provided design provided for the ESD diode is for general purposes and cannot be used for all applications without area compromise. For example, the foundry-provided design (Red line) offers a R_{dyn} of $\sim 40\Omega$ with a given area budget (limited to pad size (57 X 75 μm^2). This leads to V_{Clamp} much greater than $16V$ making it

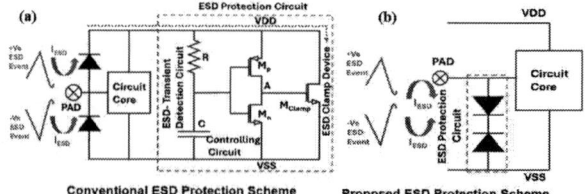

Figure 2: Illustrates two different electrostatic discharge (ESD) protection schemes. (a) The conventional ESD protection scheme is shown, which consists of an ESD detection circuit, a control unit, and an ESD clamp device. This setup typically occupies a significant amount of area. In contrast, (b) presents the proposed ESD protection scheme, which is more compact, utilizing only two back-to-back connected diodes.

unsuitable for neurostimulator application. In this work, we propose an ESD protection design (green line) that operates within the ESD design window, exhibiting R_{dyn} lower than 6Ω while having an area limited to pad size. The proposed ESD design is verified with transmission line pulse (TLP) simulations to evaluate the R_{dyn} and Human Body Model (HBM) stimulation to analyze and confirm the lower clamping voltage. Such a custom ESD solution is essential for area-constrained and high-voltage applications such as neurostimulators.

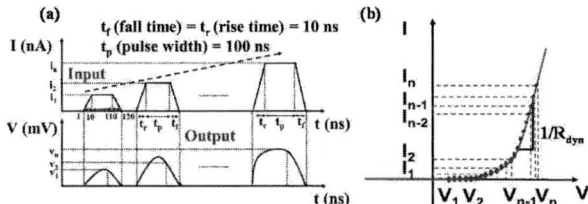

Figure 3: (a) TLP measurement for the design, current pulse of 100ns width is applied with 10ns rise and fall time and output average value of voltage is measured in the range of 80-100 ns (b) I-V plot to calculate the R_{dyn}

Methodology

The TLP simulation methodology used to evaluate R_{dyn} is shown in Fig. 3. A current pulse of increasing magnitude for a fixed pulse duration $(100ns)$ is applied to the ESD design. The clamped voltage across the diode is measured, and finally I-V plot is generated, leading to R_{dyn}.

First, The ESD protection circuit with a back-to-back diode configuration (with the foundry-provided ESD diodes pdio and ndio laid out side-by-side) is designed in a $65nm$ CMOS Technology as shown in Fig. 4(a). The P+ width of the pdio (P_p) is 25 μm, and the N+ width of the ndio (N_n) is fixed at 25 μm, resulting in a total layout area of 51 X 55 μm^2. The available PAD area for this ESD protection circuit is 57×72 μm^2. The R_{dyn} of the foundry design (Fig. 4(a)) is 31.93Ω, which exceeded the acceptable limits. The high R_{dyn} results in higher clamping voltage, which may lead to oxide breakdown during ESD events, which makes this foundry-provided device unsuitable for ESD protection. To improve the R_{dyn}, a new design of a back-to-back diode is proposed where the pdio is enclosed by the ndio instead of being placed side by side. For this design, the P+ width of the pdio (P_p) is fixed at 5 μm, while both the N+ width of the Pdio (N_p) and N+ width of the Ndio (N_n) are set to 4 μm. This modified design occupied an area of 33 X 33 μm^2 significantly smaller than the conventional design area. Furthermore, this new configuration yielded a R_{dyn} of 8.5 Ω, representing a significant improvement over the foundry-provided design. To further reduce the R_{dyn} while maintaining the structural integrity of the ESD protection, four diodes are connected in parallel, where each back-to-back diode design is represented in Fig 4(b). As four back-to-back diodes are connected in parallel, this structure distributes the current more evenly, thus reducing the total R_{dyn}. The layout of the four-diode design is shown in Fig. 4(c). The four-diode parallel structure

resulted in a R_{dyn} of 2.1 Ω, with a total design area of 66 X

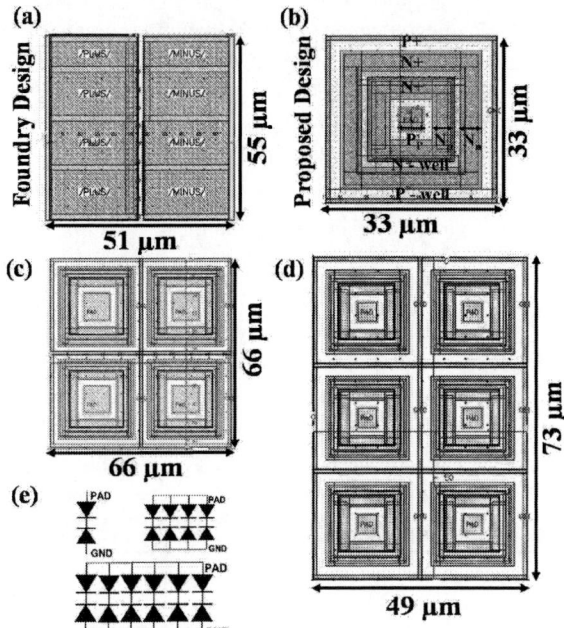

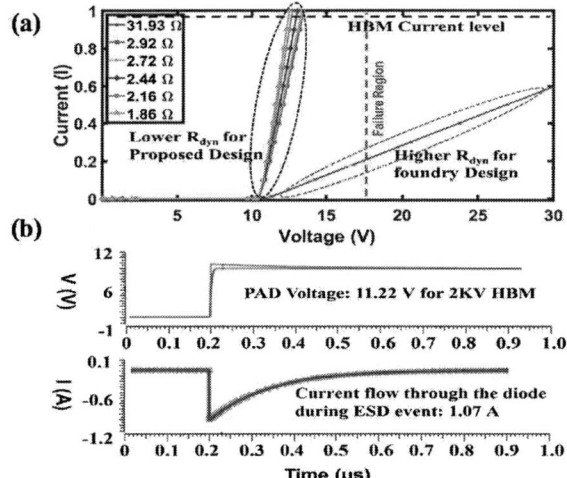

Figure 4: (a) Design of back-to-back diode provided by the foundry where Ndio and Pdio placed side by side (b) Proposed design of two diodes connected in a back-to-back configuration, design parameters are P_p, N_p, N_n (c) Design of four diodes connected in parallel, utilizing the design presented in Fig 4(a). (d) Proposed design of six parallel connected diode but asymmetric structure results in uneven current distribution (e) Schematic representation of Fig. 4(a), 4(c) and 4(d)

66 μm^2 However, this design exceeded the available PAD area of 57 X 72 μm^2; To ensure that the design fit within the pad area, the width of N_n was reduced from 4 μm to 2 μm. This modification resulted in a 57×57 μm^2 design area which is within the allowable pad dimensions. This reduction in the width increases R_{dyn} to 2.9 Ω (still much lower than the 6Ω). To optimize the design further in terms of both area and resistance, the width of N_p is reduced to 2 μm, while the width of N_n is restored to 4 μm. The width of P_p is maintained at 5

Figure 5: (a) Represents the TLP plot comparing the foundry-provided design, and the proposed design (b) represents the voltage and current level for ESD event at 2KV HBM.

979-8-3315-0417-5/25 $31.00 © 2025 IEEE

μm. This configuration resulted in an R_{dyn} of 2.4 Ω, with the design area remaining 57×57 μm^2 Since the P+ region of the Pdio is directly connected to the pad, it is susceptible to high current densities during ESD events. To reduce the current density, the P+ of the pdio (P_p) area is increased to 8 X 8 μm^2 from 4 X 4 μm^2 while the width of the N_p and N_n is maintained at 2 μm. This modified four-diode parallel structure results in a R_{dyn} of 2.72 Ω, with a total area of 52 $\times$ 52 μm^2, within the desired area limit of 57 X 75 μm^2. This design provided an optimal balance between low R_{dyn} and current distribution. To further reduce R_{dyn}, a six-diode configuration was designed by connecting six diodes in parallel, represented in Fig.4(d). This configuration further distributed the ESD current across more diodes, thus lowering the resistance, which resulted in R_{dyn} of 1.86 Ω. The total area of this design is 49×73 μm^2, which fits under the PAD. However, the asymmetricity of the structure can

cause uneven current distribution, which affects the overall reliability of the ESD device. Fig 5(a) represents the TLP plot for the foundry-provided and proposed designs. The Foundry-provided design provides larger R_{dyn} as compared to current flow through the diode during the ESD event is significantly lower than the breakdown limit of $16V$.

Results and Discussion
Table 1 compares the foundry design and various proposed designs of diodes, focusing on parameters such as R_{dyn} and design area. The foundry design exhibits a significantly higher R_{dyn}, i.e. 31.93Ω, with a design area of 51×55 μm^2. At the same time, the proposed designs with four parallel diodes show substantially reduced R_{dyn}. For instance, the proposed design with an N+ width of 4 µm achieves R_{dyn} of 2.1Ω, representing a 93.42% reduction in R_{dyn} compared to the foundry design. However, due to the larger area of

Table 1:   Comparison table for   Foundry provided design and proposed design.

	Width (μm)			PAD Area Restriction (μm^2)	ESD Diode Area (μm^2)	R_{dyn} (Ω)	V_{clamp}	Current Flow
	N_p	N_n	P_p					
Foundry Design				57 X 75	51 X 55	31.93	25 V	Non-uniform (One-direction)
Proposed Designs								
Uniform N_p, N_n, P_P (4D)	4	4	5	57 X 75	66 X 66	2.1	12 V	Uniform
Small N_n (4D)	4	2	5	57 X 75	57 X 57	2.9	12 V	Uniform
Small N_p (4D)	2	4	5	57 X 75	57 X 57	2.4	12 V	Uniform
Large P_p (4D)	2	2	8	57 X 75	52 X 52	2.72	12 V	Uniform
Uniform N_p, N_n, P_p (6D)	2	2	5	57 X 75	49 X 73	1.86	12 V	Non-uniform (Pad to diode)

66×66 μm^2, this design does not fit within the PAD area. Other proposed designs fit under the PAD with smaller design areas, improving the current density but compromising with R_{dyn}. These designs highlight the trade-off between R_{dyn} and the design area to fit under the PAD.

Conclusion
In conclusion, a custom ESD design is proposed for efficient ESD protection without compromising on the area. The proposed design provides significantly lower clamping voltage due to its lower dynamic resistance ($<3\Omega$), which is $> 10X$ lower than the foundry-provided design. Such an ESD protection solution is critical development considering the low area requirements for applications such as neurostimulators.

References
[1]   E. Noorsal, K. Sooksood, H. Xu, R. Hornig, J. Becker, and M. Ortmanns, "A neural stimulator frontend with high-voltage compliance and programmable pulse shape for epiretinal implants," *IEEE J Solid-State Circuits*, vol. 47, no. 1, pp. 244–256, Jan. 2012, doi: 10.1109/JSSC.2011.2164667.

[2]   I. Williams and T. G. Constandinou, "An energy-efficient, dynamic voltage scaling neural stimulator for a proprioceptive prosthesis," *IEEE Trans Biomed Circuits Syst*, vol. 7, no. 2, pp. 129–139, 2013, doi: 10.1109/TBCAS.2013.2256906.

[3]   Z. Luo and M. D. Ker, "A High-Voltage-Tolerant and Precise Charge-Balanced Neuro-Stimulator in Low Voltage CMOS Process," *IEEE Trans Biomed Circuits Syst*, vol. 10, no. 6, pp. 1087–1099, Dec. 2016, doi: 10.1109/TBCAS.2015.2512443.

[4]   J. A. Salcedo, J. J. Liou, and J. C. Bernier, "Design and integration of novel SCR-based devices for ESD protection in CMOS/BiCMOS technologies," *IEEE Trans Electron Devices*, vol. 52, no. 12, pp. 2682–2689, Dec. 2005, doi: 10.1109/TED.2005.859662.

[5]   S. Martinez, F. Veirano, T. G. Constandinou, and F. Silveira, "Trends in Volumetric-Energy Efficiency of Implantable Neurostimulators: A Review From a Circuits and Systems Perspective," *IEEE Trans Biomed Circuits Syst*, vol. 17, no. 1, pp. 2–20, Feb. 2023, doi: 10.1109/TBCAS.2022.3228895.

[6]   K. Bazaka and M. V. Jacob, "Implantable devices: Issues and challenges," Dec. 21, 2012, *MDPI*. doi: 10.3390/electronics2010001.

[7]   M. Katebi, A. Erfanian, M. Azim Karami, and M. Sawan, "Challenges and Trends of Implantable Functional Electrical Neural Stimulators: System Architecture and Parameters," *IEEE Access*, vol. 12, pp. 103203–103236, 2024, doi: 10.1109/ACCESS.2024.3432611.

A novel transistor free design of SOT-MRAM written by unipolar current with record bit cell size ($15F^2$)

Meiyin Yang[1,2,3], Lei Zhao[4], Bowen Yang[1,2,3], Bowen Shen[1,2,3], Yanru Li[1,2,3], Peiyue Yu[1,2,3], Jianfeng Gao[1,2], Ruipeng Shi[5], Zhuangzhuang Ye[1,2,3], Shuo Xu[1,2,3], Yan Cui[1,2,3], Xiaolei Yang[4], Ming Wang[4], Shikun He[4], Kaiming Cai[5] and Jun Luo[1,2,3]

1. Key Laboratory of Fabrication Technologies for Integrated Circuits, Chinese Academy of Sciences, Beijing 100029;2. Institute of Microelectronics, Chinese Academy of Sciences, Beijing 100029, China;3. School of Integrated Circuits, University of Chinese Academy of Sciences, Beijing 100049, China; 4. Zhejiang Hikstor Technology Co. LTD., Zhejiang 311305, China; 5. Huazhong University of Science and Technology, Wuhan 430074, China. *Email: luojun@ime.ac.cn, kmcai@hust.edu.cn

Abstract— Spin-orbit torque (SOT)-MRAM bit cell takes up a large area (~$75F^2$) due to the two access transistors architecture. To address this problem, our innovative spin-orbit torque magnetic tunnel junction (SOT-MTJ) device, which operates through unipolar current writing, facilitates the replacement of transistors with diodes in a single bit cell (2D-1M), resulting in a remarkably compact size of ~$24F^2$. In addition, we designed a 5D-4M structure to further reduce the area to $15F^2$. Four MTJs of the 5D-4M structure are on a shared SOT track and written simultaneously by unipolar current due to voltage controlled magnetic anisotropy (VCMA) effect, leading to energy consumption reduction of 37.5%.

Introduction. SOT-MRAM is very advantageous in cache and logic-in-memory computation due to its fast speed, high endurance compared with STT-MRAM. However, this three-terminal device requires two access transistors per bit-cell to perform the read and write operation as shown in Fig.1(a) [1], which results in area overhead ($75F^2$) and limits its application in high density memory. Many works focusing on minimizing the SOT-MRAM bit-cell area have been reported. For example, multi-MTJs can be arranged on the same bottom electrode (BE) to share the writing transistor (Fig.1) [2-4]. However, this approach increases the energy consumption during write operations due to the higher resistance of loner SOT track. Until now, the high integration capability of SOT-MRAM without increasing energy consumption is still under debate.

Here, we propose a SOT-MRAM bit cell with two diodes instead of two transistors to reduce the cell size to $24F^2$ (2D-1M in Fig.2). This is achieved by novel SOT-MTJ devices which are written by unipolar current. Furthermore, we designed a 5D-4M multi-bits structure further reduced the cell size to $15F^2$ with decreased energy consumption by 37.5%.

The transistor-free SOT-MTJ device. The TEM image of the SOT-MTJ device is shown in Fig. 3(a) with the hysteresis loop in Fig. 3(b). In Fig.3(c), we show a typical SOT switching loop by bipolar currents under B_x=18 mT. As shown in Fig. 3(d), the switching can also be obtained with the pulse width down to 0.8 ns. The bipolar and ultrafast switching behavior verifies the contribution of SOT effect. Interestingly, the MTJ could also be switched by a unipolar current without in-plane B_x, as shown in Fig.4(a). For a low current density J_{SOT} (indicated as "Hold" zone), the state of MTJ will not be changed, in which the MTJ could be read out safely. For an intermediate J_{SOT} (P zone), the

P state is preferred, and for higher J_{SOT} (AP zone), the AP state is preferred. Pulse sequences with high and low amplitude are applied to test the switching probability (P_{sw}) of AP to P and P to AP. Three distinct regions are obtained for specific operations, as shown in Fig.4(b). Moreover, Fig.4(c) has demonstrated the reliable reading and writing using unipolar current. Because of the unipolar writing behavior, diodes can be used as the selector to replace transistors as shown in Fig.5.

Mechanism. We proposed a model to explain the unipolar writing phenomenon as shown in Fig. 6. The coupling energy ΔE of FL/MgO/RL structure with interface roughness can be tuned from negative to positive by the K_u, and can even change sign (Fig. 6(b)). This indicates that both FM and AFM coupling can be realized by varied K_u. Experiments shown that the K_u of FL can be decreased by rising temperature (Fig.6(c)). We can utilize the current pulses to increase the temperature of the SOT-MTJ as simulated in Fig. 7. To prove this assumption, we fabricated SOT-MTJ devices with different interface roughness and measured the hysteresis loops in Fig.8(a). B_{offset} extracted from Fig.8(a) first increases and then decreases as the temperature is rising in Fig.8(b) for large roughness device (large h). B_{offset} monotonously decrease with the temperature, which is due to the reduced stray field from the SAF. Thus, the manipulation of coupling field is due to the interface roughness, which is consistent with the model in Fig. 6. The larger (smaller) B_{offset} means the magnetization of FL prefers (does not prefers) to align with the RL. Thus, the temperature can vary the coupling field in the SOT-MTJ. As a result, the final state of the SOT-MTJ is decided due to the coupling strength.

Multi-cell operations by VCMA. Next, we evaluate the impact of gate voltage (V_G) for VCMA-induced B_C and B_K modulations. We extracted the B_C and B_K from hysteresis loops of MTJs with different V_G in Fig.9(a). The V_G lowers the B_C and B_K of the FL linearly in Fig.9(b) and (c). The V_G also changed the critical switching current density J_{sw} as shown in Fig.9. The P_{sw} of P to AP operation with different V_G is presented in Fig.10(a). The P_{sw} curve shifted to left (right) at -V_G (+V_G), leaving a window for both AP and P states at the same current pulse. A change of ±12.5% of J_{sw} can be obtained for V_G=±0.8V in Fig.10(b). As shown in Fig.11(a), the R-J_{SOT} curves shows different switching current density under different V_G due to the VCMA effect. The window enables the independent operations of AP and P states in different MTJs simultaneously by the same current density with the selection

979-8-3315-0417-5/25 $31.00 © 2025 IEEE

by V_G. Therefore, an architecture with multiple cells on a shared SOT track is also expected. In Fig.11(b), we exhibit the connection of four SOT-MTJ devices with a write diode and four read diodes (5D-4M). The operation of multi-cell writing of the 5D-4M are shown in Fig.11(c). Four MTJs can be written at the same current pulse by VCMA, which can significantly increase the density and reduce the energy consumption.

The layout of 2T-1M, 2D-1M and 5D-4M SOT-MRAM cells were designed in Fig.12(a), (b) and (c). Table I lists the simulation results of different bit-cells. Compared with the 5T-4M SOT-MRAM, the area of 2D-1M and 5D-4M cell is reduced from 0.0101μm² to 0.004μm² and 0.003μm². The write energy per bit of the 5D-4M is lowered down by 75% compared with 5T-4MTJ structure due to the 4 cells writing at one sequence.

Conclusion. We propose a transistor-free SOT-MTJ device written by unipolar current to significantly reduce the bit cell area to 24F². On the top of that, 5D-4M structure was designed, which can further reduce the area to 15F². Adopting the VCMA to the 5D-4M design, multi-cell writing of SOT-MRAM structure leads to reduced energy consumption. Compared to the 5T-4M SOT-MRAM structure, the area per bit and energy consumption have been reduced by 70.3% and 75%, respectively. Our work provides a promising solution to solve the density issue for SOT-MRAM.

Reference
[1] M. Gupta et al, IEEE IEDM, 24.5.1-24.5.4 (2020) [2] H. Wang et al., IEEE TED, 68, 4494 (2021). [3] J. Doevenspeck et al., IEEE VLSI, 1-2(2020). [4] M. Gupta et al, IEEE IEDM, 1-4 (2023). [5] K. Cai et al, IEEE VLSI, 375-376 (2022).

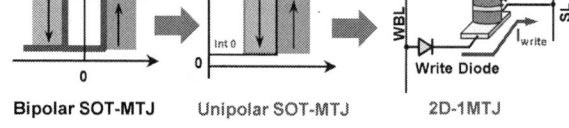

Fig. 1: Schematics of conventional bipolar SOT-MRAM unit cell with 2T-1MTJ and multi-bits on a shared SOT track as 2T-4MTJ, the averaged cell size can be reduced.

Fig. 2: The proposed unipolar SOT-MTJ, which can enable transistor free design of 2D-1M structure.

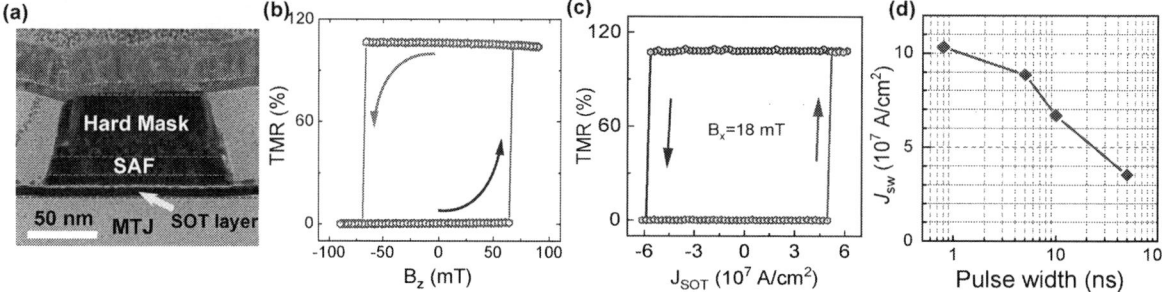

Fig. 3: (a) TEM image of an 100nm SOT-MTJ. (b) Hysteresis loop of the SOT-MTJ. (c)SOT current-induced switching under magnetic field B_x=18mT. (d) Critical current density J_{sw} as a function of pulse width.

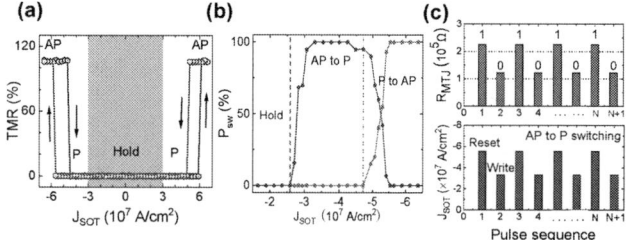

Fig. 4: (a) Switching curve without in-plane magnetic field. AP(P) state can be written by current density at AP(P) zone. The hold zone at low J can read the MTJ safely. (b) Switching probability P_{sw} of AP to P and P to AP operations by unipolar current pulses. (c)The writing of SOT-MTJ by cycled unipolar current pulse sequences.

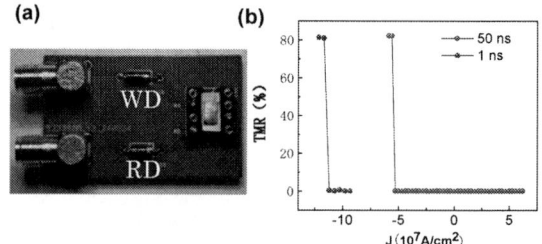

Fig. 5: (a) (b) PCB test system of the 2D1M structure. Two transistors are connected to the SOT-MTJ. (b) Current induced switching curve of the 2D1M structure.

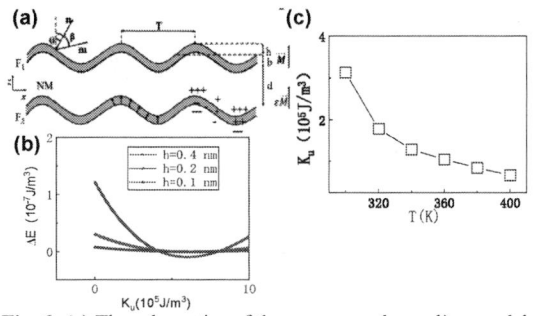

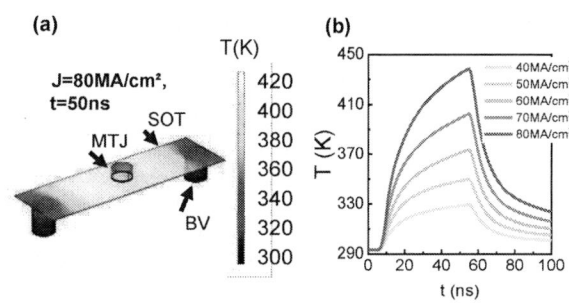

Fig. 6: (a) The schematics of the orange peel coupling model. (b) the coupling energy difference(AF-AFM) vs. Ku. (c) the dependence of Ku on temperature.

Fig. 7: (a) Simulation of a self-heated model of unipolar SOT-MTJ operated by a current pulse with J=80 MA/cm² at 50 ns. (b) Simulation curves of device temperature versus pulse time.

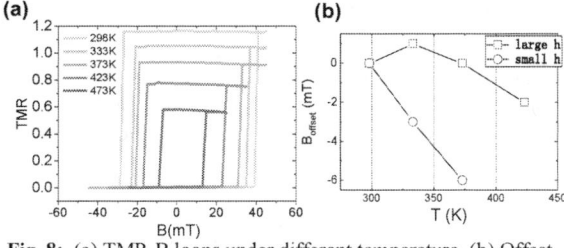

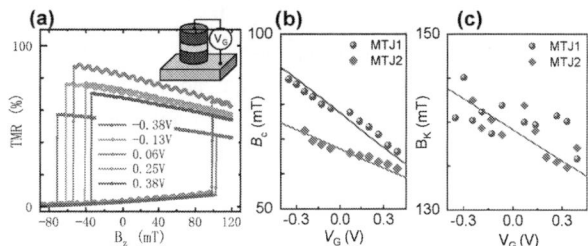

Fig. 8: (a) TMR-B loops under different temperature. (b) Offset field vs. temperature abtracted from (a).

Fig. 9: (a) TMR-B loops under voltage. (b) & (c) B_c and B_k vs. V_G of two MTJs.

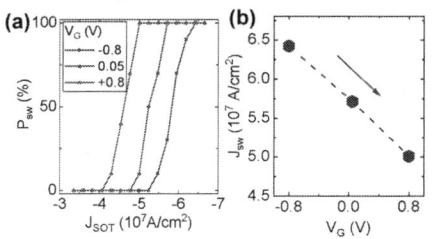

Fig. 10: (a) P_{sw} of P to AP switching with different V_G. (b) J_{sw} (@P_{sw} =50%) for different V_G.

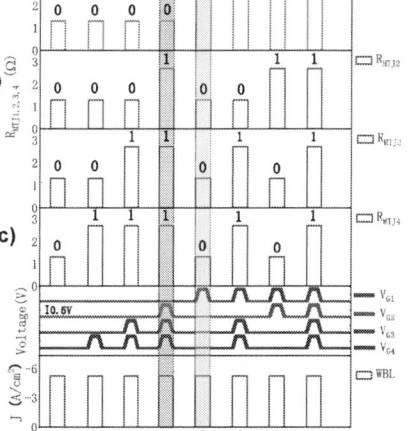

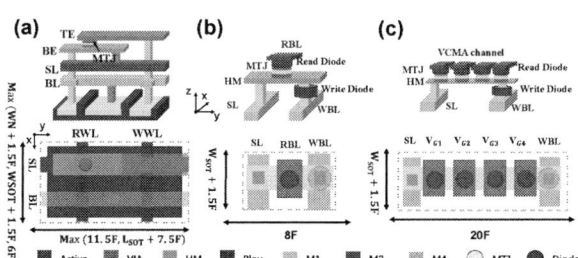

Fig. 11: (a) R_{MTJ}-J_{SOT} loops of MTJ by different V_G (b) The PCB of 5D-4M structure. (c) Experimental demonstration of simultaneous writing of the 5D-4M structure.

Fig. 12: layout of (a) 2T-1M (b)2D-1M (c)5D-4M SOT-MRAM.

Table 1. Benchmark table of different SOT-MRAM architectures.

SOT-MRAM	2T-1MTJ [5]	5T-4MTJ [5]	2D-1M [This work]	5D-4M [This work]
Parallel Write(0/1)	0	0	0	1
metal lines/bit	5	2.75	3	1.75
Transistors/cell	2	5	0	0
Write energy (a.u.)	1*	2.5	1*	0.625
bits/sequence(write)	1	4	1	4
Area/bit(um²)	0.0162 (75F²)	0.0101 (47F²)	0.004 (24F²)	0.003 (15F²)

- 1* represents the energy value for writing a 2T1M or 2D1M cell.
- F=14nm

A Physics-based Compact Model for Ambipolar Schottky-barrier CNTFETs

Rui Zhan[1], Bin Zhou[1], Zilin Teng[1], Yiheng Xue[1], Panpan Zhang[1*], Jianhua Jiang[2*]

[1]School of Integrated Circuits, Beijing University of Posts and Telecommunications，
Beijing 100876, China.

[2] Key Laboratory for the Physics and Chemistry of Nanodevices and Center for Carbon-based Electronics, School of Electronics, Peking University,
Beijing 100871, China

(*email：tanji_ic@bupt.edu.cn; jhjiang@pku.edu.cn)

Abstract

The presence of Schottky-barrier (SB) at the contact of carbon nanotube field-effect transistors (CNTFETs) is crucial in determining the transport properties. Here, by solving the Landauer equation, we can arrive at a physics-based compact model for ballistic CNTFETs that incorporates the thermionic emission and tunneling current, which is also capable of capturing the inherent ambipolar behavior of SBFETs. The excellent agreement between simulated and measured results of the fabricated 5-nm CNTFET confirms the validity of the model.

Keywords: Carbon nanotube field-effect transistor (CNTFET), Schottky-barrier (SB), Compact model, Ambipolar, TCAD.

Introduction

Carbon nanotubes (CNTs) hold great promise of extending Moore's law as alternative channel materials for high-performant transistors due to their remarkable electrostatic control over the channel and excellent immunity to short channel effects (SCEs). A physics-based compact model for CNTFETs is of great value to allow performance assessment at the application level. Extensive efforts have been made to develop physics-based and computation-efficient models, among which the virtual-source (VS) model released by Stanford University excels [3] [4]. However, the VS model failed to capture the inherent Schottky barrier contact properties of CNTFETs, such as ambipolar behavior.

In this work, we demonstrate a closed-form analytical solution of the Landauer equation (n-type CNTFETs are discussed throughout this paper), taking into account the specific SB shape and its dependency on the 2-D electrostatics. The surface potential extracted from a data-calibrated TCAD model (Sentaurus used here) was fed into the determination of tunneling probability during the derivation. Moreover, an explicit current model for holes is proposed to account for the ambipolar behavior, rather than merely mirroring the other branch [6]. Finally, the compact model shows excellent agreement with the fabricated 5-nm CNTFET [3], confirming its effectiveness.

The manuscript is organized as follows. We first present a general modeling framework for ballistic CNTFETs based on the Landauer equation, including the specific Schottky barrier potential energy shape at the contact. Note that the model is capable of dealing with electrons hoping over and tunneling through the SB in and above the subthreshold region respectively. Following, emphasis of this paper is focused on ambipolar transport and the determination of surface potential due to 2-D electrostatics. Finally, we demonstrate the accuracy of the model by comparing it with the fabricated 5-nm ballistic CNTFET.

Device Modeling

Fig. 1(a) illustrates the CNTFET device structure we investigate in this work, where single semiconducting carbon nanotube serves as the conducting channel. The schottky barrier arising from metal-CNT contacts allows electrons to hop over or tunneling through, yielding thermionic emission and tunneling current component respectively. Generally, our modeling approach starts with deriving the Landauer equation (see Eq. 1) to describe the ballistic transport of CNTFETs inspired by the reported work [1]. The key knob for this approach is figuring out the energy-dependent transmission factor (T(W)) with regard to the particular SB shape at the source as denoted in Fig. 1(b). Fig. 2 elaborates the general flow for the modeling setup, which highlights our main contributions: (1) First, we modify the expression for the surface potential to include its dependence on the drain voltage (calibrated with TCAD simulation), which will significantly penetrate into the source barrier for sub-10 nm channel length. (2) Second, we introduce a flat-band voltage parameter V_{FBP} for holes instead of the previous V_{FB}, as well as a hole barrier correction factor h_B, to accurately capture the ambipolar behavior, rather than merely mirroring the other branch [6]. Due to space limitations, more details can be found in Ref [1]. Following emphasis is focused on elucidating the above-mentioned fresh features. The Landauer equation reads as follows:

979-8-3315-0417-5/25 $31.00 © 2025 IEEE

$$I = \frac{4q}{h} \int T(W)[f_S(W) - f_D(W)]dW \qquad (1)$$

where q represents the elementary charge, and h denotes Planck's constant, I represents the current, W is the carrier energy, $T(W)$ is the energy-dependent transmission probability, and $f_S(W)$ and $f_D(W)$ are the Fermi distributions of the source and drain electrodes, respectively.

(a)

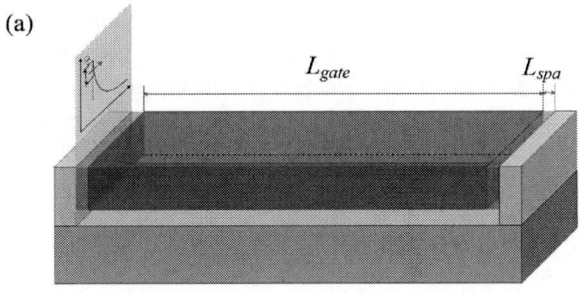

(b)

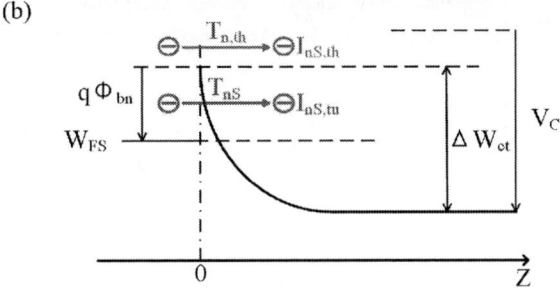

Fig.1 (a) Illustration of the device structure based on single carbon nanotube with the presence of Schottky barrier at the source and drain. (b) Schematic of band diagram at the source of the channel visualizing relevant current components across the SB. $I_{nS,th}$ and $I_{nS,tu}$ are the thermionic current and tunneling current respectively.

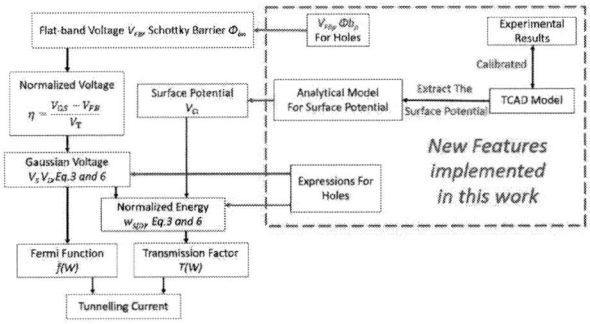

Fig.2 The general flow of the CNTFET modeling with highlighting new features implemented in this work.

A. Ambipolar Transport

To capture the inherent ambipolar transport properties in SBFETs, we introduce a flat-band voltage V_{FBP} instead of V_{FB} to determine the SB height and width, the conduction band edge profile and thus tunnelling probability for holes. This approach yields expressions for both thermionic emission and tunnelling currents of holes. Note that the tunnelling current of holes is split into source-side and drain-side components respectively.

For hole current, the drain is treated as the effective source, and V_{DS} is excluded from the drain-side hole current expression. For the source-side hole current, however, the V_{DS} bias is included, resulting in a form similar to the electron drain-side current. The derived drain expressions are as follows:

$$\frac{\Delta W_{CtD}}{q} = \Delta V_{CtD} = -V_{Ct} + \Phi_{bp} \qquad (2)$$

$$w_D = \frac{V_{Ct} - V_D}{\Delta V_{CtD}} \qquad (3)$$

$$\Phi_{bp} = h_B(E_g - \Phi_{bn}) \qquad (4)$$

the hole at the source:

$$\frac{\Delta W_{Ct}}{q} = \Delta V_{Ct} = -V_{Ct} + \Phi_{bp} - V_{DS} \qquad (5)$$

$$w_S = \frac{V_{Ct} - V_S - V_{DS}}{\Delta V_{Ct}} \qquad (6)$$

ΔW_{Ct} and ΔW_{CtD} represents the difference in conduction band edge energy for source and drain, V_{Ct} represents the device surface potential, $V_{S(D)}$ denotes the potential form of the peak energy of the Gaussian function for source and drain, as presented in [1], Φ_{bn} and Φ_{bp} represents the electron and hole barrier height, $w_{S(D)}$ are the normalized energy for source and drain, h_B is the hole barrier correction factor and E_g is the bandgap energy. The modified expression for the total intrinsic current reads as follows:

$$I_{DS} = (I_{nS} - I_{nD}) + \text{An}(I_{pD} - I_{pS}) \qquad (7)$$

with An denoting the hole current suppression factor.

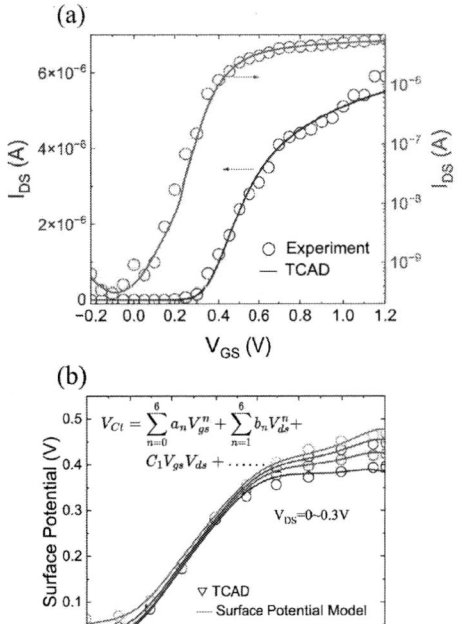

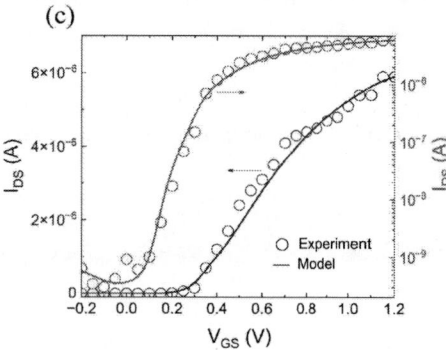
(c)

W_s	3.8	eV	source workfunction
W_d	3.8	eV	drain workfunction
W_g	3.95	eV	gate workfunction

Conclusion

The physics-based compact model presented here captures the inherent ambipolar characteristics of ballistic CNTFETs with the presence of SB at the contact. Furthermore, by taking into account the surface potential dependence on both the gate and drain voltages, the model shows excellent agreement with the fabricated 5-nm CNTFETs. This work provides a general modelling framework for SBFETs and allows the performance evaluation and design space exploration of CNTFETs in the application level.

References

[1] M. Annamalai and M. Schröter, "A Physics-Based Compact Model for the Static Drain Current in Heterojunction Barrier CNTFETs—Part I: Barrier-Related Current," in *IEEE Transactions on Electron Devices*, vol. 71, no. 1, pp. 23-29, Jan. 2024, doi: 10.1109/TED.2023.3327030.

[2] M. Annamalai and M. Schröter, "A Physics-Based Compact Model for the Static Drain Current in Heterojunction Barrier CNTFETs—Part II: Scattering, High-Field Effects, and Model Verification," in *IEEE Transactions on Electron Devices*, vol. 71, no. 1, pp. 30-36, Jan. 2024, doi: 10.1109/TED.2023.3327040.

[3] Chenguang Qiu *et al.*, Scaling carbon nanotube complementary transistors to 5-nm gate lengths. *Science* **355**, 271-276(2017). DOI: 10.1126/science.aaj1628.

[4] C. -S. Lee, E. Pop, A. D. Franklin, W. Haensch and H. -S. P. Wong, "A Compact Virtual-Source Model for Carbon Nanotube FETs in the Sub-10-nm Regime—Part I: Intrinsic Elements," in *IEEE Transactions on Electron Devices*, vol. 62, no. 9, pp. 3061-3069, Sept. 2015, doi: 10.1109/TED.2015.2457453.

[5] C. -S. Lee, E. Pop, A. D. Franklin, W. Haensch and H. -S. P. Wong, "A Compact Virtual-Source Model for Carbon Nanotube FETs in the Sub-10-nm Regime — Part II: Extrinsic Elements, Performance Assessment, and Design Optimization," in IEEE Transactions on Electron Devices, vol. 62, no. 9, pp. 3070-3078, Sept. 2015, doi: 10.1109/TED.2015.2457424.

[6] I. Bejenari, M. Schröter and M. Claus, "Analytical Drain Current Model of 1-D Ballistic Schottky-Barrier Transistors," in *IEEE Transactions on Electron Devices*, vol. 64, no. 9, pp. 3904-3911, Sept. 2017, doi: 10.1109/TED.2017.2721540.

Fig.3 (a) Comparison of transfer curves at V_{DS}=0.1 V from TCAD simulated (solid lines) and measured results (symbols) for a 5-nm CNTFET. (b) Comparison of surface potential (versus gate voltage) from the calibrated TCAD model (symbols) and the proposed analytical model (solid lines) at V_{DS}=0, 0.1, 0.2, 0.3V. (c) Comparison of transfer curves at V_{DS}=0.1 V from measured results and the proposed current model in this work for a 5-nm CNTFET.

B. Surface Potential Extraction

For ultra-scaled transistors operating in the ballistic domain, the source barrier is much more sensitive to the drain electric field, which cannot be ignored in shaping the SB profile. Therefore, 2-D electrostatics must be involved when determining the surface potential.

A TCAD model was first developed for a 5-nm CNTFET with tailoring the structural and material parameters (see Table I) to calibrate with the measured transfer curves. Good agreement can be reached as shown in Fig. 3(a). With the establishment of the baseline, the surface potential along the tube can be readily extracted as shown in Fig. 3(b). An explicit surface potential model is then setup (see Eq in Fig. 3(b)) to capture both the gate and drain voltage dependence. Following, the sophisticated surface potential model is fed into the calculation of transmission probability ($V_{Ct(D)}$, $\Delta V_{Ct(D)}$, $w_{S(D)}$), which finally yields a current model that coincides well with the measured characteristics for the fabricated 5 nm CNTFET as shown in Fig. 3(c).

Table 1: Material parameters implemented in the TCAD model for the 5-nm CNTFET.

Parameters	Value	Unit	Description
d_{CNT}	1.49	nm	CNT diameter
E_g	0.58	eV	Bandgap of CNT
E_{cnt}	3.7	eV	electron affinity of CNT
DOS	1.5×10^{10}	$eV^{-1}cm^{-2}$	Density of states
t_{ox}	3	nm	oxide layer thickness
ε_{ox}	22	1	HfO$_2$ dielectric constant
Φ_{bn}	0.1	eV	barrier height

979-8-3315-0417-5/25 $31.00 © 2025 IEEE

Wavelength-dependent reconfigurable photo memory enabled by organic-gated transistor

Xiaokun Guo[1], Yaoqiang Zhou[1,2*], Jianbin Xu[1*]

[1] Department of Electronic Engineering and Materials Science and Technology Research Center, The Chinese University of Hong Kong, Hong Kong SAR, China

[2] Department of Micro- and Nanosciences, Aalto University, Finland

*Corresponding author: yaoqiang.zhou@aalto.fi, jbxu@ee.cuhk.edu.hk

Abstract

We presented a reconfigurable two-dimensional (2D) phototransistor, fabricated by integrating a J-aggregate PTCDI-C_{13} charge-trapping. This phototransistor exhibited switchable volatile and non-volatile photo response behaviors, controlled by the tunable drain voltage (V_{ds}). At a negative V_{ds}, the device operates as a photodetector with a volatile photo response, while at positive V_{ds}, the photo transistor served as a photo memory achieving a high resistance ratio of approximately 10^3. The device demonstrates selectively retaining memory within a narrow wavelength, consistent with the absorption peak of the monolayer J-aggregate PTCDI-C_{13}, showing the potential for wavelength-specific, reconfigurable photo-memory applications.

Keywords: Reconfigurable photo-memory, Wavelength-dependent, 2D materials, J-aggregate.

Introduction

Reconfigurable phototransistor memory is gaining significant interest for adaptive visuomorphic computing, as it integrates efficient sensing, memory, and processing capabilities into a single device [1-3]. However, achieving reconfigurable phototransistor memory remains challenging, primarily due to the difficulty of implementing an electrically controlled, real-time swiching between volatile and non-volatile photo responses [4]. Schottky transistors based on 2D metal and semiconductors, due to their dopant-free nature, possess real-time reconfigurability [5], making them promising candidates for constructing reconfigurable optoelectronic memory devices [6]. By integrating 2D organic thin films with photosensitive properties into reconfigurable 2D transistor devices, reconfigurable phototransistors can exhibit wavelength-dependent optoelectronic responses [7]. For example, utilizing photoactive materials with narrowband absorption is a highly attractive strategy for filter-free narrowband photo memory [8], which is benefited for applications in spectral sensing and imaging.

In this work, we presented a reconfigurable phototransistor, fabricated by combing the PdSe$_2$-contacted WSe$_2$ transistors and photoactive PTCDI-C_{13} floating-gate.

This phototransistor exhibited switchable volatile and non-volatile photo response behaviors. with a negative V_{ds}, the device exhibited a volatile photo response, while at positive V_{ds}, the photo transistor showed a nonvolatile photo response with a high resistance ratio of approximately 10^3. Additionally, the device possessed wavelength-dependent photo-memory behavior due to narrowband absorption J-aggregated PTCDI-C_{13}. This wavelength-dependent reconfigurable photo memory device will enhance multifunctionality of photodetector and is expected to further reduce the complexity of adaptive visuomorphic computing systems.

Methods

The 2D materials, including PdSe$_2$, WSe$_2$, and h-BN (HQ-graphene, Inc.), were mechanically exfoliated and stacked on a silicon wafer with a 300 nm SiO$_2$ layer using a dry-transfer technique. Electrodes were patterned by electron beam lithography, followed by Ti/Au (5 nm /40 nm) deposition through electron beam evaporation. The growth of J-aggregated PTCDI-C_{13} was substrate-sensitive, with the PTCDI-C_{13} monolayer self-assembling on the h-BN surface via physical vapor transport (PVT) at 230°C for 120 minutes under vacuum. Structural and optical properties were analyzed through Raman and photoluminescence (PL) spectroscopy. Electrical and photoelectrical characteristics were measured using a Keithley 4200A SCS semiconductor analyzer, equipped with a probe stage and vacuum chamber. Illumination was provided by a 532 nm laser and a 300 W Xe lamp with a monochromator and electronic shutter for controlled light exposure.

Result and discussion

The schematic structure and optical image of the reconfigurable phototransistor are shown in Fig. 1(a) and (b). Staggered PdSe$_2$ layers were used as contact electrodes to

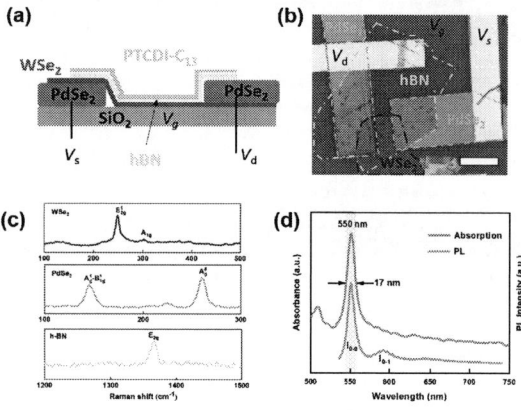

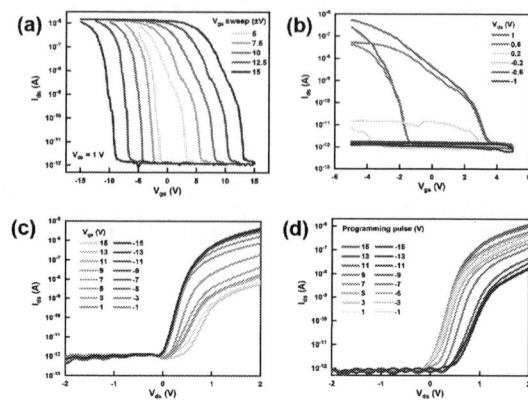

Fig. 1 (a) Schematic structure and (b) optical image of the device (scale bar: 5 μm). (c) Raman spectra of WSe₂, PdSe₂ and h-BN layers. (d) Absorption and photoluminescence (PL) spectra of PTCDI grown on h-BN.

Fig. 2 (a)-(b) Transfer characteristic curves at different V_{gs} sweeps (a) and V_{ds} bias levels (b) of the device. (c)-(d) Output characteristic curves at constant V_{gs} (c) and after programming pulses (d).

avoid the Fermi-level pining effect and form a p-type contact with the WSe₂ channel, which has a length of 2 μm and is fully covered by h-BN across the channel region. The J-aggregated PTCDI-C₁₃ was uniformly grown on the h-BN surface, serving as a light-sensitive charge-trapping layer. The Raman spectra (Fig. 1(c)) display characteristic peaks for the h-BN, WSe₂, and PdSe₂ layers, confirming the van der Waals stacking of the phototransistor. From the PL spectra (Fig. 1(d)), we observe that PTCDI-C₁₃ deposited on h-BN exhibits strong, narrow-band photoluminescence and absorption at approximately 550 nm, due to the radiative 0-0 transition of coherent Frenkel excitons in the J-aggregated structure. The narrow full-width at half maximum of ~17 nm for the absorption peak enhances its suitability as the light-sensitive gate with high spectral selectivity.

The electrical memory performance of the J-aggregated PTCDI-C₁₃ floating-gate transistor was characterized through electrical measurements. As shown in Fig. 2(a), the device exhibits fully unipolar p-type transfer characteristics with a high on-off ratio of 10^7 (at $V_{ds} = 1$ V), attributed to the close work function alignment between PdSe₂ and the valence band of WSe₂. Due to the charge-trapping effect of PTCDI-C₁₃, the transistor demonstrates a large, symmetric hysteresis loop in the transfer curves under dual sweeping of gate voltage V_{gs} dual swept. The memory window expands with increasing V_{gs} sweep range, reaching approximately 23 V when V_{gs} swept between ±17.5 V. The memory performance varies with drain voltage V_{ds}, with the window narrowing at decreasing positive bias and disappearing entirely at low negative bias, as shown in Fig. 2(b). Fig. 2(c) presents the output characteristics under various V_{gs} conditions, showing that the tunable rectification behavior with a maximum rectification ratio of 10^6. Additionally, by applying programming voltage spikes of V_{gs} ranging from -

15 V to 15 V, the device demonstrates similar tunable rectification behavior. This suggests that the PTCDI-C₁₃ floating-gate provides effective and stable electrostatic doping.

To investigate the reconfigurable photo response behaviors of the transistor, photocurrent (I_{photo}) measurements were conducted under 532 nm laser illumination at varying V_{ds}. At negative V_{ds}, the device exhibits a volatile photo response, with a stable light-to-dark current ratio (I_{photo}/I_{dark}) of approximately 10^2, as shown in Fig. 3(a) and (b), likely due to the reversed Schottky barrier formed at the PdSe₂/WSe₂ interface. Conversely, at positive V_{ds}, the device demonstrates a non-volatile photoelectric response, as shown in Fig. 3(c). As V_{ds} increases from 1 V to 2 V, the photo memory effect is enhanced, characterized by a higher high-resistance state (HRS) to low-resistance state (LRS) ratio of around 10^3 and extended retention times, as shown in Fig. 3(d)), owing to the photoinduced charge tunneling from the PTCDI-C₁₃ into the WSe₂ channel. These findings suggest a V_{ds}-dependent volatile/nonvolatile memory behavior in the phototransistor.

The wavelength-dependent non-volatile photo response of the phototransistor was investigated, revealing that the narrow-band absorption of J-aggregated PTCDI-C₁₃ enables wavelength-selective photo-memory. As shown in Fig. 4(a), the non-volatile photoelectric response occurs only when the wavelength exceeds 565 nm at $V_{ds} = 2$ V. As the wavelength increases to 700 nm, the non-volatile response fully transitions to a volatile photoelectric response, as depicted in Fig. 4(b), aligning with the absorption edge of J-aggregated PTCDI-C₁₃ (see Fig. 1(c)). This demonstrates a reconfigurable, wavelength-dependent switch between volatile and non-volatile photo response in the device. The narrow-band non-volatile photo response, matching the absorption edge of J-aggregated PTCDI-C₁₃, further

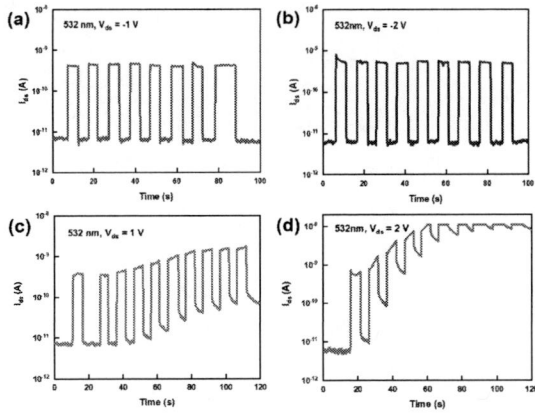

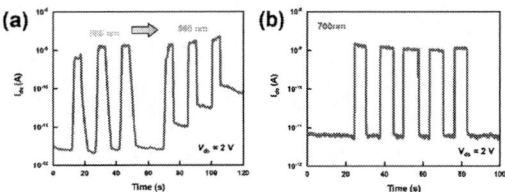

Fig. 4 (a) Output current under 700 nm light pulse at $V_{ds} = 2$ V, showing no apparent photo memory effect. (b) Output current variation at $V_{ds} = 2$V as the wavelength shifts from 565 nm to 560 nm, where the memory effect begins to emerge.

Fig. 3 (a)-(d) The output current I_{ds} as a function of time under 532 nm light pulses at the bias voltage of -1 V (a), -2 V (b), 1V (c) and 2 V (d), revealing a V_{ds} dependent photo memory behavior.

underscores the effective charge-trapping effect of the PTCDI-C$_{13}$ layer.

Conclusion

In conclusion, we developed a novel phototransistor by integrating PdSe$_2$-contacted WSe$_2$ transistor with photoactive PTCDI-C$_{13}$ monolayer floating-gate. This phototransistor exhibits unipolar p-type conductivity with a high on-off ratio of 10^7 and a broad memory window, attributed to charge-trapping effect of PTCDI-C$_{13}$. Under light illumination, we observed switchable optoelectronic behaviors from volatile to non-volatile photo responses. Specifically, at a negative V_{ds}, the phototransistor served as a photoreactor with volatile photo response, while at positive V_{ds}, it showed a nonvolatile photo-memory response with a high resistance ratio of around 10^3. Furthermore, the phototransistor exhibited wavelength-dependent photo-memory behavior around 560 nm, corresponding to the narrowband absorption of J-aggregated PTCDI-C$_{13}$. This approach of integrating 2D reconfigurable transistor with organic materials offers a promising pathway for developing novel reconfigurable memory devices.

Acknowledgments

The work is in part supported by the Research Grants Council of Hong Kong, particularly, via Grant AoE/P-701/20, 14206721, N_CUHK438/18, CUHK Group Research Scheme, CUHK Fund for Joint Research Labs.

References

[1] M. Tsai, C. Huang, C. Lin, M. Lee, F. Yang, M. Li, Y. Chang, K. Watanabe, T. Taniguchi, C. Ho, W. Wu, M. Yamamoto, J. Wu, P. Chiu and Y. Lin, "A reconfigurable transistor and memory based on a two-dimensional heterostructure and photoinduced trapping", Nat. Electro., 6, pp. 755–764 (2023).

[2] Y. Liu, W. Huang, X. Wang, R. Liang, J. Wang, B. Yu, T. Ren and J. Xu, "A Hybrid Phototransistor Neuromorphic Synapse", IEEE J. Electron. Dev., 7, pp. 13-17 (2019).

[3] S. Luo, C. Fu, C. Zhang, Z. Li, Z. Zhu, J. Wang, Y. Wang, G. He, L. Luo and F. Liang, "IGZO/InHfO$_x$ Nanowires/IGZO Phototransistor with Persistent Photoconductivity Effect for Intelligent Visual Perception Application", IEEE Trans. Electron. Devices, 71, pp. 4745-4750 (2024).

[4] H. Shao, Y. Li, W. Yang, X. He, L. Wang, J. Fu, M. Fu, H. Ling, P. Gkoupidenis, F. Yan, L. Xie and W. Huang, "A Reconfigurable Optoelectronic Synaptic Transistor with Stable Zr-CsPbI$_3$ Nanocrystals for Visuomorphic Computing", Adv. Mater., 35, pp. 2208497 (2023).

[5] Y. Zhou, L. Tong, Z. Chen, L. Tao, Y. Pang and J-B. Xu, "Contact-engineered reconfigurable two-dimensional Schottky junction field-effect transistor with low leakage currents", Nat. Commun., 14, pp. 4270 (2023).

[6] P. Y, Z. Y, T. L and J-B. Xu. "2D Dual Gate Field-Effect Transistor Enabled Versatile Functions", Small, 20, pp. 2304173 (2024).

[7] Y. Zhou, L. Tong, Z. Chen, L. Tao, H. Li, Y. Pang, J-B. Xu. "Vertical Nonvolatile Schottky-Barrier-Field-Effect Transistor with Self-Gating Semimetal Contact", Adv. Funct. Mater., 33, pp. 2213254 (2023).

[8] B. Sun, G. Zhou, Y. Wang, X. Xu, L. Tao, N. Zhao, H. Tsang, X. Wang, Z. Chen, J-B. Xu, "Ultra-narrowband photodetector with high responsivity enabled by integrating monolayer J-aggregate organic crystal with graphene", Adv. Opt. Mater., 9, pp. 2100158 (2021).

Deeply Scaled Gate Field Plate to Suppress Drain-Induced Dynamic Threshold Voltage Instability in Schottky-Type p-GaN Gate HEMT

Chen Wang, Xin Wang, Junjie Yang, Jiayin He, Ju Gao, Chengkang Ao, Ziheng Liu,
Hongjie Peng, Wenbo Xia, Jin Wei* and Jinyan Wang*

School of Integrated Circuits, Peking University, Beijing 100871, China

*Corresponding authors: jin.wei@pku.edu.cn; wangjinyan@pku.edu.cn

Abstract

This letter proposes a Schottky-type p-GaN gate HEMT with deeply scaled gate field plate (GFP) to suppress drain-induced dynamic threshold voltage (V_{th}) instability. Under high V_{DS}, the channel beneath GFP pinches off and screens capacitive coupling between drain and p-GaN, resulting in alleviated dynamic V_{th} shift, which is verified by TCAD simulations and experiments. Design and optimizations are fully investigated by simulations. Proposed structure also exhibits decent transfer, breakdown and capacitance characteristics, showing feasibility of this technology.

Keywords: p-GaN gate HEMT, V_{th} instability, gate field plate

Introduction

Thanks to the superior properties such as fast switching and low on-resistance, GaN-based high-electron-mobility transistors (HEMTs) have been widely adopted in power electronics. Currently, Schottky-type p-GaN gate HEMT, as a strong contender among all enhancement-mode technologies, has attained remarkable success in consumer electronics market [1]. However, dynamic V_{th} instabilities especially the drain-induced ones still hinder further development [2-6].

Due to the "floating" nature of p-GaN, Schottky-type p-GaN gate HEMTs suffers from gate-drain coupled barrier lowering (GDCBL) effect [3] and charge storage [4]. Under drain bias, the electron barrier under p-GaN gate is lowered, leading to negative V_{th} shift and risk of false turn-on. Meantime, holes flow out of p-GaN through the forward-biased Schottky-type gate junction. When device is turned on, the reverse-biased gate junction prevents timely restoration of holes in p-GaN, resulting in negative charge storage and positive V_{th} shift, which lowers overdrive voltage and increases on-resistance. Significant efforts have been made to address dynamic V_{th} instability [7-11]. However, improved dynamic V_{th} stability usually comes with complexity of fabrication/circuit and narrowed safety margin [12].

GFP, as a common technique to suppress dynamic R_{on}, extends from gate towards drain and modulates channel depletion region between gate and drain, which potentially interferes the interplay between drain and p-GaN.

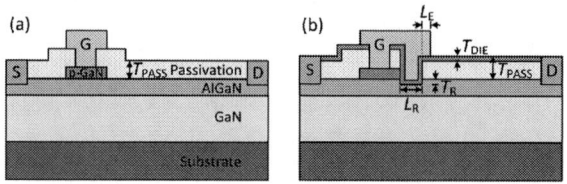

Fig. 1 Schematic cross-sectional structure of (a) conventional p-GaN gate HEMT (Conv-HEMT) and (b) proposed device (with GFP).

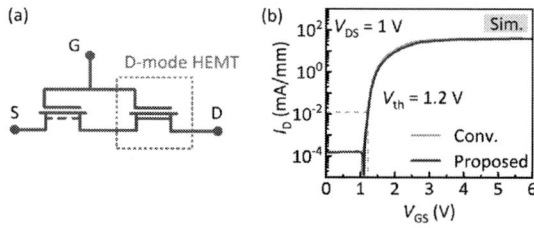

Fig. 2 (a) Equivalent circuit of the proposed structure. (b) Simulated transfer curves of conventional and proposed device. $T_{PASS}/T_{DIE}/T_R/L_R/L_E$ = 60 nm/30 nm/4 nm/2 μm/1 μm.

In this paper, we propose a Schottky-type p-GaN gate HEMT with a GFP that is deeply scaled in vertical direction for improved dynamic V_{th} stability. Suppression of drain-induced negative V_{th} shift is investigated by TCAD simulations. Design optimizations are explored while transfer, off-state breakdown as well as C-V characteristics are evaluated. Experiments further verifies effectivity of proposed structure in suppressing dynamic V_{th} shift through pulsed I-V tests.

Device Principles

Fig. 1 illustrates the device structures. Compared with the conventional structure (Fig. 1(a)), the proposed structure has an extended GFP, which is vertically scaled by replacing passivation under GFP with a thin dielectric layer as well as partially recessing barrier layer.

The vertically scaled GFP in the proposed structure functions as a depletion-mode (D-mode) HEMT (Fig. 2(a)) and will not influence forward conduction. Under high V_{DS}, the channel under GFP pinches off and clamps the potential of underneath channel. Thus, the electron barrier under p-GaN will no longer be lowered by V_{DS}, and dynamic V_{th} shift will be suppressed.

TCAD simulations calibrated by experiments are carried

979-8-3315-0417-5/25 $31.00 © 2025 IEEE

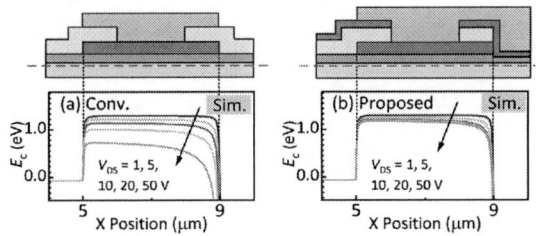

Fig. 3 Simulated E_c along the channel (extracted 3 nm below AlGaN/GaN interface) under p-GaN for (a) conventional and (b) proposed device. $T_{PASS}/T_{DIE}/T_R/L_R/L_E$ = 60 nm/30 nm/4 nm/2 μm/1 μm.

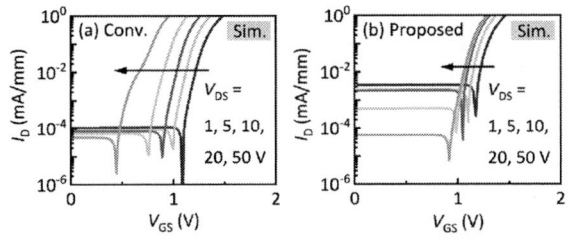

Fig. 4 Simulated transfer curves of (a) conventional and (b) proposed device. Device dimensions are the same as Fig. 3.

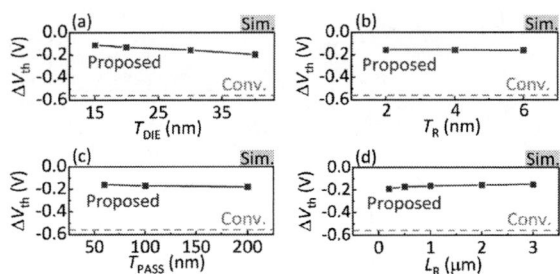

Fig. 5 ΔV_{th} of the proposed structure with different (a) T_{PASS} (b) T_{DIE} (c) T_R and (d) L_R under 50-V V_{DS}. Device dimensions are the same as Fig. 3 (except the designated variable).

out to comprehensively study the mechanism and optimizations. Main dimensions of simulated device structures are the same ($L_{GS}/L_G/L_{GD}$ = 3/4/10 μm) throughout the article. As shown in Fig. 2(b), proposed device exhibits similar forward conduction capability as conventional device.

Demonstration and Optimizations

A. Suppression of Dynamic V_{th} Instability

Fig. 3 plots the conduction band minimum (E_c) along the channel of both devices. As shown in Fig. 3(a), electron barrier under p-GaN in conventional device is continuously lowered by drain bias while the proposed device shows suppressed barrier lowering (Fig. 3(b)), indicating alleviated negative V_{th} shift [3].

Fig. 4 presents the drain-induced dynamic V_{th} shifts extracted from simulated transfer curves for conventional and proposed structure with the same dimensions. As shown in Fig. 4(a), V_{th} continuously shifts negatively with the increase of V_{DS} for conventional device, showing a negative shift of -0.56 V under 50-V V_{DS}. While negative V_{th} shift in proposed structure is greatly suppressed, with a negative shift of -0.16 V under 50 V, demonstrating an improvement of 71.4 % compared to conventional device.

B. Flexibility of Design

Detailed simulations are carried out to fully study the design flexibility. Fig. 5 presents the simulated dynamic V_{th} shift (ΔV_{th}) of proposed device with different structural parameters under 50-V V_{DS}, with ΔV_{th} of conventional device under 50-V V_{DS} plotted for comparison (dash line).

As shown in Fig. 5(a), ΔV_{th} increases about 0.09 V when the thickness of dielectric layer (T_{DIE}) increases from 15 nm to 40 nm. Thin dielectric layer leads to strong electrostatic control of the GFP towards channel, resulting in pinch-off of channel under GFP at lower V_{DS}. Thus, T_{DIE} needs to be scaled down for effective suppression of dynamic V_{th} shift. Still, T_{DIE} needs to thick enough for adequate gate swing.

ΔV_{th} as a function of the length of GFP (L_R) is plotted in Fig. 5(b). ΔV_{th} remains unchanged for L_R larger than 1 μm and slightly deteriorates (~0.04 V) for L_R < 1 μm, which is due to the weakened screening effect of the shortened GFP.

As shown in Fig. 5(c), ΔV_{th} barely changes with the increase of thickness of passivation (T_{PASS}). Variation of ΔV_{th} between T_{PASS} = 60 nm and T_{PASS} = 200 nm is only 0.02 V. Thanks to the deeply scaled GFP, the dielectric layer under GFP and passivation at other regions can be designed separately, providing more flexibility of design compared to conventional device with normal field plate (simply extending the gate metal).

Fig. 5(d) shows that ΔV_t is insensitive to the etching depth of AlGaN barrier (T_R) within the range of simulation, with a variation of 0.02 V between T_R = 2 nm and T_R = 6 nm. However, too deep recess causes depletion of 2DEG [13] and degrades forward conduction, which needs to be avoided.

Overall, changing the structural parameters affects ΔV_{th} to a much lesser extent (up to 0.09 V) compared with the difference between conventional and proposed structure (~0.4 V). Above analysis demonstrates high flexibility of design and wide process window of the proposed structure.

C. Breakdown and Capacitance Characteristics

Breakdown and switching characteristics are numerically investigated to further evaluate the feasibility of proposed device. Fig. 6 depicts the $|E|$ distribution along the channel for conventional device and proposed device. $|E|$ peak is lower for proposed device, suggesting that the deeply scaled GFP will not deteriorate off-state breakdown voltage.

C-V curves of both devices are plotted in Fig. 7. Proposed device exhibits higher capacitances than conventional device at low V_{DS}, but the difference diminishes with V_{DS} and nearly invisible at 40-V V_{DS}. Thus, the influence of increased capacitances in proposed device is limited.

Two distinct turning points in Fig. 7(b) are associated with the channel pinch-off of under the recessed and non-recessed part of GFP, which can be move towards lower V_{DS} by further optimization (e.g., decrease T_{DIE}) to reduce Q_G and Q_{oss}.

Noting that, the notorious false turn-on caused by miller

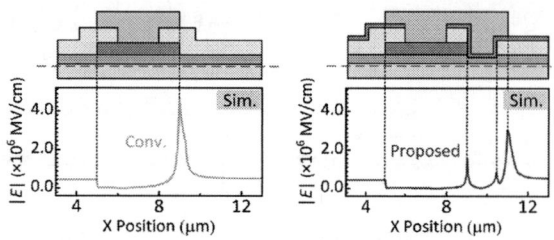

Fig. 6 $|E|$ distribution along the channel (extracted 3 nm below AlGaN/GaN interface) under 100-V V_{DS} for (a) conventional and (b) proposed device.

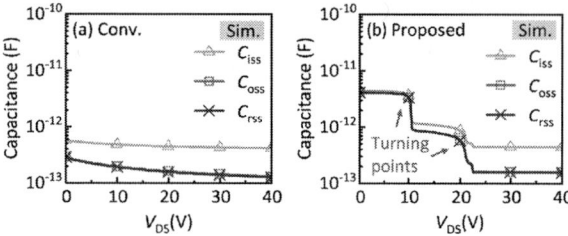

Fig. 7 C-V characteristics of (a) conventional and (b) proposed device, where $T_{PASS}/T_{DIE}/T_R/L_R/L_E$ = 60 nm/30 nm/4 nm/1.5 μm/0.5 μm.

feedback occurs during device off-state (high V_{DS}). Thus, increased C_{rss} only at low V_{DS} will not cause additional risk of false turn-on.

Experimental Verification

Conventional and proposed devices are fabricated on the same wafer to experimentally investigate the drain-induced dynamic V_{th} shift. The GaN heterostructure was grown by MOCVD on silicon substrate, which consists of p-GaN/Al$_{0.23}$Ga$_{0.77}$N/GaN (70 nm/12 nm/300 nm) and a high resistive GaN buffer layer. Mg doping concentration in p-GaN is about 10^{19} cm^{-3}. The fabrication process starts with dry etching to form the p-GaN gate. Then, the device is passivated by atomic layer deposition (ALD)-grown Al$_2$O$_3$. The source/drain ohmic contacts were fabricated by opening contact window, evaporating Ti/Al/Ni/Au stack and rapid thermal annealing. Ion implantations were then implemented for device isolation. And the fabrication is finished by gate window opening and Ni/Au evaporation.

Fig. 8 presents the drain-induced dynamic V_{th} shifts of both devices extracted from pulsed transfer curves [3]. Conventional device exhibits a negative shit of −0.15 V under 50-V V_{DS} while dynamic V_{th} shift in proposed device is invisible throughout test range. Measured dynamic V_{th} shifts are lower than the simulated results, which is caused by measurement delay and higher gate current due to non-ideal fabrication. Nonetheless, the effectivity of proposed device in suppressing dynamic V_{th} instability is verified.

Conclusion

GFP which is deeply scaled in vertical direction is proposed to suppress drain-induced dynamic V_{th} instability in Schottky-type p-GaN gate HEMT. Comprehensive TCAD simulations are carried out to study working principles and optimizations. Breakdown and C-V characteristics are

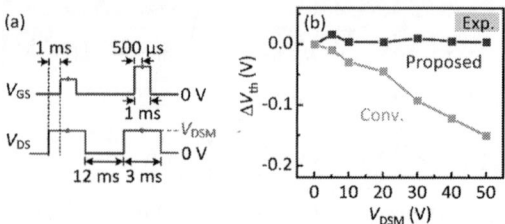

Fig. 8(a) Schematic waveforms of pulsed I-V test. (b) ΔV_{th} under varied V_{DSM} of conventional and proposed device extracted from pulsed I-V curves.

discussed. Effectivity of the proposed device is then verified by experiments. Proposed device suppresses dynamic V_{th} instability while providing high flexibility of design, demonstrating the feasibility of this technology.

References

[1] T. McDonald, "Power Conversion Semiconductor and Circuit Trends and Challenges for a Sustainable Energy Future," *IEDM Tech. Dig.*, pp. 1-4 (2023).

[2] Z. Fan *et al.*, "Analysis of Drain-Dependent Threshold Voltage and False Turn-On of Schottky-Type p-GaN Gate HEMT in Bridge-Leg Circuit," *IEEE Trans. on Power Electronics*, vol. 39, no. 2, pp. 2351-2359 (2024).

[3] M. Nuo *et al.*, "Gate/Drain Coupled Barrier Lowering Effect and Negative Threshold Voltage Shift in Schottky-Type p-GaN Gate HEMT," *IEEE Trans. on Electron Devices*, vol. 69, no. 7, pp. 3630-3635 (2022).

[4] J. Wei *et al.*, "Charge Storage Mechanism of Drain Induced Dynamic Threshold Voltage Shift in *p*-GaN Gate HEMTs," *IEEE Electron Device Lett.*, vol. 40, no. 4, pp. 526-529 (2019).

[5] X. Tang *et al.*, "Mechanism of Threshold Voltage Shift in *p* -GaN Gate AlGaN/GaN Transistors," *IEEE Electron Device Lett.*, vol. 39, no. 8, pp. 1145-1148 (2018).

[6] J. P. Kozak *et al.*, "Stability, Reliability, and Robustness of GaN Power Devices: A Review," *IEEE Trans. on Power Electronics*, vol. 38, no. 7, pp. 8442-8471 (2023).

[7] C. Zhang *et al.*, "Hybrid Gate p-GaN Power HEMTs Technology for Enhanced V_{th} Stability," *IEDM Tech. Dig.*, pp. 35.4.1-35.4.4 (2022).

[8] J. Yang *et al.*, "Simultaneously Achieving Large Gate Swing and Enhanced Threshold Voltage Stability in Metal/Insulator/n-GaN Gate HEMT," *IEDM Tech. Dig.*, pp. 1-4 (2023).

[9] X. Dai *et al.*, "Island-ohmic-PGaN Gate HEMT: Toward Steep Subthreshold Swing and Enhanced Threshold Stability," *IEEE Electron Device Letters*, vol. 45, no. 6, pp. 988-991 (2024).

[10] M. Hua *et al.*, "E-mode *p*-GaN Gate HEMT with *p*-FET Bridge for Higher V_{TH} and Enhanced V_{TH} Stability," *IEDM Tech. Dig.*, pp. 23.1.1-23.1.4 (2020).

[11] H. Zhou *et al.*, "A Gate Driver with a Negative Turn Off Bias Voltage for GaN HEMTs," *9th IPEMC2020-ECCE Asia*, pp. 1083-1086 (2020).

[12] J. Wei *et al.*, "GaN Power Integration Technology and Its Future Prospects," *IEEE Trans. on Electron Devices*, vol. 71, no. 3, pp. 1365-1382 (2024).

[13] J. P. Ibbetson *et al.*, "Polarization effects, surface states, and the source of electrons in AlGaN/GaN heterostructure field effect transistors," *Applied Physics Lett.*, vol. 77, no. 2, pp. 250-252 (2000).

Investigation of Cryogenic Ultra-Low V_{TH} MOSFETs and BEOL for Power Efficiency Enhancement

Yuanke Zhang[1], Yuefeng Chen[2], Hengxu Guo[2], Jun Xu[1], Guoping Guo[1,2,3], and Chao Luo[1*]

[1]Department of Physics, University of Science and Technology of China, China
[2]Hefei National Laboratory, University of Science and Technology of China, China
[3]Suzhou Institute for Advanced Research, University of Science and Technology of China, China
(*Email: lc0121@ustc.edu.cn)

Abstract

Power consumption in MOSFETs has emerged as a significant challenge for CMOS technology operating at cryogenic temperatures in novel applications such as quantum computing. In this paper, we present a comprehensive study on the cryogenic performance of commercial 55-nm MOSFETs with ultra-low threshold voltage (V_{TH}) for supply voltage (V_{DD}) scaling. In addition, the characteristics of back-end-of-line (BEOL) interconnects are also investigated to evaluate the power efficiency of CMOS devices at low temperatures. As the temperature decreases, power efficiency can be greatly enhanced through V_{TH} and V_{DD} scaling, resulting in a significant reduction in power consumption of CMOS circuits.

Keywords: Cryogenic CMOS, ultra-low V_{TH} MOSFETs, BEOL, power efficiency

Introduction

The low-temperature operation of CMOS transistors has been widely studied to achieve high-performance computing or implement cryogenic quantum computing control systems in recent years [1-2]. In addition to the excellent performance of the cryogenic CMOS (cryo-CMOS) transistors in switching speed, on/off ratio, and G_m/I_D ratio, the quantum transport behavior in cryo-CMOS also shows its potential application in quantum information processing (QIP). However, the dilution refrigerators that create cryogenic environments have very limited cooling power: several hundreds of microwatts at millikelvin and ~1 W at 3 K [3]. The ultra-low power consumption devices and integrated circuits (ICs) become particularly important to meet the demanding specifications for cryo-CMOS operation.

Thanks to the steeper subthreshold swing and the smaller gate-induced drain leakage (GIDL) current, the off-state leakage current (I_{OFF}) and the static power can be negligible in cryogenic MOSFETs [2]. The dynamic power consumption is proportional to V_{DD}^2, thus the low V_{TH} devices provide a solution to reduce V_{DD} and reduce the power consumption at low temperatures.

In this study, we present the characteristics of 55-nm MOSFETs with various V_{TH} over the temperature range from 300 K to 6 K. The electrical characteristics, including low-temperature electrical parameters, reliability, and interconnection capacitance and resistance are characterized and discussed. Through V_{TH} and V_{DD} scaling, the power efficiency can be greatly enhanced at cryogenic temperatures, leading to a substantial reduction in power consumption. Finally, we propose a compact model for ultra-low V_{TH} MOSFETs to facilitate the design of low power consumption cryo-CMOS circuits.

Cryogenic Ultra-Low V_{TH} MOSFETs Characteristics

The room temperature and cryogenic temperature transfer curves of commercial 55-nm MOSFETs with different V_{TH} (standard V_{TH} (SVT), low V_{TH} (LVT), and ultra-low V_{TH}) are shown in Fig. 1(a). Due to the limitation of subthreshold swing (SS, SS ≥ 60 mV/dec@300 K), the off-state current (I_{OFF}) increases exponentially with V_{TH} decrease at 300 K. Differently, the steep cryogenic SS offers a favorable V_{TH} versus I_{OFF} trade-off. Due to the reduction in thermionic transport, an improved SS ≈ 16 mV/dec and the negligible I_{OFF} are observed in both SVT, LVT, and ultra-low V_{TH} MOSFETs at 6 K. And the ultra-low V_{TH} MOSFETs provides a favorable condition for reducing V_{DD} to 0.5 V or less, thus reducing the dynamic power consumption. Thus, we carry out a comprehensive study on ultra-low V_{TH} MOSFETs.

The DC characteristics of 55-nm ultra-low V_{TH} MOSFETs are shown in Fig. 2(a)-(b). As temperature decreases, V_{TH} increase due to the raise in the substrate Fermi potential. Clear zero-temperature-coefficient (ZTC) points can be observed in both figures due to the balance between the influence of increased V_{TH} and enhanced mobility as temperature decreases. As shown in Fig. 3(a), the low-field mobility (μ_0) in long-channel transistor increases to ~3.7× at 6 K due to reduced phonon scattering. In shorter channel devices, due to the reverse short channel effect (RSCE), the higher effective substrate doping concentration results in more impurity scattering centers in the channel and thus limits the improvement of μ_0. The SS at 6 K improves to ~0.18× that at 300 K. It shows nearly ideal SS scaling proportional to k_BT/q above 77 K, with saturation occurring at lower temperatures due to the band tail effect [4]. Fig. 3(d) shows the hot carrier degradation reliability measurement results. Under hot carrier stress, I_{Dsat} degradation is slightly more pronounced at 6 K compared to 300 K due to increased mobility. But the lifetime shows no significant difference at two

979-8-3315-0417-5/25 $31.00 © 2025 IEEE

temperatures based on the typical lifetime definition (i.e., 10% I_{Dsat} degradation). Additionally, the cryogenic operation of ultra-low V_{TH} MOSFETs enables V_{DD} scaling without compromising performance, which can greatly enhance reliability. Thus, it is expected that the hot carrier reliability of ultra-low V_{TH} MOSFETs will improve at low temperatures while maintaining the same device performance by V_{DD} scaling.

BEOL Characteristics and Power Efficiency

We also evaluate the interconnects in BEOL. The test structure for interconnection resistors and capacitors is shown in Fig. 4(a). The sheet resistances of the Cu lines also reduced with temperature since phonon scattering reduces with temperature (Fig. 4(b)) in both bottom layer (M1), middle layer (M2), and top layer (UY1) metal. The wider line shows a greater reduction (Fig. 4(c)) because the narrower line experiences higher residual resistance due to the electron scattering at the sidewalls and top/bottom interfaces. The same-layer interconnect metal line capacitances and MOM capacitances value at various temperatures are presented in Fig. 4(d) and (e). There is no significant change in the interconnect metal line capacitance values down to 6 K and the MOM capacitance values decrease only slightly with temperature.

In a circuit with complex routing, the loading capacitance is dominated by the wire loading, which remains almost constant at low temperatures (Fig. 4(d)-(e)). Therefore, the normalized device currents (i.e., I_{DS}/V_{DD}), which is inversely proportional to the device operation delay time, can be further used to validate the circuit-level frequency gain at low temperatures. Due to the negligible static power at low temperatures, the electrical power is determined by the dynamic power, which is proportional to $I_{DS} \times V_{DD}$. To further evaluate the relative frequency and electrical power relationships (i.e., the relative relationship between I_{DS}/V_{DD} and $I_{DS} \times V_{DD}$) at different temperatures, we measured transfer curves under various V_{DS} biases and Ts (Fig. 5(a)). As shown in Fig. 5(b), in LVT MOSFETs, at the same device operation frequency, power is reduced by 47% at 6 K when V_{DD} is scaled to 0.8 V. At the same electrical power level, frequency increases by 41% at 80 K and 46% at 6 K, indicating improved power efficiency at cryogenic temperatures. As expected, compared to LVT MOSFETs, the ultra-low V_{TH} MOSFETs achieve a further 47% increase in frequency with reducing power by 65% when V_{DD} is scaled down to 0.6 V (Fig. 5(c)). By scaling V_{TH} and V_{DD} together while lowering the temperature, it can meet both low power and high-frequency requirements, highlighting the advantages of ultra-low V_{TH} transistors at cryogenic temperatures.

Cryogenic Modeling

The industry-standard compact models face limitations at cryogenic temperatures and do not accurately predict mobility behavior in ultra-low V_{TH} MOSFETs [6]. Thus, we modify the mobility equation in the industry-standard model as follows:

$$1/\mu_{eff} = A_1/(1+q_{inv}/q_0)^2 + A_2[(\eta \cdot q_{inv} + q_{dep})/\varepsilon_s]^2 + A_3(q_{inv} + q_{dep})^{1/3}$$

where $0 < \eta < 1$, ε_s denotes the dielectric permittivity of silicon, and q_{inv} and q_{dep} is the inversion and the depletion charge, respectively. A_1, A_2, A_3, and q_0 are temperature-dependent model parameters. The three terms in the equation represent the contributions of Coulomb, surface roughness, and lattice scattering to the effective mobility, enabling accurate modeling of I-V behavior across various temperatures. As shown in Fig. 6(a)-(c), our model shows an excellent agreement with the measurement data of the 55-nm ultra-low V_{TH} MOSFETs down to 6 K.

Conclusion

We demonstrated the electrical characteristics of commercial 55-nm MOSFETs with various V_{TH} and BEOL interconnects from 300 K to 6 K. At 6 K, the improved SS $\approx$ 16 mV/dec, cryogenic $V_{TH} <$ 0.2 V, and negligible I_{OFF} are observed in ultra-low V_{TH} MOSFETs, which can facilitate V_{DD} scaling and greatly enhance power efficiency. Based on the mobility-corrected model, we proposed a compact model of 55-nm ultra-low V_{TH} MOSFETs to facilitate the design and simulation of low power consumption cryo-CMOS circuits for high-performance computing and quantum computing.

Acknowledgments

This work was supported by the NSFC (Grant No. 12034018), Anhui Provincial Natural Science Foundation (Grant No.2408085QF190), and Innovation Program for Quantum Science and Technology (Grant No. 2021ZD0302300).

References

[1] S. J. Pauka et al., "A cryogenic CMOS chip for generating control signals for multiple qubits", Nat. Electron., 4, p. 64 (2021).

[2] M. F. Gonzalez-Zalba et al., "Scaling silicon-based quantum computing using CMOS technology" Nat. Electron., 4, p. 872 (2021).

[3] B. Patra et al., "A Scalable Cryo-CMOS 2-to-20GHz Digitally Intensive Controller for 4×32 Frequency Multiplexed Spin Qubits/Transmons in 22nm FinFET Technology for Quantum Computers" IEEE International Solid-State Circuits Conference (ISSCC), p. 304 (2020).

[4] A. Beckers et al., "Theoretical Limit of Low Temperature Subthreshold Swing in Field-Effect Transistors", IEEE Electron Device Lett., 41, pp. 276-279 (2020).

[5] M. Tada et al., "A 65nm Cryogenic CMOS Design and Performance at 4.2K for Quantum State Controller Application", IEEE J. Electron Devices Soc., 12, pp. 28-33 (2024).

[6] Y. Zhang et al., "Characterization and modeling of native MOSFETs down to 4.2 K", IEEE Trans. Electron Devices, 68, pp. 4267-4273 (2021).

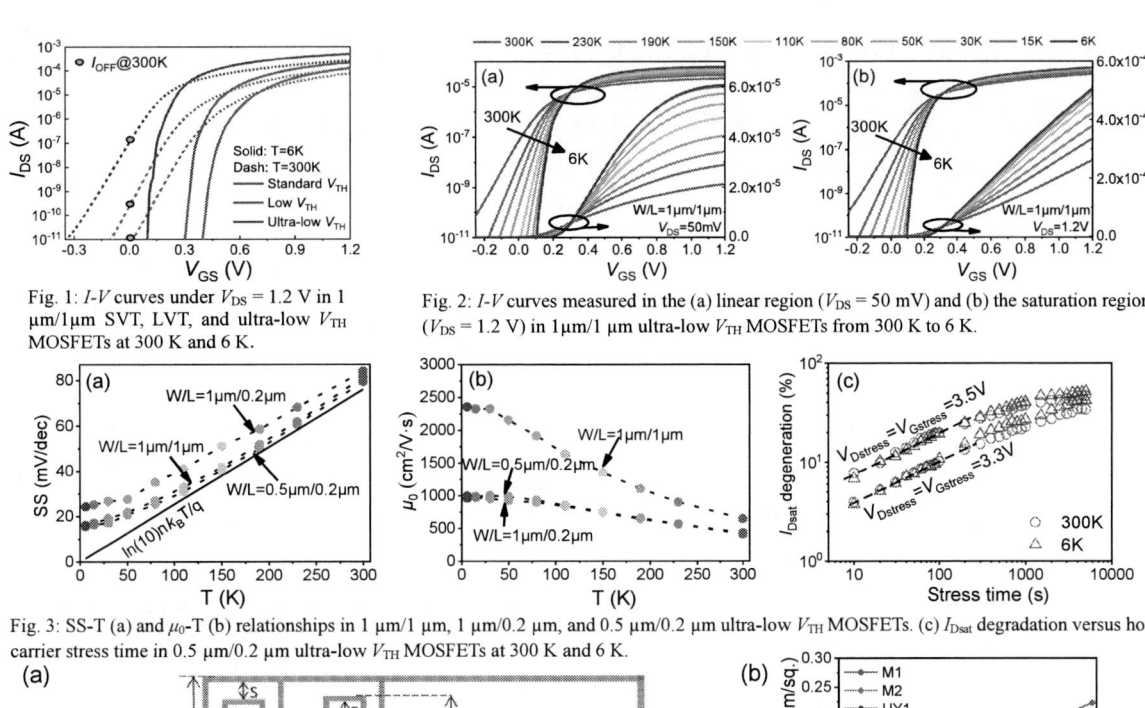

Fig. 1: I-V curves under V_{DS} = 1.2 V in 1 μm/1μm SVT, LVT, and ultra-low V_{TH} MOSFETs at 300 K and 6 K.

Fig. 2: I-V curves measured in the (a) linear region (V_{DS} = 50 mV) and (b) the saturation region (V_{DS} = 1.2 V) in 1μm/1 μm ultra-low V_{TH} MOSFETs from 300 K to 6 K.

Fig. 3: SS-T (a) and μ_0-T (b) relationships in 1 μm/1 μm, 1 μm/0.2 μm, and 0.5 μm/0.2 μm ultra-low V_{TH} MOSFETs. (c) I_{Dsat} degradation versus hot carrier stress time in 0.5 μm/0.2 μm ultra-low V_{TH} MOSFETs at 300 K and 6 K.

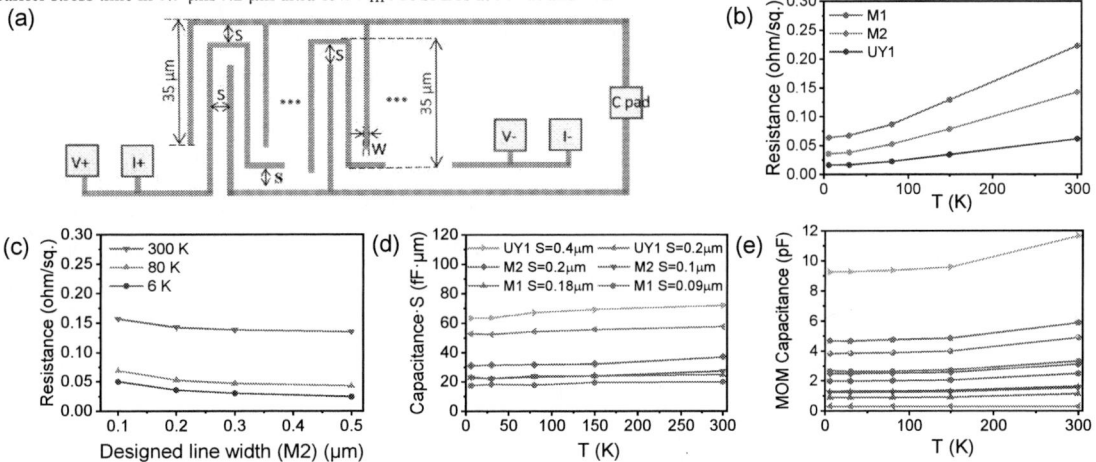

Fig. 4: (a) Layout structure for BEOL resistance and capacitance measurement. BEOL metal line resistance (b) at various temperatures and (c) with different line width. (d) Same-layer interconnect metal line capacitances and (e) ten different MOM capacitances at various temperatures.

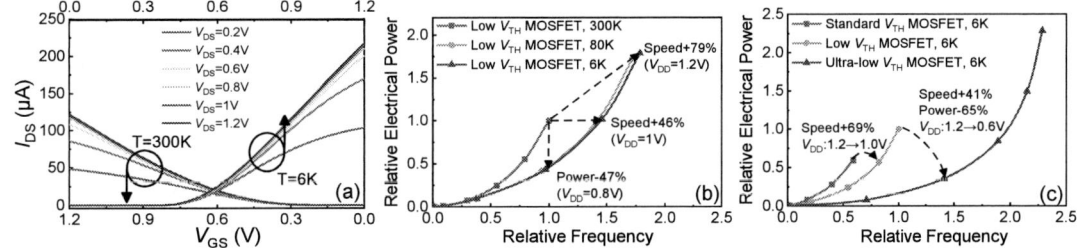

Fig. 5: (a) I-V curves in 1 μm/1 μm LVT MOSFETs under various V_{DS} biases. Relations of the relative electrical power and relative frequency in (b) LVT MOSFETs at 300 K, 80 K, and 6 K, and (c) in SVT, LVT, and ultra-low V_{TH} MOSFETs at 6 K.

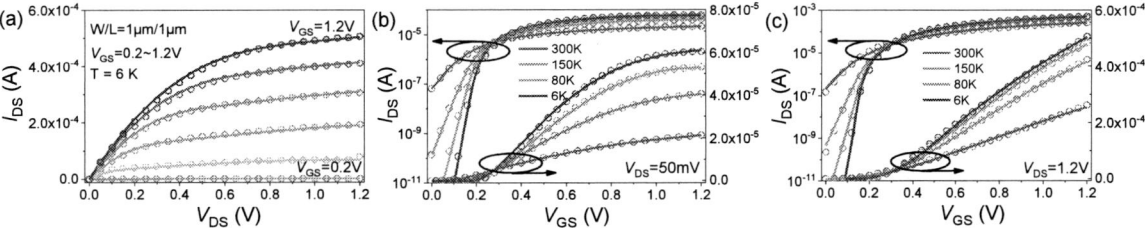

Fig. 6: Measured (symbol) and the model calculated (solid line) I-V curves of ultra-low V_{TH} MOSFETs at various temperatures.

979-8-3315-0417-5/25 $31.00 © 2025 IEEE

Low-Temperature Trench Ohmic Contact Suitable for Self-Aligned P-GaN HEMT

Zhiqiang Xue[1], Mao Jia[1], Qian Xiao[1], Bin Hou[1], Ling Yang[1], Xiaohua Ma[1], Yue Hao[1]

1 State Key Discipline Laboratory of Wide Band-Gap Semiconductor Technology, Xidian University, China

Abstract

This study explores a low-temperature ohmic contact approach suitable for self-aligned P-GaN trench etching. Through simulations, the impact of Mg diffusion on ohmic contact and trench etching improvement was investigated, followed by experimental verification. Different depths of Ti-based and Ta-based metal etching were explored. The final result achieved was Ta-based metal annealed at 550 °C for 5 minutes, yielding a low contact resistance of $3.53 \times 10^{-5}\,\Omega\cdot cm^2$ and $1.25\,\Omega\cdot mm$. Atomic Force Microscopy (AFM) revealed a smooth surface.

Keywords: P-GaN HEMT, self-aligned gate, Mg diffusion, ohmic trench, low temperature

Introduction

Enhancement-mode high electron mobility transistors (HEMTs) with P-GaN gate technology have gained favor for mass production and commercial applications due to their stable threshold voltage, high reliability, process controllability, and good reproducibility [1]. To fully exploit the potential of P-GaN HEMTs in power applications, addressing issues such as gate length reduction, process stability, and reduced on-resistance is crucial [2]. The self-aligning process effectively resolves the issues of high on-resistance and gate length scaling caused by the traditional gate metal electrode's smaller size compared to the gate area. In the self-aligned process, the gate is formed before the ohmic contact. Conventional 860 °C Ti-based metals can degrade the gate, resulting in increased gate leakage current and reduced reliability [3]. The presence of the P-GaN layer introduces Mg diffusion issues influenced by growth temperature, time, Mg concentration in the P-GaN layer, and activation temperature. Mg impurities can render the AlGaN layer weakly p-type, affecting the previously developed ohmic system. Therefore, it is necessary to study self-aligned low-temperature ohmic processes suitable for P-GaN.

Theoretical Details

Under ideal conditions, the distribution of Mg-doped impurities in the P-GaN layer should approximate a rectangular profile, as shown in Figure 1(a), where the Mg impurity concentration in the P-GaN layer is relatively high, while the Mg impurity concentration in the AlGaN layer approaches zero. However, in practical applications, due to the high-temperature growth of heterostructures and subsequent high-temperature processing, the Mg doping concentration can diffuse from the P-GaN layer into the underlying AlGaN layer, rendering the AlGaN layer p-type [4].

Simulations and experimental were employed to analyze the impact of Mg impurity diffusion on the ohmic contact and electrical characteristics of the devices. The material parameters were sourced from relevant research articles [5], as shown in Figure 2. Simulations were performed using the Silvaco TCAD software. To simplify the model, based on the simplified analysis in the literature, the Mg concentration in the P-GaN layer was set to a uniform distribution, and the Mg diffusion profile in the AlGaN layer was approximated by a Gaussian distribution, as shown in Figure 3.

Results And Discussion

Band structure simulations were conducted for the AlGaN layer in the ohmic regions of HEMT heterostructures with Mg diffusion models, and a comparative study was performed with devices without Mg diffusion models. In Figure 3(a), the comparison of band diagrams reveals that Mg diffusion increases the barrier of the AlGaN layer. To address this problem, a recessed etching scheme for the source/drain regions of P-GaN HEMTs is proposed. Since the Mg diffusion peak is at the interface and decreases rapidly with depth, recessed etching of the source/drain regions effectively reduces Mg impurities. This approach is compatible with conventional deep ohmic trenches, increasing electron transport paths and reducing contact resistance.

Band structures under the gate region were also simulated

979-8-3315-0417-5/25 $31.00 © 2025 IEEE

and compared, as shown in Figure 3(c). Mg diffusion raises the AlGaN layer's band structure, increasing the HEMT's threshold voltage, as depicted in Figure 4. The threshold voltage was extracted at a current density of 1 mA/mm, measuring 1.7 V for HEMT without Mg diffusion, 2.1 V for HEMT with diffused, and 2.2 V for recessed etched HEMT. The threshold voltage for both Devices with Mg diffusion models increased by approximately 0.3 V compared to the device without diffused model. However, recessed etching lowered the metal/semiconductor barrier, reducing contact resistance, resulting in slightly higher current at the same voltage compared to non-etched devices.

Based on the simulation work, low-temperature ohmic contact experiments suitable for self-aligned processes were conducted. Experimental results indicate that annealing the Ti/Al/Ni/Au metal stack at 500°C for 5 minutes in an N_2 atmosphere yields the best performance. As shown in Figure 5(a), the devices exhibits Schottky characteristics at different etching depths under annealing at 500 °C. The formation of ohmic contacts with Ti-based metals is primarily attributed to the reaction between Ti and the underlying AlGaN, forming TiN. TiN creates nitrogen vacancies within the barrier layer, facilitating carrier tunneling. However, the formation of TiN requires specific temperature conditions, and experimental results indicate that low temperatures around 500 °C may hinder this reaction.

Studies have shown that Ta-based metal stack can achieve better ohmic contacts at lower temperatures [6], as Ta has a lower work function compared to Ti, Facilitating the formation of ohmic contacts with barriers that have Mg-diffused. Therefore, experiments were conducted using Ta-based metals. Figure 5 (b) shows the groove-etching experiment of the Ta/Al/Ni/Au stack. Combined with deep ohmic trenches, the stack is annealed at 550°C in an N_2 environment for 5 minutes, resulting in low-resistance ohmic contacts (3.53×10^{-5} Ω·cm², 1.25 Ω·mm), as shown in Figure 6. The excellent ohmic properties are attributed to the combination of TaN and recessed etching. At the low temperature of 550 °C, TaN forms stably, creating high concentrations of nitrogen vacancies near the interface, rendering AlGaN heavily

n-doped. Trench etching effectively removes diffused Mg impurities and reduces barrier thickness, promoting electron tunneling. Atomic force microscopy (AFM) revealed smooth surface morphology with a root mean square (RMS) roughness of 25 nm (Figure 7). Compared to the conventional 860°C Ti-based system, the metal surface was smoother [7].

Conclusion

Simulation results indicate that the diffusion of Mg in the P-GaN layer increases the barrier height of the AlGaN layer, adversely affecting the ohmic contact in the source/drain regions of the devices. By exploring the low-temperature performance of Ti and Ta metal systems and combining recessed etching with Ta-based metals, a low-temperature ohmic contact solution suitable for self-aligned processes was developed. This approach not only mitigates the impact of Mg impurities but also achieves low ohmic contact resistance (3.53×10^{-5} Ω·cm², 1.25 Ω·mm).

Acknowledgments

This work was supported by the National Natural Science Foundation of China Grant Nos. 62474135.

References

[1] S. Kumar et al. "1.2 kV Enhancement-Mode p-GaN Gate HEMTs on 200 mm Engineered Substrates." Electron Device Letters, vol. 45, no. 4, pp. 657-660, 2024,

[2] G. Lükens et al. "Self-Aligned Process for Selectively Etched p-GaN-Gated AlGaN/GaN-on-Si HFETs," Transactions on Electron Devices, vol. 65, no. 9, pp. 3732-3738, 2018,

[3] Anthony Calzolaro et al. "Material investigations for improving stability of Au free Ta/Al-based ohmic contacts annealed at low temperature for AlGaN/GaN heterostructures." Semicond. Sci. Technol. 35 075011, 2020

[4] N. E. Posthuma et al. "Impact of Mg out-diffusion and activation on the p-GaN gate HEMT device performance." 2016 28th International Symposium on Power Semiconductor Devices and ICs, pp. 95-98, 2016

[5] M. Jia et al. "High VTH and Improved Gate Reliability in P-GaN Gate HEMTs With Oxidation Interlayer," Electron Device Letters, vol. 44, no. 9, pp. 1404-1407, 2023

[6] A Malmros et al. "Electrical properties, microstructure, and thermal stability of Ta-based ohmic contacts annealed at low temperature for GaN HEMTs." Semiconductor Science and Technology 26, (7), 2011

[7] Ki Hong Kim et al. "Investigation of Ta/Ti/Al/Ni/Au ohmic contact to AlGaN/GaN heterostructure field-effect transistor". J. Vac. Sci. Technol. B 1 January 23 (1): 322–326, 2005.

979-8-3315-0417-5/25 $31.00 © 2025 IEEE

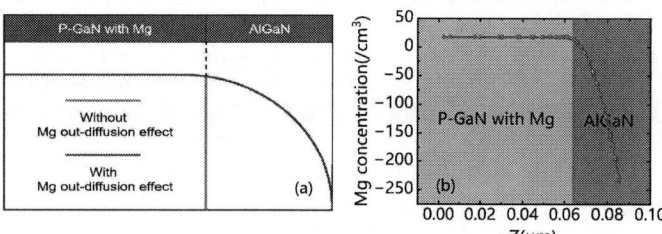

Fig. 1: (a) Distribution of Mg impurities in P-GaN/AlGaN materials

(b) Mg distribution extracted by simulation

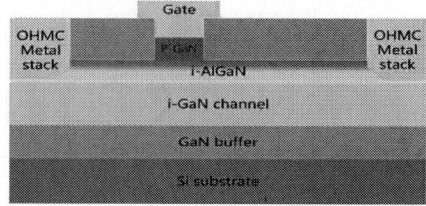

Fig. 2: Device structure with Mg diffusion

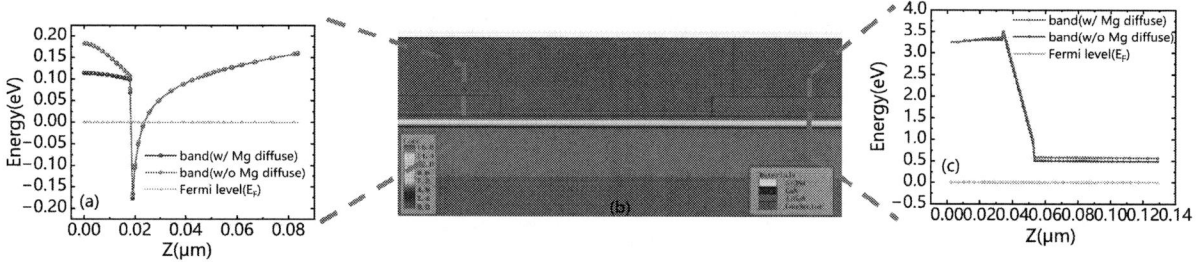

Fig. 3: (a) band diagram of ohm with Mg diffusion (b) Mg impurity distribution simulation (c) band diagram of gate Mg diffusion

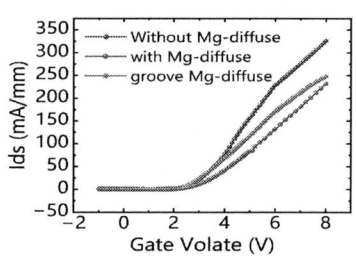

Fig. 4: simulation of transfer current

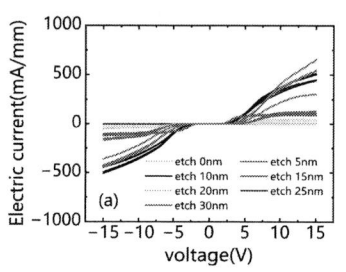

Fig. 5: Current curves at different etching depths(a)Ti base(b)Ta base

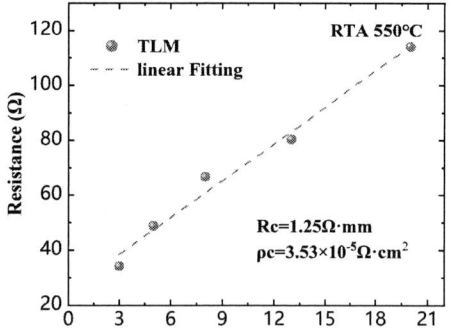

Fig. 6: 550°C 5min Ta/Al/Ni/Au TLM

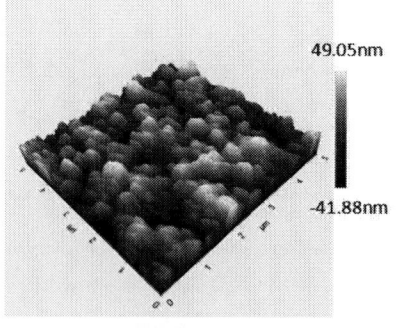

Fig. 7: AFM surface roughness measurement

979-8-3315-0417-5/25 $31.00 © 2025 IEEE   446

Enhancing Heat Dissipation in GaN-on-Diamond HEMTs through Device-First Transfer Bonding

Shiming Li, [1,] * Mei Wu, [1] Xiaohan He, [1] Haolun Sun, [1] Ling Yang, [1] Xiaohua Ma, [1] and Yue Hao [1]

[1] *School of Microelectronics, Xidian University, Xi'an 710071, China*

*E-mail: smli@stu.xidian.edu.cn

Abstract

In this work, GaN-on-SiC HEMTs were transferred and bonded to single-crystal diamond substrate using room-temperature surface-activated bonding (SAB), resulting in GaN-on-Diamond HEMTs. Due to the enhanced heat dissipation capacity, the GaN-on-Diamond HEMTs exhibited outstanding RF characteristics. With an increase in the drain bias voltage, the P_{out} of the GaN-on-Diamond HEMTs increased linearly up to 13.1 W/mm, with the PAE consistently exceeding 60%. This work provides an effective thermal management solution for high-power RF GaN HEMTs.

Keywords: GaN HEMTs, Diamond, bonding, heat dissipation

Introduction

In the past two decades, RF GaN HEMTs have rapidly advanced, with output power density (P_{out}) exceeding 30 W/mm [1], [2]. However, as the P_{out} increases, the self-heating effects of the devices become more severe. This is particularly true for multi-gate devices (gate width greater than 1 mm) used in MMICs, where the increased heat sources lead to more significant thermal accumulation. Gerrer et al. demonstrated through simulations that even with commercial high thermal conductivity SiC substrate, a DC power density (P_{DC}) consumption of only 8 W/mm resulted in peak temperatures exceeding 300 °C [3].

Diamond, the material with the highest thermal conductivity in nature, is considered the optimal passive cooling solution to address the thermal bottlenecks in GaN HEMTs. However, there are significant thermal and lattice mismatches between GaN and diamond, leading to poor performance of GaN-on-Diamond HEMTs when directly grown [4]. An alternative solution, the device-first approach, effectively avoids material mismatch issues [5]. In this method, GaN HEMTs are first fabricated, and after thinning the supporting substrate, the devices are transferred and bonded to diamond substrate using room-temperature surface-activated bonding (SAB) techniques, resulting in GaN-on-Diamond HEMTs.

In 2017, Liu et al. achieved 3-inch GaN-on-Diamond HEMTs using the device-first approach, but the introduced interface thermal resistance exceeded 50 m²K/GW [6]. Furthermore, this work completely removed the supporting substrate without reporting its impact on key electrical parameters (breakdown voltage, off-state leakage, threshold voltage, etc.) of GaN HEMTs. The thickness of the supporting substrate significantly affects the electrical characteristics and reliability of the devices [7]. In 2021, Ohki et al. transferred GaN HEMTs to single-crystal diamond substrates using the device-first method, achieving P_{out} exceeding 20 W/mm in the S-band [8]. By retaining a 50 μm thick supporting substrate, the breakdown voltage (BV) and threshold voltage (V_{th}) of the devices showed degradation-free. Thus, maintaining a certain thickness of the supporting substrate can maximize thermal benefits without degrading device performance.

In this work, we thinned the supporting substrate to 15 μm and successfully realized GaN-on-Diamond HEMTs using device-first method. The 15 μm thick supporting substrate effectively protected the heterojunction, and the electrical characteristics such as off-state leakage, V_{th}, and BV remained unchanged after transfer. Moreover, due to the ultra-thin supporting substrate and the low interface thermal resistance (TBR), the GaN-on-Diamond HEMTs demonstrated excellent heat dissipation capacity. The P_{out} of the GaN-on-Diamond HEMTs showed a linear increase with increasing drain bias (V_{DS}) resulting from improved thermal dissipation capability. When V_{DS} = 60 V, the P_{out} and power-added efficiency (PAE) of GaN-on-Diamond HEMTs were 1.8 W/mm and 11% higher than those of GaN-on-SiC HEMTs, respectively.

Device structure and fabrication

This work employed the device-first approach to transfer and bond GaN HEMTs to single-crystal diamond substrate at room temperature, as illustrated in Fig. 1. The GaN/AlGaN heterojunction was grown on a 3-inch SiC substrate. The device fabrication followed the standard GaN process established at Xidian University [9]. After device fabrication, a temporary supporting substrate protected the front of the devices while mechanical polishing removed the SiC supporting substrate. Chemical mechanical polishing (CMP) was then applied to reduce the roughness of the SiC substrate surface to below 1 nm. The thinned 3-inch the GaN-on-SiC HEMTs were diced into 10 mm × 10 mm dies. In a SAB chamber with a vacuum of 1×10^{-5} Pa, a layer of Al_2O_3 with a thickness of less than 5 nm was deposited on the polished surfaces of both samples. The two samples were then aligned and bonded under a pressure of 10 MPa for 5 minutes, resulting in the GaN-on-Diamond HEMTs.

979-8-3315-0417-5/25 $31.00 © 2025 IEEE

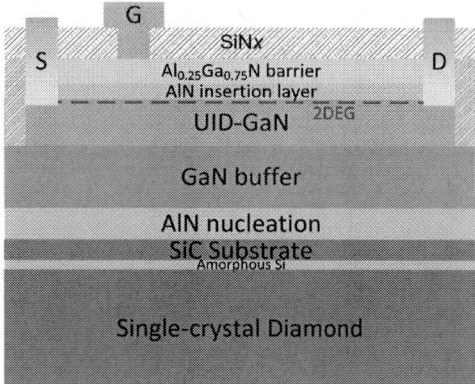

Fig. 1: Schematic structure of GaN-on-Diamond HEMTs.

The interface scanning electron microscope (SEM) images of the transferred the GaN-on-Diamond HEMTs are shown in Fig. 2. Due to the thickness of bonding interlayer less than 10 nm, making it nearly invisible at a magnification of 15,000 in the SEM. The nanoscale bonding interlayer resulted in an TBR of less than 10 m²K/GW, facilitating rapid heat dissipation in the devices [10]. The devices had a gate length (L_G) of 0.5 μm and a source-drain spacing (L_{SD}) of 6 μm, with multiple gate widths (W_G): 2×100 μm, 4×100 μm, 8×100 μm, 10×125 μm, and 12×150 μm. The same device before and after transfer was positioned for electrical and RF performance testing.

Results and discussion

The transfer characteristics and breakdown characteristics of the devices were tested at 1500 A before and after transfer bonding. Fig. 3 (a) and 3 (b) show the transfer characteristics of the same device before and after transfer. The V_{th}, peak transconductance (G_m), and $I_{on/off}$ of the GaN-on-SiC HEMTs were -3.3 V, 251 mS/mm, and ~1.6×10⁻⁶, respectively. The V_{th}, peak G_m, and $I_{on/off}$ of the GaN-on-Diamond HEMTs were -3.3 V, 260 mS/mm, and ~1.5×10⁻⁶, respectively. The breakdown characteristics of the same device before and after transfer are shown in Fig. 3 (c) and 3 (d). Both devices did not exhibit avalanche breakdown even when the devices

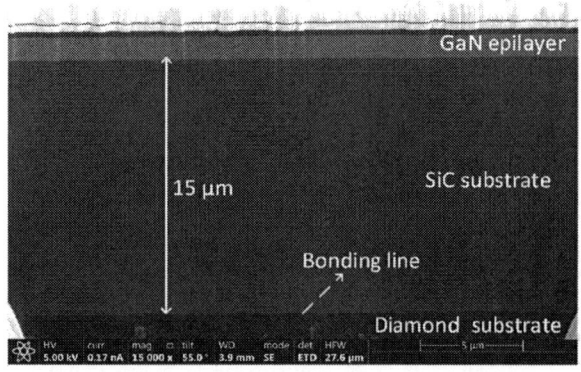

Fig. 2: Cross-sectional SEM image of the GaN-on-Diamond HEMTs.

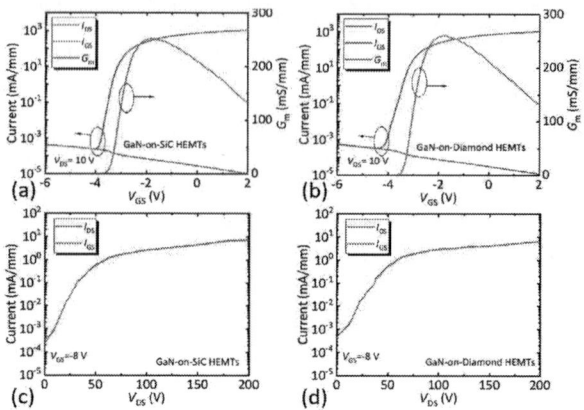

Fig. 3: Transfer characteristics of (a) the GaN-on-SiC HEMTs and (b) the GaN-on-Diamond HEMTs. Three-terminal breakdown characteristics of (c) the GaN-on- SiC HEMTs and (d) the GaN-on-Diamond HEMTs at V_{GS} = -8 V. The same device was tested before and after the transfer with a W_G of 2 × 100 μm.

reached the limits of the testing equipment (±200 V). Thus, thinning the substrate to 15 μm effectively supports the GaN/AlGaN heterojunction without degrading the performance of GaN HEMTs.

The output characteristics of the same device before and after transfer were characterized with a W_G of 2×100 μm, as shown in Fig. 4 (a) and 4 (b). V_{DS} increased from 0 V to 20 V in 0.1 V steps, while V_{GS} increased from -4 V to 2 V in 1 V steps. As the P_{DC} increased, the cooling effect became more apparent. At V_{GS} = 2 V, the saturation current (I_{sat}) of the device increased by 75 mA/mm, and the corresponding on-resistance (R_{on}) decreased to 2.6 Ω·mm. This enhancement benefits GaN HEMTs by providing higher P_{out} and PAE. Additionally, the output characteristics of devices with different W_G at V_{GS} = 0 V were characterized, as shown in Fig. 4 (c). Although the P_{DC} was the same, the degradation of I_{DS}

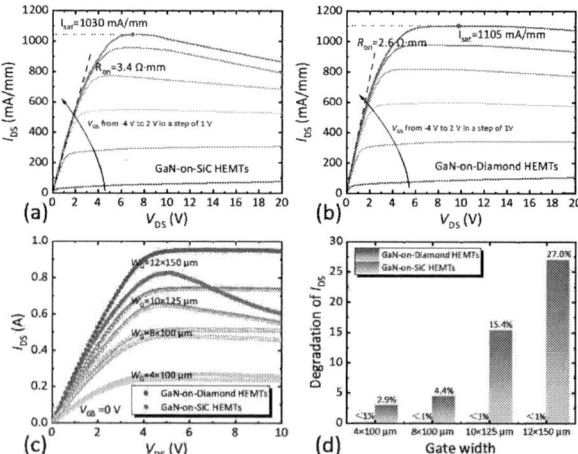

Fig. 4: Output characteristics of (a) the GaN-on-SiC HEMTs and (b) the GaN-on-Diamond HEMTs with a W_G of 2 × 100 um. I-V characteristics of two samples with different W_G at V_{GS} =0 V. Comparison of the degradation of I_{DS} for the two samples.

979-8-3315-0417-5/25 $31.00 © 2025 IEEE

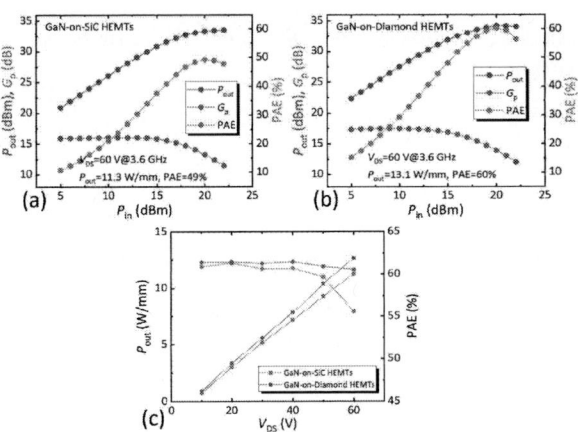

Fig. 5: Large-signal characteristics of (a) the GaN-on-SiC HEMTs and (b) the GaN-on-Diamond HEMTs at V_{DS}=60 V. (c) P_{out} and PAE versus V_{DS} for both samples.

caused by self-heating effects became more pronounced with increasing W_G. The degradation of I_{DS} was defined as: (I_{sat} - $I_{DS}@10V)/I_{sat}$) ×100%, and the degradation of I_{DS} devices for different W_G was summarized in Fig. 4 (d). It was observed that the degradation of the GaN-on-SiC HEMTs with a total W_G of 1.8 mm (12×150 μm) reached as high as 27%, while the the GaN-on-Diamond HEMTs exhibited almost no significant degradation. These results not only demonstrate the importance of the diamond substrate for heat dissipation in GaN HEMTs but also confirm that the 15 μm ultra-thin supporting substrate does not affect the thermal benefits of the GaN-on-Diamond structure.

Furthermore, large-signal characteristics of both samples were characterized through continuous wave (CW) carrier testing in the 3.6 GHz band. The GaN-on-SiC HEMTs and the GaN-on-Diamond HEMTs tested were the same device with a W_G of 2×100 μm. When V_{DS} = 60 V, the P_{out} and PAE of the GaN-on-SiC HEMTs were 11.3 W/mm and 49% (Fig. 5 (a)), while those of the GaN-on-Diamond HEMTs were 13.1 W/mm and 60% (Fig. 5 (b)). The high thermal conductivity of the diamond substrate significantly enhanced the P_{out} of the GaN-on-Diamond HEMTs, and the PAE showed no apparent degradation trend. The trends of P_{out} and PAE as V_{DS} varied before and after transfer bonding are shown in Fig. 5 (c). The P_{out} of the GaN-on-Diamond HEMTs, with enhancing heat dissipation capability, increased linearly with V_{DS}, showing degradation-free trend, and the PAE remained above 60%. These results demonstrate that the GaN-on-Diamond HEMTs with ultra-thin supporting substrates can effectively address the thermal accumulation issues present in high-power devices.

Conclusion

In conclusion, this work successfully demonstrates the fabrication of the GaN-on-Diamond HEMTs using a device-first approach with a 15 μm thick supporting substrate. The retained substrate effectively protects the heterojunction,

maintaining key electrical characteristics post-transfer. The implementation of room-temperature SAB resulted in low TBR, significantly enhancing heat dissipation capability. Consequently, the GaN-on-Diamond HEMTs exhibited improved P_{out} and PAE, addressing thermal accumulation issues common in high-power devices. This approach offers a promising solution for the development of high-performance RF GaN HEMTs with excellent heat dissipation capability.

Acknowledgments

Project supported by the National Natural Science Foundation of China (Grant Nos. 62234009, 62090014, 62104184, 62188102, 62104178, and 62104179)

References

[1] Y. -f. Wu, M. Moore, A. Saxler, T. Wisleder, and P. Parikh, "40-W/mm Double Field-plated GaN HEMTs," in *2006 64th Device Research Conference*, Jun. 2006, pp. 151–152. doi: 10.1109/DRC.2006.305162.

[2] L. Yang *et al.*, "Record Power Performance of 33.1 W/mm with 62.9% PAE at X-band and 14.4 W/mm at Ka-band from AlGaN/GaN/AlN:Fe Heterostucture," in *2023 International Electron Devices Meeting (IEDM)*, Dec. 2023, pp. 1–4. doi: 10.1109/IEDM45741.2023.10413780.

[3] T. Gerrer *et al.*, "Thermal Design Rules of AlGaN/GaN-Based Microwave Transistors on Diamond," *IEEE Trans. Electron Devices*, vol. 68, no. 4, pp. 1530–1536, Apr. 2021, doi: 10.1109/TED.2021.3061319.

[4] J. G. Felbinger *et al.*, "Comparison of GaN HEMTs on Diamond and SiC Substrates," *IEEE Electron Device Lett.*, vol. 28, no. 11, pp. 948–950, Nov. 2007, doi: 10.1109/LED.2007.908490.

[5] J. Liang *et al.*, "Fabrication of GaN/Diamond Heterointerface and Interfacial Chemical Bonding State for Highly Efficient Device Design," *Adv. Mater.*, vol. 33, no. 43, p. 2104564, Oct. 2021, doi: 10.1002/adma.202104564.

[6] T. Liu *et al.*, "3-inch GaN-on-Diamond HEMTs With Device-First Transfer Technology," *IEEE Electron Device Lett.*, vol. 38, no. 10, pp. 1417–1420, Oct. 2017, doi: 10.1109/LED.2017.2737526.

[7] L. Heuken *et al.*, "Temperature dependent lateral and vertical conduction mechanisms in AlGaN/GaN HEMT on thinned silicon substrate," *Jpn. J. Appl. Phys.*, vol. 58, no. SC, p. SCCD11, Jun. 2019, doi: 10.7567/1347-4065/ab0406.

[8] T. Ohki *et al.*, "An Over 20-W/mm S-Band InAlGaN/GaN HEMT With SiC/Diamond-Bonded Heat Spreader," *IEEE Electron Device Lett.*, vol. 40, no. 2, pp. 287–290, Feb. 2019, doi: 10.1109/LED.2018.2884918.

[9] S. Li *et al.*, "Enhanced Performance of GaN HEMTs in X-band Applications Using SixN/Si3N4 Bilayer Passivation Technique," *Phys. Status Solid A*, p. 2400047, May 2024, doi: 10.1002/pssa.202400047.

[10] Z. Cheng, F. Mu, L. Yates, T. Suga, and S. Graham, "Interfacial Thermal Conductance across Room-Temperature-Bonded GaN/Diamond Interfaces for GaN-on-Diamond Devices," *ACS Appl. Mater. Interfaces*, vol. 12, no. 7, pp. 8376–8384, Feb. 2020, doi: 10.1021/acsami.9b16959.

5-bit High-Linearity UV-Stimulated Synaptic Device Based on MoS$_2$/GaN Heterostructure

Zijia Su[1,2], Yong Yan[1], Haiding Sun[1], Chengjie Zuo[1]

email: haiding@ustc.edu.cn & czuo@ustc.edu.cn

[1]School of Microelectronics, University of Science and Technology of China, Hefei, Anhui, China

[2]Institute of Microelectronics & Beijing National Research Center for Information Science and Technology and Department of Electronic Engineering, Tsinghua University, Beijing 100084, China

Abstract

This study introduces a MoS$_2$/GaN heterostructure functioning as an optoelectronic synapse device, demonstrating significant state current enhancement under ultraviolet (UV) light stimulation. The device achieves remarkably low nonlinearity with a factor of 0.086 and supports up to 32 distinct states. These characteristics make the MoS$_2$/GaN heterostructure an ideal candidate for neuromorphic computing applications, particularly in systems requiring precise, UV-induced state modulation and high state density.

Keywords: Heterostructure, Optoelectronic synapse, Low nonlinearity

Introduction

The continuous advancement in semiconductor technology has spurred the exploration of new materials and device architectures capable of delivering superior performance and novel functionalities. Among the emerging materials, GaN and MoS$_2$ stand out due to their unique electronic properties, making them ideal candidates for advanced optoelectronic devices. GaN, with its wide bandgap, is renowned for its excellent thermal stability, high electron mobility, and strong resistance to radiation [1]. MoS$_2$, on the other hand, is a two-dimensional (2D) material that offers exceptional optical and electrical properties, including a direct bandgap in monolayer form, high on/off current ratios, and strong photodetection capabilities [2]. In the context of non-volatile memory and neuromorphic computing, combining GaN and MoS$_2$ in a heterostructure structure presents several advantages over traditional devices. Compared to purely 2D floating-gate memory devices, the MoS$_2$/GaN heterostructure offers a simpler fabrication process. The deposition techniques for GaN and MoS$_2$ are well-established, and the heterostructure formation does not require complex layer stacking or intricate control of interlayer interactions, as often necessary in purely 2D materials. This simplicity in fabrication translates to higher yield and scalability, making the MoS$_2$/GaN heterostructure device a more practical option for large-scale production. Moreover, when compared to bulk materials, the MoS$_2$/GaN

heterostructure device exhibits significantly more sensitive photoelectric characteristics. The 2D nature of MoS$_2$, combined with the high surface area-to-volume ratio, leads to enhanced interaction with light, making the device more responsive to UV excitation. This increased sensitivity is crucial for the operation of the device under UV light, where the generation and separation of electron-hole pairs within the GaN layer are key to its functionality.

One of the most significant advantages of the MoS$_2$/GaN heterostructure device is its high linearity. Linearity is a critical parameter in neuromorphic computing architectures, where analog computing often relies on the precise modulation of current or voltage levels to represent different states [3]. High linearity ensures that each state can be distinctly defined and manipulated without significant cross-talk or non-linear distortion. This is particularly important in neural network implementations, where the accuracy of weight updates (analogous to memory states in this device) directly impacts the learning efficiency and performance of the network [4].

In neuromorphic systems, where computations are performed in a manner analogous to biological neural networks, the ability to store and process information with high linearity directly correlates with the system's ability to emulate the brain's complex signal processing [5]. The MoS$_2$/GaN device, with its 32 distinct states and high linearity, is thus well-suited for such applications. It can effectively serve as a memory element or synapse within a neuromorphic architecture, offering both the storage density and precision needed for advanced computational tasks.

This work aims to demonstrate the MoS$_2$/GaN heterostructure device's potential in these emerging fields, particularly focusing on its enhanced photoelectric characteristics, simplified fabrication process, and superior linearity. The experimental results presented in this paper underscore the device's viability as a non-volatile memory component and its applicability in neuromorphic computing, highlighting its role in the future of integrated computing architectures.

979-8-3315-0417-5/25 $31.00 © 2025 IEEE

Result

A. Basic Characterization

A 3D schematic diagram and an optical image of the fabricated device are presented in Fig. 1a-b. Current-voltage (I-V) measurements were conducted at room temperature to assess the heterostructure's electrical performance. Raman spectroscopy revealed predominant vibrational modes, specifically E_{2g}^1 and A_g^1, corresponding to in-plane and out-of-plane vibrations, respectively (as shown in Fig. 1c). These modes were observed at frequencies of 383.8 cm^{-1} and 408.3 cm^{-1}. The separation of 24.5 cm^{-1} between these Raman peaks indicates the presence of multilayered MoS$_2$ [6].

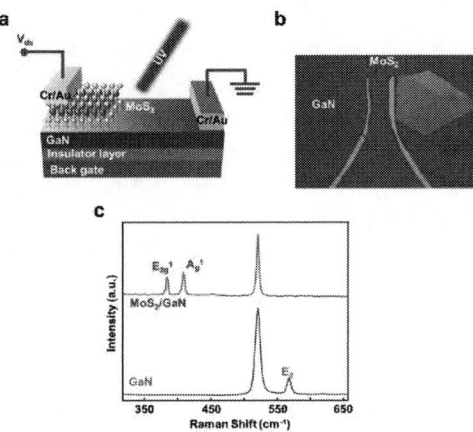

Fig. 1: (a) Schematic diagram (b) Optical image and (c) Raman spectra of MoS$_2$/GaN heterostructure.

B. Rectification Characterization

Fig. 2 illustrates the rectification characteristics of the MoS$_2$/GaN heterostructure device. In Fig. 2(a), the contour plot shows the variation of drain-source current (I_{ds}) as a function of both back-gate voltage (V_{BG}) and drain-source voltage (V_{ds}). The current increases significantly with positive V_{ds} under various gate voltages, indicating strong rectification. Fig. 2(b) presents the I-V curves for different UV wavelengths, where shorter wavelengths lead to higher I_{ds}, demonstrating the device's sensitivity and enhanced rectification under UV illumination.

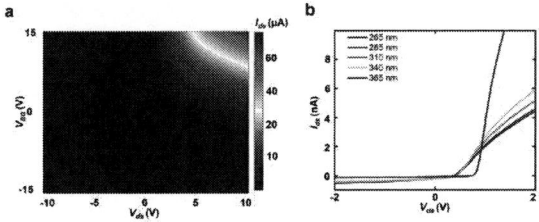

Fig. 2: (a) Contour plot of the drain-source current I_{ds} as a function of back-gate voltage V_{BG} and drain-source voltage V_{ds}. (b) I-V characteristics of the MoS$_2$/GaN heterostructure device under different UV wavelengths, demonstrating the wavelength-dependent rectification behavior.

C. Optoelectronic Characterization

To evaluate the storage capability of the heterostructure device under various wavelengths of light stimulation, a bias of 500 mV was applied to the MoS$_2$/GaN heterostructure. Photocontrol was achieved by modulating the optical shutter using spike pulse generator unit (SPGU) pulses, where the shutter's on and off states controlled the exposure. The irradiation time was set to 0.5 seconds, and light sources with wavelengths of 655, 532, 405, and 310 nm were employed with varying intensities. At the same time, a voltage pulse (+5 V, 100 ms) is applied to the back gate. The device's current response to the light pulses was measured, and the relationships between photocurrent, pulse count, wavelength, and light intensity are plotted in Fig. 3(a) to 3(d).

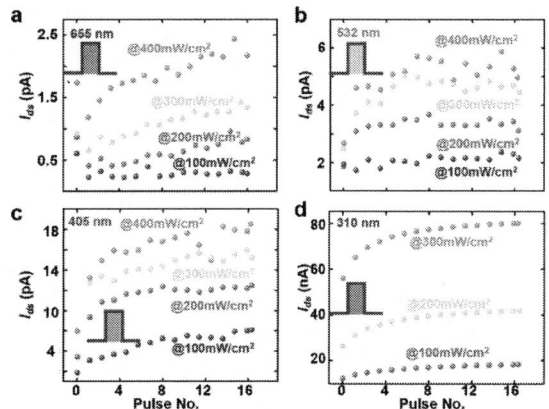

Fig. 3: Drain-source current I_{ds} response of the GaN/MoS$_2$ heterostructure device to light pulses at different wavelengths: (a) 655 nm, (b) 532 nm, (c) 405 nm, and (d) 310 nm.

Under visible light irradiation (655, 532, 405 nm), the photocurrent generally increased with the number of light pulses, though not strictly monotonically. Instances of current decrease with an increase in pulse count were observed, leading to unclear state distinctions, which could compromise the device's accuracy as a memory storage unit. Conversely, under 310 nm ultraviolet light, the state current exhibited a strictly monotonic increase with the number of pulses and eventually reached saturation. Moreover, the state current significantly increased with rising light intensity.

This behavior can be explained by the band mechanism depicted in Fig. 4. Under ultraviolet (UV) illumination, a large number of electron-hole pairs are generated on the surface of GaN. When a positive gate pulse is applied, holes move vertically towards the MoS$_2$ layer and recombine with the electrons induced on the MoS$_2$ surface, as depicted in the band diagram of Fig. 4(b) (left). After the UV light and pulse are removed, a substantial amount of residual charge

979-8-3315-0417-5/25 $31.00 © 2025 IEEE

accumulates on the GaN surface, which tends to diffuse towards regions with lower concentration, resulting in a higher channel potential. In the band diagram (Fig. 4(b), right), the accumulated residual charge modulates the barrier height and width at the heterostructure, facilitating electron tunneling. This method of modulating the channel current through the accumulation of residual charges at the heterostructure interface, controlled by UV light and voltage pulses, is promising for neuromorphic applications, where the changes in source-drain current correspond to postsynaptic excitation.

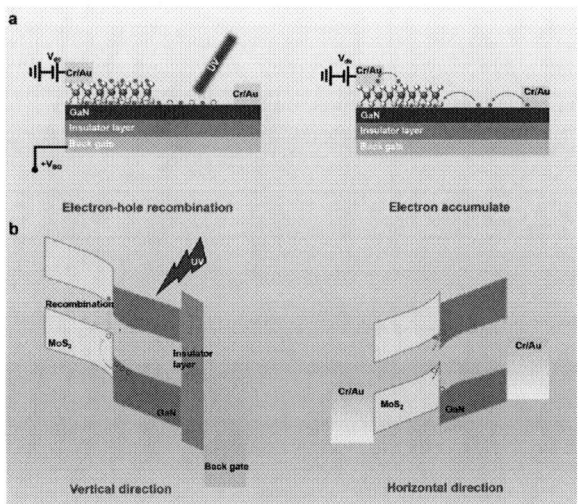

Fig. 4: (a) Carrier transport diagram: the left shows carrier dynamics during pulse stimulation, and the right shows the situation after the pulse is removed. (b) Band diagrams: UV-induced recombination (left) and residual charge modulation of the heterostructure for enhanced tunneling (right).

To enhance the linearity of the state current, the device's response was fine-tuned by reducing the light intensity and exposure duration, ensuring operation below the saturation threshold. By decreasing the exposure time to 100 milliseconds and setting the light intensity to 50 mW/cm², the device was subjected to a sequence of 32 pulses. The resulting photocurrent is shown in Fig. 5(a). Following the pulses, the state current stabilized at 0.65 nA. The relationship between the postsynaptic current (PSC) and the number of pulses can be fitted using the following equation, plotted in Fig. 5(b) [7].

$$G_p = B[1-exp(-p/v)]+G_{min} \qquad (1)$$

Where G_p, G_{min}, P, B, and v represent the conductance in the potentiation, minimum conductance, normalized number of pulses, coefficient, and nonlinearity, respectively. By fitting Eq. (1), we extracted a nonlinearity of 0.086 in the 5-bit (32 states) memory state.

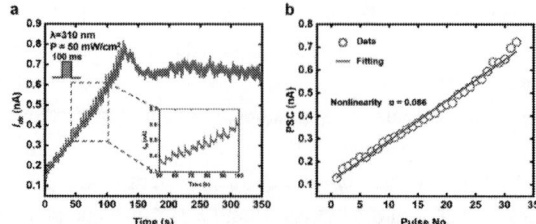

Fig. 5: (a) Time-resolved photocurrent I_{ds} response of the GaN/MoS$_2$ heterostructure device under continuous UV light stimuli @310nm&50mW/cm⁻¹. (b) Relationship between the photocurrent and the number of UV light pulses.

Conclusion

The MoS$_2$/GaN heterostructure device, acting as an optoelectronic synapse, exhibits excellent performance under UV light stimulation, with a low nonlinearity factor reaching 0.086 and the capability to differentiate 32 distinct states (5-bit). This high linearity and state density highlight the device's potential for neuromorphic computing, where accurate and reliable state control is essential. The findings suggest that this heterostructure is a promising platform for developing advanced optoelectronic synaptic devices.

Acknowledgments

This work was supported in part by the National Key Research and Development Program of China under Grant 2023YFB3610500, in part by the National Natural Science Foundation of China (Grant No. 62231023, 62322410), and in part by the USTC Center for Micro and Nanoscale Research and Fabrication.

References

[1] Z. Y. Zheng et al., "Gallium nitride-based complementary logic integrated circuits", Nat Electron, vol. 4, no. 8, pp. 595-603 (2021).

[2] B. Radisavljevic, A. Radenovic, J. Brivio, V. Giacometti, and A. Kis, "Single-layer MoS$_2$ transistors", Nat Nanotechnol, vol. 6, no. 3, pp. 147-150 (2011).

[3] I. Boybat et al., "Neuromorphic computing with multi-memristive synapses", Nat Commun, vol. 9, no. 2514 (2018).

[4] S. Ambrogio et al., "Equivalent-accuracy accelerated neural-network training using analogue memory", Nature, vol. 558, no. 7708, pp. 60-67 (2018).

[5] S. Oh, J. J. Lee, S. Seo, G. Yoo, and J. H. Park, "Photoelectroactive artificial synapse and its application to biosignal pattern recognition", Npj 2d Mater Appl, vol. 5, no. 95 (2021).

[6] B. J. Akeredolu et al., "Improved liquid phase exfoliation technique for the fabrication of MoS2/graphene heterostructure-based photodetector", Heliyon, vol. 10, no. 3 (2024).

[7] Y. Chen, Z. Wang, J. Du, C. Si, C. Jiang, and S. Yang, "Wrinkled Rhenium Disulfide for Anisotropic Nonvolatile Memory and Multiple Artificial Neuromorphic Synapses," ACS Nano, 2024/10/21 (2024)

Low Thermal Budget Ultrathin Ti Silicide for Advanced Backside Contact of Backside Power Delivery Network (BSPDN)

Hongxu Liao[1], Xijun Zhou[1], Fangze Liu[2], Lanyi Xie[3], Haixia Li[1], Jieyin Zhang[2], Jianjun Zhang[2], Xiaoyan Xu[1,4], Xia An[1,4], Heng Wu[1,4], Ru Huang[1,4], and Ming Li[1,4]*

[1]School of Integrated Circuits, Peking University, Beijing, 100871, China

[2]Beijing National Laboratory for Condensed Matter Physics and Institute of Physics, Chinese Academy of Sciences, Beijing, People's Republic of China

[3]Department of Energy and Resources Engineering, Peking University, Beijing 100871, China

[4]Beijing Advanced Innovation Center for Integrated Circuits, E-mail: liming.ime@pku.edu.cn

Abstract

The application of nanosecond laser annealing (NLA) for low-thermal-budget Ti silicide ($TiSi_x$) backside contact in Backside Power Delivery Network (BSPDN) is investigated. Silicon pre-amorphization reduces surface roughness by 90% and decreases energy density by 30%, enabling high-quality $TiSi_x$ film by NLA. Through precisely controlling the amorphous silicon thickness and energy density, self-limited ultrathin Ti silicide is achieved. Multiphysical simulations show that the NLA-induced $TiSi_x$ process can maintain Cu/low-k interconnect temperature below 400°C for BSPDN. Keywords: BSPDN, Ti silicide, Nanosecond laser annealing

Introduction

The Backside Power Delivery Network (BSPDN) is becoming essential as semiconductor technology approaches its physical scaling limitation [1-2]. By shortening power delivery paths, BSPDN reduces voltage drops and enhances signal performance, resulting in improved power efficiency and reduced interference [3]. However, BSPDN faces significant challenges by thermal constraints. When the frontside Cu/low-k interconnect is already formed, the backside interconnect processes must adhere to a stringent thermal budget (TB) (<400°C) to prevent damage to the frontside interconnect layers. This thermal limitation renders low-temperature backside source/drain (S/D) silicide contacts a critical bottleneck for BSPDN integration [4], presenting one of the major challenges to its implementation. Recently, nickel-based silicide has emerged as a potential solution for backside contacts within thermal budget limits. However, the severe diffusion of nickel in silicon presents significant reliability challenges for devices. While the incorporation of platinum aims to alleviate the diffusion issue, inevitable defects remain a concern [5]. In contrast, titanium-based silicide addresses the diffusion problem while maintaining low contact resistance. However, its formation temperature, which exceeds 450 °C, limits its applicability for the integration of BSPDN [6-7].

In this study, we investigate the feasibility of green nanosecond laser annealing (NLA) for low-TB Ti silicide ($TiSi_x$) S/D contacts in a BSPDN. By employing silicon pre-amorphization techniques, we successfully achieve $TiSi_x$

films with favorable morphology. Additionally, we design and validate a self-limited ultrathin Ti silicidation process by leveraging the free energy gap between crystalline silicon (c-Si) and amorphous silicon (a-Si). The heat penetration through the substrate and its impact on the Cu/low-k interconnects are investigated by COMSOL simulations, providing deep insights into the future optimization for BPSDN.

Experiment and simulation

In this study, the a-Si is introduced by ion implantation and molecular beam epitaxy (MBE) for comparison and verification of the self-limited silicidation.

Starting with a bare wafer, arsenic implantation at 33 keV with a dose of 4E14 cm^{-2} is firstly implemented to amorphize the substrate surface, followed by phosphorus implantation at 33 keV with a dose of 9.6E15 cm^{-2} for the contact doping. After removing the native oxide in diluted HF, a 15 nm Ti film is deposited using physical vapor deposition (PVD). In parallel, to serve as a comparative reference, an ultrathin a-Si film of 2 nm is deposited by MBE at room temperature to replace the ion implantation step. Green NLA ($\lambda = 527$ nm) is subsequently projected onto the surface of Ti film with energy density between 1.2 and 3.0 J/cm^2 and a spot size of 350×350 μm². After that, the unreacted Ti is removed by $H_2SO_4:H_2O_2$ (4:1) solution for further physical and electrical characterizations. For comparison, arsenic and phosphorus co-implanted wafers without Ti deposition, as well as bare wafers coated with 15 nm film Ti, are also subjected to the same laser annealing conditions.

To investigate the thermal impact of NLA induced Ti silicidation process on backside interconnect, which is critical for the BPSDN integration. Heat simulations of the full process are carried out with COMSOL, factors such as laser reflection, solid heat transfer, and laser absorption are taken into account, as shown in Fig. 2.

Result and discussion

A. Silicide formation by green NLA

Figure 3 illustrates the variation of sheet resistance and the content of Ti with the energy density of the Ti/c-Si system. The threshold energy density of silicide formation is

---

979-8-3315-0417-5/25 $31.00 © 2025 IEEE   453

identified to be between 1.6 and 1.8 J/cm², within which the sheet resistance decreases rapidly. Additionally, the Ti content increases in the energy density range for the annealed Ti/c-Si samples. At energy density of 2 J/cm², the sheet resistance reaches its minimum of 55 Ω/□. However, physical damage to the silicide occurs at higher energy densities, which results in increased sheet resistance. Figure 4 displays the TEM image of the Ti/c-Si sample at 2 J/cm², showing that the TiSi$_x$ film is 30 nm thick, discontinuous and droplet-shaped, rendering it unsuitable for three-dimensional devices like FinFET and gate-all-around (GAA) devices. Moreover, the surface roughness of 4.88 nm is unacceptable. The rough surface is attributed to the high formation energy of the reaction between titanium and c-Si, which results in a limited number of nucleation sites and their discontinuity. This leads to insufficient island expansion during short laser pulses.

B. Morphology improve by silicon amorphization

Since the Si-Si bonds in a-Si are generally weaker than c-Si, the formation energy for TiSi$_x$ can be reduced. Together with the increasing number of nucleation sites in a-Si, the continuous TiSi$_x$ film can then be formed by NLA. In the first pre-amorphization experiment, arsenic and phosphorus co-implantation is utilized to introduce a-Si for silicide formation in Ti/a-Si system.

Figure 5 illustrates the relationship between sheet resistance and energy density for the Ti/a-Si system and doped Si. At an energy density of 1.2 J/cm², the pure sheet resistance of doped Si as large as 2000 Ω/□ suggests the incomplete dopant activation while the sheet resistance of Ti/a-Si is only 130 Ω/□, which indicates that TiSi$_x$ is already formed. TEM images in Figure 6 confirms the continuous TiSi$_x$ formation at 1.2 J/cm² with thickness from 11 nm to 13 nm. In addition, its surface roughness of 0.353 nm, which is 90% lower than that of the Ti/c-Si system. Figure 7 shows 0.6 J/cm² gap in the formation energy between Ti/c-Si and Ti/a-Si.

C. Ultrathin Silicide design and verification

Utilizing the difference in formation energy, it is possible to obtain self-limited ultrathin Ti silicide through careful control of energy density and a-Si thickness (see Figure 8).

To validate this approach, we use MBE to grow 2 nm a-Si at room temperature and sputter 15 nm Ti film by PVD. The NLA energy density is 1.5 J/cm². Figure 9 presents the TEM image of the ultrathin TiSi$_x$ of 2.7 nm on c-Si, proving the successful self-limitation by a-Si thickness. Additionally, the TiSi$_x$/c-Si interface is exceptionally smooth, indicating minimal surface damage which is beneficial for S/D contact in advanced FinFET and GAA transistors.

D. Feasibility for backside contact application

Based on COMSOL simulations, we explore the feasibility of laser-induced Ti silicide for backside contact in BSPDN. The simulation structure is half an SRAM (1 PMOS + 2 NMOS), using a contact gate pitch of 54 nm and a cell height of 100 nm. Detailed parameters are shown in Table 1. Different green NLA pulse widths (60~300 ns) and energy densities are applied to maintain TiSi$_x$ formation temperature. As shown in Figure 10, the structure cools rapidly. When the pulse width reaches 90 ns, the peak temperature of Cu/low-k interconnect is below 400 °C while that of the TiSi$_x$ formation area exceeds 1100 °C. This balance ensures high-performance silicide formation without damaging backside interconnects, making it suitable for BSPDN applications.

Conclusion

We confirm the feasibility of green NLA for low TB backside contact in BSPDN by both experiments and simulations. By silicon amorphization, the surface roughness and the energy density with a high quality TiSi$_x$ film can be reduced by 90% and 30%, respectively. By controlling the thickness of a-Si, a 2.7 nm ultrathin TiSi$_x$ film is achieved. With proper isolation structure and NLA pulse condition, the effective temperature of Cu/low-k in BSPDN can be controlled well below 400 °C.

Acknowledgments

This work was partly supported by the National Key Research and Development Program of China (No. 2023YFB4402200), NSFC project (61927901) and Beijing Frontier Innovation Project (No. QYJS-2023-2302-B).

References

[1] A. Veloso, A. Jourdain, G. Hiblot, F. Schleicher and K. D'have et al., "Enabling Logic with Backside Connectivity via n-TSVs and its Potential as a Scaling Booster," IEEE Symposium on VLSI Technology, pp. 1-2 (2021).

[2] A. Veloso, B. Vermeersch, R. Chen, P. Matagne and M. Garcia Bardon et al., "Backside Power Delivery: Game Changer and Key Enabler of Advanced Logic Scaling and New STCO Opportunities," International Electron Devices Meeting (IEDM), pp. 1-4 (2023).

[3] B. -S. Kim, S. Choi, J. H. Lee, K. Lee and J. Park et al., "Expanding Design Technology Co-Optimization Potentials with Back-Side Interconnect Innovation," IEEE Symposium on VLSI Technology and Circuits, pp. 1-2 (2024).

[4] J. Park, J. Park, K. Hwang and J. Yun et al., "Highly manufacturable Self-Aigned Direct Backside Contact (SA-DBC) and Backside Gate Contact (BGC) for 3-dimensional Stacked FET at 48nm gate pitch," IEEE Symposium on VLSI Technology and Circuits, pp. 1-2 (2024).

[5] C. Lavoie, P Adusumilli, AV Carr, JSJ Sweet and AS Ozcan et al., "Contacts in advanced CMOS: History and emerging challenges," vol. 77, no. 5, p. 59 (2017).

[6] H. Xu et al., "Ultra-low Specific Contact Resistivity (3.2×10⁻¹⁰ Ω/cm²) of Ti/Si$_{0.5}$Ge$_{0.5}$ Contact: Deep Insights into the Role of Interface Reaction and Ga Co-doping," Symposium on VLSI Technology, pp. 1-2 (2021).

[7] C. Porret, L. Everaert, M. Schaekers, A. Ragnarsson and A. Hikavyy et al., "Low temperature source/drain epitaxy and functional silicides: essentials for ultimate contact scaling," International Electron Devices Meeting (IEDM), pp. 34.1.1-34.1.4(2022).

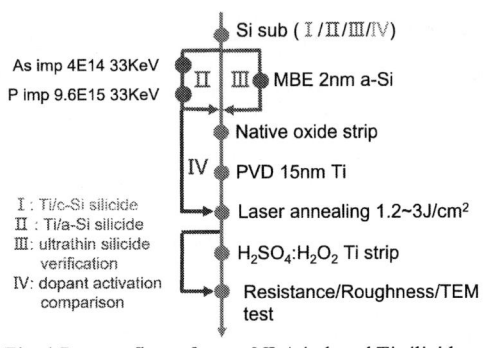

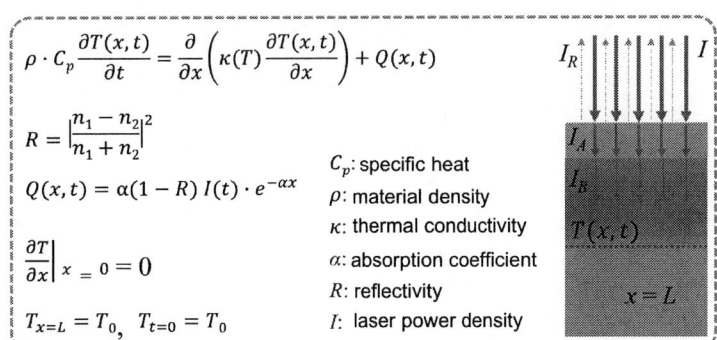

Fig. 1 Process flow of green NLA induced Ti silicide formation and dopant activation comparison.

Fig. 2 Laser heating model in this work, including laser reflection, solid heat transfer and laser absorption.

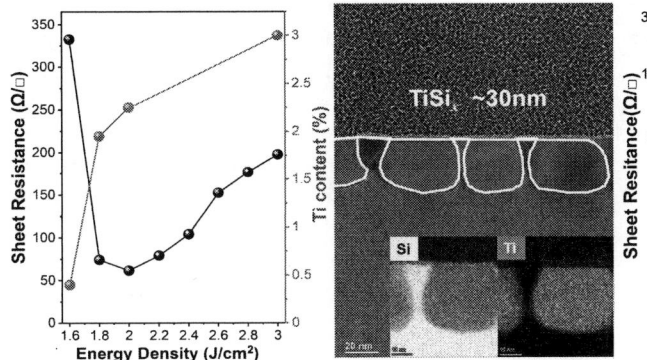

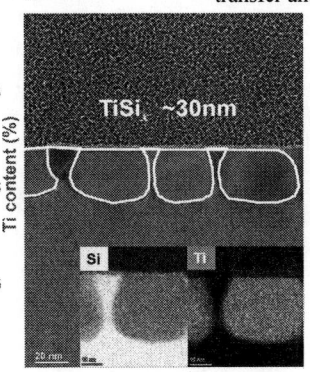

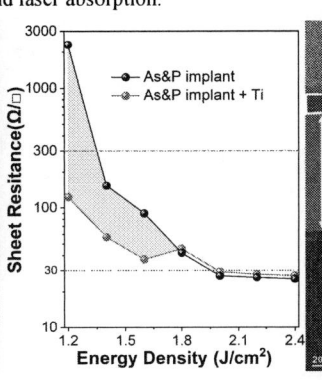

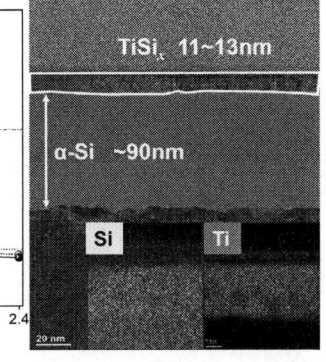

Fig. 3 Ti/c-Si sample sheet resistance and Ti content varying with energy densities.

Fig. 4 TEM of 1.8J/cm² Ti/c-Si sample, the Si and Ti mapping results are shown in the insets.

Fig. 5 As&P implant and As&P implant + Ti samples sheet resistance varying with energy densities.

Fig. 6 TEM of 1.2J/cm² Ti/a-Si sample, the Si and Ti mapping results are shown in the insets.

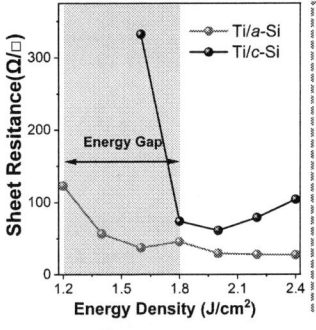

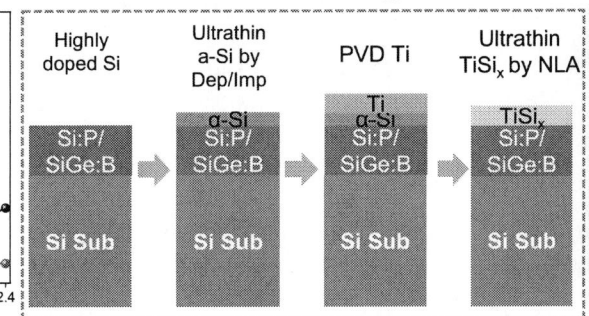

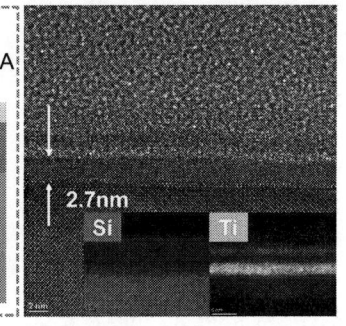

Fig. 7 The TiSiₓ formation energy gap between the Ti/c-Si and Ti/a-Si system.

Fig. 8 Ultrathin TiSiₓ contact process flow, ultrathin a-Si can be achieved by deposition or ion implantation.

Fig. 9 2.7nm Ti silicide formed by MBE a-Si & NLA, inserts are the Si and Ti mapping results.

Table1. Key parameter of BSPDN structure

Parameter	Size
Gate pitch	54 nm
Gate height	55 nm
Gate length	19 nm
Cell height	108 nm
Nanosheet width	27 nm
M0 pitch	28 nm
ILD0 thickness	60 nm
M1/M2 pitch	33 nm
ILD1/ILD2 thickness	60 nm
M3 pitch	44 nm
ILD3 thickness	67 nm

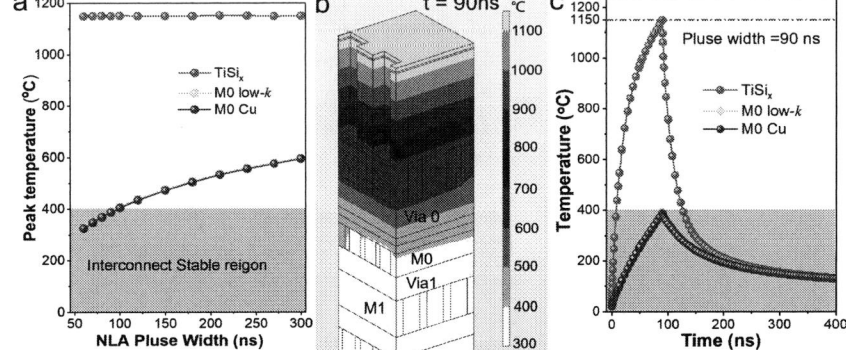

Fig. 10 (a) Peak temperatures of the interconnect and TiSiₓ layer with varying laser pulse widths; (b) The temperature distribution in a BSPDN structure (pulse width = 90 ns); (c) The TiSiₓ/M0 temperature variation with time, M0 peak temperature is below 400°C, indicating that NLA is suitable for backside contact of BSPDN.

979-8-3315-0417-5/25 $31.00 © 2025 IEEE

HRS Retention of 28 nm BEOL integrated ReRAM

Stefan Wiefels[1], Nils Kopperberg[2], and Stephan Menzel[1]

[1]PGI-7, Forschungszentrum Jülich GmbH, [2]IWE II, RWTH Aachen University

Abstract

This paper investigates the retention characteristics of resistive switching random access memory (ReRAM) integrated BEOL into 28 nm CMOS. It analyzes the high-resistance state (HRS) degradation on short and long time scales considering Mbit statistics. It combines accelerated life testing (ALT) and 3D Kinetic Monte Carlo simulations. An evaluation method based on the broadening of the observed log-normal read current distributions is proposed. Via this method, an activation energy of approximately 2 eV is estimated for oxygen vacancy migration.

Keywords: ReRAM, Retention, HRS, KMC

Introduction

Resistive switching random access memory (ReRAM) has gained significant attention as a promising non-volatile memory (NVM) for data storage solutions and neuromorphic computing, where it can mimic artificial synapses [1]. ReRAM-based in-memory computing could enhance energy efficiency in tasks like vector-matrix multiplications when directly executed in the array [2]. In particular, devices based on the valence change mechanism have been widely studied due to their scalable, fast, and low-power characteristics [3].

One of their key reliability aspects is a sufficient data retention, where typical requirements are 5-10 years of stability at 85 °C to 125 °C [3]. Numerous studies have examined VCM ReRAM retention, however, typically focusing on the decreasing read current of LRS [4]–[6]. In contrast, the HRS state exhibits more complex degradation characteristics.

This study reviews our previous works on HRS retention, providing a comprehensive analysis, ranging from the physical origin of HRS degradation to an accurate evaluation method for extrapolation of accelerated life testing (ALT) experiments towards retention at operating conditions. It is based on experimental data using 28 nm back-end-of-line (BEOL) integrated VCM ReRAM on Mbit-scale. We present ALT experiments at temperatures up to 250 °C as well as room temperature retention over nearly four years of storage. The study is supported by 3D Kinetic Monte Carlo (KMC) simulations which demonstrate the physical mechanism of degradation and provide validation for the proposed evaluation method.

I. Short and Long Term Effects

Before understanding the long-term stability, short-term effects have to be analyzed. Fig. 1 demonstrates the instability of the HRS state, typically observed on short time scales. Between two reads, the read current representing the state of the cell may fluctuate up to 100 % of the initial value as illustrated by the galaxy plot in Fig. 1, (a). This data is displayed as normalized distribution in Fig. 1, (b). The linearity on a $\log(I)$ scale indicates log-normal statistics which recover quickly even after removing tail bits above a defined threshold current [7]. This emphasizes that the HRS state follows log-normal statistics which are remarkably stable despite high fluctuations within the distribution.

On longer time scales, these intrinsic statistics remain intact. However, the distribution gradually shifts and tilts as depicted in Fig. 2 for storage at (a) elevated temperatures and (b) room temperature over nearly four years. This degradation can be interpreted as changes in the characteristic parameters of the log-normal distribution, being a shift in median (μ) and an increase in standard deviation (σ). The latter is the basis of our evaluation method discussed in section II. Both graphs show the same qualitative degradation despite different treatment. It may be noted that the 36 M devices in Fig. 2, (b) have been reprogrammed after nearly four years of storage using the same algorithm as for the initial distribution. Here, the initial and reprogrammed distributions perfectly coincide, proving that no irreversible degradation occurred.

In order to understand the physical origin of the observed short and long term behaviour, a consistent model has been developed in a previous study [9]. The 3D Kinetic Monte Carlo (KMC) model considers oxygen vacancies as defect states for trap-assisted tunneling. Diffusion limiting domains have been introduced as depicted in Fig. 3, (a). Within the boxes, defects can migrate facing a low activation energy which results in the observed short-term instability. At elevated temperatures or on long timescales, defects surpass the confinements implemented as higher activation energy accounting for the long term degradation. Fig. 3, (b) shows experimental read noise measurements (grey) as a representation of the short term instability. It is shown that the simulated curves (colors) exhibit very similar characteristics. Fig. 3, (c) displays the simulated long term degradation which shows both tilting and shifting as observed in experiment. The simulation is performed

---

979-8-3315-0417-5/25 $31.00 © 2025 IEEE

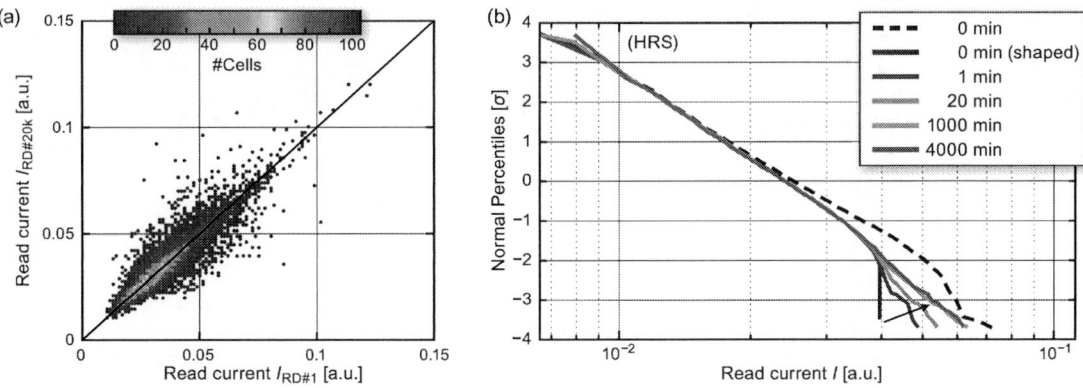

Fig. 1. Read to Read variability of embedded ReRAM (a) Galaxy plot comparing the read current of 10 k devices at the first read to the last read after 20000 s. The color marks the number of devices at the respective data point. Most devices lie close to the black diagonal, representing no change between the two reads. However, current changes of up to 0.06 a.u. (> 100 %) are observed. (b) HRS read current distributions containing 10 k devices in normal percentiles. The linear shape on logarithmic current scale reveals log-normal statistics. From the initial distribution (dashed) devices with current above a threshold are removed from the data set. Subsequent reads relax back to the log-normal distribution. Adapted from [7].

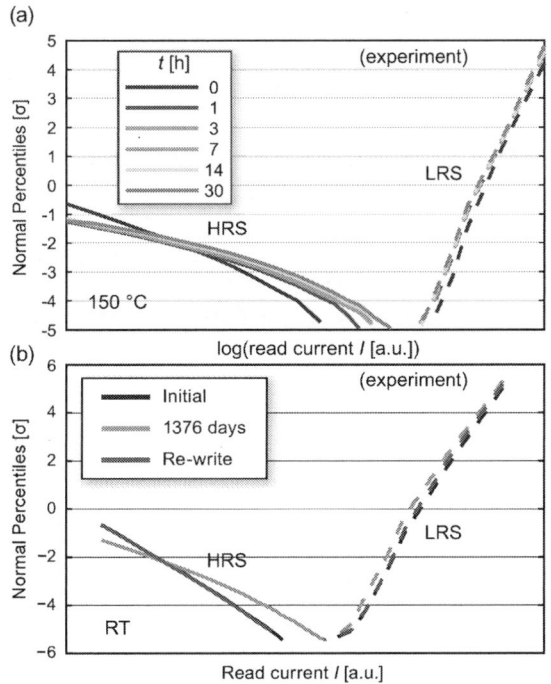

Fig. 2. (a) Normalized cumulative distributions of exemplary retention results. 2.8 M devices are programmed into HRS (solid lines) and LRS (dashed lines) each. Subsequently, the die is baked at 150 °C for up to 30 h as specified in the legend. The LRS only shows a slide shift towards lower read current. The HRS shows a complex degradation comprising shifting and tilting. Adapted from Kopperberg et al. [8]. (b) Retention of 36 M devices over 1376 days (nearly 4 years) at room temperature. The degradation is in good qualitative agreement with the ALT experiment. After read out, all devices were reprogrammed with the same algorithm. Initial and re-written distributions are identical. Adapted from Wiefels et al. [7].

with very high acceleration at 1000 K in order to limit the computation time [9].

II. EVALUATION METHOD

As discussed above, the HRS degradation is best described by a change in the parameters (μ and σ) of the intrinsic log-normal statistics. Here, the broadening of the distribution is most pronounced and has the largest impact on the read window. Thus, an evaluation method tracing the increase in σ has been developed [8]. Fig. 4, (a) exemplarily shows the log-normal fits applied to experimental HRS retention data where the degradation is accelerated by heating to 200 °C. It may be noted that due to limitations of the measurement scheme, only data above a minimum current is displayed and thus the distribution is cut-off at approx. $-0.5\,\sigma$. A comparison with the full distribution in Fig. 1, (b) shows that the applied fits are reasonable. From these fits, μ and σ are extracted and the degradation of σ is plotted over $\log(t)$ in Fig. 4, (b). The parallel lines for different temperatures indicate the validity of the Arrhenius approach, leading to the plot in Fig. 4, (c). The evaluation results in an extracted activation energy of $E_A \approx 2\,\text{eV}$. The validity of this method has been proven via 3D KMC simulations by Kopperberg et al. where it could be attributed to the migration of oxygen vacancies [8].

CONCLUSION

In this overview paper, we analyzed the stability of the HRS state, from short to long time scales, including a consistent model based on the migration of defects. This migration leads to intrinsic log-normal statistics on short time scales. The respective distributions shift and broaden on long time scales. Based on this broadening an evaluation method has been developed by which an activation energy of $E_A \approx 2\,\text{eV}$ for migration of oxygen vacancies could be extracted.

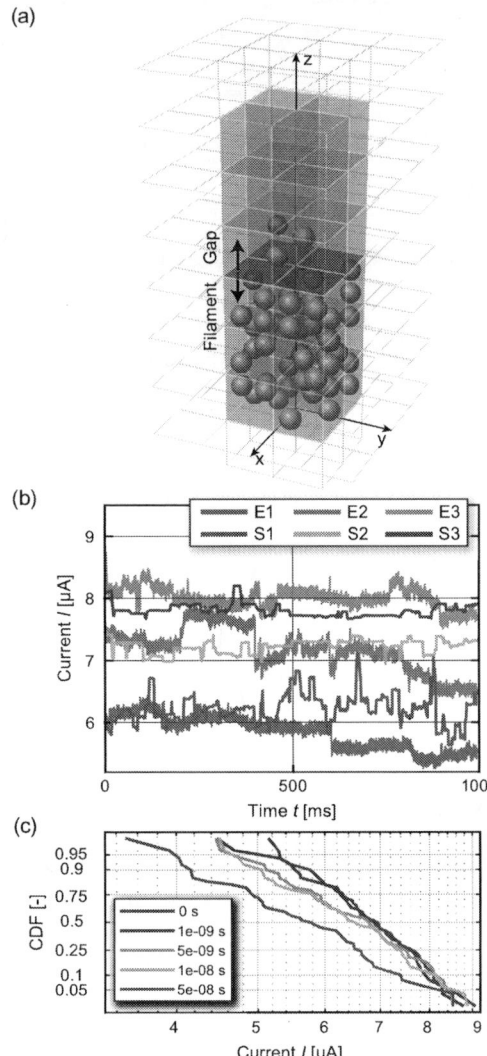

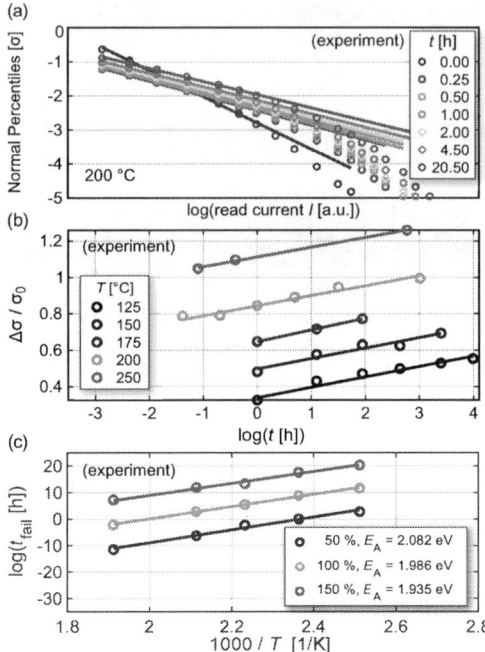

Fig. 4. (a) Linear fit of the log-normal HRS distribution demonstrated exemplarily for 200 °C. (b) The normalized degradation of the standard deviation $\Delta\sigma/\sigma_0$ is extracted from the fits in (a) and plotted over $\log(t)$. The resulting parallel lines indicate the validity of the Arrhenius approach. (c) Extracted failure time for different levels of σ degradation in Arrhenius plot, resulting in excellent fits and an extracted $E_A \approx 2$ eV. Adapted from Kopperberg et al. [8].

Fig. 3. (a) Sketch of HRS oxygen vacancy distribution in 3D KMC model with diffusion limiting domains. The filament and gap region is highlighted in blue. (b) Current evolution over time during read pulse of exemplary experimental (gray) and simulated (colored) VCM devices programmed to the HRS. The experimental results are obtained on $Pt/ZrO_2/Ta$ VCM devices with all details given in Ref. [9]. (c) Evolution of the simulated current distribution of 50 cells during baking at 1000 K. The high temperature is required for limiting simulation time. Adapted from Kopperberg et al. [9]

ACKNOWLEDGMENT

This work was supported by the Federal Ministry of Education and Research (BMBF, Germany) through the project NEUROTEC under Grant 16ME0398K and 16ME0399. We gratefully acknowledge the contributions by Dr. Karl Hofmann and Dr. Jan Otterstedt and the computing time on the supercomputer JURECA [10] at FZ Jülich under grant no. 27525.

REFERENCES

[1] D. Ielmini et al., "Brain-inspired computing via memory device physics," APL Materials, vol. 9, no. 5, p. 050 702, 2021.

[2] W. Wan et al., "A compute-in-memory chip based on resistive random-access memory," Nature, vol. 608, no. 7923, pp. 504–512, 2022.

[3] R. Dittmann et al., "Nanoionic memristive phenomena in metal oxides: The valence change mechanism," Advances in Physics, vol. 70, no. 2, pp. 155–349, 2021.

[4] E. Perez et al., "Data retention investigation in Al:HfO2-based resistive random access memory arrays by using high-temperature accelerated tests," Journal of Vacuum Science & Technology B, vol. 37, no. 1, p. 012 202, 2019.

[5] Y. Wang et al., "Algorithm-enhanced retention based on megabit array of Cu_xSi_yO RRAM," IEEE Electron Device Lett., vol. 33, no. 10, pp. 1408–1410, 2012.

[6] T. Ninomiya et al., "Improvement of data retention during long-term use by suppressing conductive filament expansion in TaOx bipolar-ReRAM," IEEE Electron Device Lett., vol. 34, no. 6, pp. 762–764, 2013.

[7] S. Wiefels et al., "Reliability Aspects of 28nm BEOL-Integrated Resistive Switching Random Access Memory," physica status solidi (a), p. 2 300 401, 2023. DOI: 10.1002/pssa.202300401.

[8] N. Kopperberg et al., "Accurate Evaluation Method for HRS Retention of VCM ReRAM," APL Materials, vol. 12, no. 3, p. 031 112, 2024. DOI: 10.1063/5.0188573.

[9] N. Kopperberg et al., "A consistent model for short-term instability and long-term retention in filamentary oxide-based memristive devices," ACS Appl. Mater. Interfaces, vol. 13, no. 48, pp. 58 066–58 075, 2021.

[10] JSC, "JURECA: Data Centric and Booster Modules implementing the Modular Supercomputing Architecture at Jülich Supercomputing Centre," Journal of large-scale research facilities, vol. 7, no. A182, 2021. DOI: 10.17815/jlsrf-7-182.

On the Evaluation of Remnant Polarization in 3D Cylindrical Hafnia-based Ferroelectric Capacitors

Yishan Wu[1], Puyang Cai[2], Junwei Guo[1], Haobo Lin[1], Xuepei Wang[1], Jinhao Liu[1], Boyao Cui[1], Yichen Wen[1], Maokun Wu[1], Runsheng Wang[2], Sheng Ye[1], Haibao Chen[1], Pengpeng Ren[1], Zhigang Ji[1]*, and Ru Huang[2]

[1]National Key Laboratory of Advanced Micro and Nano Manufacture Technology, Shanghai Jiao Tong University, Shanghai, 200240, China; [2]School of Integrated Circuits, Peking University, Beijing 100871, China.

(*E-mail: zhigangji@sjtu.edu.cn)

Abstract

In this study, we establish a phase field model for the 3D cylindrical hafnia-based ferroelectric capacitors (FeCAPs) to capture the ferroelectric (FE) characteristics of the cross-section. Based on this, we show how the remnant polarization (P_r) can be evaluated with minimum error. We further carry out an in-depth analysis on how geometric factors influence the FeCAP performance. Our results indicate that cylindrical FeCAPs with a smaller inner radius exhibit lower P_r due to intensified depolarization field. With increasing inner radius, P_r of cylindrical FeCAP approaches that of planar FeCAP. This model provides a valuable tool for assessing memory capacity and optimizing the structure of 3D FeRAMs.

Keywords: Ferroelectric, 3D cylindrical capacitor, Phase field method, Geometric optimization

Introduction

The discovery of hafnia-based ferroelectrics, which garnered significant attention for their excellent scalability and compatibility with CMOS technology, has positioned FeRAM as a critical candidate for next-generation memory devices [1, 2]. Faced with the storage density limitation, similar to DRAM, a cylindrical structure has been introduced to FeCAP to improve area efficiency [3, 4, 5], as shown in Fig. 1 (a). However, this structure introduces challenges in accurately evaluating P_r due to the identical charges on the inner and outer electrodes but different surface areas [6], as illustrated in Fig. 1 (b). Prior research calculated the P_r of cylindrical FeCAP based on the outer surface area [7]. Also, the screening surface area of cylindrical FeCAP has been used for memory window evaluation [8], which is not applicable for experiments because the position where the electric field equals 0 V cannot be precisely determined. The absence of a reliable method for P_r evaluation in cylindrical FeCAP hinders the exploration on the physical similarities and differences with planar devices and therefore requires tackled urgently.

In this work, we present a phase field model for 3D cylindrical FeCAP to compute the polarization charge density in the cross-section, enabling direct determination of P_r without specific surface selection. Next, we derive a formula for calculating an equivalent radius for P_r evaluation in experiments. Furthermore, we investigate the impact of geometric parameters on the electric properties of cylindrical FeCAPs by analyzing the ferroelectric (FE) characteristics across varying inner radius and FE layer thickness, offering insights for optimizing cylindrical FeCAP performance.

Phase Field Model for 3D Cylindrical FeCAPs

To illustrate the method for evaluating P_r, we first demonstrate the phase field simulation on the planar FeCAP using COMSOL Multiphysics [9]. As depicted in Fig. 2 (a), the 2D time-dependent Landau-Ginzburg (TDLG) equation and Poisson's equation are self-consistently solved in the x-z plane by finite difference method to track polarization evolution under applied voltage [10]. By treating polarization P as a fundamental field variable, the TDGL equation directly solves for polarization in the FE layer, without current integration for total charge. Thus, unlike in experiments, simulated P_r can be obtained without area selection. In this simulation, the total polarization charge density P_{avg} equals the average value of $(P + \varepsilon_0 \varepsilon_{FE} E_z)$ across the entire FE region. Experimentally, P_{avg} is similarly represented as the average charge density at the FE/electrode interface. In Fig. 2 (b), the P_{avg}-V loops from both methods align closely. For cylindrical FeCAP, we adopt the average value across the FE region for P_{avg} calculation to address the area selection issue.

In the phase field simulation of the cylindrical FeCAP, the polarization direction is assumed to be oriented radially, as shown in Fig. 3 (a-b). Account for the unique geometry of the cylindrical structure, we propose a modified phase field model for cylindrical FeCAP, as illustrated in Fig. 3 (c), where P refers to the polarization; ρ is the viscosity coefficient; α, β, and γ are Landau coefficients for the free energy; g is the gradient energy coefficient; U is the electric potential; θ is the angle between the radical line from the origin and the x-axis. Dirichlet boundary conditions are used at the FE/electrode interfaces. In Fig. 3 (d), the simulated P_{avg}-V loop is calibrated against experimental data, with relevant parameters listed. This model enables further investigation of the electric characteristics within the cross-section of the cylindrical FeCAP.

Results and Discussion

The phase field method enables the calculation of P_{avg} of a cylindrical FeCAP in simulations. However, directly obtaining P_{avg} across the cross-section of a cylindrical FeCAP using the same approach is not feasible. Therefore, based on the simulation results, an equivalent radius (R_{eq}) can be derived for P_r evaluation in experimental settings.

979-8-3315-0417-5/25 $31.00 © 2025 IEEE

A. Equivalent Radius for 3D Cylindrical FeCAPs

The extraction process for R_{eq} is illustrated in Fig. 3 (e). According to charge conservation, the equivalent radius (R_{eq}) should satisfy the following equation:

$$Q_{in} = Q_{out} = P_{avg} * 2\pi R_{eq} \qquad (1)$$

where Q_{in} and Q_{out} represent the polarization charges along the inner and outer surface circumference of the FE layer in the cylindrical FeCAP. To derive a general formula for R_{eq}, we demonstrate the proposed phase field simulation on the cylindrical FeCAPs with varying R_{in} and T_{FE}. Fig. 4 illustrates the trend in P_r as a function of R_{in} for different T_{FE}. For a fixed T_{FE}, R_{eq} varies linearly with R_{in}, expressed as:

$$R_{eq} = R_{in} + \frac{T_{FE}}{2} \qquad (2)$$

This relationship closely aligns with the simulation results, indicating that R_{eq} can be accurately derived from R_{in} and T_{FE}, both of which can be determined in experiments. The P_{avg} calculated using this approach is more accurate than that obtained using the inner or outer diameters alone. Fig. 5 shows the error of P_{in} calculated using R_{in} and P_{out} calculated using R_{out} compared to the accurately calculated P_{avg}. As the radius decreases, the error in P_r calculated using the inner or outer surface increases. Thus, R_{eq} is crucial for precise P_r evaluation in small-scale cylindrical FeCAPs experimentally.

B. Geometric Impact on FE Properties

The geometry-induced variation in FE properties is examined through phase field simulations. As shown in Fig. 6, the distributions of electric characteristics at 0 V across the cross-sections of both planar and cylindrical FeCAPs are compared. Unlike the planar FeCAP, the cylindrical FeCAP exhibits a radial gradient in polarization, potential, and electric field distributions. Polarization progressively decreases along the radial direction, from the inner to the outer diameter. For the electric field in the cross-section, a minimum value is observed along the radial direction, consistent with the electric field value in planar FeCAP.

The subsequent analysis examines the impact of R_{in} and T_{FE} on the FE properties of cylindrical FeCAPs. In Fig. 7 (a), the positive and negative P_r for cylindrical FeCAPs with varying R_{in} under different T_{FE} are shown. With increasing R_{in} at a fixed T_{FE}, P_r gradually increases and eventually reaches saturation. Compared to the fluctuation in E_c exhibited in Fig. 7 (b), the influence of radius on P_r is more pronounced. The corresponding P-V loops are detailed in Fig. 7 (c-e), with the P-V loop of planar FeCAP at the same T_{FE}. As R_{in} increases, P_r and E_c of the cylindrical FeCAP gradually approach those of the planar FeCAP at the same T_{FE}, with the P-V curves becoming increasingly consistent.

To further elucidate the impact of geometry on P_r, we examine the electric field distribution contours at 0V for cylindrical FeCAPs with different R_{in} with $T_{FE} = 8$ nm, as depicted in Fig. 8 (a-f). With increasing R_{in}, the radial gradient of the electric field distribution gradually flattens,

causing the near-zero minimum field to shift outward from the inner surface toward the outer surface. Fig. 8 (g) presents the absolute values of the electric field along the radial direction for different R_{in}, demonstrating that a smaller R_{in} leads to a higher electric field near the inner surface. The electric field distribution provides further insight into the observed trend of P_r with varying R_{in}. Owing to the unique cylindrical geometry, the depolarization field dominates in proximity to the inner electrode. A reduction in R_{in} leads to a substantial increase in the depolarization field. Consequently, part of polarization cannot be maintained when the external voltage is at 0 V, leading cylindrical FeCAPs with smaller R_{in} exhibit lower P_r values. As R_{in} increases, the disparity between the planar and cylindrical structures diminishes, resulting in similar FE properties.

Conclusion

In summary, this study employs a phase field simulation to investigate the electrical properties of 3D cylindrical FeCAPs, detailing polarization, potential, and electric field distributions within the FE layer. A novel method for evaluating P_r is proposed, including a formula for calculating R_{eq}, which enhances the accuracy of P_r calculation in practical tests. The impact of geometry on FE properties is extensively analyzed, revealing that smaller R_{in} in cylindrical structures leads to a lower P_r due to the intensified depolarization field near the inner electrode. This work delivers essential insights into the role of geometry in improving FE performance, offering valuable guidelines for optimizing 3D FeRAMs structural design in future applications.

Acknowledgments

This work is financially supported by the National Key Research and Development Program of China under grant 2019YFB2205005 and the National Natural Science Foundation of China (Nos. 62027818, 61874034, and 11974320).

References

[1] M.H. Park, *et al.*, *MRS Commun.*, vol. 8, no. 3, pp. 795-808, 2018. [2] U. Schroeder *et al.*, *Nat. Rev. Mater.*, vol. 7, no. 8, pp. 653-669, 2022. [3] P. Polakowski, *et al.*, *IEEE 6th International Memory Workshop (IMW)*, pp. 1-4, 2014. [4] J. Okuno, *et al.*, *IEEE Symposium on VLSI Technology*, pp. 1-2, 2020. [5] N. Ramaswamy, *et al.*, *2023 International Electron Devices Meeting (IEDM)*, pp. 1-4, 2023. [6] M. Fan, *et al.*, *IEEE Trans. Electron Devices*, vol. 67, no. 12, pp. 5810-5814, 2020. [7] J. Okuno, *et al.*, *2023 International Electron Devices Meeting (IEDM)*, pp. 1-4, 2023. [8] M. Deng, *et al.*, *7th IEEE Electron Devices Technology & Manufacturing Conference (EDTM)*, pp. 1-3, 2023. [9] *COMSOL Multiphysics®* v. 6.1. cn.comsol.com. COMSOL AB, Stockholm, Sweden. [10] A. K. Saha, *et al.*, *Appl. Phys. Lett.*, vol. 119, no. 12, 2021.

979-8-3315-0417-5/25 $31.00 © 2025 IEEE

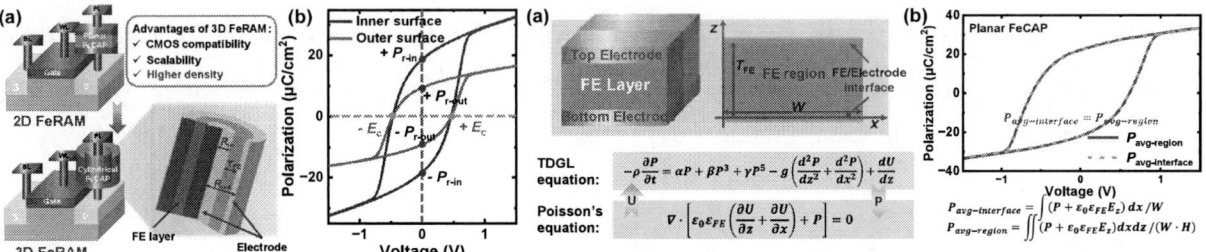

Fig. 1 (a) Schematic of 2D to 3D structural evolution in FeRAM with one transistor and one FeCAP. (b) The *P-V* loops of cylindrical FeCAP normalized to the outer and inner surfaces. Different P_r are obtained depending on the surface considered.

Fig. 2 (a) Schematic of a planar FeCAP and the phase field simulation method based on TDGL and Poisson's equations. (b) Simulated *P-V* loops of planar FeCAPs. $P_{avg\text{-}interface}$-*V* loop obtained by integrating the charge along the FE/Electrode interface, and $P_{avg\text{-}region}$-*V* loop calculated as the average polarization across the FE region, are identical.

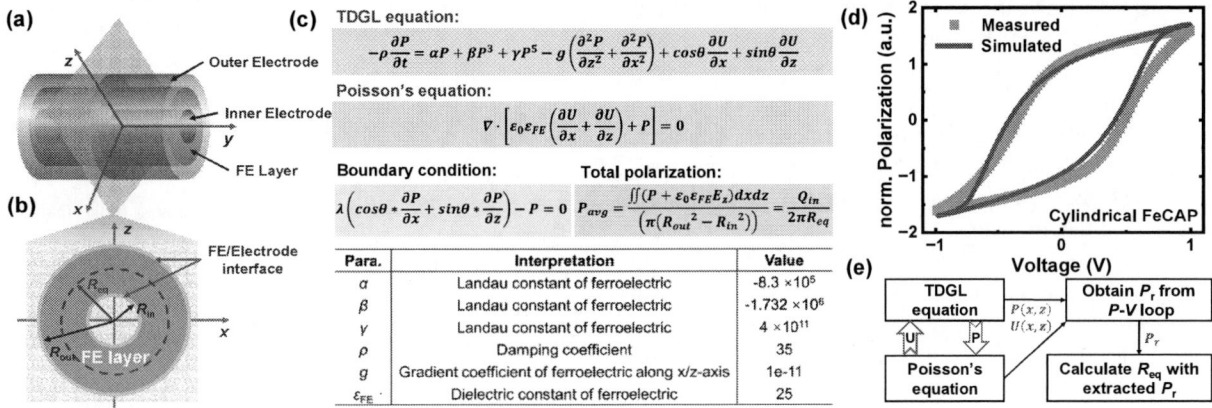

Fig. 3 (a) Schematic of cylindrical FeCAP with its cross-section shown in (b). (c) Proposed phase field simulation model of cylindrical FeCAP based on the TDGL equation coupled with Poisson's equation, with parameter listed in the table. (d) Simulated *P-V* loop calibrated with experimental data from cylindrical FeCAP. (e) Simulation flow for R_{eq} calculation.

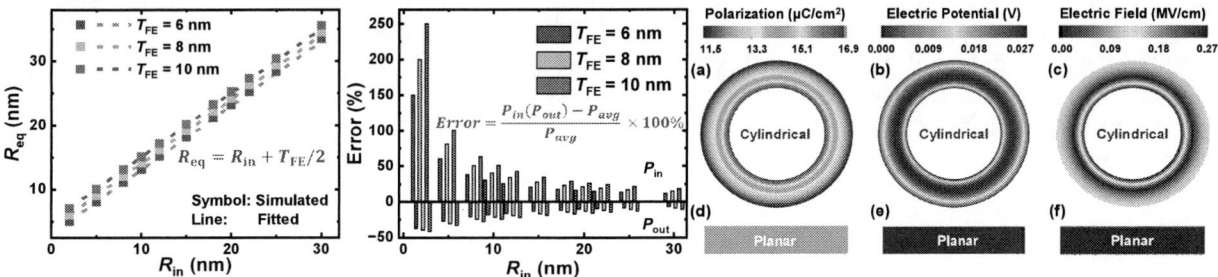

Fig. 4 Correlation between R_{in} and R_{eq} of cylindrical FeCAPs based on the proposed phase field simulation model. Symbols represent simulated R_{eq}, while the line indicates the fitting result.

Fig. 5 Error in P_{in} (evaluated using the inner surface area) and P_{out} (evaluated using the outer surface area) relative to the average *P* for different R_{in} with T_{FE} = 6, 8, 10 nm.

Fig. 6 Absolute value distribution of polarization, electric potential, and electric field in cylindrical FeCAP (a-c) and planar FeCAP (d-f) with T_{FE} = 6 nm at 0 V. In cylindrical capacitors, the distribution follows a radial gradient, whereas in planar capacitors, it remains relatively uniform.

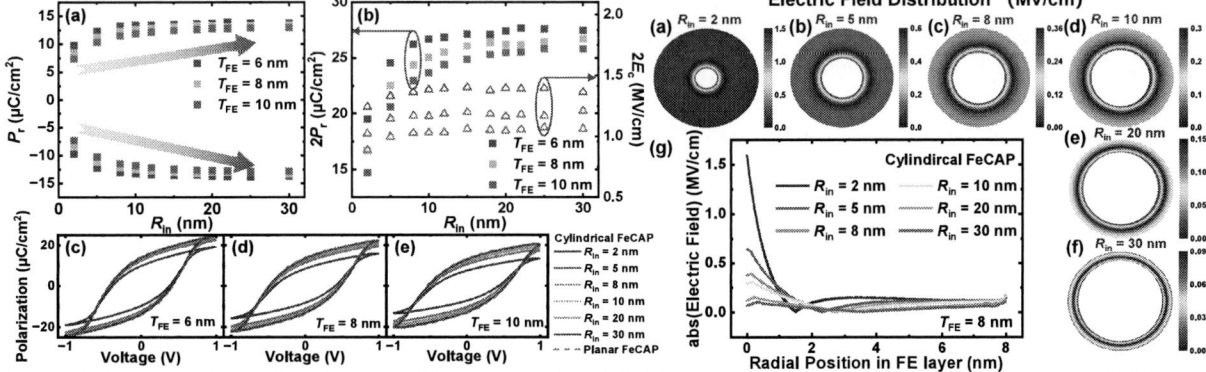

Fig. 7 (a) P_r for different R_{in} with T_{FE} = 6, 8, 10 nm. (b) Trend comparison of $2P_r$ and E_c for cylindrical FeCAPs with varying R_{in} under different T_{FE}. (c-e) *P-V* loops of planar and cylindrical FeCAPs with various R_{in} under T_{FE} = 6, 8, 10 nm.

Fig. 8 (a-f) Contours of the electric field in cylindrical FeCAPs with different R_{in} under T_{FE} = 8 nm. (g) Absolute value of electric field along the radial direction of cylindrical FeCAPs.

979-8-3315-0417-5/25 $31.00 © 2025 IEEE

Compact Modeling of Kink Effect in BULK MOSFETs at Cryogenic Temperatures

Nisha Manzoor, Wajid Manzoor, Debashish Nandi, Aloke K. Dutta, and Yogesh Singh Chauhan

Department of Electrical Engineering, Indian Institute of Technology Kanpur, India

Email: {mnisha, wajid, debashish21, aloke, chauhan} @iitk.ac.in

ABSTRACT

In this study, we examine the kink effect in output characteristics of planar bulk NMOS transistors at cryogenic temperatures, caused by incomplete ionization and increased substrate resistance. These factors lead to forward biasing of the source-body junction and activation of the parasitic bipolar transistor, which increases the drain current, creating a kink in the output characteristics. A SPICE-compatible model is proposed using the BSIM-BULK compact model framework, capturing these physical parasitic effects. The model accurately represents the kink effect at cryogenic temperatures and is validated against experimental data, showing excellent correlation. This advancement enhances the precision of simulations for bulk MOSFETs in low-temperature IC design.

Keywords: BSIM, Bulk MOSFET, Compact Modeling, Cryogenic, Incomplete ionization, Kink Effect.

INTRODUCTION

Initially used in space and defense applications [1], [2], cryogenic CMOS (Cryo-CMOS) technology has become important for quantum chip readout and control circuits [3] due to improved electrical performance at cryogenic temperatures, such as steep switching, better subthreshold behavior, reduced leakage, and lower noise. However, at cryogenic temperatures, nanoscale devices are prone to parasitic effects like the kink effect in planar bulk MOSFET's output characteristics, caused by incomplete dopant ionization, increased substrate resistance, higher built-in source-body junction potential, and activation of parasitic bipolar action, all of which can degrade performance such as lowering of an amplifier's inherent gain [4], [5].

An accurate and efficient model of these parasitic effects is imperative for low-temperature IC design. The existing models do not accurately capture the kink effect in planar bulk MOSFETs at cryogenic regime [6]–[8]. This study aims to elucidate the kink observed in drain current (I_D) as a function of drain-to-source voltage (V_{DS}) at cryogenic temperatures and proposes a model, developed using the BSIM-BULK framework [6], [7], that effectively captures this kink effect. The model has been thoroughly validated against experimental data.

KINK EFFECT AT CRYOGENIC TEMPERATURE

The kink effect is a well-known phenomenon in floating-body partially depleted silicon-on-insulator MOSFETs at room temperature [9]. It is generally attributed to the impact ionization near the drain-end of

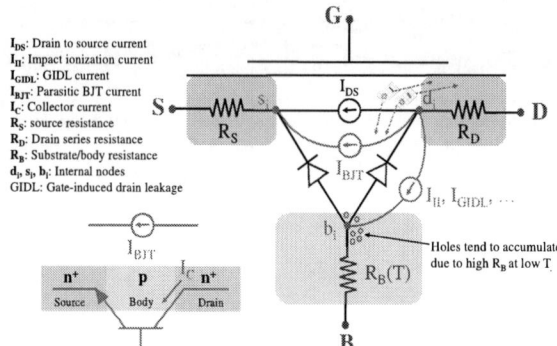

Fig. 1. Schematic representation of a typical planar bulk NMOS at cryogenic temperature.

the channel at high V_{DS}, while being more prominent in nanoscale short-channel devices. The high electric field near the drain creates electron-hole pairs in the drain side depletion region, from where the electrons move towards the drain and the holes towards the body. These holes tend to accumulate at the body, raising the body potential and lowering of the threshold voltage, which translates into an increase in drain current, and thus, a kink in the output characteristics of the device. For a planar bulk MOSFET, however, at room temperature, the holes can leave the body region through the substrate terminal (B), thus causing a body current to flow, and therefore, no such kink is generally observed.

The scenario is quite complex for short-channel planar bulk MOSFETs operating at cryogenic temperatures. A schematic representation of a typical planar bulk NMOS is presented in Fig.1. At these low temperatures, the freeze-out of carriers in the substrate/body region affects the behavior of the source-body diode. Due to the freeze-out effect, the ionized acceptor concentration is reduced below the total concentration, i.e., incomplete ionization prevails. This results in an increase in the substrate resistance, which prohibits the holes from flowing out of the body terminal, as shown in Fig.1, and increases the body potential, i.e., V_{b_i}. Therefore, a reduction in the threshold voltage and an increase in the I_D in the form of a kink in the output characteristics are observed, which is assisted by forward biasing of the source-body junction and activation of parasitic bipolar current.

MODEL FORMULATION

The effective model of the kink effect is developed using a modified BSIM-BULK framework given in [6]. The impact ionization current shown in Fig. 1, I_{II}, is modeled as [7], [10]:

979-8-3315-0417-5/25 $31.00 © 2025 IEEE

$$I_{II} = \alpha_0 \cdot I_{DS} \cdot \Delta V_{DS} \cdot \exp\left(-\frac{\beta_t}{\Delta V_{DS}}\right) \quad (1)$$

$$\Delta V_{DS} = V_{DS} - V_{DSAT} \quad (2)$$

where $\beta_t = \beta_0 \cdot (T/T_{nom})^{II_T}$. T is the operational temperature of the device and T_{nom} is the nominal reference temperature. V_{DSAT} is the drain saturation voltage. The detailed list of parameters used in the equations is provided in the Table I. Following the established methodology in BSIM-SOI [10], an enhanced model is proposed for parasitic BJT current (I_{BJT}) in BSIM-BULK model, given by:

$$I_{BJT} = \frac{\beta_{bjt1} + \beta_{bjt2} \cdot L_{eff}}{L_{eff}} \cdot I_C \cdot (V_{bci} - V_{BD})$$
$$\cdot \exp\left[-\alpha_{bjt1}(V_{bci} - V_{BD})^{(\alpha_{bjt2}-1)}\right] \quad (3)$$

where L_{eff} is the effective gate length of the device, V_{BD} is the body-drain potential. V_{bci} is the internal base-collector built-in potential of the BJT and is given by [10]:

$$V_{bci} = V_{bci_{Tnom}}\left[1 + V_{bcit} \cdot \left(\frac{T}{T_{nom}} - 1\right)\right] \quad (4)$$

$V_{bci_{Tnom}}$ is the internal base-collector built-in potential of the BJT at room temperature. The BJT collector current is modeled as [10]:

$$I_C = \alpha_B \cdot I_E\left[\exp\left(\frac{V_{BS}}{n_d \cdot V_t}\right) - \exp\left(\frac{V_{BD}}{n_d \cdot V_t}\right)\right] \cdot \frac{1}{E_{2nd}} \quad (5)$$

where E_{2nd} accounts for the 2nd order effects of BJT, n_d is the non-ideality factor, and V_t is the thermal voltage, and V_{BS} is the body-source potential. I_E is the reverse saturation current of the emitter junction of the parasitic BJT. The parameter α_B is the BJT transport factor that represents the fraction of carriers injected by the emitter that reach the collector [10].

Substrate resistance ($R_B = \rho_B \cdot W_{eff}/A_{sub}$, where W_{eff} is the effective width of the device and A_{sub} denotes the effective area of the substrate) increases with temperature due to the effective reduction of the majority carrier concentration in the substrate owing to incomplete ionization of the dopant atoms, and it is modeled as:

$$R_B = R_{B_{Tnom}} + R_{B1T} \cdot \left(1 - \frac{T}{T_{nom}}\right)^{R_{B2T}} \quad (6)$$

where $R_{B_{Tnom}}$ is substrate resistance at T_{nom}.

MODEL VALIDATION AND DISCUSSION

The proposed model is validated using experimental data for two device geometries [5], [8]. Figures 2 and 3 show baseline validation at T = 300 K, where transfer and output characteristics for Device-1 ($L = 320$ nm, $W = 2$ μm) are presented. These figures indicate the absence of a kink and demonstrate the continuity of derivatives at 300 K. Figures 4 and 5 illustrate similar validation at T = 4 K. Notably, in Fig. 5(a), the model captures the kink in the output characteristics accurately, further confirmed by the validation of higher-order derivatives of output conductance in Figs. 5(b)-5(d),

crucial for analog IC design and non-linearity analysis. Additionally, the kink-voltage shifts to higher V_{DS} as gate-source voltage (V_{GS}) increases due to the rise in V_{DSAT} (to the first order, $V_{DSAT} = V_{GS} - V_T$, where V_T is the threshold voltage), as seen in (1) - (2). Figure 6 validates the output characteristics of Device-2 ($L = 120$ nm, $W = 2$ μm) across temperatures (T = [20, 77, 300] K), with strong agreement between the model and experimental data.

TABLE I: List of Model Parameters

Symbol	Description	Unit
α_0	Impact ionization current prefactor	1/V
β_0	V_{DS} Impact ionization coefficient	V
II_T	Temperature coefficient of β_0	-
β_{bjt1}	Length scaling factors for I_{BJT}	m/V
β_{bjt2}	Length scaling factors for I_{BJT}	1/V
α_{bjt1}	Exponent factor for avalanche current	1/V
α_{bjt2}	Base-collector grading coefficient	-
V_{bcit}	Temperature coefficient of V_{bci}	
R_{B1T}	Temperature prefactor of R_B	Ω
R_{B2T}	Temperature exponent of R_B	-

CONCLUSION

This study presents a SPICE-compatible model based on the BSIM-BULK framework to capture the kink effect observed in output characteristics in planar bulk MOSFETs at cryogenic temperatures. This work incorporates the impact of temperature on ionization of dopants, substrate resistance, impact ionization, and parasitic bipolar transistor influencing the MOSFET's output characteristics. To accurately capture the kink in I_D - V_{DS} and extend the validity of model at cryogenic temperatures, the impact of parasitic bipolar action, and substrate resistance are integrated into the BSIM-BULK model. The model is validated with the experimental data and an excellent match is found between the two.

REFERENCES

[1] R.L. Patterson et al., "Assessment of electronics for cryogenic space exploration missions", in Cryogenics, vol. 46, iss. 2–3, pp. 231-6, 2006.

[2] T. Hirao et al., "Cryogenic readout electronics with silicon P-MOSFETS for the infrared astronomical satellite, ASTRO-F", in Advances in Space Research vol. 30, iss. 9, pp. 2117-22, 2002.

[3] B. Patra et al, "Cryo-CMOS Circuits and Systems for Quantum Computing Applications", in IEEE JSSC, vol. 53, no. 1, pp. 309-21, 2018.

[4] L. Deferm et al., "The importance of the internal bulk-source potential on the low temperature kink in NMOSTs", in IEEE TED, vol. 38, no. 6, pp. 1459-66, 1991.

[5] M. Koyanagi et al., "Impact ionization phenomenon in 0.1 μm MOSFET at low temperature and low voltage", in IEEE IEDM, pp. 341-4, 1993.

[6] W. Manzoor et al., "Extending Standard BSIM-BULK Model to Cryogenic Temperatures," in IEEE TED, vol. 71, no. 8, pp. 452024.

[7] BSIM-BULK 107.1.0 user manual. [Online]. Available: www.bsim.berkeley.edu/models/bsimbulk/.

[8] R. M. Incandela et al., "Characterization and Compact Modeling of Nanometer CMOS Transistors at Deep-Cryogenic Temperatures", in IEEE JEDS, vol. 6, pp. 996-1006, 2018.

[9] C. K. Dhabi et al., "Symmetric BSIM-SOI—Part II: A Compact Model for Partially Depleted SOI MOSFETs," in IEEE TED, vol. 71, no. 4, pp. 2293-300, 2024.

[10] BSIM-SOI 100.1.0 Technical manual. [Online]. Available: www.bsim.berkeley.edu/models/bsimsoi/.

979-8-3315-0417-5/25 $31.00 © 2025 IEEE

Model Validation with Device-1 at 300 K

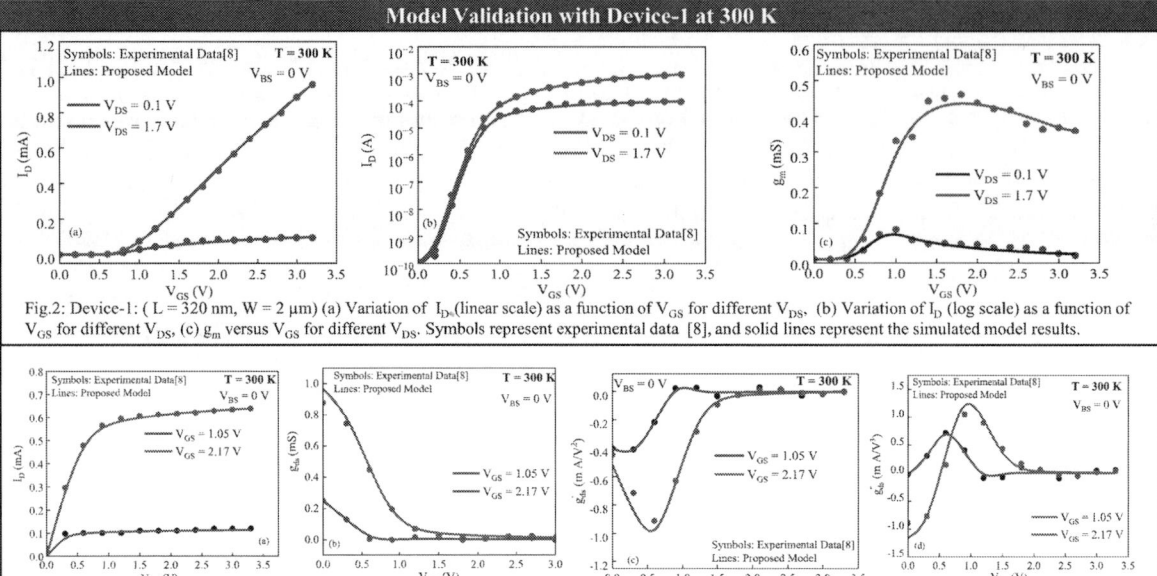

Fig.2: Device-1: (L = 320 nm, W = 2 μm) (a) Variation of I_{Ds} (linear scale) as a function of V_{GS} for different V_{DS}, (b) Variation of I_D (log scale) as a function of V_{GS} for different V_{DS}, (c) g_m versus V_{GS} for different V_{DS}. Symbols represent experimental data [8], and solid lines represent the simulated model results.

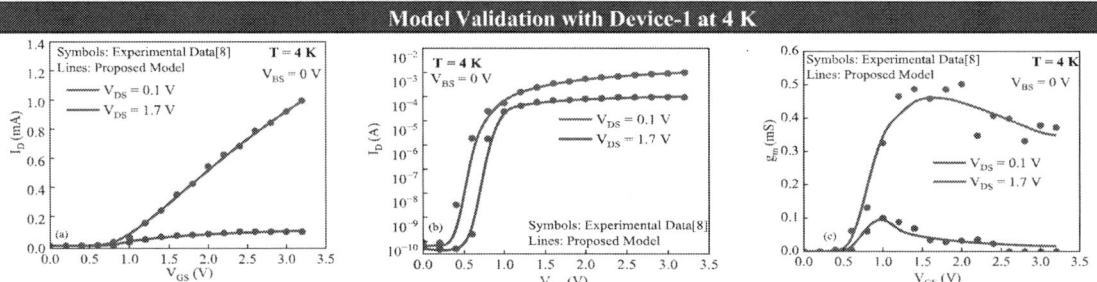

Fig.3. Device-1: (L= 320 nm, W = 2 μm) (a) Variation of I_D as a function of V_{DS} for different V_{GS},(b) Variation of g_{ds} as a function of V_{DS} for different V_{GS}, (c) g_{ds}' versus V_{DS} for different V_{GS} (d) g_{ds}'' versus V_{DS} for different V_{GS}. Symbols represent experimental data [9], and solid lines represent the simulated model results.

Model Validation with Device-1 at 4 K

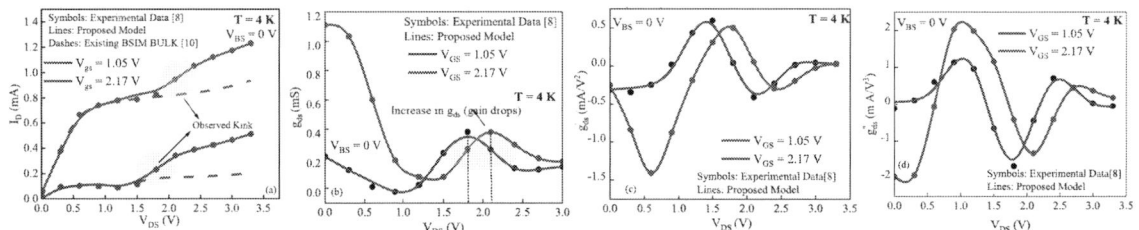

Fig.4. Device-1: (L = 320 nm, W = 2 μm) (a) Variation of I_D (linear scale) as a function of V_{GS} for different V_{DS} at (V_{BS} = 0 V), (b) Variation of I_D (log scale) as a function of V_{GS} for different V_{DS}, (c) g_m versus V_{GS} for different V_{DS}. Symbols represent experimental data [8], and solid lines represent the simulated model results.

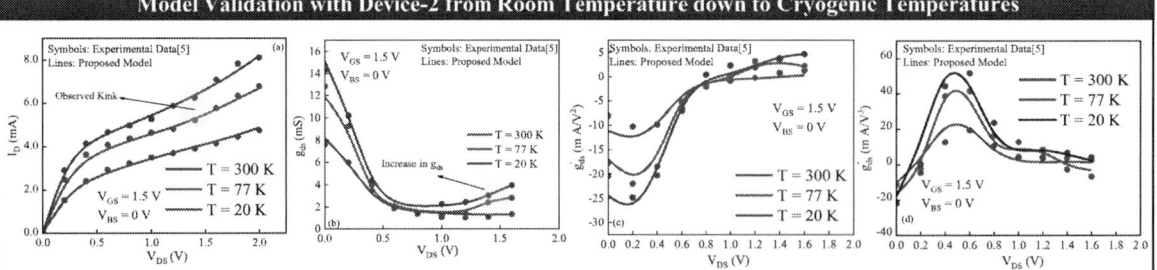

Fig.5. Device-1: (L= 320 nm, W = 2 μm) (a) Variation of I_D as a function of V_{DS} for different V_{GS}, (b) Variation of g_{ds} as a function of V_{DS} for different V_{GS}, (c) g_{ds}' versus V_{DS} for different V_{GS} (d) g_{ds}'' versus V_{DS} for different V_{GS}. Symbols represent experimental data [8], and solid lines represent the simulated model results. Comparison with existing BSIM-BULK is not shown for derivatives of current to maintain brevity.

Model Validation with Device-2 from Room Temperature down to Cryogenic Temperatures

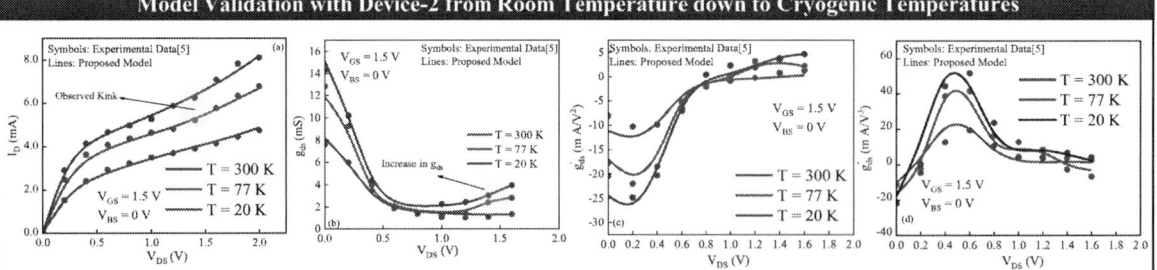

Fig.6 . Device-2: (L= 120 nm, W=2 μm) (a) Variation of I_D as a function of V_{DS} for different V_{GS}, (b) Variation of g_{ds} as a function of V_{DS} for different V_{GS}, (c) g_{ds}' versus V_{DS} for different V_{GS}, (d) g_{ds}'' versus V_{DS} for different V_{GS}. Symbols represent experimental data [5], and solid lines represent the simulated model results.

Perovskite based Artificial Vision System for Geometric Shape Recognition

Shivam Kumar[1], Swapnadeep Poddar[1], Zhenghao Long[1], Zhiyong Fan[1]

[1] Dept. of ECE, The Hong Kong University of Science and Technology,
Clear water bay, Hong Kong

Abstract

The human visual system serves as an inspiration for an efficient image sensor for applications in robotics, sensing and computer vision. Inspired by the retina, we present a perovskite nanowire based artificial vision system for integrated sensing and data preprocessing. The sensor shows stable response to different learning and forgetting visual stimuli. We demonstrate the capabilities of the artificial vision system with a crossbar array and integration with perovskite-based memory for different shape recognition.

Keywords: Perovskite, Vision system, Biomimetics

Introduction

Vision is one of the most vital sensory processes for humans, providing up to 80% of the information we perceive [1]. Humans have tried to replicate the functions of the eyes through modern day CCD and CMOS image sensors. However, these artificial image sensors lag their biological counterparts on many figures of merit [2]. The conventional frame-based image sensor always requires an image processing unit. This creates a huge computational load on the modern-day CPU. The human retina on the other hand performs simple preprocessing tasks like contrast enhancement, edge detection, etc. within itself before sending it to the visual cortex for final processing [3]. This process substantially alleviates the cognitive load on the brain and enhances energy efficiency. This type of vision system is particularly advantageous for computer vision applications, such as those utilized in autonomous robots [4].

Inspired by the human retina, herein we demonstrate an artificial vision system made of vertically aligned all-inorganic perovskite nanowire array in porous aluminum oxide membrane (PAM). The CsPbI$_3$ device shows very high stability, and the aluminum oxide membrane provides protection against moisture and stops degradation [5]. The artificial vision system can work both in the frame-based imaging mode and a neuromorphic mode owing to its unique band alignment [6]. We demonstrate imaging and simple processing tasks like contrast enhancement, learning and forgetting by just using a 10 by 10 cross bar array device. We also integrate another 6 by 6 photonic synapse device array with a perovskite based R-RAM array for different geometric shapes' recognition. The device also has the potential to be hemispherical, structurally mimicking the human retina

which provides additional benefits like a wider field of view, simplified optics and low aberration [7].

Device Fabrication and Characterization

Fig. 1(a) shows the schematic of the human retina, which includes its different parts like the photoreceptor cells, bipolar cells and ganglion cells. The photo receptor cells convert photons into electrical impulses. These electrical signals are pre-processed by the bipolar and the ganglion cells. Processes like contrast enhancement, edge detection and motion detection are performed in this layer after which it is passed to the visual cortex for final processing. Analogous to the vertical structure of the photo sensitive cells, we fabricate vertically aligned perovskite nanowires in porous aluminum oxide membrane (PAM) (Fig. 1(b)). The PAM membrane is made by two step anodization of electrochemically polished Aluminum chips [8]. Pb nanowires are then deposited on the nanochannels through the electro deposition process. A thin layer of SnO$_2$, as electron transporting layer, is conformally deposited along the side walls of the PAM through atomic layer deposition. The perovskite nanowire is then grown by the chemical vapor deposition process. The top and the bottom cross bar ITO electrodes are deposited by RF Sputtering using a shadow mask (Fig. 1(c)). The whole system is packaged using a UV Epoxy on a PCB. Through this process we can obtain high quality single crystalline perovskite nanowires. The PAM also passivates the semiconductor layer thus reducing the surface trap states and surface recombination. Figure 1(d) shows the schematic of the NI PXI 2530B multiplexer used to measure the response of the crossbar array device.

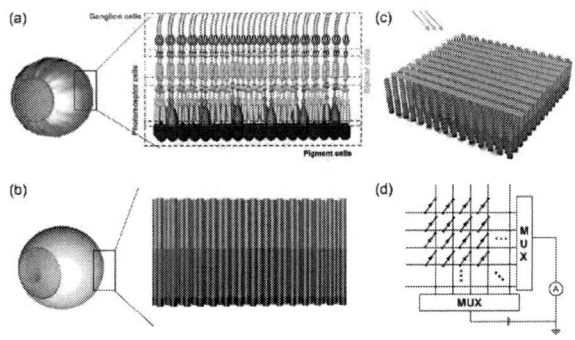

Fig 1: (a) Schematic of the human retina. (b) Schematic of Artificial Vision system with nanowire array device. (c) Crossbar array image sensor. (d) Read out circuit with multiplexer.

To properly understand the working mechanism of the device, we utilize the band diagram for the whole metal (Pb) – semiconductor (CsPbI₃) – ETL (SnO₂) – metal (ITO) interfaces (Fig. 2). When a very small voltage of 0.01V is applied across the electrodes the device works in standard imaging mode. As can be seen in figure 2(a), there is a small Schottky junction formed at the metal – semiconductor contact. When photons are absorbed by the semiconductor, the generated holes are trapped at this Schottky junction (Fig. 2(b)). Gradually with more holes reaching at this interface, the semiconductor becomes heavily p doped. This lowers its bands and reverses the formation of the Schottky contact (Fig. 2(c)). This explains the sudden spike in the photo current, which then stabilizes to a lower value (Fig. 3(a)). However, if we apply some significant bias across the electrodes, the band bending at the Schottky contact will be larger (Fig. 2(d)). Due to this, more and more photo generated holes will get trapped in the semiconductor, thus increasing its conductivity (Fig. 2(e, f)). This gradual increase in the conductivity culminates in the neuromorphic behavior of the device (Fig. 3(b)).

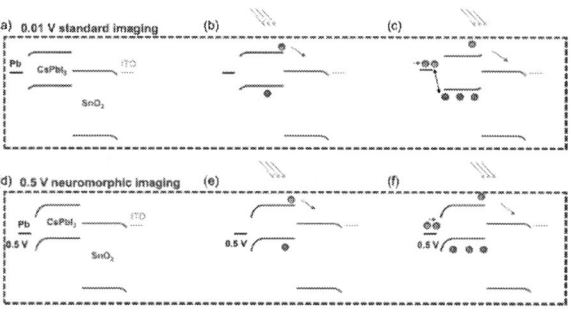

Fig 2: (a)–(c) Working mechanism of standard mode. (a) Energy band diagram at 0.01 V bias. (b) Transient after photon absorption. (c) Steady state after photon absorption. (d)–(f) Working mechanism of neuromorphic mode. (d) Energy band diagram at 0.5 V bias. (e) Transient after photon absorption. (f) Steady state after photon absorption.

Fig. 3(c) shows the photon induced neuromorphic behavior of the device for 100 spikes of photons. The uniformity of the response can be attributed to the highly stable and defect free CsPbI₃ nanowires. Figure 3(d) and 3(e) shows the response to the optical paired pulse facilitation (PPF). The PPF response decreases with the increase in the interval between the two pulses. The response to the PPF is equivalent to the neuronal synapse behavior observed in the retina for learning. All the optical measurements were performed for a light intensity of 872 µWcm⁻¹. The perovskite nanowire device in PAM shows less than 10% degradation even after 14 months of usage (Fig. 3(f)). This unique behavior contrasts with its thin film counterpart and is facilitated by the insulative properties of the membrane. Finally, we show the imaging and computational power of the device using a 10 by 10 cross bar array. As can be seen in Fig. 3(g), when the image is

continuously projected on the device, the current gradually increases. This increases the brightness of the obtained image and is similar to the learning process in humans. When the image is removed, the process of forgetting takes place and the current slowly decays.

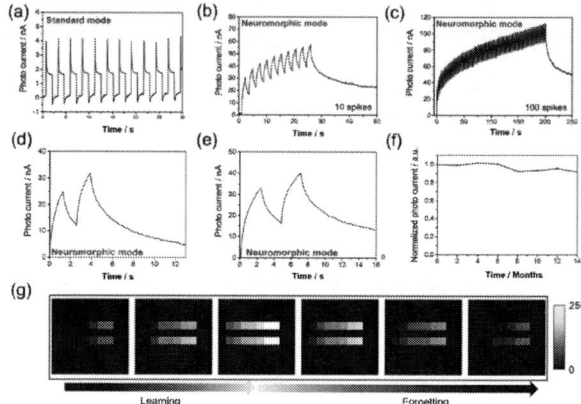

Fig 3: (a) Response of the device in standard imaging mode to 10 pulses of photons. (b) Response of device in neuromorphic mode to 10 pulses of photons. (c) Neuromorphic response to 100 pulses of photons. (d) PPF of device for 1.5s pulse interval. (e) PPF for 2.5s pulse interval. (f) Long term stability of photo response. (g) Learning and forgetting imaging in the 10 by 10 crossbar device.

Shape Classification

To further examine the advantages of the artificial vision system, we integrate the output of a 6 by 6 neuromorphic device array with a 36 by 4 electronic synapse. The electronic synapse array is fabricated using the perovskite R-RAM cross bar structure [9]. The weights of the R-RAM layer are first calculated through online training and then transferred to the R-RAM cells through resistance values. The array takes 36 inputs from the image sensor layer and produces 4 outputs calculated by the nonlinear activation function given in the equation (1).

$$output = \sum_{i=0}^{36}(pixel_i * weight) \qquad (1)$$

In total, the electronic synapse array comprises of 36 rows and 4 columns. The sum of current of each column serves as the output of a particular shape (triangle, square, circle and parallelogram). The outputs of the 4 columns are normalized to calculate the recognition probabilities of the 4 shapes. Overall, the system contains 36 outputs in the perception layer (6 by 6 array), 36 rows as the input layer and 4 rows as the output layer (36 by 4 array).

Figure 4(a) shows the recognition probabilities when different shapes are projected on the artificial vision system for a long duration of time. We observed that the recognition probabilities were always higher than 99% with false positives and false negatives always being less than 1% for all the 4 shapes. To emphasize further on the benefit of the artificial vision system we add some randomly generated noise in the projected images as shown in Fig 4(b). The noisy images are projected both on the neuromorphic sensor device (photonic synapse) and the event-based image sensor (without photonic synapse) before being fed to the electronic synapse. We observe that the training loss in the neuromorphic sensor device converges much faster (Fig. 4(c)), reaching 100% accuracy after 66 training epochs compared to 2348 epochs required by the event-based sensor system. Fig. 4(d) shows the accuracy of classification obtained against the incident light intensity for the projected shapes. We observe that higher incident light intensity facilitates faster convergence due to faster learning process.

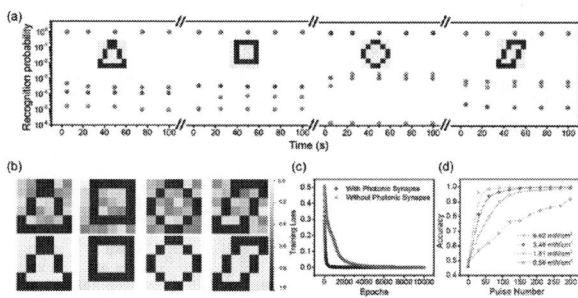

Fig 4: (a) Recognition probability of different shapes performed by the artificial vision system. (b) Classification result of geometric shapes with and without random noise. (c) Effect of photonic synapse on training loss. (d) Effect of different intensity of light on accuracy.

Conclusion

In this work, we present a human retina inspired artificial vision sensor using a very high density and vertically aligned perovskite nanowire array. The all-inorganic device shows highly stable photo response, retaining 90% of its performance even after 14 months of usage. The device exhibits synaptic response at a higher applied bias which is beneficial for data pre-processing, reducing the computation load on the image processor. We fabricate a 10 by 10 cross bar array of the device where it exhibits consistent response to different optical signals like learning and forgetting. We further highlight the full potential of the artificial vision system by integrating an array of perovskite nanowire-based photonic synapse with a perovskite nanowire-based R-RAM array, utilizing it for geometric shape classification. We obtain a very high accuracy of 99% which is considerably higher when compared to an event-based image sensor in a similar setup after fewer number of epochs. The whole work provides a pathway for the development of more sophisticated and efficient artificial vision systems.

Acknowledgments

The authors would like to thank Mr. Peter Nam and Mr. Jacob Ho from the dept. of ECE, HKUST for assistance in purchasing and setting up the equipment used in the experiments.

References

[1] F. Hutmacher, "Why Is There so Much More Research on Vision than on Any Other Sensory Modality?", Frontiers in Psychology, vol. 10, no. 2246, Oct. 2019.

[2] J. Zhang, S. Dai, Y. Zhao, J. Zhang, and J. Huang, "Recent Progress in Photonic Synapses for Neuromorphic Systems", Advanced Intelligent Systems, vol. 2, no. 3, p. 1900136, Jan. 2020.

[3] G. D. Hildebrand and A. R. Fielder, "Anatomy and Physiology of the Retina", Pediatric Retina, pp. 39–65, Aug. 2010.

[4] S. Poddar et al., "Geometric Shape Recognition with an Ultra-High Density Perovskite Nanowire Array-Based Artificial Vision System", ACS Applied Materials & Interfaces, vol. 16, no. 4, pp. 5028–5035, Jan. 2024.

[5] A. Waleed et al., "Lead-Free Perovskite Nanowire Array Photodetectors with Drastically Improved Stability in Nanoengineering Templates", vol. 17, no. 1, pp. 523–530, Dec. 2016.

[6] Z. Long, Y. Ding, X. Qiu, Y. Zhou, S. Kumar, and Z. Fan, "A dual-mode image sensor using an all-inorganic perovskite nanowire array for standard and neuromorphic imaging", Journal of Semiconductors, vol. 44, no. 9, p. 092604, Sep. 2023.

[7] L. Gu et al., "A biomimetic eye with a hemispherical perovskite nanowire array retina", Nature, vol. 581, no. 7808, pp. 278–282, May 2020.

[8] L. Gu et al., "3D Arrays of 1024-Pixel Image Sensors based on Lead Halide Perovskite Nanowires", Advanced Materials, vol. 28, no. 44, pp. 9713–9721, Sep. 2016.

[9] S. Poddar et al., "Down-Scalable and Ultra-fast Memristors with Ultra-high Density Three-Dimensional Arrays of Perovskite Quantum Wires", Nano Letters, vol. 21, no. 12, pp. 5036–5044, Jun. 2021.

Fully Tunable In-Memory Eligibility Traces Based on Ferroelectric-Semiconductor Field-Effect Transistors

Junling Liu[1], Shuo Liu[1], Xinrui Guo[1], Yu Zhu[1], Ru Huang[1], and Ming He[1,2,*]

[1] Beijing Advanced Innovation Center for Integrated Circuits, School of Integrated Circuits, Peking University, Beijing 100871, China; [2] Frontiers Science Center for Nano-optoelectronics, Peking University, Beijing 100871, China.
*E-mail: minghe@pku.edu.cn

Abstract

Traditional implementations of eligibility traces have struggled with the tunability due to fixed relaxation characteristics, usually necessitating energy-intensive auxiliary circuitry. In this work, we introduce a Bi_2O_2Se ferroelectric-semiconductor field-effect transistor-based reinforcement learning system that offers a fully tunable eligibility trace ranging from 0.001 to 0.999, spanning three orders of magnitude. Benefiting from the ferroelectric properties of Bi_2O_2Se, our device functions as a non-volatile 7-bit memory, featuring an impressive memory window of 10.5 V. In a demonstration of the Taxi Problem, our system achieves an average reward of 8.39 over 100 epochs while realizing a remarkable 99.68% reduction in energy consumption compared to conventional GPU implementations.

Keywords: Ferroelectric semiconductor, Eligibility trace, Reinforcement learning

Introduction

Reinforcement learning has emerged as a pivotal approach for enabling adaptive decision-making, allowing agents to learn through trial and error in uncertain environments[1]. Within this framework, eligibility traces serve as a critical mechanism that facilitate the distribution of rewards or penalties from the environment to relevant states with diminishing significance, largely accelerating the convergence rate during the training process[2, 3]. Previous efforts have sought to develop hardware implementations of eligibility traces to enhance the energy efficiency of reinforcement learning systems[4-6]. However, the intrinsic relaxation behavior that affected by fabrication processes and material properties poses significant challenges to achieving tunability in eligibility traces. As a consequence, additional auxiliary peripheral circuitry that facilitates the tuning of eligibility traces becomes indispensable, but it inevitably increases the system's energy consumption.

In this work, we manufacture a double-gate Bi_2O_2Se ferroelectric-semiconductor field-effect transistor that offers exceptional eligibility trace tunability from 0.001 to 0.999. This extensive tunability is present to enhance the capacity of reinforcement learning for diverse tasks. Exploiting the

ferroelectric properties of Bi_2O_2Se, the top-gate control further facilitates the achievement of 7-bit non-volatile memory with a remarkable memory window of 10.5 V. This advanced memory functionality supports the effective storage of learned policy. In a demonstration involving the Taxi Problem, our system reaches an average reward of 8.39 over 100 epochs, while achieving a remarkable 99.68% reduction in energy consumption compared to traditional GPU implementations.

Device Fabrication and Characterization

The double-gate Bi_2O_2Se field-effect transistor (FET) is depicted in **Fig. 1a**. CVD-grown Bi_2O_2Se films are transferred onto a silicon substrate coated with a 20 nm layer of HfO_2[7]. The Pd/Au electrodes are fabricated through a combination of electron-beam lithography, electron-beam evaporation, and lift-off techniques. Then, a 10 nm HfO_2 top dielectric layer is deposited, followed by the top-gate electrode fabricated using the electron-beam lithography, electron-beam evaporation, and lift-off processes. The high-resolution transmission electron microscopy (HRTEM) image in **Fig. 1b** reveals the crystal structure of Bi_2O_2Se, while the selected area electron diffraction (SAED) pattern in the inset confirms its high quality and single crystallinity. **Figures 1c-d** display the transfer and output characteristics of the device under back-gate control, illustrating a high on-off ratio of 10^6.

Results and Discussion

By optimizing the concentration of hydrofluoric acid during the transfer process, we deliberately introduce trap states at the back-gate channel-dielectric interface. Gradual charging and discharging behavior of these defects under successive back-gate voltage (V_{bg}) enables the implementation of tunable eligibility traces. We investigate the time-varying drain-source currents (I_{ds}) of the Bi_2O_2Se FETs at the V_{bg} from -1.5 V to -0.5 V across various drain-source voltages (V_{ds}). The resultant time-dependent I_{ds} curves under specific combinations of V_{ds} and V_{bg} are shown in **Fig. 2a**. Fitting analyses indicate that these curves correspond well to exponential functions of form $y = y_0 + y_1 \cdot a^t$. Notably, as a value decreases, the nonlinearity of the curve becomes increasingly pronounced.

979-8-3315-0417-5/25 $31.00 © 2025 IEEE

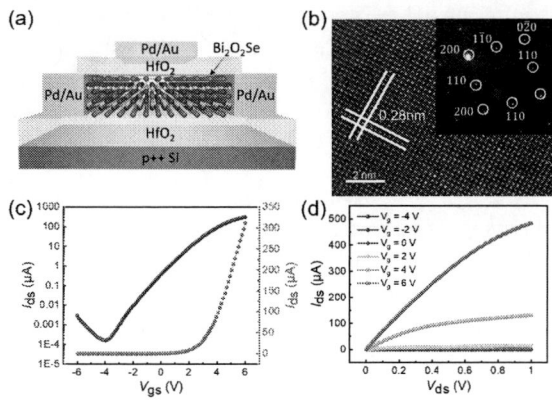

Figure 1. (a) Schematic of the double-gate Bi_2O_2Se field-effect transistor (FET). (b) High-resolution TEM image of as-synthesized Bi_2O_2Se. (c) Transfer curve of the back-gated Bi_2O_2Se FET. (d) Output curves of the Bi_2O_2Se FET with V_{ds} in the range of -0.05~0.05 V.

To demonstrate the continuous and comprehensive tunability of eligibility traces enabled by our devices, the normalized time-dependent I_{ds} fitted with a values ranging from 0.001 to 0.999 is present in **Fig. 2b**. Additionally, **Fig. 2c** illustrates the correlation between a value and the applied V_{ds} and V_{bg}, indicating that a value elevates with V_{ds} or V_{bg} decreasing. The storage of policy maps can be realized by the intrinsic ferroelectric property of Bi_2O_2Se semiconductor. Structural distortions arising from factors such as strain, doping or thickness reduction can induce symmetry breaking in Bi_2O_2Se, leading to a pronounced out-of-plane ferroelectricity[8]. As shown in **Fig. 3a**, the dual-sweep transfer curve of the top-gate Bi_2O_2Se FET exhibits a distinct counter-clockwise hysteresis indicative of ferroelectric behavior. It shows a substantial memory window that reaches up to 10.5 V. The top-gate FET constructed on PMMA-protected surface of Bi_2O_2Se during transfer suppresses etching defects, and thus enables complete exhibition of its ferroelectric semiconductor properties. Extracted from **Fig 3a,** trends of memory window and on/off ratio over cycling periods, exhibit no significant degradation over time, as shown in **Fig. 3b**. The state retention capability is revealed in **Fig. 3c**, wherein each state maintains stably for over 100 s. **Fig. 3d** displays the long-term potentiation (LTP) and long-term depression (LTD) characteristics of the top-gate device, exhibiting 7-bit states, thereby highlighting its potential for high-density and reliable information storage.

We employ the Taxi Problem as a framework to assess the efficacy of our proposed approach (**Fig. 4a**). In this scenario, a taxi driver navigates a 5×5 city grid to pick up passengers that appear randomly at one of four taxi stands (R, G, Y, B), and transports them to a destination at one of the other three stands. The taxi driver has six available actions: "down," "up," "right," "left," "pickup," and "drop off." Blocked paths are impassable. Successfully delivering each

passenger earns the taxi a reward of 20, while each step incurs a penalty of -1. Incorrect pickup or drop-off actions result in a penalty of -10. The task epoch concludes once the passenger reaches their destination, with the objective of maximizing rewards while minimizing penalties.

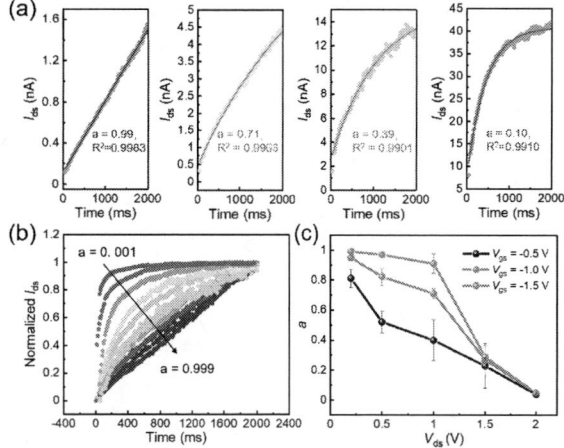

Figure 2. (a) Time-dependent I_{ds} of the back-gate Bi_2O_2Se FET under varied gate voltages and drain voltages, fitted by exponential functions. (b) Normalized time-dependent I_{ds} with tunable a from 0.001 to 0.999. (c) Relationship between a and drain voltages under varied gate voltages.

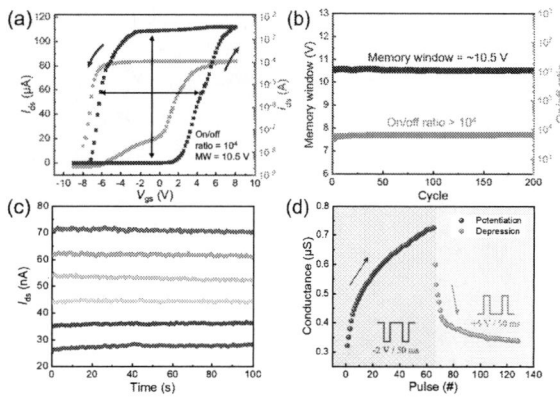

Figure 3. (a) Dual-sweep transfer curve of the top-gate Bi_2O_2Se FET. (b) Extracted trends of memory window and on/off ratio over cycling periods. (c) State retention of six specific analogue states of the top-gate Bi_2O_2Se FET. (d) Pulse-induced LTP/LTD characteristic of the device with 128 states.

We implement the tunable eligibility traces of Bi_2O_2Se ferroelectric-semiconductor FETs in conjunction with the SARSA(λ) algorithm for this task. **Fig. 4b** depicts the accumulated rewards and exploration steps throughout each training epoch, both of which converge after 1500 epochs. **Fig. 4c** illustrates the impact of eligibility trace rate (λ) on the task success rate and average reward across 100 evaluated epochs. The maximum average reward occurs at λ = 0.4, accompanied by a median failure rate, indicating that while this setting may excel in specific scenarios, it can also lead to failures in others. Conversely, setting λ to 0.8 achieves the lowest failure rate, suggesting greater universality across varied conditions. **Fig. 4d** presents the learned policy map,

979-8-3315-0417-5/25 $31.00 © 2025 IEEE

which attains an average reward of 8.39 over 100 epochs, highlighting optimal driving strategies for successful deliveries. As illustrated in **Figs. 4e-f**, we extracted a partial policy map from **Fig. 4d** to demonstrate a scenario in which a passenger begins their journey at point G and aims to reach destination R. The learned policy indicates that, regardless of the taxi's initial location, it is capable of efficiently locating and picking up the passenger, followed by successfully transporting them to their intended destination. Finally, we compare the energy consumption of executing the task on various platforms, including CPUs, GPUs, PCMs, and our system. The energy consumption for CPU, GPU, PCM and our system is estimated to be 3.34×10^{-5} J, 2.35×10^{-6} J, 1.19×10^{-8} J, and 7.54×10^{-9} J, respectively. The results indicate that our methods exhibit lower energy consumption than PCMs and a remarkable 99.68% reduction in energy consumption compared to GPUs.

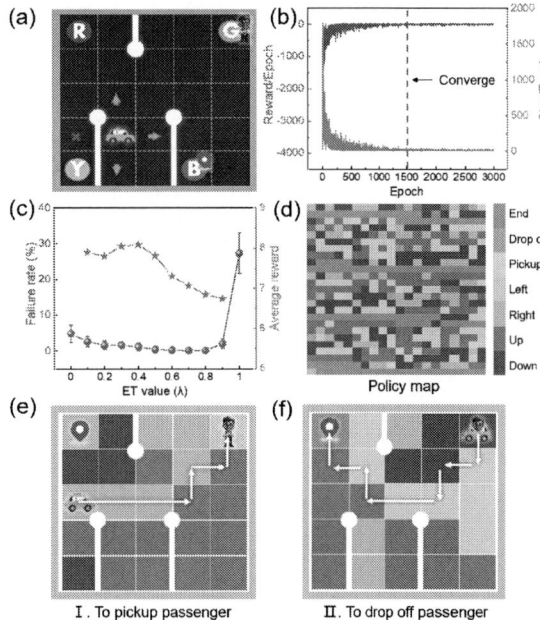

Figure 4. (a) Schematic of the environment in the Taxi Problem. (b) Reward and exploration steps in each epoch during training with SARSA(λ) method. (c) Success rate and average reward of 100 evaluated epochs with varied λ extracted from the experiment data. (d) Policy map of 500 states generated from the Q-table after training. (e-f) Partial policy map and example trajectory of the taxi moving towards the passenger location at the (e) pickup phase, and to drop off the passenger at the destination at the (f) drop-off phase.

Conclusion

In conclusion, we present a double-gate Bi_2O_2Se ferroelectric-semiconductor field-effect transistor-based reinforcement learning system that exhibit exceptional eligibility trace tunability exceeding three orders of magnitude. By adopting the ferroelectric-semiconductor feature of Bi_2O_2Se, our top-gate control realizes a 7-bit non-volatile memory with a 10.5 V memory window supporting

for efficient policy storages. The system effectiveness is validated through the resolution of the Taxi Problem, achieving an average reward of 8.39 over 100 epochs while realizing a significant 99.68% reduction in power consumption compared to conventional GPUs.

Acknowledgments

This work was supported by the National Key R&D Program of China 2022YFB4400100, the Natural Science Foundation of China (92164205, 62074004, 61927901), and 111 Project (B18001).

References

[1] L. P. Kaelbling et al. Reinforcement learning: a survey. *J. Artif. Int. Res.* **4**, 237–285 (1996).

[2] S. P. Singh et al. Reinforcement learning with replacing eligibility traces. *Mach. Learn.* **22**, 123–158 (1996).

[3] D. Precup et al. Eligibility Traces for Off-Policy Policy Evaluation. *Proceedings of the Seventeenth International Conference on Machine Learning*, 759–766 (2000).

[4] Y. Lu et al. In-Memory Realization of Eligibility Traces Based on Conductance Drift of Phase Change Memory for Energy-Efficient Reinforcement Learning. *Advanced Materials* **34**, 2107811 (2022).

[5] Y. Zhou et al. A Reconfigurable Two-WSe$_2$-Transistor Synaptic Cell for Reinforcement Learning. *Advanced Materials* **34**, 2107754 (2022).

[6] S. Oh et al. The Impact of Resistance Drift of Phase Change Memory (PCM) Synaptic Devices on Artificial Neural Network Performance. *IEEE Electron Device Letters* **40**, 1325-1328 (2019).

[7] M.He et al. Ultrasensitive dim-light neuromorphic vision sensing via momentum-conserved reconfigurable van der Waals heterostructure. *Nature Communications* **15**, 9011 (2024).

[8] M. He et al. Achieving Over 2800% Superadditive Visual-Audio Multisensory Integration in-situ Ferroelectric-Semiconducting Transistor for Fuzzy Subject Detection. *2024 International Electron Devices Meeting (IEDM)*, 27.26.21-27.26.24 (2024).

Low-Voltage Multi-Level Flash Memory Based on Intra-Float-Gate Charge Transfer

Yifan Chen[†,1], Haixia Li[†,2], Qing Wang[2], Zongwei Shang[2], Mingmin Shi[2], Xijun Zhou[2], Xiaoyan Xu[2,3], Xia An[2,3], Ru Huang[2,3], and Ming Li[2,3*]

[1] School of Software & Microelectronics, Peking University, Beijing, 102600, China,

[2] School of Integrated Circuits, Peking University, Beijing, 100871, China

[3] Beijing Advanced Innovation Center for Integrated Circuits, E-mail: liming.ime@pku.edu.cn

Abstract

Low-voltage multi-level flash memory devices with two (2FG) and four (4FG) metal floating gate layers have been successfully demonstrated experimentally. The intra-float-gate charge transfer and storage mechanism is proposed to achieve multi-level storage under sequential low-voltage pulse train. The 2FG and 4FG devices show 3-bit storage capability at 5V and 6V, respectively. The retention up to 10 years and endurance of about 10^5 cycles are also verified. (Keywords: FLASH, Multi-floating gate, Low voltage)

Introduction

The storage and processing of massive data have placed higher demands on multi-value storage technology in the big data and artificial intelligence era. Traditional NAND flash memory operations based on Multi-Level Cell (MLC) are conducted using Incremental Step Pulse Programming (ISPP) [1],[2]. However, to achieve efficient programming and erasing, the high Fowler-Nordheim (FN) tunneling current have to be obtained under high operation voltage > 10 V, which results in high power dissipation and does not suit for the on-chip application. Moreover, if the tunnel oxide layer is reduced to lower the operation voltage, the stress-induced leakage current (SILC) may be exacerbated [3]. Charge storaged in the floating gate layer may be subject to a one-time leakage when defects appear in the tunneling layer, severely affecting the reliability of the storage cell. Therefore, lowering the operation voltage while still ensuring a sustainable memory window has become crucial to high-capacity flash memory. To achieve higher density multi-level storage, many crucial material innovations to the charge storage layer have been made to increase the density of states or change the work function and carrier transport mechanism, such as the evolution from polycrystalline silicon to charge trapping layer, semiconductor nanocrystals or even metals and their nanocrystals [4],[5].

In this work, we successfully fabricated n-type multi-float-gate transistors with similar gate stack structure in our previous work [6]. Nanowire channel is used to enhance the tunneling electric field, improving the channel injection efficiency. Devices demonstrate 3-bit storage capability under 6V and 5V for 2FG and 4FG structures, respectively, which is attributed to the intra-float-gate charge transfer mechanism between floating gates, providing a solution for low-power multi-level NAND-flash applications.

Experiments

The fabrication process of the multi-floating gate flash memory (Multi-FGF) is shown in Fig. 1. Different stacking conditions are set up for single-layer floating gate (1FG), double-layer floating gate (2FG), and quadruple-layer floating gate (4FG). The silicon nanowire channel is formed on SOI wafer and sequentially followed by deposition of SiO_2 tunneling layer, TiN floating gate storage layers, Al_2O_3 barrier and blocking layer, and TiN metal control gate layer (CG). For different multi-float-gate structures, the TiN metal floating gate and Al_2O_3 barrier layer were sequentially deposited and repeated. The TiN CG was deposited by PVD. Fig. 2 displays the structure of gate stack of 2FG. Fig. 3 shows the cross-sectional TEM images of the storage gate stacks in 2FG and 4FG Multi-FGF devices. The total thickness of the TiN multi-floating gate layers is 8 and 9 nm for 2FG and 4FG respectively in order to maintain the similar total density of electron states. Al_2O_3 blocking layer is thicker than 7nm to avoid the charge loss from the top control gate [6]. About 3.4~3.9 nm SiO_2 is used as tunneling layer (TL) to ensure a large electric field across, thus a high channel injection efficiency can be obtained. For consideration of growing high-quality dielectric, 2 nm Al_2O_3 was deposited between every two metal floating gates as the barrier layer. Although the high dielectric constant may not be beneficial to increase the electric field across the BL itself, the thickness of about 2nm still allows sufficient direct tunneling probability of electrons through BL where the layered charge transfer takes place.

Results and Discussions

A. Layered Charge Transfer Mechanisms

Fig. 4 shows the basic I_d-V_g curves of 2FG and 4FG devices, of which the threshold voltages are 0.3 V and 0.7 V, respectively. Fig. 5 illustrates the layered charge transfer mechanism between the floating gates during program/erase operations. Assuming the electrical neutral state with flat band condition, as the control gate is biased at programming voltage, electrons will enter the bottom floating gate through FN tunneling from channel (green arrows) and transfer to the high-level floating gate layers through direct tunneling (red arrows) simultaneously, as shown in Fig. 5 (a). As the charge is stored in the top floating gate further away from the channel, its screening effect on the electric field across TL will be suppressed thus hence more electrons can be injected into the floating gate stacks. As a result, larger memory window can be created with more floating gates and at lower stress voltages. Conversely, under a negative pulse (erased), electrons will transfer toward channel (see in Fig. 5 (b)) and V_{th} shifts negatively. Similarly, charge transfer between floating gates in 4FG following the same tunneling mechanism.

B. Multi-level storage

Fig. 6 compares the program/erase (P/E) window of FGFs under different pulse amplitude with the pulse intervals of 100 ms. Each program pulse is applied after complete erase pulses and similar operation is also done to characterize erasing window. It is worth noting that, compared to the 4FG structure, the 2FG has a larger initial electric field across the tunneling layer under the same voltage due to fewer stacked oxide layers, leading to greater initial channel injection efficiency, as shown in Fig. 6. However, 4FG shows larger saturated memory window than 2FG. Even though 2FG shows better low voltage P/E capability than 1FG, the saturated P/E window is not much different from 1FG under high voltage because of the lack of capacity for electric field screening when a large amount of charge is stored. Fig. 7 compares the multi-value P/E capability of 4FG and 1FG by pulse sequences. The V_{th} increasement with pulse number of 4FG structure shows better linearity than 1FG, which is attributed to more total charge and layered charge transfer. 4FG also shows faster program speed than 1FG especially under low gate bias in order to achieve better performance in low-voltage multi-level storage application.

Fig. 8 and Fig.9 show the I_d-V_g curves of 4FG and of 2FG with different programming pulse numbers under amplitude voltage of 6V and 5V, respectively. The clear V_{th} shift caused by accumulated pulses indicates at least 3-bit storage capability for both structures. The V_{th} shift with the programming pulse number is depicted in Fig. 10 to show almost linear V_{th} step. Thanks to more charge injection efficiency at lower voltage, 2FG structure can achieve the multi-bit storage at lower pulse amplitude and narrower pulse width. But 4FG shows much larger memory window than 2FG also with good linearity.

C. Reliability

Fig. 11 demonstrates the retention of Multi-FGF, which can maintain four different storage states (2-bit) after being placed at 85°C for one hour and keep the four states with difference of at least 0.1V among one another for ten years extrapolated. Fig. 12 compares the retention characteristics of 4FG and 1FG to show significantly better V_{th} retention in Multi-FGF. Figure. 13 represents the endurance of Multi-FGF. To induce stronger stress effect on the dielectric, a larger program voltage (8V) has been adopted here. After 10^5

cycles of P/E operations, the V_{th} is only degraded by 4.8% and 12.1%, respectively, satisfying the requirement for NAND flash memory application.

Conclusion

A novel flash device, Multi-FGF, has been proposed and successfully fabricated, featuring low-voltage multi-bit storage capability. Based on the layered charge transfer and storage, 4FG and 2FG structures demonstrated 3-bit storage under 5V (50ms) and 6V (100ms) respectively. With more floating gates, the larger the memory window. Multi-FGF device also exhibit retention up to 10 years under 85°C and endurance of about 10^5 cycles under 8V, offering potential solution for future low-power multi-level NAND flash applications.

Acknowledgments

Y.C. and H.L. contributed equally to this work. This work was partly supported by the National Key Research and Development Program of China (No. 2023YFB4402200), NSFC project (61927901) and Beijing Frontier Innovation Project (No. QYJS-2021-0500-B).

References

[1] Kang-Deog Suh et al., 'A 3.3 V 32 Mb NAND flash memory with incremental step pulse programming scheme', IEEE J. Solid-State Circuits, vol. 30, no. 11, pp. 1149–1156, Nov. 1995, doi: 10.1109/4.475701.

[2] K.-T. Park et al., 'Dynamic Vpass ISPP scheme and optimized erase Vth control for high program inhibition in MLC NAND flash memories'.

[3] Y. Okuyama et al., 'Monte Carlo simulation of stress-induced leakage current by hopping conduction via multi-traps in oxide', in International Electron Devices Meeting 1998. Technical Digest (Cat. No.98CH36217), San Francisco, CA, USA: IEEE, 1998, pp. 905–908. doi: 10.1109/IEDM.1998.746501

[4] R. Rajput and R. Vaid, 'Flash memory devices with metal floating gate/metal nanocrystals as the charge storage layer: A status review', Facta Univ Electron Energ, vol. 33, no. 2, pp. 155–167, 2020, doi: 10.2298/FUEE2002155R.

[5] X. Yu et al., '3D NAND Flash Memory Based on Double-Layer NC-Si Floating Gate with High Density of Multilevel Storage', Nanomaterials, vol. 12, no. 14, p. 2459, Jul. 2022, doi: 10.3390/nano12142459.

[6] H. Li, H. Liao, B. Zhang, R. Bi, R. Huang, and M. Li, 'Dual-Float-Gate Capacitor for Low-Voltage Multi-Level Nonvolatile Memory with Enhanced Retention', in 2023 7th IEEE Electron Devices Technology & Manufacturing Conference (EDTM), Seoul, Korea, Republic of: IEEE, Mar. 2023, pp. 1–3. doi: 10.1109/EDTM55494.2023.10103085.

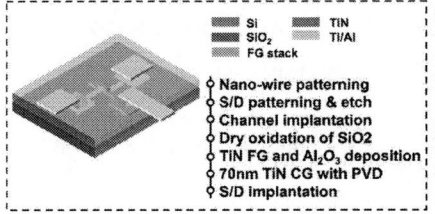

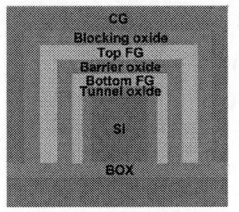

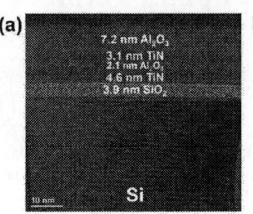

Fig. 1. Fabrication process of the floating gate transistors. Each floating storage layer is sequentially deposited and repeated.

Fig. 2. The schematic of gate stack structure. Nanowire channel is used to enhance the couple of CG.

Fig. 3. TEM images of flash devices with (a) 2FG (b) 4FG structure. The thickness of each layer is labeled on the picture.

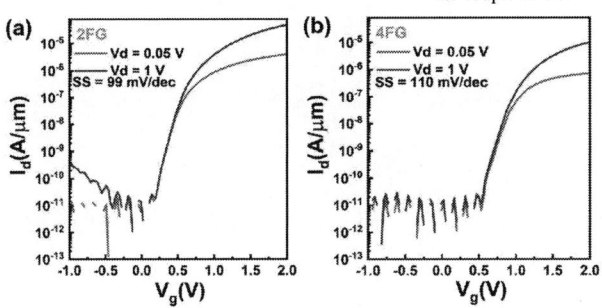

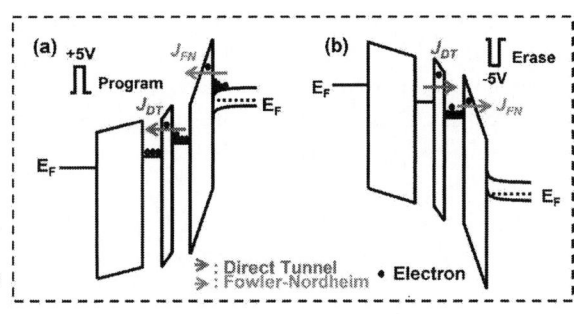

Fig. 4. Basic Id-Vg curves of (a) 2-FG (b) 4-FG. Vd has been set to 0.05V and 1V to ensure device work in linear and saturation region, respectively.

Fig. 5. Schematic of charge transfer mechanism of floating gate transistors (2FG). (a) after a programed pulse and (b) after an erased pulse. Electrons inject into metal floating gate through F-N tunneling and direct tunneling.

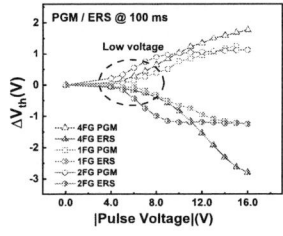

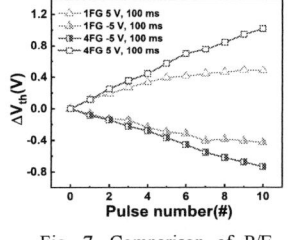

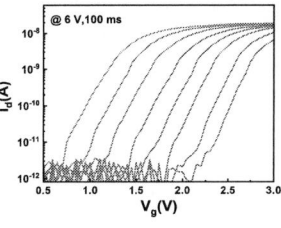

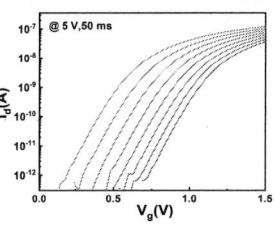

Fig. 6. Comparison of P/E window of FGFs under different pulse voltages.

Fig. 7. Comparison of P/E window of FGFs under pulses sequences.

Fig. 8. 3-bit memory of 4-FG under program pulses of 6V with pulse width of 100ms.

Fig. 9. 3-bit memory of 2-FG under program pulses of 5V with pulse width of 50ms.

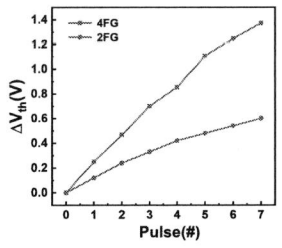

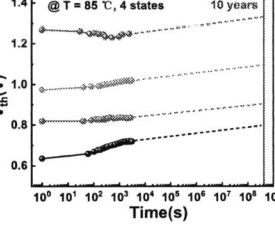

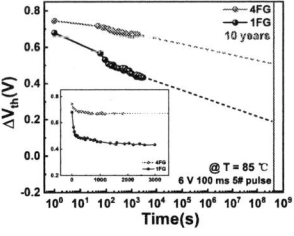

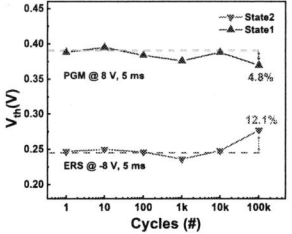

Fig. 10. The shift of Vth after continuous pulses. Both 2FG and 4FG show linearity modulation. 4FG has a larger memory window due to charge further stored.

Fig. 11. Retention characteristics of 4 states at 85°C, which can be extrapolated to 10 years.

Fig. 12. The comparison of retention characteristic of 1FG and 4FG at 85°C.

Fig. 13. Endurance of 4FG with high program voltage (8V) to verify better reliability requirements by multi-level flash memory.

First Demonstration of Tunnel FET-based Physical Unclonable Function with Independent Entropy Source through Ambipolar Current Modulation

Kaifeng Wang[1], Yingxi Zhou[1], Rundong Jia[1], Hongyan Han[2], Weihai Bu[2], Qianqian Huang[1,3*] and Ru Huang[1,3*]

[1]School of Integrated Circuits, Peking University, Beijing 100871, China; [2]Semiconductor Technology Innovation Center (Beijing), Beijing 100176, China; [3]Beijing Advanced Innovation Center for Integrated Circuits, Beijing 100871, China. (*E-mail: hqq@pku.edu.cn, ruhuang@pku.edu.cn)

Abstract

In this work, for the first time, a novel tunnel FET (TFET) with modulated asymmetric ambipolar current is proposed for physical unclonable function (PUF) applications with the independent entropy source at the drain side. In the proposed TFET, an underlap region is introduced to enlarge the variation of drain tunnel current for PUF while maintaining the relatively small variation at the source tunnel junction for logic or memory applications. This designed asymmetrical and independent variability of novel TFET devices are experimentally demonstrated on 300mm CMOS baseline platform. An AND-type TFET array as weak PUF is also designed and experimentally demonstrated with ~50% inter-PUF hamming distance. The worst raw BER at 85℃ after 4000 evaluations is only 0.46%, demonstrating the great potential of TFET for robust hardware security.
Keywords: TFET, variability, and PUF

Introduction

With the rapid development of Artificial Intelligence of Things (AIoT) applications, localized security platforms with robust, lightweight, and energy-efficient requirements are more urgent than ever. Hardware-intrinsic security primitives such as physical unclonable functions (PUFs) are promising solutions to counter possible security threats by leveraging unpredictable hardware fingerprints from physical randomness [1]. However, optimizing CMOS-based PUFs for robust and secure operation suffers from considerable area and power consumption issues for resource-limited AIoT [2]. Although alternatives, such as RRAM- and MRAM-based PUFs by employing the intrinsic device randomness from conductance filaments and freelayer switching, have attracted lots of attention [3-4], the entropy sources are still the same as the physical entities of information storage as memory functions, resulting in the intrinsic device optimization conflict for PUF and memory applications.

As one of the most promising low-power devices, tunnel FET (TFET) with gate-controlled p-i-n structure can achieve much lower leakage current and smaller VDD than MOSFET due to its band-to-band tunneling (BTBT) mechanism at source junction [5-6]. It can reduce the static power consumption of the logic circuits as well as enhance the retention time of DRAM, which has been experimentally

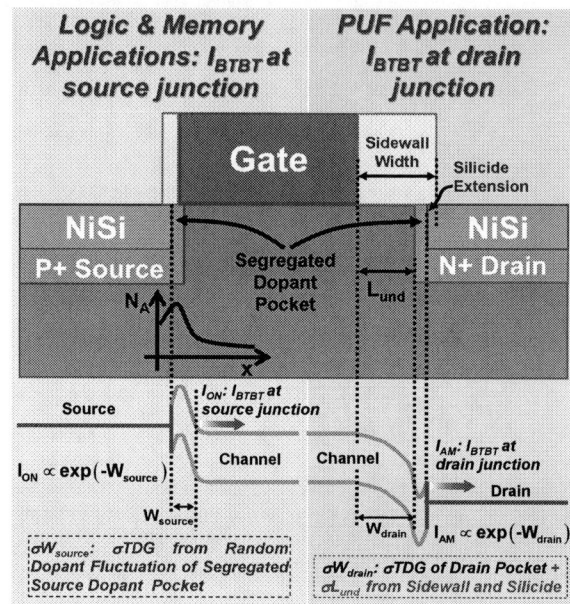

Fig. 1. Schematic of DS-TFET and the physical mechanism of I_{BTBT} at both source and drain junctions. I_{BTBT} at drain junction has an additional variation source induced by L_{und}.

verified in our recently demonstrated TFET-MCU with sub-100nA ultra-low leakage [7] and TFET-eDRAM with 3.9s retention time [8] on standard 55nm CMOS technology node.

In this work, different from the conduction current at the source junction for logic and memory applications, the band-to-band tunneling current (I_{BTBT}) at the drain junction of TFET is proposed as the entropy source of PUF. The novel dopant-segregated TFET (DS-TFET) with asymmetrical source and drain junction design is proposed and experimentally presented to enable the co-optimization for both PUF and other applications on the same wafer, showing the significant potential of TFET for robust hardware security for advanced power-dieting applications.

DS-TFET WITH INDEPENDENT ENTROPY SOURCES FOR PUF APPLICATION

As shown in Fig.1, the DS-TFET is designed with self-aligned dopant segregated source junction for high I_{ON} and drain underlap region for low I_{OFF} through asymmetrical gate spacer [9]. Taking n-type DS-TFET as an example, positive gate voltage is usually applied to open the BTBT window at source junction in logic and memory applications [7][10]. The BTBT window at drain junction is opened when the negative gate voltage is applied and this ambipolar current is

979-8-3315-0417-5/25 $31.00 © 2025 IEEE

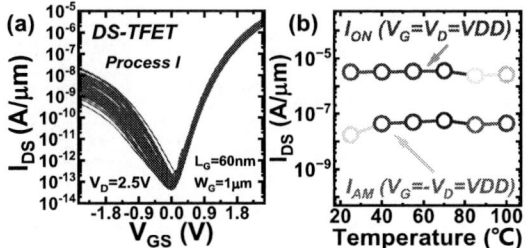

Fig. 2 **Measured** (a) transfer curves of DS-TFET across 300mm wafer, (b) conduction currents at both junctions under different temperatures.

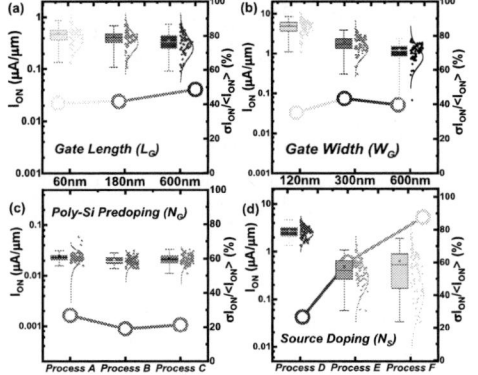

Fig. 3 **Measured** I_{ON} distribution of DS-TFETs with different (a) L_G, (b) W_G, (c) N_G and (d) N_S.

usually the dominant component of I_{OFF}. N-type DS-TFET fabricated on 55nm hybrid TFET-CMOS baseline platform is measured in Fig. 2. Although there is a underlap region with the width of L_{und}, I_{BTBT} of about 10nA/μm can still be obtained at the drain side when V_{GD}=-2VDD and its variability is larger than the I_{BTBT} at source side as shown in Fig. 2. Conduction current at V_{GS}=V_{DS}=VDD is defined as I_{ON} and conduction current at V_{GS}=-V_{DS}=-VDD is defined as I_{AM} here. The temperature-independent characteristics of I_{ON} and I_{AM} further verified the BTBT mechanism of both conduction currents at source and drain junctions.

To analyze the dominant variation source, DS-TFETs with different gate lengths (L_G), gate widths (W_G), predoping concentrations of poly-Si gate (N_G) and source doping conditions (N_S) are fabricated and compared in Fig. 3.

$$I_{DS} = W_G T_{Sieff} E_G \frac{1}{W_{t,min}^2} \frac{A_{kane}}{B_{kane}} \exp\left(-W_{t,min} B_{kane} E_G^{1/2}\right) \quad (1)$$

As equation (1) shows, the exponential I-V relationship of BTBT current makes the physical entities which are directly related to tunneling barrier become the primary variation sources of TFET [11]. Unlike MOSFETs, the variation of BTBT current can hardly be affected by L_G, W_G, and N_G. N_S shows the most significant impacts on both average and variation values of BTBT current, indicating its decisive role in TFET performance. Therefore, the dominant variation source of DS-TFET at the source tunnel junction is mainly contributed by the random dopant fluctuation of segregated source dopant pocket (σTDG). For drain junction, as shown

in Fig. 1, due to the introduction of underlap length (L_{und}), the variation of tunneling barrier width is jointly determined by both σL_{und} and segregated dopant distribution.

By modulating the source and drain doping process, the transfer curves of DS-TFETs modified for PUF across whole 300mm wafer are measured in Fig. 4. The experimental result demonstrates that the modified DS-TFETs have much larger I_{AM} variation than I_{ON}, verifying the designed asymmetrical variation characteristics. In addition, Fig. 5 demonstrates the comparison results of two device samples in mismatch testkeys across the wafer. The experimental results further reveal that variability of I_{BTBT} at both source and drain junctions are random and uncorrelated with each other. Therefore, by utilizing the asymmetrical ambipolar behavior of DS-TFET, the entropy source for PUF can leverage the variation from drain junction side while the source junction

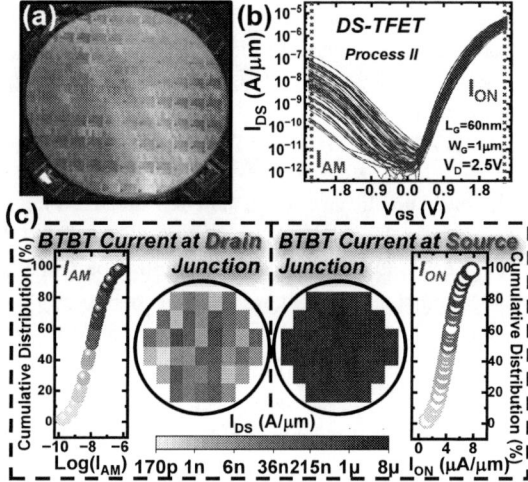

Fig.4. (a) Photo of fabricated 300mm wafer. (b) **Measured** transfer curves of DS-TFETs with process B across wafer. (c) Wafer mapping and the cumulative distribution of I_{ON} and I_{AM}.

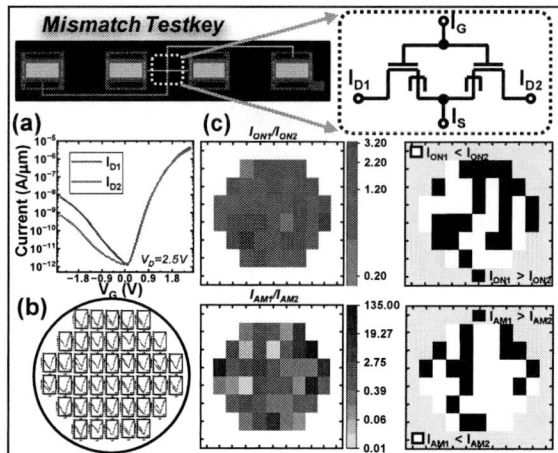

Fig.5. (a) **Measured** transfer curve of one mismatch of DS-TFETs and (b) all testkeys across wafer. (c) Current ratio and the digitalized results of I_{ON} and I_{AM} from mismatch testkeys.

979-8-3315-0417-5/25 $31.00 © 2025 IEEE   475

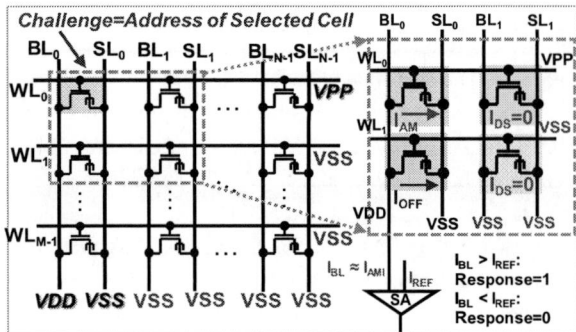

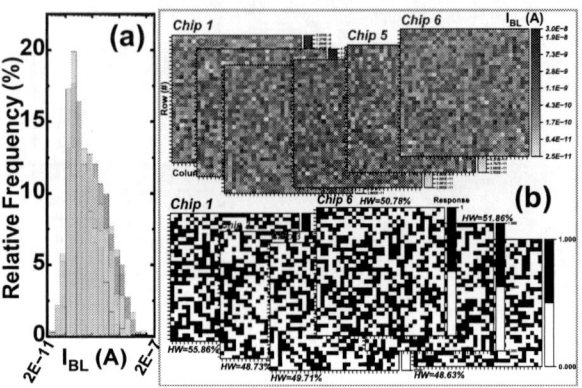

Fig.6. (a) Schematic of the proposed weak PUF based on AND-type DS-TFET array. (b) Schematic of the selected cell, unselected cell, half selected cell in weak PUF and the CRP generation mechanism.

Fig.7. Measured (a) I_{BL} distributions, (b) I_{BL} and the corresponding responses of the proposed TFET-based weak PUF from six different chips, HW of each chip is nearly 50%.

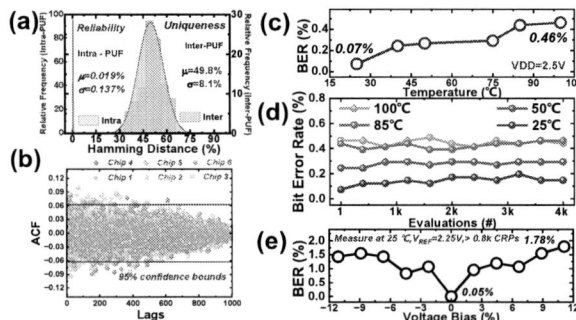

Fig.8. (a) Uniqueness, reliability and (b) auto correlation test results of TFET weak PUFs. Measured raw BER of TFET PUF at different (c) temperatures, (d) evaluations and (e) voltage biases.

is utilized for logic or memory applications. The additional variation source σL_{und} at drain junction not only enlarges the variability of I_{BTBT} at drain junction but also introduces an independent optimization dimension for PUF, enabling co-optimization for PUF and other functions on the same wafer. Furthermore, the temperature-independent characteristics of I_{BTBT} show the great noise immunity potential for high-reliability PUF.

ARRAY-LEVEL DEMONSTRATION OF TFET-BASED WEAK PUF

By utilizing I_{BTBT} at drain junction of DS-TFET with large randomness as the entropy source, a novel DS-TFET-based weak PUF is designed and experimentally demonstrated in this section. Fig. 6 demonstrates the proposed weak PUF design based on an AND-type TFET array. The challenge is the address of the selected cell. The selected WL is -VDD and the selected BL is VDD to open the BTBT window at drain junction of DS-TFET. I_{AM} of the selected DS-TFET is then readout through the bitline and compared with the reference current to generate a 1-bit response. The bitline currents (I_{BL}) of 1024 cells in six chips are measured, and the experimental results are demonstrated in Fig. 7. The measured I_{BL} in each chip show large distribution of almost 4 decades and the corresponding responses with nearly 50% hamming weight (HW) are generated using the above method. The inter-PUF hamming distance (HD) of ~50% demonstrates good uniqueness among different chips and the autocorrelation test (ACF) is also conducted to verify its randomness inside each chip as shown in Fig. 8(a-b). Moreover, benefiting from the much better device reliability of DS-TFET than conventional TFET [12], low raw BER is therefore obtained under all measured conditions as shown in Fig. 8(c-e). The worst raw BER at 85℃ after 4000 evaluations is only 0.46%, which is much smaller than CMOS-based weak PUFs reported in other works [13-14], indicating high robustness of proposed DS-TFET-based PUF.

Conclusion

DS-TFET with asymmetrical and independent variability at source and drain junctions is proposed and experimentally demonstrated for PUF applications based on 55nm hybrid TFET-CMOS platform. An AND-type DS-TFET array with I_{BTBT} at drain junction with large variability is proposed as weak PUF, and is experimentally demonstrated on 300mm baseline technology with ~50% inter-PUF HD and ~0% intra-PUF HD, showing both good uniqueness and reliability for cutting-edge power-dieting applications.

Acknowledgments

The authors would gratefully acknowledge Semiconductor Manufacturing North Corporation (SMNC) for providing the 55nm CMOS foundry platform. This work was supported by NSFC (62374009, 61927901) and 111 Project (B18001).

References

[1] Gao.Y., et al., Nat Electron (2020). [2] L. Lu, et al., TCAS-I, vol. 69, no. 6, pp. 2542-2552. [3] B. Lin et al., JSSC, vol. 56, no. 5, pp. 1641-1650. [4] Chiu.YC., et al., Nat Electron (2023). [5] Ionescu.A., et al., Nature, 329–337 (2011). [6] Liang.Z., et al., Sci. China Inf. Sci., 169406 (2023). [7] Y. Hou et al., 2023IEDM, pp. 37.6.1-37.6.4. [8] K. Wang et al., 2024ESSERC, pp. 13-16. [9] K. Wang et al., 2022ESSDERC, pp. 360-363. [10] K. Wang et al., 2023ESSDERC, pp. 13-16. [11] Q. Huang et al., 2015IEDM, pp. 22.2.1-22.2.4. [12] Y. Tang et al., 2023CSTIC, pp. 1-3. [13] S. Taneja, et al., 2021ISSCC, pp. 498-500. [14] J. Song et al., SSCL, vol. 5, pp. 58-61, 2022.

Reconfigurable and nonvolatile graphene photodetector integrated onto photonic crystal waveguide

Ruijuan Tian[1,2], Yu Zhang[1], Zhipei Sun*[2], Xuetao Gan*[1]

[1] School of Physical Science and Technology, Northwestern Polytechnical University, Xi'an, P. R. China,
[2] Department of Electronics and Nanoengineering, Aalto University, Espoo, Finland
*Email: xuetaogan@nwpu.edu.cn; zhipei.sun@aalto.fi

Abstract

A reconfigurable and nonvolatile graphene photodetector integrated graphene and ferroelectric layer onto the air-slotted photonic crystal (PC) waveguide was fabricated and characterized. The device demonstrated nonvolatile positive and negative photocurrents in reconfigurable graphene *p-i-n* and *n-i-p* homojunction configurations under telecommunication band laser illumination. This design concept can extend to other two-dimensional semiconductors, offering potential for on-chip optoelectronic synapses, in-memory sensing, computing, and neuromorphic applications. Keywords: graphene, ferroelectricity, waveguide, homojunction

Introduction

On-chip reconfigurable and nonvolatile photodetector are essential for addressing the rapid growth in telecom, datacom, computing, and sensing, etc[1-3]. Graphene stands out among Optical-to-electrical materials for on-chip photodetectors due to its excellent merits. Recently, on-chip graphene PTE photodetectors have been demonstrated by integrating them with photonic or plasmonic waveguides and resonators[4-5]. However, these strategies have different drawbacks such as inevitable metal insertion/absorption loss, complex operation and electrical circuits, and so on. Our approach uses an air-slotted PC waveguide as two separate silicon back gates, integrating graphene with a ferroelectric P (VDF-TrFE) dielectric layer to achieve a reconfigurable, nonvolatile graphene *p-i-n* homojunction photodetector.

Experiments

A. Device Fabrication

The PC waveguides as well as the electrical isolation air-slots, were fabricated via electron beam lithography and inductively coupled plasma etching. P (VDF-TrFE) (70:30 in mol %) was spin-coated on silicon PC waveguides. The graphene layer, mechanically exfoliated onto a polydimethylsiloxane (PDMS, Gel-Pak) stamp, was then dry transferred onto the P(VDF-TrFE) film using a precise alignment system[7].

B. Photoelectric Performance Characterization

Electrical and optoelectronic measurements were conducted at room temperature using a semiconductor parameter analyzer (PDA FSpro), with a telecom-band tunable laser (TUNICS T100S-HP) as the light source. Impulse responses were measured with an RF-GS probe (MPI T26A GS150), electrical amplifier (Mini-circuits, ZKLNA-020GHz), and the oscilloscope (Lecroy_SDA 825Zi-A).

Results and Discussions

To achieve the reconfigurable and nonvolatile photoresponse, a hybrid device structure was constructed, as shown in Fig. 1(a). In this device structure, the PC waveguide, split by two air slots into three electrically isolated regions, functions as two separated back-gate electrodes (G1, G2). A ferroelectric P (VDF-TrFE) layer serves as the gate dielectric. Top graphene acts as the channel material, which is contacted with drain and source electrodes (D, S). As shown in Fig. 1(b), by applying a gate voltage between the left (or right) silicon

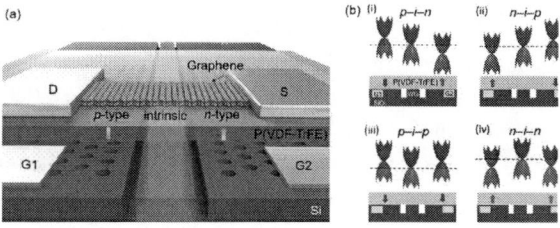

Fig. 1 (a) Schematic of the device. (b) Schematic of reconfigurable and nonvolatile graphene *p-i-n*, *n-i-p*, *p-i-p*, *n-i-n* homojunction configuration.

PC region and the top graphene layer through the P (VDF-TrFE) layer, the corresponding graphene section would be doped into *p*- or *n*-type depending on the direction of the gate voltage. Without voltage on the central waveguide, the central graphene remains intrinsic (*i*-type), forming reconfigurable *p-i-n*, *n-i-p*, *p-i-p* and *n-i-n* homojunction among graphene channel, as shown in Fig. 1(b).

The output curve of the device shows excellent Ohmic

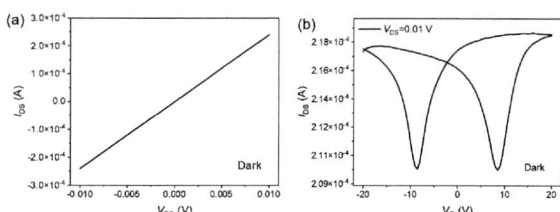

Fig. 2 (a) Output curve (I_{DS}–V_{DS}) of the device at $V_{G1}=V_{G2}=0$ V. (b) Transfer curve of the graphene channel with the global gate voltages V_G = V_{G1} = V_{G2}, and V_{DS} = 0.01 V.

contact, as indicated in Fig. 2(a). Figure 2(b) presents the transfer curve (I_{DS}–V_G) when the two bottom silicon gates are connected as a global back gate ($V_{G1}=V_{G2}=V_G$). Due to the nonvolatile polarization field of the P (VDF-TrFE) film, a clear hysteresis window appears when sweeping the gate voltages from $V_{G1}=V_{G2}=-30$ to 30 V and back to –30 V.

To confirm the reconfigurable and nonvolatile response of our device, the photoresponses were measured under different configurations with light illumination. Fig. 3(a) shows photocurrents at zero bias ($V_{DS}=0$ V) across different configurations of gate voltages, where the clearly visible 6-fold pattern indicates the PTE effect as dominant conversion

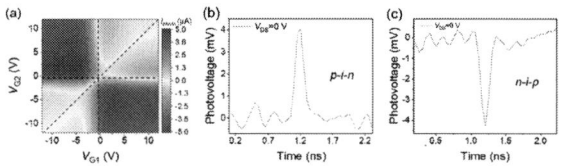

Fig. 3 (a) Photocurrent map at zero-bias with different gate voltages on G1 and G2. (b) Impulse response at zero-bias after pulsing $V_{G1}=-12$ V, $V_{G2}=12$ V. (c) Impulse response at zero-bias after pulsing $V_{G1}=12$ V, $V_{G2}=-12$ V.

process. Figure 3(b) and 3(c) display the impulse responses of the device at zero bias after pulsing gate $V_{G1}=-12$ V, $V_{G2}=12$ V and $V_{G1}=12$ V, $V_{G2}=-12$ V, producing positive and negative photocurrents in nonvolatile graphene *p-i-n* and *n-i-p* homojunction configuration. Unlike previous photodetectors that require sustained gate voltages, this reconfigurable graphene homojunction consums no energy durng photodetection. Additionally, a pulse relaxation time of ~ps exhibits high dynamic response bandwidth of ~GHz in our device.

Conclusion

We fabricated a reconfigurable, nonvolatile photodetector by integrating graphene and ferroelectric P (VDF-TrFE) layer onto air-slotted PC waveguide. It achieves positive and negative photocurrent in the *p-i-n* and *n-i-p* homojunction with zero energy consumption and high dynamic response bandwidth. This design concept is generic for other two dimensional semiconductors, provide a platform to expand on-chip capabilities for optoelectronic synapses, in-memory sensing and computing, and neuromorphic processing.

Acknowledgments

This project was primarily supported by the Key Research and Development Program (2022YFA1404800), National Natural Science Foundation of China (12374359, 62375225), Shaanxi Fundamental Science Research Project for Mathematics and Physics (22JSY004), Xi'an Science and Technology Plan Project (2023JH-ZCGJ-0023). The author also gratefully acknowledge the assistance of the Analytical & Testing Center of NPU to device fabrication and characterization.

References

[1] Romagnoli, M. et al. "Graphene-based integrated photonics for next-generation datacom and telecom", *Nat. Rev. Mater.* 3, pp. 392–414 (2018).

[2] Wu, G. et al. "Programmable transition metal dichalcogenide homojunctions controlled by nonvolatile ferroelectric domains", *Nat. Electron.* 3, pp. 43–50 (2020).

[3] Shastri, B. J. et al. "Photonics for artificial intelligence and neuromorphic computing", *Nat. Photonics* 15, pp. 102–114 (2021).

[4] Gan, X. et al. "Chip-integrated ultrafast graphene photodetector with high responsivity", *Nat. Photonics* 7, pp. 883–887 (2013).

[5] Schuler, S. et al. "Graphene photodetector integrated on a photonic crystal defect waveguide", *ACS Photonics* 5, pp. 4758–4763 (2018).

[6] Muench, J. E. et al. "Waveguide-integrated, plasmonic enhanced graphene photodetectors", Nano Lett. 19, pp. 7632–7644 (2019).

[7] Tian, R. et al. "Chip-integrated van der Waals PN heterojunction photodetector with low dark current and high responsivity", *Light Sci. Appl.* 11, pp. 101 (2022).

Back-End-of-Line Integration of Organic Thin-Film Transistor Active-Matrix on III-V Micro-LED Array for Video-Rate High-Resolution Displays

R. Shi[1,2], S. Ogier[3], J. Li[1,2], W. Tang[1,2], L. Deng[1,2], S. Li[1,2] and X. Guo[1,2,*]

[1]Department of Electronic Engineering, Shanghai Jiao Tong University, China,

[2]National Key Laboratory of Science and Technology on Micro/Nano Fabrication, Shanghai Jiao Tong University, China,

[3]SmartKem Ltd., UK

[*]E-mail: x.guo@sjtu.edu.cn

Abstract

Abstract—This work develops a back-end-of-line (BEOL) integration process for directly fabricating organic thin-film transistor (OTFT) active-matrix (AM) on top of III-V micro-size light-emitting diode (micro-LED) wafer. Through buffer layer introduction and contact interface engineering, high-performance short-channel OTFTs (mA-level driving current and 5.9×10^{11} ON/OFF ratio) are integrated on top of the micro-LED surface. A 508 PPI AM micro-LED display with a brightness of 27,200 nits is finally demonstrated for video displaying.

Keywords: Micro light emitting diode display, Back-end-of-line, Organic thin-film transistor, Low temperature, High-resolution

I. INTRODUCTION

The micro-LED has drawn much attention as a promising technology of choice for next generation displays owing to high brightness, low power consumption and long lifetime [1]. Because of high temperature strict III-V semiconductor processes, it is not able to directly integrate LED arrays onto the thin-film transistor (TFT) backplane for high-resolution AM displays like that of the organic LED. A common approach adopted in the industry is mass transfer of LED dies onto the TFT backplane [2]. However, the whole fabrication process is complex, and needs to go through three different fabs (Fig. 1 (a)). Improvement of the manufacturing efficiency and yield for lower cost is still a grand challenge [3]. Developing a TFT backplane technology that can be directly processed on top of the LED layer would enable all processes in one fab and provide a cost-effective route to fabricate AM micro-LED displays with high yield (Fig. 1 (b)). For that, the processing temperature with the TFT needs to be low enough to ensure no influence to the underneath LED layer and the metal interconnects [4-5]. However, for low temperature solution-processed OTFTs integrated in heterogeneous surface, achieving both a high driving current for micro-LEDs and a low OFF-state current for power efficiency remains a challenge.

In this work, a BEOL integration process was proposed

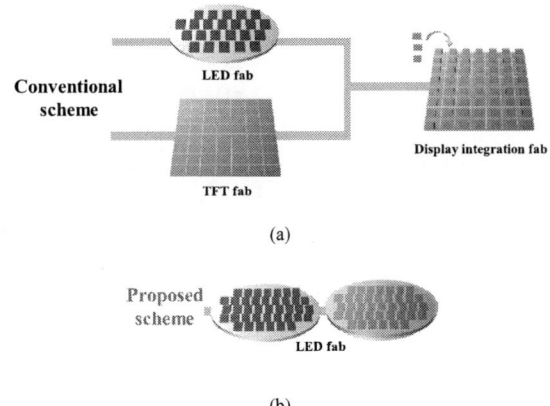

Fig.1. Schematic diagram of (a) the complex conventional micro-LED display integration scheme and (b) the proposed scheme, enabling all processes in one fab.

for the low temperature process of the AM array integrated onto the surface of the micro-LED wafer. The short-channel OTFTs with high driving current and ultrahigh 5.9×10^{11} ON/OFF ratio are fabricated after the buffer layer introduction and contact interface engineering. Lastly, a 508 PPI AM micro-LED display with a luminance of 27,200 nits has been successfully demonstrated for video displaying.

II. METHOD

The complete process flow for BEOL monolithic integration of the OTFT onto a micro-LED wafer is illustrated in Fig. 2. The used blue micro-LED array was fabricated on a 4-inch sapphire substrate in LED fab. It consists of μ-GaN buffer, n-type GaN, InGaN/GaN multiple quantum well and p-type GaN layers, which were all grown by metal-organic chemical vapor deposition (MOCVD). The micro-LED mesa structure was formed using photolithography and inductively coupled plasma (ICP) dry etching. After the growth of the ohmic contact layer indium tin oxide (ITO) using physical vapor deposition (PVD), rapid thermal annealing (RTA) was applied. A SiO_2 passivation layer was deposited using plasma-enhanced chemical vapor deposition (PECVD) to suppress sidewall leakage of small-sized micro-LEDs, and the via holes for interconnection were formed using photolithography and wet etching. (Fig.2 (a))

979-8-3315-0417-5/25 $31.00 © 2025 IEEE

Below are the steps for the BEOL integration process (Fig.2 (b-f)). The electrode of micro-LEDs and signal line in array were lithographically patterned and sputtered using a Mo/Al/Mo stack layer. To cover the high steps of micro-LED mesa structure and provide uniform surface for organic semiconductor (OSC) deposition, the planarization layer and buffer layer TRUFLEX™ were formed by spin coating (Fig.2 (b)). After that, the top-gate bottom-contact (TGBC) device structure was chosen for high resolution and large carrier injection area of source contacts. Gold (Au) electrodes of source/drain were lithographically patterned and deposited by sputtering (Fig.2 (c)), followed by self-assembled monolayer (SAM), OSC (consisting of small molecule (triethyl(2-{1,4,8,11-tetramethyl−13-[2-(triethylsilyl)ethynyl]pentacen-6-yl}ethynyl)silane) and polymer semiconductor binder) and organic gate insulator (OGI, consisting of CYTOP and sputter resistant layer) deposition using spin coating (Fig.2 (d)). Subsequently, the Mo/Al/Mo top gate metal was lithographically patterned and sputtered. Dry etching was performed to etch away the regions not covered by the top gate metal, continuing until stopping at buffer layer, thereby achieving the patterning of the OSC and OGI (Fig.2 (e)). Finally, the organic passivation layer was deposited by spin coating, followed by opening the via holes to form the opening of the electrodes of micro-LED and OTFTs for interconnection in AM array. The interconnects were ultimately deposited using the same method as previously employed for the electrode deposition (Fig. 2 (f)).

The integration process for OTFT is fully compatible with photolithography and dry etching process in standard LED fabrication procedures, with a maximum process temperature below 150 °C, which ensures the performance of

the underlying LED layers and electrical interconnects unaffected.

III. RESULTS AND DISCUSSIONS

A. OTFT device performance

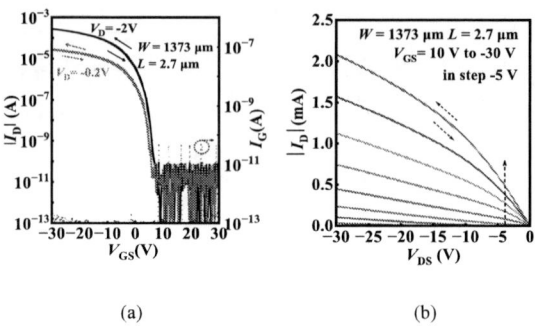

(a)   (b)

Fig.3. Measured (a) transfer and (b) output characteristics of short-channel length ($L = 2.7$ μm) OTFT on micro-LED wafer.

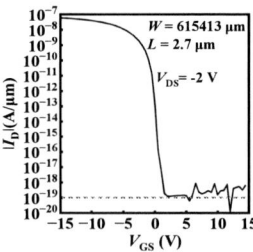

Fig.4. Typical transfer characteristics of an ultrawide short-channel OTFT device.

This work employed SAM 3-F, 4-MeOBT molecules, which possess a strong dipole interfacial layer, to modify the metal-semiconductor interface, resulting in an enhancement of the ON-state current and suppression of the OFF-state current [6].

Typical transfer and output characteristics of the 2.7 μm channel length TGBC structure OTFT are illustrated in Fig. 3 (a-b). The extracted mobility is 2.54 cm^{2}V^{-1}s^{-1} with a driving current exceeding 2 mA, which meets the requirement of the driver transistor within a finite pixel area.

The transfer characteristic of the ultra-wide OTFT, measured to extract the actual leakage current, is shown in Fig. 4. A low leakage current density of 10^{-19} A/μm, along with a high ON/OFF ratio of 5.9×10^{11} has been achieved, indicating ideal performance as a switch transistor.

B. Pixel circuit performance

A 508 PPI AM micro-LED display array with 2T1C pixel circuit (pixel size: 50 μm × 50 μm) was successfully fabricated, as shown in the optical micrograph in Fig 5 (a). Due to the decent driving capability of the short channel OTFT, the micro-LED achieved a high brightness of 27,200 nits as determined through calculation based on the pixel driving current measurement in Fig. 5 (b) and the luminance-

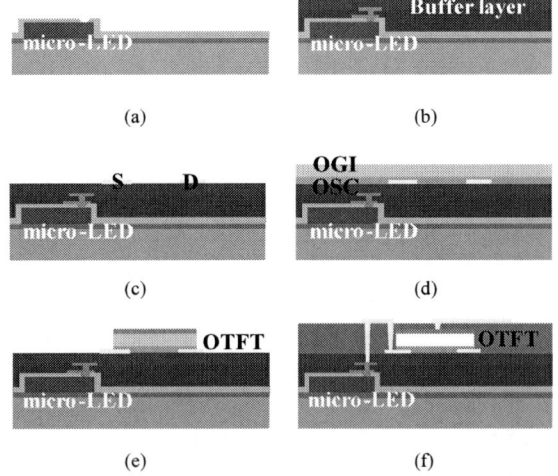

(a)   (b)

(c)   (d)

(e)   (f)

Fig.2. The complete process flow for BEOL monolithic integration of the OTFT onto a micro-LED array wafer, which is fully compatible with photolithography and dry etching process in standard LED fabrication procedures.

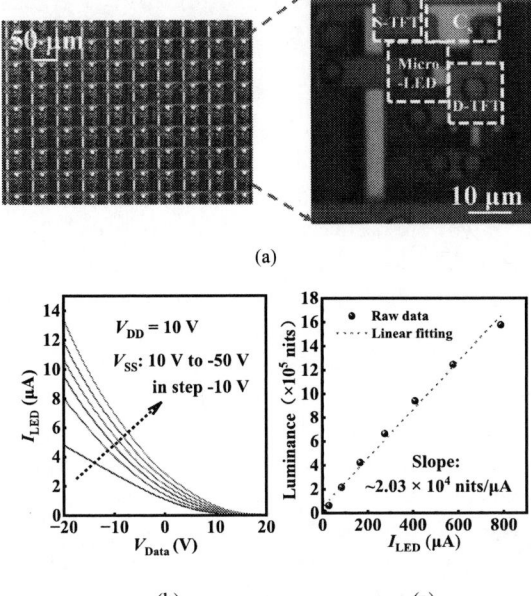

(a)

(b)

(c)

Fig.5. (a) Optical micrograph of the 508 PPI AM micro-LED array and 2T1C pixel circuit. (b) Measured *I-V* curves of the 2T1C pixel circuit modulated by various V_{Data} and V_{DD}-V_{SS}. (c) Measured luminance and linear fitting result of the micro-LED under different driving current.

current conversion efficiency measurement in Fig. 5 (c).

C. Demonstrator of AM micro-LED display

The 508 PPI high-resolution AM micro-LED display, consisting of 96 × 96 pixels, was driven by a driver system connected via a flexible printed circuit (FPC). Fig. 6 (a-f) presents clear greyscale video screenshots of the display panel operating at a refresh rate of 60 Hz. Compared with other reported AM displays based on 2T1C pixel circuit listed in Table 1, this work shows the highest resolution and brightness.

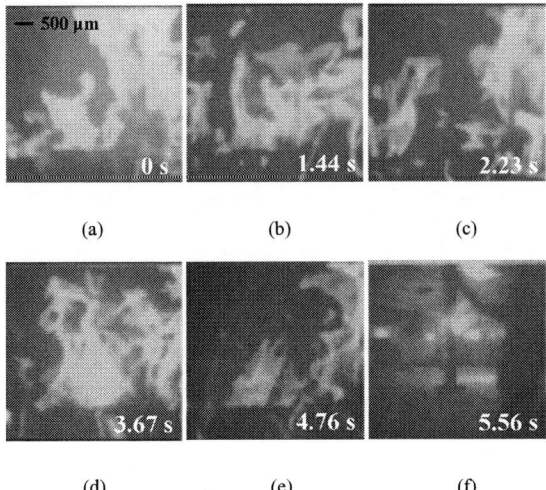

(a)   (b)   (c)

(d)   (e)   (f)

Fig.6. (a-f) Screenshots from the video displayed on the 96×96 508 PPI AM micro-LED display.

Table 1. Comparation of state-of-the-art AM display based on 2T1C pixel circuit.

TFT technology	Display technology	Highest process temperature	Integration scheme	PPI	Highest Luminance (nits)	Reference
IGZO TFT	micro-LED	300℃	Monolithic integration	78.4	1559	[7]
IGZO TFT	micro-LED	250℃	Flip chip	100	630	[8]
LTPS TFT	micro-LED	≥450℃	Mass transfer	318	5000	[9]
OTFT	OLED	≤150℃	OLED on OTFT	121	200	[10]
OTFT	OLED	150℃	OLED on OTFT	300	150	[11]
OTFT	micro-LED	150℃	Monolithic integration	508	27200	This work

Conclusion

In summary, a BEOL integration process was utilized to directly fabricate ultrahigh ON/OFF ratio ($5.9×10^{11}$) OTFT pixel array on the micro-LED wafer. A 508 PPI AM micro-LED display capable of video-rate operation was finally demonstrated. This BEOL process is considered to have significant potential for cost-effective and high-yield micro-LED display manufacturing.

Acknowledgments

The authors gratefully acknowledge funding support through the National Key R&D Program of China (Grant No. 2022YFB3603804), the National Science Fund for Excellent Young Scholars (Grant No. 61922057) and the National Natural Science Foundation of China (Grant No. 62104143). The authors gratefully acknowledge Enkris Semiconductor, Inc. for providing epitaxial wafers and LED manufacturing process.

References

[1] K. Behrman, I. Kymissi, Nat. Electron., vol. 5, no. 9, pp. 564-573, 2022.

[2] K. Ding, V. Avrutin, N. Izyumskaya, Ü. Özgür, H. Morkoç, Appl. Sci., vol. 9, no. 6, pp. 1206, 2019.

[3] L. Han, S. Ogier, J. Li, D. Sharkey, X. Yin, A. Baker, A. Carreras, F. Chang, K. Cheng, X. Guo, Nat. Commun., vol. 14, no. 1, pp. 6985, 2023.

[4] C. Youn, T. Jeong, M. Han, J. Yang, K. Lim, H. Yu, J. Cryst. Growth, vol. 250, no. 3-4, pp. 331-338, 2003.

[5] B. Tull, N. Twu, Y. Hsu, S. Leblebici, I. Kymissis, V. Lee, SID Symp. Dig. Tech., vol. 48, no. 1, pp. 246-248, 2017.

[6] L. Han, J. Li, S. Ogier, Z. Liu, L. Deng, Y. Cao, T. Shan, D. Sharkey, L. Feng, A. Guo, X. Li, J. Zhang, X. Guo, Adv. Electron Mater., vol. 8, no. 9, pp. 2200014, 2022.

[7] O. Durnan, V. Kumar, R. Alshanbari, M. Noga, I. Kymissis, J. Soc. Inf. Display, vol. 32, no. 5, pp. 350-359, 2024.

[8] J. Um, D. Jeong, Y. Jung, J. Moon, Y. Jung, S. Kim, S. Kim, J. Lee, J. Jang, Adv. Electron Mater., vol. 5, no. 3, pp. 1800617, 2019.

[9] M. Zhu, H. Ng, S. Ganapathiappan, Z. Li, N. Patibandla, SID Symp. Dig. Tech., vol. 54, no. 1, pp. 526-529, 2023.

[10] K. Zhang, D. Peng, K. Lau, Z. Liu, SID Symp. Dig. Tech., vol. 48, no. 1, pp. 357-361, 2017.

[11] K. Nomoto, M. Noda, N. Kobayashi, M. Katsuhara, A. Yumoto, S. Ushikura, R. Yasuda, N. Hirai, G. Yukawa, I. Yagi, SID Symp. Dig. Tech., vol. 42, no. 1, pp. 488-491, 201

Physics-Informed Neural Network for Predicting Out-of-Training-Range TCAD Solution with Minimized Domain Expertise

Albert Lu[1], Yu Foon Chau[2], Hiu Yung Wong[1*]

[1]San Jose State University, USA, [2]University of California, Irvine, USA, [*]hiuyung.wong@sjsu.edu

Abstract

In this paper, a Si nanowire transistor is used to demonstrate the possibility of using a physics-informed neural network to predict out-of-training-range TCAD solutions without accessing internal solvers and with minimal domain expertise. The machine can predict a 10 times larger range than the training data and also predict the inversion region behavior with only subthreshold region training data. The physics-informed module is trained without human-coded differential equations making this extendable to more sophisticated systems.

Keywords: Physics Informed Neural Networks (PINN), ML, TCAD, Nanowire, Out-of-training-range prediction

Introduction

Machine learning (ML) is promising in assisting technology computer-aided design (TCAD) simulations to alleviate difficulty in convergence and prolonged simulation time. Roughly, ML is used in TCAD in three approaches. In the first approach, ML is used to improve solver performance by providing an initial guess [1]-[2]. This requires access to the internal solver of a TCAD tool and significant domain expertise by coding physical equations (including differential operators) in the loss functions. The second approach is to use TCAD to generate terminal data such as currents and voltages and then use ML to perform reverse engineering [3]-[6] or simulation emulation [7]-[9]. This does not require access to the internal solver and has been proven to be able to predict out-of-training range data [9]. If an appropriate machine learning method is used, it can also minimize the requirement for domain expertise [7]-[9]. The third approach is to solve the TCAD problem by training a machine with the spatial distribution of physical quantities such as electron density and potential [10][11] without using device physics (as the solver is not accessible) and mostly can only predict the physical quantities within the training range.

Physics-informed neural network (PINN), in which the physics of the problem is incorporated into the NN, is expected to improve predictability [12]. In this paper, we study a new approach in which a machine is trained by the spatial distribution of physical quantities with PINN and minimal domain expertise *without accessing the solver of the tool* using a Si nanowire (NW) transistor as an example.

Data Generation

An n-type Si NW transistor is studied using TCAD Sentaurus. Half of the domain was simulated due to its symmetry. Fig. 1 shows the structure with a radius of 3 nm, a gate length of 18 nm, and an oxide thickness of 1 nm. The source/drain and body doping are 10^{20} cm^{-3} (n-type) and 10^{10} cm^{-3} (p-type), respectively. Tensor mesh is used ($129 \times 17 = 2193$ mesh points). The electrostatic potential (ϕ), electron density (n), and net space charge after device simulation at each point are extracted using a TDX script in TCAD Sentaurus. The extracted quantities are verified with an in-house Python-coded Poisson equation solver (Fig. 2).

The structure is then simulated in SDevice with a setup that includes Fermi-Dirac statistics. The Poisson equation is solved with the drain and source voltages at 0 V. The gate voltage, V_G, is ramped from 0 to 0.75 V, and snapshots of the electrostatic potential and electron density are taken for every 7.5 mV for a total of 101 snapshots. Note that each 'snapshot' (profile) records ϕ and n distributions in the 2193-point mesh.

Physics-Informing Module

Firstly, a physics-informing module machine (not PINN yet) is created. This machine predicts the ϕ profile for a given n profile. This is equivalent to solving the Poisson equation for a given space charge (ρ) profile in the NW,

$$\nabla \cdot (\varepsilon \nabla \phi) = -\rho \qquad (1)$$

where ε is the permittivity. Hole and acceptor concentrations are ignored due to their small values in this problem.

The module is trained using linear regression (LR) and the first 40 snapshots (*only* from $V_G = 0V$ to 0.3V). The input is the n profile and the output is the ϕ profile. The electron density is normalized by adding 10^{10} cm^{-3} and then dividing by 10^{19} cm^{-3}. Adding 10^{10} cm^{-3}, which is the noise level in this problem, is to avoid division by zero in the oxide region. Fig. 3 plots ϕ predicted by the machine against the ϕ calculated by TCAD for every mesh point from $V_G = 0V$ to 0.75V for the corresponding n profiles from TCAD and they agree with each other. *Note again the machine has not seen the profiles for $V_G > 0.3V$.* This demonstrates the ability of the machine to learn sophisticated physics where ϕ in the channel vary significantly and non-linearly while those in the source and drain stay almost constant. Moreover, the training data are in the subthreshold regime which has a very different V_G dependency than the inversion regime. LR is sufficient because when the n profile is given, Eq. (1) is just a linear matrix multiplication after discretization. Therefore, the machine models the inverse of the matrix at the left-hand side in Eq. (1). Systems of differential equations are generally discretized and linearized. Therefore, this methodology is expected to be applicable to more sophisticated problems.

PINN Architecture

A PINN is then built. It consists of a convolutional neural network (CNN) and the aforementioned physics-informing module. Fig. 4 shows the architecture. The input is the desired V_G and the outputs are the n and ϕ profiles. For any given V_G, the n profile is predicted by a CNN. The output layer of the CNN is chosen to be the exponential linear unit (ELU). It will then be passed to the physics-informing LR module which will calculate the corresponding ϕ profile, from which the gate voltage, V_G', can be extracted. If the solution is correct, $V_G' = V_G$. This comparison forms the first loss function and enforces the gate boundary condition.

The predicted ϕ profile is then further used to predict the n profile using the Fermi-Dirac (FD) module in Fig. 3, which is also a physics-informing module but much more straightforward. It calculates the n profile based on the ϕ profile through the approximation of the Fermi Integral of order ½ for Fermi-Dirac statistics [13] and is compared to the n profile predicted by the CNN. This forms the second constraint and is used to ensure that the predicted electron density obtained through the approximate Fermi Integral matches the predicted electron density from the CNN. This essentially enforced the Fermi-Dirac statistics. Fig. 5 shows that the approximate Fermi Integral calculation is correct by plotting the approximation against the actual value obtained from TCAD. In the loss calculation, the logarithmic value of n is taken so that low-density values also have enough weight.

For each V_G of interest, the two losses will be minimized using the Adam optimizer. To speed up the process, the learning rate is initially 10^{-3} but as it plateaus it is changed adaptively until a minimum of 10^{-5} is reached. This process can be treated as a replacement for TCAD solvers. This can also be regarded as *self-supervised learning*. Note again that, in the whole process, only the LR module has been trained with the first 40 V_G snapshots. Moreover, *unlike regular TCAD simulation, one can calculate $V_G = 0.75V$ directly without ramping V_G.*

Results

Fig. 6 and Fig. 7 show the predicted ϕ profile and n profile compared to TCAD simulation results for $V_G = 0.75V$, respectively. The maximum error is less than 0.3% for the ϕ profile and less than 0.6% for the n profile in the logarithmic scale. Fig. 8 shows the prediction of ϕ by PINN at a point (see Fig. 1) in the middle of the structure under the gate region across all V_G. This point was chosen because it experiences the transition from the depletion to the inversion regime. The result matches the TCAD simulation well. This demonstrates the ability of the PINN to learn the physics of this device and predict out-of-training-range results and is not based on simple extrapolation. Fig. 9 further shows the predicted vs. actual ϕ and n of every mesh point across all

V_G. The n comparison is in log scale and shows how the PINN is able to learn across several orders of magnitude successfully. The number of epochs in the previous figures is 200,000. Fig. 10 shows that <30,000 epochs are enough for most applications.

The out-of-range predictability is further studied using the PINN (trained up to $V_G = 0.3V$). Fig. 11 shows that it can predict up to $V_G = 3V$ well with 20,000 epochs.

Conclusion

A novel framework based on a PINN is demonstrated to have effectively learned the underlying physics by using specific physics-based loss functions. It can predict out-of-training-range simulation results as it has learned the underlying physics. A machine trained by the TCAD data from $V_G = 0V$ to 0.3V (below the subthreshold region) is shown to be able to predict the ϕ and n profiles for any given V_G (up to 3V).

Acknowledgments

Part of the work was supported by the National Science Foundation under Grant No. 2046220.

References

[1] S. -C. Han et al., "Deep Neural Network for Generation of the Initial Electrostatic Potential Profile," 2019 SISPAD, pp. 1-4.

[2] S. -C. Han et al., "Acceleration of Semiconductor Device Simulation With Approximate Solutions Predicted by Trained Neural Networks," in IEEE TED, vol. 68, no. 11, pp. 5483-5489, Nov. 2021.

[3] C. -W. Teo et al., "TCAD-Enabled Machine Learning Defect Prediction to Accelerate Advanced Semiconductor Device Failure Analysis," 2019 SISPAD, pp. 1-4.

[4] Y. S. Bankapalli and H. Y. Wong, "TCAD Augmented Machine Learning for Semiconductor Device Failure Troubleshooting and Reverse Engineering," 2019 SISPAD, pp. 1-4.

[5] H. Y. Wong et al., "TCAD-Machine Learning Framework for Device Variation and Operating Temperature Analysis With Experimental Demonstration," in IEEE JEDS, vol. 8, pp. 992-1000, 2020.

[6] K. Mehta and H. -Y. Wong, "Prediction of FinFET Current-Voltage and Capacitance-Voltage Curves Using Machine Learning With Autoencoder," in IEEE EDL, vol. 42, no. 2, pp. 136-139, Feb. 2021..

[7] H. Dhillon et al., "TCAD-Augmented Machine Learning With and Without Domain Expertise," in IEEE Transactions on Electron Devices, vol. 68, no. 11, pp. 5498-5503, Nov. 2021.

[8] K. Mehta et al., "Improvement of TCAD Augmented Machine Learning Using Autoencoder for Semiconductor Variation Identification and Inverse Design," in IEEE Access, vol. 8, pp. 143519-143529, 2020.

[9] V. Eranki et al., "Out-of-Training-Range Synthetic FinFET and Inverter Data Generation Using a Modified Generative Adversarial Network," in IEEE EDL, vol. 43, no. 11, pp. 1810-1813, Nov. 2022.

[10] J. Lee et al., "Device simulations with a U-Net model predicting physical quantities in two-dimensional landscapes," Sci. Rep., vol. 13, no. 1, p. 731, Jan. 2023.

[11] W. Jang et al., "TCAD Device Simulation With Graph Neural Network," in IEEE Electron Device Letters, vol. 44, no. 8, pp. 1368-1371, Aug. 2023, doi: 10.1109/LED.2023.3290930.

[12] B. Kim and M. Shin, "A Novel Neural-Network Device Modeling Based on Physics-Informed Machine Learning," in IEEE Transactions on Electron Devices, vol. 70, pp. 6021-6025, Nov. 2023.

[13] D. Bednarczyk, et al., "The approximation of the Fermi-Dirac integral $F_{½}(\eta)$," Physics Letters A, 64(4):409-410 (1978)

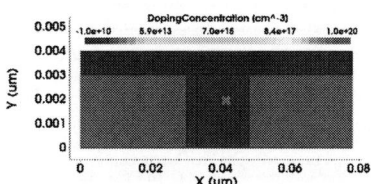

Fig. 1: Si NW structure used in this study. The cross shows the location used in Fig. 8.

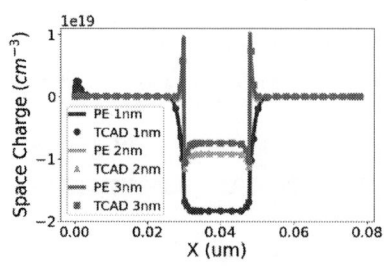

Fig. 2. Comparison between TCAD TDX extracted and in-house Poisson equation (PE) solved space charge at horizontal cutlines at different distances from SiO_2/Si interface (Fig. 1).

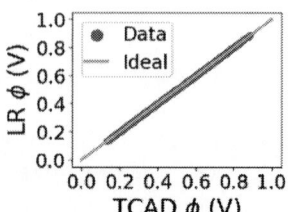

Fig. 3. Comparison between the electrostatic potential predicted using the physics-informing module and the electrostatic potential from TCAD. The blue points consist of all mesh points at all V_G from 0V to 0.75V.

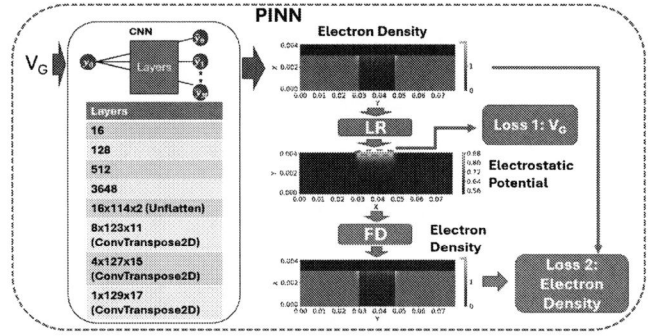

Fig. 4. The PINN used in this study consists of a CNN and 2 loss functions. V_G is input to the PINN and ϕ and n are the outputs. The physics-informing module is trained only using snapshots from $V_G = 0V$ to 0.3V. The predicted n is passed through the LR machine to obtain ϕ. The gate contact ϕ matches the input V_G through loss function 1. The predicted ϕ is also passed through Fermi Integral to obtain n. Loss function 2 is used to obtain accurate n.

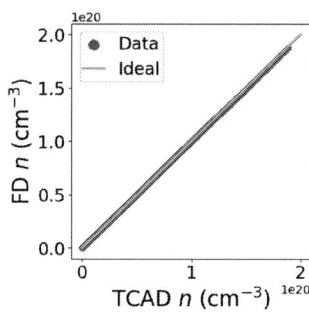

Fig. 5. Comparison between the electron density predicted using FD module and TCAD. The blue points consist of all mesh points at all V_G from 0V to 0.75V.

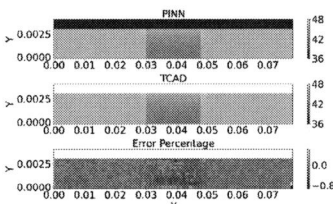

Fig. 6. Predicted electrostatic potential profile using the PINN (top) and the electrostatic potential from TCAD (middle) when V_G=0.75V. The max error percentage between them is less than 0.3% (bottom).

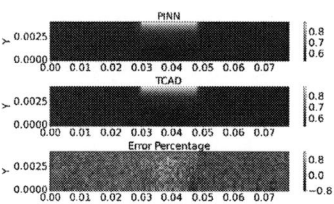

Fig. 7. Predicted electron density profile using the PINN (top) and the electron density from TCAD in log scale (bottom) when V_G=0.75V. The max error percentage between them is less than 0.6% (bottom).

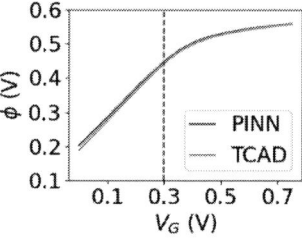

Fig. 8. Comparison of ϕ at x=40.5nm and y = 2nm (Fig. 1) predicted by PINN and TCAD at various V_G. The PINN was only trained by data before the green dotted line ($V_G < 0.3V$).

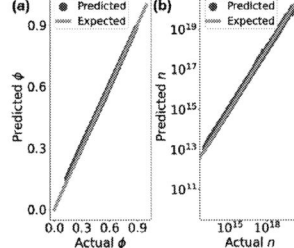

Fig. 9. PINN-predicted vs. TCAD (a) ϕ and (b) n. The blue points consist of all mesh points at all V_G from 0V to 0.75V.

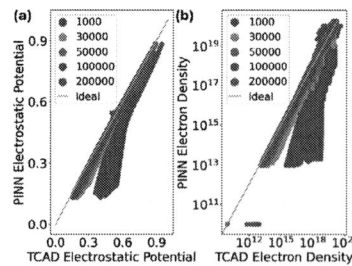

Fig. 10. PINN-predicted vs. TCAD (a) ϕ and (b) n with respect to the number of epochs.

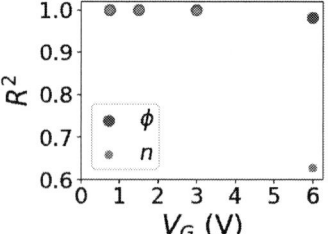

Fig. 11. R^2 values of ϕ and n against V_G for a machine trained with data only up to 0.3V.

979-8-3315-0417-5/25 $31.00 © 2025 IEEE

Optimization of RRAM Read Performance and Area Efficiency: A Large-Scale and Low-Parasitic Array with Novel Interconnection Schemes

Shengyu Bao [1†], Yuhang Yang [1†], Zongwei Wang[1,2*], Linbo Shan[1], Qishen Wang[1],

Yimao Cai[1,2*], Ru Huang[1,2]

[1]School of Integrated Circuits, Peking University, Beijing, China
[2]Beijing Advanced Innovation Center for Integrated Circuits, Beijing, China
[*]Email: wangzongwei@pku.edu.cn

Abstract

This study presents an optimization design for RRAM arrays, highlighting the trade-offs in power consumption, read speed, area efficiency, and array scalability of 1T1R arrays. To address these limitations, we propose a novel large-scale and low-parasitic array design with shared SLs across multiple rows. This design effectively reduces read power and latency while enhancing array scalability and area efficiency. Implemented in a 28nm process, it achieves an 8.6% improvement in area efficiency and a 76% enhancement in array scale compared to the conventional design, while maintaining advantages in read speed and power consumption.

Keywords: RRAM, Area efficiency, Power, Read time

Introduction

Resistance random access memory (RRAM) is a promising solution for advanced node embedded non-volatile memory, which has drawn widespread attention due to its simple metal-insulator-metal (MIM) structure and simple fabrication process, high reliability, and low operational power consumption [1], [2]. Many semiconductor companies, including TSMC and Infineon, have integrated RRAM into their technology roadmaps [3], [4].

To solve sneak path issues and write disturbances in RRAM arrays, a well-established approach involves using a MOSFET to gate individual RRAM cells, forming a one transistor one resistor (1T1R) array. Typically, four such 1T1R arrays share peripheral circuitry within a single bank, as illustrated in Fig. 1. Due to the significant proportion of area occupied by these peripheral circuits, the area efficiency of the 1T1R array is relatively low. Although larger 1T1R arrays can effectively reuse peripheral circuits to enhance area efficiency (Fig. 2), parasitic resistance and capacitance within these arrays also increase with scale. As depicted in Fig. 3, the parasitic resistance and capacitance of word lines (WL) grow rapidly with the number of array rows, leading to prolonged WL pre-charge times (Fig. 4), decreased read speed, and increased dynamic power consumption for WL charging. Additionally, as the number of columns increases, the parasitic resistance of bit lines (BL) and source lines (SL) rises. The resulting IR drop reduces the read margin of the device, which can slow down Sense Amplifier (SA) readout times or necessitate higher read voltages, further increasing

power consumption. These parasitic effects restrict the scalability of large 1T1R arrays, typically constraining their design to 0.5Mb for applications requiring high read speeds [5], [6], thereby limiting area efficiency improvements. While some studies have enhanced 1T1R area efficiency through peripheral circuit optimizations [7], optimizing the RRAM array remains a significant challenge, which has the potential to achieve even greater enhancements.

In this work, we propose a novel array structure that innovatively addresses scale-related issues at both the cell and wiring levels. Through synergistic optimization, we significantly enhance array scale and area efficiency while maintaining improvements in read speed and power consumption over conventional designs.

Array Structure

Fig. 5 compares the conventional 1T1R array (left) with a novel array (right) based on our previously developed dynamic RRAM cell [8]. The dynamic RRAM cell offers flexibility in adjusting the number of activated transistors based on specific requirements. During the write operation, all three transistors in the dynamic cell are activated simultaneously, delivering a saturation current equivalent to that in a 1T1R array. Consequently, even with the channel width reduced to two-thirds of the 1T1R, the dynamic cell still meets the switching current demands of RRAM. Under the same operating voltage, the distribution of read current for the reduced-channel-width dynamic cell and the conventional 1T1R is shown in Fig. 6. The smaller channel width of the dynamic cell reduces the array height, lowering it by 24% compared to an equivalent number of rows in a 1T1R array, thus shortening the word line (WL) length by 24%, as depicted in Fig. 7. During the read process, the dynamic cell activates only one WL, as illustrated in Fig. 8. The reduced WL length results in lower parasitic capacitance and resistance, which shortens pre-charge time for reading and enhances read speed. Alternatively, for the same read time requirement, the dynamic cell configuration supports a greater number of rows compared to the 1T1R.

Additionally, we employed a multi-row shared SL design, as shown in Figs. 5 and 8. By utilizing the space originally occupied by narrower multi-row SLs, we implemented two wider SLs in parallel, significantly reducing SL parasitic resistance, which will help to reduce the IR-drop while the

---

979-8-3315-0417-5/25 $31.00 © 2025 IEEE   485

column scale increasing. The lower SL parasitic resistance also mitigates read window degradation, reducing read voltage and power consumption. Table 1 provides a detailed comparison of the parasitic parameters between the 3T2R array and the conventional 1T1R array. Additionally, the shared SL design significantly reduces the area occupied by SL driving modules, further enhancing area efficiency.

Read Speed and Power Optimization

As illustrated in Fig. 4, the read time comprises pre-charge time, discharge time, and SA enabling time. The impact of the RRAM array on read time is primarily due to the WL pre-charge time and the read current window. With the same pull-up circuit design, the rise time in a dynamic cell array is approximately 0.6 ns shorter compared to a conventional design of the same scale, as shown in Fig. 9, indicating that the dynamic cell array achieves lower parasitic resistance and capacitance in WL, and can support a greater row scale under the same read time requirement. Fig. 10 illustrates the comparison of WL rise times between conventional and the proposed designs with varying numbers of array rows. For the same read speed, the number of rows in the 3T2R array is about a quarter more than that of the conventional design. Notably, the parasitic capacitance of the BL has a negligible impact on read time, as illustrated in Fig. 11.

SA enabling time is constrained by the read margin; the smaller the read margin, the slower the SA enabling time. Fig. 12 shows the effect of parasitic resistance on the resistance window. The series resistance of the SL and BL causes the low resistance state (LRS) distribution of remote-end devices in the array to shift higher, reducing the read margin under a certain read voltage. Fig. 13 depicts the relationship between the read current margin and the number of array columns. The low resistance SL design slows the degradation of read margin in the proposed array.

To prevent a slowdown in read speed, the read voltage needs to be increased to compensate for the reduction in the read margin caused by the enlarged column scale. However, this raises the power consumption associated with BL charging and discharging. It further increases the leakage of half-selected devices, which also exacerbates with the expansion of column scale as illustrated in Fig. 13, resulting in higher power consumption. Fig. 14 compares the necessary read voltage for different column scales between the proposed arrays and conventional 1T1R arrays. Due to the lower SL parasitic resistance, the required read voltage is smaller in the proposed design, offering power consumption advantages.

Fig. 15 depicts the one-bit read energy in conventional 1T1R arrays with different column scales and the proportion of power consumption in each part. As the column scale increases, both the BL capacitance and read voltage increase, which enlarge the BL charging and discharging energy consumption. Higher read voltage increases the power

consumption of near-end devices, while the leakage of half-selected devices also grows with the column scale, further increasing read power consumption. Meanwhile, larger row scale leads to a longer WL pre-charge time which increases DC power consumption, while the WL pre-charge energy is also increased with the WL parasitic capacitance. Nevertheless, with smaller critical parasitic parameters, the proposed design achieves a significant reduction in read energy compared to the same scale conventional array, as illustrated in Fig. 16.

By balancing read energy, read time, and array scale, the RRAM bank with the proposed array achieves an improvement of 8.6% in area efficiency and of 76% in array scale compared to the conventional 1T1R, as shown in Fig. 17, while decreasing the read time and read power consumption, as detailed in Table 2.

Conclusion

This study optimizes the area efficiency, array scale, performance, and power consumption of RRAM banks by the large-scale and low-Parasitic array with multi-row shared SL design. Our approach demonstrates an 8.6% increase in area efficiency and 76% enhancement in array scale over the conventional design with read time and read energy decreased, offering an insightful path for advanced RRAM bank development with superior efficiency and scalability.

Acknowledgments

This work was supported by National Natural Science Foundation of China under Grant 62322401, 62025401, 62341407, and 61927901, and in part by Beijing Nova Program under Grant 20220484113, "111" Project under grant B18001.

References

[1] Y. Chen, "ReRAM: History, Status, and Future," IEEE TED (2020)

[2] Z. Wang and Y. Cai, "ReRAM: Memory Technology: Development, Fundamentals, and Future Trends," Advanced Memory Technology: Functional Materials and Devices 1, 1-36 (2023)

[3] A. Grossi et al.,"28nm Data Memory with Embedded RRAM Technology in Automotive Microcontrollers," IMW (2023)

[4] C. Wu et al., "Emerging Memory RRAM Embedded in 12FFC FinFET Technology for industrial Applications," IEDM (2023)

[5] C. Chou et al., "A 22nm 96KX144 RRAM Macro with a Self-Tracking Reference and a Low Ripple Charge Pump to Achieve a Configurable Read Window and a Wide Operating Voltage Range," VLSI (2020)

[6] Y. Huang et al., "15.7 A 32Mb RRAM in a 12nm FinFet Technology with a 0.0249μm2 Bit-Cell, a 3.2GB/S Read Throughput, a 10KCycle Write Endurance and a 10-Year Retention at 105°C," ISSCC (2024)

[7] A. Levy et al., "EMBER: Efficient Multiple-Bits-Per-Cell Embedded RRAM Macro for High-Density Digital Storage," JSSC (2024)

[8] Q. Wang et al., "A Logic-Process Compatible RRAM with 15.43 Mb/mm2 Density and 10years@150°C retention using STI-less Dynamic-Gate and Self-Passivation Sidewall," IEDM (2023)

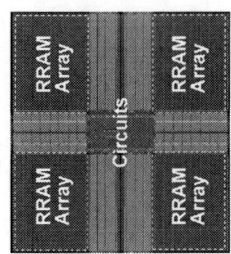

Fig. 1: Layout of our RRAM bank. The proportion of circuits area requires optimization.

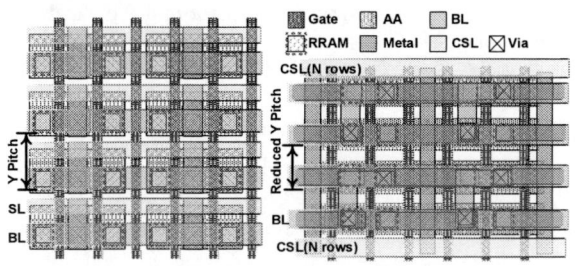

Fig. 2: Relationship between bank size, area efficiency, and array Scale

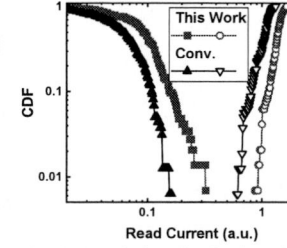

Fig. 3: Parasitic parameters limiting array scaling: WL constrains rows, BL/SL constrains columns

Fig. 4: Components of RRAM bank read time, dominated by WL pre-charge time

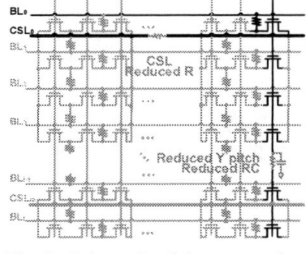

Fig. 5: Layout comparison of conventional (left) and the proposed (right) arrays: reduced channel width lowers the y pitch and WL length in our design

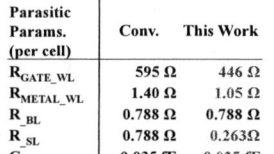

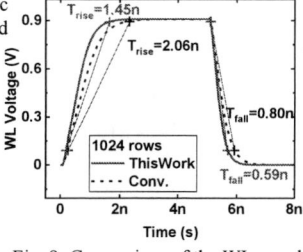

Fig. 6: Cumulative distribution of read currents for the proposed and 1T1R cells under the same operating voltage

Fig. 7: Comparison of array row count and height, showing a 24% reduction in array height achieved in this work

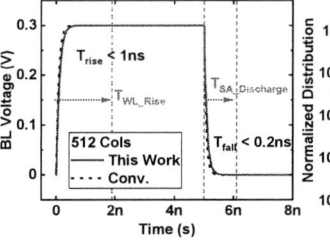

Fig. 8 : Schematic of the proposed array with multi-row shared SL design

Table 1: Comparison of parasitic parameters between this work and the conventional array

Parasitic Params. (per cell)	Conv.	This Work
R_{GATE_WL}	595 Ω	446 Ω
R_{METAL_WL}	1.40 Ω	1.05 Ω
R_{BL}	0.788 Ω	0.788 Ω
R_{SL}	0.788 Ω	0.263Ω
C_{GATE_WL}	0.035 fF	0.025 fF
C_{METAL_WL}	0.045 fF	0.036 fF
C_{BL}	0.120 fF	0.140fF
C_{SL}	0.115 fF	0.035fF

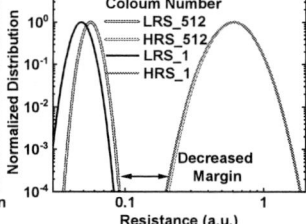

Fig. 9: Comparison of the WL pre-charge rise and fall times for conventional arrays and this work at the same scale

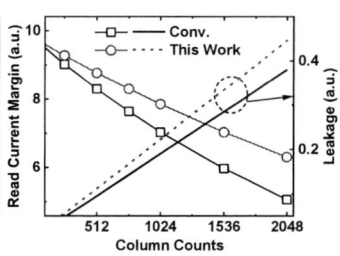

Fig. 10: Relationship between the array row count and the WL pre-charge rise time

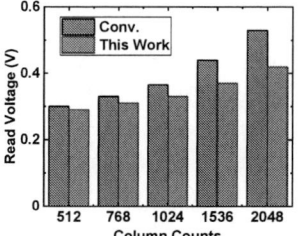

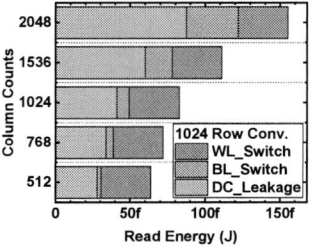

Fig. 11: Comparison of the BL pre-charge rise and fall times for conventional arrays and this work

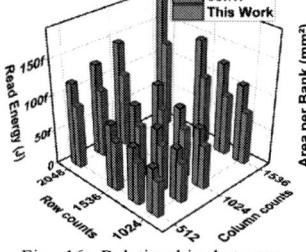

Fig. 12: Impact of BL/SL parasitic resistance on resistance distribution

Fig. 13: Relationships between array read window and column count, and between device leakage and column count

Fig. 14: Relationship between array column count and read voltage

Fig. 15: Relationship between read power consumption and array column count in conventional arrays

Fig. 16: Relationship between array read power consumption and array scale

Fig. 17: Comparison of area efficiency between this work and conventional approach

Table 2: Comparison of overall parameters between this work and conventional approach

Params	Conv. 1024 ROWs ×512 COLs	This Work 1260 ROWs ×734 COLs
Type	1T1R	Proposed
Cell Size	0.0476 um²	0.0357 um²
t_{rise_WL}	0.206 ns	0.192ns
V_{read}	0.3	0.3
E_{read}	0.0637 pJ	0.0622 pJ
Scale	0.50 Mb	0.88 Mb
Area Efficiency	31.1%	39.7%

979-8-3315-0417-5/25 $31.00 © 2025 IEEE   487

Multimode Transistors based on Ion-dynamic Capacitance

Xiaoci Liang[1], Yiyang Luo[1], Yanli Pei[1], Mengye Wang[2], Chuan Liu[1*]

[1]School of Electronics and Information technology, Sun Yat-Sen University, Guangzhou, China; [2]School of Materials, Sun Yat-Sen University, Shenzhen, China. *Email: liuchuan5@mail.sysu.edu.cn.

Abstract

Electrolyte-gated transistors can function as switching elements, artificial synapses and memristive systems. However, insight into such devices, including the ion dynamics and transient capacitances, remains limited. Here we report a concise model for the transient ion-dynamic capacitance in electrolyte-gated transistors. The model predicts that plasticity, high apparent mobility, sharp subthreshold swing, and memristive conductance can be achieved by programming interfacial ion concentrations or scan speeds. We demonstrate such multimode transistors and different capabilities experimentally.

Keywords: Electrolyte-gated transistor, Multimode transistor, Dynamic capacitance.

Introduction

Electrolyte-gated transistors offer low operational voltages and high output currents, as well as a range of other attractive features. A variety of artificial synapse and memristive systems have been developed using such devices and by exploiting the spatiotemporal dependency of ionic motion[1]. As ions in the dielectrics are driven by gate biases to approach or move away from the semiconductor/dielectric interface, the transistors exhibit transient currents that can emulate various synapse-like behaviors including short-term memory and long-term memory[2]. The accumulation of ions can also induce large capacitances due to electric double layers (EDLs) and pseudocapacitance[3]. By using these effects, combined with devices based on two-dimensional materials, the turn-on properties of transistors have been pushed to below the thermionic limit[4]: that is, a reversed subthreshold slope below 60 mV/dec at room temperature.

Despite this progress, insight into the underlying physics of the devices, and their various capabilities, remains limited. In contrast to a static system, the ion-dynamic capacitance of the electrolyte is not a constant but a function of bias and time. The electrostatic potential also does not drop linearly within the dielectric but mainly drops at the electrolyte-semiconductor interface[5], which increase the complexity of the spatiotemporal evolution of ion concentrations, electrostatic potential, effective capacitance, and charge accumulation in the channel.

Here, we report a compact theory for time- and gate-dependent current in transistors with solid-state electrolytes. The theory predicts that the features of plasticity, high apparent mobility, sharp subthreshold swing (SS), and

memristive conductance could all appear on demand in a single transistor by programming the ion distribution or sweeping speeds. We validate this with numerical simulations. Such multimode transistor and the capabilities are confirmed by using the AlO_x/InO_x electrolyte-gated transistor.

Models of Ion-dynamic Capacitance

In bottom-gate top-contact transistor, the density of ions $p(x,t)$ (cation, cm^{-3}), and the electrostatic potential $V(x,t)$ change under constant bias from the initial to the transient and the steady state. In quantity, p and V are spatiotemporally coupled in the drift-diffusion equation based on the drift flux density and diffusion flux density. The analytical solution of drift-diffusion equation and Poisson equation are complex. For simplicity, we mainly consider the time-dependent ion density at the semiconductor-dielectric interface $p(0,t) = p_i$, the dimensionless potential drop u, the induced capacitance C, and the total charges Q_i. We refer to the typical dispersive process (with the energy in Boltzmann distribution) and the EDL theory[6], and develop a set of approximated solutions for them:

$$\frac{\partial p_i}{\partial t} = -(p_i - p_{i\infty})\frac{\beta}{\tau}\left(\frac{t}{\tau}\right)^{\beta-1} \tag{1}$$

$$u = \ln\left(\frac{\gamma_1 - 1 - \sqrt{1 + 2\gamma_1\gamma_2 - 2\gamma_1}}{\gamma_1 - 2\gamma_2}\right) \tag{2}$$

$$C = C_0 \frac{\cosh\left(\frac{u}{2}\right)}{1 + 2\gamma_1\sinh^2\left(\frac{u}{2}\right)} \times \sqrt{\frac{2\gamma_1\sinh^2\left(\frac{u}{2}\right)}{\ln\left[1 + 2\gamma_1\sinh^2\left(\frac{u}{2}\right)\right]}} \tag{3}$$

$$Q_i = (p_i - n_i)et_E \tag{4}$$

In Eq. (1), the time constant τ, the power factor β and $p_{i\infty}$ are described by: $\tau = \eta/(|V_{ox}|D_i)$, $\beta = \beta_0|V_{ox}/1|^\alpha$, $p_{i\infty} = p_{i,max}[1 - A\exp(-V_{ox}/V_0)]$, where η, β_0, α, $p_{i,max}$, A, and V_0 are constants, because the process is driven by the voltage V_{ox} and the rate is limited by diffusion coefficient D_i. In Eq. (2-3), γ_1 is the initial-to-steady density ratio $\gamma_1 = (p_{io} + n_{io})/p_{i,max}$, γ_2 is the initial-to-transient density ratio $\gamma_2 = p_{io}/p_i$, and C_0 is the Debye capacitance $C_0 = \sqrt{\varepsilon_0\varepsilon_r e^2(p_{io} + n_{io})/(4\pi k_B T)}$. When applied with constant bias, Eq. (1) turns into an integral form. The possible ion transfer from electrolytes into semiconductors is considered by a transferred ratio λ_1 ($0 \le \lambda_1 < 1$). The p_i, u, and C with various V_{ox} or D_i are calculated by Eq. (1-3) in the case without interfacial ion transfer ($\lambda_1 = 0$). The compact model Eq. (1-4) could be combined with the FET equations

979-8-3315-0417-5/25 $31.00 © 2025 IEEE

to give *I-V* relations, where the charges of ions Q_i also cause shift of the flat-band voltage V_{FB} against its initial value V_{FB0}, i.e. $V_{FB} = V_{FB0} - Q_i/C$. The theory can be extended by using the non-zero transferred ratio λ_1 to describe that some ions transferred from the solid-state electrolytes into semiconductors (e.g., Faraday charge transfer) dope semiconductors and affect the conductivity. The theory predicts that multimode-operation could be obtained in a single transistor with ionic dielectric and will be presented together with experiments in the following.

Multimode Transistors

We now study the theoretically predicted and experimentally measured transistor characteristics at different operation modes. The theory (Eq. 1-4) predicts multimode-operation could be obtained by presetting the ion concentration p_i (e.g., with V_g-programming), varying the diffusion rate D_i, or changing the sweep rate $s = \partial V_g / \partial t$, as exemplified in Fig. 1a-d and validated by numerical simulation (Fig. 1e-h). Upon potentiation with different amplitudes or widths of pulses, the transistor shows short- or long-time plasticity (STP or LTP, Fig. 1a,e); or an enhanced current and apparent mobility after a high gate bias, i.e. the ultrahigh apparent-μ mode (Fig. 1b,f). By further increasing s (or the sweep range of V_g), the hysteresis curve shows an ultra-sharp reversed subthreshold-slope *SS* of about 30 mV/dec in both forward and backward curves, i.e. the sub-thermionic *SS* mode (Fig. 1c,g); or even the hysteresis with negative trans-conductance, i.e. the memristive mode (Fig. 1d,h). In experiments, we used thin semiconductor InO_x film (10 nm) and the solid-state electrolyte AlO_x film (40 nm) to fabricate bottom-gate, top-contact FETs. The maximum of leakage current is lower than 7% of I_d. As shown in Fig. 1i-l, all the above-mentioned modes could be obtained by presetting V_g or changing s. The consistent simulation and experimental results validate the theoretical predictions and the applicability of the strategy.

The complex transistor modes are summarized in the p_i-t coordinates in Fig. 2a and the evolution of electrostatic potential and charges are illustrated in Fig. 2b-d, as discussed below.

(I) Tunable plasticity mode. It occurs when ions are *gradually accumulated* and Δu gently increases with $\Delta u \ll u_M$ (the purple segment in Fig. 2a). As C gently increases, the electrostatic potential $V(x)$ and interfacial ions p_i gently increase (Fig. 2b-d, mode I). The number of ions can be controlled by adjusting the amplitude and number of V_g pulses. With a small (or large) preset p_i or large (or small) D_i, potentiation is short (or long) enabling STP (or LTP) in synaptic behavior.

(II) Ultrahigh apparent μ mode. When ions *increase synchronously* with V_g-sweep, Δu significantly increases toward u_M (the orange segment in Fig. 2a), C sharply increases, and substantial carriers are induced with large potential drop near the interface (mode II in Fig. 2b-d).

(III) Sub-thermionic *SS* mode. When ions are *released synchronously* with V_g-sweep, u increases and then decreases (the green segment in Fig. 2a). As the ions accumulated at the interface are released to the bulk with decreased p_i and potential V (mode III in Fig. 2b-d), current changes sharply and, if the rate of ion-releasing matches the rate of V_g-sweep, a negative $\partial V_{ox}/\partial V_g$ would appear with $SS < (kT/e)\ln 10$.

(IV) Memristive mode. When ions are *post-saturated*, Δu could exceed u_M (the pink segment in Fig. 2a). As C reaches the maximum and then decreases due to the steric effects of ions, the potential V near the interface decreases, resulting decreased I_d (mode IV in Fig. 2d-f).

Conclusion

We have reported the development of multimode transistors based on ion-dynamic capacitance. The ion dynamic capacitance could be described by a compact theory validated by numerical simulations. This showed that the characteristics of tunable synaptic weight, high apparent mobility, sharp subthreshold swing, and memristive conductance could all appear in a single transistor by programming the interfacial ion concentrations or matching the scan speed with ion motions. We then fabricated such multimode transistors using common solid-state electrolyte films, and experimentally confirmed this range of behaviors.

Acknowledgments

The authors gratefully acknowledge the financial support of the project from the National Natural Science Foundation of China (61922090).

References

[1] H. Ling, D. A. Koutsouras, S. Kazemzadeh, Y. van de Burgt, F. Yan, and P. Gkoupidenis, "Electrolyte-gated transistors for synaptic electronics, neuromorphic computing, and adaptable biointerfacing," *Applied Physics Reviews,* vol. 7, no. 1, p. 011307, (2020)

[2] C. S. Yang *et al.*, "All‐Solid‐State Synaptic Transistor with Ultralow Conductance for Neuromorphic Computing," *Advanced Functional Materials,* vol. 28, no. 42, p. 1804170, (2018)

[3] X. Liang, L. Liu, G. Cai, P. Yang, Y. Pei, and C. Liu, "Evidence for Pseudocapacitance and Faradaic Charge Transfer in High-Mobility Thin-Film Transistors with Solution-Processed Oxide Dielectrics," *The Journal of Physical Chemistry Letters,* vol. 11, no. 7, pp. 2765-2771, 2020/04/02 (2020)

[4] Y. Zhao, S. Bertolazzi, M. S. Maglione, C. Rovira, M. Mas‐Torrent, and P. Samorì, "Molecular Approach to Electrochemically Switchable Monolayer MoS2 Transistors," *Advanced Materials,* vol. 32, no. 19, p. 2000740, (2020)

[5] H. Du, X. Lin, Z. Xu, and D. Chu, "Electric double-layer transistors: a review of recent progress," *Journal of Materials Science,* vol. 50, no. 17, pp. 5641-5673, (2015)

[6] A. A. Kornyshev, "Double-Layer in Ionic Liquids: Paradigm Change?," *The Journal of Physical Chemistry B,* vol. 111, no. 20, pp. 5545-5557, 2007/05/01 (2007)

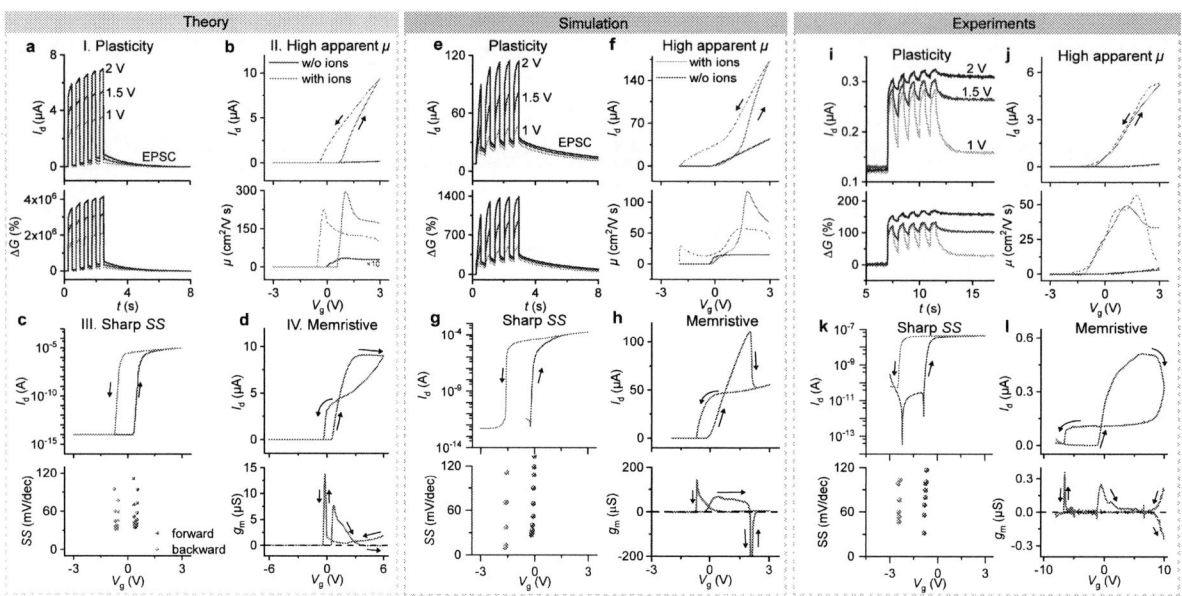

Fig. 1: Multimode transistors in the theory, simulations and experiments. (a)-(d), Calculated I-V characteristics (V_d=0.1 V) from the model with varying the sweep rate s (b: s = 2.4 V/s; c: s = 1.2 V/s; d: s = 0.24 V/s): (a) I_d-t curves under the stimulation of pulses and the relative change in conductance $\Delta G\%$; After the stimuli, the excitatory postsynaptic current is observed. (b) Transfer characteristics and the extracted apparent mobility μ; (c) Transfer characteristics and the reversed subthreshold slope SS for forward or backward curves; (d) Transfer characteristics and the trans-conductance g_m. e-h Numerical simulations by solving drift-diffusion equation and Poisson equation (V_d = 0.1 V) with varying s (f: s = 2.5 V/s; g: s = 1.0 V/s; h: s = 0.1 V/s). i-l Experimental results of the AlO$_x$/InO$_x$ transistor with the corresponding modes.

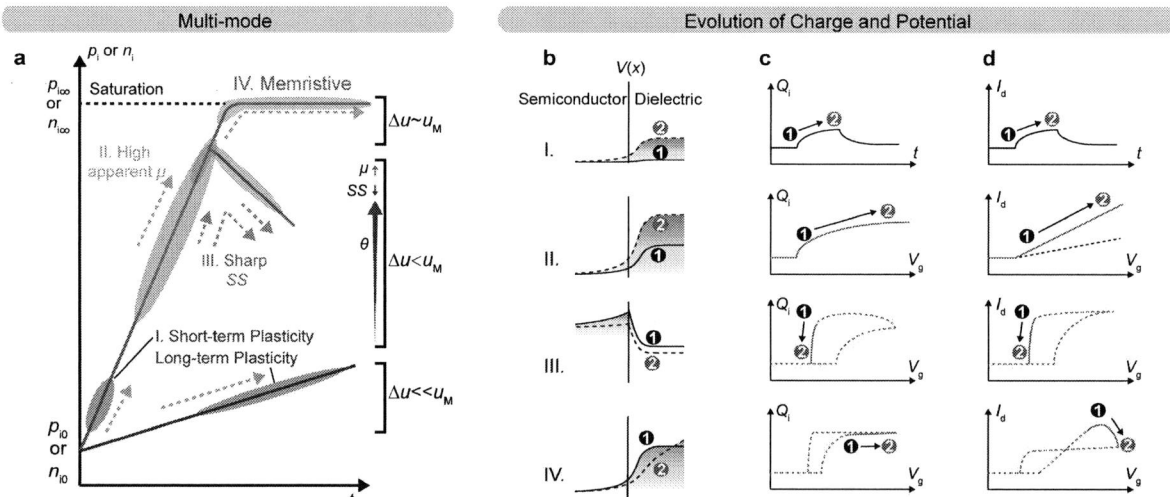

Fig. 2: Mechanisms and parameters of multimode transistors. (a) A summary of the multimode within the p_i-t coordinates. (b) Schemes of the potential near the semiconductor/dielectric interface for different modes. (c) The corresponding diagrams of Q_i-V or Q_i-t curves. (d) The corresponding diagrams of I-V or I-t curves.

979-8-3315-0417-5/25 $31.00 © 2025 IEEE   490

Pixel-to-Pixel Variance in Graphene-Silicon Photodetector Arrays

Muhammad Abid Anwar[1,3], Muhammad Malik[1,3], Srikrishna Chanakya Bodepudi[1*], Xiaolei Ding[2], Yance Chen[1], Zongwen Li[1], Zhi-Xiang Zhang[1], Wenzhang Fang[1], Huan Hu[1,2], Bin Yu[1], Yang Xu[1,2*]

[1]College of Integrated Circuits, State Key Laboratory of Silicon and Advanced Semiconductor Materials, Zhejiang University, ZJU-HIC, Hangzhou, 310027, China; *Email: bodepudi@zju.edu.cn, yangxu-isee@zju.edu.cn

[2]ZJU-UIUC Institute, International Campus, Zhejiang University, Haining, 314400, China; [3]Authors contributed equally

Abstract

Understanding pixel-to-pixel (P-P) variance of dark current and photoresponse from individual pixels in image sensors is crucial to address artifacts, noise, and photoresponse inaccuracies in the overall imaging process. Specifically, well-calibrated pixel-to-pixel variance and specific photoresponse statistics of the raw output data help to identify the unavoidable error contributions from individual pixels, thus enabling compensation through computational image processing or readout circuits. This is particularly vital for emerging image sensors based on 2D materials integrated with silicon, which often utilize readout circuits distinct from traditional CCD and CMOS sensors. This work investigates pixel-to-pixel variance in graphene-silicon Schottky photodiodes, assessing the feasibility of large-scale fabrication, uniformity of dark current, and photoresponse in small-scale arrays. The array-based pixel statistics investigated in this study have potential implications for designing in-sensor processing and developing accurate statistical estimation methods for imaging in astronomy, biomedical, and spectroscopy.

Keywords: Pixel-to-pixel variance, low dark current, coefficient of variance, graphene-silicon photodiode

Introduction

The design of large-scale logic, memory, and sensing systems requires the identical performance of multiple devices with closely matched parameters. Miniaturization and increasing device density for higher computing power, data communication, and high-resolution imaging demand uniform performance in a large number of devices. However, fabrication and readout processes introduce uncontrollable factors, leading to device-to-device variability. Understanding this variability is crucial when introducing new materials or processes. Recently, there has been an increasing interest in integrating 2D materials (2DM) with CMOS and other conventional systems to exploit the intriguing electronic, sensing, and non-volatile memory properties of 2D materials to improve the performance or introduce new functionalities to the existing device architectures[1], [2], [3], [4], [5]. In this work, we focus on the pixel-to-pixel variance in small-scale graphene-silicon photodetector arrays (6×6 & 8×8) that provide a reasonable understanding of uniformity in dark current and photoresponse as well as the feasibility of the fabrication method for large-scale systems. This study also provides preliminary evidence on van der Waals (vdW) interface engineering to integrate 2DMs and layered systems with silicon technology for reduced device-to-device variability and stability with direct implications in heterostructure and vertically integrated image sensors.

The parameters influenced by the device-to-device variance are based on specific functionality and targeted applications[1], [2]. For example, channel material quality, dielectric interface scattering, and contact resistance in transistors significantly impact the variability in integrated logic circuits.[4], [6] Many recent studies have investigated these parameters in large sets of 2DM transistors and logic circuits. The variance in these studies mainly depends on defects, impurities, and environment on in-plane charge transport in 2D materials and thus cannot be directly correlated with devices based on out-of-plane charge transport.[4], [5], [7] While out-of-plane, in 2D material-based memristors, these studies need more focus on the reliability or cycle-to-cycle variability of individual devices[2], as the memristor functionality depends on forming conductive paths and sustaining the hysteresis within the desired operational window. On the other hand, the variance studies on low-dimensional material-based image sensors must focus on comparing different devices in an array or matrix, as the intra-pixel variance alone cannot determine the stability and uniformity of the overall imaging process.[8] To be specific, dark current, read noise, quantum efficiency, gain, form factor, and photoresponse non-uniformities between pixels require special attention in emerging 2DM-based image sensors.[8], [9] This study is essential for realizing the large-scale integration of 2DMs with CMOS circuits for image sensing, as it provides required raw data for designing accurate readout circuits and developing maximum likelihood estimation methods often required for optimizing the performance of scientific CMOS (sCMOS) sensors.[10].

The variability analysis of electrical and optoelectronic parameters of graphene-Si photodetector arrays in this work demonstrates a basic variability study approach for 2D-3D hybrid image sensors that can be extended to stability, reliability, and yield to estimate and correct the predictable pixel performance and photoresponse inaccuracies with either dedicated readout circuits (ROIC) or via computational compensation techniques. This study also leverages the potential of vdW interfacial engineering to realize energy-efficient integration of 2D material and their bulk layered systems for broadband photodetection, large-array memory, and logic circuits.

Fabrication of Gr-Si Photodetector Arrays

An n-type low-doped Si/SiO$_2$ wafer is used to fabricate the Graphene-Silicon photodetector array. The fabrication involves (1) the deposition of the metal electrodes, (2) buffer oxide etchant (BOE) to remove oxide to expose the Si window, (3) Graphene is then transferred over the Si window by using the standard wet transfer technique and patterned, as shown in Fig 1 (a). The I$_{2D}$/I$_G$ of ~2 and weak I$_D$ in Raman spectra indicate low-defect monolayer graphene (Fig 1 (b)). We performed atomic force microscopy (AFM) to ensure the uniformity and roughness

979-8-3315-0417-5/25 $31.00 © 2025 IEEE

of the interface. AFM data shows very low surface roughness (257 pm) and uniform graphene-silicon interface. Furthermore, energy-dispersive X-ray microscopy represents the distinctive presence of each element such as Au, C, O, and Si in the device's active area (Fig 1 (d-g)).

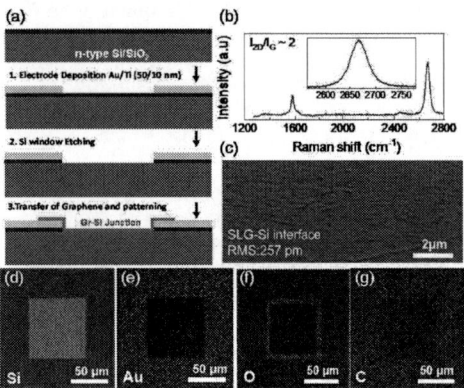

Fig. 1:(a) Fabrication steps of Gr-Si Schottky array. (b) Raman spectrum of Gr-Si interface. (c) Atomic Force Microscopy (AFM) of Gr-Si interface with a low surface roughness value (RMS = 257 pm). (d-g) Energy dispersive X-ray spectroscopy (EDS) of the active pixel.

Variance in Optoelectronic Properties of Gr-Si Arrays

The inter-pixel variance data of an image sensor is crucial in designing the post-image processing unit to effectively compensate for photoresponse inaccuracies before analog-to-digital conversion. This additional computational load for the processing unit can be minimized by reducing the inter-pixel variation in the raw electrical and optical response of the as-fabricated photodetector array. The dark current and rectification ratio of our Gr-Si photodetector array is illustrated in Fig 2(a) for a pixel size of 100×100 μm^2. We observed a low pixel-to-pixel (P-P) deviation from the estimated mean value throughout the scan range in all pixels of the 6×6 array. The inset of Fig 2(a) shows the dark current varying from ~15 to 55pA. The dark current map of 6×6 array at -1.5V is displayed in Fig 2(b). This exceptionally low spatial variation in dark current ensures uniform responsivity across all pixels in the array [8], a critical factor for achieving high-resolution imaging.

To investigate the P-P variance in photoresponse, it is important to illuminate each pixel with highly uniform light to avoid any invariance from incident light. We collected P-P variance in photoresponse under a highly uniform 532nm laser at different incident power values shown in Fig 2(c). A linear increase in photocurrent is evident with increasing optical power from 50 to 400nW. The photomapping at 50nW illuminated power is displayed in Fig. 2(d). The proportional increase in both short-circuit current and open-circuit voltage demonstrates its suitability for photovoltaic applications. [11] We evaluated the coeffct of variance (C_V) for electrical and optical parameters shown in Fig 2(e-m) to analyze pixel-to-pixel variability. We observed C_V as low as 0.33, 0.27, .052, and 0.015 for dark current, series resistance,

ideality factor, and Schottky barrier height, respectively, confirming the low deviation of pixels from the mean values of all basic Schottky diode parameters (Fig 2(e-h)). Similarly, the uniform photocurrent of all pixels in an array at specific optical power further indicates the potential of Gr-Si photodiodes for low-noise imaging. For this, we observed C_V as small as < 0.026 for all pixels at various intensities shown in Fig 2 (i-m). The raw data clearly shows the optical power dependence of photocurrent from each pixel. Based on the observed C_V of our small-scale array, a dedicated ROIC for Gr-Si Schottky photodiodes can be designed, or computational corrections can be implemented. This approach can be extended to other 2D-3D hybrid image sensors for efficient integration with CMOS readout circuits.

The geometry and the individual pixel dimensions can also contribute to the variability in the photoresponse. To understand this, we fabricated the Gr-Si photodetectors with two different pixel sizes ($100 \times 00\mu m^2$ and $500\times500\mu m^2$) and array sizes (6×6 and 8×8). The optical image of 8×8 array is displayed in Fig 3(a). We performed similar measurements for both arrays.

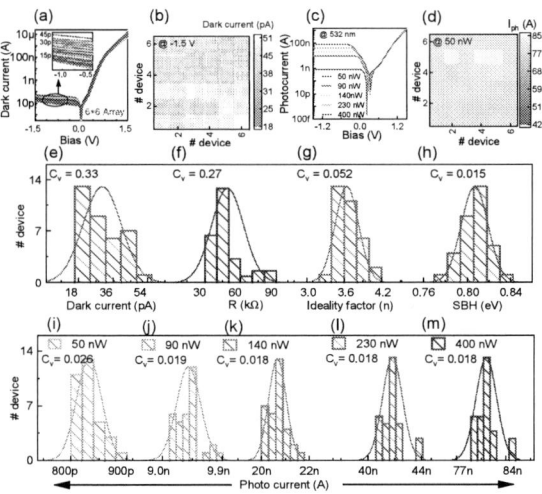

Fig. 2: (a) Current-voltage (I-V) characteristics array pixel (Array size:6×6). (b) The grid plot shows the low dark current P-P variation (~18 to 52pA). (c). Photocurrent measurement of Gr-Si pixel under various illumination power of 532nm. (d) The uniform photocurrent distribution at 50nW optical power. (e-h) Histogram of Dark current, Series resistance, Ideality factor (n), Schottky barrier height (SBH). (i-m) Photocurrent measurement of array device with optical powers.

The I-V characteristics in Fig 3(b) represent the usual pixel-to-pixel variations in Gr-Si photodiodes with an average dark current of 0.8nA. The dark current mapping for the array at a bias voltage of -2V is shown in Fig 3(c), indicating a nearly uniform functionality with a narrow variation in dark current in all 64 devices. Similarly, Fig 3(d-g) shows histograms of electrical parameters with low values of C_V such as 0.22, 0.147, 0.08, and 0.019 for dark current, series resistance, ideality factor, and Schottky barrier height, respectively.

979-8-3315-0417-5/25 $31.00 © 2025 IEEE

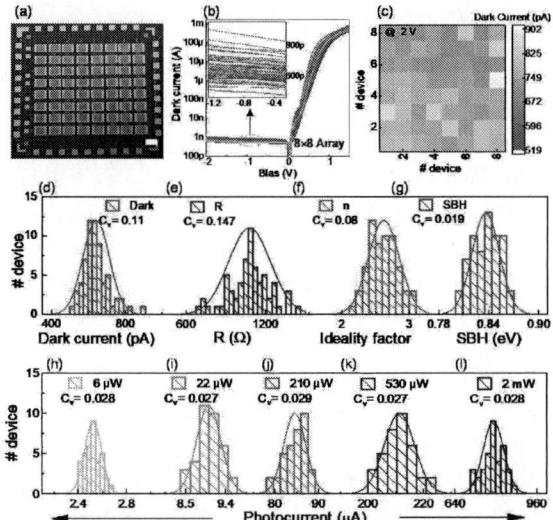

Fig. 3:(a) Optical image of Gr-Si array (Array size: 8×8, Pixel size: 500×500 μm², Scale bar: 500μm). (b) Current-voltage (I-V) plot of Gr-Si Schottky array. The inset shows the zoomed-in distribution of the dark current. (c) The grid plot shows the low P-P variation of dark current. (d-g) Histogram plots of Dark current, Series resistance, Ideality factor (n), and Schottky barrier height (SBH). (h-l) Photocurrent measurement of array device with optical powers.

Furthermore, the variance in photocurrent is much narrower than the dark current with C_V values below 0.03 at various intensities of optical powers shown in Fig 3(h-l). Such narrow deviation from the mean photoresponse can be correlated to the linear dependence of the photocurrent with light intensity and uniform laser illumination of individual pixels. The low inter-pixel deviation in electrical and optoelectronic responses demonstrates the potential of the Gr-Si interface as a basic buffer interface to integrate various 2D layered systems with Si and other conventional semiconductors.

Graphene-Schottky as a Buffer Interface for Reduced Variability

To verify the role of the Gr-Schottky interface as a buffer layer for reduced variability, we integrated macroscopic assembled graphene (MAG) via the Gr-Si interface. Our previous studies demonstrated the broadband photoresponse of MAG-Si Schottky junctions [12]. However, a large leakage current limits its responsivity. We observed that the dark current of the MAG-Si interface can be suppressed to a few orders of magnitude by decoupling MAG with the Gr-Si interface, as shown in Fig. 4(a). A narrow distribution of dark current for a 4×4 array of MAG/Gr-Si photodetector array is shown in Fig. 4(b). Photocurrent and transient absorption spectroscopy studies are required to investigate further the role of the Gr-Si interface in these hybrid systems.

The transfer of functional 2D films expands the scope of the interface and de-coupling graphene ensures low pixel-pixel or device-device variation as seen in Figure 4(b) in which de-coupling graphene enables the low variations in MAG-Si interface with reduced dark current. The current investigative study is still underway and will be expanded over various 2D-3D systems for more comprehensive results.

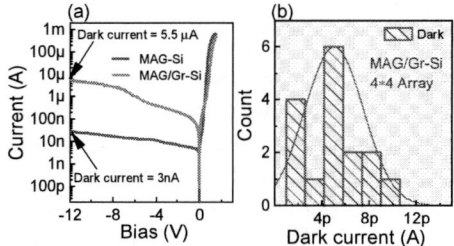

Fig. 4: (a) Current-voltage characteristics of MAG-Si and MAG/Gr-Si device. (b) Histogram showing P-P variation of dark current.

Conclusion

Statistical analysis of graphene-silicon photodetector arrays, based on raw data, revealed low coefficients of variation (C_V) for both electrical and optoelectronic responses across various array sizes and pixel dimensions. This demonstrates the potential of variability analysis to accurately design and reduce the computational load of readout circuits and post-processing units for image sensors aiming for large-scale 2D-3D integration technology.

Acknowledgments

We thank Chao Gao for providing MAG samples. This work is funded by the start-up funding of ZJU-HIC Centre (04010000-K02013007, 02170000-K3F113024) and the National Key Research and Development Program of China (2022YFA1204900).

References

[1] T. M. I. Băjenescu, et al., Rel Cha Elec Ele Sys, 2015 pp. 63–82.

[2] M. Lanza, et al., Nat Comm, 2020 vol. 11, pp. 1–5.

[3] K. K. H. Smithe, et al., ACS Nano, 2017, no. 8, pp. 8456–8463.

[4] S. Das et al., Nat Electronics 2021, vol. 4, pp. 786–799.

[5] A. Sebastian, et al., Nat Comm, 2021, vol. 12, pp. 1–12.

[6] F. Schwierz, et al., Nanoscale, vol. 7, pp. 8261–8283.

[7] P. V Pham et al., Chem Rev, 2022, vol 122, pp. 6514–6613.

[8] S. Ansari et al., Small Methods, vol. 8, p. 2300595.

[9] G. Konstantatos, et al., Nat Comm, 2018, vol. 9, pp. 1–3.

[10] F. Huang et al., Nat Methods, 2013, vol. 10, pp. 653–658.

[11] Dong Pu, M.A. Anwar, et al., App Phys Lett, 2023, vol. 122, p. 041102.

[12] Peng L, Liu L, Du S, et al., InfoMat 2022, vol. 4, p. e12309.

979-8-3315-0417-5/25 $31.00 © 2025 IEEE

50 nm GaN HEMTs Technology with High Frequency, low Noise, and High JFoM

Yun Zhang[1], Jiaheng He[12], Changxin Mi[12], Xuankun Wu[12], Shujie Xie[12], Zhe Cheng[12], Lian Zhang[1]

[1]Institute of Semiconductors, Chinese Academy of Sciences, China, [2]University of Chinese Academy of Sciences, China, yzhang34@semi.ac.cn

Abstract

Our 50 nm GaN HEMTs with source-to-drain length (L_{sd}) of 200 nm reached cut-off frequencies f_T/f_{max} of 230/525 GHz. W-band load-pull measurements indicate a 7.4 dB linear gain, 15.8 dBm output power, and 15% efficiency. A low noise figure of 1 dB is achieved from 25-35 GHz. Devices with 5 μm L_{sd} exhibited a breakdown voltage of 180 V and a f_T of 117 GHz, resulting in a JFoM of 21 THz·V.

Keywords: GaN HEMTs, device scaling and sapphire substrate

Introduction

Gallium Nitride (GaN) high electron mobility transistors (HEMTs) have emerged as critical components for 6G/terahertz (100 GHz to 10 THz) power and low-noise amplifiers due to their high two-dimensional electron gas (2DEG) concentration, elevated electron saturation velocity, and strong breakdown electric field [1]. To meet the strategic demands of future high-performance 6G/terahertz communications, GaN HEMTs must synergize high-frequency and high-power capabilities [2] with cost-efficiency [3]. Enhancing the power cutoff frequency (f_{max}) is essential for power amplifiers, while maintaining a high current gain cutoff frequency (f_T) and minimizing parasitic resistances are crucial for low-noise performance. By leveraging advanced down-scaling techniques, GaN HEMTs on sapphire substrates can achieve remarkable radio-frequency (RF) performance, W-band power performance, low noise figures, and an outstanding Johnson's figure of merit (JFoM). These attributes make them highly suitable for various 6G/Terahertz applications. Recent studies have demonstrated that GaN-on-sapphire HEMTs can deliver high operating frequencies, substantial power output, and significant cost advantages [3]–[9]. In this work, we highlight the high-frequency, low-noise, and high JFoM performances of GaN HEMTs on sapphire, underscoring their potential as a solution for next-generation communication systems.

Device Design and Fabrication

In this study, we utilized sapphire substrates and metal-organic chemical vapor deposition (MOCVD) to grow an ultra-thin 4 nm InAlN polarization layer atop a 300 nm GaN channel layer. Two sets of HEMTs were fabricated using electron beam lithography (EBL), featuring source-to-drain distances (L_{sd}) of 200 nm and 5 μm. The devices with a 200 nm L_{sd} were engineered to minimize parasitic and intrinsic delays, thereby achieving ultra-high RF performance, exceptional W-band load-pull characteristics, and low noise figures. In contrast, the devices with a 5 μm L_{sd} focused on balancing RF performance with terahertz operation (>100 GHz) while ensuring ultra-high breakdown voltage. This approach aimed to deliver high-frequency and high-power performance under high-voltage conditions, along with an outstanding Johnson's figure of merit (JFoM). Selective area growth (SAG) of n⁺-GaN source/drain electrodes was implemented using MOCVD, achieving an electron concentration of 8×10^{19} cm⁻³ [10]. Titanium/gold (20/200 nm) layers were deposited via electron beam evaporation to form source/drain ohmic contacts without the need for annealing. The SAG n⁺-GaN electrodes effectively reduced access resistance and minimized parasitic delays, resulting in lower noise figures and enhanced breakdown voltages. Additionally, high-quality 50 nm T-shaped gates made of Ni/Au (20/250 nm), with a gate width of 2×20 μm, were fabricated using EBL.

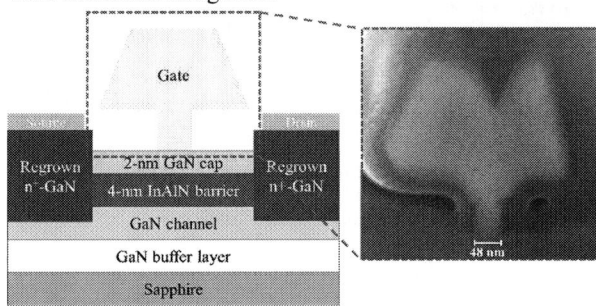

Fig. 1. Scheme of 50 nm GaN HEMT and SEM image of T-gate.

Results and Discussion

The direct current characteristics of our GaN HEMTs were evaluated using a Keithley 4200A-SCS parameter analyzer. Transfer characteristics were measured with a drain-source voltage (V_{ds}) = 2 V, while the gate-source voltage (V_{gs}) was varied from −3 V to 2 V. For HEMTs with a L_{sd} of 200 nm, the threshold voltage was determined to be -1.5V at a current density of 1 mA/mm, with a peak extrinsic transconductance (g_m) of 625 mS/mm. Output characteristics were evaluated at $V_{ds} = 5$ V, with V_{gs} varying from -3 V to 2 V in increments of 0.5 V. Under $V_{gs} = 2$ V, the devices exhibited a saturation output current density of 1.8 A/mm and an on-resistance (R_{on}) of 0.8 Ω·mm. HEMTs with a L_{sd} of 5 μm exhibited a threshold voltage of −2.0 V at a current density of 1 mA/mm, with a g_m of 545 mS/mm; at $V_{gs} = 2$ V, the devices demonstrated a saturated output current density of 1.3 A/mm and a R_{on} of 1.9 Ω·mm.

To evaluate the RF performance of our 50 nm GaN HEMTs, small-signal measurements were conducted using a Keysight N5247B network analyzer. The bias-dependent S-parameters were measured over a frequency range from 1 GHz to 67 GHz, employing standard open-load-short-through (SOLT) calibration at the probe tips. To accurately account for parasitic effects introduced by the device pads, on-wafer open and short structures were utilized to de-embed these parasitic elements from the measured S-parameters.

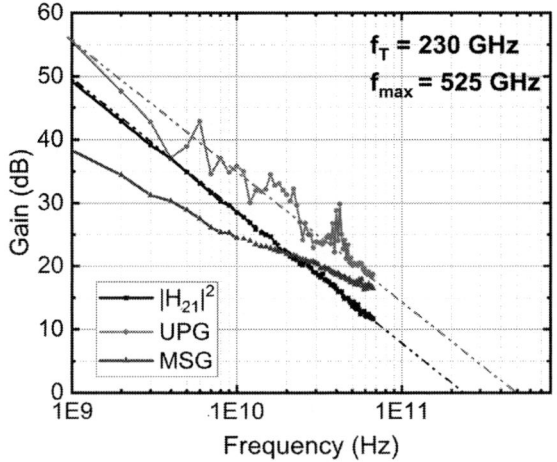

Fig. 2. RF characteristics of the 50 nm GaN HEMT with 200 nm L_{sd}.

As shown in Fig. 2, the frequency-dependent characteristics of the current gain ($|H_{21}|^2$) and Mason's unilateral gain (upg) were analyzed under the peak f_{max} bias condition, with V_{ds} set at 9 V and V_{gs} at -0.8 V for the HEMTs with a L_{sd} of 200 nm. Extrapolation of $|H_{21}|^2$ and upg with a -20 dB/dec slope yielded the f_T/f_{max} equal to 230/525 GHz, respectively. For the HEMTs with a L_{sd} of 5 μm, the measured f_T/f_{max} were 117/240 GHz, respectively. These results underscore the excellent RF performance of the devices, highlighting the impact of device scaling on key frequency metrics.

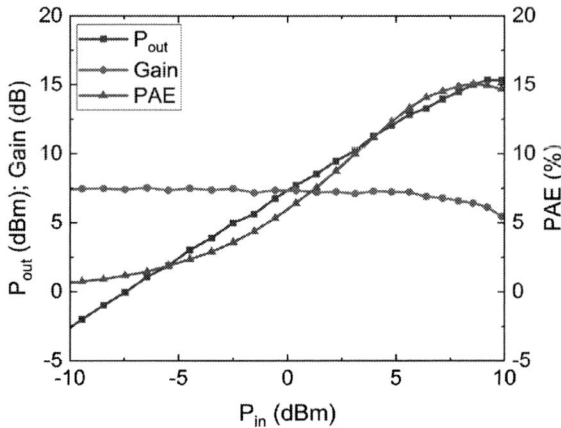

Fig. 3. 94GHz loadpull of the 50 nm GaN HEMT with 200 nm Lsd.

Additionally, as shown in Fig. 3, the results of the W-band 94 GHz continuous-wave large-signal load-pull measurement indicate that for HEMTs with a L_{sd} of 200 nm, the maximum

achievable linear gain is 7.4 dB. At an input power (P_{in}) of 10.0 dBm, the output power (P_{out}) reaches 15.8 dBm, with a peak power-added efficiency (PAE) of 15%. The combined high f_T/f_{max} values and favorable 94 GHz load-pull performance highlight the suitability of our GaN HEMTs as competitive candidates for 6G/ terahertz power amplifiers.

The noise performance of a device is a key indicator for evaluating its potential for low-noise applications. The noise figure expression relation of a transistor is defined as

$$F = F_{min} + \frac{4R_n}{Z_0}\frac{|\Gamma_s - \Gamma_{opt}|^2}{(1-\Gamma_s)^2|1+\Gamma_{opt}|^2} \quad (1)$$

where F_{min} is the minimum noise factor, R_n is the equivalent noise resistance, and Γ_{opt} is the optimal noise reflection coefficient.

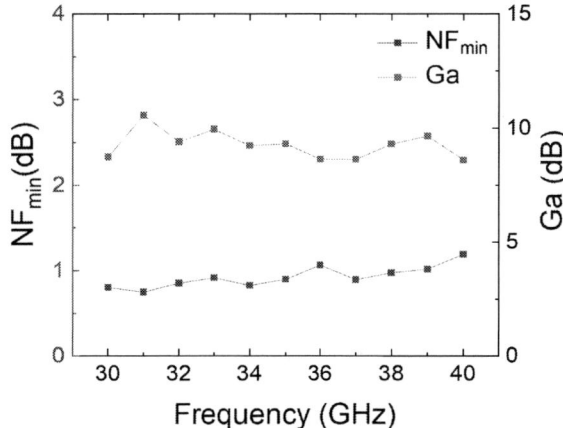

Fig. 4. Minimum noise figure (NF_{min}) and the associated gain (Ga) versus frequency for the device at V_{gs} = -1.2V, V_{ds} = 3 V.

Fig. 4 presents the minimum noise figure (NF_{min}) and associated gain (Ga) as functions of frequency under the optimal noise reflection coefficient condition. For devices with a L_{sd} of 200 nm, NF_{min} remains consistently under 1.0 dB across the frequency range of 25 GHz to 35 GHz. Concurrently, Ga exceeds 8.5 dB throughout this frequency band. The exceptional NF_{min} performance is attributed to enhancement of f_T achieved by scaling down the devices and reductions in parasitic resistances, including gate and contact resistances.

The three-terminal off-state breakdown voltages (BV_{off}) for both sets of HEMTs were measured using a Keithley 4200A-SCS parameter analyzer, with the V_g set to –4 V. To protect the testing equipment, the threshold drain current (I_d) was limited to 0.1 mA/mm, which is one-tenth of the conventional value of 1 mA/mm. The implementation of SAG effectively alleviates the density of electric field lines from the gate region to the drain, resulting in an impressive BV_{off} of 180 V for HEMTs with a source-to-drain distance L_{sd} of 5 μm, as shown in Fig. 5. The JFoM is calculated as $f_T \times BV_{off}$ = 117 GHz × 180 V = 21.06 Thz·V.

979-8-3315-0417-5/25 $31.00 © 2025 IEEE

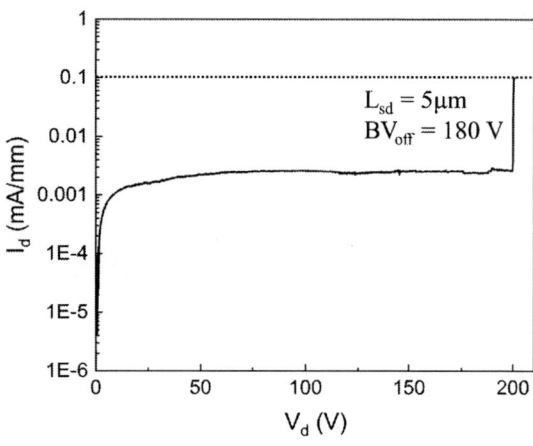

Fig. 5. BV_{off} measurement of the 50 nm GaN HEMT with 5 μm L_{sd}.

Conclusion

This work demonstrates that advanced down-scaling techniques enable 50 nm GaN HEMTs on sapphire substrates to achieve remarkable RF performance. Our devices with a L_{sd} of 200 nm exhibited outstanding f_T and f_{max} of 230 GHz and 525 GHz, respectively. In a 94 GHz load-pull test, these HEMTs achieved a linear gain of 7.4 dB, a peak output power of 15.8 dBm, and a PAE of 15%, highlighting their suitability for power amplifier applications. Additionally, the NF_{min} remained under 1.0 dB from 25 GHz to 30 GHz, indicating strong low-noise potential for sapphire-based devices. For devices with L_{sd} = 5 μm the focus on achieving high breakdown voltage while maintaining f_T > 100 GHz was validated by measurements showing f_T = 117 GHz, an off-state breakdown voltage of 180 V, and an exceptional JFoM of 21 THz·V. These attributes make them strong candidates for high-power, high-frequency applications. Overall, our results suggest that 50 nm GaN HEMTs on sapphire substrates, with their cost-effectiveness and strong performance metrics, are viable options for a range of high-performance millimeter-wave and terahertz applications.

Acknowledgments

This work was supported by the CAS Project for Young Scientists in Basic Research under Grant YSBR-064.

References

[1] K. Shinohara, D. Regan, Y. Tang, A. L. Corrion, D. F. Brown, J. C. Wong, J. F. Robinson, H. H. Fung, A. Schmitz, T. C. Oh, S. J. Kim, P. S. Chen, R. G. Nagele, A. D. Margomenos and M. Micovic, "Scaling of GaN HEMTs and Schottky Diodes for Submillimeter-Wave MMIC Applications," IEEE Transactions on Electron Devices, 60, pp. 2982-2996 (2013).

[2] M. Micovic, D. Brown, D. Regan, J. Wong, Y. Tang, F. Herrault, D. Santos, S. Burnham, J. Tai, E. Prophet, I. Khalaf, C. McGuire, H. Bracamontes, H. Fung, A, Kurdoghlian amd A, Schmitz, "High frequency GaN HEMTs for RF MMIC applications," 2016 IEEE International Electron Devices Meeting (IEDM) (2016).

[3] J. He, Z. Cheng, S. Xie, X. Wu, C. Mi, L. Zhang and Y. Zhang, "Advanced Down-Scaling Technology and its Physical Mechanism for 515 GHz GaN HEMT," 2023 20th China International Forum on Solid State Lighting & 2023 9th International Forum on Wide Bandgap Semiconductors (SSLCHINA: IFWS) (2023).

[4] D. Denninghoff, J. Lu, E. Ahmadi, S. Keller and U. K. Mishra, "N-polar GaN/InAlN/AlGaN MIS-HEMTs with 1.89 S/mm extrinsic transconductance, 4 A/mm drain current, 204 GHz fT and 405 GHz fmax," 71st Device Research Conference (DRC 2013) (2013).

[5] Y. He, L. Zhang, Z. Cheng, C. Li, J. He, S. Xie, X. Wu, C. Wu and Y. Zhang, "Scaled InAlN/GaN HEMT on Sapphire With fT/fmax of 190/301 GHz," IEEE Transactions on Electron Devices, vol. 70, pp. 3001-3004 (2023).

[6] J. He, Z. Cheng, C. Li, S. Xie, X. Wu, L. Zhang, C. Wu and Y. Zhang, "Demonstration of 420GHz highly scaled InAlN/GaN HEMTs by Electron Beam Lithography," 2023 International Workshop on Advanced Patterning Solution (IWAPS) (2023).

[7] X. Zheng, M. Guidry, H. Li, E. Ahmadi, K. Hestroffer, B. Romanczyk, S. Wienecke, S. Keller and U. Mishra, "N-Polar GaN MIS-HEMTs on Sapphire With High Combination of Power Gain Cutoff Frequency and Three-Terminal Breakdown Voltage," IEEE Electron Device Lett., 37, pp. 77-80 (2016).

[8] X. Zheng, H. Li, M. Guidry, B. Romanczyk,E. Ahmadi, K. Hestroffer, S. Wienecke, S. Keller and U. Mishra, "Analysis of MOCVD SiNx Passivated N-Polar GaN MIS-HEMTs on Sapphire With High fmax·VDS,Q," IEEE Electron Device Lett., 39, pp. 409-412 (2018).

[9] W. Li, B. Romanczyk, E. Akso, M. Guidry, N. Hatui, C. Wurm, W. Liu, P. Shrestha, H. Collins and C. Clymore, "Record RF Power Performance at 94 GHz From Millimeter-Wave N-Polar GaN-on-Sapphire Deep-Recess HEMTs," IEEE Transactions on Electron Devices, vol. 70, no. 4, pp. 2075-2080 (2023).

[10] L. Zhang, Z. Cheng, Y. W. He, J. X. Xu, L. F. Jia, X. Y. Wang, S. Y. Zhang, W. Tan and Y. Zhang, "Optimization of selective-area regrown n-GaN via MOCVD for high-frequency HEMT," Appl. Phys. Lett., 119, pp. 6 (2021).

979-8-3315-0417-5/25 $31.00 © 2025 IEEE

Enhanced Threshold Switching Devices based on Conductive Filaments in SiO$_x$ through Vertically Aligned MoS$_2$ Layers

Jimin Lee[1], Sofía Cruces[1], Dennis Braun[1], Lukas Völkel[1], Ke Ran[2,3,4], Joachim Mayer[3,4], Alwin Daus[1,5], and Max C. Lemme[1,2]

[1]Chair of Electronic Devices, RWTH Aachen University, Aachen, Germany, [2]Advanced Microelectronic Center Aachen, AMO GmbH, Aachen, Germany, [3]Central Facility for Electron Microscopy, RWTH Aachen University, Aachen, Germany, [4]Ernst Ruska-Centre for Microscopy and Spectroscopy with Electrons, Forschungszentrum Jülich GmbH, Jülich, Germany, [5]Institute of Semiconductor Engineering, University of Stuttgart, Stuttgart, Germany

E-mail: jimin.lee@eld.rwth-aachen.de

Abstract

In this work, we investigate for the first time the resistive switching (RS) behavior in a bilayer stack of SiO$_x$ and vertically aligned MoS$_2$ (VAMoS$_2$). We demonstrate threshold switching by forming silver (Ag) conductive filaments (CFs) in an amorphous SiO$_x$ layer combined with VAMoS$_2$ obtained through molybdenum (Mo) sulfurization. The SiO$_x$/VAMoS$_2$ devices exhibit repeatable RS with faster switching times and higher hold voltages compared to SiO$_x$ devices without VAMoS$_2$. The RS enhancement through VAMoS$_2$ layers shows promise for compact vertical device architectures for emerging memories and neuromorphic computing applications.

Keywords: Molybdenum disulfide (MoS$_2$), Threshold switching, Volatile resistive switching, Memristive devices, Conductive filaments, Neuromorphic computing

Introduction

Memristive devices based on the formation and dissolution of silver (Ag) or copper (Cu) conductive filaments (CFs) have shown great potential for neuromorphic computing applications [1,2]. The formation of thick metal CFs upon applying an electric field enables non-volatile memory operation, while thin CFs spontaneously dissolve when the applied voltage is removed, resulting in volatile switching [3,4].

The formation and rupture of CFs are stochastic processes influenced by multiple factors beyond the electric field, including temperature, defects in the switching layer, and the electrode interface, which can alter the stability and morphology of the CFs [2-4]. The self-rupture of filaments occurs spontaneously due to surface energy minimization or thermal diffusion but can also depend on the properties of the CFs and the switching layer [2-4].

Molybdenum disulfide (MoS$_2$), a layered transition metal dichalcogenide (TMD), has been extensively studied for resistive switching (RS) devices for neuromorphic computing applications [5,6]. RS in MoS$_2$ can be attributed to ion movement along its van der Waals (vdW) gaps [7,8] or grain boundaries [9], or phase change mechanisms [10]. RS in vertically oriented MoS$_2$, is promising for compact, small-footprint vertical device architectures for memristive devices [7,8,11].

Previously, non-volatile RS in vertically grown MoS$_2$ layers on silicon (Si) with chromium (Cr)/gold (Au) electrodes was demonstrated [7]. An interfacial silicon oxide (SiO$_x$) layer was observed between the Si substrate and the vertically aligned MoS$_2$ (VAMoS$_2$) after the sulfurization process. Additionally, non-volatile RS in MoS$_2$ on graphene with the deposition of SiO$_x$ dielectrics and nickel (Ni) electrodes was achieved [12]. However, the study did not include how this

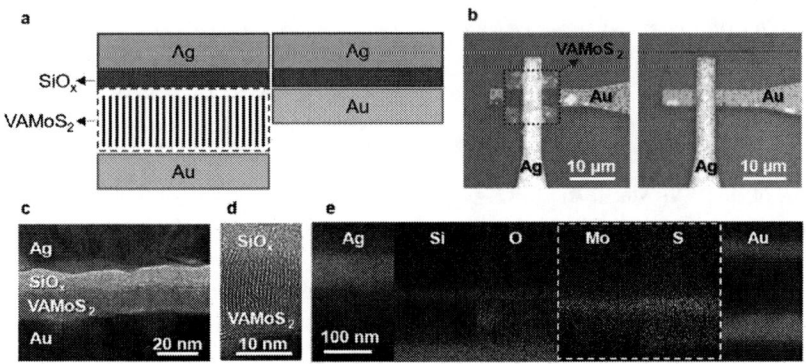

Fig. 1: Device structure and materials characterization. (a) Schematic of a RS cross-point device with SiO$_x$/VAMoS$_2$ (left) and only with SiO$_x$ (right). (b) Top-view optical microscopy images of the fabricated devices with SiO$_x$/VAMoS$_2$ (left) and SiO$_x$-only (right) corresponding to the schematics in Fig. 1a. (c) Cross-sectional TEM image of the SiO$_x$/VAMoS$_2$-based device. (d) HRTEM image showing the vertical alignment of the MoS$_2$ layers. (e) EDX elemental mapping for Ag, Si, O, Mo, S, and Au.

979-8-3315-0417-5/25 $31.00 © 2025 IEEE

stack influences the volatile RS behavior.

Here, we present Ag-CFs-based threshold switching in devices combining an amorphous SiO_x layer with VAMoS$_2$. Our Ag/SiO$_x$/VAMoS$_2$/Au devices exhibit repeatable, fast switching time of $t_{on} = 356$ ns and feature higher hold voltages than devices without VAMoS$_2$ (Ag/SiO$_x$/Au). The higher hold voltages indicate the formation of thin Ag filaments, allowing faster switching. The switching dynamics were statistically analyzed to gain insight into the RS mechanism.

Device Fabrication

Fig. 1a shows a cross-section schematic of the devices consisting of two different types of switching layers: SiO$_x$ with VAMoS$_2$ (SiO$_x$/VAMoS$_2$) and SiO$_x$ without VAMoS$_2$ (SiO$_x$-only). The switching layers are sandwiched between an active Ag top electrode (TE) and an inert Au bottom electrode (BE). The BEs and TEs were structured by photolithography, electron-beam evaporation, and lift-off. A molybdenum (Mo) thin film was deposited on the pre-patterned Au BEs via photolithography, direct-current (DC) sputtering, and lift-off. The VAMoS$_2$ was synthesized by thermally assisted conversion of the Mo thin films in a sulfur atmosphere at 800 °C for 30 min under a 20 sccm argon (Ar) flow [13]. An amorphous SiO$_x$ layer was formed during this sulfurization process from the 300 nm SiO$_2$/Si substrate by atomic diffusion of Si and O at the high process temperature [14], potentially enhanced along the vdW gaps in the VAMoS$_2$ layer. Fig. 1b shows optical microscope images of the fabricated cross-point devices, corresponding to the schematics in Fig. 1a. Fig. 1c shows a cross-sectional transmission electron microscopy (TEM) image of a SiO$_x$ layer on VAMoS$_2$ sandwiched between a 30 nm Ag TE and a 50 nm Au BE. The vertical alignment of the MoS$_2$ layers was confirmed by high-resolution TEM (HRTEM, Fig. 1d). The presence of SiO$_x$ and MoS$_2$ after sulfurization was verified by the energy-dispersive x-ray spectroscopy (EDX) elemental mapping (Fig. 1e).

Results and Discussion

We conducted direct-current (dc) current-voltage (*I-V*) and pulsed voltage measurements and evaluated how the RS behavior is influenced in devices with both SiO$_x$/VAMoS$_2$ and SiO$_x$-only configurations.

A. DC Characterization

The *I-V* characteristics of SiO$_x$/VAMoS$_2$ and SiO$_x$-only devices were recorded over 20 dc voltage sweeps from 0 V to 1.5 V and back to 0 V with a current compliance (I_{cc}) of 1 µA (Figs. 2a,b). Both devices transition from the initial high resistance state (HRS) to the low resistance state (LRS) during the forward sweep labeled "1" at a certain on-threshold voltage ($V_{t,on}$). The reverse transition from LRS to HRS occurs below a hold voltage (V_{hold}) during the backward sweep labeled "2". The voltage sweep rate was 0.3 V/s and

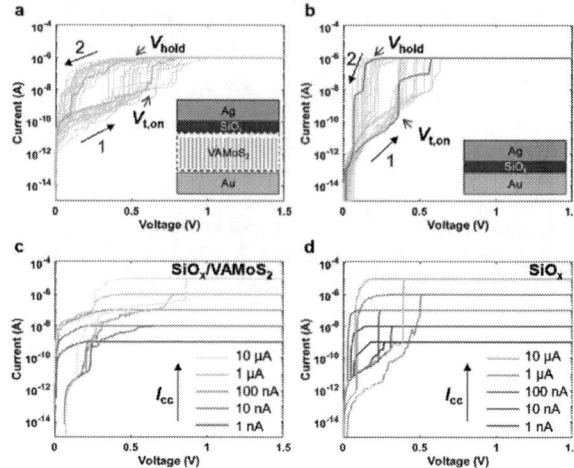

Fig. 2: Direct-current (dc) *I-V* characteristics. (a) 20 *I-V* sweeps for the SiO$_x$/VAMoS$_2$ device at $I_{cc} = 1$ µA, with the first sweep marked in orange. (b) 20 *I-V* sweeps for the SiO$_x$-only device at $I_{cc} = 1$ µA, with the first sweep marked in green. I_{cc} dependent typical *I-V* characteristics of the SiO$_x$/VAMoS$_2$ device (c) and the SiO$_x$-only device (d).

the sweep step size was 0.01 V. The SiO$_x$/VAMoS$_2$ device shows a higher average $V_{t,on} = 0.62$ V, compared to $V_{t,on} = 0.46$ V of the SiO$_x$-only device. This indicates that higher voltages are required for the Ag ions to reach the BEs through the dual material stack. Additionally, the SiO$_x$/VAMoS$_2$ device exhibited a significantly higher average $V_{hold} = 0.51$ V, compared to $V_{hold} = 0.26$ V of the SiO$_x$-only device. This suggests that the SiO$_x$/VAMoS$_2$ structure promotes the formation of thinner and weaker filaments, whereas the SiO$_x$-only structure allows the formation of thicker filaments due to unrestricted Ag ion movement.

Figs. 2c and d display typical *I-V* sweep characteristics for the SiO$_x$/VAMoS$_2$ and SiO$_x$-only devices, respectively. Both data sets show volatile switching over a wide range of I_{cc} values from 1 nA to 10 µA. Notably, the SiO$_x$/VAMoS$_2$ device exhibited a distinct stepwise switching that remained volatile for I_{cc} up to 100 µA (not shown here), making it particularly interesting for selector applications that require high on-state currents [15].

B. Pulsed Characterization

Pulsed voltage measurements were performed by applying a voltage pulse of $V_{pulse} = 4$ V for 2 µs over 30 cycles to the SiO$_x$/VAMoS$_2$ and SiO$_x$-only devices (Figs. 3a,b). The SiO$_x$/VAMoS$_2$ device showed significantly more stable and repeatable transient current responses, with faster average switching times and lower variability compared to that of the SiO$_x$-only device. The switching time (t_{on}) was defined as the time needed to reach 90% of the ON-current (I_{on}), as shown in Fig. 3c. The inset illustrates the voltage pulse waveform. The V_{pulse} was applied to transition the device to the LRS within t_{on}, after which the device returned to the HRS, and its state was read out using a 0.3 V read pulse (V_{read}) for 3 µs. Fig. 3d shows the distribution of t_{on} for both devices, with a

979-8-3315-0417-5/25 $31.00 © 2025 IEEE

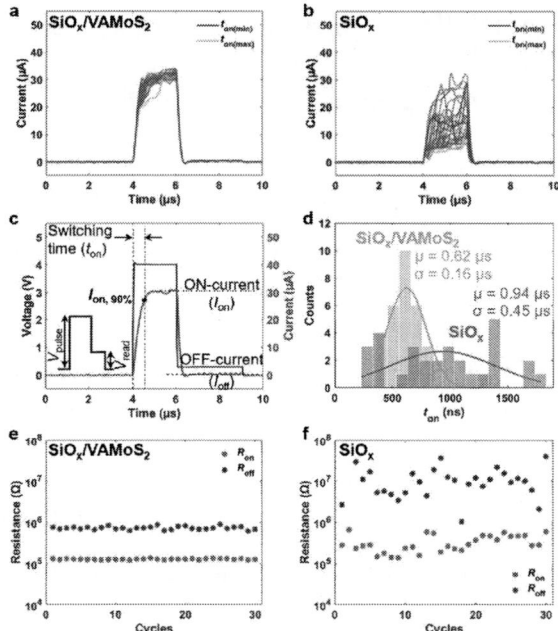

Fig. 3: Pulsed voltage characteristics. Measured current waveforms over 30 consecutive cycles under identical applied voltage pulses (4.0 V/2 µs) for the SiO_x/VAMoS$_2$ device (a) and the SiO_x-only device (b), respectively. (c) Definition of parameters for current response over time under a voltage pulse of 4.0 V/2 µs, followed by a read pulse of 0.3 V/3 µs. The inset shows the applied pulsed voltage waveform. (d) Distribution of t_{on} derived from 30 pulses for the SiO_x/VAMoS$_2$ and the SiO_x-only devices. Extracted R_{on} and R_{off} values from pulsed measurements corresponding for the SiO_x/VAMoS$_2$ device (e) and for the SiO_x-only device (f).

lower mean value of 0.62 ± 0.16 µs for SiO_x/VAMoS$_2$ compared to 0.94 ± 0.45 µs for SiO_x-only. We attribute the faster t_{on} and reduced variability to the formation of thin filaments and a self-rupture process [3] due to the restricted random generation of CFs and their limited overgrowth in the VAMoS$_2$ layer.

Finally, Figs. 3e and f display the extracted ON-state resistance (R_{on}) and OFF-state resistance (R_{off}) values, derived from the I_{on} and OFF-current (I_{off}) over 30 consecutive pulses for both devices. Details of the on-switching and off-switching characteristics are described in ref. [16]. The data shown in these figures are obtained from the measured current responses presented in Figs. 3a and b. We observed a significantly lower cycle-to-cycle variability in the SiO_x/VAMoS$_2$ device compared to the SiO_x-only device. This enhanced consistency supports that the VAMoS$_2$ layer leads to more stable volatile RS behavior and faster switching compared to devices with only the SiO_x layer.

The interlayer distance of the VAMoS$_2$ layers, i.e., the vdW gap, is approximately 6.5 Å [8], which is larger than the diameter of Ag ions (2.52 Å). Yet, it can still restrict a high flux of Ag ions by channeling their movement through the vdW gaps. Thus, Ag ions diffuse through the SiO_x layer and along the vdW gaps in the VAMoS$_2$ layer, forming thin CFs from the TE to the BE. This aligns with the literature, where the limited transport of Ag ions through atomic defects in

graphene reduces the number of ions entering the switching layer, resulting in thin and unstable filaments [4].

Conclusion

We fabricated two types of Ag-based threshold switching devices using CVD-grown SiO_x/VAMoS$_2$ and SiO_x as the switching layers. We compared the RS behavior of both device types and performed a statistical analysis of the switching dynamics to evaluate the impact of the VAMoS$_2$ layer. The SiO_x/VAMoS$_2$ device exhibited stable threshold switching, with a short minimum switching time of 356 ns for a 4 V pulsed voltage. It also showed an increased hold voltage V_{hold}. The presence of VAMoS$_2$ facilitates robust volatile switching by suppressing the generation of thick Ag conductive filaments, channeling ion movement through the vdW gaps. This also enables volatile switching at higher currents than SiO_x-only devices. Further investigations are necessary to precisely observe the morphology of the Ag-CFs and the movement of the Ag ions and to better understand the filamentary RS behavior on an atomic scale.

Acknowledgments

We acknowledge funding from the German Ministry of Education and Research (BMBF) through NEUROTEC 2 (16ME0399 and 16ME0400) and NeuroSys (03ZU1106AA) and from the European Union's Horizon 2020 research and innovation program under the project ENERGIZE (101194458).

References

[1] O. Alharbi *et al. Mater. Sci. Eng. R: Rep.* **161**, 100837, (2024).

[2] S.S. T. Nibhanupudi *et al. Nat. Commun.* **15**, 2334 (2024).

[3] S. A. Chekol *et al. Adv. Funct. Mater.* **32**, 2111242 (2022).

[4] R. D. Nikam, H. Hwang. *Adv. Funct. Mater.* **32**, 2201749 (2022).

[5] M. Naqi *et al. npj 2D Mater. Appl.* **6**, 53 (2022).

[6] W. Huh *et al. Adv. Mater.* **32**, 2002092 (2020).

[7] M. Belete *et al. Adv. Electron. Mater.* **6**, 1900892 (2020).

[8] K. Ranganathan *et al. Adv. Funct. Mater.* **30**, 2005718 (2020).

[9] R. Xu *et al. Nano Lett.* **19** (4), 2411-2417 (2019).

[10] P. Cheng *et al. Nano Lett.* **16** (1), 572-576 (2016).

[11] D. Dev *et al. IEEE Electron Device Lett.* **41** (6), 936-939 (2020).

[12] A. Krishnaprasad *et al. npj 2D Mater Appl.* **7**, 22 (2023).

[13] M. Belete *et al. ACS Appl. Nano Mater.* **1** (11), 6197-6204 (2018).

[14] H. Liu *et al. Mater. Lett.* **160**, 491-495 (2015)

[15] W. Wang *et al. IEEE Trans. on Electron Devices* **66** (9), 3795 (2019).

[16] S. Cruces *et al.* arXiv preprint arXiv:2408.09780 (2024).

Vertical Channel Gate-all-around(VCG) CMOS Transistors with MBE in-situ Doping Channel and TiN/HfO$_2$ Gate Stacks

Ran Bi[1], Haoran Zhao[1], Mingmin Shi[1], Jianhuan Wang[4], Jianjun Zhang[3], Xiaoyan Xu[1,2], Xia An[1,2], Heng Wu[1,2], Ru Huang[1,2], and Ming Li[1,2*]

[1] School of Integrated Circuits, Peking University, Beijing 100871, China

[2] Beijing Advanced Innovation Center for Integrated Circuits, Email: liming.ime@pku.edu.cn

[3] Beijing National Laboratory for Condensed Matter Physics and Institute of Physics, Chinese Academy of Sciences

[4] Beijing Academy of Quantum Information Sciences

Abstract

A vertical channel gate-all-around (VCG) device process was proposed on the CMOS platform. As the key technology for integrated process, the active area stacks by MBE and dopants distribution in the vertical direction was designed. Combined with TiN/HfO$_2$ gate stacks, both NFET and PFET were finally fabricated and obtained symmetrical Id-Vg curve. The electrical asymmetry was then analyzed and the optimization scheme was proposed for advanced logic CMOS application.

Keywords: Vertical channel, Gate-all-around, Technology and CMOS.

Introduction

At advanced technology nodes beyond 3nm, compared to lateral channel devices, vertical channel devices decouple the footprint area from the gate length, significantly mitigating short-channel effects and pressure on lithography technique at the same node[1-2]. The gate-all-around (GAA) structure further enhances gate control, meeting the demands for high-performance and low-power applications[3-4]. However, the vertical channel in VCG devices makes their process fundamentally different from traditional lateral channel devices[5-6]. The key technology challenges of the VCG process are addressed here: (1) vertical active area doping, (2) spacer formation transitioning from vertical to parallel to the wafer surface, and (3) formation and alignment of the vertical gate with the channel.

In this work, a top-down process flow was developed to create a VCG device integration scheme compatible with CMOS platform. Key process modules were sequentially developed and addressed. Ultimately, both n-type and p-type vertical channel gate-all-around devices were successfully fabricated and characterized physically and electrically.

Device Fabrication

The process flow of VCG is proposed and demonstrated in the Fig. 1. The active area (AA) stacks are grown on (100) Si wafer by molecular beam epitaxy (MBE), and the vertical nanosheet is then patterned. After the bottom spacer formation and deposition of HKMG, the gate metal is recessed to achieve alignment with the channel. Subsequently, the top source/drain, gate, and bottom source/drain contact are formed.

In the process flow, the key technologies including doping profile of AA stacks, bottom spacer formation and the recess of gate aligning to channel are developed in this work and provide support for the integrated process.

A. AA Stack Epitaxy and Patterning

As shown in Fig. 2, the doped channel-source-drain AA stacks were grown before patterning by MBE in-situ doping technology. Considering the vertical structure of VCG devices, 40nm bottom S/D layer, 30nm channel layer and 50nm top S/D layer were deposited successively on the wafer surface, with an inserted 10nm pocket doping layers between S/D and channel.

Secondary ion mass spectroscopy (SIMS) was applied to inspect the distribution of dopants in the AA stacks. In order to suppress the surface segregation effect during MBE growth, a series of measures were implemented to optimize the doping profiles in AA stacks of NFETs and PFETs. For NFETs structure, the linearly doping profile from 1E20cm^{-3} to 5E18cm^{-3} was applied instead of the constant doping in S/D layers, which can reduce the concentration gradient from S/D to channel layer to suppress the as-deposition dopant segregation. At the same time, in order to decrease the phosphorus diffusion into channel, the pocket doping of 3E18 cm^{-3} boron was introduced between S/D and channel as shown in Fig. 3(a). In contrast, for PFETs structure, an intrinsic extension layer between S/D and channel is more effective to suppress the diffusion of boron from S/D to channel, as shown in Fig. 3(b). The optimized design of AA stack dopants profile is listed in Table 1.

The epitaxial AA stack was then patterned by e-beam lithography and inductively couple plasma (ICP) dry etching, which stops at the surface of heavily doped bottom S/D layer. As shown in Fig. 4, vertical nanosheet was formed with about 22 nm width and 130 nm height.

B. Bottom Spacer

Due to the vertical structure of VCG devices, the spacer of gate from bottom S/D is formed in parallel to the wafer surface. Thus, the recess process of bottom spacer is required

979-8-3315-0417-5/25 $31.00 © 2025 IEEE

to be quite precise.

The process flow of bottom spacer formation is demonstrated in Fig. 5(a). SiO_2 was deposited by LPCVD for better conformality, and followed by CMP process. To avoid damage to the vertical nanosheet, SiO_2 was then selectively recessed by $HF:H_2O$ mixtures with controllable etching rate. Fig. 5(b) shows that a uniform SiO_2 bottom spacer was successfully formed to provide reliable isolation between gate and bottom S/D.

C. HKMG Deposition and Recess

Distinguished from lateral channel devices, the gate length is defined by the height of gate film for VCG devices. Therefore, the recess depth of the gate material is a critical parameter that determines the gate length of devices.

Fig. 6(a) displays the process flow of HKMG deposition and recess. Firstly, the gate dielectric HfO_2 and gate metal TiN film were grown by ALD, followed by the deposition of SiO_2 as protective layer. The protective layer was then polished by CMP and recessed by dry etching which stopped at TiN surface. Base on the exposing top interface of metal, the gate metal was etched back to realize the alignment to channel by $NH_4OH:H_2O_2:H_2O$ mixtures.

The TiN etching depth at different time is displayed in Fig. 6(b), and the mixtures etching rate of TiN can be calculated to be about 11.6nm/min with good linearity. Compared with dry etching process, wet etching by $NH_4OH:H_2O_2:H_2O$ mixtures provides a higher selectivity to dielectric materials, and the slower etching rate makes the recess depth controllable, which is confirmed by the TEM images in the inset of Fig.6(b).

Results and Discussion

Based on the key process module development, VCG CMOS devices were successfully fabricated and the final structure and material components were characterized by TEM and EDS mapping in Fig. 7.

The DC electrical characteristics of the CMOS devices were analyzed and the Id-Vg curves of the N-type and P-type VCG devices are shown in Fig. 8 when the top contact acts as drain in devices. NFET and PEFT devices show symmetric Id-Vg characteristics about Vg=0V. The electrical parameters of the device were extracted and listed in Table 2. Fundamentally, both devices suffer from parasitic resistance issues due to lack of epitaxy of S/D, which are electrically contacted directly with metal. As expected, the short channel effect of PFET is better than that of NFET possibly due to the inserted pocket structure. On the other hand, the gradually doped S/D in NFET doesn't cause more parasitic resistance but mitigate the issue somehow. Such results provide a guidance for the further optimization of VCG doping profile.

Exchanging the drain from top contact to bottom contact, the Id-Vg curves are measured again and shown in Fig. 9. It's found that the drive currents of both NFET and PFET are significantly decreased but the off-state leakage currents are also reduced by orders. It implies the top and bottom part of vertical nanosheet have very different resistance, which may be the most critical challenge in the future for VCG.

However, from the perspective of integrated process, the asymmetry can be optimized by reducing etching damage, heavily doped top S/D epitaxial, and better contact surface. On the other hand, from the perspective of DTCO, the asymmetric S/D doping and the alignment between gate and channel can be also utilized to balance the performance of VCG devices in SoC application such as the N/P ratio of SRAM cell and special loading circuits.

Conclusion

This work explored a top-down process integration scheme for vertical channel gate-all-around (VCG) CMOS devices based on MBE constructed substrate. The key process modules were developed and the doping profile is specially designed and implemented by low-temperature in-situ doping MBE. Both n-type and p-type VCG devices were successfully fabricated with good Vth symmetry. Additionally, the asymmetry of VCG devices was characterized and analyzed from the viewpoint of process and device physics.

Acknowledgments

This work was partly supported by the National Key Research and Development Program of China (No. 2023YFB4402200), NSFC project (61927901) and Beijing Frontier Innovation Project (No. QYJS-2021-0500-B).

References

[1] Veloso A , Altamirano-Sanchez E , Brus S , et al. "Vertical Nanowire FET Integration and Device Aspects," ECS Transactions, vol. 72. pp. 31-42 (2016).

[2] Capodieci L, Cain J P, Huynh-Bao T, et al. "Toward the 5nm technology: layout optimization and performance benchmark for logic/SRAMs using lateral and vertical GAA FETs," Spie Advanced Lithography, pp. 978102 (2016).

[3] Ryckaert J, Na MH , et al. "Enabling Sub-5nm CMOS Technology Scaling Thinner and Taller!" 2019 IEEE International Electron Devices Meeting (IEDM) (2019).

[4] Ryckaert J , et al. "3D integration for density and functionality," VLSI, (2019).

[5] Veloso A , Hikavyy A , Loo R ,et al. "Vertical Nanowire and Nanosheet FETs: Device Features, Novel Schemes for Improved Process Control and Enhanced Mobility, Potential for Faster & More Energy Efficient Circuits," 2019 IEEE International Electron Devices Meeting (IEDM) (2019).

[6] Veloso, A. , et al. "Insights and Opportunities for Junctionless Gate-All-Around Lateral and Vertical Nanowire FETs." 2017 International Conference on Solid State Devices and Materials (2017).

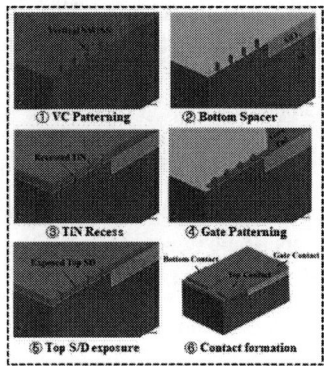

① VC Patterning ② Bottom Spacer
③ TiN Recess ④ Gate Patterning
⑤ Top S/D exposure ⑥ Contact formation

Fig. 1: Process flow and key steps

Table 1: Summary of doping profile in stacks.

Layer	NFET		PFET	
	Dopant	Concentration (cm-3)	Dopant	Concentration (cm-3)
Top SD	P	1E20~1E19	B	1E20
ext_top	B	3E18	-	0
Channel	B	1E18	P	1E18
ext_bottom	B	3E18	-	0
Bottom SD	P	1E19~1E20	B	1E20
Buffer	B	1E17	P	1E17
Intrinsic				

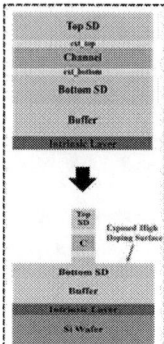

Fig. 2: AA stacks

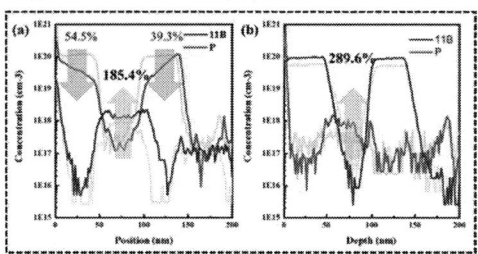

Fig. 3: Doping profile in the MBE AA stacks by SIMS (a) NFET and (b) PFET

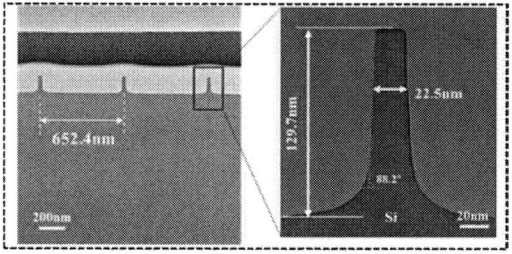

Fig. 4: TEM of vertical nanosheets after etching.

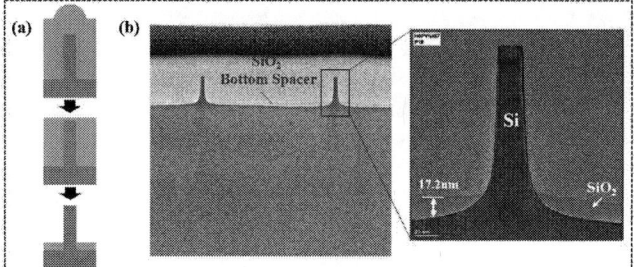

Fig. 5: (a) Process of bottom spacer and (b) TEM of SiO₂ recess cross-section

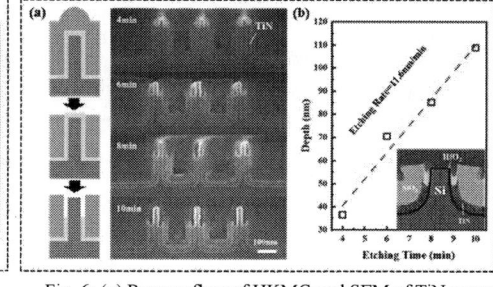

Fig. 6: (a) Process flow of HKMG and SEM of TiN recess and(b)the recess rate curve

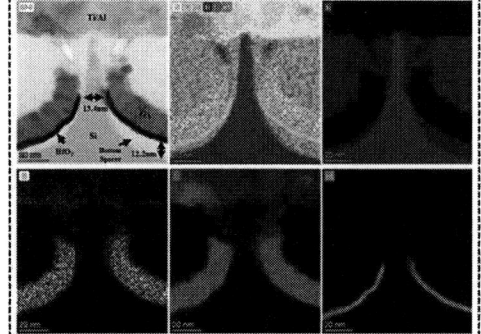

Fig. 7: TEM and EDS mapping of VCG devices

Table 2: Summary of doping profile in stacks

parameter	NFET	PFET
V_th	0.31V	-0.25V
I_on	9.64×10⁻⁶A/μm	2.95×10⁻⁶A/μm
I_off	7.43×10⁻⁹A/μm	8.90×10⁻¹⁰A/μm
ss	149.0mV/dec.	73.1 mV/dec.
Weff	0.14μm	0.1μm

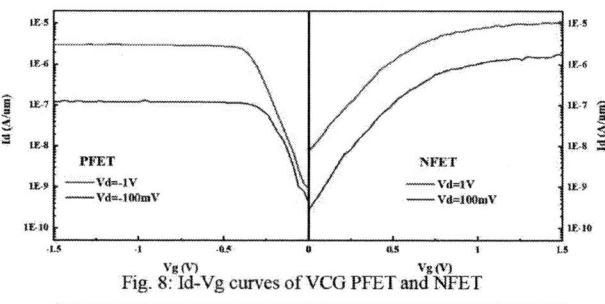

Fig. 8: Id-Vg curves of VCG PFET and NFET

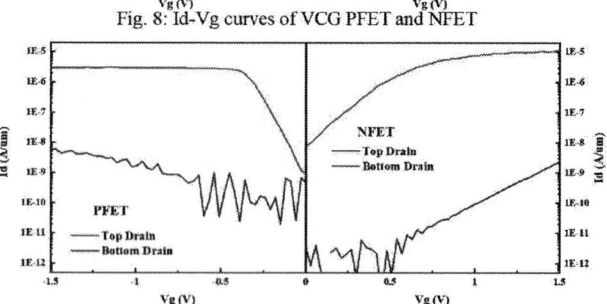

Fig. 9: Id-Vg curves of VCG PFET and NFET when changing the drain from top contact to the bottom

979-8-3315-0417-5/25 $31.00 © 2025 IEEE

Dependence of Dislocation Density Distribution on Radial Temperature Gradient in 300mm Si Wafer during IGBT High Thermal Budget Process

Jiuyang Yuan, Bozhou Cai, Yoshiji Miyamura, Wataru Saito, and Shin-ichi Nishizawa

Kyushu University, Japan

Abstract

A new experiment method was developed to investigate the relationship between dislocation propagation and temperature distribution in 300mm Si wafer. Dislocation propagation may occur due to thermal stress caused by temperature non-uniformity in Si wafer during high thermal budget process, potentially leading to wafer crystal quality degradation. In this study, we clarified that the higher radial temperature gradient can lead to more dislocation propagation at elevated temperature.

Keywords: Si wafer, RTA furnace, High thermal budget process, Thermal stress and Dislocation density

Introduction

Currently, there is a demand for high-performance and low-cost power devices. The key technologies are to enhance wafer quality and to enlarge wafer diameter simultaneously. However, in 300mm Si-IGBT fabrication, the deterioration of wafer quality like dislocation propagation may be more pronounced than small diameter wafer due to large temperature non-uniformity during high thermal budget process. Therefore, it is important to understand how the temperature non-uniformity specifically affect the dislocation propagation in Si wafer.

In previous research, we modeled a vertical furnace for 200mm Si wafer [1]. We discovered that dislocation propagation is suppressed in low-temperature process, as the wafer experience both lower temperature and stress compared to high-temperature process. However, the mechanism of dislocation propagation during high thermal budget process has not yet been demonstrated, as it is extremely difficult to experimentally measure the distribution of temperature, stress and dislocation density.

In this paper, we use the rapid thermal annealing (RTA) furnace to develop an experiment method which can create temperature distribution in 300mm wafer during high thermal budget process. Additionally, we model the RTA furnace and analyze temperature, stress and dislocation density in wafer. Finally, we demonstrate how the radial temperature gradient and actual temperature affect the dislocation propagation in wafer by comparing the results of experiment and simulation.

Experiment Method and Simulation Model

A. Experiment method

We used the rapid thermal annealing furnace (VPO-1000-300) [2] to conduct N_2 anneal process for 300mm Si wafer. The wafer is n-type, which is used for power device. When the ramping-up process, the wafer is heated by lamps, and when the cooling-down process, the wafer is cooled by gas. Fig. 1 shows the interior layout of RTA furnace. Since the lamps are designed to heat wafer to uniform temperature, we place graphite ring plate between wafer and lamp to prevent heat radiation. The inner diameter and outer diameter of graphite ring plate are 200mm and 330mm, respectively. The thickness of graphite is 2mm, and the distance between wafer and graphite is 10mm.

The temperature of wafer center is raised from room temperature to 900°C by lamps at a target heating rate of 20°C/s, kept at 900°C for 3 minutes. The lamps power is then reduced to 0, and the wafer temperature is cooled by gas. Since the significant temperature distribution only happens when the wafer temperature is controlled by lamps, this paper focuses exclusively on the ramping-up process.

B. RTA simulation model

We model the RTA furnace by using COMSOL Multiphysics [3] to calculate temperature and stress in Si wafer during high thermal budget process as shown in Fig. 1. The dimensions of the chamber, graphite, and wafer are set to the same as in the experiment. We control the power of lamps which can make wafer center to get same temperature history with experiment.

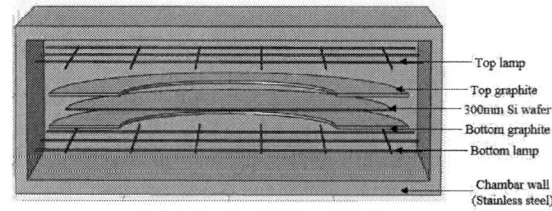

Fig. 1: Model of 300mm RTA furnace

C. Dislocation density analysis model

Based on the result of temperature and stress in wafer calculated by RTA furnace model, a three-dimensional numerical analysis of dislocation density is performed using the Haasen-Alexander-Sumino (HAS) model [4].

A silicon single crystal has a total of 12 slip systems. If the plastic strain of a specific slip system α is $\varepsilon^{c(\alpha)}$, then it can be expressed by Eq. (1).

$$\frac{d\varepsilon^{c(\alpha)}}{dt} = N_m^{(\alpha)} v^{(\alpha)} b \tag{1}$$

979-8-3315-0417-5/25 $31.00 © 2025 IEEE   503

Where N_m is the mobile dislocation density, b is the Burgers vector, and the dislocation velocity is given by Eq. (2).

$$v^{(\alpha)} = v_0 \left(\frac{\tau_{eff}^{(\alpha)}}{\tau_0}\right)^m exp\left(-\frac{U}{k_b T}\right) \qquad (2)$$

Here, τ_{eff} is the effective stress, $v_0=5000ms^{-1}$, $\tau_0=1MPa$, $m=1$, $U=2.2ev$ are used for crystal silicon. The increasing rate of mobile dislocation density in each slip system α is expressed by Eq. (3).

$$\frac{dN_m^{(\alpha)}}{dt} = KN_m^{(\alpha)} v^{(\alpha)} \tau_{eff}^{(\alpha)} - 2r_c N_m^{(\alpha)} N_m^{(\alpha)} v^{(\alpha)} \qquad (3)$$

Here, K is the dislocation multiplication coefficient, r_c is the interaction distance between dislocations. When the initial value of dislocation density is very small, the second term on the right-hand side of Eq. (3) can be ignored. Hence, a variable "f", representing the dislocation propagation rate, is introduced, and its relationship with the increasing rate of mobile dislocation density can be expressed by Eq. (4). The "f" depends on temperature and stress in the wafer, as shown in Eq. (5) [5].

$$\frac{dN_m^{(\alpha)}}{dt} = f \cdot N_m^{(\alpha)} \qquad (4)$$

$$f = Kv_0 \left(\frac{\tau_{eff}^{(\alpha)} \tau_{eff}^{(\alpha)}}{\tau_0}\right)^m exp\left(-\frac{U}{k_b T}\right) \qquad (5)$$

Fig. 2 shows exponential function curve between temperature and stress for "f" values of 0.001, 0.01, 0.1, 0.2, 0.5. Smaller "f" value indicates the lower propagation rate, and larger "f" value corresponds to higher propagation rate. It can be observed that in high-temperature regions, even a small stress can induce dislocation propagation.

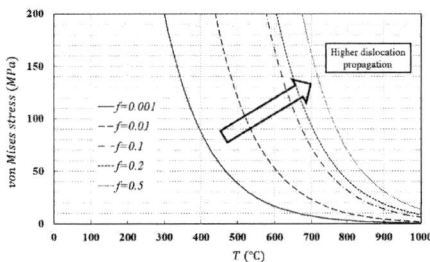

Fig. 2: Relation between temperature and stress in wafer at the same dislocation propagation rate "f"

Result and Discussion

Fig. 3 shows the wafer center target temperature and the wafer temperature history at radial positions of 0mm (wafer center), 40mm, 60mm, 80mm, 120mm, and 140mm from 0s (the beginning of ramping-up process) to 200s. During the ramping-up process, a temperature non-uniformity emerged, due to the graphite ring plate preventing heat radiation between the lamps and wafer. Once the target temperature reached 900°C, the temperature non-uniformity decreased, eventually reaching thermal equilibrium. The temperature differences relative to the wafer center, as shown in Fig. 4, indicate a trend of increasing with radius, with a maximum of around 430°C near the periphery.

Fig. 5 shows the temperature distribution in the wafer at 10s, 40s, 45s and 95s during the process. The temperature distribution is lower in the radius ranges of 0mm to 60mm and 120mm to 140mm, whereas it is higher in the region between 60mm and 120mm, which is closer to the inner diameter of the graphite ring plates.

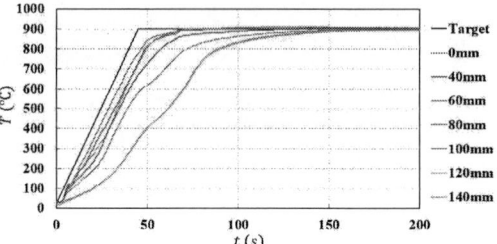

Fig. 3: Target temperature of wafer center and temperature histories at various wafer locations

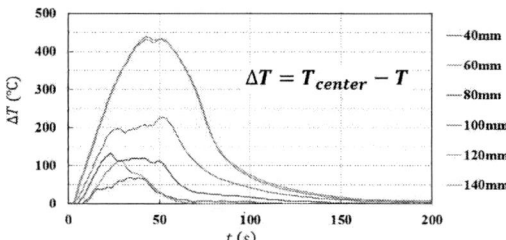

Fig. 4: Temperature differences at various wafer locations

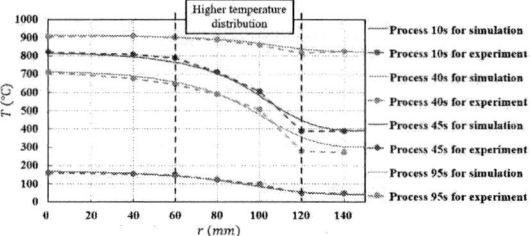

Fig. 5: Temperature distribution in wafer at various process times

Thermal stress was generated due to temperature distribution within the wafer. Fig. 6 shows the von Mises stress distribution in wafer, as calculated by the RTA model. Here, the radial temperature gradient (dT/dr) is introduced, which represents the rate of change in temperature T within a small radial range r. Fig. 7 shows the temperature gradient along the radial, calculated based on the simulation results in Fig. 5. Although thermal stress is a complex factor and isn't solely determined by the radial temperature gradient, the stress distribution follows a similar trend to the radial temperature gradient distribution. Regions with a higher radial

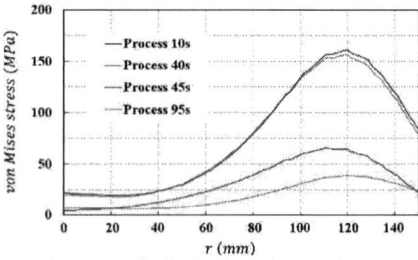

Fig. 6: von Mises stress distribution in wafer at various process times

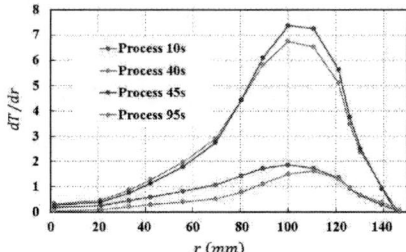

Fig. 7: Radial temperature gradient in wafer at various process times

temperature gradient also exhibit higher stress, indicating that reducing the radial temperature gradient during high thermal process may suppress the stress in wafer.

Once the process was completed, slip dislocations were predominantly observed in the radial range of approximately 60mm to 100mm. The result observed by X-ray topography is shown in Fig. 8. The inevitable diffraction lines in Fig. 8 are attributed to the optical setup, not to any defects present in the wafer.

The temperature and stress results from the RTA model shown in Fig. 5 and Fig. 6 were used to calculate the dislocation density distribution using the dislocation density analysis model. The result, as shown in Fig. 9, reveals that the larger dislocation propagation occurred in the radial range of 70mm to 100mm, which closely aligns with the slip dislocation regions in Fig. 8.

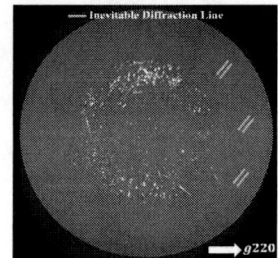

Fig. 8: Slip dislocations in the (100) plane of Si wafer after high thermal process by experiment

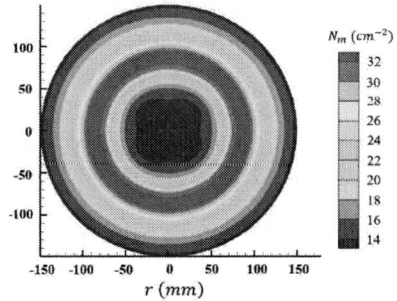

Fig. 9: Dislocation density distribution in the (100) plane of Si wafer after high thermal process by simulation

Fig. 10 shows the curves of temperature and von Mises stress at wafer radial positions from 40mm to 140mm during the ramping-up process and their relationship with the dislocation propagation rate "f". While the maximum stress was observed at 120mm, this occurred at a lower temperature (about 500°C to 600°C), where the "f" is minimal, suggesting that dislocation propagation is difficult to occur. However, at

higher temperature (about 800°C to 900°C), positions at 100mm, 80mm and 60mm displayed higher stress and "f", which can lead to dislocation propagation. This indicates that dislocation propagation is dependent not only on the stress but also on the actual temperature at which the stress occurs. Consequently, to suppress dislocation propagation during high thermal budget processes, it is critical to reduce the stress at higher temperatures, which can be achieved by maintaining a lower radial temperature gradient in the wafer.

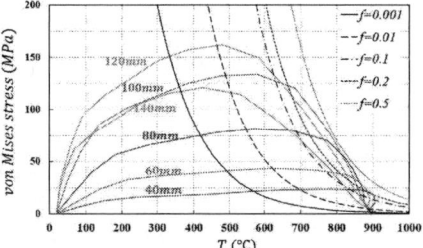

Fig. 10: von Mises stresses at different temperatures from 40mm to 140mm in the wafer

Conclusion

An experiment using the RTA furnace and the graphite ring plates was conducted to generate a temperature distribution in the 300mm Si wafer during the high thermal budget process. The dislocation propagation was observed after the process was completed. Additionally, by using the RTA model and the dislocation density analysis model, we calculated the temperature, von Mises stress, and dislocation density during process. The comparison of experimental and simulation results indicates that at higher temperature, regions with higher stress or radial temperature gradient experience more dislocation propagation. Therefore, it suggests that maintaining a small radial temperature gradient at high temperatures can effectively suppress dislocation propagation during the high thermal budget process.

Acknowledgments

This work was supported by JST SPRING, Grant Number JPMJSP2136.

This work was partially supported by NEDO (JPNP21009).

References

[1] J. Yuan et al., "The Study of Dislocation Propagation in Si Wafer during IGBT High Thermal Budget Process", IEEE Electron Devices Technology & Manufacturing Conference (EDTM), pp. 1-3, 2023

[2] https://www.unitemp.jp/rapid-thermal-process-oven/

[3] https://www.comsol.com/comsol-multiphysics

[4] B. Gao et al, "Applicability of the three-dimensional Alexander-Haasen mode for the analysis of dislocation distributions in single-crystal silicon", J. Cryst. Growth, Volume 411, pp. 49-55, 2015

[5] R. Sato et al, "Dislocation Propagation in Si 300 mm Wafer during High Thermal Budget Process and Its Optimization", International Symposium on Power Semiconductor Devices and ICs (ISPSD), pp. 494-497, 2020

Yield and Reliability Optimization of Analog RRAM for In-Memory Computing on a 28nm CMOS Platform

Siyao Yang, Bin Gao*

School of Integrated Circuits, Beijing Advanced Innovation Center for Integrated Circuits, Tsinghua University, Beijing 100084, China

*E-mail: gaob1@tsinghua.edu.cn

Abstract

RRAM is considered as one of the promising candidates for computing-in-memory (CIM) application. In pursuit of higher integration density, RRAM is gradually evolving to be integrated at increasingly advanced technology nodes. At the same time, RRAM also faces many new challenges. In this work, we improved the device yield by optimizing the material stacks and the device reliability by optimizing the operation schemes.

Keywords: RRAM, computing-in-memory (CIM)

Introduction

Resistive Random Access Memory (RRAM), characterized by its non-volatility, low energy consumption, simple structure, and compatibility with CMOS process, has garnered substantial interest in both academic and industrial fields[1]-[5]. Analog RRAM's ability to fine-tune conductance and program to various resistive states positions it as a potential device for CIM application. In analog RRAM arrays, each RRAM device's conductance acts as an analog weight within neural networks, as depicted in **Fig. 1**. By Ohm's law, the current through each RRAM device is the product of its conductance and the applied bias voltage. According to Kirchhoff's current law, the total column current is the cumulative sum of individual device currents. Consequently, complex Matrix-Vector Multiplication (MVM) can be executed through straightforward read operations.

In recent years, RRAM has been progressively evolving to be integrated at increasingly advanced technology nodes , increasing integration density and confronting a variety of new challenges[6]. At the device level, device reliability is an important issue. At the circuit level, setting current compliance during the write process is problematic, as it can lead to significant mismatch and voltage division[7]. In this research, we have improved yield through process optimization and enhanced device reliability via optimized operation schemes on a commercial 28nm Si CMOS platform. Our devices are capable of achieving 4 bits/cell multi-level programming, along

with superior retention and endurance characteristics.

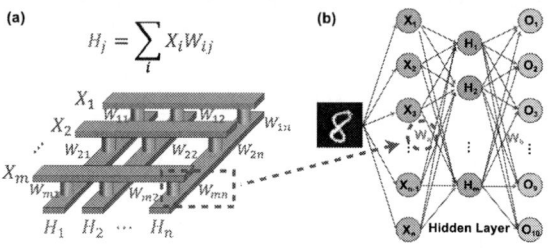

Fig.1 (a) The schematic diagram of an analog RRAM crossbar array, where MVM can be implemented. (b) The schematic diagram of a two-layer neural network, in which analog RRAM acts as the synaptic weight.

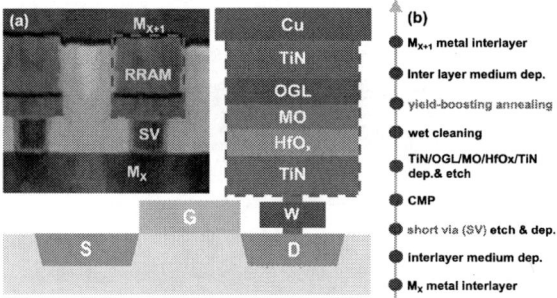

Fig.2 (a) The schematic diagram of device structure and cross-sectional TEM image. (b) The key steps of the process flow.

Device Technology

The RRAM crossbar cell with a size below 200 nm was fabricated by standard 28 nm CMOS foundry. The RRAM structure is TiN/OGL/MO/HfOx/TiN, and is fabricated between M_X metal layer and the M_{X+1} metal layer on the drain of the transistor (1T-1R structure). The transmission electron microscope (TEM) image of RRAM cell and key steps of the process-flow are shown in **Fig.2**. The metal oxidation layers MO/HfOx were fabricated using the atomic layer deposition (ALD) technique. These layers were placed between the TiN bottom electrode (BE) and the TiN/OGL top electrode (TE), which were achieved through physical vapor deposition (PVD). The role of short via (SV) is to liberate RRAM from the physical layout constraints of the Mx metal layer, enhancing the integration density.

RRAM devices that exhibit low resistance before forming operation are designated as initial low resistance (ILR) cells. These ILR cells are difficult to reset to high resistance, thereby degrading yield. We address the problem of ILR through two approaches. The first approach is inserting an interface modulation layer (IML) between the OGL layer and the RSL layer. The IML acts as a barrier, preventing the oxygen scavenging effect of the OGL from directly impacting the RSL layer, thereby reducing the probability of ILR cell, as shown in **Fig. 3a**. The second approach introduces an annealing step after the deposition and etching of the stacked materials TiN/OGL/IML/HfOx/TiN. This annealing process ensures that oxygen vacancies formed on the sidewalls of the RRAM during etching are sufficiently oxidized, further reducing the probability of ILR, as shown in **Fig. 3b**. And yield is improved to >99.99% through IML and YBA.

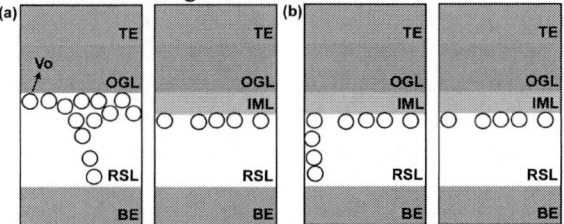

Fig.3 (a) The illustration of RRAM device without and with Interface Modulation Layer (IML). (b) The illustration of RRAM device without and with Yield-Boosting Annealing (YBA). ILR cells are eliminated and yield is improved to >99.99% through IML and YBA.

Operation Schemes

As the size of RRAM scales down, device reliability is increasingly becoming an important challenge, particularly the retention characteristic. After programming to the target states, some RRAM cells exhibit conductance drift in a short time, leading to a deviation between actual conductance and target conductance. The conductance drift shows strong randomness between cycle-to-cycle and cell-to-cell, which seriously affects the calculation accuracy of the neural network. For RRAM cells programmed to the same resistive state, the conductance distribution broadens and the center of the distribution also drift, which initially converges near the target resistive state. The phenomenon of conductance drift can be mitigated by optimizing the operation scheme. However, the optimization methods that have been proposed, including delay-verify and REWS[8], can only improve the broadening problem of the conductance distribution, but cannot solve the mean shift, as shown in **Fig. 4a**.

According to the mechanism analysis on RRAM-

array test results, the mean shift problem is related to the programming mode. In devices primarily programmed using set pulses, the conductive filaments within the RSL are relatively fragile and prone to resistance drift. Conversely, in devices primarily programmed using reset pulses, the conductive filaments are stronger and less susceptible to resistance drift, as shown in **Fig. 4b**. To ensure that the devices are predominantly programmed through reset operations, we have adopted two optimized operation schemes.

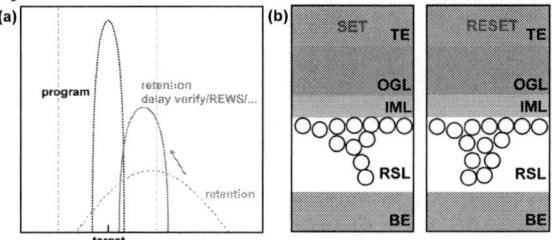

Fig.4 (a) The schematic diagram of RRAM conductance distribution of program and retention result. The current optimization operation scheme proposed can only improve the convergence degree of the distribution, but cannot improve the mean shift problem. (b) Schematic diagram of RRAM programmed mainly with set pulse and reset pulse, RRAM programmed by reset pulse has stronger conductive filaments and is less prone to conductance drift.

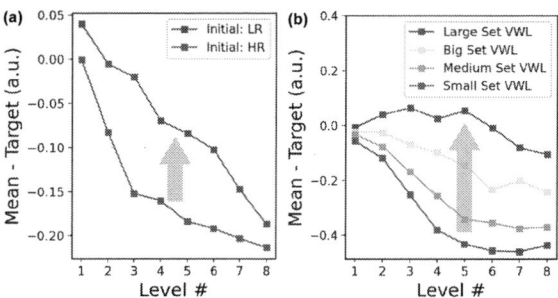

Fig.5 Short-term retention results of RRAM at 3bit. (a) The mean shift phenomenon is significantly improved after adjusting the initial resistance state to LR. (b) Increasing the initial VWL of Set significantly improves the mean-shift phenomenon.

For the electrical measurement, we utilized the Speedcury ST2500. In the programming test, we employed incremental step pulse programming (ISPP) as a two-sided verification method. In the ISPP programming process, first step VBL/VSL from the minimum value to the maximum value, and then step VWL from the minimum value to the maximum value, until programming to the target target range. If the conductance is above the target upper limit, the reset operation is performed starting from the minimum reset VSL and VWL voltages. If the conductance is below the lower target limit, the set operation is

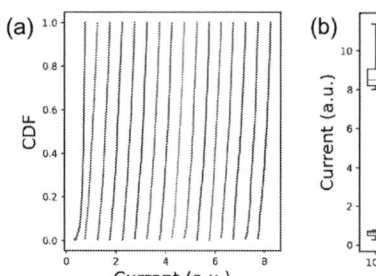 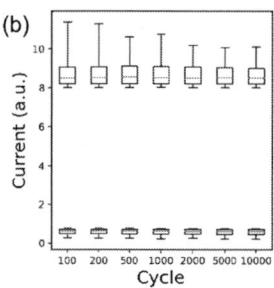 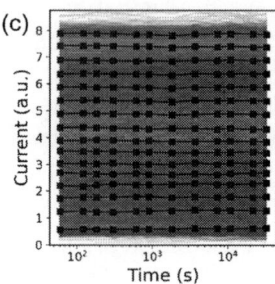

Fig.6 RRAM characteristics. (a) Programming: 4k cells were programmed to 16 different conductance states respectively. (b) Endurance: 10^4 cycle can be achieved. (c) Retention: the mean of the conductance distribution remains stable value.

performed, starting from the minimum set VBL and VWL voltages.

The first method is setting the initial state of the device to a low-resistance state before programming. This approach has significantly mitigated the problem of mean resistance shift, as shown in **Fig. 5a**.

The second method is to raise the initial set voltage. When the RRAM is below the lower limit of the target range through the reset pulse, the raised initial set voltage is used to make the RRAM reach the conductance state above the upper limit of the target range after one set pulse, and then reset operation is performed. Thus, the device can reach the target conductance state through the reset pulse. This is done by elevating the VWL of the set. This method can effectively reduce the problem of mean shift of the conductance distribution, as shown in **Fig. 5b**. On the other hand, VWL must not be increased too much, because after the transistor current limiting condition is too wide, the device will be operated to a too low resistance state, which will affect it reset to a low conductance state.

In order to evaluate the effect of these two operation schemes, we employed an identical convolutional neural network (CNN), specifically Conv3-8-Conv3-12-FC, to recognize handwritten digital images from a modified National Institute of Standards and Technology (MNIST) database. The network weight values will be affected by the conductance drift under different programming conditions. The conductance in the RRAM array was used to implement the network weights, which are susceptible to non-ideal factors that can decrease network accuracy. Simulation results reveal that the network accuracy is improved by 5% under the first method. The second method improves the network accuracy by 7%.

Device Characteristics

In the programming test, 4k cells were programmed to 16 different conductance states respectively, and the programming success rate is 100%, as shown in **Fig. 6a**.

In the endurance test, each set/reset process was verified to ensure that each cycle can achieve the highest conductance state or the lowest conductance state. 1k cells were tested, and each cell can reach 10^4 cycles, showing excellent yield andendurance, as shown in **Fig. 6b**. Due to time constraints, the endurance test only reaches 10^4 cycles, and the actual endurance of our RRAM cells may be higher.

In the retention test, 4k cells were programmed to 16 different conductance states respectively, and the conductance states were read after 24 hours at room temperature, as shown in **Fig. 6c**. With the above two optimized operation schemes, the mean of the conductance distribution remains stable.

Conclusion

In this study, we have focused on optimizing the yield and device characteristics of RRAM at a 28nm CMOS platform. We achieved a significant yield improvement through two strategies: IML and YBA. Furthermore, by refining the programming method, we effectively mitigated the issue of mean shift in conductance distribution during retention. Our devices are capable of being programmed to achieve 4 bits/cell multi-level, demonstrating both excellent retention and endurance.

Acknowledgments

This work was supported in part by the National Natural Science Foundation of China (92064001), STIC Project QYJS-2023-2701-B, and Xiamen Industrial Technology Research Institute.

References

[1] P. Yao et al., Nature, 2020. [2] W. Zhang et al., Science, 2023. [3] X. Li et al., IEDM, 2022. [4] Y. Du et al., Adv.Mater. 2024. [5] Y. Zhang et al., IEDM, 2023. [6] J. Yang et al., VLSI, 2020. [7] J. Zhang et al, IEDM, 2024. [8] Z. Jiang et al., IEDM, 2023.

979-8-3315-0417-5/25 $31.00 © 2025 IEEE

Development of an on-Chip Millimeter-Wave Antenna

Ziqi Mei[1†], Chao Liang[1†], Enze Zhou[1], Yan Zhang[2], Ji Li[3], Rongbo Xie[1], Bingbai Li[1], Xiayu Wang[1], Chi Zhang[1], Rui You[4], and Xiaoguang Zhao[1,5,6*]

[†]Equal Distribution, [1]Department of Precision Instrument, Tsinghua University, China
[2] School of Information Science and Engineering, Southeast University, China
[3]Key Laboratory of MEMS of the Ministry of Education, Southeast University, China
[4]School of Instrument Science and Opto-Electronics Engineering, Beijing Information Science and Technology University, China
[5]State Key Laboratory of Precision Measurement Technology and Instruments, China
[6]Beijing Advanced Innovation Center for Integrated Circuits, China
*Email: zhaoxg@mail.tsinghua.edu.cn

Abstract

This paper reports an on-chip antenna (OCA) operating in millimeter-wave (MM-Wave) band based on high-resistivity silicon substrate, with a small size of $2.0 \times 2.0 \times 0.3$ mm³. The proposed OCA offered an impendence bandwidth from 78.76 GHz to 86.02 GHz for S_{11} less than -10 dB. The OCA designed was fabricated using complementary metal oxide semiconductor (CMOS) process. Utilizing gold wire bonding, the OCA was fed through a substrate integrated waveguide-to-coplanar waveguide transition interfacing with a rectangular WR10 waveguide, and the measurement was performed in a microwave anechoic chamber. The OCA showed a maximum simulated gain of 3.204 dBi at 79 GHz and a maximum simulated radiation efficiency of 74.61% at 89 GHz. The demonstration of the OCA potentiates short range wireless communication like inter-chip wireless interconnections.

Keywords: OCA, MM-Wave, CMOS

Introduction

The semiconductor industry has consistently achieved remarkable advancements in the density of integrated circuits. During recent years, there has been a notable surge in the interest of system-on-chip (SoC), due to its low power, low cost and miniaturization [1]. The progress in SoC has made it possible for researchers to design on-chip antenna (OCA). Compared with traditional off-chip antennas, OCAs demonstrate numerous advantages. Firstly, OCA based systems consume less power in terms of dielectric losses and reflections, decreasing the requirements of the system power [2]. Secondly, the impedance matching network between the OCA and the RF front-end section is not necessary, so that the size of the overall system can be reduced [3]. Thirdly, they provide a co-design chance of the antenna and other transceiver components, thus offering extra flexibility and a shorter design cycle to designers [4].

The millimeter wave (MM-Wave) frequency spectrum ranging from 30 GHz to 300 GHz possesses a huge bandwidth, so MM-Wave technology has been considered as a critical part of the fifth generation (5G) wireless networks and Internet of Thing (IoT) [5]. There has been a rising trend in the use of MM-

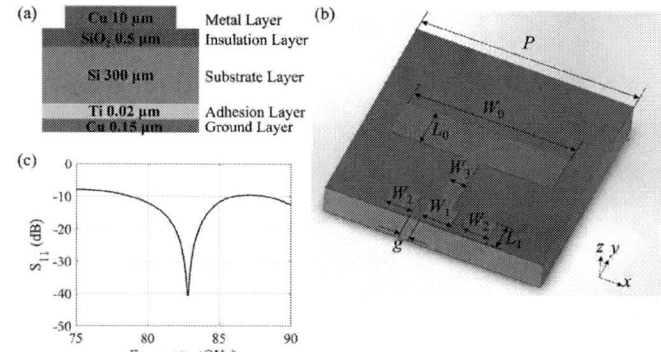

Figure 1. (a) Cross section of the OCA. (b) 3-D structure of the OCA. (c) Simulated reflection coefficient of the OCA.

Wave band for the design of OCAs, because OCAs working in MM-Wave band have small sizes and consume less chip area [6]. Multiple attractive applications using MM-Wave OCAs have been explored, including 5G wireless broadband communication, energy harvesting [7], MM-Wave radars [8], and so on.

Despite of significant advantages as well as promising applications, the challenges OCAs offer cannot be ignored. First of all, owing to the relatively low resistivity and high permittivity, electromagnetic waves propagate in the lossy silicon substrate, causing the degradation in the performance of OCAs [3]. Moreover, the layout and design of OCAs are sometimes incompatible with the specific process design rules [9]. Additionally, the characterizations of OCAs are faced with significant difficulties, as traditional off-chip antenna measurement setups are not suitable [6].

In this research, we implemented an on-chip MM-wave antenna based on high-resistivity silicon substrate, with a small size of $2.0 \times 2.0 \times 0.3$ mm³. The proposed OCA offered an impendence bandwidth from 78.76 GHz to 86.02 GHz for S_{11} less than -10 dB. Utilizing complementary metal oxide semiconductor (CMOS) process, the OCA designed was fabricated. With gold wire bonding, the OCA was fed through a substrate integrated waveguide-to-coplanar waveguide (SIW-

979-8-3315-0417-5/25 $31.00 © 2025 IEEE

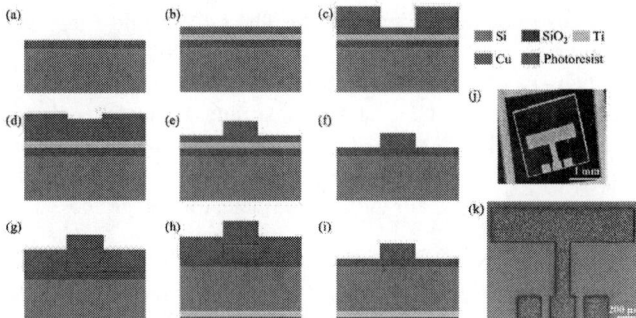

Figure 2. (a)-(i) Fabrication process flow of the OCA. (j) Photograph of the OCA. (k) Microscope image of the OCA.

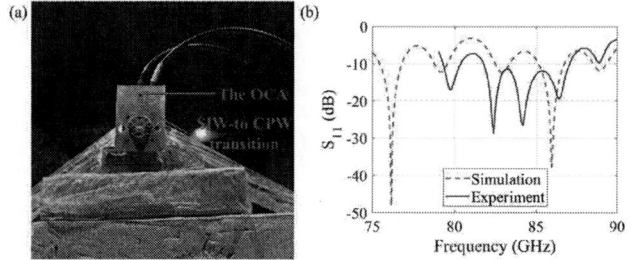

Figure 3. (a) The measurement setup of the OCA. (b) Comparison of reflection coefficient between simulation and experiment with SIW-to CPW transition.

to-CPW) transition interfacing with a rectangular WR10 waveguide, and the measurement was performed in a microwave anechoic chamber. The OCA showed a maximum simulated gain of 3.204 dBi at 79 GHz and a maximum simulated radiation efficiency of 74.61% at 89 GHz.

Methods

A. Design and Modelling

The configuration of the proposed OCA is shown in Fig. 1(a). The thickness of the silicon substrate (t_{Si}) is 300 μm, and its resistivity is 1000 Ω • cm. On the front side is 500-nm-thick SiO$_2$ as insulation layer. The metal layer, 10-μm-thick electroplated copper, is on the insulation layer. 20-nm-thick Ti and 150-nm-thick copper are on the back side of the substrate, performing as the adhesion layer and the ground layer, respectively. The structural parameters are depicted in Fig. 1(b), and the in-plane dimensions are as follows: P = 2000 μm, W_0 = 1493 μm, W_1 = 270 μm, W_2 = 250 μm, W_3 = 150 μm, L_0 = 367 μm, L_1 = 250 μm, and g = 90 μm.

We used time domain solve to calculate the responses of the proposed OCA under incidence in the y-direction. The simulation revealed that the OCA offered an impendence bandwidth from 78.76 GHz to 86.02 GHz for reflection coefficient (S_{11}) less than -10 dB, as shown in Fig. 1(c). At 82.84 GHz, the minimum S11 was -40.57 dB.

The skin depth (δ) of copper is given by

$$\delta = \frac{1}{\sqrt{\pi f \mu \sigma}}, \qquad (1)$$

where the conductivity of copper is σ = 5.96 × 10^7 S/m and the permeability of copper is μ = 4π × 10^{-7} H/m. According to (1), we calculated that at 75 GHz, δ = 0.238 μm << 10 μm, which indicated that slight change of the thickness of metal layer had almost no impact on the performance of the OCA, providing space for fabrication error.

B. Fabrication

We fabricated the OCA using surface machining processes as shown in Figs. 2(a)-(i). Firstly, 500-nm-thick SiO$_2$ was deposited on the substrate using PECVD [Fig. 2(a)]. Then, 20 nm of Ti and 150 nm of Cu were vaporized on the insulating

layer, as the seed layers for electroplating [Fig. 2(b)]. A 15-μm-thick photoresist of AZ 12XT-20PL-10 (Merck & Co., Inc.) was spin-coated and photo-defined [Fig. 2(c)]. Next, 10 μm of copper was electroplated, acting as the metal layer of the OCA [Fig. 2(d)]. Then, acetone was used to remove the photoresist [Fig. 2(e)]. Also, the seed layers, Cu and Ti, were removed by corresponding corrosion solution [Fig. 2(f)]. The following was spin-coating 15-μm-thick photoresist on the front side to protect the metal layer [Fig. 2(g)]. Then, 20 nm of Ti and 150 nm of Cu were vaporized on the back side of the substrate, as the adhesion layer and the ground layer [Fig. 2(h)]. Finally, the photoresist was removed by acetone [Fig. 2(i)].

The fabricated OCA had a small size of 2.0 × 2.0 × 0.3 mm³, as shown in Figs. 2(j)(k).

Results

As previously stated, it is hard to characterize of the OCA directly. Therefore, we referred to traditional ways of measuring off-chip antennas. The OCA was mounted through gold wire bonding and was fed through a SIW-to-CPW transition, which interfaced with a rectangular WR10 waveguide. This design ensured compatibility with the WR10 waveguide flange structure. The measurement was performed in a microwave anechoic chamber, using a network analyzer (Agilent N5245A) with a pair of extenders (V10VNA2-TIR-A-RLA) so that the responses ranging from 79 GHz to 90 GHz can be demonstrated. The measurement setup is as depicted in Fig. 3(a).

Actually, the SIW-to-CPW transition brought a large margin of error. We added the transition in the simulation and compared the results with measured ones, as shown in Fig. 3(b).

The transition introduced some troughs, while only one of them represented the resonance of the OCA itself, which was indicated in Fig. 1(c). Though S_{11} was slightly messy, we could still find out that the simulation results and the experimental results basically matched.

We analyzed gain and radiation efficiency of the OCA, as shown in Fig. 4(a) and Fig. 4(b), respectively. The OCA showed a maximum simulated gain of 3.204 dBi at 79 GHz and a maximum simulated radiation efficiency of 74.61% at 89 GHz.

979-8-3315-0417-5/25 $31.00 © 2025 IEEE

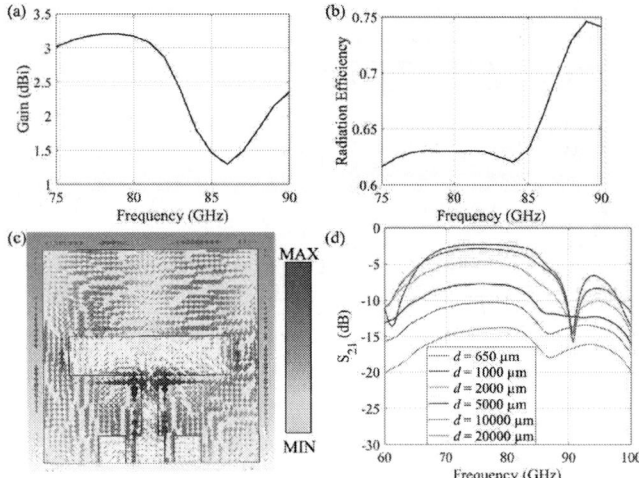

Figure 4. (a) Simulated gain of the OCA. (b) Simulated radiation efficiency of the OCA. (c) Surface current distribution of the OCA at 82.84 GHz. (d) Simulated transmission coefficient of the OCA.

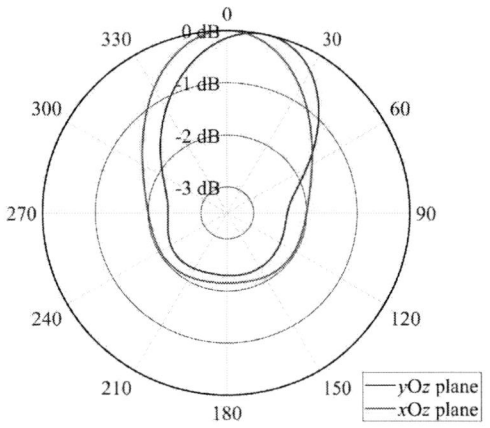

Figure 5. Simulated radiation patterns of the OCA at 82.84 GHz in the yOz plane and the xOz plane.

The surface current distribution of the OCA at 82.84 GHz was also calculated in Fig. 4(c). It can be observed that the current propagated along the copper patch and feedline on the surface of the OCA, oscillating vigorously and thereby radiating signals outward.

Additionally, the transmission performance of the OCA was also explored. Two OCAs were placed face to face with a distance of d. Fig. 4(d) showed transmission coefficient (S_{21}) under different d. It can be found out that the power transmission efficiency reached 35% when $d = 650$ μm.

Finally, the simulated radiation patterns of the OCA at 82.84 GHz in the yOz plane and the xOz plane were given in Fig. 5.

Summary

In summary, we presented the design, fabrication and characterization of an on-chip MM-Wave antenna based on high-resistivity silicon technology. The OCA could achieve an impendence bandwidth from 78.76 GHz to 86.02 GHz for S_{11} less than -10 dB. The OCA offered a maximum simulated gain of 3.204 dBi at 79 GHz and a maximum simulated radiation efficiency of 74.61% at 89 GHz. Since the fabrication process is compatible to standard CMOS process, the OCA proposed can be positioned as an attractive candidate for integrated circuit applications.

Additionally, the transmission efficiency could reach 35% when the distance between the OCAs was 650 μm. Designing OCA arrays can be effective in improving transmission efficiency. The design paves the way for short range wireless communication, promising for applications including near field communication (NFC), wireless charging and inter-chip wireless interconnections [10].

Acknowledgements

This work is supported by National Key R&D Program of China (Grant No. 2023YFB3211200) and National Science Foundation of China (Grant No. U21A6003). X. Z. acknowledges the startup funding from Tsinghua University. R. Y. acknowledges the support from the United Science Foundation of Ministry of Education of China (Grant No. 8091B032115). The authors would like to thank the Center of Nanofabrication, Tsinghua University for fabrication technical support.

References

[1] R. Saleh et al., "System-on-Chip: Reuse and Integration," Proceedings of the IEEE, vol. 94, no. 6, pp. 1050–1069, Jun. 2006.

[2] A. Barakat, A. Allam, H. Elsadek, H. Kanaya, and R. K. Pokharel, "Small size 60 GHz CMOS Antenna-on-Chip: Gain and efficiency enhancement using asymmetric Artificial Magnetic Conductor," in 2014 44th European Microwave Conference, Oct. 2014, pp. 104–107.

[3] H. M. Cheema and A. Shamim, "The last barrier: on-chip antennas," IEEE Microwave Magazine, vol. 14, no. 1, pp. 79–91, Jan. 2013.

[4] X.-D. Deng, Y. Li, C. Liu, W. Wu, and Y.-Z. Xiong, "340 GHz On-Chip 3-D Antenna With 10 dBi Gain and 80% Radiation Efficiency," IEEE Transactions on Terahertz Science and Technology, vol. 5, no. 4, pp. 619–627, Jul. 2015.

[5] Y. Niu, Y. Li, D. Jin, L. Su, and A. V. Vasilakos, "A survey of millimeter wave communications (mmWave) for 5G: opportunities and challenges," Wireless Netw, vol. 21, no. 8, pp. 2657–2676, Nov. 2015.

[6] M. R. Karim, X. Yang, and M. F. Shafique, "On Chip Antenna Measurement: A Survey of Challenges and Recent Trends," IEEE Access, vol. 6, pp. 20320–20333, 2018.

[7] H. Le, N. Fong, and H. C. Luong, "RF energy harvesting circuit with on-chip antenna for biomedical applications," in International Conference on Communications and Electronics 2010, Aug. 2010, pp. 115–117.

[8] S. Yuan, A. Trasser, and H. Schumacher, "56 GHz bandwidth FMCW radar sensor with on-chip antennas in SiGe BiCMOS," in 2014 IEEE MTT-S International Microwave Symposium (IMS2014), Jun. 2014, pp. 1–4.

[9] R. Karim, A. Iftikhar, B. Ijaz, and I. Ben Mabrouk, "The Potentials, Challenges, and Future Directions of On-Chip-Antennas for Emerging Wireless Applications—A Comprehensive Survey," IEEE Access, vol. 7, pp. 173897–173934, 2019.

[10] R. S. Narde, N. Mansoor, A. Ganguly, and J. Venkataraman, "On-Chip Antennas for Inter-Chip Wireless Interconnections: Challenges and Opportunities," 12th European Conference on Antennas and Propagation (EuCAP 2018), p. 600.

Investigation of Enhanced Robustness Against Floating-Substrate-Induced Dynamic R_{ON} Degradation in 900-V p-GaN Gate HEMT Using Virtual-Body Technology

Hao Chang[1][#], Junjie Yang[1][#], Jingjing Yu[1], Jiawei Cui[1], Youyi Yin[1], Han Yang[2], Xuelin Yang[2], Jinyan Wang[1], Maojun Wang[1], Bo Shen[2], and Jin Wei[1][*]

[1]School of Integrated Circuits, Peking University, China. [2]School of Physics, Peking University, China.
[#]Contributed equally to this work. *Corresponding author: jin.wei@pku.edu.cn.

Abstract

In this work, the enhanced robustness against floating-substrate-induced dynamic R_{ON} degradation in 900-V p-GaN gate HEMT with virtual body (VB-HEMT) is investigated. By adopting floating Si substrate, the breakdown voltage is boosted to 1779 V. Hole injection and buried AlGaN are found both essential for the formation of virtual body to suppress floating-substrate-induced negative buffer trapping effects. The VB-HEMT shows a superior dynamic R_{ON}/static R_{ON} ratio of 1.43 after 900-V V_{DS-OFF} stress with floating substrate.

Keywords: Virtual body, GaN-on-Si, floating substrate, high voltage, dynamic R_{ON}

Introduction

GaN power transistors are promising candidates for next-generation power switches owing to the favorable trade-off between ON-resistance (R_{ON}), switching speed and breakdown voltage (BV) [1]. Among all the GaN power transistors, the HEMTs based on AlGaN/GaN heterojunction grown on heterogeneous substrate, such as Si, SiC and sapphire, have gained significant attention. At present, the availability of low-cost and large-size wafer, together with the portability of existing silicon fabrication facilities, positions GaN-on-Si technology as the mainstream solution for mass production [2]. E-mode p-GaN gate HEMTs based on GaN-on-Si technology with voltage level up to 650 V have already been commercialized and gained market share [3].

For applications requiring high output power density such as electric vehicles and industrial motors, there is an urgent demand for GaN power devices with higher operation voltage [4, 5]. In commonly utilized GaN-on-Si power devices, the substrate is typically connected to the source for enhanced dynamic stability, therefore the breakdown voltage is determined by vertical drain-to-substrate leakage [6, 7]. The utilization of a thicker buffer layer is one prominent approach to improve the breakdown voltage. However, it complicates stress control during growth, leading to wafer bowing and crack formation [8].

Using floating Si substrate is another method to extend breakdown voltage of GaN-on-Si devices, which has obtained widespread attention. However, the potential of floating-substrate fluctuates across different operational states, resulting in significant dynamic R_{ON} degradation [9].

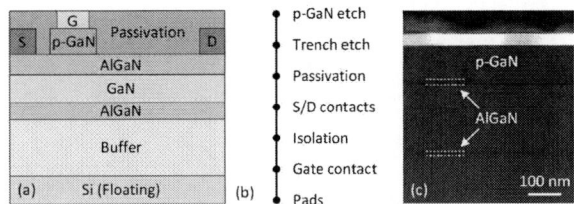

Fig. 1. (a) Schematic of the VB-HEMT. The VB-HEMT features a buried AlGaN layer. (b) Fabrication process of VB-HEMT. (c) TEM image of VB-HEMT at the gate region.

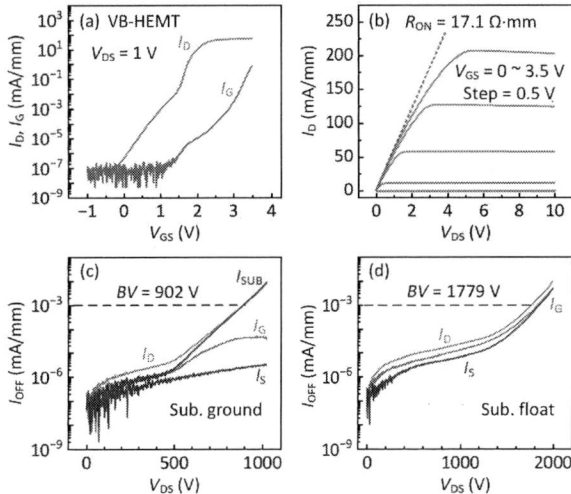

Fig. 2. I-V characteristic of VB-HEMT. The breakdown voltage of VB-HEMT extends from 902 V to 1779 V with a floating substrate.

Actively modulating the Si substrate between floating and grounded status is reported to mitigate the floating-substrate effects in 900-V applications [10]. However, the floating substrate also induces severe negative buffer trapping in the OFF-state, leading to the depletion of 2DEG and increase in dynamic R_{ON} [11]. A suitable solution to suppress the floating-substrate-induced buffer trapping and dynamic R_{ON} degradation remains elusive.

This work investigates the virtual-body technology in E-mode p-GaN gate HEMT (VB-HEMT) towards 900-V applications, which incorporates a floating Si substrate and a buried AlGaN layer. The holes injected from p-GaN gate spread along the buried AlGaN, forming a virtual body that screens the floating-substrate-induced buffer charges [12]. The dynamic R_{ON} performance is investigated under different

979-8-3315-0417-5/25 $31.00 © 2025 IEEE

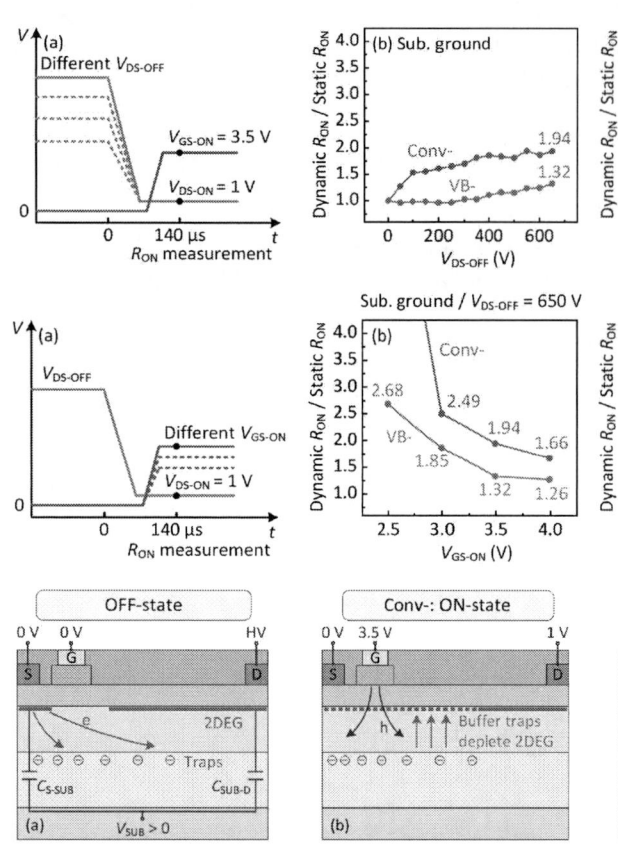

Fig. 3. (a) The testing setup of dynamic R_{ON} under different V_{DS-OFF} stress. The stress time is 1 s and the measurement delay time is 140 µs. Dynamic R_{ON}/static R_{ON} ratio of the Conv-HEMT and VB-HEMT with (b) grounded substrate and (c) floating substrate. The Conv-HEMT with a floating Si substrate exhibits pronounced dynamic R_{ON} compared to substrate-grounded case.

Fig. 4. (a) The testing setup of dynamic R_{ON} with different V_{GS-ON}. Measured dynamic R_{ON}/static R_{ON} ratio with varying V_{GS-ON} when (b) substrate is grounded and (c) substrate is floating. With a floating substrate configuration, Conv-HEMT shows severe dynamic R_{ON} degradation, while the VB-HEMT exhibits well-controlled dynamic R_{ON} performance when $V_{GS-ON} > 3$ V.

Fig. 5. (a) Negative buffer trapping as a result of positive substrate potential during OFF-state. (b) For Conv-HEMT, 2DEG partially depleted by the negative buffer charges in ON-state. (c) Negative buffer charges screened by virtual body in ON-state. Sufficient hole injection from p-GaN gate is necessary for the formation of virtual body.

ON-state gate bias (V_{GS-ON}), revealing that hole injection plays a significant role in screening negative buffer charges. A specially designed test structure is fabricated on both Conv- and VB- platforms to monitor hole current through virtual body. It is observed that the buried AlGaN layer is essential for the formation of virtual body.

Device Static Characteristics

Fig. 1(a) shows the schematic cross-sectional structure of the VB-HEMT. An energy well for holes is generated above the buried AlGaN due to negative polarization charges and energy band offset provided by the GaN/AlGaN heterojunction. Holes injected from p-GaN gate would spread along the GaN/AlGaN interface, forming the virtual body. The fabrication process of VB-HEMT is shown in Fig. 1(b), similar to that described in [12]. Fig. 1(c) presents a TEM image of a fabricated device at the gate region.

Fig. 2 presents the static characteristics of a VB-HEMT with $L_{GS}/L_G/L_{GD} = 3/5/22$ µm. The device exhibits a threshold voltage (V_{TH}) of 1.4 V and an R_{ON} of 17.1 Ω·mm. As shown in Fig. 2(c), with grounded substrate, the device exhibits a breakdown voltage (BV) of 902 V defined at $I_D = 1$ µA/mm. Fig. 2(d) shows the OFF-state I-V characteristic of the device with a floating substrate, the BV is enhanced to 1779 V.

Dynamic Characteristic and Discussion

Fig. 3 shows dynamic R_{ON} of conventional p-GaN gate HEMT without virtual body (Conv-HEMT) and VB-HEMT with different substrate configurations. The Conv-HEMT

exhibits more pronounced R_{ON} degradation with a floating substrate compared to the substrate-grounded case. In contrast, VB-HEMT presents similar and decent dynamic R_{ON} performance for both substrate-grounded and substrate-floating configurations. Moreover, the VB-HEMT presents a low dynamic R_{ON}/static R_{ON} ratio of 1.43 stressed by 900-V V_{DS-OFF} with floating substrate.

To investigate the mechanisms behind the distinct dynamic performance of Conv-HEMT and VB-HEMT, dynamic R_{ON} measurement with varying on-state gate bias (V_{GS-ON}) is conducted. Fig. 4(a) illustrates the test setup. For both devices, the dynamic R_{ON} performance gradually improves with increased V_{GS-ON} [13, 14]. However, for Conv-HEMT with floating substrate, the device presents a severe dynamic R_{ON} degradation even with $V_{GS-ON} = 4$ V due to the floating-substrate-induced negative buffer charges [11]. In contrast, with $V_{GS-ON} > 3$ V, VB-HEMT with floating substrate exhibits a low dynamic R_{ON}/static R_{ON} ratio similar as that in substrate-grounded case, indicating the hole injection is a dominant factor in suppressing floating-substrate-induced degradation.

Fig. 5 illustrates the mechanisms responsible for the dynamic performance of both devices. During the OFF-state, the potential of the floating substrate is elevated to a positive value due to capacitance coupling effect [9-11]. This positive substrate potential pulls electrons down into buffer layer, where they are captured by buffer traps, resulting in accumulation of negative buffer charges. In the ON-state, as

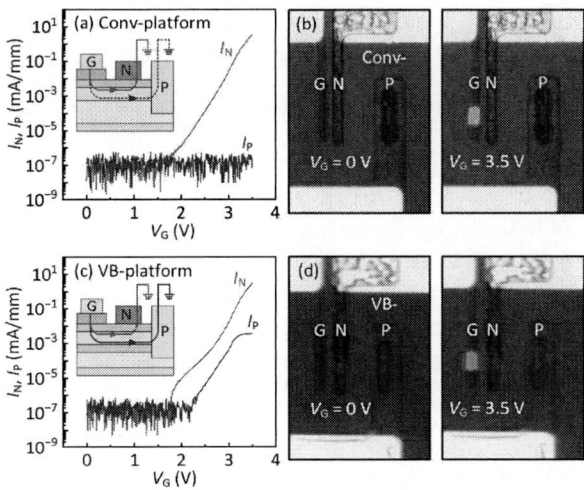

Fig. 6. (a) Measured I_N and I_P in Conv-platform. (b) Light emission in gate region on Conv-platform. (c) Measured I_N and I_P in VB-platform. (d) Light emission in gate region on VB-platform.

shown in Fig. 5(b), these negative buffer charges deplete 2DEG, leading to severe dynamic R_{ON} degradation in Conv-HEMT [7]. Moreover, in Conv-HEMT, the hole injection mainly happens at the gate region, which is insufficient to counteract the negative buffer charges under the whole active region. In VB-HEMT, holes injected from gate are blocked and spread along the GaN/AlGaN interface, forming a virtual body. The virtual body fully screens buffer negative charges, eliminating the dynamic R_{ON} degradation induced by floating substrate [12]. Furthermore, a large V_{GS-ON} together with sufficient hole injection is required for the formation of virtual body and the enhancement of dynamic performance in VB-HEMT, which is supported by test result in Fig. 4(c).

To further investigate the formation mechanism of the virtual body, a test structure is fabricated on both Conv- and VB- platform as shown in Fig. 6. The test structure features a contact on p-GaN to realize hole injection (G), a contact to 2DEG (N), and a sidewall contact (P). For test devices on both platforms, apparent light emission is captured by a CCD camera with an integration time of 15 s and the current I_N is detected, indicating sufficient hole injection from p-GaN gate However, the hole current (I_P) through the virtual body can only be detected in VB-platform, indicating the buried AlGaN is essential for the formation of virtual body.

Conclusion

In this work, the virtual-body technology demonstrates enhanced robustness against floating-substrate-induced degradation in 900-V E-mode p-GaN gate HEMT fabricated on bulk Si substrate. By incorporating a floating Si substrate, the breakdown voltage of device is improved from 902 V to 1779 V. A virtual body layer is formed by holes injected from p-GaN gate spreading along the buried AlGaN interface to screen negative buffer charges. Thus, the VB-HEMT exhibits a low dynamic R_{ON}/static R_{ON} ratio of 1.43 after 900-V

V_{DS-OFF} stress with floating substrate, compared to 3.42 in Conv-HEMT. Sufficient hole injection from gate and the buried AlGaN layer are found significant for the formation of virtual body. The results demonstrate that the virtual body is an efficient solution to enhance the dynamic stability of GaN-on-Si device with floating substrate, providing the potential to expand the voltage rating of GaN-on-Si power devices.

Acknowledgments

This work was supported by the National Natural Science Foundation of China under Grant 62174003.

References

[1] H. Amano et al., "The 2018 GaN power electronics roadmap," *J. Phys. Appl. Phys,* vol. 51, no. 16, Art. no. 163001 (2018).

[2] J. Wei et al., "GaN Power Integration Technology and Its Future Prospects," *IEEE Trans. Electron Devices,* vol. 71, no. 3, pp. 1365-1382 (2024).

[3] K. J. Chen et al., "GaN-on-Si Power Technology: Devices and Applications," *IEEE Trans. Electron Devices,* vol. 64, no. 3, pp. 779-795 (2017).

[4] S. Li et al., "1200V E-mode GaN Monolithic Integration Platform on Sapphire with Ultra-thin Buffer Technology," in *IEDM Tech. Dig.,* San Francisco, CA, USA, sec. 9-5 (2023).

[5] J. Cui et al., "High-Voltage E-Mode p-GaN Gate HEMT on Sapphire With Gate Termination Extension," *IEEE Trans. Electron Devices,* vol. 71, no. 3, pp. 1-6 (2024).

[6] C. Tsai et al., "Smart GaN platform: Performance & challenges," in *IEDM Tech. Dig.,* San Francisco, CA, USA, pp. 737-740 (2017).

[7] S. Yang et al., "Impact of Substrate Bias Polarity on Buffer-Related Current Collapse in AlGaN/GaN-on-Si Power Devices," *IEEE Trans. Electron Devices,* vol. 64, no. 12, pp. 5048-5056 (2017).

[8] A. Vohra et al., "Epitaxial buffer structures grown on 200 mm engineering substrates for 1200 V E-mode HEMT application," *Appl. Phys. Lett.,* vol. 120, no. 26 (2022).

[9] G. Tang et al., "Dynamic R_{ON} of GaN-on-Si lateral power devices with a floating substrate termination," *IEEE Electron Device Lett.,* vol. 38, no. 7, pp. 937-940 (2017).

[10] B. Li et al., "900V Normally-OFF GaN-on-Si Transistors Achieved by Substrate Potential Modulation (SPM)," in *Proc. ISPSD,* Bremen, Germany, pp. 156-159 (2024).

[11] H. Zhang et al., "GaN-on-Si lateral power devices with symmetric vertical leakage: The impact of floating substrate," in *Proc. ISPSD,* Chicago, IL, USA, pp. 100-103 (2018).

[12] J. Yang et al., "650-V GaN-on-Si Power Integration Platform Using Virtual-Body p-GaN Gate HEMT to Screen Substrate-Induced Crosstalk," in *IEDM Tech. Dig.,* San Francisco, CA, USA, sec. 9-6 (2023).

[13] M. Meneghini et al., "Time- and Field-Dependent Trapping in GaN-Based Enhancement-Mode Transistors With p-Gate," *IEEE Electron Device Lett.,* vol. 33, no. 3, pp. 375-377 (2012).

[14] Y. Wu et al., "Suppression of Buffer Trapping Effect in GaN-on-Si Active-Passivation p-GaN Gate HEMT via Light/Hole Pumping," *IEEE Trans. Electron Devices,* vol. 71, no. 1, pp. 484-489 (2024).

979-8-3315-0417-5/25 $31.00 © 2025 IEEE

A Comparative Analysis of Direct Leakage Current Compensation and Positive-Up-Negative-Down in the Characterization of Leaky Ferroelectric Structures

Tiang Teck Tan[1, *], Tian-Li Wu[2, 3, ♣], Hsien-Yang Liu[2], Chen-Yu Yu[3], Laurent Grenouillet[4], Paolo La Torraca[5], Andrea Padovani[6], Francesco Maria Puglisi[6], Nagarajan Raghavan[1], Kin Leong Pey[1, 7]

[1] Engineering and Product Development, Singapore University of Technology and Design, Singapore. [2] Institute of Electronics, National Yang Ming Chiao Tung University, Taiwan. [3] International College of Semiconductor Technology, National Yang Ming Chiao Tung University, Taiwan. [4] CEA-Leti, Univ. Grenoble Alpes, France. [5] Tyndall National Institute, University College Cork, Ireland [6] University of Modena and Reggio-Emilia, Italy. [7] NEOM University and Education, Research and Innovation Foundation, Saudi Arabia. E-mail: *tiangteck_tan@mymail.sutd.edu.sg, ♣tlwu@nycu.edu.tw

Abstract— Techniques such as the Positive-Up-Negative-Down (PUND) and Dynamic Leakage Current Compensation (DLCC) are often used to isolate the Ferroelectric (FE) switching component from other current artifacts during the characterization of FE memory devices, with DLCC being able to compensate for a wider range of measurement artifacts. An evaluation of both techniques was carried out in the context of leaky and non-leaky samples. However, it was observed that both PUND and DLCC were unable to fully compensate for large amounts of leakage currents in very leaky samples (<1 A/cm² at 3 MV/cm applied voltage). In doing so, this work aims to showcase some of the resultant signatures of inadequate compensation in the *I-V* plots and *P-V* hysteresis loops, highlighting the need for more sophisticated current compensation methodologies.

Keywords: HZO, Dynamic leakage current compensation, Positive-Up-Negative-Down, Ferroelectric, Artifacts, Characterization,

I. INTRODUCTION

Hafnia-based ferroelectric (FE) devices are one of the primary candidates for an emerging non-volatile memory (eNVM) due to their CMOS compatibility [1], scalability, simple 1T memory cell architecture and low write energy [2]. Characterization of FE properties is commonly carried out by the application of voltage pulses and collection of the transient current response [3]-[5]. The current response is integrated over time to provide a measure of the polarization in the FE material, commonly represented as a Polarization-Voltage (*P-V*) hysteresis curve.

However, the convolution of FE switching current signals with leakage and dielectric polarization contributions presents artifacts to be corrected for during characterization of any FE material system. The Positive-Up-Negative-Down (PUND) waveform is one common way of circumventing this issue [3] wherein a subtraction between current responses to two voltage pulses of similar polarity allows for the isolation of the FE switching component of the current response. However, the PUND waveform is limited by its inability to fully compensate for capacitive displacement current, antiferroelectricity and leakage current when one or more of these components heavily depend on the FE polarization state [4]-[5].

Dynamic Leakage Current Compensation (DLCC) [4]-[5] is an alternative to the PUND waveform involving multiple pulsed voltage measurements of the current response at varying voltage frequencies. While DLCC has several advantages over the much more commonly used PUND waveform, artifacts such as excessively large leakage currents which prevent accurate characterization of FE properties can still exist. In this paper, we demonstrate the limitations of both characterization methods in compensating for excessively large leakage currents.

II. TEST STRUCTURE AND ELECTRICAL CHARACTERIZATION

Leaky MFIS MOSCAP structures were fabricated on n-doped Si wafers. A schematic of the device stack, consisting of doped Si, 1 nm SiO₂, 5 nm HZO and 80 nm TiN is shown in Figure 1(a). For comparison of the efficacy of DLCC on non-leaky devices, capacitor structures consisting of doped Si and 6nm HZO between 10nm TiN top and bottom electrodes (Figure 1(b)) were also fabricated. The devices were stressed with 10^4 cycles of bipolar switching pulses (Figure 2(a)) to wake-up the devices before being characterized using PUND and DLCC.

III. PHYSICS OF PUND CHARACTERIZATION

The PUND waveform (Figure 2(b)) consists of two pairs of voltage pulses of opposite polarity. The current response to the second pulse of each pair is subtracted from that of the first pulse ($I_{Pulse\ 1} - I_{Pulse\ 2}$) to obtain the FE switching current without the influence of artifacts associated with characterization of FE devices based on current measurements. This FE switching current is then integrated over time to obtain the amount of switched polarization in the device, allowing for the generation of the FE *P-V* hysteresis loop [3].

IV. PHYSICS OF DLCC CHARACTERIZATION

The DLCC waveform (Figure 2(c)) is a simple bipolar triangular pulse. The current responses to multiple separate pulses of varying frequency are collected and a subtraction is carried out to eliminate the contributions of leakage current to obtain the ferroelectric switching current. The frequencies used are selected such that a frequency-dependent shift of the coercive field of the FE material is absent and differs between material systems. An integration of this current is carried out in a process similar to that of the PUND method to obtain *P-V* hysteresis loop [4]. The magnitude of the applied electric fields during field cycling and each of the characterization waveforms are sufficiently high to induce FE switching in the HZO layer. All voltages were applied to the top electrode, while the Si wafer was kept grounded.

The current response of a leaky FE sample can be modelled as: 1) the leakage current of a resistive element, 2) displacement current of a linear capacitive element and 3) a voltage-controlled FE switching current. DLCC eliminates leakage current contributions to the current. response by relying on the frequency independence of the resistive element. The compensated current, which includes only the frequency dependent components of the current (i.e. the dielectric and the ferroelectric components), was derived by Meyer et al. [3]:

$$i_{comp}(\omega) = \frac{\omega}{\omega_2 - \omega_1}[i(\omega_2) - i(\omega_1)] \qquad (1)$$

where i_{comp} represents the compensated current as a function of an arbitrary measurement frequency ω. $i(\omega_2)$ and $i(\omega_1)$ represent the measured current response of the device to voltage pulses applied at two selected frequencies, where $\omega_2 > \omega_1$. Since the compensated current includes both a capacitive and a ferroelectric component, integrating it over time results in a polarization loop which includes the effects of both. The resultant P-V hysteresis loop, is therefore the sum of a P-V loop obtained using PUND and the linear dielectric response, generating a diagonally slanted hysteresis loop.

V. DEMONSTRATION OF DLCC ON A NON-LEAKY SAMPLE

Figure 3(a) demonstrates a successful application of DLCC in eliminating leakage on the MIM structures from Figure 1(b). ω and ω_1 were both fixed to 10 kHz. The I-V traces all contain large current peaks around ±1V of applied voltage, representing the FE switching current transients. The circled current peaks at both ends of the voltage range in the uncompensated I-V trace indicates the presence of leakage currents that were absent in the coloured traces where DLCC was applied. Figure 3(b) shows the measured I-V traces used for the compensation process in Figure 3(a). The traces in Figure 3(b) exhibit leakage, while maintaining the FE switching transients, showing the efficacy of the DLCC compensation from Figure 3(a) in eliminating leakage currents that are significantly smaller than the transient FE signal.

VI. APPLICATION OF DLCC ON LEAKY SAMPLES

Figure 4(a) shows the I-V characteristics of a leaky MFIS device (Figure 1(a)). As shown in the circled section, there is significant leakage at the edge of the positive voltage range for both the uncompensated trace at 50 kHz and the compensated traces at various ω_2. Despite the significantly reduced leakage in the compensated traces compared to the uncompensated trace, the P-V loops in Figure 4(b) still contain gaps (indicated by arrows) between the positive and negative half loops as well as rounded edges (circled) at the peak positive voltage, features common to P-V loops in highly leaky devices.

VII. COMPARISON OF PUND VS DLCC ON LEAKY SAMPLES

Figure 5(a) shows the P-V loops of a leaky MFIS device, with various voltage magnitudes. DLCC was applied to traces obtained at lower voltages to assess its applicability at reduced leakage levels. Even at a significantly lower voltage of 3V compared to the 4V applied in Figure 4(a), it can be observed that the P-V loop still exhibits a rounded edge typical of leaky devices (circled). From Figure 5(b), it can be observed that, compared to the traces at higher voltages, the positive edge of the 3V I-V trace is approaching the FE switching current peak. However, since the signatures of leakage are still present in the 3V P-V loop, the DLCC technique is unable to produce a FE response in the MFIS device free from leakage current artifacts. From Figure 5(c), it can be observed that a curvature

in the positive edge (circled) as well as a gap (indicated by arrows) were present in the P-V loop. We can therefore conclude that PUND was also unable to produce a P-V loop free from artifacts.

VIII. CONCLUSION

DLCC and PUND were applied to leaky devices to evaluate their efficacy in eliminating leakage currents from the measured current response. Although DLCC was proposed for further compensation of artifacts compared to PUND, it is still unable to fully compensate for excessively large leakage currents. This results in P-V loops that exhibit behaviours typical of leaky devices. One way to eliminate this current after applying DLCC or PUND in P-V loops is simply to perform a graphical estimation based on two principles: 1) the top and bottom edges of the P-V loop should be horizontal. 2) The same edges should be of equal magnitude. However, this methodology merely provides an estimation rather than proper compensation. Furthermore, DLCC requires multiple measurements compared to a single pulse train in PUND (which also implies the application of multiple switching cycles). This makes it less suitable for characterization of pristine (not woken-up) samples, as the amount of switched polarization can change drastically in the first few stressing cycles from the pristine state. However, DLCC is more suitable for the disambiguation of antiferroelectric effects and dealing with FE polarization polarity dependent current. More effective means of isolating the FE switching component in the current response are required to obtain accurate and quantifiable measures of the switchable polarization in leaky FE devices.

ACKNOWLEDGEMENT

Tiang Teck Tan would like to acknowledge the research student scholarship for PhD at Singapore University of Technology and Design (SUTD), provided by the Ministry of Education (MOE), Singapore for 2020-2024. The authors would also like to acknowledge Prof. Pey's research surplus fund (RS-INSUR-00025-A3501-F00) at SUTD for the financial support. Andrea Padovani acknowledges the FAR 2023-2024 project of the "Enzo Ferrari" Engineering Department of the University of Modena and Reggio Emilia, Italy, for financial support.

REFERENCES

[1] T. S. Böscke, J. Müller, D. Bräuhaus, et al. "Ferroelectricity in hafnium oxide thin films". In: Applied Physics Letters 99.10 (2011). ISSN: 00036951. DOI: 10.1063/1. 3634052

[2] International Roadmap for Devices and Systems (IRDS) Semiconductor Industry Association, Beyond CMOS Chapter. 2021.

[3] Scott, J. F., Araujo, C. A., Meadows, H. B., McMillan, L. D., & Shawabkeh, A. (1989). Radiation effects on ferroelectric thin-film memories: Retention failure mechanisms. Journal of Applied Physics, 66(3), 1444–1453. https://doi.org/10.1063/1.344419

[4] Meyer, R., Waser, R., Prume, K., Schmitz, T., & Tiedke, S. (2005). Dynamic leakage current compensation in ferroelectric thin-film capacitor structures. Applied Physics Letters, 86(14), 1–3. https://doi.org/10.1063/1.1897425

[5] T. Schenk, U. Schroeder and T. Mikolajick, "Correspondence - Dynamic leakage current compensation revisited," in IEEE Transactions on Ultrasonics, Ferroelectrics, and Frequency Control, vol. 62, no. 3, pp. 596-599, March 2015, doi: 10.1109/TUFFC.2014.006774.

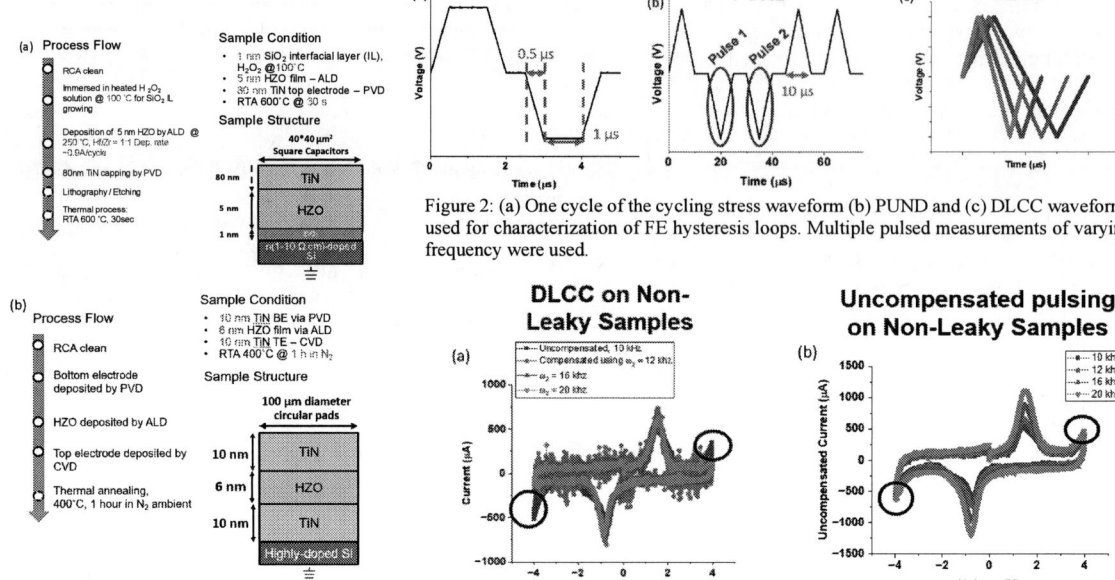

Figure 2: (a) One cycle of the cycling stress waveform (b) PUND and (c) DLCC waveforms used for characterization of FE hysteresis loops. Multiple pulsed measurements of varying frequency were used.

Figure 1: (a) Device stack of the HZO MFIS device and (b) device stack of the HZO MIM device.

Figure 3: (a) Comparison of *I-V* characteristics of a non-leaky FE device in response to the DLCC waveform, where leakage current was successfully compensated at various frequencies. (b) Comparison of *I-V* characteristics of a non-leaky FE device in response to the DLCC waveform at various frequencies, without applying compensation.

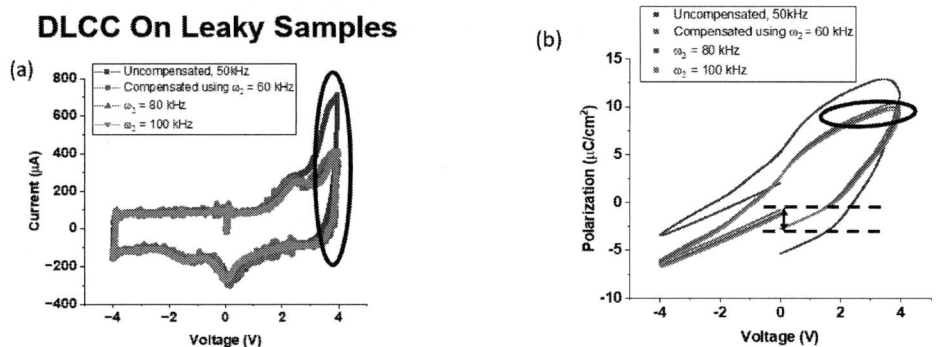

Figure 4: (a) Comparison of *I-V* characteristics of a leaky FE device across different compensation frequencies. (b) *P-V* loops obtained by integrating the currents over time.

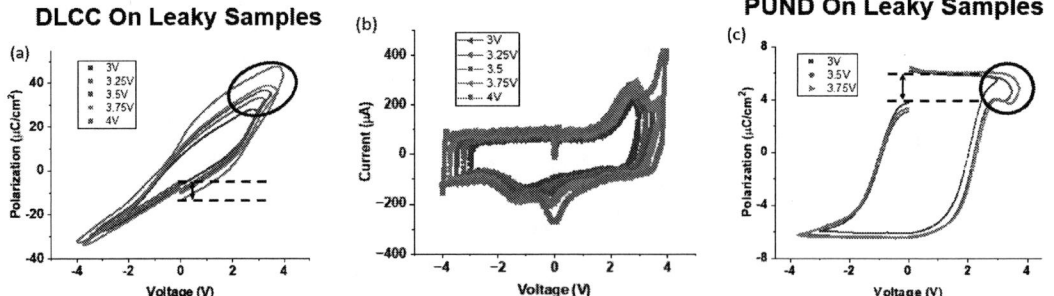

Figure 5: (a) *P-V* loops of the leaky device obtained using DLCC, at various voltage magnitudes and at ω_1 = 50 kHz, ω_2 = 100 kHz. (b) *I-V* plots of the leaky device compensated using DLCC. (c) *P-V* loops of the leaky device obtained using PUND. The devices tested all exhibit properties typical of leaky currents despite the use of both current compensation techniques, across different voltages. The polarization magnitudes in Figures 5(a) and 5(c) appear different because DLCC does not eliminate dielectric polarization.

979-8-3315-0417-5/25 $31.00 © 2025 IEEE

Efficiency Analysis of Large-Area β-Ga₂O₃ Schottky Barrier Diodes for DC-DC Converter

Yuru Lai[1], Chenxi Li[2], Shengliang Cheng[1], Huaxing Jiang[2], Leidang Zhou[3], Zimin Chen[1], Yanli Pei[1], Gang Wang[1] and Xing Lu[1*]

[1]State Key Laboratory of Optoelectronic Materials and Technologies, School of Electronics and Information Technology, Sun Yat-sen University, Guangzhou 510275, China;
[2]School of Microelectronics, South China University of Technology, Guangzhou 510641, China;
[3]School of Microelectronics, Xi'an Jiaotong University, Xi'an 710049, China;
*Emails: lux86@mail.sysu.edu.cn

Abstract

In this work, we fabricated 800 V-class, large-area (1×1 mm²) β-Ga₂O₃ Schottky barrier diodes (SBDs) and assessed their switching performance for DC-DC converter applications. Double-pulse testing revealed a reverse recovery time of 8.6 ns, with switching to 15 A at slew rates over 1125 V/µs. Additionally, β-Ga₂O₃ SBDs packaged in TO257 were evaluated as freewheeling diodes in a boost converter, delivering an efficiency of 97.64 % at a 400 V output. Their performance was comparable to SiC SBDs, highlighting the potential of β-Ga₂O₃ SBDs for power applications.

Keywords: β-Ga₂O₃ diode, double-pulse test, DC-DC boost converter

Introduction

Gallium oxide (β-Ga₂O₃) semiconductor, characterized by its 4.8 eV ultra-wide bandgap and a critical electric field of up to 8 MV/cm, is a promising candidate for next generation high-power electronics [1], [2]. Currently, small-area β-Ga₂O₃ Schottky barrier diodes (SBDs) with a Baliga figure of merit surpassing the unipolar limit of SiC have been demonstrated [3]. Nevertheless, commercial β-Ga₂O₃ devices have yet to be introduced to the market. It is crucial to fabricate ampere-level β-Ga₂O₃ SBDs to validate the circuit conversion efficiency. However, studies on Ga₂O₃ devices in DC-DC converters are still limited [4]-[6], indicating a need for further exploration of the potential of β-Ga₂O₃ SBDs.

This work presents the fabrication of large-area (1×1 mm²) β-Ga₂O₃ SBDs rated for 5 A and 800 V. The reverse recovery characteristics of the fabricated β-Ga₂O₃ SBDs were evaluated through a double-pulse test and compared against commercial SiC SBDs and Si fast recovery diodes (FRDs). Additionally, a testing platform for a DC-DC boost converter circuit has been established, enabling efficiency analysis of these devices as freewheeling diodes across varied operating conditions.

Device Static Characterization

Fig. 1(a) shows the schematic cross-section and key fabrication steps of the β-Ga₂O₃ SBDs. The β-Ga₂O₃ epi-wafer consists of a 10 µm thick n-type drift layer grown on a highly conductive (001) β-Ga₂O₃ substrate by halide vapor phase epitaxy (HVPE). After organic and acid cleaning, a layer of SiO₂ was deposited on the sample by plasma-enhanced chemical vapor deposition (PECVD) as an implantation mask. N-ion was implanted into the sample at multiple energies to form a 0.5~0.6 µm deep guard ring (GR). Thermal annealing was performed at 900 °C for 30 min in an O₂ gas atmosphere to recover the implantation damage and activate the implanted N atoms. A 360-nm-thick SiO₂ layer was deposited by PECVD and patterned. Then, the metal stack with 50 nm/100 nm Ti/Au was deposited onto the backside of the sample and annealed to form ohmic contacts. A Ni/Au metal stack (50/1000 nm) was deposited to form the Schottky contact. The Schottky electrode extends over SiO₂ and N-ion GR, forming a field plate (FP) with L_{FP} = 30 µm and L_{GR} = 50 µm. The fabricated large-area (1×1 mm²) device was bonded on a kovar substrate, and the device anode was connected to the lead frame by 32-um-diameter Al bonding wires, which were ultimately encapsulated in the silicone-based potting compounds for the TO257 package, as shown in Fig. 1(b).

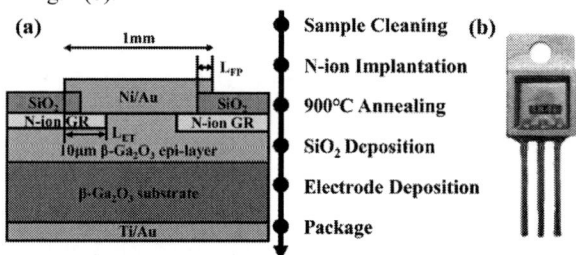

Fig. 1: (a) Schematic cross-section and key fabrication steps of the β-Ga₂O₃ SBDs. (b) Photograph of the TO257 packaged β-Ga₂O₃ SBD.

Fig. 2(a) presents the static J-V characteristics of the β-Ga₂O₃ SBDs, revealing a turn-on voltage of 0.83 V and a specific on-resistance of 34.7 mΩ·cm². The devices exhibit an ideality factor of 1.01, indicating the formation of a high-quality Schottky contact interface. The forward conduction characteristics of the devices are shown in Fig. 2(b), with a

979-8-3315-0417-5/25 $31.00 © 2025 IEEE

forward current of 5 A achieved at 6 V under a 50 μs pulse width and 1 % duty cycle. Benefiting from the composite terminal structures, the β-Ga₂O₃ SBDs achieve a breakdown voltage of 800 V, as shown in Fig. 2(c). Additionally, Fig. 2(d) demonstrates the stable reverse breakdown performance, showing that the packaging process minimally impacts the reverse characteristics of the fabricated devices.

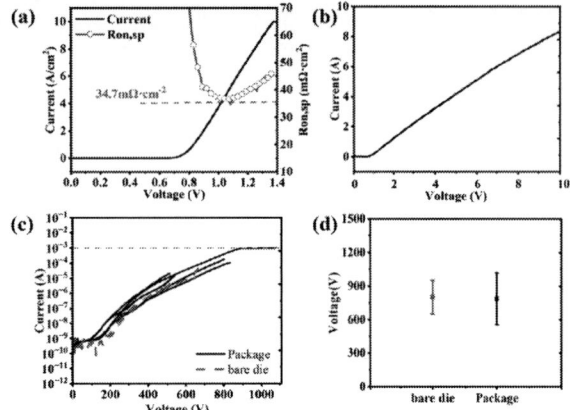

Fig. 2: (a) Static J-V and (b) pulsed I–V characteristics of the Ga₂O₃ SBDs. (c) (d) Reverse characteristics of bare die and packaged devices.

Dynamic Characterization and Conversion Efficiency

In DC-DC boost circuits, conversion losses predominantly stem from losses in both active and passive components. Active device losses contribute significantly to overall losses, as high-frequency switching supports the miniaturization of passive components[7]. In DC-DC converters, freewheeling diode losses are dominated by conduction and switching losses. Therefore, performing double-pulse tests and extracting reverse recovery parameters from the diodes are crucial for precise loss analysis.

A. Reverse Recovery Performance of the Devices

Fig. 3(a) presents a comparison of the β-Ga₂O₃ SBD with commercially mature devices, including a SiC SBD (SCS302AH, [650V/2A]) and a Si FRD (RFN20TJ6SGC9, [600V/20A]), at a forward current level of 15 A and a turn-off speed of 1125 A/μs. The Si FRD displays a classic tail current characteristic of bipolar reverse recovery behavior, contrasting with the unipolar response of the β-Ga₂O₃ SBD. The reverse recovery charge in the β-Ga₂O₃ SBD arises primarily from space charge capacitance and is independent of current levels. Fig. 3(b) further corroborates this, showing

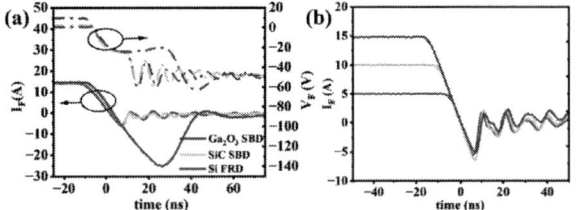

Fig. 3: (a) Diode turn-off waveforms of the β-Ga₂O₃ SBD, SiC SBD, and Si FRD at 50V and 15A. (b) Reverse recovery waveforms at different current levels (5 A, 10 A, 15 A) of the β-Ga₂O₃ SBDs.

that the reverse recovery current waveforms of the β-Ga₂O₃ SBD remain unaffected by variations in turn-off current magnitude.

Table I lists the key reverse recovery parameters, including reverse recovery time (t_{rr}), reverse current (I_{rm}), reverse storage charge (Q_c), and switching energy (E_{rr}). As shown in Table I, the switching energy of the β-Ga₂O₃ SBD is slightly lower than that of the SiC SBDs.

Table 1: Parameters of reverse recovery characteristics.

Type	t_{rr}(ns)	I_{rm}(A)	Q_c(nC)	E_{rr}(uJ)
Ga₂O₃ SBD	8.64	6.1	31.48	1.51
SiC SBD	7.36	6.21	26.80	1.55
Si FRD	38.08	25.31	570.81	15.85

B. DC-DC Boost Converter

Among various converter topologies, the DC-DC boost converter is widely recognized as a fundamental circuit, providing a straightforward platform for directly assessing device switching performance. Fig. 4(a) shows the basic configuration of the DC-DC circuit, incorporating an Infineon 650 V/83 mΩ discrete MOSFET (IMW65R083M1H) driven by an Infineon gate driver (1EDN7550B) with an 18 V drive voltage for nanosecond-level fast switching. Fig. 4(b) shows the β-Ga₂O₃ SBD-based DC-DC converter and the testing platform. The current through the active devices is measured using a Hall probe (Tektronix TCPA300), while a 50 MHz differential probe (Tektronix P5200A) captures the voltage changes across the diodes.

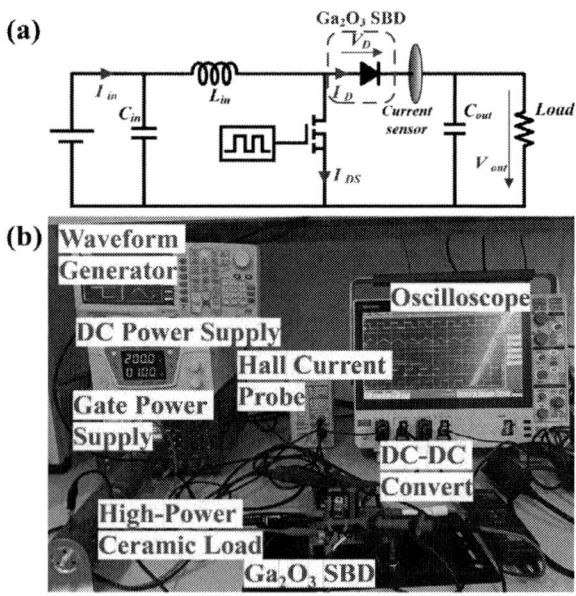

Fig. 4: (a) Schematic drawing of the 200V input to 400V output boost converter circuit. (b) Photograph of the β-Ga₂O₃ SBD-based DC-DC converter and testing platform.

To mitigate potential voltage overshoot during dynamic testing, the output voltage was limited to 400 V, providing a safety margin relative to the 800 V breakdown voltage of the β-Ga$_2$O$_3$ SBD. The primary test conditions for the boost circuit include an input voltage of V_{in} = 200 V, an output voltage of V_{out} = 400 V, a switching frequency of f = 100 kHz, a load resistance of R = 1 kΩ, a duty cycle of D = 50 %, and a power of P = 160 W. To facilitate a comprehensive comparison between the β-Ga$_2$O$_3$ SBD and the other commercial devices, four varying parameter conditions are established: input voltage, switching frequency, load resistance, and duty cycle.

As shown in Fig. 5(a), the conversion efficiency of the circuit increases with the input voltage, with the β-Ga$_2$O$_3$ SBD achieving the highest efficiency of 97.64 % at V_{in} = 200 V. However, the relationship between conversion efficiency and switching frequency exhibits the opposite trend. Fig. 5(b) indicates that under high-frequency switching conditions, the Si FRDs experience the most significant drop in conversion efficiency, decreasing by 17.3 %. In contrast, the β-Ga$_2$O$_3$ and SiC SBDs show more modest reductions of only 6.1 % and 6.3 %, respectively, maintaining the conversion efficiency consistently above 91 %, which is consistent with the earlier reverse recovery test results. Furthermore, as the load transitions from heavy to light, the conversion efficiency of all three devices decreases linearly with decreasing load resistance, as illustrated in Fig. 5(c). The Si FRDs exhibit the most substantial efficiency decline, averaging -1.79 % per 1 kΩ, while the β-Ga$_2$O$_3$ and SiC SBDs show linear decline rates of approximately -1.4 % per 1 kΩ. Fig. 5(d) illustrates the conversion efficiency across varying duty cycles, where the Si FRDs experience a 3.6 % efficiency drop at a 50 % duty cycle compared to 10 %. In contrast, the β-Ga$_2$O$_3$ and SiC SBDs exhibit efficiencies decline at duty cycles below 40 %, followed by a slight increase at D = 50 %.

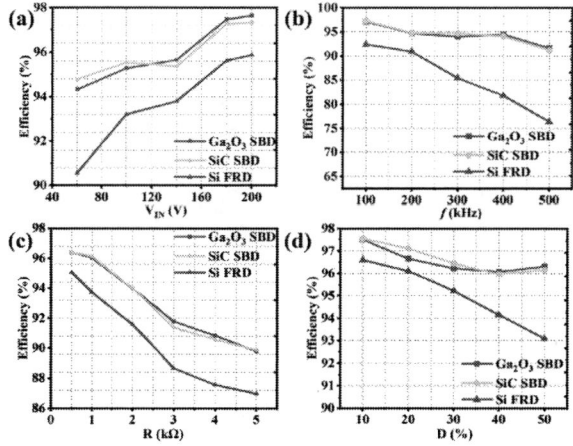

Fig. 5: Dependence of conversion efficiency on (a) input voltage, (b)switching frequency, (c) load resistance, and (d) duty cycle.

Conclusion

In this work, the ampere-level β-Ga$_2$O$_3$ SBDs with an 800 V voltage rating were fabricated, and their performance was evaluated through static and dynamic conversion efficiency testing. Despite the early stage of development, the power loop efficiency of the boost converter, especially in the high-frequency and high-power regime, is significantly higher with Ga$_2$O$_3$ SBDs than with the Si FRDs. The efficiency value of the Ga$_2$O$_3$ SBDs is almost consistent with that of the SiC SBDs. Future enhancements to β-Ga$_2$O$_3$ SBDs, including reductions in on-resistance and implementing low-inductance packaging to mitigate switching oscillations, are anticipated to further improve conversion efficiency. These results underscore the substantial potential of β-Ga$_2$O$_3$ power devices and circuits for advanced power applications.

References

[1] Y. He, F. Zhao, B. Huang, T. Zhang, and H. Zhu, "A Review of β-Ga$_2$O$_3$ Power Diodes," *Materials*, vol. 17, no. 8, Art. no. 8, Jan. 2024, doi: 10.3390/ma17081870.

[2] Y. Qin, Z. Wang, K. Sasaki, J. Ye, and Y. Zhang, "Recent progress of Ga$_2$O$_3$ power technology: large-area devices, packaging, and applications," *Jpn. J. Appl. Phys.*, vol. 62, no. SF, p. SF0801, Feb. 2023, doi: 10.35848/1347-4065/acb3d3.

[3] P. Dong *et al.*, "6 kV/3.4 mΩ·cm^2 Vertical β-Ga$_2$O$_3$ Schottky Barrier Diode With BV^2/R$_{on, sp}$ Performance Exceeding 1-D Unipolar Limit of GaN and SiC," *IEEE Electron Device Lett.*, vol. 43, no. 5, pp. 765–768, May 2022, doi: 10.1109/LED.2022.3160366.

[4] W. Guo *et al.*, "β-Ga$_2$O$_3$ Field Plate Schottky Barrier Diode With Superb Reverse Recovery for High-Efficiency DC-DC Converter," *IEEE J. Electron Devices Soc.*, vol. 10, pp. 933–941, 2022, doi: 10.1109/JEDS.2022.3212368.

[5] F. Zhou *et al.*, "1.95-kV Beveled-Mesa NiO/β-Ga$_2$O$_3$ Heterojunction Diode With 98.5 % Conversion Efficiency and Over Million-Times Overvoltage Ruggedness," *IEEE Trans. Power Electron.*, vol. 37, no. 2, pp. 1223–1227, Feb. 2022, doi: 10.1109/TPEL.2021.3108780.

[6] F. Wilhelmi *et al.*, "Switching Properties of 600 V Ga$_2$O$_3$ Diodes With Different Chip Sizes and Thicknesses," *IEEE Trans. Power Electron.*, vol. 38, no. 7, pp. 8406–8418, Jul. 2023, doi: 10.1109/TPEL.2023.3260023.

[7] D. Han, J. Noppakunkajorn, and B. Sarlioglu, "Efficiency comparison of SiC and Si-based bidirectional DC-DC converters," in *2013 IEEE Transportation Electrification Conference and Expo (ITEC)*, Jun. 2013, pp. 1–7. doi: 10.1109/ITEC.2013.6574511.

A Universal Method to Regulate Contact Resistance in Thin Film Transistors

Yanzhuo Wei[1,2], Guohui Li[2], Yanxia Cui[2], Hongwei Hao[1,3], Chen Chen[1], Dongdong Li[1,3], Shan-Ting Zhang[1*]

[1]Zhangjiang Laboratory, Shanghai, China. [2]Taiyuan University of Technology, Taiyuan, China. [3]Shanghai Jiao Tong University, Shanghai, China. (*Email: zhangst@zjlab.ac.cn)

Abstract

An atomic layer deposited ultrathin Al_2O_3 film by using different oxygen precursors (H_2O or O_3) is demonstrated to improve the contact properties of a-IGZO/Mo through different mechanisms. Combining the different processes serves as an easy-to-compliment and universal method to custom design the contact properties between channel and source/drain electrodes for oxide semiconductor based thin film transistors.

Keywords: oxide semiconductor TFTs, contact properties, negative fixed charge

Introduction

Thin film transistors (TFTs) based on oxide semiconductors are regarded as a promising technology for complementary metal-oxide-semiconductor (CMOS) back-end-of-line (BEOL) compatible ($\leq 500\,°C$) applications such as monolithic three-dimensional (3D) integration [1]. The contact properties of source/drain (S/D) electrodes to the channel material are essential to achieve high-performance TFTs [2, 3], particularly for devices made at advanced technology nodes. A low contact resistance (R_c) is highly desirable as it is responsible for achieving high electrical properties such as transconductance (hence mobility), switching properties etc. [4].

Reducing the contact resistance R_c can be achieved by increasing the surface doping concentration (N_D) and/or lowering the Schottky barrier height (Φ_B) of metal-semiconductor contact [3, 5]. One commonly reported strategy is to heavily dope the S/D region to lower R_c [6, 7]. The high-density free carriers, however, would bear the risk to diffuse into the channel layer during the subsequent annealing process, inducing unwanted V_{th} shift [8]. A proper choice of the metal electrode with negligible Schottky barrier height is essential to reduce R_c [9, 10]. For instance, metals such as aluminum (Al) and titanium (Ti) show a proper band alignment with the conduction band edge of amorphous InGaZnO (a-IGZO) thin film [11]. However, oxygen diffusion from a-IGZO to Al and Ti occurs leaving behind oxygen vacancies (V_O) at the interface, which could act as charge traps leading to unstable electrical operation. Moreover, unintentional oxidation at the interface would occur forming an additional potential barrier [12]. In comparison, molybdenum (Mo) appears fairly stable in contact with a-IGZO but the contact resistance remains high.

In this work, we propose an easy-to-implement method that can be universally applied to regulate the contact resistance of the thin film transistors. By simply modifying the oxygen precursor (H_2O or O_3) during the atomic layer deposition (ALD) growth of an ultra-thin (~ 1 nm) Al_2O_3 film on a-IGZO, the amount of negative fixed charge can be tuned at the interface of a-IGZO/Mo, leading to a custom designed modification of the effective Schottky barrier and thus the contact properties.

Experimental Details

The silicon wafer covered with a thermally oxidized SiO_2 layer of 300 nm was ultrasonic cleaned with acetone and isopropanol. A thin layer of about 40 nm tungsten (W) metal was used as the bottom gate electrode. It was patterned by photolithography followed by magnetron sputtering deposition. Subsequently, a layer of about 42 nm Al_2O_3 was deposited as the gate dielectric using atomic layer deposition (ALD) technique, using trimethylaluminum (TMA) as a precursor and H_2O as a reactant. After that, 32 nm a-IGZO was magnetron sputtered (pressure=0.6 Pa, O_2/Ar ratio=4:36). Some of the a-IGZO thin film was further deposited with an ultrathin 1 nm ALD-Al_2O_3 interlayer grown either by TMA+H_2O or TMA+O_3. All the samples were then patterned by photolithography and wet-etched. Finally, a layer of 150 nm Mo was magnetron sputtered as source/drain electrodes. Finally, the as-prepared transistors were annealed at 400 °C for 10 min. The detailed process flow and device schematics were given in **Fig. 1**.

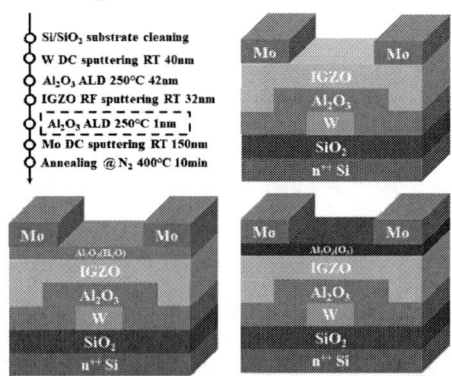

Fig. 1. The detailed process flow and devices schematics.

The physio-chemical properties of the thin films were measured by X-ray photoelectron spectroscopy (XPS) with the Thermo Scientific K-Alpha instrument and ultra-violet photoelectron spectroscopy (UPS) (model Thermo Fisher Scientific ESCALAB XI+). The transfer and output characteristics were obtained using Keithley 4200A-SCS semiconductor analyzer. According to the above process, transistors of different lengths (5,10,15,20,40,50 μm) were fabricated, and their output curves were tested, and the contact resistance was calculated using the transmission line method (TLM) [11].

Results and Discussions

As is seen from the transfer curves in **Fig. 2**, both the transistors employing a single layer of a-IGZO and those treated by an ultra-thin Al_2O_3 film show reasonably high on/off ratio ($\geq 10^9$) with low leakage current (I_g) in the order of 10^{-11} A, suggesting high response speed and low power consumption of the devices[13, 14]. The field-effect mobility (μ_{FE}) of the transistor employing a single-layer IGZO is calculated to be 10.52 cm²/V·s, which increases to 12.21 cm²/V·s and 13.1 cm²/V·s after adding an ultra-thin (~ 1 nm) layer of Al_2O_3 grown by TMA+H_2O and TMA+O_3, respectively.

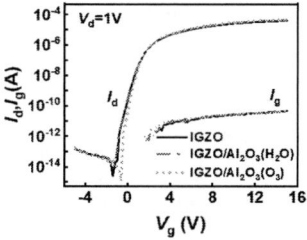

Fig. 2. Transfer characteristic curves along with the leakage current (I_g) of three types of transistors employing a single layer of a-IGZO and those treated by an ultra-thin Al_2O_3 film grown by TMA+H_2O and TMA+O_3, respectively.

Meanwhile, adding the Al_2O_3 ultrathin film greatly enhances the output current (I_d) particularly at high drain voltage (V_d), with the effect being most prominent for $Al_2O_3(O_3)$ (**Fig. 3a**). Concomitantly, one sees from **Fig. 3b** that the contact resistance R_c is effectively reduced with the lowest value shown up for $Al_2O_3(O_3)$. Without doubt, the presence of the ultrathin Al_2O_3 interlayer ameliorates the contact properties between Mo and a-IGZO. A probable reason responsible for the reduced R_c could be an increase in N_D, which is proved to be the case for a-IGZO covered by $Al_2O_3(H_2O)$. As characterized by X-ray photoelectron spectroscopy (XPS) spectra in **Fig. 4a**, the energy difference ($\Delta E_{In3d,\ VBM}$) between In3$d_{5/2}$ core level and valence band maximum (VBM) for the bare IGZO thin film is determined as 442.06 eV, comparable to the values reported by Chen et al. [15]. According to the Kraut's method[16], by measuring the In3$d_{5/2}$ core level of a-IGZO treated by $Al_2O_3(H_2O)$ (**Fig. 4b**), one calculates the VBM energy (i.e. E_F-E_V) to be 2.71 eV, higher than that of bare IGZO (E_F-E_V = 2.57 eV, **Fig. 4a**) by 0.14 eV. This indicates that depositing an ultrathin Al_2O_3

by TMA+H_2O practically increases the carrier concentration in a-IGZO. This is consistent with the experimental observation where top-gate-structured oxide semiconductor TFTs utilizing ALD-A_2O_3 grown by TMA+H_2O often exhibit a negative V_{th} shift [17]. It is suspected that H-related impurities originating from the ALD reaction of TMA+H_2O contribute to the excessive doping [18]. In our case, due to the $Al_2O_3(H_2O)$ being ultrathin of ~ 1 nm, the carrier concentration increment in a-IGZO channel seems not yet sufficient to induce any noticeable V_{th} shift. Nevertheless, the higher N_D narrows the depletion width, rendering the electrons to tunnel more easily through the barrier [6]. In other words, the ultrathin $Al_2O_3(H_2O)$ interlayer between Mo and a-IGZO enhances the probability of electrons tunneling through the contact barrier, thus reducing R_c with elevated I_d.

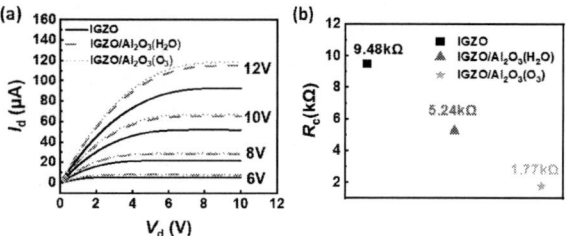

Fig. 3. (a) Output characteristic curves and (b) contact resistance R_c of three types of transistors employing a single layer of a-IGZO and those treated by an ultra-thin Al_2O_3 film grown by TMA+H_2O and TMA+O_3, respectively.

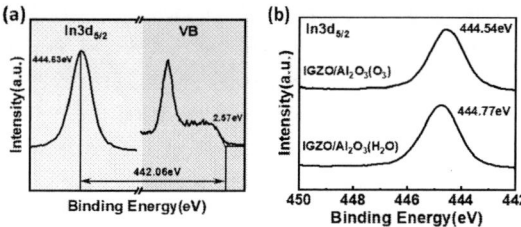

Fig. 4. X-ray photoelectron spectroscopy (XPS) spectra of (a) In 3$d_{5/2}$ core level and valence band (VB) of bare a-IGZO thin fil. (b) In 3$d_{5/2}$ core levels of a-IGZO/$Al_2O_3(H_2O)$ and IGZO/$Al_2O_3(O_3)$ thin films.

By contrast, the VBM for the a-IGZO treated by $Al_2O_3(O_3)$ is reduced by about 0.1 eV, suggesting the defects seem passivated during the ALD growth of $Al_2O_3(O_3)$. Meanwhile, the work function of the $Al_2O_3(O_3)$-passivated a-IGZO increases by ~ 0.3 eV (**Fig. 5**). The fact that the amount of work function increment is more than the VBM reduction suggests that negative fixed charges are most likely present at the a-IGZO/$Al_2O_3(O_3)$ interface. The introduction of negative fixed charges by ALD-deposited Al_2O_3 thin films are widely acknowledged[19]. Upon contact with metallic Mo, the negative fixed charges introduced by the ultra-thin $Al_2O_3(O_3)$ interlayer induce screening charges in the surface of Mo. A sheet of positive charges – the so-called "image charge" - can be thought to exist in Mo [9], which essentially lower the average potential energy in the metal with respect to its value in a-IGZO. In other words, the effective Schottky barrier Φ_B is lowered thus the contact resistance R_c reduced. On the contrary, one notes that the negative fixed charge does not seem to be present in $Al_2O_3(H_2O)$ since its work function reduction (0.17 eV) is practically the same amount as the VBM increment (0.14 eV).

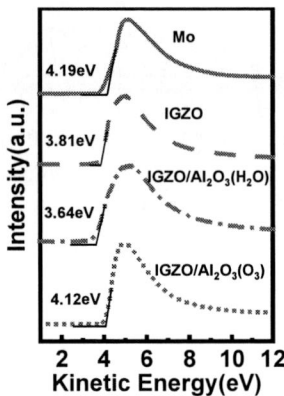

Fig. 5. Secondary electron cutoff edge for Mo, IGZO, a-IGZO/Al_2O_3(H_2O), and a-IGZO/Al_2O_3(O_3) measured by ultraviolet photoelectron spectroscopy (UPS), in which the work function is directly indicated for each sample

It is demonstrated that by simply modifying the oxygen precursor (H_2O or O_3) during the ALD process, one is capable of tuning the amount of negative fixed charges upon depositing Al_2O_3 on a-IGZO. The polarity and number of fixed charges strongly affect the contact properties between the channel and S/D electrodes. Passivating a-IGZO by an ultrathin Al_2O_3 film grown by TMA+O_3 introduces appreciable negative fixed charges and lowers the effective Schottky barrier height. Whereas the Al_2O_3 film grown by TMA+ H_2O does not show such effect, but instead reduces the contact resistance through increased carrier concentration. Care must be taken to the employment of Al_2O_3 thin film grown by TMA+ H_2O so as not to affect the V_{th} stability. The deposition of ALD-Al_2O_3 by either TMA+ H_2O or TMA+O_3 are both mature processes adopted in semiconductor industry, rendering it an easy-to-implement strategy to custom design the contact properties between channel and S/D electrodes for oxide semiconductor based TFTs.

Conclusion

Insertion of an ultrathin ALD-Al_2O_3 film is shown to effectually improve the contact properties of a-IGZO/Mo. The ALD-Al_2O_3 deposited by either TMA+ H_2O or TMA+O_3 reduces the contact resistance through different mechanisms. The former increases the carrier concentration in a-IGZO likely due to the introduction of H-related impurities, thus narrowing the depletion width and enhancing the probability of electrons tunneling through the contact barrier; The latter, on the other hand, introduces negative fixed charges at the interface which lowers the effective Schottky barrier. With both processes being widely adopted in semiconductor industry, a combination of the different ALD processes can be foreseen to stand out as an easy-to-compliment method to custom design the contact properties tailored for designated applications.

References

[1] M. Si et al., "Why In_2O_3 Can Make 0.7 nm Atomic Layer Thin Transistors," Nano Letters, vol. 21, no. 1, pp. 500-506, 2021.

[2] Z. A. Lamport et al., "A simple and robust approach to reducing contact resistance in organic transistors," Nature Communications, vol. 9, no. 1, 2018.

[3] J. H. Jeong et al., "Specific contact resistivity reduction in amorphous IGZO thin-film transistors through a TiN/IGTO heterogeneous interlayer," Sci Rep, vol. 14, no. 1, p. 10953, 2024.

[4] L. Cao, J. Wei, X. Li, S. Wang, and G. Qin, "Enhancing the Performance of MoS_2 Field-Effect Transistors Using Self-Assembled Monolayers: A Promising Strategy to Alleviate Dielectric Layer Scattering and Improve Device Performance," Molecules, vol. 29, no. 17, 2024.

[5] L. Jelver, D. Stradi, K. Stokbro, and K. W. Jacobsen, "Schottky barrier lowering due to interface states in 2D heterophase devices," Nanoscale Adv, vol. 3, no. 2, pp. 567-574, 2021.

[6] A. Y. C. Yu, "Electron tunneling and contact resistance of metal-silicon contact barriers," Solid-State Electronics, vol. 13, no. 2, pp. 239-247, 1970.

[7] A. Agrawal et al., "Fermi level depinning and contact resistivity reduction using a reduced titania interlayer in n-silicon metal-insulator-semiconductor ohmic contacts," Applied Physics Letters, vol. 104, no. 11, 2014.

[8] G. W. Mattson, K. T. Vogt, J. F. Wager, and M. W. Graham, "Hydrogen incorporation into amorphous indium gallium zinc oxide thin-film transistors," Journal of Applied Physics, vol. 131, no. 10, 2022.

[9] C. Berthod, N. Binggeli, and A. Baldereschi, "Schottky barrier heights at polar metal/semiconductor interfaces," Physical Review B, vol. 68, no. 8, 2003.

[10] C. Niu et al., "Surface Accumulation Induced Negative Schottky Barrier and Ultralow Contact Resistance in Atomic-Layer-Deposited In_2O_3 Thin-Film Transistors," IEEE Transactions on Electron Devices, vol. 71, no. 5, pp. 3403-3410, 2024.

[11] H. Park et al., "Enhancing the Contact between a-IGZO and Metal by Hydrogen Plasma Treatment for a High-Speed Varactor (>30 GHz)," ACS Applied Electronic Materials, vol. 4, no. 4, pp. 1769-1775, 2022.

[12] J.-R. Yim et al., "Effects of Metal Electrode on the Electrical Performance of Amorphous In–Ga–Zn–O Thin Film Transistor," Japanese Journal of Applied Physics, vol. 51, no. 1R, 2011.

[13] Y. Feng, K. Lee, H. Farhat, and J. Kong, "Current on/off ratio enhancement of field effect transistors with bundled carbon nanotubes," Journal of Applied Physics, vol. 106, no. 10, 2009.

[14] W. Wu et al., "High mobility and high on/off ratio field-effect transistors based on chemical vapor deposited single-crystal MoS_2 grains," Applied Physics Letters, vol. 102, no. 14, 2013.

[15] X. F. Chen et al., "Modification of band offsets of $InGaZnO_4$/Si heterojunction through nitrogenation treatment," Journal of Alloys and Compounds, vol. 647, pp. 1035-1039, 2015.

[16] S.-T. Zhang et al., "Polymorphism of the Blocking TiO_2 Layer Deposited on F:SnO_2 and Its Influence on the Interfacial Energetic Alignment," The Journal of Physical Chemistry C, vol. 121, no. 32, pp. 17305-17313, 2017.

[17] J.-M. Lee, I.-T. Cho, J.-H. Lee, W.-S. Cheong, C.-S. Hwang, and H.-I. Kwon, "Comparative study of electrical instabilities in top-gate InGaZnO thin film transistors with Al_2O_3 and Al_2O_3/SiNx gate dielectrics," Applied Physics Letters, vol. 94, no. 22, 2009.

[18] D. Baek, S.-H. Lee, S.-Y. Bak, H. Jang, J. Lee, and M. Yi, "Control of Threshold Voltage in ZnO/Al_2O_3 Thin-Film Transistors through Al2O3 Growth Temperature," Electronics, vol. 13, no. 8, 2024.

[19] J. J. H. Gielis, B. Hoex, M. C. M. van de Sanden, and W. M. M. Kessels, "Negative charge and charging dynamics in Al_2O_3 films on Si characterized by second-harmonic generation," Journal of Applied Physics, vol. 104, no. 7, 2008

Breakthroughs in Undoped HfO₂ Ferroelectric Capacitors Achieved through Enhanced Nanocrystallite Seeding in As-Deposited Films via O₂-Plasma ALD

Zongwei Shang[1†], Changqing Ye[2†], Hao Li[1], Xing Wu[2#], Runsheng Wang[1], Ming Li[1*], Ru Huang[1]

[1]School of IC, Peking University, Beijing, China, (*email: liming.ime@pku.edu.cn).

[2]School of IC, East China Normal University, Shanghai, China, (#email: xwu@cee.ecnu.edu.cn). [†]Equal contribution

Abstract

In this work, undoped HfO₂ ferroelectric capacitors with excellent material properties are fabricated by O₂-plasma atomic layer deposition (ALD). Record high ferroelectric performance is demonstrated among all reported undoped HfO₂ results, with *high endurance* (> 10^{11}) and *short switching time* (< 2 ns). An average *large single-crystal size* (> 50 nm) and an orthorhombic phase ratio of over 40% are confirmed by transmission electron microscopy (TEM). Furthermore, the impacts of oxygen vacancy (V_O) and H impurities concentration on the ferroelectric properties are elucidated.

Keywords: undoped HfO₂, seed crystallite, ultra-fast speeds

Introduction

HfO₂-based ferroelectrics have attracted significant attention nowadays due to their excellent CMOS compatibility and scalability, featuring great potential to overcome the application barriers in traditional ferroelectric materials for memory and logic devices [1]. In the past decade, various dopants such as Si, Zr, Al and La have been employed to improve the ferroelectric properties of HfO₂ [2]. However, introducing dopants by the ALD cross flow profile leads to variations of the ferroelectric film properties (thickness and doping) [3], significantly degrading design margins. Therefore, it is highly preferred to develop a undoped HfO₂ with decent ferroelectric performance. However, previously reported undoped HfO₂ devices have not yet reached the desired targets.

In this work, we fabricated undoped HfO₂ ferroelectric capacitors by increasing seed crystallite ratio in as-deposited films via plasma enhance atomic layer deposition (PEALD). Large single grain size (>50 nm) and high orthorhombic (Pca2₁, O-) phase ratio (>40%) observed. The endurance test confirms a superior potential endurance of more than 10^{11}. Furthermore, high switching speed is realized, with saturated polarization of less than 2 ns. Our results demonstrate the potential of undoped HfO₂ as a strong candidate for ferroelectrics and provide new insights into the O-phase formation without other dopants.

Experiment

Fig. 1 shows the detailed fabrication flow of the Metal-Ferroelectric-Metal (MFM) stack. The process started with cleaning the Si/SiO₂ substrate by H_2SO_4/H_2O_2 and then the bottom TiN electrode layer was sputtered by physical vapor deposition (PVD). Then, 10 nm HfO₂ films were grown by PEALD using O₂-plasma and TEMAH in optimized power and oxygen flow rates. Finally, 100 nm TiN top electrode was deposited via sputtering and patterning, and for the formation of ferroelectric phase, the rapid thermal annealing (RTA) was carried out at 600 °C for 90 s in N₂ ambient.

Results and Discussions

The scanning transmission electron microscopy (STEM) high-angle annular dark field (HAADF) cross-sectional image of the HfO₂ film is shown in **Fig. 2(a)**. The magnified image in **Fig. 2(b)** reveals the atomic arrangement, and corresponding STEM annular bright-field (ABF) image in **Fig. 2(c)**. **Fig. 2(d)** is the typical diffraction pattern of the O-phase. **Fig. 2(c)** and **Fig. 2(d)** confirms the O-phase Pca2₁ structure. Furthermore, **Fig. 2(e)** illustrates statistical results of grain size according to multiple high-resolution TEM (HRTEM) images, proving an average grain size exceeding 50 nm.

We have investigated the impact of oxygen plasma power and partial pressure on the formation of seed crystallites in the as-deposited HfO₂ films. It's found that with increased oxygen supply, the ratio of seed crystallites in the film will generally increase but reach a maximum at optimized process conditions. **Fig. 3(a)** shows the wide field-of-view cross-sectional HRTEM of the as-deposited HfO₂ films under optimal condition to clearly show the presence of seed crystallites with size of tens nanometers. The corresponding FFT images of M-phase are shown in **Fig. 3(b)**. **Fig. 3(c)** exhibits a 40% of seed crystallites in the undoped HfO₂ as shown in Fig. 3a, which leads to the crystalline ferroelectric HfO₂ film without any dopant assistance as will be shown later. The undoped HfO₂ films are then annealed in N₂ with capping TiN metal electrode to drive the growth of crystalline grain. **Fig. 4(a)** illustrates the distribution of o-/t-/m-phases and presents evidence that the size of the single crystal exceeds 50 nm, which is believed to be the largest reported for undoped HfO₂. The statistical proportions of o-/t-/m-phases in length are 44%, 47%, and 9%, respectively (**Fig. 4(c)**).

We prepared as-deposited HAO and HZO thin films separately using H₂O as the oxygen source to investigate how doping affects the formation of seed crystallites within the films. As shown in **Fig. 5(a)**, the as-deposited HZO thin film clearly exhibits 15 nm seed crystallites in the cross-sectional HRTEE image, and the corresponding FFT image (**Fig. 5(c)**) confirms that these seeds are of the M-phase. To promote grain growth and phase transformation, the HZO thin films

979-8-3315-0417-5/25 $31.00 © 2025 IEEE

with capping TiN metal electrodes were then annealed in a N_2 environment. **Fig. 5(d)** shows grains larger than 20 nm, and the corresponding FFT image **(Fig. 5(f))** confirms that these grains are of the O-phase. In contrast, the as-deposited HAO thin films introduce fewer and smaller seed crystallites compare to as-deposited HZO films, as shown in **Fig. (6)**. After annealing, O-phase grains also appeared in the HAO thin films. However, the HRTEM images reveal a typical interpenetrating region structure between different phases **(Fig. 6(d))**, which is not continuous like the lattice observed in HZO films. These results indicate that, whether through doping or O_2-plasma to form nano crystallites during the ALD process, these nanocrystalline particles facilitate the subsequent formation of the polar orthorhombic phase after RTA. However, using H_2O and O_2-plasma as oxygen sources has different effects on the crystallization of ferroelectric films, as will be demonstrated later.

The *I-V* curves and *P-V* loops obtained from Positive-Up–Negative-Down (PUND) measurement after different cycles at 3.5 V and 100 kHz are shown in **Fig. 7(a)**, indicating a coercive field of ~1.1 MV/cm and a remnant polarization $(2P_r)$ of ~28 $\mu C/cm^2$. As illustrated in **Fig. 7(b)**, undoped HfO_2 capacitor presents a $2P_r$ of 10$\mu C/cm^2$ even after 10^{11} cycles with the stress cycles of 3 V (*P-V* voltage) and 2MHz (pulse frequency). For transient ferroelectric polarization switching current measurement, the current is obtained by subtracting the non-switching current from the switching current. The polarization-time relation can be determined by integrating the corresponding net switching current as seen in **Fig. 8**. **Table. I** show the equations of the nucleation-limited switching (NLS) model and the extracted parameters [4], which can well replicate the measured *P-t* curves in Fig. 14. Our device exhibits a high switching speed represented by the extracted low τ_∞ value of 1.8 ns on 4μm^2 devices with 10 nm HfO_2. The benchmark in **Table. II** compares the key properties of this work with other reported properties of ferroelectric thin films.

Fig. 9(a) reveals that the hydrogen impurity concentration in the films grown using H_2O as an oxygen source are higher than that in O_2-plasma. The schematic illustration in **Fig. 9(b)** shows high hydrogen impurity concentrations due to the formation of an OH-terminated surface during the reaction between H_2O and Hf precursor. High concentration of hydrogen impurities may reduce the crystallinity and increases the leakage current [5], thus limiting device switching speed and endurance.

Oxygen plasma can promote the formation of oxygen vacancies and nucleation for seed crystallites, which is crucial for the ferroelectric phase formation in undoped HfO_2. **Fig. 10** illustrates the energy gap for varying concentrations of V_O in the monoclinic M- and O-phase. DFT calculations are performed by using the Quantum ATK atomic-scale modelling platform with the linear combination of atomic orbitals method for expanding the electron wave functions [6]. The results indicate that an appropriate concentration of V_O reduces the energy gap between the metastable O-phase and the ground state M-phase, thereby promoting the transition to ferroelectric O-phase. Therefore, by controlling the concentration of V_O, it is possible to optimize the metastable state of the O-phase to improve the ferroelectric properties in undoped HfO_2.

Conclusion

In summary, we successfully fabricated undoped HfO_2 ferroelectric films by adjusting the seed crystallites ratio in the as-deposited film, with optimizations in the power and the O_2 flow rate. Devices with excellent performance were demonstrated, featuring high endurance ($>10^{11}$), high switching speed (< 2 ns) and large grain size (> 50 nm). In addition, the effects of H impurities and V_O on the ferroelectric properties were investigated. This work offers new insights in the ferroelectricity of undoped HfO_2 and reveals its great potential for future non-volatile memory applications.

Acknowledgments

This work was supported by NSFC (61927901, 62125401) and the 111 Project (B18001).

References

[1] M. Saitoh et al., "HfO₂-based FeFET and FTJ for Ferroelectric-Memory Centric 3D LSI towards Low-Power and High-Density Storage and AI Applications," IEDM, pp.18.1.1-18.1.4, 2020.

[2] T. Fu, et al., "Record-high $2P_r$ = 60 $\mu C/cm^2$ by Sub-5ns Switching Pulse in Ferroelectric Lanthanum-doped HfO₂ with Large Single Grain of Orthorhombic Phase >38 nm," IEDM, pp. 6.5.1-6.5.4, 2022.

[3] L. Grenouillet et al., "Nanosecond Laser Anneal (NLA) for Si-Implanted HfO₂ Ferroelectric Memories Integrated in Back-End of Line (BEOL)," VLSI, pp. 1-2, 2020.

[4] N. Gong, et al., "Nucleation limited switching (NLS) model for HfO₂-based metal-ferroelectric-metal (MFM) capacitors: Switching kinetics and retention characteristics," Appl. Phys. Lett. 112, 262903 2018.

[5] S.J. Kim, et al., "Effect of hydrogen derived from oxygen source on low-temperature ferroelectric TiN/Hf₀.₅Zr₀.₅O₂/TiN capacitors," Appl. Phys. Lett. 115.18, 2019.

[6] H.J. Lee, et al. "Scale-free ferroelectricity induced by flat phonon bands in HfO₂," Science 369.6509, 2020.

[7] J. Hur et al., "Interplay of Switching Characteristics, Cycling Endurance and Multilevel Retention of Ferroelectric Capacitor," IEDM, pp. 39.5.1-39.5.4, 2020.

[8] T. Yamaguchi et al., "Highly Reliable Ferroelectric Hf₀.₅Zr₀.₅O₂ Film with Al Nanoclusters Embedded by Novel Sub-Monolayer Doping Technique," IEDM, pp. 7.5.1-7.5.4, 2018.

[9] J.D. Luo et al., "Ferroelectric Undoped HfOₓ Capacitor with Symmetric Synaptic for Neural Network Accelerator," in IEEE Trans. Electron Devices, 68, 3, 2021.

Growth and Structural Characterization

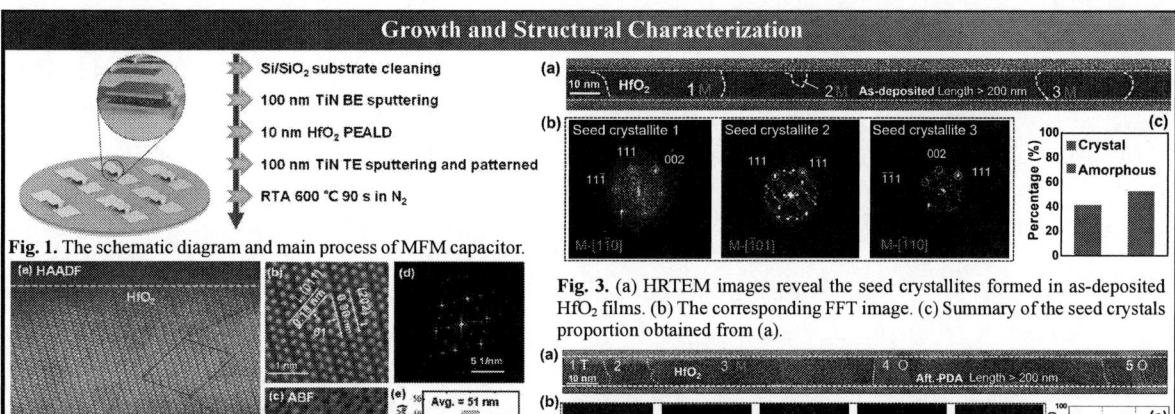

Fig. 1. The schematic diagram and main process of MFM capacitor.

Process steps:
- Si/SiO₂ substrate cleaning
- 100 nm TiN BE sputtering
- 10 nm HfO₂ PEALD
- 100 nm TiN TE sputtering and patterned
- RTA 600 °C 90 s in N₂

Fig. 2. (a) STEM-HADDF image of the HfO₂ layer. (b) Amplified view of partial Figure (a). (c) STEM-ABF images of the O-phase Pca2₁ structure lattice. (d) The corresponding FFT image. (e) The grain size distribution obtained through HRTEM.

Fig. 3. (a) HRTEM images reveal the seed crystallites formed in as-deposited HfO₂ films. (b) The corresponding FFT image. (c) Summary of the seed crystals proportion obtained from (a).

Fig. 4. (a) The cross-sectional HRTEM image reveals a distribution of o/m/t phases. (b) The corresponding FFT image of the o/t/m phases. (c) The statistical proportions of o/t/m phases.

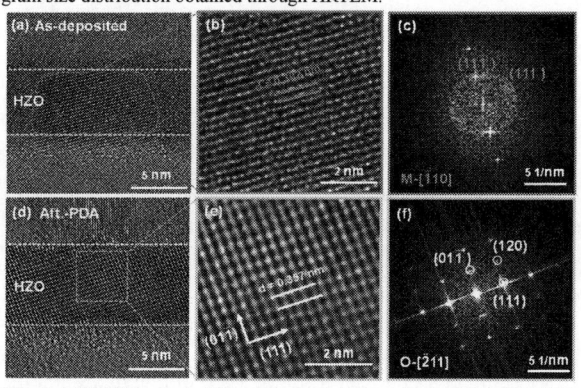

Fig. 5. HRTEM images of the as-deposited (a) and after PDA (d) HZO layers. (b) and (e), Amplified view of partial Figure (a) and (d). (c) and (f) The corresponding FFT image of (b) and (e), respectively.

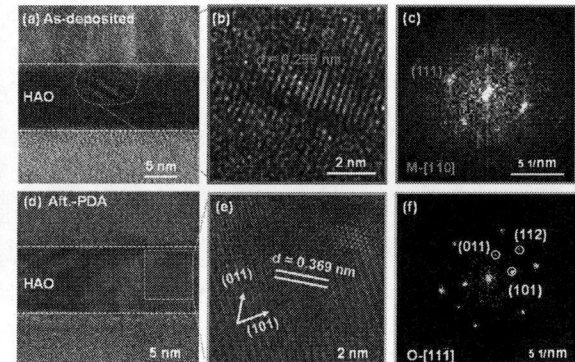

Fig. 6. HRTEM images of the as-deposited (a) and after PDA (d) HAO layers. (b) and (e), Amplified view of partial Figures (a) and (d). (c) and (f)The corresponding FFT image of (b) and (e), respectively.

High-Performance Undoped HfO₂ MFM Capacitors

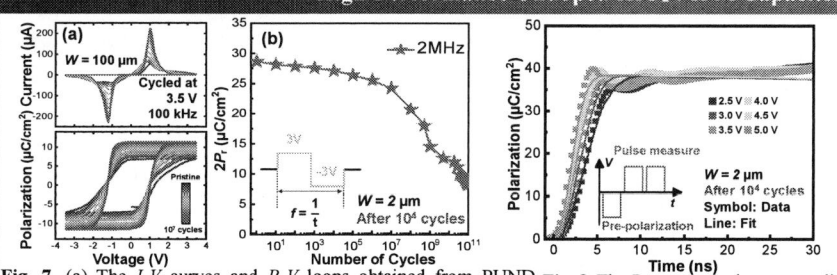

Fig. 7. (a) The I-V curves and P-V loops obtained from PUND measurement. (b) Cycling behavior with the stressing voltages of 3 V.

Fig. 8. The P-t curves and corresponding simulation lines of NLS model.

Equations of NLS model

Time constant	$\tau(E_a, E) = \tau_\infty \exp[(\frac{E_a}{E})^\alpha]$
Cumulative probability	$P(t, E, E_a) = 1 - \exp[-(\frac{t}{\tau(E, E_a)})^\beta]$

Extracted Parameters

Fitting Parameters (α, β)	$\alpha = 2$, $\beta = 2.0$
Time constant of infinite applied Field (τ_∞)	1.8 ns
Activation Field (E_a)	2.3 MV/cm
Standard Deviation for E_a	0.6 MV/cm
Remanent Polarization (P_r)	19 μC/cm²

Table I. Equations of NLS model and the extracted parameters. Assuming E_a follows a normal distribution.

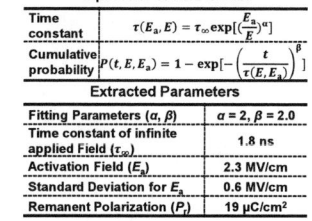

Fig. 9. (a) The depth profiles of H impurity in FE layer deposited using O₂-plasma/TEMAH or H₂O/TDMAH are obtained by SIMS. (b) Schematic of change in surface termination in HfO₂ film upon exposure to H₂O or O₂-plasma.

Fig. 10. The difference in free energy relative to M-phase under varying V_O concentrations. The lowest point is located at 6.25%.

Materials	t (nm)	$2P_r$ (μC/cm²)	Endurance	t_{sw} (ns)	Grain Size (nm)	Ref.
HfO₂:Si	10	19	~10⁸	NA	24	[3]
HfO₂:Zr	10	~24	~10⁶	> 100	NA	[7]
HfO₂:La	11	60	>10⁸	5	>38	[2]
HfO₂:Zr Al	10	~23	~10⁸	NA	~42	[8]
HfO₂	8	26	>10⁸	NA	NA	[9]
HfO₂	10	24	>10¹¹	1.8	>50	This work

Table II. The benchmark of device properties in this work and other previous studies.

979-8-3315-0417-5/25 $31.00 © 2025 IEEE

High-Performance Broadband (300-1600 nm) Si-Based Photodetector Enabled by 2.5D Out-of-Plane Architecture Metasurface Enhancement

Yunfei Xie[1†], *Member, IEEE EDS*, Jin He[2†], Zongwen Li[1], Zhi-Xiang Zhang[1,*], Xiaochen Wang[1], Feng Tian[1], Qianqian Zhang[1], Srikrishna Chanakya Bodepudi[1], Yuan Ma[1], Zijian Pan[1], Muhammad Abid Anwar[1], Bin Yu[1,*], *Fellow, IEEE*, Liaoyong Wen[2,*], Yang Xu[1,*], *Senior Member, IEEE*

[1]Institute for Microelectronics College of Integrated Circuits, State Key Laboratory of Silicon and Advanced Semiconductor Materials, Zhejiang University, ZJU-HIC, Hangzhou, 310027, China.

[2]Key Laboratory of 3D Micro/Nano Fabrication and Characterization of Zhejiang Province, School of Engineering, Westlake University, Hangzhou 310030, China.

†Equal Contribution: Yunfei Xie, Jing He.

*Email: yangxu-isee@zju.edu.cn; wenliaoyong@westlake.edu.cn; yu-bin@zju.edu.cn; zhangzhixiang@zju.edu.cn;

Abstract

Bound states in the continuum (BIC) are highly effective at confining electromagnetic waves, significantly enhancing light-matter interactions. Here, we propose unique plasmonic bound states in the continuum metasurface based on 2.5D metal/Si nanocone arrays with out-of-plane architectures, achieving broadband (300-1600 nm) absorption capabilities. At 1550 nm, the absorption of our device is nearly 100%. Under ultraviolet to visible light irradiation, our detector demonstrates an external quantum efficiency close to unity. (Keywords: 2.5D, Si, Metasurface, Graphene, Broadband photodetector)

Introduction

Broadband photodetectors are capable of detecting light across the ultraviolet (UV) to near-infrared (NIR) wavelength spectrum, rendering them indispensable in fields such as optical communications, environmental monitoring, and medical imaging [1-2]. Si is a prominent material for the development of photodetectors, owing to its abundance, cost-effectiveness, and compatibility with standard CMOS technology. Although the indirect bandgap of Si limits its intrinsic absorption in the NIR range, advanced techniques have been developed to enhance its NIR detection capabilities. These include leveraging localized surface plasmon resonances, integrating Si with two-dimensional materials (such as MoS_2 and graphene), and employing nanostructured architectures like metasurfaces. These innovations enable Si-based photodetectors to achieve improved responsivity to the NIR light [3-4].

In this work, we designed unique Au-Si 2.5D nanocone arrays, which demonstrate broadband absorption due to the bound states in the continuum (BIC) effect. Recognizing the exceptional absorption properties and significant potential of graphene (Gr) in broadband detection, we utilized it to enhance photon absorption, response speed, and photoelectric conversion efficiency. Building on these advances, we developed a Gr-Au-Si 2.5D structure to meet the growing demands for high resolution, miniaturization, and low power consumption in Si-based broadband optoelectronic integrated systems.

Fabrication

The entire fabrication process of the Gr-Au-Si 2.5D Schottky junction photodetector is shown in Fig 1i-iv. We first utilized photolithography and hydrofluoric acid etching to create a 300×300 μm² exposed Si window. The Au electrodes were fabricated using photolithography and magnetron sputtering. Next, we fabricated the special Au-Si 2.5D structure, as shown in Fig. 1A-F. First, a Ni film with periodic nanopillars was used to imprint electropolished aluminum foil under a pressure of 15 kN/cm² for 3 minutes, resulting in an array of nanopits with a spacing of 400 nm on the surface of the aluminum foil. Next, the imprinted aluminum foil was anodized at 30°C under 160 V for 15 minutes in the solution of 5 wt% phosphoric acid, forming an array of nanopores (A-pores). The A-pores were subsequently enlarged through a 15-minute treatment in 5 wt% phosphoric acid solution. Following this, a thin layer of TiO_2 was deposited inside the A-pores using an atomic layer deposition (ALD) system to protect the A-pores from chemical etching. Subsequently, an ion beam etching (IBE) system was employed to mill the top surface of the AAO template, ensuring the complete opening of the B-pores. A layer of PMMA solution was then coated on the surface of the AAO template, and the unoxidized aluminum was removed using a mixed solution of 1.5 wt% copper chloride and 53.2 wt% hydrochloric acid. The PMMA layer was dissolved in acetone. Finally, the AAO template was immersed in 0.1 mol/L NaOH solution at room temperature, and the pore size was adjusted by controlling the immersion time, leading to the formation of a set of B-pores at the fourfold junctions of the A-pores. The AAO template was then placed onto a Si substrate, and the first etching step was performed using inductively coupled plasma (ICP) etching to create pore structures in the Si. The barrier layer of the A-pores was subsequently milled off using IBE, followed by physical vapor deposition (PVD) of Cr (2 nm)/Au (20 nm).

979-8-3315-0417-5/25 $31.00 © 2025 IEEE

The Cr/Au layers served as a protective mask for the AB pores. After the AAO template was removed, the second etching step was performed, resulting in the structures with varying heights. The third drilling and etching step is carried out to etch the Si pillar transversely to obtain the final structure of this work. Finally, we transferred the Gr and patterned it, the Gr-Au-Si 2.5D Schottky junction photodetector was performed.

Results

Fig. 2a presents a schematic illustration of the device, which incorporates a Schottky junction made of Gr, Au-Si 2.5D. Fig. 2b and 2c show the scanning electron microscope and optical microscope images of the device, respectively. The device exhibits an efficient light-trapping effect, with some parts of the structure reflecting almost no light. Due to the unique preparation process of the device, we can integrate it with the Si-based CMOS back-end of-line process. Figure 2d illustrates the schematic diagram of the wafer-level preparation results of the device. Figure 1f displays the absorption spectral curve of the device. The device shows broad absorption characteristics over a wavelength range from 300 to 1600 nm, with absorption peaks at 375, 532, 808, 1064, and 1550 nm, respectively. The Finite-Difference Time-Domain simulations of the 2.5D structure (Fig. 3a) optimize the system by breaking the in-plane C_2 symmetry through alterations in the diameter of the Au/Si columns and adjusting their height differences to disrupt the out-of-plane symmetry, as shown in Fig. 3b-e. At visible light, surface plasmon resonance effectively localizes and amplifies the light field, stimulating thermoelectric charge carriers through non-radiative processes such as Landau damping, and facilitating energy transfer to the electronic system. Within the Gr-Au-Si 2.5D composite metasurface, the strong electric field generated by reverse bias voltage significantly enhances electron transport efficiency by effectively capturing excited charge carriers before they can undergo recombination, scattering, or phonon dissipation. At 1550 nm, the loss of symmetry protection results in the coupling of BIC with radiative modes, leading to the formation of quasi-bound states and achieving nearly 100 % absorption at 1550 nm. Due to the conical structure of the Si nanocones (indicated by the red arrow), the thermal carriers reaching the Schottky junction interface are directed within a conical region, which gives them specific kinetic energy and further facilitates the effective injection of thermal carriers.

To accurately evaluate the photoresponse of the device, the responsivity (R) and external quantum efficiency (EQE) were calculated [2]. Fig. 4a-d shows a comparison of the R between our detector and the device fabricated using planar Si under different light irradiance. At -2 V, the detector demonstrates an R of up to 285 mA/W and an EQE of 94.1% under 375 nm light; 409 mA/W, and 95.4% under 532 nm light; 640 mA/W and 70.4% under 1064 nm light; and 2.46 mA/W and 2.1‰ under 1550 nm light. We compared the performance of our photodetector with the reported devices [3-10], and it can be seen that our device has a broadband photoresponse and exhibits good R and EQE at 1550 nm.

Conclusion

In this study, we developed a high-performance Gr-Au-Si 2.5D Schottky junction photodetector with impressive broadband photoresponse from 300 nm to 1600 nm. The innovative fabrication process ensures compatibility with Si-based CMOS technologies, positioning our device as a strong contender for next-generation optoelectronic systems.

Acknowledgments

This research received funding from several sources, including the National Key R&D Program of China 2022YFA1204304, National Natural Science Foundation of China (62104200, 92164106, U22A2076), the Natural Science Foundation of Zhejiang Province (grant LDT23F04013F04).

References

[1] N. Na, et al., "Room Temperature Operation of Germanium–Silicon Single-photon Avalanche Diode", Nature, 627, 295-300(2024).

[2] Y. F. Xie, et al., "First Demonstration of 2.5D Out-of-Plane-Based Hybrid Stacked Super-Bionic Compound Eye CMOS Chip with Broadband (300-1600 nm) and Wide-Angle (170°) Photodetection", IEEE International Electron Devices Meeting (IEDM), 41.4 (2024).

[3] Y. Zou, et al., "Flexible, active-matrix flat-panel image sensor for low dose X-ray detection enabled by integration of perovskite photodiode and oxide thin film transistor", IEEE International Electron Devices Meeting (IEDM), 8.5 (2019).

[4] J. Liu, et al., "Flexible and Broadband Colloidal Quantum Dots Photodiode Array for Pixel-level X-ray to Near-infrared Image Fusion", Nat. Commun.,14, 5352 (2023).

[5] C. H. Xu, et al., "Bias-Selectable Si Nanowires/PbS Nanocrystalline Film n–n Heterojunction for NIR/SWIR Dual-Band Photodetection", Adv. Funct. Mater., 33, 2214996 (2023).

[6] L. Lin, et al., "An Electrically Modulated Single-Color/Dual-Color Imaging Photodetector", Adv. Mater., 32, 1907257 (2020).

[7] H. B. Yu, et al., "Highly responsive broadband (250-1000 nm) DUV-NIR Photodetector and Tunable Emitter Enabled by III-V Nanowire on Silicon for Integrated Photonics", IEEE International Electron Devices Meeting (IEDM), 20.4 (2023).

[8] Z. J. Lu, et al., "Ultrahigh Speed and Broadband Few-Layer MoTe₂/Si 2D–3D Heterojunction-Based Photodiodes Fabricated by Pulsed Laser Deposition", Adv. Funct. Mater., 30, 1907951 (2020).

[9] Y. Zhou, et al., "An Ultrawide Field-of-view Pinhole Compound Eye Using Hemispherical Nanowire Srray for Robot Vision", Sci. Robot., 9, 8666 (2024).

[10] X. P. Feng, et al., "Spray-coated perovskite hemispherical photodetector featuring narrow-band and wide-angle imaging", Nat. Commun.,13, 6106 (2022).

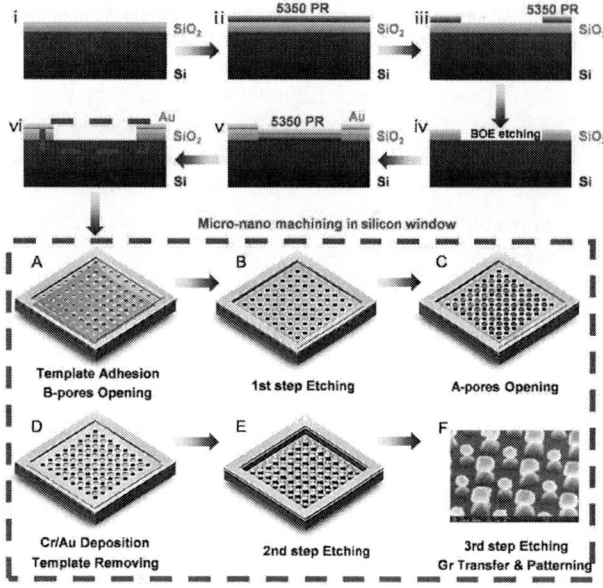

Fig. 1. The fabrication flow of the device.

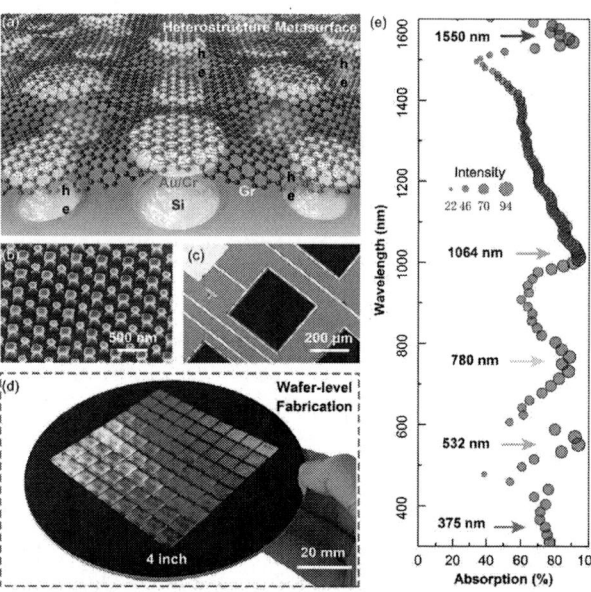

Fig. 2. Schematic illustrate of the device, measured absorption spectra, and wafer-level microfabrication fabrication.

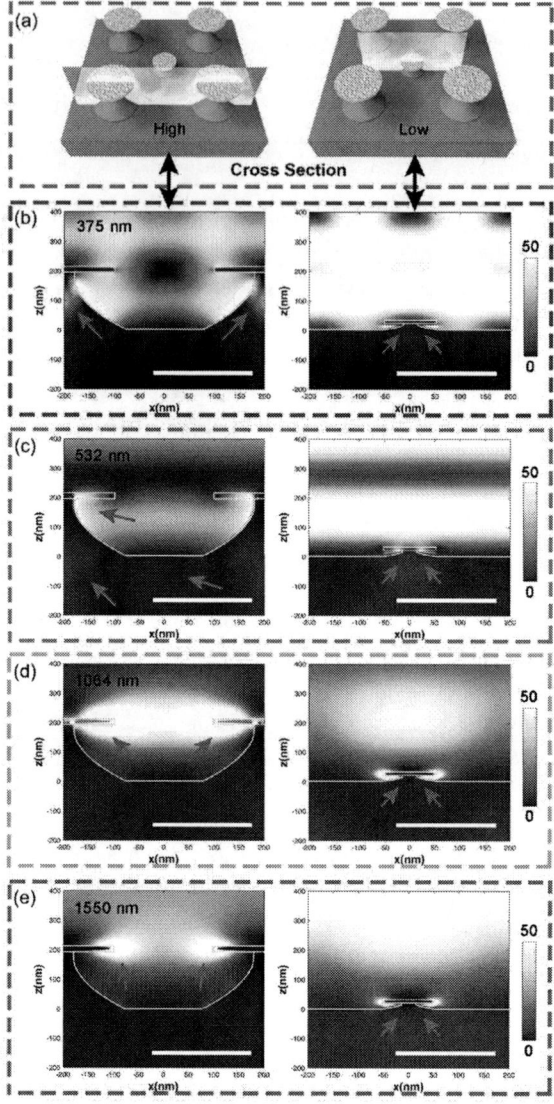

Fig. 3. The distribution of electromagnetic field strength on Au-Si 2.5D metasurface with different cross-sections. All scale bars represent 200 nm.

Device Structure	Range (nm)	R (mA/W) @ λ(nm)	EQE (%) @ λ(nm)	Reference
2D Graphene/ 2.5D Out-of-Plane Architecyures Metasurface	300-1600	285@375 409@532 604@1064 2.46@1550	94.1%@375 95.4%@532 70.4%@1064 2.1%@1550	This work
TFT/Perovskite	X-ray, 400-900	200@520	57%@520	[3] IEDM (2019).
PbS CQDs	X-ray, 400-1200	380@620 340@970	76.6%@620 43.4%@970	[4] Nat. Commun. (2023).
PbS/Si	400-2000	25@400	7.75%@400	[5] Adv. Funct. Mater. (2023).
Perovskite	530-800	50@530 100@800	11%@530 15.5%@800	[6] Adv. Mater. (2021).
AlGaN/Si	254-1000	350@254 180@1000	22.3%@1000	[7] IEDM (2023).
MoTe₂/Si	300-1800	190@980	24%@980	[8] Adv. Funct. Mater. (2020).
Perovskite Nanowire Array	400-798	123.6@798	N/A	[9] Sci. Robot. (2024).
Perovskite Hemispherical	500-900	13.8@600	4%@600	[10] Nat. Commun. (2022).

Table 1. Comparison of the broadband detection performance of different devices based on silicon, III-V materials, and frontend materials.

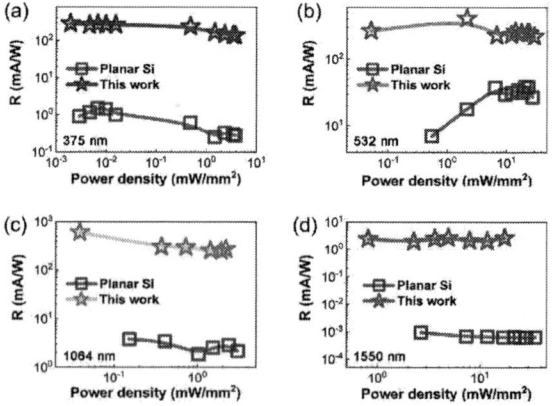

Fig. 4. The graphs of the R of the device experimentally measured under different wavelengths of excitation.

979-8-3315-0417-5/25 $31.00 © 2025 IEEE

A Bionic eye with color vision, environmental adaptivity and neuromorphic signal processing functions

Zhenghao Long[1], Xiao Qiu[1], Chak Lam Jonathan Chan[1], Zhibo Sun[1], Zhengnan Yuan[1], Zhiyong Fan[1]

[1]The Hong Kong University of Science and Technology, Hong Kong SAR, China,

zlongab@connect.ust.hk

Abstract

Here we present a novel design for a bionic eye that incorporates adaptive optics and features a hemispherical nanowire array retina. The retina's color-sensitive bidirectional photo-response enables this unique filter-free color imaging capability. Additionally, the retina's synaptic behavior-based neuromorphic preprocessing, combined with its self-powered operation, significantly lowers the system's energy consumption. Furthermore, the integration of adaptive optics in our bionic eye provides a tunable focal length and an expanded dynamic range.

Keywords: Image sensors, Bionic eye and Neuromorphic optoelectronics.

Introduction

Artificial vision devices are crucial in intelligent systems, including surveillance, autonomous vehicles, and robotics, where monitoring a broad environmental range necessitates devices with a wide field of view (FoV). However, conventional Complementary Metal-Oxide-Semiconductor (CMOS) devices face significant challenges, such as severe optical aberrations and the burden of processing large volumes of frame-based data. Bionic eye devices, featuring hemispherical detector geometry, offer promising solutions for low-aberration, wide FoV sensing[1]–[3]. Despite this, their capabilities in spectrum detection, environmental adaptivity, and in-sensor data processing[4] remain limited.

To overcome these challenges, here we show a pioneering bionic eye design. This device primarily relies on a hemispherical all-inorganic CsPbI$_3$ nanowire array for optical sensing. This array generates photocurrent without external bias, enabling a self-powered mode of operation. Remarkably, we have identified an electrolyte-assisted color-dependent bidirectional synaptic photo-response in a carefully engineered hybrid nanostructure. This response provides the retina with a unique filter-free color imaging capability. Additionally, the retina's synaptic behavior-based neuromorphic preprocessing, combined with its self-powered nature, significantly reduces the system's energy consumption. We have demonstrated that this device can reconstruct color images with high fidelity, enhancing convolutional neural network (CNN) classifications. Furthermore, our bionic eye integrates adaptive optics[5], [6], featuring an artificial crystalline lens and an electronic iris

based on liquid crystals. The artificial crystalline lens can adjust its focal length to detect objects at various distances, while the electronic iris controls the light intensity reaching the retina, thus improving the dynamic range.

Experiments

Fig. 1a schematically illustrates the human retina, while Fig. 1b details the structure of an individual cone cell unit. In the human retina, dense photoreceptor arrays, comprised of rods and cones, are vertically aligned in a hemispherical formation. Upon receiving optical stimuli through the cornea and crystalline lens, these cells absorb light to generate neuro-electric pulses. These pulses are then relayed to the retina's neural network for initial processing, where key features such as color, brightness, and the speed of moving objects are extracted and encoded[7]. Subsequently, this processed information is forwarded to the visual cortex. The efficient opto-electric conversion and integrated sensing-computation of our model enhance its effectiveness as a data acquisition system. Drawing inspiration from the intricate design and functionality of the human retina, we have engineered a unique hybrid nanostructure, depicted schematically in Fig. 1c. We utilize a dense, hemispherical CsPbI$_3$ nanowire array ($>10^9$ cm^{-2}) to emulate photoreceptors. Fig. 1d shows the schematic of the hemispherical nanowire array, which is assembled within a hemispherical porous aluminum membrane (PAM). This structure not only supports the nanowires but also isolates them to prevent crosstalk. The length of the CsPbI$_3$ nanowires is critical for optimizing photodetection performance. Additionally, our design incorporates electrically tunable optics to enhance optical adaptivity. Fig. 1e and f display the overall structure and a magnified view of our bionic eye, respectively.

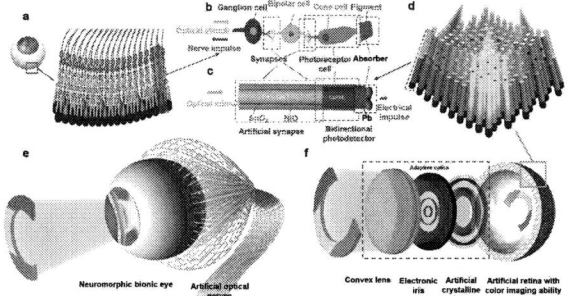

Figure 1 (a-b) Retinal neurons of human eye. (c-d) The nanowire array in the bionic eye device. (e-f) The structure of the bionic eye device.

Fig. 2 illustrates the pixel-level photo-response of our device. To minimize power consumption to as low as 0 W, we have specifically assessed the device's performance under 0 V bias voltage. Fig. 2a–c display the self-powered photocurrent responses under red, green, and blue light pulses, all at the same light intensity of 11 mW cm^{-2}. The device generates a positive current under blue light, but a negative current under red and green light. Notably, the current amplitude under green light exceeds that under red light. This differential response facilitates a unique filter-free color recognition capability: blue light is identifiable by the polarity of the current, while red and green light can be distinguished by the amplitude of the current.

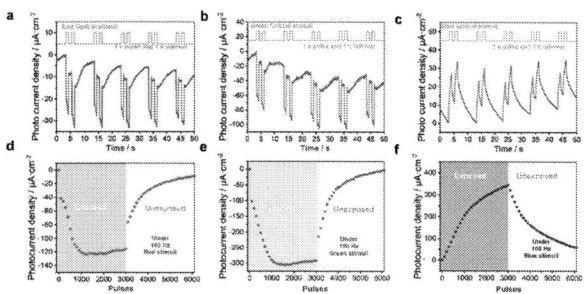

Figure 2 (a-c) Pixel level photo response under (a) red, (b) green, and (c) blue light. (d-f) Synaptic plasticity under (d) red, (e) green, and (f) blue light.

In addition to the bidirectional photo-response, the device exhibits artificial synaptic plasticity. The photocurrent gradually increases when exposed to light and decreases when the light is removed, mimicking the biological process of synaptic strengthening and weakening in the human brain. When subjected to two consecutive optical pulses, the response to the second pulse is notably stronger than to the first, suggesting an artificial paired-pulse facilitation. To further explore the bidirectional synaptic photo-response, we have examined the optical potentiation and natural depression of the device. Fig. 2d–f illustrates the photocurrent behavior under 3,000 optical pulses at a frequency of 100 Hz, followed by a depression phase in darkness. The progressive increase in photocurrent with each additional pulse demonstrates the device's potential for neuromorphic preprocessing applications.

Fig. 3 presents the pattern reconstruction and in-sensor signal processing capabilities of the bionic eye device. Fig. 3a compares the field of view (FoV) between the hemispherical detector geometry of our device and its planar counterpart. Our device achieves an ultra-wide FoV of 140°, significantly exceeding the 86° FoV of the planar model. Fig. 3b displays a photograph of the device, while Fig. 3c details its pixel distribution. The device uses nickel microneedles to emulate visual nerves, where each pixel occupies a footprint of approximately 7.85×10^{-5} cm^2, defined by the cross-sectional area of a microneedle. Mirroring the human eye, the nanowire array retina of our device features photoreceptors with a polar distribution; the density of these photoreceptors decreases from the fovea towards the periphery. This gradient in pixel density helps achieve an excellent balance between the number of pixels and the visual range. Notably, the device's polar pixel distribution allows for a recognition accuracy comparable to that of traditional matrix distributions, particularly for images centered in the visual field. The area-dependent pixel density is further illustrated in Fig. 3c.

Fig. 3d demonstrates the color pattern reconstruction capabilities of the device, showing images reconstructed with high fidelity. Fig. 3e tracks the detection and progressive clarity of a letter "C" under extended illumination times, showcasing the device's neuromorphic contrast enhancement capabilities. This enhancement suggests that carriers generated by previous stimuli can be partially stored within the device, improving lateral response. The preprocessing functions of the device, such as in-device noise filtering, significantly enhance recognition accuracy in real-life applications. These capabilities are demonstrated in Fig. 3f and g.

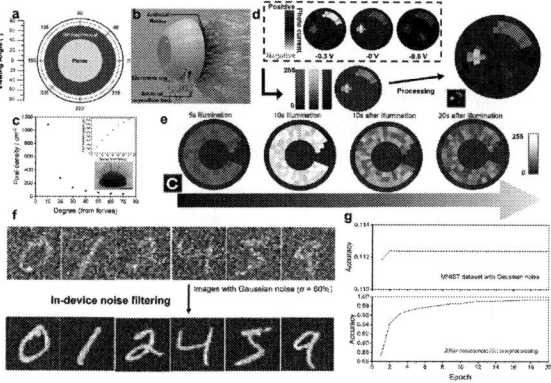

Figure 3 (a) FoV of the hemispherical retina and its planar counterpart. (b) Photograph of the bionic eye device. (c) Pixel distribution. (d) Demonstration of color pattern reconstruction. (e) Demonstration of in-sensor contrast enhancement. (f) Demonstration of in-sensor noise filtering.

In addition to its spectrum detection and signal processing capabilities, the environmental adaptability of vision devices is crucial. To enhance its versatility across different application scenes, we have integrated liquid crystal-based tunable optics, as demonstrated in Fig. 4. This integration includes both an artificial crystalline lens and an artificial iris. The artificial crystalline lens, constructed as a liquid crystal Pancharatnam-Berry lens, can quickly switch (within 16 ms) between light refraction mode, with a focal length of 15 cm, and a normally transparent mode. This rapid switching allows the system's focal length to vary between infinity and 25 cm. Such functionality enables the device to detect objects at various distances, thereby expanding its depth of field. The functional demonstration of this feature is shown in Fig. 4a. Additionally, we have incorporated an electronic iris, designed with five concentric rings covering a total area of 78.5 mm². The transparency of each ring can be individually controlled from 0% to 95%, allowing the overall aperture size to be adjusted between 3.14 mm² and 78.50 mm². This feature facilitates fast control (within 13 ms) of the light intensity reaching the retina, similar to the function of a human iris. The dynamic range of the device is thus significantly expanded, as demonstrated in Fig. 4b.

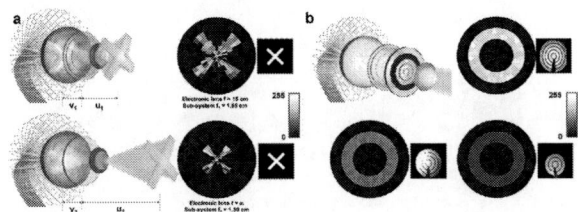

Figure 4 (a) Pattern reconstruction from various object distances. (b) Pattern reconstruction under various brightness.

Conclusion

In this study, we developed a novel hemispherical bionic retina and spherical eye device, addressing previous shortcomings in color vision, optical adaptivity, and in-sensor signal processing. Our bionic eyes, equipped with filter-free color vision, neuromorphic processing, and adaptive optics, represent a breakthrough in artificial vision systems. By emulating the structural and functional characteristics of biological Seyes, this new paradigm opens considerable opportunities for further enhancements and investigations into biomimetic functionalities. These advancements are poised to significantly boost the capabilities of machine vision and robotics, marking a substantial step forward in the field.

Acknowledgments

This work was supported by the Science and Technology Plan of Shen Zhen (JCYJ20170818114107730, JCYJ20180306174923335), The General Research Fund (projects 16205321, 16214619) from the Hong Kong Research Grant Council, Innovation Technology Fund (GHP/014/19SZ), Guangdong-Hong Kong-Macao Intelligent Micro-Nano Optoelectronic Technology Joint Laboratory (project 2020B1212030010), and Foshan Innovative and Entrepreneurial Research Team Program (2018IT100031). We also acknowledge the support from the Center for 1D/2D Quantum Materials and the State Key Laboratory of Advanced Displays and Optoelectronics Technologies at HKUST.

References

[1] Z. Long *et al.*, "A neuromorphic bionic eye with filter-free color vision using hemispherical perovskite nanowire array retina," *Nat. Commun.*, vol. 14, no. 1, p. 1972, Apr. 2023, doi: 10.1038/s41467-023-37581-y.

[2] H. C. Ko *et al.*, "A hemispherical electronic eye camera based on compressible silicon optoelectronics," *Nature*, vol. 454, no. 7205, pp. 748–753, Aug. 2008, doi: 10.1038/nature07113.

[3] L. Gu *et al.*, "A biomimetic eye with a hemispherical perovskite nanowire array retina," *Nature*, vol. 581, no. 7808, pp. 278–282, May 2020, doi: 10.1038/s41586-020-2285-x.

[4] F. Zhou and Y. Chai, "Near-sensor and in-sensor computing," *Nat. Electron.*, vol. 3, no. 11, pp. 664–671, Nov. 2020, doi: 10.1038/s41928-020-00501-9.

[5] Z.-N. Yuan *et al.*, "42.5: Fast Switchable Multi-Focus Ferroelectric Liquid Crystal Lenses for Virtual Reality," *SID Symp. Dig. Tech. Pap.*, vol. 52, no. S2, pp. 532–532, Aug. 2021, doi: 10.1002/sdtp.15186.

[6] L. Li, C. Liu, H. Ren, and Q.-H. Wang, "Adaptive liquid iris based on electrowetting," *Opt. Lett.*, vol. 38, no. 13, p. 2336, Jul. 2013, doi: 10.1364/OL.38.002336.

[7] F. M. Toates, "Accommodation function of the human eye," *Physiol. Rev.*, vol. 52, no. 4, pp. 828–863, 1972, doi: 10.1152/physrev.1972.52.4.828.

Positive and Negative Photoresponses in MoS₂ Flakes Photogated by PN Junction Diode

Yumeng Liu, Zhengfang Fan, Jianyong Wei, Yizhuo Wang, Zhijuan Su[*], Yaping Dan[*]

University of Michigan – Shanghai Jiao Tong University Joint Institute, Shanghai Jiao Tong University, Shanghai, 200240 China, Email: yumengliu@sjtu.edu.cn

Abstract

2D materials like molybdenum disulfide (MoS₂) have shown a significant potential in photodetection due to their tunable bandgap and strong light-matter interactions. In this study, we have developed a silicon-based PN junction phototransistor integrated with few-layer MoS₂ to achieve tunable positive and negative photoconductivity. The device modulates photocurrent through a photo-induced voltage from the PN junction, and photocurrent measurements reveal a logarithmic dependence on light intensity.

Keywords: MoS₂, positive photoconductivity, negative photoconductivity

Introduction

Phototransistors based on two-dimensional (2D) materials, such as molybdenum disulfide (MoS₂), have garnered significant attention for their remarkable electronic and optoelectronic properties, making them well-suited for photodetection applications [1], [2], [3], [4]. MoS₂, in particular, stands out due to its high absorption coefficient and layer-dependent tunable bandgap, allowing for strong light-matter interactions that are essential in sensitive photodetectors [5], [6], [7].

While there have been reports of both positive and negative photoconductivity in MoS₂, these findings often arise from the intrinsic properties of the material itself [8], [9], [10]. This focus on material characteristics presents challenges in consistently replicating results and integrating them into large-scale applications. Such variability complicates the development of reliable photodetectors capable of operating effectively in diverse environments.

In this paper, we report a novel approach to achieving both positive and negative photoconductivity in a phototransistor by integrating few-layer molybdenum disulfide (MoS₂) with a silicon-based PN junction photodiode. We leverage the orientation of the PN junction to control and modulate the photoconductive response. This dual photoconductive functionality represents a significant advancement, enabling controlled switching between positive and negative photoconductivity under varying conditions. Such tunability holds promise for advanced applications in optical sensing, imaging, and adaptable photodetection across diverse lighting and environmental conditions, where dynamic control over photoconductive behavior is essential for effective performance.

Result and discussion

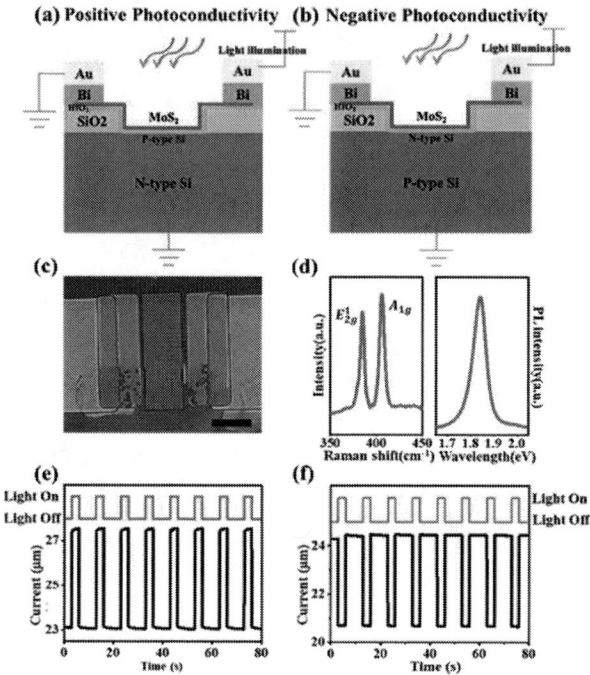

Figure 1 (a) Schematic of the device with a PN junction for positive photoconductivity and (b) Schematic of the device with a NP junction for negative photoconductivity. (c) Optical microscopy image of a fabricated device. The reference bar is 50μm. (d) Raman spectroscopy of MoS₂ (left) and Photoluminescence spectrum of MoS₂ (right) under green laser excitation ($\lambda = 532$ nm). Current transients of the (e) positive and (negative) photoconductivity device at a bias of 1 V, where illumination at $\lambda = 780$ nm is pulsed ON/OFF periodically.

Fig. 1(a) and (b) illustrate the schematic design of our device structure, comprising a silicon-based PN or NP junction diode with a few layers of MoS₂ deposited on top. A 5 nm layer of hafnium oxide (HfO₂) serves as the gate dielectric on the PN junction, effectively separating it from the MoS₂ layer. Upon illumination, the PN junction diode generates a photo-induced voltage, which is regulated by the HfO₂ gate layer to control the gate of the MoS₂ transistor. By altering the orientation of the PN junction in the substrate, we can achieve either positive or negative photoconductivity. Fig. 1(c) provides an optical microscope image of a representative device, where the gray area denotes the highly doped region of the PN junction. Devices with positive and negative photoconductivity are distinguished only by the direction of the PN junction at the center, resulting in similar optical micrographs for both.

The device fabrication process - is as follows: First, a highly doped inversion region is created on a substrate with lower doping concentration through ion implantation to form the PN junction. Subsequently, HfO_2 is deposited via atomic layer deposition. A few layers of MoS_2 are then transferred onto the structure using a gold-assisted transfer technique[11], followed by the application of a 100 nm Au/ 20 nm Bi electrode to establish contact with the MoS_2 layer. These electrodes form Ohmic contacts with MoS_2, as bismuth, being a semimetal with a work function closely matching the minimum conduction band of MoS_2, minimizes metal-induced interstitial states[12], [13].

To evaluate the material quality of the MoS_2 flakes, Raman and photoluminescence (PL) spectroscopy were conducted on the sample. The Raman spectrum, presented on the left of Fig. 1(d), shows two distinct peaks associated with the E2g and A1g vibrational modes of MoS_2, respectively. The separation between these peaks is measured at 21.56 cm^{-1}, indicating that the sample is few-layer MoS_2. The PL spectrum, acquired with green laser excitation ($\lambda = 532$ nm), is displayed on the right of Fig. 1(d), showing a prominent peak at 1.86 eV, characteristic of the MoS_2 crystal[14], [15]. Fig. 1(e)and (f) illustrate the positive and negative photoconductivity characteristics of the device, biased at a fixed source-drain voltage of 1V with the substrate grounded. A 780 nm LED is periodically cycled on and off at a 30% On / 70% Off ratio to illuminate the device. As shown, the two devices demonstrate distinct positive and negative photoconductivity effects depending on their structural orientation and exhibit rapid response to changes in illumination conditions.

The photocurrent in the MoS_2 channel can be expressed as shown in Eq. (1).

$$I_{ph} = I_{th}\ln\left(\frac{P_{light}}{P_{lig}^s} + 1\right) \qquad (1)$$

, where P_{light}^s is the critical light intensity, which indicates the performance of the PN junction as a photovoltage generator upon illumination. The threshold current I_{th} is a parameter showing - the MoS_2 flake's response in current to the photogate voltage.

Fig. 2(a) and (b) show the variation of photocurrent in positive and negative photoconductive devices with changes in light intensity and temperature. It can be seen that as the light intensity decreases, the photocurrent also gradually decreases. Excluding the influence of temperature on the dark current baseline, it is seen that as the temperature decreases, the light-to-dark current ratio increases, indicating that the photodetector sensitivity improves as the temperature decreases. We extracted the relationship between photocurrent and light intensity and compared them with Eq.

(1). As shown in Fig. 2(c) and (d), the logarithmic relationship between photocurrent and light intensity is linear, and the data fit the equation well.

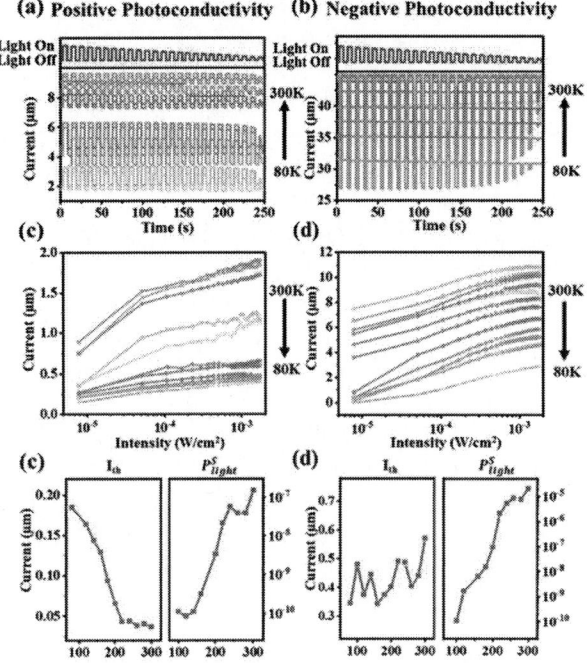

Figure 2 (a) positive and (b) negative photoconductivity device photocurrent with decrease illumination intensity, for illumination at $\lambda =780$ nm and at different temperatures. Measured photocurrent vs illumination intensity for (c) positive and (d) negative photoconductivity device. Extract threshold current (left) and extracted critical light intensity (right) for (c) positive and (d) negative photoconductivity device as a function of temperature, for illumination at different wavelengths.

We further extracted P_{light}^s and I_{th} for both positive and negative photoconductive devices and plotted them as functions of temperature, presented on the left and right sides of Fig. 2(e) and (f), respectively. Changes in I_{th} are notably complex due to various influencing factors, making accurate predictions challenging. Meanwhile, P_{light}^s decreases with temperature reduction, as the intrinsic carrier concentration n_i of the semiconductor decreases at lower temperatures. This corroborates that the device exhibits heightened sensitivity to lower light intensities in low-temperature environments.

Conclusion

This study successfully demonstrates a MoS_2-based phototransistor on a silicon PN junction substrate, exhibiting tunable positive and negative photoconductivity controlled by the orientation of the PN junction. Experimental results show that the device's photocurrent has a logarithmic dependence on light intensity, aligning well with the proposed theoretical model. Additionally, the influence of temperature further highlights the device's sensitivity in low-temperature environments. These findings underscore the potential of MoS_2-based phototransistors for highly sensitive applications.

Acknowledgments

This work was financially supported by the National Science Foundation of China (NSFC) (No. 62304131 and No. 92065103), the Oceanic Interdisciplinary Program of Shanghai Jiao Tong University (No. SL2022ZD107), the Shanghai Jiao Tong University Scientific and Technological Innovation Funds (No. 2020QY05), and National Key Laboratory of Infrared Detection Technologies (No. IRDT-23-10). The devices were fabricated at the Center for Advanced Electronic Materials and Devices (AEMD), and Raman and PL measurements were conducted at the Instrumental Analysis Center (IAC), Shanghai Jiao Tong University.

[1] R. Wadhwa, A. V. Agrawal, and M. Kumar, "A strategic review of recent progress, prospects and challenges of MoS 2 -based photodetectors," J. Phys. D: Appl. Phys., vol. 55, no. 6, p. 063002, Feb. 2022, doi: 10.1088/1361-6463/ac2d60.

[2] F. Yu, M. Hu, F. Kang, and R. Lv, "Flexible photodetector based on large-area few-layer MoS2," Progress in Natural Science: Materials International, vol. 28, no. 5, pp. 563–568, Oct. 2018, doi: 10.1016/j.pnsc.2018.08.007.

[3] P. Gant, P. Huang, D. Pérez De Lara, D. Guo, R. Frisenda, and A. Castellanos-Gomez, "A strain tunable single-layer MoS2 photodetector," Materials Today, vol. 27, pp. 8–13, Jul. 2019, doi: 10.1016/j.mattod.2019.04.019.

[4] H. S. Nalwa, "A review of molybdenum disulfide (MoS 2) based photodetectors: from ultra-broadband, self-powered to flexible devices," RSC Adv., vol. 10, no. 51, pp. 30529–30602, 2020, doi: 10.1039/D0RA03183F.

[5] L.-R. Zou et al., "Research progress of optoelectronic devices based on two-dimensional MoS2 materials," Rare Met., vol. 42, no. 1, pp. 17–38, Jan. 2023, doi: 10.1007/s12598-022-02113-y.

[6] Y. Liu and F. Gu, "A wafer-scale synthesis of monolayer MoS2 and their field-effect transistors toward practical applications," Nanoscale Adv., vol. 3, no. 8, pp. 2117–2138, Apr. 2021, doi: 10.1039/D0NA01043J.

[7] L.-R. Zou et al., "Research progress of optoelectronic devices based on two-dimensional MoS2 materials," Rare Met., vol. 42, no. 1, pp. 17–38, Jan. 2023, doi: 10.1007/s12598-022-02113-y.

[8] Y. Sun et al., "Ambipolar MoS2 Field Effect Transistors with Negative Photoconductivity and High Responsivity Using an Ultrathin Epitaxial Ferroelectric Gate," Advanced Functional Materials, vol. 34, no. 37, p. 2402185, 2024, doi: 10.1002/adfm.202402185.

[9] X. Xiao, J. Li, J. Wu, D. Lu, and C. Tang, "Negative photoconductivity observed in polycrystalline monolayer molybdenum disulfide prepared by chemical vapor deposition," Appl. Phys. A, vol. 125, no. 11, p. 765, Oct. 2019, doi: 10.1007/s00339-019-3054-2.

[10] K. Cho et al., "Gate-bias stress-dependent photoconductive characteristics of multi-layer MoS2 field-effect transistors," Nanotechnology, vol. 25, no. 15, p. 155201, Mar. 2014, doi: 10.1088/0957-4484/25/15/155201.

[11] S. B. Desai et al., "Gold-Mediated Exfoliation of Ultralarge Optoelectronically-Perfect Monolayers," Advanced Materials, vol. 28, no. 21, pp. 4053–4058, Jun. 2016, doi: 10.1002/adma.201506171.

[12] W. Li et al., "Approaching the quantum limit in two-dimensional semiconductor contacts," Nature, vol. 613, no. 7943, pp. 274–279, Jan. 2023, doi: 10.1038/s41586-022-05431-4.

[13] P.-C. Shen et al., "Ultralow contact resistance between semimetal and monolayer semiconductors," Nature, vol. 593, no. 7858, pp. 211–217, May 2021, doi: 10.1038/s41586-021-03472-9.

[14] K. F. Mak, C. Lee, J. Hone, J. Shan, and T. F. Heinz, "Atomically Thin MoS 2 : A New Direct-Gap Semiconductor," Phys. Rev. Lett., vol. 105, no. 13, p. 136805, Sep. 2010, doi: 10.1103/PhysRevLett.105.136805.

[15] G. Eda, H. Yamaguchi, D. Voiry, T. Fujita, M. Chen, and M. Chhowalla, "Photoluminescence from Chemically Exfoliated MoS2," Nano Lett., vol. 11, no. 12, pp. 5111–5116, Dec. 2011, doi: 10.1021/nl201874w.

The ultra-thin Al₂O₃ oxide layer via ALD repaired Si Channel interface and optimizied subthreshold characteristics and reduced leakage current by 96.2%

Renjie Jiang[1,2,3], Peng Wang[1,2,3], Lianlian Li[1,2,3], Qinkun Li[1,3], Zhongrui Wang[1,2,3], Hang Zhang[1,2,3], Huaxiang Yin[1,2,3] and Qingzhu Zhang[1,2]

[1]Integrated Circuit Advanced Process R&D Center, Institute of Microelectronics of the Chinese Academy of Sciences, Beijing 100029, China

[2]Key Laboratory of Fabrication Technologies for Integrated Circuits, Chinese Academy of Sciences, Beijing 100029, China

[3]School of Integrated Circuits, University of Chinese Academy of Sciences, Beijing 100049, China

*Corresponding Email: yinhuaxiang@ime.ac.cn , zhangqingzhu@ime.ac.cn

Abstract

This study proposes a p-type GAA NSFET with an interfacial Al_2O_3/HfO_2 stacked dielectric to achieve an excellent oxide interface, the electrical and interface characteristics were investigated. Adding one layer of Al_2O_3 results in a threshold voltage shift and a 29% reduction in interface state density. The subthreshold swing (SS) of the device is reduced from 69.6 to 63.1 mV/dec. Furthermore, the leakage current is decreased by 96.2%, significantly enhancing the electrical characteristics of the device.

Keywords: GAA, PMOS, interfacial Al_2O_3 layer, SS, leakage current.

Introduction

Stacked nanosheets gate-all-around (GAA) MOS transistors not only has excellent short-channel control characteristics and performance advantages but is also highly compatible with the existing mainstream FinFET process. Therefore, GAA devices are considered the most promising core transistor structure following FinFET at the 3 nm and below technology nodes [1-4]. However, the interface states on the Si channel surface affect the subthreshold characteristics of the device, thereby affecting the device's drive, on-off ratio, and other electrical characteristics. With the continuous scaling of device sizes, the thinner SiO_2 layer dielectric layer will cause a sharp increase in gate tunneling current, so it is replaced by a high-k dielectric. However, obtaining a good interface between the high-k dielectric and HfO_2 is very challenging. To improve the interface quality and electrical performance of MOSFET devices, various interface engineering process methods have been developed. The Atomic Layer Deposition (ALD) is a widely used method for depositing high-k dielectric thin films for MOSFETs. Its self-limited gas-solid reaction process achieves excellent atomic layer control over film thickness, superior film quality over large areas and lower growth temperatures. Due to its excellent thermal stability and larger bandgap, Al_2O_3 is also considered a promising interface passivation material, making it a promising interface passivation layer (IPL) [5]. Although the dielectric constant (κ) of aluminum oxide is lower than that of HfO_2, this paper considers combining the two to ensure both the higher κ value and interface optimization. This study investigates the interface characteristics of Al_2O_3/HfO_2. First, the impact of incorporating aluminum oxide films of varying thicknesses on the interface state density was verified on MOSCAP, and then the influence on the electrical performance of GAA NS devices was studied [6,7].

Fabrication Process Flow

The CMOS GAAFETs were designed and fabricated on 200 mm p-type bulk-Si (100) wafers with a resistivity of 8-12 $\Omega\cdot cm$, the process integration flow and the key fabrication steps are shown in Fig. 1. The CMOS ground plane (GP) doping was carried out with B/P impurities, respectively. Then, the multi-layer $Si_{0.7}Ge_{0.3}$/Si epitaxy was performed using a reduced pressure chemical vapor deposition approach. The SiNx spacer hardmasks (HMs) were formed by the Self-aligned Double Patterning (SADP) technique for the formation of fin arrays. Next, the shallow trench isolation (STI), dummy gate and spacer modules were finished similar to those of the traditional GAAFETs [8,9]. Then the SD regions were formed by ion implantation and a 850℃ spike anneal. After removal of the α-Si dummy gate, the stacked Si channels was formed by selective removal of SiGe in the gate trench. After the interface layer (IL), an ALD process is used to apply a layer of aluminum oxide with a thickness of 8A. Finally, the CMOS-type high-κ metal gate (HK/MG), W plug contacts and back-end-of-linc (BEOL) processes for transistor formations were performed.

Fabrication Process Flow

As shown in Fig. 2, compared to the reference, Al_2O_3 with varying deposition thicknesses (4, 8, 12 Å) is integrated with the primary HfO_2 dielectric. Following this, the initial TiN capping layer is deposited via ALD, succeeded by the deposition of a TaN metal layer by ALD, which serves as an etch stop layer (ESL). A layer of TiN is then deposited by CVD to function as the WFM. To conclude the process, metal W is filled in by ALD. The C-V curves demonstrate the impact of channel interface quality. These capacitors were electrically compared to untreated ones. The

- P (100) Si wafer
- GP engineering
- Stacked $Si_{0.7}Ge_{0.3}$ Epi.
- Fin STI formation
- Dummy gate formation
- Spacer formation
- CMOS S/D formation
- Dummy gate removal
- Channel release
- CMOS HK/MG formation by ALD
  - HK: ALD 8Å Al_2O_3
  - HK: ALD 4.5nm HfO_2
  - ......
- MOL/BEOL
- FGA 450 °C, 30 min

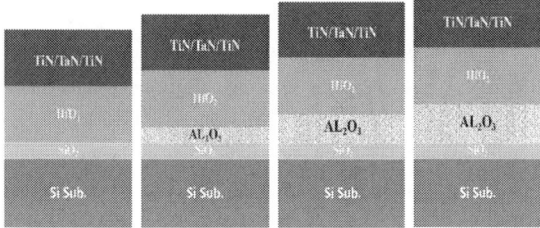

Fig. 1: The integrated process flows of PMOS with Al_2O_3 layer and the structure images of GAA FETs cut across NS top.

Fig. 2: Schematic of pMOSCAP device with the Al_2O_3 deposition in HfO2/TiN/TaN/TiN stacked structures.

relationship between ω and Gp/ω in the multi-frequency C-V curves are utilized to determine the dielectric constant of the capacitor, as indicated by the maximum value of [Gp/ω]. The interface trap density (Dit) formulate is [9]:

$$D_{it} = \frac{2.5}{A \cdot q}\left(\frac{G_P}{\omega}\right)max \qquad (1)$$

Fig. 3 shows the interface trap densit (Dit) of pMOSCAP devices with stacked layers of Al_2O_3 interfaces of different thicknesses combined with HfO_2. It is evident that the Dit has decreased by 28.7% compared to the reference devices. It may be because the incorporation of Al can reduce the trapped electron charges in the HfO_2 dielectric [10]. In this article, we found that only a small amount of Al_2O_3 (less than 1nm) decreases the trapping, which is attributed to a reduction of oxygen vacancy-related traps in HfO_2.

The typical transfer curves (I_d-V_{gs}) of 180-nm-L_g CMOS devices with and without Al_2O_3 layer is shown in Fig. 3(a). The results indicate that the SSs have achieved significant improvement from 69.6 mV/dec to 63.1 mV/dec for device w. and w./o. Al_2O_3 layer, respectively. Compared to the untreated devices, the SS_{sat} are reduced by 9.3% for PFET, resulting in a significant improvement in subthreshold characteristics. In addition, the peak values of the transconductances (G_{ms}) of the PFET are also obviously enhanced is shown in Fig. 3(b).

The statistical SS_{sat} of thePFET w. and w./o. Al_2O_3 layer is compared in Fig. 4(a). The SS is reduced by 2.9 mV/dec for PFET and reaches the median value in the statistics of 64.18 mV/dec. Additionally, with the incorporation of Al_2O_3 and

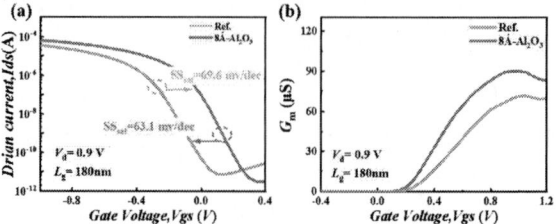

Fig. 3: Experimental electrical characteristics of the 180-nm-L_g GAAFETs with Al_2O_3 layer; (a) transfer curvess; (b) G_m.

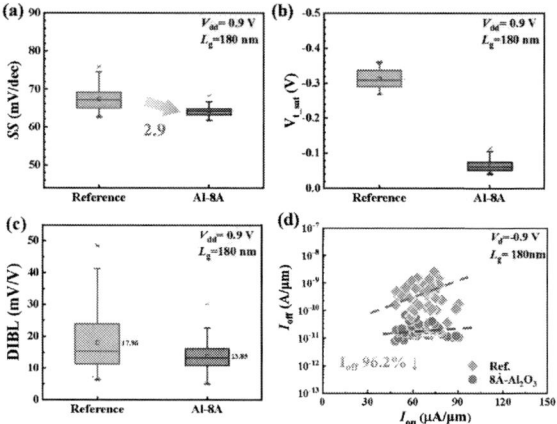

Fig. 4: The SS_{sat} for PFET GAA SiNS devices with Al_2O_3 layer. There is a significant change in V_{t-sat} (c). The devices w. Al_2O_3 layer have smaller DIBLs (d) and Ion-Ioff mapping .

the increase in the thickness of the high-κ dielectric, the number of dipoles and the VFB (Flatband Voltage) shift also increase, leading to a V_{t_sat} (Threshold Voltage) shift. Therefore, ultra-thin oxides of Al or La are commonly used to modify the EWF (Effective Work Function) of the gate electrode to meet the V_t requirements of MOSFETs [11]. Additionally, due to the repair of the oxide layer interface and the reduction of interface defects, the DIBL leakage has seen some improvement. From the Fig. 4(d), compared with reference device, the off current decreased 96.2%. The reduction in leakage current for the Al_2O_3/HfO_2 high-k laminated stacked film is likely attributed to trapped charges, with evidence suggesting that these charges are captured in deep traps [12]. This may be due to the fact that the Al-doped HfO_2 film reduces the concentration of oxygen vacancies and carbon within the film, which act as electrical defects or traps. Additionally, the doping increases the film's band gap, consequently leading to a decrease in gate leakage current [13].

Conclusion

In this paper, by adding an Al_2O_3 layer via ALD, we systematically investigated the technology of the interfacial Al_2O_3/HfO_2 stacked layers. We achieved a better oxide interface, realizing a 28.7% reduction in Dit, with the Al_2O_3 film reducing oxygen vacancies and repairing interface

979-8-3315-0417-5/25 $31.00 © 2025 IEEE

defects, thereby improving the device's interface quality. Furthermore, the SS was reduced from 69.6 mV/dec to 63.1 mV/dec, decreasing by 9.3%, and the leakage current was significantly lowered by 96.2%. The proposed approaches had enhanced the device's electrical characteristics. This is one of the important technologies for enhancing the performance of continuously scaled p MOSFET devices in the future.

Acknowledgments

This work was supported in part by the National Natural Science Foundation of China under Grant 62374183 (Corresponding authors: Qingzhu Zhang, Huaxiang Yin.)

References

[1] N. Loubet, T. Hook and P. Montanini, "Stacked nanosheet gate-all-around transistor to enable scaling beyond FinFET," in Proc. Symp. VLSI Technol., Kyoto, Japan, Jun. 2017, pp. T230−T231, doi: 10.23919/VLSIT.2017.7998183.

[2] Q. Z. Zhang, Y. K. Zhang, Y. N. Luo and H. X. Yin, "New structure transistors for advanced technology node CMOS ICs," National Science Review, vol. 11, no. 3, Mar. 2024, doi: 10.1093/nsr/nwae008.

[3] H. Mertens, R. Ritzenthaler and A. Chasin, "Vertically stacked gate-all-around Si nanowire CMOS transistors with dual work function metal gates," in IEDM Tech. Dig., San Francisco, CA, USA, 2016, pp. 19.7.1-19.7.4, doi: 10.1109/IEDM.2016.7838456.

[4] Zhang J, Frougier J, Greene A, et al. Full bottom dielec-tric isolation to enable stacked nanosheet transistor for low power and high performance applications[C]//2019 IEEE International Electron Devices Meeting. IEEE,2019.

[5] L.-F. Wu, Y.-M. Zhang, H.-L. Lu, and Y. Zhang, "Characterization of HfO2/Al2O3 gate dielectric nanometer-stacks grown by atomic layer deposition on InAlAs substrates," in Proc. 13th IEEE Int. Conf. SolidState Integr. Circuit Technol. (ICSICT), Oct. 2016, pp. 978–980, doi: 10.1109/ICSICT.2016.7998624.

[6] J. Seo and C. Shin, "Experimental study of interface traps in MOS

capacitor with Al-doped HfO2," Semicond. Sci. Technol., vol. 35, no. 8, pp. 2–7, 2020, doi: 10.1088/1361-6641/ab9847.

[7] M. Azzaz et al., "Improvement of performances HfO2-based RRAM from elementary cell to 16 kb demonstrator by introduction of thin layer of Al2O3," Solid-State Electron., vol. 125, pp. 182–188, Nov. 2016, doi: 10.1016/j.sse.2016.07.007.

[8] Q. Zhang, H. Tu, H. Yin, F. Wei, J. Li, L. Meng, Z. Zhang, J. Yan, H. Zhao, T. Ma, Z. Zhou, Y. Fan, J. Du, "Influence of the hard masks profiles on formation of nanometer Si scalloped fins arrays," Microelectron. Eng., vol. 198, pp. 48 – 54, Oct. 2018, doi: 10.1016/j.mee.2018.07.001.

[9] Z.H. Zhang, Y.N. Luo, G.B. Xu, J.X. Yao, Z. Wu, H. Zhao, Q. Zhang, H. Yin, J. Luo, W. Wang, H. Tu, "Performance improvements in complementary metal oxide semiconductor devices and circuits based on fin field-effect transistors using 3-nm ferroelectric Hf0.5Zr0.5O2," Rare Met. 2024, https://doi.org/10.1007/s12598-024-02674-0.

[10] H.-C. Wen and J. J. Chambers, "Gate contact materials in Si channel devices," MRS Bull., vol. 36, no. 2, pp. 101–105, Feb. 2011, doi: 10.1557/mrs.2011.8.

[11] T. Nabatame et al., "What is the essence of VFB shifts in high-k gate stack?" ECS Trans., vol. 11, no. 4, pp. 543–555, Sep. 2007, doi: 10.1149/1.2779589.

[12] T. J. Park et al., "Reduction of electrical defects in atomic layer deposited HfO2 films by Al doping," Chem. Mater., vol. 22, no. 14, pp. 4175–4184, Jul. 2010, doi: 10.1021/cm100620x

[13] A. Paskaleva, M. Rommel, A. Hutzler, D. Spassov, and A. J. Bauer, "Tailoring the electrical properties of HfO2 MOS-devices by aluminum doping," ACS Appl. Mater. Interfaces, vol. 7, no. 31, pp. 17032–17043, Aug. 2015, doi: 10.1021/acsami.5b03071.

Investigation of Al_2O_3/SiO_2 Interface Charge for the Feasibility Study of Charge Sheet Super Junction

Swadhin Kumar Jena[1], Chiranjibi Padhee[1], Akshay K[2] and Parlapalli Venkata Satyam[1]

[1]School of Basic Sciences, Indian Institute of Technology Bhubaneswar, Odisha, India

[2]School of Electrical Sciences, Indian Institute of Technology Bhubaneswar, Odisha, India

Email: akshay@iitbbs.ac.in, satyam@iitbbs.ac.in, s22ph09002@iitbbs.ac.in

Abstract

Charge sheet super junction (CSSJ) is a novel drift layer structure proposed as a viable alternative to superjunction (SJs). So far, studies were limited to TCAD simulations. Here, we present the fabrication, characterization, and simulation of the Al_2O_3/SiO_2 bi-layer interface on Si, which could potentially replace the p-pillar in SJs. Energy dispersive X-ray spectroscopy (EDX) shows an excess atomic percentage of oxygen possibly due to the oxygen-rich interface between Al_2O_3 and SiO_2/Si. High-frequency capacitance-voltage measurement of fabricated metal oxide semiconductor structures in combination with TCAD simulations are used to detect and quantify the negative interface charge, crucial for CSSJ operation. These results establish the preliminary feasibility of CSSJ indicating possible future applications.

Keywords: CSSJ, MOS capacitor, C-V measurement,TCAD, and fixed negative charge

Introduction

Power semiconductor devices play an important role in all power electronic applications. The primary requirements from a power device are high breakdown voltage (V_B) and low specific on-resistance ($R_{sp,on}$). Although several efforts have been made to reduce $R_{sp,on}$ of conventional junctions [1,2], these are limited by the trade-off between R_{sp-on} and V_B i.e. $R_{sp,on} \propto V_B^2$ [3] referred to as the *silicon limit*. The best trade-off between $R_{sp,on}$ and V_B is obtained by optimization of doping concentration and thickness of voltage sustaining layer(VSL) shown in Fig.1(a) in conventional power MOSFETs. Super junctions (SJs) were introduced to overcome the Silicon limit (see Fig. 1(b)). For SJs, the $R_{sp,on}$ varies linearly with V_B, i.e. $R_{sp,on} \propto V_B$ thereby enabling the reduction of $R_{sp,on}$ below the *silicon limit*. However, fabrication of high aspect ratio defect-free p-pillars [4] and activation of the dopants for achieving charge balance [5] are technological challenges. A new SJ structure called Charge Sheet SJ (CSSJ) was proposed using TCAD simulations that could potentially solve these challenges in Si [6], and later in SiC materials [7]. In CSSJ, a negative interface charge is expected to be formed at the Al_2O_3/SiO_2 interface and play the role of p-pillar in SJs. The magnitude of this negative charge may be controlled by the deposition temperature in

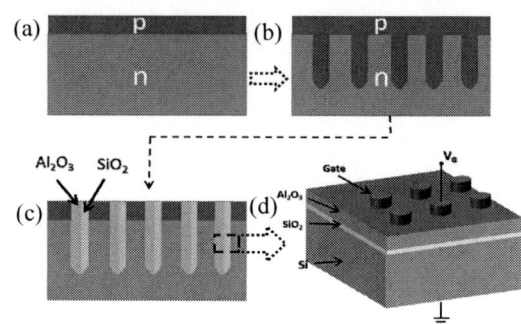

Fig. 1. (a) Conventional p-n junction (b) Super junction (c) Charge sheet super junction (d) Fabricated MOS capacitors

ALD [8]. So far no experimental attempt was made to study the practicability of the CSSJ. Therefore, in this work, we present a first attempt to investigate CSSJ by detecting negative charge at the interface Al_2O_3/SiO_2 on Si through both experiment and TCAD simulation and build motivation for realization of the device.

Device structure and Fabrication

A p-type <100> silicon substrate of resistivity 0.01-0.02 Ωcm and thickness 300 μm is used for fabricating metal-oxide-semiconductor(MOS) capacitors as shown in Fig.1(d).The substrates were cleaned with standard RCA procedure. For deposition of Al_2O_3 thin film a thermal ALD reactor is used where trimethylaluminum [TMA, $Al(CH_3)_3$] and H_2O were used as aluminum and oxygen precursor respectively as shown in fig. (2) . Al_2O_3 deposition was done at 200 °C temperature for 150 cycles and Ar gas was used as carrier gas. Spectroscopic ellipsometry was performed for measuring the thickness of Al_2O_3 film at three distinct angles (65°, 70°, and 75°) in the 200–1600 nm wavelength range. An interfacial native SiO_2 layer was considered in the analysis of the

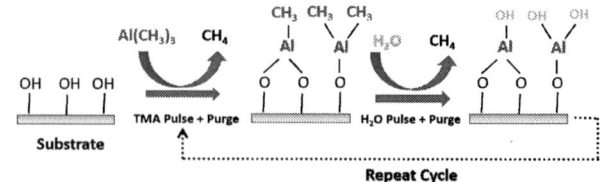

Fig. 2. Illustration of the ALD Cycle

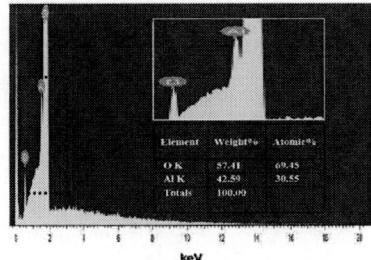

Fig.3 EDX spectrum of Al_2O_3 film.

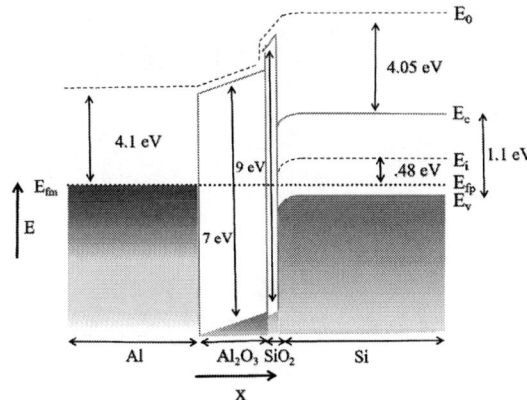

Fig.4. Band diagram of the fabricated MOS structure

findings. The thickness of the Al_2O_3 film was found as 15.78 nm. For front contact, a gate area of 5.4×10^{-7} m^2 and 200 nm thick Al pads were deposited using thermal evaporation through a shadow mask, and a 200 nm blanket Al was deposited for back contact. Morphology and stoichiometry of the deposited film were examined using field emission scanning electron microscopy (FESEM) and Energy dispersive X-ray spectroscopy (EDX) techniques respectively.

Result and Discussion
Material characterization
A smooth surface of a deposited film plays an important role in the formation of good contact between the film and the material deposited on it. FE-SEM image (not shown here) has shown smooth surface of Al_2O_3 film. EDX analysis of fig.(3) shows the atomic percentage of the deposited Al_2O_3 film is Al:O::3:7. The larger percentage of oxygen detected in the spectrum is possibly due to the presence of an oxygen-rich layer at the interface of Al_2O_3/SiO_2. The O/Al value varies between 2.5 and 3.6 because of the contribution of oxygen ions from precursor O2 [9]. So in our result, the excess oxygen is probably contributed by the OH group trapped at the interface [10], [11] during the initial cycles of ALD shown in fig.(2).

Device characterization and Parameter Extraction
Capacitance-voltage (C-V) measurement was done using a B1500A semiconductor device parameter. A high frequency (1MHz) C-V measurement was performed in order to extract the magnitude of interface fixed charge (Q_f) and trap densities (D_{it}). For accurate measurement of Q_f and D_{it}, it is important to first match the result with the properties of the film. From the calculation average value of the dielectric constant ($\varepsilon_{ox} = C_{acc} t_{Al_2O_3}/\varepsilon_0 A$) of the sample was found to be ~ 10 [12]. Here ε_0 permittivity in a vacuum, C_{acc} is the accumulation capacitance, $t_{Al_2O_3}$ is the thickness of Al_2O_3, and A is the area of the device. The extracted Si's doping concentration (N_a) is 9.4×10^{18} atoms/cm^3, which is in accordance with the expected doping concentration. Flat-band capacitance (C_{FB}) for the sample was calculated from the formula [13].

$$C_{FB} = \varepsilon_{ox} A \left\{ t_{Al_2O_3} + \frac{\varepsilon_{ox}}{\varepsilon_s}\left(\frac{KT\varepsilon_s}{q^2 N_a}\right)^{1/2} \right\}^{-1} \qquad (1)$$

where ε_s dielectric constant of silicon. The value of C_{FB} was found to be 2.3×10^{-9} F. Using C_{FB}, the flat band voltage (V_{FB}) was estimated to be .46 eV, which is comparable with the reported value [14]. The expressions used for calculating fixed insulator charge density (Q_f/q) are given by

$$\frac{Q_f}{q} = \frac{C_{ox}}{Aq}(\varphi_{ms} - V_{FB}) \qquad (2)$$

Where φ_{ms} is the work-function difference between metal and semiconductor which is calculated using band diagram at thermal equilibrium shown in fig.3. In the band diagram, E_{fm} represents the Fermi level of the metal, while E_{fp} denotes the Fermi level of the P-type silicon. The calculation shows that E_{fp} is 0.48 eV lower than the intrinsic Fermi level (E_i). Hence φ_{ms} value comes out to be -1.38 eV

By combining the characteristics of single-frequency capacitance voltage (C-V) and conductance voltage (G-V), the interface trap density is calculated using Hill's approach [15]. The expression used for the interface trap density is given by,

$$D_{it} = \frac{(2/qA)(G_{max}/w)}{\left[(G_{max}/wC_{ox})^2 + (1 - (C_m/C_{ox}))^2\right]} \qquad (3)$$

where G_{max} is the maximum conductance in the G–V plot (not shown here) with its corresponding capacitance (C_m), C_{ox} the oxide capacitance, and ω the angular frequency. The extracted values of the Q_f/q and D_{it} are found to be -6.8×10^{12}/cm^2 and 5.34×10^{13}/cm^{2}eV^{-1} respectively. Hence our experiment confirmed the presence of negative fixed charge at the interface [11], [16], [17].

979-8-3315-0417-5/25 $31.00 © 2025 IEEE

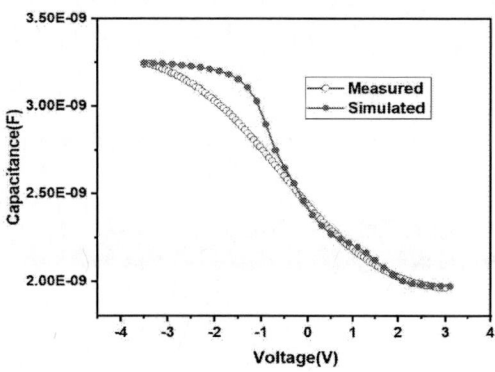

Fig.5 Comparison between measured and simulated C-V characteristics of MOS capacitors

TCAD simulations are also performed to confirm these findings. We have simulated a cylindrical MOS structure by taking radius of the gate metal and assuming uniform doping density throughout the thickness of the semiconductor. For the TCAD simulation, we used parameters extracted from the experimental C-V curve of the MOS device. From the Simulated C-V curve V_{FB} and Q_f/q was found to be 0.40 eV and $-6.6 \times 10^{12}/cm^2$ respectively. For getting best fit both donor and accept interface trap densities are included. It is interesting to note that in the Fig.5 the simulation and experimental C-V characteristics align well in both accumulation and inversion modes. However, there is a discrepancy between the C-V curves in depletion mode. This minor divergence may be attributed to the incomplete ionization of dopant atoms (N_a) which mainly contribute to the edge of the deep depletion [18].

Conclusion

The ALD technique gives a smooth and controlled Al_2O_3 layer, which is suitable for deposition in the trench of a CSSJ. EDX data shows that the presence of oxygen-rich layer at the interface of Al_2O_3 and SiO_2 is the possible reason for higher atomic percentage of oxygen in the spectrum. C-V measurement confirms the presence of negative interface fixed charge of magnitude $6.8 \times 10^{12}/cm^2$; this is in the expected range and can be used as a negative charge sheet that can play the role of p-pillars in SJ. TCAD simulation results agree with the experimental C-V data. Hence, these results should motivate further investigation for the practical realization of CSSJ.

Acknowledgment

A part of the reported work (fabrication/characterization) was carried out at the IITBNF, IITB under INUP which is sponsored by MeitY, MCIT, Government of India.

References

[1] C. Hu., "A Parametric Study Of Power Mosfets, Rec. Power Electronics Specialists Conf.(IEEE)",p.385, (1979).

[2] P.L.Hower et al., "Optimum Design of Power MOSFETS",Tech. Dig. Int.ElectronDevice Meet.(IEEE),p.87, (1983).

[3] Tatsuhiko Fujihira, "Theory of Semiconductor Superjunction Devices", Jpn. J. Appl. Phys., 36 6254, (1997).

[4] Run Tian et al, "A review of manufacturing technologies for silicon carbide superjunction devices", J. Semicond. 42 061801, (2021).

[5] H. M. Zhong, et al., "Practical superjunction MOSFET device performance under given process thermal cycles", Semicond. Sci. Technol., vol. 19, no. 8, pp. 987–996, (2004).

[6] S. Srikanth and S. KARMALKAR, "On the Charge Sheet Superjunction (CSSJ) MOSFET", IEEE Trans. Electron Devices, vol. 55, No. 12, pp. 3562-3568, (2008).

[7] K. Akshay and S. KARMALKAR, "Charge Sheet Super Junction in 4H-Silicon Carbide: Practicability, Modeling and Design", JEDS,Vol 8, (2020).

[8] J. Buckley et al., "Reduction of fixed charges in atomic layer deposited Al_2O_3 dielectrics", Microelectron. Eng., vol. 80, no. 5, pp. 210–213, (2005).

[9] T. O. K̈ari̇nen and D. C. Cameron,"Plasma-assisted atomiclayer deposition of Al2O3 at room temperature," Plasma Pro-cesses and Polymers, vol. 6, no. 1, pp. S237–S241, (2009).

[10] L.G. Gosset et.al,"Interface and material characterization of thin Al2O3 layers deposited by ALD using TMA/H2O", Journal of Non-Crystalline Solids 303,17–23, (2002).

[11] Volker Naumann et.al, "Chemical and structural study of electrically passivating Al_2O_3/Si interfaces prepared by atomic layer deposition", J. Vac. Sci. Technol. A 30, 04D106, (2012).

[12] E. Ghiraldelli et al. "ALD growth, thermal treatments and characterization of Al_2O_3 layers",Thin Solid Films , 434–436, (2008).

[13] S. M. Sze, "Physics of Semiconductor Devices", Wiley, London,pp. 425-504, (2002) .

[14] V. V. Afanas'ev et al., "Band alignments in metal–oxide–silicon structures with atomic-layer deposited Al_2O_3 and ZrO_2" . Appl. Phys. Lett. 78, 3073, (2001).

[15] W.A. Hill and C.C. Coleman, "A single frequency approximation for interface-state density determination", Solid State Electron. 23, 987–993, (1980).

[16] S. Dueñas et al., "Influence of single and double deposition temperatures on the interface quality of atomic layer deposited Al_2O_3 dielectric thin films on silicon", JOURNAL OF APPLIED PHYSICS 99, 054902, (2006).

[17] J. J. H. Gielis et al., "Negative charge and charging dynamics in films on Si characterized by second-harmonic generation", J. Appl. Phys. 104, 073701, (2008).

[18] Rejaiba Omar et al., "Effects of series and parallel resistances on the C-V characteristics of silicon-based metal oxide semiconductor (MOS) devices", Eur. Phys. J. Plus, 130, (2015).

979-8-3315-0417-5/25 $31.00 © 2025 IEEE

RRAM-Based Isotropic CNNs with High Robustness and Resource Utilization Rate

Wenyong Zhou[1,†], Yuan Ren[1,†], Jiajun Zhou[1], Chenchen Ding[1], Zhengwu Liu[1,*], and Ngai Wong[1,*]

[1]Department of Electrical and Electronic Engineering, The University of Hong Kong, Hong Kong
[†]Equal contribution, [*]Corresponding authors: {zwliu, nwong}@eee.hku.hk

Abstract

Resistive random-access memory (RRAM)-based compute-in-memory (CIM) systems show great potential for accelerating convolutional neural networks (CNNs). However, classical RRAM-based CNNs suffer from performance degradation from weight quantization and device non-idealities, and resource under-utilization due to mismatches between weight matrices and crossbar arrays. In this work, we propose RRAM-based isotropic CNNs that enhance model robustness and improve resource utilization concurrently. Extensive simulations demonstrate that the isotropic CNNs yield up to 3.74% and 12.61% accuracy improvement under quantization and non-idealities, respectively. Moreover, they increase the utilization rate by 5.2-13.4% in typical network architectures. These results make our design a highly robust and efficient RRAM-based CNN solution.

Keywords: Isotropic architecture, robustness, resource utilization

Introduction

Resistive random-access memory (RRAM)-based compute-in-memory (CIM) systems have emerged as a promising solution for accelerating convolutional neural networks (CNNs), offering significant advantages in terms of energy efficiency and processing speed, over classical computer with von Neumann architecture [1]. The reason behind it is that RRAM-based systems leverage the inherent parallelism and in situ analog computing advantages of RRAM array to perform both memory and computation within the same unit, significantly reducing data movement for the large amount of CNN weights [2].

However, despite RRAM-based CNNs have been experimentally demonstrated [3,4], achieving RRAM-based CNNs with less performance degradation and high resource utilization rate is still difficult: (1) The limited precision of weight quantization, which reduces the precision of weights and activations to lower bit representations (e.g., INT8 or INT4), can introduce computational errors and degrade CNN accuracy. Besides, inherent non-idealities of RRAM cells, such as device variations, resistance drift, and stuck-at-faults, cause weights deviations from expected values, significantly degrading model performance [5]. (2) Classical RRAM-based CNNs typically adopt a pyramidal architecture, where the number of channels progressively increases with each subsequent layer. While this design enhances feature extraction and performance in traditional digital platforms, the varying channel dimensions across layers lead to under-utilization of hardware resources. Early layers, with fewer channels relative to the crossbar size, lead to resource wastage.

In this work, we propose RRAM-based isotropic CNNs, which not only improve the model robustness against weight quantization and device non-idealities, but also optimize resource utilization on RRAM hardware. Furthermore, we conduct a comprehensive comparison between RRAM-based isotropic CNNs and their pyramidal counterparts. Through extensive experiments, we demonstrate that isotropic ones outperform pyramidal ones in robustness under various noisy conditions with higher utilization rate for lower hardware costs, validating its suitability for RRAM-based systems.

Methodology

Unlike the pyramidal architecture with varying channel numbers, the isotropic architecture maintains a uniform channel number across all layers (**Fig. 1**), making it easier to accommodate the fixed crossbar size. An intuitive approach is setting the number of channels in isotropic CNNs to a multiple of the crossbar size. For example, for 32x32 crossbars, the alternatives of channel dimension could be 32 or 64. The isotropic design ensures a more consistent workload distribution, reducing the risk of hardware under-utilization.

In addition, the isotropic architecture offers significant advantages in terms of robustness, particularly against both quantization errors and inherent RRAM non-idealities. In isotropic CNNs, the abundant convolutional kernels in early layers introduces redundancy in the weight parameters and feature maps, and enables subsequent layers to recover disturbed intermediate representations, thus reducing the network's reliance on individual weight values. This makes the network more resilient to perturbations from quantization and RRAM errors.

Results and Discussion

To verify the advantages of RRAM-based isotropic CNNs, we compare RRAM-based pyramidal and isotropic CNNs in terms of accuracies and hardware costs under different conditions on classical models.

A. High Robustness Against Quantization

We first compare the accuracy of two RRAM-based

979-8-3315-0417-5/25 $31.00 © 2025 IEEE   543

CNNs under two classical models (VGGs [6] and ResNets [7]), three types of quantization data formats (INT8, INT4, and INT3) on three datasets (CIFAR-10, CIFAR-100, and ImageNet datasets), as shown in **Fig. 2**. Although accuracy decreases with lower bit widths, the isotropic models consistently outperform their pyramidal counterparts across all datasets and quantization levels. For example, the isotropic ResNet achieves higher accuracy on CIFAR-10 (94.06% at INT8 and 93.16% at INT4) compared to the classical ResNet-18 (93.21% and 92.04%, respectively). This increased accuracy retention suggests that isotropic CNNs are well-suited for RRAM crossbars operating at various precision levels.

B. High Robustness Against RRAM Non-Idealities

We evaluate the performance of pyramidal and isotropic CNNs under various RRAM-induced noise conditions, as shown in **Fig. 3**. Following common practice, device variation was simulated as lognormal noise, while stuck-at-fault was modeled as binary noise [8]. Isotropic CNNs consistently outperformed pyramidal CNNs under both separate and simultaneous noise conditions. For instance, the accuracy of pyramidal VGG dropped from 87.91% to 48.69%, while isotropic VGG decreased from 88.25% to 61.30% under device variation. Furthermore, we simulated scenarios where both non-idealities coexist, reflecting a more realistic environment. The results demonstrate that our method exhibits better robustness against RRAM non-idealities.

C. High Utilization Rate for Lower Hardware Costs

We then simulate the hardware costs of pyramidal and isotropic CNNs using the DNN+NeuroSim workflow [9]. **Fig. 4** compares the resource utilization, energy and area efficiencies. The isotropic CNNs significantly outperform the pyramidal CNNs in resource utilization by 13.4% and 5.2% for VGG and ResNet, respectively. For energy efficiency, isotropic models demonstrate superiority, with isotropic VGG and ResNet achieving 6.79 TOPS/W and 10.92 TOPS/W, respectively, both higher than that for pyramidal networks. Area efficiency further reinforces this trend, with isotropic ResNet achieving the highest area efficiency (0.0219 TOPS/mm^2), followed by isotropic VGG (0.0189 TOPS/mm^2), both of which significantly surpass their pyramidal counterparts. These results highlight the significant improvements in the hardware efficiency of isotropic CNNs, primarily due to the improvement of resource utilization.

Fig. 5 compares the latency, energy consumption, and chip area. The isotropic architectures exhibit faster processing speeds, with isotropic VGG achieving 0.32 ms and isotropic ResNet achieving 1.06 ms, compared to their pyramidal counterparts at 0.37 ms for VGG and 1.56 ms for

ResNet. This pattern extends to energy consumption, where isotropic models prove more efficient: 1.97 mJ for isotropic VGG and 8.09 mJ for isotropic ResNet, significantly outperforming the pyramidal VGG (3.36 mJ) and pyramidal ResNet (12.77 mJ). In addition, isotropic architectures are also more compact, with isotropic VGG and ResNet requiring 8.60 mm^2 and 3.96 mm^2 respectively, compared to 11.32 mm^2 for pyramidal VGG and 5.69 mm^2 for pyramidal ResNet. These results suggest that isotropic CNNs achieve comparable or superior performance with fewer parameters and reduced computation burden than their pyramidal counterparts.

Conclusion

In this study, we propose RRAM-based isotropic CNNs for CIM systems with both high robustness and high resource utilization rate. The isotropic design preserves a consistent number of channels across all layers. Experimental results show that these CNNs achieve up to 3.74% and 12.61% higher accuracy in various noisy scenarios. Moreover, they offer resource utilization rate improvement of up to 13.4% and 5.2% for VGG and ResNet, respectively.

Acknowledgments

This work was supported in part by the Theme-based Research Scheme (TRS) project T45-701/22-R, National Natural Science Foundation of China (62404187) and ACCESS – AI Chip Center for Emerging Smart Systems, sponsored by InnoHK funding, Hong Kong SAR.

References

[1] Q. Zhang *et al.*, "On-Chip Write & Verify and Endurance Enhancer Circuits towards Multi-level RRAM Array," in *Proceedings of the EDTM*, 2024.

[2] J. Sakhuja *et al.*, " Enhancement in Bipolar Conductance Linearity by 1T1R cell with Non-Filamentary PCMO-RRAM as Synapse for Neural Networks," in *Proceedings of the EDTM*, 2023

[3] W. Wan *et al.*, "A Compute-in-Memory Chip Based on Resistive Random-Access Memory," in *Nature*, 2022.

[4] Z. Wang *et al.*, "In Situ Training of Feed-Forward and Recurrent Convolutional Memristor Network," in *Nature Machine Intelligence*, 2019.

[5] W. Zhou *et al.*, "A Time- and Energy-Efficient CNN with Dense Connections on Memristor-based Chips," in *Proceedings of the ASICON*, 2023

[6] K. Simonyan *et al.*, "Very Deep Convolutional Networks for Large-Scale Image Recognition," in *Proceedings of the ICLR*, 2014.

[7] K. He *et al.*, "Deep Residual Learning for Image Recognition," in *Proceedings of the CVPR*, 2016.

[8] T. Ketkar *et al.*, "Impact of Non-Idealities in RRAMs on Hardware Spiking Neural Networks," in *Proceedings of the EDTM*, 2021

[9] X. Peng *et al.*, "DNN+NeuroSim: An End-to-End Benchmarking Framework for Compute-in-Memory Accelerators with Versatile Device Technologies," in *Proceedings of the IEEE IEDM*,2019

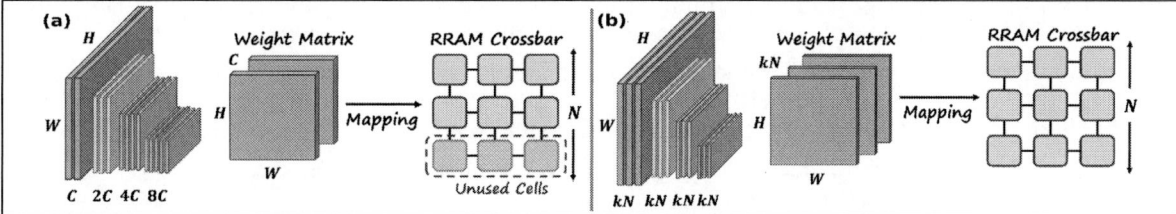

Fig. 1: Comparison between RRAM-based pyramidal and isotropic CNNs. (a) Mismatch between weight matrices (C/2C/4C/8C) and RRAM crossbars (N) result in resource under-utilization for pyramidal CNNs. (b) Isotropic CNNs ensure high resource utilization by accommodating the channel dimension (kN) to crossbars (N).

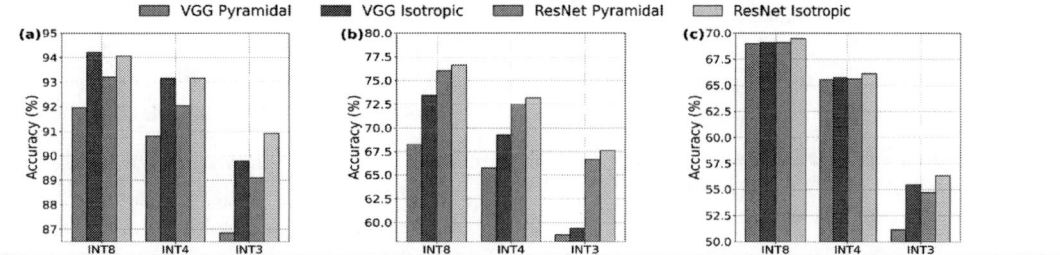

Fig. 2: Impact of quantization on RRAM-based pyramidal and isotropic CNNs. Different quantization precision, including INT8, INT4, and INT3, are evaluated on various datasets: (a) CIFAR-10, (b) CIFAR-100, and (c) ImageNet.

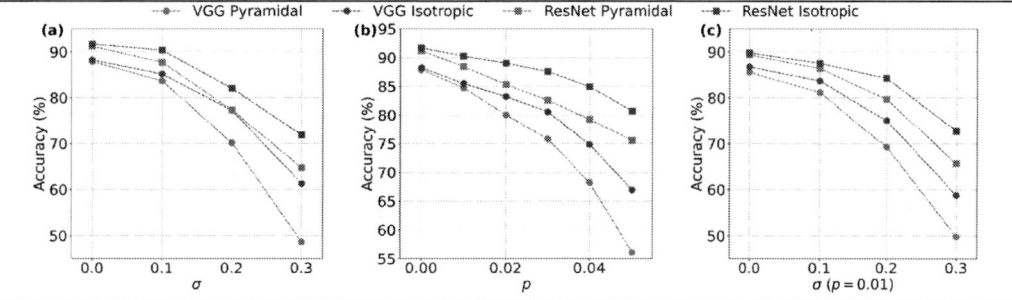

Fig. 3: Impact of non-idealities on RRAM-based pyramidal and isotropic CNNs. Two classical non-idealities, (a) device variations and (b) stuck-at-fault, are simulated. The situation where (c) they co-exist is also simulated.

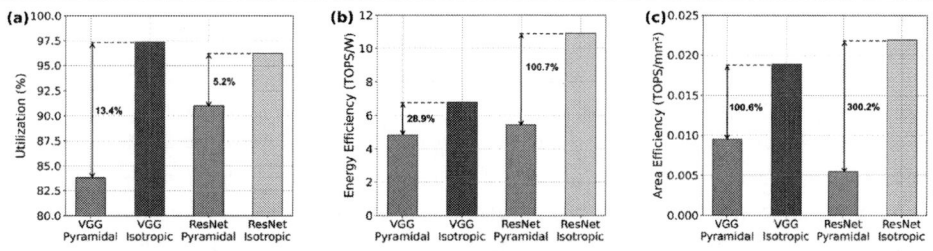

Fig. 4: Hardware efficiencies of RRAM-based pyramidal and isotropic CNNs. The (a) resource utilization (%), (b) energy efficiency (TOPS/W), and (c) area efficiency (TOPS/mm^2) are evaluated on 32×32 RRAM crossbars.

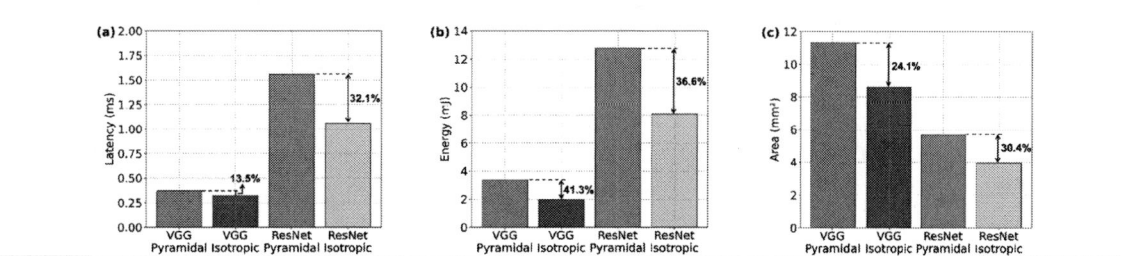

Fig. 5: Hardware costs of RRAM-based pyramidal and isotropic CNNs. The (a) latency (ms), (b) energy consumption (mJ), and (c) chip area (mm^2) for different CNNs are evaluated on 32×32 RRAM crossbars.

Laser-Induced Low Temperature Dopant Segregation Schottky Barrier MOSFET for Monolithic-3D

Feixiong Wang[1,2,3], Yadong Zhang[1,2,3]*, Jinbiao Liu[2], Yunjiao Bao[1,2,3], Zhiyao Wang[1,2,3], Shuang Liu[1,2,3], Mingzheng Ding[2], Zhaohao Zhang[1,2], Qingzhu Zhang[1,2,3], Huaxiang Yin[1,2,3]*

[1]Key Laboratory of Fabrication Technologies for Integrated Circuits, Chinese Academy of Sciences, Beijing 100029; [2]Institute of Microelectronics, Chinese Academy of Sciences, Beijing 100029, China; [3]School of Integrated Circuits, University of Chinese Academy of Sciences, Beijing, China.
*Email: zhangyadong@ime.ac.cn; yinhuaxiang@ime.ac.cn

Abstract

In this work, we present a BEOL-compatible low-temperature (Low-T, $\leq 500°C$) laser-induced silicide dopant segregation source/drain (LSDS S/D) technology for Monolithic-3D integrations. A nanosecond pulse green laser was used to induce dopant segregation of Schottky S/D, therefore a sharpened interface with high Schottky Barrier and a high impurity concentration are obtained. High performance CMOS devices are achieved with $I_{ON} > 100$ $\mu A/\mu m$ at $L_G = 500$ nm. We also demonstrate inverters, ring oscillators and logic gates, presenting an alternative fabrication scheme suitable for M3D circuit applications.
Keywords: MOSFET, M3D, Schottky Barrier, and SOI

Introduction

The emergence of generative AI, specially ChatGPT, demands computing resource requirements that surpass any existing environment[1]. Therefore, it's urgency to pursue a satisfactory system configuration in terms of energy efficiency and computing performance[2]. Follow this trend, Monolithic-3D (M3D) integration, maximizing the utilization of 3D space compared to traditional 2D ICs, has been introduced as a breakthrough technology for flexible 3D stacking of different functional circuits to achieve improvement in power, performance, area, and cost (PPAC) [3]. To sequentially stack devices upon fabricated bottom-tier devices, low thermal budget BEOL transistors such as poly-Si, Germanium, amorphous oxide semiconductors (AOS), 2D materials and Carbon Nanotube (CNT) have been extensively explored and integrated as multi-functional layer[4], [5], [6]. However, currently, no suitable candidate meets industrial criteria. Monocrystalline silicon as channel materials has optimal characteristic compared with BEOL transistors, but transistor application needs a useful way to activate S/D dopants at low temperature. SPER and Low-T epitaxy[7] were developed to achieve low thermal budget process, notably the dopant activation of S/D, but it's still challenging to activate S/D with compatible, high-yielding and cost-effective processes.

In this work, we report top-tier transistors using a BEOL-compatible laser-induced silicide dopant segregation source/drain (LSDS S/D) technology. Long-pulse (150 ns) green laser ($\lambda = 527$ nm) annealing was utilized to induce dopant segregation after silicide S/D formation and dopant implantation. The electronic performance of fabricated devices with this technique is systematically investigated. Furthermore, experimental inverters, ring oscillators (RO), and logic gate (NAND, NOR) are demonstrated to show our promising solution for future M3D AI circuit application.

Device Fabrication

Fig.1 (b) and (c) illustrate the process flow for the fabrication and schematic diagram of the device. 8-inch SOI wafer with a Si film thickness of 55 nm and BOX of 145 nm was used as start wafer. After 3 μm width active region formation, α-Si was deposited as replacement gate. After that, the dielectric of S/D region was etched by diluted-HF and $NiPt_{5\%}$ was deposited to form silicide S/D, followed by silicidation annealing to form Ni-rich silicide. Then, unreacted $NiPt_{5\%}$ was etched and second annealing was adopted to form NiSi phase. Next, the S/D regions were doped by B and P implantations for NMOS and PMOS, respectively. Laser annealing was subsequently proceeded with a long-pulse (150 ns) green laser ($\lambda = 527$ nm) to induce dopant segregation and Schottky Barrier. After silicide S/D formation, replacement metal gate was adopted and the multi-layer high-κ/metal-gate (HK/MG) films were deposited. Finally, the fabrication flow was completed with W contact plug and Al test pad metallization. Fig. 1(d) displays a cross-sectional transmission electron microscope (TEM) image of the fabricated device with gate length of 500 nm. Additionally, Fig. 1(e) presents the TEM and EDS mapping of silicide S/D and HKMG, showing the conformity and uniformity of the fabrication process.

Results And Discussion

Fig.2(a) displays the cross-section TEM and EDS mapping of LSDS S/D. 10 nm silicide was formed after totally three-step annealing and a sharpened silicide/silicon interface was achieved. The 1:1 element ratio of Ni to Si indicated the formation of low resistance NiSi phase. Secondary ion mass spectroscopy (SIMS) of the laser annealed sample is shown in Fig.2 (b). Highly segregated dopant of B > 5E20 cm^{-3} at the silicide/silicon surface was observed. Furthermore, the results show a strong Pt peak underneath the surface by 5 nm. The behave of Pt redistribution might be due to laser-induced increase in diffusion coefficient[8]. The dopants are restricted

979-8-3315-0417-5/25 $31.00 © 2025 IEEE

and sandwiched between the surface and aggregated Pt, and therefore, the dopants are segregated. To further understand the modulation of laser annealing, Schottky diode of different laser energy density was fabricated and the measured I-V characteristics is illustrated in Fig. 2(c). The effective Schottky Barrier Height (SBH) is extracted by thermal emission theory, and 0.9 eV Φ_{bn} is achieved, showing the effectivity of laser annealing to induce dopant segregation.

Fig. 3(a) shows the I_{ds}-V_{gs} characteristics of the device with a L_g of 500 nm at V_{ds} values of 0.1V and 1.2V with 1 J/cm² laser energy density. The MOSFET demonstrated exceptional performance, exhibiting DIBL values of 4.4 mV/V, 10.2 mV/V; SS values of 66.9 mV/dec, 67.6 mV/dec for PMOS and NMOS, respectively. This can be attributed to strong modulation of S/D SBH, thereby enhancing the source-side carrier injection. The V_{th}s were extracted as 0.24 V and -0.14 V for NMOS and PMOS at saturation region, which would be further modulated by work function layer tuning. Furthermore, Fig. 3(b) illustrates the output characteristics of transistor, presenting V_{gs} and V_{ds} values up to 1.2V and demonstrating excellent saturation characteristics. The transistor achieved high I_{on} of 112 µA/µm and 164 µA/µm at $V_{gs} = V_{ds} = 1.2$ V for PMOS and NMOS, respectively. Fig. 3(c) provides the I_{on}-I_{off} mappings of different laser energy density with respective to compositions for the RTA annealed reference transistor at 500 nm L_g. It demonstrates competitive I_{on}-I_{off} characteristics compared with Ref. NMOS with 33% increased I_{on} and 2 orders lower I_{off} on average values. Additionally, the mappings of laser annealed transistors exhibit a more uniform distribution compared to those of RTA annealed. In Fig. 3(d), the average values, 68.3 mV/Dec, 68.9 mV/Dec, 70.1 mV/Dec and 70.3 mV/Dec of different laser densities, show a lower and more concentrated subthreshold property compared with 76.7 mV/Dec of Ref. MOSFET.

To further investigate the potential for M3D integration, we designed several simple circuits to analysis the multi-function. Fig. 4 (a) plots the voltage transfer characteristic (VTC) curves of a fabricated inverter with 1 J/cm² laser density, showing abrupt switching from $V_{DD} = 0.15$ V to 1.2V. Almost a constant voltage gain over 38 V/V was achieved at V_{DD} in the range of 0.6 V to 1.2 V suggesting a great potential of Low-T inverter in low-power and high-performance application. Fig. 4 (b) plots the output voltage waveforms of CMOS 51-stages ring oscillators (RO) under supply voltage V_{DD} at 1.2 V, featuring output frequency f_O at 30.7 MHz. Finally, basic CMOS circuit blocks were measured to verify the proper logic function. Fig. 4 (c) shows the measured proper function of CFET based NAND and NOR logic gates.

Table I presents a comparative assessment of our LSDS S/D MOSFET in this study with other Low-T top-tier transistor schemes. The laser-induced silicide dopant segregation source/drain transistors exhibit prominent performance, low thermal budget and circuits demonstration compared to other works.

Conclusion

In this work, we introduced a BEOL-compatible LSDS S/D technology to achieve Low-T CMOS fabrication. This is accomplished by using nanosecond laser induced dopant segregation to modulate Schottky barrier with the minimum thermal budget. Experimental circuits, including inverters, ROs, and logic gate, utilizing this technology are also demonstrated. This approach provides a low-cost pathway toward low-temperature, high-performance full Si M3D circuits for AI application in the future.

Acknowledgements

The authors would like to thanks the support of Beijing Natural Science Foundation of China (No. 4222082) and the Ministry of Science and Technology (No. 2021YFA1200502).

References

[1] Y. Tanurhan, P. Paulin, and T. Michiels, "Generative AI on a Budget: Processing Transformer- based Neural Networks at the Edge," IEDM Tech. Dig., p. 15.1.1-15.1.4 (2023).

[2] S. Naffziger, "Innovations For Energy Efficient Generative AI," IEDM Tech. Dig., p. 15.4.1-15.4.4 (2023).

[3] T. Mota-Frutuoso et al., "3D sequential integration with Si CMOS stacked on 28nm industrial FDSOI with Cu-ULK iBEOL featuring RO and HDR pixel," IEDM Tech. Dig., p. 29.3.1-29.3.4 (2023).

[4] R. An et al., "A Hybrid Computing-In-Memory Architecture by Monolithic 3D Integration of BEOL CNT/IGZO-based CFET Logic and Analog RRAM," IEDM Tech. Dig., p. 18.1.1-18.1.4 (2022).

[5] C.-C. Yang et al., "Enabling low power BEOL compatible monolithic 3D+ nanoelectronics for IoTs using local and selective far-infrared ray laser anneal technology," IEDM Tech. Dig., p. 8.7.1-8.7.4 (2015)

[6] X. Xiong et al., "Top-Gate CVD WSe2 pFETs with Record-High Id~594 µA/µm, Gm~244 µS/µm and WSe2/MoS2 CFET based Half-adder Circuit Using Monolithic 3D Integration," IEDM Tech. Dig., p. 20.6.1-20.6.4 (2022).

[7] C. Porret et al., "Low temperature source/drain epitaxy and functional silicides: essentials for ultimate contact scaling," IEDM Tech. Dig., p. 34.1.1-34.1.4 (2022).

[8] S.-M. Koh et al., "Schottky barrier height tuning of silicides on p-type Si (100) by aluminum implantation and pulsed excimer laser anneal," Journal of Applied Physics, vol. 110, no. 7, p. 073703 (2012).

[9] A. Vandooren et al., "Demonstration of 3D sequential FD-SOI on CMOS FinFET stacking featuring low temperature Si layer transfer and top tier device fabrication with tier interconnections," VLSI Technology and Circuits, pp. 330–331 (2022).

[10] B.-J. Shih et al., "3DIC with Stacked FinFET, Inter-Level Metal, and Field-Size (25x33mm²) Single-Crystalline Si on SiO2 by Elevated-Epi," VLSI Technology and Circuits, pp. 11.5.1-11.5.2 (2024).

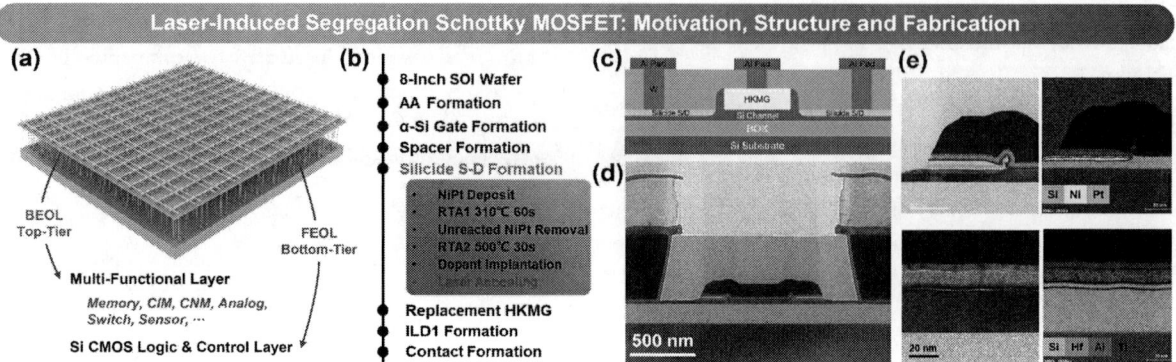

Fig. 1. (a) Motivation of this work: Achieve Low-T Top-Tier Si MOSFET for multi-function of M3D integration. (b) Fabrication process flow of simulated Top-Tier Schottky Barrier MOSFET, and laser annealing was utilized to segregate dopants. (c) Schematic diagram of Top-Tier Schottky Barrier MOSFET. (d) Cross-Section TEM of Schottky Barrier MOSFET with gate length of 500nm. (e) Cross-Section TEM and EDS of Silicide S/D and HMKG deposited channel, respectively.

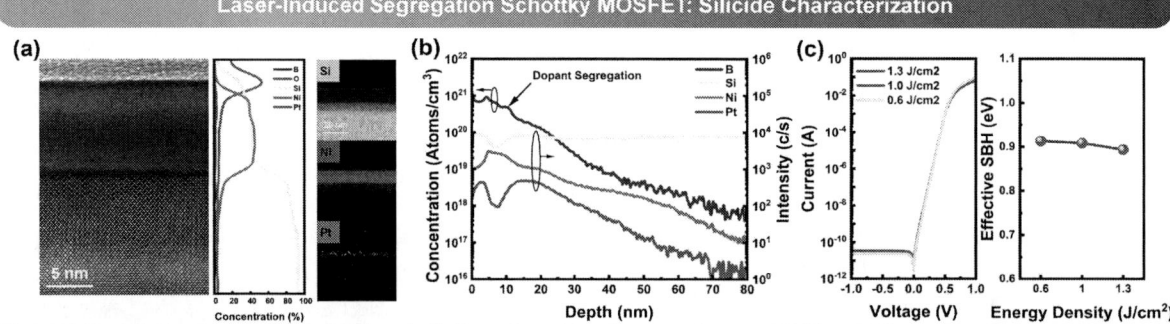

Fig. 2. (a) Cross-Section TEM and EDS of fabricated silicide. A sharped interface occurs between Silicide and Silicon. (b) SIMS of fabricated silicide. A high concentration of dopant occurs under the silicide surface. (c) Measured I-V character and extracted SBH of the Schottky Diode with different laser energy density.

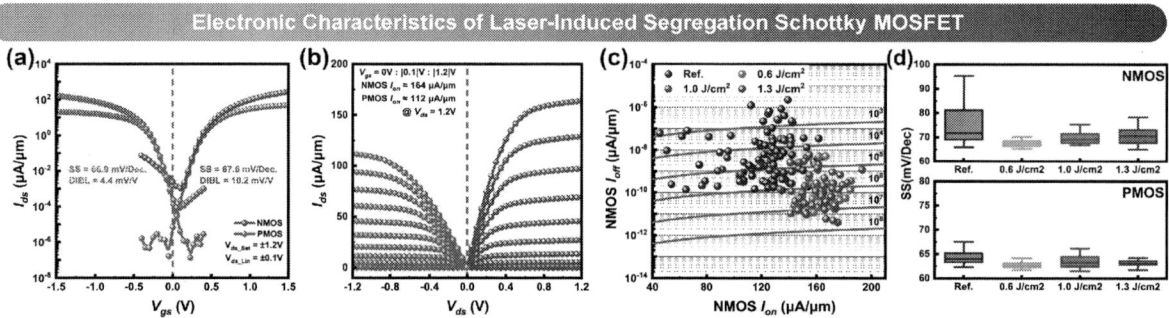

Fig. 3. (a) and (b) Measured I_{ds}-V_{gs} and I_{ds}-V_{ds} of fabricated MOSFET with 1 J/cm2 laser energy density, respectively. (c) I_{on}-I_{off} Mapping of NMOS with different laser energy density, showing a better uniformity in contrast to Ref. NMOS. (d) The measured SS of CMOS with different laser energy density.

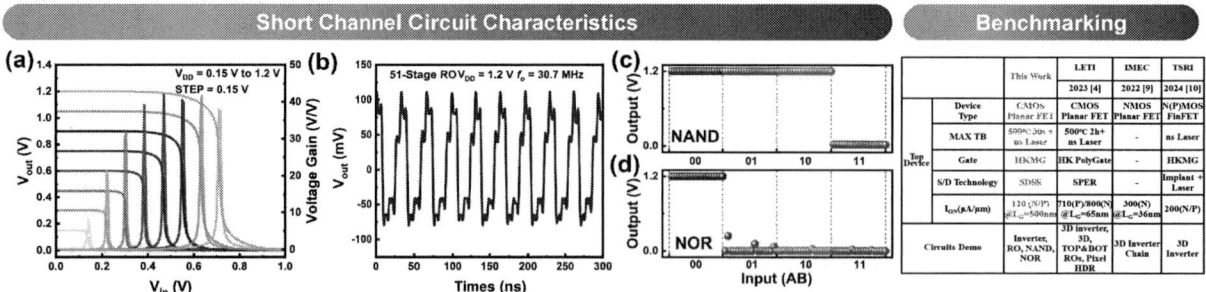

Fig. 4. (a) Voltage transfer characteristics (VTC) and Voltage Gain of a fabricated inverter. (b) Output voltage waveform of a typical 5-stage CFET RO with V_{DD} of 1.2 V. (c) and (d) Electrical measurement of NAND and NOR logic gates, respectively.

Table 1. Benchmark of Low-T process devices with previous studies.

	This Work	LETI 2023 [4]	IMEC 2022 [9]	TSRI 2024 [10]
Device Type	CMOS Planar FET	CMOS Planar FET	NMOS Planar FET	N(P)MOS FinFET
MAX TB	500°C 30s + ns Laser	500°C 2h+ ns Laser	-	ns Laser
Gate	HKMG	HK PolyGate	-	HKMG
S/D Technology	SDSS	SPER	-	Implant + Laser
I_{on}(kA/μm)	120 (N/P) @L$_g$=500nm	710(P)/800(N) @L$_g$=65nm	300(N) @L$_g$=36nm	200(N/P)
Circuits Demo	Inverter, RO, NAND, NOR	3D inverter, 3D, TOP&BOT ROs, Pixel HDR	3D Inverter Chain	3D Inverter

979-8-3315-0417-5/25 $31.00 © 2025 IEEE   548

A Heterogeneous FTJ-Based Computing-in-Memory Architecture for Vision Transformer Acceleration via Hardware and Algorithm Co-Design

Letian Wang[1†], Pinfeng Jiang[1†], Zichong Zhang[1], Yifan Yang[1], Yilong Fang[1], Yi Wang[1], Mingde Zhu[1]
Xiangshui Miao[1], Xingsheng Wang[1*]

[1]School of Integrated Circuits & Wuhan National Lab. for Optoelectronics, Huazhong University of Science and Technology, Wuhan, China.

†These authors contributed equally   *Corresponding author e-mail: xswang@hust.edu.cn

Abstract

Vision Transformers (ViT) have set new benchmarks in computer vision tasks but are often limited by their high computational and storage requirements. This work presents HEFA, a heterogeneous Computing-in-Memory architecture based on Ferroelectric Tunnel Junctions (FTJ), designed to accelerate ViT. By integrating an FTJ-based CIM Engine and a Digital Reconfigurable Engine with dynamic pruning, HEFA achieves $1.36\times\sim35.7\times$ speedup and $1.96\times\sim34.1\times$ energy efficiency improvements compared to GPU and state-of-the-art accelerators.

Keywords: Computing-in-memory, FTJ, Vision Transformer

Introduction

Transformer is a deep learning model widely adopted across various AI domains [1]. Recently, Vision Transformer (ViT) has achieved state-of-the-art performance in computer vision but faces challenges on resource-constrained devices due to its intensive feed-forward network (FFN) and attention mechanism with quadratic complexity [2].

To overcome these challenges, many works have proposed ASICs based on sparse algorithms to enhance the efficiency and speed of attention [3], but they overlook the FFN computation, which is significantly affected by memory bandwidth limitations. Additionally, computing-in-memory (CIM) architectures [4,5] have been explored to address memory bandwidth bottlenecks and accelerate Transformers, but they are limited by the non-idealities of devices and cannot fully exploit the sparsity in attention.

Inspired by the above insights, this work proposes HEFA, a heterogeneous CIM architecture based on Ferroelectric Tunnel Junction (FTJ) to accelerate ViT, comprising FTJ-based CIM Engine (FCE) and Digital Reconfigurable Engine (DRE). Hafnium-zirconium-oxide (HZO) FTJ devices exhibit superior scalability, ultra-low power, in-memory computing capabilities, and high computational parallelism [6,7]. At algorithm level, we employ Efficient Additive Attention to achieve linear complexity computation, simplifying attention by leveraging sparsity. At hardware level, our heterogeneous accelerator utilizes FCE to handle convolution and linear layer with high computational frequency and low write operations, while DRE dynamically allocates computational resources based on workload sparsity for attention layers involving frequent read-write activities. Through hardware and algorithm co-design, we enhance the efficiency and speed of ViT processing.

Experiments validate HEFA achieves a $1.36\times\sim35.7\times$ speedup and $1.96\times\sim34.1\times$ energy efficiency improvement over GPU and state-of-the-art Transformer accelerators.

Heterogeneous FTJ-based CIM Accelerator

A. HZO FTJ Device

Our previous work proposed an $Hf_{0.5}Zr_{0.5}O_2$-based FTJ device [6], which was fabricated and further enhanced by a novel "thermal rewake-up" (TR) operation at various elevated temperatures, as illustrated in Fig 1 (a) and (b). The electrical properties of the FTJ are excellent for CIM applications as shown in Fig. 1. Fig. 1(c) illustrates the impact of writing pulse amplitude on resistance when the pulse width is fixed at 1 μs. Fig. 1(d) shows the endurance of HRS and LRS created by full polarization reversal and the great ON/OFF states up to 10^5 cycles without degradation. Furthermore, Fig. 1(e) displays the excellent retention behavior of the non-volatile HRS and LRS, where the initially measured resistance is maintained for more than 10^4 s without significant loss. The FTJs with these superior features provides reliable hardware for ViT accelerator.

B. Simplification and Sparse Algorithm Design

As shown in Fig. 2(a), the original ViT relies on self-attention, which suffers from high computational complexity. Inspired by [8], we propose a hardware-friendly sparse ViT model, illustrated in Fig. 2(b) and(c), to address the limitations of the original ViT. This model incorporates sparse optimizations based on Efficient Additive Attention (EAA), replacing matrix multiplication with element-wise multiplication to achieve linear complexity. Additionally, the high sparsity significantly reduces hardware overhead. The core algorithm of EAA is outlined as follows. First, Query matrix Q is multiplied by weight vector $w_\alpha \in \mathbb{R}^d$ to attain the attention query vector $\alpha \in \mathbb{R}^n$, defined as: $\alpha = Q \cdot w_\alpha/\sqrt{d}$. Then, Q and α are multiplied element-wise to obtain the global query vector $q \in \mathbb{R}^d$ as follows: $q_j = \sum_{i=1}^n \alpha_i * Q_{ij}$. Finally, q is multiplied and concatenated with the Key matrix K to produce the final output.

During inference, the Q map exhibits significant sparsity, as shown in Fig. 2(c). To enhance the efficiency of EAA, we first apply an adaptive threshold to prune the Q map, retaining only values above this threshold. We then perform a split and reorder operation on the Q map based on the cumulative values of each column, separating it into sparse and dense parts, thereby improving hardware utilization.

979-8-3315-0417-5/25 $31.00 © 2025 IEEE

C. Heterogeneous FTJ CIM accelerator design

1) Accelerator Overview: Fig. 3 illustrates the proposed heterogeneous FTJ-based CIM accelerator, HEFA. HEFA comprises FCE, DRE, top control, SRAM, and global buffer. The top control schedules data from the SRAM to the global buffer, which in turn dispatches data to the two engines to complete the network computation. As shown in Fig. 4, traditional dataflow experiences high latency and low efficiency due to delays in intermediate transfers. HEFA uses pipeline dataflow to overlap layer computations and utilizes a global buffer for efficient data management, thereby improving overall efficiency.

2) FTJ-based CIM Engine: FCE uses a 2D mesh Network-on-Chip (NoC) for on-chip interconnect, as shown in Fig. 3(b) and (c). FCE includes multiple processing elements (PEs) of varying sizes for different mapping needs. Each PE contains multiple FTJ arrays for parallel Matrix-Vector Multiplication (MVM) in the analog domain according to Kirchhoff's law and Ohm's law. During inference, PEs read data from the input buffer, convert it to voltages using DAC, and then convert the result back to digital using ADC. Due to limited conductance levels in FTJ devices, we use bit-slicing and bit-streaming techniques for higher-precision weights and activations. The shift-and-add module and adder tree support configurable multi-precision computation. As shown in Fig. 5(a), FCE stores all weights locally to avoid frequent data interactions and achieve high-throughput computation.

3) Digital Reconfigurable Engine: Due to the lower endurance of FTJ devices compared to CMOS logic, using FTJ-CIM for frequent read-write activities is challenging. To address this, DRE was designed to accelerate the attention mechanism in ViT. As shown in Fig. 3(a), DRE consists of a reconfigurable array, sparse scheduler, addition unit, and buffer. The reconfigurable array includes multiple processing elements (PEs), each with four Calculation Units (CUs), and each CU containing four multipliers. A single multiplier can perform INT4 multiplication, and through shift and accumulation, the PE supports INT4/INT8/INT16 precision, allowing flexible configuration.

During EAA computation, DRE allocates resources dynamically based on mapping sparsity through a sparse scheduler. In dense mode, both input and weight are fed into the PE simultaneously, generating results at the terminal PE. In sparse mode, input and weight are fed sequentially into each PE based on the indices in Q map, with partial results accumulated over time. As shown in Fig. 5(b), during EAA computation for the attention query, the reconfigurable array operates entirely in dense mode, with the fixed map Q and w_α both fed into each PE, producing results at the terminal PE. During global query computation, DRE dynamically allocates PEs in sparse and dense part based on the sparsity of Q map, with the fixed Q as input and α as weight, fed accordingly based on the sparsity pattern.

Results and Discussion

To validate the proposed HEFA accelerator and its associated algorithms, we conducted a comprehensive evaluation using the classical ImageNet-1k dataset. The experimental results indicate that HEFA achieves an accuracy of 81.9% under a Q-map sparsity level of 80%. For hardware implementation, the FCE in HEFA was simulated using MNSIM [9] with hardware configuration as shown in Table 1, while the DRE and other digital circuits were implemented in Verilog and synthesized using Synopsys Design Compiler under TSMC 28nm technology to obtain the area and power consumption, with a design frequency of 500MHz. We built a cycle-accurate simulator for performance evaluation. For performance evaluation, we compared HEFA against CPU (AMD EPYC 9654), GPU (Nvidia RTX 4090) and other accelerators in terms of computation speed and energy efficiency. As shown in Fig. 6, HEFA achieved a speedup ranging from 1.36× to 35.7× compared to the GPU and other accelerators. Additionally, shown in Table 2, HEFA demonstrated significant advantages over several Transformer accelerators, with improvements in throughput by 1.73× to 13.9× and in energy efficiency by 1.96× to 4.51×.

Conclusion

This work introduces HEFA, a heterogeneous FTJ-based CIM accelerator for ViT. Through hardware and algorithm co-design, HEFA integrates an FTJ-based CIM Engine and a Digital Reconfigurable Engine to accelerate multiple computations by using processing elements of different sizes and dynamically reallocating resources between sparse and dense modes. Experimental results indicate that HEFA accelerates ViT inference with high accuracy, achieving significant improvements in throughput and energy efficiency compared to CPU, GPU, and other Transformer accelerators. These results demonstrate the effectiveness of HEFA in accelerating ViT inference.

Acknowledgments

This work is supported in part by National Natural Science Foundation of China under Grants U2341221 and 62274070, Hubei Province Science and Technology Major Project under Grant 2022AEA001, Interdisciplinary Research Program of HUST (2023JCYJ042), and Hubei Key Laboratory of Advanced Memories.

References

[1] Alec Radford et al., *International Conference on Machine Learning (ICML)*, PMLR 139:8748-8763, 2021.
[2] Alexey Dosovitskiy et al., *arXiv preprint arXiv:2010.11929*, 2020.
[3] Jyotikrishna Dass et al., *IEEE HPCA*, pp.415-428, 2023.
[4] Xiaoxuan Yang et al., *IEEE ICCAD*, no: 92, pp.1-9, 2020.
[5] Qilin Zheng et al., *IEEE Design Automation Conference (DAC)*, 2023.
[6] Yifan Yang et al., *IEEE Electron Device Letters*, vol.45, no. 1, 2023.
[7] T. Wu et al., *International Electron Devices Meeting (IEDM)*, 2019.
[8] A. Shaker, et al., *IEEE ICCV*, pp.17425-17436, 2023.
[9] Zhenhua Zhu, et al., *GLSVLSI*, pp.83-88, 2020.

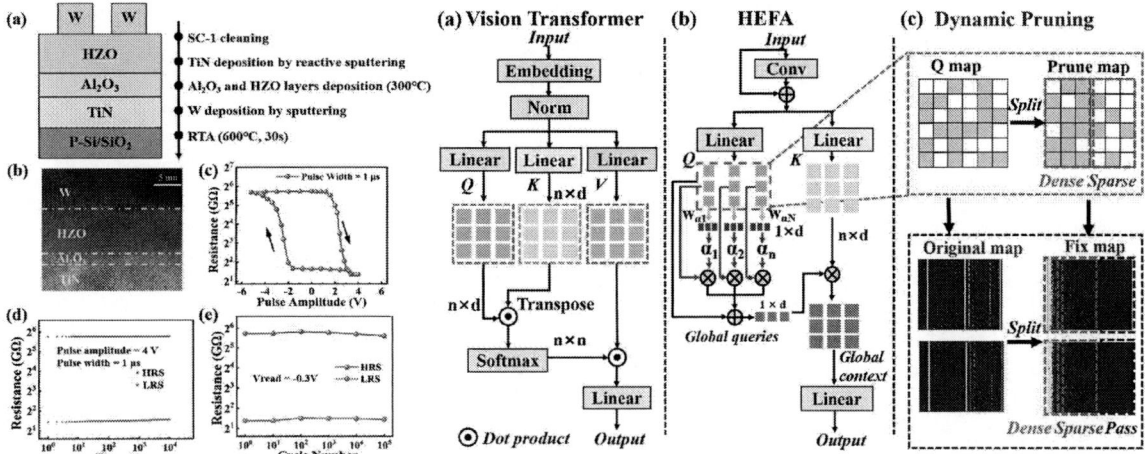

Fig. 1. (a)The device schematic and fabrication process of FTJ. (b)Cross-sectional HR-TEM image of FTJ. (c)The R-V loop. The endurance (d) and retention (e) characteristics of FTJ.

Fig. 2. (a)The architecture of Vision Transformer. (b) The Algorithm architecture of HEFA. (c)Q map mask before and after applying the split and redistribution algorithm.

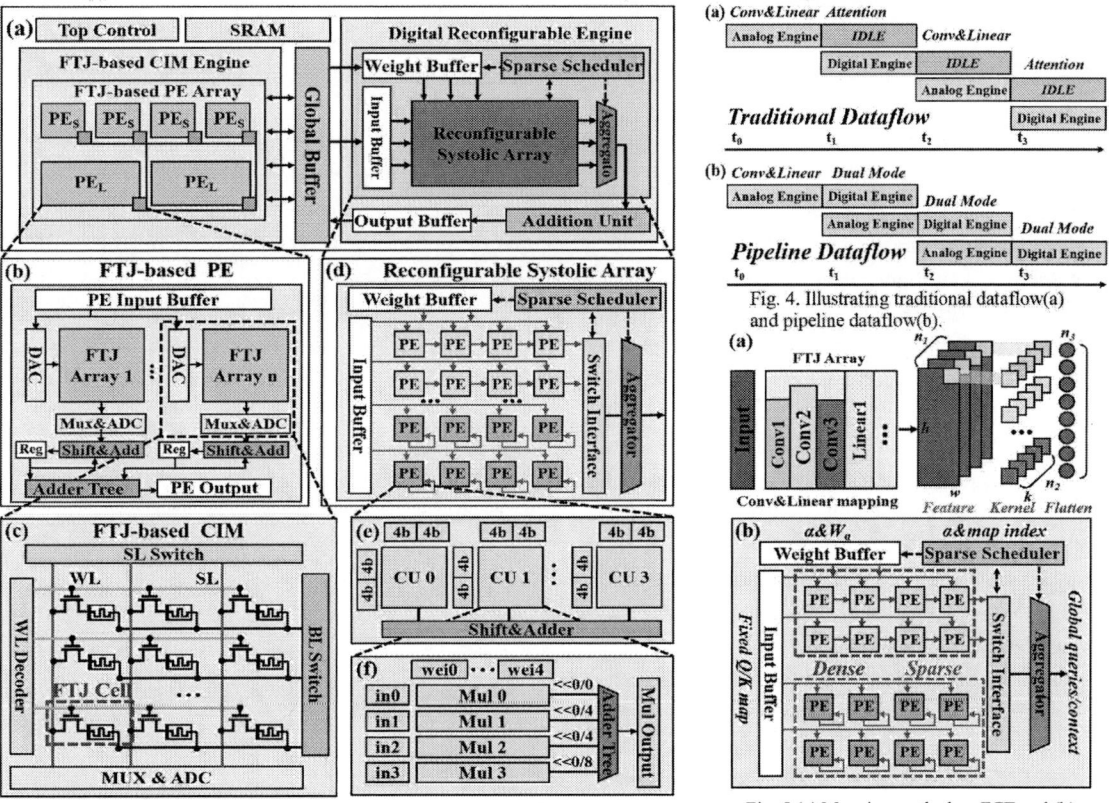

Fig. 3. Illustrating the micro-architecture details of HEFA accelerator.

Fig. 4. Illustrating traditional dataflow(a) and pipeline dataflow(b).

Fig. 5.(a) Mapping method on FCE and (b) Reconfigurable mode in DRE.

Table.1. Hardware configurations

FTJ-Based CIM Engine	
FTJ Array size	256×256,64×64
Read latency	3.4ns
ADC	8bit,1.4Gsps,1/16 cols
DAC	4bit,1.5Gsps
Digital Reconfigurable Engine	
Frequency	500MHZ
PE array size	16×16
Data Precision	INT4/8/16
Memory	200KB

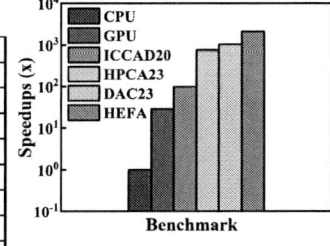

Fig. 6. Speedup of HEFA over CPU, GPU and other accelerators

Table 2.Performance comparing to other works

	HPCA23 [3]	ICCAD20 [4]	DAC23 [5]	HEFA
Technology(nm)	28	32	28	28
Implementation	ASIC	ACIM	ACIM	ACIM/CMOS
Precision	INT16	INT8	INT8	INT4/8/16
Frequency(MHz)	500	-	1200	50-500
Throughput(GOPS)	-	81.85	658	1140
Energy Efficiency(GOPS/W)	864	467.68	1074.1	2109.8

979-8-3315-0417-5/25 $31.00 © 2025 IEEE

Boosting 2D Transistor Performance via TiS$_2$ van der Waals Contact Engineering

Jialei Miao[1], Heng Zhang[1], Zheng Bian[1], Tianjiao Zhang[1] and Yuda Zhao[1*]

[1] College of Integrated Circuits, ZJU-Hangzhou Global Scientific and Technological Innovation Centre, Zhejiang University, Hangzhou, 310027, China (*Email: yudazhao@zju.edu.cn)

Abstract

In this work, we employ the 2D semimetal TiS$_2$ to construct high-quality metal-semiconductor contact for WSe$_2$ p-channel transistors and reduce subthreshold swing (SS) for MoS$_2$ n-channel transistors. The use of 2D high-work-function semimetal TiS$_2$ can effectively induce the degenerate p-doping at the contact regions with a weak fermi-level pinning effect. The contact resistance of WSe$_2$ transistors can be reduced from 303 kΩ to 6.5 kΩ. The van der Waals semimetal contact technique can improve 2D transistor performance.

Keywords: 2D materials, contact resistance, and carrier doping

Introduction

Two-dimensional (2D) materials with atomic thickness and non-dangling-bond interface are suitable channel candidates for technology nodes beyond sub-3 nm [1]. However, the charge transport in field-effect transistors (FETs) based on 2D semiconductor channels has not fully displayed superior performance, limited by the high contact resistance, interface scattering, and defects. Efficient charge carrier injection from the contacts to the 2D semiconductor channel is essential for high-performance operation, particularly in short-channel and high-frequency transistors [2].

In traditional silicon-based technology, ion implantation enables degenerate doping, effectively reducing contact resistance and lowering the Schottky barrier to enhance carrier injection. However, ion implantation is incompatible with ultrathin 2D materials because it introduces substantial defects. Nevertheless, employing degenerate doping specifically at the contact region is a viable strategy for optimizing contact engineering in 2D FETs. Semimetals such as Bi and Sb have been proposed to heavily dope n-type MoS$_2$ and reach the low contact resistance close to the quantum limit [3, 4]. However, P-type transistors with low contact resistance based on 2D materials are still difficult to realize due to the Fermi level pinning near the conduction band.

Here, we proposed contact engineering on the WSe$_2$ p-channel transistors by using the layered high-work-function semimetal TiS$_2$ (6.6 eV) as source/drain metal contact with a low R$_c$ of 6.5 kΩ [5]. Furthermore, TiS$_2$ contact in the MoS$_2$ shows subthreshold swing (SS) reduction due to the sharp band bending at the contact region. The SS of the TiS$_2$-contacted MoS$_2$ transistor has been reduced by 70.2% compared with the Au-contact MoS$_2$ transistor.

Result and discussion

The Schottky-Mott model gives incorrect predictions for the Schottky barrier height of 2D transistors due to the Fermi level pinning [2]. The Fermi level pinning originates from the gap states, such as those induced by defects, incomplete covalent bonds, and atomically sharp discontinuity between metal and semiconductor. Metal transfer technology has been proposed as an effective method to realize atomically clean contact on 2D materials [2]. The WSe$_2$ flakes in our work were mechanically exfoliated from bulk WSe$_2$ crystals and transferred to the BN/SiO$_2$/Si substrate (oxide thickness 285 nm). Figure 1a schematically shows the structure of the WSe$_2$ transistor with high-work-function TiS$_2$ (6.6 eV) contact. The TiS$_2$ flakes were mechanically exfoliated from bulk TiS$_2$ crystals and transferred in the glovebox (inert gas environment to avoid contact surface oxidation). Figure 1b shows the optical microscope image of WSe$_2$ transistor with TiS$_2$ contact.

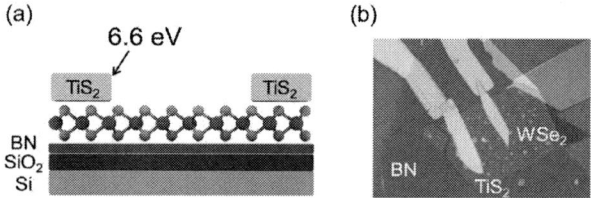

Fig. 1: a, schematic of WSe$_2$ transistor with TiS$_2$ contact. b, optical microscope image of the WSe$_2$ transistor with TiS$_2$ contact, scale bar 10 μm.

To demonstrate the potential of the van der Waals contact engineering by high work function semimetal. We fabricated the WSe$_2$ transistor with Au contact as a contrast. The Au electrodes are deposited by thermal evaporation and the thickness is 40 nm. With minimized interface disorder at the metal-semiconductor contact region, the van der Waals TiS$_2$-contacted WSe$_2$ transistor exhibits high-efficiency hole transport. The transfer curve is shown in Figure 2a. The TiS$_2$-contacted WSe$_2$ transistor shows the typically bipolar transport behavior with the dominant p transport. The mobility reaches as high as 111.5 cm^{2}V^{-1}s^{-1} due to a clean contact interface and low carrier scattering. The on-state current reaches 10.8 μA/μm (V$_d$ = 1 V). The output curves are shown in Figure 2b. The curve shows high linearity at V_g = 60 V, which represents the ohmic contact. For comparison, we also characterized the Au-contacted WSe₂ transistor with the conventional evaporated metal. The transfer curve of the Au-contacted WSe₂ transistor is shown in Figure 2c. Au-contacted WSe₂ transistor shows limited on-current due to the poor metal-semiconductor contact interface. The metal evaporation-induced trap states result in the carrier scattering and limit the electric performance of WSe₂. The output curve of the Au-contacted WSe₂ transistor is shown in Figure 2d. The on-state current is limited by the high contact resistance and Schottky barrier height.

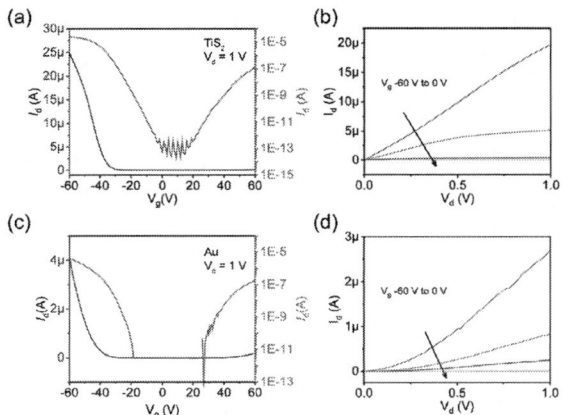

Fig. 2: a, transfer curves of WSe₂ transistor with TiS₂ contact. b, Output curves of WSe₂ transistor with TiS₂ contact. c, transfer curves of WSe₂ transistor with Au contact. d, Output curves of WSe₂ transistor with Au contact.

To character the contact quality in TiS₂-contacted and Au-contacted WSe₂ transistors. We employ the Y-function method to analyze the contact resistance [6]. The Y-function method requires only one transfer curve I_d-V_g by the linear regime with applying a small source-drain voltage and large gate voltage. I_d can be expressed as the following equation:

$$I_d = \frac{\mu_0}{1+\theta_0(V_g-V_{th})}C_i\frac{W}{L}(V_g-V_{th})V_d \qquad (1)$$

Where, θ_0 and V_{th} denote the intrinsic mobility in the linear regime, first-order mobility attenuation coefficient, and the threshold voltage.

Y-function was defined as

$$Y = \frac{I_d}{\sqrt{g_m}} = \sqrt{\mu_0 C_i V_d \frac{W}{L}(V_g-V_{th})} \qquad (2)$$

The value s_1 can be extracted from the slope of the Y-function versus V_g. The value s_2 can be extracted from the slope of $1/g_m^{0.5}$ versus V_g. The R_c follow the equation

$$R_c = \frac{S_2}{S_1}V_d \qquad (2)$$

Figure 3 shows the contact resistance (R_c) of the WSe₂ transistors. The extracted R_c values are 6.5 kΩ and 303 kΩ for TiS₂-contact and Au-contact WSe₂ transistors, respectively. Compared with Au-contact WSe₂ transistors, the contact resistance of TiS₂-contact WSe₂ transistors is greatly reduced by 45 times.

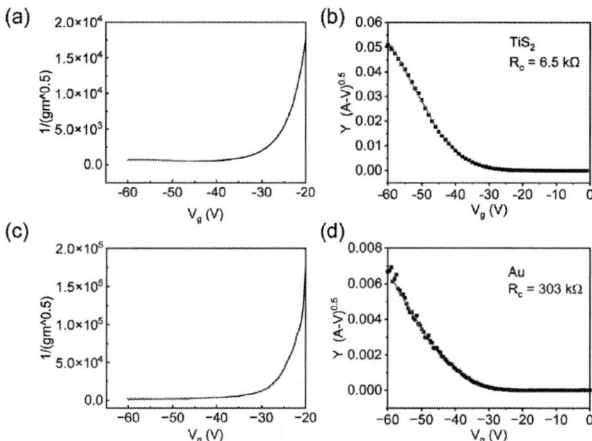

Fig. 3: a, transfer curves of WSe₂ transistor with TiS₂ contact. b, Output curves of WSe₂ transistor with TiS₂ contact. c, transfer curves of WSe₂ transistor with Au contact. d, Output curves of WSe₂ transistor with Au contact.

We character the TiS₂-contacted WSe₂ transistor at the ultralow temperature of 7 K. The transistor still shows favorable bipolar transport in Figure 5. This means that impurity scattering is insignificant in the TiS₂-contacted WSe₂ transistor. These findings are consistent with the trends of on-state current. The high-work-function semimetal TiS₂ can induce negligible defects on the channel during the fabrication process and effectively improve the hole injection in WSe₂ transistors.

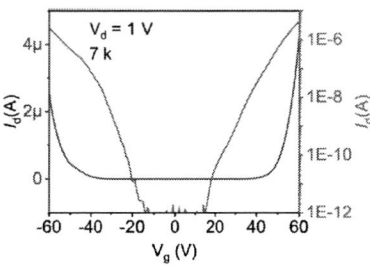

Fig. 4: Transfer curves of WSe₂ transistor with TiS₂ contact at the low temperature of 7 k.

The high-work-function TiS₂ semimetal can not only

optimize the contact resistance of p-channel 2D materials-based transistors by lowering the Schottky barrier height but also reduce SS of MoS_2 transistors by sharp band bending and reducing tunneling barrier width.

The carrier injection mechanism in MOSFET typically includes thermionic emission and tunneling, and thermionic emission plays an important role in 2D materials transistors at room temperature [7]. However, in the sub-threshold regime, the availability of electrons in the source is constrained by the high-energy tail of the Fermi-Dirac distribution in the conduction band, which ultimately restricts the SS value of the device.

To showcase the potential of low SS in n-type transistors by using TiS_2. We fabricated TiS_2-contacted MoS_2 transistors and used Au contacts for comparison. The Au electrodes were deposited via thermal evaporation and have a thickness of 40 nm. Figure 5 shows the optical microscope image of the TiS_2-contacted MoS_2 transistors.

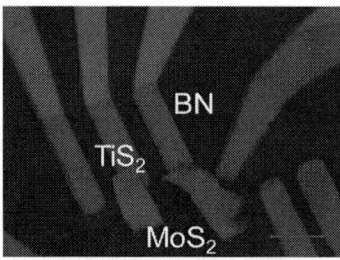

Fig. 5: Optical microscope image of MoS_2 transistor with TiS_2 contact, scale bar 10 μm.

We character the transfer curves of the TiS_2-contacted and Au-contacted MoS_2 transistors. SS can be extracted from the subthreshold region during the switching from off-state to on-state. The SS of TiS_2-contacted and Au-contacted MoS_2 transistors are 1.06 V/dec and 3.55 V/dec, respectively. The SS in TiS_2-contacted MoS_2 transistor is decreased by 70% and we believe the value is not the limitation of this engineering due to the BN and 285-nm-thick SiO_2 dielectric. When employing high-k dielectric, the TiS_2-contacted MoS_2 transistors have the potential to approach the fundamental limit of 60 mV/decade at room temperature and lower the power consumption in the 2D materials transistor-based integrated circuits.

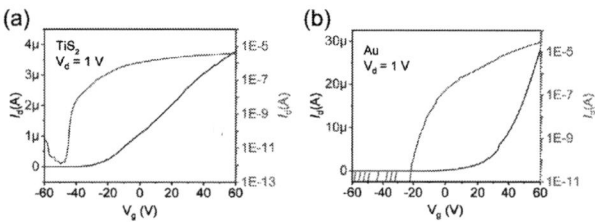

Fig. 6: a, transfer curves of MoS_2 transistor with TiS_2 contact. b, transfer

curves of MoS_2 transistor with Au contact.

Conclusion

In this work, we focused on transferring the high-work-function semimetal TiS_2 (6.6 eV) on the WSe_2 and MoS_2 transistors as the metal-semiconductor contact. By lowering the Schottky barrier height of hole transport, the p-type transport of the WSe_2 transistor has been significantly improved. In addition, TiS_2 has also been employed as the contact metal of MoS_2 transistors to lower SS by reducing the tunneling barrier width.

Acknowledgments

The project was primarily supported by the National Natural Science Foundation of China (62090034, 62104214, 62261160574, 62090030), the National Key R&D Program of China (2022YFA1204303), the Young Elite Scientists Sponsorship Program by CAST (2021QNRC001), and Kun-Peng Program of Zhejiang Province (H. W.). We thank ZJU Micro-Nano Fabrication Center and ZJU-Hangzhou Global Scientific and Technological Innovation Center for their support.

References

[1] Y. Zhao, M. Gobbi, L. E. Hueso, and P. Samorì, "Molecular Approach to Engineer Two-Dimensional Devices for CMOS and beyond-CMOS Applications," *Chemical Reviews*, vol. 122, no. 1, pp. 50-131, 2022.

[2] Y. Liu et al., "Approaching the Schottky–Mott limit in van der Waals metal–semiconductor junctions," *Nature*, vol. 557, no. 7707, pp. 696-700, 2018.

[3] W. Li et al., "Approaching the quantum limit in two-dimensional semiconductor contacts," *Nature*, vol. 613, no. 7943, pp. 274-279, 2023.

[4] P.-C. Shen et al., "Ultralow contact resistance between semimetal and monolayer semiconductors," *Nature*, vol. 593, no. 7858, pp. 211-217, 2021.

[5] T. Zhang et al., "2D semimetal with ultrahigh work function for sub-0.1 V threshold voltage operation of metal-semiconductor field-effect transistors," *Materials & Design*, vol. 231, p. 112035, 2023.

[6] H.-Y. Chang, W. Zhu, and D. Akinwande, "On the mobility and contact resistance evaluation for transistors based on MoS2 or two-dimensional semiconducting atomic crystals," *Applied Physics Letters*, vol. 104, no. 11, 2014.

[7] S. Kanungo, G. Ahmad, P. Sahatiya, A. Mukhopadhyay, and S. Chattopadhyay, "2D materials-based nanoscale tunneling field effect transistors: current developments and future prospects," *npj 2D Materials and Applications*, vol. 6, no. 1, p. 83, 2022.

Memristor-Based Approximate Adders for Efficient In-Memory Computing in Image Processing

Zhouchao Gan[†], Mengjie Li[†], Fan Yang, Fang Cheng, Xiangshui Miao, and Xingsheng Wang*

School of Integrated Circuits & Wuhan National Laboratory of Optoelectronics, Huazhong University of Science and Technology, Wuhan, China
*Correspondence e-mail：xswang@hust.edu.cn

Abstract

This paper presents two novel memristor-based approximate adder designs using the V/R-R logic principle to enhance in-memory computing (IMC) efficiency while balancing accuracy. These adders leverage approximate computation to improve the performance tolerant of minor precision losses in applications such as image processing. Experimental and simulation results demonstrate significant improvements in computational speed and area efficiency over conventional designs. Integrated into a ripple carry adder (RCA), the proposed designs maintain acceptable image quality for image addition, with peak signal-to-noise ratio (PSNR) exceeding 30 dB in most cases.

Keywords: Memristor, in-memory computing, approximate computing, full-adder, V/R-R

Introduction

Executing logic operations within memristors is an effective approach to address the von Neumann bottleneck by facilitating in-memory computing (IMC) [1]. Among the various memristor-based logic primitives, the earliest proposed IMPLY logic remains the most extensively studied [2], [3], [4], [5]. IMPLY logic integration of approximate computing can significantly enhance computational efficiency at the slight cost of accuracy, making it particularly advantageous for applications such as image processing [4], [5]. Our previous work introduced the V/R-R (Voltage/Resistance Inputs, Resistance Output) logic principle, which implements all 16 Boolean functions efficiently [6]. However, the approximation calculation using the V/R-R schemes is not yet well studied. This study investigates two approximate adder designs based on V/R-R logic, validated through experiments and simulations, demonstrateing the superior computational efficiency and area consumption over the IMPLY approximation scheme. Integrated into an RCA, they maintain acceptable image quality in most approximate cases, with PSNR exceeding 30 dB.

Implementation of Logic and Addition Functions

A. Design of V/R-R type Memristor Logic-In-Memory

Our previous work has demonstrated a reconfigurable design that can achieve 16 Boolean logic functions in the circuit structure shown in Fig. 1(a) by encoding port voltage and resistance of the first memristor as logic inputs, while the second memristor resistance represents logic output (V/R-R) [6]. Fig. 1(b) illustrates the port configuration for XOR logic, with detailed logic definitions provided in Fig. 1(c). The input variable V correlates with the port voltages V_{T1} and V_{T3}, while another input variable R is represented by the resistance of memristor M_1. The output is denoted by the resistance of memristor M_2. For an in-depth understanding of the logic design principle, please refer to [6]. Fig. 1(d) summarizes the port voltage configurations for AND, OR, and XOR logics, which are employed to achieve the adder function.

B. Realization of Exact and Approximate Adders

The full adder (FA) is a fundamental component in arithmetic units. A 1-bit FA has three inputs (addends a_i, b_i, and carry-in c_i) and two outputs (sum s_i and carry-out c_{i+1}). The logic expressions for exact FA are:

$$s_i = a_i \oplus b_i \oplus c_i \tag{1}$$

$$c_{i+1} = a_i \cdot b_i + (a_i \oplus b_i) \cdot c_i \tag{2}$$

This study proposes two approximate FA schemes: Scheme 1 (Sch. 1) focuses on speed by simplifying carry generation, while Scheme 2 (Sch. 2) enhances accuracy. Sch. 1 simplifies the carry generation process to achieve maximal computational speed. For a k-bit addition, the carry is set to zero for the first k-1 stages, and only the highest bit generates a carry-out, calculated as $c_{k+1} = a_k \cdot b_k$ [5], as shown in the following equation:

$$c_{i+1} = \begin{cases} 0 & i < k \\ a_i \cdot b_i & i = k \end{cases} \tag{3}$$

$$s_i = a_i + b_i \tag{4}$$

Sch. 2 refines the approximate logic expression to reduce error, which is given as:

$$c_{i+1} = a_i \cdot b_i \tag{5}$$

$$s_i = a_i \oplus b_i + c_i \tag{6}$$

Fig. 2(a) and (b) illustrates the circuit diagram for Sch. 1, including detailed operational steps and port voltage configuration. Similarly, Sch. 2 is shown in Fig. 2(c) and (d). Notably, Sch. 2 writes s_i by overwriting the value of c_i, which helps reduce memristor consumption. The truth tables for both schemes are shown in Fig. 3, where incorrect outputs are marked in red. It is evident that Sch. 2 yields less error at the cost of more complex logic expressions.

To implement complex applications, constructing multi-bit adders is crucial. Multi-bit adders can adopt the serial or

---

[†]These authors contributed equally. *Corresponding author

979-8-3315-0417-5/25 $31.00 © 2025 IEEE

parallel architectures using memristors. In Fig. 4(a), the serial topology for Sch. 1 uses $2n + 1$ memristors to perform n-bit approximate addition in series. Each logical cycle computes c_{n+1} and $s_1 - s_n$, requiring $n + 2$ steps. In the parallel architecture (Fig. 4(c)), each bit is computed in different rows, requiring only 3 steps and $2n + 1$ memristors. For Sch. 2, the serial topology requires $2n + 2$ memristors and $2n + 1$ steps, while the parallel version involves 4 steps and $2n + 1$ memristors (Fig. 4(b) and (d)). Sch. 2 shows a slight disadvantage in speed compared to Sch. 1, reflecting a trade-off between delay and accuracy.

C. Experimental and Simulation Verification

Key manufacturing steps and a TEM image of the cross-sectional cut of the 1-transistor-1-memristor (1T1R) unit are shown in Fig. 5(a). Fig. 5(b) presents the typical DC I-V curve of the 1T1R cell, while Fig. 5(c) displays the electron microscope image of a 32×32 1T1R array. More device characteristics can be found in [6], [7]. The test platform in Fig. 6 validates the two approximate adder schemes, with results shown in Fig. 7, indicating that both adders function correctly across all input combinations. Circuit-level simulations in Cadence Virtuoso are conducted for a 4-bit serial adder, with waveforms shown in Fig. 8(a), (b), and (c) for the exact adder, approximate Sch. 1, and Sch. 2. For the test case "0011 + 0101", all results align with expectations.

D. Comparison and Discussion

To evaluate the performance of the proposed approximate adders, a comparison is made with the exact adder based on V/R-R and the state-of-the-art (SoA) memristor-based approximate adder. Fig. 9 shows a comparison of the overhead of n-bit adders using different algorithms. The exact adder requires significantly more steps and memristors than the proposed designs, indicating improved delay and area consumption. Compared with the SoA IMPLY-based approximate adders, the proposed designs enhance both accuracy and efficiency due to the efficient V/R-R logic.

Application in Image Addition

A. Error Analysis

Approximate computing is well-suited for applications that are error-tolerant, most commonly such as image processing [8]. This paper focuses on image addition, a fundamental task for operations such as image masking. Image addition adds pixel values at corresponding positions in two images. For example, adding two 512×512 8-bit grayscale images can be accomplished with an 8-bit adder. As shown in Fig. 10, we employ a mixed-precision approach: an approximate Ripple Carry Adder (RCA) processes the lower n bits, while an exact RCA handles the upper m bits (where $m + n = 8$). This configuration is denoted as "m Exa. + n Approx.". Common error metrics, including Mean Error Distance (MED), Normalized Mean Error Distance (NMED), and Mean Relative Error Distance (MRED) [5], are used to evaluate these approximate adders. A comprehensive evaluation of all possible addend combinations is conducted using Python, with results summarized in Fig. 11. As expected, Sch. 2 consistently shows lower error than Sch. 1, making it more suitable for applications with stringent error tolerance.

B. Application-Level Simulation

To assess the quality of approximate calculations in image processing, we use metrics such as Peak Signal-to-Noise Ratio (PSNR), Structural Similarity Index (SSIM), and Mean Structural Similarity Index (MSSIM). PSNR indicates noise levels in the image, with values above 30 dB considered acceptable [8]. SSIM measures similarities and differences between images. Fig. 12 illustrates image addition using both exact and mixed-precision methods based on Sch 2. Image quality metrics are computed via Python and summarized in Fig. 13. Results show that even when processing the lower 6 bits of the 8-bit RCA with an approximate adder, it still meets acceptable quality standards (PSNR > 30 dB).

Conclusion

This paper presents two memristor-based approximate adder designs utilizing the V/R-R logic approach, aimed at enhancing computational efficiency and reducing area consumption. The proposed designs, validated through experiments and simulations, demonstrate high performance with acceptable accuracy trade-offs, especially in image processing applications where PSNR remains above 30 dB. The results indicate that the proposed adders outperform conventional methods in terms of both delay and area efficiency, providing an effective solution for error-tolerant applications within IMC systems.

Acknowledgments

This work is supported in part by National Natural Science Foundation of China under Grants U2341221 and 62274070, Hubei Province Science and Technology Major Project under Grant 2022AEA001, Interdisciplinary Research Program of HUST (2023JCYJ042), and Hubei Key Laboratory of Advanced Memories.

References

[1] Z. Sun et al., Nat. Electron., pp. 1–13, 2023.

[2] J. Borghetti et al., Nature, vol. 464, no. 7290, pp. 873–876, 2010.

[3] D. Radakovits et al., IEEE Trans. Circuits Syst. Regul. Pap., vol. 67, no. 5, pp. 1495–1506, 2020.

[4] S. E. Fatemieh et al., IEEE J. Emerg. Sel. Top. Circuits Syst., vol. 13, no. 1, pp. 175–188, 2023.

[5] F. Seiler and N. TaheriNejad, IEEE Trans. Circuits Syst. Regul. Pap., pp. 1–14, 2024.

[6] Y. Song et al., Adv. Sci., vol. 9, no. 15, p. 2200036, 2022.

[7] Z. Gan et al., 2024 8th IEEE Electron Devices Technology & Manufacturing Conference (EDTM), pp. 1–3, 2024.

[8] S. Mittal, ACM Comput. Surv., vol. 48, no. 4, pp. 1–33, 2016.

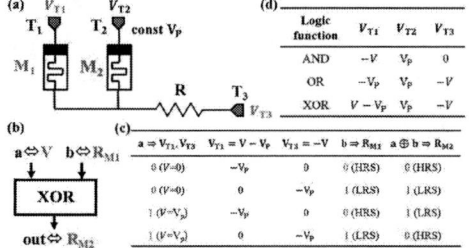

Fig. 1. (a) Circuit diagram of 16 Boolean logic implementations. (b) Example of an XOR logic gate with port configuration. (c) Truth table of XOR logic, detailing voltage mappings. (d) Port voltage configurations for various logic functions.

Fig. 2. (a) Circuit diagram of Approximate Adder Scheme 1, and (b) its detailed operation steps and voltage configuration. (c) Circuit diagram of Approximate Adder Scheme 2, and (d) its corresponding operation steps and voltage configuration.

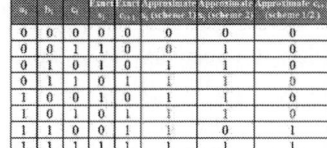

Fig. 3. Truth tables for the two proposed approximate adder schemes.

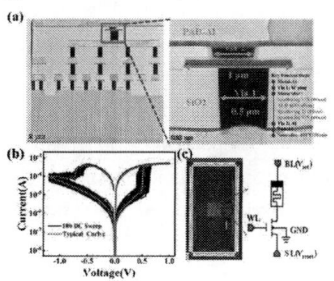

Fig. 5. (a) TEM image of the cross-sectional cut and key manufacturing steps of the 1T1R integrated unit. (b) Measured resistive switching behavior in DC IV sweeping mode. (c) Metallographic electron microscopy image of the 32×32 1T1R array.

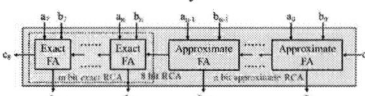

Fig. 10. Schematic diagram of the 8-bit RCA structure with varying degrees of mixed exact and approximate addition.

Scheme	m Exa. : n Approx. (m, n)	8-bit RCA MED	NMED	MRED
Sch. 1	(7, 1)	0.25	0.00049	0.00135
Sch. 2		0	0	0
Sch. 1	(6, 2)	0.62	0.00122	0.00335
Sch. 2		0.25	0.00049	0.00135
Sch. 1	(5, 3)	1.38	0.00269	0.00728
Sch. 2		0.75	0.00147	0.00399
Sch. 1	(4, 4)	2.88	0.00563	0.01489
Sch. 2		1.75	0.00342	0.00913
Sch. 1	(3, 5)	5.88	0.0115	0.0293
Sch. 2		3.75	0.00734	0.01893
Sch. 1	(2, 6)	11.88	0.02324	0.05566
Sch. 2		7.75	0.01517	0.03697
Sch. 1	(1, 7)	23.88	0.04672	0.10123
Sch. 2		15.75	0.03082	0.06843
Sch. 1	(0, 8)	47.88	0.09369	0.17387
Sch. 2		31.75	0.06213	0.11893

Fig. 11. Error metrics of the 8-bit RCA for different degrees of approximation.

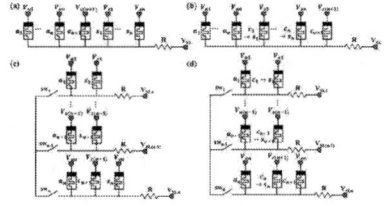

Fig. 4. (a) Serial adder architecture for Scheme 1 and (b) for Scheme 2 with n-bit inputs. (c) Parallel adder architecture for Scheme 1 and (d) for Scheme 2.

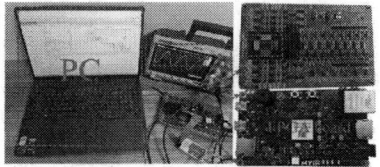

Fig. 6. Experimental test platform.

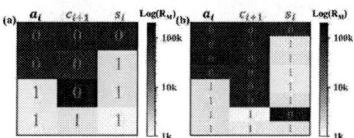

Fig. 7. Experimental results for (a) 1-bit Approximate Adder Scheme 1 and (b) Approximate Adder Scheme 2.

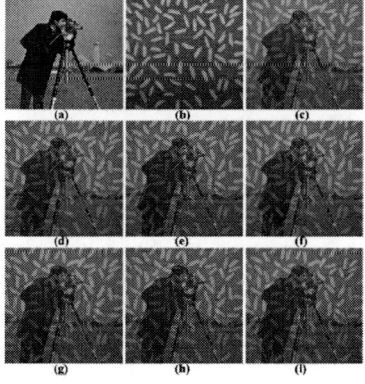

Fig. 12. Examples of image addition: (a) camera image, (b) masking, (c) exact image addition, (d-i) increasing approximation from "7 Exa. + 1 Approx." to "2 Exa. + 6 Approx.".

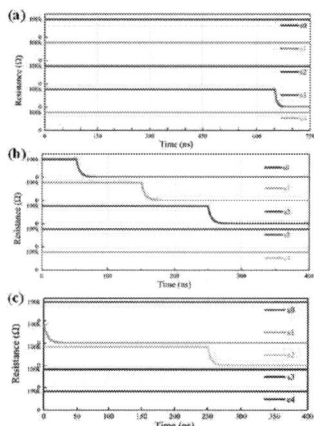

Figure 8. Simulation waveforms of (a) 4-bit exact adder, (b) Approximate Adder Scheme 1, and (c) Approximate Adder Scheme 2, taking "0011 + 0101" as an example.

Logic type	Ref	Algorithm	No. of Memristors	No. of Steps	accuracy ranking
IMPLY	[5]	SINC (serial)	2n+1	3n	4
		PINC (parallel)	3	3n	4
		SINC+ (serial)	2n+2	3n+3	3
		PINC+ (parallel)	6	3n+1	3
XOR/AND/OR	[6]	Exact Serial	6n	2n+4	1
		Exact Parallel	4n+1	5n+1	1
	This work	Approximate Serial Sch. 1	n+2	2n+1	3
		Approximate Parallel Sch. 1	3	2n+1	3
		Approximate Serial Sch. 2	2n+2	2n+1	2
		Approximate Parallel Sch. 2	4	2n+1	2

Fig. 9. Comparison of the overhead of executing n-bit adders by different algorithms.

(m, n): m Exa. +n Approx.	PSNR (dB)	SSIM	MSSIM
(7, 1)	inf	1	1
(6, 2)	54.85	0.9988	0.9988
(5, 3)	47.95	0.9949	0.9949
(4, 4)	41.57	0.9805	0.9805
(3, 5)	35.34	0.9394	0.9394
(2, 6)	30.74	0.8787	0.8787
(1, 7)	20.54	0.7329	0.7329

Fig. 13. Image quality metrics for image addition using Scheme 2 at different degrees of approximation.

979-8-3315-0417-5/25 $31.00 © 2025 IEEE

Fluorescent nanodiamond encapsulated liposomes for quantum sensing in nematode worms

Takaki Arakawa[1], F. Kamada[1], K. Kinjo[1], K. Oshimi[1], M. Sara[1], M. Maeki[2], M. Tokeshi[2], M. Fujiwara[1]

[1] Department of Chemistry, Okayama University, Okayama 700-8530, Japan.

Tel: +81-(0)86-251-7834. E-mail: pqhi08zq@s.okayama-u.ac.jp

[2] Graduate School of Engineering, Hokkaido University, Sapporo 060-8628, Japan.

Abstract

Fluorescent nanodiamonds (FNDs) are promising quantum nanosensors that can probe inside living organisms such as nematode worms. Introduction of FNDs into nematode worms is a first but critical step for the quantum nanosensors to probe inside the worms. In this paper, we report on the synthesis of FND-encapsulated liposomes (FND-Liposomes) using a microfluidic device and the introduction of FND-Liposomes to C. elegans. We successfully confirmed the fluorescence and ODMR of FND-Liposomes in C. elegans.

Keywords: Microfluidics, Fluorescent nanodiamond and ODMR

Introduction

Fluorescent nanodiamonds contain nitrogen-vacancy (NV) centers, allowing their temperature to be measured through the frequency shift in optical detection magnetic resonance (ODMR). Their ultra-low toxicity makes them suitable for use in living organisms, such as cells and nematodes [1]. To measure the internal temperature of an organism, fluorescent nanodiamonds must be introduced into its body. However, when administered orally to nematodes, these nanodiamonds are quickly expelled from the body (specifically intestine) due to the strong pumping power and do not accumulate in the intestine. Therefore, microinjection method has been used to introduce fluorescent nanodiamonds, but this method requires a skilled technique and more importantly invasive to the organism. For these reasons, a simpler and non-invasive method is needed.

We considered using liposomes, which are commonly used as carriers in drug delivery systems (DDS) and can encapsulate various substances. Liposomes are highly biocompatible since they are made from lipids contained in cells. Thus, we believe liposomes could serve as effective carriers for transporting nanodiamonds into organisms. The thin-film method is the primary technique for preparing liposomes, as it allows for the production of large quantities. However, the size of the liposomes needs to be adjusted using an extruder, etc., because reproducibility can be low, leading to size variation across samples even with the same procedure. Additionally, we needed to improve the fabrication method

to effectively encapsulate nanodiamonds within liposomes. To address these issues, we utilized the iLiNP (invasive lipid nanoparticle production) device developed by Tokeshi et al [2]. This microfluidic device enables the continuous production of nanoparticle formulations from lipid (or amphiphilic polymer) solutions and water, using a baffle flow path designed based on an original theory (Fig. 1) [3]. This baffle structure allows for control over liposome size by adjusting the speed and flow rate of the sample as it passes through the channel. Moreover, encapsulated liposomes can be easily created by adding the desired materials to the solution before it flows through the channel.

In this study, we prepared FND-liposomes using the iLiNP device (Maeki et al., in preparation) to easily and noninvasively introduce nanodiamonds into nematodes. The samples were administered orally to the worms, and ODMR measurements were conducted.

Methods

Preparation of FND-Liposomes

The iLiNP device produces liposomes by simultaneously flowing lipid and aqueous solutions into a channel using a syringe pump (Fig. 2). For this process, we used 300 µL of a lipid solution at a total concentration of 8 mM, composed of a molar ratio of DPPC (Dipalmitylphosphatidylcholine), cholesterol, and DOTAP (1,2-dioleoyl-3-trimethylammonium-propane) in a 45:45:10 ratio. The solvent was 99% ethanol. For the aqueous solution, we prepared 900 µL containing a diamond concentration of 500 µg/mL. This consisted of 450 µL of a 25 mM acetate buffer mixed with 450 µL of a 1 mg/mL solution of 100 nm fluorescent nanodiamonds (Adamas Nanotechnologies). Before use, the microfluidic device must be cleaned. We filled a syringe with more than 600 µL of 99% ethanol and flowed it through the device at a rate of 1000 µL/min using the syringe pump. Next, ultrapure water was introduced under the same conditions, and then the water in the channel was drained by pushing air through at the same rate using an empty syringe. After cleaning, the lipid and aqueous solutions were set up in the syringe pump and flowed simultaneously into the microfluidic device at a rate of 100 µL/min. Once the channel was filled with solution, the lipid solution was pushed out at 500 µL/min, while the aqueous solution was pushed out at 1500 µL/min. The resulting liposome solution was collected in a microtube. Finally, the liposome samples were dialyzed overnight using cellulose tubes.

Characterization of FND-Liposomes
We measured the size distribution and zeta potential of the prepared FND-liposomes using dynamic light scattering (DLS) and electrophoretic light scattering (ELS). For DLS, we placed 1 mL of a 100-fold diluted sample in a disposable cell. For ELS, 800 µL of the 100-fold diluted solution was placed in an Omega cuvette. Both measurements were conducted using a Litesizer DLS 500 (Anton Paar).

Feeding of FND-Liposomes to nematode worms and ODMR measurements
To prepare the nematodes, we used an egg breaching. The breaching solution consisted of 5 mL of water, 300 µL of sodium hypochlorite, and a grain of sodium hydroxide. Forty-eight hours after breaching, the nematodes were transferred to 15 mL tubes, washed, and incubated at 20°C overnight with 200 µL of FND-liposomes at a diamond concentration of 1 mg/mL. For the ODMR measurements, samples were prepared as shown in Fig. 3. We utilized a notch-shaped antenna developed by Oshimi et al [4]. To prevent movement during measurement, the nematodes were placed in a 2% sodium azide solution diluted in M9 buffer for a few minutes before being transferred to an agar pad.

Results
The characterization of FND-Liposomes
The size distribution and zeta potential of the prepared FND-liposomes were measured using dynamic light scattering (DLS) and electrophoretic light scattering (ELS) (Fig. 4). The size distribution was found to be 126 ± 29.9 nm, with a zeta potential of $+43.2 \pm 0.73$ mV. In comparison, the nanodiamonds alone had a size of 106 ± 17.0 nm and a zeta potential of -38.9 ± 1.68 mV. This indicates that the FND-liposomes increased in size by about 20 nm, and the zeta potential shifted to a significantly positive value. The increase in size is attributed to the formation of a liposomal film around the nanodiamonds. The positive zeta potential of the prepared sample is due to the use of the cationic lipid DOTAP in the liposome formulation. The absence of a peak corresponding to the nanodiamonds suggests they are encapsulated within the liposomes, with the liposomal charge masking the nanodiamond potential.

Feeding of FND-Liposomes into nematode worms and ODMR measurements
After administering FND-liposomes to the nematodes, we observed the worms using a fluorescence microscope. The results showed fluorescence from the nanodiamonds in the pharynx and intestine of the nematodes (Fig. 5). This indicates that the FND-liposomes were not expelled but rather accumulated in the pharynx and intestinal tract. According to a study by Zou et al. [5], positively charged nanoparticles tend to accumulate more in nematodes, highlighting the importance of using positively charged liposomes. ODMR measurements of the nanodiamonds within the nematodes, fed with the same sample, produced a clear ODMR spectrum (Fig. 6). This suggests that FND-Liposomes does not adversely affect the ODMR measurements.

Conclusion
In summary, we successfully prepared FND-liposomes using the iLiNP device and characterized them based on size distribution and zeta potential. The nanodiamonds accumulated in the pharynx and intestine after oral administration to nematode worms. ODMR measurements conducted on these diamonds yielded clear spectrum, confirming that the encapsulation process did not hinder their performance.

Acknowledgments
We thank M. Tokeshi and M. Maeki for providing the iLiNP device. This work was in part supported by the following grants: AMED (JP23zf0127004), JSPS-KAKENHI (20H00335, 20KK0317, 21H05599), JST-ASPIRE (JPMJAP2339), NEDO (JPNP20004), JST (JPMJMI21G1), RSK Sanyo Foundation, Asahi Glass Foundation.

References
[1] M. Fujiwara et al., "Real-time nanodiamond thermometry probing in vivo thermogenic responses", Sci. Adv. 2020, 6, eaba9636 (2020).

[2] N. Kimura, M. Maeki et al., "Development of the iLiNP Device: Fine Tuning the Lipid Nanoparticle Size within 10 nm for Drug Delivery", ACS Omega, 3, 5, 5044–5051 (2018).

[3] https://www.lilacpharma.com/proprietary_technology/ilinp/

[4] K. Oshimi et al., "Glass-patternable notch-shaped microwave architecture for on-chip spin detection in biological samples", Lab Chip, 22, 2519-2530 (2022).

[5] Y. Zou et al., "Size, polyglycerol grafting, and net surface charge of iron oxide nanoparticles determine their interaction and toxicity in Caenorhabditis elegans", Chemosphere, 358, 142060 (2024).

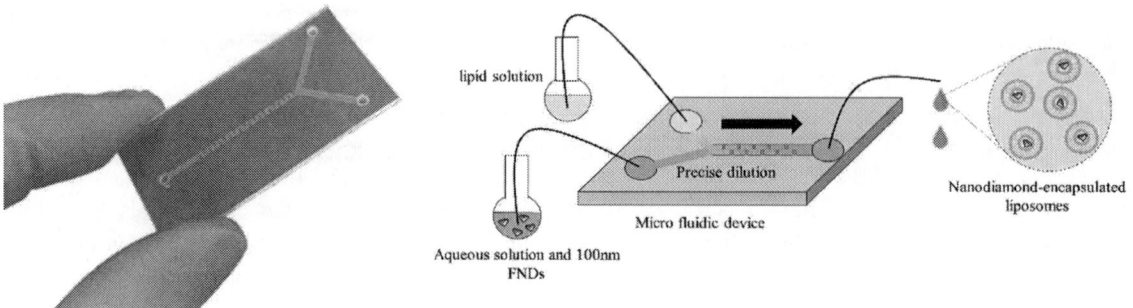

Fig. 1: The image of iLiNP device from reference [3].

Fig. 2: Schematic figure of liposome preparation method using iLiNP device

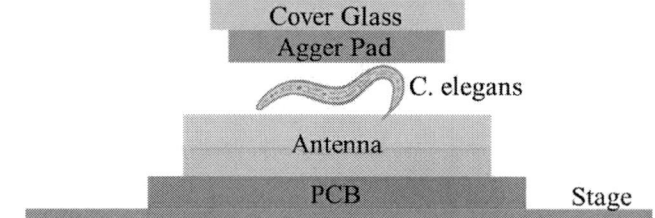

Fig. 3: Sample of ODMR measurements.

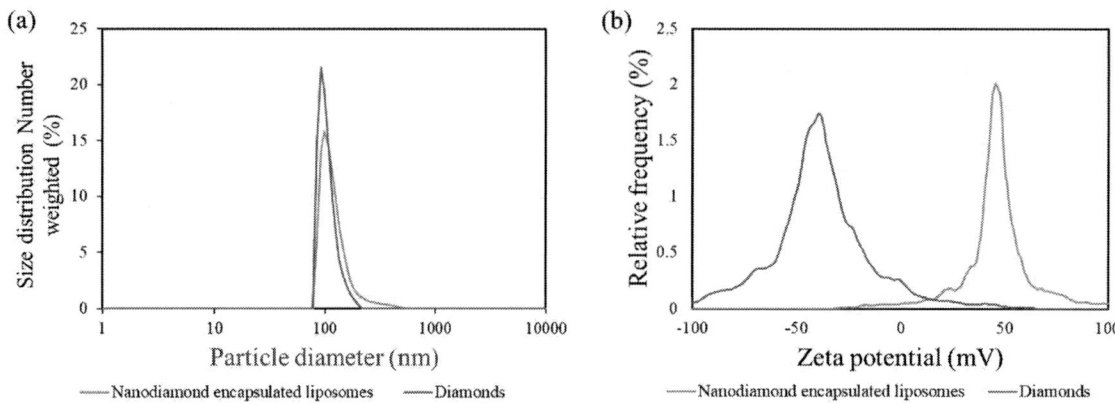

Fig. 4: (a) Size distribution of FND-liposomes (Orange) and pure diamonds (Blue). (b) Zeta potential of FND-liposomes (Orange) and pure diamonds (Blue).

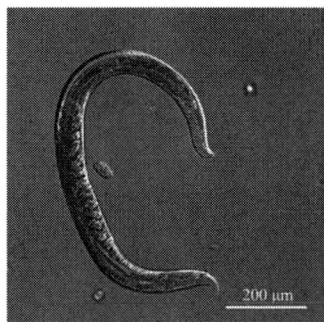

Fig. 5: Figure of a nematode feeding nanodiamond-encapsulated liposomes. Red is fluorescent nanodiamond

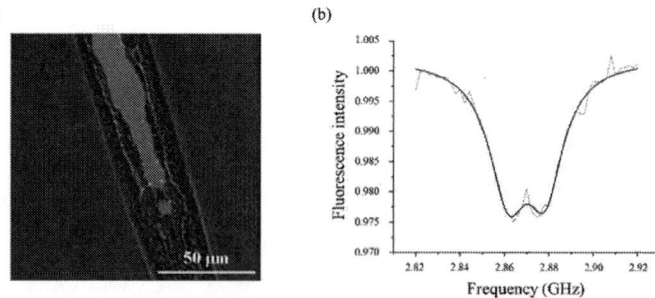

Fig. 6: nematodes feeding nanodiamond-encapsulated liposomes and ODMR spectrum.

Enhancing CAD Workflows: Heterogeneous Graph Attention Networks for Efficient Routing Congestion Prediction in Chip Design

Qingyuan Yang[1], Mingkun Xu[2], Hongyi Li[1], Yu Du[1], Dahu Feng[1], and Rong Zhao[1]

[1]Center for Brain Inspired Computing Research (CBICR), Tsinghua University, 100084, Beijing, China,
[2]Guangdong Institute of Intelligence Science and Technology, 519031, Zhuhai, China

ABSTRACT

This paper presents a heterogeneous graph attention network for predicting routing congestion on the netlists, crucial for chip design. It features a two-level attention mechanism to account for netlist structure and design specifications, which previous studies overlooked. Our approach achieves faster and more accurate congestion prediction, with up to 19% increase in accuracy and up to 5.22× speedup. It is the first work that integrates design specifications, enabling a comprehensive prediction of congestion.

Keywords: Routing congestion, Heterogeneous graph attention network and Design specifications.

INTRODUCTION

Routing congestion negatively impacts unit utilization and increases chip area and line length, ultimately decreasing chip performance while raising the cost [1]. To improve design quality, designers have to repeatedly detect congestion until all congestion is removed. However, acquiring congestion feedback necessitates several days or weeks, and it becomes a bottleneck as the number of transistors on the chip increases [2], which leads to a growing trend to employ machine learning algorithms for expeditious congestion prediction to speed up production.

Predicting on netlist is the most time-efficient as it emerges as the initial output of the chip design process. Authors in [3] apply embedding learning through matrix factorization for prediction. Other works use graph neural networks to predict congestion [4], [5], and CongestionNet [6] conceptualizes the netlist as a homogeneous graph and uses Graph Attention Network (GAT) [7]. However, these works overlook the influence of design specifications, which serve as constraints during the placement and significantly impact the routing process. As shown in Fig. 1, for the same netlist, congestion occurs at different locations under different design specifications, illustrating the limitations of relying solely on netlist structure for congestion prediction.

In this work, we formulate the netlist and design specifications as a heterogeneous graph and propose a prediction model with a specialized attention mechanism. This model accurately predicts congestion across diverse design specifications, expediting design cycles and elevating design quality.

METHODOLOGY

The overall flow of our prediction method is shown in Fig. 2. We develop a heterogeneous graph called the Spec-Circuit Graph to represent the netlist structure and the design specifications. Furtherly, we propose Spec-GAT, a heterogeneous graph neural network containing an efficient two-level attention-based message-passing mechanism, which accurately predicts congestion based on the Spec-Circuit graph.

A. Spec-Circuit Graph

In the netlist, the cell denotes the circuit components and the net denotes the connections between them. A netlist can be represented as a bi-directional homogeneous graph where the cells and the nets are the nodes and edges in the graph, respectively.

To represent the impact of design specifications, we introduce a new type of node called the **spec node**. For ease of distinction, we refer to nodes and edges in a netlist as **cell nodes** and **net edges**. The spec node establishes connections with all cell nodes through a distinct type of edge named **logic edges**, representing the impact of design specifications.

The Spec-Circuit graph is defined as $G = \{V, S, N, L, X^c, X^s\}$, where V represents the set of cells nodes, while the set of spec node is represented by S. The sets of net edges and logic edges are represented by N and L respectively. The characteristics of the two types of nodes are represented by X^c and X^s respectively.

B. SpecGAT

The cell nodes and spec nodes are connected via different types of edges, and the significance of neighbor nodes, when linked by these distinct edge types, varies in their contribution to the task of congestion prediction. To solve this challenge, SpecGAT utilizes a two-level attention mechanism including **type-level attention** and **semantic-level attention**.

1) Type-Level Attention: Type-level attention is designed to learn the importance of neighbor nodes based on different types of edges. We introduce two attention functions ($attention_{type}^{net}$, $attention_{type}^{logic}$) considering two connectivity relationships, namely net and logic. The importance of neighbors of cell node i can be formulated as follows:

$$e_{j \to i}^{c} = attention_{type}^{net}\left(x_j^c, x_i^c\right), \qquad (1)$$

$$e_i^s = attention_{type}^{logic}\left(x^s, x_i^c\right), \qquad (2)$$

where $e_{j \to i}^{c}$ and e_i^s denote the influence of cell node j and the spec node on cell node i, respectively. $attention_{type}^{net}$ and $attention_{type}^{logic}$ are implemented by deep neural networks, with identical $attention_{type}^{net}$ shared among different pairs of cell nodes.

After obtaining the influence weights of connected nodes, the embedding of cell node i can be obtained by aggregating its neighbor node features:

$$z_i^{net} = \sigma\left(\sum_{j \in N_i} e_{j \to i}^c \cdot x_i^c\right), \qquad (3)$$

$$z_i^{logic} = \sigma e_i^s \cdot x^s, \qquad (4)$$

where σ is a weighting factor, and N_i represents the cell nodes connected to node i. z_i^{net} and z_i^{logic} are net-path and logic-path embedding of cell node i.

2) Semantic-Level Attention: The cell node forges connections through two distinct types of edges, and semantic-level attention is proposed to automatically learn their importance.

$$\left(\beta^{net}, \beta^{logic}\right) = attention_{semantic}\left(z^{net}, z^{logic}\right), \qquad (5)$$

$attention_{semantic}$ is realized by deep neural networks utilized for capturing semantic-level importance.

Final representation of the cell node i is performed by the fusion of embeddings at the semantic level.

$$Z = \beta^{net} \cdot z^{net} + \beta^{logic} \cdot z^{logic}, \qquad (6)$$

where Z is the final representation to predict the congestion value of each cell node.

RESULTS AND DISSCUSION

We take the netlist "RISCY-FPU-a-2-c20" as an example, and Fig. 3 shows Pearson correlation coefficients between the congestion of this netlist and design specifications. The results show that the extent and percentage of cell nodes positively or negatively influenced by Macro placement and power mesh setting are fundamentally comparable. The impact of these two parameters is relatively well-balanced, mainly attributed to the strategic placement of macros. Primarily, the power mesh influences the layout planning without substantially affecting the design routing conditions. In contrast, utilizations and filler insertion have notable impacts on overall congestion.

We conduct comparative analysis between three distinct variants of Graph Convolutional Network (GCN) [8] and GAT [7]. The first category of GCN and GAT lack the integration of spectral information, and the second category enhance each node's feature set by incorporating specification information, while the third category comprises heterogeneous networks combining diverse features and structures. Table 1 summarizes the comparison results. SpecGAT achieves an MSE score of 0.0157, and an MAE score of 0.0642, suggesting a highly accurate prediction of congestion.

To validate the necessity of two-level attention mechanism and design specification, we conduct two ablation studies, as shown in Table 2. Removing attentions leads to a decrease in MSE and MAE, illustrating the importance of design specification information. Then, we compare the speed of our model with the baselines, as shown in Table 3. SpecGAT demonstrates superior inference speed, achieving up to 5.22× faster than the baselines.

CONCLUSION

This study proposes a congestion prediction method at logic synthesis stage, which can be seamlessly integrated into commercial EDA tools. By pioneering the integration of design specifications, the proposed method achieves up to 19% reduction in MAE and 7% decrease in MSE, and a substantial 5.22× increase in inference speed, leading to substantial improvements in quality and a marked acceleration in the chip design.

REFERENCES

[1] J. Lou, S. Krishnamoorthy, and H. S. Sheng, "Estimating routing congestion using probabilistic analysis," in Proc. of the 2001 Int. Symp. on Phys. Design, pp. 112–117 (2001).

[2] Y. Shin, "Ai-eda: Toward a holistic approach to AI-powered EDA," in Proc. of the 2023 ACM/IEEE 5th Workshop on Machine Learning for CAD (MLCAD), 2023, pp. 1-3 (2023).

[3] A. Ghose, V. Zhang, Y. Zhang, D. Li, W. Liu, and M. J. Coates, "Generalizable cross-graph embedding for gnn-based congestion prediction," in Proc. of the 2021 IEEE/ACM International Conference On Computer Aided Design (ICCAD), pp. 1–9 (2021).

[4] K. Min, S. Kwon, S.-Y. Lee, D. Kim, S. Park, and S. Kang, "Clusternet: Routing congestion prediction and optimization using netlist clustering and graph neural networks," in Proc. of the 2023 IEEE/ACM International Conference on Computer Aided Design (ICCAD), pp. 1–9 (2023).

[5] B.-L. Wang, G. Shen, D. Li, J. Hao, W. Liu, Y. Huang, H. Wu, Y. Lin, G. Chen, and P.-A. Heng, "Lhnn: lattice hypergraph neural network for vlsi congestion prediction," in Proc. of the 59th ACM/IEEE Design Automation Conference (DAC), pp. 1–9 (2022).

[6] R. Kirby, S. Godil, R. Roy, and B. Catanzaro, "Congestionnet: Routing congestion prediction using deep graph neural networks," in Proc. of the 2019 IFIP/IEEE 27th International Conference on Very Large Scale Integration (VLSI-SoC), pp. 217–222 (2019).

[7] P. Velickovic, G. Cucurull, A. Casanova, A. Romero, P. Lio', and Y. Bengio, "Graph attention networks," ArXiv, vol. abs/1710.10903 (2017).

[8] T. Kipf and M. Welling, "Semi-supervised lassification with graph convolutional networks," ArXiv, vol. abs/1609.02907 (2016).

979-8-3315-0417-5/25 $31.00 © 2025 IEEE

TABLE I: Performance Comparison

	Methods	RISCY-a	RISCY-b	RISCY-FPU-a	RISCY-FPU-b	zero-riscy-a	zero-riscy-b	**Overall**
MSE	GCN(w/o.spec)	0.0134	0.0107	0.0383	0.0145	0.0146	0.0099	0.0169
	GAT(w/o.spec)	0.0126	0.0119	0.0370	0.0135	0.0143	0.0097	0.0164
	GCN(w.spec)	0.0127	0.0122	0.0361	0.0130	0.0145	0.0097	0.0163
	GAT(w.spec)	0.0129	0.0139	**0.0328**	**0.0116**	0.0149	0.0100	0.0160
	GCN(Hetero)	0.0126	0.0118	0.0364	0.0134	0.0140	0.0098	0.0163
	GAT(Hetero)	0.0123	0.0112	0.0364	0.0131	0.0137	0.0093	0.0160
	SpecGAT	**0.0120**	**0.0103**	0.0370	0.0129	**0.0136**	**0.0089**	**0.0157**
MAE	GCN(w/o.spec)	0.0678	**0.0574**	0.0887	0.0705	**0.0653**	0.0628	0.0687
	GAT(w/o.spec)	0.0663	0.0741	0.0847	0.0715	0.0786	0.0673	0.0737
	GCN(w.spec)	0.0692	0.0713	0.0873	0.0666	0.0837	0.0636	0.0736
	GAT(w.spec)	0.077	0.0832	0.0906	0.0659	0.0918	0.0693	0.0796
	GCN(Hetero)	0.0708	0.0733	0.0897	0.0712	0.0796	0.0676	0.0753
	GAT(Hetero)	0.0674	0.0681	0.0872	0.0677	0.0754	0.0633	0.0715
	SpecGAT	**0.0594**	0.0593	**0.0798**	**0.0631**	0.0683	**0.0574**	**0.0642**

TABLE II: Result of Ablation Studies

Ablation	MSE	MAE	ΔMSE	ΔMAE
Type-level att.	0.0165	0.0743	-5.10%	-15.73%
Semantic-level att.	0.0162	0.0743	-3.18%	-15.73%
Two-level att.	0.0164	0.0749	-4.46%	-16.66%
Spec node	0.0164	0.0758	-4.46%	-18.07%

TABLE III: Result of Inference Time

Methods	Inference Time (s)	Time Spent
GCN(w/o.spec)	138.18	1.54×
GAT(w/o.spec)	140.57	1.56×
GCN(w.spec)	469.72	5.22×
GAT(w.spec)	122.53	1.36×
GCN(Hetero)	354.24	3.94×
GAT(Hetero)	129.93	1.44×
SpecGAT	89.94	/

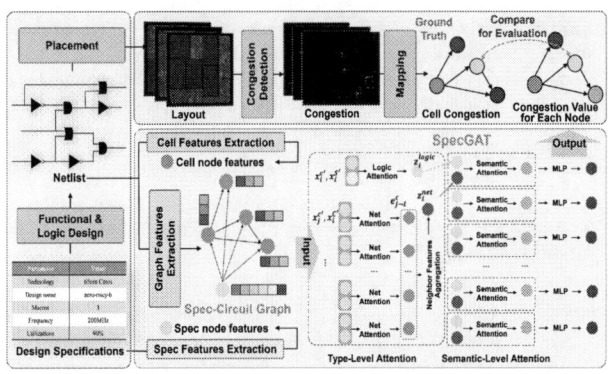

Fig. 2: The overall flow of the proposed method. The gray background represents detecting real congestion after completing the chip design. The green background represents extracting cell, graph and specification features to predict congestion using SpecGAT.

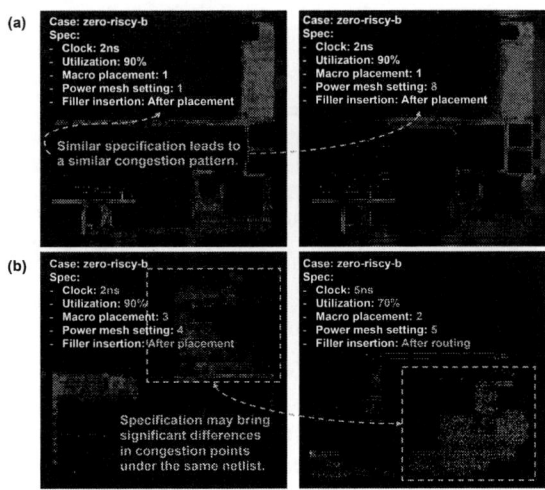

Fig. 1: Congestion maps of a netlist under different design specifications. (a) Only one parameter is different between the two congestion maps. (b) All parameters are different between the two congestion maps.

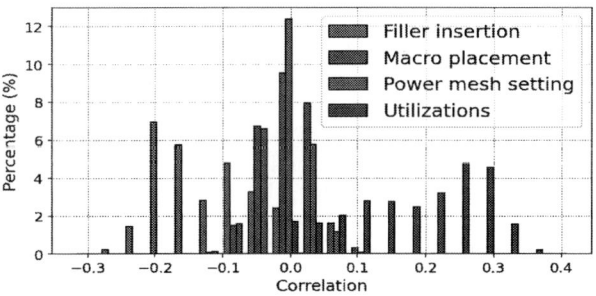

Fig. 3: Correlation of four key design specifications (Filler insertion, Macro placement, Power mesh setting) with node congestion. The y-axis represents the proportion of nodes impacted by the respective degree of influence relative to the total number of nodes, while the x-axis represents the value of the Pearson correlation, which ranges from -0.3 to 0.4.

Enabling Floating Body Effect in Bulk-Si Transistor for Area and Energy-Efficient Spiking Neuron

Shubham Patil[1], Hemant Hajare[1], Abhishek Kadam[1], Jay Sonawane[1], Shreyas Deshmukh[1], Veeresh Deshpande[1] and Udayan Ganguly[1]

[1]Department of Electrical Engineering, Indian Institute of Technology Bombay, India (e-mail: shubhampatil2.107@gmail.com, udayan@ee.iitb.ac.in)

Abstract— Ultra-compact and energy-efficient circuits are crucial for edge application. In the literature, an area and energy-efficient circuit based on band-to-band tunneling (BTBT) using partially depleted Silicon-on-Insulator (PD-SOI) MOSFETs with standard SOI technology has been proposed. However, PD-SOI technology is costly due to the expensive "smart-cut" method to fabricate SOI wafers. Therefore, in this work, we propose a bulk MOSFET design with separate body contact and laminated well technique to enable BTBT's current collection. This is followed by the implementation of a Leaky integrate and Fire (LIF) neuron using mixed-mode simulation in TCAD with a well-calibrated deck to demonstrate spiking neuron. The simulation confirms that the laminated wells successfully enable hole storage in the body, facilitating effective LIF neuron operation in bulk technology.

Keywords— Silicon-on-Insulator, Band-to-Band Tunneling, Leaky integrate and Fire neuron, TCAD, mixed-mode simulation.

I. INTRODUCTION

With the rapid advancement of the Internet of Things (IoT), there is an increasing demand for energy-efficient circuits and systems to handle complex tasks [1], [2]. Devices operating in the band-to-band tunneling (BTBT) regime in Silicon-on-Insulator technology (SOI) offer a promising solution to cater to the low-power needs of edge applications [3]. Recently, researchers have shown hardware demonstrations of multiple low-power applications of transistors operating in the BTBT regime, e.g., filters [4], oscillators [5] and Leaky integrate and Fire (LIF) neurons [3]for large-scale spiking neural network (SNN) applications [2]. SNNs offer low-power parallel computing by emulating electrical neurons and synapses (Fig. 1), making them suitable for applications such as autonomous vehicles, robotics, and medical diagnosis. These neural network applications require large-scale SNNs capable of sophisticated processing. Therefore, developing ultra-compact and energy-efficient neurons is crucial for the scalable deployment of large SNN networks.

Recently, our group demonstrated a quantum tunneling-based neuron using partially depleted (PD) SOI MOSFETs with standard SOI technology to enhance energy efficiency [3]. The lower current in the BTBT regime minimizes energy consumption while using a body capacitor for leaky integration, which allows for area-efficient LIF neuron implementation. This approach was successfully demonstrated in a 36-neuron reservoir circuit of a liquid state machine for spoken digit classification [2]. Although the proposed BTBT neuron is compact, energy-efficient, and based on mature technology that enables large-scale integration, PD-SOI technology is costly due to the expensive "smart-cut" method to fabricate SOI wafers [6]. Despite the

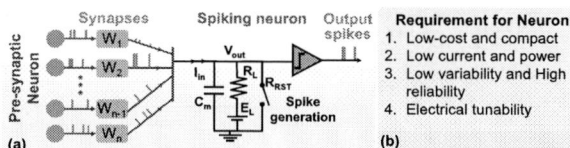

Fig. 1. **Motivation.** (a) SNN implementation and key requirement to enable a robust SNN.

low power benefit of BTBT-biased circuits, the reliance on SOI technology presents a distinct disadvantage in the fabrication cost of circuits based on BTBT-biased transistors. Facilitating band tunneling devices in bulk technology will blend the low power benefits of BTBT-based circuits with the cost efficiency of bulk technology.

Therefore, in this work, we proposed a bulk MOSFET design with separate body contact and laminated well to enable the BTBT current collection without the need for costly SOI technology. We further show the implementation of a LIF neuron using mixed-mode simulation in TCAD with a well-calibrated TCAD deck.

II. EXPERIMENTAL DETAILS

The 32nm PD-SOI MOSFET with gate length (L_G)/width (W) of 40/450 nm is used in this work. The detailed fabrication details can be found in [7]. The measurements are performed by Keysight B1500A Semiconductor parameter analyzer. A well-calibrated Sentaurus TCAD deck is used to understand the tunneling mechanism's physics and the LIF neuron implementation and analysis.

III. RESULT AND DISCUSSION

A. SOI transistor experimental characterization

The measured drain (I_D), source (I_S), gate (I_G), and body current (I_B) of the PDSOI transistor as a function of gate bias (V_{GS}) for a range of drain bias (V_{DS}) is shown in Fig. 2(a). The presence of I_B confirms the drain-body tunneling with negligible contribution from I_G. Among the low-power operating regimes, i.e., BTBT and subthreshold (SS), the BTBT regime is preferred as it offers lower variability [8]. The band diagram at the channel-drain junction with high negative V_{GS} and high positive V_{DS} bias (BTBT regime biasing) is shown for the visualization of the BTBT (Fig. 2(b)). Experimental I_D-V_{GS} and I_S-V_{GS} at fixed V_{DS} of 0.9 V for temperatures 27 °C, 85 °C, and 125 °C. With temperature, a linear increase in the BTBT current, an exponential increase in the SS current, and a linear decrease in the ON current are observed [8], [9].

B. TCAD modeling of SOI transistor

The SOI transistor schematic with dimensions modeled in TCAD is shown in Fig. 2(d). To develop a calibrated TCAD deck, the following device physics models are utilized: (1)

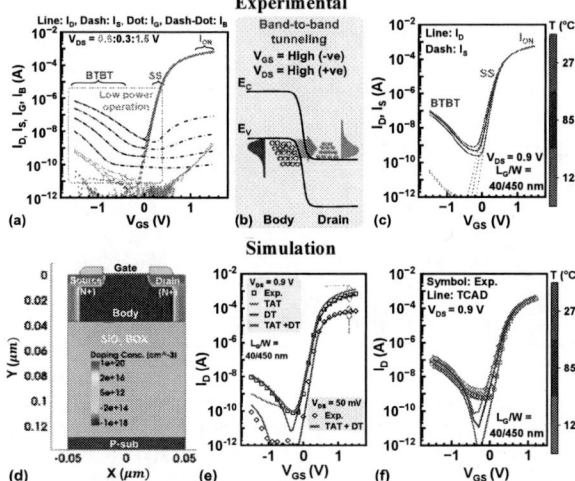

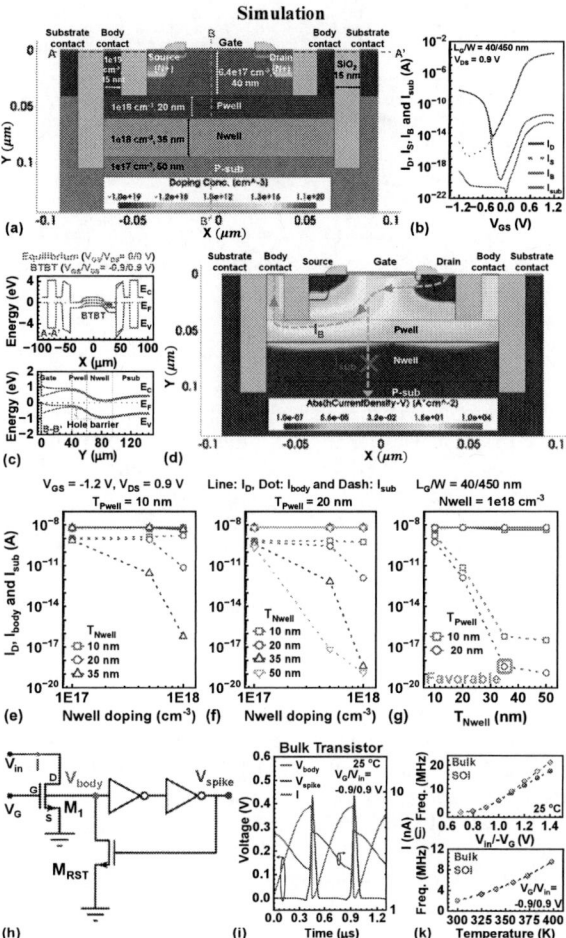

Fig. 2. BTBT experimental characterization and SOI TCAD modelling.
(a) Measured drain (I_D), source (I_S). gate (I_G) and body current (I_B) of the PDSOI transistor as a function of gate bias (V_{GS}) for a range of drain bias (V_{DS}). The presence of I_B confirms the drain-body tunneling with negligible contribution from I_G. For low power operation, BTBT and SS regime are preferred. (b) The band diagram at channel-drain junction with high negative V_{GS} and high positive V_{DS} bias (BTBT regime biasing) is shown for the visualization of the BTBT. (c) Experimental I_D-V_{GS} and I_S-V_{GS} at fixed V_{DS} of 0.9 V for temperature 27 °C, 85 °C and 125 °C. With temperature, a linear increase in the BTBT current, exponential increase in the SS current and a linear decrease in the ON- current is observed. (d) SOI MOSFET structure with dimension used in TCAD. (e) Calibrated (I_D) current of the SOI transistor as a function of V_{GS} for drain biases of 50 mV and 0.9 V at RT (25 °C). (f) TCAD calibration at higher temperature (85 °C and 125 °C).

the drift-diffusion model for capturing current transport, (2) doping-dependent mobility and the Lombardi model for capturing mobility degradation, (3) the extended Canali model for accounting high field saturation, and (4) SRH and Auger models for depicting recombination-generation processes. The dynamic non-local path (NLP) tunneling (dominant at higher negative V_{GS}) and the Hurkx model (dominant at lower negative V_{GS}) are used to describe the tunneling phenomenon. The detailed explanation can be found in our earlier work [10]. Calibration of the TCAD deck against experimental data with these models demonstrates excellent agreement for the nominal device at V_{DS} of 0.05 V and 0.9 V (Fig. 2(e)). The TCAD deck is further calibrated at elevated temperatures of 85 °C and 125 °C, showing reasonable agreement with the experimental data (Fig. 2(f)).

C. Proposed bulk transistor and LIF neuron implementation

The proposed bulk MOSFET design with a separate body contact and laminated well is illustrated in Fig. 3(a). To prevent BTBT current leakage into the substrate, an n-buried layer is incorporated into the design [11]. The calibrated TCAD deck (Fig. 2(e-f)) is employed to analyze the proposed transistor, and the mixed-mode simulations are used to evaluate its LIF neuron performance. The I_D, I_S, I_B, and substrate current (I_{sub}) as functions of V_{GS} at fixed $V_{DS} = 0.9$ V are depicted in Fig. 3(b). Fig. 3(c) shows the energy band diagram along the AA' and BB' cutline in equilibrium ($V_{GS}/V_{DS} = 0/0$ V) and BTBT biased regime ($V_{GS}/V_{DS}=-0.9/0.9$ V). Fig. 3(d) shows the hole current density in the proposed transistor. These results demonstrate that the BTBT

Fig. 3. Bulk MOSFET TCAD modelling. (a) Bulk MOSFET structure with dimension used in TCAD. (b) The I_D, I_B and I_{sub} current of the bulk transistor as a function of V_{GS} for drain biases of 0.9 V at RT (25 °C). (c) Energy band diagram along the AA' and BB' cutline in equilibrium and BTBT biased regime ($V_{GS}/V_{DS}=-0.9/0.9$ V). (d) Hole current density flow in the MOSFET. (e-g) Impact of T_{Pwell}, T_{Nwell} and Nwell doping concentration on I_D, I_B and I_{sub}. $T_{Pwell} = 20$ nm, $T_{Nwell} = 35$ nm and Nwell doping = 1e18 cm^{-3} create a sufficient hole barrier and hence a low substrate leakage. (h) LIF neuron circuit implemented in TCAD. (i) Voltage and current output characteristics vs. Time obtained using mixed-mode simulation. (j-k) Extracted LIF neuron frequency as a function of V_{DD} (25 °C) and temperature ($V_{GS}/V_{DS}=-0.9/0.9$ V). The extracted frequency of LIF neuron implemented using PDSOI transistor is given as reference.

current is collected at the body contact, resulting in negligible current flow through the substrate contact due to the hole barrier offered by the Nwell (Fig. 3(c)).

We optimized the device by conducting a design space study. The parameters considered include Pwell thickness (T_{Pwell}), Nwell thickness (T_{Nwell}), and Nwell doping concentration (Fig. 3(e-g)). Our findings indicate that a T_{Pwell} of 20 nm, T_{Nwell} of 35 nm, and Nwell doping concentration of 1e18 cm^{-3} provide an adequate hole barrier, thus minimizing substrate leakage.

Finally, the LIF neuron (Fig. 3(h)) is implemented using the proposed bulk transistor in TCAD, and its voltage and current output characteristics are obtained using mixed-mode simulation (Fig. 3(i)). The transistor M_1 serves as a leak integrator, and an inverter chain with an M_{RST} transistor is used as fire and reset circuitry. When the input voltage pulse

979-8-3315-0417-5/25 $31.00 © 2025 IEEE

TABLE II
BENCHMARKING WITH THE STATE-OF-THE-ART

Reference	OperatingRegime	Type	Cap.	Tech. (Cost)	Area μm^2 $(F^2)^*$	Energy/Spike (fJ)
A. Joubert [14]	ON	Simulation	External	Bulk (Low)	538 (127× 10^3)	4.13 × 10^4
G. Indiveri [1]	SS	Experimental	External	Bulk (Low)	2573 (21× 10^3)	9 × 10^5
S. Dutta [15]	II	Simulation	Body	PDSOI (High)	1.8 (1767)	3.5 × 10^4
Tanmay [3]	BTBT	Simulation	Body	PDSOI (High)	0.8** (784)	3.22
Ajay [2]	BTBT	Experimental	Body	PDSOI (High)	40 (19753)	8.2
This work	**BTBT**	Simulation	**Body**	Bulk (Low)	0.27** (264)	3.5

*F is the minimum feature size, ** Projected area at the 32nm node, Cap.- Capacitor, Freq.- Frequency, Tech.-Technology

(V_{in}) is applied at the drain of transistor M_1, biased in the BTBT regime with a $V_{GS} = -V_{DD}$, the minority electrons in the body tunnel into the drain, leaving behind holes, resulting in a current that integrates to charge up the body, increasing V_{body} potential. When the V_{body} exceeds a threshold (V_{th}), the V_{spike} becomes high, and the neuron is said to fire [3]. Fig. 3(j) shows the LIF neuron frequency as a function of different V_{DD} at room temperature. To evaluate the temperature dependence of the LIF neuron, mixed-mode simulations are performed at temperatures ranging from 25 °C to 125 °C. Fig. 3(k) depicts the LIF neuron frequency increases with an increase in temperature as $I_{D,BTBT}$ shows a positive temperature coefficient. [8], [9]. The extracted LIF neuron frequency implemented using the PDSOI transistor is given as a reference (Fig. 3(j-k)). These results confirm that the laminated wells successfully enable hole storage in the body, facilitating effective LIF neuron operation.

In Table II, we compare various silicon-based neuron implementations. BTBT-based neurons using PD-SOI MOSFETs with standard SOI technology [3] show the lowest energy/spike with the lowest silicon footprint compared to other state-of-the-art neuron implementations. The energy is calculated by integrating the power consumed during the time period between two consecutive spikes. However, the necessity of SOI technology undermines cost-effectiveness. In this work, we replicated the BTBT-based neuron with a lower area and similar energy/spike in bulk technology, which can significantly reduce the fabrication cost. Apart from assisting BTBT-based circuits in bulk technology, the proposed transistor can be used for analog applications such as buffers, where body transconductance ($g_{mb} = \partial i_d / \partial v_{SB}$) is undesired by making $V_{SB} = 0\ V$ [12].

D. Generalization

This study enables the implementation of BTBT-based neurons using bulk technology. Additionally, the proposed device architecture can be extended to other BTBT-based low-power circuits, such as BTBT-based low-pass filters [4], true random number generators (TRNG) [13], and temperature sensors [5], making it possible to utilize bulk technology instead of SOI technology.

IV. CONCLUSION

We propose a bulk MOSFET design with a separate body contact to collect the BTBT current as an alternative to costly SOI technology—the proposed MOSFET functions as a leaky integrator. Mixed-mode simulations in TCAD, utilizing a well-calibrated deck, are performed to demonstrate the neuronal behavior under varying voltage and temperature conditions. This design facilitates low-cost, low-area, and high-energy-efficient SNN implementations for edge applications with constraints on area, power, and cost.

ACKNOWLEDGMENT

This work was partly supported by the DST, IITBNF, CEN IIT Bombay, and GlobalFoundries.

REFERENCES

[1] G. Indiveri et al., "Neuromorphic silicon neuron circuits," 2011. doi: 10.3389/fnins.2011.00073.

[2] A. K. Singh et al., "Quantum Tunneling Based Ultra-Compact and Energy Efficient Spiking Neuron Enables Hardware SNN," IEEE TCAS I: Regular Papers, vol. 69, no. 8, pp. 3212–3224, Aug. 2022, doi: 10.1109/TCSI.2022.3172176.

[3] T. Chavan et al., "Band-to-Band Tunneling Based Ultra-Energy-Efficient Silicon Neuron," IEEE TED, vol. 67, no. 6, pp. 2614–2620, Jun. 2020, doi: 10.1109/TED.2020.2985167.

[4] A. A. Kadam et al., "A Compact and Ultra-Low-Power Low-Pass Filter Based on Band-to-Band Tunneling Effect," IEEE TCAS II: Express Briefs, vol. 70, no. 9, pp. 3298–3302, Sep. 2023, doi: 10.1109/TCSII.2023.3273621.

[5] A. A. Kadam et al., "A 42.3um2 Band to Band Tunneling-Based Oscillator Enabled Temperature to Digital Converter With Resolution FoM of 0.16 pJK2 for Embedded Temperature Sensing," IEEE SSCL, 2024, doi: 10.1109/LSSC.2024.3433610.

[6] R. Dargis et al., "Monolithic integration of rare-earth oxides and semiconductors for on-silicon technology," Journal of Vacuum Science & Technology A: Vacuum, Surfaces, and Films, vol. 32, no. 4, p. 041506, Jul. 2014, doi: 10.1116/1.4882173.

[7] B. Greene et al., "High Performance 32nm SOI CMOS with High-k/Metal Gate and 0.149μm2 SRAM and Ultra Low-k Back End with Eleven Levels of Copper," in 2009 Symposium on VLSI Technology, 2009, pp. 140–141.

[8] S. Patil et al., "Process-Voltage-Temperature Variability Estimation of Tunneling Current for Band-to-Band-Tunneling-Based Neuron," IEEE TED, pp. 1–7, Dec. 2023, doi: 10.1109/ted.2023.3331660.

[9] Y. Taur and T. H. Ning, "Fundamentals of Modern VLSI Devices, 3rd ed. Cambridge," Cambridge University Press, 2021.

[10] J. Sonawane et al., "Design Space and Variability Analysis of SOI MOSFET for Ultra-Low Power Band-to-Band Tunneling Neurons", doi: 10.48550/arXiv.2311.18577.

[11] K. Wang et al., "First Foundry Platform Demonstration of Hybrid Tunnel FET and MOSFET Circuits Based on a Novel Laminated Well Isolation Technology," in European Solid-State Device Research Conference, Editions Frontieres, 2023, pp. 13–16.

[12] Behzad Razavi, Design of Analog CMOS Integrated Circuits, 1st ed. New York, NY, USA : McGraw-Hill, 2001.

[13] A. Kadam et al., "Quantum Tunneling Noise-based Entropy Core for True Random Number Generation," 2024, doi: 10.21203/rs.3.rs-4363886/v1.

[14] A. Joubert et al., "Hardware spiking neurons design: Analog or digital?," in Proceedings of the IJCNN, 2012.

[15] S. Dutta, V. Kumar, A. Shukla, N. R. Mohapatra, and U. Ganguly, "Leaky Integrate and Fire Neuron by Charge-Discharge Dynamics in Floating-Body MOSFET," Sci Rep, vol. 7, no. 1, Dec. 2017, doi: 10.1038/s41598-017-07418-y.

Carbon Nanotube-Based High-Performance Bioelectronics

Youfan Hu

Key Laboratory for the Physics and Chemistry of Nanodevices, School of Electronics and Center for Carbon-Based Electronics, Peking University, Beijing 100871, China.

youfanhu@pku.edu.cn

Abstract

There is growing interest in ultrathin flexible devices that can provide excellent conformability toward advanced laminated bioelectronics. In addition to achieving ultrathin physical morphology, to realize the superiority of ultrathin electronics, the performance of the electronic devices is crucial to support the desired advanced functions. In this paper, a selection of the key techniques and progresses that support carbon nanotube-based flexible thin-film transistors, circuits and integrated systems toward advanced bioelectronics are discussed.

Keywords: Carbon nanotube, Ultrathin, High performance, Bioelectronics

Introduction

Since the compliance and adhesion of a film depend on its bending stiffness, which is proportional to the cubic thickness of the film [1], reducing the thickness of a flexible device can significantly improve the compliance and adhesion of the device. This characteristic makes the construction of ultrathin flexible devices very attractive for achieving an intimate interface between electronics and organisms with complex-shaped biological surfaces for high-fidelity information recording, which is very important for applications in health assessment, physical activity tracking, and personalized therapy. High-performance electronics to support advanced functions are also key enablers of these applications. Carbon nanotubes (CNTs) have been extensively explored in advanced flexible electronics due to their extraordinary intrinsic properties, including ultrathin body, high carrier mobility, excellent mechanical flexibility, and solution processability [2-4]. Extending the excellent performance of CNT thin-film transistors (TFTs) to ultrathin morphology need to address many challenges, including developing compatible fabrication strategies with fragile substrates, additional mechanical structure design to enhance mechanical stability under deformation, meeting special requirements to design interface circuits for handling bio signals, etc. We will discuss these issues one by one in the following.

Manufacturing strategies for ultrathin flexible devices

To handle the fragile ultrathin substrates during microfabrication process, electronic systems on ultrathin polymer films are generally processed with rigid supporting substrates during fabrication, followed by delamination and transfer to the targeted working areas. The manufacturing strategy we developed for ultrathin flexible devices is schematically shown in Figure 1a.

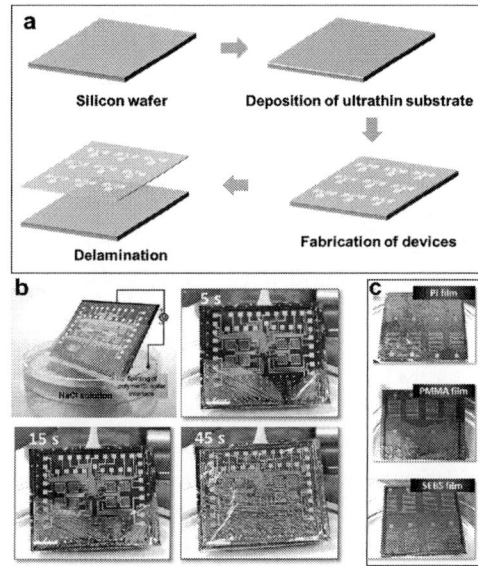

Fig. 1. (a) Manufacturing strategy for ultrathin flexible devices. (b) Schematic diagram of the CAED experimental setup, and photographs showing the progress of delaminating an ultrathin parylene substrate with electronic devices. Scale bars, 1 cm. [5] (c) Photographs showing delamination of PI, PMMA, and SEBS films deposited with Au electrical wires. [5]

A heavily doped silicon wafer works as a supporting substrate during fabrication. An ultrathin polymer film is then deposited on the silicon wafer, followed by the fabrication of devices on it. Parylene films and polyimide (PI) films are used in our case, which are deposited by chemical vapor deposition method at room temperature and spin-coating of precursor solution, respectively. Before device fabrication, a thin layer of HfO_2 was deposited on the substrate to promote the adhesion of the metal electrode on the plastic substrate. The final step is to delaminate the polymer substrate with devices from the supporting substrate. In this strategy, both the device fabrication and substrate delamination procedures play crucial roles in the obtained device performance. The former determines the ideal upper limit of the performance, while the latter determines how to maintain this upper limit at the end of the process. A technique, named capillary-assisted electrochemical delamination (CAED) [5] (Fig. 1b) to handle this issue, which can provide a high delamination efficiency at a large scale and is applicable to various ultrathin polymer substrates (Fig. 1c). When a potential is applied between the silicon wafer and the electrolyte solution of NaCl, the silicon wafer with low resistance serves as the anode to launch an electrochemical reaction at its surface, $Si(s) - 8e^- + 8OH^- \rightarrow H_2SiO_3(s) + 3H_2O + 2O_2(g)$, which introduces the etching of the silicon wafer to detach the ultrathin polymer film. Because all electronic components are on the top surface of the polymer film, these components remain dry, clean, and intact during the process, and no chemical contamination is introduced into the system as well.

Integrated sensor systems for sweat monitoring

Based on the aforementioned manufacturing strategy, a frequency-modulation quasidigital sensor system for sweat monitoring was constructed on a 2-μm-thick parylene film [6], including a resistive humidity sensor and a three-stage complementary metal-oxide-semiconductor (CMOS) ring oscillator (RO), as shown in Fig. 2a and 2b. When humidity increases, the output frequency of the system increases (Fig. 2c). During sweat monitoring, when the skin is dry, the frequency signal generated by the system is 11.4 kHz was generated by the system under dry skin conditions (blue stars in Fig. 2c), while it increases to 304.6 kHz when sweating (red stars in Fig. 2c). This humidity sensing platform demonstrated an in-situ signal processing capability that can covert relative humidity information on the skin into quasi-digital frequency signals, improving noise immunity and being ready for the next step of wireless signal transmission. In addition, the power consumption of the system is very low, which is only 1.3 and 55.4 μW when the skin is dry and sweating, respectively. Also, as shown in Fig. 2c, in fact, the system includes eight identical sensor systems, revealing the integration capability with multiple sensors for large area monitoring or diverse information collection.

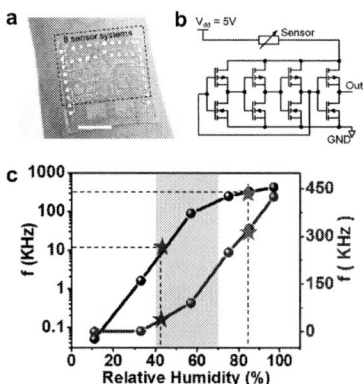

Fig. 2. (a) Photograph of the integrated sensor systems attached onto skin for sweat monitoring. Scale bar, 1 cm. (b) Circuit diagram and (c) frequency-humidity curves of the integrated sensor system. [6]

An integrated system for physiological information acquisition, processing, and storage

To provide more functions, flexible sensors, including humidity, temperature, and electrocardiogram (ECG) sensors, sensor interface circuits, which is high-performance differential amplifiers based on CNT TFTs, and an integrated flexible flash memory array of 24 bit by 16 bit are integrated on a 2-μm-thick parylene film for physiological information recording [7], as shown in Fig. 3a and 3b. The amplifier can provide great amplification capability with a voltage gain of 27 dB, a common-mode rejection ratio (CMRR) of >43 dB, and a large gain bandwidth product (GBWP) >22 kHz. The CNT-based flash memories present a long retention time, which is projected to be approximately 10^8 s, low program/erase voltages of ±2 V, and good endurance that is projected to be over 10^6 cycles, which are on par with industrial requirements. A system-level demonstration of physiological information acquisition, processing, storage was realized, and the retrieval of stored data was demonstrated, as shown in Fig. 2c and 2d, revealing reliable information retention characteristics that allow the data to serve as evidence/reference for later clinical analysis and diagnosis.

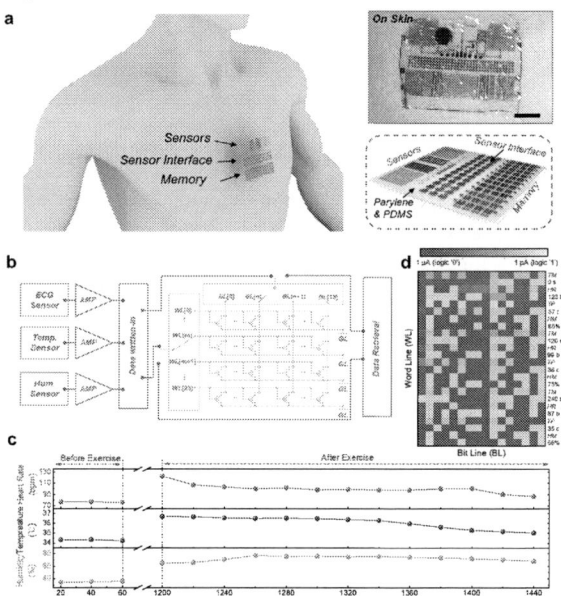

Fig. 3. (a) Schematic illustration and photograph of an electronic system conformally laminated on the skin. Scale bar, 1 cm. Bottom right: Spatial layout of the whole system. ystems attached onto skin for sweat monitoring. Scale bar, 1 cm. (b) Block diagram of the integrated system for physiological signal recording and (c) physiological information extracted from real-time monitoring results of the three sensors. (d) Data retrieved from the memory array 1 day after writing. [7]

Sub–180-nanometer-thick ultraconformable integrated systems

As mentioned before that bending stiffness of a thin film is proportional to the cubic thickness of the film, which makes the construction of ultrathin flexible devices with thicknesses of less than 1 μm very attractive for achieving ultraconformal contacts and even self-adhesive, clean, and stable interfaces with biological objects. To achieve this goal, first, we prepared PI substrate with a thickness of ~125 nm (inset of Fig. 4a) [8] by diluting the solid content of the PI precursor in an N-methyl-2- pyrrolidone solution with optimization to achieve the thinnest thickness while maintaining good mechanical robustness and integrity and minimal surface roughness. It should be mentioned that we used oxygen plasma to treat the silicon wafer surface before spin coating of the PI solution, which enhance the adhesion of the diluted PI solution and thus ensure continuity of the film at a greatly reduced thickness. And, the aforementioned CAED method makes handling this ultrathin film possible.

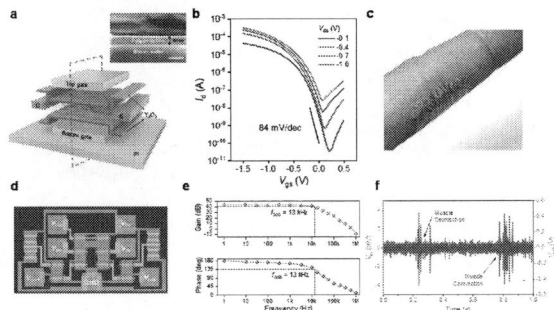

Fig. 4. (a) Schematic illustration of a dual-gate CNT TFT on a PI film. Inset: Cross-sectional SEM image of a 125-nm-thick PI film 25 nm. Scale bar, 200 nm. (b) Transfer characteristics of a typical dual-gate CNT-TFT with a channel length of 5 μm and a channel width of 50 μm. (c) Photographs of an ultrathin flexible device laminated on a human finger joint. Scale bars, 1 cm. (d) Optical microscope image of an ultrathin differential amplifier. Scale bar, 100 μm. (e) Bode plots of the differential amplifier for common-mode input signals. (f) Input-output characteristics of the circuit for amplifying EMG signals captured from the surface of the forearm skin of an adult male. [8]

To guarantee excellent gate control efficiency and mechanical stability of the ultrathin CNT TFTs, a dual-gate structure is introduced as schematically shown in Fig. 4a, which includes bottom gate layer (Pd, 8 nm), top (Pd, 18 nm) gate layers, dielectric layers of yttrium oxide (Y_2O_3) (10 nm per layer) grown by thermal oxidization. With the well-designed thickness of different layers in the stacked structure, the mechanical neutral plane of the system is situated close to the channel (located ~1 nm below the top Y_2O_3 dielectric layer). The obtained ultrathin devices exhibit high transconductance (8.96 microsiemens per micrometer), steep subthreshold swing (84 millivolts per decade) (Fig. 4b), and can sustain a bending radius of curvature of <10 micrometers. With a total thickness of less than 180 nm, the ultrathin film can be attached at finger joint with ultraconformal contact, as shown in Fig. 4c. Finally, differential amplifier constructed based on these ultrathin TFTs achieves demonstrates a low-frequency voltage gain of 141 (~43 dB) and an f_{3dB} of ~13 kHz, resulting in a maximum GBWP (GBWP = open-loop gain × f_{3dB}) of ~1.83 MHz. which is the highest gain-bandwidth product among flexible differential amplifiers, enabling higher-gain amplification of weak signals over an extended frequency spectrum that is demonstrated by amplification of electromyography signals in situ.

Conclusion

In conclusion, achieving ultrathin morphology and high performance simultaneously is very important for advanced laminated bioelectronics. Other key functional modules, such as analog-to-digital converters, filters, etc., are highly needed to further improve the information processing capabilities and achieve more comprehensive ultrathin integrated systems.

Acknowledgments

These works were supported by the National Key R&D Program of China (grant nos. 2022YFB4401603 and 2021YFA1202904).

References

[1] R. A. Nawrocki, N. Matsuhisa, T. Yokota, and T. Someya, "300-nm imperceptible, ultraflexible, and biocompatible e-skin fit with tactile sensors and organic transistors", Adv. Electron. Mater. 2, 1500452 (2016).

[2] Q. Cao, H.-S. Kim, N. Pimparkar, J. P. Kulkarni, C. Wang, M. Shim, K. Roy, M. A. Alam, and J. A. Rogers, "Medium-scale carbon nanotube thin-film integrated circuits on flexible plastic substrates", Nature, 454. 495-500 (2019).

[3] J. Tang, Q. Cao, G. Tulevski, K. A. Jenkins, L. Nela, D. B. Farmer, and S.-J. Han, "Flexible CMOS integrated circuits based on carbon nanotubes with sub-10 ns stage delays", Nat. Electron. 1, 195-196 (2018).

[4] A. D. Franklin, M. C. Hersam, and H.-S. Philip, "Carbon nanotube transistors: Making electronics from molecules", Science, 378, 726-732 (2022).

[5] H. Zhang, Y. Liu, C. Yang, L. Xiang, Y. Hu, and L.-M. Peng, "Wafer-scale fabrication of ultrathin flexible electronic systems via capillary-assisted electrochemical delamination", Adv. Mater. 30, 1805408 (2018).

[6] H. Zhang, L. Xiang, Y. Yang, M. Xiao, J. Han, L. Ding, Z. Zhang, Y. Hu and L.-M. Peng, "High-performance carbon nanotube complementary electronics and integrated sensor systems on ultrathin plastic foil", ACS Nano, 12, 2773 (2018).

[7] L. Xiang, Y. Wang, F. Xia, F. Liu, D. He, G. Long, X. Zeng, X. Liang, C. Jin, Y. Wang, A. Pan, L.-M. Peng, and Y. Hu, "An epidermal electronic system for physiological information acquisition, processing and storage with an integrated flash memory array", Sci. Adv. 8, eabp8075 (2022).

[8] Y. Wang, T. Wang, L. Xiang, R. Huang, G. Long, W. Wang, M. Xi, J. Tian, W. Li, X. Deng, Q. Gong, T. Bai, Y. Chen, H. Liu, Y. Xia, X. Liang, Q. Chen, L.-M. Peng, and Y. Hu, "Sub–180-nanometer-thick ultraconformable high-performance carbon nanotube–based dual-gate transistors and differential amplifiers", Sci. Adv. 10, eadq6022 (2024).

979-8-3315-0417-5/25 $31.00 © 2025 IEEE

Controlling the Clamping Voltage in Punch-Through Diodes via N+ Well and Contact Design for Low Voltage System Level ESD Protection

Praful Likhitkar, Navin Maheshwari, Sandip Lashkare

Electrical Engineering, IIT Gandhinagar, 382355, India

Abstract

A low voltage Electrostatic Discharge protection device is essential for low-voltage interfaces such as low-voltage MDIOs, Next-gen USB, and Thunderbolt interfaces. Here, a four-layer ($n^{++}p^+p^-n^+$) punch through diode is studied comprehensively, emphasizing lowering clamping voltage (V_{clamp}) by reducing dynamic resistance (R_{DYN}). Further, two advanced designs with multi-contact & n+ well design are proposed to reduce the R_{DYN}. The n+ well design lowers the R_{DYN} by ~ 30%, lowering the V_{clamp} and enhancing IC protection.

Keywords: Dynamic Resistance, Clamping Voltage

Introduction

Electrostatic discharge (ESD) events are unwanted voltage or current events (which occur during manufacturing, human handling) that can potentially damage the integrated circuits (ICs). Hence, an on-chip and off-chip (system level) ESD protection design is used to protect the ICs from such ESD events. ESD protection device requirement changes with respect to the applications and technology being used.

The I-V characteristic of the ESD protection device is shown in Fig. 1. Here, the ESD design window can be observed. This design window consists of 4 regions [1]. Region 1 depicts the IC normal operation region where only the leakage current (if any) flows through the ESD protection device. Region 2 shows the ESD device operation window. The trigger point (V_{t1}, I_{t1}) gives the threshold voltage required to operate the ESD protection device. If any voltage surge occurs, it passes through the ESD device shown by region 2. Region 3 is the region of IC failure. The voltage at the beginning of region 3 represents the maximum voltage tolerable by the gate dielectric of the MOSFET. Region 4 is the region of thermal failure of the ESD protection device. It can be observed that a low clamping voltage (V_{clamp}) is essential to avoid entering the IC failure voltage level. Hence, low R_{DYN} is essential to ensure a low V_{clamp} to prevent IC failure.

Driven by the need for high performance, increased functionality, and reduced power consumption, semiconductor devices have continued to scale down. This has led ICs to operate at lower supply voltages [2]. Hence, a low-voltage ESD protection device having low breakdown and clamping voltage is essential. Zener diodes used traditionally cannot offer adequate protection in very low

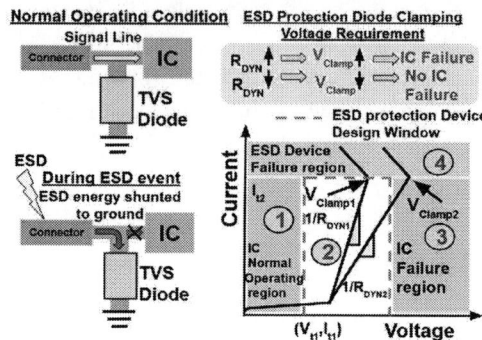

Fig. 1. shows the IV characteristic of the ESD protection device. IC operation during normal operating conditions (Region 1) and during ESD events. Region 2 shows the ESD protection diode design window. Region 3&4 respectively shows IC failure & TVS diode thermal failure regions.

voltage applications, mostly because lowering the breakdown voltage increases the leakage current [2]. A punch-through TVS structure, which is an open base BJT, always provides a trade-off between achieving low leakage current and low clamping voltage (or low R_{DYN}) [3]. Further, it was found that the 3-layer TVS had a higher V_{clamp} and larger leakage current than the 4-layer TVS due to weak punch through [3]. In $n^{++}p^+p^-n^+$ four-layer punch through diode, the p+ & p- layers of base regions are lightly doped to ensure that reverse biased p+-n+ junction will not break by avalanche breakdown [3]. Although various studies exist, a detailed analysis of R_{DYN} is not yet reported.

Here, a 4-layered low-voltage $n^{++}p^+p^-n^+$ device along with multiple n-well and contact designs is proposed to reduce the R_{DYN}, lowering clamping voltage significantly. Further, R_{DYN} dependence is also demonstrated for doping concentration and geometrical dimensions along with diffusion well and contact engineering while comparing the impact with 2D and 3D simulations. Such a study is crucial for the low-voltage electronics community which requires low clamping voltage.

Proposed Structure and Methodology

A 3D model of a 4-layered n++-p+-p--n+ vertical punch-through TVS diode is shown in Fig. 2(a). The device has four layers: n++ collector, p+ buffer, p- epi, and n+ emitter with open base. The depths of the collector and buffer regions are kept fixed at 1.28 µm & 1.32 µm respectively. When the voltage applied to the collector at cathode contact is increased it makes the collector-buffer junction more reverse biased thereby increasing the width of the depletion layer.

979-8-3315-0417-5/25 $31.00 © 2025 IEEE

Punch-through occurs when the two depletion layers of collector-buffer and epi-emitter junctions touch each other. The diffused electrons from the emitter are freely driven to the collector for the applied voltage due to barrier lowering in the open base region [1].

The R_{DYN} is evaluated through the transmission line pulse (TLP) measurement which closely follows the human body model (HBM) events. TLP simulations are performed through Sentaurus TCAD. The TVS device is stressed through a TLP pulse of 100ns duration with rise & fall time of 10ns each. By averaging the voltage values between 80 and 100 ns, the voltage created on the cathode contact is measured after each pulse. Eventually, a plot of the voltage-current data retrieved from these pulses is created and the R_{DYN} is extracted. In all these simulations the collector & emitter region doping are kept fixed at 1×10^{20} cm^{-3} & 1×10^{19} cm^{-3} respectively.

Designs and Simulation Results

A. Structural parameter Dependence

Four different designs are proposed based on 4 different structural parameters: buffer region doping (N_{Buffer}), epi region doping (N_{epi}), cathode length ($L_{cathode}$), and epi length (L_{epi}) taken one at a time, as shown in Fig. 2(a). Fig. 2(b) shows the schematic TLP pulse applied to the TVS diode. With the increase in N_{Buffer}, R_{DYN} first decreases and then

increases as shown in Fig.2(c). The initial decrease is due to the increased conductivity because of increased carriers, but later, with further increase in doping, the impurity scattering causes a hindrance to current flow. Fig.2(d) shows R_{DYN} decreases with increasing N_{epi}. R_{DYN} reduces by approximately 55% for 10 times increase in N_{epi} due to increase in charge concentration thereby increasing conductivity. Fig.2(e) shows that R_{DYN} decreases with increasing $L_{cathode}$ due to the increased area for current conduction through the device. Finally, Fig.2(f) shows R_{DYN} slightly increases and then becomes constant with L_{epi}. This happens because the electric field distribution is uniform and almost similar in all the L_{epi} cases.

B. Contact Engineering

As for the higher current, the resistance offered by all the layers, including contacts, adds to the R_{DYN}. Fig. 3(a) shows the contact engineering technique in which the cathode contact is split into multiple contact points (represented as C1, C2, C3, C4 & C5) of the same length. Here, the buffer & epi region doping is kept fixed at 1×10^{18} cm^{-3} & 2×10^{16} cm^{-3} respectively. Fig. 3(d) shows the I-V characteristics of the TVS device with variation in number of contact points. For a fixed TLP current (I_{TLP}) with an increasing number of contact points the cathode voltage almost remains constant. Fig. 3(f) shows that with increasing the number of contact

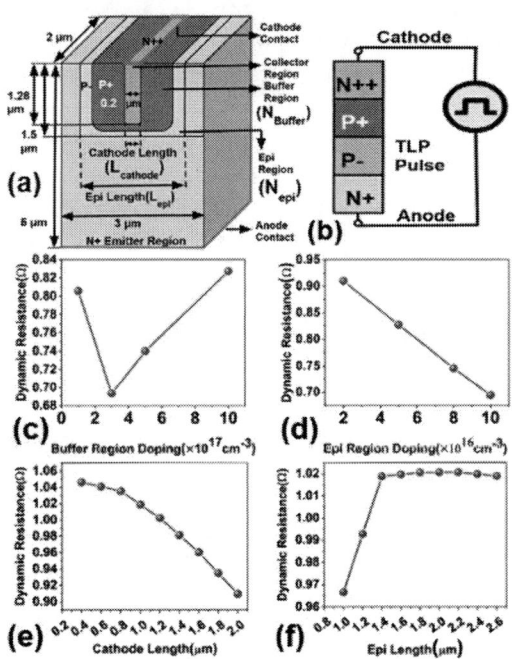

Fig. 2. (a)3D model of 4-layer n++-p+-p--n+ vertical punch-through TVS diode depicting the control parameters: N_{Buffer}, N_{epi}, $L_{cathode}$, L_{epi}. (b) Schematic showing TLP pulse applied to the TVS diode. Dynamic Resistance is calculated by scaling the device to 1 mm^2 area with (c) increasing N_{Buffer} R_{DYN} first decreases and then increases (d) increasing N_{epi}, R_{DYN} decreases (e) increasing $L_{cathode}$, R_{DYN} decreases (f) increasing L_{epi}, R_{DYN} first increases and then becomes constant.

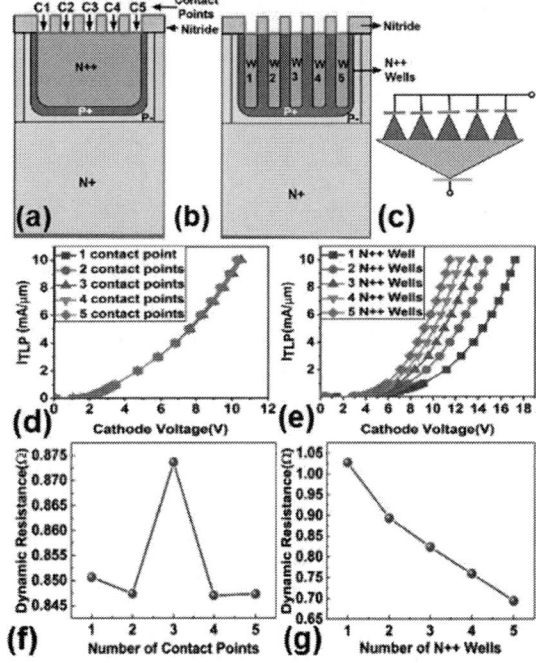

Fig. 3. 2D cross-sectional view of (a)Contact Engineering design (b)Well Engineering design (c) Schematic view of well engineering. Simulated TLP IV characteristics (d) show no change in cathode voltage with an increasing number of contact points and (e) show a reduction in punch-through threshold voltage with an increasing number of wells. R_{DYN} for 1 mm^2 area (f) shows no specific variation trend with increasing contact points (g) decreases with the increasing number of diffusion wells.

979-8-3315-0417-5/25 $31.00 © 2025 IEEE   571

points the R_{DYN} changes slightly. A slight increase in cathode voltage at fixed I_{TLP} can be attributed to the parasitic resistance of the contact points. However, as this resistance is very small compared to the TVS diode resistance, its impact is insignificant.

C. N++ Diffusion Well Engineering

Finally, the N++ well engineering is proposed to further reduce the R_{DYN}. Well engineering that divides the N++ diffusion well into several wells (denoted as W1, W2, W3, W4, and W5) is depicted in Fig. 3(b). Fig. 3(c) shows that well-engineered 4-layer punch through TVS diode can be visualized as 2 back-to-back connected p-n junction diodes with the first diode being replaced with multiple diodes of smaller junction connected in parallel. As shown in Fig. 3(e) the cathode voltage drops for a fixed I_{TLP} with the increase in number of wells. This is because of greater carrier injection. The R_{DYN} significantly drops (by more than 30%) as the number of wells increases as shown in Fig. 3(g). The current flow pathways are increased, which reduces the current crowding.

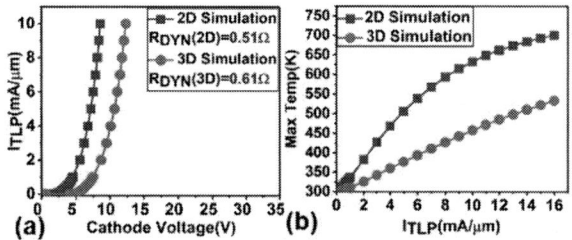

Fig. 5. Simulated TLP characteristics of 2D vs 3D TCAD simulations for single well 2D & 3D TVS structures respectively (a)IV characteristic shows threshold voltage and R_{DYN} (for 1 mm² area) is higher in 3D simulations shows (b) shows Max lattice temperature is higher in 2D.

The in-depth performance is analyzed using 3D TCAD simulation, which helps to comprehend the physical behavior of the carriers [4]. 2D TCAD considers uniform electrical and thermal response throughout the device width. TLP IV characteristics for single well 2D and 3D simulations are shown in Fig. 5(a). It is observed that the threshold voltage in 2D simulation is 61.3% lower than in 3D simulation. It happens because non-idealities like edge and corner effects are better modeled in 3D. To create depletion region consistently across the device, a larger voltage is needed to compensate for this non-uniformity. The maximum lattice temperature is lower in 3D simulations as shown in Fig. 5(b) because heat is more efficiently distributed across the device's volume. Additionally, the temperature rises linearly and uniformly with the applied current in the 3D device. In contrast incomplete edge and corner modeling in 2D simulations results in localized heating in which the temperature increases more quickly in particular areas [5]. Fig. 5(a) shows that R_{DYN} (for 1 mm² area) obtained in the 2D simulation is approx. 16% lower than in the 3D simulation.

Because 3D simulations provide a more accurate representation of parasitic resistances, including those related to contacts than 2D models.

Fig. 6(a). shows the distribution of electric field. Its gradient indicates regions where carrier acceleration takes place. Fig. 6(b) shows high current density occurs in the region of current injection. Fig. 6(c) shows localized zones of high-impact ionization critical for clamping action. Fig. 6(d) shows the lattice temperature profile which indicates areas of

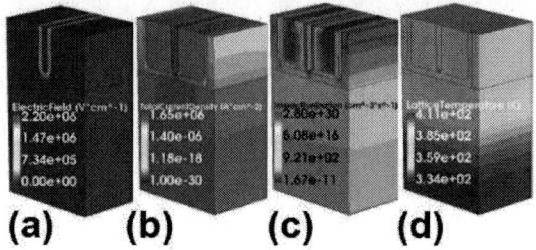

Fig. 6. 3D TCAD simulated contours for single well 3D TVS structure taken at I_{TLP} of 7 mA/μm depicting (a) Electric Field (b) Current Density (c)Impact Ionization (d) Lattice Temperature.

resistive heating.

Conclusion

Here, a low voltage 4-layer punch through system level ESD device is proposed along with a multi-N+ well design to reduce the R_{DYN} through the detailed 3D TCAD simulations. It is inferred that higher N_{epi} is advantageous as it lowers the R_{DYN} by ~55% for 10 times increase in N_{epi}. Also, R_{DYN} can be substantially lowered (over 30%) with N+ well engineering reducing clamping voltage substantially needed to prevent IC failure. It is found that threshold voltage & R_{DYN} are respectively 61.3% & 16% lower in 2D simulation compared to 3D simulations, highlighting the necessity of 3D modeling. Such a solution with detailed physical insights can be a significant development for low voltage electronic community.

References

[1] A. Dong, J. Xiong, S. Mitra, W. Liang, R. Gauthier and A. Loiseau, "Comprehensive Study of ESD Design Window Scaling Down to 7nm Technology Node," 2018 40th Electrical Overstress/Electrostatic Discharge Symposium (EOS/ESD), Reno, NV, USA, 2018, pp. 1-8.

[2] J. Urresti, S. Hidalgo, D. Flores and J. Rebollo, "Lateral Punch-Through TVS Devices: Design and Fabrication," 2009 Spanish Conference on Electron Devices, Spain, 2009, pp. 148-151.

[3] King, Ya-Chin, et al. "Punch-through diode as the transient voltage suppressor for low-voltage electronics." IEEE Transactions on Electron Devices 43.11 (1996): 2037-2040.

[4] Y. Li, Y. Wang and Y. Wang, "ESD Diode Devices Simulation and Analysis in a FinFET Technology," 2020 International EOS/ESD Symposium on Design and System (IEDS), China, 2021, pp. 1-4.

[5] P. Scharf, C. Sohrmann, S. Holland and V. Beyer, "Investigations on current filamentation in PIN diodes using TLP measurements and TCAD simulations," 49th European Solid-State Device Research Conference (ESSDERC), Cracow, Poland, 2019, pp. 226.

Demonstration of a 3D-Folded 2DEG-Channel Structure with Regrown AlGaN/GaN Heterostructures

Fuqiang Guo[1], Sen Huang[2]*, Xingyu Fu[1], Xuelin Yang[1]*, Qimeng Jiang[2], Xinguo Gao[2], Shuaiyu Chen[1], Ning Tang[1], Bo Shen[1]

[1] State Key Laboratory of Artificial Microstructure and Mesoscopic Physics, Nano-optoelectronics Frontier Center of Ministry of Education, School of Physics, Peking University, Beijing 100871, China
[2] High-Frequency High-Voltage Device and Integrated Circuits R & D Center, Institute of Microelectronics of Chinese Academy of Sciences, Beijing 100029, China

*e-mail : huangsen@ime.ac.cn; xlyang@pku.edu.cn

Abstract

A novel GaN folded channel with regrown AlGaN/GaN two-dimensional electron gas (2DEG) channel on sapphire substrates have been developed. Trench etching was performed on GaN material followed by AlGaN/GaN heterojunction regrowth, achieving the extension of the 2DEG channel from a two-dimensional plane to a three-dimensional structure. The heterojunction grown on the trench had a channel sheet resistance of ~1.2 kΩ/□ obtained through TLM model. Folding channels could be an attractive technique to optimize breakdown electric fields distribution on limited chip sizes.

Keywords: AlGaN/GaN, regrowth, fold channel, sheet resistance

Introduction

GaN materials have the advantages of large bandgap width, high breakdown voltage, and high thermal conductivity, which can achieve the high voltage and high-power technical performance of devices [1]. AlGaN/GaN heterojunction grown on C-plane GaN epitaxy was mainly used due to their high electron mobility and large electron density of the two-dimensional electron gas (2DEG) at the heterojunction interface [2,3]. At present, in the field of the medium and low voltage power devices, GaN HEMT have achieved coverage from 100 V to 1200 V [4-6]. At prensent, for the field of medium and high voltage power devices, GaN HEMT are still in the research stage [7]. Although the multi-level field plate [8] and GaN polarization super-junction structure [9] can increase the breakdown voltage to kV or even 10 kV, the main method to improve the breakdown voltage was to increase the size of the access region. Therefore, the increase in R_{on} caused by the increase in chip area seriously reduced the performance of the GaN power device.

For GaN HEMT, the channel of AlGaN/GaN heterostructures is mainly planar structure which seriously limits the distribution of the device's electric field. Recently, a novel gate design based on GaN vertical devices is realized by metal-organic chemical vapor deposition (MOCVD) regrowth of a p-GaN/AlGaN/GaN conformally over V-shaped grooves formed over the drift layer resulting in high threshold voltage [10]. It indicates that AlGaN/GaN channel over the trench sidewall is feasible. If AlGaN/GaN channel of the access region can be extended to a three-dimensional structure, it may have a significant impact on the distribution of the electric field. This may be an effective method for high-voltage GaN power devices.

In this work, we proposed a regrown AlGaN/GaN heterostructure that includes both planar and trench structures. By grooves etching on GaN material and then regrowing AlGaN/GaN heterostructures, the 2DEG channel is extended to a three-dimensional space, and the sheet resistance of the groove slope is effectively extracted by transmission line model (TLM). This is of great significance for the next step of GaN power devices with novel structures.

Experiment and method

Fig.1 shows the epitaxial structure and the regrowth process. The initial epi wafer was grown on 2-inch saphire substrate, which consists of a 2.4-μm uid-GaN. Firstly, the sample was etched in the ICP system to form a trench structure for three different channel orientations perpendicular to the m plane. Then, the AlGaN/GaN heterojunction was regrown on the etched wafer in the MOCVD, and the regrown layers were consisted of GaN cap-1nm/Al$_{25\%}$GaN-20nm/AlN-1nm/GaN-100nm from top to down. This design ensures the good property of AlGaN/GaN channel since the junction is formed without any interruption of the epitaxial growth. Next, ohmic recess and ohmic metal Ti/Al/Ni/Au deposition were performed on the regrowth layer in ohmic regions, followed by annealing at 820 °C 50 s under N_2 ambient for the wafers. Finally, mesa isolation process was done by ICP etching.

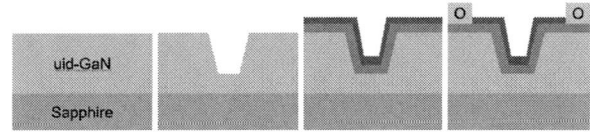

Fig. 1 Schematics of the epitaxial structure and AlGaN/GaN folded channel regrown process flow.

Four different channel structures were obtained as shown in Fig. 2. In addition，several device channel orientations varying from 0°、60° with 120° (all parallel to the m-plane

[1$\bar{1}$00]) are studied whether there are differences in the same crystal plane direction and illustrated in Fig. 3.

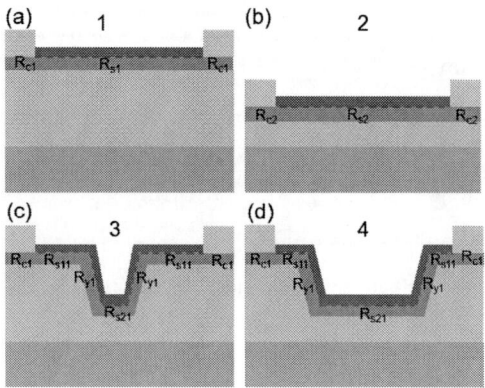

Fig. 2 Schematics of the four types regrown AlGaN/GaN structures with the resistance distribution; (a) regrown on unetched GaN surface; (b) regrown on unetched GaN surface;(c) regrown on narrow width GaN groove surface; (d) regrown on wide GaN groove surface.

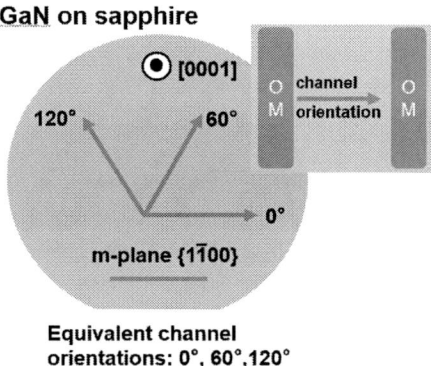

Fig. 3: GaN-on-sapphire devices structure with different channel orientations (0°、60°、120°).

Results and Discussion

SEM vertical view and cross-sectional images after etching and regrowth show the trench morphology for the 0° channel orientations in Fig. 4. GaN special V-shaped grooves formed after regrowing AlGaN/GaN heterojunctions under specific growth conditions. This is mainly because the growth rate of the sidewall is faster than that of the bottom in this growth process. There also may be slight errors in the measurement size data in SEM due to irregular sample preparation. The particles in the cross-section image also come from impurities generated during the preparation of the sample by fragmentation.

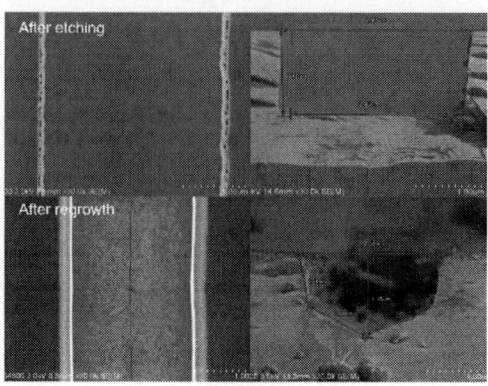

Fig. 4 SEM vertical view and cross-section images of the trench after etching and after regrowth.

The Ohmic contact properties of the regrowth heterostructure on the etched surface and the unetched surface were calculated by TLM test. Fig. 5(a) plots the linear fitting of total resistance versus spacing for the four samples. The sheet resistance (R_{sh})、 ohmic contact resistance (R_c) and sloped channel sheet resistance (R_{shy}) are extracted in Fig. 5(b) for the 0° channel orientations. The R_{sh} and R_c of regrown heterojunction on the etched and unetched GaN surface are nearly same. It indicates that the etched surface has almost no effects on the AlGaN/GaN regrowth.

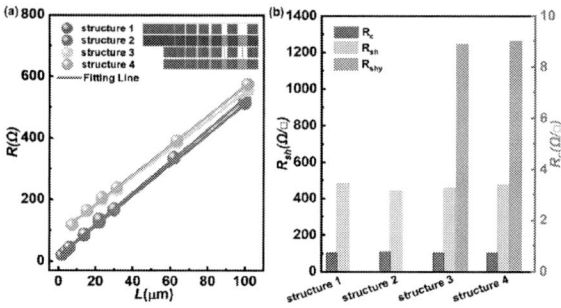

Fig. 5. (a) Linear fitting of total resistance versus spacing for different structures. (b) Comparison of R_{sh}、 R_c and R_{shy} for different structures.

Because the regrown channel in the trench is irregular inclined plane that is symmetrical on both sides, we can only extract the average sheet resistance of the sidewall channel. For channels with folded structures, the conventional TLM fitting method requires some modifications. The channel length contains planar channel length and the inclined channel length ($L_{total}=L_{s11}+ L_{s21}+ L_{s11}+2L_{y1}$). The entire channel resistance is divided into two parts of R_C、 three parts of planar R_{s1} and two parts of R_{y1}($R_{total}=R_{s11}+ R_{s21} +R_{s11}+2R_{y1}+2R_C$). Compared to the planar channel, each channel with the trench have increased $2R_{y1}$ for the channel resistance. The linear fitting formula of the TLM test was changed to:

$$R_L = \frac{R_{sh}}{W_C}L + 2R_C + 2R_{y1} \qquad (1)$$

$$R_{shy} = R_{y1}\frac{W_c}{L_y} \qquad (2)$$

L was the channel length between the two ohmic pad, W_c was the channel width, R_{sh} was the planar sheet resistance, R_c was the ohmic contact resistance, R_{shy} was the sloped surface channel sheet resistance.

According to formula 1, the planar channel sheet resistance was related to the slope of the linear fitting($R_{sh}=k×W_c$); the intercept value was related to the ohmic contact resistance and sloped channel sheet resistance. According to the R_c obtained from the sample 1, the sloped channel sheet resistance R_{shy} of the folded channel can be obtained by the formula 2.

Table 1: The results of the sheet resistance and contact resistance along three directions for the four structures.

Structure		0°	60°	120°
1	R_{sh} (Ω/□)	481	440	432
	R_c (Ω·mm)	0.74	0.97	0.96
2	R_{sh} (Ω/□)	442	428	418
	R_c (Ω·mm)	0.79	0.96	1.03
3	R_{sh} (Ω/□)	459	395	404
	R_c (Ω·mm)	0.74	0.97	0.96
	R_{shy} (Ω/□)	1245	1283	1208
4	R_{sh} (Ω/□)	475	419	423
	R_c (Ω·mm)	0.74	0.97	0.96
	R_{shy} (Ω/□)	1261	1321	1310

Table 1 summarized the results of the sheet resistance and contact resistance along three directions for the four structures. It can be seen that the planar channel R_{sh} of the regrowth AlGaN/GaN heterostructure after etching and non-etching remains at the same level for these three directions. For these channels with grooves, their planar channel R_{sh} (~450 Ω/□) is basically the same as the planar channel without folded channel. The R_{shy} of the sloped plane is calculated to be around 1.2k Ω/□. And the R_{shy} of the sloped plane are also basically the same for the different width trench along these three directions. This indicates that the transport characteristics of the regrown AlGaN/GaN on the same crystal plane direction are basically consistent. Although it has demonstrated successful conduction characteristics, the R_{on} of the folded channel is still relatively high, which requires detailed research on trench angle control and growth conformal control in the future work.

Conclusion

In summary, we developed a novel AlGaN/GaN folded channel devices with the sidewall channel R_{sh} of ~1.2 kΩ/□. AlGaN/GaN are regrown over GaN trench resulting in achieving the extension of the channel from a two-dimensional plane to a three-dimensional structure. The feasibility of three-dimensional channels may have great guiding significance for improving the key performance of GaN power devices.

Acknowledgments

This work was supported in part by National Natural Science Foundation of China under Grant 62334012, Grant 62074161, Grant 62004213, Grant U20A20208, and Grant 62304252; in part by the University of CAS; and in part by IMECAS-HKUST-Joint Laboratory of Microelectronics.

References

[1] K.J. Chen, O. Haberlen, A. Lidow, C.L. Tsai, T. Ueda, Y. Uemoto, Y. Wu, GaN-on-Si power technology: Devices and applications, IEEE Trans. Electron Devices. 64 (2017) 779–795.

[2] Y. Zhong, J. Zhang, S. Wu, L. Jia, X. Yang, Y. Liu, Y. Zhang, Q. Sun, A review on the GaN-on-Si power electronic devices, Fundam. Res. 2 (2022) 462–475.

[3] G. Greco, F. Iucolano, F. Roccaforte, Review of technology for normally-off HEMTs with p-GaN gate, Mater. Sci. Semicond. Process. 78 (2018) 96–106.

[4] R. Chu, A. Corrion, M. Chen, R. Li, D. Wong, D. Zehnder, B. Hughes, K. Boutros, 1200-V normally off GaN-on-Si field-effect transistors with low dynamic on-resistance, IEEE Electron Device Lett. 32 (2011) 632–634.

[5] A. Sharma, R. Goel, and Y. S. Chauhan, "Analysis and Modeling of OFF-state Capacitance in LDD MOSFETs", IEEE Electron Devices Technology and Manufacturing Conference (EDTM) (2023).

[6] G. Greco, F. Iucolano, F. Roccaforte, Review of technology for normally-off HEMTs with p-GaN gate, Mater. Sci. Semicond. Process. 78 (2018) 96–106.

[7] H. Ishida, D. Shibata, M. Yanagihara, Y. Uemoto, H. Matsuo, T. Ueda, T. Tanaka, D. Ueda, Unlimited high breakdown voltage by natural super junction of polarized semiconductor, IEEE Electron Device Lett. 29 (2008) 1087–1089.

[8] V. Šodan, H. Oprins, S. Stoffels, M. Baelmans, I. De Wolf, Influence of Field-Plate Configuration on Power Dissipation and Temperature Profiles in AlGaN/GaN on Silicon HEMTs, IEEE Trans. Electron Devices. 62 (2015) 2416–2422.

[9] S.W. Han, J. Song, R. Chu, Design of GaN/AlGaN/GaN Super-Heterojunction Schottky Diode, IEEE Trans. Electron Devices. 67 (2020) 69–74.

[10] D. Shibata, R. Kajitani, M. Ogawa, K. Tanaka, S. Tamura, T. Hatsuda, M. Ishida, T. Ueda, 1.7 kV/1.0 mΩcm² normally-off vertical GaN transistor on GaN substrate with regrown p-GaN/AlGaN/GaN semipolar gate structure, in: 2016 IEEE Int. Electron Devices Meet., IEEE, 2016: pp. 10.1.1-10.1.4.

[11] B. Benbakhti, M. Rousseau, A. Soltani, J.-C. De Jaeger, Analysis of Thermal Effect Influence in Gallium-Nitride-Based TLM Structures by Means of a Transport–Thermal Modeling, IEEE Trans. Electron Devices. 53 (2006) 2237–2242.

Highly stable Zn metal anodes enabled by In₂O₃ coating for high-performance flexible aqueous zinc-ion batteries

Zhongqi Liang[1], Xiaohong Tan[1], Fan Xiao[1], Yufeng Jin[1*], Guoshen Yang[2*] and Hang Zhou[1*]

[1] School of Electronic and Computer Engineering, Peking University Shenzhen Graduate School, Shenzhen, 518055, China, [2] School of Information Science and Engineering, Shandong University, Qingdao, 266237, China

E-mail: Guoshen Yang (yangguoshen@sdu.edu.cn), Hang Zhou (zhouh81@pkusz.edu.cn)

Abstract

Flexible rechargeable aqueous zinc-ion batteries (AZIBs) are a highly promising energy storage device for flexible electronic devices. This paper proposes adding a layer of indium oxide (In_2O_3) thin film to the zinc anode (In_2O_3@Zn) as an interface modification, effectively addressing issues such as zinc dendrite formation and hydrogen evolution that are inherent in zinc-ion batteries. Compared to bare Zn batteries, the In_2O_3@Zn zinc-ion batteries show significant improvements in specific capacity and cycling stability.

Keywords: aqueous zinc-ion batteries, interface modification, enhanced specific capacity, improved cycling stability

Introduction

With the rapid development of flexible and wearable devices, lightweight wearable electronics urgently require suitable energy storage solutions. Flexible zinc-ion batteries have gained significant attention due to their low cost and safety[1]. However, inevitable issues, such as zinc dendrite formation and hydrogen evolution, limit the utilization efficiency of zinc metal anodes and can even lead to mid-cycle short circuits, posing safety risks[2].

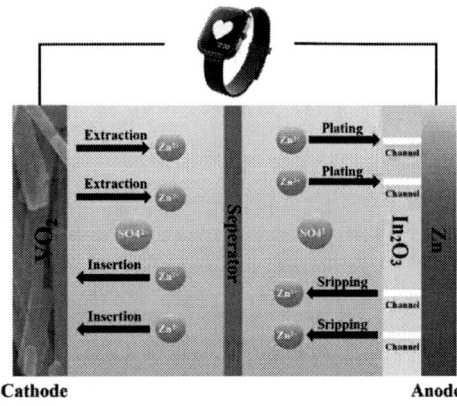

Fig. 1: Schematic illustration on the energy storage mechanism of AZIBs.

Here, we propose adding an In_2O_3 thin film to the zinc metal anode as an effective coating ensure uniform charge distribution and a rapid zinc-ion transport pathway to enable uniform Zn electroplating (Fig. 1). Symmetric In_2O_3@Zn anode systems demonstrate stable plating/stripping performance for up to 1,000 hours. Compared to bare Zn-based zinc-ion batteries, the In_2O_3@Zn batteries exhibit notable improvements in specific capacity and cycling stability, validating that the indium oxide coating strategy is a simple and effective approach for achieving long-life and deep-rechargeable flexible AZIBs.

Experimental Section

A. Preparation of vanadium dioxide (VO₂) cathode

First, 0.951 g of vanadium pentoxide, 0.473 g of $C_2H_2O_4$, and 0.07 g of cetyltrimethylammonium bromide were dissolved in 70 ml of water and stirred until fully dissolved. The mixture was then heated at 180 °C for 48 hours. Finally, the product was filtered and dried, successfully yielding vanadium dioxide nanowires[3].

B. Preparation of In2O3 thin-film coating

To prepare In_2O_3@Zn, indium oxide and polyvinylidene fluoride (PVDF) were mixed in a mass ratio of 9:1, with an appropriate amount of N-methyl-2-pyrrolidone (NMP). After stirring for three days, a uniform slurry was obtained. This slurry was spin-coated onto cleaned Zn discs at 500 rpm for 30 seconds, then dried at 70 °C for 12 hours to evaporate the NMP solvent.

C. Electrochemical characterization

Cyclic voltammetry (CV) was conducted using a commercial CHI660 electrochemical workstation (Shanghai CH Instruments Co., Ltd). Galvanostatic charge-discharge (GCD) tests were performed on a Neware BTSDA battery testing system (Shenzhen Neware Electronics Co., Ltd).

Result and Discussion

Inspired by interface materials such as Nb_2O_5 and TiO_2, which serve as n-type semiconductors and can function as electron transport layers in optoelectronic devices to optimize energy level alignment at the interface. These materials can also be used in zinc-ion batteries as interface modification materials to improve zinc ion diffusion kinetics and suppress zinc dendrite formation[4][5]. In this paper, based on the same principle, In_2O_3 as an n-type semiconductor, holds greater potential for overall performance enhancement, making it a promising candidate for practical interface modification applications[6].

979-8-3315-0417-5/25 $31.00 © 2025 IEEE

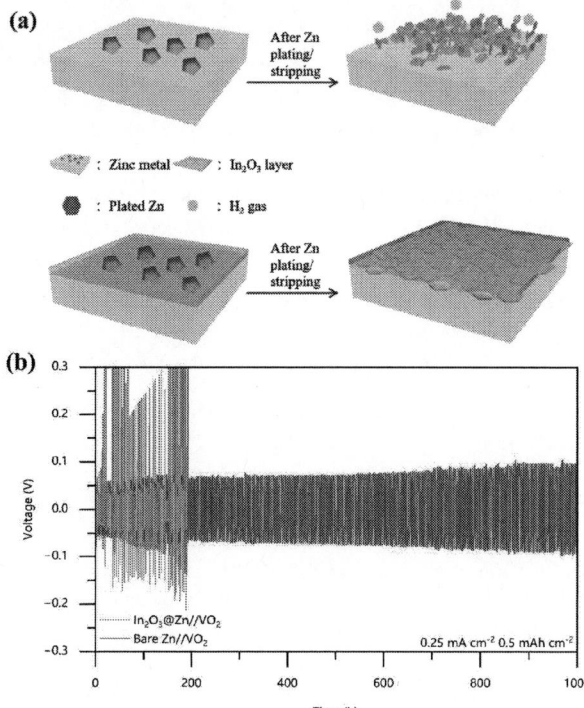

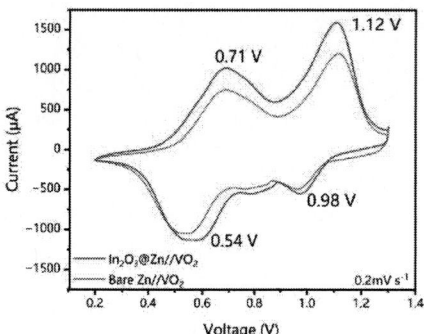

Fig. 3: Cyclic voltammetry curves of bare Zn//VO₂ and In₂O₃@Zn//VO₂ scanned at 0.2 mV s⁻¹ in range 0.2-1.3 V.

In the CV tests, it was found that both showed very similar redox peaks, with two oxidation peaks located around 0.71 V and 1.12 V, and two reduction peaks around 0.54 V and 0.98 V (Fig. 3). These separated peaks indicate the presence of multiple insertion and extraction processes of Zn^{2+} in the VO_2 material[7]. Besides, we observed that the former exhibited more pronounced electrochemical activity than the latter, indicating enhanced electrochemical reaction activity. The potential intervals of the redox peaks were slightly reduced, suggesting accelerated electrode kinetics and improved reaction reversibility.

Fig. 2: (a) Schematic illustrating Zn electrodeposition behavior on bare Zn and In₂O₃@Zn. (b) Cycling lifespan of bare Zn and In₂O₃@Zn symmetric cells at 0.25 mA cm⁻² and 0.5 mAh cm⁻² in mildly acidic aqueous 2M ZnSO₄ electrolyte.

When cycling bare zinc symmetric cells and indium oxide-coated zinc symmetric cells at 0.25 mA cm⁻² and 0.5 mAh cm⁻², respectively, to evaluate the stability of zinc metal during constant-current plating and stripping, the bare zinc symmetric cell showed highly unstable and fluctuating voltage curves with excessive polarization voltage. This confirmed that the lack of a coating to provide uniform ion channels made the zinc foil more susceptible to significant polarization, which increases the likelihood of zinc dendrite formation and hydrogen evolution (Fig. 2a). In contrast, the zinc surface with an indium oxide coating allows for uniform zinc ion transport and induces uniform growth of zinc dendrites. Besides, the zinc symmetric cell with indium oxide coating exhibited highly stable polarization voltage; for instance, the initial polarization voltage was as low as 65 mV, increasing slightly to only 80 mV after 1,000 hours of prolonged cycling (Fig. 2b). It is evident that the indium oxide coating provides a certain degree of anode protection, effectively suppressing the inevitable formation of zinc dendrites and hydrogen evolution, thereby preventing significant polarization of the zinc metal. This ultimately impacts the performance of the entire energy storage device, resulting in reduced utilization efficiency of the zinc metal anode, and may even precipitate serious electrical incidents due to short circuits.

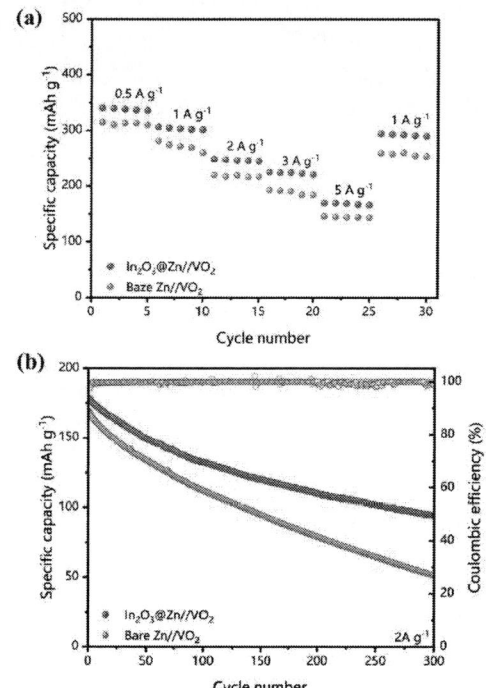

Fig. 4: (a) Rate capabilities obtained at various current densities. (b) Galvanostatic discharge/charge capacity profiles obtained at 2.0 A g⁻¹.

The In₂O₃ coating, as an interface modification material, provides a fast and uniform ionic channel for ion transport, thereby enhancing the overall specific capacity of zinc-ion batteries. Furthermore, the protective properties of the coating enhance cycling stability to a certain degree. In terms of performance testing, we conducted comparative tests between In$_2$O$_3$@Zn//VO$_2$ batteries and Bare Zn//VO$_2$ batteries. Firstly, in rate capability tests, the In$_2$O$_3$@Zn//VO$_2$ battery demonstrated increased specific capacity across various current densities compared to the Bare Zn//VO$_2$ battery (Fig. 4a). For example, at a current density of 0.5 A g^{-1}, the battery's specific capacity increased from 314.54 mAh g^{-1} to 340.33 mAh g^{-1}. Secondly, in the cycling stability test, at a current density of 2 A g^{-1}, the initial capacity of the In$_2$O$_3$@Zn//VO$_2$ battery increased from 169.06 mAh g^{-1} to 180.85 mAh g^{-1}, and after 300 charge-discharge cycles, the capacity retention improved significantly from 29.52% to 52.02%, demonstrating a marked improvement in cycling stability (Fig. 4b). On one hand, the indium oxide coating itself has high corrosion resistance and demonstrates good chemical stability in environments with relatively high humidity, providing a degree of protection to the zinc anode. On the other hand, the indium oxide coating contains a certain number of Lewis sites, which enhance the hydrophilicity on the zinc metal surface. This effectively promotes the desolvation of (Zn[(H$_2$O)$_6$]$^{2+}$), allowing it to pass rapidly through In$_2$O$_3$, thereby reducing the formation of side reaction layers.

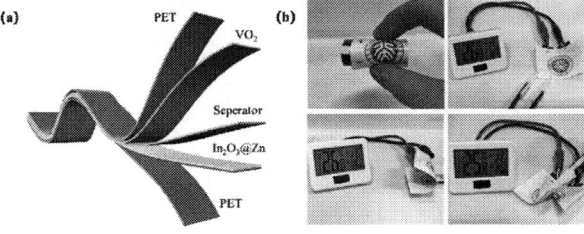

Fig 5: (a) Schematic diagram of the flexible AZIBs. (b) Optical photographs of the flexible AZIBs.

To validate the feasibility of flexible AZIBs devices, a pouch cell with a size of 4 cm × 4 cm was fabricated, in which flexible PET, In$_2$O$_3$@Zn, ZnSO$_4$, VO$_2$, and PET are used as the packaging material, anode, electrolyte, cathode, and packaging material, respectively (Fig. 5a). In the assembled flexible AZIBs, it could be bent at nearly 160 degrees and successfully provide stable power for an electronic desktop thermometer in the bent state, suggesting that the device possesses satisfactory mechanical flexibility (Fig. 5b)[8]. To assess the safety of the device, the pouch cell was subjected to bending and shearing tests. Intriguingly, the pouch cell did not malfunction throughout the entire testing process, demonstrating the outstanding reliability and safety of the device.

Conclusion

In summary, we report that indium oxide as a coating, forms a protective layer on the zinc anode and provides a stable ionic channel to suppress the corrosion and side reactions of the zinc anode, thereby stabilizing the plating/stripping of zinc ions and enhancing the cycling stability and specific capacity of the overall energy storage device. The improved zinc-ion battery from this work offers a new perspective for controlling the solid electrolyte interface for stable zinc anodes, providing a practical strategy for enhancing the performance of scalable and flexible pouch batteries. Together with improvements in the electrolyte, this can further enhance the performance of flexible energy storage devices, ultimately optimizing the practical use of flexible electronic devices.

Acknowledgments

This work is supported by Shenzhen Science and Technology Innovation Committee (KJZD20230923113759002, KJZD20230923115005009, GJHZ20240218113959009)

Reference

[1] C Li, P Li, S Yang and C Y Zhi, "Recently advances in flexible zinc ion batteries", J. Semicond., 42(10): 101603 (2021).

[2] Shin, Jaeho, Jimin Lee, Youngbin Park, and Jang Wook Choi, "Aqueous Zinc Ion Batteries: Focus on Zinc Metal Anodes", Chem. Sci. 11, no. 8 2028–44 (2020).

[3] Guoshen Yang , Xianqi Xu, Gangrui Qu, Jie Deng, Yachao Zhu, Chi Fang, Olivier Fontaine, Pritesh Hiralal, Jiaxin Zheng, and Hang Zhou, "An Aqueous Magnesium-Ion Battery Working at -50 ℃ Enabled by Modulating Electrolyte Structure", Chemical Engineering Journal 455 :140806 (2023).

[4] Kangning Zhao, Chenxu Wang, Yanhao Yu, Mengyu Yan, Qiulong Wei, Pan He, Yi fan Dong, Ziyi Zhang, Xudong Wang, and Liqiang Mai, "Ultrathin Surface Coating Enables Stabilized Zinc Metal Anode", Advanced materials, Volume 5, Issue 16 (2018).

[5] So, Seongjoon, Yong Nam Ahn, Jaewook Ko, Il Tae Kim, and Jaehyun Hur, "Uniform and Oriented Zinc Deposition Induced by Artificial Nb$_2$O$_5$ Layer for Highly Reversible Zn Anode in Aqueous Zinc Ion Batteries", Energy Storage Materials 52 : 40–51 (2022).

[6] Wenchao Huang, Bowen Zhu, Sheng-Yung Chang, Shuanglin Zhu, Pei Cheng, Yao-Tsung Hsieh and Lei Meng, "High Mobility Indium Oxide Electron Transport Layer for an Efficient Charge Extraction and Optimized Nanomorphology in Organic Photovoltaics.", Nano Letters 18, no. 9 (2018).

[7] Ding, Junwei, Zhiguo Du, Linqing Gu, Bin Li, Lizhen Wang, Shiwen Wang, Yongji Gong, and Shubin Yang, "Ultrafast Zn^{2+} Intercalation and Deintercalation in Vanadium Dioxide.", Advanced Materials 30, no. 26: 1800762 (2018).

[8] Guoqun Zhang, Lulu Fu, Yuan Chen, Kun Fan, Chenyang Zhang, Huichao Dai, Linnan Guan, Haoyu Guo, Minglei Mao, and Chengliang Wang, "Constructing Quasi-single Ion Conductors by a β-Cyclodextrin Polymer to Stabilize Zn Anode", Angewandte Chemie International Edition, e202412173 (2024).

Hardware Virtualization Technology for Multicore Brain-inspired Chip

Dahu Feng[1,2], Rong Zhao[1]

[1]Center for Brain-Inspired Computing Research (CBICR), Tsinghua University, Beijing, China
[2] Email: fengdh21@mails.tsinghua.edu.cn

ABSTRACT

With the rapid development of artificial intelligence (AI) applications, a new type of AI accelerators, known as the Brain-inspired Chip (BIC), has emerged in Non-von Neumann computing environments, such as SpiNNaker, Loihi, TrueNorth and Tianjic. However, contemporary BICs often exhibit low resource utilization due to the imbalance in hardware resource allocation across various models, which is exacerbated by the lack of virtualization support. Although prior efforts have explored (para-) virtualization techniques to share resources of other AI accelerators such as GPU, these attempts have failed to account for key characteristics of BIC such as interconnection between multiple BIC cores and scratchpad-centric memory, leading to suboptimal performance.

This paper presents vBIC, a comprehensive BIC virtualization solution featuring two key architectural extensions: **(1) BIC memory virtualization** which reduces TLB misses and table walking times for scratchpad-centric memory access; and **(2) BIC interconnection virtualization** which establishes a virtual topology among multiple BIC cores. We have implemented a prototype of vBIC on an FPGA platform. Evaluation results show that vBIC maintains end-to-end performance across various AI workloads. Furthermore, when compared with alternative approaches such as unified virtual memory, vBIC achieves a 2x performance improvement in transformer-based models.

Keywords: Brain-inspired Chip, Neuromorphic Computing and Virtualization

INTRODUCTION

With the increasing popularity of AI applications, such as ChatGPT, image recognition, and artificial general intelligence, machine learning has been widely adopted. To address the challenges and enhance AI computing performance, hardware designers have introduced specialized accelerators known as Brain-Inspired Chips (BICs). Notable examples include TrueNorth [1], SpiNNaker [2], Loihi [3], and Tianjic [4].

However, designing a solution for BIC virtualization is non-trivial. Although there have been numerous efforts to design virtualization for other accelerators like GPUs [5], these approaches are not feasible for BICs due to fundamental differences in hardware architectures.

First, GPUs employ the Single-Instruction-Multiple-Threads (SIMT) architecture, featuring thousands of homogeneous hardware threads with unified virtual memory. In contrast, BICs utilize a Multiple-Instruction-Multiple-Data (MIMD) architecture tailored to optimize data flow for AI workloads. Developers can control each BIC core using a dedicated instruction space and need to be aware of the BIC topology, which facilitates direct data transfers between BIC cores rather than relying on a unified virtual memory system.

Second, GPUs follow a classic memory hierarchy comprising multi-level caches along with main memory. Conversely, BICs adopt an SRAM/scratchpad-centric memory system, where data is managed by software without hardware mechanisms like cache coherence and address translation.

This paper presents the comprehensive BIC virtualization design named vBIC, which considers the strong isolation between multi-tenants and topology architecture for neuromorphic computing. vBIC proposes two key architectural extensions to enhance BIC virtualization:

- **BIC Memory Virtualization:** A unique feature of BIC (compared with GPU) is its scratchpad-centric memory (data cannot bypass the scratchpad) with large off-chip memory (HBM/DRAM). This paper proposes new hardware extensions, including *vSRAM* and *range table*, to achieve efficient memory virtualization for BICs.

- **BIC Interconnection Virtualization:** BICs possess the capability of direct data transfer between BIC cores, utilizing technologies such as network-on-chip (NoC) or inter-core interconnection (ICI). In the context of BIC virtualization, the system must also provide a virtual topology for virtual BIC cores to facilitate direct data transfers.

DESIGN

A. BIC Memory Virtualization (vMem)

The memory hierarchy in BICs: The performance of ML tasks is highly dependent on the BIC's memory bandwidth. Consequently, current BIC chips incorporate a substantial amount of SRAM and HBM/DRAM memory to store the ML model, KV cache, and activations.

BIC memory virtualization: Memory virtualization for BICs presents new challenges due to the significantly higher memory bandwidth of SRAM and HBM compared to traditional CPU-side DRAM. Current memory

979-8-3315-0417-5/25 $31.00 © 2025 IEEE

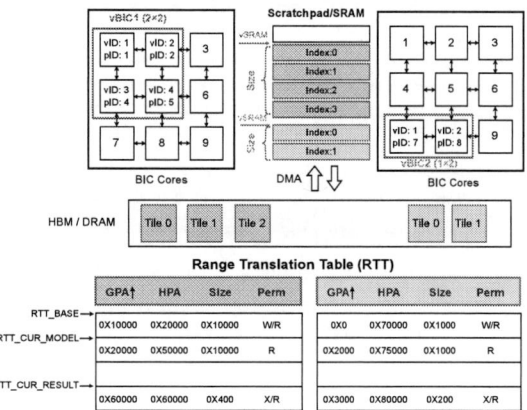

Fig. 1. **BIC memory virtualization:** vBIC virtualizes both on-chip SRAM and off-chip HBM/DRAM.

virtualization on the CPU side utilizes two-stage and four-level page tables. However, this design can result in up to 24 memory accesses for address translation in the worst-case scenario. Additionally, the page sizes in the page table are usually fixed (e.g., 4KB or 2MB) and require a large number of TLB entries to accelerate address translation.

For scratchpad (vSRAM) virtualization, since it is directly managed by software, we cannot presume that software running in guest VMs will adhere to the isolation rules. Therefore, hardware virtualization becomes indispensable for controlling scratchpad access. vBIC introduces a simple yet effective mechanism by adding a virtual offset (v_offset) and performing a boundary check for each scratchpad entry access, as illustrated in the top half of Figure 1. For instance, if an BIC core tends to load a tensor from scratchpad entry with index i, the actual scratchpad index accessed should be $(i + v\_offset)$.

For vHBM/DRAM virtualization, we fully leverage the coarse-grained and non-fragmented HBM/DRAM allocation for virtual BICs to minimize the overhead of stage-2 address translation. Unlike the memory fragmentation observed in CPU-side VMs, where VMs are usually running alongside various small-sized host applications, BICs are solely dedicated to ML/NN tasks and do not run other applications concurrently with virtual BICs. Moreover, virtual BICs require substantial amounts of HBM/DRAM memory, but only in a coarse granularity.

Therefore, we transition from the *fixed page size translation* to *range-based translation*, which stores guest physical addresses, host physical addresses, sizes, and permissions in the TLB. We also design a new translation table called the Range Translation Table (RTT), as illustrated in the bottom half of Figure 1.

B. BIC Interconnection Virtualization (vRouter)

Data flow between BIC cores: Current ML tasks are structured as computing graphs, enabling BICs to leverage the predefined data flow to accelerate these tasks efficiently. A BIC chip can be segmented into multiple pipeline stages, each equipped with several BIC cores. Each stage is exclusively dedicated to processing a single layer of the ML task, while different stages handle distinct layers.

To minimize unnecessary memory loads and stores, modern BICs employ Network-on-Chip (NoC) or Inter-Chip Interconnection (ICI) among multiple BIC cores. These interconnection mechanisms offer higher data transfer bandwidth and lower latency compared to using unified memory. Compared to other processing units, BICs can fully leverage the data flow between the multiple layers in ML tasks.

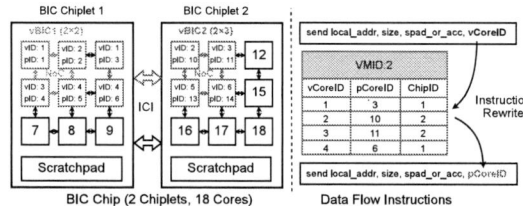

Fig. 2. **BIC interconnection virtualization:** vBIC virtualizes the interconnection between multiple BIC cores using instruction rewriting.

BIC topology virtualization: For instance, although an entire BIC chip may comprise 16 cores arranged in a 2D mesh, a user might only require 4 cores. In this case, vBIC needs to orchestrate a virtual BIC with a 2×2 topology. However, there are no existing virtualization solutions for BIC's interconnection, primarily due to their high bandwidth requirements.

In the vBIC design, we do not virtualize the interconnection between BIC cores during runtime. Instead, we employ a more traditional virtualization mechanism: instruction rewriting, as shown in Figure 2. Unlike instruction rewriting for CPUs, which often results in sub-optimal code, rewriting for BIC interconnections is much simpler, merely modifying the target node ID in send/receive instructions. The key benefit of this design is the elimination of virtualization costs from the critical path of data transfer. The instruction rewriting mechanism also depends on an additional routing table, where the virtual BIC core ID is mapped to the physical BIC core ID.

EVALUATION

C. Experimental Setup

We implement the hardware prototype of vBIC on FPGA. As for software components, we opt Linux-6.2

as the host kernel, and modify the default KVM module in Linux kernel to manage all virtual BIC resources. The configuration is shown in Table I.

TABLE I: BIC configurations used in the evaluation

Parameter	Value
Systolic array dimension (per tile)	16
Scratchpad size (per tile)	256KB
# of accelerator tiles	8
NoC topology	2D Mesh
Shared L2 size	2MB
Shared L2 banks	8
DRAM bandwidth	16GB/s

D. BIC Memory Virtualization

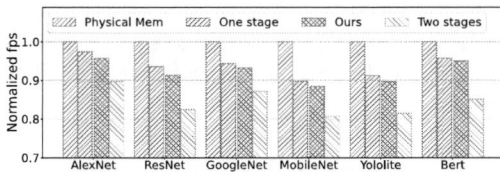

Fig. 3. The normalized performance of ML workloads with different memory virtualization methods.

We evaluate the translation overhead of various ML models with different configurations: bare metal, one-stage page translation, first-stage page translation with second-stage range translation (our solution), and two-stage page translation. To minimize the page translation overhead, we implemented huge pages for both the stage-one and stage-two translation schemes. Compared to the ideal performance(bare metal), the one-stage page translation only incurs an average of 7% overhead across six different models.However, enabling nested page translation for virtual BIC amplifies the translation overhead to 16%. In contrast, vBIC adopts the range translation in the stage-two translation scheme, and achieves an average of 1.3 times memory access for one TLB miss. Therefore, our solution achieves the similar performance across all workloads as the one-stage translation.

E. BIC Interconnection Virtualization

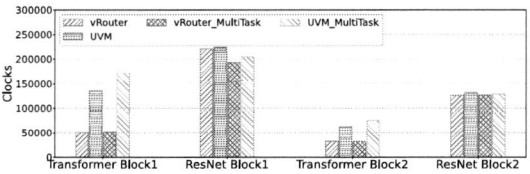

Fig. 4. **ML benchmark:** Performance of ML workloads in the single/multi-task scenarios.

In order to test the performance of vRouter mechanism in real-world scenarios, we evaluate the end-to-end performance of different ML workloads shown in

Figure 4. In the single-task scenario, we utilize four BIC cores to run a single NN block. While, in the multi-task scenario, we employ eight BIC cores to execute a ResNet block task along with a Transformer block task, and each task is running on four cores. Block1 and Block2 represent two network configurations with differet channels (i.e., 32 in Block1 and 64 in Block2). Our experimental results show that in vRouter architecture, the Transformer block achieves a 2x performance improvement compared to UVM, while the performance improvement for the ResNet block is not such significant. This is because the varying structures between layers in the ResNet block create many bubbles in the dataflow, leading to performance degradation.

As for the multi-task parallel scenarios, without the vRouter support, tasks can only run in the UVM mode, which exacerbates memory access contention between different tasks, thereby affecting overall performance. By utilizing vRouter approach, tasks can still utilize the capability of inter-core communication without any performance loss compared to the single-task scenario. Compared to the UVM, vRouter achieves an average of 52% and 29% performance improvement for two model configurations, respectively.

I. CONCLUSION

This paper presents the first comprehensive virtualization design for Brain-inspired Chip named vBIC. vBIC focuses on virtualizing specialized hardware structures for BIC, including multi-core architecture, scratchpad-centric memory, and interconnections, and proposes two hardware extensions: BIC vMem and vRouter.

REFERENCES

[1] P. A. Merolla, J. V. Arthur, R. Alvarez-Icaza, A. S. Cassidy, J. Sawada, F. Akopyan, B. L. Jackson, N. Imam, C. Guo, Y. Nakamura *et al.*, "A million spiking-neuron integrated circuit with a scalable communication network and interface," *Science*, vol. 345, no. 6197, pp. 668–673, 2014.

[2] S. Furber and P. Bogdan, *Spinnaker-a spiking neural network architecture.* Now publishers, 2020.

[3] M. Davies, N. Srinivasa, T.-H. Lin, G. Chinya, Y. Cao, S. H. Choday, G. Dimou, P. Joshi, N. Imam, S. Jain *et al.*, "Loihi: A neuromorphic manycore processor with on-chip learning," *Ieee Micro*, vol. 38, no. 1, pp. 82–99, 2018.

[4] J. Pei, L. Deng, S. Song, M. Zhao, Y. Zhang, S. Wu, G. Wang, Z. Zou, Z. Wu, and W. He, "Towards artificial general intelligence with hybrid tianjic chip architecture," *Nature*, vol. 572, no. 7767, pp. 106–111, 2019.

[5] M. Dowty and J. Sugerman, "Gpu virtualization on vmware's hosted i/o architecture," *ACM SIGOPS Oper. Syst. Rev.*, vol. 43, pp. 73–82, 2008. [Online]. Available: https://api.semanticscholar.org/CorpusID:228328

Improved linearity by Double-Channel GaN HEMTs with the ultra-thin AlN barrier layer

Long Zhang[1], Qian Yu[1], Ling Yang[1] *Member IEEE*, Meng Zhang[1], Chunzhou Shi[2], Xu Zou[1], Wenze Gao[1], and Bin Hou[1]

[1]School of Microelectronics, Xidian University, Xi'an, China
[2]School of Advanced Materials and Nanotechnology, Xidian University, Xi'an, China

Abstract

In this article, the Gallium Nitride High Electron Mobility Transistors (GaN HEMTs) with the ultra-thin AlN barrier layer is introduced to improve the linearity. Due to the ultra-thin AlN sub-barrier layer, the coupling effect between the channels is improved. Compared to the conventional AlGaN/GaN single-channel (CSC) HEMTs, the ultra-thin AlN bottom barrier layer double-channel (UBD) HEMTs exhibit a wider gate voltage swing (GVS) of 4.9 V. In comparison with traditional double-channel HEMTs, the difference between g_m peaks and valleys of UBD HEMTs is smaller. At 3.6 GHz, the higher maximum output current density of 11.3 W/mm is realized by UBD HEMTs, with a maximum power-added efficiency (PAE) of 57%. These results indicate that the double-channel structure with an ultra-thin barrier layer holds great potential in the design of high-performance GaN HEMTs.

Keywords: high linearity; ultra-thin barrier; double-channel; GaN HEMTs.

Introduction

Due to wide bandgap, high electron mobility, and excellent thermal conductivity, GaN has become a primary material for next generation semiconductor devices, particularly excelling in radar and communication systems [1] [2]. Traditional GaN HEMTs have been successfully applied in the 5th-Generation Mobile Communication Technology (5G) base stations, offering high frequency and power handling capabilities [3].

However, existing GaN HEMTs still face significant nonlinearity at higher frequency. On the one hand, one major source of the nonlinearity is the decrease in transconductance (g_m) at high gate voltages. In the g_m characteristics of CSC HEMTs, g_m drops sharply after reaching its peak. This leads to a narrow gate voltage swing (GVS), thereby introducing nonlinearity in large-signal operations [4]. On the other hand, one of the reasons for the g_m drop-off is the lower saturation velocity (v_{sat}) at high current density [5] [6]. To solve this problem, the double channel GaN HEMTs has been developed in the past few decades [7] [8]. Nevertheless, there is a deep valley in the g_m curve of the double channel GaN HEMTs, which limits the linearity of double channel GaN HEMTs [9].

To elevate the valley of the g_m curve of the double channel GaN HEMTs, and achieve a wider GVS, the double channel GaN HEMTs with the ultra-thin AlN barrier layer were fabricated in this article. Then, the DC characteristics and large-signal characteristics were tested. The comparison results show that UBD HEMTs have better saturation current and breakdown voltage than CSC HEMTs. Additionally, the g_m curve of UBD

HEMTs is flatter, providing a wider GVS. In the large-signal test, UBD HEMTs exhibit high PAE and excellent P_{out}, which means that the devices mentioned have better linearity in handling large signals and reduces distortion. From this, it can be seen that the UBD HEMTs have great potential in improving the linearity, and meeting the demands for high frequency applications.

The device fabrication and results

A. Device structure for the Fabrication

The CSC and UBD HEMTs were both grown on 3-inch semi-insulating 4H-SiC substrates by the metal organic chemical vapor deposition (MOCVD) technology. The epitaxial layer structure, consists of a 1 μm GaN buffer layer, a 400 nm GaN channel layer, a 10 nm AlN barrier layer, a 10 nm GaN channel layer, and a 20 nm AlGaN barrier layer (from bottom to top).

The schematic cross-sectional diagrams of the two HEMT devices are presented in Fig. 1 (a) and Fig. 1 (b), respectively. The fabrication processes of the devices with these two structures were completely the same. Firstly, an ohmic metal layer was deposited using Ti/Al/Ni/Au, and ohmic contact was achieved by annealing at 860 °C in a nitrogen environment for 60 seconds. Subsequently, electrical isolation was accomplished by nitrogen ion implantation technology. Then, a 0.5 μm gate foot and a 1.3 μm gate cap were formed through photolithography and CF_4-based plasma etching. Finally, the metal stack of Ni/Au was evaporated to form the Schottky contact of the gate. The source-drain spacing (L_d), gate length (L_g), gate-source spacing, and gate width of the devices are 5 μm, 0.5 μm, 1.5 μm, and 100 μm, respectively.

B. The DC characteristics

From the transfer curves of UBD HEMTs, as shown in Fig. 2 (a), it can be observed that the UBD HEMTs exhibit two transconductance peaks. This occurs because the distances from the top and bottom channels to the gate differ, resulting in varying gate control capabilities over the two channels. Consequently, the channels turn on sequentially. In contrast, Fig. 2 (b) shows that the CSC HEMTs present only one transconductance peak. Calculated by the definition of GVS, the GVS of UBD HEMTs is 4.9 V, which is higher than that of CSC with a GVS of 3.1 V. Thus, a wider transconductance range is achieved, enhancing the device's linearity. In terms of current, at $V_{gs} = 2$ V, the maximum output current of UBD HEMTs is

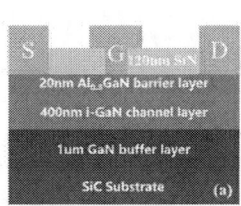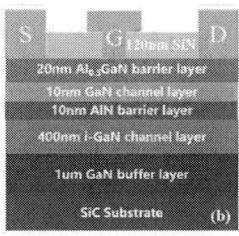

Fig. 1. The schematic diagram of the epitaxial structure of the (a) CSC and (b) UBD HEMTs.

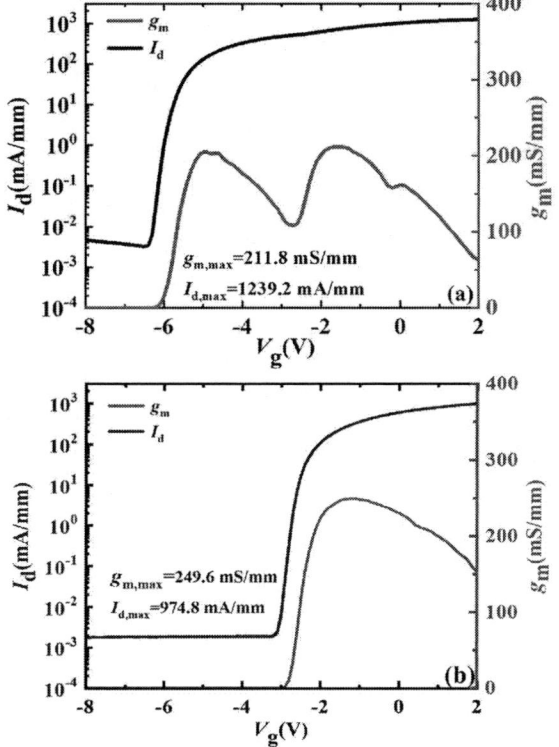

Fig. 2. The transfer curves of the (a) UBD and (b) CSC HEMTs.

1239 mA/mm, higher than that of 974 mA/mm for CSC HEMTs, highlighting its advantage in high current output.

As illustrated in Fig. 3 (a), the output characteristics were evaluated. It is observed that at $V_{gs} = 2$ V, the saturation drain current ($I_{d,max}$) for the CSC HEMTs is 1147 mA/mm, whereas for the UBD HEMTs, it is 1247 mA/mm. Notably, the curve does not exhibit complete saturation at $V_{ds} = 2$ V, suggesting that the saturation current for the UBD structure is potentially much higher than that of the CSC HEMTs. And the 3 V of knee voltage is achieved by the CSC HEMTs, which is lower than the 5 V of the UBD HEMTs. The curves of breakdown characteristics were shown in Fig. 3 (b). It demonstrates that the

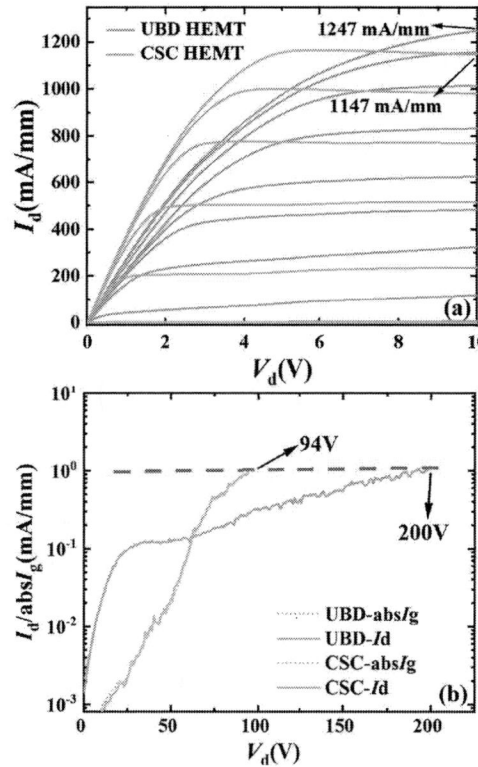

Fig. 3. The (a) output and (b) breakdown curves of the UBD and CSC HEMTs.

CSC HEMTs undergo breakdown at a drain voltage of 94 V, while the UBD HEMTs show superior breakdown characteristics, with breakdown occurring at a drain voltage of 200 V. This enhanced breakdown performance is attributed to the double-channel structure, which effectively distributes the electric field and increases the device's voltage tolerance.

C. The LARGE-SIGNAL characteristics

The load-pull measurement of the CSC and UBD HEMTs was tested at 3.6 GHz by the Maury test system. As can be seen in Fig. 4 (b), When the static drain bias voltage V_d is 48 V, the maximum output power (P_{out}) of CSC HEMTs amounts to 28.38 dBm. The 6.82 W/mm of converted maximum output power density, and the PAE of 55.7% is achieved by CSC HEMTs. Fig. 4 (a) shows while the static drain bias voltage is 48 V, the P_{out} of 33.18 dBm is achieved by UBD HEMTs. The converted maximum output power density is 11.3 W/mm, and the PAE of UBD HEMTs is 57%, which is better than that of CSC HEMTs. For single-channel devises, electrons are easily trapped in the buffer layer, resulting in poor PAE characteristics compared to dual-channel devices. According to the curve, after the 15 dBm of input power, P_{out} begins to decrease for CSC HEMTs, leading to power gain compression and degraded linearity, while UBD HEMTs exhibit great linearity.

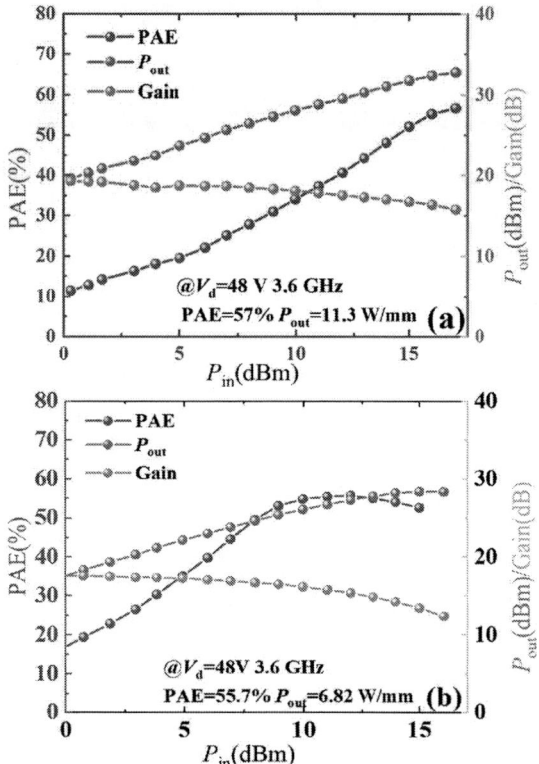

Fig. 4. Large-signal characteristics of the (a) UBD and (b) CSC HEMTs.

Conclusion

In this article, the HEMTs with the ultra-thin barrier layer accomplishing channel coupling are reported, which not only exhibit exceptional transconductance characteristics, but show high gain linearity. Compared to conventional single-channel HEMTs, the UBD HEMTs demonstrate notably a wider GVS, a flatter transconductance curve and a higher $I_{d,max}$. At 3.6 GHz with a static drain bias voltage of 48 V, the devices achieve a power-added efficiency of 57% and an output power of 11.3 W/mm. These results highlight the advantages of UBD HEMTs in high output power density and high power-added efficiency, while significantly enhancing linearity.

References

[1] V. Paidi et al., "High linearity and high efficiency of class-B power amplifiers in GaN HEMT technology," in IEEE Transactions on Microwave Theory and Techniques, vol. 51, no. 2, pp. 643-652, Feb. 2003, doi: 10.1109/TMTT.2002.807682.

[2] Yamaki F, Sano S. Mass-Production of High Reliability GaN HEMT for Wireless Communication[C]//CS MANTECH Conference. 2018.

[3] S. Nakajima, "GaN HEMTs for 5G Base Station Applications," 2018 IEEE International Electron Devices Meeting (IEDM), San Francisco, CA, USA, 2018, pp. 14.2.1-14.2.4, doi: 10.1109/IEDM.2018.8614588.

[4] X. D. Tong, S. Y. Zhang, P. H. Zheng, J. X. Xu, and X. Y. Shi "Influence of the Dynamic Access Resistance in the gm and fT Linearity of AlGaN/GaN HEMTs" *2018 22nd International Microwave and Radar Conference (MIKON)*, 2018.

[5] J.B. Khurgin, S. Bajaj, and S. Rajan, "Amplified spontaneous emission of phonons as a likely mechanism for density-dependent velocity saturation in GaN transistors," Appl. Phys. Exp., vol. 9, no. 9, 2016, Art. no. 094101, doi: 10.7567/apex.9.094101.

[6] T. Palacios, S. Rajan, A. Chakraborty, and S. Heikman, "Influence of the dynamic access resistance in the gm and fT linearity of AlGaN/GaN HEMTs," IEEE Trans. Electron Devices, vol.

[7] Palacios T, Chini A, Buttari D, et al. Use of double-channel heterostructures to improve the access resistance and linearity in GaN-based HEMTs[J]. IEEE transactions on electron devices, 2006,53(3): 562-565.

[8] C. Shi et al., "High-Efficiency AlGaN/GaN/Graded-AlGaN/GaN Double-Channel HEMTs for Sub-6G Power Amplifier Applications," in IEEE Transactions on Electron Devices, vol. 70, no. 5, pp. 2241-2246, May 2023, doi: 10.1109/TED.2023.3260809.

[9] Tilak V, Green B, Kaper V, et al. Influence of barrier thickness on the high-power performance of AlGaN/GaN HEMTs[J]. IEEE Electron Device Letters, 2001, 22(11): 504-506.

Simulation Optimization of Embedded Microchannel Heat Sink for GaN HEMTs

Bowen Yang[1], Shiming Li[1], Mei Wu[1], Ling Yang[1], Hao Lu[1], Bin Hou[1], Meng Zhang[1], Xiaohua Ma[1], and Yue Hao[1]

[1]State Key Discipline Laboratory of Wide Band-gap Semiconductor Technology, School of Microelectronics

Abstract

As the power density of Gallium Nitride (GaN)-based devices increases, heat dissipation becomes a major challenge. Embedded micro-channel heat sink have attracted much attention because of their high heat dissipation performance. In this paper, a GaN High-Electron-Mobility Transistor (HEMTs) model is established, and the effects of micro-channel geometry parameters (depth, width, spacing) and cooling fluid rate on heat dissipation performance are investigated by COMSOL Multiphysics. The pressure drop across the coolant inlet and outlet needs to be balanced while achieving effective cooling. With this trade-off, an optimal combination of microfluidic channel dimensions was obtained: depth of 200 μm, width of 100 μm, spacing of 80 μm. An increase in the flow rate of the cooling fluid generally corresponds with enhanced heat dissipation capabilities. Again, there is a trade-off between coolant pressure drop and flow rate to achieve high heat dissipation performance. Finally, in conjunction with the optimality conditions obtained from the simulation, the maximum temperature of the device is 135.5 °C, which is 45.5 °C lower than that of the device without embedded microchannel structure. In addition, the effects of different substrates on the temperature of the device are simulated. Embedded microchannels based on diamond substrates maximize the potential of liquid cooling. This paper provides theoretical and experimental basis for the design and optimization of embedded microchannel cooling structure, which is helpful to improve the heat dissipation performance of GaN devices.

Keywords: Gallium Nitride (GaN), High-Electron-Mobility Transistor (HEMTs), embedded microchannel heat sink, simulation, heat dissipation performance.

Introduction

With the increase of power density, operating frequency and circuit complexity of Gallium Nitride (GaN)-based High-Electron-Mobility Transistor (HEMTs), the current density of the device increases exponentially, resulting in an increasing heat flux in the device, and the thermal management of GaN devices faces challenges[1].

Microchannel heat sink was first proposed by Tuckerman and Pease, which has the advantages of high heat dissipation, small size and low energy consumption [2]. A large number of experiments and simulation studies have proved that substrate materials, coolant and microchannel structure are the main factors affecting the heat dissipation efficiency of microchannel heat sink [3]. Calame et al. experimentally demonstrated that the heat dissipation level of the diamond-on-SiC hybrid is about 30% higher than that of the SiC alone [4]. Qi et al. made an all-diamond microchannel heat sink with a heat transfer coefficient of up to 11447.2 W/m²K, which is 173% of the aluminum material [5]. In recent years, various microchannel structures have been proposed, such as parallel straight microchannels [6], micro pin fin arrays [7]and

manifold microchannels [8]. Feng et al. proposed an embedded cooling structure based on gradient distributed microneedle fin array, and simulated it to prove that the structure has good temperature uniformity [9]. Embedded cooling structures are designed in combination with manifolds, such as co-designed microfluidic and electronics schemes [10] and embedded manifold microchannel cooling structure [11], to enhance heat dissipation and reduce energy consumption. However, the above studies lack the analysis of the influence of the design of embedded microchannel geometric parameters on the heat dissipation performance of the GaN HEMTs.

In this work, we establish a 3D model of GaN HEMTs to optimize the geometric parameters of embedded microchannels with the goal of reducing the maximum operating temperature of the device. A set of optimal conditions is obtained considering the balance between thermal performance enhancement and coolant pressure drop. The maximum temperature of the device is 135.5°C, which is 45.5°C lower than the maximum temperature of the device without embedded microchannel structure. Meanwhile, the simulation results found that the use of diamond substrate can maximise the potential of embedded microchannel heat dissipation in GaN HEMTs.

Modeling and Simulation

A. Simulation Model

In this work, the modelling and parameter optimisation was done through the COMSOL Multiphysics 6.0. Fig. 1 shows the GaN-on-SiC HEMTs model with embedded microchannel heat dissipation structure established in this experiment. The device was placed on a copper cold plate, and the bottom temperature of the copper cold plate was maintained at 300 K. TABLE I shows the specific parameters of the 3D model of the device. Due to the thin thickness of the top GaN layer and the gate metal region, fine meshing was used to ensure the accuracy of the simulation results, and conventional meshing was used to partition the remaining regions. The total number of cells in the grid was 1,055,324.

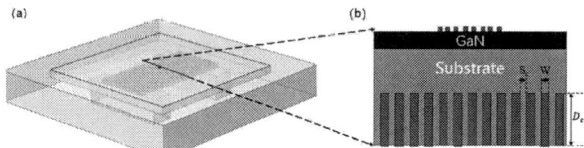

Fig. 1.(a) Establishment of the simulation device model. (b) Device cross-section diagram

TABLE I. Device model specific parameters

Simulation parameter	Value	Unit
Copper cold plate	10×10×1.5	mm×mm×mm
SiC substrate	7×7×0.35	mm×mm×mm
GaN layer	7×7×0.0015	mm×mm×mm
Microchannel size	5×0.1×0.2	mm×mm×mm
Gate	5×5×200	μm×μm×μm
Gate spacing	20	μm

979-8-3315-0417-5/25 $31.00 © 2025 IEEE

Microchannel spacing	100	μm
Number of microchannels	13	N./A.
Heat source power	15	W

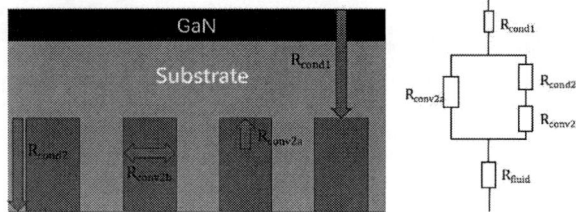

Fig. 2. The model of thermal resistance.

B. Thermal Resistance Model

Fig. 2 shows the thermal resistance model of embedded microchannel cooling structure, which mainly consists of three parts: substrate conduction thermal resistance (R_{cond1}), microchannel conduction and convection thermal resistance (R_2), and cooling zone thermal resistance (R_{fluid}).

Substrate conduction thermal resistance (R_{cond1}) is given by Fourier's heat law:

$$R_{cond1} = \frac{\delta_{GaN}}{K_{GaN}A_{heat}} + \frac{\delta_{base}}{K_{SiC}A_{heat}} \quad (1)$$

$$\delta_{base} = D_{SiC} - D_c \quad (2)$$

The δ_{GaN} is the thickness of GaN layer, δ_{base} is the thickness of the region without microchannel, D_{SiC} is SiC substrate thickness, and D_c is depth of the microchannel respectively.

The microchannel conduction and convection thermal resistance (R_2) in the microchannel structure is composed of three parts: the thermal resistance of heat transfer between the device substrate and the cooling fluid (R_{conv2a}), the thermal resistance of heat conduction (R_{cond2}) in the fin, and the thermal resistance of heat transfer between the fin wall and the cooling fluid (R_{conv2b}). The formula for calculating these thermal resistances is as follows:

$$R_{conv2a} = \frac{1}{NhWL} \quad (3)$$

$$R_{cond2} = \frac{D_c}{K_{SiC}A_{fin}} \quad (4)$$

$$R_{conv2b} = \frac{1}{2NhD_CL} \quad (5)$$

Where, N is the number of microchannels, W and L are the width and length of microchannels, A_{fin} is the total upper surface area of the fin in the microchannel structure, and h is the convective heat transfer coefficient of the microchannel.

In conclusion, The microchannel conduction and convection thermal resistance (R_2) in the microchannel structure is shown as follows:

$$R_2 = R_{conv2a} // (R_{cond2} + R_{conv2b}) \quad (6)$$

When fluid passes through the microchannel, it absorbs heat and heats up., the heat resistance generated is R_{fluid}, which is related to the density of the fluid ρ, the specific heat capacity C_p and the flow rate f, and can be expressed as:

$$R_{fluid} = \frac{1}{2} \cdot \frac{1}{\rho C_p f} \quad (7)$$

The total thermal resistance can be expressed as:

$$R_{total} = R_{cond1} + R_2 + R_{fluid} \quad (8)$$

II. RESULTS AND DISCUSSION

In order to explore the effects of various geometric parameters on the heat dissipation performance of embedded microchannel heat sink, we simulated the changes of depth, width and spacing of microchannel in turn. During the simulation, the volume flow rate of the water was fixed at 100 ml/min. Prior to this, a device model without embedded microchannel structure under the same conditions was simulated, and the maximum temperature of the device was 181 °C.

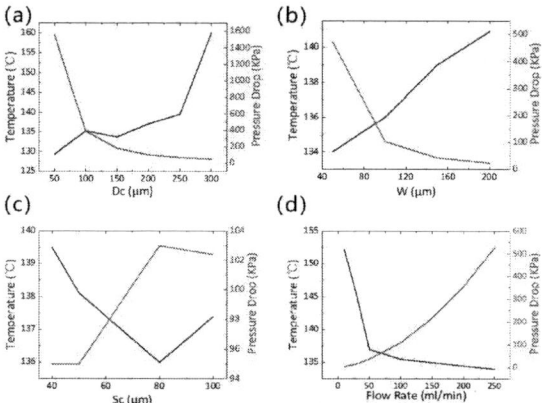

Fig. 3. The effect of individual parameter on maximum temperature and pressure drop: (a) depth of the microchannel (Dc), (b) width of the microchannel (W), (c) space between microchannels (Sc), (d) fluid flow rate.

Fig. 3 (a) shows that as the microchannel depth increased from 50 μm to 300 μm, the maximum temperature of the device gradually increased and the pressure drop decreased. In the range of 100 to 200 μm, the temperature changed little, only 3.3 °C. Microchannel deepening reduces R_{cond1}, but increases R_2, resulting in the overall increase of temperature curve with slight fluctuation. Although the device temperature was lowest at 50 μm, it brought the pressure drop that the device cannot withstand. Therefore, the optimal depth was determined to be 200 μm.

After obtaining a suitable microchannel depth, the effect of different microchannel widths on the heat dissipation performance was investigated. Fig. 3 (b) shows that when the width of the microchannel increased from 50 to 200 μm, the device temperature presented a linear rising trend, and the pressure drop decreased greatly when the width is 50~100 μm, and then slowly decreased with the increase of the width. With the increase of the width of the microchannel, the average convective heat transfer coefficient decreases and the maximum temperature of the device increases slightly. It also reduces the severe shrinkage and expansion pressure loss in the fluid area between the manifold and the microchannel, reducing the pressure drop. Through comprehensive consideration of temperature and pressure drop, the width of the microchannel was fixed at 100 μm.

After the depth and width of the microchannel were fixed, the influence of the microchannel spacing on the heat dissipation performance was studied. Fig. 3 (c) shows that when the microchannel spacing increased from 40 to 100 μm, the maximum temperature of the device first dropped and then rose, showed an inflection point, and the pressure drop curve also had an inflection point. The variation of channel spacing affects fluid flow, thermal boundary layer thickness and thermal interaction between microchannels. Under the combined action of these factors, the device temperature is lowest when the spacing was 80 μm, which is the best spacing for the structure.

The flow rate of the coolant is also a factor of attention, which largely determines the cooling efficiency. According to Fig. 3 (d), with the increased of fluid rate, the maximum temperature of the device decreased and the pressure drop increased linearly. With the increase of volume flow, the fluid boundary layer becomes thinner and the convective thermal resistance decreases. However, it will lead to increased stagnation pressure when the fluid enters and exits the microchannel, resulting in increased pressure drop. At a fluid rate of 100 ml/min, a lower device temperature was obtained while the pressure drop remained within the tolerable range.

Through the simulation of geometric factors and fluid rate, the changes of temperature and pressure drop are considered comprehensively. When the etching width of the microchannel is 100 μm, the spacing is 80 μm, the depth is 200 μm, and the fluid rate is 100 ml/min, the device's maximum temperature is 135.5 °C, and the pressure drop generated is 110 kPa.

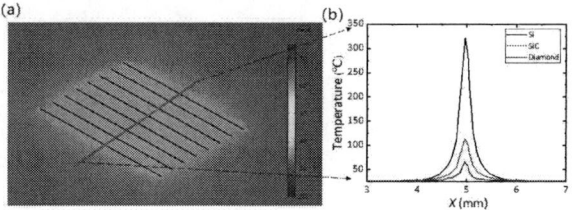

Fig. 4. (a) Temperature distribution in GaN-on-diamond gate region. (b) Device temperature curve under gate for different substrates.

Because the thermal conductivity of substrate material can affect the thermal resistance of embedded heat dissipation structure, different substrate materials are studied. Three types of substrates, Si, SiC and diamond, are used for simulation studies. In Fig. 4, the highest temperature of GaN-on-diamond HEMTs was 72.4 °C, and the highest temperature under the gate was only 20.2% of that of Si substrate and 64.8% of SiC substrate, which fully proved the high heat dissipation capacity of diamond due to high thermal conductivity.

Conclusion

In this work, we established a GaN HEMTs model with embedded microchannel cooling structure and the thermal resistance model by COMSOL Multiphysics 6.0. The effects of microchannel geometry parameters, cooling fluid rate, and different substrates on the heat dissipation performance of the device are studied. When the etching width of the microchannel is 100 μm, the spacing is 80 μm, the depth is 200 μm, and the fluid rate is 100 ml/min, the maximum

temperature of the device is 135.5 °C, which is 33% lower than the maximum temperature of the device without embedded microchannel structure under the same dissipation power. In addition, the use of diamond substrates with higher thermal conductivity can further exploit the advantages of embedded liquid cooling.

Above results provide deeply understand the influence of geometric parameters on the embedded microchannel cooling structure. This is beneficial for the thermal management of GaN devices in the future.

Acknowledgment

This work was supported in part by the National Natural Science Foundation of China under Grant 62234009, 62188102, 62090014, 62104178, 62104179, and 62104184, in part by the China Postdoctoral Science Foundation under Grant 2022T150505 and 2021M692499, in part by the Fundamental Research Funds for the Central Universities of China under Grant XJSJ23056.

References

[1] M. Wu et al., 'Integration of polycrystalline diamond heat spreader with AlGaN/GaN HEMTs using a dry/wet combined etching process', Diam. Relat. Mater., vol. 132, p. 109676, Feb. 2023, doi: 10.1016/j.diamond.2023.109676.

[2] D. B. Tuckerman and R. F. W. Pease, 'High-performance heat sinking for VLSI', IEEE Electron Device Lett., vol. 2, no. 5, pp. 126–129, May 1981, doi: 10.1109/EDL.1981.25367.

[3] Z.-H. Wang, X.-D. Wang, W.-M. Yan, Y.-Y. Duan, D.-J. Lee, and J.-L. Xu, 'Multi-parameters optimization for microchannel heat sink using inverse problem method', Int. J. Heat Mass Transf., vol. 54, no. 13–14, pp. 2811–2819, Jun. 2011, doi: 10.1016/j.ijheatmasstransfer.2011.01.029.

[4] J. P. Calame, R. E. Myers, S. C. Binari, F. N. Wood, and M. Garven, 'Experimental investigation of microchannel coolers for the high heat flux thermal management of GaN-on-SiC semiconductor devices', Int. J. Heat Mass Transf., vol. 50, no. 23–24, pp. 4767–4779, Nov. 2007, doi: 10.1016/j.ijheatmasstransfer.2007.03.013.

[5] Z. Qi et al., 'An ultra-thick all-diamond microchannel heat sink for single-phase heat transmission efficiency enhancement', Vacuum, vol. 177, p. 109377, Jul. 2020, doi: 10.1016/j.vacuum.2020.109377.

[6] H. Huang, L. Pan, and R. Yan, 'Flow characteristics and instability analysis of pressure drop in parallel multiple microchannels', Appl. Therm. Eng., vol. 142, pp. 184–193, Sep. 2018, doi: 10.1016/j.applthermaleng.2018.06.083.

[7] C. S. Sharma, S. Zimmermann, M. K. Tiwari, B. Michel, and D. Poulikakos, 'Optimal thermal operation of liquid-cooled electronic chips', Int. J. Heat Mass Transf., vol. 55, no. 7–8, pp. 1957–1969, Mar. 2012, doi: 10.1016/j.ijheatmasstransfer.2011.11.052.

[8] N. Gilmore, V. Timchenko, and C. Menictas, 'Manifold microchannel heat sink topology optimisation', Int. J. Heat Mass Transf., vol. 170, p. 121025, May 2021, doi: 10.1016/j.ijheatmasstransfer.2021.121025.

[9] S. Feng, Y. Yan, H. Li, Z. Yang, L. Li, and L. Zhang, 'Theoretical and numerical investigation of embedded microfluidic thermal management using gradient distribution micro pin fin arrays', Appl. Therm. Eng., vol. 153, pp. 748–760, May 2019, doi: 10.1016/j.applthermaleng.2019.03.017.

[10] R. Van Erp, R. Soleimanzadeh, L. Nela, G. Kampitsis, and E. Matioli, 'Co-designing electronics with microfluidics for more sustainable cooling', Nature, vol. 585, no. 7824, pp. 211–216, Sep. 2020, doi: 10.1038/s41586-020-2666-1.

[11] H. Zhang and Z. Guo, 'Near-junction microfluidic cooling for GaN HEMT with capped diamond heat spreader', Int. J. Heat Mass Transf., vol. 186, p. 122476, May 2022, doi: 10.1016/j.ijheatmasstransfer.2021.122476.

High performance pixel development for thin-film based image sensors

Jiwon Lee[1,2] and Minhyun Jin[1,2]

[1]POSTECH, South Korea, [2]Imec, Belgium

Abstract

This paper presents a high-performance pixel design for thin film image sensors based on organic, colloidal quantum dot and perovskite photodiodes. The development of a simple yet effective pixel structure reduces pixel noise to enable high quality image sensors while maintaining high resolution, or adds functionality to enable global shutters or three-dimensional image sensors. Some representative examples of the pixel structures of the developed thin-film based image sensor are presented together with image demonstrations to show their effectiveness.

Keywords: Image sensor, Thin-film image sensor, Pixel

Introduction

Thin-film-based image sensors offer the advantages of high absorption coefficients, light absorption in the short-wave (1-2 μm) or mid-wave infrared (2-4 μm) and even x-ray bands, and integration in a low-cost monolithic process on silicon readout IC (ROIC) substrates [1][2]. These unique advantages of thin-film-based image sensors enable operations that are not possible with conventional silicon image sensors, enabling a wide range of optical imaging information to be obtained [3]. In fact, active research in this area continues, with image sensors based on perovskite-based photodiodes that absorb X-rays directly, high-resolution micropixel image sensors based on organic photodiodes being developed to a commercial level, and quantum dot-based short-wave infrared image sensors on the verge of commercialization [4]. However, there has been relatively little research into pixel design in this area, which has relied on legacy pixel designs that are not optimized for thin film pixel structures; passive pixel sensor (PPS) structure [5], three-transistor active pixel sensor (3T APS) structure [6] or charge transimpedance amplifiers (CTIAs) [7]. This paper presents new pixel structures and operations that are developed for thin-film image sensors.

Conventional pixel structures

Fig 1. shows the conventional pixel structure that has been widely used for various image sensors. PPS has the advantage of a simple pixel structure with only one transistor per pixel. The transistor sequentially connects the pixel to the

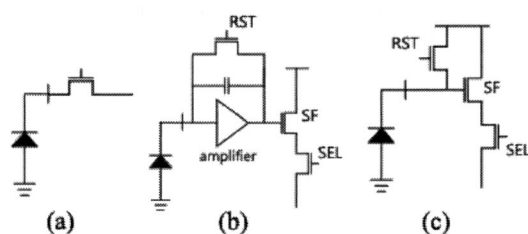

Fig. 1. Conventional pixel designs (a) PPS, (b) CTIA structure, (c) 3T APS.

readout circuit while resetting the photodiode (Fig. 1(a)). As a simple structure, the PPS structure has been widely used for the large area image sensors based on thin film transistor (TFT) technology, because TFT technologies normally takes large critical dimension therefore transistor size is large [8][9].

The CTIA pixel structure (Fig. 1(b)) has also been widely used for some image sensor applications due to the advantages of fixed photodiode bias during integration, higher conversion gain (defined by feedback capacitance). However, the presence of gain amplifiers in the pixel complicates the pixel design and consumes static power, making it difficult to implement a low power, high resolution image sensor in a small pixel pitch [10].

3T APS has dedicated reset (RST) and select (SEL) transistor while having in pixel source follower (SF) which buffer photodiode signal to the readout circuit (Fig. 1(c)). Thanks to the in-pixel buffer structure, the charge-to-voltage conversion (CG) gain is typically higher than PPS, resulting in a lower noise floor and high image quality. For thin film photodiode on silicon ROIC image sensors, 3T's APS structures have been the main workhorse in the development of high-resolution image sensors with small pixel pitch. However, the photodiode bias change during photoelectron integration and the kTC noise generated when resetting the photodiode have been the major drawbacks in implementing high quality image sensors [11].

Low noise thin-film pixel structures

Fig 2. shows the developed photogate (PG) pixel structure, which fixes the photodiode bias while increasing the conversion gain compared to the conventional 3T APS based

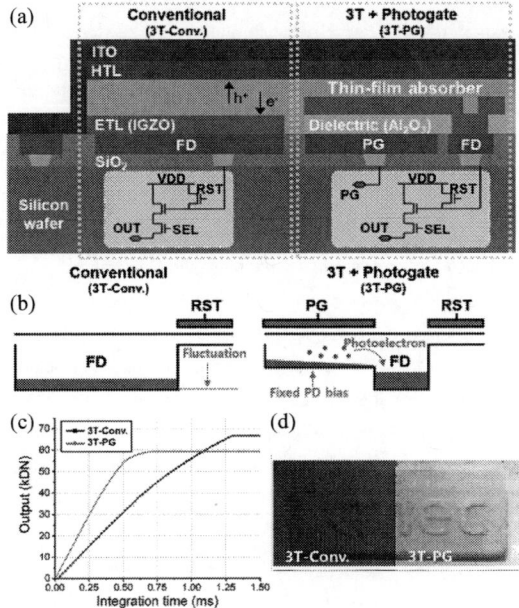

Fig 2. Thin-film photogate (a) pixel structure compared with conventional 3T pixel, (b) operation principle, (c) output characteristic, and (d) captured image comparison under same condition.

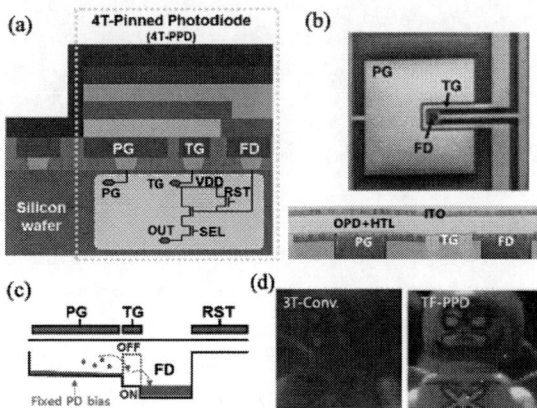

Fig 3. Thin-film pinned photodiode based 4T (a) pixel structure, (b) processed pixel images, (b) operation principle, and (d) captured image comparison under same condition.

pixel structure [12]. The applied PG voltage effectively fixes the photodiode bias remotely by vertical field effect, while the generated photoelectrons are collected in the FD region by means of the horizontal electric field by resetting FD with high voltage. Until the FD bias becomes equal to the PD potential on top of PG, photoelectrons are only filled in FD, therefore high conversion is achieved. Also, higher linearity has been achieve than 3T pixel structure thanks to the fixed photodiode bias.

Meanwhile, 4T APS with its unique pinned photodiode (PPD) structure has become the main pixel structure for most commercial CMOS image sensors (CIS) [13][14]. The PPD structure, which is fully discharged during reset, effectively suppresses kTC noise while providing high conversion gain by separating the photodiode from the charge-to-voltage conversion node (floating diffusion, FD). In addition, the separation of the photodiode from the defective FD and the surface region dramatically reduces the dark current, enabling high quality imaging. More recently, a 4T pixel structure based on thin-film PPD (TF-PPD) has been developed by integrating a TFT structure with an organic or colloidal quantum dot photodiode monolithically on silicon ROIC (Fig. 3(a)) [4]. The source of the TFT also acts as the cathode region of the thin-film photodiode and the drain is directly connected to the gate of the SF of the Si-ROIC, i.e. the drain region is configured as an FD node which converts the photoelectrons into voltage. The gate of the TFT is controlled by an external bias (TG) to connect and disconnect

the photodiode and FD node. In a TF-PPD structure, the PG bias ensures that the collected photoelectrons are fully transferred to the FD node along the lateral potential difference when the TG is turned on. This TF-PPD structure allows correlated double sampling (CDS) to be performed by fully transferring the charge from the PD to the FD node, thus maintaining the absorption capacity of the thin film photodiode while reducing kTC noise, increasing CG and reducing dark current.

Fig. 3(d) presents example images captured by developed novel image sensors in comparison with conventional 3T based image sensor. The developed image sensors present brighter images with higher image qualities.

Thin-film PD based 3D image sensor

Despite to the higher absorption coefficient in visible or even infrared light, the photodiode speed is limited due to the high exciton binding energy of organic or colloidal quantum dot photodiode; typically, mobility range of $10^{-1} \sim 10^{-5}$ cm^2/V [15]. This reduced mobility, despite to the thin photodiode structure, hinders the application of the organic and colloidal quantum photodiode to the 3D image sensors, where high speed photon travel speed needs to be measured.

Perovskite photodiode has the advantage of high mobility compared to the other thin-film photodiode. In addition, the high absorption coefficient makes it possible to achieve a thin photodiode structure, so that high-efficiency, high-speed photodiode can be achieved. Taking advantage of this, a 3D image sensor based on indirect time-of-flight (i-ToF) operation using perovskite photodiode has been developed (Fig. 4) [16]. A new i-ToF operation is proposed for the halide perovskite photodiode-based image sensor, which drives the top transparent electrode (ITO in this case) to demodulate the illuminated light pulse. The FACsPbIBr perovskite

979-8-3315-0417-5/25 $31.00 © 2025 IEEE

(a)   (b)

Fig. 4. Halide perovskite based i-ToF image sensor (a) fabricated image sensor cross-sectional image, (b) captured 2D and 3D images.

photodiode is successfully and reliably integrated on silicon ROIC to demonstrate the proposed 3D i-ToF imaging operation as well as high resolution 2D imaging (Fig. 4).

Conclusion

This paper presents a novel pixel design and operation for the thin film photodiode based high resolution image sensors. Organic or colloidal quantum dot based low noise, high quality image sensors have been successfully developed with the proposed thin film photogate and pinned photodiode pixel structure to enable high performance yet low cost infrared image sensors. A high-resolution perovskite-based i-ToF 3D image sensor is developed for the first time, to the authors' knowledge, by monolithically integrating a reliable halide perovskite photodiode on silicon ROIC. The authors believe that the developed high performance thin film based image sensors should be able to successfully subsidise image sensing technology to expand image sensing capabilities.

References

[1] J. Lee et al., "Imaging in Short-Wave Infrared with 1.82 μm Pixel Pitch Quantum Dot Image Sensor," in IEEE International Electron Devices Meeting (IEDM), Dec. 2020, pp. 16.5.1–16.5.4

[2] Ackerman, Matthew M., Xin Tang, and Philippe Guyot-Sionnest. "Fast and sensitive colloidal quantum dot mid-wave infrared photodetectors." *ACS nano* 12.7 (2018): 7264-7271.

[3] Lee, J., Georgitzikis, E., Hermans, Y. *et al.* Thin-film image sensors with a pinned photodiode structure. *Nat Electron* **6**, 590–598 (2023).

[4] Li, Z. et al. Halide perovskites for high-performance X-ray detector. Mater. Today 48, 155–175 (2021)

[5] Street, R. A., et al. "Two dimensional amorphous silicon image sensor arrays." *MRS Online Proceedings Library* 377 (1995): 757-766.

[6] Fossum, Eric R. "Active pixel sensors: Are CCDs dinosaurs?." Charge-Coupled Devices and Solid State Optical Sensors III. Vol. 1900. SPIE, 1993.

[7] Gregory, C., Hilton, A., Violette, K. & Klem, E. J. D. 66-3: invited paper: colloidal quantum dot photodetectors for large format NIR, SWIR, and eSWIR imaging arrays. SID Symp. Dig. Tech. Pap. 52, 982–986 (2021)

[8] Antonuk, Larry E., et al. "Large-area 97-um pitch indirect-detection active-matrix flat-panel imager (AMFPI)." *Medical Imaging 1998:*
Physics of Medical Imaging. Vol. 3336. SPIE, 1998.

[9] Karim, Karim S., Arokia Nathan, and John A. Rowlands. "Alternate pixel architectures for large-area medical imaging." *Medical Imaging 2001: Physics of Medical Imaging.* Vol. 4320. SPIE, 2001.

[10] Murari, Kartikeya, et al. "A CMOS in-pixel CTIA high-sensitivity fluorescence imager." *IEEE transactions on biomedical circuits and systems* 5.5 (2011): 449-458.

[11] P. E. Malinowski et al., "Miniaturization of NIR/SWIR image sensors enabled by thin-film photodiode monolithic integration," in Optical Architectures for Displays and Sensing in Augmented, Virtual, and Mixed Reality (AR, VR, MR) II, Mar. 2021, pp. 207–212.

[12] M. Jin, et al. "Thin-Film Photogate Pixel With Fixed Photodiode Bias for Near-Infrared Imaging." IEEE Electron Device Letters (2023).

[13] Teranishi, N., Kohono, A., Ishihara, Y., Oda, E. & Arai, K. No image lag photodiode structure in the interline CCD image sensor. In 1982 International Electron Devices Meeting 324–327 (IRE, 1982).

[14] Fossum, E. R. CMOS image sensors: electronic camera-on-a-chip. IEEE Trans. Electron Devices 44, 1689–1698 (1997).

[15] Joo Hyoung Kim, et al., "Detailed Characterization of Short-Wave Infrared Colloidal Quantum Dot Image Sensors", IEEE Transactions on Electron Devices, vol. 69, no.6, pp. 2900-2906, 2022

[16] Wenya Song, et al., "Halide Perovskite Photodiode Integrated CMOS Imager", ACS Nano, under review.

Spatial microwave magnetic field distribution mapped by the Permalloy/Ta bilayer sensor

Peiwen Luo[1], Guanjun Zhang[1], Bin Peng[1] and Wenxu Zhang[1]

1 National key laboratory of electronic thin films and integrated devices, University of Electronic Science and Technology of China, China

Abstract

In this work, a microwave (MW) magnetic field sensor based on a Permalloy (Py)/ Tantalum (Ta) bilayer thin film is developed. By determining the Lorentzian components of the inverse spin Hall and spin rectification voltage of the bilayer film, the MW magnetic field aptitude distributions in the three-dimensional space near a shorted MW probe were obtained. The detection limit and sensitivity of the system are obtained to be 2.1×10^{-4} Oe and 7.7 $\mu V/Oe^2$, respectively. The method provides a convenient way for characterizing the MW magnetic field in small devices and complex structures, which can be used to improve the performances of microwave integrated circuit.

Keywords: Microwave detecting, Spin rectification effect and Spin pumping effect

Introduction

Microwave (MW) technology, as one of the most widely used technologies in the twenty first century, plays an indispensable role in wireless communications, navigations, disease diagnoses and other systems. In most of these systems, the microwave components are integrated onto a single chip and the integrity is increasing with the development of semiconductor technology [1]. However, the electromagnetic interference will become prominent along with the increase of the integrity and frequency. It is demanding for the spatial image of MW field in the monolithic microwave integrated circuit in order to improve the performances of the chips. For the traditional detecting method based on the Faraday's law, it is quite a challenge to scale down the detecting sensors because its sensitivity scales down with the area of the loop. Thus, based on the needs of high-precision measurement of MW field, people have developed several other approaches such as near-field scanning MW microscopy [2], spintronic sensor [3-5] and quantum magnetic field detection [6, 7] to measure the MW electric or magnetic field. Among them, the sensors based on spintronics have received much attention because of their high sensitivity and convenience. Hu et al [4, 8] proposed a detector made by a single permalloy (Py) thin film, a great advantage of this sensor lies that the vector MW magnetic field can be detected by the angle dependence between the spin rectified (SRE) voltage of the Py film and the external DC magnetic field [5, 9]. In addition to this application, the

local MW magnetic field measurement reveals the inhomogeneity of the field during the characterization of the inverse spin Hall effects (ISHE), which may explain some of the controversies of measurements as shown by Hoffman et al [10].

So far, the spintronic sensor of MW magnetic field is mainly based on the SRE of a single Py layer while the ISHE, which is found in heavy metals, has not played its role. However, the SRE and the ISHE are mixed in a ferromagnetic/nonmagnetic metal bilayer and the separations have long been a hot topic in this field [11-13]. Even in monolayer of Py, the so-called self-induced ISHE was reported [14]. Fortunately, the two effects are both linearly dependent on the local microwave power density. They can thus be added up to enhance the proportionality of the photon voltage to the microwave power by proper choice of the directions of the DC magnetic field.

In this paper, we develop a MW magnetic field amplitude sensor based on the ISHE and SRE of a Py/Ta bilayer film. The measuring system is simple in structure and provides relatively high detection sensitivity attribute to the sensitive response of the magnetic moment to the MW field. With this sub-millimeter probe, the distribution of the MW magnetic field of an end shorted MW coaxial in three dimensions can be obtained. Our method can be used for MW integrated circuit optimization and fault diagnosis, as well as spintronics and quantum information processing where the MW field is involved.

Experiments

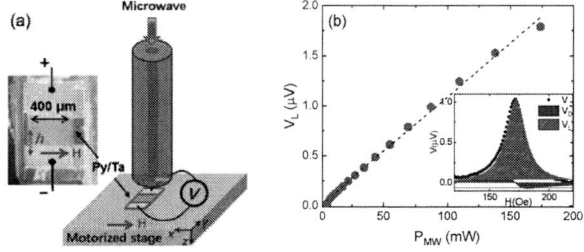

Fig. 1. (a) The schematic drawing of the setup for the measurement of the shorted MW probe, where the dimension of the bilayer film is 400 μm × 400 μm. (b) the MW power dependence of the Lorentz symmetric component of voltage (VL). The inset shows the Lorentz component (VL) and anti-Lorentz component (VD) of voltage at the power of 70 mW.

Fig. 1(a) shows the setup for the measurement of the

shorted MW probe, which is used to provide the microwave field for the high throughput characterization of the ISHE [15, 16]. The setup consists of a semirigid coaxial cable, a DC magnetic field source (not shown here), a motorized stage and a lock-in amplifier. The diameters of the inner and outer conductor of the coaxial cable are respectively 2 mm and 6.35 mm. The MW field sensor is a Py (15 nm)/Ta (20 nm) bilayer film grown by the magnetron sputtering, and has been annealed at the temperature of 350° for 3 hours after deposition. The Py/Ta bilayer film can move in the X, Y, and Z directions with the resolution of 1 μm guaranteed by the motorized stage. In the shorted-end plane of the coaxial cable where the bilayer film is mounted, the h is approximatively linearly polarized along the Y direction, so that we can expect h ≈ hy ≫ hx, hz. In order to inject the spin from the Py layer into the Ta layer with high efficiency, the shorted probe is mounted in an electromagnet with DC magnetic field (H) applied parallel to the direction of the shorted-end line, i.e., the H is perpendicular to h. The injected spins are converted into charge currents due to the ISHE of Ta layer [17, 18], which varies with H forming a symmetric Lorentzian line shape. Meanwhile, a rectified voltage in the superposition of an antisymmetric and a symmetric Lorentzian component will be generated due to the SRE in the Py layer. As discussed in the Ref. [19, 20], the total symmetric Lorentzian component (V_L) generated by ISHE and SRE is found to be:

$$V_L = V_{ISHE} + V_{SRE}^L = V_{ISHE} + (V_{AMR}^L + V_{AHE}^L) \quad (1)$$

where, the V_{AMR}^L and V_{AHE}^L are the symmetric component of the rectification voltage related, respectively, to the anisotropic magneto-resistive (AMR) and anomalous Hall effect (AHE). In our experimental setup, the H is perpendicular to h, so the expression of V_{AMR}^L and V_{AHE}^L can be written as:

$$V_{AMR}^L = \frac{A_{xx} A \Delta R \sin\emptyset}{8\pi M_S} j_e h, \quad V_{AHE}^L = \frac{A_{xy} R_{AHE} l \cos\emptyset}{2} j_e h \quad (2)$$

with

$$A_{xy} = \frac{-4\pi M_S}{\alpha(2H + 4\pi M_S)}, \quad A_{xx} = \frac{-A_{xy}\gamma(H + 4\pi M_S)}{\omega}$$

where A, ΔR, Φ, M_S, R_{AHE}, l, α, γ denote the cross-sectional area of Py layer, the change in resistance due to AMR, the relative phase between MW electric and magnetic field, the saturation magnetization, the anomalous Hall coefficients, the length of the thin film, the Gilbert damping constant and the gyromagnetic ratio respectively. The j_e is the induced MW current in the Py layer given by $j_e = (\sqrt{\omega\mu\sigma_F})h$ [21], where the μ and σ_F represent the permeability and conductivity of the Py layer. At the same time, the ISHE voltage (V_{ISHE}) of the Py/Ta bilayer film is proportional to the square of the MW field [19]:

$$V_{ISHE} = \left[\frac{2el\theta_{SH}\lambda_N \tanh(d_N/2\lambda_N)}{(d_N\sigma_N + d_F\sigma_F)} \cdot \right.$$

$$\left. \frac{g_{\uparrow\downarrow}\gamma^2(4\pi M_S\gamma + \sqrt{(4\pi M_S)^2\gamma^2 + 4\omega^2})}{8\pi\alpha^2[(4\pi M_S)^2\gamma^2 + 4\omega^2]} \right] h^2, \quad (3)$$

where e, θ_{SH}, λ_N, σ_N, d_N, d_F, $g_{\uparrow\downarrow}$ denote the electron charge, the spin Hall angle, spin diffusion length, conductivity and thickness of Ta layer, the thickness of Py layer, the spin mixing conductance respectively. The measurement is performed with a small MW power, which lies in the regime of linear region of magnetic moment response, so that both components (V_{ISHE} and V_{SRE}) are proportional to the square of h according to Eq. (2) and (3). The linear response to the MW power is demonstrated by the result of Fig. 1(b), each data is acquired at a fixed frequency while sweeping the H around the resonance conditions. Therefore, the magnitude of h can be obtained by measuring the V_L of the Py/Ta bilayer sensor at different positions.

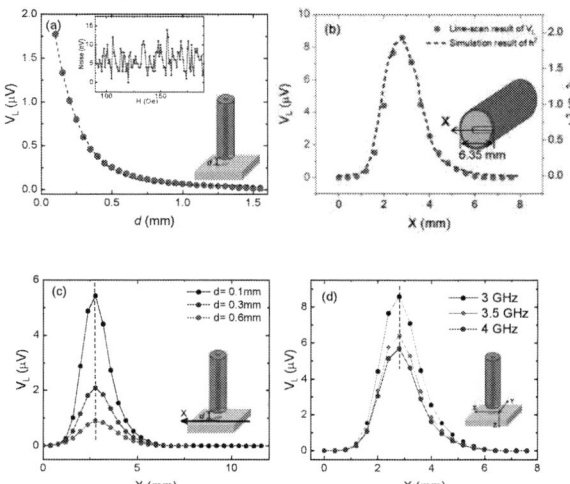

Fig. 2. (a) The dependence of V_L on d, the inset is the noise curve of the system. (b) The line scan along the central axis of the shorted MW probe at $d = 50$ μm. (c) the linear scanning results at f = 3 GHz, 3.5 GHz and 4 GHz where $d = 50$ μm. (d) the coordinate system of the scan images.

In order to determine the detection limit of the measuring system, we measured the V_L variation at different heights (d) shown in Fig. 2(a), where the step size of the scan is 50 μm. It can be observed that the V_L decreases exponentially with the increase of d and reaches detection limit (~15 nV) of our system when d increases to 1.5 mm. Then using Equation (1) and substituting relevant parameters [15] as $V_L = 15$ nV, $\theta_{SH} = -0.014$, $\lambda_N = 4.8$ nm, $\sigma_N = 7.56\times10^5 \ \Omega^{-1}m^{-1}$, $\sigma_F = 1.8\times10^6 \ \Omega^{-1}m^{-1}$, $\alpha = 0.014$, $g_{\uparrow\downarrow} = 1.1\times10^{19}$ m^{-2}, $\mu = 4.8\pi\times10^{-4}$ H/m, $\Delta R = 0.28 \ \Omega$, $\Phi = 1.8°$, $4\pi M_S = 10480$ Oe, $H = 160$ Oe, $R_{AHE} = 1.6\times10^{-9} \ \Omega \cdot m \cdot T^{-1}$ [22], we obtain the detection limit and the sensitivity of the system as 2.1×10^{-4} Oe and 7.7 μV/ Oe^2, while the system noise as measured by the Root-Mean-Square of the voltage is 6.8 nV [inset of Fig. 2(a)].

A line profile of the magnetic field was obtained as shown in the Fig. 2(b), which shows the line-scan result along the center of the shorted-end at $d = 50$ μm, where the total

979-8-3315-0417-5/25 $31.00 © 2025 IEEE

displacement and the step size are 8 mm and 0.4 mm respectively. As expected, the V_L reaches the maximum at the center of the shorted-end (x = 2.4 mm), then decreases rapidly to 0 outside the shorted probe. The measured data agrees well with the finite element method (FEM) simulation as shown by the blue curve in the figure.

To investigate the influence of MW frequency on the detection of the sensor, we measured the line-scan voltages at different frequencies under the same input power, the results are shown in Fig. 2(c). One can see that all the line-scan images have nearly the same variation tendency, i.e., the V_L reaches the maximum at x=2.4 mm and have the same half-peak width (1.8 mm). The reciprocities in the measured voltages are most likely due to the MW propagation characterization of the shorted MW probe. Specifically, for a TEM mode propagating along the axial direction of the shorted MW probe (Z axis), the variation of MW frequency only influences the distribution of h along the z axis, and the h is always on the wave peak at the shorted-end plane (XY plane), hence the line-scan voltages of the X axis at different frequencies only differ in amplitude. The frequency-independent distribution of h is helpful to demonstrate the reliability of scan images in Fig. 3. The scan results also reveal that the MW sensor can operate in a wide frequency range. In fact, the sensor can work at any frequency as long as it meets the condition of the magnetic resonance.

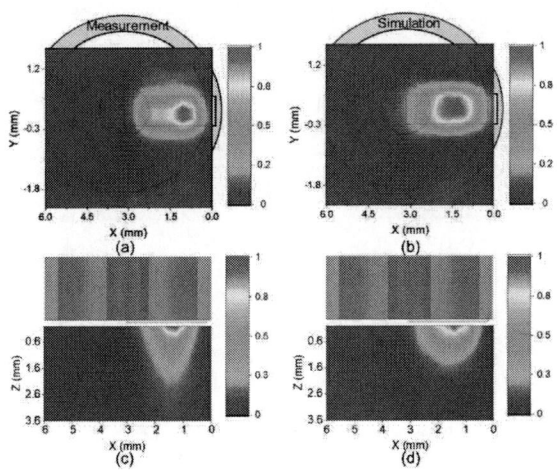

Fig. 3. the scan images (a), (c) and the simulation images (b), (d) of h2 in XY plane and XZ plane of the shorted MW probe. All the images have been normalized with its maximum value.

In the three-dimensional space measurement, we define the plane 50 µm below the shorted-end as the XY plane, the coordinate system of the scan images is shown in Fig. 2(d). Fig. 3(a) and (c) are the scan images of the area in the range of 4 mm × 6 mm in XY plane and 3.6 mm × 6 mm in XZ plane respectively, both the step size and the MW frequency are 0.2 mm and 4 GHz. For a better comparison with the simulation images [Fig. 3(b) and (d)], all the images have been normalized with its maximum value. Compared to the FEM simulation results in the XY plane, there is a weaker field intensity for the second half of the shorted-end in the actual scan image. The reason for this is that the first half of the shorted-end is closer to the Py/Ta sensor than the second half, which is caused by the manufacturing error since the

shorted-end plane is polished by hand. The lower height, according to the result of Fig. 2(a), will induce a stronger voltage response in the measuring results. For the XZ plane, the amplitude distribution of scan image is obviously "thinner" as compared to the simulation image, which illustrates the decay rate of h in the real space is slower than expectation. These experimental uncertainties, however, is difficult to reproduce in simulation quantitatively. The scan images of XZ and YZ plane demonstrate the imaging method can realize the MW magnetic field characterization in the three-dimensional space.

We have seen that the MW magnetic field in 3D space can be measured by our method. The results are dependent on the electromagnetic parameters as well as the structure of the components. The resolutions of the method can be as high as micrometers with a smaller scanning step, the method can thus be used in fault monitor or detector in MW chips. It must be noted that, in practices, due to the existence of the SRE, there is a long testing time for a large-area imaging since we have to measure the ferromagnetic resonance curve to identify the V_L. But the SRE can be effectively suppressed when the aspect ratio of the thin film is greater than 150, according to the previous report [23, 24]. At this time only the V_{ISHE} exists so that the external field H can be fixed at the resonant condition during the measurement. The scanning time for a single point will be reduced from dozens of seconds to milliseconds.

Conclusion

In conclusion, we have shown that the Py/Ta bilayer film can be used as a sensitive probe for the MW magnetic field, where the spatial resolution can be increased by scaling down the lateral dimensions of the films. The detection limit of our setup can be as low as 2×10^{-4} Oe while the system noise measured of the signal is 6.8 nV. The measured magnetic field distributions agree well with that from FEM simulations, proving the feasibility of the method. The imaging method is simple in structure and high in accuracy compared to the conventional local MW field sensing approach. Therefore, the method provides a convenient way for the characterization of MW field at the sub-wavelength scale, which can be utilized in spintronics and quantum information processing where the MW field is involved.

References

[1] Cao W, Bu H, Vinet M, Cao M, Takagi S, Hwang S, et al. The future transistors. Nature. 2023;620(7974):501-15.

[2] Torrezan AC, Alegre TPM, Medeiros-Ribeiro G. Microstrip resonators for electron paramagnetic resonance experiments. Review of Scientific Instruments. 2009;80(7).

[3] Cao ZX, Lu W, Fu L, Gui YS, Hu CM. Spintronic microwave imaging. Applied Physics a-Materials Science & Processing. 2013;111(2):329-37.

[4] Bai LH, Gui YS, Wirthmann A, Recksiedler E, Mecking N, Hu CM, et al. The rf magnetic-field vector detector based on the spin rectification effect. Applied Physics Letters. 2008;92(3).

[5] Cao ZX, Harder M, Fu L, Zhang B, Lu W, Bridges GE, et al. Nondestructive two-dimensional phase imaging of embedded defects via on-chip spintronic sensor. Applied Physics Letters. 2012;100(25).

[6] Yang B, Dong Y, Hu Z-Z, Liu G-Q, Wang Y-J, Du G-X. Noninvasive Imaging Method of Microwave Near Field Based on Solid-State Quantum Sensing. Ieee Transactions on Microwave Theory and

Techniques. 2018;66(5):2276-83.

[7] Boehi P, Treutlein P. Simple microwave field imaging technique using hot atomic vapor cells. Applied Physics Letters. 2012;101(18).

[8] Mecking N, Gui YS, Hu CM. Microwave photovoltage and photoresistance effects in ferromagnetic microstrips. Physical Review B. 2007;76(22).

[9] Fu L, Lu W, Herrera DR, Tapia DF, Gui YS, Pistorius S, Hu CM. Microwave radar imaging using a solid state spintronic microwave sensor. Applied Physics Letters. 2014;105(12).

[10] Vlaminck V, Schultheiss H, Pearson JE, Fradin FY, Bader SD, Hoffmann A. Mapping microwave field distributions via the spin Hall effect. Applied Physics Letters. 2012;101(25).

[11] Zhang W, Peng B, Han F, Wang Q, Soh WT, Ong CK, Zhang W. Separating inverse spin Hall voltage and spin rectification voltage by inverting spin injection direction. Applied Physics Letters. 2016;108(10).

[12] Bai L, Hyde P, Gui YS, Hu CM, Vlaminck V, Pearson JE, et al. Universal Method for Separating Spin Pumping from Spin Rectification Voltage of Ferromagnetic Resonance. Physical Review Letters. 2013;111(21).

[13] He K, Cheng J, Yang M, Zhang Y, Yu L, Liu Q, et al. Spin rectification effect induced by planar Hall effect and its strong impact on spin-pumping measurements. Physical Review B. 2022;105(10).

[14] Tsukahara A, Ando Y, Kitamura Y, Emoto H, Shikoh E, Delmo MP, et al. Self-induced inverse spin Hall effect in permalloy at room temperature. Physical Review B. 2014;89(23).

[15] Luo P, Wu Z, Huang F, Peng B, Zhang W. Scanning inverse spin Hall effect spectrometer by shorted coaxial probes. Review of Scientific Instruments. 2023;94(3).

[16] Soh WT, Peng B, Ong CK. Localized excitation of magnetostatic surface spin waves in yttrium iron garnet by shorted coaxial probe detected via spin pumping and rectification effect. Journal of Applied Physics. 2015;117(15).

[17] Hahn C, de Loubens G, Klein O, Viret M, Naletov VV, Ben Youssef J. Comparative measurements of inverse spin Hall effects and magnetoresistance in YIG/Pt and YIG/Ta. Physical Review B. 2013;87(17).

[18] Nakayama H, Ando K, Harii K, Yoshino T, Takahashi R, Kajiwara Y, et al. Geometry dependence on inverse spin Hall effect induced by spin pumping in Ni81Fe19/Pt films. Physical Review B. 2012;85(14).

[19] Ando K, Takahashi S, Ieda J, Kajiwara Y, Nakayama H, Yoshino T, et al. Inverse spin-Hall effect induced by spin pumping in metallic system. Journal of Applied Physics. 2011;109(10).

[20] Soh WT, Peng B, Ong CK. An angular analysis to separate spin pumping-induced inverse spin Hall effect from spin rectification in a Py/Pt bilayer. Journal of Physics D-Applied Physics. 2014;47(28).

[21] David J. Griffiths CI. Introduction to electrodynamics. American Journal of Physics. 2005;73(6): 393-5.

[22] Zhang YQ, Sun NY, Shan R, Zhang JW, Zhou SM, Shi Z, Guo GY. Anomalous Hall effect in epitaxial permalloy thin films. Journal of Applied Physics. 2013;114(16).

[23] Huang F, Zhang W, Peng B, Zhang W. Dependence of the inverse spin Hall and spin rectification voltage on the aspect ratio of thin films. Journal of Magnetism and Magnetic Materials. 2019;492.

[24] Feng Z, Hu J, Sun L, You B, Wu D, Du J, et al. Spin Hall angle quantification from spin pumping and microwave photoresistance. Physical Review B. 2012;85(21).

Characterization of a 1T-Floating Body DRAM Cell in Bulk Silicon MOSFETs for Cryogenic Memory Applications

Hengxu Guo[1†], Yuanke Zhang[1†*], Yuefeng Chen[1], Haoyu Sheng[1], Chi Fang[1], Guoping Guo[1,2,3], and Chao Luo[1*]

[1] Department of Physics, University of Science and Technology of China, Hefei 230026, China.

[2]Hefei National Laboratory, University of Science and Technology of China, China

[3]Suzhou Institute for Advanced Research, University of Science and Technology of China, China

(*Email: zyk315@mail.ustc.edu.cn, lc0121@ustc.edu.cn) [†]Co-first author

Abstract

In this work, we characterized the SMIC 180 nm bulk silicon MOSFETs down to 6 K and experimentally validated their application in cryogenic capacitor-less floating body dynamic random-access memory (DRAM). At 6 K with a drain voltage of 1.8 V, the device demonstrates a hysteresis loop with a width exceeding 0.2 V, a high-to-low drain current ratio of 10^9. Additionally, it exhibits a sense margin that exceeds 650 µA, a long retention time (>10^3 s), and a high-speed memory operation (50 ns). These characteristics make it a promising candidate for a compact, capacitor-less, single-transistor memory.

Keywords: 1T-DRAM, cryogenic memory, bulk silicon MOSFETs.

Introduction

Cryogenic CMOS (cryo-CMOS) technology has received extensive attention in recent years due to its applications in deep aerospace, and quantum computing systems [1]. In quantum computing systems, it is essential to design hierarchical memory levels located at different temperature regions within the dilution refrigerator. Quantum algorithms require substantial memory capacity to store quantum instructions. The Quantum Error Correction (QEC) controller necessitates a substantial memory capacity for storing pulse envelopes and instruction lists, further demanding enhanced storage capabilities in cryogenic memory systems [2]. To meet these requirements, various memory solutions have been proposed, including Josephson junction-based memory [3], resistive random-access memory (ReRAM) [4], and one-transistor-one-capacitor (1T1C) DRAM configurations [5]. These solutions aim to improve storage density and reduce latency. Here, we propose a cryogenic capacitor-less 1T-DRAM cell based on commercial bulk silicon MOSFETs. This memory cell demonstrates a 650 µA sense margin, a retention time of more than 10^3 s, and a 50 ns programming speed at 6 K, showing its potential application in quantum computing systems and high-performance cryo-computing systems.

Cryogenic DC Characteristics

The devices under test (DUTs) are n-type bulk silicon MOSFETs with W / L = 10 um / 180 nm. The $I_D - V_G$ curves with V_D = 50 mV of DUTs at various temperatures are shown in Fig. 1(a). As the temperature decreases, the

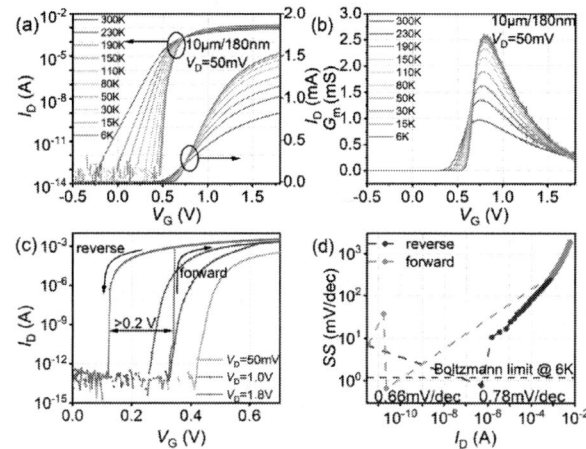

Fig. 1: (a) Transfer characteristics of the device from 300 K to 6 K, (b) transconductance of the device from 300 K to 6 K, (c) forward scan and reverse scan of I_D-V_G curves at various V_D at 6 K, (d) the subthreshold swing (SS) extracted from the hysteresis curve at V_D = 1.8 V.

following observations can be obtained: (i) the threshold voltage (V_{TH}) increases due to Fermi-Dirac scaling and bandgap widening [6]; (ii) the I_{ON}/I_{OFF} ratio improves as a result of enhanced carrier mobility and reduced gate-induced drain leakage current [7]; (iii) the subthreshold swing (SS) becomes steeper, influenced by the temperature dependence of the Boltzmann limit [8]. Fig. 1(b) shows an enhanced transconductance, which is due to reduced phonon scattering at cryogenic temperatures. Fig. 1(c) illustrates the I_D-V_G characteristics, obtained from both forward and reverse scans, with the substrate of the DUTs kept floating. At V_D = 1.8 V, a distinct hysteresis window (>0.2 V) is observed. This is attributed to the accumulation of holes in the substrate caused by carrier impact ionization, which results in a negative shift in threshold voltage. Consequently, when V_G is swept backward from a high value, the transistor cannot be turned off until it reaches the lower substrate-influenced V_{TH}. At V_D = 50 mV, the impact ionization is minimal, and thus the forward and reverse scans of the I_D-V_G curves overlap. Notably, at V_D = 1.8 V, there is a significant rise in I_D, accompanied by an ultra-steep subthreshold swing (SS). This SS reaches a minimum value of approximately 0.66 mV/dec, which is well below the Boltzmann limit of about 1.2 mV/dec at 6 K (Fig. 1(d)). At high V_D conditions, once V_G exceeds V_{TH}, a conductive channel is established, initiating carrier impact ionization. The resultant holes accumulate in the

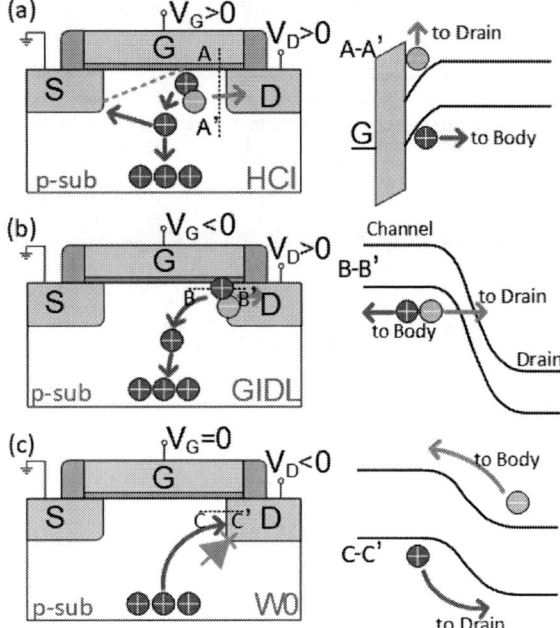

Fig. 2: Schematic diagrams of the W1 and W0 operations in MOSFETs under low-temperature conditions. (a) HCI-based W1 operation, (b) GIDL-based W1 operation, (c) W0 operation.

freeze-out substrate, increasing the floating body potential and lowering V_{TH}, which in turn boosts I_D. This increase in I_D further amplifies impact ionization, establishing a positive feedback loop that precipitates a sharp increase in I_D, and I_D reaches its final value when the channel is in strong inversion.

Memory Operation

The hysteresis phenomenon observed in bulk silicon MOSFETs at cryogenic temperatures indicates a memory effect. Controlling the presence or absence of holes in the substrate can regulate the magnitude of the device's readout current. Holes can be injected into the body through Hot Carrier Injection (HCI) and gate-induced drain leakage (GIDL) for Write "1" operations (W1). As shown in Fig. 2(a), when a relatively high V_D is applied while the device is in inversion mode ($V_D > V_G > V_{TH}$), a strong lateral electric field along the channel is established in the pinch-off region. Carriers in this region acquire high velocity and energy, and subsequently generate secondary electron-hole pairs through impact ionization. The electrons produced by impact ionization follow the channel electrons into the drain, and the holes move towards the substrate and source. The holes generated by GIDL arise from the gate-to-drain overlap region (Fig. 2(b)). A negative V_G creates an accumulation of holes at the channel surface, and an equivalent p$^+$-n$^+$ junction is formed near the overlap area. A positive V_D results in an over-bent energy band at the depletion area in this p$^+$-n$^+$ junction and a large enough local electric field. In this case, a considerable number of electrons have a certain probability to tunnel from valence band to conduction band,

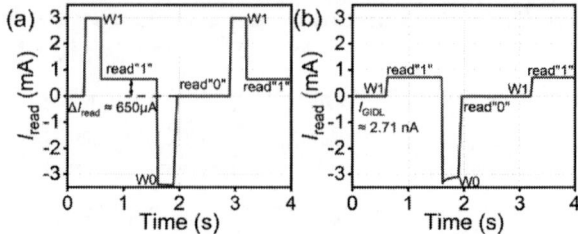

Fig. 3: The I_{read} - t characteristics at 6 K, (a) W1 operation using HCI method (W1: V_D = 1.8 V, V_G = 0.7 V, W0: V_D = -1 V, V_G = 0 V), (b) W1 operation using GIDL method (W1: V_D = 2.5 V, V_G = -0.5 V, W0: V_D = -1 V, V_G = 0 V).

Table 1: 50 ns Memory Operation Pulse Scheme.

Operation	V_D	V_G	V_S	T_{pulse}
W1(HCI)	3.0 V	1 V	0 V	50 ns
W1(GIDL)	3.5 V	-3.5 V	0 V	50 ns
W0	-3 V	0 V	0 V	50 ns
READ	1.8 V	0.3 V	0 V	5 µs

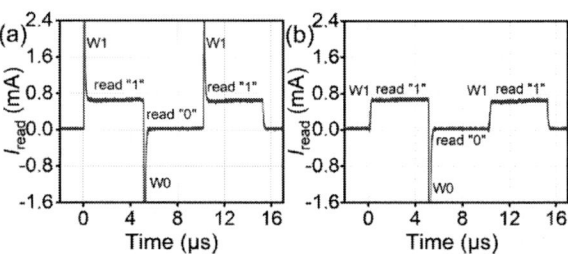

Fig. 4: Transient modulation in the device read current after W1 and W0 operations. (a) 50 ns HCI-based W1 operation, (b) GIDL-based W1 operation.

and thus the holes can be collected by the substrate. The Write "0" operation (W0) is achieved by applying a forward bias to the substrate-to-drain p-n junction (V_G = 0, V_D < 0), causing holes to migrate from the substrate to the drain (Fig. 2(c)).

Fig. 3 shows memory operations on MOSFETs using HCI and GIDL methods. The "1" state, with holes in the substrate, or the "0" state, without holes, leads to a significant current difference ($\Delta I_{read} \approx$ 650 µA) after W1 / W0 operations. It is evident that both HCI and GIDL effects can perform the W1 operation on MOSFETs at 6 K. During W1 operations, an $I_{GIDL} \approx$ 2.71 nA is observed, significantly lower than the conventional HCI method (~ 3 mA). Therefore, the GIDL method can significantly decrease the power consumption for memory operations. Fig. 4 shows the transient modulation in the device read current after W1 and W0 at 6 K, with corresponding voltages listed in Table 1. The nanosecond-level memory operations of the DUTs suggest its suitability for use in main memory and high-speed cache in low-temperature, high-performance computing systems.

Fig. 5 shows retention characteristics at various read voltages (V_D = 1.8 V, V_G variation) after W1 operation and W0 operation. According to the Shockley-Read-Hall (SRH) theory, the recombination rate sharply decreases when the

979-8-3315-0417-5/25 $31.00 © 2025 IEEE

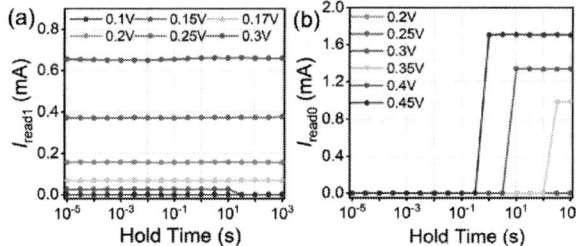
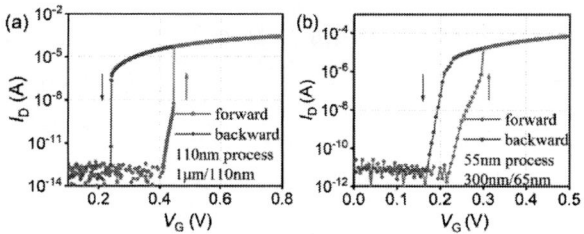

Fig. 5: The DUTs retention characteristics at 6 K, (a) the "1" state retention time after a W1 operation at various read voltages, (b) the "0" state retention time after a W0 operation at different read voltages.

Fig. 6: Low-temperature hysteresis behavior in (a) 110 nm, and (b) 55 nm commercial MOSFETs at V_D = 1.8 V at 6 K.

temperature drops owing to their exponential relationship and thermally generated holes can be negligible at 6 K. The retention time of the "1" state and "0" state will be significantly enhanced at low temperatures [9]. For the "1" state, the failure is due to hole recombination in the substrate, which results in a positive shift of the threshold voltage. Conversely, for the "0" state, holes generation via thermal excitation leads to a negative shift of the threshold voltage. As time approaches infinity, the threshold voltages of both states are anticipated to converge towards an equilibrium value, with the longest retention time being achieved when the read gate voltage approaches this equilibrium point. As shown in Fig. 5(a), at V_G = 0.17 V, the retention time improves by over two orders of magnitude compared to V_G = 0.15 V. For V_G > 0.17 V, the "1" state retention time notably exceeds 10^3 s. Similarly, the "0" state retention time at V_G = 0.35 V is 1.5 orders of magnitude better than at V_G = 0.4 V. When V_G is below 0.3 V, the "0" state retention time also far surpasses 10^3 s. Besides, an avalanche-like increase in I_{read0} can be observed in Fig. 5(b). The failure of the "0" state traditionally leads to a gradual decrease in V_{TH}. When the read voltage is applied to the DUTs, the channel is formed, and then the carrier impact ionization occurs due to the high read voltage V_D. The holes generated by this impact ionization flow towards and accumulate in the freeze-out substrate. This phenomenon is similar to the previously mentioned ultra-low SS effect and results in an increase in I_{read}. Fig. 6 shows the hysteresis phenomenon on two other commercial processes: the 110 nm technology (Fig. 6(a)), and the 55 nm technology (Fig. 6(b)). Each of these samples displayed hysteresis with V_D = 1.8 V at 6 K. These results underscore the adaptability of single-transistor memory based on bulk silicon MOSFETs in low-temperature environments. Furthermore, the simplicity and low cost of bulk silicon CMOS technology, coupled with its high compatibility with silicon-based quantum computing, make it a strong candidate for quantum computing systems.

Conclusion

This study investigates the characteristics of 1T memory devices based on SMIC 180 nm bulk silicon MOSFETs at low temperatures. Due to the floating body and the carrier

freeze-out effect, sharp I_D jumps, hysteretic loops, and a subthreshold swing as low as 0.66 mV/dec are observed at large V_D. By employing the HCI and GIDL method for the W1 operation, we have successfully achieved a 50 ns operation in bulk silicon 1T-DRAM cells, with a large sensing window of 650 μA and a retention time exceeding 10^3 s. These characteristics render bulk silicon MOSFETs promising candidates for capacitor-less, high-density memory in quantum computing systems and high-performance cryo-computing systems.

Acknowledgments

This work was supported by the NSFC (Grant No. 12034018), and Innovation Program for Quantum Science and Technology (Grant No. 2021ZD0302300).

References

[1] Y. Liu et al., "Cryogenic Characteristics of Multinanoscales Field-Effect Transistors," IEEE Trans. Electron Devices, 68, pp. 456–463, (2021).

[2] J.-H. Bae et al., "Characterization of a capacitorless DRAM cell for cryogenic memory applications," IEEE Electron Device Lett., 40, pp. 1614–1617, (2019).

[3] B. Baek et al., "Hybrid superconducting-magnetic memory device using competing order parameters," Nature Commun., 5, p. 3888, (2014).

[4] R. Fang et al., "Low-temperature characteristics of HfOx-based resistive random-access memory," IEEE Electron Device Lett., 36, pp. 567–569, (2015).

[5] B. Patra et al., "Cryo-CMOS circuits and systems for quantum computing applications," IEEE J. Solid-State Circuits, 53, pp. 309–321, (2018).

[6] A. Beckers et al., "Cryogenic MOSFET threshold voltage model," in Proc. 49th Eur. Solid-State Device Res. Conf. (ESSDERC), pp. 94–97, (2019).

[7] Y. Zhang et al., "Characterization and modeling of native MOSFETs down to 4.2 K," IEEE Trans. Electron Devices, 68, pp. 4267–4273, (2021).

[8] T.-Y. Yang et al., "Quantum transport in 40-nm MOSFETs at deepcryogenic temperatures," IEEE Electron Device Lett., 41, pp. 981–984, (2020).

[9] Y. Zhang et al., "Cryogenic Hysteresis in 110 nm Bulk Silicon MOSFETs for Capacitorless Memory Applications," IEEE Electron Device Lett., 44, pp. 1543–1546, (2023)

Double-Gate Cu-MIC Poly-Ge$_{1-x}$Sn$_x$ TFTs on Glass Substrates via the Gate-Last Process

Daiki Goshima, Akito Kurihara, Akito Hara

Tohoku Gakuin University, Japan   e-mail: akito@mail.tohoku-gakuin.ac.jp

Abstract

Double-gate (DG) polycrystalline germanium tin (poly-Ge$_{1-x}$Sn$_x$) thin-film transistors (TFTs) with different channel thickness (17, 18, and 19 nm) were fabricated on glass substrates using the gate-last process and metal induced crystallization using copper (Cu-MIC) at a maximum process temperature of 500 °C. A Cu-MIC DG poly-Ge$_{1-x}$Sn$_x$ TFT with a thickness of 19 nm demonstrated a nominal high mobility of 43 cm^2/Vs, which is two times of that of Cu-MIC DG poly-Ge TFTs.

Keywords: poly-Ge$_{1-x}$Sn$_x$, Cu-MIC, TFT and double gate

Introduction

In recent years, the miniaturization of field-effect transistors (FETs) has approached a physical limit. Thus, vertical stacking technology has attracted attention. Si is generally used as a semiconductor material. However, when vertically integrating a FET, the high process temperature of Si may have an adverse effect on FET of bottom-layer. Therefore, poly-Ge based TFTs, which can be fabricated at a low temperature and have higher mobility than that of Si, are suitable for use in upper-layer FETs [1]. In addition, germanium-tin (Ge$_{1-x}$Sn$_x$) has been reported to have higher mobility than that of Ge and can be crystallized at lower temperatures [2, 3]. The upper Ge layer should be thin to reduce the stress onto the bottom layer. However, it is difficult to crystallize thin amorphous Ge and Ge$_{1-x}$Sn$_x$ with a decreasing film thickness at low temperatures. In addition, the generation of high hole concentrations in thin poly-Ge and Ge$_{1-x}$Sn$_x$ thin films on the order of 10^{19} cm^{-3} have been reported.

We employed metal-induced crystallization using Cu (Cu-MIC) to enable the crystallization of thin films at low temperatures and developed a junctionless p-ch poly-Ge TFT [4-7]. A double-gate (DG) structure was employed to reduce the off-current and increase the on-current [4-7]. Our previous research used the SD-last process, in which SD formation was performed using lateral atomic exchange between electrode Al and Ge as a final TFT fabrication step [4-7]. However, SD formation varies greatly due to uncontrollable parameters.

In this study, we used the gate-last process to reduce the SD parasitic resistance. By combining this process with Cu-MIC Ge$_{1-x}$Sn$_x$ to achieve high mobility and a DG structure to achieve a high I$_{on}$/I$_{off}$ ratio, we developed a Cu-MIC DG poly-Ge$_{1-x}$Sn$_x$ TFT on a glass substrate via gate-last process.

Experimental methods

Figure 1 illustrates the process flow. Quartz glass was used as a substrate, and molybdenum (Mo) was deposited on a glass substrate as a bottom gate (BG) metal. Next, SiO$_2$ with a thickness of 40 nm was deposited as a BG insulating film via PECVD, and this was followed by the deposition of a-Ge$_{1-x}$Sn$_x$/Cu/a-Ge$_{1-x}$Sn$_x$ layers using a Ge$_{1-x}$Sn$_x$ (Sn = 7%) target without breaking the vacuum, resulting in Ge$_{1-x}$Sn$_x$ film thicknesses of 17, 18, and 19 nm. The film thickness was controlled by varying the deposition time from the premeasured deposition rate. The Cu deposition time was fixed at 3 s. Previous experiments have shown the Sn concentration in poly-Ge$_{1-x}$Sn$_x$ to be 3% [7]. The next step was the deposition of 20 nm thick SiO$_2$ and, then, Cu-MIC was performed by annealing at 500 °C for 5 h in N$_2$. This was the maximum temperature achieved during the TFT fabrication process. Next, taking advantage of the fact that the substrate was transparent glass, dummy gates made up of a positive resist were formed using the BG metal as a mask via exposure from the backside. Subsequently, the SiO$_2$ in the source drain (SD) area was etched using RIE. Aluminum (Al) is then sputtered, and both the Al and resist were removed using a lift-off process. Next, a heat treatment at 300 °C was performed for 10 min to reduce the high resistivity regions in the SD through the solid-state reaction of Al and Ge. Then, the excess Al existing outside of SD region was removed, and an additional 20 nm of SiO$_2$ was deposited via PECVD as the TG insulating film. Therefore, the gate SiO$_2$ in the channel region becomes 40 nm thick for both TG and BG. The bottom-gate contact holes were then opened by RIE, and the Mo of the top-gate metal (TG) was deposited. The BG and TG metals were connected at this point. After the formation of the TG electrode, an interlayer insulator SiO$_2$ with a thickness of 100 nm was deposited via PECVD, and this was followed by RIE to open the contact holes. Finally, Al was deposited to form an electrode. The gate length and width were 15 and 10 μm, respectively.

Results and discussion

Figure 2 shows the TEM and TEM-EDX of points A and B of Cu-MIC DG poly-Ge TFT (Sn = 0%) formed using the same process as that in Fig. 1. Points A and B correspond to the channel and SD areas, respectively. In the channel region, Cu is uniformly distributed along the thickness direction. However, at point B (the SD region), Al and Ge were directly contacted. Thus, solid state reaction occurred between Al and Ge during heating at 300 °C for 10 min. However, for a short duration of 10 min at 300 °C, the solid state reaction between Al and Cu does not occur. Thus, Cu catalysis in the poly-Ge is aggregated at the interface between Ge and Al. The formation of Ohmic contacts between metals and p-type Ge has been reported. Thus, a low-resistance SD region is expected.

Figures 3(a), (b), and (c) show the transfer characteristics of poly-$Ge_{1-x}Sn_x$ TFTs with film thicknesses of 17, 18, and 19 nm, respectively. Figure 4 shows the output characteristics of each TFT. Figures 5(a), (b), and (c) show the I_{on}/I_{off} ratio, nominal mobility, and V_{th} for each TFT. The nominal mobility is calculated from the maximum g_m value. Thus, the actual mobility is approximately half of the nominal mobility, considering the existence of the upper and lower channels owing to the DG mode. The threshold voltage was obtained using the linear extrapolation method.

The I_{on}/I_{off} ratio tended to decrease with increasing film thickness. In the junctionless structure, the off-state is realized by blocking the channel owing to the expansion of the depletion layer. However, in polycrystalline Ge-based thin films, the concentration of holes tends to approximately 10^{19} cm^{-3}. Therefore, when the film thickness is thick, the depletion layer does not fill the entire channel region, which leads to a high leakage current. This resulted in a low I_{on}/I_{off} ratio with an increase in the poly-$Ge_{1-x}Sn_x$ thickness. This is consistent with the phenomenon in which V_{th} shifts in the positive direction with increasing film thickness.

The nominal mobility increased with an increasing poly-$Ge_{1-x}Sn_x$ film thickness. As the film thickness increased, the current component flowing inside the semiconductor increased. However, with a decreasing poly-$Ge_{1-x}Sn_x$ film thickness, the interface scattering ratio increased, resulting in a reduction in mobility.

Conclusion

The Cu-MIC DG poly-$Ge_{1-x}Sn_x$ TFT was developed using a gate-last process to reduce the SD resistance, Cu-MIC for the crystallization of the thin $Ge_{1-x}Sn_x$ film at low temperatures, and a DG structure to achieve a high I_{on}/I_{off} ratio. Consequently, a nominal mobility of 43 cm^2/Vs and an I_{on}/I_{off} ratio of 430 were achieved for the 19 nm-thick poly-$Ge_{1-x}Sn_x$ thin film.

Acknowledgement

This work was supported by Grant-in-Aid for Scientific Research (c) 22K04247.

References

[1] Y. Kamata, M. Koike, E. Kurosawa, M. Kurosawa, H. Ota, O. Nakatsuka, S. Zaima and T. Tezuka, "Operation of inverter and ring oscillator of ultrathin-body poly-Ge CMOS", Appl. Phys. Express 7, 121302 (2014).

[2] K. Moto, K. Yamamoto, T. Imajo, T. Suemasu, H. Nakashima and K. Toko, "Sn Concentration Effects on Polycrystalline GeSn Thin Film Transistors", IEEE Electron Device Lett. 42, 1735 (2021).

[3] T. Nagano, R. Hara, K. Moto, K, Yamamoto and T. Sadoh, "Improved carrier mobility of Sn-doped Ge thin films (≤20 nm) on insulator by interface-modulated solid-phase crystallization combined with surface passivation", Mater. Sci. Semicond. Process. 165, 107692 (2023).

[4] S. Suzuki and A. Hara, "Evaluation of extremely thin polycrystalline germanium films and their TFT performance fabricated at 400 °C by Cu-induced crystallization on a glass substrate", Jpn. J. Appl. Phys. 63, 051001 (2024).

[5] A. Hara, Y. Nishimura and H. Ohsawa, "Self-aligned metal double-gate junctionless p-channel low-temperature polycrystalline-germanium thin-film transistor with thin germanium film on glass substrate", Jpn. J. Appl. Phys. 56, 03BB01 (2017).

[6] H. Utsumi, N. Nishiguchi, R. Miyazaki, H. Suzuki, K. Kitahara and A. Hara, "Double-gate polycrystalline-germanium thin-film transistors using copper-induced crystallization on flexible plastic substrate", Jpn. J. Appl. Phys. 58, 046501 (2019).

[7] R. Miyazaki and A. Hara, "Four-terminal polycrystalline $Ge_{1-x}Sn_x$ thin-film transistors using copper-induced crystallization on glass substrates and their application to enhancement/depletion inverters", Jpn. J. Appl. Phys. 59, 051008 (2020).

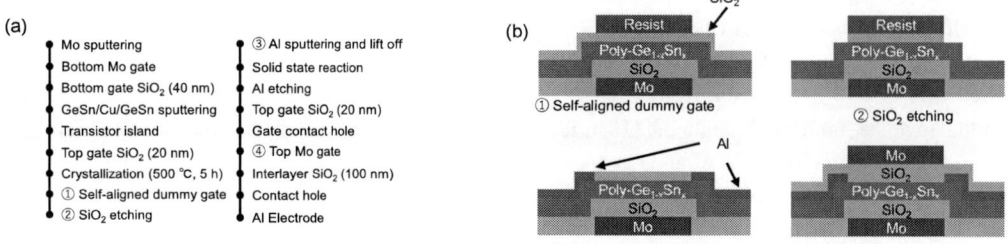

Fig.1.(a)Process flow of the gate-last process (b) Cross-sectional schematic diagram of the gate-last process

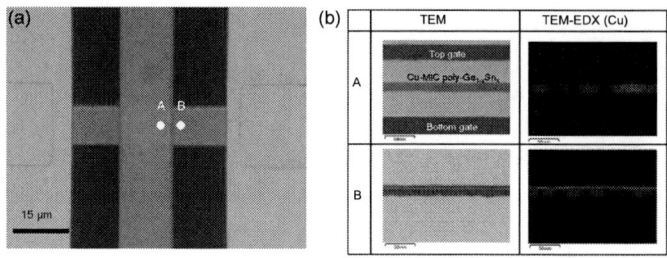

Fig. 2. (a) A and B correspond to TEM analysis region (b) TEM and TEM-EDX (Cu) of region A and B

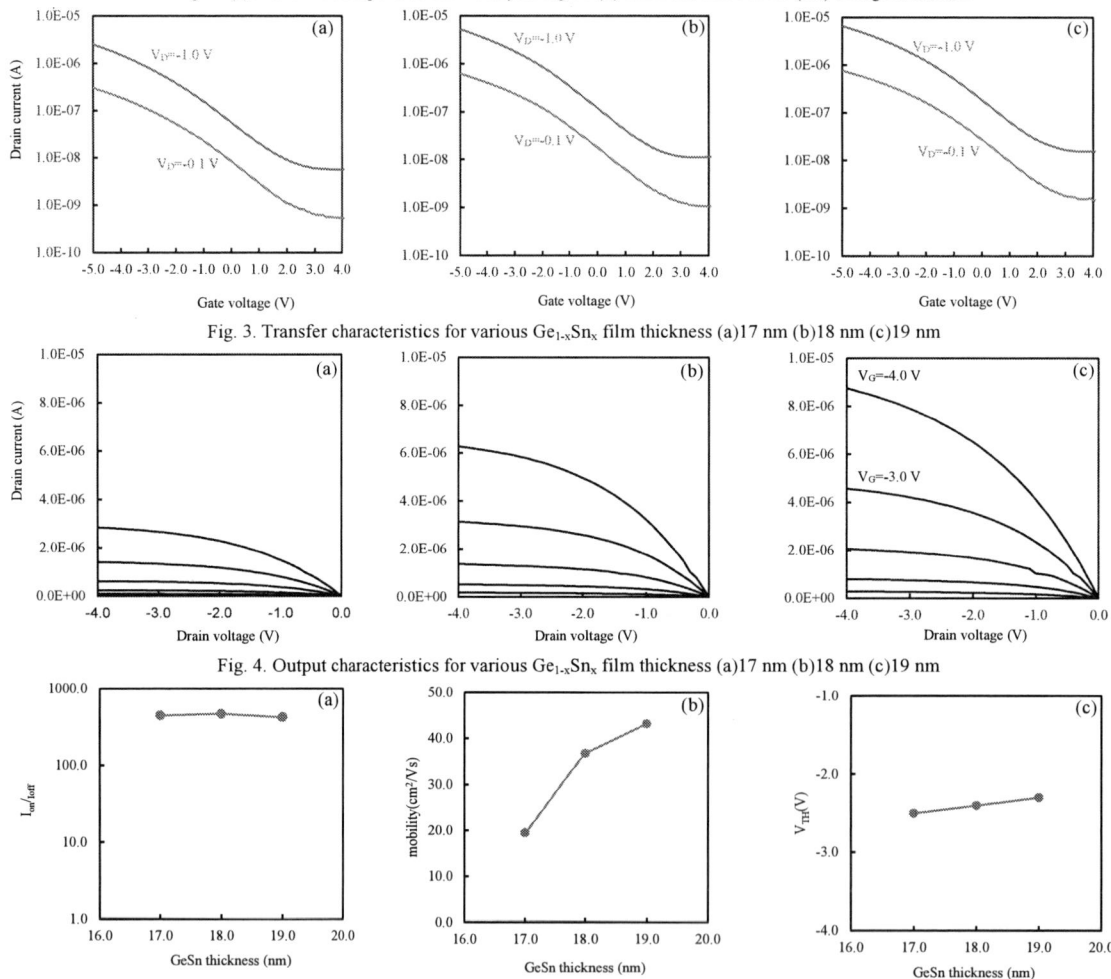

Fig. 3. Transfer characteristics for various $Ge_{1-x}Sn_x$ film thickness (a)17 nm (b)18 nm (c)19 nm

Fig. 4. Output characteristics for various $Ge_{1-x}Sn_x$ film thickness (a)17 nm (b)18 nm (c)19 nm

Fig. 5. (a) I_{on}/I_{off} (b) mobility and (c) V_{TH} for various $Ge_{1-x}Sn_x$ film thickness

From Flip FET to Flip 3D Integration (F3D): Maximizing the Scaling Potential of Wafer Both Sides Beyond Conventional 3D Integration

Heng Wu†, Haoran Lu[+], Wanyue Peng[+], Ziqiao Xu, Yanbang Chu, Jiacheng Sun, Falong Zhou, Jack Wu, Lijie Zhang, Weihai Bu, Jin Kang, Ming Li, Yibo Lin, Runsheng Wang, Xin Zhang, Ru Huang

School of Integrated Circuits, Peking University, Beijing 100871, China, †email: hengwu@pku.edu.cn

[+]These authors contribute equally

Abstract

In this work, we proposed a new 3D integration technology: the Flip 3D integration (F3D), consisting of the 3D transistor stacking, the 3D dual-sided interconnects, the 3D die-to-die stacking and the dual-sided Monolithic 3D (M3D). Based on a 32-bit FFET RISCV core, besides the scaling benefits of the Flip FET (FFET), the dual-sided signal routing shows even more routing flexibility with 6.8% area reduction and 5.9% EDP improvement. Novel concepts of Multi-Flipping processes (Double Flips and Triple Flips) were proposed to relax the thermal budget constraints in the F3D and thus support the dual-sided M3D in the F3D. The core's EDP and frequency are improved by up to 3.2% and 2.3% respectively, after BEOL optimizations based on the Triple Flips compared with unoptimized ones.

Keywords: Flip FET, Flip 3D, wafer bonding, dual-sided interconnects, 3D packaging, Monolithic 3D

Introduction

As the conventional Moore's Law coming to an end, the semiconductor industry is thriving for new methods to continuously enhance the chip integration density, mainly following two technical routes. One is pushing the scaling of advanced logic technology by DTCO, trending towards 3D transistor stacking such as the Complimentary FET (CFET) [1-3] and backside (BS) interconnects [4-9]. We have proposed the FFET previously [10] as a combination of 3D integration of 3D stacked channel FETs and dual-sided interconnects, acting as a great candidate in the 2D & 3D transistor integration roadmap as illustrated in Fig. 1. The other one is the 3D IC including the 3D die stacking enabled by the 3D packaging [4,11] or the 3D tier stacking such as the Monolithic 3D integration (M3D) [12-14].

In this work, we proposed a new 3D integration technology called the **_Flip 3D_** integration (F3D) named after the repeated usage of wafer flipping process as an extension of the FFET. The F3D is the **first** to unite the 3D transistor stacking, 3D dual-sided interconnects, 3D face-to-face/back-to-back/face-to-back die stacking and the dual-sided M3D. For below, we will first review the roadmap of dual-sided interconnects and show the block-level power-performance-area (PPA) benefits, taking FFET as an example. Then we will discuss the F3D's unique Multi-Flipping processes, with much relaxed thermal budget constraint. At last, the future blueprint of F3D will be given, validating the great potentials combining the benefits of existing 3D integration technologies for the future.

Dual-sided Interconnects

Dual-sided Interconnects (DSI) originates from the BS power delivery network (BSPDN) [4-8]. BS metals with much larger pitches are suitable not only for the power to reduce IR drop [4-7], but also for delivering timing-sensitive signals (such as the clock) to reduce the delay [7]. In **_DSI 1.0_**, std. cells only have pins on the FS. Therefore, backside signal routing should be implemented by the Signal Transfer Cell (STC) [8,9] or the nTSV [7] which transfer signals to the other side but with area penalties, which is shown in Fig. 2 (a). Considering the ultra-scaled std. cells, routing resources on the frontside (FS) become increasingly insufficient. Thus, the std. cell design needs not only BS power but also BS signals [2,3,8,13]. In **_DSI 1.5_**, std. cells are designed with BS intra-cell routing, so pins are naturally located on both FS and BS. However, in DSI 1.5, when the output pin of the current stage and the input pin of the next stage are on different sides (Fig. 2(b)), STCs or nTSVs are required to connect them. This leads to huge area loss because this kind of FS-BS interconnection occurs frequently in DSI 1.5. To solve this, in **_DSI 2.0_** we proposed the dual-sided output pin [10] existing in **_both FS and BS_** in the FFET which is composed of the common drain of nFETs and pFETs enabled by the Drain Merge structure given in Fig. 2(c). Thus, the FS signal routing and BS signal routing can be implemented independently, free of the usage of STCs or nTSVs. This results in significant area benefits w.r.t DSI 1.5. Moreover, STCs can still be added in P&R flow selectively to reduce redundant net length as illustrated in Fig. 2(c). Benchmark of these three DSI are listed in Table 1.

Physical implementation with DSI 2.0 and PPA analysis were carried out on a FFET 32-bit RISCV core (no cache, only computing core) based on the design rules listed in Table 2. Fig. 3 shows the post-P&R core layouts comparing the FS-only signal routing in FFET with FP_1BP_0 (100% FS pins and 0% BS pin: all pins set on the FS) & $FM_{0-8}BM_{0-1}$ (FM1-FM8 for inter-cell routing, M0 & M1 for intra-cell routing, etc.) and the DSI 2.0 in FFET with $FP_{0.5}BP_{0.5}$ (50% FS pins and 50% BS pins: input pins distributed evenly on both sides) & $FM_{0-4}BM_{0-4}$ at respective maximum utilization, the latter giving the area reduction of 6.8% thanks to the better routability by using the DSI 2.0. Fig. 4 shows the BS metal layer number can be reduced to 3 (even with BM1 & BM2 pitches enlarged by ×1.5) or even 2 by redistributing the input pin density, with some EDP degradation when the BS metal pitch increases or the BS metal layer number reduces. Thus, this trade-off between manufacturability and design space in the FFET should be carefully balanced. The successful implementation of the DSI 2.0 paves the way to set the I/O bumps on both sides, enabling the data communication between the face-to-face/back-to-back/face-to-back stacked dies in the F3D on both sides.

Multi-Flipping Processes

In the FFET process flow proposed previously [10], the FS RMG is formed before the BS epitaxy with single use of the wafer flipping process (***Single Flip***) for conceptual demonstration purpose. However, it is not compatible with the gate-last process in the advanced logic nodes. Though FS gate-first devices or low temperature BS epitaxy [15] can relax the thermal budget constraint, the device performance and reliability would degrade inevitably. To solve this, we proposed the ***Double Flips*** process, as illustrated in Fig, 5(a). Different from the Single Flip, the first wafer flipping is done right after the FS S/D epitaxy formation in the Double Flips. After the BS processes, the second wafer flipping is conducted, followed by the FS RMG process. Thus, both FS and BS metal gates are formed after the high temperature epitaxy. However, the BS MOL & BEOL processes in Double Flips still suffers from the high temperature FS RMG formation which is done after them. Metals and dielectrics with better high temperature stability (such as W with higher ρ and SiO_2 with higher κ) are required. This can be solved by adding another wafer flipping process (***Triple Flips***) in which both FS and BS MOL & BEOL are formed after all FEOL process, as given in Fig. 5(b). Detailed benchmarks of these 3 types of processes are listed in Table 3.

To investigate the performance difference between the Double and Triple Flips, 3 types of Double Flips (I-III) and 2 types of Triple Flips (IV-V) were studied at the block level, considering different BEOL metal and dielectric combos, which are shown in Table 4, taking Double Flips I as the baseline. Note that we assumed the same intrinsic device performance in both the Double & Triple Flips. For the Double Flips, the EDP slightly decreases by 0.2% and 0.5% (Fig. 6 (a)) and the frequency slightly increases by 0.1% and 0.7% (Fig. 6(b)) after reducing the BS metal resistance by replacing W with dual damascene Ru and subtractive Ru [16], respectively. In the Triple Flips IV, performance improves distinctly with the EDP decreases by 2.0% and the frequency increases by 1.1%. These metrics moves up to 3.2% and 2.3% respectively in the Triple Flips V for the lower via resistivity in the subtractive Ru interconnects.

101-stage ROs with fan-out 3 and distributed FS and BS BEOL loads were also simulated (Fig. 7(a)). The BEOL net length and via number are extracted from the P&R critical path statistics [17] of the FFET for each side separately, as shown in Fig. 7(b-c) (BS data are not shown). Taking Double Flips I as the baseline, the RO frequency in the Triple Flips V increases by 7.04% and 9.91% @ Vdd = 0.7 V with 1 um and 3 um BEOL loads, respectively.

By using Multi-Flipping processes, the F3D features more manufacturing-friendly runpath with clear performance gains.

Flip 3D Integration

The nowadays advanced integrated circuits mainly focus on two topics. One is the advanced CMOS, trending towards 3D integration of 3D FETs and 3D interconnects, as given in Fig. 8(a). The CFET, though a possible solution to 3D transistor stacking, faces great challenges of high aspect ratio (AR) process. The other one is the 3D IC, enabled by advanced packaging or the M3D, as depicted in Fig. 8(b). However, the advanced packaging is limited by the single-sided hybrid bonding processes to enable further stacking due to the lack of I/O bumps on the other side of dies. While, the M3D technology only explores the FS usage of the wafer.

The F3D technology could act as the ultimate form of these 3D integration technologies on the roadmap (Fig. 8(c)) as following. Firstly, the FFET (demonstrated in Die I) is more manufacturing-friendly for realizing the 3D integration of 3D FETs than the CFET because of the lower AR processes [10] thanks to the independent process on each side of the wafer. Secondly, the F3D supports the dual-sided I/O bumps thanks to the DSI 2.0. Thus, both FS and BS of chip dies support the hybrid bonding, enabling the free choice of face-to-face/back-to-back/face-to-back bonding. It could further remove the TSVs used in the 3D packaging applications such as the HBM [11] (demonstrated in Die II). Thirdly, with the help of the Multi-Flipping processes, the F3D also supports the M3D on both FS and BS of the wafer (demonstrated in Die III) with much relaxed thermal budget constraints.

Overall, the F3D fully utilizes both sides of the wafer, pushing the logic/memory density and their co-integration to the limits, and extends the boundary of the 3D packaging, enabling even broader chip stacking space for future.

Conclusion

With the manufacturing-friendly 3D stacked transistor architecture (such as FFET) supported by Multi-Flipping processes, the 3D interconnects enabled by the DSI 2.0 and the dual-sided 3D die stacking enabled by the dual-sided hybrid bonding, the F3D shows great potential of opening up new era for the next-generation 3D integration.

Acknowledgments

This work was supported by the National Key R&D Program of China under Grant 2023YFB4402200; and in part by the 111 Project under Grant 8201702520.

References

[1] J. Ryckaert, et al., *VLSI 2018*.
[2] J. Park, et al., *VLSI 2024*.
[3] S. Liao, et al., *IEDM 2024*.
[4] R. Chen, et al., *IEDM 2022*.
[5] G. Sisto, et al., *VLSI 2023*.
[6] A. Veloso, et al., *IEDM 2023*.
[7] P. Vanna-iampikul, et al., *VLSI 2024*.
[8] B. Kim, et al., *VLSI 2024*.
[9] T. Lin, et al., *ASP-DAC 2024*.
[10] H. Lu, et al., *VLSI 2024*.
[11] K. Moon, et al., *IEDM 2023*.
[12] M. Liu, et al., *VLSI 2024*.
[13] T. Srimani, et al., *IEDM 2023*.
[14] M. M. Sabry Aly, et al., *IEEE Computer 2015*.
[15] C. Porret, et al., *IEDM 2022*.
[16] C. Penny, et al., *IEDM 2022*.
[17] A. Farokhnejad, et al., *IITC 2022*.
[18] D. Gall, *VLSI-TSA 2020*.
[19] E. Milosevic, et al., *ANTS 2018*.

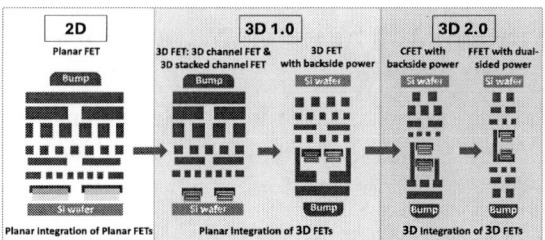

Fig. 1 The roadmap of 2D & 3D transistor integration. The 3D transistor integration began with the 3D FET (**_3D 1.0_**) and is now trending towards BS interconnects to achieve further scaling. The **_3D 2.0_** stands for 3D integration of 3D FETs including 3D stacked FETs and dual-sided interconnects. The FFET [10] acts as a great candidate for 3D transistor stacking technology beyond the CFET [1-3].

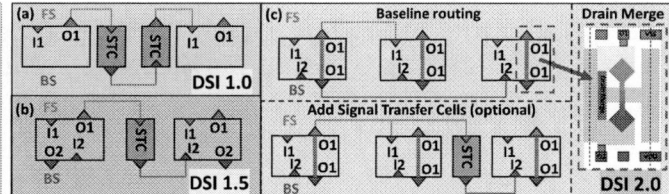

Fig. 2 (a) In DSI 1.0, FS signals can be transferred to the BS and then back to the FS through the STCs. (b) In DSI 1.5, substantial STCs should be inserted to connect the pins on different sides. (c) In DSI 2.0, the dual-sided output pin enables independent dual-sided routing, making the STC an optimizing but not necessary option.

	DSI 1.0 [7,9]	DSI 1.5 [2,3,8,13]	DSI 2.0 [10]
Std. cell pin location	FS	FS and BS	FS and BS
Output pin category	FS	Either FS or BS	Both FS and BS
BS intra-cell routing	No	Yes	Yes
nTSV / Signal Transfer Cell	Necessary	Necessary	Optional
Area waste due to inter-side routing	Moderate	Large	Small

Table 1 Benchmarks of the dual-sided interconnects of the three generations.

Layers	Pitch
FM8-FM5	76 nm
FM4-FM3	42 nm
FM2	30 nm
FM1	34 nm
FM0	28 nm
Poly	51 nm
BM0	28 nm
BM1	34 nm
BM2	30 nm
BM3-BM4	42 nm
BM5-BM6	76 nm

Table 2 FEOL and 2.0 in the FFET $FP_{0.5}BP_{0.5}$ & $FM_{0.4}BM_{0.4}$. BEOL design rules used in this work.

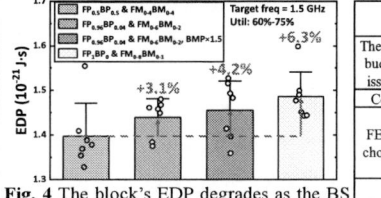

Fig. 3 Post-P&R minimum core layouts comparing the FS-only signal routing in FFET FP_1BP_0 & $FM_{0.8}BM_{0.1}$ and the DSI 2.0 in the FFET $FP_{0.5}BP_{0.5}$ & $FM_{0.4}BM_{0.4}$. The DSI 2.0 gained area reduction of 6.8% over the FS-only signal routing.

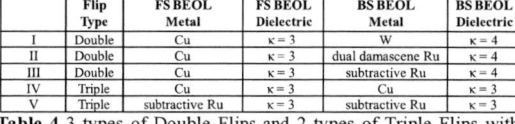

Fig. 4 The block's EDP degrades as the BS metal layer number reduces to 2 by allocating BS input pins to the FS. BS metal pitch could be enlarged with slight EDP degradation.

	Single Flip	Double Flips	Triple Flips
Thermal budget issues	FS Gate, FS BEOL	BS BEOL	None
Cost	Low	Medium	High
FEOL choices	FS gate-first devices / BS epitaxy with low thermal budget [15]	No limitation	No limitation
BEOL choices	FS: W/Ru + Higher κ dielectric BS: No limitation	FS: No limitation BS: W/Ru + Higher κ dielectric	No limitation

Table 3 Benchmarks of Multi-Flipping processes.

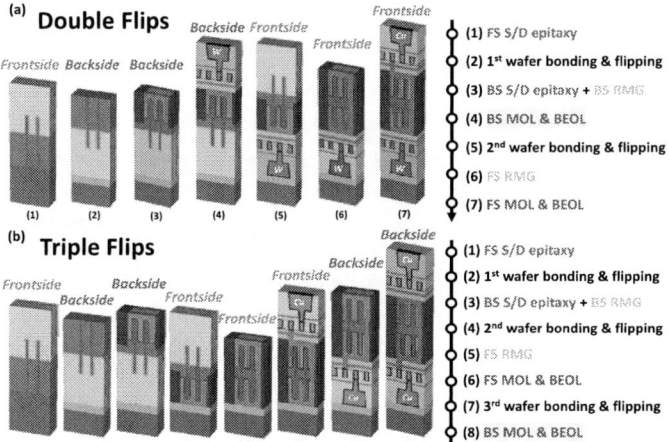

Fig. 5 (a) Process flow of the Double Flips. FS and BS metal gates are both formed after the high temperature epitaxy. (b) Process flow of the Triple Flips. The BS MOL & BEOL are formed at last by flipping the wafer right after the BS RMG in step (3).

	Flip Type	FS BEOL Metal	FS BEOL Dielectric	BS BEOL Metal	BS BEOL Dielectric
I	Double	Cu	κ = 3	W	κ = 4
II	Double	Cu	κ = 3	dual damascene Ru	κ = 4
III	Double	Cu	κ = 3	subtractive Ru	κ = 4
IV	Triple	Cu	κ = 3	Cu	κ = 3
V	Triple	subtractive Ru	κ = 3	subtractive Ru	κ = 3

Table 4 3 types of Double Flips and 2 types of Triple Flips with different process assumptions. Metal resistivity is based on [18,19]. The subtractive Ru has lower via resistance than the dual damascene Ru due to the subtractive top via with reduced liner thickness [16].

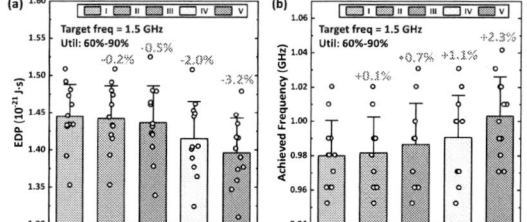

Fig. 6 The block's (a) EDP and (b) achieved frequency under the 5 types of experiments listed in Table 4. All experiments used DSI 2.0 in FFET $FP_{0.5}BP_{0.5}$ $FM_{0.4}BM_{0.4}$. The FSPDN for the Double Flips and the BSPDN for Triple Flips were removed for the fair comparison.

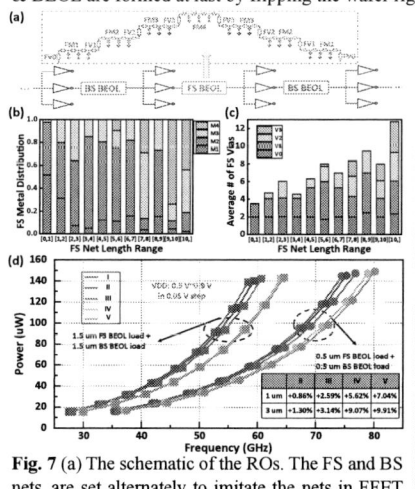

Fig. 7 (a) The schematic of the ROs. The FS and BS nets are set alternately to imitate the nets in FFET $FP_{0.5}BP_{0.5}$ & $FM_{0.4}BM_{0.4}$. (b) FS net length statistics. (c) FS via statistics. (d) Power-freq plot with 2 BEOL loads. Table inset is the frequency gain w.r.t case I.

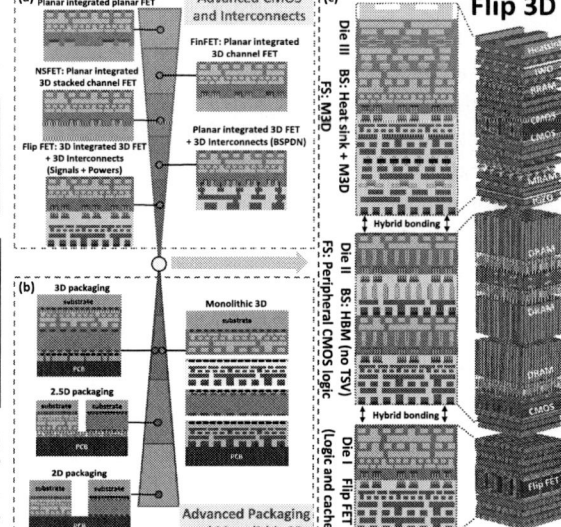

Fig. 8 (a) The roadmap of the advanced CMOS and interconnects trending towards the 3D transistor integration and 3D interconnects integration. (b) The roadmap of the advanced packaging and the Monolithic 3D which are both approaches to the 3D IC. (c) 2D schematic and 3D detailed structure demo of the F3D. The three stacked dies are fabricated separately by Multi-Flipping processes before being face-to-face/back-to-back/face-to-back stacked on each other by the hybrid bonding. The F3D is the combination of the 3D transistor stacking (Die I), the 3D die stacking (Die II) and the dual-sided M3D (Die III) and the DSI on the FS and BS of all the dies.

979-8-3315-0417-5/25 $31.00 © 2025 IEEE

Optimization of Scaling-Down Performance in Sub-100 nm AlGaN/GaN HFETs Based on Electron velocity modulation

Mingyan Wang[1], Yuanjie Lv[2], Heng Zhou[3], Chao Liu[3], Peng Cui[3], Zhaojun Lin[3], and Sen Huang[1]

[1] Institute of Microelectronics, Chinese Academy of Sciences, China, [2] Hebei Semiconductor Research Institute, China, [3] the School of Integrated Circuits, Shandong University, China

Email: linzj@sdu.edu.cn, huangsen@ime.ac.cn

Abstract

In the present study, an investigation as conducted into the effects of electron velocity modulation ($\Delta v_e/\Delta V_{gs}$) and the short-channel effect (SCE) on high-frequency and DC performance for Sub-100 nm AlGaN/GaN HFETs. Experimental and simulation methods were employed to elucidate electron and heat transport in Sub-100 nm AlGaN/GaN HFETs. Devices with gate length (L_g) varying from 500 to 80 nm were fabricated. The devices exhibited maximum transconductance (g_m) when the aspect ratio (L_g/d) fell within the range of 3.84-8.56. Additionally, L_g/d exceeding 6.85 was found to effectively regulate the channel current for high-frequency applications, mitigating SCE while maintaining other device characteristics such as threshold voltage shift (ΔV_{th}). As L_g scaled down from 350 to 80 nm, the larger $\Delta v_e/\Delta V_{gs}$ resulting from increased Polarization Coulomb field (PCF) scattering continued to enhance the g_m.

Keywords: Short channel effect, AlGaN/GaN, Delay time analysis, Monte Carlo, electron velocity, transconductance.

Introduction

AlGaN/GaN heterojunction FETs (HFETs) have demonstrated exceptional performance when deployed in high-power and high-frequency applications [1]. In recent years, a number of studies have explored the downscaling of the device L_g [2,3]. As a result of the growing severity of SCE in the progress, which include ΔV_{th} and increased subthreshold swing, the anticipated enhancement in device performance is considerably constrained. Jessen [4] et al. proposed a minimum L_g/d of 15 to mitigate the SCE of the T-gate AlGaN/GaN HFETs. Guerra [5] proposed a L_g/d of 15 for negligible short-channel effects and 10 for reduced SCE. Nevertheless, prior research on SCE primarily relies on DC and RF data obtained from TCAD simulations [4-5]. Only Jessen has prepared different devices and tested DC and RF performance. However, detailed analysis of SCEs through electron transport has not been a focus of his research. Although thermal transport of AlGaN/GaN HFETs has been explored through MC simulations [6], there is a limited number of studies that investigated thermal transport for sub-100 nm AlGaN/GaN HFETs. Moreover, as L_g scales down, it is uncertain whether the v_e and $\Delta v_e/\Delta V_{gs}$ in sub-100 nm AlGaN/GaN HFETs have an effect on high frequency performance.

In the present study, findings were made that devices with the highest g_m had a L_g/d value in the range of 3.84-8.56. In addition, it was found that as L_g scaled down to 80 nm, PCF

scattering was enhanced, and the $\Delta v_e/\Delta V_{gs}$ contributed to g_m. The utilization of extracted v_e can aid in acquiring more precise T_L profile through the integration of TCAD coupled with MC simulation. This approach offers valuable insights to guide the enhancement of HFET designs.

Experiment And Monte Carlo Simulation

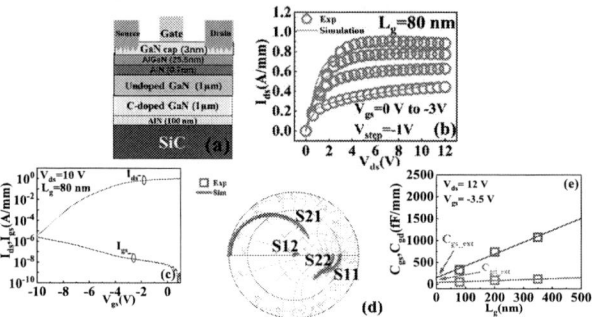

Fig. 1. (a) Schematic cross-section of AlGaN/GaN HFETs. (b)The simulated and tested output curves for the 80 nm L_g HFET on wafer A.(c)The transfer and gate current characteristic of the 80nm L_g HFET on Wafer A.(d)The simulated and measured S-parameters at V_{ds}=12V, V_{gs}=-3.5V for the 80 L_g HFET on wafer A.(e) The gate-source capacitance (C_{gs}) and gate-drain capacitance (C_{gd}) vs. L_g at V_{ds}=12V, V_{gs}=-3.5V.

Fig. 1(a) shows a schematic cross-section of the $Al_{0.21}Ga_{0.79}N$/GaN HFET. Room temperature Hall tests showed an electron density of 8.4×10^{12} cm^{-2} and electron mobility of 2060 cm²/(V·s). The source-gate spacing (L_{gs}) for the HFET was equal to the gate-drain spacing. Devices with the same gate width (W) and L_{gs} were fabricated on the same wafer. Wafer A to C have a W of 40μm and L_{gs} of 1,1.5, 2μm. Wafer D to F have a W of 20μm and L_{gs} of 1,1.5, 2μm. The L_g for Wafers A to F were as follows: 500, 350, 250, 200, 150, 100, and 80 nm, respectively. Fig. 1(b) shows the DC output of the 80 nm Lg HFET on Wafer A, while Fig. 1(c) depicts its transfer and gate current characteristics. The RF characteristics(0.1-50GHz) were obtained by Keysight N5247A vector network analyzer. Open and short structures on the wafer were used to de-embed parasitic resistances and capacitances. To extract the small-signal equivalent circuit parameters, the S-parameters of the device were measured while varying the V_{gs} from -3.5 to 0 V in increments of 0.5 V, at both 0 and 12 V V_{ds}. Subsequently, utilizing the S-parameter data, the small-signal equivalent circuit parameters were extracted at each bias point [7]. As shown in Fig. 1(d), the simulated S-parameters matched well with the measured S-parameters.v_e was extracted at different values of V_{gs} with V_{ds} of 12 V by $v_e=L_g/\tau_t$, where τ_t is the intrinsic delay time. τ_t can be obtained by $\tau_t=(C_{gs\_int}+C_{gd\_int})/g_m$ [7], where C_{gs_int} and C_{gd_int} are obtained by removing the effect of parasitic capacitance from the S-parameters of the different L_g devices. The C_{gs_ext} and C_{gd_ex} can be determined from the L_g-C_{gs} and L_g-C_{gd} relationship, where they intersect the y-axis at x = 0 in Fig. 1(e). Then, the n_s can be obtained by $n_s = I_{ds}/qv_e$. Electron transport and self-heating effect in sub-100 nm AlGaN/GaN HFETs were investigated by a simulation method coupled with a drift-diffusion solver and MC simulation [7]. Firstly, the initial n_s, T_L and electric field (E) distribution were calculated using APSYS software. Then, the v_e-E relationship was obtained by MC simulation. The v_e profile was employed as a new high-field transport model for conducting self-consistent iterative calculations. The device structure utilized in the APSYS is illustrated in Fig. 1(a), with simulation incorporating scatterings, such as optical phonon (POP), PCF [7], inter-valley phonon, acoustic deformation potential, piezoelectric, and interface roughness scattering. In the APSYS simulation, ohmic contacts are modeled as high n-type doping in the contact/GaN interface. When considering the self-heating effect, the thermal conductivity of SiC were set to 490 W/(K.m). Various heating mechanisms are considered, including Joule heating, Peltier effect, and Thomson effect.

Result and discussion

The experimental $g_{m,max}$ as a function of L_g/d is shown in Fig. 2(a), where $g_{m,max}$ is the maximum g_m for a HFET at $V_{ds} = 12$ V, for different V_{gs}. Since the g_m depends on multiple variables, it was necessary to normalize the g_m to compare $g_{m,max}$ versus L_g/d for devices of different dimensions. The peak $g_{m,max}$ between the wafers A to F varied from 209 to 234.5 mS/mm.

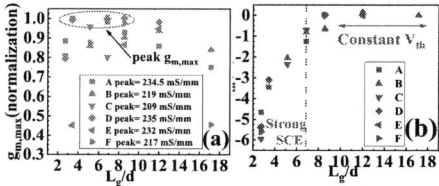

Fig 2. For each wafer (a) $g_{m,max}$ versus L_g/d.(b) ΔV_{th} versus L_g/d.

Fig. 2(a) shows that g_m increased with decreasing L_g when L_g/d was at a relatively high level. However, when L_g/d was decreased within a specific range, the gate's capacity for modulating the n_s was compromised due to the onset of SCE. Consequently, g_m decreased as L_g decreased. The highest $g_{m,max}$ for all devices occurred within the range of L_g/d from 3.84 to 8.56, marking a distinct departure from previous findings reported in the literature, which typically ranged from 5 to 9[4]. Fig. 2(b) illustrates ΔV_{th} as a function of L_g/d. The benchmark for the change in V_{th} in Fig. 2(b) is V_{th} of the 500 nm L_g device. The shorter L_g, the smaller L_g/d, the larger ΔV_{th} of the device, stemming from the weakening of the gate's ability to modulate n_s. When L_g/d =6.85, V_{th} only shifted by -1±0.2V, and the SCE was not significant. When L_g/d=5.13, V_{th} was significantly shifted by -2.15±0.1 V. Therefore, L_g/d=6.85 was identified as a critical node, where SCE was

not strong, V_{th} shifted by a small amount, and the g_m was still maintained at a relatively high value and continued to increase with decreasing L_g. Notably, in Fig. 2, the V_{th} shifted by -3 V at L_g/d=3.84, but the g_m still improved as L_g decreased. There is a scarcity of studies reporting that the g_m of FET continues to increase as L_g decreases in the presence of a sharp negative shift in V_{th}. Since the phenomenon of $\Delta v_e/\Delta V_{gs}$ has been found to have an enhancement on the GaN HFET g_m in previous research [7], it was hypothesized that $\Delta v_e/\Delta V_{gs}$ remains a crucial parameter, particularly as L_g becomes shorter (L_g/d =3.84). In such scenarios, where n_s modulation weakens, $\Delta v_e/\Delta V_{gs}$ continues to play a significant role in enhancing g_m despite the L_g decreased.

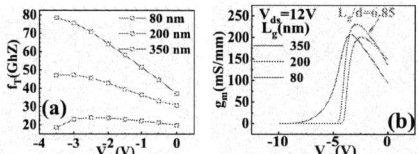

Fig3. At V_{ds} = 12V.(a) the f_T as a function of V_{gs} for three L_g devices. (b) the g_m as a function of V_{gs} for three L_g devices.

The V_{gs} as function of g_m and f_T for three L_g devices are shown in Fig. 3. In order to verify that $\Delta v_e/\Delta V_{gs}$ has a facilitating effect on high-frequency performance in sub-100 nm AlGaN/GaN HFETs, the v_e of three L_g, 80, 200, and 350 nm in Wafer A were extracted and simulated. Fig. 4(a) shows that the extracted v_e increased from 0.9×10⁷ cm/s at 0 V to 1.3×10⁷ cm/s at -1.5 V, and then decreased to 0.65 × 10⁷ cm/s at -3.5 V in 80 nm L_g device. The peak v_e values of the devices all appeared near $n_s = 3.2×10^{12}$ cm⁻² in Fig. 4(b). The v_e distribution for 80 nm L_g GaN HFET is shown in Fig. 4(c). In order to analyze the g_m of three L_g devices, V_{gs} was divided into two parts: Region I (0 V≥V_{gs}>-1.5 V), and II (-1.5 V≥V_{gs}≥-3.5 V). g_m can be expressed by

$$g_m = \frac{\Delta I_{ds}}{\Delta V_{gs}} = \left(\frac{\Delta n_s}{\Delta V_{gs}}\right)ev_e + \left(\frac{\Delta v_e}{\Delta V_{gs}}\right)en_s \quad (1)$$

As shown in Fig. 4(d), when L_g was scaled down from 350 to 200 nm, the first term of Eq. 1 was increased from 152 to 160 mS/mm and the second term was increased from 44 to 58.6 mS/mm. According to the data in Table I, with L_g scaling down, $\Delta n_s/\Delta V_{gs}$ decreased. However, the average v_e for Region II increased from 0.62×10⁷ to 0.83 ×10⁷ cm/s, leading to a slight increase in the first term. Further findings were made that as L_g scaled down, the smaller $\Delta n_s/\Delta V_{gs}$ led to a larger n_s in Region II. These two parameters together led to an increase in the second term. When L_g scaled down from 200 to 80 nm, $\Delta n_s/\Delta V_{gs}$ decreased drastically, leading to a decrease in the first term. Despite the rise in n_s and $\Delta v_e/\Delta V_{gs}$, which drove up the second term in Eq. 1, the overall g_m trended downward due to the predominant effect of $\Delta n_s/\Delta V_{gs}$. As such, the 200 nm L_g device exhibited the highest g_m among the three devices. Such findings imply that $\Delta v_e/\Delta V_{gs}$ continued to contribute to g_m in sub-100 nm GaN HFETs.The frequency characteristics of three L_g devices were analyzed using $\tau=1/2\pi f_T=\tau_t+\tau_{ext}+\tau_{par}=(C_{gs\_int}+C_{gd\_int})/g_m+(C_{gs\_ext}+C_{gd\_ext})/g_m+C_{gd}(R_s+R_d)[1+(1+C_{gs}/C_{gd})g_0/g_m]$. τ_{par} is the parasitic effect delay and τ_{ext} is the parasitic charging delay. As L_g scaled to 80 nm, $f_{T,max}$ boosted to 78.5 GHz. Fig. 4(e) illustrates the three delay times as a function of L_g. When L_g scaled from 350 to 200 nm, the increase of $f_{T,max}$ could be attributed to the decrease in τ_t due to the enhancement of v_e and the reduction of L_g, and the decrease in τ_{ext} due to the enhancement of g_m. When L_g scaled from 200 to 80 nm, τ_t decreased in accordance with the aforementioned factors, and the decrease in g_m due to the SCE caused an increase in τ_{ext}. However, since the decrease in τ_t was much more pronounced, it still contributed to an enhancement in f_T. The results show that the combined τ_{ext} and τ_{par} accounted for 54% of the 80 nm L_g device, which greatly limited the further optimization of f_T for HFETs with L_g scaling.

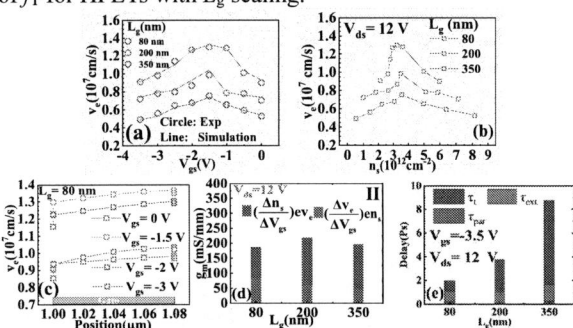

Fig. 4 (a) Simulated v_e and extracted v_e versus V_{gs}. (b) Extracted v_e versus n_s. (c) The simulated channel v_e profile at V_{ds}=12V with different V_{gs} for the 80 nm L_g device. (d) In Region **II**, the average experimental g_m components versus L_g. (e) Extracted Delay components versus L_g.

Fig. 5 shows T_L and power density (P_d) distribution at V_{ds}=12 V, V_{gs}= 0 V for 80 nm L_g HFET. The peak T_L (T_{max}) was 476 K and the hot spot was located in gate-drain spacing in Fig. 5(a). Fig. 5(b) shows the P_d along the y-axis at the rightmost of the gate. The higher heat generation was observed in the GaN Channel, while the lower heat generation was observed in the AlGaN barrier layer and GaN buffer layer because the high potential barrier formed by the C doped GaN buffer layer and AlGaN layer prevented the electrons from transferring to the GaN buffer and AlGaN layer in Fig. 5(b).

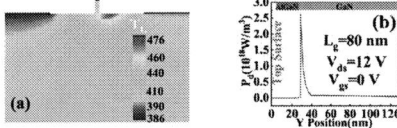

Fig. 5. V_{ds}=12V, V_{gs}= 0V for the 80 nm L_g device (a) T_L distribution(b) P_d versus Y position.

The channel T_L mainly reflected the effect on the degradation of electrical features for the HFETs. As T_L rises, increased POP scattering reduces v_e, degrading I_{ds}. T_{max} is closely related to the reliability of device. As shown in Fig. 6(b)-(c), the average T_L and T_{max} in 80 nm L_g device with the same bias were significantly higher than those of long channel devices. Besides, T_{max} had a strong dependence on V_{gs} in short-channel HFETs, while the average T_L had a weaker dependence on V_{gs}. Compared with other related research on self-heating, the simulated current and v_e were all well-matched with the

experiments shown in Fig. 1and 4, suggesting that the simulated n_s was also more accurate. Since the channel T_L is closely associated with the v_e, the extraction of v_e offers additional constraints that can facilitate the acquisition of more accurate T_{max} and average T_L in the present simulation efforts. Since T_{max}, T_L and P_d are critical for thermal optimization of sub-100 nm AlGaN/GaN HFETs, accurate thermal transport information based on the present simulations can help optimize device performance. Given that T_{max} and average T_L significantly influence reliability and performance of the device, it is essential to consider both of these physical quantities together when designing the HFET. The nonlinear v_e-n_s phenomenon in Fig.4(b) has been explained by POP and PCF scattering in previous work [7]. When n_s<3.2×10^{12} cm^{-2}, PCF scattering dominated the electron transport, and v_e decreased as V_{gs} decreased. For n_s >3.2×10^{12} cm^{-2}, POP scattering dominated, causing v_e to rise as V_{gs} decreases. The 80 nm L_g HFET had a higher v_e than the other two devices in Fig. 4(a) because of the larger E in the 80 nm L_g device. The reduced L_g and larger E accelerated v_e, consequently boosting both the g_m and f_T. However, the lowering of the channel barrier led to a decrease in $\Delta n_s/\Delta V_{gs}$, which reduced the g_m of the device [7]. Further, $\Delta v_e/\Delta V_{gs}$ was larger in Region II in the short L_g AlGaN/GaN HFETs in Table I. The main reason is that PCF scattering dominated electron transport in Region II. As L_g scaled down, the APCs would be closer to the electrons in the gate region, which would result in stronger PCF scattering to the electrons. As a result, $\Delta v_e/\Delta V_{gs}$ increased. $\Delta v_e/\Delta V_{gs}$ is a valid parameter for sub-100 nm GaN HFETs to enhance the gate-to-current modulation, providing researchers with a new avenue to suppress SCE.

TABLE I.

The average extracted $\triangle v_e/\triangle V_{gs}$, v_e, $\triangle n_s/\triangle V_{gs}$,n_s of all samples in Region **II**

L_g (nm)	80	200	350
$\triangle v_e/\triangle V_{gs}$(ms^{-1}.V^{-1})	1.93×10^4	1.49×10^4	1.3×10^4
v_e (cm/s)	1.12×10^7	0.83×10^7	0.62×10^7
$\triangle n_s/\triangle V_{gs}$(cm^{-2}.V^{-1})	0.57×10^{12}	1.21×10^{12}	1.52×10^{12}
n_s (cm^{-2})	2.76×10^{12}	2.45×10^{12}	2.12×10^{12}

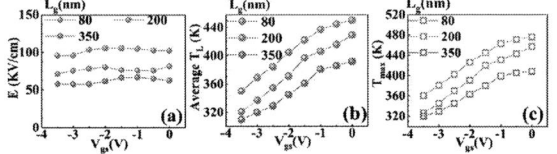

Fig. 6. V_{ds}=12V(a) Average E versus V_{gs}. (b) Average T_L versus V_{gs}. (c) T_{max} versus V_{gs}.

Conclusion

The optimal L_g/d range for achieving the $g_{m,max}$ in AlGaN/GaN HFETs has been identified in this work, along with the underlying electron transport mechanisms. This work offers a physical analysis to inform design principles for GaN HFETs under scaling down.

Acknowledgments

This work was supported by CAS-Croucher Funding Scheme under Grant No. CAS22801.

References

[1] Q. Yu *et al*, *IEEE Electron Device Letters*, vol. 44, p. 582.2023.[2] J.W. *et al*, *IEEE Electron Device Letters*, vol.31, p.19, 2010.[3] J. Moon. *et al Electronics Letters*, vol.56, p.678, 2020.[4] G. Jessen *et al*, *IEEE Transactions On Electron Devices*, vol. 54, p. 2589, 2007.[5] D. Guerra *et al*, *IEEE Electron Device Letters*, vol. 31, pp.1217, 2010, [6] A. Ashok, *IEEE Transactions On Electron Devices*, vol.57, p. 562, 2010.[7] M. Wang *et al*, *IEEE Electron Device Letters*, vol. 45, p. 160, 2024.

Monolithic Integration of GaN μLED and Normally-off p-GaN HEMT by Flip Chip Bonding

Jinxia Jiang[1,2,†], Ang Li[1,2,†], Guohao Yu[1,2,*], Han Yue[3,*], Zhongming Zeng[1,2], Baoshun Zhang[1,2,*]

[1] School of Nano Technology and Nano Bionics, University of Science and Technology of China, Hefei, China, 230026

[2] Nanofabrication facility, Suzhou Institute of Nano-Tech and Nano-Bionics, Chinese Academy of Sciences (CAS), Suzhou, China, 215123

[3] Suzhou Shinju Semiconductor Ltd., Suzhou, China

[†] J. Jiang and A. Li contributed equally to this work

*Email: ghyu2009@sinano.ac.cn, yuehan@shinju.cn, bszhang2006@sinano.ac.cn

Abstract

This work reports monolithic integration of GaN micro light emitting diode (μLED) and GaN H-treated high-electron mobility transistor (HEMT) by flip chip bonding. GaN HEMT ($W_G/L_G = 10/3$ μm) has the threshold voltage of 2.2 V, on-resistance of 8.3 Ω·mm, and output current of 0.9 mA. The turn on voltage of μLED (40×50 μm²) is 2.45 V. The current of integrated configuration is modulated by the gate voltage. GaN HEMT provides high injection current for μLED, benefiting in LED displays for high brightness and low power consumption.

Keywords: GaN HEMT, μLED, normally-off, monolithic integration, flip chip bonding

Introduction

μLED is recognized as the next-generation display technology due to its high brightness, fast response time, and small size [1], [2]. μLED can be driven by TFTs in voltage-controlled mode to realize high-precision display [3]. TFTs are limited by low electron mobility (<50 cm²·V⁻¹·s⁻¹), while GaN HEMTs exhibit significantly higher mobility (~2000 cm²·V⁻¹·s⁻¹), resulting in enhanced efficiency and reduced power consumption [4]. GaN HEMT and LED integration by selective epitaxial removal and selective epitaxial growth method has been proposed [5], [6]. Flip chip bonding is also a feasible integration.

Integrated HEMTs are typically normally-on, while normally-off operation can simplify the driving design. Among normally-off, p-GaN structure is more suitable for large scale fabrication and gained commercial [7]. H-treated technology used in p-GaN HEMT demonstrates to avoid etching damage [8].

In this work, integration of H-treated HEMT and μLED is proposed by the method based on flip chip bonding. The forward current of μLED is modulated by gate voltage of GaN device. This integrated configuration is a promising candidate for smart lighting and display systems.

Device Design and Fabrication

GaN HEMT was fabricated on Si substrate. Fig. 1(a) shows the schematic structure and fabrication process of H-treated device. The epitaxial structure from bottle to top consists of the GaN buffer layer (1.8 μm), GaN channel layer (150 nm), AlN spacer layer (1 nm), $Al_{0.2}Ga_{0.8}N$ barrier layer (20 nm), and p-GaN layer (70 nm). The devices fabrication started with isolation by fluorine ion implantation. Subsequently, using Cl_2/BCl_3-based ICP dry etching to remove the p-GaN layer above both the source and drain and depositing Ti/Al/Ni/Au (20/130/50/50 nm) electrodes followed by 875 °C 30s rapid thermal annealing (RTA). Following this, the p-GaN was treated with hydrogen plasma produced in ICP system, with a radio frequency (RF) power of 2 W and an ICP power of 100 W for 5 min. Gate metal is Ni/Au (50/150 nm). RTA at 350 °C for 5 minutes was to repair hydrogen plasma damage. 200 nm SiN_x layer was by PECVD deposited as a passivation layer. Finally, a 1.5 μm Sn was deposited for bonding.

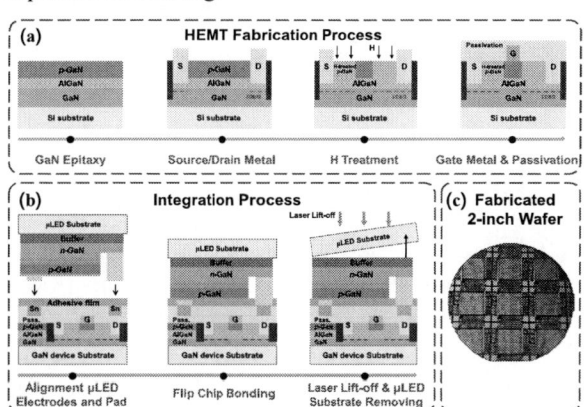

Fig. 1 (a) Schematic and fabrication process of GaN HEMT. (b) Integration process of μLED and GaN device. (c) Photograph of the proposed 2-inch GaN wafer.

μLED as commercial chip was fabricated on sapphire substrate. The epitaxial structure of μLED from bottle to top consists of AlN layer (10 nm), u-GaN layer (4 μm), n-GaN layer (2.3 μm), MQW layer (0.2 μm), and p-GaN layer (0.2 μm).

The integration process is shown in Fig. 1(b). After alignment the LED electrodes and Sn PADs, the LED was flip-chip bonded to GaN device. Following this, μLED substrate was removed with laser lift-off process.

Results and Discussion

Fig. 1(c) shows the image of the wafer integrated the μLED and GaN device. The size of the LED is 40 μm×50 μm and the size of the device is $L_{GS}/L_G/L_{GD}/W_G$ = 3.5/3/3.5/10 μm. The μLED characteristic after bonding is shown in Fig. 2. The turn-on voltage is ~ 2.45 V, indicating no damage to LED during the bonding and laser lift-off process.

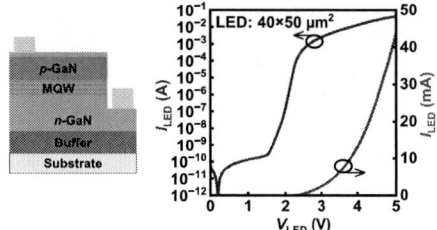

Fig. 2. μLED structure and I-V characteristic. The turn on voltage is 2.45 V.

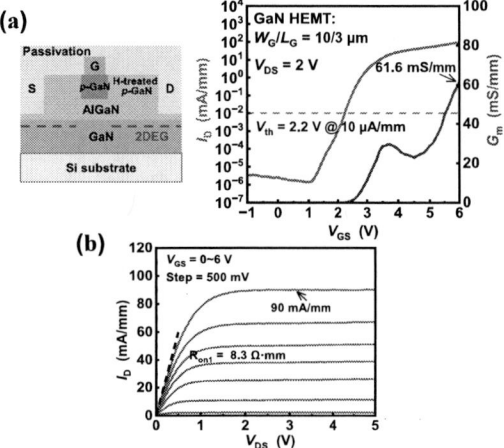

Fig. 3 (a) p-GaN device structure and transfer characteristic of GaN device (W_G/L_G = 10/3 μm). (b) Output characteristic.

Fig. 3(a) illustrates the transfer characteristic of the GaN device. Drain to source voltage (V_{DS}) is set as 2 V. The threshold voltage (V_{th}) is 2.2 V extracted at I_D = 10 μA/mm, the maximum transconductance (G_m) is 61.6 mS/mm. ON/OFF ratio (I_{on}/I_{off}) of the device is 6.8×10^7. Output characteristic is shown in Fig. 3(b). The maximum output current ($I_{D,max}$) and on-resistance (R_{on1}) are 90 mA/mm and 8.3 Ω·mm at V_{GS} = 6 V.

Fig. 4 plots the load lines of μLEDs with the GaN device output curves at supply voltage (V_{DD}) of 5 V. The current crossing points between the I_{LED} and I_D represent the driving condition for the μLED driven by GaN device.

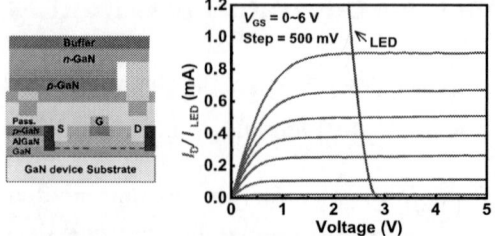

Fig. 4 Schematic diagram of integrated configuration and load line.

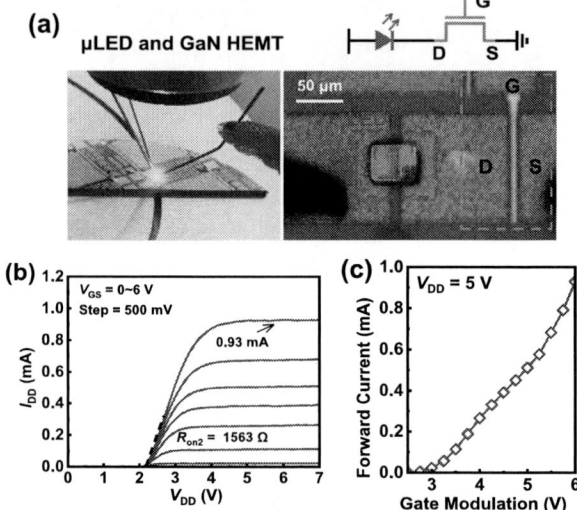

Fig. 5 (a) The emission images of μLED. (b) I-V characteristic of integrated μLED and device. (c) The forward current of the configuration modulated with gate biases.

The emission image of μLED driven by GaN device is shown in Fig.5(a) and the integrated characteristic is shown in Fig. 5(b). The current is limited by μLED before it turns on, and by the saturation current of GaN device afterwards. The configuration shows a R_{on2} of 1563 Ω at V_{GS} = 6 V. Higher on-resistance is due to the series of μLED and GaN device. The I_{DD} is 0.93 mA at V_{GS}= 6 V. Fig. 5(c) shows LED current can be modulated by the gate voltage of HEMT.

Conclusion

This work shows a feasible way to integration of GaN HEMT and μLED by flip chip bonding. HEMT exhibit a normally off operation and milliampere driving current. μLED exhibited controllable modulation in injection current by GaN device. Monolithically integrated HEMT driving μLED can possibly be one of the significant blocks for smart lighting systems.

Acknowledgments

This work was supported in part by National Natural Science Foundation of China under Grant 92163204; and in part by the Key Research and Development Program of Jiangsu Province under Grant BE20220571.

References

[1] K. Behrman and I. Kymissis, "Micro light-emitting diodes," *Nat Electron*, vol. 5, no. 9, pp. 564–573, Sep. 2022, doi: 10.1038/s41928-022-00828-5.

[2] L. Qi, P. Li, X. Zhang, K. M. Wong, and K. M. Lau, "Monolithic full-color active-matrix micro-LED micro-display using InGaN/AlGaInP heterogeneous integration," *Light Sci Appl*, vol. 12, no. 1, p. 258, Oct. 2023, doi: 10.1038/s41377-023-01298-w.

[3] L. Han *et al.*, "Wafer-scale organic-on-III-V monolithic heterogeneous integration for active-matrix micro-LED displays," *Nat Commun*, vol. 14, no. 1, p. 6985, Nov. 2023, doi: 10.1038/s41467-023-42443-8.

[4] T. Kyoung Kim *et al.*, "Realization of high-power dimmable GaN-based LEDs by hybrid integration with AlGaN/GaN HFETs," *Jpn. J. Appl. Phys.*, vol. 58, no. SC, p. SCCC12, Jun. 2019, doi: 10.7567/1347-4065/ab124a.

[5] Z. Li, J. Waldron, T. Detchprohm, C. Wetzel, R. F. Karlicek, and T. P. Chow, "Monolithic integration of light-emitting diodes and power metal-oxide-semiconductor channel high-electron-mobility transistors for light-emitting power integrated circuits in GaN on sapphire substrate," *Applied Physics Letters*, vol. 102, no. 19, p. 192107, May 2013, doi: 10.1063/1.4807125.

[6] C. Liu, Y. Cai, X. Zou, and K. M. Lau, "Low-Leakage High-Breakdown Laterally Integrated HEMT-LED via n-GaN Electrode," *IEEE Photon. Technol. Lett.*, vol. 28, no. 10, pp. 1130–1133, May 2016, doi: 10.1109/LPT.2016.2532338.

[7] H. Amano *et al.*, "The 2018 GaN power electronics roadmap," *J. Phys. D: Appl. Phys.*, vol. 51, no. 16, p. 163001, Apr. 2018, doi: 10.1088/1361-6463/aaaf9d.

[8] A. Li *et al.*, "Monolithically Integrated Power Converter Based on Etching-Free p-GaN HEMTs by Hydrogen Plasma Treatment Technology," in *2024 36th International Symposium on Power Semiconductor Devices and ICs (ISPSD)*, Bremen, Germany: IEEE, Jun. 2024, pp. 374–377. doi: 10.1109/ISPSD59661.2024.10579682.

High density 3D integration of 2D transistors via van der Waals lamination

Donglin Lu[1], Quanyao Tao[1], Yuan Liu[1,*]

[1] School of Physics and Electronics, Hunan University, Changsha 410082, China.

Abstract

In this work, we review two approaches for realizing high density three-dimensional (3D) integration using two-dimensional (2D) transistors via van der Waals (vdW) limitation, including low temperature tier-by-tier lamination, as well as integrating the planar devices onto the silicon sidewall. The two approaches provide alternatives routes for realizing high density vertical 2D transistors as well as 3D integrated system.

Keywords: van der Waals integration, 2D transistor, 3D integration

Introduction

3D integration, which involves the sequential fabrication of multiple stacked tiers on a single wafer through the deposition of upper tiers, has recently garnered significant attention [1, 2]. 2D semiconductors show significant promise for 3D integration [3-7]. With dangling-bond-free surface, 2D semiconductors can be pre-fabricated at high temperatures and subsequently transferred at lower temperatures [8]. This process addresses the previous limitation of thermal budget while ensuring the device performance of the lower tiers. Furthermore, the atomic thinness of 2D semiconductors significantly reduces short-channel effects, and off-state leakage currents, and the resultant self-heating—issues that become critical when multiple devices are stacked vertically in 3D integration [9]. However, high-density 3D systems are of great challenge using 2D semiconductor.

In this work, we review two approaches from our group, for realizing high density 3D integration using 2D transistors via vdW limitation. Within first technique, all necessary device and circuits components, including inter-tire dielectric (ITD), source/drain/gate electrodes, gate dielectric, interconnects and inter-tier-via (ITV), are pre-fabricated on a sacrificial wafer via conventional high-energy processes. These components are then mechanically released as a complete circuit tier and laminated onto the 2D semiconductor at low processing temperatures [10]. Within second approaches, lateral transistors can be pre-fabricated on planar substrate using traditional batch processes in large scale, and then dry-released and further laminated onto vertical substrates using custom-designed T-shaped. This approach effectively overcomes the incompatibility between planar processes and vertical structures, and avoided the previous complex layer-by-layer transfer. Importantly, owing to the low strain generated in the process, the dry lamination technique enables the transistors to make intimate contact with the vertical substrate without structural damage or performance degradation, resulting in high-performance vertical transistors with good uniformity [11].

High density 3D integration

Fig. 1 schematically illustrate the fabrication processes of one-step vdW 3D integration within first technique [10]. Initially, all the necessary component layers of a standard circuit tier are pre-fabricated on a sacrificial wafer in the following order: source-drain electrodes and lateral interconnects layer; low-κ ITD layer; inter-tier holes of ITD; inter-tier vertical connections of ITD layer; gate electrode and lateral interconnects layer; high-κ gate dielectric layer; inter-tier holes of gate dielectric; and inter-tier vertical connections of gate dielectric layer (Fig. 1a). Subsequently, the prefabricated circuit stack is mechanically detached from the sacrificial wafer (Fig. 1b), and physically laminated onto the target 2D wafer (Fig. 1c). MoS_2 here is utilized as a representative example of 2D semiconductors due to its scalability in production. Meanwhile, the lamination process is conducted in a dry process, free from any solutions, with a low temperature of 120 °C. This ensures the physical contact between the 2D semiconductor and other circuit components, which is important for preserving the integrity of the 2D lattice without conventional damage or degradation. Thanks to the low-energy and low-temperature vdW integration process, 3D systems with 10 circuit tiers could be constructed by repeatedly laminating 2D semiconductors and vdW circuit tiers (Fig. 2, 3).

Within this first approach, CPVA is important as an ITD layer and is essential in ne-step vdW 3D integration. This is largely because the CPVA has low adhesion to the substrate and can be mechanically peeled-off directly for successful lamination. In addition, CPVA has good flexibility and low dielectric constant, which can withstand large strains during mechanical peeling and reduce the electrical coupling between circuit tiers.

Fig. 4 schematically illustrates our second approaches of integrating 2D transistors on the silicon sidewall. The planar MoS_2 transistors are first pre-fabricated on a sacrificial wafer by conventional lithography and thermal deposition processes. The entire device layer, consisting of MoS_2

channel and Au electrodes, can then be physically detached from the sacrificial substrate using a polymer capping layer. It is important to note that the dry release method is crucial for avoiding the use of solutions, thereby preserving the intrinsic properties of the MoS$_2$ channel and metal contacts. Subsequently, the deep vertical trench is formed in the silicon substrate through wet etching process, creating the vertical substrate (Fig. 4a). The device layer is then transferred onto the vertical trench using a dry-alignment transfer technique, as shown in Fig. 4b. In traditional lamination methods, the transferred device layer typically forms conformal contact with flat planar substrate, integrating the devices onto the deep vertical substrate may cause the device layer to suspend over the silicon sidewall. To address this issue, a T-shaped stamp is designed to precisely push the suspended MoS$_2$ transistors onto the vertical substrate, forming good contact between the device layer and the substrate (Fig. 4c, d). Note that the T-shaped stamp is created from the same vertical trench used as the mold, ensuring that it matches the structure of the silicon trench perfectly. When the micro-sized T-shaped stamp mechanically inserts into the trench, the device layer could be pushed onto the vertical sidewall to form conformal contact (Fig. 4d). This is validated by the scanning electron microscopy (SEM) and scanning transmission electron microscopy (STEM) images (Fig. 5), which show the device layer is in close contact with the vertical substrate, free of any contamination or bubbles. Importantly, both the device release and vertical lamination processes are fully dry methods that eliminate the use of solvents and mechanical strain. In contrast, traditional wet-transfer process applied to non-flat or vertical substrates relies on capillary forces during solvent evaporation, resulting in significant stretching forces and distortion of the device layer (Fig. 6).

Conclusion

In this work, we reviewed two approaches for realizing high density 3D integration using 2D transistors via vdW limitation. In the first method, a low-temperature 3D technique is demonstrated by vdW laminating the entire prefabricated circuit tier, while maintaining a processing temperature of 120 °C. By repeating the vdW lamination process tier-by-tier, a 3D integrated system with 10 vertical circuit tiers is achieved, effectively overcoming previous thermal budget limitation. In the second approach, lateral transistors are pre-fabricated on planar substrates using traditional batch processes, then dry-released and laminated onto vertical substrates via custom-designed T-shaped stamp, successfully addressing the incompatibility between planar processes and vertical structures. The above research provides two alternatives routes for realizing high density vertical 2D transistors as well as 3D integrated system.

Acknowledgments

The authors acknowledge the financial support from National Key R&D Program of China (No. 2021YFA1200503) and from the National Natural Science Foundation of China (Grant Nos. 51991340, 51991341, 61874041, 62404078, 62304075)

References

[1] M. Bishop, H. Wong, S. Mitra, and M. M. Shulaker, "Monolithic 3-D integration," IEEE Micro, 39, pp. 16-27, (2019).

[2] M. M. Shulaker, G. Hills, R. S. Park, R. T. Howe, K. Saraswat, H. P. Wong, and S. Mitra, "Three-dimensional integration of nanotechnologies for computing and data storage on a single chip," Nature, 547, pp. 74-78, (2017).

[3] Y. Guo, J. Li, X. Zhan, C. Wang, M. Li, B. Zhang, Z. Wang, Y. Liu, K. Yang, H. Wang, W. Li, P. Gu, Z. Luo, Y. Liu, P. Liu, B. Chen, K. Watanabe, T. Taniguchi, X. Q. Chen, C. Qin, J. Chen, D. Sun, J. Zhang, R. Wang, J. Liu, Y. Ye, X. Li, Y. Hou, W. Zhou, H. Wang, and Z. Han, "Van der Waals polarity-engineered 3D integration of 2D complementary logic," Nature, 630, pp. 346-352, (2024).

[4] D. Jayachandran, R. Pendurthi, M. U. K. Sadaf, N. U. Sakib, A. Pannone, C. Chen, Y. Han, N. Trainor, S. Kumari, T. V. Mc Knight, J. M. Redwing, Y. Yang, and S. Das, "Three-dimensional integration of two-dimensional field-effect transistors," Nature, 625, pp. 276-281, (2024).

[5] L. Tong, J. Wan, K. Xiao, J. Liu, J. Ma, X. Guo, L. Zhou, X. Chen, Y. Xia, S. Dai, Z. Xu, W. Bao, and P. Zhou, "Heterogeneous complementary field-effect transistors based on silicon and molybdenum disulfide," Nat. Electron., 6, pp. 37-44, (2022).

[6] J.-H. Kang, H. Shin, K. S. Kim, M.-K. Song, D. Lee, Y. Meng, C. Choi, J. M. Suh, B. J. Kim, H. Kim, A. T. Hoang, B.-I. Park, G. Zhou, S. Sundaram, P. Vuong, J. Shin, J. Choe, Z. Xu, R. Younas, J. S. Kim, S. Han, S. Lee, S. O. Kim, B. Kang, S. Seo, H. Ahn, S. Seo, K. Reidy, E. Park, S. Mun, M.-C. Park, S. Lee, H.-J. Kim, H. S. Kum, P. Lin, C. Hinkle, A. Ougazzaden, J.-H. Ahn, J. Kim, and S.-H. Bae, "Monolithic 3D integration of 2D materials-based electronics towards ultimate edge computing solutions," Nat. Mater., 22, pp. 1470-1477, (2023).

[7] W. Meng, F. Xu, Z. Yu, T. Tao, L. Shao, L. Liu, T. Li, K. Wen, J. Wang, L. He, L. Sun, W. Li, H. Ning, N. Dai, F. Qin, X. Tu, D. Pan, S. He, D. Li, Y. Zheng, Y. Lu, B. Liu, R. Zhang, Y. Shi, and X. Wang, "Three-dimensional monolithic micro-LED display driven by atomically thin transistor matrix," Nat. Nanotechnol., 16, pp. 1231-1236, (2021).

[8] Y. Liu, Y. Huang, and X. Duan, "Van der Waals integration before and beyond two-dimensional materials," Nature, 567, pp. 323-333, (2019).

[9] Y. Liu, X. Duan, H. J. Shin, S. Park, Y. Huang, and X. Duan, "Promises and prospects of two-dimensional transistors," Nature, 591, pp. 43-53, (2021).

[10] D. Lu, Y. Chen, Z. Lu, L. Ma, Q. Tao, Z. Li, L. Kong, L. Liu, X. Yang, S. Ding, X. Liu, Y. Li, R. Wu, Y. Wang, Y. Hu, X. Duan, L. Liao, and Y. Liu, "Monolithic three-dimensional tier-by-tier integration via van der Waals lamination," Nature, 630, pp. 340-345, (2024).

[11] Q. Tao, R. Wu, X. Zou, Y. Chen, W. Li, Z. Lu, L. Ma, L. Kong, D. Lu, X. Yang, W. Song, W. Li, L. Liu, S. Ding, X. Liu, X. Duan, L. Liao, and Y. Liu, "High-density vertical sidewall MoS$_2$ transistors through T-shape vertical lamination," Nat. Commun., 15, p. 5774, (2024).

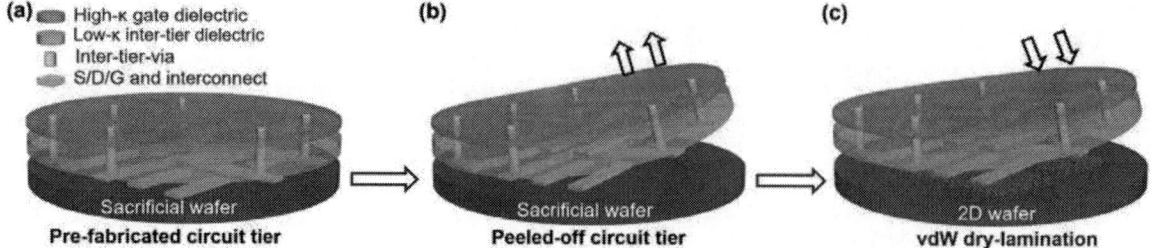

Fig. 1. Schematics of one-step vdW 3D integration processes. (a) Pre-fabrication of circuit tier with all essential components, (b) circuit tier physically peeled-off, (c) circuit tier dry laminated onto target 2D wafer. Figure adapted from [10].

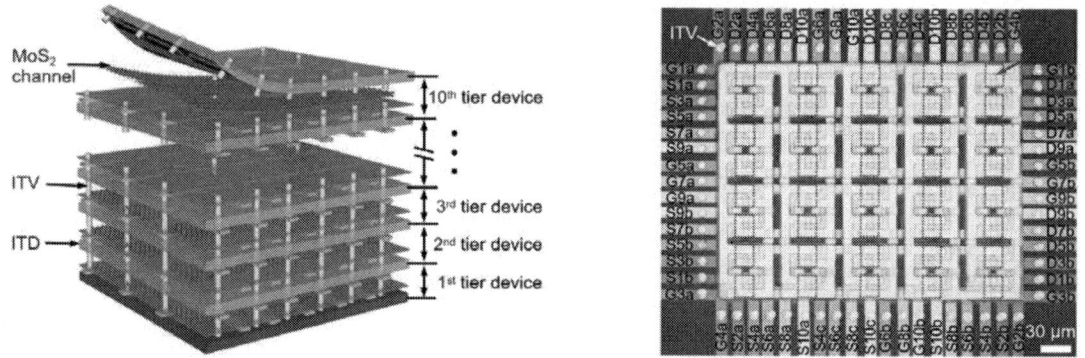

Fig. 2. Schematic diagram of 10 tiers 3D systems by multi-cycles vdW lamination. Figure adapted from [10].

Fig. 3. Optical image 10 tiers 3D systems with the MoS$_2$ arrays channels. Figure adapted from [10].

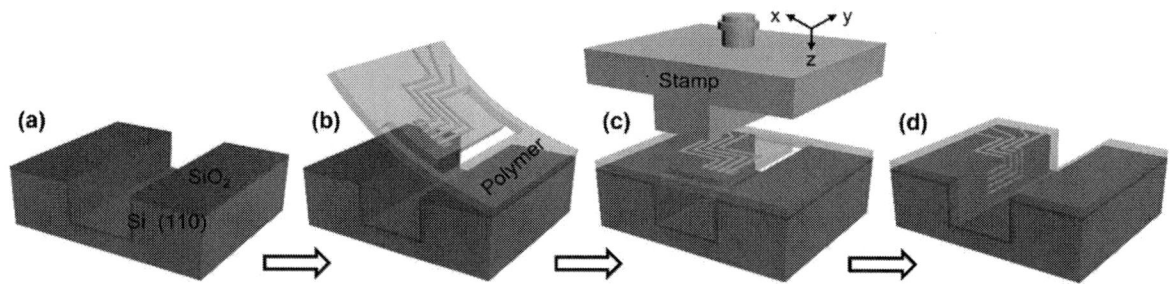

Fig. 4. Vertical lamination processes. (a) Fabrication of vertical silicon trench through etching, (b) dry-transfer of MoS$_2$ transistors , (c) T-shape stamp laminated and pushed into the trench, (d) MoS$_2$ vertical transistors after lamination. Figure adapted from [11].

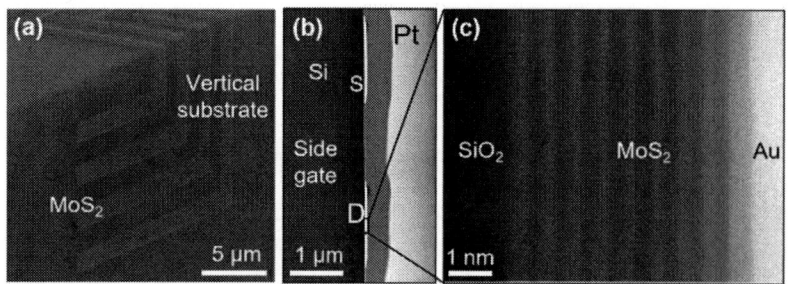

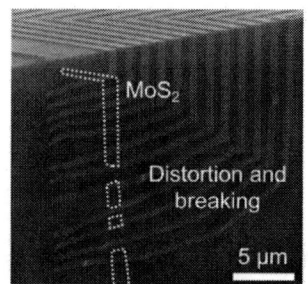

Fig. 5. (a) SEM image of the fabricated MoS$_2$ transistors on vertical substrate. (b)–(c) Cross-sectional SEM image of the vertical transistors. Figure adapted from [11].

Fig. 6. SEM image of wet-transferred devices. Figure adapted from [11].

979-8-3315-0417-5/25 $31.00 © 2025 IEEE   612

Optimizing ALD-deposited IGZO TFT Thermal Stability through Compositional Adjustments

Jianting Wu[1], Huajian Zheng[1], Min Guo[1], Yi Huang[1], Xiaoci Liang[1], Qian Wu[2,*], Chuan Liu[1,*]

[1]State Key Laboratory of Optoelectronic Materials and Technologies, School of Electronics and Information Technology, Sun Yat-sen University, Guangzhou 510275, China (*email: liuchuan5@mail.sysu.edu.cn), [2]School of Computer and Information Engineering, Guangdong Polytechnic of Industry and Commerce, Guangzhou 510510, China (*email: wuqian1427@gdgm.edu.cn)

Abstract

The integration of metal oxide TFTs is challenged by high BEOL temperatures. Our study evaluated ALD-deposited IGZO TFTs with different composition ratio. These TFTs exhibited mobilities ranging from 30.24 to 81.60 $cm^2V^{-1}s^{-1}$. The optimized-composition $In_{0.25}Ga_{0.26}Zn_{0.49}O$ TFT without encapsulation, maintained performance after 450°C annealing process and exhibited enhanced thermal stability, reducing the threshold voltage shift from over 15 V to -1.5 V. This enhancement is attributed to the stabilization of the amorphous structure, optimization of the coordination number, and reduction of defect states.

(Keyword: Thin film transistor, IGZO, thermal stability, atomic layer deposition)

Introduction

Oxide semiconductors, such as In-based materials like IGZO, have become prevalent in thin-film transistors (TFTs) for their high mobility and low-cost manufacturing, particularly in large-area AMOLED displays[1]. However, the high-temperature demands of 3D back-end-of-line (BEOL) integration and the multi-layer stacking in devices pose thermal stability challenges[2, 3]. The weak In-O bonds in IGZO are susceptible to thermal stress, leading to oxygen vacancy (V_O) defects and temperature instability. Doping with elements such as Y, Al, Si, or employing passivation layers has shown promise in enhancing stability[4]. However, IGZO's inherent V_O, which cannot be entirely eliminated, may function as donor carriers or electron traps when ionized and hydrogen-doped[5]. Minimizing V_O's impact is crucial for balancing mobility and stability.

This study deposited a series of IGZO films with varying metal cation ratios by atomic layer deposition (ALD), focusing on their impact on device performance and stability, particularly thermal stability. We report an $In_{0.25}Ga_{0.26}Zn_{0.49}O$ TFT with enhanced stability and suppressed V_O defects, attributing to the control of the amorphous state and alleviation of the overstressed condition in the film.

Experiment Details

Fig. 1a illustrates the device structure of the fabricated IGZO TFTs, which feature a bottom-gate, top-contact configuration. The fabrication process is depicted in Fig. 1b. Initially, the semiconductor active layer (IGZO) was deposited using ALD technology at a substrate temperature of 300°C on a Si/SiO₂ substrate. Ozone served as the oxygen source. As shown in Fig.1c, by adjusting the subcycle counts of In_2O_3, Ga_2O_3, and ZnO, the elemental composition was precisely controlled. Following photolithography and wet etching of the active layer, a 150 nm thick ITO film was deposited by sputtering for the source and drain and patterned using lift-off technology. The channel length and width were 350 μm and 1000 μm. Post annealing at 350°C for 2 hours in air was applied to the devices to optimize their performance.

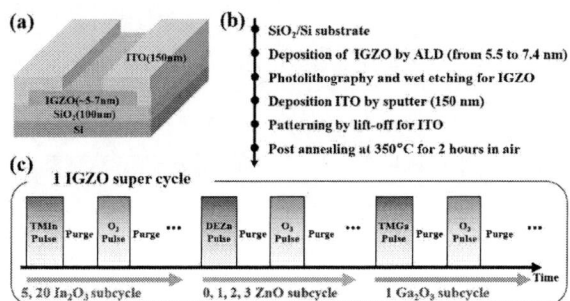

Fig. 1. (a) Device schematic of IGZO TFTs; (b) Fabrication flow of IGZO TFTs; (c) Illustration of ALD growth for per super cycle of IGZO channel.

Results and Discussion

The atomic ratio of IGZO films under various subcycle parameters was analyzed using X-ray photoelectron spectroscopy (XPS), as shown in Fig. 2a, and the samples were labeled according to their atomic ratios as $In_{0.75}Ga_{0.25}Zn_0O$, $In_{0.60}Ga_{0.40}Zn_0O$, $In_{0.41}Ga_{0.35}Zn_{0.24}O$, $In_{0.31}Ga_{0.28}Zn_{0.41}O$ and $In_{0.25}Ga_{0.26}Zn_{0.49}O$. Atomic force microscopy determined the film thicknesses, which ranged from 5.5 to 7.4 nm (Fig. 2b), indicating a good linear dependence of ALD IGZO film thickness on the number of cycles. From the thickness of IGZO film and the cycle number of Zn, a ZnO growth rate of 0.33 Å/cycle was determined.

Fig. 3a-e shows transfer characteristics of IGZO TFTs under V_D of 0.1 V and 30 V. The extracted electrical parameters are summarized in Fig. 3f. The $In_{0.75}Ga_{0.25}Zn_0O$ TFT achieved a peak mobility (μ) of 81.60 $cm^2V^{-1}s^{-1}$, which declined to 45.76 cm²V⁻¹s⁻¹ in the In$_{0.60}$Ga$_{0.40}$Zn$_0$O TFT with higher Ga content. Elevating the atomic percentage of Zn shifted the negative V$_{th}$ to 0 V and sustaining mobility over 30 cm²V⁻¹s⁻¹. The In$_{0.41}$Ga$_{0.35}$Zn$_{0.24}$O, In$_{0.31}$Ga$_{0.28}$Zn$_{0.41}$O and In$_{0.25}$Ga$_{0.26}$Zn$_{0.49}$O TFTs exhibited μ of 37.50, 32.53, and 30.24 cm²V⁻¹s⁻¹ and SS of 0.15, 0.13, and 0.09 V/dec, respectively. After post-annealing at 450°C, the In$_{0.25}$Ga$_{0.26}$Zn$_{0.49}$O TFT retained stable V$_{th}$, SS, μ, and I$_{on/off}$ (Fig. 3e).

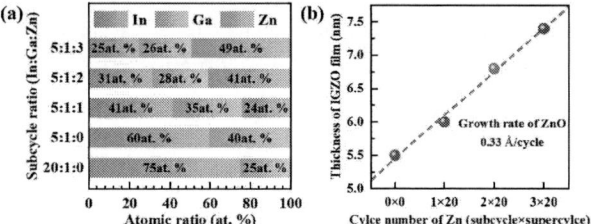

Fig. 2. (a) The atomic ratio of IGZO films with different subcycle parameters from XPS spectrum; (b) Thickness of IGZO film varies with cycle number of Zn and the growth rate of ZnO.

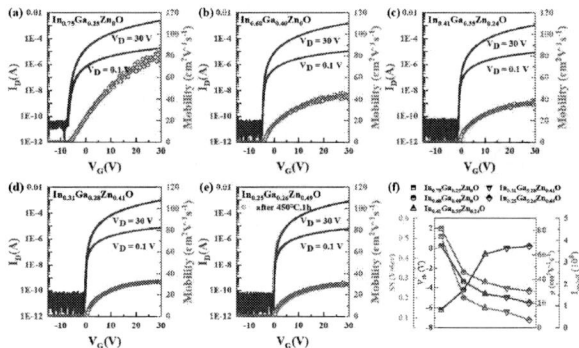

Fig. 3. (a-e) Transfer characteristics and mobilities of IGZO TFTs with varying cation ratios (The yellow marks are the transfer characteristic and mobility measured again after annealing at 450°C in air for 1h); (f) Electrical parameters of IGZO TFTs with different cation ratios.

Fig. 4a-d details the thermal stability assessment of IGZO TFTs across a temperature range incrementally increased from -25°C to 105°C. The impact of temperature on the V$_{th}$, SS, and μ of IGZO TFTs with diverse compositions is illustrated in Fig. 4e-g. An observed negative V$_{th}$ shift was mitigated by increased Zn content, with the shift ranging from over -15 V to -1.5 V. The SS increase was more substantial in devices with lower Zn content, indicating that Zn-enriched IGZO TFTs generate fewer electronic defects as the temperature rises. Mobility showed less temperature dependence, aligning with percolation transport theory[6]. These findings underscore the thermal stability of Zn-enriched IGZO TFTs.

The total defect density near the conduction band minimum of IGZO TFTs was analyzed through temperature-dependent tests and high-low frequency capacitance voltage (CV) tests,

denoted as DOS_{tt} and DOS_{CV}, derived from Eq. (1) and (2), respectively. As shown in Fig. 5, consistent results from both tests indicate that the total defect density in IGZO TFTs gradually decreases with increasing Zn content.

$$DOS_{tt} = - \varepsilon_i / [qd_i(\partial E_a / \partial V_G)] \tag{1}$$
$$DOS_{CV} = C_{INS}[C_{lf}/(C_{INS}-C_{lf}) - C_{hf}/(C_{INS}-C_{hf})]/q^2 \tag{2}$$

where ε_i and d_i correspond to the dielectric constant and thickness of the gate dielectric, respectively, q is the element charge, C_{INS} is the maximum capacitance and C_{lf} and C_{hf} are the measured capacitance per unit area at low and high frequencies, respectively.

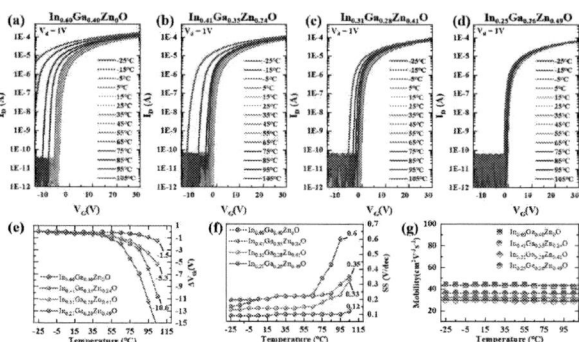

Fig. 4. (a-d) The transfer characteristic curves of IGZO TFTs with different composition ratios at different temperatures (-25°C~105°C) and the variations in (e) V$_{th}$, (f) SS, and (g) μ.

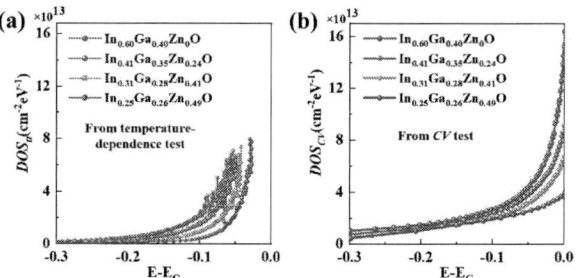

Fig. 5. Total density of defect states of IGZO TFTs extracted from (a) temperature-dependence test and (b) capacitance voltage (CV) measurements.

Besides thermal stability, the bias stability of devices was also investigated under ±20 V biases for 3600 s, and the results are shown in Fig. 6. The NBS induced threshold shifts of -4.4 V, -0.4 V, -0.2 V, and 0 V for In$_{0.60}$Ga$_{0.40}$Zn$_0$O, In$_{0.41}$Ga$_{0.35}$Zn$_{0.24}$O, In$_{0.31}$Ga$_{0.28}$Zn$_{0.41}$O and In$_{0.25}$Ga$_{0.26}$Zn$_{0.49}$O, respectively, due to V$_O$ transitioning to V$_O{}^{2+}$ and releasing electrons. The corresponding positive shift in threshold under positive bias stress (PBS) are 3.1 V, 1.8 V, 1.0 V, and 0.8 V. The instability is attributed to acceptor-like defects capturing channel electrons, with no change in SS or μ, which excludes gate insulator charge injection or the creation of new deep traps. These stability trends align with temperature dependence, confirming that the Zn composition reduces both donor-like and acceptor-like defect states.

Defects in IGZO are primarily oxygen-related, including metal-metal bonds, oxygen vacancies, and weakly bonded oxygen[7, 8]. XPS analysis, as shown in Fig. 7, deconvoluted the O 1s peak into three peaks at 529.8±0.1 eV (M-O), 530.8±0.1 eV (V_O), and 532.1±0.1 eV (M-OH). The increase in Zn content enhances the M-O state and reduces the V_O and M-OH states, as indicated by pie charts. The $In_{0.60}Ga_{0.40}Zn_0O$ has 17.30% V_O and 40.51% M-OH states, while $In_{0.25}Ga_{0.26}Zn_{0.49}O$ only has 11.17% V_O and 35.64% M-OH states. This suggests that Zn incorporation can regulate oxygen-related defects, thereby reducing electron traps and enhancing device stability..

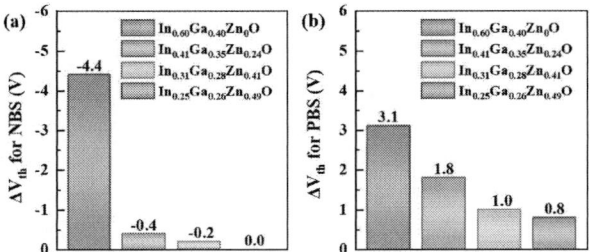

Fig. 6. The threshold voltage shifts of IGZO TFTs with different component ratios under (a) NBS and (b) PBS test

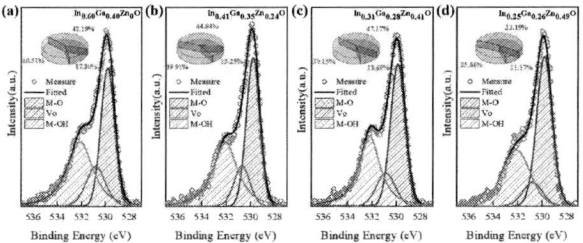

Fig. 7. (a-d) O1s XPS spectra of different composition ratios IGZO films.

GIXRD analysis reveals that increased Zn content in IGZO enhances its amorphous character. Zn's 4-coordinate structure (compared to the coordination number of 6 for In_2O_3 and Ga_2O_3) stabilizes the amorphous M-O network and inhibits crystallization. Consequently, this leads to a reduction in grain boundaries, an increase in surface roughness, and a decrease in defect states. Moreover, according to Phillips' theory[9], an ideal state is achieved with a balance between atomic freedom and interatomic constraints, quantified by the effective coordination number (ECN). The low defect formation energy in IGZO, attributed to the relaxation of metastable states, indicates an over-constrained system[10]. The incorporation of Zn not only promotes stable amorphous structures, suppresses the increase in grain boundaries and surface roughness, but also lowers IGZO's ECN, reducing over-stress, atomic relaxation, and preventing defect formation, such as V_O, which contributes to the improved stability (Fig. 8).

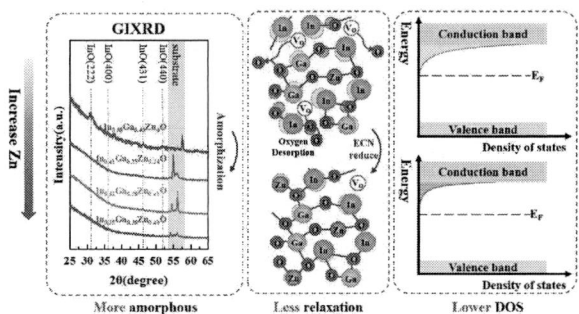

Fig. 8. A schematic diagram of the increased degree of amorphization, the reduction of over-stressed state atomic relaxation, and the decrease in DOS caused by the increase of Zn.

Conclusion

Our study on ultra-thin IGZO films, prepared via ALD, revealed that adjusting metal cation ratios significantly influences TFT performance and stability. The optimized TFTs showcased a mobility of 30.24 $cm^2V^{-1}s^{-1}$ and demonstrated superior bias stability (NBS = 0 V, PBS = 0.8 V) along with temperature stability (ΔV_{th} = -1.5 V over -25 to 105°C). Elevated Zn levels enhance IGZO's amorphous nature, alleviate stress, lower defect state density, and bolster stability.

Acknowledgments

This work was supported by the National Key Research and Development Program of China (2022YFB3603901).

References

[1] T. Kamiya, K. Nomura, and H. Hosono, "Present status of amorphous In-Ga-Zn-O thin-film transistors.," *Science and Technology of Advanced Materials.* vol. 11, no. 4, p. 044305, 2010.

[2] K.N. Chen, C.S. Tan, A. Fan, and R. Reif, "Morphology and bond strength of copper wafer bonding.," *Electrochemical and Solid-State Letters.* vol. 7, no. 1, p. G14, 2004.

[3] N. Zhang, W. Zhao, C. Yao, et al., "Transparent Multi-Level NAND Flash Memory and Circuits Based on ZnO Thin Film Transistor.," *IEEE Electron Device Letters.* vol. 44, no. 4, pp. 610–613, 2023.

[4] Y. Kim and C. Kim, "Enhancement of electrical stability of metal oxide thin-film transistors against various stresses.," *Journal of Materials Chemistry C.* vol. 11, no. 22, pp. 7121–7143, 2023.

[5] A. De Jamblinne De Meux, A. Bhoolokam, G. Pourtois, J. Genoe, and P. Heremans, "Oxygen vacancies effects in a-IGZO: Formation mechanisms, hysteresis, and negative bias stress effects.," *physica status solidi (a).* vol. 214, no. 6, p. 1600889, 2017.

[6] S. Lee, K. Ghaffarzadeh, A. Nathan, et al., "Trap-limited and percolation conduction mechanisms in amorphous oxide semiconductor thin film transistors.," *Applied Physics Letters.* vol. 98, no. 20, p. 203508, 2011.

[7] K. Ide, Y. Kikuchi, K. Nomura, M. Kimura, T. Kamiya, and H. Hosono, "Effects of excess oxygen on operation characteristics of amorphous In-Ga-Zn-O thin-film transistors.," *Applied Physics Letters.* vol. 99, no. 9, p. 093507, 2011.

[8] W. Ko, "Origin of subgap states in amorphous In-Ga-Zn-O.," *J. Appl. Phys.* p. 2024.

[9] J.C. Phillips, "Topology of covalent non-crystalline solids I: Short-range order in chalcogenide alloys.," *Journal of Non-Crystalline Solids.* vol. 34, no. 2, pp. 153–181, 1979.

[10] A. De Jamblinne De Meux, G. Pourtois, J. Genoe, and P. Heremans, "Defects in Amorphous Semiconductors: The Case of Amorphous Indium Gallium Zinc Oxide.," *Physical Review Applied.* vol. 9, no. 5, p. 054039, 2018.

Fully Printed Polymer Gas Sensors Based on Machine Learning for Calibration-free Mobile Sensing

Siying Li*, Sujie Chen, Qiuqi Zhang, Yuying Si, Xiaojun Guo*

Department of Electronic Engineering, School of Electronic Information and Electrical
Engineering, Shanghai Jiao Tong University, Shanghai 200240, China
(e-mail: lsysj2012@sjtu.edu.cn; x.guo@sjtu.edu.cn).

Abstract

Full printing processes were developed based on a composite of poly(3, 4-ethylene-dioxythiophene): poly(styrenesulfonate) (PEDOT:PSS) and silver nanowires (AgNWs) for potential of low cost manufacturing ammonia (NH_3) sensors. The sensors fabricated in the same batch exhibited good device-to-device uniformity. A calibration-free approach was developed based on early transient response characteristics of a small number of device samples. The obtained model was then implemented in a mobile phone to build a mobile sensing system. Fast and accurate detection was demonstrated by applying the system to other fresh gas sensors in the same batch without calibration.

Keywords: conductive polymer, gas sensor, printing, calibration-free, neural network, mobile sensing.

Introduction

Polymer functional materials and composite structures have been intensively studied for various gas/vapor sensors attributed to their wide-range tunable sensitivity and selectivity of electrical properties upon targeted analytes [1]. Possibility of being processed by solution printing at low temperature would also enable fabrication of sensors at low cost with ease of customization and short design-to-product time [2]. For practical use, calibration is needed for each sensor to build a relationship between the measured quantities and the actual concentration. However, for those sensors designed for one-time use or disposable tests, such calibration might not be applicable. Moreover, calibration for large number of sensors individually also consumes lots of time and resource. Calibration transfer, a technique aiming to reduce calibration cost by avoiding calibrating each sensor completely, has been explored [3]. With a calibration model obtained based on data collected from the reference sensors, only a smaller dataset transfer samples are needed to update the model and then transfer it to new sensors. However, without the uniformity of sensors, these works still require some transfer samples from new sensors. Due to material and processing limitations, printed polymer gas sensors normally suffer notable device-to-device variations, making it difficult to adopt transfer calibration. Besides, they mainly used the steady-state response value for calibration.

The response of most of the gas/vapor sensors to the analyte needs long time before reaching the steady state. Therefore, if using steady state values for calibration, it will take long time for the procedure, resulting in low efficiency. Moreover, real measurements based on such calibrated models will also be time-consuming, which is not acceptable for many mobile sensing applications. Instead of using the steady state, methods by predicting the concentration from the earlier transient response have been investigated for gas sensors to shorten the measurement time [4]. It would be ideal to develop calibration-free approach based on earlier transient response to enable one-time use or disposable tests with high efficiency in both calibration and test, for widespread adoption of those printed polymer gas sensors.

In this work, the large area, fully printed polymer gas sensors with good batch uniformity were developed based on a composite of conducting polymer (PEDOT:PSS) and silver nanowires(AgNWs). With AgNWs of optimized concentration being incorporated into the PEDOT:PSS film, the sensitivity of gas sensors were able to be significantly improved [2]. The fabricated sensors in the same batch exhibited good device-to-device uniformity. A calibration-free approach was thus developed based on early transient response characteristics of a small number of sensor samples. The obtained model was then implemented in a mobile phone to build a sensing system. Fast and accurate detection was demonstrated by applying the system to other sensors in the same batch without calibration.

Method

The fabrication processes and structure of the resistive NH_3 sensors are illustrated in Fig. 1(a). On a 125 μm thick pre-cleaned polyethylene naphihalate (PEN) substrate, AgNW dispersion of 0.50 mg/mL was bar-coated at 10 mm/s, followed by annealing for 10 minutes at 100 ℃, to form a uniform network below percolation threshold. An aqueous solution of DMSO-doped PEDOT:PSS was then deposited on top through bar coating. After a 120 °C drying process for 20 minutes, the sensing layer was formed. Details of the sensing layer design for improved sensitivity were described in previous work [2]. Afterwards, resin (JC-JY302) was screen printed as the isolation to define the sensing areas. Finally, electrodes were formed through screen printing Ag paste. Fig. 1(b) gives a photograph of a batch of 50 sensor devices on a 10 cm × 10 cm substrate, and individual devices can be cut from it for use.

A circuit board is built to read out the sensed signals, and communicate with a mobile phone (Fig. 1(c)). A resistor (R_0) is connected in series with the sensor (R_{NH3}) to convert its resistance change to voltage output (V_{OUT}). Four interface channels are designed with connection to a 12-bit analog-to-digital convertor (ADC) in the microcontroller (MCU, STM32F103C8T6) via a multiplexer (MUX). The digitalized data is transmitted to the mobile phone through Bluetooth (BLE, CC2541). A Li-ion battery with power management circuit is used to provide a 3.3 V voltage output.

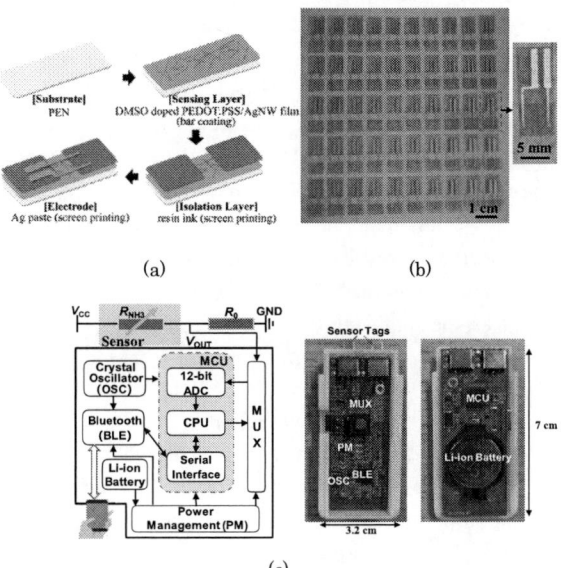

(a)   (b)

(c)

Fig. 1. (a) The fabrication processes and structure of the resistive NH₃ sensors based on the composite of PEDOT:PSS/AgNWs. (b) A photograph of a batch of 50 fabricated sensors on a 10 cm × 10 cm substrate. (c) Circuit diagram and photo images of the data acquisition (DAQ) circuit board built to read out the signals, and communicate with a smart phone via Bluetooth.

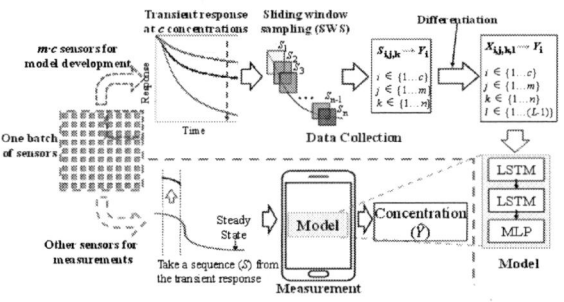

Fig. 2. Illustration of the developed calibration-free approach by using a small number of sensors in the batch for model development. The developed model based on neural network is implemented in the mobile phone for on-site measurement using the remaining sensors in the batch.

The developed calibration-free approach is illustrated in Fig. 2. A small number (m) of devices are taken from the batch to measure their transient response curves upon exposure to NH₃ of each concentration (totally c concentrations). Sliding window sampling (SWS) is used to sample a series of sequence data sets from each curve. In a detectable range (DR), the sampling window has a fixed size equal to the length of the sequence (L). By shifting its starting point forward from the beginning of the DR, a number of sequence data sets (n) is able to be obtained based on each curve. These

sequence data is denoted as $S_{i,j,k}$, standing for the kth sequence on the curve of the jth sensor measured at the ith concentration (i from 1 to c, j from 1 to m and k from 1 to n). To remove dependence on absolute values, differentiation is applied to $S_{i,j,k}$ to obtain a new sequence $X_{i,j,k,l}$ as:

$$X_{i,j,k,l} = S_{i,j,k,l+1} - S_{i,j,k,l}, \ l \in \mathbb{Z} \cap [1, L\text{-}1] \qquad (1)$$

where l represents the lth value in a sequence. $X_{i,j,k,l}$ and its actual concentration Y_i (i from 1 to c) as the label are used to train a neural network model, which consists of a two-layer stacked long short-term memory (LSTM) and a four-layer multilayer perceptron (MLP). 12 hidden neurons are used in LSTM layers. Each MLP layer has 20 neurons with the activation function *tanh*. The loss function of the model is the mean absolute error (MAE). The model is constructed, trained and evaluated using Keras, an open-source neural network framework [5]. The obtained model is implemented in an Android APP on a mobile phone to process the received data, and display the results. The remaining fresh sensors in the batch are taken for further tests through the mobile sensing system.

Results and Discussion

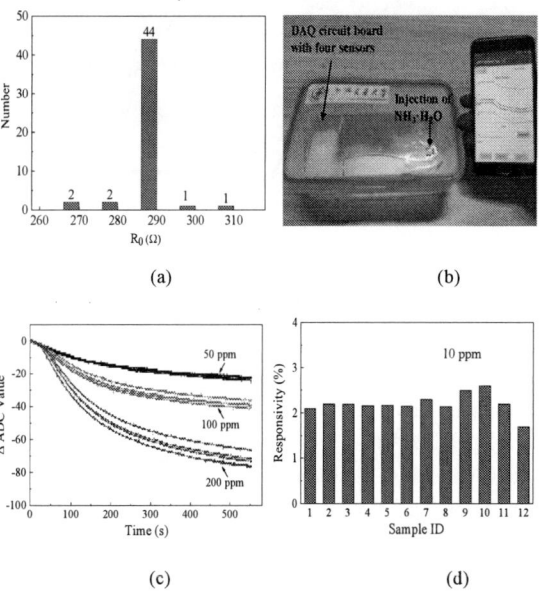

(a)   (b)

(c)   (d)

Fig. 3. (a) Measured initial resistances of all 50 sensors on a batch using a Keithley 2400 source meter. (b) Experimental setup for measuring the response of sensors to NH₃ (c) Measured relative changes of the output values from the 12-bit ADC (△ADC) over time for 12 sensors. (d) Measured responsivity to NH₃ of a relatively low concentration (10 ppm) for 12 sensors.

The initial resistances of all 50 sensors on a batch were measured as shown in Fig. 3(a), indicating good uniformity of the fabricated structures. To measure the response to NH₃, the circuit board and four sensors being connected were placed in a 1 L plastic container (Fig. 3(b)). The four sensors were measured simultaneously, and the data were wirelessly transmitted to the mobile phone. A certain volume of

NH$_3$·H$_2$O was injected into the container through a small hole in the lid to create the NH$_3$ test environment, and the concentration was estimated based on the volume of the injected NH$_3$·H$_2$O and the container [2]. Fig. 3(c) shows the measured relative changes of the output values from the 12-bit ADC ($\triangle$ADC) over time under 3 different concentrations for 12 sensors, which were randomly taken from the same batch. The results exhibit good uniformity of the transient response upon the same concentration. It can also be seen that the response requires nearly 10 minutes to reach the steady state. For more characterization of the uniformity, another 12 sensors were taken from the batch to measure the responsivity to NH$_3$ of a relatively low concentration (10 ppm). The responsivity is obtained as below:

$$Responsivity = \frac{R - R_0}{R_0} \times 100\% \qquad (2)$$

where R_0 is the measured initial resistance of the sensor, and R is its resistance after exposure to NH$_3$ for 4 minutes. As shown in Fig. 3(d), the responsivity of the 12 sensors also exhibited good uniformity.

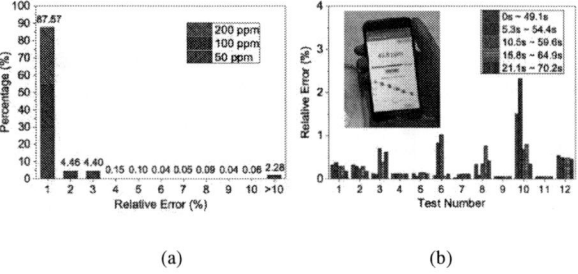

(a)　　　　　　　　(b)

Fig. 4. (a) The statistical distribution of the relative error (RE) of the prediction results using the developed model on the measured data in Fig. 3(c). (b) The RE values of the measurement results compared to the real concentrations for 12 fresh sensors using the built mobile sensing system.

Based on the uniform device-to-device characteristics, the proposed calibration-free approach was able to be used. The 12 curves in Fig. 3(c) were divided into 4 groups and each group contains curves of the three concentrations. By choosing any three groups for training and the other one for test, 4 different combinations of data were able to be obtained for 4-fold cross validation. The DR and the L for SWS were determined from the measured curves to be 147.4 s and 49.1 s, respectively. After the SWS process, 12609 sequence samples were obtained for training and 4203 samples for validation. The statistical distribution of the relative error (RE) of the prediction results using the developed model on the measured data is shown in Fig. 4(a). RE is defined as:

$$RE = \left\| \frac{\hat{y}_i - y_i}{y_i} \right\| \qquad (3)$$

where y_i and $\hat{y}_i$ are the actual and predicted concentration of the i_{th} sample, respectively. It can be seen that, for different concentrations, about 87.57 % of the prediction

results have a RE less than 1% and 96.4% of the prediction results have a RE less than 3%, proving very good accuracy with the developed model.

The obtained model was implemented in the Android APP to obtain a mobile sensing system. Another 12 fresh sensors were tested under different NH$_3$ concentrations of 50 ppm, 100 ppm and 200 ppm. The measured concentration values can be directly obtained and displayed on the APP interface without calibration as shown in the inset of Fig. 4(b). The same DR and L settings were used for the test. For each sensor test, five continuous measurements were performed, by taking the different periods in the DR of the transient response curve. The RE values of the measurement results compared to the real concentrations are shown in Fig. 4(b). All measurement results have a RE less than 3%, proving the good enough accuracy with the calibration-free approach. Moreover, the single test time only needs 49.1 s (L), which is much shorter than the required time for reaching the steady state (about 10 mins).

Conclusion

Full printed NH$_3$ sensors based on PEDOT:PSS/AgNW composite were fabricated. Based on the exhibited uniform sensing properties, a calibration-free approach is developed. Early transient response characteristics of a small number of device samples are used to generate data sets for building a neural network model. The obtained model was implemented in a mobile phone to build a sensing system. Fast and accurate detection was demonstrated by applying the system to other gas sensors in the same batch without calibration. This approach would assist widespread adoption of those printed polymer gas sensors by enabling one-time use or disposable tests with high calibration and test efficiency.

Acknowledgments

The authors gratefully acknowledge funding support by the National Natural Science Foundation of China under Grant No. 62104143 and Grant No. 62174106.

References

[1] S. Li, W. Tang, S. Chen, Y. Si, R. Liu, and X. Guo. "Flexible organic polymer gas sensor and system integration for smart packaging, " Adv. Sens. Res., 2(11): 2300030 (2023).

[2] S. Li, S. Chen, B. Zhuo, Q. Li, W. Liu, X. Guo, "Flexible ammonia sensor based on PEDOT: PSS/silver nanowire composite film for meat freshness monitoring," IEEE Electron Device Lett., 38(7), pp. 975-978(2017).

[3] L. Fernandez, S. Guney, A. Gutiérrez-Gálvez, and S. Marco, "Calibration transfer in temperature modulated gas sensor arrays," Sens. Actuators B, Chem., 231, pp. 276-284(2016).

[4] Q. Zhang, S. Li, W. Tang, and X. Guo, "Fast measurement with chemical sensors based on sliding window sampling and mixed-feature extraction, " IEEE Sens. J., 20(15), pp. 8740-8745 (2020).

[5] A. Gulli and S. Pal, "Deep learning with Keras," Birmingham, UK: Packt Publishing Ltd (2017).

Re-examination of Uniaxial Stress Effects in Ultra-scaled GAAFETs

Yusi Zhao[1], Huawei Tang[1], Rongzheng Ding[1], Yudong Lv[1], Yanbo Tang[1], Shaofeng Yu[1*]

[1]School of Microelectronics, Fudan University, Shanghai 200433, China
(*Email: shaofeng_yu@fudan.edu.cn)

Abstract

This work investigates the effects of uniaxial stress on the performance of ultra-scaled gate-all-around field-effect transistors (GAAFETs) through multi-subband Boltzmann transport equation (MSBTE)-based simulations. Both scattering-limited and ballistic on-state saturation currents are calculated to explore the effectiveness of uniaxial stress in enhancing device performance. Although uniaxial stress (within the experimentally achievable range of values) still yields performance gains, the underlying physical mechanisms differ from those observed in previous technology nodes. By extracting key parameters, such as ballistic injection velocity, quantum gate capacitance and backscattering coefficient, we provide fundamental physical insight into the impact of stress in ultra-scaled GAAFETs.

Keywords: Backscattering coefficient, Gate-all-around field-effect transistors (GAAFETs), Quantum confinement, Quasi-ballistic transport, Uniaxial stress engineering.

Introduction

Uniaxial stress engineering has been widely accepted as a performance booster, enhancing mobility at the 90 nm technology node and beyond [1]. Gate-all-around field-effect transistors (GAAFETs) have emerged as the leading candidates for future technology nodes beyond FinFETs [2]. To maintain strong electrostatic control, the cross-sectional dimensions of GAAFETs must scale down in tandem with channel length reduction. As the channel length decreases, the transport mechanism transitions from drift-diffusion to ballistic transport, where scattering-limited mobility loses physical meaning [3]. Simultaneously, aggressive cross-sectional scaling induces significant quantum confinement effects, degrading device performance [4]. Given these emerging physical mechanisms in ultra-scaled GAAFETs, the continued effectiveness of uniaxial stress engineering merits thorough investigation. Instead of simply attributing stress-induced enhancements to mobility gains, the fundamental physical effects of uniaxial stress should be carefully re-evaluated for ultra-scaled GAAFETs.

Simulation Methodology

An advanced simulation framework, multi-subband Boltzmann transport equation (MSBTE)-based numerical simulation tailored for devices at future technology nodes is adopted in this work. The MSBTE-based TCAD [5] provides a self-consistent solution of coupled 1-D Boltzmann transport, 2-D Schrödinger, and 3-D Poisson equations.

We carry out the well-calibrated MSBTE simulations for GAAFETs with 2-D cross sections of 5 nm × 10 nm. This work focuses on the n-type GAAFETs, where uniaxial tensile stress is applied along the channel direction. Stress effects are physically captured in the confined k·p Schrödinger solver by converting the stress tensor to the strain tensor based on material elasticity, and then integrating the H_{strain} term—whose elements are functions of the strain tensor—into the Schrödinger equation's Hamiltonian. For the conduction band, three pairs of ellipsoidal Delta valleys ($\Delta_{<100>}$, $\Delta_{<010>}$, $\Delta_{<001>}$, each distinguished by its longitudinal axis orientation) are computed using the 2 k·p method. Phonon and surface roughness scattering models are also integrated into the simulation.

Results and Discussion

For GAAFETs with (100) surface and <100> channel orientation, the six conduction band valleys divide into two groups with distinct band curvature. As uniaxial tensile stress increases, the light valley Δ_{Light} (comprising $\Delta_{<010>}$ and $\Delta_{<001>}$) shifts downward relative to the heavy valley Δ_{Heavy} ($\Delta_{<100>}$), prompting more electrons to occupy Δ_{Light}, as shown in Fig. 1. Consequently, the average effective mass decreases. Once all electrons reside in Δ_{Light}, the occupancy-weighted average transport effective mass saturates to the mass value of Δ_{Light}, as verified by Fig. 2. In contrast, in (100)/<110> GAAFETs, where $\Delta_{<100>}$ and $\Delta_{<010>}$ are Δ_{Heavy} while $\Delta_{<001>}$ is Δ_{Light}, the reduction in average effective mass continues even after all electrons occupy Δ_{Light}. The energy surface of Δ_{Light} is warped due to uniaxial <110> stress [6], leading to a reduced mass of Δ_{Light}, as shown in Fig. 2(a). A lighter transport effective mass m^*_{tran} results in higher scattering-limited low-field mobility μ_{low} and larger thermal injection velocity v_T, as illustrated in Fig. 3. μ_{low} enhancement induced by stress is always the main focus when exploring stress technology. However, μ_{low} is valid only for drift-diffusion transport, necessitating an alternative performance metric for ballistic transport.

Fig. 4(a) shows the distribution function along the channel (z-direction) for on-state ballistic GAAFETs, excluding all scattering models. Contributed by both source and drain carrier injection, the ballistic distribution function partially satisfies the thermal equilibrium distribution at the source, while the remaining portion adheres to the thermal equilibrium distribution at the drain, resulting in a discontinuity in k_z-space. At the position of the top of the barrier (ToB), source injection fills all positive k_z states while

979-8-3315-0417-5/25 $31.00 © 2025 IEEE

the negative k_z states are only occupied by the drain. As V_{DS} increases (Fig. 4(b)), the drain injection becomes suppressed leaving the source injection unbalanced, so the current starts to flow. At high V_{DS}, all carriers are contributed by the source injection, causing the shape of the distribution becomes hemi-Fermi-Dirac, indicating that the velocity v_{ToB} at the ToB is exactly the source injection thermal velocity v_T. This velocity is defined as the ballistic velocity v_{bal}, a pivotal indicator of ballistic on-state performance.

However, as shown in Fig. 5, the enhancement of the ballistic saturation current (I_{bal}^{on} with fixed overdrive voltage $V_{GS} - V_T = 0.2$ V) is less than the v_{bal} gain induced by stress. This is because uniaxial stress not only modifies the transport properties but also affects quantum confinement in ultra-scaled GAAFETs. The capacitor within the 2-D confined semiconductor well can be represented as a series combination of density-of-state (DOS) capacitance C_{DOS} and centroid capacitance C_{cent} [7]. C_{DOS} is proportional to the DOS at the Fermi level while C_{cent} is related to the average distance of the inversion charge from the interface, i.e. inversion-layer centroid. Increasing tensile stress leads to a degradation of effective gate capacitance C_g, due to reductions in both C_{DOS} and C_{cent}. Fig. 6 depicts the variations of the extracted quantum centroid X_{qm} and lumped DOS mass for the first subband as stress level increases. Higher tensile stress exacerbates quantum confinement, pushing electrons further away from the interface (resulting in a larger X_{qm}) and consequently reducing C_{cent}. The C_{DOS} degradation is due to the reduction of lumped m_{DOS}^*, which is caused by the greater subband separation and lighter band curvature at higher uniaxial tensile stress.

We incorporate scattering-related models and parameters to simulate scattering events that are likely to occur in GAAFETs. Fig. 7 illustrates the variation of the scattering-limited saturation current I_{scat}^{on} under different channel lengths and stress levels. It is evident that uniaxial tensile stress continues to enhance the performance of ultra-scaled GAAFETs. This improvement partially originates from the increased injection velocity and also from the rising ballistic ratio (BR). BR is defined as the ratio of scattering-limited current to ballistic current. As the channel length decreases, approaching or even becoming shorter than the mean free path (MFP), BR increases, as validated by Fig. 8. When BR=1, ideal ballistic transport is achieved.

The enhanced BR resulting from uniaxial stress is primarily attributed to the reduction in the backscattering coefficient r. We extract r under high V_{DS} conditions as the ratio of two directional currents (forward and backward) at the ToB. In high-field conditions, where carriers injected by the drain are unable to overcome the high potential barrier to reach the ToB (as validated by Fig. 4), the backward current arises from a portion of the carriers injected by the source that are redirected due to backscattering. Fig. 9 shows the

backscattering coefficient in saturation regime. A decrease in channel length and an increase in stress level can hinder backscattering, thereby enhancing the device's ballisticity.

The backscattering coefficient r_{sat} for the saturation (high-field) regime can be analytically expressed as $r_{sat} = L_{crit}/(\lambda_{MFP} + L_{crit})$ [8], where λ_{MFP} is the scattering-determined MFP and L_{crit} denotes the critical length of the narrow region around the ToB, as illustrated in Fig. 11. Once carriers cross this region, no scattering can effectively redirect them back to the source, ensuring their eventual collection at the drain contact. Fig. 10 shows the variation of λ_{MFP} and L_{crit} with changing stress levels and channel lengths. An increase in stress level leads to a larger λ_{MFP}, because of the μ_{low} enhancement. μ_{low} may still play an important rule for quasi-ballistic transport, indirectly influencing device performance. The slight variation in L_{crit} can be attributed to stress-induced changes in the channel potential profile. In contrast, the reduction of r_{sat} with decreasing channel length is primarily due to the decrease in L_{crit}, which is proportional to the channel length.

Finally, Fig. 11 summarizes the impact of uniaxial stress on key device parameters introduced and analyzed in this paper. The transport-related parameters, μ_{low} for scattering-limited transport and v_{bal} for ballistic transport, are improved attributed to stress-induced band modulation. However, the exacerbated quantum confinement due to stress degrades gate electrostatic control. The backscattering coefficient also plays a dominate role in determining the current drive and can be changed by uniaxial stress through the modulation of λ_{MFP} and L_{crit}.

Conclusion

The effects of uniaxial stress on the performance of ultra-scaled GAAFETs are studied based on MSBTE simulations. By applying experimentally achievable uniaxial tensile stress along the channel of n-type GAAFETs, device performance can be enhanced through v_{bal} improvement and r_{sat} reduction, albeit with the cost of C_g degradation. Although uniaxial stress continues to boost performance in ultra-scaled GAAFETs, the physical mechanisms underlying the stress effects merit re-examination.

References

[1] M. Chu, "Strain: A Solution for Higher Carrier Mobility in Nanoscale MOSFETs," Annu Rev Mater Res, vol. 39, no. 1, pp. 203–229, 2009
[2] G. Bae et al., "3nm GAA Technology featuring Multi-Bridge-Channel FET for Low Power and High Performance Applications," 2018 IEEE IEDM
[3] S. Cristoloveanu et al, "Intrinsic Mechanism of Mobility Collapse in Short MOSFETs," IEEE TED, vol. 68, no. 10, pp. 5090–5094, 2021
[4] A. Dasgupta et al., "BSIM Compact Model of Quantum Confinement in Advanced Nanosheet FETs," IEEE TED, vol. 67, no. 2, pp. 730–737, 2020
[5] "Sentaurus Device QTX User Guide," Dec. 2019.
[6] K. Uchida et al, "Physical Mechanisms of Electron Mobility Enhancement in Uniaxial Stressed MOSFETs and Impact of Uniaxial Stress Engineering in Ballistic Regime," 2005 IEEE IEDM
[7] R. Granzner et al, "Quantum Effects on the Gate Capacitance of Trigate SOI MOSFETs," IEEE TED, vol. 57, no. 12, pp. 3231–3238, 2010
[8] M. Lundstrom et al, "Essential Physics of Carrier Transport in Nanoscale Mosfets," IEEE TED, vol. 49, no. 1, pp. 133–141, 2002

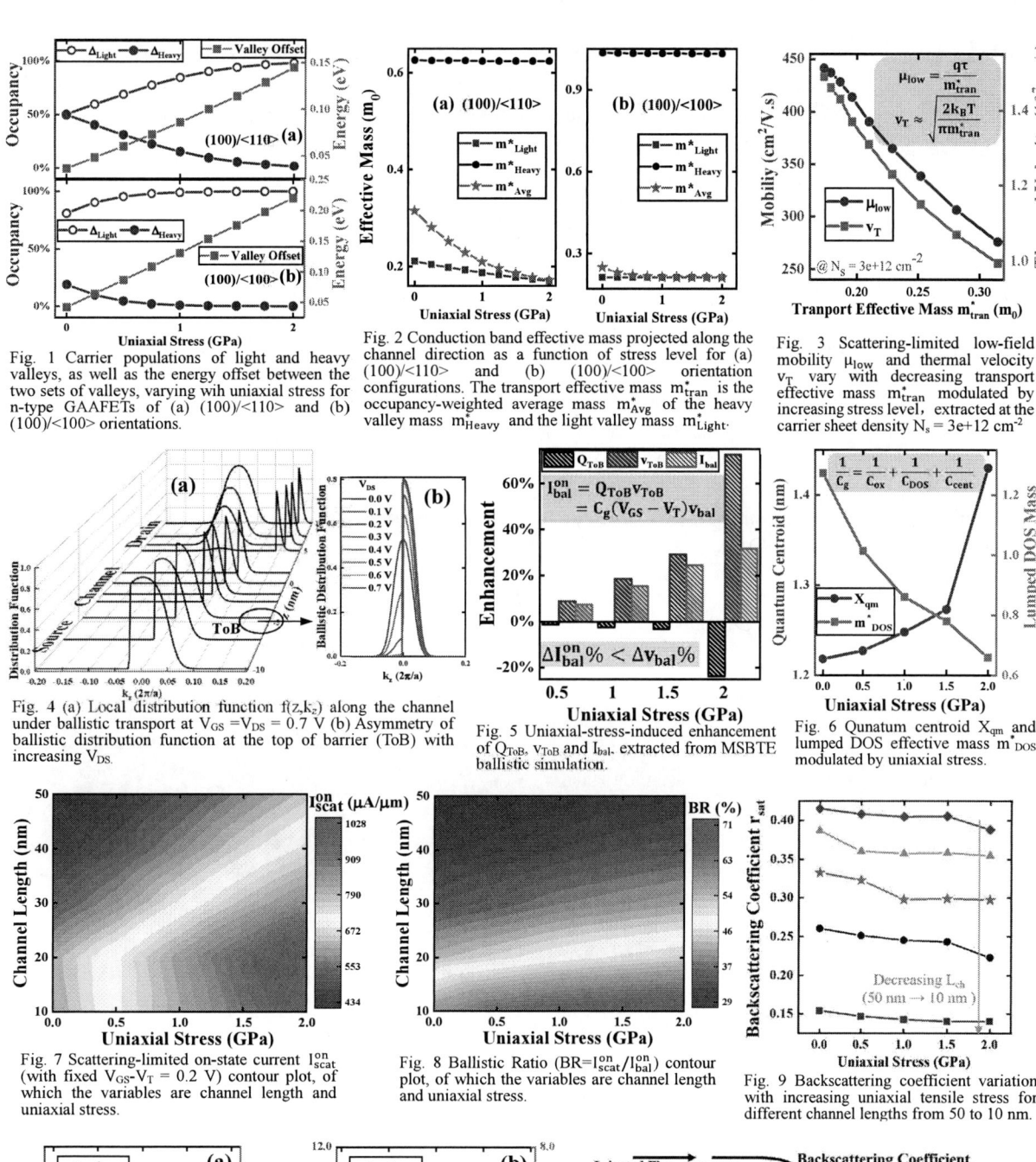

Fig. 1 Carrier populations of light and heavy valleys, as well as the energy offset between the two sets of valleys, varying wih uniaxial stress for n-type GAAFETs of (a) (100)/<110> and (b) (100)/<100> orientations.

Fig. 2 Conduction band effective mass projected along the channel direction as a function of stress level for (a) (100)/<110> and (b) (100)/<100> orientation configurations. The transport effective mass m^*_{tran} is the occupancy-weighted average mass m^*_{Avg} of the heavy valley mass m^*_{Heavy} and the light valley mass m^*_{Light}.

Fig. 3 Scattering-limited low-field mobility μ_{low} and thermal velocity v_T vary with decreasing transport effective mass m^*_{tran} modulated by increasing stress level, extracted at the carrier sheet density $N_s = 3e+12\ cm^{-2}$

Fig. 4 (a) Local distribution function $f(z,k_z)$ along the channel under ballistic transport at $V_{GS} = V_{DS} = 0.7$ V (b) Asymmetry of ballistic distribution function at the top of barrier (ToB) with increasing V_{DS}.

Fig. 5 Uniaxial-stress-induced enhancement of Q_{ToB}, v_{ToB} and I_{bal}. extracted from MSBTE ballistic simulation.

Fig. 6 Qunatum centroid X_{qm} and lumped DOS effective mass m^*_{DOS} modulated by uniaxial stress.

Fig. 7 Scattering-limited on-state current I^{on}_{scat} (with fixed $V_{GS}-V_T = 0.2$ V) contour plot, of which the variables are channel length and uniaxial stress.

Fig. 8 Ballistic Ratio (BR=$I^{on}_{scat}/I^{on}_{bal}$) contour plot, of which the variables are channel length and uniaxial stress.

Fig. 9 Backscattering coefficient variation with increasing uniaxial tensile stress for different channel lengths from 50 to 10 nm.

Fig. 10 Mean free path λ_{MFP} and Critical Length L_{crit} for (a) different stress levels at $L_{ch} = 30$ nm and (b) different channel lengths without stress.

Fig. 11 Schematic diagram of carrier channel backscattering at high V_{DS}. Uniaxial stress effects in ultra-scaled GAAFETs are summarized.

GIT-Based Bipolar *p*-FET with Enhanced Conduction Capability on *E*-mode GaN-on-Si HEMT Platform

Chengcai Wang, Jinjin Tang, Junting Chen, Mengyuan Hua*

Department of EEE, Southern University of Science and Technology, Shenzhen, China

*Email: huamy@sustech.edu.cn

Abstract

A GIT-based bipolar *p*-FET (G-BiPFET) structure is introduced to improve the conduction performance of GaN-based *p*-channel transistors. In G-BiPFET, a GaN-based gate injection transistor is cascaded with a conventional *p*-FET, boosting the conduction current by utilizing electrons as the majority carriers. The drain current density of the G-BiPFET significantly increases to 77 mA/mm, approximately 77 times higher than that of the conventional *p*-FET. Additionally, the G-BiPFET maintains a similar I_{ON}/I_{OFF} ratio and gate leakage as the conventional *p*-FET.

Keywords: GaN, HEMT, *p*-GaN, *p*-FET

Introduction

Enhancement-mode (*E*-mode) *p*-GaN gate high-electron-mobility transistors (HEMTs) have outstanding efficiency in commercial power applications [1]-[2]. The planar heterojunction-based structure naturally favors high-density all-GaN power integrated circuits (ICs), which is ideal for improving the on-chip performance [3]. As a crucial element of complementary logic, *p*-channel field-effect transistors (*p*-FETs) have been successfully implemented on the existing *E*-mode GaN-on-Si HEMT platform, further emphasizing the potential of GaN power ICs [4]-[6].

However, the limited conduction capability of *p*-FETs poses a significant challenge to the development of GaN-based complementary logic circuits, primarily due to the low hole mobility and magnesium activation ratio in *p*-GaN. One approach to improve conduction capability is to significantly downscale *p*-FETs to sizes smaller than 200 nm [7]. Another method involves designing epitaxial structures to increase the density and/or mobility of holes [8]-[9]. Although these strategies have achieved commendable current density, the intrinsic limitations of hole transport continue to restrict the current density of *p*-FETs. To address the limitations caused by low hole mobility, a novel bipolar *p*-FET (BiPFET) has recently been proposed to enhance the conduction capability of *p*-FETs on the existing *p*-GaN gate HEMT platform [10]. This BiPFET integrates an *n*-/*p*-/*n*-GaN (NPN) bipolar stack on the drain side of the traditional *p*-FET. The NPN stack can operate as a bipolar-junction transistor (BJT) to amplify the conduction current by utilizing electrons as the

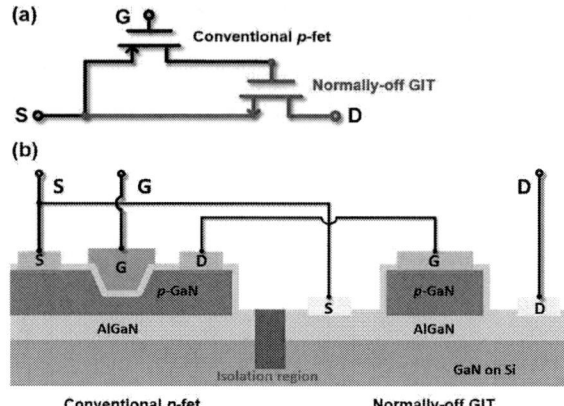

Fig. 1. (a) The equivalent circuits of G-BiPFET. (b) 3D schematic cross-sectional view of *p*-FET and GIT on the *E*-mode GaN-on-Si platform.

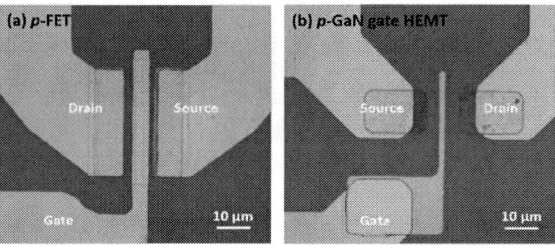

Fig. 2. Microscope images of (a) the *p*-FET and (b) *p*-GaN gate HEMT.

majority carriers. This innovative BiPFET results in 17 times increase in current density, reaching 120 mA/mm. However, the NPN stack requires selective-area growth (SAG) by metal-organic chemical vapor deposition (MOCVD), which could increase costs and complicates the manufacturing process. Therefore, developing a BiPFET that does not require SAG is an attractive solution. Fortunately, on the existing *p*-GaN gate HEMT platform, gate injection transistors (GITs) could be a replacement for BJTs because they can both be regarded as current-driven amplifiers [11].

In this work, a GIT-based bipolar *p*-FET (G-BiPFET) is proposed to enhance the conduction capability of *p*-channel devices. On the existing *p*-GaN gate HEMT platform, the G-BiPFET is formed by directly cascading a GaN-based GIT with a conventional *p*-FET, eliminating the need for additional SAG. In the G-BiPFET, the drain current density significantly increases to 77 mA/mm, which is approximately 77 times higher than that of the conventional *p*-FET. Furthermore, the G-BiPFET

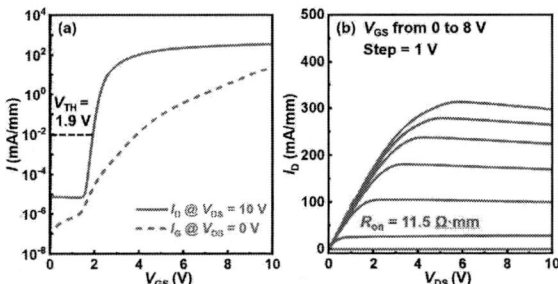

Fig. 3. (a) Transfer, gate leakage and (b) output characteristics of E-mode p-GaN gate HEMT. (Device dimension: $L_{GS}/L_G/W_G/L_{GD}$ = 3 μm/4 μm/ 20 μm/4 μm).

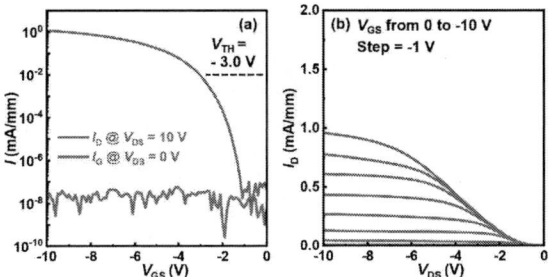

Fig. 4. (a) Transfer, gate leakage and (b) output characteristics of E-mode p-FET. (Device dimension: $L_{GS}/L_G/W_G/L_{GD}$ =3.5 μm/3 μm/40 μm/4 μm).

maintains E-mode operation, a high I_{ON}/I_{OFF} ratio and a low gate leakage current.

Device Structure and working mechanism

Fig. 1. illustrates the equivalent circuit and device structure of the proposed G-BiPFET to demonstrate its operational principle. The GIT is cascaded with the p-FET by connecting the drain of the p-FET to the gate of the GIT. The drain of the GIT serves as the drain of the G-BiPFET, while the source of the GIT is shorted with the source of the p-FET, acting as an electron source for current amplification.

As a current-driven device, the GIT can control the conduction of the 2DEG channel by adjusting the gate leakage current, enabling current amplification. When the p-FET is turned off, the gate leakage current of the GIT is very small, thus keeping the 2DEG channel off, which enables the G-BiPFET to maintain a low off-state leakage (I_{OFF}). When the p-FET is turned on, the gate leakage current of the GIT becomes sufficiently large to turn on the 2DEG channel, making electrons the majority carriers, which permits high current density flow.

Fabrication process

The p-GaN gate HEMT and p-FET device are both fabricated on a commercial 6-inch GaN-on-Si wafer, which features a 4.2-μm GaN buffer/transition layer, a

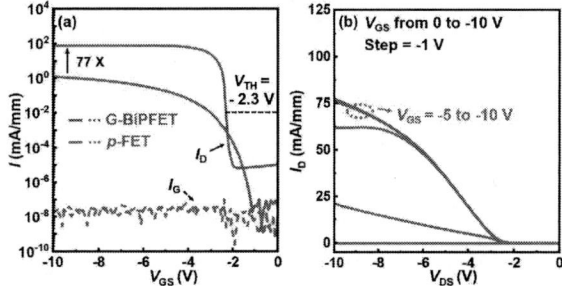

Fig. 5. (a) Transfer, gate leakage and (b) output characteristics of G-BiPFET.

300-nm undoped GaN channel layer, a 15-nm $Al_{0.15}Ga_{0.85}N$ barrier layer, and a 100-nm p-GaN layer with Mg-doping concentration of 4×10^{19} cm^{-3}. The device fabrication commenced with the removal of p-GaN outside the gate region of p-GaN gate HEMTs using the low damage BCl_3/Cl_2-based inductively coupled plasma (ICP). Then, the source/drain ohmic contacts were formed by depositing Ti/Al/Ti/Au (20/110/40/50 nm) metal stack with e-beam evaporator, followed by thermal annealing in N_2 ambient at 830 °C for 30 s.

Afterwards, p-ohmic contact regions of p-FETs were defined by photolithography, followed by soaking in buffered oxide etchant (BOE) for 5 mins. A Ni/Au/Ni (30/50/30 nm) metal stack was then e-beam evaporated, followed by lift-off and thermal annealing in N_2/O_2 = 1:1 ambient at 500 °C for 5 mins. Then PECVD-SiN_x was deposited as both passivation layer and hard-mask for p-FET gate region recess, which was also conducted by low damage BCl_3/Cl_2-based ICP dry etch. The remaining thickness of p-GaN after gate recess is 30 nm. An oxygen plasma surface treatment was conducted to form a buried p-channel [5]. Subsequently, a 20-nm Al_2O_3 layer was deposited using ALD at 300 °C as gate dielectric. After planar isolation, Ni/Au stack was evaporated to form gate electrode of p-FET and p-GaN gate HEMT. The microscope images of the final devices are shown in Fig. 2.

Results and discussion

Fig. 3 presents the transfer, gate leakage, and output characteristics of the p-GaN gate HEMT. The device operates in a normally-off mode with a threshold voltage (V_{TH}) of 1.9 V defined at I_D = 0.01 mA/mm, and also shows a large gate leakage current of 23 mA/mm at V_{GS} = 10 V (Fig. 3(a)), which is characteristic of GIT. The device delivers a low on-resistance (R_{ON}) of 16 Ω·mm and a high maximum drain current density (I_{DMAX}) of 313 mA/mm (Fig. 3 (b)). Owing to the application of oxygen plasma treatment, the p-FET shows a more negative V_{TH} of −3 V

Table 1.
Benchmark of E-mode p-FET on p-GaN gate HEMTs platform

Refs	L_{SD} (µm)	I_{DMAX} (mA/mm)	I_{ON}/I_{OFF}	V_{TH}^{*} (V)
HRL [12]	4.5	1.65	10^6	-0.36
MIT [4]	0.2	100	10^4	0
TUOS [14]	3	4.1	10^7	-0.73
ASU [13]	18	0.2	10^7	-0.6
HKUST [5]	10	6.1	10^7	-1.7
HKUST [6]	10	8	10^7	-1.7
SUSTECH [10]	8	120	10^6	-2
This work	**10.5**	**77**	**10^7**	**-2.3**

*Extracted at $|I_D| = 0.01$ mA/mm unless otherwise specified.

(Fig. 4 (a)), even though the remaining thickness of p-GaN layer is 30 nm. At $V_{DS} = V_{GS} = -10$ V, the p-FETs deliver a I_{DMAX} of approximately 1 mA/mm (Fig. 4 (b)), which is sufficient to drive the GIT.

Fig. 5 shows the transfer and output characteristics of the G-BiPFET. The drain current density of the G-BiPFET is normalized using the combined gate width (W_G) of p-FET and p-GaN gate HEMT. Compared to a conventional p-FET, the drain current density of the G-BiPFET increases by approximately 77 times (Fig. 5 (a)), reaching 77 mA/mm at a V_{GS} of -10 V (Fig. 5 (b)). Notably, if we optimize the I_{DMAX} of the p-FET in this work to match the reported value with a similar L_{SD} in [5]-[6], the W_G of the p-FET could be reduced by about 6 times while maintaining the same absolute GIT drive current. This would allow the I_{DMAX} of the G-BiPFET to increase to 174 mA/mm, indicating significant room for improvement in the G-BiPFET. The G- BiPFET also exhibits a low gate leakage and E-mode operation with a negative V_{TH} of -2.3 V. In the off-state, the G-BiPFET has a higher I_{OFF} than the conventional p-FET, which is mainly due to the drain-to-source I_{OFF} of the p-GaN gate HEMT. Despite the higher I_{OFF}, the G-BiPFET still exhibits a comparable I_{ON}/I_{OFF} ratio of 10^7.

Table 1 compares the performance of G-BiPFETs with other reported E-mode p-FETs on p-GaN gate HEMT platform [4]-[6], [12]-[14]. The G-BiPFETs exhibit a combination of E-mode operation, a high I_{ON}/I_{OFF} ratio, and a high I_{DMAX}.

Conclusion

A G-BiPFET structure is demonstrated to improve the current density of GaN-based p-FETs. The G-BiPFET stack results in approximately 77 times amplification of the conventional p-FET current density to 77 mA/mm, while preserving the control logic, low gate leakage current and high I_{ON}/I_{OFF} ratio. The G-BiPFET provides a simple and low-cost approach to enhancing the conduction capability of GaN-based p-FET by using electrons to participate in the conduction.

Acknowledgments

This work was supported in part by National Natural Science Foundation of China under Grant No. 62304097, Guang Dong Basic and Applied Basic Research Foundation under Grant 2024A1515030224，Shenzhen Fundamental Research Program under Grant No. JCYJ20220530114615035 and Grant No. 2023112115707001. We acknowledge the SUSTech Core Research Facilities (SCRF) for the device fabrication and material characterization.

References

[1] K. J. Chen *et al.*, "GaN-on-Si Power Technology: Devices and Applications," *IEEE Transactions on Electron Devices*, vol. 64, no. 3, pp. 779-795, 2017.

[2] R. Rupp *et al.*, "Application Specific Trade-offs for WBG SiC, GaN and High end Si Power Switch Technologies," in *2014 IEEE International Electron Devices Meeting*, pp. 2.3.1-2.3.4.

[3] K. J. Chen *et al.*, "Planar GaN Power Integration – The World is Flat," in *2020 IEEE International Electron Devices Meeting (IEDM)*, pp. 27.1.1-27.1.4.

[4] N. Chowdhury *et al.*, "Field-induced Acceptor Ionization in Enhancement-mode GaN p-MOSFETs," in *2020 IEEE International Electron Devices Meeting (IEDM)*, pp. 5.5.1-5.5.4.

[5] Z. Zheng *et al.*, "High I_{ON} and I_{ON} / I_{OFF} Ratio Enhancement-Mode Buried p-Channel GaN MOSFETs on p-GaN Gate Power HEMT Platform," *IEEE Electron Device Letters*, vol. 41, no. 1, pp. 26-29, 2020.

[6] L. Zhang *et al.*, "SiN/in-situ-GaON Staggered Gate Stack on p-GaN for Enhanced Stability in Buried-Channel GaN p-FETs," in *2021 IEEE International Electron Devices Meeting (IEDM)*, pp. 5.3.1-5.3.4.

[7] Q. Xie *et al.*, "Highly-Scaled Self-Aligned GaN Complementary Technology on a GaN-on-Si Platform," in *2022 International Electron Devices Meeting (IEDM)*, pp. 35.3.1-35.3.4.

[8] A. Raj *et al.*, "Demonstration of a GaN/AlGaN Superlattice Based p-channel FinFET with High On-current," *IEEE Electron Device Letters*, vol. PP, no. 99, pp. 1-1, 2020.

[9] S. J. Bader *et al.*, "GaN/AlN Schottky-gate p-channel HFETs with InGaN Contacts and 100 mA/mm On-current," in *2019 IEEE International Electron Devices Meeting (IEDM)*, pp. 4.5.1-4.5.4.

[10] J. Tang *et al.*, "Bipolar p-FET with Enhanced Conduction Capability on E-mode GaN-on-Si HEMT Platform," *2023 International Electron Devices Meeting (IEDM)*, pp. 1-4, 2023.

[11] M. H. Y. Uemoto, H. Ueno, H. Matsuo, H. Ishida, M. Yanagihara, T. Ueda, T. Tanaka, D. Ueda, "Gate Injection Transistor (GIT)—A Normally-Off AlGaN/GaN Power Transistor Using Conductivity Modulation," *IEEE Transactions on Electron Devices*, vol. 54, no. 12, pp. 3393-3399, 2007.

[12] R. Chu *et al.*, "An Experimental Demonstration of GaN CMOS Technology," *IEEE Electron Device Letters*, vol. 37, no. 3, pp. 269-271, 2016.

[13] C. Yang *et al.*, "Enhancement-Mode Gate-Recess-Free GaN-Based p-Channel Heterojunction Field-Effect Transistor With Ultra-Low Subthreshold Swing," *IEEE Electron Device Letters*, vol. 42, no. 8, pp. 1128-1131, 2021.

[14] Y. Yin *et al.*, "High-Performance Enhancement-Mode p-Channel GaN MISFETs With Steep Subthreshold Swing," *IEEE Electron Device Letters*, vol. 43, no. 4, pp. 533-536, 2022.

93.6 cm² V⁻¹·s⁻¹ Homostructure a-IGZO Thin-Film Transistor with High-k Gate Dielectric Fabricated at Room Temperature

Heng Yue Gong,[1] Jia Cheng Li,[1] Yang Hui Xia,[1] Ya Dong Zhou,[1] Hui Xia Yang,[1] Yuan Xiao Ma,[1,*] Ye Liang Wang[1,*]

[1]The School of Integrated Circuits and Electronics, Beijing Institute of Technology, Beijing 100081, China

*Author to whom correspondence should be addressed: yxma@bit.edu.cn and yeliang.wang@bit.edu.cn

Abstract

In this work, amorphous indium gallium zinc oxide (a-IGZO) thin film transistors (TFTs) with HfLaO gate dielectric have been fabricated at room temperature. The carrier mobility of the TFTs can be significantly improved by adopting homostructure a-IGZO layers with various stoichiometric ratios. This improvement can be attributed to the raised carrier density in indium-rich a-IGZO because more donor-like oxygen vacancies are induced by the content of indium oxide, reducing the source/drain electrode contact resistance. Moreover, multiple channels can be formed in the homostructure, simultaneously contributing to carrier transport. As a result, high-performance TFTs with a a-$I_{1.0}G_{2.9}Z_{0.2}O_y$/a-$I_{1.0}G_{3.1}Z_{0.2}O_y$/a-$I_{1.0}G_{2.9}Z_{0.2}O_y$ homostructure have been obtained with an ultra-high mobility of 93.6 cm² V⁻¹·s⁻¹, a low subthreshold swing of 0.12 V dec⁻¹, a low threshold voltage of 2.5 V, and an impressive I_{ON}/I_{OFF} ratio of 3.4×10^7. Besides, the intrinsic three-terminal structure of the homostructure TFTs was further exploited to concurrently mimic the biological behaviors of neurotransmitters and neuromodulators, achieving synaptic behaviors of long-term potentiation (LTP) and depression (LTD).

Keywords: a-IGZO TFTs, homostructure channel, high mobility, room-temperature fabrication

Introduction

In past decades, amorphous oxide semiconductors (AOSs) have been extensively explored due to its exceptional features of high visible transparency, high carrier mobility, excellent uniformity, and low-temperature growth process [1]. In AOSs, electron carriers can be usually generated by oxygen vacancies, which can serve as electron donors [2]. Moreover, this oxygen-vacancies-assisted carrier mechanism has been extensively studied for conductance modulation to obtain resistive-switching behavior, which is crucial for memristive devices and synaptic functions [1]. As inspired by the initial memristor demonstration using amorphous TiOₓ, AOSs TFTs have been extensively studied to mimic synaptic devices for neuromorphic applications, for which neurotransmitters and neuromodulators can be emulated by channel and gate dielectric layer, respectively [3].

As a famous AOS, a-IGZO and related TFTs have attracted much attention since its discovery in 2004 [4]. However, low-temperature-processed a-IGZO inevitably possesses inherent oxide defects of dangling bonds, rough surfaces, nanoscale voids, and interfacial traps, which can severely limit carrier mobility for next-generation high-frame-rate displays and other applications. One effective way is to construct homostructure channel layer, which can improve electrode/channel contact and carrier transport at the homostructure interface [5]. Moreover, power consumption of a-IGZO TFTs requires further reduction, prompting the study on high-k oxide dielectrics to replace traditional SiO₂. The mixture of high-k dielectric is an effective method to improve dielectric quality, allowing them complement with each other [6].

In this study, a-IGZO synaptic TFTs with HfLaO as gate dielectric have been fabricated at room temperature. Homostructure channel was obtained by depositing a-IGZO with various stoichiometric ratios, which was achieved by varying the co-sputtering power at the indium metal target. X-ray photoelectron spectroscopy (XPS) and atomic force microscopy (AFM) were conducted on the a-IGZO films to investigate the films properties. A probe station with Keithley 4200A semiconductor analyzer was used to measure the electrical characteristics of the TFTs and high-k capacitance, which were conducted at room temperature in air.

Experimental

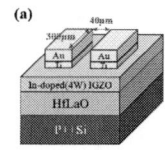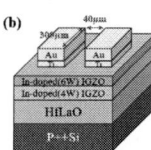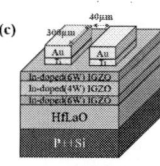

Figure 1. Schematic representation of the a-IGZO TFTs: (a) sample A, (b) sample B, and (c) sample C.

Firstly, heavily-doped p-type Si substrates (0.001–0.005 Ωcm) were ultrasonically cleaned in ethanol and deionized water each for 15 minutes. Next, HfLaO thin film was deposited on the substrates by co-sputtering a lanthanum (La) metal target at a 45 W DC power and a hafnium (Hf) metal target at a 12 W DC power, during which the ambience was Ar₂: N₂: O₂ (24: 1: 6 sccm) under a pressure of 1 Pa. Subsequently, a-$I_1G_{3.1}Z_{0.2}O_y$ thin film was deposited by co-sputtering an indium target (4W, DC) and a ceramic IGZO target (60W, RF) in an Ar₂: O₂ (30: 0.7 sccm) ambience at 1 Pa. Moreover, indium-rich a-$I_1G_{2.9}Z_{0.2}O_y$ was deposited by raising the co-sputtering power at the indium target to 6 W. As a result, channel layers of a-$I_1G_{3.1}Z_{0.2}O_y$, a-$I_1G_{2.9}Z_{0.2}O_y$/a-$I_1G_{3.1}Z_{0.2}O_y$ and a-$I_1G_{2.9}Z_{0.2}O_y$/a-$I_1G_{3.1}Z_{0.2}O_y$/ a-$I_1G_{2.9}Z_{0.2}O_y$

were deposited on the HfLaO dielectric, which were named as sample A, sample B and sample C, respectively. Finally, 10-nm Ti and 50-nm Au were deposited on all samples by e-beam evaporation to obtain source/drain electrodes, which was followed by a lift-off process to form channels with 40-μm length and 300-μm width. Figure 1 shows the schematic structure of the as-fabricated a-IGZO TFTs.

Additionally, Ti/HfLaO/Si MOS capacitors were simultaneously fabricated during the deposition of HfLaO on Si substrate to extract the capacitance per unit area (C_{ox}) of the gate dielectric, which was obtained by measuring the capacitance-voltage (C-V) characteristics.

Results and discussions

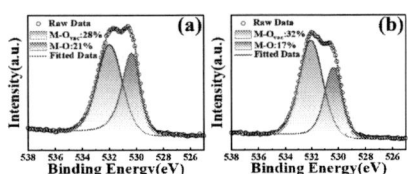

Figure 2. XPS images of (a) a-$I_1G_{3.1}Z_{0.2}O_y$ film and (b) a-$I_1G_{2.9}Z_{0.2}O_y$ film.

Figure 2 shows the XPS O $1s$ spectra of the a-$I_1G_{3.1}Z_{0.2}O_y$ and a-$I_1G_{2.9}Z_{0.2}O_y$ films. The curve can be decomposed of two subpeaks: the one around 530 eV is for metal-oxygen (M-O) bonds, and another one around 532 eV is for oxygen vacancies (M-O_{vac}). Notably, the content ratio of oxygen vacancies increases from 28% to 32%, which indicates a rise of carrier density because oxygen vacancies always act electron donors in AOSs [2].

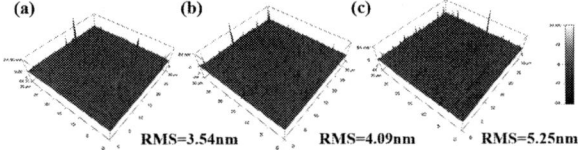

Figure 3. AFM images of a-IGZO film surface: (a) sample A with a-$I_1G_{3.1}Z_{0.2}O_y$, (b) sample B with a-$I_1G_{2.9}Z_{0.2}O_y$/a-$I_1G_{3.1}Z_{0.2}O_y$, and (c) sample C with a-$I_1G_{2.9}Z_{0.2}O_y$/a-$I_1G_{3.1}Z_{0.2}O_y$/a-$I_1G_{2.9}Z_{0.2}O_y$.

Figure 3 shows AFM characterization on the as-deposited a-IGZO thin films, and the root-mean-square (RMS) values of the film roughness were extracted. The surface roughness rises from 3.54 nm of sample A with a-$I_1G_{3.1}Z_{0.2}O_y$ to 4.09 nm of sample B with a-$I_1G_{2.9}Z_{0.2}O_y$/a-$I_1G_{3.1}Z_{0.2}O_y$, which should arise from the severe bombardment effect with higher co-sputtering power at 6 W for indium target [5]. Moreover, the roughest surface is observed for the sample C with a-$I_1G_{2.9}Z_{0.2}O_y$/a-$I_1G_{3.1}Z_{0.2}O_y$/a-$I_1G_{2.9}Z_{0.2}O_y$ due to rough a-$I_1G_{2.9}Z_{0.2}O_y$ as the bottom layer.

Figure 4(a) shows transfer characteristics, where the field-effect carrier mobility (μ) is extracted in the saturation region by linearly fitting $\sqrt{I_D}$ versus V_{GS}. As shown in Figure 4(b), the C-V characteristic of the dummy Ti/HfLaO/Si MOS capacitor was also obtained by sweeping

V_{GS} from -2 V to 6 V, demonstrating a C_{ox} of 0.095 μF cm⁻². The mobility (μ) is determined as [7]:

$$\mu = \frac{2L}{WC_{ox}}\left(\frac{\partial\sqrt{I_D}}{\partial V_{GS}}\right)^2 \qquad (1)$$

where L is the channel length; W is the channel width and I_D is the saturation drain current. Besides, the subthreshold swing (SS) can be extracted from [7]:

$$SS = \frac{\partial V_{GS}}{\partial\log I_D} \qquad (2)$$

Figure 4. (a) Transfer characteristics of the a-IZGO TFTs (b) C-V characteristic of the dummy Ti/HfLaO/Si MOS capacitor.

Table 1. Main electrical parameters of the a-IGZO TFTs.

Sample No.	A	B	C
Gate dielectric	HfLaO	HfLaO	HfLaO
μ (cm²V⁻¹·s⁻¹)	64.5	84.9	93.6
V_{th} (V)	2.4	2.7	2.5
SS (V dec⁻¹)	0.01	0.17	0.12
I_{ON}/I_{OFF} ratio	1.2×10^8	2.2×10^7	3.4×10^7
R_C (kΩ)	104	35	25

The main electrical parameters extracted from the samples are listed in Table 1. The field-effect carrier mobility is significantly improved from 64.5 cm² V⁻¹·s⁻¹ (sample A) to 84.9 cm² V⁻¹·s⁻¹ (sample B). This should arise from reduced electrode contact resistance (R_c) because more carriers can be induced in the indium-rich a-IGZO film. Another reason is the energy band bending at the a-$I_{1.0}G_{2.9}Z_{0.2}O_y$/a-$I_{1.0}G_{3.1}Z_{0.2}O_y$ interface, confining partial carriers in the potential well to preferably move along the interface direction [8]. Remarkably, sample C with a three-layer homostructure further boosts mobility to 93.6 cm² V⁻¹·s⁻¹, as carriers can be effectively accumulated at the indium-rich $I_{1.0}G_{2.9}Z_{0.2}O_y$/HfLaO interface by electric field in gate dielectric due to high carrier concentration from additional indium doping. Besides, all samples present low threshold voltages (V_{th}) (~ 2.5 V) due to the adoption of high-k HfLaO gate dielectric, indicating low power consumption.

The output characteristics are shown in Figure 5(a-c). Extracted by linearly fitting the turn-on curve at low V_{DS} (<1 V), the R_c decreases from 104 kΩ (sample A) to 35 kΩ (sample B) due to increased carriers in indium-rich a-$I_{1.0}G_{2.9}Z_{0.2}O_y$. Moreover, the R_c is further reduced to 25 kΩ (sample C) with the three-layer homostructure, which should be due to the effective carrier accumulation and multiple channels. Besides, all samples demonstrate superior stability under bias-stress tests (V_{DS} = 6 V, 10 minutes) as shown in Figure 5(d-f).

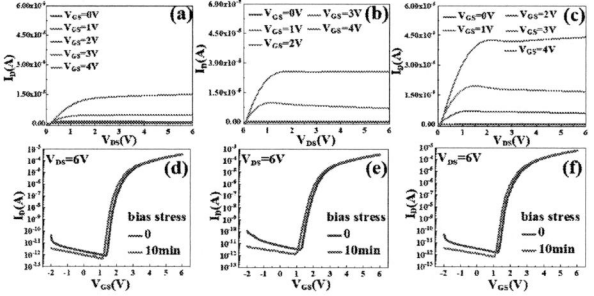

Figure 5. Output characteristics of the a-IZGO TFTs: (a) sample A, (b) sample B, and (c) sample C; Transfer characteristics after a bias stress V_{DS}=6 V for 0 min and 10mins: (d) sample A, (e) sample B and (f) sample C.

Finally, sample C with the three-layer homostructure is additionally explored as a synaptic device due to its high performances. As displayed in Figure 6(a), the source and drain electrodes are treated as a pre-neuron and a post-neuron respectively, the a-IGZO between which emulates the synaptic cleft for information transmission by neurotransmitters. Meanwhile, the gate dielectric mimics the effect of neuromodulators on activation of synaptic neurotransmitters [9]. As depicted in Figure 6(b-c), the drain current of the device maintains constant under continuous V_{DS} sweeping without V_{GS}. However, the drain current can gradually increase with V_{GS} = 1 V to manifest a gradual resistive switching, which is essential for synaptic emulation.

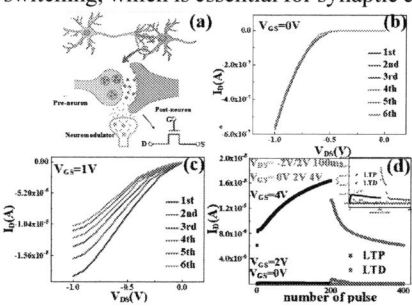

Figure 6. (a) Schematic illustrations of the artificial synapse; (b) and (c) *I-V* characteristics during subsequent V_{DS} sweep for V_{GS}=0 V and 1 V; (d) the LTP and LTD characteristics with drain spiking pulse under V_{GS}=0 V, 2 V and 4 V.

Moreover, pulses (100 ms, 2 V) were applied at the drain electrode to mimic synaptic spiking stimuli, after each of which the drain current was read at V_{DS} = 0.1 V. As shown in Figure 6(d), the drain current hardly changes in the absence of V_{GS}, which means that no synaptic behaviors are observed in the spiking pulse measurement. With increasing V_{GS}, typical LTP and LTD characteristics are observed, which successfully emulates the behaviors of neurotransmitter affected by neuromodulator. This should be originated from the migration of oxygen ions that repelled by V_{GS}, leaving

more oxygen vacancies in the a-IGZO film.

Conclusion

Adopting homostructure channel, a-IGZO TFTs with high-k HfLaO as gate dielectric have been successfully fabricated at room temperature. The carrier mobility can be significantly improved by using homostructure channel with indium-rich a-IGZO layer. Consequently, high-performance three-layer homostructure a-IGZO TFTs are obtained by room-temperature fabrication, demonstrating an ultra-high mobility of 93.6 cm^2 V^{-1}·s^{-1}, a low *SS* of 0.12 V dec^{-1}, a low threshold voltage of 2.5 V, and a high I$_{ON}$/I$_{OFF}$ ratio of 3.4×10^7. Finally, the three-layer homostructure TFT has been further explored as synaptic device to demonstrate synaptic behaviors of LTP and LTD.

Acknowledgments

We would like to acknowledge the National Key R&D Program of China (No.2023YFB3611700) and National Natural Science Foundation of China (62101044, 92163206).

References

[1] N. C. Su, S. J. Wang and A. Chin, "High-Performance InGaZnO Thin-Film Transistors Using HfLaO Gate Dielectric", IEEE Electron Device Letters, 30 (12), pp. 1317-1319 (2009)

[2] J. Yao, N. Xu, S. Deng, J. Chen, J. She, H. D. Shieh, P. Liu and Y. Huang, "Electrical and Photosensitive Characteristics of a-IGZO TFTs Related to Oxygen Vacancy", IEEE Transactions on Electron Devices, 58(4), pp. 1121-1126 (2011)

[3] D. Strukov, G. Snider, D. Stewart and R. S. Williams, "The missing memristor found", Nature 453, pp. 80-83 (2008)

[4] K. Nomura, H. Ohta, A. Takagi, T. Kamiya, M. Hirano and H. Hosono, "Room-temperature fabrication of transparent flexible thin-film transistors using amorphous oxide semiconductors", Nature, 432, pp. 488-492 (2004)

[5] R. Ye, M. Baba, Y. Oishi, K. Mori and K. Suzuki, "Air-stable ambipolar organic thin-film transistors based on an organic homostructure", Applied Physics Letters, 86, pp. 253505 (2005)

[6] J. S. Lee, S. Chang, S. -M. Koo and S. Y. Lee, "High-Performance a-IGZO TFT With ZrO$_2$ Gate Dielectric Fabricated at Room Temperature", IEEE Electron Device Letters, 31(3), pp. 225-227 (2010)

[7] J. K. Jeong, J. H. Jeong, H. W. Yang, J. -S. Park, Y. -G. Mo and H. D. Kim. "High performance thin film transistors with cosputtered amorphous indium gallium zinc oxide channel", Applied Physics Letters, 91 (11), pp. 113505 (2007)

[8] X. Ji, Y. Yuan, X. Yin, S. Yan, Q. Xin and A. Song, "High-Performance Thin-Film Transistors With Sputtered IGZO/Ga$_2$O$_3$ Heterojunction", IEEE Transactions on Electron Devices, 69(12), pp. 6783-6788 (2022)

[9] C. Henneberger, T. Papouin, S. Oliet and D. Rusakov, "Long-term potentiation depends on release of D-serine from astrocytes", Nature, 463, pp. 232-236 (2010)

Impact of Cross-Sectional Current Crowding on Electromigration in Interconnects

Yichen Wen[1], Shuying Wang[1], Xiaoman Yang[1], Hai-Bao Chen[1], Maokun Wu[1,+], Runsheng Wang[2], Zhigang Ji[1,*], and Ru Huang[2]

[1] Departure of Micro/Nano Electronics, Shanghai Jiao Tong University, Shanghai, China
[2]School of Integrated Circuits, Peking University, Beijing, China
([+]e-mail: maokunwu@sjtu.edu.cn) (*email: zhigangji@sjtu.edu.cn)

Abstract

As the critical dimensions of interconnect scaling, the enhancement of surface scattering increases the resistivity, contributing to the reduction in electromigration lifetime. Besides, the enhanced surface scattering can lower surface conductivity, resulting in cross-sectional current crowding and potentially decreasing the interconnect's electromigration resistance. Through finite element simulation, this work demonstrates the appearance of this extra decrement when the positive feedback between the Joule heat and resistivity is enhanced. As the metal's electric and thermal conductivity degrade with scaling, the cross-sectional crowding effect becomes significant, resulting in the electromigration stress increment by 10%, 33%, and 7% for Cu, Co, and Ru compared to the uniform current case at 9 nm thickness. This work reveals a potential influence on EM lifetime at advanced technology nodes and highlights Ru as an alternative interconnect metal that can effectively resist this adverse effect.

Keywords: Scaling, Electromigration, Current Crowding

Introduction

The interconnect's electromigration (EM) lifetime decreases substantially with reduced feature sizes [1]. One of the primary reasons for this degradation is the rapid increase in resistivity associated with dimensional scaling [2], which can be ascribed to intensified surface scattering [3]. Enhanced surface scattering elevates resistivity near the interconnect's interfaces, which could lead the current density to redistribute and concentrate in the interior, resulting in current crowding within the cross-section of the interconnect.

Current crowding can impact EM significantly. For instance, a higher current density is generated at vias where the current direction changes, making the via more susceptible to EM-induced failure [4]. However, most current crowding studies focus on the case that the current density is non-uniform along the length of the wire, while less attention has been given to the cross-sectional non-uniform case[4-5].

This work investigates the potential effect of the cross-section current crowding by implementing finite element simulation coupling the multiphysic fields. The influence on stress and temperature distribution in Cu is examined by including the variation of thermal conductivity and resistivity under aggressive scaling. Moreover, two promising alternative interconnect materials, Ru and Co, are also studied, and a comparison of these three metals is established.

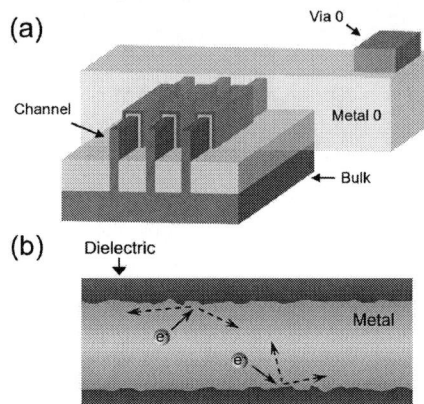

Fig. 1: (a). Illustration of the metal 0 layer. (b). Schematic for the surface scattering of the electrons in metal lines.

Simulation Methods

The simulations are conducted in COMSOL Multiphysics 6.2. The metal 0 layer (illustrated in Fig. 1a) is studied, and a 15*50 μm rectangle geometry is set for simplicity. The heat transfer boundary (shown in Fig. 3a) is as follows: three sides of the interconnect are fixed to the environment temperature, while the top side is thermally insulated, corresponding to the cap layer. As illustrated in Fig. 1b and 3b, we study the case of the current crowding in the interior and compare it with a uniform one. The cross-section is taken at 3/4 along the length of the wire to mitigate the influence of isothermal boundary conditions. The current density is 1e10 A/m2, and the simulation time is set to 1e8 s to obtain a long-term observation of EM stress.

The EM-induced vacancies flux j_v can be represented as [6]:

$$j_v = \frac{c_v D_v}{kT}\left(\frac{\nabla c_v}{c_v}\cdot kT - jeZ^*\rho - Q^*\frac{\nabla T}{T} - f\Omega\nabla\sigma\right) \quad (1)$$

where c_v is the vacancy concentration and can be written as [6]:

$$c_v = c_{v0}\exp\left(\frac{(1-f)\sigma\Omega}{kT}\right) \quad (2)$$

the evolution of c_v with time is given by [6]:

$$\frac{\partial c_v}{\partial t} = -\nabla\cdot j_v + G = -\nabla\cdot j_v + c_v - c_{v,eq} \quad (3)$$

979-8-3315-0417-5/25 $31.00 © 2025 IEEE

and the evolution of strain ε_{ij} is given by [6]:

$$\frac{\partial \epsilon_{ij}}{\partial t} = \frac{1}{3}\Omega[f \cdot \nabla j_v + (1-f)G]\delta_{ij} \qquad (4)$$

In the above equations, δ_{ij} is the Kronecker delta, k is the Boltzmann's constant, T is the absolute temperature, j is the current density and is set to 1e10 A/cm² on average, e is the elementary charge, and ρ is the resistivity. The rest of the symbols, along with the value for the case of Cu, are listed in Table I.

Table I: Parameters used in simulation for the case of Cu.

Term	Value	Description
Z^*	10	Effective charge [7]
Ω	8.78e-30 m³	Atomic volume [7]
c_{v0}	6.02e-21 m⁻³	Initial vacancy content [7]
D_v	9.13e-13 m²/s	Vacancy diffusivity [6][7]
f	0.6	Vacancy relaxation ratio [6]
τ	1.8e-3 s	Vacancy relaxation time [6]
Q^*	0.94 meV	Heat of transport [6]
T_0	473 K	Environment temperature

For Co and Ru, the vacancy diffusion coefficient was set to 4.0e-12 m²/s, calculated with an activation energy of 1.04 eV [6][8]. The effective charge for Co is 1.6, while for Ru, due to lack of data, the value was set to be the same as its group element Fe, which is 2 [9]. The evolutions of resistivity and thermal conductivity with thickness are shown in Fig. 2, where the values at 1000 nm are similar to bulk and, therefore, are set as the bulk ones [3][10]. For the thermal conductivities of Ru and Co, owing to the absence of data, we assume that the degradations are proportional to the increment of resistivity, and the proportion is the same as that for Cu. According to the Wiedemann-Franz law [11], the thermal conductivity of metals is proportional to their electrical conductivity, making this assumption reasonable.

Results and Discussions

We first study the effect of cross-sectional current crowding on Cu interconnect. Thermal conductivity and resistivity values at a 9 nm thickness are chosen to simulate the situation in advanced technology nodes. As shown in Fig. 3c, the cross-section stress gradually increases from x=0.0 (the bottom side in Fig. 3a) to a maximum at x=1.0 (the top side in Fig. 3a) for a uniform current density. The cross-section temperature (Fig. 3d) under the uniform case follows a similar trend, which can be attributed to heat accumulation caused by the thermal insulation boundary. The higher local temperature in Fig. 3d leads to more significant local stress in Fig. 3c since EM is a thermally activated process.

In the case of cross-section current crowding, the EM stress elevates from the uniform case significantly, with the maximum occurring at the center of the metal (Fig. 3c).

Correspondingly, the cross-section temperature shows a local hotspot at the same x position. This phenomenon can be attributed to localized Joule heat caused by current crowding. While copper is usually a good conductor of heat, the degraded thermal conductivity at small dimensions prevents heat dissipation and leads to more severe temperature non-uniform, as shown in Fig. 3d.

The evolution of EM stress and stress difference in the Cu interconnect is shown in Fig. 3e. As EM stress increases over time, the stress difference exhibits non-monotonic changes, which may raised by numerical errors at small values. Nevertheless, the stress difference increases during the primary period and reaches approximately 10% at 1e8 s. The change in stress difference suggests that the cross-section current crowding has a potential effect on the long-term reliability of Cu interconnect in advanced technology nodes.

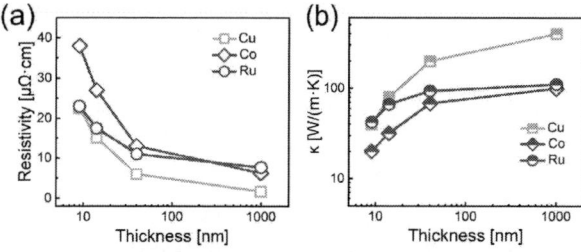

Fig. 2: The (a) resistivities and (b) thermal conductivities of Cu, Co and Ru at various thicknesses.

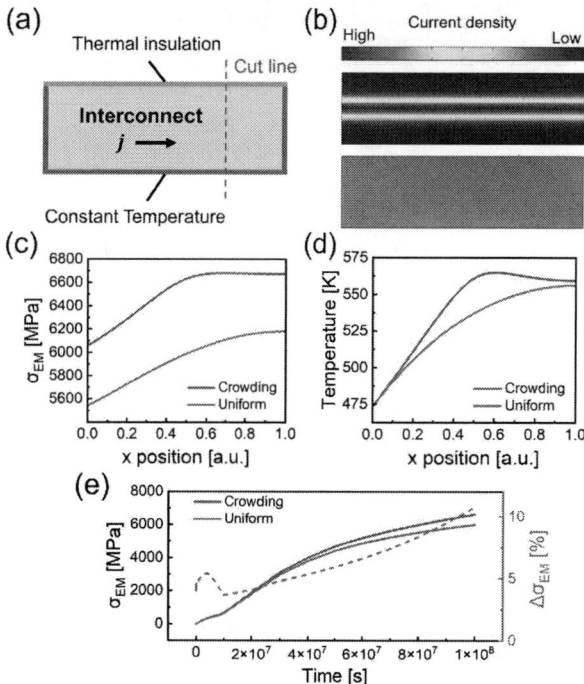

Fig. 3: (a) Thermal boundary conditions set in the simulation. (b) The current density distribution of the crowding (up) and uniform (down) case . (c) The cross-section distribution of EM stress and (d) temperature for the case of Cu. (e) Evolution of the EM stress and stress difference with time.

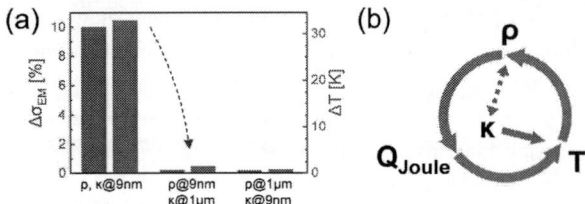

Fig. 4: (a) The change of EM stress and temperature resulted from crowding at various resistivity (ρ) and thermal conductivity (κ). (b) Illustrations of the coupling effects bewteen the ρ, κ, the joule heat Q_{Joule} and T.

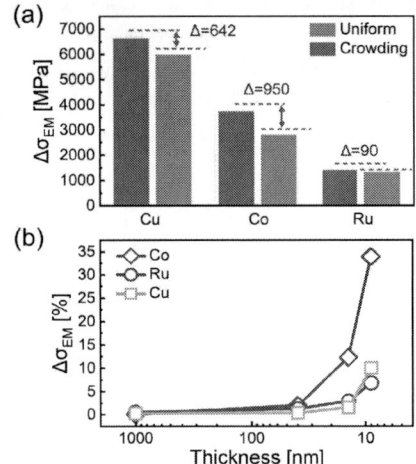

Fig. 5: (a) Comparision of the EM stress between Cu, Co, and Ru when the ρ and κ are from the value of 9nm. (b) The change of EM stress difference at various thicknesses.

Notably, neither resistivity increase nor thermal conductivity degradation alone can lead to a remarkable difference in EM stress and temperature, as shown in Fig. 4a. Such a phenomenon can be recognized as excellent conductivities of bulk Cu leading to few Joule heat or temperature gradients. Considering the positive resistivity-temperature coefficient, a positive feedback between the resistivity, the temperature, and the Joule heat is present, as shown in Fig. 4b. Note that thermal conductivity degradation further elevates local temperatures and accelerates this feedback. Consequently, the effect of cross-section current crowding becomes prominent due to the coupling of electrical and thermal conduction degeneration. Such a coupling is expected in advanced technology nodes, as the intrinsic link between electrical and thermal conductivity causes both to degrade as dimensions are scaled down [11].

We also investigated the impact of cross-section current crowding on alternative metals Co and Ru. The EM stress for each metal is calculated and compared in Fig. 5a when the values of ρ and κ at 9 nm thickness. The magnitude of stress exhibits the order Cu > Co > Ru, which is consistent with these metals' electromigration resistance. All three metals show enlarged stress under cross-section current crowding. The impact in Co is more pronounced than in Cu, attributed to Co's higher resistivity. As shown in Fig. 2a, the growth of Co's resistivity with scaling is slower than that of Cu but is

not enough to compensate for Co's higher bulk resistivity.

To achieve a comprehensive comparison, the EM stress differences at various thicknesses (reflected through variant resistivity and thermal conductivity) of these metals are tested and displayed in Fig. 5b. All the metals show the same trend for negligible cross-section current crowding effects at the large thickness and become significant when dimensions scaling, suggesting that such crowding is potential to affect interconnects in advanced technology nodes. Remarkably, Ru always exhibited the least EM stress and influence of cross-section current crowding.

Conclusions

This work employs finite element simulations to investigate the impact of cross-section current crowding on electromigration in Cu, Co, and Ru. When the resistivity and thermal conductivity reduce with scaling, a localized hotspot forms at the center of the Cu line, which leads to more remarkable EM stress than the uniform case. Under an average current density of 1e10 A/m2, the cross-section current crowding results in a temperature increase of around 33 K and a 10% rise in EM stress. Nevertheless, such an effect can be clearly observed only when both electric and thermal conduction degradation are considered. For Cu, Co, and Ru, the cross-section crowding effect is negligible at relatively thick metal lines but becomes significant as thickness shrinks to 9 nm. Co is the most affected by cross-section current crowding among the three metals despite exhibiting less EM stress than Cu. In contrast, Ru displays the most negligible impact of current crowding, making it an ideal candidate for alternative interconnect metal.

Acknowledgments

This work was supported in part by the NNSFC (T2293704, T2293700) and in part by the National Key R&D Program of China (2023YFB4402204).

References

[1] Sheldon X.-D Tan et al, *INTEGRATION the VLSI journal*, 2018, 60: 132-152.

[2] Jun Hwan Moon et al, *Advanced Science*, 2023, 10: 2207321.

[3] Daniel Gall, *Journal of Applied Physics*, 2020, 127: 050901.

[4] K. N. Tu et al, *Applied Physics Reviews*, 2017, 4: 011101.

[5] Young-Joon Park et al, *IEEE International Reliability Physics Symposium*, 2019, pp. 1-6.

[6] Hua Ye et al, *IEEE Transactions on Components and Packaging Technologies*, 2003, 26(3): 673-681.

[7] Hai-Bao Chen et al, *IEEE Transactions on Computer-Aided Design of Integrated Circuits and Systems*, 2016, 35(11): 1811-1824.

[8] Sofie Beyne et al, *IEEE Transactions on Electron Devices*, 2019, 66(12): 5278-5283.

[9] Yu-chen Liu et al, *MRS Communications*, 2019, 9, 567-575.

[10] Md. Rafiqul Islam et al, *Nature Communications*, 2024, 15: 9167.

[11] Marius Burkle et al, *Nano Letters*, 2018, 18: 7358-7361.

Effective Mass Engineering in Ultra-scaled GAAFETs

Yusi Zhao[1], Huawei Tang[1], Rongzheng Ding[1], Yudong Lv[1], Yanbo Tang[1], Shaofeng Yu[1*]

[1]School of Microelectronics, Fudan University, Shanghai 200433, China
(*Email: shaofeng_yu@fudan.edu.cn)

Abstract

This work analyzes and optimizes the performance of ultra-scaled gate-all-around field-effect transistors (GAAFETs) from the perspective of effective mass. To comprehensively understand the impact of effective mass, three types are identified: density-of-states effective mass, confinement effective mass, and transport effective mass. We introduce crystal orientation, strain/stress, and material engineering to modify the effective mass in ultra-scaled GAAFETs. The enhancements and limitations induced by these techniques are simulated and interpreted based on the effective mass theory, providing fundamental physical insights into device behavior.

Keywords: Crystal orientation, Effective mass, Gate-all-around field-effect transistors (GAAFETs), SiGe, Strain/stress technology

Introduction

The concept of effective mass was initially introduced for semiconductors as a renormalization of electron and hole masses to account for the influence of the crystal's periodic potential [1]. Effective mass directly affects the conduction charge amount and carrier transport in MOSFETs. Therefore, comprehensive understanding and engineering of effective mass are essential for optimizing and predicting the performance of leading-edge technology nodes [2], i.e. ultra-scaled gate-all-around field-effect transistors (GAAFETs). This work aims to provide fundamental physical insights into device behavior from the perspective of effective mass, guiding device engineers in interpreting experiments and developing new technologies.

Simulation Methodology

Rather than relying on classical drift-diffusion-based commercial TCAD tools, we adopt the advanced multi-subband Boltzmann transport equation (MSBTE) device simulation methodology [3]. It provides a self-consistent solution of coupled 1-D Boltzmann transport, 2-D Schrödinger, and 3-D Poisson equations. The iteration procedure is shown in Fig. 1. Detailed calibration of bandstructure and scattering related parameters has been conducted to match the experimental data [4], as validated in Fig. 2. All calculation results in this paper, unless specified, are based on the MSBTE simulations for GAAFETs with dimensions of $T_{ch} \times W_{ch} \times L_{ch}$ = 5 nm × 10 nm × 11 nm.

Types of Effective Mass in Semiconductor Devices

A. DOS Effective Mass m^*_{DOS}

We calculate the density-of-states (DOS) effective mass m^*_{DOS} based on the conventional assumption of single band occupation. The extracted m^*_{DOS} lumps the states of multiple simulated subbands into a single representative subband. A heavier m^*_{DOS} is advantageous for devices because the conduction charge is proportional to the density of states available for carrier occupation. Moreover, a larger m^*_{DOS} can help mitigate the DOS bottleneck [5].

B. Confinement Effective Mass m^*_{conf}

The nanoscale cross sections of GAAFETs induce significant quantum confinement effects, leading to threshold voltage shift, effective gate capacitance reduction, electrostatic dimensionality change and subband separation [6], all of which deteriorate device performance. In the 2-D cross sections of a 3-D GAAFET, both the effective mass projected along the thickness and width directions influence the degree of quantum confinement. The lighter the particle, the more pronounced the quantum effects. Consequently, to mitigate quantum confinement, a heavier confinement effective mass m^*_{conf} is preferred.

C. Transport Effective Mass m^*_{tran}

For dirft-diffusion transport, the scattering-limited low-field mobility μ_{low} is a key parameter for predicting device performance. As the channel length approaches the scattering-determined mean free path, carrier transport in ultra-scaled transistors is often described as quasi-ballistic. In this regime, the ballistic velocity v_{bal}, representing the injection velocity at the top of barrier (ToB) near the source in pure ballistic transport, serves as a better performance predictor [7]. A lighter transport effective mass m^*_{tran} leads to a higher μ_{low} and a larger v_{bal}, improving transistor performance in both drift-diffusion and ballistic regimes. However, for extremely short channels, quantum transport phenomena must be considered [8]. Source-to-drain tunneling (STDT), which involves current flow beneath the potential barrier, can be hindered by a heavier m^*_{tran}.

Techniques for Effective Mass Engineering

A. Crystal-orientation engineering

As depicted in Fig. 3(a), significantly anisotropic effects are observed in the carrier density profiles. Notably, quantum confinement results in the inversion charge centroid being displaced from the interface, thereby weakening effective gate control. The closer the quantum centroid is to zero, the less degradation occurs in effective gate capacitance. Fig. 3(b) demonstrates the degradation of gate capacitance. Among all

979-8-3315-0417-5/25 $31.00 © 2025 IEEE

surface orientations, the (100) and (110) surfaces exhibit the least degradation in gate capacitance for NMOS and PMOS, respectively, consistent with the results of extracted quantum centroid. The fundamental reason is that the (100) surface features the heaviest m_{conf}^* for electrons while the (110) surface has the heaviest m_{conf}^* for holes.

The simulated band dispersion E(k_z) projected on the channel direction is presented in Fig. 4, with the extracted m_{tran}^* and m_{DOS}^* at carrier sheet density $n_s = 3 \times 10^{12}\ cm^{-2}$ indicated in the figure. Lightest m_{tran}^* and heaviest m_{DOS}^* cannot be achieved simultaneously across different orientations. Effective mass varies with carrier density, especially in degeneracy limit, as shown in Fig. 5. (100)/<100> and (110)/<111> demonstrate the best transport properties for NMOS and PMOS respectively, because of the lightest m_{tran}^* at all carrier concentration.

We carry out MSBTE simulations for ultra-scaled GAAFETs on (100)-oriented and (110)-oriented wafers. All channel orientations on the same surface-orientated wafer (rotating from 0° to 360° relative to the notch direction) are calculated. The simulated results in Fig. 7 indicate that whether for ballistic or scattering-limited performance, (100)/<100> and (110)/<111> are the optimum choices for n-type and p-type ultra-scaled GAAFETs, respectively.

B. Stress/Strain engineering

Fig. 6 shows the variation of m_{tran}^* as the uniaxial stress level increases. For electrons, the improvement in m_{tran}^* is mainly attributed to carrier repopulation from the heavy valley to the light valley. Fig. 4 clearly distinguishes two sets of valleys based on their band curvatures. At extremely high stress levels, the enhancement saturates, because all electrons reside in the light valley. For holes, the stress-induced enhancement in transport properties primarily arises from the modulation of band curvature and changes in the relative occupancy of different subbands.

The reduction in m_{tran}^* certainly benefits the carrier transport above the potential barrier. However, a lighter m_{tran}^* can lead to severe tunneling when the channel enters the quantum regime. The current below the white dashed lines in Fig. 8 represents the quantum tunneling current, which degrades the off-state device performance. It is evident that the lighter m_{tran}^* is, the greater the STDT. Thus, in the regime of quantum transport, a trade-off arises in determining whether a lighter or heavier m_{tran}^* is optimal for device overall performance.

C. Material engineering

To mitigate the strong mismatch in driving strengths between PMOS and NMOS, Si_xGe_{1-x} channel emerges as a strong candidate for p-type GAAFETs[9]. We calculate the Si_xGe_{1-x} channels with x ranging from 0 to 1.

As x decrease, m_{conf}^* of Si_xGe_{1-x} becomes lighter, leading to more pronounced quantum confinement effects, as confirmed by the simulated results in Fig. 9. With increasing Ge%, the subband energy separation and the quantum centroid increase. On the other hand, as illustrated in Fig. 10, with increasing Ge%, m_{tran}^* becomes lighter, resulting in improved transport properties. However, m_{DOS}^* also reduces, leading to less density-of-states available for hole occupation.

Finally, we calculate the ballistic on-state currents for Si_xGe_{1-x} channels and extract the charge and velocity at the ToB along the channel. Fig. 11 benchmarks the ballistic performance of ultra-scaled GAAFETs having Si_xGe_{1-x} channels with respect to the Si channel. Although increasing Ge% leads to a higher v_{bal}, the amount of conduction charge at ToB is sacrificed. This reduction in conduction charge is due to lighter m_{conf}^* and lighter m_{DOS}^*. Because of this trade-off, the peak of ballistic on-state current occurs at $Si_{0.4}Ge_{0.6}$. From the perspective of effective mass, it can be concluded that merely increasing the mole fraction of Ge in the channel does not necessarily yield a sustained improvement in the performance of PMOS devices.

Conclusion

We introduce orientation, stress, and material engineering in ultra-scaled GAAFETs, evaluating the advantages and limitations of these techniques from the perspective of effective mass (m_{DOS}^*, m_{conf}^* and m_{tran}^*). (100)/<100> and (110)/<111> orientation configurations demonstrate optimum overall performance for n-type and p-type GAAFETs respectively, because of the heaviest m_{conf}^* and lightest m_{tran}^*. Uniaxial tensile/compressive stress applied along channel direction reduces m_{tran}^* and m_{DOS}^* for n-type/p-type FETs. The reduction in m_{tran}^* benefits classical/semi-classical transport, but induces increased tunneling leakage current under quantum transport. For p-type GAAFETs with SiGe channels, increasing Ge% leads to improved transport properties (lighter m_{tran}^*) but less density of states (lighter m_{DOS}^*) and more severe quantum confinement (lighter m_{conf}^*). Considering this trade-off, the optimum ballistic overall performance for p-type GAAFETs is achieved with the $Si_{0.4}Ge_{0.6}$ channel.

References

[1] S. M. Sze and K. K. Ng, "Physics of Semiconductor Devices," 2023,
[2] G. Bae et al., "3nm GAA Technology featuring Multi-Bridge-Channel FET for Low Power and High Performance Applications," 2018 IEEE IEDM
[3] "Sentaurus Device QTX User Guide," Dec. 2019.
[4] S. Mochizuki et al., "Evaluation of (110) versus (001) Channel Orientation for Improved nFET/pFET Device Performance Trade-Off in Gate-All-Around Nanosheet Technology," 2023 IEEE IEDM
[5] M. V. Fischetti et al., "Simulation of Electron Transport in HEMTs: Density of States Bottleneck and Source Starvation," 2007 IEEE IEDM
[6] A. Dasgupta et al., "BSIM Compact Model of Quantum Confinement in Advanced Nanosheet FETs," IEEE TED, vol. 67, no. 2, pp. 730–737, 2020
[7] J. T. Teherani, "A Comprehensive Theoretical Analysis of Hole Ballistic Velocity in Si, SiGe, and Ge: Effect of Uniaxial Strain, Crystallographic Orientation, Body Thickness, and Gate Architecture," IEEE TED, vol. 64, no. 8, pp. 3316–3323, 2017
[8] S. R. Mehrotra et al "Engineering Nanowire n-MOSFETs at Lg<8 nm," IEEE TED, vol. 60, no. 7, pp. 2171–2177, 2013
[9] S. Mochizuki et al., "Stacked Gate-All-Around Nanosheet pFET with Highly Compressive Strained Si1-xGex Channel," 2020 IEEE IEDM

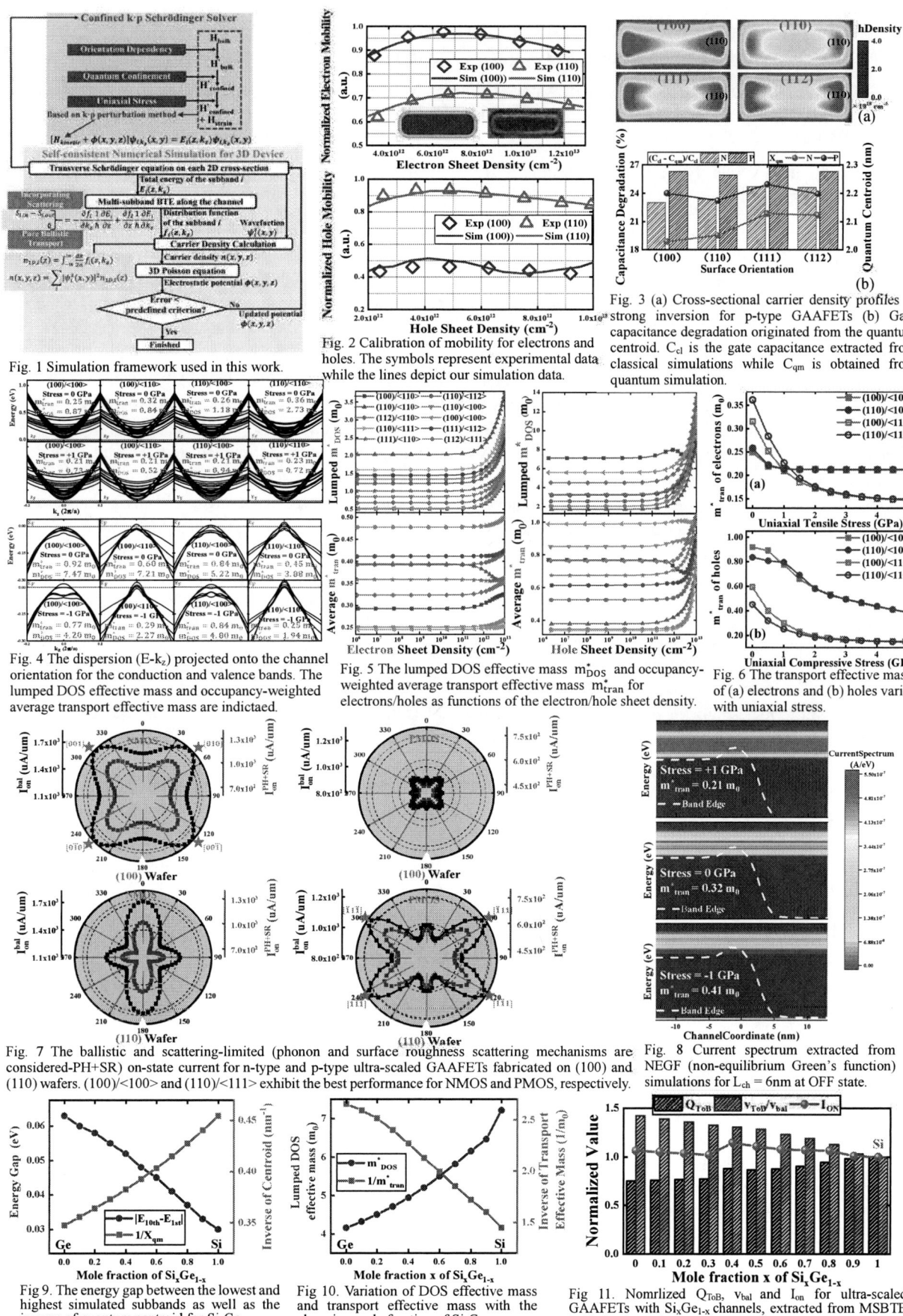

Fig. 1 Simulation framework used in this work.

Fig. 2 Calibration of mobility for electrons and holes. The symbols represent experimental data while the lines depict our simulation data.

Fig. 3 (a) Cross-sectional carrier density profiles in strong inversion for p-type GAAFETs (b) Gate capacitance degradation originated from the quantum centroid. C_{cl} is the gate capacitance extracted from classical simulations while C_{qm} is obtained from quantum simulation.

Fig. 4 The dispersion (E-k_z) projected onto the channel orientation for the conduction and valence bands. The lumped DOS effective mass and occupancy-weighted average transport effective mass are indictaed.

Fig. 5 The lumped DOS effective mass m^*_{DOS} and occupancy-weighted average transport effective mass m^*_{tran} for electrons/holes as functions of the electron/hole sheet density.

Fig. 6 The transport effective mass of (a) electrons and (b) holes varies with uniaxial stress.

Fig. 7 The ballistic and scattering-limited (phonon and surface roughness scattering mechanisms are considered-PH+SR) on-state current for n-type and p-type ultra-scaled GAAFETs fabricated on (100) and (110) wafers. (100)/<100> and (110)/<111> exhibit the best performance for NMOS and PMOS, respectively.

Fig. 8 Current spectrum extracted from NEGF (non-equilibrium Green's function) simulations for L_{ch} = 6nm at OFF state.

Fig 9. The energy gap between the lowest and highest simulated subbands as well as the inverse of quantum centroid for Si_xGe_{1-x}.

Fig. 10. Variation of DOS effective mass and transport effective mass with the changing mole fraction of Si_xGe_{1-x}.

Fig. 11. Nomrlized Q_{ToB}, v_{bal} and I_{on} for ultra-scaled GAAFETs with Si_xGe_{1-x} channels, extracted from MSBTE ballistic simulations.

Unveiling the Role of Oxygen Vacancy Inhomogeneity in Enhancing the Reliability of Ferroelectric HfO₂/ZrO₂ Superlattice Structures

Boyao Cui[1], Maokun Wu[1,*], Sheng Ye[1], Xuepei Wang[1], Yuchun Li[2], Yishan Wu[1], Yichen Wen[1], Jinhao Liu[1], Zhigang Ji[1,*], Hongliang Lu[2], David Wei Zhang[2], Runsheng Wang[3], Ru Huang[3]

[1]Department of Micro/Nano Electronics, SEIEE, Shanghai Jiaotong University, China (*email: maokunwu@sjtu.edu.cn; zhigangji@sjtu.edu.cn), [2]School of Microelectronics, Fudan University, China, [3]School of Integrated Circuits, Peking University, China

Abstract

The HfO₂/ZrO₂ ferroelectric superlattice (SL) structure is a promising material for Back-End-of-Line-compatible ferroelectric memory applications, though questions remain regarding its reliability mechanisms. In this work, for the first time, we have provided physical evidence of the uneven oxygen vacancy distribution within the SL structure. This inhomogeneity distribution not only mitigates the wake-up effect but also significantly enhances reliability, such as improved breakdown voltage and endurance. Our findings offer valuable insights for advancing SL ferroelectric thin films in memory technology.

Keywords: superlattice, ferroelectricity, reliability

Introduction

Ferroelectric hafnium oxide (HfO₂)-based materials have demonstrated significant potential for applications in memory, logic devices, and neuromorphic computing since the discovery of ferroelectricity in 2011 [1, 2]. In particular, doped hafnium oxide has gained widespread attention in non-volatile memory (NVM) applications due to its low power consumption, high opearting speed, and compatibility with complementary metal-oxide-semiconductor (CMOS) technologies [3]. For example, a high-density 3D memory architecture using zirconium oxide (ZrO₂)-doped hafnium oxide (HZO) exhibits a 10 ns program pulse, an endurance of 10^{10} cycles, and a lifetime of 10 years [4]. However, recent studies reveal that HfO₂-based ferroelectric materials require high annealing temperatures (500°C-600°C), which poses a challenge for integrating ferroelectric memory within the Back-End-of-Line (BEOL) process.

Therefore, finding materials that combine high polarization with compatibility for lower annealing temperatures has become a primary task. The HfO₂-ZrO₂ stacked superlattice (SL) not only maintains high polarization at lower annealing temperatures but also demonstrates significant reliability enhancements [5-6]. For instance, SL exhibits nearly double the remnant polarization (Pr) than that of HZO sample, along with a two-order-of-magnitude increase in endurance metrics, all within a thermal budget compatible with the BEOL process [5]. Additionally, SL also shows reduced leakage and improved fatigue performance [7].

Despite the promising potential of SL structures for future Very-Large-Scale Integration (VLSI) applications, challenges remain due to unclear mechanisms, which inhibit the further optimization of the superlattice structures. It is currently understood that increased stress within SL enhances polarization; however, the reliability mechanism behind this structure is not yet fully understood. To explore this, we have calculated the formation energy of oxygen vacancy (Vo), revealing an uneven distribution across layers within SL. Further, we have, for the first time, confirmed these differences by Angle-resolved X-ray Photoelectron Spectroscopy (ARXPS), showing higher Vo formation energy and concentration in ZrO₂ layers, with lower levels in HfO₂ layers. This uneven distribution helps regulate breakdown and affects endurance, the wake-up effect, and leakage, paving the way for innovations in ferroelectric storage technology.

Experiments

The ferroelectric capacitors, shown in Fig. 1(a), consist of top and bottom electrodes deposited by magnetron sputtering and a ferroelectric layer formed via Atomic Layer Deposition (ALD). Each metal-ferroelectric-metal (MFM) capacitor, defined using a lift-off patterning process, has an area of 100 μm × 100 μm. The 8 nm SL ferroelectric film, depicted in Fig. 1(b), includes three stacks of alternating 1.3 nm layers of HfO₂ and ZrO₂. A solid solution (SS) HZO film of the same thickness serves as a reference. The TDMAHf and TDMAZr were used as precursor gases for hafnium and zirconium, respectively. All samples were annealed in N₂ at 400°C for 30 seconds. Transmission Electron Microscopy (TEM) images reveal that the SL film exhibits a distinct layered stacking structure, whereas the SS film forms a uniform solid solution.

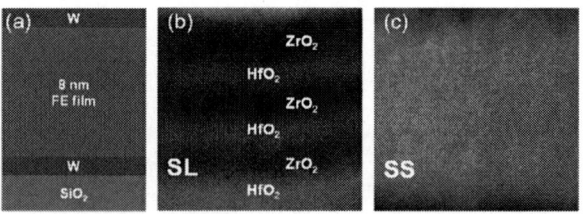

Fig. 1. (a) The schematic diagram of metal-ferroelectric-metal (MFM) capacitors. The TEM image of (b) SL and (c) SS. The total thickness of the FE film is kept constant at 8 nm.

979-8-3315-0417-5/25 $31.00 © 2025 IEEE

Results and discussion

Recognizing Vo as a key factor in regulating reliability, we used first-principles calculations to examine Vo formation energy between SL and SS structures. In SS, the formation energy of four coordinated oxygen vacancies is low and prone to occur. While in the SL structure, ZrO_2 layers has significantly lower formation energies, suggesting higher Vo concentration. Conversely, that in exhibit higher formation energies, indicating less likelihood of Vo formation.

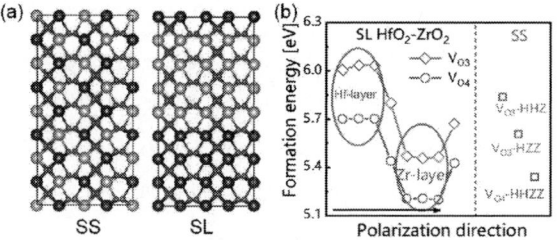

Fig. 2. (a) The structural configurations of SS and SL in DFT. (b) The formation energies of Vo in SL and SS films. The formation of four-coordinated and three-coordinated Vo is represented by green and purple symbols, respectively.

To validate the calculated Vo distribution, we examined the depth profile using XPS, which detects binding energy information from the surface to about 30 angstroms, covering at least one layer of HfO_2 and ZrO_2 and their interface within the SL [8]. By probing at different angles, we analyze Vo concentrations at various depths. Previous studies suggest that non-lattice oxygen at 531.7 eV can be used to infer Vo concentration [9].

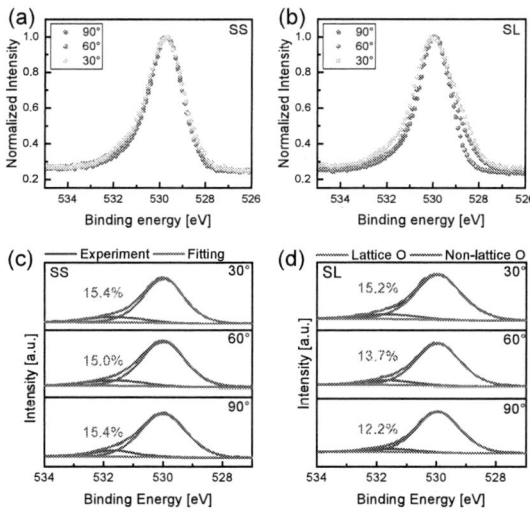

Fig. 3. Angle-resolved XPS O 1s spectrum at different angles for (a) SS and (b) SL, along with the non-lattice O analysis for (c) SS and (d) SL.

We first compare the normalized O1s spectra of SL and SS films across multiple angles. The XPS curves for the SS film are nearly identical across all angles, indicating a uniform Vo distribution with a consistent concentration of about ~15.4% across all angles, as shown in Fig. 3(a) and (c). In contrast, in Fig. 3(b) and (d)，the SL film displays notable spectral

differences, suggesting a depth-dependent variation in Vo concentration, with values of 15.2%, 13.7%, and 12.2% at angles of 30°, 60°, and 90°, respectively, confirming a depth-dependent distribution. Specifically, the 30° measurement probes the surface ZrO_2 layer, where the Vo formation energy is low, resulting in a high Vo concentration. This concentration is higher than that at the HfO_2/ZrO_2 interface measured at 60°, and significantly higher than at 90°, where the measurement primarily captures the HfO_2 layer, which has a low Vo concentration due to high Vo formation energy and therefore. Furthermore, this non-uniform distribution also results in a significantly lower overall Vo concentration in SL compared to SS.

Due to the reduced concentration and optimized distribution of oxygen vacancies, SL devices demonstrate robust performance in breakdown characteristics. As shown in Fig. 4(a), the Transient-Dielectric-Breakdown (TZDB) test indicates that SL-MFM exhibits a higher breakdown voltage and lower leakage current density than SS. Additionally, Fig. 4(b) presents the Time-Dependent-Dielectric-Breakdown (TDDB) test results, displaying the current-time curves for 20 devices various stress electric field. Notably, the breakdown duration of SL at 4.00 MV/cm is longer than that of SS at 3.81 MV/cm, highlighting superior breakdown characteristics of SL structure. Fig. 4(c) illustrates the breakdown times at 4.12 MV/cm for SL and 3.94 MV/cm for SS. These experimental data follow a classical Weibull distribution. Within this model, T_{63} serves as a key parameter characterizing the breakdown time at 63.2% failure. SL exhibits a longer breakdown time, even under high electric fields. Furthermore, the predicted ten-year lifetime at 63% failure is presented in Fig. 4(d). By fitting a power-law model, the stress field corresponding to a ten-year lifetime at 63.2% failure for SL is extrapolated to be 3.31 MV/cm, surpassing the value of SS sample (2.95 MV/cm).

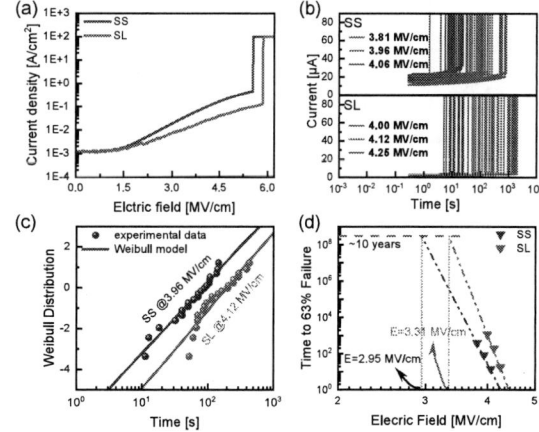

Fig. 4. (a) TZDB and (b) TDDB test of SL and SS. (c) Weibull distributions of TDDB. (d) Projected operating electric fields for ten-year lifetime prediction at 63% failure.

Reflecting on performance, SL devices also exhibit excellent behavior in wake-up and endurance cycles. This can be ascribed to the suppression of Vo due to the stacked structure in SL. The I-V curves in Fig. 5(a) indicate that SL has a higher polarization current intensity than SS, which can be attributed to the greater stress within the SL, as discussed in prior literature [6]. The dual-peak current observed in the SS device results from Vo pinning at the electrode/ferroelectric interface, a phenomenon that is less pronounced in SL where the structure effectively suppresses Vo pinning. The reduced Vo pinning in SL stems from its lower Vo concentration. This pinning effect intensifies the wake-up phenomenon in SS devices in Fig. 5(b), as demonstrated by endurance testing conducted at 3.75 MV/cm and 1 MHz, a frequency sufficient for complete polarization switching. By calculating $\Delta Pr/Pr$, we find the polarization variation to be 25% for SS and only 13% for SL, supported by statistical data. In addition, Vo decrease also achieves to an approximately two-order-of-magnitude improvement in endurance cycles. As shown in Fig. 5(c), for SL sample, over 10^9 cycles are achieved at 2.5 MV/cm, while more than 10^{10} cycles are recorded at 2 MV/cm. These results demonstrate the outstanding performance of the SL structure under conditions compatible with Back-End-of-Line processes, as shown in Fig. 5(d).

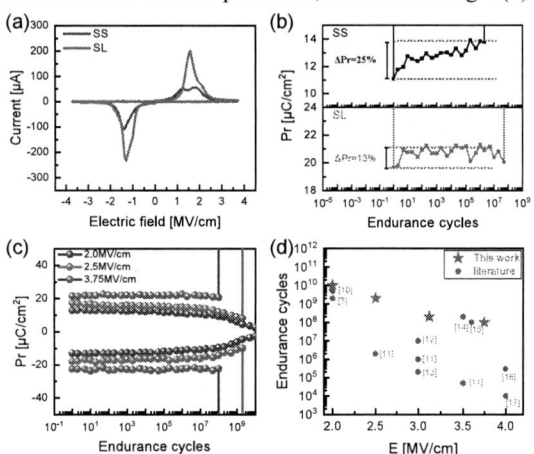

Fig. 5. (a) I-V curves of SS and SL. (b) Comparison of the wake-up effect between SS and SL. (c) Endurance cycles of SL at various electric fields. (d) Comparison of endurance results with previous literature compatible with Back-End-of-Line processes [3, 10-17].

Conclusion

In this work, we have identified experimentally and theoretically that the oxygen vacancy inhomogeneity is the fundamental factor for the enhanced reliability of superlattice structure. This leads to a significant enhancement in the breakdown characteristics of SL compared to SS, resulting in a two-order-of-magnitude increase in endurance. Additionally, the decrease in Vo concentration reduces Vo pinning, thereby mitigating the wake-up effect. These findings offer valuable insights for applying SL ferroelectric thin films in ferroelectric memory technology.

Acknowledgments

This work was supported by the National Natural Science Foundation of China (Nos. 62027818, 61874034, 62304136, and 11974320) and Startup Fund for Young Faculty at SJTU (SFYF at SJTU).

References

[1] T. S. Böscke, J. Müller and D. Bräuhaus, et.al., "Ferroelectricity in hafnium oxide thin films," Appl. Phys. Lett., vol. 99, no. 10, Art no. 102903 (2011).

[2] J. Hur, P. Wang and Z. Wang, et.al., "Interplay of switching characteristics, cycling endurance and multilevel retention of ferroelectric capacitor," IEDM Tech. Dig., pp. 39.5.1-39.5.4 (2020).

[3] Y. Wang, Y. Yang and P. Jiang, et.al., "Precrystallization engineering of Hf$_{0.5}$Zr$_{0.5}$O$_2$ film in back-end-of-line compatible ferroelectric device for enhanced remnant polarization and endurance," IEEE Electron Device Lett., vol. 44, no. 3, pp. 396-399 (2023).

[4] E. R. Hsieh, J. K. Chang and T. Y. Tang, et.al., "NVDimm-FE: a high-density 3D architecture of 3-bit/c 2TnCFE to break great memory wall with 10 ns of PGM-pulse, 10^{10} cycles of endurance, and decade lifetime at 103 °C," VLSI Technology and Circuits, pp. 359-360 (2022).

[5] B. Cui, X. Wang and Y. Li, et.al., "Back-end-of-line compatible HfO$_2$/ZrO$_2$ superlattice ferroelectric capacitor with high endurance and remnant polarization," IEEE Electron Device Lett., vol. 44, no. 6, pp. 1011-1014 (2023).

[6] Y. Peng, W. Xiao and Y. Liu, et.al., "HfO$_2$-ZrO$_2$ superlattice ferroelectric capacitor with improved endurance performance and higher fatigue recovery capability," IEEE Electron Device Lett., vol. 43, no. 2, pp. 216-219 (2022).

[7] K. Li, Y. Peng and W. Xiao, et.al., "A comparative study on the polarization, reliability, and switching dynamics of HfO$_2$-ZrO$_2$-HfO$_2$ and ZrO$_2$-HfO$_2$-ZrO$_2$ superlattice ferroelectric films," IEEE Trans. Electron Devices, vol. 70, no. 4, pp. 1802-1807 (2023).

[8] J. H. Swartz, J. M. Ammons and M. Kovac, et.al., "The adsorption of water on mietallic packages," International Reliability Physics Symposium (IRPS), pp. pp. 52-59 (1983).

[9] Y.-Y. Peng, T.-E. Hsieh, and C.-H. Hsu, "White-light emitting ZnO–SiO$_2$ nanocomposite thin films prepared by the target-attached sputtering method," Nanotechnology, vol. 17, no. 1, pp. 174-180 (2006).

[10] Y. C. Liu, J. N. Yang and Y. C. Li, et.al., "Back-end of line compatible Hf$_{0.5}$Zr$_{0.5}$O$_2$/ZrO$_2$/Hf$_{0.5}$Zr$_{0.5}$O$_2$ stack achieving 2Pr of 39.6 μC/cm^2 and endurance exceeding 10^{10} cycles under low-voltage operation," IEEE Electron Device Lett., vol. 45, no. 3, pp. 388-391 (2024).

[11] D. Lehninger, R. Olivo and T. Ali, et.al., "Back-end-of-line compatible low-temperature furnace anneal for ferroelectric hafnium zirconium oxide formation," Phys. Status Solidi A, vol. 217, no. 8, 1900840 (2020).

[12] W. Wei, W. Zhang and F. Wang, et.al., "Deep insights into the failure mechanisms in field-cycled ferroelectric Hf$_{0.5}$Zr$_{0.5}$O$_2$ thin film: TDDB characterizations and first-principles calculations," IEDM Tech. Dig., pp. 39.6.1-39.6.4 (2020).

[13] M. Yadav, A. Kashir and S. Oh, et.al., "High polarization and wake-up free ferroelectric characteristics in ultrathin Hf$_{0.5}$Zr$_{0.5}$O$_2$ devices by control of oxygen-deficient layer," Nanotechnology, vol. 33, no. 8, Art no. 085206 (2021).

[14] V. Gaddam, D. Das and T. Jung, et.al., "Ferroelectricity enhancement in Hf$_{0.5}$Zr$_{0.5}$O$_2$ based tri-layer capacitors at low-temperature (350 °C) annealing process," IEEE Electron Device Lett., vol. 42, no. 6, pp. 812-815 (2021).

[15] H. K. Peng, T. C. Lai and Y. C. Kao, et.al., "Improved reliability for back-end-of-line compatible ferroelectric capacitor with 3 bits/cell storage capability by interface engineering and post deposition annealing," IEEE Electron Device Lett., vol. 43, no. 12, pp. 2180-2183 (2022).

[16] T. Onaya, T. Nabatame and M. Inoue, er.al., "Wake-up-free properties and high fatigue resistance of Hf$_x$Zr$_{1-x}$O$_2$-based metal–ferroelectric–semiconductor using top ZrO$_2$ nucleation layer at low thermal budget (300 °C)," APL Mater., vol. 10, no. 5, Art no. 051110 (2022).

[17] T. J. Chang, Y. S. Jiang and S. H. Yi, et.al., "Atomic tailoring of low-thermal-budget and nearly wake-up-free ferroelectric Hf$_{0.5}$Zr$_{0.5}$O$_2$ nanoscale thin films by atomic layer annealing," Appl. Surf. Sci., vol. 591, Art no. 153110 (2022).

BEOL Electro-Biological Interface for 1024-Channel TFT Neurostimulator with Cultured DRG Neurons

Haobin Zhou[1], Bowen Liu[1], Taoming Guo[1], Hanbin Ma[2,3] and Chen Jiang[1*]

[1]Department of Electronic Engineering, Tsinghua University, Beijing, China, [2]CAS Key Laboratory of Bio Medical Diagnostics, Suzhou Institute of Biomedical Engineering and Technology, Suzhou, China, [3]Guangdong ACXEL Micro Nano Tech Company Ltd., Foshan, China.

*E-mail: chenjiang@tsinghua.edu.cn

Abstract

The demand for high-quality neurostimulation, driven by the development of brain-computer interfaces, has outpaced the capabilities of passive microelectrode-arrays, which are limited by channel-count and biocompatibility. This work proposes a back-end-of-line (BEOL) process for 1024-channel stimulator with bioelectrodes and waterproof encapsulation to stimulate dorsal root ganglion neurons. We introduce an active-matrix neurostimulator based on n-type low-temperature poly-silicon thin-film transistor, adding PEDOT:PSS and SU-8 as bioelectrodes and encapsulation. This enables precise stimulation of DRG neurons, addressing key challenges in neurostimulation systems.

Keywords: BEOL, Neurostimulator, Thin-film transistor and Bio-electrode.

Introduction

Due to the advances in brain-computer interfaces [1], neurostimulation therapy [2], and related fields, the interaction between electronics to biological tissues is becoming more important. However, the conventional CMOS silicon-based electronics is highly integrated and rigid, which is difficult to achieve a flexible system for biological tissues. Thin film transistors (TFTs) have the advantage of mechanical flexibility and can be designed with high-channel-count neurostimulators to adhere to biological tissue over a large area, such as brain cortex. The circuit design of TFTs with four transistors and one capacitor (4T1C) enables high-channel-count active-matrix arrays for independent programming and simultaneous stimulation of 1024 channels [3, 4]. Additionally, this fabricated chip exhibits high yield, and it has the potential to achieve a fully flexible stimulation array. However, the chips fabricated from inorganic materials can be severely degraded in solution environments, and the interfaces of the chips are not suitable for biological tissues.

To address these issues, we introduce a BEOL process to fabricate a bioelectrode layer suitable for cell attachment along with a waterproof encapsulation layer to protect chip from degradation. A bioelectrode layer formed by the poly (3,4-ethylenedioxythiophene):polystyrene sulfonate (PEDOT:PSS) is spin-coated and photolithographically patterned on the chip to enhance the biocompatibility of the electro-biological interface [5, 6]. In addition, SU-8, a polymer material commonly employed for bioencapsulation, which possesses high biocompatibility, favorable mechanical properties, and superior waterproof performance, is used as a waterproof encapsulation layer [7]. Combined with the high-channel-count active array, this system enables neurostimulation for 100 μm resolution of dorsal root ganglion (DRG) neurons.

LTPS Fabrication and Proposed Pixel Circuit

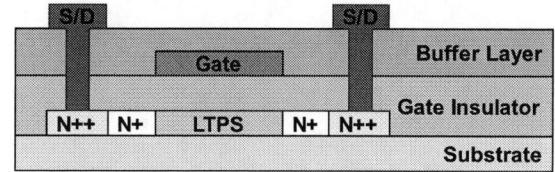

Fig. 1: Cross-sectional view of the fabricated n-type LTPS TFTs.

The proposed n-type low-temperature polycrystalline silicon thin-film transistors (LTPS TFTs) employ a top-gate structure, as illustrated in Fig. 1. The fabrication process started with the deposition of a poly-Si layer on a glass substrate. Following this, a channel-doping process was performed on the channel region, while the source and drain regions underwent an n-type doping process. Then, Mo was deposited to form the gate electrode, which was then patterned using standard photolithographic techniques. An SiO_x buffer layer was deposited, followed by the etching of vias to establish connections to the source and drain metal Finally, a trilayer of Ti/Al/Ti was deposited to serve as the metal contacts for the source and drain electrodes.

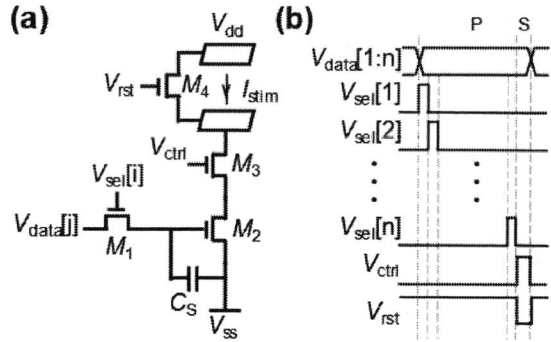

Fig. 2: (a) Schematic and (b) timing diagram of the proposed 4T1C stimulation pixel circuit. P and S stand for programming phase and stimulating phase in one cycle.

979-8-3315-0417-5/25 $31.00 © 2025 IEEE

The neurostimulation pixel circuit includes 4T1C as shown in Fig. 2(b), in the programming phase, V_{sel} changes to high voltage in sequence to turn on M_1 and V_{data} changes to corresponding voltage. M_1 acts as a switch to program the voltages on data lines to the capacitor C_s when the V_{sel} is pulled up, and C_s stores V_{data} until the next programming phase. In the stimulating phase, M_2 transistor acts as a driving transistor to supply stimulation current to the electrodes according to the voltage stored in C_s. Both M_3 and M_4 transistors act as switches to turn on the corresponding branch circuit to perform the stimulation and discharge, respectively, when V_{ctrl} and V_{rst} are pulled up.

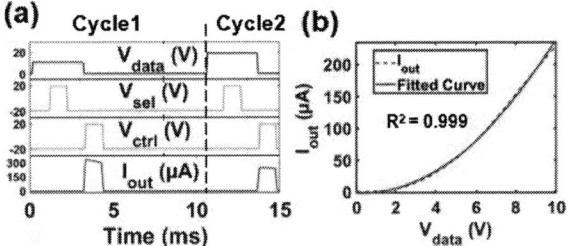

Fig. 3: (a) Measured output current I_{out} and control voltage V_{data}, V_{sel} and V_{ctrl} of the pixel circuit for neurostimulation function and (b) Measured I_{out} under different voltage of V_{data}.

We verified the neurostimulation function of the fabricated circuit, as shown in Fig. 3(a). It illustrates the outputs of the circuit, in which V_{dd} was set at 20 V, V_{sel} and V_{ctrl} toggled between –20 V and 20 V. The capacitor C_s was programmed with two different V_{data} of 10 V and 19 V, respectively. Two output currents of 315 μA and 245 μA were observed, which are sufficient to evoke action potentials. I_{out} is controlled by V_{ctrl} and the amplitude can be modulated by V_{data} effectively. Further, the I_{out}-V_{data} relation is obtained by sweeping V_{data} from 0 to 10 V, as shown in Fig. 3(b). The relation is quadratic, which corresponds to the voltage-controlled current relationship of a transistor in the saturation region.

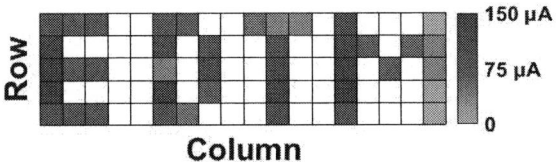

Fig. 4: The stimulation currents of 90 electrodes in the test region.

To apply patterned stimulation to DRG neurons, an area with 90 electrodes on the chip were selected, consisting of 41 electrodes programmed as stimulation pixels and the other 49 electrodes as non-stimulation pixels to form a pattern of 'EDTM'. The results of the outputs are depicted in Fig. 4, where the color intensity of each square represents the magnitude of the stimulation current. All the 41 stimulation electrodes successfully delivered stimulation currents, while all the non-stimulation electrodes did not exhibit current pulses, achieving a 100% yield. These results demonstrate the capability for precise stimulation at the targeted area.

BEOL Process for Electro-Biological Interface

The BEOL process includes a PEDOT:PSS layer at the electrode sites and a waterproof encapsulation, as shown in Fig. 5(c). The PEDOT:PSS layer transformed the indium tin oxide (ITO) electrodes into bio-electrodes to improve biocompatibility [8]. The SU-8 encapsulation layer protects the electrical circuits and further enhance the biocompatibility of the surface. The bio-electrodes and encapsulation allow neurons to adhere and grow on the surface without additional coating process, as shown in Fig. 5(a).

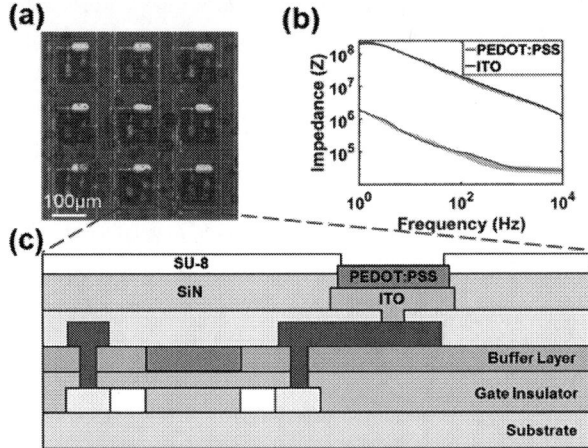

Fig. 5: (a) Cell adhesion and growth on the electro-biological interface. (b) Electrochemical impedance spectroscopy for ITO and PEDOT:PSS electrodes. (c) Cross-sectional view of the fabricated bio-electrodes and encapsulation.

Consequently, this process significantly reduced the contact resistance at the electrode interface. The electrodes are square-shaped with a side length of 100 μm. Electrochemical impedance spectroscopy of the bio-electrodes and the ITO electrodes (i.e., with and without PEDOT:PSS layer) is illustrated in Fig. 5(b). The impedance of the electrodes demonstrates typical frequency-dependent characteristics, i.e., decreasing with the increase of frequency. It is notable that the impedance of PEDOT:PSS/ITO electrodes are 1 to 2 orders in magnitude lower than that of ITO electrodes.

Precise Neurostimulation of DRG Neurons

The DRG neurons were isolated from the spine of mice and cultured on a neurostimulation chip for 12 hours. Cells were stained with Fluo-4 AM calcium ion indicators during adherent growth and patterned programmed electrical stimulation [9]. It was possible to observe fluorescent images of calcium ions with a fluorescence imaging microscope, and thus the firing process of the neurons, as Fig.6 shown.

979-8-3315-0417-5/25 $31.00 © 2025 IEEE

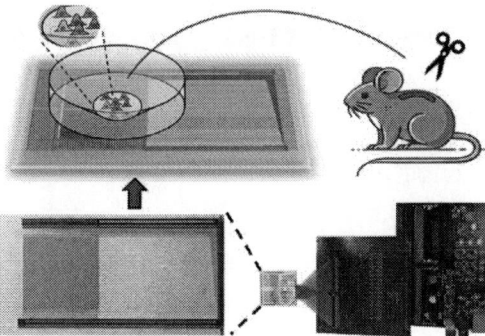

Fig. 6: Conceptual diagram of a system for neurostimulation of DRG neurons.

As Fig. 7(a) illustrated, the left two columns are stimulation electrodes, with stimulation applied from 5 to 30 seconds. In the stimulated region, the fluorescence images of neurons adhered to the electrodes exhibited a marked increase in brightness of calcium ion indicators during stimulation, followed by a decrease in intensity after the stimulation ended.

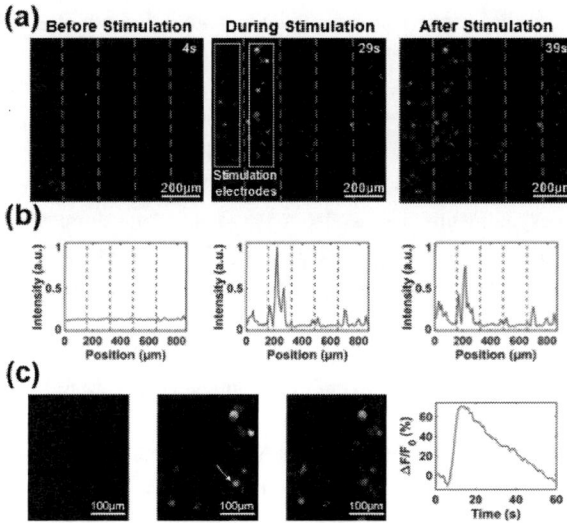

Fig. 7: Patterned electrical stimulation-induced action potentials observed through fluorescent calcium imaging. (a) Fluorescence images of neurons in regions which are programmed for current stimulation and no-current stimulation, with stimulation from the 5s to 30s. The dashed lines indicate electrode gap regions; The left two columns are stimulation electrodes, while the right side is non-stimulation region. (b) Lateral fluorescence intensity distribution across the fluorescence image. The dashed lines indicate electrode gap regions, aligned with Fig. 8(a). (c) Fluorescent images of a typical DRG neuron with 25s stimulation and calcium signal traces ($\Delta F/F_0$) over time.

Furthermore, the lateral fluorescence intensity distribution across the area in Fig. 7(a) was calculated and the results are shown in Fig. 7(b), indicating a significant increase in fluorescence intensity within the programmed stimulation region during the stimulation period, with values notably higher than those before and after stimulation. We selected a representative DRG neuron within the stimulated region and displayed its calcium signal traces ($\Delta F/F_0$) over time, as shown in Fig. 7(c).

Conclusion

In conclusion, we performed BEOL on the high-channel-count neurostimulation array, adding a PEDOT:PSS layer to construct bio-electrodes and SU-8 as the waterproof encapsulation. This addresses the pressing needs in neurostimulation for biocompatibility and reliable signal transmission, providing high-channel-count and precise stimulation to neurons. We verified that the neurostimulation system enables patterned stimulation of DRG neurons and it holds potential for high-information-transfer brain-computer interface and electrical stimulation training of neural tissue. It may also be integrated with sensor arrays to form a closed-loop sensing and stimulation system.

Acknowledgments

This work was supported in part by the National Natural Science Foundation of China (82151305 and 62374102). The authors gratefully acknowledge the assistance of TIANMA Microelectronics Corp. for manufacturing the chips.

References

[1] E. Lowet et al., "Deep brain stimulation creates informational lesion through membrane depolarization in mouse hippocampus," Nature Communications, vol. 13, no. 1, p. 7709, 2022.

[2] K. K. Sellers et al., "Closed-loop neurostimulation for the treatment of psychiatric disorders," Neuropsychopharmacology, vol. 49, no. 1, pp. 163-178, 2024.

[3] T. Guo et al., "A Low-Temperature Poly-Silicon Thin Film Transistor Pixel Circuit for Active-Matrix Simultaneous Neurostimulation," IEEE Journal of the Electron Devices Society, vol. 11, pp. 695-699, 2023.

[4] B. Liu et al., "A 1024-channel neurostimulation system enabled by photolithographic organic thin-film transistors with high uniformity," in 2024 IEEE International Symposium on Circuits and Systems (ISCAS), 2024: IEEE, pp. 1-5.

[5] C. M. Proctor, J. Rivnay, and G. G. Malliaras, "Understanding volumetric capacitance in conducting polymers," vol. 54, ed: Wiley Online Library, 2016, pp. 1433-1436.

[6] M. N. Gueye, A. Carella, J. Faure-Vincent, R. Demadrille, and J.-P. Simonato, "Progress in understanding structure and transport properties of PEDOT-based materials: A critical review," Progress in Materials Science, vol. 108, p. 100616, 2020.

[7] Z. Chen and J.-B. Lee, "Biocompatibility of su-8 and its biomedical device applications," Micromachines, vol. 12, no. 7, p. 794, 2021.

[8] E. B. Aydın and M. K. Sezgintürk, "Indium tin oxide (ITO): A promising material in biosensing technology," TrAC Trends in Analytical Chemistry, vol. 97, pp. 309-315, 2017.

[9] J. Liao, D. Patel, Q. Zhao, R. Peng, H. Guo, and Z. Diwu, "A novel Ca2+ indicator for long-term tracking of intracellular calcium flux," Biotechniques, vol. 70, no. 5, pp. 271-277, 2021.

Dual Functionality of MoS$_2$ in nFET and pFET through Contact Metal Selection

Kwok-Ho WONG* and Mansun CHAN

Dept. of Electronic and Computer Engineering, The Hong Kong University of Science and Technology,
Clear water bay, Kowloon, Hong Kong

*Email: khwongby@connect.ust.hk

Abstract

Traditional transistor fabrication requires separate doping processes for both n-type and p-type field-effect transistors (FETs), which becomes increasingly challenging as device dimensions shrink. This paper explores the dual functionality of MoS$_2$ in nFETs and pFETs through the selection of contact metals. N-type characteristics can be achieved in p-type MoS$_2$ by using a low workfunction semimetal, allowing for the fabrication of both nFETs and pFETs on the same channel material without additional doping. Our approach simplifies the fabrication process and minimizes inter-device gaps.

Keywords: Two-dimensional materials, field-effect transistor, contact metal

Introduction

Transistor scaling reduces dimensions, including device size and inter-device gaps, enabling more transistors to fit into a smaller area [1 - 3]. Traditionally, doping is required to fabricate nFETs and pFETs separately by forming N-wells and P-wells. However, as transistor dimensions shrink, maintaining precise control over doping profiles becomes increasingly challenging, further hindering efforts to minimize gaps [4].

Two-dimensional (2D) materials are promising for overcoming the scaling limits of traditional transistor technology [5]. Research has focused on enhancing the performance of both nFETs and pFETs [6-10]. Typically, n-type MoS$_2$ and p-type MoS$_2$ are used for fabricating nFETs and pFETs, respectively. Experiments show that the metal selection cannot change the device characteristics due to Fermi level pinning [11]. Despite using a high workfunction metal (Pt), n-type transfer and output characteristics are still exhibited on undoped MoS$_2$ devices, where is n-type material. Therefore, to fabricate an inverter using 2D materials, although only one type of metal is deposited as the contact for nFETs and pFETs, two differently doped flakes are needed in separate regions [12].

In this work, we show a n-type characteristic on a p-type MoS$_2$ by using a low workfunction semimetal and demonstrate the fabrication of a nFET and a pFET using the same channel material by selecting appropriate contact metals. This approach eliminates the need for doping in the fabrication of complementary 2DFETs, thereby enabling further gap minimization in 2DFET technology.

Fabrication of pFET and nFET on the same MoS$_2$

This study employs a standard 2D transistor structure with a global bottom gate setup to simplify the fabrication process and match most existing data. MoS$_2$-based FETs were fabricated with the cross-section presented in Fig. 1(a) and the optical image shown in Fig. 1(b). As described in Fig. 1(c), the fabrication of back-gated MoS$_2$ FETs starts with the deposition of a 30nm Al$_2$O$_3$, which serves as the bottom gate dielectric. Nb-doped MoS$_2$ flakes are then mechanically exfoliated onto the substrate. To achieve linear output characteristics, Bi is strategically selected as the contact metal for the pFET device, with the nature of semimetal, low workfunction, and a reported contact resistance of 123Ω on n-type MoS$_2$ [13]. As there is no semimetal with high workfunction, Pt is traditionally selected as the contact metal for the pFET device, with a reported contact resistance of 2000kΩ and a hole injection barrier of 130meV [11]. As two types of metals are needed on two different source/drain (S/D) regions for the fabrication of nFET and pFET on the same channel material, contact layers are deposited sequentially. Pt/Au (10nm/30nm) S/D contacts are first patterned on the MoS$_2$ channel, forming pFET. After the high workfunction metal evaporation and characterization, Bi/Au (10nm/30nm) S/D contacts are then deposited nearby to fabricate the device. To ensure the same contact area of both nFET and pFET and avoid asymmetrical metal contact, there is no overlap between Bi/Au and Pt/Au. As such, the device gap is defined as the distance between the low workfunction metal (Bi) and high workfunction metal (Pt), as circled in Fig. 1(b).

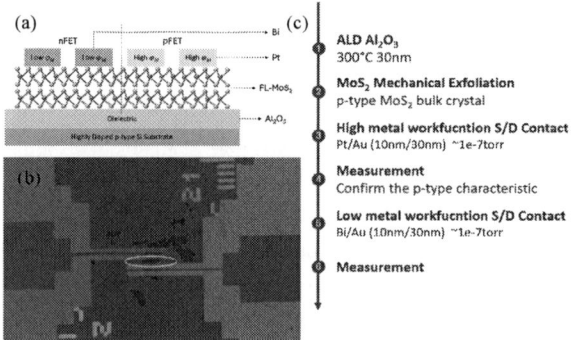

Fig. 1: (a) Device structure of back-gated 2D pFET and nFET; (b) Optical image of 2D pFET and nFET fabricated on a MoS$_2$ flake and (c) Fabrication process flow of 2D pFET and nFET with Pt and Bi contacts.

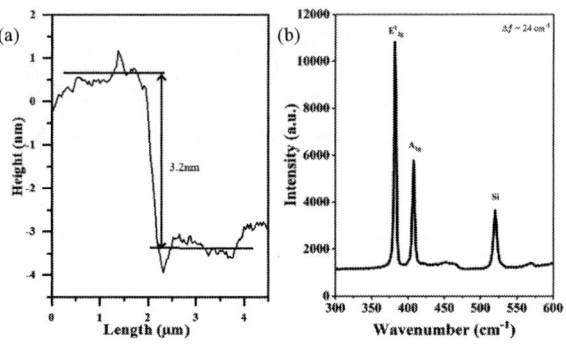

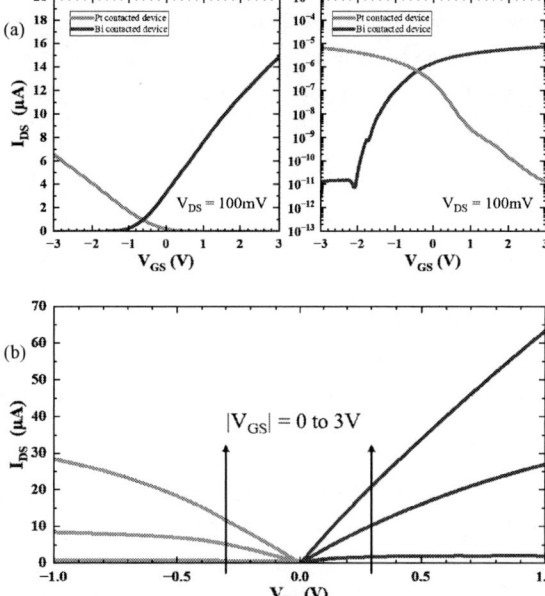

Fig. 2: (a) An AFM image of the mechanically exfoliated MoS₂ and (b) Raman spectra of p-type MoS₂.

To analyze the MoS₂ flakes, atomic force microscopy (AFM) and Raman spectroscopy were used. The AFM image in Fig. 2(a) shows a clean surface of the transferred MoS₂ flakes, with no chemical residues and identifying 4 to 5 layers. The Raman spectrum in Fig. 2(b) displays a frequency difference of 21 cm⁻¹ between the E₂g and A₁g Raman peaks at 383 cm⁻¹ and 404 cm⁻¹, respectively, which is characteristic of MoS₂.

pFET and nFET Characterization and Mechanism

The measured transfer (I_{DS}-V_{GS}) and output (I_{DS}-V_{DS}) characteristics are shown in Fig. 3(a) and Fig. 3(b) respectively. To confirm that the MoS₂ flake is p-type, the device with high metal workfunction was firstly characterized. As seen in Fig. 3(a), it exhibits p-type conduction with an ON/OFF current ratio (I_{ON}/I_{OFF}) of 10^6. As shown in Fig. 3(b), the device with a width of 9.2 μm and length of 1.2 μm achieves the drain current of 30 μA at $|V_D|$ of 1V and $|V_G|$ of 3V. The linear output characteristics of the pFET are consistent with previous reports [11, 12].

After verifying the p-type characteristic, a Bi contacted p-type MoS₂ device was measured. Initially, tunneling contact and p-type conduction were expected in p-type MoS₂ device with low metal workfunction to reduce the contact resistance since the Fermi level of the p-type channel is near the valence band, creating a high hole injection barrier but a narrow barrier width. However, as shown in Fig. 3(a), the p-type MoS₂ device with Bi contact exhibits n-type conduction with an ON/OFF current ratio (I_{ON}/I_{OFF}) of 10^6. In Fig. 3(b), linear output characteristics at low bias are indicative of ohmic contact behavior, achieving the drain current of 60 μA at $|V_D|$ of 1V and $|V_G|$ of 3V. Notably, while the subthreshold slope of the pFET is 400mV, that of the nFET is 250mV. The threshold voltages of the nFET and pFET are -1.1V and 0.1V respectively.

With this method, nFETs and pFETs are fabricated without additional doping process, resulting in both the simplification of fabrication process and the reduction of defects. By removing steps, like ion implantation and annealing, the overall fabrication process may become more streamlined, reducing the time and cost associated with the production.

Fig. 3: (a) I_D-V_G characteristics of MoS₂ pFET and nFET at V_D = 100mV; (b) I_D-V_D characteristics of MoS₂ pFET and nFET.

With the advantage of the dual functionality in 2D materials, inter-device gap can be minimized and the fabrication process is simplified as no additional doping process is involved when fabricating 2D nFET and pFET. In this experiment, the inter-device gap is measured to be around 2 μm. In such way, nFET and pFET is determined by selecting a proper contact metal and inter-device gap is only limited to the alignment in photolithography process, rather than the doping profile in the channel materials. Therefore, these devices might be easier to scale down to smaller sizes, addressing challenges in miniaturization.

To explain the dual functionality of the MoS₂ device, Fig. 4 shows the band diagram of the MoS₂ FET in different gate bias conditions. With the nature of p-type doping, the Fermi level of MoS₂ is close to the valence band. As Pt is a high workfunction metal, the Pt-MoS₂ interface forms a small hole injection barrier. On the other side, due to the Fermi level depinning and the matched band energy with the electron affinity, the interface between Bi and the MoS₂ forms a small electron injection barrier.

When negative V_{GS} is applied, holes are attracted from the p-type substrate to the oxide-semiconductor interface, increasing the hole concentration near the surface. The band diagram in Fig. 4(a) shows the bending of the valence band closer to the Fermi level at the surface, indicating an accumulation of holes. Device with small hole injection barrier conducts current under negative V_{GS} since the small hole injection barrier at the Pt-MoS₂ interface facilitates efficient hole injection. Furthermore, the thermionic emission process, which dominates the hole transport, is influenced by the barrier height and the thermal energy of the

979-8-3315-0417-5/25 $31.00 © 2025 IEEE

carriers. The small barrier height at the Pt-MoS₂ interface ensures that holes can easily overcome the barrier and contribute to the current flow. In contrast, devices with a small electron injection barrier cannot conduct electrons from the source to the drain due to hole accumulation forming a high barrier that blocks electron flow. The absence of p-type conduction through tunneling contact is probably attributed to the doping concentration in the 2D channel. For tunneling to occur, the barrier width must be extremely narrow, which requires a high doping concentration. In this case, the doping concentration may not be sufficient to achieve the necessary barrier width, thus preventing tunneling-based p-type conduction.

On the other hand, as the positive gate voltage increases further, electrons from the p-type substrate are attracted to the oxide-semiconductor interface, forming an inversion layer. The band diagram in Fig. 4(b) shows the conduction band bending downwards towards the Fermi level, indicating the presence of an inversion layer of electrons. When applying drain voltage, the small electron injection barrier at the Bi-MoS₂ interface enables electrons to flow. Similar to hole transport, electron transport is governed by thermionic emission, where electrons with sufficient thermal energy can overcome the barrier and flow through the device. The matched band energy and Fermi level depinning at the Bi-MoS₂ interface ensure efficient electron injection, resulting in current conduction.

Conclusion

In this study, the dual functionality of MoS₂ is demonstrated by fabricating both nFET and pFET using the same channel material through selection of contact metals. Our findings reveal that by utilizing a low workfunction semimetal, n-type characteristics can be achieved in p-type MoS₂, eliminating the need for traditional doping processes. This innovative approach not only simplifies the fabrication of complementary 2DFETs but also allows for significant reduction in inter-device gaps, which is essential for enhancing device density in integrated circuits. However, more simulation work should be conducted to understand the mechanism of the dual functionality. Overall, this work paves the way for more efficient and scalable transistor technologies based on 2D materials, contributing to the ongoing advancements in semiconductor design and fabrication.

Acknowledgments

This work was supported by the General Research Fund (GRF) 16201223 from the Research Grants Council (RGC) of Hong Kong.

References

[1] S. Borkar et al., "Design challenges of technology scaling." *IEEE Micro*, 19 (1999): 23-29.

[2] P. M. Zeitzoff, "MOSFET scaling trends and challenges through the end of the roadmap," in *Proceedings of the IEEE 2004 Custom Integrated Circuits Conference (IEEE Cat. No.04CH37571)*, Piscataway NJ: IEEE, 2004, pp. 233–240.

[3] R. K. Ratnesh et al., "Advancement and challenges in MOSFET scaling." *Materials Science in Semiconductor Processing*, 134 (2021): 106002.

[4] M. G. K. Alabdullah et al., "Scaling Challenges of Nanosheet Field-Effect Transistors Into Sub-2 nm Nodes," *IEEE Journal of the Electron Devices Society*, vol. 12, pp. 479-485, 2024

[5] D. Akinwande et al., "Graphene and two-dimensional materials for silicon technology". *Nature*, 573, 507–518 (2019).

[6] Y. T. Ren et al., "p-Type ohmic contact to MoS₂ via binary compound electrodes," *Journal of materials chemistry. C, Materials for optical and electronic devices*, vol. 11, no. 8, pp. 3119–3126, 2023

[7] S. Zhang, et al., "Improved contacts to p-type MoS₂ transistors by charge-transfer doping and contact engineering," *Applied physics letters*, vol. 115, no. 7, 2019

[8] A.-S. Chou *et al.*, "High-Performance Monolayer WSe₂ p/n FETs via Antimony-Platinum Modulated Contact Technology towards 2D CMOS Electronics," in *2022 International Electron Devices Meeting (IEDM)*, IEEE, 2022, p. 7.2.1-7.2.4.

[9] A. S. Chou et al., "High On-State Current in Chemical Vapor Deposited Monolayer MoS₂ nFETs With Sn Ohmic Contacts," *IEEE electron device letters*, vol. 42, no. 2, pp. 272–275, 2021

[10] A.-S. Chou *et al.*, "Antimony Semimetal Contact with Enhanced Thermal Stability for High Performance 2D Electronics," in *2021 IEEE International Electron Devices Meeting (IEDM)*, IEEE, 2021, p. 7.2.1-7.2.4.

[11] C. J. Estrada, Z. Ma, and M. Chan, "Complementary Two-Dimensional (2-D) MoS₂ FET Technology," in *European Solid-State Device Research Conference*, 2021, pp. 219–222.

[12] C. J. Estrada, Z. Ma, and M. Chan, "Complementary Two-Dimensional (2-D) FET Technology With MoS₂/hBN/Graphene Stack," *IEEE electron device letters*, vol. 42, no. 12, pp. 1890–1893, 2021

[13] P. C. Shen *et al.*, "Ultralow contact resistance between semimetal and monolayer semiconductors," *Nature (London)*, vol. 593, no. 7858, pp. 211–217, 2021

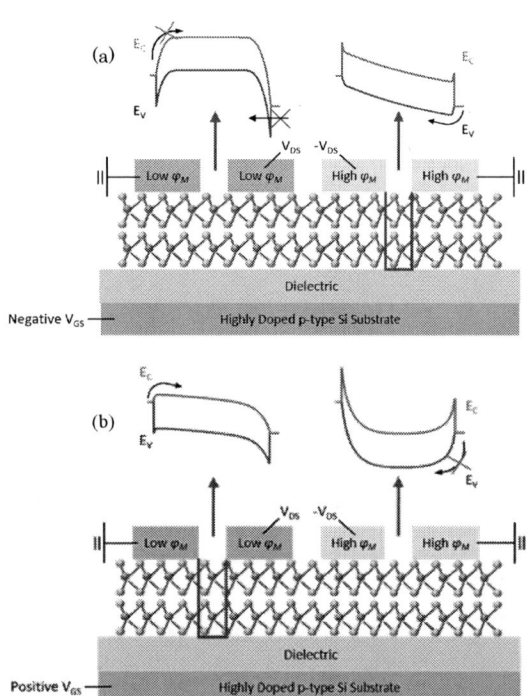

Fig. 4: Mechanism of dual functionality. (a) Band diagram of MoS₂ nFETs and pFET with low gate bias; (b) Band diagram of MoS₂ nFET and pFET with high gate bias;

An Efficient Simulation-time aware Data-Driven Automatic Design Method for Analog Circuit

Shun-Qi DAI, Yuan LEI, Bei-Ping YAN

AECS Department, Hong Kong Applied Science and Technology Research Institute (ASTRI), Hong Kong, P.R. China

Abstract

This paper presents a novel efficient data-driven approach for automated analog circuit design that considers simulation efficiency. Our method accelerates the design process by reducing both the number of design iterations and time-consuming simulations through three techniques: key target selection, elite acceptance criterion and targets classification. We validated this new approach by successfully implementing it on two fundamental analog building blocks, achieving all design specifications with significantly reduced execution time.

Keywords: Data-driven design, Automatic design method, Simulation-time aware, and Analog circuit

Introduction

Analog circuits are essential components of mixed-signal integrated circuits. While digital circuit design has become largely automated, analog circuit design still relies heavily on manual processes, creating a bottleneck in overall system chip development. Two main approaches address this challenge: analytical methods and data-driven methods. Analytical methods use approximate equations for each circuit topology to replace circuit simulators and improve design efficiency. However, they often lack accuracy and flexibility [1-3]. Data-driven methods integrate circuit simulators with optimization algorithms, achieving better accuracy and flexibility while suffering from computational inefficiency [4-5]. To overcome this limitation, we propose an efficient data-driven method that considers the simulation time difference between various performance metrics.

Efficient Simulation-time aware Data-Driven Automatic Design Method

A. Problem of Analog Circuit Design

In analog circuit design, the process begins with topology selection followed by determining design variables based on device models to meet specifications. This paper focuses specifically on the automated determination of design variables, assuming a pre-selected circuit topology. Thus, analog circuit design problem can therefore be formulated as a constrained multi-objective black-box optimization problem:

$$\text{minimize} \quad F_i(x)$$
$$s.t. \quad G_i(x) \geq p_j \quad (1)$$

where x, $f_i(x)$ and $g_i(x)$ denotes design variable, the ith

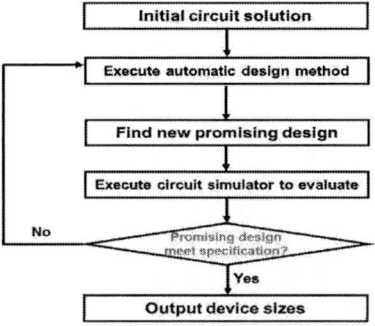

Fig. 1: A general workflow of automatic design method for analog circuits

objective and constrain, p_j represents specification of ith performance metrics. To handle the multiple objectives and constraints, weighted sum technique is used to convert a multi-objective black-box optimization problem to a single objective black-box optimization problem, which is formulated as follows:

$$\text{minimize} \quad L(x) = \sum[w_i \times F_i(x)] + \sum[w_j \times -ReLU(G_j(x) - p_j)] \quad (2)$$

where w_i is the weight factor and $L(x)$ represents the new cost function.

B. Basic Concept of Automatic Design Method

Figure 1 illustrates a general workflow of automatic design method for analog circuits. Once the circuit topology is established by circuit designer, an initial solution will be created and evaluated using a circuit simulator. If this initial solution achieves the design specifications, the device size will be output. However, if the initial netlist does not meet the specifications, the algorithm will continue to automatically seek alternative solutions. These new solutions will undergo evaluation through the circuit simulator to ascertain their compliance with the design specifications. If a solution falls to meet the specifications, the algorithm will persist in its search for a feasible solution.

C. Proposed Efficient Data-Driven Automatic Design Method

One of the challenges of the automatic design method is to reduce the overall execution time of the design procedure. The execution time T_{total} of the overall optimization algorithm is determined by two factors: the number of iterations (namely, generations) N_{iter} and running time of the circuit simulator for each solution evaluation t_{sim}:

$$T_{total} \propto N_{iter} \times t_{sim} \quad (3)$$

To enhance algorithm efficiency, the first focus is improving

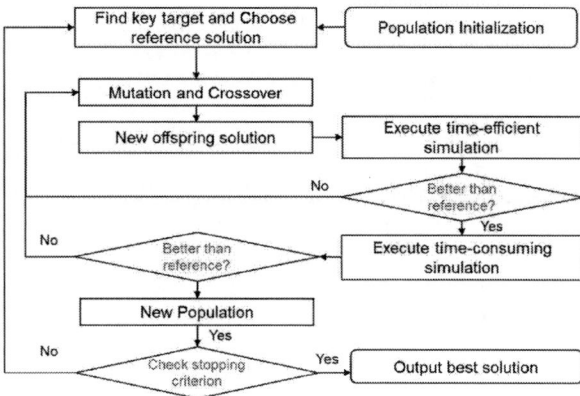

Fig. 2: A workflow of our proposed method DECT

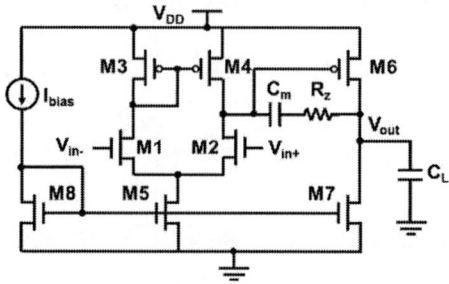

Fig. 3: Schematic of two-stage operational amplifier

TABLE I. Result of two-stage operational amplifier

Method	Specification	DECT	DE	NSGA3
DC Gain [dB]	80	81.5	81.8	86.5
GBW [MHz]	30	31.6	28.6	28.0
Phase Margin [deg]	60	60.5	9.2	60.1
CMRR [dB]	80	86.4	93.4	88.5
PSRR [dB]	80	83.7	81.5	80.1
Output Swing [V]	2.7	2.9	3	3
Positive Slewrate [V/us]	10	12.7	17	12.8
Negative Slewrate [V/us]	10	12.6	11.8	10.7
best_fitness		0.000	0.893	0.067
number of generation		10	20	50
time-efficient simulation number		938	480	2000
time-consuming simulation number		439	480	2000
run time [sec]		2171.9	2255.4	9434.2

the convergence rate of the algorithm. Improving convergence can be achieved by reducing the iteration number, which accelerates the algorithm's progress towards an optimal solution. Two methods have been proposed to achieve this goal:

Method I Key Target Selection: This method emulates human designer decision-making by identifying the most critical target in every iteration based on historical data and then ranking solution populations to determine reference solutions. This targeted approach allows for more efficient optimization of key design parameters.

Method II Elite Acceptance Criterion: This method implements a strict selection process where only offspring solutions that outperform the reference solution advance to the next iteration, ensuring consistent progress toward design improvement.

The second focus is to improve each solution's evaluation efficiency, namely minimizing the circuit simulator's execution time during solution evaluation. Performance metrics evaluation time vary significantly in their simulation durations, with certain metrics requiring substantially more computation time than others and thus dominating the overall evaluation time. Therefore, to improve evaluation efficiency, we need to minimize the number of these time-consuming simulations. One method has been outlined for this purpose:

Method III Targets Classification: This method categorizes evaluation targets into two groups: time-consuming and time-efficient targets. By firstly evaluating solutions using time-efficient targets, the algorithm can eliminate unsuitable candidates before conducting resource-intensive circuit simulations, thereby improving overall search efficiency.

These three methods are integrated into a new device sizing algorithm called Differential Evolution with Classifying Targets (DECT), which achieves both faster convergence and reduced simulation numbers, resulting in significantly shorter execution times. The detail workflow of this method is shown in Figure 2.

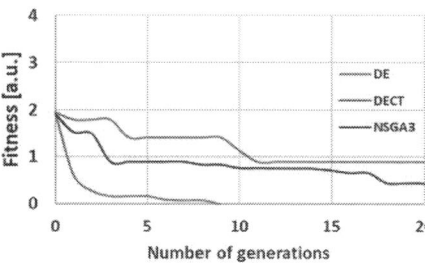

Fig. 4: Convergence behavior of two-stage operational amplifier with

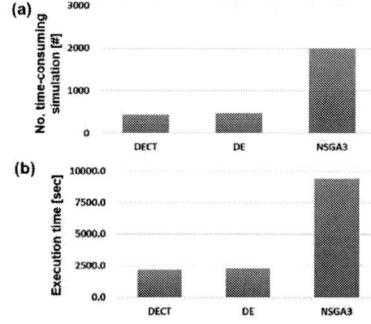

various optimization method

Fig. 5: (a) Number of time-consuming simulation and (b) execution time of two-stage operational amplifier with various optimization method.

Experimental Results

To validate the effectiveness of our design method, we apply our method to design two analog circuits, two-stage operational amplifier and class AB power amplifier, implemented in a commercial 0.18-μm CMOS mixed-signal technology. All circuit simulations were performed using a commercial circuit simulator.

979-8-3315-0417-5/25 $31.00 © 2025 IEEE

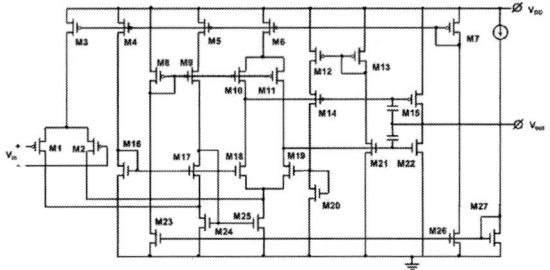

Fig. 6: Schematic of class AB power amplifier [7]

TABLE II. Result of class AB power amplifier

Method	Specification	DECT	DE	NSGA3
DC Gain [dB]	86.0	103.2	87.3	86.2
GBW [MHz]	5.0	12.0	9.0	12.0
Phase Margin [deg]	70.0	70.4	71.9	70.0
CMRR [dB]	78.1	134.8	131.4	122.3
PSRR [dB]	78.1	78.1	78.5	79.3
Output Swing [V]	2.9	3.0	2.9	3.1
Positive Slewrate [V/us]	5.0	7.2	5.7	16.3
Negative Slewrate [V/us]	5.0	10.7	5.2	5.0
best fitness		0.0	0.0	0.0
number of generation		14	90	100
time-efficient simulation number		1184	1890	4000
time-cosuming simulation number		583	1890	4000
run time [sec]		3932	7956	16960

A two-stage operational amplifier, which contains 8 transistors and 2 passive devices, serves as the first test circuit for the proposed methodology, as shown in Figure 3. The design space contains 19 variables, including 16 transistor dimensions, a bias curren, a Miller compensation capacitor, and a null resistor. Due to their computational intensity, 6 positive and negative slew rates are designated as time-consuming targets, while other targets are classified as time-efficient targets.

The convergence comparison between different optimization algorithms of two-stage operational amplifier design is shown in Figure 4. Compared with the Differential Evolution (DE) and NSGA3 methods [6], only our proposed method achieves all design specifications within 10 generations while reducing the number of time-consuming simulations to 439. The overall execution time can be reduced within 2172 seconds, which is 4-5 times faster than NSGA3, as shown in Figure 5.

The second test circuit of our method is a class AB power amplifier, which contains 27 transistors and 2 capacitors, as illustrated in Figure 6 [7]. This circuit encompasses a design space with 57 design variables including 54 transistor dimensions, 2 Miller compensation capacitors, and a bias current. As with the initial test circuit, we designate positive and negative slew rates as time-consuming targets, while other metrics are categorized as time-efficient targets.

In Figure 7, the convergence of various optimization algorithms for a class AB power amplifier is depicted. The proposed method outperforms both DE and NSGA3. DECT can successfully meet all design requirements within just 14 generations and significantly decreases the number of time-consuming simulations to 583. This optimization results in an

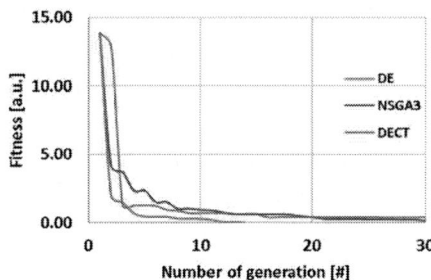

Fig.7: Convergence behavior of class AB power amplifier with various optimization method

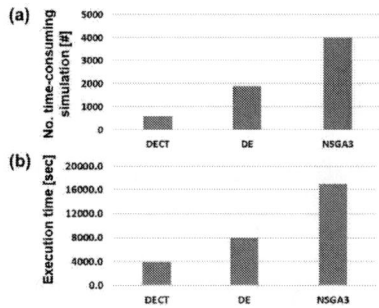

Fig. 8: (a) Number of time-consuming simulation and (b) execution time of class AB power amplifier with various optimization method.

overall execution time reduction to 3932 seconds, making it 4-5 times faster than NSGA3, as illustrated in Figure 8.

Conclusion

This work presents Differential Evolution with Classifying Targets (DECT), a new simulation-time aware data-driven approach for automated analog circuit design. We demonstrated DECT's effectiveness by implementing two widely used analog circuit blocks in a commercial 0.18-μm CMOS mixed-signal technology. Our method successfully designs an 8-transistor two-stage operational amplifier and a 27-transistor class AB power amplifier, meeting all design specifications in 2,172 and 3,932 seconds, respectively. These execution times represent a 4 to 5 times improvement in efficiency over NSGA3, demonstrating DECT's potential as an efficient solution for automated analog circuit design.

References

[1] R. Martins, N. Lourenço and N. Horta, Analog Integrated Circuit Design Automation. Cham, Switzerland: Springer, 2017.

[2] Y. Wang, M. Orshansky, C. Caramanis, "Enabling efficient analog synthesis by coupling sparse regression and polynomial optimization," IEEE Design Automation Conference, pp. 1-6, 2014.

[3] A. Sayed, A. Mohieldin, M. Mahroos, "A fast and accurate geometric programming technique for analog circuits sizing," IEEE International Conference on Microelectronics (ICM), pp. 316, 2019.

[4] B. Manuel, J. Guilherme, and N. Horta, "Analog circuits optimization based on evolutionary computation techniques," in Integration, vol. 43(1), pp. 136-155, 2011.

[5] C. Federico, and E. J. Carmona. "Automatic design of analog electronic circuits using grammatical evolution." Applied Soft Computing, 62, pp. 1003-1018, 2018.

[6] B. Julian, K. Deb, and P. C. Roy, "Investigating the normalization procedure of NSGA-III." International Conference on Evolutionary Multi-Criterion Optimization, 2019.

[7] K. Langen and J. Huijsing, "Compact low-voltage power-efficient operational amplifier cells for VLSI," IEEE JSSC, 33, pp. 1482-1496, 1998

CMOS-compatible Si₃N₄ Optical Waveguides: An Electromagnetic Study for Fabrication Considerations of SiO₂ Cladding and Silicon Wafer Choice

Wenli Zhou[1], Rui (Ray) Yao[2], and Sang Lam[1*]

[1] Department of Electrical and Electronic Engineering, School of Advanced Technology, Xi'an Jiaotong-Liverpool University, Suzhou, China; [2] Department of Electrical Engineering and Electronics, The University of Liverpool, Liverpool, United Kingdom (*email: _s.lam.cn@ieee.org_)

Abstract

We report computational electromagnetic (EM) investigation of silicon nitride (Si₃N₄) optical waveguides with CMOS-compatible manufacturing for ultimate optoelectronic integration on Si. By computing the electric field profiles of the fundamental mode of the Si₃N₄ optical waveguides, it is found about the fabrication need of 4 μm or thicker SiO₂ under-cladding layer, so as to minimise EM power loss to the resistive Si substrate. Low- or medium-resistivity (down to 0.5 Ω·cm) Si wafers are suitable choices.

Keywords: Silicon nitride optical waveguides, integrated silicon microphotonics, silicon dioxide cladding, silicon base wafer, substrate conductivity

I. Introduction

Silicon nitride (Si₃N₄) integrated photonics [1]-[2] is promising for making various on-chip optical devices [3]-[7] in the development of photonic circuits for data communications [8] and even quantum information processing [9]. This is owing to its CMOS compatibility and desirable optical properties such as allowed very broad wavelength from visible spectrum [4],[7] to mid-infrared [4]. Regardless of the complexity of photonic devices and circuits for optical information processing and other applications, the planar optical waveguide is a fundamental structure as an inherent building block in integrated photonics. There have been plentiful research works reporting Si₃N₄ optical waveguides with very low loss [10]-[13], commonly with fabrication on a Si substrate [3]-[8] and with silicon dioxide (SiO₂) cladding by thermal oxidation. Surprisingly however, there has been almost no research explicitly reporting the practical considerations of the resistive Si substrate and the SiO₂ cladding layers in making Si₃N₄ optical waveguides. Varied thickness from 1.5 μm to 15 μm [3]-[5],[14]-[15] have been adopted for the SiO₂ under-cladding. To form such a thick SiO₂ layer, the wet thermal oxidation can take 10 hours or longer [15], requiring also high temperature condition (>1000 °C). As for the substrate, there can be varied choices of Si base wafers for fabricating Si₃N₄ optical waveguides, from which complex optoelectronic devices and circuits are realised on the same wafer.

In this work, we conduct computational EM investigation into the influence of the SiO₂ cladding and of the resistive Si substrate on the performance of Si₃N₄ optical waveguides. In addition to the propagation loss, we also examine the mode profiles that might be impacted by different cladding layer thickness and different Si substrate conductivity. Our comprehensive EM study will save extensive fabrication resources in finding out the engineering considerations of the SiO₂ cladding and the resistive Si substrate for the design and chip realisation of Si₃N₄ optical waveguides, as well as their derived photonic devices and circuits.

II. Planar Si₃N₄ Optical Waveguide for EM Study

In our computational work, we constructed a device structure shown in Fig. 1. The physical structure is comprised of four layers: a rectangular dielectric core of Si₃N₄ in the middle, a resistive Si substrate layer at the very bottom and a SiO₂ under-cladding sequentially built above, following with a SiO₂ over-cladding layer at the top and besides the Si₃N₄ core. Both the Si₃N₄ layer and SiO₂ over-cladding can be formed by low-pressure chemical vapour deposition (LPCVD) [14], but the SiO₂ under-cladding by thermal oxidation [15].

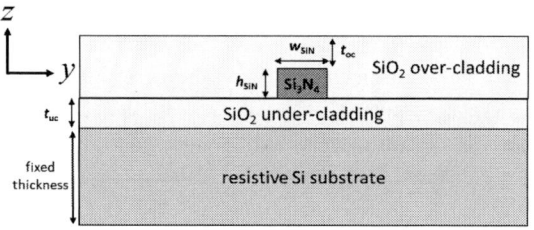

Fig. 1. A schematic cross-sectional diagram showing the device structure for investigating Si₃N₄ optical waveguides built on a Si substrate (with the height h_{SiN} = 600 nm and width w_{SiN} = 1000 nm for the Si₃N₄ core)

III. 3D EM Simulations by Finite Element Method

The computational EM investigation was carried out using mainly a commercial full-wave EM simulation software program, Ansys HFSS, with supplementary EM simulations by COMSOL. Both EM simulation programs are based on the finite element method (FEM). In the EM simulations at optical frequencies (> 300 THz) using HFSS, we manually set up the electrical properties of the materials as summarised in the following: ε_r = 3.965 and σ = 0 S/m are assigned to the Si₃N₄ core, resulting in its refractive index $n = \sqrt{\varepsilon_r}$ = 1.99 at wavelength λ = 900 nm; ε_r = 2.108 and σ = 0 S/m are assigned to the SiO₂ cladding resulting in its refractive index n = 1.45 also at λ = 900 nm; while ε_r = 13.0 (at λ = 900 nm) and σ = 2×10³ S/m are initially assigned to the resistive Si substrate at the bottom. To account for the dielectric damping loss due to polarisation of bound electrons [16] in Si as a dielectric material, the dielectric loss tangent is set $tan(\delta_c)$ = $\varepsilon''/\varepsilon'$, where ε'' = 0.0152 and ε' = 13.03 at λ = 900 nm [17]. Wave ports were used in the EM simulation with the total voltage set at 1.0 V at the excitation port (port 1) at one end of the optical waveguide, and the detection port (port 2) at the other end. Fig. 2 shows the electric field intensity distribution (in the y-z plane) in a colour scale obtained from three-dimensional (3D) EM simulations for the fundamental mode of transverse electric (TE) and transverse magnetic (TM) polarisations. It can be seen about the electric field is mostly

979-8-3315-0417-5/25 $31.00 © 2025 IEEE

confined within the Si₃N₄ core and becomes weaker and weaker at positions away from the middle axis of the core. The electric field of the TM mode seems to penetrate slightly more into the cladding layer. The computed EM fields are used to calculate the voltage and current of the signal and then the S-parameters. Among the S-parameters, $|S_{21}|$ can reveal the propagation loss of the optical waveguide for each propagation mode. In this work, we focus on the fundamental mode, while higher order modes do exist for the optical waveguide of wavelength-scale sizes ($0.6\ \mu m^2 \approx 3(\lambda/n_{SiN})^2$). The chosen dimensions of the Si₃N₄ core are based on the practical considerations of fabrication capability (especially photolithography) of typical university cleanroom facilities.

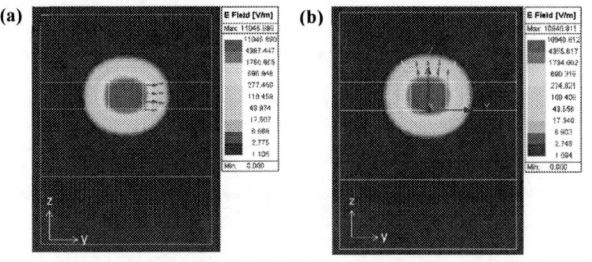

Fig. 2. Electric field intensity distribution of the fundamental mode (TE in (a) and TM in (b)) of the Si₃N₄ optical waveguides built on a Si substrate (with 2-μm SiO₂ cladding). Note the electric field is plotted in log scale to cover a large range; and the arrows show the electric field direction.

Fig. 3 shows detailed electric field (E-field) magnitude profiles of the TE fundamental mode for a Si₃N₄ optical waveguide with 2-μm SiO₂ cladding and the Si substrate at the bottom. The TM mode is almost the same and not shown.

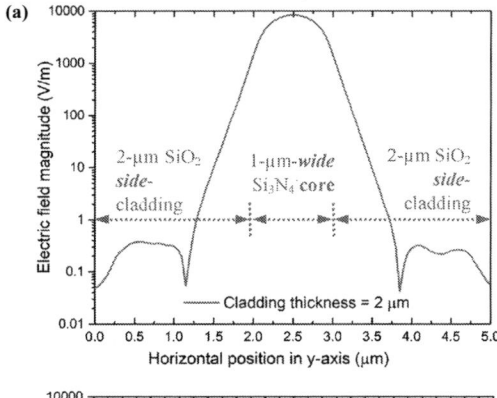

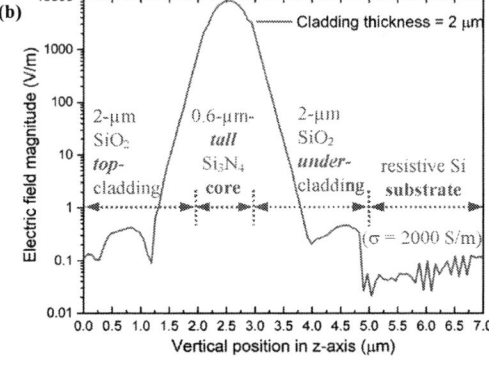

Fig. 3. Detailed electric field magnitude profiles (in horizontal direction in (a) and in vertical direction in (b)) of the TE fundamental mode of the Si₃N₄ optical waveguide (with 2-μm SiO₂ cladding), showing evanscent field magnitude < 1 V/m in the SiO₂ cladding and < 0.1 V/m in the Si substrate.

It can be seen that the E-field profile is basically symmetrical along the horizontal direction (y-axis in Fig. 1 and Fig. 2). The E-field magnitude within the 2-μm SiO₂ side-cladding drops rapidly to well below 1 V/m. Along the vertical direction (z-axis in Fig. 1 and Fig. 2), the E-field magnitude within the 2-μm SiO₂ cladding (the upper and bottom sides) behaves almost the same symmetrically; but the E-field magnitude drops further to below 0.1 V/m in the resistive Si substrate. Similar E-field profiles were also obtained for the Si₃N₄ optical waveguides with other SiO₂ under-cladding thickness. The difference lies in the evanescent E-field in the resistive Si substrate (Table 1).

Table 1: Evanescent E-field (magnitude) in the Si substrate for the Si₃N₄ waveguides of different cladding thickness

SiO₂ thickness	0.5 μm	1.0 μm	2.0 μm	4.0 μm
Evanescent E-field	7 V/m	0.3 V/m	0.03 V/m	0.02 V/m

IV. Propagation Loss Influenced by SiO₂ Cladding Thickness & Si Substrate Conductivity

With the computed EM fields, S-parameters can be determined by regarding the optical waveguide as a two-port network. Among the S-parameters, $|S_{21}|$ can reveal the propagation loss of the Si₃N₄ optical waveguides with varying SiO₂ cladding thickness and different conductivity values of the Si substrate. With the wave port settings in the EM simulations, the signal loss due to the coupling from the excitation (or detection) port to the optical waveguide is very minimal (as indicated by a small $|S_{11}|$ and $|S_{22}|$). So, $|S_{21}|$ can be used to approximate the propagation loss of the Si₃N₄ optical waveguides. With the ideal construction of the optical waveguide in EM simulations, there is neither surface roughness nor other physical defects. As a result, the computed $|S_{21}|$ and hence the propagation loss of the Si₃N₄ optical waveguides can be exceptionally low. For this reason, the absolute levels of the propagation loss can only be regarded as the theoretical best possible values. Nevertheless, the changes in the propagation loss for varying cladding thickness and different Si substrate conductivity can tell the performance trend and behaviour. Fig. 4 shows the results of the propagation loss (at $\lambda = 900$ nm) of the wavelength-scale Si₃N₄ optical waveguides for varying cladding thickness.

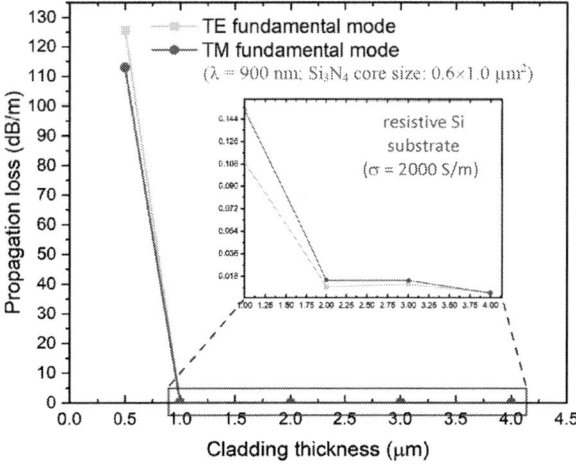

Fig. 4. Propagation loss of the Si₃N₄ optical waveguides with varying SiO₂ cladding thickness (and the Si substrate conductivity being 2000 S/m)

The propagation loss of the Si_3N_4 optical waveguides remains at very low levels (< 0.018 dB/m) for SiO_2 cladding no thinner than 2 µm. With 1-µm-thin SiO_2 cladding, the propagation loss is almost 10 times worse. With 0.5-µm SiO_2 cladding, the propagation loss is about three order of magnitude worse than those cases with thicker SiO_2 cladding. These results are generally consistent with the computed evanescent E-field magnitude (respectively 7 V/m and 0.3 V/m as in Table 1) for 0.5-µm and 1-µm SiO_2 cladding.

Fig. 5 shows the results of the propagation loss of the Si_3N_4 optical waveguides for different Si substrate conductivity (while the SiO_2 cladding is 2 µm thick). The Si substrate is connected to ground as it should be for electronic circuits integrated on the same chip. The propagation loss remains at steadily low levels when the Si substrate conductivity is less than about 20 S/m. The propagation loss changes when the Si substrate conductivity is close to 100 S/m and higher. This is likely because the Si substrate no longer behaves as a low-loss dielectric but as quasi-conducting at optical frequencies [16] corresponding to λ = 900 nm. When the conductivity is 5×10^4 S/m or higher, the Si substrate behaves as a good conductor at 333 THz (corresponding to λ = 900 nm).

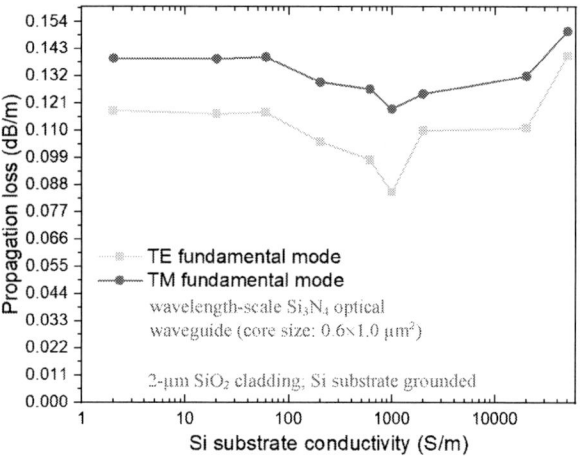

Fig. 5. Propagation loss of the Si_3N_4 optical waveguide with different Si substrate conductivity but the substrate being connected to ground (as to model the real situation that the wafer base or the chip is grounded).

V. Conclusion

We have reported a comprehensive EM study of the CMOS-compatible Si_3N_4 optical waveguides of the wavelength scale. In particular, we have used FEM-based 3D EM simulation to investigate the influence of the SiO_2 cladding and the resistive Si substrate. With the wavelength-scale Si_3N_4 core (sizes $\approx 3(\lambda/n)^2$ in this work), the evanescent electric field penetrates through the SiO_2 cladding into the Si substrate. In the fabrication of planar optical waveguides on a Si wafer, the SiO_2 cladding needs to be thick enough (>2 µm). Only when the SiO_2 under-cladding is 4 µm or thicker, the influence of the Si substrate would be minimised. The on-chip optical waveguides would have steadily low propagation loss, regardless of the Si substrate conductivity σ over a large range (preferably with σ below 200 S/m). This implies much flexibility of the Si wafer choice. Low-resistivity Si wafers, which are of lower cost, can be used for realisation of Si_3N_4 optical waveguides and other optical devices of photonic integrated circuits.

Acknowledgments

This work is supported in part by PGRS funding (FOSA2406036) of XJTLU. Both R. Yao and S. Lam acknowledge various help from Prof. Alex M. H. Wong of City University of Hong Kong, especially for R. Yao's training in the use and access to Ansys HFSS within CRAE Laboratory. S. Lam has his much appreciation to Professor Jensen T. H. Li in Department of Physics at The Hong Kong University of Science and Technology (HKUST), for the very fruitful visit to Center for Metamaterials. The discussions with him and his PhD student, Mr. Youyi Zhou, were helpful for starting the research reported in this paper.

References

[1] D. J. Blumenthal, R. Heideman, D. Geuzebroek, A. Leinse, and C. Roeloffzen, "Silicon nitride in silicon photonics," *Proceedings of the IEEE*, vol. 106, no. 12, pp. 2209-2231, December 2018.

[2] T. D. Bucio, C. Lacava, M. Clementi *et al.*, "Silicon nitride photonics for the near-infrared," *IEEE Journal of Selected Topics in Quantum Electronics*, vol. 26, no. 2, pp. 8200613-8200613, March-April 2020.

[3] J. F. Bauters, M. J. R. Heck, D. Dai, J. S. Barton, D. J. Blumenthal, and J. E. Bowers, "Ultralow-loss planar Si_3N_4 waveguide polarizers," *IEEE Photonics Journal*, 5(1), art. no. 6600207, Feb. 2013.

[4] P. T. Lin, V. Singh, L. Kimerling, and A. M. Agarwal, "Planar silicon nitride mid-infrared devices," *Apl. Phy. Lett.*, 102(251121), Jun. 2013.

[5] C. G. H. Roeloffzen, M. Hoekman, E. J. Klein *et al.*, "Low-loss Si_3N_4 TriPleX optical waveguides: technology and applications overview," *IEEE Journal of Selected Topics in Quantum Electronics*, vol. 24, no. 4, art. no. 4400321, July/August 2018.

[6] J. Liu, G. Huang, R. N. Wang *et al.*, "High-yield, wafer-scale fabrication of ultralow-loss, dispersion-engineered silicon nitride photonic circuits," *Nature Communications*, 12(2236), 16 April 2021.

[7] Y. Lin, Z. Yong, X. Luo *et al.*, "Monolithically integrated, broadband, high-efficiency silicon nitride-on-silicon waveguide photodetectors in a visible-light integrated photonics platform," *Nature Communications*, 13(6362), 26 October 2022.

[8] A. Caut, V. Shekhawat, V. Torres-Company, and M. Karlsson, "Polarization-insensitive silicon nitride photonic receiver at 1 µm for optical interconnects," *IEEE Photonics J.*, 16(3), art. no. 2200507, June 2024.

[9] C. Taballione, T. A. W. Wolterink, J. Lugani *et al.*, "8×8 reconfigurable quantum photonic processor based on silicon nitride waveguides," *Optics Express*, vol. 27, no. 19, pp. 26842–26857, September 2019.

[10] S. C. Mao, S. H. Tao, Y. L. Xu, X. W. Sun, M. B. Yu, G. Q. Lo, and D. L. Kwong, "Low propagation loss SiN optical waveguide prepared by optimal low-hydrogen module," *Optics Express*, vol. 16, no. 25, pp. 20809-20816, 8 December 2008.

[11] M.-C. Tien, J. F. Bauters, M. J. R. Heck, D. J. Blumenthal, and J. E. Bowers, "Ultra-low loss Si_3N_4 waveguides with low nonlinearity and high power handling capability," *Optics Express*, vol. 18, no. 23, pp. 23562–23568, 8 November 2010.

[12] J. F. Bauters, M. J. R. Heck, D. John *et al.*, "Ultra-low-loss high-aspect-ratio (Si_3N_4) waveguides," *Optics Express*, vol. 19, no. 4, pp. 3163–3174, 14 February 2011.

[13] H. El Dirani, L. Youssef, C. Petit-Etienne *et al.*, "Ultralow-loss tightly confining Si_3N_4 waveguides and high-Q microresonators," *Optics Express*, vol. 27, no. 21, pp. 30726-30740, 9 October 2019.

[14] R. Kou, N. Yamamoto, G. Fujii *et al.*, "Spectrometric analysis of silicon nitride films deposited by low-temperature liquid-source CVD." *Journal of Applied Physics*, 126, 133101, October 2019.

[15] P.-H. Wang, H.-Y. Zheng, Y.-H. Liu, and C.-M. Wang, "High-Q Silicon Nitride Waveguide Resonators by Nanoimprint Lithography," *2024 IEEE Silicon Photonics Conference (SiPhotonics)*, Tokyo Bay, Japan, April 2024, pp. 1-2.

[16] U. S. Inan, A. S. Inan and R. K. Said, "Waves in an Unbounded Medium" in *Engineering Electromagnetics and Waves*, 2nd edition, Upper Saddle River, New Jersey, USA: Pearson Education, 2015, ch. 8, sec. 8.3, pp. 615–635.

[17] Mikhail Polyanskiy, "Optical constants of Si (Silicon) - Schinke et al. 2015, n, k 0.25–1.45 µm", Refractiveindex.info. https://refractiveindex.info/?shelf=main&book=Si&page=Schinke (accessed 16th July 2024)

AUTHOR INDEX

Abraham, Nithin 1128
Agarwal, Harshit 793, 799, 802
Agarwal, Tarun 925
Aharonovich, Igor 998
Ahmad, Naef 426
Ahmed, Faisal 986
Ahn, Jong-Hyun.. 1225
Ahn, P.-H. 291
Ahn, S. J. 1189
Ahn, Yonghwan 652
Ai, Hao 1003
Akshay, K 540
Alant, Johan 1195
Alchalabi, Mustafa 1261
Amara, Selma 288
Amaram, Ashutosh Krishna 925
Amrouch, Hussam 1107
An, Jiayi..786
An, Xia 453, 471, 500
Anders, Jens.. 1086
Andia, Luis 1122
Ando, Koji 897
Andreev, Sergei 1261
Ang, Kah-Wee 257
Anjali, A. 1057
Anwar, Muhammad Abid............ 285, 491, 528, 868
Ao, Chengkang 369, 438
Ao, Mingrui 1059, 1062, 1065
Arabhavi, A. M. 7
Arakawa, Takaki 558
Ashraf, Syed 1222
Bae, J. H. 1189
Baek, Rock-Hyun 652, 691
Bagga, Navjeet 706, 796
Bahl, Sandeep 1246
Bahrami, Mina 1192
Bai, Jing 357
Bai, Mingkai 276
Bai, Rognxu 1252
Bai, Wubin 1198
Bao, Lin 841, 1237
Bao, Shengyu 485, 1237
Bao, Wenzhong 1059, 1062, 1065
Bao, X. 977
Bao, Yunjiao 336, 546
Barbot, Justine 399
Ben, Jianwei...360
Benini, Luca.. 1261

Benkhelifa, Mahdi.. 1107
Bernert, Kerstin 399
Bestelink, Eva 242
Bhat, Aditya K... 1057
Bhattacharjee, N. 28
Bi, Ran 500, 744
Bian, Zheng 552
Biswas, Anmol 965
Bodepudi, Srikrishna Chanakya 285, 491, 528, 868
Bolognesi, C. R. 7
Braun, Dennis 497
Brinson, Mike 1261
Bu, Saiyu 130
Bu, Weihai 474, 601
Bucher, Matthias 1261
Burgt, Yoeri Van De 759
Butler, Keith T. 895
Cai, Bozhou 503
Cai, Hao 366, 661
Cai, Hecheng 303
Cai, Jian... 980
Cai, Jing 688
Cai, Kaiming 429
Cai, Puyang 459
Cai, Weiwei 986
Cai, Yimao 485, 841, 1021, 1024, 1237
Cao, Bingyang 175
Cao, Shengjie 934, 956, 1158
Cao, Xianmao 157
Cao, Yingnan 139
Cao, Zhenyuan 1092
Chai, Junshuai 34, 40, 103, 267, 276
Chai, Y. J. 214
Chai, Yang.............. 172, 239, 333, 339, 396, 721, 783, 903
Chakarov, Ivan 181
Chan, Chak Lam Jonathan 531
Chan, Henry 1114
Chan, Kaiman.. 1195
Chan, Mansun 309, 414, 640, 989
Chan, Paddy K. L. 139
Chand, Rakesh 121
Chandra, Anirban 1114
Chang, Chin-Yu... 121
Chang, Fu-Shen 771
Chang, Hao 236, 513
Chang, Hsiang-Hung 943
Chang, Jonathan 151
Chang, Yii-Tay 771

Chang, Yu-Ming.................................257
Charan, Vanjari Sai 1057
Charbon, Edoardo.............................838
Chatterjee, Payel.............................718
Chau, Yu Foon 482
Chauhan, Amit Kumar Singh.................330
Chauhan, Yogesh Singh 462, 1080, 1107, 1137
Chen, Aobei 52
Chen, Bing 390, 1249
Chen, Chih 943
Chen, Chun-Zhang...........................82
Chen, Danyang 378, 962
Chen, Gang 865
Chen, H.-M.................................900
Chen, Haibao 459
Chen, Hai-Bao 628
Chen, Han 130
Chen, Hao 411
Chen, Haojie 1059, 1062, 1065
Chen, Hongda 372
Chen, Hongzhen 372
Chen, Jen-Hao 55
Chen, Jiajia 1249
Chen, Jianjun 423
Chen, Jianlin 1119
Chen, Jie Wei 172
Chen, Jiewei................................721
Chen, Jiezhi 85
Chen, Jingyang 679
Chen, Jun 937
Chen, Junting 300, 622
Chen, Kai 321
Chen, Kuan-Ting 121
Chen, Kun 817
Chen, Liang 956
Chen, Long 82
Chen, Nanbo 679
Chen, Peng 859, 1143
Chen, Qi 184
Chen, Qiang 780
Chen, Qingming.............................37
Chen, Rongsheng 22
Chen, Ruiqi 1003
Chen, Shuai 883
Chen, Shuaiyu 573
Chen, Sihao 823
Chen, Siyu 133, 205
Chen, Sujie 616
Chen, T.-Y. 151
Chen, Tiwei 64
Chen, Wanjun 124, 865
Chen, Weiliang..............................233

Chen, Xiaojin 912
Chen, Xin 411
Chen, Xin-Ru 10
Chen, Xuanqi 405
Chen, Y.-C. 880
Chen, Yance 285, 491
Chen, Yang 774
Chen, Yifan 471
Chen, Yitong 88, 260
Chen, Yiyang 1003
Chen, Yuefeng 441, 595
Chen, Yung-Hsiang 121
Chen, Zeqi 34, 103
Chen, Zequan 1057
Chen, Zerui 1045
Chen, Zheng 1027
Chen, Zhenyu 1030
Chen, Zhongming 109
Chen, Zhuoya 1024
Chen, Zimin 519
Chen, ...522
Cheng, Che-Chi 771
Cheng, Fang 555
Cheng, Guoyao..............................309
Cheng, Jiangong............................411
Cheng, Kai Wen 151
Cheng, Ran 390, 1249
Cheng, Shengliang 519
Cheng, Yu Jian 903
Cheng, Zengguang 408
Cheng, Zhe 494
Chhaperwal, Mayank 1128
Chi, X. 1213
Chi, Yaqing................................423
Chih, Yu-Der 151
Chiu, C.-Y. 886
Chiu, Wei-Lan 943
Cho, Byung Jin.............................1134
Cho, D. G. 1189
Cho, Kyeongrae 652
Cho, M. H. 1189
Choi, Shinhyun.............................1117
Chong, Chen 76
Chowbhury, S. 977
Chu, Wen-Ting 151
Chu, Yanbang 601, 747
Chun, Suk Yeop 1186
Chung, Jerry 151
Chunxia, Li 226
Ciabattini, F. 7
Cong, Yuqing 1101
Cruces, Sofia 1279

Cruces, Sofía ... 497
Cui, Boyao ... 459, 634, 1048
Cui, Jiawei ... 236, 513
Cui, Jie ... 393
Cui, Peng ... 604
Cui, Tianning ... 962
Cui, Xiaoqi ... 986
Cui, Yan ... 429
Cui, Yanxia ... 522
Cui, Zhi-Li ... 694
Cunman, Liang ... 217
Cüppers, Felix ... 832
Dai, Jie-Ni ... 820
Dai, Saifei ... 267
Dai, Shun-Qi ... 643
Dai, Xianqi ... 64
Dai, Xinyue ... 208
Dai, Yue ... 285
Dai, Zhongbin ... 384
Dan, Yaping ... 534
Das, Saptarshi ... 97
Das, Susobhan ... 986
Das, Tanay ... 426
Dasgupta, S. ... 706, 796
Datsuk, Anton ... 1261
Daus, Alwin ... 497, 1131
Davoudi, Mohammad Rasool ... 1192
Dayte, I. ... 977
De, Sourav ... 1173
Deen, M. Jamal ... 4
Deng, Chenkai ... 297
Deng, L. ... 479
Deng, Meng ... 163
Deng, Minyue ... 956
Deshmukh, Shreyas ... 564, 965
Deshpande, Veeresh ... 564
Devi, Reshma ... 895
Dijken, Sebastiaan Van ... 1222
Ding, Chenchen ... 543
Ding, Guanglong ... 348
Ding, Jiaxin ... 1042, 1264
Ding, Mingzheng ... 546
Ding, Rongzheng ... 619, 631
Ding, Shenlei ... 387
Ding, Xiaolei ... 285, 491
Ding, Yajing ... 103
Ding, Yi ... 847
Ding, Yian ... 390
Ding, Yichong ... 871
Dinh, T. V. ... 1207
Divyanshu, ... 288
Dixit, Ankit ... 796

Dong, Daoyi ... 727
Dong, Junchen ... 229
Dong, Xiangqi ... 1059, 1062, 1065
Dong, Yulong ... 378, 962
Đonko, Dženana ... 670
Dou, Xiaoyu ... 85
Drescher, M. ... 28
Du, Fangzhou ... 58, 67, 1243
Du, Fengyu ... 928
Du, Hanghai ... 912
Du, Haoran ... 661
Du, Peiran ... 282
Du, Peiyuan ... 390
Du, Pengbo ... 1027
Du, Yiwei ... 70
Du, Yu ... 561
Du, Zhiyuan ... 139
Duan, Jiahui ... 267
Dutta, Aloke K. ... 462, 1137
Ebrahimi, M. ... 7
Elrifai, Tarek ... 1086
Enz, Christian ... 838
Eom, Seungjoon ... 691
Ervin, Joseph ... 181
Esseni, David ... 829
Estrada, Cristine Jin ... 989
Fa, Yuan ... 1279
Fan, Shi Quan ... 342
Fan, Xuemeng ... 73, 127
Fan, Yijia ... 70
Fan, Yutong ... 312
Fan, Yuyan ... 378
Fan, Zhengfang ... 534
Fan, Zhiyong ... 465, 531
Fang, Chi ... 595
Fang, Cize ... 906
Fang, Haotian ... 196
Fang, Ryan ... 1195
Fang, Shengli ... 342
Fang, Tong ... 732
Fang, Wencheng ... 715
Fang, Wenzhang ... 285, 491
Fang, Wuqing ... 196
Fang, Yilong ... 549
Fariborzi, Hossein ... 288
Fatin, Mohammad Ajmain ... 756
Feng, Dahu ... 561, 579
Feng, Ning ... 835, 1036
Feng, Peng ... 679
Feng, Qian ... 345
Feng, Yulin ... 1003
Feng, ... 16

Figuet, C. 1216
Florica, Camelia 288
Fraser, Lachlan 1155
Fu, Chun 1243
Fu, Jianghao 193
Fu, Jun 393
Fu, Lan 727
Fu, Sulei 765, 862
Fu, Xingyu 573
Fu, Yunyi 46, 1024
Fu, Yu-Yang 682
Fu, Zhiyuan 805, 934, 1045, 1158
Fujiwara, M. 558
Fujiwara, Masazumi 115
Galderisi, G. 28
Gan, Jinghan 1167
Gan, Lin 1219
Gan, Xuetao 477
Gan, Zhouchao 555, 947
Ganguly, Udayan 564, 965
Gao, Bin 70, 133, 205, 506, 729, 847
Gao, Chao 762, 811, 937
Gao, Dawei 19, 127, 1149
Gao, Ge 52
Gao, Guoyun 139
Gao, Jiajun 765, 862
Gao, Jianfeng 429, 753
Gao, Ju 369, 438
Gao, Wenze 582
Gao, Xinguo 573
Gao, Yi 841
Gao, Yufeng 1167
Geng, Di 874
Geng, Li 342
Georgiev, Vihar 796
Geum, Dae-Myeong 1039
Geum, Daemyeong 1134
Goel, Sanket 187
Gong, Heng Yue 625
Gong, Xiao 405, 850
Gopalakrishnan, Sai Gautam 895
Goshima, D. 712
Goshima, Daiki 598
Gou, Saifei 1059, 1062, 1065
Grabinski, Wladek 1261
Grasser, Tibor 1192, 1282
Grenouillet, Laurent 516
Grothe, M. 28
Gu, Chen 874
Gu, Denshun 768
Gu, Jiangmin 387
Guan, Shan 1089

Guo, D. Y. 664
Guo, Dengyao 928
Guo, Fuqiang 573, 786
Guo, Gaofu 64
Guo, Guoping 441, 595
Guo, Haowen 919
Guo, Hengxu 441, 595
Guo, J. K. 664
Guo, Jian-Bin 10
Guo, Jianmiao 333
Guo, Jingkai 928
Guo, Junwei 459
Guo, Min 613
Guo, Qijun 381
Guo, Qinan 124
Guo, Qinhua 1018
Guo, Rui 747
Guo, Taoming 637, 953
Guo, X. 479
Guo, Xiaojun 616, 1152
Guo, Xiaokun 435
Guo, Xinrui 100, 270, 468
Guo, Yeye 983, 1012
Guo, Zhanfeng 735
Guo, Zihan 1021
Gupta, Manish 892
Gurkaynak, Frank K. 1261
Ha, Daewon 223, 1189
Hajare, Hemant 564, 965
Hamzeloui, S. 7
Han, Chuan Yu 342
Han, Cong 992
Han, Dedong 229
Han, Genquan 390, 906, 1249
Han, Hongyan 474
Han, Hung-Chi 838
Han, Jiachen 411
Han, Kaizhen 405
Han, Pengfei 303
Han, Runhao 34, 103
Han, Songjia 697
Han, Su-Ting 303, 348
Han, Yanjun 1219
Hannula, Sebastian 1222
Hao, Hongwei 522
Hao, Yue 312, 345, 372, 390, 444, 447, 585, 667,
 .... 871, 906, 912
Hao, Zhibiao 1219
Hara, A. 712
Hara, Akito 598
Hasan, Tawfique 986
Hashimoto, S. 315

Havel, V. 28
He, Gufeng 1033
He, Jiaheng 494
He, Jiayin 369, 438
He, Jin 528
He, Ming 100, 270, 468, 1071
He, Minghao 82
He, Nan 883
He, Shikun 381, 429
He, Xi 1119
He, Xiaohan 372, 447
He, Xinliu 1059, 1062, 1065
He, Y. 28
He, Yang 853
He, Yaoyu 85
He, Yongjie 1252
He, Youshui 868
He, Yuhui 184, 339, 396
He, Zhaolong 321
He, Zong Rui 903
Herfurth, Norbert 1261
Herman, Krzysztof 1261
Hikake, Kaito 1255
Hiramoto, Toshiro 1255
Ho, Yenshih 121
Hoentschel, J. 28
Hoffmann-Eifert, Susanne 832
Hong, M. 1189
Hong, Peiyan 417
Hong, S.-M. 291
Horiike, Ryota 31
Hossain, Mainul 756
Hou, Bin 444, 582, 585
Hou, Jiahao 306
Hou, Yilin 85
Hsiang, Kuo-Yu 771
Hsieh, E Ray 151
Hsieh, Y.-S. 151
Hsiung, M.-H. 880
Hsu, J.-F. 977
Hu, C. 61
Hu, Chenming 55
Hu, Chun 52
Hu, Guohua 655
Hu, Haodong 906
Hu, Hengyi 220
Hu, Huan 285, 491
Hu, K.-K. 977
Hu, Qianlan 1077
Hu, Runyu 408
Hu, Ruofei 133, 205
Hu, Tao 34, 103, 276

Hu, Tong 709
Hu, V. P.-H. 886
Hu, Vita Pi-Ho 892
Hu, Weida 959
Hu, Yan 1059, 1062, 1065
Hu, Yilin 1006
Hu, Youfan 567
Hu, Yuqing 312
Hu, Zheyuan 64
Hu, Ziyang 679
Hua, Mengyuan 300, 622
Hua, Qilin 109
Huang, Feixuan 1119
Huang, Haoxin 112
Huang, Heyi 1293
Huang, Jing-Kai 94, 257
Huang, Peng 1003
Huang, Qi 88, 199, 260
Huang, Qianqian 279, 474, 805, 934, 956, 1045, 1158
Huang, Ru 91, 100, 254, 270, 273, 279, 309, 453,
......459, 468, 471, 474, 485, 500, 525, 601, 628, 634, 676, 744,
747, 805, 823, 835, 841, 934, 956, 1006, 1021, 1024, 1036,
1045, 1048, 1051, 1071, 1158, 1161, 1237
Huang, S.-T. 900
Huang, Sen 573, 604, 786
Huang, Shan 16
Huang, Tengyan 43
Huang, Tzu-Yun 118
Huang, Wei 22, 814
Huang, Xiao-Di 703
Huang, Xingyu 1255
Huang, Yi 613
Huang, Yilong 133, 205
Hui, Fei 1009
Huo, Wenju 1143
Huo, Yikang 106
Hwang, Jason C. C. 94
Hyun, S. 1189
Illarionov, Yu. Yu. 214
Im, Changhyeok 166
Ishikawa, R. 1213
Islam, Muhammad Saif Ul 1155
Ito, Y. 712
Iwata, J. 315
Jain, Khushi 796
James, Jerry Joseph 658
Jang, G.-T. 291
Jang, M.-S. 291
Jansen, S. 28
Jena, Swadhin Kumar 540
Jeong, Jaejoong 1134
Jeong, Jaeyong 1039, 1134

Jhang, J.-H.	977
Ji, Botao	260
Ji, Haotian	387
Ji, Ning	835
Ji, Zhenghui	381
Ji, Zhigang	91, 459, 628, 634, 673, 1006, 1048
Ji, Zhongchen	786
Jia, Mao	444
Jia, Rundong	474
Jia, Xiaole	906
Jia, Xinpei J	103
Jia, Xinpei	34, 267, 276
Jia, Yueyang	750, 856
Jia, Yuping	360
Jian, Ming	685
Jiang, Biyi	49
Jiang, Bowei	811
Jiang, Chen	233, 637, 953
Jiang, Chunsheng	106
Jiang, Hao	139
Jiang, Haodong	916, 919
Jiang, Huaxing	519
Jiang, Jianbo	324
Jiang, Jianhua	432
Jiang, Jianpeng	688
Jiang, Jingru	676
Jiang, Jinxia	607
Jiang, Ke	360, 732
Jiang, Mingrui	139
Jiang, Pinfeng	549
Jiang, Qimeng	208, 573, 786
Jiang, Renjie	537
Jiang, Songyi	753
Jiang, Wei	4
Jiang, Wenfeng	874
Jiang, Xixi	172
Jiang, Yang	58, 67
Jiang, Yi	321
Jiang, Yuelin	133, 205
Jiang, Yunzhou	387
Jiang, Zhou	169
Jin, Chengji	390, 1249
Jin, Minhyun	588
Jin, Xin	762, 811
Jin, Yufeng	576
Joh, Jungwoo	1246
Jumabekov, Askhat N.	1270
Jung, S.-W.	291
Kadam, Abhishek	564, 965
Kado, M.	315
Kamada, F.	558
Kämpfe, Thomas	399

Kaneko, M.	1213
Kang, Bryan	688
Kang, J. F.	1095
Kang, Jiachen	1033
Kang, Jin	601
Kang, Jinfeng	354, 1003
Kang, Junzhe	19
Kang, Kaixiang	163
Kang, Song	937
Kang, Yuye	405
Kang, Zhuodong	729
Kar, Anirban	1107
Karasawa, Hajime	31
Karl, Alexander	1192
Kasperovich, A.	977
Kaufmann, Maik Peter	1246
Ke, Mengnan	390
Ke, Xiaoyu	40, 267
Kern, Michal	1086
Keskin, Batuhan	838
Khakbaz, Pedram	1192
Khan, Asif	1276
Khan, Imtiyaz Ahmad	330
Ki, Dong-Keun	139, 257
Kim, Bong Ho	1039
Kim, Bongho	1134
Kim, Cheol-Joo	1170
Kim, Choul-Young	1039
Kim, I. K.	291
Kim, J.	977
Kim, Ji Eun	1186
Kim, Jongmin	1039
Kim, Joon Pyo	1039
Kim, Joonpyo	1134
Kim, Kihyun	1039
Kim, Minchan	652
Kim, Mingyu	223
Kim, Sanghyeon	1039, 1134
Kim, Seong Kwang	1039
Kim, Seongkwang	1134
Kim, Sunghun	1255
Kim, Sungju	166
Kim, Younghyun	1134
Kimoto, T.	1213
Kinjo, K.	558
Knobloch, Theresia	1192, 1282
Kobayashi, Masaharu	1255
Kondo, T.	315
Kong, Shuai	1180
Kopperberg, Nils	456
Kotsugi, Masato	1273
Krivic, Senka	670

Kuang, Renfei 37
Kuang, Zhipeng 729
Kuball, Martin 1057
Kuh, B. J. 1189
Kumar, A. 28
Kumar, Harshvardhan 1176
Kumar, Naveen 796
Kumar, Sandeep 706, 796
Kumar, Shivam 465
Kunal, ...330
Kurihara, A. 712
Kurihara, Akito 598
Kwok, Hoi Sing 160
Laber, Andreas 670
Lai, Yuru 519
Lam, Sang 646
Lan, Gongpeng 196
Lao, Yunhong 236
Lashkare, Sandip 426, 570, 965
Lee, Chan Jik 1039
Lee, Dong Seup 1246
Lee, Jia-Yang 771
Lee, Jimin 497, 1279
Lee, Jiwon 588
Lee, Jongwon 1039
Lee, Jooseok 1039
Lee, Jooyoung 25
Lee, Junjong..652
Lee, K. 1189
Lee, K.-W. 291
Lee, Kyunghwan 223
Lee, Min-Hung 771
Lee, S. M. 1189
Lee, S. 1189
Lee, Sanguk 652, 691
Lee, Seunghwan 652
Lee, T. H 1189
Lee, Won-Chul...1039
Lee, Wonsok 1189
Lee, Y. 1189
Lee, Zhi-Qiang..121
Lehninger, David 399
Lei, Yuan..643
Lemme, Max C. 497, 1279
Leng, Yan-Bing..303
Leroux, Nathan 909
Li, Ang 607
Li, Bingbai 510
Li, Bochang 906
Li, Can 139
Li, Chenxi 519
Li, Chenyang 257

Li, Chunyang 688
Li, Dabing 360, 732, 774
Li, Dan 360
Li, Dapeng 52
Li, Dingwei 13, 88, 260
Li, Dongdong 522, 1030, 1180
Li, Dongya 390
Li, Fanfan 13, 260
Li, Guohui 522
Li, Haixia 453, 471, 744
Li, Haiyang 208
Li, Hang 348
Li, Hao 525, 835
Li, Haolin 877
Li, Hongsheng 1234
Li, Hongtao 1219
Li, Hongyi 561
Li, J. 479
Li, Ji 510
Li, Jia Cheng 625
Li, Jian-Cong 682, 703
Li, Jiancong 709
Li, Jiaqi 163
Li, Jie 768
Li, Jilin 160
Li, Jinghua 1074
Li, Jinxian 408
Li, Jiye 136
Li, Juan 282
Li, Junfeng 777
Li, Junjie 753
Li, Junkang 321
Li, K. 28
Li, Lain-Jong 139, 257
Li, Lianlian 537
Li, Ling 874
Li, Lingqi 257
Li, Mengdi 871
Li, Mengjie 555
Li, Ming 22, 453, 471, 500, 525, 601, 676, 744, 747
Li, Minghua 741
Li, Ming-Huang 118, 121
Li, Muchan 46
Li, Mujun 82
Li, Peihong 859
Li, Pengtao 73
Li, Qingxiu 381
Li, Qinkun 537
Li, S. Z. 1095
Li, S. 479
Li, Sean 94, 257
Li, Sheng 402

Li, Shiming	372, 447, 585
Li, Shiyu	983, 1012
Li, Shuang	700
Li, Siying	616
Li, Taige	1015
Li, Tiaoyang	844
Li, Tiefu	649
Li, Xi	715, 1231
Li, Xiangdong	912
Li, Xianming	937
Li, Xiao	43
Li, Xiaopeng	85
Li, Xin	342
Li, Xiuyan	378, 962
Li, Xuefei	417
Li, Xueqing	233
Li, Xufan	874
Li, Xun	768
Li, Xunyu	1288
Li, Yan	294, 375
Li, Yang	193
Li, Yanqin	1293
Li, Yanru	429
Li, Yi	220, 682, 694, 703, 709
Li, Yida	202
Li, Yongliang	294, 375
Li, Yu	1036
Li, Yuanbiao	1068, 1140
Li, Yuanzhe	169
Li, Yuchun	634
Li, Yun	94, 971
Li, Yunlong	321, 1149
Li, Zhe	727
Li, Zhengyue	853
Li, Zhenyuan	912
Li, Zhiyuan	248
Li, Zhonghui	1285
Li, Zhucheng	64
Li, Zhuo	1255
Li, Zi	883
Li, Zichun	16
Li, Zongwen	285, 491, 528, 868
Liang, Bin	423
Liang, Caijing	950
Liang, Chao	510
Liang, Cunman	169
Liang, Jiaqiao	245
Liang, Jing	688
Liang, Lei	318
Liang, Li-Wei	826
Liang, Renrong	423
Liang, Shijie	37

Liang, Tian	1264
Liang, Xiaoci	239, 488, 613
Liang, Xifa	37
Liang, Zhengyu	324
Liang, Zhongqi	576
Liang, Zhongxin	254
Liao, Chun-Yu	771
Liao, Hongxu	453, 744
Liao, Lihang	1119
Liao, Meiyong	145
Liao, Min	40, 267
Liao, S. Sandy	977
Liapis, Andreas C.	986
Likhitkar, Praful	570
Lim, Hyeongrak	1134
Lim, Jeong-Taek	1039
Lin, Cheng-Chien	121
Lin, Ching-Ju	151
Lin, Haobo	459
Lin, Huai-En	943
Lin, Huamao	741
Lin, J.-P.	977
Lin, Peng	859, 1143
Lin, T.-Y.	886
Lin, Xi	94
Lin, Yibo	601
Lin, Yu-Cheng	151
Lin, Yudeng	729
Lin, Yu-Hsien	151
Lin, Yuxin	1045
Lin, Zhaojun	604
Lin, Zhiyu	154, 178, 378, 1030
Lin, Ziyuan	333
Ling, Haifeng	814
Lipsanen, Harri	986
Liu, Bo	661
Liu, Bowen	637
Liu, Can	306
Liu, Chao	604
Liu, Chen	741
Liu, Cheng-Hong	771
Liu, Chenyu	906
Liu, Chuan	239, 488, 613, 697
Liu, Chunsen	1092
Liu, Dingyao	357
Liu, Enlong	381
Liu, Fangze	453
Liu, Fanyu	375
Liu, Fei	46
Liu, Fucai	318
Liu, Guolei	13, 260
Liu, Guozhu	79

Liu, Haoyan	294
Liu, Hongchao	1240
Liu, Hongxia	76
Liu, Hsien-Yang	516
Liu, Huan	390
Liu, Huanan	847
Liu, Jiacheng	889
Liu, Jian	257, 679
Liu, Jinbiao	546
Liu, Jingquan	378, 962
Liu, Jinhao	459, 634, 1048
Liu, Jun	193
Liu, Junling	100, 270, 468, 1071
Liu, Kai	130
Liu, Kaimeng	133, 205
Liu, Lifeng	229, 1003
Liu, Liwei	263
Liu, Liyuan	679
Liu, Meijie	79
Liu, Ming	1299
Liu, Mingrui	360
Liu, Mingxu	688
Liu, Peisen	765, 862
Liu, Peisong	1009
Liu, Peng	741
Liu, Pengyu	655
Liu, Qi	937
Liu, Qianqian	753, 777
Liu, Qihan	139
Liu, Qihui	411
Liu, Qing	318
Liu, Shangming	1015
Liu, Shanjing	282
Liu, Shuang	546, 1293
Liu, Shuiren	1009
Liu, Shuo	100, 270, 468, 1071
Liu, Si Rui	903
Liu, Sihang	236
Liu, Siyang	402
Liu, Siyuan	327
Liu, Weihua	342
Liu, Wen	387
Liu, Wenjun	983, 1012
Liu, Wenyuan	1054, 1164
Liu, X. Y.	1095
Liu, Xiaohua	832
Liu, Xiaolin	762, 811
Liu, Xiaosen	236
Liu, Xiaoyan	354, 420, 877, 931, 1003
Liu, Xin	1009
Liu, Xinghui	780
Liu, Xinyu	786

Liu, Xiwen	414, 1018
Liu, Xuewen	1234
Liu, Yan	390, 700, 906, 1249
Liu, Yang	1089
Liu, Yanming	148, 685
Liu, Yanyan	393
Liu, Yi	251, 992
Liu, Yibo	16
Liu, Yichen	411
Liu, Yichun	142
Liu, Yinchi	983, 1012
Liu, Yizhan	420
Liu, Yu	747
Liu, Yuan	610
Liu, Yumeng	534
Liu, Yuzhuo	856
Liu, Zhaojun	16
Liu, Zhengwu	543
Liu, Zhihong	312, 912
Liu, Ziheng	369, 438
Liu, Zuheng	790
Lizzit, Daniel	829
Lone, Aijaz H.	288
Long, Shibing	1068, 1140
Long, Yanxi	685
Long, Yinfeng	130
Long, Zhenghao	465, 531
Lotfi, Hadi	1086
Lou, Qiang	724
Lovell, Nigel	1155
Low, Kain Lu	387
Lu, Albert	482
Lu, Di	1068, 1140
Lu, Donglin	610
Lu, Hao	585
Lu, Haoran	601, 676, 747
Lu, Honghao	1243
Lu, Hongliang	634
Lu, Lei	43, 136, 160
Lu, Xing	519
Lu, Xun	983, 1012
Lu, Yipeng	1167
Lu, Yupeng	336
Lu, Yuyao	729
Luo, Chao	441, 595
Luo, Deng	423
Luo, Huaizhi	375
Luo, Jie	336
Luo, Jin	805, 1045
Luo, Jun	267, 429, 753, 777, 1293
Luo, Jun-Wei	1089
Luo, Peiwen	591

Luo, Wenpu	1161
Luo, Xiao	318
Luo, Xin	411
Luo, Yi	1219
Luo, Yiyang	488
Luo, Zhao-Feng	771
Lv, Bingchen	774
Lv, Dan	1030, 1180
Lv, Shunpeng	360
Lv, Y. Z.	214
Lv, Yuanjie	604
Lv, Yudong	619, 631
Lv, Zhong-Peng	1222
Lv, Ziyu	303
Lyu, Haiyuan	327
Ma, Awang	729
Ma, Chen	1119
Ma, Chengxiang	133, 205
Ma, Hanbin	637
Ma, Ruiqi	1180
Ma, Xiaohua	372, 444, 447, 585, 871
Ma, Yanfeng	402
Ma, Yiming	1119
Ma, Yuan Xiao	625
Ma, Yuan	528, 868
Ma, Yunfei	124
Ma, Zichao	989
Ma, Zihao	1104
Ma, Zongmin	193
Machhiwar, Yogendra	799, 802
Maeki, M.	558
Mahapatra, Souvik	718
Maheshwari, Navin	570
Mähne, Hannes	399
Mai, Zhancheng	363, 780
Majumdar, Kausik	1128
Malakoutian, M.	977
Malik, Muhammad	285, 491, 868
Manea, Paul-Philipp	909
Manhas, Sanjeev Kumar	330
Manna, Sukriti	1114
Manzoor, Nisha	462, 1137
Manzoor, Wajid	462, 1137
Mao, Jiayi	199
Mao, Peiyu	912
Mao, Ruibin	139
Martinie, Sebastien	1261
Matsui, Chihiro	897, 940
Mayer, Joachim	497
Mei, Zikang	1152
Mei, Ziqi	510
Memon, Muhammad Hunain	1146

Meng, Lingxian	1009
Menon, P. Susthitha	1176
Menzel, Stephan	456
Merkin, Tim	1195, 1246
Metze, C.	28
Mevic, Amina	670
Mi, Changxin	494
Miao, Feng	1291
Miao, Jialei	552
Miao, Songming	1068, 1140
Miao, Xiang-Shui	184, 682, 694, 703
Miao, Xiangshui	220, 248, 339, 396, 549, 555, 709, 947
Miao, Yechen	351
Mikami, K.	1213
Mikolajick, T.	28
Misawa, Naoko	897, 940
Miyamura, Yoshiji	503
Miyano, S.	315
Mizutani, Tomoko	1255
Mohapatra, Nihar Ranjan	965
Moon, K. J.	1189
Morroni, Jeffrey	1246
Mou, Xing	729
Nagahashi, Tomoya	31
Nandi, Debashish	462, 1137
Nazir, Mohammad Sajid	1080
Neve, Cesar Roda	1288
Ng, Annie	1270
Ni, Siyuan	324
Ni, Zhao	217
Nielinger, Dennis	832
Nigmatulin, Fedor	986
Nigmetova, Gaukhar	1270
Ning, Jing	667
Nishizawa, Shin-Ichi	503
Nomura, M.	977
Nugraha, Ferris Prima	889
Ogier, S.	479
Oh, Saeroonter	1
Oh, T.	291
Ootera, Y.	315
Orio, Roberto	1267
Oshimi, K.	558
Oshiyama, Atsushi	31
Ostinelli, O.	7
Ota, Takashi	897
Ou, Xin	853, 916, 919, 1042, 1264, 1285
Ouyang, Bangsen	783
Pachkawade, Vinayak	1261
Padhee, Chiranjibi	540
Padiyal, Adil	940

Padovani, Andrea	516
Pahwa, Girish	802
Pai, C.-S.	900
Pampori, Ahtisham	55, 1080
Pan, Feng	765, 862
Pan, Gang	859, 1143
Pan, Wen	1219
Pan, Zhoujie	148
Pan, Zijian	528
Park, Jaehyun	223
Park, K.	1189
Park, Minsik	1039
Park, S. W.	1189
Park, Sungil	223
Park, Youngkeun	1134
Parkhomenko, Hryhorii	1270
Patil, Deven H.	706, 796
Patil, Shubham	564, 965
Pei, Huai-Zhi	703
Pei, Mengjiao	971
Pei, Yanli	488, 519
Peller, Markus	399
Peng, Baokang	309, 1051
Peng, Bin	591
Peng, Bo	1222
Peng, Hao	136
Peng, Hongjie	369, 438
Peng, Pei	46
Peng, Wanyue	601, 676, 747
Peng, Xiao	411
Pey, Kin Leong	516
Pham, Tri T.	1270
Phùng, Qu?nh Th?	1131
Pijper, R. M. T.	1207
Plakhotnik, Taras	1000
Poddar, Swapnadeep	465
Pomeroy, James	1057
Pop, Eric	1131
Pradeep, Yelchanka Ramachandramurthy	121
Pretl, Harald	1261
Puglisi, Francesco Maria	516
Pun, Kong-Pang	655
Qi, Dianyu	1149
Qi, Yihong	811
Qi, Yixin	169
Qian, Haoji	1249
Qian, He	70, 133, 205, 729, 847
Qian, Jiang	1180
Qian, Xuanyu	4
Qian, You	741
Qian, Yuchen	871
Qiao, Bing	1285

Qin, Haiming	251
Qin, Huajun	1258
Qin, Lingjie	871
Qin, Nan	1054, 1164
Qin, Peijun	1155
Qin, Yuwen	1104
Qiu, Guoxiu	381
Qiu, Jiajun	835
Qiu, Liwen	724
Qiu, Shirong	169
Qiu, Xiao	531
Qu, Hanbin	1027
Qu, Junle	263
Qu, Yuwei	85
Qu, Zhenyu	916, 919
Quan, Dechang	387
Quinsat, M.	315
Radhakrishna, Ujwal	1195
Radu, Ionut	1288
Raghavan, Nagarajan	516, 658
Rahimi, Daniel N.	288
Rai, S.	28
Raja, Danish	799, 802
Raju, Harsh	330
Ramya, K	187
Ran, Ke	497
Rathore, Sunil	706, 796
Ren, Huihui	88, 260
Ren, Liming	46
Ren, Pengpeng	91, 459, 673, 1006, 1048
Ren, Qinghua	1119
Ren, Sheng-Guang	682, 694
Ren, Tian-Ling	735, 826
Ren, Xuanhui	193
Ren, Yuan	543
Ren, Zhipeng	1027
Ren, Zhongyang	46
Rheem, Nahyun	1039
Robert, Isaac Emanuel	399
Rossi, Chiara	829
Ruma, S R	892
Ruttloff, K.	28
Saha, Arpan	756
Saito, Wataru	503
Saitoh, M.	315
Sakai, Kota	1255
Sakhuja, Jayatika	965
Salahuddin, S.	61
Salahuddin, Sayeef	55
Sandner, Christoph	1261
Sang, Pengpeng	85
Sankaranarayanan, Subramanian K. R. S.	1114

Sanpui, Dipayan	1114
Sara, M.	558
Saraya, Takuya	1255
Sattari-Esfahlan, Seyed Mehdi	1192
Satyam, Parlapalli Venkata	540
Scholten, A. J.	1207
Scholz, Rene	1261
Seidel, A.-S.	28
Seidel, Konrad	399
Seong, Minkyoung	1039
Sessi, V.	28
Setti, Gianluca	288
Shafi, Abde Mayeen	986
Shahriar, Md.	756
Shakir, Mohd.	796
Shan, Linbo	485, 1021, 1024
Shang, Jieya	1059, 1062, 1065
Shang, Zongwei	471, 525
Shao, Feng	688
Shao, Hanyong	1161
Shao, He	814
Shao, Qiming	889, 1228
Shao, Rui	405
Shao, Xianzhou	267
Shayoub, Mir Mohammad	1080
Shen, Bo	236, 513, 573
Shen, Bowen	429, 847
Shen, Guozhen	109
Shen, Jiabin	408
Shen, Minliang	856
Shen, Rui	1285
Sheng, Bowen	1167
Sheng, Chuming	1065
Sheng, Haoyu	595
Shi, Chunzhou	582
Shi, Luping	1101
Shi, Mingcheng	70, 847
Shi, Mingmin	471, 500
Shi, R.	479
Shi, Ruipeng	429
Shi, Runxiao	160
Shi, Yuanyuan	196
Shi, Yunfei	777
Shi, Zhiming	360
Shibata, Koki	897
Shih, C.-C.	977
Shim, Joonsup	1039
Shimomura, N.	315
Shin, Hyungcheol	25, 166, 223
Shiraishi, Kenji	31
Shuyan, Zhu	226
Si, Mengwei	154, 178, 378, 962, 1030
Si, Yuying	616
Simon, M.	28
Sin, Stanislav	1
Singh, Ajay Kumar	965
Singh, Harshita	793
Singhal, Anant	793
Sinha, Advaita	1176
Slesazeck, S.-S.	28
Smith, Matthew D.	1057
Soh, Keunho	1186
Soman, R.	977
Somappa, Laxmeesha	426
Sonawane, Jay	564
Song, Changming	980
Song, Cheng	765, 862
Song, J.	1189
Song, Lekai	655
Song, Peixuan	282
Song, Q. W.	664
Song, Qingwen	928
Song, X. J.	1095
Song, Xujin	354
Song, Yichen	1296
Song, Yixian	19
Song, Yufei	1059, 1062, 1065
Song, Zhitang	715, 1231, 1296
Sporea, Radu A.	242
Stampfer, Bernhard	1267
Strachan, John Paul	909
Su, C.-J.	886
Su, Chang	956
Su, Dongyue	239
Su, Huan-Hsiang	151
Su, Jiaqi	94
Su, P.	886
Su, Pin	820
Su, Qiyi	780
Su, Xiangwei	950
Su, Yahui	282
Su, Yanbo	70, 847
Su, Yongquan	411
Su, Zhijuan	534
Su, Zijia	450
Su, Zi-Jia	826
Suh, Chang Soo	1246
Suh, Yoon-Je	1039
Sui, Nianzi	163
Sui, Zhiyuan	79
Sul, Woo-Suk	1039
Sun, Changzheng	1219
Sun, D. J.	1095
Sun, Dijiang	354

Sun, Dongdong	196
Sun, Haiding	450, 1146
Sun, Hanhan	423
Sun, Haolun	447
Sun, Huarui	853
Sun, Jiacheng	601, 676, 747
Sun, Jiahao	750
Sun, Jialiang	1042, 1264
Sun, Jia-Yi	694
Sun, Jingwei	1237
Sun, Ke	351
Sun, L. J.	664
Sun, Lejia	928
Sun, Lifei	181
Sun, Linfeng	1098
Sun, Liuyang	324
Sun, Longyu	294
Sun, Maojun	724
Sun, Mingyang	753, 777
Sun, Qicheng	1059, 1062, 1065
Sun, Qingqing	172, 1252
Sun, Ruize	124, 865
Sun, Weifeng	402
Sun, Wen	847
Sun, Xiaojuan	360, 732, 774
Sun, Xiaoqing	34, 40, 103, 267, 276
Sun, Yi	351
Sun, Yiyuan	850
Sun, Yuhua	64
Sun, Zhengjie	1062
Sun, Zhengzong	1065
Sun, Zhibo	531
Sun, Zhipei	477, 986
Sünbül, Ayse	399
Sung, C.	1189
Suzuki, Madoka	1083
Svejda, Jan Taro	1261
Szedmak, Sandor	670
Tachiki, K.	1213
Takahashi, Takanori	1255
Takeuchi, Ken	897, 940
Tan, Chaoliang	112
Tan, Hongwei	1222
Tan, Lin	980
Tan, Ping-Heng	1183
Tan, Tiang Teck	516
Tan, Xiaohong	576
Tan, Xiaojun	1062
Tang, Chuying	1243
Tang, Huansong	49
Tang, Huawei	619, 631
Tang, Jianshi	70, 133, 205, 729, 847

Tang, Jinjin	622
Tang, Jinpu	1219
Tang, Jinyao	139
Tang, Kechao	254, 273, 1161
Tang, Meng	288
Tang, Nan	1003
Tang, Ning	573
Tang, W.	479
Tang, Wei	248, 1033
Tang, X. Y.	664
Tang, Xiaoyan	928
Tang, Xinyi	58, 67, 1068, 1140, 1243
Tang, Xiyuan	1024
Tang, Y.-T.	880, 900
Tang, Yanbo	619, 631
Tang, Yingjie	88, 260
Tang, Younian	239
Tang, Zhidong	70
Tang, Zhikai	1246
Tao, Lu-Qi	826
Tao, Quanyao	610
Tao, Ran	1149
Tao, Tiger H.	324, 351, 1054, 1164
Tao, Yaoyu	1231
Teng, Qiao	19
Teng, Zilin	432
Terai, M.	1189
Tewari, Mohit	925
Thakor, Karansingh	718
Thiem, Steffen	399
Tian, Feng	528, 868
Tian, Fengbin	267
Tian, He	148, 685, 735
Tian, Jiaojiao	46
Tian, Jing	703
Tian, Qiaoling	142
Tian, Ruijuan	477
Tian, Xinyu	357
Tian, Yuchen	1059, 1062, 1065
Tilegen, Meruyert	1270
Toda, Masaya	211
Tokeshi, M.	558
Tokuhira, H.	315
Tong, Anyu	1077
Tong, Hao	184
Tong, Yi	251, 883, 992
Torraca, Paolo La	516
Tripathi, Karunesh Kumar	799, 802
Trommer, J.	28
Tsai, C.-Y.	880
Tsai, David	1155
Tsai, Y.-T.	880, 900

Tseng, Jen-Chou 151
Tu, Quanyi 414
Tung, C. T. 61
Tung, Chien-Ting 55
Uddin, Md Gius 986
Ueda, Y. 315
Uenuma, Mutsunori 1255
Umetsu, N. 315
Uraoka, Yukiharu 1255
Vater, Frank 1261
Vaziri, S. 977
Veliadis, Victor 974
Venkatesan, Prasanna 1276
Vincent, Benjamin 181
Völkel, Lukas 497, 1279
Wahid, Sumaiya 1131
Waldhoer, Dominic 1192, 1282
Waltl, Michael 1267
Wan, Changjin 971
Wan, Tianqing 333
Wan, Yi 257
Wan, Yuxi 208, 865
Wan, Yu-Xi 700
Wan, Ziqi 715
Wang, Albert 1288
Wang, Bingxiang 732
Wang, Chang-Hao 10
Wang, Chen 154, 178, 369, 438, 817
Wang, Chengcai 622
Wang, Chenguang 263
Wang, Chien-Fan 151
Wang, Chunxiu 245
Wang, Cong 333
Wang, Cuimei 1237
Wang, Di 390
Wang, Fang 351
Wang, Feixiong 546, 1293
Wang, Gang 519, 679
Wang, Guilei 688
Wang, Han 139
Wang, Hongyue 369
Wang, Huan 220
Wang, Hui 193, 282
Wang, Jiachuang 1054, 1164
Wang, Jiahao 1059
Wang, Jialiang 239, 783
Wang, Jian 1219
Wang, Jianhuan 500
Wang, Jingli 172
Wang, Jinyan 369, 438, 513
Wang, Junhao 1167
Wang, Kai 363, 762, 780, 811, 937

Wang, Kaifeng 474
Wang, Kuan 184
Wang, Lai 1219
Wang, Le 814
Wang, Lei 883, 986
Wang, Letian 549
Wang, Lihao 411
Wang, Lin 130
Wang, Lingfei 874
Wang, Long 345
Wang, Lu 1180
Wang, Lun 184
Wang, Maojun 236, 513
Wang, Mengye 239, 488
Wang, Ming 429
Wang, Mingyan 604
Wang, Nan 411, 1119
Wang, Peiran 297
Wang, Peng 336, 537, 1015
Wang, Qian 980
Wang, Qing 58, 67, 82, 297, 471, 744, 1243
Wang, Qingpeng 181
Wang, Qishen 485, 1024, 1237
Wang, Rui 765, 862
Wang, Runsheng 309, 459, 525, 601, 628, 634, 676,
.... 747, 753, 823, 835, 1006, 1036, 1048, 1051
Wang, Ruobing 715, 1296
Wang, Ruqi 37
Wang, Sen 1059
Wang, Shuang 233
Wang, Shuying 91, 628, 673
Wang, Shuyu 661
Wang, Tao 940
Wang, Tong 912
Wang, Weibiao 765, 862
Wang, Weisheng 387
Wang, Wen 661
Wang, Wenwu 34, 40, 103, 267, 276, 753, 777
Wang, Wenxiao 721
Wang, Wenyu 94
Wang, Xiangsheng 688
Wang, Xiaochen 528
Wang, Xiaohui 82
Wang, Xiaolei 34, 40, 85, 103, 267, 276, 777, 1293
Wang, Xiaoming 865
Wang, Xiaoping 208, 865
Wang, Xiayu 510
Wang, Xile 1231
Wang, Xin 369, 438
Wang, Xinghua 741
Wang, Xingsheng 549, 555, 947
Wang, Xinhe 79

Wang, Xinhua	786
Wang, Xinpeng	251, 883, 992
Wang, Xinwei	43
Wang, Xuemei	318
Wang, Xuepei	459, 634, 1048
Wang, Xueying	324
Wang, Yan	88, 260, 865
Wang, Yang	808, 959
Wang, Yao	345
Wang, Yaoyi	983, 1012
Wang, Yasai	339, 396
Wang, Ye Liang	625
Wang, Yeliang	916, 919
Wang, Yi	549
Wang, Yibo	906
Wang, Yikang	306
Wang, Yimeng	747
Wang, Yiru	814
Wang, Yizhuo	534
Wang, Yu	922, 1111
Wang, Yunda	1018
Wang, Yuxuan	405
Wang, Yuyan	847
Wang, Zebin	163
Wang, Zexi	76
Wang, Zhen	49, 73
Wang, Zhenxin	983, 1012
Wang, Zhiyao	546
Wang, Zhiyu	1077, 1119
Wang, Zhiyuan	1119, 1234
Wang, Zhongju	727
Wang, Zhongqiang	142
Wang, Zhongrui	58, 537
Wang, Zhuming	817
Wang, Zifeng	1125
Wang, Ziheng	378
Wang, Zijian	73, 127
Wang, Ziyang	58, 67
Wang, Ziyi	94
Wang, Zongwei	485, 841, 1021, 1024, 1237
Wanghe, Jianhui	1033
Waser, Rainer	832
Wei, H.-K.	977
Wei, Hu	912
Wei, Jianyong	534
Wei, Jin	236, 300, 438, 513
Wei, Jinghe	79
Wei, Jingxuan	1252
Wei, Ke	786
Wei, Lan	1195
Wei, Shaojun	688
Wei, Tiantian	688

Wei, Xiaoling	324
Wei, Xiaoyu	774
Wei, Yanzhao	922, 1111
Wei, Yanzhuo	522
Wei, Yidan	79
Wei, Yingqiang	79
Wen, Bo	139
Wen, Kangyao	58
Wen, Liaoyong	528
Wen, Xinyu	184
Wen, Yichen	459, 628, 634, 1006, 1048
Weng, Xiaoyu	263
Wiefels, Stefan	456, 832
Wijvliet, M.	28
Wittke, Christian	1261
Wong, Hiu Yung	482
Wong, Kwok-Ho	640
Wong, Man Hoi	16, 160
Wong, Ngai	543
Woon, W.-Y.	977
Wu, C.-H.	886
Wu, Chunlei	983, 1012
Wu, Daixuan	148, 685
Wu, Dong	70, 133, 205
Wu, F.	886
Wu, Haidi	667
Wu, Heng	453, 500, 601, 676, 747
Wu, Honglin	823
Wu, Hongzhao	950
Wu, Huaqiang	70, 133, 205, 729, 847
Wu, Jack	601
Wu, Jianting	613
Wu, Jixuan	85
Wu, Maokun	459, 628, 634, 1006, 1048
Wu, Mei	372, 447, 585
Wu, Nanjian	679
Wu, Qian	613
Wu, Renjie	124
Wu, Shenghao	685
Wu, Tian-Li	516
Wu, Xiaopeng	306
Wu, Xing	525
Wu, Xuankun	494
Wu, Y.	977
Wu, Ya Fei	903
Wu, Yanqing	844, 1077
Wu, Yishan	459, 634, 1048
Wu, Yiting	157
Wu, Yiyang	906
Wu, Yongbo	980
Wu, Yongyu	1149
Wu, Yuchen	375

Wu, Zhenyu	411
Wu, Zhida	1027
Wu, Zimeng	1237
Xi, Bin	239
Xi, Peng	263
Xi, Qi	79
Xia, Chenhao	46
Xia, Jiao	1167
Xia, Wenbo	438
Xia, Yang Hui	625
Xia, Yun	865
Xiang, Jinjuan	40
Xiang, Ke	768
Xiao, Boyuan	765, 862
Xiao, Fan	576
Xiao, Kai	738
Xiao, Qian	444
Xiao, Zhihua	1228
Xie, Haorong	937
Xie, Jiawei	405
Xie, Lanyi	453
Xie, Li	715
Xie, Maosong	750, 856
Xie, Qin	106
Xie, Rongbo	510
Xie, Shujie	494
Xie, Yinfei	853
Xie, Yunfei	285, 528, 868
Xie, Zhi-Fei	826
Xing, Shengpeng	73
Xing, Sicheng	1198
Xing, Weichuan	912
Xiong, Bing	1219
Xiong, Weiwei	339, 396
Xiong, Yifeng	947
Xiong, Yuwen	1104
Xiu, Huixin	239
Xu, Daiyao	1195
Xu, Dengqin	229
Xu, Gaobo	336
Xu, Guangwei	1068, 1140
Xu, Haiyang	142
Xu, Hang	10
Xu, Hao	34, 40, 103, 267, 276
Xu, Jianbin	435
Xu, Jiaqiang	109
Xu, Jinghan	931
Xu, Jun	441
Xu, Juyan	79
Xu, Kai	19, 1149
Xu, Lei	1071
Xu, Ming	220

Xu, Mingkun	561
Xu, Qiufeng	765, 862
Xu, Shaodi	934
Xu, Shuo	429
Xu, Weikai	279
Xu, Wenhui	853, 916, 919, 1285
Xu, Wenyang	768
Xu, Xiaoyan	453, 471, 500
Xu, Yang	285, 491, 528, 868
Xu, Yi	1104
Xu, Ying	850
Xu, Yitong	780
Xu, Yu	768
Xu, Yunsong	387
Xu, Zhengde	190
Xu, Zheqi	980
Xu, Zhiping	1210
Xu, Zihan	1065
Xu, Ziqiao	601, 747
Xue, Hongxia	139
Xue, Renhao	414
Xue, Yi-Bai	694, 703
Xue, Yibai	709
Xue, Yiheng	432
Xue, Yongkang	91, 673
Xue, Zhipeng	962
Xue, Zhiqiang	444
Yamane, Daisuke	995
Yan, Bei-Ping	643
Yan, Jianmin	333, 339
Yan, Longhao	1231
Yan, Weikang	937
Yan, Yong	450
Yang, Benjamin	847
Yang, Bowen	429, 585
Yang, Changhui	679
Yang, Chunzhen	239
Yang, Dandan	381
Yang, Dongliang	1098
Yang, Fan	318, 555, 947
Yang, Fengyuan	1119
Yang, Guanhua	874
Yang, Guoshen	576
Yang, Han	513
Yang, Haozhang	1003
Yang, Heng	351
Yang, Hong	753, 777
Yang, Huan	43, 136
Yang, Huazhong	233
Yang, Hui Xia	625
Yang, Huiran	324
Yang, Jia	34, 103

Yang, Jining 983, 1012, 1252
Yang, Junjie 236, 438, 513
Yang, Ling 444, 447, 582, 585
Yang, Mei 22
Yang, Meiyin 429
Yang, Ni 139, 257
Yang, Qing 1086
Yang, Qingyuan 561
Yang, Rui 248, 750, 790, 856
Yang, Shangyu 1264
Yang, Sheng 91, 673
Yang, Shengjie 43
Yang, Shuai 753
Yang, Siyao 506
Yang, Tao 777
Yang, Wenlong 381
Yang, Xiaolei 429
Yang, Xiaoman 628
Yang, Xuelin 236, 513, 573
Yang, Yafen 10
Yang, Yanyu 336
Yang, Yi 826
Yang, Yifan 549
Yang, Yingjie 193
Yang, Yintang 306
Yang, Yuchao 1231
Yang, Yuhang 485, 1009, 1237
Yang, Zongyin 986
Yao, Jiaping 248
Yao, Jiaxin 922, 1111
Yao, Peng 729
Yao, Rui Ray 646
Yao, Zikang 962
Ye, Changqing 525
Ye, Chenglin 420
Ye, Jiandong 1285
Ye, Ran 402
Ye, Sheng 459, 634, 1006, 1048
Ye, Tianchun 34, 103, 276
Ye, Zhuangzhuang 429
Yelzhanova, Zhuldyz 1270
Yeon, Deuk Ho 223
Yeung, Fion Sze Yan 160
Yi, Tingchen 229
Yin, Huaxiang 327, 336, 537, 546, 777, 922, 1111,
....1293
Yin, Youyi 513
Yip, Pak San 139
Yoo, Jinil 25
Yoo, K. 1189
Yoo, S. 1189
Yoon, Hoon Hahn 986

Yoon, Jung Ho 1186
Yoshikawa, M. 315
You, Rui 510
You, Tiangui 853, 916, 919, 1042, 1264
Yu, Bin 285, 491, 528, 868
Yu, C. X. 1095
Yu, Chen-Yu 516
Yu, Chunxiao 906
Yu, Guofang 393, 423
Yu, Guohao 607
Yu, Hao 351
Yu, Haozhe 82
Yu, Hongyu 58, 67, 82, 297, 1243
Yu, Huabin 1146
Yu, Jiexun 980
Yu, Jingjing 236, 513
Yu, Peiyue 429
Yu, Qian 582
Yu, Shaofeng 619, 631
Yu, Xiao 390, 1249
Yu, Xinxin 1285
Yu, Ying-Jie 682
Yu, Yingjie 709
Yu, Yiwen 983, 1012
Yu, Yong 688
Yu, Yue 934
Yu, Zongguang 79
Yuan, Jiahui 369
Yuan, Jian 847
Yuan, Jiuyang 503
Yuan, Mengqiang 312
Yuan, Miaojia 1006
Yuan, Shuai 790
Yuan, Wenhan 124
Yuan, Zhengnan 531
Yue, Han 607
Yue, Yuanyuan 774
Yun, P. 1189
Zang, Hang 360
Zarkob, Yawar Hayat 1137
Zeng, Chunhong 64
Zeng, Fei 765, 862
Zeng, Lang 1125
Zeng, Liang 251
Zeng, Min 1077
Zeng, Ruli 983, 1012
Zeng, Shicheng 1065
Zeng, Wei 348
Zeng, Xinlong 950
Zeng, Zhongming 607
Zeun, A. 28
Zha, Jiajia 112

Zha, Xian-Hu 700
Zhai, Yongbiao 303
Zhan, Rui 432
Zhan, Xuepeng 85
Zhang, A. Ping 1015
Zhang, Bailin 679
Zhang, Baoshun 64, 607
Zhang, Baotong 744
Zhang, Beining 248
Zhang, Bin 369
Zhang, Bo 124
Zhang, Bosen 724
Zhang, Bowen 871
Zhang, Boyang 727
Zhang, Changcheng 1027
Zhang, Chenyang 91, 673
Zhang, Chenyuan 1167
Zhang, Chi 510
Zhang, Conghui 1009
Zhang, Dao Hua 700
Zhang, David Wei 10, 634, 817
Zhang, Guanjun 591
Zhang, Guobin 127
Zhang, Guohua 303
Zhang, Haisu 1024
Zhang, Hang 537
Zhang, Hao 883, 983, 992, 1012
Zhang, Haochen 1146
Zhang, Heng 552
Zhang, Heng-Feng 694
Zhang, Hongjie 738
Zhang, Hongrui 390
Zhang, J. J. 1095
Zhang, Ji 257
Zhang, Jiajun 1018
Zhang, Jianjun 453, 500
Zhang, Jiayi 294
Zhang, Jieyin 453
Zhang, Jin 774
Zhang, Jincheng 312, 345, 667, 912
Zhang, Jingyang 16
Zhang, Jinshu 1059, 1062, 1065
Zhang, Li 1027
Zhang, Lian 494
Zhang, Lijie 601
Zhang, Lining 309, 823, 835, 989, 1036, 1051
Zhang, Long 582
Zhang, Meng 582, 585
Zhang, Miaocheng 992
Zhang, Min 245, 968
Zhang, Mingchen 871
Zhang, Panpan 157, 432

Zhang, Pengcheng 790
Zhang, Pu-Yi 682
Zhang, Qianqian 528, 868
Zhang, Qingxin 741
Zhang, Qingzhu 537, 546, 777, 922, 1111
Zhang, Qirui 318
Zhang, Qiuqi 616
Zhang, Renlong 263
Zhang, Rong-Jun 700
Zhang, Rui 321
Zhang, Shanting 1180
Zhang, Shan-Ting 522
Zhang, Shengdong 43, 136, 160
Zhang, Shiming 357
Zhang, Shiyi 378
Zhang, Shiyu 130
Zhang, Shuai 765, 862
Zhang, Shukui 172
Zhang, Tianjiao 552
Zhang, Ting 1006
Zhang, Wei 1101
Zhang, Weihang 312, 912
Zhang, Weizhe 679
Zhang, Wenjing 697
Zhang, Wenxu 591
Zhang, Xi 294
Zhang, Xiaodong 64
Zhang, Xiaohan 300
Zhang, Xin 601
Zhang, Xing 229, 1021
Zhang, Xue 190
Zhang, Xuning 983, 1012
Zhang, Y. M. 664
Zhang, Yachao 312, 345
Zhang, Yadong 546
Zhang, Yan 510
Zhang, Yewei 91, 673
Zhang, Yibei 70
Zhang, Yijian 750, 856
Zhang, Ying 741
Zhang, Yishu 73, 127, 1149
Zhang, Yu 477, 694
Zhang, Yuanke 441, 595
Zhang, Yuhan 136, 160
Zhang, Yuhang 43
Zhang, Yuming 928
Zhang, Yun 494
Zhang, Yu-Qi 303
Zhang, Zeyu 279
Zhang, Zhaohao 327, 546
Zhang, Zhejia 1059, 1062, 1065
Zhang, Zhekang 37

Zhang, Zhi-Xiang	285, 491, 528, 868
Zhang, Zichong	549
Zhang, Zihan	148
Zhang, Zijing	375
Zhao, Chao	688
Zhao, Dengrui	64
Zhao, Fangyu	1054, 1164
Zhao, Haolin	363, 780
Zhao, Haoran	500
Zhao, Jiaming	417
Zhao, Jianwen	163
Zhao, Jinxiu	154, 178
Zhao, Jiyu	348
Zhao, Junlei	300
Zhao, Lei	429, 1167
Zhao, Liang	1180
Zhao, Lu	393
Zhao, Ni	169
Zhao, Rong	561, 579
Zhao, Shujing	342
Zhao, Tiancheng	916, 919
Zhao, Wei	79
Zhao, Wenjie	1077
Zhao, Xiaoguang	510
Zhao, Xiaoning	142
Zhao, Yapeng	844
Zhao, Yuanyuan	40
Zhao, Yuda	552, 868, 950
Zhao, Yue	393, 874
Zhao, Yusi	619, 631
Zheng, Dezhi	52
Zheng, Doudou	193
Zheng, Fangyuan	257
Zheng, Gerui	405
Zheng, Hao	934
Zheng, Huajian	613
Zheng, Jun	906
Zheng, Qianze	133, 205
Zheng, Siying	1249
Zheng, Xuefeng	1027
Zheng, Yaojie	953
Zheng, Zijie	405, 850
Zhong, Kun	327
Zhong, Yujia	181
Zhou, Bin	432
Zhou, Changjian	968
Zhou, Dayu	251
Zhou, Enze	510
Zhou, Falong	601
Zhou, Feichi	202
Zhou, Guangdong	768
Zhou, Guopei	324

Zhou, Hang	576, 724
Zhou, Haobin	637, 953
Zhou, Haoxian	263
Zhou, Heng	604
Zhou, Jiajun	543
Zhou, Jiang	217
Zhou, Jin	912
Zhou, Jingjie	1062
Zhou, Jingyi	49
Zhou, Jiuren	1249
Zhou, Kaiyuan	381
Zhou, Leidang	519
Zhou, Menglong	688
Zhou, Min	916, 919
Zhou, Peng	408, 959
Zhou, Wenli	646
Zhou, Wenyong	543
Zhou, Xijun	453, 471
Zhou, Xilin	715, 1296
Zhou, Xinchen	765, 862
Zhou, Xinlong	1012
Zhou, Y.	664
Zhou, Ya Dong	625
Zhou, Yaoqiang	435
Zhou, Ye	303, 348
Zhou, Yiming	729
Zhou, Yingxi	474
Zhou, Yu	928
Zhou, Yue	339, 721
Zhou, Yuejia	254, 273, 1161
Zhou, Yuxi	871
Zhou, Zheng	229, 420, 877, 931, 1003
Zhou, Zhitao	324
Zhou,	762, 811
Zhu, Bowen	13, 88, 199, 260
Zhu, Chaoyi	339
Zhu, Fangchen	808
Zhu, Handong	983, 1012
Zhu, Hao	1252
Zhu, Jiayan	956
Zhu, Jiefei	968
Zhu, Jiejie	871
Zhu, Mingde	549
Zhu, Qihang	220
Zhu, Quanhai	366
Zhu, Runteng	1161
Zhu, Shirui	303
Zhu, Yanda	94
Zhu, Yao	741
Zhu, Yaochen	411
Zhu, Yu	100, 270, 468
Zhu, Yuxuan	1059, 1062, 1065

Zhu, Yuzhe 1077
Zhu, Zhifeng 190
Zhu, Ziyi 324
Zhuang, Kai 363, 780
Zier, M. 28
Zou, Dujuan 324
Zou, Xiangyu 1119
Zou, Xinbo 916, 919
Zou, Xu 582
Zou, Yating 838
Zou, Zhili 64
Zuo, Chengjie 384, 450
Zuo, Wen-Bin 694
Zuo, Wenbin 709

IEEE
445 Hoes Lane
Piscataway, NJ 08854-4141

ISBN 979-8-3315-0417-5

2025 9th IEEE Electron Devices Technology & Manufacturing Conference (EDTM 2025)

Hong Kong
9-12 March 2025

Pages 649-1301

IEEE Catalog Number: CFP25J58-POD
ISBN: 979-8-3315-0417-5

2025 9th IEEE Electron Devices Technology & Manufacturing Conference (EDTM 2025)

Hong Kong
9-12 March 2025

Pages 649-1301

IEEE Catalog Number: CFP25J58-POD
ISBN: 979-8-3315-0417-5

**Copyright © 2025 by the Institute of Electrical and Electronics Engineers, Inc.
All Rights Reserved**

Copyright and Reprint Permissions: Abstracting is permitted with credit to the source. Libraries are permitted to photocopy beyond the limit of U.S. copyright law for private use of patrons those articles in this volume that carry a code at the bottom of the first page, provided the per-copy fee indicated in the code is paid through Copyright Clearance Center, 222 Rosewood Drive, Danvers, MA 01923.

For other copying, reprint or republication permission, write to IEEE Copyrights Manager, IEEE Service Center, 445 Hoes Lane, Piscataway, NJ 08854. All rights reserved.

* This is a print representation of what appears in the IEEE Digital Library. Some format issues inherent in the e-media version may also appear in this print version.

IEEE Catalog Number: CFP25J58-POD
ISBN (Print-On-Demand):   979-8-3315-0417-5
ISBN (Online):   979-8-3315-0416-8

Additional Copies of This Publication Are Available From:

Curran Associates, Inc
57 Morehouse Lane
Red Hook, NY 12571 USA
Phone: (845) 758-0400
Fax: (845) 758-2633
E-mail: curran@proceedings.com
Web: www.proceedings.com

TABLE OF CONTENTS

Energy-Efficient Voltage-Induced Self-Regulated Precessional MRAM with Low Write Error Rate $<10^{-9}$ 1
Stanislav Sin, Saeroonter Oh

A 2-D Noise Model for CMOS Single Photon Avalanche Diodes 4
Xuanyu Qian, Wei Jiang, M. Jamal Deen

InP/GaAsSb DHBT Emitter Etching Process Optimization with a Simultaneous fT/FMAX =451/914 GHz and 86% Device Yield 7
M. Ebrahimi, S. Hamzeloui, F. Ciabattini, A. M. Arabhavi, O. Ostinelli, C. R. Bolognesi

A Novel Mixed Collector Structure IGBT Operating at High Temperature 10
Chang-Hao Wang, Hang Xu, Jian-Bin Guo, Xin-Ru Chen, Yafen Yang, David Wei Zhang

Controllable Oxygen Vacancies in NbOx MOTT Memristor for Tunable Spiking Neurons 13
Guolei Liu, Dingwei Li, Fanfan Li, Bowen Zhu

Study of the Characteristics of GaN Substrate-Based Microleds with Different Epitaxial Structures 16
Shan Huang, Yibo Liu, Feng Feng, Jingyang Zhang, Zichun Li, Man Hoi Wong, Zhaojun Liu

Impact of Oxide Quality in Self-Aligned Block Region on Hot Carrier Degradation in N-Type CFP-LDMOS with 0.18 μm Bipolar-CMOS-DMOS Technology 19
Qiao Teng, Yixian Song, Junzhe Kang, Kai Xu, Dawei Gao

Investigation of High-Sensitivity Pressure Sensor Integrated with InSnZnO Thin-Film Transistors 22
Mei Yang, Ming Li, Wei Huang, Rongsheng Chen

Investigation of Reliability and Optimization of Reprogram Process in 3D NAND Flash Memory Based on Physical Model 25
Jooyoung Lee, Jinil Yoo, Hyungcheol Shin

Circuit Polymorphism Enabled by RFET Devices Processed on Industrial FDSOI 28
N. Bhattacharjee, G. Galderisi, Y. He, V. Sessi, M. Drescher, V. Havel, M. Zier, M. Simon, K. Ruttloff, K. Li, A. Zeun, A.-S. Seidel, C. Metze, M. Grothe, S. Jansen, M. Wijvliet, S. Rai, A. Kumar, S. Slesazeck, J. Hoentschel, T. Mikolajick, J. Trommer

Optimizing SiN Composition for Enhanced Charge-Trapping in Next-Generation 3D NAND Flash Memories 31
Tomoya Nagahashi, Hajime Karasawa, Ryota Horiike, Atsushi Oshiyama, Kenji Shiraishi

Effect of Top Al2O3 Interlayer Thickness on the Memory Window of Fefets with TiN/Al2O3/Hf0.5Zr0.5O2 /SiOx /Si (MIFIS) Gate Structure 34
Tao Hu, Runhao Han, Xinpei Jia, Jia Yang, Zeqi Chen, Xiaoqing Sun, Junshuai Chai, Hao Xu, Xiaolei Wang, Wenwu Wang, Tianchun Ye

Erbium-Doped Aluminium Oxide Based Distributed Feedback Waveguide Laser in Thin Film Lithium Niobate Platform 37
Zhekang Zhang, Ruqi Wang, Renfei Kuang, Xifa Liang, Shijie Liang, Qingming Chen

Comparison of Border Trap Density in MIFM Capacitors with Different Interlayers Using the C−ln(f) Method 40
Min Liao, Hao Xu, Xiaoyu Ke, Yuanyuan Zhao, Junshuai Chai, Xiaoqing Sun, Jinjuan Xiang, Xiaolei Wang, Wenwu Wang

Boosting Mobility of Oxide TFTs Via PVD and ALD Hybrid Process 43
Yuhang Zhang, Xiao Li, Shengjie Yang, Huan Yang, Tengyan Huang, Xinwei Wang, Lei Lu, Shengdong Zhang

High-Performance Flexible Graphene Field-Effect Transistors and Multistage Inverter Chains for Signal Amplification 46
Zhongyang Ren, Pei Peng, Jiaojiao Tian, Chenhao Xia, Muchan Li, Liming Ren, Fei Liu, Yunyi Fu

Oxide-Based Optical Synapses with Low Consumption for In-Sensor Reservoir Computing 49
Jingyi Zhou, Biyi Jiang, Huansong Tang, Zhen Wang

A SAW Pressure Sensor Based on AlN Thin Film with a Novel Differential Structure: Balancing Sensitivity Enhancement and Temperature Decoupling 52
Aobei Chen, Dapeng Li, Ge Gao, Chun Hu, Dezhi Zheng

A BSIM Compact Model of Two-Dimensional Semiconductor Field Effect Transistors 55
Jen-Hao Chen, Ahtisham Pampori, Chien-Ting Tung, Sayeef Salahuddin, Chenming Hu

Improved Ohmic Contact Resistance and DC Performance of InAlN/GaN HEMTs with Si-Incorporated Contact Scheme 58
Yang Jiang, Fangzhou Du, Xinyi Tang, Ziyang Wang, Kangyao Wen, Zhongrui Wang, Qing Wang, Hongyu Yu

BSIM-NN: A Machine Learning Compact Model for Fast IC Simulation 61
C. T. Tung, S. Salahuddin, C. Hu

975V/4.3M Ω·cm2 Enhancement-Mode (001) β−Ga2O3 Vertical Multi-Fin Power Transistors 64
Gaofu Guo, Xiaodong Zhang, Chunhong Zeng, Tiwei Chen, Zhili Zou, Zheyuan Hu, Zhucheng Li, Dengrui Zhao, Yuhua Sun, Xianqi Dai, Baoshun Zhang

An Atomic Layer Etching Technique for MOCVD in-Situ SiNx 67
Fangzhou Du, Yang Jiang, Ziyang Wang, Xinyi Tang, Qing Wang, Hongyu Yu

Reliability-Aware Device and Programming Scheme Optimization for PBS/NBS-Immune IGZO-Based 2T0C DRAM 70
Zhidong Tang, Yanbo Su, Jianshi Tang, Yijia Fan, Yiwei Du, Mingcheng Shi, Yibei Zhang, Dong Wu, Bin Gao, He Qian, Huaqiang Wu

A 2T2R TCAM Based on RRAM and Accurate Compact Model for a Kilobit Word 73
Zhen Wang, Pengtao Li, Zijian Wang, Xuemeng Fan, Shengpeng Xing, Yishu Zhang

Prediction of Single Particle Output Response in the CTFET Inverter Based on Deep Learning Algorithm 76
Chen Chong, Hongxia Liu, Zexi Wang

Structure Design and Characteristics of 3T Sense-Switch Pflash for Computing-Inmemory 79
Wei Zhao, Jinghe Wei, Guozhu Liu, Yidan Wei, Yingqiang Wei, Zhiyuan Sui, Meijie Liu, Qi Xi, Xinhe Wang, Zongguang Yu, Juyan Xu

High-Performance Cu2O/Ga2O3 Heterojunction Diodes for Power Electronics......................... 82
 Xiaohui Wang, Mujun Li, Minghao He, Chun-Zhang Chen, Haozhe Yu, Long Chen, Qing Wang, Hongyu Yu

Comprehensive Analysis of Oxidant Effects During ALD Process of Hf0.5Zr0.5O2 Ferroelectric Thin Films 85
 Xiaopeng Li, Yilin Hou, Xiaoyu Dou, Yaoyu He, Yuwei Qu, Pengpeng Sang, Xuepeng Zhan, Xiaolei Wang, Jixuan Wu, Jiezhi Chen

Interface Engineering on Inorganic-Organic Hybrid Phototransistors for High-Responsive Near-Infrared Sensing 88
 Dingwei Li, Huihui Ren, Yingjie Tang, Yitong Chen, Yan Wang, Qi Huang, Bowen Zhu

The Comprehensive Study of Difference Between Cryogenic Planar MOSFET and FinFET Variation and Band Tail States Assessment 91
 Chenyang Zhang, Yewei Zhang, Yongkang Xue, Shuying Wang, Sheng Yang, Pengpeng Ren, Zhigang Ji, Ru Huang

Contact Engineering of 1L-WSe2 p-FETs: Straightforward WOx Doping and Thickness Dependence 94
 Wenyu Wang, Yanda Zhu, Ziyi Wang, Yun Li, Jason C. C. Hwang, Xi Lin, Jiaqi Su, Jing-Kai Huang, Sean Li

Monolithic 3D Integration of 2D Devices 97
 Saptarshi Das

Broadband Vis-NIR Neuromorphic Photodetector Based on PdSe2/Bi2O2Se Heterotransistor for Motion Detection... 100
 Yu Zhu, Xinrui Guo, Shuo Liu, Junling Liu, Ru Huang, Ming He

Study of Threshold Voltage Degradation Mechanism of Ferroelectric Field-EffectTransistors (FeFETs) with TiN/SiO2/Hf0.5Zr0.5O2/SiOx/Si (MIFIS) GateStacks 103
 Zeqi Chen, Tao Hu, Xinpei J Jia, Runhao Han, Jia Yang, Yajing Ding, Xiaoqing Sun, Junshuai Chai, Hao Xu, Xiaolei Wang, Wenwu Wang, Tianchun Ye

Modeling of the Switching Characteristics of Ag/HfO2-Based Volatile Memristors 106
 Yikang Huo, Chunsheng Jiang, Qin Xie

A Submersible Soft Robot with Ultrasonic Echolocation Capabilities.. 109
 Zhongming Chen, Qilin Hua, Jiaqiang Xu, Guozhen Shen

Recent Progress on Electronics and Optoelectronics Based on 2D Tellurium 112
 Jiajia Zha, Haoxin Huang, Chaoliang Tan

Nanodiamond Quantum Thermometers and Their Biological Applications...................................115
 Masazumi Fujiwara

Enhancing Transduction Efficiency in CMOS-MEMS CMUTs Through Atomic Layer Deposition: A Preliminary Study 118
 Tzu-Yun Huang, Ming-Huang Li

A 3.98 Ghz Aluminum Nitride Overmoded Bulk Acoustic Wave Resonator for Temperature Sensing Applications 121
 Zhi-Qiang Lee, Kuan-Ting Chen, Chin-Yu Chang, Cheng-Chien Lin, Yung-Hsiang Chen, Yelehanka Ramachandramurthy Pradeep, Rakesh Chand, Yenshih Ho, Ming-Huang Li

Effect of Gate Voltage Rise Time on Gate Charges for GaN HEMTs with Partially Depleted P-GaN Cap Layer 124

Ruize Sun, Renjie Wu, Yunfei Ma, Wenhan Yuan, Qinan Guo, Wanjun Chen, Bo Zhang

Self-Rectifying Memristors with High Rectification Ratio for Security Primitives with Ultra-Low Bit-Errors 127

Guobin Zhang, Zijian Wang, Xuemeng Fan, Dawei Gao, Yishu Zhang

Competing Ferroelectric Polarization and Defect Migration Induced Resistive Switching in Van Der Waals β-In2Se3 130

Yinfeng Long, Saiyu Bu, Han Chen, Kai Liu, Shiyu Zhang, Lin Wang

Improving the Reliability of 40nm RRAM Chip by Pre-Cycle Operation 133

Ruofei Hu, Yilong Huang, Chengxiang Ma, Qianze Zheng, Kaimeng Liu, Yuelin Jiang, Siyu Chen, Jianshi Tang, Dong Wu, Bin Gao, He Qian, Huaqiang Wu

Investigation of Passivation Layers for Self-Aligned Top-Gate Amorphous InGaZnO Thin-Film Transistors with Metal-Reacted Low-Resistance Source/Drain 136

Yuhan Zhang, Jiye Li, Hao Peng, Huan Yang, Lei Lu, Shengdong Zhang

Enabling Highly-Efficient, Low-Latency Analog CAM Operations with Optimized MoS2 Flash Memory Devices 139

Guoyun Gao, Bo Wen, Ni Yang, Zhiyuan Du, Mingrui Jiang, Ruibin Mao, Yingnan Cao, Hongxia Xue, Pak San Yip, Qihan Liu, Dong-Keun Ki, Jinyao Tang, Paddy K. L. Chan, Hao Jiang, Han Wang, Lain-Jong Li, Can Li

An Artificial Thermoresponsive Nociceptor Based on Amorphous Carbon Threshold Memristor 142

Qiaoling Tian, Xiaoning Zhao, Zhongqiang Wang, Haiyang Xu, Yichun Liu

Single-Crystal Diamond MEMS for Extreme Sensors 145

Meiyong Liao

A Variability-Aware Data Preprocessing Method for Data-Driven Memristive Device Inverse Modeling 148

Zhoujie Pan, Yanming Liu, Daixuan Wu, Zihan Zhang, He Tian

4.6-Bits-Per-Cell Resistive Probabilistic-Bit Computing for High-Efficient Evaluation of Bio-Genomic Evolution Achieving 6.25x Acceleration of Data-Operations 151

Kai Wen Cheng, Huan-Hsiang Su, Yu-Hsien Lin, Jerry Chung, Y.-S. Hsieh, T.-Y. Chen, E Ray Hsieh, Yu-Cheng Lin, Ching-Ju Lin, Jen-Chou Tseng, Chien-Fan Wang, Wen-Ting Chu, Yu-Der Chih, Jonathan Chang

Passivation Effect on Atomic-Layer-Deposited Indium-Gallium-Zinc-Oxide Transistors 154

Zhiyu Lin, Jinxiu Zhao, Chen Wang, Mengwei Si

Characteristic Length of Transition-Metal Dichalcogenides Based Complementary Field-Effect Transistors 157

Xianmao Cao, Yiting Wu, Panpan Zhang

Dual Gate-Enhanced Mechanical-Electrical Stability of Flexible InGaZnO TFTs 160

Jilin Li, Yuhan Zhang, Runxiao Shi, Fion Sze Yan Yeung, Man Hoi Wong, Hoi Sing Kwok, Shengdong Zhang, Lei Lu

Flexible Carbon Nanotube Optoelectronic Neuromorphic Devices and Irradiation-Resistance Logic Circuits with Low-Work-Function Gate Electrodes 163

Meng Deng, Kaixiang Kang, Nianzi Sui, Jiaqi Li, Zebin Wang, Jianwen Zhao

A Compact Model for GIDL-Assisted Erase Transients of 3D MONOS Charge-Trap NAND Flash Memories.. 166

Changhyeok Im, Sungju Kim, Hyungcheol Shin

From Monitoring to Modulation: Intelligent Wearable Devices for Health Sensing and Drug Delivery 169

Yuanzhe Li, Shirong Qiu, Zhou Jiang, Cunman Liang, Yixin Qi, Ni Zhao

Ultrafast Photoresponse of Vertical Diodes Utilizing WSe2/ITO Schottky Junctions.................... 172

Xixi Jiang, Jingli Wang, Shukui Zhang, Qingqing Sun, Jie Wei Chen, Yang Chai

TDA (Thermal Design Automation) for Multiscale Thermal Managements of GaN HEMTs.......................... 175

Bingyang Cao

The Observation of 2D Electron Gas at Ga2O3/IGZO Interface.. 178

Jinxiu Zhao, Zhiyu Lin, Chen Wang, Mengwei Si

Embracing SemiverseTM Solutions: Semiconductor Virtual Fabrication and Its Applications 181

Qingpeng Wang, Yujia Zhong, Lifei Sun, Benjamin Vincent, Ivan Chakarov, Joseph Ervin

GeTe9 Ovonic Threshold Switch Memristor-Centered 1S1T1R Cell as Bio-Realistic Stochastic Synapse.. 184

Xinyu Wen, Kuan Wang, Lun Wang, Qi Chen, Hao Tong, Xiang-Shui Miao, Yuhui He

Development of an Inkjet-Printed Sensor Using Ink Locked Food-Based Nano Conductive Paste for Electrochemical Sensing of Insulin: A First Attempt 187

Ramya K, Sanket Goel

Chirality Reversal Related Spin-Orbit Torque Switching in Nanoscale Perpendicular Ferromagnet.............. 190

Xue Zhang, Zhengde Xu, Zhifeng Zhu

Portable NV (Nitrogen Vacancy) Vector Magnetometer with High Sensitivity and Wide Dynamic Range... 193

Yingjie Yang, Jianghao Fu, Xuanhui Ren, Doudou Zheng, Yang Li, Hui Wang, Zongmin Ma, Jun Liu

MoS2 Transistors Based 2TOC DRAM Optimized with Optical Modulation and Read Pulse Compensation.. 196

Gongpeng Lan, Dongdong Sun, Haotian Fang, Wuqing Fang, Yuanyuan Shi

Solution-Processed Oxide Semiconductors-Based Enhancement-Mode Thin-Film Transistor Circuits for Artificial Spiking Neurons... 199

Jiayi Mao, Qi Huang, Bowen Zhu

CMOS BEOL Compatible Metal-Oxides Logics and Memories for New Paradigm Computing in the Post-Moore Era.. 202

Yida Li, Feichi Zhou

Endurance Optimization of 40nm RRAM Towards 106 Cycles by Tuning the Stoichiometry of TiN Bottom Electrode.. 205

Chengxiang Ma, Qianze Zheng, Ruofei Hu, Yilong Huang, Kaimeng Liu, Yuelin Jiang, Siyu Chen, Jianshi Tang, Dong Wu, Bin Gao, He Qian, Huaqiang Wu

Optimized Electric Field Distribution and Dynamic Performance in GaN HEMTs Using Segmented-Extended $\boldsymbol{p}-\mathbf{GaN}$ Gate Structures... 208

Xinyue Dai, Haiyang Li, Qimeng Jiang, Xiaoping Wang, Yuxi Wan

Piezoresistive Internal Stress Sensing and Gas Detection Via Polymer Swelling 211
Masaya Toda

Understanding the Impact of Discrete Trap Positions and Mapping the Hysteresis Dynamics in MoS_{2} FETs by Advanced TCAD Modeling 214
Y. Z. Lv, Y. J. Chai, Yu. Yu. Illarionov

Flexible Pulse Waveform Sensor Array for Cuffless PWV Measurement.. 217
Jiang Zhou, Liang Cunman, Zhao Ni

Physical Unclonable Function in Spiking Neural Network Based on in-Te Ovonic Threshold Switching Devices... 220
Huan Wang, Qihang Zhu, Hengyi Hu, Yi Li, Xiangshui Miao, Ming Xu

Direct Backside Contact Impact on 3-Dimensional Stacked FET SRAM Beyond 1nm Node....................... 223
Mingyu Kim, Jaehyun Park, Sungil Park, Kyunghwan Lee, Deuk Ho Yeon, Daewon Ha, Hyungcheol Shin

Highly Sensitive and Polarization Selective Au Serrate Nanogratings Fabricated by Nanoimprint Lithography 226
Li Chunxia, Zhu Shuyan

High-Performance InSnO/ZnO Heterojunction Transistors and Inverters 229
Dengqin Xu, Tingchen Yi, Junchen Dong, Lifeng Liu, Zheng Zhou, Dedong Han, Xing Zhang

A 1.8TB/S HBM Heterogeneously Integrated GPU Design Exploring 2.5D Packaging Technology.............. 233
Shuang Wang, Weiliang Chen, Xueqing Li, Chen Jiang, Huazhong Yang

9-KV p-GaN Gate HEMT with Gate Termination Extension Demonstrated on Sapphire Substrate for Improved Breakdown Voltage 236
Jingjing Yu, Junjie Yang, Jiawei Cui, Hao Chang, Sihang Liu, Yunhong Lao, Xuelin Yang, Xiaosen Liu, Maojun Wang, Bo Shen, Jin Wei

Rapid and Extensive Conductance Modulation in MoOx Based Electrochemical Random-Access Memory for Spiking Neuromorphic Systems... 239
Xiaoci Liang, Dongyue Su, Younian Tang, Bin Xi, Chunzhen Yang, Huixin Xiu, Jialiang Wang, Chuan Liu, Mengye Wang, Yang Chai

TCAD and Mixed-Mode Simulation Supporting the Development of Contact-Controlled Thin-Film Transistors and Circuits 242
Eva Bestelink, Radu A. Sporea

Dynamic Clocked Comparator Based on Flexible LTPO Technology 245
Chunxiu Wang, Jiaqiao Liang, Min Zhang

VO2 Memristor-Based Adaptive Neurons for Electromyography Signal Processing 248
Zhiyuan Li, Jiaping Yao, Beining Zhang, Wei Tang, Xiangshui Miao, Rui Yang

Fabrication and Research of Wide-Bandgap Semiconductor AlN-Based Unipolar Memristors 251
Haiming Qin, Xinpeng Wang, Dayu Zhou, Liang Zeng, Yi Liu, Yi Tong

Reliability Optimization in Hafnium Oxide Based Ferroelectric Field-Effect Transistors (FeFETs).............. 254
Kechao Tang, Yuejia Zhou, Zhongxin Liang, Ru Huang

High-Performance 2D Fets with Single-Crystal Anatase TiO2 High-κ Dielectric 257
Ni Yang, Ji Zhang, Yu-Ming Chang, Fangyuan Zheng, Lingqi Li, Chenyang Li, Jian Liu, Dong-Keun Ki, Yi Wan, Sean Li, Kah-Wee Ang, Jing-Kai Huang, Lain-Jong Li

All-Optical Modulated Artificial Synapse Based on Quantum Dots/Oxide Heterojunction for
Neuromorphic Visual Simulation 260
Yan Wang, Yingjie Tang, Yitong Chen, Dingwei Li, Huihui Ren, Guolei Liu, Fanfan Li, Qi Huang, Botao Ji, Bowen Zhu

Volumetric Super-Resolution Imaging Based on Dual Bessel Beams Sted Microscopy 263
Renlong Zhang, Haoxian Zhou, Chenguang Wang, Xiaoyu Weng, Liwei Liu, Peng Xi, Junle Qu

Direct Evidence of Oxygen Vacancy Generation in Whole Gate Stacks Through Multiple Electrical
and Atomic-Scale Physical Methods as the Cause of Endurance Failure in FeFETs 267
Xianzhou Shao, Hao Xu, Saifei Dai, Fengbin Tian, Xiaoyu Ke, Jiahui Duan, Min Liao, Xinpei Jia, Xiaoqing Sun, Junshuai Chai, Jun Luo, Wenwu Wang, Xiaolei Wang

Selective-Attention Neuromorphic Vision Classifications Based on α-In2Se3Bi2O2Se
Ferroelectric-Semiconductor Heterotransistors 270
Xinrui Guo, Yu Zhu, Shuo Liu, Junling Liu, Ru Huang, Ming He

Comprehensive Investigation of the Disturb and Retention Issues in Scaled FeNAND Arrays....................... 273
Yuejia Zhou, Ru Huang, Kechao Tang

Prediction of Endurance Characteristics in Si FeFET with Ferroelectric Hf0.5Zr0.5O2 276
Xinpei Jia, Tao Hu, Mingkai Bai, Xiaoqing Sun, Junshuai Chai, Hao Xu, Xiaolei Wang, Wenwu Wang, Tianchun Ye

A Novel FeFET-Based Multibit Content Addressable Memory Through Thermometer Encoding for
Manhattan Distance Metric with High Area- And Energy-Efficiency 279
Weikai Xu, Zeyu Zhang, Qianqian Huang, Ru Huang

Novel High-Efficiency Large-Area Optical-To-Optical Conversion Integrated Device for Optical
Signal Processing 282
Yahui Su, Shanjing Liu, Peixuan Song, Peiran Du, Hui Wang, Juan Li

Van Der Waals Interfacial Engineering for High-Performance Macroscopic Assembled Graphene
(MAG)-Silicon Schottky Photodiodes.. 285
Srikrishna Chanakya Bodepudi, Muhammad Abid Anwar, Muhammad Malik, Xiaolei Ding, Yance Chen, Yue Dai, Zongwen Li, Zhi-Xiang Zhang, Yunfei Xie, Wenzhang Fang, Huan Hu, Bin Yu, Yang Xu

Spintronic Stochastic Neuron-Based Deep Belief Networks for Image Classification 288
Aijaz H. Lone, Meng Tang, Daniel N. Rahimi, Divyanshu Divyanshu, Camelia Florica, Selma Amara, Hossein Fariborzi, Gianluca Setti

G-Universe: A Collection of In-House Technology Computer-Aided Design Simulators as a
Platform for Developing New Simulation Capabilities 291
S.-M. Hong, I. K. Kim, S.-W. Jung, M.-S. Jang, P.-H. Ahn, T. Oh, G.-T. Jang, K.-W. Lee

Impact of Load-Si Thickness and Punch-Through Stopper Doping on the Electrical Performance of
SOI Gate-All-Around Field-Effect Transistors... 294
Longyu Sun, Jiayi Zhang, Yan Li, Haoyan Liu, Xi Zhang, Yongliang Li

20 nm Gate-Length Normally-Off AlGaN/GaN HEMTs Enabled by SiNx Stress-Engineered Technique 297
    Chenkai Deng, Peiran Wang, Qing Wang, Hongyu Yu

RESURF Ga2O3-On-SiC Field Effect Transistors for Enhanced Breakdown Voltage 300
    Junting Chen, Junlei Zhao, Xiaohan Zhang, Jin Wei, Mengyuan Hua

Polyoxometalate-Doped Memristor with Redox Dynamics for Reliable Single-Component Artificial Neuron 303
    Shirui Zhu, Yan-Bing Leng, Guohua Zhang, Yu-Qi Zhang, Pengfei Han, Hecheng Cai, Ziyu Lv, Yongbiao Zhai, Ye Zhou, Su-Ting Han

Solder Fatigue Life Prediction Method Considering Intermetallic Compound Growth 306
    Yikang Wang, Xiaopeng Wu, Jiahao Hou, Can Liu, Yintang Yang

A Model-Driven Design Technology Co-Optimization (DTCO) with Multi-Objective Bayesian Algorithm for Advanced Technology 309
    Baokang Peng, Guoyao Cheng, Runsheng Wang, Ru Huang, Mansun Chan, Lining Zhang

Reverse Blocking Gan on Sic Hemts with Ultralow Dynamic Ron for 1200 V Power Applications 312
    Yutong Fan, Mengqiang Yuan, Yuqing Hu, Weihang Zhang, Yachao Zhang, Zhihong Liu, Yue Hao, Jincheng Zhang

Demonstration of Reliable Magnetic Shift Register Reading Using 50 nm MTJs on CMOS IC Towards 3D Ultra-High Density Memory 315
    M. Quinsat, Y. Ueda, N. Shimomura, S. Hashimoto, N. Umetsu, Y. Ootera, J. Iwata, H. Tokuhira, S. Miyano, M. Yoshikawa, T. Kondo, M. Saitoh, M. Kado

Artificial Hodgkin–Huxley Neurons Based on Ferro-Ionic CuInP2S6 318
    Fan Yang, Lei Liang, Qirui Zhang, Xuemei Wang, Qing Liu, Xiao Luo, Fucai Liu

High Mobility and Improved Subthreshold Characteristics of Ultra-Thin Channel IGZTO TFTs Down to 3 nm 321
    Kai Chen, Yi Jiang, Zhaolong He, Rui Zhang, Junkang Li, Yunlong Li

Design and Application of a Multi-Mode Signal Codetection System for Brain Signals 324
    Jianbo Jiang, Xueying Wang, Huiran Yang, Ziyi Zhu, Dujuan Zou, Siyuan Ni, Zhengyu Liang, Guopei Zhou, Zhitao Zhou, Liuyang Sun, Tiger H. Tao, Xiaoling Wei

Demonstration of EOT-Scaled Lg~25 nm FinFET Based on Thickness-Proportion Controlled HfO2-ZrO2-HfO2 Superlattice Gate Stacks 327
    Kun Zhong, Zhaohao Zhang, Siyuan Liu, Haiyuan Lyu, Huaxiang Yin

Global Stress Analysis in Fin Patterned Si/SiGe Multilayer Nanosheets for Nanosheet-Based CMOS Device Technology 330
    Amit Kumar Singh Chauhan, Imtiyaz Ahmad Khan, Kunal, Harsh Raju, Sanjeev Kumar Manhas

Van Der Waals Dielectrics and Electrodes in 2D Transistors for 2T0C DRAM 333
    Jianmiao Guo, Ziyuan Lin, Cong Wang, Tianqing Wan, Jianmin Yan, Yang Chai

Analysis of High Performance IGZO Thin-Film Transistors Under High-Gate and Drain Bias Stress 336
    Yupeng Lu, Yanyu Yang, Peng Wang, Jie Luo, Yunjiao Bao, Gaobo Xu, Huaxiang Yin

Ferroelectric In2Se3 Transistors with Multi-Terminal Plasticity for Highly Efficient Hardware Implementation of Reinforcement Learning 339
  Yasai Wang, Weiwei Xiong, Jianmin Yan, Yue Zhou, Chaoyi Zhu, Xiangshui Miao, Yuhui He, Yang Chai

Enabling Artificial Spiking Sensory Neurons with a Single Flexible VO2 Mott Memristor for Neuromorphic Sensing 342
  Chuan Yu Han, Shujing Zhao, Shengli Fang, Shi Quan Fan, Weihua Liu, Xin Li, Li Geng

Elemental Sn Promotes the Formation of Single-Crystal ε - Ga2O3 Films in MOCVD 345
  Long Wang, Yao Wang, Qian Feng, Yachao Zhang, Jincheng Zhang, Yue Hao

Two-Dimensional ReSe2 Based Optoelectronic Synaptic Transistor 348
  Wei Zeng, Jiyu Zhao, Hang Li, Guanglong Ding, Ye Zhou, Su-Ting Han

Capacitive Length-Extension Mode Resonators with Stress-Induced Gap-Closing Electrodes for Motional Resistance Reduction 351
  Hao Yu, Yechen Miao, Fang Wang, Ke Sun, Yi Sun, Tiger H. Tao, Heng Yang

Thermal Stability of TiO2 Channel FE-VNAND: From Fabrication to High-Temperature Operation 354
  Xujin Song, Dijiang Sun, Xiaoyan Liu, Jinfeng Kang

Rapid Customizing of Flexible Oects Arrays for Low-Cost Biosensing and Biocomputing 357
  Xinyu Tian, Jing Bai, Dingyao Liu, Shiming Zhang

Realization of Wafer-Scale AlScN Ferroelectric Films and the Investigation of AlScN/N-GaN Ferroelectric Memristors 360
  Mingrui Liu, Hang Zang, Shunpeng Lv, Zhiming Shi, Yuping Jia, Ke Jiang, Jianwei Ben, Dan Li, Xiaojuan Sun, Dabing Li

A Multi-Ion Sensing System on a Chip with Edge Computing Capability 363
  Haolin Zhao, Zhancheng Mai, Kai Zhuang, Kai Wang

Surrogate MTJ Model for Early-Stage MRAM Macro Reliability Analysis 366
  Quanhai Zhu, Hao Cai

Self-Heating and Bias Temperature Instability in Recessed-Gate GaN MIS-HEMTs with AlN/SiNx Bilayer Dielectric 369
  Xin Wang, Hongyue Wang, Chen Wang, Jiayin He, Ju Gao, Bin Zhang, Ziheng Liu, Hongjie Peng, Chengkang Ao, Jiahui Yuan, Jinyan Wang

Enhancing Heat Dissipation of GaN HEMTs Based on Finite Element Simulation 372
  Hongda Chen, Hongzhen Chen, Xiaohan He, Shiming Li, Mei Wu, Xiaohua Ma, Yue Hao

Fabrication and Performance Assessment of High-Mobility Sige Channel P-Type SOI FinFET Transistor 375
  Zijing Zhang, Yuchen Wu, Yan Li, Huaizhi Luo, Fanyu Liu, Yongliang Li

BEOL-Compatible Photosensors by Atomic-Layer-Deposited ZnO Semiconductor Transistors and Their Monolithic 3D Integration 378
  Yuyan Fan, Ziheng Wang, Yulong Dong, Zhiyu Lin, Danyang Chen, Shiyi Zhang, Jingquan Liu, Mengwei Si, Xiuyan Li

Optimizing Etch Processes for Enhanced Yield and Performance in SOT-MRAM Devices on 300 mm Wafers 381
  Zhenghui Ji, Wenlong Yang, Guoxiu Qiu, Dandan Yang, Kaiyuan Zhou, Qingxiu Li, Qijun Guo, Enlong Liu, Shikun He

High-Frequency and Wideband RF Filters for 6G and Wi-Fi 7 (Invited Paper).. 384
Chengjie Zuo, Zhongbin Dai

Dynamic Performance Analysis of Ultra-Fast Inverter Based on Tri-Gate AlGaN/GaN MIS-HEMTs............ 387
*Yunsong Xu, Weisheng Wang, Dechang Quan, Haotian Ji, Shenlei Ding, Yunzhou Jiang,
Kain Lu Low, Jiangmin Gu, Wen Liu*

Achieving High Endurance Ferroelectricity in Hf0.5Zr0.5O2 Thin Films on Ge Substrate Through
Helium Ion Doping Engineering 390
*Peiyuan Du, Huan Liu, Dongya Li, Chengji Jin, Hongrui Zhang, Di Wang, Yian Ding,
Bing Chen, Ran Cheng, Mengnan Ke, Xiao Yu, Yan Liu, Yue Hao, Genquan Han*

Wide Temperature Behavior Analyses of SiGe HBTs Based on Small-Signal Parameter Extraction
Method 393
Lu Zhao, Guofang Yu, Jie Cui, Yue Zhao, Jun Fu, Yanyan Liu

In2Se3 FET-Based Neural Network Circuit Design for Complete Associative Learning 396
Weiwei Xiong, Yasai Wang, Xiangshui Miao, Yang Chai, Yuhui He

Data Retention in Co-Doped Hzo Fecaps: Roles of Fe Thickness and Thermal Budget............................... 399
*Justine Barbot, Markus Peller, Isaac Emanuel Robert, Kerstin Bernert, Hannes Mähne,
Steffen Thiem, David Lehninger, Ayse Sünbül, Konrad Seidel, Thomas Kämpfe*

Technologies of GaN Power Integration and Modeling 402
Sheng Li, Yanfeng Ma, Ran Ye, Siyang Liu, Weifeng Sun

Indium-Tin-Oxide Transistor with Maximum Transconductance Over 1100µS/µm and Cut-Off
Frequency of 23 GHz 405
*Yuxuan Wang, Jiawei Xie, Zijie Zheng, Yuye Kang, Xuanqi Chen, Gerui Zheng, Rui Shao,
Kaizhen Han, Xiao Gong*

Optoelectronic Multiplication Emulator Based on FPGA 408
Jinxian Li, Runyu Hu, Jiabin Shen, Zengguang Cheng, Peng Zhou

Integrated MEMS Diamond Quantum Magnetometer with Active Laser Noise Suppression.........................411
*Nan Wang, Xiao Peng, Yaochen Zhu, Qihui Liu, Jiachen Han, Xin Chen, Xin Luo,
Yongquan Su, Lihao Wang, Yichen Liu, Hao Chen, Jiangong Cheng, Zhenyu Wu*

A Transistor-Free Analog Content Addressable Memory with High Bit Density 414
Renhao Xue, Quanyi Tu, Mansun Chan, Xiwen Liu

Effect of the Channel Thickness on the Pbs Reliability and 1/F Noise of the ALD Ultrathin ITO
Field-Effect Transistor... 417
Jiaming Zhao, Peiyan Hong, Xuefei Li

A Physics-Based Compact Model for Electromigration Failure Prediction and Dynamic IR-Drop
Evaluation.. 420
Chenglin Ye, Yizhan Liu, Zheng Zhou, Xiaoyan Liu

Multi-Mode Resonance and Temperature Behavior of GaN/SiC SAW Resonators 423
*Guofang Yu, Renrong Liang, Deng Luo, Jianjun Chen, Yaqing Chi, Hanhan Sun, Bin
Liang*

Enhanced ESD Protection Techniques for 10V Neurostimulator Circuits in 65nm CMOS
Technology 426
Tanay Das, Naef Ahmad, Laxmeesha Somappa, Sandip Lashkare

A Novel Transistor Free Design of SOT-MRAM Written by Unipolar Current with Record Bit Cell Size (15F2) 429

    Meiyin Yang, Lei Zhao, Bowen Yang, Bowen Shen, Yanru Li, Peiyue Yu, Jianfeng Gao, Ruipeng Shi, Zhuangzhuang Ye, Shuo Xu, Yan Cui, Xiaolei Yang, Ming Wang, Shikun He, Kaiming Cai, Jun Luo

A Physics-Based Compact Model for Ambipolar Schottky-Barrier CNTFETs................................ 432

    Rui Zhan, Bin Zhou, Zilin Teng, Yiheng Xue, Panpan Zhang, Jianhua Jiang

Wavelength-Dependent Reconfigurable Photo Memory Enabled by Organic-Gated Transistor..................... 435

    Xiaokun Guo, Yaoqiang Zhou, Jianbin Xu

Deeply Scaled Gate Field Plate to Suppress Drain-Induced Dynamic Threshold Voltage Instability in Schottky-Type p-GaN Gate HEMT.. 438

    Chen Wang, Xin Wang, Junjie Yang, Jiayin He, Ju Gao, Chengkang Ao, Ziheng Liu, Hongjie Peng, Wenbo Xia, Jin Wei, Jinyan Wang

Investigation of Cryogenic Ultra-Low V_{TH} Mosfets and Beol for Power Efficiency Enhancement 441

    Yuanke Zhang, Yuefeng Chen, Hengxu Guo, Jun Xu, Guoping Guo, Chao Luo

Low-Temperature Trench Ohmic Contact Suitable for Self-Aligned P-GaN HEMT.................................. 444

    Zhiqiang Xue, Mao Jia, Qian Xiao, Bin Hou, Ling Yang, Xiaohua Ma, Yue Hao

Enhancing Heat Dissipation in GaN-On-Diamond HEMTs Through Device-First Transfer Bonding............ 447

    Shiming Li, Mei Wu, Xiaohan He, Haolun Sun, Ling Yang, Xiaohua Ma, Yue Hao

5-Bit High-Linearity UV-Stimulated Synaptic Device Based on MoS2/GaN Heterostructure 450

    Zijia Su, Yong Yan, Haiding Sun, Chengjie Zuo

Low Thermal Budget Ultrathin Ti Silicide for Advanced Backside Contact of Backside Power Delivery Network (BSPDN).. 453

    Hongxu Liao, Xijun Zhou, Fangze Liu, Lanyi Xie, Haixia Li, Jieyin Zhang, Jianjun Zhang, Xiaoyan Xu, Xia An, Heng Wu, Ru Huang, Ming Li

HRS Retention of 28 nm BEOL Integrated ReRAM... 456

    Stefan Wiefels, Nils Kopperberg, Stephan Menzel

On the Evaluation of Remnant Polarization in 3D Cylindrical Hafnia-Based Ferroelectric Capacitors.. 459

    Yishan Wu, Puyang Cai, Junwei Guo, Haobo Lin, Xuepei Wang, Jinhao Liu, Boyao Cui, Yichen Wen, Maokun Wu, Runsheng Wang, Sheng Ye, Haibao Chen, Pengpeng Ren, Zhigang Ji, Ru Huang

Compact Modeling of Kink Effect in BULK MOSFETS at Cryogenic Temperatures................................... 462

    Nisha Manzoor, Wajid Manzoor, Debashish Nandi, Aloke K. Dutta, Yogesh Singh Chauhan

Perovskite Based Artificial Vision System for Geometric Shape Recognition.. 465

    Shivam Kumar, Swapnadeep Poddar, Zhenghao Long, Zhiyong Fan

Fully Tunable In-Memory Eligibility Traces Based on Ferroelectric Semiconductor Field-Effect Transistors 468

    Junling Liu, Shuo Liu, Xinrui Guo, Yu Zhu, Ru Huang, Ming He

Low-Voltage Multi-Level Flash Memory Based on Intra-Float-Gate Charge Transfer................................... 471

    Yifan Chen, Haixia Li, Qing Wang, Zongwei Shang, Mingmin Shi, Xijun Zhou, Xiaoyan Xu, Xia An, Ru Huang, Ming Li

First Demonstration of Tunnel Fet-Based Physical Unclonable Function with Independent Entropy Source Through Ambipolar Current Modulation...474

    Kaifeng Wang, Yingxi Zhou, Rundong Jia, Hongyan Han, Weihai Bu, Qianqian Huang, Ru Huang

Reconfigurable and Nonvolatile Graphene Photodetector Integrated onto Photonic Crystal Waveguide...477

    Ruijuan Tian, Yu Zhang, Zhipei Sun, Xuetao Gan

Back-End-Of-Line Integration of Organic Thin-Film Transistor Active-Matrix on III-V Micro-LED Array for Video-Rate High-Resolution Displays...479

    R. Shi, S. Ogier, J. Li, W. Tang, L. Deng, S. Li, X. Guo

Physics-Informed Neural Network for Predicting Out-Of-Training-Range TCAD Solution with Minimized Domain Expertise...482

    Albert Lu, Yu Foon Chau, Hiu Yung Wong

Optimization of RRAM Read Performance and Area Efficiency: A Large-Scale and Low-Parasitic Array with Novel Interconnection Schemes...485

    Shengyu Bao, Yuhang Yang, Zongwei Wang, Linbo Shan, Qishen Wang, Yimao Cai, Ru Huang

Multimode Transistors Based on Ion-Dynamic Capacitance...488

    Xiaoci Liang, Yiyang Luo, Yanli Pei, Mengye Wang, Chuan Liu

Pixel-To-Pixel Variance in Graphene-Silicon Photodetector Arrays...491

    Muhammad Abid Anwar, Muhammad Malik, Srikrishna Chanakya Bodepudi, Xiaolei Ding, Yance Chen, Zongwen Li, Zhi-Xiang Zhang, Wenzhang Fang, Huan Hu, Bin Yu, Yang Xu

50 nm Gan Hemts Technology with High Frequency, Low Noise, and High JFoM...494

    Yun Zhang, Jiaheng He, Changxin Mi, Xuankun Wu, Shujie Xie, Zhe Cheng, Lian Zhang

Enhanced Threshold Switching Devices Based on Conductive Filaments in SiOx Through Vertically Aligned MoS2 Layers...497

    Jimin Lee, Sofia Cruces, Dennis Braun, Lukas Völkel, Ke Ran, Joachim Mayer, Alwin Daus, Max C. Lemme

Vertical Channel Gate-All-Around(VCG) CMOS Transistors with MBE in-Situ Doping Channel and TiN/HfO2 Gate Stacks...500

    Ran Bi, Haoran Zhao, Mingmin Shi, Jianhuan Wang, Jianjun Zhang, Xiaoyan Xu, Xia An, Heng Wu, Ru Huang, Ming Li

Dependence of Dislocation Density Distribution on Radial Temperature Gradient in 300MM Si Wafer During IGBT High Thermal Budget Process...503

    Jiuyang Yuan, Bozhou Cai, Yoshiji Miyamura, Wataru Saito, Shin-Ichi Nishizawa

Yield and Reliability Optimization of Analog RRAM for In-Memory Computing on a 28nm CMOS Platform...506

    Siyao Yang, Bin Gao

Development of an On-Chip Millimeter-Wave Antenna...510

    Ziqi Mei, Chao Liang, Enze Zhou, Yan Zhang, Ji Li, Rongbo Xie, Bingbai Li, Xiayu Wang, Chi Zhang, Rui You, Xiaoguang Zhao

Investigation of Enhanced Robustness Against Floating-Substrate-Induced Dynamic RON Degradation in 900-V p-GaN Gate HEMT Using Virtual-Body Technology 513
Hao Chang, Junjie Yang, Jingjing Yu, Jiawei Cui, Youyi Yin, Han Yang, Xuelin Yang, Jinyan Wang, Maojun Wang, Bo Shen, Jin Wei

A Comparative Analysis of Direct Leakage Current Compensation and Positive-Up-Negative-Down in the Characterization of Leaky Ferroelectric Structures 516
Tiang Teck Tan, Tian-Li Wu, Hsien-Yang Liu, Chen-Yu Yu, Laurent Grenouillet, Paolo La Torraca, Andrea Padovani, Francesco Maria Puglisi, Nagarajan Raghavan, Kin Leong Pey

Efficiency Analysis of Large-Area $\boldsymbol{\beta-\text{Ga}_{2}} \mathrm{O}_{3}$ Schottky Barrier Diodes for Dc-Dc Converter 519
Yuru Lai, Chenxi Li, Shengliang Cheng, Huaxing Jiang, Leidang Zhou, Zimin Chen, Yanli Pei, Gang Wang, Xing Lu

A Universal Method to Regulate Contact Resistance in Thin Film Transistors 522
Yanzhuo Wei, Guohui Li, Yanxia Cui, Hongwei Hao, Chen Chen, Dongdong Li, Shan-Ting Zhang

Breakthroughs in Undoped HfO_{2} Ferroelectric Capacitors Achieved Through Enhanced Nanocrystallite Seeding in as-Deposited Films Via $\mathbf{O}_{\mathbf{2}}$-Plasma Ald 525
Zongwei Shang, Changqing Ye, Hao Li, Xing Wu, Runsheng Wang, Ming Li, Ru Huang

High-Performance Broadband (300-1600 Nm) Si-Based Photodetector Enabled by 2.5D Out-Of-Plane Architecture Metasurface Enhancement 528
Yunfei Xie, Jin He, Zongwen Li, Zhi-Xiang Zhang, Xiaochen Wang, Feng Tian, Qianqian Zhang, Srikrishna Chanakya Bodepudi, Yuan Ma, Zijian Pan, Muhammad Abid Anwar, Bin Yu, Liaoyong Wen, Yang Xu

A Bionic Eye with Color Vision, Environmental Adaptivity and Neuromorphic Signal Processing Functions 531
Zhenghao Long, Xiao Qiu, Chak Lam Jonathan Chan, Zhibo Sun, Zhengnan Yuan, Zhiyong Fan

Positive and Negative Photoresponses in MoS2 Flakes Photogated by PN Junction Diode 534
Yumeng Liu, Zhengfang Fan, Jianyong Wei, Yizhuo Wang, Zhijuan Su, Yaping Dan

The Ultra-Thin Al2O3 Oxide Layer Via ALD Repaired Si Channel Interface and Optimizied Subthreshold Characteristics and Reduced Leakage Current by 96.2% 537
Renjie Jiang, Peng Wang, Lianlian Li, Qinkun Li, Zhongrui Wang, Hang Zhang, Huaxiang Yin, Qingzhu Zhang

Investigation of Al2O3/SiO2 Interface Charge for the Feasibility Study of Charge Sheet Super Junction 540
Swadhin Kumar Jena, Chiranjibi Padhee, Akshay K, Parlapalli Venkata Satyam

RRAM-Based Isotropic CNNs with High Robustness and Resource Utilization Rate 543
Wenyong Zhou, Yuan Ren, Jiajun Zhou, Chenchen Ding, Zhengwu Liu, Ngai Wong

Laser-Induced Low Temperature Dopant Segregation Schottky Barrier Mosfet for Monolithic-3D 546
Feixiong Wang, Yadong Zhang, Jinbiao Liu, Yunjiao Bao, Zhiyao Wang, Shuang Liu, Mingzheng Ding, Zhaohao Zhang, Qingzhu Zhang, Huaxiang Yin

A Heterogeneous FTJ-Based Computing-In-Memory Architecture for Vision Transformer Acceleration Via Hardware and Algorithm Co-Design 549

Letian Wang, Pinfeng Jiang, Zichong Zhang, Yifan Yang, Yilong Fang, Yi Wang, Mingde Zhu, Xiangshui Miao, Xingsheng Wang

Boosting 2D Transistor Performance Via TiS2 Van Der Waals Contact Engineering 552

Jialei Miao, Heng Zhang, Zheng Bian, Tianjiao Zhang, Yuda Zhao

Memristor-Based Approximate Adders for Efficient In-Memory Computing in Image Processing 555

Zhouchao Gan, Mengjie Li, Fan Yang, Fang Cheng, Xiangshui Miao, Xingsheng Wang

Fluorescent Nanodiamond Encapsulated Liposomes for Quantum Sensing in Nematode Worms 558

Takaki Arakawa, F. Kamada, K. Kinjo, K. Oshimi, M. Sara, M. Maeki, M. Tokeshi, M. Fujiwara

Enhancing Cad Workflows: Heterogeneous Graph Attention Networks for Efficient Routing Congestion Prediction in Chip Design 561

Qingyuan Yang, Mingkun Xu, Hongyi Li, Yu Du, Dahu Feng, Rong Zhao

Enabling Floating Body Effect in Bulk-Si Transistor for Area and Energy-Efficient Spiking Neuron 564

Shubham Patil, Hemant Hajare, Abhishek Kadam, Jay Sonawane, Shreyas Deshmukh, Veeresh Deshpande, Udayan Ganguly

Carbon Nanotube-Based High-Performance Bioelectronics 567

Youfan Hu

Controlling the Clamping Voltage in Punch-Through Diodes Via N+ Well and Contact Design for Low Voltage System Level ESD Protection 570

Praful Likhitkar, Navin Maheshwari, Sandip Lashkare

Demonstration of a 3D-Folded 2DEG-Channel Structure with Regrown AlGaN/GaN Heterostructures 573

Fuqiang Guo, Sen Huang, Xingyu Fu, Xuelin Yang, Qimeng Jiang, Xinguo Gao, Shuaiyu Chen, Ning Tang, Bo Shen

Highly Stable Zn Metal Anodes Enabled by In2O3 Coating for High-Performance Flexible Aqueous Zinc-Ion Batteries 576

Zhongqi Liang, Xiaohong Tan, Fan Xiao, Yufeng Jin, Guoshen Yang, Hang Zhou

Hardware Virtualization Technology for Multicore Brain-Inspired Chip 579

Dahu Feng, Rong Zhao

Improved Linearity by Double-Channel GaN HEMTs with the Ultra-Thin AlN Barrier Layer 582

Long Zhang, Qian Yu, Ling Yang, Meng Zhang, Chunzhou Shi, Xu Zou, Wenze Gao, Bin Hou

Simulation Optimization of Embedded Microchannel Heat Sink for GaN HEMTs 585

Bowen Yang, Shiming Li, Mei Wu, Ling Yang, Hao Lu, Bin Hou, Meng Zhang, Xiaohua Ma, Yue Hao

High Performance Pixel Development for Thin-Film Based Image Sensors 588

Jiwon Lee, Minhyun Jin

Spatial Microwave Magnetic Field Distribution Mapped by the Permalloy/Ta Bilayer Sensor 591

Peiwen Luo, Guanjun Zhang, Bin Peng, Wenxu Zhang

Characterization of a 1T-Floating Body DRAM Cell in Bulk Silicon MOSFETS for Cryogenic Memory Applications 595

Hengxu Guo, Yuanke Zhang, Yuefeng Chen, Haoyu Sheng, Chi Fang, Guoping Guo, Chao Luo

Double-Gate Cu-MIC Poly-Ge1-XSnx TFTs on Glass Substrates Via the Gate-Last Process 598

Daiki Goshima, Akito Kurihara, Akito Hara

From Flip FET to Flip 3D Integration (F3D): Maximizing the Scaling Potential of Wafer Both Sides Beyond Conventional 3D Integration 601

Heng Wu, Haoran Lu, Wanyue Peng, Ziqiao Xu, Yanbang Chu, Jiacheng Sun, Falong Zhou, Jack Wu, Lijie Zhang, Weihai Bu, Jin Kang, Ming Li, Yibo Lin, Runsheng Wang, Xin Zhang, Ru Huang

Optimization of Scaling-Down Performance in Sub-100 nm AlGaN/GaN HFETs Based on Electron Velocity Modulation 604

Mingyan Wang, Yuanjie Lv, Heng Zhou, Chao Liu, Peng Cui, Zhaojun Lin, Sen Huang

Monolithic Integration of GaN μ LED and Normally-Off P-GaN HEMT by Flip Chip Bonding.................... 607

Jinxia Jiang, Ang Li, Guohao Yu, Han Yue, Zhongming Zeng, Baoshun Zhang

High Density 3D Integration of 2D Transistors Via Van Der Waals Lamination... 610

Donglin Lu, Quanyao Tao, Yuan Liu

Optimizing ALD-Deposited IGZO TFT Thermal Stability Through Compositional Adjustments 613

Jianting Wu, Huajian Zheng, Min Guo, Yi Huang, Xiaoci Liang, Qian Wu, Chuan Liu

Fully Printed Polymer Gas Sensors Based on Machine Learning for Calibration-Free Mobile Sensing 616

Siying Li, Sujie Chen, Qiuqi Zhang, Yuying Si, Xiaojun Guo

Re-Examination of Uniaxial Stress Effects in Ultra-Scaled GAAFETs.. 619

Yusi Zhao, Huawei Tang, Rongzheng Ding, Yudong Lv, Yanbo Tang, Shaofeng Yu

GIT-Based Bipolar p-FET with Enhanced Conduction Capability on E-Mode GaN-On-Si HEMT Platform 622

Chengcai Wang, Jinjin Tang, Junting Chen, Mengyuan Hua

93.6 cm2 V−1·s−1Homostructure a-IGZO Thin-Film Transistor with High-K Gate Dielectric Fabricated at Room Temperature... 625

Heng Yue Gong, Jia Cheng Li, Yang Hui Xia, Ya Dong Zhou, Hui Xia Yang, Yuan Xiao Ma, Ye Liang Wang

Impact of Cross-Sectional Current Crowding on Electromigration in Interconnects 628

Yichen Wen, Shuying Wang, Xiaoman Yang, Hai-Bao Chen, Maokun Wu, Runsheng Wang, Zhigang Ji, Ru Huang

Effective Mass Engineering in Ultra-Scaled GAAFETs.. 631

Yusi Zhao, Huawei Tang, Rongzheng Ding, Yudong Lv, Yanbo Tang, Shaofeng Yu

Unveiling the Role of Oxygen Vacancy Inhomogeneity in Enhancing the Reliability of Ferroelectric HfO2 ZrO2 Superlattice Structures.. 634

Boyao Cui, Maokun Wu, Sheng Ye, Xuepei Wang, Yuchun Li, Yishan Wu, Yichen Wen, Jinhao Liu, Zhigang Ji, Hongliang Lu, David Wei Zhang, Runsheng Wang, Ru Huang

BEOL Electro-Biological Interface for 1024-Channel TFT Neurostimulator with Cultured DRG Neurons 637
    Haobin Zhou, Bowen Liu, Taoming Guo, Hanbin Ma, Chen Jiang

Dual Functionality of MoS2 in nFET and pFET Through Contact Metal Selection 640
    Kwok-Ho Wong, Mansun Chan

An Efficient Simulation-Time Aware Data-Driven Automatic Design Method for Analog Circuit 643
    Shun-Qi Dai, Yuan Lei, Bei-Ping Yan

CMOS-Compatible Si3N4 Optical Waveguides: An Electromagnetic Study for Fabrication Considerations of SiO2 Cladding and Silicon Wafer Choice 646
    Wenli Zhou, Rui Ray Yao, Sang Lam

Microwave Coherent Storage Based on the Long Lifetime Cavity Electromechanical System 649
    Tiefu Li

Performance Analysis of N-Type Stacked-Vertical FET for Enhanced in Advanced CMOS Applications 652
    Yonghwan Ahn, Junjong Lee, Seunghwan Lee, Sanguk Lee, Kyeongrae Cho, Minchan Kim, Rock-Hyun Baek

Threshold Switching Memristor-Based Spiking Neuron Modeling and Simulation 655
    Pengyu Liu, Lekai Song, Kong-Pang Pun, Guohua Hu

Enhanced Gate Stack Reliability Test Framework for Gan Hemt in High-Stress Environments Leveraging on Ramp and Constant Voltage Stress Protocols 658
    Jerry Joseph James, Nagarajan Raghavan

Energy Efficient Ground Enhanced Scheme for Large-Scale SOT-MRAM Arrays 661
    Wen Wang, Haoran Du, Shuyu Wang, Bo Liu, Hao Cai

Modeling and Simulation of Prepulse and Pulse Front for 4H-SiC Drift Step Recovery Diode 664
    D. Y. Guo, Y. Zhou, J. K. Guo, X. Y. Tang, L. J. Sun, Y. M. Zhang, Q. W. Song

Van Der Waals Epitaxy h-BN/AlN Back Barrier with Controllable Boron-Diffusion for High-Erformance AlGaN/GaN HEMTs 667
    Haidi Wu, Jing Ning, Jincheng Zhang, Yue Hao

Multi-Output Virtual Metrology for Physical Vapor Deposition Using Projective Selection Algorithm 670
    Amina Mevic, Andreas Laber, Sandor Szedmak, Dženana Đonko, Senka Krivic

On the Understanding of Temperature Dependence of Flicker Noise in Advanced FinFET Technology 673
    Sheng Yang, Shuying Wang, Chenyang Zhang, Yewei Zhang, Yongkang Xue, Pengpeng Ren, Zhigang Ji

Overlay-Aware Variation Study of Flip FET and Benchmark with CFET 676
    Wanyue Peng, Haoran Lu, Jingru Jiang, Jiacheng Sun, Ming Li, Runsheng Wang, Heng Wu, Ru Huang

A 12-Bit Fully Differential SAR/SS ADC Architecture Based on Scale Reference for High-Speed CMOS Image Sensors 679
    Changhui Yang, Bailin Zhang, Weizhe Zhang, Nanbo Chen, Jingyang Chen, Ziyang Hu, Gang Wang, Peng Feng, Jian Liu, Nanjian Wu, Liyuan Liu

A High-Throughput Parasitic TRNG in Self-Rectifying Memristor Based CIM for Edge Secure Computing 682
    Ying-Jie Yu, Sheng-Guang Ren, Yu-Yang Fu, Jian-Cong Li, Pu-Yi Zhang, Yi Li, Xiang-Shui Miao

Low-Complexity Method for Shortest Path Optimization Problems Based on Nanowire Memristor Network 685
    Yanming Liu, Shenghao Wu, Ming Jian, Daixuan Wu, Yanxi Long, He Tian

Enhancing 3D DRAM Double-Gate Device Performance by Leveraging the Floating Body Effect 688
    Jing Liang, Yong Yu, Feng Shao, Menglong Zhou, Tiantian Wei, Chunyang Li, Jianpeng Jiang, Jing Cai, Bryan Kang, Mingxu Liu, Xiangsheng Wang, Guilei Wang, Chao Zhao, Shaojun Wei

Categorization of Stacking Faults and Their Effects on I-V Characteristics in 2NM P-Type GAA Nanosheet Transistors 691
    Seungjoon Eom, Sanguk Lee, Rock-Hyun Baek

A Low-Programming-Variation and High-Yield 2-Bit Programmable 1-Kb Self-Rectifying Memristor Crossbar Array 694
    Jia-Yi Sun, Sheng-Guang Ren, Heng-Feng Zhang, Zhi-Li Cui, Yi-Bai Xue, Yu Zhang, Wen-Bin Zuo, Yi Li, Xiang-Shui Miao

Based on Vertical Structures for N-Type Organic Electrochemical Transistors 697
    Wenjing Zhang, Songjia Han, Chuan Liu

Recent Research on Ultrawide Bandgap Semiconductor Ga2O3 and AlN 700
    Dao Hua Zhang, Xian-Hu Zha, Rong-Jun Zhang, Shuang Li, Yan Liu, Yu-Xi Wan

Scalable In-Memory Walsh-Hadamard Transform for Image Compression 703
    Jing Tian, Huai-Zhi Pei, Jian-Cong Li, Xiao-Di Huang, Yi-Bai Xue, Yi Li, Xiang-Shui Miao

A Novel De-Mirroring Approach for Bias-Dependent Capacitance Extraction in Nanosheet Fet Using Conformal Mapping 706
    Deven H Patil, Sandeep Kumar, Sunil Rathore, S. Dasgupta, Navjeet Bagga

4F2/Bit Memristive Multi-Bit Content Addressable Memory Enabled by Nonlinear Encoding for in-Memory Similarity Search 709
    Tong Hu, Yibai Xue, Yingjie Yu, Wenbin Zuo, Jiancong Li, Yi Li, Xiangshui Miao

Monolithic and Heterogeneous CTFT on Glass Substrate Using Ni-MIC Poly-Si TFT and Cu-MIC DG Poly-Ge TFT 712
    Y. Ito, D. Goshima, A. Kurihara, A. Hara

High Endurance Nanoscale TiN Bottom Heater for Phase Change Memory 715
    Ziqi Wan, Wencheng Fang, Li Xie, Xi Li, Ruobing Wang, Zhitang Song, Xilin Zhou

A Device to Circuit BTI and HCD Aging Analysis Framework: (Invited Paper) 718
    Payel Chatterjee, Karansingh Thakor, Souvik Mahapatra

Valley Transistors as Graded Neurons for Accurate Action Recognition 721
    Jiewei Chen, Wenxiao Wang, Yue Zhou, Yang Chai

High-Sensitivity Flexible X-Ray Detectors Based on Drop Casting 2D/3D Perovskite Thick Film 724
    Liwen Qiu, Bosen Zhang, Qiang Lou, Maojun Sun, Hang Zhou

Modeling Nanowire Single-Photon Avalanche Detectors Via Deep Neural Networks 727
Boyang Zhang, Zhe Li, Zhongju Wang, Daoyi Dong, Lan Fu

Multi-Scale Thermal Modeling of 3D-Heterogeneous Integrated Processing-Near-Memory Chip for
Edge Large Language Model Inference.. 729
*Awang Ma, Bin Gao, Yuyao Lu, Yudeng Lin, Yiming Zhou, Xing Mou, Zhuodong Kang,
Zhipeng Kuang, Peng Yao, Jianshi Tang, He Qian, Huaqiang Wu*

Polarization Regulation in Algan Solar-Blind Ultraviolet Photodetectors 732
Bingxiang Wang, Ke Jiang, Tong Fang, Xiaojuan Sun, Dabing Li

Comprehensive Performance Evaluation of 18 Perfluoroelastomers O-Ring Models for
Semiconductor Manufacturing 735
Zhanfeng Guo, He Tian, Tian-Ling Ren

A Bio-Inspired Ionic Retina... 738
Hongjie Zhang, Kai Xiao

ScAlN-Based Bulk Acoustic Wave Technology for 5G Filtering Applications................................... 741
*Chen Liu, Xinghua Wang, You Qian, Ying Zhang, Minghua Li, Peng Liu, Huamao Lin,
Qingxin Zhang, Yao Zhu*

Dual Charge-Trapping Nanowire Flash Transistor for Low-Voltage High-Precision Biomimetic
Synaptic Device with 1024 States 744
Qing Wang, Haixia Li, Ran Bi, Hongxu Liao, Baotong Zhang, Ru Huang, Ming Li

Consideration of VFET for Ultimate Logic Scaling: A Design Perspective....................................... 747
*Yimeng Wang, Yanbang Chu, Ziqiao Xu, Yu Liu, Rui Guo, Jiacheng Sun, Wanyue Peng,
Haoran Lu, Ming Li, Runsheng Wang, Heng Wu, Ru Huang*

Robot Collision Detection Acceleration with In-Memory Search Based on the Monolithic 3D
Integration of 2D Transistors and Vertical RRAMs.. 750
Yijian Zhang, Jiahao Sun, Maosong Xie, Yueyang Jia, Rui Yang

Enhanced NBTI Characteristics in HfO2/TiAlC RMG Stacks Via Low-Temperature H* Remote
Plasma Treatment After Interfacial Layer Growth 753
*Songyi Jiang, Qianqian Liu, Hong Yang, Mingyang Sun, Junjie Li, Jianfeng Gao, Shuai
Yang, Runsheng Wang, Jun Luo, Wenwu Wang*

Modeling Multi-Junction Tandem Solar Cell with High Bifacial Gain... 756
Mohammad Ajmain Fatin, Arpan Saha, Md. Shahriar, Mainul Hossain

Recent Advances in on-Chip Learning with Organic Neuromorphic Circuits.................................... 759
Yoeri Van De Burgt

Pixelated Germanium-On-Silicon Photodetector with High Responsivity for Short Wavelength
Infrared Imaging.. 762
Zhou Zhou, Chao Gao, Xin Jin, Xiaolin Liu, Kai Wang

Towards Spectrum Spurious-Free and Wideband SAW Devices Based on LN/AT-Quartz Layered
Structure 765
*Peisen Liu, Sulei Fu, Boyuan Xiao, Xinchen Zhou, Qiufeng Xu, Jiajun Gao, Shuai Zhang,
Rui Wang, Cheng Song, Fei Zeng, Weibiao Wang, Feng Pan*

Optoelectronic Switching Memory Device Based on WO3 Semiconductor Film Enables High-
Efficiency In-Sensing Reservoir Computing for Speech Recognition 768
Xun Li, Jie Li, Ke Xiang, Wenyang Xu, Yu Xu, Denshun Gu, Guangdong Zhou

High Dielectric Constant of HfO2 Technology for Memory Applications 771
    Min-Hung Lee, Zhao-Feng Luo, Chun-Yu Liao, Kuo-Yu Hsiang, Jia-Yang Lee, Fu-Shen Chang, Yii-Tay Chang, Cheng-Hong Liu, Che-Chi Cheng

Optoelectronic Artificial Synaptic Device Based on Graphene-AlGaN Van Der Waals Junction 774
    Yang Chen, Yuanyuan Yue, Bingchen Lv, Jin Zhang, Xiaoyu Wei, Xiaojuan Sun, Dabing Li

Impact of Titanium Nitride (TiN) Thickness Uniformity on the Reliability of n-FinFETs: A Comparative Study of ALD and PVD TiN Techniques 777
    Mingyang Sun, Yunfei Shi, Hong Yang, Qianqian Liu, Qingzhu Zhang, Tao Yang, Junfeng Li, Huaxiang Yin, Xiaolei Wang, Jun Luo, Wenwu Wang

Dual-Gate Thin-Film Transistor-Based Multi-Parameter Sensor for Comprehensive Water Quality Monitoring in Aquatic Environments 780
    Qiang Chen, Qiyi Su, Haolin Zhao, Zhancheng Mai, Kai Zhuang, Yitong Xu, Xinghui Liu, Kai Wang

Impact of Sb2Se3 Annealing on the Photoresponse of TiO2/Sb2Se3/Si Back-To-Back Photodiodes 783
    Bangsen Ouyang, Jialiang Wang, Yang Chai

Design and TCAD Simulation of a Surface Super-Junction Based GaN HEMT with Enhanced Breakdown Voltage 786
    Jiayi An, Fuqiang Guo, Zhongchen Ji, Sen Huang, Qimeng Jiang, Xinhua Wang, Ke Wei, Xinyu Liu

Joule Heating Effect on Quality Factor and Frequency Tuning of 2D MoS2 NEMS Resonators 790
    Shuai Yuan, Zuheng Liu, Pengcheng Zhang, Rui Yang

Improved Cmos-Based Noise-Immune Sigmoid Activation Function for Neural Networks 793
    Harshita Singh, Anant Singhal, Harshit Agarwal

Self-Heating and Process Induced Performance Barrier on Complementary Field Effect Transistor: A Reliability Perspective 796
    Sandeep Kumar, Deven H Patil, Khushi Jain, Ankit Dixit, Sunil Rathore, Mohd. Shakir, Naveen Kumar, Vihar Georgiev, S. Dasgupta, Navjeet Bagga

Compact Modeling of Silicon Carbide (SiC) Power Fets 799
    Karunesh Kumar Tripathi, Yogendra Machhiwar, Danish Raja, Harshit Agarwal

Performance Analysis of Advanced Ferroelectric $\text{HfO}_{2}\text{-}\text{ZrO}_{2}$ Superlattice Gate Stack Transistor with Multi-Phase Ferroelectric Order 802
    Danish Raja, Yogendra Machhiwar, Karunesh Kumar Tripathi, Girish Pahwa, Harshit Agarwal

Novel Ferroelectric Tunnel Fet-Based Computing-In-Memory with In-Situ XOR Cipher-Encrypted and-Type Multiply-Accumulate for Secure Edge AI 805
    Jin Luo, Zhiyuan Fu, Qianqian Huang, Ru Huang

Photoresponse Improvement in the Near-Infrared Region of Organic Photodetectors by Introducing a Trap Layer 808
    Fangchen Zhu, Yang Wang

CMOS Compatible Spike Vision Sensor 811
    Xiaolin Liu, Chao Gao, Bowei Jiang, Xin Jin, Zhou Zhou, Yihong Qi, Kai Wang

Dynamic Adaptive Visuomorphic Electronics 814
    Le Wang, Yiru Wang, He Shao, Wei Huang, Haifeng Ling

Enhanced Thermal Stability of Ru Interconnects Using h-BN as Barrier Layers 817
Kun Chen, Zhuming Wang, Chen Wang, David Wei Zhang

Leveraging Ferroelectric Negative-Capacitance Effect for Energy Efficient Electronics 820
Jie-Ni Dai, Pin Su

Thermal Crosstalk Analysis in Advanced CMOS Circuits: Insights from 3nm Gate-All-Around
Transistor Technology 823
Sihao Chen, Honglin Wu, Runsheng Wang, Ru Huang, Lining Zhang

Flexible and Skin-Compatible rGO/PVDF Composite Sensor for Multi-Functional on-Skin
Applications.. 826
Zi-Jia Su, Lu-Qi Tao, Li-Wei Liang, Zhi-Fei Xie, Yi Yang, Tian-Ling Ren

Modelling and Design of Short Channel Ferroelectric FETs with a Metal Interlayer Easing the
Multilevel Operation 829
Chiara Rossi, Daniel Lizzit, David Esseni

Effect of Pulse Schemes on Multi-Level Switching and Short-Term Instability in 1T1r
Configuration.. 832
*Xiaohua Liu, Felix Cüppers, Dennis Nielinger, Susanne Hoffmann-Eifert, Rainer Waser,
Stefan Wiefels*

Comprehensive Modeling of Ferroelectric Tunnel Junctions: Variability Analysis and Device
Design.. 835
Jiajun Qiu, Ning Ji, Hao Li, Ning Feng, Runsheng Wang, Ru Huang, Lining Zhang

Cryogenic sEKV Compact Model Applied to 22 nm FDSOI Enabling Low-Temperature Circuit
Simulation 838
Hung-Chi Han, Yating Zou, Batuhan Keskin, Edoardo Charbon, Christian Enz

An Rram-Based Multi-Mode and Pipelined Pooling Scheme in Computing-In-Memory for
Convolutional Neural Networks.. 841
Yi Gao, Zongwei Wang, Lin Bao, Yimao Cai, Ru Huang

Temperature Dependent Back-Hopping in Spin Transfer Torque Switching of Perpendicular
Magnetic Tunnel Junctions... 844
Yapeng Zhao, Tiaoyang Li, Yanqing Wu

Selective-Ferroelectricity-Defining Method Utilizing Directional Oxygen Plasma Treatment for
Optimizations of Hf0.5Zr0.5O2-Based Memory Devices.. 847
*Yi Ding, Mingcheng Shi, Yuyan Wang, Bowen Shen, Wen Sun, Benjamin Yang, Yanbo Su,
Huanan Liu, Jian Yuan, Jianshi Tang, Bin Gao, He Qian, Huaqiang Wu*

Beol-Compatible Multi-Layer ITO-ZnO-ITO Channel FETs Achieving Enhanced Mobility,
Positive V_{TH} Shift, and Improved PBTI.. 850
Ying Xu, Yiyuan Sun, Zijie Zheng, Xiao Gong

Synergistic Optimization of Thermal and Electrical Performances in Hetero-Integrated β-Ga2 O3
SBDs.. 853
Yinfei Xie, Yang He, Zhengyue Li, Wenhui Xu, Tiangui You, Xin Ou, Huarui Sun

Selecting Device, In-Memory Search, and Monolithic 3D Integration of RRAMs and 2D Devices
Towards High Reliability and High Efficiency: (Invited Paper) 856
Rui Yang, Yueyang Jia, Maosong Xie, Minliang Shen, Yijian Zhang, Yuzhuo Liu

Adaptive Update Precision with Reduced Iterative Write Cycles for Efficient Training Neural Networks on ECRAM Arrays 859

Peihong Li, Peng Chen, Peng Lin, Gang Pan

Twisted-Placed Multilayer Stack for Inherent Suppression of Transverse Modes on Layered SAW Devices 862

Boyuan Xiao, Sulei Fu, Peisen Liu, Xinchen Zhou, Qiufeng Xu, Jiajun Gao, Shuai Zhang, Rui Wang, Cheng Song, Fei Zeng, Weibiao Wang, Feng Pan

A Novel Low Loss Superjunction MOSFET with Hybrid Conduction Modes 865

Yun Xia, Gang Chen, Yuxi Wan, Wanjun Chen, Ruize Sun, Xiaoming Wang, Yan Wang, Xiaoping Wang

Graphene/Silicon Pixel Array with Integrated Imaging and Readout System 868

Yuan Ma, Zongwen Li, Youshui He, Qianqian Zhang, Yunfei Xie, Zhi-Xiang Zhang, Feng Tian, Muhammad Abid Anwar, Muhammad Malik, Srikrishna Chanakya Bodepudi, Bin Yu, Yuda Zhao, Yang Xu

Effect of Fin Dimensions on the Performance of Gan-On-Si Fin-HEMTs 871

Mengdi Li, Jiejie Zhu, Lingjie Qin, Bowen Zhang, Yuxi Zhou, Mingchen Zhang, Yichong Ding, Yuchen Qian, Xiaohua Ma, Yue Hao

Surface Potential-Based Compact Model for IGZO-DRAM Enables the TCAD-To-SPICE Framework for Reliability-Aware DTCO Flow 874

Xufan Li, Wenfeng Jiang, Chen Gu, Yue Zhao, Di Geng, Guanhua Yang, Lingfei Wang, Ling Li

Optimization of the Read Transistor in Hybrid 2T0C DRAM by Co-Modeling the Drain Current of a-IGZO and Low Temperature Poly-Silicon TFTs 877

Haolin Li, Zheng Zhou, Xiaoyan Liu

Enhanced Performance of P-FeFETs with TiN:2.5nm/Mo/TiN Gate Stacks for 3-Bit-Per-Cell Operation, 3.5V Read-After-Write, and High Endurance (109 Cycles) in Compute-In-Memory Applications 880

C.-Y. Tsai, M.-H. Hsiung, Y.-C. Chen, Y.-T. Tsai, Y.-T. Tang

Dual-Functional Volatile and Nonvolatile Resistive Switching Characteristics of $\mathbf{C U} / \mathbf{G A}_{\mathbf{2}} \mathbf{O}_{\mathbf{3}} / \mathbf{P T}$ Memristor Deposited Via Electron Beam Evaporation 883

Nan He, Zi Li, Shuai Chen, Hao Zhang, Xinpeng Wang, Lei Wang, Yi Tong

Enabling Broader Memory Windows by Double-Gate Nanosheet Ferroelectric FETS for Next-Generation Non-Volatile Memory Storage 886

F. Wu, C.-Y. Chiu, T.-Y. Lin, C.-H. Wu, V. P.-H. Hu, P. Su, C.-J. Su

On-Chip Photon-Mediated Magnon-Superconducting Qubit System and Its Quantum Application 889

Jiacheng Liu, Ferris Prima Nugraha, Qiming Shao

Reduced Process Induced Threshold Voltage Variability in Bulk Negative Capacitance Junctionless Transistors 892

Ruma S R, Vita Pi-Ho Hu, Manish Gupta

Optimal Transfer Learning Strategies for Property Predictions in Materials Science 895

Reshma Devi, Keith T. Butler, Sai Gautam Gopalakrishnan

Invertible Prediction Model for Si3N4 Wet Etching Using DHF 897

Koki Shibata, Takashi Ota, Koji Ando, Naoko Misawa, Chihiro Matsui, Ken Takeuchi

The Demonstration of Scalable-HZO/ZrO2 FeFET with Large Memory Window of 2.3V for 3Bit-Per-Cell, Immediate Read After Write, High Endurance of 109 Cycles, and the High Accuracy of 92% for Machine Learning.. 900
    S.-T. Huang, H.-M. Chen, Y.-T. Tsai, C.-S. Pai, Y.-T. Tang

A Fan-Out Wafer-Level Packaging-Based THz Communication Transceiver System for High-Speed Chip-To-Chip Wireless Interconnect Applications 903
    Si Rui Liu, Ya Fei Wu, Zong Rui He, Yu Jian Cheng, Yang Chai

Temperature-Dependent Characteristics of Field-Effect Mobility in MOCVD-Grown β-Ga2O3MOSFETs on Sapphire Substrate.. 906
    Chunxiao Yu, Yibo Wang, Yiyang Wu, Jun Zheng, Chenyu Liu, Xiaole Jia, Bochang Li, Haodong Hu, Cize Fang, Yan Liu, Yue Hao, Genquan Han

Content Addressable Memory Hierarchies for Computing in Memory... 909
    Paul-Philipp Manea, Nathan Leroux, John Paul Strachan

6-Inch GaN-On-Si Gold-Free Fabrication Technolgies for Monolithic Microwave Integrated Circuits (MMICs) 912
    Xiaojin Chen, Jin Zhou, Peiyu Mao, Tong Wang, Zhenyuan Li, Hu Wei, Hanghai Du, Weichuan Xing, Weihang Zhang, Xiangdong Li, Zhihong Liu, Jincheng Zhang, Yue Hao

Electrical and Thermal Characterization of Hetero-Integrated β-Ga2O3-On-Diamond SBDs by Transfer Printing Technology 916
    Zhenyu Qu, Tiancheng Zhao, Wenhui Xu, Haodong Jiang, Yeliang Wang, Xinbo Zou, Min Zhou, Tiangui You, Xin Ou

Electrical Transport at n-Ga2O3/N-SiC Hetero-Interface Constructed by Hydrophilic and Surface Activated Bonding.. 919
    Zhenyu Qu, Wenhui Xu, Haodong Jiang, Tiancheng Zhao, Yeliang Wang, Haowen Guo, Xinbo Zou, Min Zhou, Tiangui You, Xin Ou

Experimental Investigation of Inserted HfO2 Impact on VFb and Interface Via ALD for La2O3 Dipole-First Multi-VT Techniques 922
    Yanzhao Wei, Jiaxin Yao, Yu Wang, Qingzhu Zhang, Huaxiang Yin

Multi-Physics Modeling of Au/MoS2/Au Memristors Combining Molecular Dynamics and Electro-Thermal Simulations 925
    Mohit Tewari, Ashutosh Krishna Amaram, Tarun Agarwal

High Performance 4H-SiC DSRD with 1.4kV Peak Voltage, 400ps Risetime and 1-MHz Continuous Repetition-Rate 928
    Yu Zhou, Jingkai Guo, Dengyao Guo, Fengyu Du, Lejia Sun, Xiaoyan Tang, Qingwen Song, Yuming Zhang

Neural Network Assisted MOSFETs Gate Dielectric Traps Extraction... 931
    Xiaoyan Liu, Jinghan Xu, Zheng Zhou

A Novel Superlattice HfO2-ZrO2 Ferroelectric Tunnel FET for Overall Improvement in Memory Window, EOT and Disturb Immunity.. 934
    Shaodi Xu, Zhiyuan Fu, Shengjie Cao, Yue Yu, Hao Zheng, Qianqian Huang, Ru Huang

A 512×256 TFT-Based Image Array Sensor with High Sensitivity, High Frame Rate and Wide Dynamic Range for Industrial Soft X-Ray Detection.. 937
    Weikang Yan, Haorong Xie, Xianming Li, Chao Gao, Qi Liu, Song Kang, Jun Chen, Kai Wang

Investigation of Effects of Non-Volatile Memory-Based Computation-In-Memory Non-Idealities and Model Size on Performance and Robustness of Small Language Model During Inference Phase 940
Adil Padiyal, Tao Wang, Naoko Misawa, Chihiro Matsui, Ken Takeuchi

Effect of Cu Microstructures on Cu/SiO2 Hybrid Bonding for 3D IC Heterogeneous Integration 943
Chih Chen, Huai-En Lin, Wei-Lan Chiu, Hsiang-Hung Chang

Efficient Implementation of 16 Reconfigurable Boolean Logics Based on Memristors and Their Application in Image Edge Detection 947
Zhouchao Gan, Yifeng Xiong, Fan Yang, Xiangshui Miao, Xingsheng Wang

A CMOS-Compatible MoS2 Transistor on Silicon-Rich Silicon Nitride as Multifunctional Neuromorphic Device 950
Xiangwei Su, Hongzhao Wu, Caijing Liang, Xinlong Zeng, Yuda Zhao

A Sensing-Computing System Based on High Uniformity Photolithographic Organic Thin Film Transistor 953
Yaojie Zheng, Taoming Guo, Haobin Zhou, Chen Jiang

Physics-Based Circuit-Compatible Model of Polycrystalline Hafnia-Based 3D Ferroelectric Capacitor for High-Density Memory Applications 956
Minyue Deng, Chang Su, Jiayan Zhu, Liang Chen, Shengjie Cao, Qianqian Huang, Ru Huang

Van Der Waals Infrared Photon Detectors for Standard Blackbody Characterization 959
Yang Wang, Weida Hu, Peng Zhou

Remote Effect of Extra Metal Layer on TiN Electrode on the Ferroelectric Properties of Hf0.5Zr0.5O2 Thin Films 962
Zhipeng Xue, Danyang Chen, Zikang Yao, Tianning Cui, Yulong Dong, Jingquan Liu, Mengwei Si, Xiuyan Li

Electrical Tunability in Band-To-Band-Tunneling Based Neuron for Low Power Neuromorphic Computing 965
Shubham Patil, Jayatika Sakhuja, Anmol Biswas, Hemant Hajare, Abhishek Kadam, Shreyas Deshmukh, Ajay Kumar Singh, Sandip Lashkare, Nihar Ranjan Mohapatra, Udayan Ganguly

Direct 3D Force Mapping Enabled by Flexible Single-Crystal Piezoelectric Sensor Array 968
Jiefei Zhu, Changjian Zhou, Min Zhang

Hydrogel-Based Bipolar Synaptic Device for Artificial Neural Networks 971
Mengjiao Pei, Yun Li, Changjin Wan

Challenges in Accelerating Power SiC Device Commercialization 974
Victor Veliadis

Materials for Thermal Dissipation Applications in Stacked Devices 977
W.-Y. Woon, J.-H. Jhang, S. Vaziri, M. Malakoutian, K.-K. Hu, I. Dayte, C.-C. Shih, J.-F. Hsu, J.-P. Lin, Y. Wu, A. Kasperovich, R. Soman, J. Kim, H.-K. Wei, X. Bao, M. Nomura, S. Chowbhury, S. Sandy Liao

Thermal Analysis of Multi-Chiplet Heterogeneous Integration Based on the Lidless Fan-Out Package 980
Yongbo Wu, Jiexun Yu, Changming Song, Zheqi Xu, Lin Tan, Qian Wang, Jian Cai

Toward Low-Thermal-Budget Processing and Low-Voltage Operating in Ferroelectric Stack Films by Introducing ZrO2 Middle Layer 983

Yinchi Liu, Hao Zhang, Jining Yang, Xun Lu, Shiyu Li, Yeye Guo, Chunlei Wu, Handong Zhu, Xuning Zhang, Zhenxin Wang, Yaoyi Wang, Ruli Zeng, Yiwen Yu, Wenjun Liu

Broadband Miniaturized Spectrometers with Van Der Waals Junctions 986

Md Gius Uddin, Susobhan Das, Abde Mayeen Shafi, Lei Wang, Xiaoqi Cui, Fedor Nigmatulin, Faisal Ahmed, Andreas C. Liapis, Weiwei Cai, Zongyin Yang, Harri Lipsanen, Tawfique Hasan, Hoon Hahn Yoon, Zhipei Sun

2-D FET Modeling: Why Incorporating the Gate- And Drain-Dependent Source Tunneling Barriers Matter 989

Cristine Jin Estrada, Zichao Ma, Lining Zhang, Mansun Chan

Interface Engineering Induced Digital to Analog Switching Transition in Hafnium Oxide-Based Memory Device 992

Cong Han, Miaocheng Zhang, Yi Liu, Xinpeng Wang, Hao Zhang, Yi Tong

MEMS Integrated with Self-Assembled Electrets 995

Daisuke Yamane

"Quantum Nanophotonics with Hexagonal Boron Nitride" 998

Igor Aharonovich

Operation Principles and Applications of Ultra-Sensitive Optical Detectors at Nanoscale: Facts and Artifacts 1000

Taras Plakhotnik

Ag:SiOx-Based Volatile Memristors for Dendritic Computations 1003

Ruiqi Chen, Yulin Feng, Nan Tang, Yiyang Chen, Hao Ai, Haozhang Yang, Zheng Zhou, Lifeng Liu, Xiaoyan Liu, Jinfeng Kang, Peng Huang

Synergetic Effect of Doping and Oxygen Vacancies to Realize Higher Permittivity and Lower Leakage for DRAM Capacitors: A First-Principles Study 1006

Ting Zhang, Maokun Wu, Miaojia Yuan, Yichen Wen, Yilin Hu, Pengpeng Ren, Sheng Ye, Runsheng Wang, Zhigang Ji, Ru Huang

Wafer-Scale Fabrication of Janus-MXene Films and Its Based Flexible Artificial Synapse 1009

Xin Liu, Conghui Zhang, Yuhang Yang, Peisong Liu, Shuiren Liu, Lingxian Meng, Fei Hui

Improved Breakdown Voltage and Leakage Current in βGa2O3 Schottky Barrier Diode Realized by N Ion-Implantation Edge Termination 1012

Hao Zhang, Xinlong Zhou, Yinchi Liu, Jining Yang, Handong Zhu, Xun Lu, Shiyu Li, Yeye Guo, Chunlei Wu, Xuning Zhang, Zhenxin Wang, Yaoyi Wang, Ruli Zeng, Yiwen Yu, Wenjun Liu

Miniature Optical Fiber Fabry–Pérot Interferometric Acoustic Sensor with 3D Micro-Printed Ortho-Planar Springs 1015

Shangming Liu, Peng Wang, Taige Li, A. Ping Zhang

Diode Microheaters for Scalable Actuation in Micro-Transfer Printing 1018

Jiajun Zhang, Qinhua Guo, Xiwen Liu, Yunda Wang

Spatiotemporal Encoding Based on Mott Spiking Neurons for Sound Localization 1021

Zihan Guo, Linbo Shan, Zongwei Wang, Xing Zhang, Yimao Cai, Ru Huang

Energy-Efficient Temperature-Calibration Readout Circuits with Thermal-State Sensible Sampling and Zoom Window Switch Scheme for RRAM-Based Analog Computing-In-Memory 1024

Zhuoya Chen, Zongwei Wang, Haisu Zhang, Linbo Shan, Qishen Wang, Xiyuan Tang, Yunyi Fu, Yimao Cai, Ru Huang

S-Band Internally Matched GaN Power Amplifiers 1027

Li Zhang, Xuefeng Zheng, Zheng Chen, Changcheng Zhang, Zhida Wu, Zhipeng Ren, Pengbo Du, Hanbin Qu

The Impact of DC Stress on the Recoverable Tetragonal-To-Orthorhombic Phase Transition in Hafnium Zirconium Oxide Capacitors 1030

Zhenyu Chen, Dan Lv, Zhiyu Lin, Dongdong Li, Mengwei Si

A Particle Swarm Optimization Algorithm Based Parameters Extraction Technique to Model Organic Light-Emitting Diode for Flexible Displays 1033

Jianhui Wanghe, Jiachen Kang, Wei Tang, Gufeng He

Modeling Dynamics-Rich Devices with the Dynamic Time Evolution Method (Invited) 1036

Yu Li, Ning Feng, Runsheng Wang, Ru Huang, Lining Zhang

Heterogeneous and Monolithic 3D (HM3D) Integration of III-V and CMOS for Next-Generation Wireless Communications 1039

Jaeyong Jeong, Yoon-Je Suh, Nahyun Rheem, Chan Jik Lee, Seong Kwang Kim, Bong Ho Kim, Joon Pyo Kim, Joonsup Shim, Minsik Park, Jeong-Taek Lim, Minkyoung Seong, Jooseok Lee, Kihyun Kim, Dae-Myeong Geum, Jongmin Kim, Woo-Suk Sul, Won-Chul Lee, Choul-Young Kim, Jongwon Lee, Sanghyeon Kim

First Demonstration of 4-Inch $\mathbf{G a N}$ on $\mathbf{S I O}_{\mathbf{2}} / \mathbf{S I}(\mathbf{1 0 0})$ Monolithic Integration Materials by Ion-Cutting Technique with Hydrophilic Wafer Bonding at Elevated Temperature 1042

Jiaxin Ding, Jialiang Sun, Tiangui You, Xin Ou

Novel Hybrid Gate Ferroelectric Transistor-Based Weight Device with High Linearity and Symmetry for On-Chip Learning 1045

Yuxin Lin, Jin Luo, Zerui Chen, Zhiyuan Fu, Qianqian Huang, Ru Huang

Oxygen Vacancy-Zr Content Synergy for Morphotropic Phase Boundary Towards High-Performance DRAM Applications 1048

Jinhao Liu, Xuepei Wang, Maokun Wu, Boyao Cui, Yichen Wen, Yishan Wu, Sheng Ye, Pengpeng Ren, Runsheng Wang, Zhigang Ji, Ru Huang

Recent Advances in Compact Modeling for Advanced Semiconductor Technology 1051

Runsheng Wang, Baokang Peng, Lining Zhang, Ru Huang

Multi-Channel Intelligent Electronic Nose for Rapid Identification of Complex Hazardous Gases 1054

Wenyuan Liu, Jiachuang Wang, Fangyu Zhao, Nan Qin, Tiger H. Tao

Opportunities for Wide and Ultrawide Bandgap Devices with Heterogenous Integration 1057

Vanjari Sai Charan, Aditya K. Bhat, A. Anjali, Zequan Chen, Matthew D. Smith, James Pomeroy, Martin Kuball

Optimization of Short-Channel Top-Gate MoS2 FETs Via a Non-Transfer Fabrication 1059

Haojie Chen, Xinliu He, Jinshu Zhang, Jiahao Wang, Sen Wang, Yuchen Tian, Saifei Gou, Xiangqi Dong, Mingrui Ao, Qicheng Sun, Zhejia Zhang, Yan Hu, Jieya Shang, Yufei Song, Yuxuan Zhu, Wenzhong Bao

A Controlled Metal Doping Method Based on MoS2 Top-Gate Transistor.. 1062

Zhejia Zhang, Jingjie Zhou, Saifei Gou, Yuxuan Zhu, Xiangqi Dong, Mingrui Ao, Qicheng Sun, Yuchen Tian, Jinshu Zhang, Yan Hu, Xinliu He, Haojie Chen, Yufei Song, Jieya Shang, Zhengjie Sun, Xiaojun Tan, Wenzhong Bao

A Controllable and CMOS Compatible Doping Process for 2D Integrated Circuits................................... 1065

Yuchen Tian, Yan Hu, Shicheng Zeng, Yuxuan Zhu, Saifei Gou, Xiangqi Dong, Zhejia Zhang, Jinshu Zhang, Qicheng Sun, Mingrui Ao, Xinliu He, Haojie Chen, Yufei Song, Jieya Shang, Zihan Xu, Chuming Sheng, Zhengzong Sun, Wenzhong Bao

Research on Oxidizer Engineering of ALD for Industrial Production of ZrO2 Capacitor in Dram 1068

Xinyi Tang, Songming Miao, Yuanbiao Li, Guangwei Xu, Di Lu, Shibing Long

In-Material Multimodal Physical Computing for Multisensory Integration... 1071

Ming He, Shuo Liu, Junling Liu, Lei Xu, Ru Huang

Biofuel Cell-Inspired Chemical Sensors for Monitoring Glutamate in Mammalian Central Nervous System 1074

Jinghua Li

Boosted Performance of Atomic-Layer-Deposited Dual-Gate Indium-Gallium-Zinc-Oxide Transistors 1077

Anyu Tong, Qianlan Hu, Min Zeng, Yuzhe Zhu, Wenjie Zhao, Zhiyu Wang, Yanqing Wu

Compact Modeling of GaN Based RF Switches: Invited Paper 1080

Yogesh Singh Chauhan, Mir Mohammad Shayoub, Ahtisham Pampori, Mohammad Sajid Nazir

A Fusion of Optical Microscopy and Functional Nanomaterials for Subcellular-Scale Thermodynamic Control of Muscle Contraction.. 1083

Madoka Suzuki

An Energy-Efficient Microwave Magnetic Field Generator for NV Center Quantum Magnetometers Based on an Array of Four Injection-Locked VCOs.. 1086

Hadi Lotfi, Qing Yang, Tarek Elrifai, Michal Kern, Jens Anders

Theoretical Design of Silicon-Based Nanostructures for Spin Qubits 1089

Yang Liu, Shan Guan, Jun-Wei Luo

Integration of 2D Ultrafast Flash Memory: From Device to Chip 1092

Zhenyuan Cao, Chunsen Liu

Robust OS-FeFETs with Crystallized Anatase-TiO2 Channel-Hafnia Ferroelectric Layer Stack for Integration of 3D Memory Applications.. 1095

J. F. Kang, X. J. Song, C. X. Yu, D. J. Sun, J. J. Zhang, S. Z. Li, X. Y. Liu

2D Novel Antiferroelectric Materials for Neuromorphic Computing 1098

Dongliang Yang, Linfeng Sun

Dual Driven Approaches for General Purposed Brain Inspired Computing... 1101

Luping Shi, Yuqing Cong, Wei Zhang

Precise Transmission Matrix Measurement of a Multimode Fiber and Its Applications 1104

Yuwen Xiong, Zihao Ma, Yi Xu, Yuwen Qin

Performance Evaluation of 6T-Sram in Sub-3 nm Complementary Fet 1107

Anirban Kar, Mahdi Benkhelifa, Yogesh Singh Chauhan, Hussam Amrouch

Experiment Investigation on La2O 3 Dipole-Last Cap-Less VFB Tuning Technology Based on Nitrogen Atmosphere..1111

Yu Wang, Jiaxin Yao, Yanzhao Wei, Qingzhu Zhang, Huaxiang Yin

A Digital Twin for Advanced Manufacturing of Materials...1114

Dipayan Sanpui, Anirban Chandra, Sukriti Manna, Henry Chan, Subramanian K. R. S. Sankaranarayanan

Strategies for Reliable Emerging Memories and Their Applications 1117

Shinhyun Choi

A Hybrid Design Method of Lamb Wave Mode Filter Based on Machine Learning and COM Model 1119

Lihang Liao, Chen Ma, Zhiyu Wang, Xiangyu Zou, Zhiyuan Wang, Xi He, Feixuan Huang, Qinghua Ren, Fengyuan Yang, Yiming Ma, Jianlin Chen, Nan Wang

Engineered Substrates for 3D RF Front Ends..1122

Luis Andia

Calculation Optimization of Double-Free-Layer Magnetic Tunnel Junction 1125

Zifeng Wang, Lang Zeng

Single Photon Devices Using Layered Materials 1128

Mayank Chhaperwal, Nithin Abraham, Kausik Majumdar

Impact of Dielectrics on Hysteresis and Bias Stress Stability in Oxide Semiconductor and 2D-Material Field-Effect Transistors..1131

Alwin Daus, Sumaiya Wahid, Qu?nh Th? Phùng, Eric Pop

Heterogeneous 3D CFET with Hybrid Channel Configuration..1134

Sanghyeon Kim, Seongkwang Kim, Hyeongrak Lim, Jaeyong Jeong, Youngkeun Park, Jaejoong Jeong, Joonpyo Kim, Bongho Kim, Daemyeong Geum, Younghyun Kim, Byung Jin Cho

Benchmarking of the BSIM-BULK for Cryo-CMOS Design 1137

Wajid Manzoor, Nisha Manzoor, Yawar Hayat Zarkob, Debashish Nandi, Aloke K. Dutta, Yogesh Singh Chauhan

Oxidation of TiN Interface and Improvement of AlN Intercalation of ZrO2 Capacitor in DRAM 1140

Songming Miao, Xinyi Tang, Yuanbiao Li, Guangwei Xu, Di Lu, Shibing Long

Heterogeneously Integrated Intelligent System for Learning at the Edge 1143

Wenju Huo, Peng Chen, Peng Lin, Gang Pan

Novel GaN Integrated Photonics for Advanced Optical Communication and Imaging 1146

Muhammad Hunain Memon, Huabin Yu, Haochen Zhang, Haiding Sun

Leveraging Mature Chip Manufacturing Techniques for Innovative Technology Development....................1149

Yunlong Li, Kai Xu, Dianyu Qi, Yishu Zhang, Ran Tao, Yongyu Wu, Dawei Gao

Ultra-Low Temperature Solution Processed Organic Thin-Film Transistor for Flexible Integration 1152

Zikang Mei, Xiaojun Guo

A Compact 256-Channel CMOS Brain Surface Recording and Stimulation Array with Soft Electrodes 1155

Lachlan Fraser, Muhammad Saif Ul Islam, Peijun Qin, Nigel Lovell, David Tsai

Hafnia-Based XP-FeRAM: A Novel High-Speed and Low-Power Cross-Point Ferroelectric Memory for Data-Intensive Applications 1158
Qianqian Huang, Shengjie Cao, Zhiyuan Fu, Ru Huang

Decoupling Polarization and Trap Charges by Direct Vmid Measurement for Insights into Dynamic Mechanisms of MFMIS-FeFET 1161
Wenpu Luo, Runteng Zhu, Hanyong Shao, Yuejia Zhou, Ru Huang, Kechao Tang

Multi-Functional Flexible Intelligent Glove for Gesture Recognition and Combustible Detection 1164
Jiachuang Wang, Fangyu Zhao, Wenyuan Liu, Nan Qin, Tiger H. Tao

Piezoelectric Micromachined Ultrasonic Transducers for Advanced Sensing Applications 1167
Bowen Sheng, Lei Zhao, Jinghan Gan, Yufeng Gao, Jiao Xia, Junhao Wang, Chenyuan Zhang, Yipeng Lu

Epitaxial Growth of Stacking Faults-Free Hexagonal Bilayer MoS2 1170
Cheol-Joo Kim

Exploring the Potential of Hafnium Oxide-Based Ferroelectric Memories for Next-Generation Storage Class Memories 1173
Sourav De

Simulation of Germanium-Tin-Based $\boldsymbol{n}^{+} / \boldsymbol{i}$-Well Dot Single-Photon Avalanche Diode for Fiber-Optic Telecommunication Networks 1176
Harshvardhan Kumar, Advaita Sinha, P. Susthitha Menon

High-Frequency Capacitance Measurement Techniques and Their Applications in Memory Technology Development 1180
Jiang Qian, Dan Lv, Lu Wang, Ruiqi Ma, Shuai Kong, Shanting Zhang, Dongdong Li, Liang Zhao

Characterizing Building Blocks for Optoelectronic Devices Based on Two-Dimensional Materials by Resonant Raman Spectroscopy 1183
Ping-Heng Tan

Nanorods-Based Memristors: Advancing Bio-Inspired System and Neuromorphic Computing 1186
Ji Eun Kim, Suk Yeop Chun, Keunho Soh, Jung Ho Yoon

Novel Three-Dimensional DRAM Cell Architectures with IGZO-Channel and Key Technologies Toward Sub-10nm and Beyond 1189
Wonsok Lee, Daewon Ha, Y. Lee, S. Yoo, M. H. Cho, K. Yoo, S. M. Lee, S. Lee, M. Terai, T. H Lee, J. H. Bae, K. J. Moon, C. Sung, M. Hong, D. G. Cho, K. Lee, S. W. Park, K. Park, B. J. Kuh, P. Yun, S. Hyun, S. J. Ahn, J. Song

Evaluation of Insulator Candidates for Nanoelectronics Based on 2D Materials 1192
Mina Bahrami, Theresia Knobloch, Pedram Khakbaz, Mohammad Rasool Davoudi, Alexander Karl, Seyed Mehdi Sattari-Esfahlan, Dominic Waldhoer, Tibor Grasser

Physics of Operation of GaN Power Devices: Modeling Device and Circuit Effects Using Mit Virtual Source GaNFET (MVSG) Model 1195
Ujwal Radhakrishna, Daiyao Xu, Kaiman Chan, Tim Merkin, Ryan Fang, Johan Alant, Lan Wei

Magnetic Resonance Based Soft Electronic Implant for Wireless Electrotherapy and Thermal Ablation 1198
Sicheng Xing, Wubin Bai

Characterization of Self-Heating Using the AC Conductance Method (Invited Paper) 1207
  A. J. Scholten, R. M. T. Pijper, T. V. Dinh

Reliability in Heterogeneous Integration: A Theoretical View (Invited) 1210
  Zhiping Xu

Fundamentals and Future Challenges of SiC Power Devices 1213
  T. Kimoto, R. Ishikawa, K. Tachiki, X. Chi, K. Mikami, M. Kaneko

How Semiconductor Industry New Challenges Will Foster Engineered Substrates? 1216
  C. Figuet

Advanced Electronic Devices Empowering Opto-Sensors for Imaging and Perception 1219
  *Wen Pan, Lai Wang, Jinpu Tang, Zhibiao Hao, Changzheng Sun, Bing Xiong, Jian Wang,
  Yanjun Han, Hongtao Li, Lin Gan, Yi Luo*

Neuromorphic Multisensory Numerosity Perception Enhanced by a Tactile Glove 1222
  *Hongwei Tan, Syed Ashraf, Sebastian Hannula, Zhong-Peng Lv, Bo Peng, Sebastiaan Van
  Dijken*

2D MoS2 Thin-Film Transistors for Large-Area, Flexible Electronics 1225
  Jong-Hyun Ahn

Spintronic Foundation Cells for Scalable Unconventional Computing 1228
  Zhihua Xiao, Qiming Shao

An Efficient Pipeline Programming Scheme Based on 40nm PCM Compute-In-Memory Chip for
CNNs 1231
  Xile Wang, Longhao Yan, Xi Li, Yaoyu Tao, Zhitang Song, Yuchao Yang

Investigation on the Effect of Self-Clocking in MEMS Gyroscope 1234
  Xuewen Liu, Zhiyuan Wang, Hongsheng Li

High Density and High Reliability (H2DR) RRAM for Advanced Memory Technology 1237
  *Zongwei Wang, Zimeng Wu, Lin Bao, Qishen Wang, Yuhang Yang, Shengyu Bao, Jingwei
  Sun, Cuimei Wang, Yimao Cai, Ru Huang*

4H-SiC Semiconductor for EV and Beyond 1240
  Hongchao Liu

The Atomic Layer Etching Technique with Low Damage for p-GaN/AlGaN/GaN Structure 1243
  Xinyi Tang, Honghao Lu, Chun Fu, Chuying Tang, Fangzhou Du, Qing Wang, Hongyu Yu

Device Considerations for GaN Power Switching Transistors from Application and Reliability
Perspectives (Invited) 1246
  *Zhikai Tang, Maik Peter Kaufmann, Sandeep Bahl, Chang Soo Suh, Jungwoo Joh, Dong
  Seup Lee, Tim Merkin, Jeffrey Morroni*

HfO2-Based Ferroelectric Field-Effect Transistors for Next-Generation Storage and In-Memory
Computing Applications 1249
  *Genquan Han, Chengji Jin, Jiajia Chen, Xiao Yu, Jiuren Zhou, Siying Zheng, Haoji Qian,
  Ran Cheng, Bing Chen, Yan Liu*

Silicon Doping in Amorphous Gallium Oxide Films by Plasma-Enhanced Atomic Layer Deposition
for Dielectric and Optoelectronic Applications 1252
  Yongjie He, Jingxuan Wei, Jining Yang, Rognxu Bai, Hao Zhu, Qingqing Sun

On the Scalability of Nanosheet Oxide Semiconductor Transistors 1255
    Masaharu Kobayashi, Kaito Hikake, Xingyu Huang, Sunghun Kim, Kota Sakai, Zhuo Li,
    Tomoko Mizutani, Takuya Saraya, Toshiro Hiramoto, Takanori Takahashi, Mutsunori
    Uenuma, Yukiharu Uraoka

Reconfigurable Magnonic Devices for Spin-Wave Manipulation on the Nanoscale 1258
    Huajun Qin

The IHP OpenPDK Initiative: The Status and Roadmap 1261
    Wladek Grabinski, Mustafa Alchalabi, Sergei Andreev, Luca Benini, Mike Brinson,
    Matthias Bucher, Anton Datsuk, Frank K. Gurkaynak, Norbert Herfurth, Krzysztof Herman,
    Sebastien Martinie, Vinayak Pachkawade, Harald Pretl, Christoph Sandner, Rene Scholz,
    Jan Taro Svejda, Frank Vater, Christian Wittke

Heterogeneous Integration of Compound Semiconductor Materials and Devices by Ion-Cutting
Technique 1264
    Tiangui You, Tian Liang, Jiaxin Ding, Jialiang Sun, Shangyu Yang, Xin Ou

Impact of Charge Trapping at Defects on the Robustness of Electronic Circuits 1267
    Michael Waltl, Bernhard Stampfer, Roberto Orio

Solution-Processed Reduced-Dimensional Cesium Lead Halide Perovskites 1270
    Gaukhar Nigmetova, Zhuldyz Yelzhanova, Hryhorii Parkhomenko, Meruyert Tilegen,
    Askhat N. Jumabekov, Tri T. Pham, Annie Ng

Knowledge Discovery from Microscopic Image Data Using an Explainable AI "Extended Free
Energy Model" 1273
    Masato Kotsugi

Ferroelectric 3D NAND Storage (Invited) 1276
    Prasanna Venkatesan, Asif Khan

2D Materials for Neuromorphic Computing Devices 1279
    Max C. Lemme, Lukas Völkel, Sofia Cruces, Jimin Lee, Yuan Fa

Insulators for Devices Based on 2D Materials 1282
    Tibor Grasser, Dominic Waldhoer, Theresia Knobloch

Heterointegrated Ga2O3-On-SiC RF MOSFETS 1285
    Xinxin Yu, Wenhui Xu, Rui Shen, Bing Qiao, Zhonghui Li, Xin Ou, Jiandong Ye

Benefits of Using High-Resistivity Substrates for RF ICs 1288
    Xunyu Li, Cesar Roda Neve, Ionut Radu, Albert Wang

2D Materials for Future Physical Computing 1291
    Feng Miao

CMOS BEOL-Compatible Three-Dimensional Heterogeneous Integration with Emerging Devices
for Advanced Information Processing Systems 1293
    Heyi Huang, Feixiong Wang, Shuang Liu, Yanqin Li, Huaxiang Yin, Xiaolei Wang, Jun
    Luo

Phase Change Memory: From Technological Challenges to Materials Science 1296
    Ruobing Wang, Yichen Song, Xilin Zhou, Zhitang Song

Advancing Emerging Device and Architecture Innovations in the AI Era 1299
    Ming Liu

Author Index

Microwave coherent storage based on the long lifetime cavity electromechanical system

Tiefu Li[1]

[1]School of Integrated Circuits, Tsinghua University, Beijing, China

Abstract

Cavity optomechanical system holds potential for applications in ultrasensitive detection and serves as a fundamental element in the development of general quantum networks. Here we experimentally present the coherent storage of microwave signal utilizing a cavity electromechanical system comprising a high-quality factor SiN membrane resonator and superconducting cavity. It demonstrates that the coherent storage time is high to 50 ms, with acquisition of less than one quantum noise during this timeframe. These results indicate that long-coherent-time phonons could serve as ideal candidates for the development of quantum memories in quantum computers.

Keywords: Cavity optomechanics, SiN membrane resonator, mechanical quantum state, microwave storage

Introduction

Cavity optomechanical systems focus on the interaction between light radiation pressure and mechanical displacement, showcasing promising applications in detection, signal processing, and the construction of hybrid systems that leverage the strengths of each subsystem. Remarkable advancements have been made by enhancing the intensity of optical fields through high-reflection cavities and high-quality factor resonators, especially with the aid of advancements in modern micro- and nanofabrication techniques. In cavity optomechanical systems, the frequency shift and dissipation rate of the mechanical resonator are regulated by the light field. Effects such as mechanical oscillator bistability [1], optical spring effect [2], and the preparation of the quantum ground state have been experimentally realized [3]. This system exhibits promising application prospects in quantum storage and quantum interface due to the beam-splitter-like interaction between the resonator and light field.

In this study, we demonstrate the coherent storage of microwave signals in the gigahertz range using a cavity electromechanical system consisting of a high-quality SiN membrane resonator and a superconducting cavity. The membrane resonator was first cooled to its quantum ground state, where the phonon occupation number is 0.79. Subsequently, through the operations of writing, storing, and restoring, we demonstrate the coherent storage of microwave signals, which is manifested in the precise reproduction of the amplitude and phase of the written signal. The experimental results show a coherent storage time of 55 milliseconds, which means that less one thermal noise phonon is thermalized during this interval. This suggests that cavity electromechanical systems with high-quality factors hold significant potential for application in quantum storage devices for quantum computers.

Experimental Setup

The experiments were conducted on the stage at a temperature of 10 mK within the dilution refrigerator (Bluefors LD400). The device mainly comprised of a SiN membrane resonator and a LC circuits plane superconducting cavity. The membrane resonator was attached to the LC superconducting cavity via the flip-chip bonding method, with separations of approximately 300 nm. Figure 1(a) shows the optical image of the device, a 3D electromechanical capacitor is established between the lower capacitor of the LC circuits and the metallized SiN membrane above it. It clearly illustrates the bonding lines utilized for signal transmission, grounding, and heat conduction. The dimensions of the SiN membrane were 500 μm $\times$ 500 μm $\times$ 50 nm. It was metallized with niobium (Nb) materials through magnetron sputtering, resulting in a circular shape with a diameter of 120 μm and a thickness of 50 nm. Figure 1(b) illustrates the capacitor configuration of the planar LC cavity, which appears as a "pac-man" electrode with a diameter of 250 μm in a large circle and 80 μm in a small circle [4].

Results and Discussion

We investigate the storage performance of a microwave signal in a cavity electromechanical device at 10 mK. First, the mechanical properties of the SiN membrane are characterized. Next, the SiN membrane resonator is prepared in its quantum ground state. Finally, the microwave signal storage is implemented on the prepared mechanical quantum ground state. The layout of the measurement lines is shown in Fig. 1(c). The arbitrary waveform generator (AWG), precision network analyzer (PNA) and the real-time signal analyzer (Tektronix RSA5126B) are employed for tone generation, transmission measurement, and spectrum acquisition, respectively.

979-8-3315-0417-5/25 $31.00 © 2025 IEEE

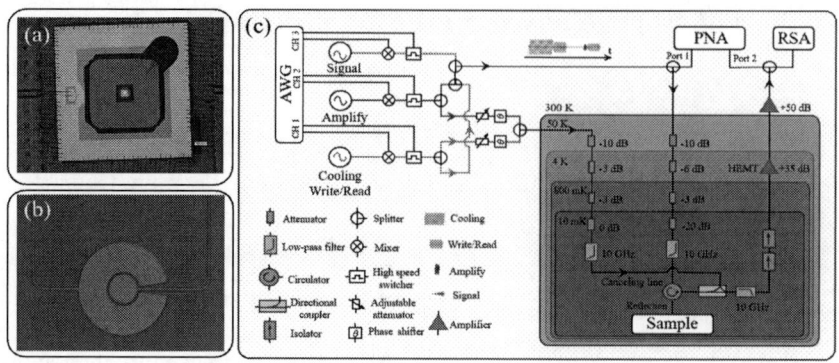

Fig. 1. (a) The optical image of the electromechanical cavity system, which is mainly comprised of a metallized SiN membrane resonator and a planar cavity of LC circuits. As a result, a 3D mechanical capacitor is formed by the metallized SiN membrane and the planar capacitor of the LC circuits beneath it. (b) Image of the "pack-man" capacitor of the LC circuits. (c) The layout of the measurement lines for this experiment.

The properties of the membrane resonator were characterized by the energy ring-down method, which was utilized by initially excited the resonator and then recording the time dependence of the vibration amplitude. Figure 2(a) shows frequency denotation used in the experiment, which includes the red sideband (ω_r), cavity frequency (ω_c), and blue sideband (ω_b), where the ω_m represents the membrane resonance angular frequency. The time sequences of tones applied in our ring-down method is depicted in Fig. 2(b), and the corresponding results are shown in Fig. 2(c). It shows that the SiN membrane resonator has the resonance frequency of $\omega_m/2\pi = 720.5$ kHz and the dissipation rate $\gamma_m/2\pi$ is as low as 10 mHz, resulting in a quality factor of $Q = 7.2 \times 10^7$. Comparing with the cavity frequency $\omega_c/2\pi = 5.315$ GHz and linewidth $\Gamma_c = 300.2$ kHz, the device works within the resolved-sideband regime.

The optomechanical coupling can be evidenced by the transmission characteristics of the cavity. It will manifest that with increasing pump power of the red sideband tone, the transmission spectrum will experience a transition from optomechanically induced absorption to optomechanically induced transparency. Accompanied by the constructive and destructive interference effects between the probe signals and the sideband signals, the phase of the transmission signal evolves a shift. Based on the diverse control methods in the cavity optomechanical system, we have also theoretically studied the unidirectional and chiral photon transfer in an multimode optomechanical system [5], and experimentally realized the hybridized combs and the anti-lasing with infinite group delay [6], [7].

By the sideband cooling, we prepared the quantum ground state of the membrane resonator. The occupancies dependence of the phonons and microwave cavity photons with the sideband cooling power is shown in Fig. 3(a). It indicates that the thermal occupancies of the phonons decrease from 112 to 0.79 as the cooling power increase from -20 dBm to 14 dBm. With the increase in cooling power, the thermal occupancies of the microwave cavity bath rise with sideband pump power. The occupation number of 0.79 means the membrane has been prepared to its quantum ground state.

The storage operation was carried out on the mechanical quantum state. A complete process of microwave signal transfer includes three steps: state capture, storage, and retrieval. It is emphasized that the signal tones are set at the resonance frequency of cavity (ω_c), the amplified tones at the blue sideband (ω_b), and the cooling, write and the read tones are set at red sideband (ω_r). The storage capabilities of the cavity electromechanical system were assessed by analyzing the spectrum after various storage durations. Experimental results show an explicit amplitudes after 80 seconds of free evolution from its quantum ground state. Via the state tomography of the mechanical oscillator, we can distinguish the coherent and thermal components of the storage signals. Figure 3(b) illustrates the variation of the coherent and thermal occupancies as a function of the storage time interval. It shows that the thermal occupancies increase with increased free evolution interval while the coherent components of the storing signal decrease over time. A coherent storage time of 55 ms has been obtained by fitting the experimental results with the theoretical model.

CONCLUSION

In conclusion, we have experimentally demonstrated the long-duration coherent storage of microwave signals based on a cavity electromechanical system. It presents that the mechanical resonator can be cooled to its quantum ground state, and the coherent storage time extends up to 55 ms, implying that less than one quantum noise can be acquired within this interval.

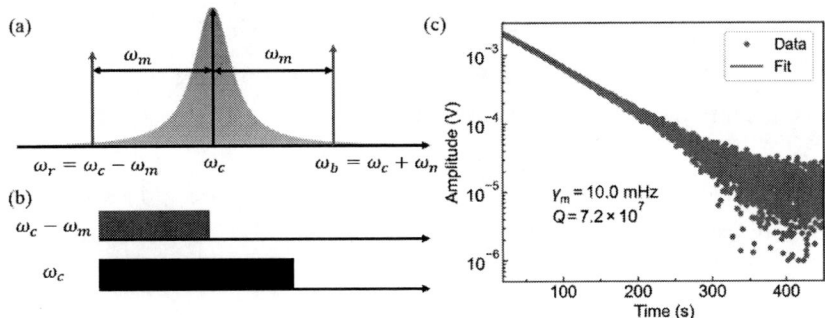

Fig. 2. (a) The frequency representation of different tones. (b) The temporal sequences of the tones used in the energy ring-down measurement. (c) Dependence of the vibration amplitude of the SiN membrane resonator on the free decay time.

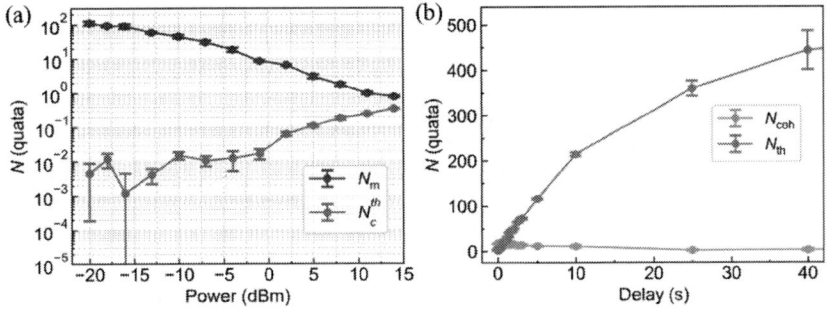

Fig. 3. (a) Variation of thermal occupancies of the phonons and microwave cavity photons with the sideband cooling power. It indicates that the quantum ground state of the membrane resonator is prepared with an occupancy number of 0.79. (b) Dependence of the coherent and thermal occupancies of the mechanical resonator on the free evolves interval.

These results indicate that long-lifetime phonons have potential applications for advancing quantum memories in quantum technology.

ACKNOWLEDGMENT

This work is supported by the Beijing Municipal Science and Technology Commission (Grant No. Z221100002722011), the National Key Research and Development Program of China (Grant No. 2022YFA1405200), and the National Natural Science Foundation of China (Grants No. 12004044, and No. 62074091).

REFERENCES

[1] A. Gozzini, F. Maccarrone, F. Mango, I. Longo, and S. Barbarino, "Light-pressure bistability at microwave frequencies," *J. Opt. Soc. Am. B*, vol. 2, no. 11, pp. 1841–1845, Nov 1985.

[2] A. Dorsel, J. D. McCullen, P. Meystre, E. Vignes, and H. Walther, "Optical bistability and mirror confinement induced by radiation pressure," *Phys. Rev. Lett.*, vol. 51, pp. 1550–1553, Oct 1983.

[3] J. D. Teufel, J. W. Harlow, C. A. Regal, and K. W. Lehnert, "Dynamical backaction of microwave fields on a nanomechanical oscillator," *Phys. Rev. Lett.*, vol. 101, p. 197203, Nov 2008.

[4] Y. Liu, Q. Liu, H. Sun, M. Chen, S. Wang, and T. Li, "Coherent memory for microwave photons based on long-lived mechanical excitations," *npj Quantum Information*, vol. 9, no. 1, p. 80, Aug 2023.

[5] Z. Chen, Q. Liu, J. Zhou, P. Zhao, H. Yu, T. Li, and Y. Liu, "Parity-dependent unidirectional and chiral photon transfer in reversed-dissipation cavity optomechanics," *Fundamental Research*, vol. 3, no. 1, pp. 21–29, 2023.

[6] S. Wu, Y. Liu, Q. Liu, S.-P. Wang, Z. Chen, and T. Li, "Hybridized frequency combs in multimode cavity electromechanical system," *Phys. Rev. Lett.*, vol. 128, p. 153901, Apr 2022.

[7] Y. Liu, Q. Liu, S. Wang, Z. Chen, M. A. Sillanpää, and T. Li, "Optomechanical anti-lasing with infinite group delay at a phase singularity," *Phys. Rev. Lett.*, vol. 127, no. 27, p. 273603, 2021.

Performance Analysis of n-type Stacked-Vertical FET for Enhanced in Advanced CMOS Applications

Yonghwan Ahn, Junjong Lee, Seunghwan Lee, Sanguk Lee, Kyeongrae Cho, Minchan Kim, and Rock-Hyun Baek*

Department of Electrical Engineering, Pohang University of Science and Technology (POSTECH),
77 Cheongam-Ro, Nam-Gu, Pohang, Gyeongbuk, 37673, Republic of Korea
Phone: +82-84-279-2220, E-mail: yonghwan05@postech.ac.kr, rh.baek@postech.ac.kr

Abstract

Novel stacked-VFET ($VFET_{ST}$) structure was proposed to overcome contact poly pitch scaling limitation. This study examined the DC/AC performance of $VFET_{ST}$ for source-drain-source (SDS) and drain-source-drain (DSD) schemes. As a result, SDS outperformed DSD because of DC coupling by the low source potential. Although RC delay was slightly degraded by 11.5% in SDS compared to VFETs with lateral connection, SDS achieved a significant 36.8% area reduction. These results demonstrate the potential for next-node logic applications.

Keywords: Vertical FET, Nanosheet FET, DC/AC, TCAD

Introduction

Nanosheet field-effect transistor (NSFET) with a three-dimensional gate structure has been successfully scaled down to the 3-nm node [1]. However, despite good gate controllability, NSFET faces the limitation of physical scaling due to performance degradation caused by the short channel effect (SCE) [2]. Vertical FET (VFET) with vertically stacked source/drain (S/D) is mentioned as a promising future CMOS because of controllable contact poly pitch (CPP) scaling with vertical direction [2], [3], [4] and small capacitance [4].

VFET also offers a reduced footprint compared to NSFET in SRAM applications. In conventional SRAM design, NSFET typically shares a source or drain, which is not feasible with VFET. We propose a novel stacked-VFET ($VFET_{ST}$) structure that enables source or drain sharing in VFETs, achieving area reduction with enhanced design flexibility. However, no research has analyzed the DC/AC performance of $VFET_{ST}$ at the device level and the DC coupling induced by their asymmetrical structure.

In this study, we comprehensively analyzed the proposed $VFET_{ST}$ as a function of S/D stacking order compared with the VFET. First, the $VFET_{ST}$ of DC/AC performance is analyzed for each upper and lower device's operation compared to 2 VFETs with lateral connection. Second, analyzed $VFET_{ST}$, which operates simultaneously. Lastly, we analyzed the DC coupling effects of $VFET_{ST}$ impact under specific operational conditions.

Device Structure and Simulation Methods

The electrical characteristics of the devices were analyzed using Sentaurus Technology Computer-Aided Design (TCAD). All devices were fully calibrated to experimental data from fabricated structures [5], employing TCAD simulation models as outlined in our previous work [6]. Fig. 1 presents the VFET and $VFET_{ST}$ structures, alongside the $VFET_{ST}$ process flow. Two different $VFET_{ST}$ S/D schemes with source-drain-source (SDS) and drain-source-drain (DSD) can be fabricated by varying the S/D stacking order from bottom to top epitaxy. DC/AC performance were evaluated using an equivalent number of vertical channels for each device.

Fig. 2 displays a top-view schematic layout of both VFET and $VFET_{ST}$. Each design includes four vertical channels with a channel width (W_{ch}) of 30 nm and a thickness (T_{ch}) of 5 nm. The CPP measures 44 nm, the L_g is 16 nm, the spacer length (L_{sp}) is 5 nm, and the S/D length (L_{sd}) is 22 nm. The other devices parameters used for simulations are indicated in Table 1. Additionally, a mole fraction of 2 % carbon enhances mobility but no conclusive evidence supports strain engineering application in the VFET [3]. Thus, we assume no mobility enhancements due to strain in the VFETs. The operation voltage (V_{DD}) is set to 0.7 V, with the metal work function adjusted to yield an off-state current (I_{on}) of 1 nA [7]. The on-state current (I_{on}) is the drain current at the V_{DD} when I_{off} is 1 nA.

Results and Discussion

A. Single VFET device performance

Fig. 3a shows the I_{on} for individual operations of both the VFET and $VFET_{ST}$. When comparing the upper and lower devices within the DSD and SDS, I_{on} consistently exceeds the single VFET, except for the lower device in the SDS. The resistance of each nFET component plays a critical role in determining I_{on}. Fig. 3b illustrates resistance components for $VFET_{ST}$. The varying dimensions of the bottom, middle, and top epitaxy layers result in differing resistances. In MOSFETs, the source that the reference for potential is more influential on I_{on} than the drain potential, thus with larger source resistance leading to a notable decrease in I_{on}. In the SDS, where a drain is shared, sources are positioned at the top and bottom, and the I_{on} of the lower device with a larger bottom source exhibits significant reduction (Fig. 3a). In contrast, the DSD shares a common source, resulting in uniform source resistance and allowing I_{on} to be primarily influenced by the remaining resistances. Since drain resistance has a lesser impact than source resistance, the difference in I_{on} between the upper and lower devices is less pronounced.

Fig. 4 shows the parasitic capacitance (C_{para}) and intrinsic capacitance (C_{int}), both of which contribute to the on-state capacitance (C_{gg_on}) in each device. C_{para} is highlighted with a red box, showing that the lower device (C_{para_L}) has a higher C_{para} than the upper device (C_{para_U}) due to the larger S/D region. Consequently, the C_{gg_on} in the VFET$_{ST}$ is increased by the presence of C_{para} from the non-operating device.

B. Analysis VFET$_{ST}$ with DC coupling

Fig. 5a presents the I_{on} versus area, where the performance of the VFET$_{ST}$ is benchmarked against 2 single VFETs, as VFET$_{ST}$ merged with two independent VFETs. The results show that 2 VFETs and the DSD exhibit nearly identical I_{on} per unit area (I_{on_A}). In contrast, SDS achieves a notably higher I_{on_A} due to its higher 7.9% I_{on} output despite occupying a smaller area than 2 VFETs. We deeply analyzed the I_{on} when the upper and lower devices operate simultaneously (Fig. 5b) for a deeper understanding of the discrepancy in I_{on} between DSD and SDS. The SDS produces significantly larger I_{on} than DSD, with each device showing a higher current. Unlike DSD, where the upper and lower devices have identical I_{on} when operate concurrently. SDS shows a larger I_{on} in the upper device, primarily due to source resistance differences discussed in Section A.

Further analysis of the I_{on} differences between DSD and SDS was conducted by computing the mobility (μ), electric field (E), and electron concentration (n) at each gate of the channel. These values, normalized to SDS, are presented in Fig. 6. Current density (J) was then calculated by multiplying μ, E, n, and the electron charge (q), with results also displayed in Fig. 6. Although the electric field is stronger in DSD, the significantly higher n in SDS dominates J, resulting in a larger I_{on} for SDS.

We also analyzed the DC coupling that occurs when the upper or lower device operates while the opposite device is operating already. Fig. 7a and 7b illustrate the I_{on} characteristic in the DSD, and Fig. 7c and 7d represent the SDS. In the DSD, one device is turned on when the other is turned on uniquely influences the other device's I_{on}. Due to the point silicide in VFETs, the contact area with the epitaxy is reduced, limiting the region and lowering the source potential, which acts as the reference potential. This effect is particularly significant in DSD since the central source is connected to both drains; a lowered source potential leads to a smaller potential difference between both drains. This results in DC coupling, which diminishes current density in the DSD, reducing I_{on} by 36.6%. Conversely, DC coupling is

not observed in the SDS. Although the source potential is similarly reduced, the two sources flow the current to a shared drain, preserving the potential difference to the drain and maintaining a low potential barrier. As illustrated in Fig. 7c, the I_{on} of the upper and lower devices crossover under specific operating conditions. This characteristic enables device designs with equal I_{on} values when applying an appropriate gate voltage.

Fig. 8 presents the RC delay and area for each device configuration. DSD experiences significant RC delay degradation when both devices operate simultaneously due to its low I_{on}. In contrast, despite a considerable increase in capacitance, the SDS benefits from a substantial increase in I_{on} with significant area reduction, resulting in the RC delay that outperforms DSD. However, the SDS also shows an 11.5% degradation over 2 VFETs. While stacking VFETs leads to an increase in capacitance, it simultaneously achieves a notable area reduction. Specifically, VFET$_{ST}$ enables a 36.8% area reduction compared to the 2 VFETs. Given the substantial area savings and enhanced DC performance in SDS, our new VFET structure will be a strong candidate for next-node logic applications.

Conclusion

This study comprehensively analyzed a VFET$_{ST}$ structure, enabling effective area reduction with high design flexibility compared to conventional VFETs. The SDS provided higher I_{on} and reduced RC delay over DSD when all the stacked devices were turned on. Especially, DSD occured DC coupling because of one source and point silicide. Furthermore, compared to two individual VFETs, SDS showed 11.5% degradation in RC delay but achieved 7.9% enhanced I_{on} with 36.8% area reductions. The VFET$_{ST}$ structure offers significant promise as a scalable and high-performance candidate for future nodes, enabling compact, flexible designs with improved current handling and RC characteristics.

Acknowledgments

This work was supported in part by the National Research Foundation of Korea (NRF)–2022R1C1C1004925 and in part by the Ministry of Trade, Industry, and Energy (20020265, and 20019450).

References

[1] G. Bae et al., IEDM Tech. Dig., pp. 28, 2018.

[2] A. V-Y Thean et al., VLSI Tech. Dig., pp. 26, 2015

[3] A. Veloso et al., IEDM Tech. Dig., pp. 11-1, 2019.

[4] H. Jagannathan et al., IEDM Tech. Dig., pp. 26.1, 2021.

[5] N. Loubet et al., VLSI Tech. Dig., pp. 230, 2017.

[6] S. Lee et al., IEEE TED, pp. 6151, 2023.

[7] J. Yoon et al., IEEE JEDS, pp. 942, 2018.

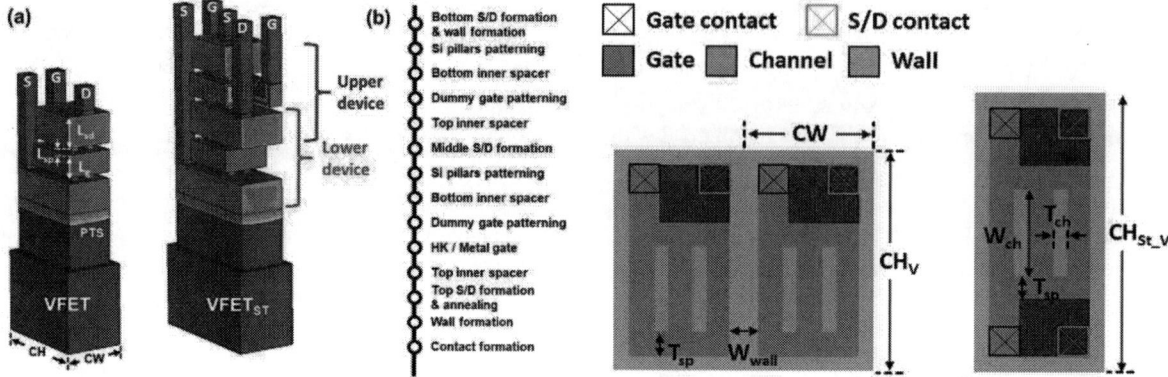

Fig. 1. (a) Difference device structures of VFET and VFET$_{ST}$. (b) The process flow to make VFET$_{ST}$.

Fig. 2. Top-view schematic layout of 2 VFETs and VFET$_{ST}$ devices. VFET$_{ST}$ has a longer CH than VFETs due to the additional gate and S/D contacts. However, the adjustable contact positions allow for design flexibility.

Table. 1. Devices parameters

Parameters	Value [nm]
Contact poly pitch (CGP)	44
Cell height (CH$_V$/CH$_{St\_V}$)	76/96
Cell width (CW)	44
Gate length (L$_g$)	16
Spacer length (L$_{sp}$)	5
S/D length (L$_{sd}$)	22
Channel thickness (T$_{ch}$)	5
Spacer thickness (T$_{ch}$)	8
Channel width (W$_{ch}$)	30
Wall width (W$_{wall}$)	10

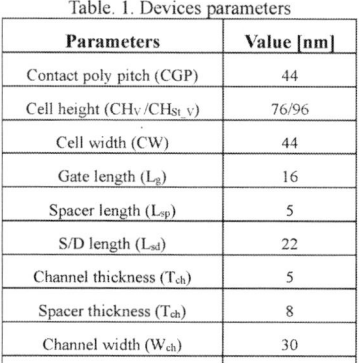

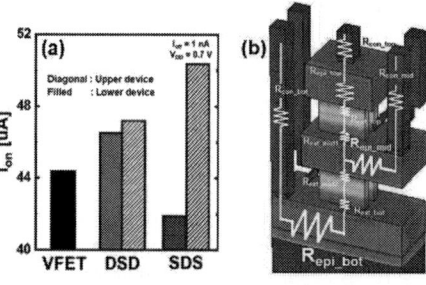

Fig. 3. (a) I$_{on}$ of each device for 1 transistor. The source resistance is the first consideration for large I$_{on}$, and the remaining resistance. (b) The resistance schematic of VFET$_{ST}$. The order of large resistance is bottom, middle, and top.

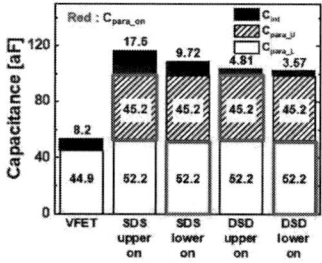

Fig. 4. Capacitance of each operation condition. C$_{para}$ of operation device is highlight with red box.

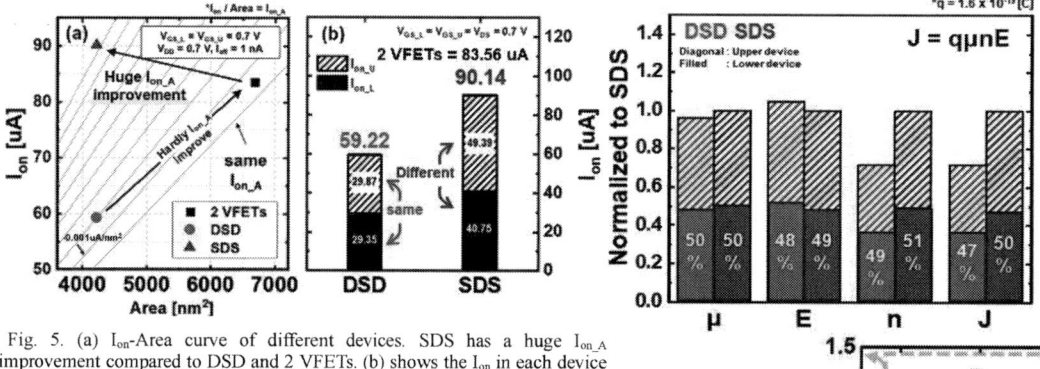

Fig. 6. μ, E, n, J which is normalized to SDS. J is proportional to μ, E, n and can be calculated by multiplying by a constant q. Although E is high in DSD, the difference in n dominantly affects J.

Fig. 5. (a) I$_{on}$-Area curve of different devices. SDS has a huge I$_{on\_A}$ improvement compared to DSD and 2 VFETs. (b) shows the I$_{on}$ in each device when the upper and lower of the VFET$_{ST}$ are operated simultaneously. The total current is reduced in the DSD, but both devices flow the same I$_{on}$. Otherwise, SDS is almost identical I$_{on}$ when operated separately.

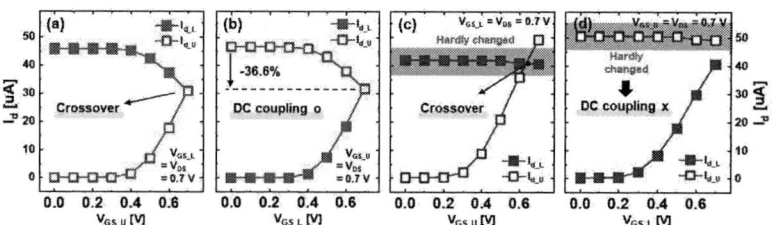

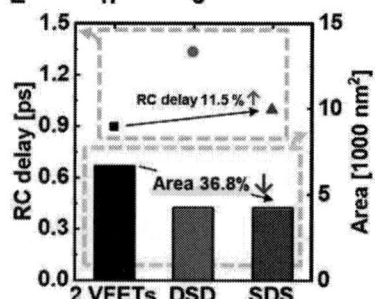

Fig. 8. RC delay and area according to each device. The SDS shows an 11.5 % increase in RC delay compared to 2 VFETs, but a ridiculous area reduction of 36.8 %.

Fig. 7. DSD (a) I$_d$-V$_{GS\_U}$ curve at V$_{GS\_L}$ = 0.7 V, (b) I$_d$-V$_{GS\_L}$ curve at V$_{GS\_U}$ = 0.7 V, SDS (c) I$_d$-V$_{GS\_U}$ curve at V$_{GS\_L}$ = 0.7 V, (d) I$_d$-V$_{GS\_L}$ curve at V$_{GS\_U}$ = 0.7 V (red is DSD, blue is SDS). DSD has DC coupling, but SDS does not.

979-8-3315-0417-5/25 $31.00 © 2025 IEEE

Threshold Switching Memristor-Based Spiking Neuron Modeling and Simulation

Pengyu Liu[1, *], Lekai Song[1, *], Kong-Pang Pun[1], *Senior Member, IEEE*, and Guohua Hu[1]

[1]Department of Electronic Engineering, The Chinese University of Hong Kong, Shatin, Hong Kong SAR, China, email: ghhu@ee.cuhk.edu.hk

Abstract

Spiking neurons are key building blocks in neuromorphic computing for information encoding into spikes. Emergent threshold switching memristors (TSMs) with unique self-rest threshold switching show great promise to facilitate compact spiking neuron designs towards VLSI. In this work, we model the TSMs in standardized Verilog-A and design compact *Leaky Integrate-and-Fire* (LIF) neurons on Cadence Virtuoso using the Verilog-A model. The neurons are proved capable of processing both digital and analog signals for spike-based neuromorphic computing.

Keywords: Threshold switching memristor, spiking neuron, neuromorphic computing

Introduction

Spike-based neuromorphic computing emerges promising for high computing efficiency. The key lies in spiking neurons that are capable of processing the information into sparse, asynchronous spikes [1]. As spiking neuron models develop, *Leaky Integrate-and-Fire* (LIF) neurons become widely used in analog-digital-hybrid very large scale integration (VLSI) circuit designs in implementing neuromorphic chips [2], given their concise design and CMOS compatibility. Figure 1a presents a widely used LIF neuron design in VLSI, termed *Axon-hillock circuit* [3]. Though promising, the spiking neurons including the *Axon-hillock* neurons with threshold switching and resetting feedback circuity often suffer from large and complex circuity and high power consumption [4].

Low-power, compact threshold switching and resetting are requested towards VLSI neuron circuit designs. As the memory technology develops, threshold switching memristors (TSMs) emerge promising for spiking neuron design owing to their unique self-rest threshold switching behavior (Fig. 1b). Different from the general memristors with memory-switching (inset in Fig. 1b), the TSMs switch on when the applied bias exceeds the threshold voltage V_{th}, and automatically switch off once the bias drops below the holding voltage V_{hold}. This self-reset threshold switching behavior fits the thresholding and resetting feedback circuity of the spiking neurons, thereby allowing for integration and compact neuron designs towards VLSI circuits. Otherwise, peripheral thresholding or resetting feedback circuity can be required for spike generation and neuron resetting.

In this work, we model the TSMs in standardized Verilog-A and design a TSM-based LIF neuron on Cadence Virtuoso with the model. The designed neuron has compact and simplified circuits to achieve low power consumption, and can generate spikes at high frequency (up to 2 MHz). We show the neurons can perform spike-based processing of both digital and analog signals for neuromorphic computing.

Spiking Neuron Design

To facilitate the integration of TSMs in VLSI neuron circuits, it is necessary to model the TSMs and study neuron design on a universal platform, such as Cadence Virtuoso. To this end, we first model the TSMs using a standardized Verilog-A language, following our previous study [5], and adapt it in the LIF neuron on Cadence Virtuoso (Fig. 1c).

As shown in Fig. 2a, based on the TSM Verilog-A modeling, our neuron specifies the spike generation in three key stages: *integrate*, *fire*, and *reset*. Specifically, as the input current I_{in} flows directly into the capacitor, the potential V_{mem} across the capacitor accumulates until V_{th}. This is the *integrate* stage and no spike is generated (Fig. 2b). When V_{mem} exceeds V_{th}, the TSM switches on and enters the low-resistance state R_{on}. The neuron at this *fire* stage fires a spike and V_{mem} gradually decreases (Fig. 2c). Due to the current flowing through the TSM, a voltage is generated on the load resistor R_L. This voltage switches on the NMOS, generates a leakage current, and accelerates the reduction of V_{mem}. The on-state NMOS at this stage acts as a resistor for current leakage. When V_{mem} is lower than V_{hold}, the TSM switches off and returns to the high-resistance state R_{off}. This is the *reset* stage. This reduces the voltage on R_L to nearly zero and causes the NMOS to switch off, thereby eliminating the leakage current. Therefore, V_{mem} returns to the *integrate* stage, awaiting the next spike generation. The neuron continuously integrates, fires, and resets as the input continues, leading to a spike train (Fig. 2d).

Spike-based Information Processing

We study our spiking neuron for processing both digital and analog signals on Cadence Virtuoso. Digital and analog signals with distinct characteristics are both widely used in current computing paradigms [6]. Specifically, digital signals with high noise immunity and compatibility are used in digital computing, while analog signals with high real-time responsiveness and signal fidelity are used in analog computing. Given the merits of the digital and analog signals, encoding the signals into sparse, asynchronous spikes and implementing spike-based neuromorphic computing using spiking neurons holds great promise [7].

In terms of digital signals, pulse trains serve as the input to our neuron. Figure 3a-c shows the spike generation under varying input intensities. We summarize the relationship

979-8-3315-0417-5/25 $31.00 © 2025 IEEE

between the output spike frequency f_{out} and the input intensity I_{in} in Fig. 3d. As shown, f_{out} increases almost linearly at lower intensities ($\leqslant 10$ μA) until saturation (>10 μA). The linear increase is due to the fact that the higher the input intensity, the faster V_{mem} accumulates. However, saturation occurs as the switching speed of the TSM and NMOS cannot keep up with the fast V_{mem} accumulation caused by a higher input intensity. The saturation can be increased by using TSM and NMOS with a higher switching speed or capacitor with a larger capacitance (so as to increase the accumulation time). Therefore, our neuron is proved capable of processing digital signals with the flexibility to adjust the electrical components for accommodating varied input frequencies.

In terms of analog signals, sinusoidal signals serve as the input to our neuron. Sinusoidal signals are a key form of information carriers in electromagnetic and audio signals, and electronic testing and calibration [8]. Similarly, Fig. 4a-c shows the spike generation under varying input intensities. The relationship between the output spike frequency f_{out} and the input intensity I_{in} is plotted in Fig. 4d. The results show that f_{out} increases almost linearly at lower intensities ($\leqslant 15$ μA) until saturation (>15 μA). Under sinusoidal signals, spikes are generated and concentrated at the positive peak region of the sinusoidal wave. This verifies the positive correlation between spike generation and input intensity. Therefore, our neuron is again proved capable of processing analog signals.

The above investigations demonstrate the capability of our neuron to process both digital and analog signals. However, to simplify VLSI neuron circuits, spiking neurons are desired to process multi-channel inputs simultaneously for spike-based information processing, mimicking *neural summation* in biology [9]. With this consideration, we investigate the neural summation capability of our neuron, as illustrated in Fig. 5a. This neuron network consists of two-layer neurons – the first-layer neurons encode digital and analog signals into spikes and then pass the spikes to the second-layer neuron. Figure 5b presents the spike generation of all the neurons involved, verifying the ability of our neurons to perform neural summation. As demonstrated, our neuron (specifically the second-layer neuron) proves the capability of summating multi-channel spikes as received. Therefore, the investigation shows that our neuron can encode multi-input signals into spikes and perform neural summation, promising for the design and implementation of VLSI neuron circuits.

Hardware Implementation
To verify the feasibility of practical implementation of our neuron, we fabricate TSMs based on hexagonal boron nitride (hBN) (Fig. 6a), following our previous study [10], and use the TSMs to realize the neuron. Figure 6b presents the cross-sectional microscopic image of a typical TSM device. We integrate the TSM into our neuron circuit (Fig. 1c) to build a spiking neuron. We study the spiking behavior the practical hardware neuron for processing pulsed and sinusoidal signals. Figure 6c and d demonstrate the experimental results where spiking is observed. Encoded by the neuron, the spikes contain and convey spatiotemporal information with the spiking frequency and intensity and as such, the spikes can be transmitted for further processing and computing. Therefore, the investigation proves the feasibility of our neuron design for implementing practical neurons, and that the neuron can process practical digital and analog signals.

Conclusion
In this work, we have designed a TSM-based LIF neuron for sparse, asynchronous spike-based information processing. The neuron with a compact circuit design owing to the unique self-reset threshold switching behavior of the TSMs proves the capability for processing digital and analog signals as well as performing neural summation of multi-inputs. Particularly, the neuron design is highly feasible for hardware implementation towards VLSI neuron circuit realization for spike-based neuromorphic computing.

Acknowledgments
GHH acknowledges support from RGC (24200521) and CUHK (4055227).

References

[1] K. Li and J. C. Príncipe, "Biologically-inspired pulse signal processing for intelligence at the edge", Front. Artif. Intell., 4, p. 568384 (2021).

[2] W. Zhang et al., "Neuro-inspired computing chips", Nat. Electron., 3, pp. 371-382 (2020).

[3] C. Mead, "Analog VLSI and neutral systems" Reading, MA: Addison-Wesley, ISBN: 0201059924 (1989).

[4] G. Indiveri et al., "Neuromorphic silicon neuron circuits", Front. Neurosci., 5, p. 73 (2011).

[5] P. Liu, L. Song, K.-P. Pun, and G. Hu, "Threshold switching memristor modelling for spiking neuron design", IEEE Electron Device Lett., 45, pp. 1649 - 1652 (2024).

[6] B. Murmann, "Mixed-signal computing for deep neural network inference", IEEE Trans. Very Large Scale Integr. (VLSI) Syst., 29, pp. 3-13 (2020).

[7] F. Ponulak and A. Kasinski, "Introduction to spiking neural networks: Information processing, learning and applications", Acta Neurobiol. Exp., 71, pp. 409-433 (2011).

[8] K. M. Cuomo and A. V. Oppenheim, "Chaotic signals and systems for communications. " in 1993 IEEE international conference on acoustics, speech, and signal processing (ICASSP), pp. 137-140, 1993.

[9] C. Molnar and J. Gair, "16.2 how neurons communicate," Concepts of Biology-1st Canadian Edition (2013).

[10] L. Song et al., "Spiking Neurons with Neural Dynamics Implemented Using Stochastic Memristors", Adv. Electron. Mater., 10, p. 2300564 (2024).

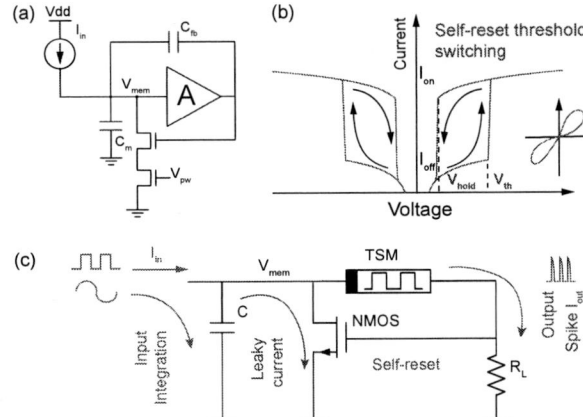

Fig. 1. (a) Axon-hillock circuit, a original silicon neuron circuit. (b) Schematic switching profile of TSMs, showing dual-self-reset threshold switching. The inset shows the memory switching of the memory switching memristors. (c) Schematic TSM-based LIF neuron circuit. The setting: V_{th}=0.5 V, V_{hold}=0.1 V, R_{on}=5 MΩ, R_{off}=1 GΩ, R_L=1 MΩ, C=1 pF, NMOS transistor: gate length of L = 180 nm and width W = 720 nm.

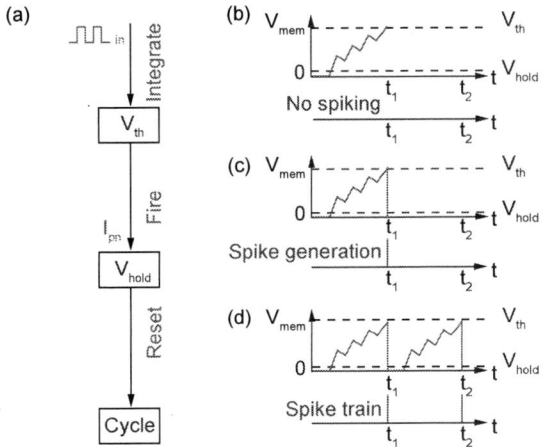

Fig. 2. Spking behavior: (a) Verilog-A modeling workflow of a TSM device in an LIF neuron. (b-d) Spike generation of the neuron. The neuron output is configured as current spiking signals.

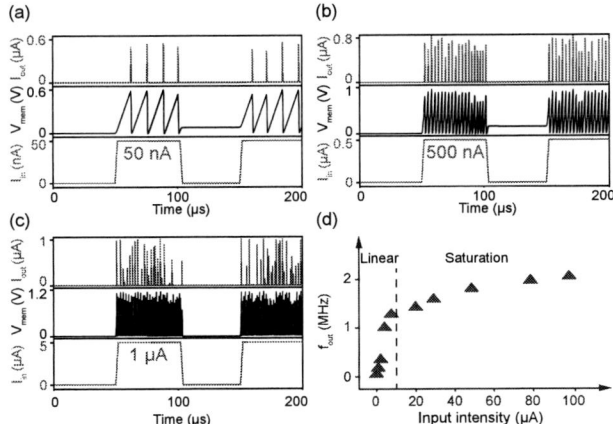

Fig. 3. Digital signal processing: (a-c) Spiking behavior of the LIF neuron at pulsed signal input with a varying current intensities at 50 nA, 500 nA and 1 µA. (d) The relation between the spiking frequency and the pulsed signal current intensity of the LIF neuron.

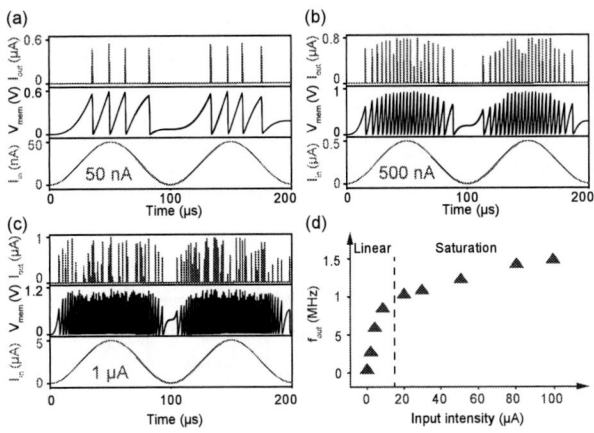

Fig. 4. Analog signal processing: (a-c) Spiking behavior of the LIF neuron at sinusoidal signal input with a varying current intensities at 50 nA, 500 nA and 1 µA. (d) The relation between the spiking frequency and the pulsed signal current intensity of the LIF neuron.

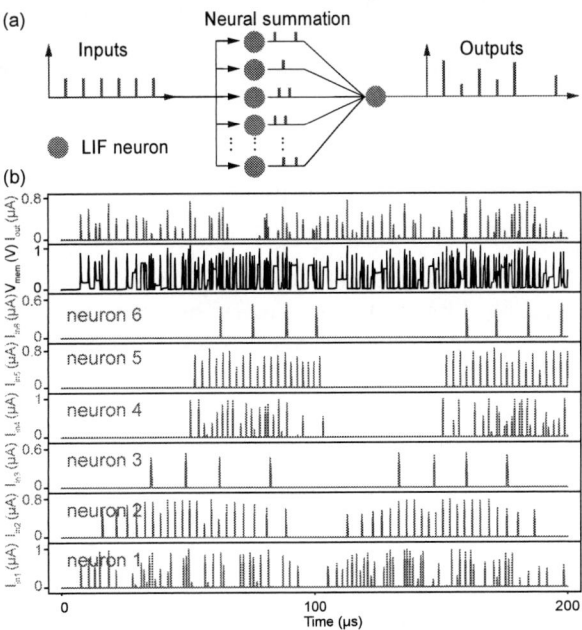

Fig. 5. Multi-input signal processing: (a) Neural summation network with 6 first-layer neurons and 1 second-layer neuron. (b) Spiking behavior of the LIF neuron at summation of spike signals.

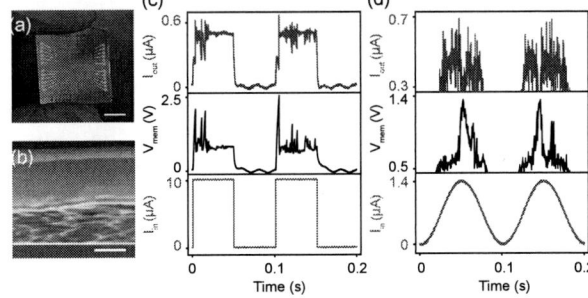

Fig. 6. Hardware implementation: (a) 12 × 12 hBN memristor array by photolithography. A typical device area is ≈20 µm × 20 µm. (b) Cross-sectional transmission electron microscopic image of a typical device. Spiking behavior of the LIF neuron at (c) pulsed signal input and (d)sinusoidal signal input. Scalebars–(a) 5 mm, and (b) 100 nm.

979-8-3315-0417-5/25 $31.00 © 2025 IEEE   657

Enhanced Gate Stack Reliability Test Framework for GaN HEMT in High-Stress Environments Leveraging on Ramp and Constant Voltage Stress Protocols

Jerry Joseph James[1,2], and Nagarajan Raghavan[1]

[1]*Engineering Product Development, Singapore University of Technology and Design (SUTD), S - 487372*
[2]*GLOBALFOUNDRIES Singapore Pte. Ltd., Singapore 738406.*

Abstract

This study proposes an enhanced reliability testing framework for GaN transistors, utilizing two distinct reliability test protocols. The first protocol, a successive multi-stage compliance ramp voltage sweep (MSC-RVS), incrementally increases the voltage to the gate breakdown threshold with varying current compliance, allowing a detailed analysis of failure progression in GaN devices. MSC-RVS enables modeling of the physics of failure mechanisms as functions of temperature and voltage through stage-specific electrical and physical failure analysis. The second protocol employs a constant voltage stress (CVS) scheme, applying fixed current compliance while varying voltage levels. CVS testing under different voltage and temperature conditions supports parameter fitting of the physics of failure models, calibration, and subsequent device lifetime predictions (extrapolation). This study, conducted at room temperature, highlights a multimodal failure mechanism within the device, revealing different regions susceptible to degradation. Three primary failure regions were identified: Schottky, Passivation, and PiN, each associated with unique failure modes: (A) Schottky-only breakdown, (B) Schottky with passivation breakdown, and (C) combined Schottky with PiN/passivation breakdown. The findings provide valuable insights into the multimodal failure pathways in p-GaN gate HEMTs, advancing the understanding of their reliability under varied stress conditions.

(*Keywords*: p-GaN, Forward Gate stress, Multi-Stage Compliance Ramp Voltage Sweep Methodology, Constant Voltage Stress, Schottky breakdown, Passivation breakdown, PiN breakdown, High electron mobility transistor (HEMT))

Introduction

Research on gate breakdown (V_{GBD}) in p-GaN HEMTs highlights mechanisms like avalanche multiplication, high electric fields in the p-GaN layer, and percolation paths in the p-GaN/AlGaN structure [1]-[5]. While these mechanisms have been speculated, the exact causes of gate breakdown remain under discussion. Traditional testing methods, such as DC sweeps and constant voltage stress, often result in catastrophic failure, making it difficult to analyze specific failure mechanisms and their sequence of occurrence. To address this, we propose two experimental approaches:

an MSC-RVS test up to gate breakdown voltage with varying current compliance, and a constant voltage stress (CVS) experiment with fixed current compliance. The combination of these two test protocols offers deeper insights into p-GaN HEMT degradation.

Experimental Details

The MSC-RVS [6] and CVS stress protocols are each structured into three phases: pre-stress, stress, and post-stress. During the pre- and post-stress phases, I_G-V_{GS} (with V_D = 0V, V_G sweeping from -4V to 4V) and I_{DSS}-V_{DS} (V_D sweeping from 0V to 200V) measurements are conducted to monitor gate leakage.

In the MSC-RVS experiment, where we tested 5 samples, an I_G-V_{GS} sweep is performed during the stress phase, ranging from -4V up to the gate breakdown voltage (V_{GBD}), while current compliance is gradually increased from 0.1mA to 10mA [6]. Using 30 samples, the CVS experiment applies a constant gate voltage with all other terminals grounded, continuously monitoring gate leakage until it reaches a 1mA compliance limit.

In our proposed enhanced reliability testing framework, MSC-RVS serves the first stage to examine failure progression and model the underlying failure mechanisms, followed by the CVS protocol for physics of failure model fitting, parameter calibration, and, ultimately, lifetime prediction (not shown). The flow chart shown in Fig.1 shows the individual advantages of each methodology and how we can potentially leverage both to build a combined reliability test framework to enhance our understanding of GaN HEMT reliability and robustness.

Results and Discussions

A. MSC-RVS Analysis

At low compliance levels, gate leakage remains stable when the gate voltage (V_G) is below the threshold voltage (V_{TH}), indicating that the PiN and passivation regions are unaffected. However, when V_G exceeds V_{TH}, an increase in gate current (I_G) occurs in the forward region where the Schottky diode is reverse-biased and the PiN region is forward-biased. Despite this, the OFF-state gate leakage (I_{GSS}) remains minimal, demonstrating the PiN region's effectiveness in blocking leakage. This behavior suggests Schottky breakdown (SCH only), where the Schottky diode becomes leakier under low compliance stress (Fig.2, BD1). At medium compliance, the I_{GS} trend mirrors that of I_G, forming a leakage path

979-8-3315-0417-5/25 $31.00 © 2025 IEEE

from the gate to the source along the passivation or through the p-GaN sidewall; also, the OFF-state gate leakage (I_{GSS}) remains unchanged, indicating intact PiN integrity (Fig.2, BD2). At high compliance levels, a significant increase in I_G-V_{GS} leakage at both negative and positive V_{GS} values, along with an increase in I_{GSS}, points to PiN breakdown (Fig.2, BD3). This is likely due to electric field crowding at the p-GaN gate edges or damage introduced during over-etching in the AlGaN stack. Similarly, at higher compliance levels, there is a higher chance of both passivation and PiN breakdown occurring simultaneously, leading to the hard breakdown of the devices.

B. CVS Analysis

A constant voltage stress (CVS) test was conducted at 10V, 10.25V, and 10.5V, as shown in Fig. 3(a). Pre- and post-stress data were categorized based on insights from MSC-RVS, with I_G (-4V, 0V, and 4V) and I_{GSS} (200V) results organized in scatter plots (Fig.4). These plots reveal three distinct failure combinations: BD1 shows increased gate leakage (I_G) in the forward region with stable OFF-state leakage (I_{GSS}), indicating a Schottky-only (SCH) failure; BD2 presents significant I_G increases at both biases with unchanged I_{GSS}, representing a Schottky + Passivation (SCH+PASS) failure; and BD3 displays elevated I_G and I_{GSS}, suggesting a Schottky + PiN/Passivation (SCH+PiN/PASS) breakdown.

Weibull analysis (Fig. 3(b)) indicated shape factors (β) of 1.2 at 10V and 1.7–1.8 at higher voltages. The absence of a single shape factor and the high concavity in the Weibull plot signal the presence of multimodal failure mechanisms in the devices under stress.

C. TCAD Simulation & Electrical Failure Analysis

We conducted TCAD simulations (Fig.5) to analyze electric field distribution in the gate region of devices under high forward gate bias, revealing four key areas with intense fields: the Schottky metal/p-GaN interface (~4-5η), gate metal/passivation (2-3η), p-GaN sidewall/passivation (~2η), and p-GaN edge/AlGaN over-etch region (~2η). These high-field regions correspond to different breakdown spots: avalanche multiplication at the SCH region, electron injection at the PASS and subsequent percolation path, and E-field crowding at the p-GaN edge and percolation at PiN edge as the main contributors to breakdown, guiding GaN HEMT optimization for improved reliability.

We conducted an electrical failure analysis (EFA) using the Thermal-Induced Voltage Alteration (TiVA) technique to evaluate faults in devices from MSC-RVS and CVS experiments. Samples with similar failure patterns (SCH + PiN/PASS) from MSC-RVS and CVS were analyzed with front-side illumination. A gate bias of 3-5V was applied to locate hot spots, primarily observed in the drain drift region (gate edge to the drain

(Fig.6)). Hot spots in the source drift region, and Schottky regions were obscured by field plates, preventing detection. A wafer backside illumination setup is needed for a more comprehensive assessment. Additionally, similar failure patterns observed in EFA indicate that both tests reveal the same failure location.

Conclusion

This study introduces a comprehensive reliability framework for GaN HEMTs, utilizing MSC-RVS and CVS protocols. The analysis of failure modes in p-GaN HEMT devices under these stress tests identified three primary failure regions: Schottky (SCH), passivation (PASS), and PiN. Observed failure combinations included SCH-only, SCH + PASS, and SCH + PiN/PASS, revealing a multimodal failure mechanism.

Statistical analysis of time-to-failure (TTF) data through Weibull modeling demonstrated significant concavity, indicative of the device's multimodal failure behavior. TCAD simulations further provided qualitative insights into electric field distribution, helping to explain the observed failure patterns. This robust framework enhances the understanding of failure mechanisms in p-GaN HEMT devices under diverse stress conditions.

Acknowledgments

The authors would like to acknowledge the support from the Industry Postgraduate Program (IPP) of GLOBALFOUNDRIES® Singapore and the Economic Development Board (EDB), Singapore under Grant No. IGIPGF2002 as well as the Research Surplus Funds from Grant No. RGSUR08.

References

[1] T.-L. Wu, D. Marcon, S. You, N. Posthuma, B. Bakeroot, S. Stoffels, M. Van Hove, G. Groeseneken, and S. Decoutere, "Forward bias Gate breakdown mechanism in enhancement-mode p-GaN Gate AlGaN/GaN high-electron mobility transistors," *IEEE Electron Device Lett.*, vol. 36, no. 10, pp. 1001–1003, Aug. 2015, doi: 10.1109/LED.2015. 2465137.

[2] M. Tapajna, O. Hilt, E. Bahat-Treidel, J. Würfl, and J. Kuzmík, "Gate reliability investigation in normally-off p-type-GaN cap/AlGaN/GaN HEMTs under forward bias stress," *IEEE Electron Device Lett.*, vol. 37, no. 4, pp. 385–388, Apr. 2016, doi: 10.1109/LED.2016.2535133.

[3] I. Rossetto, M. Meneghini, O. Hilt, E. Bahat-Treidel, C. De Santi, S. Dalcanale, J. Wuerfl, E. Zanoni, and G. Meneghesso, "Time dependent failure of GaN-on-Si power HEMTs with p-GaN Gate," *IEEE Trans. Electron Devices*, vol. 63, no. 6, pp. 2334–2339, Jun. 2016, doi: 10.1109/TED.2016.2553721.

[4] M. Meneghini *et al.*, "Degradation of GaN-HEMTs with p-GaN Gate: Dependence on temperature and on geometry," 2017 *IEEE International Reliability Physics Symposium (IRPS)*, Monterey, CA, USA, 2017, pp. 4B-5.1-4B-5.5, doi: 10.1109/IRPS.2017.7936311.

[5] A. Stockman *et al.*, "Gate Conduction Mechanisms and Lifetime Modeling of p-Gate AlGaN/GaN High-Electron-Mobility Transistors," *IEEE Trans. Electron Devices*, vol. 65, no. 12, pp. 5365–5372, Dec. 2018, doi: 10.1109/TED.2018.2877262.

[6] J. J. James and N. Raghavan, "Successive Multi-Stage Compliance - Ramp Voltage Sweep Methodology (MSC-RVS) for Understanding Evolution of Gate Breakdown Mechanisms in p-GaN HEMTs," in *IEEE Access*, doi: 10.1109/ACCESS.2024.3450472.

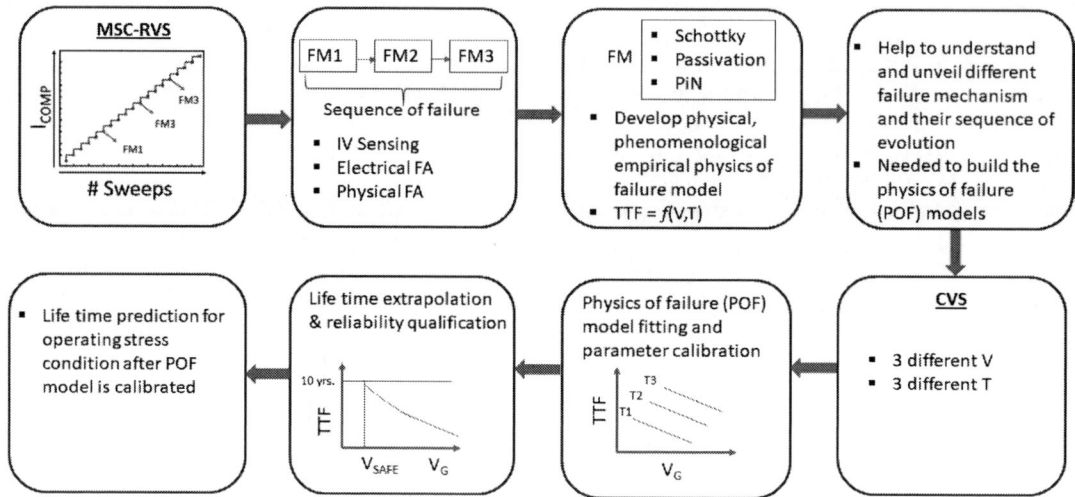

Fig. 1: Flow Chart for Enhanced Gate Stack Reliability Test Framework for GaN Transistors in High-Stress Environments.

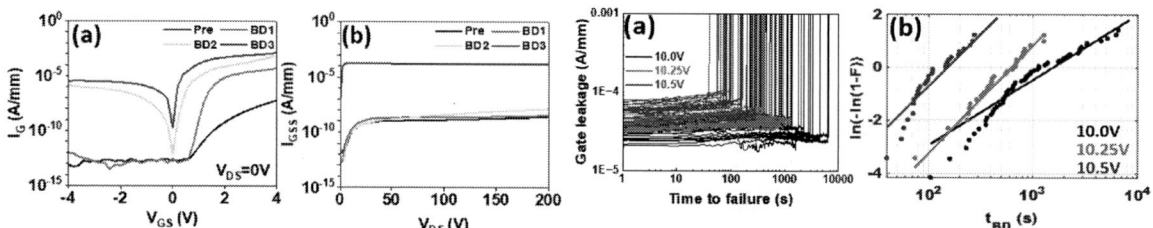

Fig. 2: - (a) I_G-V_{GS} (b) I_{GSS}-V_{DS} curves, pre-stress (black) and post-stress (red, green, blue), for typical devices during MSC-RVS. Note: I_{GSS} refers to gate leakage component from OFF state leakage data.

Fig. 3: - (a) Typical gate leakage versus stress time trends during CVS at 10V, 10.25V, and 10.5 V, at T = 25°C (b) Weibull plot of TTF distribution at 25°C for all the devices consolidated.

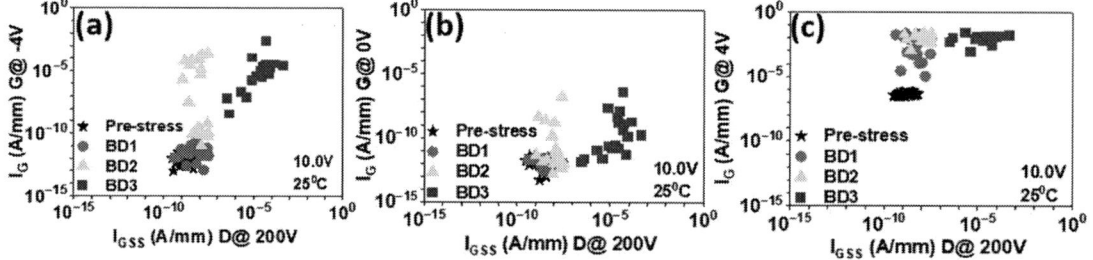

Fig. 4: Measured Pre/Post gate leakage data at 25°C for - (a) G@-4V vs. D@200V (10V stress), (b) G@0V vs. D@200V (10V stress), and (c) G@4V vs. D@200V (10V stress) for the CVS stress protocol.

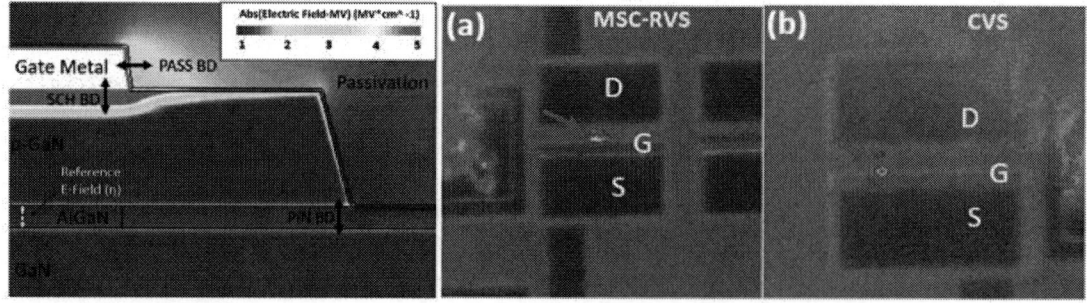

Fig. 5: TCAD E-Field analysis at $V_G < V_{GBD}$ for SCH BD, PASS BD and PiN BD region.

Fig. 6: Failure analysis data from TiVA for SCH+PiN/PASS (a) MSC-RVS, and (b) CVS indicating similar region of failure for hard breakdown.

Energy Efficient Ground Enhanced Scheme for Large-Scale SOT-MRAM Arrays

Wen Wang[1], Haoran Du[1], Shuyu Wang[1], Bo Liu[1], and Hao Cai[1]

[1]School of Integrated Circuits, Southeast University

ABSTRACT

With the development of the emerging edge computing, Spin Orbit Torque Magnetic Random Access Memory (SOT-MRAM) becomes a promising choice to replace the traditional memories. SOT-MRAM faces challenges like IR-drop and power consumption in large-scale arrays. This work proposes a ground enhanced scheme to improve write/read operation voltage at far-side bit-cells. Simulation results demonstrate successful magneto-state switching within 2ns in 512×512 array using nominal 1.2V supply voltage, write energy dissipation can be reduced by maximum to 57.5% and sensing margin is enhanced by 20%.

Keywords: SOT-MRAM, far-side, write power reduction

INTRODUCTION

With the scaling down of process technology the static power consumption of traditional volatile memory technologies such as SRAM and DRAM has become significant [1], therefore the non-volatile memory technologies enabling system to retain data are proposed. Among them, MRAM exhibits pronounced advantages as a promising alternative for caches [2, 3]. In comparison with Spin Transfer Torque MRAM (STT-MRAM) which suffers higher potential error rates and endurance risk due to its shared read/write path [4], SOT-MRAM benefits from decoupled paths. Write current does not pass through the MTJ (see Fig.1(a)), enhancing durability and enabling memory access at ns or sub-ns writing latency [5]. However, SOT-MRAM faces challenges as array size increases, includes: (1) process scaling and larger array size result in increased line resistance [6]. This reduces voltage division at far-side bit-cells and makes writing more difficult, even causing switching failures. Previous dual-side pull-down circuit could not address the problem of differing line resistances between layers [7]; (2) write current differs when writing '1' or '0' due to source degeneration in universal 2T-1M bit-cell [8]. 3T-1M bit-cell is proposed but requires higher driving capability [9].

To better leverage the power and speed advantages of SOT-MRAM in large-scale arrays, this paper proposes a dual-mode ground enhanced scheme including local-VSS to reduce IR-drop caused by excessive parasitic resistance of source-line (SL) during write-0 operation and negative voltage to increase the operating voltage

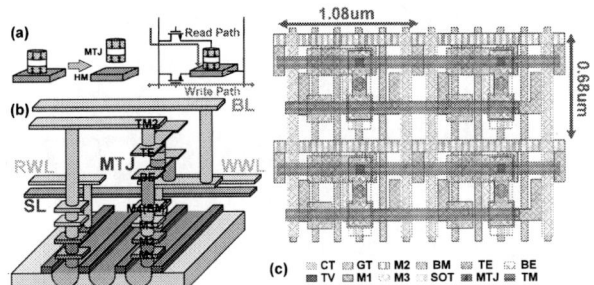

Fig. 1. (a) Schematic and (b) 3D view of SOT-MRAM and conventional 2T-1M bit-cell; (c) planar layout of 2×2 SOT-MRAM bit-cells.

during write-1 process. The design is implemented with a 40-nm CMOS process and a MTJ compact model.

PROPOSED CIRCUIT

A. Pull-Down Circuit for Far-side SL

The overview of the 2T-1M bit-cell is illustrated in Fig.1(b), where low metal layers (M1/M2) are used for the SL. Both metal layers exhibit relatively high sheet resistance. SOT-MRAM typically has a larger bit-cell area over $1\mu m^2$, in Fig.1(c). In 512×512 array, SL extends to $\sim 600\mu m$ with parasitic resistance of approximately 800Ω comparable to resistance of Heavy Metal (HM). This leads increased IR-drop, making writing operations difficult for far-side bit-cells. The parasitic resistance can be estimated using Eq. 1.

$$R_{SL} = \left(\frac{W_{bitcell}}{L_{metal}}\right) \times N_{array} \times (\rho_{M1} \parallel \rho_{M2}) \qquad (1)$$

In conventional single transmission gate (TG) write scheme, during write-0 operation, the voltage on SL at far-end which should be grounded still retains at 0.34V, as shown in Fig.2(a). Pull-down circuit is designed as local-VSS at the far-side of SL (see Fig.2(d)). This optimization can significantly shorten the write-0 path for far-side bit-cells, eliminating IR-drop and power loss caused by parasitic resistance of SL while also reducing the voltage division across TG. Notice that a 0.31V additional margin of the voltage on SL is guaranteed for successful write-0 operation. The layout with pull-down NMOS in 512×512 array is shown in Fig.3.

B. Negative Voltage Generation (NEG) Circuit

Due to the presence of MTJ, SOT-MRAM involves multiple metal layers, TM2 is used for bit line (BL) whose sheet resistance is nearly ten times lower than

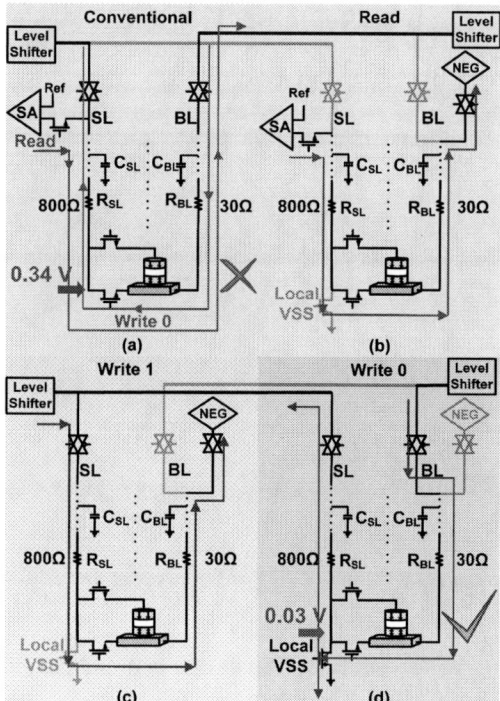

Fig. 2. Simplified schematic of (a) conventional write and read path; (b) proposed read path; (c) proposed write-1 and (d) write-0 path for columns of SOT-MRAM array.

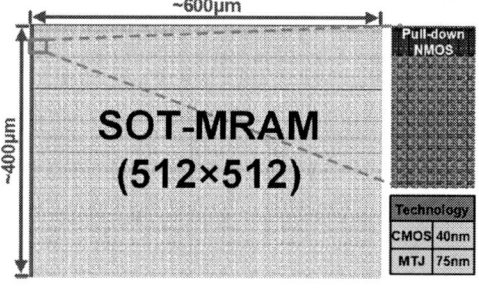

Fig. 3. 512×512 SOT-MRAM array layout with pull-down NMOS.

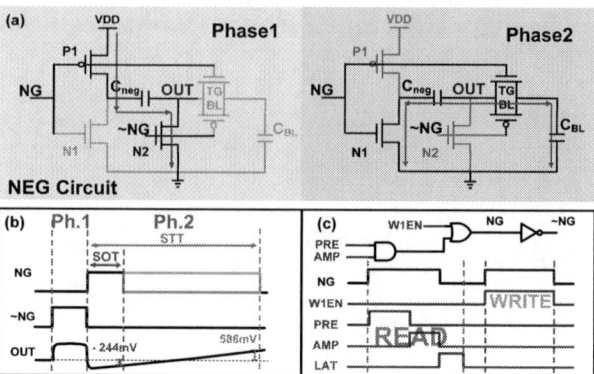

Fig. 4. (a) Schematic of NEG circuit; (b) waveform compared between STT and SOT; (c) timing diagram of signals of NEG circuit.

performed, the TG on BL selecting lower voltage is disabled, allowing charge maintain continuously without additional current. When the corresponding operation begins, the circuit enter Phase2. C_{neg} discharges and provides negative voltage. The voltage of OUT is determined by the ratio of parasitic capacitance C_{BL} and C_{neg}. Benefiting from faster write speed compared to STT-MRAM, the OUT consistently supply negative voltage throughout write-1 operation for SOT-MRAM (see Fig.4(b)). The relationship between control signals and operations for NEG circuit is illustrated in Fig.4(c).

C. Writing Path Transistor Sizing Scheme

As digital logic voltage fails to meet the requirements for writing, level shifter (LS) circuit is introduced to provide operating voltage, as shown in Fig.5(a) . During write-1 operation, supply voltage from the same voltage domain facilitates the access transistor A1 enter into saturation. To satisfy the write requirements, the saturation current must exceed the threshold switching current I_C. Conversely, in write-0 operation A1 remains in linear region and functions as a variable resistor, with

that of M1/M2 for SL. Dual-side pull-down circuit during write-1 operation reduces the division by only about 30Ω, which is insufficient to achieve P→AP switching. Therefore, we propose NEG circuit for BL, raising operating voltage without increasing driving capability. Moreover, NEG circuit also assists read operations. OCCS-SA used in this study operates in three phases: PRE, AMP, and LAT [10]. Introducing negative voltage during PRE and AMP phases could provides larger current difference $\triangle I_{SA}$. The read/write path is shown in Fig. 2(b)-(d).

The NEG circuit proposed can respond quickly without pre-charge operation, as shown in Fig.4(a). Upon initial power-up, the circuit enters Phase1. P1 and N2 are activated by NG and ∼NG respectively, forming a loop to charge C_{neg}. When no write-1 operation is

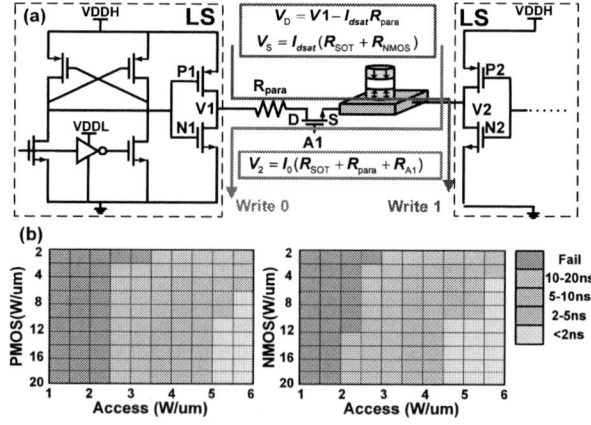

Fig. 5. (a) Level Shifter circuit and related node voltages; (b) PMOS/NMOS in LS versus Access transistor sizing in writing path.

979-8-3315-0417-5/25 $31.00 © 2025 IEEE

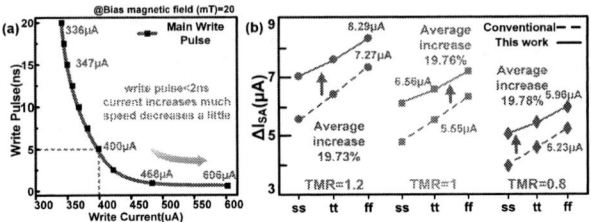

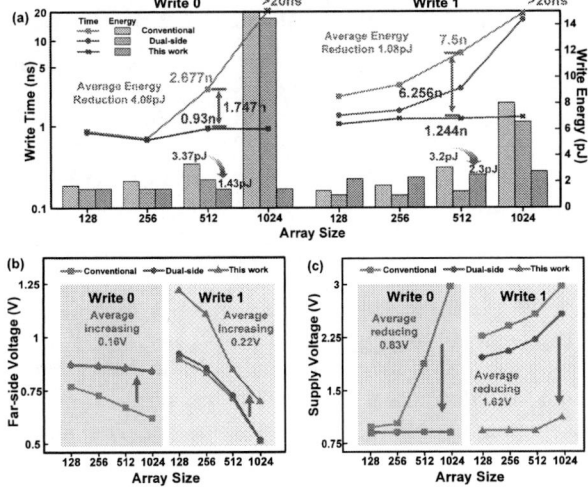

Fig. 6. (a) Relationship between write pulse length and write current; (b) comparison of read current margin under different process corners and TMR.

driving capability predominantly determined by the size of PMOS within LS, as demonstrated in Eq. 2.

$$
\begin{cases}
V_{\text{GS(A1)}} - V_{\text{TH}} \leq V_{\text{D}} - V_{\text{S}} \\
I_{\text{dsat}} = \frac{k}{2}\left(\frac{W}{L}\right)_{A1}(V_G - V_S - V_{\text{TH}})^2 \geq I_C \\
k\left(\frac{W}{L}\right)_{P1}\left(\frac{V_{\text{DDH}}}{2} - V_{\text{TH}} + \frac{V2}{2}\right)(V_{\text{DDH}} - V2) \geq I_C
\end{cases}
\tag{2}
$$

By analyzing the relationships between node voltages and current in operational regions, an improved transistor sizing scheme of write path is proposed to enhance writing speed.

SIMULATION RESULTS

The relationship between write speed and current of SOT-MRAM is shown in Fig.6(a). The improvement of read margin with average current margin increases of approximately 20% is shown in Fig.6(b). Fig.7(a) illustrates comparison of write speed and energy for far-side bit-cell across different array sizes. Write pulse length used for calculating the write power consumption is selected from the longest duration between the write-1 and write-0 operations. As the array size increases, the performance improvement for far-side bit-cells provided by the proposed circuit becomes more significant. The comparisons of operating voltages and supply voltages required under 2ns write pulse are presented in Fig.7(b) and (c), respectively. In 1024×1024 array, traditional solution fails to achieve successful switch within 20ns while this work realizes the write switching operation within 2ns, with almost minimal increase in power consumption compared to smaller arrays. Correspondingly, improving voltage division between bit-cell based on intrinsic resistance of HM increases switching speed.

Table I summarizes the performance comparison of related write operations of MRAM. Although [11] exhibits better performance, which only evaluates a single bit-cell outside an array, while our work includes a comprehensive analysis of array-level effects. Compared to [12] and [13], the proposed scheme demonstrates competitive performance in terms of write speed and power efficiency.

CONCLUSION

In this work, we proposed a ground enhanced scheme for SOT-MRAM to compensate the effect of parasitic resistance. Simulations show successful switching

Fig. 7. Comparison under different array sizes of (a) write speed and energy (b) operating voltage and (c) supply voltage at 2ns write pulse for far-side bit-cell.

TABLE I: Comparison With Previous Works

	[10]	[11]	[12]	Proposed
Technology	/	28nm	65nm	40nm
MRAM Type	SOT	STT	SOT	SOT
eCD	50nm	40nm	/	75nm
Single Device	Yes	No	No	No
Array Size	/	1024	/	512
Write Voltage	0.65V	1V	1V	1.2V
Write Pulse	1ns	10ns	10ns	2ns
Energy	~1.2pJ	~3pJ	~2pJ	1.43pJ

within 2ns without increasing supply voltage, reduction in write power by maximum to 57.5% and 20% read margin enhanced. The proposed method contributes to improve write energy utilization and efficiency in large-scale arrays.

REFERENCES

[1] S. Salahuddin et al., Nature electronics, vol. 1, no. 8, pp. 442–450, 2018.

[2] A. Hirohata, Nature Electronics, vol. 5, no. 12, pp. 832–833, 2022.

[3] Q. Shao et al., in IEEE IEDM, 2018, pp. 36.3.1–36.3.4.

[4] J. J. Kan et al., IEEE TED, vol. 64, no. 9, pp. 3639–3646, 2017.

[5] K. Garello et al., in IEEE VLSI. IEEE, 2018, pp. 81–82.

[6] P. Kumar et al., IEEE Journal on Exploratory Solid-State Computational Devices and Circuits, vol. PP, pp. 1–1, 01 2024.

[7] K. F. Lin et al., in IEEE ISSCC, vol. 67. IEEE, 2024, pp. 292–294.

[8] G. Kang et al., IEEE T-VLSI, vol. 27, no. 6, pp. 1343–1352, 2019.

[9] R. Bishnoi et al., in IEEE GLSVLSI, 2016, pp. 409–414.

[10] T. Na et al., IEEE JSSC, vol. 52, no. 2, pp. 496–504, 2016.

[11] S. Couet et al., in IEEE VLSI. IEEE, 2021, pp. 1–2.

[12] G. Patrigeon et al., IEEE Access, vol. 7, pp. 58 085–58 093, 2019.

[13] T. Kim et al., in IEEE ISCAS. IEEE, 2020, pp. 1–5.

Modeling and Simulation of Prepulse and Pulse Front for 4H-SiC Drift Step Recovery Diode

D. Y. Guo[1], Y. Zhou[1,2], J. K. Guo[1], X. Y. Tang[1], L. J. Sun[1], Y. M. Zhang[1], and Q. W. Song[1]

[1]School of Microelectronics, Xidian University, [2]Xidian-Wuhu Research Institute

ABSTRACT

This paper presents the physics-based model to capture the prepulse and pulse front of the 4H-SiC Drift Step Recovery Diode output pulse. Mathematical equations are developed to describe the physical process of prepulse and pulse front formation. The analytical model is obtained by solving the mathematical equations based on reasonable assumptions. Simulation shows that the model can be used to predict the output pulse shape when the DSRD drive parameters and structural parameters are determined.

Keywords: Pulse Power Devices, Silicon Carbide, and Nanosecond Pulse Generation

INTRODUCTION

Pulsed power technology plays a pivotal role in various fields, including environmental protection, scientific research, industrial production, and biomedical applications. At the core of this technology are pulsed power switches, which have evolved towards semiconductor plasma switches due to their notable advantages, such as high reliability and rapid switching speeds. Among these, the Drift Step Recovery Diode (DSRD) has emerged as one of the most prominent plasma switches, favored for its fast switching capabilities, high efficiency, and straightforward fabrication process [1], [2].

4H-SiC material offers distinct advantages in power device applications, including a high critical breakdown field strength, elevated carrier saturation drift velocity, and superior thermal conductivity. The 4H-SiC DSRD effectively combines these benefits, making it a promising candidate in the realm of pulsed power applications [3]–[5]. However, the approach to matching circuits with 4H-SiC DSRD has been largely empirical due to the lack of reliable models.

In this paper, we establish differential equations to accurately describe the physical operation processes of the 4H-SiC DSRD, including the formation of the prepulse region and the device turn-off under both Non-Punched-Through (NPT) and Punched-Through (PT) conditions. By solving these differential equations, we derive an analytical model that captures the output pulse characteristics of the DSRD. TCAD simulations were performed and the simulation results were compared with the model results. This model aims to alleviate the challenges associated with DSRD and drive circuit

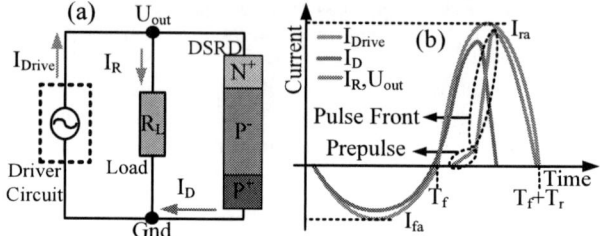

Fig. 1. (a) Simplified DSRD operating circuit and (b) 4H-SiC DSRD pulse output procedure.

matching, providing a foundational framework for the development of a DSRD SPICE model.

MODEL FORMULATION

The working mechanism of the 4H-SiC DSRD is analyzed and the prepulse and front of its output pulses are modeled.

A. Modeling Foundation

The simplified DSRD operating circuit is shown in Fig.1 (a). The driver circuit, load and DSRD are connected in parallel with each other. The DSRD operating process is divided into three stages. In the first stage, the driver circuit supplies forward current to the DSRD to inject excess carriers inside its base region. This stage is similar to the forward operation process of a common diode. In the second stage the drive circuit current (I_{drive}) is reversed and the excess carriers injected into the base region are extracted. Due to the existing of excess carriers this stage the DSRD is still in conduction. As shown in the Fig.1 (b). the output pulse voltage (U_{out}) is slightly raised during this process, this raise is called prepulse. In the third stage, the equilibrium carriers within the DSRD base region are extracted and the DSRD is rapidly turned off. The major portion of the drive current (I_{drive}) is transferred from the DSRD current (I_D) to the load current (I_R). The load voltage (U_{out}) rises rapidly and this rise is called the pulse front.

B. Modeling of Prepulse

The prepulse region is a common part of the DSRD output pulse, and the Fig.2 (a) shows how it is formed. When the electrons within the DSRD base region are extracted, the hole concentration returns to the equilibrium state, which is manifested as a quasi-neutral state. Here the region of quasi-neutral state is called quasi-neutral region. When the conduction current ($I_{Conduct}$) that the quasi-neutral region can withstand does not satisfy the value of the external extraction current (I_{Drive}), a displacement current ($I_{Displace}$) is induced at the ends of the quasi-neutral region. Displacement currents will charge in the quasi-neutral region, thus raising the electric field in the quasi-neutral region resulting in an increase in the DSRD voltage. Since the DSRD is in series with the load resistor, the rising voltage causes the current on the load to rise, shunting the drive current. As shown in the Fig.2 (b), the platform region establishment process can be equated to the charging process of a capacitor and two resistors connected in parallel. The drive current flows in three parts; the first part flows from the quasi-neutral region; the second part is used for quasi-neutral region charging; and the third part flows from the load resistance. The prepulse voltage of the 4H-SiC DSRD is usually low, so the load current caused by it is ignored.

The physical process can be described by the following equation:

$$I_{Drive} - I_{Conduct} = C_{Pre}(t)\frac{dU_{pp}}{dt} \quad (1)$$

where C_{Pre} is the quasi-neutral region equivalent capacitance and U_{pp} is the prepulse voltage. I_{Drive}, I_{drive}, and I_R can be expressed as:

$$\begin{aligned} I_{Drive} &= K_c t \\ I_{Conduct} &= \eta N_b q S v_p \\ I_R &= \frac{U_{pp}}{R} \end{aligned} \quad (2)$$

where K_c is the rate of rise of the extraction current, η is the ionization rate obtained from the literature [6], N_b is the doping concentration in the base region, q is the electron charge, S is the area of the active region of the device, and v_p is the average drift velocity of holes in the quasi-neutral region.

The equivalent capacitance in the quasi-neutral region can be expressed as:

$$C_{Pre}(t) = \frac{\varepsilon_r \varepsilon_0 S}{d} \quad (3)$$

where ε_r is the relative permittivity, ε_0 is the vacuum permittivity, d is the quasi-neutral region width, and d is bounded by the following equation:

$$dSN_p q = \int I_{Drive} dt \quad (4)$$

where N_p is the concentration of electrons injected into the base region. In order that an analytical solution can be obtained, it is assumed that all of the drive current during prepulse formation is being used to extract excess electrons.

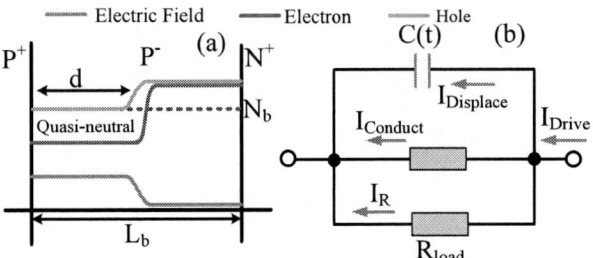

Fig. 2. (a) 4H-SiC DSRD prepulse voltage build-up process and (b) 4H-SiC DSRD prepulse stage equivalent circuit.

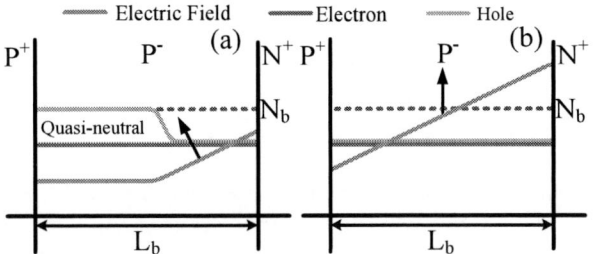

Fig. 3. Process of electric field establishment in the base region during DSRD turn-off. (a) NPT phase and (b) PT phase

v_p can be estimated by the following equation:

$$v_p = \frac{\mu_h U_{pp}}{d} \quad (5)$$

where μ_h is the mobility of the hole in the quasi-neutral region.

Solving for Equations (1), (2), (3), (4), and (5) yields:

$$U_{pp}(t) = \frac{Bt^3}{A} - \frac{Bt^2}{A^2} + \frac{6Bt}{A^3} - \frac{6B}{A^4} + \frac{6B}{A^4}e^{-At} \quad (6)$$

where A and B are:

$$\begin{aligned} A &= \frac{\mu_h \eta N_b q}{\varepsilon_r \varepsilon_0} \\ B &= \frac{K_c^2}{2N_p q S^2 \varepsilon_r \varepsilon_0} \end{aligned} \quad (7)$$

C. Modeling of Pulse Front

The physical processes during the pulse front stage are the extraction of equilibrium holes, the expansion of the depletion region, and the increase of the device voltage. Current is shifted from the DSRD to the load. Due to the short duration of the pulse front stage, the drive current I_{Drive} is assumed to be constant during this process. As shown in Fig.3, the pulse front forming process needs to be discussed in the NPT phase and the PT phase.

1) Non-Punch-Through:
The output voltage and extraction current of the NPT phase satisfy the following relationship.

$$U_{out} = \frac{Q^2}{2qN_b\varepsilon_r\varepsilon_0 S^2} = \frac{\left(\int (I_{Drive} - U_{out}/R_L)dt\right)^2}{2qN_b\varepsilon_r\varepsilon_0 S^2} \quad (8)$$

The solution to the above equation is:

$$U_{out} = \frac{\left(-\frac{tU_{out}}{R_L} + I_{Drive}t\right)^2}{2N_b\varepsilon_r\varepsilon_0 qS} \quad (9)$$

2) Punch-Through:

The output voltage and extraction current of the PT phase satisfy the following relationship.

$$U_{out} = \frac{\int \left(I_{Drive} - U_{out}/R_L \right) dt L_b}{\varepsilon_r \varepsilon_0 S} - \frac{L_b{}^2 N_b q}{2\varepsilon_r \varepsilon_0} \quad (10)$$

The solution to the above equation is:

$$U_{out} = \frac{R_L}{\varepsilon_r \varepsilon_0 S R_L + L_b t} \left(I_{Drive} \cdot L_b \cdot t - \frac{L_b{}^2 N_b q S}{2} \right) \quad (11)$$

The peak of the output voltage (U_{out}) is taken to be less than the DSRD breakdown voltage as well as the product of I_{Drive} and R_L.

SIMULATION CONDITION

The circuit used for the simulation is shown in Fig. 1 (a), where a current source is used as the drive signal for DSRD operation. The waveform of the drive current is shown in Fig. 1 (b), and two sinusoidal signals are used for DSRD current injection and extraction. I_{fa} and T_f denote the forward injection current amplitude and duration, and I_{ra} and T_r denote the reverse draw current amplitude and duration. The software used for the simulation is Synopsys Sentaurus TCAD. High field velocity saturation, SRH recombination, avalanche breakdown, and incomplete ionization models are turned on in the simulations. The devices used for simulation are $P^+P^-N^+$ structures.

RESULTS AND DISCUSSION

The modeling and simulation results for Device-I and Device-II are shown in Fig. 4, where the modeled and simulated pulse waveforms are basically the same. Both reflect that the prepulse decreases and the pulse front increases when the device area increases. As shown in the Fig. 5 and 6, the proposed model can reflect the influence of the device base region parameters as well as the driving parameters on the output pulse. These impact trends are consistent with the simulation results. Both simulation and modeling show that the prepulse and pulse amplitude increase with the length of the base region as well as with the increase of the reverse draw current amplitude when other parameters are fixed.

CONCLUSION

The prepulse and pulse front of the output pulse are modeled based on the working mechanism of 4H-SiC DSRD. The output pulses described by this model are highly consistent with those obtained from TCAD simulations. In addition, the model can accurately reflect the influence of driving conditions as well as device structural parameters on the output voltage pulse.

ACKNOWLEDGMENT

This work was supported by the National Natural Science Foundation of China (Grant No.62174123) and the Natural Science Basic Research Program of Shaanxi (Program No.2024JC-YBQN-0647).

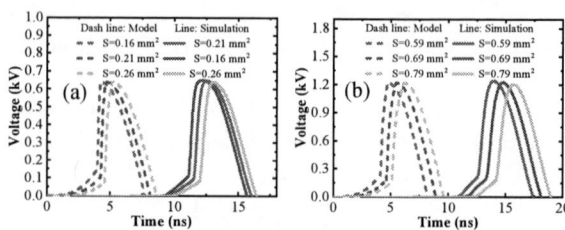

Fig. 4. Comparison of modeled and simulated output voltage pulses for different devices area. (a) Device-I L_b=6 μm, N_b=2×10^{16} cm^{-3}, I_{fa} = 20 A, T_f = 20 ns, I_{ra}= 14 ns, T_r = 7 ns; (b) Device-II L_b=9 μm, N_b=7×10^{15} cm^{-3}, I_{fa} = 20 A, T_f = 20 ns, I_{ra}= 26 ns, T_r = 7 ns

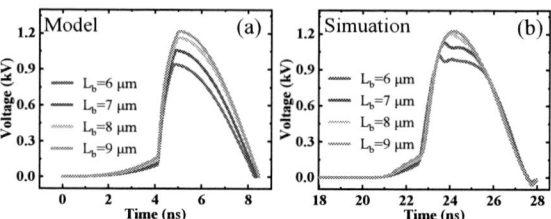

Fig. 5. Output pulses of 4H-SiC DSRD with different base region lengths. N_b = 7×10^{15} cm^{-3}, S = 0.69 mm^2, I_{fa} = 20 A, T_f = 20 ns, I_{ra}= 26 A, T_r = 7 ns. (a) Model results and (b) Simulation results.

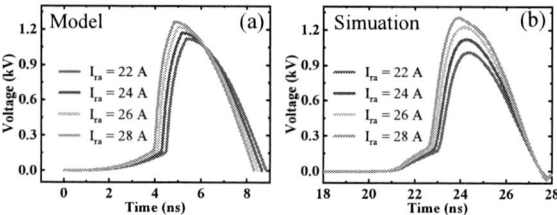

Fig. 6. Output pulses of 4H-SiC DSRD at different reverse draw current amplitudes. L_b = 9 μm, N_b = 7×10^{15} cm^{-3}, S = 0.69 mm^2, I_{fa} = 20 A, T_f = 20 ns, T_r = 7 ns. (a) Model results and (b) Simulation results.

REFERENCES

[1] B. V. Ivanov, A. A. Smirnov, S. A. Shevchenko, A. V. Afanasyev, and V. A. Ilyin, "High voltage subnanosecond silicon carbide opening switch", International Scientific Conference on Power and Electrical Engineering of Riga Technical University (RTUCON), pp. 1-4, (2016).

[2] A. G. Lyublinsky, A. F. Kardo-Sysoev, M. N. Cherenev, and M. I. Vexler, "Influence of DSRD Operation Cycle on the Output Pulse Parameters", Ieee T Power Electr, Article vol. 37, no. 6, pp. 6271-6274, 2022.

[3] P. A. Ivanov, O. I. Kon'kov, T. P. Samsonova, and A. S. Potapov, "4H-SiC Based Subnanosecond (150 ps) High-Voltage (1600 V) Current Breakers", Tech Phys Lett+, vol. 44, no. 2, pp. 87-89, 2018.

[4] R. Sun et al., "10-kV 4H-SiC Drift Step Recovery Diodes (DSRDs) for Compact High-repetition Rate Nanosecond HV Pulse Generator," International Symposium on Power Semiconductor Devices and ICs (ISPSD), pp. 62-65, (2020).

[5] X. Yan, L. Liang, Z. Yang, and H. Shang, "Investigation of Prepulse of SiC Drift Step Recovery Diode in Fast Interruption Process," Ieee T Electron Dev, pp. 1-7, 2024.

[6] C. Darmody and N. Goldsman, "Incomplete ionization in aluminum-doped 4H-silicon carbide", J Appl Phys, vol. 126, no. 14, 2019.

Van der Waals epitaxy h-BN/AlN back barrier with controllable boron-diffusion for high-erformance AlGaN/GaN HEMTs

Haidi Wu[1,2], Jing Ning[1,2]*, Jincheng Zhang[1,2], Yue Hao[1,2]

[1] State Key Laboratory of Wide-Bandgap Semiconductor Devices and Integrated Technology, Faculty of Integrated Circuit, Xidian University, Xi'an 710071, China

[2]Shaanxi Joint Key Laboratory of Graphene Faculty of Integrated Circuit, Xidian University, Xi'an 710071, China

*Corresponding author email: ningj@xidian.edu.cn

Abstract

We report a novel growth of nitride heterostructures directly on h-BN by MOCVD, achieving high-quality AlGaN/GaN HEMTs. By using aluminum ion implantation to pretreat the h-BN intermediate layer, creating a surface rich in unsaturated bonds as nucleation sites for highly oriented nitride growth. The 2DEG concentration of h-BN/AlN back barrier AlGaN/GaN HEMT by van der Waals epitaxy reached 8.80×10^{12} cm^{-2} and the electron mobility was 2091.3 cm^2/V·s, with a high On/Off ratio of 10^{10}.

Keywords: III-Nitride epitaxy, Van der Waals epitaxy, AlGaN/GaN HEMT

Introduction

In the field of semiconductor materials, third-generation nitrides stand out due to their remarkable attributes, including wide bandgaps, rapid electron saturation velocities, and superior energy conversion efficiencies, making them ideal choices for advanced high-frequency and high-power electronic devices [1-8]. However, conventional heteroepitaxial growth on mismatched substrates poses significant challenges, such as lattice mismatch and differences in thermal expansion coefficients, leading to substantial compressive stress and increased dislocation density, which severely impact device performance.

To address these issues, various epitaxial techniques have been employed to compensate for the tensile stress during growth and cooling processes, thereby improving the crystal quality of GaN-based epitaxial layers, for instance, AlN buffer layers[9], low-temperature AlN interlayers[10], and AlGaN buffer layers[11]. In addition to conventional heteroepitaxial techniques, recent years have seen the proposal of utilizing van der Waals forces to integrate two-dimensional materials with nitride films to achieve high-quality nitride films. [12-15] Notably, these van der Waals interactions are typically one to two orders of magnitude weaker than covalent bonds, presenting a significant challenge for the precise nucleation and growth of single-crystalline Group III nitride films solely via van der Waals forces on BN substrates[16].

Against this backdrop, recent advancements have led to the emergence of quasi-van der Waals (quasi-vdW) epitaxy. Quasi-vdW epitaxy innovatively integrates covalent bonding at the interface, thereby introducing a pivotal modification. By capitalizing on the inherent surface defects of BN, this method transforms these defects into beneficial features, providing unsaturated dangling bonds as preferred nucleation sites. Consequently, the targeted manipulation of these surface defects stimulates a highly oriented nucleation mechanism, transcending the limitations imposed by the inherent weakness of van der Waals bonding. As a result, quasi-vdW epitaxy charts a novel course for advancing Group III nitride technologies on BN platforms.

Furthermore, the pre-treatment of BN films of varying thicknesses during epitaxial growth leads to the upward diffusion of boron (B). If this diffusion can be controlled within a certain concentration range, it could positively influence the material quality of the GaN epitaxial layers and the control of the two-dimensional electron gas (2DEG) in the AlGaN/GaN heterojunction channels.

Experiments and methods

In the present study, all experimental samples were synthesized on 2-inch silicon carbide (SiC) substrates utilizing metalorganic chemical vapor deposition (MOCVD). Trimethylaluminum (TMAl), trimethylgallium (TMGa), and ammonia (NH₃) served as the respective precursors for aluminum, gallium, and nitrogen, while hydrogen (H₂) functioned as the carrier gas for all organometallic sources. Prior to reactor loading, variable layers of boron nitride (BN) were wet-transferred onto the SiC substrate, followed by nitrogen plasma pretreatment.

Subsequent steps in the growth protocol entailed: the deposition of a GaN buffer layer at 1050°C and 75 Torr; the growth of a 1-nm-thick AlN nanolayer (NL) spacer at identical conditions; the formation of 25-nm-thick Al$_{0.25}$Ga$_{0.75}$N barrier layers at 1010°C and 75 Torr; and the capping with a 2-nm GaN layer at 940°C and 200 Torr. Based on the varying thicknesses of the inserted BN layers (1 BN, 3 BN, and 10 BN), the samples were designated as B, C and D, respectively. A control sample devoid of the BN insertion was also prepared and labeled as Sample A.

After RCA cleaning, the epitaxial wafers underwent

isolation etching via ICP. A dual-layer adhesion technique was used to define the ohmic contact regions with 1 μm LSD, followed by Ti/Al/Ni/Au metallization through EBE. This was then annealed at 850°C for 35 seconds in nitrogen to form robust contacts. A 120-nm SiN layer was deposited by PECVD, and the gate region was defined using e-beam lithography. ICP-RIE was employed for the gate electrode opening, followed by the evaporation of Ni/Au as the gate metal, and the delineation of the external metal completed the process.

Results and discussion

The morphology and strain in GaN thin films on samples A, B, and C were investigated using X-ray diffraction (XRD) and Raman spectroscopy. As shown in Fig. 1(a-c), Raman analysis focused on the E_2 phonon mode, revealing the strain conditions; GaN on few-layer h-BN exhibited minimal peak shift compared to strain-free GaN, indicating reduced strain. The other two samples displayed E2 peak shifts corresponding to film stresses of 0.345 GPa and 0.138 GPa, suggesting that h-BN promotes stress relaxation. As shown in Fig. 1(d-e), XRD-ω scans further evaluated the crystal quality. The (002) full width at half maximum (FWHM) values for samples A, B, and D were 205, 342, and 381 arcsec, respectively, while the (102) FWHM values were 482, 551, and 566 arcsec, respectively. The incorporation of h-BN, particularly in few-layer form, reduced the FWHM values, thereby decreasing the dislocation density—screw dislocations decreased from 1.41×10^8 cm^{-2} to 4.08×10^7 cm^{-2}, and edge dislocations decreased from 8.23×10^8 cm^{-2} to 5.96×10^8 cm^{-2}. This indicates the effectiveness of h-BN in enhancing film integrity and mitigating lattice mismatch effects.

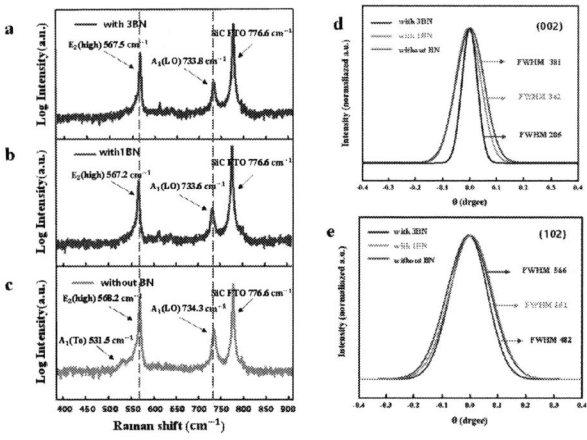

Fig. 1. Raman spectroscopy results and 2θ-HR-XRD scanning curve of the samples.

As shown in Fig. 2(a), we present a growth structure of SiC/h-BN/AlN/GaN/AlGaN, with AlGaN constituting 25% of the design. Notably, the TEM image on the left side of Fig. 2(c) reveals a dislocation annihilation depth of 171 nm within the GaN film grown on AlN. The samples grown on pre-

treated h-BN exhibit minimal basal plane dislocation extension at the AlN/GaN interface. This is attributed to vertical dislocations transforming into lateral stacking faults under quasi-van der Waals epitaxial conditions, effectively inhibiting their upward propagation. The high-resolution transmission electron microscopy (HRTEM) image and selected area electron diffraction (SAED) pattern on the right side of Fig. 2(c) depict the well-ordered lattice of GaN at different magnifications, confirming the high quality of the epitaxial GaN layer in the channel.

As shown in Fig. 2b, SIMS measurements were conducted on epitaxial heterojunctions grown on different BN layers. Comparisons reveal that the thickness of the BN layer is positively correlated with the diffusion depth and concentration. It can be observed that in the cases of 1 and 3 BN layers, B diffusion primarily concentrates within the AlN nucleation layer, with the concentration in the GaN epitaxial layer maintained at approximately 10^{15} to 10^{16} atoms/cm^3. In contrast, with 10 BN layers, the B diffusion into the channel layer reaches concentrations of 10^{18} to 10^{19} atoms/cm^3. Such high doping concentrations have a significant impact on the epitaxial quality of the layers, which is also a critical factor contributing to the degradation of the epitaxial layer quality on the 10 BN surface.

As shown in Fig. 2(d), the epitaxial layer in the region of the GaN/AlN interface where B diffusion is relatively concentrated exhibits a crystal orientation twist.

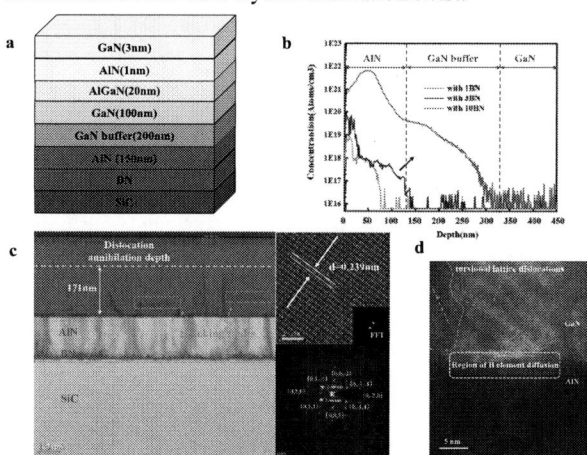

Fig. 2. TEM and SIMS characterization analysis of the HEMT layer structure.

As shown in the right-side view of Fig. 3(c), it illustrates the cross-sectional view of the device, demonstrating the alloying of the source and drain Ohmic contacts to the AlGaN barrier layer through high-temperature annealing. To evaluate the performance of the fabricated Ohmic contacts, the transmission line model (TLM) method was employed, resulting in a measured contact resistance (Rc) of 0.28 Ω-mm. The reduction in Ohmic contact resistance can be attributed to a decrease in dislocations within the GaN thin film layer, which is a result of introducing an h-BN interlayer.

As depicted in Fig. 3 (a) and (b), the output and transfer characteristics of the GaN high electron mobility transistors (HEMTs) are shown, aligning with the device simulation results, and the device exhibits an Idsat of 1396 mA/mm at VDS = 2V. The left curve in Fig. 3 (c) indicates that the HEMT devices achieve an I_{ON}/I_{OFF} ratio as high as 10^{10} by incorporating the h-BN interlayer, due to enhanced boron diffusion increasing the backside barrier of the channel, thereby improving control over the two-dimensional electron gas (2DEG) and reducing its capture into the underlying epitaxial layers. Compared to devices fabricated without the pre-processing h-BN layer under the same process conditions, all device metrics were enhanced. As shown in Fig. 3(d), the HEMT devices fabricated in this work achieved industry-leading switching ratio performance, along with superior electron transport properties. These findings highlight the effectiveness of h-BN films in enhancing the quality of nitride films and device performance via boron diffusion modulation and quasi-van der Waals epitaxy.

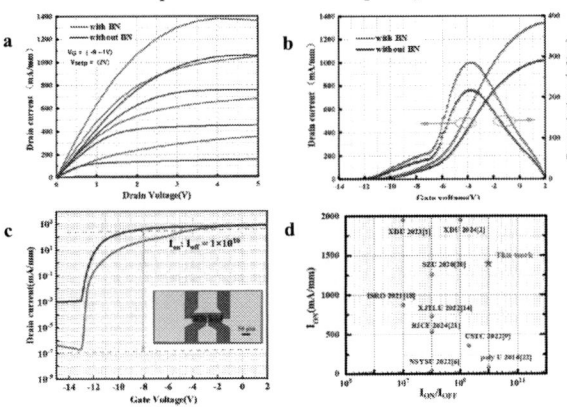

Fig. 3. The output characteristic curve and the DC transfer characteristic curve of the device.

Conclusion

In the conclusions, based on quasi-van der Waals epitaxy and B diffusion mechanisms, we have developed a novel AlGaN/GaN high-electron-mobility transistor (HEMT). During the experimental process, we successfully fabricated high-performance GaN heterojunction structures on a SiC substrate, which exhibited a low screw dislocation density of 4.08×10^7 cm^{-2} and an edge dislocation density of 5.97×10^8 cm^{-2}, alongside a two-dimensional electron gas concentration of 8.51×10^{12} cm^{-2} and a mobility of 2091.3 cm^2/V. The resulting device achieved a channel resistance (Rc) of 0.28 Ω-mm, a maximum device saturation current of 1396 mA/mm, and a high ON/OFF current ratio of 10^{10}, all without the need for complex processing steps. This demonstrates the effectiveness of our material growth method for the development of high-performance GaN radio-frequency devices.

Acknowledgments

The work was supported by General Program of Natural Science Foundation of China (Grant No: 62274134); The National Key Research and Development Program (Grant Nos: 2023YFB3609900 and 2021YFA0716400); The National Science Fund for Distinguished Young Scholars (Grant No: 61925404).

References

[1] J. Shin, H. Kim, S. Sundaram, Vertical full-colour micro-LEDs via 2D materials-based layer transfer, Nature 614(7946) 81-87. (2023)

[2] H. Kum, D. Lee, Epitaxial growth and layer-transfer techniques for heterogeneous integration of materials for electronic and photonic devices, Nat. Electron 2(10) 439-450 (2019).

[3] C. S. Chang, K. S. Kim, Remote epitaxial interaction through graphene, Sci. Adv. 9(42) 5379 (2023).

[4] Q. Chen, K. Yang , Principles for 2D-Material-Assisted Nitrides Epitaxial Growth, Adv Mater. 35(18) 2211075 (2023).

[5] K. Dang, Z. Qiu, S. Huo, P. Zhan, H. Liu, Y. Zhang, J. Ning, H. Zhou, J. Zhang, Current collapse suppressed GaN diodes with 38 Watts high RF power rectifier capability, Sci. China Inform. Sci. 67(2) 1-2 (2024).

[6] H. Wu, J. Ning, J. Zhang, Y. Zeng, Y. Jia, J. Zhao, L. Bai, Y. Wang, S. Li, D. Wang, Y. Hao, High quality AlN film assisted by graphene/sputtered AlN buffer layer for deep-ultraviolet-LED, Nanotechnology 34(29) 295202 (2023).

[7] H. Baek, C.H. Lee, Epitaxial GaN microdisk lasers grown on graphene microdots, Nano Lett. 13(6) 2782-2785 (2013).

[8] Y. Kim, S. S. Cruz, Remote epitaxy through graphene enables two-dimensional material-based layer transfer, Nature 544 340-343 (2017).

[9] S. Arulkumaran, T. Egawa, S. Matsui, H. Ishikawa, Enhancement of breakdown voltage by AlN buffer layer thickness in AlGaNGaN high-electron-mobility transistors on 4 in. diameter silicon, Appl. Phys. Lett. 86 123503 (2005).

[10] A. Dadgar, M. Poschenrieder, crack-free blue light-emitting diodes on Si(111) using low-temperature AlN interlayers and in situ SixNy masking, Appl. Phys.Lett. 80 3670 (2002).

[11] X. Yu, J. Ni, Z. Li, J. Zhou, C. Kong, Reduction in leakage current in AlGaN/GaN HEMT with three Al-containing step-graded AlGaN buffer layers on silicon, Jpn. J. Appl. Phys. 53 051001 (2014).

[12] Q. Chen, X.-Y. Liu, S. B. Biner, Solute and dislocation junction interactions, Acta Mater 56 (2008) 2937-2947.

[13] H. Zhang, F. Liang, K. Song, C. Xing, D. Wang, H. Yu, C. Huang, Y. Sun, L. Yang, X. Zhao, H. Sun, S. Long, Demonstration of AlGaN/GaN-based ultraviolet phototransistor with a record high responsivity over 3.6×107 A/W, Appl. Phys. Lett. 118 242105 (2021).

[14] W. Wu, C. Liu, L. Han, X. Wang, J. Li, Wafer-scale high sensitive UV photodetectors based on novel AlGaN/n-GaN/p-GaN heterostructure HEMT, Appl. Surf. Sci. 618 156618 (2023).

[15] H. You, X. Sun, High-responsivity and fast-response ultraviolet phototransistors based on enhanced p-GaN/AlGaN/GaN HEMTs, ACS Photonics 9 2040-2045 (2022).

[16] J. Yu, L. Wang, Z. Hao, Van der Waals epitaxy of III-nitride semiconductors based on 2D materials for flexible applications, Adv. Mater. 32 1903407 (2020).

Multi-output Virtual Metrology for Physical Vapor Deposition using Projective Selection Algorithm

Amina Mević [1,2], Andreas Laber[2], Sandor Szedmak[3], Dženana Đonko[1], and Senka Krivić[1]

[1]Faculty of Electrical Engineering, University of Sarajevo, [2]Infineon Technologies AG, [3]Aalto University

ABSTRACT

Technologies such as virtual metrology (VM), which monitors fabrication processes and predict product properties without physical measurements have numerous positive impacts. In this paper, we propose a VM system that predicts multiple physical properties of metal layers after the physical vapor deposition. We employ the Projective Selection (ProjSe) algorithm, which is suitable for variable selection in multi-output problems, to investigate the relationship between process parameters and layer properties. The effectiveness of the feature selection process combined with different regression models is demonstrated on real-world datasets collected from semiconductor manufacturer Infineon Technologies AG.
Keywords: Manufacturing, virtual metrology and physical vapor deposition

INTRODUCTION

The rising demand for digitalization and decarbonization has led to the increasing production of chips (9.2 billion chips by Infineon Technologies AG in 2023 [1]), highlighting the need for faster, more reliable, and efficient semiconductor manufacturing with minimal waste. It is often impossible to fully inspect products during manufacturing due to the destructive nature of some testing techniques [2]. The industry relies on random sampling and inspection, which does not always ensure comprehensive quality control or optimal yields. With the growing use of artificial intelligence (AI), there is potential to predict product properties using data from existing monitoring systems such as Advanced Process Control (APC) and Statistical Process Control (SPC).

Based on AI and machine learning (ML) techniques [2], virtual metrology (VM), introduced in 2005 in the semiconductor manufacturing industry [3], estimates the quality of a product directly from production process data without physically measuring it [2] and in this way reduces production times and costs. The VM workflow is demonstrated in Fig. 1. Following the previous efforts in this direction [2], we focus on creating a VM system to predict the properties of a thin film produced in the physical vapor deposition (PVD) process. PVD is one of the main steps in the production process, and it is used to create thin metal layers by depositing metal vapor onto a substrate [4]. The important physical properties of the film, such as thickness, resistance, and resistivity, depend on factors like deposition time, power, voltage, pressure, and temperature in the chamber.

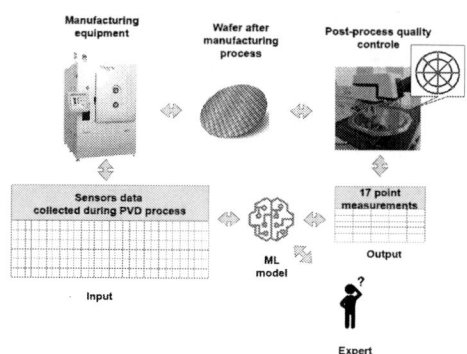

Fig. 1. An example of using VM to replace post-process measuring of wafer properties at 17 characteristic points. VM utilizes ML models to predict product quality and equipment reliability from sensor data. Experts use these predictions to adjust process parameters and schedule maintenance.

Sensors monitor these parameters, and sensor signals are collected from production equipment, aggregated, and integrated into the APC system. After a process, another procedure, metrology, measures the product's physical properties at 17 different characteristic points. Measured values are processed and analyzed within the SPC system (Fig. 1). A highly accurate and reliable prediction system can replace post-process metrology and data analysis, resulting in faster production and significant cost reductions. Therefore, in this paper, we propose building such VM system (Fig. 2).

Chen at al. [5] used the APC dataset to build prediction models using a tree-based ensemble model to estimate a single PVD process output value. They use the XGBoost variable selection method to improve their ensemble method. However, the existing works in VM do not examine the prediction problems as multiple-output predictions but rather independently predict thickness, resistance, and resistivity. Since these parameters are consequences of the same physical process, we treat this problem as a multiple-output prediction problem.

We utilize Projective Selection (ProjSe) [6], the state-of-the-art method for multiple-output feature selection. Moreover, we predict all statistical variables of the measurement along with the thickness, resistance, and resistivity, making our model more reliable. The best-performing prediction model is selected based on model performance metrics. The number of necessary features for predictions is again examined with respect to the prediction model, and the smallest subset of features is determined by evaluating the model's performance.

979-8-3315-0417-5/25 $31.00 © 2025 IEEE

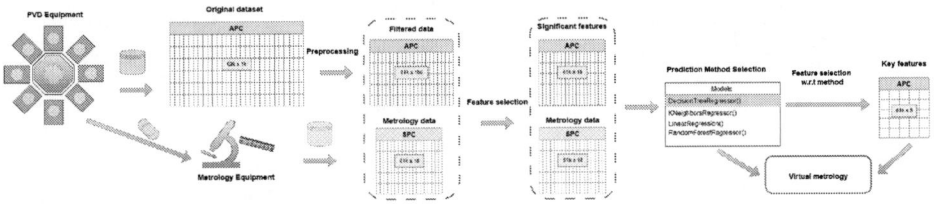

Fig. 2. Steps of the VM design include data collection and preprocessing, feature selection, prediction model selection, and the prediction system finetuning.

METHODOLOGY

Dataset. The data used in this work was collected from 16 chambers of six PVD machines at the Infineon Technologies AG fab from 2021 to 2023. We consider each of the 17 measured points as individual samples characterized by 104 APC parameters. These parameters are considered as features for the simultaneous prediction of 18 SPC parameters, which include measurements and 5 aggregated statistical measures (EWMA-Mean, EWMA-S, EWMA-R, MA and MS) for each physical property: *resistivity, resistance, and thickness.* After preprocessing, measurements from 3598 products are included in the dataset, resulting in a total of 61165 samples.

Feature Selection. As a first step in designing a VM system, we interviewed industry experts in PVD equipment and processes. They identified *10 process parameters* (*voltage, current, power in target, a lifetime of target, a lifetime of shielding, deposition time, direct current bias, pressure in a chamber, speed of a magnetron, and gas*) that have the greatest influence on the physical properties of metal layers. We use feature selection and importance algorithm ProjSe [6] to select and rank process parameters that are the most important for the prediction of *resistivity, resistance, and thickness* of the wafer. ProjSe is the state-of-the-art approach for variable selection for multi-output learning problems based on projection operators and their algebra. The method uses a kernel-based representation to capture complex relationships between variables. The algorithm chooses iteratively the input variable that has the highest correlation with the outputs while being as uncorrelated as possible with the inputs already selected. This ensures each new variable adds relevant information to the prediction model without redundancy.

Prediction models. The next step after selecting features is to choose the best prediction model. Following the previous work [3] and the thorough data analysis, we use four regression models for multiple output prediction: Linear Regressor (LR) [7], K Neighbors Regressor (KNN) [8], Random Forest Regressor (RFR) and Decision Tree Regressor (DTR) [9].

Feature tuning for the prediction model. Once the model is selected, it is possible to check which features are most relevant to the prediction with respect to the model itself. We use the best-performing prediction model to identify the minimal subset of features selected by ProjSe necessary for accurate prediction. Features with the highest importance are added iteratively to the input space until the difference in model performance metrics falls below a specified threshold.

RESULTS

Minimal subset of relevant features. Using a linear kernel to select features with ProjSe relevant to predicting 18 output variables, we identified the 18 most important features for prediction. Three of these 18 features coincided with the ones chosen by experts, and four were statistical metrics of the selected features. We compared the prediction results of the least-squares regression model [10] using features selected by ProjSe with those selected randomly. The data are centralized and normalized to ensure unbiased evaluation. Prediction accuracy is measured using Pearson correlation between actual Y and predicted outputs Y_p: $\rho_{Y,Y_p} = \frac{\text{Cov}(Y,Y_p)}{\sigma_Y \sigma_{Y_p}} = \frac{\sum_{i=1}^{n}(y_i - \bar{y})(y_{p_i} - \bar{y}_p)}{\sqrt{\sum_{i=1}^{n}(y_i - \bar{y})^2}\sqrt{\sum_{i=1}^{n}(y_{p_i} - \bar{y}_p)^2}}$ and iteratively recalculated as features are added.

Fig. 3 demonstrates how the number of selected features enhances prediction accuracy compared to random selection. The projective operator-based selection yields higher Pearson correlation coefficients, indicating a correlation between selected features and outputs. After 14 features, the cumulative correlation does not increase. The 14 top-ranked features have a high correlation with the output variables and minimal correlation with each other.

Evaluation of prediction models. In the second experiment, we evaluate the performance of prediction models concerning selected features. The performances of prediction algorithms LR, KNN, FRF, and DTR are evaluated on three datasets: `DS1` - dataset consisted of all process parameters

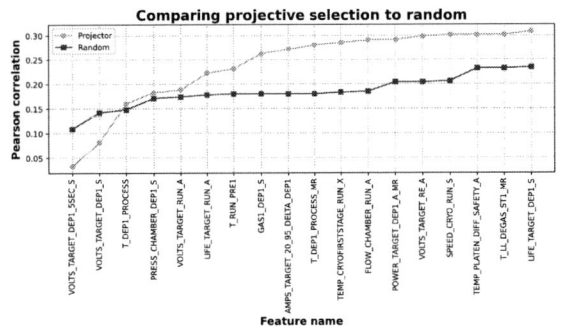

Fig. 3. The blue line shows how prediction accuracy changes when variables are added randomly, whereas the orange line demonstrates the performance when variables are chosen using the projective selection method.

979-8-3315-0417-5/25 $31.00 © 2025 IEEE

TABLE I: Performance Metrics for Predictive Models

Dataset	Model	MSE	MAE	MAPE (%)	RMSE	R2 Score
DS1	LR	62.74	2.82	$6.3 \cdot 10^{14}$	3.96	0.26
DS1	**DTR**	**37.68**	**1.22**	**0.28**	**1.45**	**0.90**
DS1	KNN	65.41	1.53	0.34	1.91	0.84
DS1	RFR	37.69	1.22	0.28	1.45	0.90
DS2	LR	72.14	3.01	$9.1 \cdot 10^{14}$	4.33	0.10
DS2	**DTR**	**37.68**	**1.22**	**0.28**	**1.45**	**0.90**
DS2	KNN	54.76	1.41	0.32	1.74	0.86
DS2	RFR	37.70	1.22	0.28	1.45	0.90
DS3	LR	76.68	3.12	$7.5 \cdot 10^{14}$	4.42	0.16
DS3	**DTR**	**37.68**	**1.22**	**0.28**	**1.45**	**0.90**
DS3	KNN	65.41	1.53	0.34	1.91	0.84
DS3	RFR	37.70	1.22	0.28	1.44	0.90

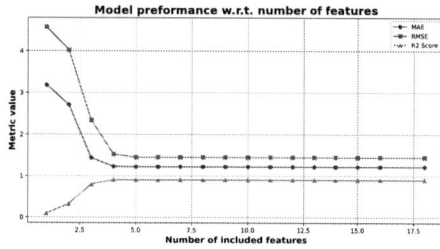

Fig. 4. Performance metrics of the DTR model regarding the number of features selected and ranked in descending order.

(104 feature values and 18 output values per measurement point), DS2 - dataset of features selected by experts (10 feature values and 18 output values per measurement point) and DS3 - a dataset of features selected by ProjSe (18 feature values and 18 output values per measurement point). We employed 10-fold cross-validation to evaluate model performance across various data subsets comprehensively and utilized a suite of performance metrics, including MSE, MAE, MAPE, RMSE, and R-squared score. Results are given in Table I. There is no difference between the datasets, which implies that ProjSe selected enough informative features. For all datasets, the best performance has DTR, but RFR demonstrates comparable results. Our results confirm the findings in earlier work [5].

Feature tuning. In the third experiment, we examine feature importance with respect to the Decision Tree model, which was selected for prediction. We examine how features ranked by ProjSe impact the prediction success rate. For this purpose, we expanded the input space of the prediction model, adding features one by one and ranking them in descending order. For each input space created in this way, we performed a 10-fold-cross validation test and observed changes in prediction metrics. Fig. 4 shows model performance results for each combination of features, starting from single, the most significant variable selected by ProjSe. We can notice a decrease in MAE and RMSE with increased input feature numbers and an increase in R2 scores. After creating input space with the five most significant features, we observe the metrics have constant values. Among the 5 most relevant features three were selected by experts: average voltage in target during deposition, time of deposition, and standard deviation of pressure in the chamber.

CONCLUSION

This study used the ProjSe algorithm to identify features relevant for predicting multiple variables and developed an algorithm to select the minimal subset of features concerning model performance. The subset of the most relevant features differs from the 10 features identified by semiconductor industry experts. While the selected sensors remain consistent, varying statistical metrics have a greater impact on predicting resistivity, resistance, and thickness compared to the initially selected average values. Additionally, ProjSe highlights more process parameters as influential for prediction, going beyond the considerations of experts. Our prediction methods exhibit high accuracy, showing no significant differences in prediction accuracy whether we use parameters chosen by experts, those selected by ProjSe, or the complete set of parameters.

ACKNOWLEDGMENT

The work is linked to the FID activities of the IPCEI on ME (Important Project of Common European Interest on Microelectronics), funded by national authorities from Germany, France, Italy, UK and Austria.

REFERENCES

[1] https://www.infineon.com/cms/austria/en/press/GJ2324/Bilanz-Geschaeftsjahr_23.html.

[2] V. Maitra, Y. Su, and J. Shi, "Virtual metrology in semiconductor manufacturing: Current status and future prospects," *Expert Systems with Applications*, 2024.

[3] P. Chen, S. Wu, J. Lin, F. Ko, H. Lo, J. Wang, C. Yu, and M. Liang, "Virtual metrology: A solution for wafer to wafer advanced process control," in *ISSM 2005, IEEE International Symposium on Semiconductor Manufacturing, 2005.*, 2005.

[4] R. A. Powell and S. M. Rossnagel, *PVD for microelectronics: sputter deposition applied to semiconductor manufacturing*, 1999.

[5] C.-H. Chen, W.-D. Zhao, T. Pang, and Y.-Z. Lin, "Virtual metrology of semiconductor pvd process based on combination of tree-based ensemble model," *ISA transactions*, 2020.

[6] S. Szedmak, R. Huusari, T. H. Duong Le, and J. Rousu, "Scalable variable selection for two-view learning tasks with projection operators," *Machine Learning*, 2023.

[7] X. Su, X. Yan, and C.-L. Tsai, "Linear regression," *Wiley Interdisciplinary Reviews: Computational Statistics*, 2012.

[8] Y.-P. Mack, "Local properties of k-nn regression estimates," *SIAM Journal on Algebraic Discrete Methods*, 1981.

[9] W.-Y. Loh, "Classification and regression trees," *Wiley interdisciplinary reviews: data mining and knowledge discovery*, 2011.

[10] R. W. Farebrother, *Linear least squares computations*. Routledge, 2018.

979-8-3315-0417-5/25 $31.00 © 2025 IEEE

On the Understanding of Temperature Dependence of Flicker Noise in Advanced FinFET Technology

Sheng Yang [1,2], Shuying Wang [1,2], Chenyang Zhang [1,2], Yewei Zhang [1,2], Yongkang Xue [1,2]

Pengpeng Ren [*1,2] and Zhigang Ji [*1,2]

[1] National Key Laboratory of Science and Technology on Micro/Nano Fabrication, Shanghai Jiao Tong university, Shanghai, China [2] Department of Micro/Nano Electronics, School of Electronic Information and Electrical Engineering, Shanghai Jiao Tong university, Shanghai, China

*Email: pengpengren@sjtu.edu.cn, zhigangji@sjtu.edu.cn

Abstract

Temperature Dependence will complicate the noise problem of FinFET devices. Our experiments show that Flicker noise is obviously affected by high temperature, and shows different temperature dependence under different gate bias. That is, at low gate bias, high temperature makes the flicker noise of the device lower than that at room temperature, while at high gate bias, high temperature makes the flicker noise of the device higher than that at room temperature. A new physical-based model is used to analyze the temperature dependence of flicker noise. The decrease of high temperature noise at low bias is mainly affected by the decrease of Hooge parameter, and the increase of high temperature noise at high bias is mainly affected by the access resistance induced noise. This increase in noise due to high temperatures is likely to be more significant in future GAAFET and CFET devices.

Keywords: FinFET, flicker noise , high temperature

Introduction

Due to the increasing demand for analog/mixed-signal circuits in mobile and artificial intelligence applications, flicker (1/f) noise presents a pressing problem in the scaling of CMOS technology [1,2]. However, because of high temperature environment due to environmental reasons or thermal effects, the flicker noise of the devices may be different from room temperature conditions. Therefore, the main objective of this paper is to use recently developed flicker noise separation method to model and characterize the temperature dependence of flicker noise in advanced nodes, elucidate the physical origin of the associated noise, and evaluate its impact on future devices.

Device and characterization

In this study, an industrial-grade P-FinFETs was used to investigate temperature dependence of flicker noise. The device has a channel width and length of 338 nm (10 Fins and 4 fingers) and 20 nm, respectively. We measured the noise of the device at different gate voltages at the ambient temperature of 300K, 365K and 425K.

Result and discussion

A. Temperature Dependence of flicker noise

As shown in Fig.1a to 1c, when the gate voltage V_g=0.25V, the normalized current noise of the device decreases with the increase of the temperature of the device; when the gate voltage V_g =0.45V, the normalized current noise of the device at high temperature could increase and could decreases compared with room temperature; and when the higher gate voltage V_g =0.75V, the normalized current noise obviously increases with the increase of the temperature of the device. The above experimental results have similar experimental results in the literature, but there is no in-depth analysis of the reasons behind [3-5]. Fig.2 shows the statistical result of the original noise spectrum of the device at different temperatures at 10Hz. It is obvious that the noise of devices has opposite high temperature dependence at low gate bias and high gate bias.

B. Noise separation method

It is well known that flicker noise or 1/ f noise can be derived from both carrier number fluctuations (CNF) and mobility fluctuations [2], [10]. In the CNF model, 1/ f noise is assumed to be a superposition of a large number of trap induced noise sources. However, the traditional noise model cannot explain the new phenomenon of noise on nanodevices. Therefore, we believe flicker noise can originate from more than one sources: (de-)trapping in the dielectric layer for the oxide trap (OT)-induced noise; scattering of carriers during transport for channel scattering (CS)-induced noise; nonnegligible source/drain access resistance (AR)-induced noise [2,6]. The total flicker noise consists of oxide trap-induced noise (i.e. random telegraph noise,RTN) caused by defects in the gate oxide dielectric layer and mobility noise(i.e. $S_{id} \propto 1/f$), as described in Eq. 1-3. Mobility noise consists of channel scattering noise(CS noise) and access resistance noise(AR noise), as described in Eq. 4-6.

C. Oxide trap (OT)-induced noise

Figure 3 shows the comparison of noise spectrum changes at different temperatures under high and low gate bias. Fig. 3a& c show the changes of RTN noise in the noise spectrum at different temperatures under gate bias of 0.25V and 0.7V respectively. It can be clearly seen that high temperature causes the corner frequency of RTN noise of the device to move towards high frequency under low bias and high bias.

This is mainly because the high temperature makes the carrier trapping speed of the oxide trap faster and the time constant becomes smaller. As shown in Fig. 3b&d, the amplitude of RTN of the device is relatively small at low bias, so the influence of RTN on the noise spectrum of the device is relatively small at high temperature. It can be observed that the RTN amplitude at high bias has little change compared with that at normal temperature, while the RTN amplitude at low bias decreases with the increase of temperature, which is caused by the difference in current path at high and low bias. According to the penetration model [9], the current path of the device at low bias is elongated and curved, so the oxidation trap rapidly captures charge carriers at high temperature. It is difficult to save carriers in low-density carrier paths, so the RTN noise current phase amplitude decreases at high temperature, while the current path of the device at high bias is wide and straight, and the trapping speed of the trap does not affect the extent to which the trap fills the carrier, because there are enough carriers to trap the carrier for the oxide trap. Therefore, the RTN amplitude of the device hardly changes with temperature at high bias.

D. Channel Scattering (CS)-induced noise

Through the above noise separation method, we get the component of the channel scattering noise spectrum and the channel mobility noise coefficient a_H. As shown in Fig. 5a &b, it can be found that the channel scattered noise of the device decreases with the increase of temperature under both high and low bias conditions. As shown in Fig.5c, the extracted Hooge parameter decreases with the increase of temperature. By using the Y-function method, we extract the mobility changes at different temperatures, as shown in Fig.5d. This is mainly because with the increase of temperature, the lattice scattering and impurity scattering of the device become stronger, resulting in a decrease in the mobility of the device, and the Hooge parameter is related to the mobility, so the Hooge parameter decreases with the increase of temperature. As shown in Eq.4, m is the fitting parameter of mobility change with temperature

E. Access resistance (AR)-induced noise

Note that Fig.6a shows the series resistance noise at different temperatures extracted by the noise separation method. It can be clearly seen that the series resistance noise increases with the increase of temperature. Fig.6a&b show the extracted series resistance and series resistance noise coefficient Kr change with temperature. It can be found that both series resistance and Kr increase with the increase of temperature, which is consistent with the results in reference [10]. As shown in Eq.5, n is the fitting parameter for the variation of access resistance noise with temperature. Hooge parameter dominates the change of mobility noise, which can explain

that at low bias, the noise at high temperature becomes smaller than that at normal temperature, while at high bias, high temperature makes the device noise increase, which indicates that the series resistance noise may dominate the device noise at high temperature and high bias compared with that at normal temperature.

Because at low bias, the device channel current is small, the total on-resistance is particularly large, and the series resistance ratio is particularly small, it can be ignored at low bias. However, with the increase of channel current, the on-resistance of the device becomes larger and larger, and gradually dominates the mobility noise, while the mobility noise caused by channel scattering decreases. Fig.6d shows the change of the series resistance noise ratio with the gate voltage, and it can be seen that at high bias, the proportion of series resistance at high temperature becomes larger than that at room temperature. This is because the total on-resistance of the device becomes smaller at high temperature, the series resistance becomes larger, and the series resistance noise coefficient Kr becomes larger, causing the proportion of series resistance noise to increase at high temperature. This can also explain the increase of the total noise spectrum of the device at high temperature and high bias.

Fig. 7&8 shows the simulation diagram of the structure and heat distribution of FinFET, GAAFET and CFET devices. As shown in Table 1, the self-heating temperature and series resistance ratio of advanced node devices are compared. It can be obviously seen that the self-heating temperature of GAAFET and CFET devices is higher than that of FinFET, and the proportion of series resistance is larger [8]. Therefore, GAAFET and CFET devices may face greater 1/f noise than FinFET devices at high temperature and high bias. This has important reference significance for our future integrated circuit design and reliability.

Conclusion

In this paper, the high temperature flicker noise in FinFETs of advanced nodes is studied. The physical model based on noise separation clarified the physical reason behind the phenomenon of high temperature noise. We believe that the impact of high temperature on the noise of new devices (GAAFET and CFET) in the future will be more obvious, providing guidance for designers to improve the reliability of products.

References

[1] P. Kushwaha., et al., EDL 2019. [2] J. Wu, et al., IEDM 2023. [3] A. R. Molto, Y., et al., SBMicro 2016. [4] Srinivasan, T., et al., ECS Meeting 2010. [5] Dong, D, et al., Micromachines 2023. [6] P. Ren, et al., TED 2024. [7] J. Wu, et al., TED 2023. [8] S. Wang, et al., IRPS 2017. [9] W. H. Lin, et al., EDL 2003. [10] L. K. J. Vandamme, et al., TED 2008

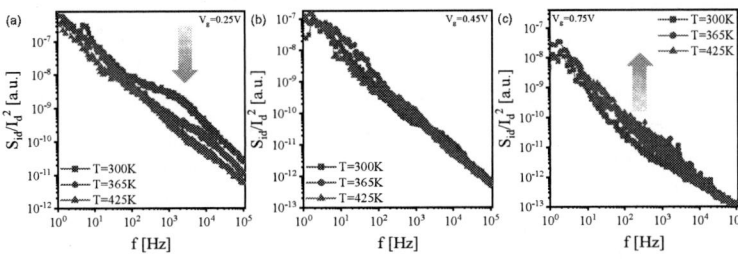

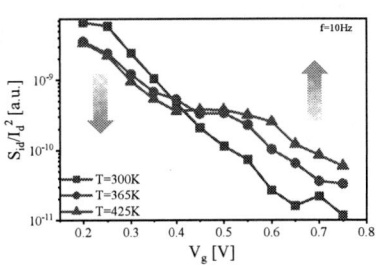

Fig. 1: Comparison of normalized noise currents S_{id}/I_d^2 at 300K,365K and 425K based on a) Vg=0.2V; b) Vg=0.45V; c) Vg=0.7V, the results show opposite high temperature dependence of flicker noise at low gate bias and high gate bias in advanced technology.

Fig. 2: Comparison of the trend of the mean noise of the original noise spectrum S_{id}/I_d^2 with Vg at different temperature

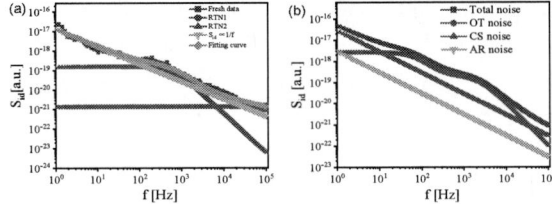

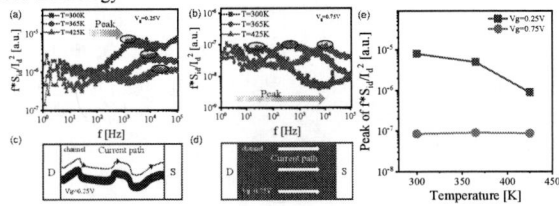

Fig. 3: a) Illustration of separation of RTN noise from total flicker noise; b) Separation method of different noise components ; flicker noise be divided into three parts: OT noise, CS noise and AR noise.

Fig. 4: The change of Sid * f /Id2 with different temperature based on a) Vg=0.25V; b) Vg=0.75V; e) Comparison of 0.2V and 0.75V; Different current paths at d) high and c) low voltages lead to different RTN amplitude

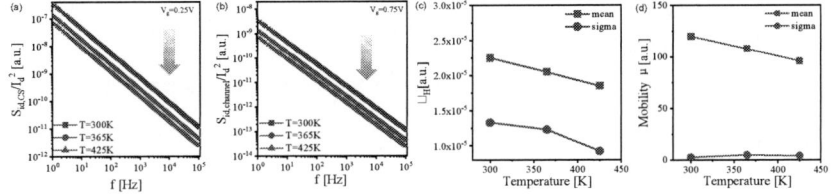

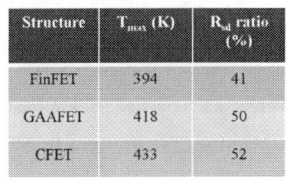

Fig. 5: Comparison of CS noise at 300K,365K and 425K based on a) Vg=0.25V; b) Vg=0.75V; Mean and sigma values of c) extracted Hooge parameter and d) mobility μ with temperature; the Hooge parameter is related to the mobility, so the Hooge parameter decreases with the increase of temperature

Table I: Comparison of self-heating temperature and access resistance ratio of advanced devices

Structure	T_{max} (K)	R_{sd} ratio (%)
FinFET	394	41
GAAFET	418	50
CFET	433	52

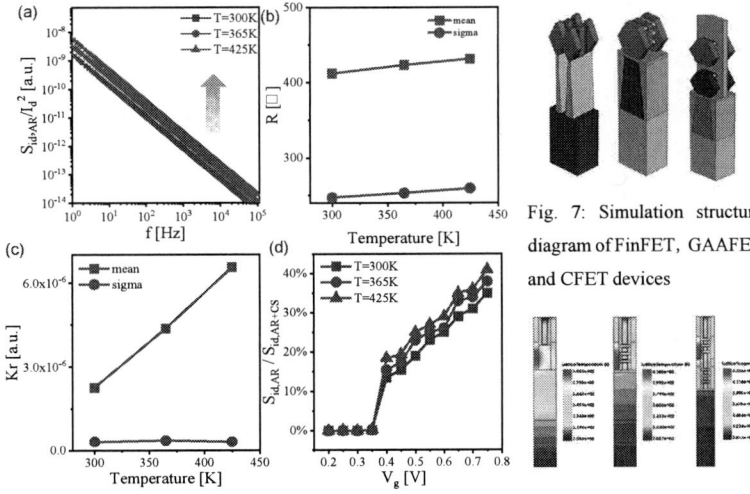

Fig. 6: a) Comparison of AR noise at 300K,365K and 425K; Mean and sigma values of extracted b) R_{access} and c) Kr with temperature ; d) Comparison of the trend of the mean ratio of AR noise to the mobility noise at 300K,365K and 425K

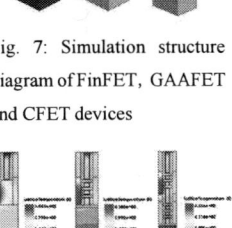

Fig. 7: Simulation structure diagram of FinFET, GAAFET and CFET devices

Fig. 8: Simulation diagram of the sectional channel heat distribution of devices with three structures

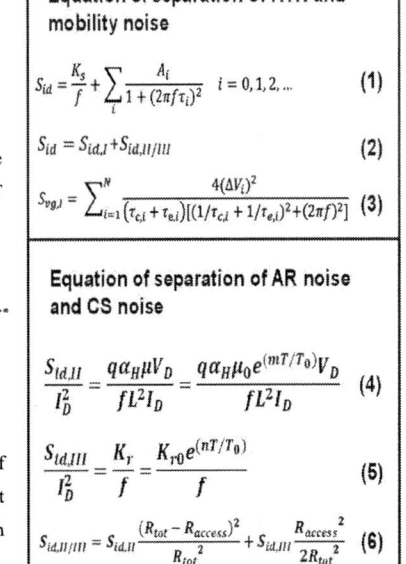

Equation of separation of RTN and mobility noise

$$S_{id} = \frac{K_s}{f} + \sum_i \frac{A_i}{1+(2\pi f \tau_i)^2} \quad i=0,1,2,\dots \quad (1)$$

$$S_{id} = S_{id,I} + S_{id,II/III} \quad (2)$$

$$S_{vg,I} = \sum_{i=1}^{N} \frac{4(\Delta V_i)^2}{(\tau_{c,i}+\tau_{e,i})[(1/\tau_{c,i}+1/\tau_{e,i})^2+(2\pi f)^2]} \quad (3)$$

Equation of separation of AR noise and CS noise

$$\frac{S_{id,II}}{I_D^2} = \frac{q\alpha_H \mu V_D}{fL^2 I_D} = \frac{q\alpha_H \mu_0 e^{(mT/T_0)} V_D}{fL^2 I_D} \quad (4)$$

$$\frac{S_{id,III}}{I_D^2} = \frac{K_r}{f} = \frac{K_{r0} e^{(nT/T_0)}}{f} \quad (5)$$

$$S_{id,II/III} = S_{id,II}\frac{(R_{tot}-R_{access})^2}{R_{tot}^2} + S_{id,III}\frac{R_{access}^2}{2R_{tot}^2} \quad (6)$$

979-8-3315-0417-5/25 $31.00 © 2025 IEEE

Overlay-aware Variation Study of Flip FET and Benchmark with CFET

Wanyue Peng, Haoran Lu, Jingru Jiang, Jiacheng Sun, Ming Li, Runsheng Wang, Heng Wu†, Ru Huang

School of Integrated Circuits, Peking University, Beijing 100871, China, †email: hengwu@pku.edu.cn

Abstract

In this work, we carried out an overlay-aware variation study on Flip FET (FFET) considering the impact on RC parasitics induced by the lithography misalignment in backside processes, and benchmarked it with CFET in terms of the power-performance (PP) and variation sources. The iso-leakage frequency degrades up to 2.20% with layout misalignment of 4 nm. It's found that the Drain Merge resistance degrades significantly with misalignment increasing and is identified as the major variation source. Through careful DTCO with design rule optimization, the variation can be greatly suppressed, while the resistance fluctuation of the DM also drops substantially. Monte Carlo random experiments were also conducted, validating the variation reduction. Comparing with the CFET featuring self-aligned gate and much less overlay induced misalignment, fortunately, FFET's PP is still better except when misalignment reaches 8 nm, which is out of spec and nearly impossible. Considering the variabilities induced by the high aspect ratio processes, CFET still faces big challenges compared with FFET.

Keywords: Flip FET, CFET, Misalignment, Variation, Monte Carlo, Design Technology Co-optimization (DTCO)

Introduction

As the traditional scaling coming to an end, the Flip FET (FFET) was promoted as a promising solution for future logic technology [1]. However, normal FFET only features self-aligned active (Fig. 1(a)), leaving the backside (BS) layers possibly misaligned to frontside (FS) ones, which is different from CFET [2][3] with naturally self-aligned gate, as shown in Fig. 1(b).

Design of experiments (DoEs) based on misalignment assumption has been conducted to investigate how the misalignment affects, followed by careful variation source validation and DTCO. Further power-performance (PP) analysis between the FFET and the CFET proves that FFET could still be a better solution than the CFET despite its misalignment-induced PP degradation, while CFET has more variation sources due to high aspect ratio processes. Note that the overlay impact on the intrinsic device performance is not yet considered in this work, deserving further investigation.

Variation Sources Exploration

The FFET [1] is composed of back-to-back-stacked transistors with only the active formed in a self-aligned manner. As the first lithography on the backside is the BS-Gate, it's assumed that misalignment mainly occurred between BS-Gate and FS-Gate, while other BS masks would align to BS-Gate. In this case, the intrinsic performance of FS and BS devices keeps unchanged in this work.

To describe the misalignment distance and directions more precisely, we introduce the Misalignment Vector along which the BS masks shift relative to the FS layers, as illustrated on Fig. 2. Based on backside EUV lithography experiment [4], $0 \sim 4$ nm is assumed to be the most reasonable misalignment range between FS and BS layers. Under each misalignment condition, a 15-stage ring-oscillator with fan-out 3 was used for PP analyses at iso-leakage @VDD = 0.7 V, and the results are displayed on Fig. 3. Only frequency variation is shown as the power variation is too small compared with the frequency. The maximum frequency degradation is up to 2.20% when the misaligned BS layers move along the Y-axis to a very negative place.

To lower the variation, the main source of frequency fluctuation should be identified first. Specially, the Drain Merge (DM) and the Gate Merge (GM) are formed on the BS S/D region and gate region respectively to connect BS and FS transistors. Once there is a misalignment, the DM and GM cannot directly land on frontside transistor's metalized drain (MD) and gate metal precisely (Fig. 4). Therefore, the DM and the GM have the largest structural variations, which influence their parasitic resistances a lot. It's assumed that the frequency variation mainly derives from the resistance variation of the DM and the GM and developed two models. One is the DM-only model, which only considers the resistance variation of the DM and replaces the DM resistance sub-netlist in the baseline Inverter (INV) netlist with DM resistance sub-netlist under various misalignment conditions. The other one is the GM-only model which follows a similar approach. After comparing different converted frequency with experimental frequency, it's validated that the resistance of the DM contributes to the majority of the frequency variation (Fig. 5).

The reason why the variation in GM resistance (ΔR_{GM}) has a smaller influence compared to DM is that the ΔR_{GM} is very small relative to the total resistance of the gate (less than 10%), so the impact of ΔR_{GM} is minimal. Additionally, it should be acknowledged that the parasitic variation under misalignment varies with different design rules. In the baseline design rule, the DM is mainly buried in the Gate Cut and the overlay between the DM and the gate is small (as shown in Fig. 6(a)), so the DM contributes little to C_{gd}. As a result, the ΔC_{gd} matters less than ΔR_{DM}.

979-8-3315-0417-5/25 $31.00 © 2025 IEEE

DTCO and Monte Carlo Validation

Previous work has confirmed that the DM determines the frequency variation. So to reduce the variation, lowering the variation of the DM resistance (ΔR_{DM}) is needed. Further investigation into the baseline design rule reveals that the DM would be blocked by the S/D epi (Fig. 6(a), taking INV as an example), thus the size of DM varies significantly. To leave more margin between the Drain Merge and the active, the power rail can be moved to other side of the layout thanks to the flexibility of the FFET layout design, making the Drain Merge farther away from the active, as shown in Fig. 6(b). The MD was also enlarged at the same time while all the other structures kept the same. The ΔR_{DM} under the baseline design rule and the optimized design rule were also extracted. The standard deviation of ΔR_{DM} decreases 97.2% compared with the baseline one, which matches the expectation. Misalignment-induced frequency variation is suppressed evidently after the design rule optimization (Fig. 8) compared to the baseline in Fig. 3. The maximum frequency degradation is optimized from 2.20% to 1.30% due to the ΔR_{DM} improvement.

In the experiments above, the probabilities of different Misalignment Vectors occurring across the wafer is not considered, which actually is not the same. To mimic the real case, It's assumed that the probability of the Misalignment Vector decreases as the misalignment distance increases and most of the Vectors exist within the range of 0-4 nm. The misalignment distribution assumption is depicted in Fig. 9. Based on this distribution, 10,000 Monte Carlo simulations were conducted, randomly selecting a Vector each time and calculating the corresponding frequency. Then the misalignment-induced frequency distributions under the baseline design rule and the optimized design rule were extracted, as shown in Fig. 10(a) & (b), respectively. The standard deviation of frequency after DTCO decreases by 19.7% compared with the baseline one.

All the elements above should be considered into new design rule enablement, and the variation of C_{gd} should also be considered.

Benchmark with CFET

In the previous work [1], it's proved that the 3.5T dual-fin FFET outperforms the 4T dual-fin CFET by using RO simulation. For here, the power-performance degradation derived from the overlay in the FFET is evaluated, with worst cases under each misalignment range selected. The results in Fig.11 show that the PP of the FFET is weaker than that of the CFET only when the misalignment distance reaches 8 nm, while the situation is considered impossible if the EUV lithography is used[4]. Both FFET and CFET are under tight design rules (with short distance between vias to active, small gate extension [4], etc.). So even though there is some misalignment-induced degradation in the FFET, its performance can still outperform the CFET.

Misalignment between frontside FEOL layers and backside FEOL layers is a unique source of variation in FFET. But CFET is free of this issue because its structures are mostly formed on the frontside of the wafer in a self-aligned manner. However, the unique high aspect ratio processes in CFET could also introduce significant variations [4-8], as shown in Fig. 12.

Table. 1 gives an analysis of these variations. The wafer flip process is considered the reason for patterning misalignment. But it also greatly reduced the numbers of high aspect ratios (HAR). The CFET can also replace some HAR processes at the cost of introducing flip process, like the bottom MD formation and BPR formation can be substituted by direct backside contact [9-13]. As the influence of patterning-induced fluctuation of FFET is limited while the high aspect ratio processes matter significantly, it's indicated that the FFET can still exceed the CFET even in variation aspect.

Conclusion

In this work, we conducted an overlay-aware variation study on FFET using advanced DTCO flow, in a parasitic perspective. A Precise model was established to verify the DM resistance as the main variation source. Based on it, new design rule was proposed and validated with much lower the frequency variation. Furthermore, to comprehensively study the frequency distribution, Mento Carlo random experiments were also conducted. Power-performance comparison and the trade-off between flip-induced patterning misalignment and high aspect ratio processes variations were analyzed, implying that FFET retains great advantages over CFET.

Acknowledgments

All the authors gratefully acknowledge the funding support from National Key Research and Development Program of China under Grant 2023YFB4402200; and in part by the 111 Project under Grant 8201702520.

References

[1]   H. Lu, et al., *VLSI 2024*.

[2]   J. Ryckaert, et al., *VLSI 2018*.

[3]   N. Horiguchi et al., *IEDM 2023*.

[4]   S. Demuynck *et al.*, *VLSI 2024*.

[5]   S. Liao et al., *IEDM 2023*.

[6]   M. Radosavljević et al., *IEDM 2023*.

[7]   V. Vega-Gonzalez et al., *VLSI 2023*.

[8]   A. Pal et al., *VLSI 2024*.

[9]   A. Veloso *et al.*, *IEDM 2023*.

[10]   R. Xie *et al.*, *VLSI 2024*.

[11]   J. Park, et al., *VLSI 2024*.

[12]   S. Liao, et al., *IEDM 2024*.

[13]   B. Kim, et al., VLSI 2024.

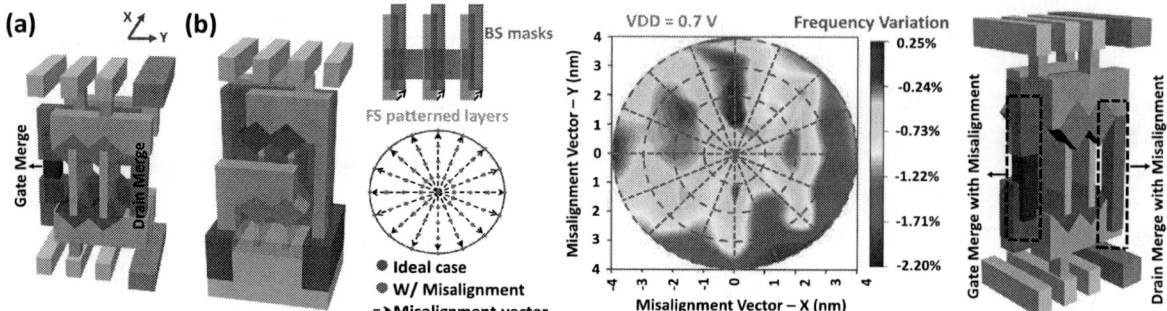

Fig. 1 (a) 3D schematic of the 3.5T dual-fin FFET. (b) 3D schematic of the 4T dual-fin CFET with BPR.

Fig. 2 The Misalignment of the FFET. BS masks shift along the Misalignment Vectors relative to the FS layers.

Fig. 3 Misalignment-induced FFET frequency variation distributed along all directions with the maximum possible displacement assumed to be 4 nm.

Fig. 4 The FFET with misalignment where the DM and the GM have obvious mismatches.

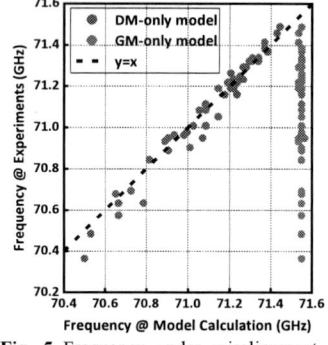

Fig. 5 Frequency under misalignment experiments vs. frequency converted by replacing netlists without misalignment with DM or GM resistance sub-netlists.

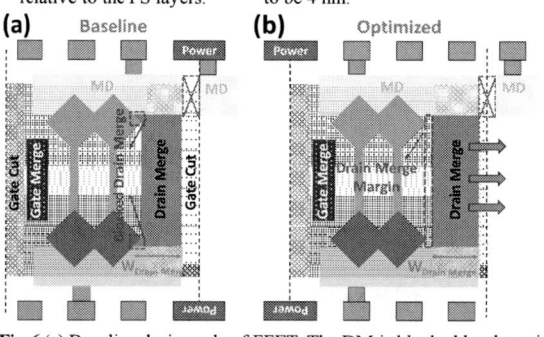

Fig. 6 (a) Baseline design rule of FFET. The DM is blocked by the epi with misalignment. (b) The optimized design rule with better margins between the DM and the active. The MD can be closer to the boundary after modifying the power rail, thus the same for the DM.

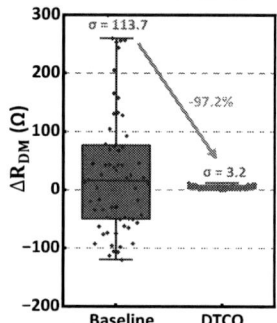

Fig. 7 ΔR_{DM} under the baseline and optimized design rule. The standard deviation of ΔR_{DM} decreases by 97.2%.

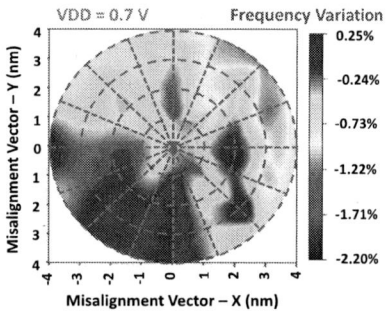

Fig. 8 Misalignment-induced FFET frequency variation distributed along all directions after DTCO. Compared to baseline case in Fig. 3, the frequency variation is suppressed evidently.

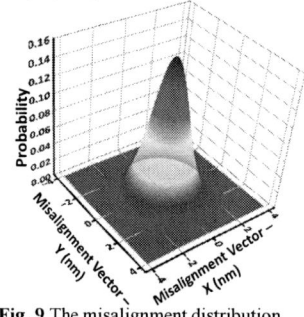

Fig. 9 The misalignment distribution is assumed to follow a two-dimensional normal distribution. The cumulative probability within the range of 0-4 nm is 99.7%.

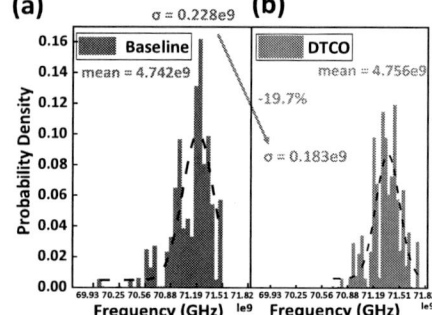

Fig. 10 Monte Carlo experiments on frequency occurrence were performed 10,000 times, for baseline (a) and DTCO (b), based on the misalignment vectors distribution in Fig. 6. The standard deviation decreases by 19.7%.

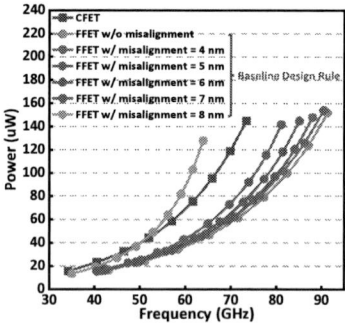

Fig. 11 Power-Performance curves of the CFET's and the FFET's baselines and worst cases of different misalignment conditions.

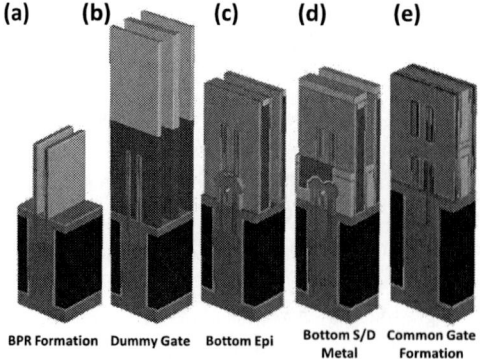

Fig. 12 High aspect ratio processes of the CFET.

Variation Sources	FFET	CFET	Description
Lithography	Yes	No	Wafer flip induced misalignment in the FFET.
BPR formation	No	Yes	A deep trench should be etched on the wafer frontside in the CFET.
Dummy gate	No	Yes	Dummy gate is formed on wafer frontside for both top and bottom FETs in the CFET.
Epi formation	No	Yes	Bottom epi & top epi & metal gate & MD are all formed on the frontside of the wafer in the CFET while they are not in the FFET.
RMG	No	Yes	
MD formation	No	Yes	

Table 1 Variation sources from lithography and HAR processes in FFET and CFET.

979-8-3315-0417-5/25 $31.00 © 2025 IEEE

A 12-bit Fully Differential SAR/SS ADC Architecture Based on Scale Reference for High-speed CMOS Image Sensors

Changhui Yang[1,2], Bailin Zhang[1,2], Weizhe Zhang[1], Nanbo Chen[1,2], Jingyang Chen[1,2], Ziyang Hu[2], Gang Wang[2], Peng Feng[1,3]*, Jian Liu[1,3], Nanjian Wu[1,3], Liyuan Liu[1,3]

[1] Institute of Semiconductors, Chinese Academy of Sciences, Beijing, China

[2] School of Physical Science and Technology, Ningbo University, Ningbo Zhejiang, China

[3] Center of Materials Science and Optoelectronics Engineering, University of Chinese Academy of Sciences, Beijing, China

Abstract

This paper presents a 12-bit fully differential SAR/SS ADC designed for high-speed CMOS image sensors, that uses SAR ADC with a scaled reference CDAC for coarse quantization and SS ADC with a current-steering ramp generator for fine quantization. The circuit, designed in a 110 nm 1P4M CMOS process, saves at least 80% of chip area and achieves a favorable speed-area trade-off. Simulation results indicate a conversion time of 3.9μs at a clock frequency of 20MHz.
Keywords: CMOS image sensors, SAR/SS ADC, scaled reference, current-steering ramp generator.

Introduction

Compared to CCD image sensors, CMOS image sensors (CIS) have notable advantages in integration, power consumption, and processing speed. However, meeting the demands for higher frame rates, larger dynamic ranges, and smaller chip sizes remains a challenge for high-speed CIS. The analog-to-digital converter (ADC) in high-speed CIS is a key component that determines the performance and quality of CIS. CIS primarily uses column-parallel ADCs, which offer a better trade-off between area, power consumption, and sampling rate. Traditional column-parallel ADCs use smaller-area, higher-linearity SS ADCs. However, SS ADCs require 2^N clock cycles to complete quantization, making it difficult to meet high-speed design requirements. In contrast, SAR ADCs complete quantization in just N clock cycles, significantly enhancing the ADC's quantization rate. Nevertheless, the area of the capacitor digital-to-analog (DAC) array in SAR ADCs increases exponentially with resolution, making it challenging to apply them in CIS while maintaining high precision [1]. This paper proposes a 12-bit fully differential SAR/SS ADC by analyzing and comparing the two types of ADCs and leveraging their structural advantages. The higher 8 bits (H8B) are quantized by the SAR ADC while the lower 4 bits (L4B) are quantized by the SS ADC. This design reduces the number of unit capacitors by 80%, saves chip area, and has a faster conversion cycle than a 12-bit SS ADC.

Proposed SAR/SS ADC Structure

The ADC employs a differential structure, which offers superior noise suppression [2]. The block diagram of the proposed hybrid fully differential SAR/SS ADC is depicted in Fig.1. The H8B of the ADC is quantized by a scale reference SAR ADC, while the L4B is quantized by a current-steering SS ADC. This approach mitigates excessive area from the SAR ADC and enhances SS ADC conversion speed, achieving a favorable trade-off between speed and area. Furthermore, there are two pairs of reference sources, specifically V_{refn1} and V_{refp1}, as well as V_{refn2} and V_{refp2} [3]:

$$V_{FS} = V_{refp1} - V_{refn1} \quad (1)$$
$$V_{refp2} = V_{refp1} + V_{FS}/2^4, \quad V_{refn2} = V_{refn1} + V_{FS}/2^4 \quad (2)$$

Where V_{refn1} is 2.6V, V_{refp1} is 0.6V, and V_{FS} is the quantization range. In this architecture, the scale reference, which means another pair of references V_{refn2} and V_{refp2} is used to greatly reduce the number of capacitors in SAR ADC.

Fig. 2 shows the timing diagram of the SAR/SS ADC. During the sampling period, both comparator input voltage V_P and V_N equal to the common mode voltage V_{cm}. Then, the input terminals of the comparator are disconnected with V_{cm}. Before comparison, without quantization. Fig.3 shows the upper 4-bit (U4B) conversion, the SAR ADC determines U4B by switching between V_{refn1} and V_{refp1}. Take the Most Significant Bit (MSB) as an example. The b_{11} in U4B, which means the twelfth highest bit in SAR/SS ADC, is initially set to "1" with the rest as "0". The bottom plate of C_3 in the positive capacitor array is switched from V_{refn1} to V_{refp1}, while the rest remain unchanged. Similarly, switch the C_3 bottom plate in the negative capacitor array from V_{refp1} to V_{refn1}. The comparator then determines b_{11} based on the magnitude of V_P and V_N. If V_P is less than V_N, the output of the comparator is "0", b_{11} is "1", and the voltage of bottom plate of C_3 remains unchanged. If V_P is greater than V_N, the output of the comparator is "1", b_{11} is "0", and the switches of C_3 are set to initial state. Meanwhile, the voltage of bottom plate of C_2 switched between V_{refn1} and V_{refp1}. The remaining U4B values are determined sequentially.

During the T1 period in Fig.1, the reference voltage of the

979-8-3315-0417-5/25 $31.00 © 2025 IEEE

capacitor array must be switched to V_{refp2} or V_{refn2}, which increases the negative capacitor array's bottom plate reference voltage by $V_{FS}/16$. During the M4B period in Fig.1, as shown in Fig.4, the voltage of the bottom plate of the positive capacitor array remains unchanged. The voltage of the bottom plate of the dummy capacitor in the negative capacitor array switches from V_{refp1} to V_{refn2}, while the voltage of the bottom plate of the other capacitors switches from V_{refp1} or V_{refn1} to their corresponding scaled reference V_{refp2} or V_{refn2}. The V_N change dV_{N_1} caused by dummy capacitor switching in the negative capacitor array and the V_N change dV_{N_2} caused by other capacitors switching in the negative capacitor array are as follows [4]:

$$dV_{N\_1} = (V_{refp2} - V_{refp1})/16 = -15V_{FS}/256 \qquad (3)$$
$$dV_{N\_2} = (V_{FS}/16) \cdot 15/16 = 15 \cdot V_{FS}/256 \qquad (4)$$

The change in V_N caused by dV_{N_1} and dV_{N_2} is zero, ensuring V_N remains unchanged. When performing M4B comparison, the bottom plate of the positive capacitor array can switch between V_{refp1} or V_{refn1} and V_{refp2} or V_{refn2}, and the same applies to the negative capacitor array. The remaining values of M4B are determined by this switching.

In Fig.5, the ramp generator uses a current steering DAC structure to produce a rising and falling ramp with differential characteristics. It has a current source array that maintains current stability and safeguards against source jitter. After quantizing the 8th bit, the comparator output determines if V_N and V_P remain constant or if the process is repeated to ensure V_N is greater than V_P. The voltage difference between V_P and V_N is the residual voltage V_{res}. According to the principle of charge redistribution, the output V_{sig} voltage of the capacitor array can be expressed as [5-6]:

$$V_{sig} = V_{res} + \sum_{i=1}^{8} \left[\frac{V_{FS}}{2^i} \cdot (2D_{CMP\_SAR}[i] - 1) \right] \qquad (5)$$

Where $D_{CMP\_SAR}[i]$ is the i-th output of the comparator.

It is necessary to provide a period T2 as shown in Fig.1, before L4B conversion of the SS ADC. The length of T2 depends on the voltage of V_N and V_P in the 8th comparison results. During the T2 conversion period as shown in Fig.1, V_{rise} and V_{fall} are connected to the bottom plate of the dummy capacitor. Then the quantization of L4B begins, with each step in V_P and V_N having a height of $V_{FS}/2^5$ and a step size of $V_{FS}/2^4$. When the output of the comparator changes, the four digital codes are the quantized results of L4B. Upon completion of the L4B conversion, the quantization result of the H8B, which has been quantized by the SAR ADC and stored in a register, is combined with the quantization result of the L4B quantized by the SS ADC at time T3. This yields the final quantization result of the 12-bit SAR/SS ADC, which is outputted immediately before the next quantization cycle begins.

Simulation Results

The circuit is implemented in 110nm 1P4M CMOS process and the power supply voltage is 3.3V. Simulation results for both coarse and fine quantization SAR/SS ADCs are presented and discussed.

The coarse quantization simulation results for the SAR section are depicted in Fig.6. At 20MHz clock rate, during the sampling period, $V_P = V_N = V_{cm}$, and V_{ip} and V_{in} are sampled to the lower plate of the capacitor. Then, the SAR logic sequentially switches the capacitor connection mode through the switch to alter the values of V_P and V_N. As shown in Fig.6, with each switch, the values of V_P and V_N gradually converge to the middle value until 8-bit quantization is complete. The quantization period of SAR ADC is 9 clock cycles, which is 0.45us. Fig.7 shows simulation results for two current references, I_1 and I_2, in the current source array. Each reference is 1.0019µA and 4.0062µA, meeting current accuracy requirements. After the SAR ADC finishes high 8-bit quantization, the SS ADC waits one to two clock cycles. Then, quantization begins, and 1-bit requires 16 clock cycles, which is 0.8us, while L4B requires a total of 3.2us. As shown in Fig.7, the output signal stabilizes within 1/4 LSB by 13ns, meeting fast establishment requirements, and each step value is 0.0625V, meeting design requirements.

Conclusion

The fully differential SAR/SS ADC suitable for high-speed CIS was designed. Which uses a scaled reference DAC instead of a traditional DAC in the H8B SAR ADC, reducing DAC area by 93.75%. The L4B SS ADC improved the linearity of the ramp signal by optimizing the current steering DAC, reducing the use of unit capacitors. This method saves chip area and achieves a conversion time of 3.9us.

Acknowledgments

This work is supported by the National Key Research and Development Program of China (2022YFB2804402), National Natural Science Foundation of China (62134004), and Basic Frontier Scientific Research Program of the Chinese Academy of Sciences (ZDBS-LY-JSC008).

References

[1] M. K. Kim et al, "An Area-Efficient and Low-Power 12-b SAR/SS ADC Without Calibration Method for CMOS Image Sensors", in IEEE Transactions on Electron Devices, pp. 3599-3604 (2016).

[2] Galvez A M. "Design of a fully-differential dual-slope analog-to-digital converter". California State University. (2019).

[3] Zhang, Xiaowei, Wei Fan, et al. "14-Bit Fully Differential SAR ADC with PGA Used in Readout Circuit of CMOS Image Sensor". Sensors 2021, pp. 1-17 (2021).

[4] Li, Z et al. "A 14-bit column-parallel two-step SA ADC with digital calibration based on scaled references and redundancy for CMOS image sensors". Analog Inte Circ Sig Process 100, pp. 295–309 (2019).

[5] G. Liu. "Research on Column level Fully Differential SAR/SS ADC in CMOS Image Sensor". Xi'an University of Technology. (2019).

[6] G. Liu et al. "A Fully Differential SAR/Single-Slope ADC for CMOS Imager Sensor", IEEE of 2019 EDSSC, Xi'an, China, pp. 1-3 (2019).

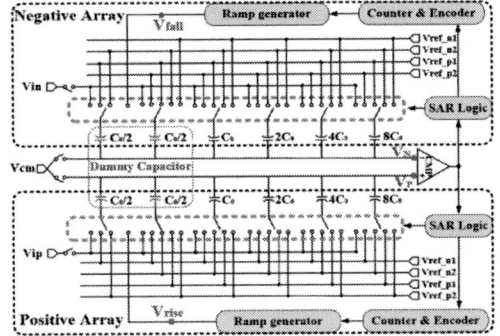

Fig.1: The structure block diagram of the proposed SAR/SS

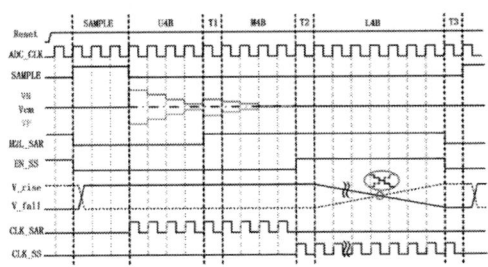

Fig.2: The working timing of SAR/SS ADC

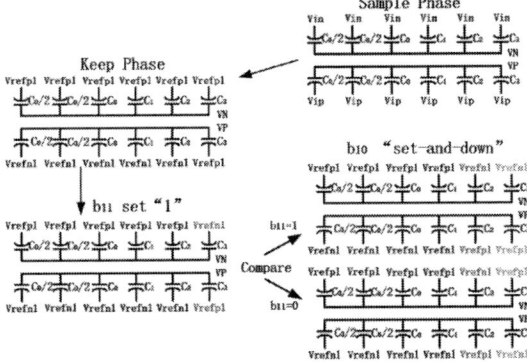

Fig.3: The switch scheme for U4B

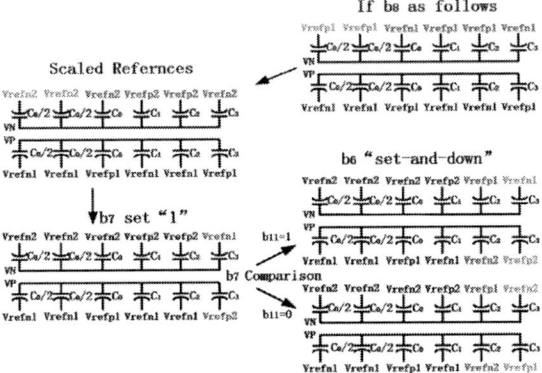

Fig.4: The switch scheme for M4B

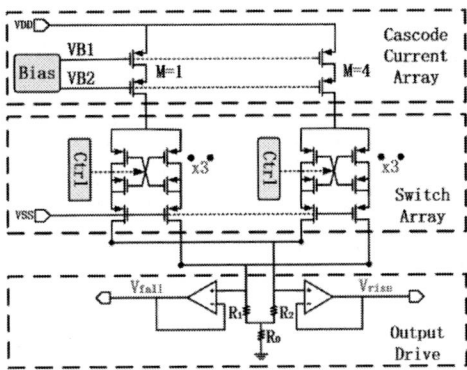

Fig.5: Current steering DAC block diagram

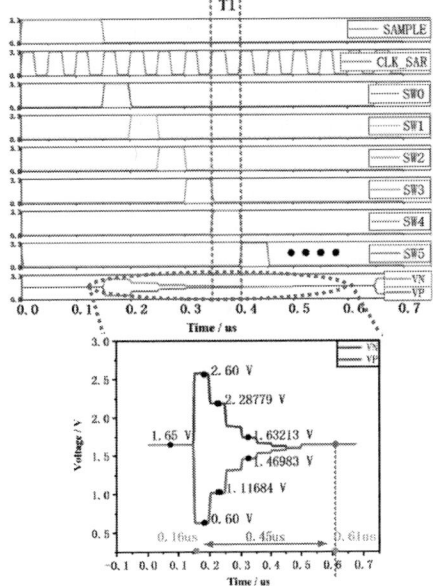

Fig.6: Simulation of SAR ADC

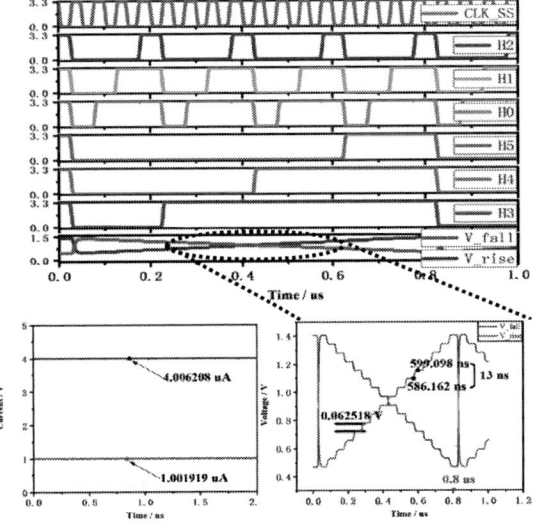

Fig.7: Current steering DAC simulation results

979-8-3315-0417-5/25 $31.00 © 2025 IEEE   681

A High-Throughput Parasitic TRNG in Self-Rectifying Memristor based CIM for Edge Secure Computing

Ying-Jie Yu[1], Sheng-Guang Ren[1], Yu-Yang Fu[1], Jian-Cong Li[1], Pu-Yi Zhang[1],
Yi Li[1, *] and Xiang-Shui Miao[1]

[1]School of Integrated Circuits, Huazhong University of Science and Technology, Wuhan 430074, China
*Email: liyi@hust.edu.cn

Abstract

Edge computing systems are resource-constrained and security-critical, necessitating on-chip TRNG with high area efficiency. For the first time, we demonstrated a parasitic TRNG in a self-rectifying memristor (SRM) based CIM array by sharing the same array and peripheral circuit with CIM. The resistance fluctuations at low read voltage are utilized to generate entropy. Thanks to the high parallelism, the proposed TRNG achieves a high throughput of 2.56 Gb/s, meeting the demands for frequent cryptographic key generation of edge computing system. The parasitic-in-CIM method presented in this work can be extended to other devices exhibiting resistance fluctuations. Our research offers an area-efficient and high-throughput solution for future CIM-based edge secure computing systems.

Keywords: True random number generator (TRNG), computing-in-memory (CIM) and edge secure computing

Introduction

Edge computing systems are resource-constrained and security-critical, requiring hardware that offers high computing efficiency and robust privacy protection [1]. On one hand, computing efficiency is limited by the separation of storage and computation inherent in conventional von Neumann architecture. Non-volatile computing-in-memory (CIM) technology can help overcome this efficiency bottleneck by enabling analog in-situ parallel multiply-and-accumulate (MAC) operations. Non-volatile CIM chips based on resistive random-access memory (RRAM) have demonstrated effectiveness in edge computing applications, such as image processing [2] and artificial intelligence (AI) [3]. On the other hand, true random number generators (TRNGs) play a crucial role in ensuring data privacy by generating true random numbers that are unpredictable through computation, relying on physically random phenomena. Traditional CMOS-based TRNGs require a large circuit area for random number generation and calibration. Recently, several TRNGs based on non-volatile devices, such as RRAM [4-5] and magnetic random-access memory (MRAM) [6], have gained significant attention due to their low power consumption and reduced area overhead. However, the aforementioned TRNGs all necessitate additional array and readout circuits, which impose a considerable burden on resource-constrained edge computing systems.

In this work, for the first time, a self-rectifying memristor (SRM) based parasitic-in-CIM TRNG was designed and demonstrated for CIM-based edge computing system. This innovative TRNG utilizes the same array and readout circuit as the CIM architecture. The entropy source comes from the resistance fluctuation at low read voltage. Low energy consumption and high throughput are achieved thanks to the high parallelism of the SRM array and high-precision readout of the analog-to-digital converter (ADC) for CIM. This parasitic-in-CIM TRNG departs from traditional methods of implementing TRNGs that rely on specialized circuits, thereby minimizing area overhead in non-volatile CIM-based edge computing systems (Fig. 1).

Design of parasitic-in-CIM TRNG

A. Working Principle

The SRM is constructed by Pt/HfO$_2$/TaO$_x$/Ta structure, which has been reported in our previous work [9]. Fig. 2 shows the 10 consecutive I-V curves of the SRM. We select 2 V as the read voltage for CIM (V$_{read-CIM}$), at which the high and low resistance states (HRS, LRS) can be clearly distinguished, as shown in Fig. 3. However, when reading the SRM at 1 V (V$_{read-TRNG}$), whether the SRM is in HRS or LRS, only randomly varying fluctuating currents at the pA level are obtained, as demonstrated in Fig. 4. This phenomenon, which we call nonlinear behavior, is mainly caused by the barrier effect at the HfO$_2$/TaO$_x$ interface which interferes with electron conduction. Detailed physical modeling analysis can be referred in Ref. [10].

B. The circuit implementation of parasitic-in-CIM TRNG

Thanks to the different phenomena exhibited by the SRM at two read voltages (V$_{read-CIM}$ and V$_{read-TRNG}$), both the CIM and TRNG functions can be implemented within the same circuit by simply adjusting the input read voltage and the amplification of the readout circuit. The proposed circuit design is illustrated in Fig. 5, which consists of digital-to-analog converters (DACs), a SRM array, transimpedance amplifiers (TIAs), and ADCs. The DAC is employed to apply a read voltage to the SRM array based on the function being executed. For the SRM arrays, different current magnitudes are detected when performing different functions (~1 pA for TRNG and ~100 pA for CIM). Consequently, the TIA linearly converts the current to voltage for readout, with the conversion multiplier (100 MΩ and 10 GΩ) being adjustable and determined by the specific functions. A differential readout scheme is implemented to eliminate bias, enhancing

979-8-3315-0417-5/25 $31.00 © 2025 IEEE

the accuracy of computation results for CIM. In the case of TRNG, high-quality random numbers can be generated without the need for additional corrections. It is important to note that both the sign and amplitude of the differential current readout in TRNG mode are random. Therefore, each bit of the ADC's digital readout can be regarded as a binary random number.

In practical applications, varying environmental conditions can cause the average weight of binary random numbers to deviate by 0.5. To enhance the robustness of TRNG, an XOR-based calibration scheme is proposed (Fig. 7). This method requires only two read operations and a simple XOR gate. By implementing this straightforward calibration scheme, we can ensure high-quality random number output, even when the weights of the output random bit streams are skewed.

Result and Discussion

A 4×1 SRM array and its peripheral circuit have been fabricated to validate the CIM and TRNG functions (Fig. 6). The HRS and LRS can be distinguished, and the random fluctuations of the read current in TRNG mode can be observed. Based on the HRS and LRS distributions shown in Fig. 3, the current outputs with different MAC results for 16×8 arrays have been simulated, taking device variations into account. In CIM mode, the MAC values (-8~8) and output current are positively correlated, as demonstrated in Fig. 8. A 4-bit ADC is used to sense the converted output voltage, allowing for the acquisition of four binary random numbers per read operation in TRNG mode. Utilizing the circuit depicted in Fig. 5, the converted differential output voltage obtained from 4.5 million read operations is presented in Fig. 9. The voltage distribution shows a Gaussian distribution ($\mu = 0$ mV, $\sigma = 32.42$ mV) with a range of approximately -100 mV to 100 mV. The resolution of the ADC is set to 14 mV. After ADC conversion, the bitmaps of four random bitstreams obtained are shown in Fig. 10. To evaluate the randomness, 2.25 million bits sequence per bit were used in the standard NIST test, as shown in Table 1. All four bits passed the statistical tests, confirming their independent randomness and suitability as generation keys for the edge computing security system.

We have also benchmarked the performance of the proposed TRNG implementation against state-of-the-art TRNGs, as shown in Table 2. Due to the parasitic-in-CIM-based SRM array and the direct utilization of the device's resistance fluctuations during read operations, the proposed TRNG does not require any programming of the device or additional specialized circuit. Thanks to the high parallelism of the read operation and high ADC precision, our TRNG achieves a throughput of 2.56 Gb/s and exhibits low energy consumption of 1.6 pJ/bit.

The parasitic TRNG scheme proposed in this work can be extended to any non-volatile devices with resistance fluctuations. This scheme parasitizes the TRNG within the CIM architecture and shares the same array and peripheral circuit, significantly reducing the area burden on resource-limited CIM-based edge computing system and allowing for more space for computation. Our study explores the possibility of a non-specialized circuit implementation of TRNG for future CIM-based edge secure computing system.

Conclusion

In summary, we have developed a novel methodology that leverages the combination of CIM and TRNG to maximize the area efficiency of the edge security system. An XOR-based calibration scheme with minimal circuit overhead was proposed for robust randomness. Benefiting from the high parallelism of array reads, our TRNG can achieve a high throughput to cope with frequent key generation. The illustrated idea can also be extended to other devices with feature of resistance fluctuations. Our study offers an area-efficient and high-throughput solution for future CIM-based edge secure computing systems.

Acknowledgments

This work was supported by the STI 2030—Major Projects (Grant No. 2021ZD0201201), the National Key Research and Development Plan of MOST of China (2022YFB4500101), the Fundamental Research Funds for the Central Universities (HUST: 5003190012), and the Natural Science Foundation of Hubei Province under Grant (2024AFA043).

References

[1] S. Taneja, M. Alioto, "Fully Synthesizable Unified True Random Number Generator and Cryptographic Core", IEEE Journal of Solid-State Circuits, 56(10), pp. 3049-3061 (2021).

[2] C. Li, M. Hu, Y. Li et al, "Analogue signal and image processing with large memristor crossbars", Nature Electronics, 1, pp. 52-59 (2018).

[3] C. Xue, Y. Chiu, T. Liu et al, "A CMOS-integrated compute-in-memory macro based on resistive random-access memory for AI edge devices", Nature Electronics, 4, pp. 81-90 (2021).

[4] B, Lin, B. Gao, Y. Pang, et al, "A High-Speed and High-Reliability TRNG Based on Analog RRAM for IoT Security Application", IEEE International Electron Devices Meeting (IEDM), pp. 14.8.1-14.8.4 (2019).

[5] Q. Ding, H. Jiang, J. Li, et al, "Unified 0.75pJ/Bit TRNG and Attack Resilient 2F²/Bit PUF for Robust Hardware Security Solutions with 4-layer Stacking 3D NbOx Threshold Switching Array", IEEE International Electron Devices Meeting (IEDM), pp. 39.2.1-39.2.4 (2021).

[6] K. Yang, Q. Dong, Z. Wang, "A 28NM Integrated True Random Number Generator Harvesting Entropy from MRAM", IEEE Symposium on VLSI Circuits, pp. 171-172 (2018).

[7] S. Taneja, V. K. Rajanna and M. Alioto, "36.1 Unified In-Memory Dynamic TRNG and Multi-Bit Static PUF Entropy Generation for Ubiquitous Hardware Security", IEEE International Solid-State Circuits Conference (ISSCC), pp. 498-500 (2021).

[8] M. S. Equbal, T. Ketkar, S. Sahay, "Hybrid CMOS-RRAM True Random Number Generator Exploiting Coupled Entropy Sources", IEEE Trans. on Electron Devices, 70(3), pp. 1061-1066 (2023).

[9] J. Li, S. Ren, Y. Li, et al, "Sparse matrix multiplication in a record-low power self-rectifying memristor array for scientific computing", Science Advances, 9(25): eadf7474 (2023).

[10] S. Ren, G. Mao, Y. Xue, et al, "Interface modeling analysis using density functional theory in highly reliable Pt/HfO2/TaOx/Ta self-rectifying memristor", Applied Physics Letters, 125(12), pp: 123503 (2024).

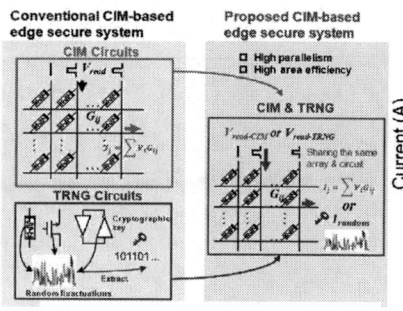

Fig 1: Illustration of Conventional CIM-based edge secure system and proposed system.

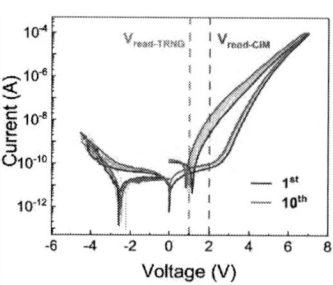

Fig 2: Typical I-V curves of the SRM.

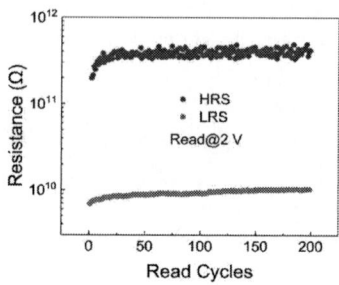

Fig 3: HRS and LRS distributions under $V_{read-CIM}$=2 V.

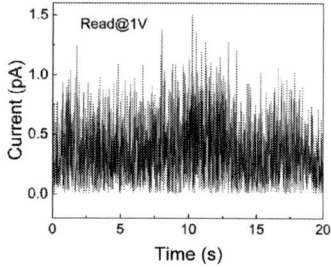

Fig 4: Current fluctuation of SRM under $V_{read-TRNG}$=1 V.

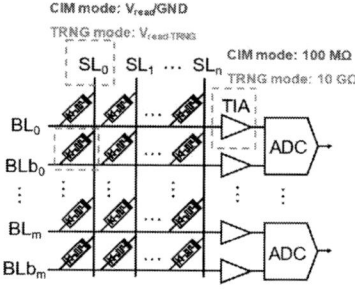

Fig 5: Schematic illustration of the parasitic TRNG in CIM system.

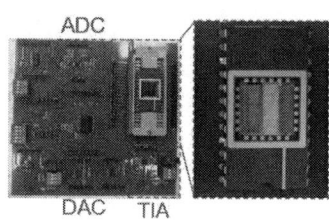

Fig 6: The hardware test platform for 4×1 SRM array.

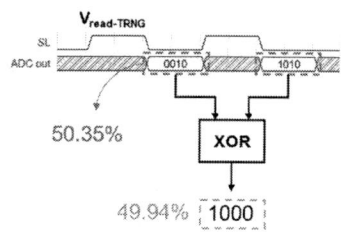

Fig 7: The XOR-based calibration method.

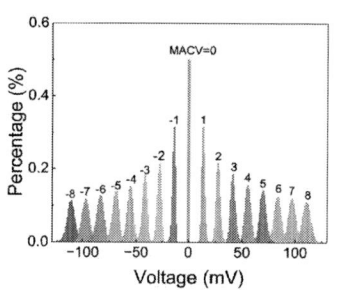

Fig 8: Simulated CIM functions with 16×8 SRM array.

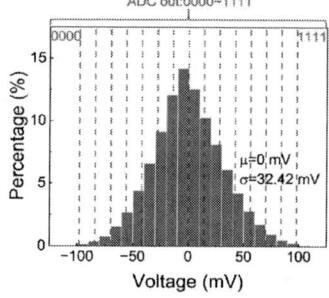

Fig 9: Voltage distribution after TIA conversion under TRNG mode.

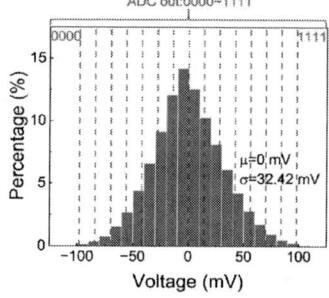

Fig 10: 1500×1500 bitmaps of four bitstream based on the 4-bit ADC readout.

Table 2: Performance Comparison of the-state-of-the-art TRNGs.

Device	This work	ISSCC 2021[7]	IEDM 2019[4]	TED 2023[8]	IEDM 2021[5]
	RRAM	SRAM	RRAM	CMOS-RRAM	TS
Entropy Source	Resistance Fluctuations	Leakage Noise Jitter	Pulse Number	SR-Latch	Leakage Mismatch
Throughput	2.56 Gb/s	3.6 Mb/s	12 Mb/s	10 Mb/s	-
Specialized Circuit?	No (CIM+TRNG)	No (SRAM+PUF+TRNG)	Yes	Yes	Yes
NIST test passed?	All	All	All	All	All
Energy efficiency	1.6 pJ/bit	9.6 pJ/bit	3.72 pJ/bit	-	750 fJ/bit

Table 1: NIST SP800-22 randomness test results.

NIST test	p-value				If pass?
	bit3	bit2	bit1	bit0	
Frequency	0.03	0.625	0.655	0.851	Pass
Block Frequency	0.144	0.5	0.842	0.452	Pass
Cumulative Sum-1	0.011	0.209	0.704	0.657	Pass
Cumulative Sum-2	0.046	0.122	0.429	0.493	Pass
Runs	0.596	0.386	0.823	0.408	Pass
Longest Runs	0.642	0.259	0.206	0.613	Pass
FFT	0.883	0.646	0.849	0.23	Pass
Rank	0.175	0.693	0.072	0.784	Pass
Universal	0.125	0.86	0.669	0.469	Pass
Approximate Entropy	0.457	0.247	0.241	0.19	Pass
Non Overlapping	147/148	139/148	145/148	143/148	Pass
Overlapping	0.865	0.207	0.049	0.202	Pass
Random Excusions	8/8	8/8	8/8	8/8	Pass
Random Excusions Variant	18/18	18/18	18/18	18/18	Pass
Linear Complexity	0.38	0.317	0.214	0.296	Pass

Low-Complexity Method for Shortest Path Optimization Problems based on Nanowire Memristor Network

Yanming Liu[#1], Shenghao Wu[#2], Ming Jian[#3], Daixuan Wu[#1], Yanxi Long[2], He Tian[*1] IEEE Senior Member

[1] School of Integrated Circuits, Tsinghua University, China; [2] WeiYang College, Tsinghua University, China; [3] XingJian College, Tsinghua University, China

(Email: tianhe88@tsinghua.edu.cn)

Abstract

The shortest path problem is a cornerstone in graph theory, with traditional algorithms often incurring a computational complexity of $O(n^2)$. In recent years, Ag nanowire memristor networks have emerged as a promising candidate for next-generation hardware due to their excellent random full-interconnectivity and memristive properties. This work introduces a novel approach to implement the shortest path problem using Ag nanowire network hardware, which significantly reduces the computational complexity from $O(n^2)$ to $O(1)$, thereby enhancing computational efficiency dramatically. Our method not only offers a substantial improvement in performance but also paves the way for future algorithmic deployment on nanowire networks.

Keywords: Ag nanowire network, memristor and shortest path optimization.

Introduction

The shortest path problem stands as a cornerstone in graph theory, with applications ranging from transportation networks to network routing in computer science [1]. Traditional algorithms, such as Dijkstra's or Bellman-Ford, often incur a computational complexity of $O(n^2)$, which can be prohibitive for large-scale problems [2, 3]. In recent years, there has been a significant shift towards exploring novel hardware architectures that can offer more efficient solutions to this classic problem.

One such promising development is the use of Ag nanowire memristor networks. These networks have emerged as a candidate for next-generation hardware due to their inherent random full-interconnectivity and memristive properties, which are crucial for neuromorphic computing [4-6]. The potential of Ag nanowire networks in computational tasks has been highlighted in various studies. For instance, Zdenka Kuncic et al. demonstrated the self-organization capabilities of inhomogeneous memristive hardware for handwritten recognition, showcasing the adaptability of these networks for complex computational tasks [7]. Similarly, Carlo Ricciardi et al. proposed a graph theory for computing machines that leverage the unique properties offered by various physical phenomena, including memristive systems [8].

In this work, we proposed a method to implement the shortest path problem using Ag nanowire network hardware. This approach aims to reduce the computational complexity

from $O(n^2)$ to $O(1)$, thereby significantly enhancing the efficiency of solving this problem. We fabricated the circuits and tested the basic algorithm. By harnessing the random full-interconnectivity and memristive properties of Ag nanowire networks, we aim to not only improve the performance of the shortest path problem but also lay the groundwork for future algorithmic deployment on similar nanowire networks.

Device Structure and Behaviors

In this section, we detail the construction and fundamental behaviors of the random Ag nanowire memristor network. We utilized a silver nanowire (AgNW) solution to fabricate the entire network, the basic structure of which is depicted in **Fig.1(a)**. The core structure is shown in **Fig.1(b)**. Each AgNW is coated with a layer of polyvinylpyrrolidone (PVP) approximately 1-2 nm thick, resulting in a typical memristive structure when two nanowires overlap, forming an Ag/PVP/Ag trilayer.

Under current-limiting conditions, the fundamental behavior of these memristors is illustrated in **Fig.1(c)**. Initially, there is no conductive filament present. Upon the application of voltage, Ag ions gradually drift from the anode to the cathode, eventually forming a conductive filament. However, due to the current-limiting nature, once the voltage is removed, the conductive filament spontaneously breaks, reverting the device back to a high-resistance state. This is the fundamental principle of the nanowire memristor network.

We completed the basic device fabrication by spin-coating AgNW solution onto a oxide silicon wafer, allowing it to dry, and then depositing metal electrodes. **Fig.2(a)** presents the results of 150 DC measurement curves for the same device. It can be observed that under a 10^{-8}A current-limiting protection, the device exhibits the characteristics of a volatile memristor. Moreover, due to the complexity of the nanowire network, the device demonstrates a self-healing property. After a clear breakdown phenomenon, it can still recover the memristor window after multiple set operations. This can be observed in the switching voltage statistical graph extracted from **Fig.2(b)**. The device shows a distinct breakdown behavior with reversal points at very low voltages, but after multiple DC cycles, it remains a good resistance switching window. Therefore, it can be concluded that the system possesses excellent stability and robustness, making it a suitable hardware foundation for constructing algorithms.

979-8-3315-0417-5/25 $31.00 © 2025 IEEE

Algorithm Construction and Simulation

Following the successful fabrication of AgNW device, we discovered that the silver nanowires exhibit stable volatile memristive characteristics under low current-limiting conditions. Leveraging the unique, nearly uniform random distribution characteristic of the nanowire network, we proposed a model for constructing an optimal path algorithm using the nanowire network. Our fundamental research approach is to utilize the similarity between memristor networks and ant colony optimization (ACO) algorithms to build the entire architecture.

As illustrated in **Fig.3**, which depicts the ant colony model, the path on the left is closer compared to the path on the right. In terms of memristor circuits, this analogy means that the memristor on the left is smaller, while the memristor on the right is equivalent to a series of more memristors. For the ACO, this implies that the ant colony's path is continuously amplified. For the nanowire memristor network, this means that avalanche breakdown occurs on the shorter paths, allowing the shorter path to be instantly selected.

We prepared the chips and conducted tests, confirming that the nanowire network at the centimeter scale still possesses the characteristics we require. **Fig.4** displays the I-V curves from our testing, which clearly show volatile memristive characteristics. The inset in **Fig.4** shows the wafer we prepared, along with scanning electron microscope (SEM) and optical microscope images of the nanowires.

Fig.5 illustrates our overall research approach. By segmenting the nanowire network and programming interconnections, we can reconfigure all paths to deploy shortest path algorithm. We established an emulating model to illustrate the formation of the conductive filament distribution. We completed the current distribution simulation of the nanowire network through meticulous grid simulation. The darker the color, the greater the current. **Fig.5(a)** shows the basic structure diagram. We change the structure interconnections by routing on a PCB, as shown in **Fig.5(b)**, which demonstrates the connection relationships. **Fig.5(c)** is the current distribution diagram after voltage is applied. It can be observed that the shortest path has the greatest current. **Fig.5(d)** shows that, based on the current results, we have selected the final shortest path, proving the effectiveness of our design.

Since our algorithm can find the optimal path in a single pulse, its complexity is only $O(1)$, which is lower than that of traditional algorithms. The Table.1 compared with traditional algorithm for shortest path problem and showed better efficiency. The stark contrast in complexity not only implies a substantial improvement in processing speed but also suggests that our algorithm is less resource-intensive, making it more suitable for real-time applications where rapid decision-making is crucial.

Conclusion

This work introduced an innovative approach to solving the shortest path problem using Ag nanowire memristor networks, achieving a remarkable reduction in computational complexity from $O(n^2)$ to $O(1)$. Our findings underscore the potential of these networks for neuromorphic computing, offering a significant leap in efficiency for graph-based problems. The successful implementation and simulation of our algorithm on Ag nanowire networks pave the way for further exploration in hardware-based graph algorithms. Future work will focus on scalability, integration with existing technologies, and applications to other complex computational tasks, further solidifying the role of Ag nanowire memristor networks in advancing computational hardware.

Acknowledgments

This work was supported in part by STI 2030—Major Projects under Grant 2022ZD0209200, in part by Beijing Natural Science Foundation-Xiaomi Innovation Joint Fund (L233009), in part by Beijing Natural Science Foundation (20231190052), in part by National Natural Science Foundation of China under Grant No. 62374099, in part by Independent Research Program of School of Integrated Circuits, Tsinghua University.

References

[1] Kitsak, M., Ganin, A., Elmokashfi, A. et al. Finding shortest and nearly shortest path nodes in large substantially incomplete networks by hyperbolic mapping. Nat Commun 14, 186 (2023).

[2] A. Candra, M. A. Budiman and K. Hartanto, "Dijkstra's and A-Star in Finding the Shortest Path: a Tutorial," 2020 International Conference on Data Science, Artificial Intelligence, and Business Analytics (DATABIA), Medan, Indonesia, pp. 28-32, (2020).

[3] R. Abousleiman and O. Rawashdeh, "A Bellman-Ford approach to energy efficient routing of electric vehicles," 2015 IEEE Transportation Electrification Conference and Expo (ITEC), Dearborn, MI, USA, pp. 1-4, (2015).

[4] Hochstetter, J., Zhu, R., Loeffler, A. et al. Avalanches and edge-of-chaos learning in neuromorphic nanowire networks. Nat Commun 12, 4008 (2021).

[5] Milano, G., Pedretti, G., Montano, K. et al. In materia reservoir computing with a fully memristive architecture based on self-organizing nanowire networks. Nat. Mater. 21, 195–202 (2022).

[6] Milano, G., Montano, K., & Ricciardi, C., In materia implementation strategies of physical reservoir computing with memristive nanonetworks. Journal of Physics D: Applied Physics, 56(8), 084005, (2023).

[7] Zhu, R., Lilak, S., Loeffler, A. et al. Online dynamical learning and sequence memory with neuromorphic nanowire networks. Nat Commun 14, 6697 (2023).

[8] Milano, G., Miranda, E., & Ricciardi, C., Connectome of memristive nanowire networks through graph theory. Neural Networks, 150, 137-148, (2022).

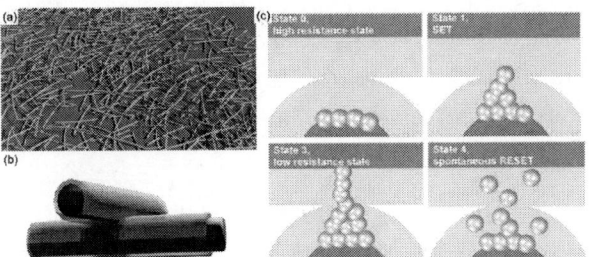

Fig. 1: (a) Schematic of the nanowire network stochastic structure. (b) The Ag/PVP/Ag memristor structure formed by nanowire junction. (c) Schematic representation of the resistive switching mechanism occurring at the nanowire junctions under low current compliance, where the conductivity can be modulated by the formation/rupture of a metallic Ag conductive path across the nanowire shell layer, under the action of the applied electric field and the current compliance.

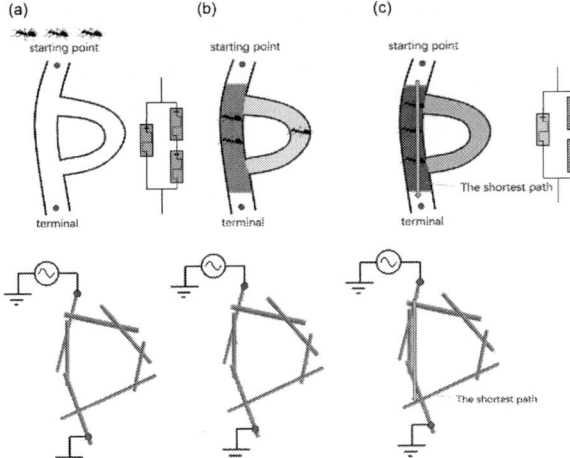

Fig. 3: (a) Schematic shows the representation of the relationship and similarity between path distribution, memristor distribution, and the layout of the nanowire network. (b) Schematic illustrates the analogous effects produced by the ant colony algorithm and the nanowire network, despite being based on different principles. (c) Schematic demonstrates that both the ant colony algorithm, from the perspective of path accumulation, and the nanowire network, from the perspective of current accumulation, arrive at the same final results.

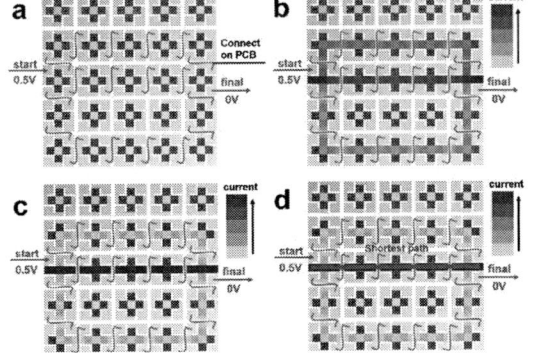

Fig. 5: (a) Schematic of the experimental setup for the nanowire network. (b) Initial current state when voltage is applied. (c) Final current distribution after algorithm execution. (d) The identified shortest path from the simulation results.

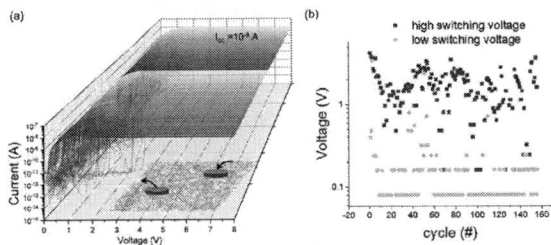

Fig. 2: (a) I-V curves of the nanowire network memristor with a schematic of the two terminal testing shown in the inset. The 150 I-V curves show great stability. (b) The figure shows extracted switching voltage during the DC voltage applying. The nanowire network memristor shows volatile behaviors.

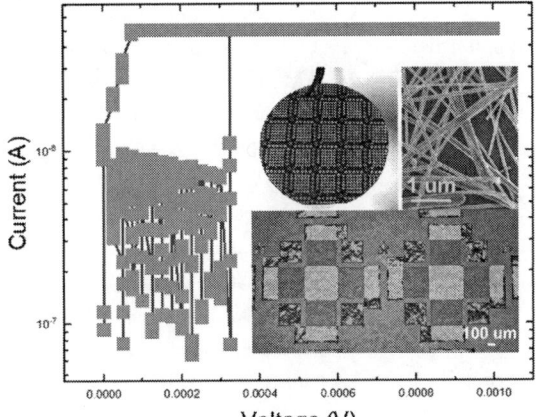

Fig. 4: The figure displays the DC mode I-V curves of long-distance silver nanowires. The reduction in switching voltage is attributed to the use of oxygen plasma etching, which leads to the accumulation of charges that break down some memristors. In the inset figure, the entire wafer image, scanning electron microscope (SEM) image of the nanowires, and the optical microscope image of the nanowire network are presented.

Algorithm	Time Complexity
Dijkstra's Algorithm	$O(V^2)$ simple arrays
	$O(E + V \log V)$ a binary heap
Bellman-Ford Algorithm	$O(VE)$
Floyd.Warshall Algorithm	$O(V^3)$
A* Search Algorithm	$O(E)$ average case
	$O(b^d)$ worst case
Johnson's Algorithm	$O(V^2 \log V + VE)$
Bidireetional Seareh	$O(b^{d/2})$
Yen's K-Shortest Paths Algorithm	$O(K(VE + V \log V))$
Bhandar''s Algorithm	$O(K(VE))$
AgNW Hardware Implementation (This work)	$O(1)$

Table. 1: Comparative Analysis of Shortest Path Algorithms Based on Time Complexity This table provides a detailed comparison of various shortest path algorithms, highlighting their time complexity in terms of the number of edges (E) and vertices (V) within a graph. The time complexity is crucial for understanding the efficiency of each algorithm, especially when dealing with large-scale graph data.

979-8-3315-0417-5/25 $31.00 © 2025 IEEE

Enhancing 3D DRAM Double-Gate Device Performance by Leveraging the Floating Body Effect

Jing Liang[1], Yong Yu[1,2*], Feng Shao[1], Menglong Zhou[1], Tiantian Wei[1], Chunyang Li[1], Jianpeng Jiang[1], Jing Cai[1], Bryan Kang[1], Mingxu Liu[1], Xiangsheng Wang[1], Guilei Wang[1], Chao Zhao[1], Shaojun Wei[2*]

[1]Beijing Superstring Academy of Memory Technology, China

[2]School of Integrated Circuits, Tsinghua University, China

*Email: yong.yu@bjsamt.org.cn, wsj@tsinghua.edu.cn

Abstract

Three-dimensional stackable dynamic random access memory (3D DRAM), featuring horizontally stacked transistors with capacitors, is highlighted as a promising approach to continue DRAM scaling beyond planar architecture. However, the floating body effect (FBE) of this structure is not well understood. Its potential to enhance device performance represents a new direction for the future development of 3D DRAM. This study systematically investigates the physical mechanisms of the FBE and the effect of process parameters on the electrical characteristics of the access transistor. The results show that channel length and doping are not sensitive to leakage current, suggesting that the FBE can be leveraged to reduce the cell area, thereby providing design guidelines for high-density 3D DRAM applications. Furthermore, a novel structure is proposed that perfectly validates the theoretical feasibility.

Keywords: 3D DRAM, Floating Body Effect, BTBT

Introduction

The continuous scaling of DRAM is now facing developmental constraints, primarily due to the mismatch between the reduced capacitor and the increasing off-current. Existing approaches roughly fall into two categories. In the first approach, 4F² DRAM [1] with vertical channels reduces the die area by 30% compared with the 6F² DRAM at the same technology node [2], but this only alleviates the scaling challenges for several generations. The second approach introduces another dimension, taking planar DRAM into a 3D stackable architecture, which has attracted more attention due to its advantages for high-density integration without the need for extremely high-resolution patterning equipment.

The innovative 3D stackable multi-layer cell (MLC) architecture shown in **Fig. 1(a)** features a double-gate structure with a silicon channel [3], offering advantages such as reduced area and easier interconnection compared with the Channel-All-Around (CAA) architecture shown in **Fig. 1(b)**, which uses amorphous semiconductor oxide channel [4]. However, this double-gate structure requires a high tolerance to process variations and easily induces FBE by the absence of body connections.

Thus, investigating the effect of the FBE and its potential applications will be crucial for the future development of 3D DRAM. We analyzed the mechanisms of the FBE in double-gate devices of 3D DRAM and the effect of device parameters on leakage current in a systematic way. The corresponding structure was further proposed to lower leakage current in 3D DRAM access transistors by leveraging the FBE.

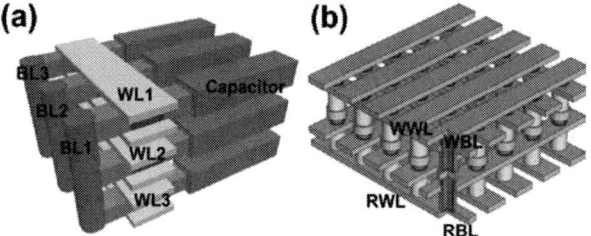

Fig. 1. Schematic of two innovative 3D DRAM architectures: **(a)** MLC architecture [3] and **(b)** stacked CAA 2T0C DRAM architecture [4].

Simulation Results and Discussion

A. Physical mechanism of the FBE in DRAM double-gate devices

We calibrated the Technology Computer-Aided Design model parameters based on the experimental data from the gate-all-around silicon device to improve the simulation accuracy [1,5,6].

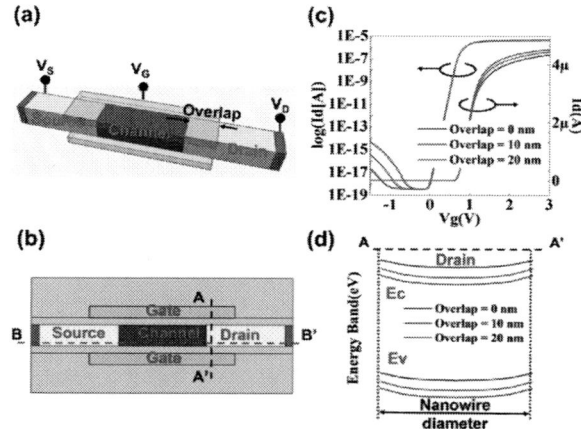

Fig. 2(a) Three-dimensional view, **(b)** two-dimensional schematic, **(c)** transfer curves of the MLC memory cell, and **(d)** energy band profile in the off-state (@Vg = −1.5 V) of transistor along the longitudinal direction (cutline A – A'), indicating transverse band-to-band tunneling (T-BTBT).

Fig. 2(a, b) shows the MLC memory cell. **Fig. 2(c)** presents the transfer curves with the overlap region being 0nm/10nm/20nm. The result fits the well-known trend that

979-8-3315-0417-5/25 $31.00 © 2025 IEEE   688

the gate-induced drain leakage (GIDL) current increases as the overlap increases. The energy band along the longitudinal direction as shown in **Fig. 2(d)** confirms that no transverse band-to-band tunneling (T-BTBT) has occurred [7].

Fig. 3 shows the energy band diagram along the lateral direction, describing the leakage mechanism induced by the FBE in the MLC device.

1) $-0.6\,V < Vg < 0\,V$: The L-BTBT probability is low due to the wide barrier. However, as the gate voltage (Vg) increases, the barrier height rises while the width narrows, leading to stronger tunneling.

2) $Vg < -0.6\,V$: When Vg is relatively negative, L-BTBT generates electron-hole pairs, where electrons flow toward the drain, and holes accumulate in the channel, thereby reducing the overall barrier height. The lightly doped drain (LDD) near the channel is transitioning into an inverted state, which causes tunneling in the LDD close to the drain. Hence, as Vg increases, the tunneling width decreases while the probability increases, resulting in increased leakage current.

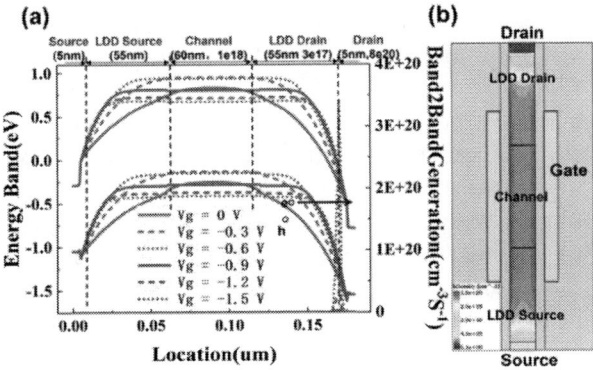

Fig. 3(a) Energy band profile as the Vg become more negative along the lateral direction (cutline B – B' of **Fig. 2(b)**), indicating the L-BTBT gradually worsens. **(b)** Hole density when $Vg = -1.5\,V$ shows the LDD has been inversed.

Fig. 4 shows the electron and hole concentration as well as the generation-recombination rate distribution in the channel center and surface under different negative gate biases. When $Vg = 0\,V$, the LDD region is depleted, which reduces the vertical electric field. At this point, the gate-drain field is insufficient to induce significant BTBT current, making the overall leakage primarily governed by Shockley-Read-Hall (SRH) generation-recombination current. As Vg becomes more negative, L-BTBT increases rapidly, causing the depletion region to be narrower and shift toward the source and drain, thereby reducing SRH. The LDD region undergoes inversion, resulting in a rapid increase in the lateral electric field along the channel direction, with BTBT current occupying the leakage current [8].

Analysises validate the FBE mechanism and indicate that the FBE extends the channel length, which may serve as a pivotal point for MLC 3D DRAM devices.

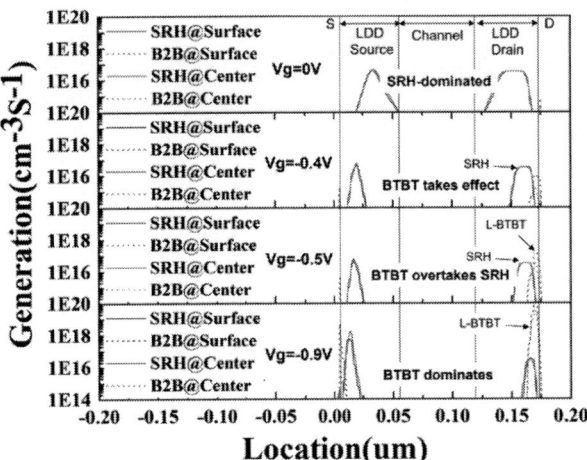

Fig. 4. Distribution of electron and hole generation rates at the channel center and surface under different gate biases.

B. Effect of factors on leakage current

Since the primary role of the access transistor is to minimize capacitor leakage in DRAM 1T1C structure, I_{GIDL} at $Vg = -1.5\,V$ and I_{off} at $Vg = -0.4\,V$ were used as evaluation parameters in our analysis to investigate the effect of the FBE.

Fig. 5 shows the effect of channel doping (Nch), where I_{GIDL} and I_{off} remain consistent under different doping levels, while I_{on} and V_{th} follow conventional trends. The primary cause of the leakage behavior is the hole accumulation due to the FBE, which extends into the LDD depletion region, effectively lengthening the channel. At this stage, channel doping no longer contributes to the current leakage, thereby reducing the need for stringent control over channel doping. This presents a favorable method to regulate I_{on} and threshold voltage.

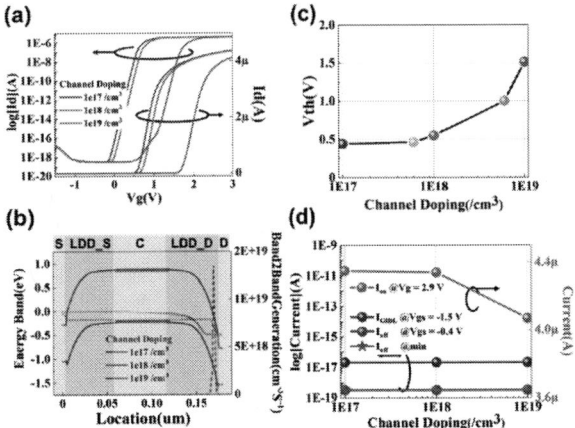

Fig. 5(a) Transfer characteristics and **(b)** energy band along the lateral direction under various channel doping levels, indicating L-BTBT. Trends of **(c)** V_{th} and **(d)** I_{on}, I_{GIDL}, and I_{off} for different Nch levels.

Fig. 6 shows the effect of the channel length (Lch). As the channel length decreases, I_{GIDL} remains nearly constant, while V_{th} slightly decreases. Benefiting from the FBE, a part of the LDD region is inverted into the effective channel, making the short-channel effect inconspicuous. Due to the channel extension, the short-channel design of inversion-mode MLC devices is feasible, providing a direction for further scaling.

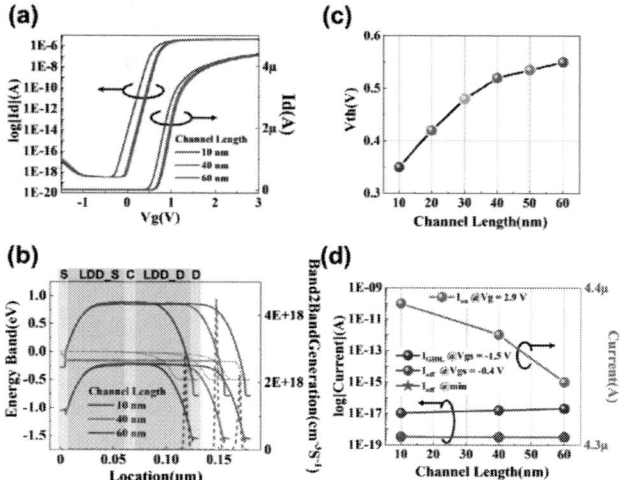

Fig. 6(a) Transfer characteristics and **(b)** energy band along the lateral direction under various channel length (Lch) levels, indicating L-BTBT. Variations in **(c)** Vth and **(d)** I_{on}, I_{GIDL}, and I_{off} for different Lch levels.

In addition to the aforementioned parameters, other relevant factors were also simulated, and **Table 1** ranks the parameter sensitivity of the device characteristic. The following 4 factors are most sensitive to I_{GIDL}: ldd_length, channel_thickness (CH_THK), overlap_length (Lov), and ldd_doping. Lch and Nch show little sensitivity to gate leakage, indicating that the FBE can reduce device cell area and simplify processing difficulties, which is significant for the technology node advancement of 3D DRAM.

C. Proposed structure

A structure based on the FBE mechanism is proposed, as shown in **Fig. 7**, which increases the lateral area of the LDD regions to suppress the depletion region caused by highly doped source and drain. This structure effectively extends the channel length, reduces LDD resistance, and ultimately enhances the on/off ratio. The transfer characteristics and electric field distribution for different LDD widths (H) are also illustrated.

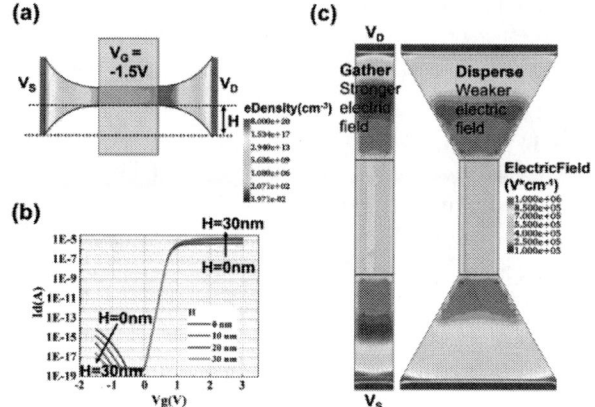

Fig. 7(a) Proposed structure and **(b)** corresponding transfer curves, where I_{on} increases and I_{off} decreases as H increases. Increasing the LDD width suppresses the depletion region extension to the source/drain region, which has **(a)** a higher electron concentration and **(c)** a weaker electric field.

Conclusion

We investigated the physical mechanisms of the FBE in 3D DRAM double-gate devices and analyzed the influence of process parameters on the leakage current and other electrical characteristics systematically. The simulation result showed the insensitivity of channel length and doping to leakage current. This suggests that the FBE can be used to suppress the short-channel effect, reduce memory cell area, and ultimately simplify process complexity. Furthermore, based on the aforementioned conclusions, a structure has been proposed, which introduces a new strategy for reducing leakage current in 3D DRAM access transistors.

Acknowledgments

This work was supported by the National Key R&D Program of China (2022YFB3606900), and the Beijing Science and Technology Plan (Z221100007722025).

References

[1] D. Feng et al., IMW, pp. 1-4, 2023. [2] Y. Cho et al., TED, vol. 65, no. 8, pp. 3237-3242, 2018. [3] M. Huang et al., IMW, pp. 1-4, 2023. [4] J. Liang et al., JJAP, vol. 63, no. 6, 2024. [5] W. Wang et al., ICSICT, pp. 1-3, 2022. [6] A. Yoo et al., IEDM, pp. 1-4, 2023. [7] Sahay et al., IEEE Press Series on Microelectronic Systems, 2019. [8] M. Shi et al., IC

Table 1: Parameter sensitivity of the device characteristic.

Formula of SX	Sx	Vth	Ioff	Ioff_min	I_GIDL	Ion
	(X=	(@1nA)	(Vgs=-0.4V)		(Vgs=-1.5V)	(Vgs=2.9V)
$\dfrac{2*(Xmax-Xmin)}{Xmax+Xmin}$	CH_THK	18.55%	64.74%	68.72%	199.98%	9.63%
	Ldd_length	3.43%	199.99%	198.31%	200.00%	18.66%
	CH_length	46.67%	10.46%	54.98%	63.18%	1.08%
	N_channel	112.71%	10.46%	11.61%	40.85%	6.54%
	Ldd_dopping	6.18%	200.00%	167.55%	188.76%	2.71%
	Lov	1.10%	21.60%	5.50%	198.40%	4.30%

Categorization of Stacking Faults and Their Effects on I-V Characteristics in 2nm p-Type GAA Nanosheet Transistors

Seungjoon Eom[1], Sanguk Lee[1], and Rock-Hyun Baek[1]

[1]Pohang University of Science and Technology (POSTECH), Electrical Engineering, 77 Cheongam-Ro, Nam-Gu, Pohang, Gyeongbuk, 37673, Republic of Korea

Abstract

Gate-All-Around (GAA) nanosheet field-effect transistors (NSFETs) face challenges with stacking faults (SFs) that disrupt stress transmission, affecting device performance. This study categorizes these faults and examines their impact on stress, carrier mobility, and I-V characteristics using technology computer-aided design (TCAD) simulations. Results demonstrate that specific fault types significantly alter stress distribution and electrical characteristics, leading to variations in boron diffusion rates and on/off-current, thereby emphasizing the importance of managing SFs to enhance p-type NSFET performance.

Keywords: Stacking Fault, Nanosheet FET, S/D stressor

Introduction

FinFET technology has been crucial in scaling semiconductor devices beyond planar Si-MOSFETs [1]. However, as device dimensions shrink, FinFETs face challenges in electrostatic control, scalability, and increased variability. To address these issues, gate-all-around (GAA) nanosheet FETs (NSFETs) have been proposed as promising successors to FinFETs [2]. However, the effectiveness of source/drain (S/D) stressors in GAA structures diminishes due to epitaxial growth on multiple, separated silicon surfaces, leading to incomplete merging and stacking faults (SFs) in the S/D that hinder stress transmission [3], [4]. Previous studies on SFs in NSFETs primarily focused on stress effects, with less emphasis of their impact on I-V characteristics [5]. Based on our previous work [6], we used lattice kinetic Monte Carlo (LKMC) simulations to extract the epitaxy merging interface, allowing us to classify SF types in p-type NSFETs and assess their effects on stress, hole mobility, and dopant diffusion across each sheet. Our analysis explores the influence of these faults on I-V characteristics and also examines how partial non-stressor S/D impacts device performance.

TCAD Structure & SF Simulation

We used LKMC simulations (Fig. 1(a)) to identify where the S/D $Si_{0.5}Ge_{0.5}$ epitaxial layers grown from separate Si nanosheets meet each other. In an ideal scenario where no SFs form at the merge interface, the S/D stressor induces strong compressive stress in the channel due to lattice mismatch with the silicon substrate (Fig. 1(b)). This compressive stress enhances hole mobility, thereby boosting the pFET current. However, SFs at the merge interface can disrupt this effect: vertical SFs (V-SFs) can cause stress relaxation in S/D stressors, while horizontal SFs (H-SFs) can hinder the SiGe above from merging with the silicon substrate (Fig. 1(c), (d)). To evaluate their impact, we simulated 2nm node p-type Si-NSFETs using Synopsys Sentaurus technology computer-aided design (TCAD) (Fig. 2), calibrated with hardware data, and scaled to meet IRDS specifications (Table 1) [7], [8]. Process simulations included doping profiles and strain effects. Carrier mobility was calculated using Lombardi and thin-layer mobility models; recombination mechanisms included Shockley-Read-Hall, Auger, and Hurkx band-to-band tunneling models. To account for the stress relaxation at SFs, we inserted a $Si_{0.5}Ge_{0.5}$ buffer layer with its bulk modulus (a measure of material stiffness) drastically reduced to 10^{-5} times its original value. Therefore, strain is hardly converted into stress in the buffer layer, which weakens the functionality of the adjacent S/D stressors. Additionally, the SiGe layers separated from the silicon substrate by H-SFs and bottom SF (B-SF) are set to not act as stressors, as they do not experience lattice mismatch. For example, when H2 occurs, the middle and top layer SiGe are set to not function as stressors. To simplify the scenarios, we assumed identical SFs in both S/D SiGe.

Results and Discussion

Fig. 3 shows the impact of SFs on the I-V characteristics, deviating from the ideal case with no SFs present. SFs lead to increased off current (I_{off}), shifts in threshold voltage (V_{th}), and decreased on current (I_{on}). As shown in Fig. 4, locations of V-SFs significantly affect the subthreshold region, leading to considerable variations in I_{off}. B-SF shows degraded I_{off} while maintaining I_{on} similar to the ideal case. Conversely, H-SFs consistently degrade I_{off} across all cases while causing greater variations in I_{on}. Fig. 5 illustrates the longitudinal compressive stress in the device's channel for each nanosheet layer (top, middle, bottom) extracted after the simulation steps. In the All-SF device, stress is minimal due to SFs in all interfaces, whereas the ideal device faces high compressive stress in all layers, making All-SF the worst-case and the ideal device the best-case for current boosting. For V-SFs, V2 only reduces middle sheet stress by 0.32GPa because both the top and bottom SiGe stressors offset the decrease. In contrast,

V3 reduces top sheet stress by 0.96GPa because it is supported solely by the middle layer S/D. In the V2&V3 configuration, top sheet stress is eliminated, and the bottom layer S/D partially compensates to moderate the middle sheet's stress relaxation. For B-SF, the stress is significantly relaxed only on the bottom sheet, while the stress levels on the other sheets remain unaffected. As is evident in V-SF, B-SF, and most H-SF scenarios, SFs generally decrease stress in the layers where they originate and do not affect stress in other layers. In contrast, the H2 scenario affects the middle and top sheet stress by preventing the SiGe in these layers from acting as a stressor. For H2&H3, although a buffer layer is added between the middle and top layers, it has minimal impact on stress levels compared to H2 alone. This suggests that the non-stressor SiGe more significantly influences the stress relaxation by H-SF than by the buffer layer.

A. On-state analysis (Vgs=Vds=Vdd= -0.65 V)
Fig.6 depicts the on-state hole velocity and density for H-SF and V-SF. Compared to the ideal case, H3 shows hole velocity reduction only in the top sheet, while H2 and H2&H3 exhibit decreases in both middle and top sheets. This occurs because the compressive stress in these sheets is reduced by SFs, as shown in Fig. 5, leading to less mobility enhancement. Consequently, H2 and H2&H3 show lower I_{on} values than H3 due to decreased hole velocity. However, the decrease in I_{on} is not directly proportional to the reduction in velocity, as the hole density increases, following an opposite trend. Therefore, the increased hole density mitigates the reduction in I_{on}. Similarly, the I_{on} of V-SFs depends on both hole velocity and density. Specifically, V2&V3 exhibit significant velocity reductions compared to V3, yet maintain similar I_{on} values due to increased hole density. The opposite trend in hole density arises from stress relaxation in each sheets which lowers V_{th}.

B. Off-state analysis (Vgs= 0 V, Vds=Vdd= -0.65 V)
To understand V_{th} lowering and I_{off} increase by SF, examining the stress levels within the channel during the annealing step is essential. Compressive stress in the channel reduces boron diffusivity, while tensile stress enhances it [9]. Therefore, SFs can influence boron diffusion into the channel by modifying channel stress. As shown in Fig. 7(a), in the ideal case, the S/D stressor applies strong compressive stress to the channel, slowing down boron diffusion. However, when SF occurs, the stress from the S/D is reduced, allowing tensile stress to become more prominent in the channel due to the lattice mismatch between the sacrificial $Si_{0.7}Ge_{0.3}$ layer (removed before gate formation) and the silicon channel. Therefore, as shown in Fig. 7 (b), V2&V3, which exhibit the highest tensile stress in the top sheet, shows the deepest boron diffusion in the top sheet. H2&H3 have moderate boron diffusion in the top sheet due to stress relaxation, and only B-SF shows significant differences in boron diffusion in the bottom sheet.

Higher boron diffusion in the channel lowers the valence band energy barrier, leading to V_{th} shift and I_{off} increase. Fig. 8 illustrates the normalized I_{off} contributions by sheet. I_{off} is evenly distributed across all layers in the ideal device, whereas in other devices, I_{off} is concentrated in the stress-relaxed layers. Thus, the type of SF influences the stress distribution to change boron diffusivity, ultimately impacting the turn-on timing and I_{off} specific to each layer.

C. Effect of partial non-stressor S/D on device performance
We also investigated the effect of partial non-stressor S/D, caused by H&B-SFs, on device performance. As shown in Fig. 9(a), direct H&B SFs involve the non-stressor SiGe S/D contacting the silicon channels, while indirect H&B SFs have this contact on the neighboring devices. Fig. 9(b) demonstrates that both types' longitudinal compressive stress is reduced. In the middle and top sheets, both direct and indirect H-SFs exhibit similar stress relaxation; however, in the bottom sheets, direct B-SF show more relaxation than indirect B-SF. According to Fig. 9(c), during annealing, direct H&B-SFs induce tensile channel stress in the bottom and middle channels, affecting boron diffusion, while indirect H&B-SFs show either compressive stress or minimal tensile stress. This stress relaxation adversely affects the I-V characteristics, as depicted in Fig. 10. Specifically, indirect H&B-SFs, due to less boron diffusion into the channel, primarily reduce I_{on}. In contrast, direct H&B-SFs induce tensile stress and deeper boron diffusion, causing a V_{th} shift and higher I_{off} by impacting the subthreshold region. Even a partial absence of lattice mismatch in the S/D epitaxial layer diminishes the induced stress, influencing the I-V characteristics of neighboring devices.

Conclusion
Our study provides a comprehensive analysis of the impact of various SFs on p-type Si-NSFETs at the 2nm node using TCAD simulations. We classified different SF types—vertical, horizontal, and bottom—and evaluated their effects on stress distribution, carrier mobility, and dopant diffusion across each nanosheet layer. Our results reveal that SFs degrade device performance by altering the stress-induced enhancements in hole mobility and affecting boron diffusion into the channel. V-SFs mainly influence subthreshold behavior, while H-SFs show consistent I_{off} increase and reduced I_{on}. Additionally, partial non-stressor S/D regions due to SFs further compromise device performance. These findings underscore the importance of controlling SFs to optimize NSFET performance at advanced technology nodes.

Acknowledgments
This work was supported in part by the National Research Foundation of Korea (NRF)–2022R1C1C1004925 and in part by the Ministry of Trade, Industry, and Energy (20020265, and 20019450).

979-8-3315-0417-5/25 $31.00 © 2025 IEEE

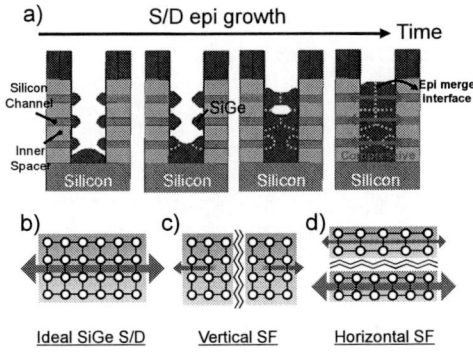

Fig. 1. (a) S/D epitaxy merge interface obtained via LKMC simulation, (b) Ideal S/D structure. (c) S/D with vertical SF, (d) S/D with horizontal SF.

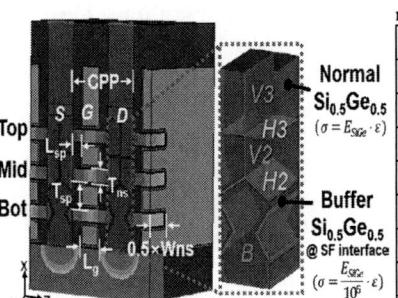

Fig. 2. Structures of NSFETs and various S/D SF-types implemented with buffer SiGe layers.

Table. 1. Key parameters for 2nm node NSFET

Parameter	Value
Gate Pitch (CPP)	45 [nm]
Gate Length (L_g)	14 [nm]
Spacer Length (L_{sp})	6 [nm]
NS Spacing (T_{ns})	12 [nm]
NS Thickness (T_{sp})	6 [nm]
NS Width (W_{ns})	30 [nm]
Spacer Dielectric (κ_{sp})	3.3
Supply Voltage (V_{DD})	0.65 [V]

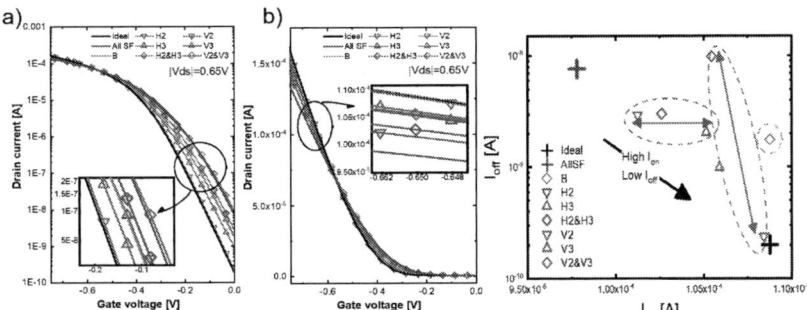

Fig. 3. Transfer curves of p-type NSFETs with various S/D SF types. (a) logarithmic scale and (b) linear scale.

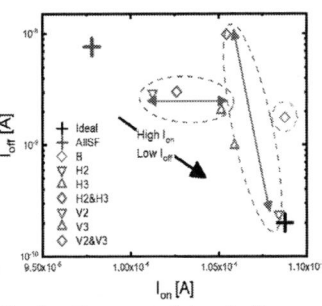

Fig. 4. On current (I_{on}) and off current (I_{off}) distributions according to SF types

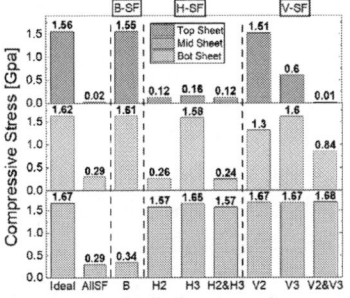

Fig. 5. Longitudinal compressive stress applied to top, bottom, and middle channels based on SF type.

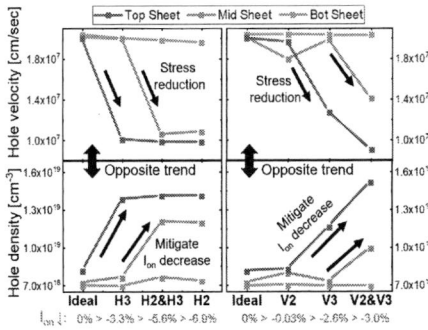

Fig. 6. Average hole velocity and hole density at the top, middle, and bottom sheet for H-SF and V-SF extracted at on-state.

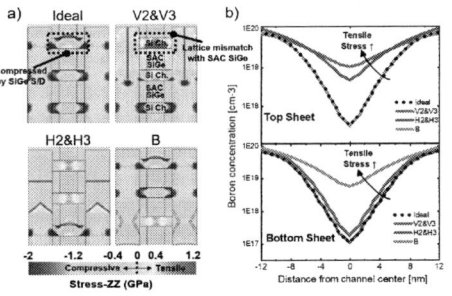

Fig. 7. (a) Longitudinal channel stress during annealing process after S/D epitaxy step for each SF types (b) Boron concentration in the channel center after annealing.

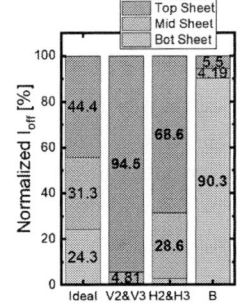

Fig. 8. Contribution of each sheet to the I_{off} for each SF type.

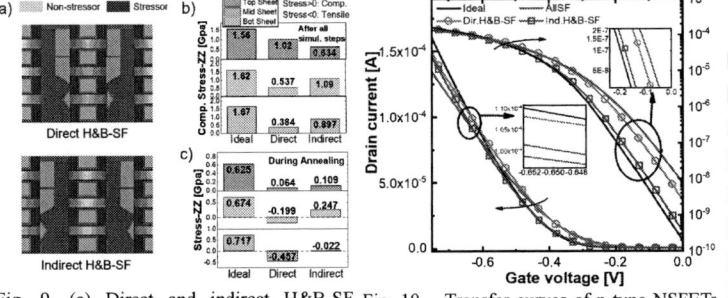

Fig. 9. (a) Direct and indirect H&B-SF NSFETs. Longitudinal channel stress (b) after simulation steps. (c) during annealing step.

Fig. 10. Transfer curves of p-type NSFETs with direct and indirect H&B-SF.

References

[1] B. Yu et al., IEDM Tech. Dig., pp. 100, 2002.

[2] N. Loubet et al., VLSI Tech.Dig, pp. T230, 2017.

[3] J. Kavalieros et al., VLSI Tech.Dig, pp. 50, 2006.

[4] A. Veloso et al., IEEE ICICDT, pp. 51, 2022.

[5] J. Yoon et al., IEEE Access, pp. 22032, 2022.

[6] G. Eneman et al., IEEE TED, pp. 5380, 2021.

[7] J. Jeong et al., IEEE TED, pp. 396, 2023.

[8] https://irds.ieee.org/editions/2023

[9] N. R. Zangenberg et al., J. Appl. Phys., pp. 3883, 2003.

A Low-programming-variation and High-Yield 2-bit Programmable 1-Kb Self-Rectifying Memristor Crossbar Array

Jia-Yi Sun[1], Sheng-Guang Ren[1], Heng-Feng Zhang[1], Zhi-Li Cui[1], Yi-Bai Xue[1], Yu Zhang[1], Wen-Bin Zuo[1], Yi Li[1*] and Xiang-Shui Miao[1]

[1]School of Integrated Circuits, Huazhong University of Science and Technology, China

Email: liyi@hust.edu.cn

Abstract

Self-rectifying memristor (SRM) array with multi-bit programming ability has been proven to be an ideal device prototype for energy-efficient in-memory computing (IMC). In this work, we fabricated Pt/HfO$_2$/TaO$_x$/Ta SRM-based 1-Kb (32×32) crossbar arrays with high uniformity. Using the write and verify method, we successfully modulated the array-level 2-bit low resistance states (60 MΩ, 89 MΩ, 173 MΩ, and 3 GΩ) with a >99% programming yield. These 2-bit states can well be retained at both 25 °C and 85 °C. The array-level 2-bit programming errors and inter-array variations are less than 6% and 2%, respectively. Our work provides a viable solution to optimizing the multi-bit programming ability for SRM, thereby further improving the energy efficiency of SRM-based IMC towards the AI era.

Keywords: Self-rectifying memristor, Multi-bit and In-memory computing

Introduction

In-memory computing (IMC) has been considered a promising solution for bridging the energy efficiency gap in the AI era [1] The self-rectifying memristor (SRM) with its ability to suppress the sneak paths in crossbar array provides an ideal device prototype for large-scale even high-density three-dimensional (3D) IMC [2].There are two typical ideas for improving the energy efficiency of IMC. One is to adopt devices with multi-bit characteristics, and the other is to increase the number of devices to compensate for the devices with only binary characteristics [3]. Obviously, the former demonstrates a higher area efficiency.

But few previous researches on SRM have mentioned its multi-bit programming characteristics [4-8]. This is probably because the self-rectifying behavior of most SRMs is realized by introducing an interfacial barrier layer with high resistance [4-6]. This scheme was indeed proven to exhibit better rectification characteristics [4,5,8]. However, in contrast to the conductive filament type devices, this interface effect type device is difficult to pulse modulate for multi-bit programming because the overall resistance is dominated by the tough barrier layer [2].

Here, we fabricated a 1-Kb crossbar array based on an interface effect type SRM. To enhance the multi-bit programming characteristics of the array, we have improved both the preparation process and the programming method at two levels. And we have successfully implemented a 2-bit programmable 1-Kb SRM array with good retention (>20000 s @85 °C), low programming variation (<6%), low inter-array variation (<2%), and high yield (>99%).

Experiments

Fig. 1 presents the optical image of the 1-Kb (32×32) SRM array. The bottom electrodes of the Ta (100nm) layer were patterned by ultraviolet lithography followed by direct-current (DC) sputtering. A 13 nm TaO$_x$ resistive switching (RS) layer is deposited by radio frequency sputtering using a Ta$_2$O$_5$ target. A 7 nm HfO$_2$ barrier layer was deposited onto the TaO$_x$ layer by atomic layer deposition (ALD). Finally, the 100 nm Pt as top electrodes were patterned by ultraviolet lithography and were grown by DC sputtering. All electrical measurements were performed by Keysight B1500A semiconductor parameter analyzer and Lakeshore PS-100 probe station.

Results and Discussions

Fig. 2 shows the well-defined crossbar structure captured using scanning electron microscopy (SEM). In Fig. 3, the transmission electron microscopy (TEM) image reveals the obvious HfO$_2$ and TaO$_x$ layers. Fig. 4 and 5 show energy-dispersive spectroscopy (EDS) mapping and EDS line scan images of the Pt/HfO$_2$/TaO$_x$/Ta bilayer SRM, confirming the 7 and 13 nm thicknesses of the HfO$_2$ and TaO$_x$ layers. Compared to our previous work [7], we have adjusted the occupancy ratio of TaO$_x$ and HfO$_2$ layers in this work. The thickness of the TaO$_x$ RS layer is significantly increased while ensuring that the rectification characteristics are not diminished. This contributes to the enhanced control of the applied bias on electron conduction with the assistance of oxygen vacancy defects [8]. Thus, the possibility of multi-bit programming of our SRM is further improved.

As illustrated in Fig. 6, DC voltage sweeps were performed on 12 randomly selected SRMs from the 1-Kb array (10 cycles per SRM), demonstrating good cycle-to-cycle uniformity. Fig .7 presents a series of *I-V* curves measured at gradual Set voltages from 7 V to 8 V with a fixed interval of 0.1 V. Under a constant reset voltage, the incremental increase of the Set voltage yields a continuously tunable low resistance state (LRS) in the device. As illustrated in Fig. 8, a series of *I-V* curves were obtained by systematically varying the Reset voltage from -1 V to -4.5 V in increments of 0.5 V. Analogously, the high resistance state (HRS) demonstrates a voltage-dependent tunability as the reset voltage is incrementally varied. This phenomenon presents a viable strategy for realizing multi-state resistance modulation in subsequent applications.

As depicted in Fig. 9, the DC write-and-verify method entails a sequential process of repetitive, gradual DC Set and DC Reset operations, enabling precise control over the device's resistance state. An initial resistance (R_i) measurement was conducted on the device. A Set operation was subsequently implemented using a 7.5 V voltage. A comparative analysis was then performed to determine the relationship between the measured R_i and the pre-defined target resistance R_{target}. If R_i exceeded the R_{target}, the Set voltage was systematically incremented from 7 V to 8 V. Conversely, if R_i was less than the R_{target}, the Reset voltage was systematically incremented from -1 V to -4.5 V. This iterative process of measurement, comparison, and voltage adjustment continued until the criteria for successful DC write verification were satisfied.

As a result of the programming method mentioned above, Fig. 10 and 11 exhibit the retention characteristics of the conductance equidistant 2-bit LRS (60 MΩ, 89 MΩ, 173 MΩ, and 3 GΩ) at 25 °C and 85 °C, respectively. All these four states can be well retained exceeding 20000 s even at 85 °C. This phenomenon can be attributed to the amorphous interfacial characteristics of the ALD-deposited HfO_2 layer, where oxygen vacancies drift homogeneously across the entire layer rather than forming localized conductive filaments. Under applied electric fields, this uniform defect distribution facilitates bulk switching mechanisms instead of conventional filamentary switching [9]. Although we have reduced the percentage of HfO_2 to enhance multi-bit programmability, it still brings good robustness to the SRM.

For comparative analysis, the pulse endurance characteristics were also investigated, as demonstrated in Fig. 12. The SRM underwent 10^7 switching cycles with a Set pulse voltage of 7.3 V/10 μs and a Reset pulse voltage of -2.8 V/2 μs. The switching energies of Set and Reset are calculated to be 6.59 fJ and 235.3 fJ, respectively, rivaling the state-of-the-art SRMs. This phenomenon arises from the overshoot effects in pulse signals, which introduce instability in resistance modulation, thereby compromising programming precision and degrading the multi-bit characteristics of the array [10]. The comparative analysis demonstrates that DC write verification yields significantly improved resistance state stability and enables stable modulation of the 2-bit low-resistance state.

In Fig. 13, we demonstrate the implementation of array-level 2-bit resistance state programming results. In Fig. 13(a), the target resistance was configured to 60 MΩ, resulting in a programming error rate of 3.47% and an inter-array variation of 1.39%. In Fig. 13(b), the target resistance was set to 89 MΩ, yielding a programming error rate of 3.48% and an inter-array variation of 1.69%. In Fig. 13(c), the target resistance was adjusted to 173 MΩ, with a programming error rate of 5.06% and an inter-array variation of 1.77%. In Fig. 13(d), the target resistance was established at 3 GΩ, with a programming error rate of 5.23% and an inter-array variation of 1.98%. During the whole DC write-and-verify process, 10 SRMs in the 1-Kb array experienced breakdown, indicating a programming success rate exceeding 99%. These results showcase the superiority of the DC write-and-verify in SRM-based array-level resistance modulation.

Conclusion

In summary, we proposed a 2-bit programmable 1-Kb crossbar array based on Pt/HfO2/TaOx/Ta SRM structure. To improve the multi-bit programmability of our SRM, in the preparation process, we increase the relative thickness of the RS TaO_x layer to enhance the ability of voltage-controlled electronic conduction. And during the measurement, we use DC write-and-verify method to avoid overshoot effect. As a result, the array-level programming yield exceeds 99% with programming variation below 6% and inter-array variations below 2%, indicating the superior reliability and stability of our solutions. We believe this work provides ideas for the refinement of multi-bit programmability of SRM arrays, which can further improve the energy efficiency of IMC to meet the demands of AI scenarios.

Acknowledgments

This work was supported by the National Key Research and Development Plan of MOST of China (2022YFB4500101), the STI 2030—Major Projects (Grant No. 2021ZD0201201), the Fundamental Research Funds for the Central Universities (HUST: 5003190012), and the Natural Science Foundation of Hubei Province under Grant (2024AFA043).

References

[1] W. Wan, et al., 'A compute-in-memory chip based on resistive random-access memory', Nature, 608, pp. 504–512, 2022.

[2] S.-G. Ren, et al., 'Self-Rectifying Memristors for Three-Dimensional In-Memory Computing', Adv. Mater., 36, p. 2307218, 2024.

[3] D. Ielmini, et al., 'Device and Circuit Architectures for In-Memory Computing', Adv. Intell. Syst., 2, p. 2000040, 2020.

[4] K.-M. Kim, et al., 'Low-Power, Self-Rectifying, and Forming-Free Memristor with an Asymmetric Programing Voltage for a High-Density Crossbar Application', Nano Lett., 16, pp. 6724–6732, 2016.

[5] J. Zhou, et al., 'Very Low-Programming-Current RRAM With Self-Rectifying Characteristics', IEEE Electron Device Lett., 37, pp. 404–407, 2016.

[6] C.-W. Hsu, et al., 'Bipolar Ni/TiO2/HfO2/ Ni RRAM With Multilevel States and Self-Rectifying Characteristics', IEEE Electron Device Lett., 34, pp. 885–887, 2013.

[7] J. Li, et al., 'Sparse matrix multiplication in a record-low power self-rectifying memristor array for scientific computing', Sci. Adv.9, eadf7474, 2023.

[8] S.-G. Ren, et al., 'Interface modeling analysis using density functional theory in highly reliable Pt/HfO2/TaOx/Ta self-rectifying memristor', Appl. Phys. Lett., 125, p. 123503, 2024.

[9] J. Park, et al., 'Multi-level, forming and filament free, bulk switching trilayer RRAM for neuromorphic computing at the edge', Nat Commun, 15, p. 3492, 2024.

[10] P.-R. Shrestha, et al., "Analysis and Control of RRAM Overshoot Current," in IEEE Trans. Electron Devices, 6, pp. 108-114, 2018.

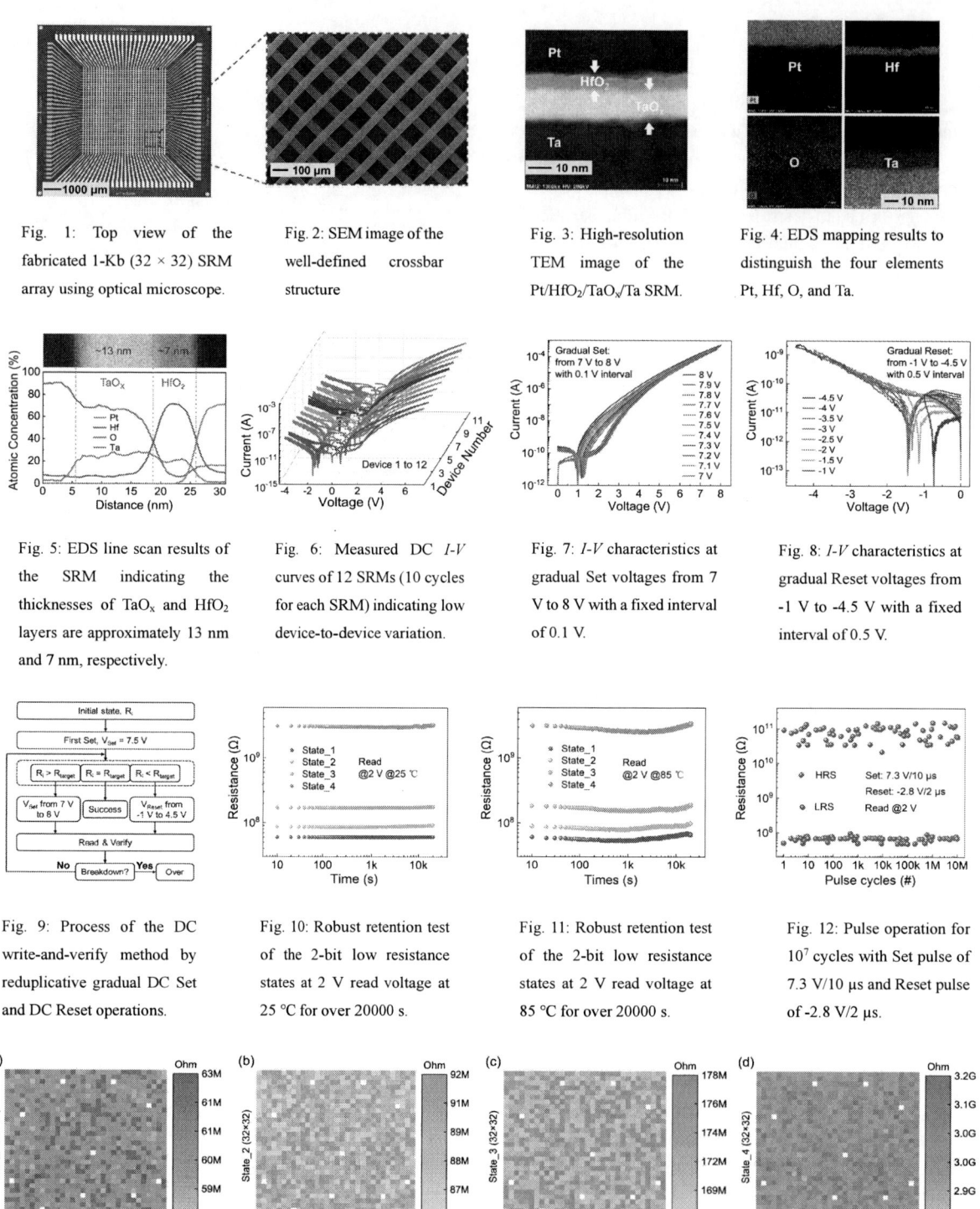

Fig. 1: Top view of the fabricated 1-Kb (32 × 32) SRM array using optical microscope.

Fig. 2: SEM image of the well-defined crossbar structure

Fig. 3: High-resolution TEM image of the Pt/HfO2/TaOx/Ta SRM.

Fig. 4: EDS mapping results to distinguish the four elements Pt, Hf, O, and Ta.

Fig. 5: EDS line scan results of the SRM indicating the thicknesses of TaOx and HfO2 layers are approximately 13 nm and 7 nm, respectively.

Fig. 6: Measured DC *I-V* curves of 12 SRMs (10 cycles for each SRM) indicating low device-to-device variation.

Fig. 7: *I-V* characteristics at gradual Set voltages from 7 V to 8 V with a fixed interval of 0.1 V.

Fig. 8: *I-V* characteristics at gradual Reset voltages from -1 V to -4.5 V with a fixed interval of 0.5 V.

Fig. 9: Process of the DC write-and-verify method by reduplicative gradual DC Set and DC Reset operations.

Fig. 10: Robust retention test of the 2-bit low resistance states at 2 V read voltage at 25 °C for over 20000 s.

Fig. 11: Robust retention test of the 2-bit low resistance states at 2 V read voltage at 85 °C for over 20000 s.

Fig. 12: Pulse operation for 10^7 cycles with Set pulse of 7.3 V/10 μs and Reset pulse of -2.8 V/2 μs.

Fig. 13: Array-level 2-bit resistance states programming. (a) State_1 (target: 60 MΩ) with a 3.47% programming error and a 1.39% array variation, (b) State_2 (target: 89 MΩ) with a 3.48% programming error and a 1.69% variation, (c) State_3 (target: 173 MΩ) with a 5.06% programming error and a 1.77% variation, and (d) State_4 (target: 3 GΩ) with a 5.23% programming error and a 1.98% variation. 10 SRMs break down in DC write-verify programming indicating high programming success rate for over 99%.

979-8-3315-0417-5/25 $31.00 © 2025 IEEE

Based on Vertical Structures for N-Type Organic Electrochemical Transistors

Wenjing Zhang[1], Songjia Han[2*], Chuan Liu[1*]

[1]State Key Laboratory of Optoelectronic Materials and Technologies, School of Electronics and Information Technology, Sun Yat-sen University, Guangzhou 510275, China (*email: liuchuan5@mail.sysu.edu.cn),[2]College of Electronic Engineering, College of Artificial Intelligence, South China Agricultural University, Guangzhou 510642, China

Abstract

A novel vertical structure for n-type organic electrochemical transistors (OECTs) employing side-chain-free planar rigid conjugated polymer polybenzobisimidazole dibenzofuran (BBL) as the semiconductor layer and a cellulose gel with ionic liquid as the electrolyte. The vertical OECT structure offers a significant improvement in transconductance, nearly 83-fold higher compared to conventional planar designs. The n-type OECT demonstrates non-volatile behavior during single-pulse operation, and short-term plasticity can be modulated by varying the pulse count. Furthermore, the device maintains robust cycling stability, with a maximum current loss of less than 9% after 400 cycles.

(Keywords: BBL, OECT and Vertical structure)

Introduction

At present, there is an imbalance in the development of semiconductor channel materials, with research on p-type conjugated polymers being relatively mature. In contrast, n-type polymers have developed more slowly due to low electron mobility[1-3], poor environmental stability, and incompatibility with integration. The performance mismatch between p-type and n-type materials limits practical applications such as complementary inverters[4,5]. In addition, there are few reports on vertical structure gel-based organic electrochemical transistors. Some researchers propose that the directionality of charge transport in these structures differs from that of planar structures, primarily due to the edge orientation of n-type semiconductor microcrystal structures or the interaction between the semiconductor layer and the electrolyte, which restricts ion doping and dedoping activities[6,7].However, the planar rigid conjugated structure without side chains, polybenzimidazole dibenzofuropyrrole (BBL) [8-10], may limit the development of this theory. This study selects BBL as the semiconductor layer to investigate the electrical performance of organic electrochemical transistors with the same electrolyte in different structures, to explore the main reasons for the performance differences between planar and vertical structures, and to conduct research on artificial neural synaptic responses.

Device Structure and Mechanism

The fabrication processes for the planar and vertical n-type BBL organic electrochemical transistors were shown in Figure 1(a) and (b). In the planar OECT (p-OECT) structure, the channel width between the source and drain was 50 μm, which was obtained by thermally evaporating electrodes on a flexible PEN substrate, followed by spin-coating a BBL semiconductor film and photolithographic patterning of the cellulose gel electrolyte. Unlike the planar OECT electrodes, the source and drain electrodes of the vertical OECT (v-OECT) were prepared by double thermal evaporation, and the gel electrolyte was patterned to cover the channel. The v-OECT had a shorter channel width compared to the p-OECT, with the source, semiconductor, and drain vertically stacked onto the substrate surface. Ions from the gate electrolyte injected into the shorter channel to modulate the channel resistance, sensing and driving the current along the vertical direction, thus offering advantages such as low operating voltage and high current density.

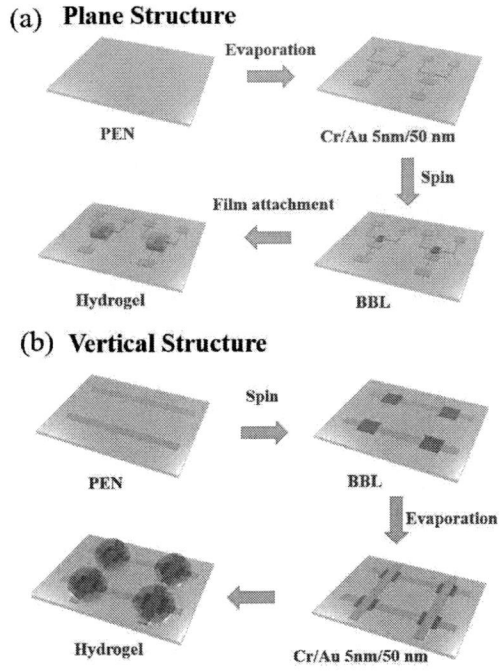

Figure 1. Planar and Vertical Structure OECT Process Flowchart.

Results and Discussion

The transfer characteristics and transconductance of p-OECT and v-OECT were discussed in Figures 2 and 3. As shown in Figure 2(a), with the increase of the source-drain voltage, the source-drain current of the p-OECT also increased. At the optimal drain voltage $V_d=0.6V$, the maximum current for the planar structure channel of 50 μm reached 1.6×10^{-3} A, and the on-off ratio reached 10^3. Figure 3(a) showed that the source-drain current of the v-OECT also increased with the voltage. At its optimal drain voltage $V_d=0.1V$, the maximum current reached 1.05×10^{-2} A, and the on-off ratio reached 10^4.

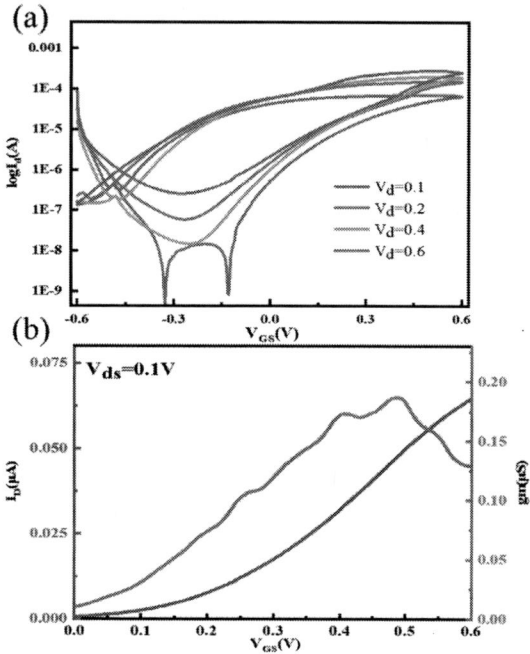

Figure 2. (a) p-OECT Transfer Characteristics (b) Transconductance.

Transconductance characterized the transistor's ability to control the drain current under different gate voltages, with the maximum transconductance indicating the highest current conversion efficiency achieved by the organic electrochemical transistor at that gate voltage. Under the same source-drain voltage $V_{DS}=0.1V$ condition, Figure 2(b) showed that the maximum transconductance of the p-OECT was 0.18 mS, which occurred at $V_G = 0.5$ V, with the current reaching 0.05 μA. Figure 3(b) demonstrated that the maximum transconductance of the v-OECT, g_{max}, was 15 mS, and it occurred at $V_G = 0.35$ V, with the current reaching 2.5 μA. The experimental results indicated that, with the same semiconductor and electrolyte materials, the v-OECT operated at a lower voltage, with increased working current and an 83-fold increase in transconductance compared to the p-OECT. This might have surpassed the performance differences caused by variations in the edge orientation and face orientation of microcrystal structures in the semiconductor material or the interaction between the semiconductor layer and the electrolyte, indicating that the

different charge transport directions in planar and vertical structures were the main reasons for the significant performance changes.

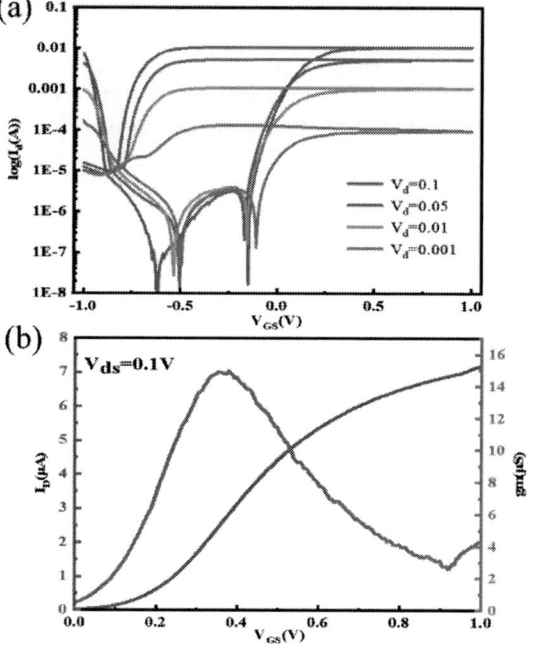

Figure 3. (a) v-OECT Transfer Characteristics (b) Transconductance.

The transfer characteristics of OECTs based on BBL semiconductor material exhibited an enhancement-type change in current, which could serve as the fundamental effect for artificial neural synapses, that is, the synapse generated excitatory postsynaptic current (EPSC) after being stimulated by a nerve impulse.

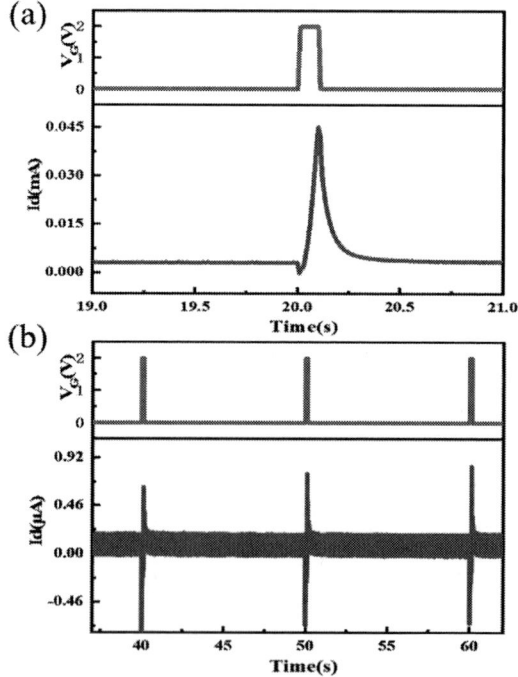

Figure 4. (a) 100ms Single Pulse Response (b) Multi-Pulse Regulation.

EPSC was often related to the input presynaptic action potential, which was a common form of short-term synaptic plasticity. From Figure 4(a), it could be seen that a response occurred at 100 ms under a single pulse, exhibiting non-volatility, where the spike-and-tail curve was related to the relaxation process of ions. Figure 4(b) similarly set different numbers of gate voltage pulses, resulting in varying degrees of amplification of EPSC, thus n-type organic electrochemical transistors could achieve synaptic current response by regulating the number of pulses. Figure 5 evaluated the cyclic stability of the OECT, under high-frequency continuous stimulation with a 0.5 s pulse period, after 400 cycles, the maximum current loss was less than 9%, indicating good stability.

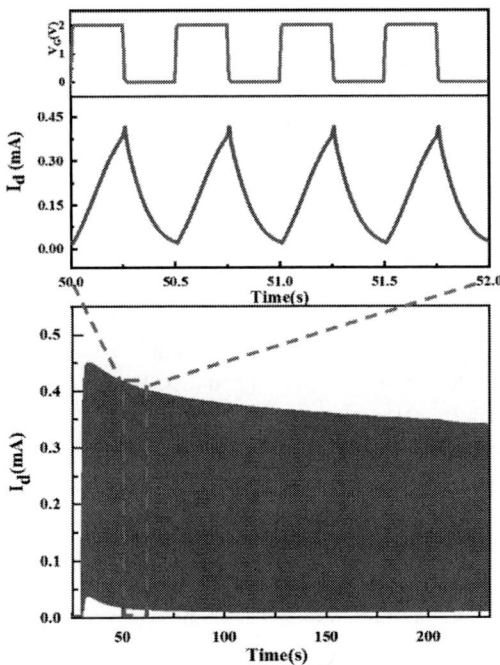

Figure 5. High-frequency pulse cyclic stability.

Conclusion

In the study, the performance of planar and vertical structures of organic electrochemical transistors with the same electrolyte using poly(benzimidazole dibenzofuropyrrole) as the semiconductor layer was explored. The significant difference in transconductance suggested that the direction of charge transport might have been the primary cause of the discrepancy. Additionally, pulse modulation of synaptic characteristics in n-type OECTs was conducted, and good cyclic stability under high-frequency continuous pulse stimulation was assessed. This was of significant importance for downsizing device dimensions, developing large-scale arrays, and enhancing response speeds.

Acknowledgments

The authors gratefully acknowledge the financial support of the National Natural Science Foundation of China (No. 62104262).

References

[1] Sophie Griggs, "n-Type organic semiconducting polymers: stability limitations, design considerations and applications" J. Mater. Chem. C, 9, 8099-8128 (2021).

[2] Pushpa Raj Paudel, Joshua Tropp, Vikash Kaphle, Jason David Azoulay, "Organic electrochemical transistors – from device models to a targeted design of materials", J. Mater. Chem. C, 9, 9761-9790 (2021).

[3] Dafei Yuan, "Efficient and air-stable n-type doping in organic semiconductors", Chem. Soc. Rev., 52, 3842-3872 (2023).

[4] Cindy G. Tang, " A Universal Biocompatible and Multifunctional Solid Electrolyte in p-Type and n-Type Organic Electrochemical Transistors for Complementary Circuits and Bioelectronic Interfaces", Adv. Mater., 2405556 (2024).

[5] Chi-Yuan Yang, "Low-Power/High-Gain Flexible Complementary Circuits Based on Printed Organic Electrochemical Transistors" Adv. Electron. Mater., 2100907 (2022)

[6] B. Wang, Y. Kong, S. Zhang, Z. Wu, S. Wang. " Face-on Orientation Matches Vertical Organic Electrochemical Transistors for High Transconductance and Superior Non-Volatility " Adv. Funct.Mater.,34, 2312822 (2024).

[7] Wang, M., " Exceptionally high charge mobility in phthalocyanine-based poly(benzimidazobenzophenanthroline)-ladder-type two-dimensional conjugated polymers", Nat. Mater., 880–887 (2023).

[8] Amit Babel and Samson A. Jenek he "Electron Transport in Thin-Film Transistors from an n-Type Conjugated Polymer" Adv. Mater. 14 (2002).

[9] Benjamin D. Naab, Xiaodan Gu, "Role of Polymer Structure on the Conductivity of N-Doped Polymers," Adv. Electron. Mater., 1600004 (2016).

[10] H.Meng, "π-Extended Poly(benzimidazoanthradiisoquinolinedione) Ladder-type Conjugated Polymer", ACS Macro Lett.,1136−1141 (2022).

Recent research on ultrawide bandgap semiconductor Ga₂O₃ and AlN

Dao Hua ZHANG[1*+], Xian-Hu ZHA[1+], Rong-Jun ZHANG[1], Shuang LI[2], Yan LIU[1] and Yu-Xi WAN[1]

[1]Shenzhen Pinghu Laboratory, Shenzhen, 518111, China.
[2]Nanjing University of Science and Technology, Nanjing, 210094, China.
*Email: zhangdaohua@phlab.com.cn.
+These authors contribute equally.

Abstract

Ga₂O₃ and AlN have the characteristics of ultrawide bandgap, making them a promising choice for applications in power devices. In this talk, we will primarily present our recent work on Ga₂O₃ and AlN, including bandgap engineering by incorporating different metal elements into Ga₂O₃ to realize new ternary compound semiconductors and easier realization of p-type doping, photoluminescence characterization of 6.2 eV bandgap AlN epi-layers under bias voltage and various temperatures.

Keywords: Ga₂O₃, valence band maximum, AlN.

Introduction

High-power semiconductor devices have gained wide attention in electric transportation and smart grids. Baliga's figure of merit (BFOM) normally depicts a semiconductor device's power performance, which critically depends on the semiconductor bandgap [1]. Ga₂O₃ and AlN with ultrawide bandgaps are at the forefront of power semiconductor materials. Ga₂O₃ has a bandgap of 4.9 eV [2], and can be grown by various melt-growth methods. One of the challenges faced is the lack of p-type doping because of the relatively low energy and flat band dispersion near the valence band maximum (VBM) of Ga₂O₃ [3]. Most dopants show deep acceptor levels higher than 1 eV [4]. AlN has the widest bandgap of 6.2 eV and the highest critical electric field of 15.4 MV/cm [5]. Due to the ultrawide bandgap, it is still challenging to characterize the semiconducting properties of AlN at different temperatures and bias voltages. Photoluminescence characterization is an effective approach to measuring the bandgaps and defect levels. In addition, UV photoexcitation can be used to reduce carbon contamination for n-type doping of AlN [5].

Regarding the difficulty of p-type doing, the valence band of Ga₂O₃ is engineered by incorporating different metal elements to realize new ternary compounds in our recent works [6, 7]. Forty-nine different transition metals (TMs) alloyed β-$(TM_{0.125}Ga_{0.875})_2O_3$ are studied. Further, the β-$(Rh_xGa_{1-x})_2O_3$ with x from 0 to 0.5 are elaborated from first-principles calculations. An optical measurement system is designed to characterize the ultrawide-bandgap AlN. The photoluminescence characterization of AlN epi-layers under bias voltage and various temperatures is performed.

Experimental and computational details

First-principles density functional calculations were implemented in the plane-wave VASP code [8]. The GGA-PBE form was employed for the exchange-correlation functional. To obtain more precise bandgaps, the SCAN functional was used to relax the structural models, and the hybrid functional HSE was adopted to correct the bandgap based on the SCAN structure.

Our designed optical measurement system was implemented by the Zolix Company. Photoluminescence spectra and minority carrier lifetime of the AlN epi-layer under 83 K and 173 K were measured.

Results and discussion

A. Ga₂O₃

To engineer the valence band of β-Ga₂O₃, its configuration and electronic structure are studied first. The crystal structure of β-Ga₂O₃ belongs to the C2/m space group (No. 12). The structure has two crystallographic nonequivalent Ga sites denoted as Ga-I and Ga-II respectively. The Ga-I is six-coordinated, and the Ga-II is four-coordinated as shown in the inset of Figure 1. The band structure and density of states (DOS) of β-Ga₂O₃ based on its primitive cell are also provided. The configuration has an indirect bandgap, with its VBM located between the high-symmetry points I and L. The valence bands are mainly contributed by the O 2p orbitals. After the HSE correction, the bandgap is increased to 4.81 eV. Based on the flat dispersion near the VBM, the conduction average hole mass is calculated to be 4.24 m_e.

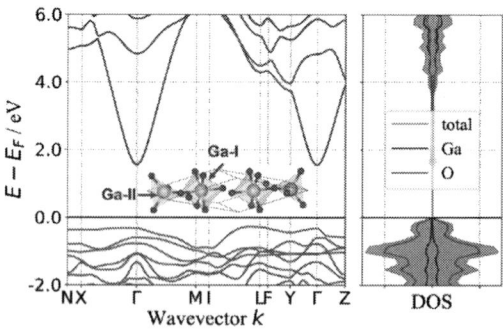

Fig. 1: Electronic band structure and density of states (DOS) of β-Ga₂O₃ based on its primitive cell and the GGA functional [6]. The inset shows the primitive cell of β-Ga₂O₃.

979-8-3315-0417-5/25 $31.00 © 2025 IEEE

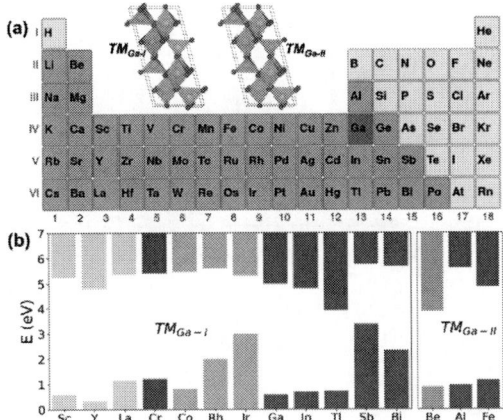

Fig.2: (a) TM_{Ga-I}: a Ga-I atom and TM_{Ga-II}: a Ga-II atom replaced by a TM based on the β-Ga_2O_3 unit cell; all the investigated TMs are denoted in gold in the periodic table of elements; (b) Energy-band alignment for the TM_{Ga-I} (TM=Sc, Y, La, Cr, Co, Rh, Ir, Ga, In, Tl, Sb, and Bi) and TM_{Ga-II} (TM=Be, Al, and Fe) based on the HSE functional [7].

To promote p-type doping, it is required to increase the energy level and band-dispersion curvature at the β-Ga_2O_3's VBM [4]. Various TMs are attempted to engineer the valence band. Based on the β-Ga_2O_3 unit cell, the β-$(TM_{0.125}Ga_{0.875})_2O_3$ configurations are constructed by replacing a Ga atom with a TM. Since two nonequivalent Ga sites exist, the configurations with a TM respectively replaced the Ga at the Ga-I and Ga-II are considered. The corresponding structure models are provided in Figure 2a, denoted as TM_{Ga-I} and TM_{Ga-II} to facilitate elaboration. All the studied TMs are also shown in Figure 2a. Most TMs except for TMs of Li, Be, Na, Al, K, Fe, Zn, Ge, Ba, and Po stabilize at the Ga-I site. Based on the stable configurations, the electronic band structures are studied. The electronic properties are significantly dependent on the group number of TM. The β-$(TM_{0.125}Ga_{0.875})_2O_3$ alloys with TMs in groups 3, 9, 13, and 15, and the TMs of Be, Cr, and Fe are semiconductors, and their energy band alignments are presented in Figure 2b. Compared to the VBM of β-Ga_2O_3, the VBMs' energy levels in the TM_{Ga-I} (TMs=Sc, La, Cr, Co, In, and Tl) and the TM_{Ga-II} with TMs of Be, Al, and Fe are slightly higher. The energy difference is below 0.6 eV, much less than the acceptor levels higher than 1 eV in the β-Ga_2O_3. The alloys with TM_{Ga-I} (TM=Rh, Ir, Sb, and Bi) are the four members with their VBMs' energy levels much higher, mainly contributed by the electronic orbitals of TM and neighboring O atoms. Noteworthily, the alloys with TM_{Ga-I} (TM= Ir and Sb) are

not wide bandgap semiconductors as their bandgaps are only 2.32 and 2.40 eV, while the alloys with TM_{Ga-I} (TM=Rh and Bi) still have wide bandgaps larger than that of commercial silicon carbide. Since the Bi-alloyed β-Ga_2O_3 has been well-studied previously [9], the Rh-alloyed configuration is discussed further.

The β-$(Rh_xGa_{1-x})_2O_3$ with various x from 0 to 0.5 are studied based on the 1×2×2 β-Ga_2O_3 supercell. The Rh atoms are found to occupy the Ga-I site in the x range studied. The stabilities of β-$(Rh_xGa_{1-x})_2O_3$ alloys are tested based on their mixing enthalpy. Because the atomic radius of Rh is close to that of Ga, the β-$(Rh_xGa_{1-x})_2O_3$ are quite stable. Specifically, the mixing enthalpy of β-$(Rh_{0.5}Ga_{0.5})_2O_3$ is approximately 0.14 eV/cation. This low mixing enthalpy implies that the β-$(Rh_xGa_{1-x})_2O_3$ could exist with x up to 0.5. The electronic band structures, energy-band alignments, and conduction average hole masses of the β-$(Rh_xGa_{1-x})_2O_3$ are studied and compared to those of β-Ga_2O_3. As shown in Figure 3, the VBM of β-$(Rh_{0.0625}Ga_{0.9375})_2O_3$ shows an abrupt increase compared to that of β-Ga_2O_3, and the energy difference is 1.35 eV. Then, the VBM energy increases uniformly with x, and the VBM of β-$(Rh_{0.5}Ga_{0.5})_2O_3$ is 2.95 eV higher than that of β-Ga_2O_3. Based on the significantly increased VBMs, it might be easy to realize p-type doping in the β-$(Rh_xGa_{1-x})_2O_3$ alloys. Noteworthily, all these alloys are ultrawide bandgap semiconductors, with their bandgaps ranging from 3.77 to 4.10 eV. The nonlinear relationship between the bandgap and x is because the bandgap shows a direct-indirect transformation. These large bandgaps imply that the β-$(Rh_xGa_{1-x})_2O_3$ alloys could have high power performances. Additionally, the conduction average hole mass decreases in most β-$(Rh_xGa_{1-x})_2O_3$ alloys. The hole mass of β-$(Rh_{0.25}Ga_{0.75})_2O_3$ is only 52.4% of that of the β-Ga_2O_3, which is beneficial to p-type conduction.

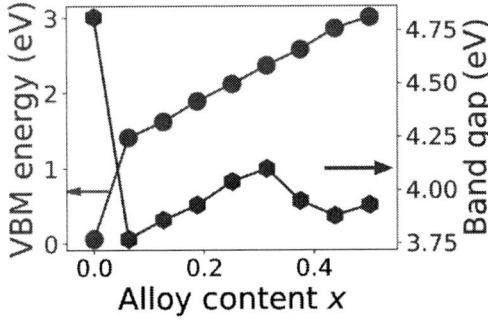

Fig. 3: VBM energy levels and bandgaps of β-$(Rh_xGa_{1-x})_2O_3$ with x from 0 to 0.5. The left and right y-axis respectively show the VBM energy levels and bandgaps.

979-8-3315-0417-5/25 $31.00 © 2025 IEEE

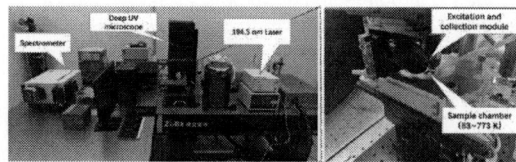

Fig. 4: Our optical characterization system.

B. AlN

Regarding the characterization of AlN, an optical characterization system is built and shown in Figure 4. It includes a vacuum sample chamber, a first UV objective, an excitation and collection, and coupled optical, and fluorescence processing modules. The vacuum sample chamber possesses a sample holder. and its temperature can be adjusted from 83 to 773 K. Moreover, the sample can be biased up to a maximum bias voltage of 800 V. The excitation and collection module comprises a laser and a long-wave-pass filter. The laser wavelength is 194.5 nm and the photoluminescence can be collected from both top and side. Based on our optical characterization system, the photoluminescence spectra of AlN samples at different temperatures are measured. As shown in Figure 5, obvious band-edge emission peaks are observed. The band-edge peaks show a redshift with the increasing temperature. The band-edge peaks at 83 K and 173 K are located at 198.2 nm and 200.2 nm, respectively. The minority carrier lifetime at 198.2 nm is determined to be about 4.27 ns.

Conclusion

Our recent works on the ultrawide bandgap semiconductors Ga_2O_3 and AlN are introduced. The valence bands of β-Ga_2O_3 are engineered by incorporating forty-nine different TMs. The β-$(TM_{0.125}Ga_{0.875})_2O_3$ with TMs in groups 3, 9, 13, and 15, and the TMs of Be, Cr, and Fe are semiconductors. The β-$(TM_{0.125}Ga_{0.875})_2O_3$ with TMs of Rh and Bi could be promising for power devices due to their significantly enhanced VBMs and ultrawide bandgaps. Furthermore, the β-$(Rh_xGa_{1-x})_2O_3$ with x ranging from 0 to 0.5 are studied. Their VBMs' energies increase with increased x, and the increased magnitudes range from 1.35 to 2.95 eV compared to that of β-Ga_2O_3. All the Rh-alloyed configurations present ultrawide bandgaps larger than 3.77 eV. The Rh-alloyed β-Ga_2O_3 could have high power performances with p-type conduction. An optical characterization system has been built to characterize the ultrawide-bandgap AlN and the photoluminescence spectra under different temperatures.

Acknowledgments

The authors gratefully acknowledge the financial support

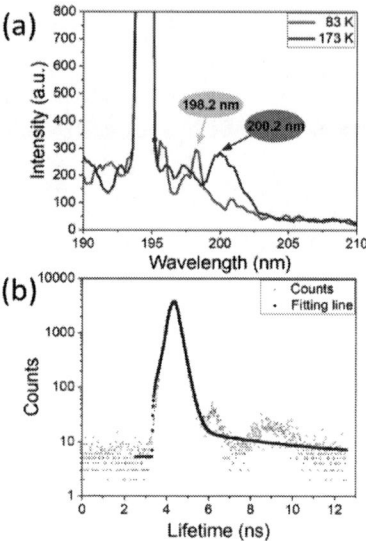

Fig. 5: (a) band-edge peaks of AlN at different temperatures; (b) minority carrier lifetime of AlN at 83 K.

from the Shenzhen Pinghu Laboratory Projects (Grant Nos. 224120, 224160, and 224210).

References

[1] B.J. Baliga, "Semiconductors for high-voltage, vertical channel field-effect transistors", J. Appl. Phys. 53(3): p. 1759-1764 (1982).

[2] T. Matsumoto, et al., "Absorption and reflection of vapor grown single crystal platelets of β-Ga_2O_3," Jpn. J. Appl. Phys. 13(10): p. 1578 (1974).

[3] A.T. Neal, et al., "Donors and deep acceptors in β-Ga_2O_3", Appl. Phys. Lett. 113(6): p. 062101 (2018).

[4] J.L. Lyons, "A survey of acceptor dopants for β-Ga_2O_3", Semicond. Sci. Technol. 33(5): p. 05LT02 (2018).

[5] W.A. Doolittle, et al., "Prospectives for AlN electronics and optoelectronics and the important role of alternative synthesis", Appl. Phys. Lett. 123(7): p. 070501 (2023).

[6] X.-H. Zha, et al., "Rhodium-alloyed beta gallium oxide materials: New type ternary ultra-wide bandgap semiconductors", Adv. Electron. Mater. 2400547 (2024). https://doi.org/10.1002/aelm.202400547.

[7] X.-H. Zha, et al., "Engineering the valence band of β-Ga_2O_3 via alloying transition metals: A first-principles study", Available at SSRN: http://dx.doi.org/10.2139/ssrn.4973092.

[8] G. Kresse, and J. Furthmüller, "Efficient iterative schemes for ab initio total-energy calculations using a plane-wave basis set", Phys. Rev. B 54(16): p. 11169-11186 (1996).

[9] X. Cai, et al., "Approach to achieving a p-type transparent conducting oxide: Doping of bismuth-alloyed Ga_2O_3 with a strongly correlated band edge state", Phys. Rev. B 103(11): p. 115205 (2021).

Scalable in-memory Walsh-Hadamard Transform for image compression

Jing Tian, Huai-Zhi Pei, Jian-Cong Li, Xiao-Di Huang, Yi-Bai Xue, Yi Li*, and Xiang-Shui Miao

School of Integrated Circuits, Huazhong University of Science and Technology, Wuhan 430074, China
(e-mail: liyi@hust.edu.cn).

Abstract

Image compression is of great significance for improving the efficiency of image transmission and storage. In this work, we proposed a scalable in-memory Walsh-Hadamard Transform (WHT) using self-selective memristors for image compression. With the self-selective memristors and scalable in-memory WHT method, arbitrarily-size WHT computing can be realized with a fixed-size memristor array, and image compression is achieved with a small result error that the Peak Signal-to-Noise Ratio (PSNR) of the compressed result is 32 dB. In addition, non-ideal characteristics in devices and arrays are analyzed and compared with Discrete Cosine Transform (DCT), which shows that in-memory WHT has higher robustness.

Keywords: In-memory computing, Image compression, Walsh-Hadamard transform, Self-selective memristor

Introduction

One of the key technologies in digital image processing is the image compression algorithm, which is usually used to reduce the image storage overhead [1]. Image compression faces von Neumann's bottleneck when implemented using traditional computational paradigms. Then, in-memory computing (IMC) is introduced to accelerate image compression, which can implement in-situ computation [2].

Discrete Cosine Transform (DCT) and Walsh-Hadamard Transform (WHT) are both widely used image compression operators [3]. The binary WHT matrix is easy to map on the memristors compared to DCT. For different task scales, the size of the WHT operator needs to be changed, while the scale of the memristor chip is usually fixed after fabrication. Therefore, it is necessary to consider using a fixed size memristor array to achieve WHT operations of different sizes.

In this work, we proposed an in-memory scalable Walsh-Hadamard Transform method using self-selective memristor for image compression. With this approach, arbitrarily WHT computing can be realized with a fixed-size memristor array, and image compression is achieved with a small result error that the PSNR of the compressed result is 32 dB. Additionally, the impact of non-ideal characteristics in memristor devices and arrays is discussed in detail and compared with DCT based on IMC.

In-memory WHT Method

WHT is one of the most important image compressions, which includes forward transformations, quantization, and inverse transformation, as shown in Fig. 1(a). The mathematical essence of forward and inverse transform is orthogonal transformation, which involves a large number of Matrix-Vector Multiplication (MVM) [4].

The WHT matrix size can only be to the power of 2. The 4×4 WHT implemented by 4×4 WHT matrix is shown in Fig. 1(b). The calculation consists of 16 multiplication operations and 16 addition operations. According to the definition of matrix multiplication and the characteristics of WHT matrix values, the 4×4 WHT can also be realized by 2×2 WHT matrix, as shown in Fig. 1(c). The calculation of this computation method consists of 8 multiplication operations and 12 addition operations, which is 40% less computational than the previous method. According to the proposed calculation method, WHT operations of 8×8, 16×16, or even larger sizes can be completed based on a 2×2 WHT matrix, which can be used to achieve arbitrary WHT operations. And the value of the WHT matrix only contains "-1" and "1", which can be mapped on two binary memristors for each value, as shown in Fig. 1(d). LRS is used to map "1" and HRS is used to map "0". The mapping of "1" and "-1" in a memristor array is achieved by applying positive and negative voltages to HRS and LRS in different ways.

Device Fabrication and Characterization

Fig. 2 illustrates a fabricated 4×4 self-selective memristor array, where each device at the crosspoint is constructed within a 250 nm via-hole structure, employing a $V/AlO_x/Al_2O_3/Pt$ vertical stacking configuration. This configuration has been corroborated by the transmission electron microscopy (TEM) result in Fig. 3. Typical one-selector-one-memristor (1S1R) characteristics can be observed in this device by using direct current sweeping within a range of (-3 V, 3 V), as shown in Fig. 4. The results indicate highly uniform switching with forming-free characteristics.

The 1S1R characteristics are due to the formation of the naturally oxidized interfacial VO_x layer at the interface between V and AlO_x layers. The VOx layer functions as a built-in selector, exhibiting a volatile threshold-switching behavior. Due to the built-in selector structure, the low-resistance states (LRS) when the selector off were enhanced by approximately 100 times to a sub-high resistance region, as shown in Fig. 5. This improved LRS mitigates practical line-resistance-induced current errors and reduces power consumption, thus leading to substantially improved energy efficiency compared with selector-less memristors.

Additionally, this self-selective device offers several features that enhance efficient large-scale MvM. First, this device exhibited an extraordinary switching endurance of up to 10^{10} cycles while maintaining a memory window of ~10^2, as shown in Fig. 6. The device consumes 122.5 fJ for SET operations and 2.4 pJ for RESET operations, which are beneficial for efficient mapping operations on a memristor array. Second, the fast-switching speed (< 20 ns) of the device contributes to high computational speeds within the system, as shown in Fig. 7. Finally, high-precision programming was realized using a write-with-verify approach. As shown in Fig. 8, the programming error was minimized to approximately 2% within 40 programming cycles, which is critical for reducing analog computing error and attaining higher accuracy in image compression. Fig. 9 presents the programmed binary stabilized conductance states including LRS when the selector off and HRS in the fabricated array for computational demonstration, where the average programming error was reduced to ~0.4%.

Results of in-memory WHT

Using the 4×4 memristor array, we mapped the 2×2 WHT matrix and successfully computed the 4×4 WHT. To verify the feasibility of the in-memory WHT calculation method, we first performed a WHT calculation. The input and ideal versus IMC-based output results are presented in Fig. 10, with an average relative error of 0.014.

Then, the in-memory WHT calculation is expanded to realize image compression. The input image was firstly divided into 4 × 4 image blocks and then performed a 2D WHT operation, which is equivalent to two 1D WHT operations. Then, the values of the smaller 90% portion of the WHT result matrix were set to 0. Last, the WHT result matrix was subjected to Inverse WHT (IWHT) operation and obtained a compressed image. Fig. 11 shows the original image and IMC-based image compression results, and the Peak Signal-to-Noise Ratio (PSNR) of the compression result is 32 dB [5]. When the PSNR is greater than 30db, it is difficult for the human eye to perceive the difference between the compressed and original images.

Non-ideal Effects Discussion

Various non-ideal factors are analyzed for the influence on in-memory WHT compression quality. First, the influence of wire resistance should be considered at the array level, which causes voltage loss. The lower the memristor resistance, the greater the influence of wire resistance [6]. Therefore, the worst condition is that memristors are all in low resistance. As shown in Fig. 12, supposing the ratio of the low resistance of memristor to wire resistance is 25000 in 65nm node [7], the voltage at every memristor array node is calculated. It can be found that the voltage drop is smaller than 1% when the array size is 16×16 or smaller, leading to a negligible effect on the image compression. However, when the array size is 128×128 or larger, the voltage drops more than 20%, which will greatly influence the accuracy of the calculation. In Fig. 13, the result of the read accuracy is

analyzed. The read accuracy is defined as I_{real}/I_{ideal}, which is still more than 99% when the array size is 16 × 16 or smaller. However, it is less than 80% when the array size is 128 × 128 or larger. Thus, to minimize the effect of wire resistance, memristor array size should be limited to 16 × 16.

Then, the impact of stuck issues is discussed In Fig. 14(a) and Fig. 14(b) [8]. The value of PSNR drops sharply as the ratio of stuck devices increases. It can be found that the impact of stuck-on devices is similar to the impact of stuck-off devices. When the ratio of stuck on/off is more than 1%, the PSNR value will be less than 30, which greatly affects the image compression result. Thus, the yield of memristor arrays should be more than 99%.

Comparison of WHT and DCT

To further discuss the robustness of the in-memory image compression system based on WHT, the compression results of WHT and DCT are discussed. The DCT matrix values are quantized as 4-bit data and then mapped into memristor arrays. In Fig. 15(a), the PSNR value of DCT decreases faster than that of WHT as the variation becomes larger, meaning WHT has a better anti-variation capability than DCT [9]. The compressed image using WHT has a better compression effect than the ones using DCT. As shown in Fig. 15(b), WHT has a superior performance than DCT in terms of stuck-on issues. The results suggest that in-memory image compression based on WHT has better robustness than the one based on DCT.

Conclusion

In this work, a scalable in-memory WHT method using self-selective memristors is proposed for image compression. With this approach, arbitrarily WHT computing can be realized with a fixed-size memristor array, and image compression is achieved with a small result error that the PSNR of compressed result is 32 dB. Additionally, the in-memory WHT is more robust than the DCT. Thus, this study has good applicability and is expected to be widely used.

Acknowledgments

This work was supported in part by the National Key R&D Plan of China (2022YFB4500101), the STI 2030—Major Projects (Grant No. 2021ZD0201201), the Fundamental Research Funds for the Central Universities (HUST: 5003190012), and the Natural Science Foundation of Hubei Province under Grant (2024AFA043).

References

[1] Xu, Mai, et al. IEEE J-STSP 14.1 (2020): 5-26.
[2] B. Zhang, et al. in 2020 DATE., 2020: IEEE, pp. 1594-1597.
[3] H. Kekre, et al. in IJCA 2th ICWET., 2011:
[4] Jayathilake, et al. IJRIT Int. J. Res. Inf. Technol 1 (2013): 80-89.
[5] Z. Wang, et al. IEEE SPL, vol. 9, no. 3, pp. 81-84, 2002.
[6] X. Cong, et al. in IEEE HPCA., 2015.
[7] J. Woo, et al. IEEE VLSI systems., no. 27-9, 2019.
[8] Yeo, Injune, et al. IEEE TED 66.7 (2019): 2937-2945.
[9] T. Dongale et al., Mater. Sci. Semicond. Process., vol. 35, pp. 174-180, 2015.

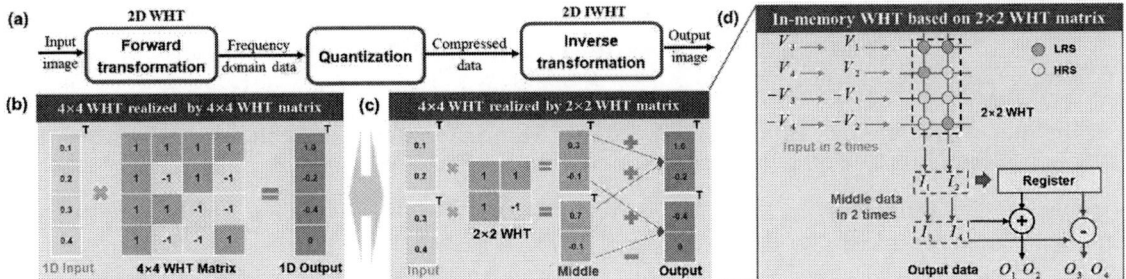

Fig. 1. (a) Flowchart of WHT-based image compression. (b) Schematic diagram of 4×4 WHT based on 4×4 WHT matrix. (c) Schematic diagram of 4×4 WHT based on 2×2 WHT matrix. (d) Realization of in-memory WHT based on 2×2 WHT matrix.

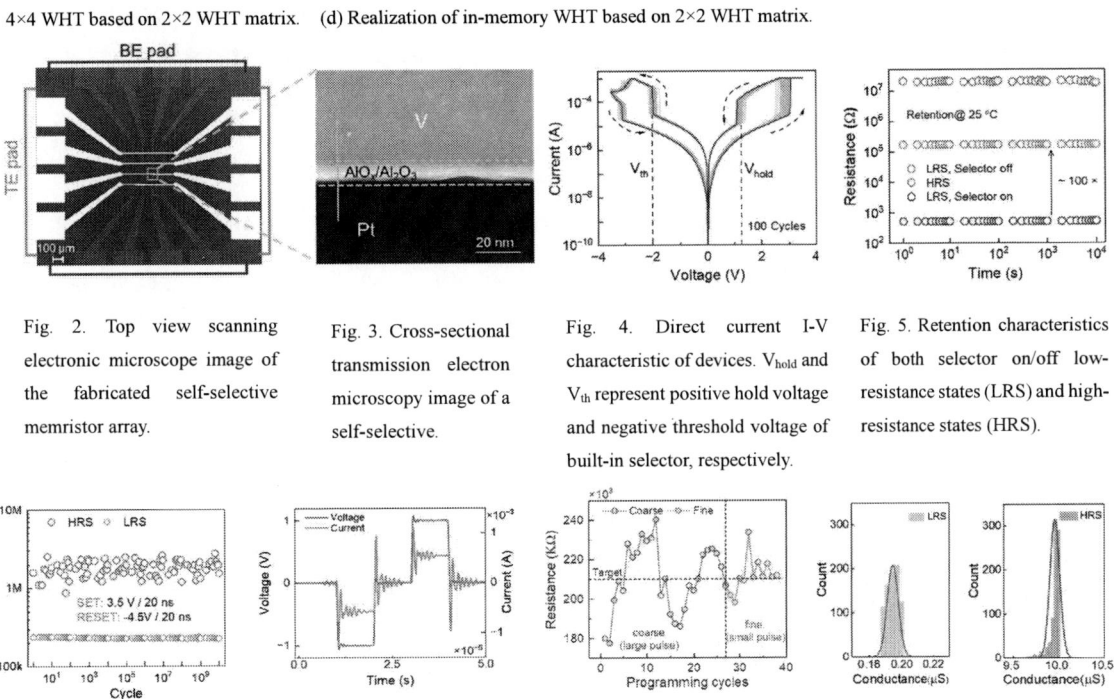

Fig. 2. Top view scanning electronic microscope image of the fabricated self-selective memristor array.

Fig. 3. Cross-sectional transmission electron microscopy image of a self-selective.

Fig. 4. Direct current I-V characteristic of devices. V_{hold} and V_{th} represent positive hold voltage and negative threshold voltage of built-in selector, respectively.

Fig. 5. Retention characteristics of both selector on/off low-resistance states (LRS) and high-resistance states (HRS).

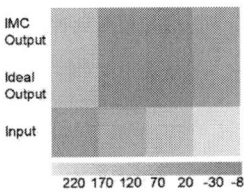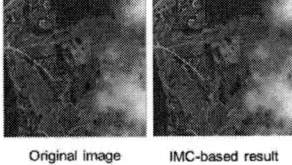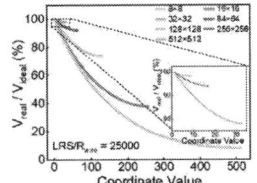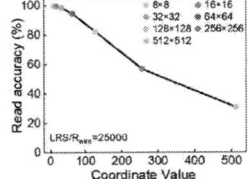

Fig. 6. Programming endurance test over 10^{10} cycles of a device.

Fig. 7. Fast switching speed of the VO_x-based self-selective layer.

Fig. 8. A programming example using write-with-verify method.

Fig. 9. C2C and D2D device variations of LRS when the selector off and HRS.

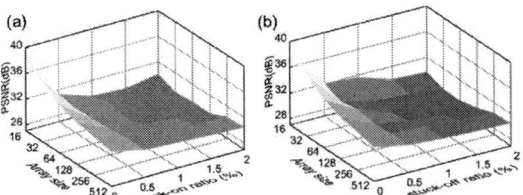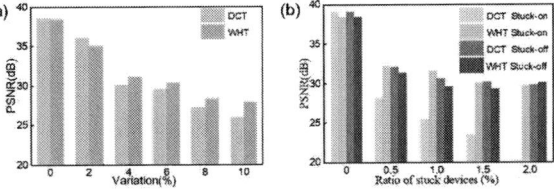

Fig. 10. 1D WHT input, ideal and IMC-based output results.

Fig. 11. Original input image and IMC-based result.

Fig. 12. Impact of wire resistance on the voltage at different nodes.

Fig. 13. Impact of wire resistance on reading accuracy.

Fig. 14. (a) Impact of different ratio of stuck-on to image compression. (b) Impact of different ratio of stuck-off respectively to image compression.

Fig. 15. (a) Impact of different variations in image compression between DCT and WHT. (b) Impact of different ratio of stuck devices in image compression between DCT and WHT.

A Novel De-Mirroring Approach for Bias-dependent Capacitance Extraction in Nanosheet FET using Conformal Mapping

Deven H Patil[1], Sandeep Kumar[1], Sunil Rathore[2], S. Dasgupta[3], Navjeet Bagga[1*]

[1] Indian Institute of Technology Bhubaneswar, India, [2] Manipal Institute of Technology, Manipal, India.
[3] Indian Institute of Technology Roorkee, India. (Email: navjeet@iitbbs.ac.in)

Abstract

The sheets/channels are vertically stacked in Nanosheet FET (NSFET), which forms a complex capacitive network. Till date, to the best of our knowledge, the only bias-independent capacitance model for NSFET has been presented by considering that the device is mirrored symmetric around the center of the channel. This mirroring approach does not precisely address the nominal biasing conditions at the source and drain; thus, we need to model the two halves separately. In this paper, we propose a *bias-dependent* (i.e., drain voltage (V_{DS}) dependent) capacitance extraction of an NSFET using a de-mirroring approach. The electric field coupling from the drain to the source impacts the capacitances. The obtained results reveal that the mirroring approach overestimates the capacitances by ~15%, which is accurately predicted by our proposed de-mirroring approach.

(Keywords: Nanosheet FET, Capacitance Modeling, De-mirroring, Parasitic, Intersheet Capacitance)

Introduction

The technological advancement of emerging semiconductor devices paves the way to achieve exceptional performance, energy, economy, and scalability. The miniaturization of the devices leads to reduced contacted poly pitch (CPP), metal pitch, fin pitch, etc., thereby resulting in complex non-planar geometries with improved gate-electrostatic integrity [1]-[2]. Unlike planar MOSFET, the non-planar architectures (e.g. FinFETs, Nanosheets, etc.) comprise various capacitances, which require proper attention as capacitance extraction is essential to explore the dynamic behavior of the device. For instance, the parasitic source/drain (S/D) capacitance becomes a major problem, which restricts the drive current (I_{ON}) at high voltages, when gate length and CPP get smaller. A modified transmission line approach has been proposed for N-stack Nanosheet FETs to handle this problem by offering a better optimized parasitic capacitance model to fairly depict the non-uniform current density in the S/D extension regions (L_{EXT}) [3]. Other than this, several approaches have been proposed for capacitance extraction, like conformal mapping, elliptical integration, Poisson integration, etc. [4]. However, in all such literature, a mirroring approach has been followed, which states that the device is symmetric if the drain voltage (V_{DS}) is not being applied. In the mirroring approach, the one-half of the device being modeled and then the overall capacitance would be considered as double the value. For instance, extract all the possible capacitance at the source-channel (S-C) side and consider that they are identical at the drain-channel (D-C) side. Though mirroring is a simple approach, this is not an accurate one because the bias-dependency needs to be included in the model, to accurately predict the dynamic behavior of the NSFET. Therefore, in this work, we propose a novel *de-mirroring* approach to extract the bias-dependent capacitance by considering the S-C & D-C sides separately. While considering the V_{DS} dependency in our proposed de-mirroring approach, a significant contribution of gate-to-drain capacitance (C_{gd}) is observed in the overall gate capacitance (C_{gg}), which was unaddressed in the mirroring technique. Further, the role of varying gate length (L_g), L_{EXT}, and sheet thickness (T_{ch}) in the dynamic performance of NSFET has been explored by V_{DS}-dependent capacitance extraction. Finally, using a simple RC network, we conclude that the proposed de-mirroring approach predicts an accurate delay performance of the device.

Device Structure and Simulation Setup

A three-sheet Nanosheet of 5nm thick is opted as a baseline reference (Fig.1a) to extract all the possible capacitances (Fig.2). A stacked gate-oxide of SiO_2-HfO_2 layer is employed to get an effective oxide thickness (EOT) of 0.9nm. The source/drain (S/D) are uniformly doped whereas the S/D extensions have Gaussian doping to avoid the abrupt field breakdown at S-C and D-C junctions and to mitigate the effects of random dopant fluctuations (RDF). The Si_3N_4 is used as a spacer over the extension regions. *Sentaurus* TCAD [5] is employed for the device simulation and MATLAB [6] is used for analytical modeling. The conventional drift-diffusion model with Poisson solver is invoked for governing the carrier transport. The electrostatic phenomenon and spatial quantum confinement effect have been incorporated using the MLDA model. To capture the mobility degradation, the inversion accumulation (IAL-Mob) mobility model, thin layer, and high-field saturation models are included in the TCAD framework [7]. The Lombardi model is included to capture the dipole phonon and remote coulomb scattering. The SRH and Auger models have been included for considering the carrier generation recombination. The transfer characteristic (I_{DS}-V_{GS}) is well-calibrated (Fig.1c) against the experimental data [8]. Table I comprises the default values of all the parameters unless stated otherwise.

Results and Discussion

In this paper, we propose a novel de-mirroring approach to extract all the capacitances of an NSFET, as shown in Fig.2.

Though the parasitic capacitance is explicitly dependent on physical components, its dependency on any bias should not be neglected. We employed conformal mapping and elliptical integration in this work to frame out the analytical model, which generates the complex plane where plates of the capacitors are kept at an angle of θ to some complex plane where both the plates are parallel to each other (Fig.3a-b). Eq. (1-4) governs the underlying mechanism of conformal mapping and transmission line method used for capacitance calculation. The given equation uses an electric field between two planes to calculate the capacitance. The model captures the different components of capacitances, such as inner and outer fringing field capacitances with an accuracy of ~1.63% than simulation data. Fig.3c shows a schematic of NSFET showing the traditional mirroring technique, i.e., capacitance extraction considering the S-C and D-C to be identical. Therefore, the drain and source voltages are kept grounded, and only gate voltage (V_{GS}) is applied at any instance. Thereby, no impact of V_{DS} comes into the picture during capacitance calculation. Though this is a simple approach, yet not a correct way, as it does not include the impact of V_{DS}. The lateral electric field from the drain side (Fig.3d) results in a wider depletion layer at the D-C junction, which offers a voltage-controlled transition capacitance, unlike the case in the mirroring technique. This reduces the overall gate-to-drain capacitance in the de-mirroring approach [$C_{D(dM)}$] than the mirroring approach [$C_{D(M)}$], as shown in Fig. 4(a). The respective capacitance, i.e., C_S and C_D for de-mirroring (dM) cases with varying V_{DS} is plotted in Fig. 4b, showing that C_D is significantly smaller than C_S, which would be the same as $C_{S(M)}$ or $C_{D(M)}$. The calculated value of $C_{D(M)}$ is 1.4 times that of $C_{D(dM)}$. Thus, the overall gate capacitance (C_{gg}) is ~15% smaller in the proposed de-mirroring approach (Fig.4c). The model also predicts the physical device parametric variation, such as L_{EXT}, L_g, etc. The increase in the extension region results in lowering all of the spacer (S_1 & D_1) and extension capacitances, i.e., S_2 and D_2. Thus, in turn, reduces the C_{gg} with varying L_{EXT} (Fig.5a). Further, the increase in L_g reduces the impact of the lateral field (inset: Fig.5b), and due to an increase in the longitudinal field, the intersheet capacitances increase. Further, an increase in L_g increases the overlap capacitances. This, in turn, increases the difference in capacitances between mirroring and de-mirroring approaches (Fig. 5b). Similarly, the intersheet capacitance will be reduced in the de-mirroring case (Fig.5c). The fabrication non-idealities cause overlap (OL) and underlap (UL) regions at the edge of S-C and D-C junctions, where these UL and OL results in the additional capacitances. Thus, the overall gate capacitance gets modulated with UL and OL regions (Fig.6). An increase in sheet thickness (T_{ch}) causes more charges to reside in the channel, and thus, the capacitance increases. This is further modulated as the increase in T_{ch} increases the impact of the lateral field which

significantly modulates the de-mirroring capacitance, whereas, the mirroring capacitance stays almost constant (Fig.6). Overall impact of increasing OL (Fig.6a) and UL (Fig.6b) with T_{ch} is plotted for all three cases, i.e., (i) mirroring: $V_{GS}=V_{DD}$ and $V_{DS}=0V$; (ii) de-mirroring: $V_{GS}=V_{DS}=V_{DD}$; and (iii) parasitic capacitance: $V_{GS}=V_{DS}=0$. At last, to evaluate the delay performance, we designed a simple RC network by considering NSFET as a two-terminal capacitor under mirroring and de-mirroring cases (Fig.7a), where results reveal that the proposed de-mirroring approach accurately predicts the capacitances. The delay performance is further evaluated by defining a figure of merit (FoM), i.e., I_{ON}/C_{gg}. The negative gradient in Fig.7(b-c) shows that the impact of C_{gg} variation is dominant in addressing the FoM. In Fig. 8(a), the contribution of individual capacitance (shown in Fig.2) is plotted, where the inner circle is for the de-mirroring case. This shows that the overlap capacitance has the highest contribution, i.e., 48%, and the inter-sheet has the lowest contribution. The error bar chart (Fig.8b) shows the percentage error in model and simulation data for individual capacitance extraction.

Conclusion

In this work, we proposed a novel de-mirroring approach to extract the overall capacitance of a Nanosheet FET, which comprises a complex capacitive network. Unlike the mirroring technique, here in the de-mirroring approach, the channel is not considered symmetrical around the center owing to applied V_{DS}. Thus, bias-dependent governance should be included for accurate capacitance prediction, in turn, offers a valid delay-performance merit. The results reveal that the traditional mirroring approach overestimates the overall capacitance by ~15%, a significant value for scaled devices. We further evaluated the role of the individual source side and drain side capacitances on overall gate capacitance (C_{gg}) with the varying channel length, S/D extension length, underlap, and overlap capacitance with mirroring and de-mirroring approaches. Thus, the proposed study is worth exploring to define a proper design guideline in terms of accurate capacitance prediction.

References

[1] IRDS 2021 Roadmap. Accessed: May 26, 2021. [Online]. Available: https://irds.ieee.org/editions/2021.

[2] S. Rathore et al., "Demonstration of a Nanosheet FET with High Thermal Conductivity Material as Buried Oxide: Mitigation of Self-Heating Effect, IEEE TED, vol. 70, no. 4, pp 1970-1976, Apr. 2023.

[3] S. B. Cohn, "Shielded Coupled-Strip Transmission Line," in *IRE Transactions on Microwave Theory and Techniques*,1955.

[4] Yumin Xiang," The electrostatic capacitance of an inclined plate capacitor", *Journal of Electrostatics*, 2006

[5] Sentaurus Device User Guide Version T-2022.03, *Synopsys, Inc.*, 2022.

[6] The MathWorks Inc., *MATLAB* (Version 9.15.0 R2024a), Natick, Massachusetts, United States, 2024. [Software]. Available: https://www.mathworks.com

[7] S. Banchhor, et. al., "A New Insight into the Saturation Phenomenon in Nanosheet Transistor: A Device Optimization Perspective," IEEE EDTM, 2023.

[8] N. Loubet et al., "Stacked nanosheet gate-all-around transistor to enable scaling beyond FinFET", *IEEE SVLSI*, pp.230,2017.

979-8-3315-0417-5/25 $31.00 © 2025 IEEE

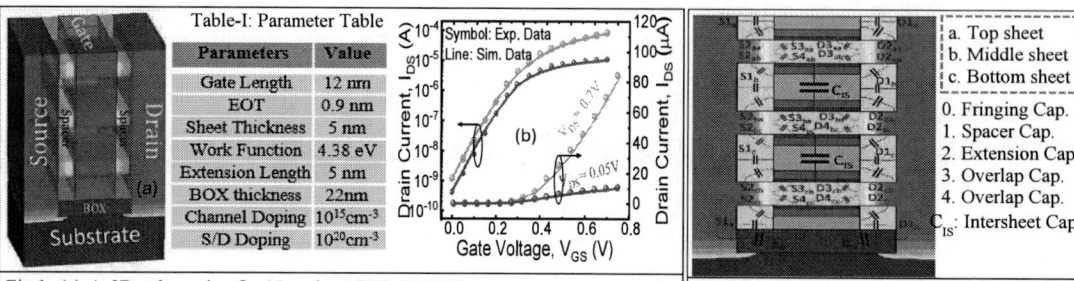

Table-I: Parameter Table

Parameters	Value
Gate Length	12 nm
EOT	0.9 nm
Sheet Thickness	5 nm
Work Function	4.38 eV
Extension Length	5 nm
BOX thickness	22nm
Channel Doping	$10^{15} cm^{-3}$
S/D Doping	$10^{20} cm^{-3}$

a. Top sheet
b. Middle sheet
c. Bottom sheet

0. Fringing Cap.
1. Spacer Cap.
2. Extension Cap.
3. Overlap Cap.
4. Overlap Cap.
C_{IS}: Intersheet Cap.

Fig.1. (a) A 3D schematic of a Nanosheet FET (NSFET), considered as a baseline reference; (b) the calibrated I_{DS}-V_{GS} characteristics with the experimental data [8]. Table I: comprises all the parameters used in the TCAD simulation.

Fig.2. A complete capacitive network of NSFET. Here, 'S' stands for source, and 'D' stands for drain side.

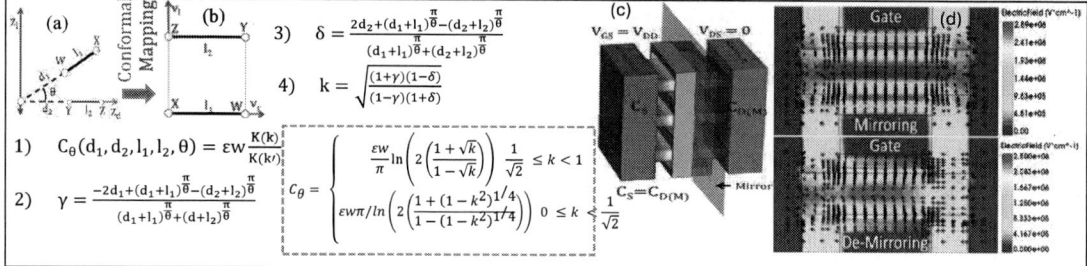

3) $\delta = \dfrac{2d_2+(d_1+l_1)^{\frac{\pi}{\theta}}-(d_2+l_2)^{\frac{\pi}{\theta}}}{(d_1+l_1)^{\frac{\pi}{\theta}}+(d_2+l_2)^{\frac{\pi}{\theta}}}$

4) $k = \sqrt{\dfrac{(1+\gamma)(1-\delta)}{(1-\gamma)(1+\delta)}}$

1) $C_\theta(d_1,d_2,l_1,l_2,\theta) = \varepsilon w \dfrac{K(k)}{K(k')}$

2) $\gamma = \dfrac{-2d_1+(d_1+l_1)^{\frac{\pi}{\theta}}-(d_2+l_2)^{\frac{\pi}{\theta}}}{(d_1+l_1)^{\frac{\pi}{\theta}}+(d_1+l_2)^{\frac{\pi}{\theta}}}$

$C_\theta = \begin{cases} \dfrac{\varepsilon w}{\pi}\ln\left(2\dfrac{1+\sqrt{k}}{1-\sqrt{k}}\right) \dfrac{1}{\sqrt{2}} \le k < 1 \\ \varepsilon w\pi/ln\left(2\dfrac{1+(1-k^2)^{1/4}}{1-(1-k^2)^{1/4}}\right) 0 \le k \le \dfrac{1}{\sqrt{2}} \end{cases}$

Fig.3. (a) inclined plate capacitor in z-plane, considered as a reference for conformal mapping technique; (b) conformal mapped parallel plate capacitor in 'v' plane using Schwarz-Christoffel transformation; (c) traditional mirroring approach used for capacitance calculation, where the source-channel and drain-channel are considered to be symmetric under V_{DS}=0V condition; (d) de-mirroring approach, which results in lateral field, thereby, affects the capacitance extraction. C_θ is the overall capacitance calculated using conformal mapping (Eq. 2-3) & coupled transmission line (Eq.1), where k is coupling factor (Eq.4).

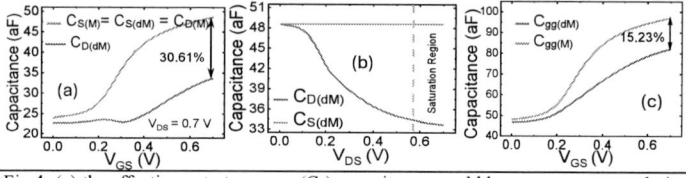

Fig.4. (a) the effective gate-to-source (C_S) capacitance would be same as gate-to-drain capacitance (C_D) in mirroring approach. However, in de-mirroring, the applied V_{DS} results in ~30% reduction in C_D, while keeping the same C_S; (b) C_S & C_D for de-mirroring with varying V_{DS}; (c) overall gate capacitance (C_{gg}) in mirroring & de-mirroring techniques, revealing accurate prediction of C_{gg} by de-mirroring approach.

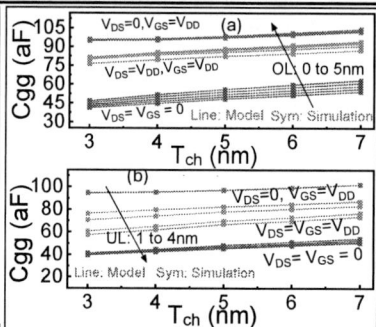

Fig.6. C_{gg} with varying sheet thickness (T_{ch}) for (a) overlap length (OL) of 0-5nm; and (b) underlap length (UL) of 1-4nm. The OL and UL require additional capacitance extraction.

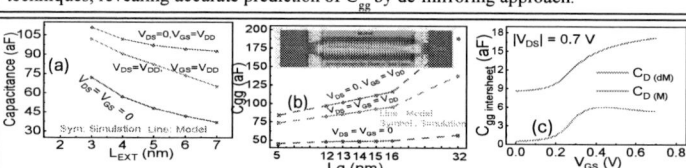

Fig.5. Total gate capacitance with varying (a) extension length (L_{EXT}); (b) gate length (L_g). Inset of (b) shows lateral field component reduction for L_g=32nm. Here, the mirroring technique (V_{GS}=V_{DD}, V_{DS}=0V) shows a higher value than the de-mirroring approach (V_{GS}=V_{DS}=V_{DD}). Parasitic capacitance is extracted at V_{GS}=V_{DS}=0V; (c) Inter-sheet capacitance is also reduced in the de-mirroring technique.

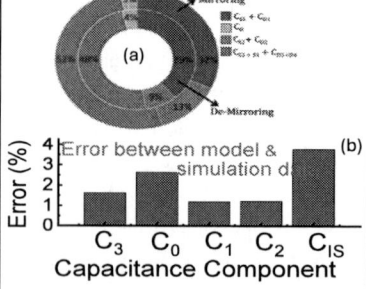

Fig.8. (a) the contribution of individual capacitances, where the symbols have known meaning as explained in Fig.2; (b) percentage error while calculating the respective capacitances using modeling and simulation.

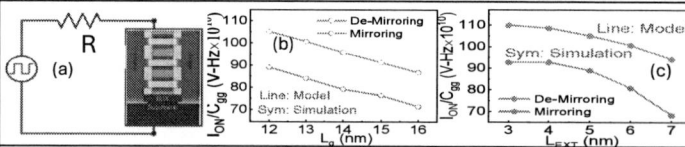

Fig.7. (a) a simple RC network, where NSFET is considered as a two-terminal capacitor for both mirroring and de-mirroring cases. I_{ON}/C_{gg}, defined as a figure-of-merit (FoM) is plotted with varying (b) L_g and (c) L_{EXT}. The proposed de-mirroring approach offers an accurate prediction of NSFET capacitances.

979-8-3315-0417-5/25 $31.00 © 2025 IEEE

4F^2/bit Memristive Multi-bit Content Addressable Memory Enabled by Nonlinear Encoding for In-Memory Similarity Search

Tong Hu[1], Yibai Xue[1], Yingjie Yu[1], Wenbin Zuo[1], Jiancong Li[1], Yi Li[1*], and Xiangshui Miao[1]

[1] School of Integrated Circuits, Huazhong University of Science and Technology, Wuhan 430074, China

Email: liyi@hust.edu.cn

Abstract

Search density and energy efficiency are critical metrics for similarity search, particularly in resource-limited edge AI applications. In this study, we present an innovative nonlinear encoding scheme for Content Addressable Memory (CAM) to enhance density and energy efficiency. Utilizing a 2R memristor CAM, this work leverages multi-level programmable states and similarity-based matching within a single CAM cell by a Euclidean-like distance function. The design achieves a high density of 4F^2/bit and an energy consumption of 0.88 fJ/bit. Experimental results demonstrate software-comparable accuracy in few-shot learning tasks, offering 213x improvement in energy efficiency compared with GPU, highlighting its potential for scalable, low-power edge AI applications.

Keywords: Content addressable memory, Associative memory, Memristor and few-shot learning

Introduction

The human brain's innate ability to associate unfamiliar objects with prior knowledge has inspired the advancement in associative memory (AM), which emulates this cognitive function by associating new query with past memory based on similarity (Fig. 1(a)). Content addressable memory (CAM), especially those based on non-volatile memories (NVM), has emerged in recent years as a hardware implementation of AM [1],[2], owing to its aptness of efficient similarity search between query and memory.

Fig. 1(b) shows the basic type of CAM. Traditional binary (BCAM) and ternary (TCAM) CAM structures utilize Hamming distance calculations, offering basic similarity matching capabilities. However, these structures are constrained by limited data states, since only two states (except the 'X' state) are stored in CAM. This approach also suffers from accuracy loss when applying binarization methods by locality-sensitive hashing (LSH), unless the vector dimensionality is exceptionally high. To further exploit multi-level or analog tuning ability of NVMs, multi-bit CAM (MCAM) and analog CAM (ACAM) are trending in recent years, offering excellent density and energy benefits [3-6]. However, most of these designs only output exact match results within a single cell, lacking efficient distance-based matching capabilities required for optimal nearest neighbor (NN) search.

In this work, we propose and experimentally validate a multi-bit CAM design enabled by nonlinear encoding scheme that incorporates similarity-based matching. The 2R architecture integrates two memristors per CAM cell to calculate the Euclidean-like distance between vectors, outputting a similarity measure via match line (ML) current. This compact design achieves a high density of 4F^2/bit, showing high area efficiency, as shown in Fig. 1(c). Based on the design, we validated our encoding scheme in few-shot learning and achieved software-comparable accuracy with robustness to resistance variation and high energy efficiency.

Design and Experiment of Multibit CAM

A. Device structure of memristor based CAM

As shown in Fig. 2(a), the proposed MCAM is composed of two W/Al$_2$O$_{3-x}$N$_x$/Pt memristors. To prevent the crosstalk between cells, we deposited a SiO$_2$ isolation layer by plasma-enhanced chemical vapor deposition, followed by ultraviolet lithography and inductively coupled plasma etching. This process ensures that the resistive switching layer is isolated within the via-holes, as shown in Fig. 2(b). The device's transmission electron microscopy (TEM) image is presented in Fig. 2(c), while Fig. 2(d) shows 100 continuous I-V curves without an electrical forming process, indicating great cycle-to-cycle uniformity.

B. Proposed MCAM circuit and nonlinear encoding

The proposed MCAM circuit is revealed in Fig. 1(c). It consists of 2 memristors, where the top electrodes are connected to two search lines SL and SLB, respectively. The other electrodes of two memristors are connected to the same ML. Search line voltage V_1, V_2 and conductance of memristors G_1, G_2 represent input data k and stored data j, respectively ($0 \leqslant k, j \leqslant 1$). Here, our design utilizes a nonlinear encoding scheme, where input voltage and storage conductance are regulated by rules in Eq. (1)-(4):

$$V_1 = V_S \cos(k\pi/2) \quad (1)$$
$$V_2 = V_S \sin(k\pi/2) \quad (2)$$
$$G_1 = G_H \cos(j\pi/2) \quad (3)$$
$$G_2 = G_H \sin(j\pi/2) \quad (4)$$

The output current I_{ML_i} of a single cell and match line current I_{ML_total} can be obtained as in Eq. (5) and (6):

$$I_{ML\_i} = V_1 G_1 + V_2 G_2 = \alpha \cos\left(\frac{\pi}{2}(k_i \text{-} j_i)\right) \quad (5)$$

$$I_{ML\_total} = \sum_i \alpha \cos\left(\frac{\pi}{2}(k_i \text{-} j_i)\right) \quad (6)$$

979-8-3315-0417-5/25 $31.00 © 2025 IEEE

where $\alpha = V_S G_H$. V_S and G_H are maximum search voltage and memristor conductance, respectively. In an array shown in Fig. 2(e), where multiple MCAM cells are connected to one ML, I_{ML_i} in each cell are accumulated according to Kirchhoff's laws. The accumulated current, I_{ML_total}, is sent to winner-take-all (WTA) circuits, where the highest ML current is selected among all match lines, indicating the stored data that shares the highest similarity with the input. Fig. 2(f) demonstrates the array- level distance function on a ML connected with 2 MCAM cells. The proposed distance function in Eq. (6) has a similar trend to Euclidean distance, proving suitable for NN search.

C. Multibit encoding with 2R MCAM

To enhance robustness against resistance variation in memristors and relieve analog encoding burden, we experimentally split data into 4 quantized states. Fig. 2(g) illustrates the ideal distance function across these states. Here, k, j are set at (0, 1/3, 2/3, 1) corresponding to 2-bit state (00, 01, 10, 11). It is noted that the encoded states are experimentally configured to prevent zero-conductance settings by assigning a low conductance value G_L when $j=0$ or $j=1$. Fig. 3(a) illustrates the operating waveform of search in MCAM, while Fig. 3(b), (c) show the measured current response in fabricated memristors. Results suggest that I_{ML} is positively correlated to the degree of match, indicating that our MCAM can accurately compute the similarity for NN search.

Evaluation of MCAM in Few-shot Learning Task

Few-shot learning is vital for AI applications where adapting to new classes from limited data is essential. Memory-augmented neural networks (MANNs) [9] address this challenge by incorporating associative memory (AM) to store feature vectors as prior knowledge for retrieval. As illustrated in Fig. 4(a), MANN utilizes CNN to extract features and store new class features directly within CAM for similarity-based retrieval when encountering new inputs.

We evaluated the proposed MCAM with non-linear encoding in a 20-way, 1-shot few-shot learning task using the Omniglot dataset. Feature vector was extracted with a pre-trained 4-layer CNN, and weights were fixed during inference. Fig. 4(b), (c) shows the test results. Our nonlinear encoding scheme achieves an accuracy of 87.72% with 2-bit encoding, only 4.03% and 2.15% below Euclidean distance results for full-precision and 2-bit quantized cases, respectively. Our 2-bit MCAM outperforms LSH+TCAM approach by 15.59% at the same word length ($n=64$), since LSH requires much higher dimension (>256) for optimal feature representation. Testing revealed minimal accuracy degradation with device-to-device (D2D) variation under 5%, and accuracy drops less than 6% when $\sigma = 25\%$ in 5way, 1shot task, suggesting our MCAM has a high variation tolerance. Finally, a 213x energy

efficiency improvement over Nvidia H100 GPU was observed in Fig. 4(d), confirming our MCAM's robustness and energy efficiency.

Conclusion

In this work, we reported a novel nonlinear encoding scheme for CAM, featuring a compact design with 2R structure, which is validated on fabricated memristors. The general encoding approach is adaptable to other CAM designs with complementary structures. By leveraging multi-level states within a single MCAM cell, the design achieves a density of $4F^2$/bit and an energy efficiency of 0.88fJ/bit, as shown in Table 1. Furthermore, the proposed MCAM demonstrates software-comparable accuracy and high energy efficiency in few-shot learning tasks, showing 213x improvement in energy efficiency over GPU and high tolerance to device variability, reinforcing its viability for resource-limited edge AI applications.

Acknowledgments

This work was supported by the STI 2030—Major Projects (Grant No. 2021ZD0201201), the National Key Research and Development Plan of MOST of China (2022YFB4500101), the Fundamental Research Funds for the Central Universities (HUST: 5003190012), and the Natural Science Foundation of Hubei Province under Grant (2024AFA043).

References

[1] K. Ni *et al.*, "Ferroelectric ternary content-addressable memory for one-shot learning," *Nat Electron.*, 2(11), pp. 521–529 (2019).

[2] Y. Zhang *et al.*, "Semantic memory–based dynamic neural network using memristive ternary CIM and CAM for 2D and 3D vision," *Science Advances*, 10(33), p. eado1058 (2024).

[3] C. Li *et al.*, "Analog content-addressable memories with memristors," *Nat Commun.*, 11(1), p.1638 (2020).

[4] X. Yin *et al.*, "FeCAM: A Universal Compact Digital and Analog Content Addressable Memory Using Ferroelectric," *IEEE Transactions on Electron Devices*, 67(7), pp. 2785–2792 (2020).

[5] J. Luo *et al.*, "A Novel Ambipolar Ferroelectric Tunnel FinFET based Content Addressable Memory with Ultra-low Hardware Cost and High Energy Efficiency for Machine Learning," in *2022 IEEE Symposium on VLSI Technology and Circuits (VLSI Technology and Circuits)*, pp. 226–227 (2022).

[6] X. S. Hu *et al.*, "In-Memory Computing with Associative Memories: A Cross-Layer Perspective," in *2021 IEEE International Electron Devices Meeting (IEDM)*, p. 25.2.1-25.2.4 (2021).

[7] L.-Y. Huang *et al.*, "ReRAM-based 4T2R nonvolatile TCAM with 7x NVM-stress reduction, and 4x improvement in speed-wordlength-capacity for normally-off instant-on filter-based search engines used in big-data processing," in *2014 Symposium on VLSI Circuits Digest of Technical Papers*, pp. 1–2 (2014).

[8] J. Li, R. K. Montoye, M. Ishii, and L. Chang, "1 Mb 0.41 μm² 2T-2R Cell Nonvolatile TCAM With Two-Bit Encoding and Clocked Self-Referenced Sensing," *IEEE Journal of Solid-State Circuits*, 49(4), pp. 896–907 (2014).

[9] H. Li *et al.*, "One-Shot Learning with Memory-Augmented Neural Networks Using a 64-kbit, 118 GOPS/W RRAM-Based Non-Volatile Associative Memory," in *2021 Symposium on VLSI Technology*, pp. 1–2 (2021).

[10] I. Hayashi *et al.*, "A 250-MHz 18-Mb Full Ternary CAM With Low-Voltage Matchline Sensing Scheme in 65-nm CMOS," *IEEE Journal of Solid-State Circuits*, 48(11), pp. 2671–2680 (2013).

979-8-3315-0417-5/25 $31.00 © 2025 IEEE

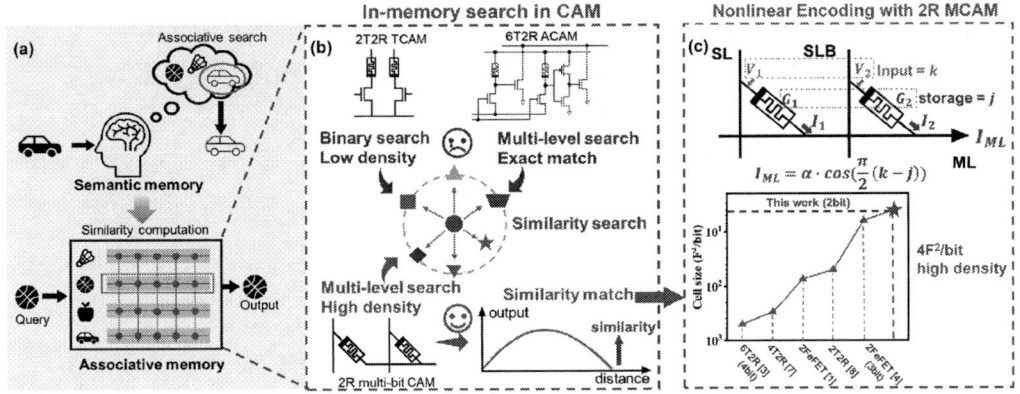

Fig. 1: (a) Comparison of human learning and associative memory. (b)Associative memory realized by CAM. Exact match can only tell match and mismatch, while the proposed MCAM can realize similarity-based match with high density. (c) 2R MCAM structure with nonlinear encoding and density comparison with other works in [1],[3],[4],[7],[8].

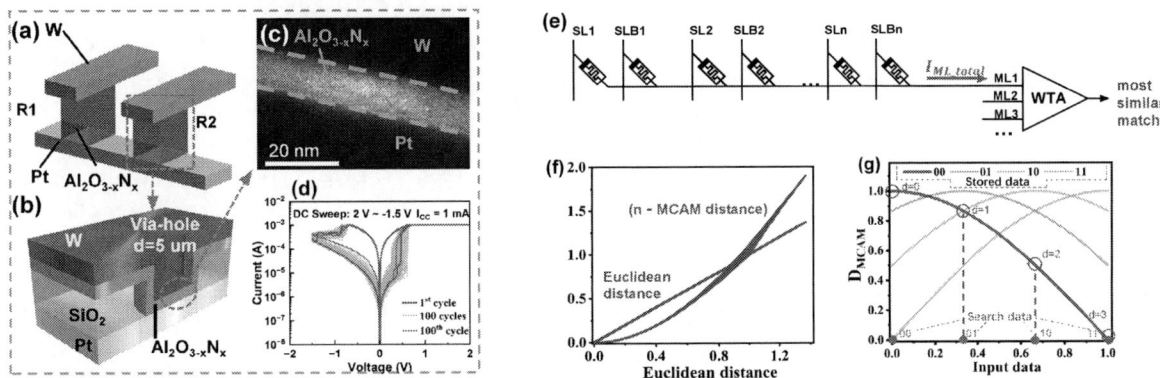

Fig. 2: (a) Schematic of the 2R CAM cell. (b) Detailed structure of W/Al$_2$O$_{3-x}$N$_x$/Pt memristor with 5 μm width via-hole. (c) Cross-sectional TEM image of the memristor (d) DC I-V characteristic with 100 continuous cycles. (e) MCAM array. (f) Distance functions on one row has similar trend with Euclidean distance. The MCAM distance here is set to (2-d$_{MCAM}$) for the ease of comparison. (g) Distance function of MCAM in 2bit encoding

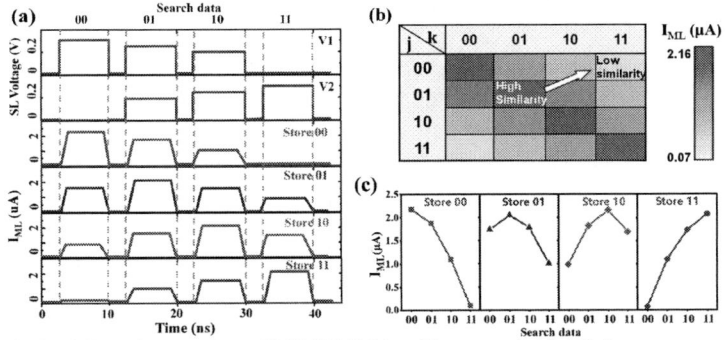

Table 1: Benchmark of proposed MCAM

Reference	[10]	[8]	[4]	[3]	This work
CAM cell	16T	2T2R	2FeFET	6T2R	2R
Encoding	binary	binary	multi-bit	analog	multi-bit
Match type	exact	exact	exact	exact	**similarity**
Search energy (fJ/bit)	1.98	0.75	0.069	/	0.88
Density (F^2/bit)	400	50	6.23	506	4

Fig. 3: (a) Operation waveform of 2-bit MCAM. k and j are set at 0, 1/3, 2/3, 1.
(b), (c) **Measured** current of search operation. ML current is positive with the degree of match.
(V$_S$ is set at 0.2V, and G$_H$ = 10μS, G$_L$ = 0.33μS)

Fig. 4: Results of few-shot learning. (a) MANN structure. TCAM uses LSH for binary quantization while MCAM directly applies 2-bit quantization. (b) Accuracy of 20way, 1shot task. Results of MCAM is based on measured data (c) Accuracy under different resistance variation. (d) Energy efficiency comparison with Nvidia H100 GPU.

979-8-3315-0417-5/25 $31.00 © 2025 IEEE

Monolithic and heterogeneous CTFT on glass substrate using Ni-MIC poly-Si TFT and Cu-MIC DG poly-Ge TFT

Y. Ito, D. Goshima, A. Kurihara, A. Hara

Tohoku Gakuin University, Japan, e-mail: akito@mail.tohoku-gakuin.ac.jp

Abstract

We realized a heterogeneous complementary TFT (CTFT) fabricated using a monolithic process consisting of an n-channel poly-Si TFT and a p-channel double-gate (DG) poly-Ge TFT on a glass substrate. The lower layer poly-Si was crystallized at temperatures below 600°C using a metal induced crystallization method that involves Ni-MIC. The DG poly-Ge TFT on the upper layer was crystallized at temperatures below 500°C by Cu-MIC. The dynamic characteristics of the inverters are also evaluated.
Keywords: TFT, complementary TFT, poly-Si and poly-Ge

Introduction

Complementary FETs (CFETs), a three-dimensional integration technology, have garnered considerable attention in recent years [1-13]. For example, p-ch FETs are formed on the upper layer of the n-ch FETs and connected to form a three-dimensional integrated circuit. This helps reduce the overall footprint and achieve higher speed and integration. The semiconductors used are not limited to conventional silicon (Si) but may also include different semiconductors such as germanium (Ge), which possesses a higher bulk hole mobility compared to Si and a higher affinity for Si process. In recent years, several attempts have been undertaken to fabricate hetero semiconducting CTFTs consisting of polycrystalline materials [14-16].

We have aiming to improve the performance of low-temperature (LT) poly-Si TFTs and LT poly-Ge TFTs on glass substrates, and we have already developed and reported complementary TFTs (CTFTs) [17] on glass substrates using CLC [18] for the crystallization of the lower n-ch poly-Si layer and Cu-MIC for the upper DG poly-Ge TFT [19].

In this study, we realized a CTFT inverter on a glass substrate using Ni-MIC [20], which utilizes Ni as a catalyst for the crystallization of poly-Si in the lower layer. We further improved the gate length to achieve an equal on-current between the n-ch Ni-MIC poly-Si TFT and p-ch Cu-MIC DG poly-Ge TFTs. The dynamic characteristics of the CTFT inverter are also evaluated.

Experimental methods

DG poly-Ge TFTs were used as the upper layer of the CTFT because they can be formed at low temperatures. Figure 1 shows a photograph of the fabricated device. Figure 2 shows the cross-sectional images of the CTFT and the circuit of the CMOS inverter. The lower-layer poly-Si TFT was formed in the order shown in Fig. 3-1. After growing 50 nm of amorphous Si (a-Si) on a glass substrate, a poly-Si thin film was formed using Ni-MIC. In Ni-MIC, a-Si is immersed in a solution containing Ni for 3 h, followed by crystallization by heat treatment at 580 °C for 12 h. Next, transistor islands are formed by RIE, and SiO_2 is grown as the gate insulator (SO-1) of the lower n-ch poly-Si TFT by plasma enhanced CVD (PECVD) at a thickness of 30 nm, followed by a 4-h heat treatment at 550 °C in an N_2 atmosphere to improve the quality of the SiO_2 gate insulator. Then, molybdenum (Mo) was sputtered as a bottom gate (BG) metal. Then, the BG electrode (G-1) was formed, and after etching the SiO_2 in the SD region by RIE using the gate (G-1) as a mask, phosphorus (P) ion implantation was performed in a self-aligned manner on the SD of the poly-Si TFT. Then, SiO_2 of thickness 30 nm is grown by PECVD (SO-2), and the SD is heat-treated for activation at 550 °C for 6 h in an N_2 atmosphere. The gate metal (G-1) of the lower-layer poly-Si TFT acts as the lower gate metal (G-1) of the upper-layer DG poly-Ge TFT.

Next, the upper-layer Cu-MIC DG poly-Ge TFT was fabricated using the process shown in Fig. 3-2. A Ge/Cu/Ge layer with a Ge thickness of 14.4 nm is deposited. The upper gate SiO_2 with thickness 30 nm (SO-3) was deposited, and Cu-MIC crystallization is performed by heat treatment at 500 °C for 5 h in N_2 atmosphere. To operate the upper-layer poly-Ge TFT in the DG mode, the upper and lower metals (G-1 and G-2) were connected. This was achieved by forming contact holes in the SiO_2 film on BG (G-1) and then sputtering Mo to form TG (G-2). At this point, the two metal gates were connected. The lower gate metal (G-1) acted as the gate for the poly-Si TFT and the lower gate for the DG poly-Ge TFT. Finally, a 100-nm SiO_2 interlayer insulator was formed through PECVD. Contact holes were subsequently formed through RIE, and aluminum (Al) is deposited to form V_{DD}, V_{IN}, V_{OUT}, and ground electrodes. Finally, the SD region of the poly-Ge TFT was changed into low-resistance Al through heat-treating at 300 °C in an $N_2 + H_2$ atmosphere and by performing atomic exchange between Ge and Al. Metals and p-type semiconductors can achieve ohmic contact, enabling the low-resistance SD of poly-Ge TFTs.

The gate length (L) of the lower layer poly-Si TFT is 10 μm. From the CTFT photograph in Fig. 1, it can be

979-8-3315-0417-5/25 $31.00 © 2025 IEEE

confirmed under an optical microscope that the exchange between Al and Ge occurred in the channel region of the upper Ge layer. Based on this, for the upper layer p-ch DG poly-Ge TFT, the value of L was determined to be 5 μm.

Results and discussion

The transfer characteristics of the lower-layer n-ch poly-Si TFT and the upper-layer p-ch DG poly-Ge TFT, which constitute the CTFT, are shown in Figure 4. The threshold voltages of n- and p-ch TFTs are 1.9 V and -1.2 V, and the mobilities are 32.5 cm^2/Vs and 9.6 cm^2/Vs, respectively. From a comparison of the respective TFTs, it was observed that the on- and off-currents of both TFTs were not significantly different, and their I_{on}/I_{off} ratios were almost equal. Figure 5 shows the output characteristics of the two TFTs. A comparison of the maximum values of the on-currents confirmed that they were almost identical. The mobility, threshold voltage, and s.s. values of the n-ch poly-Si and p-ch DG poly-Ge TFT used in this CTFT are summarized in Table 1.

Figure 6 shows the performance of the static inverter with $V_{DD} = 5$V. The operation of the inverter was confirmed, and the transition threshold was 2.5 V. The theoretical value of the transition threshold was 2.6 V, which is roughly consistent with the measured value. This may be attributed to the fact that the on-current of the p-ch DG poly-Ge TFT and n-ch poly-Si TFT are identical, and an ideal CTFT has been successfully fabricated on a glass substrate.

The CTFT used in this study possesses a structure wherein the upper-layer p-ch DG poly-Ge TFT and the lower-layer n-ch poly-Si TFT share part of the gate, but each TFT has a different effective gate length. The effective gate length of the upper-layer p-ch DG poly-Ge TFT is reduced to 5 μm by replacing Ge by Al in the channel region, and the SD is made up of metallic Al with low resistance. On the other hand, the gate length of the lower-layer n-ch poly-Si TFT is 10 μm. Therefore, the difference in mobility between the TFTs was offset, and a similar on-current was achieved.

Figure 7 shows the dynamic characteristics of the CTFT inverter. The dynamic characteristics were measured at 100 Hz, 1 kHz, and 10 kHz. Consequently, normal inverter dynamic characteristics were confirmed at 100 Hz and 1 kHz. The pull-up and pull-down delay times were almost equal, indicating that the on-current was balanced. It was observed that the performance at 10 kHz does not provide a sufficient output signal. The proposed CTFT possesses a large gate length and insufficient mobility. Moreover, the parasitic capacitance is large. Therefore, the output signal is insufficient at a frequency of 10 kHz.

To improve the performance further, the implementation of germanium tin (GeSn), which is reported to possess a higher mobility than Ge, as a channel material, and HfO$_2$, which is a high-k gate dielectric, are under consideration. In addition, the integration of CLC poly-Si TFTs with high mobility is considered.

Conclusion

We used Ni-MIC and Cu-MIC for Si and Ge crystallization, respectively, to form a heterogeneous semiconductor CTFT consisting of an n-ch Ni-MIC poly-Si TFT in the lower layer and a p-ch Cu-MIC DG poly-Ge TFT in the upper layer on a glass substrate, and we operated it as an inverter at $V_{DD} = 5.0$ V. The effective gate length of the upper layer DG poly-Ge TFT was reduced to 5 μm by replacing Ge in channel region to Al, and the SD of this TFT is made of metallic Al with lower resistance. On the other hand, the lower-layer n-ch poly-Si TFT possesses a gate length of 10 μm. Therefore, the difference in mobility between the two TFTs was offset, and an identical on-current was achieved. The dynamic characteristics of the fabricated CTFT inverters were evaluated.

Acknowledgement

This work was supported by Grant-in-Aid for Scientific Research (c) 22K04247.

References

[1] J. Ryckaert et al., Tech. Dig. VLSI symp., p.141 (2018).

[2] S.-W. Chang et al., Tech. Dig. IEDM., p.254 (2019).

[3] S. Subramanian et al., Tech. Dig. VLSI Symp., TH3.1 (2020).

[4] P.-J. Sung et al., IEEE Trans. Electron Devices 67, 3504 (2020).

[5] S.-G. Jung et al., IEEE J. Electron Devices Soc. 10, 78 (2021).

[6] X. Yang et al., IEEE Trans. Electron Devices 69, 4029 (2022).

[7] X.-R. Yu et al., Tech. Dig. IEDM., p.487 (2022).

[8] H. Mertens et al., Tech. Dig. VLSI Symp., T1-3 (2023).

[9] C.-T. Tu et al., Tech. Dig. IEDM., 29-5 (2023).

[10] S.K. Kim et al., IEEE Trans. Electron Devices 71, 393 (2024),

[11] S.K. Kim et al., Tech. Dig. VLSI Symp., T5.3 (2024).

[12] T. Chiarella et al., Tech. Dig. EDTM., 4A-4 (2024).

[13] C.-L. Chu et al., ACS Appl. Electron. Mater. 6, 4304 (2024).

[14] W. Tang et al., Tech. Dig. IEDM., p.483 (2022).

[15] Z. Wang et al., IEEE Trans. Electron Devices 71, 4664 (2024).

[16] C. Lee et al., Mater. Sci. Semicond. Process. 185, 108871 (2025).

[17] Y. Ito et al., Ext. Abst. of the 2024 SSDM, K-7-04 (2024).

[18] A. Hara et al., Jpn. J. Appl. Phys. 43, 2004, 1269 (2004).

[19] S. Suzuki et al., Jpn. J. Appl. Phys. 63, 051001 (2024).

[20] S.-W. Lee et al., IEEE Electron Device Lett. 17, 160 (1996).

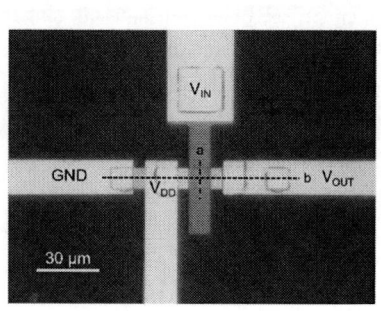

Fig.1. Photograph of CTFT

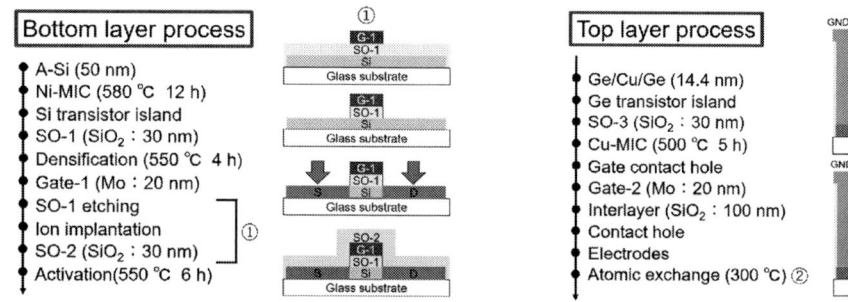

Fig. 2. (a) Cross sectional image along line-a in Fig. 1 (b) Cross sectional image along line-b in Fig. 1 (c) CMOS inverter circuit in this experiment

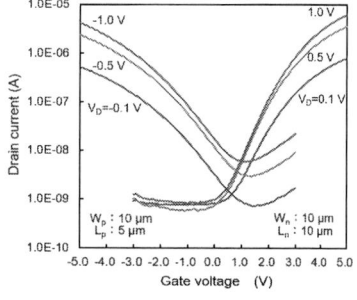

Bottom layer process

- A-Si (50 nm)
- Ni-MIC (580 ℃ 12 h)
- Si transistor island
- SO-1 (SiO₂：30 nm)
- Densification (550 ℃ 4 h)
- Gate-1 (Mo：20 nm)
- SO-1 etching ⎫
- Ion implantation ⎬ ①
- SO-2 (SiO₂：30 nm) ⎭
- Activation(550 ℃ 6 h)

Fig.3-1. TFT fabrication process of bottom n-ch poly-Si TFT

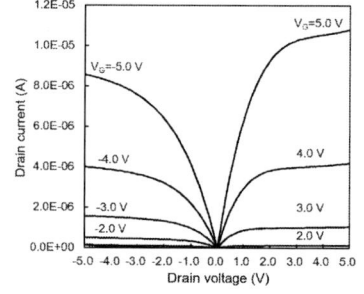

Top layer process

- Ge/Cu/Ge (14.4 nm)
- Ge transistor island
- SO-3 (SiO₂：30 nm)
- Cu-MIC (500 ℃ 5 h)
- Gate contact hole
- Gate-2 (Mo：20 nm)
- Interlayer (SiO₂：100 nm)
- Contact hole
- Electrodes
- Atomic exchange (300 ℃) ②

Fig.3-2. TFT fabrication process of top p-ch DG poly-Ge TFT

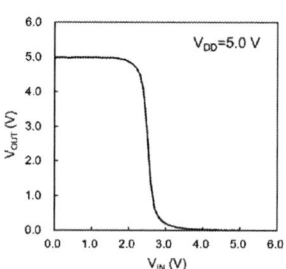

Fig. 4. Transfer characteristic of n- and p-TFTs

Fig. 5. Output characteristic of n- and p-TFTs

Table I. TFT performance

	Mobility (cm²/Vs)	V_th (V)	s.s. (V/dec)
Bottom layer Poly Si	32.5	1.9	0.8
Top layer Poly Ge	9.6	-1.2	1.5

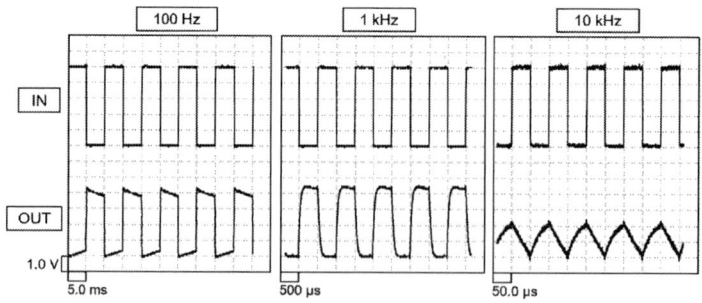

Fig. 6. Inverter characteristic

Fig. 7. Dynamic characteristic of CMOS inverter

979-8-3315-0417-5/25 $31.00 © 2025 IEEE

High Endurance Nanoscale TiN Bottom Heater for Phase Change Memory

Ziqi Wan, Wencheng Fang, Li Xie, Xi Li, Ruobing Wang*, Zhitang Song*, and Xilin Zhou*

State Key Laboratory of Materials for Integrated Circuits, Shanghai Institute of Microsystem and Information Technology, Chinese Academy of Sciences, Shanghai 200050 China.

*Email: wrb@mail.sim.ac.cn; ztsong@mail.sim.ac.cn; xilinzhou@mail.sim.ac.cn

Abstract

In this work a highly reliable blade-type TiN bottom heater is prepared with the contact size down to 3 nm × 150 nm which contributes to the improved heating efficiency and high switching performance of phase change memory (PCM). The TiN thin film is deposited by atomic layer deposition (ALD) on 12-inch wafer. It shows good thickness uniformity and deposition conformality. The TiN bottom heater exhibits high endurance performance over 1E12 cycles. A cyclic switching endurance of more than 1E7 cycles is demonstrated in the TiN bottom heater enabled PCM cells.

Keywords: phase change memory, TiN heater, ALD

Introduction

Phase change memory (PCM) has become a representative of emerging non-volatile memory due to its good scalability, excellent endurance, fast operating speed, and high thermal stability [1]-[4]. The high switching current is one of the main challenges of PCM to be widely adopted as storage class memory (SCM) [5].

The bottom heater that locates between the bottom electrode contact (BEC) and phase change layer (PCL) is an important part to transfer the Joule heat to phase change material which is crucial for the melting-quench process of PCM [6]. The blade-type bottom heater has been proposed to improve the heating efficiency of PCM by minimizing the contact area between the heater and phase change material. The schematic diagram of PCM cell with nanoscale bottom heater is shown in Fig. 1. The critical dimension (CD) of the blade-type heater is typically smaller than 10 nm because the smaller the contact area the higher the heating efficiency of PCM [7]. Additionally, the blade-type heater does not add process complexity in the fabrication of PCM as the CD of the bottom heater can be simply reduced by decreasing the thickness of the film which is not limited , by the lithography process [8]. The blade-type bottom heater is fabricated by depositing thin film in a patterned groove structure which is followed by the subtractive etching. The prepare of square wave-shaped bottom heater puts high requirement in the deposition process in terms of thickness uniformity and step coverage of the thin film across the 300 mm wafer. Atomic layer deposition (ALD) shows significant advantages over the physical vapor deposition (PVD) method for the deposition of the bottom heater material due to its precise thickness control, excellent conformality, and high film uniformity [9]. Thus, ALD is

employed to fabricate the sub-10nm thick blade-type TiN bottom heater in this work.

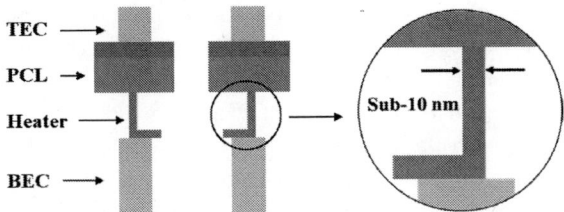

Fig. 1 Schematic diagram of the typical mushroom-type PCM cells with blade-type bottom heater.

Compared to tungsten, TiN is more suitable as a bottom heater material for PCM due to its lower thermal conductivity that is beneficial for reducing the thermal dissipation from the heater [10]. In this work, the TiN film deposited by the ALD process show good thickness uniformity and step coverage on 12-inch wafer. The 3 nm thick TiN bottom heater with small contact area exhibits high endurance after 1E12 operation cycles, showing a good reliability for the low power PCM devices. The PCM devices with 3 nm TiN heater shows high endurance switching of more than 7.5E7 cycles.

Atomic Layer Deposition of TiN Bottom Heater

The TiN thin film is deposited by ALD on 12-inch wafer at an elevated temperature. Figure 2 illustrates the ALD reaction purging sequence to grow TiN film. $TiCl_4$ and NH_3 are used as precursors with high purity N_2 as the purging gas. In each ALD cycle, the $TiCl_4$ gas is introduced into the reactor for 0.05 s first which is followed by a N_2 gas purging step. The NH_3 gas is then introduced into the reactor for 2 s and a reaction with $TiCl_4$ that has been adsorbed on the substrate is expected. Finally, a N_2 gas purging is performed again to remove the remaining precursors and by-products.

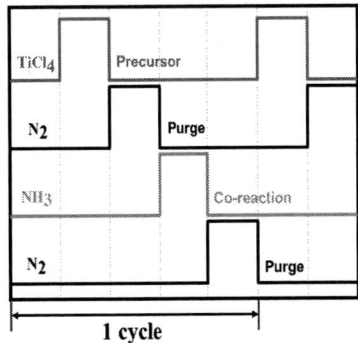

Fig. 2 Reaction purging sequence in the atomic layer deposition of TiN film.

The deposition temperature is set at 450°C which facilitates the decomposition of the reaction by-product of NH₄Cl, thus reduces the chlorine contamination of the TiN film [11].

Figure 3(a) and 3(b) shows the thickness and resistance distribution of the TiN film, respectively, that is measured after 60 ALD cycles. The thickness of film is measured using spectroscopic ellipsometry and the resistance is measured by the four-point probe method. The measurement is performed on 49 different positions of the 12-inch wafer. The average values is labeled in each figure. The average thickness of the TiN film is 3.19 nm across the wafer. The thickness in the edge region is slightly thicker than the middle area. The variation in the film thickness across the whole wafer is smaller than 0.12 nm. The high uniformity in the thickness of the film is attributed to the surface saturation of precursors and the self-limiting reaction, which is the essential feature of the ALD process with only one layer deposited in each cycle [12].

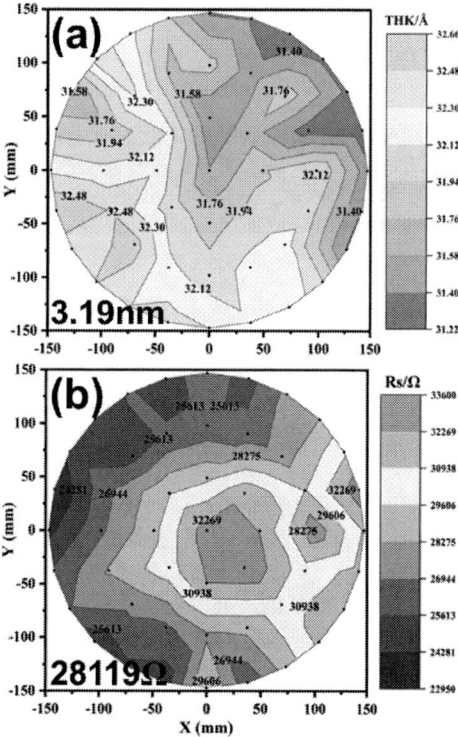

Fig. 3 The distribution of (a) film thickness and (b) sheet resistance of the ALD TiN film on 12-inch wafer.

The distribution of sheet resistance across the wafer shows different trends to the film thickness where the sheet resistance in the middle region is larger than that at the wafer edge which is due to the fact that the sheet resistance of the film is inversely proportional to its thickness. It is therefore concluded that the 3 nm thick TiN film deposited by ALD exhibit good uniformity in film thickness and sheet resistance, which will benefit the reliability and cell-to-cell variation of the bottom heater in the PCM cells. What's more, the average sheet resistance of the 3 nm thick TiN films is lower than 3 kΩ. It is smaller than the resistance of PCM cell in the SET state [13], which ensure a small contact resistance and high heating efficiency without affecting the performance of PCM devices.

Fabrication and Electrical Performance of 3 nm Thick TiN Bottom Heater

The cross-sectional transmission electron microscopy (TEM) images of the 3 nm thick TiN heater deposited in the groove structure by ALD are shown in Fig. 4. Good conformality and step coverage is displayed without voids or interfacial peeling in the TiN film. The magnified TEM image of the TiN film in the sidewall and bottom of the groove is shown in Fig. 4(b) and 4(c), respectively, in which the excellent uniformity in the film thickness of around 3 nm is observed. The uniform TiN film not only ensures the effective and reliable contact between the BEC and PCL, but also benefits the consistent heat transfer among different cells across the 12-inch wafer.

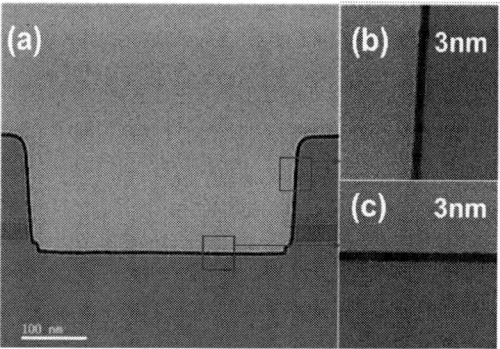

Fig. 4 The cross-sectional TEM images of the 3 nm thick ALD TiN film in the (a) groove, (b) sidewall, and (c) bottom of the groove.

The PCM cell with blade-type TiN bottom heater is fabricated on 12-inch wafer. The nanoscale TiN heater is deposited by ALD which is patterned by plasma etching with a contact size of 3 nm × 150 nm to the PCL. The deposition of phase change material film is finished using sputtering tool. A TiN adhesive layer is deposited on the top of PCL by PVD. The tungsten top electrode contact (TEC) is prepared by chemical vapor deposition. The control devices without otherwise PCL are also fabricated to investigate the electrical performance of the TiN bottom heater. The SET pulse of 400 μA and 440 ns and RESET pulse of 1000 μA and 15 ns are used in the electrical measurement.

Figure 5(a) shows the endurance characteristic of the device without the PCL. It is noted that the resistance of TiN heater keeps stable up to 1E12 cycles which gives rise to excellent stability and reliability of the electrical performance of PCM cells. The switching endurance performance of more than

1E7 cycles is demonstrated with resistance ratio larger than one order of magnitude in the PCM cells adopting 3 nm TiN heater, as shown in Fig. 5(b). These results indicate that the 3 nm TiN bottom heater not only exhibits a small contact size to improve the heating efficiency, but also demonstrates a long endurance cycle to ensure a good switching performance of the PCM device.

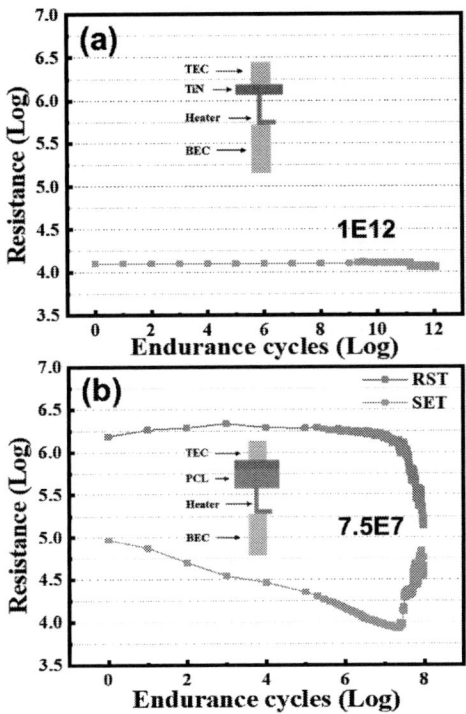

Fig. 5 The endurance characteristics of the PCM cells adopting 3nm TiN bottom heater: (a) without and (b) with PCL.

Conclusion

In this work, the TiN film deposited by ALD tool is investigated to function as the bottom heater of the PCM cell. By precisely controlling the film thickness, the TiN film shows high uniformity in thickness across the 12-inch wafer. A good conformality of TiN film is achieved in the patterned groove substrate with a consistent film thickness and morphology in both the sidewall and bottom region. The electrical measurment indicates that the 3 nm thick TiN heater exhibits a high endurance characteristic of up to 1E12 cycles. The PCM device with 3 nm TiN bottom heater demonstrates a good switching endurance performance of more than 1E7 cycles.

Acknowledgments

This work is supported by the National Key Research and Development Program of China (2023YFB4404500, 2023YFB4502903, 2023YFB4502202), National Natural Science Foundation of China (62174168, 92164302), Strategic Priority Research Program of the Chinese Academy of Sciences (XDB44010200, XDB0670000), Science and Technology Council of Shanghai (23XD1404700, 23JC1400900, 22DZ2229009), China Postdoctoral Science Foundation (2023TQ0363,2024M753370), Postdoctoral Fellowship Program of CPSF (GZC20232839).

References

[1] X. Zhou et al., "Understanding Phase-Change Behaviors of Carbon-Doped Ge2Sb2Te5 for Phase-Change Memory Application," *ACS Appl. Mater. Interfaces*, vol. 6, no. 16, pp. 14207–14214, Aug. 2014.

[2] Chen, "A review of emerging non-volatile memory (NVM) technologies and applications," *Solid-State Electronics*, vol. 125, pp. 25–38, Nov. 2016.

[3] Y. Xue et al., "Phase change memory based on Ta–Sb–Te alloy – Towards a universal memory," *Materials Today Physics*, vol. 15, p. 100266, Dec. 2020.

[4] R. Wang et al., "Phase-change memory based on matched Ge-Te, Sb-Te and In-Te octahedrons: Improved electrical performances and robust thermal stability," *InfoMat*, vol. 3, pp. 1008–1015, Sep. 2021.

[5] H.-S. P. Wong et al., "Phase Change Memory," *Proc. IEEE*, vol. 98, no. 12, pp. 2201–2227, Dec. 2010.

[6] J. Y. Wu et al., "A low power phase change memory using thermally confined TaN/TiN bottom electrode," in *2011 International Electron Devices Meeting*, IEEE, Dec. 2011, p. 3.2.1-3.2.4.

[7] Y. Y. Lu et al., "The Impact of the Electrode Performance on the Endurance Properties of the Phase Change Memory Device," *IEEE Transactions on Device and Materials Reliability*, vol. 19, no. 1, pp. 164–168, Mar. 2019.

[8] Z. T. Song et al., "High Endurance Phase Change Memory Chip Implemented based on Carbon-doped Ge2Sb2Te5 in 40 nm Node for Embedded Application," in *2018 IEEE International Electron Devices Meeting (IEDM)*, IEEE, Dec. 2018, p. 27.5.1-27.5.4.

[9] J. Kim, "Properties including step coverage of TiN thin films prepared by atomic layer deposition," *Applied Surface Science*, vol. 210, no. 3–4, pp. 231–239, Apr. 2003.

[10] Z.-J. Cui, D.-L. Cai, Y. Li, C.-X. Li, and Z.-T. Song, "WN coating of TiN electrode to improve the reliability of phase change memory," *Materials Science in Semiconductor Processing*, vol. 138, p. 106273, Feb. 2022.

[11] K.-E. Elers et al., "Diffusion Barrier Deposition on a Copper Surface by Atomic Layer Deposition," *Chem. Vap. Deposition*, vol. 8, no. 4, p. 149, Jul. 2002.

[12] S. Xie et al., "Properties and morphology of TiN films deposited by atomic layer deposition," *Tsinghua Science and Technology*, vol. 19, no. 2, pp. 144–149, Apr. 2014.

[13] X. Zhou et al., "Carbon-doped Ge2Sb2Te5 phase change material: A candidate for high-density phase change memory application," *Applied Physics Letters*, vol. 101, no. 14, p. 142104, Oct. 2012

A Device to Circuit BTI and HCD Aging Analysis Framework

Payel Chatterjee[#], Karansingh Thakor[#], and Souvik Mahapatra[*]

Department of Electrical Engineering, Indian Institute of Technology Bombay, Mumbai 400076, India
([#]Equal contributions, * Corresponding and Presenting author, Email: souvik@iitb.ac.in)

(Invited Paper)

Abstract

Physics-based frameworks are used to model BTI and HCD stress-recovery time kinetics in GAA-SNS FETs. Validation is done using measured data for BTI at multiple V_G, T and HCD at multiple V_G, V_D. A cycle-by-cycle circuit simulator utilizes these physical frameworks to study the impact of random input activity / mission profiles and DVFS. Results are compared to effective duty approach and fixed V_{DD} simulations respectively to highlight the significance of the device-to-circuit platform.

Introduction

Device parametric shift due to Bias Temperature Instability (BTI) and Hot Carrier Degradation (HCD) is serious reliability concern for Gate-All-Around Stacked Nano Sheet (GAA-SNS) FETs [1]-[3]. During circuit operation under dynamic mode, BTI impacts the on and off phases while HCD impacts the on-to-off transition phases, resulting in timing degradation [4]-[6]. It is often necessary to simulate the impact of device parametric drift on circuit timing degradation, as circuit test structures are not always available during early technology development. It is now well-known that BTI recovers while HCD does not recover after removal of stress. BTI recovery is difficult to handle via a compact model, and circuit simulations often use an effective-duty approach [7]. HCD does not recover, and hence it is often calculated at a particular gate (V_G)-to drain (V_D) bias ratio with a multiplier in the time scale to account for the ratio of transition time to total pulse period [5]. However, these approaches are not compatible with random activity and Dynamic Voltage and Frequency Scaling (DVFS) or turbo/throttle bias profiles.

We have developed physics-based BTI Analysis Tool (BAT) and Hot Carrier Analysis Tool (HCAT) to respectively model the BTI and HCD stress-recovery time kinetics [1]-[3], [8]-[11], and Circuit Aging Reliability Analysis Tool (CARAT), a cycle-by-cycle simulation framework that uses BAT and HCAT and is compatible with random input activity and DVFS [8]-[15]. In this paper, some of the salient capabilities of BAT, HCAT and CARAT are reviewed from our previous works.

Device Degradation Framework

The BAT framework, Fig.1, calculates threshold voltage shift (ΔV_T) during and after BTI stress by uncorrelated contributions from generated interface (ΔV_{IT}) and bulk (ΔV_{OT}) gate insulator traps and trapping of electrons (ΔV_{ET}) or holes (ΔV_{HT}) in pre-existing traps. The Reaction-Diffusion (RD) model, along with the Transient Trap Occupancy Model (TTOM) is used for ΔV_{IT}. The Reaction-Diffusion Drift (RDD) model is used for ΔV_{OT}, Fig.2 (a), the RD model is a subset of the RDD model having no ions. The Activated Barrier Double Well Thermionic (ABDWT)

model is used for TTOM and ΔV_{TP}, Fig.2 (b). All these models are triggered by V_G and temperature (T). HCAT is based on the RDD model, where the initial trigger depends on V_G, V_D and T.

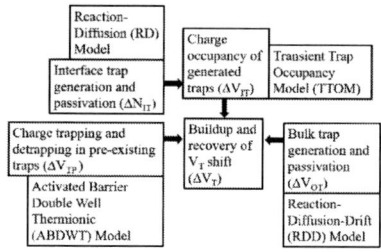

Fig 1. Schematic of BAT framework ΔV_{TP} is ΔV_{HT} for NBTI and ΔV_{ET} for PBTI) [8] implemented in TCAD and standalone versions.

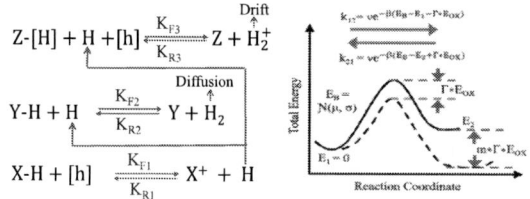

Fig.2. Schematic of (a) RDD and (b) ABDWT models, used respectively for time kinetics of trap generation and trapping [8] in TCAD and standalone versions.

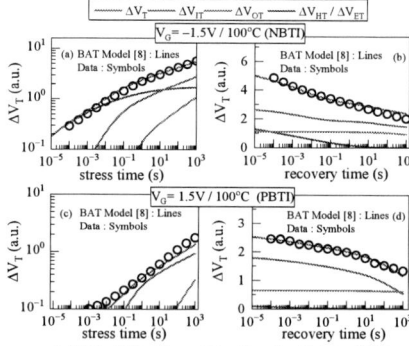

Fig. 3. Measured BTI ΔV_T time kinetics in GAA SNS p-FET (above) and n-FET (below) during (a, c) stress and (b, d) recovery, modeled using BAT framework with contributions from uncorrelated subcomponents.

Fig.3 plots the BAT modelling of measured ΔV_T during and after BTI stress along with the subcomponents. ΔV_{HT} (or ΔV_{ET}) saturates and recovers fast. ΔV_{IT} and ΔV_{OT} increase gradually with long-time slope of ~1/6 and ~1/4 respectively, the former recovers over extended timescale but recovery is negligible for the latter. The V_G and T dependencies of measured and modeled ΔV_T with the subcomponents after fixed-time BTI stress, Fig.4, show that Voltage Acceleration Factor (VAF) and T activation energy (E_A) are highest for ΔV_{OT} and lowest for ΔV_{HT} (or ΔV_{ET}),

ΔV_{IT} has intermediate values. Modeling of a few measured ΔV_T stress-recovery kinetics is shown in Fig.5.

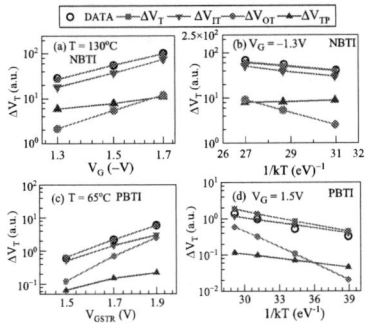

Fig. 4. Measured (symbols) and BAT modelled (line + symbols) fixed time ΔV_T versus (a) stress V_G and (b) stress T during NBTI (a,b) and PBTI (c,d) stress [8].

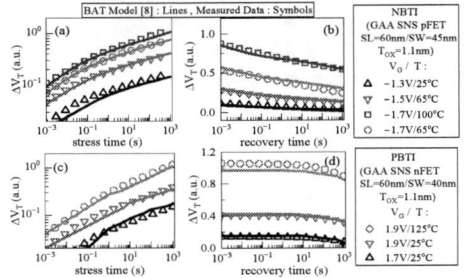

Fig.5. Measured ΔV_T time kinetics in GAA SNS p-FET (above) and n-FET (below), during stress (a, c), and recovery (b, d) modelled using BAT framework.

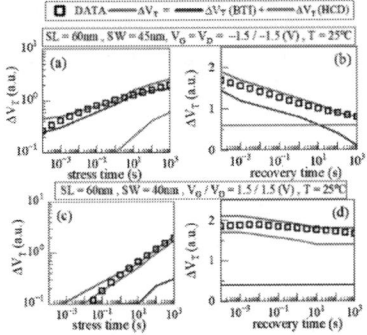

Fig. 6. Modeling of measured HCD ΔV_T time kinetics (symbols), in a GAA SNS (above) p-FET and (below) n-FET during (a, c) and after (b, d) HCD stress with the underlying BTI and pure HCD components [9].

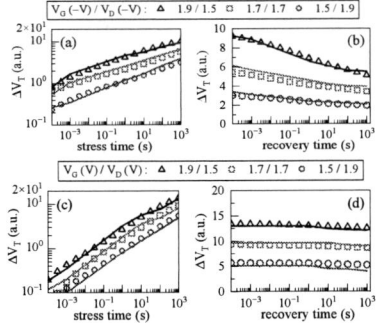

Fig. 7 Measured GAA SNS p-FET (above) and n-FET (below) HCD ΔV_T time kinetics (symbols) under multiple V_G/V_D ($V_G >$, $=$, $< V_D$) conditions modeled during (a,c) and after (b,d) stress (line+symbols) with the calibrated standalone BAT and HCAT modules simultaneously turned on as explained in [9].

Fig.6 plots the modeling of measured ΔV_T during and after HCD stress and underlying BTI and pure-HCD contributions. BTI dominates p-FET while HCD dominates n-FET; BTI under non-zero V_D is different from BTI under zero V_D due to lateral field reduction near drain and self-heating related T increase. In Fig.7, BTI-HCD co-modeling is shown with measured ΔV_T for few different V_G-to-V_D stress combinations. The extrapolated EOL ΔV_T under BTI and HCD stress is shown in Fig.8.

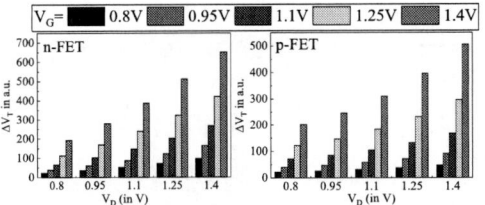

Fig. 8. EOL ΔV_T in a.u. due to different V_G/V_D, at T= 65°C for (Left) n-FET and (Right) p-FET [10]

Circuit Aging Flow

CARAT is fully automated and is governed by the Control-Framework (CF), Fig.9. CF runs a short-time SPICE simulation (*e.g.*, 100ns) on user-defined circuit Netlist, using fresh Model-Cards (MC's) for p- and n- FETs, and input waveforms during stress. After transient analysis, the Waveform Grabber (WG) captures the resulting waveforms at all terminals of all FETs in the Netlist, which are then processed by the Pulse-Shaper (PS) to prepare them for aging analysis, by converting trapezoidal to rectangular shapes for BTI and linear to stepped transitions for HCD. BTI and HCD related drift of all FETs is computed using BAT and HCAT respectively, with parameters pre-calibrated using device data. ΔV_T for all FETs in the circuit is calculated in a cycle-by-cycle basis. The short-time ΔV_T is then passed to the Degradation Extrapolator (DE), which estimates its End of Life (EOL) value by time-shifting a DC reference simulation. ΔV_T at EOL is used in the Modelcard Modifier (MM) module to update the FET MC's. SPICE simulation is run in sense mode with fresh and aged MC's to obtain the timing degradation of the circuit and input stress waveforms under consideration.

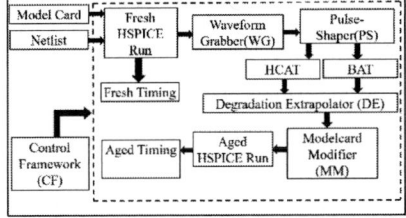

Fig. 9 Schematic of CARAT simulation flow. The calibrated HCAT module for HCD and calibrated BAT module for BTI are used here [10].

Circuit Simulation – DVFS

An Inverter-based 21 stage Ring Oscillator (RO), Fig.10 (a), is analyzed using CARAT under fixed V_{DD} and with stepped V_{DD} subjected to DVFS waveform, Fig.10 (b). The frequency degradation (%Δf) values are shown in Fig.10 (c), to study the impact of only Negative BTI (NBTI) in p-FET, total BTI, total BTI + only n-FET HCD, and total BTI + total HCD, while the fixed V_{DD} values are chosen equal to V_{MIN}, V_{AVG} and V_{MAX} of the DVFS waveform. NBTI has the largest impact, while the n-

HCD impact shows up at higher V_{DD}. DVFS results cannot be replicated using fixed V_{DD} simulations.

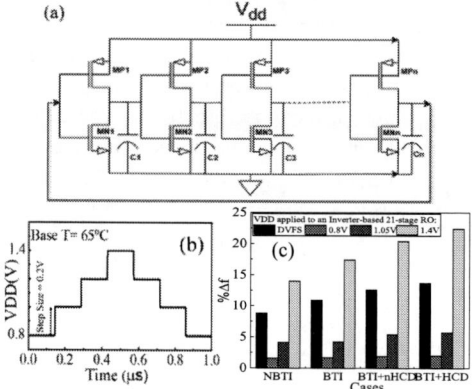

Fig. 10: (a) Schematic of Inverter-based RO, (b) DVFS waveform (c) EOL $\Delta f / f$ (a.u.) for 21 stage Inverter-based RO using DVFS case compared to V_{MIN}, V_{AVG}, and V_{MAX} [11].

Circuit Simulation – Random Activity

2-input XOR gate is implemented by different combinations of (a) NAND and (b, c) NOR gates, Fig.11, and is subjected to different combinations of random input waveforms, Fig.11 (d), all having 50% effective duty. Changes in rise ($\Delta\tau_{PD\text{-}RISE}$) and fall ($\Delta\tau_{PD\text{-}FALL}$) side propagation delays obtained from CARAT are listed in Table-I, for only total BTI and both total BTI + total HCD. The results are compared with RO output (regular 50% duty pattern).

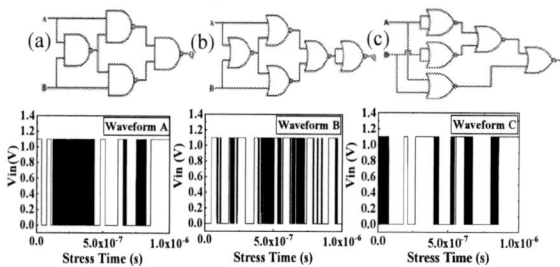

Fig. 11: Schematic of XOR using (a) NAND, (b) NOR1, and (c) NOR2, (d) Activity waveforms with effective duty 50%

Cases	Propagation Delay degradation, $\Delta\tau/\tau$ (in a.u.) for XOR with NAND/NOR1/NOR2 for the following cases			
	$\Delta\tau_{RISE\text{-}BTI}$	$\Delta\tau_{FALL\text{-}BTI}$	$\Delta\tau_{RISE\text{-}BTI+HCD}$	$\Delta\tau_{FALL\text{-}BTI+HCD}$
AA	21/19/20	15/27/24	23/19/21	16/30/30
AB	28/29/25	11/18/17	28/31/26	14/22/24
AC	27/28/25	12/19/18	29/30/26	14/23/26
BA	27/29/26	11/18/18	29/31/27	14/22/25
BB	22/18/20	15/29/24	23/20/21	16/29/32
BC	28/29/26	11/18/18	29/32/27	15/22/24
CA	27/29/25	12/20/18	28/30/26	15/24/25
CB	28/29/26	11/18/17	28/31/27	15/22/24
CC	21/19/20	16/28/24	22/19/21	17/29/31
RO-like	21/16/21	16/28/25	23/18/22	16/29/34

Table-I: Listing $\Delta\tau_{PD\text{-}RISE}$ and $\Delta\tau_{PD\text{-}FALL}$ due to BTI and BTI+HCD with different combinations of waveforms (Fig. 11d) as input and compared against a pulse from inverter-based RO [14].

The impact of BTI dominates all cases, the impact of HCD is more prominent for cases with more on/off transitions. The application of symmetric waveforms to both inputs (like AA) can be replicated by RO input for both $\Delta\tau_{PD\text{-}RISE}$ and $\Delta\tau_{PD\text{-}FALL}$. However, $\Delta\tau_{PD\text{-}RISE}$ under asymmetric waveforms (such as AB) is higher compared to symmetric and RO-like, while the exact opposite is seen for $\Delta\tau_{PD\text{-}FALL}$. The above clearly show that BTI and HCD related aging in a circuit is history dependent, and the use of effective-duty approach cannot replicate random input or mission profile results.

Conclusion

The use of physical BAT and HCAT models inside CARAT is key in making the circuit simulator compatible with random input activity / mission profiles and DVFS. The limitations of effective duty approach and fixed V_{DD} simulations respectively in handling complex input waveforms and DVFS patterns are demonstrated. The proposed device-to-circuit flow would help with incorporation of aging as a metric in Design Technology Co-Optimization (DTCO).

Acknowledgement

Industrial partners for device data and SPICE license.

References

[1] N. Choudhury and S. Mahapatra, TED, pp. 3535, July 2022, doi: 10.1109/TED. 2022.3172055.

[2] N. Choudhury and S. Mahapatra, TED, pp. 6576, Dec. 2022, doi: 10.1109/TED.2022.3217714.

[3] R. Saikia, N. Choudhury, A. S. Bisht and S. Mahapatra, TED, pp. 6499, Nov. 2024, doi: 10.1109/TED.2024.3459867.

[4] S. Mahapatra, V. Huard, A. Kerber, V. Reddy, S. Kalpat and A. Haggag, IRPS, 2014, pp. 3B.1.1, doi: 10.1109/IRPS.2014.6860615.

[5] A. Kerber, T. Nigam, P. Paliwoda and F. Guarin, TED, pp. 230, June 2020, doi: 10.1109/TDMR.2020.2981010.

[6] S. Mukhopadhyay et al., IRPS, 2023, pp. 1, doi: 10.1109/IRPS48203.2023.10117914.

[7] S. Mishra at al., TED, pp. 271-278, Jan 2019, doi. 10.1109/TED.2018.2875813

[8] S. Mahapatra et al., TED, pp. 114, Jan. 2024, doi: 10.1109/TED.2023.3291333

[9] S. Mahapatra et al., TED, pp. 126, Jan. 2024, doi: 10.1109/TED.2023.3308526.

[10] K. Thakor, P. Chatterjee and S. Mahapatra, IRPS, 2024, pp. 1, doi: 10.1109/IRPS48228.2024.10529395.

[11] P. Chatterjee, K. S. Thakor and S. Mahapatra, SISPAD 2024 (accepted).

[12] P. Gholve, et al., Solid-State Electronics, pp: 108586, 2023, doi: 10.1016/j.sse.2022.108586

[13] P. Chatterjee, K. Thakor, A. S. Bisht, A. Ansari, S. Samaga and S. Mahapatra, IIRW,2023,pp. 1, doi: 10.1109/IIRW59383.2023.10477643.

[14] P. Chatterjee, K. S. Thakor and S. Mahapatra, EDTM, 2024, pp. 1, doi: 10.1109/EDTM58488.2024.10511488.

[15] J. Sonawane, P. Chatterjee, and S. Mahapatra, IIRW 2024 (accepted)

Valley Transistors as Graded Neurons for Accurate Action Recognition

Jiewei Chen[1*], Wenxiao Wang[2], Yue Zhou[1], and Yang Chai[1*]

[1]Department of Applied Physics, The Hong Kong Polytechnic University, Hong Kong, P. R. China

[2]Future Institute of Technology, School of Optoelectronic Science and Engineering, South China Normal University, P. R. China

e-mail: jie-wei.chen@connect.polyu.hk ; ychai@polyu.edu.hk

Abstract

Neuromorphic computing aims to mimic the neural architecture of the biology to process information efficiently. Valley transistors, with their unique low-power information carrier properties, offer a promising approach for implementing bio-inspired computing systems. This study investigates the use of nonlocal valley transistors as graded neuron. After integrating the graded response properties in processing tempoeral signals, we have achieved high classification accuracy (95%) of dynamic icons.

Keywords: neuromorphic computing, valley transistors, graded neuron, action recognition

Introduction

The growing demand for computational power necessitates the development of energy-efficient electronics[1]. As conventional charge-based electronics are approaching fundamental limits, valley (a quantum degree of freedom) provides an information carrier (Table 1) with low-power features due to its low-dissipation transport characteristics[2, 3]. Electric fields, magnetic fields, and light are capable of generating and manipulating valley states. Specifically, circularly polarized light can localize electrons in MoS_2 within specific momentum valleys[4]. Nevertheless, realizing fundamental computing functions based on valley states at room temperature remains a considerable challenge.

Neuromorphic computing endeavors to replicate the neural architecture of biological systems to achieve efficient information processing[5]. A significant challenge within this domain is the energy-intensive task of recognizing and processing temporal information, which has spurred considerable research activity[6]. Graded neurons with volatile reponse (Figure 1) present a promising approach to addressing this challenge, as they facilitate more efficient handling of temporal data[7]. By emulating the nuanced, variable responses of biological neurons, graded neurons are well-positioned to tackle tasks involving time-dependent pattern recognition, thereby potentially reducing the energy consumption typically associated with these processes.

In this work, we proposed using nonlocal valley transistors as graded neuron for action recognition. We demonstrate the high nonlinearity and short-term memory characteristics of valley transistors, which can act as the graded neurons for transforming time-series data into spatiotemporal states. Our experiments show that valley transistors as graded neurons can effectively process temporal signals with low readout power, achieving high classification accuracy (95%) of 2020 Tokyo Olympics dynamic icons even in the presence of noise (10%).

Gate-tunable nonlocal valley transistors for emulating the graded neuron

We have prepared valley transistors with tellurium (Te), a typical Weyl semiconductor[8, 9]. Figure 2 presents both the schematic and the scanning electron microscope image of the nonlocal tellurium (Te) device. To achieve the functionality of a graded neuron, we have employed a $PEO/LiClO_4$ electrolyte to modulate the valley signal effectively. Our design involves crafting valley transistors by patterning the Te flake into a Hall bar configuration, which includes a strategically placed side gate electrode, as illustrated in Figure 3. The unique combination of electrical double layer formation and ion intercalation mechanisms within these valley transistors enables the tuning of response curves, thereby facilitating a graded-neuron-like response.

The application of a gating voltage plays a crucial role in this process, as it can efficiently adjust the Fermi level (E_F) either closer to or further from the Weyl point, consequently influencing the Berry curvature, as depicted in Figure 4. This capability allows the valley transistors to produce an output curve that mirrors the behavior of conventional transistors. By manipulating the amplitude and duration of the gate voltage, as well as the timing, we can precisely control multiple volatile states within the valley transistors. This adaptability underscores the potential of valley transistors in applications requiring nuanced control of electronic states, paving the way for advancements in neuromorphic computing.

To effectively encode time-series data, we have developed a nonlinear decaying curve that exhibits a graded-neuron-like response following pulse gating, as illustrated in Figure 5. The device's response is primarily driven by the processes of ion adsorption and relaxation, which dynamically modulate the E_F. This modulation is essential for achieving the desired temporal encoding and processing capabilities. In our system, the signal "1" is defined as an applied stimulation pulse with a specific amplitude of 4 V and a duration of 1 ms. This precise control over the input parameters allows for the nuanced manipulation of the valley FET's response, enabling it to handle complex temporal patterns effectively.

The graded-neuron-like response is crucial for mimicking the temporal dynamics observed in biological neural systems. Figure 6 demonstrates the representative modulation capabilities of our dynamic valley transistor when subjected to a sequence of four distinct input signals: (10010), (11011), (01110), and (10101).

Recognizing dynamic scenes with the valley transistors

Figure 7 illustrates the schematic of our neuromorphic computing network, which incorporates graded neurons to enhance processing capabilities. This network employs pulse-coding techniques to effectively process dynamic icons from the Tokyo Olympic Games, as depicted in Figure 8. Each dynamic icon represents distinct motions, requiring sophisticated processing to capture their unique characteristics. In our approach, every five pixels within a motion sequence are grouped and processed nonlinearly using a single valley field-effect transistor (FET). For instance, the pixel sequence for the "golf" icon is encoded as (1100111111100001000111111), which is subsequently divided into segments: (11001), (11111), (10000), (10001), and (11111).

The states of graded neurons corresponding to the rhythmic gymnastics icon are experimentally measured and presented in Figure 9. These measurements are integral to the training and inference processes within our reservoir computing framework. Remarkably, the classification accuracy for dynamic icons, even with the introduction of 10% noise, reaches an impressive 95% after 65 training epochs, as shown in Figure 10a. This high level of accuracy underscores the capability of our valley transistors to effectively classify dynamic data, despite the presence of diverse time-series information.

Furthermore, the confusion matrix in Figure 10b provides a detailed overview of the classification results for the five dynamic icons. It demonstrates that our neuromorphic computing system, based on valley transistors, can accurately classify nearly all dynamic icons. During the training process, we intentionally introduced 10% noise into the dynamic figures to test the robustness of our system. The results confirm that our approach is resilient and capable of maintaining high classification accuracy, highlighting the

potential of valley transistors in advanced neuromorphic computing applications.

Conclusion

Our study highlights the promising potential of valley transistors in advancing bio-inspired computing, particularly for precise action recognition tasks. These devices emulate the graded response of neurons, making them exceptionally well-suited for processing dynamic information with remarkable resilience to noise. By leveraging the unique properties of valley transistors, graded neurons can efficiently handle temporal signals with minimal readout power, approximately in the femtowatt range, while maintaining impressive classification accuracy of 95%, even amidst a noise level of 10%. This underscores their viability as a robust and energy-efficient solution for future computing applications.

Acknowledgments

This work is supported by MOST National Key Technologies R&D Programme (2022YFA1203804), National Natural Science Foundation of China (NSFC) for Distinguished Young Scholars (62425405), Research Grant Council of Hong Kong (15301023 and CRS_PolyU502/22), and the Hong Kong Polytechnic University (WZ4X and ZE1T).

References

[1] Y. Chai, "In-sensor computing for machine vision," *Nature,* vol. 579, pp. 32-33, 2020.

[2] K. F. Mak, D. Xiao, and J. Shan, "Light–valley interactions in 2D semiconductors," *Nature Photonics,* vol. 12, no. 8, pp. 451-460, 2018.

[3] J. Chen *et al.*, "Room-temperature valley transistors for low-power neuromorphic computing," *Nature Communications,* vol. 13, no. 1, p. 7758, 2022.

[4] L. Li *et al.*, "Room-temperature valleytronic transistor," *Nature Nanotechnology,* vol. 15, no. 9, pp. 743-749, 2020/09/01 2020.

[5] H. Ning *et al.*, "An in-memory computing architecture based on a duplex two-dimensional material structure for in situ machine learning," *Nature nanotechnology,* vol. 18, no. 5, pp. 493-500, 2023.

[6] J. Chen, Y. Zhou, J.-H. Ahn, and Y. Chai, "Bioinspired in-sensor vision processing network for action recognition," in *2024 8th IEEE Electron Devices Technology & Manufacturing Conference (EDTM)*, 2024: IEEE, pp. 1-3.

[7] J. Chen *et al.*, "Optoelectronic graded neurons for bioinspired in-sensor motion perception," *Nature Nanotechnology,* vol. 18, no. 8, pp. 882-888, 2023.

[8] N. Zhang *et al.*, "Magnetotransport signatures of Weyl physics and discrete scale invariance in the elemental semiconductor tellurium," *Proceedings of the National Academy of Sciences,* vol. 117, no. 21, pp. 11337-11343, 2020.

[9] J. Chen *et al.*, "Topological phase change transistors based on tellurium Weyl semiconductor," *Science advances,* vol. 8, no. 23, p. eabn3837, 2022.

Table 1. Comparision of charge and valley as information carriers

	Charge	Valley
Properties	a. Easy to generate, control and detect b. Short mean free path c. Unavoidable Joule heating d. Work at room temperature	a. No direct Joule heating b. Berry curvature sensitive c. Electric and magnetic fields for controlling Berry curvature d. Long diffusion distance e. Hard to work at room temperature

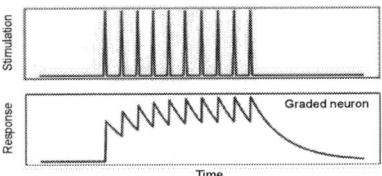

Fig. 1. Response of graded neurons.

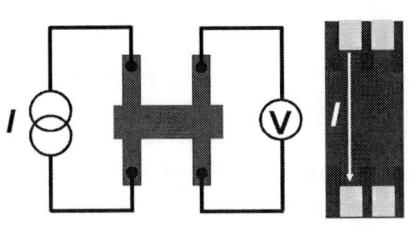

Fig. 2 Schematic and scanning electron microscope image of testing the nonlocal Te device.

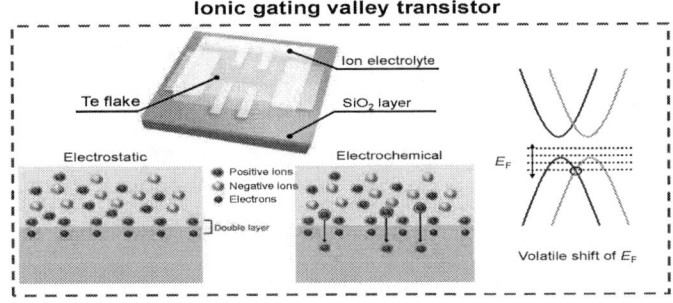

Fig. 3. Ionic gating valley transistors along with the volatile shift of E_F.

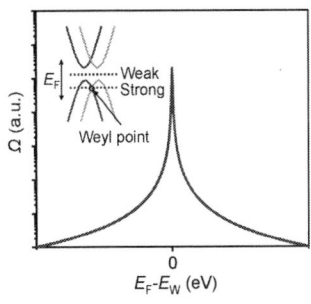

Fig. 4. The relationship between the Berry curvature strength (Ω) and Fermi level (E_F). The Weyl point is located at the valence band of Te.

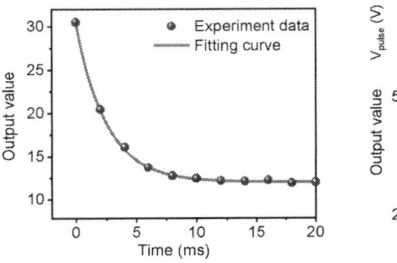

Fig. 5. The resposne of nonlocal valley transistor to pulse stimulation.

Fig. 6. Typical output values based on valley transistors.

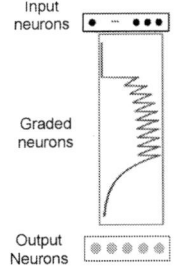

Fig. 7. Graded neuron with valley transistors for temporal encoding.

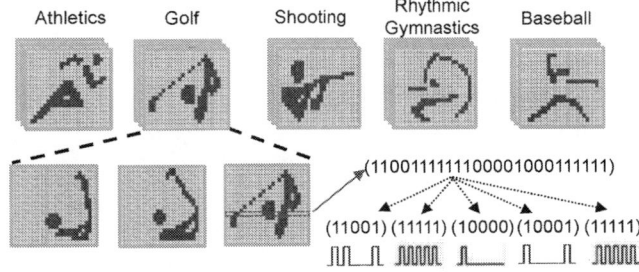

Fig. 8. Demonstration of encoding temporal images.

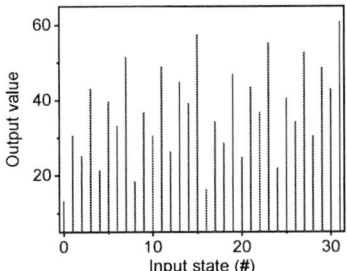

Fig. 9. Summarized output states for different input states varying from state 0 (00000) to state 31 (11111).

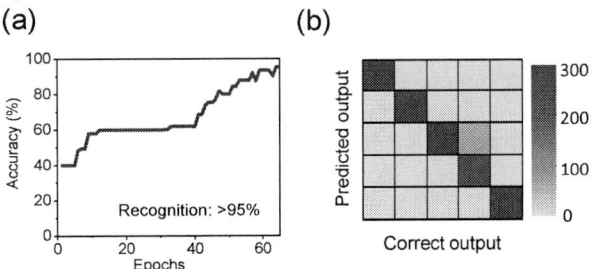

Fig. 10. The response of MoS_2 phottransistor to the optical stimulation. (a) Accuracy of classifying the dynamic ions with 10% noise based on the reservoir computing network. (b) Confusion matrix for classifying the five dynamic icons of the test set.

High-Sensitivity Flexible X-ray Detectors Based on Drop Casting 2D/3D Perovskite Thick Film

Liwen Qiu[1], Bosen Zhang[2], Qiang Lou[2], Maojun Sun[2], Hang Zhou[2*]

[1]Science department, faculty of chemistry, National University of Singapore, Singapore

[2*]School of Electronic and Computer Engineering, Peking University Shenzhen Graduate School, Shenzhen, 518055, China

Abstract

Flexible X-ray detectors hold great promise for applications in wearable devices, smart wall systems, and safety monitoring technologies. This study presents a large-area, flexible perovskite X-ray detector fabricated using the drop-casting method to produce 2D/3D perovskite thick films. Optimized low-temperature deposition and annealing processes ensure compatibility with flexible substrates. The devices demonstrate high sensitivity ($232\ \mu CGy^{-1}_{air}cm^{-2}$) and excellent mechanical durability (120° bending angle), underscoring their suitability for cost-effective and efficient X-ray detection in diverse applications.

Keywords: 2D/3D perovskite heterojunction, Direct X-ray detection, Flexible electronics

Introduction

Flexible X-ray detectors have attracted considerable attention for their potential applications in advanced fields such as wearable electronics, industrial and security screening, heritage conservation, and space exploration [1]. Unlike rigid detectors, flexible ones can conform to irregular surfaces, enabling robust imaging for non-planar objects and complex shapes [2]. This adaptability is particularly beneficial in medical diagnostics, where close-skin adherence enhances imaging accuracy and patient comfort. Additionally, the increasing demand for lightweight, compact, and versatile electronics has driven efforts to develop flexible photodetectors with improved performance under bending stress.

Recent research on flexible X-ray detectors has focused on developing materials and structures that integrate mechanical flexibility with high sensitivity and stability [3]. Metal halide perovskites (MHP) have emerged as a promising candidate due to their high X-ray absorption coefficients, low-temperature solution processability, and tunable bandgaps [4]. However, optimizing the performance of perovskite-based flexible X-ray detectors remains challenging. Issues such as insufficient film thickness, poor adhesion to flexible substrates, and increased dark current noise negatively affect device sensitivity and stability. Traditional rigid detectors, such as those made from a-Se and CdTe, are less suitable for flexible applications due to their brittleness and complex fabrication processes, underscoring the importance of transitioning to perovskite-based flexible detectors.

To address these challenges, we present a novel fabrication approach utilizing a 2D/3D perovskite heterostructure optimized for flexible substrates. We successfully fabricated a dense, thick perovskite film with excellent adhesion to polyimide (PI) substrates through a refined drop-casting process. This structure enhances the detector's X-ray absorption capabilities and ensures reliable performance even under extreme bending conditions. Furthermore, the proposed device structure effectively reduces noise current, resulting in improved sensitivity and durability.

Experiments, Results and Discussion

A flexible large-area perovskite X-ray detector was constructed by combining 3D MAPbI$_3$ as the X-ray absorption layer with 2D PEA$_2$PbI$_4$, forming a 2D/3D heterostructure. This configuration was utilized to fabricate a p-i-n type direct X-ray detector, comprising Mo/ITO/SnO$_2$/MAPbI$_3$/PEA$_2$PbI$_4$/CNT film, as illustrated in the structural diagram and energy band alignment in Fig.1a and Fig.1b. In the p-i-n structure, the energy level alignment between the electron transport layer SnO$_2$ and the hole transport layer (2D MHP) enhances the separation and extraction of photogenerated carriers.

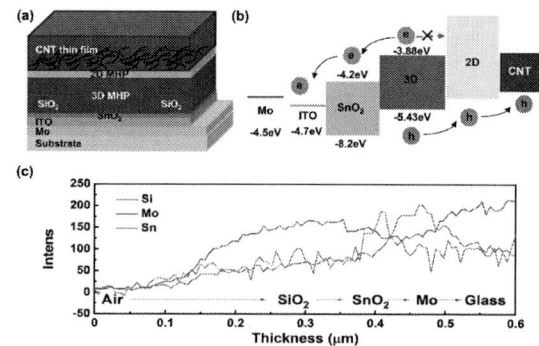

Fig.1: Fabrication of the X-ray detector via drop-casting process: (a) device architecture diagram, (b) device energy band structure, (c) Mo/ITO/SnO$_2$/SiO$_2$ structure fabricated on flexible PI substrate, and (d) EDX line scan analysis.

To prevent short-circuiting caused by the large-area CNT film electrode, a 200nm-thick SiO$_2$ insulating layer was sputtered onto the substrate outside the device area. The patterned SiO$_2$ layer also serves to confine the perovskite droplet coating area, helping ensure uniform film thickness.

The SnO₂ layer [5], 2D PEA₂PbI₄ perovskite [6], and CNT thin-film electrode [7] were grown and prepared using reported method. In general, the Mo/ITO/SnO₂/SiO₂ structure was fabricated on a flexible PI substrate. EDX spectrum analysis (Fig.1c) reveals the peak distribution variations of Si, Sn, and Mo elements from the SiO₂ layer to the glass substrate.

To enhance the spread ability of the drop-cast solution on the hole transport layer, polyvinylpyrrolidone (PVP) was added to the MAPbI₃ precursor to increase hydrophilicity without affecting carrier transport. Additionally, N-methylpyrrolidone (NMP) was introduced to achieve a more uniform MAPbI₃ MHP layer. NMP forms a stable complex with PbI₂ (PbI₂·NMP), controlling nucleation [8]. The addition of an equimolar amount of NMP to PbI₂ significantly improves film morphology, suppresses needle-like crystal growth, and promotes spherical nuclei, resulting in a high-purity perovskite thin film.

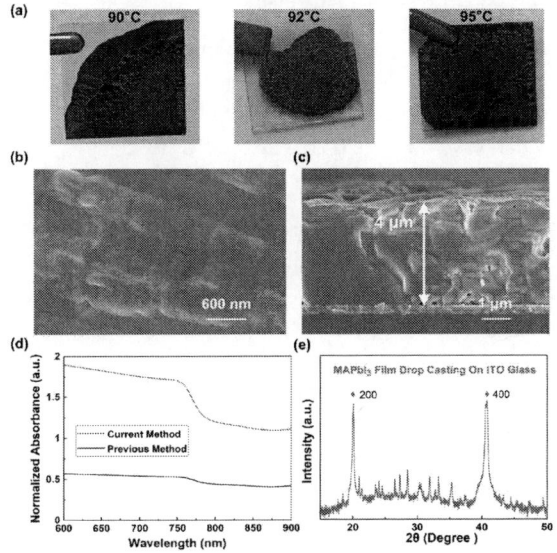

Fig.2: (a) Morphology of perovskite films at various annealing temperature conditions. (b) SEM image of the optimized MAPbI₃ MHP film at 30,000× magnification. (c) SEM cross-sectional image of the optimized MAPbI₃ MHP film at 2,000× magnification. (d) Comparison of absorption peaks of perovskite films fabricated by previous methods and the method used in this study. (e) XRD results of the optimized MAPbI₃ MHP film.

After optimizing the MAPbI₃ precursor solution formulation, we investigated the optimal annealing temperature (Fig.2a) by studying the crystallization morphology and quality of drop-cast perovskite films at temperatures ranging from 90°C to 95°C. At 90°C, black-phase growth from both edges and center caused a notable coffee ring effect. At 92°C, black-phase growth from the center minimized the coffee ring effect and resulted in a smoother surface. However, at 95°C, the higher temperature led to surface roughness. Thus, 92°C was identified as the optimal annealing temperature. Fig.2b and Fig.2c show the SEM analysis results for the optimized

MAPbI₃ film. At 30,000× magnification (Fig.2b), the MAPbI₃ film prepared with NMP in the precursor solution and annealed at 92°C shows densely packed spherical grains along the growth path. Cross-sectional analysis (Fig.2c) reveals a well-compacted film with a thin thickness of approximately 4μm.

Analysis of absorption peaks (Fig.2d) indicates that the film annealed at 150°C without NMP exhibits reduced absorbance due to low coverage by needle-like crystals, displaying a weaker perovskite absorption peak at 750 nm. In contrast, the film prepared with NMP and annealed at 92°C exhibits a significantly enhanced absorption peak at the same wavelength, indicating that the addition of NMP contributes to the formation of a higher-quality perovskite layer. XRD result (Fig.2e) reveals that the perovskite film annealed at 92°C displays two main peaks at 19.8° and 40.5°, corresponding to the (200) and (400) planes of MAPbI₃ respectively.

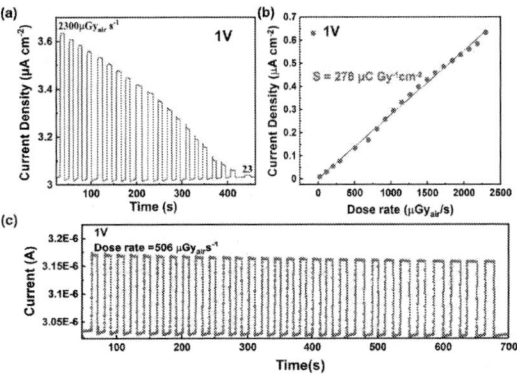

Fig.3: X-ray response characteristics of the 2D/3D perovskite X-ray detector on a glass substrate: (a) Photocurrent response under X-ray irradiation, (b) Photocurrent density as a function of dose rate, and (c) Continuous X-ray photocurrent response under a 40 kV X-ray source.

The performance of a p-i-n type X-ray detector fabricated with a Mo/ITO/SnO₂/MAPbI₃/PEA₂PbI₄/CNT film structure on a glass substrate was evaluated. Under intermittent 40 kV X-ray irradiation, the photocurrent response effectively tracked dose variations from 2300 $\mu Gy_{air}S^{-1}$ to 23 $\mu Gy_{air}S^{-1}$ over 400 seconds, as shown in Fig.3a. The detector demonstrated excellent baseline current stability under a constant 1V bias. Upon X-ray exposure, photocurrent increased rapidly, showing a linear relationship with the dose. Based on this linear growth (Fig.3b), detection sensitivity was calculated as 278 $\mu CGy_{air}^{-1}cm^{-2}$. Additionally, the performance stability of the 2D/3D MHP thick film X-ray detector was confirmed, with continuous photocurrent tests under a 40 kV X-ray source at 506 $\mu Gy_{air}S^{-1}$. As depicted in Fig.3c, the detector maintained a stable response to pulsed X-ray exposure over 700 seconds, without performance degradation.

Subsequently, a p-i-n type X-ray detector with a Mo/ITO/SnO₂/MAPbI₃/PEA₂PbI₄/CNT film structure on a flexible PI substrate was constructed. The dark current behavior was analyzed, showing a standard photodiode mode in the I-V characteristic curve within the voltage range of -1V to 1V. Under a forward bias of 1V, the device exhibited a dark current density of 3 $\mu A\,cm^{-2}$, significantly lower than the 0.32 $mA\,cm^{-2}$ reported for MAPbI$_3$-based detectors under a 0.125 $V\mu m^{-1}$ electric field [9]. This highlights the effectiveness of the 2D/3D perovskite heterostructure in reducing dark current density. Under a 120° bending condition (Fig.4a), a 0.5-fold increase in dark current was observed, which returned to its initial value once the device was flattened. This phenomenon was attributed to nanoscale microstructural changes in the polycrystalline layers caused by bending in flexible electronic devices.

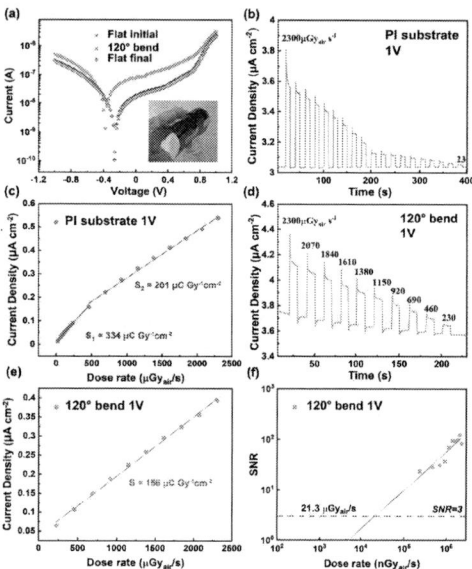

Fig.4: Electrical and X-ray detection performance of the device: (a) I-V characteristics in dark before (orange dots), during (green dots), and after (purple dots) bending; (b) Photocurrent response under X-ray irradiation at 1 V bias; (c) Photocurrent density versus dose rate; Performance under 120° bending: (d) Photocurrent response under X-ray irradiation at 1 V bias; (e) Photocurrent density versus dose rate; (f) Dose rate-dependent signal-to-noise ratio (SNR).

As depicted in Fig.4b, the X-ray response of the flexible detector was measured in a flat state. Under intermittent irradiation by a 40 kV X-ray source, the photocurrent decreased from 2300 $\mu Gy_{air}S^{-1}$ to 23 $\mu Gy_{air}S^{-1}$ over 400 seconds. At a 1V bias, excellent stability of the detector's baseline current was maintained, showing a linear relationship between the photocurrent and dose. The sensitivity was calculated to be 232 $\mu CGy_{air}^{-1}cm^{-2}$. Detailed tests in a 120° bent state (Fig.4d) showed that over 230 seconds, the baseline current drifted from 3.75 μAcm^{-2} to 3.57 μAcm^{-2}. Despite the dark current shift and reduced photocurrent signal under bending, the sensitivity remained largely unchanged at 156 $\mu CGy_{air}^{-1}cm^{-2}$. The limit of detection (LoD) was 21.3 $\mu Gy_{air}S^{-1}$, with sensitivity

dropping by 32% compared to the flat state.

Conclusion

In summary, a flexible large-area perovskite X-ray detector was developed through meticulous fabrication and performance optimization. Precise control of deposition processes for electrodes and electron transport layers, alongside refinement of the drop-casting technique for 2D/3D MHP thick films, resulted in excellent energy level alignment and uniform film thickness. The Mo/ITO/SnO$_2$ structure on flexible PI substrates, developed via sputtering and chemical bath deposition, enhanced device reliability. Including PVP and NMP in the precursor solution improved film morphology, while optimal annealing at 92°C produced high-purity MAPbI$_3$ layers. The detectors exhibited impressive sensitivity and low detection limits, maintaining performance under mechanical stress. These findings highlight the potential of flexible perovskite detectors for advanced X-ray sensing applications.

Acknowledgments

L.Q. and B.Z. contributed equally to this work. This work is supported by Shenzhen Science and Technology Innovation Committee (GJHZ20240218113959009).

References

[1] X. Ou, X. Qin, B. Huang, et al, "High-resolution X-ray luminescence extension imaging", Nature, 590, pp. 410-415 (2021).

[2] W. Jang, B. G. Kim, S. Seo, et al., "Strong dark current suppression in flexible organic photodetectors by carbon nanotube transparent electrodes", Nano Today, 37, pp. 10181 (2021).

[3] X. Zhang, X. Liu, Y. Huang, et al, "Review on flexible perovskite photodetector: processing and applications", Frontiers of Mechanical Engineering, 18, article no. 33 (2023).

[4] B. Q. Wang, X. Yang, S. Chen, et al., "Flexible perovskite scintillators and detectors for X-ray detection", iScience, 25, pp. 105593 (2022)

[5] R. Zhi, C.-Q. Yang, M. U. Rothmann, et al., "Direct Observation of Intragrain Defect Elimination in FAPbI$_3$ Perovskite Solar Cells by Two-Dimensional PEA$_2$PbI$_4$", ACS Energy Lett., 8, pp. 6 (2023).

[6] K. Lee, J. Kim, H. Yu, et al., "A highly stable and efficient carbon electrode-based perovskite solar cell achieved via interfacial growth of 2D PEA$_2$PbI$_4$ perovskite", Journal of Materials Chemistry A, 6, pp. 24560-24568 (2018).

[7] L. Qiu, M. Wang, T. Sun, et al., "An interfacial toughening strategy for high stability 2D/3D perovskite X-ray detectors with a carbon nanotube thin film electrode", Nanoscale, 15, pp. 14574-14583 (2023).

[8] T. Bu, J. Li, H. Li, et al., "Lead halide–templated crystallization of methylamine-free perovskite for efficient photovoltaic modules", Science, 372, pp. 1327-1332 (2021).

[9] W. Qian, X. Xu, J. Wang, et al., "An aerosol-liquid-solid process for the general synthesis of halide perovskite thick films for direct-conversion X-ray detectors," Matter, 4, pp. 942-954 (2021).

Modeling Nanowire Single-Photon Avalanche Detectors via Deep Neural Networks

Boyang Zhang*, Zhe Li*, Zhongju Wang‡, Daoyi Dong†, and Lan Fu*

*Australian Research Council Centre of Excellence for Transformative Meta-Optical Systems,
Research School of Physics, The Australian National University, Acton, ACT 2601, Australia
†School of Engineering, The Australian National University, Acton, ACT 2601, Australia
‡School of Systems and Computing, The University of New South Wales, Campbell, ACT 2612, Australia
Correspondence: zhe.li@anu.edu.au, lan.fu@anu.edu.au

Abstract—Recently, single-photon avalanche detectors based on III-V compound semiconductor nanowires (NWs) have shown promising performance in emerging photonic quantum information technologies. However, their design and fabrication are expensive and time-consuming. Predicting key device performance metrics for different nanowire structures before experimental prototyping is critical. We introduce a deep learning-based modelling framework named NW-SPAD-Net that predicts SPAD performance for given nanowire configurations several orders of magnitude faster than the conventional drift-diffusion modelling approach.

Keywords: single-photon avalanche detectors, III-V compound semiconductor nanowires, scientific machine learning, deep learning.

I. INTRODUCTION

Single-photon avalanche detectors (SPADs) are crucial in cutting-edge applications such as quantum communications and light detection and ranging (LiDAR) [1]. These devices harness the avalanche effect, where a single photon hitting the detector and generating an electron-hole pair sets off a chain reaction of electron multiplication. The avalanche process amplifies the signal to a level large enough to be registered by the external electronics.

Traditional near-infrared single-photon avalanche detectors often rely on planar InGaAs/InP and Ge-on-Si structures, for which mitigating the performance trade-off between dark count rate and photon detection efficiency remains a challenging issue. Recent advancements have highlighted the potential of employing III-V compound semiconductor nanowires for single-photon avalanche detectors to minimise the trade-off between photon detection efficiency and dark count rate [3]. However, nanowires' complicated material growth and device fabrication processes require precise predictive modeling to optimize designs before experiments begin. Traditionally, SPAD modeling has been a slow and resource-intensive process. It relies on methods such as Monte-Carlo simulations or steady-state drift-diffusion models, where the former often require extensive random trials to accurately predict outcomes, and the latter constantly struggles with numerical convergence beyond avalanche breakdown [3].

This paper introduces a deep learning-based model that substantially speeds up SPAD modeling, reducing computation time from hours to seconds. Using advanced artificial neural network architectures, our model efficiently predicts key performance metrics such as photon detection efficiency (PDE), dark count rate (DCR), avalanche triggering probability (ATP) and quantum efficiency (QE). With the proposed workflow, users can generate a detailed mapping between the nanowire's properties (e.g., nanowire lengths and doping) and PDE and DCR; coupled with general optimization algorithms, one can easily find a group of optimal candidate structures with minimized performance trade-offs. These short-listed candidate structures will significantly accelerate the time-consuming material growth and device fabrication cycles.

II. METHODOLOGY

The proposed deep-learning-based framework is named nanowire single-photon avalanche detector prediction net (NW-SPAD-Net), based on a full-connection network, where artificial neurons between each layer are fully connected. NW-SPAD-Net consists of two sub-networks: i) the electric field prediction network is specifically tailored to accurately predict the electric field within the nanowires, a critical parameter that can be used to calculate photon detection efficiency and dark count rate, and ii) the quantum efficiency prediction network is used to predict the intrinsic light absorption efficiency within a nanowire. These networks are designed to handle the experimentally accessible input parameters that characterize a nanowire's physical and electrical properties. This includes the length of each segment, doping concentrations, and applied voltage. The NW-SPAD-Net is constructed assuming that nanowires have the most common axial p-i-n device structure grown by the metal-organic chemical-vapor deposition (MOCVD) technique [2]. The model has seven input parameters, including p-/i-/n-region lengths, doping concentrations for each region, and the applied voltage. The electric field prediction network takes all parameters, while the quantum efficiency prediction network works without the applied voltage. The development of two separate sub-networks is to ensure that different physics models involved in predicting electric fields and quantum efficiency can be learned by the neural networks more accurately. NW-SPAD-Net operates as follows: 1) given a vector of seven nanowire parameters, the electric field prediction sub-network calculates a spatial distribution of electric field within a nanowire, which is then used to

979-8-3315-0417-5/25 $31.00 © 2025 IEEE

determine the avalanche triggering probability (ATP); 2) the quantum efficiency prediction sub-network also generates a spatial quantum efficiency distribution; 3) photon detection efficiency can be computed based on ATP and quantum efficiency; 4) extra information on nanowire's bulk/interface defect densities, together with ATP, can be used to compute dark count rate.

III. Results

We conducted the simulations for SPADs based on indium phosphide (InP) nanowires operated at 150 K. For demonstration, only the nanowire's segment length was varied while the nanowire's total length and doping concentrations were fixed. The p-region is fixed at 1 μm, and the i-region length is varied from 600 to 1800 nm, with the n-region length changed correspondingly. The p-/n-region doping is 10^{18} cm^{-3}, and the i-region is slightly n-doped (10^{16} cm^{-3}) due to background doping commonly observed in undoped InP nanowires grown by selective-area MOCVD [2]. The training data was generated by the time-dependent drift-diffusion model [3] with the Semiconductor Module, COMSOL Multiphysics [4]. The electric field prediction sub-network achieved a mean squared error (MSE) of 0.00077 for training, 0.00084 for validation, and 0.00108 for testing. The quantum efficiency prediction sub-network achieved an MSE of 0.00096 for training, 0.00264 for validation, and 0.00466 for testing. The training process was completed in only 1 minute and 37 seconds, allowing for predictions of spatial distributions of electric field and quantum efficiency in less than a second per input nanowire structure.

Figures 1 and 2 show a mapping of photon detection efficiency and dark count rate, respectively, as a function of the i-region length and excess bias (i.e., the bias above the individual breakdown voltage). The overlap between high PDE and low DCR regions in these two mappings gives the optimal nanowire structures and operating conditions. A bulk defect density of 10^{12} cm^{-3} was used for the dark count rate calculation. Such a density typically indicates poor material-quality nanowires grown by MOCVD. Nevertheless, there is still a large overlap of high PDE and low DCR, featuring larger i-region lengths and higher operating excess biases. Such performance mapping, coupled with any generic optimization algorithms, can assist researchers in locating the optimal device designs with minimal trial-and-error efforts.

IV. Conclusion

We proposed a deep-learning based NW-SPAD-Net that can efficiently predict photon detection efficiency and dark count rate for nanowire single-photon avalanche detectors with high accuracy and minimal computation time. The demonstrated PDE and DCR mappings generated by the NW-SPAD-Net reveal the great potential of nanowire SPADs that can effectively mitigate the trade-offs between key performance metrics. The proposed NW-SPAD-Net greatly advances the high-throughput, highly efficient SPAD design workflow and toolkits.

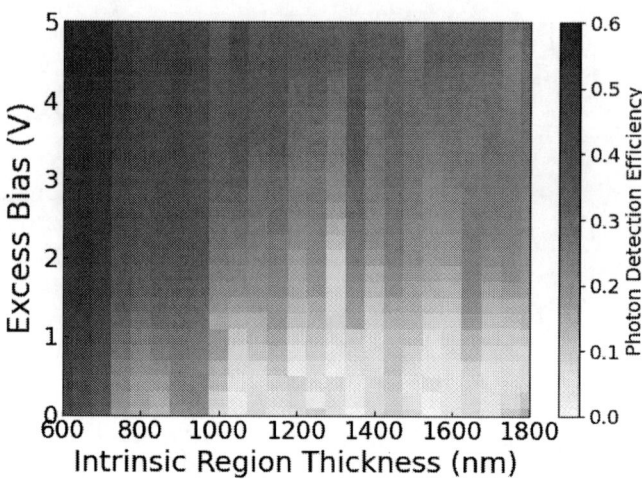

Fig. 1. Photon detection efficiency mapping as a function of the i-region length (from 600 nm to 1800 nm) and applied excess bias (from 0 to 25 V).

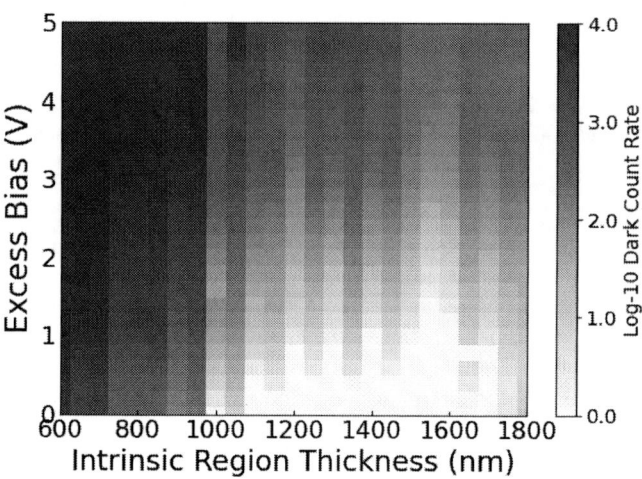

Fig. 2. Dark count rate mapping as a function of the i-region length and excess bias. A defect density of 10^{12} cm^{-3} was used.

Acknowledgment

The authors acknowledge the financial support from the Australian Research Council. D.D acknowledges the support of an Australian Research Council Future Fellowship through project FT220100656.

References

[1] Xuanyu Qian et al., *"Modeling for single-photon avalanche diodes: State-of-the-art and research challenges."*, Sensors (2023), 23(7), 3412.

[2] Ziyuan Li et al., *"Review on III-V Semiconductor Single Nanowire-Based Room Temperature Infrared Photodetectors."*, Materials (2020), 13, 1400.

[3] Zhe Li et al., *"An efficient modeling workflow for high-performance nanowire single-photon avalanche detector."*, Nanotechnology (2024), 35, 175209.

[4] COMSOL Multiphysics v5.5. www.comsol.com. COMSOL AB, Stockholm, Sweden.

Multi-Scale Thermal Modeling of 3D-Heterogeneous Integrated Processing-Near-Memory Chip for Edge Large Language Model Inference

Awang Ma, Bin Gao*, Yuyao Lu, Yudeng Lin, Yiming Zhou, Xing Mou, Zhuodong Kang, Zhipeng Kuang, Peng Yao, Jianshi Tang, He Qian, and Huaqiang Wu

School of Integrated Circuits, Beijing National Research Center for Information Science and Technology (BNRist), Tsinghua University, Beijing, China. *Email: gaob1@tsinghua.edu.cn

Abstract

In this paper, we demonstrated a multi-scale thermal modeling work for 3D-heterogeneous integrated processing-near-memory (PNM) chips. To achieve this, We extracted thermal conductivity using a compact model, reducing simulation time by 95% combining with FEM. Furthermore, we explored solutions for running LLAMA2 on 3D DRAM PNM chips, and proposed a bank-level refresh scheme to lower temperatures and reduce refresh power by 34%. We proposed a distributed DRAM solution to further optimize the maximum operating temperature.

Keywords: multi-scale thermal modeling, 3D DRAM PNM chip, edge large language model.

Introduction

3D chiplet integration technologies, including High Bandwidth Memory (HBM), have significantly enhanced memory density and data transfer rates, and thus are widely used for high performance AI computing hardware[1-3], like GPU. However, thermal management issues are becoming increasingly critical due to the 3D architecture. Recent developments in edge Large Language Model (LLM) inference hardware have garnered significant attention. For an edge LLM inference chip capable of processing over 20 tokens per second, 3D heterogeneous integration and processing-near-memory technologies are vital. However, thermal dissipation poses greater challenges in confined environments like smartphones and robots. Thus, there is a pressing need for rapid and accurate thermal simulation and efficient heat dissipation techniques, particularly for upcoming DRAM dies and computing die stacking technologies.

Computing dies and DRAM dies are integrated in a three-dimensional (3D) structure using TSV and micro-bump or hybrid bonding technology. Heat from the bottom computing die must transfer to the heat sink through the DRAM dies. This 3D solution has a higher power consumption density than traditional 2.5D GPUs. Key challenges in thermal modeling include: 1) Over 100,000 Through-Silicon Vias (TSVs) in DRAM dies, which critically affect thermal performance. Traditional Finite Element Method (FEM) models face long computation times and inefficiency with large-scale TSV networks. 2) Anisotropic thermal conductivity in TSV structures complicates accurate thermal analysis[6].

To address these challenges, this work realized multi-scale thermal modeling with a compact model to assess the anisotropic equivalent thermal conductivity of TSV region. Thermal simulations examine the effects of DRAM layers, power consumption of DRAM dies and processing elements (PEs), and so on. We also explore solutions for running LLAMA2 on 3D DRAM PNM. A fine-grained bank-level refresh scheme is proposed to lower temperatures in DRAM dies within thermal budgets. Additionally, a distributed DRAM-based PNM architecture is optimized for better thermal performance and higher computing throughput than traditional centralized transfer designs.

Multi-Scale Thermal Modeling

Fig. 1 shows the thermal modeling framework for the 3D DRAM PNM chips. Model parameters are extracted firstly. Then, the FEM solves temperature distributions of the 3D PNM chip. After that, decode_speed and maximum temperature will be evaluated. Finally, solutions and optimization guidelines are proposed.

The 3D DRAM PNM chip in our research includs 12 DRAM dies and one computing die. TSVs have a great impact on thermal simulation time. **Fig. 2** illustrates the extraction of thermal conductivity in TSV regions, where thermal conductivity is anisotropic, to speed up 3D thermal simulation. The proposed compact model shown in **Fig. 3**. The thermal conductivity calculated by the model and finite element method are consistent. Thermal simulation results with or without using the proposed model have little difference in temperature distribution that can be ignored, but saving the simulation time by 95% as presented in **Fig. 4**. As shown in **Fig. 5**, the anisotropic thermal conductivity of the TSV region changes with the TSV diameter in both the x-direction and the z-direction.

Fig. 6 presents the chip architecture and power density map of DRAM dies and the computing die. The main parameters for these work are specified in **Table 1**, including thermal conductivity, ambient temperature, power consumption and so on [7-8].The temperature maps of the top DRAM die and the computing die at full load are shown in **Fig. 7**. The maximum temperatures of each DRAM die and the computing die when dies are presented in **Fig. 8**. Obviously, the maximum temperature of the hot spot has been over 130 °C, far exceeding safe temperature limitation. **Fig. 9** shows the contribution of each part to the maximum

979-8-3315-0417-5/25 $31.00 © 2025 IEEE

temperature, when all dies are working at full load. To reduce hotspot temperatures, in addition to operating in low-temperature environments, but also we can reduce the power consumption of DRAM dies and computing die.

Analysis of Edge Large Language Model Inference

Edge LLMs are gaining significant attention. We assessed the temperature and decode_speed of a 3D DRAM PNM running LLAMA2. **Fig. 10** illustrates the LLAMA2 structure and one parameter mapping scheme across all 12 layers of DRAM. We conducted a series of temperature evaluations, with **Fig. 11** showing the maximum temperature relative to the power of DRAM dies and computing units. **Fig. 12** presents the maximum and average temperature variations based on the number of stacked DRAM layers. As depicted in **Fig. 13**, with a maximum temperature limit of 95 °C, the decode speed of LLAMA2 on the 3D DRAM PNM reaches 82 tokens/s at FP16, with a computing power of 15 TOPS under thermal constraints. In contrast, at FP8, the decode speed is 90 tokens/s, and the computing power is 65 TOPS.

To mitigate hotspot temperatures, we explored two approaches: reducing power consumption and minimizing temperature differences. **Fig. 14** shows the refresh frequency map for a DRAM die based on temperature distribution and refresh requests (double refresh rate for T > 95°C and double again for T > 105°C). Notably, bank-level refresh saves 34% refresh power compared to die-level refresh (**Fig. 15**). The centralized DRAM configuration, with TSVs grouped together, leads to local hotspots, while the distributed DRAM solution in **Fig. 16** effectively addresses this issue. **Fig. 17** illustrates that, under full load, the distributed DRAM solution results in a 6.9 °C lower temperature difference and a 3.4 °C lower maximum temperature compared to the centralized DRAM solution..

Conclusion

This work presents a multi-scale thermal modeling approach for 3D DRAM PNM chips. We extracted thermal conductivity using a compact model, reducing simulation time by 95% combining with FEM. Additionally, we investigated viable solutions for executing LLAMA2 on 3D DRAM PNM chips, and we proposed an innovative bank-level refresh scheme that reduces refresh power by 34%, and a distributed DRAM solution to further optimize the maximum operating temperature.

Acknowledgments

Acknowledgment: This work is supported in part by the MOST of China (2021ZD0201200), the NSFC (92064015), Beijing Advanced Innovation Center for Integrated Circuits.

References

[1] P. Yao et al., Nature 2020, 577(7792): 641-646.

[2] W. Zhang et al., Science 2023, 381(6663): 1205-1211.

[3] Z. Liu et al., Nature communications 2020, 11(1): 4234.

[4] A. Ma et al., IEDM 2022, pp. 15.5.1-15.5.4.

[5] A. Ma et al., IRPS 2023, pp. 1-6,

[6] T. Kim et al., iTherm 2022, pp. 1-5.

[7] K. Son et al., EDAPS 2023, pp. 1-3.

[8] K. Son et al., TCPMT 2022, vol. 12, no. 9, pp. 1542-1556.

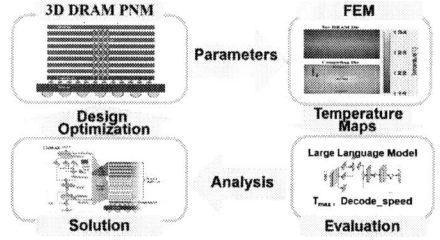

Fig. 1. The thermal modeling framework for the 3D DRAM PNM chips. Model parameters are extracted firstly. Then, the FEM solves temperature distributions of the 3D PNM chip. After that, decode_speed and maximum temperature will be evaluated. Finally, solutions and optimization guidelines are proposed.

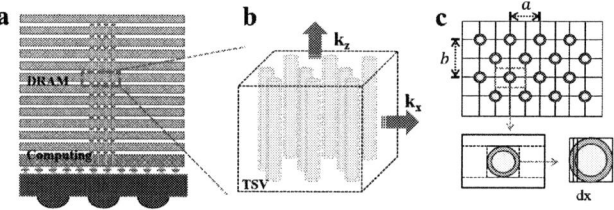

Fig. 2. Extraction of thermal conductivity in TSV regions to speed up 3D thermal simulation, a) the cross-sectional schematic of 3D DRAM PNM chips, including 12 DRAM dies and one computing die, b) schematic of the TSV region where thermal conductivity is anisotropic. c) the TSV arrangement is regular, and repeating units can be used to extract thermal conductivity.

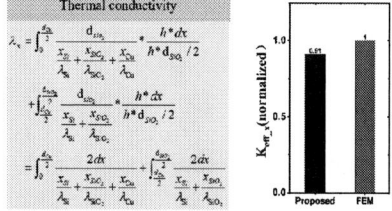

Fig. 3. The thermal conductivity is extracted through compact model and FEM simulation. The figure on the left shows the model proposed. The figure on the right compares the thermal conductivity calculated by the model and finite element method.

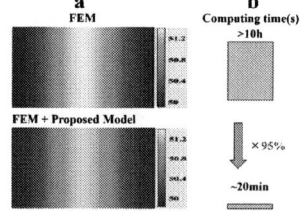

Fig. 4. Thermal simulation results with or without using the proposed model. a) temperature distribution in TSV area. b) simulation time required for 3D DRAM PNM.

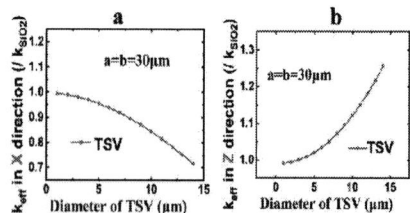

Fig. 5. Based on the proposed model, the thermal conductivity of the TSV region changes with the TSV diameter, a) thermal conductivity in x direction. b) thermal conductivity in z direction.

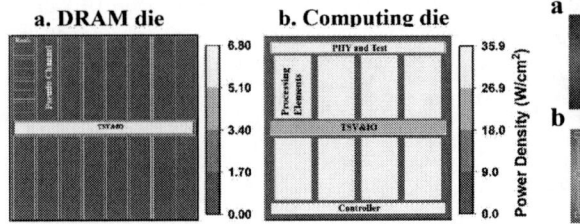

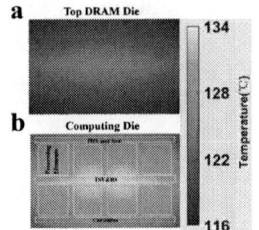

<table>
<tr><td>Parameter</td><td>Description</td><td>Value</td></tr>
<tr><td>k_{Cu}</td><td>Thermal conductivity of Cu</td><td>401 W/m·K</td></tr>
<tr><td>k_{SiO2}</td><td>Thermal conductivity of SiO2</td><td>1.4 W/m·K</td></tr>
<tr><td>$k_{underfill}$</td><td>Thermal conductivity of underfill</td><td>0.35 W/m·K</td></tr>
<tr><td>k_{Si}</td><td>Thermal conductivity of Si</td><td>149 W/m·K</td></tr>
<tr><td>R_{up}</td><td>Thermal resistance upperside</td><td>1.69 cm²·K/W</td></tr>
<tr><td>T_o</td><td>Environment temperature</td><td>45°C</td></tr>
<tr><td>T_{limit}</td><td>limit of working temperature</td><td>95°C</td></tr>
<tr><td>P_{Bank}</td><td>Power of DRAM die bank(per die)</td><td>1.5W</td></tr>
<tr><td>$P_{TSV, DRAM}$</td><td>Power of DRAM die TSV(per die)</td><td>0.5W</td></tr>
<tr><td>$P_{TSV, GPU}$</td><td>Power of computing die TSV</td><td>2W</td></tr>
<tr><td>P_{PE}</td><td>Power of one Computing Unit</td><td>3W</td></tr>
</table>

Fig. 6. Power density maps, a) the DRAM die based on the typical DRAM architecture. b) the computing die including Processing Elements(PEs), TSV, controller and so on.

Fig. 7. The temperature maps, a) the top DRAM die, b) the computing die.

Table. 1. The main parameters used in this study include thermal conductivity, ambient temperature, power consumption and so on.

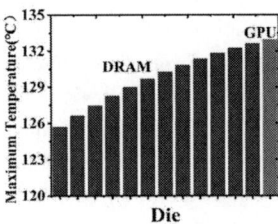

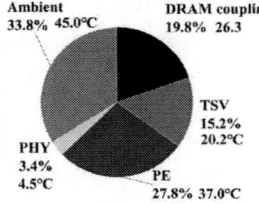

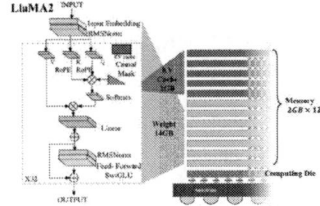

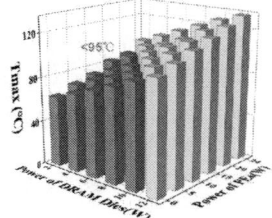

Fig. 8. The maximum temperatures of each DRAM die and the computing die when dies are operating at full load.

Fig. 9. The contribution of each part to the maximum temperature, when all dies are working at full load.

Fig. 10. The structure of the large language model LLAMA2 and one scheme of its parameters mapping onto all 12-layer DRAM.

Fig. 11. Evaluation of the maximum temperature changes with power of DRAM dies and PEs of computing die.

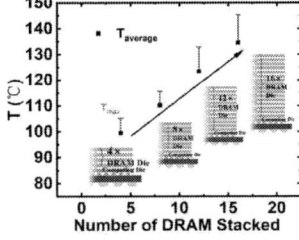

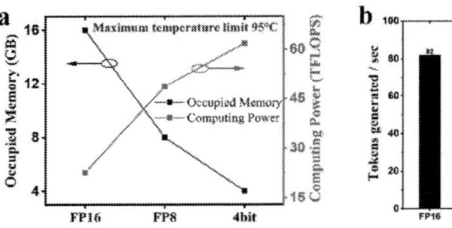

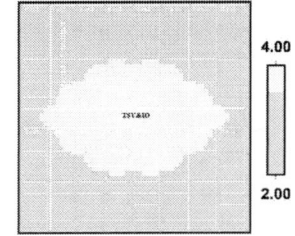

Fig. 12. Evaluation of maximum temperature and average temperature variation with number of DRAM stacking layers.

Fig. 13. Evaluation of LLAMA2 running on 3D DRAM PNM, under maximum temperature limitation at 95°C, a) the memory required and computing power allowed at precision of 4-bit, FP8, and FP16. b) decoding speed of LLAMA2 at precision of FP8, and FP16.

Fig. 14. The refresh frequency map of one DRAM die according to temperature distribution and refresh request.

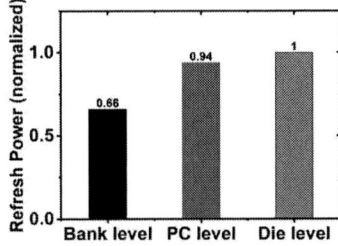

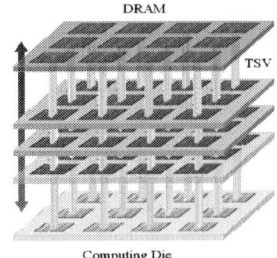

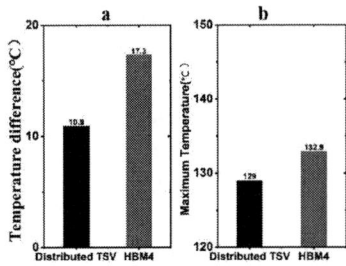

Fig. 15. DRAM refresh power changes with refresh granularity at bank level、PC level and Die level, according to the refresh frequency map of one DRAM die.

Fig. 16. Distributed DRAM solution: for example, each 1Gb DRAM is one unit, connected layer by layer through TSV, and finally connected to the computing die.

Fig. 17. Comparison between distributed DRAM solution and centralized DRAM solution at full load, a) difference between T_{max} and T_{mini}, b) maximum temperature.

Polarization regulation in AlGaN solar-blind ultraviolet photodetectors

Bingxiang Wang,[1] Ke Jiang,[1,2,*] Tong Fang,[1,2] Xiaojuan Sun,[1,2] and Dabing Li[1,2,*]

[1]Key Laboratory of Luminescence Science and Technology, Chinese Academy of Sciences & State Key Laboratory of Luminescence and Applications, Changchun Institute of Optics, Fine Mechanics and Physics, Chinese Academy of Sciences, China,

[2]Center of Materials Science and Optoelectronics Engineering, University of Chinese Academy of Sciences, Beijing, China

Abstract

This paper presents the impact of polarization effect on AlGaN-based solar-blind photodetectors. The impact of polarization heterojunction on charge distribution, electric field modulation and device performance are investigated for both planar and vertical device structure. Results shows that polarization heterojunction enhanced the channel conductance and carrier separation efficiency of planar structure device. For vertical structure, polarization charge present an enhanced carrier transportation under higher voltage, while the carrier screening effect in vertical transportation regime suppress the polarization enhancement under low bias.

Keywords: AlGaN, solar-blind photodetection, polarization

Introduction

AlGaN shows great potential in solar-blind ultraviolet (SBUV) photodetection due to its tunable direct wide bandgap and efficient absorption in the SBUV region, showing great potential in military and civil applications. The strong spontaneous and piezoelectric polarization in AlGaN heterojunctions have significant effects on the performance of AlGaN-based optoelectronic devices, which usually reflected in the regulation of the distribution of charges and electric fields [1-2]. Utilization of AlGaN-based polarization heterojunction has proven to be effective on improving nitride-based device performance [3-6].

This article demonstrates that in planar structures with horizontal carrier transport, polarized heterojunctions effectively enhance the separation efficiency of photo-generated carriers. Additionally, the triangular potential well at the heterojunction increases the electron concentration in the transverse electron transport channel, thereby significantly improving device responsivity. Furthermore, incorporating a high electron barrier AlN spacer layer into the heterojunction effectively suppresses the dark current of the planar device while simultaneously increasing its photocurrent. Conversely, in the vertical structures, the accumulation of high concentrations of interface carriers induced by polarized heterojunctions significantly diminishes the enhancement effect of polarization electric fields under low bias (the interface carrier shielding effect). Under higher external bias, the carrier gatherde at the heterojunction interface is depleted, achieving a more substantial enhancement of the active region electric field.

Methods

Standard semiconductor device fabrication processes are used to fabricate the devices. The dielectric and passivation layer are deposited through plasma enhanced chemical vapor deposition (PECVD). Photolithography and inductively coupled plasma (ICP) etching are conducted to fabricate device mesa and isolation. E-beam evaporation and rapid thermal annealing are used to form proper device contact electrodes. The APSYS software is employed to simulate the AlGaN-based SBUV detectors. The hole and electron mobilities are set to be 5 and 1000 cm2/(V·s), respectively. The conduction and valance band offset ratio for the AlGaN heterojunction is set to be 1:1. The polarization level is set as 30%. The monochromatic incident light of 280 nm with a power density of 500 W/m2 is used to offer a SBUV illumination condition. The optical absorption coefficients for the AlGaN layers are determined as $1.30 \times 10^4 \mathrm{cm}^{-1}$ for all AlGaN layers to exclude the impact on the device performance difference.

Results and Discussion

A. Polarization regulation in planar photodetector

The non-centrosymmetric crystal structure of AlGaN generates an obvious spontaneous polarization (Fig. 1). As the AlGaN heterojunction is formed, the strain can lead to an additional piezoelectric polarization, generating the fix charges at the heterojunction interface. The polarized charge at the heterojunction interface of device S_A induces the generation of high-concentration electron channels (Fig 2a), whereas there is no corresponding process in device S_B (Fig 2b). Consequently, device SA exhibits superior lateral current conduction capability. Fig 2c illustrates the measured characteristic curve of device S_A. The difference between the photocurrent and dark current of device S_A is substantially larger than that of device S_B, aligning with our simulation expectations and indicating that the response of the polarization-enhanced detector is superior to that of the non-polarization-enhanced detector. Fig 2d presents the response spectra of devices S_A and S_B under varying biases. The response peak of device SA is located at 257 nm, with peak responsivities of 1.18 A/W and 1.42 A/W at 5 V and 10 V, respectively. In comparison, the response peak of device S_B is at 255 nm, with peak responsivities of 0.022 A/W and

979-8-3315-0417-5/25 $31.00 © 2025 IEEE

0.0258 A/W at 5 V and 10 V, respectively. These results confirm our analysis that polarized heterojunctions enhance the efficiency of photogenerated carrier separation in horizontally structured detectors while generating highly conductive carrier channels, significantly improving the device's photo response intensity.

B. Suppressing photocurrent with AlN spacer

Although polarized heterojunctions enhance carrier separation efficiency and lateral current conduction capability, high-concentration electron accumulation at the interface can also lead to increased device dark current [4]. To address this, we proposed a polarization-enhanced structure with an aluminum nitride (AlN) insertion layer (Fig. 3a). Device S_C incorporates a high potential barrier AlN insertion layer on top of the conventional polarization-enhanced heterojunction, whereas the control device S_D lacks an AlN layer. Fig. 3b displays the experimentally measured dark state I-V curves of devices S_C and S_D, as well as the 250 nm photocurrent curve at 5 $\mu W/cm^2$. Both devices exhibit a dark current intensity of approximately 1 pA within ±5 V (the instrument's detection limit), while under illumination, device S_C shows an increase in photocurrent intensity by two orders of magnitude compared to device S_D. Device S_C achieved a photodetector current ratio (PDCR) of up to 10^7 at 5 V, whereas device S_D had a PDCR of 10^5 at 5 V. Fig. 3c and 3d illustrate the response spectrum curves of devices S_C and S_D, respectively, both exhibiting good solar-blind characteristics with their response spectrum cutoff at 270 nm. Under a 5 V bias at 234 nm, the responsivities of device S_C and device S_D are 1.2×10^6 A/W and 6.8×10^3 A/W, respectively. These results demonstrate that the AlN insertion layer can further enhance the polarization effect in AlGaN heterojunctions, resulting in more efficient carrier separation, higher electron concentration in the optical state, and increased photocurrent. In the dark state, the high potential barrier introduced by the AlN insertion layer effectively suppresses the device's dark current by limiting the electron transport process across the top potential barrier. Finally, to demonstrate the potential application of the proposed structure in SBUV band detection, we set up a single-pixel scanning imaging system, results are shown in Fig. 4.

C. Polarization regulation in vertical transport regime

The performance control mechanism of the polarization effect on vertical carrier transport is crucial in designing AlGaN-based UV avalanche photodiode (APD) structures. To explore this, we designed AlGaN-based p-i-n type devices with various polarization configurations, as depicted in Fig. 5a-d. These configurations achieved different polarization heterojunctions by adjusting the heterojunction structure at the i-n junction. Device S1 features a conventional p-i-n type control structure without polarized heterojunctions, while the other devices, labeled as S2 (n-i heterojunctions x=0.45, y=0.3), S3 (i-i heterojunctions with a lower i-$Al_{0.3}Ga_{0.7}N$

layer of 10 nm, y=0.3), and S4 ("Al grading" heterojunction with graded n-AlGaN layers ranging from Al content 0.3 to 0.45 within 10 nm, y=0.3), incorporate varied heterojunction structures.

We calculated the light-dark I-V characteristics and gain curves of the four structures at 0-160 V (Fig. 6a). The critical avalanche voltages of devices S1-S4 are 142.85 V, 135.60 V, 130.38 V, and 139.79 V, corresponding to avalanche gains of 3.85×10^4, 5.43×10^4, 5.76×10^4, and 4.51×10^4, respectively. Notably, device S3 exhibits the lowest critical voltage and highest avalanche gain. Fig. 6b displays the electric field intensity distribution under a 100 V bias voltage, with the depletion region electric field intensity decreasing in the order of S3, S2, S4, and S1. Due to a relatively low electron concentration on both sides of device S3's heterojunction, its electron concentration at the interface is lower than that of device S2. The polarization charges shielding effect caused by electron aggregation in device S3 is weaker, and the reverse peak electric field strength at its heterojunction interface is less than that of device S2, supporting this conclusion.

We further analyzed the low-voltage characteristics of devices with different thicknesses (10 nm, 20 nm, 30 nm) of i-$Al_{0.3}Ga_{0.7}N$ layers based on the S3 structure with the best performance (labeled as IP-1~IP-3). As the heterojunction interface moves away from the highly doped n-type layer (Fig. 6c), the accumulated electron concentration in the heterojunction interface area decreases, thus the photocurrent increases. Comparing device S3 with device S1, which lacks polarization heterojunctions, reveals that the photocurrent of device S3 is weaker than that of device S1 below 10 V (Fig. 6d). This difference is primarily due to the shielding effect of interface electrons on polarized charges, generating an electric field opposite to the built-in field direction in the i-$Al_{0.3}Ga_{0.7}N$ layer at low voltages, causing electron deceleration.

Conclusion

The polarization modulation mechanisms in AlGaN-based photodetectors are thoroughly investigated. The effects of polarization heterojunctions on charge distribution, electric field modulation, and device performance are investigated for both planar and vertical device structures. The results show that polarization heterojunctions enhance the channel conductance and carrier separation efficiency of planar devices. In vertical structures, polarization charges enhance carrier transportation under higher voltages, while the carrier screening effect in the vertical transportation regime suppresses polarization enhancement at low bias.

Acknowledgments

This work was supported by National Key R&D Program of China [2022YFB3604903]; National Natural Science Foundation of China [62121005, 62004196, 61827813,

U22A2084]; Youth Innovation Promotion Association of CAS [2023223]; Young Elite Scientist Sponsorship Program by CAST [YESS20200182]; the Natural Science Foundation of Jilin Province [20230101345JC, 20230101360JC, SKL202302026].

References

[1] D. Li, K. Jiang and X. Sun, et al., "AlGaN photonics: recent advances in materials and ultraviolet devices", Adv. Opt. Photonics, 2018, 10, 43.

[2] Q. Cai, H. You and H. Guo, et al., "Progress on AlGaN-based solar-blind ultraviolet photodetectors and focal plane arrays", Light: Sci. Appl., 2021, 10, 94.

[3] Guo, Long, et al. "Multiple-quantum-well-induced unipolar carrier transport multiplication in AlGaN solar-blind ultraviolet photodiode." Photonics Research 9.10 (2021): 1907-1915.

[4] Jiang, Ke, et al. "Polarization-enhanced AlGaN solar-blind ultraviolet detectors." Photonics Research 8.7 (2020): 1243-1252.

[5] Z. G. Shao, D. J. Chen, Y. L. Liu, H. Lu, R. Zhang, Y. D. Zheng, L.

Li, and K. X. Dong, "Significant performance improvement in AlGaN solar-blind avalanche photodiodes by exploiting the built-in polarization electric field," IEEE J. Sel. Top. Quantum Electron 20(6), 187-192 (2014).

[6] Z. G. Shao, X. F. Yang, H. F. You, D. J. Chen, H. Lu, R. Zhang, Y. D. Zheng, and K. X. Dong, "Ionization-enhanced AlGaN heterostructure avalanche photodiodes," IEEE Electron Device Lett. 38(4), 485-488 (2017).

[7] J. F. Muth, J. D. Brown, M. A. L. Johnson, Z. H. Yu, R. M. Kolbas, J. W. Cook Jr. and J. F. Schetzina, "Absorption coefficient and refractive index of GaN, AlN and AlGaN alloys," Mater. Res. Soc. Internet J. Nitride Semicond. 4, 502–507 (1999).

[8] D. Brunner, H. Angerer, E. Bustarret, F. Freudenberg, R. Höpler, R. Dimitrov, O. Ambacher, and M. Stutzmann, "Optical constants of epitaxial AlGaN films and their temperature dependence," J. Appl. Phys. 82 (10): 5090–5096 (1997).

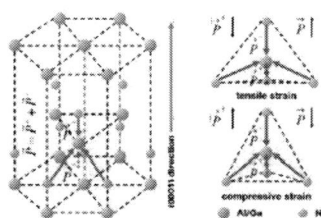

Fig. 1: Lattice diagram and origination of AlGaN polarization.

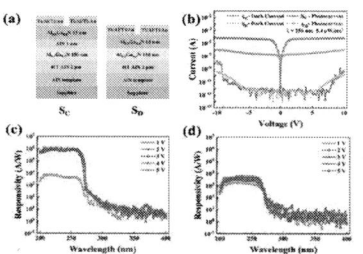

Fig. 3: Structure of Device (a) S_C and S_D. (b) IV characteristics of Device S_C and S_D. Spectral response of device (c) S_C and (d) S_D, respectively.

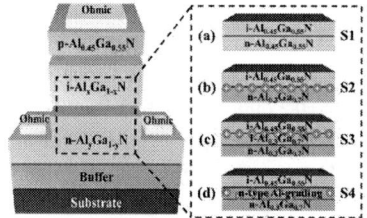

Fig. 5: Vertical structure with different polarization heterojunction. (a) Conventional pin structure. (b) "n-i" polarization. (c) "i-i" polarization. (d) "Al-grading" polarization.

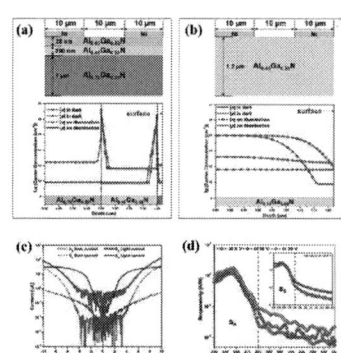

Fig. 2: Structure and simulated carrier distribution of Device (a) S_A and (b) S_B. (c) IV characteristics and (d) spectral response of Device S_A and S_B.

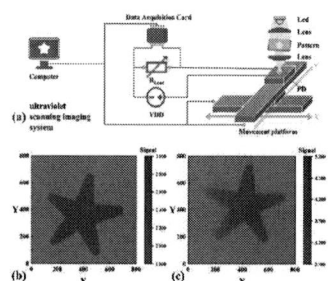

Fig. 4: (a) Schematic diagram of the single-pixel SBUV scanning imaging system. Resulting images for devices (b) S_C and (c) S_D, respectively.

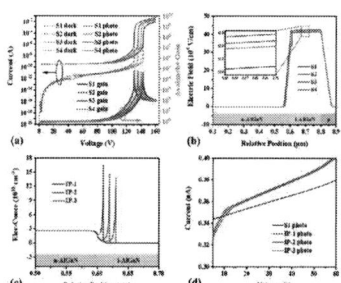

Fig. 6: (a) IV characteristics of divec S1-S4 within 0-160 V. (b) Electric field distribution of device S4-S1. (c) Electron distribution profile and (d) photocurrent of device S3 with different i-Al$_{0.3}$Ga$_{0.7}$N thickness. thickness.

Comprehensive Performance Evaluation of 18 Perfluoroelastomers O-Ring Models for Semiconductor Manufacturing

Zhanfeng Guo[1], He Tian[1,*], Tian-Ling Ren[1,*]

[1] School of Integrated Circuits & Beijing National Research Center for Information Science and Technology (BNRist), Tsinghua University, Beijing 100084, China

e-mail: RenTL@tsinghua.edu.cn; tianhe88@tsinghua.edu.cn

Abstract

Perfluoroelastomers (FFKM) O-rings are crucial in semiconductor processes because of their excellent resistance to high temperatures, plasma and corrosion. However, semiconductor manufacturers often face confusion during procurement due to the multiplicity of parameters and varying evaluation standards across suppliers. This study evaluated 18 FFKM O-ring models from 8 suppliers, focusing on compression set, O_2 plasma, ozone, and steam resistance tests, to establish a reliable reference for O-ring selection in the semiconductor industry.
Keywords: FFKM, O-ring, Comparison

Introduction

Semiconductor processes involve a lot of vacuum environments, and often operate at high temperatures, plasma radiation and other harsh conditions. Perfluoroelastomers (FFKM) is widely used in advanced semiconductor equipment because of its excellent cleanliness, high temperature resistance, plasma resistance and corrosion resistance [1]. For semiconductor foundries, the performance, service life and cost of FFKM O-rings are important factors in reducing costs and risks. Previous work usually focused on the difference between FFKM and fluoroelastomer(FKM)[2], [3], or some companies have tested the performance of their own FFKM O-rings[4], [5]. In this paper, a series of comparative tests were carried out on FFKM O-rings commonly used in China market, including high temperature resistance, plasma resistance, ozone resistance and steam resistance, which provided important and reliable reference for semiconductor manufacturers when choosing FFKM O-rings.

Methodology

18 O-ring models involved in the tests are listed in Table 1. Most of the models are sized AS568A-214. However, due to procurement challenges and delivery times, some models were purchased as AS568A-226 and AS568A-211, which differ in inner diameter (ID) but maintain the same cross-sectional diameter (CS). At the same time, due to the delivery delays, two models—Valque's F-plus and Greene Tweed's 547—were not fully received in time, limiting their use to only part of the experiments.

The performance evaluation of FFKM O-rings under various conditions was conducted using a series of standardized tests to simulate real-world semiconductor industry environments. The tests included compression set tests at high temperatures, plasma resistance test, ozone resistance test and steam resistance test.

Table 1: List of the FFKM models.

Type	Supplier	Brand	Model	Size (AS568A)
Heat resistance	Daikin	DUPRA	DU551	214
	Dupont	Kalrez	K7075	211
	Dupont	Kalrez	K8900	226
	Greene Tweed	Chemraz	XCD	226
	IC SEAL	Enduraz	XM02	214
	Is Sealing	PERFREZ	6301	214
	Valque	FLUORITZ	F-plus	214
	Jishang	JST	TFC-87512	214
Plasma resistance	Daikin	DUPRA	DU341	214
	Dupont	Kalrez	K9100	214
	Dupont	Kalrez	K9500	226
	Greene Tweed	Chemraz	XRZ	226
	Greene Tweed	Chemraz	547	226
	EKK	Superior	301F	214
	IC SEAL	Enduraz	TS02	214
	Jishang	JST	TFC-87520	214
	Is Sealing	PERFREZ	6133D	214
	Valque	FLUORITZ	F-TR	214

1. Compression Set Tests (250°C and 300°C)

The compression set test is widely regarded as a reliable method to evaluate the thermal stability of FFKM[6]. In this work, the compression set tests were performed at 250°C and 300°C using an oven (ESPEC PHH-102M) and fixtures (Chengsheng CS-6024). Samples were prepared by cutting to specified sizes, followed by thickness measurements. The samples were compressed by 25% in the fixtures and heated in the oven for 72 hours and 168 hours. After heating, samples were removed, cooled, and measured for thickness.

2. Plasma Resistance Test

Plasma resistance was evaluated under an O_2 plasma environment by microwave plasma etcher (Alpha Plasma Q235HT). Samples were exposed to plasma with an O_2 flow rate of 300 sccm, at 500W power for 60 minutes. After plasma exposure, the samples were ultrasonically cleaned, dried and weighed to calculate weight change rate. Additionally, a 72-hour compression set test at 250°C was also performed to assess whether the plasma treatment affected the thermal resistance of the O-rings.

979-8-3315-0417-5/25 $31.00 © 2025 IEEE

3. Ozone Resistance Test

The ozone resistance test was conducted using an ozone generator (CY50G) and an aging test chamber (QYL-100). Samples were exposed to a 10% (100000ppm) ozone concentration at 50°C for 50 hours. Same as the previous, a 72-hour compression set test at 250°C was also performed after ozone resistance test.

4. Steam Resistance Test

Steam resistance was tested using a high-pressure reactor (SL-KH). Due to the limited volume of the reactor, custom fixture matching the shape of the chamber was designed to accommodate the samples. Samples were heated in the reactor with deionized water at 200°C and a pressure of 1.5-1.6 MPa for 72 hours. After cooling, samples were measured for changes dimensions to assess their durability against steam.

These tests provided a comprehensive evaluation of the FFKM O-rings' performance under various challenging conditions. The combination of compression set, plasma, ozone, and steam resistance tests ensures that the selected O-rings meet most of the key requirements of semiconductor manufacturing environments. The obtained data will assist semiconductor manufacturers in model selection, enable suppliers to optimize product design, and enhance the reliability of FFKM O-rings in semiconductor applications.

Results and discussion

1. Compression Set Tests

The compression set test results reveal notable differences in material performance under high temperatures. The results in Figure 1 show a consistent pattern across both 250°C and 300°C, with models that perform well at 250°C also showing better resistance at 300°C. However, some models have poor tolerance to high temperature of 300°C, 6301 and XM02 had two out of three specimens break, while TS02 experienced complete failure, with all three specimens breaking.

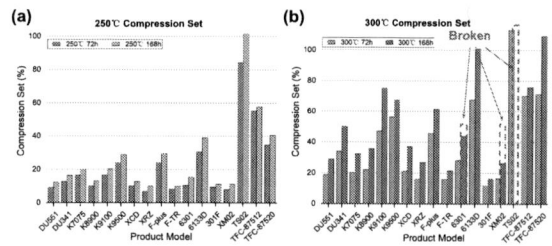

Fig. 1: Compression set of 250°C and 300°C.

On the other hand, Greene Tweed's XRZ, Valque's F-TR, EKK's 301F and Daikin's DU551 maintained relatively stable performance, highlighting their suitability for high-temperature applications.

2. Plasma Resistance Test

Figure 2(a) presents the percentage of weight change for various O-ring models after exposure to O₂ plasma. Most models show a weight loss between 0.1% and 0.3%, reflecting degradation of the material in the plasma. Models such as Dupont's K9100, Daikin's DU341, and IC SEAL's TS02 show relatively low weight change, indicating better stability and resistance in plasma environments. These results highlight the varying capabilities of the materials, with some more suitable for plasma-intensive semiconductor applications than others.

Figure 2(b) shows the compression set values of the O-rings before and after plasma exposure. The result indicates that for all samples, the compression set values after plasma exposure are generally consistent with those without plasma treatment. This suggests that although plasma exposure may cause surface degradation and weight loss, it does not significantly impact the thermal stability of the O-rings.

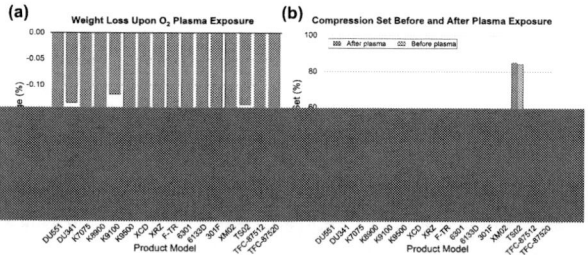

Fig. 2: Weight change and compression set comparison of plasma resistance test.

It is worth noting that the chamber temperature of O2 plasma test is relatively low (<100°C), while in actual semiconductor operations, such as PVD processes, high temperatures often accompany plasma exposure, need to take into account the resistance to plasma and high-temperature performance to make the equipment selection. Therefore, model selection for such applications must consider both plasma resistance and thermal stability to ensure long-term performance and reliability.

3. Ozone Resistance Test

Figure 3 presents the results of ozone resistance tests for various FFKM O-ring models, focusing on weight change rates and compression set values before and after ozone exposure.

Among the tested models, Dupont's K7075 and Is sealing's 6301 stand out with superior performance, maintaining lower weight loss after ozone exposure. In the meantime, it can be seen that the heat-resistant models from the same company also typically exhibit better ozone resistance. This consistency may because both ozone and high-temperature resistance in FFKM O-rings are primarily determined by the strength of the C-F bonds in their molecular structure. Therefore, models designed for high-temperature resistance generally exhibit lower weight change rates after ozone exposure.

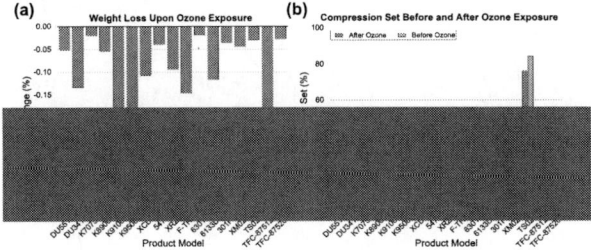

Fig. 3: Weight change and compression set comparison of ozone resistance test.

Similar to the results observed after plasma treatment, ozone exposure does not significantly affect the compression set values of the samples, indicating that ozone exposure under the tested conditions did not have a noticeable impact on the material's heat resistance. Nonetheless, if the ozone concentration and exposure duration are increased, the oxidative stress could disrupt weak cross-links or introduce surface-level degradation, which might eventually affect the O-ring's thermal resistance.

In semiconductor foundries, while ozone is indeed required in high-temperature environments like oxidation processes (aligning with the need for high-temperature and ozone resistance), it is also commonly used as a cleaning gas in plasma-related processes. These processes demand materials that are both plasma-resistant and ozone-resistant. Therefore, depending on the specific operating conditions, different test results can be referenced to select the appropriate O-rings.

4. Steam Resistance Test

Figure 4 presents the compression set values for various FFKM O-ring models exposed to high-pressure steam at 200°C. Many samples demonstrate significant permanent deformation, with compression set values nearing 100%. This contrasts with the results at 250 °C, where the compression set values are typically much lower. Among all the models, Daikin's DU551, Greene Tweed's XCD and XRZ showed the best resistance to steam at 200°C.

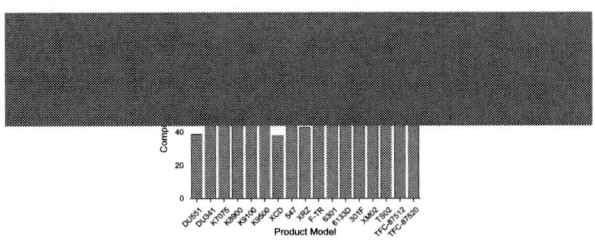

Fig. 4: compression set after 200°C steam test.

The significant difference in compression set under two conditions can be attributed to the combined effects of temperature and the environment of high-pressure steam.

The steam can infiltrate the material, causing hydrothermal degradation, thus weakening the polymer network and leading to higher deformation. Furthermore, the presence of water under high pressure can act as a plasticizer and soften the material, making it easier to permanently deform compared with dry heat environment. This combination of factors results in the much higher compression set values seen in the 200°C steam condition.

Conclusion

In this work, targeting the typical working conditions of semiconductor foundries, we carried out comprehensive tests on 18 FFKM O-ring models from 8 suppliers, including tests on high-temperature compression set, plasma resistance, ozone resistance, and high-pressure steam resistance. This work provides a systematic assessment of the suitability of FFKM O-rings for use in semiconductor equipment, alleviating the selection challenges faced by semiconductor manufacturers and motivating suppliers to further optimize their products. With these tests results, we provide a reliable reference for reducing uncertainty in O-ring selection, improving the operational reliability and efficiency of semiconductor lines.

Acknowledgments

This work was supported in part by Beijing Natural Science Foundation-Xiaomi Innovation Joint Fund (L233009), in part by National Natural Science Foundation of China under Grant No. 62374099. This work was also supported by the Daikin Tsinghua Union Program.

References

[1] S. T. Mhaske, J. D. Mohanty, and K. W. Chugh, "1 - Fluoropolymers: brief history, fundamental chemistry, processing, structure, properties, and applications," in Advanced Fluoropolymer Nanocomposites, K. Deshmukh and C. M. Hussain, Eds., in Woodhead Publishing Series in Composites Science and Engineering. , Woodhead Publishing, 2023, pp. 1–27.

[2] T. Goto, S. Obara, T. Shimizu, T. Inagaki, Y. Shirai, and S. Sugawa, "Study on CF4/O2 plasma resistance of O-ring elastomer materials," Journal of Vacuum Science & Technology A, vol. 38, no. 1, p. 013002, Dec. 2019.

[3] S. Wang and J. M. Legare, "Perfluoroelastomer and fluoroelastomer seals for semiconductor wafer processing equipment," Journal of Fluorine Chemistry, vol. 122, no. 1, pp. 113–119, Jul. 2003.

[4] T. S. Reger and G. J. Reichl, "Science of Sealing: Advanced Materials for High-Temperature Applications," in 2020 31st Annual SEMI Advanced Semiconductor Manufacturing Conference (ASMC), Aug. 2020, pp. 1–5.

[5] M. J. Heller, S. Sogo, J. Chen, and J. Legare, "Outgassing characterization of elastomeric seals used in semiconductor wafer processing," in 2010 IEEE/SEMI Advanced Semiconductor Manufacturing Conference (ASMC), Jul. 2010, pp. 68–71.

[6] M. Heller, J. Legare, S. Wang, and S. Fukuhara, "Thermal stability

A bio-inspired ionic retina

Hongjie Zhang[1,2] and Kai Xiao[1,*]

[1]Department of Biomedical Engineering, Guangdong Provincial Key Laboratory of Advanced Biomaterials, Institute of Innovative Materials, Southern University of Science and Technology, Shenzhen 518055, P.R. China.
[2]State Key Laboratory of Quality Research in Chinese Medicine, Institute of Chinese Medical Sciences, University of Macau, Taipa, Macau SAR 999078, China.
*Correspondence to xiaok3@sustech.edu.cn

Abstract

Herein, we demonstrate a bio-inspired retina by developing inhibitory and excitatory artificial synapses. By fine-tuning the ionic hydrogel structures to confine ion transport, these synapses achieve negative or positive modulation of optical signals without additional input. Integrating these synapses enables advanced tasks such as image recognition, motion analysis, and obstacle avoidance. This work offers a new approach for constructing bio-inspired retinas by precisely regulating ion transport, bringing it closer to the functionality of biological retinas.
Keywords: Ionic hydrogel, Artificial synapses, Bio-inspired retina

Introduction

Biological retina is the foundation of visual perception, relying on the collaboration of inhibitory and excitatory synapses to regulate neuronal activity [1]. Inhibitory synapses decrease postsynaptic potentials (PSP) by opening chloride channels, resulting in inhibitory postsynaptic potentials (IPSP), which attenuate noisy signals and modulate signal transmission. In contrast, excitatory synapses increase PSP through sodium and calcium channels, leading to excitatory postsynaptic potentials (EPSP) that enhance image contrast and sharpness. The diverse and tunable ion transport behaviors allow the visual system to precisely distinguish changes in light, facilitating efficient visual information processing (Fig. 1a). Inspired by the biological retina, significant progress has been made in neuromorphic visual processing platforms. These platforms utilize electrons as carriers, enabling functions such as feature extraction and image processing [2, 3]. However, compared to biological systems that use ions as charge carriers, these methods exhibit less regulatory diversity and may face challenges related to signal mismatch [4, 5].

Hydrogels, known for their excellent ion transport properties, are considered ideal materials for constructing flexible and biocompatible visual platforms. Herein, we developed a bio-inspired retina using ionic hydrogel

synapse devices that achieve both inhibitory and excitatory functions. These synapse devices effectively process optical signals and enable the execution of both image and motion tasks, offering a novel solution for intelligent bio-inspired retina systems.

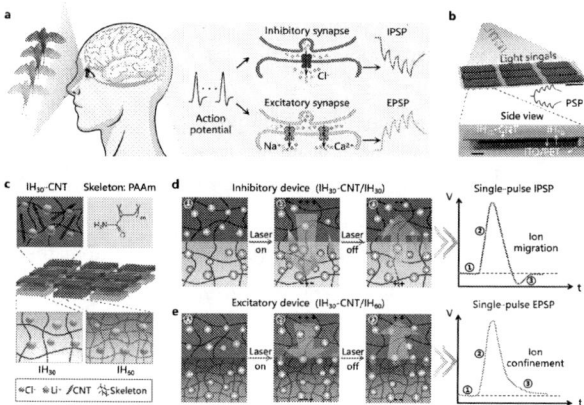

Figure 1. (a) Schematic of human visual system for processing image and motion information. (b) Photographs of bio-inspired retina with inhibitory and excitatory hydrogel artificial synapses for optical signal processing. In the figure above, the scale bar represents 0.3 cm. The scale of the enlarged side view is 0.5 mm. (c) Chemical structures of the polymer skeletons. (d, e) Schematic of the ion transport in inhibitory devices (IH_{30}-CNT/IH_{30}, d) and excitatory devices (IH_{30}-CNT/IH_{60}, e).

The hydrogel artificial synapses consist of a bilayer ionic hydrogel containing 0.1 M lithium chloride (LiCl), covered by two ITO/PET electrodes on the hydrogel's top and bottom surfaces (Fig. 1b). The structure of the ionic hydrogel can be adjusted by varying the solid content, denoted as IH_x, where x represents the mass percentage of polyacrylamide (Fig. 1c). The ionic hydrogel possesses soft elasticity and high water content, facilitating the movement of lithium ions (Li^+) and chloride ions (Cl^-). Introducing the light-responsive carbon nanotubes (CNT) into the upper hydrogel layer (IH_{30}-CNT) enables excellent photothermal conversion capabilities, generating a temperature gradient in the heterogeneous hydrogel under light exposure, which drives ion migration via the Soret effect [6]. By regulating the structure of the bottom hydrogel, inhibitory devices (IH$_{30}$-CNT/IH$_{30}$) and excitatory devices (IH$_{30}$-CNT/IH$_{60}$) can be fabricated. Under laser irradiation, due to the differing migration rates of anions and cations [7], Cl$^-$ rapidly migrates to the bottom layer, creating an uneven ion distribution and generating an ionic potential. When the laser is turned off, Cl$^-$ in the inhibitory device diffuses quickly back to the IH$_{30}$-CNT layer along the concentration gradient, while the slower return of Li$^+$ results in a potential lower than initial value. This process produces biosimilar IPSP, enabling negative signal regulation (Fig. 1d). In the excitatory device, the dense IH$_{60}$ layer confines ion migration. When the laser is turned off, both anions and cations gradually return to their original positions across the interface, restoring the ionic potential to its initial level. This process generates biosimilar EPSP, achieving positive signal regulation (Fig. 1e).

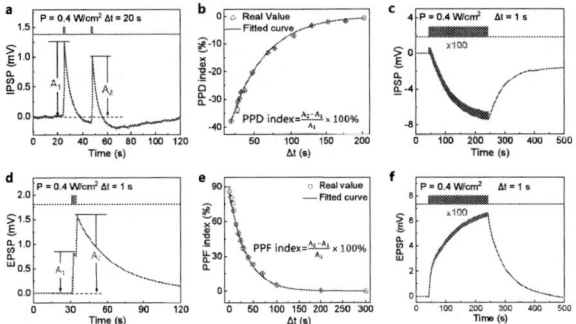

the experimental results, indicating that varying bottom structures lead to different fluxes of Cl$^-$ and Li$^+$ across the interface, consequently resulting in different changes in ion potential.

Figure 3. (a) Paired-pulse depression (PPD) effect of the inhibitory device. (b) Relationship between PPD index and the time interval (Δt) between two stimuli, where the blue curve is the fitted bi-exponential curve. (c) Long-term depression (LTD) effect of the inhibitory device. (d) Paired-pulse facilitation (PPF) effect of excitatory device. (e) Relationship between PPF index and Δt, where the red curve is the fitted bi-exponential curve. (f) Long-term potentiation (LTP) effect of excitatory device.

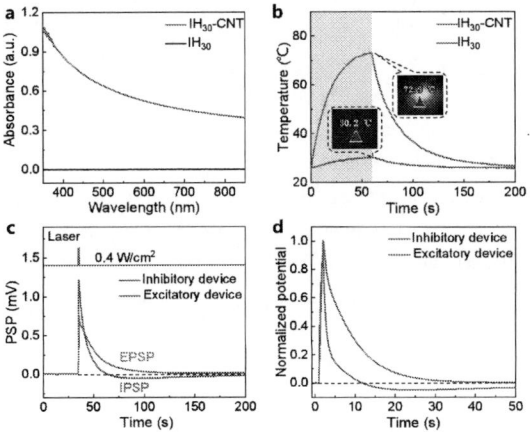

Figure 2. (a) Ultraviolet-visible-near infrared (UV-vis-NIR) absorption spectroscopy of IH$_{30}$-CNT and IH$_{30}$. (b). Heating and cooling curves of IH$_{30}$-CNT and IH$_{30}$ under 808 nm laser irradiation. (c) PSP generated by the synapse devices exposed to 808 nm laser irradiation. (d) The normalized potential signals of synapse devices induced by 808 nm laser irradiation, simulated by finite element method (FEM).

Figure 2a and 2b show that IH$_{30}$-CNT has a broadband absorption in the UV-vis-NIR region and can effectively convert laser irradiation into heat. When a single light pulse was applied, both inhibitory and excitatory synapse devices generated positive potentials. At the end of laser irradiation, the inhibitory device exhibits IPSP; whereas the excitatory device shows EPSP (Fig. 2c). Additionally, both devices exhibited a relaxation time after laser irradiation, providing time-related information. FEM simulations are used to explore the mechanism. The simulation results indicated that the temperature gradient generated by laser irradiation in the hydrogel facilitates ion migration. Figure 2d shows the simulated potential-time curve, which closely matches

Further experiments revealed the PPD effect, as shown in Fig. 3a, by applying a pair of laser pulses. When the second pulse arrives before the full recovery of the IPSP induced by the first pulse, the cumulative effect leads to a more negative IPSP. By adjusting Δt from 20 s to 200 s, the PPD index decreased with increasing Δt. Furthermore, the decay curves could be well fitted with a double exponential function (Fig. 3b), consistent with biological synapses [8]. In addition, under 100 consecutive laser pulses, the IPSP showed significant memory behavior, taking over 200 seconds to return to the initial value (Fig. 3c), indicative of LTD. The PPF and LTP effects can also be observed in excitatory devices (Fig. 3d-f).

By adjusting the hydrogel structure in the synaptic devices, convolutional kernels with varying weights can be implemented, successfully enabling image processing tasks like edge detection, sharpening, and denoising (Fig. 4a and 4b). The image recognition task was further demonstrated. A visual processing platform was constructed to preprocess noising images, featuring a kernel denoising operator simulated by a 3 × 3 synaptic device array. The preprocessing significantly improved image quality, achieving an MNIST dataset classification accuracy of 83% (Fig. 4c). The processing of dynamic tasks was further demonstrated, leveraging the relaxation properties of hydrogel artificial synapses to encode temporal information. Figure 4d shows the EPSP characteristics of excitatory devices trained with 20 consecutive light pulses. Each single pulse can be regarded as a frame of motion, indicating that the device can deduce the relationship between motion position and the time axis based on the spatiotemporal differences in relaxation properties. Subsequently, a 6 × 6 array of excitatory devices was used to analyze moving targets, demonstrating the recognition of object motion direction (Fig. 4e). By fitting the relaxation segments of the curves in Fig. 4d, we obtained a function curve representing EPSP changes over time, simulating a 28 × 28 synaptic array where each synapse corresponds to a pixel. After 20 training epochs, the model achieved a classification accuracy of 94.03% (Fig. 4f). Further path analysis was explored using a simulated array of excitatory devices. Initially, we captured the current signal state by sampling the signals generated as a block moved across the array. By analyzing these signals with relaxation functions and a set sampling rate, we were able to infer the block's movement path (Fig. 4g), which was then used to guide the robotic vehicle for obstacle avoidance (Fig. 4h).

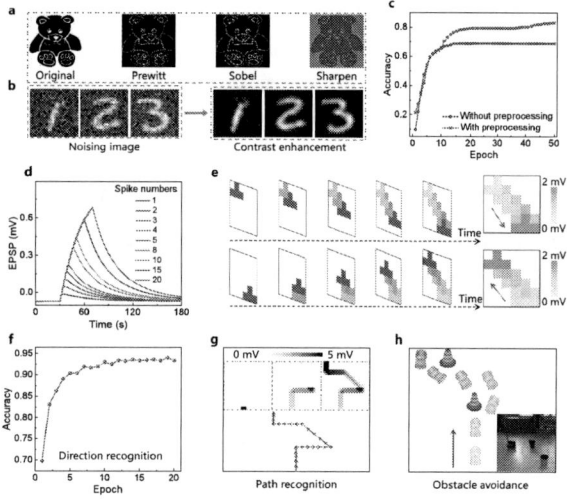

Figure 4. (a) Demonstration of various convolutional processing operations. (b) Nosing image and contrast enhancement. (c) Recognition accuracies without and with preprocessing. (d) Spike-number dependent plasticity of the excitatory hydrogel artificial synapse. (e) Recognition of the motion direction on the device array. (f) The direction recognition accuracy. (g) Demonstration of specific path recognition. (h) Schematic of a robot vehicle avoiding obstacles, with an inset image showing the optical image of a robot vehicle in motion.

Conclusion

In conclusion, ion transport can be effectively controlled by regulating the ionic interfacial energy barrier of hydrogel structures, allowing for both negative and positive modulation of optical signals. The artificial hydrogel synapses demonstrate broadband optical responses and tunable synaptic plasticity, facilitating the optimization of signal processing akin to the perceptual mechanisms of the human retina. Utilizing these synapses, we enhanced image recognition accuracy, motion direction detection, path analysis, and robotic vehicle obstacle avoidance. This bio-inspired retina, achieved through ion-confined transport, represents a new paradigm for neuromorphic visual information processing.

Acknowledgments

This work was financially supported by the National Key Technologies R&D Program of China (Grant No. 2023YFC2415900), the National Natural Science Foundation of China (No.22275079), Shenzhen Science and Innovation Committee (20220815164834003).

References

[1] F. Liao, Z. Zhou, B. J. Kim, J. Chen, J. Wang, T. Wan, Y. Zhou, A. T. Hoang, C. Wang, J. Kang, J.-H. Ahn, and Y. Chai, "Bioinspired in-sensor visual adaptation for accurate perception," *Nat. Electron.*, 5, pp. 84-91, (2022).

[2] L. Mennel, J. Symonowicz, S. Wachter, D. K. Polyushkin, A. J. Molina-Mendoza, and T. Mueller, "Ultrafast machine vision with 2D material neural network image sensors," *Nature*, 579, pp. 62-66, (2020).

[3] G. Zhou, J. Li, Q. Song, L. Wang, Z. Ren, B. Sun, X. Hu, W. Wang, G. Xu, X. Chen, L. Cheng, F. Zhou, and S. Duan, "Full hardware implementation of neuromorphic visual system based on multimodal optoelectronic resistive memory arrays for versatile image processing," *Nat. Commun.*, 14, p. 8489, (2023).

[4] Y. Hou, Y. Ling, Y. Wang, M. Wang, Y. Chen, X. Li, and X. Hou, "Learning from the Brain: Bioinspired Nanofluidics," *J. Phys. Chem. Lett.*, 14, pp. 2891-2900, (2023).

[5] W. Chen, L. Zhai, S. Zhang, Z. Zhao, Y. Hu, Y. Xiang, H. Liu, Z. Xu, L. Jiang, and L. Wen, "Cascade-heterogated biphasic gel iontronics for electronic-to-multi-ionic signal transmission," *Science*, 382, pp. 559-565, (2023).

[6] L. Talbot, R. K. Cheng, R. W. Schefer, and D. R. Willis, "Thermophoresis of particles in a heated boundary layer," *J. Fluid Mech.*, 101, pp. 737-758, (1979).

[7] E. R. Nightingale, "Phenomenological Theory of Ion Solvation. Effective Radii of Hydrated Ions," *J. Phys. Chem.*, 63, pp. 1381-1387, (2002).

[8] S. Zhao, W. Ran, Z. Lou, L. Li, S. Poddar, L. Wang, Z. Fan, and G. Shen, "Neuromorphic-computing-based adaptive learning using ion dynamics in flexible energy storage devices," *Natl. Sci. Rev.*, 9, p. nwac158, (2022).

ScAlN-based Bulk Acoustic Wave Technology for 5G Filtering Applications

Chen Liu, Xinghua Wang, You Qian, Ying Zhang, Minghua Li, Peng Liu, Huamao Lin, Qingxin Zhang, Yao Zhu

Institute of Microelectronics (IME), Agency for Science, Technology and Research (A*STAR), 2 Fusionopolis Way, Innovis #08-02, Singapore 138634, Republic of Singapore

Abstract

This paper briefly introduces the research efforts of leveraging the superior piezoelectric properties of Sc doped AlN (ScAlN) thin films to address the current challenges of bulk acoustic wave (BAW) technology for 5G filtering applications. The film bulk acoustic resonators (FBAR) based on $Sc_{0.3}Al_{0.7}N$ films can boost the bandwidth of the filters, scaling up the frequency up to mmWave range, and even enable the frequency reconfigurability of the FBAR resonators and filters.

Keywords: ScAlN, piezoelectric, RFMEMS, BAW

Introduction

Acoustic resonators and filters are essential in wireless communication, selecting specific frequency bands. They convert electrical signals into acoustic waves using piezoelectric materials like $LiNbO_3$ and AlN. Unlike surface acoustic wave (SAW) resonators, bulk acoustic wave (BAW) resonators use a metal/piezoelectric/metal sandwich structure and operate in thickness-extensional (TE) mode. To prevent acoustic waves from entering the substrate, an air cavity or Bragg mirror is placed beneath the device. The resonant frequency of BAW resonators mainly depends on stack thickness and acoustic phase velocity.

A typical frequency spectrum of the impedance of a BAW resonator is shown in Figure 2 (a). The resonator has the lowest impedance at the series frequency (f_s) and the highest impedance at the parallel frequency (f_p). BAW filters commonly employ the ladder-type filter topology (Figure 2(b)). As shown in Fig. 2 (c), all series resonators have a similar resonant frequency while all shunt resonators have another resonant frequency, normally lower than the series one. The bandwidth of the filter depends on the effective coupling coefficient (k^2_{eff}) of the resonator, while the quality factor (Q factor) determines the insertion loss and slope of the passband's skirt.

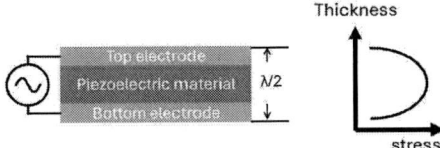

Fig.1. Working mechanism of BAW resonator

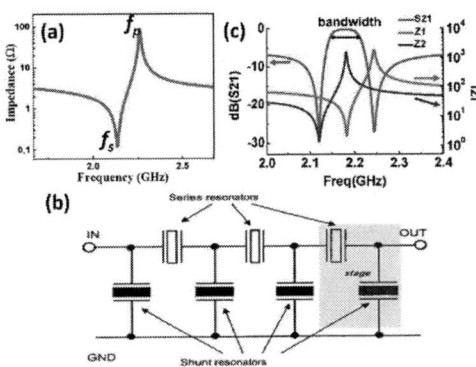

Fig.2. (a) Typical spectrum of a BAW resonator; (b) Topology of a ladder filter; (c) Typical spectrum of a filter.

Although BAW filters have achieved significant success in the commercial 4G filter market, ranging from 2.5 GHz to sub-6 GHz, they face challenges as the industry transitions to 5G. These challenges include limited bandwidth that cannot meet the new requirements, difficulties in achieving frequencies suitable for mmWave bands, and the constraint that each filter can only accommodate one frequency band. This article will outline our research endeavors to address the aforementioned challenges utilizing ScAlN-based film bulk acoustic wave resonator (FBAR) technology.

Boosting k^2_{eff} with higher Sc%

In 2009, researchers discovered that adding Scandium (Sc) to AlN, a conventional piezoelectric material for BAW resonators, enhances the intrinsic longitudinal piezoelectric coefficient (e_{33}) [1]. However, increasing Sc concentration also raises the dielectric constant of ScAlN, which reduces material's coupling coefficient (K_t^2) as following equation:

$$K_t^2 = \frac{e_{33}^2}{c_{33}\varepsilon_{33}}$$

Our experimental results in Fig. 3 (a) show that for ScAlN-based FBAR devices, with Sc% increasing from 0% to 30%, the k^2_{eff} of the resonators is significantly improved from 8% to around 20% [2]. The results indicate the efficiency of doping more Sc into AlN. Wide bands in 5G standards, such as n79 (4.4 to 5 GHz) and n78 (3.3 to 3.8 GHz), will benefit from the development of ScAlN thin films.

In the other hand, manufacturing highly Sc doped films induces challenges including crystallinity, stress control and high density of the abnormal grains [3], especially when scaling down the film thickness for higher frequency bands. These result in the degradation of Q factor with increasing the doping level of Sc in ScAlN [4]. With advanced deposition equipments and recipes, we have demonstrated 6 GHz FBAR resonators with stack shown in Fig. 3 (b) [5]. Performances of the fabricated devices are shown in Fig. 4, recording a k^2_{eff} of over 20% and Q factor over 600.

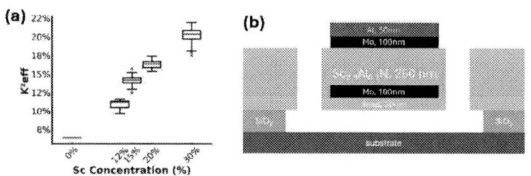

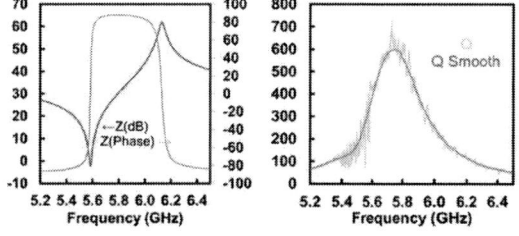

Fig.3. (a) The k^2_{eff} of FBAR resonators based on ScAlN with different Sc%. (b) Cross-sectional schematic of an FBAR device.

Fig. 4. Spectrum of fabricated 6 GHz $Sc_{0.3}Al_{0.7}N$-based FBAR. Left: Impedance and phase; Right: Bode Q factor.

Scaling frequency with thinner film

mmWave is a critical part of 5G communication featuring a high data speed and an ultralow latency. While holding the benefits of miniaturized size and CMOS integration compatibility, scaling BAW filter into mmWave frequency range requires a much thinner film stack and smaller lateral dimensions, which introduces extra process and design difficulties, as shown in Fig.5 (a) [7].

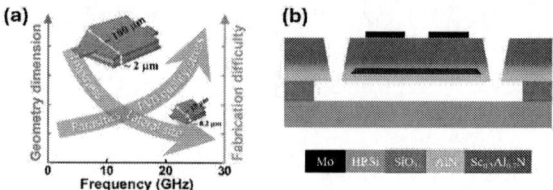

Fig. 5. (a) Challenges of BAW resonator for mmWave frequency; (b) Cross-sectional schematic of two series FBAR devices.

Recently, we showcased 21 GHz FBAR devices featuring a stack of 20 nm Mo/45 nm $Sc_{0.3}Al_{0.7}N$/50 nm Mo, as depicted in Fig. 5 (b). By optimizing the seed layer, we achieved a thin film with dielectric loss below 0.01. Consequently, as illustrated in Fig. 6, the fabricated devices exhibit a high k^2_{eff} of approximately 13.4% and a Q factor of around 300, resulting in a best-in-class FOM ($Q*k^2_{eff}*f$) for all reported acoustic resonators within this frequency range.

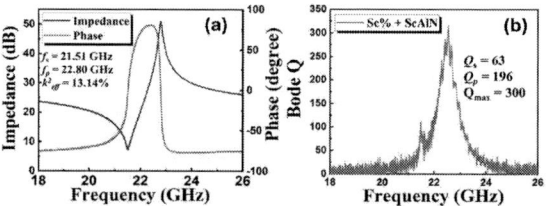

Fig. 6. Spectrum of fabricated 20 GHz $Sc_{0.3}Al_{0.7}N$-based FBAR: (a) Impedance and phase; (b) Bode Q factor.

We further designed and fabricated ladder filters. The form factor of this mmWave filter is compacted to 0.3x0.2 mm², as illustrated in Fig. 7(a). Fig. 7(b) showcases the measured insertion and return loss of the filter, which demonstrates impressive performance with a center frequency of approximately 21 GHz, a fractional 3-dB bandwidth near 11%, and a minimal insertion loss in the passband around 3.2 dB, suggesting that ScAlN-based FBAR is a promising option for 5G mmWave filters.

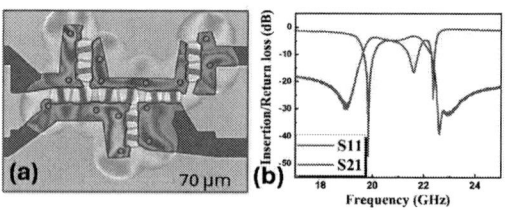

Fig. 7. (a) Micro photo of fabricated 20 GHz ladder filter. (b) Insertion and return loss of the fabricated filter.

Enabling switchable frequencies by ferroelectricity

Entering the 5G era, the expanding RF spectrum places a greater burden on the mobile phone's circuit board, necessitating over 50 filters per device, each dedicated to a specific frequency band. A promising solution to this challenge is to reduce the total number of filters by using reconfigurable resonators.

With increased Sc concentration to over 20%, ferroelectricity emerges due to lowered energy barrier of polarization switching [6]. This results in electrically-controlled polarization direction of ScAlN thin films, which can be utilized to design frequency-switchable FBAR resonators and filters, although controlling as-deposited polarity of sputtered film ScAlN remains challenging. The working mechanism of a bilayer $Sc_{0.3}Al_{0.7}N/AlN$ switchable FBAR is shown in Fig.8. Leveraging the gap of coercive field between AlN (~7.5 MV/cm) and $Sc_{0.3}Al_{0.7}N$ (~3.5 MV/cm), the polarization of ScAlN layer can be switched without in-between electrode layer.

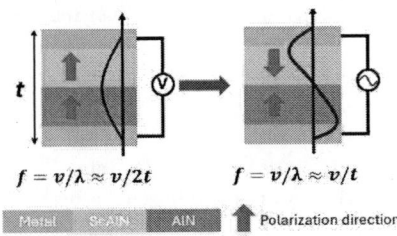

Fig. 8. Mechanism of bilayer frequency-switchable FBAR resonator.

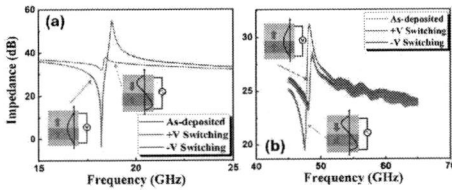

Fig. 9. Impedance spectrums of the as-deposited FBAR, after positive and negative DC pulses. (a) Low-frequency range; (b) High-frequency range.

The functionality is examined on fabricated bilayer FBAR devices, and is shown in Fig. 9. For the as-deposited device, both layers are N-polar and the f_s is at 18.8 GHz with a Q-factor of around 550 and a k^2_{eff} reduced to ~7.6%. After the $Sc_{0.3}Al_{0.7}N$ layer is switched to Al-polar by a negative pulse, the observed mode is weakened significantly to a k^2_{eff} of only 2%. Then applying a positive pulse can restore the resonance back to original state. Regarding the high frequency spectrum in Fig. 9 (b), switching the $Sc_{0.3}Al_{0.7}N$ layer from N-polar to Al-polar results in an increase of k^2_{eff} and R_p/R_s ratio of the resonance at ~ 48 GHz, reversible with a positive pulse.

Conclusion

This paper provides a concise overview of our research into using ScAlN thin films to tackle the existing issues in 5G filtering applications. Experimental results indicate that ScAlN-based FBAR technology holds great potential for next-generation 5G filters, offering high bandwidth, capability of reaching mmWave frequencies and electrically switchable frequencies.

Acknowledgments

This research is supported by A*STAR under its Industry Alignment Fund – Industry Collaboration Projects (IAF-ICP), with Grant No. I2301E0027, Project Title: Piezo Specialty Lab-in-Fab 2.0 (LiF 2.0) and the Science and Engineering Research Council of A*STAR Singapore, under Grant No. A20G9b0135.

References

[1] Akiyama, M., Kamohara, T., et al., "Enhancement of piezoelectric response in scandium aluminum nitride alloy thin films prepared by dual reactive cosputtering", Advanced Materials, 21(5), 593–596.

[2] Qian, Y. et al, "Acoustic filters for RF communication", IEEE Nanotechnol. Mag. 18, 23–33 (2024).

[3] Bogner, A., Timme, H. J., Bauder, R., et al., "Impact of High Sc Content on Crystal Morphology and RF Performance of Sputtered Al1-xScxN SMR BAW". IEEE International Ultrasonics Symposium, IUS, 2019-Octob, 706–709.

[4] Aigner, R., Fattinger, G., Schaefer, M., Karnati, K., Rothemund, R., & Dumont, F. "BAW Filters for 5G Bands", International Electron Devices Meeting (IEDM), 2018-Decem, 14.5.1-14.5.4.

[5] Y. Qian, X. Wang, Y. Zhang, C. Liu and Y. Zhu, "Demonstration of 8-Inch Thin-Film $Sc_{0.3}Al_{0.7}N$ BAW Resonator with High Electromechanical Coupling Coefficient," 2024 IEEE MTT-S International Conference on Microwave Acoustics & Mechanics (IC-MAM), Chengdu, China, 2024, pp. 5-8

[6] L. Chen et al., "Scandium-doped aluminum nitride for acoustic wave resonators, filters, and ferroelectric memory applications," ACS Appl. Electron. Mater., vol. 5, pp. 612–622, 2023.

[7] Wang, X. et al. "Thin Sc0.2Al0.8N Film Based 15 GHz Wideband Filter: Towards mmWave Acoustic Filters", International Electron Devices Meeting (IEDM), 2023

Dual Charge-Trapping Nanowire Flash Transistor for Low-Voltage High-Precision Biomimetic Synaptic Device with 1024 States

Qing Wang[†,1], Haixia Li[†,1], Ran Bi[1], Hongxu Liao[1], Baotong Zhang[1], Ru Huang[1,2], and Ming Li[1,2]*

[1]School of Integrated Circuits, Peking University, Beijing, 100871, China
[2]Beijing Advanced Innovation Center for Integrated Circuits, E-mail: liming.ime@pku.edu.cn

Abstract

In this work, a novel longitudinal coupled flash structure using Si_3N_4/HfO_2 as dual charge trapping layers (Dual-CTFs) is proposed to expand the conventional flash devices into synapse application. The fabricated Si nanowire Dual-CTFs features superior memory and synaptic characteristics, exhibiting remarkable long-term plasticity with low pulse voltage (-2.3 V/3.7 V). Through the field-assisted charge hopping, effectively 1024 continuous conduction states are realized in a single device with good linearity and low power (~ 0.5 fJ/spike).

Keywords: Nanowire, Dual-Charge-Trapping, Long-Term Plasticity (LTP), Synaptic Transistor.

Introduction

Neuromorphic computing has attracted much attention as a promising technique for next-generation computing architectures [1]. The development of high-performance low-power biomimetic synaptic devices to simulate biological plasticity is crucial for the construction of efficient neuromorphic computing systems. The Si flash memory is a strong candidate for biomimetic synaptic device hardware because of its high-density integration and multi-value storage characteristics [2]. Especially, charge trapping device exhibits superior linearity, symmetry and reliability, which makes it widely used in synapse applications [3]. At present, the main problems of Si flash devices for neuromorphic computing are high operation voltage (~ 10 V) and high power consumption, which hinder the further development of its neuromorphic applications.

In this work, we propose a nanowire Dual-CTF memory with $p\text{-}Si/SiO_2/Si_3N_4/HfO_2/Al_2O_3/TiN$ stack as synapses device to realize low power linear and multi-level weight updates. Noticeable linearity and training accuracy of Dual-CTF as a biomimetic synaptic device can be obtained. Compared to previously reported CTF with dual capture layers (Si_3N_4/HfO_2) but without tunneling oxide ($V_{PGM} = 4.8$ V for [4] and $V_{PGM} = 4.5$ V for [5]), this work achieves superior LTP/LTD characteristics at lower operating voltages (- 2.3 V/3.7 V). It also exemplifies the ability of linear multi-level weight modulation and high accuracy in the MNIST dataset. This architecture shows a promising potential in low-power neuromorphic application.

Experiment Method

Fig. 1(a) shows a schematic of the Dual-CTF. The devices were fabricated via conventional silicon-based processes as exhibited in Fig. 1(b). Firstly, the p-type silicon on insulator (SOI) substrate was thermally oxidized and thinned down to 50 nm. After the nanowire pattern formation, the whole device was doped by As^+ implantation to form junctionless structure. The channel and S/D were doped and activated at 1×10^{18} cm^{-3} and 1×10^{20} cm^{-3}, respectively. The $SiO_2/Si_3N_4/HfO_2/Al_2O_3$ stack (2nm/3nm/5nm/4nm) was then deposited to serve as the tunneling layer, first charge trap layer, second charge trap layer and the blocking layer, respectively. The total EOT was calculated to be 6.2 nm. After that, metal gate was formed by 61 nm TiN deposition and patterning. Fig. 2(a) shows a cross-sectional transmission electron microscopy (TEM) image and the corresponding energy dispersive spectrometer (EDS) mapping image along a direction perpendicular to the channel is depicted in Fig. 2(b).

Results and Discussion

Compared with the traditional CTF using single Si_3N_4 trapping layer, the programming process of Dual-CTF contains a transfer behavior of injected charges between the separated traps of the Si_3N_4/HfO_2 double layers. Fig. 3 exhibits the cross-section schematic and energy band diagram including charges hopping process and current components perpendicular to the channel of Dual-CTF. Under continuous low-voltage programming(PGM) pulses, the charges can be tunneled and migrated longitudinally within Si_3N_4/HfO_2 double charge-trapping layers by Trap-Assisted Tunneling (TAT) and Poole-Frenkel (PF) mechanism. The threshold voltage is varied accordingly to realize multi-value design and weight modulation. The expressions of the threshold voltage (V_{th}) in the initial and programmed states as follows:

Initial state：

$$V_{th0} = \left(\phi_{ms} - \frac{Q_f}{C_{tot}}\right) + 2\phi_B + \frac{\sqrt{4q\varepsilon_s N_A \phi_B}}{C_{tot}} \tag{1}$$

PGM state 1 (charges injected to Si_3N_4):

$$V_{th1} = \left(\phi_{ms} - \frac{Q_f}{C_{tot1}} + \frac{Q_1}{C_{tot1}}\right) + 2\phi_B + \frac{\sqrt{4q\varepsilon_s N_A \phi_B}}{C_{tot1}} \tag{2}$$

PGM state 2 (charges injected from Si_3N_4 to HfO_2):

$$V_{th2} = \left(\phi_{ms} - \frac{Q_f}{C_{tot2}} + \frac{Q_1}{C_{tot2}} + \frac{Q_2}{C_{tot2}}\right) + 2\phi_B + \frac{\sqrt{4q\varepsilon_s N_A \phi_B}}{C_{tot2}} \tag{3}$$

To investigate the low-voltage operation of the Dual-CTF, programming (PGM) and erasing (ERS) operations using a single pulse were performed. The P/E speed of the device is characterized by extracting ΔV_{th} under excitation. Fig. 4 illustrates that low-voltage amplitude (5.7 V and - 6.3 V) with short-pulse width (~ ns) operations of the Dual-CTF can be realized. Fig. 5 shows that the programming efficiency is significantly improved when the programming voltage is increased from 5 V to 5.7 V, which suggests that a voltage of 5.7 V for ns-scale stimulation can be used for fast writing, whereas voltages lower than 5 V are suitable for achieving graded linear synaptic weight modulation. The tri-gate structure and thin tunneling oxide layer (2 nm) can further provide the capability to reduce the operating voltage. Meanwhile, the introduction of trap-dense HfO_2 greatly increases the storage window and actually enables the device to achieve a slowly varying linear synaptic weight modulation.

In the following section, we demonstrate a synaptic strength-enhancing effect in the proposed gate stack. We characterize and compare the excitatory post-synaptic currents (EPSC) with different voltage amplitudes (V_{PGM} = 3.7 V to 7.7 V for Dual-CTF; V_{PGM} = 7 V to 11 V for Single-CTF) at pulse width W_{PGM} = 100 ns and 100 μs respectively, as shown in Fig. 6(a) and (b). It's shown that when the pulse amplitude is 7 V~11 V for Single-CTF, the postsynaptic current changes smaller than 100 nA and presents a noticeable decreasing trend even at 100 μs pulse width, which are much smaller than that of Dual-CTF with shorter pulse width (100 ns) and lower pulse voltages (3.7 V/- 3.3 V). With ultra-low voltages at wider pulse widths of 10 ms and 5 ms, it demonstrates typical LTP/LTD characteristics as in Fig. 7. The additional introduced HfO_2 trapping layer has sufficient trap density and deeper trap energy level to strongly prison the charges, thus playing a critical role in charge retention [6], [7].

To further investigate the application of the Dual-CTF in simulating learning and memory, the excitatory/inhibitory postsynaptic currents under continuous stimuli with low voltage long pulse (~ ms) are used to simulate the learning and re-learning processes. The Dual-CTF exhibits sustained LTP/LTD properties during the process transition as shown in Fig. 8.

Next, an investigation was conducted into the linearity and training accuracy of Dual-CTF as a biomimetic synaptic device in neural networks. Good linearity and symmetry with successive pulse scheme can be obtained. In Fig. 9 and Fig. 10, high linearity and symmetry of long-term potentiation and depression are demonstrated with (α_{pot}, α_{dep}) = (0.69, -1.09), which results a high accuracy of 97.48% after 1500 training iterations in the MNIST dataset (Fig. 11). Thanks to the continuous charge migration mechanism inside the gate stack, more likely field-assisted charge hopping, a 1024 continuous conductance states can be obtained linearly with

the pulse number. The calculated power consumption is ~ 0.5 fJ/spike, which is much lower than the level of normal biological synapse devices (~ 10 fJ/spike) [8].

In Fig. 12, the retention measurements were carried out for the proposed stack at 85 °C after applying a PGM pulse of 5 V for 5 ms. The threshold voltage of Dual-CTF can maintain higher stability than single CTF, and its retention lifetime can be extrapolated to 10 years. Fig. 13 shows the endurance test results of the Dual-CTF. The results show that Dual-CTF maintains a stable window after 100 k P/E cycles.

Conclusion

In this work, we have proposed to use Si_3N_4/HfO_2 as Dual-CTL in Si CTFs as low-voltage high-precision synaptic devices. The Dual-CTL significantly enlarges the memory window and improves P/E speed because it has deeper trap distribution and higher trap density in the second trapping layer HfO_2. It can also work at low pulse voltages, short pulse widths, and excellent weight retention characteristics. 1024 continuous states are achieved with high linearity. We believe our proposed Dual-CTF is the one with great potential for the synaptic device in low-power high-precision neuromorphic application.

Acknowledgments

Q.W. and H.L. contributed equally to this work. This work was partly supported by the NSFC project 61927901.

References

[1] Y. LeCun, Y. Bengio, G. Hinton, "Deep learning", Nature, 521, pp. 436-444 (2015).

[2] C.-Y. Lu, "Future prospects of NAND flash memory technology—the evolution from floating gate to charge trapping to 3D stacking", Journal of nanoscience and nanotechnology, 12, pp. 7604-7618 (2012).

[3] C. Wang, H. Liu, L. Chen, H. Zhu, L. Ji, Q.-Q. Sun, and D. W. Zhang, "Ultralow-Power Synaptic Transistor Based on Wafer-Scale MoS2 Thin Film for Neuromorphic Application", IEEE Electron Device Lett., 42, pp. 1555-1558 (2021).

[4] M. S. Lee, J. W. Lee, C. H. Kim, B. G. Park, and J. H. Lee, "Implementation of Short-Term Plasticity and Long-Term Potentiation in a Synapse Using Si-Based Type of Charge-Trap Memory", IEEE Trans. Electron Devices, 62, pp. 569-573 (2015).

[5] Y. T. Seo, M. S. Lee, C. H. Kim, S. Y. Woo, J. H. Bae, B. G. Park, and J. H. Lee, "Si-Based FET-Type Synaptic Device With Short-Term and Long-Term Plasticity Using High- κ Gate-Stack", IEEE Trans. Electron Devices, 66, pp. 917-923 (2019).

[6] V. Gritsenko, S. Nekrashevich, V. Vasilev, and A. Shaposhnikov, "Electronic structure of memory traps in silicon nitride", Microelectronic Eng., 86, pp. 1866-1869 (2009).

[7] Y. Lu, "Electrical characterization of deep trap properties in high-k thin-film HfO_2", Chin. Physics Lett., 27, p. 077102 (2010).

[8] T.-Y. Wang, J.-L. Meng, Z.-Y. He, L. Chen, H. Zhu, Q.-Q. Sun, S.-J. Ding, P. Zhou, and D. W. Zhang, "Ultralow Power Wearable Heterosynapse with Photoelectric Synergistic Modulation", Adv. Sci., 7, p. 1903480 (2020).

979-8-3315-0417-5/25 $31.00 © 2025 IEEE

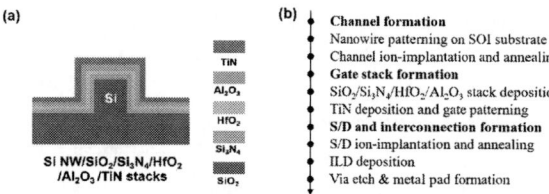 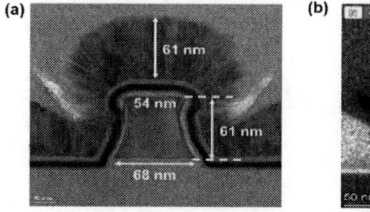 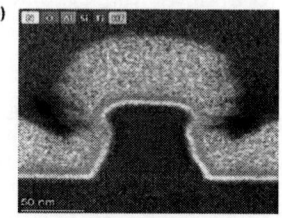

Fig. 1. (a) Schematic of the Dual-CTF. (b) Simplified fabrication flow for the Dual-CTF.

Fig. 2. (a) Cross-sectional TEM image and (b) EDS image along a direction perpendicular to the channel.

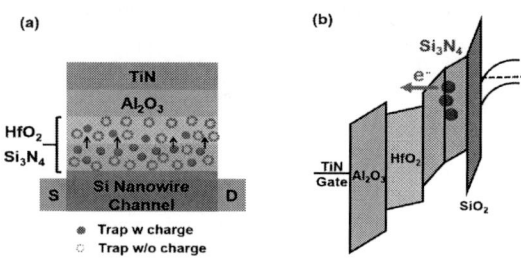 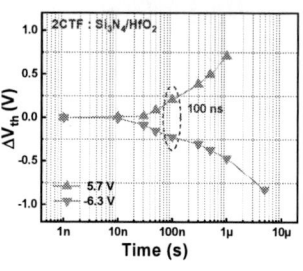 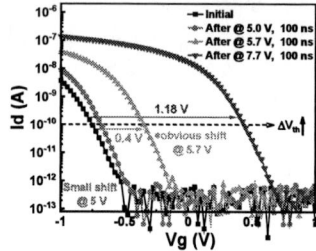

Fig. 3. (a) Cross-sectional schematic and (b) energy band diagram of Si (bottom) /SiO$_2$/Si$_3$N$_4$/HfO$_2$/Al$_2$O$_3$ (top) stack of Dual-CTF.

Fig. 4. P/E speed of Dual-CTF (V_{PGM} = 5.7 V, V_{ERS} = - 6.3 V.

Fig. 5. The drift of the I_d-V_g curve after stimulating the Dual-CTF.

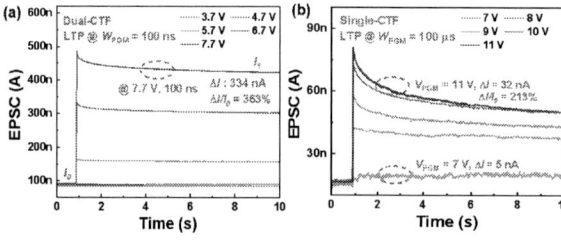 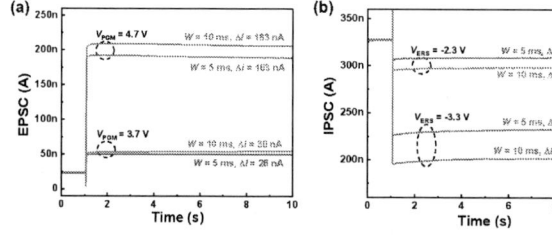

Fig. 6. LTP characteristics of (a) Dual-CTF (V_{PGM} = 3.7 V ~ 7.7 V, W_{PGM} = 100 ns) and (b) Single-CTF (V_{PGM} = 7 V ~ 11 V, W_{PGM} = 100 ms).

Fig. 7. (a) LTP characteristics (V_{PGM} = 3.7 V/ 4.7 V, W_{PGM} = 5 ms/10 ms) and (b) LTD characteristics (V_{PGM} = - 3.3 V/- 2.3 V, W_{PGM} = 5 ms/10 ms) of Dual-CTF.

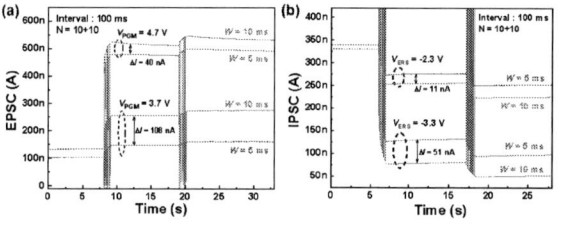 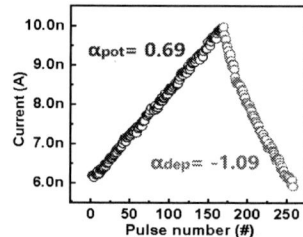 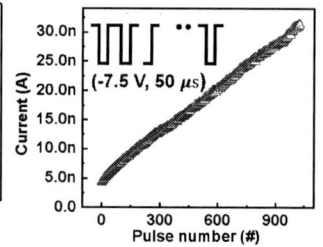

Fig. 8. LTP and LTD characteristics of Dual-CTF under continuous low voltage long pulses (~ ms) stimulation (a) V_{PGM} = 3.7 V/4.7 V, W_{PGM} = 5 ms/10 ms; (b) V_{ERS} = - 2.3 V/- 3.3 V, W_{ERS} = 5 ms/10 ms.

Fig. 9. Conductance update curves (V_{PGM} = - 7.5 V, W_{PGM} = 50 μs; V_{ERS} = 5 V, W_{ERS} = 50 μs).

Fig. 10. 1024 continuous conduction states of Dual-CTF (V_{PGM} = - 7.5 V, W_{PGM} = 50 μs).

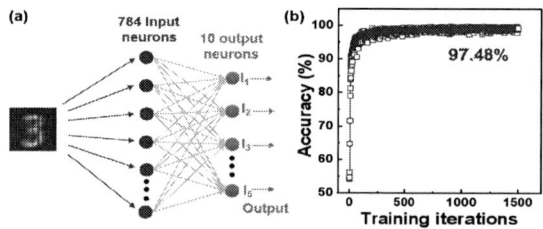 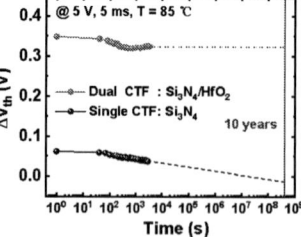 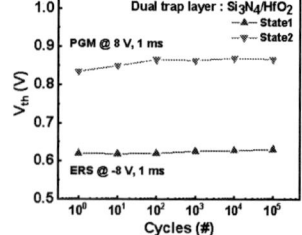

Fig. 11. (a) Schematic of the single-layer perceptron network; (b) Accuracy of network training used Dual-CTF with the WTA algorithm, which can reach 97.48% after 1500 training iterations.

Fig. 12. Comparison of retention characteristics of Single-CTF and Dual-CTF at 85 °C.

Fig. 13. Endurance test of Dual-CTF.

979-8-3315-0417-5/25 $31.00 © 2025 IEEE

Consideration of VFET for Ultimate Logic Scaling: A Design Perspective

Yimeng Wang, Yanbang Chu, Ziqiao Xu, Yu Liu, Rui Guo, Jiacheng Sun, Wanyue Peng, Haoran Lu,
Ming Li, Runsheng Wang, Heng Wu*, Ru Huang

Peking University, Beijing, China, *Email: hengwu@pku.edu.cn

Abstract

In this work, the design of Vertical FET (VFET) for logic applications is comprehensively studied by the DTCO (Design Technology Co-Optimization) methodology. The nanosheet channel placement orientation and intrinsic device asymmetry are thoroughly investigated in the perspective of circuit design for the first time. For the nanosheet channel placement orientation, the Horizontal (along x-axis) Nanosheet has great layout flexibility, while the Vertical (along y-axis) Nanosheet design outperforms its counterpart by ~7% performance gain @ iso-power, with ~33% reduction of cell area. For the device asymmetry, significant Ieff improvement (~20%) and Cgd reduction (~30%) in the Forward mode were identified, leading to ~68% better performance @ iso-power in circuit simulation. A novel mixed design of Forward and Reverse mode is proposed for complex stdcell for the vertical nanosheets placement. This study brings new insights into the implementation of VFETs.
Keywords: VFET, layout, DTCO, circuit level PPA, nanosheet channel placement, device asymmetry

Introduction

In the post-Moore era, as the device scaling slows down, some new device structures have been proposed beyond FinFET, such as L-GAA [1] (Lateral Gate-All-Around), VFET [2] (Vertical Field Effect Transistor), CFET [3] (Complementary FET), and FFET [4] (Flip FET). L-GAA enhances the gate control capability of lateral-transport devices for larger Weff (effective gate width) in smaller footprint, but the area gain is limited for the CGP (Contacted Gate Pitch) [2]. CFET shows process complexity due to high-AR (aspect ratio) requirements [3],[4]. VFET, proposed with vertical channel direction, shows significant area scaling by stacking the Source/Gate/Drain vertically. The Lg (gate length), top/bottom spacer thickness and contact size are no longer limited by CGP in the lateral devices [2],[5], delivering significant flexibility in device performance optimization. In addition, VFET can greatly relax the AR requirement for processes. Overall, VFET is very a valuable research topic for advanced technology nodes.

In the view of layout coordinate system, the VFET nanosheets can be placed along the x-axis or y-axis, which are respectively defined as Horizontal Nanosheet (HNS) design and Vertical Nanosheet (VNS) design (as in Fig. 3). The design perspective for these two placement strategies is critical for the VFET implementation.

Fig. 1(a) shows the three-dimensional structure of the VFET, with the current conducting along the white dash line [6]. Fig. 1(b) shows two cross-sections of the VFET [6], where the BSD and TSD are the abbreviations for the bottom and top source/drain, the CTB, CTT, and CTG are the metal contacts of BSD, TSD, and Gate respectively. Note that the CTB and CTG are placed on either side of the CTT. The B-Spacer and T-Spacer are the dielectrics deposited between metal gate and source/drain epitaxy. Because the top and bottom epitaxies have different morphologies and positions relative to the channel, the VFET is an asymmetric device. This asymmetry results in inconsistent electrical behavior when exchanging the source and drain. Thus, there are two operating modes of VFET, the Forward (FWD) mode (top electrode as drain) and the Reverse (REV) mode (bottom electrode as drain). The influence of this asymmetry on layout design and PPA (Power-Performance-Area) is also investigated in this work.

This paper will first evaluate various VFET layout design schemes in view of layout flexibility. Then the PPA comparison of each layout design scheme will be given. What's more, the complex stdcell (like AOI21) layout design strategy of VFET will also be discussed. Finally, conclusions about VFET layout design will be presented.

Design and Evaluation Methodology

The DTCO methodology used in this work is shown in Fig. 2. Two SPICE models are extracted for the FWD/REV modes through transistor modelling, and two parasitic netlists are extracted for the different nanosheet channel placement orientations through the parasitic extraction (PEX) flow. Finally, the circuit simulation of the 15-stage Ring Osculator with Fanout3 and typical BEOL loads [7] is used to obtain the power-performance for three schemes, including HNS FWD scheme, HNS REV scheme and VNS FWD scheme. However, the VNS REV scheme is excluded, and the reason will be given later.

The layouts discussed in this paper are based on key dimensional assumptions as shown in Table. 1, and are all based on 2-Nanosheet and 6-Track design. Both Horizontal and Vertical Nanosheet designs follow these rules.

The nanosheet channel placement orientation is critical in the VFET layout design. As shown in Fig. 3(a), the CTB and CTT in Horizontal Nanosheet design has the same capability of connecting to the power rail. However, in the Vertical Nanosheet design, only the CTB is accessible for the power rail on M0 level because it can be formed just underneath the power rail, while the CTT must go through

979-8-3315-0417-5/25 $31.00 © 2025 IEEE

M1 metal to access the power rail with a big resistance penalty. As a result, Horizontal Nanosheet design is more flexible in the layout design due to its better connectivity to power rail.

Layout Considerations and Implementation

Considering the asymmetry of the VFET, the two operating modes should first be investigated. Fig. 4(a) gives the device Id-Vg curves based on the TCAD simulation, showing clear difference of drive current between the two modes. Furthermore, Fig. 4(b) compares the Ieff (effective current) and the Cgd0 (gate to drain capacitance @ Vg=0) between the Forward and Reverse modes. Resulted from the asymmetry of the top and bottom junctions/SDs/contacts, the FWD mode is much stronger than REV mode as indicated by the larger effective current (~20%) and lower Cgd (~30%). The fittings of extracted SPICE models to TCAD data are shown in Fig. 5, indicating good accuracy of the fitting based on VFETs' Id-Vg and Cgg-Vg (Cgg is the gate capacitance).

Inverter layouts based on Horizontal and Vertical Nanosheet design in the Forward or Reverse mode are displayed in Figs. 6(a-e). The HNS FWD, HNS REV and VNS FWD layout can be easily designed for the inverters, as shown in Figs. 6(a-c).

However, the Vertical Nanosheet design with REV mode shows clear difficulty in the stdcell design, as shown in Figs. 6(d-e). The blockage zone (distance needed to separate the electrodes to gate metals) from the nanosheet channel restricts the CTG and CTB from being placed directly underneath the M0 tracks. Therefore, both CTBs and CTGs fail to directly access the M0s through vias without using vertical-horizontal-vertical (VHV) routings [8]. Note that enlarging the M0 pitch to have M0s directly on top of CTBs/CTGs, with clear area penalties, is still a poor solution for the reasons below. Firstly, a long M1 metal connection between two CTBs (Fig. 6(d)) for the OUTPUT or two CTGs (Fig. 6(e)) for the INPUT are required, with huge RC degradations. Secondly, the CTTs also need extra M1 metals to reach the power rails, degrading the power connection resistance. Thirdly, in the ultra-dense layout design of VNS, only two M1s are allowed in the x-axis direction. However, two M1s are already occupied, leaving no room to connect the INPUT CTG (Fig. 6(d)) or OUTPUT CTB (Fig. 6(e)).

The cell areas of HNS and VNS layouts are compared in Fig. 6, indicating a clear area reduction of ~33% by using VNS design. This is due to the smaller cell width of VNS layout. The power-performance curves of the three schemes are given in Fig. 7, further revealing the performance of the FWD mode is stronger than that of the REV mode, with ~68% higher frequency @ iso-power, and the performance of the VNS FWD design is the best among all the three schemes, with ~7% higher frequency @ iso-power compared to the HNS FWD design. These can be attributed to the larger Ieff

and lower Cgd of the FWD mode (as in Fig. 4(a)) and less BEOL loads from a smaller layout area with VNS design.

Other Crucial Aspects

The design of complex stdcell layout should also be considered and investigated for implementation of various circuit functions. Therefore, AOI21 (AND-OR-Inverter) is evaluated as an example, as a widely used complex logic gate. The layouts of the AOI21 with HNS design circuit are given in Figs. 8(a-b), showing that both FWD mode and REV mode layouts can be flexibly designed. However, VNS design features great routing difficulties due to the poor connectivity of CTBs to the signal M0s shown in Figs 6(d). Thus, a novel mixed mode scheme was proposed, with both FWD and REV modes devices used in the layout. In this scheme, The BSDs of VFETs are not only used to connect the power rails but also used as connections between two adjacent transistors. However, inevitably, this mixed mode faces some challenges of the source-drain asymmetry in VFETs, which may degrade the performance and induce more variation in stdcell due to the worse performance of the REV mode. This issue can be mitigated by optimizing the process in the top/bottom spacer module, or by growing symmetric SD epitaxies.

Conclusion

Four layout design schemes for VFET have been detailly investigated for the first time, considering the nanosheet channel placement orientation and the device asymmetry. The VFET in FWD mode has ~20% larger Ieff and ~30% lower Cgd, leading to ~68% better PP than those of the REV mode. It's found that the HNS design offers great flexibility but wastes areas. The VNS design in FWD mode can greatly reduce the layout area by ~33% with the best PP, outperforming the HNS design in FWD mode by ~7% frequency gain @ iso-power, with great scaling potential. However, the VNS design sustains some design constraints, causing that the CTB is preferred to be connected to the power rail. A novel mixed mode design was proposed on VNS design with both FWD and REV devices, to keep the area reduction benefits. However, it may still suffer from the degradation of PP and more variation in stdcell. More efforts are needed to improve the source-drain symmetry of VFETs for better performances with aggressive area scaling.

Acknowledgement

This work is supported by the National Key R&D Program of China under Grant 2023YFB4402200; and in part by the 111 Project under Grant 8201702520.

References

[1] N. Loubet et al., *VLSI 2017*. [2] H. Jagannathan et al., *IEDM, 2021*. [3] S. Subramanian et al., *VLSI 2020*. [4] H. Lu et al., *VLSI 2024*. [5] B. Senapati et al., *SPIE 2023*. [6] G Tsutsui et al., *IEDM 2022*. [7] S. Yang et al., *IEDM 2021*. [8] B. Chehab et al., *IITC 2021*.

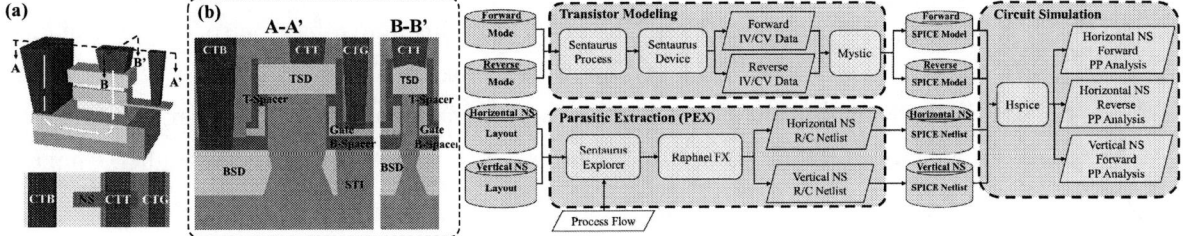

Fig. 1: (a) 3D schematic of the VFET device. The top-down view shows the positions of the vertical nanosheet channel and three contacts. (b) Cross sections along A-A' and B-B' of the VFET structure.

Fig. 2: The DTCO methodology with three main procedures in blue blocks for evaluating Power-Performance (PP). Four layout design schemes and their SPICE models or netlists are highlighted in red. The software used is marked in yellow.

Table. 1: Design rule assumptions.

Parameter	Value
Track Number	6
M2 CD	20nm
M2 Pitch	35nm
M1 CD	24nm
M1 Pitch	42nm
M0 CD	18nm
M0 Pitch	28nm
Nanosheet (NS) CD	6nm
NS Height	40nm
NS Length	30nm
NS Pitch	42nm
Lg	20nm
CTB CD	24nm
CTT CD	20nm
CTG CD	14nm

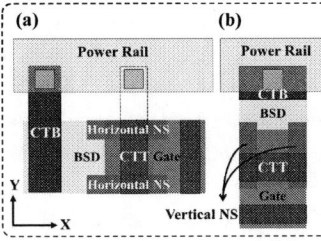

Fig. 3: The capability of connecting to the power rail is different in the (a) Horizontal design and (b) Vertical design. The CTT needs an M1 metal (not shown in the figure) to reach the power rail in the VNS design.

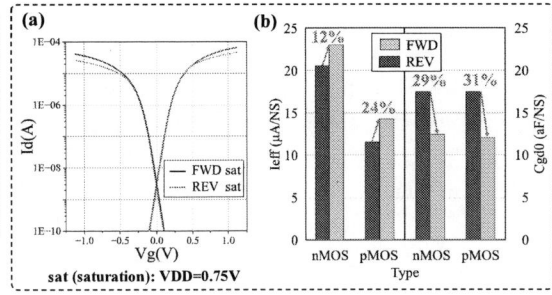

sat (saturation): VDD=0.75V

Fig. 4: (a) Id-Vg curves of the Forward and Reverse mode from TCAD simulation. (b) The Forward mode has ~20% larger effective current and ~30% lower Cgd than the Reverse mode, resulted from the asymmetry of the VFET structure.

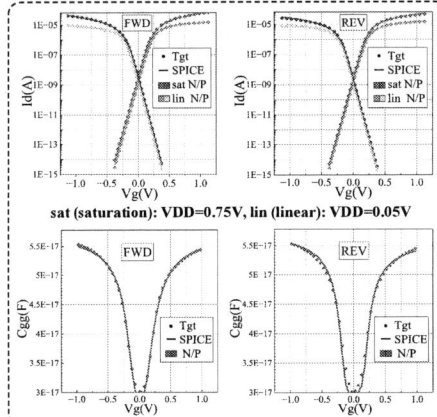

sat (saturation): VDD=0.75V, lin (linear): VDD=0.05V

Fig. 5: The fit of the SPICE model about Id-Vg curves at VDD=0.05V and VDD=0.75V, as well as Cgg-Vg curves. The dots represent the target data obtained by TCAD, and the lines represent the fitting of the SPICE model.

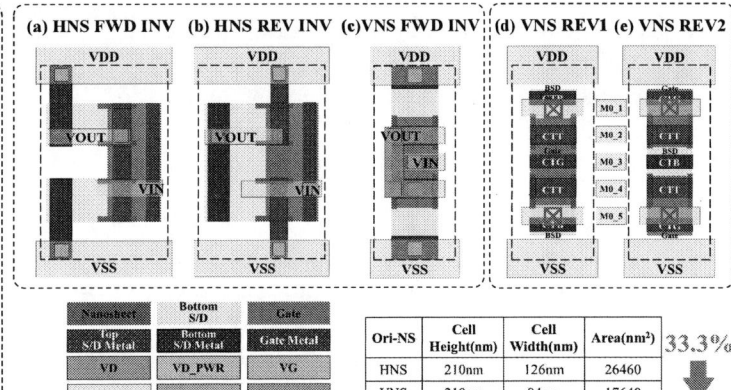

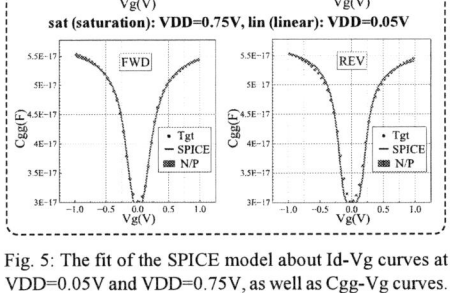

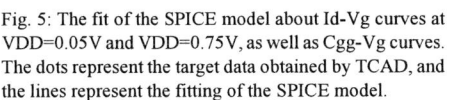

Ori-NS	Cell Height(nm)	Cell Width(nm)	Area(nm²)
HNS	210nm	126nm	26460
VNS	210nm	84nm	17640

33.3%

Fig. 6: The inventor layout comparison between (a) Horizontal Nanosheet design in Forward mode, (b) HNS design in Reverse mode and (c) Vertical Nanosheet design in FWD mode. The area calculation shows the advantage of VNS design in terms of area. (d) and (e) are two possible VNS designs in REV mode but the CTB and CTG cannot connect to the M0_1 and M0_5 tracks.

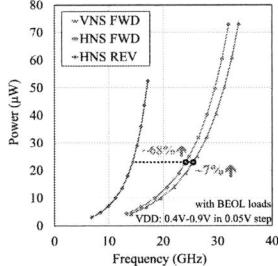

Fig. 7: Power-frequency plots of the HNS design in FWD mode, the HNS design in REV mode and the VNS design in FWD mode.

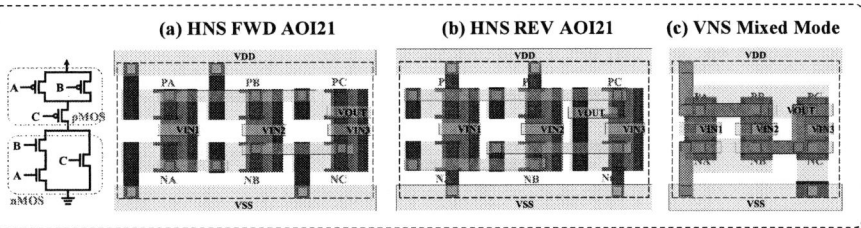

Fig. 8: AND-OR-Inverter21 (AOI21) based on (a) Horizontal Nanosheet design in Forward mode, (b) Horizontal Nanosheet design in Reverse mode, and (c) Vertical Nanosheet design in mixed mode. The circuit diagram is placed on the left. The HNS layout has two designs with larger cell width, while the VNS layout only has the mixed mode due to the poor connectivity of CTB to the signal M0s, where the BSDs connect to the power rails or act as bottom connection of the two adjacent transistors.

979-8-3315-0417-5/25 $31.00 © 2025 IEEE   749

Robot Collision Detection Acceleration with In-Memory Search based on the Monolithic 3D Integration of 2D Transistors and Vertical RRAMs

Yijian Zhang[1], Jiahao Sun[1], Maosong Xie[1], Yueyang Jia[1], Rui Yang[1,2,3,*]

[1]University of Michigan – Shanghai Jiao Tong University Joint Institute, Shanghai Jiao Tong University (SJTU), Shanghai, China

[2]School of Electronic Information and Electrical Engineering, SJTU, Shanghai, China

[3]State Key Laboratory of Radio Frequency Heterogeneous Integration, SJTU, Shanghai, China

[*]Email: rui.yang@sjtu.edu.cn

Abstract

Collision detection consumes over 90% of the total computation time in robot motion planning, making its optimization crucial. Previous hardware accelerators have struggled to meet the time requirement (<1 ms) in dynamic environment while supporting the data storage for large roadmaps. In this work, we design and fabricate the monolithic 3D integrated structure based on 2D transistors and vertical resistive random-access memories (VRRAMs) to enhance memory density, and accelerate collision detection based on in-memory search.

Keywords: Collison detection, in-memory search, RRAM, TCAM, monolithic 3D integration

Introduction

Motion planning is one of the key technologies in robot motion navigation, and includes three major components: roadmap construction, collision detection and graph search (Fig. 1a). Among them, collision detection consumes more than 90% of the total robot motion planning time, which is the key to ensuring a safe path for robots [1]. However, due to the space complexity of high-degree-of-freedom (DOF) robotic arms, GPU-based and CPU-based collision detection can take a long time of more than hundreds of milliseconds [2], which cannot meet the requirements for real-time collision detection (<1 ms) in a dynamic environment. Various accelerator designs based on application-specific integrated circuits (ASICs) and field-programmable gate arrays (FPGAs) have been proposed, which can reduce the computation time to below a millisecond [3,4], but can only store small roadmaps due to the limited storage capacity and memory bandwidth. Processing-in-memory (PIM) accelerator based on dynamic random-access memory (DRAM) has also been demonstrated for collision detection [5], but it faces the problem of large area overhead for supporting the PIM logic and the high energy consumption of DRAM arrays. Furthermore, the data storage is not efficient due to the large amount of data generated from robot roadmaps. Therefore, it is urgent to improve the memory density, speed, and energy efficiency for collision detection, by optimizing both memory devices and architectures.

Resistive random-access memory (RRAM) is a non-volatile memory with high potential for in-memory computing due to its high density and low power consumption [6]. Furthermore, 3D VRRAMs can further increase the memory density by vertical stacking, which make them highly promising towards large-scale on-chip memory and in-memory computing. From the architecture perspective, in-memory search offers the advantages of reduced size, fast search speed, and high energy efficiency for performing look-up table function [7]. Meanwhile, it can also perform high-parallelism comparison between stored data and query data, which is similar with the collision detection logic.

In this work, we design and fabricate the one-transistor-four-VRRAMs (1T–4R) structure based on monolithic 3D integration of 2D molybdenum disulfide (MoS$_2$) field-effect transistors (FETs) and HfO$_x$-based 3D VRRAMs for high memory density. We further implement the in-memory search operation and realize collision detection acceleration with increased memory capacity on chip (Fig. 1).

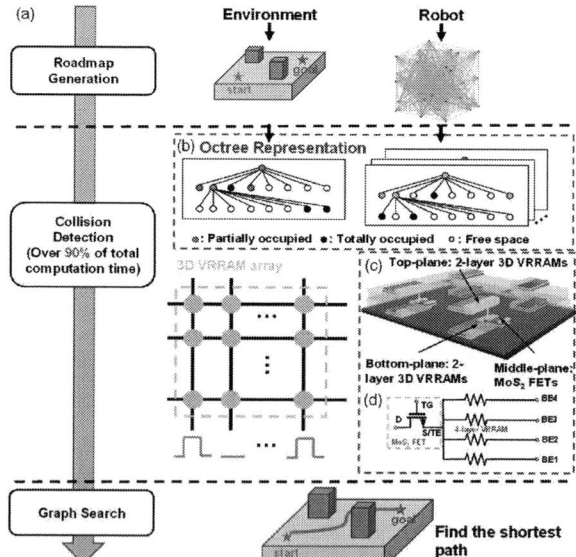

Fig. 1: (a) Illustration of motion planning procedure. Stage 1 is for recognizing the obstacle in the environment and generating the roadmap of the robot. In Stage 2, collision detection is performed to determine the collision-free regions, which is the focus of this work. In Stage 3, graph search is performed to find the shortest path for controlling the robot motion. (b) Octree representation to encode the data. (c) The monolithic 3D integrated 1T–4R structure and (d) Equivalent circuit of the 1T–4R structure.

979-8-3315-0417-5/25 $31.00 © 2025 IEEE

1T-4R Device Design, Fabrication and Measurement

To support the storage of large roadmaps, 3D VRRAMs are further stacked to enhance the memory density, thanks to the low fabrication temperature that enables monolithic 3D integration. Therefore, we fabricate the monolithic 3D integrated 1T-4R structure [8], as shown in Fig. 1c-1d. The VRRAMs with layer 1 (L1) and layer 2 (L2) are fabricated on the bottom plane, followed by an isolation oxide layer deposition. The VRRAM switching regions are formed by the titanium nitride (TiN)/hafnium oxide (HfO_x)/platinum (Pt) structure. Next, the 2D MoS_2 transistors with local back gates are patterned on the middle plane, to provide the access transistors for the VRRAMs during search operation and for providing the drive current when programming the VRRAMs. This is enabled by the low-temperature transfer of 2D MoS_2, which is different from Si transistors that require high temperature fabrication. To further increase the memory density within the same device area, after another isolation oxide layer deposition, we repeat the processes for fabricating the two-layer 3D VRRAMs, and obtain the top-plane layer-3 (L3) and layer-4 (L4) VRRAMs. As such, we obtain the monolithic 3D integrated structure with a 2D MoS_2 transistor driving four VRRAMs.

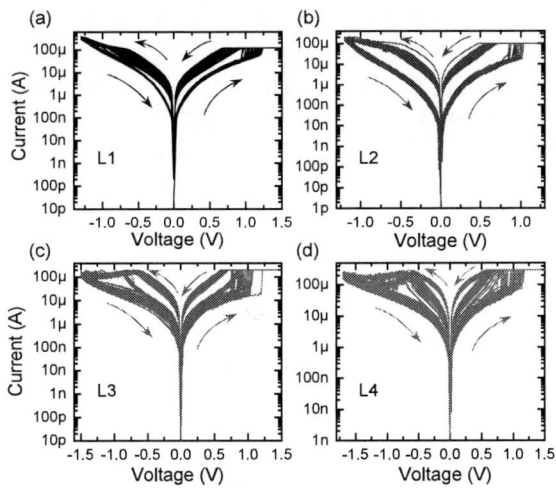

Fig. 2: The DC I–V sweeps of the four-layer VRRAMs, for (a) L1, (b) L2, (c) L3 and (d) L4 VRRAMs.

We then characterize the electrical properties for each of the four layers of 3D VRRAMs. Fig. 2 shows the DC I–V sweeps with 50 cycles of switching for each layer, which shows high uniformity among the four layers of VRRAMs. Fig. 3 illustrates the resistance distributions for the high resistance state (HRS) and low-resistance state (LRS) for the four VRRAM layers. It shows On/Off ratio up to 33 with small variations (σ/μ) of 25% and 18.94% for LRS and HRS, respectively. Furthermore, we measure the retention property at 85 °C for all four layers of VRRAMs. Our device shows a negligible drift both in HRS and LRS up to 10,000 s (Fig. 4). Such property can well prepare them for the massively parallel in-memory search operations for collision detection.

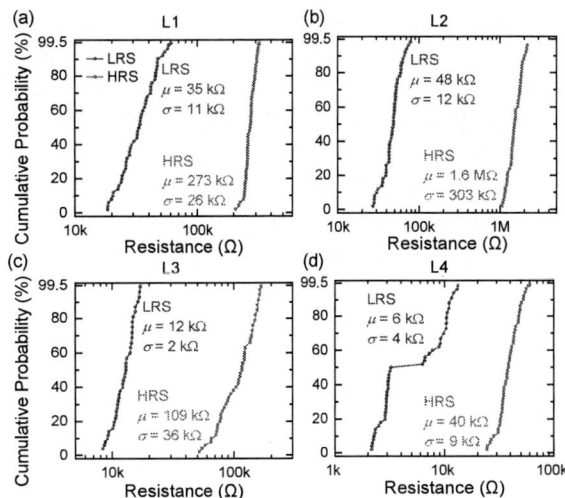

Fig. 3: Extracted LRS and HRS resistance distributions from the I–V sweeps in Fig. 2, for (a) L1, (b) L2, (c) L3 and (d) L4 VRRAMs.

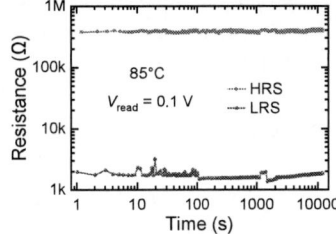

Fig. 4: Retention measurements at 85 °C for the VRRAM.

In-Memory-Search Collision Detection with VRRAMs

A. Overview

For robots, the environment and the roadmaps are all 3D models. In this work, we use octree to partition the 3D models into sub-cubes [9], as shown in Fig. 1b. Each finest-resolution sub-cube is named as an element. There are three possible states for an element: totally occupied, partially occupied, or free space. The collision detection process can be regarded as a comparison between elements of the environment and all possible routes in the roadmap (also called edges) within the same volume. Here, we use 2 bits to encode an element and our encoding scheme is listed as follows: 0/0 for free space elements, 1/0 for partially occupied elements and 1/1 for totally occupied elements.

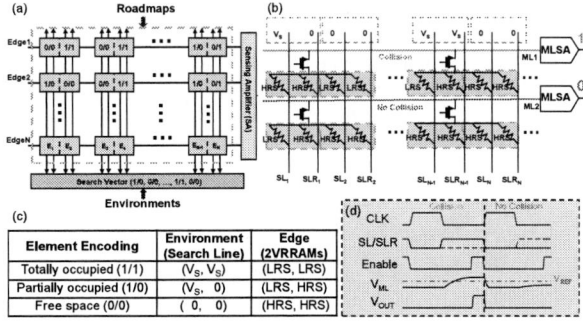

Fig. 5: (a) Structure of the proposed 3D VRRAM-based collision detection accelerator. (b) Circuit diagram of the 2-VRRAM cell architecture. (c) State definitions. (d) Timing of collision detection operation.

Here we propose to use the 3D VRRAM array to implement collision detection. The architecture of the collision detection accelerator is shown Fig. 5a. For each 1T–4R device, we use two of the VRRAMs as one cell to store the elements of one edge. Each row in the array stores one edge of the robot. The environment element inputs are fed into the search lines (SLs), and then in-memory search is performed to obtain the collision detection results.

B. In-Memory Collision Detection in 2-VRRAMs Cell

The detailed schematic of the 1T-4R array is shown in Fig. 5b. The roadmap data are represented by the resistance states of the 2-VRRAM cells, while the environment data are represented by two voltage pulses (0 or 0.2 V search voltage, V_S) applied to the search lines (SL/SLR), as shown in Fig. 5c. For each 1T-4R structure (represented in one row), they share a match line (ML), which is connected with a sensing amplifier (SA) to output the collision detection results [10]. When the environment and the roadmap are collided, the V_S is applied to the LRS device with the HRS device grounded, so the ML voltage is charged to a high value that is higher than the reference voltage (V_{REF}), outputting logic high "1" (Fig. 5d). Otherwise, the V_S is applied to the HRS device, and the ML is charged slowly and remains at a voltage that is lower than V_{REF} and the SA outputs logic low "0" (no collision). One exception exists for collision detection between two partially occupied elements, where the result needs further recursive checking. Here we detect the state as a collision state, which does not have much effect on finding the collision-free path.

To further investigate the robustness of the collision detection, we analyze the worst-case sense margin (collision-free *vs.* one collision element case) with device variation. For a 4×4 1T-4R crossbar array, the SPICE simulation shows a 0.9 mV sense margin (Fig. 6a). The maximum word length is also simulated, as shown in Fig. 6b. In this work, we can compare 16 elements of environment and roadmap in parallel. If the On/Off ratio increases, the parallelism will be even higher.

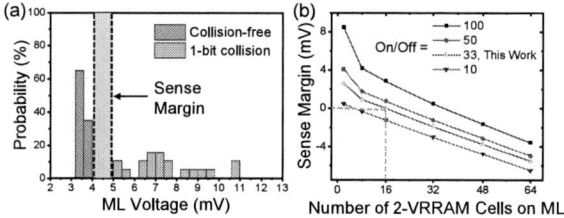

Fig. 6: (a) SPICE simulation of ML voltage distribution in a 4×4 1T-4R crossbar array for collision-free and 1-bit collision cases. (b) Relationship of the sense margin and word length for different HRS/LRS (On/Off) ratios.

C. Latency, Area and Power Estimation

The search delay of the proposed 3D VRRAM-based collision detection function is 2 ns. Here, we set the clock frequency at 250 MHz. For a traditional 6-level octree representation (8^6 elements), the 16-bit 2-VRRAM cells can achieve a 65.54 μs collision detection latency, which satisfies

the requirement for a dynamic environment. By exploiting the low-power and compact 1T–4R devices, the proposed 2-VRRAM cells show low search energy of 0.3 fJ/bit/search and a shared cell area of 24 F^2. Compared with collision detection accelerator based on 2T2R RRAM-TCAMs [11], when performing collision detection on the same-size roadmap, the proposed 1T–4R VRRAM can achieve a 4× area reduction and 2.3× lower energy consumption, and far exceeds those metrics for CPU or GPU based accelerators.

Conclusion

In summary, we design the in-memory search collision detection accelerator based on the experimentally demonstrated monolithic 3D integration of 2D MoS₂ FETs and 3D VRRAMs, as 1T–4R structure with five device layers. Based on the high-density and low-power on-chip memory, we achieve collision detection with 65.54 μs latency. Compared with a 2T2R-based collision detection accelerator, it achieves 4× less area and 2.3× energy reduction.

Acknowledgments

The authors thank the support from the National Natural Science Foundation of China (NSFC) (Grants 92364107, 62104140), Science and Technology Commission of Shanghai Municipality (STCSM) (Grant 23QA1405300), Natural Science Foundation of Chongqing (CSTB2022NSCQ-MSX1095), State Key Laboratory of Radio Frequency Heterogeneous Integration (Shenzhen University), Lingang Laboratory Open Research Fund (Grant LG-QS-202202-11), and the Center for Advanced Electronic Materials and Devices (AEMD).

References

[1] J. Bialkowski et al., "Massively Parallelizing the RRT and the RRT*," in IROS, pp. 3513-3518, 2011.

[2] A. Hermann et al., "Unified GPU Voxel Collision Detection for Mobile Manipulation Planning," in IROS, pp. 4154-4160, 2014,

[3] S. Lian et al., "Dadu-P: A Scalable Accelerator for Robot Motion Planning in A Dynamic Environment," in ACM/IEEE Design Automation Conf. (DAC), pp. 1-6, 2018.

[4] S. Murray et al., "The Microarchitecture of A Real-Time Robot Motion Planning Accelerator," in MICRO, pp. 1-12, 2016.

[5] Y. Yang et al., "Dadu-CD: Fast and Efficient Processing-in-Memory Accelerator for Collision Detection," in ACM/IEEE Design Automation Conf. (DAC), pp. 1-6, 2020.

[6] S.-T. Wei et al., "Trends and Challenges in the Circuit and Macro of RRAM-based Computing-in-Memory Systems," Chip, 1, pp. 100004 (2022).

[7] K.-J. Zhou et al., "The Trend of Emerging Non-Volatile TCAM for Parallel Search and AI Applications," Chip, 1, pp. 100012 (2022).

[8] M. Xie et al., "Monolithic 3D Integration of 2D Transistors and Vertical RRAMs in 1T–4R Structure for High-Density Memory," Nature Comm., 14, pp. 5952 (2023).

[9] A. Hornung et al., "OctoMap: An Efficient Probabilistic 3D Mapping Framework Based on Octrees," Autonomous robots, 34, pp. 189-206 (2013).

[10] C. C. Lin et al., "A 256b-Wordlength ReRAM-based TCAM with 1ns Search-time and 14× Improvement in Wordlength-Energyefficiency-Density Product Using 2.5T1R Cell," in IEEE Int. Solid-State Circuits Conference (ISSCC), pp. 136-137, 2016.

[11] J. Sun et al., "RTSA: An RRAM-TCAM based In-Memory-Search Accelerator for Sub-100 μs Collision Detection," In DATE, pp. 1-2, 2024.

Enhanced NBTI Characteristics in HfO₂/TiAlC RMG Stacks via Low-Temperature H* Remote Plasma Treatment after Interfacial Layer Growth

Songyi Jiang[1,2,4], Qianqian Liu[1,2], Hong Yang[1,2,4*], Mingyang Sun[1,2,4], Junjie Li[1,2,4*], Jianfeng Gao[1,2], Shuai Yang[1,2], Runsheng Wang[3], Jun Luo[1,2,4] and Wenwu Wang[1,2,4]

[1] Key Laboratory of Fabrication Technologies for Integrated Circuits, Chinese Academy of Sciences, Beijing 100029, China.

[2] Institute of Microelectronics, Chinese Academy of Sciences, Beijing 100029, China.

[3] School of Integrated Circuits, Peking University, Beijing 100871, China

[4] the School of Integrated Circuits, University of Chinese Academy of Sciences, Beijing 100049, China.

Abstract

The remote hydrogen plasma technique at 250°C after interfacial layer (IL) formation in HfO₂/TiAlC RMG is studied. It shows that the equivalent oxide thickness (EOT) increases about 0.31nm by H* treatment, which is related to the interfacial layer re-growth. The threshold voltage shift of H* treated-device in NBTI with -1.0V and 1000s at 125°C, is improved about 90%. It's proved that the H* treatment enhances H concentration in IL and HK layer, and reduces interface state density, effectively passivating interface dangling bonds and oxide hole traps.

Introduction

In CMOS integration processes, high-temperature rapid annealing methods, such as spike annealing and laser annealing, are commonly employed to enhance device reliability effectively[1]. However, for advanced CMOS device, especially sequential CFET (Complementary FET) fabrication, where devices are stacked from the bottom up, high-temperature annealing could impact the performance of underlying devices. In monolithic CFET integration, the high-temperature reliability annealing for replacement metal gate (RMG) processes can degrade contact performance [2].Therefore, a low thermal budget annealing approach is required to protect the bottom device layers in these integration schemes [3].

However, low thermal budget processing may induce numerous hydroxyl-E' interface defects within the interfacial layer (IL). Low-temperature H* treatment on the MOS capacitor's IL layer has shown potential to repair these interface defects, enhancing fundamental electrical and NBTI characteristics[4, 5]. In this paper, low-temperature remote H* treatment after IL formation to improve IL quality significantly reduces the interfacial state density and enables IL regrowth, while high-temperature post HK PDA provides minimal improvement in performance. These findings provide valuable insights for optimizing the continuous 3D integration process.

Device Fabrication

The MOS capacitor (MOSCAP) is fabricated on a 200 mm p-type (100) silicon wafer, where stress accumulation occurs in the p-well. The manufacturing process follows a conventional MOSCAP fabrication method and employs a simplified Replacement Metal Gate (RMG) process, as illustrated in Figure 1. The interfacial layer is formed by ozone oxidation of Si, while the high-k gate stack consists of a SiO₂ interfacial layer, a 2.7 nm HfO₂ high-k layer, an 8 nm TiAlC/2 nm TiN work function metal layer, and a 100 nm W metal layer.

The samples treated with H* were exposed to remote plasma H* at 250°C for 10 seconds immediately after the formation of the chemical oxide layer(IL), and underwent a 450°C annealing process for 15 seconds immediately after the formation of the high-k layer. In contrast, the PDA samples only underwent the 450°C annealing for 15 seconds after the formation of the high-k layer. Both were compared with the baseline devices, which followed the standard manufacturing process.

Unless otherwise specified, all measurements were conducted at 125°C. Electrical measurements were performed using the Keysight B1500A. EOT and Vfb were estimated from the measured C-V curves using the CVC-Hauser tool [6], while the interface state density was calculated using the conductance method based on multifrequency C-V results.

Experimental and Results

Electrical Characteristics:

Figure 2 compares the C-V characteristics of devices treated with H*, post-HK PDA, and baseline samples. An obviously rightward shift in the C-V curve for H* treated devices is observed. The V_{fb} shift of -0.13V for H*-treated MOSCAPs, compared to a slight positive shift of +0.04V for post-HK PDA devices. It means that H* treatment effectively suppress defect levels within the Si bandgap. However, EOT increased by 0.35 nm after H* treatment, while post-HK PDA showed limited EOT growth.

In Figure 3(a), H* treatment reduced gate leakage current by a factor of 10 when EOT increased by 2 Å. Considering the typical scaling behavior with SiO₂ thickness, the relationship between gate leakage current density (Jg) and EOT in Figure 3(b) confirms that H* treatment increases the physical thickness of the IL. Cross-sectional TEM analysis of the high-k/metal gate stack further quantifies this, as shown

979-8-3315-0417-5/25 $31.00 © 2025 IEEE

in Figure 4, with the IL thickness increasing by approximately 3 Å after H* treatment, while PDA treatment yielded negligible IL thickness increase. In other words, the H* treatment induces the regrowth of interfacial layer of device, which may helpful for reliability.

Reliability:

Figure 5 illustrates the V_{fb} shift (ΔV_{fb}) and extracted voltage acceleration factor (VAE) of devices under different Vov stresses ranged from -0.8V to -1.5V with 1000 seconds. In Figure 5 (a), following H* treatment, a substantial reduction in ΔV_{fb} is observed, indicating that H* treatment effectively mitigates NBTI (90% at V_{ov}=-1V). In addition, the VAE of MOSCAP after H* treatment increased from 1.96 to 5.22 compared to baseline, which may be attributed to the annealing process that repaired the interfacial defects and thus reduced the initial interfacial state density. In order to eliminated the influence of different EOT, the NBTI under constant electrical field is shown in Figure 5(b). Here, equivalent electric fields is defined as $E_{ox}=V_{ov}/(CET \equiv EOT+0.4nm)$. It shows a similar gain in NBTI reliability, further suggesting that H* treatment passivates dangling bonds on the SiO_2 surface and neutralizes hole traps within the oxide layer. This dual effect enhances NBTI reliability and increases the FAE. Figure 6 depicts the temperature dependence of ΔV_{fb} under identical stress voltages across three processing conditions, allowing extraction of activation energies. The results indicate that the activation energy is significantly elevated in devices treated with H*, while the activation energy shows minimal change after post-HK PDA.

Physical Understanding

Multi-frequency conductance measurements were conducted to investigate the effects of H* treatment, post-HK PDA and baseline on the initial interface characteristics of MOSCAPs, as well as Dit changes before and after stress application. As shown in Figure 7(a), H* treatment reduces interface trap density by 86%, improving initial interface properties. While post-HK PDA devices also show a reduction in initial interface trap density, the effect is less pronounced than that of H* treatment. Furthermore, Figure 7(b) illustrates that MOSCAPs treated with H* exhibit relatively minor changes in interface trap density under stress conditions. We attribute this reduction to the passivation effect of H* on hole traps within the SiO_2 and dangling bonds on the SiO_2 surface[6].

To further investigate the underlying physical mechanisms of H*-dependent properties, SIMS analysis was performed to examine the distribution of impurity elements within the gate stack, as shown in Figure 8. The results indicate that H concentrations in the IL and HK layers of H*-treated devices are significantly higher than those in PDA-treated and baseline devices, with a broader O peak, corroborating the increased IL thickness.

Conclusion

This study systematically investigated the effects of H* treatment and post-HK annealing on the interfacial layer (IL) quality and negative bias temperature instability (NBTI) of HfO2/TiAlC/TiN RMG structures. The results show that the equivalent oxide thickness (EOT) of the H*-treated device increases by ~0.3 nm, and V_{fb} shifts by -0.17 V. In contrast, post-HK annealed devices showed minimal EOT change and a slight positive V_{fb} shift of +0.04 V. The current degradation behavior of H*-treated MOSCAPs aligned with the SiO_2 equivalent field reference, indicating SiO_2 IL regrowth, supported by TEM images. Furthermore, H* treatment significantly improved NBTI characteristics (90%), especially under low V_{ov} conditions critical for device lifetime. This improvement is attributed to the enhanced quality of the SiO_2 insulator layer, with reduced Dit (86%) and oxide hole traps due to hydrogen passivation of dangling bonds in H*-treated devices. In conclusion, atomic H* treatment after IL deposition, performed with a low thermal budget, effectively enhances IL quality and NBTI by suppressing interface and oxide traps.

Acknowledgments

This work is financially supported by National Natural Science Foundation of China under Grant 62374177 and 61927901.

References

[1] A. T. Fiory, "Recent developments in rapid thermal processing," Journal of Electronic Materials, vol. 31, no. 10, pp. 981-987, 2002.

[2] S. Subramanian et al., "First Monolithic Integration of 3D Complementary FET (CFET) on 300mm Wafers," in 2020 IEEE Symposium on VLSI Technology, 16-19 June 2020, pp. 1-2.

[3] J. Franco et al., "Low-temperature atomic and molecular hydrogen anneals for enhanced chemical SiO_2 IL quality in low thermal budget RMG stacks," in 2021 IEEE International Electron Devices Meeting (IEDM), 11-16 Dec. 2021, pp. 31.4.1-31.4.4.

[4] T. Grasser et al., "On the microscopic structure of hole traps in pMOSFETs," in 2014 IEEE International Electron Devices Meeting, 15-17 Dec. 2014, pp. 21.1.1-21.1.

[5] J. Franco et al., "Atomic Hydrogen Exposure to Enable High-Quality Low-Temperature SiO_2 with Excellent pMOS NBTI Reliability Compatible with 3D Sequential Tier Stacking," in 2020 IEEE International Electron Devices Meeting (IEDM), 12-18 Dec. 2020, pp. 31.2.1-31.2.4.

[6] J. R. Hauser and K. Ahmed, "Characterization of ultra-thin oxides using electrical C-V and I-V measurements," AIP Conference Proceedings, vol. 449, no. 1, pp. 235-239, 1998.

[7] A. M. El-Sayed, Y. Wimmer, W. Goes, T. Grasser, V. V. Afanas'ev, and A. L. Shluger, "Theoretical models of hydrogen-induced defects in amorphous silicon dioxide", PHYSICAL REVIEW B, vol. 92, no. 1, JUL 10 2015, Art no. 014107.

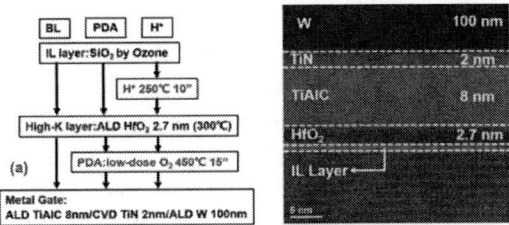

Fig. 1. (a) Gate stack process flow and split process conditions (b) TEM of MOSCAP's gate stack (HfO$_2$/TiAlC/TiN/W)

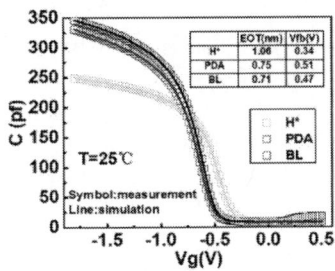

Fig. 2. The C-V curve was fitted using the CVC Hauser tool to estimate the EOT and V$_{fb}$.

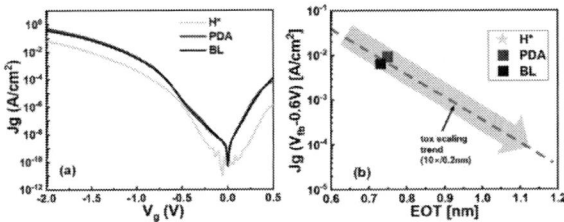

Fig. 3. (a)Ig-Vg curves of MOSCAPs under three different fabrication conditions. (b) The gate leakage current density (Vfb · 0.6 V) attenuation with EOT follow the SiO$_2$ equivalent field reference line.

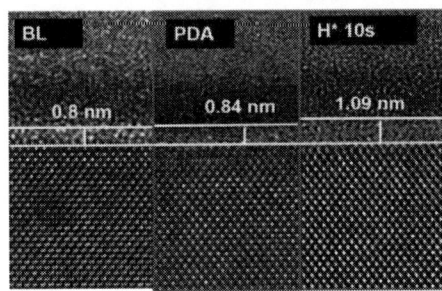

Fig. 4. TEM images of the chemical oxide IL layer between the MOSCAPs.

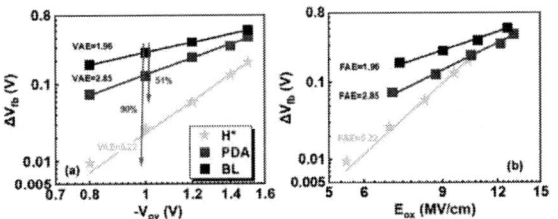

Fig. 5. The ΔV$_{fb}$ and extracted (a)VAE and (b)FAE values of the MOSCAPs were measured under various V$_{ov}$ conditions after 1000 s of stress at 125 °C.

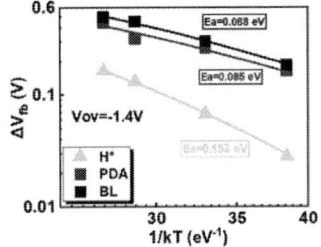

Fig. 6. Temperature-dependent characterization of ΔVfb measured at the same stress voltage for three MOSCAPs with different fabrication conditions..

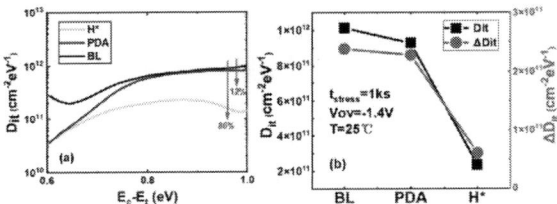

Fig. 7.(a) The energy band profile of inintial Dit extracted using multi-frequency conductivity.(b) Comparison of the initial Dit and the amount of change in Dit before and after stress for MOSCAPs with three different fabrication conditions

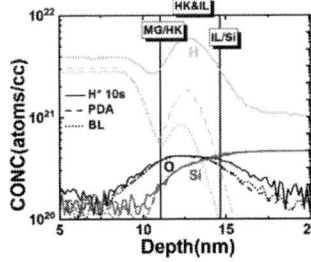

Fig. 9. Frontside SIMS test results of MOSCAPs for different fabrication conditions. The IL layer and HK layer in MOSCAP treated with H* contain significantly more H elements than the other two devices.

979-8-3315-0417-5/25 $31.00 © 2025 IEEE

Modeling Multi-Junction Tandem Solar Cell with High Bifacial Gain

Mohammad Ajmain Fatin[1], Arpan Saha[1], Md. Shahriar[1], and Mainul Hossain[1]

[1]Department of Electrical and Electronic Engineering, University of Dhaka, Dhaka-1000, Bangladesh

Abstract

Exploiting albedo light at the rear, bifacial solar cells provide a better energy yield than their mono-facial counterparts. This work exhibits a novel 4-T bifacial solar cell with soil as the reflecting surface. Two mechanically stacked 2-T tandem solar cells function as a novel 4-T bifacial tandem solar cell to maximize effective light absorption from both the front and rear surface of the photovoltaic device. An enhanced photon absorption in the rear surface of the bifacial solar cell demonstrates a bifacial gain of 22.25% indicating a 22.75 percent increase in photo conversion efficiency compared to the corresponding mono-facial configuration.

Keywords: Bifacial solar cell, Albedo, Bifacial Gain, PCE, Perovskite, WBH, NBH.

Introduction

Bifacial solar cells are designed to enhance power conversion efficiency (PCE) by enabling the absorption of albedo light in the rear surface thus ensuring maximum utilization the AM1.5G solar spectrum per unit area [1]. The PCE of mono-junction bifacial solar cells can be further enhanced by using multijunction architectures in bifacial configuration. Asadpour *et al.* demonstrated a multijunction Perovskite-Silicon bifacial solar cell with a PCE of 33% whereas it's mono-facial architecture shows a PCE of 25% [2]. Perovskite based multijunction solar cells have gained popularity in recent times due to their cost effectiveness, decent conversion efficiency, low temperature processing, bandgap tunability and bifacial design potential [3]. In our work we demonstrate two perovskite based 2-T tandem solar cells mechanically stacked to form a 4-T tandem solar cell used in bifacial configuration. The top 2-T tandem cell consists of a wideband halide (WBH) perovskite with a bandgap of 1.6 eV as the top sub-cell and a silicon heterojunction bottom sub-cell with a bandgap of 1.12 eV. The top 2-T tandem is illuminated with the standard AM1.5G solar spectrum. The bottom 2-T tandem is an all-perovskite tandem cell which consists of the same 1.6 eV wide band halide in conjunction with a narrow band halide (NBH) of 1.2 eV. The bottom 2-T tandem cell absorbs the albedo reflected from a soil surface. Fig. 1 depicts the arrangement of both the top and bottom 2-T tandem cells used in a bifacial configuration. Fig. 2 shows the reflection percentage of standard AM1.5G from a soil surface.

The performance of 2-T tandem solar cells is limited by the lowest short circuit current (J_{sc}) which demands a current matched condition to exhibit maximum PCE and mitigate power loss. 4-T tandem cells, on the contrary, optically couples the top and bottom sub-cells by mechanically stacking them on top of the other. Hence 4-T tandems require no current matching and ensures maximum PCE [4]. We have independently current matched and optimized the top and

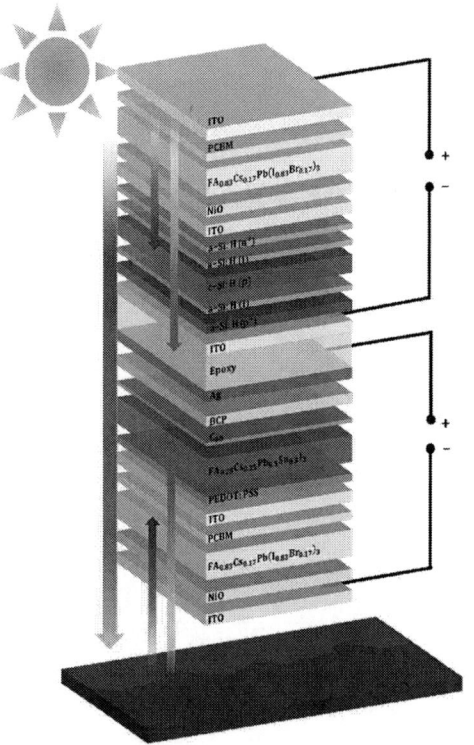

Fig. 1. Schematic of a dual tandem bifacial solar cell using soil as albedo reflecting surface

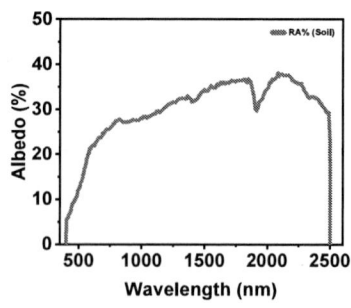

Fig. 2. Reflection percentage of AM1.5G from soil surface

bottom 2-T tandems by tailoring the thickness of the absorber layers. Our bifacial 4-T tandem solar cell shows a staggering PCE of 33% when positioned over a soil surface. The results show great promise for agri-voltaic applications.

Device Structure and Modeling

In SCAPS-1D software, three set of equations named Poisson, continuity and transport are used to solve for properties of carriers and determine the current-voltage characteristics. Our proposed structure consists of a 4-T configuration with two separate 2-T tandems. The top 2-T tandem comprises a perovskite top sub-cell and silicon bottom sub-cell. The top cell has a n-i-p configuration featuring a 350nm, 1.6eV metal halide perovskite absorber layer of $FA_{0.83}Cs_{0.17}Pb(I_{0.83}Br_{0.17})_3$ composition with PCBM and NiO functioning as the ETL and HTL respectively. The bottom cell of this tandem is a planar n-i-p silicon cell of 230 μm thickness and 1.12 eV bandgap to absorb lower energy photons. The bottom 2-T tandem is all-perovskite configuration, where the top perovskite sub-cell is is identical to the one in the top tandem. The bottom cell has a p-i-n structure comprising of a 210nm thick $FA_{0.75}Cs_{0.25}Pb_{0.5}SnO_{0.5}I_3$ metal halide perovskite and uses PEDOT:PSS and C_{60} as HTL and ETL respectively.

Results and Discussions

A. Calibration of the standalone top and bottom sub-cells

The simulation model is calibrated by matching the simulated current density voltage (J-V) curves of the three standalone cells used in our work with experimentally determined characteristics reported in literature [5], [6]. Figure 3 shows the simulated results for the standalone cells with experimental data.

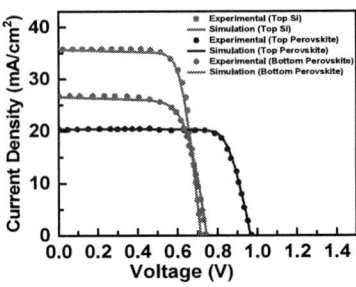

Fig. 3. Calibration of the sub-cells of the 4-T tandem

B. Output characteristics of the proposed 2T tandem cells

Fig. 4(a)-(b) shows the J-V characteristics of our top and bottom tandems in current unmatched condition. The top tandem is illuminated with AM1.5G solar spectrum while the bottom tandem is illuminated with the albedo from the soil surface. The top tandem achieved a J_{sc} of 17.41 mA/cm², V_{oc}

of 1.66V, FF of 82.25% and PCE of 23.82%. The bottom tandem shows a J_{sc} of 2.93 mA/cm², V_{oc} of 1.58 V, FF of 85% and PCE of 13.52% when illuminated by the soil albedo of 282 W/m². The lower short circuit currents for each of the tandems act as the limiting current.

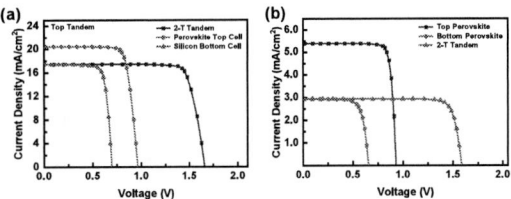

Fig. 4. J-V curves of (a) Top Tandem (b) Bottom Tandem without current matching

C. Current Matching of the Independent 2-T Tandem Cells

Fig 5(a)-(b) depicts the current matching of the independent top and bottom 2-T tandem cells. For the top tandem the silicon bottom sub-cell thickness is fixed at a value of 400 μm. The thickness of the perovskite top sub-cell is varied from 50nm-500nm. The top 2-T tandem achieves a current matched condition at 306nm WBH perovskite absorber layer thickness when the bottom silicon sub-cell is fixed at 400 μm. The bottom tandem, on the other hand, is current matched keeping the thickness of bottom NBH perovskite at 500nm. The bottom tandem achieves a current matched condition at 258nm WBH perovskite layer thickness. After current matching the top tandem achieves a PCE of 27% and the bottom tandem achieves a PCE of 21.33% under albedo illumination from the soil.

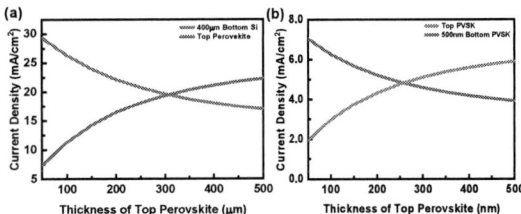

Fig. 5. Current Matching of (a) Top Tandem (b) Bottom Tandem

D. PCE and Bifacial Gain for Various Reflection Percentage of the Soil Albedo

The intensity of albedo affects the PCE of 4-T tandem solar cell based on the incident angle of the solar irradiance on the reflecting soil surface. Fig. 6(a) shows how the PCE varies with soil surface reflection percentage (RA%) from 20% to 100%. The PCE of the top tandem cell is unaffected by RA(%) while the PCE of the bottom tandem changes with

RA(%), affecting the overall 4-T tandem PCE The efficiency of a 4-T tandem solar cell is given by:

$$PCE_{4\text{-}T} = PCE_{Top\text{-}2T} + PCE_{Bot\text{-}2T} \quad (1)$$

The equivalent efficiency of bifacial solar cell is given by:

$$PCE_{eq} = PCE_{Front} + PCE_{Rear}\,\chi \quad (2)$$

with χ=0.27 for soil. Increasing RA(%) from 20% to 100% raises PCE_{eq} from 28% to 33%.

Bifacial gain is another parameter to determine the increase in PCE of a bifacial solar cell compared to it's mono-facial architecture [7]. The bifacial gain of a bifacial solar cell is given by:

$$BG(\%) = PCE_{rear}\,\chi / PCE_{Front} \times 100\% \quad (3)$$

Fig. 6(b) shows the BG(%) of the bifacial tandem cell. At normal incidence BG(%) reaches 22.75% reflecting a 22.75% PCE increase over its mono-facial configuration.

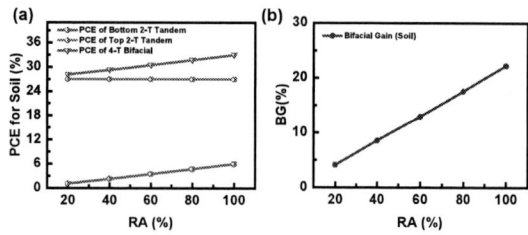

Fig. 6. (a) PCE and (b) Bifacial Gain of the 4-T bifacial tandem

Conclusion

We presented a novel 4-T bifacial tandem solar cell comprising of two separate 2-T tandems. Increase in PCE and BG(%) was investigated while positioned on soil surface. Furthermore, we investigated the effect of varying incident angles on the parameters demonstrating a similar pattern as we increase RA(%). The results open up new avenues for further advancing bifacial solar cell technology.

References

[1] M. De Bastiani et al., "Efficient bifacial monolithic perovskite/silicon tandem solar cells via bandgap engineering," Nat. Energy, vol. 6, no. 2, pp. 167–175, 2021, doi: 10.1038/s41560-020-00756-8.

[2] R. Asadpour, R. V. K. Chavali, M. Ryyan Khan, and M. A. Alam, "Bifacial Si heterojunction-perovskite organic-inorganic tandem to produce highly efficient (ηT* ~33%) solar cell," Appl. Phys. Lett., vol. 106, no. 24, 2015, doi: 10.1063/1.4922375.

[3] Z. Zhang, Z. Li, L. Meng, S. Y. Lien, and P. Gao, "Perovskite-Based Tandem Solar Cells: Get the Most Out of the Sun," Adv. Funct. Mater., vol. 30, no. 38, 2020, doi: 10.1002/adfm.202001904.

[4] S. Sarker et al., "A SCAPS simulation investigation of non-toxic MAGeI3-on-Si tandem solar device utilizing monolithically integrated (2-T) and mechanically stacked (4-T) configurations," Sol. Energy, vol. 225, no. July, pp. 471–485, 2021, doi: 10.1016/j.solener.2021.07.057.

[5] G. E. Eperon et al., "Perovskite-perovskite tandem photovoltaics with optimized band gaps," Science (80-.)., vol. 9717, no. 6314, pp. 1–10, 2016.

[6] A. Descoeudres, Z. C. Holman, L. Barraud, S. Morel, S. De Wolf, and C. Ballif, ">21% Efficient Silicon Heterojunction Solar Cells on N-and P-Type Wafers Compared," IEEE J. Photovoltaics, vol. 3, no. 1, pp. 83–89, 2013, doi: 10.1109/JPHOTOV.2012.2209407.

Recent advances in on-chip learning with organic neuromorphic circuits

Yoeri van de Burgt[1,2,3]

[1] Department of Mechanical Engineering, Eindhoven University of Technology, Netherlands
[2] Eindhoven Hendrik Casimir Institute (EHCI), Eindhoven University of Technology, Netherlands
[2] Institute for Complex Molecular Systems (ICMS), Eindhoven University of Technology, Netherlands

Abstract

Artificial neural networks power AI applications but require substantial computing resources. Neuromorphic engineering, inspired by the brain, uses organic materials that emulate neuroplasticity, offering low-energy operation and adaptability. Organic electrochemical transistors can be used as neuromorphic devices (ECRAM) to enable efficient edge computing with local, adaptive learning capabilities. We have developed compact neuromorphic circuits that learn through reinforcement and supervised learning, advancing low-power, on-chip training suitable for autonomous systems.

Keywords: Organic Electronics, Neuromorphic Engineering, on-chip learning.

Introduction

Artificial neural networks (ANNs) form the backbone of many artificial intelligence (AI) applications, including language translation, image classification, and the advancement of autonomous driving. However, these algorithms demand significant computing power and energy resources[1], [2]. Neuromorphic engineering, by contrast, draws on the brain's efficiency, implementing core concepts like neurons and synapses directly and efficiently in hardware.

Organic materials have recently shown promise as foundational elements for neural processing, exhibiting basic forms of neuroplasticity that enable device-level emulation of brain-like functions[3]. Compared to inorganic materials, organic electronic materials offer notable advantages such as low-energy operation and easy tunability, making them excellent candidates for efficient neuromorphic computing systems[4], [5]. Our research group pioneered the use of these materials in a single synaptic device (ECRAM)[6], which is based on the organic electrochemical transistor[7] and demonstrates remarkable performance characteristics. These tuneable materials are particularly well-suited for edge-computing systems, where part of the neural network circuit can be embedded directly in the electronics, allowing localized training without external systems or networks. This allows for low power and autonomously adaptive circuits that can be efficiently trained locally and on-chip using supervised, reinforcement or associative learning without external computers.

In this mini-review, I highlight recent advancements from our group and collaborators in on-chip learning with organic neuromorphic systems and circuits, demonstrating that adaptive systems can be developed with a small number of neuromorphic devices and simple circuits, while exploring pathways toward greater complexity with larger-scale device integration.

Organic Electrochemical Resistive Access Memory

The conductivity of mixed ionic/electronic conductors such as PEDOT:PSS can be tuned by the controlled injection of ions[7]. The mixed ionic/electronic nature of organic electrochemical devices (OECT) make them highly sensitive to changes in ionic concentration due to a volumetric response in capacitance. Benefitting from these characteristics, OECTs and related materials have been shown successful for applications in bioelectronics because of their biocompatibility and a variety of low-cost fabrication methods compatible with a wide range of substrates. The volumetric response in electrical conduction to ion injection (and high on/off ratios) make OECTs interesting for neuromorphic applications as well [3].

We have demonstrated a non-volatile electrochemical neuromorphic organic device, based on a working mechanism in which the ionic charges are trapped in the channel by preventing the electronic charges in the outer circuit to flow [6], see Fig. 1. This device was shown to not only demonstrate a high number of analogue conductance states but operating it also requires low energy while maintaining long-term non-volatility (retention) due to the working mechanism that resembles a supercapacitor. Similar to an OECT, the conductance states can be modified by the

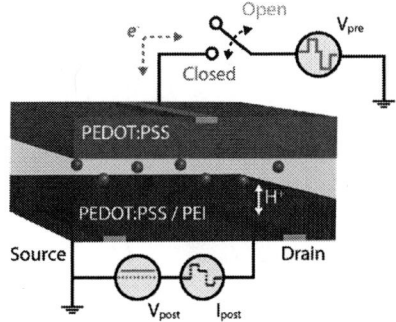

Fig. 1. Sketch of the device structure. Pre- and postsynaptic layers are separated by an electrolyte layer transporting ions/protons (red spheres). Reproduced from [3].

injection of protons in the film. The conductance of the PEDOT:PSS/PEI channel represents the synaptic weight of the connection between two neurons, an essential property of an neuromorphic device to be used in in-memory accelerators.

Sensorimotor integration and reinforcement learning in robotics using a simple neuromorphic circuit

In the brain communication and cooperation across different regions enable advanced cognitive functions such as learning, motor control, decision-making, pattern recognition, and predicting future events based on experience. In motor control, the sensory and motor systems work together as a cohesive unit. During this sensorimotor integration, sensory processing occurs alongside motor behaviors, while simultaneously motor actions are continuously informed by sensory input. We have emulated this sensorimotor integration in a behavioural mobility task using a simple, low-voltage organic neuromorphic circuit based on a single (non-volatile) neuromorphic device combined with a single (volatile) transistor device. Here the non-volatile device represented the long-term learning and plasticity while the volatile OECT represented the short-term plasticity. A standalone robot was equipped with this circuit and learned to navigate a two-dimensional maze by following a predetermined path to the exit after training its neuromorphic circuit with direct, real-time feedback from its sensorimotor system [8], see Fig. 3. The circuit underwent online learning within the sensorimotor loop, allowing it to establish associations between sensory inputs and motor responses—an essential link for completing the navigation task. Visual stimuli served as navigation cues, triggering specific motor actions. This integration, occurring locally in the analog domain, guided the robot toward the maze exit. The robot, along with its sensors, actuators, and neuromorphic circuit, was battery-powered and operated autonomously. These results showcased the potential of organic neuromorphic electronics as localized, decentralized learning systems for mobile applications in energy-constrained environments.

The straightforwardness of locally integrating low-power organic neuromorphic circuits with sensory inputs also allows for more advanced robotic systems using multimodal sensory inputs to explore and interact with a real-world environment in real time, adapting dynamically through bio-inspired mechanisms. We created a standalone robotic system

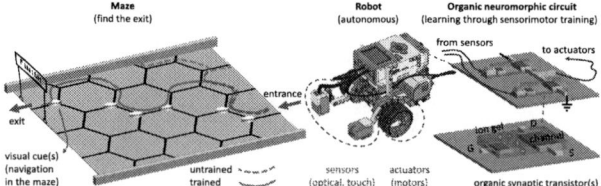

Fig. 3. Path-planning robot with an organic neuromorphic circuit for sensorimotor integration where an autonomous robot gradually learns to navigate a 2D-maze by following navigation cues. Reproduced from [8].

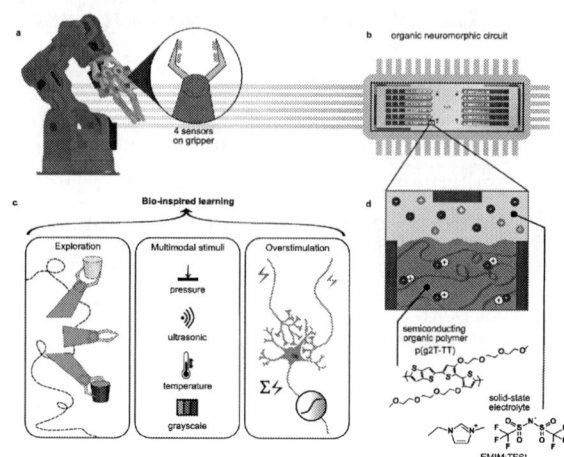

Fig. 2. **a** Robotic manipulator with a custom-made gripper is equipped with four multimodal sensors. **b** The robot employs the following bio-inspired principles for learning: exploration of its environment through random movement, collection of multimodal sensory inputs and adaptive processing leading to behavioral conditioning. **c** The robotic system is connected to a local organic neuromorphic circuit emulating neuronal processing. Reproduced from [9].

that intelligently interacts with a changing environment by presenting it with cups that it had to pick up [9]. By incorporating an organic neuromorphic circuit, the system adapts its behavior in response to multimodal sensory feedback from environmental cues and thereby learns that it should pick up the cups presented to it but refrain from picking up the hot ones. The synaptic devices within the circuit thus enabled associative learning, allowing for both Pavlovian conditioning and more advanced operant conditioning, in which overstimulation introduced noxious behavior and trained the robot agent to link positive and negative outcomes with different stimuli. This demonstrated its adaptability and ability to distinguish between safe and potentially harmful objects.

On-chip supervised learning with error backpropagation

Most current artificial intelligence systems and algorithms are based on large-scale artificial neural networks that are trained using error backpropagation with gradient descent[10]. Using the highly predicable, linear and easy to tune organic ECRAM devices it is possible to move the programming to the chip itself [11]. We have shown the versatility of these systems in an application with on-chip learning where low-power, autonomous and *in-situ* training is important such as point-of-care devices [12]. We demonstrated a modular smart biosensor chip that enabled software-free learning with sensors and neuromorphic hardware emulating a single layer perceptron [13]. The modular biosensor consisted of a sensor input layer, a linear array of organic neuromorphic devices as the tunable

979-8-3315-0417-5/25 $31.00 © 2025 IEEE

synaptic weights and an output classification layer. We used the chip to classify the disease cystic fibrosis using modified donor sweat and ion-selective sensors. Backpropagation was done using the error signal to modulate the conductance of the organic neuromorphic devices.

Beyond single layer neural networks it is more complex to implement gradient descent in hardware due to the required calculation of partial derivatives and the related digital storage of local error gradients and weight values [14], [15]. We have demonstrated the potential of a novel *in situ* backpropagation strategy for multi-layer neural networks in which all training is done in hardware and on-chip [16]. No digital storage of information was necessary because the partial derivative of the error was calculated locally in order to apply an outer product weight update step[14]. The method we proposed allowed us to circumvent this restriction by progressively updating each layer. The method only works under the assumption that $w_{ij}^L \approx w_{ij}^L + \Delta w_{ij}^L$ but this condition can be satisfied in a straightforward manner by optimising the hyperparameters, and the learning rate. Our progressive backpropagation method allows for *in situ* training of multilayer neural networks which marks significant progress in intelligent computing, especially for applications requiring continuous, online learning.

Conclusion

In this paper I reviewed recent advances in on-chip learning with organic neuromorphic systems. Using simple organic neuromorphic circuits, we have shown that we can use real world information and signals in combination with reinforcement or supervised learning to achieve local decentralised on-chip learning and adaptability, showcasing

the potential for organic neuromorphic hardware in adaptive, and intelligent systems for real-world applications as well as efficient large-scale neural network training.

Acknowledgments

The author gratefully acknowledges all contributions from his research group, students, postdocs and collaborators, as well as financial support from The European Union's Horizon Europe Research and Innovation Programme, grant agreement no. 101125598.

References

[1] P. Dhar, 'The carbon impact of artificial intelligence', *Nat. Mach. Intell.*, vol. 2, no. 8, pp. 423–425, Aug. 2020,

[2] A. Mehonic and A. J. Kenyon, 'Brain-inspired computing needs a master plan', *Nature*, vol. 604, no. 7905, Art. no. 7905, Apr. 2022,

[3] P. Gkoupidenis, N. Schaefer, B. Garlan, and G. G. Malliaras, 'Neuromorphic Functions in PEDOT:PSS Organic Electrochemical Transistors', *Adv. Mater.*, vol. 27, no. 44, pp. 7176–7180, 2015,

[4] Y. van de Burgt, A. Melianas, S. T. Keene, G. Malliaras, and A. Salleo, 'Organic electronics for neuromorphic computing', *Nat. Electron.*, vol. 1, no. 7, pp. 386–397, Jul. 2018,

[5] P. Gkoupidenis *et al.*, 'Organic mixed conductors for bioinspired electronics', *Nat. Rev. Mater.*, pp. 1–16, Dec. 2023,

[6] Y. van de Burgt *et al.*, 'A non-volatile organic electrochemical device as a low-voltage artificial synapse for neuromorphic computing', *Nat. Mater.*, vol. 16, no. 4, pp. 414–418, Apr. 2017,

[7] J. Rivnay, S. Inal, A. Salleo, R. M. Owens, M. Berggren, and G. G. Malliaras, 'Organic electrochemical transistors', *Nat. Rev. Mater.*, vol. 3, p. 17086, Jan. 2018,

[8] I. Krauhausen *et al.*, 'Organic neuromorphic electronics for sensorimotor integration and learning in robotics', *Sci. Adv.*, vol. 7, no. 50, p. eabl5068, Dec. 2021,

[9] I. Krauhausen, S. Griggs, I. McCulloch, J. M. J. den Toonder, P. Gkoupidenis, and Y. van de Burgt, 'Bio-inspired multimodal learning with organic neuromorphic electronics for behavioral conditioning in robotics', *Nat. Commun.*, vol. 15, no. 1, p. 4765, Jun. 2024,

[10] Y. LeCun, Y. Bengio, and G. Hinton, 'Deep learning', *Nature*, vol. 521, no. 7553, pp. 436–444, May 2015,

[11] E. J. Fuller *et al.*, 'Parallel programming of an ionic floating-gate memory array for scalable neuromorphic computing', *Science*, vol. 364, no. 6440, pp. 570–574, May 2019,

[12] I. Kurt, I. Krauhausen, S. Spolaor, and Y. van de Burgt, 'Predicting Blood Glucose Levels with Organic Neuromorphic Micro-Networks', *Adv. Sci.*, vol. 11, no. 27, p. 2308261, 2024,

[13] E. R. W. van Doremaele, X. Ji, J. Rivnay, and Y. van de Burgt, 'A retrainable neuromorphic biosensor for on-chip learning and classification', *Nat. Electron.*, vol. 6, no. 10, pp. 765–770, Oct. 2023,

[14] S. Ambrogio *et al.*, 'Equivalent-accuracy accelerated neural-network training using analogue memory', *Nature*, vol. 558, no. 7708, pp. 60–67, Jun. 2018,

[15] G. W. Burr *et al.*, 'Experimental Demonstration and Tolerancing of a Large-Scale Neural Network (165 000 Synapses) Using Phase-Change Memory as the Synaptic Weight Element', *IEEE Trans. Electron Devices*, vol. 62, no. 11, pp. 3498–3507, Nov. 2015,

[16] E. R. W. van Doremaele, T. Stevens, S. Ringeling, S. Spolaor, M. Fattori, and Y. van de Burgt, 'Hardware implementation of backpropagation using progressive gradient descent for in situ training of multilayer neural networks', *Sci. Adv.*, vol. 10, no. 28, p. eado8999, Jul. 2024,

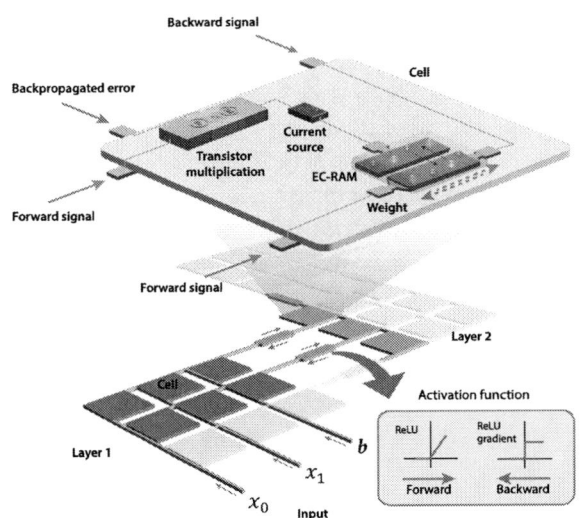

Fig. 4. illustration of multiple crossbars including the cells with the neuromorphic ECRAM devices, transistor multiplication, and current source and ReLU activation function, representing a hardware neural network. Reproduced from [16].

Pixelated Germanium-on-Silicon Photodetector with High Responsivity for Short Wavelength Infrared Imaging

Zhou Zhou[1], Chao Gao[1], Xin Jin[1], Xiaolin Liu[1] and Kai Wang[1*]

[1]Guangdong Province Key Laboratory of Display Material and Technology, State Key Laboratory of Optoelectronic Materials and Technologies, School of Electronics and Information Technology, Sun Yat-sen University, Guangzhou, China.

Abstract

We present a novel short-wavelength infrared detector, the Ge-on-Si photodiode-body-biased MOSFET (GePD-MOS), which offers high sensitivity, a broad spectral range, and CMOS compatibility. This detector achieves responsivities of 10^3 A/W at 1310 nm and 8×10^2 A/W at 1550 nm, with a spectral response extending to 1.9 μm due to its unique internal amplification mechanism. The GePD-MOS demonstrates adjustable responsivity and maintains low dark current, highlighting its potential for various applications in SWIR imaging.

Keywords: SWIR, Ge-on-Si photodetector, High sensitivity and Broadband Spectrum

Introduction

Imaging in short-wavelength infrared (SWIR) light has garnered increasing attention for its potential applications in autonomous driving, security, drone, night vision imaging, space remote sensing, medical diagnostics, and other fields [1-2]. The pixel formed by a photodetector plays a critical role in SWIR imaging. Currently, the most popular photodetectors operated in the SWIR range are InGaAs/InP-based devices; however, they are relatively expensive and not fully compatible with CMOS processes, necessitating three-dimensional heterogeneous integration. Nevertheless, the absorption coefficient of germanium (Ge) extends significantly beyond 1 μm into the SWIR range, and its preparation process is fully compatible with CMOS technology, making it a promising alternative to SWIR imaging. Ge photodetectors have been the subject of investigation for decades, particularly with the development of direct epitaxial growth of Ge on Si [3]. The common types of Ge-on-Si photodetectors are p-i-n photodiodes (PIN) and avalanche photodiode (APD) [4-5]. The PIN has a relatively simple structure; however, its low responsivity is insufficient for low-level light exposure. The APD exhibits high internal gain; however, its high power consumption and significant shot noise limit its applicability.

Since 2019, we have developed a new type of photodetector, termed as photodiode-body-biased MOSFET (PD-MOS), which offers both high gain and a wide dynamic range [6]. However, limited by the absorption cut-off wavelength of Si, its response is limited at 1.3μm. The introduction of materials sensitive to short-wavelength infrared light, such as Ge, as the absorption layer is likely to enable the development of an ultra-high responsivity SWIR detector. Furthermore, in this work, we propose a novel Ge-on-Si photodetector termed GePD-MOS (Ge-on-Si PD-MOS) to achieve high responsivity and a broadband spectrum based on the PD-MOS architecture, which is compatible with CMOS technology.

Design and Simulation of Ge-on-Si PD

Fig. 1(a) presents a schematic diagram of the proposed G-PDMOS, while the corresponding equivalent circuit is illustrated in Fig. 1(b). As depicted in the figure, one side of the device features a p-i-n type Ge-on-Si photodetector (GePD) composed of heavily doped p-Ge, native Ge, and heavily doped n-Si; this configuration is then connected to an adjacent MOSFET through the silicon substrate to form an integrated device. When Ge absorbs SWIR radiation to generate electron-hole pairs, the electrons drift to the silicon substrate under the influence of the built-in electric field, resulting in photocurrent amplification in the MOSFET via the body-effect. The specific physical mechanism can be referenced in [6].

The primary photoelectric conversion process of the G-PD-MOS occurs within the GePD. The conversion efficiency of this component is critical to the overall performance of the detector. We first optimize and design the GePD using Technology Computer-Aided Design (TCAD) tools. The photodetector typically operates in a reverse bias state to enhance the built-in electric field and improve the collection efficiency of photo-generated electrons and holes. In this scenario, the dark current (I_D) of the GePD is primarily composed of the minority carrier diffusion current near the barrier region and the majority carrier drift current generated by defect centers in the barrier region, which can be expressed as follows:

$$I_D = qn_i^2 A \left(\frac{1}{N_A} \sqrt{\frac{V_T \mu_n}{\tau_n}} + \frac{1}{N_D} \sqrt{\frac{V_T \mu_p}{\tau_p}} \right) + qn_i A \frac{W_D}{2\tau_0} \tag{1}$$

where A represents the junction area of the GePD, $N_{A/D}$ is the impurity concentration, V_T is the thermal voltage, $\mu_{n/p}$ and $\tau_{n/p}$ refer to the carrier mobility and lifetime of Si, τ_0 is the minority carrier lifetime of the intrinsic Ge. Although the process of epitaxial growth of germanium on silicon has advanced rapidly, threading dislocations caused by lattice mismatch still persist in large quantities in Ge, serving as defect centers that provide minority carrier drift current. Previous studies have indicated that threading dislocation density (TDD) is dependent on minority carrier lifetime [7]. In this study, we utilize TCAD to simulate the dark current characteristics of the GePD under varying TDDs based on this relationship. The results are presented in Fig. 2. Reducing TDD effectively decreases the dark current, aligning with industry understanding and reflecting the ongoing efforts in the Ge epitaxial growth process. Furthermore, we compared the performance of the GePD with n-i-p (n-i Ge/p Si) and p-i-n (p-i Ge/n Si) structures. As illustrated in Fig. 3, the p-i-n structure exhibits higher responsivity. This disparity can be attributed to the variation in band structure. As shown in Fig. 4, a barrier exists in the Ge/Si conduction band that impedes the movement of holes to p-Si, resulting in reduced photocurrent.

GePD-MOS Simulation and Modelling

Based on the optimization results of the GePD, the structure and associated parameters of the GePD-MOS were established, and its optoelectronic performance was validated using TCAD. The primary parameter settings are presented in Table 1. Fig.5 shows the transfer characteristics of the GePD-MOS at typical SWIR wavelengths of 1550 nm. The curve exhibits a continuous rightward shift with increasing light intensity, thereby confirming the underlying physical mechanism of the photo-induced body-effect. By adjusting the gate voltage, the operational range of the GePD-MOS can be toggled between the off-state region and the subthreshold region, corresponding to different responsivities, thereby demonstrating its adjustable responsivity, as illustrated in Fig. 6. Specifically, we analyzed the relationship between dark current and responsivity under varying operating conditions. As illustrated in the inset of Fig. 6, the device retains a high responsivity of 10 A/W while maintaining a low dark current density of 0.1 mA/cm^2. Fig. 7 illustrates the spectral response of the GePD-MOS operating in the subthreshold region. The responsivity achieves values of 10^3 A/W at 1310 nm and 8×10^2 A/W at 1550 nm. Owing to its high gain characteristics, the detector maintains a responsivity of 0.6 A/W at 1.9 μm, whereas ordinary Ge photodetectors exhibit essentially no response in this wavelength range.

Fig. 8 shows the measured result of GePD-MOS at 1310nm. The GePD-MOS is realized by integrating a commercial germanium photodetector with a MOSFET substrate. Even when operating in the off-state region, the GePD-MOS exhibits a prominent response at 1310 nm, whereas the response of the silicon-based PD-MOS is nearly imperceptible. This observation preliminarily verifies the superiority and feasibility of the GePD-MOS. Furthermore,

we designed a 5T-APS pixel circuit based on the GePD-MOS (Fig. 9a) to meet the imaging requirements for SWIR applications. Given the linear relationship between photocurrent and light intensity, we utilize the photocurrent corresponding to varying light intensities to simulate the light signal and obtain a series of output voltage values. The final transient simulation results of the pixel circuit are presented in Fig. 9(b). The results presented above demonstrate the unique advantages of the GePD-MOS in SWIR detection and imaging, including high sensitivity and a broad spectral range.

Conclusion

Considering the current challenges posed by traditional detectors in the SWIR imaging, which are price-sensitive, along with the insufficient performance of new devices, we have proposed a novel SWIR detector characterized by high sensitivity, a broad spectral range, and CMOS compatibility, exhibiting a responsivity of 10^3 A/W at 1310 nm and 8×10^2 A/W at 1550 nm. Owing to its unique internal amplification mechanism, the detector's spectral response range extends to 1.9 μm, significantly enhancing its application potential and prospects in the SWIR imaging.

Acknowledgments

This work is supported in part by the National Key Research and Development Program of China under Grant 2022YFA1204202 and in part by the National Natural Science Foundation of China under Grant 62274186.

References

[1] Hansen, Marc P., and Douglas S. Malchow. "Overview of SWIR detectors, cameras, and applications." Proc. SPIE, 6939 (2008).

[2] Thimsen, Elijah, Bryce Sadtler, and Mikhail Y. Berezin. "Shortwave-infrared (SWIR) emitters for biological imaging: a review of challenges and opportunities." Nanophotonics, 6, pp.1043 (2017).

[3] Michel, Jurgen, Jifeng Liu, and Lionel C. Kimerling. "High-performance Ge-on-Si photodetectors." Nature photonics, 4, pp. 527-534 (2010).

[4] F Thorburn, X Yi, ZM Greener, J Kirdoda, RW Millar, LL Huddleston, DJ Paul and GS Buller. "Ge-on-Si single-photon avalanche diode detectors for short-wave infrared wavelengths." Journal of Physics: Photonics 4, 012001 (2021).

[5] G Chen, Y Yu, Y Shi, N Li, W Luo, L Cao, AJ Danner, AQ Liu and X Zhang. "High‐Speed Photodetectors on Silicon Photonics Platform for Optical Interconnect." Laser & Photonics Reviews, 16, 2200117 (2022).

[6] Y Qi, X Liu, Z Feng, Q Li, K Su, X Zhou, J Guo and K Wang. "Human Retinal Photoreceptor-Inspired Sensor with Adjustable Gain from 0.1–10 6 and Wide Dynamic Range Over 140dB." IEEE International Electron Devices Meeting (IEDM) (2021).

[7] B Son, Y Lin, KH Lee, Q Chen and CS Tan. "Dark current analysis of germanium-on-insulator vertical pin photodetectors with varying threading dislocation density." Journal of Applied Physics, 20, (2020)

979-8-3315-0417-5/25 $31.00 © 2025 IEEE

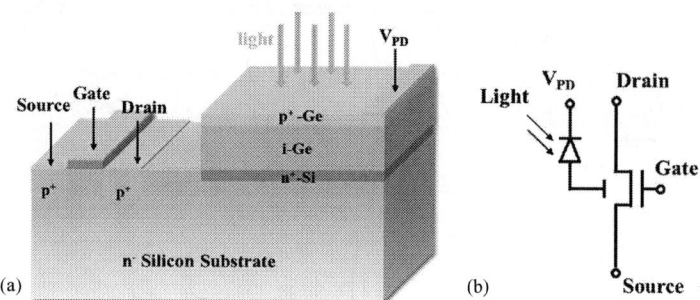

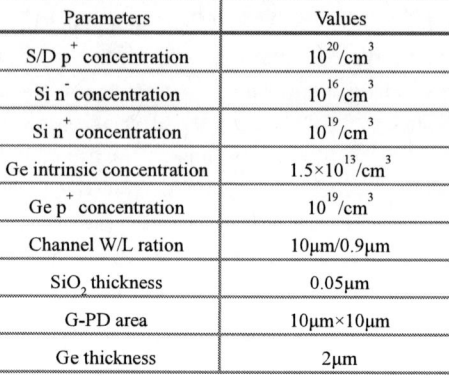

Table 1: Major Parameters Used in Simulation.

Parameters	Values
S/D p^+ concentration	$10^{20}/cm^3$
Si n^- concentration	$10^{16}/cm^3$
Si n^+ concentration	$10^{19}/cm^3$
Ge intrinsic concentration	$1.5\times10^{13}/cm^3$
Ge p^+ concentration	$10^{19}/cm^3$
Channel W/L ration	$10\mu m/0.9\mu m$
SiO$_2$ thickness	$0.05\mu m$
G-PD area	$10\mu m\times10\mu m$
Ge thickness	$2\mu m$

Fig. 1: (a) the schematic diagram of the GePD-MOS and (b) its equivalent circuit.

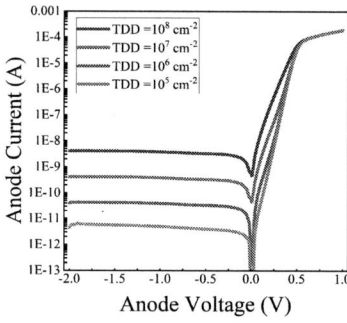

Fig. 2: Dark current characteristics of the GePD for different TDDs.

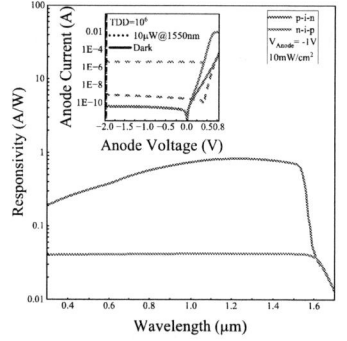

Fig. 3: Spectral responsivity of the GePD with n-i-p and p-i-n structures. The inset shows the comparison of I-V curve.

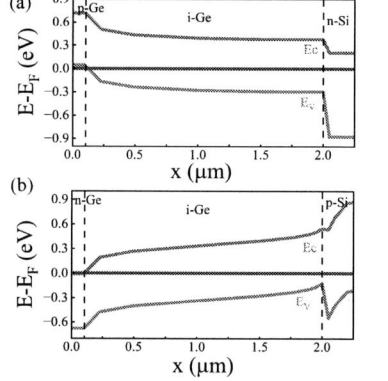

Fig. 4: Band diagram of (a) p-i-n and (b) n-i-p structures.

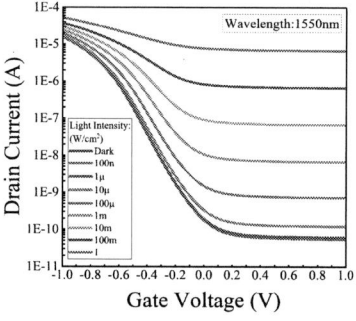

Fig. 5: Transfer characteristics under various light intensity.

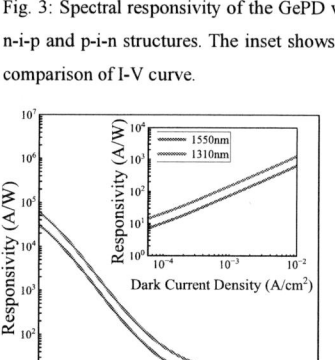

Fig. 6: Tunable responsivity versus V_{gs}. The inset shows adjustable dark current with high responsivity.

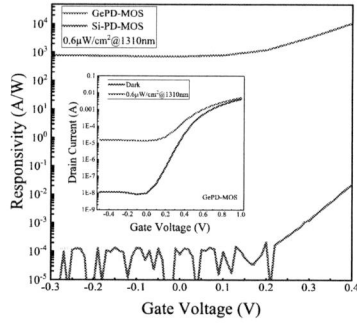

Fig. 8: Responsivity of GePD-MOS and silicon based PD-MOS measured at 1310nm. The inset shows the I-V curve of GePD-MOS.

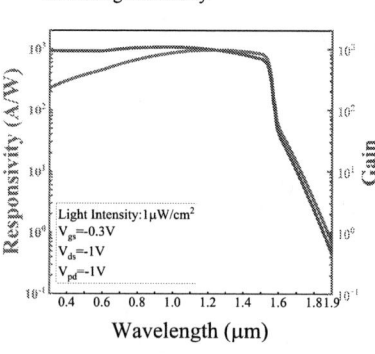

Fig. 7: Spectral responsivity and gain of the GePD-MOS in subthreshold regions.

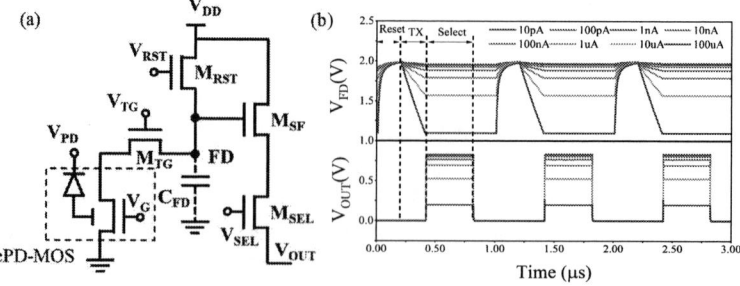

Fig. 9: 5T-APS based on GePD-MOS. (a) Circuit schematic. (b) Transient simulation result.

979-8-3315-0417-5/25 $31.00 © 2025 IEEE

Towards Spectrum Spurious-free and Wideband SAW Devices Based on LN/AT-Quartz Layered Structure

Peisen Liu[1], Sulei Fu[1,*], Boyuan Xiao[1], Xinchen Zhou[1], Qiufeng Xu[1], Jiajun Gao[1], Shuai Zhang[2], Rui Wang[1], Cheng Song[1], Fei Zeng[1], Weibiao Wang[2], and Feng Pan[1,*]

[1] Key Laboratory of Advanced Materials (MOE), School of Materials Science and Engineering, Tsinghua University, Beijing 100084, China, [2] Shoulder Electronics Ltd., Wuxi 214124, China.
*E-mail: suleifu@163.com; panf@mail.tsinghua.edu.cn

Abstract

This work reports on a novel lithium niobate thin-film on AT-Quartz layered structure with inherent spurious suppression for shear-horizontal surface acoustic wave (SH-SAW) wideband devices. The dispersion curves are twisted vertically by incorporating AT-Quartz substrate with strong concave shear horizontal slowness. We compared the proposed LN/AT-Quartz platform with typical LN/SiO$_2$/SiC structure through device measurements. One port spurious-free LN/AT-Quartz resonators without transverse and high-order modes are experimentally confirmed, while maintaining a high coupling factor and admittance ratio.

Keywords: Surface acoustic wave, lithium niobate and spurious suppression

Introduction

In next-generation mobile communication systems, the radio frequency (RF) front-ends based on layered shear-horizontal surface acoustic wave (SH-SAW) using lithium-niobate-on-insulator (LNOI) have achieved notable improvements in fabricating high-performance SAW filters with low loss and large bandwidth [1], [2]. Another key challenge for device usability is to maintain spectral responses free from spurious interference [3].

The LNOI platform generally employs high-velocity substrates such as Si [1-3] and SiC [4], which easily excite high-order mode above SH-SAW, as annotated (iii) in **Fig. 1**. Due to the non-vertical slowness curve in the interdigital transducer (IDT) region, another severe spurious signal namely transverse modes, inevitably arise around the main SH-SAW resonance, as annotated (ii) in **Fig. 1**. While the high-order modes can be eliminated through layer optimization [3] or specific Si orientation [5], suppressing transverse resonances remains a major bottleneck as typically requiring re-design of IDT layout but with trade-offs in electromechanical coupling (k_{eff}^2) and quality factor (Q) [6]. Recently, layered SAW built on quartz substrate has shown giant potential in delivering spurious-free resonators [7-8]. The moderate acoustic velocity of quartz allows high-order modes leak into substrate while not impacting main mode performance. More importantly, its rich anisotropy significantly expands the design space for manipulating slowness curve, as demonstrated in the structure of lithium tantalate (LT) on 69°Y-90°X quartz, where transverse

modes are inherently suppressed without deliberate IDT redesigns or undermining performance metrics [9]. However, these findings based on LT thin films cannot be directly applied to the LNOI platform, as LN with larger k_{eff}^2 exhibits stronger transverse modes, and intrinsic suppression of transverse modes in LN wideband systems has not yet been achieved [10-11].

In this work, we explore an alternative LN/AT-Quartz platform towards the inherent suppression of transverse resonance and high-order modes in SAW wideband devices. The slowness profiles are optimized and the spectrum purity of SH-SAW resonators on LN/AT-Quartz are compared with typical LN/SiO$_2$/SiC structure. Fabricated LN/AT-Quartz resonators exhibit excellent spurious-free responses as well as giant k_{eff}^2 over 20% and admittance ratio (AR) of 75 dB.

Design and Simulation

Fig. 2 presents the slowness profile of a piezoelectric layer of 32°Y-X LN, a dielectric layer of SiO$_2$ and two handling substrates of SiC and AT-Quartz, showcasing the SH, shear vertical (SV), and longitudinal (L) components. To assess the transverse mode suppression capabilities of each material, we calculated the curvature (γ) of SH waves along S_x as a measure of convexity, as also shown in **Fig. 2**. A more negative γ indicates greater concavity, correlating with more efficient suppression of transverse modes [9]. Notably, 32°Y-X LN exhibits a slightly concave slowness, while both SiC and SiO$_2$, with $\gamma=1$, show distinctly convex profiles, which explains reported LN/SiO$_2$/SiC SH-SAW devices suffer from transverse modes. In contrast, AT-Quartz offers a much more negative γ than SiC, suggesting that direct bonding of LN thin-film to AT-Quartz in a bi-layered structure could enable intrinsic suppression of transverse modes.

Building on the material slowness analysis, FEM simulations were performed to investigate the slowness characteristics of the 32°YX-LN/AT-Quartz layered structure. **Fig. 3(a)** illustrates the model, which employs periodic boundary conditions along the x-axis to simulate an infinite array of IDTs and applies Floquet boundary conditions along the y-axis to capture transverse resonance effects. A perfect match layer (PML) was implemented to minimize acoustic reflections. In the simulation, a wavelength (λ) of 4 μm was used. The resulting dispersion curves are shown in **Fig. 3(b)** and **(c)**, where the LN thickness was varied from 100 to 500

979-8-3315-0417-5/25 $31.00 © 2025 IEEE

nm, and the Al thickness was varied from 80 to 240 nm. The shape of the dispersion curve is strongly influenced by the LN and Al layer thicknesses: thin LN and Al layers yield a concave dispersion profile, while thicker layers produce a convex shape. A vertical dispersion profile is achieved at h_{LN} = 300 nm and h_{Al} = 160 nm.

Fabrication

Using smart-cut technology, 4-inch single-crystalline 32°YX-LN films were successfully transferred onto AT-quartz and 4H-SiC substrates, which are both commercially available. **Fig. 4(a)** shows the high-resolution X-ray diffraction (XRD) rocking curve for the 32°YX-LN (300) plane, with a full width at half maximum (FWHM) of 120 arcsec, confirming the high single-crystal quality of LN thin-film. **Fig. 4(b)** displays a cross-sectional transmission electron microscope (TEM) image of the bilayer substrate, demonstrating that the LN film and Al electrode thicknesses match the specified values. Additionally, **Fig. 4(c)** presents optical microscope images of the resonator utilizing standard photolithography and lift-off processes. The IDTs with standard layout are well patterned with sharp boundaries.

Results and Discussion

Fig. 5 shows the measured admittance and conductance of corresponding LN/SiO$_2$/SiC and LN/AT-Quartz SH-SAW resonators. Under the layout of standard IDT, numerous severe transverse modes resonating between resonant frequency (f_r) and anti-resonant frequency (f_a) can be clearly identified for LN/SiO$_2$/SiC resonator, which will deteriorate passband response in the filter application. On the contrary, the optimized LN/AT-Quartz resonator can inherently suppress the transverse spurious modes, while retaining a larger k_{eff}^2 and AR than LN/SiO$_2$/SiC, as supported by experimental metrics in **Fig. 5**. The transverse modes are amplified due to the waveguide effect and can be affected by the anisotropy of the substrate. Therefore, the slowness profile of layered structure is twisted vertically by incorporating AT-Quartz and optimizing layer thickness, hence eliminating transverse modes inherently.

Fig. 6 (a) and **(b)** depict the measured admittance of SH-SAW resonators based on LN/AT-Quartz and LN/SiO$_2$/SiC structures, with resonant frequencies spanning 1 to 2 GHz. In the case of the LN/AT-Quartz resonator, all devices exhibit clean spectral responses between f_r and f_a, validating the effectiveness of proposed design in suppressing transverse modes. The resonators with λ=2.5 and 3 μm display very minimal transverse modes, which are negligible for practical application. This remarkable robustness in suppressing transverse modes across different λ or relative h_{Al} values offers potential for further enabling the fabrication of spurious-free LN/AT-Quartz wideband filters. Furthermore, the spurious high-order and Rayleigh modes are effectively mitigated in the proposed LN/AT-Quartz resonators,

achieving a wideband spurious-free spectrum to enable an excellent far-end rejection. The LN/AT-Quartz resonators demonstrate k_{eff}^2 from 18.7% to 24.8% and Q_{max} values between 596 and 865. However, the fabricated LN/SiO$_2$/SiC resonators suffer from severe in-band transverse and Rayleigh modes, requiring additional suppression method that could compromise performance metrics. Some devices with smaller λ even show spurious high-order modes.

Conclusion

In this work, we inherently eliminated transverse and high-order modes in LN-film wideband devices without any trade offs. The fabricated spurious-free LN/AT-Quartz SH-SAW resonators exhibit high k_{eff}^2 of 20.8 and AR of 75 dB. This prototype shows enormous potential for enabling wideband, spurious-free, and compact RF devices.

Acknowledgments

This work was supported in part by the National Key Research and Development Program of China (Grant No. 2022YFB3606700), Natural Science Foundation of Beijing Municipality (Grant No. JQ20010), and National Natural Science Foundation of China (Grant No. 52002205).

References

[1] T.-H. Hsu, K.-J. Tseng, M.-H. Li, IEEE Electron Device Letters, 41, pp. 1825-1828 (2020).

[2] R. Su, S. Fu, Z. Lu, J. Shen, H. Xu, P. Liu, Z. Xu, H. Wang, S. Zhang, R. Wang, C. Song, F. Zeng, W. Wang, F. Pan, International Electron Devices Meeting (IEDM), p. 4.2.1-4.2.4 (2022)

[3] Y. Yang, L. Gao, S. Gong, IEEE Transactions on Microwave Theory and Techniques, 70, pp. 5185–5194 (2022).

[4] S. Zhang, R. Lu, H. Zhou, S. Link, Y. Yang, Z. Li, K. Huang, X. Ou, S. Gong, IEEE Transactions on Microwave Theory and Techniques, 68, pp. 3653–3666 (2020).

[5] H. Xu, S. Fu, R. Su, P. Liu, B. Xiao, S. Zhang, R. Wang, C. Song, F. Zeng, W. Wang, F. Pan, Journal of Microelectromechanical Systems, 33, pp. 163–173 (2024).

[6] T.-H. Hsu, C.-H. Tsai, S.-S. Tung, M.-H. Li, IEEE Electron Device Letters, 44, pp. 1200–1203 (2023).

[7] X. Ke, J. Wu, Y. Chen, S. Zhang, X. Zhao, P. Zheng, M. Zhou, K. Huang, X. Ou, IEEE Transactions on Electron Devices, 71, pp. 6343–6349 (2024).

[8] B. Xiao, S. Fu, R. Su, H. Xu, P. Liu, X. Zhou, S. Zhang, R. Wang, C. Song, F. Zeng, W. Wang, F. Pan, IEEE Transactions on Electron Devices, 71, pp. 1266–1273 (2024).

[9] S. Inoue, M. Solal, IEEE International Ultrasonics Symposium (IUS), pp. 1–4 (2020).

[10] P. Liu, X. Zhou, S. Fu, R. Su, H. Xu, B. Xiao, Q. Xu, C. Song, F. Zeng, W. Wang, F. Pan, IEEE Electron Device Letters, 44, pp. 1796–1799 (2023).

[11] Y. Guo, M. Kadota, S. Tanaka, IEEE Transactions on Ultrasonics, Ferroelectrics, and Frequency Control, 71, pp. 182–190 (2024).

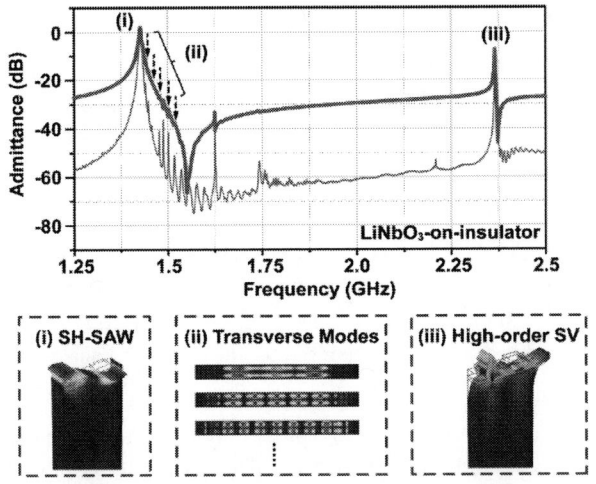

Fig. 1. Typical admittance curve of SAW resonator on LNOI and displacement contours of SH-SAW, transverse and High-order modes.

Fig. 2. The calculated slowness curves: (a) 32° Y-X LiNbO₃, (b) SiO₂, (c) 4H-SiC, and (d) AT-Quartz.

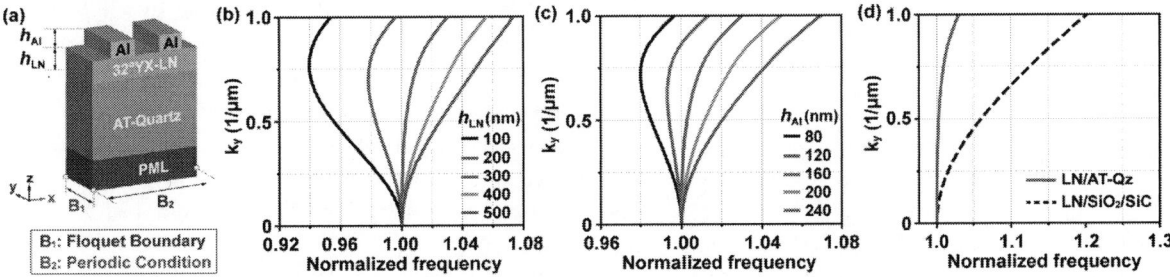

Fig. 3. (a) 3-D unit cell model of simulated 32°YX-LN/AT-Quartz resonator. Calculated slowness under different of (b) h_{LN} and (c) h_{Al}. (d) Comparison of slowness between LN/AT-Qz and LN/SiO₂/SiC structures under the same h_{LN} of 300 nm and h_{Al} of 160 nm.

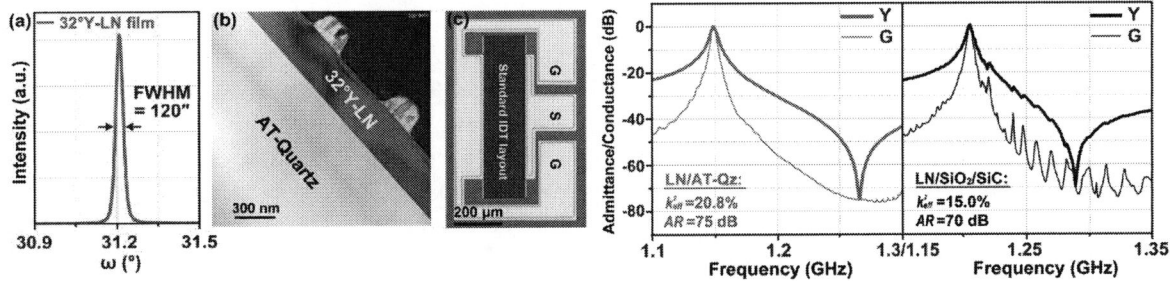

Fig. 4. (a) XRD rocking curves of 32°Y-LiNbO₃ thin film. (b) Cross-sectional TEM image of LN/AT-Quartz structure. (c) Optical microscope image of fabricated SH-SAW resonator with standard IDTs.

Fig. 5. Measured admittance and conductance responses of SH-SAW resonators based on LN/AT-Quartz and LN/SiO₂/SiC layered structures, with k_{eff}^2 and AR shown in the figure.

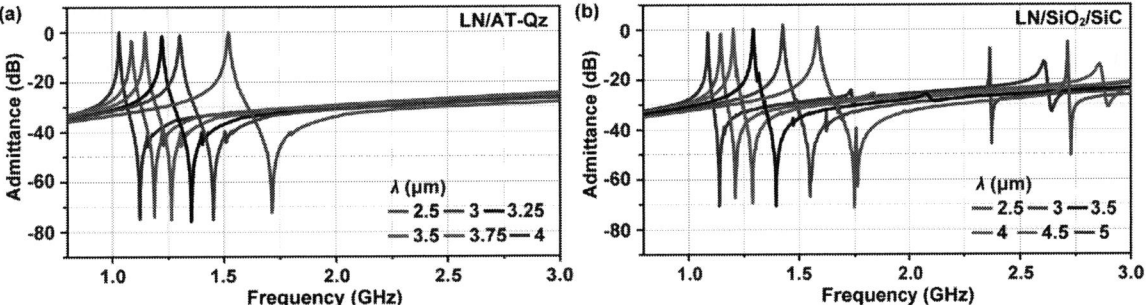

Fig. 6. Measured admittance responses of the (a) LN/AT-Quartz and (b) LN/SiO₂/SiC SAW resonators with different λ.

979-8-3315-0417-5/25 $31.00 © 2025 IEEE

Optoelectronic Switching Memory Device based on WO_3 Semiconductor Film enables High-efficiency in-sensing Reservoir Computing for Speech Recognition

Xun Li[1†], Jie Li[1†], Ke Xiang[1], Wenyang Xu[1], Yu Xu[1*], Denshun Gu[1*], Guangdong Zhou[1*]

[1]College of Artificial Intelligence, Southwest University, Chongqing, 400715, China.

[†]The authors equally contribute to this work.

[*]Corresponding Author Email: hazylight@163.com; gds15156501712@email.swu.edu.cn; zhougd@swu.edu.cn (lead contact)

Abstract

Optic in-sensing computing system exhibits great potential in advanced computing. An emerging optoelectronic synapse device with the structure of $Au/WO_3/Au$ is prepared by sputtering, showing an analogue switching memory behavior and positive photoconductance effect-driven short-term synaptic plasticity (STP), long-term synaptic plasticity (LTP), paired pulse facilitation (PPF). This optic synapse device can provide high data transmit rate ($\sim$ 9496.62 bit s^{-1}), response on 1000 Hz 405-nm light, and execute reservoir computing for speech processing with 95.8% accuracy.

Keywords: WO_3 Semiconductor, Optoelectronic Switching Memory Device, Reservoir Computing

Introduction

Optic in-sensing computing as a new computing platform shows great potential in analogue signal processing requires the optic switching memory device to have high data transmit rate, synapse mimicry, and highly-sensitive light response [1]. In term of the device levels, emerging optic devices including optoelectronic transistor and two-terminal optoelectronic switching memory device based on oxide semiconductors (e.g. HfO_2, MoO_3) [2], two-dimension materials (e.g. MoS_2, WSe_2) [3], and even natural organic materials (e.g. Fibroin protein, egg albumen) [4]. In the function level, the conductance set or reset through by light stimuli and then reset or set by electric stimuli is developed for conductance weight update in in-sensing neuromorphic computing for image preprocessing, or utilizing positive photoconductance effect and negative photoconductance effect to realize fully-optical weight update for complex information processing including image recognition and motion detection [5-6]. In the large-scale manufacturing and CMOS compatibility, oxide semiconductors such as the WO_3 are advantaged by the large-scale preparation and matured technology node. Recently, these emerging optoelectronic devices are severed as physical node for reservoir computing (RC) for complex analogue signal processing. An ultrafast light response and high data transmit rate are desired for the RC system for complex signal processing.

Herein, a symmetrical two-terminal device with structure of $Au/WO_3/Au$ is developed by sputtering, exhibits typical photoelectric synaptic properties, ultrafast light response, and high data transmit rate and then is operated for RC system for speech processing, yielding over 95% recognition accuracy. A physical mode is proposed for the observed behaviors.

Results and Discussion

Figure 1a shows the optoelectronic switching memory device with 268-nm WO_3 semiconductor film sandwiched between 30-nm Au top electrode and 98-nm Au bottom electrode. Typical current-voltage (I-V) curve illustrates that the WO_3 device exhibits an analogue switching behaviors under darkness and positive photoconductance effect under 405-nm UV light illumination (Fig. 1b). Figure 1c-d show the analogue switching behaviors that can be maintained for 200 cycles and good device-to-device stability characterized by the resistance ratio distribution in 10×10 crossbar arrays.

Figure 1d exhibits physical dynamic for this positive photoconductance effect observed in the $Au/WO_3/Au$ device. The work function of Au electrode and WO_3 semiconductor film is 5.1 eV and 1.2 eV, respectively. Under this case ($\Phi_m > \Phi_s$), a wider depletion layer and potential barrier are formed due to the metal-semiconductor contact (the stage i of Figure 1d). The electrons in valence band of the WO_3 semiconductor are excited by external 405-nm UV light and then hoping to defeat energy levels formed by oxygen vacancy and finally enter conduction band (the stage ii of Figure 1d). Under number of light pulses, the excited photoelectrons lead to a thin depletion layer and finally realize conduction (the stage iii of Figure 1d).

Figure 2 shows photoelectric synaptic properties of the $Au/WO_3/Au$ device. When operating single light pulse with intensity of 47.8 mW/μm^2 and width of 0.5 seconds, the device presents typical short-term synaptic plasticity (STP) while increasing the light intensity to 78.8 mW/μm^2 and width of 2 seconds, the device exhibits long-term synaptic plasticity (LTP), suggesting that the device can work as the manner of synapse to process information. Figure 2c is the light intensity-dependent LTP features, indicating that the LTP features can be linearly enhanced through modulating light intensities from 4.75 to 19.70 mW/μm^2. Similarly, the LTP features can be enhanced after increasing the light pulse idth from 1.0 to 7.0 seconds (Figure 2d). To further investigate the learning and forgetting processes the WO_3-

based optic synapse, the light pulse with 13 mW/μm^2 and width of 1 seconds is operated at room temperature. It can be seen that the increasing photocurrent is acquired as increase in the time of pulse, as shown in Figure 2e. In other words, the learning capacity is gradually accumulated through giving cycle period training. Figure 2f is the photocurrent response of the prepared device under paired pulses, showing an obviously increasing transient photocurrent. According to the intensity of transient photocurrent under the first pulse (A$_1$) and second pulse (A$_2$), the paired pulse facilitation (PPF) can be defined as A$_2$-A$_1$/A$_1$, as shown in the inset of Figure 2f. After operating the paired pulse with light intensity of 31 mW/μm^2 and an increasing interval time ranged from 50 to 3500 ms, the temporary photocurrent distribution can be well fitted by index function, indicating that the WO$_3$-based optoelectronic device can execute the information transmission as the manner of nerve cell, as shown Figure 2g. To further investigate the speed of the light response, the light stimuli with a constant light intensity of 12.8 mW/μm^2 and the variety of frequency ranged from 1.0 to 1000.0 Hz are operated on Au/WO$_3$/Au device, indicating that our device is highly-sensitive to the high frequency light of 1000 Hz, as shown in Figure 2h. According to the photocurrent response under different frequence of 0~1000Hz, the response trajectory function is described as follow:

$$S(t) = (S_1(t)+S_2(t) +S_3(t) +S_4(t) +S_5(t))/5 \qquad (1)$$

And the noise track function can be described as follow:

$$N_n(t) = S_n(t)-S(t) \qquad (2)$$

where the n denotes the average value. The signal to noise ratio (SNR) can be acquired:

$$SNR = |S(f)|^2/|N(f)|^2 \qquad (3)$$

After that, the channel capacity (C) can be further obtained as follow:

$$C = B \times log_2(1+SNR) \qquad (4)$$

The B is the bandwidth (Hz). Therefore, the device response on different frequency light under specific intensity, the channel capacity can be obtained. Figure 2i gives the channel capacity of 9496.62 bit s^{-1} under the bandwidth of 0~1000Hz, indicating that this optoelectronic switching memory device holds large information transmission.

This study utilized the NIST TI-46 public dataset to train and test a speech recognition system. The TI-46 dataset consists of audio waveforms of spoken digits (0 ~ 9 in English pronunciation) from five different female speakers. The goal of the speech signal recognition is to distinguish between different audio signals of digits while disregarding speaker variations. The overall architecture of the analog signal processing system based on Au/WO$_x$/Au optoelectronic

switching memory device is schematically shown in Figure 3a.

The raw digital speech signals are used as input, processed using Lyon's passive ear model to generate preprocessed input signals. After dimensional conversion, each row of the feature matrix is treated as an input vector with a duration of τ. This decodes into a series of spectral signals, reducing unnecessary information and improving system processing speed. Figure 3a on the left shows a schematic diagram of the preprocessed input signal after masking. Figure 3b illustrates how the Au/WO$_3$/Au memristor responds to 16 types of encoding modulation from "0000" to "1111" with light stimuli. After modulating the 16 encoding states, the output can be described as the different photocurrents. The Au/WO$_3$/Au optoelectronic switching memory device, therefore, can conduct the encoding task for the speech recognition. Figure 3c gives the encoding current of the "0001" and "1111", indicating that the encoding photocurrent can be completely separated, suggesting the availability of the WO$_3$ device in information encoding for RC system.

The system employs a parallel reservoir system (integrating masks, matrix operations, and other complex functionalities) for deep training. Within the parallel reservoir system, multiple different mask matrices are used to repetitively mask and record the same input signal, totaling N times. This process aims to capture the dynamic responses of each memory resistor within each time interval τ, and combine these responses into reservoir states for use in subsequent MLP classification tasks. Linear regression is employed to calculate weight out (W_{out}). To verify the reliability and accuracy of the model, eight-fold cross-validation strategy is used during the speech signal prediction. Figure 3d exhibits the 500 random photocurrent values and the prediction through the RC system. It can be seen that the random photocurrent encodings are well predicted. Based upon results, digit speech recognition tasks are demonstrated using the Au/WO$_x$/Au-based RC system, yielding an accuracy 95.8%.

Conclusion

An emerging photoelectronic switching memory device based on WO$_3$ semiconductor is developed by sputtering method. This device presents typical synaptic plasticity such as STP, LTP and PPF, is sensitive to the light stimuli with high frequency of 0~1000Hz, enabling high data transmission of 9496.62 bit s^{-1}. An in-sensing optic RC system based on the rich optic switching dynamic of the WO$_3$ device is demonstrated in speech recognition.

Acknowledgments

This work was supported by the Key Project of Chongqing Natural Science Foundation Joint Fund [CSTB2023 NSCQ-LZX0103, (G.Z.)], Chongqing Natural Science Foundation [CSTB2024NSCQ-MSX0012, (C.L.)], Fundamental

979-8-3315-0417-5/25 $31.00 © 2025 IEEE

Research Funds for the Central Universities [SWU: ZLPY03, (G.Z.)], and Fundamental Research Funds for the Central Universities [SWU:020019,(G.Z.)], [SWU:XDJH202319,(G.Z.)].

References

[1] Y. Chai. "In-sensor computing for machine vision." Nature, 579, pp. 32-33.

[2] F. Zhou, Z. Zhou, J. Chen, T. H. Choy, J. Wang, N. Zhang, and Y. Chai, "Optoelectronic resistive random access memory for neuromorphic vision sensors" Nature Nanotechnology, 14, pp. 776-782 (2019).

[3] Z. Wang, T. Wan, S. Ma, and Y. Chai, "Multidimensional vision sensors for information processing" Nature Nanotechnology, 19, pp.

919-930 (2024).

[4] G. Zhou, Z. Ren, L. Wang, B. Sun, S. Duan and Q. Song, "Artificial and wearable albumen protein memristor arrays with integrated memory logic gate functionality", Materials Horizon, 6, pp. 1877-1882 (2019).

[5] G. Zhou, J. Li, Q. Song, L. Wang, Z. Ren, B. Sun, X. Hu, W. Wang, G. Xu, X. Chen, L. Cheng, F. Zhou, and S. Duan, "Full hardware implementation of neuromorphic visual system based on multimodal optoelectronic resistive memory arrays for versatile image processing", Nature Communications, 14, pp. 8489 (2023).

[6] Z. Zhang, S. Wang, W. Hu, and P. Zhou, "All-in-one two-dimensional retinomorphic hardware device for motion detection and recognition. Nature Nanotechnology, 17, pp. 27-32 (2022).

Figures

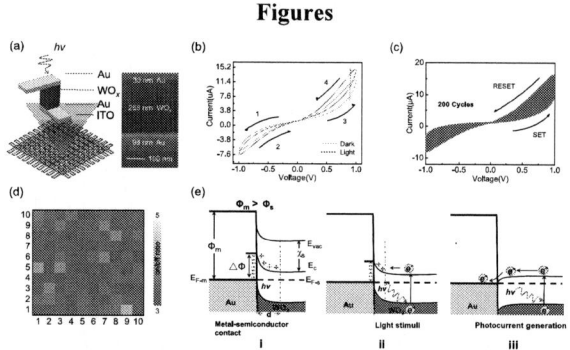

Fig.1: Simulated Photoresistive Memristor Based on Au/WO₃/Au. a, Schematic diagram of Au/WO₃/Au memristor array. The right figure shows the FE-SEM image of the memristor cross-section. **b,** *I-V* characteristic curves of memristors under light and dark conditions. **c,** The cycle endurance testing of the memristor. **d,** Device-to-device. **e,** Energy band model of Au/WO₃/Au memristors.

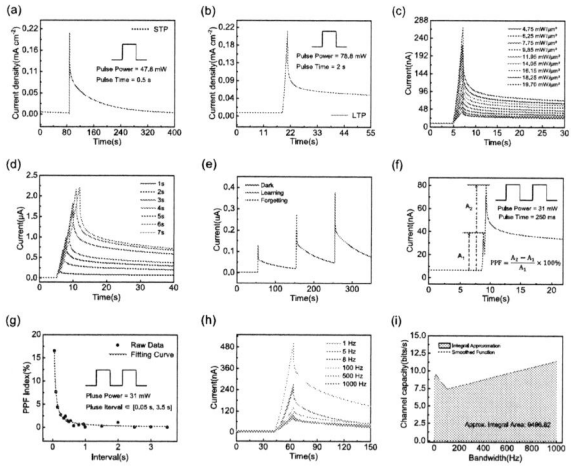

Fig.2: Optical property testing based on Au/WO₃/Au and its reflected synaptic properties. a-b, The simulated memristor exhibits short-term plasticity similar to synapses. Further adjustment of the light pulse (47.8mW,

1s) can achieve a transition from STP to LTP, laying the foundation for neuromorphic computing. **c,** The response of Au/WO₃/Au memristor under different power laser. **d,** The response of Au/WO₃/Au memristor under different pulse width laser. **e,** The optomemristor-based graded neurons exhibit a learning-forgetting property. **f-g,** The phenomenon of double-pulse facilitation in memristors. **h,** The response of Au/WOₓ/Au memristor under different frequencies laser. **i,** The total channel capacity at frequencies of 1-1000 Hz.

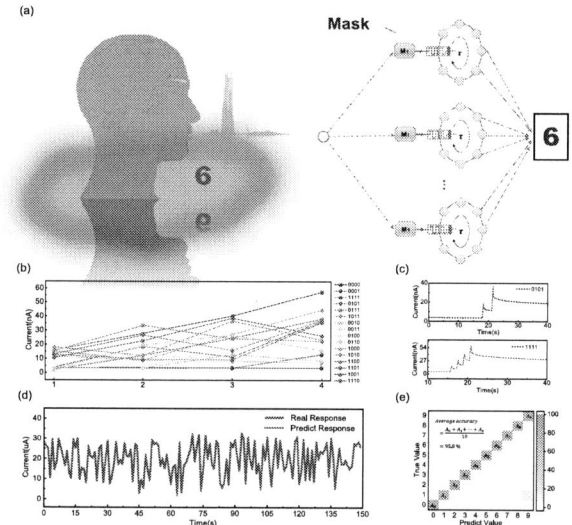

Fig.3: Analog signal recognition system based on photo resistive memory Au/WO₃/Au. a, For the real speech time series sample (speech signal 6), a parallel RC system based on the Au/WO₃/Au dynamic memristor is proposed. **b,** Comparison of real response and predictive response of digital speech recognition system based on parallel reserve pool calculation. **c,** The prediction results of the RC system based on Au/WO₃/Au memristors are presented using a confusion matrix, achieving an accuracy of 95.8%.

High Dielectric Constant of HfO$_2$ Technology for Memory Applications

Min-Hung Lee[1,*], Zhao-Feng Luo[1], Chun-Yu Liao[2], Kuo-Yu Hsiang[3], Jia-Yang Lee[1], Fu-Shen Chang[1], Yii-Tay Chang[1], Cheng-Hong Liu[1], and Che-Chi Cheng[1]

[1] Program for Semiconductor Devices, Materials, and Hetero-integration (DMHI), Graduate School of Advanced Technology (GSAT), National Taiwan University, Taipei, Taiwan

[2] Institute Program of Electro-Optical Engineering, National Taiwan Normal University, Taipei, Taiwan

[3] Institute of Electronics Engineering, National Yang Ming Chiao Tung University, Hsinchu, Taiwan

*Phone: +886-2-33669925 *E-mail: minhunglee@ntu.edu.tw

Abstract

HfO$_2$ with ferroelectricity has been widely investigated for memory and logic applications, including doped by Zr, Si, La, …etc. The superlamination (SL) technique exhibits higher capacitance and dielectric constant as compared to solid-solution (SS). The maximum dielectric constant achieves as high as 46 for HZZ at [Zr] = 66% with morphotropic phase boundary (MPB) effect. Ferroelectric-based memory of ferroelectric capacitive memory (FCM) has gained significant attention due to the charge transfer concept, which is adopted with the SL technique. The SL technique exhibits higher capacitance and applicable remnant polarization (P$_r$) benefits for FCM application to demonstrate a C_{HCS}/C_{LCS} ratio of 245x. The feasible concept of coupling the MPB SL Hf$_{1-x}$Zr$_x$O$_2$ is practicable into emerging memory/synapse.

Keywords: ferroelectric, ferroelectric capacitive memory (FCM), morphotropic phase boundary (MPB)

Introduction

HfO$_2$ with ferroelectricity has attracted lots of attention for memory applications due to BEOL-compatible, high conformal deposition for 3D architecture, <10nm ability by ALD for high-density requirement, HKMG process, and Lead-free [1]. For the CMOS evolution following Moore's Law, the requirement of scaling down supply voltage V$_{DD}$ and the power consumption for low-power devices meets the bottleneck and slows down the scaling rate. In the past decade, negative capacitance (NC) has been widely investigated [2]-[6]. However, hysteresis is one of the inevitable issues. **Fig. 1** shows the load line of FET and MFM to lead from hysteresis of positive capacitance to the NC region [7]. Recently, the superlattice of ferroelectric (FE) HfO$_2$ and antiferroelectric (AFE) ZrO$_2$ as gate stack was reported for the dielectric constant of 52 by NC effect [8]. The dopant concentration and annealing temperature would lead the phase transition occurrence from orthorhombic to tetragonal by curie temperature and results in antiferroelectric (AFE) characteristics. **Fig. 2** shows the recent study on superlattice

(SL) and nanolamination (NL) of HfO$_2$ and ZrO$_2$ for drain current enhancement with dielectric constant increasing [9]. In this study, the [Zr] concentration will be modulated with multi-lamination technology to achieve the morphotropic phase boundary (MPB) in order to further enhance the dielectric constant. The applications of the technology will be introduced for memory in this study.

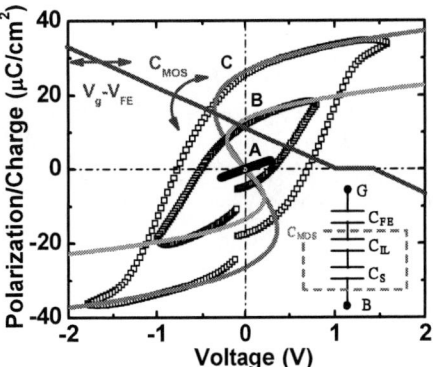

Fig. 1. The polarization hysteresis-loop of MFM is employed to match the FET load line and balance the charge from positive capacitance (PC) to NC with increasing applied bias. The adjustment for charge of FET load line benefits for reducing NC onset voltage [7].

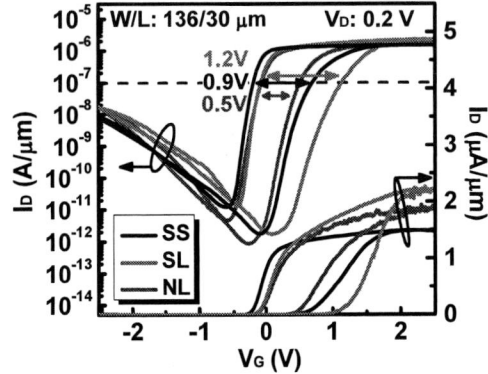

Fig. 1. I$_D$-V$_G$ of the SS-, SL-, and NL-HZO. Both SL- and NL-HZOs exhibit higher currents due to higher dielectric constant [9].

Super-lamination Process

Due to the multi-lamination integrated with the MPB effect in this work, the super-lamination (SL) process is named. The deposition rates of the HfO_2 and ZrO_2 films were 0.96 Å/cycle and 0.92 Å/cycle, respectively. The 3.62 nm HfO_2 or HZO/3.27 nm ZrO_2/3.62 nm HfO_2 or HZO for SL or MPB-SL were deposited, respectively, and the cycle numbers were determined from the above deposition rates by ALD. The supercycles of HfO_2-ZrO_2 were performed for SS-HZO and comparison. Note that the HZOs were kept at a physical thickness of 10 nm.

For superlattice, Moreover, the interplanar spacing of o(111) with $Pca2_1$ group and/or t(011) with $P4_2/nmc$ group is smaller than the equilibrium values by a distorted lattice of SL growth [10][11] as shown in **Fig. 3** [12]. For MPB effect, around [Zr] = 70% is the coexistence of both orthorhombic (o-phase) and tetragonal (t-phase) phases. The MPB near this concentration reveals the vanishing of the energy barrier separating the two phases due to the increased dipole interaction with reducing the elastic energy [13] as shown in **Fig. 3** [12].

The SL technique exhibits higher capacitance and dielectric constant as compared to SS, as shown in **Fig. 4**. The maximum dielectric constant achieves as high as 46 for HZZ at [Zr] = 66% with MPB effect. This agrees with MPB around [Zr] = 70% [14]. The partial t-phase is necessary for MPB. However, over high [Zr] may lead to AFE characteristic due to high t-phase fraction with narrowing P_r. This would be against memory application.

Memory Applications

Ferroelectric-based memory has ferroelectric random-access memory (FeRAM), ferroelectric field-effect transistor (FeFET), and ferroelectric tunneling junction (FTJ), as shown in **Fig. 5**. The high demands of three kinds of memory are large polarization. However, the high dielectric constant is not necessary. Recently, the ferroelectric capacitive memory (FCM) has gained significant attention due to the charge transfer concept as shown in **Fig. 6**, which has been investigated to realize non-destructive read operation (NDRO) [15][16] due to non-current based operation.

The proposed FCM is presented by the pulse-drive program (PG) and erase (ER) in write pulse voltages and widths as compared to HZO FCM. It is clear to observe that a higher C_{HCS}/C_{LCS} ratio for proposed FCM indicates reduced voltage and/or pulse width to keep a similar ratio. The proposed FCM not only enhances the on/off ratio but also reduces access voltage and time to improve energy efficiency. Note that the extraction ratio is set at 0 V for standby without power consumption. The benchmark shows the outstanding C_{HCS}/C_{LCS} of proposed FCM 245x, compared with prior works in **Fig. 7**.

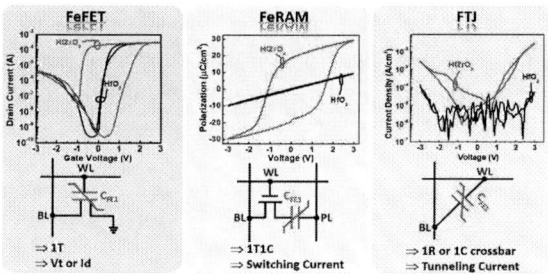

Fig. 5. Ferroelectric-based memory has ferroelectric random-access memory (FeRAM), ferroelectric field-effect transistor (FeFET), and ferroelectric tunneling junction (FTJ).

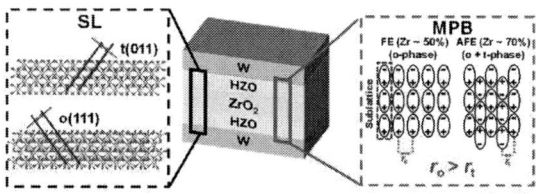

Fig. 3. Schematic diagram of the mechanisms of SL and MPB to improve EOT [12].

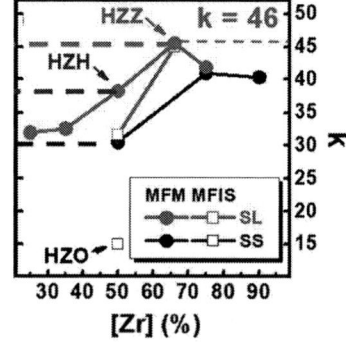

Fig. 4. Summarized the capacitance and dielectric constant with various [Zr] for SL and SS techniques [12].

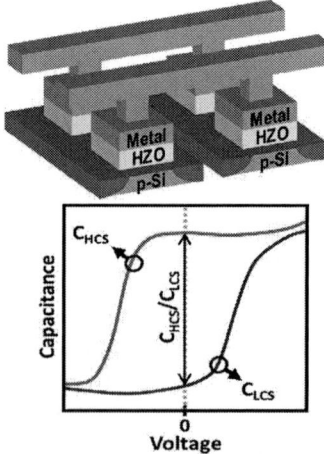

Fig. 6. ferroelectric capacitive memory (FCM) has gained significant attention due to the charge transfer concept.

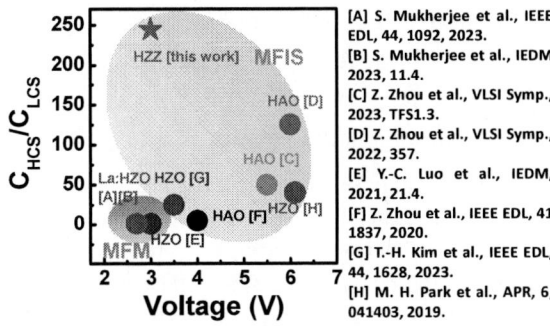

Fig. 7. Benchmark of the ratio vs. applied voltage for FCMs, including MF(I)S and MFM. This work exhibits the highest ratio, 245x.

Conclusion

The proposed SL technology with MPB effect enhances dielectric constant and keeps large 2Pr to realize for FCM. The SL technique exhibits higher capacitance and dielectric constant as compared to SS. The maximum dielectric constant achieves as high as 46 for HZZ. The promising high-quality ferroelectric layer is feasible for emerging NVM and logic applications.

Acknowledgments

This work was supported in part by the Semiconductor Research Corporation (SRC) 2024-LM-3236, the Ministry of Science and Technology (NSTC) 113-2640-E-002-004, 113-2218-E-A49-022-MBK, 112-2221-E-002-252-MY3, 111-2221-E-002-203-MY3, MATek 2024-T-013, the Powerchip Semiconductor Manufacturing Corporation (PSMC) 113H1004-C09, 113H1004-C10, and the National Taiwan University NTU-CC-113L893406. Processes were supported by the Taiwan Semiconductor Research Institute (TSRI) and the Nano Facility Center (NFC), Taiwan.

References

[1] M. Hoffmann *et al.*, "Unveiling the double-well energy landscape in a ferroelectric layer," *Nature*, vol. 565, pp. 464-467, Jan. 2019, doi: 10.1038/s41586-018-0854-z.

[2] M. H. Lee *et al.*, "Prospects for Ferroelectric HfZrOx FETs with Experimentally CET=0.98nm, SSfor=42mV/dec, SSrev=28mV/dec, Switch-OFF <0.2V, and Hysteresis-Free Strategies, " IEDM, pp. 616-619, 2015.

[3] M. H. Lee *et al.*, "Physical Thickness 1.x nm Ferroelectric HfZrOx Negative Capacitance FETs, " IEDM, pp. 306-309, 2016

[4] M. H. Lee *et al.*, "Ferroelectric Al:HfO$_2$ Negative Capacitance FETs, " IEDM, pp. 565-568, 2017.

[5] M. H. Lee *et al.*, "Extremely Steep Switch of Negative-Capacitance Nanosheet GAA-FETs and FinFETs, " IEDM, pp. 735-738, 2018.

[6] M. H. Lee *et al.*, "Bi-directional Sub-60mV/dec, Hysteresis-Free, Reducing Onset Voltage and High Speed Response of Ferroelectric-AntiFerroelectric Hf$_{0.25}$Zr$_{0.75}$O$_2$ Negative Capacitance FETs, " IEDM, pp. 447-450, 2019.

[7] K.-T. Chen *et al.*, "Ferroelectric HfZrO2 FETs for Steep Switch Onset, " *Microelectronic Engineering*, vol. 215, 110991, 2019.

[8] S. S. Cheema et al., "Ultrathin ferroic HfO$_2$–ZrO$_2$ superlattice gate stack for advanced transistors", Nature, vol. 604, pp. 65–71, 2022.

[9] C.-Y. Liao *et al.*, "Superlattice HfO$_2$-ZrO$_2$ based Ferro-Stack HfZrO$_2$ FeFETs: Homogeneous-Domain Merits Ultra-Low Error, Low Programming Voltage 4 V and Robust Endurance 10^9 cycles for Multibit NVM, " IEDM, pp. 878-881, 2022.

[10] Shinji Migita et al., "Accelerated ferroelectric phase transformation in HfO$_2$/ZrO$_2$ nanolaminates" 2021 Appl. Phys. Express 14 051006.

[11] S. L. Weeks et al., "Engineering of Ferroelectric HfO2-ZrO2 Nanolaminates" ACS Appl. Mater. Interfaces, vol. 9, no. 15, pp. 13440-13447, 2017.

[12] Z.-F. Lou et al., "Super-lamination HZO/ZrO2/HZO of Ferroelectric Memcapacitors with Morphotropic Phase Boundary (MPB) for High Capacitive Ratio and Non-destructive Readout, " accepted by *IEEE Electron Device Letter*, 2024.

[13] M. Jung, et al., "A review on morphotropic phase boundary in fluorite-structure hafnia towards DRAM technology", Nano Convergence, vol. 9, no. 44, 2022.

[14] K. Ni et al., "Equivalent Oxide Thickness (EOT) Scaling With Hafnium Zirconium Oxide High-κ Dielectric Near Morphotropic Phase Boundary," IEDM, 2019, pp. 7.4.1-7.4.4.

[15] S. Mukherjee et al., "Capacitive Memory Window With Non-Destructive Read in Ferroelectric Capacitors," in IEEE Electron Device Letters, vol. 44, no. 7, pp. 1092-1095, July 2023.

[16] S. Mukherjee et al., "Pulse-Based Capacitive Memory Window with High Non-Destructive Read Endurance in Fully BEOL Compatible Ferroelectric Capacitors," 2023 International Electron Devices Meeting (IEDM), San Francisco, CA, USA, 2023,11.4.1-11.4.4.

Optoelectronic Artificial Synaptic Device Based on Graphene-AlGaN van der Waals Junction

Yang Chen[1], Yuanyuan Yue[2], Bingchen Lv[1], Jin Zhang[1], Xiaoyu Wei[1], Xiaojuan Sun[1*], Dabing Li[1*]

[1]Changchun Institute of Optics, Fine Mechanics and Physics, Chinese Academy of Sciences, China

[2]Jilin University of Finance and Economics, China

*Corresponding Authors: sunxj@ciomp.ac.cn, lidb@ciomp.ac.cn

Abstract

The graphene-AlGaN van der Waals junction and their array were prepared by a novel patterned graphene growth method, in which the photosensitive polymer was served as the carbon source. Apart from the high quality and uniformity of the as-grown graphene patterns, this graphene-AlGaN junction was applied to an optoelectronic artificial synaptic device, and it has successfully emulated the functionalities of a biological synapse, the long-term memory time could sustain to 10 min.

Keywords: graphene, AlGaN, and optoelectronic device

Introduction

To break through the limitation of von Neumann's computing configuration, the neuron-like computing style was proposed, which benefits from the compact inner integration of storage and possessing units [1]. In comparison to the early artificial synaptic device based on electrical stimuli, the total light-modulated device is featured with large input bandwidth, high possessing rate, and small energy consumption [2,3]. Considering that the light contains plentiful information and could be easily collected through humans' eyes, then making the development of optoelectronic artificial synaptic device more meaningful.

The synaptic weight is one of the key parameters to evaluate the performance of optoelectronic artificial synaptic device, and it is known that the light-matter interaction and photo-carrier behavior control are crucial for the enhancement of synaptic weight [4]. For a double material system, such as two-dimensional (2D) graphene and three-dimensional (3D) AlGaN (Gr-AlGaN), the interfacial junction electric field could promote the carrier separation and collection by graphene, which has the high carrier transportation mobility [5,6]. On the other hand, the graphene layer acts as a highly "transparent window" for the ultraviolet light, and thus the underneath AlGaN will fully utilize the incident light signal. Hence, it could be expected that the optoelectronic artificial synaptic device with the van der Waals junction may trigger a great advance.

The valuable application of optoelectronic artificial synaptic device is visual learning like that of human brain, but requires the mass production of Gr-AlGaN van der Waals junction array. Herein, we demonstrate a novel strategy for the easy fabrication of Gr-AlGaN van der Waals junctions, in which patterned graphene was directly grown on AlGaN by using photosensitive polymer as the carbon source. The as-grown graphene on the AlGaN exhibits excellent crystalline quality and layer uniformity. With no need of a gate controller, the optoelectronic artificial synaptic device has emulated most functionalities of a biological synapse, such as the short-term plasticity (STP), long-term plasticity (LTP), and spike-rate-dependent plasticity (SRDP). By applying the proper external stimulation, this device reveals a long-term memory ability with the persistent time up to 10 min.

Results and Discussion

Fig. 1a-d illustrate the main growth procedure of patterned graphene on the non-catalytic substrate (e.g. sapphire), and the photosensitive polymer (PSP) was applied as the carbon source. At first, the PSP solution was spin-coated onto target substrates, and the standard photoetching technique was used to produce the PSP pattern array. Then, the Ni catalyst layer and SiO_2 capping layer were deposited before the high-temperature annealing conversion process, then, the polymer could be directly converted into patterned graphene.

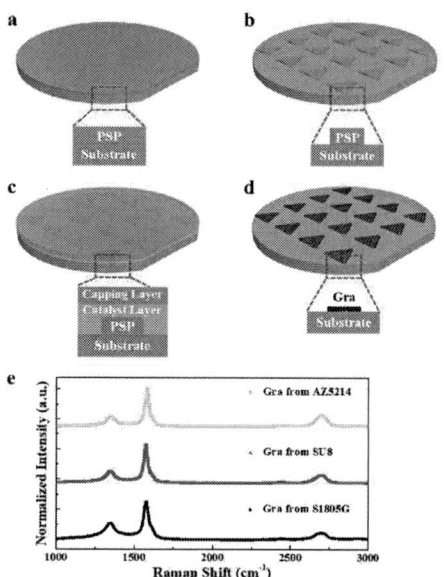

Fig. 1: a-c Schematic diagram of the direct growth of patterned graphene on the non-catalytic substrate. **e** Raman spectra of the patterned graphene grown from different photosensitive polymers.

The Raman spectra in Fig. 1e shows the typical G and 2D band of graphene, demonstrating that this growth method has successfully converted the polymer into graphene. Besides, the intensity of defect-related D band is relatively small, corresponding to a high crystalline quality.

The optical image and Raman mapping of patterned graphene in triangle, square, and circle are shown in Fig. 2a-c. The graphene patterns are featured with high graphic accuracy, and the graphic characteristics (e.g. shape and size) could be easily controlled during the PSD photoetching process. The transmission electron microscope (TEM) has been applied to investigate the graphene properties. In Fig. 2d, the peeled graphene flake has a continuous membrane feature. Based on the enlarged TEM image at graphene boundary, the layered structure is apparent, and the layer number of graphene is estimated as 5-6 (Fig. 2e). The high-resolution TEM image exhibits clear hexagonal lattice of graphene, and the lattice constant (0.239 nm along the zigzag side) is consistent with pervious works. In addition, the hexagonal pattern in selected area electron diffraction (SAED) of Fig. 2g further identifies the single-crystal quality of graphene.

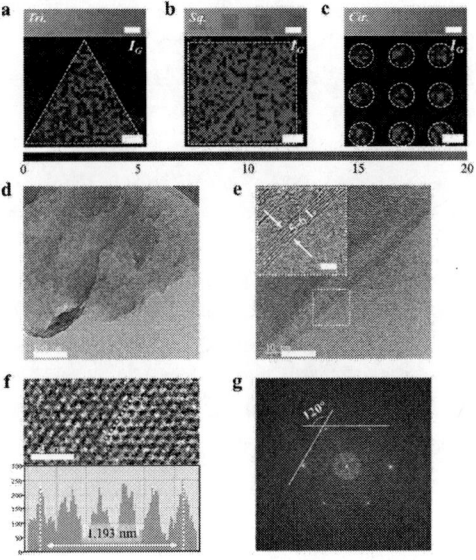

Fig. 2: **a-c** Raman mapping of the G band intensity for patterned graphene in triangle, square, and circle. **d-e** TEM images of the peeled graphene flakes and enlarged image at the flake boundary. **f** High-resolution TEM image of the graphene lattice. **g** The SAED pattern.

The optoelectronic artificial synaptic device was fabricated by directly growing patterned graphene on AlGaN epilayer, which forms a Gr-AlGaN van der Waals junction, as shown in Figure 3a. Here, the graphene was served as the carrier transportation layer, and AlGaN as the light absorber. When the light was absorbed by AlGaN, the photon-induced carrier could be separated by the van der Waals junction field, and the electric signal could be further collected via the paired metal electrodes on the graphene. In contrast to a biological synapse, the pulse light, AlGaN, photon-induced carrier, and graphene could be recognized as the external stimulation, pre-synapse, neurotransmitter, and post-synapse. As shown in Fig. 3b, the graphene possesses the high transparency over the whole measured wavelength, confirming its capability as a "transparent window" for the underneath AlGaN. The PL peak of AlGaN is well matched to its absorption band edge, it reveals an Al content of ~0.39.

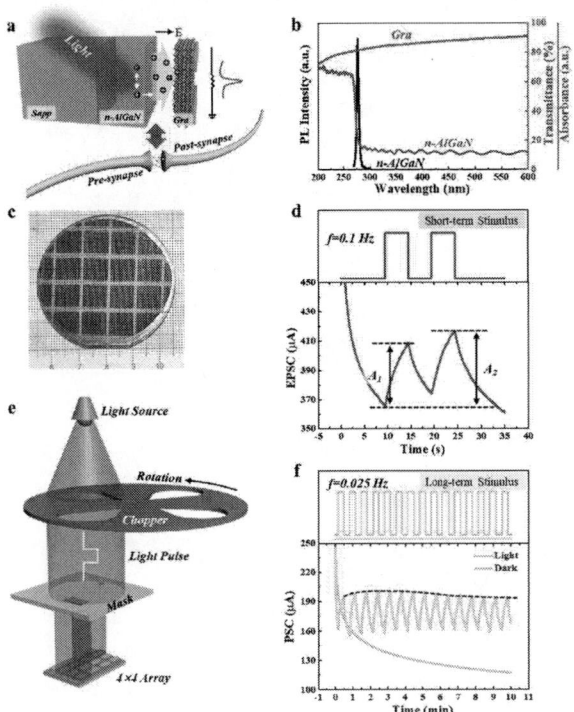

Fig. 3: **a** Schematic diagram of the optoelectronic artificial synaptic device and the comparison to a biological synapse. **b** optical measurement of the Gr-AlGaN junction. **c** Photography of the synaptic device array in the 2" scale. **e** Schematic diagram of the light pulse generation and modulation of the synaptic functionalities. **d** The PSC variation as it illuminated with paired light pulse and **f** long-term light pulse.

Benefiting from the direct growth of graphene on AlGaN, the 2" Gr-AlGaN van der Waals junction array is easily obtained, as shown in Fig. 3c. With the illumination of pulsed light, the post-synaptic current (PSC) of this optoelectronic artificial synaptic device could be collected. In Fig. 3d, the paired light pulse with the frequency of 0.1 Hz triggers the excitatory PSC, it is obvious that the PSC induced by the second light pulse is enhanced comparing to the first one. In Fig. 3f, the LTP of this device is evaluated by a long-term pulsed light, whose frequency is 0.025 Hz. After the light stimulation for 10 min, it still maintains 100% of the initial PSC, showing the ability for long-term memory (LTM).

The pulsed light with a graphic of "L" is illuminated onto the 4×4 optoelectronic artificial synaptic device array, which can mimic the visual learning like that in a human brain. In Fig. 4a, it shows the real imaging of letter "L" along with the dynamic acquisition time at 5 s, 25 s, 55 s, and 105 s. Initially, the imaging contrast of letter "L" is enhanced within 5-45 s, and it tends to be stable for the longer time (45-105 s), which is consistent with the LTM behavior shown in Fig. 3f. These trends could be observed more visualized via the difference between two optoelectronic artificial synaptic devices with the maximum and the minimum PSC, as shown in Fig. 4b. It perfectly matches with the visual imaging results. We believe that this optoelectronic artificial synaptic device may serve as a crucial part for the development of future biomimetic vision system.

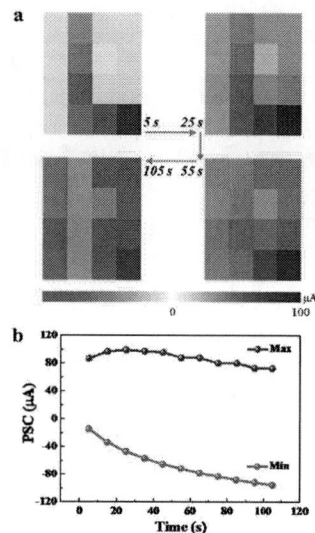

Fig. 4: a Visual imaging of "L" on a 4×4 device array. **b** Comparison of the maximum and minimum PSC as a function of pulsed light illumination time.

Conclusion

In summary, the wafer-scale fabrication of Gr-AlGaN van der Waals junction was achieved by the direct growth method of patterned graphene, and the key novelty is the utilization of photosensitive polymer as the carbon source. Based on the Gr-AlGaN junction, the optoelectronic artificial synaptic device was prepared. It has emulated different functionalities of a biological synapse, such as STP, LTP, and SRDP. Even better, the 4×4 device array achieves the visual learning like that of a human brain, and the LTM can retain for a long time up to 10 min. This construction strategy for van der Waals junctions would guide the design trend of future artificial synaptic devices, and further contributing to the advances of vision imaging and learning products.

Acknowledgments

This work was supported by the National Natural Science Foundation of China (62121005, 62374165, 61827813, 62074147, and 61922078), the Young Elite Scientists Sponsorship Program by CAST (2023QNRC001), the Youth Innovation Promotion Association of the Chinese Academy of Sciences.

References

[1] M. M. Waldrop, "The Semiconductor Industry Will Soon Abandon Its Pursuit of Moore's Law. Now Things Could Get a Lot More Interesting", Nature, 530, pp. 144-147 (2016).

[2] X. Han, Z. Xu, W. Wu, X. Liu, P. Yan, C. Pan, "Recent Progress in Optoelectronic Synapses for Artificial Visual-Perception System", Small Struct., 1, pp. 2000029 (2020).

[3] N. Li, S. Zhang, Y. Peng, X. Li, Y. Zhang, C. He, G. Zhang, "2D Semiconductor-Based Optoelectronics for Artificial Vision", Adv. Funct. Mater., 33, pp. 2305589 (2023).

[4] X. Chen, B. Chen, B. Jiang, T. Gao, G. Shang, S.-T. Han, C.-C. Kuo, V. A. L. Roy, Y. Zhou, "Nanowires for UV-vis-IR Optoelectronic Synaptic Devices", Adv. Funct. Mater., 33, pp. 2208807 (2023).

[5] C. Fan, X. Sun, Z. Shi, B. Lü, Y. Chen, S. Li, J.-M. Liu, "Wafer-Scale Fabrication of Graphene-Based Plasmonic Photodetector with Polarization-Sensitive, Broadband, and Enhanced Response", Adv. Opt. Mater., 11, pp. 2202860 (2023).

[6] R. Dutta, A. Bala, A. Sen, M. R. Spinazze, H. Park, W. Choi, Y. Yoon, S. Kim, "Optical Enhancement of Indirect Bandgap 2D Transition Metal Dichalcogenides for Multi-Functional Optoelectronic Sensors", Adv. Mater., 35, pp. 2303272 (2023).

Impact of Titanium Nitride (TiN) Thickness Uniformity on the Reliability of n-FinFETs: A Comparative Study of ALD and PVD TiN Techniques

Mingyang Sun[1,2,3], Yunfei Shi[1,2,3], Hong Yang[1,2,3*], Qianqian Liu[1,2], Qingzhu Zhang[1,2,3], Tao Yang[1,2,3], Junfeng Li[1,2,3], Huaxiang Yin[1,2,3], Xiaolei Wang[1,2,3], Jun Luo[1,2,3] and Wenwu Wang

[1] Key Laboratory of Fabrication Technologies for Integrated Circuits, Chinese Academy of Sciences, Beijing 100029, China.

[2] Institute of Microelectronics, Chinese Academy of Sciences, Beijing 100029, China.

[3] the School of Integrated Circuits, University of Chinese Academy of Sciences, Beijing 100049, China.

Abstract

To study the thickness uniformity of TiN in FinFETs, the electrical characteristics and reliability of n-FinFETs with ALD TiN and PVD TiN as barrier layer are investigated. Despite almost similar threshold voltages (~24mV), PVD TiN-based devices exhibit poorer BTI and HCD reliability (11% and 24%). TCAD simulation shows that the thick-TiN in top-Fin with lower workfunction (WF) induces the large electrical field. Therefore, initial threshold voltage and reliability of n-FinFET is sensitive to the thickness variation of TiN barrier layer.

Introduction

TiN is widely used as a gate electrode material in advanced semiconductor devices due to its excellent electrical and thermal properties[1-2]. As an effective work-function-determining metal (WFM), the thickness of TiN can be tuned to modulate the effective work function[3], thereby optimizing device performance. Additionally, the thickness of the TiN layer plays a crucial role in device reliability. It is proposed that the use of TiN stacks influence trap behavior and physical mechanisms underlying degradation.[4]. Additionally, previous studies have demonstrated that thicker TiN layers generally enhance the Bias Temperature Instability (BTI) and Time-Dependent Dielectric Breakdown (TDDB) reliability of the devices[5-6]. Furthermore, variations in TiN thickness have been found to significantly impact device performance[7]. Despite these findings, the impact of TiN thickness variation across the fin on device reliability remains insufficiently explored. This study aims to address this gap by comparing the BTI and Hot Carrier Degradation (HCD) reliability of n-FinFETs fabricated using physical vapor deposition (PVD) or Atomic Layer Deposition (ALD) TiN. Through comprehensive electrical characterizations and TCAD simulations, we aim to elucidate the mechanisms underlying the influence of TiN thickness uniformity on device degradation and reliability.

Device and Experiment

A. Devices

The replacement metal gate Si n-FinFETs are fabricated utilizing a full-gate-last replacement-metal-gate (RMG) process. As shown in Fig.1(a), the gate stack is composed of a chemical oxide-based interfacial layer, a HfO2-based high-k layer, a TiAlC/TiN-based multilayer work function metal, and a W-based filling metal. Two different processes are employed for the growth of TiN. In Fig. 1(b), TiN layer is grown using ALD, while in Fig. 1(c), TiN layer is deposited via PVD. As can be observed from TEM image, thickness of the ALD TiN is more uniform. In contrast, the PVD TiN exhibits a thinner deposition on the fin sidewalls and a thicker deposition on the top of the fin. To further compare their electrical performance, corresponding MOSCAP are also fabricated using the same two processes.

B. Experiment

Unless otherwise stated, all measurements are conducted at 125°C. The equivalent oxide thickness (EOT) and flat band voltage (V_{fb}) are estimated from the measured C-V curves utilizing the CVC-Hauser tool on MOSCAPs with the Keysight B1500A. BTI and HCD tests are conducted on devices with a channel length of 500 nm for a duration of 1000 seconds. The time-dependent changes in the threshold voltage (ΔV_{th}) are characterized using the measure-stress-measure method. An Agilent B1500A equipped with a waveform generator/fast measurement unit (WFGMU) is utilized for these measurements. The threshold voltage (V_{th}) is extracted employing the constant current method. The Discharge-based Multiple Pulse (DMP) experiments are also performed with Keysight B1530 to investigate the energy distribution of generated oxide traps.

Results and Discussion

A. Electrical Characterization

A comparative analysis of the C-V curves between the ALD and PVD-TiN based MOSCAPs is presented in Fig. 2(a). Results obtained through fitting using the CVC-Hauser tool reveal that they exhibit similar EOT. However, V_{fb} of MOSCAP with PVD-TiN is found to be 190 mV lower than that of MOSCAP with ALD-TiN. This indicates a 0.19eV difference in the work function between the TiN layers fabricated using two different processes.

In Fig. 2(b), it is evident that the gate leakage of MOSCAP with PVD-TiN is increased by an order of magnitude compared to MOSCAP with ALD-TiN.

In Fig. 3(a), I_d-V_g curves is tested on n-FinFET devices fabricated using the two different processes. The cumulative distribution of V_{th} is compared in Fig. 3(b). It is observed that the V_{th} distribution of n-FinFET with PVD-TiN is more uneven and is 24 mV lower than that of n-FinFET with ALD-TiN on average. This finding appears to contradict the previous conclusion that the work function difference between MOSCAPs with PVD and ALD-based TiN is 0.19 eV. This discrepancy can be attributed to the non-uniform deposition of PVD TiN. The work function of the thicker PVD TiN deposited on the top of the fin is found to be closer to that of the corresponding MOSCAPs, whereas the work function of the thinner TiN on the fin sidewalls is observed to be higher. Consequently, the V_{th} of FinFETs with PVD and ALD-based TiN are ultimately more similar.

B. Reliability Characterization

As shown in Fig. 4(a), BTI tests are conducted on FinFETs under various overdrive voltages ($V_{ov}=1.0V/1.2V/1.4V$). In Fig. 4(b), It is found that the average degradation of the FinFET with PVD-TiN is 11% higher than that of the FinFET with ALD-TiN. In Fig. 5(a), HCD tests are also performed under different V_{ov}, revealing that the average degradation of the FinFET with PVD-TiN is 24% higher. These results indicate that FinFETs with PVD-TiN exhibit poorer reliability compared to FinFETs with ALD-TiN.

Subsequently, a comparison of the degradation due to interface and oxide traps is conducted. In Fig. 6, the change in subthreshold swing (SS) is used to compare the number of newly generated interface traps on FinFETs after BTI and HCD. It is observed that FinFETs with PVD-TiN exhibit a slightly higher number of new interface traps compared to FinFETs with ALD-TiN after 1000 seconds at $V_{ov} = 1.4$ V.

In Fig. 7, the energy level distribution of newly generated oxide traps after BTI and HCD stress is compared using DMP. After BTI stress, it is found that the density of traps near the valence band in FinFETs with PVD-TiN increased by 13% at higher V_{ov} compared to FinFETs with ALD-TiN. After HCD stress, density of traps near the valence band in FinFETs with PVD-TiN increases by 24% at higher V_{ov}. This further demonstrates the poorer reliability of FinFETs with PVD-TiN.

C. TCAD Simulation

To elucidate the cause of different reliability observed in FinFETs with PVD-TiN and ALD-TiN, TCAD simulations are performed. FinFETs with ALD-TiN is simulated using device with a contact with uniform work function, as seen in Fig.8(a). In contrast, FinFET with PVD-TiN is simulated using n-FinFET whose the work function at the top is 0.19 eV lower than that on the sidewalls, as seen in Fig.8(c). The FinFET structure is shown in Fig. 9(a). By adjusting the distribution of contact with different work function, V_{th} of the

FinFET with PVD-TiN is tuned to differ by 24 mV from that of FinFET with ALD-TiN, as illustrated in Fig. 9(b).

As shown in Fig. 10, the electric field in oxide layer reveals that, compared to FinFET with ALD-TiN, a stronger localized electric field exists at the top of the fin on FinFET with PVD-TiN. This is because the uneven thickness of the PVD TiN results in a non-uniform work function, which in turn leads to the stronger localized electric field. Consequently, the reliability of FinFET with PVD-TiN is significantly poorer than that with ALD-TiN.

Conclusion

By comparing the electrical performance and reliability of n-MOSCAPs and n-FinFETs fabricated using PVD or ALD TiN as barrier layer, it is found that the PVD TiN exhibits non-uniform thickness across the Fin (thick top-Fin and thin side-Fin). Although the devices prepared by both processes exhibit similar V_{th}, PVD TiN-based devices demonstrate inferior reliability. This is attributed to the variations in the work function across different regions of the fin, which leads to stronger localized electric fields, particularly at the top of the fin. Our findings highlight the critical role of thickness uniformity of TiN deposition in achieving reliable and high-performance n-FinFETs.

Acknowledgments

This work is financially supported by National Natural Science Foundation of China under Grant 62374177.

References

[1] C. Kang et al., "Effects of ALD TiN Metal Gate Thickness on Metal Gate /High-k Dielectric SOI FinFET Characteristics," in 2006 IEEE international SOI Conferencee Proceedings, Niagara Falls, NY, USA: IEEE, Oct. 2006, pp. 135–136.

[2] W. Zhang, J. Cai, D. Wang and Q. Wang, "Properties of TiN films deposited by atomic layer deposition for through silicon via applications," in *2010 11th International Conference on Electronic Packaging Technology & High Density Packaging*, Aug. 2010, pp. 7–11.

[3] M. Kadoshima et al., "Effective-Work-Function Control by Varying the TiN Thickness in Poly-Si/TiN Gate Electrodes for Scaled High-k CMOSFETs," IEEE Electron Device Lett., vol. 30, no. 5, pp. 466–468, May 2009.

[4] T.-S. Jang et al., "Study on the Vt variation and bias temperature instability characteristics of TiN/W and TiN metal buried-gate transistor in DRAM application," in 2014 IEEE International Reliability Physics Symposium, Jun. 2014, p. 2E.4.1-2E.4.4.

[5] C.-L. Chen, "TiN Metal Gate Electrode Thickness Effect on BTI and Dielectric Breakdown in HfSiON-Based MOSFETs," IEEE Trans. Electron Devices, vol. 58, no. 11, pp. 3736–3742, Nov. 2011.

[6] C. L. Chen, "TiN Thickness Impact on BTI Performance," IEEE Electron Device Lett., vol. 32, no. 6, pp. 707–709, Jun. 2011.

[7] Singanamalla et al., "On the impact of TiN film thickness variations on the effective work function of poly-Si/TiN/SiO/sub 2/ and poly-Si/TiN/HfSiON gate stacks," IEEE Electron Device Lett., vol. 27, no. 5, pp. 332–334, May 2006.

979-8-3315-0417-5/25 $31.00 © 2025 IEEE

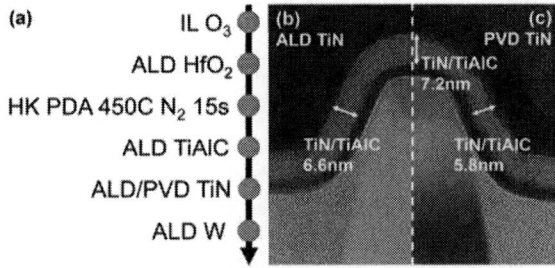

Fig. 1:(a) Schematic illustration of the n-FinFET fabrication process, (b) Transmission electron microscopy (TEM) image of the TiN layer deposited by ALD, (c) TEM image of the TiN layer deposited by PVD.

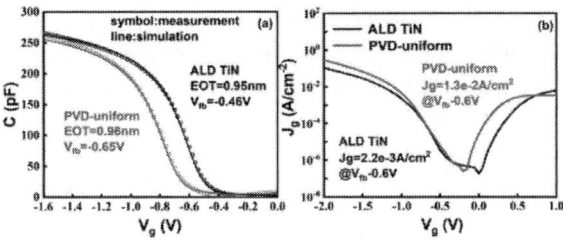

Fig. 2: (a)C-V curves of MOSCAPs with ALD or PVD-TiN. A linear model is applied to characterize the C-V curves for estimating the EOT and V_{fb}. (b) I_g-V_g curve of MOSCAPs with ALD or PVD-TiN.

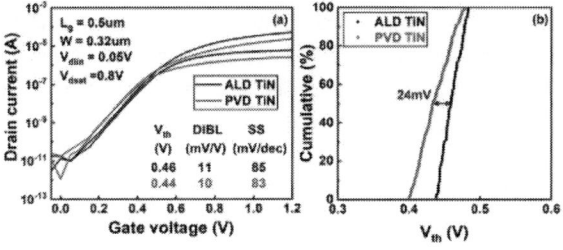

Fig. 3: (a)I_d-V_g curves of n-FinFETs with ALD or PVD-TiN. (b) Cumulative distribution of V_{th} of all n-FinFETs with ALD or PVD-TiN.

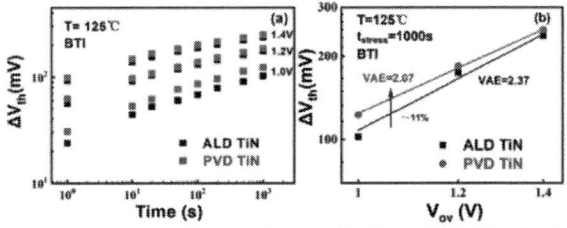

Fig. 4: (a) BTI degradation of n-FinFETs with ALD or PVD-TiN. (b) BTI degradation after 1000s stress. VAE of PVD TiN-based n-FinFET is lower.

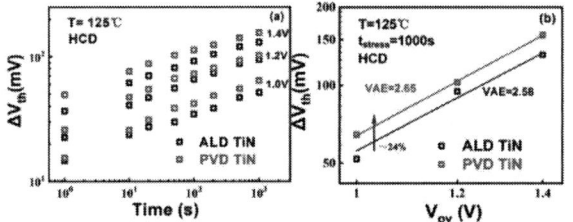

Fig. 5: (a)HCD degradation of n-FinFETs with ALD or PVD-TiN. (b) HCD degradation after 1000s stress. VAE of n-FinFETs with ALD or PVD-TiN is more consistent.

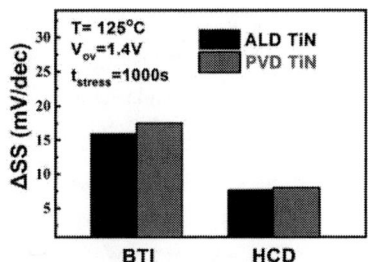

Fig. 6: Variation in SS is extracted to compare newly grown interface traps. PVD TiN-based n-FinFET is observed to grow more interface traps.

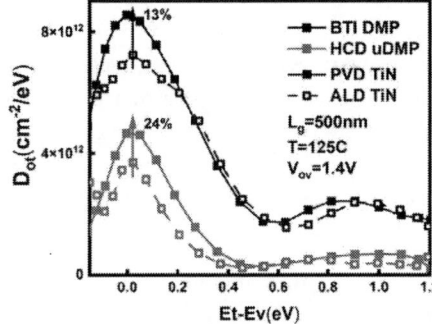

Fig. 7: Density of traps is extracted through DMP tests. PVD TiN-based n-FinFET is observed to grow more oxide layer traps.

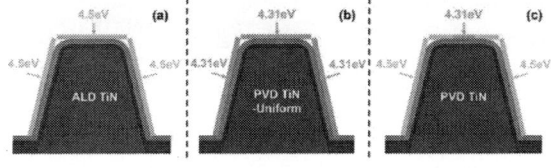

Fig.8: Simulation structure of n-FinFET with(a)ALD TiN, (c)PVD TiN and (b) PVD TiN with uniform work function.

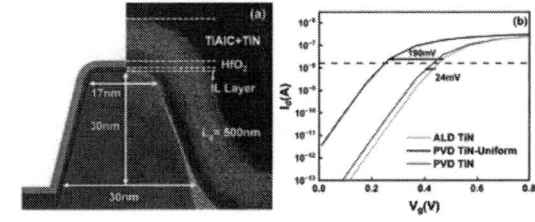

Fig.9: (a)Simulation structure of n-FinFET (b) I_g-V_g curve of n-FinFET with ALD TiN, PVD TiN and PVD TiN with uniform work function.

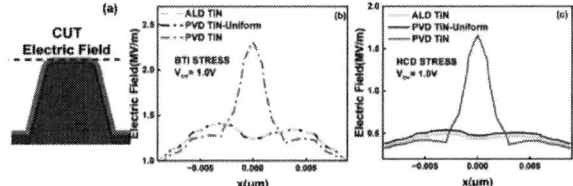

Fig.10: Electric field at the top of the fin in oxide layer (b)at BTI stress when $V_g=V_{ov}= 1.0V$ (c) at HCD stress when $V_g=V_d= V_{ov}=1.0V$.

Dual-Gate Thin-Film Transistor-Based Multi-Parameter Sensor for Comprehensive Water Quality Monitoring in Aquatic Environments

Qiang Chen[1], Qiyi Su[1], Haolin Zhao[1], Zhancheng Mai[1], Kai Zhuang[1], Yitong Xu[1], Xinghui Liu[2], and Kai Wang[1*]

[1]Guangdong Province Key Laboratory of Display Material and Technology, State Key Laboratory of Optoelectronic Materials and Technologies, School of Electronics and Information Technology, Sun Yat-sen University, Guangzhou, China

[2]Guangke Chipwey Sensing Technologies Co. Ltd., Foshan, China

*Corresponding author (wangkai23@mail.sysu.edu.cn)

Abstract

This work presents a multi-parameter water quality sensor system based on dual-gate thin-film transistor (DG-TFT) for accurate aquatic monitoring, targeting pH, dissolved oxygen (DO), and temperature. The sensor chip achieves low-cost, high-sensitivity performance, offering a compact alternative to traditional discrete modules that are bulky and costly. Integrated with temperature compensation algorithms, the final circuit system ensures precise measurements in dynamic environments.

（Keywords: Water quality monitoring,Dual-gate thin-film transistor, Multi-parameter sensor, On-chip integration）

Introduction

Effective water quality monitoring is critical for maintaining healthy aquatic environments, particularly in applications such as aquaculture, where parameters like DO, pH, and temperature are essential for sustaining aquatic ecosystems[1].Traditional multi-parameter sensors, which often rely on discrete, standalone components for each parameter, present challenges in terms of bulkiness, high power consumption, and limited integration potential. These limitations reduce their effectiveness in modern monitoring systems that demand compact, real-time, and energy-efficient solutions[2].

The sensor system's design enables significant miniaturization while enhancing measurement accuracy and stability through advanced features such as temperature compensation and signal amplification. The DG-TFT, a key component of the sensor chip, facilitates precise control of sensor responses, enabling high sensitivity and improved signal processing for reliable operation in dynamic aquatic environments.The primary innovation lies in integrating multiple sensing functions into a unified platform, overcoming traditional water quality monitoring challenges. Leveraging DG-TFT technology, this system offers a scalable, cost-effective solution to meet the growing demand for efficient, high-performance water quality monitoring across various applications.

Sensor Circuit Design

The equivalent circuit model of the dual-gate thin-film transistor is shown in Fig. 1(a), offering independent control of the top and bottom gates. This configuration enhances flexibility in modulating electrical properties, providing greater precision for multi-parameter sensing applications[3]. Fig. 1(b) shows the DG-TFT-based amplification circuit for the hydrogen ion-selective electrode. The Working Electrode (WE) transfers the hydrogen ion potential to the circuit voltage. The common-source configuration offers high input impedance, minimizing readout circuit impact on sensitive front-end elements and reducing interference from the readout circuit on the sensor.Fig. 1(c) shows a temperature sensing circuit based on the gate-source voltage and temperature relationship of the DG-TFT. Two DG-TFTs with identical width-to-length ratios share their bottom gates and drains, each connected to different resistors. This differential design controls the drain-source currents (I_{DS1} and I_{DS2}) by adjusting resistor values, enabling fine-tuning of output voltage sensitivity.DO sensing uses a potentiostatic method for signal readout. Due to the low open-loop gain of the TFT amplifier, an external circuit is required for driving and reading signals. Fig. 1(d) shows this external circuit, which stabilizes the potential between the Reference Electrode (RE) and the WE in a three-electrode system. The RE provides a stable reference potential, while the Counter Electrode (CE) completes the current loop with the WE, ensuring accurate current measurement through feedback control.

Chip Design and Characterization

Fig. 2 shows the non-crystalline silicon dual-gate thin-film transistor glass chip we designed and fabricated, along with microscopic images of some components. This chip integrates sensing units for dissolved oxygen, pH, temperature, and conductivity, each specifically optimized to achieve high sensitivity and accuracy for monitoring aquatic environmental parameters. Additionally, the chip incorporates amplification circuits and a dedicated testing unit to enhance signal fidelity and facilitate precise calibration, ensuring the reliability of real-time data acquisition.

To meet sensor circuit design requirements, this work employs DG-TFTs with dimensions of 1000 μm/2.8 μm and 5 μm/2.8 μm. Fig. 3 shows their transfer characteristics, revealing a strong on/off current ratio and low off-state

979-8-3315-0417-5/25 $31.00 © 2025 IEEE

current, indicating reliable switching. The top-gate bias provides precise control over the operating region, making DG-TFTs ideal for applications needing accurate sensor parameter tuning.

Subsequent characterization focused on the three primary sensing parameters of the chip, beginning with the pH sensing circuit. To optimize sensor performance, the top-gate bias was set to 5 V, ensuring the sensor operated in the amplification region. In this configuration, a PVC membrane-based hydrogen ion carrier I was used as the ion-sensitive layer, replacing the previously tested low-response silicon nitride membrane[4]. Samples with varying pH values were prepared by diluting hydrochloric acid and sodium hydroxide solutions. The test results, shown in Fig. 4, indicate a high sensitivity of 72 mV/pH over a response range of pH 4 to 9.

During the characterization of the temperature sensing circuit, with V_{DD} fixed at 5 V, an optimal resistor pair of 100 MΩ and 1 MΩ was selected for the external differential resistors. Using this configuration, the optimal V_{DD} value was determined to be 15 V. Temperature tests were then conducted with top-gate biases set at -10 V, -5 V, 0 V, 5 V, and 10 V. As shown in Fig. 5, the response sensitivity remained nearly constant under negative top-gate biases. When the top-gate bias was positive, however, sensitivity increased with rising temperature, reaching -17 mV/C at a top-gate bias of 10 V. This demonstrates that positive top-gate biases enhance the temperature response of the sensor.

The DO sensor was characterized using an electrochemical workstation[5] with a startup voltage set at -1.2 V. DO concentration levels were adjusted by introducing nitrogen gas into deionized water, thereby reducing the oxygen concentration. The DO concentration in the samples was calibrated using a fluorescence-based dissolved oxygen probe, ensuring accuracy across different concentration levels. The fabricated glass chip was then immersed in the water samples, and the output current was measured. The results, shown in Fig. 6, indicate that the current response of the DO sensor circuit is in the sub-microampere range, achieving a sensitivity of -13.5 nA/(mg/L).

System Integration and Temperature Compensation

In the previous section, the response sensitivity parameters of each sensor in the multi-parameter sensing chip were obtained through individual characterization. As shown in Fig.7, a hardware circuit based on an STM32 microcontroller was designed to support the multi-parameter sensing chip. This circuit integrates the readout circuitry and control functions for sensor operation and driving adjustments. Additionally, a LabVIEW-based upper computer system was developed for real-time visualization and data acquisition, enabling continuous monitoring and efficient data handling.

The output signals of the dissolved oxygen and pH sensors exhibit a dependency on temperature parameters, necessitating temperature compensation to ensure the accuracy of the sensor outputs[6]. Temperature data is incorporated into the system to correct the sensor readings dynamically, enabling improved precision in the measurement results. The results after applying temperature compensation to the system-tested data are presented in Table 1, illustrating the effectiveness of the compensation algorithm in refining the sensor's output signals for reliable environmental monitoring.

Conclusion

The DG-TFT sensor chip exhibits high response sensitivities of 72 mV/pH for pH, -13.5 nA/(mg/L) for DO, and -17 mV/°C for temperature. With integrated temperature compensation, the system achieves precise and stable measurements across varying conditions. The demonstrated capabilities suggest this system can replace bulky and costly traditional discrete modules, offering a compact, efficient solution for continuous and accurate aquatic ecosystem monitoring.

Acknowledgments

This work was funded by Foshan National Hi-tech Industrial Development Zone Industrialization Entrepreneurial Teams Program (2120197000110).

References

[1] Islam, Md Jakiul, Andreas Kunzmann, and Matthew James Slater. "Responses of aquaculture fish to climate change - induced extreme temperatures: A review." Journal of the World Aquaculture Society 53.2 (2022): 314-366

[2] Yang, Haibo, et al. "A review of remote sensing for water quality retrieval: progress and challenges." Remote Sensing 14.8 (2022): 1770.

[3] M. M. Billah, M. H. Rabbi, C. Park and J. Jang, "Highly Sensitive Temperature Sensor Using Low-Temperature Polysilicon Oxide Thin-Film Transistors," in IEEE Electron Device Letters, vol. 42, no. 12, pp. 1864-1867, Dec. 2021

[4] Shojaei Baghini, Mahdieh, et al. "Ultra - thin ISFET - based sensing systems." Electrochemical Science Advances 2.6 (2022): e2100202.

[5] Lv H, Pan Q, Song Y, et al. A review on nano-/microstructured materials constructed by electrochemical technologies for supercapacitors[J]. Nano-Micro Letters, 2020, 12 : 1-56

[6] Datar, Rishikesh, and Gautam Bacher. "Influence of gate material, geometry, and temperature on ISFET performance in pH sensing applications." Silicon 15.12 (2023): 5393-5405.

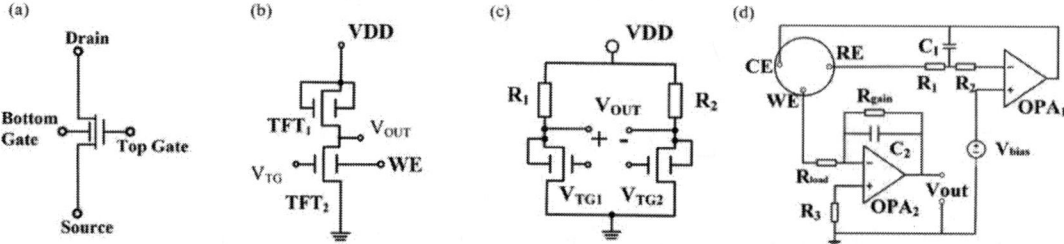

Fig. 1: (a) Device circuit model (b) PH sensing signal amplification circuit (c) Temperature sensing circuit
(d) External DO sensing drive and readout circuit

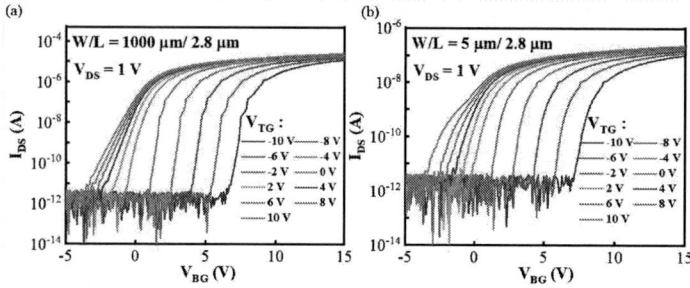

Fig. 3: (a) Transfer characteristics curve of a 1000 μm/2.8 μm DG-TFT
(b) Transfer characteristics curve of a 5 μm/2.8 μm DG-TFT

Fig. 4: Response curve of pH sensor

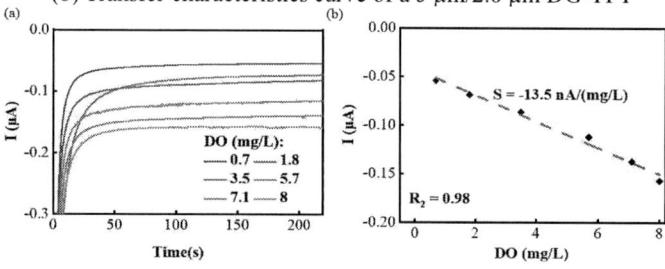

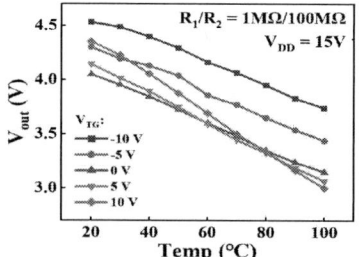

Fig. 6: (a) Current-time curve under different DO concentrations
(b) Dissolved oxygen concentration-current fitting curve

Fig. 5: Temperature-output voltage curves under different top-gate Biases

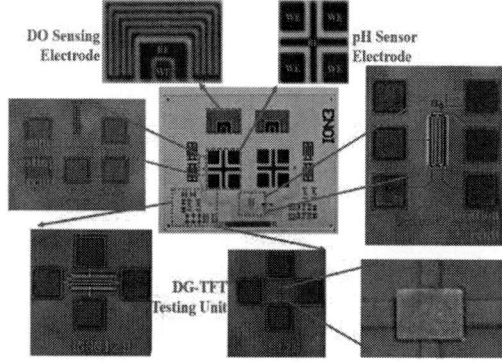

Fig. 2: Chip and partial microscopic

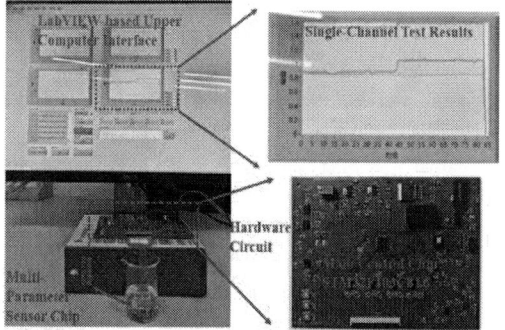

Fig. 7: Hardware circuit testing environment

	Sample 1	Sample 2	Sample 3	Sample 4	Sample 5
pH(Actual)	7	5.7	8.2	6.3	4.4
pH(Test)	6.91	5.64	8.18	6.22	4.43
DO(Actual)(mg/L)	7.8	5.6	4.3	2.2	3.4
DO(Test)(mg/L)	7.77	5.63	4.35	2.26	3.36
Temp(Actual)(°C)	23	20	28	15	11
Temp(Test)(°C)	23.13	20.07	27.96	14.98	11.6

Table. 1: Multi-parameter sensing test results after temperature compensation

979-8-3315-0417-5/25 $31.00 © 2025 IEEE

Impact of Sb$_2$Se$_3$ annealing on the photoresponse of TiO$_2$/Sb$_2$Se$_3$/Si back-to-back photodiodes

Bangsen Ouyang[1,2,#], Jialiang Wang[1,2,#] and Yang Chai[1,2,*]

[1]Department of Applied Physics, The Hong Kong Polytechnic University, Hong Kong, China

[2] Joint Research Center of Microelectronics, The Hong Kong Polytechnic University, Hong Kong, China

\# These authors contribute equally

e-mail: ychai@polyu.edu.hk

Abstract

Vision sensors utilizing TiO$_2$/Sb$_2$Se$_3$/Si back-to-back photodiodes demonstrate the ability to rapidly enhance image quality. However, the performance of the back-to-back photodiodes still requires improvement. Here, we show that raising the annealing temperature of Sb$_2$Se$_3$ from 250 to 350 °C results in a 67-fold increase in photocurrent density. Results indicate that the reduced series resistance in the photodiodes contributes to this enhancement. This work offers a practical approach to enhance the performance of back-to-back photodiodes.

Keywords: Sb$_2$Se$_3$ photodiode, vision sensor, and post-deposition annealing

Introduction

Light spectra encode rich visual information[1-3]. Conventional vision sensors with fixed spectral response only work for the specific waveband, which hinders the comprehensive perception of the visual information distributed among various wavebands in the environment[4]. Recently, a spectra-adapted vision sensor utilizing titanium dioxide (TiO$_2$)/ antimony selenide (Sb$_2$Se$_3$)/ silicon (Si) back-to-back photodiodes has been reported[5]. This sensor can operate effectively across two different wavebands, offering a promising solution for improving machine vision in challenging lighting conditions.

Previous research has shown that optimizing the annealing temperature can improve the crystallinity and carrier concentration of Sb$_2$Se$_3$ thin films[6-8]. Nevertheless, the effect of the post-deposition annealing on the back-to-back photodiode structure remained unexplored. In this work, we have studied the effects of annealing temperature on the performance of the back-to-back photodiodes, for strengthening the understanding of the proposed device structure and potentially improving the device performance.

Experimental details

Fig. 1 depicts the spectra-adapted vision sensor with a 5-layer-stack structure. Thick (~1μm) antimony selenide (Sb$_2$Se$_3$) deposited on n-Silicon (0.05 - 0.20 Ω cm) wafer by thermal evaporation served the light absorber. To study the effects of post-deposition annealing on the device, the Sb$_2$Se$_3$ was annealed at various temperatures in flowing argon, before 200 nm titanium oxide (TiO$_2$) was sputtered onto the Sb$_2$Se$_3$.

The two back-to-back heterojunctions formed at two ends of the absorber through TiO$_2$ / Sb$_2$Se$_3$ and Sb$_2$Se$_3$ / n-Si contacts. The activated top or bottom junctions collect short- or long-wavelength photoexcitation mostly generated at the top- and bottom-end of the Sb$_2$Se$_3$ absorber. Through this charge-collecting modulation mechanism, the sensor achieves distinctive spectral responses vertically integrated into one footprint, without the participation of additional optical filters. More detailed device physics is reported in our previous paper[5].

Results and discussion

Fig. 2 shows the current-voltage curves of the back-to-back photodiodes with the Sb$_2$Se$_3$ layer annealed at temperatures of 250, 300, and 350 °C. The dark current density for all three groups of back-to-back photodiodes remains low, consistently below 6×10^{-2} mA/cm^2, regardless of whether a negative or positive bias is applied (Fig. 2a-c). This is attributed to at least one of the photodiodes in the back-to-back configuration being reverse-biased. The current density under light illumination increases with higher annealing temperatures of the Sb$_2$Se$_3$ (Fig. 2a-c). To facilitate a clearer comparison, the photocurrent density of the back-to-back photodiodes was derived by subtracting the dark current density from the current density recorded under 808 nm light illumination (Figs. 2d,e). At a bias voltage of -1 V (Figs. 2d), the photocurrent density increases tenfold with annealing temperatures rising from 250 to 350 °C, while at 1 V (Figs. 2e), it increases by 67 times. The increase in photocurrent density indicates an enhancement in the light-to-electricity conversion efficiency of both the top and bottom photodiodes in the back-to-back configuration.

Since the current density for the forward-biased top or bottom photodiode cannot be directly obtained from the back-to-back photodiodes, we prepared single-photodiode devices of TiO$_2$/Sb$_2$Se$_3$ and Sb$_2$Se$_3$/Si. This approach allows us to investigate the effect of the annealing temperature of Sb$_2$Se$_3$ on the series resistance generated by the forward-biased photodiodes.

979-8-3315-0417-5/25 $31.00 © 2025 IEEE

Fig. 3a illustrates the current density-voltage curves of the TiO_2/Sb_2Se_3 single-photodiode devices, measured under dark conditions, with annealing temperatures of Sb_2Se_3 ranging from 250 °C to 350 °C. The dark current density of all three devices at a reverse bias (corresponding to positive voltages) closely matches that of the back-to-back photodiodes at the corresponding annealing temperatures. This result verified the proposed mechanism of the back-to-back photodiodes, indicating that the current density at positive voltages is primarily governed by the top junction. The TiO_2/Sb_2Se_3 single-photodiode annealed at relatively low temperatures (250 °C and 300 °C) shows poor rectify behavior, with a rectification ratio (R) at ±1 V of only 1.9 and 11.5, respectively. These two photodiodes show low current density at both reverse and forward bias, probably because of the high bulk resistance of the poorly crystallized Sb_2Se_3 layer. As the annealing temperature of Sb_2Se_3 rises to 350 °C, the current density of the forward-biased TiO_2/Sb_2Se_3 photodiode significantly increases, while the reverse saturation current density remains unchanged at high positive voltages. Consequently, the high-temperature TiO_2/Sb_2Se_3 photodiode demonstrates improved rectifying behavior, achieving an R of 1100. The reverse saturation current density at low voltage also increases noticeably, resulting in a current density-voltage curve at the reverse bias that resembles a leaky-saturation shape, characteristic of the current density-voltage response in heterojunctions.

Figure 3b shows the current density-voltage curves of the Sb_2Se_3/Si single-photodiode devices. The reverse saturation current density (corresponding to negative voltages) of these photodiodes also aligns well with the current density of back-to-back photodiodes (corresponding to negative voltages) with corresponding annealing temperatures. This observation indicates that the current density-voltage response at negative voltages is primarily influenced by the bottom photodiode.

The forward bias current density of the control devices increases with the annealing temperature from 200 °C to 300 °C. This result aligns with the trend observed in the TiO_2/Sb_2Se_3 top photodiode, suggesting that the bulk resistance of the Sb_2Se_3 layer decreases as the annealing temperature increases. The forward bias current of the Sb_2Se_3/Si single-photodiode density annealed at 350 °C noticeably decreased, resulting in the loss of the rectifying behavior. This deviation is likely due to the Fermi level of the Sb_2Se_3 annealed at 350 °C falling below that of the sputtered ITO, leading to the formation of a Schottky contact between the Sb_2Se_3 and the ITO electrode.

The test results of single-photodiode control devices reveal that increasing annealing temperature can enhance the rectification contact between the Sb_2Se_3 absorption layer with TiO_2 and Si. The enhanced internal electrical field at reverse bias for activated junction facilitates electron-hole splitting, while large forward current for inactive junction indicates reduced series resistance in the back-to-back photodiodes. Subsequently, a high annealing temperature of 350 °C yields optimum device characteristics as shown in Fig. 2.

Fig. 4a shows the photoresponsivity of the back-to-back photodiodes with Sb_2Se_3 annealed at 350 °C, when a positive bias is applied, the operational spectral range is broad and covers the visible spectra; when a negative bias is applied, the operational spectral range is narrowly distributed within the near-infrared spectra. The maximum photoresponsivity reaches 0.4 A/W at a bias of 2 V and − 0.4 A/W at a bias of − 2 V. The specific detectivity of the device is shown in Fig. 4b, the maximum specific detectivity reaches $3.5×10^{11}$ Jones. The photoresponsivity and specific detectivity of the back-to-back photodiodes are comparable to those of high-performance Sb_2Se_3-based photodetectors[9], highlighting their potential for practical applications.

Conclusion

We fabricated back-to-back photodiodes based on a $TiO_2/Sb_2Se_3/Si$ architecture. Increasing the annealing temperature of Sb_2Se_3 from 250 to 350 °C results in a 67-fold enhancement in photocurrent density. The forward bias current density of the fabricated TiO_2/Sb_2Se_3 and Sb_2Se_3/Si photodiodes increases by two orders of magnitude as the annealing temperature rises from 250 to 300 °C. This indicates that higher annealing temperatures lower the series resistance from the forward-biased photodiodes in the back-to-back configuration, thereby contributing to the enhancement of output photocurrent. The maximum photoresponsivity of the back-to-back photodiodes reaches 0.4 A/W showing potential for practical applications.

Acknowledgments

This work is supported by MOST National Key Technologies R&D Programme (2022YFA1203804), National Natural Science Foundation of China (NSFC) for Distinguished Young Scholars (62425405), Research Grant Council of Hong Kong (15301023 and CRS_PolyU502/22), and the Hong Kong Polytechnic University (WZ4X and ZE1T).

References

[1] S. Yuan *et al.*, "Geometric deep optical sensing," *Science*, vol. 379, no. 6637, p. eade1220, 2023.

[2] Z. Wang, T. Wan, S. Ma, and Y. Chai, "Multidimensional vision sensors for information processing," *Nat. Nanotechnol.*, vol. 19, no. 7, pp. 919-930, 2024.

[3] X. Tang, M. M. Ackerman, M. L. Chen, and P. Guyot-Sionnest, "Dual-band infrared imaging using stacked colloidal quantum dot photodiodes," *Nat. Photonics*, vol. 13, no. 4, pp. 277-282, 2019.

[4] S. Z. Zang, M. Ding, D. Smith, P. Tyler, T. Rakotoarivelo, and M. A. Kaafar, "The Impact of Adverse Weather Conditions on Autonomous Vehicles: How Rain, Snow, Fog, and Hail Affect the Performance of a Self-Driving Car," *IEEE Veh. Technol. Mag.*, Article vol. 14, no. 2, pp. 103-111, 2019.

[5] B. S. Ouyang *et al.*, "Bioinspired in-sensor spectral adaptation for perceiving spectrally distinctive features," *Nat. Electron.*, vol. 7, no. 8, pp. 705-713, 2024.

[6] L. Zhang *et al.*, "Scalable Low-Band-Gap Sb₂Se₃ Thin-Film Photocathodes for Efficient Visible-Near-Infrared Solar Hydrogen Evolution," *ACS Nano*, vol. 11, no. 12, pp. 12753-12763, 2017.

[7] R. Tang *et al.*, "Highly efficient and stable planar heterojunction solar cell based on sputtered and post-selenized Sb₂Se₃ thin film," *Nano Energy*, vol. 64, 2019.

[8] X. Liu *et al.*, "Thermal evaporation and characterization of Sb2Se3 thin film for substrate Sb₂Se₃/CdS solar cells," *ACS Appl. Mater. Interfaces*, vol. 6, no. 13, pp. 10687-95, 2014.

[9] S. Chen *et al.*, "Carrier recombination suppression and transport enhancement enable high-performance self-powered broadband Sb₂Se₃ photodetectors," *InfoMat*, vol. 5, no. 4, 2023.

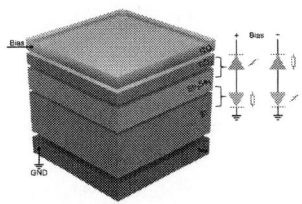

Fig. 1. Schematic device structure of the back-to-back photodiodes.

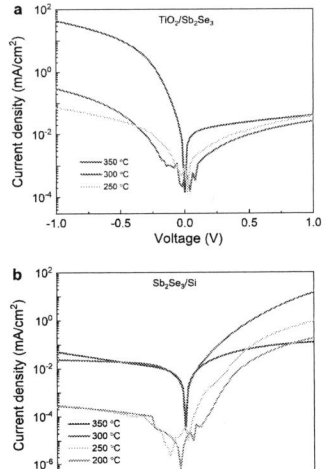

Fig. 3. Impact of annealing temperature of Sb₂Se₃ on the current density of single-photodiode devices. (a,b) Current density-voltage curves of TiO₂/Sb₂Se₃ (a) and Sb₂Se₃/Si (b) single-photodiode under dark conditions, utilizing Sb₂Se₃ annealed at various temperatures.

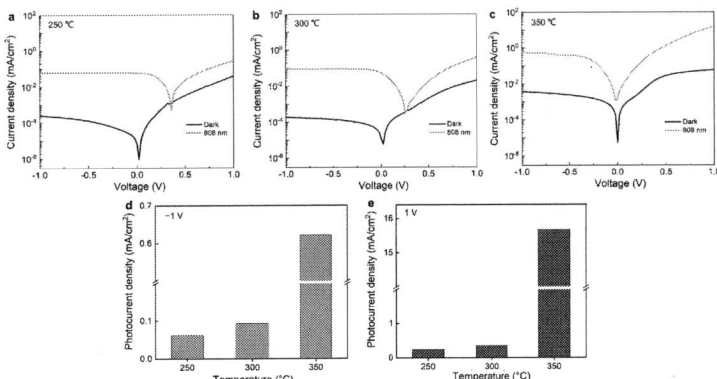

Fig. 2. Influence of annealing temperature of Sb₂Se₃ thin film on photoresponse of the back-to-back photodiode. (a-c) Current density-voltage curves of the back-to-back photodiode in the dark condition and under light illumination when the annealing temperature of Sb₂Se₃ thin film is 250 °C (a), 300 °C (b), and 350 °C (c). (d-e) photocurrent of the back-to-back photodiode extracted from (a-c) under bias voltage of − 1V (d) and 1V (e).

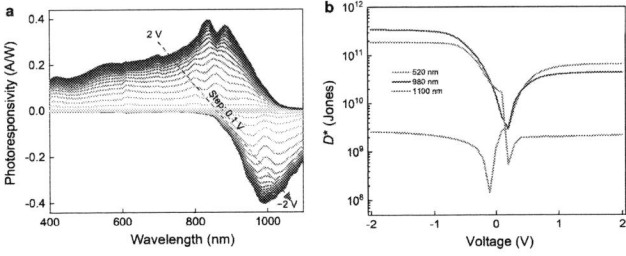

Fig. 4. Photoelectric performance of the optimized back-to-back photodiodes based on an Sb₂Se₃ thin film annealed at 350 °C. (a) Photoresponsivity of the back-to-back photodiode under different bias voltages. (b) The specific detectivity of the back-to-back photodiode.

Design and TCAD Simulation of a Surface Super-Junction Based GaN HEMT with Enhanced Breakdown Voltage

Jiayi An[1], Fuqiang Guo[2], Zhongchen Ji[1], Sen Huang[1], Qimeng Jiang[1], Xinhua Wang[1], Ke Wei[1], Xinyu Liu[1]

[1] Institute of Microelectronics, University of Chinese Academy of Science, Beijing 100049, China

[2] State Key Laboratory of Artificial Microstructure and Mesoscopic Physics, Nano-optoelectronics Frontier Center of Ministry of Education, School of Physics, Peking University, Beijing 100871, China

Email: huangsen@ime.ac.cn

Abstract

A surface super-junction effect based AlGaN/GaN-on-Sapphire HEMT is discussed. The structure featuring an ohmic contacted p-GaN gate region extending to the drain area is called a p-string (PST), which is capable of regulating the off-state channel electric field. The impact of doping concentration and length of PST is thoroughly discussed. Additionally, the dynamic depletion and injection behavior of holes in PST are firstly demonstrated in GaN-based power devices. PST acts as a self-adjusting gate at the on-state, thus avoiding deteriorating conduction characteristics of the device. Through TCAD simulation, a blocking voltage (BV) of 930 V under L_{GD}=7 μm is achieved, enhanced by about 7 times compared with traditional-HEMT. While the device possesses R_{on}=10.2 mΩ·mm and I_{sat}=208 mA/mm at V_{GS}=5 V remaining unaffected. The dynamic turning-on time of ~20 ns is also obtained.

Keywords: GaN HEMT with p-string structure (PST-HEMT), surface super-junction, mobile holes, electric field modulation.

Introduction

Gallium Nitride (GaN), as a type of wide-band-gap semiconductor, possesses about 10 times higher material breakdown voltage (3×10^6 V/cm) than silicon. GaN-based high-electron-mobility-transistor (HEMT) features a high-density two-dimensional electron gas (2DEG) at the hetero-junction interface with a mobility of over 1200 cm²/Vs, making GaN HEMT device a promising candidate for application occasions of radio frequency and high voltage[1].

Nevertheless, due to the lack of a built-in p-n junction in the AlGaN/GaN channel, the off-state electric field tends to crowd at the gate edge, which limits exploiting its full potential of blocking voltage. The high electric field peak could not only lead the device to breakdown in advance, but also exacerbate the so-called current collapse effect, deteriorating the stability and power consumption of the device.

In order to flatten the channel off-state electric field, several efforts have been made to induce the so-called super-junction (SJ) structure into GaN HEMTs. Nakajima et al. are the first to fulfill SJ-HEMT, by means of introducing a string-shaped p-doped GaN area (PST) covering the surface[2]. However, the epitaxial thickness of AlGaN barrier layer there must be increased then; otherwise the on-state 2DEG density would be depleted by the surface P-type region. Cui et al. have proposed an ohmic-gate termination extension (GTE) with constant length and doping concentration to optimize R_{on} [3]. Furthermore, the detailed discussion about the impact of PST doping and size on the characteristics of the device, as well as its specific dynamic behaviour is still needed to be clarified for further research.

As key parameters for PST-HEMT, the doping concentration (N_{st}) and length (L_{st}) of PST will be thoroughly discussed in the paper. TCAD simulation is operated to obtain static and dynamic electronic characteristics of MIS-HEMT with key parameters varying. The mechanism behind PST about hole moving is also clearly demonstrated.

Structure and Simulation

The cross-section schematic figure of PST-HEMT is shown in Fig.1, with electrode distance of (L_{GS}/L_G/L_{GD}, 2/1/7 μm). The heterostructure is based on MOCVD-grown AlGaN barrier/GaN channel/GaN buffer on a sapphire substrate. The device features a string-like p-doped cap layer called PST under gate electrode to form an enhance-mode device. PST layer overhangs from under the gate by a length of L_{st}. The top layer of PST is heavily doped with 3×10^{19} cm⁻³ and junction depth of 10 nm to form a p-ohmic contact.

During the off-state, holes in PST are completely or partly depleted by V_{DS} and removed from ohmic-gate. Therefore, the remaining ionized acceptors in PST act as a field-plate to modulate the electric field. A 50nm undoped GaN cap is set under PST to improve interface quality. In the on-state, PST filled with holes will form a two-dimensional hole gas (2DHG) at the uGaN/AlGaN interface.

The key parameters of PST-HEMT are listed in Table 1.

Table 1: Device Parameters Specification

Parameters	Notations	Values
N_{st}/cm⁻³	PST Mg doping	1×10^{17}~3×10^{19}
L_{st}/μm	PST length	2/4/6
T/μm	PST thickness	0.1
T_{AlGaN}/nm	AlGaN barrier thickness	10
$T_{Channel}$/nm	GaN channel thickness	100
T_{Buffer}/μm	GaN buffer thickness	4
N_{AlGaN}/cm⁻³	AlGaN barrier Si doping	1×10^{18}
N_{uid}/cm⁻³	Background Si doping	1×10^{15}
N_{Buffer}/cm⁻³	GaN Buffer C/O doping	1×10^{19}/1×10^{18}
N_{ohm}/cm⁻³	S/D contact Si doping	5×10^{20}
T_{SiN}/μm	SiN$_x$ layer thickness	0.5

In TCAD simulation, several necessary physical models are induced. The Piezoelectric Polarization model is set between hetero-junctions. Low doping efficiency of Mg in GaN is fulfilled by the incomplete ionization model, with ionization energy of acceptor magnesium (Mg) in GaN of 200 meV. Then, hole carrier density in PST is calculated to be 3×10^{15} cm⁻³ ~8×10^{17} cm⁻³ respectively[4].

Source and drain electrodes are evaporated as a Ti/Al/Ni/Au stack with a work-function of ~3.5 eV, while gate electrode as Ni/Au with a work-function of ~4.5 eV. To simulate the actual schottky and ohmic contact, the thermionic emission (TE) model and the trap-assist-tunneling (TAT) model are utilized[4].

The passivation layer is deposited as silicon-nitride. Donor-like traps of 5×10^{13} cm^{-3} with 1.6 eV from the conduction band are set at the SiN$_x$/AlGaN interface to provide a source of channel 2DEG, while deep-acceptor-like traps of 1×10^{13} cm^{-3} with 0.89 eV from valance band are set at the buffer layer to simulate carbon-induced C$_N$ impurities. Simulation results and data measured will be discussed then.

Results And Discussion

A. On-state characteristics

The influence of N$_{st}$ and L$_{st}$ on the output I–V characteristics of PST-HEMT is shown in Fig.2(a). When L$_{st}$ is set at 2/4/6 µm, the saturation current I$_{sat}$ is shown to be 469,304,208 mA/mm respectively, while I$_{sat}$ is almost not affected by N$_{st}$. Fig.2(a) indicates that I$_{sat}$ is negatively correlated to longer L$_{st}$, because the pinch-off point is moved to the PST edge further rather than gate edge, resulting in lower I$_{sat}$. The result indicates that PST-HEMT with longer L$_{st}$ is more beneficial for the short-circuit capability, which is quite important for robustness in high-voltage application [5].

We can also conclude from Fig.2(a) that R$_{on}$ ($\sim$10.2m$\Omega\cdot$mm) is even lower with longer L$_{st}$. That's because the p-ohmic contact injects plentiful holes into PST during the on-state, turning the electrical static potential of PST to be equal with the gate. In other words, PST is capable to enhance 2DEG density as an extended gate in the on-state.

Another conclusion from Fig.2(a) is that both I$_{sat}$ and R$_{on}$ are not obviously affected by N$_{st}$. That's because the p-ohmic gate obtains a high-level injection in the on-state, therefore the intrinsic doping concentration can be negligible.

Fig.3(a) displayed the gate leakage current I$_{GSS}$ characteristics. From Fig.3 we can see that I$_{GSS}$ reaches 0.1 mA/mm^2 as V$_{GS}$ rises to 9.6 V, when gate breakdown occurs. A V$_{GS}$ swing of over 7 V is identical for an ohmic gate.

B. Blocking-Voltage characteristics

The following discussion is about the impact of N$_{st}$ and L$_{st}$ on BV at off-state, while BV is sensitive to both of the parameters, shown in Fig.4. Then length of L is set to be 6 µm constant, covering areas from X=2 to 8 µm. After the device turned off, a static state is reached, then electric field and carrier distributions in the PST-HEMT are obtained.

Fig.6 is shown at a low doping of N$_{st}$=1×10^{17} cm^{-3}. When V$_{DS}$ rises from 0 to 719 V (breakdown voltage), the electric field shows a peak at the gate-edge like traditional HEMTs do, while other parts of channel remain a low level of electric field. In Fig.6(c), the field peak rises sharply when V$_{DS}$ reaches above 700 V. That's because PST had already been fully depleted by drain voltage at 200 V, thanks to the relatively low N$_{st}$ concentration. Therefore, the fixed positive charge provided by acceptors of PST is not enough to withstand the high electric field. Then electric field lines induced by excessive V$_{DS}$ point to the gate corner like Fig.5 .

Experiment in Fig.7 features N$_{st}$ as 5×10^{17} cm^{-3}. When V$_{DS}$ rises from 0 to 930 V, the depletion region length of PST expands from 0 to 3.5 µm synchronously with electric field. During V$_{DS}$ rises, the shape of channel electric field below PST is like a trapezoidal region that continuously lengthens its bottom and upper bottom, until its upper bottom widens to 3.5 µm, similar to GTE-HEMT in [3] does. When V$_{DS}$

reaches 930 V, the channel region under the depleted part of PST from X=4.5 to 8 µm are affected by superjunction, so that the maximum off-state electric field in the region forms flatly without reaching 3×10^6 V/cm. That's an ideal channel field thanks to balanced charges between drain and PST.

If the doping concentration N$_{st}$ reaches 5×10^{18} cm^{-3} or higher, then PST is hardly depleted by V$_{DS}$, shown in Fig.8. Even at V$_{DS}$ raises 800 V, there's only a short depletion region formed at the PST edge (X=8 µm). The small depletion region would attract a large number of electric field lines at high V$_{DS}$, thus forming a new electric field peak there, with a high risk of breakdown. Moreover, owing to the electric field peak closer to the drain electrode, the drain induced barrier lowering (DIBL) effect near drain region would be much more severe, thus deteriorating buffer leakage current.

As for buffer leakage current paths, electrons always start from the drain 2DEG, then step over the depletion region of the channel and inject into the source 2DEG. Due to DIBL effect, Fig.8(c) with highest PST doping concentration possesses the most remarkable leakage current at V$_{DS}$=200 V.

C. Dynamic characteristics

By utilizing Masetti and Canali model in simulation, the hole mobility in p-GaN is calculated to be $\sim$7 to 11 cm^2/(V·s) with N$_{st}$ varying, thus the depletion of PST might be retarded. DUT is tested under the mono-pulse circuit in Fig.9. When turning on, although Fig.9(a) shows the voltage drop time t$_{ON}$ is mainly positively correlated with L$_{st}$, t$_{ON}$ still stands quite fast from 10 to 20 ns, which is capable of a 10 MHz switching frequency. The hole extraction time about devices turning off is quite longer and needs to be further optimized.

Conclusion

This study focuses on the doping, size and dynamic behavior of PST area upon HEMT. The PST structure is set to modulate the off-state electric field, which behaves like a gate or a field-plate during the on or off-state, thanks to the p-ohmic gate contact extracting and injecting holes towards it. By selecting N$_{st}$=1×10^{17} cm^{-3} and L$_{st}$=6 µm we can enhance the BV to 930 V, while obtaining R$_{on}$=10.2 m$\Omega\cdot$mm and I$_{sat}$=208 mA/mm at V$_{GS}$=5 V. The dynamic turning-on time is measured to be $\sim$20 ns in the mono-pulse test circuit.

Acknowledgments

This work was supported in part by National Natural Science Foundation of China under Grant 62334012, Grant 62074161, Grant 62004213, Grant U20A20208, and Grant 62304252; in part by the University of CAS; and in part by IMECAS-HKUST-Joint Laboratory of Microelectronics.

References

[1] D. Piedra et al., "Advanced power electronic devices based on Gallium Nitride (GaN)," in 2015 IEEE International Electron Devices Meeting (IEDM), Washington, DC, USA, Dec. 2015, p. 16.6.1-16.6.4.

[2] A. Nakajima et al.,"GaN-Based Super Heterojunction Field Effect Transistors Using the Polarization Junction Concept," in IEEE Electron Device Letters, vol. 32, no. 4, pp. 542-544, April 2011.

[3] J. Cui et al., "High-Voltage E-Mode p-GaN Gate HEMT on Sapphire with Gate Termination Extension," in IEEE Transactions on Electron Devices, vol. 71, no. 3, pp. 1592-1597, March 2024.

[4] A. N. Tallarico et al., "TCAD modeling of the dynamic VTH hysteresis under fast sweeping characterization in p-GaN gate HEMTs," IEEE Trans. Electron Devices, vol. 69, no. 2, pp. 507–513, Feb. 2022.

[5] Y.H.Ng et al., "Distribution and transport of holes in the p-GaN/AlGaN/GaN

979-8-3315-0417-5/25 $31.00 © 2025 IEEE

heterostructure," Appl. Phys. Lett., vol. 123, no. 14, Oct. 2023.

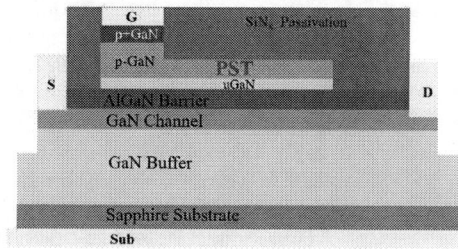

Fig. 1: Schematic cross-sectional structure of PST-HEMT.

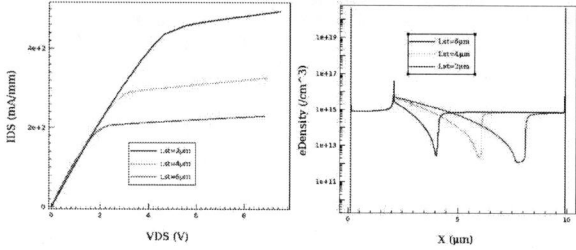

Fig. 2: (a) Output I-V characteristics of PST-HEMT at V_{GS}=5 V,
(b) Channel electron distribution of PST-HEMT at V_{GS}=5 V, V_{DS}=5 V.

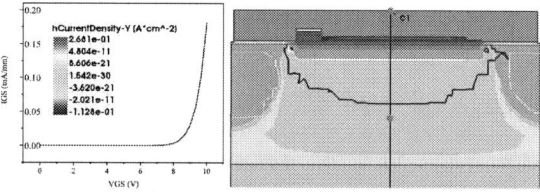

Fig. 3: (a) Gate leakage current I_{GS}-V_{GS} of PST-HEMT at V_{DS}=2 V,
(b) Simulated hole distribution in PST-HEMT when gate is broken.

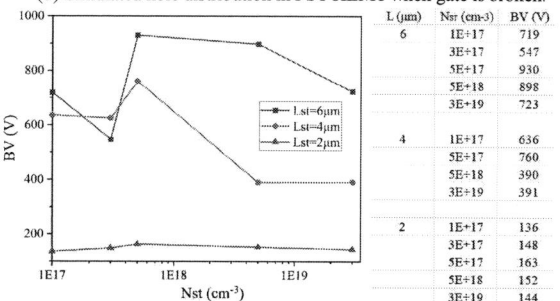

L (μm)	N_{st} (cm-3)	BV (V)
6	1E+17	719
	3E+17	547
	5E+17	930
	5E+18	898
	3E+19	723
4	1E+17	636
	5E+17	760
	5E+18	390
	3E+19	391
2	1E+17	136
	3E+17	148
	5E+17	163
	5E+18	152
	3E+19	144

Fig. 4: Impact of N_{st} and L_{st} on BV for the PST-HEMT (with data in table).

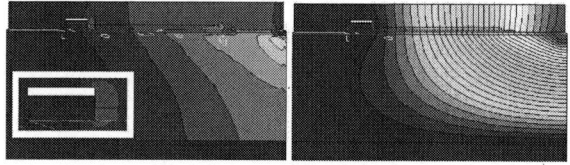

Fig. 5: Simulated results in N_{st}=1×10^{17} cm^{-3} and L_{st}=6 μm about (a) Electric
field distribution (index: gate corner electric field), (b) Electrostatic potential
distribution under bias of V_{GS}=0, V_{DS}=719 V.

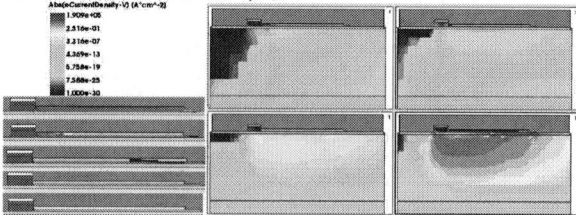

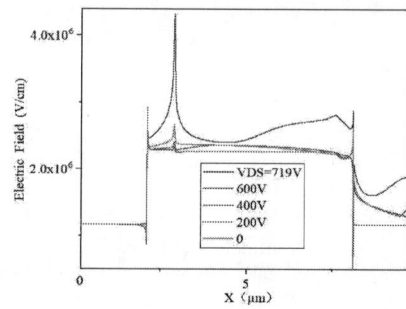

Fig.6: Simulated results in N_{st}=1×10^{17} cm^{-3} and L_{st}=6 μm about (a) hole
depletion behavior of PST, (b) Buffer leakage current behavior. V_{GS}=0, V_{DS}
rises from 0 to 200,400,600,719 V, (c) Channel electric field distribution.

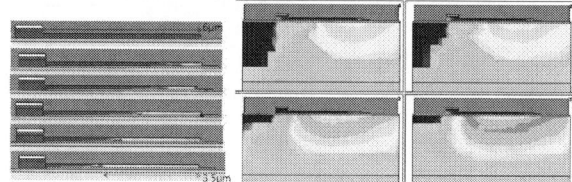

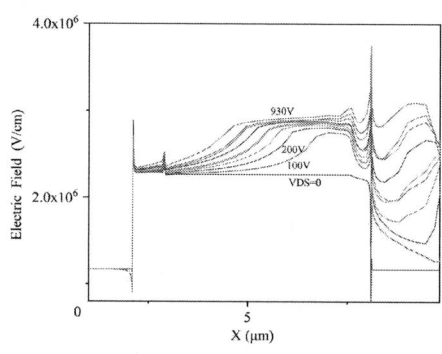

Fig.7: Simulated results in N_{st}=5×10^{17} cm^{-3} and L_{st}=6 μm about (a) Hole
depletion behavior of PST, (b) Buffer leakage current behavior. V_{GS}=0, V_{DS}
rises from 0 to 200,500,800,930 V, (c) Channel electric field distribution.

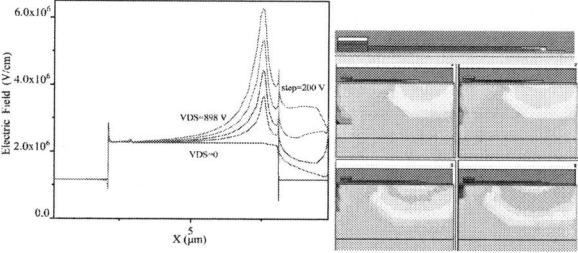

Fig.8: Simulated results in N_{st}=5×10^{18} cm^{-3} and Lst=6 μm about (a) Channel
electric field distribution, (b) hole depletion behavior of PST, (c) Buffer
leakage current behavior. V_{GS}=0, V_{DS} rises from 0 to 200,400,600,898 V.

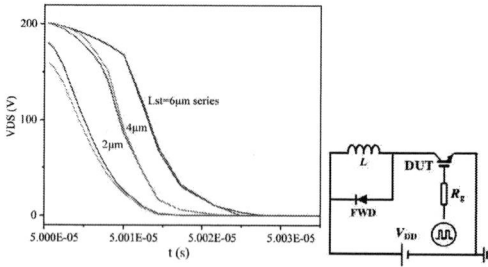

Fig.9: Simulated voltage dropping time during the device turning on (Index:
Schematic of mono-pulse test circuit in the spice model, where V_{DD}=200 V,
L=25 nH, R_g=5 Ω. Vin pulse rises from 0 to 15 V at t = 5×10^{-5} s).

979-8-3315-0417-5/25 $31.00 © 2025 IEEE

Joule Heating Effect on Quality Factor and Frequency Tuning of 2D MoS₂ NEMS Resonators

Shuai Yuan[1], Zuheng Liu[1], Pengcheng Zhang[1,4,*], Rui Yang[1,2,3,*]

[1]University of Michigan–Shanghai Jiao Tong University Joint Institute, Shanghai Jiao Tong University (SJTU), Shanghai, China

[2]School of Electronic Information and Electrical Engineering, SJTU, Shanghai, China

[3]State Key Laboratory of Radio Frequency Heterogeneous Integration, SJTU, Shanghai, China

[4]Micro and Nanosystems, Department of Mechanical and Process Engineering,
ETH Zürich, Switzerland

*Corresponding authors. Emails: zpengchen@ethz.ch, rui.yang@sjtu.edu.cn

Abstract

This work mainly studies the frequency and quality factor (Q) tuning effect and mechanism by Joule heating in two-dimensional (2D) molybdenum disulfide (MoS₂) nanoelectromechanical systems (NEMS). We demonstrate decreased frequency and Q with a larger drain voltage which contributes to stronger Joule heating. The measured tunability of Joule heating effect on Q is approximately four times the tunability of gate voltage effect on Q. The results provide insights for dissipation mechanisms in 2D NEMS resonators.

Keywords: 2D NEMS resonator, Joule heating, quality factor, dissipation, frequency tuning

Introduction

Two-dimensional materials, such as MoS₂ and graphene, have unique properties, including variable bandgaps depending on thickness, high strain limit, high mechanical flexibility, and tunable thermal and electronic properties, making them ideal for NEMS. Due to their high sensitivity, low power, and high tunability, 2D NEMS resonators are promising for many potential applications, including sensors [1], memories and computing devices [2,3], and quantum engineering [4]. Previous studies have demonstrated tuning of frequency and Q through gate voltage or external strain [5]. However, when drain voltage is applied, the Joule heating effect generated by the 2D transistor due to current across drain and source electrodes can also effectively control the resonator's frequency and Q [6]. Therefore, the mechanisms of Q tuning by Joule heating require further exploration.

In this study, we explore Joule heating effects on tuning the resonance characteristics of 2D MoS₂ NEMS resonators, by examining its influence on both frequency and Q at varying drain and gate voltages. The devices demonstrate strong frequency and Q modulation, highlighting the role of drain-voltage-induced Joule heating for enhancing device tunability beyond gate voltage alone. The results provide important guidelines for controlling the frequency and dissipation in 2D NEMS resonators.

Measurement System

The resonance signal of devices is measured by our custom-built laser interferometry system (Fig. 1), under vacuum at room temperature. To facilitate electrical excitation and optical measurements, DC gate voltage V_{GS} and radio-frequency voltage v_{RF} are applied *via* a bias-tee to the Si back gate with the source grounded. The laser interferometry signal is converted through a photodetector before entering the network analyzer. Drain voltage is applied by a DAQ card output to provide Joule heating.

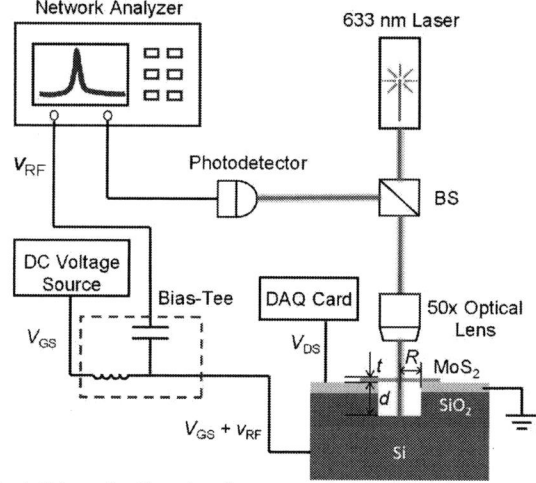

Fig. 1: Schematic of laser interferometry system for 2D MoS₂ resonators. The detection laser is a He-Ne laser with wavelength of 632.8 nm. The V_{GS} and v_{RF} signals are generated by a DC voltage source and a network analyzer, respectively. The drain voltage V_{DS} is applied by a data acquisition (DAQ) card output.

We use the dry-transfer process to transfer the 2D MoS₂ onto a pre-patterned substrate with electrodes and microtrenches [7], forming circular fully-clamped resonators, as shown in Fig. 2a. A basic mechanical resonance signal at around 45 MHz is shown in Fig. 2b.

Joule Heating Effect Measurement and Discussion

The structure of the resonator is analogous to that of a suspended-channel MoS₂ transistor. The Joule heating effect is closely related to the device's DC current–voltage (*I–V*) characteristics, which is shown in Fig. 2c and 2d. The

979-8-3315-0417-5/25 $31.00 © 2025 IEEE

MoS$_2$ transistor shows n-type transistor behavior with clear gate tuning of the current, and relatively linear I_D–V_{DS} relationship. In the following measurements, we follow the upward V_{GS} sweep, and within the applied voltage range, we assume that the transistor resistance remains constant for a fixed gate voltage. It is important to note that within the V_{DS} range of -10 to 10 V, the transistor current is in the microampere range, producing Joule heating power on the order of several microwatts. This power is sufficient to induce localized heating effects in the resonator. Specifically, the thermal effect may reduce surface stress, leading to softening of the membrane spring. Additionally, the dissipation induced by Joule heating contributes to a further decline in the Q of the resonator.

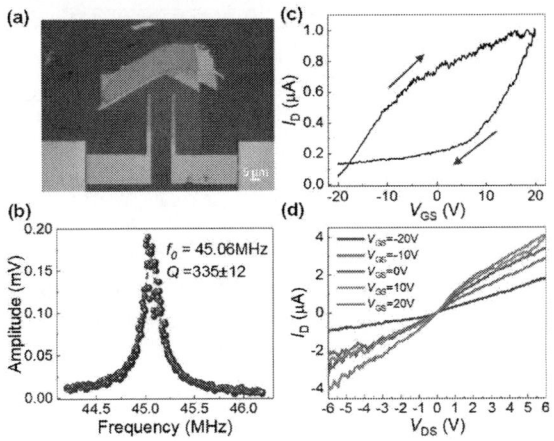

Fig. 2: DC I–V and mechanical resonance measurements. (a) Optical image of a 2D MoS$_2$ resonator. *Scale bar*: 5 μm. (b) Mechanical resonance signal of a MoS$_2$ membrane near 45 MHz with V_{GS}=5 V, v_{RF}=50 mV, and V_{DS}=0 V. Fitting results are presented as f_0=45.06 MHz and Q=335. (c-d) Electrical characteristics of the MoS$_2$ transistor with suspended channel. (c) Dual-sweep I_D-V_{GS} curve at V_{DS}=1 V, and (d) I_D-V_{DS} curve at different V_{GS}. The SiO$_2$ dielectric layer thickness is 300 nm.

For a fully-clamped resonator, the frequency of the fundamental mode can be expressed as follows [6]:

$$f = \frac{1}{2\pi}\sqrt{f_m^2 + f_B^2 + f_z^2 - f_s^2}. \tag{1}$$

The four parts of f can be defined as:

$$f_m^2 = 2.405^2 \frac{\gamma_0 + \gamma(T)}{R^2 \rho t}, \tag{2}$$

$$f_B^2 = 2.405^4 \frac{Et^3}{12(1-v^2)R^4\rho}, \tag{3}$$

$$f_z^2 = 2.405^2 \frac{12}{3} \frac{Etz^2}{(1-v^2)R^4\rho}, \tag{4}$$

$$f_s^2 = 1.23 \frac{\epsilon_0 V_g^2}{d^3\rho}, \tag{5}$$

where γ is the mechanical tension of the membrane; ϵ_0 is vacuum permittivity; E, v, R, d, t, ρ, and z are the Young's modulus, Poisson ratio, radius, initial vacuum gap, MoS$_2$ thickness, density, and vibration amplitude, respectively. The Joule heating effect primarily influences the f_m term.

$\gamma(T)$ can be estimated by $\gamma(T) = -Et \int_{T_0}^{T} \alpha\left(T'\right) dT'$, where α is the thermal expansion coefficient, which can be treated as a constant. Temperature can be regarded as a function of Joule heating power ($P=V_{DS} \times I_D$), allowing us to derive the frequency tuning curve of the resonator.

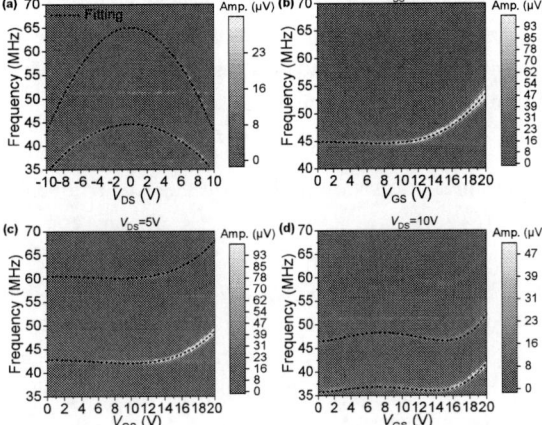

Fig. 3: Frequency tuning characteristic by varying V_{DS} and V_{GS}. (a) Resonance amplitude shown in color scale when sweeping V_{DS} from -10 V to 10 V with a fixed v_{RF} of 50 mV and a fixed V_{GS} of 0 V. (b-d) Resonance amplitude shown in color scale when sweeping V_{GS} from 0 to 20 V with a fixed v_{RF} of 50 mV at different V_{DS} of (b) V_{DS}=0 V, (c) V_{DS}=5 V and (d) V_{DS}=10 V. The black dashed lines in each subfigure show the fitting results using the frequency tuning model.

With a gate voltage V_{GS}=0 V, we measure the frequency response of the resonator across a V_{DS} range from -10 to 10 V (Fig. 3a). The results reveal that Joule heating softens the membrane in both two modes, enabling a frequency tuning range of up to 55%. We further characterize the resonator frequency response for V_{GS} from 0 to 10 V at V_{DS}=0, 0.5 and 10 V (Fig. 3b-3d). At V_{DS}=5 V, Joule heating effect makes the resonance frequency lower than that when V_{DS}=0 V. Note that the increased V_{DS} also means a smaller effective gate voltage drop V_{GD}, which leads to a smaller gate tuning range. Intriguingly, at V_{DS}=10 V, the frequency response adopts an 'S'-shaped tuning curve (Fig. 3d), where the resonance frequency first increases, and then decreases, and then increases again with a larger V_{GS}. This is different from the previous frequency tuning by the strain induced by V_{GS}, where the frequency should increase with V_{GS}. Even with capacitive softening, the frequency tuning usually first decreases and then increases with V_{GS}. The measured frequency tuning curve could be related to the fact that the larger gate voltage also increases the transistor's suspended channel conductance, which then increases the Joule heating and decreases the resonance frequency. Therefore, there is competition between the strain effect, capacitive softening effect, and Joule heating effect, leading to such frequency tuning curve. The results offer a versatile approach for resonator frequency tuning.

979-8-3315-0417-5/25 $31.00 © 2025 IEEE

The thermoelastic dissipation can be described by [5]:

$$Q_{TED}^{-1} = \eta \frac{5.576 \delta z^2 (v_{RF}, V_{GS})}{\varepsilon_r (V_{GS}) R^2 (1-v^2)} \delta_{elong}, \qquad (6)$$

where η, δz and ε_r are the mode parameter, vibration amplitude, and the total strain of the membrane, respectively. The loss angles due to elongational loss is $\delta_{elong} = \Delta E_{elong}/(2\pi E_{elong})$. Additionally, we introduce an electrical loss term in the total dissipation to account for the impact of Joule heating on Q of the resonator.

$$Q_{total}^{-1} = Q_{TED}^{-1} + Q_{elec,loss}^{-1}. \qquad (7)$$

The channel resistance of MoS$_2$ can be determined by $R_d = (\beta V_{GS} + G_0)^{-1}$, where β and G_0 are constants depending on the material. The electrical loss term, $Q_{elec,loss}^{-1} = Q_{AC,loss}^{-1} + Q_{DC,loss}^{-1}$, can be viewed as two components: the AC Joule heating, primarily driven by the capacitive gate oscillating current, and the DC Joule heating, which is mainly caused by the DC drain voltage. The total gate capacitance is $C = C_g + \delta C_g \sin(\omega t)$, and the electrical drive is $V = V_{GS} + \delta V_g \sin(\omega t)$ [8]. The two damping terms are given by the following equations:

$$Q_{AC,loss}^{-1} = \frac{1}{E_{stored}} \frac{\pi \omega}{\beta V_{GS}+G_0} [V_{GS}^2 (\delta C_g)^2 + 2 C_g V_{GS} \delta C_g \delta V_g + C_g^2 (\delta V_g)^2], \qquad (8)$$

$$Q_{DC,loss}^{-1} = \frac{2\pi V_{DS}^2 (\beta V_{GS}+G_0)}{\omega E_{stored}}. \qquad (9)$$

We measure the variation of Q under different gate and drain voltages (Fig. 4). The results indicate that the Q decreases with both a larger $|V_{GS}|$ and $|V_{DS}|$. When the DC Joule heating effect is not considered, the tuning effect of the gate voltage on Q is approximately 2%/V (Fig. 4a). Conversely, when the tensile strain induced by the gate voltage is disregarded, the tuning effect of the drain voltage on Q is around 8%/V, approximately four times that of the gate voltage, for the measured 2D MoS$_2$ NEMS resonator in this study (Fig. 4b). This indicates that energy dissipation due to the DC Joule heating effect has a significant effect on the dissipation and Q of 2D NEMS.

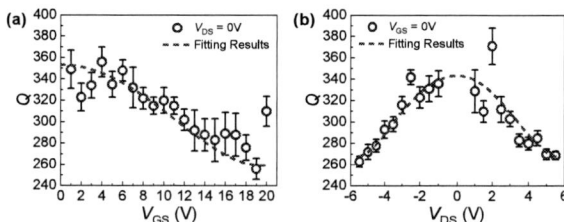

Fig. 4: Q factor tuning characteristic as a function of V_{DS} and V_{GS}. (a) Measured Q by sweeping V_{GS} from 0 to 20 V without DC Joule heating effect. (b) Measured Q by sweeping V_{DS} from -10 to 10 V due to Joule-heating-induced damping. The black dash line in each subfigure shows the fitting to the Q tuning model, showing decreased Q with a larger V_{GS} and V_{DS}, which is consistent with the measured data.

Conclusion

In conclusion, this study demonstrates the significant impact of Joule heating on frequency and Q in 2D MoS$_2$ NEMS resonators. By applying a drain voltage, we achieve a substantial frequency softening with tuning range of up to 55%. This effect becomes even more pronounced compared to gate voltage control alone, with the drain voltage showing strong influence on dissipation and Q. These findings highlight the Joule heating as a potential mechanism for enhancing the resonator tunability, expanding potential applications in nanoscale sensing, memory, and computing.

Acknowledgments

The authors thank the support from the National Natural Science Foundation of China (NSFC) (Grants U21A20505, 92364107, 62104140, 62250073), Science and Technology Commission of Shanghai Municipality (STCSM) (Grants 23QA1405300, 21ZR1433800), Natural Science Foundation of Chongqing (CSTB2022NSCQ-MSX1095), State Key Laboratory of Radio Frequency Heterogeneous Integration (Shenzhen University), and Lingang Laboratory Open Research Fund (Grant LG-QS-202202-11). The authors thank the Center for Advanced Electronic Materials and Devices (AEMD) for the support in device fabrication and characterization.

References

[1] P. Weber, J. Güttinger, A. Noury, et al., "Force sensitivity of multilayer graphene optomechanical devices", Nat. Commun., 7, 12496 (2016).

[2] P. Zhang, Y. Wang, Z. Li, and X. Chen, "Nanoelectromechanical memories based on nonlinear 2D MoS$_2$ resonators", Int. Conf. Micro Electro Mech. Syst., pp. 208-211, Tokyo (2022).

[3] L. Wang, P. Zhang, et al., "On-chip mechanical computing: status, challenges, and opportunities", Chip, 2, 100038 (2023).

[4] B. Xu, P. Zhang, et al., "Nanomechanical resonators: toward atomic scale", ACS Nano, 16 (10), 15545-15585 (2022).

[5] P. Zhang, et al., "Strain-modulated dissipation in two-dimensional molybdenum disulfide nanoelectromechanical resonators", ACS Nano, 16, 2261-2270 (2022).

[6] A. Chiout, C. Brochard-Richard, L. Marty, et al., "Extreme mechanical tunability in suspended MoS$_2$ resonator controlled by Joule heating", npj 2D Mater. Appl., 7, 20 (2023).

[7] R. Yang, et al., "Multilayer MoS$_2$ transistors enabled by a facile dry-transfer technique and thermal annealing", J. Vac. Sci. Technol. B, 32, 061203 (2014).

[8] J. Lee, et al., "Electrically tunable single- and few-layer MoS$_2$ nanoelectromechanical systems with broad dynamic range", Sci. Adv., 4, eaao6653 (2018).

979-8-3315-0417-5/25 $31.00 © 2025 IEEE

Improved CMOS-Based Noise-Immune Sigmoid Activation Function for Neural Networks

Harshita Singh*, Anant Singhal*, and Harshit Agarwal

Nano Devices and Application Lab, Indian Institute of Technology Jodhpur, Rajasthan, India

ABSTRACT

This work discusses an improved CMOS-based, noise-immune nonlinear sigmoid activation function designed for enhanced neural network performance. The proposed architecture provides a precise, variability-resistant approximation of the sigmoid function, significantly boosting network efficiency. Implemented using the UMC $180nm$ technology node PDK, this compact design offers substantial performance improvements for neural networks, achieving both noise immunity and compactness with a trade-off of only a minimal 1% increase in power consumption.

Keywords: Activation Function, CMOS, Neural Networks, Sigmoid Function.

INTRODUCTION

Artificial Neural Networks (ANNs) are paving their way into the semiconductor industry, enabling innovative solutions across a range of applications, from compact modeling [1] to analog circuit optimization [2]. Despite the potential advantages of hardware implementations, the semiconductor industry continues to predominantly focus on software-based approaches for integrating ANNs [2]–[4]. This reliance on software limits the full exploitation of the inherent parallelism and efficiency offered by neural networks. Implementing ANNs in hardware presents unique challenges, such as complexity in circuit design, power consumption, and noise sensitivity. As demand for faster and more efficient electronic devices grows, developing effective hardware solutions for neural networks and activation functions is critical to harnessing their full capabilities and optimizing performance in semiconductor applications [5]–[8].

The sigmoid activation function is a crucial component in the design of ANNs, and its implementation can significantly impact overall performance. Various approaches have been explored, including CMOS-based designs [5], current-mode circuits, and voltage-mode configurations [8], each aiming to optimize accuracy and efficiency. Authors in [5] have proposed a simplistic sigmoid function for the first time; however, the circuit is not immune to noise due to its simplistic nature. Authors in [8] have further improved the circuit using a voltage-to-current converter to gain better control

*Equal contribution.

over the input signal. However, the circuit has become more complex due to large number of transistors and is computationally inefficient.

In this work, we introduce an improved CMOS-based, noise-immune nonlinear sigmoid activation function specifically designed to enhance the performance of ANNs. Our architecture provides accurate and variability-resistant approximation of the sigmoid function owing to the presence of additional driver transistors, addressing the limitations of existing approaches. By utilizing the driver transistors, the proposed design can be operated for a wider input range as well as flexible towards the actual sigmoid function. It is implemented using the $180nm$ technology node and offers substantial performance improvements, achieving both noise immunity with only a marginal increase in power consumption. This approach not only enhances the reliability of neural networks but also paves the way for more efficient hardware implementations in semiconductor applications. Rest of the paper is organized as follows: Section II details the methodology employed, Section III presents the results, followed by conclusion in Section IV.

METHODOLOGY

Fig. 1 shows the proposed circuit designed to approximate the sigmoid function which plays crucial role as an activation function in ANNs. Equation of the sigmoid fuction is represented by

$$f(x) = \frac{1}{1 + e^{-\gamma x}}$$

The parameter γ determines the slope, which in turn controls the sensitivity of the output node to the input signal. This sigmoid behavior enables the circuit to perform non-linear transformations, essential for learning complex patterns during ANN training. Authors in [5] proposed a simplistic six-transistor configuration for the sigmoid activation function. Building upon this simplistic circuit, we have introduced two additional driver transistors ($M7$ and $M8$) for adjustable sigmoid function by adjusting the biasing voltages V_{bias1} and V_{bias2}. In the proposed circuit, transistors $M5$ and $M6$ are responsible for generating a stable bias voltage close to $V_{DD}/2$. The core sigmoid function is realized through transistors $M1$, $M2$, $M3$, $M4$, $M7$, and $M8$, which together shapes the output response. For the ease of calculation, the threshold voltage of NMOS and PMOS are considered equal. Operation of sigmoid function is designed in such way that when the input current $I_{sig} = 0$, V_{sig} must be equal to $V_{DD}/2$. When I_{sig} becomes negative, $M2$ and $M4$ will be in cutoff region and triode region respectively and other two transistors $M1$ and $M3$ will be in saturation region which is responsible for the voltage drop at V_{sig}. Now when I_{sig} starts increasing and starts to move towards zero, it makes $M4$ to operate in saturation region and $M1$ and $M2$ to operate in cutoff region. On further increasing the I_{sig}, $M3$ goes into deep triode region which results in increase in the output voltage V_{sig}.

To address the limitations of previous simplistic configurations, the proposed circuit introduces two additional driver transistors, $M7$ and $M8$, which operate in the triode region, effectively functioning as voltage-modulated resistors. By incorporating these driver transistors, we achieve precise control over the steepness and flexibility of the sigmoid function through external bias voltages, V_{bias1} and V_{bias2}. These adjustable biasing voltages control the channel resistance of $M7$ and $M8$, which in turn dynamically alters the slope parameter γ in the sigmoid function, allowing for finer control of the activation response. This adaptability enables the circuit to accommodate a range of ANN architectures, enhancing its versatility for various hardware implementations.

RESULTS AND DISCUSSION

To validate the proposed circuit (see Fig. 1), input current I_{sig} is mapped to the theoretical sigmoid function within a range of $-75\mu A$ to $75\mu A$, confirming the circuit's applicability. Fig. 3 demonstrates that the simulated output aligns well with the theoretical sigmoid function for different γ values, highlighting the model's flexibility. This tuning of sigmoid behavior across various γ values is achieved by adjusting transistor aspect ratios (W/L), as detailed in Table I. Furthermore, Fig. 4 shows the derivative of the simulated circuit's output alongside the theoretical function, accurately overlapping with a maximum error of only 0.52% for $\gamma = 1$, giving lower error than existing circuits (see Fig. 5).

To check the noise immunity of the proposed design, the input-referred noise (IRN) is compared with the simplistic configuration given in [5]. Due to the additional control of the voltage-modulated resistors, the proposed circuit shows a 7.78% reduction in IRN as shown in Fig. 6. The voltage-controlled resistors formed by these transistors enable fine-tuning of the sigmoid function while also stabilizing biasing conditions, effectively reducing signal fluctuations. Additionally, Monte Carlo analysis for V_{bias1} and V_{bias2} has been carried out to check the control of the biasing resulting in a 24.68% reduction in the IRN as shown in Fig. 7.

To further affirm the flexibility of the proposed circuit, parametric analysis for the controlling parameters is carried out. Fig. 8 shows the output characteristics for a wide range of V_{bias1} and V_{bias2} (ensuring $M7$ and $M8$ always remain in the triode region). Similar parametric analysis is carried out for the aspect ratio of the load transistors $M3$ and $M5$ to validate the control over the steepness of the sigmoid function. Fig. 9 shows a good control in the slope by varying the width of the load transistors. Finally, transient analysis is carried out for the proposed design. Fig. 10 and 11 shows the applied current pulses and the corresponding voltage response V_{sig} respectively, showing smooth output pulses without kinks, validating the circuit's reliability. Table II compares the proposed circuit's performance metrics, achieving superior accuracy and robustness with only a 1% increase in power consumption over existing designs [5], [8].

CONCLUSION

This work proposes an improved CMOS-based nonlinear sigmoid activation function with improved noise immunity that enhances the performance of artificial neural networks. The design offers accurate and variability-resistant approximation of the sigmoid function by utilizing additional driver transistors, allowing for a broader input range while maintaining compactness. Validation through simulations using the $180nm$ technology node confirms significant improvements with only a 1% increase in power consumption.

ACKNOWLEDGMENT

This work is partly supported by PMRF, GoI.

REFERENCES

[1] C. -T. Tung and C. Hu, "Neural Network-Based BSIM Transistor Model Framework: Currents, Charges, Variability, and Circuit Simulation," in IEEE Trans on Electron Devices, vol. 70, no. 4, pp. 2157-2160, (2023).

[2] A. Singhal, P. Goyal and H. Agarwal, "Artificial Neural Network Driven Optimization for Analog Circuit Performance," 2024 IEEE Latin American Electron Devices Conference (LAEDC), pp. 1-4, (2024).

[3] K. Hakhamaneshi, M. Nassar, M. Phielipp, P. Abbeel and V. Stojanovic, "Pretraining Graph Neural Networks for Few-Shot Analog Circuit Modeling and Design," in IEEE Trans on CAD of Integrated Circuits and Systems, vol. 42, no. 7, pp. 2163-2173, (2023).

[4] S. Jain, A. Sengupta, K. Roy and A. Raghunathan, "RxNN: A Framework for Evaluating Deep Neural Networks on Resistive Crossbars," in IEEE Trans on CAD of Integrated Circuits and Systems, vol. 40, no. 2, pp. 326-338, (2021).

[5] G. Khodabandehloo, M. Mirhassani and M. Ahmadi, "Analog Implementation of a Novel Resistive-Type Sigmoidal Neuron," IEEE Trans on VLSI Systems, vol. 20, no. 4, (2012).

[6] J. Shamsi, A. Amirsoleimani, S. Mirzakuchaki, A. Ahmade, S.Alirezaee and M. Ahmadi, "Hyperbolic tangent passive resistive-type neuron," 2015 IEEE International Symp on Circuits and Systems (ISCAS), pp. 581-584, (2015).

[7] Shakiba, F.M. and Zhou, M., Novel analog implementation of a hyperbolic tangent neuron in artificial neural networks. IEEE Trans on Industrial Electronics, 68(11), pp.10856-10867, (2020).

[8] Moposita, T., Trojman, L., Crupi, F., Lanuzza, M. and Vladimirescu, A., Voltage-to-voltage sigmoid neuron activation function design for artificial neural networks. In 2022 IEEE 13th Latin America Symposium on Circuits and System (LASCAS) (pp. 1-4), 2022.

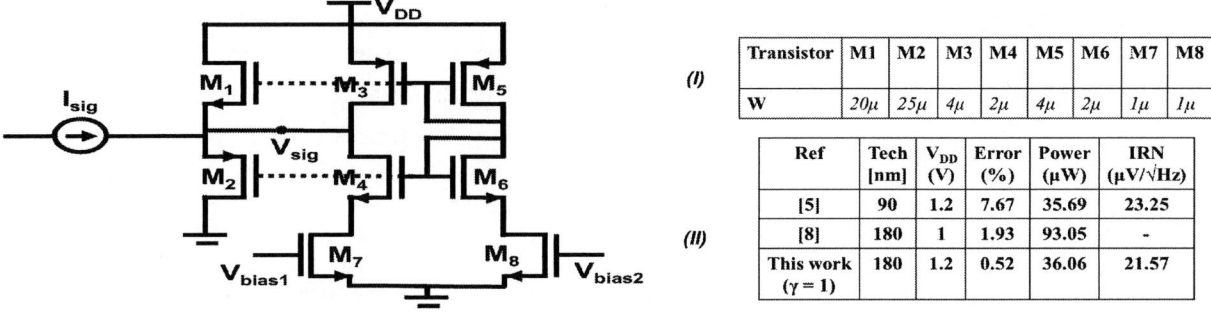

Transistor	M1	M2	M3	M4	M5	M6	M7	M8
W	20µ	25µ	4µ	2µ	4µ	2µ	1µ	1µ

(I)

Ref	Tech [nm]	V_{DD} (V)	Error (%)	Power (µW)	IRN (µV/√Hz)
[5]	90	1.2	7.67	35.69	23.25
[8]	180	1	1.93	93.05	-
This work ($\gamma = 1$)	180	1.2	0.52	36.06	21.57

(II)

Fig. 1. Schematic of the proposed CMOS-based sigmoid activation function: The driver transistors $M7$ and $M8$ helps in increasing the noise immunity owing to the voltage control i.e. V_{bias1} and V_{bias2}.

Fig. 2. Table I: (W/L) used for each of the transistor. Table II: Comparison of figure of Merits with the existing circuits.

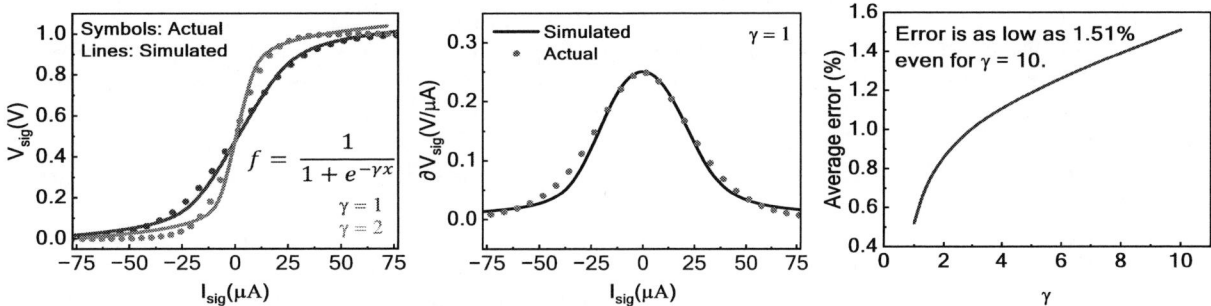

Fig. 3. I_{sig} vs V_{sig}: The corresponding current has been mapped between -5 to 5 to fit the actual sigmoid function.

Fig. 4. Derivative of the sigmoid function. The derivative is continuous and smooth showing robustness of the circuit.

Fig. 5. Error for different values of γ. The error is calculated at the mid the supply voltage V_{DD}. The error is as low as 1.51%

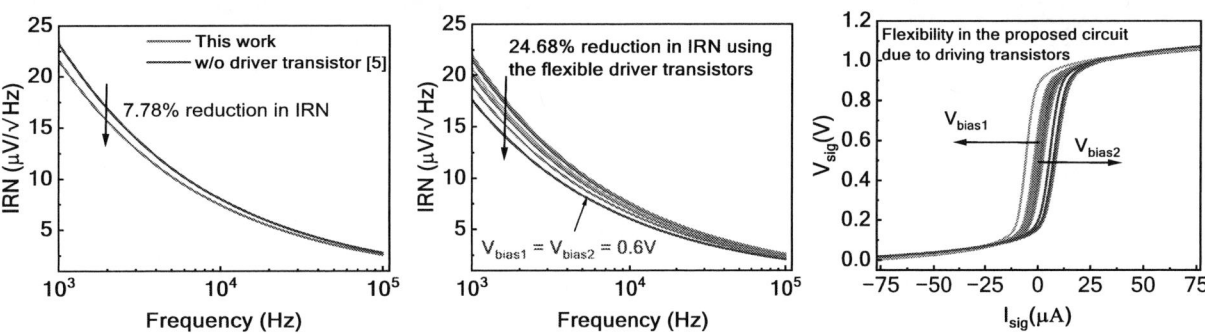

Fig. 6. Input referred noise (IRN): Proposed circuit shows better noise performance.

Fig. 7. Monte Carlo Analysis for the bias control of driving transistors.

Fig. 8. Flexibility in fitting the sigmoid function owing to driving transistors.

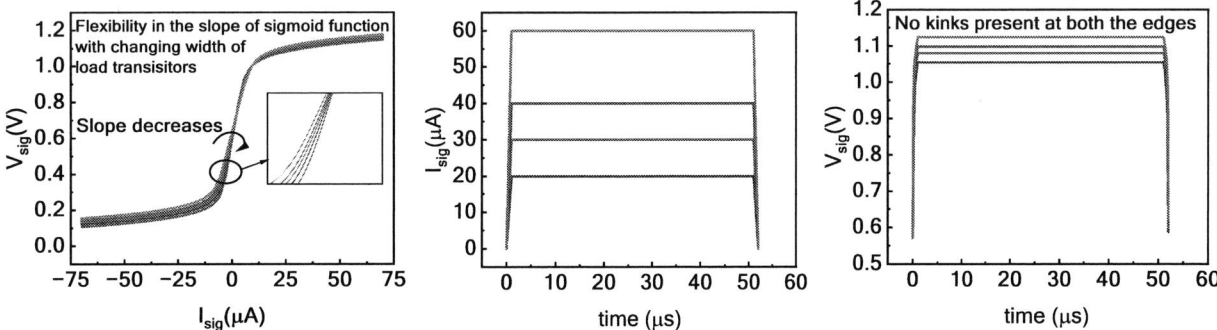

Fig. 9. Flexibility in capturing the control using the (W/L) of the load transistors.

Fig. 10. Applied current pulse I_{sig} vs time. Rise time and fall time are taken as $1\mu s$.

Fig. 11. Output voltage response: There is no ringings at the edges, hence reliable.

Self-heating and Process Induced Performance Barrier on Complementary Field Effect Transistor: A Reliability Perspective

Sandeep Kumar[1], Deven H Patil[1], Khushi Jain[1], Ankit Dixit[2], Sunil Rathore[3], Mohd. Shakir[3], Naveen Kumar[2], Vihar Georgiev[2], S. Dasgupta[3], Navjeet Bagga[1*]

[1]IIT Bhubaneswar, India, [2]University of Glasgow, UK, [3]IIT Roorkee, India. (Email: navjeet@iitbbs.ac.in)

Abstract

The vertical nanosheet (channels) stacking and aligned in such a way that nFET is kept over pFET, or vice-versa, raises severe reliability concerns in Complementary FET (CFET). In this paper, using well-calibrated TCAD models, all the related reliability issues are being analyzed, such as: (i) the role of dielectric separation wall (D_{SW}) in electrical and thermal cross-talk from nFET to pFET and vice-versa; (ii) the impact of self-heating effect (SHE) on self- and the other side of the D_{SW}; (iii) impact of random dopant fluctuations (RDF) on threshold voltage (V_{th}); (iv) impact of line-edge roughness (LER) on I_{ON} and V_{th}; (v) effect of metal grain granularities (MGG) and the ratio of grain size to gate area (RGG) on device merits, viz I_{ON} and V_{th}; and finally (vi) the device aging is predicated using the 'shift in V_{th}' by ±50mV. Thus, the proposed analysis benchmarks a reliable CFET design.

(Keywords: Complementary FET, Self-heating, Metal grain granularity, Line-edge roughness, Dielectric separation wall)

Introduction

Achieving the effective area scaling of the device footprint is difficult while maintaining better gate-electrostatic integrity, efficient power consumption, and a reliable cell design. Thus, the complementary field effect transistor (CFET) is proposed as an emerging architecture that is realized by vertical stacking of nFET over pFET, or vice-versa. In turn, a CFET utilizes the scaled footprint, which was previously allocated for nFET devices only, and offers 50% structural scaling [1]-[2]. Furthermore, the vertical stacking of N & P simplifies the transistor terminal access, which mitigates the poly pitch (PP) and metal pitch (MP) scaling restrictions. Despite CFET's advantages in terms of area and performance, the ultra-scaled dimension and stacking raise severe reliability concerns, which require proper attention. In CFET, the N and P are separated by a dielectric separation wall (D_{SW}) whose transparency/opacity causes the electrical and thermal crosstalk across the N and P, or vice-versa. Thus, the thickness and the permittivity of D_{SW} play a significant role in CFET reliability. Further, the sheets/channels are wrapped by a stacked SiO_2-HfO_2 gate oxide, which hinders the heat flow from the active region of the device. Therefore, the lattice temperature (T_L) of the device increases, which in turn, reduces the effective mobility of the charge carriers and results in self-heating effect (SHE) [3]-[4]. In addition to SHE, the intercoupling heat flux from N to P (or vice-versa) severely influences the device's operation. Considering the external process variations, the RDF, LER, and MGG contribute significantly as unreliability agents. Therefore, a thorough investigation is needed to predict the early aging of the CFET under the influence of these internal and external variabilities. ***The key contributions are:*** (a) analyzing how the D_{SW} affects the thermal-electrical coupling in CFET; (b) investigating the impact of SHE on one side (N or P) of D_{SW} and also at the other side; (c) evaluating the impact of RDF, LER, and MGG on device merits; and finally (d) predicting the early aging in consideration to all reliability barriers.

Device Structure and Simulation Setup

A 3-D schematic and a cross-section view of a dual-sheet Complementary FET (CFET) are shown in Fig. 1, considered as a baseline reference for analysing the performance barrier viz self-heating and process-induced reliability concerns. The source/drain (S/D) regions are uniformly doped, whereas the Gaussian doping is used in the extension region to mimic the realistic scenario, which avoids the abrupt change in the electric field at the S/D-channel junction. *Sentaurus* TCAD [5] and Garand VE [6] are used for simulating the electrical, thermal, and process-induced randomization characteristics of the CFET. The drift-diffusion model, Poisson solver, and quantum potential model are invoked in the TCAD setup to capture the charge transport. The carrier-carrier scattering and phonon scattering are incorporated using the Phillips Unified Mobility (Phu Mob) model, and high-field saturation model as drift velocity is no longer proportional to the electric field, and IAL Mob models are utilized to encapsulate the mobility degradation. The Enormal model is used to capture the effect of surface roughness at the interface. Doping and temperature dependence SRH model and Auger recombination model are inserted to capture the generation and recombination. Thermodynamic and hydrodynamic models are invoked for capturing the SHE-based thermal degradation. For process variability, the statistical impedance field method is used, which treats randomness as a perturbation of a reference device using Green's function-based approach. The calibrated I_{DS}-V_{GS} curve against experimental data [7] is shown in Fig.2a, and Table I comprises all the relevant parameters used in the simulation.

Results and Discussion

This section provides a comprehensive analysis of CFET performance while addressing the role of SHE and process-induced reliability barriers such as metal grain granularities (MGG), line edge roughness (LER), and random dopant fluctuation (RDF). We also explained the electrical and thermal coupling causes due to separation wall (D_{SW}). The section is divided into the following sub-sections:

(a) ***SHE-induced Performance Evaluation:*** The confined geometry, vertical stacking, and phonon boundary scattering, result in the lattice temperature increasing to ~396K in nFET and ~384K in pFET (Fig.3a), as the channel/sheets are wrapped by low thermal conductivity materials (SiO_2 and

HfO$_2$), which obstructs the heat flow from the active channel region. Due to mobility degradation with increased lattice temperature (T$_L$), the SHE results in ~8% I$_{ON}$ degradation in nFET and ~5% in pFET (Fig.3b). This is because when the nFET is activated, the heat transfer to the substrate is hindered by the pFET. Conversely, when the pFET is activated, the heat transfer to the back-end-of-line (BEOL) is hindered by the nFET, resulting in the temperature rise. Further, the choice of D$_{SW}$ material plays a crucial role as it affects the heat and electrical coupling. Smaller D$_{SW}$ results in reduced I$_{ON}$ due to electrical coupling (Fig.2b). Using different materials (Al$_2$O$_3$, HfO$_2$, Si$_3$N$_4$, SiO$_2$) in D$_{SW}$ manifests slight variation in thermal resistance as depicted in the power vs temperature curve but shows a homogeneous effect and can be sustainable (Fig.3c). The intra-device crosstalk is also a key issue in CFET, which requires proper attention. The induced SHE does not only impact the self-side (i.e., either N or P) of the active device in CFET but also, affects the attributes caused by the heat flux transfer on the other side by Dsw. This thermal cross-coupling causes the temperature of the pFET to rise when the nFET is turned on. As Dsw lowers from 35 nm to 20 nm, the temperature on pFET (nFET) rises by 2.38% (2.18%) due to increased thermal cross-coupling on the other side (Fig. 3d).

(b) Role of RDF, LER, and MGG on CFET performance:
The device miniaturization raises various process-induced reliability concerns, such as random dopant fluctuation (RDF), line edge roughness (LER), metal grain granularities (MGG), etc. The discrete nature of dopants can result in considerable fluctuations in electrical properties, including carrier concentration and mobility. The influence of RDF on I$_{DS}$-V$_{GS}$ is investigated with 200 samples (Fig.4a) of CFET, revealing a standard deviation in threshold voltage (σVth) of 5mV for nFET and 4mV for pFET, respectively. The comparable σVth is justified by the fact that it exhibits a similar variation with T$_{ch}$, as evidenced by the normal quantile plot (Fig. 4b). With an increase in T$_{ch}$, the more pronounced variation in quantile plot becomes evident as a result of random dopants. The increase in channel doping leads to greater fluctuations in dopants, resulting in a higher variation in σVth and σLog$_{10}$I$_{OFF}$ (Fig.4c). LER denotes the inconsistencies and fluctuations observed along the edges of patterns created during the photolithography process. In this work, Si/SiO$_2$ and SiO$_2$/HfO$_2$ interface are considered for LER process variation. The I$_{ON}$ vs I$_{OFF}$ correlation plot indicates a stronger positive correlation between I$_{ON}$ and I$_{OFF}$, which is attributed to the fact that maximal velocity saturation occurs at higher V$_{DS}$ (Fig.5a). The histogram illustrates the variation in threshold and the number of samples, indicating that the nFET exhibits higher distribution from the reference V$_{th}$, resulting in increased deviation compared to the pFET. The histogram plot for the linear region shows the V$_{th}$ distribution at a higher value compared to the saturation region (Fig.5b). Further, the variability in V$_{th}$ is caused by MGG, a dominant source of variability in transistors utilizing the confined metal gate stacks. This is due to the non-uniform distribution of work functions (WF)

over the metal. Grain size (G$_r$) determines the average number of grains in a region, which is used as the expectation value for a Poisson distribution. Each grain is assigned to a WF depending on the likelihood of each grain orientation for the gate material (Fig.6a). In TiN, grains have a 60% chance of having a WF=4.6 eV and a 40% chance of 4.4 eV. For an average G$_r$ (5nm), the plotted I$_{DS}$-V$_{GS}$ (Fig.6b) with 200 samples, exhibits σVth of 24mV for nFET and 23mV for pFET, respectively. For the scaled gate area, effective grain size (G$_{reff}$) and ratio of G$_r$ to gate area (i.e., RGG) are considered. Due to randomness and non-uniformity in the electrical properties of the metal gate, the statistical variance in local WF as G$_r$ increases, leading to higher V$_{th}$ variation with RGG (Fig.6c). As the G$_r$ enlarges, the σV$_{th}$ escalates, resulting in higher statistical fluctuations in the I$_{ON}$ (Fig. 6d).

(c) Early Aging Prediction:
In practice, the V$_{th}$ shift is regarded as a criterion for forecasting device longevity, i.e., early aging. The fluctuation in threshold voltage (ΔV$_{th}$) of ±50mV serves as a criterion for determining the aging prediction [8]. The smaller D$_{SW}$ results in reduced I$_{ON}$, thereby, showing a lesser variation of I$_{ON}$ and I$_{OFF}$ as plotted in Fig.7(a) with considering all the process-induced effects and SHE. ΔV$_{th}$=50mV reaches earlier in the T$_L$ scale if D$_{SW}$ is small (Fig.7b), resembling SHE-dependent early aging. Finally, the variation in L$_g$ shows that lower channel length CFET is more prone to early aging (Fig.7c).

Conclusion

The crux of the paper is to analyze the self-heating (SHE) and process-induced reliability issues and electrical-thermal coupling in Complementary FET. Using well-calibrated TCAD models, the imperative role of the dielectric separation wall (D$_{SW}$) is being analyzed, which causes the electrical-thermal intercoupling between nFET and pFET. Despite SHE, the observed cross-talk temperature rise on pFET (when nFET is at ON state) is ~336K, which significantly affects the normal operation. The correlation plot, quantile plot, and histograms have been drawn for analyzing the impact of process-induced variation, like RDF, LER, and MGG. At last, the early aging of the CFET under such variations has been evaluated based on threshold voltage shift, i.e., ΔV$_{th}$ = ±50mV. Thus, the proposed analysis is worth needed for a reliable CFET design.

References

[1] J. Ryckaert et al., "The Complementary FET (CFET) for CMOS scaling beyond N3," IEEE Symp. VLSI Tech, pp. 141-142, 2018.

[2] S. Subramanian et al., "First Monolithic Integration of 3D Complementary FET (CFET) on 300mm Wafers," IEEE Symp. VLSI Tech, pp. 1-2, 2020.

[3] I. Myeong et al., "Analysis of Self-Heating Effect in DC/AC Mode in Multi-Channel GAA-Field Effect Transistor," IEEE TED, 66, pp. 4631-4637, 2019.

[4] S. Zhao, et. al, "Self-Heating and Thermal Network Model for Complementary FET," IEEE TED, 69, pp. 11-16, 2022.

[5] Synopsys TCAD, "Sentaurus Device User Manual, 2022.

[6] Garand User Guide, Synopsys Inc.

[7] P. Schuddinck et al., "PPAC of sheet-based CFET configurations for 4 track design with 16nm metal pitch," 2022 IEEE Symp. VLSI Tech. and Circuits, pp. 365-366, 2022.

[8] S. Rathore et al., "Self-Heating Aware Threshold Voltage Modulation Conforming to Process and Ambient Temperature Variation for Reliable Nanosheet FET", *IEEE IRPS*, pp.1-5, 2023.

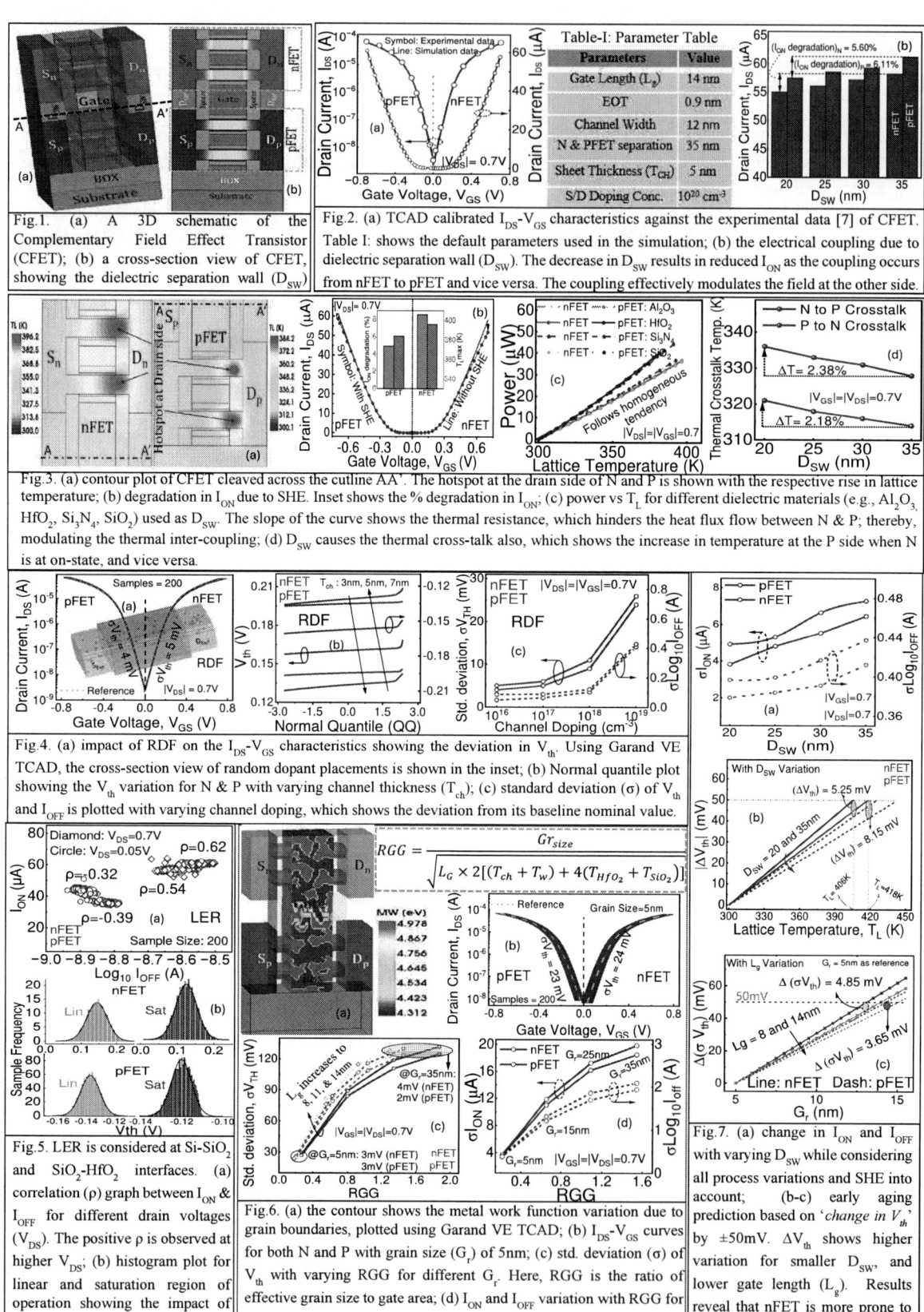

Fig.1. (a) A 3D schematic of the Complementary Field Effect Transistor (CFET); (b) a cross-section view of CFET, showing the dielectric separation wall (D_{SW})

Fig.2. (a) TCAD calibrated I_{DS}-V_{GS} characteristics against the experimental data [7] of CFET. Table I: shows the default parameters used in the simulation; (b) the electrical coupling due to dielectric separation wall (D_{SW}). The decrease in D_{SW} results in reduced I_{ON} as the coupling occurs from nFET to pFET and vice versa. The coupling effectively modulates the field at the other side.

Table-I: Parameter Table

Parameters	Value
Gate Length (L_g)	14 nm
EOT	0.9 nm
Channel Width	12 nm
N & PFET separation	35 nm
Sheet Thickness (T_{CH})	5 nm
S/D Doping Conc.	10^{20} cm^{-3}

Fig.3. (a) contour plot of CFET cleaved across the cutline AA'. The hotspot at the drain side of N and P is shown with the respective rise in lattice temperature; (b) degradation in I_{ON} due to SHE. Inset shows the % degradation in I_{ON}; (c) power vs T_L for different dielectric materials (e.g., Al$_2$O$_3$, HfO$_2$, Si$_3$N$_4$, SiO$_2$) used as D_{SW}. The slope of the curve shows the thermal resistance, which hinders the heat flux flow between N & P; thereby, modulating the thermal inter-coupling; (d) D_{SW} causes the thermal cross-talk also, which shows the increase in temperature at the P side when N is at on-state, and vice versa.

Fig.4. (a) impact of RDF on the I_{DS}-V_{GS} characteristics showing the deviation in V_{th}. Using Garand VE TCAD, the cross-section view of random dopant placements is shown in the inset; (b) Normal quantile plot showing the V_{th} variation for N & P with varying channel thickness (T_{ch}); (c) standard deviation (σ) of V_{th} and I_{OFF} is plotted with varying channel doping, which shows the deviation from its baseline nominal value.

$$RGG = \frac{Gr_{size}}{\sqrt{L_G \times 2[(T_{ch} + T_w) + 4(T_{Hfo_2} + T_{SiO_2})]}}$$

Fig.5. LER is considered at Si-SiO$_2$ and SiO$_2$-HfO$_2$ interfaces. (a) correlation (ρ) graph between I_{ON} & I_{OFF} for different drain voltages (V_{DS}). The positive ρ is observed at higher V_{DS}; (b) histogram plot for linear and saturation region of operation showing the impact of LER.

Fig.6. (a) the contour shows the metal work function variation due to grain boundaries, plotted using Garand VE TCAD; (b) I_{DS}-V_{GS} curves for both N and P with grain size (G_r) of 5nm; (c) std. deviation (σ) of V_{th} with varying RGG for different G_r. Here, RGG is the ratio of effective grain size to gate area; (d) I_{ON} and I_{OFF} variation with RGG for different grain sizes.

Fig.7. (a) change in I_{ON} and I_{OFF} with varying D_{SW} while considering all process variations and SHE into account; (b-c) early aging prediction based on 'change in V_{th}' by ±50mV. ΔV_{th} shows higher variation for smaller D_{SW}, and lower gate length (L_g). Results reveal that nFET is more prone to early aging issues.

Compact Modeling of Silicon Carbide (SiC) Power FETs

Karunesh Kumar Tripathi*, Yogendra Machhiwar*, Danish Raja, and Harshit Agarwal

Department of Electrical Engineering, Indian Institute of Technology Jodhpur

ABSTRACT

In this work, we discuss compact modeling strategy for Silicon Carbide (SiC) based FETs using non-linear bias dependent drift resistance in series with low-voltage MOSFET. Our results show that BSIM-BULK HV compact model can capture SiC device characteristics if the parameters are correctly extracted. We provide step-by-step parameter extraction procedure, covering the complete range of operation, from the weak inversion to the strong inversion region, including both low and high drain biases. The model is validated with numerical simulation (TCAD simulation of SiC MOSFET) as well as experimental data from different technologies.
Keywords: BSIM-BULK, Silicon Carbide (SiC), BSIM6 and power FETs.

INTRODUCTION

Power MOSFETs based on wide-bandgap (WBG) energy semiconductors, such as Silicon Carbide (SiC) and Gallium Nitride (GaN), offer superior performance compared to those based on Silicon (Si) due to their intrinsic material properties, such as higher critical electric fields and higher bandgap energy [1]. The high critical electric field allows WBG semiconductors to withstand higher electric fields, resulting in thinner drift regions and consequently lower ON-resistance compared to Si.

SiC-based power devices hold a substantial market share and find use across a range of applications, including high-voltage switches, power converters, and inverters. These devices are highly mature in terms of technology and industrialization [2]. Consequently, there is a strong and growing demand for precise, compact models of SiC power MOSFETs that can support device performance evaluation, circuit design, and system-level design.

In the field of SiC power MOSFET modeling, the model put forth in [3] is regarded as the seminal study. This offered a physics-based model with a distinct channel current expression that incorporates both the enhanced linear region transconductance and the formulation of the lower gate voltages. An improved model for medium voltage SiC MOSFET is proposed in [4],

*These authors contributed equally to this work.

the MOSFETs are modeled by introducing temperature-dependent voltage source and current source. Experimental validation presented in [4] shows that it is suitable for wide temperature range applications particularly at low temperature (-25 °C). A three layer artificial neural network based equivalent circuit simulation behavior model for SiC MOSFET is presented in [5]. Over the past two decades, numerous SiC MOSFET models have been developed [6], which have been comprehensively studied however, there is still a lack of accurate and comapct industry-standard models for SiC power MOSFETs. In [7], method to extend the industry-standard BSIM3 model to describe SiC vertical power MOSFETs is presented. The simulation validation demonstrated that the model is capable of accurately capturing the properties of SiC power MOS-FETs at various electrical and thermal operating points.

In this work, we present compact modeling of SiC FETs using industry standard BSIM-BULK HV compact model [8]–[10]. We show that the model can capture characteristics of different flavors of SiC FETs, including derivatives, provided the parameters are correctly extracted. We provide brief guidelines for the same. The parameter extraction encompasses the full range of operation, from weak inversion to strong inversion regions, covering both low and high drain biases. The extraction is validated using TCAD-simulated SiC device, as well as experimental data of 600V vertical 4H-SiC MOSFETs [11] and 600V lateral 4H-SiC MOS-FETs [12].

MODELING DETAILS AND PARAMETER EXTRACTION

SiC FET can be modeled as a low voltage transistor in series with a bias-dependent non-linear resistor (R_{drift}) as shown in Figure 1(e). While actual transistor is more complex, this simplified representation works well as we have seen with high-voltage silicon transistor cases [9]. The drift resistance is modeled as [9], [10]:

$$R_{Drift} = RDLCW \left[1 + \delta_v \left(\frac{|V_{di1,di}|}{V_c} \right)^{MDRIFT} \right]^{\frac{1}{MDRIFT}} \quad (1)$$

$$V_c = RDLCW \cdot I_{sat} \quad (2)$$

$$I_{sat} = W.q.N_{DR}.t_{dr}.VDRIFT \quad (3)$$

$$\delta_v = \frac{|V_{di1,di}|^{4-MDRIFT}}{|V_{di1,di}|^{4-MDRIFT} + \alpha.V_c^{4-MDRIFT}} \quad (4)$$

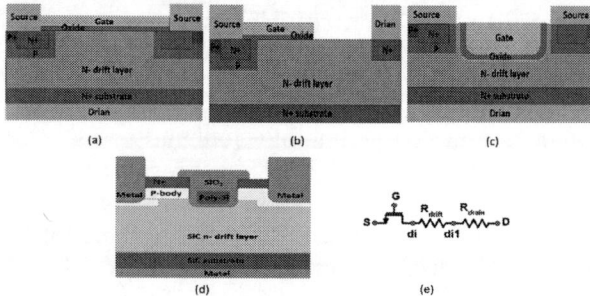

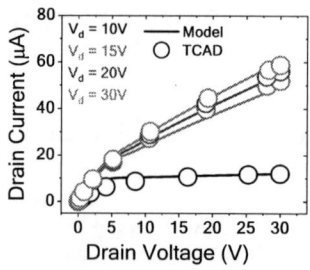

Fig. 1. Illustration of different types of SiC FET configuration : (a) Vertical SiC planar-gate MOSFET (b) Lateral SiC MOSFET (c) Vertical SiC double trench MOSFET (d) Cross-section of simulated SiC FET (e) Modeling of SiC FETs using bias-dependent non-linear drift resistance in series with low voltage transistor.

Fig. 3. Model validation with TCAD data, $I_d - V_d$ plot for $V_g =$ 10V, 15V, 20V and 30V.

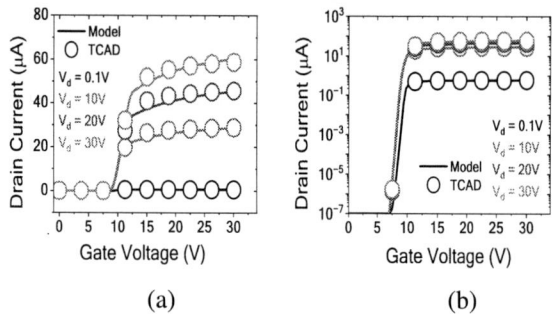

Fig. 2. Model validation with TCAD simulation for V_d = 0.1V, 10V, 20V and 30V.

Here, t_{dr} and N_{DR} are the effective drift region thickness and doping concentration. **RDLCW** is the resistance parameter of the drift region at low drain current, **MDRIFT** is a smoothing parameter for the transition of drain current from linear to saturation region in I_d-V_d characteristics. **VDRIFT** represents carrier saturation velocity in the drift and δ_v in (4) is used to ensure correctness of derivatives at $V_{di1,di} = 0$ [9]. The above model takes care of velocity saturation in the drift region. In addition, I_{sat} in the above is improved to account for gate/drain bias-dependency [8] using parameters **GADRIFT**, **RDVDS**, **PTWGHV** and **PTWGHV1**,

$$I_{sat} = I_{sat} \cdot \alpha_1 \cdot \alpha_2 \quad (5)$$
$$\alpha_1 = 1 + GADRIFT \cdot (|V_{di1,di}| - RDVDS) \quad (6)$$
$$\alpha_2 = 1 + PTWGHV \cdot (q_s - PTWGHV1) \quad (7)$$

A. $I_d - V_g$ in Sub-threshold region

To begin with, firstly process parameters are correctly initialized. These includes parameters like gate length **L** and width **W**, as well as the equivalent oxide thickness **TOXE**. Doping parameters **NSD** (source or drain doping concentration) and **NDEP** (channel doping concentration) are then set to reasonable values. Although **NDEP** can be adjusted to suit model requirements.

To enable the High Voltage (HV) module, set the parameters **HVMOD** and **RDSMOD** to value 1 [8], [13].

In the subthreshold region, at low gate voltages, the channel region resistance predominates over the drift region resistance, and henceforth the HV device characteristics are primarily governed by MOS parameters. At low drain bias (typically V_d = 100 mV), both the flat band voltage **VFB** and the subthreshold swing parameter **NFACTOR** should be extracted [8].

B. $I_d - V_g$ Above threshold region, at low drain bias

At low drain voltages, carriers do not reach velocity saturation so the drift region operates similar to a resistor. The channel region also acts like a nonlinear resistor, with its resistance being function of the channel length (L), inversion charge density, and carrier mobility. The inversion charge in the channel varies proportionally with the gate voltage [$Q_{inv} \propto (V_{gs} - V_{th})$]. In the BSIM-BULK model, effective channel mobility (and thus channel resistance) is modeled using the universal mobility model [13]. Parameters used to model the effective channel mobility are **U0** (low field mobility) **ETAMOB** (effective vertical field), **UA** and **EU** (mobility degradation at high electric field), **UD** and **UCS** (other short-channel influencing coulomb scattering effect mobility parameters).

Keeping **VDRIFT** at large value, MOS resistance and drift resistance parameters are extracted: **U0, UA, EU, ETAMOB, RDW, PRWG** etc. (for the MOS) and **RDLCW** for the drift region.

C. $I_d - V_d$ Characteristics

At low gate voltage, the primary reason behind current saturation is carrier velocity saturation in channel region. In contrast, at high gate voltage the current saturation is because of carrier velocity saturation in drift region. Further the current in HV devices saturates gradually with drain voltage at higher gate voltages. This quasi-saturation behavior is a function of drift region doping and length, and based on their structure devices may show stronger or weaker quasi-saturation effect.

979-8-3315-0417-5/25 $31.00 © 2025 IEEE

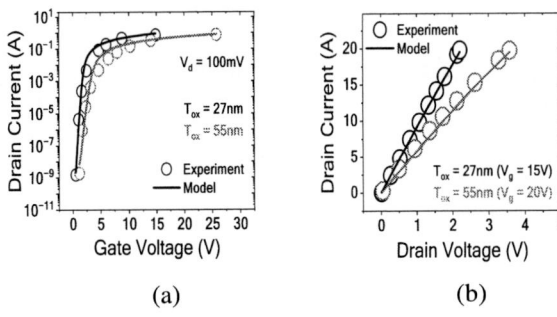

Fig. 4. Model validation with experimental data [11], 4H-SiC Vertical MOSFET (a) $I_d - V_g$ plot for oxide thickness (T_{ox}) = 27nm and 55nm at V_d = 100mV (b) $I_d - V_d$ plot for T_{ox} = 27nm and 55nm at V_g = 15V and 20V respectively.

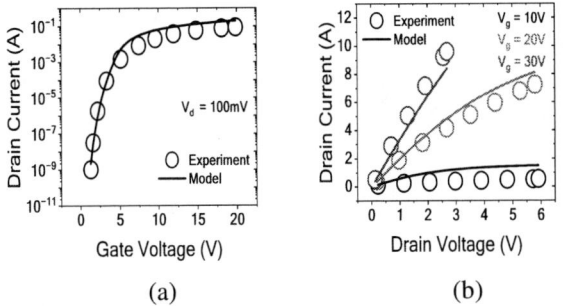

Fig. 5. Model validation with experimental data [12], 4H-SiC Lateral MOSFET (a) $I_d - V_g$ plot at V_d = 100mV (b) $I_d - V_d$ plot for V_g = 10V, 20V and 30V.

Keeping **VDRIFT** high, extract VSAT, PTWG, PSATX and other MOS parameters to match $I_{d,sat}$ level then start reducing **VDRIFT** to tune linear to saturation transition. Extract MDRIFT for smoothing this region.

RESULTS AND DISCUSSION

To validate the model, SiC Double Trench MOS-FET [14] with channel length $0.42\mu m$ and gate oxide thickness of $50nm$ was simulated in Sentaurus TCAD [15]. The cross-section of simulated device is shown in Figure 1(d). Figure 2 shows $I_d - V_g$ model validation with TCAD simulation in linear and logarithmic scale for V_d = 0.1V, 10V, 20V and 30V. The $I_d - V_d$ model validation plot for V_g = 10V, 15V, 20V and 30V is shown in Figure 3. Figure 4(a) shows $I_d - V_g$ plots for both model and 600V Vertical 4H-SiC MOSFET [11] for 27nm and 55nm oxide thickness. Figure 4(b) shows the $I_d - V_d$ plot for the same. For 600V 4H-SiC Lateral MOSFET [12] Figure 5(a) shows the $I_d - V_g$ plot for both model and device whereas Figure 5[b] shows $I_d - V_d$ plot for V_g = 10V, 20V and 30V.

CONCLUSION

We have presented compact modeling of power SiC FETs using industry standard BSIM-BULK HV compact model. A step-by-step parameter extraction strat-

egy, encompassing the full range of operation, from weak inversion to strong inversion regions and covering both low and high drain biases is detailed. The proposed parameter extraction is validated with TCAD simulation data and multiple sets of experimental data.

REFERENCES

[1] Jerry L. Hudgins, Grigory S. Simin, Enrico Santi, M. Asif Khan, "An Assessment of Wide Bandgap Semiconductors for Power Devices", IEEE Trans. Power Electron., Vol. 18, May 2003.

[2] Xu She, Alex Q. Huang, ´Oscar Luc´ıa, Burak Ozpineci, "Review of Silicon Carbide Power Devices and Their Applications", IEEE Trans. on Industrial Electronics, Vol. 64, OCTOBER 2017.

[3] Ty R. McNutt, Allen R. Hefner, Jr., H. Alan Mantooth, David Berning, Sei-Hyung Ryu, "Silicon Carbide Power MOSFET Model and Parameter Extraction Sequence", IEEE Trans. on Power Electronics, Vol. 22, March 2007.

[4] Kai Sun, Hongfei Wu, Juejing Lu, Yan Xing, Lipei Huang, "Improved Modeling of Medium Voltage SiC MOSFET Within Wide Temperature Range", IEEE Trans. on Power Electronics, Vol. 29, May 2014.

[5] Yuhao Lee, Mu Zhang, Pingjuan Niu, Pingfan Ning, Lei Liu, Shushu Lee, "Simplified Silicon Carbide MOSFET Model Based on Neural Network", Materials Science Forum, Trans Tech Publications Ltd., Vol. 954, May 2019.

[6] Homer Alan Mantooth, Kang Peng, Enrico Santi, Jerry L. Hudgins, "Modeling of Wide Bandgap Power Semiconductor Devices—Part I", IEEE Trans on Electron Devices, Vol. 62, Feb. 2015.

[7] Lixi Yan , Kanuj Sharma, Ingmar Kallfass, "A Compact Model Extending the BSIM3 Model for Silicon Carbide Power MOSFETs", IEEE Trans. on Power Electronics, Vol. 38, April 2023.

[8] (2020). BSIM-BULK107.2.0 MOSFET Compact Model. Available: https://www.bsim.berkeley.edu/models/bsimbulk/

[9] H. Agarwal, C. Gupta, R. Goel, P. Kushwaha, Y.-K. Lin, M.-Y. Kao, J.-P. Duarte, H.-L. Chang, Y. S. Chauhan, S. Salahuddin, C. Hu, "BSIM-HV: High-Voltage MOSFET Model Including Quasi-Saturation and Self-Heating Effect", IEEE Trans. on Electron Devices, Vol. 66, Oct. 2019.

[10] Girish Pahwa, Ayushi Sharma, Ravi Goel, Garima Gill, Harshit Agarwal, Yogesh Singh Chauhan, Chenming Hu, "Robust Compact Model of High-Voltage MOSFET's Drift Region" IEEE TRANSACTION ON COMPUTER-AIDED DESIGN OF INTEGRATED CIRCUITS AND SYSTEMS, Vol. 42, Jan. 2023.

[11] Aditi Agarwal, Kijeong Han, B. Jayant Baliga, "600V 4H-SiC MOS-FETs Fabricated in Commercial Foundry With Reduced Gate Oxide Thickness of 27nm to Achieve IGBT-Compatible Gate Drive of 15V", IEEE ELECTRON DEVICE LETTERS, Vol. 40, Nov. 2019.

[12] Nick Yun, Woongje Sung, "Design and Fabrication Approaches of 400–600V 4H-SiC Lateral MOSFETs for Emerging Power ICs Application", IEEE Trans. on Electron Devices, Vol. 67, Nov. 2020.

[13] Garima Gill, Yogendra Machhiwar, Girish Pahwa, Chenming Hu, Harshit Agarwal, "Comprehensive High-Voltage Parameter Extraction Strategy for BSIM-BULK HV Model", IEEE Trans. on Electron Devices, Vol. 71, Jan. 2024.

[14] T. Nakamura, Y. Nakano, M. Aketa, R. Nakamura, S. Mitani, H. Sakairi, Y.Yokotsuji, "High performance SiC trench devices with ultra-low ron", International Electron Devices Meeting, Dec. 2011.

[15] User's Manual, Synopsys Inc., Mountain View, CA, USA, 2013.

979-8-3315-0417-5/25 $31.00 © 2025 IEEE

Performance Analysis of Advanced Ferroelectric $HfO_2 - ZrO_2$ Superlattice Gate Stack Transistor with Multi-Phase Ferroelectric Order

Danish Raja[1], Yogendra Machhiwar[1], Karunesh Kumar Tripathi[1], Girish Pahwa[2], and Harshit Agarwal[1]

[1]Indian Institute of Technology Jodhpur, India, [2]ICST, NYCU, Taiwan

ABSTRACT

In this work, we investigate the integration of advanced ferroelectric $HfO_2 - ZrO_2$ (HZH) superlattice gate stacks with mixed t/o-phase ferroelectric order, onto Fully Depleted Silicon-On-Insulator (FDSOI) transistors. Our study reveals that the HZH superlattice gate stack leads to improved equivalent oxide thickness (EOT). The improvement in EOT leads to enhanced gate capacitance (C_{gg}), and better I_{ON}/I_{OFF} ratio compared to traditional high-κ oxides, while maintaining low leakage and high transconductance. Furthermore, while the location of t-phase can affect C-V behaviour, little impact on I-V is observed.

Keywords: HZH, ferroelectric, EOT, SOI, negative capacitance.

INTRODUCTION

With advancements in semiconductor technology, new material-device concepts that surpass Boltzmann's limit are essential for improved transistor performance, particularly in terms of power consumption and size. The discovery of ferroelectricity in doped HfO_2, particularly Zr-doped HfO_2 (HZO), through the negative capacitance (NC) effect, offers a promising solution for further scaling of transistors and enhance their energy efficiency. This novel concept of negative capacitance was introduced in [1], where a ferroelectric insulator is used as gate oxide to achieve voltage amplification. The integration of high-κ dielectrics with this negative capacitance effect into transistor gate stacks holds significant potential in breaking the limitations of standard scaling approaches [2]. This integration fueled the reduction of equivalent oxide thickness (EOT) of gate oxides to enhance gate capacitance (C_{gg}). The enhancement in transistor performance depends on the extent of capacitance matching between the ferroelectric and underlying MOS transistor [3]. In conventional high-κ HfO_2 gate stacks, however, EOT scaling has been constrained by the presence of an SiO_2 interfacial layer (IL), which typically limits EOT to around 9.5Å. Although, the IL scavenging techniques can be employed to reduce EOT further, these methods come with drawbacks, such as degraded electron transport and increased gate leakage current, which compromise the overall device performance.

Recent studies have demonstrated that ultrathin ferroelectric $HfO_2 - ZrO_2$ (HZH) superlattice structures offer a promising solution to this issue. These layered heterostructures enable EOT values as low as 6.5Å in Metal-Oxide-Semiconductor (MOS) capacitors, achieving a gate oxide thickness smaller than the IL itself [4]. This reduction is attributed to the mixed ferroelectric-antiferroelectric (FE-AFE) phase of the HZH stack, which allows for the flattening of the energy landscape due to large depolarization fields. As a result, the effective C_{gg} is significantly enhanced without the adverse effects on electron mobility or leakage current typically observed in conventional approaches. This enhancement, in turn, leads to improved transconductance (g_m) in both bulk and SOI n-MOSFETs [5], [6].

To optimize HZH superlattice gate stacks, advanced modeling methods like phase-field and TCAD simulations are essential. Phase-field simulation provides a microscopic view like domain-wall motion and NC stabilization by solving complex interactions, including the Ginzburg-Landau and Poisson equations [7]–[9]. In this work, we present Technology Computer-Aided Design (TCAD) simulation based study of HZH heterostructures, calibrated with experimental data [4] to evaluate the macroscopic effects of NC, including capcitance enhancement, g_m boosting and EOT reduction, in multi-layer gate stacks. The effect of varying position of the tetragonal (t-phase) fraction along the length of the ferroelectric in the gate stack is also studied.

SIMULATION METHODOLOGY

The MOS capacitor structure with HZH as ferroelectric is shown to have a mixture of t/o-phase (tetragonal/orthorhombic) with partially in-plane polarization [4]. Hence, the simulation is done in accordance with [7] as shown in Figure 1(a).

The ferroelectric material can be modeled using Ginzburg-Landau-Khalatnikov equation as follows

$$-\rho\frac{dP}{dt} = \alpha P + \beta P^3 + \gamma P^5 - g\nabla^2 P - E \quad (1)$$

where α, β and γ are Landau Coefficients and P is the ferroelctric polarization. The electric field E can be expressed as $-\frac{d\phi}{dy}$, where ϕ is the electric potential distribution in the device and is obtained by solving Poisson's equation as

$$\nabla \cdot \epsilon\nabla\Phi = \nabla \cdot P \quad (2)$$

with ϵ as the permittivity of the ferroelectric.

The TCAD simulations are carried out using physical models including constant mobility model, Shockley-Read-Hall (SRH) model for carrier generation and recombination, and high field saturation. The HZH heterostructure is first calibrated with the experimental data [4] by fine-tuning the ferroelectric parameters starting with [10] as reference. Then the superlattice stack is integrated on to the FDSOI device as shown in Figure 1(b) to study various characteristics. The parameters used for the simulations (unless stated otherwise) are given in Table 1.

RESULTS AND DISCUSSION

Firstly, the capacitance-voltage (C-V) characteristics of HZH MOS capacitor obtained from the TCAD simulations is calibrated against experimental measurements. This was done to extract the reported EOT of 6.5Å in such gate stacks. Figure 2 shows the calibration of HZH superlattice with the experimental data [4].

After the EOT calibration in MOS capacitor, the HZH superlattice is integrated as the gate stack in FDSOI MOSFET. Figure 3 shows the capacitance enhancement in this device with 1.8Å HZH + 8Å SiO_2 as gate stack. The introduction of HZH as gate stack in FDSOI leads to improvement in the EOT as shown in Figure 4. The improvement in EOT is due to underlying mixed ferroic order in the HZH gate stack. This results in **39.64% increase in the effective C_{gg}** compared to the baseline FET with 1.8nm HfO_2 + 8Å SiO_2 gate stack as shown in Figure 3. Figure 5 shows the current voltage (I-V) characteristics of FDSOI MOSFETs with HZH gate stack compared with the baseline FET. The observed increase in the ON-current (I_{ON}) is attributed primarily to the significant enhancement in C_{gg}. The advantages of EOT scaling are further reflected in the reduction of the OFF-current (I_{OFF}). Furthermore, the introduction of HZH superlattice as gate stack in FDSOI FET contributes to improvement in subthreshold swing (SS), which helps to suppress leakage currents and improves the overall switching efficiency of the device. We also obtain better transconductance efficiency (g_m/I_D) as depicted in Figure 6 which again is linked to capacitance enhancement due to EOT lowering as the carrier velocity remains unaffected.

Role of t-phase fraction location: The impact of varying the position of the t-phase fraction along the length of the gate stack from source end to drain end was also investigated. Interestingly, no significant change was observed in key parameters such as threshold voltage, I_{ON}/I_{OFF} ratio, SS and drain induced barrier lowering (DIBL) by varying the position of t-phase fraction. Figure 7 shows the variation of these parameters with the position of t-phase fraction over the oxide. However, the effect on C_{gg} is more pronounced.

The placement of the t-phase fraction closer to the drain leads to a considerable enhancement in C_{gg}. This enhancement is explained by the stronger electric fields and greater capacitive coupling near the drain, especially in short-channel devices where the electric field near the drain is more intense due to the higher potential gradient. In contrast, placing the t-phase near the source, where the electric field is weaker as shown in Figure 8, results in **8.56% reduction in C_{gg}** compared to placing the t-phase near the drain, as shown in Figure 9. The increased C_{gg} from placing the t-phase near the drain also contributes to further EOT reduction, enhancing overall device performance. While varying the position of the t-phase along the gate length does not significantly impact C_{gg}, placing the t-phase in the middle yields the most enhanced device characteristics.

CONCLUSION

This work demonstrates the experimentally observed EOT reduction in HZH superlattice gate stacks integrated on FDSOI MOSFETs through detailed TCAD simulations. Our results show that the HZH superlattice enables sub-nm EOT, resulting in a 39.6% enhancement in C_{gg}, an improved I_{ON}/I_{OFF} ratio, reduction in SS and better transconductance efficiency(g_m/I_d). Our analysis also shows that varying the t-phase position along the gate length may affect overall C_{gg}.

ACKNOWLEDGMENT

This work is supported by DST-Nanomission grant.

REFERENCES

[1] S. Salahuddin, and S. Datta. "Use of negative capacitance to provide voltage amplification for low power nanoscale devices." Nano letters 8.2, 405-410 (2008).

[2] M. M. Frank, "High-k/metal gate innovations enabling continued CMOS scaling." Proceedings of the ESSDERC,pp. 25-33 (2011).

[3] H. Agarwal et al. "Proposal for Capacitance Matching in Negative Capacitance Field-Effect Transistors." IEEE Electron Device Lett., 40, pp. 463-466 (2019).

[4] S. S. Cheema et al. "Ultrathin ferroic HfO2–ZrO2 superlattice gate stack for advanced transistors." Nature 604, 65–71 (2022).

[5] W. Li et al., "Demonstration of Low EOT Gate Stack and Record Transconductance on Lg=90 nm nFETs Using 1.8 nm Ferroic HfO2-ZrO2 Superlattice." IEEE IEDM, pp. 13.6.1-13.6.4 (2021).

[6] LC. Wang et al., "Record Transconductance in Leff 30 nm Self-Aligned Replacement Gate ETSOI nFETs Using Low EOT Negative Capacitance HfO2-ZrO2 Superlattice Gate Stack." IEEE Symposium on VLSI Technology and Circuits, pp. 1-2 (2023).

[7] M. Hoffmann, S. S. Cheema, N. Shanker, W. Li and S. Salahuddin, "Quantitative study of EOT lowering in negative capacitance HfO2-ZrO2 superlattice gate stacks." IEEE IEDM, pp. 13.2.1-13.2.4 (2022).

[8] P. Kumar, A. Nonaka, R. Jambunathan, G. Pahwa, S. Salahuddin, Z. Yao "FerroX: A GPU-accelerated, 3D phase-field simulation framework for modeling ferroelectric devices", Computer Physics Communications, 290, (2023)

[9] P. Kumar, M. Hoffmann, A. Nonaka, S. Salahuddin and Z. Yao, "3D Ferroelectric Phase Field Simulations of Polycrystalline Multi-Phase Hafnia and Zirconia Based Ultra-Thin Films." Adv. Electron. Mater., 10, 2400085 (2024).

[10] A.K.Saha, S.K.Gupta "Multi-Domain Negative Capacitance Effects in Metal-Ferroelectric-Insulator-Semiconductor/Metal Stacks": A Phase-field Simulation Based Study. Sci Rep 10, 10207 (2020).

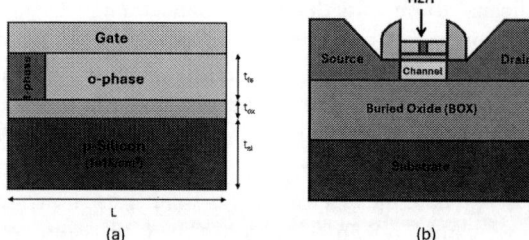

(a)　　　　　　　　(b)

Fig. 1. (a) The Metal-Oxide-Semiconductor (MOS) capacitor heterostructure HfO_2-ZrO_2 (HZH) superlattice as ferroelectric. (b) Fully Depleted Silicon On Insulator (FDSOI) structure simulated in TCAD with HZH superlattice as gate stack.

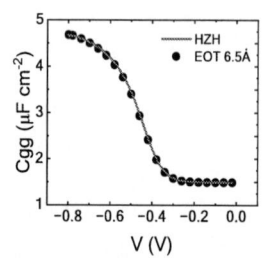

Fig. 2. The simulated Capacitance-Voltage (CV) characteristics of HZH + 8Å SiO_2 as IL calibrated with experimentally obtained 6.5Å EOT.

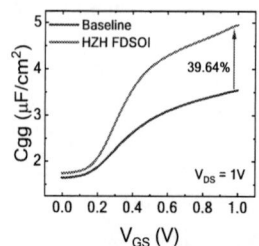

Fig. 3. Capacitance-Voltage (CV) characteristics of $L = 20nm$ FD-SOI FET with HZH + 8Å SiO_2 gate stack showing enhancement in C_{gg} compared to the Baseline FET.

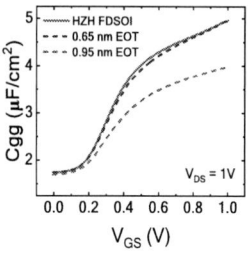

Fig. 4. EOT improvement and enhanced C_{gg} in FDSOI with the HZH gate stack compared to HfO_2 control with 9.5Å EOT.

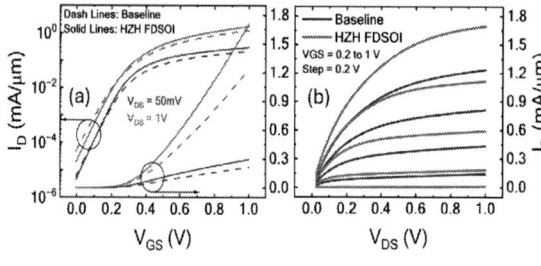

Fig. 5. The simulated current-Voltage (IV) characteristics of FDSOI FET with HZH + 8Å SiO_2 gate stack (a) $I_d - V_{GS}$ characteristics depicting considerable increase in I_{ON}/I_{OFF} ratio, SS and DIBL. (b) $I_d - V_{DS}$ characteristics.

Fig. 6. Improvement in transconductance efficiency (g_m/I_D) of $L = 20nm$ FDSOI FET with HZH + 8Å SiO_2 gate stack compared to the Baseline FET.

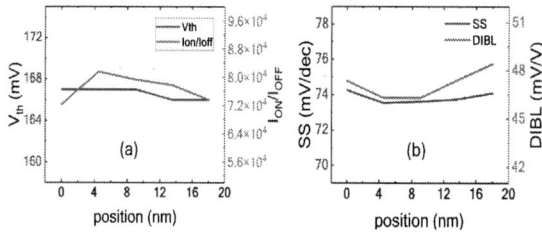

Fig. 7. Effect of position variation of t-phase fraction along the length of the ferroelectric on parameters (a) V_{th} and I_{ON}/I_{OFF} (b) SS and DIBL.

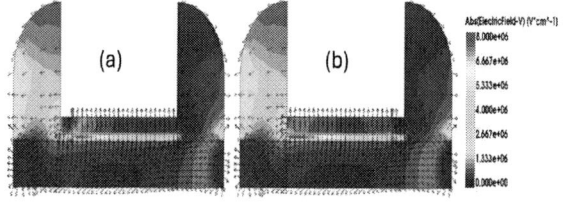

Fig. 8. Electric field contours with t-phase fraction at (a) source side (b) drain side. The decrement in electric field intensity with t-phase fraction on source side results in degradation of C_{gg} when compared to t-phase fraction position at drain side.

Fig. 9. Reduction in capacitance enhancement with t-phase fraction at Source side compared to Drain Side.

Table 1: List of Default Parameters used in Simulations.

Parameter	Value	Description
α	-1.65×10^{11} cm/F	Landau Coefficient
β	2.4×10^{20} cm^5/FC^2	Landau Coefficient
γ	1.5×10^{29} cm^9/FC^4	Landau Coefficient
g	8×10^{-5} cm^3/F	Gradient Energy Coefficient
L	20 nm	Channel Length
W	1 μm	Channel Width
t_{si}	5 nm	Channel thickness
t_{ox}	0.8 nm	Oxide thickness
t_{fe}	1.8 nm	Ferroelectric thickness
t_{BOX}	25 nm	Buried oxide thickness
L_{ex}	8 nm	Extension region length
%t	10%	t-phase fraction

979-8-3315-0417-5/25 $31.00 © 2025 IEEE

Novel Ferroelectric Tunnel FET-based Computing-in-memory with In-situ XOR Cipher-Encrypted AND-Type Multiply-Accumulate for Secure Edge AI

Jin Luo[1*], Zhiyuan Fu[1], Qianqian Huang[1,2*] and Ru Huang[1,2*]

[1]School of Integrated Circuits, Peking University, Beijing 100871, China.
[2]Beijing Advanced Innovation Center for Integrated Circuits, Beijing 100871, China.
*E-mail: ruhuang@pku.edu.cn, hqq@pku.edu.cn, luoj@pku.edu.cn

Abstract

In this work, a novel ferroelectric tunnel FET (FeTFET) based computing-in-memory (CIM) system incorporating in-situ XOR cipher-encrypted AND-type multiply-accumulate (MAC) functionality is proposed and experimentally demonstrated for the first time. Leveraging the dual-modulation effect and ambipolarity of FeTFET, the activation (x) is represented by drain-controlled band-to-band tunneling (BTBT), while XOR operations between the *Key* and stored encrypted weight (w_e) are performed by ferroelectric-gate-controlled non-monotonic BTBT at both drain and source junctions. The in-situ XOR cipher-encrypted AND-type multiplication is realized within a single FeTFET, fabricated on a 14 nm FinFET platform, thereby eliminating decryption circuits. Based on the design, encrypted neural networks are demonstrated with reduced hardware costs and high energy efficiency, highlighting its potential for secure edge AI applications. (*Keywords*: Ferroelectric tunnel FET, computing-in-memory, secure AI, XOR cipher)

Introduction

Computing in memory (CIM) architectures based on high-density non-volatile memory (NVM) have garnered significant attention [1, 2]. The raw weight matrices stored in CIM are susceptible to security breaches, presenting potential vulnerabilities (Fig. 1a) [3]. Encrypting CIM with an authorized *Key* is essential for secure edge AI. Recently, several secure CIM schemes employing in-situ XOR cipher in parallel expandable AND-type multiply-accumulate (MAC) computations have been explored [4, 5], where the *Key* and activation (x) of neural networks are applied to the CIM array concurrently (Fig. 1b). However, the encrypted weight (w_e) cell designs tend to be complicated, requiring complementary encrypted weight (w_e) storage and *Key* inputs for MAC with in-situ decryption, such as 2T-2RRAM configurations [5] or the SRAM with split word-line (WL) [4], suffering from high hardware cost and energy consumption. In our previous work [6], by algorithmically combining XNOR-type MAC with XOR cipher, and then utilizing gate-controlled ambipolar BTBT mechanism, we proposed a ferroelectric tunnel FET (FeTFET) based non-volatile CIM (nvCIM) array, which features single-transistor w_e along with row-wise XOR gates for decryption (Fig. 2a).

In this work, we further propose a AND-type MAC with in-situ XOR cipher, enhancing precision expansion capability while maintaining a lower cumulative current

compared with our previously proposed XNOR-type encrypted MAC (eMAC). By simultaneously leveraging the drain- and gate-controlled BTBT, AND-type eMAC is experimentally demonstrated within a single FeTFET device, eliminating extra decryption gates (Fig. 2b). Based on this design, encrypted neural networks for pattern recognition are demonstrated for secure edge AI with reduced hardware cost and high energy efficiency.

Design of FeTFET-based nvCIM with in-situ XOR cipher-encrypted AND-type MAC

Fig. 2b shows the proposed FeTFET-based encrypted nvCIM array design. The XOR cipher, as an effective cryptography method for neural networks, is used for row-wise weight encryption (Fig. 2c), and the *Key* is applied to the word-line (WL) shared by the gates of each row's FeTFET. The AND-type multiplication with expandable precision is employed for MAC operation, and the activation (x) is applied to the read-word-line (RWL) shared by the drain of each row.

XOR cipher-encrypted AND-type MAC: As shown in Fig. 2b, the correct partial sum of MAC is obtained by two steps. First, XOR operation is applied to the ciphertext weight ($w_{e\_i,j}$) and the authorized *Key* to retrieve the plaintext weight ($w_{i,j}$). Second, AND operation is performed between $w_{e\_i,j}$ and x_i for multiplication. Finally, the bitwise multiplication results are accumulated along source-lines (SLs) as current. Both XOR cipher and AND-type multiplication can be simultaneously implemented by a single three-terminal FeTFET device with ambipolarity. Thus, by advancing the algorithm, the row-wise XOR gates presented in our earlier work [6] can be effectively eliminated.

FeTFET-based AND operation with XOR cipher: The FeTFET structure integrates a ferroelectric (FE) layer into the gate of the tunnel FET (TFET) with gate-controlled reverse-biased P-I-N structure (Fig. 2d). When a relatively large write voltage is applied, the w_e can be stored in FeTFET with different non-volatile FE polarization states, modulating the threshold voltage (V_{TH}), and shifting the ambipolar I_D-V_G curves. The x_i of "0/1" is applied through the drain voltage (V_D) of $0V/V_{DD}$ in RWLs. The *Key* bit of "0/1" is set through the readout gate voltage ($V_{G\text{-read}}$) of V_L/V_H, allowing the FeTFET to operate in a non-destructive read mode. For $x_i = 0$ (V_D=0V), the device generally exhibits a negligible low current. For $x_i = 1$ (V_D=VDD), only when the key is correct can FeTFET displays a drain current (I_D) corresponding to the plaintext weight (w) of "0/1", indicating the XOR cipher-encrypted AND operation for multiplication.

Therefore, this approach results in a higher sparsity of MAC outcomes, which is advantageous for reducing energy consumption compared with our previous proposed XNOR-based MAC scheme [6]. Moreover, this design eliminates the need of row-wise XOR gates for decryption [6], as well as the complementary weight storage and key-input in conventional encrypted CIM schemes [4, 5].

FeTFET-based XOR-encrypted AND-type MAC

A. Characteristics of FeTFET for in-situ XOR cipher

The proposed FeTFET is experimentally constructed by connecting an FE layer to the gate of TFET. The tungsten top electrode is paired with a 10nm $Hf_{0.5}Zr_{0.5}O_2$ FE layer to optimize write- and read-disturb immunity of FeTFET array (Fig. 3a-c) [7]. The TFET structure is fabricated based on a 14nm FinFET platform, which exhibits distinct ambipolar BTBT current (Fig. 3d) due to a specialized layout without a drain underlap region. Because of the unique dual-modulation effects [8], the BTBT windows and ambipolar current are controlled by both gate and drain voltage, representing the Key and x inputs, respectively. The I_D-V_G characteristics of FeTFET device exhibits ambipolarity with typical FE hysteresis under relatively high sweep voltage (Fig. 3e), due to the FE polarization switching, indicating the w_e storage capability.

B. FeTFET-based AND-type MAC with XOR cipher

Based on the fabricated FeTFET, the XOR cipher-encrypted AND-type multiplication behavior is experimentally characterized (Fig. 4). During initialization, the encrypted weight w_e, dependent on both w and the authorized key (Fig. 4a), is programed into FeTFET with different V_{TH} states and BTBT location. During the implicitly decrypted multiplication phase, the input x_i (V_X) and key (V_{key}) are respectively applied to the drain and gate of FeTFET. When $x_i = 0$, the I_D maintains the off state (I_L). When $x_i = 1$ (Fig. 4b-d), the correct product ($x \times w$) is only represented by different drain currents (I_L/I_H) when the authorized key is applied. If an unauthorized key is used, the product results are incorrect, leading to mistake in the MAC computation, thus achieving security enhanced nvCIM. Compared with our previous XNOR-eMAC, the AND-type eMAC eliminates the decryption gates by leveraging the drain terminal.

FeTFET-based encrypted nvCIM macro

Based on the proposed FeTFET-based nvCIM with XOR cipher-encrypted AND-type MAC, encrypted neural networks are demonstrated by utilizing our developed FeTFET model, featuring ambipolar BTBT in both source and drain junctions [6] (Fig. 5). The precision of x and w in nvCIM is extended through serial bits and multiple cells

(Fig. 6a). If the key is incorrect, the accumulated current (I_{SL}) of each column shows errors (Fig. 6b), directly resulting in a significant accuracy degradation when using an unauthorized key (Fig. 7a). Moreover, the energy efficiency of FeTFET-based AND-type encrypted MAC (eMAC) is evaluated (Fig. 7b), showing improvement over SRAM-based scheme and further improvement compared with our previous FeTFET-based XNOR-type eMAC.

Conclusion

This work reports a novel FeTFET-based nvCIM with in-situ XOR cipher, and experimentally demonstrates precision-expandable AND-type MAC with implicitly decryption within a single-transistor w_e cell, enabling significantly hardware cost reduction (Table I). Based on our proposed design, the encrypted neural networks are implemented with high energy efficiency, highlighting its great potential for secure edge AI applications.

Acknowledgments

This work was supported by NSFC (62404008, 62374009, 61927901), 111 Project (B18001), and China Postdoctoral Science Foundation (2024M750098, GZC20240026).

References

[1] J. An *et al.*, "Design memristor-based computing-in-memory for AI accelerators considering the interplay between devices, circuits, and system," *Science China Information Sciences*, vol. 66, no. 8, p. 182404, 2023, doi: https://doi.org/10.1007/s11432-022-3627-8.

[2] Z. Zhang *et al.*, "Recent progress of hafnium oxide-based ferroelectric devices for advanced circuit applications," *Science China Information Sciences*, vol. 66, no. 10, p. 200405, 2023, doi: https://doi.org/10.1007/s11432-023-3780-7.

[3] P. Zuo, Y. Hua, L. Liang, X. Xie, X. Hu, and Y. Xie, "Sealing neural network models in encrypted deep learning accelerators," in *2021 58th ACM/IEEE Design Automation Conference (DAC)*, 2021: IEEE, pp. 1255-1260, doi: https://doi.org/10.1109/DAC18074.2021.9586199.

[4] S. Huang, H. Jiang, X. Peng, W. Li, and S. Yu, "XOR-CIM: Compute-in-memory SRAM architecture with embedded XOR encryption," in *Proceedings of the 39th International Conference on Computer-Aided Design*, 2020, pp. 1-6, doi: https://ieeexplore.ieee.org/document/9256595.

[5] W. Li, S. Huang, X. Sun, H. Jiang, and S. Yu, "Secure-RRAM: A 40nm 16kb compute-in-memory macro with reconfigurability, sparsity control, and embedded security," in *2021 IEEE Custom Integrated Circuits Conference (CICC)*, 2021: IEEE, pp. 1-2, doi: https://doi.org/10.1109/CICC51472.2021.9431558.

[6] J. Luo *et al.*, "Novel ferroelectric tunnel FinFET based encryption-embedded computing-in-memory for secure AI with high area-and energy-efficiency," in *2022 International Electron Devices Meeting (IEDM)*, 2022: IEEE, pp. 36.5. 1-36.5. 4, doi: https://doi.org/10.1109/IEDM45625.2022.10019387.

[7] Z. Fu, S. Cao, H. Zheng, J. Luo, Q. Huang, and R. Huang, "First demonstration of hafnia-based selector-free FeRAM with high disturb immunity through design technology co-optimization," in *2023 International Electron Devices Meeting (IEDM)*, 2023: IEEE, pp. 1-4, doi: https://doi.org/10.1109/IEDM45741.2023.10413887.

[8] C. Wu, R. Huang, Q. Huang, C. Wang, J. Wang, and Y. Wang, "An analytical surface potential model accounting for the dual-modulation effects in tunnel FETs," *IEEE Transactions on Electron Devices*, vol. 61, no. 8, pp. 2690-2696, 2014, doi: https://doi.org/10.1109/TED.2014.2329372.

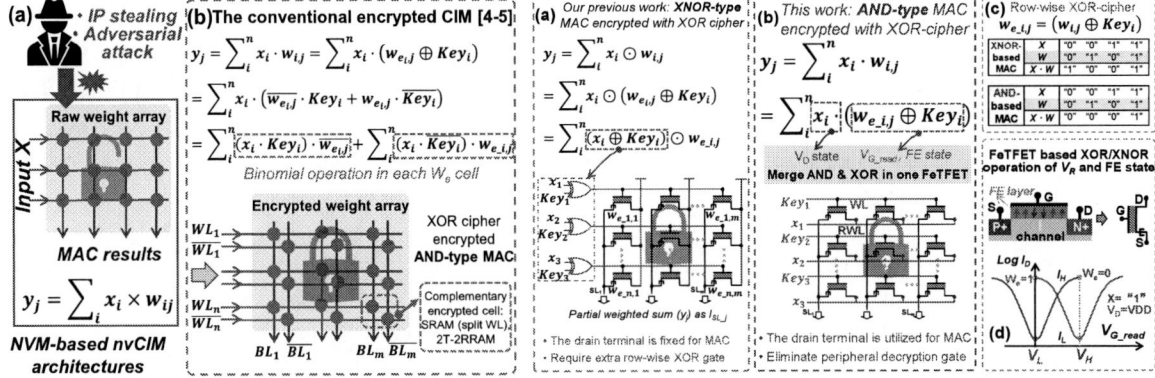

Fig. 1 (a) The security vulnerabilities of non-volatile memory (NVM) based computing-in-memory (CIM) architectures. (b) The conventional encrypted CIM schemes require complementary form for encrypted weight (w_e) storage and the Key inputs.

Fig. 2 (a) The FeTFET-based XNOR-type encrypted multiply-accumulate (eMAC) with extra XOR gate for decryption. (b) The proposed FeTFET-based AND-type eMAC in this work. (c) Truth table of XOR cipher and XNOR/AND type multiplication. (d) The FeTFET-based non-monotonic XOR/XNOR operation.

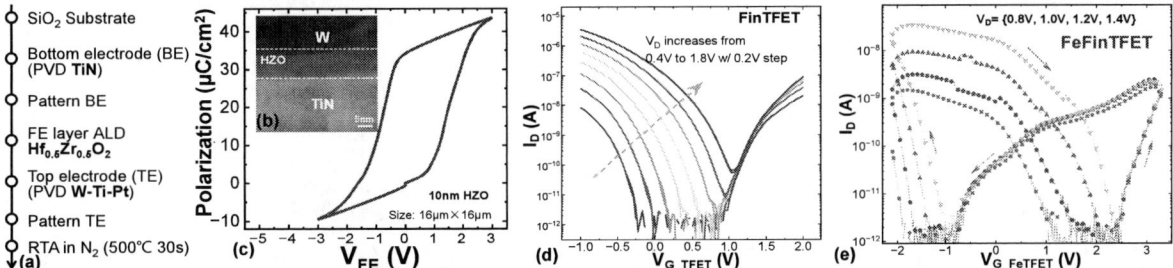

Fig. 3 Experimental characterization of FeTFET. (a) The process flow, (b) TEM image and (c) P-V loop of fabricated HZO ferroelectric (FE) layer. (d) The I_D-V_G characteristics of fabricated n-type TFET on a 14nm FinFET platform, and the drain voltage (V_D) modulated band-to-band tunneling (BTBT) behavior of FinTFET. (e) The I_D-V_G curves of FeFinTFET, which is constructed by connecting FE layer to the gate of FinTFET.

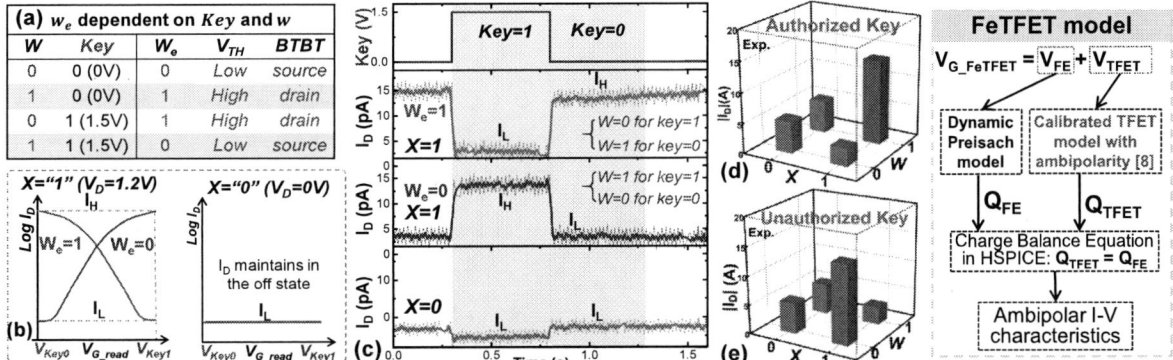

Fig. 4 (a) The encrypted weight (w_e) and FeFinTFET states are determined by the Key and plaintext weight (w). (b) The schematic diagram of FeFinTFET based XOR encrypted AND-type multiplication. (c-e) **Measured** XOR cipher encrypted AND-type MAC behavior of FeFinTFET-based nvCIM without explicit decryption.

Fig. 5 The modeling framework for FeTFET with ambipolar BTBT effect.

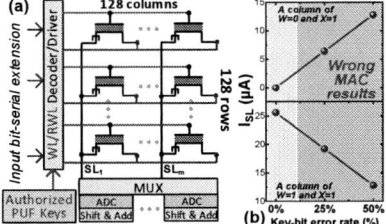

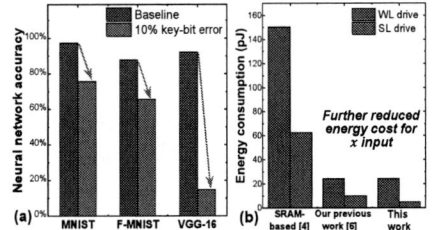

Fig. 6 (a) The schematic of FeTFET-based nvCIM with expandable precision of x and w. (b) The encrypted MAC operation in HSPICE.

Fig. 7 (a) The encrypted neural networks based on the proposed secure nvCIM. (b) The comparison of energy consumption for MAC operation with decryption.

Table I Benchmark of proposed FeTFET-based secure nvCIM in this work with other reported solutions.

	ICCAD [4]	CICC [5]	Our previous work [6]	This work
Cell type	SRAM	RRAM	FeTFET	**FeTFET**
MAC type	AND		XNOR	**AND**
W_e cell size	6T	2T-2R	1T & row-wise XOR gate	1T
Energy (pJ)[a]	212.6	123.2	34.4	**29.4pJ**

(a) Energy cost for eMAC, calculated in HSPICE based on the encrypted array of the same size (128×128) with respectively proposed encrypted weight cell.

Photoresponse improvement in the near-infrared region of organic photodetectors by introducing a trap layer

Fangchen Zhu, and Yang Wang

State Key Laboratory of Electronic Thin Films and Integrated Devices, School of Optoelectronic Science and Engineering, University of Electronic Science and Technology of China, Chengdu, China

Abstract

Visible-near-infrared organic photodetectors (Vis-NIR OPDs) show promising prospects in many fields. However, the intrinsic narrow band gap of NIR organic materials enhances the nonradiative recombination of photogenerated carriers, thereby limiting the device photoresponse. Herein, we introduced a trap layer of $P3HT:PC_{71}BM$ (0.5:100, w/w) into the typical diode-type Vis-NIR OPDs, whose photoresponse improved obviously in the NIR region as compared to its counterpart. This strategy provides a new idea for developing high-performance Vis-NIR OPDs.

Keywords: Near Infrared Organic Photodetectors, Photoresponse, Trap Layer

Introduction

Visible-near-infrared (Vis-NIR) photodetectors have a wide range of applications in night vision imaging, spectral analysis, human health monitoring, biomedicine, and other fields [1]. Current Vis-NIR photodetectors are mainly based on inorganic materials (*e.g.* silicon and indium gallium arsenide) due to their excellent optoelectronic properties, but their complicated and costly manufacturing processes and lack of flexibility leave room for emerging organic photodetectors (OPDs). Due to the low energy of photons in the near-infrared range, the band gap of organic semiconductor materials needs to be narrow enough for their ground state electrons to be excited by incident photons to generate photogenerated charges. However, narrowing the bandgap of organic semiconductors will increase the recombination of photogenerated carriers, leading to a decrease in device photoresponse.

At present, typical OPDs are diode-type, which mainly composed of two asymmetrical electrodes and an organic photoactive layer sandwiched between the electrodes. To improve the device's performance, the bulk-heterojunction (BHJ), formed by the blending of electron donor (D) and acceptor (A), is mostly used active layer structure in OPDs, which can improve the dissociation efficiency of photogenerated excitons due to a large number of D/A interfaces [2]. But the recombination efficiency of photogenerated carriers is still high due to the inherent narrow bandgap of NIR materials. Another strategy is to introduce a modified layer between the electrode and the active layer, acting as an electron transport layer (ETL) and a hole transport layer (HTL), to increase the transport efficiency of photogenerated carriers of the device by optimizing the energy level alignment [3]. However, this strategy cannot address the issue of high carrier recombination of the device in the NIR region.

In this paper, based on a typical diode-type OPD, a hole trap layer of $P3HT:PC_{71}BM$ (0.5:100, w:w) was inserted between the ETL (ZnO) and the photoactive layer (PCE10:TQpp) by transfer printing [4]. The device showed a structure of $ITO/ZnO/P3HT:PC_{71}BM$ (0.5:100)/ PCE10:TQpp/MoO$_3$/Ag, as shown in Fig. 1a. The energy level arrangement of the device is shown in Fig. 1b. Under a bias voltage of -3 V and illumination of a monochromatic light (1000 nm), the responsivity (R) of the obtained OPD reached 0.25 A/W, improved by 53% as compared to the OPD without a trap layer, and the device specific detectivity (D^*) was higher than 10^{12} Jones. Our findings open the door toward realizing high-performance Vis-NIR OPDs by ingenious trap layer design.

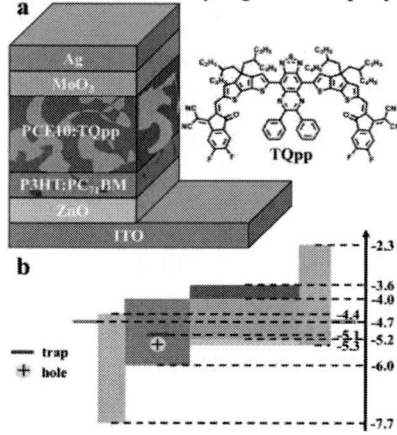

Fig. 1. a) Device structure diagram and the chemical structure of TQpp, b) Energy level arrangement diagram.

Experimental

A. Device Fabrication

The OPD was fabricated with an inverted architecture of ITO/ETL/trap layer/BHJ/HTL/Ag. The ITO substrates and glasses were sonicated in dish soap, deionized water (DI water), acetone and isopropanol first. And then dried with a nitrogen stream and treated with UV-Ozone for 20 min before use. The ETL (ZnO) precursor solution was prepared by dissolving Zn(Ac)2·2H2O and monoethanolamine (molar ratio of 1:1) in 2-methoxyethanol, to form a sol-gel solution. The ETL layer was obtained by spin-coating the precursor solution (4000 rpm, 40 s) and then annealing (200°C, 30 min) under air. For the trap layer solution, P3HT: $PC_{71}BM$ (weight ratio of 0.5:100, 10 mg/mL) was dissolved in chlorobenzene

(CB) and stirred overnight at 50°C under N2. Afterward, the substrates were transferred into the N2-filled glovebox, and the trap layer was spin-coated (1000 rpm, 40 s) using the solution and then thermally annealed (110°C, 10 min).

The preparation method of the BHJ layer was similar to that of the trap layer. The PCE10:TQpp (1:1, w:w, 20 mg/mL) solution was spin-coated (2000 rpm, 40 s) onto a piece of glass. Then the film was transferred onto the above-prepared sample by transfer printing. Briefly, the PDMS stamp was placed on top of the organic layer till they were fully attached, the entire stack was immersed in deionized water for 10 s. Then remove the stamp with attached active layer from the glass and cover it onto the trap layer, and anneal the sample for 30 s. Then remove the stamp, and the organic layer was transferred onto the trap layer.

The MoO_3 layer (~10 nm) and Ag layer (~100 nm) were sequentially thermally evaporated under 10-4 Pa, and the OPD with trap layer (named as 0.5:100-OPD) was obtained. We also prepared two control samples: one is the typical diode-type OPD without the trap layer (named wo-OPD), and the other one is the OPD with P3HT: $PC_{71}BM$ layer of weight ratio of 1:1 (named as 1:1-OPD).

B. Characterization

The film morphology was characterized by the scanning electron microscope (SEM, ZEISS Gemini 300, Germany), atomic force microscope (AFM, Bruker Dimension Icon, Germany) and transmission electron microscope (TEM, FEI Tecnai G2 F20 S-TWIN, USA). The optical transmittance was measured by SHIMADZU UV-3600. The current-voltage characteristic, R, external quantum efficiency (EQE) and D* spectra, noise spectral density (NEP) and response time were measured at room temperature by the PD-QE and PD-RS-L2 (ENLITECH, Taiwan) system.

Results and Discussion

The working mechanism of our Vis-NIR OPD (under reverse bias) is predicted as follows. i) When in the dark (Fig. 2a), the external electrons and holes are blocked by the MoO_3 and ZnO layers, respectively, which suppresses the dark current of the device; ii) Under illumination, the trap layer captures the photogenerated holes, causing the energy level of the photoactive layer to bend close to the region of the trap layer (Fig. 2b), which makes it easier for photogenerated electrons in the active layer to transfer to the trap layer. In addition, due to the existence of continuous electron transport channels in the trap layer, more photogenerated electrons can be collected by the ITO electrode. It means that the hole trap layer is beneficial to the external output of photogenerated electrons from the PCE10:TQpp layer.

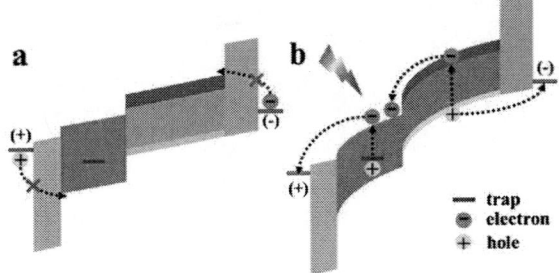

Fig. 2. Schematic diagrams illustrating the working mechanism. a) In the dark, b) Under the light.

In the cross-sectional SEM image of Fig. 3a, we can observe the bottom-up functional layers of the device. The thickness of PCE10:TQpp and P3HT:PC_{71}BM layers were about 150 nm and 60 nm, respectively. The ZnO layer and MoO_3 layer were only 20 nm and 10 nm in thickness, which cannot be clearly distinguishable in the SEM image. Fig. 3b shows the normalized Vis-NIR (300-1400 nm) absorption spectra of the PCE10:TQpp (1:1), P3HT:PC_{71}BM (0.5:100) and P3HT:PC71BM (1:1) films. In the NIR range, there was no obvious absorptivity for the two P3HT:PC_{71}BM films, indicating that the introduction of P3HT:PC_{71}BM film did not affect the NIR absorption of the photoactive layer, nor did it affect the generation of photogenerated carriers in this region. Fig. 3c and 3d are AFM phase image and TEM image of the trap layer, and obvious phase separation can be observed in both cases. As shown in Fig. 3c, there were some shadows of low phase difference against a uniform high phase difference background, indicating that there was another material existed that was different from the background material in morphology, structure, or hardness [5]. It meant that the distribution of P3HT on the surface of PC_{71}BM with a severe imbalance ratio. Fig. 3d shows a large number of lumpy white shades evenly distributed against a black background, and the white shade was clustered P3HT and the black background was PC_{71}BM [6].

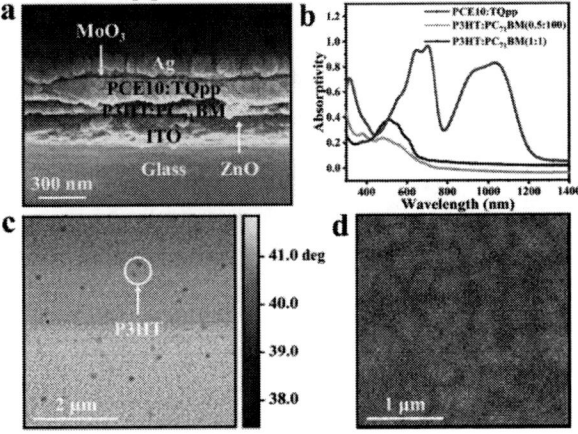

Fig. 3. a) Cross-sectional SEM image of the device, b) Normalized Vis-NIR absorption spectra of different films, c) AFM phase image and d) TEM image of the trap layer

979-8-3315-0417-5/25 $31.00 © 2025 IEEE

Table 1. Carrier mobility of the P3HT:PC$_{71}$BM films.

	electron mobility (μ_e)	hole mobility (μ_h)
1:1	1.83×10^{-3} cm^2/Vs	1.87×10^{-3} cm^2/Vs
0.5:100	9.47×10^{-4} cm^2/Vs	4.55×10^{-9} cm^2/Vs

The carrier mobility of the P3HT:PC$_{71}$BM (0.5:100) trap layer and the P3HT:PC$_{71}$BM (1:1) layer was measured by the space charge limited current (SCLC) method [7], and the calculated value was shown in Table 1. As to the P3HT:PC$_{71}$BM (1:1) layer, the electron mobility (μ_e) and hole mobility (μ_h) were 1.83×10^{-3} cm^2/Vs and 1.87×10^{-3} cm^2/Vs, respectively, the μ_h/μ_e ratio was almost 1. While for the trap layer, the hole mobility and electron mobility are 4.55×10^{-9} cm^2/Vs and 9.47×10^{-4} cm^2/Vs, leading to a μ_h/μ_e ratio of 5×10^{-4}, which further proved that hole traps exhibited in the P3HT:PC$_{71}$BM (0.5:100) layer.

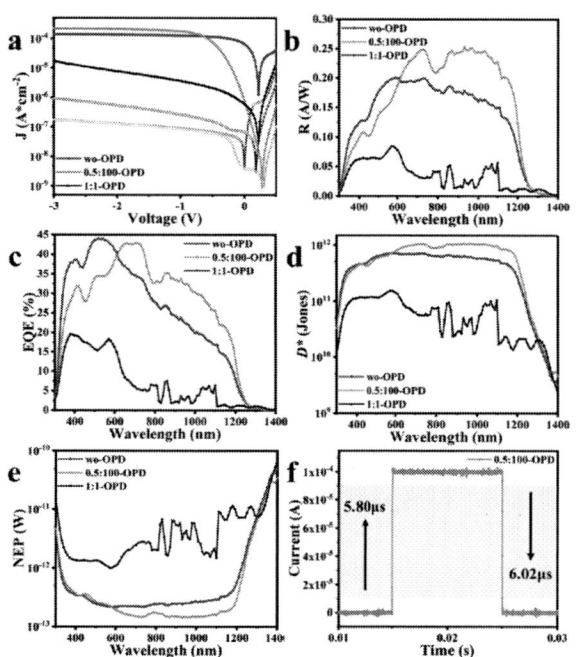

Fig. 4. Device (0.5:100-OPD, wo-OPD, 1:1-OPD) performance. a) *J-V* characteristics, b) *R*, c) EQE, d) *D**, e) NEP and f) response time.

Fig. 4 shows a performance comparison of the three devices (0.5:100-OPD, wo-OPD, 1:1-OPD). Among them, the 0.5:100-OPD showed better performance than the others. As shown in Fig. 4a, under the irradiation of 1000 nm monochromatic light, the photocurrent of the 0.5:100-OPD increased (reverse bias) as compared to the wo-OPD, while that of the 1:1-OPD was significantly reduced. Fig. 4b-4e show the *R*, EQE, *D** and NEP spectra of the three devices (under -3 V). It can be observed that the performance of the 0.5:100-OPD showed significant improvement in the NIR region as compared to the counterparts. Typically, the responsivity of the 0.5:100-OPD reached 0.25 A/W, improved by 53% compared with the 1:1-OPD. Fig. 4f shows the transient photocurrent of the 0.5:100-OPD (980 nm, -3 V). The rise time (t_r) and fall time (t_f), are defined as the duration

time from 10% to 90% and 90% to 10% of the maximal photocurrent, were 5.80 µs and 6.02 µs, respectively, which were very fast among recently-reported OPDs.

Conclusion

In this paper, we reported a trap-layer strategy that effectively improved the photoresponse in the NIR region of Vis-NIR OPDs. By introducing the P3HT:PC$_{71}$BM (0.5:100) hole trap layer between the ETL and the photoactive layer, the obtained 0.5:100-OPD showed significant improvement in R, EQE, D* and NEP performance in the NIR region, as compared to the wo-OPD, 1:1-OPD counterparts. This was due to the trap layer can capture holes, thereby inducing the outward export of photogenerated electrons from the active layer. This trap-layer strategy can be further extended to develop other NIR OPDs for next-generation application.

Acknowledgments

The authors gratefully acknowledge the contributions of Liu Yuan and Jianhua Xiao to the device fabrication.

References

[1] Q. Liu, L. Li, J. Wu, Y. Wang, L. Yuan, Z. Jiang, J. Xiao, D. Gu, W. Li, H. Tai, and Y. Jiang, "Organic photodiodes with bias-switchable photomultiplication and photovoltaic modes", Nat. Commun. 14, pp. 6935-6944 (2023).

[2] J. Halls, C. Walsh, N. Greenham, E. Marseglia, R. Friend, S. Moratti, , and A. Holmes, "Efficient photodiodes from interpenetrating polymer networks", Nature, 376, pp. 498–500 (1995).

[3] K. Wang, C. Liu, T. Meng, C. Yi, and X. Gong, "Inverted organic photovoltaic cells", Chem Soc Rev, 45, pp. 2937-2975 (2016).

[4] A. Abdellah, A. Falco, U. Schwarzenberger, G. Scarpa, and P. Lugli, "Transfer Printed P3HT/PCBM Photoactive Layers: From Material Intermixing to Device Characteristics", ACS Appl. Mater. Interfaces, 8, pp. 2644–2651 (2016).

[5] R. GarcÂõa and R. PeÂrez, "Dynamic atomic force microscopy methods", Surf. Sci. Rep., 47, pp. 197-301 (2002).

[6] K. Kaku, A. Williams, B. Mendis, and C. Groves, "Examining charge transport networks in organic bulk heterojunction photovoltaic diodes using 1/f noise spectroscopy", J. Mater. Chem. C, 3, pp. 6077-6085 (2015).

[7] J. Blakesley, F. Castro, W. Kylberg, G. Dibb, C. Arantes, R. Valaski, and J. Kim, "Towards reliable charge-mobility benchmark measurements for organic semiconductors", Org. Electron., 15, pp. 1263-1272 (2014).

979-8-3315-0417-5/25 $31.00 © 2025 IEEE

CMOS Compatible Spike Vision Sensor

Xiaolin Liu[1], Chao Gao[1], Bowei Jiang[1], Xin Jin[1], Zhou Zhou[1], Yihong Qi[2], Kai Wang[1]*†

[1] Guangdong Province Key Laboratory of Display Material and Technology, State Key Laboratory of Optoelectronic Materials and Technologies, School of Electronics and Information Technology, Sun Yat-sen University, Guangzhou, China

[2] National Innovation Center for Advanced Medical Devices, Shenzhen, China

*wangkai23@mail.sysu.edu.cn, †Keynote speaker

Abstract

Intelligent vision systems hold a unique position in the era of big data due to their high efficiency and low power consumption. Conventional intelligent vision systems with traditional image sensors are reaching the limit and not smart enough to address rapidly-growing computer vision applications. Energy-efficient vision sensors are therefore urgently needed for the next-generation vision systems. Spike vision sensors, possessing the ability to realize in-sensing computing with low power consumption, are one of promising candidates for the future intelligent vision systems.

Keywords: Intelligent vision systems, Spike vision sensors and in-sensing computing

Introduction

With the development of machine learning and artificial intelligence (ML/AI) technologies for computer vision, intelligent vision systems have been widely applied in autonomous driving, security, robotics, scientific instrumentation and others. Conventional intelligent vision systems typically employ traditional image sensors embedded with pattern recognition algorithms. However, increasingly complex ML/AI algorithms are consuming much more energy. Research indicates that by 2040, the energy demands on running ML/AI algorithms on modern computational devices may exceed the world's available energy resources [1]. More urgently, as shown in Fig. 1, the computational power demands on ML/AI algorithms are growing far beyond the Moore's Law, with the current demand doubling every two months [2]. Leveraging edge computing, which processes data close to the source, could be the future of intelligent vision systems. Inspired by biological systems, spike vision sensors process information as biological-like spike signals and may be compatible with various brain-inspired hardware and algorithms, aiming for integration of sensing, storage, and even computing. Intelligent vision systems based on spike vision sensors might effectively reduce the complexity and computational demand of centralized image recognition algorithms, potentially enabling efficient imaging and even object or motion recognition at the edge directly.

Device and circuit co-optimization

Traditional image sensors scan and store image information at fixed frame rates, recording all pixel data in each frame even if no changes occur between frames [3]. As shown in Fig. 2, this imaging mechanism generates substantial redundant data, while also limiting resolution and increasing power consumption. The spike vision sensors aim to compress visual information into spike trains similar to the biological vision sensors and achieve dynamic imaging and pattern recognition (Fig. 3).

Such differentiated imaging mechanism of spike vision sensor raises new challenges. For instance, the output from spike vision sensors cannot be converted to images directly like the traditional ones. Additionally, readout circuits based on traditional image sensors are naturally inefficient in processing visual information from spike vision sensors. As the development requires, device technology co-optimization (DTCO) or system technology co-optimization (STCO) across materials, devices, circuits, systems, and backend algorithms is required to achieve high-efficient spike vision imaging systems [4], as shown in Fig. 3.

Specifically, in terms of materials and novel semiconductors, such as bulk crystal-like perovskites, or low dimensional materials such as transition metal dichalcogenides, black phosphorus and III-V or II-VI compound quantum dots, have broad-spectrum photo-responsibility, low dark current, and high carrier mobility. Y. Chai et al. [5] have reported WSe_2 based vision sensor that can electrically programs its photo-response rate to generate spike signals of varying amplitude and polarity, thus capturing dynamic motions.

At the system level, dynamic vision sensor (DVS) is one typical spike vision sensor. T. Delbruck and his colleague [6] have designed an event-based camera enabling motion tracking in six-degrees-of-freedom (6-DOF) based on DVS. Though promising, emerging novel semiconductors are still far from mass production readiness in industry. Though mature, conventional DVSs still rely on traditional photodetectors and pixel circuits including photodiodes and phototransistors that are still power hungry. CMOS-based photosensitive device re-design may alleviate the current situation. By bridge the connection between device and circuit, novel CMOS-based photosensitive device can realize DTCO and STCO towards efficient spike vision sensor.

979-8-3315-0417-5/25 $31.00 © 2025 IEEE

Spike vision sensors by using PD-MOSs

Device-wise, existing CMOS-based photodiodes and phototransistors have relatively high-power consumption, approximately 100-time beyond the Landauer limit, restricting their applications in coherent sensing. Innovating device structures may be possible to approach or even surpass the Landauer limit. For example, subthreshold-operated dual-gate transistor tends to have a high amplification efficiency might be a solution. Inspired by these device characteristics, we proposed PD-MOS, a dual-gate device utilizing photoinduced body-bias effect (Fig. 4) [7]. By adjusting the bias voltage, we can control the circuit's operational region and gain, simulating retinal functionality as the photoreceptor. Based on the PD-MOS, we further designed a retinal parallel pathway pixel circuit to encode light signals into spikes. Comparing to conventional silicon-based retinal image sensors, the parallel pathway circuit achieves both light enhancement and light suppression within a single pixel. To simulate ON and OFF bipolar cells, we used PD-MOS and n-type MOSFETs to form a current-mirror-like structure, generating stimulation currents either positively or negatively correlated with light intensity under current bias, as shown in Fig. 5. The illumination signals are converted to corresponding spike frequencies via a simplified pull-down current-type axon-hillock circuit architecture. This parallel pathway circuit enables light enhancement and suppression within a single pixel to widen the dynamic range, allowing adjustment of pulse firing frequency through current bias, thus reserving compatibility with neuromorphic computing hardware at the system level for energy-efficient design [8].

Visual perception by spiking neural networks

As aforementioned, our PD-MOS based spike vision sensors can encode visual information into spike trains. Hardware-implemented spiking neural networks (SNNs) are designed to efficiently process these spike signals. In the SNNs, each neuron receives the spike output from n front-end neurons, through the attached synapses, as shown in Eq. 1 and Fig. 6. The specific processing operation logic can be summarized as: if the j-th spike passing through the i-th synapse to the soma causes the membrane potential to exceed the threshold, the neuron will output spike signal to the back-end neuron. After firing output spike, the membrane potential of the soma reset to the static status. The affecting weight of the synapse attached to the n-th front-end neuron is ω_n. The array of such electronic spiking neurons can be used to build the spike visual information processing circuit, as shown in Fig. 7.

$$\tau \frac{du(t)}{dt} = -[u(t) - u_0] + \sum_j \omega_j \sum_{t_j^k \in S_j}^{T_w} K(t - t_j^k) \quad (1)$$

$$s(t) = 1, u(t) = u_1, if \ u(t) \geq u_{th}$$

$$s(t) = 0, if \ u(t) \leq u_{th}$$

Where t represents the time step, τ is a constant, u and s represents membrane potential and output peak. u_0 and u_1 are the resting potential and reset potential respectively. ω_j is the weight of the j-th input synapse. t_j^k is the moment when the k-th spike of the j-th input synapse fires within the integration time window T_w. $K(\cdot)$ is a function of the delay effect, and u_{th} is the threshold voltage. As a demonstration, we designed computational units to process pulse signals [9], as shown in Fig. 8.

Conclusion

Spike vision sensor holds a promise for the future intelligent vision systems. Its dynamic information processing ability enables more efficient imaging in multiple scenarios. In addition, its low power consumption allows extended usage duration and is suitable for edge computing. Given these merits, the realization of such systems is highly needed but remains challenging. Efforts should be put towards co-optimization of the hardware architecture in different levels across materials, devices, systems, and even algorithms.

Acknowledgments

This work is supported in part by the National Key Research and Development Program of China under Grant 2022YFA1204202 and in part by the National Natural Science Foundation of China under Grant 62274186.

References

[1] Damsgaard H. J., Ometov A., Mowla M. M., et al., "Approximate computing in B5G and 6G wireless systems: A survey and future outlook", Computer Networks, 2023, 233: 109872.

[2] Mehonic A., Kenyon A. J., "Brain-inspired computing needs a master plan", Nature, 2022, 604(7905): 255-260.

[3] Tang W., Yang Q., Xu H., et al., "Review of bio-inspired image sensors for efficient machine vision", Advanced Photonics, 2024, 6(2): 02400-024001.

[4] Schuman C. D., Kulkarni S. R., Parsa M., et al., "Opportunities for neuromorphic computing algorithms and applications", Nature Computational Science, 2022, 2(1): 10-19.

[5] Zhou Y., Fu J., Chen Z., et al., "Computational event-driven vision sensors for in-sensor spiking neural networks", Nature Electronics, 2023, 6(11): 870-878.

[6] Gallego G., Lund J. E. A., Mueggler E., et al., "Event-based, 6-DOF camera tracking from photometric depth maps", IEEE transactions on pattern analysis and machine intelligence, 2018, 40(10): 2402-2412.

[7] Qi Y., Liu X., Feng Z., et al., "Human Retinal Photoreceptor-Inspired Sensor with Adjustable Gain from 0.1–10^6 and Wide Dynamic Range Over 140dB",2021 IEEE International Electron Devices Meeting (IEDM), 2021: 23.3. 1-23.3. 4.

[8] Wang K., Jiang, B., "A retinal-inspired circuit for frequency encoding of light intensity", CN202410102582.8, May 28, 2024.

[9] Wang K., Gao C., "Integrated circuit for in-memory sensing and computing", CN202410162604.X, June 14, 2024.

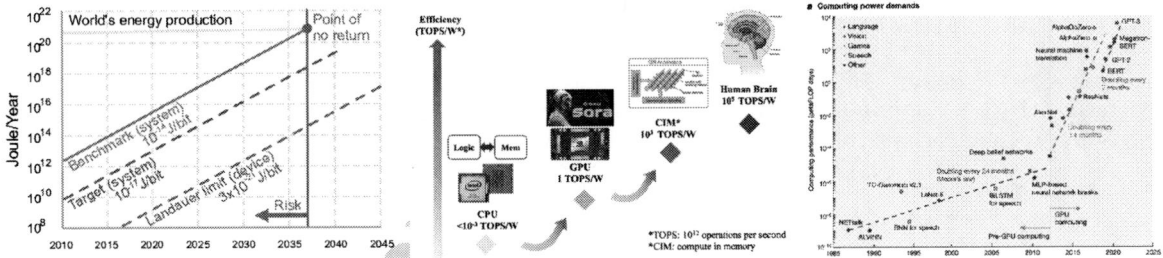

Fig.1: (a) World energy production and energy consumption in computing; (b) roadmap for neuromorphic computing;(c) computational demands are increasing rapidly [1-2].

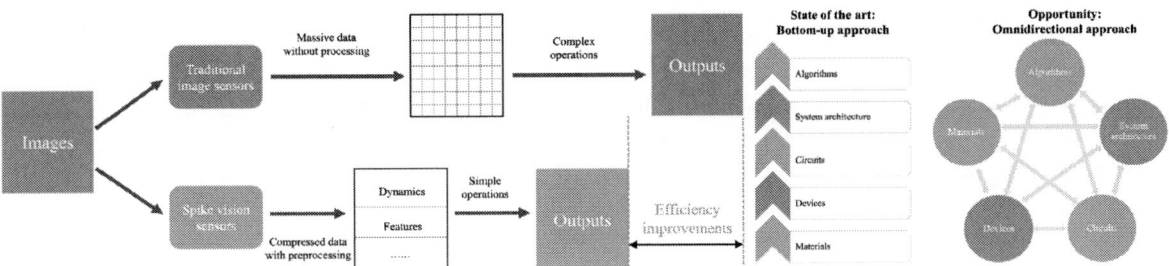

Fig.3: Comparison between traditional image sensors and spike vision sensors [3].

Fig.2: The current bottom-up approach contrasts with a potential future codesign approach, where all aspects interact directly [4].

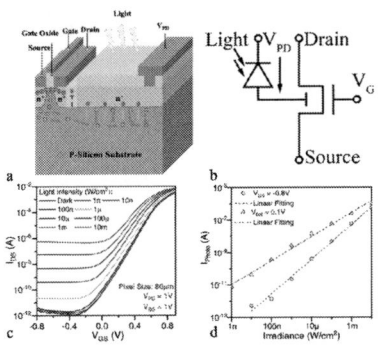

Fig.4: (a) Schematic cross-sectional view and (b) equivalent circuit of a PD-MOS. (c) measured transfer characteristics of PD-MOS. (d)measured photocurrent versus light intensity.

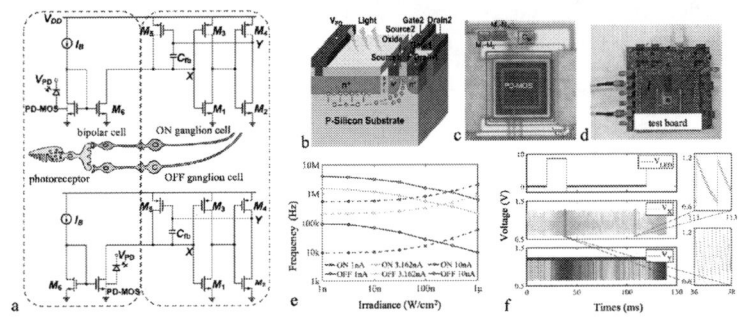

Fig.5:(a) Schematic diagram of the parallel pathway corresponding to the retinal fovea structure; (b) schematic diagram of the substrate-integrated PD-MOS structure;(c) microscopic image of the ON visual pathway circuit(d) photographs of the test board and packaged chip;(e) test results of the light intensity pulse emission rate for the parallel pathway circuit;(f) output waveform test results of the ON pathway circuit under square wave illumination [7].

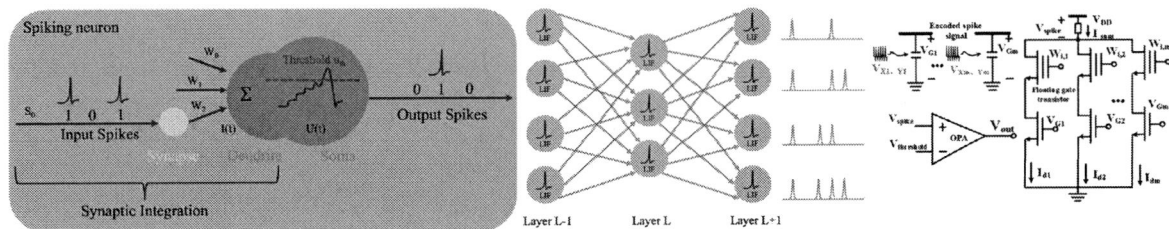

Fig. 6: Structure of spiking neuron.

Fig. 7: Topology of spiking neural network.

Fig. 8: SNN neuromorphic computing unit.

979-8-3315-0417-5/25 $31.00 © 2025 IEEE   813

Dynamic Adaptive Visuomorphic Electronics

Le Wang[1], Yiru Wang[1], He Shao[1], Wei Huang[1,2], Haifeng Ling[1]

[1] State Key Laboratory of Organic Electronics and Information Displays & Institute of Advanced Materials (IAM), Nanjing University of Posts & Telecommunications (NJUPT), China, [2] Frontiers Science Center for Flexible Electronics (FSCFE), MIIT Key Laboratory of Flexible Electronics (KLoFE), Northwestern Polytechnical University (NPU), China

Abstract

Photo-adaptive neuromorphic devices are crucial for artificial vision systems to process complex light conditions. However, challenges remain in achieving precise, fast responses across varying light environments. Expanding optical ranges and using multiple modes enhance adaptability, yet bidirectional modulation across the visible and near-infrared spectrum is difficult. We addressed this by developing organic photosensitive field-effect transistors with complementary p-n heterojunctions, nanopore patterning, and ternary organic heterostructures, enabling metaplasticity, adaptive light relaxation, and bidirectional photoresponses, foundational for efficient vision systems in complex lighting.

Keywords: Visuomorphic computing; Optoelectronic adaptation and Synaptic transistors

Introduction

Neuromorphic vision systems, which emulate biological vision to perceive and recognize information, have been developed to enhance scene understanding and response capabilities [1], [2]. However, environmental complexities such as uneven lighting and unstable visual fields can lead to suboptimal image capture. To address this, reducing hardware complexity and enhancing image recognition efficiency necessitates precise in-situ enhancement and filtering of unstructured data.

Photo-adaptive neuromorphic devices are crucial for precise perception in dynamic lighting, yet challenges in response precision and speed persist [3], [4]. Adaptive modulation of response direction, amplitude, and relaxation speed is essential. Expanding the optical response range and integrating multi-modal responses would enhance adaptability to complex environments. However, achieving bidirectional modulation across the visible and NIR spectrum in a single device remains difficult, as most devices respond positively or negatively to specific wavelengths.

To address these challenges, we investigated adaptive optoelectronic transistors (AOPTs) by constructing complementary photosensitive p-n heterojunctions, introducing nanopore patterning, and developing ternary organic heterostructures. These strategies fully realized the plasticity function, achieving light-intensity adaptive relaxation and demonstrating bidirectional photoresponses from volatile PPC to nonvolatile NPC across the visible to infrared spectrum. These advancements provide a foundation for visual systems capable of efficiently perceiving and recognizing information in complex light environments.

Artificial Visuomorphic Computing System

Fig. 1 schematically illustrates the proposed artificial visuomorphic computing system inspired by the human vision system. In the system, for the former image acquiring stage, eye fixation caused by eye movement can locate the most highlighted target object to the fovea, thus acquiring the specific image in the visual field. For the following preprocessing stage, strong background light in the obtained image might lead to low contrast in part of the sub-images and annihilate the corresponding information. The light-intensity dependent adaptive decay rate in optoelectronic neuromorphic devices was needed to suppress the too bright background and enhance the dark information, thus enhancing the contrast and improving the following recognition accuracy.

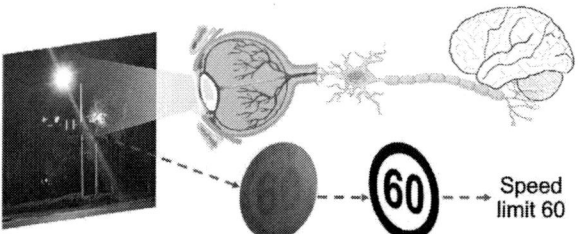

Fig. 1: Schematic diagram of the proposed visual system based on the nanoporous structured adaptive optoelectronic transistor.

Results and discussion

Optoelectronic devices with light intensity adaptation can be used for enhancing contrast, and filtering background noise of the images, thus improving the perception of scenes under complex illumination [5]. Fig. 2a depicts the structure of our as-prepared AOPT device with organic heterojunction, which consists of p-type/n-type channels with a pentacene/PTCDI-C_{13} heterojunction, a dielectric layer with stack poly (2-vinyl naphthalene) (PVN)/SiO_2, and bottom gate/top contact electrodes. Kelvin probe force microscopy (KPFM) was employed to investigate the interfacial charge distribution and photoresponse in the organic heterojunction. The surface morphology and topographic image in Fig. 2b showed the PTCDI-C$_{13}$ and pentacene/PTCDI-C$_{13}$ interfaces of the samples. The contact potential difference (CPD) in pentacene surface was higher than that of PTCDI-C$_{13}$ in dark, indicating the built-in potential would be formed at the interface.

The transfer curves before and after the stimuli of negative gate pulses were exhibited in Fig. 2c. With the amplitude of negative gate pulse increased, the transfer curve shifted gradually to the negative direction due to the charge trapping effect. The transfer curves of the device under white light illumination with different light intensity were investigated and showed in Fig. 2d. The decreased channel current under illumination implying that NPC behavior was observed in the p-type channel. Generally, in phototransistor, positive photoconductivity is the common behavior that can detect light signals, while NPC is significant for artificial vision adaption. In addition, together with the fact that LTP could be achieved by applying only the positive gate pulses, our as-prepared device can perform a threshold sliding for LTD and LTP under different light intensity (Fig. 2e), demonstrating the photo-adaptation function and endowing the device a component for in-sensor pre-processing with contrast enhancement functions.

Setting the same initial current value as about −1 nA, the synaptic weight change (ΔG_c) with different priming stimuli were examined, and Fig. 2f presented the calculated ΔG_c ((A_2−A_1)/(A_1)×100%) as a function of drain voltage amplitude. For each of the curves in Figure 3f, applying the same priming gate spike, ΔG_c showed a non-monotonic dependence with the drain voltage, the depression was enhanced at the low drain voltage region, indicating the observation of the EDE region. Thus, profited from the history-dependent plasticity and the unique U-SVDP, synaptic metaplasticity was perfectly implemented in our AOPT with organic heterojunction.

Photoadaptive neuromorphic devices are essential for artificial vision systems to accurately process complex light scenes. However, achieving adaptive optoelectronic function remains challenging due to the complexities of light-dependent charge modulation. Nano-patterning functional layers—such as nanopillars, metallic nanodots, and nanoporous semiconductors—can enhance light absorption, improve photoexciton separation, and introduce localized electric fields, conferring unique properties to AOPTs [6]. While prior studies have focused on nano-patterning techniques or performance enhancement, further exploration into nano-patterning's potential to enable optoelectronic adaptability is required.

Fig. 3a depicts the device structure of the as-prepared nanoporous structured AOPT device, consisting of a dielectric layer with stack nanoporous PVK/SiO$_2$, p−type pentacene channel, and bottom gate/top contact Copper (Cu) electrodes. We achieved the preparation for the nanoporous PVK thin film via the Breath-figure method. When moist air contacted with the cold surface of PVK solution, the moisture condenses forming water droplets with ordered patterning on the surface, sinking into the polymer film, and thus forming an ordered hole array on the surface. The surface morphology of the prepared thin films was characterized by atomic force microscopy (AFM). Following sequential deposition on the large-aperture nanoporous PVK film, the pentacene channel exhibited a nanoporous morphology. Additionally, conformal growth was observed in the Cu source/drain electrodes (Fig. 3b), confirming that the nanoporous PVK film effectively serves as a template. After the templated growth of pentacene, a significantly enhanced light absorption was observed for both PVK (232 nm) and pentacene (586, 633, and 673 nm, inset in Fig. 3c) in nanoporous structured pentacene/PVK bilayer (Fig. 3c). Moreover, muti-pulse PSCs under the stimuli of light (Figure 3d) further proved the advantages of nanoporous device in transient response and charge trapping capability.

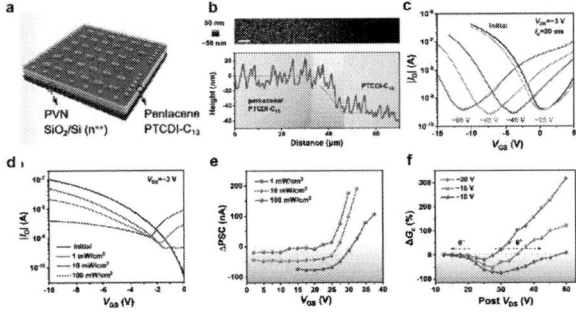

Fig. 2: Photoresponse and metaplasticity in the AOPT device with organic heterojunction. a) The device structure of the AOPT device with organic heterojunction. b) The surface morphology of the sample with PTCDI-C$_{13}$ and pentacene/PTCDI-C$_{13}$ interface. c) Transfer curves of the devices with different stimuli of negative gate pulses. d) Transfer curves of the devices under different illumination intensity. e) Photo-adaptive threshold sliding under different illumination intensity. f) Metaplasticity achieved in the AOPT device with organic heterojunction.

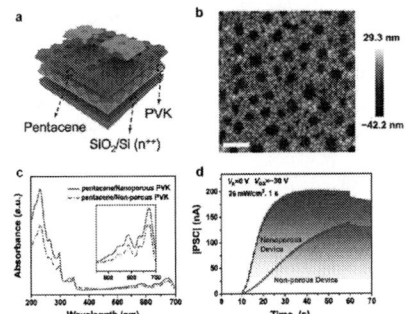

Fig. 3: a) Device structure of the nanoporous structured organic transistor. b) AFM images for Cu electrode on pentacene/PVK/SiO2/Si. Scale bars, 1 μm. c) Absorption spectra for pentacene/PVK bilayers with nanoporous and non-porous structures. d) Synaptic plasticity of nanoporous and non-porous devices triggered by multi light pulses.

Enhancing visuomorphic devices for large-scale image data

requires critical feature processing. While expanding photoresponse and integrating multi-modal responses could improve adaptability to complex environments, achieving bidirectional modulation across the visible and NIR spectrum in a single device remains challenging, as typical hardware supports only limited wavelength-specific responses [7].

An AOPT device with ternary organic heterostructure was developed, featuring a donor-acceptor bulk-heterojunction (BHJ) with a p-type semiconductor layer (Fig. 4a). A tunable 3D porous photosensitive layer was achieved through phase separation (Fig. 4b). The complementary absorption of PM6 and Y7-BO enhances visible and NIR light capture, significantly boosting the device's efficiency and sensitivity. The final phototransistor incorporates a BHJ/pentacene heterostructure, achieving enhanced optoelectronic performance.

The electrical characteristics of the phototransistor are further investigated. Fig. 4d illustrates the transfer curves of the device constructed from BHJ/pentacene under both dark conditions and varying light illuminations with different wavelengths. The device without additives shows a typical p-type behavior and a current ratio of over 10^3 for dark conditions. Under light illumination, the photoresponse transitions from NPC to PPC by applying gate voltage. The transformation of photoresponse characteristics under different gate reading voltages (V_R) is shown in Fig. 3e. At a 5 V reading voltage, the photocurrent shows a positive response upon illumination, stabilizing at a higher equilibrium after the light is off, indicating non-volatile behavior. Conversely, the device exhibits a negative response with threshold switching at −5 V, where the peak current falls below the initial state and quickly recovers post-stimulation.

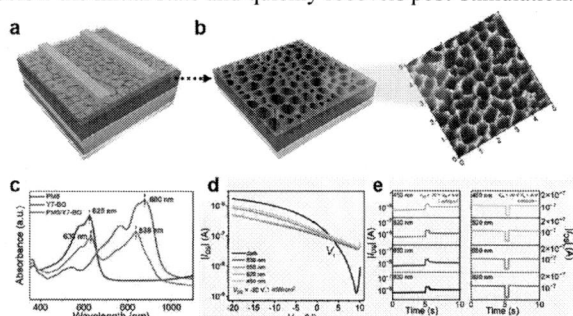

Fig. 4: a) The structure of the AOPT with the BHJ/pentacene. The BHJ layer is composed of PM6, Y7-BO, CN and DDO. b) The schematic diagram and the corresponding AFM image of the porous film. c) The absorption spectra of PM6, Y7-BO, and PM6/Y7-BO composite film. d) Transfer characteristics of the AOPT. e) PPC and NPC response behaviors exhibited by the phototransistor.

Application

Benefiting from the improved photomemory performances and the unique light intensity-dependent adaptability, the AOPT device array was used in the proposed artificial visuomorphic computing system for image acquiring and preprocessing features. As illustrated in Fig. 5a, the system integrates image acquisition, neuromorphic preprocessing, and recognition. In image acquisition, visual selective attention prioritizes specific information in complex lighting, such as bright areas. Additionally, excessive light triggers a protective reflex, causing the eyes to close. A biomimetic eye simulates human visual selective attention and self-protection (Fig. 5b). Optical signals collected by the nanoporous device array (Figure 5c) are transmitted to control eye rotation and closure (Fig. 5d, e). In preprocessing, adaptive light-intensity decay enables bright pixels to fade faster than dark ones, enhancing edge contrast and detail in high-brightness sub-images for improved feature recognition (Fig. 5f).

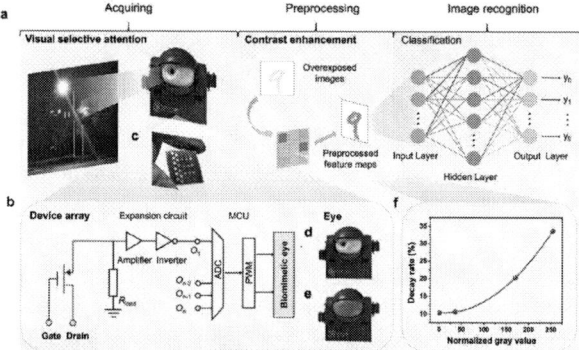

Fig. 5: a) Schematic illustration of the image acquiring, preprocessing, and recognition processes of the proposed artificial vision system. b) Circuit diagrams of the constructed artificial vision system. c) 3×3 nanoporous device array. Biomimetic eye in the d) eye rotation and e) eye closing states. f) Fitted curve of the relationship between normalized gray value and decay rate.

Conclusion

We have developed three types of AOPTs with complementary p-n heterojunctions, nanopore patterning, and ternary organic heterostructures, which respectively achieve meta plasticity, adaptive light relaxation, and bidirectional light response. These results offered a path to design photo-adaptive neuromorphic devices for precise sensing and efficient learning, providing potential for future visuomorphic computing under harsh lighting conditions.

Acknowledgments

The authors gratefully acknowledge the contributions of Wei Huang, Haifeng Ling, Linghai Xie, Yiru Wang, and He Shao.

References

[1] L. Mennel, *et al.*, Nature, *579*, p. 62, 2020.

[2] D. Jayachandran, *et al.*, Nat. Electron., 3, p. 646, 2020.

[3] B. Dang *et al.*, IEEE EDL, 41, p. 1641, 2020.

[4] A. Naseer, *et al.*, IEEE EDTM, p. 1, 2024.

[5] Y. Wang *et al.* Adv. Mater. 36, p. 2404160, 2024.

[6] Y. Wang *et al.* Adv. Funct. Mater. p. 2414546, 2024.

[7] H. Shao *et al.* Nano Energy, 130, p. 110133, 2024.

Enhanced Thermal Stability of Ru Interconnects Using h-BN As Barrier Layers

Kun Chen[1,2], Zhuming Wang[1], Chen Wang[1,2,#] David Wei Zhang[1,2]

[1]School of Microelectronics, Fudan University, China,
[2]Shanghai Integrated Circuit Manufacturing Innovation Center, China
[#]Email: chen_w@fudan.edu.cn

Abstract

As semiconductor technology advances, Ruthenium (Ru) emerges as a promising interconnect material. However, its integration into advanced Backside Power Delivery Networks (BSDPN) faces significant challenges due to the high-temperature processing requirements of front-end-of-line (FEOL) fabrication. This study demonstrates that hexagonal boron nitride (h-BN) is a viable barrier layer for Ru-based BSDPN interconnects, enabling them to withstand high-temperature processing without significant diffusion , highlighting the promising future of h-BN for next-generation integrated circuits.

Keywords: Ru interconnect, BSDPN, h-BN and barrier layer

Introduction

As CMOS transistors continue to shrink towards ever smaller dimensions, the interconnects also face significant challenges. While copper (Cu) based dual damascene process has been the dominant interconnect technology for decades, its resistivity increases substantially at nanoscale dimensions due to increased electron scattering at grain boundaries. This leads to performance degradation and increased power consumption [1][2]. Additionally, highly scaled Cu interconnects are more susceptible to electromigration, thermal management issues, signal integrity concerns, and complex manufacturing processes [3][4]. To address these multifaceted challenges, a comprehensive approach is required. This includes the introduction of low-resistivity metals like Ru, Co, and Mo [5][6], along with innovative interconnect structures such as backside power delivery networks (BSDPN) [7][8]. The adoption of these novel materials and structures also necessitates the development of a full suite of advanced integration solutions, including diffusion barrier and isolation layers.

Ruthenium (Ru), with its superior resistivity than Cu at nanoscale dimensions, as well as higher electromigration resistance. emerges as a promising candidate to replace Cu in the 3nm node and beyond. However, the integration of Ru into the upcoming novel BSDPN interconnect structures poses significant challenges. The high thermal budget required for the Front-End-of-Line (FEOL) processes can adversely affect Ru's thermal stability [9]. Moreover, the proximity of BSDPN structures to transistors makes metal diffusion a critical concern, as it can severely degrade transistor performance. Therefore, novel diffusion barrier materials and related process solutions need to be investigated to overcome these challenges.

The potential novel barrier layer for Ru needs to be very thin due to the extremely limited space in the trench advanced BSDPN interconnect structure. Meanwhile, ideal two-dimensional materials can be thin down to one atomic layer in thickness, while still maintaining a high degree of film integrity. This is highly desirable as a barrier layer. Such an advantage has already been explored, P.H. Feng et,al have studied the potential of using MoS2 as diffusion barriers in Ru interconnects[10]. Compared to MoS2, hexagonal boron nitride (h-BN) is also with excellent physical, chemical, thermal, mechanical, and dielectric properties. Its application in the CMOS area has already attracted many attending, Knobloch et. al. have assessed the performance limits of h-BN as an insulator in scaled CMOS devices [11], and Meng et. al have investigated the electrical breakdown limit of the multilayer h-BN film [12]. However, no attempt has been made to integrate h-BN into the advanced Ru interconnect structure as a barrier layer solution.

In this study, the potential of h-BN as a barrier layer for Ru-based BSDPN applications has been evaluated through electrical testing. Results for voltage breakdown (VrBD) and resistance demonstrate that Ru wires with an h-BN barrier layer can withstand high-temperature stress up to 800°C without performance degradation, unlike Ru wires without a barrier layer. This finding provides a material reference for the future incorporation of Ru into BSDPN structures.

Experiments

As illustrated in Figure 1, a planar film stack test structure was utilized to replicate the BSDPN application environment. A 50 nm high-temperature CVD SiO$_2$ layer was deposited on top of a silicon wafer to serve as the dielectric isolation layer. This thickness was selected to emulate the proximity of the silicon channel and Ru wire, while also ensuring accurate breakdown voltage measurements following high-temperature annealing. Regarding the vital h-BN barrier layer, although wafer-scale growth of polycrystalline h-BN on SiO$_2$/Si substrates has been demonstrated [13], the presence

979-8-3315-0417-5/25 $31.00 © 2025 IEEE

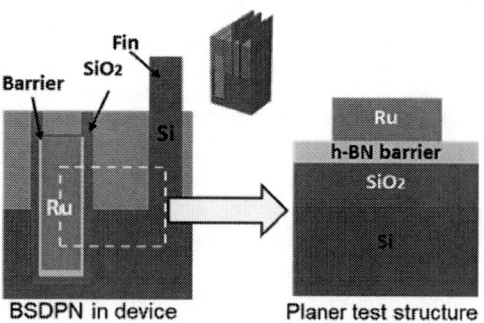

Fig. 1. Test structure design for diffusion barrier performance evaluation for BSDPN application.

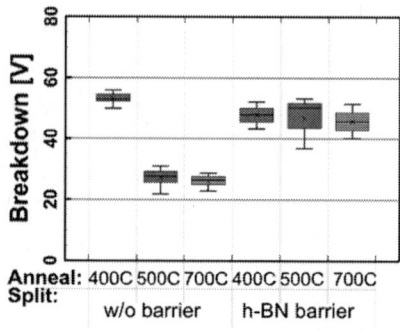

Fig. 3. VrBD breakdown voltages under different annealing temperature conditions with or without h-BN barrier layer.

of grain boundaries may affect its performance as a diffusion barrier. For this demonstration, a more traditional dry transfer technique was employed. After exfoliating from high-quality bulk crystal, few-layer h-BN nanosheets (approximately 10 nm thick) were placed onto flat polydimethylsiloxane (PDMS) stamps and subsequently transferred to the Si/SiO₂ substrate inside an argon-filled glove box, where the concentrations of oxygen and water were kept below 0.1 ppm. After that, metal wires and metal pad capacitor test structure for VrBD test were patterned using polymethyl methacrylate (PMMA) photoresist through electron beam lithography (EBL). The test structure was completed with a physical vapor deposition (PVD) of 50 nm Ru layer, followed by a lift-off process.

Results and Discussion

For barrier layers used in advanced nodes interconnect, a high degree of uniformity is very important, otherwise, it can affect the Ru metal wire filling integrity in nanoscale trench. The atomic force microscope (AFM) was performed to characterize whether the introduction of h-BN affected the deposition of Ru wires. As shown in Figure 2(a), the surface roughness of Ru wire with or without an h-BN barrier underneath is essentially the same after deposition. The adhesion and stability of the barrier layer and the metal were subsequently tested under thermal stress. Each sample was annealed in the RTP for 15 minutes at the corresponding temperature, and then the roughness of the

surface was evaluated by AFM. The result is shown on Figure 2(b), The observations indicate that the Ru film becomes partially crystallized when temperatures reach 700 °C or higher, which leads to increased surface roughness. However, there is no significant difference in surface roughness between samples with the h-BN barrier layer and those without it. This observation demonstrates that h-BN is highly stable at elevated temperatures and exhibits excellent adhesion to the Ru interface, a critical property for advanced barrier layers.

The introduction of the BSDPN architecture has resulted in a significant reduction in the distance between metal wires and transistors, now measuring just a few tens of nanometers. The close proximity raises concerns about metal contamination, which can severely impact transistor performance. To evaluate this risk, voltage breakdown (VrBD) structures based on planar capacitive simulation structure designs are employed. If Ru atoms diffuse into the SiO₂ dielectric layer at high thermal stress, it would lead to a degradation in the breakdown performance of the dielectric layer. The VrBD performance test results at different annealing temperatures are plotted on Figure 3. The breakdown voltage degradation of the Ru is significantly reduced with h-BN barrier layer. Even at 800 °C conditions, the VrBD of Ru with h-BN barrier layer has degraded by less than 7%, compared to more than 50% for Ru without the barrier layer. To obtain physical insight into the observed VrBD test result, the film structure was

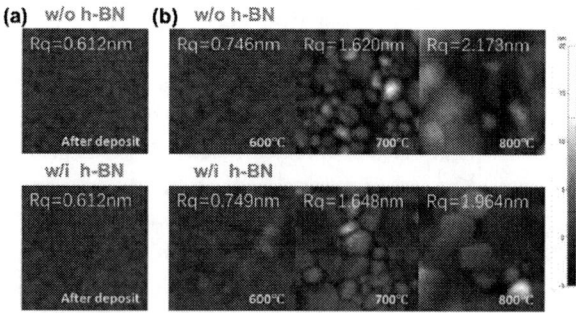

Fig. 2. Surface roughness of Ru with or without h-BN barrier layer under various annealing condition.

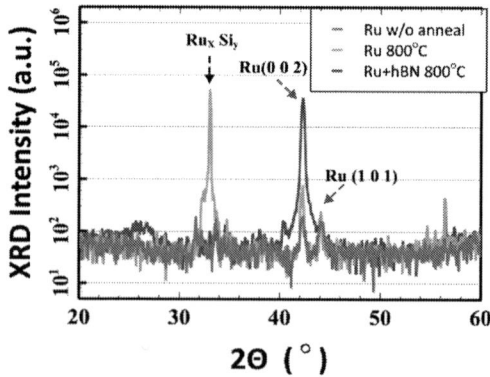

Fig. 4. Comparison of XRD spectrum. The h-BN barrier has effectively prevented the formation of Ru$_x$Si$_y$ at 800 °C.

979-8-3315-0417-5/25 $31.00 © 2025 IEEE

examined using X-ray diffraction (XRD) to investigate phase formation, as shown in Figure 4. The intensity of the Ru (002) and Ru (101) peaks increase with high-temperature annealing, indicating that Ru grain growth occurs following the annealing process. However, the h-BN barrier effectively prevents the formation of Ru_xSi_y in the film, even at temperatures as high as 800 °C. This result suggests that the h-BN barrier layer can potentially dramatically improve the compatibility of BSDPN buried Ru wire structures with high FEOL thermal budgets.

Conclusion

In this study, the potential of hexagonal boron nitride (h-BN) as a barrier layer integration solution for Ruthenium (Ru)-based BSDPN interconnects technology has been investigated. A comprehensive physical and electrical characterization, including voltage breakdown test and AFM and XRD measurements, confirmed that Ru wires with an h-BN barrier layer can withstand high-temperature processing involved in BSDPN technology without performance degradation. These findings support the development of next-generation interconnect technology with enhanced performance and reliability. However, further research and optimization of the h-BN integration process are necessary to fully realize the potential of this promising material.

References

[1] D. Prasad et al., "Buried Power Rails and Back-side Power Grids: Arm® CPU Power Delivery Network Design Beyond 5nm," in 2019 IEEE International Electron Devices Meeting (IEDM), San Francisco, CA, USA: IEEE, Dec. 2019, p. 19.1.1-19.1.4. doi: 10.1109/IEDM19573.2019.8993617.

[2] D. Gall, A. Jog, and T. Zhou, "Narrow interconnects: The most conductive metals," in 2020 IEEE International Electron Devices Meeting (IEDM), San Francisco, CA, USA: IEEE, Dec. 2020, p. 32.3.1-32.3.4. doi: 10.1109/IEDM13553.2020.9372060.

[3] J. W. McPherson, "Reliability Trends with Advanced CMOS Scaling and The Implications for Design," in 2007 IEEE Custom Integrated Circuits Conference, San Jose, CA, USA: IEEE, 2007, pp. 405–412. doi: 10.1109/CICC.2007.4405763.

[4] W. Ahn, Y.-P. Chen, and M. A. Alam, "An Analytical Transient Joule Heating Model for an Interconnect in a Modern IC: Material Selection (Cu, Co, Ru) and Cooling Strategies," in 2019 IEEE International Reliability Physics Symposium (IRPS), Monterey, CA, USA: IEEE, Mar. 2019, pp. 1–6. doi: 10.1109/IRPS.2019.8720497.

[5] A. Gupta et al., "Barrierless ALD Molybdenum for Buried Power Rail

and Via-to-Buried Power Rail metallization," in 2022 IEEE International Interconnect Technology Conference (IITC), San Jose, CA, USA: IEEE, Jun. 2022, pp. 58–60. doi: 10.1109/IITC52079.2022.9881304.

[6] A. Jourdain, M. Stucchi, G. Van Der Plas, G. Beyer, and E. Beyne, "Buried Power Rails and Nano-Scale TSV: Technology Boosters for Backside Power Delivery Network and 3D Heterogeneous Integration," in 2022 IEEE 72nd Electronic Components and Technology Conference (ECTC), San Diego, CA, USA: IEEE, May 2022, pp. 1531–1538. doi: 10.1109/ECTC51906.2022.00244.

[7] S. B. Samavedam et al., "Future Logic Scaling: Towards Atomic Channels and Deconstructed Chips," in 2020 IEEE International Electron Devices Meeting (IEDM), San Francisco, CA, USA: IEEE, Dec. 2020, p. 1.1.1-1.1.10. doi: 10.1109/IEDM13553.2020.9372023.

[8] A. Veloso et al., "Backside Power Delivery: Game Changer and Key Enabler of Advanced Logic Scaling and New STCO Opportunities," in 2023 International Electron Devices Meeting (IEDM), San Francisco, CA, USA: IEEE, Dec. 2023, pp. 1–4. doi: 10.1109/IEDM45741.2023.10413867.

[9] J.-J. Tan, X.-P. Qu, Q. Xie, Y. Zhou, and G.-P. Ru, "The properties of Ru on Ta-based barriers," Thin Solid Films, vol. 504, no. 1–2, pp. 231–234, May 2006, doi: 10.1016/j.tsf.2005.09.129.

[10] P.-H. Feng et al., "Unleashing the Power of 2D MoS 2 : In Situ TEM Study of Its Potential as Diffusion Barriers in Ru Interconnects," ACS Appl. Mater. Interfaces, vol. 15, no. 41, pp. 48543–48550, Oct. 2023, doi: 10.1021/acsami.3c10656.

[11] T. Knobloch et al., "The performance limits of hexagonal boron nitride as an insulator for scaled CMOS devices based on two-dimensional materials," Nat Electron, vol. 4, no. 2, pp. 98–108, Feb. 2021, doi: 10.1038/s41928-020-00529-x.

[12] G. Meng, Y. Cheng, D. Zhang, and G. Zhang, "Insulation performance of atomic hexogonal boron nitride film under ultra-high DC electric stress," in 2017 International Symposium on Electrical Insulating Materials (ISEIM), Toyohashi: IEEE, Sep. 2017, pp. 54–57. doi: 10.23919/ISEIM.2017.8088688.

[13] J. S. Lee et al., "Wafer-scale single-crystal hexagonal boron nitride film via self-collimated grain formation," Science, vol. 362, no. 6416, pp. 817–821, Nov. 2018, doi: 10.1126/science.aau2132.

Leveraging Ferroelectric Negative-Capacitance Effect for Energy Efficient Electronics

Jie-Ni Dai and Pin Su

Institute of Electronics, National Yang Ming Chiao Tung University, Taiwan

E-mail: jieni76688.ee10@nycu.edu.tw, pinsu@nycu.edu.tw

Abstract

This work, through experimentally calibrated Ginzburg-Landau-Khalatnikov model for an ultrathin ferroelectric (1.5 nm) with stacked-GAA FETs, investigates and analyzes the potential for V_{DD} reduction via the ferroelectric negative-capacitance (NC) effect based on the IRDS 2025-2031 nodes. Our study suggests that the ferroelectric NC effect may enable a reduction in V_{DD} to ~0.4 V for 2028 and 2031 nodes. In addition, an IL-free FE-GAA design may further reduce the V_{DD} to ~0.35 V, provided that the impact of mobility degradation can be mitigated. Our study may provide insights for future energy efficient electronics.

Keywords: Ferroelectric negative-capacitance (NC) effect, stacked-GAA FET, Ginzburg-Landau-Khalatnikov equation

Introduction

Reducing the supply voltage (V_{DD}) remains one of the most critical challenges in advancing CMOS logic technology, as it directly impacts the power consumption and energy efficiency [1][2]. Integrating a ferroelectric gate stack has shown promise in addressing this issue by enhancing the overall gate capacitance and increasing the inversion charge density, enabling V_{DD} reduction while maintaining required I_{ON} and I_{OFF} [2]-[6]. This improvement leverages the unique ferroelectric negative-capacitance (NC) effect [2]-[6], a phenomenon described by the Ginzburg-Landau-Khalatnikov (GLK) equation, applicable specifically to single-crystalline ferroelectric materials. It is possible to obtain a single-crystal-like ferroelectric by technologies such as epitaxial growth of the ferroelectric directly on the semiconductor substrate [7][8].

In this work, through experimentally calibrated GLK model for an ultrathin ferroelectric with stacked gate-all-around (GAA) FET architectures, we investigate and analyze the potential for V_{DD} reduction via the ferroelectric NC effect based on the 2025-2031 technology nodes specified by the IRDS [1].

Methodology

Fig. 1 shows the stacked GAA structures and pertinent device parameters based on the IRDS 2025 node. Fig. 2 shows the schematics of the baseline gate stack (HK-GAA), ferroelectric gate stack (FE-GAA) with interfacial layer (IL), and FE-GAA without the IL considered in this study. The ferroelectric is modeled by the 3D Ginzburg-Landau-Khalatnikov equation in TCAD. Pertinent Landau parameters

(α, β, γ) are determined based on the recent data of ultrathin ferroelectric (T_{FE} = 1.5 nm) reported in [9] (α = -1.911×10^9 Vm/C, β = 5.445×10^{11} Vm^5/C^3, γ = 3.22×10^{14} Vm^9/C^5). The g factor for the polarization gradient employed is 5×10^{-9} Vm^2/C [10].

Results and Discussion

Fig. 3 shows the electron density profiles for baseline HK-GAA, FE-GAA with IL, and FE-GAA without IL at ON state (V_{GS} = V_{DS} = V_{DD} = 0.65 V). For the baseline HK-GAA structure, the electron density in the top and bottom channels (representing the longer channel sides) is notably lower than that in the side channels (shorter channel sides). This discrepancy is attributed to the quantum confinement effect, which is stronger in the vertical direction, where the channel thickness T_{GAA} is only 6 nm, compared to the lateral direction, where the width W_{GAA} is 30 nm. When the high-k gate stack is replaced with a ferroelectric gate stack, the electron density in the top and bottom channels significantly increases, particularly near the source region (see the FE-GAA with IL column in Fig. 3). This enhancement arises from the ferroelectric layer's ability to induce a higher gate capacitance and thus a stronger electric field, which drives a greater inversion charge density in the channel. Furthermore, removing the IL from the FE-GAA boosts the channel electron density even more, as shown in the FE-GAA without the IL column in Fig. 3. This IL-free FE-GAA device allows for an even stronger ferroelectric negative capacitance effect, further enhancing electron density.

Fig. 4 shows the ferroelectric polarization distribution near the source region for the FE-GAA devices. It shows that the ferroelectric polarization in the top and bottom channels is markedly higher than that in the side channels. Additionally, the FE polarization for the FE-GAA without IL (Fig. 4(b)) is larger than that for the FE-GAA with IL (Fig. 4(a)). This increased polarization in the IL-free FE-GAA contributes to the higher electron density observed in Fig. 3, highlighting the effectiveness of direct ferroelectric epitaxial growth on the semiconductor channel in maximizing the channel charge density and potential device performance.

Fig. 5 shows the I_{DS}-V_{GS} curves of the baseline HK-GAA and the FE-GAA with/without IL at V_{DD} = 0.65 V. It indicates that the introduction of ferroelectric gate capacitance can significantly boost I_{ON} and improve the I_{ON}/I_{OFF} ratio. While a hysteresis-free operation in FE-GAA

NC devices typically relies on the depolarization field produced by the IL between the ferroelectric material and the channel [2], the IL-free FE-GAA can also remain hysteresis-free. This stability is due to the capacitance associated with the inversion layer (or $C_{Centroid}$), as depicted in Fig. 6. Compared to the FE-GAA with IL, Fig. 5 shows that the FE-GAA without IL has greater potential in I_{ON}.

The I_{DS}-V_{GS} curves of the baseline HK-GAA and the FE-GAA with/without IL under the same I_{OFF} are shown in Fig. 7. The same I_{OFF} are matched by adjusting the gate work function to align with the I_{OFF} of the baseline. Under these conditions, the FE-GAA with IL enables a reduction in V_{DD} from 0.65 V to 0.5 V while maintaining the same I_{ON} specifications as the baseline. By using the IL-free FE-GAA, V_{DD} can be further reduced to 0.43 V, demonstrating the IL-free design's potential for achieving low-power operation while still meeting the performance requirement.

Fig. 8(a) shows the C_{gg}-V_{GS} curves for baseline HK-GAA and FE-GAA with/without IL. The CET in Fig. 8(b) is calculated at $V_{GS} = 0.65$ V from Fig. 8(a). It can be seen that the ferroelectric NC effect enhances the gate capacitance, thereby reducing the CET. The FE-GAA with IL reduces the CET from 0.84 nm (baseline HK-GAA) to 0.63 nm, while the FE-GAA without IL further reduces the CET to 0.2 nm. It is noteworthy that although the absence of an IL may reduce the carrier mobility, NC devices exhibit superior electrostatic integrity [11]. It has recently been reported in [12] that the NC device, with improved electrostatics, can possess a superior carrier injection velocity.

Fig. 9 shows the comparison of I_{ON} for the baseline HK-GAA and the FE-GAA with/without IL at $V_{DD} = 0.65$ V for the 2025-2031 nodes. It shows an increasing I_{ON} enhancement for the FE-GAA structures as the gate length is scaled down from 14 nm to 12 nm, achieving approximately 1.47× improvement for FE-GAA with the IL and up to 2.51× improvement for the IL-free FE-GAA for the 2028 and 2031 nodes. This trend results from improved capacitance matching at shorter gate lengths, where the ferroelectric layer effectively boosts gate capacitance.

Importantly, the substantial I_{ON} improvement seen with the IL-free FE-GAA helps offset potential I_{ON} loss from mobility degradation, a common challenge with aggressive scaling. Table I summarizes the projected V_{DD} reductions achievable from 2025 to 2031, demonstrating the advantages of ferroelectric gate stacks in enabling lower-power operation across these future technology nodes.

Conclusion

Our study, based on the IRDS specifications and GLK theory, suggests that employing the ferroelectric NC effect in stacked-GAA FETs may enable a reduction in V_{DD} to ~0.4 V for 2028 and 2031 nodes. In addition, an IL-free FE-GAA design may further reduce the V_{DD} to ~0.35 V, provided that the impact of mobility degradation can be mitigated (by, e.g., the NC enhanced carrier injection velocity). This work underscores the potential of ferroelectric gate technology for future energy efficient electronics.

Acknowledgments

This work is supported in part by NSTC, Taiwan, under Grants 113-2634-F-A49-008 and 113-2221-E-A49-092-MY2, and in part by the "Advanced Semiconductor Technology Research Center" from the Featured Areas Research Center Program within the framework of the Higher Education Sprout Project by the Ministry of Education (MOE), Taiwan. The authors are grateful to the comments of Prof. Chenming Hu during the work.

References

[1] IRDS 2023 Edition. [https://irds.ieee.org].

[2] R. Ramesh et al. "Roadmap on low-power electronics," *APL Materials*, 12, 9 (2024).

[3] S. Salahuddin and S. Datta, "Use of negative capacitance to provide voltage amplification for low power nanoscale devices," *Nano Lett.*, 8, 2, pp. 405-410 (2008).

[4] S. S. Cheema et al., "Ultrathin ferroic HfO2-ZrO2 superlattice gate stack for advanced transistors," *Nature*, 604, 7904, pp. 65-71 (2022).

[5] M. Hoffmann et al., "Quantitative study of EOT lowering in negative capacitance HfO2-ZrO2 superlattice gate stacks," *IEDM*, pp. 13.2.1-13.2.4 (2022).

[6] N. Shanker et al., "CMOS Demonstration of Negative Capacitance HfO2-ZrO2 Superlattice Gate Stack in a Self-Aligned, Replacement Gate Process," *IEDM*, pp. 34.3.1-34.3.4 (2022).

[7] Z. Zhao et al., "Towards Epitaxial Ferroelectric HZO on n^+-Si/Ge Substrates Achieving Record $2P_r = 84$ μC/cm^2 and Endurance> 1E11," *Proc. IEEE Symp. VLSI Technol. Circuits*, pp. 1-2 (2023).

[8] W.-H. Hsieh et al., "Interfacial-Layer-Free Ge$_{0.95}$Si$_{0.05}$ Nanosheet FeFETs," *IEEE TED*, 71, 3, pp. 1758-1763 (2024).

[9] S. Jo et al., "Negative differential capacitance in ultrathin ferroelectric hafnia," *Nature Electronics*, 6, 5, pp.390-397 (2023).

[10] M.-Y. Kao, S. Salahuddin, and C. Hu, "Negative capacitance enables GAA scaling V_{DD} to 0.5 V," *Solid-State Electronics*, 181, 108010 (2021).

[11] P. Su and W.-X. You, "Electrostatic Integrity in Negative-Capacitance FETs – A Subthreshold Modeling Approach," *IEDM*, pp. 7.3.1-7.3.4 (2019).

[12] C. Garg et al. "Boost in Carrier Velocity due to Electrostatic Effects of Negative Capacitance Gate Stack," *IEEE EDL*, 45, 3, pp. 460-463 (2024).

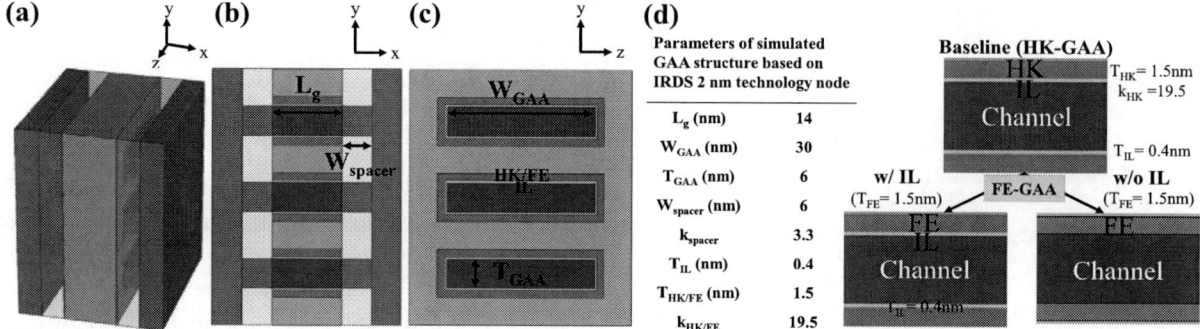

Fig. 1 (a) 3D view of GAA structure. (b) Cross section along the direction of channel length. (c) Cross section at the middle of the channel. (d) Pertinent device parameters of the GAA structure based on the IRDS 2025 technology node.

Fig. 2 Schematics of the baseline gate stack (HK-GAA), FE-GAA with IL, and FE-GAA without IL in this study.

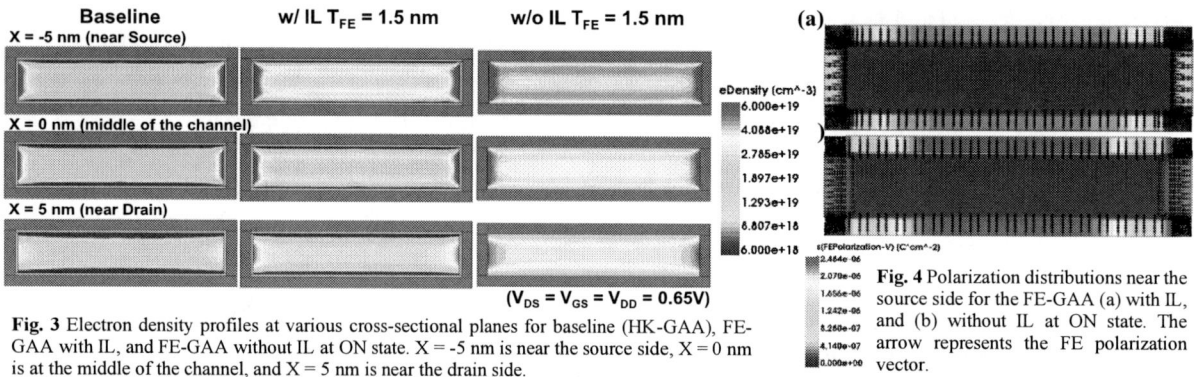

Fig. 3 Electron density profiles at various cross-sectional planes for baseline (HK-GAA), FE-GAA with IL, and FE-GAA without IL at ON state. X = -5 nm is near the source side, X = 0 nm is at the middle of the channel, and X = 5 nm is near the drain side.

Fig. 4 Polarization distributions near the source side for the FE-GAA (a) with IL, and (b) without IL at ON state. The arrow represents the FE polarization vector.

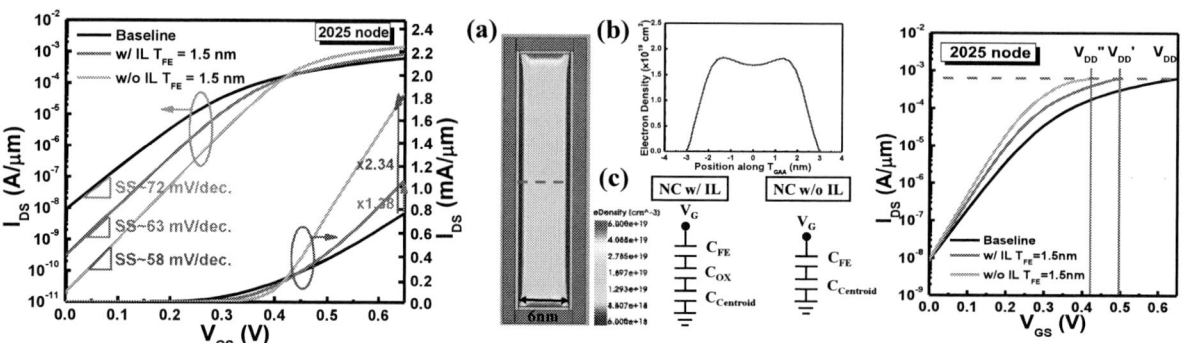

Fig. 5 I_{DS}-V_{GS} curves of the baseline HK-GAA and FE-GAA with and w/o IL at V_D = 0.65 V. The FE-GAA w/o IL has a higher I_{ON} in spite of lower carrier mobility.

Fig. 6 (a) Electron density profile at the middle of the channel. (b) Electron density distribution along the direction of T_{GAA}. (c) Equivalent capacitance models of the FE-GAA device.

Fig. 7 V_{DD} reduction for the baseline HK-GAA by FE-GAA with/without IL with the same I_{OFF}.

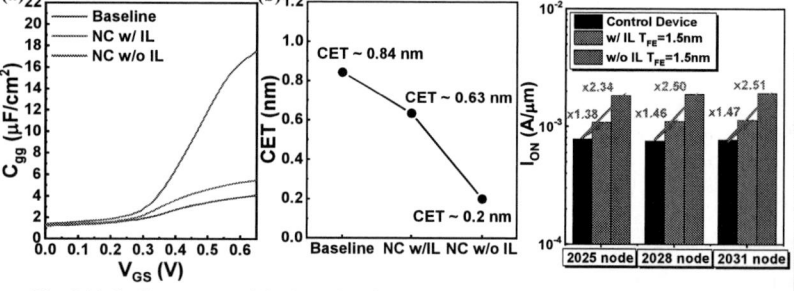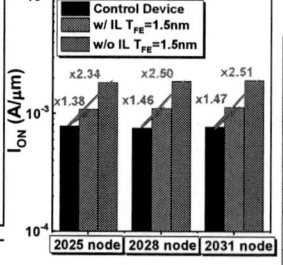

Fig. 8 (a) C_{gg}-V_{GS} curves and (b) CET of the baseline HK-GAA and FE-GAA with/without IL. (The CET value corresponds to Fig. 8(a) at V_{GS} = 0.65 V.)

Fig. 9 Comparison of I_{ON} for baseline HK-GAA and FE-GAA with/without IL at V_{DD} = 0.65 V in 2025 - 2031 nodes.

Table I Projected V_{DD} Reduction in This Work.

Year (node)	2025 ("2nm")	2028 ("1.5nm")	2031 ("10am eq")
L_g (nm)	14	12	12
W_{GAA} (nm)	30	20	15
T_{GAA} (nm)	6	6	6
V_{DD} (V)	0.65	0.60	0.60
SS (mV/dec.)	72	70	70
I_{ON} (µA/µm) at I_{eff} = 10 nA/µm	787	759	775
Required V_{DD} (V) for FE (w/ IL T_{FE}=1.5nm)	0.50	0.43	0.42
Required V_{DD} (V) for FE (w/o IL T_{FE}=1.5nm)	0.43	0.35	0.34

979-8-3315-0417-5/25 $31.00 © 2025 IEEE

Thermal Crosstalk Analysis in Advanced CMOS Circuits: Insights from 3nm Gate-All-Around Transistor Technology

Sihao Chen[1], Honglin Wu[1], Runsheng Wang[2], and Ru Huang[2], Lining Zhang[1*]

[1]School of Electronic and Computer Engineering, Peking University, China. *Email: eelnzhang@pku.edu.cn
[2]School of Integrated Circuits, Peking University, China.

Abstract

The thermal crosstalk issue in various typical circuit layouts featuring 3nm Gate-all-around transistors (GAAFET) is systematically investigated in this work. In a standard inverter with 6 tracks, thermal crosstalk from the active NMOS induces a temperature rise in the adjacent PMOS, elevating it by 7.5% above the ambient temperature. Further increasing the number of devices to three and four results in temperature increases of 11.8% and 14.9%, respectively. Moreover, the dynamic thermal response time and transient heating behavior induced by thermal crosstalk are found to be strongly dependent on device density and layout configuration, indicating that high-precision thermal modeling and thermal-aware reliability assessment urgently need to account for thermal crosstalk effects in advanced CMOS technology.

Keywords: GAAFET, Thermal Crosstalk, Self-heating, Dynamic thermal response.

Introduction

As advanced technology nodes continue to scale down beyond 3nm, thermal issues have emerged as critical challenges, significantly affecting the reliability and performance of logic and memory devices, as well as the stability of entire circuit systems [1-4]. The persistent increase in operating temperature exacerbates reliability degradation mechanisms such as hot carrier degradation (HCD) and bias temperature instability (BTI), leading to an accelerated aging process in advanced devices. Additionally, the dynamic temperature response during circuit operation has emerged as a crucial metric in assessing power-performance-reliability (PPR) metrics in advanced CMOS circuits [5, 6].

In fact, the total temperature of a single device is influenced by both self-heating effect (SHE) and partial temperature increases due to thermal crosstalk with other surrounding transistors and metal wires in the back-end-of-line (BEOL) [7]. Moreover, as chip integration density increases, the contribution of thermal crosstalk to overall device heating becomes increasingly significant. Fig. 1 illustrates four different types of thermal crosstalk phenomena, along with the vector distribution of heat flux observed in advanced CMOS circuits [8]. In Fig. 1(a), thermal crosstalk between different NSFETs within a 3 nm GAAFET structure is shown to be largely due to the introduction of lower thermal conductivity gate oxide and surrounding dielectric materials. Fig. 1(b) shows the thermal crosstalk experienced by a device in the off state due to the operation of neighboring devices, a common occurrence in both digital and analog circuits. Furthermore, the integration of power and signal metal wires during the place and route (PR) stage introduces Joule heating, resulting from high current densities in these metal wires. This thermal effect contributes to thermal crosstalk between front-end-of-line (FEOL) devices and BEOL interconnect metals, as depicted in Fig. 1(c) and (d). To address these challenges, we systematically evaluate thermal crosstalk across different circuit layouts based on 3 nm GAAFETs. The intensification of thermal crosstalk is highly dependent on device density and layout configuration, which influence hotspot formation and the dynamic thermal response of devices.

Self-heating Effect of 3nm GAAFET

The increased current density in the channel results in a pronounced self-heating effect in the 3 nm GAAFET device. Fig. 2(a) presents the transient thermal response of the 3 nm GAAFET, in which four distinct stages of the heating process can be observed. Figure 2(b) illustrates the temperature response of the device under dynamic power conditions at frequencies of 100 MHz, 1 GHz, and 10 GHz, showing that the peak temperature of the device increases as the frequency decreases. Fig. 3(a)-(d) show the temperature distribution of the GAA device at different response times, where heat flux in the channel is gradually conducted to the gate metal, source/drain epitaxial regions, high-thermal-conductivity silicon substrate, and interconnect metals, corresponding to the four stages of the heating process depicted in Fig. 1(a).

Thermal crosstalk analysis in various circuit layouts

A. Thermal Crosstalk in an Inverter Cell with 6 Tracks

Fig. 4(a) and (b) illustrate the two operating modes of the inverter in "on" and "off" states. The input signal is high, the NMOS transistor is active, while the PMOS transistor is inactive. Fig.4(c) depicts the transient temperature response of the inverter under a DC power of 30 μW for the N/P MOS in their respective on states. Additionally, it shows the approximate reduction in heating due to SHE, which results in a decrease of the steady-state temperature by 6 K compared to a single GAAFET with an equivalent substrate geometry. The inset in Fig. 4 presents the steady-state temperature distribution across the device for the three different scenarios, while Table I provides the geometrical and material parameters employed in the thermal simulations. To further elucidate the differences in peak temperature between a single GAAFET and a standard inverter, we compare the heat flux distributions of these two configurations, as shown in Fig. 5(a) and 5(b). In the single NMOS configuration, the heat flux dissipates through the interconnect metals of BEOL and the substrate. In contrast, in the inverter operating in the on state, most of the heat flux is observed to diffuse laterally through the gate metal, resulting in significant heating of the PMOS device in the off state due to thermal crosstalk from the adjacent NMOS device.

Fig. 6(a) compares the temperature variations of the device due to self-heating and thermal crosstalk induced by the inverter in different operating modes. The crosstalk coefficient, defined as $\alpha = \Delta T_{cross}/\Delta T_{SH}$, quantifies the proportion of temperature rise in neighboring devices attributable to the thermal crosstalk. When the input is "1", the temperature rise of NMOS due to self-heating is 64.8 K and the temperature rise of PMOS due to thermal crosstalk induced by NMOS is 22.4 K with a crosstalk coefficient of 34.6%. Conversely, When the input is "0", the temperature rise in the NMOS device due to thermal crosstalk from the PMOS is 20.2 K, with a corresponding crosstalk coefficient of 31.7%. The differences in the temperature rise due to SHE and thermal crosstalk coefficient between the two operating modes can be attributed to the difference in TC of the channel materials and interconnect metal density between the NMOS and PMOS devices. Fig. 6(b) illustrates the temperature response by SHE of NMOS and PMOS under dynamic power at a frequency of 1 GHz, where the peak temperature of PMOS is slightly higher than that of NMOS, consistent with the transient temperature response shown in Fig. 4(b).

B. Thermal Crosstalk in three GAAFETs with Common Gate

Fig. 7(a) presents a cross-sectional view of three NSFETs that share a common gate and separated S/D contacts. Three distinct operating scenarios are analyzed to illustrate how the intensity of thermal crosstalk varies with the operational states of the devices. The transient thermal response of NS2 is consistently extracted for each scenario, as depicted in Fig. 7(b). In Case 3, where both neighboring devices are active, NS2 experiences a temperature rise of 335 K. The peak temperature of NS2 occurs at the central NS device, as shown in the inset of Fig. 7(b), which is 13 K higher than the thermal crosstalk-induced temperature rise between the NMOS and PMOS in the inverter. When self-heating effects are considered (Case 1 and Case 2), the steady-state peak temperature of the device increases significantly, resulting in a 21% and 8.9% rise in steady-state temperature for Case 1 and Case 2, respectively, compared to Case 3. Additionally, the heating of NS2 occurs earlier than in Case 3 due to the progressive heat flux transfer from the metal gate and interconnects to the NS2 channel. Fig. 7(c) shows the dynamic thermal response of NS2 under each scenario with dynamic power at a frequency of 1 GHz and a 50% duty cycle. Fig. 7(d) illustrates the thermal crosstalk in Case 3 under dynamic power at frequencies of 100 MHz, 1 GHz, and 5 GHz.

C. Thermal Crosstalk in Typical Circuit Layouts

Thermal crosstalk issues are prevalent in various circuit layouts. To elucidate how thermal crosstalk varies with the density of on-state devices and layout configurations, we compare four distinct scenarios, as depicted in Fig. 8(a). In these diagrams, red and black boxes represent devices in the "on" and "off" states, respectively, while the blue box denotes the target device, which remains in the off state in all cases except Case 1. In Case 1, the target device reaches a peak steady-state temperature of 405.9 K due to SHE of the device. When only thermal crosstalk effect is considered, the peak steady-state temperature of the target device decreases as the number of neighboring active devices decreases. Specifically, the steady-state temperature in Case 2 decreases to 344.6 K, representing an 18% reduction compared to Case 1 due to the exclusion of SHE. Additionally, most of the heat flux is conducted through the interconnecting metals, resulting in different thermal response times for the two cases. Further reducing the number of neighboring devices, nearly identical peak temperatures and heating processes were observed in Cases 3 and 4.

Fig. 9(a) presents the dynamic temperature response of four cases under dynamic power with a frequency of 1 GHz, corresponding to the transient thermal response depicted in Fig. 8(b). Fig. 9(b) compares the steady-state temperature rise for devices with different numbers of active devices and layout configurations. For an inverter in the on state, thermal crosstalk from the PMOS device results in a 7.5% temperature increase, reaching 322.4 K above ambient temperature. With three neighboring active devices, thermal crosstalk effect results in an 11.8% increase to 335.3 K, while with four neighboring devices, this increase reaches 14.9%, bringing the temperature to 344.6 K. Overall, the temperature rise induced by thermal crosstalk is highly dependent on device density and layout configuration.

Conclusion

In summary, we systematically investigated the thermal crosstalk issues in advanced CMOS technology across various typical circuit layouts. Within a standard inverter cell with 6 tracks, the PMOS exhibits a temperature rise of 22.4K due to thermal crosstalk induced by an adjacent NMOS device, corresponding to a 7.5% increase over the ambient temperature of 300K. Further increasing the number of on-state devices to three and four across different layouts, the thermal crosstalk-induced temperature rise further escalates by 11.8% and 14.9%, respectively, indicating that this ratio is closely related to the density of on-state devices and circuit layout configurations. Consequently, these findings indicate that thermal crosstalk issues warrant special concern in thermal modeling and thermal-aware reliability assessment at the device-circuit level.

Acknowledgments

This work was supported in part by the National Key Research and Development Program of China under Grant 2023YFB4402204, the Natural Science Foundation of China under Grant 62474009, and in part by the Guangdong Basic and Applied Basic Research Foundation under Grant 2024B1515020064.

References

[1] B. Vermeersch, et al., IEDM, 2022. [2] S. Chen et al., TED, 2024. [3] F. Klemme et al., DATE, 2023. [4] U. Uttarwar, et al., SISPAD, 2023. [5] Z. Yu et al., IEDM, 2017. [6] Z. Sun et al., TED, 2023. [7] S. Zhao et al., TED, 2022. [8] S. Wang, et al., IRPS, 2024.

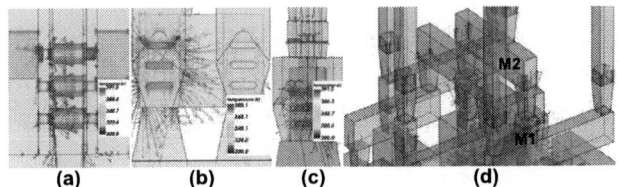

Fig. 1. Thermal crosstalk in advanced technology occurs in (a) between the nanosheets within a NSFET device; (b) between different devices; (c) between FEOL and BEOL metals; and (d) among BEOL interconnect metals for signal and power rails.

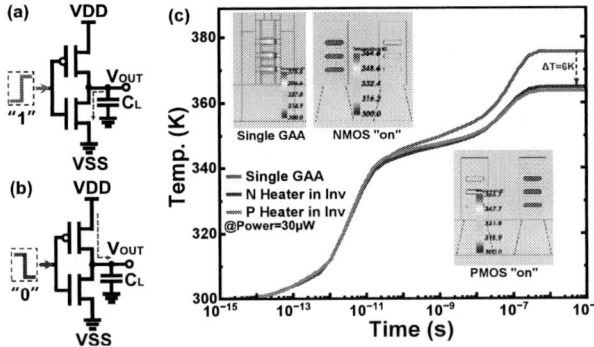

Fig. 4. CMOS inverter operating in (a) on-state and (b) off-state modes. (c) Transient thermal response with NMOS and PMOS as heat sources, respectively, compared to a single GAAFET with the same substrate geometry.

Dimensional Parameters	Value [nm]	Thermal conductivity	Value [W/K·m]
Channel Length	14	N-Channel	[22, 22, 6]
Nanosheet width	20	P-Channel	[18, 18, 4]
Thickness of Nanosheet	5	Gate HfO₂	2.3
Thickness of Gate Oxide	1.1	Gate Oxide	1.4
Thickness of SubFin	40	Copper	385
Metal2 Pitch of Inverter	21	Substrate (Si)	148

Table I: Geometric and thermal parameters for thermal simulation in this work.

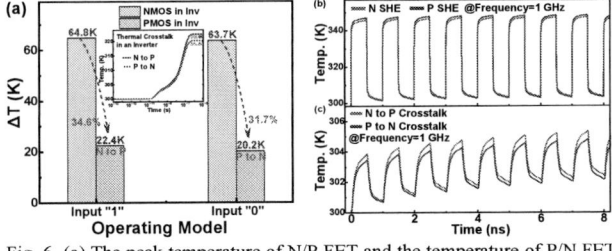

Fig. 6. (a) The peak temperature of N/P FET and the temperature of P/N FET induced by thermal crosstalk when the inverter operates at on-state (input "1") and off-state (input "0"), respectively. (b) The AC thermal response of the NFETs and PFETs and (c) crosstalk thermal response of the PFETs and NFETs when the inverter acts with dynamic power at a frequency of 1GHz.

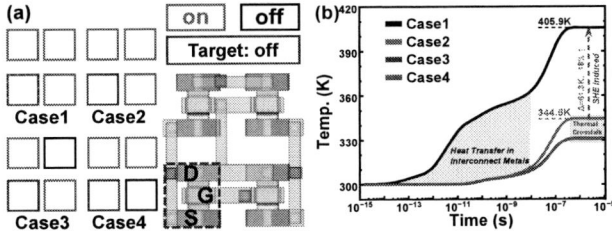

Fig. 8. (a) Four different cases in actual circuit layout where red and black squares represent devices in on-state and off-state, respectively, and the blue square indicates the target device for temperature response extraction in each case. Note that in case1, the blue square represents the target device itself in operation. (b) Transient thermal response across the four cases.

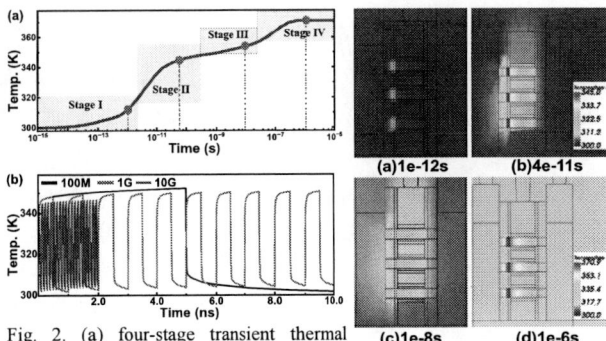

Fig. 2. (a) four-stage transient thermal response of the GAAFET device; (b) thermal response under dynamic powers conditions at different frequencies.

Fig. 3. (a)-(d) The temperature distribution of the 3nm GAAFET with different response times.

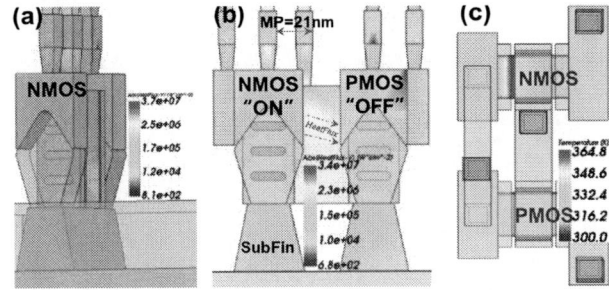

Fig. 5. (a) The temperature distribution of an inverter with the NMOS acting as a heater. The heat flux distribution of an inverter with (b) NMOS as a heater and (c) PMOS as a heater. Note that the standard inverter cell consists of 6 tracks with the metal pitch of 21nm.

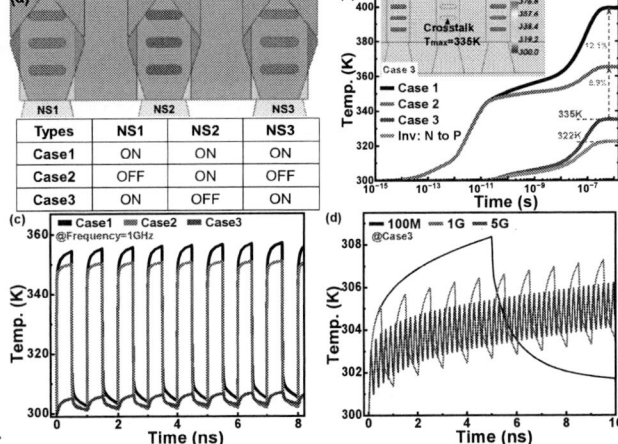

Fig. 7. (a) Schematic view of three NSFETs with a common gate, illustrating three operating models for thermal crosstalk analysis. (b) Transient thermal response of the three different operating cases, with the inset showing the temperature distribution of Case 3. (c) AC thermal response of the three cases under dynamic power at the frequency of 1GHz. (d) Thermal response for Case 3 at frequencies of 100 MHz, 1 GHz, and 5 GHz, respectively.

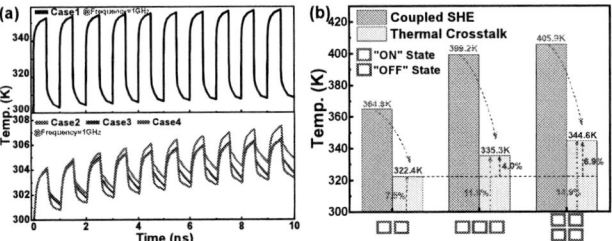

Fig. 9. (a) AC thermal response of the target device under four different cases. (b) Comparison of temperatures induced by the thermal crosstalk across three different layouts configurations in this work.

979-8-3315-0417-5/25 $31.00 © 2025 IEEE

Flexible and Skin-Compatible rGO/PVDF Composite Sensor for Multi-Functional On-Skin Applications

ZI-JIA SU[1], LU-QI TAO[1], LI-WEI LIANG[1], ZHI-FEI XIE[1], YI YANG[1], TIAN-LING REN[1]

[1]Institute of Microelectronics & Beijing National Research Center for Information Science and Technology and Department of Electronic Engineering, Tsinghua University, Beijing 100084, China

Email: taoluqi@tsinghua.edu.cn & RenTL@tsinghua.edu.cn

Abstract

This study presents an electrospinning rGO/PVDF composite sensor designed for on-skin applications, leveraging rGO's conductivity and PVDF's flexibility. Fabricated through CO_2 laser patterning and transfer onto a PVDF nanofiber network, the sensor demonstrates high sensitivity in acoustic detection, voiceprint recognition, and gesture control. It reliably captures frequencies up to 15,000 Hz and differentiates voice and gesture patterns, showcasing potential in wearable health monitoring and human-machine interaction.

Keywords: Electrospinning, Composite sensor, Acoustic and gesture recognition

Introduction

On-skin electronics are crafted to be ultra-thin, flexible, stretchable, and resilient, allowing them to seamlessly integrate with the human body for continuous, long-term monitoring of vital health metrics [1]. Among these devices, strain gauges have garnered significant attention due to their diverse applications, from human-machine interfaces to personal health diagnostics [2]. Notably, soft and highly precise strain gauges have been used for ongoing in vivo monitoring of organ function. These gauges operate on a straightforward principle: they produce consistent electrical signals in response to mechanical deformations [3]. Their compatibility with soft biological tissues makes them ideal for real-time health tracking. Essential attributes for such devices include high mechanical adaptability, flexibility, sensitivity, stretchability, durability, lightness, thinness, and biocompatibility [4].

In practical applications, strain sensors must conform closely to biological surfaces and endure significant deformation across various motions [5]. A common approach involves laser-induced patterning on polymer substrates or electrode deposition onto polymers [6, 7]. However, the flexibility of these substrate materials often limits their durability under repeated use or restricts their ability to withstand high strains [8]. Reduced graphene oxide (rGO) is a widely used piezoresistive material, but conventional rGO structures tend to wrinkle under repeated strain, which adversely affects their electrical performance [9]. This limitation highlights the need for innovative methods to fabricate rGO-based sensors.

Polyvinylidene fluoride (PVDF) electrospinning offers high flexibility, excellent skin compatibility, and a straightforward, cost-effective fabrication process, making it increasingly popular [10]. In this study, we propose a method to combine an rGO sensor with a PVDF electrospinning substrate to create a composite sensor. This approach ensures flexibility and superior electrical properties, meeting the demands of various movement patterns and enabling reliable performance in dynamic environments. We propose an rGO/PVDF composite sensor that leverages the electrical properties of rGO along with the skin-friendly and stretchable characteristics of PVDF, achieving a design that is both highly sensitive and adaptable for on-skin applications.

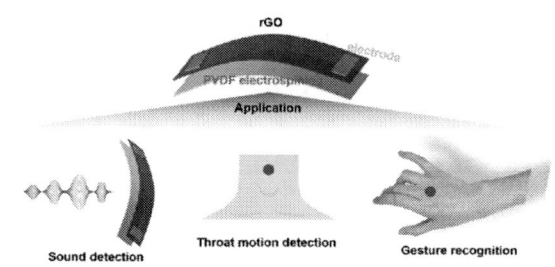

Fig. 1: Schematic illustration of the rGO/PVDF composite sensor and its applications. The sensor can be used for sound detection, throat motion sensing, and gesture recognition due to its flexibility and sensitivity.

The rGO/PVDF composite sensor shown in Fig. 1 is designed for precise monitoring of subtle mechanical forces and can be applied in various scenarios such as acoustic wave detection, throat motion sensing, and gesture recognition. Its flexibility and high sensitivity make it ideal for capturing fine deformations, enabling applications in voiceprint recognition, pulse monitoring, and human-machine interaction through gesture control, all while maintaining comfort and adaptability for on-skin use.

Results

A. Fabrication

Fig. 2 illustrates the fabrication process of the rGO/PVDF composite sensor: (1) GO solution is drop-coated onto a water-soluble substrate, forming a uniform GO dispersion layer. (2) A CO_2 laser is used to pattern and

979-8-3315-0417-5/25 $31.00 © 2025 IEEE

partially reduce the GO layer, creating a conductive rGO pattern. (3) The patterned rGO is immersed in deionized (DI) water to detach it from the soluble substrate. (4) The rGO pattern is then transferred onto a PVDF electrospinning membrane, providing a flexible support layer. This process ensures the sensor achieves high electrical sensitivity and mechanical flexibility, making it suitable for on-skin bio-monitoring applications.

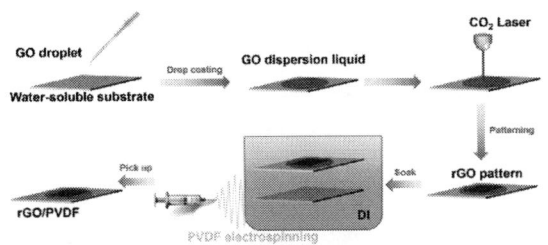

Fig. 2: Fabrication process of the rGO/PVDF composite sensor. The process includes drop-coating a GO layer onto a water-soluble substrate, laser patterning to create an rGO pattern, immersion in deionized water to release the rGO layer, and transfer onto a PVDF electrospinning membrane.

The interconnected PVDF nanofiber network acts as a strong mechanical reinforcement, allowing the rGO/PVDF composite structure to withstand greater mechanical strain without compromising its integrity. This enhanced support greatly improves the mechanical strength and flexibility of the composite, enabling it to accommodate larger strains and achieve a higher gauge factor compared to standalone rGO devices. As illustrated in Fig. 3, the rGO/PVDF composite demonstrates significantly improved stretchability and sensitivity.

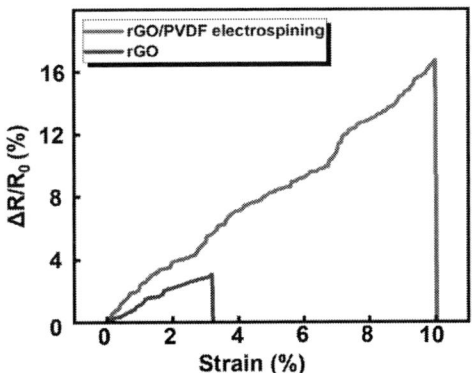

Fig. 3: Mechanical performance of the rGO/PVDF composite sensor compared to a bare rGO device.

B. Muti-motion scene identification

Based on the results of mechanical properties, the composite sensor demonstrates excellent sensitivity and reliability in sound detection across different scenarios.

Under the specified experimental conditions, the composite sensor was first positioned 5 cm away from the sound source for a frequency sweep test (Fig. 4a). The clear and continuous yellow diagonal line in the spectrogram demonstrates the sensor's capacity to reliably detect a broad frequency range (20 Hz to 15,000 Hz), making it suitable for precise acoustic monitoring applications. Next, the sensor was applied directly to a volunteer's throat to capture voiceprint patterns during speech (Fig. 4b). This setup enabled the sensor to detect unique voiceprint signatures from vocal cord vibrations, highlighting its potential for personalized voice recognition. The distinct patterns in each waveform provide a basis for differentiating spoken words. Further noise reduction can enhance these patterns by minimizing background noise and improving signal clarity. This capability supports not only high-quality voice recognition but also potential applications in silent speech interfaces and real-time health monitoring, where subtle throat vibrations can offer valuable insights into vocal health and activity.

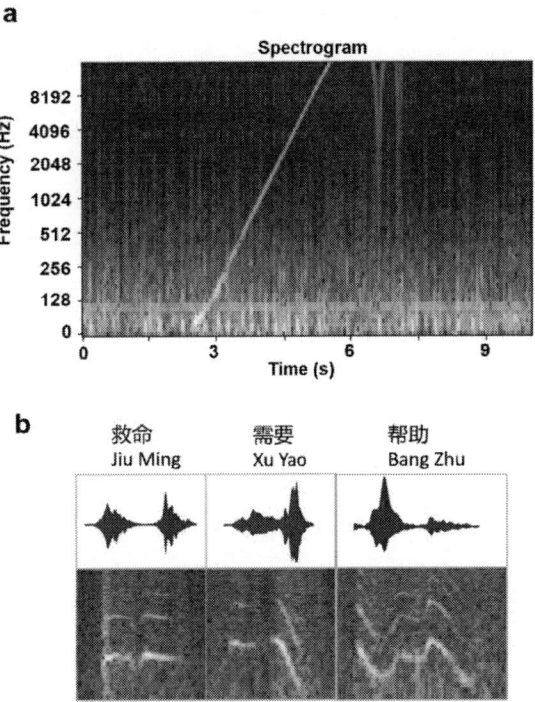

Fig. 4: Sound detection performance of the rGO/PVDF sensor. (a) Frequency sweep test with the sensor placed 5 cm from the sound source, demonstrating consistent detection across a wide frequency range. (b) Voiceprint patterns captured from a volunteer's throat, showing distinct signatures for different words.

Fig. 5 demonstrates the application of the rGO/PVDF sensor in hand gesture recognition. The sensor is attached to the back of the hand, where it detects resistance changes caused by different hand gestures, such as finger bending and extension, to achieve gesture recognition. The results indicate that the sensor can accurately distinguish between various gestures, providing a highly sensitive tool for gesture-controlled interfaces. Leveraging the device's ability to differentiate gestures, we implemented a Morse code application, encoding the word "Tsinghua" through hand motions corresponding to Morse code signals. This showcases the sensor's strong potential for advanced gesture recognition applications.

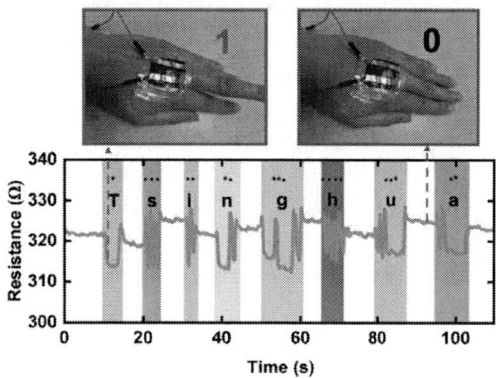

Fig. 5: Application of the rGO/PVDF sensor in gesture recognition. The sensor, attached to the back of the hand, detects resistance changes from finger movements, enabling Morse code transmission of the word "Tsinghua" through hand gestures.

The result highlights the sensor's potential for real-time communication and interaction, especially in scenarios where conventional input methods are impractical. This high sensitivity to subtle resistance changes enables the sensor to serve as an intuitive interface for applications in wearable technology, remote communication, and assistive devices for individuals with speech or mobility impairments. Such versatility underscores the sensor's potential to transform gesture recognition systems, making them more accessible and efficient in a wide range of environments.

Conclusion

This study presents an rGO/PVDF composite sensor that combines the electrical properties of rGO with the flexibility and skin compatibility of PVDF, resulting in a highly sensitive, durable, and stretchable design. The sensor demonstrated effective performance across applications such as sound detection, voiceprint recognition, and gesture control. Experimental results confirmed its ability to capture a wide frequency range for acoustic monitoring, identify voiceprint patterns for personalized recognition, and detect hand gestures, enabling Morse code encoding. Overall, the rGO/PVDF composite sensor offers a versatile and efficient solution for on-skin electronics, making it suitable for wearable technology, real-time health monitoring, and intuitive human-machine interaction. Its adaptability highlights its potential for future innovations in biosensing and interactive interfaces.

References

[1] J. A. Rogers, T. Someya, Y. Huang, "Materials and mechanics for stretchable electronics," *Science*, vol. 327, pp. 1603–1607, 2010.

[2] C. Pang, G. Y. Lee, T. I. Kim, S. M. Kim, H. N. Kim, S. H. Ahn, K. Y. Suh, "A flexible and highly sensitive strain-gauge sensor using reversible interlocking of nanofibres," *Nature Materials*, vol. 11, pp. 795–801, 2012.

[3] Y. R. Jeong, H. Park, S. W. Jin, S. Y. Hong, S. S. Lee, J. S. Ha, "Highly stretchable and sensitive strain sensors using fragmentized graphene foam," *Advanced Functional Materials*, vol. 25, pp. 4228–4236, 2015.

[4] C. Yan, J. Wang, W. Kang, M. Cui, X. Wang, C. Y. Foo, K. J. Chee, P. S. Lee, "Highly stretchable piezoresistive graphene–nanocellulose nanopaper for strain sensors," *Advanced Materials*, vol. 26, pp. 2022–2027, 2014.

[5] C. Tan, Z. Dong, Y. Li, et al., "A high-performance wearable strain sensor with advanced thermal management for motion monitoring," *Nature Communications*, vol. 11, no. 1, p. 3530, Jul. 15, 2020.

[6] J. Lin, Z. Peng, Y. Liu, et al., "Laser-induced porous graphene films from commercial polymers," *Nature Communications*, vol. 5, no. 1, p. 5714, Dec. 10, 2014, doi: 10.1038/ncomms6714.

[7] S. Lee, D. H. Ho, J. Jekal, et al., "Fabric-based lamina emergent MXene-based electrode for electrophysiological monitoring," *Nature Communications*, vol. 15, no. 1, p. 5974, Oct. 2, 2024.

[8] L. Lan, J. Ping, J. Xiong, et al., "Sustainable Natural Bio-Origin Materials for Future Flexible Devices," *Advanced Science*, vol. 9, issue 15, 2022.

[9] M. Xie, G. Qian, Y. Yu, C. Chen, H. Li, and D. Li, "High-performance flexible reduced graphene oxide/polyimide nanocomposite aerogels fabricated by double crosslinking strategy for piezoresistive sensor application," *Chemical Engineering Journal*, vol. 480, p. 148203, Jan. 15, 2024.

[10] J. Xue, T. Wu, Y. Dai, and Y. Xia, "Electrospinning and Electrospun Nanofibers: Methods, Materials, and Applications," *Chemical Reviews*, vol. 119, no. 8, pp. 5298–5415, Apr. 24, 2019.

Modelling and Design of Short Channel Ferroelectric FETs with a Metal Interlayer easing the Multilevel Operation

Chiara Rossi, Daniel Lizzit, and David Esseni

DPIA, University of Udine, Via delle Scienze 206, 33100 Udine, Italy; email: chiara.rossi@uniud.it

ABSTRACT

This work presents a simulation study of a ferroelectric field effect transistor (FeFET), which leverages a metal interlayer to achieve a multilevel operation thanks to the interplay between the ferroelectric polarization and the charge stored in the interlayer. We show that the metal interlayer can effectively stabilize the ferroelectric polarization even for a negligible charge trapping in the dielectric stack and, moreover, enable a multilevel operation even for a uniform ferroelectric polarization.

Keywords: Ferroelectric field-effect transistor, TCAD simulation, metal interlayer, charge trapping.

INTRODUCTION

Ferroelectric devices are among the most promising implementations of memristors for neuromorphic computing, thanks to their energy efficiency, scalability and CMOS compatibility [1], [2]. Unfortunately, however, in FeFETs and Ferroelectric Tunneling Junctions (FTJs) the switching of the ferroelectric (FE) polarization and thus the device operation are inextricably linked to charge trapping. In fact, charge trapping, by partly compensating the FE polarization, can reduce the ferroelectric depolarization field and stabilize the polarization, as it has been clearly observed in both FTJs [3], [4], and FeFETs [5]–[7]. The spatial and energy position of traps is difficult to engineer though [8]. In this latter respect, a metal interlayer (ML) integrated in the gate stack could be a better design option for FeFETs (see Fig. 1), as the charge injected from the semiconductor into the ML can be tuned via the programming pulses. Such a tuning of the net charge in the gate stack can enable a multilevel operation even in highly scaled devices, where the polarization exhibits a two-level or a very-few-level behavior [9]. Hence, a ML can unlock the multilevel operation from the size of FE domains [10].

The influence of a ML has been previously studied for the negative capacitance operation [11], and in FeFETs for the tuning of the ratio between the dielectric (DE) and FE capacitance. In [12], [13] a large area ratio between DE and FE (i.e. up to 40:1) has been employed to enhance charge injection through the HZO and thus enlarge the Memory Window (MW). Such a large area ratio, however, is a hyndrance to the device scaling, particularly for planar devices. Charge injection through both the DE and FE was studied in [14], but only in quasi-static conditions.

In this paper, we present numerical simulations for an aggressively scaled FeFET that integrates a ML into the gate stack. Our analysis elucidates the device operation and reports a good memristor behavior, whereby the conductance of the ML–FeFET can be potentiated and depressed over a wide range of values. We investigate the interplay between the charge in the ML and the FE polarization, and show that the FE polarization is pivotal for achieving a write operation at a <5 V voltage.

MODELING FRAMEWORK

Fig. 1 reports the structure of the device analyzed in this work. A tungsten metal interlayer lays in between a 10 nm hafnium zirconium oxide (HZO) ferroelectric layer and a 5 nm HfO$_2$ dielectric layer. The gate length is 30 nm. The gate stack sits on top of a p–doped silicon semiconductor ($N_A=10^{16}$ cm^{-3}) and the source and drain regions are $n+$ silicon wells.

Simulations are performed with Sentaurus TCAD simulator [20]. The ferroelectric switching dynamics is modeled with the phenomenological Landau-Ginzburg-Devonshire (LGD) equation, calibrated on experimental data for a metal-ferroelectric-metal (MFM) capacitor featuring the same 10 nm ferroelectric thickness [22]. The simulated total polarization (Q_{FE}), i.e. the spontaneous polarization plus the linear contribution, is plotted against experiments in Fig. 2. Simulations in Fig. 2 are performed considering 100 domains with a Gaussian distribution of coercive fields E_C with $\sigma = 0.4E_C$, as previously described in [7].

Experiments and simulations in Fig. 2 refer to MFM capacitors with large areas, corresponding to a very high number of ferroelectric domains. In the simulations of aggressively scaled FeFETs, instead, we employed a homogeneous description of the ferroelectric material, as in scaled FeFETs the polarization switching has been observed to have a two-level behavior [9]. The Landau's anisotropic constants used in FeFET simulations are the average values of the ones used in Fig. 2 and are listed in Table I. The corresponding remnant polarization P_R is 19 μC/cm^2 and the coercive field E_C is 1.1 MV/cm [22]. The leakage current through the HfO$_2$ layer is known to be dominated by Trap Assisted Tunneling (TAT) [23]. This has been simulated in Sentaurus by including bulk traps in HfO$_2$, using the parameters extracted in [23], and enabling tunneling of carriers [20], [24]. Fig. 3 reports a fairly good agreement between our simulations and experiments for a 5 nm–thick HfO$_2$ layer [23]. The

979-8-3315-0417-5/25 $31.00 © 2025 IEEE

leakage current through the HZO has been neglected, as the HZO layer is two times thicker than the HfO$_2$ layer between the ML and the transistor channel. Moreover, HZO has a lower leakage compared to HfO$_2$ at the same thickness, which has been attributed to the zirconium oxide doping [25].

SIMULATION RESULTS

We define net polarization, P_{net}, as the sum between HZO polarization, P, and the charge, Q_{ML}, in the ML.

The FeFET is first set to a well determined initial state using a *reset* pulse. This is achieved by applying a negative voltage to the gate electrode (see Fig. 4(a)) and grounding the source, drain and substrate electrodes, such that a negative P and a negative P_{net} for $t<0$ can be set.

The *write* biasing consists of a positive gate voltage pulse (see Fig. 4(a)). During the positive pulse, the P switches to positive values and electron injection into the ML is triggered from the semiconductor and through the HfO$_2$ layer. The Q_{ML} can reduce the depolarization field and stabilize the P even for a modest charge trapping in the dielectrics [4], [5]. Indeed, in our simulations trapped charge in HfO$_2$ is only $\sim$5% of P_R, but allows TAT to effectively charge/discharge the ML. Fig. 4(b) shows the P and P_{net} as a function of time. At the end of the pulse, when the applied gate voltage goes back to zero, the resulting P_{net} is $4.2\,\mu\text{C/cm}^2$. By increasing the width of the gate pulse, the polarization gets more and more compensated by the negative Q_{ML}, and the positive P_{net} is progressively reduced. This is shown in Fig. 5, where the total P and the compensated P_{net} at $V_G=0$ V are plotted versus the write pulse width. A similar scenario occurs for negative V_G pulses, whereby a negative P is stabilized by a positive Q_{ML}, and the overall P_{net} at $V_G=0$ V is negative. Different values of P_{net} result in a different threshold voltage V_T for the FeFET. After the end of the write pulse, the state of the FeFET is sensed by applying a read voltage V_G sweep from -0.6 V to 0.6 V (at $V_{DS}=50$ mV), which does not disturb the stored P_{net}. Typical simulated drain current versus gate voltage curves are reported in Fig. 6. The leftmost curve corresponds to the maximum positive $P_{net}\simeq4.2\,\mu\text{C/cm}^2$. As the write pulse width increases, the V_T shifts towards positive values because the positive P_{net} decreases. Thus, by using the leftmost curve in Fig. 6 as a *preset* state, it is possible to operate the device as a memristor and modulate its conductance, here defined at read voltages $V_G=0.2$ V and $V_{DS}=50$ mV.

A simulated depression curve is reported in Fig. 7(b), obtained by using the gate depression pulses at $V_G=+4.5$ V shown in Fig. 7(a). The simulated potentiation curve reported in Fig. 7(c), has been obtained by using the potentiation pulses at $V_G=-5$ V also

shown in Fig. 7(a). The preset state for Fig. 7(c) has been taken as the lowest conductance state in Fig. 7(b), corresponding to the longest depression pulse. Likewise, the largest conductance in Fig. 7(c) is the same as the preset conductance for the depression curve in Fig. 7(b).

Finally, we investigated the advantages of using a ferroelectric material in the gate stack. In this respect, Fig. 8 provides a comparison between the ML–FeFET and a device structure with the same gate stack as in Fig. 1 but with the HZO replaced by a non ferroelectric HfO$_2$ layer. As it can be seen in Fig. 8, in order to obtain a similar MW of $\sim$0.75 V as in the ML–FeFET a much higher 7 V write pulse is needed.

CONCLUSION

Based on calibrated TCAD simulations, we have investigated FeFETs exploiting a ML in the gate stack and, in particular, the interplay between the FE polarization and the charge injected in the ML. In such a FeFET architecture, the stabilization of the FE polarization is decoupled from charge trapping, and the multi–level operation is decoupled from the multi-domain behavior of the ferroelectric. These features are particularly attractive for the scaling of FeFETs, and for their exploitation in neuromorphic computing applications.

ACKNOWLEDGMENT

This work has received funding from the European Union's Horizon Europe research and innovation programme under the Marie Skłodowska-Curie Actions Postdoctoral Fellowships, project No. 101108023 (EASIFeT), and under FIXIT (GA No. 101135398). *

REFERENCES

[1] E. Covi *et al.*, Neuromorph. Comput. Eng., vol. 2, 012002 (2022).
[2] H. Mulaosmanovic *et al.*, Nanoscale, 10, 21755-21763 (2018).
[3] H. W. Park *et al.*, Nanoscale, 13, 2556-2572 (2021).
[4] R. Fontanini *et al.*, IEEE JEDS, 10, 593 (2022).
[5] S. K. Toprasertpong *et al.*, IEDM Tech. Dig., 23.7.1 (2019).
[6] S. Deng *et al.*, IEDM Tech. Dig., 4.4.1 (2020).
[7] D. Lizzit *et al.*, IEEE ESSDERC 2022, p. 344 (2022).
[8] R. Fontanini *et al.*, IEEE TED, 69, 7, 3694 (2022).
[9] H. Mulaosmanovic *et al.*, Nanotechnology, 32, 502002 (2021).
[10] H. Mulaosmanovic *et al.*, ACS Appl. Mater. Interfaces, 9, 3792 (2017).
[11] T. Rollo *et al.*, Nanoscale, 12, 6121-6129 (2020).
[12] Z. Zheng *et al.*, IEEE TED, 71, 3, 1827 (2024).
[13] Z. Zheng *et al.*, IEEE TED, 71, 9, 5325 (2024).
[14] X. Wang *et al.*, IEEE TED, 71, 9, 5332 (2024).
[15] V. P.-H. Hu *et al.*, IEEE Symposium on VLSI, T134 (2019).
[16] K. Seidel *et al.*, IEEE Symposium on VLSI, 355 (2022).
[17] S.-J. Yoon *et al.*, IEEE TED, 67, 2, 499 (2020).
[18] D. Kwon *et al.*, Adv. Intell. Syst., 5, 12, 2300125 (2023).
[19] D. Lehninger *et al.*, IEEE EDL, 43, 11, 1866 (2022).
[20] Sentaurus Device of Synopsys TCAD, User Guide, V-2023.12.
[21] B. Jiang *et al.*, Symposium on VLSI Technology, p.141 (1997).
[22] T. Kim,*et al.*, IEEE TED, vol. 69, no. 7, pp. 4016–4021 (2022).
[23] S. Cimino *et al.*, Microelectr. Engin., vol. 95, p. 71 (2012).
[24] F. Matteo *et al.*, IEEE ICD, p. 764 (2022).
[25] C.-C. Fan *et al.*, IEEE EDTM, p. 224 (2017).

*Views and opinions expressed are however those of the author(s) only and do not necessarily reflect those of the European Union or the European Commission. Neither the European Union nor the granting authority can be held responsible for them.

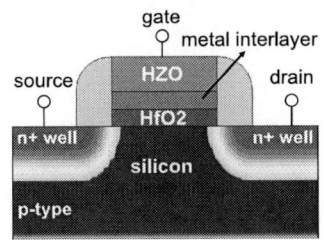

Fig. 1. Structure of the simulated FeFET with metal interlayer.

TABLE I: Parameters of the HZO layer.

Param.	Description	Value
α	Landau's	$-6.12 \cdot 10^{10}$ cm/F
β	anisotropic	$3.29 \cdot 10^{19}$ cm^5/(FC^2)
γ	constants	$9.59 \cdot 10^{28}$ cm^9/(FC^4)
ρ	resistivity	11000 Ω cm
ε_{FE}	background dielectr. const.	35

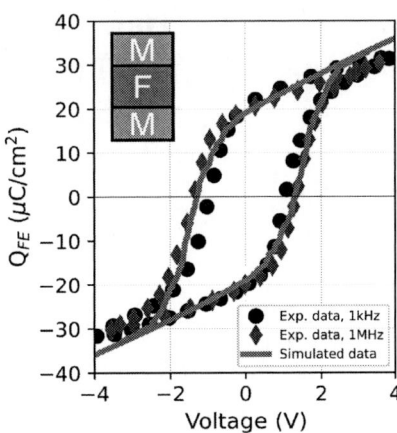

Fig. 2. Simulated total polarization Q_{FE} (solid line) compared to experimental data [22] (symbols) for a 10 nm HZO capacitor.

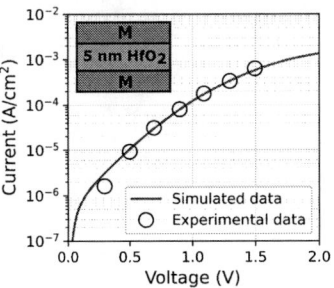

Fig. 3. TAT Leakage current through a 5 nm HfO$_2$ MIM capacitor. Simulated results (solid line) are compared to experimental data (symbols) [23]. The traps parameters are: density of $3 \cdot 10^{19}$ cm^{-3}, energy levels in the range $2.0 - 2.6$ eV below the bottom of the HfO$_2$ conduction band and cross-section of 10^{-14} cm^2 [23].

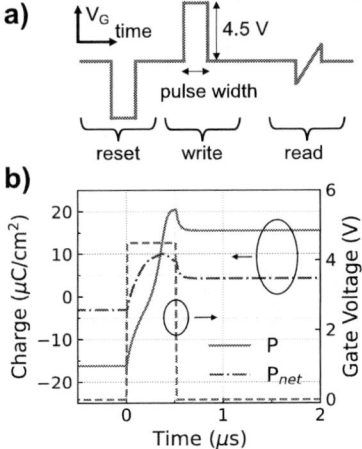

Fig. 4. a) Gate voltage pulses used for *reset*, write and read operations of the FeFET. b) P and P_{net} during and after a gate write pulse.

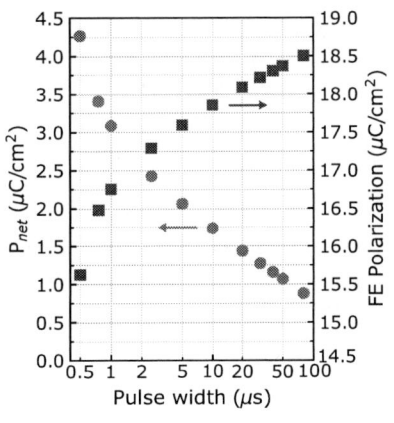

Fig. 5. P_{net} and P at $V_G=0$ V after the write pulse ($V_G=4.5$ V) versus the pulse width.

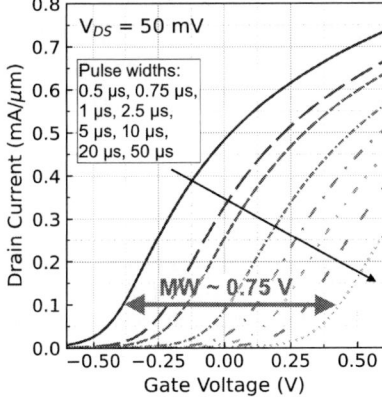

Fig. 6. I_D versus V_G characteristics of the FeFET in read conditions and after different write pulse widths. The MW is extracted for a drain current of 0.1 mA/μm.

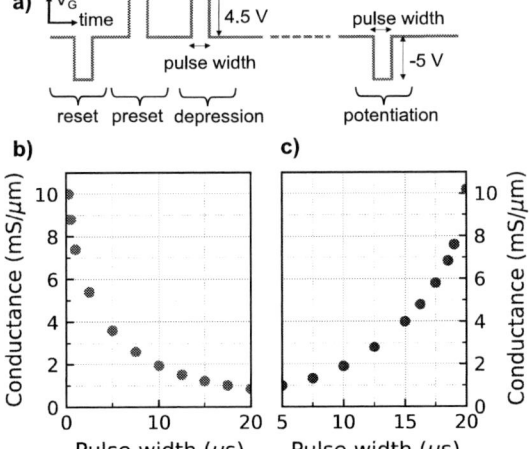

Fig. 7. Depression (b) and potentiation (c) curves of the FeFET, obtained with the pulsing scheme depicted in (a). Conductance extracted at $V_G=0.2$ V and $V_{DS}=50$ mV.

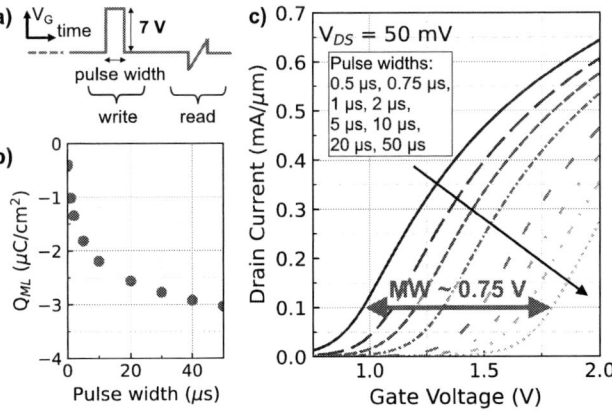

Fig. 8. a) Write pulse applied at the gate, starting from a condition with neutral ML. b) Charge Q_{ML} in the ML obtained after the write pulse at $V_G=7$ V. c) Corresponding $I_D - V_G$ characteristics. Without the ferroelectric layer, a much higher V_G is needed to have the same MW with similar pulses duration.

Effect of pulse schemes on multi-level switching and short-term instability in 1T1R configuration

Xiaohua Liu[1], Felix Cüppers[1], Dennis Nielinger[1], Susanne Hoffmann-Eifert[1], Rainer Waser[2], and Stefan Wiefels[1]

[1]Forschungszentrum Jülich, (PGI-7/10, ZEA-2), [2]RWTH Aachen University, IWE II

ABSTRACT

This study investigates the characteristics and short-term stability of multi-level storage in 1T1R structures under single-pulse and multi-pulse programming schemes with varying step-size. The findings indicate that, due to limitations in the transistor's transfer characteristics, the read current distributions become narrower as the current value increases. The multi-pulse programming scheme enables finer resistance tuning, but there is no significant advantage over the current distribution obtained by the single-pulse scheme at the same current level. Furthermore, the current under the single-pulse scheme exhibits better short-term stability. Keywords: ReRAM, 1T1R, programming schemes, multi-level switching, short-term stability

INTRODUCTION

Redox-based resistive random-access memory (ReRAM) holds significant potential for in-memory computing systems [1]. The non-volatile storage characteristics and fast switching speed make valence change mechanism (VCM)-based filamentary ReRAM stand out among various candidates. After series integration with a transistor, the one-transistor-one-resistor (1T1R) structure allows multi-level storage through gate voltage control, thereby enhancing the processing precision for in-memory computing applications.

One of the primary reliability challenges for VCM-based filamentary ReRAM is variability, which is typically observed from device-to-device, from cycle-to-cycle, and from read-to-read (R2R) [2]. Incremental step pulse programming (ISPP) is commonly used to address the first two types [3]. However, optimization strategies for the latter have been less frequently reported, highlighting the need to explore the impact of different programming schemes on the R2R variability of ReRAM in each storage state.

This study examines the effects of different programming schemes on the multi-level switching and short-term stability of 1T1R structures. Multi-pulse programming schemes with varied step sizes are designed, comparing with a baseline single-pulse programming scheme. Statistical analysis of the resulting current from different schemes indicates that, while multi-pulse programming remains the

preferred method for fine resistance tuning, the current distributions achieved through the single-pulse scheme demonstrate superior short-term stability.

EXPERIMENTS

The 1T1R structures, as shown in Fig. 1 (a), measured in this study are $Pt/TaO_x/Ta$-based VCM cells, connected in series with n-channel transistors in a $0.18\,\mu m$ CMOS technology, which are fabricated in collaboration with X-FAB semiconductor foundries GmbH. The W/L ratio of the used transistor equals 1. For the detail of the bias scheme for switching, the readers is referred to our previous study [4].

Figure. 1 (b) shows the programming schemes, taking a multi-pulse scheme as an example. Each scheme consists of two parts: the programming section for multi-level switching highlighted in blue background and the read section for the investigation of short-term instability highlighted in yellow background. The programming section involves four types of programming schemes: a single-pulse scheme (baseline, denoted as S1) and multi-pulse schemes with the V_{WL} step size of $0.4\,V$ (S2), $0.2\,V$ (S3), and $0.1\,V$ (S4). In the single-pulse scheme, a single SET pulse with a width of $1\,\mu s$ and an amplitude of $2.0\,V$ is applied at the target V_{WL}. The multi-pulse schemes, resembling typical program-

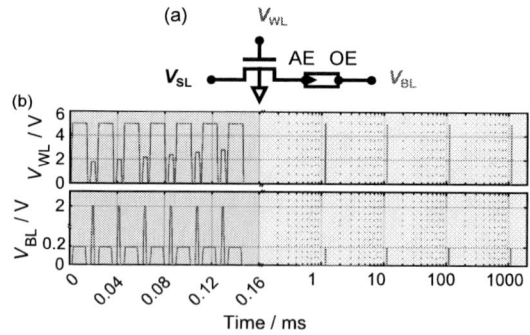

Fig. 1. (a) Sketch of the 1T1R structure. The active electrode (AE) of the ReRAM device is connected to the transistor, the ohmic electrode (OE) to V_{BL}. (b) Example for pulse scheme S3 used in this study, taking the target V_{WL} of $2.8\,V$ with $0.2\,V$ step as an example. Here, the programming section for multi-level switching and the reading section for investigating short-term instability are highlighted with blue and yellow backgrounds, respectively. The programming section is presented using a linear x-axis scale, while the read section is in a logarithmic x-axis scale. Bulk and V_{SL} are connected to ground.

979-8-3315-0417-5/25 $31.00 © 2025 IEEE

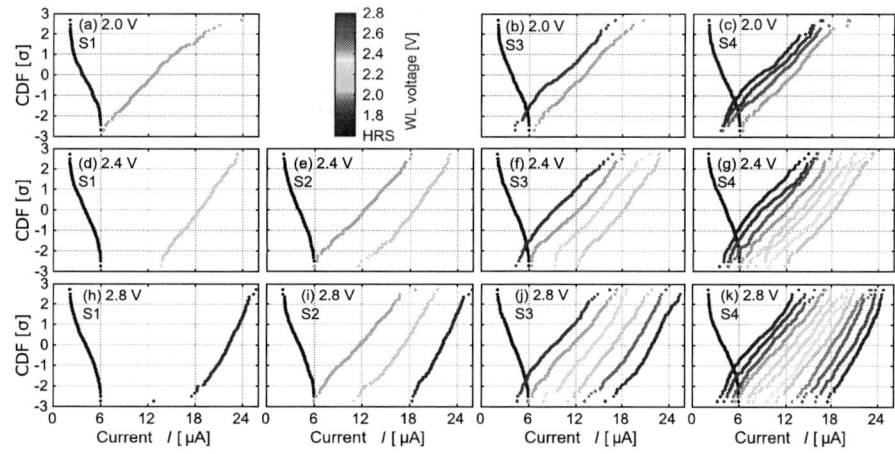

Fig. 2. Read current distributions measured after each programming pulse in different schemes. The target V_{WL} are 2.0 V (a-c), 2.4 V (d-g), and 2.8 V (h-k). For the target V_{WL} of 2.0 V, S1 and S2 are identical; thus, only the results for S1 are shown. At each target V_{WL}, the read current after each pulse is displayed using the same color gradient, as shown in the color bars. The initial HRS distribution, shown as the deep blue curve on the far left, has been filtered within a range of 2 μA to 6 μA.

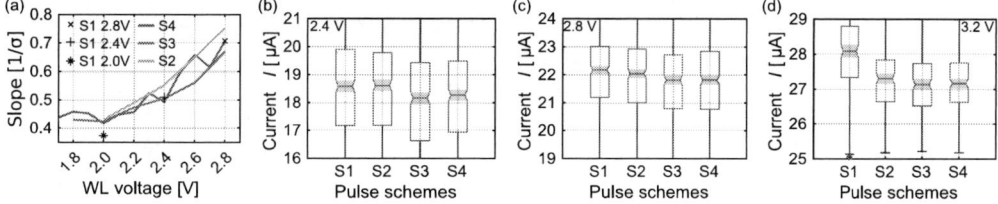

Fig. 3. (a) Slope of the current distribution as a function of V_{WL}. The yellow, red, and blue lines represent the variation in slopes for S2, S3, and S4, respectively, at a target V_{WL} of 2.8 V. Additionally, the slopes of the current distribution for S1 at target V_{WL} of 2.8 V, 2.4 V, and 2.0 V are indicated by symbols. (b-c) Box plots of the final current distribution at the target V_{WL} 2.4 V, 2.8 V, and 3.2 V.

verify algorithms, consist of a series of SET pulses with the same parameters as those in single-pulse schemes, followed by read pulses with a width of 10 μs and an amplitude of 0.2 V. During the SET process, a gradually increasing V_{WL} is applied, starting from 1.6 V up to the target V_{WL}. Immediately after the last programming pulse, a read operation is conducted. Then, 4 read pulses with log-spaced delay time ranging from 1 ms to 1 s are performed to check the short-term stability of the readout current. Each programming scheme includes 400 cycles of experiments. At the beginning of each cycle, the 1T1R structure is reset to an HRS.

RESULTS

Figures 2(a-k) present the read current distributions after each programming pulse under different programming schemes for three target V_{WL}. In each plot, the deep blue curve on the far left represents the current distribution of the device in HRS before programming, filtered within a current range of 2 μA to 6 μA. The color bar legend shows the V_{WL} corresponding to the current distributions after each programming pulse. As the V_{WL} increases, the read current distribution of the device shifts gradually to the right, demonstrating good programmability.

The experiments with different target V_{WL} were conducted on separate devices. However, when the V_{WL}

is the same, the current distribution curves are similar (e.g., the same blue curve in (a), (e), and (i), each at a V_{WL} of 2.0 V), indicating minimal device-to-device variability. Compared to the counterpart, the multi-pulse programming schemes offer greater possibility in the current distribution. Therefore, in a practical program-verify algorithm, once the target current range is reached, programming of the device will cease, resulting in a narrower final distribution. This finding suggests that the multi-pulse programming scheme is more advantageous for fine-tuning device resistance.

Since the distributions in the Fig. 2 are approximately linear, a respective fit can be used to obtain an approximate slope for each distribution. The relationship between the slope and the V_{WL} is shown in Fig. 3(a). Considering that the distributions are similar for the same V_{WL}, only the slope variations for S2, S3, and S4 at the target V_{WL} of 2.8 V are plotted. Additionally, the slope data for S1 at each of the three target V_{WL} are also indicated. As seen in Fig. 3(a), the distribution slope increases with the V_{WL}, indicating that the current distributions become narrow gradually.

According to our previous study [4], this effect occurs because, at a constant SET voltage, the transistor's operation regime transitions gradually from saturation to the linear region as the V_{WL} increases, resulting in

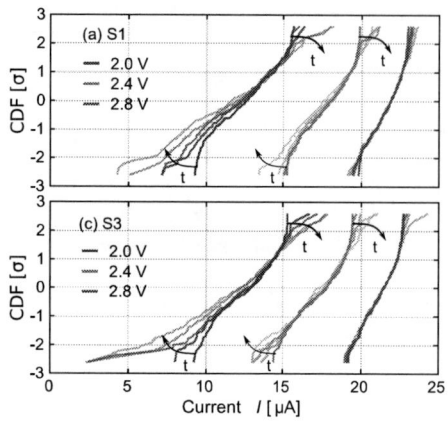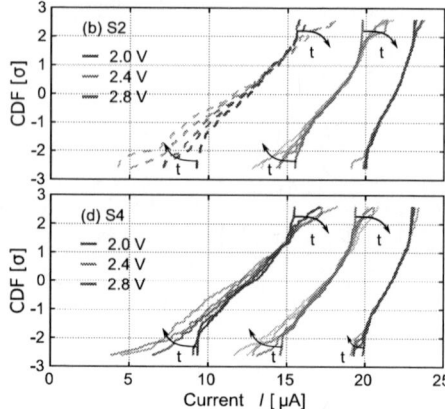

Fig. 4. (a-d) Read current distributions at different times after programming the devices with three different target V_{WL} by S1-4. The initial post-programming currents were reshaped by retaining 200-cycle data around the median. The dark blue, light blue, and red curves represent the distribution of the target V_{WL} of 2.0 V, 2.4 V, and 2.8 V, respectively. The low to high transparency of each color shows the distribution over time with applied read voltage. The data for S2 at the target V_{WL} of 2.0 V is identical to S1 and is therefore represented with a dashed line.

a progressively smaller increment in allowable current. Further increases in the V_{WL} cause the device's resistance to approach its lower limit, narrowing the current distribution.

Figures 3(b-d) show box plots of the final current distribution under different programming schemes at target V_{WL} 2.4 V, 2.8 V, and 3.2 V. The median current obtained with the large-step programming scheme (S1, S2) is significantly higher than that achieved with the small-step programming scheme (S3, S4). Since the current values in the 2.0 V experimental results are relatively low, this trend is not prominent. Therefore, we supplemented the study with experiments at higher V_{WL} (3.2 V) to further validate the existence of this trend. This is counter-intuitive as more pulses are applied if the step size is smaller. However, the large steps lead to more abrupt transitions and are thus more prone to capacitive overshoots which may result in higher read current after SET.

Figures 4(a-d) show the time-dependent changes in the read current distribution under different target V_{WL} across different programming schemes. It is evident that as the current values increase, the degree of distribution broadening decreases gradually. This suggests that the R2R variability of the device is significantly alleviated as the device resistance decreases. Additionally, it can be noted that the distribution of 2.4 V from S4 exhibits the largest degree of broadening among the other distribution at the same level, indicating the worst stability. Based on our previous research hypothesis [5], it is suggested that the multi-pulse programming scheme leads to the formation of broader, but less dense conductive filaments within the ReRAM. Consequently, lower concentrations of oxygen vacancies are more likely to jump at the interface between the conductive filaments and the electrodes, resulting in the current relaxation.

Therefore, the results in S4 (the highest number of pulses) show the least preferable short-term stability.

CONCLUSION

This study compares the effects of single-pulse and multi-pulse programming schemes with varying step sizes on the multi-level switching characteristics and stability of 1T1R structures. The multi-pulse schemes facilitate precise tuning of the device resistance, while the single-pulse scheme provides better short-term stability in the resulting current distribution. Given the importance of R2R variability, careful selection of the step size in the programming scheme design for 1T1R structures is essential.

ACKNOWLEDGMENT

This work was supported by the Federal Ministry of Education and Research (BMBF, Germany) in the projects NEUROTEC (Project Nos. 16ME0398K and 16ME0399) and NeuroSys (No. 03ZU1106AA and 03ZU1106BA). It is based on the Jülich Aachen Research Alliance (JARA-FIT).

REFERENCES

[1] A. Mehonic *et al.*, "Roadmap to neuromorphic computing with emerging technologies," *APL Materials*, vol. 12, no. 10, p. 109 201, Oct. 2024, DOI: 10.1063/5.0179424.

[2] S. Wiefels *et al.*, "Reliability aspects of 28nm BEOL-integrated resistive switching random access memory," *physica status solidi (a)*, p. 2 300 401, 2023, DOI: 10.1002/pssa.202300401.

[3] E. Pérez *et al.*, "Reduction of the cell-to-cell variability in Hf1-xAlxOy based RRAM arrays by using program algorithms," *IEEE Electron Device Lett.*, vol. 38, pp. 175–178, 2017, DOI: 10.1109/LED.2016.2646758.

[4] X. Liu *et al.*, "Effect of transistor transfer characteristics on the programming process in 1T1R configuration," *IEEE Trans. Electron Devices*, vol. 71, no. 4, pp. 2423–2430, 2024, DOI: 10.1109/TED.2024.3370536.

[5] X. Liu *et al.*, "Effect of programming schemes on short-term instability in 1T1R configuration," *IEEE Access*, submitted.

Comprehensive Modeling of Ferroelectric Tunnel Junctions: Variability Analysis and Device Design

Jiajun Qiu[1,2], Ning Ji[1], Hao Li[2], Ning Feng[1], Runsheng Wang[2*], Ru Huang[2], Lining Zhang[1*]

[1]School of Electronic and Computer Engineering, Peking University, Shenzhen, China. *E-mail: eelnzhang@pku.edu.cn

[2]School of Integrated Circuits, Peking University, Beijing, China. *E-mail: r.wang@pku.edu.cn

Abstract

In this work, we develop a model that comprehensively captures the physical mechanisms of ferroelectric tunnel junctions (FTJ) of the metal-ferroelectric-insulator-metal (MFIM) structure. This model utilizes kinetic Monte Carlo (KMC) statistical distributions to capture nucleation-limited switching of multi-domain ferroelectrics and accurately describes the I-V characteristics of MFIM across the full voltage region. Through this model, we quantify the device-to-device variation of FTJ and propose strategic approaches to mitigate this variability by refining grains and optimizing pulse waveform. Additionally, the design space of material parameters for enhancing FTJ performance under non-saturated polarization is investigated, providing design insights for neuromorphic computing applications.

Keywords: ferroelectric tunnel junctions, model, design and variability.

Introduction

The exponential increase in data storage and computational demands has prompted significant interest in emerging memory technologies among researchers. Among these, hafnium-based ferroelectric tunnel junctions (FTJ) have the advantages of non-destructive readout, fast read and write, multi-level states, low energy consumption and high-density integration, which considered to be one of the potential candidates for future non-volatile memory and neuromorphic devices [1].

Among the reported FTJs, the metal-ferroelectric-insulator-metal (MFIM) type FTJ is distinguished by its high endurance and low interface defects [1-2]. As shown in Fig. 1, MFIM-FTJ introduces an ultra-thin interlayer and forms an asymmetric tunneling barrier by regulating the polarization direction, thereby differentiating between the low (LRS) and high resistance state (HRS) and achieving a substantial tunneling electro-resistance (TER) ratio. To guide the design of FTJ in applications such as neuromorphic devices and non-volatile memory, it is essential to develop a comprehensive model that can fully elucidate the underlying physical mechanisms of FTJ.

In this paper, we construct an efficient FTJ model for MFIM structure. This model can accurately describe the multi-domain characteristics of polycrystalline hafnium-based ferroelectric films and the I-V characteristics of FTJ in the full voltage region. Based on this model, the fluctuation characteristics of MFIM are investigated, providing effective strategies for controlling device-to-device variation. Furthermore, we explored the design space for material parameter tuning TER in low-voltage ranges, which will aid in applications within neuromorphic devices.

Simulation Methodology

Fig. 2 illustrates the simulation framework and theoretical formulations of the MFIM FTJ model. Initially, charge balance is achieved through iterative calculations to determine the correct voltage drop across each layer, which is then used to update the band profile and calculate the tunneling current. The ferroelectric polarization reversal is calculated by Monte Carlo nucleation-limited switching (MC-NLS), which can characterize the multi-domain and dynamic features [3]. Tunnel Current is calculated based on Wentzel-Kramers-Brillouin (WKB) approximation, which can cover both Direct and Fowler Nordheim Tunneling. The MFIM structure comprises two insulating layers, which affords the possibility of electrons tunneling through both barriers simultaneously or through a single barrier. The WKB approximation can be utilized for various tunneling barrier shapes.

Fig. 3 shows the I-V curve calculated by the model, and the sub-figure is the correction result of the ferroelectric material P-V and experimental data [4]. The model demonstrates precise calculation of the I-V characteristics of the MFIM FTJ with full region voltage, exhibiting a high degree of correlation with TCAD simulation results. Table I lists the material parameters that are used in the simulation.

Variability analysis for MFIM-FTJ

The multi-domain characteristics of polycrystalline ferroelectric films (Fig. 4b), which are described by the E_a statistical distribution of the NLS model, result in significant device-to-device variation in FTJs. As shown in Fig. 4(a), we conducted I-V characteristic simulations for 100 devices, which results exhibited significant fluctuations. Fig. 4(c) illustrates the frequency distribution of J_{HRS}, respectively, read at -0.5 V. The mean (μ) and standard deviation (σ) are extracted by fitting a normal distribution to the current density distributions of HRS, thus enabling the assessment of the magnitude of variation through the σ/μ ratio.

From the insights of modeling, several methods to control the variation of FTJ are explored. As illustrated in Fig. 5, an increase in the number of ferroelectric domains can serve to diminish the variability of TER. The variation of LRS is

979-8-3315-0417-5/25 $31.00 © 2025 IEEE

always higher than that of HRS as shown in Fig. 5(a). In the fabrication of the device, grain refinement can be regulated by modifying the annealing temperature or the doping ratio, etc. [5-6]. Furthermore, the pulse waveform can be optimized to reduce the variation during measurement, for instance, by modifying the pulse amplitude or width. The impact of device-to-device variation on TER is investigated by altering the width (50 ns-1000 ns) and amplitude (2-5 V) of the input voltage pulses, as illustrated in Fig. 6. Through modeling, the fluctuation of TER exhibited an increasing trend with an expansion in pulse width, accompanied by a concurrent rise in the μ of TER. At narrow pulse widths, the majority of domains are unable to reach the speed of the voltage response, exhibiting reduced polarization flip and displaying characteristics more akin to an ordinary dielectric layer. Conversely, as pulse width increases, more domains become involved in polarization flip, resulting in a notable rise in the variation of ferroelectric polarization. This highlights the heightened impact of fluctuations on TER. With a fixed pulse width and increasing pulse amplitude, the variation of TER initially rises before declining, while the mean (μ) of TER continues to rise. The initial increase in TER fluctuations is attributed to the growing number of domains participating in the switching process under conditions of unsaturated polarization. Conversely, in the case of saturated polarization, the voltage drop across the ferroelectric layer is observed to decrease, which serves to reduce the tunneling barrier and contribute to a reduction in variability in the tunneling current. Given that FTJs exhibit multi-state characteristics and low power consumption, a reduction in either the pulse amplitude or pulse width may prove an effective compromise for improving variation.

Design Space for Neuromorphic FTJs

In practical applications of neuromorphic circuits, it depends more on the scale of the TER of the FTJ in the unsaturated polarization state [8]. This section concentrated on the optimization strategies for device design under conditions of unsaturated polarization. The programming voltage is ±2 V and the reading voltage is -0.5 V (material parameters list in Table I). The design space for the thickness of the Fe and IL layer is illustrated in Fig. 7(a). A larger t_{Fe} is preferable for the modulation of the tunneling barrier, which is indicated by an increase in the TER. The combination of a large t_{Fe} and a small t_{IL} is more conducive to enhancing TER. The selection of the dielectric constant is examined in Fig. 7(b), with fixed thicknesses of t_{Fe} 5.5nm and t_{IL} 1nm. The objective is to achieve the lowest possible value for ε_{Fe} and a range of 9-20 for ε_{IL}. This combination may prove beneficial in achieving a more optimal partial voltage. However, in practice, the dielectric parameters of the materials are not readily tunable, necessitating the consideration of trade-offs.

The MFIM structure has the capacity to utilize a range of dielectric oxides, with its intrinsic properties being contingent upon the characteristics of the IL material, including electron affinity and bandgap. Differences in IL materials primarily lead to variations in voltage drop, thereby affecting the tunneling barrier characteristics. To achieve the greatest possible TER in FTJs, selecting an optimal thickness for each IL material is essential. The optimal thickness design space for SiO_2 and ZrO_2 is investigated and is presented in Fig. 8(a) and 8(b). A comparison of the two dielectric materials reveals that a larger t_{Fe} corresponds to a higher TER. However, the optimal thickness for SiO_2 is approximately 1 nm, while that for ZrO_2 is nearly 3 nm. Fig. 8(c) provides a summary of the TER for five common IL materials at fixed thickness, indicating that MgO exhibits the highest TER.

Conclusion

A MFIM-FTJ model is developed on the basis of the MC-NLS framework. In conjunction with TCAD simulation, the voltage domain operations, including read and write, are investigated. Through modeling, the FTJ device-to-device variability is quantified and several methods to control the device variability are proposed. In this work, we explore the design space for enhancing the performance of the MFIM at small voltages to provide guidance for its applications in neuromorphic devices.

Acknowledgments

This work was supported in part by the Natural Science Foundation of China under Grant 62474009 and in part by the Guangdong Basic and Applied Basic Research Foundation under Grant 2024B1515020064.

References

[1] Park S H, Lee H J, Park M H, et al. "Ferroelectric tunnel junctions: promise, achievements and challenges", Journal of Physics D: Applied Physics, 2024.

[2] Hwang, Junghyeon, Youngin Goh, and Sanghun Jeon. "Physics, Structures, and Applications of Fluorite-Structured Ferroelectric Tunnel Junctions." Small, 20.9, pp. 2305271 (2024).

[3] Cai P, Li H, Ji Z, et al. "Modeling the Coupled Charge Trapping Dynamics in FeFETs for Reliability Characterizations", IEEE Trans. Electron Devices, (2024).

[4] Lee K, Kim S, Kim M, et al. "Comprehensive TCAD-based validation of interface trap-assisted ferroelectric polarization in ferroelectric-gate field-effect transistor memory." IEEE Trans. Electron Devices, 69, pp. 1048-1053 (2022).

[5] Hyuk Park M, Joon Kim H, Jin Kim Y, et al. "Evolution of phases and ferroelectric properties of thin $Hf_{0.5}Zr_{0.5}O_2$ films according to the thickness and annealing temperature", Applied Physics Letters, 102 (2013).

[6] Lederer M, Lehninger D, Abdulazhanov S, et al. "Process influences on the microstructure of BEoL integrated ferroelectric hafnium zirconium oxide" IEEE International Symposium on Applications of Ferroelectrics (ISAF), pp 1-4 (2021).

[7] Applied Materials. Ginestra? User's Guide. Accessed: Feb. 12, 2024. [Online]. Available: https://www.appliedmaterials.com/cn/zh_cn/semiconductor/ginestra-software.html.

[8] Zhao L, Fang H, Wang J, et al. "Ferroelectric artificial synapses for high-performance neuromorphic computing: Status, prospects, and challenges", Applied Physics Letters, 124, (2024).

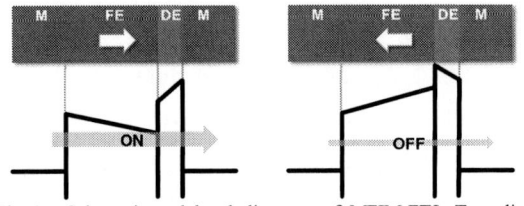

Fig 1 . Schematic and band diagrams of MFIM-FTJ: Tunneling barriers are regulated by changing the direction of polarization to distinguish between LRS(left) and HRS(right).

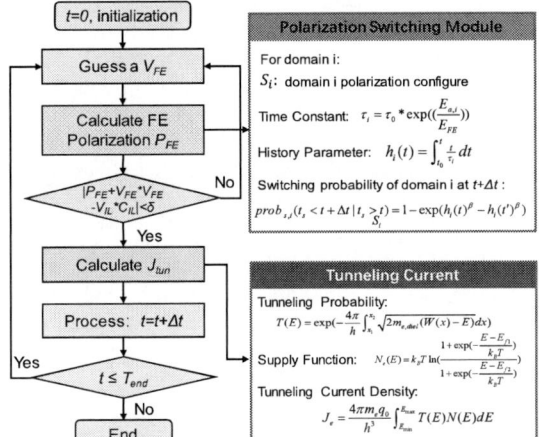

Fig 2. Framework and theoretical formulations of MFIM FTJ model. Ferroelectric polarization switching is implemented using the Monte Carlo NLS model, with the tunnel current calculated based on WKB approximation.

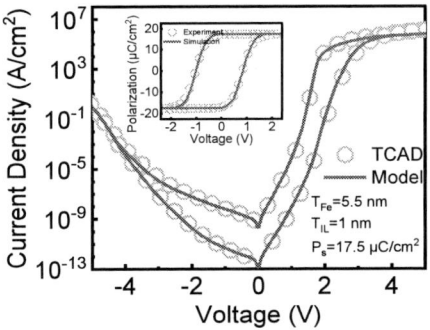

Fig 3 . I–V characteristics of a typical MFIM FTJ with full region voltage. Model verification with TCAD Ginestra [7] simulation. Insert: P-V loop calibrated with experiment data [4].

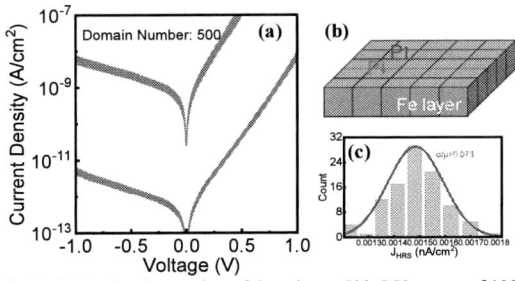

Fig 4. (a) Setting the number of domains to 500, I-V curves of 100 samples demonstrate device-to-device variation; (b) Schematic of Fe layer; (c) Frequency distribution plots in HRS, with results consistent with a normal distribution.

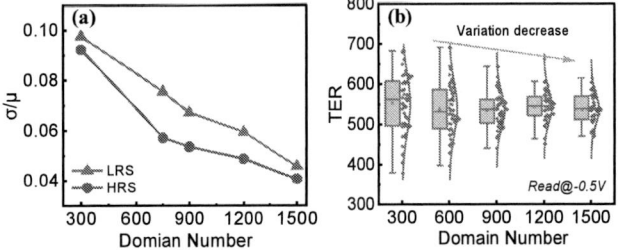

Fig 5 . Increasing the domain number: (a) Evaluation of device variation through σ/μ of current density in LRS and HRS; (b) Box plots of TER and the data distribution of TER.

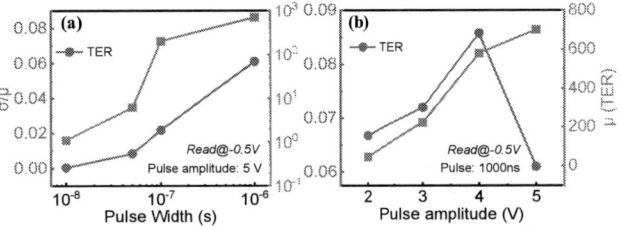

Fig 6 . The right side of the fig. shows the σ/μ of TER, and the left side shows the mean (μ) of TER. Device variation analysis with pulse width (a) and pulse amplitude (b).

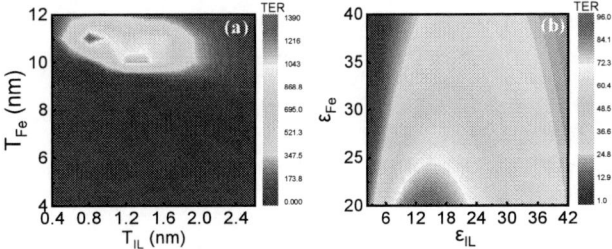

Fig 7 . Design space of the FE layer and IL layer property for tuning TER: (a) thickness and (b) dielectric constant.

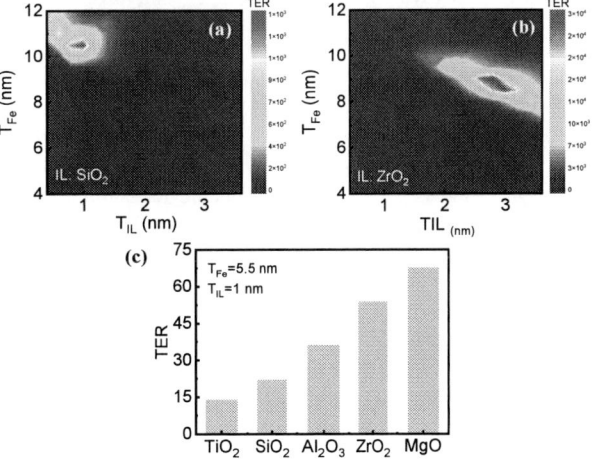

Fig 8 . Design space for thickness selection of different IL materials: (a) SiO₂; (b) ZrO₂. (c) Comparison of TER for different IL materials at a fixed thickness.

Material	Band gap(eV)	Affinity (eV)	Permittivity	mass
HZO	5.75	2.7	25	$0.2m_0$
Al₂O₃	8.7	1.35	11	$0.35m_0$

Table 1. Parameters used in this work.

Cryogenic SEKV Compact Model Applied to 22 nm FDSOI Enabling Low-temperature Circuit Simulation

Hung-Chi Han[1], Yating Zou[1], Batuhan Keskin[1], Edoardo Charbon[1], and Christian Enz[1]

[1]École Polytechnique Fédérale de Lausanne (EPFL), Lausanne, Switzerland

ABSTRACT

This paper introduces the sEKV compact model for cryogenic circuit simulation, addressing the key challenge of cryo-CMOS circuit validation. This model accurately represents the transistor and circuit DC characteristics at both room and cryogenic temperatures of a commercial 22 nm FDSOI technology using only a few parameters. It serves as a practical alternative solving the lack of PDKs currently available for the design of cryo-CMOS circuits. To reach a wider audience, the Verilog-A code of the model is available as open source.

Keywords: Compact modeling, Cryogenic, Quantum computing, FDSOI, MOSFET.

INTRODUCTION

The advances in classical and quantum computing set an increasing demand for low-temperature electronics [1], [2]. Operating transistors at cryogenic temperatures (CT) offers many benefits including faster speed, higher gain, reduced leakage, and lower power consumption [3]. Unfortunately, most PDKs provided by semiconductor foundries are not verified at CT due to numerical issues and a lack of models accounting for cryogenic physics [4], [5]. This complicates the circuit simulation and makes verification before tape-out (TO) impossible. As a result, the engineering process for developing cryo-CMOS circuits requires many re-spins and therefore becomes costly and inefficient. To address the simulation challenges at low temperatures, extensive cryogenic characterization and modeling activities have been reported, as summarized in [6]. Several solutions based on industrial compact models (CMs) have been proposed, such as the BSIM [7], [8], EKV [9], and PSP [10]. Nevertheless, extending existing CMs to 4 K may face (i) an inaccurate representation of the underlying physics due to the missing modeling of phenomena present at CT and (ii) complex parameter extraction resulting from a large number of model parameters, e.g., ~ 240 for BSIM 4. This complexity combined with limited cryogenic measurements available makes this approach too much resource demanding. Taking advantage of the temperature independence of the normalized G_m/I_D characteristic, we have recently proposed to use the G_m/I_D methodology for the design of analog circuits using the inversion coefficient IC as the main design parameter combined with the sEKV parameters extracted from measurements made at CT [11], [12]. However, the design still cannot be verified by simulation before TO. This work addresses this issue by proposing the sEKV CM using a minimal set of parameters that can be run in a circuit simulator using the Verilog-A code.

COMPACT MODELING

The proposed model stems from the EKV charge-based model [13]–[16]. It is basically made of two equations, the first relating the voltages to the charges according to

$$v_p - v_{s,d} = \frac{V_P - V_{S,D}}{U_T} = \ln(q_{s,d}) + 2q_{s,d}, \quad (1)$$

where the pinch-off voltage V_P and source/drain voltages $V_{S,D}$ are normalized to the thermal voltage $U_T = kT/q$ (k Boltzmann constant, T temperature, and q elementary charge). q_s and q_d are the source and drain inversion charge normalized to the specific charge $Q_{spec} = -2nC_{ox}U_T$ with C_{ox} the dielectric capacitance. The pinch-off voltage is approximated as $V_P = (V_G - V_T)/n$ with V_G the gate voltage and V_T the threshold voltage. Note that in this model, the terminal voltages are referenced to the substrate preserving the S/D symmetry. V_T is influenced by the *drain-induced barrier lowering* (DIBL) effect according to $V_T = V_{T0} - \sigma_d(V_D + V_S)$. The term σ_d is given by $\sigma_d = 0.5/(\cosh(\xi) - 1)$ with $\xi = L/\sqrt{\epsilon_{ch} t_d t_{ox}/\epsilon_{ox}}$ (L channel length, t_d depletion thickness, t_{ox} oxide thickness, ϵ_{ch} and ϵ_{ox} are permittivities of channel and dielectric) [17]. The second expression of the sEKV model gives the normalized drain current in terms of the charges [13]

$$i_{ds} = q_s^2 + q_s - (q_d^2 + q_d). \quad (2)$$

Note that the current i_{ds} models the thermionic transport, differing from the hopping transport that will be introduced later for modeling the subthreshold swing saturation at low temperatures. At high longitudinal field, q_d decreases to negligible value according to (1), but saturates to $q_{d_{sat}}$ in the presence of *velocity saturation* (VS). This results in a saturated current given by

$$i_{ds_{sat}} = 2 q_{d_{sat}}/\lambda_c, \quad (3)$$

with $\lambda_c = L_{sat}/L$ the VS parameter [14]. By inserting (3) into (2), $q_{d_{sat}}$ can then be expressed as

$$q_{d_{sat}} = \frac{\lambda_c (q_s^2 + q_s)}{1 + \sqrt{1 + \lambda_c^2 (q_s^2 + q_s)}}, \quad (4)$$

where the diffusion term is ignored in (2), i.e., $i_{ds} = q_s^2 + q_s - q_{d_{sat}}^2$, because VS does not appear in diffusive transport. Consequently, the minimum of q_d is defined by $q_{d_{sat}}$, as shown in Fig. 1. Then, the normalized drain saturation voltage in strong inversion can be computed by inserting $q_{d_{sat}}$ into (1), resulting in $v_{d_{sat}} = v_p - 2 q_{d_{sat}}$. Toward the weak inversion regime, $v_{d_{sat}}$ is clamped at 4, i.e., $V_{D_{sat}} = 4U_T$, as plotted in Fig. 1.

The *channel length modulation* (CLM) effect is represented by $L_{eff} = L(1 - \Delta L/L) = Lf_{clm}^{-1}$, where [18]

$$f_{clm}^{-1} = 1 - \frac{l_{clm}}{L} \ln\left(1 + \frac{v_d - v_{d_{sat}}}{v_{dclm}}\right) \quad (5)$$

with l_{clm} and v_{dclm} are characteristic length and voltage for CLM effect. Specifically, $l_{clm} = \sqrt{\epsilon_{ch} x_j t_{ox}/\epsilon_{ox}}$ with x_j the depth of junction, and $v_{dclm} = 2 l_{clm}/l_{sat}$ with l_{sat} the fitting parameter. On the other hand, effective mobility μ_{eff} with respect to operating bias gets more prominent at low temperatures mainly due to the absence of phonon scattering

979-8-3315-0417-5/25 $31.00 © 2025 IEEE

[19]. The factor $f_{mu} = \mu_{eff}/\mu_0$ is then introduced to account for the noticeable change in mobility given by [20]

$$f_{mu} = \frac{1}{1 + \theta_1 Q_i + \theta_2 Q_i^2} \qquad (6)$$

where θ_1 and θ_2 are fitting parameters and $Q_i = \frac{1}{2}(q_s + q_d)Q_{spec}$.

It is now well established that the subthreshold swing (SS), which is an important phenomenon at CT, does not follow the Boltzmann limit ($\frac{kT}{q}\ln 10$), but actually saturates to a constant value ranging from 10 to 20 mV/dec below a critical temperature of typically 50 K [21], [22]. In this model, the hopping transport through the band tail is included to capture this phenomenon. According to Eqs. (2.8) and (4.10) in [23], the hopping charge can be deduced from the thermionic charge. Therefore, similar to (1), the hopping charge can then be written as

$$\frac{V_P - V_{S,D}}{W_t/q} = \ln\left(q_{hops,d}\right), \qquad (7)$$

where q_{hop} exponentially scales with the band tail characteristic energy W_t rather than U_T. Additionally, the value of q_{hop} is clamped at a maximum value of 1 because the hopping current can be neglected compared to the thermionic current which dominates in strong inversion. Subsequently, the normalized hopping current can be written as $i_{hop} = q_{hops}^2 + q_{hops} - q_{hopd}^2 - q_{hopd}$. Finally, de-normalizing i_{ds} and i_{hop} results in

$$I_{th} = i_{ds} I_{spec} f_{clm} f_{mu}, \qquad (8)$$

$$I_{hop} = i_{hop} I_{spec} \left(\frac{W_t}{kT}\right)^2, \qquad (9)$$

with $I_{spec} = Q_{spec}\mu_0 U_T W/L$. The factor $(W_t/(kT))^2$ applied to I_{hop} stems from the difference in effective density of states between thermal and hopping charges [23]. Fig. 2 demonstrates the thermionic/hopping charge and current as functions of v_p. Note that the Verilog-A code of the proposed model is available on GitLab [24]. Fig. 3 compares the I-V measurements from a 22 nm FDSOI technology to simulations using the proposed Verilog-A model for various channel lengths at both 300 K and 3.8 K. The very good agreement between the model and the measurements at RT and CT demonstrates the validity of the proposed DC model, despite it requires only 11 parameters.

CIRCUIT SIMULATION

This section verifies the proposed model on a simple circuit made of a common-source nMOS transistor providing a transconductance G_{m1} and loaded by a common-gate nMOS stage presenting a load conductance G_{m2} (Fig. 4). If the two transistors have the same dimensions, the small-signal voltage gain G_{m1}/G_{m2} is close to unity as long as both transistors remain in saturation.

Fig. 5 shows the simulations of the large-signal transfer characteristic V_{out}-V_{in}, the current efficiency G_{m1}/I_D and the small-signal voltage gain versus the current I_D performed at RT for various gate lengths for M1 and for a supply voltage of 0.8 V. The simulation results obtained from the proposed model are compared to the results obtained from the full CM of the PDK. We can observe a very good agreement between the simple model and the full CM.

Fig. 6 presents the same transfer characteristic V_{out}-V_{in}, current efficiency G_{m1}/I_D and small-signal voltage gain but simulated with the proposed model at 3.8 K. We can observe that the simulations stop for $V_{out} < 0.3V$. This is not due to the model, but to the fact that the simulation does not converge anymore because of the extremely small current flowing when the devices are biased in deep weak inversion. However, the simulations are still valid in the bias region of interest for the designers, which often happens to be moderate inversion. We can observe that the V_{out}-V_{in} characteristic at 3.8 K is shifted to a higher V_{in} because of the significant increase of V_T at CT. Moreover, the output swing at 3.8 K is significantly lower than that at 300 K. This highlights the importance of applying the back-gate bias to reduce the V_T [25] and finally achieve a larger voltage swing. The maximum current efficiency at 3.8 K is enhanced by about a factor five compared to 300 K, but the magnitude of the enhancement is lower than expected due to the SS saturation [11]. The gain is increased at 3.8 K because of the improved mobility. However, this improvement is not noticeable for $L = 24$ and 28 nm since the mobility increase of short-channel devices is limited by temperature-independent neutral defect [26].

CONCLUSION

A cryogenic charge-based compact model that has a minimal number of parameters is presented. The model effectively captures the essential transistor DC behavior both at room and cryogenic temperatures. It includes the subthreshold swing saturation effect observed at low temperatures by accounting for the hopping transport through the band tail. The model is validated using measurements and PDK from a 22 nm FDSOI technology. Furthermore, the simulation of a common-source amplifier using the Verilog-A code of the proposed model at 3.8 K is presented. This work is a first step towards finding a solution for a more efficient design of cryo-CMOS circuits.

ACKNOWLEDGMENT

This project has received funding from the EU Horizon research and innovation programme, ARCTIC, under grant agreement No 101075725 and Pathfinder program, QUADRATURE, under grant agreement No 101099697.

REFERENCES

[1] R. Saligram et al., Chip, p. 100082, Dec. 2023.
[2] E. Charbon et al., in IEDM, San Francisco, CA, USA, Dec. 2016, pp. 13.5.1–13.5.4.
[3] H. L. Chiang et al., in VLSI. Honolulu, HI, USA: IEEE, Jun. 2020.
[4] C. Enz et al., in IEDM, San Francisco, CA, USA, Dec. 2020, pp. 25.3.1–25.3.4.
[5] A. Akturk et al., IEEE TED, vol. 57, no. 6, pp. 1334–1342, Jun. 2010.
[6] C. Enz et al., in 50th ESSERC, Bruges, Belgium, Sep. 2024.
[7] S. K. Singh et al., IEEE EDL, vol. 43, no. 5, pp. 689–692, May 2022.
[8] S. S. Parihar et al., in 7th EDTM, Seoul, Korea, Mar. 2023, pp. 1–3.
[9] Y. Zhang et al., IEEE TED, vol. 68, no. 9, pp. 4267–4273, Sep. 2021.
[10] R. M. Incandela et al., in ESSDERC, Leuven, Belgium, Sep. 2017, pp. 58–61.
[11] C. Enz and H.-C. Han, in ISCAS, Monterey, CA, USA, May 2023.
[12] H.-C. Han et al., IEEE OJCAS, vol. 3, pp. 162–167, 2022.
[13] C. C. Enz and E. A. Vittoz. Wiley, 2006.
[14] A. Mangla et al., in 18th MIXDES, Jun. 2011, pp. 85–89.
[15] C. Enz et al., IEEE SSC-M, vol. 9, no. 3, pp. 26–35, 2017.
[16] C. Enz et al., in ISCAS, Singapore, May 2024.
[17] Z.-H. Liu et al., IEEE TED, vol. 40, no. 1, pp. 86–95, 1993.
[18] N. Arora, ser. Computational Microelectronics. Springer, 1993.
[19] S. Takagi et al., IEEE TED, vol. 41, no. 12, pp. 2357–2362, 1994.
[20] F. Serra Di Santa Maria et al., SSE, vol. 186, p. 108175, 2021.
[21] A. Beckers et al., IEEE EDL, vol. 41, no. 2, pp. 276–279, 2020.
[22] H.-C. Han et al., IEEE EDL, vol. 45, no. 1, pp. 92–95, Jan. 2024.
[23] H.-C. Han, Ph.D. dissertation, EPFL, Lausanne, Switzerland, 2024.
[24] H.-C. Han. (2024) SEKV MOSFET Compact Model. [Online]. Available: https://gitlab.com/quanexum/modeling/sekv-verilog-a
[25] H.-C. Han et al., IEEE Access, vol. 11, pp. 56951–56957, 2023.
[26] A. Cros et al., in IEDM, Dec. 2006.

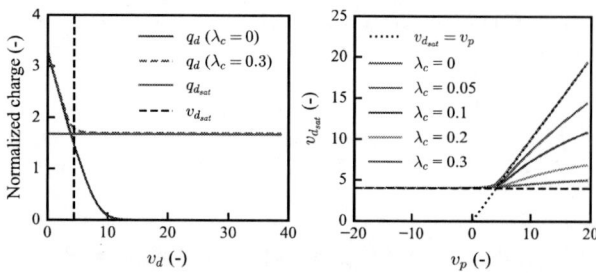

Fig. 1: Velocity saturation setting the lower boundary of q_d and impacting on $v_{d_{sat}}$. The stronger VS yields higher $q_{d_{sat}}$ and lower $v_{d_{sat}}$.

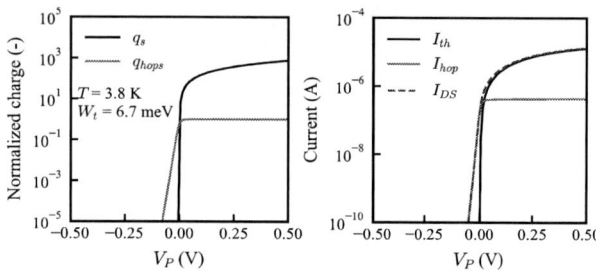

Fig. 2: Hopping transport leading to SS saturation at CT. The hopping charge exponentially scales with W_t rather than kT, which governs the subthreshold current at low temperatures.

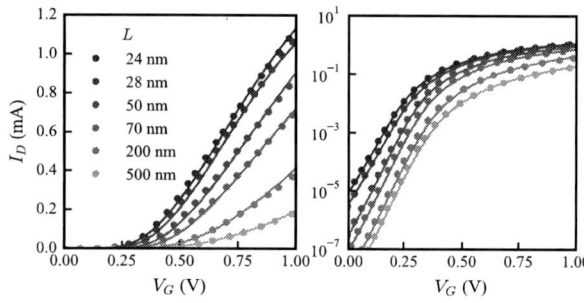

(a) I_D-V_G at 300 K and $V_{DS} = 0.8$ V, left: linear scale, right: log scale.

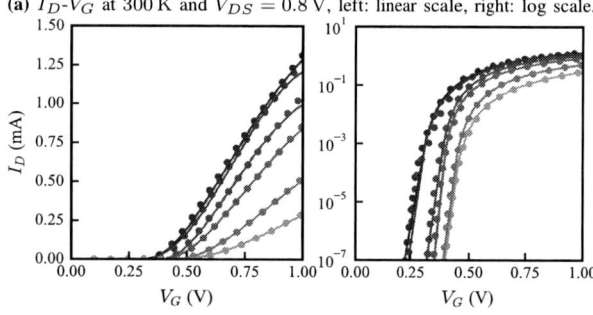

(b) I_D-V_G at 3.8 K and $V_{DS} = 0.8$ V, left: linear scale, right: log scale

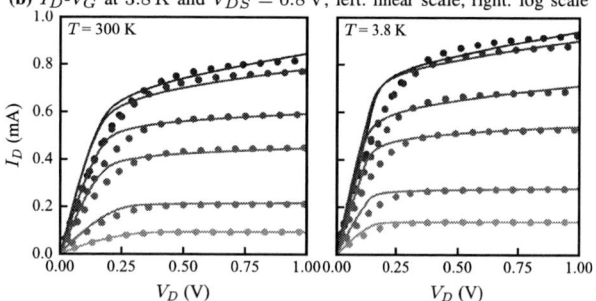

(c) I_D-V_D at $V_G = 0.8$ V, left: 300 K, right: 3.8 K

Fig. 3: Model (lines) comparing to DC measurement (markers) for FDSOI devices with various channel lengths and at 300 K and 3.8 K.

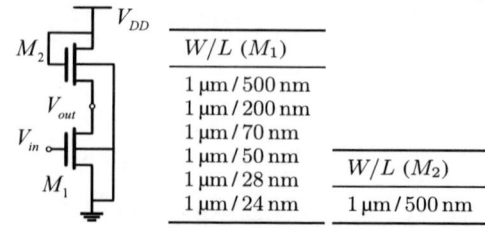

Fig. 4: Schematic of a common-source amplifier with diode-connected load and list of simulated transistor dimensions. M_1 and M_2 are nMOS.

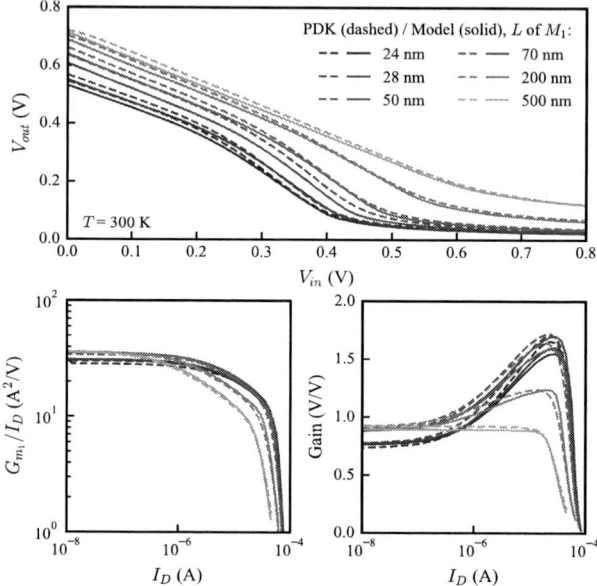

Fig. 5: Comparison between PDK and proposed model for common-source amplifiers in Fig. 4 at room temperature. The slight discrepancy between PDK and model may arise from the variability of the process.

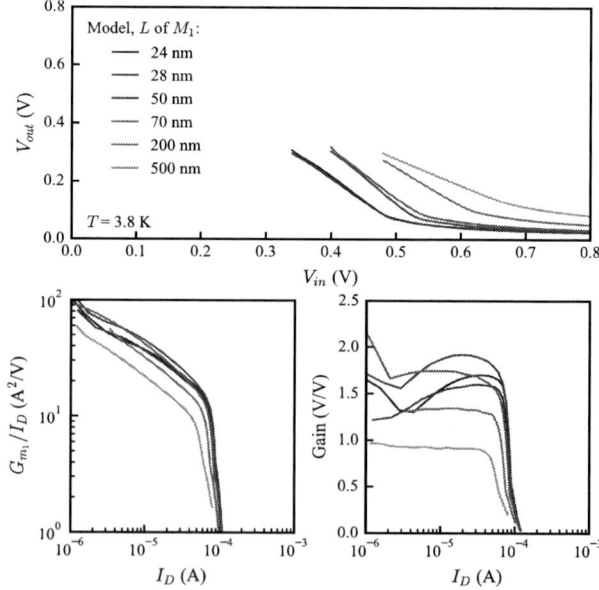

Fig. 6: Simulation of common-source amplifiers with various channel lengths for M_1 at 3.8 K. The simulation does not converge for M_1 operating in deep weak inversion, and hence, V_{in} can not be swept down to 0 V.

An RRAM-based Multi-Mode and Pipelined Pooling Scheme in Computing-in-Memory for Convolutional Neural Networks

Yi Gao[1], Zongwei Wang[1,2*], Lin Bao[1], Yimao Cai[1,2*], and Ru Huang[1,2]

[1]School of Integrated Circuits, Peking University, Beijing, China
[2]Beijing Advanced Innovation Center for Integrated Circuits, Beijing, China
*Email: {wangzongwei, caiyimao}@pku.edu.cn

Abstract

RRAM-based computing-in-memory acceleration for convolutional neural networks (CNNs) has shown promising advancements across various fields. In this work, to further accelerate the pooling operations between convolution layers, we propose an RRAM-based pooling scheme for CIM that enables in-suit pooling without extra data transfer and supports flexible switching between multi-mode (max, sum, and mean) pooling functions. We achieve pipelined operation by integrating RRAM arrays and ADCs within the CIM system. Validation using the VGG-11 architecture demonstrates a significant reduction of 10.4× in latency and 9.9× in energy consumption with negligible accuracy loss (0.04%@6 bit).

Keywords: RRAM, Computing-in-memory, Pooling, Convolutional neural networks.

Introduction

Convolutional neural networks (CNNs), such as VGG [1], play a significant role in various fields, including image recognition. CNN architectures typically consist of convolution layers, pooling layers, and fully connected layers. Resistive random-access memory (RRAM)-based computing-in-memory (CIM) can significantly enhance the latency and efficiency of convolution operations in CNNs by accelerating matrix-vector multiplication (MVM) within a cross-bar structure [2]. Many CNN architectures, such as VGG-11, rely on essential pooling operations between convolution layers to extract abstract features, as shown in Fig. 1. However, existing CIM solutions provide limited discussions on pooling solutions and predominantly depend on CMOS pooling designs [3,4], as shown in Fig. 2. The mass outputs from the convolution layers must be buffered and transferred to the CMOS pooling unit with additional hardware overhead from comparator circuits for further processing, resulting in significant data movements and latency.

Moreover, as MVM efficiency improves through CIM acceleration, the energy consumption of convolutional layers decreases, thereby amplifying the impact of pooling layers on overall latency and power consumption (Fig. 3). Although some designs have proposed dedicated pooling circuits within CIM chips, these typically support limited operations [5,6] and encounter issues with array area matching and pooling hardware reuse. Additionally, existing max-pooling schemes rely on prior comparison results, limiting the computational parallelism potential of CIM systems.

To address these limitations, we propose a multi-mode pooling scheme based on the current-driven characteristic of RRAM devices, which contains the following advantages:

1. Implementation of pooling operations within the RRAM array, eliminating frequent data transfers and reducing hardware costs;

2. Generalized pooling schemes that require only simple timing control to switch between different modes (max, sum, and mean pooling) without restrictions on input window size, enabling adaptive pooling to meet the requirements of various CNN architectures;

3. Seamless integration of pipelined operations with convolution processing and ADC sampling.

Using the RRAM chip test system, we validated the functionality of the RRAM-pooling module on CIFAR-10 recognition tasks by VGG-11, demonstrating significant potential in CIM acceleration for CNNs.

Current-Driven Characteristics for Pooling Application

The resistance of RRAM can be easily modulated under stress. Given that most RRAM computing-in-memory (CIM) systems output via current, we investigated the current memory characteristics of RRAM device to develop pooling modules. We designed a test structure for the pooling module as shown in Fig. 4, where RRAM devices R_1, R_2, and R_3 are configured to represent different resistance states. By applying a read voltage to the bit-line (BL) and a clamp voltage to the source-line (SL), we measured varying currents that simulate the computation results during CIM operations. Similarly, during computations in the RRAM CIM array, it is essential to apply a clamping voltage on the SL to ensure stable voltage division between the CIM array and the pooling device.

During the pooling period, the pooling-word-line (PWL) is activated, enabling the analog current I_{in} to flow through the pooling device and complete the sampling process. Subsequently, the RRAM array is turned off to initiate the read process, where a small read voltage is applied to the read-bit-line (RBL) to obtain the read current I_{out}. After completing the pooling and read operations, the pooling device is reset, and this process is repeated to gather statistical relationships between I_{in} and I_{out}. As shown in Fig. 5(a), the RRAM device exhibits strong current-driven characteristics within specific input ranges, which can be reliably reproduced through simple read operations.

Furthermore, in a single pooling cycle, we sequentially activated WL_1, WL_2, and WL_3 to input currents I_n, followed by reading the corresponding output currents. After reset operations, we varied the order of I_{in} to generate multiple sets of I_{out}. As illustrated in Fig. 5(b), multiple input sequences at currents of 90 µA, 110 µA, 130 µA, and 150 µA were analyzed. The results indicate that, despite variations in the order of input current, the output current depends solely on the maximum input current, reinforcing the max-pooling resistive memory characteristics of our RRAM devices. We conclude that:

$$I_{read} = \alpha * max\ (I_1, I_2, \cdots, I_n)$$

where α is determined by the fitting line in Fig. 4(a). Taking the trans-impedance amplifier as an example, the gain factor α can be adjusted by changing the R_F during ADC sampling (Fig. 6(b)). From the perspective of the intrinsic physical mechanism of RRAM devices, the thickness of the conductance filament is positively correlated with the input current magnitude. As shown in Fig. 7, when a larger current I_1 flows through the pooling device, a thicker conductance filament is formed. Conversely, with a smaller stress of current I_2, the existing conductance filament sufficiently accommodates the current, resulting in a negligible impact on the RRAM resistance, as indicated by the test results in Fig. 5(b). Ultimately, the RRAM pooling device retains the conductance filament formed by the maximum current, which can be reliably reproduced through the read process.

Multi-Mode RRAM-Pooling Circuits

Based on the previous discussion, we further develop the RRAM-based pooling circuit and CIM architecture as shown in Fig. 6(a). The RRAM-pooling module can perform various functions such as max-pooling, sum-pooling, and mean-pooling through simple timing control.

A. Max-pooling Scheme

As shown in the timing diagram of Fig. 8(a), during the pooling stage, the ADC sensing switch connected to the CIM array is first turned off while the pooling-word-line (PWL) is activated to engage the pooling device. WL_1 to WL_3 in the pooling array are sequentially activated, allowing currents from the CIM array to flow through the SL into the pooling device, which records the maximum current. During the reading phase, the input signal from the CIM array is turned off, and the read path is activated to transfer the recorded results from the pooling devices to the ADC. Specific quantization relationships are applied to the read currents during ADC sampling to obtain the final max-pooling results.

B. Sum-pooling Scheme

For the sum-pooling scheme (Fig. 8(b)), multiple WLs are activated simultaneously within a single pooling cycle, enabling convolution results from different SLs to accumulate on the RBL, as indicated by the red line in Fig. 6(a). After stabilization, the PWL is activated, and the sum-pooling results are recorded in the pooling devices.

On the one hand, compared to current accumulations on the SL of the RRAM CIM array, our solution enables sum-pooling across different columns without complex connections, enhancing compatibility with the computation and output processes of the CIM array. On the other hand, due to line width limitations, activating too many BLs can cause the current on a single SL to exceed the critical threshold. Our approach increases the current tolerance of the RBL, enabling sum-pooling for more SLs and larger pooling kernels with minimal area overhead. Moreover, currents can be automatically accumulated on the SL before being fed into the sum-pooling device, further improving pooling efficiency.

C. Mean-pooling Scheme

Similar to the sum-pooling operation, we obtain a sum result I_{sum} after pooling and a count of pooling data N_{count} during the pooling period. This count is used to configure the ADC sampling ratio (Fig. 6(b)). After conversion by the ADC, the resulting voltage value corresponding to I_{sum}/N_{count} represents the mean-pooling result.

Pipelined Pooling Design for Convolutional CIM

Convolutional calculations in the CIM array are typically performed in a row-wise and column-wise manner, but the pooling windows are usually sized at 2×2 or 3×3 (Fig. 9&10). Therefore, due to the different processing order requirements, traditional max-pooling schemes require multiple reads of the same data across different windows, resulting in significant data movements and time overhead.

The proposed pipelined design based on max-pooling scheme, as shown in Fig.11(a) and (b), allows for sequential operations within the convolution CIM array. During the convolution calculation in a row-wise and column-wise manner, when convolution devices overlap with the corresponding pooling kernels, the associated PWLs are activated to realize parallel pooling. After a single traversal of the entire convolution CIM array, all pooling records can be obtained simultaneously, requiring only read operations to achieve max-pooling results. This approach significantly reduces computation time and data access volume.

Furthermore, when pooling is not required after a convolution layer, all RWLs are turned off, allowing currents to flow directly to the ADC for quantization and subsequent computation in the next layer. The pooling module operates only when necessary, ensuring complete physical and functional compatibility with CIM arrays and ADCs for seamless integration.

Implementation of RRAM-based M²P² Scheme

An 1M RRAM chip test system (Fig. 12) is employed to assess the performance of the RRAM-based multi-mode and pipelined pooling (M²P²) scheme. We pre-trained a VGG-11 network for CIFAR-10 image recognition tasks, with weight quantization applied for RRAM-based CIM implementation. A convolution layer with a max-pooling layer was extracted from VGG-11, with the corresponding weights mapped into the 1M RRAM chip and pooling module [7]. As shown in Fig. 13, experimental results indicate that the RRAM-pooling module achieved nearly 100% accuracy in max value extraction, with minimal loss in network recognition precision (Fig. 14). Furthermore, as shown in Fig. 15, we evaluated the latency and energy consumption improvements of RRAM-based CIM with the M²P² module. The results demonstrate a promising potential for RRAM-based CIM systems in CNN network applications.

Conclusion

In this work, we proposed an in-suit RRAM-based pooling module and M²P² scheme for CNN networks acceleration, which can flexibly execute max-pooling, sum-pooling and mean-pooling through timing control adjustments. The RRAM-based pooling scheme enables pipelined operations with convolution computation results and seamless hardware integration with the RRAM-based CIM system. With negligible loss in accuracy, our solution achieved a significant reduction of 10.4× in latency and 9.9× in energy consumption.

Acknowledgement

This work was supported by the National Natural Science Foundation of China (62341407, 62025401, 62322401, 61927901), Beijing Nova Program (20220484113), and in part by "111" Project (B18001).

References

[1] Simonyan, K et al., *ICLR*, 2015, arXiv:1409.1556.
[2] Z. Wang et al., *IEEE TED*, 2020, 67(10): 4166-4171.
[3] Chi, P et al., *ACM SIGARCH Computer Architecture News*, 2016, 44(3): 27-39.
[4] D. Chen et al., *TCAS-II*, 2022, vol. 70, pp. 276-280.
[5] A. Dorzhigulov et al., *ISCAS*, 2022, pp. 1828-1832.
[6] Y. Ling et al., *EDTM*, 2021, pp. 1-3.
[7] J. Sun et al., *IRPS*, 2024, pp. 1-4.

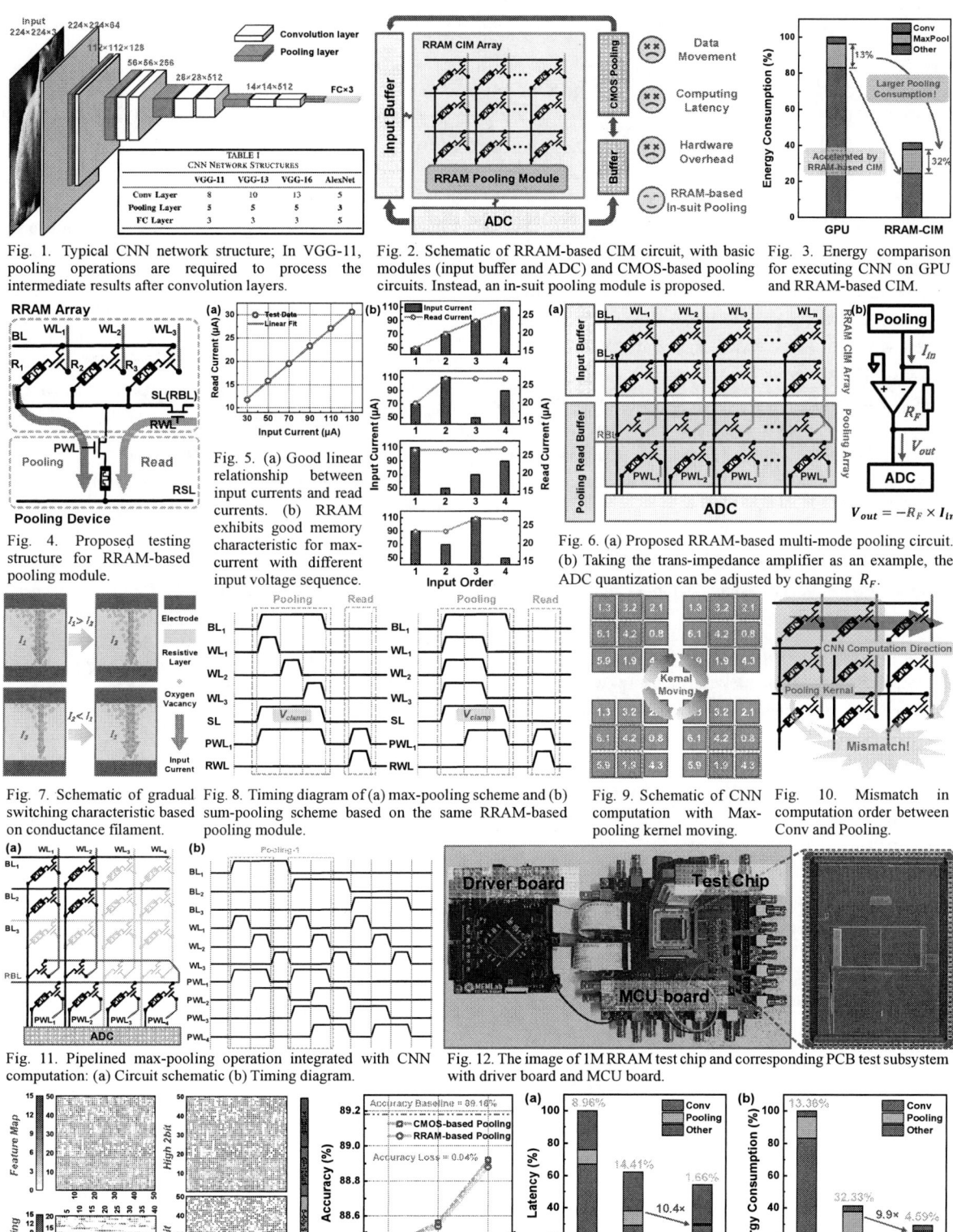

Fig. 1. Typical CNN network structure; In VGG-11, pooling operations are required to process the intermediate results after convolution layers.

Fig. 2. Schematic of RRAM-based CIM circuit, with basic modules (input buffer and ADC) and CMOS-based pooling circuits. Instead, an in-suit pooling module is proposed.

Fig. 3. Energy comparison for executing CNN on GPU and RRAM-based CIM.

Fig. 4. Proposed testing structure for RRAM-based pooling module.

Fig. 5. (a) Good linear relationship between input currents and read currents. (b) RRAM exhibits good memory characteristic for max-current with different input voltage sequence.

Fig. 6. (a) Proposed RRAM-based multi-mode pooling circuit. (b) Taking the trans-impedance amplifier as an example, the ADC quantization can be adjusted by changing R_F.

$$V_{out} = -R_F \times I_{in}$$

Fig. 7. Schematic of gradual switching characteristic based on conductance filament.

Fig. 8. Timing diagram of (a) max-pooling scheme and (b) sum-pooling scheme based on the same RRAM-based pooling module.

Fig. 9. Schematic of CNN computation with Max-pooling kernel moving.

Fig. 10. Mismatch in computation order between Conv and Pooling.

Fig. 11. Pipelined max-pooling operation integrated with CNN computation: (a) Circuit schematic (b) Timing diagram.

Fig. 12. The image of 1M RRAM test chip and corresponding PCB test subsystem with driver board and MCU board.

Fig. 13. 4-bit weight map of pre-trained VGG-11 and experimental RRAM conductance, with input and output data from max-pooling layer.

Fig. 14. Simulations of VGG-11 recognition accuracy using extracted test data.

Fig. 15. (a)Latency and (b)energy consumption improvements of RRAM-based CIM system with the RRAM-pooling module compared to CMOS-based solution.

979-8-3315-0417-5/25 $31.00 © 2025 IEEE

Temperature Dependent Back-hopping in Spin Transfer Torque Switching of Perpendicular Magnetic Tunnel Junctions

Yapeng Zhao[1], Tiaoyang Li[1*], Yanqing Wu[1,2]

[1]Fuzhou University-Jinjiang Joint Institute of Microelectronics and School of Physics, Information Engineering and Microelectronics, Fuzhou University, Fuzhou 350108, China. *Email:tyl@fzu.edu.cn
[2]School of Integrated Circuits and Beijing Advanced Innovation Center for Integrated Circuits, Peking University, Beijing 100871, China

Abstract

This work systematically investigated the magnetoresistance switching behavior and back-hopping mechanisms of p-MTJs at high temperatures (up to 360 K). Experimental results show that the TMR ratio and the switching voltage (V_C) of p-MTJs decrease with increasing temperature due to reduced magnetic anisotropy. The back-hopping phenomenon, characterized by random resistance state fluctuations, occurs with increasing pulse voltage. Systematic analysis reveals that back-hopping results from weakening antiferromagnetic coupling under high-voltage pulse excitation, leading to flipping of the reference layer. As temperature rises, the switching voltage associated with back-hopping decreases.

Keywords: Perpendicular magnetic tunnel junctions, Spin transfer torque, Back-hopping

Introduction

Perpendicular magnetic tunnel junctions (p-MTJs) with a MgO/CoFeB double-interface free layer structure exhibit excellent performance, including high tunneling magnetoresistance (TMR) ratio and good thermal stability [1]. Previous theoretical and practical studies have shown that the TMR ratio and STT switching characteristics are greatly affected by temperature [2]. As the applied voltage increases, random oscillations of the resistance state will occur, called the back-hopping phenomenon [3]. This phenomenon may be caused by the decrease of magnetic anisotropy under high voltage excitation or the instability of the RL [4]. This paper studied the dependence of the TMR ratio and STT switching characteristics of p-MTJ on temperature. We also found that the back-hopping phenomenon originates from the RL flip caused by the weakening of antiferromagnetic coupling under high-voltage pulse excitation. Then, the dependence of the back-hopping phenomenon on temperature is explored.

Device Fabrication

As illustrated in Fig. 1(a), the p-MTJ stack can be divided into three distinct layers: hard layer (HL; [Pt/Co]$_6$/Co), double MgO interface-free layer (FL; MgO/CoFeB/Ta/CoFeB/MgO), and reference layer (RL; Co/[CoPt]$_5$/Ta/CoFeB), where RL and HL form a synthetic antiferromagnetic (SAF) coupling through Ru. As illustrated in Fig. 1(b), the fabrication of the p-MTJs involved electron beam lithography (EBL) and

multistep Ar ion beam etching (IBE). After the etching process, the SiO$_2$ passivation layer was deposited through e-beam evaporation (EBE). Finally, Ti/Au top electrodes were fabricated on the p-MTJs. Fig. 1(c) shows the optical microscopy image of the final device structure. Fig. 1(d) presents the scanning electron microscope (SEM) image of p-MTJ after the IBE process, which reveals that the diameter of the p-MTJ is about 150 nm.

Results and Discussion

Fig. 2(a) shows the roof-shaped curves of p-MTJ at different temperatures from 300 K to 360 K. Fig. 2(b) shows the temperature dependence of parallel and antiparallel resistance (R_P/R_{AP}) and TMR ratio, where the R_{AP} value decreases by about 9% from 300 K to 360 K. This is because thermomagnetic disorder causes a significant decrease in R_{AP}. In contrast, the parallel resistance is mainly generated by tunneling electrons [5], so R_P is relatively stable at different temperatures.

As shown in Fig. 2(c), the temperature dependence of the magnetic switching voltage (V_C) was determined, and V_C showed a linear dependence with increasing temperature. According to the fitted linear curve of V_C dependence on temperature, the $V_c^{P(AP)}$ of p-MTJ is $-0.36/+0.27$ V at 300 K. We expresses the magnetic switching voltage (V_C) as a function of temperature [8] as follows:

$$V_C^{P(AP)} = V_{C0}^{P(AP)} \left[1 - \frac{k_B T}{E_b} ln \left(\frac{t_p}{\tau_0} \right) \right] \qquad (1)$$

Where $V_C^{P(AP)}$, $V_{C0}^{P(AP)}$, and E_b are the parameters of the initial magnetization configuration before the magnetization reversal occurs, and P and AP represent parallel and antiparallel, respectively. V_{C0} is the intrinsic critical magnetic switching voltage, t_p is the duration of the applied voltage pulse, and τ_0 is the inverse of attempt frequency (assumed to be 1 ns). On fitting the experimental data with equation (1), we got $V_{C0}^{P(AP)} = -0.72/+0.81$ V, this imbalance could be induced by excessive etching and magnetic damage to the RL [6].

To investigate the STT characteristics of p-MTJ under pulse measurements, as shown in Fig. 3(a), we applied a ramp pulse voltage waveform sweep to determine the switching of p-MTJ. Fig. 3(b) shows the switching of p-MTJ under voltage pulses. The p-MTJ state completes the R_{AP} to R_P switching at 0.35 V. However, after the STT effect-based R_P to R_{AP} switching occurred at −0.45 V, the resistance state of the device oscillated from −0.71 V to −0.78 V, which can be attributed to the back-hopping phenomenon [7].

As shown in Fig. 3(c), State 1 is the initial magnetization configuration of each functional layer of the p-MTJ. The resistance state switching from state 1 to state 3 is all due to the stable magnetization reversal of FL under the STT effect. With the further increase of the negative pulse voltage, the STT effect from FL is more significant than the antiferromagnetic coupling between RL and HL, causing the magnetization of RL to flip. Since electrons flow downward from FL to RL at this time, the magnetization of FL will also flip immediately. Meanwhile, the antiferromagnetic coupling between RL and HL is broken, and their magnetization directions are parallel. In the p-MTJ resistance reading state, because the reading voltage is not enough to affect the antiferromagnetic coupling, the magnetization directions of RL and HL will return to antiparallel in the reading state, and the resistance state at this time is R_P in state 4. As shown in Fig. 3(b), the back-hopping phenomenon does not occur under positive voltage sweeping, because under the magnetization configuration of state 2, the stray field from FL will inhibit the reversal of RL [8].

To further explore the dependence of the back-hopping phenomenon on temperature and to avoid the impact of excessive voltage on device durability, the characteristics of the p-MTJ were tested in the range of −0.9 V to 0.5 V from 300 K to 360 K. As shown in Fig. 4(a), The p-MTJ at all temperatures exhibits the back-hopping phenomenon. Fig. 4(b) shows that the critical voltage corresponding to the back-hopping phenomenon exhibits a linear dependence on temperature. As the temperature rises, the critical back-hopping voltage $V_C^{(back\text{-}hopping)}$ gradually decreases. We expresses the $V_C^{(back\text{-}hopping)}$ as a function of temperature as follows:

$$V_C^{(back\text{-}hopping)} = V_{C0}^{(back\text{-}hopping)} \left[1 - \frac{k_B T}{E_b} ln\left(\frac{t_p}{\tau_0}\right) \right] \qquad (2)$$

According to the fitted linear curve of $V_C^{(back\text{-}hopping)}$ dependence on temperature, the $V_C^{(back\text{-}hopping)}$ of the p-MTJ at 300 K is −0.7 V. On fitting the experimental data with equation (2), we got intrinsic critical back-hopping switching voltage $V_{C0}^{(back\text{-}hopping)}$ = −0.87 V. As shown in Fig. 4(c), the p-MTJ was measured with 200 consecutive pulse sequences of 200 ns pulse width at 300 K, where the voltage was −0.68 V in the back-hopping range. The resistance state of the p-MTJ showed prominent random fluctuation characteristics. This phenomenon can be applied to random number generators, neural networks, and probabilistic bits.

Conclusion

This work systematically investigated the magnetoresistance switching behavior and back-hopping mechanisms of p-MTJs at high temperatures (up to 360 K). We have systematically analyzed the magnetoresistance switching behavior, mainly focusing on the back-hopping phenomenon. Our findings reveal that the back-hopping phenomenon is attributed to the reference layer flipping due to weakened antiferromagnetic coupling under high-voltage pulse excitation, which is significantly influenced by temperature. The random fluctuation characteristics of the back-hopping phenomenon can be further applied to random number generators, neural networks, and probabilistic bits.

Acknowledgments

This work was supported by the National Science and Technology Major Project of China (Grant No. 2020AAA0109005), the National Natural Science Foundation of China (Grant No. 62204042), and the Science and Technology Major Project of Fujian Province, China (Grant No. 2021HZ021027).

References

[1] H. Sato, M. Yamanouchi, S. Ikeda, S. Fukami, F. Matsukura, and H. Ohno, "Perpendicular-anisotropy CoFeB-MgO magnetic tunnel junctions with a MgO/CoFeB/Ta/CoFeB/MgO recording structure," Appl. Phys. Lett., 101, 022414 (2012).

[2] L. Yuan, S. H. Liou, and D. Wang, "Temperature dependence of magnetoresistance in magnetic tunnel junctions with different free layer structures," Phys. Rev. B, 73, 134403 (2006).

[3] T. Min, J. Z. Sun, R. Beach, D. Tang, and P. Wang, "Back-hopping after spin torque transfer induced magnetization switching in magnetic tunneling junction cells," J. Appl. Phys., 105, 07D126 (2009).

[4] W. Kim, S. Couet, J. Swerts, T. Lin, Y. Tomczak, L, and A. Furnemont, "Experimental Observation of Back-hopping With Reference Layer Flipping by High-Voltage Pulse in Perpendicular Magnetic Tunnel Junctions," IEEE Trans. Magn., 52, pp. 1-4 (2016).

[5] H. Sato, M. Yamanouchi, S. Ikeda, S. Fukami, F. Matsukura, and H. Ohno, "MgO/CoFeB/Ta/CoFeB/MgO Recording Structure in Magnetic Tunnel Junctions With Perpendicular Easy Axis," IEEE Trans. Magn., 49, pp. 4437-4440 (2013).

[6] Y. H. Wang, S. H. Huang, D. Y. Wang, K. H. Shen, C. W. Chien, K. M. Kuo, S. Y. Yang, and D. L. Deng, "Impact of stray field on the switching properties of perpendicular MTJ for scaled MRAM", IEEE International Electron Devices Meeting (IEDM), pp. 29.2.1-29.2.4, 2012.

[7] T. Devolder, O. Bultynck, P. Bouquin, V. D. Nguyen, S. Rao, D. Wan, B. Sorée, I. P. Radu, G. S. Kar, and S. Couet, "Back hopping in spin transfer torque switching of perpendicularly magnetized tunnel junctions," Phys. Rev. B, 102, 184406 (2020).

[8] C. Yoshida, T. Tanaka, T. Ataka, and A. Furuya, "Micromagnetic Study of Edge-Damage Effects in Perpendicular CoFeB/MgO Magnetic Tunnel Junction," IEEE Trans. Magn., 55, pp. 1-5 (2019).

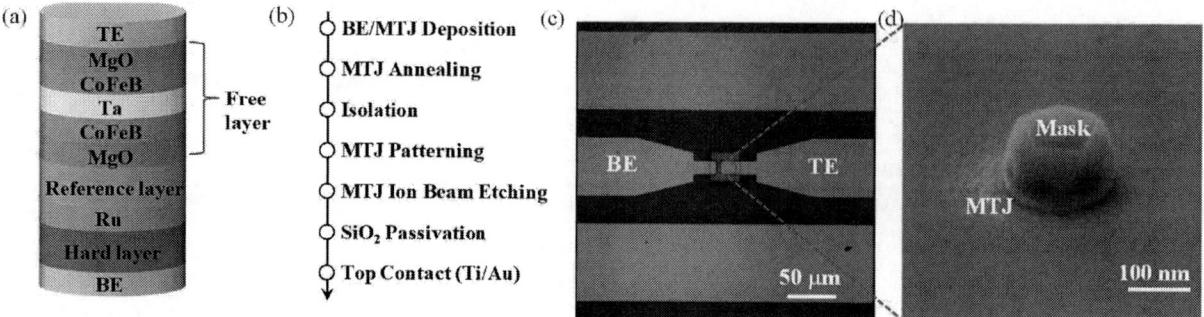

Fig. 1 (a) The double-interface free layer p-MTJ stack structure. (b) Critical fabrication process flow of the p-MTJ device. (c) Optical top view of the fabricated device. (d) Scanning electron microscope (SEM) image of p-MTJ after ion beam etch (IBE) process with photoresist as an etching mask.

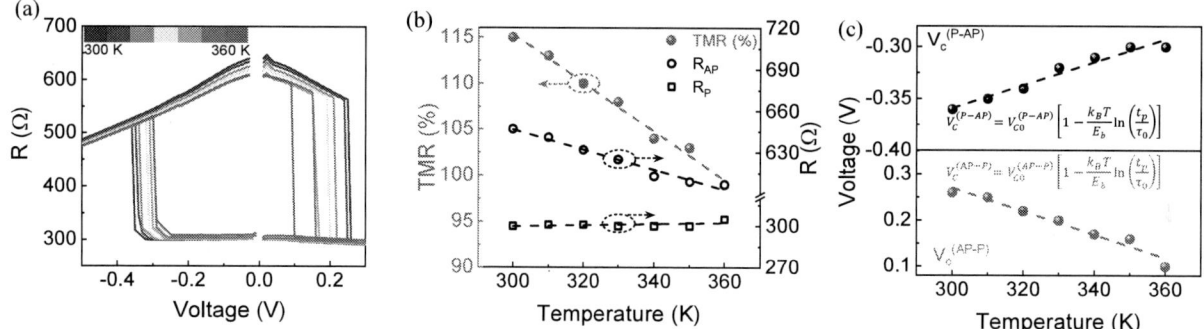

Fig. 2 (a) Resistance versus applied voltage (R-V) curves, measured at various temperatures. (b) Temperature dependence of antiparallel/parallel state resistance (R_{AP}/R_P) and tunnel magnetoresistance (TMR) ratio of p-MTJ devices, along with the theoretical fits. The magnetoresistance switching voltage V_c for the AP to P (P to AP) direction as a function of temperatures.

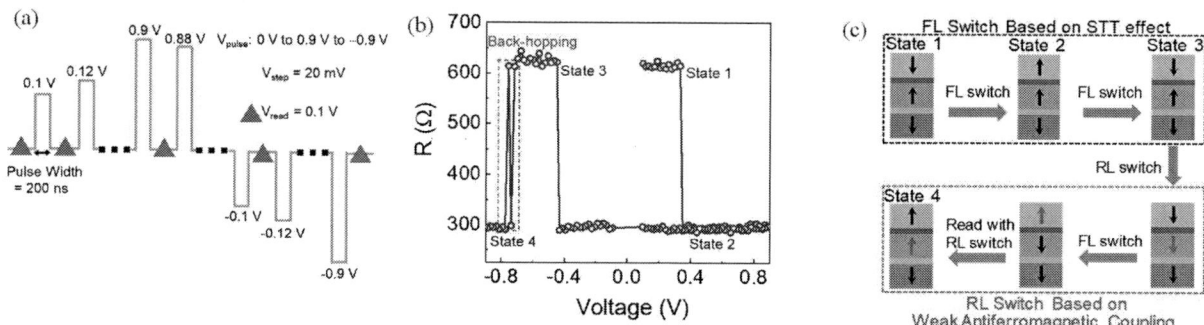

Fig. 3 (a) Staircase voltage waveform sweep. (b) Experimental measurement of the STT switching behavior of p-MTJ, showing the back-hopping phenomenon of resistance fluctuation under negative voltage sweep. (c) Schematic for magnetization configuration in p-MTJ correlated with state 1 to state 4 in (b).

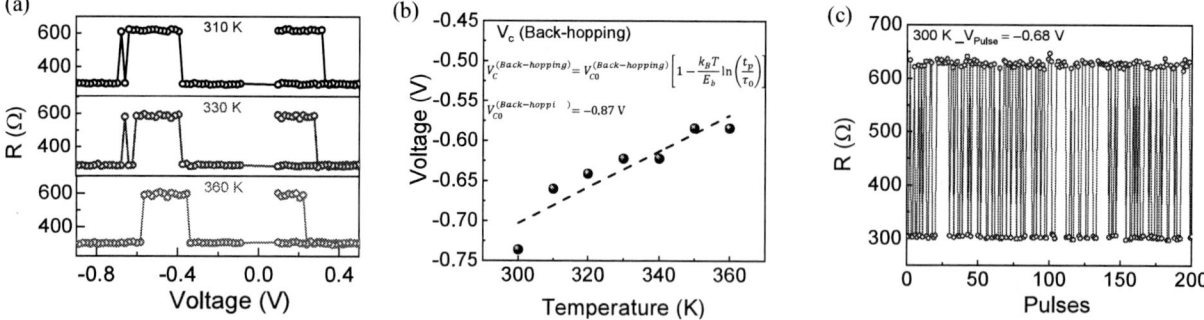

Fig. 4 (a) Experimental measurement of the STT switching behavior of a p-MTJ device at different temperatures, where the back-hopping phenomenon is observed at each temperature. (b) The dependence of the critical voltage corresponding to the back-hopping phenomenon on temperature. (c) The p-MTJ was measured with 200 consecutive pulse sequences of 200 ns pulse width at 300 K, where the voltage value was −0.68 V in the back-hopping range. The resistance state of the p-MTJ showed prominent random fluctuation characteristics.

979-8-3315-0417-5/25 $31.00 © 2025 IEEE   846

Selective-Ferroelectricity-Defining Method Utilizing Directional Oxygen Plasma Treatment for Optimizations of Hf$_{0.5}$Zr$_{0.5}$O$_2$-based Memory Devices

Yi Ding[1,†], Mingcheng Shi[1,†], Yuyan Wang[1,*], Bowen Shen[1], Wen Sun[1], Benjamin Yang[1], Yanbo Su[1], Huanan Liu[1], Jian Yuan[1], Jianshi Tang[1,*], Bin Gao[1], He Qian[1], Huaqiang Wu[1]

[1]School of Integrated Circuits, Beijing Advanced Innovation Center for Integrated Circuits, Tsinghua University, Beijing, †Equal Contribution, *Email: wangyuyan@tsinghua.edu.cn; jtang@tsinghua.edu.cn

Abstract

We present a method to control the ferroelectricity in specific regions of Hf$_{0.5}$Zr$_{0.5}$O$_2$ (HZO) through directional oxygen plasma treatment. Using this method, the ferroelectricity can be selectively eliminated in the treated areas, and leakage currents up to 10 times lower than untreated areas can be achieved after rapid thermal annealing (RTA), allowing HZO in the treated regions to be used as the ideal gate dielectric. This method streamlines the fabrication process by reducing the number of steps needed to design and prototype ferroelectric devices. In this way, it can facilitate the integration of ferroelectric devices with regular access transistors and has promising potential to simplify fabrication process and reduce cost in mass productions.

Keywords: Ferroelectric Random-Access Memory (FeRAM), Oxygen Vacancies (V$_o$), Plasma Processing Technology.

Introduction

Hafnium oxide (HfO$_2$) based ferroelectric memory has drawn extensive interest for its many intriguing properties, including low operation voltage, good CMOS compatibility, and exceptional scalability. HZO can be used in FeRAM for next-generation DRAM or FeFET for both memory and in-memory computing applications. Device density is a crucial factor for these applications, making HZO ferroelectrics a promising solution. However, designing the process flow for ferroelectric devices is challenging, as high-quality HZO growth for advanced nodes is primarily achieved through atomic layer deposition (ALD). This method typically results in HZO coverage across the entire wafer surface. Unlike stable dielectrics, HZO cannot be used as the conventional gate oxide for normal transistors. As shown in Fig. 1, HZO in this context presents unwanted ferroelectricity at the gate that must be eliminated. In this study, we demonstrate an innovative and convenient method to process HZO films after ALD to confine ferroelectricity within selected regions. As shown in Fig. 2, the proposed method can greatly optimize device design and fabrication process flows.

Experimental Procedures and Results

HZO film is an ideal candidate for ferroelectric devices, but it cannot be used as a gate dielectric for regular transistors because of unfavorable properties such as large leakage [1] and unstable control. Therefore, all process flows involving ferroelectrics and gate dielectrics for MOSFETs eventually require replacing specific ferroelectric areas with normal dielectrics. Aside from the fabrication complexity, this step could compromise the quality of the interface between the channel region and the gate oxide, adversely affecting device performance. HfO$_2$ is a standard high-κ gate dielectric widely used in advanced logic and device research; zirconium oxide (ZrO$_2$) is also a standard material for fabricating DRAM capacitors [2]. Thus, HZO has a foundation to be used as a dielectric, but the key issue is to prevent the formation of ferroelectric phase during the RTA process.

The proposed treatment aims to process the HZO film immediately after ALD growth. Positive photoresist (PR) is used to mask the intended ferroelectric areas while exposing the regions where ferroelectricity is to be eliminated. With this selective-ferroelectricity defining method, the covered and exposed portions of the film can exhibit selectively designed ferroelectricity after RTA. In theory, oxygen inductively coupled plasma (ICP) provides precise control of oxygen injection, which could reduce the concentration of V$_o$ in HZO; therefore, ferroelectricity can be precisely and selectively defined via this method.

In conventional ferroelectric device fabrication, HZO is grown using ALD as previously discussed. For the metal electrodes, both sputtered titanium nitride (TiN) and tungsten (W) were tested. A 550 °C RTA for 30 s was then conducted with RTA, which induces a ferroelectricity 2P_r close to 45 μC/cm^2. All variations of stack thicknesses and electrode combinations (including TiN/HZO/TiN, W/HZO/W, and TiN/HZO/W) yield consistent results with the proposed method, thus confirming this method's universality. This experiment uses multiple ICP powers and durations to tune the HZO ferroelectricity of specific regions and even fabricate gate dielectrics with extremely low leakage. The ICP powers tested were 300 W, 400 W, and 600 W, and the treatment times tested were 2 s, 15 s, and 45 s.

To demonstrate the effectiveness of this method, the positive-up-negative-down (PUND) measurements of samples with different treatments are shown in Fig. 3(a), and an untreated sample is also illustrated as the control group. Light treatments (such as using 300 W plasma for 2 s) retain most of the ferroelectricity, while heavy treatments using 600 W plasma can cause severe damage to the sample (not shown). Ferroelectricity diminishes with increasing plasma power and duration, and is completely eliminated when using a 400 W treatment for 15 seconds. As shown in Fig. 3(b), the I-V curves show that the 15 s 400 W sample shows no obvious flipping current, suggesting no ferroelectricity. This result is also supported by the relative permittivity measurements in Fig. 3(c). Whereas the ferroelectric samples exhibit a butterfly-shaped curve with ε_r up to 62 at the corresponding coercive field, the sample after the oxygen ICP treatment has a stable dielectric constant of 26, which is relatively high within the ε_r range between HfO_2 and ZrO_2.

Besides the elimination of unwanted ferroelectricity, HZO shows excellent dielectric properties after the treatment. In Fig. 4, the 15 s 400 W treated sample shows the lowest leakage among all groups tested (a reduction of more than 10 times compared to the control group). The 45 s 400 W O_2 plasma sample shows a large leakage due to excessive O_2 plasma dose. Endurance tests were conducted under 0.5 MHz in Fig. 5. The pristine HZO sample shows outstanding endurance performance with 10^8 pulses tested and 10^{12} projected, thanks to the highly dense and conformal tungsten electrodes used. For the post-treatment HZO dielectric, there is an ideally minimal change in ε_r after 10^8 pulses, suggesting strong reliability and potential for gate dielectric use.

Underlying Mechanism Analysis

HZO theoretically has four common phases: monoclinic, tetragonal, cubic (three dielectric phases), and orthorhombic (the ferroelectric phase). After ALD growth at 250 °C, the HZO film is amorphous, and small granules tend to be monoclinic. During the RTA process, HZO becomes tetragonal as the temperature reaches 350 °C and can transform into the cubic phase at extremely high temperatures. Ferroelectricity appears during the cooling process when portions of the HZO film transform from the tetragonal phase to the orthorhombic phase.

As illustrated in Fig. 6, high-resolution transmission electron microscopy (HRTEM) and precession electron diffraction (PED) are used to characterize the crystal structure and phase distribution. The pristine HZO film is predominantly polycrystalline orthorhombic phase (O-phase), which is consistent with its strong ferroelectricity. In contrast, the treated sample retains a large portion of amorphous HZO after RTA, indicating reduced ferroelectricity. Fundamentally, the decrease in V_o concentration is the essential cause of these changes, as V_o has a significant impact on phase transition to orthorhombic [3]. To be specific, the transformation from T-phase to O-phase is hindered by the low V_o concentration. Along with the experimental results of electrical measurements shown in the previous section, the mechanism for the proposed selective-ferroelectricity-defining method is summarized in Fig. 7.

While a high O-phase concentration can lead to a large $2P_r$, repeated switching operations are likely to generate V_o, leading to reliability degradation. High V_o concentrations can exacerbate the ferroelectric imprint, fatigue, and time-dependent dielectric breakdown (TDDB) [4]. Thus, there is a tradeoff between ferroelectricity and reliability. The optimization of HZO involves a thorough consideration of the stack structure, interface characteristics, thermodynamics, and other factors, similar to the optimization of HfO_2-based resistive random-access memory (RRAM) devices. As the proposed O_2 plasma treatment can regulate V_o concentration, it may offer a new approach to optimize this tradeoff.

Technology Applications and Designs

As shown in Fig 8(a), we further demonstrate the usage of the selective-ferroelectricity-defining method in a FeRAM cell for its practicality. This method can be further applied to other ferroelectric devices such as 2T0C gain cells and ferroelectric NAND (Fig. 8(b-c)). Aside from the aforementioned material performance demonstration, this method can also efficiently simplify the fabrication process flow and accelerate the ferroelectric device design.

Previous studies have sought to control the concentration of V_o through various methods, including the use of specialized electrodes, the incorporation of interlayers, and adjustments of oxygen pulse duration during ALD [5]. However, these efforts typically optimize ferroelectricity within a limited range and apply uniformly to the entire wafer film. In contrast, our innovative method effectively and precisely regulates the HZO film's property (dielectric/ferroelectric) within selected areas using O_2 plasma after lithography, making this selective-ferroelectricity-defining method a useful knob for design technology co-optimization (DTCO).

Conclusion

To summarize, an innovative method to selectively define ferroelectricity of HZO film is demonstrated. Undesired ferroelectricity can be conveniently removed using an oxygen ICP plasma treatment, and the reliability can be further enhanced as well. This method has two advantages: simplified fabrication process and reduced leakage. This study also provides new insights into the correlation between the V_o concentration and the ferroelectric property, offering a practical method to control V_o for various applications.

Acknowledgments

This work was in part supported by the NSFC 92264201 and 62025111.

References

[1] H. Kim *et al.*, *Nanotechnology*, vol. 32, no.5, 055703, 2021.

[2] D. I. Suh *et al.*, *2023 7th IEEE Electron Devices Technology and Manufacturing Conference (EDTM)*, pp. 1-3, 2023.

[3] T. Mittmann *et al.*, *2020 IEEE International Electron Devices Meeting (IEDM)*, pp. 18.4.1-18.4.4, 2020.

[4] A. Chavan *et al.*, *2024 IEEE International Memory Workshop (IMW)*, pp. 1-4, 2024.

[5] Y. -C. Li *et al.*, *IEEE Electron Device Letters*, vol. 45, no. 5, pp. 829-832, 2024.

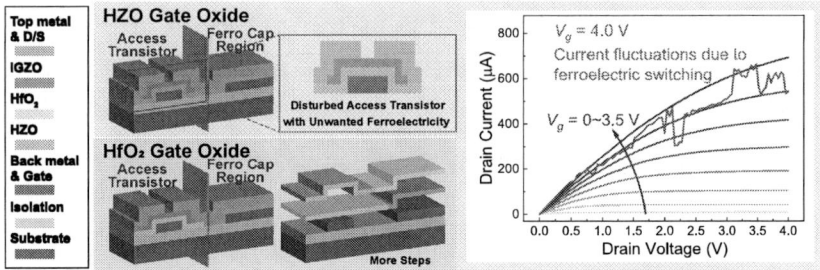

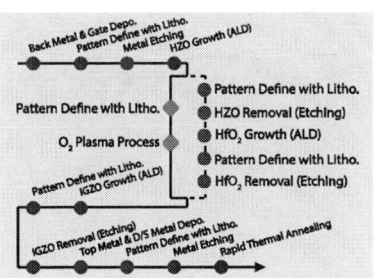

Fig. 1. Illustration of the dilemma in FeRAM fabrication caused by unwanted ferroelectricity: either suffer from unreliable transistor or incur additional fabrication costs. The I_d-V_g curve in the inset clearly shows current fluctuations due to ferroelectric switching in HZO gate dielectric, while using a normal HfO$_2$ gate dielectric would introduce extra processes, including lithography, deposition, and etching.

Fig. 2. Standard and optimized FeRAM process flows. Additional steps can be saved with the proposed selective-ferroelectricity defining method.

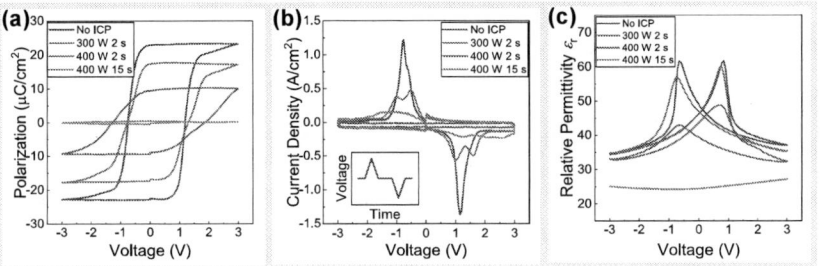

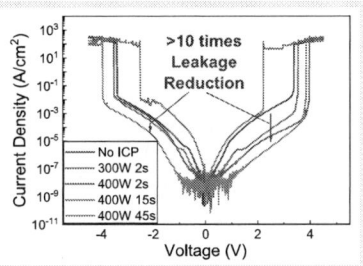

Fig. 3. (a) PUND test, (b) I-V loop, and (c) C-V loop of the differently treated samples after a wake-up process of 10^3 cycles. The $2P_r$ is reduced from 45 µC/cm^2 to less than 1 µC/cm^2 after a 400 W treatment for 15 seconds. The I-V and C-V loops of the treated samples also show capacitor-like behaviors. The samples treated with 600 W or for 45 s are not shown due to their low breakdown voltage.

Fig. 4. Leakage of samples with different oxygen plasma processing doses. A 400 W treatment for 15 seconds can reduce the leakage by more than 10 times.

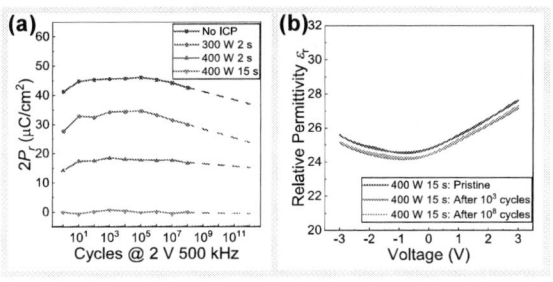

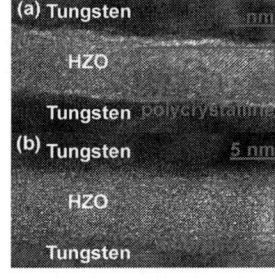

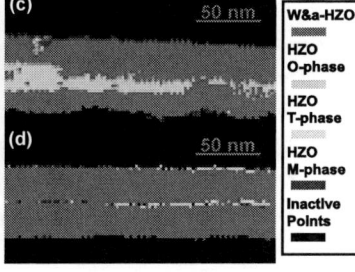

Fig. 5. Endurance tests. (a) The pristine ferroelectric sample is projected to endure beyond 10^{12} cycles, while (b) the capacitance of treated non-ferroelectric sample remains unchanged. This stability makes the latter suitable for use as the gate dielectric in normal transistors.

Fig. 6. (a, b) HRTEM images and (c, d) PED images of (a, c) pristine and (b, d) non-ferroelectric samples. More amorphous HZO (a-HZO) and less O-phase HZO can be clearly observed in the treated sample. The lower concentration of O-phase explains the absence of ferroelectricity in the treated sample.

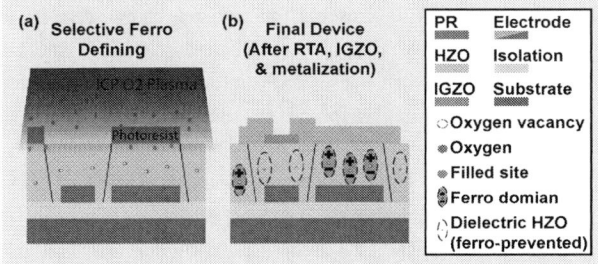

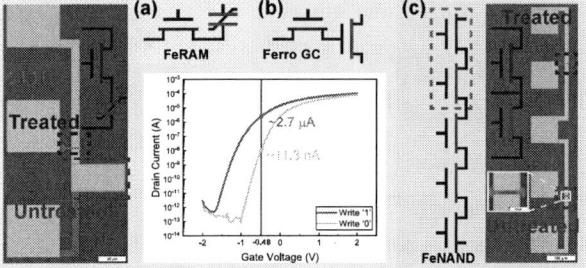

Fig. 7. Illustration of oxygen vacancy and plasma treatment. The oxygen vacancies in the target region exposed to O$_2$ plasma have been filled, reducing its ferroelectricity in the final device. The diagram is intentionally not drawn to scale for clarity.

Fig. 8. Illustrated applications of selective-ferroelectricity-defining method in (a) FeRAM, (b) 2T0C ferroelectric gain cell, and (c) ferroelectric NAND, which facilitates the integration of ferroelectrics components with regular access transistors.

979-8-3315-0417-5/25 $31.00 © 2025 IEEE

BEOL-Compatible Multi-layer ITO-ZnO-ITO Channel FETs Achieving Enhanced Mobility, Positive V_{TH} Shift, and Improved PBTI

Ying Xu[1#], Yiyuan Sun[1#], Zijie Zheng[1], and Xiao Gong[1*]

[1]Department of Electrical and Computer Engineering, National University of Singapore, 117576,

Singapore

[#]Equal contributing authors; *Phone: +65 6516-7871, E-mail: elegong@nus.edu.sg

Abstract

In this work, we present ultra-thin body bottom-gate ITO-ZnO-ITO FETs by sandwiching a thin ZnO layer between two ITO layers. Using a low thermal budget process of less than 250 °C, our long channel ITO-ZnO-ITO FETs demonstrate ~28% enhancement in field-effect mobility (μ_{eff}), improved positive bias temperature instability (PBTI), and more positive shift of the threshold voltage (V_{TH}) compared to the 2.5 nm-thick ITO FETs. We further scaled down the ITO-ZnO-ITO FET with channel length (L_{ch}) as small as 50 nm, achieving a high on-current (I_{ON}) of 652 µA/µm.

Keywords: Field-Effect Transistor, ITO-ZnO-ITO, Low Thermal Budget, PBTI

Introduction

The semiconductor industry faces significant challenges in further transistor miniaturization. A promising solution to empower Si complementary metal-oxide-semiconductor (CMOS) is the development of high-density monolithic 3D integrated circuits (M3DICs), which are considered pivotal for future semiconductor technology that will drive innovations in AI, communication, automotive, etc. [1], as depicted in Fig. 1. The success of M3DICs depends on fabricating upper-layer transistors at low temperatures (< 400 °C) to maintain the electrical performance of the underlying devices [2]. Amorphous oxide semiconductors (AOS) have emerged as a strong candidate for back-end-of-line (BEOL) compatible transistors, offering both a low thermal budget and moderate mobility [3-4]. However, achieving high performance in AOS field-effect transistors (FETs) poses a significant challenge due to trade-offs between mobility, V_{TH}, and reliability [5]. Optimizing these trade-offs is crucial to advancing AOS-based technology.

In this study, we introduce a novel ITO-ZnO-ITO multi-layer channel structure to improve both mobility and reliability, and at the same time optimize V_{TH}. Our approach achieves substantial performance improvement over the single-layer ITO devices. The ultra-thin 3 nm ITO-ZnO-ITO (1-1-1) FET, featuring a 1 nm ZnO layer sandwiched between two 1 nm ITO layers, demonstrates a high I_{ON} of 652 µA/µm and μ_{eff} of 66.1 cm²/V·s. Furthermore, we understand the improved PBTI of our ITO-ZnO-ITO FET by proposing the interaction between the hydrogen (H) and oxygen interstitials (O_i) in the ITO-ZnO-ITO stack.

Experiment

The bottom-gate multi-layer channel device was designed and fabricated as depicted in Fig. 2(a). Fig. 2(b) and (c) show the cross-section TEM image of various device layers and the top-view SEM image of device with a L_{ch} of 50 nm, respectively. Fig. 2 also summarizes the key achievements of this work.

Fig. 3 lists key process steps for realizing the ITO-ZnO-ITO FETs. The fabrication began with a substrate consisting of a 100 nm-thick SiO_2 layer on a Si substrate which was cleaned using acetone and IPA. A 45 nm-thick tungsten (W) layer was sputter-deposited as the bottom gate. Subsequently, a 10 nm HfO_2 gate dielectric was grown via atomic layer deposition (ALD) at 250 °C, using O_3 as the oxygen precursor.

Multi-layer channel structures were deposited by alternating between sputtering and ALD. The ITO layer was sputtered at room temperature, and ZnO channel layers of varying thickness were deposited using the ALD at 150 °C with H_2O as the oxygen precursor. It should be noted that, after the ITO layer in contact with HfO_2 was deposited, the devices were annealed at 200 °C in air for 10 minutes. Finally, a 35 nm-thick nickel (Ni) layer was deposited using E-beam evaporation and patterned to form the source/drain (S/D) contacts.

Results and Discussions

A. Basic Electrical Characteristics:

We first measured the devices with pure ITO films having various thicknesses from 1.5 to 3 nm. The electrical characteristics demonstrate a strong dependent on film thickness. Fig. 4(a) presents the transfer characteristics of ITO FETs with varying channel thicknesses (T_{ch}), where a notable positive shift in V_{TH} is observed as the T_{ch} is reduced. This shift can be attributed to the thickness-dependent carrier concentration and the bandgap widening caused by quantum effects [6-7]. The transconductance (G_M) of these ITO FETs, extracted at V_{DS} = 1 V, is shown in Fig. 4(b). Devices with thicker T_{ch} demonstrate higher G_M, but also exhibit more negative V_{TH}, highlighting a key trade-off when developing high-performance ITO FETs that operate in enhancement mode ($V_{TH} > 0$).

To address this challenge for high-performance AOS

979-8-3315-0417-5/25 $31.00 © 2025 IEEE

FETs, we incorporated a ZnO layer between two ITO layers to create an ITO-ZnO-ITO channel FET. Fig. 5 illustrates the transfer characteristics of both the ITO FET and the ITO-ZnO-ITO (1-1-1) FET. This ITO-ZnO-ITO structure shows obvious positive shift of V_{TH}. Fig. 6 further highlights the performance advantage of the ITO-ZnO-ITO (1-1-1) FET, showing a ~17% higher maximum transconductance ($G_{M, max}$) compared to the pure ITO FET with T_{ch} of 2.5 nm.

Fig. 7(a) and (b) display the transfer characteristics and G_M vs. V_{GS} curves for ITO-ZnO-ITO FETs with varying ZnO thicknesses. As the ZnO thickness increases from 1 nm to 3 nm, V_{TH} shifts slightly in the positive direction. However, this comes with a decrease in G_M. Fig. 8(a) provides the C-V curves of the ITO-ZnO-ITO FETs and the pure ITO FETs with a T_{ch} of 2.5 nm, measured at a frequency of 1 MHz. The μ_{eff} of these devices was extracted using the split C-V method as shown in Fig. 8(b). Fig. 9 presents the extracted peak μ_{eff} from Fig. 8(b). The ITO-ZnO-ITO (1-1-1) FETs achieve a high peak μ_{eff} of 66.1 cm^2/V·s, representing a ~28% enhancement over the pure ITO FETs with a T_{ch} of 2.5 nm, which have a peak μ_{eff} of 51.7 cm^2/V·s.

Using the ITO-ZnO-ITO (1-1-1) multi-layer structure, we further scaled the L_{ch} down to 50 nm. Fig. 10(a) presents the transfer characteristics of the ITO-ZnO-ITO (1-1-1) FETs with a L_{ch} of 50 nm at V_{DS} of 0.1 V and 1 V, demonstrating excellent transfer characteristics with a subthreshold swing (SS) of 122 mV/decade and an on/off current ratio of exceeding 8 orders of magnitude. The I_D - V_{DS} curves in Fig. 10(b) for the same ITO-ZnO-ITO (1-1-1) FET show a high I_{ON} of 652 μA/μm at V_{GS} = 3 V and V_{DS} = 2 V.

B. PBTI Characteristics:

We conducted PBTI tests on both ITO-ZnO-ITO FETs and pure ITO FETs using the Fast I-V method, which minimizes the effects of recovery. Fig. 11 illustrates the waveform of the applied V_{GS} during the PBTI measurement. During the stressing phase, the overdrive voltage (V_{OV}) was applied, defined as the difference between the gate stress voltage (V_{stress}) and V_{TH}. Fig. 12(a) presents the PBTI results for both types of FETs at the same V_{OV} of 2.5 V with stressing time (t_{str}) up to 1000 s. For pure ITO FETs, V_{TH} initially decreases significantly due to the increased carrier concentration in the channel, induced by H formation. This is followed by an increase in V_{TH}, attributed to the electron trapping effect. In contrast, the ITO-ZnO-ITO (1-1-1) FETs show only a slight shift in V_{TH}, suggesting that the excess O$_i$ in the ZnO can offset the H-induced negative V_{TH} shift, leading to improved PBTI.

Fig. 13 illustrates the interaction between the defects of H and O$_i$ in the ITO-ZnO-ITO structure for improved PBTI. In the ITO-ZnO-ITO channel, the bond dissociation energy (BDE) of the Zn-O bond (284 kJ/mol) is significantly lower than that of the In-O (360 kJ/mol) and Sn-O (548 kJ/mol) bonds [8]. This lower BDE facilitates the formation of more O$_i$. These O$_i$ can form O-O dimers with host oxygen atoms in their neutral state. However, the O-O dimer bond is easily broken upon electron capture during stressing. As the ZnO thickness and stress time increase, the impact of O$_i$ becomes more pronounced, shifting V_{TH} further in the positive direction. This effect is accelerated at a higher V_{OV}, as shown in Fig. 12(b).

Conclusion

In conclusion, we introduced an innovative ITO-ZnO-ITO channel structure that improves mobility and device reliability, while achieving a more positive V_{TH}. The ITO-ZnO-ITO (1-1-1) FETs, with a 50 nm channel length, demonstrated a high I_{ON} of 652 μA/μm. This approach shows potential for advancing emerging applications, such as high-performance 3D monolithic integrated nano-electronic systems requiring BEOL-compatible transistors.

Acknowledgments

We acknowledge funding support from Singapore Ministry of Education MOE Tier 2 (MOE-T2EP50221-0008) and MOE Tier 1 (A-8001168-00-00 and A-8002027-00-00).

References

[1] K. Dhananjay, P. Shukla, V. F. Pavlidis, A. Coskun, and E. Salman, "Monolithic 3D Integrated Circuits: Recent Trends and Future Prospects," IEEE Transactions on Circuits and Systems II: Express Briefs, vol. 68, no. 3, pp. 837-843, 2021, doi: 10.1109/tcsii.2021.3051250.

[2] S. Datta, S. Dutta, B. Grisafe, J. Smith, S. Srinivasa, and H. Ye, "Back-end-of-line compatible transistors for monolithic 3-D integration," IEEE micro, vol. 39, no. 6, pp. 8-15, 2019.

[3] K. Han, Y. Kang, X. Chen, Y. Chen, and X. Gong, "Indium-Tin-Oxide Thin-Film Transistors with High Field-Effect Mobility (129.5 cm^2/V·s) and Low Thermal Budget (150 °C)," IEEE Electron Device Letters, 2023.

[4] K. A. Aabrar et al., "Improved Reliability and Enhanced Performance in BEOL Compatible W-doped In$_2$O$_3$ Dual-Gate Transistor," in 2023 International Electron Devices Meeting (IEDM), 2023: IEEE, pp. 1-4.

[5] T. Kim et al., "Progress, challenges, and opportunities in oxide semiconductor devices: a key building block for applications ranging from display backplanes to 3D integrated semiconductor chips," Advanced Materials, vol. 35, no. 43, p. 2204663, 2023.

[6] Y. Kang, K. Han, Y. Chen, and X. Gong, "Thickness-Engineered Extremely-thin Channel High Performance ITO TFTs with Raised S/D Architecture: Record-Low R_{SD}, Highest Mobility (Sub-4 nm T_{CH} Regime), and High V_{TH} Tunability," in 2023 IEEE Symposium on VLSI Technology and Circuits (VLSI Technology and Circuits), 2023: IEEE, pp. 1-2.

[7] Y. Kang, K. Han, X. Chen, Y. Chen, and X. Gong, "BEOL-Compatible High-Performance Indium-Tin-Oxide Transistors Enabled by Quantum Confinement-Engineered Properties," IEEE Transactions on Electron Devices, vol. 71, no. 8, pp. 4692-4700, Jan. 2024, DOI: 10.1109/TED.2024.3410239.

[8] Speight, James G., ed. 2017. Lange's Handbook of Chemistry. 17th ed. New York: McGraw-Hill Education.

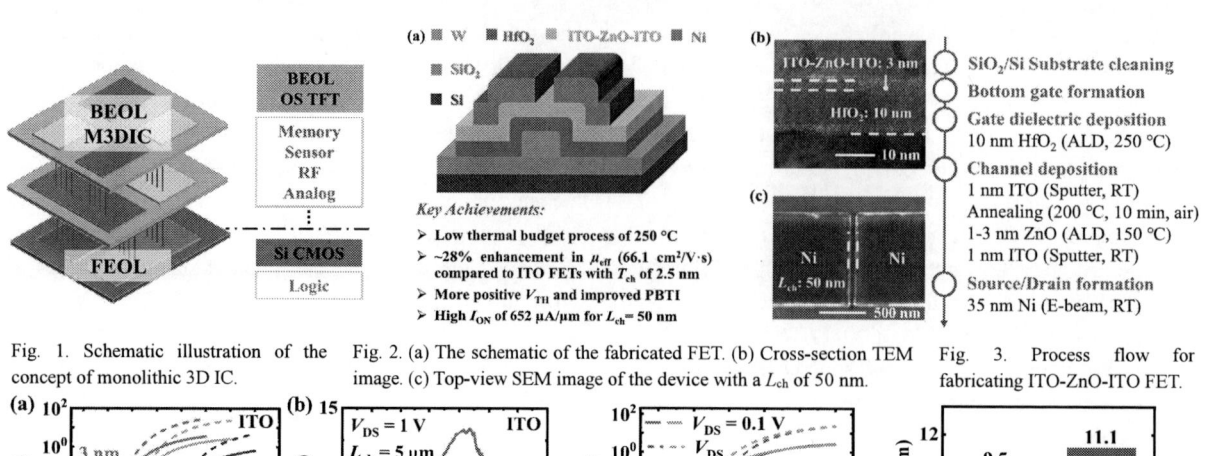

Fig. 1. Schematic illustration of the concept of monolithic 3D IC.

Fig. 2. (a) The schematic of the fabricated FET. (b) Cross-section TEM image. (c) Top-view SEM image of the device with a L_{ch} of 50 nm.

Fig. 3. Process flow for fabricating ITO-ZnO-ITO FET.

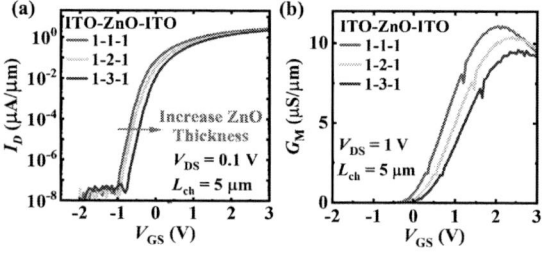

Fig. 4. (a) I_D–V_{GS} characteristics of ITO FETs with different T_{ch} at V_{DS} of 0.1 V and 1 V. (b) Extracted G_M for these ITO FETs at V_{DS} = 1 V. Devices with thicker T_{ch} demonstrate higher G_M

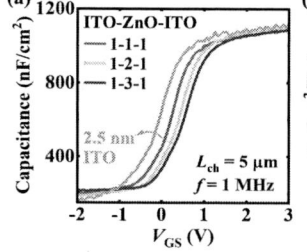

Fig. 5. The ITO-ZnO-ITO FET shows positive V_{TH} shift as compared with the ITO FET.

Fig. 6. The ITO-ZnO-ITO (1-1-1) FET gives higher G_M as compared with the ITO FET.

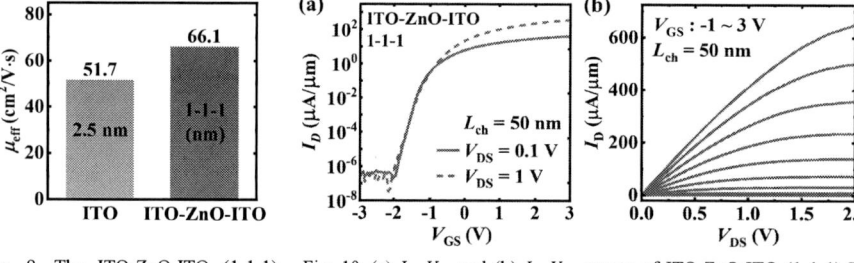

Fig. 7. (a) I_D–V_{GS} and (b) G_M vs. V_{GS} curves of ITO-ZnO-ITO FETs with varying ZnO thicknesses.

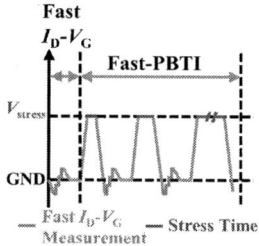

Fig. 8. (a) C-V and (b) μ_{eff} vs. V_{GS} curves of ITO-ZnO-ITO FETs and pure ITO FETs.

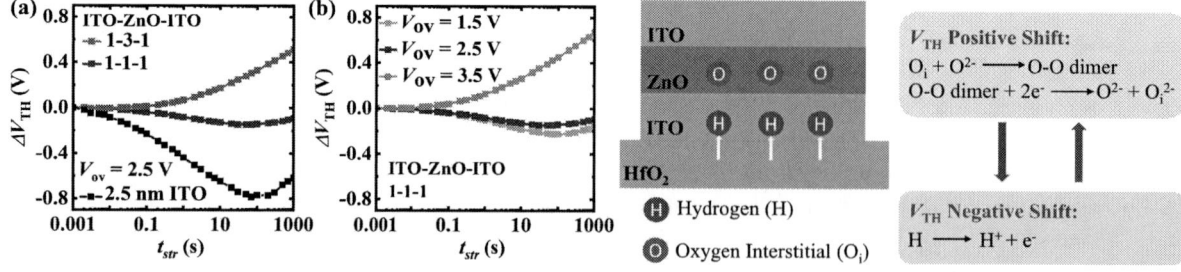

Fig. 9. The ITO-ZnO-ITO (1-1-1) FET displays higher peak μ_{eff} compared to the ITO FET.

Fig. 10. (a) I_D–V_{GS} and (b) I_D–V_{DS} curves of ITO-ZnO-ITO (1-1-1) FET with L_{ch} of 50 nm, showing a SS of 122 mV/decade and a high I_{ON} of 652 μA/μm at V_{GS} = 3 V and V_{DS} = 2 V.

Fig. 11. Schematic illustration of the V_{GS} waveform for PBTI measurement.

Fig. 12. (a) ΔV_{TH} of ITO-ZnO-ITO FETs and pure ITO FET at V_{OV} = 2.5 V. (b) ΔV_{TH} of ITO-ZnO-ITO (1-1-1) FET at V_{OV} = 1.5 V, 2.5 V and 3.5 V.

Fig. 13. The schematic illustration of the interaction between H and O_i in the multilayer structure of ITO-ZnO-ITO.

Synergistic Optimization of Thermal and Electrical Performances in Hetero-integrated β-Ga$_2$O$_3$ SBDs

Yinfei Xie[1], Yang He[1], Zhengyue Li[1], Wenhui Xu[2], Tiangui You[2], Xin Ou[2], and Huarui Sun[1,*]

[1]School of Science and Ministry of Industry and Information Technology Key Laboratory of Micro-Nano Optoelectronic Information System, Harbin Institute of Technology, Shenzhen, China

[2]National Key Laboratory of Materials for Integrated Circuits, Shanghai Institute of Microsystem and Information Technology, Chinese Academy of Sciences, Shanghai, China

Abstract

This work combines 3D Raman thermography with electrothermal simulations to study the thermal behaviors of hetero-integrated β-Ga$_2$O$_3$-on-SiC (GaOISiC) and β-Ga$_2$O$_3$ bulk Schottky barrier diodes (SBDs). It shows the benefits of hetero-integration in significantly enhancing heat dissipation. Moreover, we optimize the electrode layouts in GaOISiC SBDs by balancing electrical and thermal characteristics to improve performance. These techniques are beneficial for accurate temperature measurement and effective heat dissipation, which are crucial for β-Ga$_2$O$_3$-based devices.

Keywords: Hetero-integrated β-Ga$_2$O$_3$ Schottky barrier diodes, Raman 3D thermography, electrothermal modeling.

Introduction

β-Ga$_2$O$_3$ is a promising semiconductor for high-power electronics with an ultrawide bandgap of approximately 4.9 eV, a critical electric field of 8 MV/cm, and a high Baliga's Figure of merit (surpassing SiC and GaN) [1]–[3]. However, its low thermal conductivity (ranging from 0.1 to 0.3 W/cm·K at 300K) challenges high-power applications [4]. Heteroepitaxy of β-Ga$_2$O$_3$ on high thermal conductivity substrates such as SiC or diamond has been proposed to improve heat dissipation, but lattice mismatch problems exist [5]. The wafer-scale heterogeneous integration of β-Ga$_2$O$_3$ on SiC substrates via the ion-cutting/bonding process has been achieved. This process ensures crystalline quality and avoids lattice mismatch. Nevertheless, the thermal behavior of such devices is not fully understood [6]–[8], which is due to the difficulty in accurately characterizing the thermal properties of sub-micrometer scale ultrawide bandgap devices.

Common thermometry techniques like thermocouples and pulsed I–V have limitations in capturing junction temperature, and infrared thermography has limited spatial resolution and accuracy [9]–[11]. None of these can provide three-dimension (3D) temperature information. Intriguingly, Raman thermometry, with its submicron-scale spatial resolution and material selectivity, can address these issues [12]. In addition to thermal aspects, optimal electrode layout design is crucial for β-Ga$_2$O$_3$ devices. Current research on electrode layout separately studies electrical or thermal impacts without comprehensively evaluating the electrothermal coupling process.

In this study, 3D Raman thermography is used to characterize the peak and spatial temperature in a hetero-integrated β-Ga$_2$O$_3$-on-SiC (GaOISiC) SBD, and the thermal resistance of a β-Ga$_2$O$_3$ bulk SBD is investigated with TiO$_2$ nanoparticle Raman temperature sensors. 3D electrothermal models are developed for verification. The work also investigates on how the electrode layout design of β-Ga$_2$O$_3$ SBDs hetero-integrated on SiC affects electrical and thermal performances. Different electrode gaps are studied through experiment and modeling. Electrode layout optimization strategies are quantitatively evaluated to provide insights into current distribution and thermal management in these devices.

Experimental Details

The ion-cutting/bonding technique is utilized to fabricate SBDs on GaOISiC heterostructures. An SBD on β-Ga$_2$O$_3$ bulk with the same process was made for comparison. Six electrode gaps were studied. Circular electrode SBDs were fabricated via the circular transmission line method (CTLM). In these SBDs, the anode-cathode gap varies from 15 to 65 μm (increments of 10 μm, named C1-C6), with a constant anode radius of 100 μm. Raman measurements were carried out using a Renishaw InVia micro-Raman spectrometer (532 nm laser, 50× lens). The laser power was 5 mW, and spectra were integrated 40 times (2 s exposure time) for reliable signal quality. A Linkam THMS600 stage heated samples from 25 to 275 °C in 25 °C steps for calibration. For temperature-depth scans, the laser focus varied with depth, and for spatial temperature analysis, measurements were evenly spaced between the anode and the cathode (Fig. 1). 3D electrothermal models in Slivaco TCAD were employed to evaluate the reliability of Raman thermography. An effective thermal boundary resistance (TBR_{eff}) was set at ~7.5 m^{2}K/GW, considering multiple interfaces. The substrate bottom was maintained isothermally at 300 K, and other boundaries were adiabatic.

Results and Discussion

A. Comparison of GaOISiC and β-Ga$_2$O$_3$ bulk SBDs

Raman spectroscopy of GaOISiC SBDs records signals originating from the underlying substrate, as illustrated in Fig. 2. The spectrum displays four distinct phonon modes characteristic of the 4H-SiC substrate. Conversely, Raman signals from the thin surface layer of β-Ga$_2$O$_3$ exhibit notably reduced intensities and are nearly negligible after

979-8-3315-0417-5/25 $31.00 © 2025 IEEE

normalization. The linear temperature coefficient χ_T extracted from the $A_g(6)$ mode was $-(0.0047 \pm 0.00020)$ cm^{-1}/°C for β-Ga$_2$O$_3$ bulk SBD and $-(0.0051 \pm 0.00020)$ cm^{-1}/°C for GaOISiC SBD, respectively.

Fig. 3 shows simulated temperature distribution under 520 mW power on both SBDs, with heat near the inner electrode in the channel area, supporting experimental results and indicating self-heating. Fig. 4(a) compares lateral and depth-wise temperature near the hotspot. Simulated and measured temperatures match with a 14 °C (13.6% relative) error. Heat accumulates in the SBD with an inner anode due to current density and contact resistance, so the outer anode placement is safer. Due to material transparency, Raman spectra at different depths can be recorded for depth - wise temperature distribution. Fig. 4(b) shows depth-dependent temperature measured by adjusting the laser focus (from 0 to 200 μm). Temperature decreases with depth as heat diffuses to the heat sink, visualizing heat dissipation pathways. Good agreement between experiment and simulation shows Raman thermography's potential for 2D/3D temperature analysis.

Fig. 5 compares SBDs' peak temperature rise and thermal resistance via Raman thermography, TCAD simulation, and IR thermography. Raman-sensed temperatures in TiO$_2$/β-Ga$_2$O$_3$ SBDs are higher than IR-measured and closer to TCAD-simulated. In GaOISiC SBD, Raman-measured thermal resistance is 42% higher than IR-measured; in β-Ga$_2$O$_3$ bulk SBD, the value is 22%. IR underestimates junction temperature due to depth-averaging. Raman-sensed peak temperature can approximate junction temperature for device evaluation. Raman thermography has high lateral resolution and access to depth- and material-resolved temperature, enabling 3D temperature profiling in β-Ga$_2$O$_3$ devices with TCAD simulations.

B. Comparison of Different Electrode Gap of Circular GaOISiC SBDs

Fig. 6(a) shows peak temperatures of SBDs with different electrode gaps, with a linear correlation to input power. Thermal resistance (R_{th}) is from the linear-fit slope (Fig. 6(b)). The SBD with the smallest gap has the highest R_{th} and peak temperature per unit power. Larger gaps offer better heat dissipation and aid lateral cooling. Fig. 6 compare lateral and depth-wise temperature near the peak. The peak temperature is near the inner electrode. Reducing the gap intensifies heat accumulation. Good experiment-simulation agreement shows the reliability of 3D Raman thermography for ultrawide bandgap semiconductor thermal analysis.

In Fig. 7, anode current decreases with increasing electrode gap due to the higher resistance based on Ohm's law. I_{ON}/I_{OFF} ratios are about 10^9 and cut-off current is at pA level. Circular SBDs have 17.3% S-factor increase and 66.1% heat reduction. Larger gaps increase series resistance, making ideality factor n increase 20.4%. Therefore, widening the gap

moderately affects electrical performance but significantly improves heat dissipation, so device layout should balance electrical and thermal properties.

Conclusion

In conclusion, this study uses Raman thermography and electrothermal simulations to analyze β-Ga$_2$O$_3$-based SBDs. Raman-sensed temperature better indicates the channel junction temperature than IR thermography. In GaOISiC SBDs, the SiC substrate leads to lower thermal resistance. Different electrode geometries affect peak temperature and thermal resistance. Widening the electrode gap reduces peak temperature with some electrical performance loss. This work highlights the importance of Raman thermography and electrothermal simulations in understanding thermal behavior and of balancing electrical and thermal characteristics in β-Ga$_2$O$_3$ device layout design, especially for hetero-integrated devices.

Acknowledgments

This work was supported in part by the Guangdong Special Support Program under Grant 2021TQ06C953, in part by the Shenzhen Science and Technology Innovation Program under Grant JCYJ20190806142614541, and in part by the Open Research Fund of State Key Laboratory of Materials for Integrated Circuits under Grant SKLIC-K2024-04.

References

[1] M. Higashiwaki, K. Sasaki, A. Kuramata, T. Masui, and S. Yamakoshi, "Gallium oxide (Ga$_2$O$_3$) metal-semiconductor field-effect transistors on single-crystal β-Ga$_2$O$_3$ (010) substrates," Appl. Phys. Lett., 100, p. 013504, (2012).

[2] B. Wang, M. Xiao, X. Yan, H. Y. Wong, J. Ma, K. Sasaki, H. Wang, and Y. Zhang, "High-voltage vertical Ga$_2$O$_3$ power rectifiers operational at high temperatures up to 600 K," Appl. Phys. Lett., 115, p. 263503, (2019).

[3] C. Chen, X. Zhao, X. Hou, S. Yu, R. Chen, X. Zhou, P. Tan, Q. Liu, W. Mu, Z. Jia, G. Xu, X. Tao, and S. Long, "High-performance β-Ga$_2$O$_3$ solar-blind photodetector with extremely low working voltage," IEEE Electron Device Lett., 42, pp. 1492-1495, (2021).

[4] P. Jiang, X. Qian, X. Li, and R. Yang, "Three-dimensional anisotropic thermal conductivity tensor of single crystalline β-Ga$_2$O$_3$," Appl. Phys. Lett., 113, p. 232105, (2018).

[5] X. C. Guo, N. H. Hao, D. Y. Guo, Z. P. Wu, Y. H. An, X. L. Chu, L. H. Li, P. G. Li, M. Lei, and W. H. Tang, "β-Ga$_2$O$_3$/p-Si heterojunction solar-blind ultraviolet photodetector with enhanced photoelectric responsivity," J. Alloy. Compd., 660, pp. 136-140, (2016).

[6] W. Xu, Y. Wang, T. You, X. Ou, G. Han, H. Hu, S. Zhang, F. Mu, T. Suga, Y. Zhang, Y. Hao, and X. Wang, "First demonstration of waferscale heterogeneous integration of Ga$_2$O$_3$ MOSFETs on SiC and Si substrates by ion-cutting process," IEEE International Electron Devices Meeting (IEDM) (2019).

[7] W. Xu, T. You, Y. Wang, Z. Shen, K. Liu, L. Zhang, H. Sun, R. Qian, Z. An, F. Mu, T. Suga, G. Han, X. Ou, Y. Hao, and X. Wang, "Efficient

thermal dissipation in wafer-scale heterogeneous integration of single-crystalline β-Ga$_2$O$_3$ thin film on SiC," Fundamental Res., 1, pp. 691-696, (2021).

[8] W. Xu, Z. Shen, Z. Qu, T. Zhao, A. Yi, T. You, G. Han, and X. Ou, "Current transport mechanism of lateral Schottky barrier diodes on β-Ga$_2$O$_3$/SiC structure with atomic level interface," Appl. Phys. Lett., 124, p. 112102, (2024).

[9] T. J. Bajzek, "Thermocouples: A sensor for measuring temperature," IEEE Instrum. Meas. Mag., 8, pp. 35-40 (2005).

[10] J. Joh, J. A. d. Alamo, U. Chowdhury, T. M. Chou, H. Q. Tserng, and J. L. Jimenez, "Measurement of channel temperature in GaN high-

electron mobility transistors," IEEE Trans. Electron Devices, 56, pp. 2895-2901 (2009).

[11] A. Sarua, H. Ji, M. Kuball, M. J. Uren, T. Martin, K. P. Hilton, and R. S. Balmer, Integrated micro-Raman/infrared thermography probe for monitoring of self-heating in AlGaN/GaN transistor structures," IEEE Trans. Electron Devices, 53, pp. 2438-2447 (2006).

[12] M. Kuball, J. W. Pomeroy, "A review of Raman thermography for electronic and opto-electronic device measurement with submicron spatial and nanosecond temporal resolution," IEEE Trans. Device Mater. Reliab., 16, pp. 667-684, (2016).

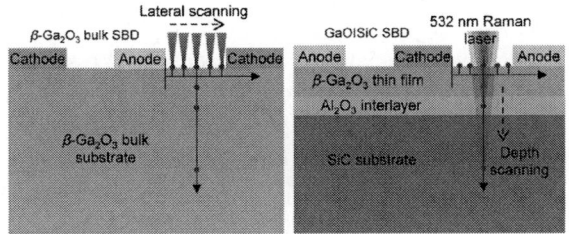

Fig. 1   Schematic illustrations of circular electrode SBDs.

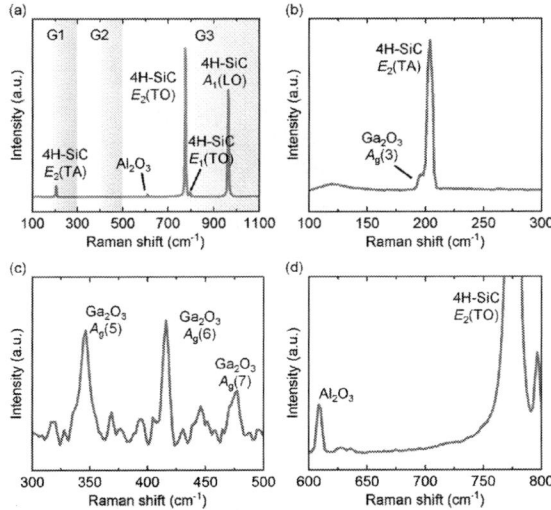

Fig. 2   Raman spectrum of GaOISiC SBD.

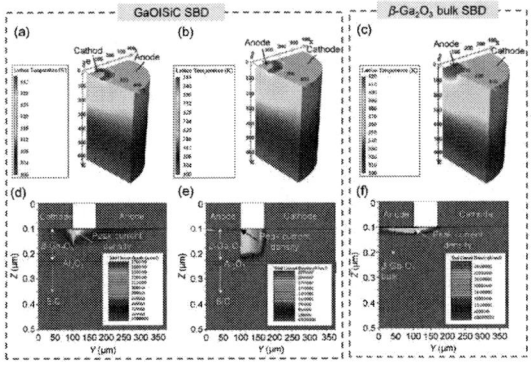

Fig. 3   Simulated temperature and electric field.

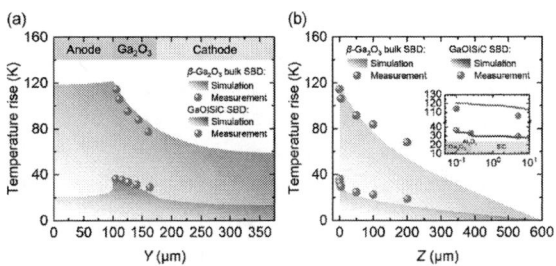

Fig. 4   Temperature profiles along lateral and depth directions.

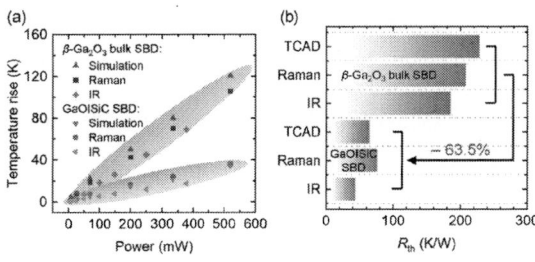

Fig. 5   Comparison of thermal resistance at different powers between Raman thermography, simulation, and IR thermography.

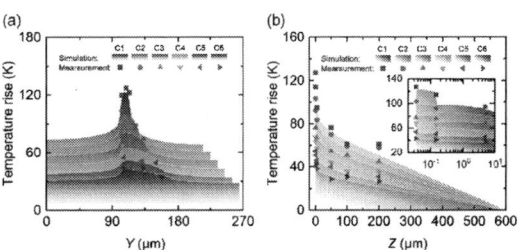

Fig. 6   Temperature profiles along lateral and depth directions.

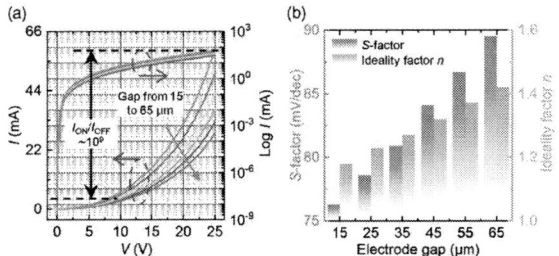

Fig. 7   Measured and simulated electrical characteristics.

Selecting Device, In-Memory Search, and Monolithic 3D Integration of RRAMs and 2D Devices Towards High Reliability and High Efficiency

Rui Yang[1,2,3,*], Yueyang Jia[1], Maosong Xie[1], Minliang Shen[1], Yijian Zhang[1], Yuzhuo Liu[1]

[1]University of Michigan–Shanghai Jiao Tong University Joint Institute, Shanghai Jiao Tong University, Shanghai, China.

[2]School of Electronic Information and Electrical Engineering, Shanghai Jiao Tong University, Shanghai, China.

[3]State Key Laboratory of Radio Frequency Heterogeneous Integration, Shanghai Jiao Tong University, Shanghai, China.

(*Corresponding Author, Email: rui.yang@sjtu.edu.cn)

(Invited Paper)

Abstract

Resistive random-access memories (RRAMs) are highly promising for high-density memory and energy-efficient in-memory computing. Towards the integration of RRAM arrays, the sneak path leakage and IR drop problem need to be alleviated and the memory density should be further enhanced. Here we develop the consistent RRAM model, demonstrate two-dimensional (2D) selectors and one-selector-one-selector (1S1R) memory cells, perform monolithic 3D integration of 2D transistors and vertical RRAMs (VRRAMs), and show the promise for in-memory search applications.

Keywords: RRAM, In-memory search, 2D devices, Monolithic 3D integration

Introduction

Towards the large-scale integration and applications based on RRAMs, further optimizations are necessary in various aspects such as more reliable device model, higher memory density, better selecting device, and more efficient in-memory computing paradigms. Previous RRAM compact models have been able to reproduce a number of RRAM switching behaviors [1,2,3]. However, a comprehensive model that can well explain and reproduce the experimentally observed gradual, abrupt, and abnormal reset transitions in unipolar and bipolar RRAMs remains elusive. Such consistent RRAM model is critical for large-scale RRAM circuit simulation, and for predictable tuning of RRAM properties. With the device model, the efficient in-memory computing scheme should be developed. In experiments, the current RRAM arrays are either 1R arrays or 1-transistor-1-resistor (1T1R) arrays [4], yet the 1R arrays usually suffer from high leakage, while 1T1R arrays sacrifice the memory density. In this work, we develop a consistent RRAM model that well explains the gradual, abrupt, and abnormal reset phenomena with experimental verification, develop the 2D selectors and 1S1R memory cells for high integration density while suppressing the leakage, design an in-memory-search collision detection accelerator for high energy efficiency in robotic motion planning tasks in a dynamic environment, and

demonstrate monolithic 3D integration system based on 2D transistors and 3D VRRAMs for ultrahigh-density and ultralow-latency on-chip memory (Fig. 1). Targeting at the key challenges for the developments of nonvolatile memories and in-memory computing, these advances hold high promises for the development of high-density, high-reliability, and energy-efficient memories and in-memory computing architectures towards applications in artificial intelligence (AI), robotics, embodied intelligence, and neuromorphic computing.

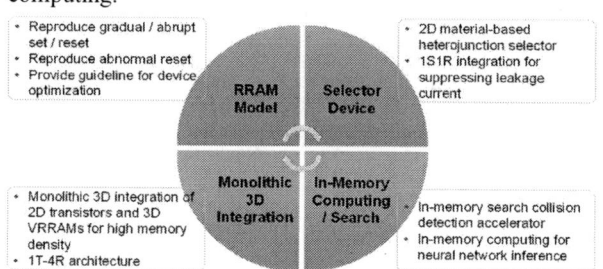

Fig. 1: The four aspects of optimization for RRAMs and the integration with 2D materials, including RRAM model, 2D selector device and 1S1R integration, RRAM-based in-memory computing and in-memory search, and monolithic 3D integration of 2D transistors and 3D VRRAMs.

Consistent RRAM Model

The RRAM compact model is necessary for further circuit simulation and integration of RRAM arrays. In the RRAM compact model, we systematically consider the programming processes of metal-oxide RRAMs as the competition between the generation and recombination, and the competition between the drift and diffusion processes of oxygen ions and oxygen vacancies [5]. We further model the energy barriers for oxygen ion migration into and out of the reservoir layer of the electrode, and the oxygen vacancy (V_O) concentration change of the filament. The oxygen vacancy generation probability is given by:

$$P_G = \Delta t \cdot f \cdot \exp\left(-\frac{E_a - k_E|E|qa_0}{k_B T}\right), \quad (1)$$

where electric field $E=V/d$, V is the applied voltage, d is the oxide thickness, k_E is field enhancement factor, the equivalent electric field in the gap region is $k_E \cdot |E|$, Δt is the operating time, f is the vibration frequency of oxygen ions (O^{2-}), E_a is

979-8-3315-0417-5/25 $31.00 © 2025 IEEE

the oxygen vacancy generation barrier, q is the unit electron charge, a_0 is the lattice constant, k_B is the Boltzmann constant, and T is the local temperature. The recombination probability of oxygen vacancy and oxygen ion is:

$$P_R = \Delta t \cdot n_{O^{2-}} \cdot f \cdot \exp(-\frac{E_r}{k_B T}), \quad (2)$$

where $n_{O^{2-}}$ is the available O^{2-} number, and E_r is the relaxation energy. The net recombination probability is given by the difference between P_R and P_G to represent the competition process, and the gap change can be obtained from the net recombination probability. Then by modeling the drift and diffusion processes, the O^{2-} number in the electrode and the oxide, and the energy barriers, we model the whole I–V relationship of the RRAMs (Fig. 2). The model can reproduce the gradual and abrupt reset phenomena well by tuning the model parameters. Based on the guidance from the model, we experimentally tune the programming voltage and current compliance, which results in controllable gradual or abrupt reset operations. Furthermore, we can tune the same RRAM between normal and abnormal reset phenomena, and obtain the corresponding model parameter change.

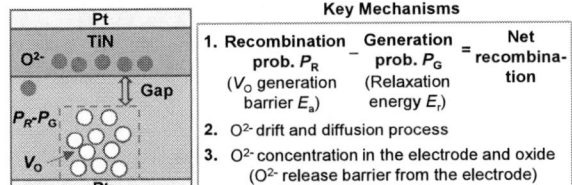

Fig. 2. Illustration of the RRAM device structure and the key mechanisms considered in the model, including the competition between the defect generation probability P_G and recombination probability P_R, the O^{2-} drift under electric field and diffusion under O^{2-} gradient, and the O^{2-} number in the electrode and oxide. Check Ref. [5] for details.

2D Selector and 1S1R Integration

With the consistent RRAM model, the RRAM arrays can be simulated in the circuit level. However, when experimentally building RRAM arrays, there are several problems. Because the RRAM array can be regarded as a resistive network, there are inevitable sneak path leakage problem and IR drop problem when voltages are applied [6]. To alleviate such problems, nonlinear components such as transistors are typically used for selecting and accessing the RRAMs, forming 1T1R structure. Yet due to the relatively large area of transistors compared with RRAMs, the integration density is usually limited. To enable ultrahigh-density integration of RRAMs while limiting the sneak path and IR drop problem, selectors are actively explored.

Here we leverage the van der Waals integration of 2D materials, and demonstrate exponential selectors based on 2D heterostructures of multilayer graphene (MG)/tungsten disulfide (WS$_2$)/Pt (Fig. 3a) [7]. Such selector is based on two Schottky-type contacts, which differs from the previous 2D selectors with hexagonal boron nitride isolation [8]. Furthermore, to promote the integration of selectors and

RRAMs, the selector should show asymmetric I–V behavior, because the set process of RRAM requires a current compliance, while the reset process requires a high enough drive current (Fig. 3b). The different MG/WS$_2$ and WS$_2$/Pt junctions provide such asymmetric behavior. Moreover, the selectors show high nonlinearity, which is defined as the current at an operating voltage (V_{op}) divided by the current at $V_{op}/2$. The nonlinearity of the 2D selector reaches 120 due to the suitable Schottky barrier height provided by the 2D heterojunction. When integrated with the RRAM, the selector can further suppress sneak path leakage current by more than 100 times, which makes it promising for low-power, high-density memory and in-memory computing applications.

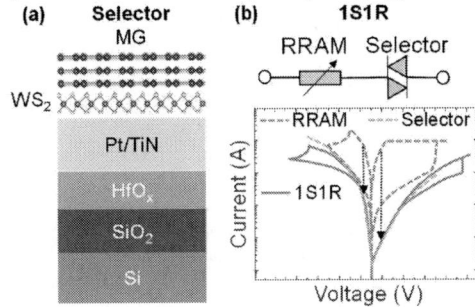

Fig. 3. (a) Illustration of the 2D selector device structure. (b) Illustrations of 1S1R integrated structure, and the effect for suppressing the leakage current compared with RRAM devices only. Check Ref. [7] for details.

In-Memory Computing and In-Memory Search

The conductance of the 1S1R structure can be programmed by applying consecutive voltage pulses, which can represent the synaptic weights in artificial neural networks (NNs). By using positive voltage pulses for the potentiation and negative pulses for depression, we achieve multiple conductance levels in one 1S1R synapse (Fig. 4a) [7]. The voltage is used as input, so the matrix-vector multiplication (MVM) which is widely used in NNs can be computed directly inside the 1S1R array by reading out the summed currents. Such in-memory computing architecture is quite energy-efficient because it avoids the frequent data transfer between the memory and logic units. We find that the inference accuracy of the VGG-8 network remain high even with large device variations (Fig. 4b), which shows high robustness of the in-memory computing using 1S1R cells.

Besides MVM, ternary content-addressable memory (TCAM) can perform look-up table function in a single clock cycle, which is suitable for in-memory searching and pattern matching applications [9]. Previously, we have demonstrated the integration of 2D molybdenum disulfide (MoS$_2$) field-effect transistors (FETs) with HfO$_x$-based RRAMs into 2-transistor-2-RRAM (2T2R) TCAM cells with high On/Off ratio between match and mismatch states and high sense margin [10]. Based on the 2T2R TCAM arrays, we further design an in-memory search robot collision detection accelerator, which can store a large roadmap with 100,000 edges and perform collision detection within 100 μs [11].

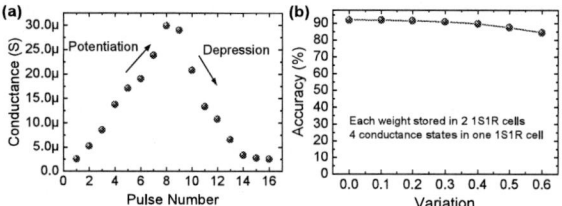

Fig. 4. (a) Potentiation and depression of a 1S1R synapse based on the 2D selector. (b) Simulated inference accuracy of VGG-8 neural network based on the1S1R cells with different device variations. Check Ref. [7] for details.

Monolithic 3D Integration of 2D FETs and VRRAMs

To further enhance the on-chip memory density, besides developing 1S1R cells, the integration of 1T–nR cells with one transistor driving n RRAMs is also a promising route. Furthermore, the memory and logic layers can be fabricated on different layers on the same chip, interconnected by inter-layer vias, forming monolithic 3D integration, which is promising for achieving high energy efficiency [12]. Based on the previous work on 1T1R integration using 2D MoS_2 FETs and RRAMs [13], we further fabricate monolithic integration structure based on 2D MoS_2 FETs and 3D VRRAMs (Fig. 5), forming 1T–4R structure [14]. Such integrated structure can achieve high memory density, low latency and high energy efficiency.

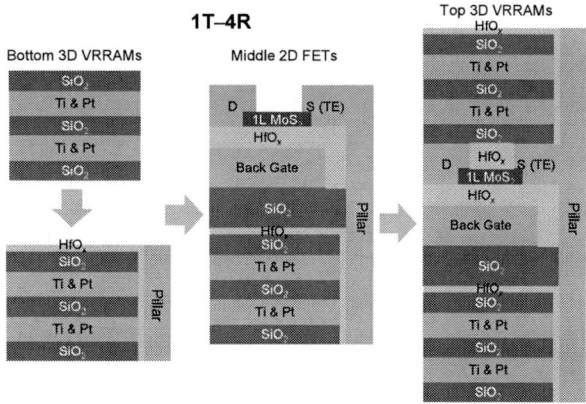

Fig. 5. Illustration of the fabrication processes for the monolithic 3D integration of 2D FETs and VRRAMs. Check Ref. [14] for details.

Conclusion

In summary, through research into modeling, selecting device, in-memory computing application, and 3D integration techniques, it is promising to achieve efficient memory and computing systems based on 2D devices and RRAMs.

Acknowledgments

The authors thank the support from the National Natural Science Foundation of China (NSFC) (Grants 92364107, 62104140), Science and Technology Commission of Shanghai Municipality (Grant 23QA1405300), State Key Laboratory of Radio Frequency Heterogeneous Integration (Shenzhen University), Natural Science Foundation of Chongqing (CSTB2022NSCQ-MSX1095), and Lingang Laboratory Open Research Fund (Grant LG-QS-202202-11).

References

[1] Z. Jiang, Y. Wu, S. Yu, L. Yang, K. Song, Z. Karim, H.-S. P. Wong, "A compact model for metal-oxide resistive random access memory with experiment verification," IEEE Trans. Electron Devices, 63, pp. 1884–1892 (2016).

[2] H. Li, P. Huang, B. Gao, B. Chen, X. Liu, and J. Kang, "A SPICE model of resistive random access memory for large-scale memory array simulation," IEEE Electron Device Lett., 35, pp. 211–213 (2014).

[3] A. Padovani, L. Larcher, O. Pirrotta, L. Vandelli, and G. Bersuker, "Microscopic modeling of HfOx RRAM operations: from forming to switching," IEEE Trans. Electron Devices, 62, pp. 1998–2006 (2015).

[4] S.-T. Wei, B. Gao, D. Wu, J.-S. Tang, H. Qian, H.-Q. Wu, "Trends and challenges in the circuit and macro of RRAM-based computing-in-memory systems," Chip, 1, 100004 (2022).

[5] Y. Jia, S. Shen, M. Xie, P. Zhang, M. Shen, R. Yang, "A consistent model for gradual, abrupt, and abnormal reset phenomena in bipolar/unipolar metal oxide RRAMs," IEEE Trans. Electron Devices, 71, pp. 3142–3149 (2024).

[6] H.-S. P. Wong, S. Salahuddin, "Memory leads the way to better computing," Nat. Nanotechnol., 10, pp. 191–194 (2015).

[7] M. Shen, S. Shen, Y. Jia, P. Zhang, M. Xie, J. Wei, R. Yang, "One-selector-one-resistor integrated memory cells based on two-dimensional heterojunction memory selectors," ACS Nano, 18, pp. 28292–28300 (2024).

[8] C. H. Wang, V. Chen, C. J. McClellan, A. Tang, S. Vaziri, L. Li, M. E. Chen, E. Pop, H.-S. P. Wong, "Ultrathin three-monolayer tunneling memory selectors," ACS Nano, 15, pp. 8484–8491 (2021).

[9] K.-J. Zhou, C. Mu, B. Wen, X.-M. Zhang, G.-J. Wu, C. Li, H. Jiang, X.-Y. Xue, S. Tang, C.-X. Chen, Q. Liu, "The trend of emerging non-volatile TCAM for parallel search and AI applications," Chip, 1, 100012 (2022).

[10] R. Yang, H. Li, K. K. H. Smithe, T. R. Kim, K. Okabe, E. Pop, J. A. Fan, H.-S. P. Wong, "Ternary content-addressable memory with MoS2 transistors for massively parallel data search," Nat. Electron., 2, pp. 108–114 (2019).

[11] J. Sun, F. Liu, Y. Zhang, L. Jiang, R. Yang, "RTSA: an RRAM-TCAM based in-memory-search accelerator for sub-100μs collision detection," IEEE Design, Automation and Test in Europe Conference (DATE 2024), DOI: 10.23919/DATE58400.2024.10546586 (2024).

[12] M. M. S. Aly, et al., "Energy-efficient abundant-data computing: the N3XT 1,000x," IEEE Computer, 48, pp. 24–33 (2015).

[13] R. Yang, H. Li, K. K. H. Smithe, T. R. Kim, K. Okabe, E. Pop, J. A. Fan, H.-S. P. Wong, "2D molybdenum disulfide (MoS2) transistors driving RRAMs with 1T1R configuration," IEEE Int. Electron Devices Meeting (IEDM 2017), pp. 477–480 (2017).

[14] M. Xie, Y. Jia, C. Nie, Z. Liu, A. Tang, S. Fan, X. Liang, L. Jiang, Z. He, R. Yang, "Monolithic 3D integration of 2D transistors and vertical RRAMs in 1T–4R structure for high-density memory," Nat. Commun., 14, 5952 (2023).

979-8-3315-0417-5/25 $31.00 © 2025 IEEE   858

Adaptive Update Precision with Reduced Iterative Write Cycles for Efficient Training Neural Networks on ECRAM arrays

Peihong Li[1], Peng Chen[1,2], Peng Lin[1,2,*], Gang Pan[1,2,3]

[1] College of Computer Science and Technology, Zhejiang University, China, [2] State Key Laboratory of Brain Machine Intelligence, China, [3]MOE Frontier Science Center for Brain Science and Brain-Machine Integration, Zhejiang University, Hangzhou, China, *corresponding to penglin@zju.edu.cn

Abstract

Training neural networks in large-scale non-volatile memory arrays suffers from stochastic conductance modulation and device variations, requiring closed-loop write iterations to achieve desired update precisions. Here we propose an adaptive algorithm combining static evaluation of network structure and real-time network activity to substantially reduce write iterations without compromising performance. Our method improves energy efficiency by up to ×81 and ×242 for VGG19 and ResNet18, while maintaining high training accuracy.

Keywords: ECRAM, Adaptive write precision, Training

Introduction

The continuous evolution of deep neural networks (DNNs) urgently demands new hardware solutions to handle the growing computation burdens in training these large-scale models. One promising direction is the use of analog non-volatile memory-based arrays to accelerate the training process. Electrochemical random-access memory (ECRAM) is a promising candidate for training neural networks[1], thanks to a linear and symmetric analog programming capability (Fig. 1a), which naturally and linearly maps the weight update values in algorithms to the stimulation conditions of each device[2]. However, achieving decent linearity and symmetry across a large ECRAM array remains challenging and has yet to be experimentally verified, Suffering from cycle-to-cycle (C2C) and device-to-device (D2D) variations. To compensate for inaccurate weight update, multiple closed-loop write cycles are required to achieve desired update precision[3], but will add significant energy consumption and programming overhead.

Reducing update precision has been proposed as a trade-off between write iterations and network performance. One limitation in existing methods is that static allocation cannot fully track the precision requirement of weights. For instance, the update precisions are typically adjusted uniformly across the entire network[4], or layer-wise[5], still leading to redundant or insufficient update precision.

In this paper, we propose a two-step mixed-precision update scheme that adaptively sets weight update precision. Initially, we allocate coarse-grained precision to each layer based on its number of parameters, ensuring that layers with fewer parameters receive higher precision. During training, we dynamically adjust precisions for each synapse based on the activation frequency of its postsynaptic neuron. Based on the write precision model derived from experimental data, we evaluated the training performance of this method. This strategy leverages both layer-wise noise robustness and neuron activities to adaptively tune update precision, demonstrating a co-optimization of the write overhead and network performance.

Write Precision

A. Definition of Equivalent Write Precision

When programming ECRAM devices, it is essential to account for the inherent programming errors that arise during the write operations. Specifically, each write operation to an ECRAM device introduces a write error (σ), defined as the difference between the actual written resistance and the target value. Equivalent write precision is defined based on the distribution of write errors observed during repeated write attempts to the target resistance state. Based on the write error distribution, we categorize resistance states more than $\pm\sigma$ as distinguishable, then the corresponding number of states N can be calculated as follows:

$$N = \frac{G_{max} - G_{min}}{2\sigma} + 1 \tag{1}$$

Fig. 2a illustrates different precision levels (2-bit, 3-bit, and 4-bit) derived from varying write errors (corresponding σ values are 8.6 μS, 3.7 μS, and 1.7 μS, respectively, G_{min} and G_{max} are 1.9 μS and 53.7 μS). Notably, narrowing margins is observed between adjacent states, showing challenges to achieve higher precisions.

B. Impact on Training Accuracy

We trained an AlexNet model on CIFAR-10 using various equivalent write precisions, as shown in **Fig. 2b** and **c**. The results demonstrated that the required write cycles grow exponentially with higher update precision. Notably, a minimum 6-bit update precision is essential in fixed-precision training to achieve convergence with acceptable accuracy, consistent with previous findings[6]. This illustrates the importance of adaptively adjusting the update precision to balance efficiency and accuracy.

Algorithmic Scheme

Our mixed-precision update scheme comprises two key precision adjustments: static and dynamic precision settings, which are conducted before and during training, respectively (as illustrated in **Fig. 3**). **Stage I**: We assign the required write precision for each weight layer based on its number of parameters. This layer-wise assignment is performed before training begins, ensuring that layers with fewer parameters receive higher precision, while layers with larger parameter scales are allocated lower precision, optimizing energy usage by concentrating high precision on critical layers. **Stage II**: During the training process, we implement fine-grained precision adjustment for each weight at the neuron level, determined by the activation frequency (AF) of its postsynaptic neuron, i.e., the percentage of nonzero iterations in each epoch. Neurons exhibiting higher AF value have stronger synaptic connections with neurons in preceding layers[7]. Therefore, the synapses connected to these active neurons are allocated increased write precision to maintain accurate weight updates, whereas synapses of less active neurons operate at reduced precision. This dynamic adjustment compensates for the limitations of Stage I by enabling more precise tracking of precision needs through real-time AF monitoring.

A. Coarse-grained Precision Presetting

The sensitivity of each neural network layer to write noise varies with the number of parameters which determines its inherent noise robustness. In an 8-layer AlexNet[8], layers with more parameters exhibit greater resilience to noise **(Fig. 4a and b)**, enabling us to exploit this property for energy-efficient training.

By assessing the parameter scale of each layer, we can allocate higher write precisions to weight-less layers and vice versa. The write precision of layer l is calculated by:

$$k_l = round((1-\alpha_l) \times K) \qquad (2)$$

where α_l denotes the proportion of the parameter scale of layer l relative to the total parameter scale, and K is a user-defined initial precision. k_l represents the write precision allocated to weights in layer l and will be rounded to the nearest integer in practice.

B. Fine-grained Precision Tuning Scheme

To further track the precision requirement for each weight, we introduced a fine-grained mixed-precision scheme that dynamically adjusts the write precision of individual weights based on the AF values of postsynaptic neurons during the training process. It operates in tandem with the coarse-grained layer-wise precision allocation established before training. The AF value of a neuron reflects its importance during training, with highly active neurons necessitating greater update precision for accurate representation of synaptic weights.

In **Fig. 5a**, we randomly select a layer and illustrate how the update precision evolved with the AF of its postsynaptic neuron during the training process. It can be seen that, the AF varies both spatially and temporally, indicating different degrees of redundancy[9]. Furthermore, the trends of increasing and decreasing AF also reflect the under-capacity

and over-capacity of the neuron, as well as its incoming synapses.

Based on this observation, the adjustment method for the write precision is derived using Eq. (3).

$$\delta^{l-1}_{(n,:)} = \begin{cases} round(|\Delta AF^l_n| \times k_l), & \Delta AF^l_n > 0 \\ -round(|\Delta AF^l_n| \times k_l), & \Delta AF^l_n \leq 0 \end{cases} \qquad (3)$$

where $\delta^{l-1}_{(n,:)}$ represents the change in write precision of all incoming synaptic weights connected to neuron n in layer l, ΔAF^l_n denotes the change in the AF of neuron n in layer l, and k_l is the corresponding write precision for layer l. This algorithm will increase the write precision when the capability needs to be improved (i.e., when the AF progressively increases) and decreases it otherwise, as shown in **Fig. 5b**.

Results

Our adaptive update scheme achieved substantial energy efficiency improvements in training VGG19 and ResNet18 architectures across CIFAR-10, CIFAR-100, and Tiny-ImageNet datasets. Specifically, our approach showed up to ×242 increases in energy efficiency of training ResNet18 model, demonstrating it as a highly efficient solution for DNNs training.

To comprehensively assess its effectiveness, we compared our scheme to a state-of-the-art mixed-precision one-time training method based on activation density (AD) [9]. We evaluated the performance of our coarse-grained and fine-grained schemes independently (denoted as Coarse I and AF II, respectively) and in combination (denoted as AF I+II), benchmarking them versus 16-bit and 8-bit fixed-precision models for reference. The comparison results, as shown in **Fig. 6** and **Table 1**, underscore the superior energy efficiency and adaptability of our adaptive weight update scheme, particularly advantageous for DNN training.

Conclusion

We present an adaptive update scheme that tracks the precision requirements for weights across different layers and neuron connections. By implementing both coarse-grained layer-wise and fine-grained neuron-level precision adjustments, our method substantially reduces write overhead without sacrificing network performance, which can be applicable to other emerging non-volatile memories.

Acknowledgments

This work was supported in part by National Key R&D Plan of China (2022YFB4500100), Natural Science Foundation of China (62404198), Major Program of Natural Science Foundation of Zhejiang Province in China (LDQ23F040001).

References

[1] Y. Van De Burgt et al., Nat. Mater., 16, 414-418, (2017). [2] P. Chen et al., Nat. Commun., 14, 6184, (2023). [3] C. Li et al., Nat. Electron., 1, 52-59, (2018). [4] Hubara et al., JMLR, 18, 1-30, (2018). [5] C. Tang et al., ECCV, 259-275, (2022). [6] J. Tang et al., IEDM, 13.1. 1-13.1. 4, (2018). [7] T. Foldy-Porto et al., ICPR, 8929-8936, (2021). [8] A. Krizhevsky et al., NIPS, 25, (2012). [9] K. Vasquez et al., DATE, 1360-1365, (2021).

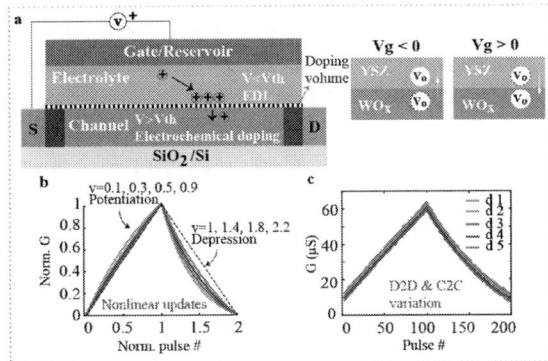

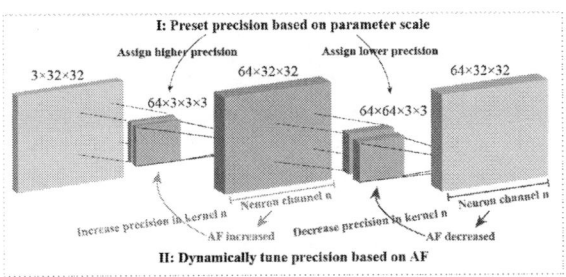

Figure 1. ECRAM device and its non-idealities in large-scale ECRAM arrays. a) Schematic illustration depicting the structure and working mechanism of ECRAM. **b)** Based on the behavior model [6], we applied different levels of nonlinearity for potentiation and depression. **c)** Illustration of device-to-device variation with the variability of 5.33% and cycle-to-cycle variation with the variability of 0.61%.

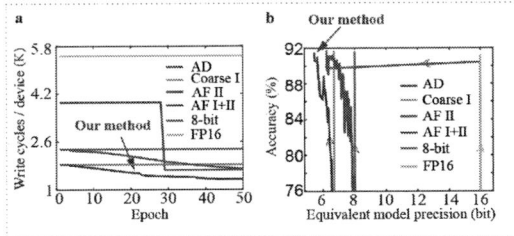

Figure 3. Overview of our proposed algorithm, composed of two stages (I and II).

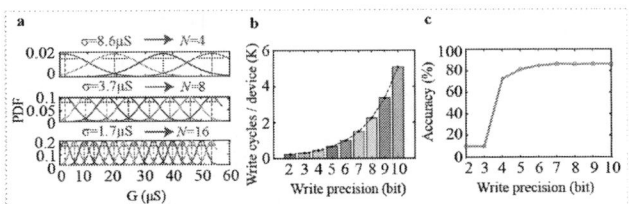

Figure 2. Modeling of equivalent write precision and its impact on training neural networks. a) Illustration of different precision levels (2-bit, 3-bit, and 4-bit) under varying noise conditions. **b)** The average number of writes per device and epoch in AlexNet trained on CIFAR-10 under different write precisions. **c)** CIFAR-10 classification accuracy obtained by AlexNet under different equivalent write precisions.

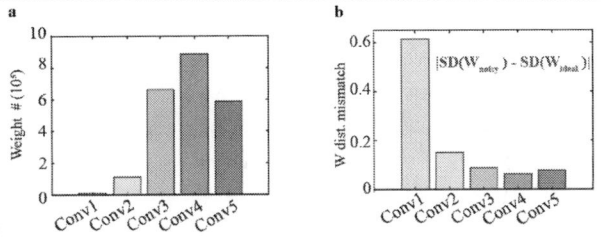

Figure 4. Relationship between layer-wise parameter scale and layer-wise robustness to noise. a) The number of weights in each layer. **b)** Weight distribution mismatch in each layer of AlexNet trained on CIFAR-10, calculated using models with and without noise. A layer with a larger number of weights results in a lower degree of mismatch, indicating that a lower write precision is required. The distribution mismatch is defined as the absolute difference in the standard deviation (SD) of weights with noise (W_{noisy}) and weights without noise (W_{ideal}).

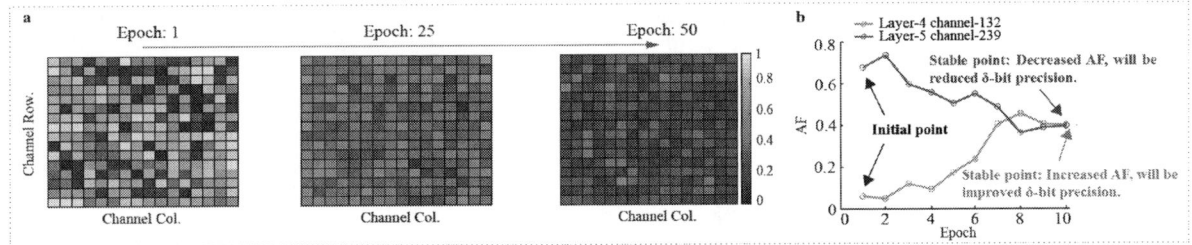

Figure 5. Motivation of the finer-grained mixed-precision update scheme and its strategy. a) Illustration of the activation frequency (AF) in selected feature map channels of the 4th layer of AlexNet on CIFAR-10 during training epochs 1, 25, and 50. In the convolutional layer, the average AF over the (batch, height, width) dimensions is calculated for each point in the heat map, as each channel of a feature map corresponds to an incoming convolutional kernel. **b)** Selected example channels from two feature maps illustrate our proposed write precision adjustment method, based on the increase or decrease of AF over a series of training epochs, with write precision increased by δ-bit and decreased by δ-bit, respectively.

Figure 6. Training efficiency and accuracy of VGG19 on CIFAR-10. a) The average number of writes per device in each epoch. Our proposed update scheme (AF I+II) shows the fewest write cycles. **b)** Evolution of accuracy and corresponding model precision over each epoch. Our method gradually decreases the equivalent model precision while maintaining a high level of accuracy.

Table 1. Accuracy and efficiency of various networks and datasets, normalized to their respective baselines. The results are averaged over the epochs corresponding to the highest accuracy achieved for each candidate, ensuring fairness in comparison.

Architecture	Test Accuracy	Write Operations Reduction	Energy Efficiency	Training Speedup	Model Precision: Best Acc Epochs
(a) VGG19 on CIFAR-10					
Baseline	92.5%	0	1×	1×	16 : 200
Fixed 8-bit	91.3%	60%	65.2×	2.64×	8 : 50
AD [9]	91.6%	40%	27.79×	1.71×	11.4 : 153
AF	**91.3%**	**73.6%**	**80.69×**	**4.22×**	**5.9 : 45**
(b) ResNet18 on CIFAR-100					
Baseline	72.6%	0	1×	1×	16 : 100
Fixed 8-bit	71.3%	60%	211.71×	2.64×	8 : 50
AD [9]	71.5%	33.3%	56.9×	1.53×	14.3 : 70
AF	**71.4%**	**69.3%**	**241.8×**	**3.55×**	**6.6 : 39**
(c) ResNet18 on Tiny-ImageNet					
Baseline	55.98%	0	1×	1×	16 : 200
Fixed 8-bit	55.6%	60%	219.26×	2.64×	8 : 42
AD [9]	55.6%	46.7%	79.79×	1.94×	9.53 : 92
AF	**55.2%**	**70.7%**	**231.17×**	**3.75×**	**6.4 : 39**

979-8-3315-0417-5/25 $31.00 © 2025 IEEE

Twisted-placed Multilayer Stack for Inherent Suppression of Transverse Modes on Layered SAW Devices

Boyuan Xiao[1], Sulei Fu[1, *], Peisen Liu[1], Xinchen Zhou[1], Qiufeng Xu[1], Jiajun Gao[1], Shuai Zhang[2], Rui Wang[1], Cheng Song[1], Fei Zeng[1], Weibiao Wang[2], and Feng Pan[1, *]

[1] Key Laboratory of Advanced Materials (MOE), School of Materials Science and Engineering, Tsinghua University, Beijing 100084, China, [2] Shoulder Electronics Ltd., Wuxi 214124, China.
*E-mail:suleifu@163.com; panf@mail.tsinghua.edu.cn

Abstract

This study explores a novel twisted-placed multilayer structure with inherent spurious transverse suppression for shear-horizontal surface acoustic wave (SH-SAW) devices. By adjusting the twisted angle of 45° between lithium tantalate (LT) and Si (100) substrate, the slowness curve of material stack was twisted vertically to enhance spectrum purity while inherently suppressing transverse modes. The fabricated resonators based on the twisted-placed multilayer demonstrated a significant reduction in transverse modes compared to traditional stacks, ensuring a cleaner spectrum without compromising performance. This method presents a promising advancement in RF filter technology for next-generation communications.

Keywords: SAW resonator; twisted-placed multilayer; transverse modes; slowness curve.

Introduction

The evolution of next-generation wireless communication has enabled an era characterized by an exponential increase in data transmission, necessitating radio frequency (RF) frontend solutions that meet demands for larger bandwidth, lower latency, and minimal interference [1], [2], [3]. Recent innovations in the integration of lithium tantalate (LT) and lithium niobate (LN) thin films onto insulating substrates has led to performance boost in the multilayered surface acoustic wave (SAW) filters, providing substantial enhancements such as increased electromechanical coupling (k^2_{eff}) and quality factors (Bode-Q) [4], [5].

However, the excellent waveguide property of piezoelectric multilayer stack is a double-edged sword, that the velocity profile inherent in multilayer configurations can inadvertently excite spurious transverse modes [6]. These unwanted modes manifest as notches or spikes within the filter passband, potentially leading to data loss and reduced system efficiency, which in turn hampers the development of heterostructure technology towards next generation applications. To suppress transverse modes in multilayered SAW devices, one common approach is to modify the interdigital transducer (IDT) design. Techniques such as tilted, repetitive X-IDT and zigzag structure have been employed to address this issue [7], [8], [9]. However, these methods often lead to increased fabrication complexity, larger device footprints, and potential performance trade-offs

in terms of quality factor (Q) and spectral purity [7].

Recently, there has been a growing interest in utilizing novel materials stack such as 69° Y-90° X quartz and Y-Z lithium tetraborate [10], [11]. The inherent concave slowness curves of such material effectively suppress transverse modes within new platforms. However, the adoption of these materials is still in its infancy, facing challenges such as difficulties in fabrication and incompatibility with existing processes. As a widely used substrate material in integrated circuit manufacturing, silicon (Si) offers favorable electrical properties, excellent thermal conductivity, and good machinability, making it an ideal candidate for advancing front-end module integration. However, there is currently a lack of methods for the intrinsic suppression of transverse modes within established I.H.P. structures.

In this work, we present twisted-placed multilayer to modulate slowness curve of the multilayer stack through twisting precisely the angle between stacked LT and substrate Si (100). The slowness profile manipulation in twisted-placed multilayer and its comparative analyses of spectrum purity with normal LT/SiO₂/Si stack were conducted by finite element method (FEM). The fabricated resonators based on the twisted-placed multilayer demonstrated a significant reduction in transverse modes compared to traditional stacks, ensuring a cleaner spectrum without compromising performance.

Design and Simulation

Fig. 1(a) illustrates a basic configuration of the 42° Y-X LT/SiO₂/Si heterostructure, which is recognized as a highly favored structure in the realm of industrial filter applications as incredible high performance (I.H.P.) SAW technology [12]. 42° Y-X LT was chosen to ensure efficient excitation of the SH main mode, while the Si substrate limits energy leakage due to its high acoustic velocity. It is well known that Si (100) serves as a commonly used type of commercial Si substrate, which is widely employed in I.H.P. heterostructure. The alignment of Si (100) wafer flat is typically along the <110> direction, resulting in an Euler angle of (90°, 90°, 45°) [12]. However, such multilayer stack often leads to a convex slowness curve, which gives rise to substantial transverse spurious modes. Herein, a twisted placed bilayer lithium tantalate structure was adopted, as shown in Fig. 1 (b), in

979-8-3315-0417-5/25 $31.00 © 2025 IEEE

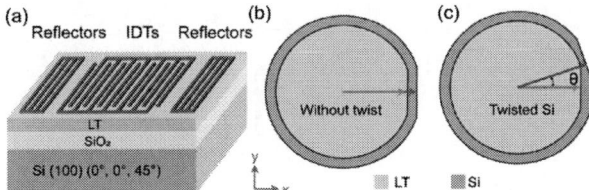

Fig. 1. Basic configuration of resonators based on LT/SiO₂/Si heterostructure. (b) Schematic of the heterostructure with a twisting angle.

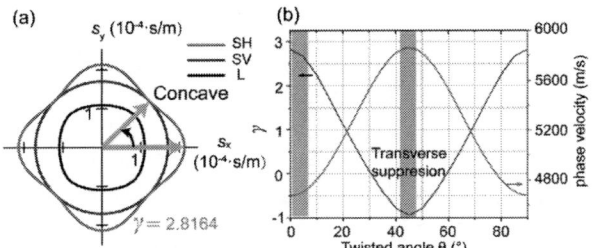

Fig. 2. (a) Calculated BAW slowness curves for Si (100). (b) Slowness curvature and phase velocity for SH wave at different twisting angles.

which the slowness curve of multilayer stack can be manipulated through changing the twisted angle θ.

Fig. 2(a) shows the BAW slowness curve of the Si (100) substrate. The capacity of the substrate to attenuate transverse modes is closely linked to the characteristics of its slowness curve; specifically, a more concave shape enhances its suppression capabilities [13]. Notably, a distinctly convex slowness curve for the SH wave is exhibited in the s_x direction.

The curvature γ is employed to evaluate the convexity and can be calculated as follows:

$$\gamma = -s_x \frac{\partial^2 s_x}{\partial s_y^2}$$

where s_x and s_y represent the slowness in the x- and y-direction, respectively. The traditional Si (100) substrate shows a significant positive curvature value in the s_x direction, which accounts for the susceptibility of reported I.H.P. devices to interference from transverse modes. **Fig. 2(b)** plots the acoustic velocity of the SH wave in the Si (100) substrate alongside the curvature γ as a function of the twisted angle θ. At a twist of 45°, the curvature reaches its minimum value, at which point this twisted multilayer structure demonstrates optimal capability for suppressing transverse modes, corresponding to a Euler angle of (90°, 90°, 0°).

To validate the influence of substrates on the regulation of transverse response, resonators based on twisted placed multilayer and normal heterostructure were simulated through FEM. As shown in **Fig. 3(a)**, a 3-D model of the unit cell was established, while Floquet boundary conditions were set along the y-axis to capture transverse resonance effects. In simulation, wavelength (λ) was set as 2 μm, and the thickness of LT, SiO₂ and Al was 600 nm, 200 nm and 120

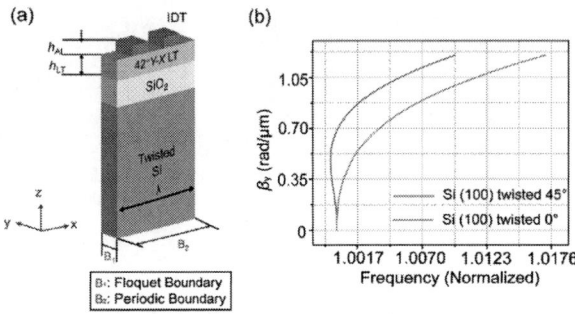

Fig. 3. (a) 3-D unit cell model of simulated resonators based on the 42°Y–X LT/SiO₂/Si structure. (b) Calculated dispersion curves under longitudinal SH resonance conditions at varying twisting angles.

nm respectively. Under the same conditions, the dispersion curves of resonators based on different platforms exhibit significant differences, as shown in **Fig. 3(b)**. The dispersion curve for the twisted multilayer device is vertical, while that of the conventional heterostructure displays a convex shape. This further demonstrates the ability of the twisted structure to regulate the overall dispersion characteristics, highlighting its considerable potential for suppressing transverse modes.

Fabrication

To verify the above simulation results, two types of substrates including the twisted-placed and normal layout of LT/SiO₂/Si were prepared. First, the SiO₂ layer of 200 nm were deposited onto the Si by chemical vapor deposition (CVD). Then, 4-inch single-crystalline 42°YX-LT films were successfully transferred onto Si substrates using smart-cut technology. Subsequently, the chemical mechanical polishing (CMP) technique was employed to trim the 42° Y-X LT film to the target thickness of 600 nm while achieving the desired surface roughness. It is worth noting that the wafer flat of Si (100) substrate was rotated by 45° relative to the 42° Y-X LT substrate for twisted-placed multilayer, as shown in **Fig. 4(a)**. In this configuration, the SH-SAW propagates along the <100> direction of the Si (100) substrate instead of <110>.

Results and Discussion

Building on the simulation results, SAW resonators of λ=2 μm with varying electrode thickness were fabricated using standard photolithography and lift-off processes on the multilayer structures. **Fig. 4(b) and (c)** shows the optical microscope image and enlarged IDT of the devices. The IDT employed a standard layout without utilizing common transverse mode suppression techniques such as tilted, weighted, or piston operations.

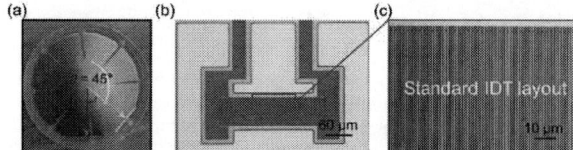

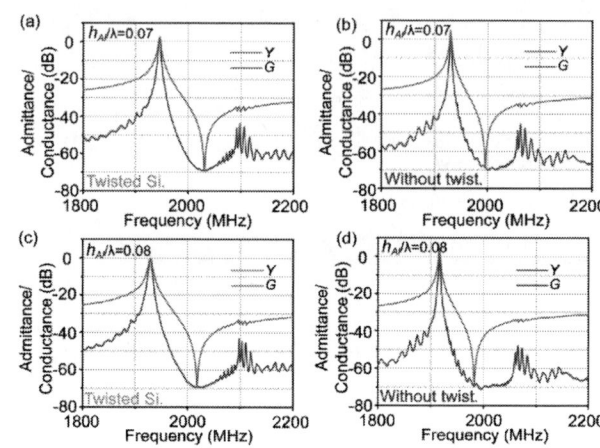

Fig. 4. Bonded LN wafer on Si substrate with twisting angle of 45°. (b) Optical microscope image and (c) enlarged images of the fabricated one-port resonator with λ = 2 μm

Fig. 5 shows the measured admittance and conductance of resonators fabricated on the twisted-placed multilayer and normal heterostructure. When the relative electrode thickness is 0.07 and 0.08, an important distinction confirmed by the conductance and admittance curves is that the transverse spurious modes can be modulated through adjusting the structure parameters such as the thickness of electrode and the twisting angle. Significant transverse modes between resonance (f_r) and anti-resonance (f_a) frequency were identified in the normal heterostructure without a twist, as shown in **Fig. 5(b) and (d)**. These transverse modes create notches or spikes in the filter passband, which may result in data loss, miscommunication, and decreased system efficiency. In contrast, the devices based on the twisted-placed multilayer exhibit clean responses, with no spurious transverse modes observed as shown in **Fig. 5(a) and(c)**. The twisted-placed multilayer leverages the substrate anisotropy to achieve a vertical twist in the stacked slowness curve. It is worth noting that the manipulation of the slowness curve through the twisted-placed multilayer structure does not adversely affect the performance of the SH main mode. Due to the higher SH bulk wave velocity of the twisted 45° Si (100), the resonant frequency of the twisted-placed devices is slightly elevated compared to the traditional structure, while both configurations demonstrate similar Q_{max} values of 1711 and 1746 for h_{Al}/λ=0.07 respectively.

Fig. 5. Measured admittance and conductance curves of resonators with different h_{Al}/λ for Si (100) with twisting angle of 45° and Si (100) without twist.

Science Foundation of China (Grant No. 52002205).

Conclusion

In this study, we achieved intrinsic suppression of transverse modes in LT/SiO$_2$/Si multilayered utilizing a twisted-placed technique through slowness modulation. The resonators fabricated on the twisted-placed multilayer showed a notable decrease in transverse modes compared to conventional stacks, delivering a cleaner spectrum while maintaining performance. Our prototype demonstrates significant potential for enabling spurious-free, and compact RF devices suitable for advanced communication systems.

Acknowledgments

This work was supported in part by the National Key Research and Development Program of China (Grant No. 2022YFB3606700), Natural Science Foundation of Beijing Municipality (Grant No. JQ20010), and National Natural

References

[1] Y. Yang, L. Gao, S. Gong, IEEE Transactions on Microwave Theory and Techniques, 70, pp. 5185–5194 (2022).

[2] R. Su, S. Fu, Z. Lu, J. Shen, H. Xu, P. Liu, Z. Xu, H. Wang, S. Zhang, R. Wang, C. Song, F. Zeng, W. Wang, F. Pan, International Electron Devices Meeting (IEDM), p. 4.2.1-4.2.4 (2022)

[3] H. Yao, P. Zheng, S. Zhang, C. Hu, X. Fang, L. Zhang, D. Ling, H. Chen, X. Ou, Nat Communication, 15, p. 5002 (2024).

[4] P. Liu, S. Fu, R. Su, H. Xu, B. Xiao, C. Song, F. Zeng, F. Pan, Applied Physics Letters, 122, p. 103502 (2023).

[5] T.-H. Hsu, K.-J. Tseng, M.-H. Li, IEEE Electron Device Letters, 41, pp. 1825-1828 (2020).

[6] F. Qian, T. F. Ho, and Y. Yang, IEEE/MTT-S International Microwave Symposium (IMS) pp. 907–910 (2023)

[7] H. Xu, S. Fu, R. Su, P. Liu, B. Xiao, S. Zhang, R. Wang, C. Song, F. Zeng, W. Wang, F. Pan, IEEE Electron Device Letters., 45, pp. 1353–1356 (2024)

[8] Y. Guo, M. Kadota, S. Tanaka, IEEE Transactions on Ultrasonics, Ferroelectrics, and Frequency Control, 71, pp. 182–190 (2024)

[9] Z.-Q. Lee, T.-H. Hsu, Y.-C. Yu, C.-C. Lin, Y.-C. Liao, S. Cho, R. Lu, M.-H. Li, IEEE Transactions on Electron Devices, vol. 71, no. 6, pp. 3880–3887, Jun. 2024.

[10] S. Inoue, M. Solal, IEEE International Ultrasonics Symposium (IUS), pp. 1–4 (2020)

[11] Y. He, T. Wu, Y.-P. Wong, T. B. Workie, J. Bao, and K. Hashimoto, IEEE Transactions on Ultrasonics, Ferroelectrics, and Frequency Control, 70, pp. 1246–1251 (2023)

[12] R. Nakagawa, M. Ozasa, A. Michigami, H. Iwamoto, IEEE International Ultrasonics Symposium (IUS), pp. 1–4 (2023)

[13] B. Xiao, S. Fu, R. Su, H. Xu, P. Liu, X. Zhou, S. Zhang, R. Wang, C. Song, F. Zeng, W. Wang, F. Pan, IEEE Transactions on Electron Devices, 71, pp. 1266–1273 (2024)

A Novel Low Loss Superjunction MOSFET with Hybrid Conduction Modes

Yun Xia[1], Gang Chen[1], Yuxi Wan[1], Wanjun Chen[2], Ruize Sun[2], Xiaoming Wang[2], Yan Wang[3], Xiaoping Wang[1]

[1] Shenzhen Pinghu Laboratory, China, [2] University of Electronic Science and Technology of China, China, [3] Tsinghua University, China,

Abstract

A novel superjunction MOSFET with hybrid conduction modes (HCM SJ-MOSFET) is proposed. The proposed HCM SJ-MOSFET can operate in unipolar conduction mode in the N- pillar and bipolar conduction mode in P- pillar, which dramatically reduces the specific on-resistance ($R_{on,sp}$) by 61.5% in comparison with the conventional SJ-MOSFET (Con SJ-MOSFET). Besides, the conductivity modulation in the P- pillar of the HCM SJ-MOSFET helps reduce its turn-off loss (E_{off}) and turn-on loss (E_{on}) by 56.6% and 33.3%, respectively.

Keywords: Superjunction, MOSFET, and IGBT

Introduction

The superjunction MOSFET (SJ-MOSFET) is widely used as a switch in switching mode power supply circuits and inverter systems [1-4]. Due to the SJ-MOSFET being a unipolar device, its specific on-resistance ($R_{on,sp}$) is greatly related to the doping of the N- pillar. However, the doping of N- pillar is restricted by the breakdown voltage (BV), and a too high N- pillar doping causes a lower tolerance of the variation of N- pillar's doping. Besides, the P- pillar narrows the forward conduction path in SJ-MOSFET which does not contribute to its $R_{on,sp}$.

In this paper, a novel SJ-MOSFET with hybrid conduction modes (HCM SJ-MOSFET) is proposed, it can operate unipolar conduction mode in the N- pillar, meanwhile, it can operate bipolar conduction mode in the P- pillar, which significantly reduces the $R_{on,sp}$ in high current density, breaking the silicon SJ limit. Besides, the HCM SJ-MOSFET reduces the turn-off loss (E_{off}) and turn-on loss (E_{on}) by 56.6% and 33.3% compared with conventional SJ-MOSFET, respectively.

Structure and Mechanism

Fig. 1(a) shows the structures and simplified circuits of the proposed HCM SJ-MOSFET. For comparison, the SJ-MOSFET with oxide pillar (OP SJ-MOSFET) and conventional SJ-MOSFET (Con SJ-MOSFET) are also studied as shown in Fig. 1(b) and 1(c), respectively. The proposed HCM SJ-MOSFET separates the P- pillar and the N- pillar with an oxide pillar, forming a P-drift IGBT (PD-IGBT) in the P- pillar side with the introduction of the N carrier storage (CS) layer, N field stop (FS) layer, and P drain [5-6]. Meanwhile. a dummy gate trench shorted to the source electrode is used to assist in the depletion of the N CS layer. When the proposed device turns on, on its N- pillar side,

operates as a regular unipolar SJ-MOSFET, while on its P-pillar side, bipolar PD-IGBT is operated [5-6]. The P drain/N FS/P- pillar/N CS operates as a PNPN diode, the hole injected from the P drain causes strong conductivity modulation in the P- pillar, which contributes to a dramatically reduced $R_{on,sp}$, E_{on}, and E_{off} in the proposed device.

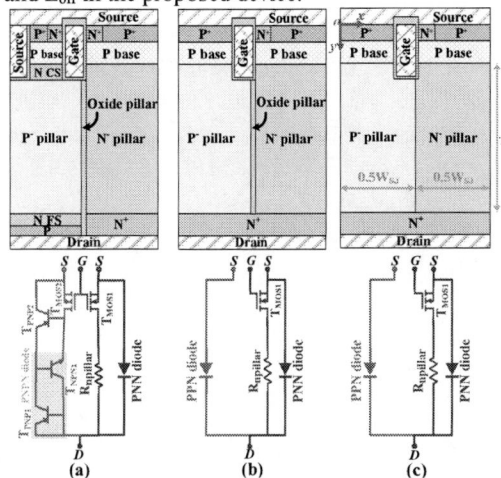

Fig. 1. Schematic views and equivalent circuits of (a) proposed HCM SJ-MOSFET, (b) OP SJ-MOSFET, and (c) Con SJ-MOSFET.

Results and Discussions

Device simulation and mixed-mode simulation results are obtained by using MEDICI TCAD. Calibrated models including IMPACT.I, CONSRH, AUGER, BGN, CONMOB, FLDMOB, ANALYTIC, and SRFMOB2 are all considered. Optimized key device parameters are listed in Table 1. If it is not specified, an active device area of 0.2 cm², and a rated current density of 200 A/cm² are set for the studied SJ-MOSFET devices.

Table 1: Optimized Key Device Parameters

Parameters	HCM SJ.	OP SJ.	Con SJ.
N⁻/P⁻ pillar thickness, T_{SJ} (μm)	40	40	40
N⁻/P⁻ pillar doping, N_{SJ} (×10¹⁵cm⁻³)	4	4	4
N⁻/P⁻ pillar width, W_{SJ} (μm)	5.9	5.9	6.0
Gate oxide thickness, T_{GOX} (μm)	0.1	0.1	0.1
Gate oxide trench depth, D_{GOX} (μm)	3	3	3
Gate oxide trench width, W_{GOX} (μm)	1	1	1
Oxide pillar width, W_{OXP} (μm)	0.1	0.1	/
N CS layer doping, (×10¹⁶ cm⁻³)	2	/	/
N CS layer thickness, (μm)	1	/	/
N FS layer doping, (×10¹⁶ cm⁻³)	2	/	/
N FS layer thickness, (μm)	1.5	/	/
P drain doping, (×10¹⁷ cm⁻³)	2	/	/
P drain thickness, (μm)	0.5	/	/
Hole/electron carrier lifetime (μs)	10	10	10

Owing to the introduction of PD-IGBT in the p-pillar part, the BV of the proposed HCM SJ-MOSFET (690 V) is slightly lower than the OP SJ-MOSFET (707 V) and the Con SJ-MOSFET (707 V) as shown in Fig. 2.

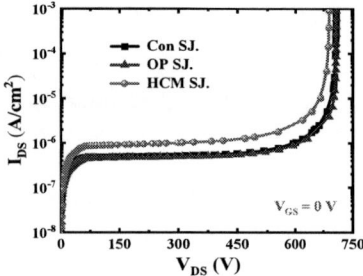

Fig. 2. The forward blocking characteristic.

Fig. 3(a) and 3(b) show the transfer characteristic and forward conduction characteristic, respectively. As shown in Fig. 3(a), these studied devices share the same threshold voltage (V_{TH}), but the proposed HCM SJ-MOSFET shows the highest transconductance and saturation current. As shown in Fig. 3(b), the proposed device achieves a low $R_{on,sp}$ of 5.31 m$\Omega \cdot$cm^2 at I_{DS} = 200 A/cm^2 and V_{GS} = 10 V, which reduces by 59.8% and 61.5% in comparison with OP SJ-MOSFET (13.2 m$\Omega \cdot$cm^2) and Con SJ-MOSFET (13.8 m$\Omega \cdot$cm^2), respectively. Owing to the existence of the oxide pillar, the current in the HCM SJ-MOSFET in the P drain and the N$^+$ drain is mutually independent as shown in Fig. 3(b).

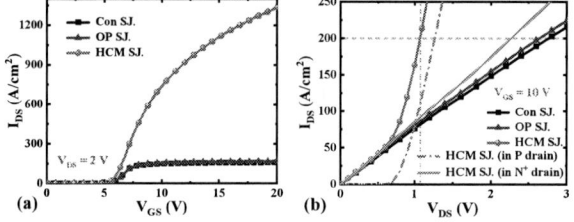

Fig. 3. (a) The transfer characteristic. (b) The forward conduction curves.

The factor that accounts for the improvement in the $R_{on,sp}$ of the proposed device is shown in Fig. 4. As depicted in Fig. 4, the current flows in the proposed HCM SJ-MOSFET indicate that bipolar conduction occurs in its P- pillar side, this makes great contributions to the reduction in $R_{on,sp}$.

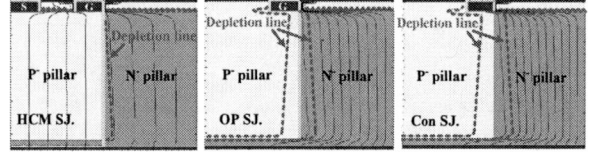

Fig. 4. The current flowlines and depletion lines distribution at V_{GS} = 10 V and I_{DS} = 200 A/cm^2.

The turn-off performance of the three studied devices is shown in Fig. 5, where the proposed device obtains an E_{off} of 0.33 mJ/cm^2, which is 50.0% and 56.6% lower than the 0.66 mJ/cm^2 in OP SJ-MOSFET and 0.76 mJ/cm^2 in the Con SJ-MOSFET, respectively. Though bipolar conduction occurs in the P-pillar of the proposed device, it obtains the lowest E_{off}. This is because the blocking junction in the P- pillar part of

the proposed device is the P- pillar/N FS junction, the electric field can be built in this junction until the on-state excess electron/hole in the P- Pillar is fully extracted as shown in Fig. 6(a), 6(b) and 6(c). As a result, the V_{DS} stays low and the V_{GS} is gradually reduced until the on-state excess hole/electron in the P- Pillar is fully extracted. Hence, the electron channel can be shut down before the V_{DS} rises. Then the I_{DS} charges both the C_{GD} and C_{DS} simultaneously as shown in Fig. 5(d), which speeds the turn-off and contributes to greatly reduced E_{off} in the proposed HCM SJ-MOSFET.

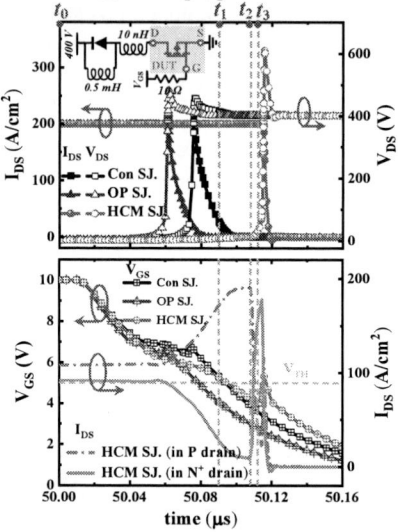

Fig. 5. The turn-off waveforms. The current waveforms in the P drain and N$^+$ drain of the proposed device are also depicted. The inset is the turn-off test circuit, where the V_{GS} is from 10 V to 0 V.

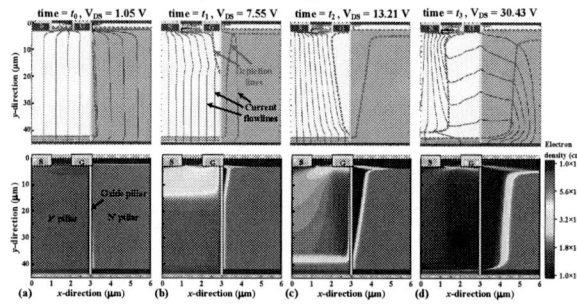

Fig. 6. The current flowlines, depletion lines and electron density distributions of the proposed HCM SJ-MOSFET at varying times during the turn-off transient.

Owing to the conductivity modulation in the P- pillar of the proposed device being less influenced by the channel resistance, the resistor in the P- pillar can reduce dramatically once the gate channel is conducted. As a result, the time for the I_{DS} rising in the turn-on transient of the proposed can be dramatically reduced as shown in Fig. 7. This causes a 33.3% decrease in the E_{on} (1.4 mJ/cm^2) compared with the OP SJ-MOSFET and Con SJ-MOSFET (2.1 mJ/cm^2).

Fig. 8 shows the reverse conduction performance of the three studied devices. Due to the P- pillar unconducted in the reverse conduction state, the proposed device shows a

979-8-3315-0417-5/25 $31.00 © 2025 IEEE

slightly higher conduction voltage (V_f). However, owing to the half-reduced active conduction area, the hole stored in the drift region during the reverse conduction state can be reduced, this causes a 50.8% reduction in reverse recovery charge (Q_{rr}) in the proposed device as shown in Fig. 8 [7].

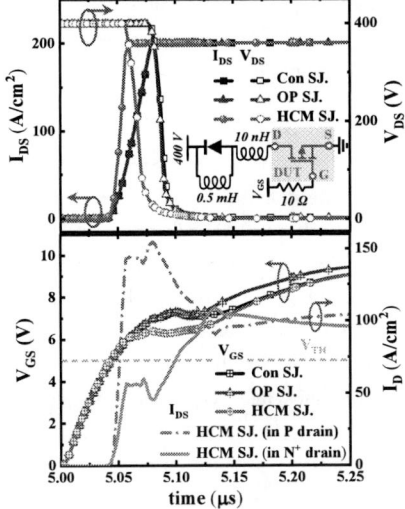

Fig. 7. The turn-on waveforms. The current waveforms in the P drain and N+ drain of the proposed device are also depicted. The inset is the turn-on test circuit, where the V_{GS} is from 0 V to 10 V. For simplicity, the diode used in the simulation is an ideal diode with no reverse recovery charge.

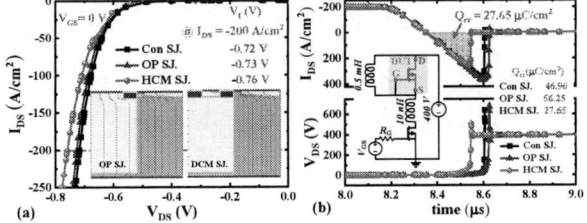

Fig. 8. (a) The reverse conduction characteristic. The inset is the current flowlines distribution at V_{GS} = 0 V and I_{DS} = -200 A/cm^2. (b) The reverse recovery characteristic. The inset is the test circuit.

The BV–$R_{on,sp}$ performance of the proposed HCM SJ-MOSFET is compared with the reported results as shown in Fig. 9 [8-11]. The proposed HCM SJ-MOSFET breaks the silicon SJ limits and achieves the best BV–$R_{on,sp}$ performance.

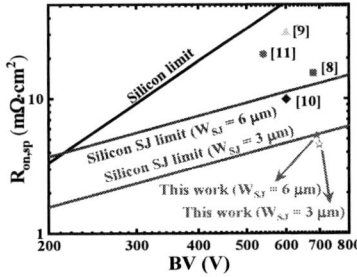

Fig. 9. The BV–$R_{on,sp}$ performance of the proposed HCM SJ-MOSFET and some reported results.

Conclusion

A novel superjunction MOSFET with hybrid conduction modes is proposed and investigated in this paper. Owing to the conductivity modulation in the P- pillar part, the proposed HCM SJ-MOSFET obtains dramatically reduced $R_{on,sp}$, which is 61.5% lower than the Con SJ-MOSFET, breaking the relationship of BV and $R_{on,sp}$ of silicon SJ limit. Besides, the conductivity modulation in the P-pillar makes its E_{on} and E_{off} reduce by 33.3% and 56.6%, respectively, compared with the Con SJ-MOSFET. Since the P- pillar of the proposed HCM SJ-MOSFET does not conduct in the reverse conduction mode, hence its Q_{rr} can be reduced by 50.8%.

References

[1] Xing-Bi Chen and J. K. O. Sin, "Optimization of the specific on-resistance of the COOLMOS^{TM}," IEEE Transactions on Electron Devices, 48, pp. 344-348 (2001).

[2] Fujihira and Y. Miyasaka, "Simulated superior performances of semiconductor super junction devices," Proc. of the ISPSD (1998).

[3] W. Saito, "Comparison of theoretical limits between superjunction and field plate structures," 2013 25th International Symposium on Power Semiconductor Devices & IC's (ISPSD) (2013).

[4] H. Kang and F. Udrea, "Analytic Model of Specific ON-State Resistance for Superjunction MOSFETs With an Oxide Pillar," IEEE Electron Device Letters, 40, pp. 761-764 (2019).

[5] X. Xu et al., "An Ultralow Loss N-channel RB-IGBT with P-drift Region," 2020 32nd International Symposium on Power Semiconductor Devices and ICs (ISPSD) (2020).

[6] X. Xu et al., "Numerical Analysis for a P-Drift Region N-IGBT With Enhanced Dynamic Electric Field Modulation Effect," IEEE Transactions on Electron Devices, 69, pp. 3277-3282 (2022).

[7] Y. Xia, W. Chen, R. Sun, C. Liu, Z. Li and B. Zhang, "A High Voltage Superjunction MOSFET with Enhanced Reverse Recovery Performance," 2021 5th IEEE Electron Devices Technology & Manufacturing Conference (EDTM) (2021).

[8] W. Saito et al., "A 15.5mΩ cm2-680V Superjunction MOSFET Reduced On-Resistance by Lateral Pitch Narrowing," 2006 IEEE International Symposium on Power Semiconductor Devices and IC's (2006).

[9] P. Nautiyal, A. Agrawal, S. Kumari, H. Sahu, A. Naugarhiya and S. Verma, "Electrical characteristic investigation of variation vertical doping superjunction UMOS," 2019 IEEE 16th India Council International Conference (INDICON) (2019).

[10] Z. Lin, H. Huang and X. Chen, "An Improved Superjunction Structure With Variation Vertical Doping Profile," IEEE Transactions on Electron Devices, 62, pp. 228-231 (2015).

[11] S. Iwamoto, K. Takahashi, H. Kuribayashi, S. Wakimoto, K. Mochizuki and H. Nakazawa, "Above 500V class Superjunction MOSFETs fabricated by deep trench etching and epitaxial growth,". The 17th International Symposium on Power Semiconductor Devices and ICs, (2005).

Graphene/Silicon Pixel Array with Integrated Imaging and Readout System

Yuan Ma[1,#], Zongwen Li[1,#], Youshui He[1,#], Qianqian Zhang[1], Yunfei Xie[1], Zhi-Xiang Zhang[1], Feng Tian[1], Muhammad Abid Anwar[1], Muhammad Malik[1], Srikrishna Chanakya Bodepudi[1], Bin Yu[1], Yuda Zhao[1], Yang Xu[1,*]

[1]College of Integrated Circuits, ZJU-Hangzhou Global Scientific and Technological Innovation Center, State Key Laboratory of Silicon and Advanced Semiconductor Materials, Zhejiang University, Hangzhou 310027, China,

*Corresponding Author, E-mail: yangxu-isee@zju.edu.cn

Abstract

Driven by advances in intelligent imaging and sensing technologies, the integration between novel materials and silicon (Si)-based optoelectronics processes is significantly expanding the functionality of optoelectronic devices. Herein, we have fabricated a state-of-the-art graphene (Gr)/Si heterojunction photodetector with 3 × 3 pixel array, demonstrating promising photoelectric detection behavior. This work provides a simplified device model for performance simulation and designed a readout circuit that enhances efficiency and reduces noise, achieving effective signal processing for Gr/Si integrated imaging and readout.

(Keywords: Gr/Si photodetector, readout circuit, imaging, integration)

Introduction

With the increasing demand for advanced imaging technologies across fields such as biomedical diagnostics, environmental monitoring, and high-resolution sensors, the integration of innovative materials into imaging and readout systems has become a focal point of research [1]-[3]. Traditional silicon (Si)-based imaging systems, despite their mature development and wide deployment, face fundamental limitations in sensitivity, resolution, and scalability when tasked with capturing complex or low-intensity signals [4]. The challenge of poor compatibility between Si and other optoelectronic materials is also present. As such, the exploration of alternative materials that can synergize with Si has become critical. Among various candidates, graphene (Gr), an atomic layer-thick material with remarkable electrical, optical, and mechanical properties, has emerged as a promising option for enhancing the performance of imaging and readout technologies [5]. Especially, the integration of Gr with Si technology leverages the established Si infrastructure while embedding Gr unique capabilities, creating a pathway toward highly efficient, and compact imaging devices [6], [7]. However, achieving a seamless integration between Gr and Si that optimizes both the material interface and the signal readout efficiency poses significant technical challenges. In this work, we have fabricated an Gr/Si photodetector array with low noise current (< 10⁻¹¹ A) and favorable photoresponse. A simplified device model and readout circuit have been developed that improve efficiency and reduce noise, enabling efficient signal processing in integrated imaging and readout.

Device Fabrication

The Gr/Si photodetector with a 3 × 3 pixel array was fabricated by using the complete Si-based optoelectronics process. The detailed fabrication processes of selected pixels are documented in Figure 1, including I) primary detector chip, II) upper cladding etching by buffered oxide etch solution, III) Gr transferring and IV) patterning. Here, polymethyl methacrylate was spin-coated onto the Gr surface, followed by wet transfer, and then the excess Gr was removed by the photolithographic mask and plasma cleaning processes.

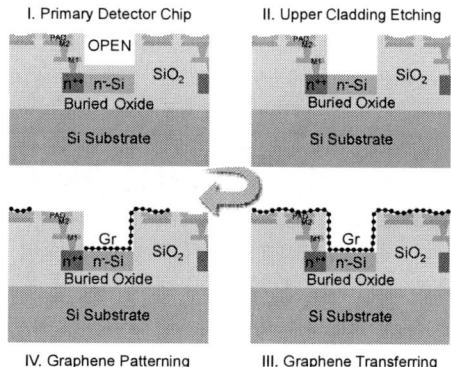

Fig. 1: Schematic diagram of Gr/Si photodetector fabrication processes.

Device Properties and Characterization

The schematic diagram of the Gr/Si photodetector with charge transport and current flow is demonstrated in Figure 2a. The scanning electron microscope (SEM) image (right image in Figure 2a) shows that Gr was tightly laminated onto the exposed Si, providing a transparent conductive layer for resistive optoelectronic sensing. The energy band diagram in Figure 2b clearly displays the operating mechanism of the Gr/Si Schottky photodetector. When the Gr is contacted with the n-type Si (n-Si), the electrons in Si will be injected into Gr since the work function of Gr is larger than n-Si, resulting in a depletion region and a Schottky barrier at the interface. Under dark conditions, the Schottky barrier at the Gr/Si interface determines charge carrier flow. In forward bias, the barrier height decreases, promoting charge carrier injection from Si to Gr, while reverse bias increases the barrier height, suppressing current flow. Upon illumination, photogenerated electron-hole pairs in the Si diffuse toward the Gr/Si interface, where the built-in electric field of the Schottky junction separates them. Electrons are driven toward the Gr layer, while holes migrate deeper into the Si substrate. Consequently, the presence of these additional photoexcited carriers results in a significant photocurrent.

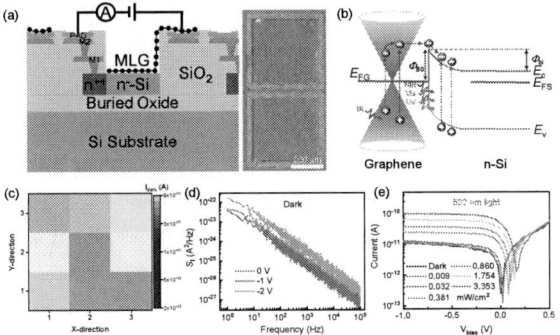

Fig. 2: (a) Schematic design of Gr/Si photodetector. Right image shows a SEM image of the pixel unit. (b) Energy band diagram of device working mechanism. (c) Dark current distributions of all array pixels at -1 V bias. (d) The NSD as a function of frequency at low reverse bias. (e) The logarithmic I–V curves of the device under 532 nm light illuminations.

The negligible fluctuation in dark current, as shown in Figure 2c, demonstrates the high homogeneity of all pixels in the Gr/Si photodetector array. The average dark current was observed to be ~10 pA. Improving the junction quality via highly conformal transfer of Gr and vacuum annealing significantly suppresses the direct influence of dark currents generated by the thermal effect from Gr due to device size. The noise-spectral-density S_I (Figure 2d) of the device, only ~10^{-23} A^2/Hz at low bias, demonstrates a comparatively high signal-to-noise ratio and a low-defect interface. Figure 2e illustrates the I–V characteristics of the Gr/Si photodetector under the dark and 532 nm light irradiation, different power densities from 0.009 to 3.353 mW/cm^2, indicating a promising photodetection response.

Integrated Imaging and Readout System

A simplified photodiode model is used to simulate various properties of our photodetector, as plotted in Figure 3a. The model consists of a parallel photocurrent source, a shunt resistor, a diode, and a series resistor. As shown in Figure 3b, the lines and circles represent the simulated and experimental data, respectively. The simulation aligns well with experimental results under different light intensity.

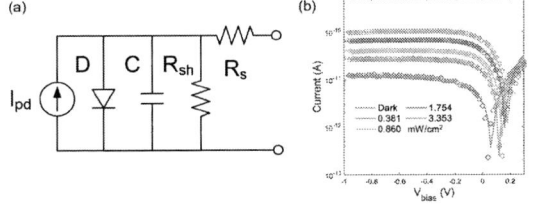

Fig. 3: Equivalent circuit of photodiode (a) and a comparison between simulation and experimental results (b).

After converting the model into Verilog-A code, it can be imported into circuit simulation software to design a customized readout circuit based on device characteristics. The circuit architecture is shown in Figure 4. Each pixel is equipped with a readout analog front-end (AFE) circuit to enhance speed and parallelism. During readout, the device currents are converted into voltages simultaneously through their respective capacitive transimpedance amplifiers (C-TIAs) [8]. The transimpedance gain of the C-TIA depends on the ratio between the integration time and the value of the feedback capacitor. To minimize the impact of analog-to-digital converter (ADC) offset on readout uniformity, a global ADC is employed for analog-to-digital conversion. The voltage output from each AFE is sequentially gated and fed into the ADC. After processing within the digital circuitry, the data is output off-chip via UART interface.

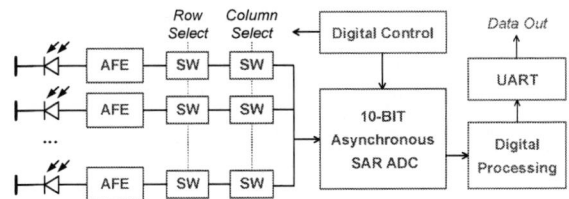

Fig. 4: Overall architecture of the readout circuit.

Circuit configuration of AFE are shown in Figure 5a. By choosing a small capacitance or long integration time, a large transimpedance gain can be generated with higher accuracy than R-TIA [9]. In this case, we achieve a current input level ranging from 10 pA to 1 nA with a specific integration time, which is compatible with the device's performance. Furthermore, we adopt correlated double sampling (CDS) to reduce noise. The implementation involves performing two exposures at the beginning and end stages, as shown in Figure 5b. Both exposures undergo analog-to-digital conversion, and the final signal is obtained by subtracting the two converted values [10].

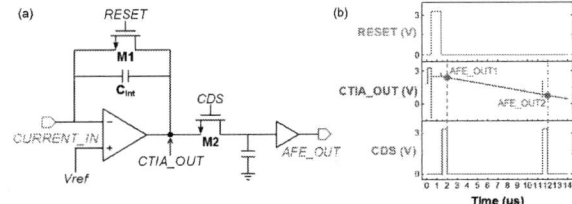

Fig. 5: Circuit configuration (a) and operation (b) of AFE.

An asynchronous successive approximation register (SAR) ADC is used to meet the conversion speed requirements [11], as shown in Figure 6. The signal is passed through the sampling and Single-ended to differential conversion module and input to capacitor analog-to-digital converter (CDAC) arrays. The comparator, composed of a preamplifier and a dynamic latch comparator, compares the voltages of two CDACs and outputs the comparison result. The clock generation module is responsible for generating the Start of Conversion (SOC) and End of Conversion (EOC) signals as well as the clocks required by the SAR logic circuit. The quantization process employs a Vcm-based switching scheme to reduce power consumption [12]. According to simulations, this ADC achieves an effective number of bits (ENOB) of 9.6 bits at a 5 MS/s sampling rate.

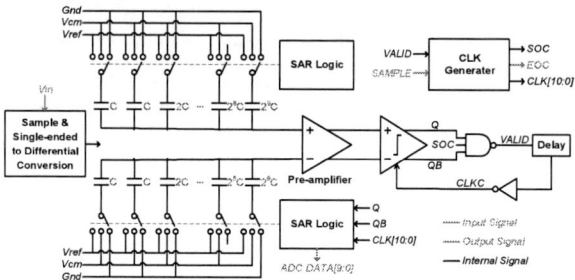

Fig. 6: 10-bit asynchronous SAR ADC circuit architecture.

The detector chip and readout chip were fabricated with Si-based optoelectronics and 55 nm CMOS technologies, respectively. The two chips were wire-bonded to a substrate and interconnected through the substrate to form an Gr/Si integrated imaging and readout (GSIIAR) system, as shown in Figure 7a. Imaging tests were conducted on the GSIIAR system. The target images and results after imaging and readout are shown in Figure 7b.

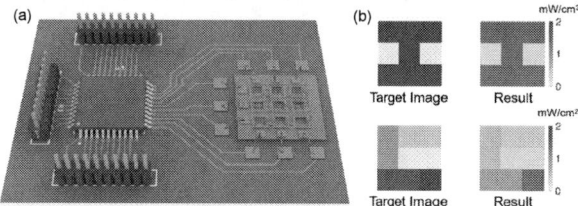

Fig. 7: Integration schematic (a) and imaging results (b) of photodetector array and readout circuits.

Conclusion

In summary, an Gr/Si heterojunction photodetector with 3 × 3 pixel array was fabricated and demonstrated low noise and promising photodetection performance. Moreover, an imaging test system, integrating the readout circuit designed via device modeling with the photodetector chip, has been successfully developed. This work shows the potential for integrated imaging applications.

Acknowledgments

#Yuan Ma, Zongwen Li and Youshui He contribute equally to this work. This work is supported by National Key R&D Program of China 92164106, National Natural Science Foundation of China (62104200, U22A2076), Natural Science Foundation of Zhejiang Province (LDT23F04013F04), ZJU Micro-Nano Fabrication Center.

References

[1] Z. Li, T. Yan, and X. Fang, "Low-dimensional wide-bandgap semiconductors for UV photodetectors", Nat. Rev. Mater., 8, pp. 587-603 (2023).

[2] S. Das, A. Sebastian, E. Pop, C. J. McClellan, A. D. Franklin, T. Grasser, T. Knobloch, Y. Illarionov, A. V. Penumatcha, J. Appenzeller, Z. Chen, W. Zhu, I. Asselberghs, L.-J. Li, U. E. Avci, N. Bhat, T. D. Anthopoulos, and R. Singh, "Transistors based on two-dimensional materials for future integrated circuits", Nat. Electron, 4, pp. 786–799 (2021).

[3] L. Mennel, J. Symonowicz, S. Wachter, D. K. Polyushkin, A. J. M.-Mendoza, and T. Mueller, "Ultrafast machine vision with 2D material neural network image sensors", Nature, 579, pp. 62-66 (2020).

[4] W. Yang, J. Chen, Y. Zhang, Y. Zhang, J.-H. He, and X. Fang, "Silicon-compatible photodetectors: trends to monolithically integrate photosensors with chip technology", Adv. Funct. Mater., 29, pp. 1808182 (2019).

[5] A. K. Geim, K. S. Novoselov, "The rise of graphene", Nat. Mater., 6, pp. 183-191 (2007).

[6] Y. Xu, L. Chen, L. Peng, W. Fang, Y. Chen, X. Wang, Y. Dong, S. C. Bodepudi, Y. Zhao, C. Gao, and Bin Yu, "Broadband Graphene-Silicon Integrated Imagers", IEEE Electron Devices Technology and Manufacturing Conference (EDTM) (2023)

[7] W. Liu, J. Lv, L. Peng, H. Guo, C. Liu, Y. Liu, W. Li, L. Li, L. Liu, P. Wang, S. C. Bodepudi, K. Shehzad, G. Hu, K. Liu, Z. Sun, T. Hasan, Y. Xu, X. Wang, C. Gao, B. Yu, and X. Duan, "Graphene charge-injection photodetectors", Nat. Electron., 5, pp. 281-288 (2022).

[8] Y. Zhuo, W. Lu, S. Yu, Y. Zhou, J. Kong, and Y. Zhang, "A Low-Noise CTIA-based pixel with CDS for SWIR Focal Plane Arrays." in International Conference on Integrated Circuits and Microsystems (ICICM), pp. 258-262, 2021.

[9] Caizzone, Antonino, Assim Boukhayma, and Christian Enz. "17.8 A 2.6 µW Monolithic CMOS Photoplethysmographic Sensor Operating with 2µW LED Power." In 2019 IEEE International Solid State Circuits Conference-(ISSCC), pp. 290-291. IEEE, 2019.

[10] N. Chen, S. Zhong, M. Zou, J. Zhang, Z. Ji, and L. Yao, "A Low-Noise CMOS Image Sensor With Digital Correlated Multiple Sampling," in IEEE Transactions on Circuits and Systems I: Regular Papers, vol. 65, no. 1, pp. 84-94, Jan. 2018

[11] C. P. Huang, J. M. Lin, Y. T. Shyu, and S. J. Chang, "A Systematic Design Methodology of Asynchronous SAR ADCs," in IEEE Transactions on Very Large Scale Integration (VLSI) Systems, vol. 24, no. 5, pp. 1835-1848, 2016.

[12] Z. Fu, X. Tang, D. Li, J. Wang, D. Basak, and K. -P. Pun, "A 10-bit 2 MS/s SAR ADC using reverse VCM-based switching scheme." in 2016 IEEE International Symposium on Circuits and Systems (ISCAS), Montreal, QC, Canada, pp. 1030-1033, 2016.

979-8-3315-0417-5/25 $31.00 © 2025 IEEE

Effect of fin dimensions on the performance of GaN-on-Si Fin-HEMTs

Mengdi Li[1], Jiejie Zhu[1], Lingjie Qin[1], Bowen Zhang[1], Yuxi Zhou[1], Mingchen Zhang[1], Yichong Ding[1],
Yuchen Qian[1], Xiaohua Ma[1], Yue Hao[1]

[1]Xidian University, China

Abstract

The influence of fin dimensions variation on characteristics the AlGaN/GaN Fin-HEMTs on Si are reported in this work. The device with the fin widths of 250 nm exhibited a maximum drain current of 1.5 A/mm, a peak transconductance of 314 mS/mm, a gate voltage swing of 3.7 V, a cut-off frequency f_T of 35 GHz, and a maximum oscillation frequency f_{max} of 78 GHz. Good performance demonstrates its potential for cost-effective microwave wave applications.

Keywords: GaN-on-Si, FinFET and fin dimension

Introduction

As the third generation of semiconductor materials, GaN has obvious advantages in breakdown field strength, electron saturation speed, electron mobility, thermal conductivity and so on compared with the first generation of semiconductor materials (Si, Ge) and the second generation of semiconductor materials (GaN). The two-dimensional electron gas (2DEG) of GaN HEMT has attracted much attention due to its high carrier density and mobility [1], [2], [3]. GaN-on-Si HEMTs are preferred for large-scale commercialization of GaN HEMTs due to their low cost, high thermal conductivity, and ability to provide wafers up to 12 inches in size [4], [5]. In addition, GaN-on-Si HEMTs can be integrated with other silicon-based devices, enabling a more compact design and higher power density.

Faced with problems such as short channel effect, the preparation of transistors must break the traditional planar structure and develop into three-dimensional structure. The invention of FinFET (Fin-Field Effect Transistor) led to the continuation of Moore's law, which is characterized by the ability of the gate to control the conducting channel to a greater extent. For AlGaN/GaN Fin-HEMTs, 2DEG channel is controlled not only by the top of the gate, at the same time also by the side wall of the gate.

In this paper, the AlGaN/GaN Fin-HEMTs with the different fin widths (W_{fin}) are fabricated to investigate the effect of the W_{fin} on device performance. The AlGaN/GaN Fin-HEMTs with W_{fin} of 0.25 μm show a high level of current of 1.5 A/mm, a maximum transconductance of 314 mS/mm, and a gate voltage swing of 3.7 V. Subsequently, the f_T/f_{max} value of 35/78 GHz is obtained for the Fin-HEMT at V_{ds} of 6 V by small signal measurement.

Device and Fabrication

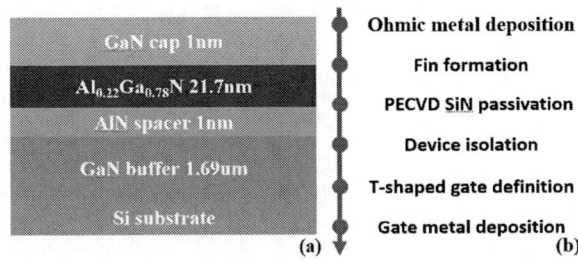

Fig. 1. (a) The epitaxial information of the AlGaN/GaN heterostructure. (b) Fabrication process flow of the AlGaN/GaN Fin-HEMTs.

The epitaxial information of the AlGaN/GaN heterostructure and the fabrication process flow of the AlGaN/GaN Fin-HEMTs is shown in Fig. 1. The heterostructure was grown on a 6-inch high-resistivity Si (111) substrate by metal-organic chemical vapor deposition (MOCVD). The epilayers include a 1-nm of GaN cap, a 21.7-nm of unintentionally doped $Al_{0.22}Ga_{0.78}N$, a 1 nm of AlN spacer, and a 1.69 μm of GaN buffer layer. A 2DEG density of 8.35×10^{12} cm^{-2}, an high electron mobility of 2226 cm^2/V·s, and a sheet resistance of 336 Ω/sq were obtained from Hall measurements.

The device fabrication process began with the deposition of the metal Ti/Al/Ni/Au stack as ohmic contact formation, followed by rapid thermal annealing at 850 °C for 60 s in N$_2$ atmosphere. Afterward, fin pattern was defined by Deep ultraviolet lithography (DUV), followed by a BCl$_3$/Cl$_2$ plasma etching to remove the 2DEG channel. Concretely, the W_{fin} are 150, 200, 250 nm with the same W_{trench} of 200 nm. A 120 nm SiN passivation layer was deposited at 250 °C by using plasma enhanced chemical vapor deposition (PECVD), followed by the device isolation enabled by nitrogen ion implantation. The ohmic contact resistances of 0.36 Ω·mm and the sheet resistance of 313 Ω/sq were measured for the devices by using on-wafer transmission line model. Then, the gate foot was defined by DUV. Subsequently, the gate foot was removed by CF$_4$/O$_2$ plasma plasma etching to remove SiN passivation. Finally, Ni/Au metal stack was deposited as the gate Schottky contact. For comparison, the planar HEMT with the same device dimension was also fabricated. A drain-source spacing (L_{sd}), a gate length (L_g), a drain-gate spacing (L_{gd}), and a gate width for the fabricated devices were 2.5, 0.2, 1.3, and 2 × 50 μm, respectively.

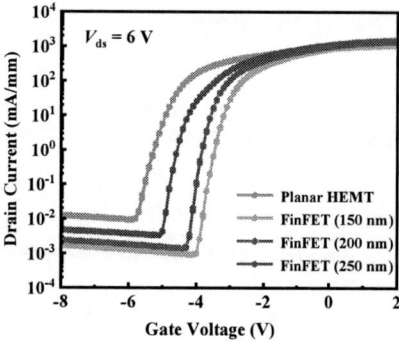

Fig. 2. The transfer characteristics of the AlGaN/GaN planar HEMTs and Fin-HEMTs.

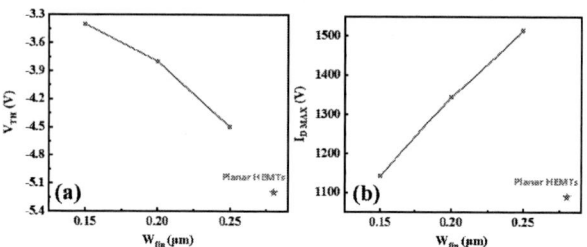

Fig. 3. Influence of the W_{fin} on the (a) V_{th} and (b) I_{ds} for the AlGaN/GaN Fin-HEMTs.

Results and Disscusion

Fig. 2 gives the transfer characteristics of the AlGaN/GaN planar HEMTs and Fin-HEMTs. The Fin-HEMTs has a lower leakage level (~1E-3 mA/mm) compared to the planar HEMTs, which is attributed to the excellent low damage etching process. The AlGaN/GaN planar HEMTs exhibits a threshold voltage (V_{th}) of -5.2 V, while the V_{th} of the Fin-HEMTs is increased due to the depletion of the gate side wall. The V_{th} of -3.4, -3.8, and -4.5 V was observed to the Fin-HEMTs with the W_{fin} of 150, 200, and 250 nm, respectively. As shown in Fig. 3(a), for the devices with the different W_{fin}, it can be found that the V_{th} of the Fin-HEMTs shifts in a positive direction as the W_{fin} decreases. The saturation output current of the Fin-HEMTs increases with the increase of the W_{fin}, as shown in Fig. 3(b). The device with W_{fin} = 250 nm exhibited a maximum drain current density (I_{ds}) of 1.51 A/mm. Fig. 4 shows the transconductance characteristics of the AlGaN/GaN planar HEMTs and Fin-HEMTs. As shown in Fig. 5(a), since the Fin-HEMTs enhances the gate control ability to the channel, the Fin-HEMTs have a higher transconductance (>300 mS/mm) compared to the planar HEMTs (236 mS/mm). The gate voltage swing as a function of W_{fin} are summarized in Fig. 5(b). It can be observed that the Fin-HEMTs have a better gate voltage swing, which reaches 3.7 V for the Fin-HEMTs with W_{fin} = 250 nm.

Output characteristics of the Fin-HEMTs with W_{fin} = 250 nm at V_{gs} = -6 ~ 2 V are shown in Fig. 6, indicating the I_{ds} of 1.53 A/mm and the knee voltage (V_{knee}) of 1.8 V. Fig. 7 presents the gate diode characteristics of the Fin-HEMTs with W_{fin} = 250 nm. The off-state gate leakage current at V_{gs} = -8 V for

the device is 4E-3 mA/mm. The breakdown voltage (V_{br}) is defined as the drain voltage where the drain current density reaches 1 mA/mm, while the device is biased at V_{gs} = -8 V. As shown in Fig. 8, the breakdown is caused by the buffer leakage for the Fin-HEMTs with the V_{br} of 53 V.

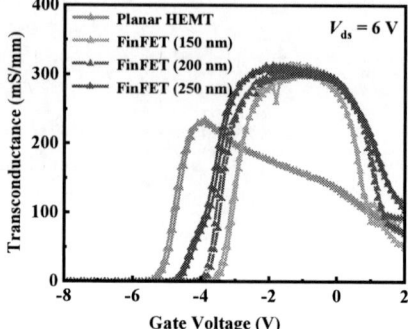

Fig. 4. The transconductance characteristics of the AlGaN/GaN planar HEMTs and Fin-HEMTs.

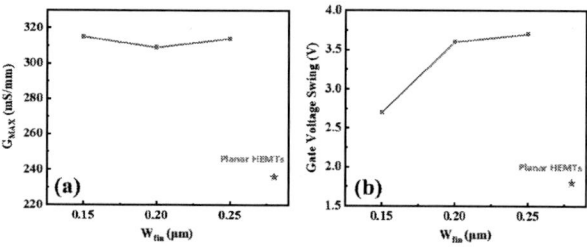

Fig. 5. Influence of the W_{fin} on the (a) transconductance and (b) gate voltage swing for the AlGaN/GaN Fin-HEMTs.

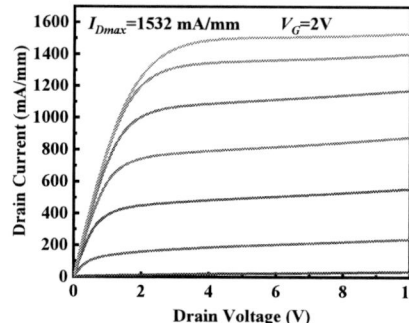

Fig. 6. The output characteristics of the AlGaN/GaN Fin-HEMTs with W_{fin} = 250 nm.

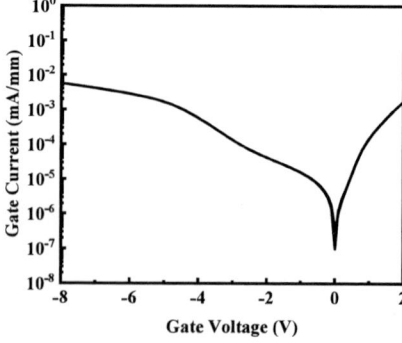

Fig. 7. The gate leakage of the AlGaN/GaN Fin-HEMTs with W_{fin} = 250 nm.

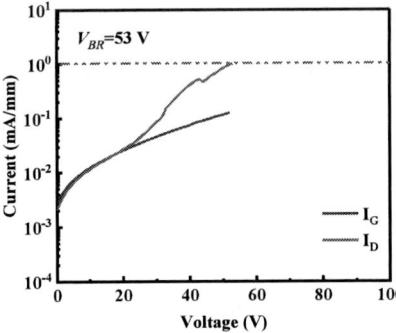

Fig. 8. The breakdown characteristics of the Fin-HEMTs with W_{fin} = 250 nm.

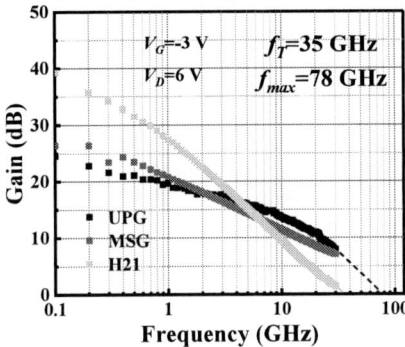

Fig. 9. RF performance of the AlGaN/GaN Fin-HEMTs with W_{fin} = 250 nm.

S-parameters were measured in the frequency ranging from 1 to 40 GHz using Agilent 8363B network analyzer calibrated with a short-open through calibration standard. The small signal characteristics were measured in the frequency ranging from 1 to 40 GHz using Agilent E8363B PNA network analyzer calibrated with the through-reflect-line calibration standard. Fig. 10 illustrates the microwave performances of the Fin-HEMTs with W_{fin} = 250 nm, biased V_{ds} = 6 V and V_{gs} = -3 V. The Fin-HEMTs deliver a current gain cutoff frequency (f_T) of 35 GHz and a maximum oscillation frequency (f_{max}) of 78 GHz.

Conclusion

In summary, the AlGaN/GaN Fin-HEMTs on Si with W_{trench} = 150, 200, 250 nm are fabricated to make a comparison and find out the influence of W_{trench} on performance including V_{th}, I_{ds}, transconductance and gate voltage swing. The performance shows the great potentials of GaN-on-Si Fin-HEMTs for microwave applications.

References

[1] S. Nakajima, "GaN HEMTs for 5G Base Station Applications," *2018 IEEE International Electron Devices Meeting (IEDM)*, San Francisco, CA, USA, 2018, pp. 14.2.1-14.2.4, doi: 10.1109/IEDM.2018.8614588.

[2] H. W. Then et al., "3D heterogeneous integration of high performance high-K metal gate GaN NMOS and Si PMOS transistors on 300mm high-resistivity Si substrate for energy-efficient and compact power delivery, RF (5G and beyond) and SoC applications," *2019 IEEE International Electron Devices Meeting (IEDM)*, San Francisco, CA, USA, 2019, pp. 17.3.1-17.3.4, doi: 10.1109/IEDM19573.2019.8993583.

[3] Y. Hao et al., "High-Performance Microwave Gate-Recessed AlGaN/AlN/GaN MOS-HEMT With 73% Power-Added Efficiency," in *IEEE Electron Device Letters*, vol. 32, no. 5, pp. 626-628, May 2011, doi: 10.1109/LED.2011.2118736.

[4] K.-Y. R. Wong et al., "A next generation CMOS-compatible GaN-on-Si transistors for high efficiency energy systems," *2015 IEEE International Electron Devices Meeting (IEDM)*, Washington, DC, USA, 2015, pp. 9.5.1-9.5.4, doi: 10.1109/IEDM.2015.7409663.

[5] D. Marti, S. Tirelli, A. R. Alt, J. Roberts and C. R. Bolognesi, "150-GHz Cutoff Frequencies and 2-W/mm Output Power at 40 GHz in a Millimeter-Wave AlGaN/GaN HEMT Technology on Silicon," in *IEEE Electron Device Letters*, vol. 33, no. 10, pp. 1372-1374, Oct. 2012, doi: 10.1109/LED.2012.2204855.

[6] M. Sun, Y. Zhang, X. Gao and T. Palacios, "High-Performance GaN Vertical Fin Power Transistors on Bulk GaN Substrates," in *IEEE Electron Device Letters*, vol. 38, no. 4, pp. 509-512, April 2017, doi: 10.1109/LED.2017.2670925.

[7] K. Zhao et al., "P-GaN/AlGaN/GaN Fin-HEMT with High Saturation Current and Enhanced VTH Stability," in *IEEE Electron Device Letters*, doi: 10.1109/LED.2024.3466955.

[8] A. U. H. Pampori, S. A. Ahsan and Y. S. Chauhan, "Modeling the Impact of Dynamic Fin-Width on the I– V, C– V and RF Characteristics of GaN Fin–HEMTs," in *IEEE Transactions on Electron Devices*, vol. 69, no. 5, pp. 2275-2281, May 2022, doi: 10.1109/TED.2022.3156966.

[9] H. -S. Zhang et al., "Influence of Different Fin Configurations on Small-Signal Performance and Linearity for AlGaN/GaN Fin-HEMTs," in *IEEE Transactions on Electron Devices*, vol. 66, no. 8, pp. 3302-3309, Aug. 2019, doi: 10.1109/TED.2019.2921445.

[10] S. Takashima, Z. Li and T. P. Chow, "Sidewall Dominated Characteristics on Fin-Gate AlGaN/GaN MOS-Channel-HEMTs," in *IEEE Transactions on Electron Devices*, vol. 60, no. 10, pp. 3025-3031, Oct. 2013, doi: 10.1109/TED.2013.2278185.

979-8-3315-0417-5/25 $31.00 © 2025 IEEE

Surface Potential-based Compact Model for IGZO-DRAM Enables the TCAD-to-SPICE Framework for Reliability-aware DTCO Flow

Xufan Li[1,2], Wenfeng Jiang[1,2], Chen Gu[1,2], Yue Zhao[1,2], Di Geng[1,2], Guanhua Yang[1,2], Lingfei Wang[1,2*], Ling Li[1,2*]

[1]Key Lab of Fabrication Technologies for Integrated Circuits, CAS, Beijing, 100029, China,
[2]Lab of Microelectronics Devices and Integrated Technology, IMECAS, Beijing, 100029, China
emails: wanglingfei@ime.ac.cn, lingli@ime.ac.cn

Abstract

With the Design-Technology Co-Optimization (DTCO) being increasingly significant in IGZO-DRAM, a surface potential-based compact model is highly required. For better understanding the disorder effects, an in-house TCAD tool is proposed to study sub-threshold properties. On this basis, a unified surface potential based compact is provided with underlying transport mechanisms for various structures. By addressing critical reliability issues, such as D2D and C2C variability, it enables the TCAD-to-SPICE framework for realizing a reliability-aware DTCO flow of IGZO-DRAM.
Keywords: DTCO, Compact model, Surface potential, IGZO-DRAM

Introduction

Monolithic 3D (M3D) DRAM has garnered significant attention in response to the increasing industry demand for enhanced DRAM capacity with the rapid advancement of artificial intelligence. Amorphous In-Ga-Zn-O (a-IGZO) has emerged as a promising candidate for 3D DRAM integration[1] due to its excellent uniformity, extremely low leakage current, and inherent compatibility with back-end-of-line (BEOL) processing. However, charge transport mechanisms in a-IGZO are inherently complex owing to its material disorder. Therefore, the development of reliable compact models is crucial to address the challenges of precise performance characterization of transistors (AC or DC) associated with device scaling, structural variations in IGZO-based DRAM.

Considering the ongoing scaling and technology compatibility of IGZO FETs in very-large scale integration (VLSI) circuit design, compact models should be adapted to accommodate complex device structures, including single-gate[2], dual-gate, independent dual-gate[3], channel-all-around (CAA)[4], as shown in Fig. 1(a). In this work, a surface potential-based compact model incorporated critical reliability factors, such as device-to-device (D2D) and cycle-to-cycle (C2C) variation, has been developed for a-IGZO FETs, enabling intrinsic potential profiling and improved reliability-awareness.

In-House Built TCAD Tool

Prior to compact modeling, technology computer aided design (TCAD) is a necessary tool to investigate the device properties. Particularly, in the presence of structural disorders in transistors, it's key to interpret the disorder effects for device modeling due to the amorphous active layer. Such disorders (e.g., oxide vacancies) contribute to the localized density of states, influencing the carrier density distribution, surface potential profile and mobility etc. Therefore, an in-house built TCAD tool has been developed well incorporated the disordered effects on performances of IGZO-DRAM. Fig. 2 shows the simulation results for an a-IGZO FET. The trap states within the band gap of a-IGZO are characterized through a general method that models the density distribution by categorizing it into four bands: a donor-like valence band, an acceptor-like conduction band, and two Gaussian-distributed deep-level bands. The donor-like and acceptor-like bands are described by exponential tail distributions, while the deep-level bands are represented by Gaussian distributions. The assembly of these trap states in the Poisson equation and transport equations is achieved by discretizing the energy levels within the band gap. The localized and free carrier density can be included into the Poisson equations. Amorphous semiconductors inherently possess a significant number of trap states, making their characterization crucial for simulating amorphous semiconductor devices. A more precise description can also be assembled in this in-house TCAD tool, enabling simulations to more accurately reflect the device characteristics.

Surface Potential-Based Compact Model

Based on the analysis of disorder effects by the TCAD, a compact model framework can be provided to paves the path to SPICE as shown in Fig. 3. It starts with DoS sampling and the localized states are generally expressed in Gaussian or exponential forms. Combining the Fermi-Dirac distribution, the carrier density can be numerically solved. Moreover, with Poisson-equation and boundary conditions, a complicated relation between density and potential is obtained, and it is hard to derive a close-form equation. Thus, we have developed a unified surface potential calculation methodology to simplify the procedure. An intermediate equation for surface potential can be obtained via approximation approaches, such as Taylor expansion. Then the initial solutions are corrected by the Schoder Series.

Provided by surface potential, transport mechanisms can be described more accurately due to their carrier density and temperature dependences. By varying the gate voltage, the effective mobility is dominated by the different transport mechanisms, corresponding to the position of Fermi-level. At a microscopic level, carrier transport is influenced by the energy and spatial distribution of defects. Generally, such

---

979-8-3315-0417-5/25 $31.00 © 2025 IEEE

mechanisms include multiple-trapping and releasing[5], percolation transport[6, 7], trap-limited transport[7] and hopping transport[8, 9]. Furthermore, the finite-size effects will emerge on the transport mechanism in nanoscale of the thin-film transistors[10-12], and the mobility is dimension-related. This finite-size correction method is important for future ultra-scaled transistor modeling. With an in-depth insight into physics for different materials or structures, analytical models of surface potential and mobility will be incorporated in the modeling framework for current expression.

This framework has been adopted to several structures including single-gate, dual-gate and vertical-gate and the compact model exhibits a good convergence. The dual-gate coupling effect can be described accurately, and the independent-gate control of the threshold-shift is incorporated[3]. It is important for a-IGZO transistor in multi-threshold voltage applications; in particular, a compensation scheme has been proposed with dual-gate coupling. Multi-nanoscale compact model is provided in [12], and the transport is corrected due to finite-size effects. Such models also describe short channel effects with serious contact issues. For high density design, vertical transistors, such as IGZO CAA-FETs, are demanded. Accurate surface potential solutions are given in cylindrical coordinates for both inner and outer surface of the ultra-thin channel around 5 nm-thickness. Due to the difficulty of the process, the asymmetry of the drain/source electrode and gate-overlap are investigated for the future manufactured device uniformity.

Reliability-Aware DTCO Flow

Reliability issues are critical in circuit simulation. It includes the D2D and C2C variations. C2C variations are mainly induced by bias temperature instability, which is important for DRAM cells. Particularly, the anomalous PBTI phenomenon induced by hydrogen has been discovered. To Accurate characterization of BTI effects is key in compact model for prediction of the device lifetime. D2D variations are mainly induced by process variations, showing difference of material or structure in fabrication. Particularly, in multicomponent oxide materials, the composition is relatively complex and exhibits a certain degree of randomness. Thus, the energy and spatial distribution of defects will be slightly different among devices. By modulating DoS parameters, the statistical effect can be included. To take the IGZO-DRAM as an example, the model is calibrated to the experimental data as shown in Fig. 4(a) and (b). Then, the disorder information can be extracted based on the physical model. Considering the statistical effect on the sub-threshold region of DRAM cells, the statistical distribution of retention time can be obtained as shown in Fig. 4(c) and (d). Moreover, the variation will be aggregated in a larger timescale. Thus, for future circuit design, a reliability-aware DTCO enabled by compact model is quite significant.

By incorporating the BTI or other stress-induced degradation, the lifetime prediction can be also achieved.

Conclusion

In this paper, the framework comprising surface potential-based compact model cooperated with an in-house TCAD tool is outlined. The surface-potential-based compact model serves as practical tool in DTCO flow, assisting in a high-density IGZO DRAM simulation and continuing to support the future ultra-scaling and 3D integration.

Acknowledge

This work was supported in part by the Strategic Priority Research Program of Chinese Academy of Sciences (Grant No. XDA0330401), the National Natural Science Foundation of China (Grant Nos. 62274178, 92264204), CAS Interdisciplinary Innovation Team [JCTD-2022-07].

References

[1] Chen C, et al., First Demonstration of Stacked 2T0C-DRAM Bit-Cell Constructed by Two-Layers of Vertical Channel-All-Around IGZO FETs Realizing 4F 2 Area Cost, 2023 International Electron Devices Meeting (IEDM)

[2] Liu M, et al., First Demonstration of Monolithic Three-Dimensional Integration of Ultra-High Density Hybrid IGZO/Si SRAM and IGZO 2T0C DRAM Achieving Record-Low Latency (< 10ns), Record-Low Energy (< 10fJ) of Data Transfer and Ultra-Long Data Retention (> 5000s), 2024 IEEE Symposium on VLSI Technology and Circuits (VLSI Technology and Circuits)

[3] Xu L, et al., Reliability-Aware Ultra-Scaled IDG-InGaZnO-FET Compact Model to Enable Cross-layer Co-design for Highly Efficient Analog Computing in 2T0C-DRAM, 2023 International Electron Devices Meeting (IEDM)

[4] Duan X, et al., Novel Vertical Channel-All-Around (CAA) IGZO FETs for 2T0C DRAM with High Density beyond 4F2 by Monolithic Stacking, 2021 IEEE International Electron Devices Meeting (IEDM)

[5] Zong Z, et al., A new surface potential-based compact model for a-IGZO TFTs in RFID applications, 2014 IEEE International Electron Devices Meeting

[6] Lu N, et al., Universal carrier thermoelectric-transport model based on percolation theory in organic semiconductors, Physical Review B, 2015

[7] Guo J, et al., A new surface potential based compact model for independent dual gate a-IGZO TFT: Experimental verification and circuit demonstration, 2020 IEEE International Electron Devices Meeting (IEDM)

[8] Lu N, et al., Charge carrier relaxation model in disordered organic semiconductors, AIP Advances, 2013

[9] Xu L H, et al., A Surface Potential Based Compact Model for Ferroelectric a-InGaZnO-TFTs Toward Temperature Dependent Device Characterization, IEEE Electron Device Letters, 2023

[10] Wang J, et al., Collective transport for nonlinear current-voltage characteristics of doped conducting polymers, Physical Review Letters, 2023

[11] Guo J, et al., Compact Modeling of IGZO-based CAA-FETs with Time-zero-instability and BTI Impact on Device and Capacitor-less DRAM Retention Reliability, 2022 IEEE Symposium on VLSI Technology and Circuits (VLSI Technology and Circuits)

[12] Guo J, et al., A New Surface Potential and Physics Based Compact Model for a-IGZO TFTs at Multinanoscale for High Retention and Low-Power DRAM Application, 2021 IEEE International Electron Devices Meeting (IEDM)

DTCO Enabled Density Increasing of IGZO-DRAM

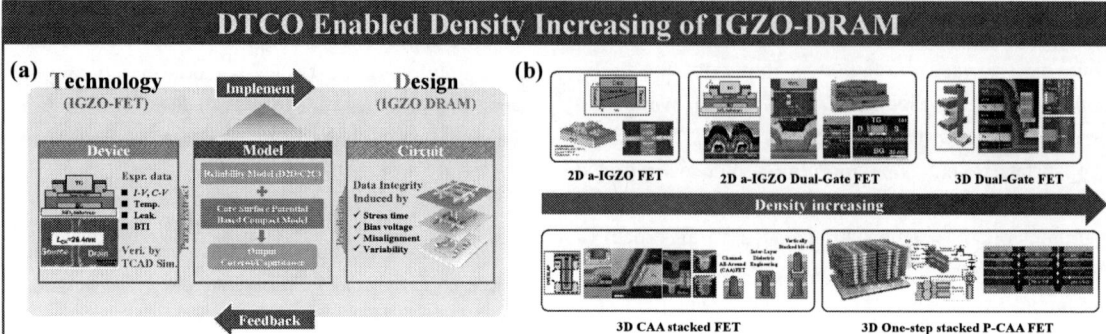

Fig. 1. (a) The DTCO flow utilizing the surface potential-based compact model as a bridge between design and devices enables (b) continuous density increases in IGZO-DRAM along the two primary pathways: dual-gate and channel-all-around (CAA).

In-House Built TCAD for Describing Disorder-Effects

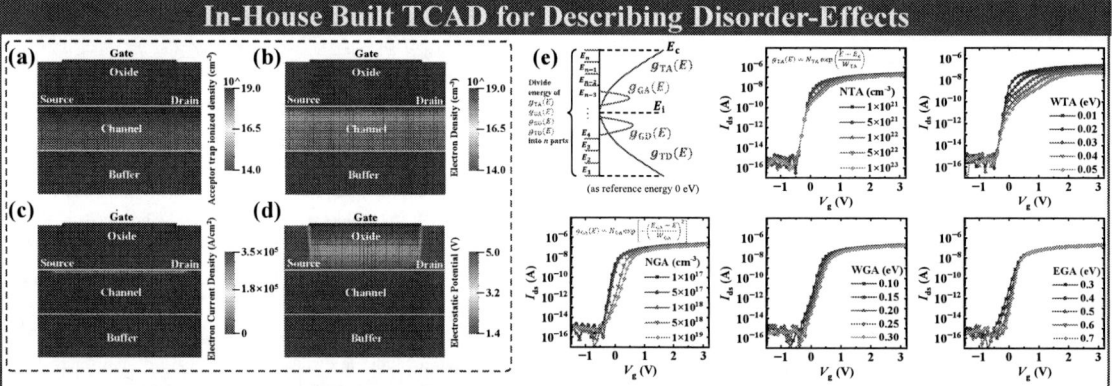

Fig. 2. In-house TCAD simulation of IGZO-FET. Spatial distribution of (a) acceptor-like traps ionized density, (b) electron density, (c) electron current density and (d) electrostatic potential @V_{ds} = 0.1 V, V_{gs} = 5 V (x-axis scaled by 10^{-3}). (e) Effects of trap DoS on the electrical characteristics, tail acceptor-like trap states and Gaussian acceptor-like trap states are considered here.

Framework of Surface Potential-based Compact Model for TCAD-to-SPICE

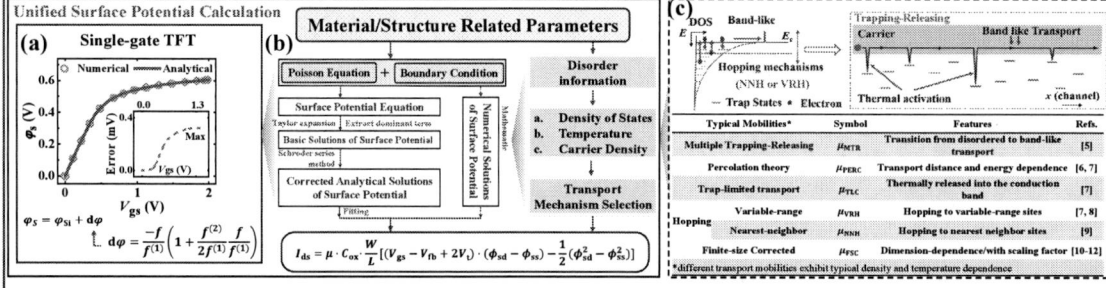

Fig. 3. (a) Good agreement between analytical and numerical results for surface potential with error in the inset. (b) A flow diagram of surface potential-based compact model for unified surface potential calculation. (c) Typical transport mobilities utilized in compact model.

TCAD-SPICE Platform for IGZO-DRAM based Circuit Prediction

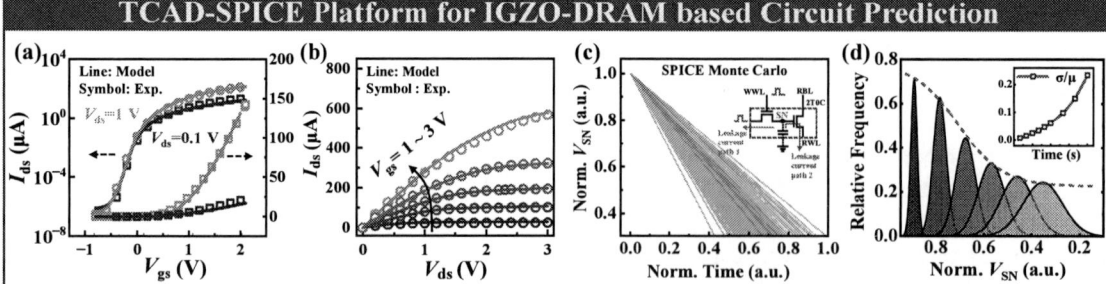

Fig. 4. Agreement between surface potential-based compact model and experiments for (a) transfer and (b) output characteristics under different V_{ds} and V_{gs} respectively. (c) Circuit simulation of 2T0C eDRAM to predict the retention time with Monte Carlo method and extract bit distribution. (d) Gaussian fitting of bit distribution, with σ/μ in the inset.

979-8-3315-0417-5/25 $31.00 © 2025 IEEE

Optimization of the Read Transistor in Hybrid 2T0C DRAM by Co-modeling the Drain Current of a-IGZO and Low Temperature Poly-Silicon TFTs

Haolin Li[1], Zheng Zhou[1,2], Xiaoyan Liu*[1,2]

[1] School of Integrated Circuits, Peking University, Beijing 100871, China,

[2] Beijing Advanced Innovation Center for Integrated Circuits, Beijing 100871, China.

* Email: liuxiaoyan@pku.edu.cn

Abstract

A hybrid 2T0C DRAM cell with an n-type write transistor and a p-type read transistor is analyzed by modeling the characteristics of a-IGZO and LTPS-TFTs respectively. Both models are based on surface potential calculation and verified with experiments. Dynamic trap effect is involved as well. The simulation results demonstrate the advantage of hybrid 2T0C DRAM over full n-type DRAM in sense margin. The trade-off between memory window, retention and write/read speed has to be taken into account to design the read transistor size. Our work proposes a guideline for 2T0C DRAM design.

Keywords: 2T0C DRAM, TFT, Modeling and Simulation

Introduction

Amorphous Indium-Gallium-Zinc-Oxide (a-IGZO) thin film transistor (TFT) has been considered as a promising candidate for future application in system on panel (SoP), which is attributed to its extremely low leakage current, relatively high mobility, on-off ratio and uniformity [1, 2]. With the development of low temperature polycrystalline oxide (LTPO) technology, a-IGZO TFTs have been implemented to the fabrication of flat panel display (FPD) to suppress the flicker effect, especially in active-matrix organic light emitting diode (AMOLED) panels. Meanwhile, the back-end of line (BEOL) compatible IGZO-based two-transistor-zero-capacitor (2T0C) dynamic random access memory (DRAM) has been put forward to achieve memory with higher density [3]. To optimize the design of 2T0C DRAM cell and emulate the impact of various device parameters, it is necessary to establish an a-IGZO TFT model which can accurately describe its electrical characteristics with prominent trap behaviors.

In this work, a-IGZO TFT and low temperature poly-silicon (LTPS) TFT drain current (Ids) model are established to analyze the behavior of 2T0C DRAM. A hybrid 2T0C DRAM with n-type a-IGZO write transistor (WTR) and p-type LTPS read transistor (RTR) is simulated. Factors including RTR equivalent width, hysteresis and WTR leakage are investigated to figure out the impact on hybrid 2T0C performance. The results are benefit for 2T0C design and optimization [4, 5].

TFT Models

Fig.1 shows the schematic of a 2T0C DRAM cell. The WTR is first switched on to charge the storage node (SN) during the write period. The signal input by Write Bit Line (WBL) is stored by gate storage capacitance (C_{ST}) of RTR and detected by Read Bit Line (RBL) in read period. In the case of writing "1", V(SN) will be pulled down after WWL is disabled due to the parasitic capacitance between Write Word Line (WWL) and SN ($C_{WWL,SN}$). The capacitance between SN and RBL ($C_{RBL,SN}$) may also make a difference for fully n-type transistor based 2T0C DRAM[5, 6]. To address this, p-type TFT is implemented as RTR to lift V(SN) by higher RWL/RBL signals. Feasible for complementary metal-oxide-semiconductor (CMOS) logic and mature in FPD fabrication, p-type LTPS-TFT is utilized in this work for further analysis.

To design and optimize the hybrid 2T0C DRAM, both LTPS-TFT and a-IGZO TFT models are necessary. The cross-sectional structure diagram of a common self-aligned top gate TFT is shown in Fig.2, where surface potential (ϕ_s) based model is adopted thanks to its explicit physics and good continuity in all operation regions. The core of this model is to connect ϕ_s with input signals and device parameters. Ids can be calculated based on the gradual channel approximation (GCA) and the drift-diffusion model.

Due to different operation mode and transport mechanisms, a-IGZO and LTPS-TFT models are established respectively. The basic quasi-static Ids modeling flowchart is shown in Table 1. in detail. For LTPS-TFT switching on at inversion mode, both majority and minority carriers are taken into account as in the case of MOS and charged traps are involved [7]. Semi-empirical model is also established for the leakage Ids and mobility degradation effect in LTPS-TFTs. For a-IGZO TFT working in the accumulation mode, the charge density along x direction perpendicular to the channel is dominated by electrons in localized and extended states, which increases with the rise of Fermi level (E_F). Additional trap limited conduction model is adopted in the mobility term of a-IGZO model.

Our models are transferred into Verilog-A version and verified with the experiment data [8], by HSPICE simulator. Main modeling parameters are listed in Table 2. The storage and parasitic capacitors are also added and modulated to evaluate their impact on 2T0C cell performance, where data retention, memory window and write/read speed are focused.

Dynamic trap capture/emission behavior is also combined in our model. By calculating the trapped charge density by a time-dependent differential process, calculated ϕ_s and Ids

979-8-3315-0417-5/25 $31.00 © 2025 IEEE

will change accordingly and the impact of bias temperature instability (BTI) induced by trap can be further evaluated.

Results and Discussion

A. Device-level Verification

As is shown in Fig.3 (a) and (b), calculated transfer curves have a good agreement with the experiment data in both LTPS-TFT and a-IGZO TFT case, which verifies the accuracy of our models. Note that the switch-off Ids of LTPS-TFT enhances with higher drain-source voltage (Vds), while for a-IGZO TFT its leakage Ids maintains a lower level. The results show the feasibility for a-IGZO TFT to act as the WTR in DRAM.

Trap induced hysteresis is also calculated in Fig.4. A negative Vth shift is observed in LTPS-TFT during the forward sweep, which may decrease Ids and lower the read speed compared to the pristine state. If negative BTI induced Ids degradation exceeds the preset level, a logic error may arise while reading.

B. Hybrid 2T0C DRAM Cell Behavior

The input signals of read and write process for the hybrid 2T0C DRAM cell with n-type a-IGZO TFT and p-type LTPS-TFT are shown in Fig. 5(a). Note that for p-type TFTs Ids declines with higher gate voltage. Limited by the relatively negative threshold voltage (Vth), RWL and RBL have to be biased at a higher read voltage to guarantee the sense margin. The 1:10 ratio of $C_{WWL-SN}:C_{ST}$ is originally set.

As is shown in Fig.5 (b), V(SN) drop arises when WWL is disabled especially when writing "1", which leads to an undesirable lift of the read current in Fig. 5 (c), narrowing the memory window and sense margin. Hybrid 2T0C suppresses the V(SN) degradation during the read operation, which is shown in Fig. 6. Suppose that the prospective Ids and gate bias (Vgs) of p/n-type RTR are the same, a further V(SN) drop is observed in n-type RTR, leading to a smaller sensing current. For p-type RTR, although "0" state read current declines, low-current readout level corresponding to read "1" state remains unchanged. Our simulation verifies the reliability of hybrid 2T0C DRAM compared to full n-type DRAM memory.

C. Impact of the W/L for LTPS-TFT

Acting as C_{ST}, the design of RTR may have a significant impact on the performance of hybrid 2T0C DRAM, including memory window/sense margin, retention and write/read speed. Process of writing "1" is mainly focused due to the severe V(SN) drop. The V(SN) response right after write operation with different RTR channel width is shown in Fig.7. As $C_{WWL-SN}:C_{ST}$ decreases with larger C_{ST}, the gap between V(SN) and V(WWL) becomes larger to reach charge conservation in the capacitor network, leading to less severe V(SN) drop. However, larger C_{ST} also leads to longer writing time, which is in proportion to the RTR channel width.

For the retention, due to the suppression of WTR leakage, V(SN) level of hybrid 2T0C DRAM can be maintained in C_{ST} in longer time compared to conventional 1T1C structure and can reduce the refresh rate. For simplicity, the retention time is extracted when V(SN) reaches 20% drop. Simulated retention of hybrid 2T0C DRAM with different WTR leakage current are shown in Fig. 8 (a) and (b). It is obvious that retention time (RT) is inversely proportional to WTR Ids leakage and the simulation result shows the same trend. The impact of W/L for RTR is also plotted in Fig.8 (c) and (d). Attributed to the steady low leakage current, longer retention is prospected to discharge larger storage capacitor.

DRAM access time is another important spec of DRAMs. Wider LTPS-TFT with larger C_{ST} will extend the charge time while lowering the read time due to large RTR Ids. The simulation result of write/read speed on RTR width is shown in Fig.9.

Conclusion

LTPS-TFT and a-IGZO TFT models are established to improve the performance of hybrid 2T0C DRAM. Focusing on the RTR optimization, the memory window, write/read speed, and retention under different equivalent channel width are investigated. Larger LTPS-TFT width is required with the trade-off of read speed. Our work paves a way for further 2T0C DRAM design and optimization.

References

[1] A. Yan, C. Wang, J. Yan, et al., "Thin-Film Transistors for Integrated Circuits: Fundamentals and Recent Progress," Advanced Functional Materials, 34, 2023.

[2] S. Yan, Z. Cong, N. Lu, et al., "Recent progress in InGaZnO FETs for high-density 2T0C DRAM applications," Science China Information Sciences, 66, 2023.

[3] A. Belmonte, H. Oh, N. Rassoul, et al., "Capacitor-less, Long-Retention (>400s) DRAM Cell Paving the Way towards Low-Power and High-Density Monolithic 3D DRAM," 2020 IEEE International Electron Devices Meeting (IEDM), pp. 28.2.1-28.2.4 (2020)

[4] Z. Wang, L. Zheng, Z. Lin, et al., "CMOS Logic and Capacitorless DRAM by Stacked Oxide Semiconductor and Poly-Si Transistors for Monolithic 3-D Integration," IEEE Transactions on Electron Devices, 71, pp. 4664-4669, 2024.

[5] S. Liu, S. Li, Q. Lin, et al., "Hybrid 2T nMOS/pMOS Gain Cell Memory With Indium-Tin-Oxide and Carbon Nanotube MOSFETs for Counteracting Capacitive Coupling," IEEE Electron Device Letters, 45, pp. 188-191, 2024.

[6] C. H. Jeon, K. J. Kwon, S. K. Hong, et al., "Study on Residual Image in Low-Temperature Poly-Si Oxide TFT-Based OLED Display on Polyimide Substrate," IEEE Transactions on Electron Devices, 69, pp. 4958-4961, 2022.

[7] H. Li, Z. Zhou, and X. Liu, "Modeling the Transient Characteristics with Trap Behaviors in LTPS-TFTs," 2024 IEEE 17th International Conference on Solid-State and Integrated Circuit Technology (ICSICT) (2024)

[8] C. H. Jeon, K. J. Kwon, S. K. Hong, et al., "Study on Residual Image in Low-Temperature Poly-Si Oxide TFT-Based OLED Display on Polyimide Substrate," IEEE Transactions on Electron Devices, 69, pp. 4958-4961, 2022.

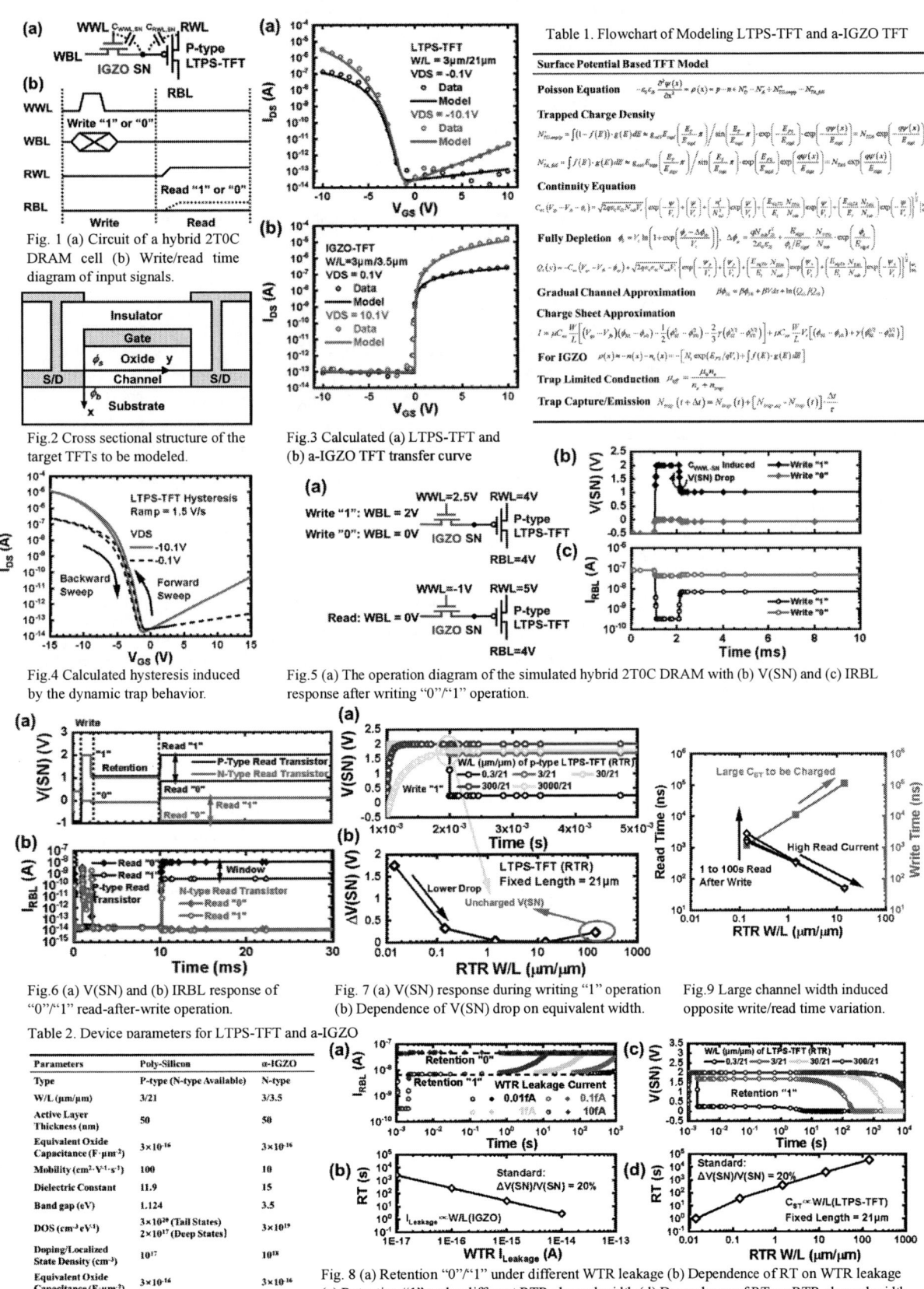

Fig. 1 (a) Circuit of a hybrid 2T0C DRAM cell (b) Write/read time diagram of input signals.

Fig.2 Cross sectional structure of the target TFTs to be modeled.

Fig.3 Calculated (a) LTPS-TFT and (b) a-IGZO TFT transfer curve

Fig.4 Calculated hysteresis induced by the dynamic trap behavior.

Fig.5 (a) The operation diagram of the simulated hybrid 2T0C DRAM with (b) V(SN) and (c) IRBL response after writing "0"/"1" operation.

Table 1. Flowchart of Modeling LTPS-TFT and a-IGZO TFT

Fig.6 (a) V(SN) and (b) IRBL response of "0"/"1" read-after-write operation.

Fig. 7 (a) V(SN) response during writing "1" operation (b) Dependence of V(SN) drop on equivalent width.

Fig.9 Large channel width induced opposite write/read time variation.

Table 2. Device parameters for LTPS-TFT and a-IGZO

Parameters	Poly-Silicon	α-IGZO
Type	P-type (N-type Available)	N-type
W/L (μm/μm)	3/21	3/3.5
Active Layer Thickness (nm)	50	50
Equivalent Oxide Capacitance (F·μm⁻²)	3×10⁻¹⁶	3×10⁻¹⁶
Mobility (cm²·V⁻¹·s⁻¹)	100	10
Dielectric Constant	11.9	15
Band gap (eV)	1.124	3.5
DOS (cm⁻³·eV⁻¹)	3×10²⁰ (Tail States) 2×10¹⁷ (Deep States)	3×10¹⁹
Doping/Localized State Density (cm⁻³)	10¹⁷	10¹⁸
Equivalent Oxide Capacitance (F·μm⁻²)	3×10⁻¹⁶	3×10⁻¹⁶

Fig. 8 (a) Retention "0"/"1" under different WTR leakage (b) Dependence of RT on WTR leakage (c) Retention "1" under different RTR channel width (d) Dependence of RT on RTR channel width

Enhanced Performance of P-FeFETs with TiN:2.5nm/Mo/TiN Gate Stacks for 3-bit-per-cell Operation, 3.5V Read-after-Write, and High Endurance (10^9 Cycles) in Compute-in-Memory Applications

C.-Y. Tsai, M.-H. Hsiung, Y.-C. Chen, Y.-T. Tsai and Y.-T. Tang*

Dept of Electrical Engineering, National Central University, Taoyuan City, Taiwan,

email: yttang@ee.ncu.edu.tw

Abstract

This study examines the performance of P-type FeFETs with various gate stacks, focusing on the TiN:2.5nm/Mo/TiN electrode. XPS and interface trap density (Dits) analyses confirm that ultra-thin barrier of TiN (b-TiN) suppresses Mo oxidation and carrier trapping in HZO, enhancing the memory window and reducing Vth variability. The sandwiched structure improves write speed, switching time, and low-voltage read-after-write operations. In neural network applications, b-TiN/Mo/TiN ensures more linear conductance and faster accuracy convergence, driving computational efficiency for CIM technologies. This work addresses the potential of b-TiN/Mo/TiN in enhancing FeFET reliability, endurance, and performance for neuromorphic and CIM applications.

Introduction

In recent years, ferroelectric field-effect transistors (FEFETs) have emerged as a promising embedded non-volatile memory technology for data-centric computing, attracting significant interest due to their CMOS compatibility, high-density integration potential, low operating voltages, ultra-fast write speeds, and low energy requirements. However, one of the primary challenges limiting their performance is the read-after-write delay (td), which is crucial for restoring the memory window after each write operation, especially in real-time applications demanding low read latencies.[1] [2]

Most FeFETs face difficulties achieving immediate read capability post-write due to parasitic charge trapping. While recent studies have shown that p-type Si-based FeFETs (p-FeFETs) can exhibit immediate read-after-write capabilities, the impact of bipolar stress on td has not been thoroughly examined. This study aims to investigate the nature and physical origins of bipolar stress-induced degradation in td of p-FeFETs, focusing on how trap characteristics influence td and the role of temperature in read recovery time. By exploring the effects of charge trapping from electrodes, we seek to uncover the underlying mechanisms responsible for write-read delays in p-FeFETs, ultimately paving the way for enhanced performance in future computing applications.[4]

Device Fabrication

Figure 1(a)-(c) shows schematic of the stacked structures for p-FeFET with different gate configurations. The TEM image of the fabricated structure is shown in the inset. Figure 1(d) presents the fabrication process. The HZO layer is deposited using the ALD process, with HfO_2/ZrO_2 stacked in 7Å cycles. The top electrode is capped with TiN (50 nm), Mo (30 nm), and a sandwiched TiN/Mo/TiN structure, where the initial TiN layer, only 2.5 nm thick, serves as a barrier layer (b-TiN). We then patterned three different conditions before performing BF_2 doping, followed by annealing at 800 °C for 15 seconds. [3]

We then used the Keysight B1500A with the 1525 module for Id-Vg measurements, achieving cycling up to 10^9 cycles. The relationship between the Vth shift and cycling is shown in Figures 2(d)-(f). It was observed that the p-type FeFET with traditional TiN capping exhibited the smallest initial memory window (approximately 0.5V). In contrast, Mo capping showed a higher saturation current I_{ON} with a larger memory window (MW: 0.93V). With increased cycling, the program state remained stable while the erase state decayed, indicating the presence of defect-induced hole trapping. This trapping may be attributed to inter-diffusion between Mo and HZO, driven by thermal stress due to their differing thermal expansion coefficients. To mitigate this effect, b-TiN can be introduced as a barrier layer. Although the b-TiN/Mo structure may undergo oxidation during annealing, the work function of the resulting metal oxides increases from TiN (4.7 eV) / Mo (4.7 eV) to TiOxNy (5.4 eV) / MoOx (6.3 eV), reducing carrier transitions and thereby slowing the decay of the erase state.[5] [7]

Defects can be analyzed through both electrical and optical properties. Figures 3(a)-(c) illustrate the Dits for three types of gate stacks. Notably, the introduction of b-TiN between Mo and HZO reduces Dit by 22% compared to pure Mo capping. From the optical XPS spectra, we observe an 11% reduction in suboxide changes of the Hf4f peaks with b-TiN insertion, confirming that b-TiN effectively suppresses defects and prevents excessive oxidation of Mo. Compared to the metallic MoO_2, MoO_3 serves as a dielectric material ($\kappa_{MoO3} \sim 0.6\kappa_{HZO}$). With a 3V voltage, an electric field of 1.8 MV/cm is across MoOx layer (~10nm), close to the breakdown field $E_b^{MoO3} \sim 2MV/cm$. During cycling operations, Mo ion diffuse into HZO, resulting in fatigue. A comparison in Figures 3(g)-(h) shows that b-TiN suppress the proportion of MoO_3 from 82% to 75%, enhancing the electrode's stability.

FeFET is a device that supports both writing and reading functions. Here, we conducted a read-after-write test on the device, as shown in Figure 4(a)-(c). It shows that with Mo and TiN capping, a small voltage of 3.5V and 1 μs write pulse requires a 1s wait before reading. However, by adding an ultra-thin TiN, reading can occur immediately.

In addition to improved read performance, the write speed has also been significantly enhanced. Figure 5(a) shows that the switching speed of Vth is noticeably faster with the inclusion of b-TiN before capping Mo. We proceeded to extract the switching speeds based on Merz's law, with the results shown in Figures 5(a) and 5(b). It can be observed that ERS is generally faster than PGM, primarily because ERS involves fewer trapped carriers compared to PGM in the P-FeFET, corresponding band diagram were illustrated in Figures 5(c) and 5(d). As discussed earlier, the b-TiN/Mo electrode's high Schottky barriers at the TiOxNy/HZO interface reduce electron migration, resulting in fewer trapped electrons in HZO. This benefits polarization and increases the switching speed.[6]

Preventing excessive domain growth helps improve the number of Vth states. Figure 7 shows the TLC characteristics of FeFETs with different electrode structures under the same area. For conventional TiN electrode P-FeFETs, although the domains are not large, the defects inside domain boundaries (DB) and interface result in a high read error rate. In contrast, Mo-capping easily forms larger o-phase domains with fewer DBs; however, the oversized domains make it challenging to find domains that can be flipped at the corresponding voltages. The b-TiN/Mo/TiN structure, on the other hand, prevents excessive domain growth and reduces interface defect density, resulting in a uniform distribution of states.

In machine learning, we observe the relationship between conductance and pulse number. Compared to Figures 6(a) and (b), the P-FeFET shown in (c) exhibits a more linear conductance behavior, with a similar trend observed in the N-FeFET (shallow color circles). Figure 6(d) illustrates the accuracy versus epoch relationship, showing that FeFETs with the b-TiN/Mo/TiN electrode achieve high accuracy with relatively fewer epochs, making them highly competitive for future compute-in-memory (CIM) technology development.

Conclusion

This study demonstrates the potential of b-TiN/Mo electrodes in P-FeFET applications. By suppressing defects in HZO and minimizing excessive Mo oxidation, b-TiN/Mo significantly improves memory window (MW) stability and Vth endurance over cycles. Compared to conventional TiN and Mo capping, b-TiN/Mo electrodes exhibit faster write speeds, stable switching behavior, and more accurate read-after-write functionality, especially under low voltage conditions. Additionally, in neural network applications, b-TiN/Mo electrodes provide a more linear conductance response to pulse number and achieve high accuracy with fewer epochs, highlighting their promise for compute-in-memory (CIM) applications. These findings indicate that b-TiN/Mo offers significant advantages in enhancing FeFET performance, supporting the development of more reliable and durable CIM and neuromorphic computing technologies in the future.

Acknowledgments

The authors gratefully acknowledge the supports from NSTC113-2218-E-A49-022, 113-2119-M-A49-007, 112-2622-8-A49-013-SB and TSRI JDP113-Y1-035.

References

[1] Kim, Bong Ho, et al. "Oxygen Scavenging in HfZrOx-Based n/p-FeFETs for Switching Voltage Scaling and Endurance/Retention Improvement." Advanced Electronic Materials 9.5 (2023): 2201257.

[2] Aspiotis, N., Morgan, K., März, B. et al. Large-area synthesis of high electrical performance MoS2 by a commercially scalable atomic layer deposition process. npj 2D Mater Appl 7, 18 (2023).

[3] J. Müller, T. S. Böscke, U. Schröder, et al., "Ferroelectricity in Simple Binary ZrO2 and HfO2," Nano Letters, vol. 12, pp. 4318-4323, 2012.

[4] Cai, Zuocheng, et al. "HZO Scaling and Fatigue Recovery in FeFET with Low Voltage Operation:Evidence of Transition from Interface Degradation to Ferroelectric Fatigue." 2023 IEEE Symposium on VLSI Technology and Circuits (VLSI Technology and Circuits). IEEE, 2023.

[5] F. Huang; B. Saini; L. Wan; H. Lu; X. He; S. Qin. "First Observation of Ultra-high Polarization (~ 108 μC/cm²) in Nanometer Scaled High Performance Ferroelectric HZO Capacitors with Mo Electrodes." 2023 IEEE Symposium on VLSI Technology and Circuits (VLSI Technology and Circuits)

[6] Yixin Qin; Venkateswarlu Gaddam; Taeseung Jung; Sanghun Jeon. "HZO (>10 nm) Films for Achieving High-κ Near Morphotropic Phase Boundary at Low-Temperature Furnace Annealing Process." IEEE Transactions on Electron Devices (Volume: 71, Issue: 9, September 2024)

[7] Song-Hyeon Kuk, Seung-Min Han, Bong Ho Kim, Seung-Hyub Baek, Jae-Hoon Han, Associate Member, IEEE, and Sang-Hyeon Kim, Member, IEEE. "An Investigation of HZO-Based n/p-FeFET Operation Mechanism and Improved Device Performance by the Electron Detrapping Mode"

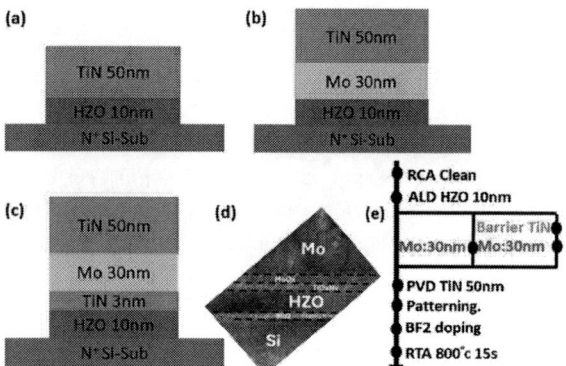

Fig1. (a)-(c)MFIS-FeFET with different gatestack structures. (d) TEM image. (e) Process flow of FeFET

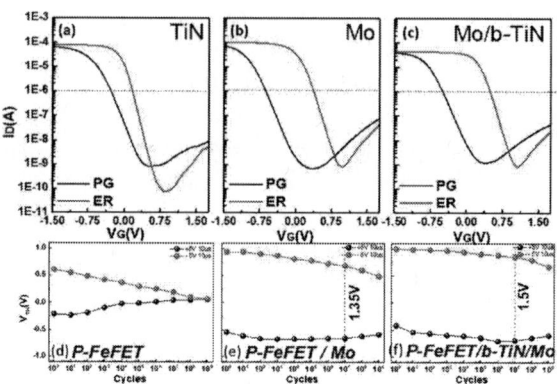

Fig2. (a)-(c) Id-Vg characteristic curve for different capping electrode with memory window 0.4V,1V and 1V. (d)-(f) Cycling number with pulse ±5V

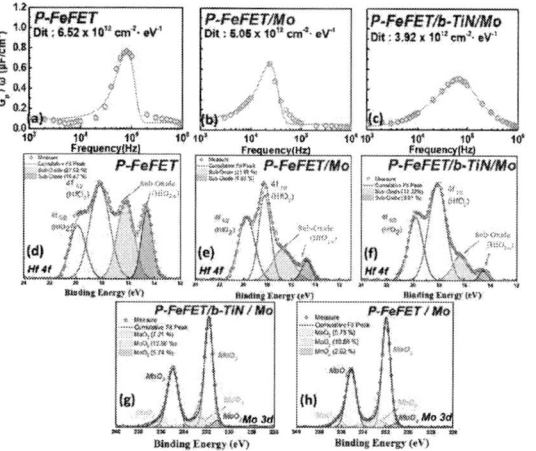

Fig3. (a)-(c) Represent interfacial trap charges, (d)-(f) show Hf4f peak variations in XPS spectra, and (g-h) illustrate Mo3d peak variations in the XPS spectra.

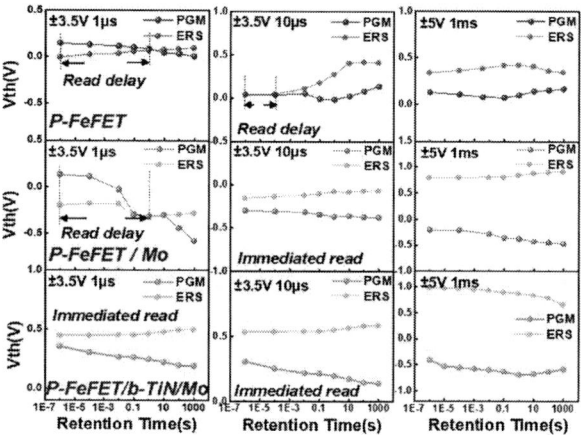

Fig4. Read after write with P-FeFET, P-FeFET/Mo and P-FeFET/b-TiN/Mo

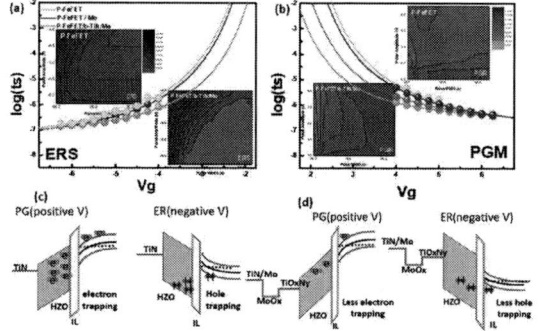

Fig5. (a)-(b)Switching time as a function of pulse amplitude VG. (c)-(d) Schematic band diagram of program and erase in FeFETs.

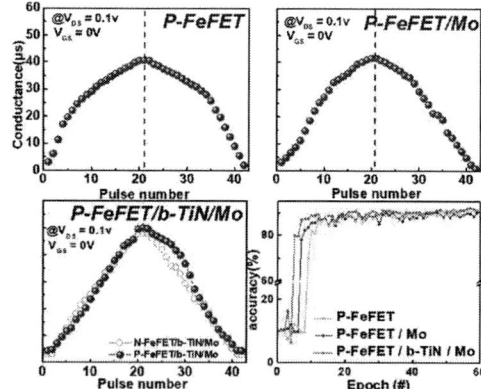

Fig6. (a)-(c)Conductance in FeFET. (d) Comparison of recognition accuracy with different capping electrode.

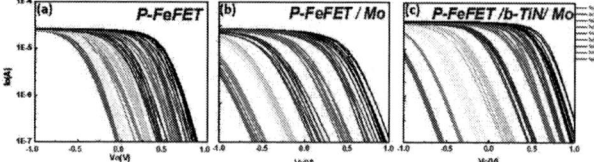

Fig7.Triple-level-cell of FeFET with different capping electrode.

	P-FeFET	P-FeFET/Mo	P-FeFET/b-TiN/Mo	[7]
MW (V)	0.14 V	1.35 V	1.5 V	1 V
Endurance (#)	10^8	10^9	10^9	10^4
Accuracy (%)	89.78	90.72	92.28	X
On/Off ratio(#)	10^4	10^6	10^6	10^5
MLC	2 bit	3 bit	3 bit	X

Fig8.Benchmarking of different type FeFET.

979-8-3315-0417-5/25 $31.00 © 2025 IEEE

Dual-Functional Volatile and Nonvolatile Resistive Switching Characteristics of Cu/Ga₂O₃/Pt Memristor Deposited via Electron Beam Evaporation

Nan He[1,2,3], Zi Li[1], Shuai Chen[1], Hao Zhang[2,3], Xinpeng Wang[2,3], Lei Wang[1,4*], and Yi Tong[2,3*]

[1]College of Integrated Circuit Science and Engineering, Nanjing University of Posts and Telecommunications, Nanjing, 210023, China, [2]Suzhou Laboratory, Suzhou, 215123, China, [3]Gusu Laboratory of Materials, Suzhou, 215123, China, [4]Nantong Institute of Nanjing University of Posts and Telecommunications, Nantong, 226021, China

*Email: LeiWang1980@njupt.edu.cn; tongyi2020@gusulab.ac.cn.

Abstract

In this work, we investigate the dual functionality of volatile and nonvolatile memory in Cu/Ga₂O₃/Pt memristor by tuning the compliance current. These devices, fabricated using electron beam evaporation, demonstrate impressive nonvolatile memory switching characteristics, including low switching voltages, 10^2 switching cycles, $>10^3$ s data retention, and $>10^2$ HRS/LRS ratio. Our work provides an effective and straightforward fabrication technique for using Ga₂O₃ in the development of advanced memristors.

Keywords: wide-bandgap semiconductor, Ga₂O₃, electron beam evaporation, dual-functional resistive switching

Introduction

Recent advancements in artificial intelligence (AI), particularly those based on ChatGPT, have attracted considerable global attention due to their remarkable capabilities in natural language processing. These systems can comprehend and produce text that closely resembles human writing [1]. However, this progress is accompanied by growing concerns regarding the substantial energy consumption associated with running ChatGPT on traditional computer hardware platforms. [2]. Such situation calls for a critical examination of the conventional von Neumann architecture, which has long been the standard in computing by distinctly separating processing and memory units [3]. In response to these challenges, there is increasing interest in exploring alternative computational paradigms, particularly memristor-based neuromorphic computing, which offers the potential for more efficient processing solutions [4]. Memristors, with their unique volatile and nonvolatile characteristics, have been developed using various dielectric materials, making them vital for multifunctional applications in electronic systems.[5] Despite these advancements, there remains a significant gap in investigating wide-bandgap semiconductors, such as Ga₂O₃, as functional layers for dual-functional memristors. Additionally, the fabrication processes for these devices have yet to be fully optimized. Addressing these gaps could result in significant progress in the development of energy-efficient, multifunctional memory systems, thereby laying the groundwork for the next generation of electronic applications.

In this work, we demonstrate dual-functional volatile and nonvolatile resistive switching characteristics in Cu/Ga₂O₃/Pt memristors fabricated via electron beam evaporation (EBE). These Cu/Ga₂O₃/Pt devices exhibit remarkable nonvolatile memory behavior, characterized by low SET and RESET voltages, 10^2 switching cycles, data retention exceeding 10^3 s, and a memory window greater than 10^2. These attributes indicate that our Cu/Ga₂O₃/Pt memristor holds significant promise as a candidate for advanced memory applications.

Experiment

The schematic representation of the fabrication process for the Cu/Ga₂O₃/Pt device featuring a crossbar structure is depicted in Fig. 1. The fabrication utilized electron beam evaporation (EBE) technology to sequentially deposit material layers from the bottom electrode (BE) to the top electrode (TE). Specifically, an 80 nm thick platinum (Pt) BE was deposited at a rate of 0.4 Å/s, followed by a 20 nm thick insulating layer of Ga₂O₃, which was deposited at a slower rate of 0.25 Å/s. The copper (Cu) TE, with a thickness of 90 nm, was prepared at an increased rate of 0.5 Å/s.

The electrical characteristics of the Cu/Ga₂O₃/Pt memristor were thoroughly evaluated using a Keithley 4200A-SCS semiconductor analyzer in conjunction with a Cascade MPS 150 probe station. All measurements were conducted at room temperature.

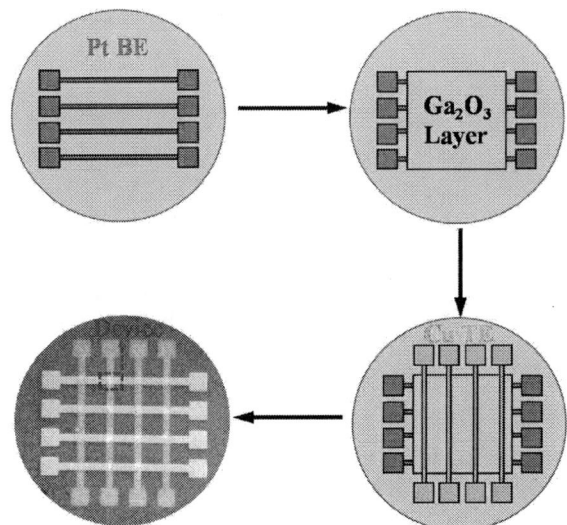

Fig. 1: Schematic of the fabrication process for the Cu/Ga₂O₃/Pt crossbar memristor.

Results and discussion

Fig. 2(a) and 2(b) show the surface morphology and the corresponding three-dimensional (3D) atomic force microscope (AFM) image of the Ga_2O_3 film, obtained from a scanning area of 10 μm × 10 μm. The measured surface roughness (Rq) of the Ga_2O_3 film is 0.27 nm, indicating a remarkably flat and smooth surface, which is highly beneficial for ensuring reliable resistive switching behaviors.

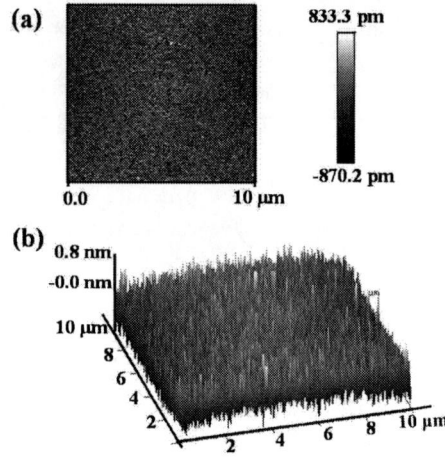

Fig. 2: AFM images of the Ga_2O_3 film. (a) flat scale. (b) 3D scale.

The Cu/Ga_2O_3/Pt devices exhibit a unique capability to operate in both volatile and nonvolatile switching modes upon controlling the applied compliance current (I_{CC}). The current−voltage ($I−V$) characteristics of the Cu/Ga_2O_3/Pt device were investigated under varying I_{CC} conditions, revealing a transition from volatile to nonvolatile behavior as the I_{CC} increases. At a low I_{CC} of 1 μA, as illustrated in Fig. 3(a), the device demonstrates distinct volatile switching characteristics. In this mode, the Cu/Ga_2O_3/Pt device transitions sharply from an insulating high resistance state (HRS) to a low resistance state (LRS, >4 × 10^4 Ω) upon application of an appropriate positive bias voltage (SET voltage). However, once the external excitation is removed, the device quickly reverts to the HRS, as evidenced by the sustained HRS throughout the entire negative voltage region.

Additionally, as depicted in Fig. 3(b), when the I_{CC} is slightly increased to 1 μA, the device exhibits a coexistence of both volatile (black curves) and nonvolatile (red curves) switching properties. It can be noted from the black curves that the Cu/Ga_2O_3/Pt device clearly maintains the HRS throughout the negative voltage sweep, confirming its volatile behavior. In contrast, during positive voltage application, the device transitions from HRS to LRS, completing the SET process. The device then retains the resulting LRS until a specific negative bias voltage (RESET voltage) is applied. This sudden drop in current prompts a transition back to the HRS, thereby completing the RESET process. This behavior aligns well with the characteristics of nonvolatile memory [6].

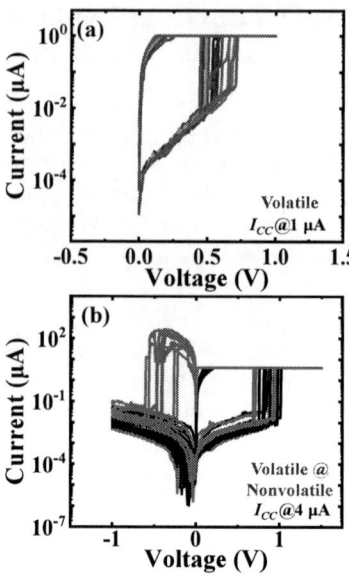

Fig. 3: (a) Volatile I-V curves of the Cu/Ga_2O_3/Pt memristor at an I_{CC} of 1 μA. (b) Coexistence of volatile and nonvolatile switching characteristics at an I_{CC} of 4 μA.

As the applied I_{CC} to the Cu/Ga_2O_3/Pt memristors increases to 10 μA, we observe exclusively bipolar nonvolatile memristive behavior, as illustrated in Fig. 4(a). Specifically, when the bias voltage starts at 0 V and increases, a sudden increase in current occurs at about 0.8 V during the SET process. This transition signifies the device's shift from a HRS of >2 × 10^6 Ω to a LRS of <2 × 10^4 Ω. It is noteworthy that the nonvolatile LRS depicted in Fig. 4(a) is lower than the volatile LRS shown in Fig. 3(a). This discrepancy can be attributed to the formation of more robust Cu conductive filaments facilitated by the higher I_{CC} [7].

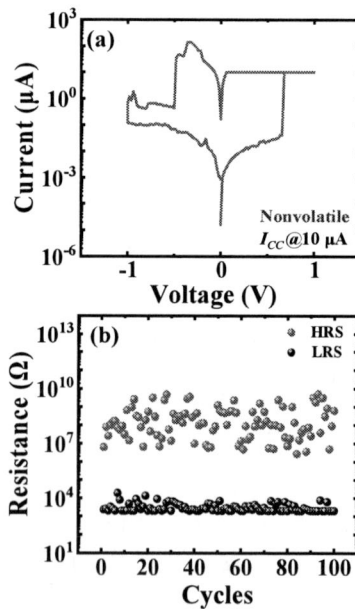

Fig. 4: (a) Nonvolatile behavior at an I_{CC} of 10 μA. (b) Cyclic endurance characteristics over 100 cycles.

The Cu/Ga$_2$O$_3$/Pt devices maintain the LRS when the voltage is swept from a maximum of 1.0 V back to 0 V. Following this, a negative RESET voltage of approximately −0.5 V initiates a RESET process, transitioning the device from the LRS to a HRS. Fig. 4(b) shows the endurance characteristics of the Cu/Ga$_2$O$_3$/Pt memristor at an I_{CC} of 10 μA, with resistance states measured at 0.1 V. The HRS/LRS ratio remains stable at around 10^2 after 100 cycles. However, it should be referred that the cycle-to-cycle variation in the HRS is significant, which indicates the need for further optimization through device engineering.

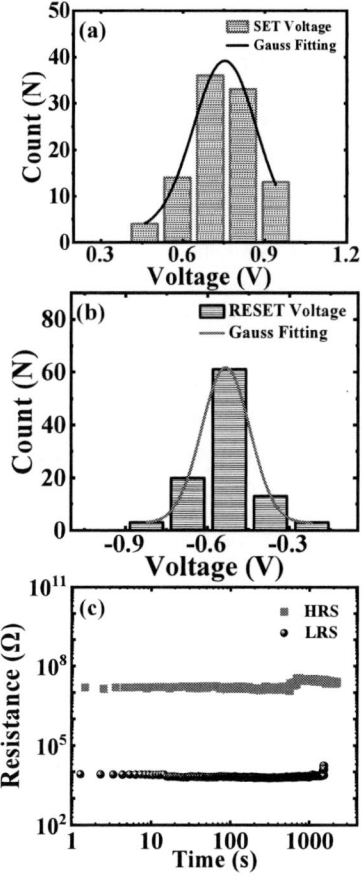

Fig. 5: Distributions of (a) SET and (b) RESET voltages of the Cu/Ga$_2$O$_3$/Pt memristor. (c) Data retention characteristics of HRS and LRS.

The histogram distributions of the SET and RESET voltages for the nonvolatile Cu/Ga$_2$O$_3$/Pt memristor are presented in Fig. 5(a) and 5(b), respectively. The SET voltage exhibits approximately a 36% probability at 0.7 V and a 33% probability at 0.82 V, as shown in Fig. 5(a). Fig. 5(b) shows that the RESET voltage has about a 61% probability of occurring at −0.52 V. The relatively narrow distributions of the SET and RESET switching voltages highlight the good uniformity of the Cu/Ga$_2$O$_3$/Pt device. Fig. 5(c) illustrates the data retention characteristics of the Cu/Ga$_2$O$_3$/Pt memristor, demonstrating that this bipolar resistive switching behavior can be consistently observed across successive voltage cycles,

with a retention time exceeding 10^3 s at room temperature, thereby confirming its nonvolatile bipolar features.

Conclusion

In conclusion, we demonstrate the dual-functional behaviors of a memristor with a Cu/Ga$_2$O$_3$/Pt structure. Notably, the Ga$_2$O$_3$-based device exhibits forming-free, volatile resistive switching when operated under low I_{CC} conditions. Conversely, at higher I_{CC} levels, it presents a unique mixed-mode switching behavior. This versatility makes it suitable for a wide range of applications. Our findings provide valuable insights into device design strategies, facilitating advanced integration and multifunctional applications within memristor technology.

Acknowledgments

This work was supported in part by 2030 Major Project of the Chinese Ministry of Science and Technology (2021ZD0201200), Research Fund of Suzhou Laboratory (SK-1202-2024-012), and Opening Project of Advanced Integrated Circuit Package and Testing Research Center of Jiangsu Province (NTIKFJJ202302).

References

[1] I. Dergaa, K. Chamari, P. Zmijewski and H. B. Saad, "From human writing to artificial intelligence generated text: examining the prospects and potential threats of ChatGPT in academic writing", Biol Sport., 40, pp. 615–622 (2023).

[2] J. Yang, H. Jin, R. Tang, X. Han, Q. Feng, H. Jiang, S. Zhong, B. Yin and X. Hu, "Harnessing the power of LLMs in practice: A survey on ChatGPT and beyond", ACM Trans. Knowl. Discov. Data, 18, p. 160 (2024).

[3] X. Y. Xu and X. M. Jin, "Integrated photonic computing beyond the von Neumann architecture", ACS Photonics, 10, pp. 1027–1036 (2023).

[4] Y. Li and K. Ang, "Hardware implementation of neuromorphic computing using large-scale memristor crossbar arrays", Adv. Intell. Syst., 3, p. 2000137 (2021).

[5] C. Yang, B. Sun, G. Zhou, H. Zhao, S. Zhu, C. Ke, Y. Zhao and H. Wang, "Evolution between volatile and nonvolatile resistive switching behaviors in Ag/TiO$_x$/CeO$_y$/F-Doped SnO$_2$ nanostructure-based memristor devices for information processing applications", ACS Appl. Nano Mater., 6, pp. 8857−8867 (2023).

[6] G. D. Zhou, Z. R. Wang, B. Sun, F. C. Zhou, L. F. Sun, H. B. Zhao, X. Hu, X. Peng, J. Yan, H. Wang, W. Wang, J. Li, B. Yan, D. Kuang, Y. Wang, L. Wang and S. K. Duan, "Volatile and nonvolatile memristive devices for neuromorphic computing", Adv. Electron. Mater., 8, p. 2101127 (2022).

[7] N. He, Q. Zhang, L. Tao, X. Chen, Q. Qin, X. Liu, X. Lian, X. Wan, E. Hu, J. Xu, F. Xu and Yi Tong, "V$_2$C-based memristor for applications of low power electronic synapse", IEEE Electron Device Lett., 42, pp. 319-322 (2021).

Enabling Broader Memory Windows by Double-Gate Nanosheet Ferroelectric FETs for Next-Generation Non-Volatile Memory Storage

F. Wu[1,#], C.-Y. Chiu[2,#], T.-Y. Lin[2], C.-H. Wu[3], V. P.-H. Hu[3], P. Su[1,*] and C.-J. Su[1,2,4,**]

[1]Institue of Electronics, National Yang Ming Chiao Tung University, Taiwan; [2]Department of Electrophysics, National Yang Ming Chiao Tung University, Taiwan; [3]Graduate Institute of Electronics Engineering, National Taiwan University, Taiwan; [4]Taiwan Semiconductor Research Institute, Taiwan

#Equal contribution; *Email: pinsu@nycu.edu.tw, **Email: cjsu@nycu.edu.tw

Abstract

In this work, we present a novel lateral electric field-driven transistor that leverages an asymmetric double-gate nanosheet (DGNS) design to optimize the balance between ferroelectric (FE) layer scaling and memory window (MW) expansion. This device architecture, implemented through precise digital etching, defines a controllable active area together with a self-aligned etching method to facilitate the formation of the NS channels, significantly improving fabrication efficiency. The unique design separates read and write functions, by taking advantages of body effect and a high surface-to-volume ratio of channel to expand the MW without compromising electrostatic integrity. These innovations provide enhanced performance and scalability, positioning the proposed device as a viable candidate for advanced applications in next-generation electronic systems and memory storage technologies.

Keywords: Memory Window, FeFETs and Nanosheet

Introduction

As the rapid advancements in artificial intelligence continue, the demand for next-generation non-volatile memories (NVMs) is rising correspondingly. The energy efficiency and operational speed of ferroelectric field-effect transistors (FeFETs) are key attributes being explored, positioning them as promising candidates for future NVM technologies. However, the relationship between the maximum MW and the ferroelectric thickness (t_{FE}) combined with the coercive field (E_c) presents a scaling challenge: reducing t_{FE} favors low-voltage operation while negatively impacts MW. To mitigate this challenge, various strategies have emerged. For instance, interfacial layer (IL)-free FeFETs are able to eliminate the influence of ILs that act as voltage dividers, thereby enhancing the MW [1,2]. Similarly, utilizing germanium as the channel material holds potential for reducing writing voltages while maintaining adequate MW due to the higher dielectric constant of germanium oxide [3]. Additionally, the asymmetric DG configuration in fully depleted silicon-on-insulator (FDSOI) structures leverages the body effect to further amplify MW without compromising the integrity of ferroelectric layer [4]. However, FDSOI is limited by its requirement for stacked structures and high fabrication cost. To this end, we propose a lateral-controlled, back-end-of-line (BEOL)-compatible architecture, which not only enhances MW through the body effect but also opens avenues for unique three-dimensional integrated circuit (3DIC) designs.

Device Fabrication

The fabrication of the developed device has been successfully demonstrated. Following the process flow illustrated in **Fig. 1(a)**, we can fabricate asymmetric DG FeFETs with thin NS channels. Conceptually, to address the challenges of achieving vertical stacks in fully FDSOI devices, we "rotate" the structure by 90 degrees and manipulate the electric field laterally to control the channel. The gate stacks, capped with a top etching stop layer, are deposited using a horizontal furnace. The side-gate (SG) is fabricated by *in-situ* doped n$^+$-poly-Si. Next, the gate stack is defined through dry etching, as depicted in **Fig. 1(b)**. A crucial step in the process is the lateral SG recess **(Fig. 1(c))**. We employ sophisticated digital etching techniques to define the space for the SG dielectric (DE) and the NS channels. After the lateral SG recess, a 6 nm layer of silicon nitride (SiN) is deposited as the SG DE (**Fig. 1(d)**). A subsequent 50 nm amorphous silicon layer is then deposited. Ion implantation is performed for source and drain (S/D) regions, followed by annealing to crystallize the amorphous silicon. The NS channels are defined through a self-aligned process using anisotropic etching, as shown in **Fig. 1(e)**. The NS is confined by the cavity created by the SG recess, allowing for precise nanometer-scale self-alignment without advanced lithography tools. Next, in **Fig. 1(f)**, atomic layer deposition (ALD) is used to deposit an 8 nm superlattice HZO layer and applied *in-situ* ALD NH₃ treatment to suppress oxygen vacancy (V_o) formation [5]. Titanium nitride (TiN) is then deposited via physical vapor deposition (PVD) as the metal gate for the top-gate (TG), which controls the ferroelectric layer. The TG is defined using anisotropic etching in **Fig 1(g)**, completing the fabrication of the DGNS FeFETs. Finally, microwave annealing (MWA) is performed to crystallize the HZO into the ferroelectric phase and activate dopants while maintaining a low thermal budget.

The successful formation of the NS and lateral DG profiles is confirmed by the Energy Dispersive X-Ray Spectroscopy (EDS) line scan shown in **Fig. 2(a)**, where the silicon peak is observed between the peaks of HZO (representing the ferroelectric layer) and silicon nitride (SiN, representing SG_DE).

979-8-3315-0417-5/25 $31.00 © 2025 IEEE

Results and Discussion

A. Ferroelectricity and Body Effect

A typical PUND (Positive Up, Negative Down) pulse train method is employed to extract the polarization-voltage (P-V) loops and the transient current response, as illustrated in **Fig. 2(b)** and **Fig. 2(c)**. These results indicate that the FE properties have been successfully obtained. To achieve maximum writable MW amplification, it is essential to analyze and quantify the body effect factor resulting from our innovative lateral-controlled asymmetric DG configuration.

The mechanism accounting for the MW amplification is the body effect [6]. The equivalent circuit of the voltage divider corresponding to our laterally controlled device is depicted in **Fig. 2(d)**. The body effect factor is defined as the ratio of the change in threshold voltage (V_T) to the various biases of the other gate. Since the MW is a function of V_T, the MW measured from the SG is proportional to the MW measured from the TG scaled by the SG body effect factor (γ_{SG}). This relationship provides a clear guideline: to enhance MW, we simply need to reduce the capacitance of the silicon nitride (C_{SiN}). The details are shown in **Fig. 2(e)**.

B. MW Amplification

The DC I_D-V_G measurements under various bias conditions from the two gates help us to extract the body effect factors γ_{TG} and γ_{SG}, as shown in **Figs. 3(a)** to **3(d)**. The results indicated $\gamma_{TG} = 0.6$ and $\gamma_{SG} = 1.32$. Ignoring the channel capacitance (C_{NS}), we obtained comparable values using the voltage divider approach. We investigated the optimal writing conditions, focusing on the switching dynamics on the TG when reading and writing solely on it [7]. Increasing pulse height is correlated with an increase in MW, attributed to the sensitivity of the low threshold voltage (LVT) to pulse width.

We then applied the same writing scheme to the TG while reading from both TG and SG. The MW amplification, illustrated in **Fig. 4(a)**, is found to be a factor of 2, with certain discrepancy to γ_{SG}, likely due to actual device variations and potential defects. To assess non-ideal DC responses and examine the body effect in transient behavior, we conducted pulse-IV measurements. The device was written under the same conditions and read at the microsecond scale with varying the SG biases. We extracted γ_{TG} values for both LVT and high threshold voltage (HVT) in **Fig. 4(b)**. By comparing the γ_{TG} extracted under DC in **Fig. 3(b)** and γ_{TG} extracted by pulsed-IV in **Fig. 4(b)**, the two values are comparable (0.6 and 0.56). This indicates that the body effect factor is independent of transient responses, as no frequency dispersion relationship exists for the FE and SG DE.

Conclusion

In this work, we outlined the fabrication process of novel DGNS FeFETs, emphasizing sophisticated lateral etching and self-aligned technology. We extracted the γ_{TG}, γ_{SG} and analyzed switching dynamics, achieving an amplified MW, showcasing the potential of these devices for the next generation of NVMs with improved endurance and energy efficiency.

Acknowledgments

This work was supported by the NSTC in Taiwan under Grants NSTC 112-2223-E-002-011-MY3, 112-2218-E-002-028, 110-2221-E-A49-165-MY3, 112-2622-8-A49-013-SB, 113-2218-E-A49-022-MBK, 113-2221-E-A49-099-MY3. The authors would like to thank TSRI and NFC for the process support.

References

[1] Sourav Dutta, Huacheng Ye, Akif A. Khandker, Sharadindu Gopal Kirtania, Abhishek Khanna, Kai Ni and Suman Datta, "Logic Compatible High-Performance Ferroelectric Transistor Memory", IEEE EDL, 43, 3 (2022).

[2] Kailiang Huang, Minglong Zhai, Xueyuan Liu, Bing Sun, Hudong Chang, Jianhua Liu, Chao Feng, and Honggang Liu, "Hf0.5Zr0.5O2 Ferroelectric Embedded Dual-Gate MoS2 Field Effect Transistors for Memory Merged Logic Applications", IEEE EDL, 41, 10 (2020).

[3] Dipjyoti Das, Prasanna Venkatesan Ravindran, Chinsung Park, Nujhat Tasneem, Zheng Wang, Hang Chen, Winston Chern, Shimeng Yu, Suman Datta and Asif Khan, "A Ge-Channel Ferroelectric Field Effect Transistor with Logic-Compatible Write Voltage", IEEE EDL, 44, 2 (2023).

[4] Zhouhang Jiang, Yi Xiao, Swetaki Chatterjee, Halid Mulaosmanovic, Stefan Duenkel, Steven Soss, Sven Beyer, Rajiv Joshi, Yogesh S. Chauhan, Hussam Amrouch, Vijaykrishnan Narayanan, and Kai Ni, "Asymmetric Double-Gate Ferroelectric FET to Decouple the Tradeoff Between Thickness Scaling and Memory Window", VLSI, 395 (2022).

[5] Yi-Jan Lin, Chih-Yu Teng, Chenming Hu, Chun-Jung Su, and Yuan-Chieh Tseng, "Impacts of surface nitridation on crystalline ferroelectric phase of Hf1-xZrxO2 and ferroelectric FET performance", Appl. Phys. Lett. 119, 192102 (2021).

[6] Halid Mulaosmanovic, Dominik Kleimaier, Stefan Dünkel, Sven Beyer, Thomas Mikolajick, and Stefan Slesazeck, "Ferroelectric transistors with asymmetric double gate for memory window exceeding 12 V and disturb-free read", Nanoscale, 13, 16258 (2021).

[7] Zijian Zhao, Shan Deng, Swetaki Chatterjee, Zhouhang Jiang, Muhammad Shaffatul Islam, Yi Xiao, Yixin Xu, Scott Meninger, Mohamed Mohamed, Rajiv Joshi, Yogesh Singh Chauhan, Halid Mulaosmanovic, Stefan Duenkel, Dominik Kleimaier, Sven Beyer, Hussam Amrouch, Vijaykrishnan Narayanan, and Kai Ni, "Powering Disturb-Free Reconfigurable Computing and Tunable Analog Electronics with Dual-Port Ferroelectric FET", ACS Appl. Mater. Interfaces, 15, 54602 (2023).

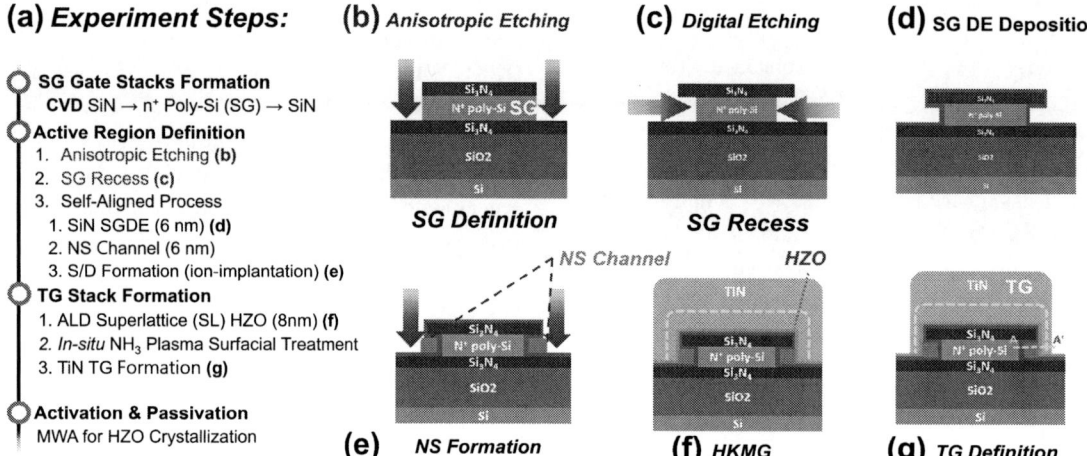

Fig. 1. (a) The key process flow for the DGNS FeFETs. (b) Formation of SG. (c) SG recess by digital etching. (d) 6 nm SiN is deposited as SG DE. (e) NS channel formation by self-aligned etching. (f) 8 nm HZO as the FE layer and TiN as the TG. (g) TG patterning, followed by MWA to recrystallize the HZO into the ferroelectric phase.

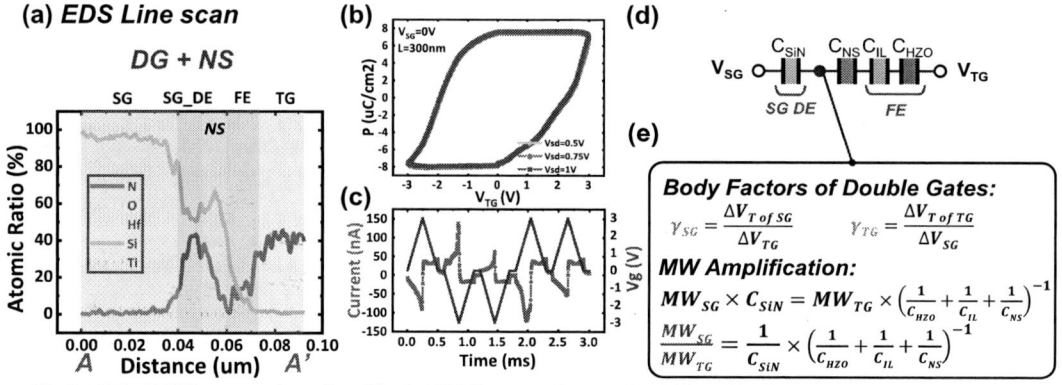

Fig. 2. (a) The DGNS structure is confirmed by the EDS line scan, along A-A' as indicated in Fig. 1(g). (b) On-device P-V loop extracted by (c) PUND method and the switching current. (d) Equivalent circuit of the voltage divider. (e) Body effect factors and the correlation to MW amplification.

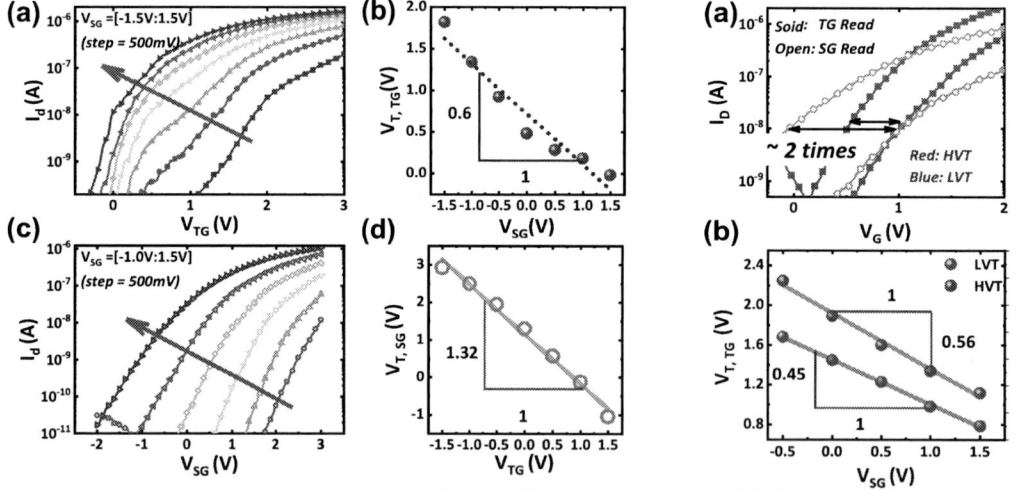

Fig. 3. (a) Transfer characteristics of TG under different SG biases. (b) γ_{TG} is extracted by the V_T dependence to V_{SG}. (c) I_D-V_G curves of the SG control. (d) γ_{SG} extracted by the slope of the V_T dependency.

Fig. 4. (a) The HVT and LVT behaviors for TG and SG read with the same TG write condition under pulsed-IV measurement, revealing an amplified MW by the SG read. (b) Consistent body effect factor in HVT and LVT as well as transient vs. DC operation (Fig. 3(b)).

979-8-3315-0417-5/25 $31.00 © 2025 IEEE

On-chip photon-mediated magnon-superconducting qubit system and its quantum application

Jiacheng LIU[1*], Ferris Prima NUGRAHA[1*], Qiming SHAO[1]

[1]Department of Electronic and Computer Engineering, The Hong Kong University of Science and Technology, Clear Water Bay, Kowloon, Hong Kong, China

*Equal contributions. Email: jliuex@connect.ust.hk , fpnugraha@connect.ust.hk

Abstract

On-chip hybrid system utilizing photon-mediated coupling between magnon and superconducting qubit is proposed in this paper. By integrating a magnetic thin film in the z-axis with superconducting circuits, quantum operations between two-level magnon modes and qubit are enabled without increasing the circuit footprint, controlled by external magnetic field. Physical parameters of the coupled system are obtained using the proposed Multiphysics model following our design framework. Finally, we demonstrated the behavior of the system and performed quantum operations based on the Hamiltonian model.

(Keywords: Hybrid system, photon mediated magnon-qubit coupling, quantum operation)

Introduction

Hybrid magnonics, an interdisciplinary field combining magnonics with coherent information processing [1], [2], has emerged as a promising platform for quantum information science. Also referred to as quantum magnonics, this field enables the exploration of quantum-entangled states and single-magnon states by coupling collective spin excitations, or magnons, to various quantum systems [2]. Magnons offer three primary advantages: they operate within the microwave bandwidth and can be tuned across a wide frequency range via magnetic fields; they exhibit strong dipolar coupling with electromagnetic waves; and their short wavelength allows for compact integration in microwave circuits. Additionally, magnetic materials are intrinsic microwave resonators with their frequency and performance almost independent of dimensions. Conversely, recent work [1] has integrated, through a 3D cavity, the hybrid magnonic system with the superconducting qubit [3] known for its mature technology and accurate control for potential quantum computing or sensing application. However, an on-chip photon mediated magnon-qubit hybrid circuit has yet to be demonstrated, and comprehensive Multiphysics simulation methods are still under development.

Here, we propose a framework for designing and analyzing an on-chip photon-mediated magnon-superconducting qubit system. By extracting parameters through Multiphysics analysis, we can use our Hamiltonian model to reveal strong coupling and demonstrate quantum operations such as natural time-evolution or state-swapping mechanism.

Model for photon-mediated magnon-qubit hybrid circuit

The proposed on-chip hybrid system model, illustrated in Fig. 1(a), primarily consists of three components: superconducting resonator (photon, shown in gold), magnetic thin film coupled via dipolar interaction (magnon, shown in green), and a capacitively coupled superconducting qubit (qubit, shown in purple). The two insets in Fig. 1(a) display detailed view of the magnon and qubit components. An overview of the hybrid system's interactions and self-term is presented in the hybrid Hamiltonian model (Fig. 2), which includes the photon, qubit, and magnon (both Kittel and PSSW) mode.

A. Dissipation rate and operation frequency obtained

The properties of superconducting resonators used in standard transmon qubits can be achieved by designing a $\lambda/4$ resonator and optimizing the capacitive coupling to achieve a high-quality factor. Fig. 3(a) presents the S21 curve and fitting results, with $f_p = 7.058\ GHz, \kappa_p = 10^{-4}\ GHz$.

Since the resonance of magnetic materials is not limited by in-plane geometric dimensions, we integrated a magnetic insulator film on top of the metal wire (Fig. 1(a)). Notably, beyond the uniform spin wave precession (Kittel mode), the skin effect generates a non-uniform dynamic magnetic field across the film thickness, breaking the symmetry in the z direction and inducing higher-order perpendicular standing spin waves (PSSW) mode [4]. Fig. 3(b) shows the spectrum of the two magnon modes, where the resonant frequency is tuned by the magnetic field. The inset illustrates the spatial distribution of spin waves in both the Kittel and PSSW modes.

B. Strong coupling between magnon and photon

Achieving strong coupling between oscillators is fundamental to quantum information processing. Magnon and photon are coupled through dipolar interaction, which could be described as an effective permeability tensor:

$$\overrightarrow{\mu_r} = \begin{bmatrix} 1+u_k & -iv_k & 0 \\ iv_k & 1+u_k & 0 \\ 0 & 0 & 1 \end{bmatrix}, u_k = \frac{\omega_{m1,2}\omega_{Ms}}{\omega_{m1,2}^2 - \omega^2}, v_k = \frac{\omega\omega_{Ms}}{\omega_{m1,2}^2 - \omega^2}$$

where $\omega_{Ms} = \gamma\mu_0 Ms$, ω_{m1}, ω_{m2} and all other physical parameter definitions are detailed in the notes below Table 1. The anti-crossing spectrum in Fig. 3(c) was obtained by numerically solving the hybrid dynamics using Maxwell and Landau-Lifshitz-Gilbert (LLG) equations. By extracting the parameters, we get the cooperativity ($C_{pm1} = g/\kappa_p \cdot g/\kappa_{m_1} = 562.5 > 1$ and $C_{pm2} = g/\kappa_p \cdot g/\kappa_{m_2} = 640 > 1$).

C. Lumped element analysis of superconducting qubit

Figure 3(d) illustrates the analysis workflow for a

979-8-3315-0417-5/25 $31.00 © 2025 IEEE

superconducting qubit, typically of the transmon type. Essentially, the qubit functions as a nonlinear LC resonator with the Josephson Junction serving as the inductor. The remaining superconductor islands form a capacitance network, which can be represented as a capacitance matrix obtainable through electrostatic analysis using ANSYS Q3D. This qubit couples to its environment—in this case the $\lambda/4$ resonator via electric dipole interaction. Utilizing the lumped oscillator model (LOM) method [5], we can convert the capacitance matrix into the effective capacitance involved in qubit frequency and coupling. Using transmon physics [3], we calculate the relevant qubit parameters for our Hamiltonian model.

All extracted physical parameters based on the proposed hybrid circuit are summarized in Table 1.

Quantum operation for magnon-qubit system

In the following section, we analyze the behavior of the magnon-qubit hybrid system based on the Hamiltonian model provided with the characteristics in Table 1.

A. Avoided crossing from strong coupling

Inspired from [6], we simulate the spectroscopy process by applying a small drive to the hybrid system through the readout resonator and monitoring the steady-state photon excitation. The frequencies of both qubit and magnon are varied by adjusting the external magnetic fields H_q and H_m, respectively. Fig. 4a shows the avoided crossing between the qubit and photon, confirming the coupling rate g_{pq} determined by the qubit lumped element analysis. By sweeping the H_m, the heatmap in Fig. 4(b) illustrates how both the Kittel and PSSW mode frequencies vary. In addition to the magnon-photon crossing seen in Fig. 3(c), we observe strong coupling between the magnon and qubit, facilitated by the direct qubit-photon and the magnon-photon couplings.

B. Time-evolution without external drive

We analyze the time-dependent behavior of the hybrid system using the Lindblad master equation. In addition to the Hamiltonian, dissipation is incorporated through collapse operators [7]. Without external drive, we initially excite the qubit, photon, magnon Kittel/PSSW modes, observing the population dynamics shown in Fig. 5. H_m is set such that the Kittel and PSSW frequencies exceed that of the qubit.

In Fig. 5(a), the qubit population decreases over time with minimal excitation in the photon and both magnon modes. This may result from the weak photon-qubit coupling, despite the dominant photon and Kittel modes due to frequency proximity. Figs. 5(b-d) show that the photon and two magnon modes coupled strongly, evidenced by their higher excitation levels compared to the qubit.

C. Demonstration of swapping mechanism

By adjusting H_m, we can vary the frequencies of the Kittel and PSSW modes to facilitate a swapping operation [8]. The

magnetic fields are applied in a trapezoid pulse form, with the flat time τ_f and edge time τ_e, and idle time τ_i on both sides. For this simulation, we use $\tau_f = 100ns$ and $\tau_i = 50$ ns, while we keep $\tau_e = 0.05ns$ to approximate a rectangular pulse.

We implemented two scenarios based on the resonance with either Kittel or PSSW mode to approach the qubit frequency. In Fig. 6(a) when the Kittel mode coincides with the qubit frequency, we observe sinusoidal exponential dynamics between the qubit and the Kittel mode. A similar pattern is evident in Fig. 6(b) for the qubit and PSSW modes when the PSSW mode aligns with the qubit frequency. The sinusoidal period is inversely proportional to the coupling strength between qubit and the respective magnon mode. The decay rate is influenced by the dissipation rate of each mode. Notably, we observe a larger maximum excitation in the PSSW mode compared to the Kittel one, likely due to the its lower dissipation rate.

Conclusion

In this work, we proposed and analyzed an on-chip hybrid system featuring photon-mediated coupling between magnons and a superconducting qubit. Using a Multiphysics framework and Hamiltonian model, we demonstrated strong coupling and quantum operations, including time evolution and state-swapping mechanisms in the hybrid system. This framework provides a foundation for advancing hybrid magnon-qubit systems, supporting future applications in quantum information processing and coherent control of quantum states.

Acknowledgments

The authors appreciate insightful discussions with W. Yu and Y. Liu. This work was supported by the National Key R&D Program of China (Grant No. 2021YFA1401500) and the RGC General Research Fund (Grant No. 16303322). Ferris acknowledges the support from HKPFS.

References

[1] Y. Tabuchi *et al.*, "Coherent coupling between a ferromagnetic magnon and a superconducting qubit," *Science*, 2015.

[2] D. D. Awschalom *et al.*, "Quantum Engineering With Hybrid Magnonic Systems and Materials," *IEEE Trans. Quantum Eng.*, 2021.

[3] A. Blais, A. L. Grimsmo, S. M. Girvin, and A. Wallraff, "Circuit quantum electrodynamics," *Rev. Mod. Phys.*, 2021.

[4] J. Liu *et al.*, "Strong magnon-magnon coupling and low dissipation rate in an all-magnetic-insulator heterostructure," *Phys. Rev. Appl.*, 2024.

[5] Z. K. Minev *et al.*, "Circuit quantum electrodynamics (cQED) with modular quasi-lumped models," 2021, [Online]. Available: http://arxiv.org/abs/2103.10344

[6] L. S. Bishop *et al.*, "Nonlinear response of the vacuum Rabi resonance," *Nat. Phys.*, 2009.

[7] J. R. Johansson et al., "QuTiP 2: A Python framework for the dynamics of open quantum systems," *Comput. Phys. Commun.*, 2013.

[8] C. Trevillian and V. Tyberkevych, "Universal Set of Magnon-Mediated Quantum Gates," *2023 IEEE Int. Magn. Conf. - Short Pap.*, 2023.

Hybrid photon-mediated magnon-qubit system & Framework

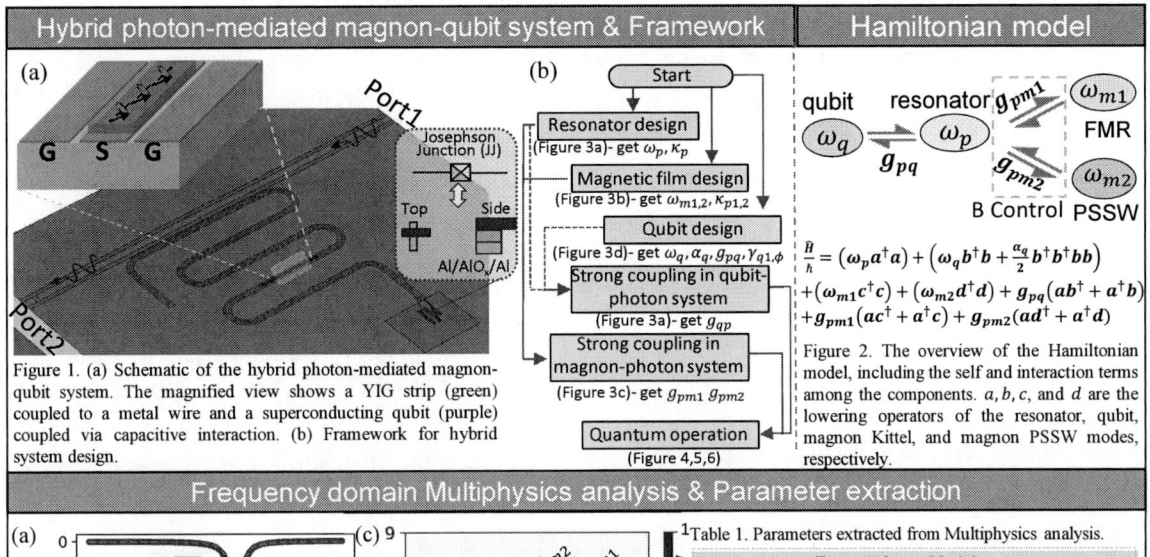

Figure 1. (a) Schematic of the hybrid photon-mediated magnon-qubit system. The magnified view shows a YIG strip (green) coupled to a metal wire and a superconducting qubit (purple) coupled via capacitive interaction. (b) Framework for hybrid system design.

Hamiltonian model

$$\frac{H}{\hbar} = \left(\omega_p a^\dagger a\right) + \left(\omega_q b^\dagger b + \frac{\alpha_q}{2} b^\dagger b^\dagger bb\right) + \left(\omega_{m1} c^\dagger c\right) + \left(\omega_{m2} d^\dagger d\right) + g_{pq}\left(ab^\dagger + a^\dagger b\right) + g_{pm1}\left(ac^\dagger + a^\dagger c\right) + g_{pm2}\left(ad^\dagger + a^\dagger d\right)$$

Figure 2. The overview of the Hamiltonian model, including the self and interaction terms among the components. a, b, c, and d are the lowering operators of the resonator, qubit, magnon Kittel, and magnon PSSW modes, respectively.

Frequency domain Multiphysics analysis & Parameter extraction

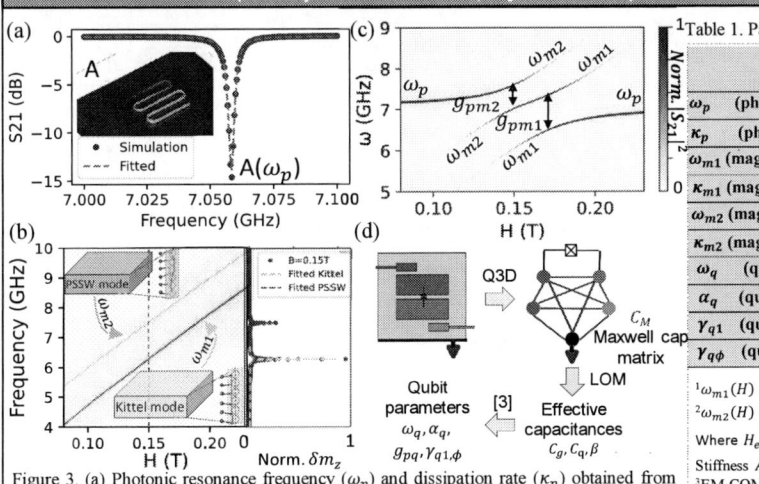

Figure 3. (a) Photonic resonance frequency (ω_p) and dissipation rate (κ_p) obtained from COMSOL. The inset shows the spatial distribution of the dynamic magnetic field at resonance A point. (b) Two level magnon frequency ($\omega_{m1,2}$) and dissipation rate ($\kappa_{m1,2}$) obtained from COMSOL. The insets show the spatial distribution of the spin wave. (c) Spectrum of magnon-photon strong coupling system obtained in COMSOL. (d) The process flow of qubit lumped element analysis.

[1]Table 1. Parameters extracted from Multiphysics analysis.

		Extracted parameter	Model Tool
ω_p	(photon)	7.058 GHz	EM-COMSOL[3]
κ_p	(photon)	0.1 MHz	EM-COMSOL[3]
ω_{m1}	(magnon)	$\omega_{m1}(H)$[1]	MM-COMSOL[4]
κ_{m1}	(magnon)	8 MHz	MM-COMSOL[4]
ω_{m2}	(magnon)	$\omega_{m2}(H)$[2]	MM-COMSOL[4]
κ_{m2}	(magnon)	5 MHz	MM-COMSOL[4]
ω_q	(qubit)	5.568 GHz	Q3D-ANSYS[5]
α_q	(qubit)	-0.32 GHz	Q3D-ANSYS[5]
γ_{q1}	(qubit)	0.31 MHz	LOM-KOCH[6]
$\gamma_{q\phi}$	(qubit)	0.28 MHz	LOM-KOCH[6]

g_{pm1}: 0.45GHz
g_{pm2}: 0.32GHz
g_{pq}: 0.11GHz

$\omega_r = 8.0 GHz$[6]
$g_{qr} = -94 MHz$[6]
[6]For readout to confirm avoided crossing

[1]$\omega_{m1}(H) = \gamma\sqrt{\mu_0 H(\mu_0 H + \mu_0 M_s)}$
[2]$\omega_{m2}(H) = \gamma\sqrt{(\mu_0(H + H_{ex})) \times (\mu_0(H + H_{ex} + M_s))}$
Where $H_{ex} = \frac{2A_{ex}}{M_s}\left(\frac{\pi}{d}\right)^2$; γ: 28GHz/T; M_s: $1.48 \times 10^5 \frac{A}{m}$;
Stiffness A_{ex}: 2.67 pJ/m; Thickness d: 100nm; Magnetic field H
[3]EM-COMSOL: Electromagnetic wave module in COMSOL
[4]MM-COMSOL: Micromagnetic module in COMSOL
[5]Q3D-ANSYS: Extracted capacitance C_q is 60.5fF, assuming JJ to have inductance of 12nH and capacitance of 2fF.
[6]LOM-KOCH: Using the lumped element analysis together with the transmon physics.

Strong coupling and quantum operation demonstration

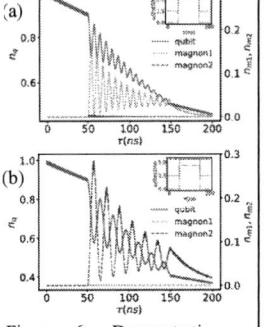

Figure 4. Avoided crossing showing strong coupling between (a) qubit-photon and (b) magnon-qubit with the Hamiltonian model of the hybrid system. The heatmap represent the excited photon number in the readout resonator given a small drive, simulating the spectroscopy process.

Figure 5. Time-evolution of the hybrid system in the absence of any external drive. The upper half represents the population of the initially excited (a) qubit, (b) photon, (c) Kittel, and (d) PSSW mode progress. The lower half shows the evolution of the remaining components.

Figure 6. Demonstrating a swapping mechanism between initially excited qubit with magnon modes, either the (a) Kittel or (b) PSSW dominating.

979-8-3315-0417-5/25 $31.00 © 2025 IEEE

Reduced Process Induced Threshold Voltage Variability in Bulk Negative Capacitance Junctionless Transistors

Ruma S R[1], Vita Pi-Ho Hu[2] and Manish Gupta[1]

[1]Birla Institute of Technology and Science Pilani, K K Birla Goa Campus, Goa, India

[2]Department of Electrical Engineering, National Taiwan University, Taiwan

[1]p20220061@goa.bits-pilani.ac.in, [1]manishg@goa.bits-pilani.ac.in

Abstract- This work provides a comprehensive analysis of threshold voltage (V_{th}) variability (σV_{th}) in negative capacitance (NC) junctionless (JL) transistors on silicon-on-insulator (SOI) and bulk substrates through well-calibrated simulations. Bulk NCJL transistors show improved performance due to a shift in the conduction channel towards the front surface, which improves the gate capacitance (C_{gg}) and enhances the negative capacitance in bulk NCJL devices. The results highlight a significant reduction in σV_{th} by ~89 %, ~91 %, ~60 %, ~80 %, and ~56 % due to the variations in gate length (L_g), silicon film thickness (T_{si}), equivalent oxide thickness (EOT), channel doping (N_{ch}), and ferroelectric thickness (T_{fe}), respectively. Additionally, the reduced V_{th} variability further translates to an enhanced noise margin (NM) and improved process-induced NM variations in inverter circuits. The results reported for the first time showcase the superior performance of bulk NCJL designs compared to SOI NCJL transistors.

Keywords: Negative capacitance, Junctionless, Threshold voltage, Variability, MOSFET.

Introduction

Junctionless (JL) devices have emerged as a promising alternative for various applications due to their simpler architecture [1]. Although JL devices are easier to fabricate, their heavily doped channel results in higher threshold voltage (V_{th}) variability, influenced by statistical fluctuations in the device geometrical parameters [1]. This variability is further exacerbated by factors like random dopant fluctuation (RDF) [2], line edge roughness (LER) [3], and work-function variability (WFV) [4]. Previous studies [5] have shown that V_{th} sensitivity towards device parameters in JL devices can be reduced through the negative capacitance (NC) effect. However, achieving low σV_{th} remains challenging, especially at shorter technology nodes. Therefore, this work systematically reports on reducing σV_{th} caused by process-induced structural parameter variations by designing NCJL devices on a bulk substrate.

Simulation Methodology

Fig. 1 shows the NCJL transistor designed on bulk (Fig. 1*a*) and SOI (Fig. 1*b*) substrates. TCAD simulations were performed with field and doping-dependent mobility modules, bandgap narrowing, quantum correction, and Shockley-Read-Hall (SRH) recombination models [6]. The mean (μ) device parameters are shown in Table. 1, were considered from the International Roadmap for Devices and Systems (IRDS) 2023 [7]. While the buried oxide layer of a thickness (T_{BOX}) 10 nm was used in designing the SOI NCJL device, a *p*-type bulk substrate doped (N_B) with 5×10^{18} cm^{-3} was for the bulk NCJL transistor. The heavily doped source/drain region with doping of 10^{20} cm^{-3} was considered to minimize the series

resistance effect. The calibrated values of the E_c and P_r, shown in our previous work [5], were used to model the NC regime offered by the Zr-doped HfO$_2$ Fe layer [5].

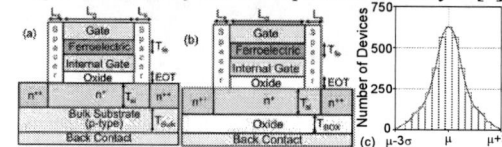

Fig. 1. Schematic view of negative capacitance (NC) junctionless (JL) transistor designed on a (a) bulk and (b) silicon-on-insulator (SOI) substrates. (c) Normal distribution of 750 devices to analyze the threshold voltage (V_{th}) variability in NCJL transistors with bulk and SOI substrates.

Parameter	μ-3σ	μ	μ+3σ
T_{si}	6.3 nm	7 nm	7.7 nm
EOT	0.91 nm	1 nm	1.09 nm
L_g	18 nm	20 nm	22 nm
N_{ch}	9×10^{18} cm^{-3}	10^{19} cm^{-3}	1.1×10^{19} cm^{-3}
T_{fe}	3.6 nm	4 nm	4.4 nm

Table I. Reference (μ) and extreme values (μ-3σ and μ+3σ) of the device parameters considered to analyze the V_{th} variability in the bulk and SOI NCJL transistors.

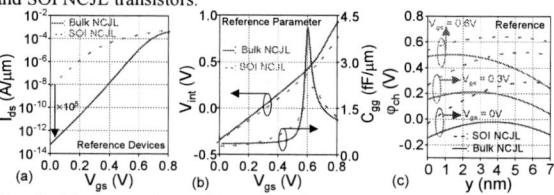

Fig. 2. (a) Comparison of I_{ds} - V_{gs} characteristics of NCJL transistor designed on the bulk and SOI substrates. Comparison of (b) internal gate voltage (V_{int}) (on the primary y-axis) and gate capacitance (C_{gg}) (on the secondary y-axis) as a function of gate voltage (V_{gs}), (c) channel potential (φ_{ch}), extracted along the vertical cut-line at the mid-gate position, at varying V_{gs}. The results are obtained with mean device parameters.

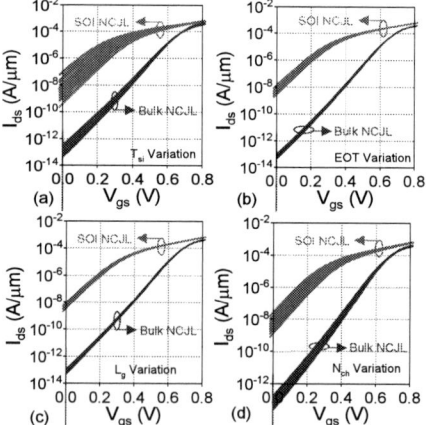

Fig. 3. Comparison of I_{ds} - V_{gs} characteristics of NCJL transistors designed on the bulk and SOI substrate. Results are obtained considering normally distributed variation in (a) T_{si}, (b) EOT, (c) L_g, and (d) N_{ch}.

The process-induced V_{th} variation was analyzed by varying each device parameter individually, including T_{si}, EOT, L_g, N_{ch}, and T_{fe} [1]. Each parameter follows a normal distribution (Fig. 1*c*) with a 3σ range equal to ±10 % of the mean (μ) value and extreme cases (i.e., μ+3σ and μ-3σ), as shown in Table 1. Statistical analysis was performed on 750 samples to evaluate σV_{th} for each parameter using the formula: $\sigma(V_{th}) = [((\partial V_{th}/\partial X) \times X)^2]^{0.5}$, where X represents any normally distributed parameter [1]. The total standard deviation of V_{th}, i.e., $(\sigma V_{th})_{Total}$, was calculated as $[\Sigma(\sigma V_{th}@X)^2]^{0.5}$ [8].

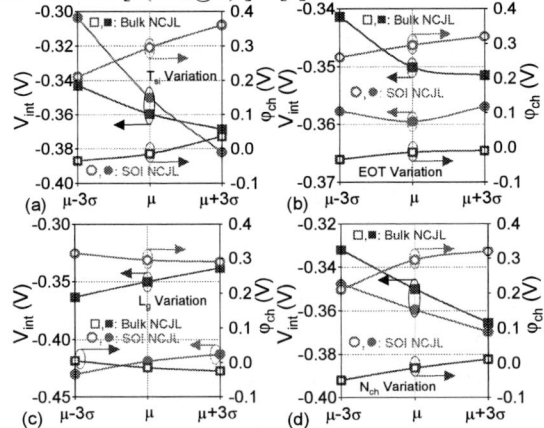

Fig. 4. The internal gate voltage (V_{int}) and channel potential (φ_{ch}) across the channel as a function of (a) T_{si}, (b) EOT, (c) L_g, and (d) N_{ch}, obtained at reference (μ) and extreme cases (i.e., $\mu+3\sigma$ and $\mu-3\sigma$) for each device parameter variations.

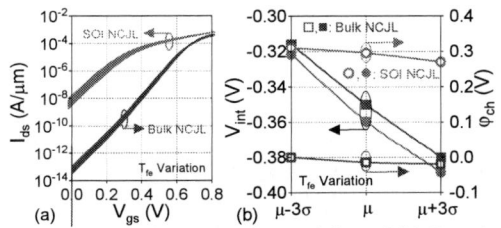

Fig. 5. Comparison of (a) I_{ds} - V_{gs} characteristics and (b) V_{int} and φ_{ch} in SOI and bulk NCJL devices considering the variation in T_{fe}.

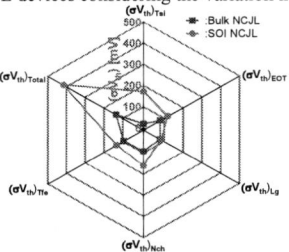

Fig. 6. Comparison of σV_{th} in bulk and SOI NCJL transistors.

Results and Discussion

A. Variability Study of Bulk and SOI NCJL Transistor

Fig. 2a compares the I_{ds}-V_{gs} characteristics of bulk and SOI NCJL devices with reference parameters. Both the transistors were realized with near mid-gap gate work function [9]. Compared to the SOI NCJL transistor, the result highlights bulk NCJL device achieves sufficiently lower off-current (I_{off}) compared to the SOI NCJL transistor along with improved subthreshold swing. To understand the difference in functionality, Fig. 2b shows the internal gate voltage (V_{int}) as a function of V_{gs}. While both the bulk and SOI NCJL transistors achieve the same negative V_{int} (Fig. 2b) at $V_{gs} = 0$ V, the sufficiently lower I_{off} is owing to the excess depletion of the electrons due to the formation of the pn junction at the back surface in bulk NCJL device. This additional depletion at the back surface improves the gate capacitance (C_{gg}) as the

conduction channel shifts towards the front surface. Fig. 2c compares the channel potential, extracted at $x = L_g/2$ along the y-axis at different V_{gs}, in SOI and bulk NCJL devices. While the higher potential is observed at the back surface in the SOI NCJL device, the higher potential (or conduction channel location) was observed towards the front surface. This shift of the conduction channel towards the front surface improves the C_{gg} in non-NC bulk JL devices, thereby enhancing the NC effect due to better C_{fe} and C_{gg} matching in bulk NCJL transistors. Fig. 2b compares the C_{gg} in bulk and SOI NCJL devices to confirm this stronger NC effect. The result highlights the enhanced C_{gg} due to better C_{fe} and C_{gg} matching in bulk NCJL devices than their conventional counterparts. Thus, this enhancement of the NC effect due to the shift of the conduction channel towards the front surface and the additional depleting field at the back surface of the bulk NCJL device is useful for lowering the process-induced geometrical device parameter variation in V_{th}. Our previous work has reported a detailed explanation of the functionality of bulk NCJL devices [5].

Fig. 3 compares the I_{ds}-V_{gs} characteristics of 750 bulk and SOI NCJL devices considering the process-induced device parameter variation in T_{si} (Fig. 3a), EOT (Fig. 3b), L_g (Fig. 3c), N_{ch} (Fig. 3d). It is evident that bulk NCJL transistor has ~88 % (T_{si}), ~91 % (EOT), ~59 % (L_g), and ~80 % (N_{ch}) reduction in ΔV_{th} (= $(V_{th})_{\mu+3\sigma} - (V_{th})_{\mu-3\sigma}$) than in SOI NCJL transistor. To understand the reason for the improved V_{th} variability in bulk NCJL transistors, Fig. 4 shows the internal gate voltage (V_{int}) and channel potential (φ_{ch}) as a function of the mean (μ) and extreme cases (i.e., $\mu+3\sigma$ and $\mu-3\sigma$) of the T_{si} (Fig. 4a), EOT (Fig. 4b), L_g (Fig. 4c), N_{ch} (Fig. 4d) at $V_{gs} = 0$ V. In addition, the improved bulk NCJL performance can also be seen with varying T_{fe} (Fig. 5), with ~ 56 % reduction in $(\Delta V_{th})_{Tfe}$. It is worth mentioning that the negative V_{int} leads to efficient channel depletion in bulk and SOI NCJL devices. However, the enhanced NC effect and additional depletion of the carriers resulting from the pn junction at the back surface lowers the change in channel potential against the fluctuation in the device parameters, which leads to the reduction of V_{th} variability in bulk NCJL devices. Results (Fig. 6) showcase that a reduction of ~89 %, ~91 %, ~60 %, ~80 %, and ~56 % in σV_{th} is observed due to the statistical variation in the individual device parameter along with ~82 % lowering in $(\sigma V_{th})_{Total}$ in bulk NCJL transistor as compare to SOI NCJL transistor.

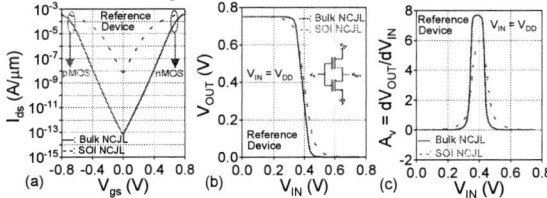

Fig. 7. (a) Comparison of I_{ds} - V_{gs} characteristics of p-type and n-type NCJL transistors designed on bulk and SOI substrates. Comparison of (b) transfer characteristics and (c) gain (A_v) as a function of input voltage (V_{IN}) for bulk and SOI NCJL inverters. The results are obtained for the inverter circuit realized with mean device parameters at $V_{DD} = 0.75$ V.

B. Variability Study of Bulk and SOI NCJL Inverter

To analyze the inverter variability, the gate work function was adjusted to achieve similar V_{th} values for n-

type and p-type NCJL transistors on bulk and SOI substrates (Fig. 7a). Fig. 7b compares the transfer characteristics (V_{OUT}-V_{IN}) of inverters designed with bulk and SOI NCJL devices using mean device parameters. While both bulk and SOI-based inverters exhibit the transition between logic '1' and logic '0' ~ at $V_{DD}/2$, the switching is steeper in the inverter designed with bulk NCJL transistor owing to NC effect-induced enhanced gate electrostatics. Fig. 7c shows the gain ($A_v = dV_{OUT}/dV_{IN}$) as a function of V_{IN} in both the inverter circuits. The higher peak magnitude of the gain further confirms the enhanced performance of an inverter designed with n- and p-type bulk NCJL devices.

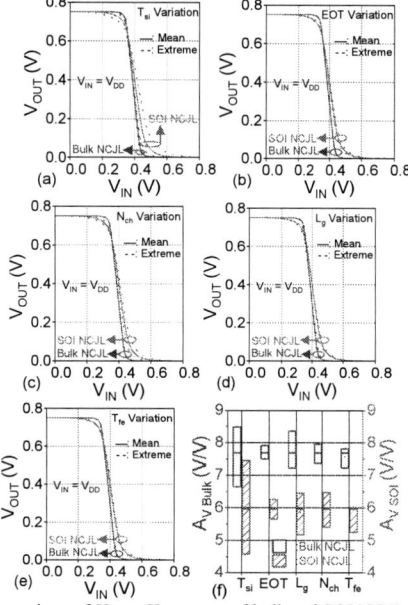

Fig. 8. Comparison of V_{OUT} - V_{IN} curves of bulk and SOI NCJL transistor inverter circuit considering the fluctuation in (a) T_{si}, (b) EOT, (c) L_g, (d) N_{ch}, and (e) T_{fe}. (f) Variation of gain (A_v) as a function of T_{si}, N_{ch}, EOT, L_g, and T_{fe} in bulk (on the primary y-axis) and SOI (on the secondary y-axis) NCJL inverters. The results were extracted at reference (μ) and extreme cases (i.e., $\mu+3\sigma$ and $\mu-3\sigma$) for each device parameter variation.

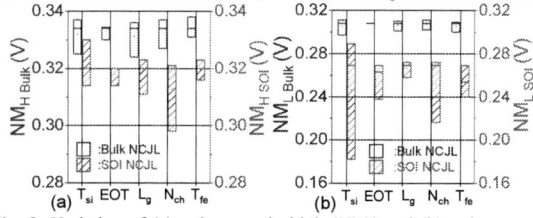

Fig. 9. Variation of (a) noise margin high (NM_H) and (b) noise margin low (NM_L) as a function of T_{si}, N_{ch}, EOT, L_g, and T_{fe} in bulk (on primary y-axis) and SOI (on secondary y-axis) NCJL inverters.

For fair assessment, the impact of process-induced device parameter variations on inverter performance, using bulk and SOI NCJL transistors, was analyzed by varying one device parameter at a time for both p-type and n-type transistors. Figs. 8(a-e) show the transfer characteristics of the mean and extreme cases considering the variation in T_{si} (Fig. 8a), EOT (Fig. 8b), L_g (Fig. 8c), N_{ch} (Fig. 8d), and T_{fe} (Fig. 8e) in both n- and p-type devices. The enhanced NC effect in the bulk NCJL device, due to the conduction channel shift towards the front surface, leads to lower variation than the SOI NCJL

transistor-based inverter. Fig. 8f compares A_v as a function of T_{si}, EOT, L_g, N_{ch}, and T_{fe} in bulk and SOI NCJL inverter circuits extracted for the mean and extreme cases. Results showcase that inverter circuits designed with bulk NCJL devices have higher gain along with lower gain variability against parameter variation.

To further evaluate the competitiveness of NCJL inverter circuits, Fig. 9 compares the noise margin (NM) of bulk and SOI NCJL inverter circuits, considering the variation in T_{si}, EOT, L_g, N_{ch}, and T_{fe}. The results are extracted for the mean and extreme cases. Figs. 7a and 7b correspond to the noise margin high ($NM_H = V_{OH} - V_{IH}$) and noise margin low ($NM_L = V_{IL} - V_{OL}$) of bulk and SOI NCJL inverter as a function of varying device parameters, respectively. The results show that bulk NCJL inverter circuits showcase a higher NM, which determines the maximum amount of noise voltage that can be added to a digital circuit, and has significantly lower variation in NM than SOI NCJL inverter circuits.

Conclusion

This study demonstrates the potential of bulk NCJL transistors in reducing process-induced σV_{th}. Enhanced channel depletion and improved NC effect in bulk NCJL devices lead to reduced channel potential variation and superior $\sigma(V_{th})$ performance compared to SOI NCJL devices. The comparison of inverter realized with bulk and SOI NCJL devices showcases the benefits of bulk NCJL transistors in achieving higher gain and NM along with lower process-induced variability. The present work underscores the suitability of bulk NCJL devices for achieving lower process-induced σV_{th} at shorter L_g.

Acknowledgment

This work was supported by Birla Institute of Technology and Science-Pilani, K K Birla Goa Campus, Goa, India through the Research Initiation Grant (RIG)- BPGC/RIG/2021-22/03-2022/02 and Additional Competitive Grant (ACG)-GOA/ACG/2022-2023/Oct/08. Authors would also like to thank ChipIN centre for providing TCAD tool under Chips to Stratup (C2S) programme by Ministry of Electronics and Information Technology, India.

References

[1] R. Trevisoli, M. A. Pavanello, R. T. Doria, C. E. Capovilla, S. Barraud, and M. de Souza, "Variability modeling in triple-gate junctionless nanowire transistors," IEEE Transactions on Electron Devices, vol. 69, pp. 4730-4736, 2022.

[2] N. D. Akhavan, G. A. Umana-Membreno, R. Gu, J. Antoszewski and L. Faraone, "Random dopant fluctuations and statistical variability in n-channel junctionless FETs," Nanotechnology, vol. 29, pp. 025203, 2017.

[3] G. Leung and C. O. Chui, "Variability of Inversion-Mode and Junctionless FinFETs due to Line Edge Roughness," IEEE Electron Device Letters, vol. 32, pp. 1489-1491, 2011.

[4] M. S. Bae and I. Yun, "Impact of process variability in junctionless FinFETs due to random dopant fluctuation, gate work function variation, and oxide thickness variation," Semiconductor Science and Technology, vol. 35, pp. 035015, 2020.

[5] S. R. Ruma and M. Gupta, "Sub-60 mV/Decade Dynamic Subthreshold Swing in Bulk Negative Capacitance Junctionless MOSFET," IEEE Transactions on Electron Devices, vol. 71, no. 11, pp. 7156-7161, Nov. 2024.

[6] Sentaurus Device User Guide, Version N-2017.09, Synopsys, Mountain View, CA, USA, Sep. 2017.

[7] [Online]. Available: https://irds.ieee.org/

[8] J. A. Croon, M. Rosmeulen, S. Decoutere, W. Sansen, and H.E. Maes, "An easy-to-use mismatch model for the MOS transistor," IEEE Journal of Solid-State Circuits, vol. 37, no. 8, pp. 1056-1064, Aug. 2002.

[9] M Gupta, and V. P. H. Hu, "Negative capacitance junctionless device with mid-gap work function for low power applications," IEEE Electron Device Letters, vol. 41, no. 3, pp. 473-476, Jan. 2020.

Optimal transfer learning strategies for property predictions in materials science

Reshma Devi[1], Keith T. Butler[2], Sai Gautam Gopalakrishnan[1]

[1]Indian Institute of Science, India, [2]University College London, United Kingdom

Abstract

Materials science is a domain characterised by 'small' datasets (i.e., < 10,000 datapoints) of critical properties that govern performance of various applications and devices. For instance, there are no large, reliable datasets available for several key 'performance determining' metrics in energy applications, such as diffusivities in battery electrodes, carrier recombination rates in photovoltaics, and molecular adsorption energies for catalysis. On the other hand, there are reasonably 'large' datasets (> 100,000 datapoints) available on some properties, such as, bulk formation enthalpies, computed band structures, and crystal structures across wide chemical spaces. Thus, if key chemical, compositional, and structural trends can be captured in available large datasets and subsequently transferred (or re-learnt), it will enable the use of deep learning and graph based neural network models in smaller datasets as well. Hence, my talk will explore the utility of current transfer learning (TL) approaches that are available for computational materials science and identify optimal ways to employ TL-based strategies. Specifically, TL involves training a neural network model on a larger dataset and subsequently retraining a fraction of the model on a smaller dataset. I will quantify the accuracy, transferability, and efficiency of TL models compared to models that have been trained from scratch. Finally, I will focus on TL models that can generalise over multi-properties during pre-training and can efficiently be re-trained on small datasets, which pave the way towards creating more general, foundational models, in the near future.

Keywords: Transfer learning, Materials Properties, Multi-property training

Description of work

In the realm of materials design and optimization, machine learning (ML) architectures have played a pivotal role given their ability to predict material properties at low computational costs. However, there are several materials properties that critically govern the efficiencies of devices which are notoriously hard to predict (and train) ML models. For example, ionic conductivity in electrolytes employed in batteries, defect formation energies of semiconductors, and surface properties of key materials used in catalytic applications. The primary limitation in training of reliable ML models on such critical material properties is the lack of availability of 'large' datasets (i.e., > 100,000 datapoints),

with currently available experimental and/or computational datasets often being 'small' (< 10,000 datapoints). Given that training of reliable deep learning (DL) models require large datasets, an alternative strategy is required to predict such material properties quickly.

Importantly, transfer learning (TL) is a promising strategy to train DL models with sparse datasets, which has been demonstrated in computer vision and in biological applications. Briefly, TL involves a DL model that is pre-trained (PT) on a larger, easily-available dataset (e.g., formation energies of materials) and subsequently fine-tuned (FT) on the target datasets that are often small (e.g., piezoelectric modulus). Thus, we identify the optimal strategies [1] to perform TL among various materials datasets using the atomistic line graph neural network (ALIGNN) architecture. Specifically, we optimize the model architecture, tune the hyperparameters, identify optimal ways to sample sparse datasets, and observe the amount of retraining required during FT for good performance. Additionally, we demonstrate the utility of the TL approach in swiftly predicting materials properties, especially against models trained from scratch. Also, we show that TL models generally learn properties (upon FT) much faster (i.e., at fewer datapoints) compared to scratch models.

In addition to identifying optimal TL strategies, we also demonstrate a pathway to create models that are generalizable over a wide range of materials properties. Specifically, we train ALIGNN models simultaneously over multiple material property datasets, by modifying the prediction head, resulting in models that are PT on multiple properties (or MPT models). Subsequently, we demonstrate that the MPT models perform better than both scratch and PT-FT models on 4/6 occasions. Moreover, we show that the MPT models generalize significantly better on an out-of-domain dataset consisting of properties of two dimensional materials, which is fully different from the three dimensional bulk properties used during PT. Thus, we demonstrate an architecture that can be used to generalize across multiple properties efficiently and can be systematically made better by including more properties during PT in the near future.

We hope that our work enables the creation of reliable TL and generalizable models, further accelerating property predictions among materials, resulting in materials discovery for novel applications.

Acknowledgments

The authors gratefully acknowledge the financial support from the Royal Society, the Engineering and Physical Sciences Research Council of the United Kingdom Research and Innovation and the Science and Research Engineering Board of the Government of India.

References

[1] https://arxiv.org/abs/2406.13142

Invertible Prediction Model for Si₃N₄ Wet Etching Using DHF

Koki Shibata[1], Takashi Ota[2], Koji Ando[2], Naoko Misawa[1], Chihiro Matsui[1], and Ken Takeuchi[1]

[1]Dept. of Electrical Engineering and Information Systems, The University of Tokyo, Japan,

[2]Application Software Engineering Operations, SCREEN Semiconductor Solutions Co., Ltd., Japan

Abstract

Optimizing semiconductor fabrication is challenging due to numerous recipe parameters. This study explores various nozzle operations and wafer rotation speeds to enable selective Si₃N₄ etching using dilute hydrogen fluoride (DHF). By using measured etching rate (ER), the invertible model dependent on the nozzle scanning period is constructed, enabling determination of etching recipes for the desired ER results. The model is sufficiently accurate for complex scanning patterns, providing a practical tool for process optimization.

Keywords: Semiconductor manufacturing, Wet etching, Process optimization

Introduction

As semiconductor devices shrink, maintaining strict uniformity on wafer in wet etching processes becomes crucial for high yields. Achieving this uniformity requires selective etching. For instance, if an over-deposition occurs in the central area of a wafer due to variations in previous processes, recipes that selectively etch more in the center are required to compensate (Fig. 1). However, the multitude of etching recipe parameters—chemical concentrations, flow rates, temperature, wafer rotation speed, and nozzle scanning patterns—makes it extremely challenging for process engineers to devise recipes that yield the desired etching results.

In this study, single-wafer wet etching experiments are conducted on Si₃N₄ using DHF (Fig. 1). The etching rate (ER), defined as the depth etched per unit time, is measured at different radial positions across the wafer surface. To collect these datasets, wafer rotation speeds and nozzle scanning patterns are systematically varied (Fig. 2). The aim of this research is to develop a model capable of accurately predicting the ER distribution across the wafer, thereby aiding in the optimization of the etching process.

Due to the characteristics of the ER dependence in nitride films—where the influence of wafer rotation speed is partial, and the scan pattern's influence is dominant—it is found that the developed model depends on the square root of the scan pattern's period (Fig. 2). Leveraging this dependency makes the proposed model attractive, as it not only enables the prediction of etching results but also allows for the direct solution of the inverse problem. In other words, when a specific ER distribution is given, the model can determine how the nozzle should be moved to achieve it, providing practical guidance for recipe formulation (Fig. 1 (c)).

Several prior studies are relevant to this research. Habuka et al. [1] uses computational fluid dynamics (CFD) to predict the etching distribution for SiO₂ etched with DHF at different radial positions on the wafer. Singh et al. [2] investigates amorphous silicon etched with SC1 at various fixed nozzle positions and used convolution on these results to predict etching outcomes for moving nozzles. While these studies are limited to estimating the ER, the proposed model not only ensures extrapolative capability in simulations but also provides a recipe for how to move the nozzle within certain constraints when a desired ER distribution is specified. In [3, 4], neural network models not only predict etching results for SiO₂ but also predict process recipes. However, these are proposed as separate models, and the introduction of neural networks has resulted in the process being treated as a black box.

The proposed approach is based on an inductive data-

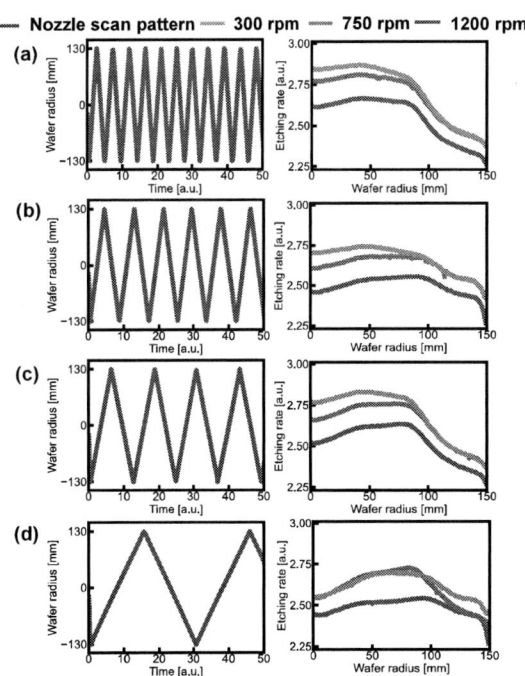

Fig. 2 Triangle wave scan patterns as training dataset and corresponding measured ER data in ascending order by scan pattern's period. (a) Scan-A. (b) Scan-B. (c) Scan-C. (d) Scan-D.

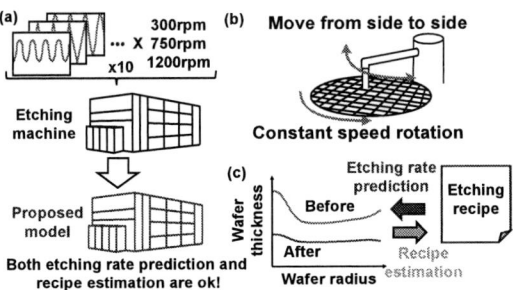

Fig. 1 (a) Model building through comprehensive data collection across multiple scan patterns and wafer rotation speeds. (b) Nozzle arm scanning and wafer rotation during single wafer wet etching process. (c) Selective etching recipe is practically useful to compensate for stable yield. While simple simulation model only achieves etching rate prediction, the proposed invertible model simultaneously enables estimation of recipe parameters. Therefore, the model has high practical value.

driven model rather than relying on top-down CFD simulations. Moreover, the model itself not only predicts the distribution of ER but also provides practical methods to achieve the desired etching results (Fig. 1).

Experimental setup

Etching is performed using DHF at wafer rotation speeds of 300 rpm, 750 rpm, and 1200 rpm (Fig. 2). Nozzle scanning patterns include fixed-point dispensing at radial positions—wafer center (0 mm), 10 mm, 20 mm, and 30 mm—with 80 mm and 130 mm as extremely large patterns. Additionally, triangular wave patterns with a fixed amplitude of 130 mm (Scan-A, Scan-B, Scan-C, Scan-D, increasing in period) are utilized, along with more complex patterns for validation data (Scan-Y and Scan-Z) (Fig. 2). Combining 10 scan patterns with 3 rotation speeds yields 30 combinations. For validation, the following three are used: 750rpm_scan-Y, 750rpm_scan-Z, and 1000rpm_scan-Z. Experiments are conducted under conditions using the same etching chamber. ER data are measured every 1 mm across the wafer radius (0 to 150 mm) using an ellipsometer.

Data analysis

Three distinct inflection points are observed at radii of approximately 40 mm, 90 mm, and 140 mm for all samples (Fig. 2). Additionally, two different etching behavior patterns ("mode 1", "mode 2") are identified, independent of both wafer rotation speed and nozzle scanning pattern (Fig. 3). Wafers in the "mode 1" group exhibit an abrupt change in etching depth near 100 mm, while those in the "mode 2" group show a more uniform etching profile. These differences are found to correlate with wafer No., suggesting that variations in film preparation preprocesses or Front Opening Unified Pod (FOUP) conditions may be responsible for the observed discrepancies. Identifying the

factors influencing such external variations is expected to be an important direction for future research.

It is observed that short-period scans, such as Scan-A, enhance the ER in the central region compared to long-period scans like Scan-D (Fig. 2, right column). This phenomenon can be attributed to the continuous supply of fresh HF in short-period patterns. In contrast, long-period patterns have extended intervals between droplet depositions when passing through the center, allowing centrifugal forces to scatter the chemicals towards the periphery, thus reducing the availability of chemicals in the central area.

At the 80 mm and 130 mm nozzle positions, etching does not occur in the central region (Fig. 4). This is due to the "inflow" effect described in [2], where the chemical solution spreads inward from the dispensing position, causing etching further inside than the drop point. Increasing the wafer rotation speed decreases the "inflow" effect (Fig. 4). While this phenomenon is proposed in [2], discrepancies in the inflow formulation suggest differences due to varying physical targets. Despite 750 rpm being the average of 300 rpm and 1200 rpm, the "inflow" effect at 750 rpm is not intermediate between those at 300 rpm and 1200 rpm, indicating that inflow is not a linear function of rotation speed.

Model framework

The model $f(r)$ is developed to predict the ER of nitride films at each 1 mm interval from 0 mm to 150 mm (Fig. 5). Since etching efficiency is most likely to progress at the fixed drop position, almost independent of the wafer rotation speed, the model $f(r)$ is designed to predict the difference in etching results when the nozzle operation is altered, compared to the ER of the central fixed drop pattern $f_{\text{fixed drop}}(r)$. Specifically, the following equation is proposed based on the observation that a constant decrease occurs below r_1, while a decrease dependent on the triangular wave period is observed below r_2. Here, T

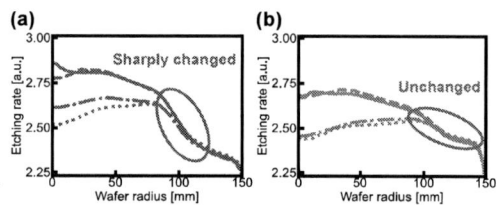

Fig. 3 Comparison of measured ER data between two modes. (a) All samples in "mode 1" display a pattern with a sharp change in ER at a radial distance of 100 mm. (b) All samples in "mode 2" show a pattern with relatively uniform ER beyond 100 mm.

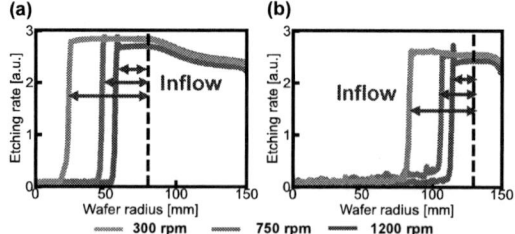

Fig. 4 Comparison of measured ER data at different wafer rotation speeds. Dotted line represents fixed nozzle position respectively. (a) 80 mm fixed. (b) 130 mm fixed.

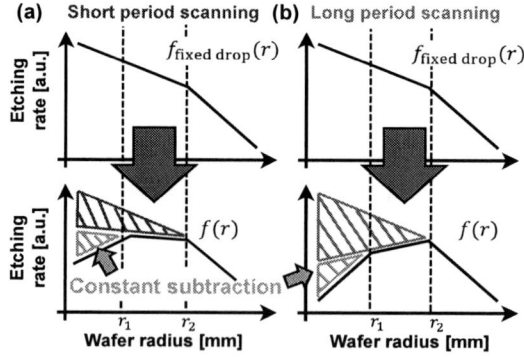

Fig. 5 The model for calculating ER by subtracting specific components from central fixed ER data $f_{\text{fixed drop}}$. The subtraction comprises two parts: one below r_2, which depends on the wave's period (blue or red), and another below r_1, which is independent of the waveform (orange). (a) A short period leads to a small subtraction. (b) A long period results in a large subtraction.

979-8-3315-0417-5/25 $31.00 © 2025 IEEE

represents the period of the triangular wave.

$$f(r) = \begin{cases} f_{\text{fixed drop}}(r), & r_2 < r \\ f_{\text{fixed drop}}(r) - b\sqrt{T}|r - r_2|, & r_1 < r \le r_2 \\ f_{\text{fixed drop}}(r) - b\sqrt{T}|r - r_2| - a|r - r_1|, & r \le r_1 \end{cases}$$

Parameters a and b are obtained using the nonlinear least-squares method. The value of r_1 is tuned in 1 mm increments between 35 mm and 45 mm, and r_2 is adjusted similarly between 90 mm and 110 mm using grid search. Using these parameters, the developed models are applied to the validation dataset (scan-Y and scan-Z). To mitigate the impact of complex waveforms differing from triangular waves, linear regression is performed on each half-cycle to approximate the wave (Fig. 6). By using the effective amplitude of the approximated wave, the difference between this effective amplitude and the training data amplitude of 130 mm is added to r_2 to dynamically adjust the model.

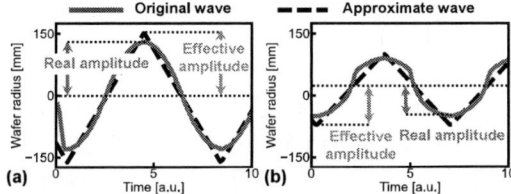

Fig.6 Nozzle scan patterns used as validation dataset and their approximations. (a) Scan-Y. (b) Scan-Z.

Results and discussion

The results of the grid search reveal that for "mode 1", the optimal parameters are $r_1 = 39$ mm and $r_2 = 103$ mm yielding $a = 2.1 \times 10^{-3}$ and $b = 5.5 \times 10^{-3}$, and a minimum mean squared error (MSE) of 9.7×10^{-3} and mean absolute error (MAE) of 6.6×10^{-2}. In contrast, for "mode 2", the optimal parameters are $r_1 = 44$ mm and $r_2 = 119$ mm, resulting in $a = 2.9 \times 10^{-3}$ and $b = 2.0 \times 10^{-3}$, with a minimum MSE of 1.0×10^{-1} and MAE of 2.8×10^{-1} (Table 1). This indicates an essential difference between "mode 1" and "mode 2", necessitating distinct model configurations.

In "mode 1", it is found that the results are consistent with the measured data in all areas (Fig. 7 (a)). Moreover, the measured data (750rpm_scan-Z and 1000rpm_scan-Z)

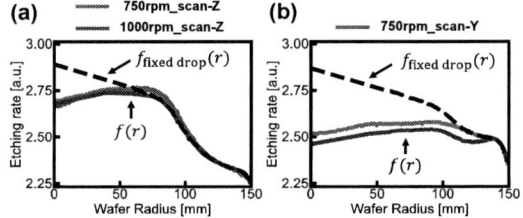

Fig. 7 Comparison between $f(r)$ (blue) and measured data. (a) For mode 1, since the rotation speed is not considered in the proposed model, the predictions $f(r)$ for two samples are consistent. (b) For mode 2, the prediction $f(r)$ is offset from the measured data.

reconfirm that the wafer rotation speed does not significantly influence the etching results (Fig. 7 (a)). In "mode 2", offsets between $f(r)$ and the measured data are observed at all radial positions; however, the predicted trends are accurate (Fig. 7 (b)). These results demonstrate that the model possesses extrapolative capability across different nozzle patterns.

This model also enables estimating process recipes from etching results—that is, it solves the inverse problem. Once the simulation model is constructed for a specific mode, it becomes possible to predict recipes that achieve arbitrary etching outcomes by determining the period and amplitude based on $f(r)$, even though the ER at the wafer edge (beyond approximately 110 mm) is determined by selecting "mode1" or "mode2". In the left column of Fig. 2, the ER increase due to lower rotation speeds in the region where $r \le r_1$ is evident; specifically, the ER in the central area at 300 rpm is higher than at 750 rpm and 1200 rpm. Therefore, changing the rotation speed can potentially adjust the ER in this region. The ability to formulate process recipes simultaneously with the simulation offers a significant practical advantage.

Table 1: MSE and MAE for each mode

	Mode1		Mode2	
	Training	Validation	Training	Validation
MSE	9.7×10^{-3}	4.1×10^{-4}	1.0×10^{-1}	1.9×10^{-3}
MAE	6.6×10^{-2}	1.7×10^{-2}	2.8×10^{-1}	4.2×10^{-2}

Conclusion

In this study, DHF is applied to Si_3N_4, ER is measured at various radial positions on the wafer, followed by detailed data analysis. The results show that ER is more influenced by nozzle scanning patterns than wafer rotation speed. Using the measured data, a model is constructed to predict deviations from central fixed-drop conditions, demonstrating extrapolative capabilities even for more complex scanning patterns than simple triangles. Unlike computationally intensive physics simulations, the model uses straightforward calculations with minimal computational overhead for Si_3N_4, enabling recipe identification to achieve arbitrary etching results by selecting the proper parameters. This model is expected to improve yield and enhance manufacturing throughput by reducing future labor costs.

Acknowledgments

The authors thank T. Mikawa, H. Takahashi, H. Horiguchi, and Y. Yoshinaga for the fruitful discussion.

References

[1] H. Habuka, S. Ohashi, and T. Kinoshita. "Numerical calculation model of a single wafer wet etcher using a swinging nozzle." Materials science in semiconductor processing 15.5, pp. 543-548 (2012).

[2] M. K. Singh, et al. "An empirical approach to accurate single wafer wet etch simulation." 2015 26th Annual SEMI Advanced Semiconductor Manufacturing Conference (ASMC) (2015).

[3] S. Yoshikiyo, et al. "Machine Learning-based Accurate Single Wafer Wet Etching Amount Prediction." 2023 Silicon Nanoelectronics Workshop (SNW) (2023).

[4] K. Shibata, et al. "Automated Recipe Creation & Verification of Single Wafer Wet Etching: Ensemble Learning with Backcasting & Forecasting AIs using Scarce Data." 2024 IEEE Silicon Nanoelectronics Workshop (SNW) (2024).

The Demonstration of Scalable-HZO/ZrO₂ FeFET with Large Memory Window of 2.3V for 3bit-per-cell, Immediate Read after Write, High Endurance of 10^9 cycles, and The High Accuracy of 92% for Machine Learning.

S.-T. Huang[+], H.-M. Chen[+], Y.-T. Tsai, C.-S. Pai1, and Y.-T. Tang*

Dept of Electrical Engineering, National Central University No. 300, Zhongda Rd., Zhongli District, Taoyuan City 320317, Taiwan,

Phone: +886-972-815-521, Email: yttang@ee.ncu.edu.tw

Abstract

In this study, we show that using HZO/ZrO₂ as a ferroelectric layer in FeFETs outperforms HZO alone. The thick HZO layer forms a low-defect-density ZrO₂ layer, enhancing the ON/OFF ratio, memory window, and reducing leakage. Interface defects for the 5/3(5nm/3nm), 7/3(7nm/3nm), and 9/3(9nm/3nm) structures were analyzed, revealing optimal performance at 9/3 due to uniform domain growth and low defect density. The 9/3 structure achieved faster polarization switching and high write speed. Despite a smaller ON/OFF ratio, it exhibited excellent Gmax/Gmin linearity, crucial for neural networks. Machine learning tests demonstrated 92% accuracy in 20 epochs, highlighting its AI application potential.

Introduction

With the rapid advancement of artificial intelligence (AI), the need for high-performance memory technologies has become critical to support extensive data processing and computation. Ferroelectric field-effect transistors (FeFETs) are a key focus for in-memory computing due to their high speed, low power consumption, and non-volatility. HZO/ZrO₂ heterostructures, known for their excellent polarization switching speed and stability, have significantly enhanced FeFET performance to meet AI demands. Over the past five years, these structures have been valued for providing superior memory windows at reduced dimensions. MPB technology [4], which involves stacking a ZrO₂ layer on HZO, has improved dielectric constants and reduced leakage, enhancing stability and efficiency. However, challenges remain in controlling phase transitions during film deposition and balancing high polarization switching speed with low power consumption while maintaining linearity, limiting large-scale commercialization.

Device fabrication

Fig. 1 illustrates the gate oxide stack structure and fabrication workflow of the Metal-Antiferro-Ferro-Insulator Silicon (MAFIS)-FeFET used in this experiment. We first deposited HZO with thicknesses of 5 nm, 7 nm, and 9 nm using ALD, followed by a 3 nm ZrO₂ layer to ensure crystallization (denoted as HZO/ZrO₂ = 5/3, 7/3, 9/3). Next, a 50 nm TiN top electrode was deposited, and patterned.

Electrical measurements were conducted using a Keysight 1500B system with a 1525 module to apply pulse voltages with varying lengths, numbers, and amplitudes. Optical XPS analysis was performed using the (X-ray photoelectron spectroscopy) system.

Result and discussion

Fig. 2 shows the Id-Vg characteristics of the MAFIS-FeFET, with the memory window values read at 10^{-6} A. The memory windows for 5/3, 7/3, and 9/3 are 1.24 V, 1.3 V, and 2.4 V, respectively. The memory windows of 5/3 and 7/3 are similar, while 9/3 is almost double compared to the former, which may be ascribed to phase transitions influenced by the bottom HZO. According to first-principles calculations of surface energy, HZO primarily exhibits the tetragonal (t-) phase when its thickness is below 7 nm. The ideal dielectric constant of κ values for t-ZrO₂ and t-HZO are 45 and 35, respectively. However, when the HZO thickness increases to 9 nm, the higher surface energy of the orthorhombic (O)-phase dominates crystalline phase [8]. Since κ_{ortho} (25) is lower than κ_{terta} (35), more voltage is applied to the HZO region, yielding the largest polarization and a rightward shift in V_{TH}. A higher voltage across the FE-HZO implies that less voltage within the t-ZrO₂ layer. Reducing the probability of weak Zr-O bonds consequently decreases leakage.

In Fig. 3, we tested the speeds of ER and PG by applying different pulse widths and amplitudes, and summarized the variations in V_{TH}. Next, we convert this into the relationship between switch time (ts) and Vg, as shown in Eq. 1. The results show that V_{TH} switching speed increases with increasing HZO thickness, suggesting a reduction in the depolarization field within the HZO.

$$\log(ts) = \log(t0) + A * \frac{1}{(Vg - V0)^2} \quad (1)$$

To identify these defects, we analyzed Dit using the conductance method [5]. Fig. 4 presents Dit for the three structures, clearly showing that the 5/3 has a higher defect density than 7/3 and 9/3. Defects can also be verified in Fig. 5. Fig. 5 presents the Hf 4f and Zr 3d spectra at the HZO/ZrO₂ interface. The defect proportion of Zr 3d for 5/3 is 42.7%, significantly higher than 27.3% for 7/3 and 20.3% for 9/3. A similar trend can be seen in the Hf 4f spectra. In summary,

979-8-3315-0417-5/25 $31.00 © 2025 IEEE

the FE-HZO/ZrO₂ heterostructure shows high write speed and a wide memory window (>2.3 V) when the FE-HZO thickness is at least 9 nm.

Next, we conducted a read-after-write test. Fig. 6 shows the variation of V_{TH} with retention under ER and PG conditions. At a low voltage of 3.5 V and a pulse width of 10 μs, memory windows are closed due to the high content of defects, which shield the polarization. Nonetheless, under an operating voltage of 5 V(FLASH), all three structures are immediate in read after write. This demonstrates that HZO/ZrO₂ still hold potential in replacing Flash. In Fig 7, it is observed that the breakdown voltage (BV) of the 9/3 reaches 6V, and exhibits the lowest leakage current, showing that the 9/3 has better performance. Fig. 8 depicts the V_{TH} distribution for 20 devices measured at different voltages. These eight Vg values correspond to 3-bit levels. Due to the reduced memory window, the 5/3 had the highest error rate (3.28×10^2). Conversely, the 9/3 with a larger memory window. Significantly reduced the read error rate (3.76×10^3).

Fig. 9 illustrates the training schematic of the neural network, divided into three layers: Input layer, Hidden layer, and Output layer. The circuit array is shown below. In this training, the weight coefficients (A, B) were obtained from the conductance of FeFET through its linearity. The conductance G(P) is presented in Eq. (2) and (3). Where P represents the pulse number and G represents the conductance. We extracted the weight coefficients LTP (PG) and LTD (ER) by using Eq. (4) [1]. The cartoon in Fig. 9 shows that out of 10 images, 9 were correctly matched, indicating an accuracy of 90%. Comparing conductance vs. pulse number for the three devices, 9/3 presents the most linear behavior, which enhances model predictability and speeds up computation. Note that the conductance extraction point is fixed at Vgs = 0.2 V, and the operating voltages for PG and ER were chosen from 2 V to 4 V and -2 V to -4 V, respectively, with a step of 0.05 V. Fig. 10 compares the conductance of 5/3 and 9/3 at different pulse numbers. It shows that the training of 9/3 is more efficient than 5/3.

$$G_{LTP} = B*(1-e^{(-\frac{P}{A})}) + G_{min} \tag{2}$$

$$G_{LTD} = -B*(1-e^{(\frac{P-Pmax}{A})}) + G_{max} \tag{3}$$

$$B = (G_{max}-G_{min})/ (1-e^{(-\frac{Pmax}{A})}) \tag{4}$$

Fig. 11 depicts the phase field model of HZO [3], where the domain grows with an increasing pulse number, and the increase in polarization (-Pr) causes the channel V_{TH} to shift rightward. The results indicate that the domain growth in the 9/3 is more uniform compared to 5/3. Fig. 12 shows the results from running machine learning using neurosim, where the 9/3 achieved the highest accuracy, reaching 92% in just 20 epochs. This performance surpasses that reported in other

literature (see Table 1), demonstrating its potential for artificial intelligence applications.

Conclusion

In this work, we demonstrate that using HZO/ZrO₂ as a ferroelectric layer significantly enhances FeFET performance compared to using HZO alone. The 9/3 with its optimal combination of low-defect density and high linearity, provides a large memory window and efficient polarization switching, making it highly suitable for neural network applications. Interface defect analysis and phase field modeling confirm that a thicker HZO layer leads to more uniform domain growth and reduces leakage. The 9/3 also achieves a high accuracy of 92% in neural network training, indicating its potential for advanced AI applications.

Acknowledgments

The authors gratefully acknowledge the supports from NSTC113-2218-E-A49-022, 113-2119-M-A49-007, 112-2622-8-A49-013-SB and TSRI JDP113-Y1-035.

References

[1] Min-Kyu Kim and Jang-Sik Lee, "Ferroelectric Analog Synaptic Transistors", (ACS) (2019).

[2] Chen Sun, Kaizhen Han, Subhranu Samanta, Qiwen Kong, Jishen Zhang, Haiwen Xu, Xinke Wang, Annie Kumar, Chengkuan Wang, Zijie Zheng, Xunzhao Yin, Kai Ni, and Xiao Gong, "First Demonstration of BEOL-Compatible Ferroelectric TCAM Featuring a-IGZO Fe-TFTs with Large Memory Window of 2.9 V, Scaled Channel Length of 40 nm, and High Endurance of 10^8 Cycles", (VLSI) (2021).

[3] K. Ni, B. Grisafe, W. Chakraborty, A. K. Saha, S. Dutta, M. Jerry, J. A. Smith, S. Gupta, and S. Datta, "In-Memory Computing Primitive for Sensor Data Fusion in 28 nm HKMG FeFET Technology", (IEDM) (2018).

[4] Venkateswarlu Gaddam, Giuk Kim, Taeho Kim, Minhyun Jung, Chaeheon Kim, and Sanghun Jeon, "Novel Approach to High κ (~59) and Low EOT (~3.8Å) near the Morphotrophic Phase Boundary with AFE/FE(ZrO2/HZO) Bilayer Heterostructures and High-Pressure Annealing", (ACS) (2022).

[5] Yiming Qu, Junkang Li, Mengwei Si, Xiao Lyu, and Peide D. Ye, "Quantitative Characterization of Interface Traps in Ferroelectric/ Dielectric Stack Using Conductance Method", (EDS) (2020).

[6] S. Dutta, H Ye, W. Chakraborty, Y.-C. Luo, M. San Jose, B. Grisafe, "Monolithic 3D Integration of High Endurance Multi-Bit Ferroelectric FET for Accelerating Compute-In-Memory", (IEDM) (2020).

[7] Fei Mo, Yusaku Tagawa, Chengji Jin, MinJu Ahu, Takuya Saraya, Toshiro Hiramoto, "Experimental Demonstration of Ferroelectric HfO2 FET with Ultrathin-body IGZO for High-Density and Low-Power Memory Application", (VLSI) (2019).

[8] R. Materlik, C. Künneth, A. Kersch, "The origin of ferroelectricity in Hf₁₋ₓZrₓO₂: A computational investigation and a surface energy model", (Journal of Applied Physics) (2015).

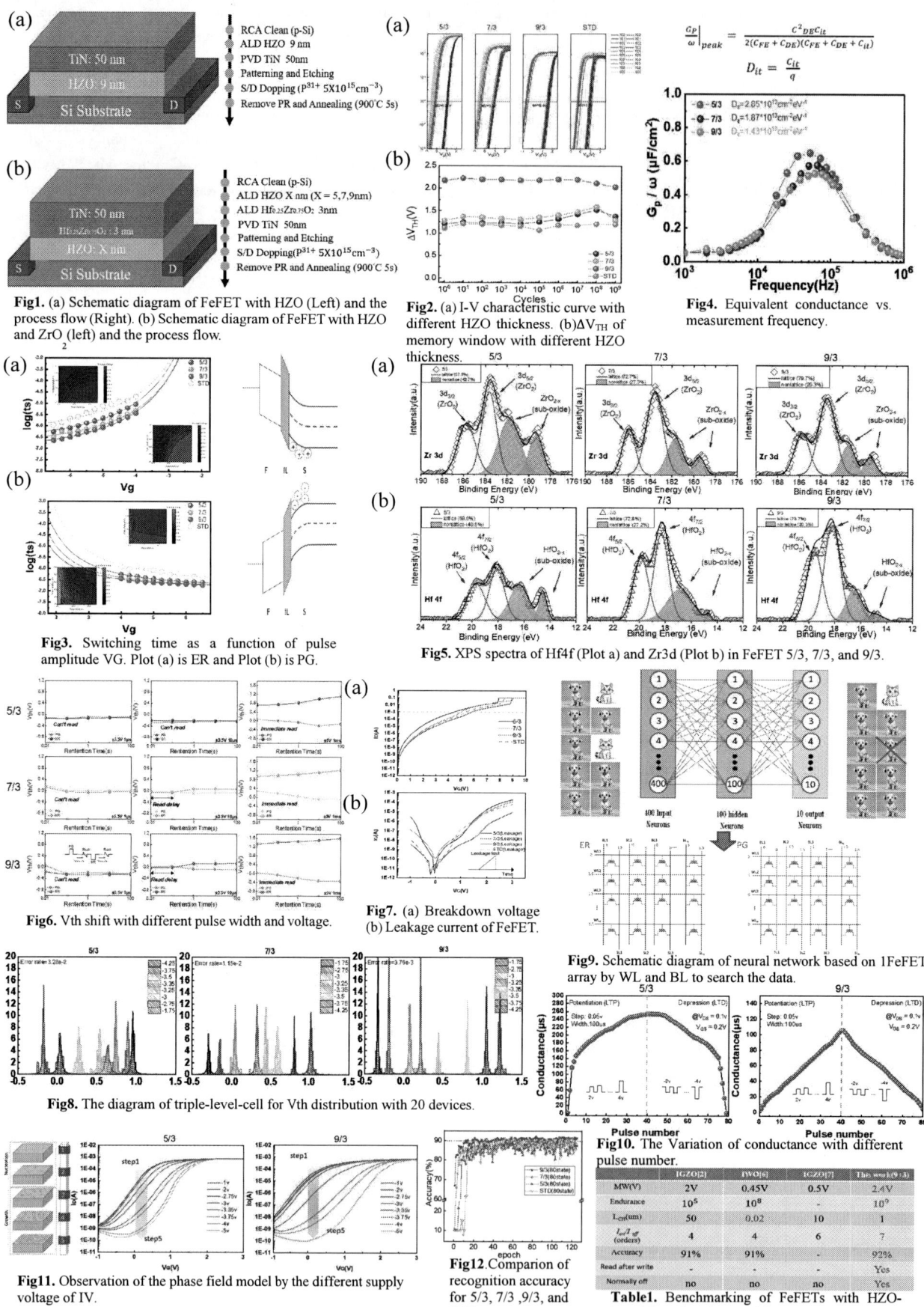

Fig1. (a) Schematic diagram of FeFET with HZO (Left) and the process flow (Right). (b) Schematic diagram of FeFET with HZO and ZrO$_2$ (left) and the process flow.

Fig2. (a) I-V characteristic curve with different HZO thickness. (b) ΔV_{TH} of memory window with different HZO thickness.

$$\left.\frac{G_P}{\omega}\right|_{peak} = \frac{C^2_{DE}C_{it}}{2(C_{FE}+C_{DE})(C_{FE}+C_{DE}+C_{it})}$$

$$D_{it} = \frac{C_{it}}{q}$$

Fig4. Equivalent conductance vs. measurement frequency.

Fig3. Switching time as a function of pulse amplitude VG. Plot (a) is ER and Plot (b) is PG.

Fig5. XPS spectra of Hf4f (Plot a) and Zr3d (Plot b) in FeFET 5/3, 7/3, and 9/3.

Fig6. Vth shift with different pulse width and voltage.

Fig7. (a) Breakdown voltage (b) Leakage current of FeFET.

Fig8. The diagram of triple-level-cell for Vth distribution with 20 devices.

Fig9. Schematic diagram of neural network based on 1FeFET array by WL and BL to search the data.

Fig10. The Variation of conductance with different pulse number.

Fig11. Observation of the phase field model by the different supply voltage of IV.

Fig12. Comparison of recognition accuracy for 5/3, 7/3 ,9/3, and STD.

Table1. Benchmarking of FeFETs with HZO-based ferroelectric material.

	IGZO[2]	IWO[5]	IGZO[7]	This work(9 : 3)
MW(V)	2V	0.45V	0.5V	2.4V
Endurance	10^5	10^8	-	10^9
L_{CH}(um)	50	0.02	10	1
I_{on}/I_{off} (orders)	4	4	6	7
Accuracy	91%	91%	-	92%
Read after write	-	-	-	Yes
Normally off	no	no	no	Yes

A Fan-Out Wafer-Level Packaging-Based THz Communication Transceiver System for High-Speed Chip-to-Chip Wireless Interconnect Applications

Si Rui Liu[1], Ya Fei Wu[1,2*], Zong Rui He[1], Yu Jian Cheng[1], Yang Chai[2]

[1]University of Electronic Science and Technology of China, Chengdu 611731, China

[2] The Hong Kong Polytechnic University, Hong Kong, China

*wuyafei@uestc.edu.cn

Abstract

This paper presents the design of a THz communication transceiver system based on fan-out wafer-level package (FOWLP) technology for high-speed chip-to-chip (C2C) wireless interconnect applications. The design focus on the transceiver (TR), antenna in package (AiP)s, and interconnect transition structures, integrating coaxial-like structures and coplanar waveguide (CPW) transmission line transitions to enable low-loss signal transmission across both vertical and horizontal pathways. Simulation results show that the proposed FOWLP structures achieve an insertion loss of less than 1 dB and a return loss greater than 10 dB within the 25-60 GHz range for the local oscillator (LO) frequency band. The AiP provides an impedance match of -15 dB within the 200-230 GHz range for the radio frequency (RF) band. Link budget calculations further verify that the proposed system enables C2C communication at data rates up to 10 Gbps, highlighting its potential for advanced interconnect applications.

Keywords: Antenna in package (AiP), THz communication, chip-to-chip (C2C) wireless interconnect, fan-out wafer-level package (FOWLP), transceiver.

Introduction

In recent years, the semiconductor industry has been seeking smaller, more segmented, and more modular IC solutions called chiplets [1]. Chiplets offer individual optimization, production with different process nodes, and reduced complexity, enhancing yield and lowering costs [2]. However, that poses unique challenges for large-scale C2C interconnects, particularly regarding high precision and low power consumption.

C2C wireless interconnection is gaining recognition for its design flexibility and low power consumption. Technologies such as printed circuit board (PCB), low-temperature cofired ceramic (LTCC), and wafer-level packaging (WLP) are used for C2C wireless interconnect. PCB technology is cost-effective and mature but introduces parasitic parameters, hindering high-frequency and miniaturization designs. LTCC technology integrates multi-layers and ensures thermal stability, suitable for high-temperature environments but less ideal for highly integrated miniaturized applications. WLP technology excels in the THz band with higher integration and improved electrical performance due to its precise processing and shorter interconnect paths.

Fan-out wafer-level packaging (FOWLP) technology has gained attention for future terahertz applications due to its superior electrical performance and design flexibility [3]. D. Schwantuschke demonstrated a GaN traveling wave amplifier embedded in an epoxy molding compound using FOWLP [4], achieving a maximum power-added efficiency of 10%. Zihao Chen integrated V-band and W-band antennas with a SPDT switch on FOWLP [5], showing 9 GHz and 7.2 GHz impedance bandwidths and peak realized gains of 12.84 dBi and 14.42 dBi, respectively. Gang Zhuang proposed an omnidirectional AiP for THz C2C wireless interconnect [6], demonstrating an impedance band of -10 dB from 216-231 GHz.

In this paper, an RF transceiver system based on FOWLP technology is designed for THz C2C wireless interconnection. The transmitting and receiving system have been verified through simulation to achieve low-loss interconnection from the chip to the micro solder bumps to the antenna, as well as from the chip to the through-molding vias (TMV) to the ball grid array (BGA) to the PCB. Furthermore, theoretical calculations have demonstrated that this system can meet the requirements for wireless transmission of 10 Gbps high-speed data, with a packaging antenna gain increased to 3 dB, a transmission power of 5 dBm, and a transmission distance of 2.5 cm.

Design Process

A. FOWLP RF Transceiver System

The RF transceiver system package structure shown in Fig. 1 consists of three layers: antenna, chip, and PCB. The top layer features a glass-based omnidirectional antenna for signal radiation. The middle layer houses a reconstituted resin-wafer-packaged RF transceiver chip, with signals transmitted via polyimide (PI) layers and coplanar waveguide (CPW) to the top antenna using micro solder bumps. The redistribution layer (RDL) connects the DC, local oscillator (LO), and intermediate frequency (IF) signals through TMV to the bottom of the package, then to the PCB via BGA. The external signal source powers the RF chip through the PCB, then the signal is transmitted via the top antenna.

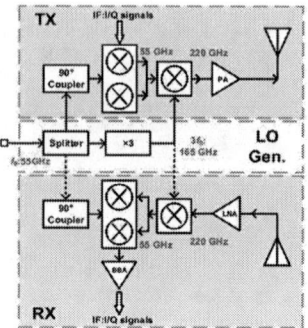

Fig. 2. Block diagrams of the proposed 220-GHz Tx and Rx.

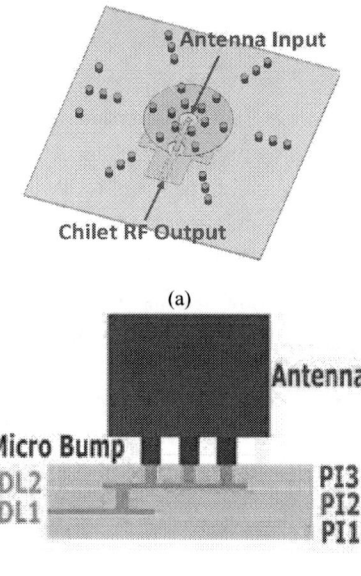

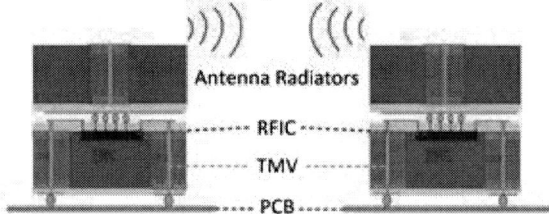

Fig. 1. Illustration of a FOWLP RF Transceiver System.

1) 220-GHz Transmitter (Tx) and Receiver (Rx) Chipset

The block diagrams of the proposed sliding-IF 220 GHz Tx and Rx are shown in Fig. 2. The LO signal of 165 GHz is generated by the first LO signal of 55 GHz via a frequency tripler. Furthermore, the frequency choice of the sliding-IF architecture keeps the image and the LO far away from the band of interest while guaranteeing sufficient IF bandwidth. For the high linear output power of the Tx, a power amplifier (PA) is employed after the mixer. A low noise amplifier (LNA) and a baseband amplifier (BBA) are adopted in the Rx to improve the noise performance and enhance the dynamic range.

2) Interconnect Transition Structures

The interconnection between the chip and the antenna is depicted in Fig. 3, comprising three PI layers, two RDL layers, and electroplated micro-convex structures. The PI layers ensure good metal adhesion and favorable electromagnetic properties, while the two-layer RDL area is slightly smaller than the antenna substrate to minimize interference. The interconnection from the PCB to the chip is shown in Fig. 4. An LO signal at 55 GHz, sourced externally, transitions through a coaxial-like structure via BGA and TMV to the RDL connected to the chip's LO input.

Fig. 3. Illustration of chip to antenna design. (a) 3D view, (b) side view schematic diagram.

B. Interconnect Structure Simulation and Communication Validation

The simulation results of the structure are presented in Fig. 5 by using commercial software HFSS. It shows that the AiP achieves an impedance bandwidth of -15 dB covering the frequency range from 200-230 GHz for the RF frequency band. The structures achieve insertion loss less than 1 dB and return loss greater than 10 dB within the 25-60 GHz frequency band for the LO frequency band.

3) Link Calculation for the Communication System

The Fries transmission formula can be used to estimate the communication system. The formula for calculating link space loss is:

$$L_s = 10\lg(\frac{4\pi d}{\lambda})^2 \tag{1}$$

where d is the transmission distance which is 2.5 cm, λ is the wavelength. Thus, the formula for calculating the signal-to-noise ratio (SNR) is as follows:

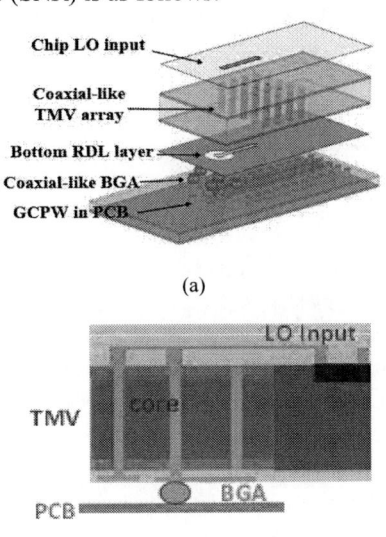

(a)

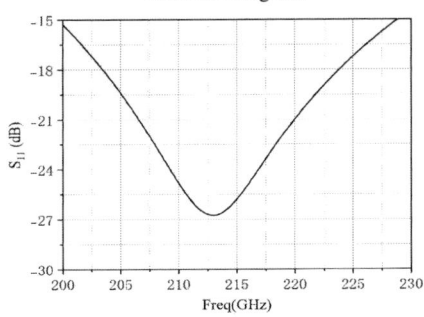

(b)

Fig. 4. Illustration of PCB to chip design. (a) 3D view, (b) side view schematic diagram.

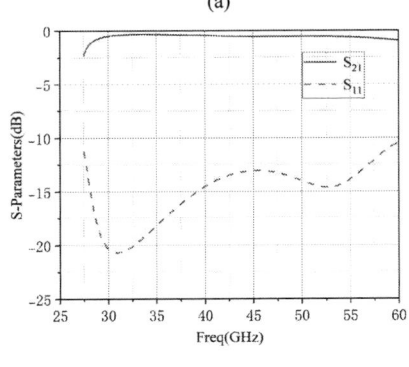

(a)

(b)

Fig. 5. Simulated performance of interconnect structures. (a) S_{11} of the AiP, (b) transmission capability from the PCB to the chip

$$SNR = P_r - P_N \qquad (2)$$

$$P_r = P_{out} - L_t + 2 \times Gain - L_t - L_s \qquad (3)$$

where P_r is the received power, P_N is the transmission power, and L_t is the chip to antenna transition loss. It can be concluded that, under the conditions of antenna gain of 3 dB, transmission power of 5 dBm, and a transmission distance of 2.5 cm, the requirements for high-speed wireless data transmission at 10 Gbps are satisfied.

Conclusion

This paper designs a communication transceiver system based on FOWLP technology, achieving a low-loss interconnect structure. The system will undergo physical testing in the next step and is expected to provide a promising solution for THz C2C wireless interconnects.

Acknowledgment

This work is supported in part by the National Natural Science Foundation of China (NSFC) under grant 62101109, in part by the Hong Kong Scholars Program under Grant XJ2022031, and in part by Young Elite Scientists Sponsorship Program by CAST under grant 2023QNRC001.

References

[1] T. N. Theis and H.-S. P. Wong, "The end of Moore's law: A new beginning for information technology," Comput. Sci. Eng., vol. 19, pp. 41–50 (2017).

[2] G. H. Loh, S. Naffziger and K. Lepak, "Understanding Chiplets Today to Anticipate Future Integration Opportunities and Limits," 2021 Design, Automation & Test in Europe Conference & Exhibition (DATE), Grenoble, France, pp. 142-145 (2021).

[3] J.-Y. Lee, J. Choi, D. Choi, Y. Youn, and W. Hong, "Broadband and wide-angle scanning capability in low-coupled mm-Wave phased-arrays incorporating ILA with HIS fabricated on FR-4 PCB", IEEE Transactions on Vehicular Technology, vol. 70, no. 3. pp. 2076-2088 (2021).

[4] D. Schwantuschke et al., "Fan-out Wafer Level Packaging of GaN Traveling Wafer Amplifier," 2022 IEEE/MTT-S International Microwave Symposium - *IMS 2022*, Denver, CO, USA, 2022, pp. 579-582 (2022).

[5] Z. Chen and L. Xu, "Integration of V-band and W-band Antennas with SPDT Switch on Fan-Out Wafer Level Packaging (FOWLP)," 2019 IEEE MTT-S International Wireless Symposium (IWS), Guangzhou, China, 2019, pp. 1-3 (2019).

[6] G. Zhuang et al., "An Omnidirectional Antenna-in-Package Based on FOWLP for THz Chip-to-Chip Wireless Interconnect Applications," 2024 15th Global Symposium on Millimeter-Waves & Terahertz (GSMM), Hong Kong, 2024, pp. 1-3 (2024).

Temperature-dependent Characteristics of Field-Effect Mobility in MOCVD-Grown β-Ga$_2$O$_3$ MOSFETs on Sapphire Substrate

Chunxiao Yu[1], Yibo Wang[1,2,*], Yiyang Wu[3], Jun Zheng[3,*], Chenyu Liu[1], Xiaole Jia[1], Bochang Li[1,2], Haodong Hu[1], Cize Fang[1,2], Yan Liu[1], Yue Hao[1], Genquan Han[1,2,*]

[1]State Key Discipline Laboratory of Wide Band Gap Semiconductor Technology, School of Microelectronics, Xidian University, Xi'an 710071, China.

[2]Emerging Device and Chip Laboratory, Hangzhou Institute of Technology, Xidian University, Hangzhou 311200, China

[3]State Key Laboratory on Integrated Optoelectronics, Institute of Semiconductors, Chinese Academy of Sciences, Beijing 100083, People's Republic of China

*e-mail: wangyibo2576@126.com; zhengjun@semi.ac.cn; gqhan@xidian.edu.cn

Abstract

β-Ga$_2$O$_3$ MOSFETs were fabricated based on the heteroepitaxial β-Ga$_2$O$_3$ film grown on the sapphire substrate by metalorganic chemical vapor deposition (MOCVD). Thanks to the high epitaxial quality and an optimized Si$^+$ implantation and activation process, a high field effect mobility (μ_{FE}) of 35 cm^2/V·s was achieved at a high carrier density(n_e) near the surface exceeding 10^{18} cm^{-3}. As the temperature rises, the density of ionized impurities increases, and Coulombic scattering becomes the dominant mechanism, resulting in the enhancement of μ_{FE} from 35 cm^2/V·s to 55 cm^2/V·s when the temperature increases from 25°C to 300°C. The μ_{FE} is the highest reported for heteroepitaxial Ga$_2$O$_3$ devices on sapphire substrates.

Keywords: High mobility, Heteroepitaxy, β-Gallium Oxide, MOSFET

Introduction

β-Ga$_2$O$_3$ is regarded as a high-performance material for next-generation power devices because of its ultrawide bandgap (UWBG) of approximately 4.8-4.9 eV [1], along with its high melting points and breakdown electric fields, making it ideal for high-temperature applications. However, the β-Ga$_2$O$_3$ homoepitaxial substrate with a high cost and low thermal conductivity (κ) of 10-25 W/(m·K) eventually limits its practicality [1]. Heteroepitaxial Ga$_2$O$_3$ on well-developed foreign substrates offers an avenue for cost reduction. An alternative, cost-effective, available sapphire substrate with a greater κ of 40.1 W/m·K is a promising choice to grow the hetero-epitaxy Ga$_2$O$_3$ conductive channels achieving power MOSFETs [2]. Studies using the metalorganic chemical vapor deposition (MOCVD) method have successfully grown Ga$_2$O$_3$ films on sapphire [3], however, the challenges related to effective doping techniques and defects in the epitaxial layers still need to be faced [4], [5].

Many reports focus on heteroepitaxial β-Ga$_2$O$_3$ transistors grown on sapphire substrates [6], [7]. Mobility, as a key parameter for characterizing the conductivity of these devices, remains at a relatively low value below 20 cm^2/V·s [3], [8]

for the reported works. Additionally, the impact of temperature on the mobility of heteroepitaxial β-Ga$_2$O$_3$ transistors is still unclear.

In this work, we achieved the reported best value of the field effect mobility (μ_{FE}) by optimizing the epitaxial quality and the implantation and activation processes in the fabricated heteroepitaxial β-Ga$_2$O$_3$ MOSFETs on sapphire. The temperature-dependent characteristics of the μ_{FE} and the underlying mechanisms have been investigated. Due to the the enhanced heat dissipation of the sapphire substate and increased ionized impurity with rising temperature, Coulombic scattering becomes the dominant mechanism in the β-Ga$_2$O$_3$ channel. As a result, the μ_{FE} shows a decent improvement with the increasing temperature.

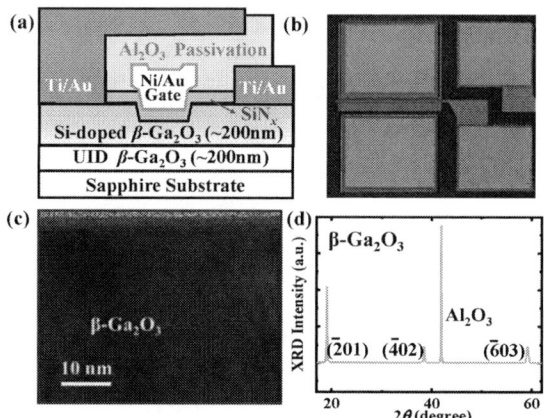

Fig. 1 (a) A schematic and (b) Top-view optical microscope image of the fabricated β-Ga$_2$O$_3$ MOSFETs on sapphire. (c) HRTEM image and (d) HRXRD 2θ - ω spectra obtained for MOCVD-grown β-Ga$_2$O$_3$ on the sapphire substrate.

Device Fabrication

β-Ga$_2$O$_3$ MOSFETs were fabricated on the heteroepitaxial β-Ga$_2$O$_3$/sapphire by MOCVD. Fig. 1(a) and (b) show the schematic structure and the top-view optical microscope image of the transistors with a ~ 400 nm heteroepitaxial β-Ga$_2$O$_3$ layer. Si$^+$ implantation with a dose of 1×10^{15} cm^{-2} was done targeting a depth of 160~180 nm, followed by rapid annealing for 20 min at 950°C under N$_2$ to improve the conductivity of the whole Ga$_2$O$_3$ channel layer. The recessed Plasma (ICP) with Cl$_2$/BCl$_3$ gases, and a wet etching solution

979-8-3315-0417-5/25 $31.00 © 2025 IEEE

of tetramethylammonium hydroxide (TMAH) at 70°C for 3 minutes was used to repair the etching surface. SiN_x (60 nm) is deposited as the gate dielectric using Plasma-Enhanced Chemical Vapor Deposition (PECVD). Ni/Au (20 nm /80 nm) gate electrodes were deposited by an electron beam deposition system, followed by a lift-off process. Source, gate-field plates, and Al_2O_3 passivation layer were also fabricated.

Fig. 1(c) shows the High-resolution TEM image of the heteroepitaxial β-Ga_2O_3 film, exhibiting the high crystalline quality. Fig. 1(d) shows the high-resolution X-ray diffraction (HRXRD) 2θ - ω spectra measured for MOCVD-grown β-Ga_2O_3 ($\bar{2}01$) on sapphire at a point. The observed three peaks indicate the single-crystal nature of the epitaxial β-Ga_2O_3 on sapphire.

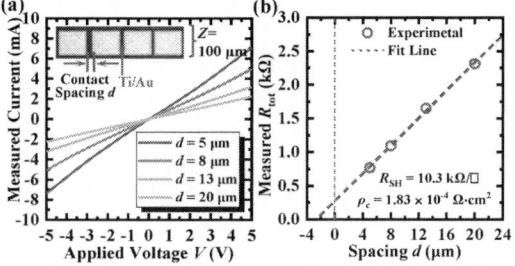

Fig. 2. (a) DC I - V curves were measured by the transmission line model (TLM) structure formed on n^+-Ga_2O_3 on sapphire. The inset shows the optical micrograph of the TLM structure and key parameters. (b) A plot depicts the total resistance between Ti/Au contacts as a function of d, showing the key values of the devices.

Result and Discussion

Fig. 2(a) illustrates the I - V curves of the fabricated transmission line model (TLM) structure with the same step of the Source/Drain Ti/Au Ohmic contacts forming. The nice I-V behavior at different d indicates the decent Ohmic performance. The inset shows the optical micrographs of the TLM structure and key parameters. Fig. 2(b) presents a plot of the total resistance between Ti/Au contacts as a function of d to extract the contact resistivity (ρ_c) and sheet resistance (R_{SH}) values, the obtained ρ_c is the lowest value for heteroepitaxial Ga_2O_3 materials [9].

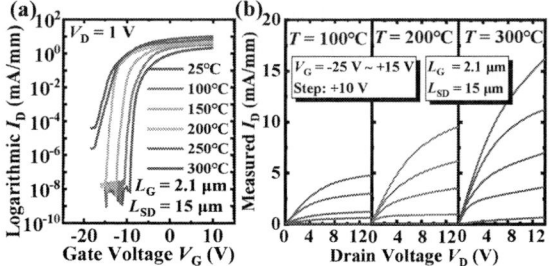

Fig. 3. (a) The Log-scale transfer of β-Ga_2O_3 MOSFETs on sapphire under a variety of temperature ranging from 25°C to 300°C. (b) The Linear-scale output characteristics of β-Ga_2O_3 MOSFETs on sapphire at T = 100°C, T = 200°C and T = 300°C.

Fig. 3(a) shows the measured transfer characteristics I_D - V_G curves at a V_D of 1 V with a measured temperature ranging from 25 °C to 300°C. The off-state current remains at a stable

value from 25°C to 200°C, and a gradual increase is observed at the temperature above 200°C. This behavior can be attributed to the enhanced heat dissipation provided by the sapphire substrate, which also suppresses Phonon scattering—a phenomenon that typically decreases carrier mobility at higher temperatures. Consequently, the maximum drain current (I_D) shows an obvious increase with temperature rising, which can be attributed to the increased activation of carriers from dopants or traps [10]. Fig. 3(b) illustrates the output curves of the β-Ga_2O_3 MOSFET on sapphire with a V_G ranging from -25 V to 15 V at T = 100°C, T = 200°C, and T = 300°C. The notable changes in the gate override (V_G - V_{TH}) observed in the output curves at different temperatures are attributed to the decreased threshold voltage V_{TH} as temperature increases, suggesting an increase in the density of activated carriers [11].

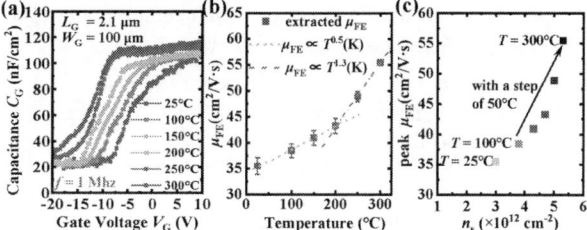

Fig. 4. (a) The capacitance-voltage C_G - V_G characteristics of β-Ga_2O_3 MOSFETs on sapphire under a variety of temperature ranging from 25°C to 300°C. (b) Extracted μ_{FE} for β-Ga_2O_3 MOSFETs on sapphire increases with temperature rising, corresponding to the Coulombic scattering. The dash lines are the fitting results. (c) Extracted peak μ_{FE} and the corresponding sheet density n_s under a variety of temperature ranging from 25°C to 300°C.

Fig. 4(a) presents the temperature-dependent capacitance-voltage (C_G - V_G) characteristics of the β-Ga_2O_3 MOSFETs on sapphire. The extracted temperature-dependent field-effect mobility (μ_{FE}) representing the average of five devices with error bars based on the I_D -V_G and the C_G - V_G curves is shown in Fig. 4(b). It was calculated by the following equation [12]:

$$\mu_{FE} = \frac{L_G \sqrt{g_m}}{W_G C_{OX} V_D} \tag{1}$$

where g_m is the transconductance, V_D is the drain voltage, and C_{OX} is gate capacitance per unit area, measured using C_G - V_G configuration, L_G and W_G are the gate length and width, respectively. The average μ_{FE} of β-Ga_2O_3 MOSFETs on sapphire increases with rising temperature, from 35 cm²/V·s at 25°C to 55 cm²/V·s at 300°C. The fitting results indicate that μ_{FE} is proportional to $T^{0.5}$ from 25°C to 200°C, while it is proportional to $T^{1.3}$ from 200°C to 300°C. These results suggest that the dominant scattering mechanism is Coulomb scattering, which results in increased mobility as temperature rises due to a reduction of interaction time, as expressed in the following equation [13]:

$$\mu_{FE} \propto \frac{T^{1.5}}{N_e m^*} \tag{2}$$

where N_e is the ionized impurity density, and m^* is the effective mass. When $25°C \leq T \leq 200°C$, the temperature-dependent coefficient ($\alpha = 0.5$) is smaller than the theoretical Coulomb scattering coefficient ($\alpha = 1.5$), this difference can be attributed to the increased density of ionized impurities as the temperature rises. When $200°C < T \leq 300°C$, the temperature dependence coefficient ($\alpha = 1.3$) approaches the theoretical value of 1.5 due to the nearly complete ionization of impurities at high temperatures ($> 200°C$), as shown in Fig. 4(c).

Fig. 4(c) illustrates the μ_{FE} vs. the sheet density n_s, derived from the V_G - I_D and C_G - V_G curves as a function of temperature. Both the peak μ_{FE} and the corresponding n_s increase linearly with the rising temperature. At $T = 25°C$, n_s and μ_{FE} are extracted at 5×10^{12} cm^{-2} and 35 cm^2/V·s, respectively. The measured value of n_s aligns with the increased activation of carriers as the temperature rises. The maximum n_s reaches approximately 5.5×10^{12} cm^{-2} at $T = 300°C$, corresponding to a μ_{FE} of 55 cm^2/V·s.

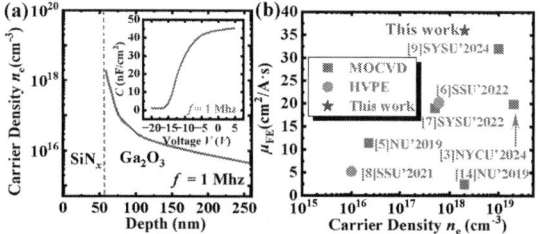

Fig. 5. (a) Carrier distribution profile extracted from the C - V measurement of the SiN$_x$/β-Ga$_2$O$_3$ MOS capacitor at 1MHz. (b) μ_{FE} versus carrier density n_e benchmark plot of MOCVD and Hydride Vapor Phase Epitaxy (HVPE)-grown Ga$_2$O$_3$ MOSFETs on sapphire.

Fig. 5 presents the carrier distribution profile and benchmarks the extracted μ_{FE} versus carrier density n_e of the fabricated β-Ga$_2$O$_3$ MOSFETs against other reported lateral Ga$_2$O$_3$ MOSFETs on sapphire substrate. Our devices exhibited a competitive μ_{FE} of 35 cm^2/V·s at room temperature with a carrier density of 2×10^{18} cm^{-3}, a record-high value for heteroepitaxial Ga$_2$O$_3$ devices on sapphire.

Conclusion

For the first time, the temperature dependence field-effect mobility in heteroepitaxy β-Ga$_2$O$_3$ MOSFETs on sapphire substrates has been investigated. A peak field effect mobility of μ_{FE} of 35 cm^2/V·s was achieved at room temperature with a $n_s = 3 \times 10^{12}$ cm^{-2}, which increased to 55 cm^2/V·s at $T = 300°C$, corresponding to $n_s = 5.5 \times 10^{12}$ cm^{-2}. The temperature dependence of field-effect mobility suggests an increase in activated carriers from dopants or traps as the temperature rises, while Coulombic scattering becomes the dominant mechanism influencing carrier transport. The transistor achieved the highest μ_{FE} comparable to that of other heteroepitaxial Ga$_2$O$_3$ devices on sapphire with similar n_e.

Acknowledgments

The authors acknowledge support from the National Natural Science Foundation of China (Grant No. 62293522, 62204255, 62234007, and 62025402) and Strategic leading science and technology project, CAS (XDB43020100).

References

[1] M. Higashiwaki et al., "Development of gallium oxide power devices," *Phys. Status Solidi A*, vol. 211, no. 1, pp. 21–26, Jan. 2014.

[2] Zhi Guo et al., "Anisotropic thermal conductivity in single crystal β-gallium oxide. *Appl. Phys. Lett.* 16 March 2015.

[3] Y.-H. Chuang et al., "Performance Improvement of Enhanced-Mode β-Ga$_2$O$_3$ MOSFETs by Partial Gate Recess Structure," *ACS Appl. Electron. Mater.*, p. acsaelm.4c00835, Sep. 2024.

[4] Y. J. Jeong et al., "Heteroepitaxial α- Ga$_2$O$_3$ MOSFETs with a 2.3 kV breakdown voltage grown by halide vapor-phase epitaxy," *Appl. Phys. Express*, vol. 15, no. 7, p. 074001, Jul. 2022.

[5] J.-H. Park et al., "MOCVD grown β-Ga$_2$O$_3$ metal-oxide-semiconductor field effect transistors on sapphire," *Appl. Phys. Express*, vol. 12, no. 9, p. 095503, Sep. 2019.

[6] S. Y. Oh et al., "A 2.8 kV Breakdown Voltage α- Ga$_2$O$_3$ MOSFET with Hybrid Schottky Drain Contact," *Micromachines*, vol. 15, no. 1, p. 133, Jan. 2024.

[7] W. Chen et al., "First demonstration of hetero-epitaxial ε-Ga$_2$O$_3$ MOSFETs by MOCVD and a F-plasma surface doping," *Applied Surface Science*, vol. 603, p. 154440, Nov. 2022.

[8] Y. J. Jeong et al., "Fluorine-based plasma treatment for hetero-epitaxial β- Ga$_2$O$_3$ MOSFETs," *Applied Surface Science*, vol. 558, p. 149936, Aug. 2021.

[9] D. Zeng et al., "Heteroepitaxial ε-Ga$_2$O$_3$ MOSFETs on a 4-inch Sapphire Substrate with a Power Figure of Merit of 0.29 GW/cm^2," in *2024 36th International Symposium on Power Semiconductor Devices and ICs (ISPSD)*, Bremen, Germany: IEEE, Jun. 2024, pp. 192–195.

[10] N. P. Sepelak et al., "First Demonstration of 500 °C Operation of β-Ga$_2$O$_3$ MOSFET in Air," in *2022 Compound Semiconductor Week (CSW)*, Ann Arbor, MI, USA: IEEE, Jun. 2022, pp. 1–2.

[11] K. Fu et al., "GaN-Based Threshold Switching Behaviors at High Temperatures Enabled by Interface Engineering for Harsh Environment Memory Applications," *IEEE Trans. Electron Devices*, vol. 71, no. 3, pp. 1641–1645, Mar. 2024.

[12] D. K. Schroder, *Semiconductor Material and Device Characterization*, 1st ed. Wiley, 2005.

[13] S. M. Sze and K. K. Ng, *Physics of semiconductor devices*, 3. ed. in Wiley-Interscience online books. Hoboken, N.J: Wiley-Interscience, 2007.

[14] J.-H. Park et al., "Ga$_2$O$_3$ metal-oxide-semiconductor field effect transistors on sapphire substrate by MOCVD," *Semicond. Sci. Technol.*, vol. 34, no. 8, p. 08LT01, Aug. 2019.

Content Addressable Memory Hierarchies for Computing in Memory

Paul-Philipp Manea[1,2], Nathan Leroux[1], and John Paul Strachan[1,2]

[1]Peter Grünberg Institute, Forschungszentrum Jülich, [2]RWTH Aachen University

ABSTRACT

Content Addressable Memories (CAMs) match input data to stored content, ideal for search-intensive tasks like network address lookup and similarity matching. This paper explores CAM hierarchies (similar to the standard RAM hierarchy), focusing on two implementation layers: non-volatile, analog Content Addressable Memory (aCAM) for static data patterns, and volatile, gain cell-based aCAM for dynamic tasks. Mitigating key non-idealities in memristor-based aCAMs increases storage from 2 to 4 bits per cell, enhancing performance, capacity, and energy efficiency for Compute in Memory (CIM) applications.

Keywords: Content Addressable Memory, Compute In Memory, Memristor

INTRODUCTION

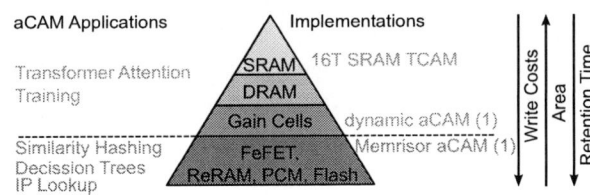

Fig. 1. Memory hierarchies in aCAM systems and their potential application

A Content Addressable Memory (CAM) is a memory type that retrieves the stored address corresponding to a given data input, rather than fetching data based on address, as in Random Access Memory (RAM). This functionality is a form of computing in memory, and valuable for search-based applications as well as in Decision Trees, Random Forests, Finite State Machines, and Hyper-dimensional Computing [1]–[3]. One special type of CAM is the aCAM [4], which can store analog intervals, acting also like a multi-bit CAM. As with conventional NOR-type memory systems, each cell of the word is connected to a shared Match Line (ML) via a pull-down transistor. A shared ML determines word matches, with non-binary outputs available to measure the distance to stored keys by analyzing mismatch timing or currents [1]. aCAMs have been implemented using various types of non-volatile memory, including memristors [4], FeFETs [5], and Flash [6]; however, dynamic and easily writable aCAMs implementations have yet to be explored, which we address in this work.

Memory technology hierarchies are crucial in modern computing systems, including Graphical Processing Units (GPUs). Non-volatile memories such as NAND Flash store long-term data and operating systems, while volatile memories like DRAM serve as working memory. The latter are also typically off-chip, thus access times are slower, adding latency. In contrast, SRAM is used on-chip, such as local cache memory, providing high-speed access for dynamic operands but offering low capacities. However, in CIM systems, memory hierarchies are less considered. In this context, we apply this paradigm to aCAM systems and describe two complementary layers: one optimized for dynamic data patterns using gain cell memory technology, and another for non-volatile computing tasks leveraging Memristors. Figure 1 highlights the advantages of aCAM systems, depending on the used memory technology.

While memristors are known for non-volatile CIM and aCAM implementations [4], capacitor gain cells offer volatility with lower programming energy/time. The potential for CIM, particularly in performing analog vector-matrix multiplications, was demonstrated in [7]. Typically, gain cells incorporate specialized write transistors using oxide-based 2D materials such as Indium Gallium Zinc Oxide (IGZO), which provide extremely low leakage currents and retention times in the range of seconds [8]. Additional advantages include 3D integration capabilities and CMOS compatibility. We develop an aCAM system with this technology targeting dynamic tasks such as transformer attention.

The remainder of the paper is organized as follows: we begin with an exploration of non-volatile, memristor-based aCAMs, demonstrating enhanced bit precision by mitigating non-idealities. Next, we present the implementation of a dynamic aCAM and its application in transformer attention. Finally, we conclude with a summary of our findings.

NON-VOLATILE MEMRISTOR-BASED ACAMS: RELIABILITY AND CAPACITY

Memristor-based aCAMs leverage the non-volatility and tunable conductance of memristive devices, which support multiple stable states, but require higher programming energy and time than SRAM. Figures 2 (a) and (b) illustrate the functional principle of an aCAM, where a match is determined by comparing an input data line voltage to a defined boundary range of analog reference voltages, encoded by specific memristor

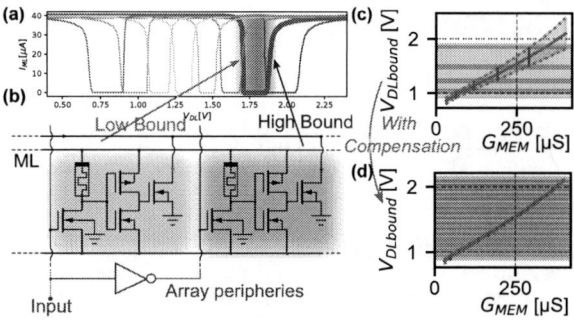

Fig. 2. **a)** Multiple aCAM windows encoding a boundary range of V_{DL}, which is marked in one example as the green area indicating a match. Low- and high-bound are encoded by each memristor comparator. **b)** Proposed 8T2M aCAM cell. **(c)** Combined effects of temperature dependence, memristor fluctuations, CMOS process variations, and read noise. The green area indicates the estimated CAM windows. **(d)** number of levels after compensation of temperature- and CMOS process variations

conductance values. The boundary voltages $V_{DLbound}$ are stored in two memristors per cell, one for the lower boundary and one for the upper boundary. If the input voltage V_{DL} falls within this range, the aCAM cell registers a match; otherwise, it returns a mismatch, resulting in a binary output for each aCAM cell. This setup allows an aCAM to store multiple intervals, effectively functioning as a multi-bit CAM. Similar to conventional NOR-type memory systems, each cell within a word is connected to a shared ML via a pull-down transistor. An external sense amplifier then determines whether an entire ML matches or not.

We have investigated [9] the inherent non-idealities that affect Memristor-based Analog Content Addressable Memories (aCAMs), as seen in many analog computing systems, including temperature dependence, CMOS process variations, and memristor read noise [10]. Effectively addressing these non-idealities is crucial to ensure the accuracy, reliability, and efficiency of aCAM systems. In previous work, we demonstrated how these three types of non-idealities could be mitigated to enhance aCAM efficiency. By incorporating these mitigation techniques, aCAM cells achieve increased reliability and data capacity, with an estimated storage potential of 4 bits (16 levels) per cell. This is illustrated in Figure 2 (c), which shows variations in the boundary voltage $V_{DLbound}$, accounting for all three non-idealities, and in Figure 2 (d), where only memristor read noise remains. The identified non-idealities and corresponding mitigation measures are as follows:

- **Temperature Dependence:** Addressed through temperature-aware transformations of input data, achieved by monitoring temperature across the array.
- **CMOS Process Variations:** Managed through a program-and-verify algorithm that directly targets

boundary voltage $V_{DLbound}$, regardless of specific memristor conductance values.
- **Memristor Read Noise:** While read noise has minimal impact on overall accuracy, it cannot be deterministically mitigated.

DYNAMIC ANALOG CONTENT ADDRESSABLE MEMORY WITH CAPACITOR GAIN CELLS

Our dynamic aCAM [11] employs capacitor gain cells to store analog intervals and utilizes strong-arm latched comparators to compare these stored values with the analog input voltage, which is illustrated in Figure 3. The dynamic aCAM cell is implemented in TSMC's 28 nm technology and has been simulated in Cadence Virtuoso. The circuit for a single dynamic aCAM cell of the proposed design is shown in Figure 3 (a). For clarity, the circuit can be divided into four stages:

1) A write enable stage, consisting of a transmission gate (P_1, N_1) that connects or disconnects the capacitor from the write word line (WL_W).
2) A storage capacitor C_1 that temporarily stores the reference voltage to which the input is compared.
3) A comparator stage (I_1, I_2), where the input voltage is compared to the reference using two strong arm latched comparators: one for testing the lower and one for testing the upper bound.
4) The ML stage (P_2, P_3), where the cell is connected to the common ML via two PMOS transistors. This stage combines the results from all cells within a single word.

The macro architecture is shown in Figure 3 (b), featuring an $N \times M$ array along with the necessary peripheral circuitry.

The dynamic aCAM receives two input signals, VDL_L and VDL_H. Unlike previous approaches that store the range within the cell and search for a fixed voltage, our method applies the range directly at the aCAM input while storing the search value within the cell, with a bias voltage V_{Range} added to or subtracted from the actual search voltage VDL, as shown in Figure 3 (c&d).

By using Monte carlo simulations capturing variability we estimated the bit precision to 8 levels (3) bits. Our system can be This distinction yields an estimated precision of 8 distinguishable levels (3 bits). In terms of operational speed, the read speed in the worst-case scenario is 6 ns, while the write speed reaches 20 ns. For energy consumption, the search operation requires an average of 3.0 fJ per cell, with an estimated energy cost per bit per search of 1.0 fJ. Write operations have a mean energy consumption of 4.8 fJ per cell.

To demonstrate the novel capabilities, we have replaced the classical Transformer's Softmax-scaled dot-product similarity (SDPS) with a similarity measure computed by this aCAM. To assess our aCAM similarity-based attention, we implemented a PyTorch model of the circuit and hardware [12] and applied it to bio-signal processing on the NinaproDB8 dataset [13] for predicting finger-joint angles from forearm Electromyographic signals. Our aCAM similarity achieved a mean absolute error of 6.35°, comparable to the conventional SDPS error of 6.33°, validating the differential aCAM's trainability and potential for attention score computation.

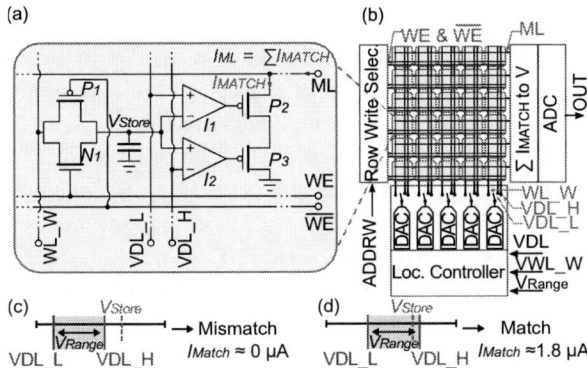

Fig. 3. **(a)** Circuit configuration of the proposed dynamic aCAM cell. **(b)** Architecture of the dynamic IMC aCAM macro including peripheral circuitry. **(c&d)** Input voltage range within the dynamic search window for a given stored key voltage V_{store} in both mismatch **(c)** and match **(d)** cases.

CONCLUSION

In this paper, we described the importance of engineering multiple layers in a memory hierarchy for CAM systems. We presented two implementation layers within this hierarchy: using non-volatile memristors for static tasks but having high programming energy and time, and another based on volatile gain cells for dynamic tasks with low programming energy and time. We developed a framework to enhance the capacity of memristor-based aCAM systems despite non-idealities from CMOS process variations, temperature-dependence, and memristor read noise. We proposed countermeasures, increasing the storage capacity from 2 to 4 bits. Furthermore, we described the gain cell-based aCAM, offering fast write speeds and capable of performing attention similarity computations for sequence-processing tasks.

REFERENCES

[1] G. Pedretti, C. E. Graves, T. Van Vaerenbergh, S. Serebryakov, M. Foltin, X. Sheng, R. Mao, C. Li, and J. P. Strachan, "Differentiable content addressable memory with memristors," *Advanced electronic materials*, vol. 8, no. 8, p. 2101198, 2022.

[2] C. E. Graves, C. Li, G. Pedretti, and J. P. Strachan, "In-memory computing with non-volatile memristor cam circuits," in *Memristor Computing Systems*. Springer, 2022, pp. 105–139.

[3] A. Kazemi, F. Müller, M. M. Sharifi, H. Errahmouni, G. Gerlach, T. Kämpfe, M. Imani, X. S. Hu, and M. Niemier, "Achieving software-equivalent accuracy for hyperdimensional computing with ferroelectric-based in-memory computing," *Scientific reports*, vol. 12, no. 1, p. 19201, 2022.

[4] C. Li, C. E. Graves, X. Sheng, D. Miller, M. Foltin, G. Pedretti, and J. P. Strachan, "Analog content-addressable memories with memristors," *Nature communications*, vol. 11, no. 1, p. 1638, 2020.

[5] X. Yin, C. Li, Q. Huang, L. Zhang, M. Niemier, X. S. Hu, C. Zhuo, and K. Ni, "Fecam: A universal compact digital and analog content addressable memory using ferroelectric," *IEEE Transactions on Electron Devices*, vol. 67, no. 7, pp. 2785–2792, 2020.

[6] A. Kazemi, S. Sahay, A. Saxena, M. M. Sharifi, M. Niemier, and X. S. Hu, "A flash-based multi-bit content-addressable memory with euclidean squared distance," in *2021 IEEE/ACM International Symposium on Low Power Electronics and Design (ISLPED)*, 2021, pp. 1–6.

[7] Y. Wang, H. Tang, Y. Xie, X. Chen, S. Ma, Z. Sun, Q. Sun, L. Chen, H. Zhu, J. Wan, Z. Xu, D. W. Zhang, P. Zhou, and W. Bao, "An in-memory computing architecture based on two-dimensional semiconductors for multiply-accumulate operations," *Nature Communications*, vol. 12, no. 1, jun 2021.

[8] A. Belmonte, S. Kundu, S. Subhechha, A. Chasin, N. Rassoul, H. Dekkers, H. Puliyalil, F. Seidel, P. Carolan, R. Delhougne, and G. S. Kar, "Lowest ioff < 3×10-21 a/μm in capacitorless dram achieved by reactive ion etch of igzo-tft," in *2023 IEEE Symposium on VLSI Technology and Circuits (VLSI Technology and Circuits)*, 2023, pp. 1–2.

[9] P.-P. Manea, C. Sudarshan, F. Cüppers, and J. P. Strachan, "Non-idealities and design solutions for analog memristor-based content-addressable memories," in *Proceedings of the 18th ACM International Symposium on Nanoscale Architectures*, 2023, pp. 1–6.

[10] S. Wiefels, C. Bengel, N. Kopperberg, K. Zhang, R. Waser, and S. Menzel, "Hrs instability in oxide-based bipolar resistive switching cells," *IEEE Transactions on Electron Devices*, vol. 67, no. 10, pp. 4208–4215, 2020.

[11] P.-P. Manea, N. Leroux, E. Neftci, and J. P. Strachan, "Gain cell-based analog content addressable memory for dynamic associative tasks in ai," 2024. [Online]. Available: https://arxiv.org/abs/2410.09755

[12] P.-P. Manea, "torchcam," https://iffgit.fz-juelich.de/manea/torchcam, 2023.

[13] A. Krasoulis, S. Vijayakumar, and K. Nazarpour, "Effect of User Practice on Prosthetic Finger Control With an Intuitive Myoelectric Decoder," *Frontiers in Neuroscience*, vol. 13, 2019. [Online]. Available: https://www.frontiersin.org/articles/10.3389/fnins.2019.00891

6-inch GaN-on-Si Gold-Free Fabrication Technolgies for Monolithic Microwave Integrated Circuits (MMICs)

Xiaojin Chen[1,2], Jin Zhou[2], Peiyu Mao[2], Tong Wang[2], Zhenyuan Li[2], Hu Wei[1,2], Hanghai Du[1,2], Weichuan Xing[2], Weihang Zhang[2], Xiangdong Li[2], Zhihong Liu[1,2,*], Jincheng Zhang[1], Yue Hao[1]

[1]1Key Laboratory of Wide Band Gap Semiconductor Materials and Devices, School of Microelectronics, Xidian University, Xi'an 710071, China

[2]Guangzhou Institute of Technology, Xidian University, Guangzhou 510555, China

*Email: liuzhihong@ieee.org

Abstract

We have fabricated GaN HEMTs with gate lengths of 0.5 μm and 0.25 μm on 6-inch silicon-based GaN substrates using a gold-free process. The 0.5 μm devices exhibit a maximum drain current density (I_{dmax}) of 680 mA/mm, a peak transconductance (G_{max}) of 230 mS/mm, and small-signal characteristics with a cutoff frequency (fT) and maximum oscillation frequency (fmax) of 19 GHz and 28 GHz, respectively. Continuous wave (CW) loadpull measurements at 3.6 GHz demonstrate an output power density of 3.1 W/mm, a peak power added efficiency (PAE) of 43.5%, and a linear gain of 20 dB. The 0.25 μm devices show a maximum drain current density (Imax) of 979 mA/mm, a peak transconductance (Gmax) of 286 mS/mm, and small-signal characteristics with a cutoff frequency (fT) and maximum oscillation frequency (fmax) of 34 GHz and 50 GHz, respectively. CW loadpull measurements at 10 GHz yield an output power density of 2.71 W/mm, a peak power added efficiency (PAE) of 44.5%, and a linear gain of 12 dB. The low and uniform ohmic contact resistance (Rc) of 0.2-0.3 Ω·mm across the wafer indicates the potential for high-performance and cost-effective 6-inch GaN with silicon CMOS compatibility for RF applications.

Keywords: Gold-Free, 6-inch, High Electron Mobility Transistors (HEMTs), CMOS compatibility

Introduction

Gallium nitride (GaN) technology is extensively utilized in the realms of radio frequency (RF) and millimeter-wave applications, such as 5G network infrastructures, satellite communications, and military and aviation electronics [1,2,3,4]. Gold-containing GaN processes have demonstrated notable advancements in both power density and efficiency. However, factors such as gold contamination, cost, and high-temperature processes have led to the incompatibility of GaN processes with Si CMOS processes, which restricts the integration of GaN with Si CMOS [5]。
The challenges associated with gold pollution and the high-temperature processes in the conventional ohmic process (Ti/Al/Ni/Au) for Si-based GaN HEMTs have hindered the integration of GaN with Si CMOS technology, thereby limiting the potential for GaN to be employed in a broader range of applications. To date, numerous research outcomes have been achieved in gold-free GaN processes. For instance, Liu et al[6] utilized a Ti/Al/Ni/Pt metallization stack and achieved a contact resistance of 0.6 Ω·mm after annealing at 975°C. Lu et al[7] employed a Si/Ti/Al/Ti/TiN multilayer stack and obtained a contact resistance of 0.12 Ω·mm following annealing at 725°C. Hyung-Seok Lee et al [8] reduced the ohmic contact resistance to 0.5 Ω·mm using a Ti/Al/W metallization stack. However, the process suffered from low repeatability, complexity, excessively high temperatures, and was not suitable for large-scale GaN integration with Si CMOS. In this work, we successfully implemented smooth, low, and uniform ohmic contacts on 6-inch silicon-based GaN using a Ti/Al bilayer metallization.

Fabrication process development

The epilayers of wafer in this work include a 1.5 μm undoped GaN buffer layer, a 150 nm GaN channel layer, a 1 nm AlN spacer layer, a 23 nm AlGaN barrier layer, and a 3 nm GaN cap layer grown on a 6-inch high-resistivity (HR) silicon substrate by a metal-organic chemical vapor (MOCVD) system. Room temperature Hall measurement shows a 2DEG density of 1×10^{13} cm^{-2} and an electron mobility of 1380 cm²/V·s.

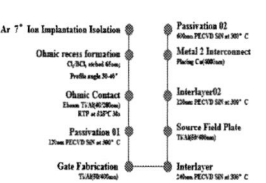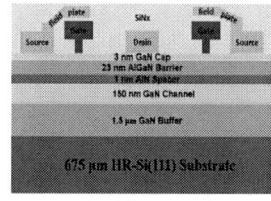

Figure 1. 6-inch GaN HEMT process flow diagram and device structure schematic.

The device fabrication commenced with Ar ion implantation for isolation. Subsequently, ohmic contact trenches were etched using a Cl2/BCl3 plasma dry etching process to a depth of 60 nm, forming a tilt angle of 30°-40° show in Figure 2. After evaporating Ti/Al=40/200 nm, rapid thermal annealing (RTA) was performed at 525°C in a nitrogen ambient for 30 seconds to form the ohmic contacts. Then, 120 nm of SiN$_x$ was grown at 300°C using PECVD to support the T-gate structure. The gate electrode was prepared through electron beam lithography, with a gate metal stack of Ti/Al=50/400 nm. Finally, source field plates, metal interconnects, and device passivation were carried out. Detailed process steps and a schematic diagram of the device structure are shown in Figure 1.

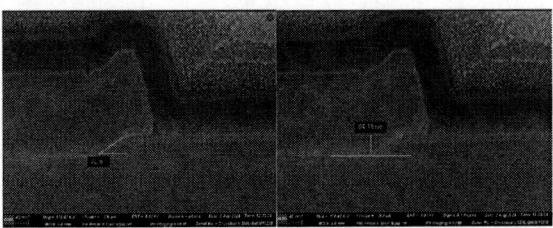

Figure 2. FIB image of ohmic contact.

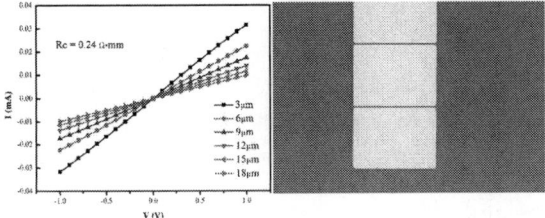

Figure 3. I-V characteristics of TLM structure and optical picture after anneal at 525°C.

Results and Discussion

Figure 3 presents the TLM (Transmission Line Method) fitting results alongside optical microscope images, indicating the lowest on-wafer ohmic contact resistance value of 0.24 $\Omega\cdot$mm. The images reveal that the metal surface is relatively smooth after annealing. Figure 4 displays the on-wafer isolation leakage (ISO pattern spacing of 3 μm), ohmic contact resistance, and saturation current test results for 3 μm electrode spacing. Benefiting from effective ion implantation isolation, the leakage current does not exceed 1 μA/mm at 200 V, the saturation current for 3 μm spacing is 900 mA/mm, and the contact resistance across the wafer is uniformly maintained between 0.2-0.3 $\Omega\cdot$mm, demonstrating good uniformity.

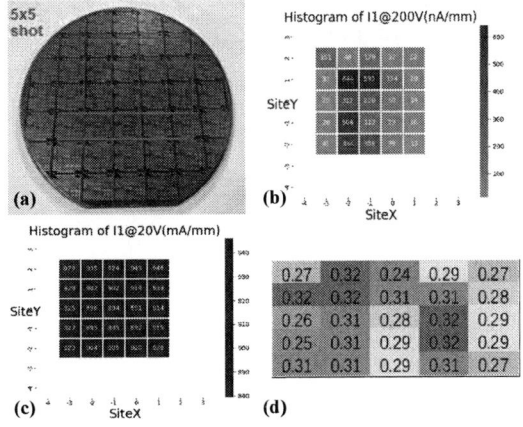

Figure 4. Photograph of the fabricated 6-inch GaN-on-Si wafer(a). Whole-Wafer leakage current(b), saturation current(c) and ohmic contact resistance(d) uniformity testing.

0.5 μm gold-free GaN HEMT

Figure 4 presents the SEM image of the core region of a gallium nitride (GaN) high electron mobility transistor (HEMT) device, with the right-hand side image showing an enlarged view of the gate area, demonstrating a well-defined T-gate morphology.

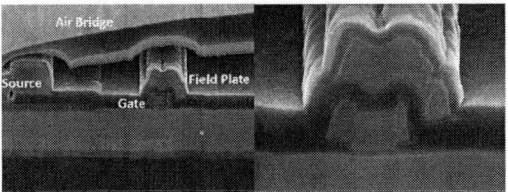

Figure 4. The SEM image of the core region of HEMT

The DC testing of the device was conducted using a Keysight B1500 semiconductor parameter analyzer system. The test results are depicted in Figure 5a and 5b, with the tested device having physical dimensions of $W_g = 2\times100$ μm and $L_{gd} = 4$ μm. The device exhibits a maximum saturation current (I_{dmax}) of 680 mA/mm, a peak transconductance (G_m) of 230 mS/mm, and an on-resistance of approximately 4.5 $\Omega\cdot$mm.

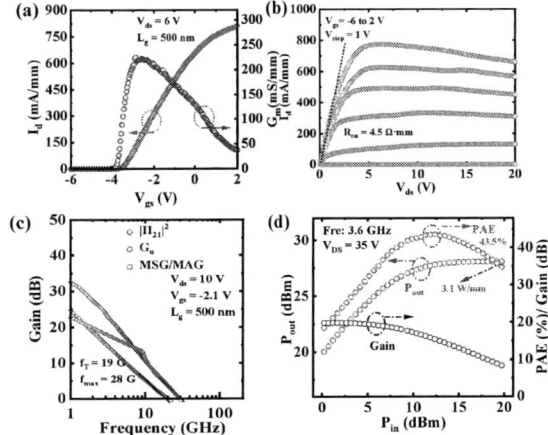

Figure 5. (a)transfer and (b)output characteristics of 0.5 μm GaN HEMT, (c)Small signal characteristics at $V_{gs} = -2.1$ V, $V_{ds} = 10$ V with the open structure de-embedded, (d)Load-pull measured power performance at CW 3.5 GHz.

Fig. 5 (c) shows the RF small-signal characteristic of the device which was carried out using an Agilent N5080B ENA network analyzer from 1 to 40 GHz. The influence of the pad capacitance was de-embedded using an on-wafer open test structure. At Vg = −2.1 V and Vd = 10 V, the fT/fmax = 19/28 GHz.

The continuous-wave (CW) large-signal power performance was measured by an on-wafer Focus load-pull system at 3.6 GHz. The device was biased in Class A amplification region, yielding a linear gain of 20 dB, a saturated output power density as high as 3.1 W/mm, and a peak power added efficiency (PAE) of 43.5%.

0.25 μm gold-free GaN HEMT

The fabrication process for the 0.25 μm gate-length HEMT is identical to that of the 0.5 μm process, and both are fabricated on the same 6-inch wafer. Figure 6 illustrates the device's DC and RF characteristics. The dimensions of the tested device are $W_g = 2 \times 100$ μm and $L_{gd} = 1.75$ μm. The device exhibits a maximum saturation current (I_{dmax}) of 979 mA/mm, a peak transconductance (G_m) of 286 mS/mm, and an on-resistance of approximately 3.3 Ω·mm.

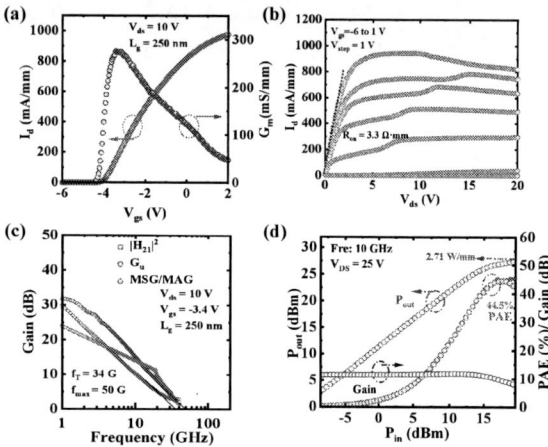

Figure 5. (a)transfer and (b)output characteristics of 0.25 μm GaN HEMT, (c)Small signal characteristics at V_{gs} = -3.4 V, V_{ds} = 10 V with the open structure de-embedded, (d)Load-pull measured power performance at CW 10 GHz.

The de-embedded small-signal characteristics with a cutoff frequency (f_T) and maximum oscillation frequency (f_{max}) of 34/50 GHz are extrapolated from the magnitude squared of the $|H_{21}|$ and the unilateral gain (G_u). Continuous wave (CW) load-pull measurements at 10 GHz demonstrate a linear gain of 12 dB, a saturated output power density of 2.71 W/mm, and a peak power added efficiency (PAE) of 44.5%.

Conclusion

We have successfully fabricated GaN HEMTs with gate lengths of 0.5 μm and 0.25 μm on 6-inch silicon-based GaN substrates using a gold-free ohmic process. Loadpull testing results indicate that they achieve power densities of 3.1 W/mm at 3.6 GHz and 2.71 W/mm at 10 GHz, respectively. The low and uniform ohmic contact resistance values of 0.2-0.3 Ω·mm, along with good uniformity, provide a guarantee for the compatibility of 6-inch silicon-based GaN with Si CMOS processes. There is reason to believe that the gold-free processes for 0.15 and 0.1 μm will further enhance device performance, promising the realization of low-cost monolithic millimeter-wave integrated circuits.

Acknowledgments

This work was supported by the NationalI Key Research and Development Program (No.2020YFB1807300), the National Natural Science Foundation of China under Grant 62074122 and JCKY research plan (No.JCKY2020110B010)

References

[1] B. Green, K. Moore, D. Hill, M. CdeBaca and J. Schultz, "GaN RF device technology and applications, present and future", 2013 IEEE Bipolar/BiCMOS Circuits and Technology Meeting (BCTM), Bordeaux, France,pp. 101-106(2013).

[2] U. K. Mishra, L. Shen, T. E. Kazior and Y. -F. Wu, "GaN-Based RF Power Devices and Amplifiers", in Proceedings of the IEEE, vol. 96, no. 2, pp. 287-305 (2008).

[3] K. Shinohara, D. C. Regan, Y. Tang, A. L. Corrion, D. F. Brown, J. C. Wong, J. F. Robinson, H. H. Fung, A. Schmitz, T. C. Oh, S. J. Kim, P. S. Chen, R. G. Nagele, A. D. Margomenos, M. Micovic, "Scaling of GaN HEMTs and Schottky Diodes for Submillimeter-Wave MMIC Applications", IEEE Transactions on Electron Devices, vol. 60, no. 10, pp. 2982-2996(2013).

[4] T. Palacios, A. Chakraborty, S. Rajan, C. Poblenz, S. Keller, S. P. DenBaars, J. S. Speck, and U. K. Mishra, "High-Power AlGaN/GaN HEMTs for Ka-Band Applications", IEEE Electron Device Letters, vol. 26, no. 11, pp. 781-783(2005)

[5] U. Peralagu, A. Alian, V. Putcha, A. Khaled, R. Rodriguez, A. Sibaja-Hernandez, S. Chang, E. Simoen, S. E. Zhao, B. De Jaeger, D. M. Fleetwood, P. Wambacq, M. Zhao, B. Parvais, N. Waldron, and N. Collaert, "CMOS-compatible GaN-based devices on 200 mm-Si for RF applications: Integration and performance", IEDM Tech. Dig., pp. 17.2.1-17.2.4,(2019).

[6] Z. Liu, M. Sun, H.-S. Lee, M. Heuken and T. Palacios, "AlGaN/AlN/GaN figh-electron-mobility transistors fabricated with Au-free technology", Appl. Phys. Exp., vol. 6, no. 9(2013).

[7] H. Lu, B. Hou, M. Zhang, L. Yang, M. Wu, L. Deng, L. Zhou, Z. Si, Q. Zhu, X. Ma, and Y. Hao, "High Performance CMOS-Compatible RF GaN-on-Silicon HEMTs With Low-Resistive and Highly-Conformal Ohmic Contacts", IEEE Transactions on Electron Devices, 71, no. 9, pp. 5218-5224(2024).

[8] H.-S. Lee, D. S. Lee, and T. Palacios, "AlGaN/GaN high electron mobility transistors fabricated through an Au-free technology", IEEE Electron Device Lett., vol. 32, no. 5, pp. 623–625(2011).

979-8-3315-0417-5/25 $31.00 © 2025 IEEE

Electrical and Thermal Characterization of Hetero-integrated β-Ga₂O₃-on-Diamond SBDs by Transfer Printing Technology

Zhenyu Qu[1,#], Tiancheng Zhao[1,#], Wenhui Xu[1,#,*], Haodong Jiang[1,2], Yeliang Wang[1], Xinbo Zou[2], Min Zhou[1], Tiangui You[1] and Xin Ou[1*]

[1] State Key Laboratory of Materials for Integrated Circuits, Shanghai Institute of Microsystem and Information Technology, Chinese Academy of Sciences, Shanghai 200050, China

[2] School of Information Science and Technology, ShanghaiTech University, Shanghai 201210, China

*E-mail: xuwh@mail.sim.ac.cn, ouxin@mail.sim.ac.cn; #Authors contributed equally to this work.

Abstract

Hetero-integrated β-Ga₂O₃-on-diamond (GaOD) SBDs were fabricated using transfer printing technology. The GaOD SBDs showed a more uniform Schottky barrier than the β-Ga₂O₃ Bulk (GaO Bulk) SBDs. Pulse I-V characteristics revealed a significantly reduced self-heating effect in GaOD SBDs compared to GaO Bulk SBDs. The device thermal resistance was as low as 16.69 K/W, only 1/11 of that of GaO Bulk SBDs. This work offers a promising solution for efficient heat dissipation in high-power devices integrated with diamond substrates.

Keywords: β-Ga₂O₃, diamond, transfer printing

Introduction

As one of the most promising ultrawide bandgap semiconductor, β-Ga₂O₃ exhibits an exceptionally high Baliga's figure of merit, making it particularly suitable for high-power applications [1]-[4]. However, its intrinsic low thermal conductivity hinders its application in high-power conditions [5],[6]. Heterogeneous integration of β-Ga₂O₃ on a diamond substrate with extremely high thermal conductivity can break the thermal dissipation bottleneck of β-Ga₂O₃ high-power devices. However, the large lattice mismatch between diamond and β-Ga₂O₃, along with the poor surface morphology of diamond, has thus far impeded the heterogeneous integration of β-Ga₂O₃ with diamond.

In this work, GaOD heterogeneous material was fabricated by transfer printing technique. The SBDs were fabricated to investigate its electrical and thermal properties. The heat dissipation capability of the device was significantly improved and a low device thermal resistance of 16.69 K/W was achieved by integrating of β-Ga₂O₃ with diamond substrate.

Materials and Device Fabrication

The fabrication process of the GaOD SBDs is illustrated in Fig. 1. A β-Ga₂O₃(200nm)/SiO₂(500nm)/Si(500μm) composite wafer was firstly fabricated by ion-cutting technique [7],[8]. Afterwards, the β-Ga₂O₃ thin film was etched into 200*200 μm square film arrays using ICP reactive ion etching (ICP-RIE) and released from the Si substrate after etching the SiO₂ layer with 40% HF. Then, the polyvinyl alcohol (PVA) film was used to transfer the β-Ga₂O₃ thin film from the Si substrate to the diamond substrate. A high-doping region was defined as the cathode through silicon ion implantation at 10 keV with a dose of 1×10^{15} cm⁻², followed by activation at 950°C for 2 min. Ti/Au (20/160 nm) cathode was deposited using electron beam evaporation (EBE) and annealed at 475°C for 2 min to form ohmic contacts. Ni/Au (20/160 nm) Schottky contact was also deposited by EBE.

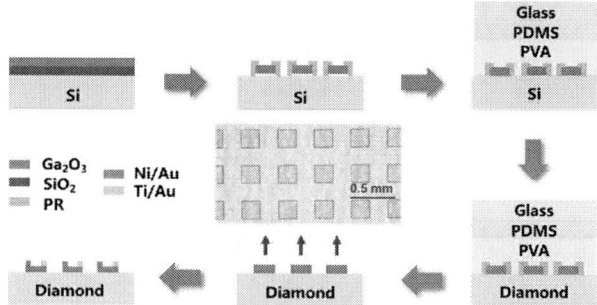

Fig. 1. The process flow of the GaOD SBDs fabricated by transfer printing technique.

Results and Discussion

Fig. 2(a) displays a cross-sectional scanning transmission electron microscopy (STEM) image of the GaOD sample. An amorphous layer with a thickness of ~ 2 nm is observed at the interface between the β-Ga₂O₃ and diamond. Fig. 2(b) and 2(c) display the STEM images of the β-Ga₂O₃ and diamond regions, respectively. The orderly and regular atomic arrangement indicates that both β-Ga₂O₃ thin film and diamond substrate maintain well single-crystal quality after the transfer printing process.

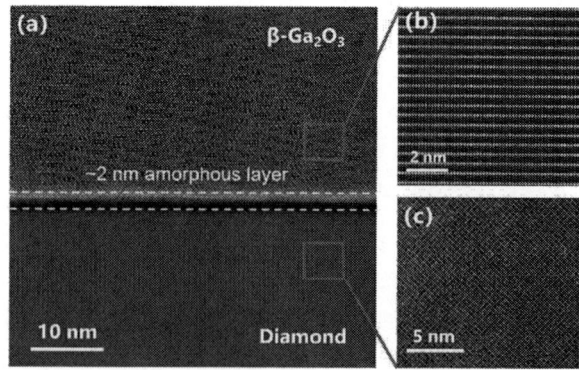

Fig. 2. (a) The cross-sectional STEM HAADF image of the GaOD sample, the STEM images of the (b) β-Ga₂O₃ and (c) diamond regions at high magnification.

Fig. 3(a) displays the temperature-dependent J-V

characteristics of the GaOD SBD from 300K to 375K. The extracted temperature dependent performance of I_{ON}/I_{OFF} ratio (±2V) and the on-state resistance (R_{on}) are shown in Fig. 3(b). At 300 K, the device exhibits an I_{ON}/I_{OFF} ratio over 10^8. As the temperature increases to 375 K, the I_{ON}/I_{OFF} ratio decreases to 10^7, which is primarily due to the leakage current rise. The R_{on} can be extracted from the slope of linear region of the I-V characteristics in Fig. 3(a), which is observed to decreases from 0.42 to 0.36 Ω·cm² as the temperature rises from 300 to 375 K. As the temperature rises, more impurities in β-Ga₂O₃ thin film are ionized, which increases the carrier concentration and consequently leads to the decrease of R_{on}.

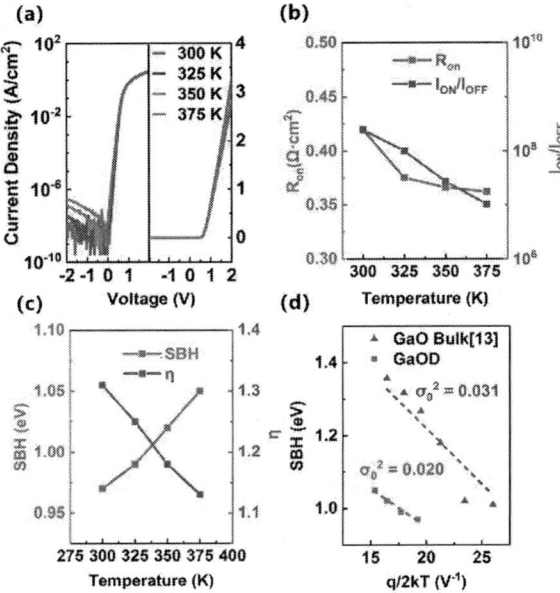

Fig. 3. The temperature dependent (a) J–V characteristics, (b) R_{on} and I_{ON}/I_{OFF} ratio, (c) SBH and η of the GaOD SBD. (d) SBH against q/2kT and the linear fits for the GaO Bulk SBD and GaOD SBD according to the Gaussian distribution of the barrier heights.

According to the thermionic emission model, the ideality factor (η) and Schottky barrier height (SBH) can be extracted by equation (1) and (2) [9], [10], respectively:

$$\eta = \frac{q}{kT}\left(\frac{dV}{d\left(ln\left(J\right)\right)}\right) \quad (1)$$

$$\varphi_b = \frac{kT}{q}ln\left(\frac{A^* T^2}{J_0}\right) \quad (2)$$

where φ_b is the SBH, J_0 is the saturation current density, k is the Boltzmann constant, T is the ambient temperature, and A* is the Richardson constant, which is 41.1 A·cm⁻²·K⁻² for β-Ga₂O₃ [11].

The temperature-dependent extracted ideality factor and SBH are depicted in Fig 3(c). As the temperature increases, the η of the device decreases, while the SBH increases. This is a typical characteristic of barrier inhomogeneities at the Schottky contact [12],[13]. Based on the Gaussian distribution model, the Schottky barrier can be expressed as equation (3) [13]:

$$\varphi_b = \varphi_{b0} - \frac{q\sigma_0^2}{2kT} \quad (3)$$

where φ_{b0} represents the mean SBH and σ_0 represents the standard deviation, which is used to characterize the dispersion of the barrier distribution. Extracted from the linear fitting of the SBH-q/2kT curve, as shown in Fig 3(d), the φ_{b0} and σ_0^2 is 1.37 eV and 0.020, respectively. The σ_0^2 is lower than the previously reported value of 0.031 for GaO Bulk SBD with Ni/Au anodes [14], which indicates a more uniform SBH in the GaOD SBD. The discrepancy in the average barrier height may be attributed to the differences in the deposition methods of β-Ga₂O₃ and Ni/Au anodes.

To demonstrate the mitigation of the self-heating effect of β-Ga₂O₃ SBD by integration of diamond substrate, the J-V characteristics of GaOD SBD were measured under direct current (DC) and pulse conditions, as shown in Fig.4. It is noteworthy that the current density of GaOD SBDs under a power density of 1800 W/cm² exhibits negligible variation across a wide range of duty cycles, including DC conditions. In contrast, GaO Bulk SBD devices exhibited a significant decrease in current density under the same DC condition compared to pulse conditions. This indicates the effective alleviation of self-heating in the GaOD SBD, demonstrating its excellent heat dissipation performance.

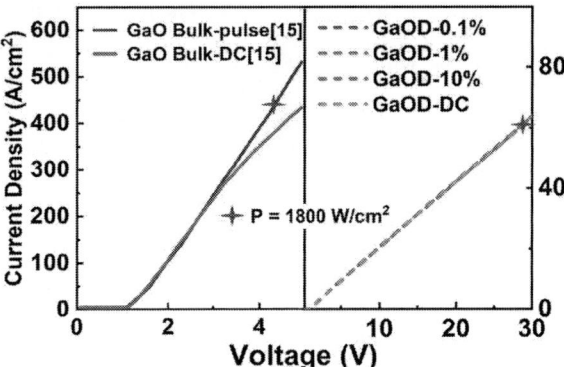

Fig. 4. The J-V characteristics of GaO Bulk SBD and GaOD SBD under DC and pulse conditions at different duty cycles.

To visually temperature distribution of GaOD SBD, QFI infrared photography was employed to map the temperature rising under various power densities. As depicted in Fig. 4(a), the surface temperature distributions of the GaOD SBD are visualized under the DC power of 0.17, 0.59, 0.88, 1.24 W. The highest temperature rising within the channel is measured to be 23.1 K under the power of 1.24 W. In addition, a slightly higher temperature was observed in the anode region compared to the cathode region, which is due to the electric field concentration at the anode edge. The abnormally high temperature on both anode and cathode is due to the contact resistance between the probe and the electrode.

The device thermal resistance of the GaOD SBD, determined by linearly fitting the peak temperature rising with applied

979-8-3315-0417-5/25 $31.00 © 2025 IEEE

power, was measured to be 16.69 K/W. This value is substantially lower than those of GaOSiC SBDs and GaO Bulk SBDs, being approximately 1/3 and 1/11, respectively [16]. This highlights the beneficial impact of incorporating diamond on the heat dissipation capabilities of β-Ga₂O₃ devices.

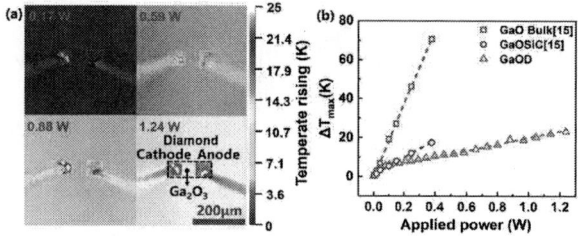

Fig. 5. (a) The surface temperature rising of GaOD SBD under different applied DC power measured by infrared photography. (b) The peak temperature rising of GaO Bulk SBD, GaOSiC SBD and GaOD SBD under the different applied power.

Conclusion

In conclusion, this work demonstrated the successful fabrication of GaOD SBDs using transfer printing technology. The electrical characteristics exhibited a more uniform Schottky contact than GaO Bulk SBDs. Moreover, the integration of β-Ga₂O₃ thin film with diamond resulted in a significantly reduced device R_{th} of 16.69 K/W. This promising approach offers an effective thermal management strategy for high power devices.

Acknowledgments

The work was supported by Science and Technology Commission of Shanghai Municipality (NO.24DP1500300), National Natural Science Foundation of China (62293520, 62293521, 6240033389).

References

[1]   M. Higashiwaki and G. H. Jessen, "Guest editorial: The dawn of gallium oxide microelectronics", Appl. Phys. Lett., 112, pp. 060401 (2018).

[2]   M. Higashiwaki, K. Sasaki, H. Murakami, Y. Kumagai, A. Koukitu, A. Kuramata, T. Masui and S. Yamakoshi, "Recent progress in Ga₂O₃ power devices", Semicond. Sci. Technol., 31, pp. 034001 (2016).

[3]   S. J. Pearton, J. Yang, P. H. Cary, F. Ren, J. Kim, M. J. Tadjer, M. A. and M. A. Mastro, "A review of Ga₂O₃ materials, processing, and devices", Appl. Phys. Rev., 5, pp. 011301 (2018).

[4]   J. Y. Tsao, S. Chowdhury, M. A. Hollis, D. Jena, N. M. Johnson, K. A. Jones, R. J. Kaplar, S. Rajan, C. G. Van de Walle, E. Bellotti, C. L. Chua, R. Collazo, M. E. Coltrin, J. A. Cooper, K. R. Evans, S. Graham, T. A. Grotjohn, E. R. Heller, M. Higashiwaki, M. S. Islam, P. W. Juodawlkis, M. A. Khan, A. D. Koehler, J. H. Leach, U. K. Mishra, R. J. Nemanich, R. C. N. Pilawa-Podgurski, J. B. Shealy, Z. Sitar, M. J. Tadjer, A. F. Witulski, M. Wraback and J. A. Simmons, "Ultrawide -

bandgap semiconductors: research opportunities and challenges", Adv. Electron. Mater., 4, pp. 1600501 (2018).

[5]   P. Jiang, X. Qian, X. Li and R. Yang, "Three-dimensional anisotropic thermal conductivity tensor of single crystalline β-Ga₂O₃", Appl. Phys. Lett., 113, pp.232105 (2018).

[6]   M. Slomski, N. Blumenschein, P. P. Paskov, J. F. Muth and T. Paskova, "Anisotropic thermal conductivity of β-Ga₂O₃ at elevated temperatures: Effect of Sn and Fe dopants", J. Appl. Phys., 121, pp. 235104 (2017).

[7]   W. Xu, Y, Wang, T. You, X. Ou, G. Han, H. Hu, S. Zhang, F. Mu, T. Suga, Y. Zhang, Y. Hao and X. Wang, "First demonstration of waferscale heterogeneous integration of Ga₂O₃ MOSFETs on SiC and Si substrates by ion-cutting process", IEEE International Electron Devices Meeting (IEDM), pp. 12-5 (2019).

[8]   Z. Shen, W. Xu, Y Chen, J. Lin, Y. Xie, K. Huang, T. You, G. Han and X. Ou, "Wafer-scale single-crystalline β-Ga₂O₃ thin film on SiC substrate by ion-cutting technique with hydrophilic wafer bonding at elevated temperatures", Sci. China Mater., 66, pp.756–763 (2023).

[9]   H. Fu, X. Huang, H. Chen, Z. Lu, I. Baranowski and Y. Zhao, "Ultra-low turn-on voltage and on-resistance vertical GaN-on-GaN Schottky power diodes with high mobility double drift layers", Appl. Phys. Lett., 111, pp. 152102 (2017).

[10]   M. K. Yadav, A. Mondal, S. K. Sharma and A. Bag, "Substrate orientation dependent current transport mechanisms in β-Ga₂O₃/Si based Schottky barrier diodes", J. Vac. Sci. Technol. A, 39, pp. 033203 (2021).

[11]   K Sasaki, M Higashiwaki, A Kuramata, T Masui and S Yamakoshi, "Ga₂O₃ Schottky Barrier Diodes Fabricated by Using Single-Crystal β-Ga₂O₃ (010) Substrates", IEEE Electron Device Lett., 34, pp. 493-495 (2013).

[12]   W. Xu, Z. Shen, Z. Qu, T. Zhao, A. Yi, T. You, G. Han and X. Ou, "Current transport mechanism of lateral Schottky barrier diodes on β-Ga₂O₃/SiC structure with atomic level interface", Appl. Phys. Lett., 124, pp. 112102 (2024).

[13]   J. H. Werner and H. H. Güttler, "Barrier inhomogeneities at Schottky contacts", J. Appl. Phys., 69, pp. 1522–1533 (1991).

[14]   H Sheoran, BR Tak, N Manikanthababu and R Singh, "Temperature-dependent electrical characteristics of Ni/Au vertical Schottky barrier diodes on β-Ga₂O₃ epilayers", ECS J. Solid State Sci. Technol., 9, pp. 055004 (2020).

[15]   W. Li, Z. Hu, K. Nomoto, Z. Zhang, J. Hsu, Q. T. Thieu, K. Sasaki, A. Kuramata, D. Jena and H. G. Xing, "1230 V β-Ga₂O₃ trench Schottky barrier diodes with an ultra-low leakage current of < 1 μA/cm²", Appl. Phys. Lett., 113, pp. 202101 (2018).

[16]   W. Xu, T. You, Y. Wang, Z. Shen, K. Liu, L. Zhang, H. Sun, R. Qian, Z. An, F. Mu, T. Suga, G. Han, X. Ou, Y. Hao and X. Wang, "Efficient thermal dissipation in wafer-scale heterogeneous integration of single-crystalline β-Ga₂O₃ thin film on SiC", Fundamental Res., 1, pp. 691-696 (2021).

Electrical Transport at n-Ga₂O₃/n-SiC Hetero-interface Constructed by Hydrophilic and Surface Activated Bonding

Zhenyu Qu[1,#], Wenhui Xu[1,#,*], Haodong Jiang[1,2], Tiancheng Zhao[1], Yeliang Wang[1], Haowen Guo[2], Xinbo Zou[2], Min Zhou[1], Tiangui You[1] and Xin Ou[1]*

[1]State Key Laboratory of Materials for Integrated Circuits, Shanghai Institute of Microsystem and Information Technology, Chinese Academy of Sciences, Shanghai 200050, China

[2]School of Information Science and Technology, ShanghaiTech University, Shanghai 201210, China

*E-mail: xuwh@mail.sim.ac.cn, ouxin@mail.sim.ac.cn; #Authors contributed equally to this work.

Abstract

The n-Ga₂O₃/n-SiC (GaOSiC) hetero-structure was achieved by ion-cutting process based on surface activated bonding (SAB) and hydrophilic bonding (HB) techniques. The HB GaOSiC hetero-interface showed an amorphous layer of only ~1 nm, contributing to 16.87% of the total on-resistance ($R_{on,total}$) for the GaOSiC structure, which is much less than that of the SAB hetero-interface. The low interface resistance (R_i) of HB GaOSiC structure provides a feasible path towards the vertical β-Ga₂O₃ power electronics with high power operating capability and acceptable heat dissipation performance.

Keywords: β-Ga₂O₃, surface activated bonding, hydrophilic bonding

Introduction

β-Ga₂O₃ has garnered significant attention due to its exceptionally high Baliga figure of merit [1]-[3]. In particular, vertical β-Ga₂O₃ devices, capable of handling high currents, have shown promising potential for high-power applications [4],[5]. However, the poor thermal conductivity of β-Ga₂O₃ presents a significant heat dissipation challenge, especially for vertical devices operating at high power densities. Hetero-integration of β-Ga₂O₃ onto SiC substrates through wafer bonding has been demonstrated as an effective approach to address thermal management issue in β-Ga₂O₃ lateral and vertical devices [6]-[8]. However, the effects of amorphous layer induced by wafer bonding on vertical current transport across the hetero-interface remain unclear.

In this work, GaOSiC hetero-structures were demonstrated by ion-cutting technique based on SAB and HB. The current-voltage (J-V) characteristics of SAB and HB GaOSiC structures were compared. Moreover, a detailed analysis of the $R_{on,total}$ of these structures was conducted to investigate the impact of the hetero-interface on current transport. The HB GaOSiC structure showed a low R_i of 16.87%, demonstrating the advantages of HB over SAB in the fabrication of high-power vertical β-Ga₂O₃ power devices.

Materials and Device Fabrication

The 200 nm/350 μm n-Ga₂O₃/n-SiC hetero-structures were fabricated using an ion-cutting technique based on SAB and HB [9],[10]. The transmission electron microscopy (TEM) micrographs of GaOSiC interface are shown in Fig. 1(a) and 1(b), which indicate the single crystal quality of β-Ga₂O₃ and

4H-SiC. The thicknesses of the amorphous layers at the hetero-interface are ~1 nm and ~4 nm in the HB and SAB GaOSiC structures, respectively. The thick amorphous layer at the SAB interface is primarily caused by the high energy Ar ion bombardment during the SAB process [11].

The β-Ga₂O₃ and SiC ohmic contacts were fabricated as illustrated in Fig. 1(c) to characterize the electrical properties across n-GaO/n-SiC hetero-interface. Firstly, the composite wafers were subjected to O₂ annealing at 900°C for 1 h to eliminate the defects from the β-Ga₂O₃ thin film. Afterwards, mesa isolation was achieved through ICP-RIE with a forming gas of BCl₃/Cl₂/Ar. Si ions with an energy of 10 kV and a dose of 1×10^{15} cm⁻² were then implanted into β-Ga₂O₃ thin film to define the n⁺-Ga₂O₃ layer. Rapid thermal annealing at 1100°C for 3 min was used to activate the implanted Si ions. Before depositing the ohmic metal stacks of β-Ga₂O₃ thin film, Ni (100 nm) layers were deposited on the SiC surface by electron beam evaporation (EBE) and annealed at 1050°C for 2 min to avoid the influence of high temperature on the β-Ga₂O₃ ohmic contacts. It is noteworthy that the whole structures were immersed in 4% HF for 5 min to remove the oxide layers on the SiC surface before the deposition of Ni electrodes. Finally, Ti/Au (20/160 nm) metal stacks were deposited using EBE followed by a 2 min annealing at 475 °C to improve the ohmic contact.

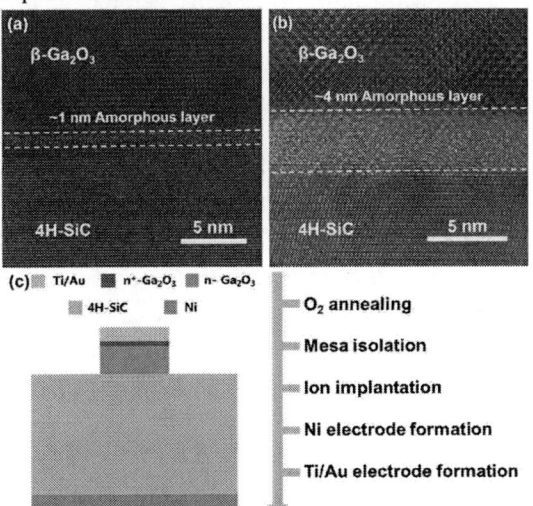

Fig. 1: The cross-sectional TEM micrographs of the GaOSiC interface constructed by (a) HB and (b) SAB technique, and (c) the device structure and process flow of GaOSiC structures.

Results and Discussion

Fig. 2(a) displays the J-V characteristics of both SAB and HB GaOSiC structures. It is observed that the current density of HB GaOSiC structure is 1.5 orders of magnitude higher than that of SAB GaOSiC structure at 8 V, which demonstrates the overwhelming advantage of HB GaOSiC structure over SAB GaOSiC structure in high-power vertical devices. The specific on-resistances ($R_{on,sp}$) of the HB and SAB GaOSiC structures are extracted from the slope of the linear fitting of J-V characteristics, which are 0.93 and 12 m$\Omega\cdot$cm^2, respectively. According to equation (1), the net carrier concentrations of β-Ga$_2$O$_3$ thin films can be extracted from the slope of $1/C^{-2}$-V characteristics.

$$N_d - N_a = \frac{-2}{q\varepsilon_s\varepsilon_0 A^2}\left[\frac{dV}{d\left(1/C^2\right)}\right] \quad (1)$$

where q is the electron charge, $\varepsilon_s\varepsilon_0$ is the dielectric constant, and A is the Schottky contact area. The measured $1/C^{-2}$-V characteristics are shown in Fig. 2(b) and the measured SBD structure is illustrated in the inset. The net carrier concentrations of β-Ga$_2$O$_3$ are extracted as 1.83×10^{18} and 1.77×10^{18} cm^{-3} for the HB and SAB GaOSiC structures, respectively. It reveals the similar electrical properties in β-Ga$_2$O$_3$ thin films fabricated by ion-cutting technique based on SAB and HB. Considering the same specification of commercial SiC substrates, the dispersion of the current density is originated from the different amorphous layer thicknesses at the GaOSiC hetero-interface.

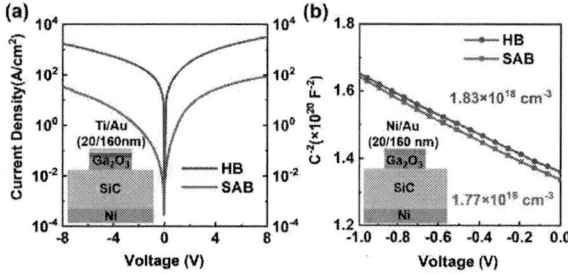

Fig. 2: (a) J-V and (b) C^{-2}-V characteristics of HB and SAB GaOSiC structures.

To quantitatively characterize the influence of amorphous layer on the $R_{on,total}$, $R_{on,total}$ is divided into five parts and the proportion of each part is analyzed, as expressed in equation (2):

$$R_{on,total} = R_{c,GaO} + R_{GaO} + R_i + R_{SiC} + R_{c,SiC} \quad (2)$$

where $R_{c,GaO}$ and $R_{c,SiC}$ are the ohmic contact resistances of the β-Ga$_2$O$_3$ and SiC while R_{GaO} and R_{SiC} are the series resistances of β-Ga$_2$O$_3$ and SiC, respectively. $R_{c,GaO}$ and R_{GaO} are measured by the transfer length model (TLM), as shown in Fig. 3(a). The sheet resistance (R_{sh}) and specific contact resistance (ρ_c) of β-Ga$_2$O$_3$ can be extracted as equation (3):

$$R = 2\frac{R_{sh}}{W}L_T + \frac{R_{sh}}{W}d \quad (3)$$

where W is the width of the electrode, d is the length of space between two electrodes and L_T is the transmission length

equal to $\sqrt{\rho_c/R_{sh}}$. The extracted $\rho_{c,GaO}$ is 1.29×10^{-6} $\Omega\cdot$cm^2. On account of the 100 μm radius of Ti/Au electrodes, $R_{c,GaO}$ is calculated to be 4.106×10^{-3} Ω, which is 0.14% and 0.01% of $R_{on,total}$ for HB and SAB GaOSiC structures. The $R_{sh,GaO}$ is extracted as 3487 Ω and R_{GaO} is 4.44×10^{-3} Ω in view of the 200 nm β-Ga$_2$O$_3$ thin film. The electron mobility in the β-Ga$_2$O$_3$ thin film is calculated to be 51 cm^2/V·s from equation (4), which is comparable to that of the previously reported ion-cutting β-Ga$_2$O$_3$ thin film [12].

$$\rho_{GaO} = 1/nq\mu_n = R_{sh,GaO}\times t_{GaO} \quad (4)$$

where ρ_{GaO} is the resistivity of β-Ga$_2$O$_3$, n is the electron concentration, μ_n is the electron mobility and t_{GaO} is the thickness of β-Ga$_2$O$_3$.

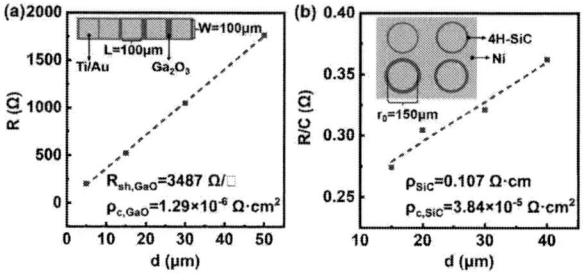

Fig. 3: (a) The R-d curves of TLM structure on the β-Ga$_2$O$_3$ thin film and (b) the R/C-d curves of CTLM structure on the SiC substrate.

Circular transmission line model (CTLM) is used to measure R_{SiC} and $\rho_{c,SiC}$, as shown in Fig. 3(b). The extracted ρ_{SiC} and $\rho_{c,SiC}$ are 0.107 $\Omega\cdot$cm and 3.84×10^{-5} $\Omega\cdot$cm^2 from equation (5) and (6) [13]:

$$R = \frac{R_{sh}}{2\pi r_0}\left(d + 2L_T\right)C \quad (5)$$

$$C = \frac{r_0}{d}\ln\left(1+\frac{d}{r_0}\right) \quad (6)$$

where r_0 is the radius of the circular electrode and C is the correction factor. Taking the lateral current spreading effect into account, R_{SiC} is calculated to be 2.438 Ω using equation (7) [14]:

$$R_{SiC} = \frac{\rho_{SiC}\tan^{-1}\left(2t_{SiC}/r\right)}{2\pi r} \quad (7)$$

where ρ_{SiC} is the resistivity of SiC, t_{SiC} is the thickness of the SiC layer and r is the radius of contact area on the top surface of SiC, which is 100 μm.

To obtain the effective contact area of the SiC ohmic electrode, the electrical characteristics of GaOSiC structures with different radius of Ni electrode are simulated by Sentaurus TCAD. As shown in Fig. 4(a), the $R_{on,total}$ is observed to be independent of Ni electrode radius when the radius is larger than 550 μm, which indicates that the length of lateral current spreading is about 450 μm. The current potential drop is concentrated in the area of 550 μm from the center, as depicted in Fig. 4(b), which proves the extracted effective contact area. Combined with $\rho_{c,SiC}$ extracted from the CTLM structure, $R_{c,SiC}$ is calculated to be 4.041×10^{-3} Ω.

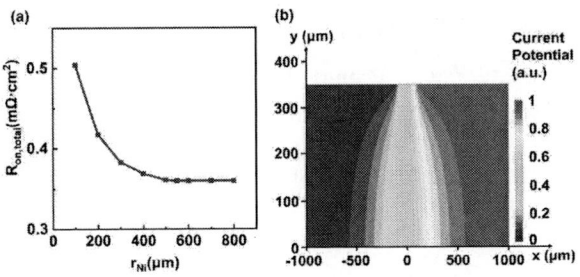

Fig. 4: (a) $R_{on,total}$ of GaOSiC structures with various radius of Ni electrodes, and (b) the current potential mapping of GaOSiC structure.

The resistance components of $R_{on,total}$ are concluded in Table 1. In the HB GaOSiC structure, the majority of $R_{on,total}$ comes from R_{SiC}, accounting for 82.7%. However, with the increase of amorphous layer thickness, R_i increases sharply in SAB GaOSiC structure and contributes the majority of $R_{on,total}$, accounting for 93.8%. The gap in R_i between HB GaOSiC and SAB GaOSiC structures indicates that HB is more suitable for the fabrication of vertical β-Ga$_2$O$_3$ power devices.

Table 1: The values and ratios of the resistance components in HB and SAB GaOSiC structures.

	Value$_{HB}$ (Ω)	Ratio$_{HB}$ (%)	Value$_{SAB}$ (Ω)	Ratio$_{SAB}$ (%)
$R_{c,GaO}$	4.106×10^{-3}	0.14	4.106×10^{-3}	0.01
R_{GaO}	4.440×10^{-3}	0.15	4.440×10^{-3}	0.01
R_i	0.497	16.87	37.045	93.80
R_{SiC}	2.438	82.70	2.438	6.17
$R_{c,SiC}$	4.041×10^{-3}	0.14	4.041×10^{-3}	0.01
$R_{on,total}$	2.948	100	39.496	100

Conclusion

In conclusion, the GaOSiC hetero-structure was fabricated by ion-cutting technique based on SAB and HB. A thin amorphous layer of ~1 nm was achieved at the HB GaOSiC interface. Furthermore, R_i contributed only 16.87% to $R_{on,total}$ of HB GaOSiC structure, significantly lower than that of the SAB GaOSiC structure with a ~4 nm amorphous layer. This work demonstrates that GaOSiC structure obtained by HB, combined with homoepitaxial growth on it as template for thick drift layer, is more suitable for the application of β-Ga$_2$O$_3$ vertical power devices with effective heat dissipation.

Acknowledgments

The work was supported by Science and Technology Commission of Shanghai Municipality (NO.24DP1500100), National Natural Science Foundation of China (62293520, 62293521, 6240033389).

References

[1] M. Higashiwaki and G. H. Jessen, "Guest editorial: The dawn of gallium oxide microelectronics", Appl. Phys. Lett., 112, pp. 060401 (2018).

[2] M. Higashiwaki, K. Sasaki, H. Murakami, Y. Kumagai, A. Koukitu, A. Kuramata, T. Masui and S. Yamakoshi, "Recent progress in Ga$_2$O$_3$ power devices", Semicond. Sci. Technol., 31, pp. 034001 (2016).

[3] S. J. Pearton, J. Yang, P. H. Cary, F. Ren, J. Kim, M. J. Tadjer, M. A. and M. A. Mastro, "A review of Ga$_2$O$_3$ materials, processing, and devices", Appl. Phys. Rev., 5, pp. 011301 (2018).

[4] M. H. Wong and Masataka Higashiwaki, "Vertical β-Ga$_2$O$_3$ power transistors: A review", IEEE Trans. Electron Devices, 67, pp. 3925-3937 (2020).

[5] P. Dong; J. Zhang; Q. Yan; Z. Liu; P. Ma; H. Zhou and Y. Hao, "6 kV/3.4 mΩ·cm^2 vertical β-Ga$_2$O$_3$ schottky barrier diode with BV^2/R$_{on,sp}$ performance exceeding 1-D unipolar limit of GaN and SiC", IEEE Electron Device Lett., 43, pp. 765-768 (2022).

[6] W. Xu, T. You, Y. Wang, Z. Shen, K. Liu, L. Zhang, H. Sun, R. Qian, Z. An, F. Mu, T. Suga, G. Han, X. Ou, Y. Hao and X. Wang, "Efficient thermal dissipation in wafer-scale heterogeneous integration of single-crystalline β-Ga$_2$O$_3$ thin film on SiC", Fundamental Research, 1, pp. 691-696 (2021).

[7] Z. Qu, Y. Xie, T. Zhao, W. Xu, Y. He, Y. Xu, H. Sun, T. You, G. Han, Y. Hao and X. Ou, "Extremely low thermal resistance of β-Ga$_2$O$_3$ MOSFETs by co-integrated design of substrate engineering and device packaging", ACS Appl. Mater. Interfaces, 16, pp. 57816–57823 (2024).

[8] C. Lin, N. Hatta, K. Konishi, S. Watanabe, A. Kuramata, K. Yagi and M. Higashiwaki, "Single-crystal-Ga2O3/polycrystalline-SiC bonded substrate with low thermal and electrical resistances at the heterointerface", Appl. Phys. Lett., 114, pp. 032103 (2019).

[9] W. Xu, Y, Wang, T. You, X. Ou, G. Han, H. Hu, S. Zhang, F. Mu, T. Suga, Y. Zhang, Y. Hao and X. Wang, "First demonstration of waferscale heterogeneous integration of Ga$_2$O$_3$ MOSFETs on SiC and Si substrates by ion-cutting process", IEEE International Electron Devices Meeting (IEDM), pp. 12-5 (2019).

[10] Z. Shen, W. Xu, Y Chen, J. Lin, Y. Xie, K. Huang, T. You, G. Han and X. Ou, "Wafer-scale single-crystalline β-Ga$_2$O$_3$ thin film on SiC substrate by ion-cutting technique with hydrophilic wafer bonding at elevated temperatures", Sci. China Mater., 66, pp.756–763 (2023).

[11] Z. Cheng, F. Mu, T. You, W. Xu, J. Shi, M. E. Liao, Y. Wang, K. Huynh, T. Suga, M. S. Goorsky, X. Ou and S. Graham, "Thermal transport across ion-cut monocrystalline β-Ga$_2$O$_3$ thin films and bonded β-Ga$_2$O$_3$-SiC interfaces", ACS Appl Mater Interfaces, 12, pp. 44943–44951 (2020).

[12] Y. Wang, W. Xu, G. Han, T. You, F. Mu, H. Hu, Y. Liu, X. Zhang, H. Huang, T. Suga, X. Ou, X. Ma and Y. Hao, "Channel properties of Ga$_2$O$_3$-on-SiC MOSFETs", IEEE Trans. Electron Devices, 68, pp. 1185-1189 (2021).

[13] S. M. Sze, Y. Li and K. K. Ng, "Physics of semiconductor devices", John wiley & sons, ISBN: 9783319030012 (2021).

[14] R. H. Cox and H. Strack, "Ohmic contacts for GaAs devices", Solid-State Electron., 10, pp. 1213-1218 (1967)

Experimental Investigation of Inserted HfO$_2$ Impact on V$_{FB}$ and Interface via ALD for La$_2$O$_3$ Dipole-first Multi-V$_T$ Techniques

Yanzhao Wei[1,2,3], Jiaxin Yao[1,2,*], Yu Wang[1,2,3], Qingzhu Zhang[1,2], and Huaxiang Yin[1,2,3,*]

[1]Integrated Circuit Advanced Process R&D Center, Institute of Microelectronics of the Chinese Academy of Sciences, Beijing 100029, China

[2]Key Laboratory of Fabrication Technologies for Integrated Circuits, Chinese Academy of Sciences, Beijing 100029, China

[3]School of Integrated Circuits, University of Chinese Academy of Sciences, Beijing 100049, China

*Corresponding Email: yaojiaxin@ime.ac.cn, yinhuaxiang@ime.ac.cn

Abstract

An inserted HfO$_2$ approach is proposed to achieve V$_{FB}$ tuning and interface improvement in La$_2$O$_3$ dipole-first gate stack via ALD. A series of V$_{FB}$ levels were achieved with 61 mV linear-tuning step and 310 mV tuning range by manipulating inserted HfO$_2$ thickness (<5Å) ahead of La$_2$O$_3$/HfO$_2$ stacks in MOSCAPs. Furthermore, the device performance and interface are improved indeed. These results indicate that the proposed approach provides a promising way for advanced NS-GAA multi-V$_T$ techniques.

Keywords: multi-V$_T$, dipole, GAA nanosheet

Introduction

To match high performance and low power consumption requirements in stacked nanosheet (NS) gate-all-around field effect transistors (GAA-FETs), a multiple threshold voltage (multi-V$_T$) technique with linear tunability is in need for state-of-the-art devices [1], [2]. With conventional method, the V$_T$ linear modulation was achieved by tuning work function metal (WFM) thickness in high-k/metal gate (HKMG) stacks [3]. However, this thickness-tuning technique is strictly challenged by the sheet-to-sheet spacing (T$_{sus}$) limitation in NS GAA-FETs **(Fig.1)** [4]. Instead, dipole engineering can modulate V$_T$ with less or even no WFM-thickness tuning requirements. In general, La-dipole is believed to be a n-type dipole technique due to its negative dipole effect for less electronegativity [5]. Compared to the classic dipole-last approach, the dipole-first approach with La-dipole layer depositing under HfO$_2$ HK layer is a promising way for advanced device integration due to its low thermal budget process [6]. While with the dipole-first integration approach, the La-dipole demonstrates a much wide V$_T$ modulation ability, where the linear tuning results is a crucial issue [7]. On the other hand, La$_2$O$_3$ is reported to easily react with SiO$_2$ interfacial layer (IL) during gate stack formation, which may cause interface issue [8].

In this paper, an inserted HfO$_2$ approach is proposed to achieve flat-band voltage (V$_{FB}$) tuning and interface improvement in metal-oxide-semiconductor capacitors (MOSCAPs) with La$_2$O$_3$ dipole-first gate stack. By depositing 1~6 atomic layer deposition (ALD) cycles HfO$_2$ ahead of La$_2$O$_3$/HfO$_2$ stacks, a series of V$_{FB}$S with 61 mV

linear tuning step in 310 mV range was achieved. Furthermore, the device uniformity, gate leakage, trap/detrap electron density (N$_{ot}$), and interface trap density (D$_{it}$) are also improved. These results indicate that the proposed approach provides a promising La$_2$O$_3$ dipole-first V$_T$ modulation ability for the development of advanced multi-V$_T$ techniques in future NS GAA-FETs.

Experiment and Fabrication

Fig.2 demonstrates the key process flow and schematic of the La$_2$O$_3$ dipole-first technique with inserted HfO$_2$. Firstly, an 8-inch p-type Si (100) wafer was used as the substrate. The standard RCA process was used to clean the silicon wafer. After that, a high-density 7-Å SiO$_2$ IL was grown by deionized water and O$_3$ (DI-O$_3$). Then, the multi-layered HKMG structure was fully deposited by ALD process. 1~6 ALD cycles HfO$_2$ were deposited on IL as the insertion layer, following with a 0.5-nm La$_2$O$_3$ dipole layer and a 3-nm HfO$_2$ HK layer. The HfO$_2$/La$_2$O$_3$/HfO$_2$ stack was deposited by *in-situ* process with no vacuum break to keep a high-quality stack. As the reference, the device with only a 3-nm HfO$_2$ HK layer was also fabricated. No drive-in annealing and PDA process was applied during and after HK laminates deposition. Next, the metal gate stack and a 100-nm gap-filling metal of W were consecutively deposited by ALD process. Then, the electrode was patterned for forming the electrical test area. A thin Al film was sputtered at the backside of the wafer to reduce the series resistance (R$_s$) in final electrical test. At last, a metallization process at 400 °C by forming gas annealing (FGA) in environment with 5% H$_2$ and 95% N$_2$ was performed for 30 min to improve interface quality.

Results and Discussions

Fig.3 presents stable high-frequency C-V characteristics of devices with different thickness of inserted HfO$_2$. A series of uniform positive C-V curve shifts are obtained by increasing inserted HfO$_2$ thickness, which demonstrates the weakened La-dipole modulation ability. The extracted V$_{FB}$ values in **Fig.4** exhibit a linear V$_{FB}$ modulation achieved by inserted HfO$_2$ thickness tuning from 1 to 6 ALD cycles with a step of 61 mV per cycle. Moreover, as shown in **Fig.5**, the

979-8-3315-0417-5/25 $31.00 © 2025 IEEE

V_{FB} uniformity is greatly improved by only depositing 1 ALD cycle inserted HfO_2, which has nearly the same uniformity as reference device with no La_2O_3. The reason for the worse V_{FB} uniformity of device without inserted HfO_2 may be the nonuniform interdiffusion of La-Si atoms, especially with no annealing process after HfO_2 HK deposition. The existence of inserted HfO_2 between SiO_2 IL and La_2O_3 restrains the La-Si interdiffusion process, which greatly improves the device performance. **Fig.6** shows the extracted equivalent oxide thickness (EOT) values of devices with different thickness of inserted HfO_2. The linear EOT increase exhibits a high-quality atomic-scale inserted HfO_2 deposition, and a maximum of 1.5 Å EOT increasing is obtained by 5 Å La_2O_3 with 6 ALD cycles inserted HfO_2, which has no obvious EOT penalty. Meanwhile, the TEM images of devices with 1/6 ALD cycles inserted HfO_2 showed in **Fig.7** exhibit good interface quality, and only a ~0.4 nm thickness increasing by 5 ALD cycles inserted HfO_2 is observed, which means that the inserted HfO_2 approach meets the need of space limitation in advanced NS GAA-FETs.

To deeply explore the inserted HfO_2 approach impact on other electrical characteristics, gate leakage current density was then studied. Considering the obvious V_{FB} difference between devices with different thickness of inserted HfO_2, the gate leakage current at the same overdrive bias of V_{FB}-1 V is presented in **Fig.8**. With the same overdrive gate voltage, the leakage current is obviously restrained by inserted HfO_2. Moreover, the inserted HfO_2 approach impacts on defects are also investigated. **Fig.9** shows the dual-sweep C-V curves of devices with 0/1/6 ALD cycles inserted HfO_2. The device with 1 cycle inserted HfO_2 shows a smaller hysteresis than devices with 0/6 cycles inserted HfO_2. The calculated N_{ot} values from V_{FB} hysteresis are demonstrated in **Fig.10** [7]. The N_{ot} decrease with first ALD cycle inserted HfO_2 indicates an improved dielectric stack. With the inserted HfO_2 thickness increasing, the N_{ot} increases due to the oxygen vacancies in additional inserted HfO_2, which is still not higher than the device without inserted HfO_2, indicating an improvement on gate stack bulk. **Fig.11** shows the D_{it} values extracted by conductance method of devices with 0/1/6 ALD cycles inserted HfO_2 [7]. The D_{it} value decreases significantly by depositing Hf-DBL between SiO_2 IL and La_2O_3, demonstrating that interface is greatly improved by inserted HfO_2 approach.

Conclusion

In this paper, an inserted HfO_2 approach is proposed to achieve V_{FB} tuning in MOSCAPs with La_2O_3 dipole-first gate stack. By depositing 1~6 ALD cycles inserted HfO_2 ahead of La_2O_3/HfO_2 stacks, a series of V_{FBS} with 61 mV linear tuning step in 310 mV range was achieved. Furthermore, the device uniformity, gate leakage, N_{ot}, and D_{it} are also improved. These results indicate that the proposed approach provides a promising La_2O_3 dipole-first V_T modulation ability for the development of advanced multi-V_T techniques in future NS GAA-FETs.

Acknowledgments

This work is supported by National Natural Science Foundation of China (No. 62304247), and Strategic Priority Research Program of Chinese Academy of Sciences (No. XDA0330302). The authors would like to thank all staff at the Integrated Circuit Advanced Process Center of IMECAS, for their support in the fabrication on 8-inch CMOS pilot line.

References

[1] J. Lee et al, "Tunable work function dual metal gate technology for bulk and non-bulk CMOS," Digest. International Electron Devices Meeting, San Francisco, CA, USA, 2002, pp. 359-362, doi: 10.1109/IEDM.2002.

[2] H. Jagannathan et al, "Engineering High Dielectric Constant Materials for Band-Edge CMOS Applications," ECS Transactions, vol. 16, no. 5, p. 19, 2008, doi: 10.1149/1.2981584.

[3] M. Kadoshima et al, "Effective-Work-Function Control by Varying the TiN Thickness in Poly-Si/TiN Gate Electrodes for Scaled High-k CMOSFETs," in IEEE Electron Device Letters, vol. 30, no. 5, pp. 466-468, May 2009, doi: 10.1109/LED.2009.2016585.

[4] Q. Zhang et al, "New structure transistors for advanced technology node CMOS ICs, " in National Science Review, Volume 11, Issue 3, March 2024, nwae008, https://doi.org/10.1093/nsr/nwae008.

[5] P. Sivasubramani et al, "Dipole Moment Model Explaining nFET Vt Tuning Utilizing La, Sc, Er, and Sr Doped HfSiON Dielectrics," 2007 IEEE Symposium on VLSI Technology, Kyoto, Japan, 2007, pp. 68-69, doi: 10.1109/VLSIT.2007.4339730.

[6] M. Zhu et al, "Study of Lanthanum Diffusion in HfO2-Based High-k Gate Stack," ECS Transactions, vol. 85, no. 8, p. 131, 2018/04/10 2018, doi: 10.1149/08508.0131ecst.

[7] Y. Wei et al, "Sub-5-Å La2O3 In Situ Dipole Technique for Large VFB Modulation With EOT Reduction and Improved Interface for HKMG Technology," in IEEE Transactions on Electron Devices, vol. 71, no. 1, pp. 746-751, Jan. 2024, doi: 10.1109/TED.2023.3335900.

[8] L. Liu et al, "Advances in La-Based High-k Dielectrics for MOS Applications," Coatings, vol. 9, no. 4, p. 217, Mar. 2019, doi: 10.3390/coatings9040217.

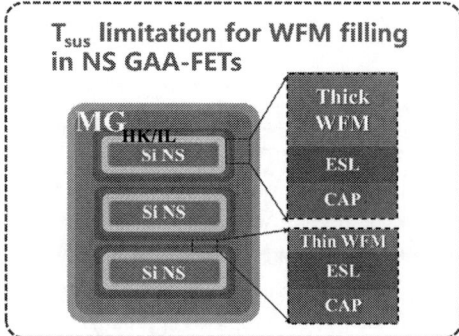

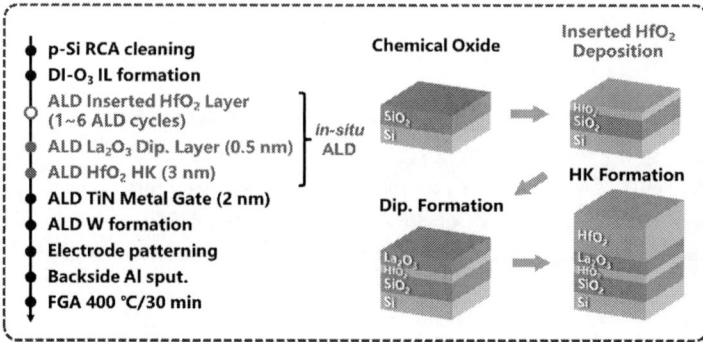

Fig.1 The conventional multi-V_T method with work function metals thickness tuning is strictly challenged by the sheet-to-sheet spacing limitation in advanced NS GAA-FETs.

Fig.2 The key process flow and schematic of the La_2O_3 dipole-first technique with inserted HfO_2 approach. An atomic-scale HfO_2 is deposited after SiO_2 IL formation, following with La_2O_3 dipole layer and HfO_2 HK layer. The $HfO_2/La_2O_3/HfO_2$ stack is deposited by in-situ ALD process with no vacuum break to keep a high-quality stack.

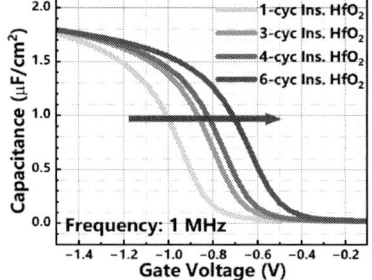

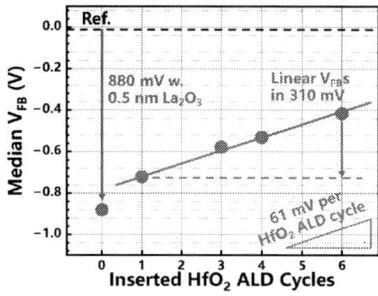

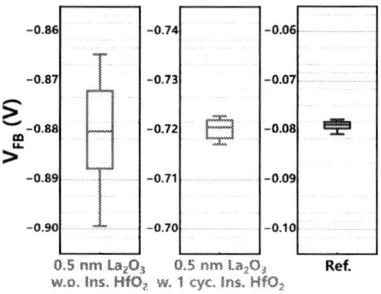

Fig.3 C-V characteristics of devices with 1/3/4/6 ALD cycles inserted HfO_2 in gate stack. A series of uniform positive shift is obtained by increasing inserted HfO_2 thickness.

Fig.4 A linear V_{FB} modulation is achieved by inserted HfO_2 thickness tuning from 1 to 6 ALD cycles with a step of 61 mV per cycle.

Fig.5 By depositing HfO_2 between SiO_2 IL and La_2O_3, the V_{FB} uniformity is greatly improved due to the restraint of La-Si interdiffusion, which is nearly the same as reference device with no La_2O_3.

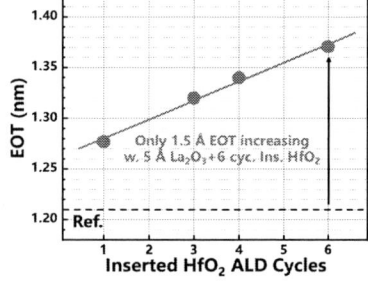

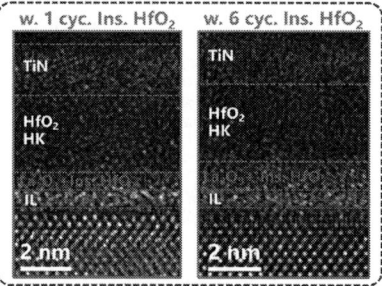

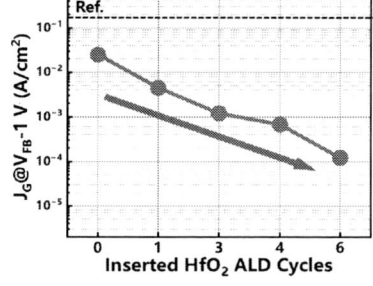

Fig.6 The linear EOT increasing shows a high-quality inserted HfO_2 deposition, and a maximum of 1.5 Å EOT increasing is obtained by 5 Å La_2O_3 with 6 ALD cycles inserted HfO_2.

Fig.7 The TEM images of devices with 1/6 ALD cycles inserted HfO_2 exhibit good interface quality. An only ~0.4 nm thickness increasing by 5 ALD cycles inserted HfO_2 is observed.

Fig.8 The gate leakage at V_{FB}-1V of devices with different inserted HfO_2 thickness. With the same overdrive voltage, the leakage current is obviously restrained by inserted HfO_2.

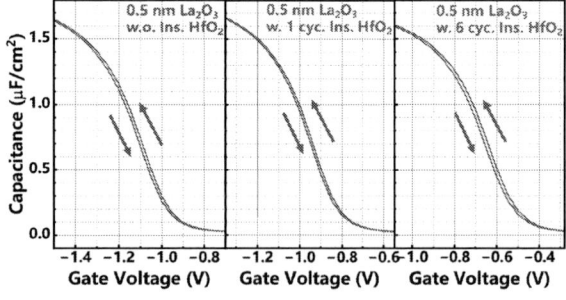

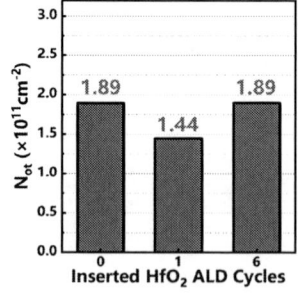

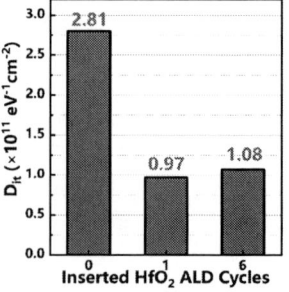

Fig.9 The dual-sweep C-V curves of devices with 0/1/6 ALD cycles ins. HfO_2. The device with 1 cycle ins. HfO_2 shows a smaller hysteresis than devices with 0/6 cycles ins. HfO_2, indicating an improved dielectric stack.

Fig.10 The N_{ot} values demonstrate a slight improvement by thinner inserted HfO_2.

Fig.11 The D_{it} of devices with 0/1/6 ALD cycles inserted HfO_2. The interface is greatly improved.

Multi-physics Modeling of Au/MoS₂/Au Memristors combining Molecular Dynamics and Electro-thermal Simulations

Mohit Tewari[1], Ashutosh Krishna Amaram[2] and Tarun Agarwal[1]

[1]Department of Electrical Engineering, IIT Gandhinagar

[2]Department of Materials Engineering, IIT Gandhinagar

Abstract

The state transition in Au/MoS₂/Au memristor across several cycles is analysed in this work using a multi-physics modeling framework that combines electro-thermal Finite Element Method (FEM) and Reactive Molecular Dynamics (MD) simulations. MD simulations provide the field-driven realistic atomistic filament structure of HRS and LRS, while temperatures from FEM simulations are fed back into MD for state transitions. The framework demonstrates that the HRS is caused by constriction of the filament near the top Au and monolayer MoS₂ interface rather than complete filament rupture.

Keywords: ReRAM, 2D materials and Reactive Molecular Dynamics, Electro-thermal Simulations

Introduction

Resistive Random Access Memory (ReRAM) is promising technology for neural network applications, due to its compact size, low power consumption, high storage density, sub-nanometer scalability, fast switching speeds, and high endurance [1]. However, ReRAM-based neural networks continue to face challenges in accuracy, primarily due to the intrinsic device variability [2]. This variability is notably reduced in memristors based on 2D materials, such as MoS₂, compared to conventional materials like HfO₂ [2]. Understanding the cause of cycle-to-cycle (C2C) variability in ReRAM using experimental and modeling techniques remains an important research topic [3]. The cause of variability reported Au/MoS₂/Au atomristors [4] however remains elusive.

In this work we present a modeling framework that aims to investigate the sources of variability in 2D material-based ReRAM, focusing on an Au/MoS2/Au structure through FEM and MD simulations. Filament geometry is the primary contributor to the C2C variability [2]. Using MD simulations, we examine atomistic filament changes over multiple cycles. An electron transport module integrated within our simulation framework allows us to compare resistance ratio of HRS and LRS over multiple cycles. Notably, we observe that localized temperature within the filament during LRS significantly impacts the LRS to HRS transition, hence HRS structure, which have been reported earlier [5]. Electro-thermal modules in COMSOL Multiphysics are employed to calculate the filament temperature in LRS, which are shown to reach 1000 K for RESET, a parameter used in subsequent MD simulations to refine our analysis. The reactive MD-FEM coupled framework is then used to assess HRS to LRS resistance ratios over ratios over multiple cycles.

Methodology

The detailed methodology is outlined in Fig.1. For macroscopic modeling of Au/MoS₂/Au memristor, we use electro-thermal modules within COMSOL Multiphysics. Here the conductivity of filament is modeled with (1) [5]

$$\sigma_{fila} = \frac{r_0 \times t_{MoS_2}}{Rlrs \times (r_0 \times A_{fila} + A_{MoS_2})} \quad (1)$$

where $t_{MoS2} = 1$ nm is the thickness between the top and bottom Au electrode corresponding to monolayer MoS₂, $R_{lrs} = 20$ Kohm [6] is the LRS resistance, A_{fila} and A_{MoS2} is the area of the filament and the non-conducting MoS₂, and $r_0 = 10$ is the fitting parameter. For two Sulfur vacancies, we adjusted the filament width within the Au/MoS₂/Au memristor structure, incorporating geometrical parameters derived from MD simulations. The macroscopic structures for electro-thermal simulation are obtained from the atomic structure obtained from MD simulations transformed as cuboids with filament geometry and modeled in COMSOL. The values of current are in the range of 10^{-6} A which agrees reasonably well with the previously reported experiment [7] and simulation [6].

The COMSOL simulations generated a detailed temperature profile (Fig. 2), which was the input for MD simulations to study LRS to HRS transitions over multiple cycles. The MD simulations were carried out using LAMMPS [8]. The interactions between the atoms are modeled using the reactive force field [9] parameters. For the simulations, we use a timestep of 0.5. Periodic boundary conditions along the x and y axes, and non-periodic boundary conditions along the z axis. Ambient pressure was set using the Noose-Hoover barostat and the temperature was fixed at 300K for SET process and 1000K for RESET process using the Berendsen thermostat. The voltage was applied across the gold electrodes using the ECHEMDID [10] package. For the set process a voltage difference of 38 volts was applied across the explicit gold electrodes and for the reset process, a voltage difference of -38 volts was applied. The voltages applied are very high because of the limitation of MD to capture effects such as filament formation and rupture at larger time and length scales. However, this approach helps capture the kinetics of the set and reset process very effectively.

979-8-3315-0417-5/25 $31.00 © 2025 IEEE

Results and Discussion

To study the C2C variation in Au/MoS$_2$/Au memristor we obtain LRS and HRS structures for consecutive three cycles from our modeling framework. Based on the experimental insights [7] we use the MoS$_2$ layer with two Sulfur vacancies where the first cycle represents the forming step. These defects are active sites for a conductive filament of gold to be formed. We observe from the FEM simulations that when the filament is formed (LRS state), the temperature of the interface between the conducting filament and gold electrodes rises up to 1400 K for the current density of 9.7×10^{11} A/m^2 as observed in Fig.2(b). Hence it becomes important to apply a raised temperature obtained from electro-thermal simulations in MD simulations while modeling the LRS to HRS transitions.

The cycle-to-cycle MD simulations indicate that the width of the filament changes with each cycle, for both LRS and HRS as shown in Fig. 3(a-f). Both HRS and LRS structures over each cycle show a small variation which can contribute to the C2C variability. We can observe that HRS structures from MD show the constriction of the filament compared to the LRS structure rather than complete rupture. These constrictions or thinning of filament can also be seen in macroscopic structure in Fig. 3(f- k) which have been reported in the conventional Conductive Bridge Random Access Memory (CBRAM) [11]. The constriction of the filament in HRS leads to an increase in the resistance that arises due to the migration of Au atom from near the top Au-MoS$_2$ interface. There it results in a reduced number of modes (i.e., number of parallel conducting paths) and thus, reduced current in the HRS structures. As shown in Table 1, resistance ratios for LRS and HRS for three consecutive SET-RESET cycles of two sulfur vacancy structure, are in the range of 1.5-3 (Fig. 4(b)) from COMSOL simulation and 3-3.7(Fig. 4(a)) from MD simulations, which are also experimentally reported from single or double defects memristors [7]. The low HRS to LRS resistance ratios are attributed to the constriction of the filament rather than a complete rupture. A rupture near the interface is expected to create a tunneling gap and thus, much higher resistances compared to the constriction of the filament.

Conclusion

In this work, we combine the MD and Electro-thermal FEM simulations to holistically assess the filament construction and constrictions over multiple cycles. The observations from our framework have been validated with experimentally reported a few defect MoS$_2$ memristors. The combined MD and FEM framework can be used to design the Au/MoS$_2$ interface to have controlled forming and breaking of the filament leading to lower C2C variability.

References

[1] X. Hong, D. J. Loy, P. A. Dananjaya, F. Tan, C. Ng, and W. Lew, "Oxide-based rram materials for neuromorphic computing," Journal of materials science, vol. 53, pp. 8720–8746, 2018.

[2] J. B. Rold´an, E. Miranda, D. Maldonado, A. N. Mikhaylov, N. V. Agudov, A. A. Dubkov, M. N. Koryazhkina, M. B. Gonz´alez, M. A. Villena, S. Pobladoret al., "Variability in resistive memories," Advanced Intelligent Systems, vol. 5, no. 6, p. 2200338, 2023

[3] W. Xu, J. Wang, and X. Yan, "Advances in memristor-based neural networks," Frontiers in Nanotechnology, vol. 3, p. 645995, 2021.

[4] R. Ge, X. Wu, M. Kim, J. Shi, S. Sonde, L. Tao, Y. Zhang, J. C. Lee, and D. Akinwande, "Atomristor: nonvolatile resistance switching in atomic sheets of transition metal dichalcogenides," Nano letters, vol. 18, no. 1, pp. 434–441, 2018.

[5] I. M. Datye, M. M. Rojo, E. Yalon, S. Deshmukh, M. J. Mleczko, and E. Pop, "Localized heating and switching in mote2-based resistive memory devices," Nano letters, vol. 20, no. 2, pp. 1461–1467, 2020.

[6] A. K. Amaram, S. Kharwar, and T. K. Agarwal, "Investigation of resistive switching in au/mos2/au using reactive molecular dynamics and ab-initio quantum transport calculations," 2024.

[7] S. M. Hus, R. Ge, P.-A. Chen, L. Liang, G. E. Donnelly, W. Ko, F. Huang, M.-H. Chiang, A.-P. Li, and D. Akinwande, "Observation of single-defect memristor in an mos2 atomic sheet," Nature Nanotechnology, vol. 16, no. 1, pp. 58–62, 2021.

[8] A. P. Thompson, H. M. Aktulga, R. Berger, D. S. Bolintineanu, W. M. Brown, P. S. Crozier, P. J. in 't Veld, A. Kohlmeyer, S. G. Moore, T. D. Nguyen, R. Shan, M. J. Stevens, J. Tranchida, C. Trott, and S. J. Plimpton, "LAMMPS - a flexible simulation tool for particle-based materials modeling at the atomic, meso, and continuum scales," *Computer Physics Communications*, vol. 271, p. 108171, Feb. 2022.

[9] Q. Mao, Y. Zhang, M. Kowalik, N. Nayir, M. Chandross, and A. C. T. van Duin, "Oxidation and hydrogenation of monolayer MoS2 with compositing agent under environmental exposure: The ReaxFF Mo/Ti/Au/O/S/H force field development and applications," *Frontiers in Nanotechnology*, vol. 4, Oct. 2022.

[10] N. Onofrio and A. Strachan, "Voltage equilibration for reactive atomistic simulations of electrochemical processes," The Journal of Chemical Physics, vol. 143, no. 5, Aug. 2015.

[11] M. N. Kozicki and H. J. Barnaby, "Conductive bridging random access memory materials, devices and applications," Semiconductor Science and Technology, vol. 31, no. 11, p. 113001, 2016.

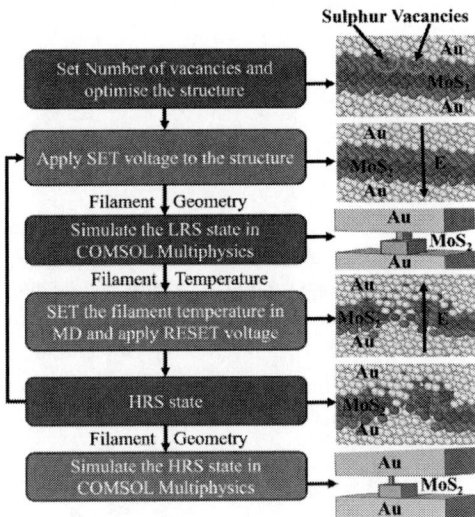

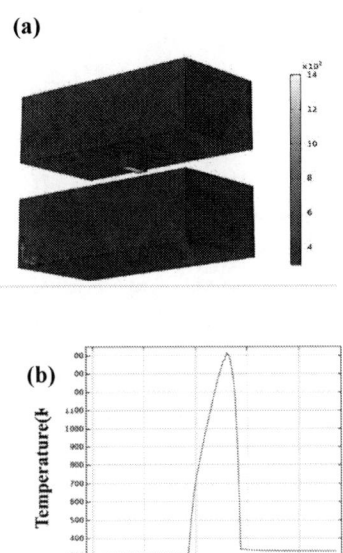

(Figure 1: Framework for multi-physics modelling combining FEM and reactive MD for C2C simulation in Au/MoS$_2$/Au memristor.)

(Figure 2 (a) Temperature profiles obtained from COMSOL. (b) The temperature profile across the filament of the device.)

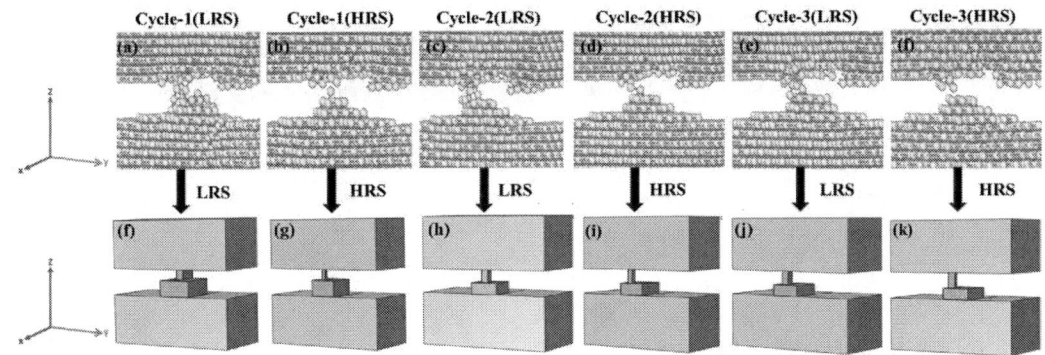

(Figure 3: (a), (c), (e) and (f), (h), (j) represents LRS of cycle 1,2 and 3 obtained from MD and COMSOL respectively. Similarly (b), (d), (f) and (g), (i), (k) represents the HRS of cycle 1,2 and 3. Here MoS$_2$ is not shown in both MD and COMSOL structure for better visualization of the

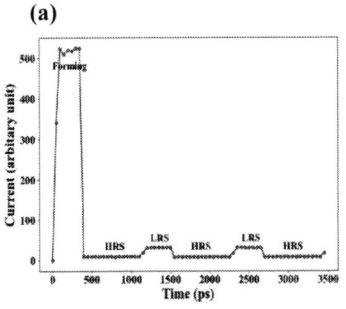

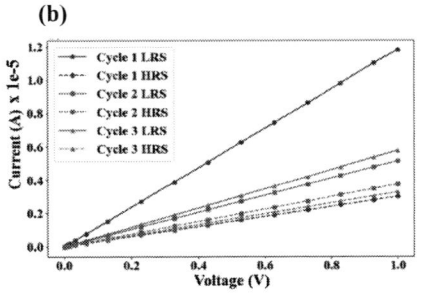

	R_{HRS}/R_{LRS}
Cycle 1	3.950
Cycle 2	1.540
Cycle 3	1.670

(Table 1. R_{HRS}/R_{LRS} ratio for consecutive three cycles.)

(Figure 4: (a) The current transient from MD simulation (b) the IV simulated using the COMSOL modelling.)

979-8-3315-0417-5/25 $31.00 © 2025 IEEE

High Performance 4H-SiC DSRD With 1.4kV peak voltage ,400ps risetime and 1-MHz Continuous Repetition-rate

Yu zhou[1,2],Jingkai Guo[1],Dengyao Guo[1], Fengyu Du[1,2], Lejia Sun[1], Xiaoyan Tang[1],
Qingwen Song[1,2], Yuming Zhang[1,2]

[1]Xidian University, China,
[2]Xidian-Wuhu Research Institute, China

Abstract

In this paper, a high-performance 4H-SiC Drift Step Recovery Diode (DSRD) is achieved. Experimental measurements demonstrate that our reported 4H-SiC DSRD could achieve a pulse peak voltage of 1.45 kV on 50 Ω load with a rise-time of 400-ps, resulting in a high voltage rise rate of 3.6 kV/ns. Then, the repetition frequency measurement result at 1 MHz is demonstrated in continuous operation mode, revealing that fabricated 4H-SiC DSRD with excellent dynamic stability.

Keywords: Sub-nanosecond, SiC DSRD, Pulsed power

Introduction

The applications of Nanosecond pulsed power source have been extensively studied in recent years, including plasma waste gas treatment [1],[2] high power microwave systems[3]. These applications require high-voltage pulse, fast pulse front and high pulse repetition frequency rate. The pulsed power system based on drift step recovery diode (DSRD) can satisfy these requirements well. Recently, DSRD based on SiC has attracted strong interest in ultrafast power pulse technology due to its excellent properties such as high breakdown electric field, high saturation electron velocity, excellent thermal conductivity, and thermal stability. These properties enable DSRD based on SiC to operate at a higher frequency, compared with Si. Importantly, given a blocking voltage, higher base doping concentration and thinner base layer, make SiC DSRD cut off larger current density with shorter time [4].

In this paper, we successfully demonstrate the robust power-pulses capability of the SiC DSRD in a HV generation circuit with continuous operation at repetition rate up to MHz range into a 50-Ω load, achieving a dynamic voltage rise rate of 3.6 kV/ns and a reverse current density exceeding 5.4 kA/cm², rise time of 400 ps (20%-90% Vpeak). Rugged reliability is validated after over 1-million times dynamic continuous operation with the 1.45-kV peak voltage.

Device structure, Fabrication and Characterization

Fig. 1(a),(b) show the simplified cross-sectional views of beveled mesa 4H-SiC DSRD and the SEM image of fabricated 4H-SiC DSRD. The 4H-SiC DSRD devices were fabricated on a multilayer 4H-SiC homoepitaxial structure.

By using chemical vapor deposition (CVD) process, a 5-μm n+ buffer layer was grown on a heavily nitrogen doped SiC substrate with donor concentration of 5×10^{18} cm^{-3}. A 9-μm p-base layer and a 1-μm p+ layer with concentration of 7×10^{15} cm^{-3} and 2×10^{19} cm^{-3} were grown on the n+ buffer layer. And the mesa structure was fabricated by our proposed improved Bosch etching process[5].

The static characteristics of the fabricated 4H-SiC DSRD have been performed by an Agilent B1505A analyzer. Fig. 1(c) shows the representative forward I-V characteristics of the device. The turn-on voltage of 4H- SiC DSRD is 2.76 V. A differential on-resistance (Ron,sp) of ~0.9 mΩ·cm² can be obtained at current density of 520 A/cm². Fig. 1(d) shows the measured reverse I–V characteristics of the fabricated device. The maximum breakdown voltage achieved 1450V, which is the upper limit of the pulsed peak voltage.

Experiment Results

A. Principle and Experimental Setup

To evaluate the dynamic pulse performance of the fabricated 4H-SiC DSRD, a pulsed power test measurements have been performed. Fig. 2(a) shows the diagram of the basic testing circuit , according to the working principle, the trigger current of the DSRD is affected by the values of inductance L_1 and L_2, capacitance C_1 and C_2, supply voltage V_1, and the on-time duration ΔT of the MOSFET. Selecting a set of capacitance and inductance parameters the maximum reverse current I_{rmax} flowing through the DSRD can be changed by only adjusting the value of supply voltage V_1. The increase in V_1 will cause the increase in the pumping current and the current flowing through L1. Larger pumping current fills the base area of the SiC DSRD with more plasma and larger current through L_1 results in faster plasma extraction. Both factors will lead to the increase of maximum reverse current I_{rmax}, which directly affects the cut off process of the SiC DSRD. The experimental platform is shown in Fig. 2(b).

D. Experimental Results

Fig. 3(a) shows the measured trigger current waveform with different supply voltage V_1, which is changed from 30 to 140 V. Consistent with the analysis results, the pumping current $i_{pump}(t)$ and the maximum reverse current I_{rmax} are increasing with the increase of V_1. Fig. 3(b) shows the waveform of the output voltage under different values of V_1, it can be seen V_{peak} is increasing with the increase of V_1, and t_r is decreasing as V_1 increases. The measured maximum V_{peak} and minimum t_r is shown in Fig. 3(c),(d). When I_{rmax} reaches about 80 A, corresponding to a current density of 5400 A/cm^2, voltage rise time reaches 400 ps with V_{peak} of 1.45kV.

A continuous operation experiment was conducted to verify the stability of fabricated SiC DSRD, the output voltage waveforms measured at 1 MHz ultra-high frequency are displayed in Fig. 4(a). It can be seen that a stable MHz high frequency short electrical pulse signal is successfully achieved, and the overall voltage fluctuation is controlled within 20V, less than 2% of V_{peak}, revealing our fabricated 4H-SiC DSRD with robust dynamic breakdown capability for high-reliability power-pulses applications. And a device temperature of about 60 °C measured by thermal imagers during operation under air-cooling conditions is also present in Fig. 4(b). Since the work principle of this circuit requires the snappy switch-off of MOSFET and DSRD, the power consumption and temperature rise of the MOSFET and DSRD is unavoidable.

Table I compares the performance of recently reported Si and SiC DSRD dies [4], [6], [7], [8]. And all the pulsed voltage is measured on the 50 Ω matched load. And this paper also demonstrates the longest operation time at 1 MHz repetition rate among recently reported devices.

Conclusion

In this paper, the 1.45 kV 4H-SiC DSRDs are designed, fabricated, and characterized. A sub-nanosecond switch performance, reaching 1.45 kV at arising time of 400 ps is achieved. The dynamic reliability of devices is also verified by over 10 million times work at 1 MHz repetition frequency and there is no obvious peak voltage fluctuation, which shows excellent rugged reliability and thermal stability of devices. Together with the advantages of high voltage ratings, superior dynamic performance, and high robustness, beveled mesa 4H-SiC DSRD developed herein are promising in all-solid, high-repetition frequency rate, and high-reliability high power applications.

Acknowledgments

This work was supported by the National Natural Science Foundation of China (Grant Nos. 61234003, 61434004, 61504141) and CAS Interdisciplinary Project (Grant No. KJZD-EW-L11-04).

References

[1] M. Scapinello, L. M. Martini, G. Dilecce, and P. Tosi, "Conversion of CH4/CO2 by a nanosecond repetitively pulsed discharge," J. Phys. D. Appl. Phys., vol. 49, no. 7, Art. no. 75602. (2016).

[2] S. V. Korotkov, Y. V. Aristov, A. K. Kozlov, D. A. Korotkov, A. G.Lyublinsky, and G. L. Spichkin, "Installation for air cleaning from

organic impurities by plasma formed by barrier discharge of nanosecond duration," Instrum. Exp. Techn., vol. 55, no. 5, pp. 605–607.(2012).

[3] A. F. Kardo-Sysoev, "New power semiconductor devices for generation of nano- and subnanosecond pulses," in Ultra-Wideband Radar Technology, J. D. Taylor, Ed. Boca Raton, FL, USA: CRC Press, pp. 214–299. (2001).

[4] A. S. Kesar, M. Wolf, A. Raizman, D. Cohen-Elias and S. Zoran, "A 1-kV, 0.84-kV/ns Epi-Si Drift-Step-Recovery-Diode," in IEEE Transactions on Plasma Science, vol. 51, no. 4, pp. 1133-1137.(2023).

[5] C. Han et al., "An Improved ICP Etching for Mesa-Terminated 4H-SiC p-i-n Diodes," in IEEE Transactions on Electron Devices, vol. 62, no. 4, pp. 1223-1229 (2015)

[6] A. G. Lyublinsky, A. F. Kardo-Sysoev, M. N. Cherenev and M. I. Vexler, "Influence of DSRD Operation Cycle on the Output Pulse Parameters," in IEEE Transactions on Power Electronics, vol. 37, no. 6, pp. 6271-6274.(2022).

[7] Ivanov, Pavel A., and Igor V. Grekhov. "Subnanosecond Semiconductor Opening Switch Based on 4H-SiC Junction Diode." Materials Science Forum, vol. 740–742, Trans Tech Publications, Ltd., pp. 865–868. (2013).

[8] X. Yan, L. Liang, Z. Yang and H. Shang, "Investigation of Prepulse of SiC Drift Step Recovery Diode in Fast Interruption Process," in IEEE Transactions on Electron Devices, vol. 71, no. 5, pp. 3102-3108, May (2024)

Table I: Comparison of reported DSRDs

Ref./Material	V_{peak} (kV)	t_r (ns)	dv/dt (kV/ns)
[4] Si 2022	0.4	0.67	0.60
[6] Si 2023	1	1.2	0.83
[7] SiC 2013	1.76	0.6	2.93
[8] SiC 2024	2.3	1.2	1.92
This work	1.45	0.4	3.62

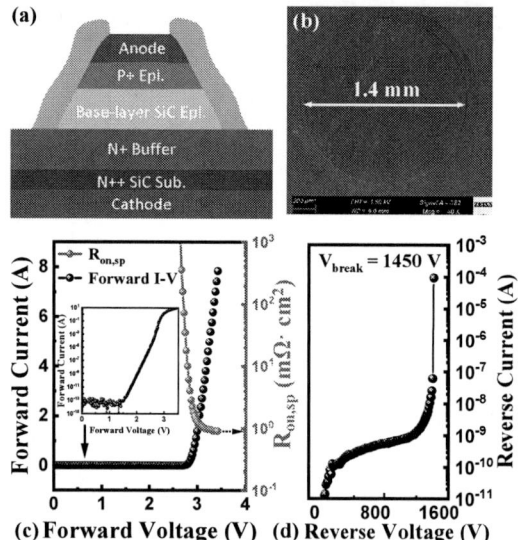

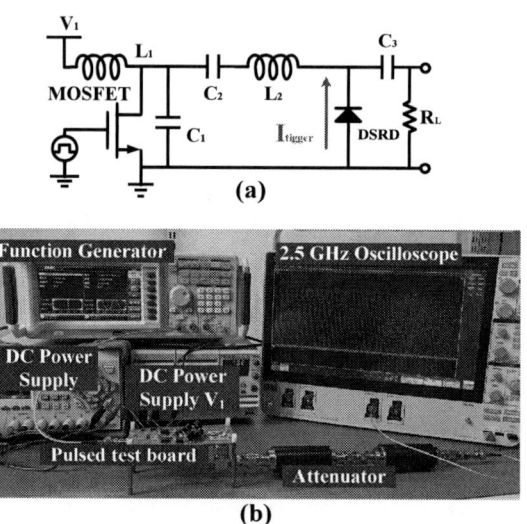

Fig. 1(a)Schematic cross-sectional views of 4H-SiC DSRD. (b) SEM image of fabricated 4H-SiC DSRD (top view). (c) forward I-V characteristics and extracted Ron,sp .(d) reverse I-V characteristics.

Fig. 2(a) Schematic diagram of pulsed test circuit on 4H-SiC DSRD. (b) Experimental setup for dynamics test

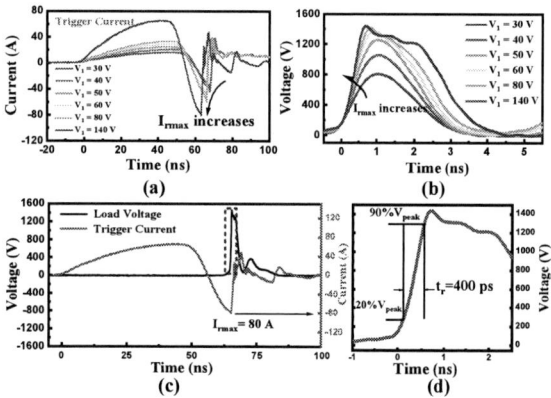

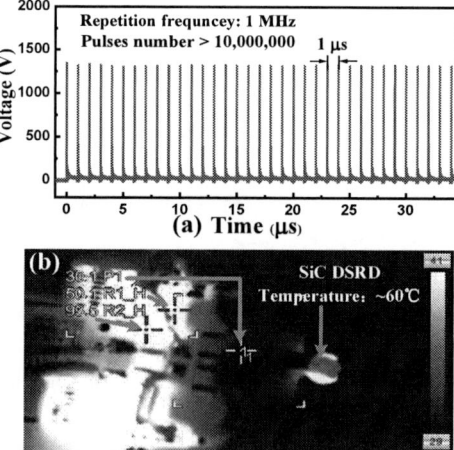

Fig. 3: (a)experiment waveforms of trigger current with V1 increase. (b)output voltage pulse measured on 50 Ω load with V1 increase. (c)output pulse and trigger current waveform @ Irmax=80 A. (d) partial enlarged detail of pulsed front

Fig. 4(a) 1 MHz continuous waveform in experiment. (b) temperature picture of test board @ 1 MHz continuous work under air-cooling condition.

979-8-3315-0417-5/25 $31.00 © 2025 IEEE

Neural Network Assisted MOSFETs Gate Dielectric Traps Extraction

Xiaoyan Liu[*], Jinghan Xu, Zheng Zhou

School of Integrated Circuits, Peking University, Beijing 100871, China.

Abstract

Neural network assisted MOSFETs gate dielectric traps extraction method as an emerging fast data-driven approach is studied. The proposed method extracts trap spatial and energetic distribution from the noise power spectral density (PSD) with two steps: 1) rough extraction using a customized multi-scale receptive fields Convolutional Neural Network (CNN), 2) refinement using Backpropagation Optimization (BO) network based on Gradient Descent (GD). The method is applied to a practical case, and successfully extracts the trap distribution.

Keywords: MOSFET, Trap, Neural network

Introduction

Traps pose a significant challenge in MOSFETs as they can induce various performance and reliability issues, such as low-frequency noise (LFN) [1]. As the gate dielectric thins, these problems worsen [2]. However, existing trap extraction methods face limitations. Simultaneously extracting trap energy levels and locations has been a long-standing challenge [3].

In recent years, neural networks have been widely applied across various fields due to their effectiveness in managing complex, nonlinear, and high-dimensional problems. Their proficiency in pattern recognition, data-driven prediction, and feature extraction from large datasets offers valuable insights into how these networks can be applied to address reliability challenges. We developed an automated, DL-assisted method for extracting trap spatial and energetic distributions from noise PSD [4]. The method comprises two steps: first, a CNN with multi-scale receptive fields is employed to roughly extract the trap distribution from the noise PSD; then, the initial rough extraction result is iteratively refined using a Backpropagation Optimization network. The efficacy of this method is validated using both ideal datasets and measurement results.

Method

A bulk n-MOSFET device, with its structure and the theoretical framework used for simulation depicted in Fig. 1. LFN originates from traps located in the gate dielectric. The voltage noise PSD (S_{Vg}) is calculated using the carrier number fluctuations theory (CNF) model [5]:

$$S_{Vg} \cong q^2 \cdot \left(\frac{1}{WL \cdot C_{ox}^2}\right) \cdot \iint S_{\Delta N} \cdot N_{bt} \cdot \left(1 - \frac{x_t}{t_{ox}}\right)^2 dE_t dx_t \quad (1)$$

where $S_{\Delta N}$ is the noise PSD due to a single trap, t_{ox} is the equivalent oxide thickness (EOT) of the gate dielectric, C_{ox} symbolize the equivalent capacitances of gate dielectric, g_m is the transconductance, and W and L are the width and length of the gate, respectively. The electron capture and emission of the traps is modeled as inelastic, phonon-assisted processes [6].

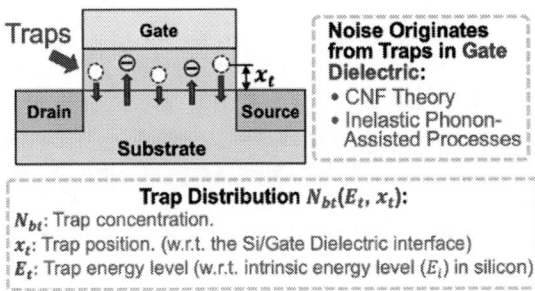

Fig. 1. Device structure and the theory of trap-induced noise in MOSFETs

Fig. 2 shows the simulation results of LFN. The local noise source (LNS) figure illustrates the noise generated by traps at discretized E_t and x_t, providing an intuitive demonstration of the noise sources. Under different Vg, traps at varying E_t and different x_t contribute to noise at different frequencies, suggesting that spatial and energetic distributions can be extracted from noise PSD.

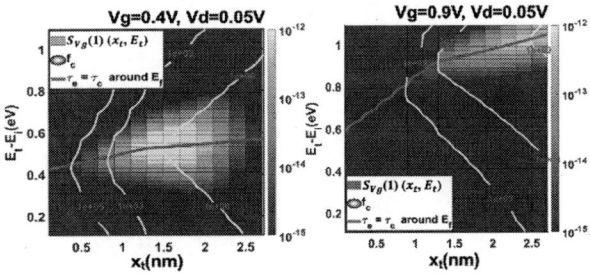

Fig. 2. LNS figure of Vg=0.4V/0.9V, containing characteristics of PSD pattern from traps at different x_t and E_t under different Vg.

We developed an automated DL-assisted method to extract trap spatial and energetic distributions from the noise PSD. The overall workflow is illustrated in Fig. 3. The network is trained on datasets generated from TCAD simulations [7] and theoretical calculations. The proposed DL-assisted trap distribution extraction method is a two-step process: 1) initial rough extraction using NW1; 2) refinement through the synergistic use of NW2 and NW3 by gradient backpropagation. The final output delivers parameters for N groups of Gaussian-distributed traps.

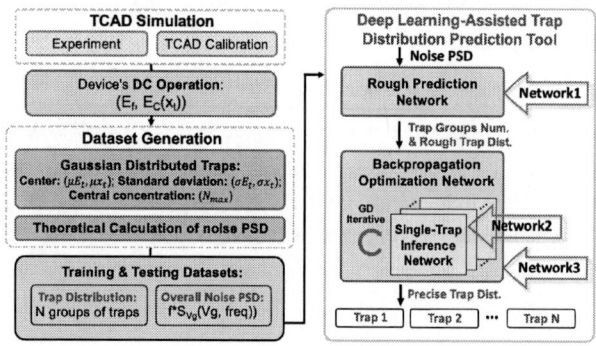

Fig. 3. Workflow of the method.

NW1 is a customized CNN, which uses multi-scale receptive fields to extract trap features from both local and overall PSD information (Fig. 4). NW1 accepts noise PSD as its input and predicts the presence probability of trap group center in six potential positions and their Gaussian distribution parameters. Within this network, we conceptualize the Noise PSD with varying Vg and frequency as analogous to images. This approach enables the extraction of trap distributions from the absolute values of individual elements and the intricate interconnections among neighboring elements. The extraction result in Fig. 5 indicates that the network can predict the number of trap groups and provide a rough guess about their distribution.

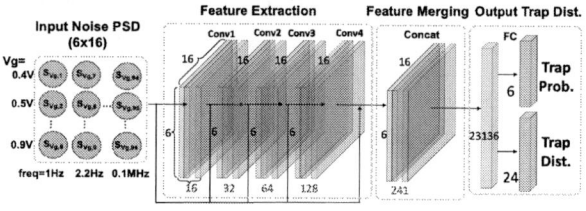

Fig. 4. Architecture of the trap distribution extraction network (NW1).

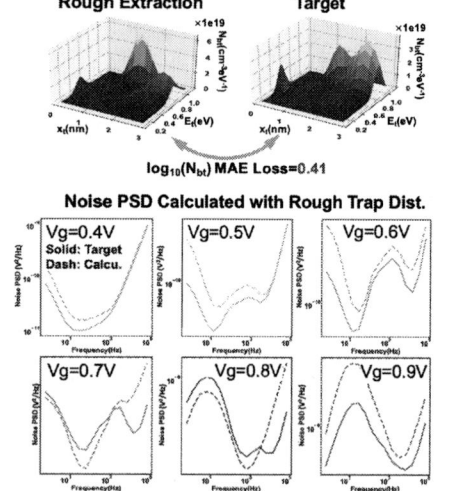

Fig. 5. Comparison between the rough extraction results with the target. (a) Trap distribution, (b) Corresponding noise PSD at various Vg.

The noise PSD calculated from the trap distribution extracted from NW1 aligns well with the input data, but there's still some discrepancies due to slight variations in trap distribution resulting in noticeable gaps in the noise PSD. We use the GD iteration algorithm to refine NW1 results. To enable gradient backpropagation, a PSD inference network (NW2), which is compatible with GD algorithm, is trained to model the relationship between single trap distribution and noise PSD. NW2 accepts a set of Gaussian parameters for a single trap group as input, and produces the corresponding noise PSD as its output, as shown in Fig. 6.

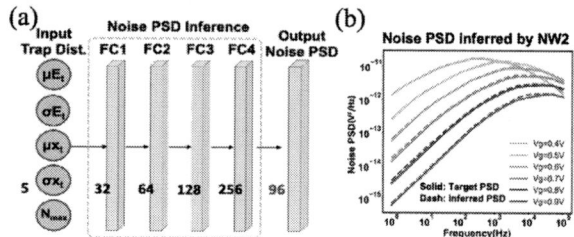

Fig. 6. Single trap noise PSD inference network (NW2) (a) Architecture, (b) Inference results.

As depicted in Fig.7, the architecture of NW3 consists of multiple independent units of NW2, and its structure is dynamically adaptable. We created a pool of networks containing six identical, pre-trained NW2 units. During operation, NW3 activates N NW2 units, each producing a predicted noise PSD. The sum of these outputs forms the total noise PSD. The initial extraction results from NW1, which include both the predicted number of trap groups (N) and their associated Gaussian distribution parameters, serve as initial inputs for NW3. The calculated total noise PSD is then compared with the target noise PSD to identify discrepancies. The discrepancies are rectified through gradient backpropagation, which sequentially updates the input parameter set for each of the N trap groups. This iterative refinement process enhances the accuracy of the extracted trap distribution.

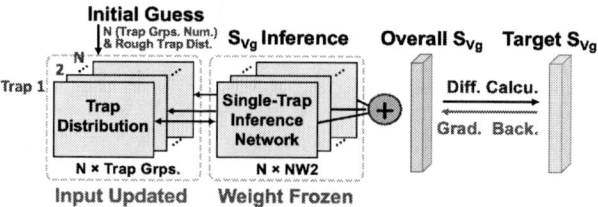

Fig. 7. Architecture of the GD algorithm refinement network (NW3).

An example in Fig. 8 shows that after several iterations, both PSD loss and trap distribution loss significantly decreased, demonstrating the effectiveness of NW2 and NW3 in refining initial extraction results.

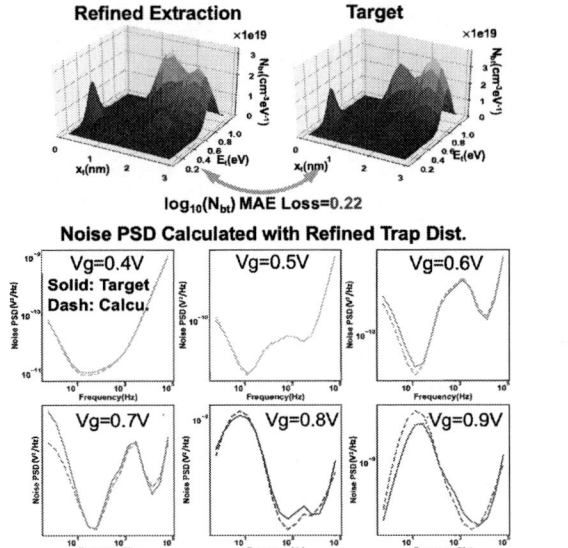

Fig. 8. Comparison between the refined extraction results with target. (a) Trap distribution, (b) Corresponding noise PSD at various Vg.

Results and Discussion

To validate the effectiveness of our methodology, we conducted extensive testing using a dataset of 6,000 samples. Figure 9 displays the evaluation results of different number of trap groups. Our statistical analysis reveals efficient optimization through the refinement steps with NW2 and NW3. As the number of trap groups increases, the loss of predicted trap distribution rises, indicating a reduction in extraction accuracy. Conversely, the loss of calculated noise PSD is more pronounced with fewer trap groups. Overall, the statistical results indicate a close match between the extracted traps and their targets, and the calculated noise PSD aligns well with the input data.

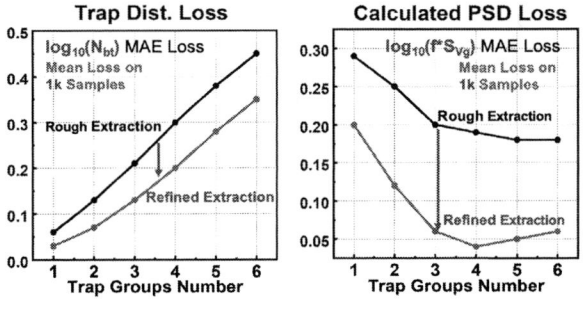

Fig9. Statistical results of 1k samples for varying trap group number.

Evaluation of (a) Trap distribution loss and (b) Calculated Noise PSD loss.

Additionally, we applied this method to an actual device scenario, as depicted in Fig. 10. The outcomes show that the noise PSD generated from the extracted trap distribution closely matches the empirical data, affirming our method's proficiency in extracting traps from measured noise PSD.

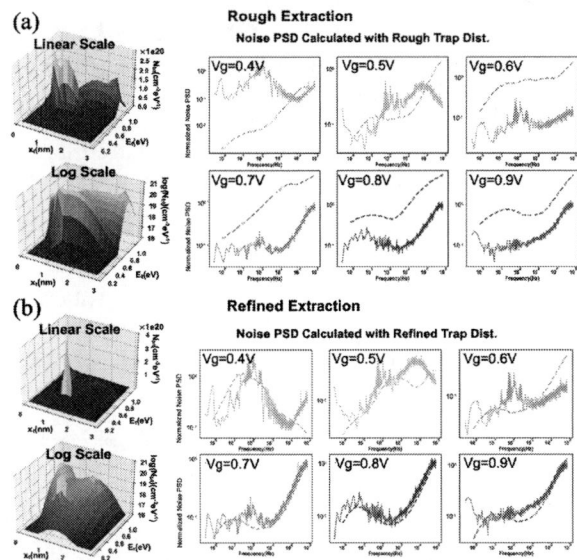

Fig10. Extracted trap distribution and calculated noise PSD using the proposed trap distribution extraction method. (a) Rough extraction, (b) Refined extraction.

Conclusion

We proposed a method that combines multiple DL techniques to automatically extract trap spatial and energetic distributions directly from MOSFET noise PSD. The effectiveness of the method is verified by both ideal dataset and actual case. This provides a novel perspective for solving complex reliability issues using deep learning techniques.

Acknowledgments

This work was supported in part by the Natural Science Foundation of China under Grant 92364104 and the National Key Research and Development 2022YFB4401704.

References

[1] E. G. Ioannidis *et al.*, "Low frequency noise variability in high-k/metal gate stack 28nm bulk and FD-SOI CMOS transistors," *International Electron Devices Meeting* (IEDM) (2011).

[2] S. Zafar, A. Callegari, E. Gusev, and M. V. Fischetti, "Charge trapping related threshold voltage instabilities in high permittivity gate dielectric stacks," *Journal of Applied Physics*, vol. 93, no. 11, pp. 9298–9303 (2003).

[3] G. Sereni, L. Vandelli, D. Veksler, and L. Larcher, "A New Physical Method Based on CV – GV Simulations for the Characterization of the Interfacial and Bulk Defect Density in High- k /III-V MOSFETs," *IEEE Transactions on Electron Devices*, 62, pp. 705–712 (2015).

[4] J. Xu, Z. Zhou, M. Fan, Z. Sun, S. Wang, Z. Tang, F. Liu and X. Liu, "Deep Learning-Assisted Trap Extraction Method from Noise Power Spectral Density for MOSFETs", IEEE Int. Rel. Phys. Symp (IRPS) (2024).

[5] R. Asanovski, P. Palestri, E. Caruso, and L. Selmi, "A Comprehensive Gate and Drain Trapping/Detrapping Noise Model and its Implications for Thin-Dielectric MOSFETs," *IEEE Transactions on Electron Devices*, vol. 68, no. 10, pp. 4826–4833 (2021).

[6] A. Palma, A. Godoy, J. A. Jiménez-Tejada, J. E. Carceller, and J. A. López-Villanueva, "Quantum two-dimensional calculation of time constants of random telegraph signals in metal-oxide--semiconductor structures," *Phys. Rev. B*, vol. 56, no. 15, pp. 9565–9574 (1997).

[7] Sentaurus Device User Guide, "Sentaurus Device User Guide," Synopsys Inc., CA, USA (2019)

979-8-3315-0417-5/25 $31.00 © 2025 IEEE

A Novel Superlattice HfO₂-ZrO₂ Ferroelectric Tunnel FET for Overall Improvement in Memory Window, EOT and Disturb Immunity

Shaodi Xu[1], Zhiyuan Fu[1], Shengjie Cao[1], Yue Yu[1], Hao Zheng[1], Qianqian Huang[1,2*] and Ru Huang[1,2*]

[1]School of Integrated Circuits, Peking University, Beijing 100871, China;
[2]Beijing Advanced Innovation Center for Integrated Circuits, Beijing 100871, China.

(*Email: hqq@pku.edu.cn, ruhuang@pku.edu.cn)

Abstract

In this paper, a novel superlattice (SL) HfO₂-ZrO₂ ferroelectric (FE) junction-modulated tunnel FET (TFET) is proposed and experimentally demonstrated with overall improvement in memory window (MW), equivalent oxide thickness (EOT) and disturb immunity. The gate stack is optimized with SL FE layer for larger MW and smaller EOT simultaneously which usually has an optimization conflict in conventional FE layer. In addition, a more abrupt tunnel junction is introduced by a striped gate stack design, which is found to further increase the MW. The fabricated novel SL FE-JTFET demonstrates a 2.6× improvement in MW along with a 19% reduction in EOT comparing to FE-TFET, which is very beneficial for practical read current margin improvement. Moreover, the SL FE layer can also mitigate the disturb issue in the memory operations, showing the great potential of proposed device for practical low-power and high-robust memory applications.

Introduction

HfO₂-based ferroelectric field-effect transistors (FE-FETs) have become one of the most promising emerging non-volatile memories due to its advantages of high-speed, low-power, CMOS compatibility and good scalability [1], [2]. For practical memory operation of FE-FETs, since the stored information is read out by sensing the read current difference, the read current margin (I_{PGM}/I_{ERS}) is an important memory metric, which is determined by both device memory window (MW) and subthreshold swing (SS) [3]. However, achieving a large MW typically requires a thick ferroelectric layer [4], which may inevitably in turn degrade the EOT and SS. Therefore, optimizing the gate stack to achieve both small EOT and high MW is essential.

Ferroelectric Tunnel FETs (FE-TFETs) with inherent SS advantages thanks to the band-to-band tunneling (BTBT) mechanism [5] can be a promising low-power memory device to achieve large read current margin. Several Si-based FE-TFETs have been reported in previous literatures [6], [7], yet further SS and MW improvement is still in urgent need.

In this work, a novel Si FE-TFET design is proposed and experimentally demonstrated. The gate stack is optimized through well-designed superlattice (SL) layer, resulting in improvement in both EOT and MW. To achieve an abrupt tunnel junction, we introduce junction-depleted modulation effect through a striped gate design [8], which is found to further increase the MW due to the increase in the electric field of tunnel junction. In addition, by using the pulse-disturb evaluation method for FE disturbance behavior [9], it is found that the SL FE layer shows better disturb immunity during memory operation, and the fabricated novel SL FE-TFET shows less disturb impacts after 1000 $V_W/2$ write disturb pulses.

Gate Stack Optimization through Superlattice HZO

MFM metal-ferroelectric-metal (MFM) capacitors are first fabricated using the process flow shown in Fig. 1. Bottom electrode (BE) TiN and top electrode (TE) TiN are deposited by DC sputtering. Three different kinds of HZO films are deposited on the BE using thermal ALD, with the same totaling 100 cycles, i.e., the same total layer thickness, but with different deposition periodicities for HfO₂ and ZrO₂, i.e., 1 cycle (H1Z1), 5 cycles (H5Z5 SL), and 10 cycles (H10Z10 SL). After TE deposition and patterning, a rapid thermal process (RTP) is performed for FE crystallization.

Fig. 2 presents the measured P-V hysteresis loops of the HZO and SL capacitors with increment sweeping voltage in steps of 0.2V. For the clearer comparison, the switched polarization charge density (P_{SW}) with the applied voltage is extracted through the positive-up-negative-down (PUND) measurement in Fig. 3(a). The H10Z10 SL shows the best ferroelectricity among the three FE films. Fig. 3(b) presents the measured C-V curves for three HZO films, both SL FE films of H5Z5 and H10Z10 show larger capacitance than the HZO film, suggesting the smaller EOT. The endurance tests in Fig. 3(c) reveal that H1Z1 FE layer is more prone to fatigue. Therefore, Fig. 3(d) summaries the key performances of the three FE layers, and H10Z10 SL is chosen as the optimal FE structure for the gate stack.

Based on the above discussion, p-type FE-FETs with H1Z1 and H10Z10 SL FE layers are further fabricated to validate the superiority of H10Z10 SL. An additional 2.5nm-thick SiO₂ layer is also grown in the gate stack for a high quality interlayer between the FE layer and Si. Fig. 4 presents the measured gate capacitance of two FE-FET devices, demonstrating 19% reduction in EOT with H10Z10 SL. The improved SS and I_{ON} in Fig. 5a further evidence the decrease in EOT. The measured pulsed IV

979-8-3315-0417-5/25 $31.00 © 2025 IEEE

curves of the FE-FETs shown in Fig. 5(b) indicate a 40% improvement in MW, resulting from the enhanced ferroelectricity of the H10Z10 FE layer.

Optimization of FE-TFETs

Gate-first process as shown in Fig. 6(a) is used for the fabrication of FE-TFETs in this work. To achieve a steep doping profile, a hard mask is used for the patterning of metal gate for an abrupt sidewall. Since the doping profile is a key factor affecting the steepness of the tunneling junction, the RTP condition is carefully selected to minimize impurity diffusion while ensuring the activation of impurities. Fig. 6(b) shows the measured distribution of Arsenic with depth by secondary ion mass spectrometry (SIMS). The diffusion of Arsenic after RTP process of 600°C 60s is negligible, and the measured sheet resistance (Fig. 6(c)) confirms that the activation rate of Arsenic is sufficient, thus we choose 600°C 60s as an optimum RTP condition. Fig. 7(a) shows the schematic illustration of FE-TFETs with H1Z1 and H10Z10 SL FE layers. The cross-section high resolution TEM (HRTEM) images of the as-deposited gate stacks are shown in Fig. 7(b), and an obvious layered structure is observed in H10Z10 SL FE layer. Fig. 8(a) shows the measured single sweep DC transfer characteristics of fabricated FE-TFETs. The I_{ON} and SS of FE-TFETs can be largely improved by the SL FE layer.

To achieve an ideally abrupt doping profile, we further introduce junction-depleted modulation effect [8] into FE-TFETs, i.e., the ferroelectric junction-modulated TFET (FE-JTFET). Fig. 9(a) shows the schematic structure of a p-type FE-JTFET which has a striped gate stack configuration. The striped gate increases the tunnel junction area for the higher current. Fig. 9(b) shows the SEM images of the fabricated FE-JTFET.

Fig. 10(a) illustrates the device measurement methods for the MW evaluation. As shown in Fig. 10(b-c) of the measured pulsed I_d-V_g curves of the fabricated FE-FETs and FE-JTFETs. Under the same gate stack, the FE-JTFET with an abrupt junction demonstrates a larger MW, and when the gate stack is optimized with SL FE layer, the MW improvement in FE-JTFET is more pronounced. This is likely due to the higher electric field at the BTBT junction of FE-JTFETs, which makes it more sensitive to reduction in EOT and polarization switching. Through combined optimizations of the gate stack and tunnel junction, the SL FE-JTFET has a large MW of 0.8V, showing a 2.6× improvement compared to the FE-TFET.

Table I compares the proposed FE-JTFET in this work with other reported Si-based FE-TFETs. Due to the optimization in both gate stacks and the tunnel junction in the FE-TFET, the fabricated FE-JTFET with H10Z10 SL gate stack shows superior MW of ~0.8V and improved EOT, showing its great potential as low-power memories with high read current margin.

Improved Disturb Immunity through Superlattice Ferroelectric Layer

The disturb behavior in FE-FETs in practical memory operation have been widely investigated, and operational methods to reduce disturbances have been proposed [10], [11]. The key objective is to minimize the amount of switched polarization during disturb. Therefore, optimizing the FE layer for a better disturb immunity is of great importance. Based on our previously proposed method for assessing the disturbance behavior of FE layer [9], we conducted FORC and pulse-disturb measurements on the H1Z1 and H10Z10 SL capacitors as shown in Fig. 11. The H10Z10 SL demonstrates a tighter E_C distribution (Fig. 11(a)), leading to better disturb immunity. Fig 11(b) shows the measured P_{sw} versus different pulse width (t_{pw}) under varied negative pulse amplitude. Extracted normalized P_{SW} under the same t_{pw} is shown in Fig. 11(c). Fig. 11(d) shows the extracted normalized P_{sw} ($P_{sw}/2Pr$). At low pulse amplitudes, the SL FE layer has less partial switched polarization, which indicates better disturbance immunity. In contrast, due to the more concentrated polarization switching in the SL FE layer, a high proportion of polarization switching is achieved at higher pulse amplitudes. By using the test pulse scheme shown in Fig. 12(a), the disturb behavior in the fabricated FE-JTFETs is further evaluated. The $V_W/2$ write disturb pulse is applied for 1000 times and then the V_{th} shift is extracted. Fig 12(b) indicates that the FE-JTFET with H10Z10 SL FE layer also demonstrates superior disturb immunity compared with conventional FE layer.

Conclusion

In this work, we reported a novel superlattice HfO$_2$-ZrO$_2$ ferroelectric junction-modulated TFET. By the gate optimization through a superlattice HZO FE layer and a striped gate configuration, the EOT and MW is simultaneously improved, and the modulated abrupt tunnel junction can further enhance the MW. In addition, the disturb immunity is improved by the superlattice HZO, making the proposed SL FE-JTFET a promising low-power device in the practical memory applications.

Acknowledgments

This work was supported by NSFC (62374009, 61927901) and 111 Project (B18001).

References

[1] A. I. Khan *et al.*, *Nature Electronics*, vol. 3, pp.588-597, 2020. [2] Z. Zhang *et al.*, *Sci. China Inf. Sci*, vol. 66, p. 200405, 2023. [3] C. Su *et al.*, *ESSDERC*, pp. 89-92, 2023. [4] H. Mulaosmanovic *et al.*, *IEEE TED*, vol. 66, pp. 3828-3833, 2019. [5] A. M. Ionescu *et al.*, *Nature*, vol. 479, pp. 329-337, 2011. [6] W. C.-Y. Ma *et al.*, *IEEE TED*, vol. 69, pp. 6072-6077, 2022. [7] M. Ryu *et al.*, *IEEE EDL*, vol. 45, pp. 144-147, 2024. [8] Q. Huang *et al.*, *IEDM*, pp. 187-190, 2012. [9] Z. Fu *et al.*, *IEDM*, pp. 11.3.1-11.3.4, 2023. [10] K. Ni *et al.*, *IEEE EDL*, vol. 9, pp. 1656–1659, 2018. [11] M. Otomo *et al.*, *VLSI*, T15.2, 2024.

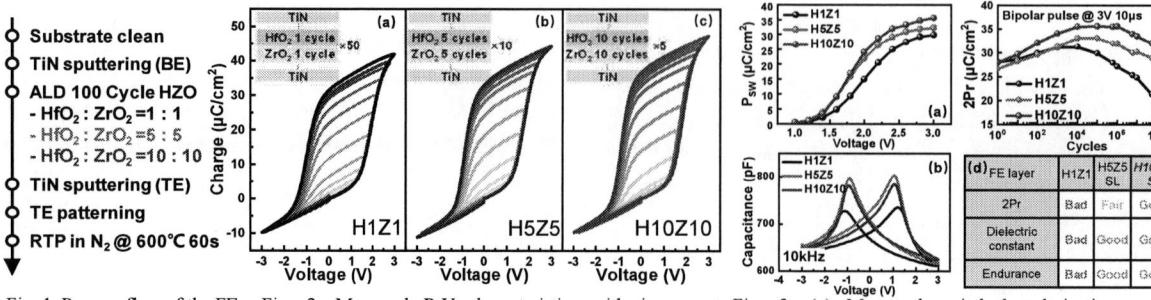

Fig. 1. Process flow of the FE capacitors. The superlattice FE layers are deposited by altering the ALD cycles.

Fig. 2. Measured P-V characteristics with increment voltage amplitude for the MFM capacitors with (a) H1Z1, (b) H5Z5 SL and (c) H10Z10 SL. Insets are the schematic illustration of capacitors with different FE layers.

Fig. 3. (a) Measured switched polarization under different voltages. (b) Small signal C-V characteristics of capacitors. (c) Endurance test of the capacitors. (d) Comparation of the three FE layers.

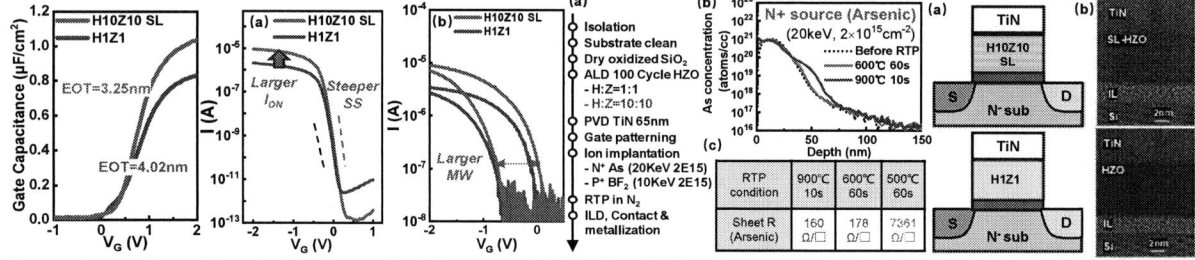

Fig. 4. Measured gate capacitance, the gate stack with SL FE layer has a lower EOT.

Fig. 5. (a) Measured single sweep DC transfer characteristics of the fabricated FE-FETs. (b) Measured pulsed I_d-V_g curves of the fabricated FE-FETs for MW evaluation.

Fig. 6. (a) Fabrication process of the FE-TFETs. (b) SIMS measurement results of As. (c) Measured sheet resistance after typical RTP conditions.

Fig. 7. (a) Illustration of the FE-TFETs with H1Z1 and H10Z10 SL FE layers. (b) TEM images of the as-deposited gate stacks.

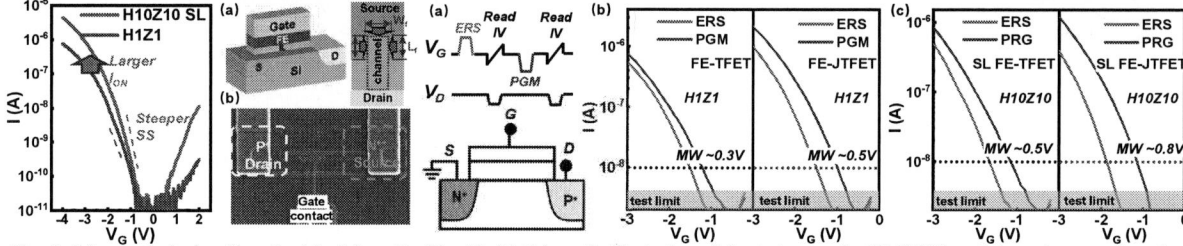

Fig. 8. Measured single sweep DC transfer characteristics of the FE-TFETs. I_{ON} and SS are improved by SL FE layer.

Fig. 9. (a) Schematic illustration of the FE-JTFET. (b) Top view SEM image of the fabricated FE-JTFET.

Fig. 10. (a) Schematic illustration of the test setup for FE-TFETs and the pulse waveform for measuring MW. Measured pulsed Id-Vg curves of the FE-TFET and FE-JTFET with H1Z1 FE layers on the same die (b) and FE-TFET and FE-JTFET with H10Z10 FE layers on the same die (c). The superlattice FE layer and an abrupt junction are both able to improve the MW. The SL FE-JTFET demonstrates the largest MW. V_{ERS} = 7 V, V_{PRG} = -6 V, pulse width = 10µs.

	Ref. [6]	Ref. [7]	This work	
Device structure	FE-TFET	FE-TFET	FE-TFET	FE-JTFET
N/P type	N-type	P-type	P-type	P-type
FE layer	10nm HZO	6nm HZO	8nm SL-HZO	8nm SL-HZO
Inter layer	-	2nm SiO₂	2.5nm SiO₂	2.5nm SiO₂
MW	~0.5V	~0.5V	~0.5V	~0.8V
EOT Opt-imization	No	No	Yes	Yes

Table I. Benchmark of the reported Si FE-TFETs. This work reports optimization in gate stack and tunnel junction. The SL FE-JTFET shows superior MW and improved EOT.

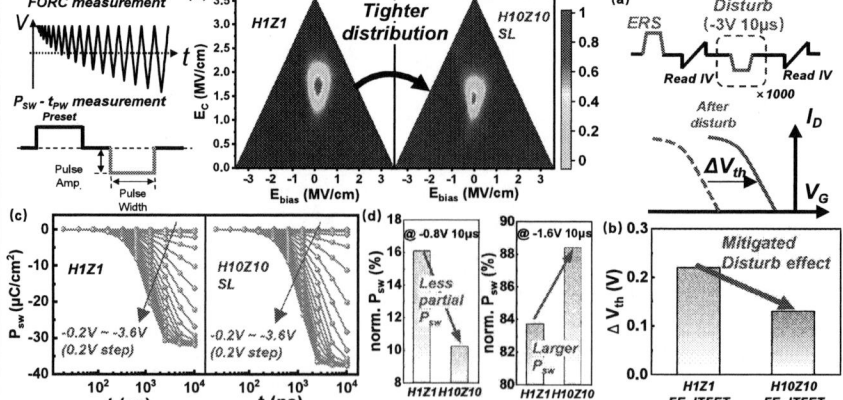

Fig. 11. (a) Schematic illustration of the measurement waveform for FORC and P_{SW}-t_{PW} tests. (b) Measured FORC diagrams. The H10Z10 SL shows tighter EC distribution. (c) Measured P_{SW}-t_{PW} curves. (d) Extracted normalized P_{SW}. The SL FE layer demonstrates less partial P_{SW} at low pulse amplitudes and larger P_{SW} at high pulse amplitudes.

Fig. 12. (a) Schematic illustration of the disturb test with V_W/2 in FE-JTFETs. (b) Extracted V_{th} shift. The SL-FEJTFET suffers less disturb.

979-8-3315-0417-5/25 $31.00 © 2025 IEEE   936

A 512×256 TFT-based image array sensor with high sensitivity, high frame rate and wide dynamic range for industrial soft X-ray detection

Weikang Yan[1], Haorong Xie[1], Xianming Li[1], Chao Gao[1], Qi Liu[1], Song Kang[1], Jun Chen[1], and Kai Wang[1*]

[1]Guangdong Province Key Laboratory of Display Material and Technology, State Key Laboratory of Optoelectronic Materials and Technologies, School of Electronics and Information Technology, Sun Yat-sen University, Guangzhou, China

Abstract

This work presents a novel active pixel sensor (APS) design for soft X-ray flat panel detectors (FPDs) intended for non-destructive testing (NDT) applications that require high sensitivity, a wide dynamic range, and fast imaging capabilities. Unlike conventional passive pixel sensor (PPS), this design employs dual-gate thin film transistor (TFT) operating in the subthreshold region to compress and amplify optical signals, thereby expanding both detection limit and range. The array features randomly-accessible capabilities, achieving a frame rate of 88fps, making it as a promising candidate for real-time in-line inspection applications.

Keywords: Active pixel sensor, high speed, X-ray detectors

Introduction

X-ray detection is an NDT technology that plays a crucial role in the field of industrial and food testing. Soft X-rays are essential for achieving optimal contrast when evaluating soft materials such as PU, apple and etc. [1]. However, current X-ray FPDs exhibit limitations in detecting low-energy X-rays within the industrial sector, primarily due to the detectors' inadequate low-dose detection capabilities and frame rates. To address these issues, two approaches have been developed, corresponding to two distinct types of X-ray FPDs: direct-conversion X-ray FPDs and indirect-conversion X-ray FPDs. The former relies on photoconductive materials that exhibit greater absorption and photoelectric effect in the presence of low energy X-rays [2]. In addition to this material dependence, the latter also involves modifying the pixel design. By incorporating a transistor with amplification functionality within the pixel, it is possible to enhance pixel gain and signal-to-noise ratio (SNR). For in-line inspection, X-ray FPDs are typically configured as line arrays, such as time delay integration (TDI) CMOS linear array camera systems, which significantly reduce data volume compared to area arrays and maintain high frame rates. Due to the limitations of the CMOS process especially over large area, the costs associated with CMOS detectors are higher than those of TFT-based FPDs. Nevertheless, due to nature of disordered thin film semiconductor, TFT-based panels often exhibit lower frame rates compared to CMOS detectors. The pixel operation of both CMOS and TFT detectors

requires timing of reset, integration and readout which in turn determine its speed. In the 1990s, the Belgian Microelectronics Research Center (IMEC) proposed a randomly-accessible logarithmic APS circuit based on CMOS technology [3], leading to a wide dynamic range and a fast random readout. However, the design has not been adopted in practice because its high fixed pattern noise and low responsivity at low-level light conditions. Since 2019, we have developed a 2T1D logarithmic-exponential APS circuit based on dual-gate TFT technology, aiming to reduce noise and lower the detection limit and hedge the disadvantages of logarithmic APS circuit.

Design

The building block of the 2T1D circuit is a dual-gate TFT, a four-terminal device. Compared with a single-gate TFT, the dual-gate TFT has both top gate and bottom gate, which enhance its performance in terms of controllability, drive ability and stability [4]. As shown in the inset of Fig. 1, by adjusting the top gate of TFT_1, it operates in the subthreshold region, enabling the logarithmic conversion of photocurrent to Vds1, as illustrated in Eq. (1).

$$V_{ds1} = V_{TH1} + \frac{\beta_1 KT}{q} \cdot ln(\frac{I_{PD}}{I_{D01}}) \qquad (1)$$

Subsequently, by adjusting the top gate of TFT_2, it is also operated in the subthreshold region, where it exhibits the function of exponential amplification. The relationship between the output current and Vds1 is illustrated in Eq. (2).

$$I_{out} = I_{D02} \cdot exp[\frac{q(V_{ds1} - V_{TH2})}{\beta_2}] \qquad (2)$$

The block diagram of the 2T1D pixel array and signal acquisition system is shown in Fig. 1. The readout chip is ADI's AD71124, and the driver chip is Himax's HX8698-D. Such a APS does not need the integration process of PD during global X-ray exposure which differs from PPS，but adopts rolling shutter exposure, as shown in Fig. 2, which can achieve higher frame rates to cope with high-speed industrial production lines [5]. The minimum line scan time for the readout chip is 22μs, and the array has 512 lines, so its maximum frame rate reaches 88.78fps.

Experimental

In this paper, a 512×256 array sensor with a pixel pitch of 300μm is fabricated utilizing commercial TFT-LCD processes. The cross-sectional view of the single pixel is shown in Fig. 3, and the schematic cross-section of circuit are shown in Fig. 4. To evaluate the performance of 2T1D, the light intensity-current curve is shown in Fig. 5. Compared with PD of the same size, the dark current of 2T1D remains unchanged, whereas photocurrent is enhanced by 2 orders of magnitude, and the light detection capability is improved from $10nw/cm^2$ to $100pw/cm^2$, and the dynamic range reaches 120dB. Then we calculate SNR according to Eq. (3).

$$SNR = 20 \cdot lg[\frac{I_{light} - I_{dark}}{I_{dark}}] \qquad (3)$$

At an illumination intensity of $10 \ nW/cm^2$, the SNR reaches 46 dB, and with each order of magnitude increase in light intensity, the SNR roughly increases by 15 dB. At an illumination intensity of $1mW/cm^2$, the SNR can reach up to 115 dB as shown in Fig. 6. Meanwhile, the external quantum efficiency (EQE) of 2T1D and a single PD under various light wavelengths are compared in Fig. 7. A single PD can respond to visible light wavelengths, its EQE is less than 100%, while the 2T1D active pixel circuit's gain can be as high as $1.5 \times 10^5 \%$ in the visible light wavelengths after the internal TFT amplification, more than three orders of magnitude higher than PD. Simultaneously, the photoelectric conversion efficiency of PD in the near ultraviolet and near-infrared wavelengths is very weak, and the gain can exceed $10^5\%$ and $10^2\%$ after the amplification of 2T1D active pixel circuit. Due to the characteristics of amorphous silicon materials, its detection ability is strongest in the visible light wavelengths. Therefore, the photoelectric detection capability of 2T1D active pixel circuit is enhanced compared to the PD, and the sensitivity of the pixel circuit is improved.

Then, an imaging system is built for the entire array. To validate its high frame rate, a coin is released in the air and dynamic images are captured at a visible light intensity of $1mw/cm^2$ in Fig. 8. Finally, we attached 405-μm-thick GOS (Gd_2O_2S) scintillator from Toray Industries Inc. to the surface of the array. GOS is chosen over CSI due to its low afterglow, which is advantageous for high-frame-rate imaging. For X-ray source, we select the IXS080BP210P396 hot-cathode tungsten target X-ray source from VJ Company. The array is positioned about 20cm from the X-ray source, with the tube voltage at 20kV, and the tube current from 0.5mA to 5mA to perform sensitivity and SNR analysis. The mean gray value and SNR at dose rates corresponding to various tube voltages and currents are calculated and calibrated in Fig. 9. At this dose, the mean gray value increases linearly with the dose rate. The SNR is calculated by referring to ASTM's international Standard E2597/E2597M-14, titled "Standard Practice for Manufacturing Characterization of Digital Detector Arrays". At a dose of 59.5μGy/s, the system maintains the SNR of 7.46. The modulation transfer function

(MTF) describes the spatial resolution properties of imaging systems. In Fig. 10, we evaluated the MTF of the sensor covered with the GOS scintillator at tube voltage of 80kV and tube current of 80μA. MTF reached 83.8% at 0.5cy/mm and 52.5% at 1cy/mm. We also take images of the resolution chart to evaluate its spatial resolution, which has a theoretical resolution of 1.67 lp/mm for 300 μm pixel. Finally, we tested the practical imaging capabilities by photographing an apple with a flawed core and an insole with bubbles. In Fig. 11, the pentagon star core inside the apple, the lines and defects inside the insole are clearly visible in low-energy X-rays.

Conclusion

In order to solve the non-destructive testing of soft substances in online industrial production, we propose to use an active pixel sensor circuit, simply 2T1D with high sensitivity and high frame rate, which's output current per pixel is 100 times that of PD with the same size. With 405-μm GOS scintillator, 88fps is achieved, with MTF reaching 83.8% at 0.5cy/mm, for in-line industrial X-ray inspection.

Acknowledgments

This work is supported in part by the National Key Research and Development Program of China under Grant 2022YFA1204202 and in part by the National Natural Science Foundation of China under Grant 62274186.

References

[1] Z. Du, Y. Hu, N. Ali Buttar, and A. Mahmood, "X‐ray computed tomography for quality inspection of agricultural products: A review," Food Science & Nutrition, vol. 7, no. 10, pp. 3146-3160, Aug. 2019.

[2] J. Zhao et al., "Perovskite-filled membranes for flexible and large-area direct-conversion X-ray detector arrays," Nature Photonics, vol. 14, no. 10, pp. 612–617, Oct. 2020.

[3] N. Ricquier and B. Dierickx, "Pixel structure with logarithmic response for intelligent and flexible imager architectures," ESSDERC '92: 22nd European Solid State Device Research conference, Leuven, Belgium, 1992, pp. 631-634.

[4] M.-J. Spijkman, K. Myny, E. C. P. Smits, P. Heremans et al., "Dual-Gate Thin-Film Transistors, Integrated Circuits and Sensors," Advanced Materials, vol. 23, no. 29, pp. 3231–3242, Jun. 2011.

[5] C. -K. Liang, L. -W. Chang and H. H. Chen, "Analysis and Compensation of Rolling Shutter Effect," in IEEE Transactions on Image Processing, vol. 17, no. 8, pp. 1323-1330, Aug. 2008.

[6] G. H. Gelinck et al., "X-Ray Detector-on-Plastic With High Sensitivity Using Low Cost, Solution-Processed Organic Photodiodes," in IEEE Transactions on Electron Devices, vol. 63, no. 1, pp. 197-204, Jan. 2016.

[7] T. Shan, J. Li, C. Zhou, F. Chang and X. Guo, "Organic Active-Matrix Imager with Ultra-low Illumination Detection Capacity for Lens-Free Optical Analysis," 2023 International Electron Devices Meeting (IEDM), San Francisco, CA, USA, 2023, pp. 1-4.

979-8-3315-0417-5/25 $31.00 © 2025 IEEE

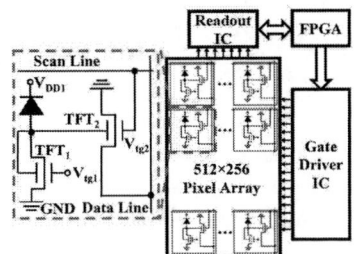

Fig. 1: 2T1D pixel array and signal acquisition system

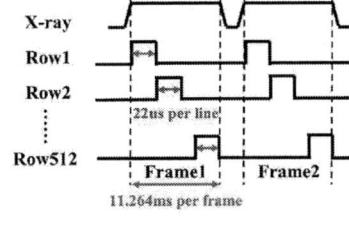

Fig. 2: Rolling shutter exposure mode

Fig. 3: Micrographic photo of the pixel

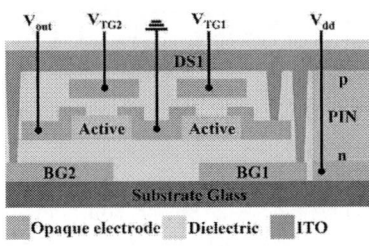

Fig. 4: Schematic cross-sectional of the circuit

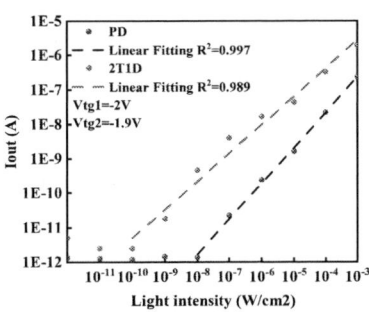

Fig. 5: Output currents of PD and 2T1D under different light intensities

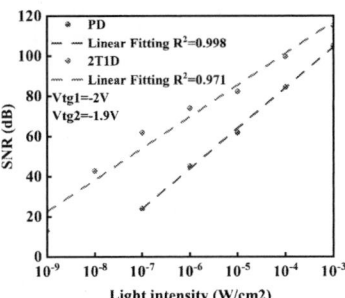

Fig. 6: SNR of the 2T1D (wavelength: 560nm)

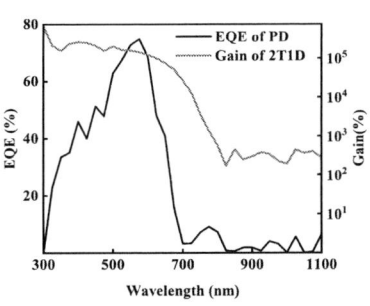

Fig. 7: Gain of the 2T1D APS and EQE of the PD as a comparison

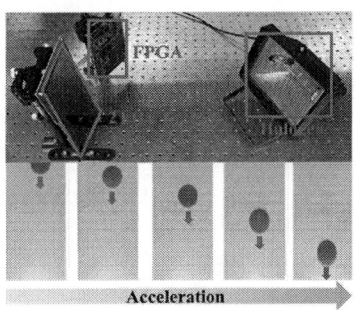

Fig. 8: Array chip dynamic characteristic test system

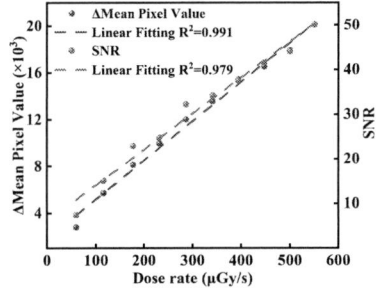

Fig. 9: Mean gray value and SNR at different dose rate

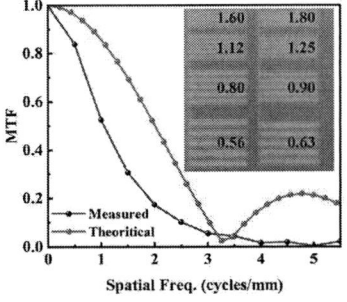

Fig. 10: MTF and resolution chart

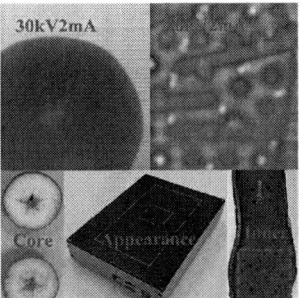

Fig. 11: Prototype, Internal structure of apple and insole

Table 1: Comparison with other works

Parameter	This work	[6]	[7]
Technology	a-Si	IGZO	a-Si
PD type	PIN	OPD	OPD
Pixel pitch(μm)	300	200	500
Pixel Resolution	512×256	160×180	64×64
Farme Rate(fps)	88	10	NA
Scintillator	GOS	CSI	CSI
EQE or gain(%)	>10^5	25	78.8
MTF@(1cy/mm)	52%	54%	37%

Investigation of Effects of Non-Volatile Memory-based Computation-in-Memory Non-Idealities and Model Size on Performance and Robustness of Small Language Model During Inference Phase

Adil Padiyal[1], Tao Wang[1], Naoko Misawa[1], Chihiro Matsui[1], Ken Takeuchi[1]

[1]Department of Electrical Engineering and Information Systems, The University of Tokyo, Japan

E-mail – adil.padiyal@co-design.t.u-tokyo.ac.jp

Abstract

This paper comprehensively evaluates the effect of employing non-volatile memory (NVM)-based computation-in-memory (CiM) for small language model acceleration. The effect of memory non-idealities on the performance of the small language models, along with the similarity between CiM-based CNNs and language models, is documented with similarities between error tolerance of large language models and convolutional neural networks. The study also establishes a trend between the type of NVM non-idealities, model performance, model size, and noise tolerance.

Keywords: Computation-in-Memory, Non-volatile Memory (NVM), Small Language Models, Edge AI

Introduction to Computation-in-Memory

A. Computation-in-Memory (CiM)

In recent times, to overcome the barrier of AI memory wall [1] as well as to mitigate the memory bandwidth and access limitations of the von Neumann architecture, interest in Computation-in-Memory (CiM) architecture-based accelerators for Neural Networks (NNs) and other machine learning models has been growing rapidly. Neural networks and LLMs consist of multiple types of computations such as matrix multiplications, batch normalization, and non-linear functions (such as softmax and GeLU). As such, the usage of CiM modules significantly speeds up the computations involving matrices and multiply-accumulate (MAC) operations in Fig. 1 at appropriate locations in various models resulting in a much more energy-efficient, and low-latency implementation compared to general purpose GPU based implementations.

The non-volatile memory (NVM) based CiM performs the MAC operations in NNs as AI accelerators with a potential application in Edge AI. This is done by CiM storing model weights in NVM devices and performing the model inference at much lower power and higher speed. This is due to the capability of CiM array, in Fig. 1, to map the input as word line voltage and add bit line current to perform accumulation operation. It should be noted however that NVM, like ReRAM and FeFET, are subject to device-to-device variations and are limited in bit precision to represent model weights which can result in a significant inference accuracy degradation due to insufficient bits for weight representation [2]. The type of variation and error respective to each type of memory is introduced in the subsequent section.

B. Non-volatile memory and its non-idealities

Using NVM in CiM brings about its set of challenges and obstacles, the non-idealities the system has to tolerate can result in a significant performance drop with a possibility of a non-functioning application. Therefore, it is important to recognize and study the effects of the non-idealities that are incurred due to the usage of NVM-CiM. The most significant non-idealities can be segregated into quantization errors and read-write and data retention variation. The quantization errors occur due to the limitation of a single cell to store a value bounded by an upper limit to bit precision, and hence the format of the stored weights from a neural network cannot be an arbitrary high bit precision.

The main focus of this study is the errors introduced in the system due to read-write variations and data retention errors. These memory non-idealities can manifest as errors in the model. The corresponding underlying physical phenomena to the conductance non-idealities are listed in Table 1. Since the said memory errors have adverse effects on the performance of the models [3], the models must be robust against such corruption, and the dependence of the said robustness on the model specifications must be studied to confirm the viability of NVM-CiM and remains the primary focus of this research.

NVM-CiM, LLMs and Small Language Models

Large Language Models are massive machine learning models aimed to generate text (and in some cases, images) for a given prompt/input (which can be either text or input). LLMs can have parameters ranging from a few billion to as much as a trillion compared to a few million for convolutional neural networks (CNNs, E.g. Fig. 2). The basic building block of a LLM is the transformer [4]. A transformer block, illustrated in Fig. 3), consists of various layers which in turn involve different linear and non-linear operations. To be able to deploy LLMs for edge AI, the most important metric is the number of parameters to be stored on the edge device. To put the size of regular LLMs in the perspective of memory, for GPT-3, the amount of memory to just store/load the parameters (with 2-byte precision) is approximately 350GB. This foreshadows an uncharacteristically high memory requirement for edge devices for LLMs. Therefore, in the study, the focus is on Small Language Models (SLMs) that can potentially fit onto high-end personal devices and quantized versions onto edge as well as personal devices with a limit of 32GB. E.g., Phi-3-mini sizes up to 7.6 GB (for 16-bit precision), while the quantized version of 4-bit can run on modern phones [5]. Due to the feasibility of deploying SLMs

979-8-3315-0417-5/25 $31.00 © 2025 IEEE

on personal and edge devices, a plausible application of CiM accelerators, the research is done from a perspective of SLMs. Since the intended area of acceleration for NVM CiM is matrix multiplications (involving MAC operations) which constitute the bulk of SLM computations and consume the most time and energy in the system, in an SLM, CiM is used in the indicated layers in Fig 3. Due to the different nature of matrix weights, other research proposes architecture such as Hetero-CiM [6] for inference phase, necessitating a need for analysis of the different error types and the respective model performance during inference.

Experiment and Simulation Environment

To emulate non-ideality of NVM CiM, the errors incurred due to the modules were introduced only in the layers where CiM is employed to accelerate the computations. For SLMs, these are the linear layers of self-attention, fully connected (FC) layers and multi-layer perceptron (MLP) layers/feed-forward layers during inference. The SLMs selected for the purpose of the research were Phi-2 (2.7 billion parameters) [7], Phi-3-mini-4k-instruct (3.8 billion parameters) [5], and LLaMA-3.1-8B-instruct (8 billion parameters) [8]. For ResNet-32, in Figs. 2 and 4, a CNN composed of various convolution layers and batch normalization layers, the errors were introduced in the conv2d and FC layers, for models trained at various gradient bit-precisions [3]. For training and evaluation of ResNet-32, the environment setup of [3] was recreated. For reference of scale of models, the total number of parameters that in ResNet-32 model is around 460,000, smaller than that of SLMs by a magnitude of 3-4.

The language models were evaluated using LLM-as-a-judge method using Prometheus 2-7B model [9] as the judge for the Big Bench Hard dataset [10] over 5 seeds each for generating responses from the model during inference and evaluating the responses using Prometheus 2. The responses were evaluated using rubrics score (1-5) to distinguish incorrect answers from garbled text as illustrated in Fig. 5. The errors introduced were zero-mean additive Gaussian noise and conductance shift errors (as seen in Table 1) with varying rates to simulate potential asymmetry in device non-idealities without further quantization [2]. The absolute standard deviation (S.D.) for Gaussian errors and delta for shift errors, added to model weights to emulate CiM, was varied from 0.0005 to 0.05. The degradation of the error-injected model is calculated w.r.t. performance of conventional models with 16-bit weights with no non-idealities.

Performance and Robustness Analysis of SLMs, CNN

The effect of noise injection in appropriate layers of language models can be observed in Figs. 6 and 7 while the sensitivity of ResNet-32 to noise can be observed in Fig 8. Figures 6 and 7 are classified into (a)-(d) by the respective percentage of error prone weights in a layer termed as rate. Figure 8 explores response of ResNet-32 to errors in different layer blocks (set of 10 layers) of the network trained for different gradient bit precision as indicated in the legend of Fig. 8. For each rate in Figs. 6 (a)-(d) and 7 (a)-(d), the degradation of the model performance increases with the size of the model for same amount of error, i.e., a larger model's performance degrades worse in presence of noise as LLaMA3.1 has a worse degradation than Phi-3 which in turn degrades more than Phi-2. This effect is possibly due to overfitting and higher dimensionality of bigger SLM models. Despite different architecture, the degradation of much smaller models such as ResNet-32 (with a degradation of -1.32% for error characteristics for Gaussian error with S.D. = 0.0005, refer to Fig. 8 (a)-(d)) is comparable to that of the Phi-2 model (with a degradation of -6.23% for same error characteristics, refer to Fig. 6 (a)-(d)) indicating a similar robustness. Due to different architecture compared to SLMs, ResNet32's robustness cannot be solely contributed to the overfitting or training quality of CNN and remains a topic of future research. The negative degradation indicates an improvement in model performance in the presence of small noise. This is owing to the fact that noise can act as a regularization phenomenon for CNNs and reduce overconfidence for SLM. Additionally, SLMs are more robust to Gaussian errors than to conductance shift errors owing to distribution of gaussian and shift errors for a similar S.D. and delta. To quantify the performance, the degradation of SLMs, averaged over rates, for a common error type and characteristics (S.D. and delta) is reported in Table 2.

Conclusions

This research establishes the effect of NVM CiM non-idealities on SLMs for deployment on edge devices as well as personal devices, illustrating the relationship between sensitivities of SLMs w.r.t. model size and error type. The degradation for different models for common error (Gaussian and constant shift) is noted in Table 2. The robustness of much smaller models such as CNNs is similar to the smaller language models. The potential effect of quantization on performance and fine-tuning the model to mitigate degradation is left up to future research.

Acknowledgements

This paper is based on results obtained from a project, JPNP23015, subsidized by the New Energy and Industrial Technology Development Organization (NEDO). The authors thank Koki Shibata for his valuable feedback.

References

[1] A. Gholami *et al.*, *IEEE Micro*, vol. 44, no. 3, pp. 33-39, 2024. [2] K. Higuchi *et al.*, *JJAP*, vol. 61, pp. SC1054, 2022. [3] A. Padiyal *et al.*, *JJAP*, vol. 63, pp. 04SP15, 2024. [4] A. Vaswani *et al.*, *NeurIPS*, pp. 5998-6008, 2017. [5] A. Marah *et al.*, *arXiv preprint arXiv:2404.14219*, 2024. [6] N. Misawa *et al.*, *IEEE IMW*, pp. 1-4, 2024. [7] M. Javaheripi et al., "Phi-2: The surprising power of small language models.", *Microsoft Research Blog*, 2023. [8] A. Dubey et al., *arXiv preprint arXiv:2407.21783*, 2024. [9] K. Seungone et al., *arXiv preprint arXiv:2405.01535*, 2024. [10] M. Suzgun et al., *arXiv preprint arXiv:2210.09261*, 2022.

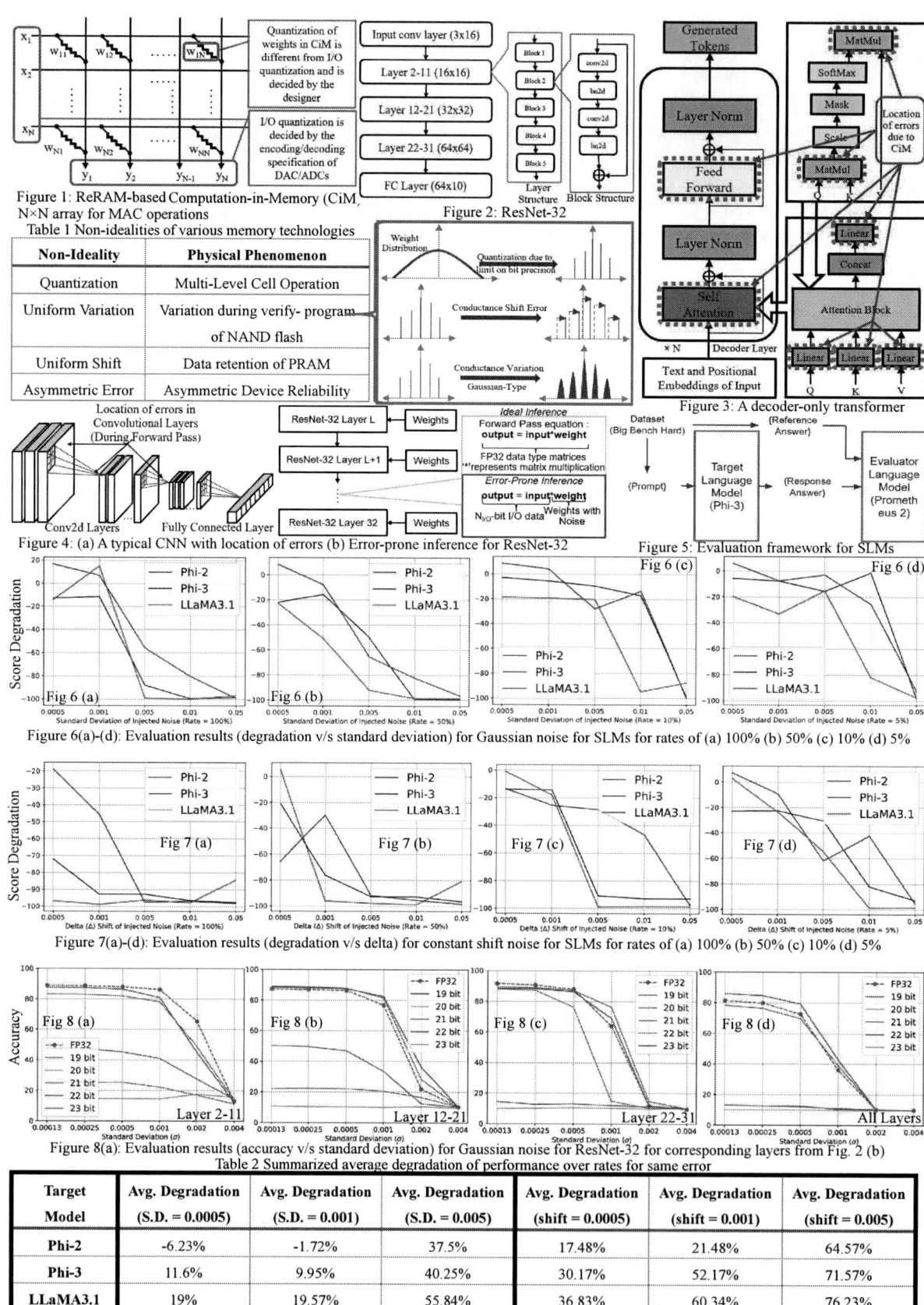

Figure 1: ReRAM-based Computation-in-Memory (CiM) N×N array for MAC operations

Figure 2: ResNet-32

Table 1 Non-idealities of various memory technologies

Non-Ideality	Physical Phenomenon
Quantization	Multi-Level Cell Operation
Uniform Variation	Variation during verify-program of NAND flash
Uniform Shift	Data retention of PRAM
Asymmetric Error	Asymmetric Device Reliability

Figure 3: A decoder-only transformer

Figure 4: (a) A typical CNN with location of errors (b) Error-prone inference for ResNet-32

Figure 5: Evaluation framework for SLMs

Figure 6(a)-(d): Evaluation results (degradation v/s standard deviation) for Gaussian noise for SLMs for rates of (a) 100% (b) 50% (c) 10% (d) 5%

Figure 7(a)-(d): Evaluation results (degradation v/s delta) for constant shift noise for SLMs for rates of (a) 100% (b) 50% (c) 10% (d) 5%

Figure 8(a): Evaluation results (accuracy v/s standard deviation) for Gaussian noise for ResNet-32 for corresponding layers from Fig. 2 (b)

Table 2 Summarized average degradation of performance over rates for same error

Target Model	Avg. Degradation (S.D. = 0.0005)	Avg. Degradation (S.D. = 0.001)	Avg. Degradation (S.D. = 0.005)	Avg. Degradation (shift = 0.0005)	Avg. Degradation (shift = 0.001)	Avg. Degradation (shift = 0.005)
Phi-2	-6.23%	-1.72%	37.5%	17.48%	21.48%	64.57%
Phi-3	11.6%	9.95%	40.25%	30.17%	52.17%	71.57%
LLaMA3.1	19%	19.57%	55.84%	36.83%	60.34%	76.23%

979-8-3315-0417-5/25 $31.00 © 2025 IEEE

Effect of Cu microstructures on Cu/SiO₂ hybrid bonding for 3D IC heterogeneous integration

Chih Chen[1], Huai-En Lin[1], Wei-Lan Chiu[2], Hsiang-Hung Chang[2]

[1]Department of Materials Science and Engineering, National Yang Ming Chiao Tung University, Hsinchu 300093, Taiwan.

[2]Electronic and Optoelectronic System Research Laboratories, Industrial Technology Research Institute (ITRI), Hsinchu 310401, Taiwan

Abstract

This paper reviews the effect of Cu microstructures on the Cu/SiO₂ hybrid bonding for 3D IC heterogeneous integration. Four Cu microstructures are discussed, including regular Cu, (111)-oriented nanotwinned Cu, fine-grained Cu, and nanocrystalline Cu. The Cu microstructures affect the amount of expansion of Cu pads, as well as the surface diffusivity during the bonding process. Therefore, it is critical to select proper Cu microstructures for different applications.

Keywords: Cu hybrid bonding, Heterogeneous integration, 3D IC packaging.

Introduction

As the dimension of the solder microbumps continue to scale down below 20 μm pitch and beyond, several problems may occur in solder joints, including solder bridging and low yield, [1] sidewall wetting induced necking, [2] and brittle joints. Even the fine-pitch solder microbumps can be fabricated successful, their high resistance would increase the RC delay and thermal dissipation. Therefore, Cu/SiO₂ hybrids have

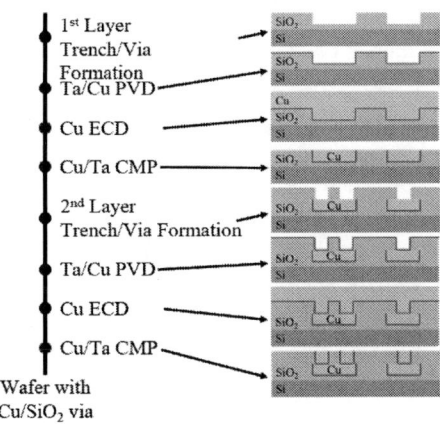

Figure 1. Schematic drawing showing the process flow for the fabrication of Cu/SiO2 hybrid structures.

been developed to meet the requirements for high performance devices, such as AI chips and high performance computing chips. [3-5]. The pitch of the Cu/SiO₂ hybrids joints can be scaled down to sub-micron. This is because the bonding schemes and processes are perfectly designed. Figure 1 shows the schematic process flow for the Cu/SiO₂ hybrids joints. First, an oxide layer is deposited by plasma-enhanced chemical vapor deposition (PECVD) on a Si wafer,

and lithography processes and oxide etching are carried out to define the patterns for Cu interconnects. A Ta diffusion barrier and a Cu seed layer are sputtered on the surfaces of the Si wafer, followed by the electrodeposition (ECD) of Cu. Usually, this Cu layer is designated as redistribution layer for signal and power delivery. Then the Cu film is planarized by chemical-mechanical planarization (CMP), so that the individual Cu lines and pads can be exposed. Then another PECVD oxide is deposited, and the above processes are repeated to form the Cu pads for bonding. The critical step for the hybrid joint fabrication is the second step CMP, which well-control the height difference between the Cu pads and the surrounding SiO₂ films in a 12" Si wafer. Typically, the heights of the Cu pads should be few nanometers lower than the surrounding SiO₂ films, which is called recessed Cu. We will explain why the few-nm recessed Cu pads are preferred later.

The bonding mechanism is shown in Figure 2 schematically. The Si wafers are cleaned and activated by plasma, and they are aligned face-to-face for Cu direct bonding, as illustrated in Figure 2(a). SiO₂-SiO₂ dielectric bonding is carried out at room temperature, as depicted in Figure 2(b). Because the Cu pads are recessed, so they do not touch each other at this stage. Then the bonded wafers are heated to temperatures ranging from 200 °C to 350 °C to form Cu-Cu bonding without external pressure, as presented in Figure 2(c). During the heating, the Cu pads expand more than the SiO₂ layer does due to that the linear coefficient of thermal expansion (CTE) of Cu and SiO₂ is 17×10^{-6} /°C and 0.5×10^{-6} /°C, respectively. They contact each other, and provide the compressive stress needed for the bonding. Through this smart design, numerous Cu-Cu pads can be bonded simultaneously by batch

(a) Alignment (b) Dielectric bonding (c) Cu bonding

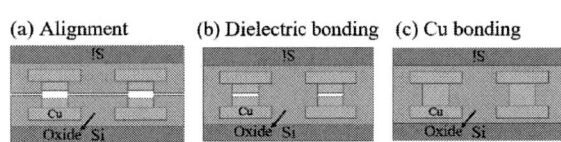

Figure 2. Schematic diagram showing the bonding processes and mechanism for CuSiO₂ hybrid bonding. (a) Alignment of top and bottom wafers. (b) Bonding of dielectric-to-dielectric layers at room temperature. (c) Increasing temperature to let the Cu pads expand to form Cu-Cu bonding.

annealing processes. Since the dielectric layers are sealed prior to Cu-Cu bonding, and the Cu-Cu is achieved in solid

979-8-3315-0417-5/25 $31.00 © 2025 IEEE

state, almost no bridging issue for the Cu/SiO₂ hybrid bonding, and it can be scaled down to sub-micron. Therefore, it has become the solutions for interconnects in 3D IC heterogeneous integration.

Several studies reported that different Cu microstructures possess different thermal expansion and surface diffusivity.[5-8] The microstructures of Cu pads may affect the bonding processes, quality, and reliabilities, and it is critical to control the microstructures of Cu pads in the Si wafers.

A.   Effect Cu Microstructures on the Thermal Expansion

The thermal expansion behavior of Cu pads in SiO₂ via is

Vertical expansion in the z-axis

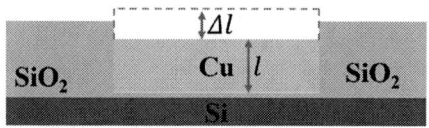

Figure 3. Schematic drawing for the thermal expansion of a Cu pad in a SiO₂ via.

quite different from that of a free-standing Cu. This is because the Cu pads adhere to the sidewalls and the bottom SiO₂ vias, so that they cannot expand freely in all direction, as shown in Figure 3. They are confined in the SiO₂ via. The SiO₂ is more rigid than the Cu, and thus the Cu expansion is limited in the X-Y planes. The Cu pad can only expand vertically toward the free surface, that is +Z axis direction. It is noteworthy to state that the SiO₂ layer may be damaged when the temperature is too high to induce high stress in the SiO₂ layer.

The Cu expansion plays in crucial role in the formation of the Cu-Cu bonding. As shown in Figures 2(b) and 2(c), if the amount of the Cu expansion is too small, the two Cu pads could not touch each other, or the induced compressive stress is not high enough to form a robust bonding. On the other hand, if the amount of the Cu expansion is too large, the Cu pads may create large tensile stress in the dielectric layers and may cause delamination in the dielectric layers. Therefore, it is very important to know the exact amount of Cu expansion at the bonding temperature. Then we can tailor the amount of Cu recess during the CMP process.

Several studies have been carried out to measure the amount of thermal expansion experimentally by atomic force microscopy (AFM) or by simulation. [6-9] The thermal expansion of Cu pads depends on the grain size, orientation, and dimension (Cu volume). In this paper, we will compare the thermal expansion of regular Cu (R-Cu), (111)-oriented nanotwinned Cu ((111) NT-Cu), fine-grained Cu (FG-Cu), and (111)-oriented nanocrystalline and nanotwinned Cu ((111) NC-NT-Cu). The grain size for the LG-Cu is few microns, and its orientation is random. The grain size of (111) NT-Cu is approximately 0.3 μm, and it is highly (111)-oriented. The FG-Cu has an average grain size of 0.15 μm with random orientation. Finally, the (111) NC-NT-Cu has a grain size of

80 nm with very high (111) preferred orientation. The thermal expansion may consist of elastic and plastic deformation. Generally speaking, the smaller the grain size is, the larger the thermal expansion. This is because creep may take place at a temperature higher than 200 °C for FG-Cu and NC-Cu. Table 1 summaries the effective thermal expansion of Cu measured by in-situ heating AFM for Cu pads with various microstructures. [7,8]

The effective CTE ($\Delta\alpha$) is defined by the Equation (1):

$$\Delta l = l \Delta\alpha \Delta T \qquad (1)$$

where Δl is the measured heigh difference at room temperature and at an elevated temperature, relative to the neighboring surface of SiO₂, as illustrated in Fig. 3, l is thickness of the Cu pad, ΔT is the temperature difference between room temperature and the elevated temperature. The results show that the (111) NC-NT-Cu with an 80 nm grain size has the largest effective CTE of 69.2 ppm/°C.[8] This

Table 1. The measured effective CTE of Cu by in-situ heating AFM for Cu pads with various microstructures.

	Diameter (μm)	Thickness (μm)	Expansion at 200 °C (nm)	Volume	α (ppm/°C)	Grain Size (μm)	Orientation
Regular-Cu	12, 8 (Two layers)	2.3	14.6	187.9	35.1	3.5	random
NT-Cu		2.3	10.9	187.9	27.6	0.3	70% of (111) Cu
Regular-Cu	3	1.5	3.7	10.6	14.7	2.5	random
FG-Cu	3	1.5	8.4	10.6	32.6	0.15	random
NC-NT-Cu	3	0.3	3.6	2.1	69.2	0.08	~100% of (111) Cu

vale exceeds the theoretic value of linear CTE of a free standing Cu. The large CTE value may be attributed two reasons. First, the Cu pads could not expand in the X-Y plane and in the -Z axis, thus all the volume expansion goes to the +Z direction, from which the deformation was measured. Second, the deformation comprised plastic deformation induced by Coble creep, which takes place along the grain boundary at elevated temperature. Since the NC-CT-Cu and FG-Cu Cu possess many grain boundaries, it is reported that more than 50% of the deformation is plastic deformation. [8] On the other hand, for regular Cu and NT-Cu, most of the deformation is elastic deformation. In addition, for regular Cu, the effective CTE of two-layer Cu pad is higher than that of single layer Cu pads. Finally, the (111) oriented Cu has the smallest value of effective CTE. [7,9,10] This may be attributed to that the (111) planes have the highest packing density.

As the dimension of the Cu/SiO₂ hybrid joints continues to scale down, the Cu pads would experience much more confinement from the SiO₂ layer, since most of the Cu volume adheres to the sidewalls and the bottom of the SiO₂ via. Therefore, it is reported the amount of Cu expansion decreases as the Cu volume decreases. [6] The process window decreases as the dimension of the Cu pads scale down. Therefore, packaging industry may adopt FG-Cu or NC-Cu for ultra-fine pitch Cu/SiO₂ hybrid joints.

B.   Effect Cu Microstructures on the Surface Diffusivity

During the bonding processes, the surfaces of the two Cu pads contact together at elevated temperatures and the two Cu pads are under compressive stress. Surface creep takes place to achieve bonding. [11] Shie et al developed a model to calculate the bonding time, and the time required to achieve the Cu-Cu bonding is described in Equation 2.[12] It shows the higher the surface diffusivity is, the shorter the bonding time. Therefore, surface diffusivity plays a key role on achieving a robust bonding interface in the Cu-Cu joints.

$$t_{bonding} \approx (7.3 \times 10^{30} m^{-3}) \times \varphi_{max} \frac{R_q^6}{6\delta D} \frac{kT}{p\Omega}.$$

The crystal of Cu is face-centered cubic, and its closest packing planes and slip planes are (111). The surface diffusivity on (111) planes are fastest among all the major planes. As listed in Table 2, the surface diffusivity on (111) planes is approximately four orders higher than those on (100) and (110). In addition, although the grain boundary serves as a fast diffusion path, its diffusivity is in the order of 10^{-10} cm^2/s at 300 °C. Therefore, (111)-oriented Cu pads can be bonded at 200 °C with a low contact resistivity. [13] On the other hand, several studies reported the Cu-Cu bonding using NC-Cu or FG-Cu, and they found Cu-Cu joints can be bonded at 250 °C, and the bonding interfaces can be eliminated. But the bonding of regular-Cu has to be higher than 300 °C. It indicates the FG-Cu and NC-Cu also facilitate the bonding. Their surfaces consist of numerous grain boundaries. Their surface diffusivity is still unknown. It is noteworthy to state that the electrodeposition of (111)-oriented NT-Cu encounters a bottleneck as the via size is smaller than 2 μm. It is difficult to electroplate highly (111) NT-Cu in a small via [14]. Therefore, FG-Cu and (111) NC-NT-Cu appear to be the candidates for Cu/SiO$_2$ hybrid joints below 4 μm pitch.

Table 2. Surface and grain boundary diffusivity at 300 °C.

Diffusivity / Temp.	(111)	(100)	(110)	Grain bourdary
300°C	1.51×10^{-5}	1.48×10^{-8}	1.55×10^{-9}	4.30 × 10^{-10}

Conclusion

This paper reviews the current status for the effect of Cu microstructures on the Cu/SiO$_2$ hybrid joints. (111)-oriented NC-Cu possesses a very high surface diffusivity and it is suitable to low temperature Cu/SiO$_2$ hybrid joints at pitches larger than 8 μm. On the other hands, FG-Cu and NC-Cu have a high ability to expand in the SiO$_2$ vias and they are promising candidates for ultra-fine pitch Cu/SiO$_2$ hybrid joints.

Acknowledgments

This work was financially supported by National Science and Technology Council, Taiwan, through the T-Star Center project "Future Semiconductor Technology Research Center", under Grant No. NSTC 113-2634-F-A49-008

References

[1] K. Oi et al., "Development of new 2.5D package with novel integrated organic interposer substrate with ultra-fine wiring and high density bumps," 2014 IEEE 64th Electronic Components and Technology Conference (ECTC), pp. 348-353 (2014).

[2] C. Chen, D. Yu, and K.N. Chen, Vertical interconnects of microbumps in 3D integration, MRS BULLETIN, 40, pp. 257-262 (2015).

[3] M.-F. Chen, F.-C. Chen, W.-C. Chiou, C. Doug, System on integrated chips (SoIC (TM) for 3D heterogeneous integration, 2019 IEEE 69th Electronic Components and Technology Conference (ECTC), IEEE, 2019, pp. 594-599.

[4] Y. Kagawa et al., Novel stacked CMOS image sensor with advanced Cu2Cu hybrid bonding, 2016 IEEE International Electron Devices Meeting (IEDM), pp. 8.4. 1-8.4. 4. (2016).

[5] E. Beyne et al., Scalable, sub 2μm pitch, Cu/SiCN to Cu/SiCN hybrid wafer-to-wafer bonding technology, 2017 IEEE International Electron Devices Meeting (IEDM), pp. 32.4. 1-32.4. 4. (2017).

[6] Y. Kim, et al., "Die to Wafer Hybrid Cu Bonding for Fine Pitch 3D-IC Applications," 2023 IEEE 73rd Electronic Components and Technology Conference (ECTC), Orlando, FL, USA, 2023, pp. 1043-1047

[7] H.E. Lin et al., In-situ measurement of thermal expansion in Cu/SiO$_2$ hybrid structures using atomic force microscopy at elevated temperatures, Applied Surface Science 662, pp.160103. (2024)

[8] H.E. Lin, D.P. Tran, W.L. Chiu, H.H. Chang, C. Chen, Enhanced thermal expansion with nanocrystalline Cu in SiO2 vias for hybrid bonding, Applied Surface Science 672, pp. 160784. (2024)

[9] B. Ayoub et al, In-situ characterization of thermomechanical behavior of copper nano-interconnect for 3D integration, Microelectronic Engineering 261, pp. 111809, (2022)

[10] J. De Messemaeker et al.,, New Cu "Bulge-Out" Mechanism Supporting SubMicron Scaling of Hybrid Wafer-to-Wafer Bonding, 2023 IEEE 73rd Electronic Components and Technology Conference (ECTC), pp. 109-113 (2023).

[11] C.M. Liu, H.W. Lin, Y.S. Huang, Y.C. Chu, C. Chen, D.R. Lyu, K.N. Chen, K.N. Tu, Low-temperature direct copper-to-copper bonding enabled by creep on (111) surfaces of nanotwinned Cu, Scientific Reports 5, pp. 09734 (2015)

[12] K.C. Shie et al., A kinetic model of copper-to-copper direct bonding under thermal compression, Journal of Materials Research and Technology, 15, 2322-2344 (2021)

[13] J.J. Ong et al., Low-Temperature Cu/SiO2 Hybrid Bonding with Low Contact Resistance Using (111)-Oriented Cu Surfaces. Materials 2022, 15, 1888.

[14] S.C. Yang et al., Periodic reverse electrodeposition of (111)-oriented nanotwinned Cu in small damascene SiO2 vias, Journal of Electroanalytical Chemistry, 935, 117328 (2023)

979-8-3315-0417-5/25 $31.00 © 2025 IEEE

Efficient Implementation of 16 Reconfigurable Boolean Logics Based on Memristors and Their Application in Image Edge Detection

Zhouchao Gan[†], Yifeng Xiong[†], Fan Yang, Xiangshui Miao, and Xingsheng Wang*

School of Integrated Circuits & Wuhan National Laboratory for Optoelectronics, Huazhong University of Science and Technology, Wuhan, China
*Correspondence e-mail: xswang@hust.edu.cn

Abstract

This paper presents a novel memristor-based logic-in-memory (LIM) design that integrates resistance input resistance output (R-R) and voltage input resistance output (V-R) principles to enhance computational efficiency. We implement 16 reconfigurable Boolean logic functions within a unified circuit framework. The design is applied to create both 1-bit and n-bit full adders, demonstrating significant improvements in speed and area efficiency. Additionally, the designs are utilized for image edge detection using both logical operations and the Sobel operator, with image results validating their effectiveness. This work highlights the potential of memristor-based logic to advance edge computing applications.

Keywords: Memristor, logic-in-memory, Boolean logic, full adder, image edge detection

Introduction

Logic-in-memory (LIM) computing using memristors is a promising non-von Neumann paradigm [1]. Based on the physical properties of input and output signals, existing schemes for memristor LIM are categorized into resistance input resistance output (R-R) [2], [3], [4], [5], [6] and voltage input resistance output (V-R) types [7], [8]. These logic types are typically realized through different circuit structures and voltage configurations, each with distinct advantages and limitations. Achieving 16 Boolean logic functions within a unified circuit framework remains challenging. In our previous work [9], we proposed a hybrid scheme combining R-R and V-R logic gates, forming a reconfigurable V/R-R logic unit with three memristors and a load resistor (3M1R). However, further exploration is required to validate its broader applicability. In this paper, we implement an efficient 1-bit full adder (FA) using V/R-R logic and extend this to an n-bit FA, demonstrating improvements in both speed and area efficiency. As a case study, the proposed logic and adder units are applied to image edge detection using both logical operations and the Sobel operator, highlighting their potential in edge computing.

Implementation of Logic and Addition Functions

A. 16 Reconfigurable Boolean Logic Functions

The proposed primitive circuit consists of two input memristors (M_1, M_2) and one output memristor (M_3) connected in parallel with a series resistor, as depicted in Fig. 1(a). The bit line (BL) connects to the top electrode, while the word line (WL) connects to the bottom electrode, with a load resistor for voltage division and current limiting. M_1 and M_2 are preset to input states a and b, while M_3 is initially set to high resistance (HRS), representing logical "0", and low resistance (LRS) represents "1". Dynamic configuration of interfaces (T_1, T_2, and T_4) allows for flexible logic operations. Fig. 1(b) illustrates the XOR logic configuration, with detailed port voltage configurations shown in Fig. 1(c), indicating that the port voltage V_{T2} depends on the value of input a. The key to logic implementation is conditionally setting memristor M_3 to low resistance ("1") when specific port voltages are applied. The value of the series resistor is $\sqrt{R_{HRS}R_{LRS}}$, and the voltage V_p must satisfy:

$$V_{set} < V_p < 1.5V_{set} \tag{1}$$

The threshold voltage for setting the memristor is denoted as V_{set}. Based on this design principle, all 16 Boolean logic functions can be implemented configurable using the 3M1R circuit by applying the corresponding voltages $V_{T1} - V_{T4}$ as shown in Fig. 2. Half of these functions require fixed voltage inputs, while the others depend on the logical input a. The proposed logic scheme features a unified circuit structure and operation steps with a limited number of required excitation voltages, facilitating efficient construction of computing units.

B. Implementation of the Adder

The full adder is a fundamental component in arithmetic units, with its logic expression as follows:

$$c_{i+1} = (a_i \oplus b_i)c_i + \overline{(a_i \oplus b_i)}a_i = P_ic_i + \overline{P_i}a_i \tag{2}$$

$$s_i = P_i \oplus c_i \tag{3}$$

Efficient logic cascading for FA computation is achieved by applying coordinated voltage sequences to the memristor crossbar array. Fig. 3(a) shows the circuit schematic of the proposed FA composed of six memristors. The 1-bit FA function is executed in five steps, with each step's computed result and port voltage configuration detailed in Fig. 3(b). In the final step, the value of P_ic_i is overwritten to obtain the carry output c_{i+1}.

Further investigations focus on implementing n-bit adders for specific applications. Fig. 4(a) and 4(b) illustrate the topologies of n-bit serial and parallel adders based on the proposed 1-bit adder design. In the serial structure, n-bit

---

[†]These authors contributed equally *corresponding author

addition is completed along a single row of memristors by sequentially executing the logic operations for each bit, requiring 5n steps and 5n+1 memristors. In contrast, the parallel adder architecture allows simultaneous computation of all $a_i \oplus b_i$, since different bits are processed in separate rows of memristors, with a delay of three logic cycles for the carry chain between adjacent bits. Thus, the n-bit parallel adder requires 3n+2 steps and $5n+1$ memristors.

C. Results and Discussion

Key fabrication steps and a TEM image of the cross-sectional cut of the 1-transistor-1-memristor (1T1R) unit are shown in Fig. 5(a). Fig. 5(b) presents the typical DC I-V curve of the 1T1R cell, while Fig. 5(c) displays the electron microscope image of a 32×32 1T1R array. Additional device characteristics are provided in [8], [10]. The test platform in Fig. 6 validates the proposed adder schemes, with results for XOR logic shown in Fig. 7, confirming correct operation across all input combinations. Circuit-level simulations for a 2-bit serial adder in Cadence Virtuoso, depicted in Fig. 8, align with expected outcomes for the test case "11+11".

Fig. 9 compares the overhead of n-bit adders using different logic primitives. The proposed n-bit adder scheme demonstrates a higher quality factor (steps × memristors) than other designs, indicating enhanced computational performance. The primary improvement is in computation speed, as repeated iterations of single logic types (such as IMPLY or MAGIC) are inefficient, whereas the proposed efficient reconfigurable logic operates effectively. However, improvements in area overhead are less significant due to the increased number of devices required by the proposed logic.

Application of Image Edge Detection

Edge detection is a key technology in image processing, used to extract important features from images. This article introduces two edge detection methods: logical operations-based edge detection and Sobel operator-based edge detection.

A. Edge Detection Based on Logical Operations

The Boolean logic-based edge detection algorithm identifies edge features in a binary image by applying shifts in four cardinal directions (left, right, up, and down), as illustrated in Fig. 10. For each direction, the shifted image is combined with the original image using a logical OR operation, followed by a logical XOR operation to determine differences and extract directional edge information. Finally, the edge data from all directions are merged using a logical OR operation to produce the complete edge detection result. Fig. 11 demonstrates this process applied to a binary ring image, with individual directional results shown in Fig. 11(b-e) and the combined outcome in Fig. 11(f).

B. Edge Detection Based on the Sobel Operator

The Sobel operator is an effective method for edge detection, approximating the gradient of image intensity in horizontal and vertical directions to achieve edge recognition [11], allowing the proposed adder unit to be seamlessly embedded. Fig. 12 compares the two edge detection methods on the "cameraman" image. Fig. 12(a) is the original grayscale image, Fig. 12(b) shows the result of the Sobel operator, highlighting fine edge details. Fig. 12(c) displays the binary image after thresholding, and Fig. 12(d) presents the edge detection result using Boolean logic, which is suitable for binary images but lacks gradient detail compared to the Sobel operator. Therefore, the Sobel operator excels in capturing detailed intensity variations in grayscale images, while the Boolean logic-based method offers simplicity and efficiency for binary image processing, making it essential to choose the appropriate method based on the specific application context.

Conclusion

In conclusion, this study demonstrates the effectiveness of a novel memristor-based LIM design that integrates R-R and V-R principles to achieve enhanced computational efficiency. By implementing 16 reconfigurable Boolean logic functions within a unified circuit framework, we have successfully developed both 1-bit and n-bit FAs that show significant improvements in speed and area efficiency. The application of these designs to image edge detection, utilizing both logical operations and the Sobel operator, confirms their capability to maintain high performance in practical scenarios. This work underscores the potential of memristor-based logic to drive advancements in edge computing applications, paving the way for further research and development in this promising area.

Acknowledgments

This work is supported in part by National Natural Science Foundation of China under Grants U2341221 and 62274070, Hubei Province Science and Technology Major Project under Grant 2022AEA001, Interdisciplinary Research Program of HUST (2023JCYJ042), and Hubei Key Laboratory of Advanced Memories.

References

[1] J. Borghetti et al., Nature, vol. 464, no. 7290, pp. 873–876, 2010.
[2] S. G. Rohani and N. TaheriNejad, in 2017 IEEE 30th Canadian Conference on Electrical and Computer Engineering (CCECE), pp. 1–4, 2017.
[3] A. Karimi and A. Rezai, J. Comput. Electron., vol. 17, no. 3, pp. 1303–1314, 2018.
[4] S. Ganjeheizadeh Rohani et al., IEEE Trans. Very Large Scale Integr. VLSI Syst., vol. 28, no. 1, pp. 297–301, 2020.
[5] N. TaheriNejad et al., in 2019 17th IEEE International New Circuits and Systems Conference (NEWCAS), pp. 1–4, 2019.
[6] N. Talati et al., IEEE Trans. Nanotechnol., vol. 15, no. 4, pp. 635–650, 2016.
[7] Z.-R. Wang et al., IEEE Trans. Electron Devices, vol. 65, no. 10, pp. 4659–4666, 2018.
[8] Y. Song et al., Adv. Sci., vol. 9, no. 15, p. 2200036, 2022.
[9] Y. Song et al., ICET, pp. 333–337, 2023..
[10] Z. Gan et al., in 2024 8th IEEE Electron Devices Technology & Manufacturing Conference (EDTM), pp. 1–3, 2024.
[11] J. Jing et al., Neurocomputing, vol. 503, pp. 259–271, 2022.

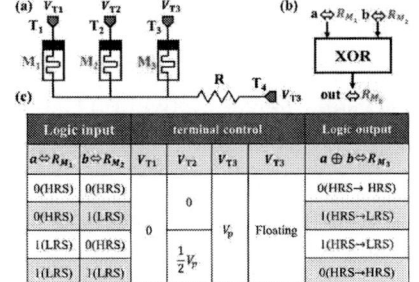

Fig. 1. (a) Circuit diagram of 16 Boolean logic implementations. (b) Example of an XOR logic gate with port configuration. (c) Truth table of XOR logic, detailing terminal voltages.

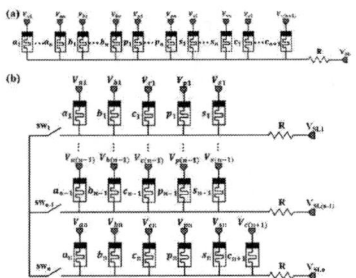

Fig. 4. (a) Serial and (b) parallel topologies of an N-bit adder.

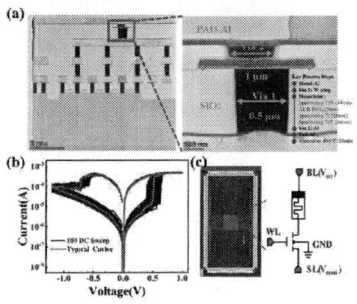

Fig. 5. (a) TEM image of the cross-sectional cut and key manufacturing steps of the 1T1R integrated unit. (b) Measured resistive switching behavior in DC IV sweeping mode. (c) Metallographic electron microscopy image of the 32×32 1T1R array.

```
Algorithm 1: Edge Detection in Binary Image using Boolean Logic
  Input: Binary image img
  Output: Final edge detection result img_edges
 1 Left Edge Detection:
 2   img_left_shifted = img(x, y-1)
 3   img_left_or = img | img_left_shifted
 4   img_edge_left = img ⊕ img_left_or
 5 Right Edge Detection:
 6   img_right_shifted = img(x, y+1)
 7   img_right_or = img | img_right_shifted
 8   img_edge_right = img ⊕ img_right_or
 9 Top Edge Detection:
10   img_top_shifted = img(x-1, y)
11   img_top_or = img | img_top_shifted
12   img_edge_top = img ⊕ img_top_or
13 Bottom Edge Detection:
14   img_bottom_shifted = img(x+1, y)
15   img_bottom_or = img | img_bottom_shifted
16   img_edge_bottom = img ⊕ img_bottom_or
17 Combine All Edges:
18   img_edges = img_edge_left | img_edge_right |
19              img_edge_top | img_edge_bottom
20 return img_edges
```

Fig. 10. Binary Image Edge Detection Algorithm Based on COPY, OR, XOR Logic.

Fig. 2. 16 types of Boolean logic control interface voltage configuration.

Logic Function		V_{T_1}	V_{T_2}	V_{T_3}	V_{T_4}
XOR	a=0	0	0	V_p	floating
	a=1	0	$\frac{1}{2}V_p$	V_p	floating
XNOR	a=0	0	$\frac{1}{2}V_p$	V_p	0
	a=1	$\frac{1}{2}V_p$	$-\frac{1}{2}V_p$	V_p	0
AND	a=0	0	$\frac{1}{2}V_p$	V_p	floating
	a=1	$\frac{1}{2}V_p$	$-\frac{1}{2}V_p$	V_p	floating
NAND	a=0	0	0	V_p	0
	a=1	0	$\frac{1}{2}V_p$	V_p	0
IMP	a=0	0	0	V_p	0
	a=1	$\frac{1}{2}V_p$	$-\frac{1}{2}V_p$	V_p	0
RIMP	a=0	0	$\frac{1}{2}V_p$	V_p	0
	a=1	0	0	V_p	floating
COPY P	a=0	0	$\frac{1}{2}V_p$	V_p	floating
	a=1	0	0	V_p	floating
COPY Q	a=0	$\frac{1}{2}V_p$	0	V_p	floating
	a=1	$\frac{1}{2}V_p$	$-\frac{1}{2}V_p$	V_p	floating
NIMP		0	$\frac{1}{2}V_p$	V_p	floating
RNIMP		$\frac{1}{2}V_p$	0	V_p	floating
OR		0	0	V_p	floating
NOR		$\frac{1}{2}V_p$	$\frac{1}{2}V_p$	V_p	0
NOT P		$\frac{1}{2}V_p$	0	V_p	0
NOT Q		0	$\frac{1}{2}V_p$	V_p	0
TRUE		0	0	V_p	0
FALSE		$\frac{1}{2}V_p$	$\frac{1}{2}V_p$	V_p	floating

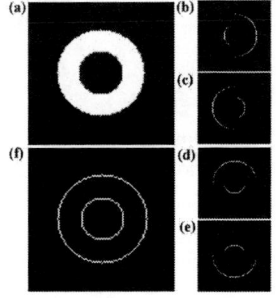

Fig. 6. Experimental test platform.

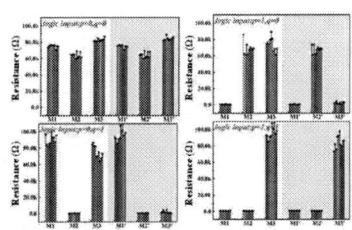

Fig. 7. Experimental results for proposed XOR logic.

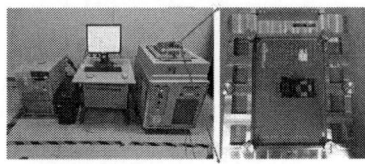

Fig. 11. Schematic diagram of a simple example of binary image edge detection. (a) Original image. (b-e) Edge detection results in four directions: left, right, top, and bottom. (f) Final result of edge detection.

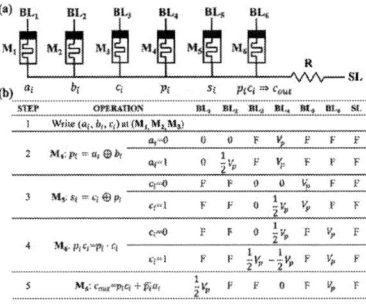

Fig. 3. (a) Circuit diagram and (b) detailed implementation steps and port voltage configuration of a 1-bit full adder.

STEP	OPERATION		BL_1	BL_2	BL_3	BL_4	BL_5	BL_6	SL
1	Write (a_i, b_i, c_i) at $(M_1 M_2 M_3)$								
2	$M_4: p_i = a_i \oplus b_i$	a_i=0	0	0	F	V_p	F	F	F
		a_i=1	0	$\frac{1}{2}V_p$	F	V_p	F	F	F
3	$M_5: s_i = c_i \oplus p_i$	c_i=0	F	F	0	0	V_p	F	F
		c_i=1	F	F	0	$\frac{1}{2}V_p$	V_p	F	F
4	$M_6: p_i c_i = p_i \cdot c_i$	c_i=0	F	F	0	V_p	F	V_p	F
		c_i=1	F	F	$\frac{1}{2}V_p$	$-\frac{1}{2}V_p$	F	V_p	F
5	$M_6: c_{out} = p_i c_i + \bar{p_i} a_i$		$\frac{1}{2}V_p$	F	F	0	F	V_p	$\frac{1}{2}V_p$

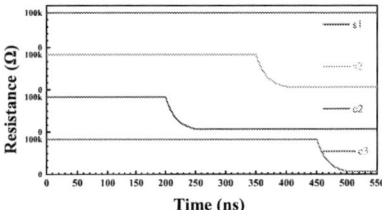

Figure 8. Simulation waveforms "11 + 11" as an example.

Ref	Logic type	Number of Memristors		Number of Steps	
		Total	N=32	Total	N=32
[2]	IMPLY Serial	2N+3	67	22N	704
[3]	IMPLY Parallel	4N+1	129	5N+16	176
[4]	IMPLY Semi-Parallel	2N+3	67	17N	544
[5]	IMPLY Semi-Serial	2N+6	70	10N+2	322
[6]	MAGIC Area Optimized	3N+7	103	15N	480
	MAGIC Latency Optimized	14N+1	449	12N+1	385
This work	Serial	5N+1	161	5N	160
This work	Parallel	5N+1	161	3N+2	98

Fig. 9. Comparison of the overhead of executing n-bit adders.

Fig. 12. (a) Original image; (b) Sobel edge detection result; (c) Binarized image; (d) Result of edge detection based on Boolean logic.

A CMOS-Compatible MoS₂ Transistor on Silicon-Rich Silicon Nitride as Multifunctional Neuromorphic Device

Xiangwei Su[1], Hongzhao Wu[1], Caijing Liang[1], Xinlong Zeng[1] and Yuda Zhao[1*]

[1] College of Integrated Circuits, ZJU-Hangzhou Global Scientific and Technological Innovation Centre, Zhejiang University, Hangzhou, 310027, China (*Email: yudazhao@zju.edu.cn)

Abstract

Neuromorphic systems process information efficiently, but scalable hardware implementation of synaptic units remains challenging. In this work, we fabricated silicon-rich silicon nitride films as the gate dielectric and leveraged charge trap interactions with MoS₂ to simulate both electrical and optoelectronic synaptic functions in a single device. This device exhibits excellent linearity with positive weights (R^2 =0.99) and demonstrates significant photo-induced short-term plasticity. Additionally, the entire fabrication process is CMOS-compatible, facilitating large-scale integration.

Keywords: Silicon-rich silicon nitride, CMOS-compatible, Neuromorphic systems, MoS₂ transistor array

Introduction

Neuromorphic computing systems integrate sensing, computing, and storage functions, mimicking the structure of brain-like computational units such as synapses and neurons to enhance processing efficiency[1]. However, replicating these units using silicon-based transistors poses challenges due to their structural complexity[2]. To address this, researchers have explored 2D materials with unique optoelectronic properties for constructing neuromorphic devices[3]. Despite this, many attempts to simulate synaptic behavior using 2D materials face integration and structural limitations, hindering scalability[4].

In this study, we used silicon-rich silicon nitride films grown by PECVD as the substrate and fabricated a 5×5 MoS₂ transistor array on top. The entire fabrication process is compatible with CMOS technology, demonstrating excellent scalability. The interaction between the charge traps in silicon nitride and electrons in MoS₂ enables the device to exhibit synaptic responses to both electrical and optical signals. When multiple electrical pulses are applied, the device shows a significant increase in conductivity to assign weights, along with excellent linearity (R^2 =0.99), which is crucial for enhancing the computational efficiency of neuromorphic systems. In terms of optical response, the device can effectively perceive multidimensional information from optical signals and displays notable short-term synaptic plasticity. The paired-pulse facilitation (PPF) reaches 134%, supporting efficient and rapid image processing capabilities in the device. These multifunctional functions present extensive application prospects in future AI hardware, especially in neuromorphic systems requiring real-time, multimodal processing.

Result and discussion

Silicon nitride (SiN$_x$), a common dielectric layer in CMOS processes, is typically grown by chemical vapor deposition (CVD) and is widely used in applications such as passivation, stress modulation, and memory dielectrics. When SiN$_x$ is used as the gate dielectric layer in contact with a thin MoS₂ channel, it can readily influence the electrical and optical behaviors of MoS₂. By adjusting the charge traps within silicon nitride, it is possible to emulate synaptic electrical and optoelectronic responses in MoS₂. Studies have shown that controlling the growth parameters can significantly affect the silicon-to-nitrogen ratio, creating dangling bonds and a high density of charge traps[5]. In this work, PECVD is used to grow approximately 100 nm SiN$_x$ film on SiO₂/p++-Si substrate, with growth conditions of SiH₄: N₂ = 40: 196, T = 623 K, time =240 s, and RF power =100 W. XPS analysis confirmed a silicon-to-nitrogen ratio of 1.17, indicating silicon-rich SiN$_x$. We then transferred a few-layer MoS₂ film via wet-transfer method and fabricated a 5×5 MoS₂ transistor array through photolithography, etching, and metal deposition techniques, as shown in Figure 1a. Figure 1b illustrates the structure of a single transistor and the schematic of electrical characterization.

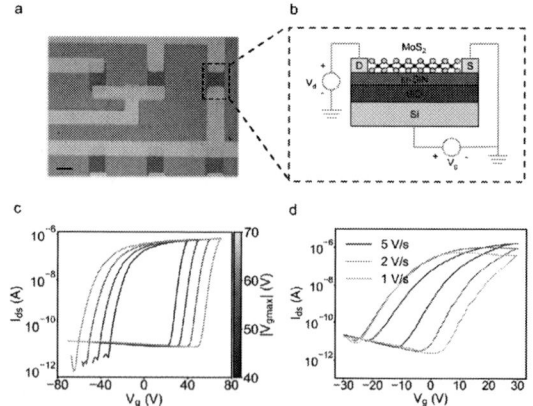

Fig. 1: Optical microscope image of MoS₂ transistor array. b, Schematic of device structure and electric measurement. c, Transfer curves under different sweeping voltage ranges. d, Transfer curves under different sweeping voltage speeds.

979-8-3315-0417-5/25 $31.00 © 2025 IEEE

To demonstrate that silicon-rich SiN_x can modulate MoS_2 to exhibit synapse-like responses, we first tested the transfer characteristics. As shown in Figure 1c, when the sweeping range gradually changes from -40 V~ 40 V to -70 V~ 70 V, the hysteresis window also increases correspondingly. This indicates that, due to the presence of charge traps, the device exhibits memory behavior, meaning that previously applied signals can influence the current state, which is analogous to synaptic behavior. Typically, the memory behavior of synapses is transient, so it is necessary to further determine whether the memory behavior exhibited by the device can dissipate over time, which requires confirming whether the traps generated in SiN_x source from shallow energy levels. By varying the sweeping speeds from 1 V/s to 5 V/s, it can be observed that as the sweeping speed increases, the hysteresis window in a clockwise direction gradually decreases (Figure 1d). This demonstrates that MoS_2 is influenced by shallow-level electron traps[6], further indicating that the behavior exhibited by the device is consistent with synaptic behavior.

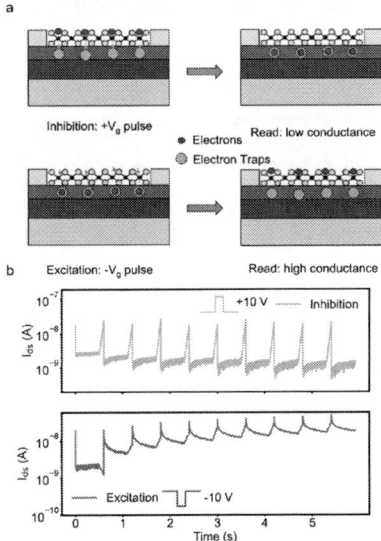

Fig. 2: a, Movement of electrons in MoS_2 under the application of positive or negative gate voltage, as well as the conductance after applying positive and negative voltages. b, The plot of I_{ds}-t under continuous +10 V, 5 ms (upper) and -10 V, 5 ms (bottom) gate pulses.

Figure 2a reveals the working mechanism of the device. There are electron traps at the interface between SiN_x and MoS_2. When a relatively large positive voltage is applied to the gate, electrons move toward the SiN_x due to the electrostatic field and get captured by the electron traps. After the positive voltage is removed, some electrons remain trapped in the charge traps, resulting in a lower conductivity state compared to the initial state. This behavior is analogous to the inhibitory behavior of synapses. Conversely, if a larger negative gate voltage is applied at this point, the electrons in

the traps are expelled and return more rapidly to the MoS_2, leading to an increase in the device conductance, which resembles the excitatory action of synapses.

Figure 2b intuitively demonstrates the process of simulating inhibitory and excitatory behaviors in biological neurons using positive and negative gate pulses. It can be observed that with multiple instances of inhibition or excitation, the current of the device gradually decreases or increases, effectively simulating the basic functions of synapses.

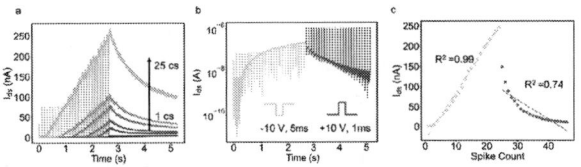

Fig. 3: a, The plot of I_{ds}-t as the number of pulses increasing from 1 to 25. b, The plot of I_{ds}-t under the continuous application of 25 negative gate pulses (width = 5 ms, amplitude = 10 V) followed by 25 positive gate pulses (width = 1 ms, amplitude =10 V). c, The relationship between the conductance of the device and the number of gate pulses.

Moreover, it is essential to investigate whether the device can simulate synaptic plasticity, which is key to achieving the transition from short-term to long-term memory. Figure 3a shows that as the number of positive gate pulses increases from 1 to 25, the device's conductance retention gradually improves, which indicates that synaptic plasticity has been achieved. Figure 3b and Figure 3c evaluate the performance metrics of the device for neuromorphic computing. Generally, the number of pulses reflects the weights, and a better linear relationship with conductivity enhances computational efficiency[7]. It can be observed that when positive weights are assigned to the device, the linearity is excellent (R^2 =0.99). In contrast, when negative weights are applied, the R^2 is 0.74. We speculate that this metric could be improved by optimizing the growth parameters of SiN_x.

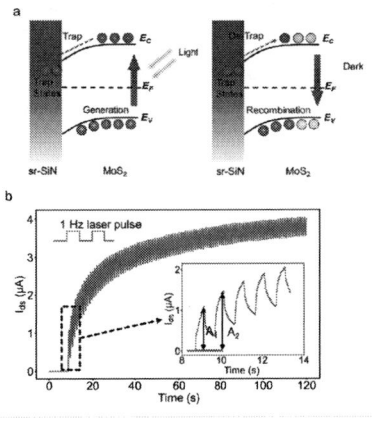

Fig. 4: a, The operation mechanism of the device depicts the capture of electrons by traps during illumination and their subsequent release upon the

cessation of illumination. b, The plot of I_{ds} under the continuous exposure of 100 laser pulses with a frequency of 1 Hz, a duty ratio of 50%, and a power intensity of 586 mW/cm².

Simulating synaptic behavior from an electrical perspective can enhance computational efficiency, while simulating synaptic behavior from an optical perspective is significant for processing image information. Figure 4a illustrates the device operating mechanism. In an ideal scenario without trap states, photogeneration and recombination in MoS₂ under laser irradiation quickly achieve dynamic equilibrium, resulting in a stable photocurrent. However, SiN$_x$ introduces trap states that capture electrons, significantly slowing down the electron-hole recombination process. This leads to a gradual increase in photocurrent and a longer time to reach equilibrium. It is also important to note the charge traps introduced by SiN$_x$ are limited, which determines the range to which the optical signal can increase the device conductance. In Figure 4b, laser pulses with a high-power intensity of 586 mW/cm² (wavelength of 532 nm, frequency of 1 Hz and laser width of 0.5 s) are applied on MoS₂ phototransistors. I_{ds} gradually increases and reaches approximately 4 μA after 100 laser pulses, under the conditions of V_{gs}=0 V and V_{ds}=1 V. This demonstrates that the traps in the SiN$_x$ we fabricated are abundant, allowing for the processing of optical signals over a wide range.

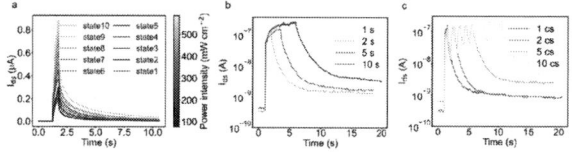

Fig. 5: a, The plot of I_{ds} when applying laser pulses with 10 different intensities (V_{ds}=1 V, V_{gs}=0 V and pulse width=0.5 s). b, The plot of I_{ds}-t under different laser pulse widths. c, The plot of I_{ds}-t under different laser pulse numbers.

The PPF was also evaluated, resulting in 134% according to Eq. (1), where A_1 and A_2 represent the response to the first and second pulses, respectively. This reflects that the device exhibits a tendency toward short-term plasticity in response to optical signals, which facilitates rapid image processing and reduces energy consumption.

$$PPF = A_2/A_1 \tag{1}$$

Figure 5a, Figure 5b, and Figure 5c evaluate the device response to varying light intensities, pulse widths, and pulse numbers, respectively. Similar to the electrical characteristics, the device demonstrates synaptic plasticity across all three dimensions of input optical information, which proves that the input optical signals can be adjusted from multiple dimensions to alter the state of device, enhancing its functionality in image information processing.

Conclusion

This work successfully realizes a MoS₂ transistor structure with silicon-rich SiN$_x$ as gate dielectric, achieving both electrical and optoelectronic synaptic functionality within a scalable design. The device exhibits high linearity and notable short-term plasticity, addressing key requirements for efficient neuromorphic computing. Its compatibility with standard CMOS fabrication paves the way for practical large-scale integration, making it a promising foundation for future neuromorphic systems that require both computational efficiency and scalability.

Acknowledgments

The project was primarily supported by the National Natural Science Foundation of China (62090034, 62104214, 62261160574, 62090030), the National Key R&D Program of China (2022YFA1204303), the Young Elite Scientists Sponsorship Program by CAST (2021QNRC001), and Kun-Peng Program of Zhejiang Province (H. W.). We thank ZJU Micro-Nano Fabrication Center and ZJU-Hangzhou Global Scientific and Technological Innovation Center for their support.

References

[1]   J. Chen, Y. Zhou, J.-H. Ahn, and Y. Chai, "Bioinspired in-sensor vision processing network for action recognition," in 8th IEEE Electron Devices Technology & Manufacturing Conference (EDTM), pp. 1–3, 2024.

[2]   P. A. Merolla et al., "A million spiking-neuron integrated circuit with a scalable communication network and interface," Science, vol. 345, no. 6197, pp. 668–673, Aug. 2014.

[3]   A. Pal et al., "An ultra energy-efficient hardware platform for neuromorphic computing enabled by 2D-TMD tunnel-FETs," Nat Commun, vol. 15, no. 1, p. 3392, Apr. 2024.

[4]   F. Zhang, "Recent progress in three-terminal artificial synapses based on 2D materials: from mechanisms to applications," 2023.

[5]   A. E. Kaloyeros, Y. Pan, J. Goff, and B. Arkles, "Review—Silicon Nitride and Silicon Nitride-Rich Thin Film Technologies: State-of-the-Art Processing Technologies, Properties, and Applications," ECS J. Solid State Sci. Technol., vol. 9, no. 6, p. 063006, Jan. 2020.

[6]   X. Su et al., "Integrating Image Perception and Time-to-First-Spike Coding in MoS2 Phototransistors for Spiking Neural Network," Adv Funct Materials, vol. 34, no. 30, p. 2315323, Jul. 2024.

[7]   J. Tang et al., "A Reliable All-2D Materials Artificial Synapse for High Energy-Efficient Neuromorphic Computing," Adv. Funct. Mater., 2021.

A sensing-computing system based on high uniformity photolithographic organic thin film transistor

Yaojie Zheng[1], Taoming Guo[1], Haobin Zhou[1], and Chen Jiang[1*]

[1]Department of Electronic Engineering, Tsinghua University, Beijing, China

Abstract

In this work, a sensor computing system based on high uniformity photolithographic organic thin film transistor (OTFTs) is presented. Different from traditional sensing computing system, this work avoids the step-down module required by the mismatch of operating voltage between OTFT and CMOS circuits by introducing capacitor weight array and compensation circuit, and realizes the recognition accuracy of 100% by taking the 5*6 system as a demonstration.

Keywords: Sensing-computing system, OTFT

Introduction

The sensing-computing circuit has been developing rapidly in recent years. With the continuous emergence of demand, the performance requirements for sensing computing circuits have gradually increased. The integration of sensing computing circuits shows great prospects in energy efficiency and data bandwidth [1,2]. In order to achieve sensing, large-area bonding and dense sensor arrays are required. A hybrid system of large-area electronics (LAE) and silicon CMOS circuits provides diverse sensing and high-performance computing/control, making this sensing platform possible [3]. Thin film transistor (TFT) technology is widely used in sensing applications due to its enormous advantages such as flexibility, transparency, large-area integration, and low-cost manufacturing. In addition, it has been proposed to use reconfigurable two dimensional (2D) semiconductor photodiode array and parallel dual gate TFT to implement the artificial neural network(ANN), but the sensing-computing circuit that meets both flexibility and edge computing functions still has challenges [4,5].

This work proposes an architecture for implementing a sensing-computing circuit using flexible OTFT. The improved Adam algorithm of gradient descent is used to determine the capacitance values of the capacitor weight array, and a compensation circuit is used to reduce the impact of V_{th} variation of OTFT on circuit performance [6,7]. At the same time, the problem of mismatch between OTFT high operating voltage and CMOS low operating voltage is solved.

Finally, accurate recognition results can be obtained by layer normalization of the circuit outputs. Thanks to this highly uniform OTFT manufacturing process, the accuracy of the sensing computing system is guaranteed.

Decive Structure and Characteristics

For OTFT fabrication, 10-nm chromium (Cr) and 40-nm gold (Au) was evaporated on a glass substrate, using photolithography and wet-etching to form the source and drain electrodes [8]. Indacenodithiophene-benzothiadiazole (IDT-BT) solution was spin-coated and annealed as the semiconductor. Then, CYTOP was spin-coated and parylene-C (par-C) was vacuum deposited as the insulation layer, followed by vapor deposition of 10-nm Cr and 40-nm Au and wet etching to form the gate electrode. The OSC and dielectric layers were patterned by reactive-ion etching (RIE) with oxygen. Then, a 20-nm Cr and 80-nm Au layer was evaporated and patterned by wet etching, serving as interconnections for the source/drain and gate electrodes. The measured transfer characteristics of a typical OTFT with a width of 300 μm and length of 10 μm are shown in Fig. 1(b).

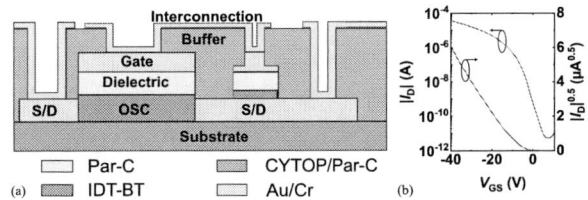

Fig. 1: (a) The cross-sectional view of the fabricated p-type OTFT. (b) The measured transfer characteristics of a typical fabricated OTFT.

Design and Operation of Sensing-computing System

A. System Framework

The entire designed system is described in Fig. 2, consisting of four parts: the pressure sensing array, the capacitor weight array, the compensation circuit, and normalization.

The pressure sensing circuit is shown in Fig. 3 (a). A pressure-sensitive resistor and a fixed resistor

979-8-3315-0417-5/25 $31.00 © 2025 IEEE

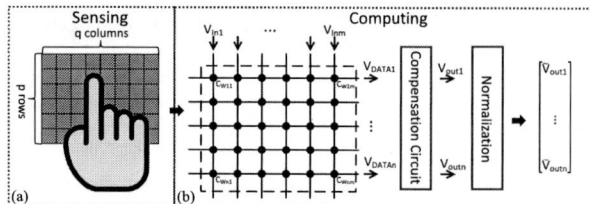

Fig. 2: Schematic of sensing-computing system, consisting of (a) a pressure sensing array and (b) a computing circuit.

to form a voltage divider. The change in pressure alters the resistance of the pressure-sensitive resistor, thereby changing the input voltage on the capacitor-weight array.

The output voltage of the buffer is connected to the capacitor-weight array. Each capacitance of the capacitor weight unit is trained by the neural network, and T_{write} & T_{read} controls the input and output of the array, respectively. The unit circuit of the array is shown in Fig. 3(b). There are three capacitors, corresponding to three recognition modes trained by the neural array. When the first mode of recognition is desired, T_{w1} of all units are turned on, and so on to the other two types.

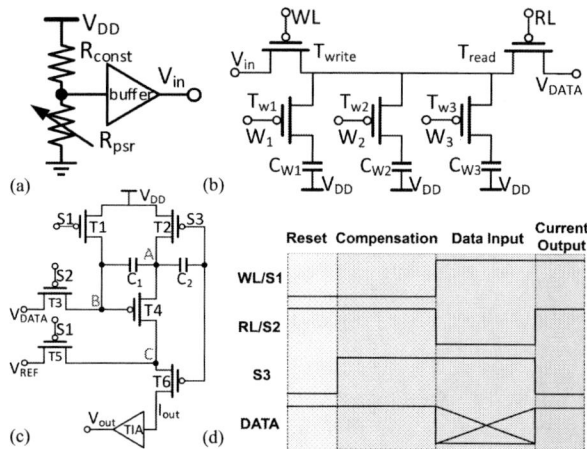

Fig. 3: Schematic of (a) sensing unit, (b) capacitor weight unit and (c) compensation circuit. (d) Control-signal timing diagram.

In this system, the input voltage signal is first stored in the capacitor-weight array, and then the voltage signal is converted into current signal by the 6T2C OTFT compensation circuit shown in Fig. 2(c), so as to avoid voltage matching through additional step-down circuits.

B. Proposed Circuit Operation

The proposed pressure recognition system can be divided into four parts, as the timing diagram shown in the Fig. 2(d). Detailed operating principles of the proposed circuit are described as follows [?].

1) Reset & Compensation Stage: At this stage, T_{write} is turned on, according to the desired recognition mode, one of the $T_{w1,w2,w3}$ is turned the input signal is written to the C_w. At the same time, reset and compensation are completed in the compensation circuit. The voltage of node A (V_A) is boosted to $V_{DD} + |V_{TH\_T4}|$.

2) Data Input Stage: T_{write} is turned off, T_3 and T_{read} is turned on, and the capacitors of the capacitor-weight array in each column and C_1 & C_2 in the compensation circuit achieve charge redistribution, thus getting the first step of calculation. S3 remains high to keep the T6 off, so no current flows out of T4.

3) Current Output Stage: During this stage, S3 goes low to turn on T6. The relationship between the output current of T4 and the voltage stored in capacitor C_1 follows a quadratic relation shown in Fig. 4(a). Therefore, the second step of calculation is completed. Figure 4(b) and (c) show the I_{out} versus the data voltages of the proposed 6T2C and conventional 4T1C. Obviously, the proposed 6T2C shows compensation effect.

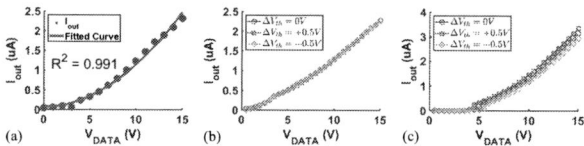

Fig. 4: (a) Output characteristics of proposed 6T2C. The output currents in (b) proposed 6T2C and (c) conventional 4T1C circuits with the consideration of device threshold-voltage variation.

4) The recognition result is obtained after the layer normalization of V_{out}.

C. Operation of a 5 × 6 Network

As a demonstration, we take a 5×6 pressure recognition system as an example. The shape to be recognized is an area composed of 2×3 squares, with 3 or 4 square areas above which pressure exists. The encoding method with or without pressure is shown in Fig. 5 (a). The four shapes of one pattern recognition in the system are shown in Fig. 5 (b). The pressure is converted into an input vector of six elements $[V_{in\alpha}]$. When there is pressure, the pressure

979-8-3315-0417-5/25 $31.00 © 2025 IEEE   954

sensing array outputs a low voltage, and vice versa, the pressure sensing array outputs a high voltage.

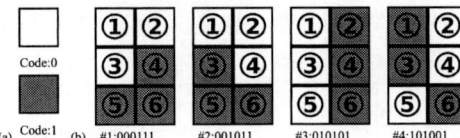

Fig. 5: Generation of (a) code word and (b) those of all four designated patterns

The entire system includes an input vector of 6 elements $[V_{in\alpha}]$, a 5×6 capacitor-weight array $[C_{W\alpha\beta}]$, and an output vector of 5 elements $[V_{out\beta}]$.

We trained the training set $[V_{in\alpha}]$ using the improved Adam algorithm of gradient descent, and encoded the expected $[V_{out\beta}]$ (represented by $[\hat{V}_{out\beta}]$) with a unique hot code. For example, the output of the first pattern to be recognized is $(V_H, 0, 0, 0, 0)$.

During training, we used $[C_{W\alpha\beta}^{\{e\}}]$ to represent the weight array, and $[V_{out\beta}^{\{e\}}]$ to represent the output vector. We used the loss function $J^{\{e\}}([V_{W\alpha\beta}^{\{e\}}])$ to denote the difference between the observed $[V_{out\beta}^{\{e\}}]$ and the expected $[\hat{V}_{out\beta}]$.

$$J^{\{e\}}([V_{W\alpha\beta}^{\{e\}}]) \equiv - \sum_{[V_{out\beta}]} p_{V_{out\beta}} log(\hat{p}_{V_{out\beta}}) \quad (1)$$

Here the loss function $[J^{\{e\}}([V_{W\alpha\beta}^{\{e\}}])]$ is the cross-entropy error function for all inputs, and $[p_{V_{out\beta}}]$ is the value of the target label in class $[V_{out\beta}]$, and $[\hat{p}_{V_{out\beta}}]$ is the prediction probability of the model for the class $[V_{out\beta}]$. If the sample belongs to class $[V_{out\beta}]$, then $[p_{V_{out\beta}}]=1$; otherwise $[p_{V_{out\beta}}]=0$. Starting with a random $[V_{W\alpha\beta}^{\{0\}}]$, $[V_{W\alpha\beta}^{\{e\}}]$ is updated through the Adam algorithm until $[J^{\{e\}}([V_{W\alpha\beta}^{\{e\}}])]$ converges to the minimum value.

The recognition system trained using four specified pattern sets is deployed to infer the classification of inference set elements containing all possible patterns composed of three or four shapes. According to the output voltage result of each line $V_{out\beta}$, a figure-of-merit Φ_β is obtained by using layer normalization method and softmax function below:

$$\Phi_\beta(\%) = \text{Softmax}(V_{out}) = \frac{e^{\widetilde{V}_{out\beta}}}{\sum_{\widetilde{V}_{out\beta}} e^{\widetilde{V}_{out\beta}}} \times 100\% \quad (2)$$

$\widetilde{V}_{out\beta}$ is the result of layer normalization of $V_{out\beta}$. The recognition accuracy is displayed in Fig. 6 in the form of a heatmap. Obviously, the recognition system can accurately distinguish between the patterns it trained to recognize.

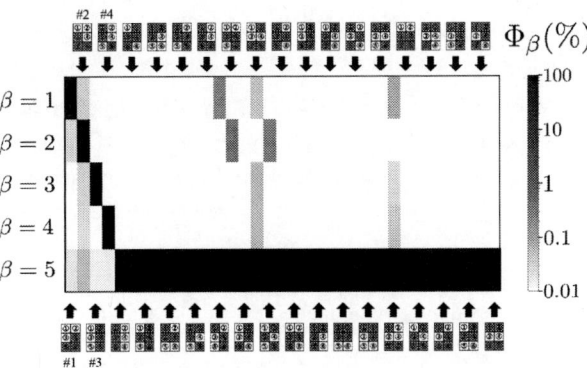

Fig. 6: The inference outcome of all patterns containing three or four solid shapes.

Conclusion

This work proposes a novel sensing-computing system based on high-uniformity photolithographic organic thin film transistors. By utilizing the Adam algorithm of neural networks, we recognized different pressure distributions with 100% accuracy, demonstrating the potential of sensing-computing systems in future large-scale flexible and wearable systems.

Acknowledgment

This work was supported in part by the National Key R&D Program of China under Grant 2019YFA0706100.

References

[1] Y. Yuan et al., "Artificial Leaky Integrate-and-Fire Sensory Neuron for In-Sensor Computing Neuromorphic Perception at the Edge," ACS Sensors 2023 8 (7), 2646-2655.

[2] E. Bestelink, H. -J. Teng and R. A. Sporea, "Contact Doping as a Design Strategy for Compact TFT-Based Temperature Sensing," in IEEE Transactions on Electron Devices, vol. 68, no. 10, pp. 4962-4965, Oct. 2021.

[3] L. E. Aygun et al., "Hybrid LAE-CMOS Force-Sensing System Employing TFT-Based Compressed Sensing for Scalability of Tactile Sensing Skins," in IEEE Transactions on Biomedical Circuits and Systems, vol. 13, no. 6, pp. 1264-1276, Dec. 2019.

[4] Mennel, L., Symonowicz, J., Wachter, S. et al., "Ultrafast machine vision with 2D material neural network image sensors," Nature 579, 62–66 (2020).

[5] Y. Hu, T. Lei, Y. Wang, F. Wang and M. Wong, "An Artificial Neural Network Implemented Using Parallel Dual-Gate Thin-Film Transistors," in IEEE Transactions on Electron Devices, vol. 69, no. 10, pp. 5574-5579, Oct. 2022.

[6] Kingma, Diederik , and J. Ba . "Adam: A Method for Stochastic Optimization." Computer Science (2014).

[7] C. -L. Lin et al., "Compensation Pixel Circuit to Improve Image Quality for Mobile AMOLED Displays," in IEEE Journal of Solid-State Circuits, vol. 54, no. 2, pp. 489-500, Feb. 2019.

[8] B. Liu et al., "A 1024-Channel Neurostimulation System Enabled by Photolithographic Organic Thin-Film Transistors with High Uniformity," IEEE International Symposium on Circuits and Systems (ISCAS) (2024).

Physics-Based Circuit-Compatible Model of Polycrystalline Hafnia-based 3D Ferroelectric Capacitor for High-Density Memory Applications

Minyue Deng[1], Chang Su[1], Jiayan Zhu[1], Liang Chen[1], Shengjie Cao[1], Qianqian Huang[1,2,3*] and Ru Huang[1,2,3*]

[1]Peking University, Beijing 100871, China; [2]Beijing Advanced Innovation Center for Integrated Circuits, Beijing 100871, China; [3]Beijing Superstring Academy of Memory Technology, Beijing 100176, China.
(*E-mail: hqq@pku.edu.cn, ruhuang@pku.edu.cn)

Abstract

In this work, based on finite element method, a physics-based equivalent circuit model for polycrystalline hafnia-based 3D ferroelectric capacitor is proposed and developed, which can capture the cylindrical geometric feature and the impacts from complex phase distribution in actual ferroelectric film for the first time. The proposed circuit-compatible model shows high accuracy compared with TCAD simulation and significantly improved computation efficiency compared with previous methods. Moreover, two typical types of phase distribution in ferroelectric capacitors are discussed based on the experimental characterization for model modification. Based on the proposed model, it is suggested to improve the ferroelectricity near inner electrode and reduce the possible interfacial dielectric layer thickness of 3D ferroelectric capacitor for high density memory application.

Keywords: ferroelectric, 3D, circuit-compatible model, grain distribution and hafnium oxide

Introduction

To realize ultra-high memory density of future nonvolatile memory applications, 3D integration featuring a cylindrical ferroelectric (FE) layer is essential for both ferroelectric random-access memories (FeRAM) and ferroelectric field-effect transistors (FeFET) (Fig.1a) [1-2]. With the changed geometry, electric characteristics of 3D cylindrical ferroelectric are significantly different from those of 2D planar FE (Fig.1b) [3-5], and cannot be simply derived from the conventional understanding based on planar FE. Moreover, the hafnia (HfO_2)-based FE is polycrystalline in nature, characterized by multi-phase coexistence [6-7]. The complicated grain distribution in film further considerably increases the analysis and computation complexity of the electric characteristics for real HfO_2-based 3D FE. In order to further promote the applications of high-density and large-scale FE memory, it is crucial to establish an in-depth analysis and an efficient accurate model for the 3D FE.

In our previous work [8], a 3D FE physical model is developed based on the proposed equivalent screen radius (R_{eq}), which can clarify the intrinsic relationship of ferroelectricity between 2D planar and 3D cylindrical FE from a physical perspective. However, the process of determining R_{eq} is quite tedious due to its nonanalytic solution and complex dependence on various parameters. Therefore, from the perspective of large-scale circuit simulation, there is an urgent demand for a more efficient model for 3D FE.

In this work, for the first time, a physics-based equivalent circuit model for the polarization switching of 3D cylindrical FE capacitors (FeCAP) is proposed based on *finite element method (FEM)* with high accuracy and high efficiency. Moreover, according to experimental grain distribution feature observed in the fabricated FeCAP, two types of FeCAP are discussed and the modified circuit-compatible model can accurately reflect impacts from complex multi-phase distribution. The modeled results show that it is important to enhance the ferroelectricity near inner electrode and eliminate the possible dielectric interface layer between FE and electrodes.

Equivalent Circuit Model of 3D FE

To model the electric characteristics of 3D FeCAP in a simple form, we develop the equivalent circuit structure inspired by concept of the *FEM* which is also reflected in our previously proposed model based on R_{eq} (Fig.2a) [8] and TCAD simulation [9].

For 3D FeCAP in ideal case, the analysis framework can be simplified to one dimension due to its radial symmetry. Then, the continuous film is discretized into multiple connected grid points along the thickness direction. Although the electric field (E) and FE polarization (P_{FE}) show a non-uniform distribution in the whole region [5], they both are discretized and take a local definite single value at each of the discrete grid points. The local P_{FE}–E relationship is described by *planar FE polarization switching model (1)* separately at each point. And the individual electric variables of discrete points are interconnected through *Gauss's Law (2)* with different radius but the same free charge (only exists on electrodes). In previous work [8], combining *(1)* and *(2)*, as well as the applied voltage on the whole film ($V_{Applied}$), FE charge can be numerically solved by value iteration of R_{eq}.

However, from the above physics, a simple and intuitive equivalent circuit structure can be abstracted. As shown in Fig.2b, the 3D cylindrical FeCAP can be converted into a series of multiple (N) planar sublayer FeCAP (Sub-cap) with the same thickness and height, but different areas. The Q-V loop of each Sub-cap is separately described by the planar analytical Preisach model [10]. To reflect the geometry featured by increased radius from inside to outside, the areas of Sub-caps increase in succession. It is noticeable that the physics formula for each Sub-cap(i) with individual ideal electrodes is completely equivalent to that of corresponding grid point with $R_i=R_{Sub-cap(i)}$ in *FEM* for continuous materials, demonstrating rationality of the equivalence. This method facilitates directly embedding the existing planar analytical FE model as an independent module into the circuit simulators (such as HSPICE) to self-consistently solve electric characteristics of FeCAP with cylindrical geometry.

Model Consideration for Polycrystalline 3D Hafnia-FE

HfO_2-based FE is polycrystalline and featured by multi-phase coexistence. To accurately reflect the impacts of phase distribution, $Hf_{0.5}Zr_{0.5}O_2$ (HZO)-based FeCAP with cylinder structure are fabricated and characterized to observe grain distribution characteristics (Fig.3). The HZO is grown by ALD on the sidewall of etched trenches, showing good crystallinity and interface quality. TEM image (Fig.4) shows that the FE and dielectric (DE) grains are irregularly and complexly distributed, but predominantly present as a combination of two kinds of simplified types: FE/DE phases laminated pattern along radius (Type A) and FE/DE phases partitioned pattern along angle (Type B). Therefore, the modeling can be initially focused on the two simplified types.

For Type A: Benefiting from the layered structure of equivalent circuit with sublayer FeCAP in series, the established model can be easily expanded to capture the laminated distribution by setting certain sublayer as a DE capacitor with $P=0$ and $D_{DE}=\varepsilon_0\varepsilon_{DE}E_{DE}$.

For Type B: To capture the impact from partitioned distribution along angle, E and P_{FE} distribution in the Type-B 3D FeCAP are first simulated and analyzed by *Sentaurus* TCAD with the novel method proposed in our previous work [5] (Fig.5). As shown in Fig.6, the E distribution in FE-phase region is nearly unchanged compared with the 3D FE with uniform materials under the same $V_{Applied}$. Considering that E along the thickness direction (dominant in determining P_{FE} and then charge on electrodes) is mostly unaffected by nearby region, the capacitor charge of Type-B 3D FeCAP can be approximated by linear superposition based on individual sector with respective proportion.

Modeling Results and Discussion

To verify feasibility and accuracy of the proposed model in this work, electric characteristics of 3D FeCAP are presented and compared by utilizing the proposed model in this work, TCAD simulation and the model based on R_{eq}.

A. Verification of proposed equivalent circuit model

For the proposed equivalent circuit model, high simulation accuracy can be achieved with the number of sublayers $N\geq$ 5 (Fig.7a). The model shows excellent agreement with TCAD simulation in Q_{in}-$V_{Applied}$ relationship (Fig. 7b), distributed E (Fig. 7c) and P_{FE} (Fig. 7d) along the radius direction, demonstrating feasibility and accuracy of the proposed model. Moreover, the calculation time can be reduced by two orders of magnitude with the proposed model compared with the other two methods, which is very suitable for large-scale circuit simulation.

B. Results of Type A and Type B polycrystalline 3D FE

With the consideration of capturing laminated feature of Type-A film, the modeled results still match closely with TCAD simulation in various electric characteristics (Fig.8a-c). Besides, different from the uniform FE film, for Type A, the introduction of DE layer escalates the incomplete screening of P_{FE}, leading to that the whole FE layer is depolarized under zero bias (Fig.8d).

Considering the non-uniformly distributed E along radius direction, it should be noted that 3D FeCAP will have spatial pattern dependence for FE-phase distributed near inside (case 1) or outside (case 2) (Fig. 9a). Compared with case 1, 3D FeCAP of case 2 will show remnant charge degradation due

to less switched P_{FE} and poorer retention due to enhanced depolarization field (Fig.9bc). Therefore, it is suggested that the ferroelectricity near inner electrode should be paid more attention during material selection and fabrication process.

Furthermore, Fig. 10 shows the impact of FE/DE thickness ratio T_{FE}/T_{DE} on the electric characteristics of Type-A 3D FeCAP. Higher FE ratio will lead to higher Q_{in} due to better ferroelectricity (Fig.10b). However, more voltage will drop on FE layer with higher T_{FE}/T_{DE}, causing worse non-uniformity of E and P_{FE} distribution (Fig.10c). Therefore, it is suggested to reduce the possible DE interfacial layer during material selection and fabrication process for higher remnant charge.

For Type B, the model result calculated by linear superposition also agrees excellently with TCAD simulation (Fig.11), demonstrating the unaffected electric characteristics by nearby grain.

C. Modeling results for complex grain distribution of FE

Based on the above discussion, a random complex grain distribution can always be simplified to a union of Type A and Type B. Then FeCAP charge Q can be calculated as (Fig.12a):

$$Q = \sum (Sector\ proportion \times Charge\ of\ individual\ film)$$

where *proportion* corresponds to the Type B distribution along the angle, *charge of individual film* covers both the uniform and Type-A condition.

Assigning a grain distribution with complexity equivalent to that of observed grains according to the TEM image in Fig.4, the proposed model can accurately describe the P-V loop of polycrystalline 3D FeCAP, which matches well with TCAD simulation (Fig.12b).

Conclusion

For the first time, this work proposed a physics-based equivalent circuit model of 3D FeCAP for efficient large-scale circuit simulation with consideration of polycrystalline feature. The spatial pattern dependence of electric characteristics across different types of FeCAP is further discussed, highlighting the importance of ferroelectricity near inner electrode. This work provides an easy-to-use model and a guidance of further material and device optimizations for 3D FE memory applications.

References

[1] G. Kim et *al.*, *IEEE Symposium on VLSI Technology and Circuits*, pp. T2.3 (2024). [2] N. Ramaswamy *et al.*, International Electron Devices Meeting (IEDM), pp. 15.7.1-15.7.4 (2023). [3] K. Florent *et al.*, *47th European Solid-State Device Research Conference (ESSDERC)*, pp. 164-167 (2017). [4] M. Fan *et al.*, *IEEE Transactions on Electron Devices*, 67, pp. 5810-5814 (2020). [5] M. Deng *et al.*, *7th IEEE Electron Devices Technology & Manufacturing Conference (EDTM)*, pp. 273-275 (2023). [6] Z. Zhang et *al.*, Sci China Inf Sci, pp. 66.200405 (2023). [7] Z. Zhao *et al.*, *IEEE Symposium on VLSI Technology and Circuits (VLSI)*, pp. T15.5 (2023). [8] C. Su et *al.*, *8th IEEE Electron Devices Technology & Manufacturing Conference (EDTM)*, pp. 373-375 (2024). [9] Sentaurus Device User Guide Version: O-2018.06, Synopsys, Mountain View, CA, USA, June. 2018. [10] A. D. Gaidhane *et al.*, *IEEE Transactions on Computer-Aided Design of Integrated Circuits and Systems*, 42, pp. 1634-1642 (2023).

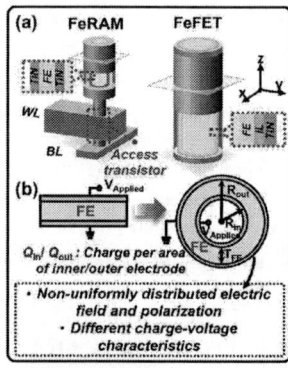

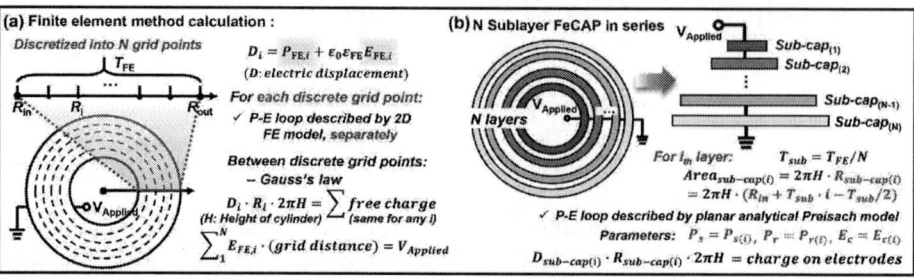

Fig.1 (a) Schematic of the FE devices for high-density memory. **(b)** Different electric characteristics between 2D and 3D FeCAP. **Fig.2 (a)** Basic concept of solving electric characteristics through finite element method for 3D FeCAP with radially symmetric cylinder geometry. **(b)** Structure of the equivalent circuit model.

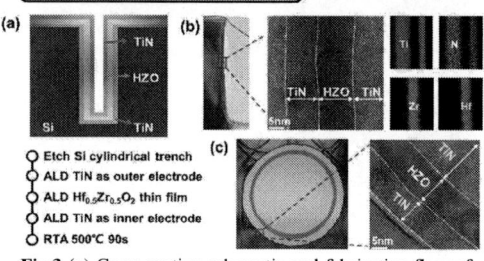

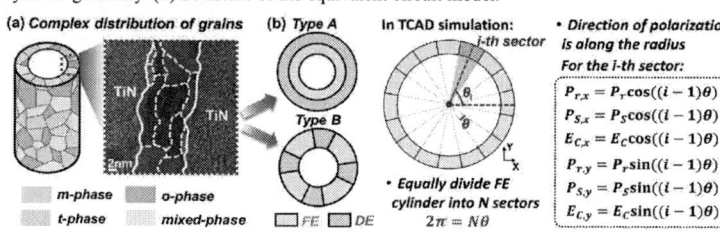

Fig.3 (a) Cross section schematic and fabrication flow of the 3D FeCAP. **(b)** TEM image and EDS mapping of the sidewall FE material. **(c)** TEM image in a plane view.

Fig.4 (a) Schematic of polycrystalline 3D FeCAP and the experimentally characterized grain distribution. **(b)** Simplification and classification of complex grain distribution into two typical types.

Fig.5 Novel TCAD simulation method for 3D FeCAP proposed in our previous work [5].

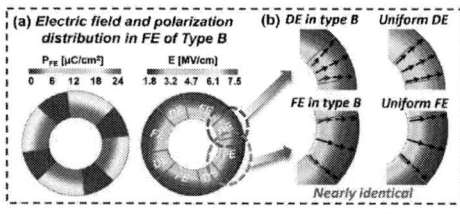

Fig.6 (a) Simulated E and P_{FE} distribution for FE/DE distribution of Type B. **(b)** Simulated comparison of E distributions between the FE (DE) region in Type B and the individual FE (DE) cylinder.

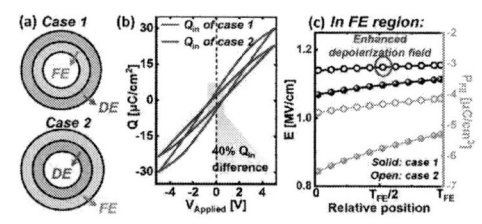

Fig.9 (a) Schematic of spatial phase distribution in Type A, including case 1 and case 2. **(b)** Comparison of the modeled Q_{in}-$V_{Applied}$ loops between case 1 and case 2. **(c)** Comparison of the modeled E_{FE} and P_{FE} distribution between case 1 and case 2.

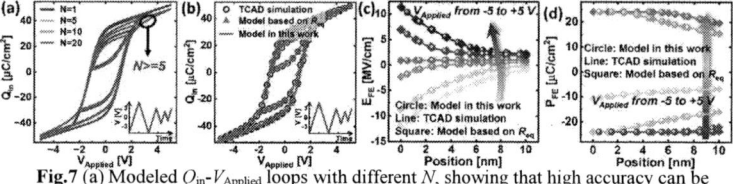

Fig.7 (a) Modeled Q_{in}-$V_{Applied}$ loops with different N, showing that high accuracy can be achieved with a small $N \geq 5$. Good agreement between the equivalent circuit model proposed in this work, the TCAD simulation and the model based on R_{eq} [8] in (b) Q_{in}-$V_{Applied}$ loop, distribution of (c) E_{FE} and (d) P_{FE} under different $V_{Applied}$.

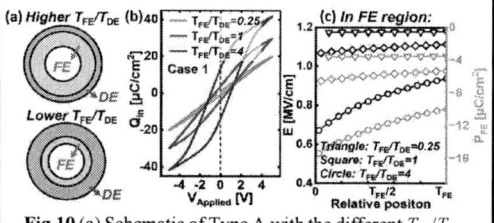

Fig.10 (a) Schematic of Type A with the different T_{FE}/T_{DE}. **(b)** Comparison of the modeled Q_{in}-$V_{Applied}$ loops for Type A with T_{FE}/T_{DE}=0.25, 1, 4, respectively. **(c)** Comparison of the modeled E_{FE} and P_{FE} distribution for Type A with T_{FE}/T_{DE}=0.25, 1, 4, respectively.

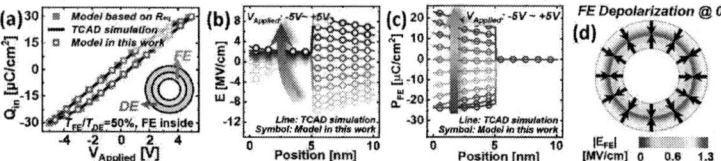

Fig.8 For Type A with T_{FE}/T_{DE}=50% and FE inside, the good agreement between the equivalent circuit model proposed in this work, TCAD simulation and model based on R_{eq} [8] in (a) Q_{in}-$V_{Applied}$ loop, distribution of (b) E_{FE} and (c) P_{FE} under different $V_{Applied}$. **(d)** Simulated depolarization field distribution in Type A under zero bias.

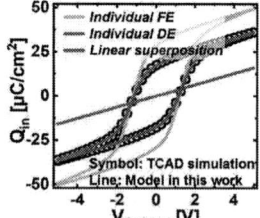

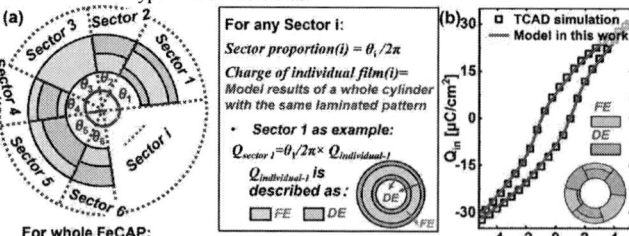

Fig.11 Linear superposition shows good agreement with TCAD simulation in description of Q_{in}-$V_{Applied}$ loop for Type B.

Fig.12 (a) Calculation method for quite complex grain distribution. **(b)** Accurate agreement between model proposed in this work and TCAD simulation in the Q_{in}-$V_{Applied}$ loops of 3D FeCAP with complex grain distribution.

Acknowledgments
This work was supported by NSFC (62374009, 61927901), Beijing SAMT Project (SAMT-BD-KT-22030101), 111 Project (B18001).

Van der Waals infrared photon detectors for standard blackbody characterization

Yang Wang[1,2,3], Weida Hu[2], Peng Zhou[1,3]

[1]1State Key Laboratory of ASIC and System, School of Microelectronics, Fudan University, Shanghai 200433, China, [2]State Key Laboratory of Infrared Physics, Shanghai Institute of Technical Physics, Chinese Academy of Sciences, Shanghai 200083, China, [3]Shaoxin Laboratory, Shaoxing 312000, China

Abstract

We will discuss the standard blackbody characterization and time response characterization techniques for infrared photon detectors. These standardized techniques provide a reliable and consistent way to evaluate the performance of infrared photon detectors and are essential for ensuring the accuracy and repeatability of device measurements. Overall, this perspective will provide valuable insights into the latest innovations in the field, and how these innovations are advancing the capabilities and reliability of these devices.

Keywords: Van der Waals, infrared photon detectors, standard blackbody characterization

Introduction

Currently, Van der Waals narrowband materials (vdWs-NMs) photon detectors have surpassed conventional thin-film material photon detectors in many performance metrics, but these test data are usually obtained by non-standard measurement methods. Due to the Gaussian distribution of energy within the small spot of the laser, the laser has a strong coherence and high power density, which makes the calculated power density misjudged by a large margin. It is easy to lead to the overestimation of the photoelectric performance of the photon detectors. Therefore, the laser as the excitation light source doesn't accurately measure the actual detection capability and key performance indicators of the photon detectors.

This paper analyzes the importance of blackbody characterization and response time measurement techniques. Also, the paper presents several challenges and potential future directions in this research field.

Standard characterization of next-generation infrared photon detectors

Blackbody radiation is mixed light, and its wavelength distribution follows Planck's formula and is only related to the blackbody temperature (Figure 1b). When the blackbody temperature decreases, the overall power decreases and the peak wavelength shifts to the right, which is a stable source of standard infrared radiation. The total incident power of blackbody radiation on the device surface can be calculated using the formula:

$$P = \frac{\alpha \varepsilon \delta (T^4 - T_0^4) A}{\pi L^4} A_n \qquad (1)$$

Where α is the modulation factor, ε is the average emissivity

of the blackbody radiation source, σ is the Stefan-Boltzmann constant. T is the blackbody radiation source temperature, T_0 is the test environment temperature, A is the blackbody radiation source area, A_n is the device area and L is the distance of the photon detector from the aperture. Therefore, with this standard algorithm, the optical power radiated by the blackbody is error-free compared to the Gaussian spot of the laser.

Therefore, we believe that the infrared band (especially in the mid-wave and long-wave infrared bands) should be dominated by blackbody testing in standard measurements. As an example, the specific detectivity (D^*) of an important indicator of infrared photon detectors is 10^8 cm Hz$^{-1/2}$ W^{-1} for commercial thermistors, and meaningful research efforts should be higher than the detectivity of commercial thermistors. The non-standard measurement methods make it impossible to accurately assess their optoelectronic performance, resulting in a direct comparison with commercial equipment. (Figure 1c)[1-8] Therefore, it is particularly important to summarize standardized methods for characterizing the quality factor of vdWs-NMs photodetectors.

The equation of responsivity ($R_{blackbody}$) under blackbody radiation is:

$$R_{blackbody} = \frac{I_{ph}}{P} \qquad (2)$$

where I_{ph} is a blackbody photocurrent.

The responsivity of the photon detector in different wavelength bands is different. Fourier transform infrared spectroscopy (FTIR) was used to obtain the relationship between the responsivity of the photon detector and the radiation wavelength ($R_{(\lambda)}$). The response spectrum obtained by FTIR is only a relative response spectrum. The response spectrum measured by FTIR is an equal power curve in which the radiated power at any wavelength is equal. However, there are more photogenerated electrons and holes in the long-wavelength region than in the short-wavelength region under the same optical power. Ideally, the responsivity of the photon detector increases with wavelength and peaks at the wavelength corresponding to the band gap, followed by an attenuation value cutoff wavelength. After calibration of the blackbody response, the responsivity of the response spectrum of the photon detector

979-8-3315-0417-5/25 $31.00 © 2025 IEEE

can be obtained. The relative responsivity spectrum $R'(\lambda)$ is obtained from the FTIR characterization:

$$R'(\lambda) = \frac{R(\lambda)}{R(\lambda_P)} \tag{3}$$

Since the blackbody radiation has a continuous spectrum and the emissivity of each wavelength is different, the blackbody light signal generated by the photon detector is the sum of the signals generated by the radiation at each wavelength. The ratio of the blackbody responsivity ($R_{blackbody}$) to the peak responsivity ($R(\lambda_p)$) of an infrared photon detector is a constant, the g-factor. The g-factor can be calculated from the tested $R'(\lambda)$ and the blackbody spectral emissivity ($(\phi(\lambda))$). Therefore, by calculating the g-factor, the peak responsivity ($R(\lambda_p)$), quantum efficiency ($QE(\lambda_p)$) and the peak detectivity ($D^*(\lambda_p)$) can be obtained:

$$R(\lambda p) = \frac{R_{blackbody}}{g} \tag{4}$$

$$QE(\lambda_P) = R(\lambda_P)\frac{hc}{e\lambda} \tag{5}$$

$$D^*(\lambda p) = \frac{(A_n \Delta f)^{1/2}}{NEP} = R(\lambda_P)\frac{(A_n \Delta f)^{1/2}}{i_n} \tag{6}$$

where h is Planck's constant, c is the speed of light in vacuum, λ is the photon wavelength, NEP is noise equivalent power and in is noise current.

When estimating the detectivity of vdWs-NMs photon detectors, the influence of different types of noise current (i_n) needs to be considered. The modulation frequency of the device has a great relationship with the noise, which determines the main noise source of the device (Figure 1d). Generally, shot noise is considered to be the main noise source in vdWs-NMs photon detectors. However, for photon detectors operating at room temperature, thermal noise and recombination noise cannot be ignored due to the long carrier lifetime, high photogain, and slow response speed, among which the light-induced recombination noise contributes greatly. In addition, low-frequency ($1/f$) noise should be studied as an important indicator for evaluating the performance of photon detectors. However, we found that the detectivity of many vdWs-NMs photon detectors exceeds the limit detectivity or is close to the limit detectivity, which is mainly caused by the non-standard measurement methods. This results in many photon detectors computing orders of magnitude higher detectivity. Therefore, we need to perform standard measurement methods and standard calculation formulas. Alleviate researchers' doubts about new vdWs-NMs photon detectors.

The response time is an important criterion for the performance of the photon detector, which is related to the carrier lifetime in the photon detector material, the sweep-out time of the photoexcited carriers, and the RC time constant of the photon detector. Bandwidth (f_{-3dB}) is defined as the frequency range corresponding to the attenuation of the frequency response of the signal within -3dB. Figure 1f shows that the correct -3dB response should first be constant and then gradually decrease. The -3dB bandwidth is given by:

$$\frac{1}{f_{-3dB}^2} = \frac{1}{f_t^2} + \frac{1}{f_{RC}^2} \tag{7}$$

where f_t and f_{RC} are the carrier transit time limited and RC limited bandwidths respectively. They are defined as:

$$f_t = \frac{3.5}{2\pi t_{tr}} \tag{8}$$

$$f_{RC} = \frac{1}{2\pi RC} \tag{9}$$

where t_{tr} is the carrier transit time, R is the total series resistance including the photodiode series resistance, sheet resistance, contact resistances and load resistances in the measurement circuit, and C is the sum of the capacitance of the device as well as the parasitic capacitance of the measurement system. The carrier transit time can be given as:

$$t_{tr} = \frac{L^2}{\mu V_{bias}} \tag{10}$$

where V_{bias} is the applied bias voltage, μ is the carrier mobility, and L is the length of the channel. Since mobility is determined by the material itself, the transport length and applied bias can be changed to further improve the carrier transit time-limited bandwidths but lead to a lower quantum efficiency.[4,9] As shown in Figure 1f, another impulse response characterization method is by modulating the incident light with a square wave to determine the time for the photocurrent to rise from 10% to 90% (rise time, t_r) or fall from 90% to 10% (fall time, t_f), the larger value between the two is taken as the response time. The response time extraction must be at equilibrium in both the dark and light states (Figure 1f). It versus signal bandwidth equation:

$$t \times f_{-3dB} = 0.35 \tag{11}$$

In actual experiments, the two methods are not significantly different from each other. The bandwidth of the preamplifier decreases as the amplification increases. When the bandwidth of the device exceeds the bandwidth of the amplifier, the response time curve of the device will be inaccurate. The response time of photoconductive photon detectors generally depends on the diffusion time of photogenerated carriers. Therefore, the response time of defective materials is relatively longer. The response time of photovoltaic photon detectors is mainly determined by the drift process of minority carriers in the junction region. Since the drift length of the junction is relatively short, this process can be very fast.

979-8-3315-0417-5/25 $31.00 © 2025 IEEE

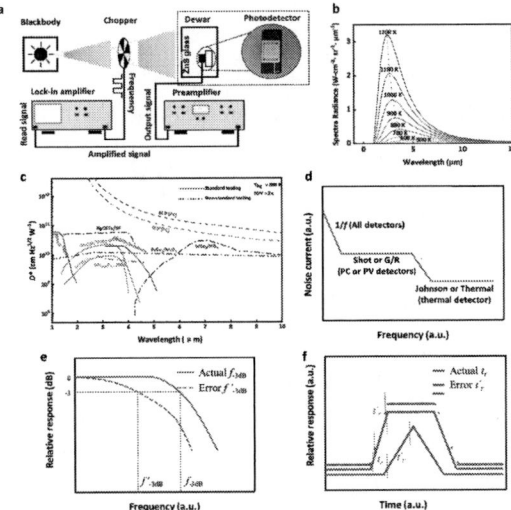

Figure 1. Standard characterization of vdWs-NMs photon detectors. a) Characterization principle diagram of the blackbody test system. b) Relationship between wavelength and radiant power at different blackbody temperatures. c) The spectral dependence of the detectivity of standard and non-standard measurements of vdWs-NMs photon detectors. d) Relationship between noise current and frequency. e) Actual and error -3db bandwidth response curves of vdWs-NMs photon detectors. f) Actual and error response time curves of vdWs-NMs photon detectors.

Conclusion

The current and future challenges to realizing vdWs-NMs photon detectors mainly come from large-scale material synthesis and doping techniques, device physics, standardizing evaluation methods and array integration techniques. However, all recent advances suggest that 2D vdW heterostructures can offer many opportunities for exploration and provide real-time and energy-efficient data for dark current regulation and standard characterization of IR photon detectors for ideal IR detection innovation prospects for analytics applications[10,11].

The significant technological potential is rapidly developing in the field of next-generation vdWs-NMs-based photon detectors. Protocols for accurate determination and validation of comparative performance data for these devices are urgently needed[12]. The broad practice of simply defining specific detectivity from responsivity and dark noise via laser characterization is simply not sufficient, that approach may overestimate specific detectivity by orders of magnitude. Therefore, we recommend determining an accurate specific detectivity by taking separate measurements of standard noise current and blackbody responsivity. The accurate response time is related to the carrier lifetime in the detector material, the sweep-out time of the photoexcited carriers, and the RC time constant of the detector. These are properties that must be considered for all reported vdWs-NMs photon detectors. Therefore, the vdWs-NMs-based photon detectors are still in the research stage and need further development and improvement to meet the needs of practical applications.

Acknowledgments

This work was supported by the National Key Research and Development Program of China (Grant Nos. 2023YFB3611400, 2021YFA1200500), National Natural Science Foundation of China (Grant Nos. 61925402, 62090032, and 62304040), Open Fund of State Key Laboratory of Infrared Physics (Grant No. SITP-NLIST-ZD-2023-01).

References

[1]   J. Bullock, M. Amani, J. Cho, Y.-Z. Chen, G. H. Ahn, V. Adinolfi, V. R. Shrestha, Y. Gao, K. B. Crozier, Y.-L. Chueh, Nat. Photon., 12, 601 (2018).

[2]   M. Long, A. Gao, P. Wang, H. Xia, C. Ott, C. Pan, Y. Fu, E. Liu, X. Chen, W. Lu, Sci. Adv., 3, e1700589 (2017).

[3]   C. Tan, M. Amani, C. Zhao, M. Hettick, X. Song, D. H. Lien, H. Li, M. Yeh, V. R. Shrestha, K. B. Crozier, M. C. Scott, A. Javey, Adv. Mater., 32, e2001329 (2020).

[4]   Y. Chen, Y. Wang, Z. Wang, Y. Gu, Y. Ye, X. Chai, J. Ye, Y. Chen, R. Xie, Y. Zhou, Nat. Electron., 4, 357 (2021).

[5]   S. Lukman, L. Ding, L. Xu, Y. Tao, A. C. Riis-Jensen, G. Zhang, Q. S. Wu, M. Yang, S. Luo, C. Hsu, Nat. Nanotechnol., 17, 220 (2022).

[6]   M. Long, Y. Wang, P. Wang, X. Zhou, H. Xia, C. Luo, S. Huang, G. Zhang, H. Yan, Z. Fan, X. Wu, X. Chen, W. Lu, W. Hu, ACS Nano, 13, 2511 (2019).

[7]   H. Jiao, X. Wang, Y. Chen, S. Guo, S. Wu, C. Song, S. Huang, X. Huang, X. Tai, T. Lin, Sci. Adv., 8, eabn1811 (2022).

[8]   Y. Wang, Y. Gu, A. Cui, Q. Li, T. He, K. Zhang, Z. Wang, Z. Li, Z. Zhang, P. Wu, Adv. Mater., 34, 2107772 (2022).

[9]   Y. Wang, P. Wu, Z. Wang, M. Luo, F. Zhong, X. Ge, K. Zhang, M. Peng, Y. Ye, Q. Li, Adv. Mater., 32, 2005037 (2020).

[10]   X. Pang, Y. Wang, Y. Zhu, Z. Zhang, D. Xiang, X. Ge, H. Wu, Y. Jiang, Z. Liu, X. Liu, C. Liu, W. Hu, P. Zhou, Nat. Commun., 15, 1613 (2024).

[11]   Y. Zhu, Y. Wang, X. Pang, Y. Jiang, X. Liu, Q. Li, Z. Wang, C. Liu, W. Hu, P. Zhou, Nat. Commun., 15, 6015 (2024).

[12]   F. Wang, T. Zhang, R. Xie, Z. Wang, W. Hu, Nat. Commun. 14 (1), 2224 (2023).

Remote effect of extra metal layer on TiN electrode on the ferroelectric properties of $Hf_{0.5}Zr_{0.5}O_2$ thin films

Zhipeng Xue[1,2], Danyang Chen[1,2], Zikang Yao[1,2], Tianning Cui[1,2], Yulong Dong[1,2] Jingquan Liu[1], Mengwei Si[1] and Xiuyan Li[1,2*]

1 National Key Lab of Micro/Nano Fabrication Technology, Shanghai Jiao Tong University, Shanghai China

2 Department of Nano/Microelectronics, Shanghai Jiao Tong University, Shanghai China

Email: xiuyanli@sjtu.edu.cn

Abstract—An extra metal cap is generally employed for improving the top electrode conductivity of TiN/$Hf_{0.5}Zr_{0.5}O_2$(HZO)/TiN ferroelectric capacitor. This work demonstrates that the extra metal cap also has a significant remote effect on polarization, endurance performance and crystalline structure of HZO films. And, based on a strain evaluation of the HZO films, the remote effect of extra metal cap is understood to be a modulation of the in-plane strain in the films.

Keywords: extra metal cap, remote strain effect, ferroelectric properties of HZO

I. Introduction

Ferroelectric (FE) $Hf_{0.5}Zr_{0.5}O_2$(HZO) thin films have been intensively studied due to their good CMOS compatibility and scalability, promising for application in future non-volatile memory devices. The FE properties of HZO films, such as remanent polarization (P_r), endurance and retention, are significantly influenced by several factors in material and process designing. Among them, electrode materials and the interface between electrode and HZO have been considered to make great influences [1]. And, TiN electrode has been generally employed to get a high P_r and endurance performance. Concerning the benefits of TiN electrodes, it has been proposed that TiN plays a role of oxygen scavenger to induce oxygen vacancy (V_O) into HZO and to suppress the further immigration of V_O due to the TiON interlayer formation, leading to a stable retention property and enhanced endurance [2]. Besides, it has been also considered that TiN electrode introduce strain into HZO films in annealing due to mismatch of thermal expansion coefficient [3]. Despite these advantages, the conductivity of top TiN electrode is easily degraded due to its easy oxidation. Therefore, an extra metal cap, such as W, is usually deposited on TiN to suppress its oxidation and to improve the conductivity.

In this work, we find that not only TiN electrode but also the extra metal has a significant effect on the FE properties of HZO thin films. An extra Mo or Au capping layer on TiN top electrode are enable to enhance P_r values and endurance performance of HZO through enhancing tetragonal(T) and/or orthorhombic(O) phase formation, in contrast to cases with extra W or without extra capping layer. Moreover, via the 2D grazing incident x-ray diffraction (GI-XRD) measurement and $\sin^2\psi$ method, we demonstrate that the remote effect of the extra metal layer intrinsically originates from the strain modulation in HZO. Our results provide new insights into engineering the strain and FE properties of HZO films.

II. Device Fabrication

Fig. 1 shows the device structure and fabrication process of HZO capacitors. Typical sandwich TiN/HZO/TiN structure with three kinds of extra metal capping layer (Au, W and Mo) on top TiN electrode were employed. 30 nm bottom TiN layer was deposited by sputtering, followed by 10 nm HZO growth by ALD at 250 °C with TDMAH and TDMAZ as precursors for Hf and Zr. After patterned by photolithography, 30 nm TiN was sputtered on HZO and 40 nm extra metal layer was grown on TiN with different methods (W and Mo via sputtering and Au via thermal evaporation) as top electrodes. The device without extra metal cap was also prepared for a contrast. Then, post-metallization annealing (PMA) was carried out for the crystallization of HZO in an N_2 ambient for 60 sec at the temperature range from 450 °C to 600 °C.

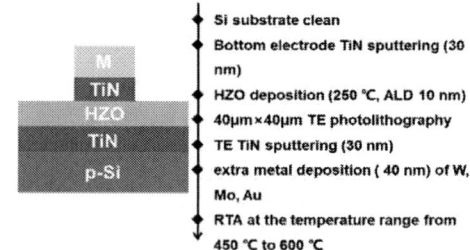

Fig.1. Structure and process flow of FE

III. Remote effect of extra metal on HZO properties

Fig. 2(a)-(d) show the polarization-electric field (P-E) characteristics of HZO capacitors with three kinds of and without extra metal capping layers. It is found that the P-E loop of four devices both before and after wake-up show an obvious difference though the basic structure and process is the same. First, P_r value of HZO capacitors with

979-8-3315-0417-5/25 $31.00 © 2025 IEEE

extra metal capping is much larger than that without. Second, devices with different extra metal capping layers show obvious difference in P_r value in both cases before and after wake-up. Particularly, the Mo and Au capped devices with all PMA conditions exhibit enhanced P_r values after wake up, while W capped one only show enhanced P_r with PMW at 600 °C. To get a clear comparison, P_r values of four kinds devices before and after wake-up are plotted as function of PMA temperature in **Fig. 2(e)** and **(f)**, respectively. Clearly, there is a gap in P_r value of devices with different remote metal, particularly, with PMA at temperature lower than 550°C. And, the Mo capped device shows the largest P_r with PMA at 50°C. Interestingly, its P_r slightly decreases when PMA temperature is increased to 600°C, while those of others increases with this trend. As a results, P_r value of devices with different extra metal capping become comparable after wake-up with a PMA temperature of 600 °C.

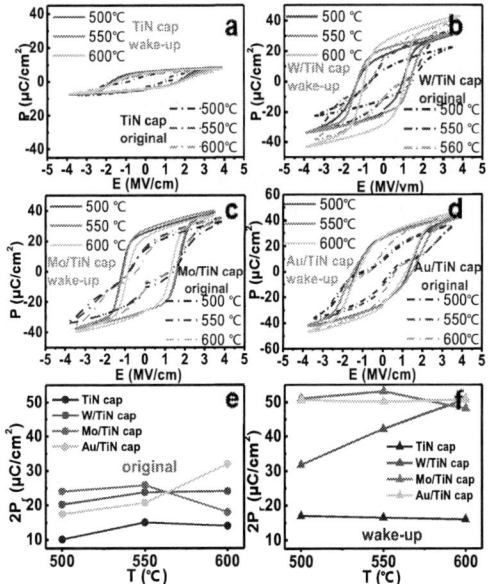

Fig.2. The PE loops of the capacitors with TiN (a), W/TiN (b), Mo/TiN (c) and Au/TiN (d) capped samples PMA at different temperature. (e) (f) The original and wake-up $2P_r$ functions of four kinds of devices at different temperature.

The endurance and retention of some capacitors were also investigated. As shown in **Fig. 3(a)**, under a program/erase voltage of 3.5V, the Mo capped devices with PMA at 550 °C exhibits higher endurance performance ($>10^9$ cycles) than W capped one (10^8) and the one without extra capping layer ($>10^6$). Besides, **Fig 3 (b)** shows a good retention of such a Mo capped device. Only a 11% degradation occurs when the retention is extrapolated to 10 years. These results directly demonstrate that the extra metal capping layer has a significant remote effect on the electric properties of HZO.

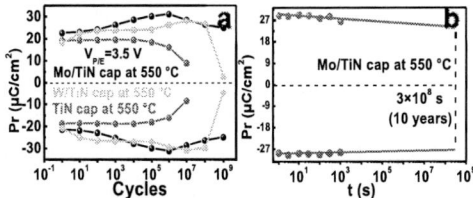

Fig.3. (a) The endurance measurements of three kinds of capped capacitors annealing at 550 °C. (b) The data retention measurement and extrapolated result to 10 years.

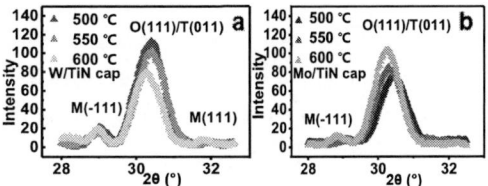

Fig.4. (a) (b) The XRD patterns of W/TiN cap and Mo/TiN capped samples under different annealing temperature.

In order to understand the remote effect of extra metal on the P_r, the crystalline structure of Mo and W capped samples with PMA at different temperatures were studied. As seen from the patterns in Fig. **3(c)** and **(d)**, the films with extra W capping show a larger intensity of monoclinic(M) phase in all PMA cases, while those with extra Mo only show obvious M-phase peak with PMA at 600 °C. Since formation of M-phase limit the content of O-phase and hence P_r value, the enhancement of Pr values of Mo capped devices should be due to suppression of M-phase content, namely enhancement of O-phase formation. Also, we have proposed that the endurance could be enhanced by enhancing O-phase content in pristine state. So, the better endurance of Mo capped sample should be also due to larger T/O phase content in HZO film [4]. In a word, it is understandable that remote effect of extra metal cap on P_r and endurance performance of HZO lies in the modulation of crystalline composition of the HZO films.

IV. Intrinsic origin of remote effect of extra metal

Next, the issue how the extra metal remotely modulates the crystalline structure is considered. We have proposed that in-plane strain in HZO intrinsically affects the crystalline structure of HZO films together with temperature [5]. Considering that different extra metal capping layer performs different thermal expansion and causes different thermal stress, the remote effect of extra metal capping is considered from a viewpoint of strain modulation.

To figure out the possible strain effect induced by the extra metal layer, 2D GIXRD characterization were carried out for the Mo and W capped devices with all PMA conditions. **Fig. 5(a)** show a typical 2D GIXRD image, where ψ is defined as the angle between the selected orientation (dash yellow line) and the normal line (solid yellow line) of sample surface. The XRD pattern is extracted from 2D image with a Δψ of 10° as shown in the marked area in Fig. 5(a). Peaking the 2θ value in the XRD pattern, the corresponding d-spacing is extractable. The d-spacing of peak position of O-phase is calculated by the Bragg's equation

$$d = \frac{n\lambda}{sin\,\theta} \qquad (1)$$

where the λ is the X-ray wavelength and the n is the diffraction order (n = 1). Meanwhile, the d-spacing has a liner relationship with $sin^2\psi$ as

$$d = k\,sin^2\,\psi + b \qquad (2)$$

where the k and b are the slope and intercept of the fitted d-$sin^2\psi$. As a result, the in-plane strain ($\varepsilon_{11} = \varepsilon_{22}$) can be calculated by

$$\varepsilon_{11} = \varepsilon_{22} = \frac{(1-v)k}{(1+v)b + 2vk} \qquad (3)$$

where υ is the Poisson's ratio to 0.29 [5]. **Fig 5(b)** shows GI-XRD patterns of Mo capped sample with PMA at 500 °C, varying ψ from 0° to 80°. **Fig. 5(c)** demonstrates the extracted linearly d-$sin^2\psi$ function and fitting parameters of slope and intercept.

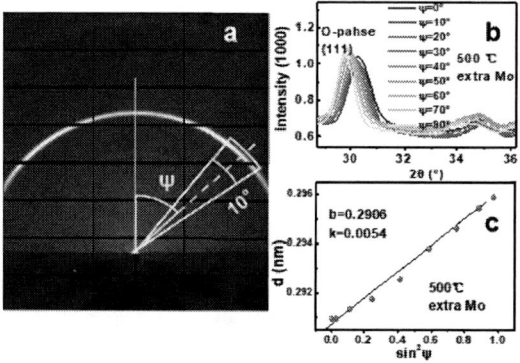

Fig.5. d-spacing with different ψ extracting process and fitted liner function fabrication of HZO film in the sample with Mo extra cap and PMA at 450 °C.

Adding the parameters extracted, the in-plane strain of HZO films with extra Mo and W capping is calculated and plotted as a function of PMA temperature as displayed in **Fig. 6(a)**. Interestingly, the in-plane strain presents a similar tendency with the 2P$_r$ after wake-up with increase of PMA temperature for both Mo and W capped films. This

indicates extra metal cap affect P$_r$ value via modulating in-plane strain in HZO. To get a clear view on the relationship between Pr and in-plane strain, 2P$_r$ in both cases before and after wake-up is plotted as function of strain in **Fig. 6(b)** and **(c)**, respectively. A universal 2P$_r$-ε relation is clearly observed in both cases independent of extra cap material and PMA temperatures. These results clearly demonstrate that the remote effect of extra metal on FE properties and crystalline structure composition intrinsically lies in the strain modulation.

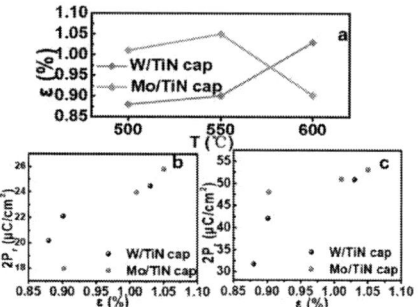

Fig.6. (a) Stain vs Temperature functions of W/TiN capped capacitors and Mo/capped. (b) The scatter plot between original 2P$_r$ and strain. (c) The scatter plot between wake-up 2P$_r$ and strain.

Conclusion

In this work, we studied several kinds of metal as the extra cap on TiN/HZO/HZO capacitors and demonstrated the remote effect of extra metal cap on the polarization, endurance and crystalline structure of HZO film. Besides, the remote effect of extra metal cap is understood to be the strain modulation due to the thermal expansion mismatch.

Reference

[1] S.J. Kim, et al. "A comprehensive study on the effect of TiN top and bottom electrodes on atomic layer deposited ferroelectric Hf$_{0.5}$Zr$_{0.5}$O$_2$ thin films" Materials, 13, pp. 1-8 (2020),

[2] S. Wang, et al. "Tunable defect engineering of Mo/TiON electrode in angstrom-laminated HfO$_2$/ZrO$_2$ ferroelectric capacitors towards long endurance and high temperature retention" Nanotechnology, 35, pp. 205704 (2024).

[3] Pešić, et al. "Physical mechanisms behind the field-cycling behavior of HfO$_2$-based ferroelectric capacitors" Adv. Funct. Mater., 26, pp. 4601-4612 (2016).

[4] D. Chen et al. "Correlation between crystal phase composition, wake-up effect, and endurance performance in ferroelectric HfxZr1−xO2 thin films" Appl. Phys. Lett. 122, pp. 212903 (2023)

[5] T. Cui et al. "Universal Temperature-Strain Phase Diagram of Hf$_x$Zr$_{1-x}$O$_2$ Films Towards to Ferroelectric Phase Engineering", International Conference on Solid State Devices and Materials (SSDM), pp. 71-72, (2024)

Electrical Tunability in Band-to-Band-Tunneling based Neuron for Low Power Neuromorphic Computing

Shubham Patil[1], Jayatika Sakhuja[1], Anmol Biswas[1], Hemant Hajare[1], Abhishek Kadam[1], Shreyas Deshmukh[1], Ajay Kumar Singh[1], Sandip Lashkare[2], Nihar Ranjan Mohapatra[2], Udayan Ganguly[1]

[1]IIT Bombay, India. [2]IIT Gandhinagar, India Email: udayan@ee.iitb.ac.in

Abstract—In this work, we show the electrical control in the ultra-energy and area-efficient BTBT-based Si-neuron and the impact on network performance. We show the control of gate bias and current threshold on the spiking threshold and frequency. Finally, we show the impact of such design space on SNN performance and a 10-layer spiking Convolutional Neural Network (CNN). The result demonstrates that neurons' post-fabrication electrical tuning capability is essential for SNN performance improvement.

I. INTRODUCTION

With the emergence of the Internet of Things, intensive computing and colossal data generation have led to the need for energy-efficient computing [1], [2], [3], [4]. The third-generation Spiking Neural Network (SNN) enables efficient computation by realizing brain functionalities [5], such as low-power parallel computing in electrical neurons and synapses (Fig. 1).

Various Silicon-based [3], [6], [7], [8], [9], [10] and non-silicon-based [11], [12], [13], [14], [15], [16] neurons are explored in the literature. Among the Silicon-based neurons, most operate in the on-regime of MOSFETs with large external capacitors for charge integration [3], [6], [7], [8], [9], [10] and dissipate higher power. In contrast, the neurons operating in the subthreshold (SS) regime [2], [17] provide power efficiency and facilitate using smaller capacitors, giving area efficiency. However, the exponential dependence of current in the SS regime makes it prone to variability. To eliminate the requirement for an external capacitor, recently, an Impact-Ionization (II) based neuron in partially depleted-Silicon on Insulator (PD-SOI) MOSFET has been demonstrated, which utilizes the body capacitor for leaky integration [18], [19]. However, the operation in II uses higher power; hence, it is unsuitable for realizing the energy-efficient SNN network (Table I).

However, to address the complexity of applications such as autonomous vehicles, robotics, medical diagnosis, and others, large-scale Spiking Neural Networks (SNN) are essential for their sophisticated processing requirements. Recently, our group has also demonstrated quantum tunnelling based neuron with the PD-SOI MOSFET using standard SOI technology to gain area and energy efficiency [5], [4]. Apart from compactness and energy efficiency, the capability to tune the electrical behavior of the neurons could be a key enabler to maximize the network performance post-fabrication according to hardware requirements for different applications. Further, using frequency control, diverse neuronal spiking patterns observed in the human brain can be implemented [20]. Hence, in this work, we characterize the BTBT-based neurons and demonstrate the effect of gate bias and current threshold parameters on the neuronal behavior (i.e., spiking threshold and spiking frequency (f) to input (V) sensitivity (df/dV)). Finally, we show the impact of such design space on SNN performance by analyzing the network learning accuracy on the Fisher's Iris classification dataset using a simple one-layer spiking network and on the CIFAR-10 image classification dataset using a 10-layer spiking Convolutional Neural Network (CNN) with the observed characteristics of the proposed spiking neuron device implemented in the simulations.

II. DEVICE DETAILS AND CHARACTERIZATION

The PD-SOI transistor used to demonstrate the LIF neuron functionality is fabricated at GlobalFoundries (Fig. 1) [21], [22], [23]. A transistor with a L_G of 40 nm and W of 450 nm is used for the experiment. The BTBT characterization is carried out by operating the SOI MOSFET in a body-grounded configuration, and during neuron demonstration, the body contact is deliberately left floating to facilitate charge integration.

III. RESULTS AND DISCUSSION

A. BTBT characterization

The experimentally measured drain current (I_D) and source current (I_S) of the SOI transistor for a range of V_{DS} is shown in Fig. 2a. An increase in I_D compared to the I_S in the tunneling regime with the increase in V_{DS} is observed, indicating the presence of drain-body tunneling (Fig. 2a). The I_B confirms the presence of the tunnelling current, and negligible gate leakage is observed (Fig. 2b-c). Fig. 2d shows the band diagram in the channel region at the channel-drain junction for visualization. Next, a measurement is performed with source-drain shorted, fixed gate bias, and body grounded to validate the tunneling current explicitly. The I_D as a function of drain-to-body voltage (V_{DB}) for different gate biases is shown in Fig. 2e.

B. Neuron electrical tunability

For neuron, the PD-SOI MOSFET is operated in the BTBT regime, enabling the low-power compact integrator. The threshold and reset circuit is implemented using conventional CMOS [4], [5]. The body terminal is the input for the threshold and reset circuit. For leaky integration, a high positive V_{DS} is applied with a negative bias on the V_{GS} with the source grounded, or a high positive drain and source bias is applied with the gate grounded. This enables the tunnelling of electrons into the drain terminal, leaving behind the holes in the body, resulting in a current that is integrated to charge the floating body. Once the body potential reaches the threshold, the external circuit issues the spike and discharges the body.

After programming the SOI MOSFET, it is difficult to read the integrated body voltage off-chip using an oscilloscope as it has a low input impedance ($M\Omega$) compared to the inverter input impedance ($T\Omega$) used in the threshold and reset circuitry [4]. To circumvent this issue, we performed a high drain current read before and after the programming to estimate the measure of charge integration. Hence, a read (R1) – Program (P) – read (R2) pulsing scheme is used [5] (Fig. 3a). In all measurements, the body contact is kept floating to enable integration. During programming, the pulse at the drain and gate terminal biases the device in the tunneling regime, enabling charge integration in the body (Q_B). The change in Q_B produces the equivalent change in the gate-body potential, reflecting the change in I_D. Hence, the difference in read current ($I_{D,R2}$) performed after programming compared to the $I_{D,R1}$, will reflect if charge integration is taking place. The read measurements are performed at fixed $V_{GS}/V_{DS} = 0.35/0.1$ V. The gate bias (V_{GS}) and threshold current (I_{th}) parameters are varied (Fig. 3b) to demonstrate neuron tunability.

C. Neuron threshold tunability

To demonstrate charge integration pulses with different T_P (at fixed bias $V_{GS}/V_{DS} = -0.7/0.9$ V) are applied. The biasing voltage is chosen systematically to represent all the regimes of operation, i.e., from no charge integration to gradual increase to saturation. The increase in $I_{D,R2}$, as compared to $I_{D,R1}$ with program duration (T_P), implies the gradual increase in the hole's integration in the body (Fig. 3c). However, after a certain T_P, the saturation in the $I_{D,R2}$ is observed, implying the saturation in the charge integration. The saturation is observed when the steady state is reached between the hole injection at the drain-body junction and hole leakage at the source-body junction. Also, with the storage of injected

979-8-3315-0417-5/25 $31.00 © 2025 IEEE

holes, the effective V_{DB} lowers, limiting the further charge injection with time for a given V_{GS} and V_{DS}. The sampled $I_{D,R2}$ vs. T_P, for different V_{DS}, is shown in Fig. 3d. The charge integration rate increases with V_{DS} for a fixed V_{GS}. The measured $I_{D,R2}$ (Fig. 3d) is compared to the threshold of 2 µA to determine the spike time, which is inverse of the spiking frequency (f). The obtained spiking frequency response versus V_{DS} for different values of V_{GS} is shown in Fig. 3e. The two key observations are (a) for the same V_{DS}, higher negative V_{GS} (at fixed V_{DS}) results in higher frequency due to a decrease in the tunneling barrier/distance at the drain-body junction, causing the tunneling of a higher number of electron/hole pairs in a shorter timescale. (b) The spiking threshold (V_{th}) shifts to lower V_{DS} at higher negative V_{GS} (Fig. 3f). The neuron's sensitivity is retained up to a gate bias from (-0.9 to -1.5 V), enabling only the threshold tuning capability. However, a lower gate bias can modulate the threshold with a reduction in neuron sensitivity (Fig. 3e).

D. Neuron sensitivity modulation

Fig. 3g shows the sampled $I_{D,R2}$ vs. T_P, for different V_{DS} at a fixed V_{GS}. The set of I_{th} (grey shaded region, 1.1 – 1.6 µA) is shown to tune the neuron's sensitivity (i.e., change in frequency vs. change in drain voltage, df/dV) without altering the neuron threshold (V_{th} = 0.7 V). The effect of I_{th} variation on f vs. V_{DS} response (Fig. 3h). The calculated sensitivity vs. I_{th} is shown in Fig. 3i. The increase in I_{th} in the range from 1.1 µA to 1.6 µA causes a decrease in neuron sensitivity with fixed neuron threshold. However, any higher I_{th} (>1.6 µA) will change neuron sensitivity and threshold.

E. SNN implementation

The impact of such designability on SNN performance is evaluated by performing the classification task on the Fisher Iris dataset. A 16×3 one-layer feed-forward SNN network is implemented with leaky integrate and fire neurons, plastic synapses, and spike time-dependent plasticity (STDP) as the learning rule. The framework used is shown in Fig. 4a [24]. The neuron characteristics (f vs. VDS) are modeled using Eqs. (1) [5], [18], [24].

$$\frac{1}{f} = R_p + \tau * \ln\left(\frac{1}{1-\frac{I_{th}}{I_{in}}}\right), I_{th} = \frac{C\,(V_{th}-E_l)}{\tau} \ \& \ g_L = \frac{C}{\tau} \quad (1)$$

Where f is the spiking frequency, R_p is the neuron refractory period, τ is the time constant, I_{th} is the threshold current, I_{in} is the input current, V_{th} is the firing threshold, E_l is the resting potential, g_L represents the conductance & C is the capacitor.

The network weights are learned by a supervised STDP rule where the supervision is provided by negative bias currents going into all of the incorrect output neurons, thus allowing the weights to learn a correlation between input features and the expected output classification by simple STDP. The weights are also directly related to conductance values in these simulations. The learning algorithm used is the same as described in [24].

Fig. 4b shows the network learning accuracy vs. the neuron firing threshold (V_{th}). A maximum software equivalent accuracy of ~96.6% is demonstrated for the experimentally obtained V_{th} compared to the ideal LIF neuron (~97.3 %) for the Fisher Iris dataset. A high or low threshold leads to a drop in learning accuracy. At a high threshold, the network ignores the smaller inputs, which results in information loss and reduces learning accuracy. However, a low threshold results in excessive spiking, leading to inappropriate training and decreased accuracy. Fig. 4c shows the impact of a change in neuron sensitivity at a fixed threshold on network learning accuracy. A maximum accuracy of ~96% is demonstrated for experimentally obtained sensitivity. A high or low sensitivity leads to a fall in network accuracy. The decrease in accuracy at high sensitivity is induced by weight saturation due to large

synaptic weight updates, while insufficient synaptic weight change at low sensitivity leads to a drop in network accuracy.

F. Spiking CNN Implementation

We have also simulated the training and testing of a larger spiking CNN consisting of 9 convolutional layers and 1 fully connected layer, as shown in Fig. 4(d). This was done (a) to perform hardware-like CIFAR-10 image classification by taking into consideration the characteristics of the proposed spiking neuron device and (b) to showcase the application of electrical tunability of these spiking neuron devices. For our simulations, we characterized the proposed spiking neuron device as a LIF neuron whose Resistance (R), Capacitance (C), and threshold voltage (V_{th}) values are chosen to replicate, as closely as possible, the experimentally observed spiking frequency vs. input current (f vs. I) curves of the proposed spiking neuron devices. Then, to anchor the temporal aspect of the simulations to realistic values, we calculate the time-constant ($\tau = RC$) of the LIF neuron, set the simulation time-step (Δt) to $\Delta t = \tau/5$, and finally, we simulate the network for 10 time-steps for training and testing, which equivalently corresponds to a sample presentation time of $T = 2\tau$ for the spiking network. For the simulations, we have assumed relatively ideal synapses, implemented as network weights and per-layer current scaling factors. Weights are quantized to 8-bit precision, and empirically determined per-layer current scaling factors are used to ensure spikes continue to propagate through the network. The current scaling factors are scaled according to the number of parameters in the layer. The network is trained using backpropagation through time (BPTT) and surrogate gradients [25]. We achieved a classification accuracy of 89.5% on the CIFAR-10 dataset using the 10-layer spiking CNN (9 conv layers + 1 fully connected layer) with the proposed spiking neuron device characteristics implemented as LIF neuron parameters and 8-bit quantized weights. This is comparable to but slightly lower than the non-spiking baseline of 91.4%, achieved using the same architecture but with ideal ReLU artificial (non-spiking) neurons and 32-bit floating point weights.

One of the primary applications of electrical tunability of the neuron parameters like V_{th} and sensitivity (incorporated in the equivalent LIF model through the capacitance value of the LIF R-C parameters) is that if there is a mismatch between the expected theoretical values of the parameters (V_{th} and C) used for training the spiking neural network and the actual ones in the fabricated devices, the mismatch can be addressed by taking advantage of the electrical tunability of the neuron devices to match the parameter values used while training, thereby not requiring the network to be retrained with the actual parameter values. To show the effect of such a mismatch between neuron parameters used for training and the actual neuron parameters extracted from fabricated devices, we simply train with one set of LIF neuron parameters (V_{th} and C) and then test with a different set of LIF neuron parameters. We find that a mismatch in the V_{th} values causes a significant drop-off in classification performance, as shown in Fig. 4(e-f), which can be recovered by using the electrical tunability properties to tune the V_{th} values of the fabricated devices. The code implementations for both the STDP-based Fisher-Iris classification and the spiking CNN training experiments (including the code for modeling spiking neuron devices as LIF neurons) have been made available online at: https://github.com/SNNalgo/device_matched_SNN_sims

G. Performance and Benchmarking

Table II presents the performance benchmarking of various novel devices-based neurons (Magnetic Tunnel Junction (MTJ), Photonic, Phase change Memory (PCM), Resistive random-access memory, and Ferroelectric Field Effect

979-8-3315-0417-5/25 $31.00 © 2025 IEEE

Transistors and standard CMOS based neuron operating in different regimes. The proposed neuron performs better compared to the novel devices-based neurons. Further, these novel devices are at various levels and add to the complexity and cost of technology integration. In standard CMOS-based neurons, the BTBT-based neuron offers ultra-compact and energy-efficient realization with higher spiking threshold variation and better sensitivity control.

IV. CONCLUSION

We experimentally demonstrated the electrical tunability in the BTBT-based neuron. Threshold tunability is achieved using gate bias, whereas sensitivity is modulated using the current threshold. Finally, the impact of such design space on the SNN performances is demonstrated. It shows that BTBT-based neurons can be effectively tuned electrically to optimize network performance for different applications. This provides a promising platform for ultra-energy-efficient and compact area SNNs.

Acknowledgment

This work is partially supported by DST and MeitY.

Reference

[1] T. Chavan *et al.*, *IEEE DRC*, pp. 1-2, 2018. [2] G. Indiveri et al., 2011. [3] A. Joubert *et al.*, *IJCNN*, 2012. [4] A. K. Singh et al., *IEEE TCASI*, 2022. [5] T. Chavan, S. Dutta *et al.*, *IEEE TED*, 2020. [6] K. M. Hynnaet al., *IEEE ISCC*, 2006. [7] J. M. Cruz-Albrecht et al., *Nanotechnology*, vol. 24, no. 38, Sep. 2013. [8] V. Ostwal et al., VLSI-TSA, 2015. [9] A. Joubert et al., *IEEE NEWCAS* 2011. [10] G. Indiveri et al., *IEEE TNN*, 2006. [11] S. Lashkare et al., *IEEE EDL*, 2018. [12] T. Tuma et al., *Nat Nanotechnol*, 2016. [13] J. Lin *et al.*, *IEDM*, 2017. [14] S. Dutta et al., *Front Neurosci*, 2020. [15] I. Chakraborty et al., *Sci Rep*, vol. 8, no. 1, Dec. 2018. [16] A. Sengupta et al., *NLM*, Aug. 29, 2017. [17] I. Sourikopoulos et al., *Front Neurosci*, vol. 11, Mar. 2017. [18] S. Dutta et al., *Solid State Electron*, vol. 160, Oct. 2019. [19] S. Dutta et al., *Sci Rep*, vol. 7, no. 1, Dec. 2017. [20] X. Fang et al., *Front Neurosci*, vol. 16, May 2022. [21] N. Butt *et al.*, *IEDM*, 2010. [22] M. Horstmann *et al.*, *CICC*, 2009. [23] B. Greene *et al.*, 2009. [24] A. Biswas et al., 2016.

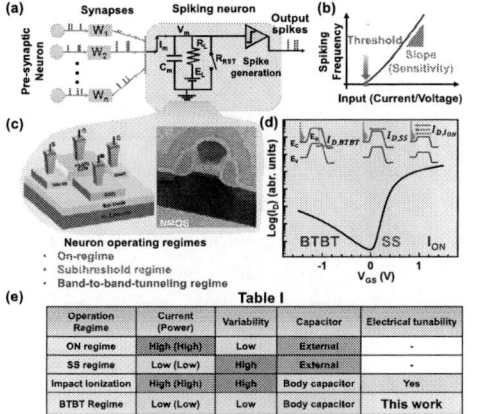

Fig. 1. **BTBT neuron for electrical tunability.** (a) Schematic of SNN with spiking neurons and plastics synapses. (b) Frequency (f) vs. input characteristics for the LIF neuron. Threshold and sensitivity are the important parameters to evaluate the neuronal response. (c) Schematic and TEM image of the PD-SOI MOSFET. (d) I_D-V_{GS} of the SOI MOSFET shows the three operating regimes (1) BTBT, (2) SS and (3) I_{ON}. (e) Benchmarking of neurons with different operating regimes.

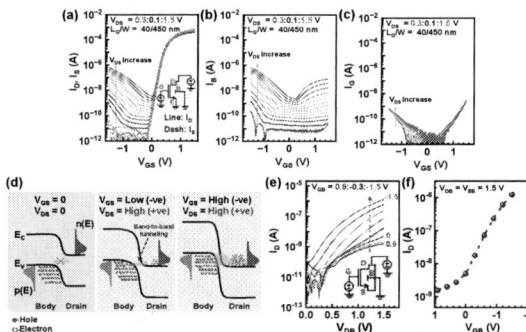

Fig. 2 **BTBT characterization.** Experimental measured (a) I_D and I_S, (b) I_B and (c) I_G of the SOI MOSFET. The increase in I_D compared to I_S shows significant drain-body tunnelling. (d) Band diagram in the channel region at the channel-drain junction. (e) I_D vs. V_{DB} for different gate bias (source-drain are shorted to avoid any leakage current contribution). (f) Extracted I_D vs. V_{GB} at source-drain bias (1.5 V).

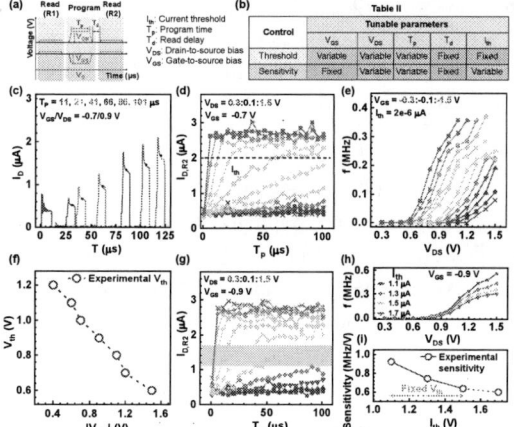

Fig. 3 **Threshold and sensitivity tunability.** (a) A read (R1) – Program – read (R2) pulsing scheme is used to capture the impact of charge integration in body. Body contact is kept floating during the measurement. (b) Usable parameters to modulate the neuronal behavior. (c) The measured I_D vs. T_p shows the gradual increase in $I_{D,R2}$ after programming due to continuous charge integration in the body (floating body enable the hole accumulation). (d) Sampled I_D vs. T_p for different V_{DS} at $V_{GS} = -0.7$ V shows the rate of integration increases with increase in the drain voltage. (e) Extracted spiking frequency (f) vs. V_{DS} for different V_{GS} shows the neuron threshold tunability electrically. (f) Extracted spiking threshold (V_{th}) vs. V_{GS}. (g) Sampled I_D vs. T_p for different V_{DS} at $V_{GS} = -0.9$ V. (h) The f vs. V_{DS} for different I_{th} (grey region, 1.1 - 1.7 µA). (i) Calculated sensitivity (i.e., change in frequency vs. change in drain voltage, df/dV) vs. I_{th}. The increase in I_{th} reduces the f as well as sensitivity of neuron at fixed threshold (V_{th}=0.7 V).

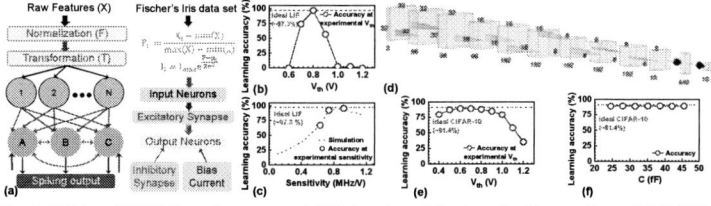

Fig. 4 **SNN and SCNN performance.** (a) Network architecture for the one-layer SNN [24]. (b-c) Effect of threshold and sensitivity on the learning accuracy of SNN. A high or low threshold/sensitivity leads to a drop in the network learning accuracy. (d) 10-layer spiking CNN architecture (9 convolutional layers + 1 fully connected layer) used for CIFAR-10 classification. (e-f) Effect of threshold and C mismatch (analogous to sensitivity in the one-layer SNN simulations) on the learning accuracy (test set).

979-8-3315-0417-5/25 $31.00 © 2025 IEEE

Direct 3D force mapping enabled by flexible single-crystal piezoelectric sensor array

Jiefei Zhu[1], Changjian Zhou[2*], and Min Zhang[1*]

[1] School of Electronic and Computer Engineering, Peking University, Shenzhen 518055, China,
[2] School of Microelectronics, South China University of Technology, Guangzhou 510640, China
*Corresponding Authors: zhoucj@scut.edu.cn; zhangm@ece.pku.edu.cn

Abstract

The ability to map 3D forces is crucial in tactile sensors. However, most traditional sensors can generally only sense forces in one direction, that is, normal or shear stress. Here, we report a flexible sensor array, which is enabled by combining single-crystal piezoelectric films with different crystal orientations for direct 3D force mapping. Both normal and shear pressure can be quickly and freely distinguished without any crosstalk. This direct 3D sensing feedback has illustrated its high potential application in intelligent robotics, human-computer interaction, medical care, and health monitoring with real-time feedback.

Keywords: Flexible Sensor, Force mapping, Single-crystal Piezoelectric thin film

Introduction

In recent years, with the rapid development of flexible electronics and the rapid popularization of intelligent terminals, flexible sensing technology has been developed rapidly, and widely used in intelligent robots, human-machine interfaces, virtual and augmented reality, medical treatment, health monitoring, and other emerging applications [1-5]. Various principles including piezoelectric, triboelectric, and capacitive sensing have been proposed to realize flexible sensors [6-8]. Among them, piezoelectric sensing has the advantages of low power consumption and fast dynamic response to external stimuli, which makes it suitable for real-time interactive force sensing applications. However, the commonly used piezoelectric materials for traditional sensors, such as ZnO, AlN, and PZT, generally exhibit a certain piezoelectric coefficient in response to normal strain, and their polarization direction is the normal direction of the thin film, which is only suitable for detecting normal force signals and cannot directly detect shear stress/force signals [9-12]. Therefore, it is necessary to design a new flexible sensor structure that can simultaneously monitor normal stress and shear stress, and achieve more accurate tactile simulation functions and high integration using a consistent large-scale array fabrication method.

Here, we report an ultra-flexible tactile sensor array based on transferred single-crystal LiNbO₃(LN) piezoelectric film, which is capable of direct mapping of real-time 3D force. It is enabled by combining single-crystal piezoelectric films with different crystal orientations for direct 3D force mapping. Single-crystal piezoelectric films with different crystal orientations can respond to normal or shear forces separately, so forces in different directions can be directly distinguished in real time. Moreover, our sensor array based on only 300 nm-thick single-crystal piezoelectric film exhibits very good flexibility. The sensors can quickly and freely distinguish normal pressure and shear pressure without any crosstalk. It can provide direct 3D sensing feedback for slip detection, interaction with fragile objects, and medical monitoring for blood and pressure and so on, illustrating its high potential application in intelligent robotics, human-computer interaction, medical care, and health detection with real-time feedback.

Device Structure and Working Principles

When a flexible piezoelectric tactile sensor comes into contact with an external object, charges are generated between the upper and lower electrodes of the single-crystal piezoelectric film. As shown in Fig. 1a, for a z-cut thin film, the charge generated between the upper and lower electrodes is D_z, and

$$D_Z = d_{33}T_{ZZ} + d_{34}T_{ZY} + d_{35}T_{ZX} \qquad (1)$$

For the single-crystal LiNbO₃ piezoelectric film, the value of d_{33} is 6 pC/N, while both d_{34} and d_{35} are zero. So

$$D_Z = d_{33}T_{ZZ} \qquad (2)$$

The flexible tactile sensor based on z-cut single-crystal LN piezoelectric film can only respond to the normal stress which corresponds to T_{ZZ} and cannot sense the shear stress which corresponds to T_{ZY} or T_{ZX}. For an X-cut thin film, as shown in Fig. 1b, the charge generated between the upper and lower electrodes is D_x, and

$$D_X = d_{11}T_{XX} + d_{15}T_{XZ} + d_{16}T_{XY} \qquad (3)$$

For the single-crystal LiNbO₃ piezoelectric film, the value of d_{11} is zero while d_{15} is 68 pC/N and d_{16} is -42 pC/N. So

$$D_X = d_{15}T_{XZ} + d_{16}T_{XY} \qquad (4)$$

Based on the same principle, the flexible tactile sensor based on X-cut single-crystal LN piezoelectric film can only respond to the shear stress which corresponds to T_{XZ} and T_{XY}, and cannot sense the normal stress corresponding to T_{XX}.

979-8-3315-0417-5/25 $31.00 © 2025 IEEE

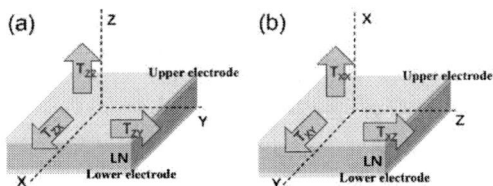

Fig. 1 Schematic diagrams of the stress notation when applying force to (a) z-cut and (b) x-cut LN piezoelectric thin films.

We innovatively propose a flexible sensor array based on single-crystal LN piezoelectric thin films with different crystalline orientations according to the above principle. The optimal single-crystal orientation, z-cut thin films for normal stress detection and x-cut thin film for shear stress, were selected and then transferred to the same flexible substrate, which can achieve single-point simultaneous non-crosstalk detection of normal stress and shear stress.

To verify the proposed device structure for force sensing of different orientations, we first constructed a multi-physics three-dimensional model for the standalone LN thin films. We simulated the piezoelectric response of 300 nm-thickness z-cut and x-cut LN films in response to normal and shear forces using COMSOL Multiphysics simulation software. As shown in figure 2, both types of forces, including normal stress and shear stress are applied to the surface of the z-cut and x-cut LN films. It is shown that when normal stress is applied, the z-cut LN thin film exhibits a large voltage, while there is no charge and voltage generated in the x-cut thin film. In contrast, the shear stress can only generate charges and build the voltage in the x-cut LN thin film. The preliminary simulation results suggest that we can take advantage of the non-crosstalk characteristics of the two kinds of crystal orientations to construct an integrated force-sensing platform.

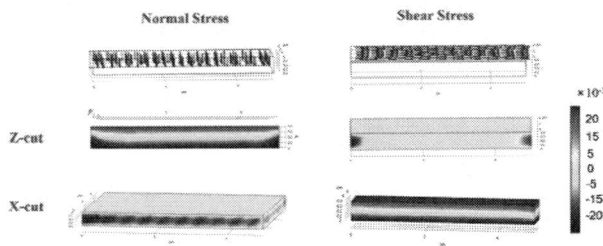

Fig. 2 Finite element simulation of the LN thin films in response to normal and shear stresses.

Device Fabrication

Fig. 3 shows the schematic process flow of preparing the transferred single-crystal thin films. The original wafer with a single-crystal thin film bonded to the substrate is prepared by ion implantation, wafer bonding, annealing and splitting, and chemical mechanical polishing (CMP) processes. The thickness of the thin film can be controlled by the ion-implantation energy and the polishing process. Up to now, the z-cut LN thin film is readily available in the market and is

provided by NANOLN company in this work. To obtain multiple small-area thin films (about 0.01 mm^2) for the subsequent transfer process, standard photolithography, electrode deposition, and lift-off process are used to first form the patterned upper electrodes. To obtain the isolated small flakes for sensing purposes, the thin film stack needs to be etched into the SiO$_2$ interlayer. After forming a hard mask by the PECVD method, the ICP process with Cl$_2$-based plasma is applied to remove the LN and also the underlying Pt/Cr bottom electrode until the SiO$_2$ is exposed.

After the etching process, the wafer is ready for the following transfer process, as shown in Figure 4. First, the exposed SiO$_2$ is removed by wet etching in a buffered HF solution. At this stage, the LN flake is still connected with the substrate as there is still SiO$_2$ underneath the LN thin film. The second step involves the formation of the anchor to the substrate by patterned photoresist stripes. Third, the SiO$_2$ under the LN thin film can be removed by using the same wet etching method, and the anchor helps to stabilize the LN flakes for the subsequent transfer process. At last, the LN flake can be picked up by the PDMS stamp and released to another flexible substrate for sensing purposes.

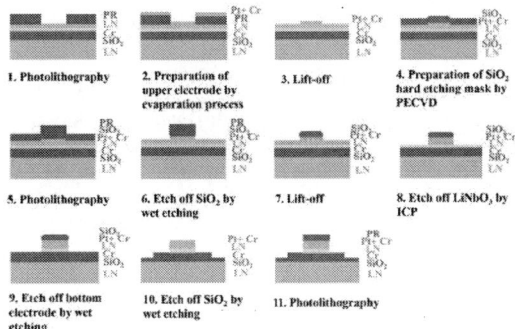

Fig. 3 Fabrication process for preparing patterned LN thin film flakes.

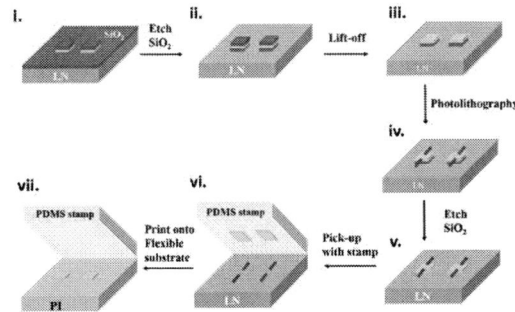

Fig. 4 Schematic diagram of the transfer process.

Electrical Measurements

Figure 5a shows the layout design for testing the force response of the transferred LN flake, and Figure 5b shows one fabricated device for electrical measurements. The large peripheral electrodes are formed on the flexible PI substrate for external electrical connections. The response to normal stress is measured with a motor stage equipped with a force meter (Mark-10). As shown in Fig. 5c, with a 5 N force applied to the center of the sample, the transferred LN sample on the flexible substrate exhibits a voltage one order higher compared with the LN sample on the original hard substrate. However, at this stage, the X-cut thin films are still not available because the bonding process is under development. A fully functional 3D force sensing system will be prepared once both two types of materials are integrated shortly.

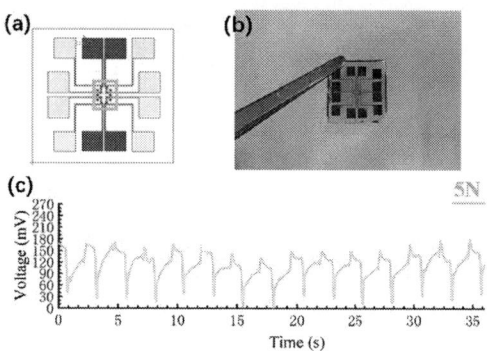

Fig. 5 (a) Circuit layout of the device with two 2*2 sensor arrays in the yellow box. (b) Optical photograph of sensor arrays. (c) Response of the z-cut LN thin film to the normal pressure.

Conclusion

In Summary, the ultra-flexible tactile sensor array based on transferred single-crystal LN film was developed. Both normal and shear pressure can be quickly and freely distinguished without any crosstalk, because of the physical properties of the $LiNbO_3$ material itself that z-cut LN film only responds to normal forces and x-cut LN film responds to shear forces. We conducted a normal piezoelectric test for z-cut LN film and verified the whole fabrication process for flexible force sensors. This direct 3D sensing feedback of LN piezoelectric film illustrated its high potential application in intelligent robotics, human-computer interaction, medical care, and health detection with real-time feedback.

Acknowledgments

This work was financially supported by National Natural Science Foundation of China (62074008), National Key Research and Development (2022YFB3603600), and the Fundamental Research Funds for the Central Universities (2023ZYGXZR107).

References

[1] S. Chun, J. S. Kim, Y. Yoo, Y. Choi, S.J. Jung, D. Jang, G. Lee, K.-Il Song, K. S. Nam, I. Youn, D. Son, C. Pang, Y. Jeong, H. Jung, Y. J. Kim, B. D. Choi, J. Kim, S.-P. Kim, W. Park and S. Park, "An artificial neural tactile sensing system", Nature Electronics, 4, pp. 429-438 (2021).

[2] G. H. Lee, H. Moon, H. Kim, G. H. Lee, W. Kwon, S. Yoo, D. Myung, S. H. Yun, Z. Bao and S. K. Hahn, "Multifunctional materials for implantable and wearable photonic healthcare devices", Nat. Rev. Mater., 5(2), pp. 149-165(2020).

[3] X. Yu, Z. Xie, Y. Yu, J. Lee, A. Vazquez-Guardado, H. Luan, J. Ruban, X. Ning, A. Akhtar, D. Li, B. Ji, Y. Liu, R. Sun, J. Cao, Q. Huo, Y. Zhong, C. Lee, S. Kim, P. Gutruf, C. Zhang, Y. Xue, Q. Guo, A. Chempakasseril, P. Tian, W. Lu, J. Jeong, Y. Yu, J. Cornman, C. Tan, B. Kim, K. Lee, X. Feng, Y. Huang and J. A. Rogers, "Skin-integrated wireless haptic interfaces for virtual and augmented reality", Nature, 575(7783), pp. 473-479(2019).

[4] S. Pyo, J. Lee, K. Bae, S. Sim and J. Kim, "Recent Progress in Flexible Tactile Sensors for Human-Interactive Systems: From Sensors to Advanced Applications", Adv. Mater., e2005902(2021).

[5] J. J. Kim, Y. Wang, H. Wang, S. Lee, T. Yokota and T. Someya, "Skin electronics: next-generation device platform for virtual and augmented reality", Adv. Funct. Mater., 31, e2009602(2021).

[6] Z. Ma, D. Kong, L. Pan and Z. Bao, "Skin-inspired electronics: emerging semiconductor devices and systems", Journal of Semiconductors, 41(2020).

[7] D. Li, K. Yao, Z. Gao, Y. Liu and X. Yu, "Recent progress of skin-integrated electronics for intelligent sensing", Light: Advanced Manufacturing, 2(1), pp. 1-20(2021).

[8] C. M Boutry, M. Negre, M. Jorda, O. Vardoulis, A. Chortos, O. Khatib and Z. Bao, "A hierarchically patterned, bioinspired e-skin able to detect the direction of applied pressure for robotics", Sci. Robot, 3(2018).

[9] M.-O. Kim, S. Pyo, Y. Oh, Y. Kang, K.-H. Cho, J. Choi, and J. Kim, "Flexible and multi-directional piezoelectric energy harvester for self-powered human motion sensor", Smart Materials and Structures, 27(2018).

[10] E. S. Nour, C. O. Chey, M. Willander, and O. Nur, "A flexible anisotropic self-powered piezoelectric direction sensor based on double sided ZnO nanowires configuration", Nanotechnology, 26(9), pp. 095502(2015).

[11] L. Jin, Z. Li, Z. Liu, B. Richardson, Y. Zheng, L. Xu, Z. Chen, H. Zhai, H. Kim, Q. Song, P. Yue, S. Q. Xie, K. J. Kim and Y. Li, "Flexible unimodal strain sensors for human motion detection and differentiation", npj Flexible Electronics, 6(2022).

[12] X. Ning, X. Yu, H. Wang, R. Sun, R. E. Corman, H. Li, C. M. Lee, Y. Xue, A. Chempakasseril, Y. Yao, Z. Zhang, H. Luan, Z. Wang, W. Xia, X. Feng, R. H. Ewoldt, Y. Huang, Y. Zhang and J. A. Rogers, "Mechanically active materials in three-dimensional mesostructures", Sci. Adv., 4(9), eaat8313(2018).

Hydrogel-based bipolar synaptic device for Artificial Neural Networks

Mengjiao Pei[1], Yun Li[1], Changjin Wan[1]

[1] School of Electronic Science and Technology, Nanjing University, Nanjing 210093, China

Abstract

We present a bipolar synaptic device based on PAA/PMMA hydrogel. Bidirectional weight modulation, including multilevel positive and negative weights, is realizes within a single device, eliminating the need for differential circuits. This synaptic device, with ultra-low energy consumption of only ~1.32 pJ per pulse, exhibits linear and symmetric weight adjustments and achieves a 91.3% recognition accuracy on the MNIST task, highlighting its potential for efficient neuromorphic computing.

Keywords: hydrogel, bipolar synaptic device, ANNs

Introduction

Neuromorphic computing, inspired by the human brain's structure and function, seeks to overcome the limitations of conventional computing architectures in handling complex, data-intensive tasks [1-2]. At the core of neuromorphic systems are synaptic devices, which emulate the signal transmission and plasticity observed in biological synapses. Among various synaptic devices, resistive switching devices and capacitive-based memristors have been widely explored for their potential in emulating biological synapses [3-4]. However, most of these devices are unipolar, meaning they require differential circuits to achieve both excitatory (positive weight) and inhibitory (negative weight) functions, adding complexity to circuit design and increasing power consumption.

In response to these limitations, hydrogel-based devices have emerged as promising candidates due to their biocompatibility, flexibility, and ability to undergo reversible physical changes [5]. The tunable ionic conductivity and stable electrochemical properties make them suitable for artificial synaptic applications. This study introduces a hydrogel-based bipolar synaptic device using poly(acrylic acid) (PAA)/poly(methyl methacrylate) (PMMA), capable of modulating positive and negative weights within a single component, thereby eliminating the need for differential circuitry. The device demonstrates low power consumption, with each pulse requiring only 1.32 pJ, making it energy-efficient compared to traditional unipolar devices. Furthermore, the device maintains linear and symmetric weight modulation, enhancing its potential for precise and stable performance in neural network applications. This study evaluates the device's performance in an artificial neural network (ANN) model using the MNIST digit recognition task, achieving an impressive 91.3% recognition accuracy. These findings underscore the device's potential in neuromorphic computing, providing a scalable and efficient approach for developing next-generation artificial neural systems.

Result and discussion

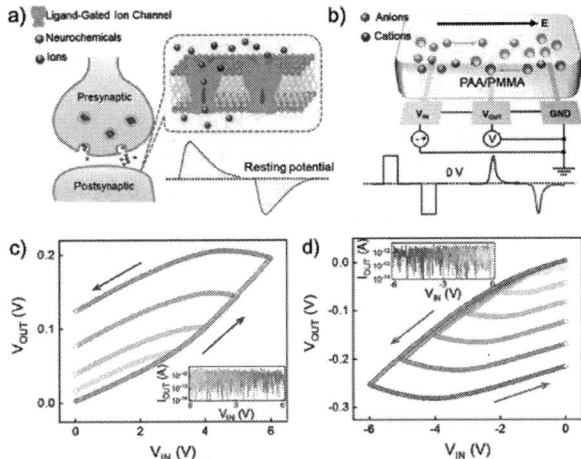

Fig.1: (a) Schematic of excitatory and inhibitory synapse with positive depolarization (purple) and hyperpolarization (blue), respectively. (b) Schematic diagram of PAA/PMMA hydrogel-based memcapacitor device and its bipolar responses. (c) VIN–VOUT hysteresis curves at different sweeping ranges with maximum voltages of +6 and (d) -6V, respectively. Inserts are the corresponding output currents.

In biological synapses, electrical signals propagate from the presynaptic terminal to the corresponding postsynaptic membrane through ion flow and the release of neurotransmitters in the synaptic cleft, resulting in the generation of postsynaptic signals. The bipolar responses include both positive depolarization and negative hyperpolarization relative to the resting potential (Fig. 1(a)). Hydrogels, also based on ionic conductance, were used to prepare our synaptic devices. As shown in Fig. 1(b), a three terminal in-plane memcapacitor is fabricated based on the PAA/PMMA hydrogels. First, the Au electrodes were deposited onto a glass substrate at a rate of 3 Å/s with the thickness of 200 nm. Subsequently, a mixture of N,N'-methylenebisacrylamide (BIS), tetramethylethylenediamine (TEMED), and potassium persulfate (KPS) is prepared in deionized water. This solution is then combined with another solution of 0.1 wt.% PMMA in anisole, and the

979-8-3315-0417-5/25 $31.00 © 2025 IEEE

resulting mixture is drop-cast onto the Au-coated substrate. The assembly is polymerized on a 70 ℃ hot plate for 20 minutes to complete the fabrication. In our design, the input was applied on left electrode (V_{IN}), which forms a circuit with the right electrode (GND). The output voltage was measured on the middle electrode (V_{OUT}). As shown in Fig. 1(c) and (d), the anticlockwise V_{IN}-V_{OUT} hysteresis curves swept at different ranges with maximum voltages of +6V and -6V, respectively. Such bipolar voltage responses arise from the migration of H^+ ions, which are released from the carboxyl groups in PAA and migrate in the direction of the applied electric fields. During these processes, PMMA forms hydrogen bonds with water molecules, creating stable "water bridge" structures that retain hydration and control ion mobility, thereby slowing H^+ migration and enhancing the hysteresis [6]. As seen in Fig. 1(c) and 1(d) insets, the operating current remains at the pA level, indicating the PAA/PMMA hydrogel's inherently low conductivity. This limited ionic migration underpins the stable hysteresis effect, reinforcing memory behavior critical for synaptic plasticity in neuromorphic applications.

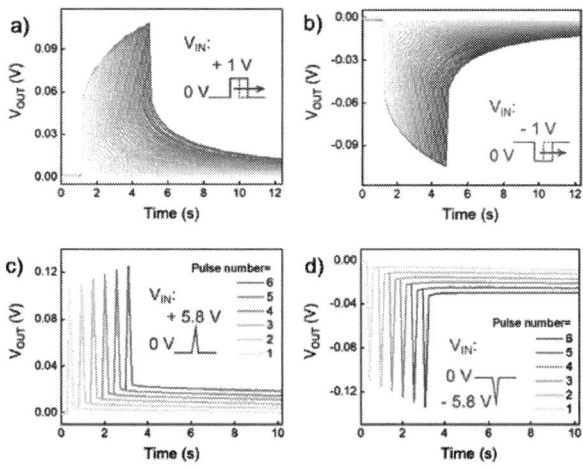

Fig.2: (a)The short-term memory triggered by electrical pulses of +1 V and (b) -1 V with different duration. (c)The short-term memory triggered by electrical pulses of +5.8 V and (d) -5.8 V with different spike numbers.

The short-term memory (STP) behavior of hydrogel synapse under electrical pulses was investigated, as shown in Fig. 2(a)-(b). As the pulse width increases, the device exhibits a typical excitatory post-synaptic potential (EPSP) and inhibitory post-synaptic potential (IPSP). The gradual decay in response after each pulse cessation reflects a transient memory effect caused by the electric double layer (EDL) effect of the hydrogel/electrode interfaces. At higher amplitudes (+5.8 V and -5.8 V), the long-term memory (LTP) behavior of hydrogel synapse was observed (Fig.

2(c)-(d)). With an increasing number of pulses, the output levels can be effectively controlled, demonstrating distinct, stable states for different pulse counts.

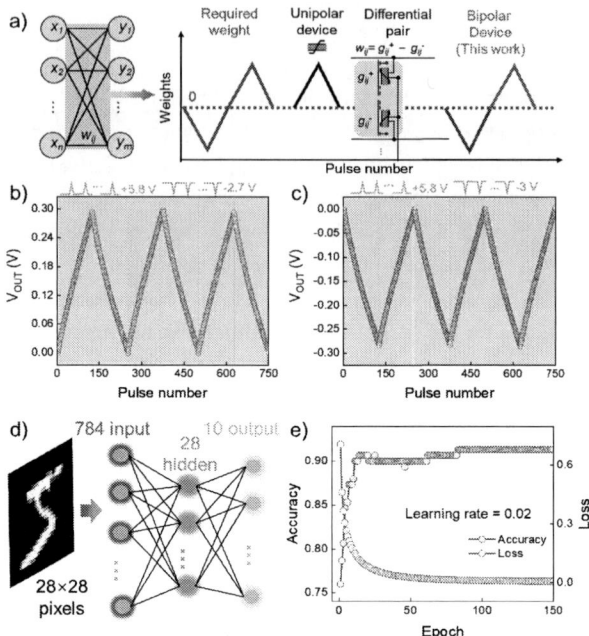

Fig.3: (a) Required weight values in ANN and weight values of unipolar and bipolar synaptic device. The inset is a differential pair, combining two unipolar devices to produce both the positive and negative conductance needed for the network. (b) Positive and (c) Negative conductance update for the potentiation and depression. (d) Schematic illustration of the simulated ANN in the MNIST recognition task. (e) Accuracy as a function of the training epoch.

Bidirectional weight updating is one of core values in the ANN (Fig. 3(a)). Considering that weight is normally represented as conductance of the synapse device which always has a positive value, a pair of unipolar synaptic device is employed to express the negative weight: named as differential pair method. Conductance values of the two synapse devices are subtracted from each other through additional circuitry to indirectly express the negative weight, which requires additional devices, larger area, higher power consumption, and increased circuit complexity. As shown in Fig.3(b), the hydrogel-based memcapacitor achieves bidirectional weight updating behavior with high linearity and symmetry. In the positive weight zone, 125 positive pulses (red, the pulse amplitude is +5.8V) and 125 negative pulses (blue, the pulse amplitude is -2.7 V) were used in potentiation and depression processes, respectively. In the negative weight zone, the pulse amplitude for depression processes is changed to -3 V for better linearity and symmetry (Fig. 3(c)). The pulse width during all measurements is fixed to 100 ms. Based on the output current observed from the device in Fig. 1(c), the maximum energy consumption per spike is calculated to be ~1.32 pJ.

The linearity of weight updates during potentiation and depression was further analyzed using conductance change equations. Nonlinearity factors for potentiation and depression were calculated to be 0.0656 ± 0.0094 and 0.1087 ± 0.0024, respectively during three repetitions [7]. These results affirm the device's capability to exhibit stable and symmetric weight adjustments, crucial for reliable synaptic behavior in neuromorphic systems. To assess the device's functionality in practical neuromorphic tasks, a pattern recognition test was conducted using the MNIST dataset [8], utilizing 3000 randomly selected images in MATLAB simulations. The simulated artificial neural network (ANN) is depicted schematically in Fig. 3(d) and consists of a three-layer structure with 768 input neurons corresponding to 28×28 pixels, a hidden layer with 28 neurons, and an output layer of 10 neurons. As shown in Fig. 3(e), the network based on our hydrogel synaptic device achieved a recognition accuracy of 91.3% after 83 training epochs. These results provide valuable insights for designing hardware-based neuromorphic systems. Future optimization of the device structure and performance could further enhance recognition accuracy and broaden the applicability of hydrogel-based ANNs.

Conclusion

In this work, we demonstrate the feasibility of a hydrogel-based bipolar synaptic device for neuromorphic computing. Capable of positive and negative weight modulation without differential circuitry, the device achieves low power consumption, linearity, and symmetry in weight adjustments. The recognition accuracy of 91.3% in the handwritten digit recognition task highlights its potential for energy-efficient, scalable artificial neural networks, positioning hydrogel materials as promising candidates for future bio-inspired computing technologies.

Acknowledgments

This work was supported by the National Natural Science Foundation of China (Grant No. 62174082).

References

[1] C. Wan et al., "Toward a Brain–Neuromorphics Interface," Adv. Mater., 36, p. 2311288 (2024).

[2] Z. Yu, A. M. Abdulghani, A. Zahid, H. Heidari, M. A. Imran, and Q. H. Abbasi, "An Overview of Neuromorphic Computing for Artificial Intelligence Enabled Hardware-Based Hopfield Neural Network," IEEE Access, 8, pp. 67085–67099 (2020).

[3] Y. Wang et al., "Optoelectronic Synaptic Devices for Neuromorphic Computing," Adv. Intell. Syst., 3, p. 2000099 (2021).

[4] G. Cao et al., "2D Material Based Synaptic Devices for Neuromorphic Computing," Adv. Funct. Mater., 31, p. 2005443 (2021).

[5] J. Yan, J. P. K. Armstrong, F. Scarpa, and A. W. Perriman, "Hydrogel-Based Artificial Synapses for Sustainable Neuromorphic Electronics," Adv. Mater., 36, p. 2403937 (2024).

[6] W. Kang, Y. Bao, H. Wang, N. Cao, and S. Cui, "Force-Induced Enhancement of Hydrophilicity of Individual Polymethyl Methacrylate Chain," Chin. J. Chem., 41, pp. 2289–2295 (2023).

[7] P.-Y. Chen, X. Peng, and S. Yu, "NeuroSim+: An integrated device-to-algorithm framework for benchmarking synaptic devices and array architectures," in Proc. IEEE Int. Electron Devices Meeting (IEDM), pp. 6.1.1–6.1.4 (2017).

[8] Y. Lecun, L. Bottou, Y. Bengio, and P. Haffner, "Gradient-based learning applied to document recognition," Proc. IEEE, 86, pp. 2278–2324 (1998).

Challenges in Accelerating Power SiC Device Commercialization

Victor Veliadis

PowerAmerica/North Carolina State University, USA

Abstract

There are several reasons behind silicon's dominance of the power electronics market. Silicon is renowned for its excellent starting material quality, its ease of processing, the opportunity for low-cost mass production, proven reliability, and circuit design legacy. However, despite significant progress, silicon devices are now approaching their operational limits. They are held back by their relatively low bandgap and low critical electric field, traits that result in high conduction and switching losses and a substandard high-temperature performance. To address these shortcomings, much effort has been directed at increasing the competitiveness of commercial SiC power devices.

Keywords: SiC, Fabrication, Processing, Manufacturing

Introduction

Transistors and diodes made with SiC have superior material properties, enabling the production of highly efficient power devices with a smaller form factor and simplified cooling management. Infineon launched the world's first commercial SiC Schottky diode in 2001. The first commercial SiC power MOSFET was released by Cree in 2011. As SiC continues to grow to a projected US$10 billion market by 2029 [1], the industry is lifting the last barriers to mass commercialization that primarily include three factors. One is that costs are higher than those for mass-produced silicon, chiefly because of the complexity of synthesizing SiC substrates and the more labor intensive SiC fab manufacturing and modules. A second significant issue is defects that limit chip yield and area, compromising long-term reliability, as well as ruggedness concerns. The third issue is that the workforce lacks expertise in integrating SiC technologies into systems – like many industries, the power electronics industry is traditionally slow to change and adapt to new technologies.

Cost Analysis

Today, the SiC wafer represents 45-65% of the overall SiC device cost [1], a consequence of its unique complex fabrication specifics. Conventional SiC substrates are primarily grown by the seeded sublimation technique at temperatures of 2300-2500 °C, which creates process control challenges. Crystal expansion is limited requiring the use of large high-material quality seeds, and the sublimation growth rates can be relatively low in the order of 0.5-2 mm/h. Furthermore, the SiC material's hardness, which is comparable to that of diamond, makes sawing and polishing SiC substrates slow and costly relative to Si.

The epitaxial layers, where SiC devices are fabricated, are grown by chemical vapor deposition in horizontal or planetary reactors at 1500-1650 °C. Pressure typically ranges from 30 to 90 Torr and growth rates can be as high as 46 mm/hr. Epitaxial growth is done on 4-degree off-cut substrates to maintain the polytype stability of the substrate. Since defects in SiC wafers limit large-area device yields, and numerous devices are paralleled in modules to increase current output, tight epitaxial doping and thickness uniformities are required, particularly as the wafer area increases.

For mass SiC commercialization, high yielding fabrication processes are required. Numerous well-established processes from silicon technology have been successfully transferred to SiC. However, SiC material properties necessitate use of specific processes not available from silicon, including wafer thinning, etching, heated implantation and anneal, and low resistivity Ohmic contact formation. SiC is inert against chemical solvents and only dry etching is practical. Furthermore, the hardness of SiC results in low photoresist selectivity and a "hard" mask, usually composed of metal or dielectrics, is required for SiC photolithographic patterning and etch. Conventional thermal diffusion is not practical in doping SiC due to its high melting point and the low diffusion constant of dopants within SiC. 600-900 °C heated ion-implantation is typically performed. After ion implantation, a 1600-1800 °C anneal assists in lattice damage recovery and dopant electrical activation. During the anneal, a protective cap layer protects the wafer surface from degradation due to the higher than carbon desorption and migration of silicon surface atoms. The high value of the SiC/metal barrier results in rectifying metal contacts and post metal deposition anneal is required for ohmic contact formation. Typically, a 50-100 nm Ni layer is blanket deposited and patterned on the wafer for the simultaneous ohmic contact formation on the n-type and p-type doped regions. Unlike Si wafers, SiC wafers are transparent. This complicates the use of "silicon" tools for inspection and metrology measurements, as the focal plane is determined with the use of an optical microscope. SiC-specific wavelength metrology/inspection tools are now available from multiple vendors. Another issue is the relative lack of flatness of SiC wafers, compared to those of Si, which can complicate photolithography. In addition, the high-temperature SiC processing can further degrade wafer flatness, occasionally rendering wafers unusable. This is particularly problematic with the thick epitaxy wafers used in

+3.3 kV device fabrication. Lastly, the poor SiC/SiO2 gate interface quality reduces inversion layer mobility and can cause threshold voltage instability. Thus, passivation techniques including annealing in nitrides are utilized to improve the SiC/SiO2 interface quality.

Overall, SiC wafer fabrication and fab processing are more complex, more time consuming, and more labor intensive than those in silicon. The result is more expensive manufacturing, contributing to higher device costs. A key part of the vertical integration occurring in today's SiC industry is securing internal substrate and epitaxy wafer capabilities to eliminate "outside purchasing" profit margins. In addition, investments in disruptive SiC wafering like engineered substrates and synthesis, boule slicing and sawing/polishing with less wasted material, etc., have high returns and are being actively sought.

SiC wafer processing savings can be introduced by shifting production from today's 150-mm to 200-mm wafers. Presently, the majority of SiC device production is on 150-mm wafers. 200-mm SiC wafers were demonstrated in 2015 and are in the process of becoming commercially available. For wide adoption, it is highly desirable that the 200-mm wafer defect-density and cost of material, per square area, are equal or lower than those of 150-mm wafers. In addition, wafer planarity should not be worse in 200-mm wafers. Due to the large overhead fab costs, the price of a fully processed wafer is, to a first approximation, independent of its area. Thus, the cost of processing a 150-mm wafer is similar to that of a 200-mm wafer, with the latter producing about 1.8 times more devices. Of course, a 200-mm SiC wafer will be more expensive than a 150-mm wafer, and that needs to be factored into the overall cost calculations. Several silicon fabs are starting to process SiC wafers, and given the plethora of mature 200-mm fabs with fully depreciated equipment, there are many companies waiting on the sidelines to enter SiC production when 200-mm wafers become widely available from multiple sources. These are companies that have established 200-mm silicon wafer production and do not want to retool to fabricate the 150 mm SiC wafers that are currently commercially available. To illustrate the economic benefits of moving to larger SiC wafer area, let's assume cost and defect density parity, per cm², between 150-mm and 200-mm wafers. Further assuming a $1500 cost for a fully processed 150-mm SiC wafer, and 60% of that coming from the "starting" wafer (thus 40% fab manufacturing cost), a back of the envelope calculation points to a 17% device cost reduction when switching to 200 mm. The same fabrication cost scenario, with now 50% of the overall cost coming from the "starting" wafer, allows for a 22% device cost reduction when switching to 200-mm production. These calculations do not include the additional processing cost reductions of streamlined mass production using the newer equipment of 200-mm volume fabs, as well as the higher yield of "edge"

devices in larger area wafers.

With respect to fabrication, the SiC industry is successfully leveraging the fully depreciated legacy silicon fab infrastructure, and is making the relatively small financial investments that allow mature silicon fabs to process SiC. Consequently, SiC chip fabrication alongside silicon has emerged as a cost-effective model that exploits silicon manufacturing economies of scale. Through repurposing mature fully depreciated 150-mm (and soon 200-mm) Si fabs, SiC power devices can be manufactured with the relatively small investments necessary to support the unique SiC processing steps. Minimizing fabrication cost by exploiting the mature Si volume production assumes the fab is loaded close to capacity with standard Si and SiC processes running in the same lines. In addition, aggregating the demand for SiC substrates and epilayers in volume fabs contributes to lower material costs. Lower fabrication costs in a fully depreciated "Si+SiC capacity loaded" fab, coupled with decreased material costs, can lead to significant price reductions for SiC devices [2]. Today, SiC manufacturing has matured and its fab infrastructure mirrors that of Si. Integrated SiC device manufactures coexist with foundries and fabless companies, and design houses provide know-how and IP that can be leveraged to accelerate entry to market.

Based on the discussion above, the higher cost of SiC devices is being lowered as mass production brings economies of scale, fully depreciated mature silicon fabs introduce SiC in their lines, and 200-mm wafers are becoming commercially available. It should be noted that in several power applications, like PV, price parity with silicon is achieved at the system level even though SiC devices cost three times more than their silicon counterparts. This is due to the reduced mass/volume of the passive components and the cooling system simplifications enabled by the efficient high-frequency SiC operation. For mass commercialization, cost savings at the system level must outweigh the higher cost of manufacturing SiC devices.

Defects, Reliability, and Ruggedness

"Killer" defects have been virtually eliminated in modern SiC wafers. Basal-Plane-Dislocations (BPDs) are the last major defect degrading device performance [3]. When bipolar current flows through a SiC device, electron–hole pair recombination at BPDs provides the energy to activate dislocation glides, which give rise to stacking faults that degrade electric performance. BPDs can propagate from the wafer substrate through the thickness of the epitaxial layers where devices are fabricated. Luckily, in commercial wafers, more than 95% of substrate BPDs propagate as relatively "benign" threading edge dislocations in epitaxial layers grown off-axis by Chemical-Vapor-Deposition [4]. BPDs can also be generated during epitaxy and during the high-temperature ion-implantation fabrication process. BPD-related degradation can be readily identified in post-fabrication production testing and is thus a yield issue; not a reliability one.

Threshold voltage instability is the main remaining reliability concern in SiC MOSFETs, which are the dominant transistors in SiC-based power electronics applications. It is primarily due to the high density of oxide traps at the SiC/gate-oxide interface. A positive shift in the MOSFET's threshold voltage increases resistance and associated conduction losses, while a negative shift is undesirable as it can unintentionally turn the device on. In addition, oxide traps reduce electron mobility leading to higher channel resistance.

SiC MOSFETs exhibit short-circuit withstand times that are well below the 8-10 μm of silicon IGBTs. Short-circuit withstand time increases with longer MOSFET channel-length and device area, but both increase resistance. SiC devices can be made more rugged by design through design trade-off optimizations. Fast gate drives, with prognostic and diagnostic functions, have been developed and provide adequate short-circuit protection within the typical 3-4 μm SiC MOSFET short circuit capability [5]. Extensive SiC reliability data have been accumulated over many years of field operation and are building confidence in the insertion of devices in power electronic systems.

Education and workforce training

A highly skilled workforce is the third challenge to creating a substantial demand for wide bandgap devices, which in turn, spurs SiC mass manufacturing and lowers overall costs. The high dV/dt and dI/dt of power SiC devices can lead to severe overshoot and ringing system issues, impacting reliability, safety, protection, efficiency, and performance. Circuit insertion of high-voltage SiC packages and modules must address the high dV/dt and related exacerbated EMI/EMC issues. Entities like PowerAmerica carry out University/industry applied collaborative projects, offer industry-driven WBG short courses and tutorials at conferences, and match students with industrial internship opportunities. These activities train the existing workforce and prepare the next generation of SiC technologists, facilitating accelerated deployment of SiC power electronics. Additional educational initiatives run by PowerAmerica include free, available-to-all monthly technical webinars, delivered by member experts. There are also: wide bandgap lectures, being added to power electronics courses across member universities; and tutorials offered at the winter and summer PowerAmerica member meetings, which provide an informal setting for learning and networking.

Conclusion

SiC chip fabrication alongside silicon has emerged as a cost-effective model that exploits silicon manufacturing economies of scale. SiC manufacturing is now mature and its fab infrastructure mirrors that of Si; integrated SiC device manufactures coexist with foundries and fabless companies. Overall, SiC wafer synthesis and fab processing are more complex, more time consuming, and more labor intensive than those in silicon. The result is more expensive devices. Shifting production from today's 150-mm to 200-mm wafers, innovations in wafering, and economies of scale are expected to reduce costs in the near future. Basal-Plane-Dislocations are the last major material defect compromising yields. Threshold voltage instability due to SiC/gate-oxide interface traps is the main remaining reliability concern in SiC MOSFETs and it is aggressively addressed. A highly skilled workforce is essential in creating a substantial demand for SiC devices, which in turn, spurs SiC mass manufacturing and lowers overall costs. Multiple initiatives across the world train students and the workforce to accelerate deployment of SiC devices and power electronics.

Acknowledgments

The author gratefully acknowledges the support of the PowerAmerica consortium members.

References

[1] Yole Group market research, https://www.yolegroup.com/

[2] V. Veliadis, Compound Semiconductor Magazine 25, vol. 36, pp. 36-42, 2019.K.

[3] V. Veliadis, H. Hearne, E. J. Stewart, M. Snook, W. Chang, J. D. Caldwell, H. C. Ha, N. El-Hinnawy, P. Borodulin, R. S. Howell, D. Urciuoli, and C. Scozzie, "Degradation and full recovery in high-voltage implanted-gate SiC JFETs subjected to bipolar-current stress," IEEE Electron Dev. Lett., Vol. 33, No. 7, pp. 952-954, 2012.

[4] T. Kimoto, Jpn. J. Appl. Phys. 54 (2015) 040103-1-27.

[5] A. Kumar, S. Parashar, S. Sabri, E. Van Brunt, S. Bhattacharya, and V. Veliadis, IEEE 30th Inter. Sym. on Power Semiconductor Devices and ICs (ISPSD), (2018) pp. 423-426.

979-8-3315-0417-5/25 $31.00 © 2025 IEEE

Materials for Thermal Dissipation Applications in Stacked Devices

W.-Y. Woon[1], J.-H. Jhang[1], S. Vaziri[1], M. Malakoutian[2], K.-K. Hu[1], I. Dayte[1], C.-C. Shih[1], J.-F. Hsu[1], J.-P. Lin[1], Y. Wu[3], A. Kasperovich[2], R. Soman[2], J. Kim[2], H.-K. Wei[1], X. Bao[1], M. Nomura[3], S. Chowbhury[2], S. Sandy Liao[1]

[1]Research and Development, Taiwan Semiconductor Manufacturing Company, Hsinchu, Taiwan, R.O.C.

[2]Deaprtment of Electrical Engineering, Stanford University, CA, U.S.

[3]Institute of Industrial Science, University of Tokyo, Tokyo, Japan

Abstract

We tackle the thermal dissipation challenges inherent in three-dimensional (3D) stacked devices and explore innovative strategies to overcome these issues through advancements in materials and processes. Back-end-of-line (BEOL) compatible materials, including aluminum nitride (AlN) and diamond are identified as promising dielectric solutions to enhance thermal management in 3D-stacked architectures. Furthermore, we propose the integration of phonon dispersion-matched bridging layers to effectively mitigate thermal boundary resistance, paving the way for improved thermal performance in stacked devices.

Keywords: Thermal dissipation, stacked devices

Introduction

The 3D stacking of high-performance integrated circuits (ICs) offers a promising solution to meet the increasing demands of computing workloads, particularly as traditional 2D scaling reaches its lithographic limitations [1]. However, as more layers of active devices are stacked, managing heat dissipation becomes increasingly challenging [2]. Specifically, thermal management from hot spots through the back-end-of-line (BEOL) routing within these 3D structures is particularly difficult with current inter-layer dielectric (ILD) materials (Fig. 1). The integration of additional tiers in 3D ICs will exacerbate the thermal resistance of these materials, ultimately impacting junction temperatures, degrading device performance, and potentially leading to premature failure [3].

To alleviate degradation issues due to self-heating of the operating devices, efficient thermal dissipation paths that mediate the heat generated from the devices to heat sinks are needed. The efficiency of thermal dissipation is related to the thermal properties of the surrounding materials and their interfaces, how and where they are integrated and placed, and how the devices are connected in a circuit. Therefore, the thermal dissipation of stacked devices is highly dependent on the materials, integration, and design architecture. In this paper we focus on the materials and their processing methods that could enable integration scheme to boost thermal dissipation performance in stacked devices.

Simulation of heat dissipation in 3D-stacked device

Fig. 2 shows an example of the simulated temperature distribution for a stacked system built with two bonded chips, akin to the chiplet packaging scenario [4]. The operating high-power device forms a hot spot and heats up the nearby low power chip, leading to leakage and reliability issues. By slightly increasing the κ of the ILD materials, significant thermal dissipation improvement can be achieved. Nevertheless, since the conventionally used mesoporous SiO_2 based ILD materials is critical to the overall RC delay in the circuit caused by the parasitic capacitance, replacing the low dielectric constant ILD materials is not a trivial task. Alternatively, by increasing the κ of the bonding materials, the hot spot temperature can also be suppressed. The temperature rising per watt in the stacked device is significantly lower by inserting a high-κ material in the film stack. The above simulation results indicate that efforts in high κ materials processing and integration could be very critical for 3D stacking of transistors and chips.

In general, high κ materials that are used as either ILD or bonding materials must concurrently satisfy the electrical and process criteria for device fabrication while improving the thermal properties. Therefore, the selected materials must be insulating, relatively low in dielectric constant, and processed at low (preferably BEOL compatible) temperature [5]. For insulators, the heat transport is mediated through collective phonon modes [6]. Thus, wide band gap crystalline materials with high Debye temperature can be promising candidates for high κ materials [7].

AlN as high thermal conductivity material

AlN has been studied as a high κ material for improvement of thermal dissipation efficiency in power devices for decades. The bulk κ value of AlN can be as high as 320 W/m/K. Nevertheless, the effective κ decreases as the film thickness decreases down to the sub-micron scale, due to phonon-boundary scattering [8]. The crystallinity and chemical stoichiometry of AlN films strongly affect its thermal properties. Here we show a transmission electron microscopy (TEM) image of a columnar AlN thin film deposited through PVD sputtering at 200 °C (Fig. 3a). The X-ray diffraction (XRD) of the AlN film indicates aligned crystal structure along the (002) orientation, and the full-width-at-half-maximum (FWHM) of the (002) peak shows a grain size of about 40 nm (Fig. 3b). Through time domain

979-8-3315-0417-5/25 $31.00 © 2025 IEEE

thermal reflectance (TDTR) measurements, we found that the good crystallinity of AlN results in high out-of-plane thermal conductivity ($\kappa_\perp$) at the thin film limit, depicted by the star data points in the benchmark figure (Fig. 4). These values are amongst the highest reported for low temperature processed nanometer-scale AlN thin films.

For AlN PVD, low RF bias condition would result in formation of triple-junction voids in the film (Fig. 5). While the columnar poly-crystalline AlN showed high $k_\perp$, its in-plane thermal conductivity ($k_\parallel$) needs to be further verified for heat spreader application. To correctly measure the $k_\parallel$ of AlN, we need to eliminate the out-of-plane heat transport. Suspended AlN cantilever structures with micron-size Au transducer are fabricated for micro-time domain thermal reflectance (μ-TDTR) measurement (Fig. 6). It is shown that $k_\parallel$ of AlN film without triple-junction voids is more than twice as high as the void-free film albeit the grain size is slightly smaller (Fig. 7). To implement AlN as heat spreader bonding layer, it is imperative to planarize the film down to sub-nanometer scale roughness for bonding. Chemical mechanical polishing (CMP) with facet insensitive polishing rate showed feasibility for the planarization (Fig. 8(a)). Note that, random pits may appear after the CMP (Fig. 8(b)) for the AlN grown under low RF bias. Proper RF bias condition is therefore vital for application of PVD AlN as bonding layer. The BEOL compatible deposition conditions and the good insulating film properties could enable application of PVD AlN in BEOL thermal management.

Diamond as high thermal conductivity material

Diamond is one of the ideal materials that promises extremely high thermal conductivity. Owing to its strong sp³ bonding and low atomic number, the Debye temperature of diamond exceeds 2200 K and the κ value of bulk single crystalline diamond is 2200 W/m/K [9]. Conventionally, the growth temperature of single crystalline diamond can be as high as 1700 °C, far exceeding the temperature limits in silicon-based semiconductor fabrication processes [10]. It has been reported that the growth temperature of diamond can be significantly reduced through hot filament [11] or plasma assisted chemical vapor deposition [12] (CVD).

Here we report growth of a high κ diamond poly-crystalline thin film through a BEOL compatible microwave plasma assisted CVD process using CH_4 and H_2 as precursor and reactant gas, respectively [13]. The nucleation of the diamond grain is assisted by pre-deposited diamond nanoparticles on the target substrate while the high-power microwave plasma provides high density radicals that break CH_4 bonds. Furthermore, to enable diamond growth at BEOL compatible temperature, low concentration oxygen is mixed into the CVD process to promote CH_4 dissociation (Fig. 9).

As larger grain size is needed to reduce phonon scattering and improve κ of the poly-crystalline diamond film, high pressure, and plasma power, along with low CH_4/H_2 ratio conditions are employed to promote grain growth and suppress nucleation density. Conventionally, the diamond grains tend to grow vertically and form columnar structures with many grain boundaries between the columns, resulting in high $\kappa_\perp$ but much lower $\kappa_\parallel$. To promote better lateral heat spreading efficiency, nearly isotropic grains are needed. Fig. 10a shows a TEM image of the diamond poly-crystalline thin film grown at 400 °C. The sp³ purity of the diamond film is high as no significant sp² peak can be identified by Raman spectroscopy [14] (as shown Fig. 10b).

TDTR measurement showed > 500 W/mK high $\kappa_\perp$, $\kappa_\parallel$, and low TBR values for the MPCVD grown diamond film [15]. To integrate diamond the heat spreader bonding layer in backside or buried power rail architectures [16], novel diamond CMP method would be needed to enable good wafer bonding [17]. The polishing rate of PCD is extremely low, but facet insensitive planarization can still be achieved using slurry that contains high-hardness nanoparticles (Fig. 11). Finally, a more subtle issue is the potentially additional TBR arising from the phonon dispersion mismatch between diamond and the bonded materials (Fig. 12) [18]. Inserting a thin layer of amorphous bridging materials with many overlapping phonon modes from both materials can effectively reduce the TBR, as shown in the simulation result (Fig. 13).

Conclusion

In summary, we present challenges in thermal dissipation for 3D stacked devices through simulation and discuss strategies to tackle the issues through innovation in materials and their processing methods. BEOL compatible AlN and diamond are evaluated and proposed as candidates for thermal dissipation applications in stacked devices. Thermal boundary resistance alleviation through insertion of bridging layers is proposed to solve the phonon dispersion mismatch issue.

References

[1] 3D System Integration, IEEE 3DIC 2009. [2] Guan et al., NPJ 2D Mater Appl 7. p. 10, 2023. [3] Lau et al., IEEE ECTC, p. 635 2009. [4] Jeng et al., IEDM p. 3.2, 2022. [5] Oates, IEDM, p. 20.7.1 2014. [6] Cahill et al., JAP 93, p. 793, 2003. [7] Watari et al., MRS Bulletin, 26, p. 440, 2001. [8] Xu et al., JAP 126, p. 185105, 2019. [9] Ward et al., PRB 80, p. 125203, 2009. [10] Liang et al., Diamond and Related Materials 18, p. 698, 2009. [11] Ahmed et al., Cryst. Growth Des. 19, p. 672, 2019. [12] Gurbuz et al., IEEE Trans. Power Electronics, 20, p. 1, 2005. [13] Malakoutian et al., Cryst. Growth Des. 21, p. 2624, 2021. [14] Osswald et al., JACS 128, p. 11635, 2006. [15] Woon et al., IEDM, P. 19.3 2023. [16] Gupta et al., IEDM, p. 20.3.1, 2020. [17] Iavoco et al., IEEE ECTC, p. 2206, 2019. [18] Swarz et. al., Rev. Mod. Phys. 61, p. 605, 1989.

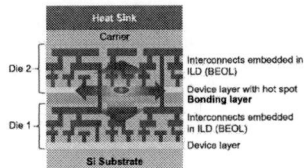

Fig. 1. Schematic of a 3D IC with two stacked dies, which consist of a device layer and interconnect layers embedded in an ILD, and a bonding layer between the dies. There is overall heating in the device layers of both dies, with a hot spot in the upper die 2 (not to scale).

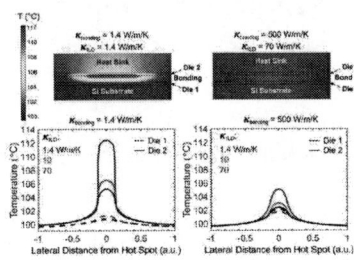

Fig. 2. Simulation showing reduced temperature of the 3D IC through increased thermal conductivity of the bonding and ILD materials in BEOL.

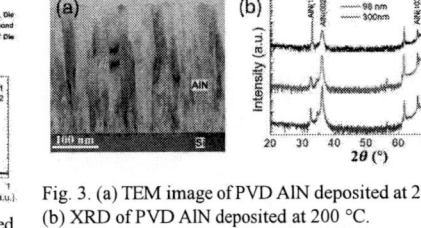

Fig. 3. (a) TEM image of PVD AlN deposited at 200 °C. (b) XRD of PVD AlN deposited at 200 °C.

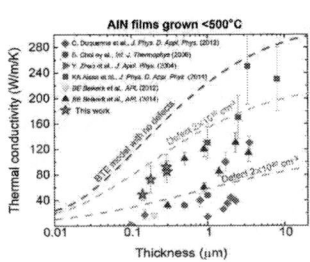

Fig. 4. Benchmark of thermal conductivity vs. thickness for AlN thin films processed at temperatures below 500 °C.

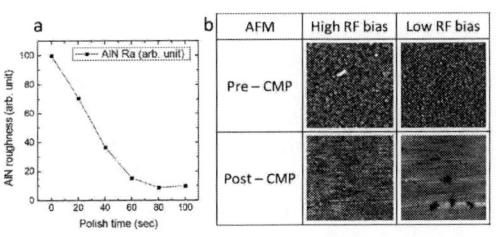

Fig. 5. (a) Cross-section-view SEM of PVD AlN with columnar grains along [0001] with the hexagonal wurtzite structure. (b) Plane-view SEM images of PVD AlN deposited under low RF bias (left), and high RF bias (right), respectively.

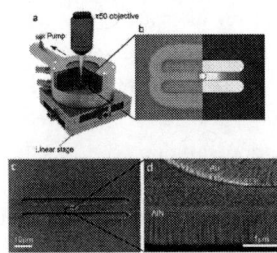

Fig. 6. (a) Schematics of micro-TDTR. (b) Optical image (left) and COMSOL simulation of temperature mapping (right) of the AlN suspended cantilever. (c) SEM images of the suspended AlN cantilever structure (d) zoom-in view of the Au transducer and AlN cantilever.

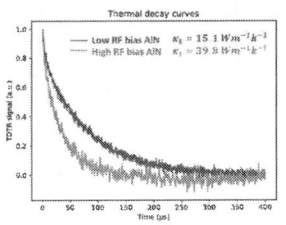

Fig. 7. Thermal decay curves for two AlN samples, showing faster decay rate for PVD AlN grown under high RF bias.

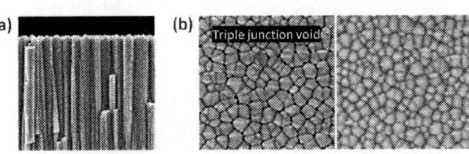

Fig. 8. (a) RMS roughness of AlN surface versus polishing time. Significant roughness reduction can be found by prolonged CMP duration. (b) AFM images of PVD AlN film deposited under high RF bias (left) and low RF bias (right) before and after CMP.

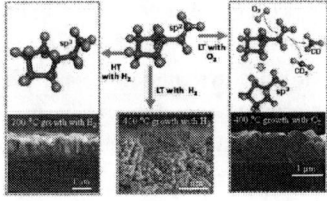

Fig. 9. CH₄ dissociation in microwave plasma assisted CVD at high and low temperatures in H₂ and O₂ showed different diamond growth quality.

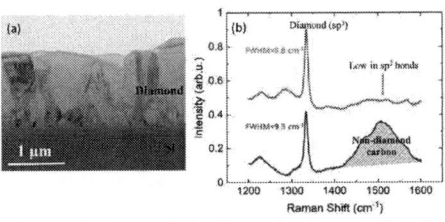

Fig. 10. (a) TEM image of the diamond poly-crystalline thin film grown at 400 °C. (b) Raman spectroscopy of high quality (upper curve) and low quality (lower curve) diamond poly-crystalline thin films grown at 400 °C.

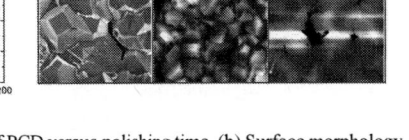

Fig. 11. (a) RMS roughness of PCD versus polishing time. (b) Surface morphology of PCD before and after CMP. Intrinsic voids in PCD exposed during CMP, resulting in a challenge to achieve an atomically flat diamond surface.

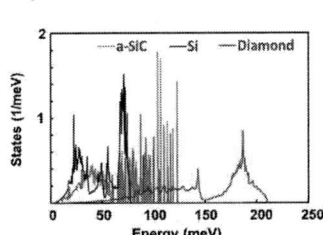

Fig. 12. Phonon dispersion of diamond (red), Si (black), and a-SiC (green).

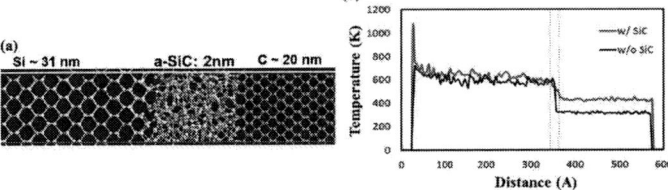

Fig. 13. (a) Simulation framework for heat transport from Si to diamond through a sandwiched 2 nm SiC bridging layer and (b) Temperature profile across diamond and Si with and without a SiC bridging layer.

979-8-3315-0417-5/25 $31.00 © 2025 IEEE

Thermal Analysis of Multi-Chiplet heterogeneous integration Based on The Lidless Fan-out Package

Yongbo Wu[1], Jiexun Yu[1], Changming Song[1], Zheqi Xu[1], Lin Tan[1], Qian Wang[1,2], Jian Cai[1,2*]

[1]School of Integrated Circuits, Tsinghua University, Beijing, China, [2]Beijing National Research Center for Information Science and Technology, Beijing, China

Abstract

The fan-out package with many distinct advantages is widely adopted for the state-of-the-art mobile applications and has great potential in the server applications. In this paper, the lidless structure with an heatsink directly integrated on the fan-out package is proposed as the efficient thermal dissipation solution with low thermal resistance at package and system level. The thermal modelling and characterization of the lidless fan-out package with SoC and Memory is carried out.

Keywords: Heterogeneous integration, Fan-out package, Thermal performance, Power envelop

Introduction

As the scaling of transistors slows down, the paradigm of heterogeneous integration of chiplets and components to accelerate performance enhancement in computing and electronic systems becomes imperative. Particularly, advanced system integration of logic and memory has become a critical focus for the mobile and server applications in the era of 5G and AI[1, 2]. The fan-out package offering numerous benefits of small form factor, large routing space, low cost, high integration flexibility, great thermal performance and superior electrical performance can meet the requirements of the applications. In the applications of server and network, RDL interposer as a high bandwidth, large area and cost-effective solution of heterogeneous integration holds significant potential to replace silicon interposer[3].

In the state-of-the-art mobile applications, Samsung has developed the first and the second generation high-density fan-out package platforms, FO-POP and FO-SIP respectively. FO-SIP with the logic and memory side by side placement achieves higher power density, better thermal performance, and higher interface bandwidth than FO-POP with the package on package placement[1, 4]. With the increasing demands of the more functional features of advanced system integration, the fan-out package with higher density is manifesting from single-chip package to multi-chiplet heterogeneous integration and from being initially designed for low-power devices to high-power devices[5, 6]. However, the power (or power density) increases with the transitions of the fan-out package. The rise of power in turn causes a sharp increase of junction temperature, which could degrade and compromise the system performance and component reliability. Thus, the critical concern is brought in the thermal management of the fan-out package.

In our previous work, the lidless fan-out package is proposed to be an effective thermal enhancement solution, and the integrated heatsink directly attached on the package [5]. The thermal characteristics of the fan-out package with one chip were investigated. In this paper, the lidless structure of the high-density fan-out package with SoC and Memory is presented for enhancing heat dissipation. The thermal performance of the high-density fan-out package with various key parameters are characterized.

Methodology

Compared with the lidded structure of the fan-out package, the lidless structure has better thermal dissipation performance owing to the removal of TIM2 and the lid and the absence of their thermal interface [7]. Therefore, the lidless structure with the additional heatsink directly integrated on the fan-out package is used for the advanced high-density system integration with SoC and Memory in this research, and the SoC and Memory are placed side by side and communicated via fine RDL. The schematic diagram of the lidless fan-out package is depicted in Fig.1. The epoxy molding compound (EMC) can protect the chip in the operating environment, control the warpage, and support the heatsink [5].

The SoC and Memory sizes are 10mm*10mm and 10mm*11mm with the same thickness, respectively. The Memory size corresponds to the high bandwidth memory (HBM) dimension[8]. The power consumption for SoC and Memory are 30W and 20W, respectively. The molded package size is 13.5mm*27.5mm. The total thickness of 3 layers RDL is 40μm[9]. The diameter and pitch of the solder ball are 0.25mm and 0.4mm, and the number of solder balls for both SoC and Memory is 1024.

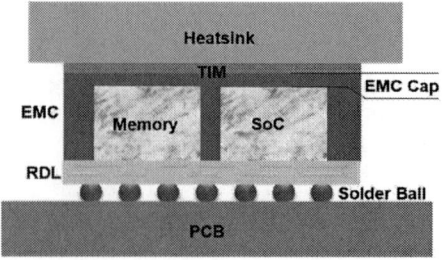

Fig. 1: The schematic diagram of the lidless high-density fan-out package

The thermal characterization of the lidless fan-out package with SoC and Memory is performed by the 3D finite element method. In order to improve the solving efficiency, some of the detailed models with high-density interconnections can be replaced by simplified models with anisotropic thermal conductivities, such as PCB, RDL and solder ball. The heat sources of SoC and Memory are set to the bottom of the SoC and Memory. The ambient temperature is set to 25 °C. The top side of the heatsink with a certain standardized size is with different convective heat transfer coefficients(h_{eff}) corresponding to different cooling environments[10], and the bottom side of PCB is with an effective convective heat transfer coefficient of 10 W/m²K, as shown in Fig. 2(b).

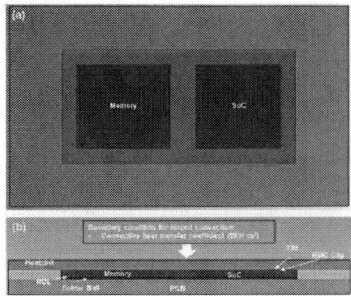

Fig. 2: The lidless fan-out package with SoC and Memory: (a) horizontal view and (b)Vertical view of the thermal simulation model

Results and Discussions

The junction temperature is positively proportional to the EMC cap thickness between chip and heatsink of the lidless fan-out package with SoC and Memory, and the junction temperature is minimized in the absence of EMC cap due to the low thermal conductivity, as shown in Fig. 3(a). The maximum junction temperature of the SoC is 149.20°C, 89.38°C, and 61.58°C with the EMC Cap thickness of 0.1mm, 0.03mm and without the EMC cap. The maximum junction temperature of the Memory is 111.15°C, 72.79°C, and 56.05°C with the EMC Cap thickness of 0.1mm, 0.03mm and without the EMC cap, respectively. Fig. 3(b) shows power envelops with EMC cap and without EMC cap of the lidless fan-out package. The maximum operating junction temperatures (T_{jmax}) of the SoC and memory are limited to 105 °C and 85 °C, respectively. The absence of EMC cap results the largest power envelop, and the power envelop shrinks with the increase of the EMC cap thickness. The temperature profiles with EMC cap and without EMC cap are shown in Fig. 4.

Nevertheless, the electrical connection of the chip and the heatsink is usually prohibited. The certain EMC cap thickness can ensure the chips at different potentials and less compromise thermal performance in the high-density fan-out package. In the subsequent analysis, the EMC Cap thickness was set to 0.03mm.

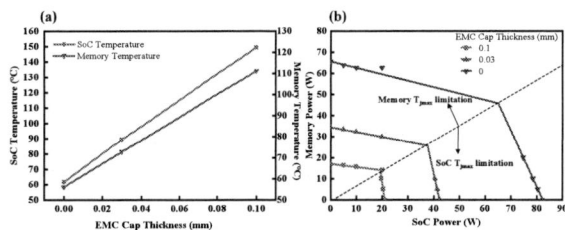

Fig. 3: (a) The junction temperatures is positively proportional to the EMC cap thickness and (b) the package power envelops

Fig. 4: The temperature profiles with EMC cap thickness of (a) 0.1mm, (a) 0.03mm and (c) without EMC cap

The thermal performance of the lidless fan-out package with different convective heat transfer coefficient on the heatsink for specific cooling methods is evaluated. The h_{eff} varies from 300W/m²K to 10000W/m²K, representing the range from forced air cooling to forced liquid cooling. From Fig. 5, it can be seen that the maximum junction temperatures of the SoC and Memory decrease with the increase of the h_{eff}. The maximum junction temperature of SoC decreases from 334.08 °C with the h_{eff} of 300W/m²K to 68.34°C with the h_{eff} of 10000W/m²K. The larger h_{eff} generates the larger package power envelop and effectively reduces convective thermal resistance of the top path and improves thermal performance.

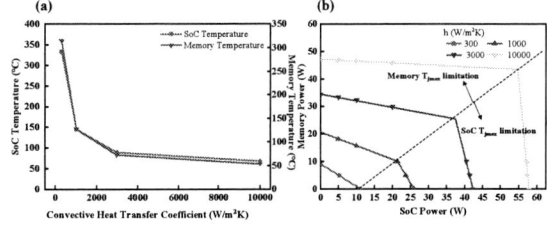

Fig. 5: (a) The junction temperatures decrease with the increase of the h_{eff} and (b) the power envelops with various h_{eff}

In addition to the convective heat transfer coefficient, the convective heat transfer area is also an important factor influencing the thermal resistance of top path. Fig. 6(a) exhibits the maximum junction temperature and the convective heat transfer area of the heatsink. As the ratio of the horizontal area of the heatsink to the package increases, the junction temperature decreases by lower convective thermal resistance, which is achieved by harnessing the space external to the package.

Fig. 6(b) shows the impact of the thermal interface material (TIM) with different thermal conductivity on the thermal performance of the lidless fan-out package. The maximum junction temperatures of SoC and Memory reduce as the thermal conductivity increases. TIM with high thermal conductivity has low thermal resistance.

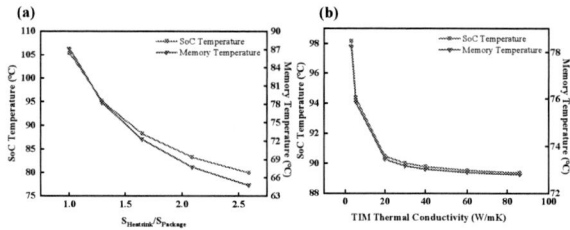

Fig. 6: The junction temperatures decrease with the increase of (a) the convective heat transfer area and (b) the TIM thermal conductivity

The junction temperature first rises then declines as the spacing of SoC and Memory the increases from 0mm to 4mm, as shown in Fig. 7(a). When the spacing is 0mm, the SoC and Memory can be regarded as merged together, and the junction temperature is further averaged by heat conduction. Fig. 7(b) shows the thermal resistance of junction to ambient (R_{J-A}) with various spacing. $R_{J-A\ SS}$ is R_{J-A} of SoC due to the power applied to SoC, and $R_{J-A\ MM}$ is R_{J-A} of Memory due to the power applied to Memory. $R_{J-A\ SM}$ and $R_{J-A\ MS}$ are R_{J-A} of Memory owning to the power applied to SoC and R_{J-A} of SoC owning to the power applied to Memory. The thermal crosstalk of the split chips attenuates as the spacing increases.

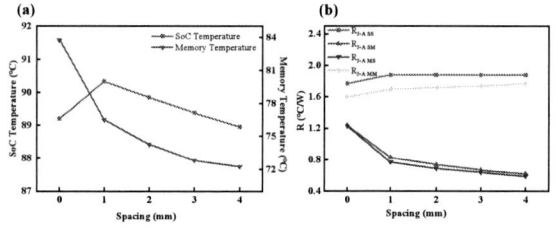

Fig. 7: (a) The relationship between the junction temperatures and the spacing, (b) the thermal crosstalk effect weakens as the spacing increases

Conclusion

In this paper, the lidless structure with the additional heatsink directly integrated on the high-density fan-out package with SoC and Memory is presented for enhancing heat dissipation.

The maximum junction temperature of the integrated system is 149.20°C, 89.38°C, and 61.58°C with the EMC Cap thickness of 0.1mm, 0.03mm and without the EMC cap. The package power envelop enlarges with the decrease of the EMC cap thickness. The thermal characteristic of the high-density fan-out package with various key parameters including the convective heat transfer coefficient, the convective heat transfer area, the thermal conductivity of TIM and the spacing of SoC and Memory are investigated.

Acknowledgments

The authors would like to appreciate the financial support under Grant No. QYJS-2022-2100-B and the Beijing Natural Science Foundation Grant No. L223005.

References

[1] S. H. You et al., "Advanced Fan-Out Package SI/PI/Thermal Performance Analysis of Novel RDL Packages," in 2018 IEEE 68th Electronic Components and Technology Conference (ECTC), 2018, pp. 1295-1301.

[2] C. T. Wang and D. Yu, "Signal and Power Integrity Analysis on Integrated Fan-Out PoP (InFO_PoP) Technology for Next Generation Mobile Applications," in 2016 IEEE 66th Electronic Components and Technology Conference (ECTC), 2016, pp. 380-385.

[3] J. g. Jang et al., "Advanced RDL Interposer PKG Technology for Heterogeneous Integration," in 2020 International Wafer Level Packaging Conference (IWLPC), 2020, pp. 1-5.

[4] J. H. Lau, "FOWLP: PoP" in Fan-out wafer-level packaging, Springer, 2018.

[5] Y. Wu, J. Cai, C. Song, L. Tan, and Q. Wang, "Simulation Analysis on Thermal Performance of a Lidless Fan-Out Package," in 2024 IEEE 10th Electronics System-Integration Technology Conference (ESTC), 2024, pp. 1-8.

[6] A. Cardoso, H. Barros, and G. Hantos, "Fabrication, performance and reliability of a thermally enhanced wafer level fan out demonstrator with integrated heatsink," in 2016 22nd International Workshop on Thermal Investigations of ICs and Systems (THERMINIC), 2016, pp. 336-344.

[7] K. Yan, P. Y. Lin, and S. L. Kuo, "Thermal Challenges for HPC 3DIC Packages and Systems," in 2022 6th IEEE Electron Devices Technology & Manufacturing Conference (EDTM), 2022, pp. 151-153.

[8] M. J. Park et al., "A 192-Gb 12-High 896-GB/s HBM3 DRAM with a TSV Auto-Calibration Scheme and Machine-Learning-Based Layout Optimization," in 2022 IEEE International Solid-State Circuits Conference (ISSCC), 2022, vol. 65, pp. 444-446.

[9] J. Lau et al., "Warpage and Thermal Characterization of Fan-Out Wafer-Level Packaging," in 2017 IEEE 67th Electronic Components and Technology Conference (ECTC), 2017, pp. 595-602.

[10] N. R. Jankowski and F. P. McCluskey, "Modeling transient thermal response of pulsed power electronic packages," in 2009 IEEE Pulsed Power Conference, 2009, pp. 820-825.

Toward Low-Thermal-Budget Processing and Low-Voltage Operating in Ferroelectric Stack Films by Introducing ZrO₂ Middle Layer

Yinchi Liu[1], Hao Zhang[1], Jining Yang[1], Xun Lu[1], Shiyu Li[1], Yeye Guo[1], Chunlei Wu[1], Handong Zhu[1], Xuning Zhang[1], Zhenxin Wang[1], Yaoyi Wang[1], Ruli Zeng[1], Yiwen Yu[1], Wenjun Liu[1,2,*]

[1]The school of Microelectronics, Fudan University Shanghai 200433, China, [2]Zhangjiang Fudan International Innovation Center, Fudan University, Shanghai 201203, China
*E-mail: wjliu@fudan.edu.cn

Abstract

In this work, four ferroelectric (FE) capacitors with thickness dependent ZrO₂ middle layer (ML) were designed and fabricated at the low annealing temperature of 350 °C. By modulating the thickness of ZrO₂ ML, it is demonstrated that the FE capacitor with 2 nm ZrO₂ ML exhibits a superior double remanent polarization ($2P_r$) of 36.7 μC/cm² under the V_{op} of 2 V. The insertion of ZrO₂ ML effectively suppresses the wakeup effect in the FE capacitor and decreases the wakeup ratio by approximately 31 % under the low operating voltage of 2 V. In addition, the FE capacitor with ML presents an excellent endurance property of the cycling characteristic. These findings show the great potential for BEOL-compatible non-volatile memory applications.

Keywords: Back end of line, ZrO₂ middle layer, reliability, low-voltage operation, ferroelectricity.

Introduction

The Zr-doped HfO₂ (Hf₀.₅Zr₀.₅O₂, HZO) ferroelectric (FE) thin films have garnered lots of attention in embedded memory toward VLSI integration owing to its high CMOS compatibility as well as scalability [1-3]. However, the poor crystallization of HZO thin films during low-temperature annealing results in excessive operating voltages (V_{op}), necessitating further reduction for meeting the power driving requirements in advanced process nodes. Additionally, the excessively high V_{op} leads to the premature hard breakdown during the erase/write cycles, limiting the reliability improvement [4-6]. A direct method to reduce V_{op} is to shrink the thickness of FE films. Nonetheless, the reduction in thickness increases the phase formation barrier, leading to a rise in the annealing temperature (> 450 °C) for maintaining a considerable remanent polarization (P_r), which is incompatible with the back-end of line (BEOL) process. On the other hand, the increasing leakage current and decreasing breakdown voltage also restrict the application in the non-volatile memory devices [7]. Therefore, middle layer strategy offers a promising means to improve the ferroelectricity and reliability under the low V_{op} and low thermal budget. Yet, systematic studies on these issues have not been reported so far.

Here, we proposed the ZrO₂ ML strategy to achieve a

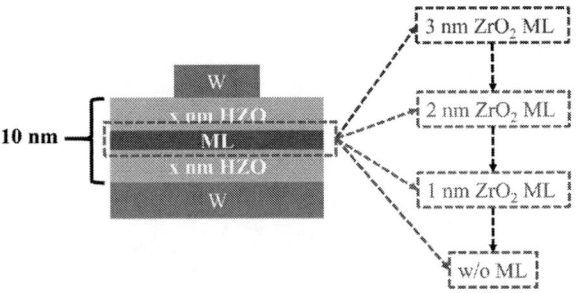

Fig. 1 The schematic of the fabricated ferroelectric capacitor with different ZrO₂ middle layer thickness. The total thickness of ferroelectric thin film is 10 nm.

superior $2P_r$ of 36.7 μC/cm² and endurance exceeding 10⁹. Specially, compared to the FE capacitor without ML, the FE capacitor with ZrO₂ ML also shows a suppressed wakeup effect under the room temperature. Furthermore, the FE capacitor with ZrO₂ ML shows an excellent reliability under different frequency and temperature, while maintaining the excellent retention time of 10⁴ s.

Experimental Details

Fig. 1 shows the schematic diagram of the fabricated capacitor with HZO/ZrO₂ ML/HZO stack. Firstly, a 50 nm tungsten (W) was deposited onto the Si/SiO₂ substrate by physical vapor deposition (PVD). Then, the 10 nm HZO/ZrO₂/HZO stack were grown by atomic layer deposition (ALD) at 280 °C, in which Hf[N(CH₃)₂]₄, Zr[N(CH₃)₂]₄, and oxygen plasma were used as Hf, Zr, and oxygen sources, respectively. Subsequently, a 50 nm W is sputtered by PVD, photo-lithography and wet etching in sequence to form electrode with an area of 80 × 80 μm². Finally, the FE capacitors were annealed at 350 °C in N₂ for 300 s using rapid thermal annealing (RTA). The electrical performance was measured by semiconductor analyzer (Agilent B1500A) and ferroelectric test system (Precision Premier II).

Results and Discussion

Fig.2(a) and (b) depict the polarization-voltage (*P-V*) and dynamic current-voltage (*I-V*) characteristics of the FE capacitors with different thickness of ZrO₂ ML under 2 and 4 MV/cm at the frequency of 1 kHz. Obviously, the insertion

979-8-3315-0417-5/25 $31.00 © 2025 IEEE

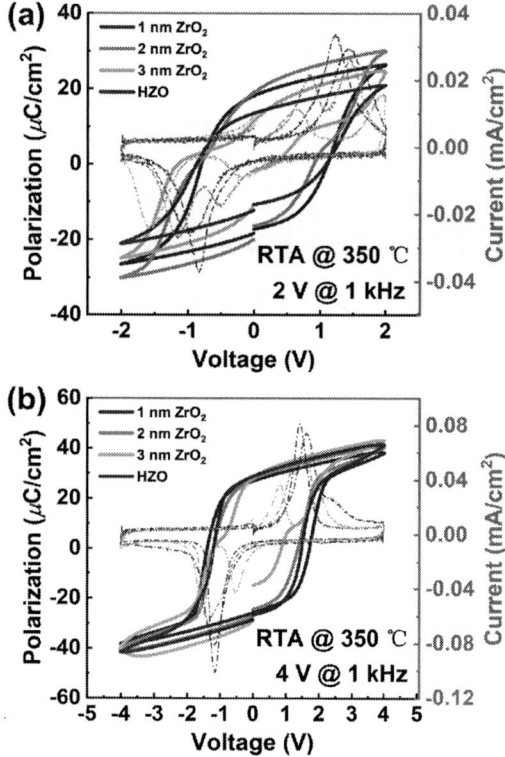

Fig. 2 the polarization-voltage (*P-V*) and dynamic current-voltage (*I-V*) characteristics of the FE capacitors with different thickness of ZrO_2 ML under (a) 2 and (b) 4 MV/cm with the frequency of 1 kHz.

of ZrO_2 ML has been shown to effectively enhance the $2P_r$ of FE thin films. In particular, when the ML thickness is 1 nm, the FE capacitor shows the $2P_r$ of up to 36.7 $\mu C/cm^2$ under the V_{op} of 2 V. However, as the ML thickness increases to 2 nm or more, a pronounced splitting of the current switching peak is observed at an V_{op} of 2 V, indicative of a noticeable wakeup effect in the FE capacitor. Under higher V_{op}, the FE capacitor with 2 nm ML displays no significant current peak splitting. This is contributed to the release of domain pinning effects induced by defects, which facilitates more uniform ferroelectric domain switching. With the ML thickness increased to 3 nm, distinct dual current switching peaks are observed at both 2 and 4 V. This phenomenon could be attributed to an increased presence of an antiferroelectric phase within the material [8].

Fig. 3(a) and (b) illustrate the $2P_r$ of FE capacitors with varying ML thicknesses, both before and after the wakeup cycling. As the operating voltage increases, the $2P_r$ of FE capacitors without ML exhibits a linear increase. Note that, FE capacitors with ZrO_2 ML show a sudden increase around 1.25 V and approach nearly complete polarization reversal at approximately 2.0 V. Such distinct behavior could be caused by the crystal structure change, since that the introduction of the ZrO_2 ML increases the proportion of the ferroelectric phase within the film [9]. Fig 3(c) and (d) present the

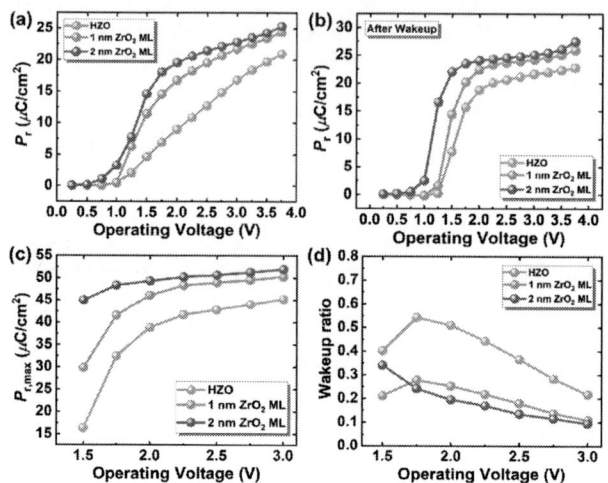

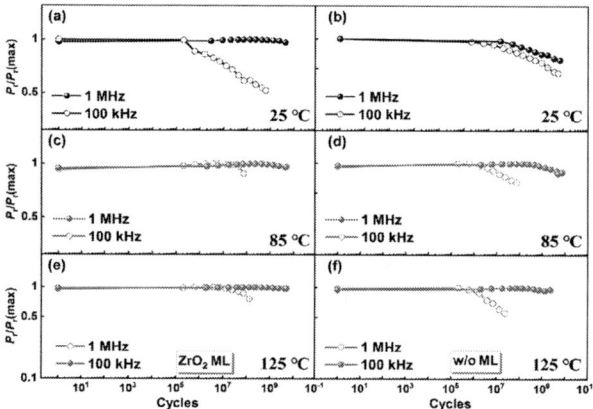

Fig. 3 The polarization as a function of the operating voltage (a) before and after (b) wakeup cycling. The (c) $P_{r,\,max}$ and (d) wakeup ratio $(1 - 2P_{r\_initial}/2P_{r\_max})$ of the FE capacitors with different ML thickness under different V_{op}.

Fig. 4 (a)-(f) The endurance characteristics of capacitors without and with ZrO_2 ML under different frequencies and temperature with the cycling voltage of 1.8 V.

saturated remanent polarization and wakeup ratio of the ferroelectric capacitors after wakeup, respectively. The results indicate that the introduction of 2 nm ML allows FE domains to switch rapidly at low operating voltages, while also reducing the wakeup ratio of the FE film. This provides an experimental basis for designing wakeup free memory devices that operate effectively at low operating voltages.

To further investigate the impact of ZrO_2 ML on the durability, the endurance cycling characteristics of the ferroelectric capacitors at different temperatures and frequencies were measured, as shown in Fig. 4. During the endurance cycling test, a cycling voltage of 1.8 V was applied to the ferroelectric capacitors, with frequencies set at 1 MHz and 100 kHz. Initially, at room temperature, the FE capacitor with ZrO_2 ML exhibited superior fatigue effect, maintaining performance up to 10^9 cycles, indicating that the inserting of ZrO_2 ML could effectively suppress the formation of internal

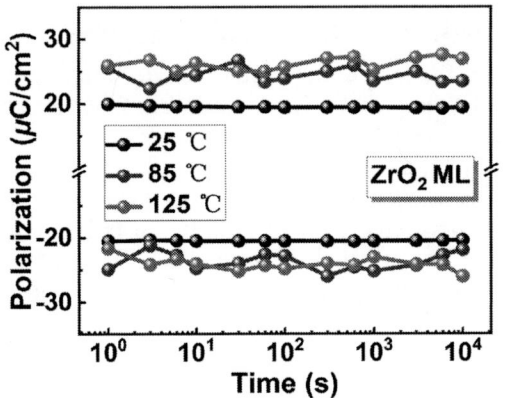

Fig. 5 The retention time at 25, 85 and 125 °C of FE capacitor with ZrO₂ ML annealed at 350 °C under 1.8 V.

defects within the FE film. Although the FE capacitor with ZrO₂ ML exhibited faster fatigue compared with the control sample at 100 kHz, its relatively high initial remanent polarization allowed it to remain suitable for non-volatile memory applications even after fatigue. As the temperature increased, the fatigue effect of both FE capacitors with and without ZrO₂ ML was mitigated. However, the FE capacitor without ML experienced breakdown at an earlier stage. In contrast, the FE capacitor with ML can sustain over 10^8 cycles at 100 kHz. Fig. 5 presents the retention time of FE capacitor with ZrO₂ ML baking at different temperatures after wakeup. The pulse schemes used for retention state measurements were consistent with the previous studies [10]. As shown in Fig. 5, stable polarization was obtained for up to 10^4 s across different temperatures.

Conclusion

In summary, we have demonstrated the ZrO₂ ML strategy in FE capacitor for low-voltage operation with annealing at 350 °C. Compared to the capacitors without ZrO₂ ML, the FE capacitor with ZrO₂ ML exhibits an enhanced $2P_r$ under the low operating voltage of 2 V. Additionally, the insertion of the ZrO₂ ML effectively suppresses the wakeup effect in the FE capacitor. The FE capacitor with ZrO₂ ML maintains the excellent retention time of 10^4 s even elevated up to 125 °C. The results indicate that the introduction of ZrO₂ ML effectively inhibit the fatigue effect of the film under various environmental conditions and testing frequencies.

Acknowledgments

This work was supported by Shanghai Municipal Science and Technology Commission under Grant 24DP1500105, 23511102300 and National Key Research and Development Program of China under Grant 2021YFB3202500.

References

[1] U. Schroeder, M. H. Park, T. Mikolajick, and C. S. Hwang, "The fundamentals and applications of ferroelectric HfO₂," Nature Rev. Mater., 7, pp. 653-669 (2022).

[2] M. H. Park, Y. H. Lee, H. J. Kim, Y. J. Kim, T. Moon, K. D. Kim, J. Muller, A. Kersch, U. Schroeder, T. Mikolajick, and C. S. Hwang, "Ferroelectricity and antiferroelectricity of doped thin HfO₂-based films," Adv. Mater., 27, pp. 1811-1831 (2015).

[3] J. Hur, Y.-C. Luo, N. Tasneem, A. I. Khan, and S. Yu, "Ferroelectric hafnium zirconium oxide compatible with back-end-of-line process,", IEEE Trans. Electron Devices, 68, pp. 3176-3180 (2021).

[4] P. Jiang, A 256 Kbit Hf₀.₅Zr₀.₅O₂-based FeRAM Chip with Scaled Film Thickness (sub-8nm), "Low Thermal Budget (350 °C), 100% Initial Chip Yield, Low Power Consumption (0.7 pJ/bit at 2V write voltage), and Prominent Endurance (>10^{12})," IEDM Tech. Dig., p.32 (2023).

[5] J. Y. Park, D. H. Lee, G. H. Park, J. Lee, Y. Lee, and M. H. Park, "A perspective on the physical scaling down of hafnia-based ferroelectrics," Nanotechnology, 34, 202001 (2023).

[6] K. Toprasertpong, K. Tahara, Y. Hikosaka, K. Nakamura, H. Saito, M. Takenaka, and S. Takagi, "Low Operating Voltage, Improved Breakdown Tolerance, and High Endurance in Hf₍₀.₅₎Zr₍₀.₅₎O₍₂₎ Ferroelectric Capacitors Achieved by Thickness Scaling Down to 4 nm for Embedded Ferroelectric Memory," ACS Appl Mater Interfaces, 14, pp.51137-51148 (2022).

[7] E. Yu, X. Lyu, M. Si, P. D. Ye, and K. Roy, "Interfacial Layer Engineering in Sub-5-nm HZO: Enabling Low-Temperature Process, Low-Voltage Operation, and High Robustness," IEEE Trans. on Electron Devices 70, pp.2962-2969 (2023).

[8] K. Xu, T. Wang, Y. Liu, J. Yu, Z. Li, J. Meng, H. Zhu, Q. Sun, D. W. Zhang, and L. Chen, "Ferroelectric and Antiferroelectric Phenomenon in Lanthanum doped Hafnium based Thin Films," IEEE Electron Device Letters, 44, pp. 1472-1475, (2023).

[9] Y.-C. Liu, J.-N. Yang, Y.-C. Li, X.-L. Zhou, K.-L. Xu, Y.-C. Chen, G.-R. Xie, H. Zhang, L. Chen, S.-J. Ding, and W. J. Liu, "Back-End of Line Compatible Hf₀.₅Zr₀.₅O₂/ZrO₂/Hf₀.₅Zr₀.₅O₂ Stack Achieving 2Pᵣ of 39.6 μC/cm² and Endurance Exceeding 10^{10} Cycles Under Low-Voltage Operation," IEEE Electron Device Letters, 45, pp. 388-391 (2024).

[10] S. Mueller, J. Muller, U. Schroeder, and T. Mikolajick, "Reliability characteristics of ferroelectric Si:HfO₂ thin films for memory applications," IEEE Trans. Device Mater. Rel., 13, pp. 93-97 (2013).

979-8-3315-0417-5/25 $31.00 © 2025 IEEE

Broadband miniaturized spectrometers with van der Waals junctions

Md Gius Uddin [1], Susobhan Das[1], Abde Mayeen Shafi[2], Lei Wang[3], Xiaoqi Cui [1,2], Fedor Nigmatulin[1],
Faisal Ahmed[1], Andreas C. Liapis[2], Weiwei Cai[3], Zongyin Yang[4], Harri Lipsanen[1], Tawfique Hasan[5],
Hoon Hahn Yoon[1,6] & Zhipei Sun[1]

[1]QTF Centre of Excellence, Aalto University, Finland. [2]Department of Electronics and Nanoengineering, Aalto University, Finland. [3]Key Lab of Education Ministry for Power Machinery and Engineering, School of Mechanical Engineering, Shanghai Jiao Tong University, China. [4]College of Information Science and Electronic Engineering and State Key Laboratory of Modern Optical Instrumentation, Zhejiang University, China. [5]Cambridge Graphene Centre, University of Cambridge, UK. [6]Department of Semiconductor Engineering, Gwangju Institute of Science and Technology, Republic of Korea

Abstract

Miniaturized spectrometers are crucial for applications in on-chip and implantable devices, requiring high spectral resolution in limited spaces. Here, we present our van der Waals heterojunction-based spectrometers that achieve tunable spectral responses through band structure engineering, offering sub-nanometer wavelength resolution and a broad operational bandwidth of ~500 to 1600 nanometers.

Keywords: Miniaturized spectrometers, van der Waals heterojunctions, Electrically tunable spectral responses, Algorithms.

Introduction

The growing demand for compact and portable spectrometers is driven by various applications [1], such as environmental monitoring, chemical analysis, biological imaging, and industrial quality control. Conventional spectrometers, though accurate, rely on bulky optical elements like gratings and interferometers, making them impractical for use in small, integrated devices, such as wearable electronics, on-chip systems, and consumer products. A truly miniaturized spectrometer with high resolution and low power requirements could vastly expand the reach of spectral sensing into mobile and remote applications, where size and energy efficiency are critical.

Two-dimensional (2D) materials, especially van der Waals heterostructures, offer an ideal platform for miniaturized spectrometers [2]. These atomically thin materials exhibit unique optoelectronic properties, such as tunable photoresponse and strong light-matter interactions, which allow for compact and highly sensitive spectral sensing. By stacking 2D materials with different bandgaps, it is possible to achieve heterostructures where the spectral response is adjustable through an external electric field, eliminating the need for traditional optical components [2,3]. This approach not only reduces device size and complexity but also enables high spectral accuracy over a broad wavelength range, making 2D materials highly promising for creating portable spectrometers.

Here, we present our recent miniaturized spectrometer results with van der Waals heterojunctions [4], achieving tunable spectral responses via band structure engineering. The devices cover a broad operational range from ~500 to 1600 nanometers, with a resolution approaching sub-nanometer levels. The device spectral response matrix is encoded during a learning phase and later used to reconstruct unknown spectra with computational algorithms. This simplified, power-efficient design is suitable for portable applications, opening avenues for spectral sensing in compact, low-power devices. Future developments will aim to extend its range further into the mid-infrared, enhancing the device versatility for a broader range of applications.

Methods

Our miniaturized spectrometer is based on van der Waals heterostructure junctions [4]. The materials are mechanically exfoliated. We used silicon substrates with a 285 nm SiO_2 capping layer for device fabrication, preparing high-quality hBN-encapsulated van der Waals heterostructures from commercially available hBN, black phosphorus (BP), and 2H-phase MoS_2 crystals via deterministic dry-transfer. The BP/MoS_2/bottom-hBN heterostructures were assembled layer-by-layer using a motorized 2D transfer stage in an argon-filled glovebox to keep oxygen and water below 0.1 ppm. After patterning with electron beam lithography, Ti/Au contacts were added by vacuum electron beam evaporation and acetone lift-off. The top hBN was transferred to protect against degradation in the glovebox, followed by an annealing process at 180°C in a high vacuum for process stability. An additional 20 nm Al_2O_3 layer was deposited at 120°C by atomic layer deposition for environmental protection. Finally, Ti/Au electrodes were wire-bonded to a printed circuit board, and the heterostructure topology was confirmed via atomic force microscopy.

The operation principle of our spectrometer relies on encoding the device's photoresponse to various known light wavelengths through an initial learning phase. During this phase, the device's response to multiple monochromatic wavelengths (e.g., from 400 to 1600 nm) is recorded under varied voltages. This process creates a spectral response

979-8-3315-0417-5/25 $31.00 © 2025 IEEE

matrix that serves as a reference for spectral reconstruction. Our approach utilizes gate or drain-source voltage adjustments to precisely tune the device's spectral response, eliminating the need for the moving parts typically required in conventional spectrometers. To measure unknown spectra, we expose the device to the target light and record its voltage-dependent photocurrent response. Using computational algorithms, we then match this response against the pre-established spectral response matrix, reconstructing the unknown spectrum through constrained least-squares optimization.

Results

A. Device scheme

The device scheme and characterization of the miniaturized spectrometer are focused on a BP/MoS_2 heterostructure fabricated as a diode-based spectrometer on a silicon substrate with a 300 nm SiO_2 capping layer (Fig.1) [4]. To ensure device stability, the structure is encapsulated with hBN and further protected by an additional Al_2O_3 layer. This construction enhances the robustness of the spectrometer and minimizes degradation from environmental exposure. The used multilayer BP and MoS_2 flakes have thicknesses of approximately 45 nm and 33 nm, respectively, which were confirmed using atomic force microscopy.

Raman spectroscopy was employed to verify the quality of the materials and to confirm the formation of a high-quality van der Waals heterojunction. The Raman spectra displayed characteristic modes for MoS_2 and BP, indicating that both materials maintained their structural integrity within the heterostructure. The electrical properties of the heterojunction were also measured, where a drain bias applied to the BP while connecting MoS_2 to the source terminal revealed efficient electron transport across the junction.

The van der Waals junction-based spectrometer's spectral response was characterized by varying the incident light wavelength, establishing a responsivity matrix. This responsivity matrix enables spectral reconstruction using

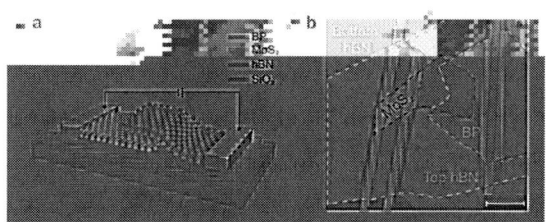

Figure 1 Device structure. a. Schematic of our BP/MoS_2 spectrometer. The junction is encapsulated by hBN passivation flakes. **b.** Corresponding optical microscope image. The black and yellow dashed lines indicate MoS_2 and BP flakes, respectively. Dashed white line marks the edges of the top hBN flake. Scale bar: 10 μm. Figures from Ref. [4].

computational algorithms and achieves a high peak wavelength accuracy of approximately 2 nm.

B. The demonstration of our spectrometer at the visible range

Upon optical irradiation of our BP/MoS_2 heterojunction, photocurrent generation is influenced by the carrier states within the absorption spectra of the constituent flakes. In the visible wavelength range, incident photons excite electrons from the valence bands of both MoS_2 and BP to their respective conduction bands.

For the spectrometer's learning phase, the BP/MoS_2 diode's response was recorded across various known light wavelengths (500–800 nm) in 10 nm steps, creating a spectral response matrix (Fig.2a). During testing, the device's bias-tunable photocurrent for an unknown spectrum is measured. Using an adaptive Tikhonov regularization method, the system then reconstructs the spectrum by solving a constrained least-squares problem based on the previously encoded matrix. Testing results (Fig.2b) show strong alignment between reconstructed spectra and reference measurements from a commercial spectrometer, with an average peak wavelength difference of ~2.5 ± 0.9 nm. A minor secondary peak at 675 nm is attributed to computational error, which can be minimized by filtering or model refinement. Additional broadband measurements further validate the device's performance, showing close agreement with commercial spectrometer data.

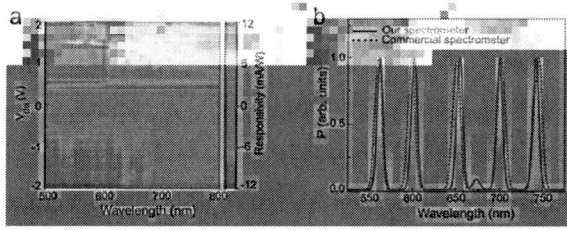

Figure 2 Our spectrometer demonstration at the visible range. a. Color contour plot of experimentally obtained spectral response matrix. **b.** Quasi-monochromatic spectra reconstructed with our spectrometer (solid curves). Dashed curves represent corresponding spectra obtained with a commercial spectrometer. Figures from Ref. [4].

C. The demonstration of our spectrometer at the near-infrared range

We also demonstrate our spectrometer's operation in the near-infrared region, where photocurrent generation is primarily due to the direct-bandgap transition in BP. Incident near-infrared photons lack the energy to excite electrons in MoS_2 from the valence to the conduction band.

To achieve high spectral resolution in the near-infrared range, we constructed a spectral response matrix (Fig.3a) using a fine learning step of approximately 0.5 nm for monochromatic lights from 1550 to 1560 nm. This responsivity matrix allows precise spectral reconstruction. Testing confirmed that our spectrometer effectively resolves closely spaced near-infrared wavelengths, successfully distinguishing peaks at ~1557 nm and ~1558 nm (Fig.3b). Larger learning steps would reduce peak-to-peak wavelength accuracy due to a lower signal-to-noise ratio.

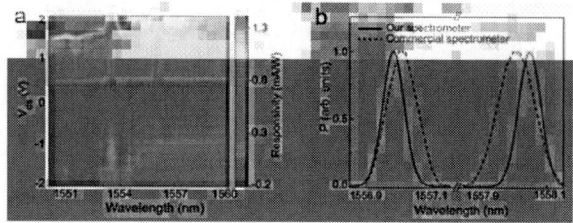

Figure 3 Our spectrometer demonstration at the near-infrared range. a. Color contour plot of spectral response matrix obtained with a small learning step of ~0.5 nm. **b.** Reconstruction of typical quasi-monochromatic spectra with our spectrometer (solid curves). Dashed curves represent corresponding spectra measured with a commercial spectrometer. Figures from Ref. [4].

C. Stability and reproducibility tests

Stability and reproducibility tests over a 10-day period showed reliable performance with minimal variation, maintaining accuracy within ~1.2 nm in the visible and ~0.02 nm in the near-infrared range. This high stability and device-to-device consistency support the spectrometer's robustness for extended use and its suitability for applications requiring precise near-infrared measurements.

Conclusion

In conclusion, we have developed miniaturized broadband spectrometers using BP/MoS$_2$ van der Waals heterostructures, achieving high spectral resolution across the visible and near-infrared regions. This compact design leverages the unique properties of 2D materials, allowing electrically tunable spectral responses without the need for traditional optical components or moving parts. The device's simple architecture and operating at low voltages - demonstrate high wavelength precision, successfully distinguishing narrow wavelength differences in both visible and near-infrared ranges. Stability tests confirmed consistent performance over extended use, highlighting the robustness and reliability of the spectrometer. This approach offers significant potential for integration into portable and on-chip applications, from environmental monitoring to bio-imaging. Future advancements could focus on expanding new bands (e.g., into the mid-infrared range [4]), new heterojunctions [5] and new methods [6], enhancing the device's versatility for a broader scope of real-world spectral sensing applications.

Acknowledgments & Notes

The authors acknowledge the funding from the Academy of Finland, the Academy of Finland Flagship Program, the EU H2020-MSCA-RISE-872049 (IPN-Bio), Business Finland (AGATE), the Jane and Aatos Erkko foundation, the Technology Industries of Finland centennial foundation (Future Makers 2022), and ERC (834742). This research was conducted at the Micronova, Nanofabrication Centre of Aalto University. Note that the results have been published as Ref. [4] under the terms of the Creative Commons CC BY license.

References

[1] Z. Yang, T. Albrow-Owen, W. Cai, T. Hasan, Science, 371 (2021), eabe0722.

[2] H.H. Yoon, H.A. Fernandez, F. Nigmatulin, W. Cai, Z. Yang, H. Cui, F. Ahmed, X. Cui, M.G. Uddin, E.D. Minot, H. Lipsanen, K. Kim, P. Hakonen, T. Hasan, Z. Sun, Science, 378 (2022), P.296-299.

[3] W. Deng, Z. Zheng, J. Li, R. Zhou, X. Chen, D. Zhang, Y. Lu, C. Wang, C. You, S. Li, L. Sun, Y. Wu, X. Li, B. An, Z. Liu, Q.j. Wang, X. Duan, Y. Zhang, Nat. Commun., 13 (2022), p. 4627.

[4] M.G. Uddin, S. Das, A.M. Shafi, L. Wang, X. Cui, F. Nigmatulin, F. Ahmed, A.C. Liapis, W. Cai, Z. Yang, H. Lipsanen, T. Hasan, H.H. Yoon, Z. Sun, Nat. Commun., 15 (2024), p. 571.

[5] . Du, M.R. Molas, Z. Huang, G. Zhang, F. Wang, Z. Sun, Science, 379 (2023), p. eadg0014.

[6] L. Du, Z. Huang, J. Zhang, F. Ye, Q. Dai, H. Deng, G. Zhang, Z. Sun, Nat. Mater., 23 (2024), pp. 1179-1192.

2-D FET Modeling: Why Incorporating the Gate- and Drain-Dependent Source Tunneling Barriers Matter

Cristine Jin Estrada[1], Zichao Ma[2], Lining Zhang[3], and Mansun Chan[4]

[1]Electronics Engineering Department, University of Santo Tomas, Philippines, Email: cdestrada@ust.edu.ph
[1]Research Center for the Natural and Applied Sciences, University of Santo Tomas, Philippines
[2]School of Microelectronics, South China University of Technology, China,
[3]School of ECE, Peking University, Shenzhen, China,
[4]Department of ECE, The Hong Kong University of Science and Technology, Hong Kong

Abstract

This paper presents the modeling of a two-dimensional (2-D) field-effect transistor (FET). By considering two tunneling regions at the source and showing how these regions vary with the gate and drain voltages, a compact model that captures even the non-linear dependence of the drain current at low drain voltages observed in fabricated devices is developed. The proposed model reproduces the results from TCAD simulations and published experimental data with close agreement. This highlights the importance of incorporating the gate- and drain-dependent source tunneling barriers in modeling 2-D FETs.

Keywords: Compact modeling, two-dimensional (2-D) materials, field-effect transistor (FET).

Introduction

Two-dimensional (2-D) materials-based field-effect transistor (FET) technology has opened new possibilities for enhanced performance in modern electronic technology. In our earlier works in [1] and [2], we addressed the major challenges in the device design, fabrication, and integration of 2-D transistors. The quality of the underlying 2-D film, the choice of the source-drain electrode, and the processing conditions for the metal evaporation played key roles in matching the output current of the complementary devices [1]. To achieve desirable threshold voltage polarity and to avoid interface states in the gate dielectric and gate electrode stack, complementary devices with van der Waals channel stack are also developed [2]. Based on these two approaches in forming 2-D transistors, CMOS inverter circuits are realized with rail-to-rail performance, gain larger than unity during switching, and a wide noise margin. We also studied the threshold voltage of 2-D FETs [3], which is an important parameter in CMOS circuits, and provided new insights into the turn-on condition for such devices.

Moving forward, compact models for 2-D FETs are needed to demonstrate further their potential, especially in building integrated circuits. While several compact models have been developed to support large circuit design and simulation of such devices, these models do not fully capture the non-linear trend on the output curve at low drain voltages typically seen in fabricated devices [1],[4]. This paper, therefore, investigates the mechanism behind this phenomenon and presents a compact model to incorporate such behavior in response to the applied terminal voltages.

The Behavior of 2-D FETs

Bottom-gated 2-D FETs with excellent device characteristics have been demonstrated [5] based on undoped transition metal dichalcogenide (TMD) 2-D films. These devices have their gate and source-drain (S-D) contacts at the opposite side of the active film, and with the gate overlapping the S-D as shown in Fig. 1(a). Hence, the gate also modulates the mechanism by which the carriers from the source contact are injected into the channel and collected in the drain. Schematically, the device can be represented by a gated diode in series with a MOSFET as shown in Fig. 1(b), which is similar to that of a tunneling FET [6].

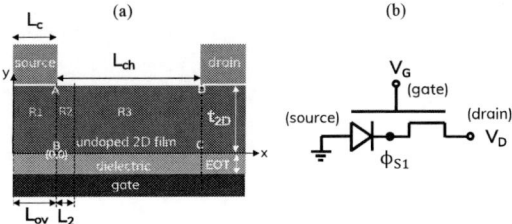

Fig. 1(a). Physical structure of a bottom-gated 2-D FET. (b) The device can be represented schematically by a gated diode in series with a MOSFET.

As presented in our previous work [3], the conduction path for such a device can be approximated by the line segment from A to D marked in the given physical structure. Based on our earlier analysis, the 2-D film body at the source and the source/channel transition region act as tunneling barriers. As such, the device can then be divided into three parts: Region 1 (R1) – 2-D film body at the source, Region 2 (R2) – source/channel transition region, and Region 3 (R3) – channel. To derive the compact model, the band diagram, specifically the conduction band edge (CBE), along the approximated path at different biasing conditions must be analyzed using TCAD, and they are shown in Fig. 2.

We will begin our analysis by first considering the gate control while the drain voltage (V_D) is set to a low value. When the gate voltage increases (from V_{G1} to V_{G5}), as shown in Fig. 2(a), carriers are injected from the source to the channel, starting with a thermionic emission-dominated (V_{G2}) to a tunneling-dominated mechanism (beyond V_{G3}). At V_{G3}, the tunneling barrier comprises the 2D film thickness, t_{2D}, in R1 and the width of the source/channel transition region,

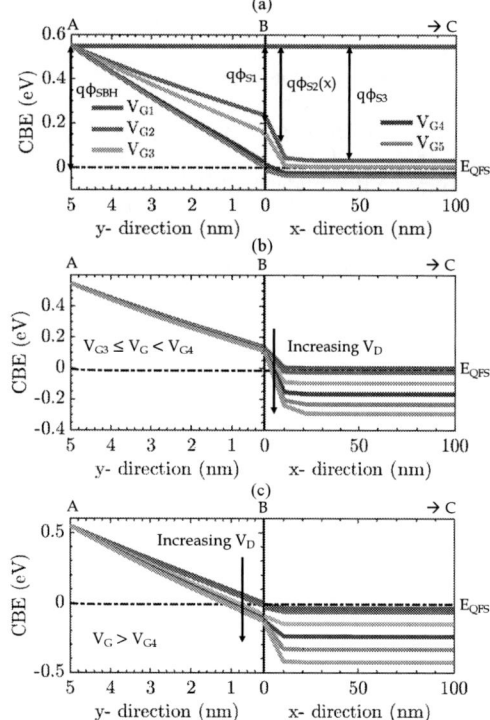

Fig. 2(a). V_G control on the tunneling regions where V_D is set to a low value. V_D control on the tunneling regions when (b) $V_{G3} \leq V_G < V_{G4}$, and (c) $V_G > V_{G4}$.

$W_{t,sc}$, which is formed when ϕ_{S2} becomes equal to the Schottky barrier height, ϕ_{SBH}, at the edge of R2. The device fully turns on at V_{G4} when the conduction band edge at the 2-D/dielectric interface (point B) aligns with the quasi-Fermi level of electrons at the source, E_{QFS}. Beyond V_{G4}, the drain current will be dominated by the narrowing of the barrier width along the 2-D film of R1.

For the drain control, when V_G is biased between V_{G3} and V_{G4}, the barrier narrowing occurs in R2 as V_D increases, as shown in Fig. 2(b). Whereas when V_G is biased beyond V_{G4}, the barrier narrowing occurs in R1, as shown in Fig. 2(c). In both cases, the channel remains flat, and a zero-field approximation in R3 can be therefore assumed.

Three inferences can be made based on these behaviors: (i) Since the dependence of the tunneling current on the tunneling width is exponential, it can be expected that the I_D-V_D curve will exhibit some non-linearity before I_D saturates. (ii) I_D saturation is expected to happen when the overall shape of the tunneling barrier becomes insensitive to V_D. (iii) The overall shape of the source tunneling barriers, which is both gate- and drain-dependent, must be considered to capture the behavior of 2-D FETs fully.

Incorporating the Gate- and Drain-Dependent Source Tunneling Barriers into the *I-V* Model

The shape of the barrier along R1 and R2 must be, therefore, known to derive the model for 2-D FETs. To achieve this, the surface potentials, ϕ_{S1} and ϕ_{S3}, must be

derived, and their boundaries at R2 must be matched to identify the minimum tunneling, $W_{t,min}$, distance and, by extension, the maximum generation rate, $G_{tun,max}$, along the tunneling regions. Using appropriate boundary conditions in solving the Poisson's equation in [7] and using a parabolic approximation to capture the potential profile in R2 [8], the surface potentials can be determined. After straightforward derivation, the minimum tunneling distance and the maximum generation rate along the tunneling barrier will also be known. The tunneling component of I_D, $I_{D(tunneling)}$, can be then found by integrating the maximum generation rate over a linearly changing tunneling distance as an approximation:

$$I_{D(tunneling)} = qG_{tun,max} \cdot (2\lambda_{tun}Wt_{2D}) \cdot N_{supply} \quad (1)$$

where $\lambda_{tun} = 1/\left(B\sqrt{\phi_{SBH}}\right)$ is the tunneling decay length, B is the tunneling coefficient, W is the device width, and N_{supply} represents the supply function that considers the difference in the supply of carriers at the interfaces of the tunneling barrier:

$$N_{supply} = \ln\left[\frac{1+e^{\frac{qV_{window}}{kT}}}{2}\right] \quad (2)$$

and V_{window} is the window voltage proportional to V_D which opens from when the CBE of the channel aligns with E_{QFS}:

$$V_{window} = V_D\left[1 - \left(1 + e^{\frac{q(V_G - V_{G3})}{\eta_1 kT}}\right)^{-1}\right] \quad (3)$$

For the thermionic component of I_D, $I_{D(thermionic)}$, it can be expressed as:

$$I_{D(thermionic)} = A_C RT^2 \cdot e^{-\frac{q\phi_{SBH,eff}}{kT}}\left(1 - e^{-\frac{q\phi_{S3}}{kT}}\right)$$
$$\times \left[1 - \left(1 + e^{\frac{q(V_G - V_{G2})}{\eta_2 kT}}\right)^{-1}\right] \quad (4)$$

where $\phi_{SBH,eff}$ is the effective barrier height at the metal/2-D junction, V_{Gn} are the onsets defined in the previous section, A_C is the cross-sectional area of the device, R is the Richardson's constant, and η_n are empirical fitting parameters. The total drain current, $I_{D(total)}$, can be then found by taking the sum of the tunneling component in (1) and the thermionic component in (4). Fig. 3 shows the verification of the transfer characteristics of the device at low V_D where a good matching between our model and simulation is achieved.

When the device is on, the main contact transport mechanism is mainly by tunneling, and the minimum tunneling distance can be approximated using $W_{t,min} = (\phi_{SBH}/\phi_{S1})t_{2D}$. With the proposed approach and as discussed in [6], $G_{tun,max}$ is non-zero even at $V_D = 0$ V, hence, a correction factor, f_{FERMI}, must also be incorporated to ensure a zero-drain current at zero drain voltage. The total drain

979-8-3315-0417-5/25 $31.00 © 2025 IEEE

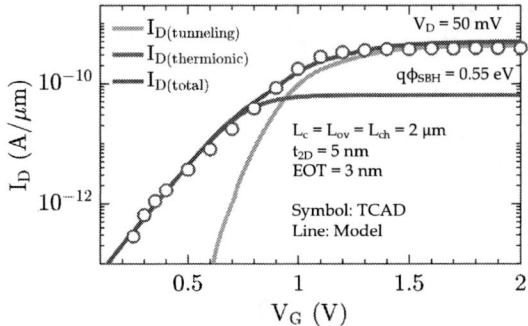

Fig. 3. Drain current model verifications using TCAD results for the transfer characteristics.

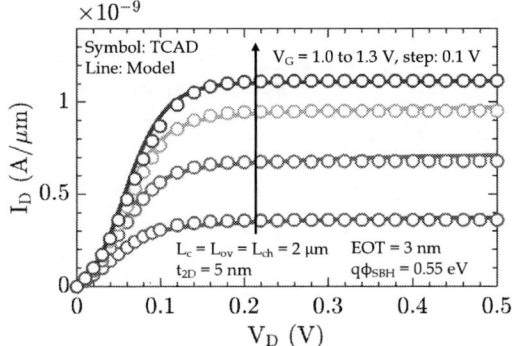

Fig. 4. Drain current model verifications using TCAD results for the output characteristics of the device.

current to capture the output characteristics of the device can be then expressed as follows:

$$I_D = qG_{tun,max} \cdot (2\lambda_{tun}Wt_{2D}) \cdot f_{FERMI} \quad (5)$$

where

$$f_{FERMI} = 2 \times \left[\frac{1}{2} - \left(1 + e^{\frac{qV_D}{f_n kT}} \right)^{-1} \right]$$
$$\times \left[1 - \left(1 + e^{\frac{q(V_D - V_{DT})}{\eta_3 kT}} \right)^{-1} \right] \quad (6)$$

f_n is an empirical fitting parameter, and V_{DT} is the drain threshold voltage, resembling that of the characteristics of tunneling FETs. Fig. 4 shows the verification of the output characteristics of the device where a good matching is again achieved. The model is also used to predict the experimental results in [4] where the critical device features are captured, as shown in Fig. 5.

Conclusion

A model for bottom-gated 2-D FETs that considers how the gate and drain control the shape of the barrier seen by the carriers from the source contact to the channel was derived. By determining the minimum tunneling distance along an approximated current conduction path, the maximum generation rate is determined, and it serves as the core in deriving the drain current model. The model is verified using TCAD results and experimental data from published

literature where a close matching is achieved. The proposed model captured the non-linear trend in the output characteristics, especially at low drain voltage, typically observed in fabricated 2-D FETs.

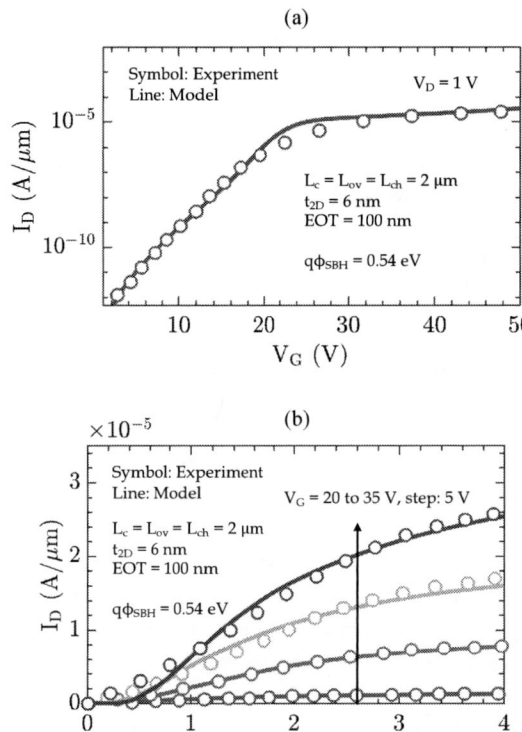

Fig. 5. Drain current model verifications using the experimental results in [4] for the (a) transfer characteristics and (b) output characteristics of the device.

Acknowledgment

This work was supported by the General Research Fund (GRF) 16201223 from the Research Grants Council (RGC) of Hong Kong.

References

[1] C. J. Estrada, Z. Ma, and M. Chan, "Complementary Two-Dimensional (2-D) MoS₂ FET Technology," *ESSDERC 2021*, Grenoble, France, 2021, pp. 219-222.

[2] C. J. Estrada, Z. Ma, and M. Chan, "Complementary Two-Dimensional 2-D) FET Technology with MoS₂/hBN/Graphene Stack," *IEEE Electron Device Lett.*, vol. 42, no. 12, pp. 1890–1893, 2021.

[3] C. J. Estrada, Z. Ma, L. Zhang and M. Chan, "Threshold Voltage Model for 2-D FETs With Undoped Body and Gated Source," in *IEEE Transactions on Electron Devices*, vol. 70, no. 5, pp. 2575-2580, 2023.

[4] S. Das and J. Appenzeller, "WSe₂ field effect transistors with enhanced ambipolar characteristics," *Appl. Phys. Lett.*, vol. 103, no. 10, 2013.

[5] P. C. Shen *et al.*, "Ultralow contact resistance between semimetal and monolayer semiconductors," *Nature*, vol. 593, no. 7858, pp. 211–217, 2021.

[6] L. Zhang and M. Chan, "SPICE modeling of double-gate tunnel-FETs including channel transports," *IEEE Trans. Electron Devices*, vol. 61, no. 2, pp. 300–307, 2014.

[7] Y. Taur, "An analytical solution to a double-gate MOSFET with undoped body," *IEEE Electron Device Lett.*, vol. 21, no. 5, pp. 245–247, 2000.

[8] X. Aymerich-Humet, F. Serra-Mestres, and J. Millán, "A generalized approximation of the Fermi-Dirac integrals," *J. Appl. Phys.*, vol. 54, no. 5, pp. 2850–2851, 1983.

979-8-3315-0417-5/25 $31.00 © 2025 IEEE

Interface engineering induced digital to analog switching transition in Hafnium Oxide-based memory device

Cong Han[1,2], Miaocheng Zhang[1,2], Yi Liu[1], Xinpeng Wang[2,3], Hao Zhang[2], Yi Tong[2,3*]

[1] College of Integrated Circuit Science and Engineering, Nanjing University of Posts and Telecommunications, China

[2] Suzhou Laboratory, China, [3] Gusu Laboratory of Materials Science, China

ABSTRACT

Oxide-based memory devices are promising candidates for future non-volatile resistive random-access memories (RRAMs). However, these devices typically exhibit digital resistive switching behavior, which constrains their applicability in neuromorphic computing. In this article, we demonstrate that the insertion of an ultra-thin by ALD Al_2O_3 layer prior to a TiN/HZO/W memristor facilitates a transition from digital to analog switching characteristics. The device exhibits good durability (10^4%) and excellent retention (>3000s) across multiple resistance states. Through an analysis of the device's interface chemical state and fitting of the I-V curve, we observe that the oxygen within the Al_2O_3 layer undergoes gradual oxidation and reduction, avoiding the abrupt formation and rupture of conductive filaments. This work provides valuable insights for the modification and optimization of oxide memristors.

Keywords: Resistive switching behavior; HZO; Al_2O_3; ALD; abrupt formation and rupture

INTRODUCTION

Memristive devices have emerged as some of the most promising non-volatile components due to their low power consumption, rapid switching speed, and high integration density[1]. Typically, memristors are structured as a sandwich, such as metal/insulator/metal or metal/semiconductor/metal. Various oxides, including TiO_2[2], SiO_2[3], and HfO_2[4], have been demonstrated as effective dielectric layers in memristors. Among these, HfO_2-based memristors exhibit exceptional resistive switching behavior attributed to the formation of conductive filaments driven by oxygen vacancies[5]. However, these devices commonly show digital switching characteristics due to the abrupt formation and rupture of conductive filaments, which restricts their applicability in certain domains, such as neuromorphic computing.

To modify the switching behavior of memristors, one effective approach involves the insertion of a barrier layer at the interface of the dielectric layer, thereby altering the conditions for the generation of conductive filaments. Al_2O_3 is a commonly utilized interfacial insertion layer with higher barrier to oxygen ion migration[6]. This increased barrier requires oxygen ions to overcome a higher energy threshold to generate oxygen ions to facilitate the formation and rupture of conductive filaments. As a result, this configuration reduces the effective distance between electrodes caused by the abrupt breakage of these filaments.

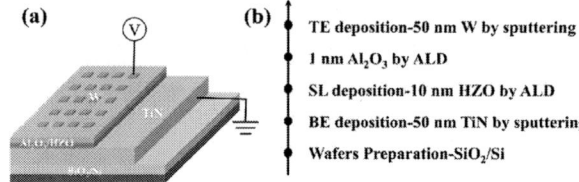

Figure 1. (a) Structure schematic of Al_2O_3/HZO device. (b) The fabrication process flow of Al_2O_3/HZO device.

In this work, we fabricated an Al_2O_3/HZO-based memory device, utilizing an ultra-thin Al_2O_3 layer to regulate the formation and rupture of internal conductive filaments. Titanium nitride (TiN) and tungsten (W) were employed as the bottom and top electrodes, respectively, due to their widespread use in foundries and excellent compatibility with CMOS technology[7]. Compared to conventional HZO-based devices, our device demonstrates analog resistive switching behavior, effectively enabling the transition between two non-volatile switching modes. Furthermore, it exhibits a good on/off ratio (10^4%) and retention(>3000s). Additionally, the device shows reliable analog switching characteristics and corresponding synaptic behaviors. Through an analysis of the interface chemical state and fitting of the current-voltage (I-V) curve, we conclude that the resistive switching mechanism is governed by conductive filament dynamics, while the conduction mechanism of the resistive state is attributed to Ohmic conduction and Schottky emission mechanisms.

METHODS

The TiN/HZO/Al_2O_3/W memristor was fabricated following the steps outlined in **Figure 1(a)** and **(b)**. Initially, a 100 nm layer of TiN was deposited on the substrate using magnetron sputtering at a power of 500 W. Subsequently, a 10 nm HZO film was deposited on the TiN bottom electrode via atomic layer deposition (ALD) method at a temperature of 250 °C, maintaining a 1:1 ratio of hafnium to zirconium through super-cycles during the deposition. An ultra-thin Al_2O_3 film, measuring 1 nm, was then deposited using the same ALD

979-8-3315-0417-5/25 $31.00 © 2025 IEEE

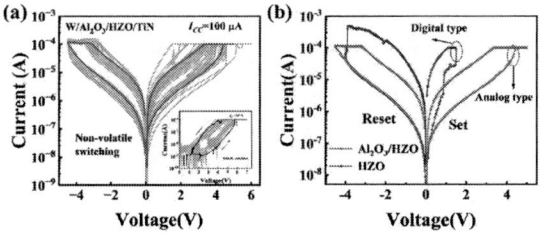
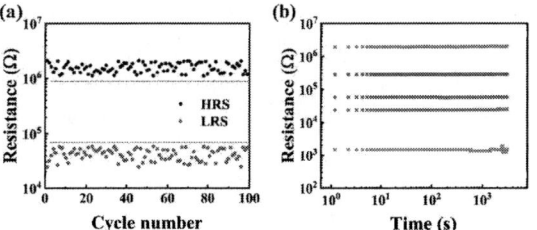

Figure 2. (a) *I-V* curves of Al₂O₃/HZO device under a I_{CC} of 100 μA. (inset shows the *I-V* curves under a low I_{CC} of 1 μA). (b) the effect of Al₂O₃ layer to the switching behavior of memristor based on HZO under equal I_{CC}.

Figure 3. (a) Endurance characteristics of Al₂O₃/HZO device; (b) Retention of multiple resistance.

method. Finally, the W top electrode was deposited using magnetron sputtering, achieving a deposition rate of 2 Å/s.The electrical properties of the device were characterized using Keithley 4200A-SCS semiconductor parameter analyzer.

RESULTS and DISCUSSION

Figure 2(a) illustrates the *I-V* measurements of the TiN/HZO/Al₂O₃/W memristor under a compliance current (I_{CC}) of 100 μA, with the inset depicting the IV curve of the device at a low I_{CC} of 1 μA. It is evident that this device can switch between non-volatile and volatile switching behaviors by varying the I_{CC}. To investigate the role of the Al₂O₃ layer, two devices were compared. As shown in **Figure 2 (b)**, in the absence of the Al₂O₃ layer, the SET and RESET processes of the HZO-based memristor exhibit abrupt transitions, as highlighted by the blue circle, confirming its digital switching behavior. Conversely, when the Al₂O₃ layer is incorporated, the current values during the SET and RESET processes change gradually rather than abruptly, as highlighted by the red circle. Notably, both devices were tested under the same I_{CC}. This observation demonstrates that the introduction of ultra-thin Al₂O₃ layer facilitates the transformation of the HZO memristor from a digital to an analog switching mode.

In order to quantitatively study the performance of Al₂O₃/HZO device, the cycle-to-cycle distributions of high resistance state (HRS) and low resistance state (LRS) have been plotted in **Figure 3(a)**. It can be seen that the on/off ratio is above 10^2, which is sufficient for application. Additionally, the retention of conductance is crucial for memristors[8], as it signifies improved long-term synaptic plasticity. In this study, we evaluated the conductance retention characteristics of the device modulated a variety of different resistance states.

As depicted in **Figure 3(b)**, each resistance state of the device can be sustained for over 3000 seconds without significant drift. This indicates that the device exhibits excellent conductance retention and a capability for multi-resistance states.

When a DC bias voltage is repeatedly applied to the device, its resistance gradually increases or decreases, which is associated with the multi-resistance characteristics previously mentioned. Continuous SET and RESET processes are achieved within voltage ranges of 2.2 V to 4.2 V and -1.9 V to -3.8 V, as illustrated in **Figures 4(a)** and **4(c)**. The current values recorded at 1.5 V and -1.5 V during this gradual switching process are presented in **Figures 4(b)** and **4(d)**. These results demonstrate that the conductance of the device can be adjusted by varying the applied voltage. This characteristic aligns with the processes of potentiation and depression observed in synaptic plasticity.

The subsequent fracture model of the conductive filament mechanism was derived by fitting the HRS data, as shown in **Figure 5(a)** and **(b)**. In the low-voltage region, the slope of the fitted line (ln I-ln V) is 1.04, suggesting that the ohmic conduction mechanism is predominant. In contrast, when a higher voltage is applied, the slope of the fitted line (ln I-V¹ᐟ²) mechanism becomes dominant.

In summary, we propose the resistive switching mechanism for the TiN/HZO/Al₂O₃/W memristor, as illustrated in **Figure 6**. For the memristor lacking an Al₂O₃ layer, the application of an electric field leads to the accumulation of oxygen ions (O^{2-}) at the interface between the upper electrode and the dielectric layer, while oxygen vacancies migrate in the opposite direction to form conductive filaments, as depicted in **Figure 6(a)**. During the RESET process, the electric field repels the oxygen ions at the interface, causing them to recombine with the conductive filaments and resulting in a sudden rupture of these filaments, as shown in **Figure 6(b)**. Consequently, the device without the Al₂O₃ layer exhibits digital resistive switching behavior.

In contrast, for Al₂O₃/HZO devices, the high oxygen formation energy of Al₂O₃ leads to a gradual generation of oxygen vacancies during the SET process, facilitating the formation of conductive filaments, as illustrated in **Figure 6(c)**. During the RESET process, oxygen ions are gradually ionized, contributing to the breaking of the conductive filaments, as illustrated in **Figure 6(d)**. As a result, this device demonstrates analog resistive switching behavior. Furthermore, the presence of the Al₂O₃ layer introduces significant internal resistance, which accounts for the relatively high SET voltage observed in the TiN/HZO/Al₂O₃/W memristor.

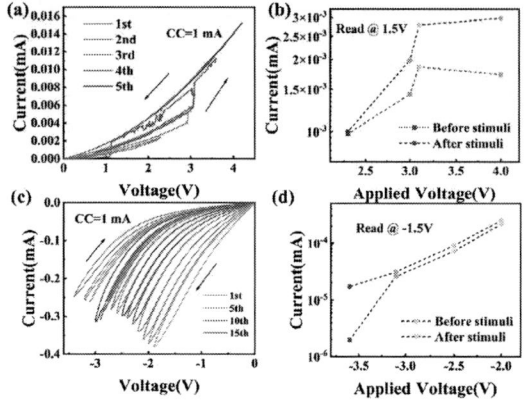

Figure 4. (a) Analog switching characteristics of Al₂O₃/HZO memristor during SET process; (b) Current values read at 1.5V under different applied voltages; (c) Analog switching characteristics during RESET process; (d) Current values read at 1.5V under different applied voltages.

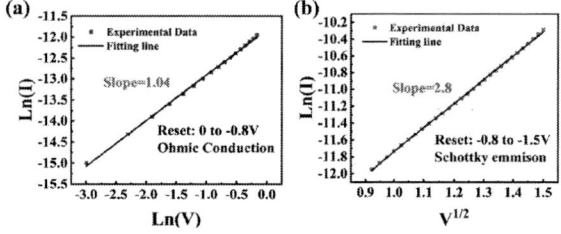

Figure 5. Fitting HRS results of Al₂O₃/HZO memristor at (a) low voltage range; (b) higher voltage range.

CONCLUSION

In summary, we have demonstrated that the insertion of an Al₂O₃ layer into the HZO memristor transforms the device's resistive switching behavior from digital to analog. The Al₂O₃/HZO device exhibits good on/off ratio (10^2%) and retention (>3000s) characteristics, along with the ability to support multiple resistance states. Through a comprehensive analysis of the device's interface chemical state and current-voltage (IV) characteristics, we conclude that the oxygen vacancies generated by the Al₂O₃ layer play a crucial role in the formation and rupture of conductive filaments. However, the introduction of the Al₂O₃ intercalation also leads to an increase in the switching voltage of the device. This work offers a viable method for regulating the performance of oxide memristors.

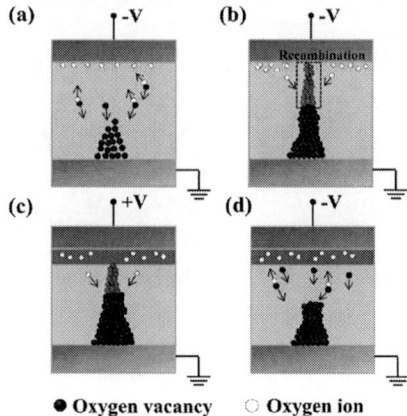

Figure 6. Formation/rupture of conductive filament in (a) (b) HZO-based and (c)(d) Al₂O₃/HZO-based memristor.

ACKNOWLEDGMENTS

This work was supported in part by the open research fund of Suzhou Laboratory (Grant No.SK-1202-2024-012), 2030-Major Project of the Chinese Ministry of Science and Technology (Grant No. 2021ZD0201200).

REFERENCES

[1] J. J. Yang, D. B. Strukov, and D. R. Stewart, "Memristive devices for computing," Nature Nanotechnology, vol. 8, no. 1, pp. 13-24, 2013/01/01 2013.

[2] S. Srivastava, J. P. Thomas, X. Guan, and K. T. Leung, "Induced Complementary Resistive Switching in Forming-Free TiO$_x$/TiO$_2$/TiO$_x$ Memristors," ACS Applied Materials & Interfaces, vol. 13, no. 36, pp. 43022-43029, 2021/09/15 2021.

[3] S. Kwon et al., "Structurally Engineered Nanoporous Ta$_2$O$_{5-x}$ Selector-Less Memristor for High Uniformity and Low Power Consumption," ACS Applied Materials & Interfaces, vol. 9, no. 39, pp. 34015-34023, 2017/10/04 2017.

[4] Z. Peng et al., "HfO2-Based Memristor as an Artificial Synapse for Neuromorphic Computing with Tri-Layer HfO$_2$/BiFeO$_3$/HfO$_2$ Design," Advanced Functional Materials, vol. 31, no. 48, p. 2107131, 2021/11/01 2021.

[5] J. Woo, A. Padovani, K. Moon, M. Kwak, L. Larcher, and H. Hwang, "Linking Conductive Filament Properties and Evolution to Synaptic Behavior of RRAM Devices for Neuromorphic Applications," IEEE Electron Device Letters, vol. 38, no. 9, pp. 1220-1223, 2017.

[6] B. Traoré et al., "Microscopic understanding of the low resistance state retention in HfO2 and HfAlO based RRAM," in 2014 IEEE International Electron Devices Meeting, 2014, pp. 21.5.1-21.5.4.

[7] E. Shahrabi, J. Sandrini, B. Attarimashalkoubeh, T. Demirci, M. Hadad, and Y. Leblebici, "Chip-level CMOS co-integration of ReRAM-based non-volatile memories," in 2016 12th Conference on Ph.D. Research in Microelectronics and Electronics (PRIME), 2016, pp. 1-4.

[8] W. Q. Pan et al., "Strategies to Improve the Accuracy of Memristor-Based Convolutional Neural Networks," IEEE Transactions on Electron Devices, vol. 67, no. 3, pp. 895-901, 2020.

MEMS Integrated with Self-Assembled Electrets

Daisuke Yamane[1,2,3]

[1]Department of Mechanical Engineering, Ritsumeikan University, Japan,
[2]Graduate School of Science and Engineering, Ritsumeikan University, Japan,
[3]Ritsumeikan Advanced Research Academy, Ritsumeikan University, Japan

Abstract

SAEs (self-assembled electrets) are recently reported electrets that can be deposited in semiconductor manufacturing processes and, unlike conventional electrets, do not require any charging process. We have developed an integration technology of SAEs with MEMS (micro-electro-mechanical systems) for the first time and successfully demonstrated SAE-based MEMS vibrational energy harvesters. This presentation introduces the integration technology of MEMS and SAEs, the characterization of SAE films formed in MEMS devices, and SAE-MEMS co-design technologies, including future prospects.

Keywords: Micro-electro-mechanical systems, self-assembled electrets, integration and energy harvesters

Introduction

Electrets are widely known as dielectric materials with quasi-permanent charges or polarizations and are used in applications such as microphones and air filters [1]. MEMS (microelectromechanical systems) is a collective term for three-dimensional electronic and mechanical devices with micro-meter dimensions manufactured using semiconductor microfabrication technology. In the MEMS field, application technologies for sensors, actuators, and energy harvesters have been developed, and electrostatic MEMS VEHs (vibrational energy harvesters) can be implemented by the combination of electrets and MEMS [2-4]. Conventional electrostatic MEMS VEHs require non-semiconductor process for electret deposition or charging, which makes it difficult to achieve mass production and integration of MEMS and ICs (integrated circuits) through semiconductor processes. There were also design constraints on the materials and structure of MEMS because of the need for robust mechanical structures that can withstand high-voltage and/or high temperature charging processes. Accordingly, further miniaturization, higher performance, and mass production of the devices can be expected by allowing electrostatic MEMS VEHs to be fabricated using only semiconductor processes.

Recently, self-assembled electrets (SAEs) have been developed using organic electroluminescent device materials that can be deposited by semiconductor processes and require no charging process [5]. SAEs are electrets that utilize the polarization state generated from the spontaneous orientation phenomenon of polar molecules, and a variety of materials have been reported. There are also reports of SAEs with surface charge densities equivalent to those of the electrets used in conventional electrostatic MEMS VEHs [5]. In addition that SAEs do not require any charging processes, they can be formed by vacuum deposition, one of the semiconductor processes, enabling the production of electrostatic MEMS VEHs using only semiconductor processes. We are currently working on the integration of MEMS and SAEs and their device applications. In this presentation, we introduce our recent research results on the integration technology of MEMS and SAEs [6], characterization of SAE films formed in MEMS devices [7, 8], and SAE-MEMS co-design technology [9, 10].

Methods and Results

Prior to our work, there were no reported cases of integrating SAE and MEMS. For example, when attempting to form MEMS structures on the same substrate after SAE deposition, the polarization of SAE is lost during the MEMS fabrication process. To solve this issue, we proposed a method of forming SAE films inside MEMS structures by creating through holes, called through holes, in some parts of the MEMS structures in advance and forming SAE films via these through holes after MEMS fabrication. This allows us to maintain the electrical properties of the electret, since no MEMS fabrication process is required after SAE deposition. Figure 1 shows a schematic image of our SAE-MEMS devices. In this device, SAE films are formed inside the

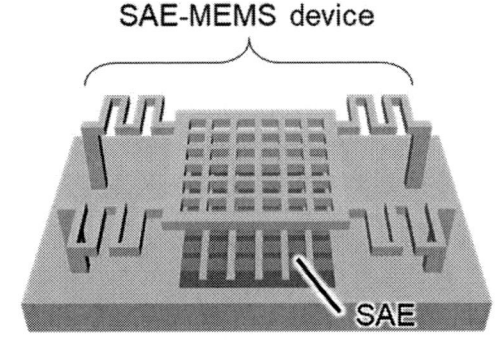

Fig. 1: Schematic image of an SAE-MEMS device.

Fig. 2: Photo of a developed SAE-MEMS device [6].

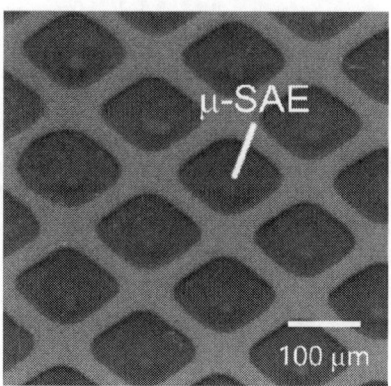

Fig. 3: Micro-patterned SAE films [7].

MEMS, under the movable electrodes, by making through holes in the MEMS movable electrodes and depositing SAE films via the through holes. Figure 2 shows a photograph of an actual SAE-MEMS VEH device with almost the same structure as shown in Fig. 1. In this experiment, we applied external vibration to the developed device and were able to measure the induced current due to the electret, confirming that the electret properties of the SAE films were maintained [6].

The proposed SAE-MEMS integration technology uses semiconductor processes and can form SAE films in various MEMS and peripheral LSIs, such as sensors and actuators, in addition to energy harvesters. When using electronic devices, it is necessary to control and design the electrical characteristics of the electrets. It has been known that the surface potential (from a few to about 30 V) is proportional to the SAE thickness for thin and flat films in the order of nm in thickness [5]. For future MEMS applications, for example, energy harvesters may require higher surface potentials (> 50 V) [4, 11]. Before our study, it was unclear whether such a high surface potential for SAE could be achieved only by increasing film thickness. Moreover, SAEs formed by this method are not simple flat films, but rather SAE films patterned on the micro-meter scale (μ-SAE), and no direct observation of the shape or surface potential of μ-SAEs has been reported. We have recently achieved an increase in SAE surface potential, as well as direct observation of the μ-SAE geometry and its surface potential.

We prepared a deposition equipment environment for thick film SAE deposition and succeeded in forming thick film SAE of 1 μm or more, achieving a surface potential of 200 V or more in the flat film state [7]. It was also experimentally clarified that the surface potential of SAE was proportional to film thickness even at film thickness of 1 μ m or more, and

that the proportionality coefficient is comparable to that of conventional thin-film SAE [7]. Furthermore, μ-SAE evaluation samples with removable through-hole structures were developed, and direct observation of μ-SAEs formed by the proposed method was successfully performed as shown in Fig. 3 [7]. Through this experiment, we can visually confirm that μ-SAEs reflecting through-hole geometries has been fabricated in MEMS structures. Also, the surface potential of μ-SAE was successfully measured directly using a Kelvin probe for the first time, and surface potentials above 50 V were realized [7, 8].

In relation to these technologies, it was also necessary to develop MEMS design techniques that take into account the characteristics of SAEs. Aiming to build a co-design environment for SAE and MEMS, we developed equivalent circuit models of SAE-MEMS devices to enable integrated analysis of mechanical behavior (mainly related to MEMS) and electrical behavior (mainly related to electrets and peripheral LSIs) on circuit simulators. As an example, an equivalent circuit of an SAE-MEMS VEH implemented on a circuit simulator, LTspice (Analog Devices, Inc.), is shown in Fig. 4 [9, 10]. Each component of the device (electrode, spring, etc.) was made into a module as an equivalent circuit, and by linking each module via voltage and current values, an equivalent circuit of an SAE-MEMS VEH was developed on LTspice. This result is useful for optimizing device design

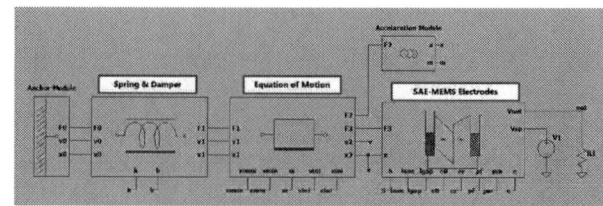

Fig. 4: Equivalent circuits of an SAE-MEMS VEH on a circuit simulator [9].

979-8-3315-0417-5/25 $31.00 © 2025 IEEE

because the SAE-MEMS VEH and its peripheral power management circuits can be analyzed simultaneously on the same simulator.

Conclusion

We have realized the integration technology of MEMS and electrets using semiconductor processes. Unlike conventional electrets, SAEs used in this technology do not require charging processes. Unlike conventional electrets, SAEs used in this technology do not require charging processes and can achieve high surface potentials required for MEMS applications. We have also developed SAE-MEMS co-design environments using circuit simulators. Electret-based MEMS are known to include sensors and actuators in addition to energy harvesters, and thus further miniaturization, higher performance, and mass production of such devices can be expected by applying this technology.

Acknowledgments

The author gratefully thanks Associate Professor Yuya Tanaka, Gunma University and Professor Hisao Ishii, Chiba University for useful discussions. This work was partly supported by the JSPS KAKENHI (Grant Nos. 22H01929 and 21H04555) , the Adaptable and Seamless Technology Transfer Program through Target-driven R&D (A-STEP) from the Japan Science and Technology Agency (JST) (Grant No. JPMJTR22R5), JST Program for co-creating startup ecosystem, Grant Number JPMJSF2315, Japan, the Samco Science and Technology Foundation, the Mitutoyo Association for Science and Technology and RARA (Ritsumeikan Advanced Research Academy), Ritsumeikan University.

References

[1] G. M. Sessler, Electrets, 3rd ed., Laplacian Press (1998).

[2] P. D. Mitcheson, E. M. Yeatman, G. K. Rao, A. S. Holmes, and T. C. Green, "Energy Harvesting from Human and Machine Motion for Wireless Electronic Devices," Proc. IEEE, 96 pp. 1457–1486 (2008).

[3] Y. Suzuki, "Recent progress in MEMS electret generator for energy harvesting," IEEJ Trans. Elec. Electron. Eng., 6, pp.101-111 (2011).

[4] H. Toshiyoshi, S. Ju, H. Honma, C.-H. Ji, and H. Fujita, "MEMS vibrational energy harvesters", Sci. Technol. Adv. Mater., 20, pp. 124-143 (2019).

[5] Y. Tanaka, N. Matsuura, and H. Ishii, "Self-Assembled Electret for Vibration-Based Power Generator," Sci. Rep., 10, 6648 (2020).

[6] D. Yamane, H. Kayaguchi, K. Kawashima, H. Ishii, and Y. Tanaka, "MEMS post-processed self- assembled electret for vibratory energy harvesters", Appl. Phys. Lett., 119, 254102 (2021).

[7] D. Yamane, K. Kawashima, R. Sugimoto, R. Li, H. Kayaguchi, K. Kurihara, H. Ishii, and Y. Tanaka, "Observation of Surface Potential of Micropatterned Self-assembled Electrets for MEMS Vibrational Energy Harvesters," Sens. Mater., 35, pp. 1985-1993 (2023).

[8] R. Li, S. Hosoi, K. Kakuno Y. Sunagawa, A. Jingu, R. Koike, R. Sugimoto, Y. Tanaka, and D. Yamane, "Investigation of the relationship between surface potential and film thickness of micro-patterned self-assembled electrets," in Proc. PowerMEMS 2023, Dec. 11-14, 2023, Abu Dhabi, UAE, pp. 45-47 (2023).

[9] K. Tokuno, S. Kinoshita, H. Kayaguchi, K. Kurihara, H. Ishii, Y. Tanaka, and D. Yamane, "Circuit Simulator Implementation of an Equivalent Circuit Model of Self-Assembled Electret Vibrational Energy Harvesters Based on an Energy Diagram", IEEJ Trans. Electr. Electron. Eng. E, 19 (2024).

[10] K. Tokuno, S. Kinoshita, F. Sugitani, T. Sono, Y. Tanaka, and D. Yamane, "An equivalent circuit model of self-assembled electret MEMS vibration energy harvesters based on an energy diagram in hardware description language," in Proc. PowerMEMS 2023, Dec. 11-14, 2023, Abu Dhabi, UAE, pp. 274-276 (2023).

[11] Z. Mao, C. Chen, Y. Zhang, K. Suzuki, and Y. Suzuki, "AI-driven discovery of amorphous fluorinated polymer electret with improved charge stability for energy harvesting," Adv. Mater. 2303827 (2023).

Title: "Quantum Nanophotonics with Hexagonal Boron Nitride"

Igor Aharonovich

[1]*School of Mathematical and Physical Sciences, Faculty of Science, University of Technology Sydney, Ultimo, NSW, 2007, Australia*
[2] *ARC Centre of Excellence for Transformative Meta-Optical Systems, Faculty of Science, University of Technology Sydney, Ultimo, NSW, 2007, Australia*

Engineering robust solid-state quantum systems is amongst the most pressing challenges to realize scalable quantum photonic circuitry. While several 3D systems (such as diamond or gallium arsenid) have been thoroughly studied, solid state emitters in two dimensional (2D) materials are still in their infancy.

In this presentation I will discuss the appeal of an emerging van der Waals crystal – hexagonal boron nitride (hBN). This unique system possesses a large bandgap of ~ 6 eV and can host single defects that can act as ultra-bright quantum light sources. In addition, some of these defects exhibit spin dependent fluorescence that can be initialised and coherently manipulated. I will discuss in details various methodologies to engineer these defects and show their peculiar properties. Furthermore, I will discuss how hBN crystals can be carefully sculpted into nanoscale photonic resonators to confine and guide light at the nanoscale. Taking advantage of the unique 2D nature of hBN, I will also show promising avenues to integrate hBN emitters with silicon nitride photonic crystal cavities.

All in all, hBN possesses all the vital constituents to become the leading platform for integrated quantum photonics. To this extent, I will highlight the challenges and opportunities in engineering hBN quantum photonic devices and will frame it more broadly in the growing interest with 2D materials nanophotonics.

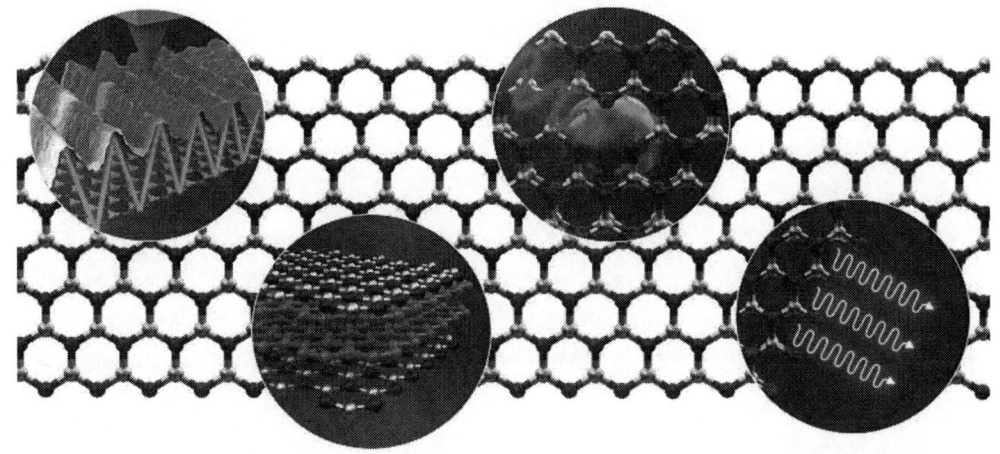

979-8-3315-0417-5/25 $31.00 © 2025 IEEE

Operation principles and applications of ultra-sensitive optical detectors at nanoscale: Facts and artifacts

Taras Plakhotnik[1]

[1]School of Mathematics and Physics, The University of Queensland, Australia

ABSTRACT

Measurements on a nanoscale are at technological limits in several factors and therefore are more complex than measurements on a macro-scale. The required sensitivity and precision critically depends on the factors which may be marginal in conventional measurements. We describe phenomena observed in scientific CMOS cameras at the level of several photoelectrons per pixel and the effect of the local refractive index on the radiative and non-radiative transition rates.

Keywords: Photo luminescence, radiative and non-radiative lifetimes, thermosensors, CMOS cameras

INTRODUCTION

Point defects (vacancies, substitutional atoms etc) are defects found on a singe crystal site. However, the optical properties of such quantum systems are affected by various external factors such as, for example, temperature of the environment. These dependencies can be used to design a nano-sensor, a tiny device capable of measuring properties at the nanoscale (size range 10-100 nm). However, at least two problems appear when this idea is put to a practical test. The first problem is that the changes are small and the second problem is that the optical signal frequently used for readout is very low. The first problem requires careful consideration of all the factors which may result in the same readouts while the second problem can be addressed by using ultra-sensitive photodetectors capable of detecting one emitted photon. In both cases, solutions of the initial problem may result in creating a different set of nontrivial questions. In this paper, we consider both aspects of sensing at the nanoscale using single photon detection with a CMOS camera and dependence of the radiative and non-radiative rates on the local refractive index of the environment as examples.

SPECTROSCOPY USING sCMOS CAMERAS AT EXTREMELY LOW LIGHT INTENSITIES

Silicon-vacancy center (SiV) is a known point defect in diamond composed of a silicon atom and a vacancy replacing two adjacent atoms of carbon in the crystal

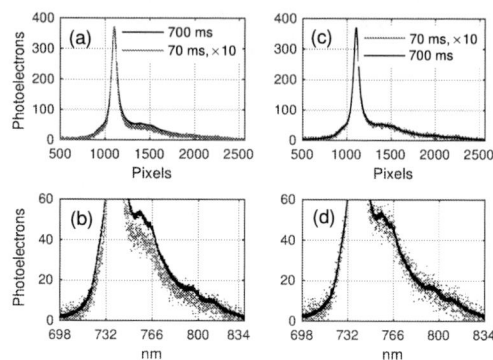

Fig. 1. Spectra of SiV centers in 100-nm diamond crystal recorded with a CMOS camera. The vertical scales show the average number of photoelectrons per pixel. Data recorded with 700-ms exposure are shown as black curves and 70-ms exposure data as red curves (vertically multiplied by 10). The relatively strong and narrow feature with a maximum at 740 nm is a zero-phonon line (ZPL). The broad feature on the right is a phonon sideband (PSB). Panels (c) and (d) show spectra after correction using the data from Fig. 2. The associated error in the estimated ZPL width without this correction, can be as high as 0.5 nm ($\approx$ 8 K if the linewidth is used for temperature measurements). Panels (b) and (d) have a smaller vertical range for better visibility of the difference in the PSB region. Adapted with permission from [1] © 2021 Optical Society of America.

lattice. These centers are characterized by strong photo emission dominated by a narrow line (of about 5 nm full width at half maximum) centered at around 740 nm (see Fig.1). The spectral position of this line and its width are sensitive to the external temperature. Their thermal susceptibilities are $0.013 \pm 0.0012 \, \mathrm{nm \, K^{-1}}$ for the position and $0.061 \pm 0.0046 \, \mathrm{nm \, K^{-1}}$ for the width [2]. This means, for example, that the width increases by 1% for a 1-K increment of the external temperature. This is a relatively small change which requires accurate measurements of the line shape.

We have investigated a property of a scientific CMOS camera that makes accurate spectroscopy at ultra-low photoelectron accumulations difficult. The problem is illustrated in Fig.1 where the results are shown for two exposures, 700 ms and 70 ms. The data recorded with 70-ms exposures are multiplied by a factor of 10 and

979-8-3315-0417-5/25 $31.00 © 2025 IEEE

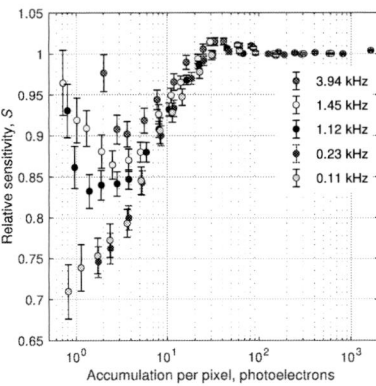

Fig. 2. Relative sensitivity of a sCMOS camera at different accumulated numbers of photoelectrons per pixel and different photon detection rates as displayed in the legend. The "accumulation per pixel" is the average number over a selected set of pixels. The selected set satisfies the condition that the difference between the smallest and the largest number of photoelectrons accumulated in each pixel at the longest exposure is less than 2% of the mean value for the set. This ensures uniformity of illumination for the selected pixels. Adapted with permission from [1] © 2021 Optical Society of America.

overlap with the data recorded with 700-ms exposure near the peak intensity at 740 nm. However, intensities at 760 nm show significant disagreement between the two sets of data. This is quite an unusual disagreement. Conventionally, it is expected that issues related to the nonlinearity in a detector response to incident light (pixel illumination) become significant only at very high intensities of light. This nonlinearity is normally referred to as saturation and largely determined by the capacity of a pixel to retain the electric charge as well as limitations of the ADC converter. Fig. 1 demonstrates nonlinearity of a different type. While the high intensity signal seems to scale proportionally to the exposure time, the detector demonstrates a smaller response at a low intensity signal (less than 50 electrons per pixel). Figure 2 shows the results of a systematic investigation of these phenomena. As a source of light, we have used a light emitting diode (LED). The intensity of light emitted by the LED has been adjusted so that the image of the LED had the electron count rate per pixel as specified in the legend. A large set of exposures has been used to record the accumulated photoelectrons. The relative sensitivity is defined as

$$S = \frac{C/\tau}{C_r/\tau_r} \qquad (1)$$

where C is the number of ADC counts recorded with exposure τ, and C_r is the number of ADC counts recorded

with a reference exposure τ_r. The reference exposure has been selected as 400 electrons per pixel. In an ideal case, $S = 1$ for any combination of the count rate and exposure. As one can see, the value of S depends on both the number of accumulated photoelectrons and the actual photoelectron count rate. This makes correction complicated but possible with proper calibration (see Fig. 2). The corrected spectra in Fig. 1 (c, d) agree with each other within the noise level.

ELECTRODYNAMICS AND ENVIRONMENTAL EFFECTS IN ULTRA-THIN RUBY FLAKES (RUBYENE)

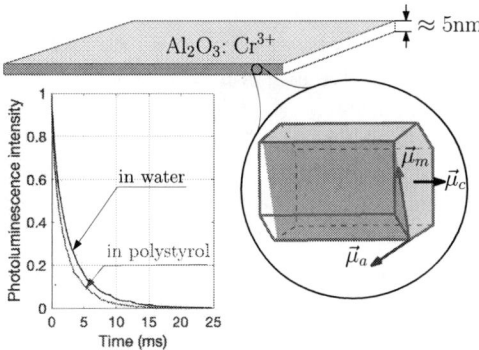

Fig. 3. Crystals of Al_2O_3 doped with chromium ions Cr^{3+} are obtained in a form of about 5-nm thin flakes. When such flakes are immersed in media with different refractive indices, the decay rate of Cr^{3+} varies and is faster if the medium has a higher refractive index. These variations are caused by the variations of both radiative and non-radiative rates. The encircled inset shows a possible orientation of the crystal. The direction of the c-axis parallel to the plane of the surface provides the best agreement with the data. $\vec{\mu}_a$, $\vec{\mu}_m$, and $\vec{\mu}_c$ are three transition dipole moments associated with each chromium ion ($|\vec{\mu}_a|^2 = |\vec{\mu}_m|^2 \approx 10|\vec{\mu}_c|^2$). © IOP

Fluorescence lifetime is another temperature dependent property which may be used for thermal sensing. The small size of the nano-sensors makes them suitable for various delicate applications including measuring temperature in complex biological environments such as, for example, intracellular space [3]. Calibration of such thermometers is frequently done at laboratory conditions when the sensors are immersed in pure water or a buffer solution. This simplifies the temperature control and the data analysis but the actual measurement may be affected by factors which are absent during the calibration procedure. Photonic effects caused by inhomogeneity of the refractive index may in some cases significantly modify the radiative lifetime and spectra of the emitting entities [4]. Such sensitivity may result in misleading interpretations of the measurements when the change

979-8-3315-0417-5/25 $31.00 © 2025 IEEE

of a certain parameter are entirely explained by the change of the temperature and the contribution of other factors are ignored. At the same time, the additional sensitivity to the nanoscale surroundings may be used as an advantage because it provides additional information about the system of interest.

As an example, we have investigated (both theoretically and experimentally) radiative and non-radiative rates in Cr^{3+} ions imbedded (see Fig. 3) in few-nm-thin flakes of ruby (Al_2O_3: Cr^{3+}) called rubyene [5]. Although the effect of the environment on the non-radiative relaxation rates is an order of magnitude smaller than its effect on the radiative rates, the sensitivity of our measurements was sufficient to observe both. Our theoretical analysis has general implications for Förster-type energy transfer in non-uniform media. The experimental results are in good agreement with the theory and establish the potential of the material in optical nanosensing applications, highlighting its sensitivity to environmental refractive index changes.

The approach to the data analysis is based on the following equation

$$I_m(t) = I_w(t) \exp(-\nu_m t) \tag{2}$$

where $I_m(t)$ is the time-dependent intensity of luminescence (decay curve) when flakes are immersed in a medium (refractive index n_m) and $I_w(t)$ is the decay curve when the medium is water (refractive index n_w) which is used as a reference. The exponent depends on the only fitting parameter for each medium, ν_m. The results of multiplication $I_w(t)$ and $\exp(-\nu_m t)$ are shown in Fig. 4 (a, b). The value of ν_m can be also obtained theoretically using appropriate assumptions about the flakes and the orientation of three transition-dipole moments of the Cr^{3+} ions. For example, in the case of a thin but large in the lateral direction flake and c-axis parallel to the surface of the flake, the expression for ν_m reads

$$\nu_m = \gamma_b \left[\frac{\frac{n_m^4}{n_r^4} + 1.1}{2.1 n_r} n_m - \frac{\frac{n_w^4}{n_r^4} + 1.1}{2.1 n_r} n_w \right] - \Delta k_m \tag{3}$$

where γ_b is the radiative rate of Cr^{3+} in a bulk crystal, n_r is the refractive index of ruby, and Δk_m is the change of the non-radiative rate when water is replaced with the medium (obtained by numerical simulations) [6].

CONCLUSION

We have demonstrated two examples of sensing at the nanoscale when factors negligible at the macro scale affect the measurements. One case is rather technical and

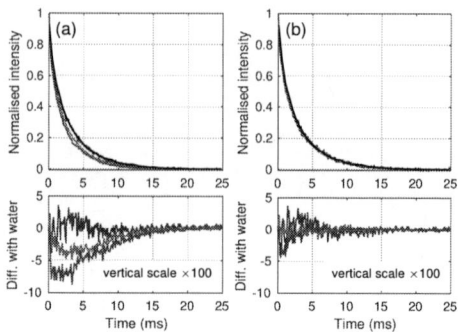

Fig. 4. Photo luminescence decay in rubyene. Panel (a): Normalized experimental curves measured when the flakes are dissolved in solvents of increasing refractive index: methanol ($n_m = 1.325$, blue curve), water ($n_m = 1.331$, black curve), DMSO ($n_m = 1.474$, red curve), and polystyrene ($n_m = 1.583$, dark red curve). The bottom panel shows the difference with water. In panel (b) the experimental curves are multiplied by the corresponding $\exp(\nu_m t)$. Note, that only one fitted parameter per curve has been used in transitioning from panel (a) to panel (b). © IOP

can be resolved by improvement in CMOS technology. The second case is related to fundamental electrodynamics. However all factors are essential for the accuracy of the results.

ACKNOWLEDGMENT

The author gratefully acknowledge the contributions of O. Matsiaka, A. Mukhamedshin, W. A. W. Razali, X. Yang, E. V. Khaydukov, J. M. Dawes and A. V. Zvyagin.

REFERENCES

[1] O. Matsiaka and T. Plakhotnik, "Accurate spectroscopy with sCMOS cameras at ultra-low intensities of light" Opt. Lett., 46, pp. 961-964 (2021).

[2] S. Choi, V. N. Agafonov, V. A. Davydov, L. F Kulikova and T. Plakhotnik, "Formation of interstitial silicon defects in Si- and Si,P-doped nanodiamonds and thermal susceptibilities of SiV$^-$ photoluminescence band", Nanotechnology, 31, 205709 (2020).

[3] K. Okabe, N. Inada, C. Gota, Y. Harada, T. Funatsu, S. Uchiyama, "Intracellular temperature mapping with a fluorescent polymeric thermometer and fluorescence lifetime imaging microscopy", Nature Commun. 3, 705 (2012).

[4] S. J. W. Vonk, Th. P. van Swieten, A. Cocina, and F. T. Rabouw, "Photonic Artifacts in Ratiometric Luminescence Nanothermometry", Nano Lett. 23 6560-6566 (2023).

[5] W. A. W. Razali, X. Yang, P. A. Demina, A. V. Atanova, E. V. Khaydukov, V. A. Semchishen, J. M. Dawes, T. Plakhotnik, and A. V. Zvyagin, "Ruby Nanoflakes (Rubyene) for Efficient 2D Förster Resonance Energy Transfer: Implications for Engineered Emitters in Multiplexed Imaging", ACS Appl. Nano Mater, 7, 11320-11329 (2024).

[6] A. Mukhamedshin, W. A. W. Razali, X. Yang, P. A. Demina, K. V. Khaydukov, V. A. Semchishen, E. V. Khaydukov, J. M. Dawes, R. C. Willson, A. V. Zvyagin, T. Plakhotnik "Electrodynamical and environmental effects in ultra-thin ruby flakes (rubyene): A promising platform for quantum technologies and optical nanosensing", J. Phys: Photonics (submitted).

Ag:SiOx-based Volatile Memristors for Dendritic Computations

Ruiqi Chen[1,2], Yulin Feng[3], Nan Tang[1,2], Yiyang Chen[1,2], Hao Ai[1,2], Haozhang Yang[1,2],
Zheng Zhou[1,2], Lifeng Liu[1,2#], Xiaoyan Liu[1,2], Jinfeng Kang[1,2], Peng Huang[1,2*]

[1]School of Integrated Circuits, Peking University, Beijing 100871, China
[2]Beijing Advanced Innovation Center for Integrated Circuits, Beijing 100871, China
[3]School of Instrument Science and Opto-Electronics Engineering, Beijing Information Science and Technology University, Beijing 100192, China
Email: lfliu@pku.edu.cn, phwang@pku.edu.cn

Abstract

Dendrites play a crucial role in the information processing of neuromorphic systems by virtue of their unique functions. However, research realizing hardware dendritic computation, especially active functions is still insufficient. Here we fabricated Ag:SiOx-based volatile memristors and demonstrated multiple dendritic functions. We experimentally achieved not only the temporal integrations but also the active function of dendrites, namely gain modulation and selectivity. Our work shows the potential of implementing versatile dendritic computation with emerging memristors.

Keywords: Memristor, Dendritic computing, Gain modulation, Selectivity

Introduction

The exponential growth of both the volume of data and demand for computing, coupled with the looming end of Moore's law and the end of Dennard scaling, has fueled interest in alternative computing primitives [1-3]. Neuromorphic computing is a compelling technology, which may yield momentous performance improvements in the rapidly growing areas of identification and classification of information buried within massive datasets [4-6]. Various electronic components have been designed for the hardware implementation of neuromorphic computing. Nevertheless, most works so far focus on mimicking the synapse and the soma of neurons [7], [8]. Less attention is paid to developing dendritic computations, especially the active functions [9-11]. Dendrites with fine morphology consist of countless proximal and distal branches as shown in **Fig. 1(a)**, which makes a single neuron a multifunctional computational unit. Passive dendrites nonlinearly transform inputs from different branches with delay and attenuation [12-15]. Active dendritic responses are triggered under coincident inputs from both proximal and distal branches, which is important for cognition and perception [16-20]. However, dendritic computations are usually simplified in widely used neuron models as shown in **Fig. 1(b)**. Several studies have demonstrated electronic dendrites based on complex CMOS circuits [21], [22]. Therefore, scalable artificial dendrites

with passive and active functions are highly desired to offer the opportunity to enrich computing capabilities and reduce the energy consumption of electronic neuromorphic computing systems. In this work, we fabricated Ag:SiOx-based volatile memristors to demonstrate both the passive and active functions of dendrites experimentally. The fabricated memristor shows temporal nonlinear integration characteristics, which are the essential function of biological neurons. Gain modulation, referring to the processing of coincident inputs received from different dendritic branches, is also demonstrated experimentally. Additionally, the proposed dendritic computation enables efficient detections of moving direction and velocity. This work paves the way towards compact and densely functional neuromorphic computing primitives and energy-efficient validation of neuroscientific models.

Results and Discussions

A. Ag:SiOx-based memristors

Fig. 2(a) shows the fabrication process of Ag:SiOx-based memristors. The device structure is illustrated in **Fig. 2(b)**, with Pt as the top/bottom electrodes and SiOx:Ag as the resistive layer. The cross-sectional TEM image is shown in **Fig. 2(c)**, in which the profile can confirm the doping of Ag nanoclusters. The typical DC I-V curves of Ag:SiOx-based memristors are shown in **Fig. 3**. When the voltage is larger than the threshold voltage (V_{th}), the device switches from a low conductance state (LGS) to a high conductance state (HGS). When the applied voltage is smaller than the hold voltage (V_h), the device relaxes back to LGS spontaneously. The distributions of V_{th} and V_h are plotted in **Fig. 4**. The I-V sweep results can indicate good uniformity of our memristors. AC switching behaviors are presented in **Fig. 5**, showing the gradual switching and relaxation behaviors. **Fig. 6** shows the excellent endurance behaviors above 10^8 cycles by using the oxygen annealing process according to the physical mechanisms based on the density of Si dangling bonds [23].

B. Temporal integration

Biological dendrites integrate inputs from different branches over correlated timescales in the form of superlinear, linear, or sublinear summation [15]. Temporal integration refers to the nonlinear responses determined by the input temporal order [15], which is represented by the frequency here. Pulse sequences (2.4 V, 20 μs) with different frequencies are applied to artificial dendrites as shown in **Fig. 7**. The response current, similarly to biological EPSP, gradually increases and then becomes saturated. With the increase of input frequency, the response current experiences superlinear, linear, and sublinear increases as shown in **Fig.8**.

C. Gain modulation

Gain modulation plays a critical role in the complex arithmetic operations of human brains [16]. **Fig. 9** shows the gain modulation of temporal integration, which refers to the phenomenon that a slight increase in distal inputs will significantly increase the gain of temporal integration curves of proximal inputs. Pulse sequences with different frequencies (f_p) are applied through synapse P as proximal inputs, and pulses with minor constant voltage (V_d) are applied through synapse D as distal inputs. The linear summation currents are converted to voltage applied on the fabricated artificial dendrites by a trans-impedance amplifier (TIA). **Fig. 9(b)** shows the transient current responses under diverse V_d. The change of saturation current (I_s) increases from 40 μA to 70 μA with a slight increase in V_d, showing an amplification of gain. I_s-f_p curves under different V_d and the corresponding gain ($\Delta I_s / \Delta f_p$) are shown in **Fig. 10**. The results indicate that the gain of proximal inputs can be modulated by the distal inputs, which is similar to the characteristics in biological Layer 5 pyramidal neurons [17].

D. Direction and velocity selectivity

Travelers can detect the motion direction of trains based on the directional selectivity of neurons, which is attributed to dendritic involvement [18] (**Fig. 11(a)**). Direction-selective neurons are triggered in the preferred direction that sequential inputs from distal dendrite to soma (IN), and are less active in the OUT direction [19], as shown in **Fig. 11(b)**. This function relies on the nonlinear integration of several synaptic inputs along the branches. Accordingly, multiple signals with different temporal correlations encoded by frequencies are integrated and applied to artificial dendrites. **Fig. 12** shows the transient current responses to signals representing IN and OUT directions. The saturation currents are around 30μA and 100μA, indicating the distinct selectivity of moving direction in 1-D space. The dendritic responses also show strong sensitivity in velocity as can be seen in **Fig. 13**, where the IN and OUT current both vary with input velocity, as is in accordance with biological phenomenon [19]. Leveraging the relationship in **Fig. 13**, we can readily deduce the velocity component in the IN-OUT axis direction based on a certain current. Furthermore, for objects moving within 2-D space, the velocity (v_0) and

direction (θ) can both be derived as long as two velocity components of different axes are obtained. This observation is exploited in **Fig. 14**, where dendrites with different preferred directions, labeled as North, South, East, and West, are used for a single neuron [20]. The four artificial dendrites pointing in orthogonal directions are used to detect direction and velocity. We verified this function by detecting an object moving (v_0=1) from southwest to northeast at a 30° angle. The statistical distributions of transient current responses are shown in **Fig. 15**. We observed a 5% error of v_y direction due to the velocity component being relatively small and causing overlap in the North/South current (**Fig. 15(b)**). Fortunately, this error is overcome by the intrinsic nature of dendrites. Dendrites with diverse geometric morphologies show different transmission delays, making them sensitive to input velocities. Therefore, v_y direction can be ascertained distinctly by slow dendrite with relatively slow signal propagation (**Fig. 15(c)**), showing adaptable detection capabilities. The v_0 and θ are finally calculated to be 0.95 and 29.8°based on the velocity sensitivity curve in **Fig. 13** and equations in **Fig. 14**, showing remarkable accuracy to ground truth. We further show the responses of the four dendrites to all directions in **Fig. 16**. The polar plot indicates comprehensive selectivity in 2-D space.

Conclusion

In this work, we fabricated Ag:SiO$_x$-based artificial dendrites to enrich the basic functions of neuromorphic computing. Key achievements include: 1) Ag:SiO$_x$-based memristor is fabricated for emulating dendritic computation with reliable uniformity and endurance. 2) Temporal integrations, along with gain modulation and selectivity are experimentally demonstrated. 3) Direction and velocity selectivity are demonstrated by the proposed artificial dendrite. This work has validated the effectiveness of dendritic computation. It also provides a promising method for implementing compact and energy-efficient neuromorphic computational elements.

Acknowledgments

This work was supported in part by the National Sci-Tech Innovation 2030 under Grant 2021ZD0201202, the National Natural Science Foundation of China Program under Grant 62474005, 62034006, the STIC under QYJS-2022-1501-B and the 111 Project Program under Grant B18001.

References

[1] N. Mirzadeh et al., *ASBD*, 2015. [2] M. Bohr, et al., *IEEE SSCS Newsletter*, 2007. [3] D. Marković, et al., *Nat. Rev. Phys.*, 2020. [4] C. Mead, *Proc. IEEE*, 1990. [5] G. W. Burr, et al., *Adv. Phys.*, 2017. [6] K. Roy, et al., *Nature*, 2019. [7] Z. Wang et al., *Nat. Mater.*, 2016. [8] Z. Wang et al., *Nat. Electron.*, 2018. [9] X. Li et al., *Nat. Nanotechnol.*, 2020.[10] X. Li, et al., *Adv. Mater.*, 2022. [11]R. Nauda, et al., *PNAS.*, 2017. [12] P. Poirazi et al., *Nat. Rev. Neurosci.*, 2020. [13] S. Antic et al., *J. Neurosci. Res.*, 2010. [14] M. Larkum et al., *Nature*, 1999. [15] M. London et al., *Annu. Rev. Neurosci.*, 2005. [16] R. Silver, *Nat. Rev. Neurosci.*, 2010. [17] M. E. Larkum, et al., *Cereb. Cortex*, 2004. [18] S. Single, et al., *Science*, 1998. [19] T. Branco et al., *Science*, 2010. [20] D. Vaney et al., *Nat. Rev. Neurosci.*, 2012. [21] J. Schemmel et al., *IJCNN*, 2017. [22] A. Bhaduri et al., *Neural. Comput.*, 2018. [23] R. Li et al., *SNW.*, 2023.

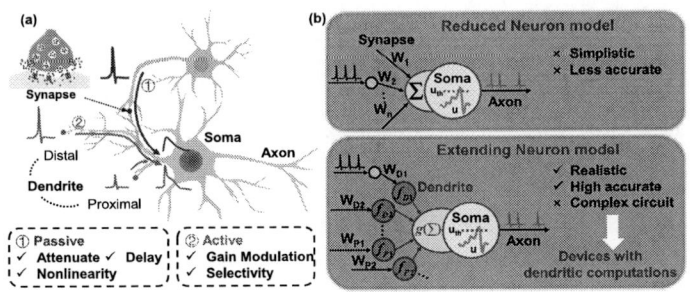

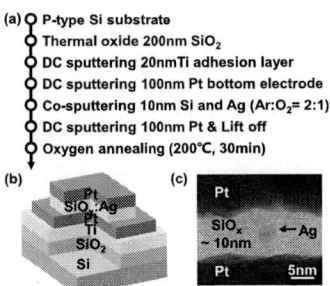

Fig. 1 Sketch of the computation functions of dendrites in neuron. (a) Passive and active dendritic computations. (b) Reduced neuron model and extending neuron model with dendrites equipped with both passive and active functions.

Fig. 2 (a) Fabrication process. (b) Device structure. (c) Cross-sectional TEM image of Ag:SiO$_x$-based memristors.

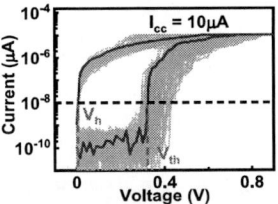

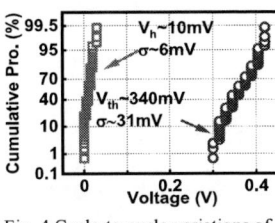

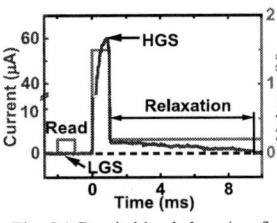

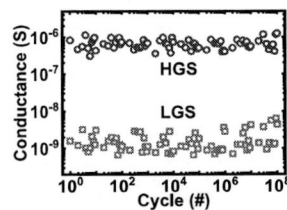

Fig. 3 Basic I-V curves of Ag:SiO$_x$-based memristors.

Fig. 4 Cycle-to-cycle variations of V_h and V_{th} under 100 I-V cycles.

Fig. 5 AC switching behaviors of Ag:SiO$_x$-based memristors.

Fig. 6 Endurance behaviors of Ag:SiO$_x$-based memristors.

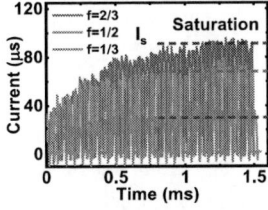

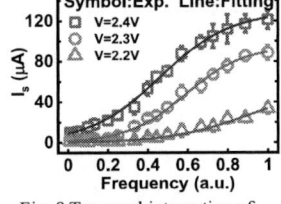

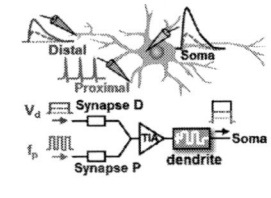

Fig. 7 Transient current responses to pulse sequences with different input frequencies (normalized by 50kHz).

Fig. 8 Temporal integration of inputs with different frequencies under diverse amplitudes.

Fig. 9 Gain modulation of temporal integration. (a) Sketch and hardware implementation. (b) Transient current responses.

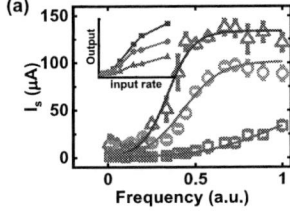

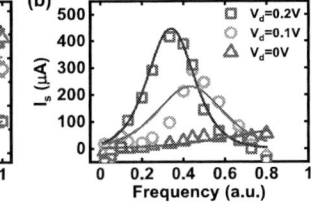

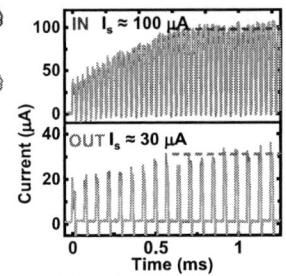

Fig. 10 Gain modulation (a) I_s-f_p curves under different V_d, along with biological results [17]. (b) Corresponding gain ($\Delta I_s/\Delta f_p$). The gain of proximal input is substantially affected by distal input.

Fig. 11 Directional selectivity. (a) Real-life cases. (b) Sketch of the direction-selective neuron.

Fig. 12 Transient current responses of the artificial dendrite to IN and OUT directions.

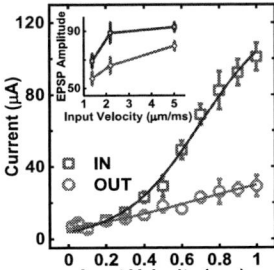

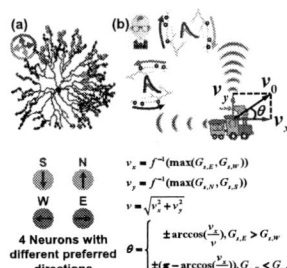

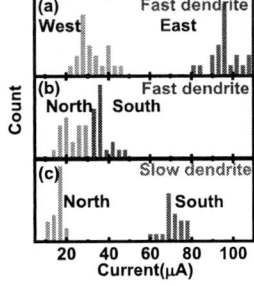

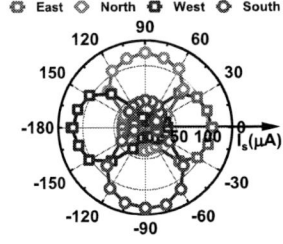

Fig. 13 Measured device responses of IN and OUT directions to different input velocities.

Fig. 14 (a) Dendrites with different preferred directions [22]. (b) Sketch and implementation principle of motion detection. directions.

Fig. 15 Statistical results of device responses of four dendrites to object moving at 30° angle when v_0 = 1.

Fig. 16 Polar plots of four dendritic responses to all directions. Comprehensive selectivity is verified

979-8-3315-0417-5/25 $31.00 © 2025 IEEE

Synergetic effect of doping and oxygen vacancies to realize higher permittivity and lower leakage for DRAM capacitors: A first-principles study

Ting Zhang[1], Maokun Wu[1,*], Miaojia Yuan[1], Yichen Wen[1], Yilin Hu[1], Pengpeng Ren[1], Sheng Ye[1], Runsheng Wang[2], Zhigang Ji[1,*], and Ru Huang[2]

[1]Departure of Micro/Nano Electronics, Shanghai Jiao Tong University, Shanghai, China
(*email: maokunwu@sjtu.edu.cn, zhigangji@sjtu.edu.cn),
[2]School of Integrated Circuits, Peking University, Beijing, China

Abstract

A dielectric material with higher permittivity and lower leakage is required to meet the demands of three-dimensional (3D) DRAM. The t-phase HfO_2 has higher permittivity but instability. Our First-principles calculations reveal that doping Y and the associated oxygen vacancies (Vo) play a critical role in phase stability. Additionally, defect states induced by Vo can be passivized at charge balance state to enable large bandgap, suppressing leakage currents. Our findings offer a feasible technical pathway for further 3D DRAM.

Keywords: 3D DRAM, Doping, Oxygen vacancy, First-principles calculations

Introduction

As devices reach their physical limitations, the three-dimensional (3D) Dynamic Random Access Memory (DRAM) becomes inevitable. Consequently, there is an urgent need for dielectric materials that possess higher permittivity and lower leakage to facilitate its progression [1-3]. HfO_2 has gained significant attention in industry due to its the compatibility with complementary metal-oxide-semiconductors technology and mature atomic layer deposition (ALD) process [4,5]. HfO_2 exists in monoclinic (m-), tetragonal (t-), and orthorhombic (o-) phases. Among these, the m-HfO_2 is the most stable but exhibits the lowest dielectric constant (k), whereas the t-HfO_2 offers higher k value but is relatively unstable [6,7]. Therefore, stabilizing the t-phase is required to elevate the overall k value of HfO_2 thin films [8,9].

Doping is an effective method to stabilize the t-HfO_2 by suppressing the formation of the m-HfO_2, thereby enhancing its dielectric properties. Govindarajan et al. found that doping Gd with 20% concentration can significantly stabilizes t-HfO_2, increasing the k value to approximately ~27 [10]. Harashima et al. reported that t-HfO_2 can be stabilized through doping 6% Si [6]. Bhanu et al. demonstrated that Mg doping can stabilize the t-ZrO_2. Despite these efforts,, its leakage current density at -1.0 V under doping strategy is on the order of 1×10^{-6} A/cm^2, which cannot satisfy the device requirement [11]. Thus further enhancement of k is necessary through achieving more t-HfO_2 fraction. Additionally, k and leakage exhibit an inverse relationship from the reported literatures. Thus how to break this trade-off between k and leakage is crucial for optimizing future 3D DRAM devices. This work systematically investigates the impact of nine dopants (Mg, Ca, Sr, Sc, Y, La, Si, Ge and Ti) on phase stability through first-principles calculations, revealing that the ratio of doping to oxygen vacancies is crucial for stabilizing the t-HfO_2. This can be attributed to the fact that the local structure distortion. Thus Y doping is suggested to further stabilize t-HfO_2. Moreover, the defect states induced by oxygen vacancies (Vo) can be shifted near the conduction band under doping systems, enabling the t-HfO_2 to maintain a large band gap and consequently inhibiting leakage current.

Computational Details

First-principles calculations based on density functional theory are adopted utilizing PWmat software. The norm-conserving pseudopotential (NCPP) from the SG15 set is employed to describe the exchange-correlation potential. The plane-wave energy cutoff is set to 50 Ry, and the k-point grid was generated with a k-point density of 2×2×2. The structures are fully optimized until the force on each atom is less than 0.03 eV/Å and the system was considered converged when the energy difference was below 10^{-5} eV.

Results and Discussion

The unit cell of various HfO_2 is shown in **Fig. 1**. For m-phase, a = 5.072 Å, b = 5.127 Å, and c = 5.261 Å (**Fig. 1a**); for the o-phase, a = 4.983 Å, b = 5.017 Å, and c = 5.209 Å (**Fig. 1b**); and for the t-phase, a = 5.013 Å, b = 5.013 Å, and c = 5.178 Å (**Fig. 1c**). The lattice constants are consistent with the reported work [12]. The 2*2*2 supercell is built to simulate doping effect.

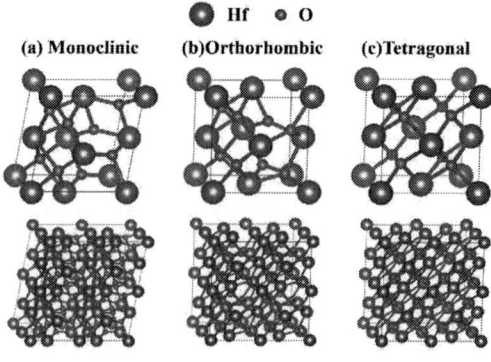

Fig. 1 Atomic structure of pristine HfO_2 (a) m-phase, (b) o-phase, (c) t-phase, with unit cell and 2*2*2 supercell.

Here, we discuss three different types of dopants based on their valence states: divalent (Mg, Ca, Sr), trivalent (Sc, Y, La), and tetravalent (Si, Ge, Ti). Initially, doping systems without Vo are considered at different doping concentration. As shown in **Fig. 2**, compared with other dopants (Mg, Ca, Sr, Sc, Y, La, Ti), Si and Ge doping exhibit significant effects in stabilizing the t-HfO$_2$. The reason for this can be attributed to the similarity between the local structure after doping and its native oxide. This point would be explained in the following part. Furthermore, with increasing concentrations of Si and Ge, the relative energy of t-HfO$_2$ further decreases (**Fig. 2g,h**).

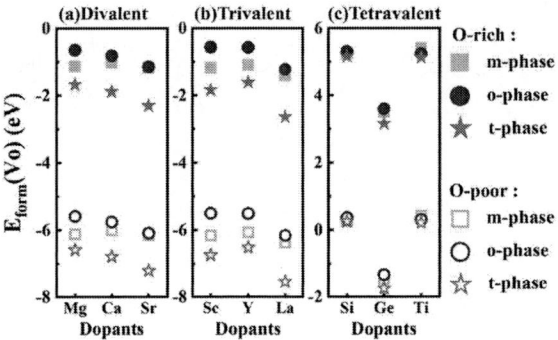

Fig. 3 Vo formation energies in doped HfO$_2$ (a) divalent dopants, (b) trivalent dopants, and (c) tetravalent dopants. Solid markers represent oxygen-rich conditions, while hollow markers represent oxygen-poor conditions.

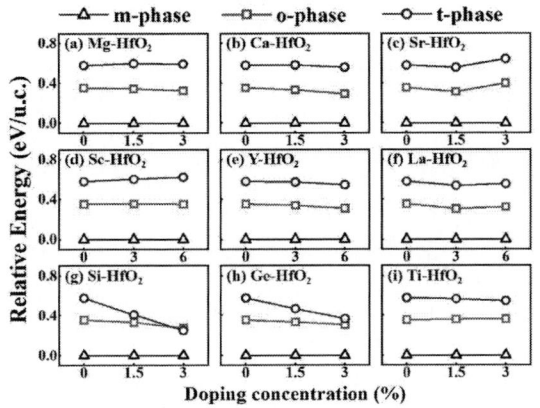

Fig. 2 Effect of doping on phase stability (in the absence of Vo) (a) Mg, (b) Ca, (c) Sr, (d) Sc, (e) Y, (f) La, (g) Si, (h) Ge, and (i) Ti doped m-, o-, and t-phase HfO$_2$. The energies shown are relative to the m-phase.

Considering the actual process, Vo are inevitable in ALD processes. To investigate the Vo behavior in doping HfO$_2$ systems, the Vo formation energies are estimated under both oxygen-rich and oxygen-poor conditions, as shown in **Fig. 3**. Remarkably, since stoichiometric group-II and group-III oxides have a metal-to-oxygen ratio of 1:1 and 1:1.5, thus the stoichiometry are further considered in Mg, Ca, Sr and Sc, Y, La doping HfO$_2$ to satisfy charge balance through removing corresponding oxygen atoms number. Under oxygen-rich conditions, the high oxygen chemical potential leads to a higher Vo formation energy, necessitating more energy for its creation. Conversely, in oxygen-poor conditions where the oxygen chemical potential is reduced, Vo formation requires less energy. This is understood that Vo formation is more favorable in oxygen-poor environments. Furthermore, for divalent (Mg, Ca, Sr) and trivalent (Sc, Y, La) dopants (**Fig. 3a,b**), the Vo formation energy is negative, indicating that Vo can form spontaneously. For tetravalent (Si, Ge, Ti) dopants (**Fig. 3c**), the Vo formation energy is generally higher compared with divalent and trivalent dopants regardless of the low chemical potential of oxygen.

Based on the doping configurations at different dopants, we further explore the phase stability for divalent dopants, trivalent dopants, and the tetravalent dopant when considering Vo effect (**Fig. 4**). It is found that the relative energy of t-HfO$_2$ tends to decrease significantly under the synergetic regulation of doping and oxygen vacancies especially for Y and La. Furthermore, as the doping concentration increases, the trend becomes more pronounced. Although La doping at a concentration of 6% can effectively stabilize the t-phase, it concurrently stabilizes the o-phase energy, which is not conducive to increasing the proportion of t-HfO$_2$. Furthermore, doping elements with smaller ion radius have a tendency to penetrate the crystal lattice structure more readily, facilitating precise control over their properties. Therefore, based on these results, doping Y is promising in stabilizing the t-HfO$_2$.

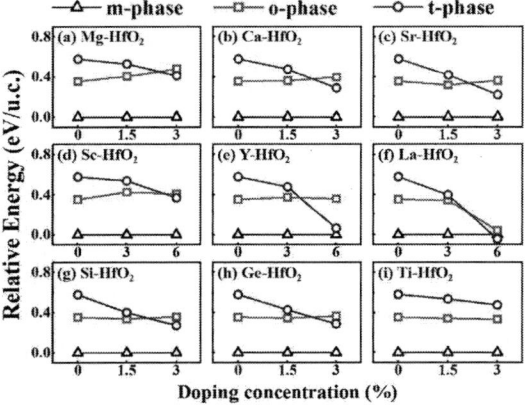

Fig. 4 Effect of doping on phase stability (with Vo present) (a) Mg, (b) Ca, (c) Sr, (d) Sc, (e) Y, (f) La, (g) Si, (h) Ge, and (i) Ti doped m-, o-, and t-phase HfO$_2$.

To understand the physical mechanism behind the doping-modulated phase stability, we further analyze local atomic structures around dopants. **Fig. 5** illustrates the local structures of doped t-HfO$_2$ without/with Vo effect. When oxygen vacancies are not considered (left side), only the local structures of Si and Ge resemble their native oxides, which explains their stabilizing effect on the t-HfO$_2$. In the presence of oxygen vacancies (right side), the local structures of divalent dopants (Mg, Ca, Sr) and trivalent dopants (Sc, Y, La) more closely resemble their native oxides, indicating a pronounced stabilizing tendency for the t-HfO$_2$ [13].

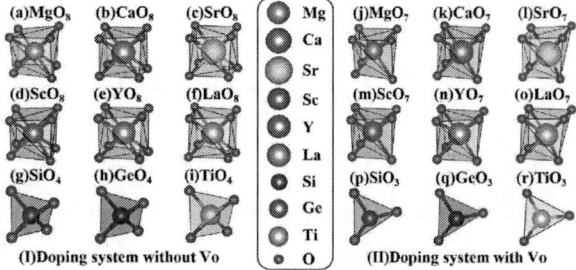

Fig. 5 The intrinsic mechanism by which doping and Vo synergistically stabilize the t-phase (a, j) Mg, (b, k) Ca, (c, l) Sr, (d, m) Sc, (e, n) Y, (f, o) La, (g, p) Si, (h, q) Ge, and (i, r) Ti sites are shown for both structures without Vo (left side) and with Vo (right side).

The electronic structures of doped t-HfO$_2$ are taken into account due to their potential role in leakage and other reliability issues. In HfO$_2$ with Vo system, the defect states corresponding to two electrons can be observed within the bandgap, contributing to electron hopping and increase the leakage currents [14]. Thus the defect states are expected to be eliminated to enable the large band gap in t-HfO$_2$ [15]. **Fig. 6** presents the DOS for cases where both dopants and Vo are considered. For divalent (Mg, Ca, and Sr) and trivalent dopants (Sc, Y, and La) doped t-HfO$_2$, the defect states merge to the conduction band minimum (CBM) (**Fig. 6a-f**), suggesting the passivized effect. This can be attributed to the charge balance effect. In contrast, the defect state is still remained within band gap in Si, Ge and Ti doped t-HfO$_2$ (**Fig. 6g-i**). Due to the inevitability of oxygen vacancies during the process, divalent and trivalent dopants are suggested in reducing leakage due to the suppression of defect states. Thus the concentration of dopants and associated Vo needs to be tuned carefully to meet charge balance, achieving low leakage current.

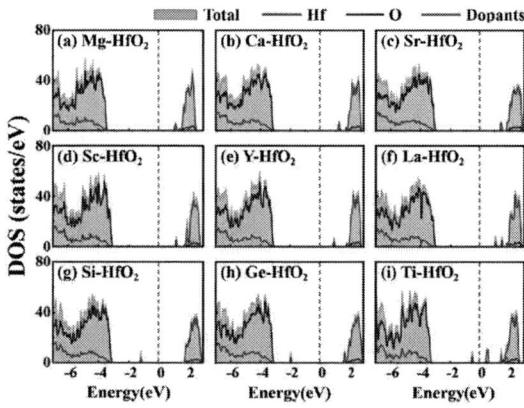

Fig. 6 DOS in doped t-HfO$_2$ (a) Mg-, (b) Ca-, (c) Sr-, (d) Sc-, (e) Y-, (f) La-, (g) Si-, (h) Ge-, and (i) Ti-doped t-HfO$_2$. The Fermi levels are indicated by grey dashed lines.

Conclusion

This study explores the effects of doping on the stability of the t-HfO$_2$ using first-principles calculations. When doping divalent and trivalent dopants, oxygen vacancies tend to occur readily and play a crucial role in phase stability. Especially for Y, a significant effect can be observed to stabilize t-HfO$_2$. Furthermore, this effect becomes more pronounced as the doping concentration increases. Additionally, dopants such as Mg, Ca, Sr, Sc, Y, and La can merge the defect states induce by Vo to the CBM, restoring the large bandgap and suppressing the leakage. Overall, Y dopants with the associated Vo effect are suggested to achieve higher-k and lower leakage, providing feasible strategy for further 3D DRAM devices.

Acknowledgments

This work was supported by the National Natural Science Foundation of China (Nos. 62304136, 62027818, 61874034, and 11974320) and Startup Fund for Young Faculty at SJTU (SFYF at SJTU).

References

[1] J. Kumar, S. Birla, and G. Agarwal, Materials Today: Proceedings 79, 297 (2023).

[2] J.-P. Locquet, C. Marchiori, M. Sousa, J. Fompeyrine, and J. W. Seo, Journal of Applied Physics 100 (5) (2006).

[3] W. Jeon, Journal of Materials Research 35 (7), 775 (2020).

[4] Y.-S. Lin, R. Puthenkovilakam, and J. Chang, Applied physics letters 81 (11), 2041 (2002).

[5] A. Nishiyama, in High Permittivity Gate Dielectric Materials (Springer, 2013), pp. 153.

[6] Y. Harashima, H. Koga, Z. Ni, T. Yonehara, M. Katouda, A. Notake, H. Matsui, T. Moriya, M. K. Si, and R. Hasunuma, Applied Physics Letters 122 (26) (2023).

[7] M. N. K. Alam, S. Clima, B. O'sullivan, B. Kaczer, G. Pourtois, M. Heyns, and J. Van Houdt, Journal of Applied Physics 129 (8) (2021).

[8] M. Kingsland, S. Lisenkov, S. Najmaei, and I. Ponomareva, Journal of Applied Physics 135 (5) (2024).

[9] D. Banerjee, C. Dey, R. Kumar, B. Modak, S. Hazra, S. Datta, B. Ghosh, S. Thakare, S. Jha, and D. Bhattacharyya, Physical Chemistry Chemical Physics 25 (32), 21479 (2023).

[10] S. Govindarajan, T. Boscke, P. Kirsch, M. Quevedo-Lopez, P. Sivasubramani, S. Song, R. Wallace, B. Gnade, P. Hung, and J. Price, 2007 International Symposium on VLSI Technology, Systems and Applications (VLSI-TSA), 2007.

[11] J. U. Bhanu, A. Amutha, G. R. Babu, B. Sundaravel, and P. Thangadurai, Materials Science in Semiconductor Processing 81, 7 (2018).

[12] R. Materlik, C. Künneth, and A. Kersch, Journal of Applied Physics 117 (13) (2015).

[13] K. Chae, A. C. Kummel, and K. Cho, ACS Applied Materials & Interfaces 14 (25), 29007 (2022).

[14] M. Wu, B. Cui, X. Wang, M. Yuan, Y. Wu, Y. Wen, J. Liu, T. Zhang, P. Ren, and S. Ye, Journal of Applied Physics 136 (14) (2024).

[15] K. Xiong, J. Robertson, M. Gibson, and S. Clark, Applied physics letters 87 (18) (2005).

Wafer-Scale Fabrication of Janus-MXene Films and Its Based Flexible Artificial Synapse

Xin Liu[1], Conghui Zhang[1], Yuhang Yang[1], Peisong Liu[2], Shuiren Liu[1], Lingxian Meng[1], and Fei Hui[1]*

[1]School of Materials Science and Engineering, Zhengzhou University, Zhengzhou, China
[2]Engineering Research Center for Nanomaterials (ERCN)
National & Local Joint Engineering Research Center for Applied Technology, Henan University, Zhengzhou, China
Phone: (0086)15370083707, Email: feihui@zzu.edu.cn

Abstract—MXenes, a member of the two-dimensional (2D) materials family, are of interest in the field of flexible electronics because of their unique mechanical, physical and chemical properties. However, the challenges of large-area preparation of MXene films have limited their development in the fields of wearable electronics. In this work, we used a surface modification method to prepare Janus MXene film with distinct hydrophilic and hydrophobic properties on each surface, and used phase-interface transfer method for preparing large-area and high-quality MXene films. Furthermore, we constructed the Ag/Janus-MXene/ITO/PET flexible memristors. The results show that a stable unipolar and bipolar resistive switching (RS) with low switching voltages (~0.27 V), high mechanical endurance (> 200 cycles), and long retention time (~2900 s). In addition, we have successfully simulated several typical synaptic behaviors, such as spike-amplitude-dependent Plasticity (SADP) and spike-number-dependent plasticity (SNDP), demonstrating its potential in the field of neuromorphic computing.

Keywords—*MXenes; Janus structure; flexible artificial synapses.*

I. INTRODUCTION

In the age of the Internet of Things (IoT), the parallel processing and storage of data are essential to address the challenges posed by the explosion of information. Traditional computers experience a rapid increase in power consumption as computing speed and storage capacity rise, while the individual device units in use are approaching their physical-size limits. Consequently, Moore's Law is increasingly at risk of failure [1]. The conventional von Neumann architectures, which separate the memory units from computation units, face developmental bottlenecks because of the significant power consumption and data latency associated with data transmission [2]. As a result, an increasing number of researchers are turning their attention to the emerging paradigm of in-memory computing. Memristors, which are nonlinear resistors capable of changing their resistances under applied voltages, hold significant promise for high-integration brain-inspired chips. The design and fabrication of low-power, multi-bit, non-volatile memristors are crucial for advancing processing-in-memory technology [3,4].

MXenes, a tunable family of 2D carbides and nitrides, have been recognized as the largest class of 2D materials because of its diverse compositions with abundant surface functional groups, which have been extensively utilized in memristors [5-9]. Given the rich functional groups on the surface of MXene, modifications can be made to enhance the performance of its-based memory devices [10-15]. For example, W. Sun et al. [6] demonstrated for the first time that octylphosphonic acid modified $Ti_3C_2T_x$ MXene used as

an active layer in memory devices and presented stable ternary memory behaviors. Afterwards, by engineering the surface structure of MXene, N. B. Mullani, et al. [7] and J. Huang et al. [8] constructed surface-modified MXene-based memristors with enhanced performance in terms of large I_{ON}/I_{OFF} ratio (>10^4) and low power consumption (~230 nJ), as well as high transmittance (>90%) and high mechanical endurance (>10^4 bending cycles).

In this work, we selected hexadecyltrimethoxysilane (HDTMS) as the surface modifier for the modification of MXene to create Janus-MXene materials and prepared homogeneous wafer-scale MXene films using the phase interface transfer method. Subsequently, we constructed memristors using Janus-MXene and evaluated their electrical RS characteristics, including endurance, retention, and variability. The results demonstrated that Ag/Janus-MXene/ITO flexible memristors exhibit typical volatile RS behavior when a surface modifier is present between the MXene and top electrodes, and non-volatile RS behavior with the modifier between the MXene and bottom electrodes, resulting in two different RS mechanisms. Additionally, the Janus-MXene-based memristors exhibit low switching voltage (~0.27 V), high retention time (~2900 s), and reliable endurance (over 200 cycles). Furthermore, the memristors display synaptic behaviors (such as PPF, SADP and SNDP), indicating promising potential for neuromorphic computing applications.

II. RESULTS AND DISCUSSION

$Ti_2C_3T_x$, a common member of the MXene family, was prepared through a bottom-up approach and possesses abundant surface terminals (-OH, -O-, -COOH, etc.) [16]. **Fig. 1** illustrates the etching process for MXene nanosheets and the modification process of Janus-MXene. To obtain $Ti_3C_2T_x$ nanosheets, Ti_2AlC_3 powder was mixed with LiF and HCl, and then stirred at 50 °C for 30 hours to remove aluminum atoms from the MAX phase. The mixture was subsequently subjected to repeated centrifugation with water to wash away the acid until the solution reached pH neutrality. Finally, single or few-layer MXene was obtained.

In the next step, we grafted HDTMS, a common silane coupling agent, onto the surface functional groups of MXene to transform its surface from hydrophilic to hydrophobic [16]. MXene nanosheets, toluene, and HDTMS were added into a bottle and stirred at 700 rpm while being heated in a water bath at 60 °C for 2 hours to obtain HDTMS-grafted MXene. During this process, HDTMS hydrolyzed to produce silanol, which then reacted with hydroxyl or carboxyl groups on the MXene surface, generating ester groups or hydrogen bonds that facilitate

979-8-3315-0417-5/25 $31.00 © 2025 IEEE

organic-inorganic hybridization. Janus MXene nanosheets were formed at the oil-water interfacial when HDTMS-MXene was dispersed.

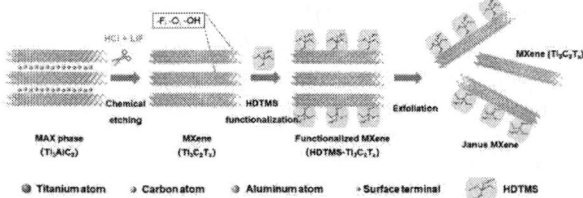

Fig. 1 Schematic representation of MXene sheets obtained from top-down etching process and Janus-MXene flakes via modification process.

To fabricate large-area, high-quality Janus-MXene films, we dispersed HDTMS-MXene into an oil-water system. Through migration and self-assembly, the HDTMS-MXene nanosheets can naturally generate a continuous, uniform MXene film at the oil-water phase interface. This film was then fished and transferred from the oil-water interface onto the target substrate to obtain the Janus-MXene film. **Fig. 2** shows the preparation process of wafer-scale Janus-MXene films.

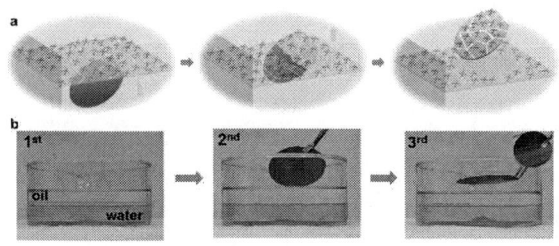

Fig. 2 (a) Schematic illustration and (b) optical images of self-assembly and wafer-scale transfer of Janus-MXene films at the oil/water two-phase interface.

Afterwards, we carried out a series of characterizations, including transmission electron microscopy (TEM), Raman spectroscopy, atomic force microscopy (AFM), among others, see **Fig. 3**. The AFM image of the Janus-MXene film demonstrates good homogeneity and with a surface roughness of ~7.0 nm (Fig. 3a). The layered structure of Janus-MXene films (~10 layers) has been verified by TEM image (Fig. 3b). Transferring the film to the surface of quartz glass, it can be found that the film obtained by the transfer method has good transparency (Fig. 3c). When Janus-MXene films are self-assembled at the interface of hexane and water, a hydrophilic substrate is required for the transfer, and the hydrophobic surface of the transferred Janus MXene films is upward; on the contrary, when Janus MXene films are self-assembled in water and chloroform, a hydrophobic substrate is required for the transfer, and the hydrophilic surface of the transferred Janus MXene films is upward.

Then, we tested the water contact angle of two different films (i.e., hydrophobic and hydrophilic top surfaces) and found that there was a significant difference in contact angle between these two Janus-MXene films, with the hydrophilic side having a contact angle of 42.9° and the hydrophobic side having a contact angle of 105.2° (Fig. 3d). From the Raman spectra, we can see that similar typical peaks (198 cm^{-1}, 302 cm^{-1}, 377 cm^{-1}, 622 cm^{-1}) were obtained by MXene films before and after modification (Fig. 3e). This indicated that the phase structure of MXene did not change after

modification. The XRD pattern further proves that the Al element in MAX was removed by LiF and HCl etching and (002) peak shifted to a smaller angle after etching and surface modifying, which indicated that the layer spacing increased (Fig. 3f).

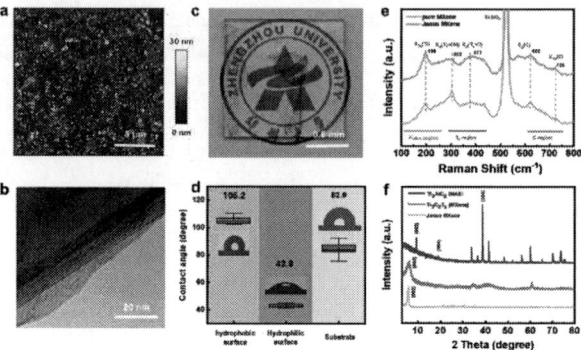

Fig. 3 Morphological and structural characterizations of Janus-MXene films. (a) AFM topographic map with a surface roughness of ~7.0 nm; (b) TEM image indicating the correct layered structure of Janus-MXene; (c) Janus-MXene transferred onto a piece of quartz glass with good transparency; (d) Contact-angle tests, (e) Raman spectra and (f) XRD patterns reveal Janus-MXene films with hydrophobic and hydrophilic surfaces after modifying.

Based on high-quality, continuous and homogeneous large-area Janus MXene films, we prepared Ag/Janus MXene/ITO memristors with the size of 100 μm×100 μm and explored its RS behaviors as shown in **Fig. 4**. All the electrical measurements were conducted by Keithley 4200 semiconductor parameter analyzer (SPA) and probe station. Fig. 4a shows the structure of the device with a rigid substrate, i.e., ITO/glass. When the hydrophobic surface of MXene faces up to top Ag electrodes, the devices show typical volatile RS behavior with low operating voltage (~0.72 V), large on/off ratio (>10^3) and reliable endurance (Fig. 4b-d). In contrast, when reversing the functional layer (i.e., hydrophilic surface faces up to Ag electrodes), the device exhibits non-volatile RS behavior with a lower set voltage (V_{set}=~0.27 V) and reset voltage (V_{reset}=~-0.31 V), large on/off ratio (>10^3), reliable endurance, and long retention time (~2900 s) (Fig. 4e-h). Herein, the RS mechanisms attributes to the formation of conductive filaments under external electric fields, and the surface-dependent RS behaviors may be attributed to the fact that the modifiers play a hindering role on the migration of silver ions, further resulting in a difference in the strength of the conductive filaments formed.

In addition, we also explored the RS behaviors on a Janus-MXene based flexible memeristors (with ITO/PET as a substrate), see Fig. 4i. The devices exhibit good mechanical robustness under in-situ bent with different bending radii (0 mm, 15 mm, 10 mm, 5 mm) (Fig. 4j). The switching ratio of the device decreased significantly after 1000 bending cycles (Fig. 4k). It is speculated that the reason for the increase in the high resistance state may be that the conductive filaments inside the device cannot completely return to the initial state after each spontaneous break; the reason for the decrease in the low resistance state is attributed to the fact that the ITO itself is not resistant to bending, and the conductivity is not as good as it used to be after bending to a certain extent. Further work need to be done to determine the causes of the device degradation and further provide strategies to improve the endurance.

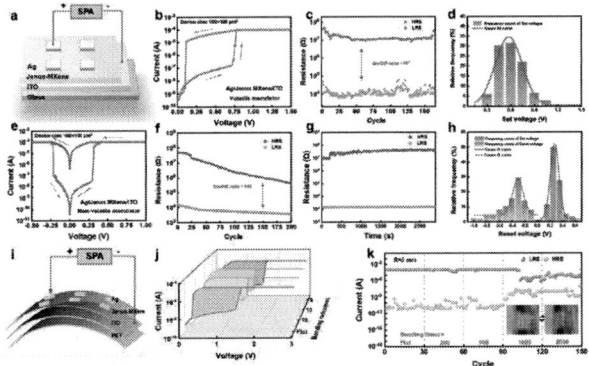

Fig. 4 Resistive switching (RS) characteristics of Ag/Janus-MXene/ITO. (a) schematic diagram of vertical Ag/Janus-MXene/ITO/Glass devices with top patterned electrodes; I-V curves indicate typical (b-d) unipolar RS and (e-h) bipolar RS, including multiple performance of (c,f) endurance, (g) retentions, (d,h) V_{set} and V_{reset} distribution; (i) schematic diagram of flexible Ag/Janus-MXene/ITO/PET devices, and (j,k) its RS behaviors under different bending radii and times, respectively.

Similar to the working principle of biological synapses, postsynaptic membrane potential changes can be simulated by changing the resistance of memristors (Fig. 5a). Based on Ag/Janus-MXene/ITO devices, we have preliminary simulated long-term plasticity (LTP), including SNDP and SADP behaviors (Fig. 5b,c), by varying the numbers and amplitudes of the pulses to regulate synaptic weights. All of the above results demonstrate the great potential of the Janus MXene memristor for neuromorphic computing.

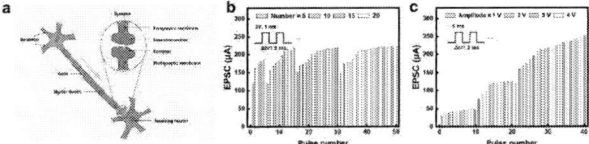

Fig. 5 Simulating the LTP behaviors of flexible artificial synapses. (a) schematic structure of a biological synapse; typical dynamic behaviors of (b) SNDP and (c) SADP.

III. CONCLUSIONS

In conclusion, we proposed a surface modification and phase interface transfer method to obtain wafer-scale Janus-MXene films, and constructed Janus-MXene based memristors. The Ag/Janus-MXene/ITO memory devices exhibited surface modification- (i.e., hydrophobic- or hydrophilic-) dependent RS behaviors with low set and reset voltages (<0.3 V), high I_{ON}/I_{OFF} ratio and high mechanical endurance. This work provides a facile and versatile methodology to prepare large-area 2D materials and presents its promising potential in flexible memristors and neuromorphic computing as an artifical synapses, which is also facilitating the development of wearable electronics.

ACKNOWLEDGMENT

This work was supported by the National Natural Science Foundation of China (grant no. 62204224); the China Postdoctoral Science Foundation (grant no. 2023M743149,

2024T170839); Provincial international cooperation project (grant no. 242102521025); Innovation and Entrepreneurship Training Program of Zhengzhou University (grant no. 2023CXCY041).

REFERENCES

[1] J. V. Neumann. "The Principles of Large-Scale Computing Machines," IEEE Ann, vol. 10, pp. 243-256. Oct.-Dec. 1988.

[2] D. S. Jeong, I. Kim, M. Zieglerb and H. Kohlstedt, "Towards artificial neurons and synapses: a materials point of view," RSC Adv., vol. 3, pp. 3169–3183, November 2013.

[3] Y. Zhu, Y. Wang, X. Pang, Y. Jiang, X. Liu, Q. Li, Z. Wang, C. Liu, W. Hu P. Zhou, "Non-volatile 2D MoS2/black phosphorus heterojunction photodiodes in the near- to mid-infrared region," Nat. Commu., vol. 15, pp. 1-10, July 2024.

[4] H. Zhou, S. Li, K. Ang, Y. Zhang, "Recent Advances in In-Memory Computing: Exploring Memristor and Memtransistor Arrays with 2D Materials," Nano-Micro Lett., vol. 16, pp.121, February 2024.

[5] B. Zeng; X. Zhang; C. Gao; Y. Zou; X. Yu; Q. Yang, T. Guo, H. Chen, "MXene-Based Memristor for Artificial Optoelectronic Neuron," IEEE Trans Electron Devices, vol. 70, pp. 1359-1362, 2023.

[6] W. Sun, Y.n Zhao, X. Cheng, J. He, J. Lu, "Surface Functionalization of Single-Layered Ti3C2Tx MXene and Its Application in Multilevel Resistive Memory," ACS Appl. Mater. Interfaces, vol. 8, pp. 9865-9871, February 2020.

[7] N. B. Mullani, D. D. Kumbhar, D. Lee, M. J. Kwon, S. Cho, N. Oh, E. Kim, T. D. Dongale, S. Y. Nam, J. H. Park, "Surface Modification of a Titanium Carbide MXene Memristor to Enhance Memory Window and Low-Power Operation," Adv. Funct. Mater., vol.33, pp. 2300343, June 2023.

[8] J. Huang, S. Yang, X. Tang, L. Yang, W. Chen, Z. Chen, X. Li, Z. Zeng, Z. Tang, X. Gui, "Flexible, Transparent, and Wafer-Scale Artificial Synapse Array Based on TiOx/Ti3C2Tx Film for Neuromorphic Computing," Adv. Mater., vol. 35, pp. 2303737, June, 2023.

[9] X. Feng, J. Huang, J. Ning, D. Wang, J. Zhang, Y. Hao, "A novel nonvolatile memory device based on oxidized Ti3C2Tx MXene for neurocomputing application," Carbon, Vol. 205, pp. 365-372, March 2023.

[10] E.C. Ahn, H.S.P. Wong, E. Pop, "Carbon nanomaterials for non-volatile memories," Nat. Rev. Mater., vol. 3, 18009, March 2018.

[11] B. Lyu, M. Kim, H. Jing, J. Kang, C. Qian, S. Lee, J. H. Cho, "Large-Area MXene Electrode Array for Flexible Electronics," ACS Nano, vol. 13, pp. 11392–11400, September 2019.

[12] K. Wang, J. Chen, X Yan, "MXene Ti3C2 memristor for neuromorphic behavior and decimal arithmetic operation applications," Nano Energy, vol. 79, pp. 105453, January 2021.

[13] H. Riazi, M. Anayee, K. Hantanasirakul, A. A. Shamsabadi, B. Anasori, Y. Gogotsi, M. Soroush, "Surface Modification of a MXene by an Aminosilane Coupling Agent," Adv. Mater. Interfaces, vol. 7, pp. 1902008, February, 2020.

[14] C. Hu, Z. Wei, L. Li, G. Shen, "Strategy Toward Semiconducting Ti3C2Tx-MXene: Phenylsulfonic Acid Groups Modified Ti3C2Tx as Photosensitive Material for Flexible Visual Sensory-Neuromorphic System," Adv. Funct. Mater., vol.33, pp. 2302188, May,2023

[15] S. Ling, C. Zhang, C. Ma, Y. Li, Q. Zhang, "Emerging MXene-Based Memristors for In-Memory, Neuromorphic Computing, and Logic Operation," Adv. Funct. Mater. vol. 33, pp.2208320, November 2022.

[16] F. Hui, C. Zhang, H. Yu, T. Han, J. Weber, Y. Shen, Y. Xiao, X. Li, Z. Zhang, P. Liu, "Self-Assembly of Janus Graphene Oxide via Chemical Breakdown for Scalable High-Performance Memristors," Adv. Func. Mater., vol. 33, pp. 2302073, April 2023.

Improved Breakdown Voltage and Leakage Current in β-Ga$_2$O$_3$ Schottky Barrier Diode Realized by N Ion-implantation Edge Termination

Hao Zhang[1], Xinlong Zhou[1], Yinchi Liu [1], Jining Yang[1], Handong Zhu[1], Xun Lu[1], Shiyu Li[1], Yeye Guo[1], Chunlei Wu[1], Xuning Zhang[1], Zhenxin Wang[1], Yaoyi Wang[1], Ruli Zeng[1], Yiwen Yu[1], Wenjun Liu[1,2, *]

[1] School of Microelectronics, Fudan University Shanghai 200433, China, [2]Zhangjiang Fudan International Innovation Center, Fudan University, Shanghai 201203, China
*E-mail: wjliu@fudan.edu.cn

Abstract

In this work, we have demonstrated high performance β-Ga$_2$O$_3$ Schottky barrier diodes (SBDs) with superior power figure-of-merit (PFOM) and breakdown voltage (V$_{br}$) by adopting the N ion-implantation edge termination (NIET) structure. It is observed that the introduction of the NIET structure significantly reduces the reverse leakage current, in the meanwhile, it can maintain the decent forward characteristic. The device with the N ion implantation at the edge of the Schottky electrode exhibited a V$_{br}$ of 1.65 kV, a specific on-resistance (R$_{on,sp}$) of 5.34 mΩ·cm^2, a PFOM of 0.509 GW/cm^2, and a reverse current density of 5.5×10^{-6} A/cm^2 @ −1000 V. These results show the great potential for the β-Ga$_2$O$_3$ SBDs with NIET structure in power electronic applications.

Keywords: β-Ga$_2$O$_3$, breakdown voltage, N ion implantation, Schottky barrier diode;

Introduction

β-Ga$_2$O$_3$, with its ultra-wide bandgap of 4.9 eV and high breakdown electric field of 8 MV/cm, has emerged as one of leading candidates for next-generation power electronics [1]. The material's unique properties, such as its ultra-wide bandgap, contribute to its high efficiency and thermal stability, making it particularly suitable for high-power devices. The advent of large-scale melt-growth techniques for β-Ga$_2$O$_3$ wafers has significantly reduced production costs, advancing widespread implementation across multiple electronic fields [2]. β-Ga$_2$O$_3$ has led to the development of high-performance Schottky barrier diodes (SBDs) and metal-oxide-semiconductor field-effect transistors (MOSFETs) [3,4]. These devices are crucial in power electronics due to their ability to handle high voltages and currents [5]. However, the practical use of β-Ga$_2$O$_3$ devices often falls short of their theoretical prediction. However, this shortfall is primarily due to material defects and the concentration of electric fields at the edges of Schottky electrodes, which limits the breakdown electric field. To address these limitations, various approaches have been explored. Field plate structures, trench structures, and different termination techniques [6-8] have been proposed to mitigate electric field concentration

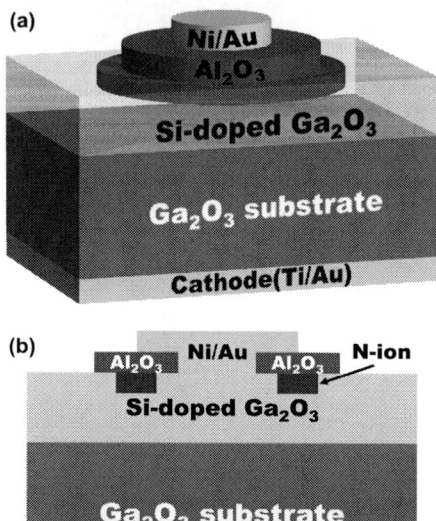

Fig. 1 (a) 3-D and (b) cross-sectional schematic of the fabricated Schottky barrier diode.

and improve device performance. Among these, ion implantation stands out as a highly effective method [9,10]. It creates high-resistance regions within the β-Ga$_2$O$_3$ material, thereby enhancing the V$_{br}$ and significantly reducing leakage current.

Here, we have demonstrated high-performance β-Ga$_2$O$_3$ Schottky barrier diodes featuring nitrogen ion-implantation edge termination structure. The device with ion implantation at the Schottky electrode edge significantly reduced the reverse leakage current from 4.1×10^{-2} A/cm^2 to 5.5×10^{-6} A/cm^2 @ -1000 V. The device exhibited a specific on-resistance of 5.34 mΩ·cm^2 and a breakdown voltage of 1.65kV.

Experimental Details

Fig. 1 depicts the cross-sectional schematic of our device. The (001) Si-doped β-Ga$_2$O$_3$ epitaxial layer with a thickness and doping concentration (N$_D$) of 10 um / 3.6×10^{16} cm^{-3} was grown by halide vapor phase epitaxy (HVPE) on the Sn-doped conductive substrate, which the thickness and doping concentration are 587 um and 5.1×10^{18} cm^{-3}. The device

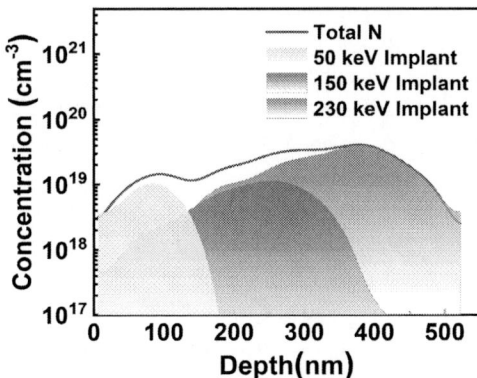

Fig. 2 Simulated N depth profile after implantation.

fabrication began with the N ion-implantation edge termination (NIET) being patterned, followed by N-ion implantation with various energy of 50, 150, and 230 keV and targeted dose of 3.5×10^{13} cm^{-2}. The doping profile, simulated using the SRIM software, displayed a mountain-like distribution extending approximately 500 nm, as illustrated in Figure 2. Then, thermal annealing was executed at 1100 ℃ for 30 minute in N$_2$ to recover the implantation damage and activate the implanted N atoms. Subsequently, The back Ohmic contact is formed by EBE growth of a Ti/Au (30/100 nm) stack, followed by annealing at 470℃ in N$_2$ atmosphere for 1 minute. Then the device was sent into the EBE chamber to deposit 100 nm of aluminum oxide as the Schottky electrode field plate. Finally, Schottky anode electrodes were prepared through the evaporation of Ni/Au (50 nm/100 nm) stack. The forward and reverse current–voltage characteristics were measured at room temperature using the Keysight B1500A and B1505A semiconductor device analyzers.

Results and Discussion

Fig. 3 (a) plots the linear forward J–V characteristics of the β-Ga$_2$O$_3$ SBDs with and without NIET. Both devices can reach the order of magnitude of 10^2 A/cm^2 @ 2V. The specific on-resistance of the SBDs with the NIET structure and without NIET structure is 5.34 mΩ·cm^2 and 4.05 mΩ·cm^2, respectively. Surface N implantation has little influence on the forward characteristics. Considering the SBDs to be turned on when the current density reaches 1 A/cm^2, the turn-on voltage of the SBDs with the NIET structure is 0.95V, while that of the control SBDs is 0.89V, respectively. Herein, the turn-on voltage of the two devices differs by only 0.06V.

To examine the impacts of the NIET structure on the reverse characteristics of Ga$_2$O$_3$ SBDs, the Weibull distributions of the breakdown voltage and leakage current are presented in Fig. 4. The measurement is carried out under the Fluorinert solution to avoid the premature breakdown caused by the air and the V$_{br}$ is extracted at the current density

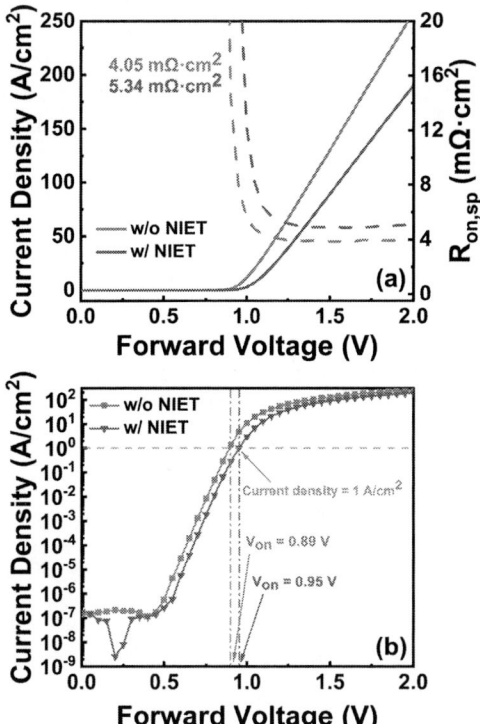

Fig. 3 (a) The forward J–V characteristics and R$_{on,sp}$. (b) semi-log scale J–V characteristics and turn-on voltage of SBDs with and without NIET structure.

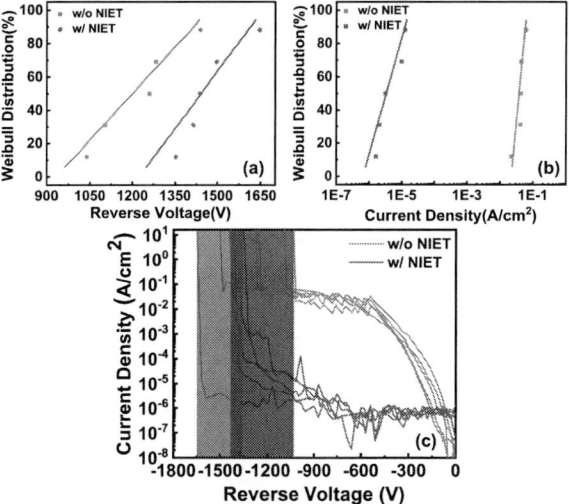

Fig. 4 The Weibull distributions of the (a) breakdown voltage and (b) leakage current for Ga$_2$O$_3$ SBDs with and without the NIET structure. (c) The reverse characteristics of Ga$_2$O$_3$ SBDs with and without the NIET structure.

of 1 A/cm^2. By introducing the NIET structure at the edge of the Schottky electrode, the average breakdown voltage increases from 1.2 kV to 1.65 kV. Additionally, it is notable that the reverse leakage current decreases from 4.1×10^{-2} A/cm^2 to 5.5×10^{-6} A/cm^2 @-1000V, a reduction of approximately four orders of magnitude. Fig.4(c) shows the reverse characteristics of two groups of devices, with each group consisting of five devices. It is seen that the performance optimization by the NIET structure is uniform and reproducible. This provides a new approach to fully leverage the excellent material properties of Ga_2O_3 in the future.

Conclusion

In summary, we have investigated the impact of the NIET structure on the performance of β-Ga_2O_3 SBDs. The NIET-structured device achieved a breakdown voltage of 1.65 kV, a specific on-resistance of 5.34 mΩ·cm², a power figure of merit of 0.509 GW/cm², and a reverse leakage current density of 5.5×10^{-6} A/cm² at −1000 V. These findings suggest that the NIET structure significantly enhances the performance of β-Ga_2O_3 power electronic devices.

Acknowledgments

The authors would like to thank Xing Lu and Yuxin Deng, Sun Yat-Sen University, Guangzhou, China, for constructive advice and conducting the ion implantation in this work. This work was supported by Shanghai Municipal Science and Technology Commission under Grant 24DP1500105, 23511102300 and National Key Research and Development Program of China under Grant 2021YFB3202500.

References

[1] S.J. Pearton, J. Yang, Cary, P. H., IV; Ren, F.; Kim, J.; Tadjer, M.J.; Mastro, M.A. "A Review of Ga_2O_3 Materials, Processing, and Devices." Appl. Phys. Rev., 5, 011301 (**2018**).

[2] A. Kuramata, K. Koshi, S. Watanabe, Y. Yamaoka, T. Masui, S Yamakoshi, "High-Quality β- Ga_2O_3 Single Crystals Grown by Edge-Defined Film-Fed Growth." Jpn. J. Appl. Phys. 55, 1202A2,(2016).

[3] Sasaki, K., Wakimoto, D., Thieu, Q.T., Koishikawa, Y., Kuramata, A., Higashiwaki, M., Yamakoshi, S. "First Demonstration of Ga_2O_3 Trench MOS-Type Schottky Barrier Diodes." IEEE Electron Device Lett. 38, 783–785,(2017).

[4] Y. Lv, , H. Liu, X. Zhou, Y. Wang, X. Song, Y. Cai, Q. Yan, C. Wang, S. Liang, J. Zhang, et al. "Lateral β- Ga_2O_3 MOSFETs with High Power Figure of Merit of 277 MW/cm2." IEEE Electron Device Lett. 41, 537–540, (2020).

[5] F. Gucmann, P. Nadazdy, K. Husekova, E. Dobrocka, J. Priesol, F. Egyenes, A. Satka, A. Rosova, M. Tapajna, "Thermal Stability of Rhombohedral α- and Monoclinic β- Ga_2O_3 Grown on Sapphire by Liquid-Injection MOCVD." Mater. Sci. Semicond. Process. 156, 107289,(2023).

[6] S. Roy, A. Bhattacharyya, C. Peterson, S. Krishnamoorthy. "2.1 kV (001) β- Ga_2O_3 Vertical Schottky Barrier Diode with High-k Oxide Field Plate." Appl. Phys. Lett. 122, 152101,(2023).

[7] W. Li, K. Nomoto, Z. Hu, D. Jena, H.G Xing. "Field-Plated Ga_2O_3 Trench Schottky Barrier DiodesWith a BV2/Ron,sp of up to 0.95 GW/cm2." IEEE Electron Device Lett. 41, 107–110,(2020).

[8] Y. Wang, Y. Lv, S. Long, X. Zhou, X. Song, S. Liang, T. Han, X. Tan, Z. Feng, S. Cai, et al. "High-Voltage (−201) β- Ga_2O_3 Vertical Schottky Barrier Diode with Thermally-Oxidized Termination." IEEE Electron Device Lett. 41, 131–134,(2020).

[9] H. Zhou,; Q. Yan, J. Zhang, Y. Lv, Z. Liu, Y. Zhang, K. Dang, P. Dong, Z. Feng, Q. Feng, et al. "High-Performance Vertical β-Ga_2O_3 Schottky Barrier Diode with Implanted Edge Termination." IEEE Electron Device Lett. 40, 1788–1791, (2019).

[10] C.H. Lin, Y. Yuda, M.H. Wong, M. Sato, N. Takekawa, K. Konishi, T. Watahiki, M. Yamamuka, H. Murakami, Y. Kumagai, et al. "Vertical Ga_2O_3 Schottky Barrier Diodes with Guard Ring Formed by Nitrogen-Ion Implantation." IEEE Electron Device Lett. 40, 1487–1490,(2019).

Miniature Optical Fiber Fabry–Pérot Interferometric Acoustic Sensor with 3D Micro-Printed Ortho-Planar Springs

Shangming Liu[1], Peng Wang[1], Taige Li[1], A. Ping Zhang[1,2]

[1] Photonics Research Institute, Department of Electrical and Electronic Engineering, The Hong Kong Polytechnic University, Kowloon, Hong Kong SAR, China.

[2] State Key Laboratory of Ultraprecision Machining Technology, The Hong Kong Polytechnic University, Kowloon, Hong Kong SAR, China.

Abstract

Optical fiber acoustic sensors have attracted remarkable attention in the detection of weak acoustic signals. To achieve both small size and high sensitivity, we present a sub-millimeter-scale optical fiber acoustic sensor based on a 3D Fabry–Pérot (FP) interferometer with ortho-planar springs. The design of ortho-planar springs can not only effectively enhance mechanical sensitivity but also reduce the total size of the sensor head. Experimental results demonstrate that its noise equivalent pressure of the fabricated sensor is 49.32 µPa/√Hz around 2 kHz and the mechanical quality factor of its fundamental vibration resonance is around 31.

Keywords: Optical fiber acoustic sensor, Ortho-planar springs, Fabry-Pérot interferometer, 3D micro-printing

Introduction

Optical fiber acoustic sensors have attracted remarkable attention due to many promising applications, such as industrial monitoring equipment [1,2], photoacoustic gas detection [3,4], and bioimaging [5-7]. Compared to conventional acoustic sensors, optical fiber acoustic sensors have advantages of compact size, electromagnetic immunity, and remote sensing. Among the various types of optical fiber acoustic sensors, optical fiber-based Fabry-Pérot interferometer (FPI) based acoustic sensors are broadly used in acoustic sensing due to their simple structure, high sensitivity, and stability. One may fabricate either a diagraph or a cantilever beam on the end-face of optical fiber as the reflector for interferometric acoustic sensing [8, 9]. However, the sensor heads of previously reported high-sensitivity optical fiber acoustic sensors have dimensions on the centimeter scale [10,11], which limits their wide use in many space-restricted applications. One of promising solutions is to directly 3D print micrometer-scale FPI on the end-face of optical fiber for acoustic sensing [12]. However, the sensitivity of optical fiber-tip acoustic sensor is relatively low, which may limit their use in many applications for detection of very weak acoustic waves. In this work, we present an optical fiber acoustic sensor based on a Fabry–Pérot interferometric that is 3D micro-printed on a fiber-optic ferrule. The design of ortho-planar springs is introduced to enhance acoustic sensitivity of the sensor. An own-built 3D micro-printing technology was applied to fabricate this sub-millimeter-scale sensor head of optical fiber acoustic sensor.

Design and Fabrication

Fig. 1(a) illustrates the schematic design of the optical fiber Fabry–Pérot interferometric acoustic sensor. The sensor consists of four parts: a three-petal base, three pillars, ortho-planar spring microbeams, and a thin-film reflector. The three-petal base is designed to increase the cavity length of FPI, and the thin-film reflector is suspended with three ortho-planar spring microbeams and pillars attached upon the three-petal base on the end-face of a ceramic fiber-optic ferrule.

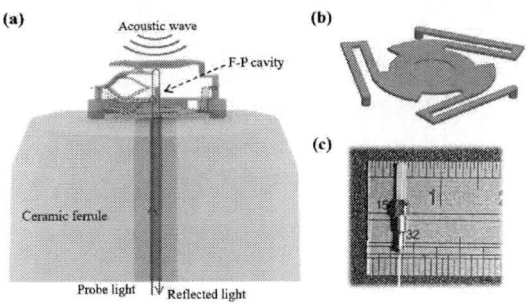

Fig. 1. (a) Schematic design of the optical fiber FPI acoustic sensor with ortho-planar springs. (b) 3D model of the thin-film reflector suspended with ortho-planar springs, (c) Photo of the sensor head fabricated using a fiber-optic ferrule.

Traditional suspended structures are typically suspended with cantilever, which suffer from unwanted tilt under vertical acoustic pressure. To overcome this issue, the supporting microbeams of our sensor are designed to be ortho-planar springs. Due to its cantilever nature, this design allows the central reflector to move up and down relative to the fiber-optic ferrule end-face. Moreover, it can avoid the twisting and flipping of the central thin-film reflector, but enhance its sensitivity to vertical acoustic pressure.

An optical 3D micro-printing technology is applied to directly print FPI sensor head on the end-face of a fiber-optic ferrule. **Fig.** 2(a) shows the schematic diagram of the in-house optical 3D micro-printing setup [14,15]. The material

---

979-8-3315-0417-5/25 $31.00 © 2025 IEEE

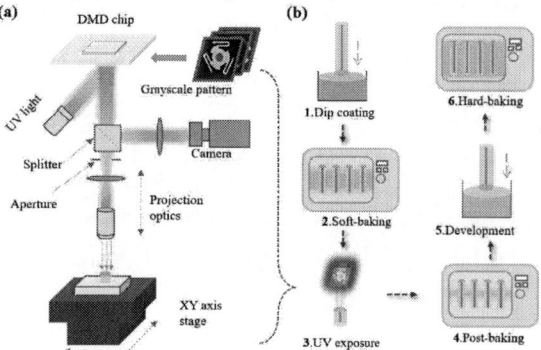

Fig. 2. (a) Schematic diagram of the optical 3D micro-printing setup. (b) Flow chart of the optical 3D micro-printing processes.

used for the fabrication of FPI sensor head is SU-8, whose elastic modulus and Poisson ratio are 3.5 GPa and 0.22, respectively. **Fig**. 2(b) shows the flow chart of 3D micro-printing processes. A cleaved optical fiber was inserted into a fiber-optic ferrule and then deposited with SU-8 on the end-face of the fiber-optic ferrule via dip coating. The sample was soft-baked in an oven at 95 °C for 15 minutes to evaporate solvent. Then dynamic UV exposure was performed using grayscale patterns. After optical exposure, the sample was post-baked at 65 °C for 60 minutes and developed with PGMEA. Finally, the sample was hard-baked in an oven at 105 °C for 15 minutes to enhance the mechanical strength of the fabricated 3D microstructure. The processes were repeated twice to fabricate the bottom layer, i.e., the three-petal base and the upper layer, including the pillars, ortho-planar spring microbeams and the thin-film reflector.

Results and Discussion

The experimental setup for acoustic response testing is shown on **Fig**. 3. A tunable laser source was used to generate a probe laser beam for signal demodulation. Its wavelength was adjusted to the quadrature point to maximize optical sensitivity. A commercial speaker was employed to generate sinusoidal sound waves for acoustic response testing. The light reflected from the sensor head was redirected to a photodiode (PD) via a circulator, where the light signal was

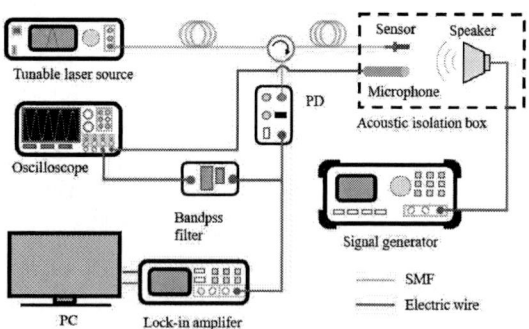

Fig. 3. Experimental setup for testing the responses of the fabricated optical fiber FPI acoustic sensor.

converted to an electrical signal and collected by a lock-in-amplifier and an oscilloscope through a bandpass filter.

Fig. 4 (a) shows the measured time-domain signal versus different applied acoustic pressures at the frequency of 2 kHz. The acoustic sensor exhibited smooth sinusoidal waveforms whose amplitude varied according to the input acoustic pressures.

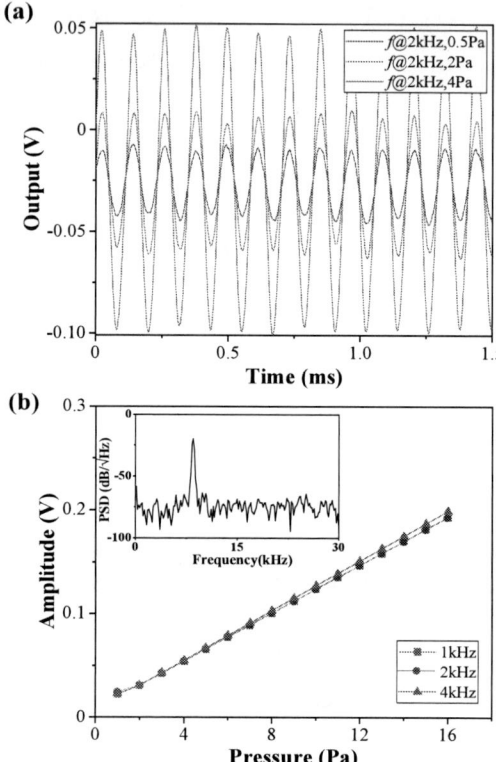

Fig. 4. (a) Measured responses of the optical fiber acoustic sensor to different acoustic pressures (i.e., 0.5 Pa, 2 Pa, 4 Pa) at the frequency of 2 kHz. (b) Measured dependances of the output magnitude on the input acoustic pressures at the frequencies of 1 kHz (red), 2 kHz (black), and 4 kHz (blue), respectively. The inset is the measured power spectral density spectrum under 0.5-Pa acoustic pressure at the frequency of 2 kHz.

Fig. 4 (b) shows the measured dependances of the fabricated optical fiber acoustic sensor on input acoustic pressures below the resonant frequency. The sensor exhibited an almost identical response for the input acoustic waves at 1 kHz and 2 kHz, while its sensitivity at 4 kHz became slightly higher due to relatively closer to resonant frequency, i.e., 8.47 kHz. The SNR of the fabricated acoustic sensor is measured when the applied acoustic pressure is 0.5 Pa at the frequency of 2 kHz. As shown in the inset, it is about 57.9 dB. As the resolution bandwidth (RBW) in the measurement is 166.67 Hz, the noise equivalent pressure (NEP) of the fabricated acoustic sensor is calculated to be 49.32 μPa/√Hz.

979-8-3315-0417-5/25 $31.00 © 2025 IEEE

The measured frequency response of the fabricated optical fiber acoustic sensor is shown in **Fig.** 5. One can see that the peak frequency of the fundamental resonant mode of the fabricated acoustic sensor is around 8.47 kHz. Its 3-dB bandwidth is 0.28 kHz, from which the mechanical quality factor is deduced to be around 31. A comparison of the measured responses of the fabricated acoustic sensor to different acoustic pressures between the resonant frequency and non-resonant frequency is shown in the inset of **Fig.** 5. One can see that the measured sensitivities of the acoustic sensor are 101.3 mV/Pa and 12.2 mV/Pa at the resonant frequency of 8.47 kHz and the non-resonant frequency of 2 kHz, respectively. The measured resonant acoustic sensitivity is 9 times higher than the sensitivity at non-resonant frequency.

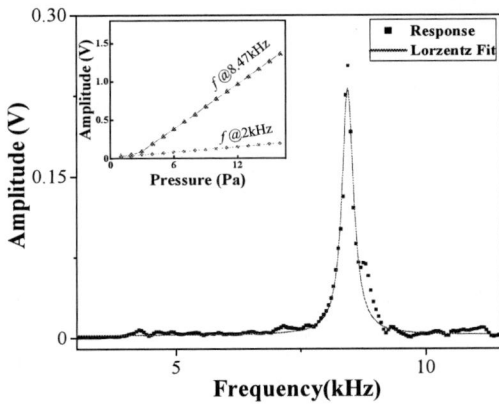

Fig. 5. Measured measures frequency response of the fabricated acoustic sensor in the range from 2 kHz to 20 kHz. The inset shows a comparison of the optical fiber acoustic sensor's responses to acoustic pressures at the resonant frequency of 8.47 kHz (blue) and the non-resonant frequency of 2 kHz (red), respectively.

Conclusion

We have presented an optical fiber acoustic sensor based on a 3D micro-printed Fabry–Pérot interferometer with ortho-planar springs. This optical fiber acoustic sensor has a sub-millimeter size, and its noise equivalent pressure can reach 49.32 μPa/√Hz around 2 kHz. Therefore, such a small-size high-sensitivity optical fiber acoustic sensor is promising in many applications such as photoacoustic gas detection and bioimaging.

Acknowledgments

The work was fully supported by a grant from the Research Grants Council of the Hong Kong SAR, China (Grant No.: 15213522).

References

[1] Fischer, Balthasar. "Optical Microphone Hears Ultrasound." *Nature Photonics*, vol. 10, no. 6, pp. 356–358, 2016.

[2] Bilaniuk, Nykolai. "Optical Microphone Transduction Techniques." *Applied Acoustics*, vol. 50, no. 1, pp. 35–63, 1997.

[3] Cao, Yingchun, et al. "Miniature Fiber-Tip Photoacoustic Spectrometer for Trace Gas Detection." *Optics Letters*, vol. 38, no. 4, pp. 434–434, 2013.

[4] Chen. Ke "Ultra-High Sensitive Fiber-Optic Fabry-Perot Cantilever Enhanced Resonant Photoacoustic Spectroscopy." *Sensors and Actuators B-Chemical*, vol. 268, pp. 205–209, 2018.

[5] Guggenheim, James A., et al. "Ultrasensitive Plano-Concave Optical Microresonators for Ultrasound Sensing." *Nature Photonics*, vol. 11, no. 11, pp. 714–719, 2017.

[6] Jathoul, Amit P., et al. "Deep in Vivo Photoacoustic Imaging of Mammalian Tissues Using a Tyrosinase-Based Genetic Reporter." *Nature Photonics*, vol. 9, no. 4, pp. 239–246, 2015.

[7] Hajireza, Parsin, et al. "Glancing Angle Deposited Nanostructured Film Fabry-Perot Etalons for Optical Detection of Ultrasound." *Optics Express*, vol. 21, no. 5, p. 6391, 2013.

[8] Ni, Wenjun, et al. "Ultrathin Graphene Diaphragm-Based Extrinsic Fabry-Perot Interferometer for Ultra-Wideband Fiber Optic Acoustic Sensing." *Optics Express*, vol. 26, no. 16, p. 20758, 2018.

[9] Iannuzzi, D, et al. "Fibre-Top Cantilevers: Design, Fabrication and Applications." *Measurement Science and Technology*, vol. 18, no. 10, pp. 3247–3252, 2007.

[10] Guo, Min, et al. "Ultrahigh Sensitivity Fiber-Optic Fabry–Perot Interferometric Acoustic Sensor Based on Silicon Cantilever." *IEEE Transactions on Instrumentation and Measurement*, pp. 1–8, 2021.

[11] Guo, Min, et al. "High-Sensitivity Silicon Cantilever-Enhanced Photoacoustic Spectroscopy Analyzer with Low Gas Consumption." *Analytical Chemistry*, vol. 94, no. 2, pp. 1151–1157, 2022.

[12] Yao, Mian, et al. "Ultracompact Optical Fiber Acoustic Sensors Based on a Fiber-Top Spirally-Suspended Optomechanical Microresonator." *Optics Letters*, vol. 45, no. 13, p. 3516, 2020.

[13] Parise, John J., et al. "Ortho-Planar Linear-Motion Springs." *Mechanism and Machine Theory*, vol. 36, pp. 1281–1299, 2001.

[14] Li, Taige, et al. "Miniature Optical Fiber Photoacoustic Spectroscopy Gas Sensor Based on a 3D Micro-Printed Planar-Spiral Spring Optomechanical Resonator." *Photoacoustics*, vol. 40, no. 11-12 p. 100657, 2024.

[15] Wu, Jushuai, et al. "Optical Fiber-Tip Fabry–Pérot Interferometric Pressure Sensor Based on an in Situ μ-Printed Air Cavity." *Journal of Lightwave Technology*, vol. 36, no. 17, pp. 3618–3623, 2018.

Diode Microheaters for Scalable Actuation in Micro-Transfer Printing

Jiajun Zhang[1], Qinhua Guo[1], Xiwen Liu[2], Yunda Wang[1]*

[1]Smart Manufacturing Thrust, The Hong Kong University of Science and Technology (Guangzhou)
Guangzhou, China

[2]Microelectronics Thrust, The Hong Kong University of Science and Technology (Guangzhou)
Guangzhou, China

*Email: ydwang@ust.hk

Abstract

Micro-transfer printing (μTP) enables precise integration of microscale devices on a single substrate. Selective μTP can be achieved using thermo-responsive shape memory polymer (SMP) stamps with resistive heater arrays as actuators. However, resistive heaters face scalability challenges. This work demonstrates 50 μm × 50 μm diode microheaters with ZnO/PEDOT: PSS Schottky contacts, achieving up to 47.8°C through a simple fabrication process, potentially suitable for scalable SMP-based μTP application.

Keywords: Micro-transfer printing, Diode microheaters and Schottky contacts

Introduction

Micro-transfer printing (μTP) is an advanced technique for transferring microscale devices or materials from a donor to a receiving substrate with high precision. It enables the heterogeneous integration of diverse materials onto a single substrate, making it particularly effective for wafer-level microsystems integration. μTP has been widely applied in silicon photonics [1], flexible electronics [2], and large-scale microLED transfer [3]. An effective approach in μTP utilizes thermo-responsive shape memory polymer (SMP) stamps [4][5]. By locally controlling the temperature of the SMP, its adhesion can be selectively adjusted, enabling precise pick-up and release of microdevices. The temperature range for tuning the adhesive properties of SMP has been reported to be between 44 °C and 90 °C [4-6].

Previous work has demonstrated that selective micro-transfer printing can be achieved using arrays of individually controlled resistive microheaters [6][7]. However, approaches based on resistive heaters encounter challenges in scalability. As shown in Fig. 1a, an individually controlled (m × n) resistive microheater array requires each heater to have a positive terminal and a shared negative terminal, resulting in (m × n + 1) wires. As the array size increases, the wiring complexity grows rapidly, making layout management challenging and limiting the scalability. The crossbar addressing scheme, as illustrated in Fig. 1b, has been used in microheater array to overcome the wiring challenge [8]. This approach reduces the wiring requirements to (m + n) for an (m × n) matrix, simplifying the layout. However, due to the lack of rectifying properties in resistive microheaters, this scheme results in current leakage. When power is applied to a specific heater, unintended current flows through neighboring heaters, leading to thermal crosstalk and preventing precise localized heating. For example, in a (5 × 5) heater array, simulations using Multisim software show that the central microheater receives only 35.97% of the total output power. In a (45 × 45) matrix, this drops to 4.39%. A promising solution to the limitations of resistive microheaters is the use of diode heaters, as illustrated in Fig. 1c and 1d. The diode-based design, shown in Fig. 1c, employs a crossbar scheme with diodes at each intersection, enabling individual control of microheaters while eliminating current leakage and thermal crosstalk. Fig. 1d shows a (3 × 3) array example, demonstrating how diode rectification directs current only to the selected microheater, isolating adjacent heaters. The equivalent circuit clarifies current flow direction.

In this work, we developed a type of 50 μm × 50 μm diode microheater based on zinc oxide (ZnO)/poly(3,4-ethylene-dioxythiophene)-poly(styrenesulfonate) (PEDOT:PSS) Schottky contacts, reaching temperatures up to 47.8 °C. These diode heaters are intended for heating SMP with phase transition temperatures below 45 °C, providing scalability for precise actuation in μTP. We are fabricated through a simple process with a maximum processing temperature below 100 °C, enabling integration on flexible substrates.

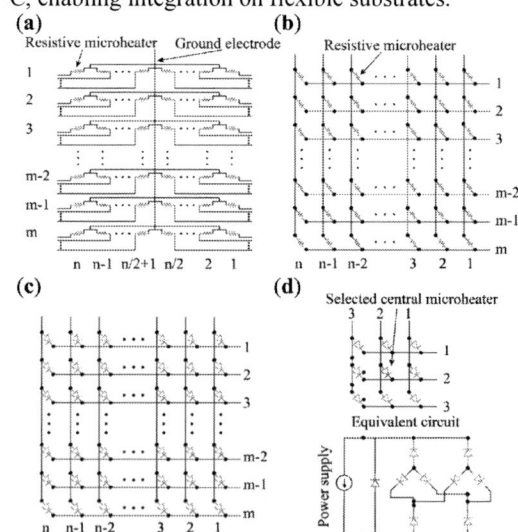

Fig. 1: Resistive microheaters array v.s. diode microheater array (a) Resistive microheater array with individual control wiring (m × n + 1) wires. (b) Crossbar addressing scheme for resistive microheater array (m + n) wires. (c) Crossbar scheme for diode microheater array (d) Diode microheater array example (3 × 3) with equivalent circuit demonstrating current direction.

979-8-3315-0417-5/25 $31.00 © 2025 IEEE

Design and Fabrication

A. Design

Fig. 2a and 2b show the schematic cross-sectional and the microscope image of a fabricated diode microheater device, respectively. The diode's rectifying function is achieved by the Schottky contact formed between ZnO and the high work-function conductive polymer PEDOT:PSS. It has been reported that spin-coating PEDOT on a ZnO (0001) single crystal substrate can form a Schottky junction, achieving a current density of nearly 80 mA/cm² at 2V [9]. Indium tin oxide (ITO) is used as the bottom electrode to form an ohmic contact with ZnO, while Au is used as the top electrode for its compatible work function with PEDOT:PSS. PEDOT:PSS also acts as a barrier, preventing Au from diffusing into ZnO. The substrate is glass to facilitate observation during micro-transfer printing.

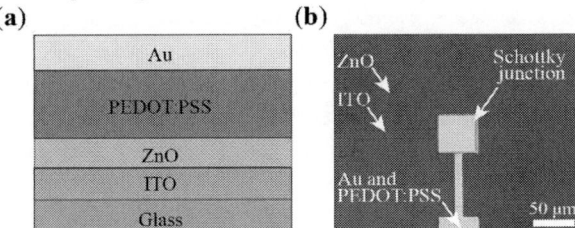

Fig. 2: Diode microheater structure and optical image (a) Cross-sectional schematic of the microheater. (b) Optical microscope image of a diode microheater (scale bar: 50 μm).

B. Fabrication Process

Commercially available ITO-coated glass wafers, with an ITO sheet resistance of about 10 Ω/sq and an ITO thickness of around 185 nm, are used as the substrate for microheater fabrication. First, a positive photoresist is applied and patterned using photolithography, followed by ion beam etching (IBE) to define the ITO bottom electrode. Next, a negative photoresist is patterned, and a 300-nm-thick ZnO layer is deposited via magnetron sputtering. The lift-off process is then employed to define the ZnO pattern. Subsequently, a 1.3-μm-thick layer of PEDOT:PSS is spin-coated from a 1.3 wt.% aqueous dispersion. The device is annealed in a vacuum furnace at 100 °C for one hour to form a Schottky contact between ZnO and PEDOT:PSS. A 120-nm-thick gold layer is then thermally evaporated to form the top electrode, followed by the application of a positive photoresist mask and IBE to pattern the top electrode. Finally, reactive ion etching (RIE) is used to pattern the PEDOT:PSS layer and remove any remaining photoresist from the microheater surface.

Results and Discussion

A. Cross-Section Imaging

A test region on the same wafer, sharing the cross-sectional structure of the diode microheaters, with dimensions of approximately 10 mm × 10 mm, was subjected to brittle fracture in liquid nitrogen, followed by scanning electron microscope (SEM) tomography imaging. Fig. 3a shows the distinct multilayer structure of the test region, with clearly defined layer boundaries, confirming the successful fabrication of the heterojunction.

B. Rectification Performance of the Diode Microheater

The electrical characterization of a fabricated diode microheater was performed using a Keithley 2636B System SourceMeter instrument connected to a probe station. The top Au electrode was connected as the anode, while the bottom ITO electrode was connected as the cathode. A voltage sweep from -5 V to 11 V was applied Fig. 3b shows that the diode microheater reaches 950 μA at 11 V, while the reverse current is only -1.73 μA at -5 V, confirming its effective rectifying behavior.

C. Surface Temperature of the Microheater

The thermal characterization of the diode microheater was performed simultaneously with the electrical characterization described above, using a Fluke TiX1060 infrared thermal imaging camera. Fig. 3c shows thermal images at different input power levels and the relationship between power and temperature. The maximum temperature reached in this test is 47.8 °C at a microheater input power of 10.9 mW.

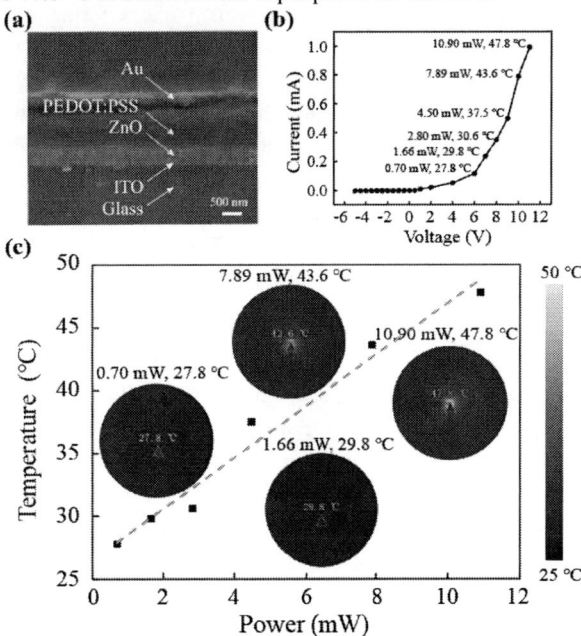

Fig. 3: Diode microheater characterization (a) Cross-sectional SEM image of the test region (scale bar: 500 nm). (b) IV curve of the diode microheater, along with the corresponding power and temperature. (c) The power v.s. temperature curve of diode microheater, along with the infrared thermal images (The triangular marker in the thermal image indicates the location of the highest temperature. A unified temperature scale is provided).

D. Degradation Issues and Potential Improvements

In a separate experiment, a diode microheater fabricated using the same process showed a sharp current increase at high voltages (15V), with subsequent tests revealing a permanent loss of rectification, yet no visible surface changes were observed. This may suggest potential internal breakdown, such as interface degradation or thermal runaway. To further increase the operating temperature, potential approaches could include introducing semiconductor materials with higher thermal stability to replace PEDOT:PSS to enhance stability under high voltage and temperature conditions.

The electrical and thermal tests shown in Fig. 3 were performed within 2 days of fabrication, with samples stored in air. We observed that diode microheaters stored in ambient air and tested after one week showed a clear decrease in both forward and reverse current. This degradation is likely due to the hygroscopic nature of PEDOT:PSS, which absorbs moisture, forming an acidic solution that reacts with ZnO [10]. Studies have shown that polydimethylsiloxane (PDMS) encapsulation can effectively prevent degradation in ZnO/PEDOT Schottky devices [11]. Further studies will focus on understanding the degradation mechanisms and exploring encapsulation methods for the diode microheater to enhance its stability.

Conclusion

This study demonstrates the fabrication and performance of 50 μm × 50 μm diode microheaters using ZnO/PEDOT Schottky contacts, achieving rectifying behavior and a maximum surface temperature of 47.8 °C. The fabrication process is simple and has a low processing temperature. These microheaters have potential applications in SMP-based micro-transfer printing and flexible electronics. Further improvements are needed to increase thermal performance and achieve long-term stability.

Acknowledgments

The authors would like to acknowledge the contributions of Jiahao Jiang and Jingyang Zhang at HKUST(GZ), and Qingming Chen, Xifa Liang, and Shijie Liang at Sun Yat-sen University for their support in equipment training and discussions. This work is partially supported by the National Natural Science Foundation of China (No. 52375580), the Guangdong Basic and Applied Basic Research Foundation (No. 2024A1515011397), the Department of Education of Guangdong Province Project (No. 2023ZDZX1036), and the Guangzhou-HKUST(GZ) Joint Funding Program (No. 2023A03J0688). ChatGPT is used to refine the English in the manuscript.

References

[1] J. Yoon, S. M. Lee, D. Kang, M. A. Meitl, C. A. Bower, and J. A. Rogers, "Heterogeneously integrated optoelectronic devices enabled by micro-transfer printing", Advanced Optical Materials, vol. 3, no. 10, pp. 1313-1335, (2015).

[2] C. Linghu, S. Zhang, C. Wang, and J. Song, "Transfer printing techniques for flexible and stretchable inorganic electronics", npj Flexible Electron., vol. 2, no. 1, pp. 26, (2018).

[3] J. Li, B. Luo, and Z. Liu, "Micro-LED mass transfer technologies", in Proc. 2020 21st Int. Conf. Electron. Packag. Technol. (ICEPT), IEEE, (2020).

[4] C. Linghu, S. Zhang, C. Wang, K. Yu, C. Li, Y. Zeng, H. Zhu, X. Jin, Z. You, and J. Song, "Universal SMP gripper with massive and selective capabilities for multiscaled, arbitrarily shaped objects", Sci. Adv., vol. 6, no. 7, pp. eaay5120, (2020).

[5] J. D. Eisenhaure, T. Xie, S. Varghese, and S. Kim, "Microstructured shape memory polymer surfaces with reversible dry adhesion", ACS Appl. Mater. Interfaces, vol. 5, no. 16, pp. 7714-7717, (2013).

[6] L. Yang, Q. Guo, J. Zhang, Y. Gan, and Y. Wang, "Selective micro-transfer printing of microspheres using adhesion-switchable stamp", in Proc. 2024 IEEE 19th Int. Conf. Nano/Micro Eng. Mol. Syst. (NEMS), IEEE, pp. 1-4, (2024).

[7] Y. Wang, S. Solberg, J. Lu, Q. Wang, N. Chang, D. Schwartz, and M. Chintapalli, "Programmable micro-transfer-printing for heterogeneous material integration", AIP Adv., vol. 12, no. 6, pp. 065110, (2022).

[8] Y. Bai, J. Tian, Z. Lin, M. You, J. Liu, and X. Wang, "Development of a high throughput micro-heater array with controllable temperature for each heating unit", Microsyst. Technol., vol. 26, pp. 787-792, (2020).

[9] M. Nakano, A. Tsukazaki, R. Y. Gunji, K. Ueno, A. Ohtomo, T. Fukumura, and M. Kawasaki, "Schottky contact on a ZnO (0001) single crystal with conducting polymer", Appl. Phys. Lett., vol. 91, no. 14, pp. 143514, (2007).

[10] N. Hernandez-Como, M. Lopez-Castillo, F. J. Hernandez-Cuevas, H. Baez-Medina, R. Baca-Arroyo, and M. Aleman, "Flexible PEDOT:PSS/ZnO Schottky diodes on polyimide substrates", Microelectron. Eng., vol. 216, 111060, (2019).

[11] M. U. Khan, G. Hassan, M. A. Raza, J. Bae, and N. P. Kobayashi, "Schottky diode based resistive switching device based on ZnO/PEDOT:PSS heterojunction to reduce sneak current problem", J. Mater. Sci.: Mater. Electron., vol. 30, pp. 4607-4617, (2019).

Spatiotemporal Encoding Based on Mott Spiking Neurons for Sound Localization

Zihan Guo[1, 2†], Linbo Shan[1†], Zongwei Wang[1*], Xing Zhang[1], Yimao Cai[1*], Ru Huang[1]

[1]School of Integrated Circuits, Beijing Advanced Innovation Center for Integrated Circuits, Peking University, Beijing, China, [2]School of Software & Microelectronics, Peking University, Beijing, China
*Email: {wangzongwei; caiyimao}@pku.edu.cn

Abstract

In this work, we propose a spatiotemporal sound localization system based on VO$_2$ neurons. Our scheme employs five parallel one-transistor-one-neuron (1T1N) units as coincidence detectors with delay lines, encoding interaural time differences into spatial information. The results demonstrate that the proposed design achieves angular localization with a resolution of $\pi/60$ rad (98% accuracy) and an average energy consumption of around 20.07 nJ. Compared to the existing method, it enhances resolution by 25% and reduces energy consumption by 7.1%, highlighting its potential for efficient, biologically inspired auditory processing.

Keywords: Sound source localization, Coincidence detectors, One-transistor-one-neuron

Introduction

Spatiotemporal sound source localization is essential in applications like robotics, hearing aids, and surveillance (Fig.1a) [1]. Traditional methods such as time-difference-of-arrival and microphone arrays are effective but require high computational power and complex hardware. In contrast, biological neurons achieve sound source localization through spatial-temporal stamps of correlated neuronal activity, such as interaural time differences (ITDs), coincidence detectors (CDs), and delays based on Jeffress model (JM). Fig.1b illustrates the mechanism of JM, which processes auditory signals using delay lines and CDs, converting temporal differences into spatial information. The axons act as delay lines, introducing precise timing shifts, while neurons respond to the coupling of pulses from the left and right [2].

Although these functionalities have been implemented in chips based on complementary metal-oxide-semiconductor (CMOS) technology, they tend to be area- and power-intensive (Fig.1c). Emerging devices like RRAM and Mott devices are considered promising candidates for emulating the rich functionalities of neural systems [4-5]. Previous studies [2] have shown that the RRAM-based CD circuit is sensitive only to the spatial order of impulse arrivals. However, they are unable to directly encode the timing sequence of inputs as an angle, meaning that achieving even a basic accuracy in angle detection demands a significant number of CD units.

In this work, we proposed a neuromorphic sound localization circuit (SLC) with VO$_2$ spiking neurons, which consists of five parallel one-transistor-one-neuron (1T1N) units that implement ITDs, CDs and delay lines (Fig.2). The system processes input through two pathways, V_{left} and V_{right} representing the left and right ear input signals. Each 1T1N unit functions as a CD, firing only when the left and right insputs align temporally. Delay modules between units introduce adjustable time offsets, representing different ITDs. The SLC achieves angle localization with 98% accuracy and an average energy consumption of 20.07 nJ.

Results and Discussions

A. VO$_2$ device and Spiking Neuron

The Au/Ti/VO$_2$/Ti/Au device fabrication process was carried out as follows: first, a 30 nm VO$_2$ layer was deposited on SiO$_2$ substrate using atomic layer deposition. Next, the planar Ti/Au electrodes were formed by beam evaporator, following electron beam lithography, and were patterned using a lift-off technique. Fig.3 shows the VO$_2$ device DC curves with stable bipolar switching over 50 cycles and scanning electron microscope (SEM) image. The transition from a high resistance state (HRS) to a low resistance state (LRS) occurs when the applied voltage exceeds the threshold voltage (V_{th}) of approximately ±1.86 V. The device reverts to HRS when the voltage drops below the holding voltage (V_{hold}) of around ±0.72 V. We also established an LTspice model [3]. The simulation results are consistent with experiment results (Fig.3).

Next, combining resistance and capacitance, we constructed a leaky integrate-and-fire (LIF) neuron, as the illustratied in Fig.4. When the voltage V_{IMT} across the VO$_2$ device is below V_{th}, the device remains in the HRS. When V_{IMT} reaches V_{th}, it switches to LRS, and voltage division causes the capacitor to discharge rapidly, generating a spike. After that, the voltage across the device decreases. When $V_{IMT} > V_{hold}$, the device returns to HRS, quickly discharging. The VO$_2$ devices then cycle between charging and discharging across V_{th} and V_{hold}, firing spikes. Figs.4a-c demonstrate that the spike frequency increases with the input amplitude. When the input voltage is below 2.1 V, it fails to trigger oscillations, as the voltage on VO$_2$ device does not reach the V_{th}. Once input amplitude exceeds 2.2 V, the neuron fires spikes, with the frequency increasing as the pulse amplitude rises, attributed to the reduced charging time.

B. Characteristics of 1T1N Unit

To implement the coincidence detection, we integrated a transistor in series with the neuron circuit to form a 1T1N unit. The 1T1N unit outputs when a signal is applied to the gate of the transistor, as shown the insect in Fig.5. The gate voltage of the transistor and the input voltage of the neuron

979-8-3315-0417-5/25 $31.00 © 2025 IEEE

circuit represent the left and right inputs of JM, respectively. Fig.5 illustrates that 1T1N will generate spike waveforms only when both inputs are present. To enhance angle encoding precision, we introduce a D flip-flop (DFF) in Fig.6. The transistor gate voltage (V_{right}) and the input signal (V_{left}) to the neuron circuit are connected to the clock (CLK) terminal and D terminal of the DFF, respectively. Since the CLK terminal is sensitive to the rising edge, the output of the DFF can indicate the sequence of V_{right} and V_{left}.

When the V_{right} signal is positioned to the right of the V_{left}, the left-right detection signal will be set high. In Fig.7, the output of the DFF divides the spatial angles into left and right sections. The overlap width between V_{right} and V_{left} encodes more precise angles. The formula for calculating the localization angle can be expressed as: $\pi/10 + \pi/5\cdot(5-m) \pm \pi/10\cdot(1-x/x_{max})$, where m is the sequence number of the excited 1T1N unit. x represents the the output spike count, and x_{max} is the maximum number of output spikes observed when V_{left} and V_{right} fully coincide. Here x_{max} is 67. The first two terms denote the center angle, while the third term indicates the angle offset, with its sign encoded by the timing of V_{right} and V_{left}. As the time difference between the input voltages increases, the number of output spikes decreases. Fig.8 illustrates the case where the encoded angle offset is -$\pi/30$. Both V_{left} and V_{right} are 150 μs pulses with a time difference of Δt = -50 μs, positioning V_{left} before V_{right}. Consequently, the left-right detection signal is set to high, producing 38 spikes. This demonstrates that the angle offset can be expressed as different configurations of input pulse signals, and is thereby encoded by a particular number of spikes and the state of the left-right detection signal.

C. Demonstration of SLC

We then cascade five 1T1N units to form a SLC, with each unit responsible for a $\pi/5$ angle range, collectively covering π rad. Delay module consists of D flip-flops, and four delay modules are evenly distributed along each delay line. The delay time of each module is equal. The V_{right} is fed into the gate terminal of the 1T1N unit's transistor and the CLK terminal of the DFF through the delay line. All output ports Q of the DFFs are connected to an OR gate to detect the order of arrival of V_{right} and V_{left}. The V_{left} is connected to the input terminal of the 1T1N neuron and the D terminal of the DFF via the delay line. All output terminals of the neurons are connected to the OR gate for precise angle encoding. The system diagram is presented in Fig.9.

The signals from the left and right sides are sequentially input into the five 1T1N units after being Delay modules. When V_{right} and V_{left} signals meet at the same neuron, the neuron is activated. The position of the activated neuron directly corresponds to the approximate direction of the sound source. Complete overlap of input pulses within a unit indicates that the sound source is precisely aligned with the unit's center detection angle. There are five such center angles: $\pi/10$, $3\pi/10$, $\pi/2$, $7\pi/10$, $9\pi/10$, and π rad.

Fig.10 presents simulation examples demonstrating pulse inputs from both sides of the SLC system. In Fig.10a, a time difference between the input signals results in partial pulse overlap, triggering the second neuron unit, which covers the $3\pi/5$ to $4\pi/5$ rad range, with a center angle of $7\pi/10$ rad. Δt is measured as -20 μs. In Fig.10b, the system generates 47 spikes, and the left_right detection signal is set high. Therefore, the localization angle can be calculated as: $\pi/10 + \pi/5\cdot(5-2) \pm \pi/10\cdot(1-47/67) = 2295\pi/3350$ (rad). Through extensive simulations, the relationship between spike count and encoded angle is captured in Fig.11, which shows the relationship between input pulse time difference Δt, angle offset, and number of spikes. These results are generally consistent with the calculations provided in Fig.10. From the above, it is evident that the system achieves an angular resolution of $\pi/60$ rad with 98% accuracy, consuming only 20.07 nJ per operation on average. Compared to [2], it improves resolution by 25% ($\pi/60$ vs. $\pi/45$) and reduces energy consumption by 7.1% (20.07 nJ vs. 21.6 nJ), while requiring just 5 CD modules instead of 40 in [2], greatly simplifying hardware design.

Conclusion

This work presents a spatiotemporal sound localization system utilizing VO₂ neurons. The design features five parallel 1T1N units functioning as coincidence detectors integrated with delay lines, effectively encoding ITDs into spatial information. The system achieves an angular resolution of $\pi/60$ rad with nearly 98% accuracy and an average energy consumption of approximately 20.07 nJ, demonstrating strong robustness.

Acknowledgments

This work was supported by the National Natural Science Foundation of China (62025401, 62341407, 62322401, and 61927901), Beijing Nova Program (20220484113), and in part by "111" Project under grant No. (B18001).

References

[1] G. Song, X. Wang, and Z. Chen, "Bionic Binaural Sound Localization Circuit Design Based on Memristor," in 2024 39th Youth Academic Annual Conference of Chinese Association of Automation (YAC), IEEE, pp. 526-531 (2024). [2] F. Moro, E. Hardy, B. Fain, et al., "Neuromorphic Object Localization Using Resistive Memories and Ultrasonic Transducers," Nature Commun., vol. 13, no. 1, p. 3506 (2022). [3] M. D. Pickett and R. S. Williams, "Sub-100 fJ and sub-nanosecond thermally driven threshold switching in niobium oxide crosspoint nanodevices," Nanotechnology, 23, p. 215202 (2012). [4] C. Ban, L. Shan, G. Yang, et al., "Artificial VO2 Spiking Neurons with Protective Mechanism for Enhancing Resilience of Spiking Neural Network Against Adversarial Attacks," 2024 IEEE Silicon Nanoelectronics Workshop (SNW), pp. 43-44. (2024) [5] L. Shan, Z. Wang, L. Bao, et al., "In Materia Neuron Spiking Plasticity for Sequential Event Processing Based on Dual-Mode Memristor," Advanced Intelligent Systems, vol. 4, no. 8, p. 2100264 (2022)

Jeffress Model: Delay lines and Coincidence detectors

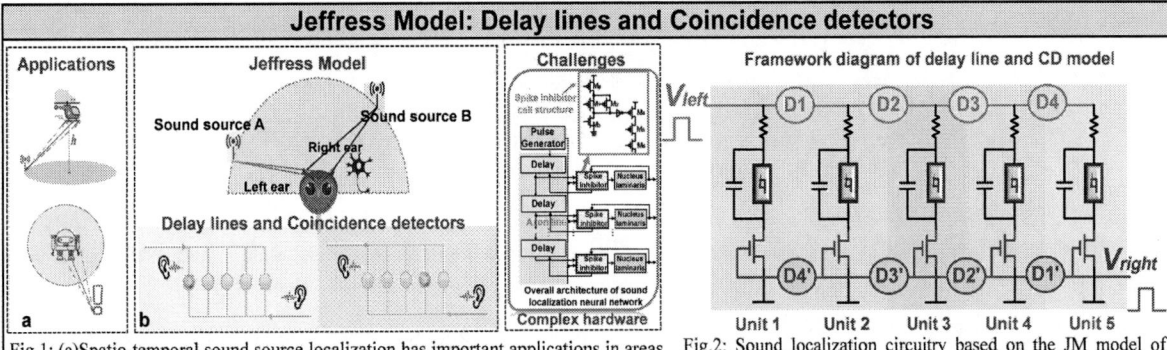

Fig.1: (a)Spatio-temporal sound source localization has important applications in areas such as robotics, hearing aids. (b) JM mechanism for processing auditory signals using delay lines and CDs. (c) Challenges of the proposed technology.

Fig.2: Sound localization circuitry based on the JM model of VO_2, which includes multi VO_2 spiking neurons, CDs and delay lines

Characteristics of the VO₂ device and spiking neuron

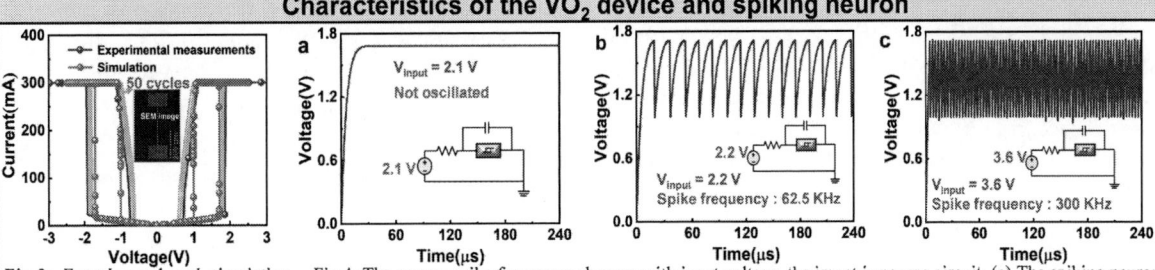

Fig.3: Experimental and simulation DC curves of the VO_2 device, both showing stable bipolar switching.

Fig.4: The neuron spike frequency changes with input voltage, the insect is neuron circuit. (a) The spiking neuron does not oscillate at an input voltage of 2.1 V. (b) At an input voltage of 2.2 V, the neuron reaches a spike frequency of 62.5 KHz. (c) Spike frequency of 300 KHz at 3.6 V.

Characteristics of 1T1N Unit

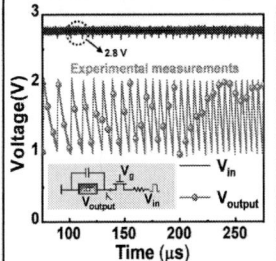 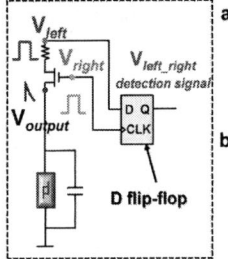 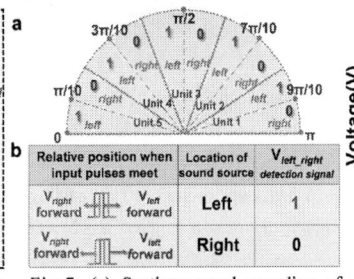 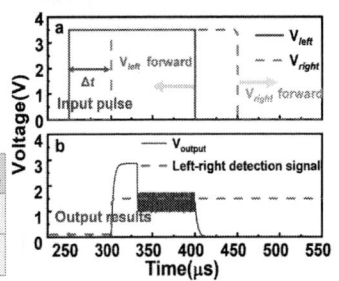

Fig.5: Only when V_g and V_{in} both inputs are present, the 1T1N generate spike, the insect is 1T1N circuit.

Fig.6: 1T1N unit structure with a E-D flip-flop used to determine the sequence of V_{right} and V_{left}

Fig.7: (a) Spatiotemporal encoding of left_right detection signal. (b) The relationship between input pulse timing, sound source location and the signal.

Fig.8: Coupling time difference Δt between V_{left} and V_{right} (V_{left} is located to the left of V_{right}) and their corresponding neuronal outputs.

Characteristics of SLC

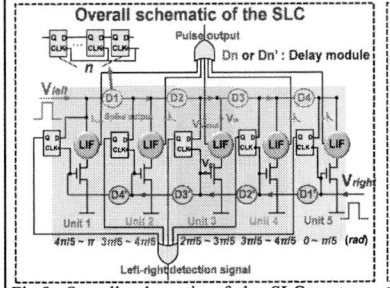 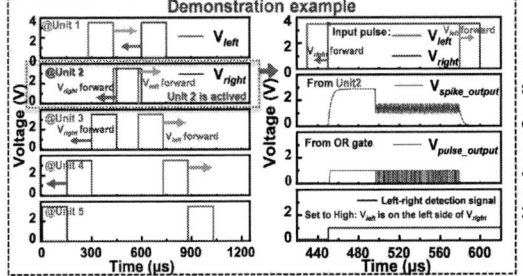 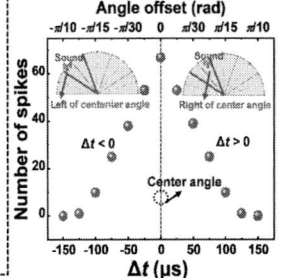

Fig.9: Overall schematic of the SLC system, demonstrating a π rad detection range with a rough $\pi/5$ rad localization, refined to $\pi/60$ accuracy within each segment.

Fig.10: Example illustrating the operating principle based on the SLC. (a) Pulse inputs from both sides of the system overlap in Unit 2. (b) Output curves of the spike, pulse, and $V_{left-right\ detection\ signal}$ when Unit 2 is activated.

Fig.11: The relationship between input pulse time difference Δt, angle offset, and number of spikes

979-8-3315-0417-5/25 $31.00 © 2025 IEEE

Energy-Efficient Temperature-Calibration Readout Circuits with Thermal-State Sensible Sampling and Zoom Window Switch Scheme for RRAM-Based Analog Computing-in-Memory

Zhuoya Chen[1], Zongwei Wang[1,2*], Haisu Zhang[1], Linbo Shan[1], Qishen Wang[1], Xiyuan Tang[1],

Yunyi Fu[1], Yimao Cai[1,2*], Ru Huang[1,2]

[1]School of Integrated Circuits, Peking University, Beijing, China
[2]Beijing Advanced Innovation Center for Integrated Circuits, Beijing, China
*Email: {wangzongwei; caiyimao}@pku.edu.cn

Abstract

In this work, we present a temperature calibration readout circuit for RRAM-based analog computing-in-memory (ACIM), which includes i) a thermal-state sensible sampling circuit with a compact in-array reference and resistance compensation column to mitigate the temperature-induced resistance shift issue, ii) a zoom window switch scheme to improve the energy efficiency. Based on 40nm foundry PDK, the implemented sampling circuit enhances recognition accuracy by 16% at 125°C. Meanwhile, zoom window switch scheme achieves an 93% reduction in average capacitor switching energy compared to traditional methods. Keywords: RRAM, Computing-in-Memory, Temperature, Reliability

Introduction

Resistive random-access memory (RRAM)-based analog computing-in-memory (ACIM) has emerged as a promising candidate for artificial intelligence (AI) applications in low-power edge computing [1-2]. However, AI edge computing systems are deployed in varied environments, requiring mobile devices such as smartphones and automotive systems to withstand temperature fluctuations due to seasonal and geographic changes. Furthermore, applications in natural gas and aviation demand robust chip performance under extreme thermal conditions [3]. Specifically, resistance of RRAM varies under different temperatures and leads to severe accuracy degradation of ACIM, necessitating sophisticated design for energy-efficient temperature-calibration readout circuits. Recent research has introduced a temperature-compensated transimpedance amplifier (TC-TIA) designed specifically to account for the drift in the low resistance state (LRS), thereby ensuring conductivity accuracy over varying temperature ranges [4]. Other studies have explored leveraging sections of the on-chip RRAM array to generate a reference voltage for energy-intensive Flash analog to digital converts (ADC) [5]. While this approach enhances accuracy, it also results in increased ADC power consumption.

This work presents an energy-efficient, temperature-calibrated readout circuit for RRAM-based ACIM, comprising a thermal-state-sensitive sampling circuit that achieves a 16% improvement in accuracy, along with a zoom

window switch SAR ADC that enhances the efficiency of capacitor array switching by 93%.

Temperature-Calibration Readout Circuit

The typical readout circuit for ACIM mainly comprises two primary components: a sampling circuit and a conversion circuit. When the RRAM array performs a computing operation, the bit lines (BLs) are clamped at a specific read voltage by operational amplifiers (OPAs) with source lines (SLs) ground. The sampling circuit is engineered to capture the results of multiply-accumulate (MAC) operations, represented as a voltage derived from the cumulative currents produced by conductance-voltage multiplications (as shown in Fig.2). As multiple word lines (WLs) are activated simultaneously, the currents accumulate on BLs and pass through a fixed resistor to generate the sampled voltage output. Next, the conversion circuit is designed to quantify the sampled voltage signal using ADC. The reliability and precision of the sampling circuit are of paramount importance to ensure the accuracy of the quantized values generated by the ADC. However, the conductance of the LRS decreases with rising temperature, while the HRS exhibits the opposite trend, leading to a decrease in the accuracy of the ACIM system. The proposed temperature-calibration readout circuit combines a thermal-state sensible sample circuit and zoom window switch to alleviate the temperature-induced accuracy degradation while simultaneously reduce the power consumption.

A. Thermal-State Sensible Sampling Circuit

The thermal-state sensible sampling circuit employ an additional HRS reference column and a LRS cell to calibrate the temperature coefficient of the HRS and LRS, respectively (as illustrated in Fig.3), based on our previous scheme [6]. The 1056×1024 ACIM array supports simultaneous activation of 32 rows, with a 16-to-1 MUX selecting columns, and every set of 16 CIM columns includes a reference column for MAC operations. During MAC operation, the current in reference column, denoted as I_{ref}, is scaled through a current mirror with a ratio of n_1:1 and subsequently subtracted from I_{BL}. In differential weight array, n_1 is set to 0.5 for uniform distribution of LRS and HRS cells in a pair of RRAM cells to estimate average HRS offset in activated column. The result current I_{out}^* is scaled again by another current mirror with a ratio of $n_2{:}1$ and then flows through an LRS RRAM (R_{ref} at room temperature) to generate the final corrected output voltage (V_{out}). Factor n_2 is determined by the number of WLs activated simultaneously and set to 1/32 to prevents read disturb during MAC operation. Assuming the temperature coefficient of the LRS is represented as $\alpha(T)$, V_{out} can be expressed as:

$$V_{out} = I_{out}{}^* \times R_{ref}{}^* = I_{out}/\alpha(T) \times R_{ref} \times \alpha(T) = I_{out} \times R_{ref}, \quad (1)$$

where I_{out} represents the current flowing through R_{ref} at room temperature while $I_{out}{}^*$ and $R_{ref}{}^*$ represent corresponding current and resistance at the operating temperature.

The functionality of the proposed circuit is validated using 40nm foundry PDK and measured RRAM resistance data under conditions from -40° to 125°C, with the read voltage set to 0.2V. Fig. 8 illustrates the relationship between the number of concurrently activated LRS cells and output voltages, highlighting a 39% increase in read margin at 125°C with calibration. Fig. 9 shows output voltage variations across temperatures for LRS counts of 15, 16, and 17. Fig. 10 demonstrates the normalized recognition accuracy of a 784×20×10 neural network, confirming a 16% improvement as temperature changes. These findings validate that the thermal-state-aware sampling circuit effectively addresses resistance drift reliability issues induced by temperature fluctuations.

B. Zoom Window Switch Scheme

The ADCs account for 86% of the energy consumption in previous CIM accelerators [7], emphasizing the need to minimize their overhead. Fig.1(b) illustrates that SAR ADCs, with their simple structure and ultra-low power consumption, are an ideal readout scheme. An energy-efficient zoom window switch scheme is proposed, optimizing comparison steps using distinct window settings that leverage the Gaussian distribution of intermediate neural network outputs (Fig. 4). The design employs a 5-bit SAR ADC, adequate for AI edge computing. With this resolution, the input starts from a midpoint of 16, defining three windows—central, side, and extreme—illustrated in Fig. 5, with the corresponding number of comparisons detailed in Table 1.

- Central Window: In 16-20 (or 12-16), adjustments are made with a +1LSB (or −1LSB) step size.
- Side Window: In 20-24 (or 8-12), adjustments are made with a +2LSB (or −2LSB) step size.
- Extreme Window: In 0-8(or 24-32), a +4 LSB (or −4 LSB) step size is first applied, followed by a binary search.

As shown in Fig.6, when V_{IN} is near the midpoint, the zoom window switch significantly reduces capacitor array switching energy and comparison cycles compared to binary search. Fig.7 shows the most complex situation for V_{IN} at 25.8,

requiring 10 cycles.

Fig. 11 presents the switching energy for all inputs using the proposed scheme, alongside several conventional SAR ADC binary search schemes. Meanwhile, Fig. 12 depicts the energy consumption based on the probability distribution of the neural network output prior to activation. The proposed energy-efficient Zoom Window Switch scheme achieves a significant reduction in capacitor array switching energy by approximately 93% and decreases the average number of comparison cycles by around 27% compared to traditional methods.

Conclusion

Targeting AI edge computing, we present an energy-efficient temperature-calibration readout circuit for the RRAM-based ACIM core. Table 2 compares overall parameters with conventional approaches. Using the thermal-state sensible sample circuit, we achieve a 39% improvement in read margin and a 16% increase in recognition accuracy at extreme temperatures. The zoom window switch SAR ADC strategy reduces power consumption, cutting capacitor array switching energy by 93% and average comparison cycles by 27%.

Acknowledgments

This work was supported by the National Natural Science Foundation of China (62025401, 62322401, 62341407, 61927901), Beijing Nova Program (20220484113), and in part by "111" Project (B18001).

References

[1] Cai Y, et al.," Technology-array-algorithm co-optimization of RRAM for storage and neuromorphic computing: Device non-idealities and thermal cross-talk," IEDM (2020)

[2] Y Ling, et al., "An isolated symmetrical 2T2R cell enabling high precision and high density for RRAM-based in-memory computing," Sci China (2024).

[3] Watson J, et al., "A review of high-temperature electronics technology and applications," J MATER SCI-MATER EL (2010).

[4] C. Lee, et al., "Improved On-chip Training Efficiency at Elevated Temperature and Excellent Inference Accuracy with Retention (> 10^8 s) of $Pr_{0.7}Ca_{0.3}MnO_{3-x}$ ECRAM Synapse Device for Hardware Neural Network," IEDM (2021).

[5] W. Li, et al., "A 40-nm MLC-RRAM compute-in-memory macro with sparsity control, on-chip write-verify, and temperature-independent ADC references," JSSC (2022).

[6] Y Ling, et al., "Temperature-dependent accuracy analysis and resistance temperature correction in RRAM-based in-memory computing," TED (2023).

[7] Sun H, et al., "An energy-efficient quantized and regularized training framework for processing-in-memory accelerators, "ASP-DAC (2020).

[8] J. L. McCreary, et al., " All-MOS charge redistribution analog-to-digital conversion techniques. I," JSSC (1975)

[9] C. Liu, et al., " A 10-bit 50-MS/s SAR ADC with a monotonic capacitor switching procedure " JSSC (2010).

[10] Y. Chang, et al., " A 8-bit 500-KS/s low power SAR ADC for bio-medical applications," ASSCC (2007).

[11] Y. Zhu, et al., " A 10-bit 100-MS/s reference-free SAR ADC in 90 nm CMOS," JSSC (2010).

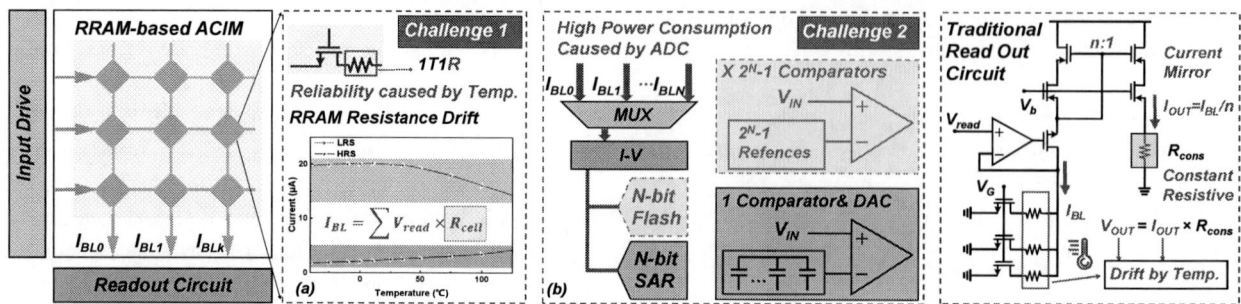

Fig.1 Challenges in RRAM-based ACIM (a) The reliability is affected by temperature variations, causing RRAM resistance drift.(b) High power consumption caused by ADCs: SAR is more energy-efficient compared to Flash.

Fig.2 The diagram of traditional read out circuit.

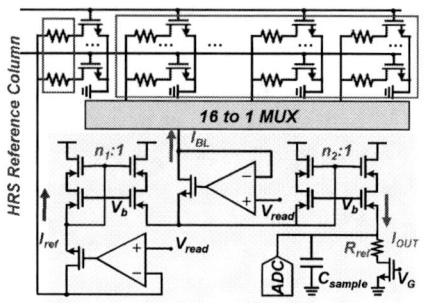

Fig.3 The placement of HRS reference column and the schematic of thermal-state sensible sample circuit.

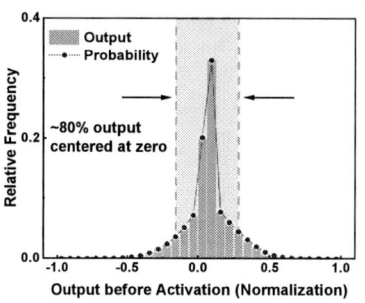

Fig.4 Relative frequency of the normalized output before activation.

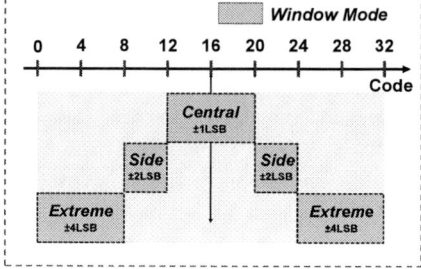

Fig.5 Zoom Window scheme: three windows are defined: the central, side and extreme window.

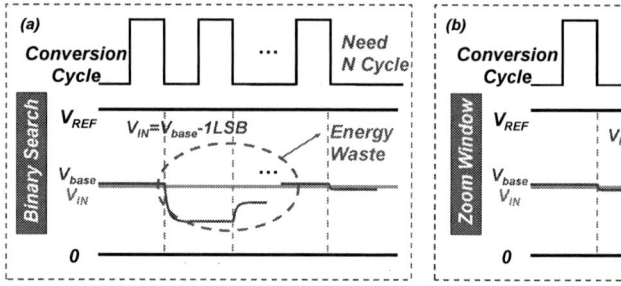

Fig.6 Capacitor array switching voltage and comparison cycle count for (a) binary search and (b) zoom window switch scheme.

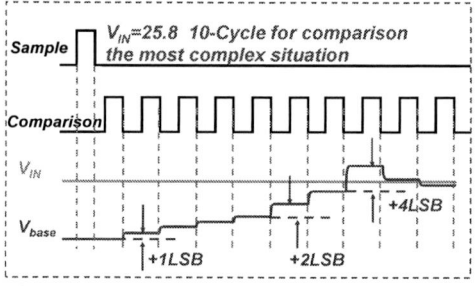

Fig.7 The most complex comparison situation occurs at V_{IN}=25.8, requiring 10 cycles for conversion.

Window	Signal	Number
Central	16<V_{IN}<17	2
	17<V_{IN}<18	3
	18<V_{IN}<19	4
	19<V_{IN}<20	5
Side	20<V_{IN}<22	7
	22<V_{IN}<24	8
Extreme	24<V_{IN}<28	10
	28<V_{IN}<32	10

Table 1: Cycle number of comparisons for different window.

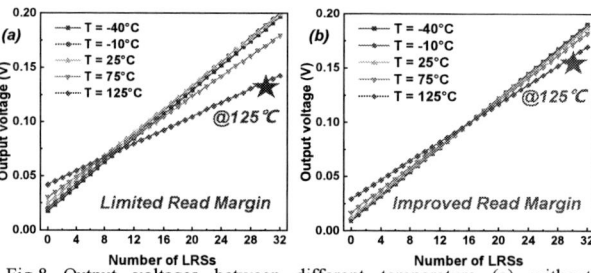

Fig.8 Output voltages between different temperature (a) without calibration and (b) with calibration.

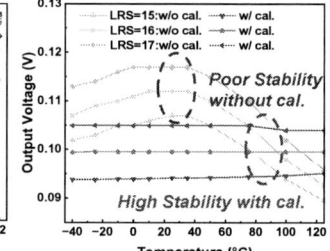

Fig.9 Output voltages for LRS counts of 15, 16, and 17.

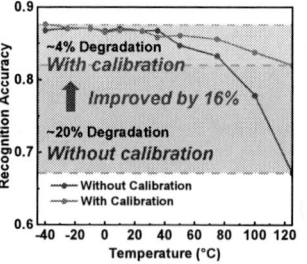

Fig.10 Recognition accuracy at different temperatures.

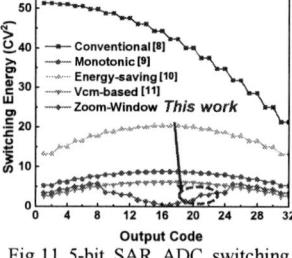

Fig.11 5-bit SAR ADC switching energy (Equal Probability Input).

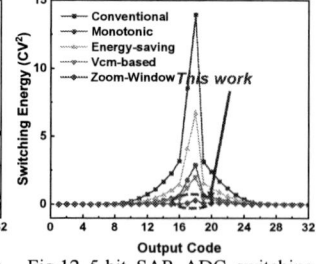

Fig.12 5-bit SAR ADC switching energy(Based on fig.4).

	Conv.	This work
Accuracy Degradation	20%	4%
Read Margin	0.101	0.141
Conversion Energy*	41.98	1.25
Conversion Latency*	5	3.67

*Only considering the energy consumption of capacitor array switch of SAR ADC.

Table 2: Comparison of overall parameters between this work and conventional approach.

979-8-3315-0417-5/25 $31.00 © 2025 IEEE

S-band Internally Matched GaN Power Amplifiers

Li Zhang[1], Xuefeng Zheng[1], Zheng Chen[2], Changcheng Zhang[2], Zhida Wu[2], Zhipeng Ren[2], Pengbo Du[1,2], Hanbin Qu[2]

[1]State Key Laboratory of Wide Bandgap Semiconductor Devices and Integrated Technology, School of Microelectronics, Xidian University, Xi'an 710071, China

[2]North-China Integrated Circuit Co. Ltd, Shijiazhuang 050200, China

Abstract

In this work, the design and measurement results of two different sizes of S-band internally matched power amplifiers (IMPAs) utilizing a single GaN HEMT bare-die with a total gate width of 4200 μm are presented. The first IMPA (PA 1) aiming for a smaller package size, only matches the impedance at the fundamental frequency. The output power and power-added efficiency (PAE) are 20W/55% and a gain variation of only ±0.2 dB within a 20% fractional bandwidth (FBW). Conversely, the second IMPA (PA 2) employs a 2nd-harmonic matching network, effectively enhancing the output power and PAE. Pulse measurement results show that PA 2 achieves a maximum PAE of 60% at a center frequency of 3 GHz, with a maximum saturated output power of 44 dBm (25 W). Compared with PA 1, the lateral dimension of the package is increased by 4.5 × 3 mm². Both IMPAs achieve an optimal trade-off between wide bandwidth, high PAE, and small size.

Keywords: S-band, internally matched power amplifier (IMPA), GaN HEMT

Introduction

Power amplifiers (PAs) have become indispensable to RF wireless communication systems [1]. Compared to other devices based on GaAs and Si, GaN-based HEMT devices are widely used due to their unique material properties, such as wide bandgap, high breakdown voltage, and high thermal conductivity [2]. A series of GaN-based products have been developed, including GaN discrete devices on SiC technology, internally matched amplifiers (IMPAs), and monolithic microwave integrated circuits (MMICs) [3].

Discrete devices often design matching networks outside the package, usually resulting in larger sizes that are not conducive to the miniaturization requirements of communication system integration [4]. In addition, due to the package's cutoff frequency, it is generally impossible to achieve optimal harmonic impedance matching [5]-[6]. Therefore, it is necessary to insert specific matching networks as close to the active chip within the package as possible. The manufacturing of MMIC is more complex and costly, and it is widely used in higher frequency bands (i.e., Ka/V/W-band).

In this work, two S-band internally matched power amplifiers (PA 1 and PA 2) are reported. PA 1 based on smaller size requirements, only matches the impedance at the fundamental frequency, and it exhibits excellent gain variation from 2.7 GHz to 3.3 GHz. To achieve higher efficiency, 2nd harmonic tuning is performed on the PA 2. The results show that 2nd harmonic tuning network can effectively enhance the power and PAE of the PA.

Design of the PA

The GaN HEMT die used in this work is fabricated on SiC technology and consists of an epitaxial structure comprising an AlN nucleation layer, a GaN buffer, an AlGaN barrier layer, and a GaN cap layer. The total gate width of the bar-die is 4200 μm, with a typical breakdown voltage greater than 90V, and gate pinch-off at -3V. The HEMT is operational with a gate voltage of -1.8 V and a drain voltage of 28 V, as the process design kit (PDK) recommended.

Firstly, load-pull analysis was performed on the HEMT in the frequency range of 2.7 GHz to 3.3 GHz, to determine its maximum output power and PAE at both the fundamental frequency f_0 and the 2nd-harmonic frequency $2f_0$. The analysis results are shown in Table 1 and Table 2. As the data indicate, performing 2nd harmonic matching on the HEMT is essential. By tuning the second harmonic at the center frequency of 3GHz, the PA's efficiency rises from 64% to 71%.

Secondly, the matching circuits are designed based on the impedance at the f_0 and the $2f_0$. The schematic diagrams of PA 1 and PA 2 are shown in Fig.1. For PA 1, the input side utilizes a low-pass LC network and a λ/4 wavelength transformer to match the source impedance to 50 Ω. On the output side, gold wires and high-low-high impedance transformers achieve power matching with the termination.

Table 1: The load-pull result at f_0.

Freq (GHz)	Pout (dBm)	PAE (%)
2.7 @ f_0	43.7	68
3 @ f_0	43.8	64
3.3 @ f_0	43.5	63

Table 2: The load-pull result at $2f_0$.

Freq (GHz)	Pout (dBm)	PAE (%)
2.7 @ $2f_0$	44.3	74
3 @ $2f_0$	44.7	71
3.3 @ $2f_0$	44.5	70

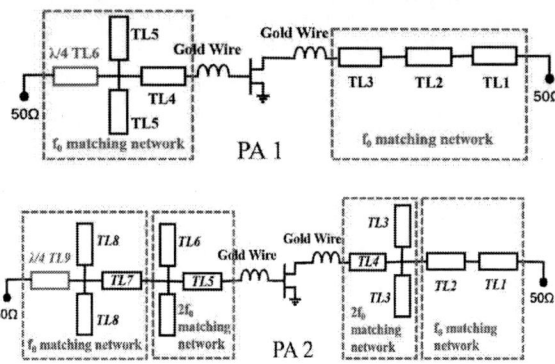

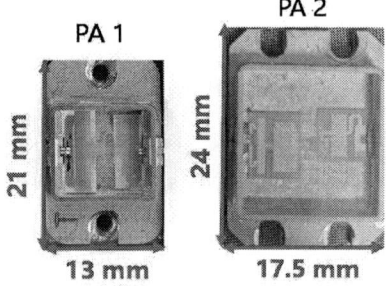

Fig.1: Schematic of the PAs.

Fig.2: The picture of the complete module.

For PA 2, the matching network consists of gold wires, a 2nd-harmonic matching network, and a fundamental matching network. The 2nd-harmonic network comprises a $\lambda/4$ wavelength microstrip line and two $\lambda/2$ wavelength open stubs [7]. The open stubs not only adjust unnecessary reactance parameters but also modify the output reactance of the HEMT, aligning the second harmonic impedance of the HEMT to a short-circuited point. For the input termination, due to the small value of the source impedance ($Zs = 4.21 + j*8.185$), a low-pass LC network is employed to expand the bandwidth of the PA. Moreover, a $\lambda/4$ wavelength transformer is used in the final stage of the input matching to achieve a real-to-real impedance transformation. It is worth mentioning that the impedances of the transmission lines (TL) with identical numbers in PA 1 and PA 2 are inconsistent.

The matching networks for PA 1 and PA 2 are realized on ceramic (Al_2O_3, $\varepsilon r = 9.9$) with a thickness of 0.381 mm. The diameter of gold wires is 25um, which is utilized to realize the equivalent inductance in the matching circuits. These matching networks are mounted in standard metal packages without a lid, as shown in Fig.2. The package size of PA 1 is 13×21 mm^2, and the PA 2 is 17.5×24 mm^2.

Measurement Results

Fig.3 presents the RF performance of PA 1 and PA 2 in the S-band, with a pulse width of 100 µs and a duty cycle of 20%. PA 2 achieved a maximum saturated output power of 44 dBm with an efficiency of 60% and a power gain flatness of ± 1 dB. Although PA 1's maximum saturated output power was 0.5 dB lower than that of PA 2, it exhibited exceptional power gain flatness, with a variation of ± 0.2 dB across the 2.7-3.3 GHz, and a peak PAE of 55%.

Fig.4(a) illustrates the S-parameter measurement results for both power amplifiers under pulsed waves. PA 1 and PA 2 exhibit nearly identical small-signal gains, exceeding 17dB. It is noteworthy that compared with PA 2, PA 1 demonstrates superior gain flatness, with a variation range of only ±0.2 dB across the 2.7 GHz to 3.3 GHz. In addition, the input reflection coefficients are less than -10 dB in both PAs.

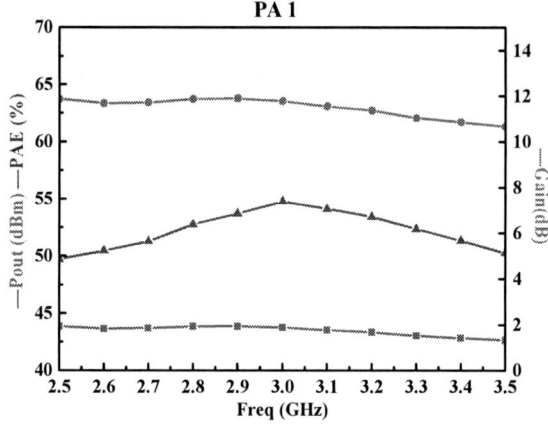

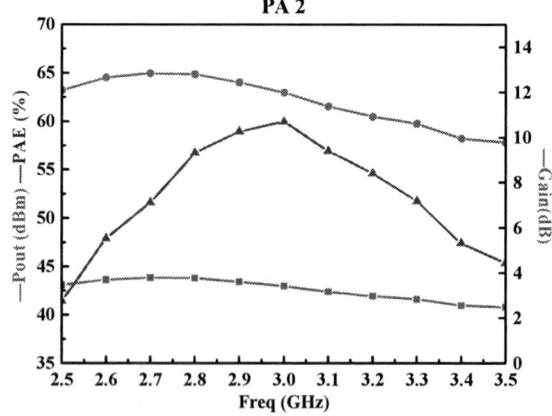

Fig.3: The measurement result of the large signal performance vs. frequency of PA 1 and PA 2.

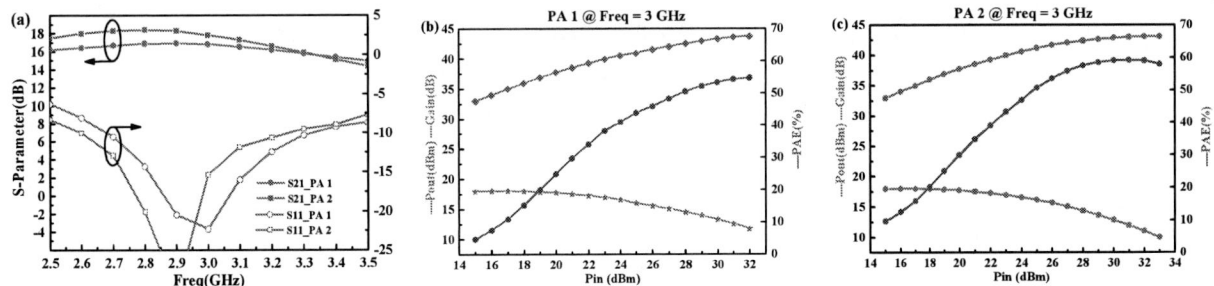

Fig.4: (a) Measured S-parameter result of proposed PAs. (b) The large signal performance vs. power sweep at 3GHz of PA 1 and (c) PA 2.

Table 3: The performance of PAs.

Freq (GHz)	Psat (dBm)	PAE (%)	S21 (dB)	Gain (dB)	Size (mm²)	No.
2.7-3.3	42.8-43.5	51-55	16.5	11.5±0.2	13 × 21	PA 1
2.7-3.3	41.6-44	52-60	17	11±1	17.5 × 24	PA 2

The results of the input power sweep are depicted in Fig.4. At the center frequency of 3 GHz, the output power of PA 2 is 43 dBm, with a PAE of 60%. Meanwhile, the output power at 1dB compression (P1dB) was P1dB = 41.2 dBm, with a PAE of 50.6%. In contrast, PA 1 demonstrated an output power and PAE of 43.5 dBm/53% at 3GHz, with a corresponding P1dB = 40.5 dBm and PAE of 40.7%. Table 3 summarizes the RF performance of PA 1 and PA 2, the differences in variation of S21 can largely be attributed to variations in the matching network and testing fixtures.

Conclusion

This work reports two S-bands internally matched power amplifiers. In the range of 2.7-3.3 GHz, the PAs with the maximum saturated output powers of 44 dBm and 43.5 dBm, respectively. The peak PAEs of 55% and 60%, respectively. The IMPAs that have been designed meet the requirements of high power, high efficiency, and compact size.

Acknowledgments

This work was supported in part by the National Natural Science Foundation of China (Grant No. U2241220), and in part by the National Key Research and Development Project of China (Grant No. 2021YFB3602404).

References

[1] J. Jeong, P. Pech, Y. Jeong, S. Lee. "Wafer-Level-Packaged X - Band Internally Matched Power Amplifier Using Silicon Interposer Technology," *IEEE Microwave and Wireless Components Letters,* vol. 29, no. 10, pp. 665-668, Oct. 2019.

[2] Rubio, J. Cheron, M. Campovecchio, D. Barataud, et al. "Wideband 50 W packaged GaN HEMT with over 60% PAE through internal harmonic control in S-band," *IEEE MTT-s Int. Microwave Symp.* QC, Canada, 2012, pp. 1-3.

[3] S. Lee et al. "Recent progresses in R&D and production of GaN HEMT based power amplifiers," *IEEE International Symposium on Radio-Frequency Integration Technology (RFIT)*, Seoul, Korea (South), pp. 83-86, 2017.

[4] J. Chéron, M. Campovecchio, D.Barataud, Reveyrand, M.Stanislawiak, P. Eudeline, D. Floriot, W. Demenitroux."Harmonic Control In Package of Power GaN Transistors for High Efficiency and Wideband Performances in S-Band," *European Microwave Integrated Circuits Conference (EuMIC).* pp.550-553, 10-11 Oct. 2011.

[5] Y. Park, J. Y. Jeong, W. Kang, M. Park, D. Kim, "A Ku-band Internally Matched 50W GaN HEMT Power Amplifier using Advanced Cu-Mo-Cu Heat Sink," *2024 IEEE/MTT-S International Microwave Symposium - IMS 2024.*Washington, DC, USA. pp. 772-775, 2024.

[6] J. Chéron *et al.* "Wideband 50W packaged GaN HEMT with over 60% PAE through internal harmonic control in S-band," *2012 IEEE/MTT-S International Microwave Symposium Digest,* Montreal, QC, Canada. pp. 1-3, 2012.

[7] K. kuroda, R. Ishikawa, and K. Honjo. "High-Efficiency GaNHEMT Class-F Amplifier Operating at 5.7 GHz," *38th European Microwave Conference 2008*, pp. 440-443, October 2008.

The impact of DC stress on the recoverable tetragonal-to-orthorhombic phase transition in hafnium zirconium oxide capacitors

Zhenyu Chen[1], Dan Lv[2], Zhiyu Lin[1], Dongdong Li[2,*] and Mengwei Si[1,*]

[1]Department of Electronic Engineering and State Key Laboratory of Radio Frequency Heterogeneous Integration, Shanghai Jiao Tong University, Shanghai, China; [2]Zhangjiang Laboratory, Shanghai, China;

Email: lidd@zjlab.ac.cn, mengwei.si@sjtu.edu.cn

Abstract

In this work, we studied the effect of DC stress at room temperature on the polarization strength and dielectric constant of $Hf_xZr_{1-x}O_2$ film capacitors, as well as the retention of this effect. We found that under DC stress, the device changes from a significant non-polarized characteristics to a gradually increased polarization in the low-voltage range. Therefore, it is understood the device exhibits a certain phase transition from the tetragonal to the orthorhombic phase. The polarization decreases after a long time of recovery, suggesting it returns to a state close to the initial state. Accordingly, under DC stress, the dielectric constant of the device gradually decreases, and after recovery, it gradually returns to a k value close to the initial state. The above results suggest the correlation between k value and crystal phases during reliability testing.

Keywords: HZO, high-k, phase transition, reliability

Introduction

Recently, there has been a surge of interest in HfO_2-based ferroelectric thin films, particularly $Hf_xZr_{1-x}O_2$ (HZO) thin films, due to their potential applications in dynamic random-access memory (DRAM), ferroelectric memory (FRAM), etc. [1-3] The advantages of this film include a large band gap (5.2–5.8 eV), and excellent compatibility with complementary metal-oxide semiconductor (CMOS) processes [4], [5]. In DRAM application, a stable high-k film is necessary for continuing scaling. The prevailing technical approach is to enhance the t-phase ratio through phase modulation, thus enhancing the device's k-value [6], [7]. However, in the complex stress environment of DRAM, the t-phase capacitor may undergo an unstable t-o-m phase transition [8-10], which presents a significant challenge to its application in DRAM. In this work, we performed unipolar DC stress and evaluated the reliability of HZO films. The mechanism of the phase change during DC stress was analysed and the associated reliability issues were studied.

Experiments

In this work, HZO films were formed by atomic layer deposition (ALD) process, and a schematic diagram of the metal-dielectric-metal capacitor is shown in Fig. 1. First, a 10 nm thick titanium nitride film was deposited on low-resistance Si by the ALD process. Then a 6.5 nm thick HZO film with a Zr:Hf ratio of 2:1 was deposited using ALD. Subsequently, a 4.5 nm thick TiN top electrode was deposited

using ALD. The next step was the growth of Mo to protect the titanium nitride and patterning of top electrode with area of 100 μm^2. The capacitor was subjected to a rapid thermal annealing process at 250°C.

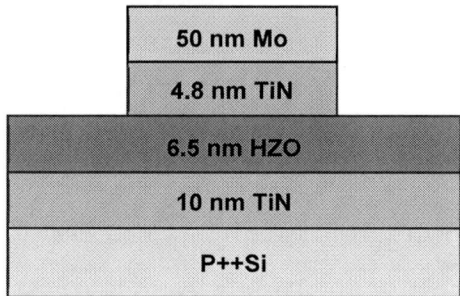

Fig. 1: Schematic diagram of a metal-insulator-metal capacitor structure.

Keysight Technologies waveform generator fast measurement unit (WGFMU) was used to perform basic electrical characterization tests on the device. The voltage was applied to the top electrode of the device, and the bottom electrode was grounded. The device was subjected to a low voltage of 0.6 V and a high voltage of 3 V for polarization-voltage measurements.

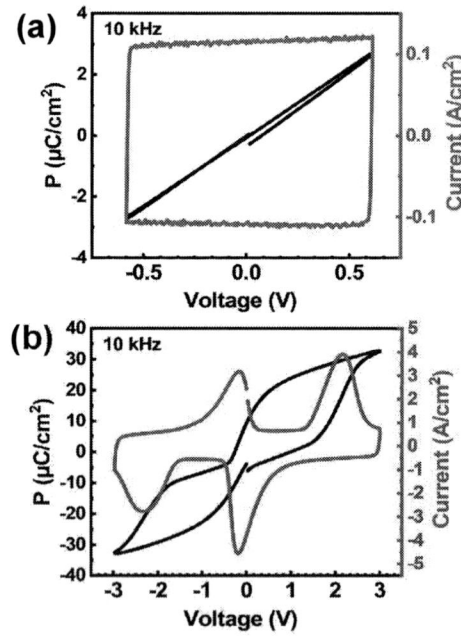

Fig. 2: Polarization-voltage measurement of a typical device at (a) a voltage of 0.6 V at 10 kHz and (b) a voltage of 3 V at 10 kHz.

In Figs. 2(a) and (b), polarization-voltage tests are performed on the device at 0.6 V and 3 V, respectively. The device shows

979-8-3315-0417-5/25 $31.00 © 2025 IEEE

no obvious ferroelectric hysteresis at the low voltage of 0.6 V. When the polarization-voltage test voltage is changed to 3 V, the device shows obvious antiferroelectric properties.

Results and Discussion

A. Effect of DC stress on device polarization
The reliability of HZO capacitors under DC stress was performed using the test waveform shown in Fig. 3. Stress and test signal are applied to the top electrode of the device, and the bottom electrode of the device is grounded. A DC stress with a stress level of V_{stress} is continuously applied to the top electrode, and a 10 kHz triangular wave is inserted during the stress application for polarization-voltage measurements at a test voltage of 1 V [11], [12].

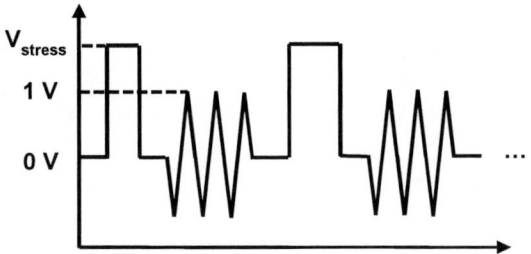

Fig. 3: Schematic diagram of the polarization-voltage test waveform under continuous unipolar DC stress.

Fig. 4 shows the impact of stress and recovery time at a V_{stress} of 3 V on the polarization of the device. As the duration of stress increases, the device changes from non-ferroelectric hysteresis loop to strong ferroelectric hysteresis loop in the 1 V low voltage range, where the remnant polarization (P_r) gradually increases. As shown in Fig. 4(a), in devices with anti-ferroelectric properties (same devices as in Fig. 2), ferroelectric hysteresis loop is not observed at low voltages without DC stress. While after the DC stress for a certain time, ferroelectric hysteresis loop appears, and the P_r of the device increases with stress time. The energy states of different crystal phases are understood to be different [13], [14]. The energy distribution at different voltages may be different. This behaviour may be attributed to the phase change of HZO thin film during the continuous high DC stress. Since the lowest energy state of each crystalline phase of the film is different under different voltage stress levels, part of the film t phase transforms to the o phase, thereby observing the gradually increasing of P_r in polarization-voltage measurements at a low voltage of 1 V. After the stress test is completed, V_{stress} is set to 0 V to capture the recovery performance. As the recovery time increases, P_r gradually decreases. In the short time domain, the polarization decreases rapidly, while in the long-time domain, the polarization intensity gradually approaches the initial state with no polarization. After a recovery for 7 days, a small polarization intensity still remains.

The change of P_r during stress and recovery process is shown in Fig. 4(c). Under DC stress the polarization gradually increases, and eventually tends to saturate. During the recover process, the polarization gradually decreases with time. This recoverable polarization change may be attributed to the mechanism that the o-phase in the HZO thin film has a lower energy than the t-phase under a high voltage stress. Consequently, part of the t-phase in the crystal is converted to the o-phase under DC stress, resulting in a significant polarization enhancement that can be observed in the P-V test in the low voltage range. At V_{stress} of 0 V or without DC stress, t phase is more stable than o phase, so that P_r decrease during recovery and eventually close to 0.

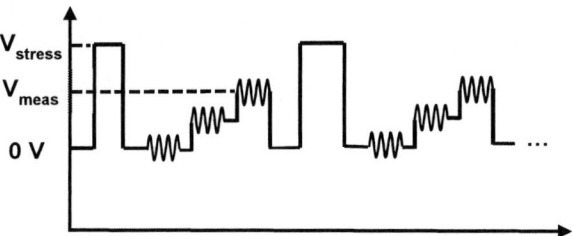

Fig. 5: Schematic diagram of the C-V test waveform under continuous unipolar DC stress.

B. Effect of DC stress on the dielectric constant
The dielectric constant is a key parameter for DRAM application. Experiments are conducted to measure the degradation and recovery of dielectric constant under DC stress. As shown in Fig. 5, a DC stress at V_{stress} is

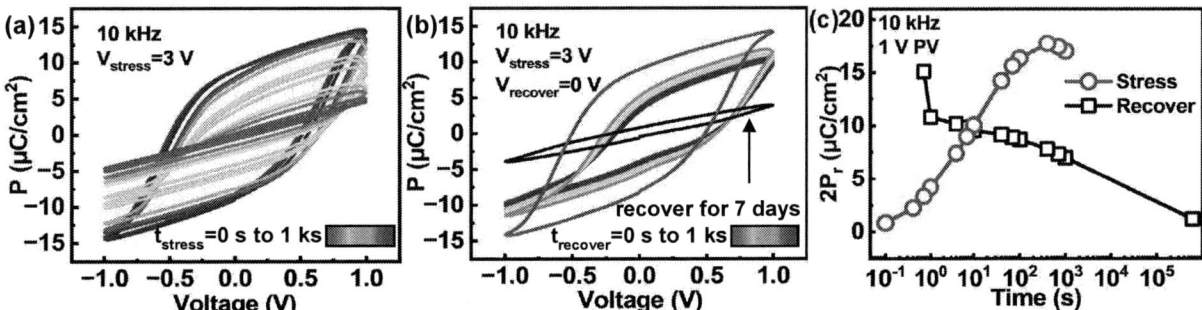

Fig. 4: (a) P-V curve at different stress times. (b) P-V curve at different recovery times. (c) $2P_r$ versus time during DC stress and recovery, extracted from (a) and (b).

continuously applied to the top electrode, and a C-V test at a low voltage of V_{meas}=0.6 V is performed without introducing a polarization switching during C-V test. The above measurement is performed by an impedance analyser. V_{stress} is set to 3 V while V_{stress} is set 0 V during the recovery phase below 1 ks and V_{stress} is floating for 7 days after recovery.

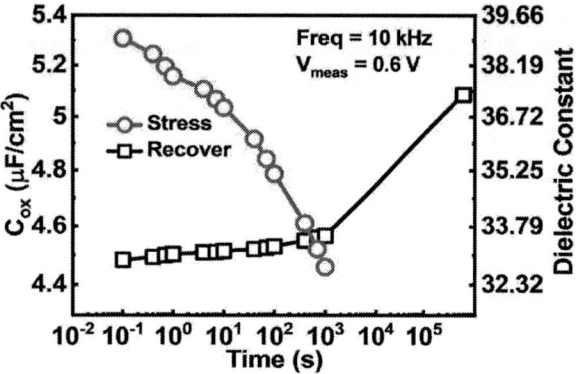

Fig. 6: C_{ox} versus time curve for stress and recovery phases.

As shown in Fig. 6, under DC stress the capacitance of the device continues to decrease. During the recovery phase, the capacitance gradually increases, and after 7 days, the capacitance value recovers to a state close to the initial state. This process is similar to the change of P_r during stress and recovery. Therefore, both the change of capacitance and k value are related with the phase change between t-phase and o-phase. However, it is worth noting that during stress, the capacitance decreases without saturation. At the same time, there was no rapid recovery of the capacitance value in the short time during the recovery phase. Therefore, there are other degradation mechanisms occurred in capacitance degradation and recovery in addition to the recoverable t-o phase transition.

Conclusion

The impact of DC stress on the polarization and capacitances of the HZO capacitor was investigated. It was observed that continuous DC stress causes a transformation from the t-phase to the o-phase. Furthermore, following the removal of the stress, the phase transition gradually returns to its initial state. This will consequently result in a decrease in the dielectric constant of the device during the stress period, followed by an increase upon the removal of the stress. However, the degradation and recovery of the dielectric constant remain coupled to other influencing mechanisms, causing the dielectric constant to continue to decrease even when the t-o phase transformation reaches saturation.

Acknowledgements

This work was supported by the National Natural Science Foundation of China under Grant 52350195. The samples used in the research were supported by Zhangjiang Laboratory.

References

[1] M. Jung, V. Gaddam, and S. Jeon, "A review on morphotropic phase boundary in fluorite-structure hafnia towards DRAM technology," Nano Convergence, vol. 9, no. 1, p. 44, 2022.

[2] H. Chen et al., "HfO₂-based ferroelectrics: From enhancing performance, material design, to applications," Applied Physics Reviews, vol. 9, no. 1, p. 011307, 2022.

[3] T. Ali et al., "Tuning Hyrbrid Ferroelectric and Antiferroelectric Stacks for Low Power FeFET and FeRAM Applications by Using Laminated HSO and HZO films," Advanced Electronic Materials, vol. 8, no. 5, p. 2100837, 2022.

[4] Z. Dou et al., "Microstructural evolution and ferroelectricity in HfO₂ films," Microstructures, vol. 2, no. 2, p. 2022007, 2022.

[5] D. Lehninger et al., "Back-End-of-Line Compatible Low-Temperature Furnace Anneal for Ferroelectric Hafnium Zirconium Oxide Formation," Physica Status Solidi A, vol. 217, no. 8, p. 1900840, 2020.

[6] V. Gaddam et al., "Low-Damage Processed and High-Pressure Annealed High-k Hafnium Zirconium Oxide Capacitors near Morphotropic Phase Boundary with Record-Low EOT of 2.4Å & high-k of 70 for DRAM Technology," 2024 IEEE Symposium on VLSI Technology and Circuits, pp. 1-2, 2024.

[7] J. Y. Kim et al., "Stabilization of Tetragonal Phase in Hafnium Zirconium Oxide by Cation Doping for High-K Dielectric Insulators," ACS Applied Materials & Interfaces, vol. 16, no. 44, pp. 60811-60818, 2024.

[8] P. Nukala et al., "Reversible oxygen migration and phase transitions in hafnia-based ferroelectric devices," Science, vol. 372, no. 6542, pp. 630-635, 2021.

[9] Z. Zhang et al., "Phase Transformation Driven by Oxygen Vacancy Redistribution as the Mechanism of Ferroelectric $Hf_{0.5}Zr_{0.5}O_2$ Fatigue," Advanced Electronic Materials, vol. 10, no. 9, p. 2300877, 2024.

[10] Z. Weng, L. Zhao, C. Lee, and Y. Zhao, "Phase Transitions and Anti-Ferroelectric Behaviours in $Hf_{1-x}Zr_xO_2$ Films," IEEE Electron Device Letters, vol. 44, no. 10, pp. 1780-1783, 2023.

[11] M. Si, X. Lyu, P. Shrestha, X. Sun, H. Wang, K. P. Cheung, and P. D. Ye, "Ultrafast Measurements of Polarization Switching Dynamics on Ferroelectric and Anti-Ferroelectric Hafnium Zirconium Oxide," Appl. Phys. Lett., vol. 115, p. 072107, 2019.

[12] X. Lyu, M. Si, X. Sun, M. A. Capano, H. Wang, and P. D. Ye, "Ferroelectric and Anti-Ferroelectric Hafnium Zirconium Oxide: Scaling Limit, Switching Speed and Record High Polarization Density," in 2019 IEEE Symposium on VLSI Technology, pp. 44-45, 2019.

[13] M. Jung, V. Gaddam, and S. Jeon, "A review on morphotropic phase boundary in fluorite-structure hafnia towards DRAM technology," Nano Convergence, vol. 9, no. 1, p. 44, 2022.

[14] S. S. Fields et al., "Phase-Exchange-Driven Wake-Up and Fatigue in Ferroelectric Hafnium Zirconium Oxide Films," ACS Applied Materials & Interfaces, vol. 12, no. 23, pp. 26577-26585, 2020.

A Particle Swarm Optimization Algorithm Based Parameters Extraction Technique to Model Organic Light-emitting Diode for Flexible Displays

Jianhui Wanghe, Jiachen Kang, Wei Tang[*], Gufeng He[*]

Department of Electronic Engineering, Shanghai Jiao Tong University, Shanghai 200240, China
[*]E-mail: gufenghe@sjtu.edu.cn, terry_tang@sjtu.edu.cn

Abstract

In this study, an automatic parameter extraction algorithm using particle swarm optimization (PSO) is introduced to model organic light-emitting diode (OLED). It is demonstrated both DC and AC characteristics are modeled with superior efficiency and accuracy when validated against experimental data and benchmarked against existing methodologies. This approach streamlines the parameter extraction process for rapid modeling of OLEDs with high precision, contributing to ongoing efforts in OLED research and development.

Keywords: Organic Light-Emitting Diode; Device Modeling; Particle Swarm Optimization Algorithm.

Introduction

Organic Light Emitting Diode (OLED) has drawn more and more attention in the display and lighting applications due to its superior properties.[1] However, the unique structural and energetic disorder of the organic molecular films in OLED devices creates certain challenges for modeling. [2] Due to the energy disorder and nonlinearity of OLEDs, establishing a versatile electrical model requires consideration of multiple parameters and segmented small-signal models. Some studies focus solely on static models or small-signal dynamic models where the equivalent capacitance is not affected.[3] Others have expanded the applicability of the model by combining the OLED characteristics across multiple regions, but they do not prioritize fitting accuracy and the complexity of parameter extraction.[4] Based on the current state of OLED research, this study aims to design an efficient, fast, and accurate OLED parameter extraction and model optimization algorithm based on the Particle Swarm Optimization (PSO) algorithm.

Experimental

A. Modeling of OLED

Fig. 1(a) shows the stacking structure of an OLED device consisting of an anode, a cathode, and multiple organic layers. The hole injection layer (HIL) and electron injection layer (EIL) ensure efficient charge carrier injection into OLED. The electron transport layer (ETL) and hole transport layer (HTL) provide pathways for electrons and holes to be transported to the emissive layer (EML), where

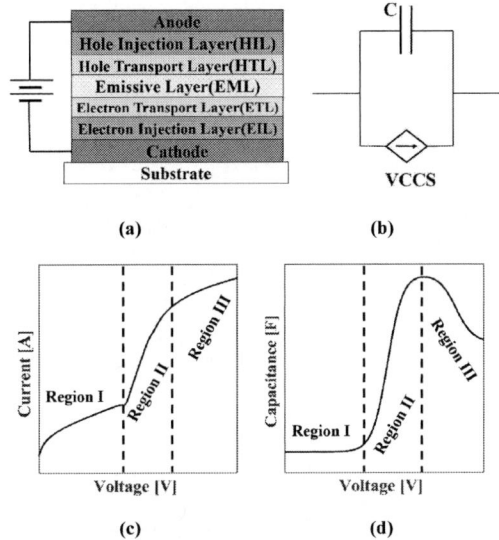

Fig. 1. (a) Structure of and (b) equivalent circuit model of OLED. Schematics of (c) current-voltage and (d) capacitance-voltage characteristics showing the dark current region (Region I), injection region (Region II), and recombination region (Region III).

they combine to generate light emission. **Fig. 1(b)** shows the widely established circuit model consisting of a capacitance and a voltage controlled current source (VCCS).[5, 6] The electrical characteristics of OLED, including current-voltage (*I-V*) and capacitance-voltage (*C-V*) performance, can be divided to three parts: the dark current region (Region I), the injection region (Region II), and the recombination region (Region III), as shown in **Fig. 1 (c)-(d)**. In Region I, electrons and holes are weakly injected into the organic film layer, with the carrier concentration following a Boltzmann distribution and a geometric capacitance. In Region II, the injection of electrons and holes is enhanced with the current follows an exponential distribution, leading to accumulation at the interface and an increase in capacitance. In Region III, further enhancement of the external voltage allows for a large injection of carriers, resulting in recombination within the film. The carrier transport is predominantly driven by drift, and the recombination reduces the accumulated charge, resulting in a rapid decrease in capacitance. [7]

Based on the above properties, the DC characteristics in different regions are depicted by using the following *I-V* **Equations (1)-(3)** while the AC characteristics is depicted by

C-V **Equations (4)**, with the definition of the used physical parameters shown in **Table 1** and **Table 2**:

$$j_{\text{region1}} = j_0(e^{\frac{qV}{n_1 kT}} - 1 + \frac{V}{V_{\text{bi}}}) \qquad V \le V_{\text{bi}} \tag{1}$$

$$j_{\text{region2}} = j_0\, exp[\frac{q(V-V_{\text{bi}})}{n_2 kT}] \qquad V_{\text{bi}} < V < V_{\text{th}} \tag{2}$$

$$j_{\text{region3}} = G_{\text{hi}}(V - V_{\text{th}})(\frac{\sqrt{V-V_{\text{th}}}}{\sqrt{V_{\text{pf}}}}) \qquad V > V_{\text{th}} \tag{3}$$

$$C(w) = C_{\text{g}} + \frac{\chi_1 C_{\text{g}}}{1+w^2 \tau_{\text{trap}}^2} - \frac{\chi_2 C_{\text{g}}}{1+w^2 \tau_{\text{r}}^2} \tag{4}$$

Hereby, the tanh function is used to connect the currents of the three regions, making them differentiable, as shown in Equation (5).

$$j = \frac{1}{2}j_{region1}(1 - tanh(b(V - V_{bi}))) +$$
$$\frac{1}{2}j_{region2}(tanh(b(V - V_{bi})) + tanh(b(V_{th} - V))) $$
$$+ \frac{1}{2}j_{region3}(1 + tanh(b(V - V_{th}))) \tag{5}$$

Table 1: DC physical parameters

Name	Description
J_0 (A/m^2)	Region1 Reverse saturation current density
n_1	Device structure and material-related ideal factors
J_{li0} (A/m^2)	Depends on the current density of HTL and ETL
n_2	Ideal factors related to device parameters
V_{bi} (V)	OLED Built-in potential
G_{hi} (S)	Average carrier concentration-related conductivity
V_{th} (V)	Threshold voltage related to confined accumulation
V_{pf} (V)	Voltage term proportional to material layer thickness

Table 2: AC physical parameters

Name	Description
χ_1	Related to the volume of carriers and τ_{trap}
χ_2	Related to the volume of carriers and τ_r
A_1 (s/K)	Related to $q/(k\sigma_s N_A n_{t0})$
A_2 (s)	Related to $\varepsilon\varepsilon_0/(q\mu_0 n_{r0})$
β_1	Correction factor, a dimensionless constant.
β_2	Correction factor, a dimensionless constant.

B. *PSO-based Parameter Extraction Technique*

Particle swarm optimization (PSO) is a simple and effective heuristic algorithm that can solve swarm optimization problems, making it well-suited for the extraction and optimization of multiple parameters in OLED applications. **Fig. 3** shows the schematic algorithm of PSO for extraction of the key parameters to model the OLED devices. **Fig. 4** shows the PSO algorithm-based flowchart for parameter extraction by employing two error calculation functions: Root Mean Square (RMS) and Mean Absolute

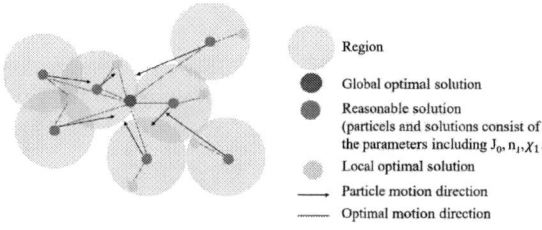

Fig. 3. PSO schematic algorithm

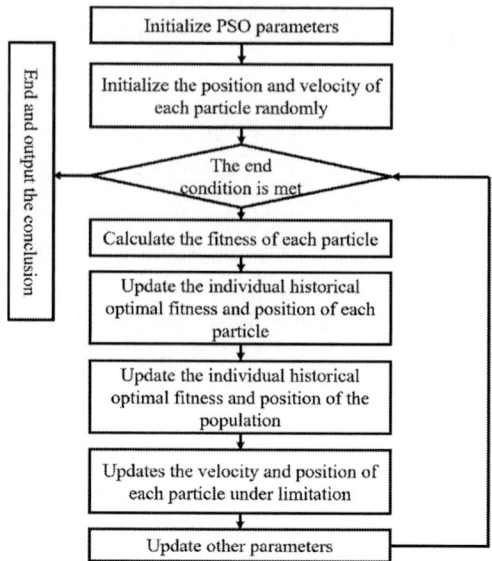

Fig. 4. PSO algorithm-based flowchart for parameter extraction.

Percentage Error (MAPE). RMS focuses more on the model accuracy in the high current/capacitance region, while MAPE emphasizes the overall model accuracy. Therefore, RMS is used as a fitness function for *I-V* model in regions I and region III, and MAPE is used as a fitness function for *I-V* model in regions II as well as for the entire *C-V* model. In particular, this algorithm incorporates a position constraint that the search space for model parameter extraction is constrained by expanding it upwards by 1-2 orders of magnitude, thereby accelerating convergence and improving results.

To conduct parameter extraction, the input parameters required by the proposed algorithm are only the measured *I-V* data, the OLED area, and the temperature. The V_{bi} and V_{th} extracted from Region II and Region III are also used as voltage segmentation points, which are input into the next step for parameter extraction. For the parameter extraction of AC characteristics to model the OLED, the capacitance under zero bias from the measured *C-V* data is first extracted as the geometric capacitance. Then, the remaining parameters are extracted by using the PSO algorithm.

Results and Discussion

Table 3 shows the extracted parameters for high-accuracy modeling the DC and AC characteristics, respectively. **Fig. 5** shows the discrepancy between the fitting curve of the DC model and the measured data. When the applied voltage exceeds the built-in potential, the *I-V* curve exhibits a high degree of fitting in the charge injection region on the right side of the *I-V* curve. However, when the applied voltage is lower than the built-in potential, the current passing through the diode itself is very small, resulting in some error in the exponential coordinate. Nevertheless, in

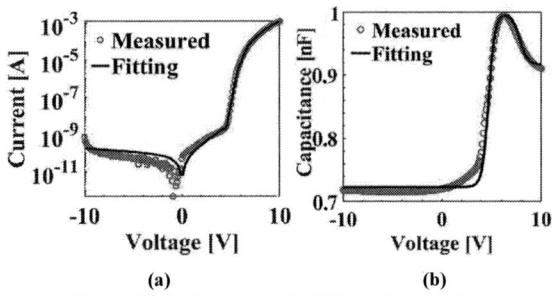

Fig. 5. The modeled (a) *I-V* and (b) *C-V* results with the parameters extracted by PSO-based algorithm.

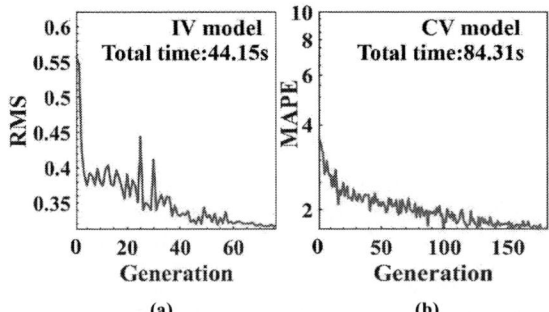

Fig. 6. Curves of fitness functions of (a) *I-V* model in Region III and (b) *C-V* model.

most voltage ranges, the modeled DC and AC characteristics remain high accuracy with RMS errors less than 0.1717% and 0.5673%, respectively. **Fig. 6** shows that with the increase of generation, the error calculation functions decrease rapidly and eventually stabilized, while the fitting curves were quickly generated in both *I-V* and *C-V* models.

Table 3: Extracted parameters at 27 °C

Name	J_0	n_1	J_{li0}	n_2	V_{bi}	G_{hi}	V_{th}
Value	2.8e-5	45.2	9.2e-4	6.1	4.5	7.5	5.9
Unit	A/m²	-	A/m²	-	V	S	V
Name	V_{pf}	χ_1	χ_2	A_1	A_2	β_1	β_2
Value	0.46	0.398	0.134	6.6 e-7	5.9e-4	0.028	0.02
Unit	V	-	-	s/K	s	-	-

Table 4: Comparison with similar studies

Reference	Device type	Algorithm	RMS
This work	OLED	PSO	0.369%
[8]	OTFT	PSO	6.6%
[9]	TFT	PSO	6.82%
[9]	TFT	MLPSO	0.37%

In [8] the OFET model was established based on various theoretical models. In [9], mutual learning (ML) PSO can find the global optimal solution but not automatically.

Compared to the methods that only utilize PSO in [8] and [9], this study demonstrates a significant improvement in accuracy for both DC and AC modeling of OLED as shown in **Table 4**, validating the accuracy of the proposed models. Although the semi-automatic algorithm based on MLPSO exhibits slightly higher accuracy than this work, it incurs a considerable increase in complexity. This indicates that the approach employed in this study achieves a good balance between accuracy and automation, showing broad

application potential in display research.

Conclusion

In summary, the electrical model of OLED characteristics has been developed. The parameters have been defined and the extraction method based on the PSO has been implemented. The results indicate that the PSO-based OLED model has made significant progress in the speed and accuracy of parameter extraction.

Acknowledgments

This work was supported by National Key R&D Program of China (Grant No. 2023YFB3611300).

References

[1]   S. C. Xia, R. C. Kwong, V. I. Adamovich, M. S. Weaver, and J. J. Brown, "OLED Device Operational Lifetime: Insights and Challenges," in *2007 IEEE Int. Reliability Phys. Symp. Proc. 45th Annual*, 15-19 April 2007, pp. 253-257.

[2]   X. Shi and J. Sun, "The Seebeck Coefficient for Disordered Organic Semiconductors," in *2018 Int. Conf. Microwave Millimeter Wave Technol. (ICMMT)*, 7-11 May 2018, pp. 1-3.

[3]   A. Haldi, A. Sharma, W. J. Potscavage, Jr., and B. Kippelen, "Equivalent Circuit Model for Organic Single-Layer Diodes," *J. Appl. Phys.*, vol. 104, no. 6, p. 064503, 2008.

[4]   A. Fischer, M. Pfalz, K. Vandewal, S. Lenk, M. Liero, A. Glitzky, and S. Reineke, "Full Electrothermal OLED Model Including Nonlinear Self-Heating Effects," *Phys. Rev. Appl.*, vol. 10, no. 1, p. 014023, 2018.

[5]   L. m. Yan and H. Wang, "Research on the Models of OLED-On-Silicon Pixel Circuits," in *2007 Int. Symp. High Density Packaging Microsystem Integr.*, 26-28 June 2007, pp. 1-4.

[6]   V. C. Bender, T. B. Marchesan, and J. M. Alonso, "Solid-State Lighting: A Concise Review of the State of the Art on LED and OLED Modeling," *IEEE Ind. Electron. Mag.*, vol. 9, no. 2, pp. 6-16, 2015.

[7]   G. Cummins, I. Underwood, and A. Walton, "Electrical Characterization and Modelling of Top-Emitting PIN-OLEDs," *J. Soc. Inf. Display*, vol. 19, no. 4, pp. 360-367, 2011.

[8]   S. Fatima, U. Rafique, U. F. Ahmed, and M. M. Ahmed, "A Global Parameters Extraction Technique to Model Organic Field Effect Transistors Output Characteristics," *Solid-State Electron.*, vol. 152, pp. 81-92, 2019.

[9]   P. Liu, B. Liu, J. Feng, Z. Wang, Q. Zhang, X. Tang, Y. Li, G. Yuan, and X. Dong, "Research on Parameter Extraction of Thin-Film Transistors Based on Swarm Intelligence," *J. Soc. Inf. Display*, vol. 31, no. 5, pp. 398-411, 2023.

Modeling Dynamics-rich Devices with the Dynamic Time Evolution Method
(Invited)

Yu Li[1], Ning Feng[1], Runsheng Wang[2], Ru Huang[2], Lining Zhang[1*]

[1]School of Electronic and Computer Engineering, Peking University, Shenzhen, China. *E-mail: eelnzhang@pku.edu.cn

[2]School of Integrated Circuits, Peking University, Beijing, China.

Abstract

One key assumption behind the compact modeling of semiconductor devices is the quasi-static approximations (QSA). However, more emerging devices or phenomena in classical devices show characteristics of rich dynamics for which a paradigm shift from QSA is required. This work reviews the representative dynamics-rich device behaviors, spanning a variety of devices from transistors to memories. For transistors, self-heating and aging are representative dynamics of short and long terms, respectively. Hysteresis due to the forward and backward physical process is also shown up in thin film transistors and negative capacitance MOSFETs. For non-volatile memories, an individual state variable is dynamically evolving with dependences on the voltage or current stimulus. Applications of the dynamic time evolution method (DTEM) to these dynamic scenarios are reviewed, for accurate and efficient circuit simulations.

Introduction

A quasi-static approximation (QSA) is assumed essentially behind state-of-the-art MOSFET modeling. It implies that the channel carrier density does not explicitly depend on time t, but depends on the space x, and bias voltages $V(t)$ which have an dependence on time t in a transient scenario [1].

$$n(x, V(t), t) = n(x, V(t)) \quad (1)$$

QSA leads to the classical expression of transient terminal current in terms of the summation of a steady-state current and a displacement current with a partitioned terminal charge. Further implications of QSA include the charge partition schemes of MOSFETs and the charge-based capacitance modeling which conserves the charge under the integration methods of SPICE simulations.

On the other hand, QSA holds under assumptions of ignorable time constants of fast surface states and dielectric relaxation, as stated in the original proposal [1]. There are occasions where the assumption is not valid. For example, the surface states in some thin film transistors are not fast enough, such that the charging and discharging take a long period. A hysteresis in the transfer characteristics is observed [2] which is attributed to the dynamics of surface states. Another example is the ferroelectric device with the ferroelectric properties involved in determining the electrical properties but with a clear relaxation characteristic [3]. In fact, QSA does not hold for MOS transistors with fast transitions where the charge distribution in the channel is not given by the above Eq. (1). The non-quasi-static (NQS) term is often used to refer to these cases, for which some excellent models [4] had been established. While the NQS effect has been discussed thoroughly in literature, this work focuses on other dynamics-rich device behaviors, including the above examples.

The dynamic time evolution method (DTEM) was first developed to capture the aging effects in circuit simulations [5]. Since then, it has been found useful in many scenarios where certain dynamics are involved in the electronic devices. In this work the typical dynamics-rich devices are reviewed first, including the transistors and memories. Applications of the DTEM are then analyzed. The model interface, currently used in the industry, is possibly extendable to cover some of the dynamic behaviors.

Dynamic Modeling of Transistors

Fig.1 shows the schematic of a MOS transistor, in which several possible differential equations (DE) are listed, including the self-heating (DE3), the interface states (DE2), and the dielectric relaxations (DE1). When one or more of them show up, the transistor becomes a dynamics-rich device.

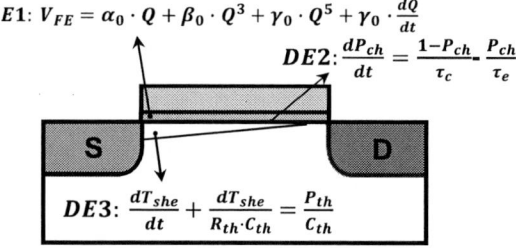

$$DE1: V_{FE} = \alpha_0 \cdot Q + \beta_0 \cdot Q^3 + \gamma_0 \cdot Q^5 + \gamma_0 \cdot \frac{dQ}{dt}$$

$$DE2: \frac{dP_{ch}}{dt} = \frac{1 - P_{ch}}{\tau_c} - \frac{P_{ch}}{\tau_e}$$

$$DE3: \frac{dT_{she}}{dt} + \frac{dT_{she}}{R_{th} \cdot C_{th}} = \frac{P_{th}}{C_{th}}$$

Fig.1 Possible dynamic processes in a transistor.

A. Self-Heating

Self-heating refers to the effect of Joule heat P_{th} in increasing the device temperature and hence changing the device characteristics. For example, the saturation current is reduced due to the temperature dependence of mobility. The Joule power as a voltage and current product is probably a dynamic term in an operational circuit. Due to the increased thermal resistance in nanoscale transistors, the temperature rise is becoming more significant.

The differential heat equation (DE3) in Fig.1, in which R_{th} and C_{th} are the thermal resistance and capacitance, resembles that of a simple RC circuit [6]. Note that a single equation is used here as an example, and an equation set is observed for nanoscale transistors like GAA MOSFETs [7]. As a result, one classical way to model the self-heating effect is to use an equivalent circuit [6]. The sub-circuit is implemented into the model with an additional node in the stamping matrix, which represents the temperature rise T_{she} from the ambient temperature. The whole circuit matrix is then increased, with one more node for each transistor in the circuit.

There is one way to remove the internal node, which is not connected to other instances, i.e. with the non-diagonal matrix entries only inside the stamping matrix. Once the stamping matrix is available, the solution of T_{she} could be formulated with the solution of terminal voltage and current by looking into the sub-circuit corresponding matrix row. By substituting this formulation into all the other matrix rows, the circuit matrix is folded. As a result, the matrix solving in the circuit simulator is done in a shorter time. The advantage is that the folded matrix is rigorously equivalent to the original matrix. The disadvantage is that the matrix folding is subject to a model compiler that directly handles the circuit matrix. Further, it becomes too complicated when there are more internal nodes for the complex heat equation set.

DTEM was first proposed [5] to trace the evolution of a certain dynamic variable in the device and circuits. Special attentions are devoted to accommodating the dynamic voltage scaling and dynamic frequency scaling in the circuit operations with the effective time t_{eff} approach. With the DTEM, the self-heating temperature is traced without turning to an internal node to retain the higher efficiency of self-heating free circuit simulations. The differential heat equation is solved first, by assuming a constant Joule power. Then the cases where the power is either increased or decreased are formulated with the t_{eff}. Fig.2 plots the schematic of dynamic Joule power and the t_{eff}. The model is then discretized for the simulator implementations. For those more complex heating equations, DTEM works well for an accurate and efficient simulation [7].

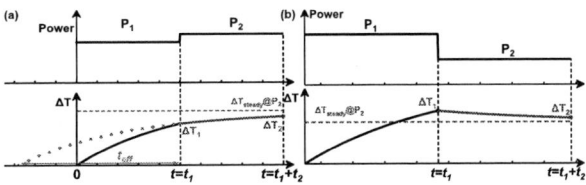

Fig.2 An illustration of the DTEM in the case of self-heating.

B. Dynamics of Interface States

Interface states are almost eliminated in state-of-the-art Si transistors of their fresh states. However, they are observed in a variety of thin film transistors due to different reasons like the device processing conditions. Depending on their physics origins, these interface states (or interface traps) may differ in energy levels, density, and cross sections. With a certain combination of these parameters, the traps will be occupied or emptied depending on the applied voltages.

The differential equation for trap dynamics (DE2) in Fig.2 also resembles that of a simple RC circuit [2], and the capture and emission time constants are voltage-dependent. It means that the amount of occupied traps changes with a dynamic relaxation process. It has been further proved that the trap dynamics lead to an important concern of hysteresis in the I-V characteristics. Similar to the self-heating effects, the trap dynamics could be modeled with the sub-circuit [2], or the DTEM. The time-dependent trap charge affects the threshold voltage hence the current, with a clear hysteresis captured.

C. Aging

Aging refers to the degradation of transistor characteristics like the threshold voltage, mobility, etc. due to, for example, the hot carrier degradation (HCD) and bias temperature instability (BTI). Recent studies [8] show that they are both caused by different kinds of traps, hence trap-based modeling is the new solution for aging prediction. On the other hand, the aging characterizations are always accelerated using higher stress which are most often constants. Any model derived under constant stress shall be reformulated to fit into the dynamic circuit stress. Recent studies of trap physics reveal the two-state or three-state dynamics of trapping and de-trapping. The two-state dynamics are similar to that in Fig.2, which is the basis for applying DTEM. It has been proven that the three-state oxide traps (those also responsible for stress and recovery) are also reproduced by DTEM with the modified t_{eff}. The time constants for the aging-related traps are too large to fulfill the requirement of QSA.

D. Dielectric Relaxation

A significant dielectric relaxation is not common among advanced CMOS but is substantial in the emerging negative capacitance transistors due to the ferroelectric gate stack. The polarization switching is described by the L-K equation with a damping factor [3]. Before the damping factor was reduced to an ignorable level due to fabrication process advancements, the relaxation showed up and affected the device behaviors. It is shown that a hysteresis by sweeping the drain voltage is also possible besides the more common gate voltage induced hysteresis.

The gate control equation becomes dynamic in this case, which in its static state is a transcendental equation. High accuracy is required in solving the gate control equation, making the sub-circuit approach not valid. It has been shown that the integration theme such as the Backward Euler method works in reproducing the relaxation and hysteresis.

Dynamic Modeling of Neuromorphic Devices

Fig.3 shows the schematic of a two-terminal memristor, in which several possible dynamic equations are listed per their operation mechanisms. Ferroelectric and phase change memories are used as two examples. Neuromorphic devices, in a wide scope, are those with state variables subject to gradual changes corresponding to synaptic properties and hence are intrinsically dynamics-rich devices.

A. Ferroelectric Polarization Switching

Ferroelectric materials favor two stable states with upward

and downward polarizations. Mathematically, the switching is a two-state dynamic process, with two transition rates between the states [9]. The external voltage adjusts the transition rates and induces polarization switching. Details involved in the process, such as the randomness in the nucleation are not fully represented in the differential equation (DE4) for the desirable compactness.

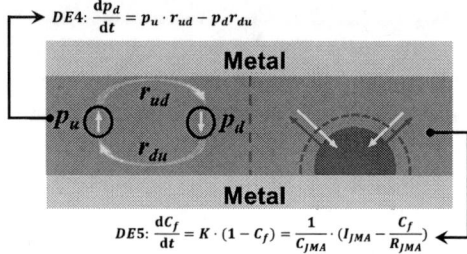

Fig.3 Possible dynamic processes in a neuromorphic device.

In the equation, the fraction of the downward polarization is noted as p_d, and the transition rates r_{ud}, r_{du} represent those inwards to and outwards from the downward polarization state. They are constrained with a detailed balanced principle. The equation is reformulated to cover the slow limit, and its final form again resembles the form of the simple RC circuit. DTEM applies to the dynamic equation if the sub-circuit approach is not desired from the perspective of simulation efficiency. With the dynamic modeling of polarization switching, emerging ferroelectric devices like the tunneling junctions [10] are characterized by frequency dependence as expected from experimental observations.

B. Dynamics of Crystallization

Phase change memory (PCM) utilizes the crystalline and amorphous states to represent the boundaries of a tunable resistance range. Joule heating is induced after a snapback, which enhances the nucleation and growth (DE5) of crystalline out of the amorphous states. From this aspect, PCM is very rich in multi-physics dynamics including thermal, nucleation, and electrical filament, posing a challenge in modeling and the associated circuit simulations.

It turns out that all the differential equations [11] follow the same mathematical formulation. Fundamentally, they all represent a phase space model with transitions, rooted in Pauli's master equation. For heating, the temperature rise varies between zero and a peak. For nucleation, the fraction varies between zero and one. For the filament, a relaxation is expected between its appearance and disappearance. As a result, the complex dynamics are solvable in a similar way, either with the sub-circuit or the DTEM. At the same time, the neuromorphic circuit simulations face a challenge in efficiency due to the large-scale network. It has been shown that with DTEM to reduce the circuit matrix size the simulation is accelerated by around four times.

A Modeling Interface Approach

From the above discussions, using a sub-circuit to solve

the differential equation is supported in the state-of-the-art SPICE framework, together with a Verilog-A compiler. On the other hand, deploying the DTEM could be supported with a modeling interface approach. Take the aging simulation as one example. The calculations of device parameter shifts are done by the DTEM with circuit simulation results at each time step. In other words, the transient simulations are run in parallel to the DTEM calculations, without the latter affecting the former at the first stage. The accumulated aging will be considered at the second stage to check its effects by including the parameter shifts in circuit simulations. This mechanism is implementable into a model interface, such as the Open Model Interface (OMI) of CMC [12].

When the dynamic variables are to be fed into the transient simulations, the model interface in principle still works, with the DTEM results of previous time points taken to the current time point. For example, the dynamics-rich PCM is modeled by an implementation of DTEM in the Verilog-A code. The convergence and acceleration of the simulations verify the idea. The model interface integrating instantaneous feedback of DTEM into the circuit matrix is expandable from its current architecture.

Conclusion

This work reviews the modeling of dynamics-rich devices and the methodology. Quasi-static approximation (QSA) is a legacy for transistor transient modeling. At the same time, there are requirements for a paradigm shift from QSA to dynamic modeling to adapt to the device mechanisms. Typical differential equations involved in the devices are reviewed, including the self-heating, the interface traps, the aging and dielectric relaxations of transistors, and the polarization relaxation, the crystallization of neuromorphic devices. The DTEM has been successfully deployed to model these dynamic phenomena, generally with significant efficiency enhancements. The modeling interface approach, currently supporting aging and self-heating simulations, is reviewed and proposed to further support the DETM in a wide scope.

Acknowledgments

This work was supported in part by the Natural Science Foundation of China under Grant 62474009, the Guangdong Basic and Applied Basic Research Foundation under Grant 2024B1515020064, and in part by the Beijing Outstanding Young Scientist Program (JWZQ20240101004).

References

[1] S. Oh, et al., TED, pp. 1571-1578, 1980. [2] Y. Li, et al., SSE, 108459,2022. [3] X. Huang, et al., TED, pp. 3665-3671, 2021. [4] M. Chan, et al., IEDM, pp. 169-172, 1994. [5] L. Zhang, et al., TED, pp. 184-190, 2019. [6] L. Zhang, et al., JEDS pp. 291-297, 2018. [7] S. Chen et al., IRPS, pp. 7C.1-1-7C.1-6, 2024. [8] Y. Li, et al., TED, pp. 206-212, 2024. [9] N. Feng, et al., EDL, pp. 390-393, 2022. [10] X. Chen, et al., TED, pp. 4404-4410, 2021. [11] N. Feng, et al., EDL, pp. 261-264, 2023. [12] OMI, Si2, 2018.

Heterogeneous and monolithic 3D (HM3D) integration of III-V and CMOS for next-generation wireless communications

Jaeyong Jeong[1], Yoon-Je Suh[1], Nahyun Rheem[1], Chan Jik Lee[1], Seong Kwang Kim[1], Bong Ho Kim[1], Joon Pyo Kim[1], Joonsup Shim[1], Minsik Park[2], Jeong-Taek Lim[2], Minkyoung Seong[3], Jooseok Lee[4], Kihyun Kim[4], Dae-Myeong Geum[5], Jongmin Kim[6], Woo-Suk Sul[3], Won-Chul Lee[3], Choul-Young Kim[2], Jongwon Lee[2], and Sanghyeon Kim[1*]

[1]School of Electrical Engineering, KAIST, Daejeon, Korea, [2]Chungnam National University, Daejeon, Korea, [3]Nano Convergence Technology Division, NNFC, Daejeon, Korea, [4]Samsung Electronics Company Ltd., Hwaseong, Korea, [5]Inha University, Incheon, Korea, [6]KANC, Suwon, Korea

[*]E-mail: shkim.ee@kaist.ac.kr

Abstract

Future wireless communication, such as 6G, demand systems that operate at high frequencies, offer multifunctionality and are energy efficient. However, individual solutions relying on Si CMOS or III−V technologies are no longer sufficient to meet these critical requirements. In this paper, we will discuss the potential of heterogeneous and monolithic 3D (HM3D) integration of III-V and CMOS to address these challenges and will present recent progress in HM3D integration from device level to circuit level.

Keywords: Heterogeneous integration, Monolithic 3D integration, HEMT, RF circuits, 6G

Introduction

High-frequency, multifunctional, and power-efficient systems are essential for upcoming next-generation wireless communication systems such as 6G and other advanced wireless technologies. However, the scaling of traditional CMOS technology presents challenges for RF applications, as shown in Fig. 1, restricting its ability to meet these emerging needs [1], [2]. On the other hand, III-V compound semiconductors demonstrate excellent capabilities for high-speed and high-frequency operations, achieving remarkable results in applications such as low-noise amplifiers (LNAs) and power amplifiers (PAs). To overcome the current technological barriers and realize next-generation wireless communications, integrating III-V semiconductors with Si CMOS in a heterogeneous and monolithic 3D presents a promising solution. Specifically, III-V materials are employed for RF operations, while Si CMOS is used for digital and analog functions, as shown in Fig. 2.

In this paper, we will discuss the challenges and potential benefits of HM3D integration of III-V on Si CMOS. Additionally, it will demonstrate the need for heterogeneous integration of III-V and CMOS, and provide a brief discussion on extending this approach from the device level to the circuit level.

Results and Discussion

A. HM3D integration of III-V on Si CMOS technology

III-V compound semiconductors, although known for their superior RF performance, are difficult to monolithically

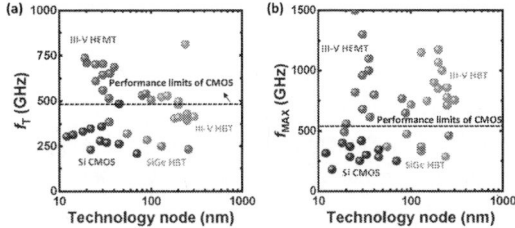

Fig. 1. The (a) f_T and (b) f_{MAX} as a function of technology node for state-of-the-art RF transistors [3].

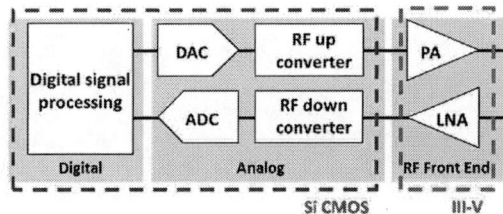

Fig. 2. The schematic of the HM3D integration technology for next-generation wireless communications [3].

integrate with Si CMOS. This difficulty arises from the significant physical disparities with Si, including lattice constant and thermal expansion coefficient differences, making direct growth challenging. To overcome these issues, several approaches have been recently demonstrated including direct wafer bonding (DWB) [4], [5], nano-ridge engineering [6], direct growth [7], and aspect ratio trapping [8]. Two major concerns in HM3D are the quality of the active layers and the temperature limitations during the fabrication of top devices. A high-quality top layer is crucial for achieving high performance in the top devices, and maintaining a low process temperature is necessary to prevent degradation of the lower interconnects and devices. Therefore, DWB with its relatively low processing temperature, is a more favorable approach compared to the other methods.

The HM3D integration of InGaAs HEMTs on Si CMOS using direct wafer bonding (DWB) was recently demonstrated as shown in Fig. 3 by our group [9]. The low-temperature DWB process ensures that the bottom Si CMOS circuits, which include ADC, can carry out analog-to-digital conversions without performance loss caused by the integration process. Moreover, the top-tier III-V RF devices successfully amplify high-frequency signals, achieving outstanding f_T and f_{MAX} of 329 GHz and 742 GHz.

However, one major challenge for 3D integration is thermal management, and the ability to implement effective heatsinks will be key. In HM3D systems, top devices are stacked over bottom devices using an interlayer dielectric like SiO2 (Fig. 4). Due to the low thermal conductivity of the interlayer dielectric, heat from the top device is not dissipated efficiently, making the structure more prone to self-heating effects compared to conventional 2-D planar devices. This results in a negative impact on both performance and reliability.

To improve this issue, we proposed the HM3D integrated InGaAs HEMTs with back metal that relaxes the self-heating of top devices as shown in Fig. 4. To evaluate this effect, we performed a thermal analysis using thermoreflectance microscopy (TRM) [10]. The TRM results in Fig.5 indicate that the back metal layer effectively alleviates the self-heating of the 3D stacked top devices.

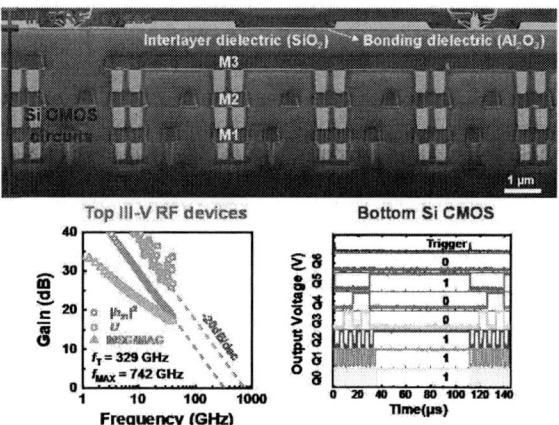

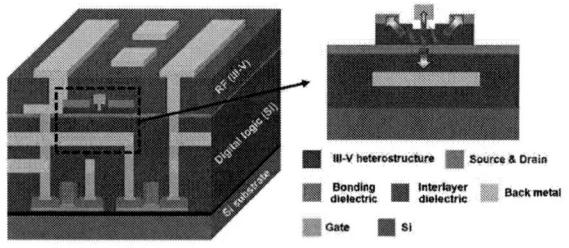

Fig. 3. (a) Cross-sectional SEM image of HM3D integration of III-V RF devices on Si CMOS. (b) Gain plot of top III-V RF devices. (c) Output waveforms of each bit of 7-bit ADC [9].

Fig. 4. Illustration of HM3D integration of III-V and Si CMOS and heat dissipation path in HM3D integration [10].

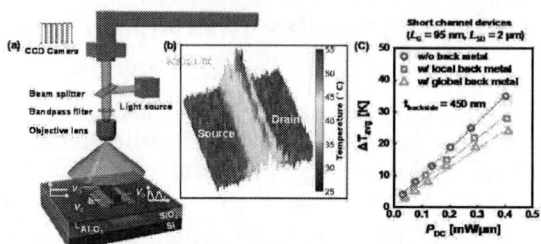

Fig. 5. (a) Schematic of TRM system. (b) 3D surface temperature image measured from TRM system. (c) ΔT as a function of P_{DC} of HM3D integrated InGaAs HEMTs with three different structures [10].

B. HM3D integrated RF circuits

Although significant progress has been made toward HM3D integrated RF systems, most studies still focus on discrete III-V-based RF devices rather than fully integrated RF circuits on Si CMOS [4], [5], [9]. Achieving HM3D integrated RF circuits requires thorough research that addresses both III-V active devices and passive components, as well as circuit configurations. Unfortunately, there is currently a lack of clear methodologies for implementing HM3D at the RF circuit level. While III-V technology offers advantages in the RF performance of active devices, it faces substantial drawbacks in terms of manufacturability and cost for passive components, mainly due to the limited wafer size and complex processing steps.

To successfully implement high-performance, multifunctional, power-efficient, highly manufacturable, and cost-effective RF systems, we proposed a novel approach involving HM3D integration of III-V-based active devices (here, InGaAs HEMTs) on CMOS-based backside integrated passive devices (BS-IPDs) as shown in Fig. 6 [3]. By doing so, we can harness the advantages of both technologies—high RF performance from III-V devices and cost-effectiveness and manufacturability from CMOS-based passive devices.

Using the developed HM3D integrated top-tier InGaAs HEMTs and BS-IPDs technologies, a 28-GHz HM3D integrated RF amplifier has been demonstrated [3]. Fig. 7 shows the chip photograph and the simplified schematic of the RF amplifier, respectively, which integrates top-tier InGaAs HEMTs with BS-IPDs. The design of the HM3D RF amplifier was centered on achieving high gain and low noise figures. The amplifier consists of two common-source stages with inductive source degeneration, which enhances performance in terms of gain, noise figure, and input matching. A low-loss CPW system at the backside and M3D interconnects are used to connect the BS-IPDs to the top-tier InGaAs HEMTs.

Fig. 8 displays the first successful circuit operation using HM3D technology within a frequency range of 10 to 40 GHz. The circuit achieves a maximum gain of 15.6 dB at 31 GHz.

979-8-3315-0417-5/25 $31.00 © 2025 IEEE

A small signal gain greater than 10 dB was observed between 20 and 38 GHz. At 28 GHz, both input return loss (S11) and output return loss (S22) are better than -10 dB. Optimizing the circuit design is expected to bring about additional performance improvement.

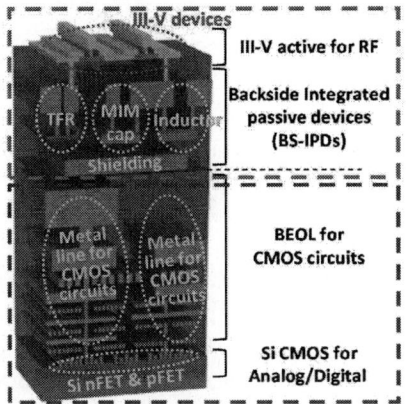

Fig. 6. The proposed schematic of the HM3D integration technology for next-generation wireless communications [3].

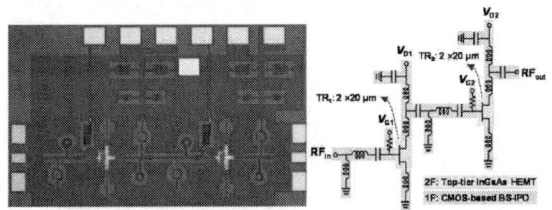

Fig. 7. Chip photograph and simplified schematic of HM3D integrated RF amplifier with top InGaAs HEMTs and CMOS-based BS-IPD [3].

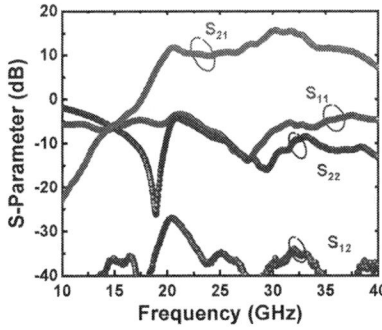

Fig. 8. S-parameters of HM3D integrated RF amplifier [3].

Conclusion

For next-generation wireless communications, the innovations from device to circuit and architecture are needed. We present the potential solution of HM3D integration, highlighting key advances from the device level to the circuit level. The results presented in this work pave the way for realizing highly integrated, high-performance, multifunctional RF systems that are essential for next-generation wireless communication technologies.

Acknowledgments

This work was partly supported by the National Research Foundation of Korea (NRF) grant (No. 2022M3F3A2A01065057, RS-2024-00407767), and BK21 FOUR.

References

[1] H. -J. Lee *et al.*, "Intel 22nm FinFET (22FFL) Process Technology for RF and mmWave Applications and Circuit Design Optimization for FinFET Technology," *IEEE International Electron Devices Meeting (IEDM)*, p. 14.1.1, 2018.

[2] J. Jeong *et al.*, "Heterogeneous and Monolithic 3D Integration Technology for Mixed-Signal ICs," *Electronics*, 11, p. 3013, 2022.

[3] J. Jeong *et al.*, "Heterogeneous and Monolithic 3D (HM3D) Integrated RF Circuits: HM3D Integration of InGaAs HEMTs on CMOS-Based Backside Passive Devices," *IEEE International Electron Devices Meeting (IEDM)*, p. 16.4.1, 2024.

[4] J. Jeong *et al.*, "Stackable InGaAs-on-Insulator HEMTs for Monolithic 3-D Integration," *IEEE Trans. Electron Devices*, 68, p. 2205, 2021.

[5] C. B. Zota *et al.*, "High performance quantum well InGaAs-on-Si MOSFETs with sub-20 nm gate length for RF applications," *IEEE International Electron Devices Meeting (IEDM)*, p. 39.4.1, 2021.

[6] A. Vais *et al.*, "First demonstration of III-V HBTs on 300 mm Si substrates using nano-ridge engineering." *IEEE International Electron Devices Meeting (IEDM)*, p. 9.1.1, 2019.

[7] M. Radosavljevic *et al.*, "Advanced high κ gate dielectric for high performance short channel In0.7Ga0.3As quantum well field effect transistors on silicon substrate for low power logic applications" *IEEE International Electron Devices Meeting (IEDM)*, p. 13.1.1, 2009.

[8] N. Waldron *et al.*, "An InGaAs/InP quantum well FinFET using the replacement fin process integrated in an RMG flow on 300 mm Si substrates" *IEEE Symposium on VLSI*, p. 1-2, 2014.

[9] J. Jeong *et al.*, "Heterogeneous and Monolithic 3D Integration of III−V-Based Radio Frequency Devices on Si CMOS Circuits, *ACS Nano*, 16, p. 9031, 2022.

[10] J. Jeong *et al.*, "Thermal Studies of 3D Stacked InGaAs HEMTs and Mitigation Strategy of Self-Heating Effect using Buried Metal Insertion", *IEEE Transactions on Electron Devices*, 71, p. 4517 2024.

979-8-3315-0417-5/25 $31.00 © 2025 IEEE

First demonstration of 4-inch GaN on SiO$_2$/Si(100) monolithic integration materials by ion-cutting technique with hydrophilic wafer bonding at elevated temperature

Jiaxin Ding[1,2], Jialiang Sun[1], Tiangui You[1,2], Xin Ou[1,2]

[1]National Key Laboratory of Materials for Integrated Circuits, Shanghai Institute of Microsystem and Information Technology, Chinese Academy of Sciences, China, [2]Center of Materials Science and Optoelectronics Engineering, University of Chinese Academy of Sciences, China

e-mail: t.you@mail.sim.ac.cn, ouxin@mail.sim.ac.cn

Abstract

In this study, we first demonstrate a 4-inch GaN single crystal thin film monolithically integrated with Si(100) substrate, forming a GaNOI structure (GaN on SiO$_2$/Si(100)), which was fabricated by ion-cutting technique with hydrophilic wafer bonding at elevated temperature. The exfoliated 4-inch GaN bulk wafer was survival which can be recycled to reduce the material cost. After the post-annealing and CMP processes, the GaNOI possessed high material quality. Our study suggests the possibility of combination of GaN-based optoelectronic devices and Si-based CMOS.

Keywords: Monolithic integration, GaNOI, Ion-cutting, Wafer bonding

Introduction

In recently years, GaN has gradually entered the fields of consumer electronics, e.g., mobile phone. It also has great potential applications in the power systems of electric vehicles.[1] In addition, thanks to the excellent performance of GaN, especially the radiation resistance, it may also have great applications in space applications, e.g., power supplies, digital-to-analog conversion. However, due to the high price of GaN bulk wafer, the large-scale application of GaN bulk wafer has been hindered, especially in the consumer electronics field.[2] To reduce the cost, hetero-epitaxially grown GaN on heterogeneous substrates, i.e., sapphire, Si(111), SiC, have been well developed and used for the device fabrications.[3-4] However, the hetero-epitaxially grown GaN thin film has high threading dislocation density (TDD) of $>10^8$ cm^{-2}, which is three orders of magnitude higher than that of GaN bulk wafer.[5] Moreover, the commonly used substrate for the hetero-epitaxial growth of GaN is CMOS incompatible. Therefore, we proposed to fabricate GaN on SiO$_2$/Si(100) (GaNOI) with ion-cutting technique which can exfoliate a high quality GaN thin film from the GaN bulk wafer and transfer it to Si(100) and even other desired substrates, which will reduce the cost of GaN wafer and provide a material platform for the integration of GaN-based optoelectronic devices and Si-based CMOS. Previously, we reported a 2-inch wafer-scale GaN film transferred onto Si(100) substrate by the ion-cutting technique.[6] In this study, we first demonstrate a 4-inch GaN single crystal thin film monolithically integrated with Si(100) substrates by ion-cutting technique with hydrophilic wafer bonding at elevated temperature.

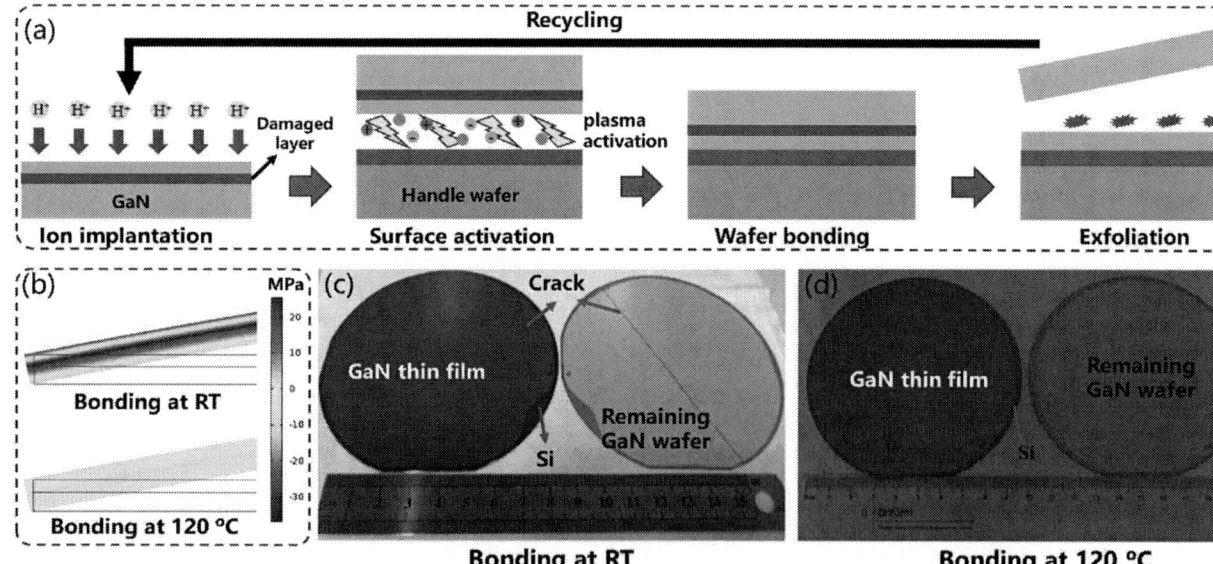

Fig. 1. (a) Fabrication process of Ion-cutting. (b) Thermal stress simulation of 4-inch GaN/Si bonding pair. (c) and (d) 4-inch GaNOI wafers fabricated by ion-cutting technique with wafer bonding at room temperature (RT) and at 120 °C.

Experimental details

The fabrication process of GaNOI was presented in Fig.1(a). Firstly, hydrogen ions (H⁺) with an energy of 160 KeV and a dose of 3.5E17 cm⁻² were implanted into 4-inch GaN bulk wafer, forming a ion implantation damage layer 1000 nm below the implanted surface. The GaN wafer were tilted by 7° from normal to minimize the ion channeling effect. After thoroughly cleaning the implanted GaN bulk wafer and the SiO$_2$/Si(100) handle substrate, they were activated by oxygen plasma with a power of 100/75W for 30 second using a low-temperature plasma-activation system. After that, they were bonded at a temperature of 120 °C to reduce the thermal stress between implanted GaN and SiO$_2$/Si(100) wafers during the annealing process. After the bonding process was finished, the GaN/SiO$_2$/Si bonding pair was cooled down slowly to room temperature, and then immediately transferred to a furnace for a further annealing process. After annealing at 450°C for 20 h, a 4-inch GaN thin film was successfully transferred to the SiO$_2$/Si substrate. The crystalline quality of the transferred GaN thin films were characterized by a Philips X' Pert X-ray diffractometer with a Cu Kα radiation source (λ=1.5418 Å). JEOL 2100F high-resolution transmission electron microscopy (HRTEM) was used to examine the material quality. The warp of wafer was measured by MicroProf® 200. And surface roughness was examined by Dimension ICON.

Results and discussions

It is worth noting that the previously reported 2-inch GaNOI was fabricated by modified surface-activated bonding (SAB) which was carried out in an ultrahigh-vacuum condition.[6] Moreover, the bonding interface defects were induced by the fast Ar ion beam bombardment during the SAB process. In this study, the 4-inch GaNOI was fabricated by hydrophilic bonding which is a more practical solution to combine different wafers under atmospheric conditions. Furthermore, to reduce the thermal stress caused by the thermal expansion mismatch of GaN and Si at the high-temperature exfoliation process, the hydrophilic bonding process of GaN and SiO$_2$/Si(100) was carried out at an elevated temperature of 120 °C. Fig. 1 (b) shows the simulation result of the thermal stress distribution which was bonded at room temperature (20°C) and 120 °C. It is clear that the maximum stress appears at the edge GaN/SiO$_2$/Si bonding interface. The maximum stress is 25.3 MPa for the wafer bonding at 120 °C which is much smaller than that for the wafer bonding at room temperature (2.01 MPa). As a result, an intact 4-inch GaNOI was fabricated with the hydrophilic wafer bonding at 120 °C, while the GaNOI was cracked when GaN and /SiO$_2$/Si was bonded at room temperature as shown in Fig. 1 (c) and (d).

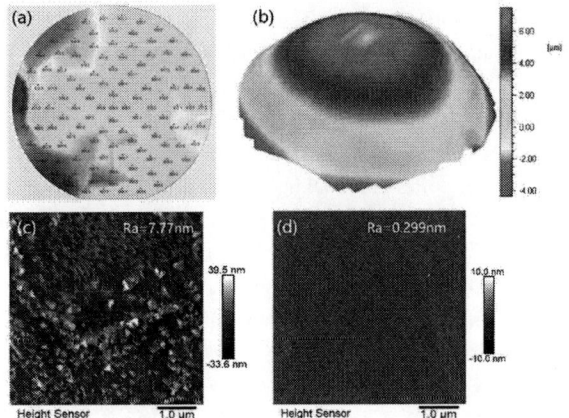

Fig. 2. (a) Thickness mapping of GaN thin film. (b) Warp of 4-inch GaNOI. AFM images of GaN thin film (c) before and (d) after CMP.

The thickness of the as-transferred GaN thin film was measured by an automated film thickness mapping system, as shown in Fig. 2 (a), which reveals an average thickness of 900 nm with a thickness deviation of ±1%. After the ion-cutting process, the wafer bow of the 4-inch GaNOI is as large as 7.387 μm convex, which was measured by the wafer bow mapping measurement, as shown in Fig. 2 (b). Fig. 2 (c) shows the AFM image of the as-transferred GaNOI. The RMS of the as-transferred GaN thin film is 7.77 nm with a scanning area of 5 μm × 5 μm. The rough surface could be attributed to the formation of implantation-induced micro-cracks layer during exfoliation process. A CMP process was carried out to smooth the surface of GaN thin film and to remove the top damaged layer induced by the ion-implantation. After CMP process, the RMS was decreased to 0.299 nm shown in Fig. 2 (d).

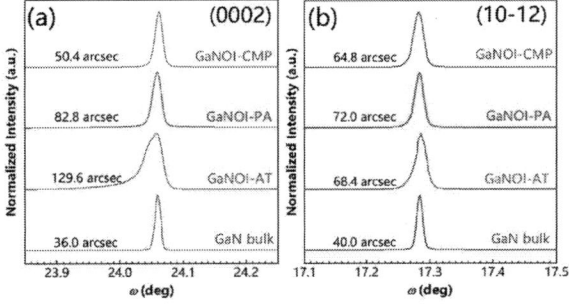

Fig. 3. X-ray rocking curves of GaNOI for (0002) and (10-12) plane

The single crystalline qualities of the transferred GaN thin films before and after the CMP process were further characterized using XRD. The normalized (0002) and (10-12) XRDs of the GaNOI and GaN bulk are shown in Fig. 3 (a) and (b), where the (0002) facet reflects the regularity of the out-of-plane lattice arrangement of the GaN lattice, while the (10-12) facet reflects the regularity of the in-plane lattice arrangement of the GaN lattice. It is obvious that the full width at half maximum (FWHM) was narrowed from 129.6 to 82.8 arcsec for (0002) plane after a post annealing process at 800°C for 2 h, which was employed to recover the implantation-induced damages. After the CMP process, the crystalline quality of GaN thin film was further improved since the top damage layer was removed, and the FWHM of (0002) plane was decreased to 50.4 arcsec, which is comparable to that of GaN bulk (36.0 arcsec). However, for the (10-12) plane, the change of the FWHM is much smaller after the post-annealing and CMP process in comparison with that of (0002) plane, which suggests that the ion implantation mainly affect the the out-of-plane lattice arrangement of the GaN lattice.

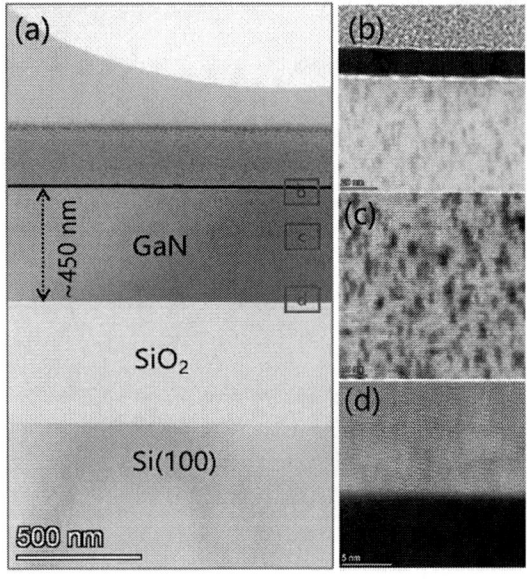

Fig. 4. XTEM images of GaNOI

The microstructure of the GaNOI after CMP process is further investigated by XTEM measurements as shown in Fig. 4 (a), which reveals a thickness of around 450 nm of GaN thin film after CMP. A clear and sharp interface of GaN/SiO₂ suggests that there is no significant interdiffusion. The high-resolution XTEM (HRTEM) images of the top, middle and bottom regions in GaN thin film are shown in Fig. 4 (c), (d) and (e). There are some residual nano-defects induced by the ion implantation in the top and middle regions of GaN thin film, while the bottom region of the GaN thin film maintains the good crystalline quality as indicated by the regular and clear GaN lattice shown in Fig. 4 (d).

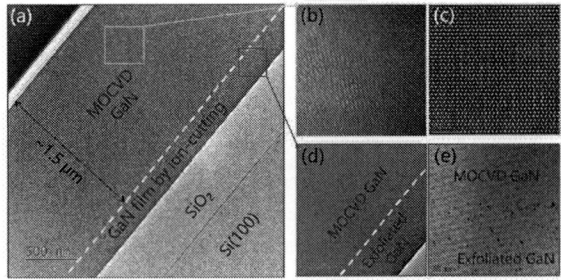

Fig. 5. XTEM images of GaN layer epitaxially grown on GaNOI by

MOCVD

A GaN layer with a thickness of around 1.5 μm was epitaxially grown on GaNOI as shown in the XTEM image in Fig. 5 (a). As shown in Fig. 5 (b) and (c), the HRTEM images suggest that the epitaxially grown GaN layer possesses good crystalline quality with clear and regular GaN lattice. Fig. 5 (d) and (e) show the HRTEM images in the area of the interface of exfoliated GaN layer and epitaxially grown GaN layer. It can be seen that there are some implantation-induced point defects in the exfoliated GaN layer. Fortunately, the point defect was not observed in the epitaxially grown GaN layer. It is suggested that the GaNOI fabricated by ion-cutting technique can be used to grow high quality GaN, which provides a platform for the integration of GaN devices on Si CMOS ICs for next-generation electronics and optoelectronics.

Conclusion

In summary, we achieved a 4-inch GaNOI by using ion-cutting technique with hydrophilic wafer bonding at elevated temperature. The fabricated GaN thin film possesses high crystalline quality as suggested by the XRD and TEM measurements. Even though there are some residual implantation-induced point defects in GaNOI, it is achievable to grow high quality GaN on GaNOI. Our study shows the potential of integration of GaN-based optoelectronic devices and Si-based CMOS.

Acknowledgments

This work was supported by the National Key Research and Development of China (No. 2021YFB3601900), National Natural Science Foundation of China (No. 62293521, 62174167), Strategic Priority Research Program of Chinese Academy of Sciences (No. XDB0670202), Shanghai Rising-Star Program (No. 22QA1410700).

References

[1] Chaudhary, O. S., Dena?, M., Refaat, S. S. & Pissanidis, G. Technology and Applications of Wide Bandgap Semiconductor Materials: Current State and Future Trends. Energies 16, 6689 (2023).

[2] Nela, L. et al. Ultra-compact, High-Frequency Power Integrated Circuits Based on GaN-on-Si Schottky Barrier Diodes. IEEE Trans. Power Electron. 36, 1269–1273 (2021).

[3] Feng, Y. et al. Epitaxy of Single - Crystalline GaN Film on CMOS - Compatible Si(100) Substrate Buffered by Graphene. Adv. Funct. Mater. 29, 1905056 (2019).

[4] Aspnes, D. E. & Ihm, J. Biatomic Steps on (001) Silicon Surfaces. Phys. Rev. Lett. 57, 3054–3057 (1986).

[5] Setera, B. & Christou, A. Threading dislocations in GaN high-voltage switches. Microelectron. Reliab. 124, 114336 (2021).

[6] Shi, H. et al. Realization of wafer-scale single-crystalline GaN film on CMOS-compatible Si(100) substrate by ion-cutting technique. Semicond. Sci. Technol. 35, 125004 (2020).

979-8-3315-0417-5/25 $31.00 © 2025 IEEE

Novel Hybrid Gate Ferroelectric Transistor-based Weight Device with High Linearity and Symmetry for On-Chip Learning

Yuxin Lin[1#], Jin Luo[2*#], Zerui Chen[1], Zhiyuan Fu[2], Qianqian Huang[2,3*] and Ru Huang[2,3*]

[1]School of Software & Microelectronics, Peking University, Beijing 102600, China. [2]School of Integrated Circuits, Peking University, Beijing 100871, China. [3]Beijing Advanced Innovation Center for Integrated Circuits, Beijing 100871, China. *E-mail: ruhuang@pku.edu.cn, hqq@pku.edu.cn, luoj@pku.edu.cn. [#]Equal contribution

Abstract

In this work, a novel hybrid gate ferroelectric (FE) transistor (HG-FeFET) based weight device is proposed and experimentally demonstrated for the first time, enabling high linearity and symmetry for enhanced on-chip learning. By leveraging the FE-thickness-dependent coercive voltage (V_C) distributions of FE layers, and the superposition effect of V_C distributions in parallel FE layers, the HG-FeFET, featuring hybrid FE layers of varying thickness connected in parallel, achieves a broadened V_C distribution. This results in a 3.6-fold improvement in linearity and symmetry under identical programming pulses, as well as higher weight precision. Based on the proposed HG-FeFET, the on-chip learning process of image recognition is demonstrated with a high accuracy of 95.33%, showing its great potential for edge AI applications.

(*Keywords*: Ferroelectric FET, computing-in-memory, on-chip learning, linearity and symmetry)

Introduction

Deep neural networks (DNNs) have been widely adopted for edge applications such as embodied intelligence and autonomous driving (Fig. 1a), where on-chip learning is crucial for real-time adaptation to diverse environments [1-2]. Computing-in-memory (CIM) architectures based on high-density non-volatile memory (NVM), capable of performing highly parallel vector-matrix multiplication (VMM) and in-situ weight updates within memory arrays, have garnered significant attention for accelerating on-chip learning in DNNs [3]. Hafnium-based ferroelectric (FE) field-effect transistors (FeFETs) are particularly promising for CIM due to their high CMOS compatibility, low energy consumption, and high on/off ratios [4]. To enhance the accuracy of on-chip learning in FeFET-based CIM, improving the linearity and symmetry of FeFET weight updates is essential. Recent advancements include superlattice gate stack-based FeFETs with improved FE domain density [5-6] and 2T-1FeFET weight cells with hybrid precision [7], both of which aim to address linearity and symmetry challenges. However, these designs often rely on complex peripheral circuits for non-identical programming pulse schemes (Fig. 1b)

or intricate weight cell architectures, which still suffer from high hardware cost.

In this work, a novel ferroelectric transistor with a hybrid gate stack (HG-FeFET) incorporating FE layers of varying thicknesses is proposed as a weight device for on-chip learning (Fig. 2). By leveraging the broadened coercive voltage (V_C) distribution of parallel FE layers, the HG-FeFET is experimentally demonstrated to achieve improvement in linearity and symmetry, along with more conductance states within a single device. Based on this design, the high-accuracy on-chip learning of DNN is achieved, highlighting its great potential for edge applications.

Characteristics of ferroelectric capacitors with hybrid layers of differing thickness

The FeFET leverages non-volatile FE polarization in the FE gate to modulate channel conductance. The linearity and symmetry of the weight update process are directly influenced by the uniformity of polarization changes in the FE layer in response to programming pulses [5-6]. The structure of the proposed HG-FeFET weight device is illustrated in Fig. 2a. The hybrid gate stack consists of two FE layers with varying thicknesses connected in parallel, designed to broaden the overall distribution of ferroelectric coercive voltage (V_C) and improve programming uniformity. To analyze this behavior, we first investigate the properties of FE layers with different thicknesses using fabricated ferroelectric capacitors (FeCaps).

A. FeCap fabrication and modeling

The FeCaps of $TiN/Hf_{0.5}Zr_{0.5}O_2/TiN$ structure (Fig. 3a) with varying FE layer thicknesses are experimentally fabricated and analyzed based on the multi-domain Preisach FE model [8] (Fig. 3b). In this model, the overall distribution of coercive field (E_C) for a large number of FE domains in the FE layer is assumed to follow a Gaussian distribution, which corresponds to the coercive voltage (V_C) distribution of the FeCaps. The V_C distribution and remanent polarization (P_r) of FeCaps with different FE layers are calibrated using measured P-V characteristics (Fig. 3c). The mean and variance of V_C distribution increase with the thickness of the FE layer, and the distribution is relatively concentrated (Fig. 4a). This

979-8-3315-0417-5/25 $31.00 © 2025 IEEE

concentration leads to non-uniform polarization switching under continuous programming pulses in FeFETs with a constant FE layer thickness.

B. *Behavior of FeCaps with hybrid thicknesses*

In this work, we propose a novel hybrid FE layer design with varying thicknesses connected in parallel to broaden the V_C distribution through the superposition effect of independent distributions. As shown in Fig. 4b, the overall V_C distribution of the hybrid FeCap, simulated using the multi-domain Preisach FE model, appears more dispersed compared to a single FeCap. Moreover, due to the thickness-dependent P_r, the V_C distribution of hybrid FeCaps can be further optimized by adjusting the area ratio of the parallel-connected FeCaps with different thicknesses (Fig. 4c). Experimentally, the hybrid FeCap is constructed by connecting FeCaps with 10 nm and 15 nm FE layers in parallel. As shown in Fig. 5, the dynamic I-V curves confirm that the hybrid FeCap broadens the V_C distribution, leading to improved linearity and symmetry of FE polarization switching under continuous identical pulses (Fig. 6).

HG-FeFET based weight update behavior

The proposed HG-FeFET and conventional FeFET based weight cells are experimentally constructed by respectively connecting the hybrid FeCap and single-thickness FeCap to the gate of a MOSFET (Fig. 7ab). The I_d-V_g curves under relatively high sweep gate voltage exhibit typical FE hysteresis due to the polarization switching (Fig. 7c), indicating the weight storage capability. As shown in Fig. 8a-c, the continuous positive and negative programming pulses are applied to the gate of weight cells. The channel conductance (G_{DS}) is measured at the interval of pulses, corresponding to the long-term potentiation and depression behavior. Benefitting from the broadened V_C distribution, the HG-FeFET weight update exhibits significantly improved linearity and symmetry compared to the conventional FeFET, highlighting its potential as a weight device for on-chip learning.

Moreover, the HG-FeFET with optimized area ratio of FE layers with different thicknesses is investigated by HSPICE model, which is established based on the multi-domain Preisach FE model and BSIM-4 MOSFET model (Fig. 9a). As shown in Fig. 9b, under identical programming pulses, the proposed

HG-FeFET with a 2:1 area ratio of 15 nm and 8 nm FE layers achieves superior linearity and symmetry, showing a 3.6-fold improvement compared to conventional FeFET weight devices. Using the proposed HG-FeFET weight cell, the on-chip learning process of a two-layer neural network is demonstrated for image recognition (Fig. 10a). The network achieves an accuracy of 95.33% after 200 iterations (Fig. 10b), matching the precision expected for an ideal 6-bit weight. These results indicate that the proposed HG-FeFET weight device holds significant promise for edge on-chip learning in DNNs.

Conclusion

In this work, a novel HG-FeFET weight device is proposed and experimentally demonstrated with broadened VC distribution, enabling a significant improvement in linearity and symmetry for weight updates, as well as an increased number of intermediate conductance states. Based on the proposed HG-FeFET, on-chip learning is successfully demonstrated for image recognition with high accuracy, highlighting its great potential for edge AI applications.

Acknowledgments

This work was supported by NSFC (62404008, 61927901, 62374009), 111 Project (B18001), and China Postdoctoral Science Foundation (2024M750098, GZC20240026).

References

[1] W. Zhang et al., "Edge learning using a fully integrated neuro-inspired memristor chip," Science, vol. 381, no. 6663, pp. 1205–1211, Sep. 2023, doi: 10.1126/science.ade3483.

[2] Z. Wang et al., "CPE: An Energy-Efficient Edge-Device Training with Multi-dimensional Compression Mechanism," in 2023 60th ACM/IEEE DAC, Jul. 2023, pp. 1–6, doi: 10.1109/DAC56929.2023.10247968.

[3] J. An et al., "Design memristor-based computing-in-memory for AI accelerators considering the interplay between devices, circuits, and system," Science China Information Sciences, vol. 66, no. 8, pp. 182404, Jul. 2023, doi: 10.1007/s11432-022-3627-8.

[4] Z. Zhang et al., "Recent progress of hafnium oxide-based ferroelectric devices for advanced circuit applications," Science China Information Sciences, vol. 66, no. 10, pp. 200405, Sep. 2023, doi: 10.1007/s11432-023-3780-7.

[5] K. A. Aabrar et al., "BEOL Compatible Superlattice FerroFET-based High Precision Analog Weight Cell with Superior Linearity and Symmetry," in 2021 IEEE IEDM, Dec. 2021, pp. 19.6.1-19.6.4, doi: 10.1109/IEDM19574.2021.9720713.

[6] Y. Zhou et al., "Hybrid-FE-Layer FeFET With High Linearity and Endurance Toward On-Chip CIM by Array Demonstration," IEEE Electron Device Letters, vol. 45, no. 2, pp. 276–279, Feb. 2024, doi: 10.1109/LED.2023.3346030.

[7] X. Sun et al., "Exploiting Hybrid Precision for Training and Inference: A 2T-1FeFET Based Analog Synaptic Weight Cell," in 2018 IEEE IEDM, Dec. 2018, pp. 3.1.1-3.1.4, doi: 10.1109/IEDM.2018.8614611.

[8] J. Luo et al., "Capacitor-less Stochastic Leaky-FeFET Neuron of Both Excitatory and Inhibitory Connections for SNN with Reduced Hardware Cost," in 2019 IEEE IEDM, Dec. 2019, pp. 6.4.1-6.4.4, doi: 10.1109/IEDM19573.2019.8993535.

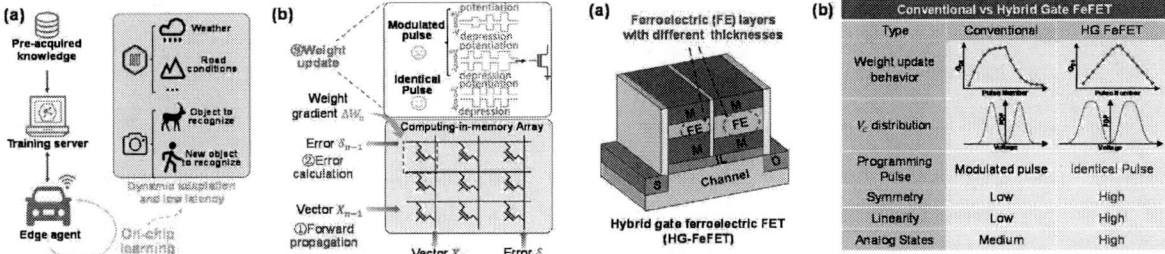

Fig. 1 (a) On-chip learning applications in autonomous driving. (b) On-chip learning process based on CIM array, and two programming pulses for weight update.

Fig. 2 (a) Schematic of hybrid gate ferroelectric FET. (b) Comparison of conventional ferroelectric and hybrid gate FeFET.

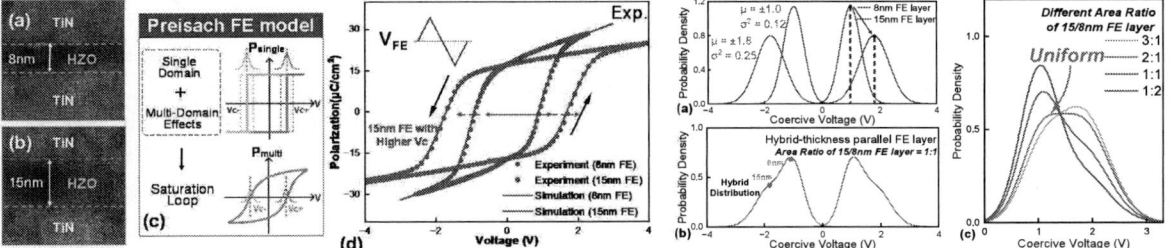

Fig. 3 (a) (b) TEM image of FE layer. (c) Multi-domain FE model based on the Preisach. (d) Measured P-V characteristics of the FeCAPs compared with the curve fitted using the Preisach model.

Fig. 4 Probability density functions of the Vc distributions for (a) 8nm and 10nm FE, (b) a hybrid FE layer, (c) hybrid FE with varying area ratios.

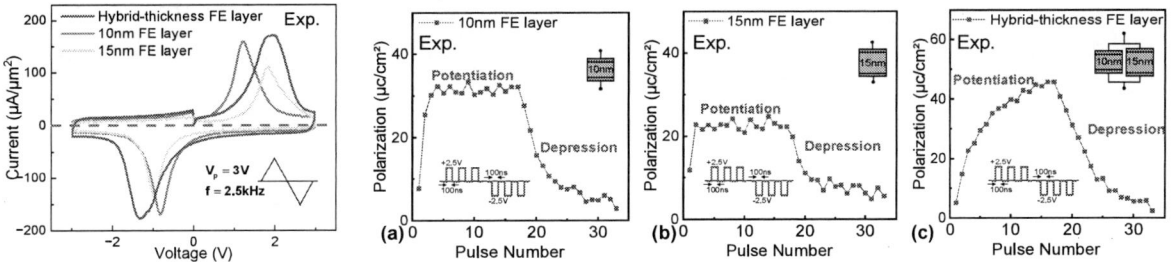

Fig. 5 Measured I-V characteristics showing stretched polarization current peaks of hybrid FE layer.

Fig. 6 The polarization versus pulse number diagrams for (a) 10nm single-thickness FE capacitor, (b) 15nm single-thickness FE capacitor, and (c) Hybrid-thickness FE capacitor (10nm and 15nm FE connected in parallel).

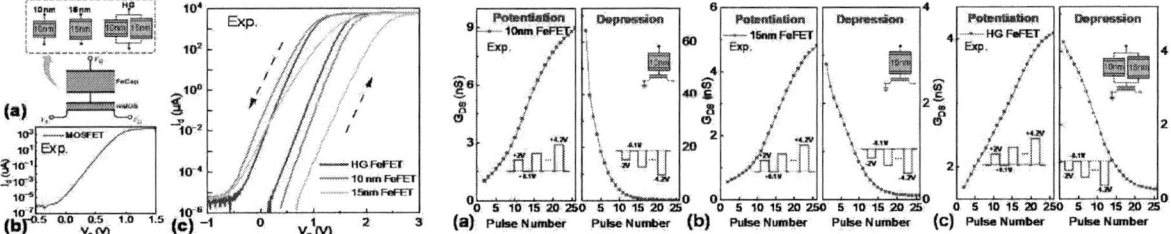

Fig. 7 (a) Experimental design for Hybrid Gate FeFET. (b) (c) Measured I-V characteristics of the MOSFET and FeFETs.

Fig. 8 Conductance versus pulse number diagrams for (a) 10nm single-thickness FeFET, (b) 15nm single-thickness FeFET, and (c) mixed-thickness FeFET (10nm and 15nm FE connected in parallel).

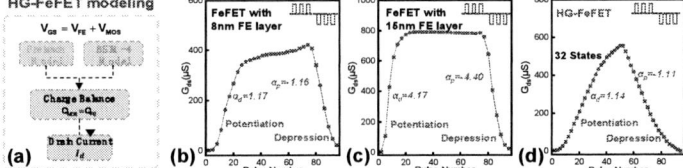

Fig. 9 (a) The modeling framework of HG-FeFET. (b) Simulated Conductance versus pulse number diagrams of FeFETs under identical pulse.

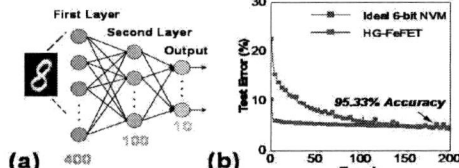

Fig. 10 (a) Diagram of neural network architecture for image recognition. (b) Simulated on-chip learning.

979-8-3315-0417-5/25 $31.00 © 2025 IEEE

Oxygen Vacancy-Zr Content Synergy for Morphotropic Phase Boundary Towards High-performance DRAM Applications

Jinhao Liu[1], Xuepei Wang[1], Maokun Wu[1,*], Boyao Cui[1], Yichen Wen[1], Yishan Wu[1], Sheng Ye[1], Pengpeng Ren[1], Runsheng Wang[2], Zhigang Ji[1,*], and Ru Huang[2]

[1]Departure of Micro/Nano Electronics, Shanghai Jiao Tong University, Shanghai, China (*email: maokunwu@sjtu.edu.cn, zhigangji@sjtu.edu.cn), [2]School of Integrated Circuits, Peking University, Beijing, China

Abstract

To advance DRAM technology, balancing high permittivity with reliability remains a critical challenge. This study achieves ultra-high permittivity (~67) and low leakage (3E-8 A/cm² at 0.8 V) in hafnium-based films through combined modulation of oxygen vacancy (V_O) and Zr content. TDDB and endurance tests confirm the superior reliability of these films. By adjusting ozone dose time during atomic layer deposition (ALD), oxygen vacancies are flexibly controlled, influencing phase ratio to induce morphotropic phase boundary and significantly enhance permittivity. The synergistic effect of Zr content further boosts permittivity while mitigating reliability issues. This co-modulation strategy overcomes traditional trade-offs between high permittivity and reliability, providing a promising approach for next-generation high-performance DRAM applications.

Introduction

As the explosive growth of AI model sizes intensifies the demand for high-performance memory, DRAM remains indispensable despite advancements in emerging memory technologies due to its unique ability to meet stringent product-level requirements for power, latency and bandwidth [1]. To increase integration density, mainstream DRAM technology is scaling toward the 10 nm node. However, reduced cell area lowers cell capacitance, underscoring the need for optimized higher-κ and more reliable dielectric materials to maintain data sensing margins [2]. Hafnium-based high-κ materials, known for their excellent CMOS compatibility, thickness scalability and relatively large bandgap (~5.5 eV), present a promising solution for next-generation DRAM capacitors [3].

Researchers have leveraged the morphotropic phase boundary (MPB) between ferroelectric (FE) and antiferroelectric (AFE) phases as an effective strategy to enhance the permittivity of hafnium-based films [4]. Recently, a record permittivity of ~68 was achieved in Al-doped HfO₂ near MPB via deposition temperature engineering [5]. However, high leakage (~1E-5 A/cm²) and precise doping requirements limit its scalability. Hafnium zirconium oxide (HZO) has attracted considerable interest for its tunable Zr content, which facilitates phase transition and broadens the design space for MPB. Both experimental evidence and theoretical predictions confirm the existence of MPB in the HZO system [6, 7]. Nevertheless, boosting permittivity via Zr-only modulation proves limited,

indicating the need for further structural and process optimizations [8-12]. Oxygen vacancies also play a crucial role in phase stability and dielectric response [13, 14], offering another enhancement pathway but require careful control of the process window to mitigate reliability trade-offs. By strategically balancing V_O and Zr content, optimal dielectric reliability and performance can be achieved.

In this study, electrical and physical characterizations confirm the effective modulation of phase ratio by V_O concentration, yielding a permittivity of 61 in oxygen-deficient $H_{0.5}Z_{0.5}O_2$ sample. Additionally, through co-adjustment of V_O and Zr content, the $Hf_{0.45}Zr_{0.55}O_2$ sample achieved ultra-high permittivity (~67), low leakage (3E-8 A/cm² at 0.8 V) and outstanding reliability. **Fig. 1** compares the performance with state-of-the-art reports, highlighting the exceptional capabilities of the fabricated devices.

Experiments and Methods

Fig. 2 presents a schematic diagram of the metal-insulator-metal (MIM) capacitors on SiO₂/Si substrate, along with the key process flow. After standard RCA cleaning of the substrate, a 50 nm tungsten (W) layer was sputtered via physical vapor deposition (PVD) as the bottom electrode (BE). Next, 10 nm thick $H_{0.5}Z_{0.5}O_2$ (HZO) films were fabricated via ALD, using TDMAH, TDMAZ, and ozone as the Hf, Zr, and oxidant precursors, respectively. The hafnium-to-zirconium ratio was fine-tuned through the number of ALD cycles. Ozone dose times, ranging from 0.1 to 10 s, were adjusted to control oxygen content. A 50 nm W layer was then deposited as the top electrode (TE), and a 50×50 μm^2 device cell was defined by the lift-off process. Finally, post-metal annealing (PMA) was conducted for 30 s in N₂ ambient at 450–650 °C to achieve crystallization.

The electrical properties were characterized using a semiconductor parameter analyzer (Keithley 4200). The dielectric response was obtained from small-signal (50 mV) capacitance measurements at 1 kHz using the LCR meter (Agilent E4980A). Physical characterization, including X-ray photoelectron spectroscopy (XPS) for chemical state analysis and grazing incidence X-ray diffraction (GIXRD) for crystallization analysis, was also conducted.

Results and Discussion

A. Oxygen Vacancy modulation for MPB via Ozone Dose

Fig. 3(a)&(b) illustrate the typical polarization-electric field (*P-E*) loops and relative permittivity-electric field (κ-*E*)

979-8-3315-0417-5/25 $31.00 © 2025 IEEE

curves of $H_{0.5}Z_{0.5}O_2$ films under different ozone dose times, reflecting their ferroelectric and dielectric responses, respectively. As the ozone dose decreases, a transition from FE to AFE-like behavior is observed, with P-E loops evolving from saturated to pinched and even showing MPB-like broken loop [15]. Correspondingly, the κ-E curves exhibit an upward shift, transitioning from FE butterfly-like shapes to merged peak shapes. Both trends indicate a significant transformation from the orthorhombic phase (o-phase) to the tetragonal phase (t-phase) driven by the modulation of ozone dose, which could be ascribed to the fact that more oxygen vacancies can stabilize the t-phase [13]. The **inset of Fig. 3(b)** shows the cumulative probability of relative permittivity at 0V ($\kappa_{@V=0}$) for 20 devices under each ozone dose condition, confirming strong uniformity and great device-to-device variation (DDV). **Fig. 4(a)** depicts the leakage characteristics of the above samples, with leakage current density (J_{leak}) at 0.8 V under oxygen-deficient conditions increasing by approximately one order of magnitude compared to oxygen-rich conditions, possibly related to V_O concentration as analyzed through the trap-assisted tunneling (TAT) mechanism [16]. These observations demonstrate the phase modulation effects induced by V_O across varied ozone doses. Specifically, as summarized in **Fig. 4(b)**, the values of both $2P_r$ and $\kappa_{@V=0}$ as a function of ozone dose time reveal that the $2P_r$ value at 3 MV/cm for the oxygen-deficient sample is less than half of that for the oxygen-rich sample (21 $\mu C/cm^2$ at 0.1 s sample, 48 $\mu C/cm^2$ at 10 s sample), while its $\kappa_{@V=0}$ value is nearly twice as high (61 at 0.1 s sample, 36 at 10 s sample).

Fig. 5(a) demonstrates the GIXRD results for HZO thin films with different oxygen conditions. With ozone dose increased, the mixed o(111)/t(101) phase peak (located at 30.5°) gradually shifts to lower 2θ values, providing direct physical evidence of the evolution from o-phase to t-phase dominance. **Fig. 5(b)** illustrates the XPS spectrum of O 1s, identifying 2 peaks corresponding to lattice oxygen and oxygen vacancy [17]. The concentration of V_O increases from ~10% in the 0.1 s ozone dose sample to ~16% in the 10 s ozone dose sample, consistent with the leakage results discussed above.

Therefore, our results reveal the profound impact of V_O on the phase modulation in HZO thin films. The FE and AFE behaviors of HZO films can be flexibly modulated by adjusting the ozone dose, leading to significantly enhanced permittivity that exceeds the κ-value of typical t-phase (~40) [6] and indicating the presence of an effective MPB formation window enabled by this V_O-driven phase modulation strategy. The accompanying leakage and reliability issues are also critical considerations to ensure their suitability for DRAM technology applications. Current MPB region requires an oxygen dose of 0.1 s, necessitating precise equipment control. Moreover, reliability issues resulting from these conditions have been corroborated in various studies [13, 16]. Previous works [6, 7] have shown that Zr content is crucial for inducing MPB. Inspired by these findings, we achieved HZO films with high-κ, low leakage and robust reliability through synergetic modulation of V_O and Zr content.

B. Synergetic modulation with Oxygen Vacancy and Zr content for MPB

We employed the V_O-Zr content synergy strategy to fabricate $H_{0.45}Z_{0.55}O_2$ samples with a 0.3 s ozone dose, aiming to induce sufficient t-phase formation and enhance permittivity through increased MPB regions. **Fig. 6** presents the electrical characterization results of these samples at different PMA temperatures. As the annealing temperature increases, the P-E loops gradually open from the highly pinched state, while the κ-E curves evolve from merged, nearly flat peaks to typical AFE double-butterfly shape, eventually developing into significantly enhanced MPB-like merged peaks. The $\kappa_{@V=0}$ value rises, reaching a maximum of ~67 at PMA temperature of 650 °C. The behaviors reflect the evolution from the dielectric monoclinic phase (DE m-phase) to AFE t-phase, and ultimately to FE o-phase dominance (PMA > 650 °C, data not shown), correlating with the energy provided by PMA to overcome the phase transition barrier [18]. The group of 0.1-10 s ozone dose $H_{0.5}Z_{0.5}O_2$ samples follows the same trend described above, with our discussion focused on the results observed near the MPB window (PMA at 550 °C). In addition to the high permittivity, leakage performance was evaluated in **Fig. 7(a)**. Although the leakage level increases with annealing temperature, it consistently meets leakage criteria for DRAM capacitors, reaching 3E-8 A/cm^2 at 0.8 V when the maximum $\kappa_{@V=0}$ value is achieved. The AC & DC reliability of $Hf_{0.45}Zr_{0.55}O_2$ sample annealed at 650 °C was further assessed. Endurance testing was conducted under an electric stress of 1 MV/cm at 1 MHz. As illustrated in **Fig. 7(b)**, $\kappa_{@V=0}$ and J_{leak} at 1 MV/cm show minimal degradation even after 10^{10} cycles. **Fig. 7(c)** presents the DC lifetime evaluation using a time-dependent dielectric breakdown (TDDB) method under constant voltage stress (CVS) mode. 60 devices were tested at 4.0-4.2 MV/cm, establishing the corresponding Weibull distribution. Based on the 63% failure time, the extrapolated 10-year lifetime operating electric field reaches 2.9 MV/cm, as depicted in **Fig. 7(d)**. Together, these results provide evidence of the outstanding reliability of our samples, highlighting the promising potential for high-performance DRAM applications.

Conclusion

This work experimentally demonstrates MPB-HZO films achievement through oxygen vacancy-Zr content synergy, resulting in high permittivity (~67) and low leakage (3E-8 A/cm^2 at 0.8 V). With excellent AC & DC reliability, including robust endurance and a high TDDB electric field, the strategy opens opportunities for future DRAM technology.

Benchmark & Fabrication

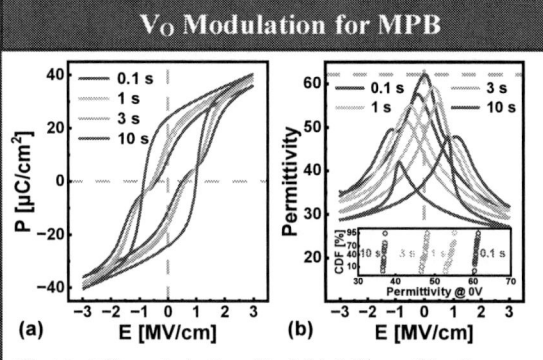

Fig. 1 Benchmark of κ values and leakage for our hafnium-based films with previous literature.

Fig. 2 Schematic diagram and key process flow for the fabrication of the MIM capacitors with ozone dose time engineering.

References

[1] N. Ramaswamy, et al., *IEDM*, 2023. [2] S. Kim, et al., *Adv. Mater. Technol.*, 2023. [3] Z. Luo, et al., *APL*, 2023. [4] M. Jung, et al., *Nano Converg.*, 2022. [5] J. Zhou, et al., *IEDM*, 2021. [6] M. Park, et al., *ACS AMI*, 2018. [7] K. Ni, et al., *IEDM*, 2019. [8] V. Gaddam, et al., *VLSI*, 2024. [9] H. Shin, et al., *PSS-RRL*, 2024. [10] A. Kashir, et al., *ACS AEM*, 2021. [11] V. Gaddam, et al., *ACS AMI*, 2022. [12] D. Das, et al., *TED*, 2022. [13] T. Mittmann, et al., *IEDM*, 2020. [14] A. Kashir, et al., *PSS-A*, 2021. [15] M. Park, et al., *Nanoscale*, 2016. [16] A. Pal, et al., *APL*, 2017. [17] X. Wang, et al., *EDL*, 2024. [18] S. Kim, et al., *EDL*, 2021.

V_O Modulation for MPB

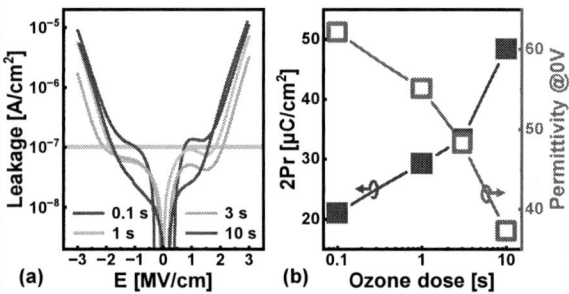

Electrical Characterization: Fig. 3 (a) *P-E* loops (b) *κ-E* curves with device-to-device variation in the inset of the $H_{0.5}Z_{0.5}O_2$ films at different ozone dose time.

Leakage Limitation & Trend Summarization: Fig. 4 (a) Leakage characteristics for the group of 0.1-10 s ozone dose samples. (b) Evolution of $2P_r$ and $κ_{@V-0}$ values with various ozone dose times.

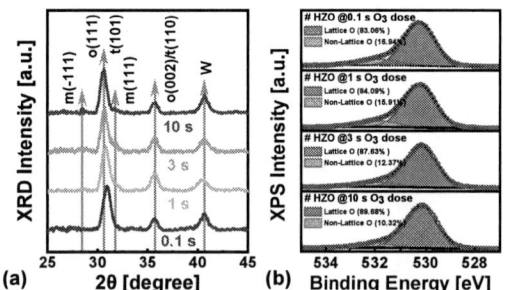

Physical Characterization: Fig. 5 (a) GIXRD result of samples across varied ozone dose times for crystal structure analysis. (b) O 1s XPS spectra for V_O concentration analysis.

V_O-Zr Content Synergy for MPB

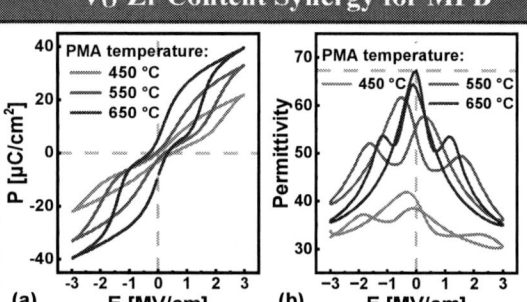

Electrical Characterization: Fig. 6 (a) *P-E* loops (b) *κ-E* curves for the 0.3 s ozone dose $H_{0.45}Z_{0.55}O_2$ films measured under different PMA temperatures.

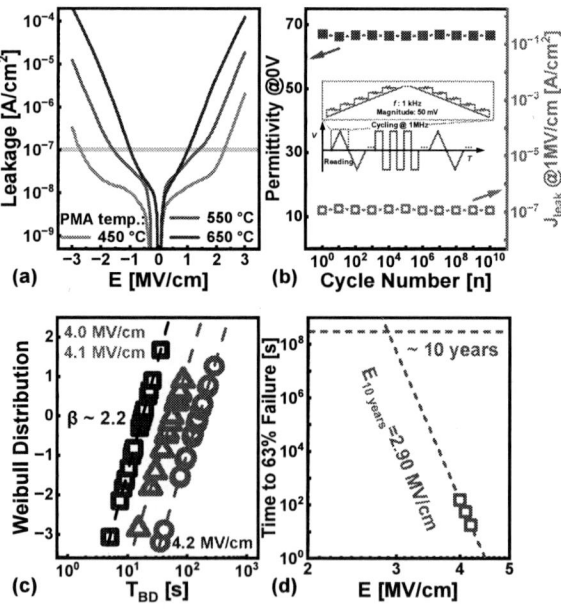

AC & DC Reliability: Fig. 7 (a) Leakage current density of the 0.3 s ozone dose sample measured at different annealing temperatures. (b-d) Results for the 0.3 s sample after 650 °C PMA: (b) Endurance test showing stable $κ_{@V-0}$ and J_{leak} at 1 MV/cm up to 10^{10} cycles; (c) Weibull distribution of the time-to-breakdown at 4.0-4.2 MV/cm; (d) extrapolated 10-year lifetime operating electric field reaching 2.9 MV/cm.

979-8-3315-0417-5/25 $31.00 © 2025 IEEE

Recent Advances in Compact Modeling for Advanced Semiconductor Technology

Runsheng Wang[1,3*], Baokang Peng[2], Lining Zhang[2,3], Ru Huang[1,3]

[1]School of Integrated Circuits, Peking University, Beijing, China.

[2]School of Electronic and Computer Engineering, Peking University, Shenzhen, China.

[3]Institute of Electronic Design Automation, Peking University, Wuxi, China.

[*]Email: r.wang@pku.edu.cn

Abstract

A thorough review of recent advances in compact modeling for advanced semiconductor technology is performed. The development of compacting modeling is based on the modeling coordinate system (Fig. 1), which is composed of model applications, modeling techniques, and modeling scenarios. From the perspective of the model application, recent developments include advanced logic devices without/with the backside power delivery network (BSPDN) and their design technology co-optimizations (DTCO), computing memories, cryogenic CMOS, etc. Along the dimension of modeling techniques, recent advancements span AI-assisted modeling and fusion modeling, besides the traditional approaches. In terms of modeling scenarios, dynamic modeling has been discussed more often recently, in addition to the classical static modeling scenarios. By reviewing the modeling coordinate system, we aim to provide a holistic picture of the recent advances for deeper understanding and alignment toward a better compact modeling community.

Keywords: compact modeling, artificial intelligence, BSPDN, advanced CMOS, computing memory.

Introduction

Compact device models play an important role in the semiconductor industry to bridge circuit designs and chip manufacturing. At the same time, the need for a compact model is becoming earlier, as part of the shifting left in technology development and circuit design automation. In the path-finding stage, or the technology optimization stage, a compact model is needed to explore the technology options and refine the parameters. To this end, the model application side poses requirements for the modeling.

There are different modeling techniques, which refer to ways to develop a model. Traditionally, a physics-based model is chosen in the community. Other techniques are also used for circuit simulations, for example, the table lookup model is a solution when simulation speed overtakes the accuracy. In general, the modeling techniques are also matching the calls from the application side.

Fig.1 shows the modeling coordinate system, in which the model application and model technique are two important dimensions. At the same time, modeling scenarios also essential concerns. There are many possible combinations in the coordinate, which deserve research and development to further push the technology. In this work, recent advances along each dimension are reviewed, and the open questions are discussed.

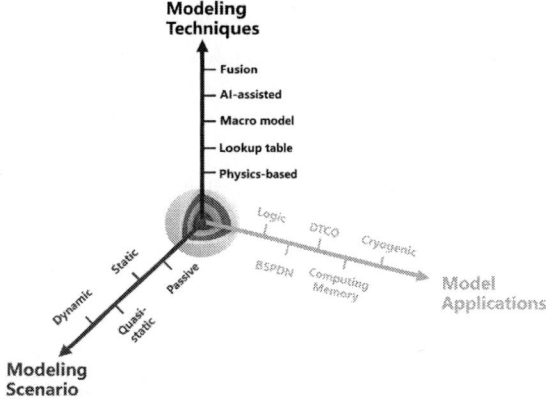

Fig. 1: The development of compacting modeling is based on the modeling coordinate system, comprising of dimensions along the model applications, modeling techniques, and modeling scenarios.

Emerging applications and model requirements

Several interesting applications require advancement in compact modeling, including the next-generation CMOS technology, analog memory suitable for neuromorphic computing, and cryogenic CMOS.

The community aligns that GAA transistors will succeed the FinFET for the 2nm and below technology nodes. Besides the complexity increments in the modeling like the nanosheet width-dependent modules, the self-heating effect becomes more significant [1,2]. The temperature rise under the same Joule power is even higher than that of the FinFET due to the increased thermal resistance. At the same time, more time constants are observed in the heating and cooling process, which is different from the classical self-heating model.

Going forward along with further scaling using stacked transistor technology, CFET may be deployed. Due to the coupling between the stacked transistors, compact modeling faces a certain challenge in terms of the electrical coupling, e.g., the threshold voltage of a transistor is tuned by the gate voltage of an adjacent transistor. At the same time, the thermal cross-talk is also more significant due to the small transistor space. Another possible stacked transistor

---

979-8-3315-0417-5/25 $31.00 © 2025 IEEE

architecture, FFET [3], is also subject to thermal coupling inside the transistor pair. How to develop a coupling model consistent with the state-of-the-art model framework is one problem to solve.

Meanwhile, DTCO [4] is the new paradigm for advanced logic technology development, which requires an efficient model generation from the test data. The models of different versions are used in the standard cell characterizations, to eventually identify an optimized device and circuit design. A model-driven design space exploration is necessary for the DTCO. In the center, how to get the model parameters from the data efficiently will be a key question to answer.

Besides of logic technology, analog memories for computing purpose (i.e., computing memory) are under extensive research and development. One main target is to support the circuit design of new paradigm computing, e.g. neuromorphic computing. One essential feature of these emerging memories is the state variable determined by the voltage or current stimulus, in a continuous manner. The possible challenge, therefore, is how to solve the multiphysics equations for an efficient memory model.

Another application that has recently been under extensive study is cryogenic CMOS [5,6]. A lot of benefits to putting the CMOS transistors under the cryogenic environment, including the steep slope, low power consumption, and higher saturation current. At the same time, the advanced effects like the band tails, and the incomplete ionizations, show up calling for modeling efforts. Numerical robustness is another key concern in the modeling due to the very small intrinsic carrier concentration, which may lead to data overflow if not handled properly.

Advances in modeling techniques

Physics-based compact modeling is the classical option in the community due to its many advantages. Understandings gained from the model are valuable for device designs as well as circuit designs. However, one main shortcoming is the model development cycle, which may take years. At the same time, parameter extraction becomes more complicated when the number of model parameters also scales up. The disparity between the requirement and status inspires the development of new modeling techniques.

AI-assisted compact modeling is one topic that attracts a lot of attention [7-10]. Within the category of data-driven modeling, in addition to the table lookup model, one essential goal of the AI model is to improve the modeling efficiency. One early work [8] dedicated to developing the artificial neural network (ANN) in the 2010s was inspired by the success of AlexNet for computer vision. There are several domain-specific issues for an ANN-based compact model. First, the data range of transistors is very large, covering several orders of magnitude. How to prepare the ANN to accommodate the data range with a proper loss function?

Second, the constrain posed by the device physics, like the zero current for zero voltage property, should be strictly satisfied by the ANN model. Third, the ANN model should be implemented into a circuit simulator to be practically useful for DTCO. These several issues have been solved with the recent advancements as reported in the literature [7-10].

Within the scope of the ANN models, the incorporation of statistical modeling and self-heating has been recently realized. For statistical modeling, the ANN model is first interpreted with sensitivity analysis, to identify the ANN parameters responsible for current and capacitance variation. The fine-tuning is performed to find the distribution of the ANN parameters [11]. In this way, the Monte Carlo simulations with the ANN model prove to be accurate and efficient. For the self-heating modeling, the classical sub-circuit is combined with the ANN model for training and deployment [12]. In this way, the key features important for advanced CMOS applications are developed.

There are also efforts towards applying AI to develop a surrogate model for analog memory [13,14]. One important concern is how to incorporate the recurrency into the NN, and to keep the prediction ability, especially in the time domain. A macro model is sometimes used for the memory, utilizing the circuit elements to construct an equivalent circuit. While it depends on the understanding of memory mechanisms, a macro model also takes a long development time.

Similar to AI-assisted transistor modeling, the memory modeling with recurrent neural network (RNN) is also proposed. It is still in the early stage for the AI-assisted memory modeling generally, but recent reports such as that using the NARX [13] show promising results. More efforts toward this direction are desired.

There are a lot of discussions around new parameter extraction techniques using the neural network for the proven physics-based model. It is based on the assumption that there is a black box mapping function from the model output to the model parameter set. As a result, parameterization with neural networks is a process of encoding, quite different from traditional nonlinear regression [14]. At present, the approach is limited to extracting a few tens of parameters. On the other hand, it makes a promising approach for reliability model parameterization [15], where only a certain parameter set is changed.

Further, some interesting ideas [7] of merging device physics and neural networks for compact modeling have been discussed, which may combine the capability of both. In fact, the self-heating module around the ANN model follows this approach. There is a lot of room in between the physics-based model and AI-assisted model, for the fusion strategy to be further optimized.

Advances in modeling scenarios

The emerging modeling techniques such as the AI-assisted

ones are effective solutions to the questions posed by the emerging applications. For example, the data-driven ANN models could accelerate the DTCO loop, and in a similar position, the RNN models seem promising for analog memory devices. At the same time, there are occasions when the fundamental model assumption should be re-examined to better answer the requirements from the application side.

Design for reliability (DfR) is becoming an important practice in advanced technology, requiring time-dependent aging model [16-21] and dynamic simulation solutions. The common approach of reliability characterizations is to use constant stress voltages to collect the parameter shifts and to further develop a continuous-time model. However, it is not compatible with circuit simulation which is a discrete-time process. The BERT or similar solutions bypass this compatibility problem for some reliability issues. However, for advanced technologies, a concurrent simulation with an aging model embedded seems necessary [16,20,22,23]. In this regard, reformulating the continuous-time model to a discrete-time one with approaches like effective time is needed.

This further leads to a possible paradigm shift from static or quasi-static modeling to dynamic modeling. The quasi-static approximation is a legacy in transistor modeling, leading to the state-of-the-art separate current and terminal charge modeling. Some devices are intrinsically dynamic, like the emerging non-volatile memory. Some device properties are also dynamic naturally like the self-heating and aging. The possible shift from quasi-static to real-time dynamic solution is an open question, which is touched on in another work [24].

Conclusion

This work reviews the recent advances in compact modeling by a coordinate system. At the application side, advanced device technologies like the GAAFET, CFET, or FFET, together with the new development strategy of DTCO, as well as the advanced computing memory technologies, are making rapid progress requiring model solutions. New modeling techniques are emerging, like the AI-assisted approach with neural networks. Modeling scenarios is an important consideration to match the requirements from the application side. A paradigm shift from quasi-static to dynamic solution is discussed.

Acknowledgments

This work was supported in part by the Natural Science Foundation of China under Grant 62125401 and 62474009, the Beijing Outstanding Young Scientist Program under Grant JWZQ20240101004 and in part by the Guangdong Basic and Applied Basic Research Foundation under Grant 2024B1515020064.

References

[1] S. Chen *et al.*, *IEEE Trans. Electron Devices*, vol. 71, no. 11, pp. 6478–6485, Nov. 2024.

[2] Z. Sun *et al.*, *IEEE Trans. Electron Devices*, vol. 70, no. 11, pp. 5528–5534, Nov. 2023.

[3] H. Lu *et al.*, *2024 IEEE Symposium on VLSI Technology and Circuits (VLSI Technology and Circuits)*, Honolulu, HI, USA: IEEE, Jun. 2024, pp. 1–2.

[4] M. Liu, *2021 IEEE International Solid-State Circuits Conference (ISSCC)*, Feb. 2021, pp. 9–16.

[5] C. Enz *et al.*, *2020 IEEE International Electron Devices Meeting (IEDM)*, San Francisco, CA, USA: IEEE, Dec. 2020, p. 25.3.1-25.3.4.

[6] R. Saligram *et al.*, *Chip*, vol. 3, no. 1, p. 100082, Mar. 2024.

[7] L. Zhang *et al.*, *Proceedings of the 2024 ACM/IEEE International Symposium on Machine Learning for CAD*, Salt Lake City UT USA: ACM, Sep. 2024, pp. 1–6.

[8] L. Zhang and M. Chan, *J Comput Electron*, vol. 16, no. 3, pp. 825–832, Sep. 2017.

[9] J. Wang *et al.*, *IEEE Trans. Electron Devices*, vol. 68, no. 3, pp. 1318–1325, Mar. 2021.

[10] C.-T. Tung and C. Hu, *IEEE Transactions on Electron Devices*, vol. 70, no. 4, pp. 2157–2160, Apr. 2023.

[11] W. Dai *et al.*, *IEEE Transactions on Computer-Aided Design of Integrated Circuits and Systems*, vol. 42, no. 12, pp. 5156–5160, Dec. 2023.

[12] H. Liu *et al.*, *2024 2nd International Symposium of Electronics Design Automation (ISEDA)*, Xi'an, China: IEEE, May 2024, pp. 89–93.

[13] Z. Rong *et al.*, *IEEE Electron Device Lett.*, vol. 44, no. 8, pp. 1272–1275, Aug. 2023.

[14] Y. Shintani *et al.*, *2024 Design, Automation & Test in Europe Conference & Exhibition (DATE)*, Valencia, Spain: IEEE, Mar. 2024, pp. 1–6.

[15] C. Shen *et al.*, *2024 8th IEEE Electron Devices Technology & Manufacturing Conference (EDTM)*, Bangalore, India: IEEE, Mar. 2024, pp. 1–3.

[16] R. Wang *et al.*, *2018 IEEE International Electron Devices Meeting (IEDM)*, Dec. 2018, p. 17.2.1-17.2.4. doi: 10.1109/IEDM.2018.8614594.

[17] R. Wang *et al.*, *2021 IEEE International Electron Devices Meeting (IEDM)*, San Francisco, CA, USA: IEEE, Dec. 2021, p. 31.2.1-31.2.4. doi: 10.1109/IEDM19574.2021.9720674.

[18] Z. Ji *et al.*, *IEEE Trans. Electron Devices*, vol. 71, no. 1, pp. 138–150, Jan. 2024, doi: 10.1109/TED.2023.3330834.

[19] Z. Sun, S. Chen, L. Zhang, R. Huang, and R. Wang, *Micromachines*, vol. 15, no. 1, p. 127, Jan. 2024, doi: 10.3390/mi15010127.

[20] R. Wang *et al.*, *2023 7th IEEE Electron Devices Technology & Manufacturing Conference (EDTM)*, Seoul, Korea, Republic of: IEEE, Mar. 2023, pp. 1–3. doi: 10.1109/EDTM55494.2023.10103061.

[21] R. Wang, Z. Zhang, Z. Sun, Z. Guo, Y. Lin, and R. Huang, *2022 IEEE 16th International Conference on Solid-State & Integrated Circuit Technology (ICSICT)*, Nangjing, China: IEEE, Oct. 2022, pp. 1–4. doi: 10.1109/ICSICT55466.2022.9963318.

[22] S. Guo *et al.*, *2017 IEEE/ACM International Conference on Computer-Aided Design (ICCAD)*, Irvine, CA: IEEE, Nov. 2017, pp. 780–785.

[23] L. Zhang *et al.*, *IEEE Trans. Electron Devices*, vol. 66, no. 1, pp. 184–190, Jan. 2019.

[24] Y. Li, *et al*, *2025 9th IEEE Electron Devices Technology & Manufacturing Conference (EDTM)*, Hong Kong: IEEE, Mar. 2025, pp. 1–3.

Multi-channel Intelligent Electronic Nose for Rapid Identification of Complex Hazardous Gases

Wenyuan Liu[1,2], Jiachuang Wang[1,2,3], Fangyu Zhao[1,2], Nan Qin[1,2],
and Tiger H. Tao[1,2,3,4,5,6,7,8]

[1]State Key Laboratory of Transducer Technology, Shanghai Institute of Microsystem and
Information Technology, Chinese Academy of Sciences, Shanghai, CHINA
[2]School of Graduate Study, University of Chinese Academy of Sciences, Beijing, CHINA
[3]2020 X-Lab, Shanghai Institute of Microsystem and Information Technology,
Chinese Academy of Sciences, Shanghai, CHINA
[4]Center of Materials Science and Optoelectronics Engineering,
University of Chinese Academy of Sciences, Beijing, CHINA
[5]Center for Excellence in Brain Science and Intelligence Technology,
Chinese Academy of Sciences, Shanghai, CHINA
[6]Neuroxess Co., Ltd. (Jiangxi), Nanchang, Jiangxi, CHINA
[7]Guangdong Institute of Intelligence Science and Technology,
Hengqin, Zhuhai, Guangdong, CHINA and
[8]Tianqiao and Chrissy Chen Institute for Translational Research, Shanghai, CHINA

Abstract

We report an intelligent electronic nose based on the 18-channel sensors (16 gas sensors, 1 humidity sensor and 1 temperature sensor), integrated with *Drosophila*-inspired neural network algorithm. This microsystem is capable of detecting toxic and harmful gases within 1 s and identifying various hazardous gases with a high accuracy of 98.5% even in harsh environments such as device damage and high humidity. It demonstrates the potential for precise and rapid gas recognition.

Keywords: Sensor array, electronic nose, hazardous gas detection

Introduction

Gas sensors have a wide range of applications in industrial production [1,2]. With the advancement and development of science and technology [3], the performance requirements for accuracy, sensitivity and repeatability are becoming increasingly high. Combining advanced microstructure design and sensing materials selection [4], multi-channel gas sensor arrays have made significant progress in the identification of various toxic and hazardous gases. Microelectromechanical system (MEMS) utilizes precision

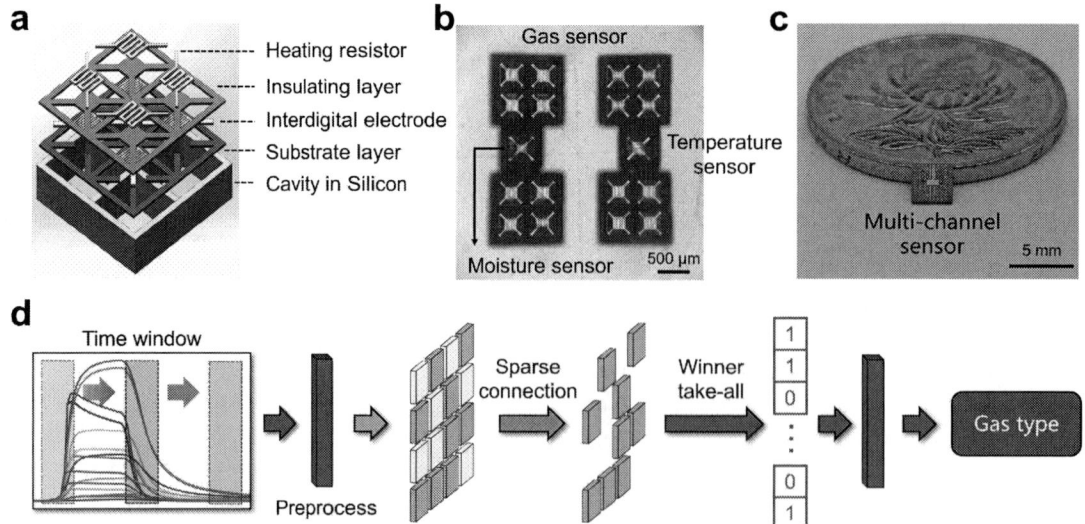

Figure 1: (a) 3D structure diagram of a 2 × 2 gas sensor array. Microscope image (b) and optical image (c) of an 18-channel gas sensor. (d) A *Drosophila*-inspired neural network algorithm model for gas type recognition.

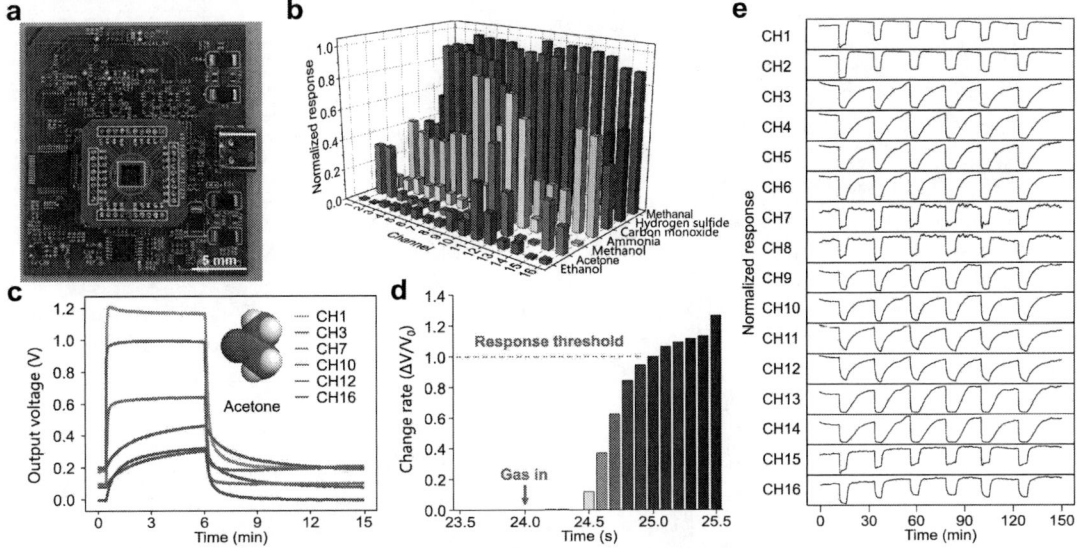

Figure 2: (a) A multi-channel sensor driver board. (b) Response spectra of gas sensor array for 7 gases. (c) Output voltage curves of six different channels to acetone gas. (d) The change rate of the device for acetone gas within 2 s. (e) Cyclic experiment of gas sensor array on 0.8 ppm hydrogen sulfide gas.

machining technology to manufacture complex and extremely small sensors, enabling gas sensing technology to achieve a great development in miniaturization, low power consumption, and intelligence.

Drosophila is considered one of the fastest reacting organisms in nature [5]. Researchers are enthusiastic about exploring the mechanisms behind its ability to quickly identify gas phenomena, and related research is constantly emerging. The combination of high-performance sensors and artificial intelligence is considered an important research direction for applying the principles of *Drosophila* neural networks to the gas sensing.

Design and Fabrication

Figure 1a shows the 3D structure of a 2 × 2 gas sensor array, including heating resistor, insulating layer, interdigital electrode, substrate layer, and cavity in silicon. The microstructure is fabricated based on MEMS technology. Firstly, a layer of silicon oxide is deposited by thermal oxidation as a protective layer, and then dry etching and deposition of metal Pt are used to pattern the forked electrodes in the sensitive area. Next, the insulation layer of the gas sensor is formed by plasma enhanced chemical vapor deposition (PECVD), and then the heating resistor is patterned by dry etching and deposition of metal Pt. Deep etching is performed on the back of the silicon wafer, followed by dry etching and anisotropic wet etching of the dielectric layer to form a suspended beam structure. Finally, the gas sensing materials are modified onto the forked electrode to complete the fabrication of the multi-channel gas sensor.

The entire device consists of 16 channel gas sensor, 1 channel temperature sensor, and 1 channel humidity sensor (Figure 1b), with a size of only 2.5 mm × 2.5 mm (Figure 1c). We designed a *Drosophila*-inspired neural network algorithm (Figure 1d) for identifying gas types, analyzing the time window of multi-channel sensor response value changes, including preprocess, sparse connection, winner take-all, and classification, to achieve rapid identification of various hazardous gases.

Experimental Results

According to the data collection requirements, we designed corresponding adapter board and driver board (Figure 2a), and further used FPGA to control the heating voltage and output channels. Figure 2b shows the specific response spectra of a 16-channel gas sensor to 7 gases (ethanol, acetone, methanol, ammonia, carbon monoxide, hydrogen sulfide, methanal), laying the foundation for intelligent olfactory sensing systems to identify multiple gases.

Acetone is flammable and volatile, so its rapid detection plays an important role in industrial production. We set the sampling rate to 10 Hz and conducted a response experiment with a duration of 15 minutes. Figure 2c shows the output voltage value of six channels with different modified materials, exhibiting a typical response recovery curve. When the response change rate exceeds the set threshold, it can be considered as a response to the test gas，so that our multi-channel gas sensor achieved rapid response to acetone gas in 1 s (Figure 2d). Hydrogen sulfide is highly toxic and corrosive. Figure 2e shows the cyclic response of the array device to 0.8 ppm hydrogen sulfide gas. There are significant differences in the response between various channels, which also reflects the stability of the device. This is crucial for the development of practical gas detection and analysis equipment.

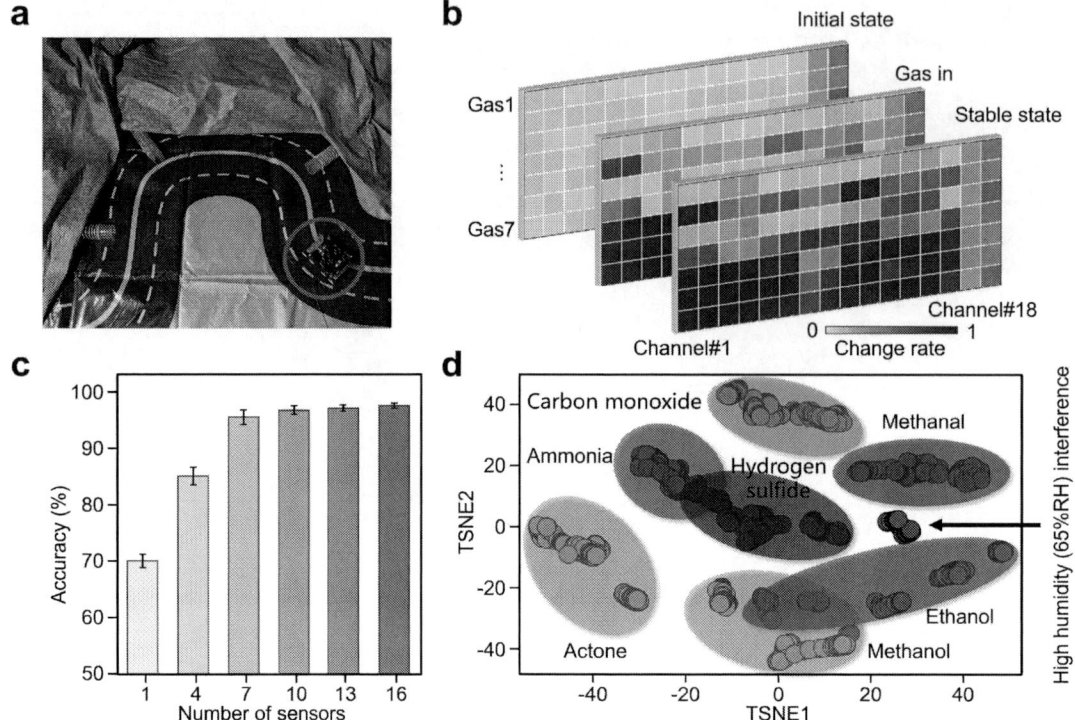

Figure 3: (a) Simulated gas leakage scenario. (b) The olfactory mappings of the device for 7 gases in the initial state, gas in, and stable state. (c) The variation of algorithm recognition accuracy with the number of gas sensors. (d) The visualization result of the feature space obtained after data dimensionality reduction, including 7gases and high humidity interference.

Finally, we constructed a simulated emergency rescue scenario for detecting gas leakage (Figure 3a). Figure 3b indicates that our device can monitor the differences in response between the initial state, gas in, and stable state, while utilizing algorithm for identification. Meanwhile, this *Drosophila*-inspired algorithm achieves a high recognition accuracy of 98.5% and maintains high performance when half of the sensors are damaged (Figure 3c). Figure 3d shows that our intelligent electronic nose is able to distinguish 7 hazardous gases even under complex conditions with the interference of high humidity.

Conclusion

In this work, we demonstrate a multi-channel and multi-dimensional intelligent electronic nose system, which uses a fruit fly simulation algorithm to correct and compensate for gas type recognition based on temperature and humidity. When facing emergency situations such as toxic gas leaks, it helps to quickly and accurately identify the type of hazardous gas, demonstrating its potential for application in emergency rescue scenarios for alarm systems.

Acknowledgments

This work was supported by National Science and Technology Major Project from the Minister of Science and Technology of China (grant no. 2018AAA0103100), National Natural Science Foundation of China (grant no. 62236005), National Science Fund for Excellent Young Scholars (grant no. 61822406), Key Research Program of Frontier Sciences, CAS (grant no. ZDBS-LY-JSC024) and Fund of Youth Innovation Promotion Association CAS (grant no. 2022234). These authors contributed equally: Wenyuan Liu, Jiachuang Wang and Fangyu Zhao.

Contact

*Tiger H. Tao, tel: +86-21-62511070, tiger@mail.sim.ac.cn
*Nan Qin, qinnan@mail.sim.ac.cn

References

[1] S.-Y. Jeong, Y. K. Moon, J. Wang & J.-H. Lee, Exclusive detection of volatile aromatic hydrocarbons using bilayer oxide chemiresistors with catalytic overlayers. Nat. Commun. 14, 233 (2023).

[2] P. H. Rogers, K. D. Benkstein & S. Semancik, Machine learning applied to chemical analysis: sensing multiple biomarkers in simulated breath using a temperature-pulsed electronic-nose. Anal. Chem. 84, 9774–9781 (2012).

[3] L. Cheng, Q.-H. Meng, A. J. Lilienthal & P.-F. Qi, Development of compact electronic noses: a review. Meas. Sci. Technol. 32, 062002 (2021).

[4] C. Wang, et al. Biomimetic olfactory chips based on large-scale monolithically integrated nanotube sensor arrays. Nat. Electron. 7, 157–167 (2024).

[5] W. Y. Peter, et al. Evolving the olfactory system with machine learning, Neuron, 23, 109 (2021).

Opportunities for Wide and Ultrawide Bandgap Devices with Heterogenous Integration

Vanjari Sai Charan, Aditya K. Bhat, A. Anjali, Zequan Chen, Matthew D. Smith,
James Pomeroy, and Martin Kuball

Centre for Device Thermography and Reliability, University of Bristol, United Kingdom

Abstract

This work reviews the latest progress in wide and ultrawide bandgap semiconductor RF and power devices, in particular selected highlights in GaN and Ga_2O_3 technologies are presented. Ga_2O_3 trench diodes with the highest reported breakdown voltage of 4kV are discussed, in addition to the benefits of heterogenous integration, for example, with diamond, are highlighted.

Keywords: GaN, Ga_2O_3, Gallium Oxide, diamond, QST, heterogeneous integration, RF, power

Introduction, Results and Discussions

RF and power electronic devices underpin a wide range of modern technologies across communications, transportation, energy generation and power distribution. Further system-level improvements depend on the continued development of devices, particularly, reliable high voltage and high power density technologies. We will focus here on GaN, and its opportunities, and Ga_2O_3 will also be touched upon. GaN-based devices are now commercialized with growing market shares, both in the power and RF field, although performance and reliability can still be limited by poor heat extraction; we discuss the advantages that heterogenous integration of GaN with diamond can offer, with the potential to dramatically reduce device temperature. Furthermore, despite major advancements in material growth technologies, GaN still features relatively high defect densities that can impact device performance, for example local variations in threshold voltages; we will discuss techniques to map these local changes in the device properties as well as the impact of device stress on long-term device performance. A good understanding of the role of Carbon impurities in GaN buffer layers is also important for high voltage applications; the substrate backbias technique, now an industry standard, enables understanding of charge transport related to Carbon impurities and their impact dynamic on-resistance, toward optimizing material growth processes. Most prior work on the backbias technique was focused on GaN-on-Si; this is extended here to GaN-on-QST. Finally, materials beyond GaN will be discussed e.g. the opportunities to utilize even wider bandgap materials such as Ga_2O_3 which can offer advantages in e.g. breakdown voltage, with us recently demonstrating >4kV devices. Commercialization of these technologies within the UK government-funded Innovation

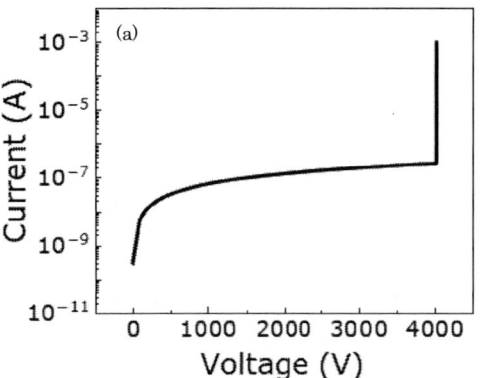

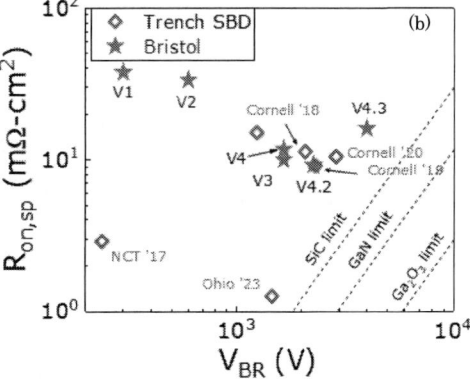

Fig. 1: (a) Reverse characteristics of Ga_2O_3 Trench Schottky Barrier Diode (TSBD), fabricated in Bristol, with 4kV reverse breakdown voltage; (b) evolution of the Ga_2O_3 TSBD technology in Bristol over the last 2 years (V1 to V4.3).

and Knowledge Centre (IKC) REWIRE will be highlighted, as will routes to become involved.

Conclusion

Wide and ultrawide bandgap semiconductors offer new opportunities for power and RF device technology, in particular, when integrated with high thermal conductivity materials such as diamond.

Acknowledgments

The authors gratefully acknowledge funding from the UKRI Innovation and Knowledge Centre REWIRE under grant number EP/Z531091/1 and from the EPSRC programme grant GaN-DaME and ULTRAlGaN under grant number EP/P00945X/1 and EP/X035360/1, respectively. M. Kuball acknowledges financial support by the Royal Academy of Engineering through the Chair in Emerging Technologies

979-8-3315-0417-5/25 $31.00 © 2025 IEEE

Scheme. We thank our various collaborators.

979-8-3315-0417-5/25 $31.00 © 2025 IEEE

Optimization of Short-channel Top-gate MoS$_2$ FETs via a Non-transfer Fabrication

Haojie Chen[1][†], Xinliu He[1][†], Jinshu Zhang[1][†], Jiahao Wang[1], Sen Wang[1], Yuchen Tian[1], Saifei Gou[1], Xiangqi Dong[1], Mingrui Ao[1], Qicheng Sun[1], Zhejia Zhang[1], Yan Hu[1], Jieya Shang[1], Yufei Song[1], Yuxuan Zhu[1*], and Wenzhong Bao[1,2*]

[1]School of Microelectronics, Fudan University, Shanghai 200433, China.

[2]Shaoxin Laboratory, Shaoxing 312000, China.

[†]Equal Contribution. [*]Corresponding author, E-mail: zhuyuxuan23@m.fudan.edu.cn；
baowz@fudan.edu.cn

Abstract

We fabricated short-channel top-gate transistor arrays based on wafer-scale MoS$_2$. The dielectric layer consisting of 8 nm HfO$_2$ is developed to meet the scaling down targets. The typical device achieves I_{ON}/I_{OFF} ratio of 1.2×10^9 and a positive V_{TH} of 0.57 V with the channel length of 400 nm. Through gate engineering and width-to-length ratio adjustment, the E/D-NMOS inverters were realized. This integration process could pave the way for scaling down of the 2D transistors.

Keywords: MoS$_2$, Short channel and Dielectric optimization.

Introduction

Among the transition metal dichalcogenides (TMDs) materials, molybdenum disulfide (MoS$_2$) is regarded as a promising channel material for scaled-down transistors due to its atomic level thickness, absence of dangling bonds on the surface and high mobility. In comparison with the conventional back-gate devices that are fabricated through transfer process, the top-gate (TG) devices present a series of advantages, including a smaller device size, higher integration, and greater compatibility with integration processes [1]. However, the channel lengths (L_{CH}) of TG MoS$_2$ devices reported so far are mostly at the micrometer scale [2], which cannot present the best performance of the material. And with the L_{CH} scaling down, the growth of dielectric for TG FETs becomes more urgent [3].

In this work, we explored the suitable oxides combination and adopted a gate-last process to fabricate short-channel (SC) TG MoS$_2$ FETs at sub-micrometer L_{CH} without transfer process. The typical device achieved threshold voltage (V_{TH}) of 0.57 V, I_{ON}/I_{OFF} ratio up to 10^9, and SS of 136.1 mV/dec. In addition, we regulated the transistors V_{TH} by using different gates and adjusting the width-to-length ratio (W/L) to achieve high-performance E/D-NMOS inverters. This work contributes to the future industrialization and integration of SC TG devices.

Short-channel Top-gate MoS$_2$ FETs Fabrication Process

Fig.1 summarizes the fabrication flow. Monolayer (1L) MoS$_2$ was initially grown on the 2-inch sapphire substrate by chemical vapor deposition (CVD). As shown in Fig.2, the frequency difference between the E$^1_{2g}$ and A$_{1g}$ vibration modes for 1L-MoS$_2$ is approximately 19 cm^{-1}, verifying the homogeneity of the CVD-grown monolayer MoS$_2$.

Electron beam lithography (EBL) was used to define the sub-micrometer L_{CH}, followed by thermal evaporation of source and drain (S/D) electrodes (35 nm Au). Then MoS$_2$ channel region was etched using reactive ion etching (RIE). To investigate the impact of the W/L on the performance of transistors, we designed two different W/L ratios (10/1 and 20/1). Next, a seeding layer was deposited by electron-beam evaporation (EBE), which can facilitate the growth of a uniform high-K dielectric [4]. To optimize the dielectric of SC TG FETs, HfO$_2$ was grown by atomic layer deposition (ALD) with three different thicknesses T_{high-K} (5 nm, 8 nm, and 12 nm). Subsequently, Au-gate (35 nm Au) and Al-gate (5 nm Al/30 nm Au) was deposited. MoS$_2$ FET arrays with different L_{CH} were fabricated without transferring throughout the entire process. The scanning electron microscopy (SEM) images exhibit the sharp edges of the S/D electrode lines and the precise channel length of 200 nm, as shown in Fig.3. Furthermore, we integrated Au-gate and Al-gate transistors for optimizing V_{TH} matching to fabricate E/D-NMOS inverters.

Electrical Characteristics of the MoS$_2$ FETs

Fig.4(a) shows the I_D-V_{TG} characteristics of the MoS$_2$ transistors (L_{CH} = 400 nm) with different T_{high-K}, representing excellent electrostatic control and I_{ON}/I_{OFF} ratios regardless of the dielectric thickness. When T_{high-K} = 5 nm, the thinner dielectric leads to the larger C$_{ox}$, improving their gate control capability and resulting higher I_{ON} at V_{TG} = 5 V. But a leakage current (I_G) rises with increasing V_{TG} also appeared during device operation when T_{high-K} = 5 nm, which is unacceptable for applications, as shown in Fig.4(b). Other two curves of 8 nm and 12 nm dielectric show a negligible I_G with pA magnitude, proving the stability of dielectric above 8 nm. Also, compared with 5 nm, I_{OFF} under 8 nm and 12 nm significantly decrease. We finally chose 8 nm HfO$_2$ as the dielectric, which meets the scaling down targets for low I_G and presents smaller EOT than 12 nm.

Fig.5 and Fig.6 show the I_D-V_{TG} and I_D-V_{DS} characteristics of a typical device with L_{CH} = 400 nm, which demonstrates the superior electrical performance of our SC TG transistors. The device exhibits an excellent I_{ON}/I_{OFF} ratio of 1.2×10^9, a

979-8-3315-0417-5/25 $31.00 © 2025 IEEE

positive V_{TH} of 0.57 V and SS of 136.1 mV/dec at V_{DS}=1.0 V. The Fig.5 inset is the double-sweep transfer characteristics at V_{DS} = 1.0 V, showing a small hysteresis about 0.1V, which is negligible in non-transfer gate-last fabrication process. The output characteristic curves show the saturation current I_{ON} of 700 µA at V_{TG} = 5.0 V in the Fig.6. As shown in Fig.6 inset, the linear output characteristics at small V_{DS} = 50 mV presents the ohmic contact between the S/D electrodes and MoS$_2$. We further fabricated SC TG transistors with different L_{CH}, ranging from 200 nm to 800 nm, as shown in Fig.7. All these transistors exhibit excellent I_{ON}/I_{OFF} ratios and gate control capability, which proves the stability of fabrication process and the prospects for future scaling down.

V_{TH} Modulation of E/D-NMOS Inverters of Short-channel Top-gate MoS$_2$ FETs

To further demonstrate the application potential of our process, we modulated the V_{TH} of the MoS$_2$ FETs and fabricated inverters. Though doping treatment can affect V_{TH}, it may damage and change MoS$_2$ due to its atomically thin thickness, resulting in a decrease in mobility [5]. Some methods fitting well on µm-length MoS$_2$ FETs may not applicable on SC FETs. Here we proposed gate selection and W/L ratio adjustment to achieve the match between the pull-down (driver) transistor and the pull-up (load) transistor.

We used a low-work-function Al (4.08 eV) gate to effectively modulated V_{TH} towards negative and also fabricated Au-gate (5.1 eV) FETs with positive V_{TH} [6]. Fig.8 plots the I_D-V_{TG} characteristics of Au/Al-gate FETs with different L_{CH} (solid line: Au-gate, dotted line: Al-gate). Al-gate FETs all show larger I_{on} at V_{DS} = 1.0V and negative shifted V_{TH}, the V_{TH} of 400 nm and 600 nm are -1.05 V and -0.49 V respectively.

E/D-NMOS inverters require enhancement-mode FETs and depletion-mode FETs with a positive and negative V_{TH} respectively [7]. We used Au-gate FET as a driver transistor and Al-gate FET as a load transistor. Besides gate engineering, we also adjusted the W/L ratios of the driver and load transistor to optimize their match. Fig.9(a) shows the voltage transfer curve (VTC) of inverters with different ratio (solid line: W/L=10, dotted line: W/L=20). The inverters at W/L=20 are unable to switch to 0 V, while the inverters at W/L=10 have improved flip-flop performance. W/L=10 is confirmed as the suitable width-to-length ratio. This optimization can be attributed to Al gate. When the ratio increasing, Al gate tuned the MoS$_2$ channel to the electron accumulation regime greatly which weaken the driver capability of Au-gate transistor, leading to the shift of V_M (Switch Threshold Voltage).

Benefiting from gate selection and the width-to-length ratio optimization, we improved the electrical performance of E-D NMOS inverters of the 600 nm and 400 nm. The inset of Fig.10(a) shows the schematic and the optical image of the inverter. And Fig.10(a) shows the VTC of a typical optimized inverter of 600 nm with gain of 5.9, V_M of 1.1 V, which satisfies the requirements of the complex circuits. Gate engineering and W/L adjustment will be more effective to V_{TH} modulation with L_{CH} scaling, which lead to superior 2D circuits performance in future.

Conclusion

In summary, we fabricated high-performance SC TG device arrays using a non-transfer gate-last process. The optimized oxide couple dielectric consisting of the seeding layer and 8 nm HfO$_2$ are suitable for scaling down. Moreover, we modulate V_{TH} through changing gate metal and adjust the W/L ratio to reach appropriate match between the Au-gate load transistor and the Al-gate driver transistor of inverters. The entire fabrication flow without any transfer process is industry-compatible and has promising prospects in the future integration and industrialization of 2D transistors.

Acknowledgments

This work was supported by the National Key Research and Development Program (Grant No.2021YFA1200500), Science and Technology Commission of Shanghai Municipality (NO. 23JC1401100), the National Natural Science Foundation of China (Grant No. 62334011) and the Shanghai Pilot Program for Basic Research – Fudan University 21TQ1400100 (23TQ008).

References

[1] C. M. Sheng, X. Q. Dong , Y. X. Zhu Y, et al. " Two - Dimensional Semiconductors: From Device Processing to Circuit Integration", Adv. Funct. Mater., 33, 2304778 (2023).

[2] Y. L. Peng, L. Li, B. Y. Huang, et al. "Gate-Last MoS$_2$ Transistors for Active-Matrix Display Driving Circuits", Adv. Funct. Mater., 33, 2304879 (2023).

[3] H. Y. Lan, J. Appenzeller, Z. Chen, "Dielectric Interface Engineering for High-Performance Monolayer MoS$_2$ Transistors via hBN Interfacial Layer and Ta Seeding", 2022 IEEE International Electron Devices Meeting (IEDM), pp. 7.7. 1-7.7. 4 (2022).

[4] Y. C. Sheng, X. Y. Chen, F. Y. Liao, et al. "Gate stack engineering in MoS$_2$ field-effect transistor for reduced channel doping and hysteresis effect", Adv. Electron. Mater., 7, 2000395 (2021).

[5] C. C. Chiang , H. Lan. Y, C. S. Pang, et al. "Air-stable P-doping in record high-performance monolayer WSe$_2$ devices", IEEE Electron Device Lett., 43, 319-322 (2021).

[6] J. Y. Ma, X. Y. Chen, X. Y. Wang, et al. "Engineering top gate stack for wafer-scale integrated circuit fabrication based on two-dimensional semiconductors", ACS Appl. Mater. Interfaces, 14, 11610-11618 (2022).

[7] D. X. Fan , W. S. Li, H Qiu, et al. "Two-dimensional semiconductor integrated circuits operating at gigahertz frequencies", Nat. Electron., 6, 879-887 (2023).

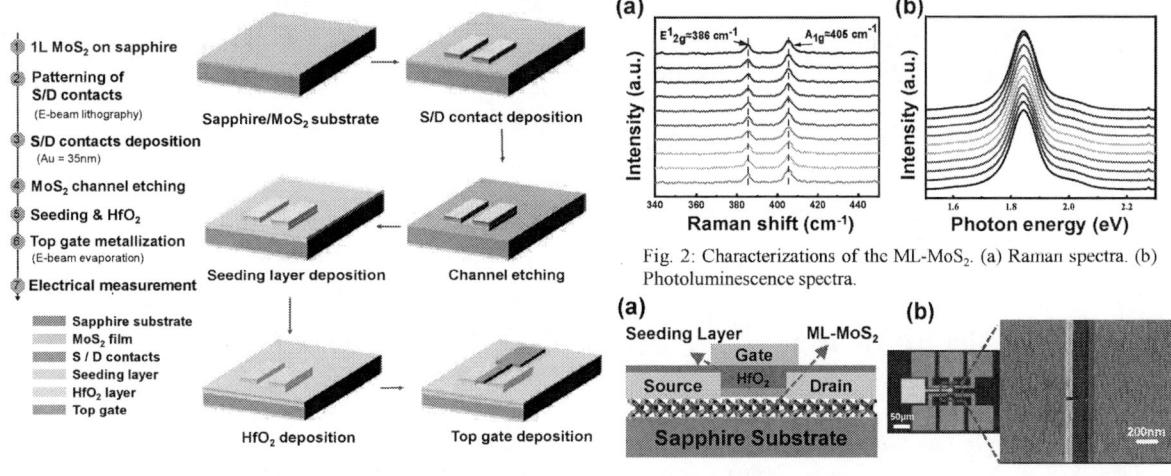

Fig. 1: Fabrication flow of the top-gated 1L-MoS₂ FETs.

Fig. 2: Characterizations of the ML-MoS₂. (a) Raman spectra. (b) Photoluminescence spectra.

Fig. 3: Characterizations of the MoS₂-FETs. (a) Schematic of the 1L-MoS₂ device structure. (b) Optical image and the SEM image of the devices. (L_{CH} = 200 nm).

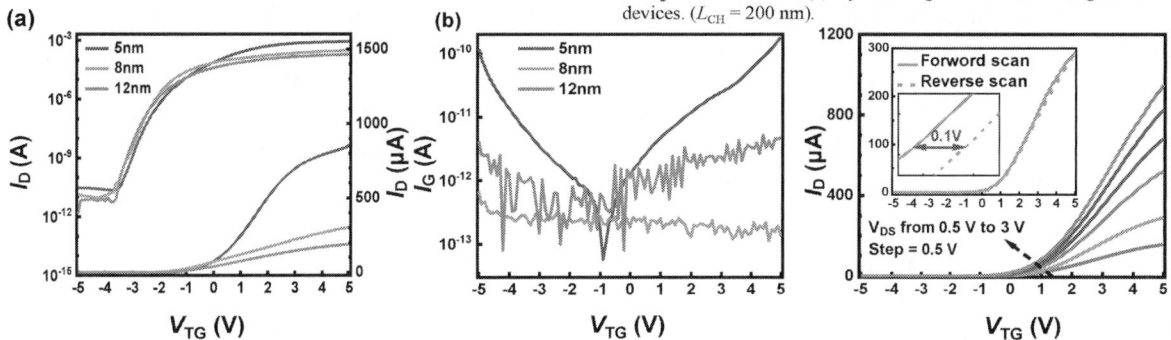

Fig. 4: Characteristics of MoS₂ FETs (L_{CH} = 400 nm) with different dielectric thickness. (a) I_D-V_{TG} characteristics. (b) I_g-V_{TG} characteristics.

Fig. 5: I_D–V_{TG} characteristics of a MoS₂ FET (L_{CH} = 400 nm) at different V_{DS}. The inset shows the characteristics at V_{DS} = 1.0V and the hysteresis behavior.

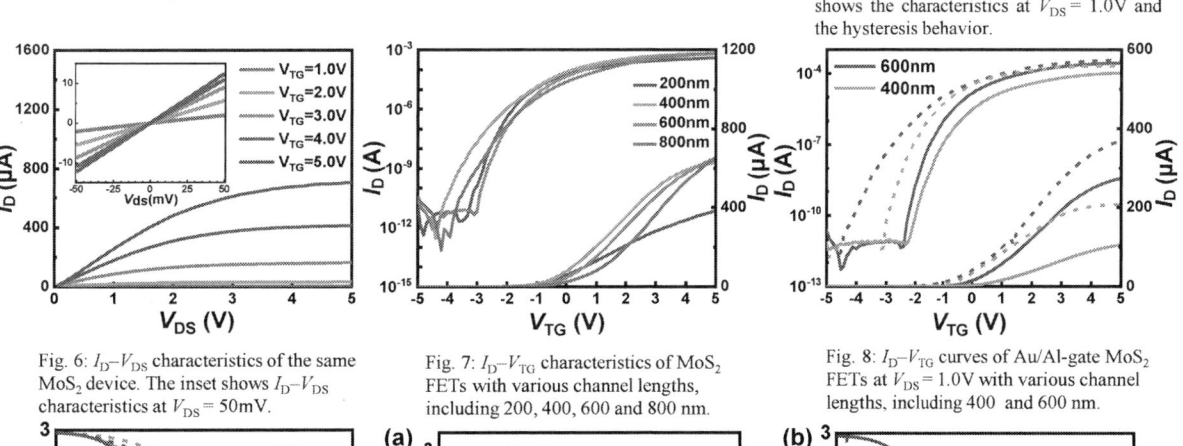

Fig. 6: I_D–V_{DS} characteristics of the same MoS₂ device. The inset shows I_D–V_{DS} characteristics at V_{DS} = 50mV.

Fig. 7: I_D–V_{TG} characteristics of MoS₂ FETs with various channel lengths, including 200, 400, 600 and 800 nm.

Fig. 8: I_D–V_{TG} curves of Au/Al-gate MoS₂ FETs at V_{DS} = 1.0V with various channel lengths, including 400 and 600 nm.

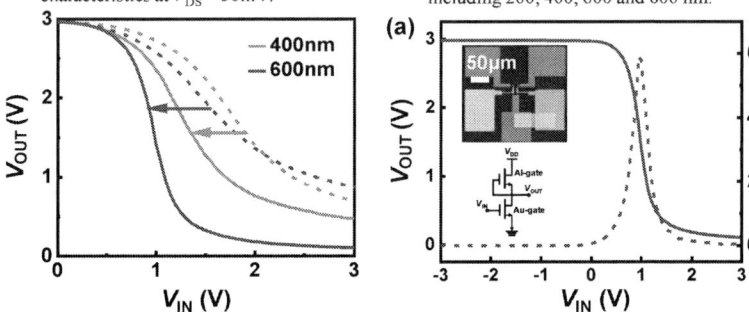

Fig. 9: Voltage transfer curve of the MoS₂ inverter at V_{DD} = 3 V with different W/L. W/L = 10 are the solid line; W/L = 20 are dotted line.

Fig. 10: Characteristics of the optimized typical inverter with L_{CH} = 600 nm. (a) Voltage transfer characteristics and the corresponding voltage gains of the inverter. The inset shows the schematic and the optical image of the inverter. (b) The original curve and the mirroring curve.

979-8-3315-0417-5/25 $31.00 © 2025 IEEE

A Controlled Metal Doping Method Based on MoS$_2$ Top-gate Transistor

Zhejia Zhang[1†], Jingjie Zhou[1†], Saifei Gou[1], Yuxuan Zhu[1], Xiangqi Dong[1], Mingrui Ao[1], Qicheng Sun[1], Yuchen Tian[1], Jinshu Zhang[1], Yan Hu[1], Xinliu He[1], Haojie Chen[1], Yufei Song[1], Jieya Shang[1], Zhengjie Sun[1], Xiaojun Tan[2*], and Wenzhong Bao[1,2*]

[1]School of Microelectronics, Fudan University, Shanghai, 200433, China

[2]Shaoxin Laboratory, Shaoxing, 312000, China

†Equal Contribution; *E-mail: 20112020031@fudan.edu.cn, baowz@fudan.edu.cn

Abstract

This study proposes a strategy for controllably tuning the threshold voltage of two-dimensional (2D) transistors. By depositing different metals with low work functions in the channel region of MoS$_2$ transistors, different doping levels are achieved. Experimental results indicate that aluminum enables controllable n-doping through adjustments in thickness without degrading mobility and subthreshold swing. By combining transistors with different threshold voltages, we successfully fabricated enhancement and depletion (E-D) mode inverters with adjustable flip-flop voltages. This research provides a pioneering methodology for achieving controllable doping in 2D semiconductors.

Keywords: MoS$_2$ transistor, metal cluster and doping

Introduction

Two-dimensional (2D) semiconductors offer significant opportunities and solutions for overcoming the bottleneck of further scaling down of devices, attributed to their atomic thickness, high carrier mobility, dangling-bonds-free, and tunable energy band structures. Recent advancements in wafer scale synthesizing of 2D semiconductors have established a robust foundation for their integration into electronic devices [1,2]. Traditional silicon-based semiconductors allow effective doping through high-energy ion implantation, but the atomic-scale thickness of 2D semiconductors requires gentler methods to prevent lattice damage.

To facilitate the large-scale integration of 2D semiconductors in the industry, it is imperative to address the challenges of controllable doping [3]. This study investigates a method for achieving controllable doping in 2D semiconductors by depositing low-work-function metals at varying thicknesses in the channel of the transistor to achieve precise n-doping. Aluminum is identified as the optimal doping material, with doping levels precisely controlled through variations in deposition thickness, allowing for performance optimization of the inverter [4]. This study not only demonstrates the controllability of surface metal deposition for doping of 2D semiconductors, but also facilitates the application of 2D semiconductors in integrated circuits.

Device Fabrication

The schematic structure of the MoS$_2$ top-gate transistors fabricated using a metal doping method and the fabrication process are illustrated in Fig. 1 and Fig. 2. First, high-quality monolayer MoS$_2$ was grown on a sapphire substrate using chemical vapor deposition (CVD). Fig. 3 presents atomic force microscopy (AFM) height profile of the CVD-grown MoS$_2$ film, indicating a thickness of approximately 0.7 nm, which demonstrates the monolayer nature of MoS$_2$. The source and drain electrodes were then defined using laser direct writing, and a 35 nm Au film was deposited by electron beam evaporation (EBE). Next, the channel was etched using inductively coupled plasma (ICP), followed by the deposition of various types and thicknesses of thin metal layers in the channel using EBE. Afterward, a seeding layer was grown through EBE, followed by the deposition of a 16 nm of hafnium oxide layer using atomic layer deposition (ALD). Finally, the top gate electrode was completed by depositing an additional 35 nm Au by EBE. This process ensured high quality and excellent electrical performance of the devices, laying a solid foundation for further research and applications.

Results and Discussion

Fig. 4 presents scanning electron microscopy (SEM) images after the deposition of a thin metal layer, revealing the formation of discontinuous metal clusters that prevent short-circuiting in the transistor. Fig. 5 shows the Raman spectra of MoS$_2$ films before and after the deposition of Ti, Ni, Al, and Mo. Notably, the A_{1g} peak of MoS$_2$ exhibits a blueshift when Ti, Ni, and Mo are deposited, indicating film damage and formation of impurities, such as metal oxides, which degrade film quality [5,6].

To further evaluate the effects of metal doping, low-work-function metals like Ti, Ni, Al, and Mo were deposited at thicknesses of 0.5, 1, and 2 nm in the transistor channel, and transfer characteristic curves were measured, as shown in Fig. 6. After the deposition of metals Ti, Ni and Mo, the MoS$_2$ top-gate transistor demonstrated markedly enhanced n-doping, while the gate control capability of the transistor is significantly compromised, and the switching ratio drops from 10^8 to ~10^2. For Ti and Ni, thicker films led to reduced current due to superior surface adhesion, which caused partial shielding of the gate voltage. In contrast, increasing Mo thickness results in higher current. In particular, after Al

979-8-3315-0417-5/25 $31.00 © 2025 IEEE

deposition, the current on/off ratio remains relatively unchanged, the threshold voltage (V_{TH}) shows negative shifts, and the source-drain current (I_{DS}) increases, indicating controlled n-doping effect compared to the MoS_2 top-gate transistor without metal deposition. The n-doping effect becomes more pronounced with increasing Al thickness. The transfer characteristic curves of transistors with 1 nm of different metals are combined for comparison, as illustrated in Fig. 7(a), allowing a more accurate assessment of the doping effects brought about by the different metals.

Fig. 7(b) shows the energy band diagram explaining the n-doping mechanism: when low-work-function metal clusters contact with MoS_2, electrons flow from the metal to MoS_2, causing the energy bands of MoS_2 to bend downward, which results in electron accumulation and can even lead to degenerate state, forming a heavily doped region locally in the channel, which provides additional carriers.

Fig. 8 presents a further statistical analysis of the electrical performance of the Al-doped transistors. As the thickness of the Al metal increased, the V_{TH} shifted leftward from 2.5 V to 0.5 V, the subthreshold swing (SS) remained below 500 mV/dec, the current switching ratios reached 10^8, and the carrier mobilities were above 13 cm²V⁻¹s⁻¹. As illustrated in Fig. 9, the 1 nm Al-doped MoS_2 top-gate transistors display remarkable homogeneity, indicating that the n-doping process of the present species is highly controllable and suitable for deployment in large-scale arrays.

Based on the aforementioned process, we fabricated an enhancement and depletion (E-D) mode inverter based on NMOS, with its optical microscope image and circuit structure illustrated in Fig. 10 [7,8]. The inverter consists of two n-type MoS_2 top-gate transistors: the pull-up transistor is a depletion-mode Al-doped MoS_2 transistor with gate-source shorting, serving as the load transistor, while the pull-down transistor is an enhancement-mode bare MoS_2 transistor acting as the driver transistor. We adjusted the electrical performance of the load transistor by controlling the deposition thickness of the aluminum metal, thereby enabling regulation and optimization of the inverter's flip-flop point. The load characteristic curve of the load transistor and the output characteristic curve of the driver transistor are presented in Fig. 11.

Fig. 12 shows that under the test condition of V_{DD} = 4 V, the flip-flop point of the inverter shifts to the right as the thickness of the deposited Al on the load transistor increases. When the Al deposition thickness is 2 nm, the inverter's logic output flips from high to low at V_{IN} = 1/2 V_{DD}. Fig. 13 shows the voltage transfer characteristic curves of ten randomly selected inverters, illustrating that each inverter successfully achieves a logic transition at V_{IN} = 1/2 V_{DD}, exhibiting excellent uniformity and repeatability, thereby confirming the potential of the proposed method for large-scale wafer-level digital circuit applications.

Conclusion

This study involves the deposition of different thicknesses of low-work-function metals channel of MoS_2 top-gate transistors to investigate their effects on transistor performance. The experimental results demonstrate that the deposition of Al enables controllable n-doping of MoS_2 top-gate transistors. Ultimately, by adjusting the deposition thickness of Al, we successfully modulated the V_{TH} of the pull-up load transistor in the inverter, aligning it with the electrical performance of the driver transistor and resulting in an inverter with a tunable flip-flop point. The findings of this research provide an effective solution to the challenge of controllable doping in two-dimensional semiconductors.

Acknowledgments

This work was supported by the National Key Research and Development Program (Grant No.2021YFA1200500), Science and Technology Commission of Shanghai Municipality (NO. 23JC1401100), the National Natural Science Foundation of China (Grant No. 62334011) and the Shanghai Pilot Program for Basic Research – Fudan University 21TQ1400100 (23TQ008).

References

[1] F. Wu, H. Tian, Y. Shen, et al., "Vertical MoS_2 transistors with sub-1-nm gate lengths," Nature, vol. 603, no. 7900, pp. 259–264, Mar. 2022.

[2] S. B. Desai, S. R. Madhvapathy, A. B. Sachid, et al., "MoS_2 transistors with 1-nanometer gate lengths," Science, vol. 354, no. 6308, pp. 99–102, Oct. 2016.

[3] H.-Y. Park, S. R. Dugasani, D.-H. Kang, et al., "n- and p-Type Doping Phenomenon by Artificial DNA and M-DNA on Two-Dimensional Transition Metal Dichalcogenides," ACS Nano, vol. 8, no. 11, pp. 11603–11613, Nov. 2014.

[4] S. Das, H.-Y. Chen, A. V. Penumatcha, et al., "High performance multilayer MoS_2 transistors with scandium contacts," Nano Lett, vol. 13, no. 1, pp. 100–105, Jan. 2013.

[5] N. H. Nickel, P. Lengsfeld, I. Sieber, et al., "Raman spectroscopy of heavily doped polycrystalline silicon thin films," Phys. Rev. B, vol. 61, no. 23, pp. 15558–15561, Jun. 2000.

[6] K. Schauble, D. Zakhidov, E. Yalon, et al., "Uncovering the Effects of Metal Contacts on Monolayer MoS_2," ACS Nano, vol. 14, no. 11, pp. 14798–14808, Nov. 2020.

[7] S. Wachter, D. K. Polyushkin, O. Bethge, et al., "A microprocessor based on a two-dimensional semiconductor," Nat Commun, vol. 8, no. 1, p. 14948, Apr. 2017.

[8] Z. Lin, Y. Liu, U. Halim, et al., "Solution-processable 2D semiconductors for high-performance large-area electronics," Nature, vol. 562, no. 7726, pp. 254–258, Oct. 2018.

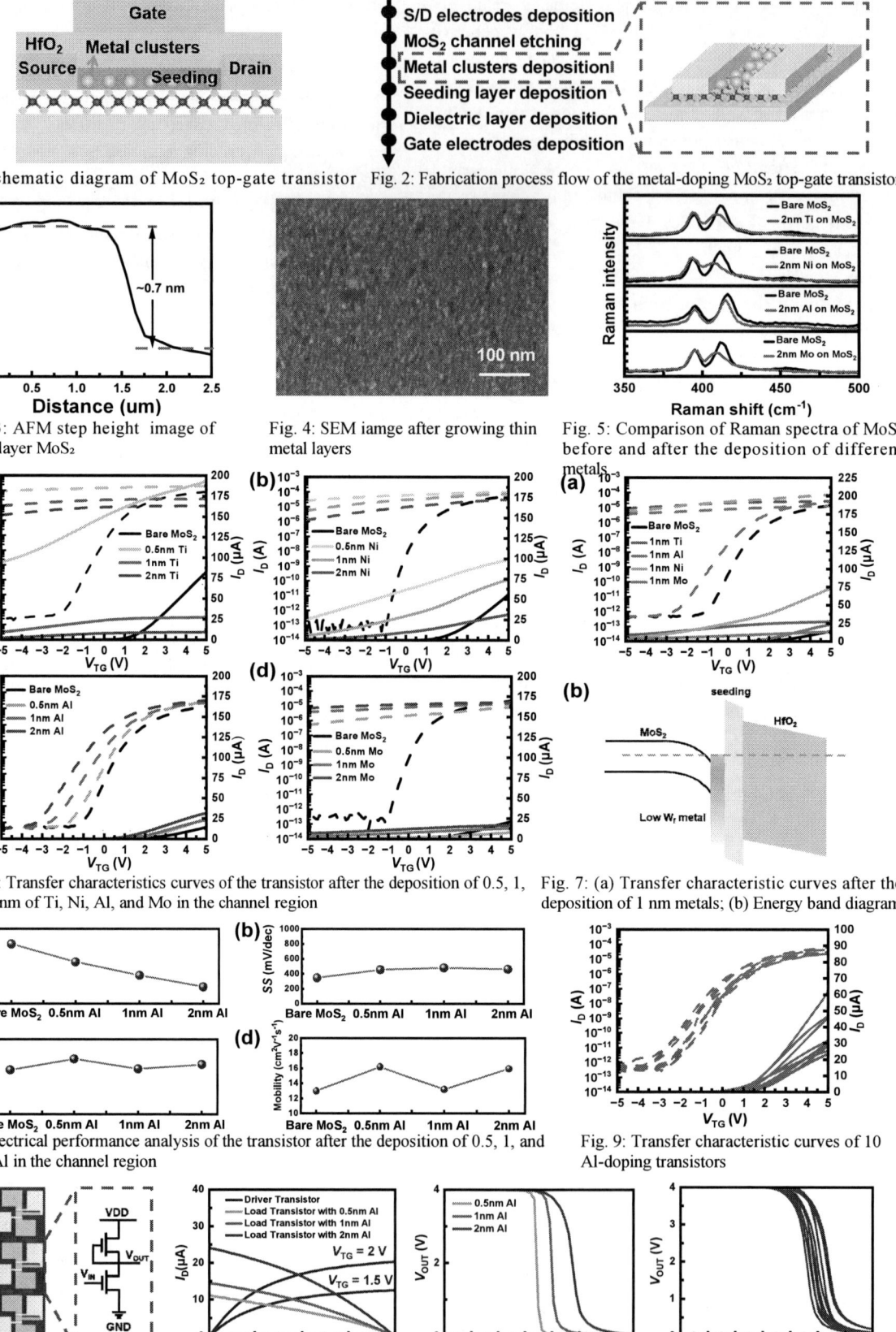

Fig. 1: Schematic diagram of MoS₂ top-gate transistor structure

Fig. 2: Fabrication process flow of the metal-doping MoS₂ top-gate transistors

Fig. 3: AFM step height image of monolayer MoS₂

Fig. 4: SEM iamge after growing thin metal layers

Fig. 5: Comparison of Raman spectra of MoS₂ before and after the deposition of different metals

Fig. 6: Transfer characteristics curves of the transistor after the deposition of 0.5, 1, and 2 nm of Ti, Ni, Al, and Mo in the channel region

Fig. 7: (a) Transfer characteristic curves after the deposition of 1 nm metals; (b) Energy band diagram

Fig. 8: Electrical performance analysis of the transistor after the deposition of 0.5, 1, and 2 nm of Al in the channel region

Fig. 9: Transfer characteristic curves of 10 Al-doping transistors

Fig. 10: Optical microscope image and circuit structure diagram of the inverter

Fig. 11: Load characteristic curve and output characteristic curve

Fig. 12: Voltage transfer characteristic curve of the inverter

Fig. 13: Voltage transfer characteristic curves of 10 randomly selected inverters

979-8-3315-0417-5/25 $31.00 © 2025 IEEE

A Controllable and CMOS Compatible Doping Process for 2D Integrated Circuits

Yuchen Tian[1†], Yan Hu[1†], Shicheng Zeng[1†], Yuxuan Zhu[1], Saifei Gou[1], Xiangqi Dong[1], Zhejia Zhang[1], Jinshu Zhang[1], Qicheng Sun[1], Mingrui Ao[1], Xinliu He[1], Haojie Chen[1], Yufei Song[1], Jieya Shang[1], Zihan Xu[2], Chuming Sheng[1*], Zhengzong Sun[1*] and Wenzhong Bao[1,3*]

[1]School of Microelectronics, Fudan University, Shanghai 200433, China.
* E-mail: baowz@fudan.edu.cn, zhengzong_sun@fudan.edu.cn, cmsheng20@fudan.edu.cn
[2]Shenzhen SixCarbon Technology, Shenzhen 518000, China.
[3]Shaoxin Laboratory, Shaoxing 312000, China.
[†]These authors contributed equally to this work

Abstract

Various methods have been investigated to modulate the electrical characteristics of MoS_2 transistors, which are essential for large scale integrated logic circuits. Here, we demonstrated that a controllable doping method, plasma treatment of source/drain contact, can effectively modulate the threshold voltage (V_{TH}) and the on-state current of top-gate transistors to optimize the performance of inverters. The adjusted inverter exhibits a controllable switching threshold voltage (V_M), ranging from -0.8 V to 1.5 V, with a voltage gain exceeding 20.

Keywords: MoS_2, plasma and inverter.

Introduction

Transition metal dichalcogenides (TMDs) have attracted widespread attention due to their unique atomic thickness, excellent electrical and optoelectronic properties. Among them, MoS_2 demonstrates significant potential in the manufacturing of logic devices and integrated circuits due to its excellent mobility and large on/off current ratio [1-4]. However, how to precisely modulate the threshold voltage (V_{TH}) of MoS_2 transistors to manipulate the switching threshold voltage (V_M) of the inverter remains a challenge [5,6]. A common method is simply adjusting the width-to-length ratio (W/L) of the transistors. But this method is only effective within a limited range and greatly increases the device area [4]. The method of covering the channel of the back-gate transistor with a global oxide layer has also been proven to be effective, but it cannot be suitable for more precise regulation of local devices, and cannot be used in large-scale integrated circuits [7].

This work proposes a stable and controllable strategy that uses remote plasma to treat the source/drain (S/D) contact region of the MoS_2 transistor. The doping level can be controlled by modulating the gas atmosphere and processing time of plasma to change the V_{TH} and significantly improve the on-state current of the transistor. Finally, we successfully utilize this strategy to modulate the V_M of E/D NMOS inverter.

Experimental Methods

The large-scale monolayer MoS_2 film in this work was grown by chemical vapor deposition (CVD) on the sapphire substrates. The specific growth details have been reported in our previous work [8]. The Raman spectrum in Figure 1 shows a Raman peak spacing (Δk) of approximately 20 cm^{-1} between E^1_{2g} and A_{1g}. Meanwhile, the photoluminescence (PL) spectrum displays an emission peak at approximately 1.88 eV, with a full width at half maximum (FWHM) of about 120 meV, as shown in Figure 2.

The fabrication processes of plasma-treated MoS_2 top-gate (TG) FETs are as follows. Firstly, the S/D region was patterned by laser direct writing, and the exposed area was treated by Ar or O_2 plasma, followed by depositing 35 nm Au as the electrodes. Subsequently, the channel regions (W/L=30 μm /10 μm) were patterned, and the non-channel regions were etched by O_2 reactive ion etching (RIE). The seeding layer was deposited by electron beam evaporation (EBE), followed by the deposition of 16 nm HfO_2 as the dielectric using atomic layer deposition (ALD). Finally, 35 nm Au was deposited as the top gate electrode of the transistor by thermal evaporation. The structural schematic diagram and optical microscope image of the MoS_2 TG transistor are shown in Figure 3. All electrical characterization was conducted by Agilent B1500A semiconductor parameter analyzer in ambient air environment.

Results and Discussion

The typical transfer characteristics of MoS_2 TG FETs, the contact regions of which were treated by different plasma，are shown respectively in Fig. 4a-c. All the devices exhibit a typical n-type transistor behavior, and the on/off current ratio reaches 10^8, but these devices also show some noticeable differences. Compared with the untreated pristine device, the device with S/D region treated by Ar plasma shows a negative shift of V_{TH} and a higher on-state current. However, the V_{TH} of the device treated by O_2 plasma shows a noticeable positive shift, and the on-state current is slightly lower than that of the pristine device. Noticeably, the subthreshold swing (SS) and on/off ratio of the device did not change conspicuously. This indicates that the Ar plasma treatment leads to a n-type doping for MoS_2, while the O_2 plasma treatment induces a slight p-type doping effect, which is consistent with the previous researches [9]. As a non-reactive gas, the main role of the Ar in plasma is to produce physical bombardment effects, that can increase the number of Sulphur vacancies on the surface of monolayer MoS_2, thereby producing an effect similar to n-type doping. O_2 plasma, a common reactive plasma, etches the sulfur (S) atom layer on the surface of MoS_2. The resulting sulfur vacancies are subsequently filled by oxygen atoms, leading to a p-type doping effect[9].

The plasma treatment for contact region is a very effective strategy to modulate the V_{TH} of MoS_2 FETs, which can be utilized to fabricate the E/D-NMOS inverter. The fabrication processes of the E/D-NMOS inverter are shown in Figure 5. The load transistor was a depletion-type FET with a negative V_{TH}, which has a relatively higher current at $V_{TG} = 0$ V. Meanwhile, the V_{TH} and on-state current of the device can be further modulated by adjusting the processing time. The driver transistor was treated by O_2 plasma to produce an enhanced FET with a relatively more positive V_{TH}.

Figure 6 shows the schematic diagram of the E/D-NMOS inverter and the modulated performance of the inverter by plasma treatment. It is obvious that whether modulating the load transistor or the driver transistor independently, the V_M of the inverter shifts significantly. With the increase of the processing time of the load transistor, the V_M shifts positively. This is due to a stronger n-doping level induced by the increased processing time. Therefore, the V_M of the inverter shifts from -0.8 V to 0.8 V, and the voltage gain increases from 10.4 to 21.6. When we individually modulated the driver transistor by using O_2 plasma to treat the S/D region of driver transistor, the V_M of the inverter shifts positively from 0 V to 1.5 V. We successfully modulate the V_M of inverter by local optimizing the S/D treatment process of load and driver transistors, which is critical for the industrial application of 2D semiconductors. Meanwhile, the noise margin and voltage gain of inverter are also improved significantly.

Conclusion

In conclusion, this work marks the first time that the plasma treatment method, equivalent to ion implantation in silicon-based technology, has been employed to precisely realize the control of n-type or p-type doping in two-dimensional semiconductor devices. This approach effectively modulates V_M of inverters. Moreover, the remote plasma treatment process is seamlessly compatible with silicon-based CMOS technology, and has the potential to be applied to other monolayer or few-layer two-dimensional materials.

Acknowledgments

This work was supported by the National Key Research and Development Program (Grant No.2021YFA1200500), Science and Technology Commission of Shanghai Municipality (NO. 23JC1401100), the National Natural Science Foundation of China (Grant No. 62334011) and the Shanghai Pilot Program for Basic Research – Fudan University 21TQ1400100 (23TQ008).

References

[1] Radisavljevic, B., Radenovic, A., Brivio, J. et al, "Single-layer MoS_2 transistors", Nature Nanotech. 6, 147–150 (2011).

[2] Chen, X., Xie, Y., Sheng, Y. et al, "Wafer-scale functional circuits based on two dimensional semiconductors with fabrication optimized by machine learning", Nat. Commun. 12, 5953 (2021).

[3] Wang, X., Chen, X., Ma, J., et al, "Pass-Transistor Logic Circuits Based on Wafer-Scale 2D Semiconductors", Adv. Mater. 34, 2202472(2022).

[4] Wachter, S., Polyushkin, D., Bethge, O. et al, "A microprocessor based on a two-dimensional semiconductor", Nat. Commun. 8, 14948 (2017).

[5] Branimir, R., Michael, B., and Andras, K., "Integrated Circuits and Logic Operations Based on Single-Layer MoS_2", ACS Nano. 5 (12), 9934-9938(2011).

[6] Sheng, Y., Chen, X., Liao, F., et al, "Gate Stack Engineering in MoS_2 Field-Effect Transistor for Reduced Channel Doping and Hysteresis Effect", Adv. Electron. Mater. 7, 2000395(2021).

[7] Fan, D., Li, W., Qiu, H. et al, "Two-dimensional semiconductor integrated circuits operating at gigahertz frequencies", Nat. Electron. 6, 879–887 (2023).

[8] Xia, Y., Chen, X., Wei, J. et al, "12-inch growth of uniform MoS_2 monolayer for integrated circuit manufacture", Nat. Mater. 22, 1324–1331 (2023).

[9] Sheng, Y., Zhang, L., Li, F., et al, "A novel contact engineering method for transistors based on two-dimensional materials", Journal of Materials Science & Technology, 69, 15-19(2021).

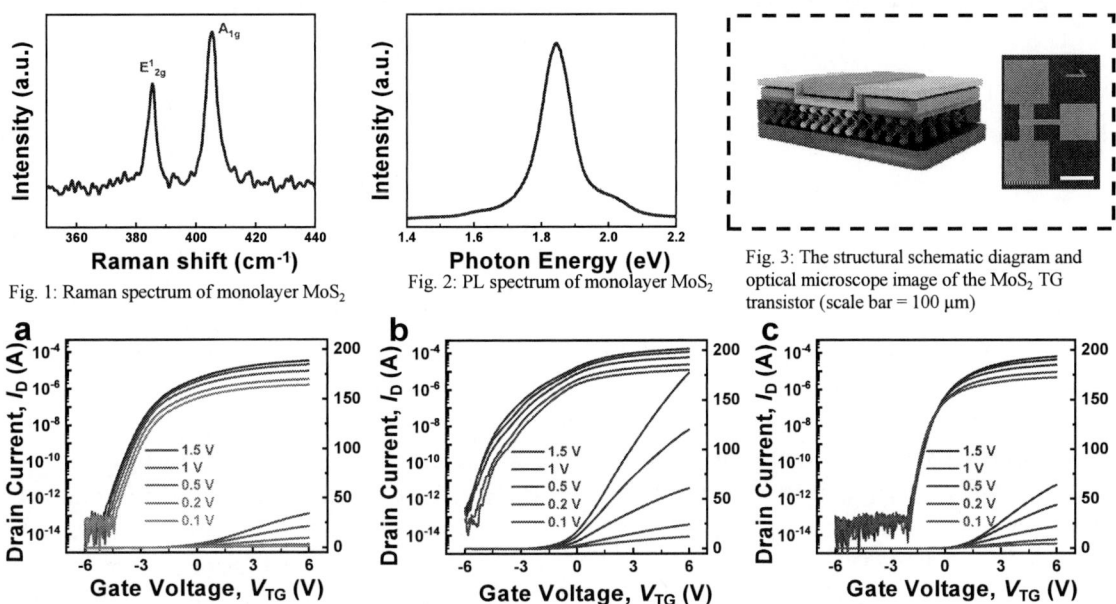

Fig. 1: Raman spectrum of monolayer MoS$_2$

Fig. 2: PL spectrum of monolayer MoS$_2$

Fig. 3: The structural schematic diagram and optical microscope image of the MoS$_2$ TG transistor (scale bar = 100 μm)

Fig. 4: The transfer characteristics of MoS$_2$ TG FETs (W/L=30 μm /10 μm) fabricated with the (a) pristine S/D contact, treated by (b) Ar plasma and (c) O$_2$ plasma contact.

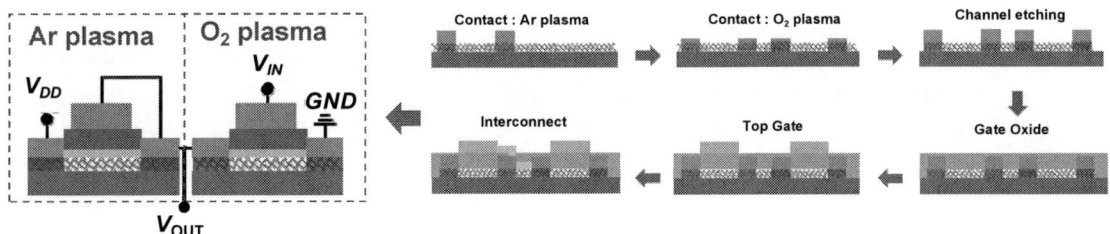

Fig. 5: The fabrication process of E/D NMOS inverters using different plasma treatments for load and driver transistors.

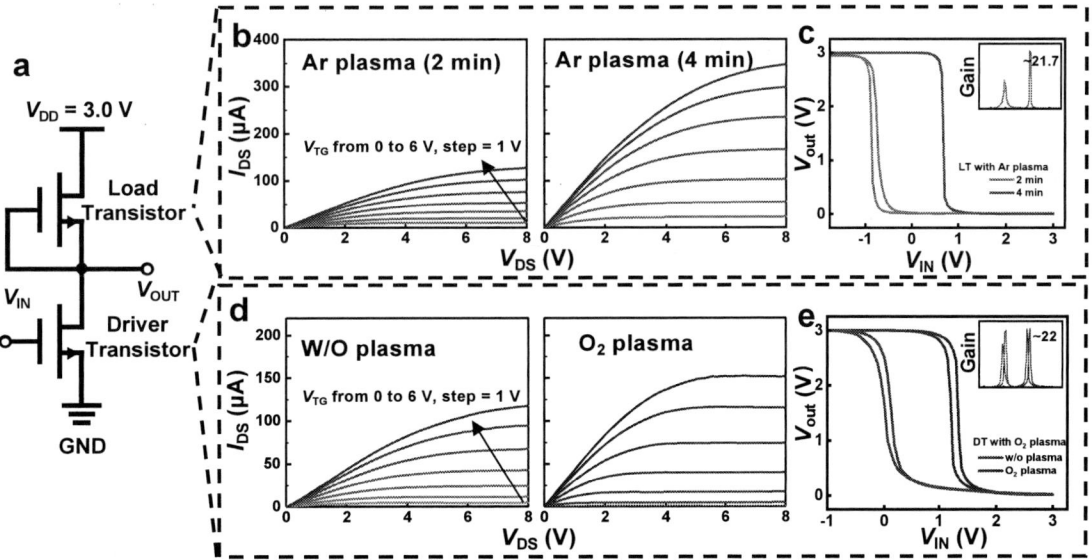

Fig. 6: Modulating the performance of E/D NMOS inverters based on S/D contact process. (a) Equivalent circuit and test diagram of inverter. (b) Output curves of load transistor treated with Ar – plasma for different times. (c) The electrical characteristics of inverters for different load transistors. The illustration shows the corresponding voltage gain. (d) Output curves of driver transistors treated with different plasma. (e) The electrical characteristics of inverters for different driver transistors. The illustration shows the corresponding voltage gain. LT: Load Transistor. DT: Driver Transistor.

979-8-3315-0417-5/25 $31.00 © 2025 IEEE

Research on Oxidizer Engineering of ALD for Industrial Production of ZrO₂ Capacitor in DRAM

Xinyi Tang*, Songming Miao*, Yuanbiao Li, Guangwei Xu, Di Lu and Shibing Long

*Contributed equally to this work; University of Science and Technology of China, Hefei 230026, China

Email: shibinglong@ustc.edu.cn

Abstract

This manuscript aims to enhance the production efficiency while maintaining its properties by optimizing the growth processes of the DRAM capacitor material ZrO₂, through oxidizer engineering by increasing the O₃ flux and using an extremely fast pulse time (1.5 s). This "short pulse - high oxidizer flux" method elevates the k value, effectively reduces leakage, and cuts off the growth time. Moreover, this method also provides high reliability and uniformity of the resulting devices.

Keywords: DRAM, ZrO₂, oxidizer engineering

Introduction

Atomic layer deposition (ALD) is one of the most time-consuming process in the fabrication of dynamic random-access memories (DRAM) [1]. Reducing the pulse time of precursors of the ALD-grown materials, such as the dielectric, could significantly boost the efficiency of production. For example, if the pulse time of the oxidizer of ZrO₂ is reduced from 3 s to 1.5 s, the ALD growth process can be shortened by about 20% (assuming cycle length 7 s in total). However, we have found the short pulse time impacts the quality of ZrO₂ (Fig. 1(a)). By greatly increasing the oxidizer flux, we have researched a "short pulse - high oxidizer flux" method to achieved high-k, low-leakage and highly reliable DRAM capacitors while reducing its growth time.

Device Structure and Fabrication

Fig. 1(b-c) shows the structure and the cross-sectional element distribution of the 6 nm ZrO₂ capacitors and Fig. 1(d) depicts the key procedures of the fabrication. The deposition temperature is 300 °C. The ALD process sequence for ZrO₂ followed a pattern of CpZr(NMe₂)₃ pulse - N₂ purge - O₃ pulse - N₂ purge (2-1.5-1.5-2 s), with saturated growth rates of 0.9 Å per cycle (Fig. 1(e)) [2]. The O₃ flux ranged from 1k sccm to 10k sccm (the maximum flux limited by the equipment). The bottom electrodes were TiN (20.0 nm), while the top electrodes were TiN (3.0 nm) and SiGe (280.0 nm). All films deposited under different O₃ flux were not annealed. The areas of the capacitors were 2.5×10^{-5} cm².

Results and Discussion

A. Electrical Characterizations of the Capacitors

Fig. 2 shows the C-V and I-V curves of the ZrO₂ capacitors by different O₃ flux, showing significant impact. As the increasing of the O₃ flux, dielectric constant (k) increases and the leakage current decreases at the same time. The ZrO₂ film grown by using 10k sccm O₃ flux has the highest dielectric constant ($k\sim47$ at 1 V). The increase of k can also be expressed as the decrease of effective oxide thicknesses (EOT), a commonly used dielectric performance indicator:

$$EOT = \frac{k_{SiO_2}}{k_{high-k}} d_{high-k}, \tag{1}$$

These are likely due to the reduction of the low-k amorphous phases and the reduction of the oxygen vacancy density. To elucidate the origin of the improved k and leakage current by oxidizer engineering, we do more analysis about the ZrO₂ films.

B. Structural Analysis of the Films using Different O₃ Flux

To explore the chemical composition and impurity concentrations of the films, we tested GIXRD data of the unpatterned ZrO₂ films by low (1k sccm) and high (10k sccm) O₃ flux, showing in Fig. 3. Due to the low thickness (6 nm) of the test sample and the limitations of the equipment, the signal peak of the ZrO₂ film is weak. Obviously, the characteristic peak of the t-phase or o-phase ($2\theta \sim 30°$) is stronger for the sample grown under 10k sccm than that for the sample grown under 1k sccm, showing better crystallization in high O₃ flux.

Cross-sectional high-resolution transmission electron microscopy (HRTEM, Fig. 4(a-d)) images also show significant difference. The HRTEM images show that the ZrO₂ film (1k sccm O₃ flux) is mostly amorphous with a small proportion of crystalline phase, confirmed by the fuzzy circle in the FFT pattern (Fig. 4(c)). As a comparison, the ZrO₂ film (10k sccm O₃ flux) is polycrystalline, confirmed by the clear array of the diffraction peaks in the FFT pattern (Fig. 4(d)). This result also indicates that high O₃ flux results in high crystallinity. Higher atomic concentration of O in the ZrO₂ films tends to promote the crystallization, leading to higher k value, while lower atomic concentration of O tends to destabilize the ZrO₂ crystal into the amorphous phase (Fig. 4(e)).

To understand the reduction of leakage by using higher O₃ flux, we counted the number of the grain boundaries in the TEM images within a width of about 55 nm. Fig. 4(f) shows the statistical data of the number of the grain boundaries in each image of the ZrO₂ films grown by using 1k and 10k sccm O₃ flux (11×2 images, about 600×2 nm in total). And XPS was conducted to investigate the chemical and electronic states of two films (Fig. 5(a-b)). The O 1s peak was decomposed into three sub-peaks: lattice oxygen of the metal oxides, oxygen vacancy and chemisorbed oxygen. The oxygen vacancy density in the ZrO_2 were significantly lower for the sample grown under high (10k sccm) O_3 flux (Fig. 5(c-d)). As the gain boundaries and excessive oxygen vacancies introduce defect states as well as leakage current [3], the improvement of leakage current by increasing the O_3 dosage can be attributed to the reduction of the grain boundary density and the bulk oxygen vacancy density.

C. Quality Characterizations of the Capacitors

Annealing typically crystallizes the dielectric and enhances k [4]. We have annealed the ZrO_2 samples in different temperature (in N_2, 60 s), however, only enhances the highest k values for the samples with larger O_3 flux, even with a high annealing temperature of 500 °C showing in Fig. 6(a). It indicates that low pulse time combined with low O_3 may cause irreversible damage to the dielectric crystals. Periodically applying voltages (2.5 V) higher than DRAM working voltage (0.5 V) examines their reliability under stress for high potential safety test (Fig. 6(b)). While maintaining stability over 10^4 cycles, all samples exhibit the enhancement of k values, especially during the initial 100 cycles. The leakage current density J and k at 0.5 V of an 8×8 array of the best devices (10k sccm O_3 flux) demonstrate excellent uniformity (Fig. 7).

Fig. 8(a) and Fig. 8(b) benchmarks the leakage current density J at 0.5 V and the ALD cycle time of our oxidizer-engineered film in 10k sccm O_3 flux against other reported ZrO_2-based dielectrics [5-17]. It should be highlighted that this "short pulse - high oxidizer flux" method produces ZrO_2 films exceeding all other reports by about 10% in terms of EOT (0.55 nm), among the best results reported for ZrO_2-based dielectrics, in particular with such a small ALD cycle time of 7 s. This excellent EOT is attributed to the ultra-thin thickness, as well as the carefully optimized ALD growth process. The leakage current density was also effectively reduced to 2×10^{-8} A/cm^2 at 0.5 V, meeting the demand for DRAM capacitors. For the double data rate (DDR) 4- or 5-DRAM technology, the leakage current should be below 10^{-7} A/cm^2 at the operation voltage (0.5 V) [18].

Summary and Conclusion

This study presents novel insights into the ALD growth of ZrO_2 films using high O_3 flux and fast pulse time conditions. We found that the "short pulse - high oxidizer flux" method can lead to the high-k t-phase in ZrO_2, reduce the leakage current by reducing the grain boundaries and the bulk oxygen vacancy density. Although short-pulse itself deteriorates the performance of the dielectric, high (10k sccm) O_3 flux can effectively improve

the highest k value by about 20% and reduce the ALD cycle time. This method is also likely applicable for advanced DRAM materials, like HZO. The minimal EOT is as low as 0.55 nm and the leakage current is at 2×10^{-8} A/cm^2 level at 0.5 V. The outstanding properties exhibited by these ZrO_2 layers highlight their significant potential in DRAM applications

Acknowledgment

This work is supported by the NSFC under Grant Nos. U20A20207 and 61925110. This work was partially carried out at the Center for Micro and Nanoscale Research and Fabrication of University of Science and Technology of China.

References

[1] X. Tang, Y. Li, S. Miao, X. Chen, G. Xu, D. Lu and S. Long, *IEEE Electron Device Letters*, vol. 45, no. 11, pp. 2114-2117, Nov. 2024, doi: 10.1109/led.2024.3455338.

[2] H. Song, H. Jeon, C. Shin, C. Shin, W. Jang, J. Park, J. Chang, J. H. Choi, Y. Kim, H. Lim, H. Seo and H. Jeon, *Thin Solid Films*, vol. 619, no. 30, pp. 317-322, Nov. 2016, doi: 10.1016/j.tsf.2016.10.044.

[3] K. McKenna, A. Shluger, V. Iglesias, M. Porti, M. Nafría, M. Lanza and G. Bersuker, *Microelectronic Engineering*, vol. 88, no. 7, pp. 1272-1275, Apr. 2011, doi: 10.1016/j.mee.2011.03.024.

[4] J. Zhou, Z. Zhou, L. Jiao, X. Wang, Y. Kang, H. Wang, K. Han, Z. Zheng, and X. Gong, *IEEE International Electron Devices Meeting*, pp. 13.4.1-13.4.4, 2021, doi: 10.1109/IEDM19574.2021.9720632.

[5] H. Song, D. Kim, Y. Kim, H. Jung, H. Lim, S. Lee and K. Yong, *Thin Solid Films*, vol. 675, no. 5, pp. 153-159, Feb. 2019, doi: 10.1016/j.tsf.2019.02.040.

[6] J.-H. Kim, V. Ignatova, P. Kücher, J. Heitmann, L. Oberbeck and U. Schröder, *Thin Solid Films*, vol. 516, no. 23, pp. 8333-8336, Jan. 2008, doi: 10.1016/j.tsf.2008.03.051.

[7] C. Y. Tsai, K. C. Chiang, S. H. Lin, K. C. Hsu, C. C. Chi and A. Chin, *IEEE Electron Device Letters*, vol. 31, no. 7, pp. 749-751, Jul. 2010, doi: 10.1109/led.2010.2049636.

[8] D.-K. Lee, H.-B. Kim, S.-H. Kwon and J.-H. Ahn, *Materials Letters*, vol. 279, p. 128490, Nov. 2020, doi: 10.1016/j.matlet.2020.128490.

[9] J. H. Lee, B.-E. Park, D. Thompson, M. Choe, Z. Lee, I.-K. Oh, W.-H. Kim and H. Kim, *Thin Solid Films*, vol. 701, no. 1, p. 137950, May 2020, doi: 10.1016/j.tsf.2020.137950.

[10] Y. Li, X. Tang, G. Xu, H. Li, S. He, X. Hu, X. Su, W. Bai, D. Lu and S. Long, *IEEE Transactions on Electron Devices*, vol. 70, no. 1, pp. 59-64, Jan. 2023, doi: 10.1109/ted.2022.3223327.

[11] H. Song, D. Kim, S. Kang, H. Jung, H. Lim and K. Yong, *Thin Solid Films*, vol. 713, no. 1, p. 138368, Nov. 2020, doi: 10.1016/j.tsf.2020.138368.

[12] W. Weinreich, A. Shariq, K. Seidel, J. Sundqvist, A. Paskaleva, M. Lemberger and A. J. Bauer, *Journal of Vacuum Science & Technology B*, vol. 31, no. 3, p. 01A109, Dec. 2013, doi: 10.1116/1.4768791.

[13] S. Knebel, U. Schroeder, D. Zhou, T. Mikolajick and G. Krautheim, *IEEE Transactions on Device and Materials Reliability*, vol. 14, no. 1, pp. 154-160, Mar. 2014, doi: 10.1109/tdmr.2012.2204058.

[14] B. Zhu, X. Wu, W. J. Liu, S. J. Ding, D. W. Zhang and Z. Fan, *Nanoscale Research Letters*, vol. 14, no. 1, p. 53, Feb. 2019, doi: 10.1186/s11671-019-2874-5.

[15] C. An, W. Lee, S. Kim, C. Cho, D. Kim, D. Kwon, S. Cho, S. Cha, J. Lim, Jeon and C. Hwang, *Phys Status Solidi-R*, vol. 13, no. 3, Mar. 2019, doi: 10.1002/pssr.201800454.

[16] W. Lee, C. H. An, S. Yoo, W. Jeon, M. J. Chung, S. H. Kim and C. S. Hwang, *Phys Status Solidi-R*, vol. 12, no. 10, p. 1800356, Aug. 2018, doi: 10.1002/pssr.201800356.

[17] Y. W. Yoo, W. Jeon, W. Lee, C. H. An, S. K. Kim and C. S. Hwang, *ACS Appl Mater Interfaces*, vol. 6, no. 24, pp. 22474-82, Dec. 2014, doi: 10.1021/am506525s.

[18] S. Kim, S. H. Lee, M. J. Kim, W. S. Hwang, H. S. Jin and B. J. Cho, *IEEE Electron Device Letters*, vol. 42, no. 4, pp. 517-520, Apr. 2021, doi: 10.1109/led.2021.3059901.

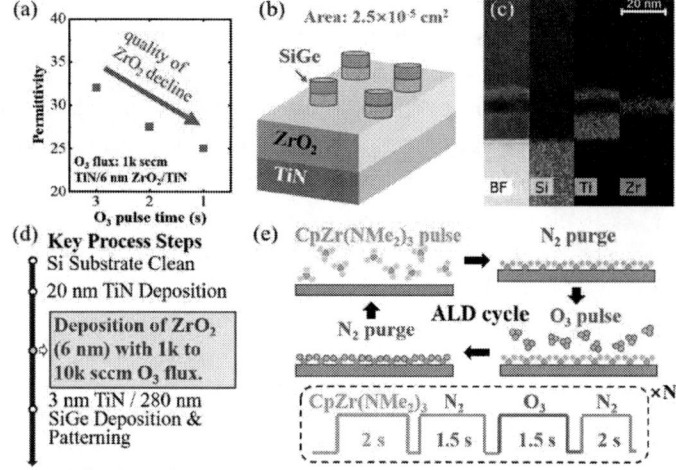

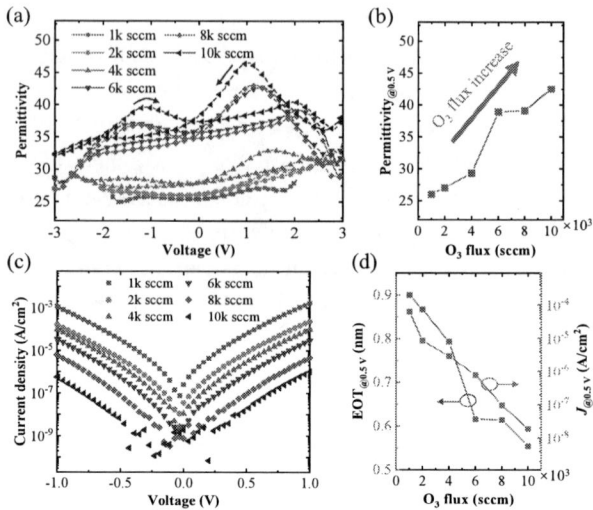

Fig. 1 (a) k values of the ALD-grown ZrO_2 as a function of the O_3 pulse time. The short pulse time destabilizes the ZrO_2 crystal (see Fig. 4). (b) Structure of the ZrO_2-based DRAM capacitors. (c) Element distribution across a ZrO_2 capacitor by TEM-EDS, showing atomic concentrations. (d) Process flow to fabricate the capacitors with O_3 flux varied from 1k to 10k sccm. (e) Schematics of the typical ALD growth cycle for ZrO_2.

Fig. 2 (a) k-V curves of the ZrO_2 films. (b) Dielectric constant at 0.5 V vs O_3 flux for the ZrO_2 films. When the O_3 flux increases, k also increases. (c) The I-V curves of the ZrO_2 films. Leakage current decreases as the O_3 flux increases. (d) $EOT_{@0.5 V}$ and $J_{@0.5 V}$ vs different O_3 flux.

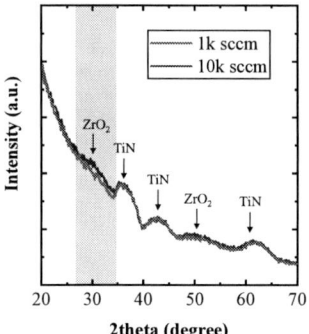

Fig. 3 Grazing incident X-ray diffraction (GIXRD) of the ZrO_2 films grown under 1k and 10k sccm O_3 flux. The ZrO_2 film with higher O_3 flux shows better crystallization with stronger characteristic peak of the t-phase or o-phase ($2\theta \sim 30°$).

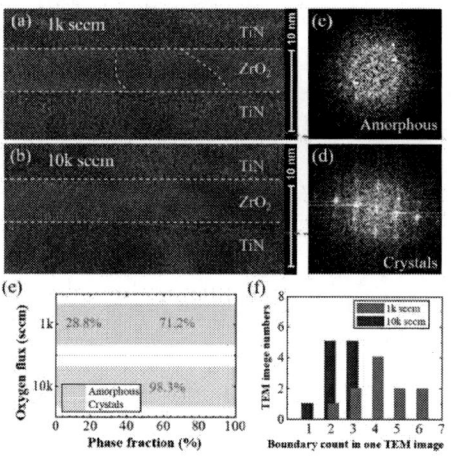

Fig. 4 Cross-sectional HRTEM image of the ZrO_2 films with (a) 1k and (b) 10k sccm O_3 flux, clearly showing partially amorphous (c, 1k) and polycrystalline (d, 10k). Crystal structures (e) and grain boundaries (f) of them.

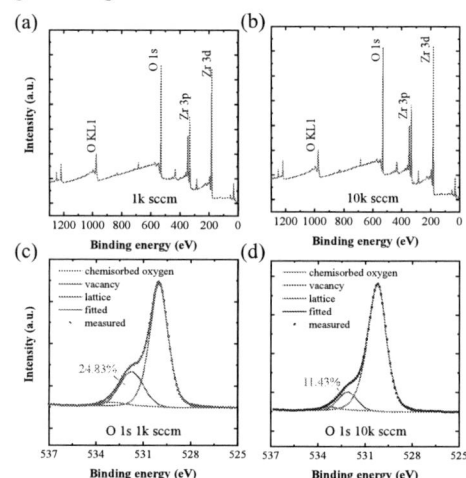

Fig. 5 Survey scan of a ZrO_2 film with a 1k (a) and 10k (b) sccm O_3 flux by X-ray photo-electron spectroscopy. (c-d) Individual high-resolution spectra of O 1s for the samples with different O_3 flux.

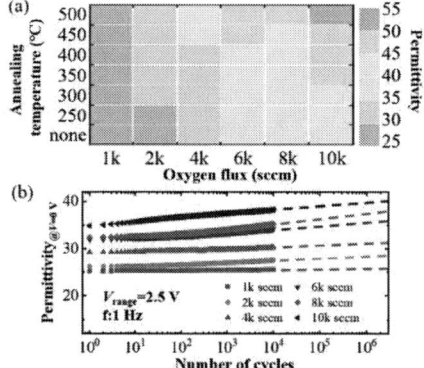

Fig. 6 (a) The highest k value of the ZrO_2 capacitors by annealing at various temperatures from 250 °C to 500 °C in N_2 for 60 s. (b) The k value as a function of number of cycles.

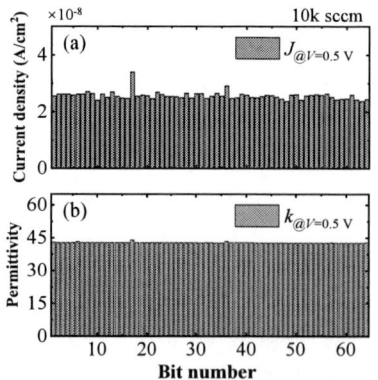

Fig. 7 The J(a) and the k value (b) at 0.5 V of the ZrO_2 capacitance array (64 devices) with 10k sccm O_3 flux. The results show that the devices are highly uniform, necessary for further applications.

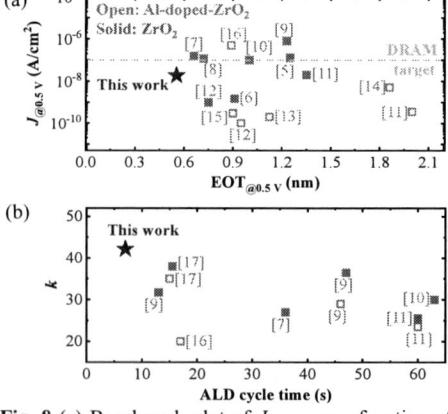

Fig. 8 (a) Benchmark plot of $J_{@0.5 V}$ as a function of EOT. (b) The highest k value vs ALD cycle time of our film and other reported capacitors.

979-8-3315-0417-5/25 $31.00 © 2025 IEEE

In-Material Multimodal Physical Computing for Multisensory Integration

Ming He[1,2,*], Shuo Liu[1], Junling Liu[1], Lei Xu[1], Ru Huang[1]

[1]School of Integrated Circuits, Beijing Advanced Innovation Center for Integrated Circuits, Peking University, Beijing 100871, China; [2]Frontiers Science Center for Nano-optoelectronics, Peking University, Beijing 100871, China. *E-mail: minghe@pku.edu.cn

Abstract

We present multisensory-integration devices by adopting the in-material multimodal physical computing of novel Bi_2O_2Se ferroelectric semiconductor, which enables the simultaneous detection and perception of visual, audio, thermal, bolometric, and electrical stimuli. The non-volatile ferroelectric characteristics achieve an impressive on/off ratio exceeding 10^6 and a memory window of approximately 4 V, demonstrating significant potential for non-volatile memory applications. We demonstrate the fusion responses for optical-thermal, optical-electrical, and optical-bolometric signals, allowing for effective information pre-processing. In addition, the stochastic resonance model is introduced to decouple the respective signals within the fusion response, addressing the challenge of high-precision signal separation in noisy environments. Our work highlights the substantial potential of this all-in-one device for intelligent perception in the era of artificial intelligence. (**Keywords:** In-Material Physical Computing, Multisensory Integration, Bi_2O_2Se Ferroelectric Semiconductor)

Introduction

Biological systems perceive their environments through a diverse array of sensory modalities: vision (83%), audition (11%), olfaction (3.5%), gustation (1%), and touch (1.5%). Each modality relies on specialized receptors that facilitate the detection and transformation of incoming stimuli. This sensory information is then transmitted to the cerebral cortex, where multisensory neurons integrate and interpret the data, allowing for nuanced understanding of the external world. Within the realm of artificial intelligence applications, such as robotics, autonomous driving, and smart furniture, the development of effective, sensitive, and stable multisensory sensors that combine advanced sensing capabilities with signal preprocessing functionality represents a significant challenge.

We have been demonstrating the simultaneous transformation of multimodal physical signals, including sound, light, heat, and electricity, based on the novel two-dimensional ferroelectric semiconductor Bi_2O_2Se. This material exhibits an exceptional carrier mobility exceeding 100 cm² V^{-1} s^{-1}, an ultra-sensitive optical response of 5×10^5 A/W across the broad wavelength range of 400 to 1500 nm, and a high thermal resolution of 10 mK [1,2]. Moreover, we have disclosed the non-volatile ferroelectric memory of Bi_2O_2Se, resulting in the on/off ratio of 10^6 and the large memory window of 4 V. We have further demonstrated the optical-thermal fusion, optical-electrical fusion, and optical-bolometric fusion, confirming fundamental principles of biological multisensory integration such as the super-additivity, the inverse effectiveness, and the temporal congruency. To facilitate the analysis of individual components within the fused signal, we have designed the stochastic resonance decoupling method based on the response frequency differences of the various signals. Finally, we have developed the multisensory integration hardware that enables reliable recognitions of practical targets.

Results and Discussion

The sensing characteristics of the as-fabricated back-gated Bi_2O_2Se field-effect transistor are presented in **Fig. 1**. The device exhibits exceptional electrical performance, exhibiting ultrasensitive photoresponse across the visible light spectrum and a broad thermal detection range spanning from 280 K to 380 K [1-3]. Notably, the device demonstrates a peak photoresponse value exceeding 5×10^5 A/W and achieves a thermal resolution superior to 2.5%/K. Collectively, these attributes position our device as an exemplary platform for multisensory optical-thermal detection [4].

To investigate the non-volatile potential of the fabricated Bi_2O_2Se transistor, we characterize piezoresponse force microscopy (PFM) as illustrated in **Fig. 2a-c**. The butterfly-shaped curve of amplitude and the hysteresis behavior of the PFM phase confirm the ferroelectric properties of the Bi_2O_2Se semiconductor channel. Subsequently, we construct the ferroelectric-channel field-effect transistor that exhibits a remarkable on/off ratio exceeding 10^6 and a large memory window near 4 V, indicating substantial memory capacity (**Fig. 2d**). To assess the stability of the polarization behavior, we applied over 10^6 set and reset pulses at the gate terminal (**Fig. 2e**). The resulting curve demonstrates no significant temporal shift, with a memory retention time exceeding 10^8 seconds (**Fig. 2f**). Collectively, these findings indicate that our device also functions as an exceptional memory device.

We conduct comprehensive investigations into the

interactions between multimodal sensory signals. **Fig. 3a-b** presents the optical-bolometric response of our multimodal device with the corresponding individual optical and bolometric responses. In fact, the device exhibits pronounced fusion response to optical-thermal and optical-sound interactions (**Fig. 3c**). To validate the principles of multisensory integration, we apply unisensory stimuli of optical, sound, and simultaneous inputs to the device (**Fig. 3d**). Remarkably, we observe the fusion response value exceeding 1, indicative of super-additivity, which surpasses the linear summation of the unisensory responses. As the optical intensity increases, the integration factor decreases, supporting the inverse effectiveness principle in multisensory integration. **Fig. 3e** illustrates the fusion response under visual-first and audio-first conditions, while the results in **Fig. 3f** reveal that the fusion response diminishes with increasing time intervals between the two modes of stimulation, thereby confirming the principle of temporal congruency. These findings indicate that by combining different stimulus modes, both the response speed and neuronal excitability are significantly enhanced, requiring only one pulse to elicit a response, compared to 20 or 5 pulses needed for individual modalities.

The multisensory decoupling method is also thoroughly examined as illustrated in **Fig. 4**. We first design a digital filter to effectively distinguish between two individual signals within the fusion response, capitalizing on the distinct characteristic frequencies of optical, thermal, and sound stimuli. Additionally, by solving the Langevin equation within the framework of a constructed nonlinear bistable potential, we enable energy transfer from noise to the effective signal through stochastic resonance. This approach employs noises to assist the particle transition between stable states, thereby addressing the challenge of accurately recognizing desired signals in a noisy background (**Fig. 4a**). **Fig. 4b** displays the original noise fusion response, which reveals an indistinguishable fusion current at the output. In contrast, **Fig. 4c-d** illustrate the processed results following one and three iterations of the corresponding stochastic resonance decoupling model. Notably, we observe a significant reduction in noise at the output of the two individual signals.

We demonstrate the capability of our fusion mode for the detection and recognition of ambiguous targets. The signal flow is depicted in **Fig. 5a**, where a video composed of visual and audio information is sequentially fed into the visual-audio integration neuron array. Through in-situ computation, the fusion signal exhibits a larger amplitude and more distinct characteristics, emerging from the fusion array. To facilitate further on-chip processing, we constructed a hardware pulse-transformation circuit that converts the current into a voltage spike once it exceeds a predetermined threshold, as shown in **Fig. 5b**. The corresponding visualized results of this framing process are illustrated in **Fig. 5c**. Subsequently, the fused images are routed to either software or hardware units for on-chip recognition, with the algorithmic structure detailed in **Fig. 5d**. Employing this multisensory integration technique, we achieve a recognition accuracy nearing 100%, significantly surpassing that of unisensory visual or audio-based neural networks (**Fig. 5e**). Moreover, our fusion mode exhibits enhanced noise tolerance compared to other models, further highlighting the potential of this multisensory integration strategy for signal dimension compression and feature enhancement [5].

Conclusion

We have developed multisensory-integration devices capable of sensitive detection across optical, thermal, sound, and electrical modalities within a single device, especially possessing exceptional ferroelectric memory properties (i.e., memory window of 4 V). The fusion responses of optical-thermal, optical-electrical, and optical-bolometric signals are demonstrated to facilitate effective information compression for front-end signal processing. We also propose a stochastic resonance decoupling model to differentiate the respective component modal responses. The fusion response is utilized for recognition tasks with heightened reliability and enhanced feature clarity. We construct both hardware preprocessing and software recognition components to achieve reliable fuzzy target detections, significantly surpassing the performance of unisensory recognition and algorithmic fusion models. Our work highlights the substantial potential of this multisensory integration strategy for applications in AI scenarios.

Acknowledgments

This work was supported by the National Key R&D Program of China 2022YFB4400100, the Natural Science Foundation of China (92164205, 62074004, 61927901), and the 111 Project (B18001).

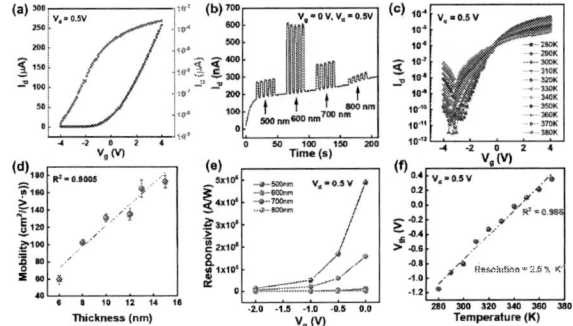

Figure 1. Multisensory response of Bi_2O_2Se transistors: (a) transfer curves, (b) time-varied photoresponse, (c) transfer curves under 280K to 380K range of the Bi_2O_2Se transistors. (d) Extracted carrier mobility for different thicknesses of Bi_2O_2Se channel. (e) Photo-responsivity for V_g ranging from -2V to 0V. (f)V_{th} change with temperature.

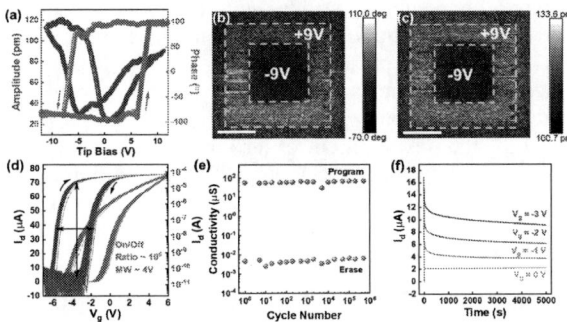

Figure 2. (a) PFM phase and height images of ferroelectric Bi_2O_2Se. (b) PFM phase image, (c) PFM height image after poling with ±9 V voltage bias. Scale bar = 2 μm. (d) Transfer curves of Bi_2O_2Se transistor under 200 cycles. (e) Program and erase states for ferroelectric polarization over 10^6 cycles. (f) Long-time memory over 10 years with single gate pulse from -1V to -3V.

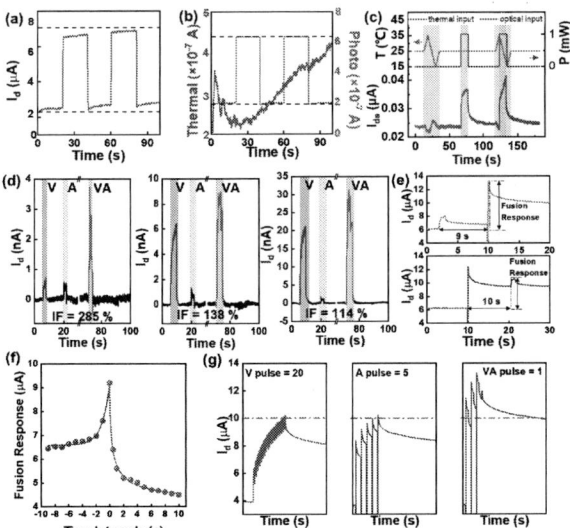

Figure 3. Time varied (a) photo-bolometric fusion response and (b) individual response. (c) light-thermal, and (d) Visual-audio fusion response. (e) Diagram demonstrating fusion response. (f) Fusion response for different V-A time intervals. (g) Number of stimuli needed to reach the threshold for V and A stimuli.

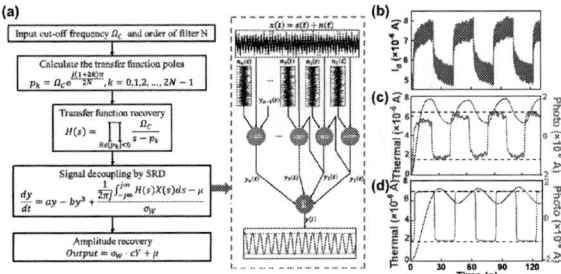

Figure 4. (a) Characteristics of paired pulse facilitation (PPF), (b) spike-interval (Δt)-dependent PPF ratio curves at different wavelengths, (c) excitatory postsynaptic current (EPSC) dynamics triggered by multi-pulse of different frequencies, and (d) spike-number-dependent EPSC variation.

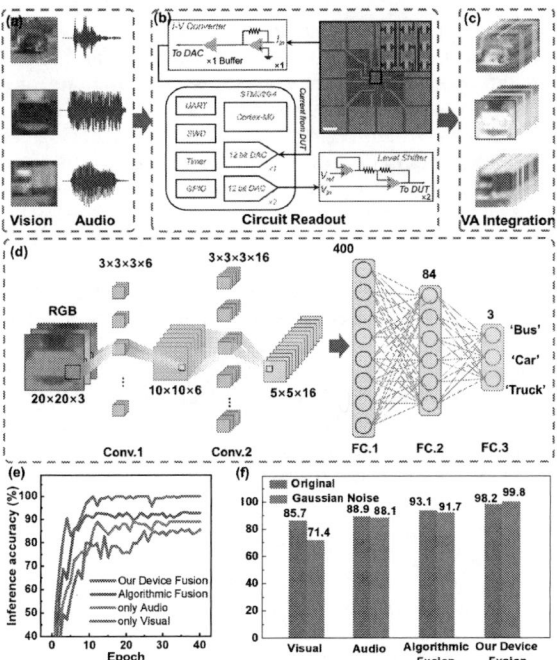

Figure 5. Diagrams for (a) original visual and audio information, (b) pulse transformation process and (c) integrated fusion response, (d) the algorism neural network structure. (e) Recognition results for different models. (f) Comparison of recognition result for noise tolerance and properties.

References

[1] He, M., et al., Sub-10mK-Resolution Thermal-Bolometric Integrated FET-Type Sensors Based on Layered Bi_2O_2Se Semiconductor Nanosheets. 2020 IEEE International Electron Devices Meeting (IEDM), San Francisco, CA, USA, 26.1.1-26.1.4 (2020).

[2] He, M., et al. Ultrasensitive Photo-Thermal Multimodal Sensory based on Self-Doping Modulation of Bi_2O_2Se Semiconductor. 8th IEEE Electron Devices Technology & Manufacturing Conference, Edt, 37-39 (2024).

[3] He, M., et al. Ultrasensitive Retinomorphic Dim-Light Vision with In-Sensor Convolutional Processing Based on Reconfigurable Perovskite- Bi_2O_2Se Heterotransistors, 2023 IEEE International Electron Devices Meeting (IEDM), San Francisco, CA, USA, 1-4 (2023).

[4] He, M., et al. Bi_2O_2Se-Perovskite Heterostructure Based Bipolar Photosensors as Reconfigurable Logic-In-Sensor Devices. 2023 7th IEEE Electron Devices Technology & Manufacturing Conference (EDTM), 10103057 (2023).

[5] He, M., et al. Ultrasensitive dim-light neuromorphic vision sensing via momentum-conserved reconfigurable van der Waals heterostructure, *Nature Communications*, 15, 9011 (2024).

979-8-3315-0417-5/25 $31.00 © 2025 IEEE

Biofuel Cell-Inspired Chemical Sensors for Monitoring Glutamate in Mammalian Central Nervous System

Jinghua Li, Department Of Materials Science and Engineering
The Ohio State University
Columbus, OH, United States
li.11017@osu.edu

Abstract— **Chemical biomarkers in the central nervous system can provide valuable quantitative measures to gain insight into the etiology and pathogenesis of neurological diseases. Glutamate, one of the most important excitatory neurotransmitters in the brain, has been found to be upregulated in various neurological disorders, such as traumatic brain injury, Alzheimer's disease, stroke, epilepsy, chronic pain, and migraines. However, quantitatively monitoring glutamate release in situ has been challenging. This work presents a novel class of flexible, miniaturized probes inspired by biofuel cells for monitoring synaptically released glutamate in the nervous system. The resulting sensors, with dimensions as low as 50 by 50 µm2, can detect real-time changes in glutamate within the biologically relevant concentration range. Experiments exploiting the hippocampal circuit in mice models demonstrate the capability of the sensors in monitoring glutamate release via electrical stimulation using acute brain slices. These advances could aid in basic neuroscience studies and translational engineering, as the sensors provide a diagnostic tool for neurological disorders. Additionally, adapting the biofuel cell design to other neurotransmitters can potentially enable the detailed study of the effect of neurotransmitter dysregulation on neuronal cell signaling pathways and revolutionize neuroscience.**

Keywords—bioimplants, neurochemical sensors, flexible electronics, glutamate, biofuel cells

I. Introduction

Biofluids contain a diverse range of chemical biomarkers that can provide insights into health and age-related conditions. These biomarkers are highly relevant to fields such as biomedical research, advanced healthcare, and clinical medicine. [1] Continuous monitoring of biomarker levels, in particular, can offer valuable evidence that enables the effective diagnosis and treatment of chronic diseases and injuries. As one example, glutamate is one of the most important excitatory neurotransmitters in the mammalian central nervous system (CNS) that is responsible for learning, memory, and communication between neurons. [2] However, excessive glutamate can lead to hyperexcitability of post-synaptic neurons, resulting in induced excitotoxicity. [3] This phenomenon can result in decreased neuronal regeneration and dendritic branching which can not only impair memory and cognition but increase the risk of the development of various neurological diseases. Accordingly, elevated glutamate concentrations show a strong correlation with many neurological disorders, such as traumatic brain injury (TBI), Alzheimer's disease, stroke,

epilepsy, chronic pain, and migraines. [4] In patients who have suffered an ischemic stroke and neurological deterioration, glutamate concentrations in the brain can increase from 15 to 200 µM, and in plasma from 100 to 300 µM. Glutamate concentrations that exceed 200 µM in plasma can serve as a predictor of neuronal damage progression at 48 h post-stroke. [5] Thus, the development of an implantable glutamate sensor that can relay real-time glutamate concentration levels to medical professionals would aid in both the diagnosis and prevention of stroke in high-risk patients (e.g., smokers, patients with high cholesterol or blood pressure).

However, the accurate, continuous, and real-time measurement of glutamate in situ remains a challenging topic. Conventional techniques, such as glutamate microdialysis and ultraperformance liquid chromatography-tandem mass spectrometry (UPLC-MS/MS), have limited capability for real-time monitoring of subtle changes in glutamate concentration with a spatial resolution. Recently developed glutamate sensor protein, iGluSnFr, enables the optical measurement of glutamate dynamics. However, it only measures glutamate semi-quantitatively without providing an absolute concentration reading and needs genetic modification of the subject. [6] Through the use of electrochemical amperometry techniques, alternative strategies have successfully leveraged biosensors to continuously monitor glutamate release in real-time and/or explore their correlations with electrophysiological signals. Despite the great advances enabled by pioneering studies, the existing approaches still involve a complex collection of on-chip hardware for signal generation, such as potentiostats, and power supply/management systems. The difficulty in minimizing the form factors of these subsystems poses substantial challenges in achieving the goal of building miniaturized and lightweight neural interfaces for understanding and intervening in various neurological diseases.

To address this issue, this work presents a class of flexible, miniaturized probes inspired by the structure and working principle of biofuel cells for monitoring synaptically released glutamate in the nervous system. The sensor exploits an enzyme functionalized sensing interface, where the anodic and cathodic reactions will spontaneously generate electrical currents proportional to the concentration of glutamate in the solution. The elimination of potentiostat simplifies the measurement setup compared to that for conventional three-electrode electrochemical cells. Systematic studies investigate the

979-8-3315-0417-5/25 $31.00 © 2025 IEEE

structure-performance interrelationship of the enzyme-functionalized interface for quantitative analysis. The resulting sensors, with dimensions as small as 50 by 50 µm2, can detect real-time changes in glutamate within the biologically relevant concentration range. Ex vivo experiments using the hippocampal circuit in mice models demonstrate the capability of the sensors in monitoring glutamate release during synaptic transmission in electrically stimulated acute brain slices. By functioning as sensing media for brain-machine interfaces, the resulting devices have the potential to improve our understanding of the underlying molecular and cellular mechanisms of structural and functional maladaptive changes in neural circuits after a variety of CNS trauma and diseases. This knowledge can shed light on the development of new therapeutic interventions for these pathological conditions.

II. METHODS AND WORKING PRINCIPLES

The previously reported biofuel cell structure in the literature inspires the design of the glutamate sensors developed in this work, where glutamate spontaneously generates electrical signals proportional to the concentration. The anode is composed of a gold layer, a rough electrically conductive layer to increase the surface area (either platinum black (Pt-black) or carbon nanotubes (CNTs)), a redox mediator layer containing tetrathiafulvalene (TTF) for electron transfer, an enzyme layer containing immobilized glutamate oxidase (GlutOx) linked to bovine serum albumin (BSA), and a protective layer of Nafion against interferent molecules. The cathode is composed of a gold layer, a rough electrically conductive surface layer of either Pt-black or CNTs, a platinized carbon (Pt-C) layer, and a layer of Nafion. The anodic and cathodic reactions generate electrical currents proportional to the concentration of the glutamate. A load resistor connecting the anode and cathode transforms the current into a voltage for signal readout using an electrochemical workstation. Figure 1 illustrates the procedures for the preparation and functionalization of the electrodes with detailed information provided in the Experimental Section. After the preparation of the conductive layer on the gold electrode, drop-casting TTF yields a thin redox mediator layer on the surface. Then, drop-casting a mixture of BSA, Nafion and GlutOx followed by drying the system at 4 °C for a day forms the enzyme coating layer. To prepare the cathode, drop-casting a solution of Pt-C and Nafion deposits the catalyst for the reduction of oxygen.

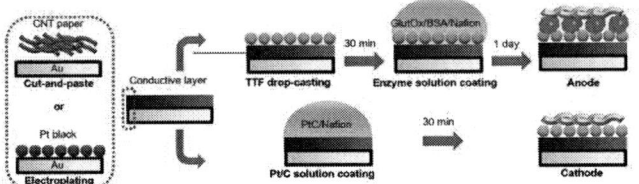

Figure 1. Schematic illustration showing the functionalization protocol of active components onto the surface of the anode and cathode.

III. PERFORMANCE VALIDATION IN BRAIN SLICES

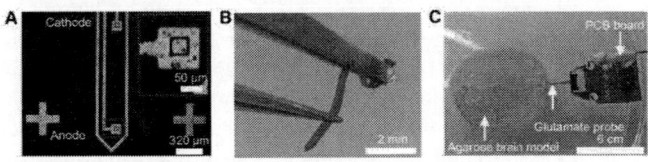

Figure 2. (A) Microscope image of a 50 X 50 µm2 probe. (B) Photograph of a miniaturized probe in a bent configuration. (C) Photograph of a probe penetrating into a 0.6% (w/w) agarose brain model.

Building on the successful development of glutamate sensors, the study leverages these probes to quantitatively analyze the release of glutamate from acute mouse brain slices. Further scaling down the form factors of the probes to nearly cellular-scale dimensions (as small as 50 by 50 µm² for the sensing area) enhances the spatial resolution and biocompatibility for future in vivo applications. This will enable real-time visualization of changes in glutamate concentration at the tissue-device interface. Figure 2A and 2B show a microscopic image and a photograph of a representative probe in a bent configuration, respectively, providing a visual representation of the size and design parameters of the probes. Figure 2C shows a mechanical test of the probe using a brain model made of a 0.6 w/w % agarose solution in DI water, which has a similar Young's modulus to a mouse brain. The Young's modulus of the probe, determined by the value of the Kapton film substrate, is estimated to be ~4.0 GPa. The photograph demonstrates that the probe can penetrate the agarose model with minimal damage to the biochemical interface and gold traces.

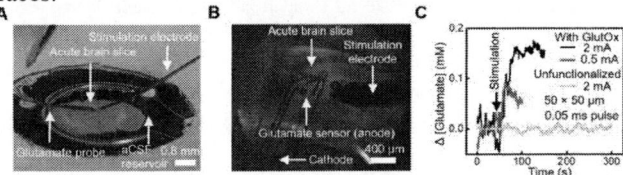

Figure 3. (A) Photograph showing the experimental setup for the ex vivo study using a glutamate probe and a stimulation electrode on a mouse brain slice (thickness: 300 µm). (B) Microscopic image of the glutamate probe and the stimulation electrode on the brain slice. The distance between the stimulation electrode and the probe is approximately 400 µm. The cathode of the glutamate probe is in the surrounding aCSF but not in direct contact with the hippocampus. (C) Change in glutamate concentration detected using a miniaturized probe following electrical stimulation with different pulse intensities (2 mA and 0.5 mA), and comparison with results obtained using an unfunctionalized probe (pulse width: 0.05 ms).

The study utilizes the hippocampal circuit in mice to evaluate the performance of the miniaturized probes in detecting synaptically released glutamate: When Schaffer collateral fibers originating from CA3 neurons are stimulated electrically using a bipolar electrode, glutamatergic synaptic vesicles undergo exocytosis in the CA1 Stratum Radiatum region. Subsequently, glutamate is released from the presynaptic neuron into the synaptic cleft via exocytosis. Excitatory amino acid transporters (EAAT) will then facilitate the uptake of glutamate and transport glutamate from the synaptic cleft back into the neurons and astrocytes, thereby reducing the overall glutamate concentration in the extracellular space and preventing the continuous firing of action potentials. Figure 3A and 3B depict the experimental setup, which includes a stimulation electrode and a glutamate probe laminated onto the surface of the CA1 region of the hippocampus. Figure 3C shows changes in local glutamate concentration obtained using the probe with and without key functional layers on the anode (i.e., TTF and GlutOx). All stimulations use a bi-phasic pulse width of 0.05 ms. The results suggest an immediate increase in the recorded signal following the stimulation, as detected using the functionalized probe, whereas the signal captured by the unfunctionalized probe remains constant due to the lack of bio-recognition elements for glutamate. The increased voltage indicates the successful detection of the synaptic release event using the glutamate probe following the hippocampal circuit model. For the same functionalized probe, increasing the intensity of the stimulation current (from 0.5 to 2 mA) results in an increased amplitude in detected glutamate concentration, possibly due to more axons being recruited by larger stimulus intensity, resulting in more synaptic release.

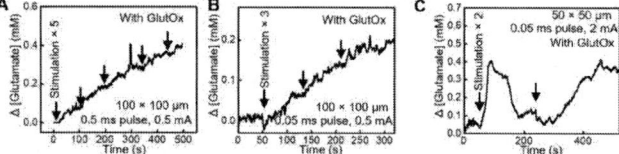

Figure 4. (A) Change in glutamate concentration with five repeated pulses (intensity: 0.5 mA, width: 0.5 ms) measured using a miniaturized probe. (B) Change in glutamate concentration with three repeated pulses (intensity: 0.5 mA, width: 0.5 mA) measured using a miniaturized probe. (C) Change in glutamate concentration with two repeated pulses (intensity: 2 mA, width: 0.05 mA) measured using a miniaturized probe showing the increase and decrease process.

The data in Figure 4A illustrates the progressive rise in glutamate concentration, as measured by a 100 by 100 µm2 probe, during five consecutive stimulations (as indicated by the arrows) with a pulse width of 0.5 ms and a current intensity of 0.5 mA. Each pulse results in the release of a similar amount of glutamate in the local area, creating a stepwise pattern. Figure 4B depicts the outcomes obtained using the same 100 by 100 µm² probe from a comparable 0.5 mA pulse stimulation, but with a stimulation period of 0.05 ms. The results indicate that in this experimental setup, increasing the pulse width does not have a significant impact on the amount of glutamate released from each stimulation. With a single pulse (width: 0.05 ms, intensity: 2 mA) applied, and no further stimulations over a period of approximately 300 s, the concentration of glutamate gradually decreases and returns to the baseline (as depicted in Figure 4C). This observation is likely due to a combined effect of glutamate uptake by transporters in the extracellular space, as well as diffusion in the solution.

IV. CONCLUSION

In summary, this study illustrates the feasibility of utilizing a biofuel cell design in a flexible and miniaturized sensing probe to enable the continuous, real-time detection of glutamate in complex biological environments. The findings indicate that the biofuel cell configuration exhibits sensitivity to glutamate, with detectable concentrations spanning from about 0.05 mM

to 1.0 mM. This research comprises comprehensive investigations of critical parameters that influence the sensing performance, as well as the necessary requirements and considerations for constructing miniaturized sensing probes. Testing the resulting sensing platform with biological samples, including 50X dilutions of mouse brain homogenates and mouse brain slices, validates the effectiveness of the system. The results obtained from experiments utilizing the hippocampal circuit in mice suggest that the miniaturized probes can detect synaptically released glutamate triggered by electrical stimulation. Future studies will aim to explore changes in glutamate concentrations in different regions of the brain, such as the cortex or the ventral tegmental area (VTA), by employing miniaturized probes.

V. REFERENCES

[1]   a) Martin, J. Kim, J. F. Kurniawan, J. R. Sempionatto, J. R. Moreto, G. Tang, A. S. Campbell, A. Shin, M. Y. Lee, Liu, X. Liu, J. Wang, ACS. Sens. 2017, 2(12), 1860-1868.; b) J. Kim, A.S. Campbell, J. Wang, Talanta 2018, 177, 163-170.; c) W. Gao, S. Emaminejad, H. Y. Y. Nyein, S. Challa, K. Chen, A. Peck, H. M. Fahad, H. Ota, H. Shiraki, D. Kiriya, D. -H. Lien, G. A. Brooks, R. W. Davis, A. Javey, Nature 2016, 529(7587), 509-514.; d)T. Arakawa, Y. Kuroke, H. Nitta, P. Chouhan, K. Toma, S.-I. Sawada, S. Takeuchi, T. Sekita, K. Akiyoshi, S. Minakuchi, K. Mitsubiyashi, Biosens. Bioelectron. 2016, 84, 106-111.; e) H. Lee, Y. J. Hong, S. Baik, T. Hyeon, D. -H. Kim, Adv. Healthc. Mater. 2018, 7(8), 1701150.; f) J. Andreu-Perez, D. R. Leff, H. M. D. Ip, G. -Z. Yang, IEEE. Trans. Biomed. Eng. 2015, 62(12), 2750-2762.; g) R. Li, H. Qi, Y. Ma, S. Liu, Y. Jie, J. Jing, J. He, X. Zhang, L. Wheatley, C. Huang, X. Sheng, M. Zhang, L. Yin, Nat. Commun. 2020, 11(1), 3207.; h) G. Rong, E. H. Kim, Y. Qiang, W. Di, Y. Zhong, X. Zhao, H. Fang, H. A. Clark, ACS. Sens. 2018, 3(12), 2499-2505.

[2]   a) W. J. McEntee, T.H. Crook, Psychopharmacology 1993, 111(4), 391-401.; b) Y. -T. Li, X. Jin, L. Tang, W. -L. Lv, M. -M. Xiao, Z. -Y. Zhang, C. Gao, G. -J. Zhang, Anal. Chem. 2019, 91(13), 8229-8236.

[3]   P. -M. Herminia, K. Tuz, Contrib. Nephrol. 2006, 152, 221-240.

[4]   a) C. Baliatas, J. Bolte, J. Yzermans, G. Kerlfkens, M. Hooiveld, E. Lebert, I. V. Kamp, Int. J. Hyg. Environ. Health. 2015, 218(3), 331-344.; b) M. -C. Haces, J. Tang, G. Acosta, J. Fernandez, R. Shi, Transl. Neurodegener. 2017, 6, 20.; c) J. Schultz, Z. Uddin, G. Singh, M. M. R. Howlader, Analyst 2020, 145, 321-347.

[5]   J. Castillo, M. I. Loza, D. Mirelman, J. Brea, M. Blanco, T. Sobrino, F. Campos, J. Celeb. Blood. Flow. Metab. 2016, 36(2), 292-301.

[6]   J. S. Marvin, B. Scholl, D. E. Wilson, K. Podgorski, A. Kazemipour, J. A. Muller, S. Schoch, F. J. U. Quiroz, N. Rebola, H. Bao, J. P. Little, A. N. Tkachuk, E. Cai, A. W. Hantman, S. S. -H. Wang, V. J. Dipiero, B. G. Borghuis, E. R. Chapman, D. Dietrich, D. A. Digregorio, D. Fitzpatrick, L. L. Looger, Nat. Methods. 2018, 15, 936-939.

Boosted Performance of Atomic-Layer-Deposited Dual-Gate Indium-Gallium-Zinc-Oxide Transistors

Anyu Tong[1], Qianlan Hu[1*], Min Zeng[2], Yuzhe Zhu[1], Wenjie Zhao[2], Zhiyu Wang[1], and Yanqing Wu[1,2,3*]

[1] School of Integrated Circuits and Beijing Advanced Innovation Center for Integrated Circuits, Peking University, Beijing 100871, China. [2] Wuhan National High Magnetic Field Center and School of Integrated Circuits, Huazhong University of Science and Technology, Wuhan 430074, China. [3] Beijing Superstring Academy of Memory Technology, Beijing 100176, China. *Email: qlhu@pku.edu.cn, yqwu@pku.edu.cn

Abstract

In this work, an oxygen-rich surface passivation process was applied to ALD IGZO transistors, effectively reducing the oxygen vacancy density within the channel while maintaining a relatively low contact resistance. The dual-gate transistors with improved electrostatic control capability demonstrate boosted performance compared to the back-gate transistors, achieving a record-high I_{on} of 2.24 mA/μm and g_m of over 1 mS/μm for 50 nm short-channel device at V_{ds} = 1 V, the highest values among IGZO transistors.

Keywords: Indium-gallium-zinc-oxide, atomic layer deposition, surface passivation, dual-gate transistor

Introduction

Indium-gallium-zinc-oxide (IGZO), a representative of amorphous oxide semiconductors (AOS), is commonly used in display industry [1]. Its excellent comprehensive performance, including low process temperature, ultra-low leakage current, and suitable mobility, makes it a potential candidate for back-end-of-line (BEOL) compatible logic and memory devices [2], [3], [4]. Atomic layer deposition (ALD) technique offers distinct obvious advantages in component modulation, precise growth rate control, complete chemical reaction, and arbitrary surface deposition, meeting the performance requirements of various device applications [5], [6], [7]. However, in the aforementioned work, high-mobility AOS transistors usually show relatively negative threshold voltage (V_{th}) with significant short-channel effects, such as degraded V_{th} roll-off, subthreshold swing (SS), and drain-induced barrier lowering (DIBL) [8], [9], [10]. Several effective methods can achieve enhancement-mode operation, such as thinning the film thickness and optimizing the metal cation composition of AOS [11], [12]. However, the percolation conductance mechanism of AOS results in a trade-off between the on-state current (I_{on}) and threshold voltage [13]. Most approaches lead to a degradation in on-state performance due to the reduced mobility and increased contact resistance [14]. Therefore, implementing separate control of the channel and source-drain contact regions is critical for achieving high-performance transistors. Additionally, adopting dual-gate structures is considered an effective method to enhance electrostatic control capability and mitigate short-channel effects.

In this work, we have developed an oxygen-rich surface passivation process that significantly improves the current on/off ratio of short-channel back-gate IGZO transistors. The electrostatic control capability is further enhanced by the dual-gate structure. The 50 nm dual-gate IGZO transistor exhibits not only a steep SS of 67 mV/dec with a low DIBL of 18 mV/V, but also a record-high I_{on} of 2.24 mA/μm and g_m of 1.024 mS/μm.

Device Fabrication

Fig. 1 shows the key fabrication process flow of the dual-gate ALD IGZO transistors in this work. First, high-resistance silicon substrates with 100 nm SiO_2 were used for electrical insulation and cleaned using the standard RCA-1 process. The back-gate area was patterned by electron beam lithography (EBL), and a nickel/platinum metal stack was deposited as the gate electrode using electron beam evaporation (EBE). Then, a 5-nm-thick HfSiO back-gate dielectric and an 8-nm-thick IGZO channel were deposited using ALD at 300 °C. The channel region was defined using EBL, followed by wet etching in dilute hydrochloric acid. Source and drain contacts were formed using an ITO interlayer deposited by sputtering, followed by the deposition of a nickel/gold metal stack by EBE to complete the fabrication of the fresh back-gate transistors. Subsequently, 5 nm HfSiO was deposited using oxygen-rich ALD as the passivation layer and the top-gate dielectric, to modulate the carrier density within the channel and enhance the electrostatic control capability [15]. Finally, the top-gate was formed using the same EBL and process with the back-gate. The thermal budget of the entire fabrication process is below 300 °C, ensuring compatibility with back-end-of-line (BEOL) requirements. The back-gate and top-gate were then connected to measure the dual-gate characteristics. The electrical performance was measured using a Keysight B1500A semiconductor parameter analyzer with the device placed into a high-vacuum (<10⁻⁴ mbar) Lakeshore TTPX probe station.

Results and Discussion

As shown in Fig. 2, an oxygen-rich surface passivation process was adopted to reduce oxygen vacancy density

979-8-3315-0417-5/25 $31.00 © 2025 IEEE

within the IGZO channel of the as-fabricated back-gate transistors [16]. Fig. 3 compares the transfer characteristics of the back-gate IGZO transistors before and after passivation, with channel length (L_{ch}) scaled from 1 μm to 100 nm at drain bias (V_{ds}) of 0.5 V, showing a positive V_{th} shift of approximately 2 V with an optimized V_{th} roll-off after passivation. The current on/off ratio of the back-gate transistors before and after passivation as a function of L_{ch} was extracted and plotted in Fig. 4. Before passivation, the back-gate transistors show a sharp decrease in the current on/off ratio as the channel length is scaled down, due to the significantly negative shift in V_{th}. In contrast, the current on/off ratio slightly increases after passivation, owing to the ultra-low off-state current being below the measurement limit and the enhanced on-state current with the scaled channel length. The 100 nm back-gate transistor achieves a more than 7 orders of magnitude improvement in the current on/off ratio after passivation. It's worth noting that the oxygen-rich surface passivation is applied to the channel region, which allows the high carrier density in the source/drain areas to be maintained, preserving low contact resistance while positively shifting the V_{th}.

To further enhance electrostatic control ability to suppress short-channel effects and improve the on-state performance, we have fabricated the dual-gate IGZO transistors by connecting the back-gate and top-gate, as shown in Fig. 5. Fig. 6 compares the transfer characteristics of separate back-gate (BG), top-gate (TG), and dual-gate (DG) IGZO transistors with L_{ch} = 1 μm. The dual-gate transistor exhibits a more positive threshold voltage and an on-state current nearly double that of the back-gate and top-gate transistors. When the channel length scaled from 1 μm to 100 nm, the dual-gate transistors show negligible V_{th} roll-off, as depicted in Fig. 8. This demonstrates a significant improvement over the back-gate transistors in Fig. 3, attributable to enhance gate control.

To investigate the suppression of short-channel effects through structural optimization, shorter-channel back-gate and dual-gate transistors with L_{ch} = 50 nm were fabricated to compare the key electrical parameters. Fig. 9 shows the transfer characteristics of back-gate and dual-gate IGZO transistors at V_{ds} = 0.05 and 0.5 V. The scaled dual-gate transistor exhibits a better immunity to short-channel effects, with a positive V_{th} shift of about 1 V, a steeper SS of 67 mV/dec and a lower DIBL of approximately 18 mV/V. In contrast, the back-gate transistor shows a SS of 97 mV/dec and a DIBL of 331 mV/V. Fig. 10 shows the extracted g_m as a function of gate bias voltage (V_{gs}) for both back-gate and dual-gate IGZO transistors at V_{ds} = 0.5 V. The dual-gate structure also improves on-state performance, increasing the peak value of gm ($g_{m,peak}$) from 365 μS/μm to 527 μS/μm, a 1.5 times enhancement. Fig. 11 compares the extracted SS as a function of drain current (I_d), where the dual-gate transistor

shows a low SS of 67 mV/dec, representing an approximately 1.5 times improvement owing to the enhanced gate control ability. Additionally, a relatively low SS level (< 80 mV/dec) can be maintained exceeding 4 orders of magnitude, which is beneficial for reducing the operating voltage. The DIBL of both back-gate and dual-gate transistors with L_{ch} varying from 1 μm to 50 nm has been extracted and plotted in Fig. 12, showing a sharp increase for back-gate transistors as channel length decreases, while remaining at a considerably low level (< 18 mV/V) for dual-gate transistors. For transistors with the shortest 50 nm channel length, adopting the dual-gate structure reduces the DIBL by 18 times.

Fig. 13 shows the transfer characteristics of 50 nm dual-gate IGZO transistors at V_{ds} = 1V, showing excellent switching performance with remarkable $g_{m,peak}$ of 1.024 mS/μm. Fig. 14 presents the corresponding output characteristics, where a maximum drain current of 2.24 mA/μm is achieved, attributed to the performance boost provided by the dual-gate structure. Fig. 15 and Fig. 16 show the benchmark of on-state current (I_{on}), extracted at V_{ds} = 1 V and V_{gs} = V_{th} + 2 V and $g_{m,peak}$, extracted at V_{ds} = 1 V, as a function of L_{ch} for reported sub-100 nm IGZO transistors [3], [11], [14], [16], [17], [18], [19], [20], [21]. Our dual-gate transistor achieves both highest on-state current of 1.34 mA/μm and highest $g_{m,peak}$ exceeding 1 mS/μm.

Conclusion

In summary, an effective passivation process has been adopted to optimize the switching characteristics of ALD IGZO back-gate transistors. Subsequently, the deposition of a top gate to form a dual-gate transistor further improves electrostatic control capability. The 50 nm short-channel dual-gate transistor demonstrated optimized electrical performance compared to back-gate transistors, achieving a record-high I_{on} of 2.24 mA/μm and g_m exceeding 1 mS/μm.

Acknowledgments

This work was supported by the National Key Research and Development Program of China (2021YFA1202903) and the National Natural Science Foundation of China (6240030623 and 62425402).

References

[1] Y. Zhu, et al., J. Semicond., 2021. [2] A. Belmonte, et al., VLSI, 2023. [3] Q. Li, et al., IEDM, 2022. [4] Q. Hu, et al., IEDM, 2022. [5] M. H. Cho, et al., Journal of Information Display, 2019. [6] J. Sheng, et al., ACS AMI, 2019. [7] M. H. Cho, et al., IEEE TED, 2019. [8] M. Si, et al., Nat. Electron., 2022. [9] S. Li, et al., IEDM, 2019. [10] S. Li, et al., IEDM, 2020. [11] K. Chen, et al., VLSI, 2022. [12] J. Zhang, et al., IEDM, 2023. [13] W. Chakraborty, et al., VLSI, 2020. [14] J. Zhang, et al., IEEE TED, 2023. [15] C. Gu, et al., IEEE EDL, 2023. [16] K. Han, et al., IEEE TED, 2023. [17] W. Lu, et al., IEDM, 2022. [18] Z. Wu, et al., IEEE EDL, 2024. [19] Q. Li, et al., IEDM, 2023. [20] S. Samanta, et al., VLSI, 2020. [21] C. Wang, et al., VLSI, 2022.

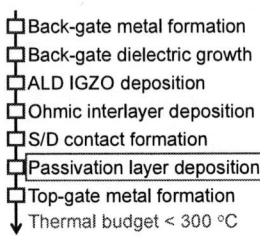
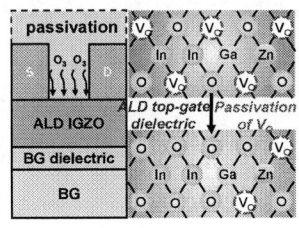
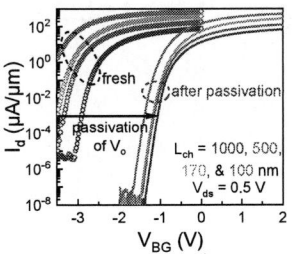

Fig. 1. Key fabrication process flow of the dual-gate ALD IGZO transistor.

Fig. 2. Mechanism schematic of the oxygen-rich surface passivation for the back-gate transistor.

Fig. 3. Comparison of the I_d-V_g curves for back-gate ALD IGZO transistors with various L_{ch} before (hollow) and after (solid) passivation.

Fig. 4. The extracted on/off ratio as a function of L_{ch} for back-gate ALD IGZO transistors before and after passivation.

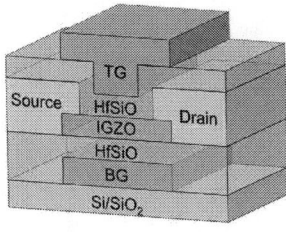
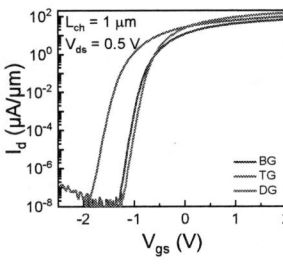
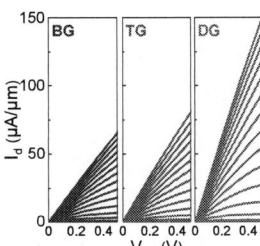
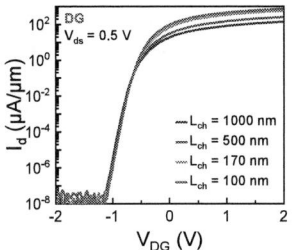

Fig. 5. Device schematic of the dual-gate ALD IGZO transistor.

Fig. 6. Comparison of I_d-V_g curves of ALD IGZO transistors with different device structures: BG, TG, and DG.

Fig. 7. Corresponding I_d-V_d curves characteristics of ALD IGZO transistors in Fig. 6.

Fig. 8. I_d-V_g curves of the dual-gate ALD IGZO transistors at $V_{ds} = 0.5$ V with various L_{ch}.

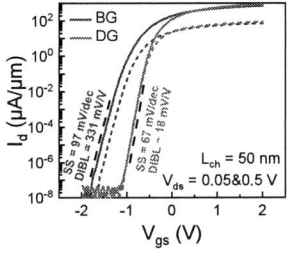
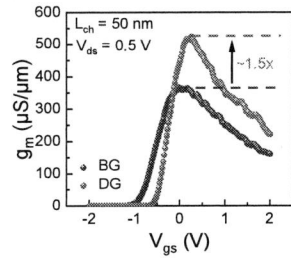
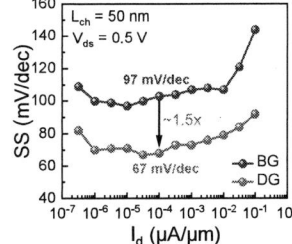
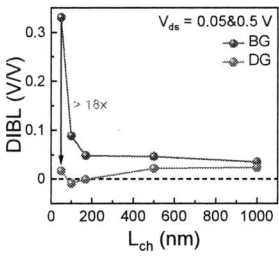

Fig. 9. Transfer characteristics comparison of the BG and DG IGZO transistors with $L_{ch} = 50$ nm at $V_{ds} = 0.05$&0.5 V.

Fig. 10. Comparison of the extracted g_m versus V_{gs} for BG and DG IGZO transistors with $L_{ch} = 50$ nm at $V_{ds} = 0.5$ V.

Fig. 11. Comparison of the extracted SS versus I_d for BG and DG IGZO transistors with $L_{ch} = 50$ nm at $V_{ds} = 0.5$ V.

Fig. 12. Comparison of the extracted DIBL versus L_{ch} between BG and DG IGZO transistors.

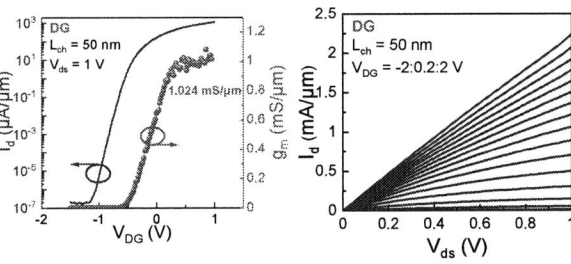
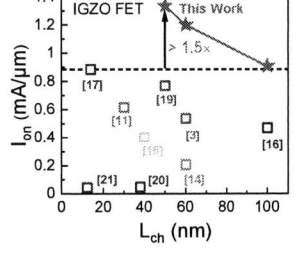
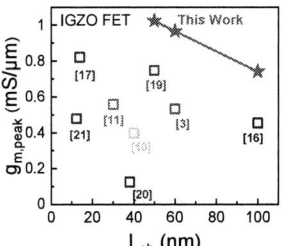

Fig. 13. I_d-V_g curve of the 50 nm dual-gate IGZO transistor at $V_{ds} = 1$ V.

Fig. 14. I_d-V_d curve of the 50 nm dual-gate IGZO transistor.

Fig. 15. Benchmark of I_{on} versus L_{ch} for sub-100 nm IGZO transistors.

Fig. 16. Benchmark of $g_{m,peak}$ versus L_{ch} for sub-100 nm IGZO transistors.

979-8-3315-0417-5/25 $31.00 © 2025 IEEE

Compact Modeling of GaN Based RF Switches

(*Invited Paper*)

Yogesh Singh Chauhan, Mir Mohammad Shayoub, Ahtisham Pampori, and Mohammad Sajid Nazir

Department of Electrical Engineering, Indian Institute of Technology Kanpur, India

Email: chauhan@iitk.ac.in

ABSTRACT

This work introduces a SPICE-compatible compact model for GaN-based dual-gate RF switches. The model is width-scalable and improves accuracy in RF, large-signal, and harmonic balance simulations by using a robust gate network to represent the distributed nature of the serpentine gate. The model is validated against experimental data for multiple device peripheries.
Keywords: Compact model, GaN HEMT, RF switch

INTRODUCTION

Gallium nitride (GaN)-based semiconductor devices produce lower power loss in electrical energy conversion because of their small device on-state resistance and inter-electrode capacitance, making them the suitable candidates for high-efficiency, high-frequency switching and high-power density applications [1]. Owing to these properties, GaN-based switches outperform the switches based on other technologies in RF switching applications [2], [3].

It is imperative for circuit designers to carefully account for switching characteristics. While existing models for GaN switches [4], [5] are primarily empirical in nature, a physics-based model is still missing in the literature. In this work, we present a physics-based model for a dual-gate RF GaN HEMT switch.

DEVICE CHARACTERIZATION

Four devices, 32×50 μm, 32×150 μm, 32×200 μm and 10×100 μm, have been considered in this study. The I-V measurements are carried out by sweeping the gate bias (from -12 to 0 V for three devices and -20 to 0 V for one device) and the drain bias (from 0 to 5 V for all the devices). The s-parameter measurements are carried out for a frequency range of 0.5 to 40 GHz, with V_{ds} = 0 V for various gate bias conditions. Large-signal characterization is performed at 2 GHz.

MODEL FORMULATION

The modeling process involves gate network, DC, and RF modeling, all of which are discussed in the following subsections.

A. Gate Network Modeling

The distributed nature of the serpentine gate is modeled by taking into consideration the transmission line effects of the gate under two assumptions [6] – (a) $L \ll |A_v|W$, where A_v is the common-source

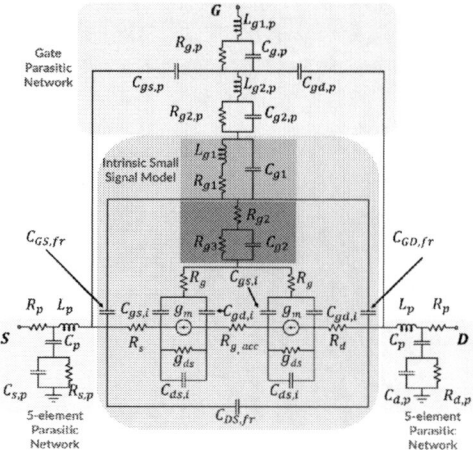

Fig. 1. Small-signal model of the GaN switch device containing the intrinsic model (shaded green) and the parasitic components around it. The region in blue represents the distributed nature of the serpentine gate. The region in red depicts the non quasi-static (NQS) effect.

voltage gain, and (b) the bias conditions across the width of the device remain uniform, implying that only small signals applied at the gate are considered. Thus, the input impedance at the gate node can be expressed as [6] $Z_{in} = (R_1'/\gamma) \coth(\gamma W)$ where $R_1' = (R_g + sL_g)/W$, W is the total gate width, R_g and L_g are the gate resistance and inductance, respectively. γ is the transmission coefficient and is given by $\gamma^2 = R_1' G_1'$ where $G_1' = sC_{gt}/W$ (C_{gt} represents the total gate capacitance, including fringing capacitance). From the above equation, we derive (1), as discussed in detail in our previous work [7]

$$Z_{\text{in}} = \frac{1}{sC_{\text{gt}}}$$
$$+ \frac{X_{R1}R_{\text{g}} + s(X_{L1}L_{\text{g}})}{1 + s(X_{R1}R_{\text{g}})(X_{C1}C_{\text{gt}}) + s^2(X_{L1}L_{\text{g}})(X_{C1}C_{\text{gt}})} \quad (1)$$

where X_{R1}, X_{L1} and X_{C1} are treated as model parameters. It represents an LCR circuit whose equivalent circuit (shaded in blue) is shown in Fig. 1. The NQS behavior is modeled in a similar way as [8], which is an RC network in series with the gate resistance as shown (shaded in red) in Fig. 1. The input impedance can be expressed as follows:

$$Z_{\text{in}} = (R_{g2} + R_{g3}) \left(\frac{1 + \frac{(1+R_{g2}R_{g3})}{(R_{g2}+R_{g3})} C_{g2}s}{1 + R_{g3}C_{g2}s} \right) \quad (2)$$

where $1/R_{g2} = X_{R2}.NF.G_{ch}$, $1/R_{g3} = X_{R3}.NF.G_{ch}$, and $C_{g2} = X_{C2}.(2(C_{gs,i} + C_{g,di}))$. G_{ch} is the conductance of the 2DEG channel and is defined as $G_{ch} = I_{ds}/V_{dseff} = \mu_{eff}\,(W/L)C_g Q_{ch}$.

B. DC Modeling

In between the two gates, the access region resistance is given by [9] as $R_{g,\mathrm{acc}} = R_{g,\mathrm{acc}0}/\left[1 - (I_D/I_{\mathrm{acc,sat}})^\gamma\right]^{(1/\gamma)}$, where $I_{\mathrm{acc,sat}} = Q_{\mathrm{acc}}.v_{\mathrm{sat}}$ is the saturation current or maximum current supported in the access region, and $R_{g,\mathrm{acc}0}$ is the low current access resistance. The parameters, including the threshold voltage, thermal resistance, and access region resistance, are scaled linearly as:

$$M_{eff} = M_0 * (1 + M_W \cdot W) \tag{3}$$

where M_{eff} is the effective parameter value, M_0 is the parameter (to be scaled) and M_W is the factor describing the width (W) dependence. The sign of M_W is positive for threshold voltage and negative for access region resistance and thermal resistance parameter. The self-heating thermal resistance parameter decreases with an increase in size. The model accurately captures the current voltage characteristics, as shown in Fig. 2.

C. RF Modeling

Manifold and substrate parasitics are modeled using a five-element parasitic network [Fig. 1] [10]. The reflection S-parameters of the de-embedded devices (de-embedded using OPEN-SHORT standard) are shown in Fig. 3, and the transfer S-parameters in Fig. 4 for both the ON-state (0 V) and OFF-state (-15 V). The parasitic elements are also scaled according to device dimensions. To accurately capture the compression characteristics, a bias-dependent drain-source fringing capacitance defined as follows is used

$$C_{ds,\mathrm{fr}} = W \cdot \mathrm{NF} \cdot \left(CDSO - CDSL \cdot \sqrt{1e^{-6} + V_{DSE}^2} \right) \tag{4}$$

where $CDSO$ is the drain-source fringing capacitance parameter, $CDSL$ is the parameter for bias dependence of parasitic drain-source capacitance, V_{DSE} is the effective drain bias, given as $V_{DSE} = V_{DS} \cdot V_{DSAT}/(V_{DS}^2 + V_{DSAT}^2)^{1/2}$ and V_{DSAT} is the drain current saturation voltage. Moreover, to consider the higher-order harmonics, we define the gate-drain fringing capacitance as follows:

$$C_{gd,\mathrm{fr}} = W \cdot \mathrm{NF} \cdot \left(CGDO - CGDL \cdot \sqrt{1e^{-6} + V_{DSE}^2} \right)$$
$$- CGDL2H \cdot V_{DSE}^{2N} - CGDL3H \cdot V_{DSE}^{3N} \tag{5}$$

where $CGDO$ is the gate-drain fringing capacitance parameter, $CGDL$ is a parameter to control the drain-bias dependence of $C_{gd,\mathrm{fr}}$, $CGDL2H$ and $CGDL3H$ are parameters to tune the even and odd harmonics, and N is a parameter to adjust the order of the exponents.

RESULTS AND DISCUSSION

The model effectively captures the device behavior and self-heating effects, as shown in Fig. 2, small-signal characteristics in both the OFF-state (-15 V) and ON-state (0 V), as shown in Fig. 3 and Fig. 4, insertion loss as shown in Fig. 5, compression characteristics as shown in Fig. 6, and the harmonic behavior as shown in Fig. 7. In the OFF-state, the devices exhibit compression [Fig. 6] whereas in the ON-state, devices exhibit negligible compression until delivered power levels of 40 dBm (10 W), which is the measurement system's upper limit. The drain-source fringing capacitance plays a crucial role in affecting the second harmonic in the OFF-state, while gate resistance is a major factor for the second harmonic in the ON-state, which aligns closely with the existing literature.

CONCLUSION

We presented a width-scalable physics-based compact model for GaN-based dual-gate RF switches. The model validation is performed against the experimental data for different device peripheries. The model includes the distributed nature of the serpentine gate and also incorporates an improved model for the gate-gate access region as well as fringing capacitances, enabling accurate harmonic simulations in the ON- and OFF states of the RF switch.

REFERENCES

[1] K. Li et al., "SiC and GaN power transistors switching energy evaluation in hard and soft switching conditions," in IEEE WWBPDA, pp. 123–8, 2016.

[2] C. F. Campbell et al., "Wideband high power GaN on SiC SPDT switch MMICs," in IEEE IMS, pp. 145–8, 2010

[3] D. Nandi et al., "Validation of Dynamically Depleted Symmetric BSIM-SOI Compact model for RF SOI T/R Switch Applications," IEEE EDTM, 2024.

[4] A. Wentzel et al., "A simplified switch-based GaN HEMT model for RF switch-mode amplifiers," in EuMIC, pp. 77–80, 2009

[5] Z. Hu, et al., "An improved compact large-signal GaN HEMT model for switch application," in IEEE TED, vol. 69, pp. 3061–7, 2022.

[6] E. Abou-Allam et al., "A small-signal MOSFET model for radio frequency IC applications," in IEEE TCDICS, vol. 16, pp. 437–47, 1997.

[7] A. Pampori et al., "A Large-Signal SPICE Model for a Dual-Gate GaN RF Switch With OFF-State Harmonic Control," in IEEE TED, vol. 71, pp. 84-90, 2024.

[8] C. Gupta et al., "Accurate and computationally efficient modeling of nonquasi static effects in MOSFETs for millimeter-wave applications," in IEEE TED, vol. 66, pp. 44–51, 2019.

[9] S. Ghosh et al., "Modeling of source/drain access resistances and their temperature dependence in GaN HEMTs," in Proc. IEEE EDSSC, pp. 247–50, 2016

[10] A. K. Sahoo et al., "Small-signal modeling of high electron mobility transistors on silicon and silicon carbide substrate with consideration of substrate loss mechanism," in SSE, vol. 115, pp. 12–6, 2016.

979-8-3315-0417-5/25 $31.00 © 2025 IEEE

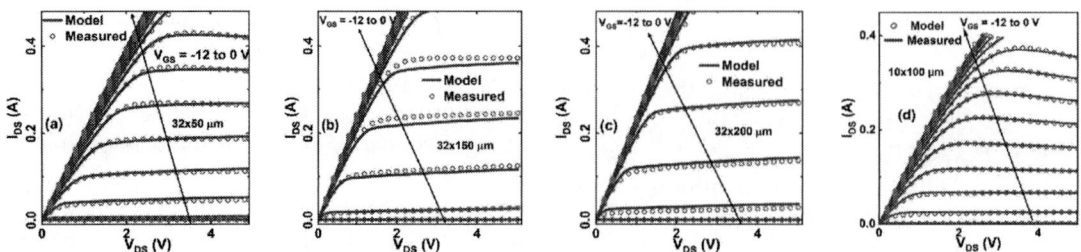

Fig. 2. Output ($I_{DS} - V_{DS}$) characteristics, with drain voltage varying from 0 to 5 V and gate voltage from -12 to 0 V for devices of size (a) 32 x 50 μm (b) 32 x 150 μm (c) 32 x 200 μm (d) 10 x 100 μm.

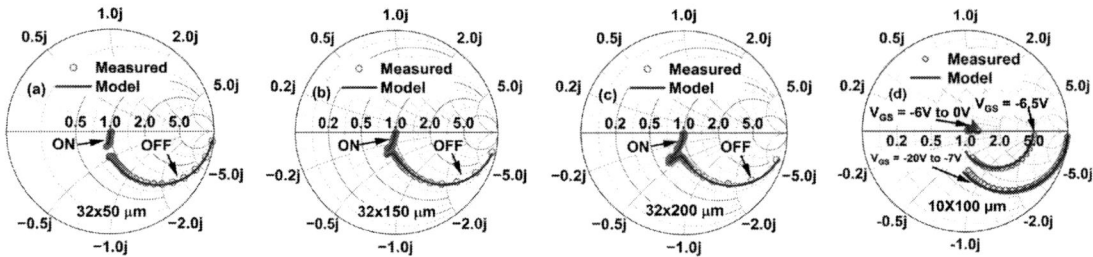

Fig. 3. Reflection S-parameter ($S_{11} = S_{22}$) for a frequency range of 0.5 - 40 GHz for devices of size (a) 32 x 50 μm (b) 32 x 150 μm (c) 32 x 200 μm (d) 10 x 100 μm. Gate bias was swept from -10V to 0V for (a), (b) and (c), and from -20 to 0V for (d).

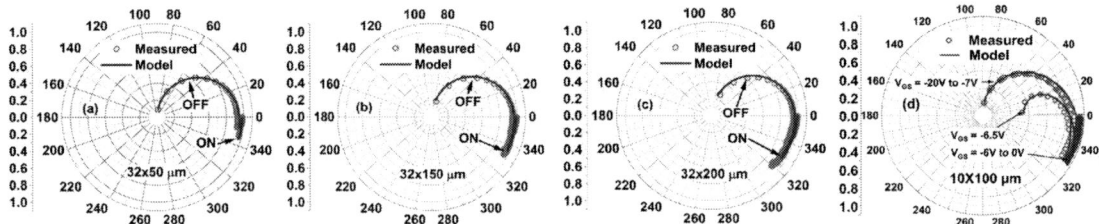

Fig. 4. Transmission S-parameter ($S_{12} = S_{21}$) for a frequency range of 0.5 - 40 GHz for devices of size (a) 32 x 50 μm (b) 32 x 150 μm (c) 32 x 200 μm (d) 10 x 100 μm. Gate bias was swept from -10V to 0V for (a), (b) and (c), and from -20 to 0V for (d).

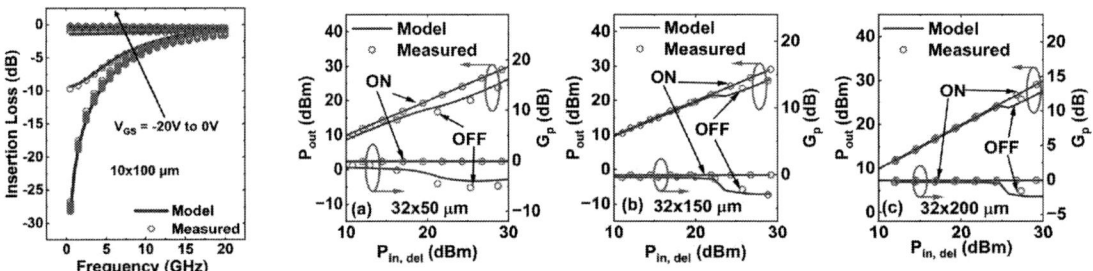

Fig. 5. Insertion Loss validation for the 10 x 100 μm device with V_{GS} swept from −20V to 0V in steps of 0.5V.

Fig. 6. Large-signal power measurements (symbols) and model (solid lines) at 2 GHz for devices of size (a) 32 x 50 μm (b) 32 x 150 μm (c) 32 x 200 μm

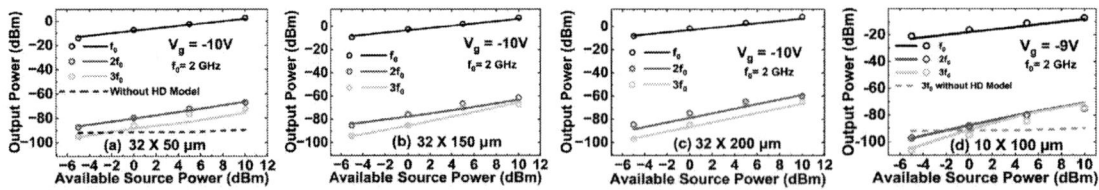

Fig. 7. Harmonic distortion measurements (symbols) and model (solid lines) for devices of size (a) 32 x 50 μm (b) 32 x 150 μm (c) 32 x 200 μm (d) 10 x 100 μm. The dotted line in (a) and (d) represents a fixed fringing gate–drain capacitance (standard ASM-HEMT model) while the solid lines represent the new harmonic distortion (HD) submodel.

A fusion of optical microscopy and functional nanomaterials for subcellular-scale thermodynamic control of muscle contraction

Madoka Suzuki[1]

[1]Institute for Protein Research, Osaka University, Japan

Abstract

This paper challenges the traditional view of heat as a mere byproduct of muscle activity. We demonstrate that heat can directly induce muscle contraction, even in the absence of calcium signaling. We introduce novel techniques based on light, fluorescent molecules, and nanoparticles, for precise heat manipulation at the microscopic scale, enabling the localized control of muscle function. These findings have implications for tissue engineering and regenerative medicine, exploring new strategies for therapeutic interventions and advanced biomaterials.

Keywords: temperature imaging, microscopic heating, fluorescence microscopy

Introduction

Our traditional understanding of skeletal muscle function has primarily focused on the process of force generation leading to heat production, or thermogenesis. This process is particularly evident during physical exercise or shivering in cold environments. The reverse process, where thermogenesis may directly influence force generation, has been much less explored. The common view of the reverse process is that the thermogenesis supports force generation indirectly by elevating overall body temperature. However, recent research has challenged this traditional view. While it's true that individual cells have low heat power that is quickly dissipated, studies have shown that localized thermogenesis can have a direct impact on cellular processes, including muscle contraction. This suggests a more intricate relationship between thermogenesis and force generation than previously considered.

Here, I highlight the limitations of the traditional view of thermogenesis as a byproduct and proposes an advanced understanding of the direct impact of heat flow in living systems, particularly in the muscle contractile systems. The paper introduces novel techniques for measuring and manipulating heat flows at the microscopic scale, that have played a major role in these discoveries [1,2].

Heat Induced Muscle Contraction

We have examined the role of temperature in regulating muscle contractions. Muscle contraction is initiated by an increase in intracellular calcium (Ca^{2+}) levels from the relaxed state where the Ca^{2+} level is maintained low at sub-μM range. The Ca^{2+} binds to the muscle regulatory proteins, triggering the switch from OFF to ON to initiate the interaction between myosin motor proteins and actin filaments, i.e., the muscle contraction. However, our studies demonstrate that heating of cardiac [3] and skeletal [4] muscle cells within the physiological range can induce contractions even in the absence of Ca^{2+} increase. These results suggest that the thermogenesis plays a crucial role in modulating muscle contractility.

The temperature manipulation with a precision of the order of 0.1 K in these studies was made possible by the combination of optical heating microscopy and the fluorescence temperature imaging. The former is based on the water-absorbable intra-red (IR) laser beam focused into the aqueous medium by the objective lens. The latter employs the thermal quenching of highly temperature-sensitive fluorescent molecules. The quantum yield is decreased by increasing the temperature.

To further investigate this heat induced muscle contraction, we conducted experiments using the *in vitro* motility assay, where the sliding motions of fluorescently labeled actin filaments over myosin molecules attached to the glass substrate are observed under the fluorescence microscope [4,5]. We constructed artificial contractile systems composed of skeletal muscle filaments interacting with cardiac myosins, and vice versa. Then, compared the thermosensitivity with those of physiologically relevant combinations. Our data demonstrated higher thermosensitivity of skeletal muscle filaments than cardiac ones, whereas cardiac myosins were more thermosensitive than skeletal counterparts. Interestingly, the combination of higher thermosensitivities, i.e., skeletal muscle filaments with cardiac myosins, demonstrated an extremely high thermosensitivity. With the right combinations, skeletal muscle system demonstrated about 1.6 times higher thermosensitivity than that of cardiac system.

The findings of this research highlight the interplay between thermogenesis and force generation in muscle contractile systems. Furthermore, the results propose the possibility of precisely controlling muscle contractions by externally applied thermal stimulations. However, stimulating the muscle tissue requires the simultaneous excitation of a large number of muscle cells in a controllable manner. In the next section, I introduce a method of simultaneously stimulating intracellular space of a multiple number of myotubes using gold nanoshells (NSs) with near-IR (NIR) radiation.

Gold Nanoshells Remotely Activating Muscles

This study proposes the use of the plasmonic properties of NIR-absorbing NSs to wirelessly stimulate and induce the contraction of C2C12-derived myotubes by heat [6]. NSs are the nanoparticles characterized by a SiO_2 core with a diameter of about 120 nm covered by a shell of Au with a thickness of about 15 nm. When exposed to NIR radiation (peak absorbance is at $\lambda = 828$ nm), NSs absorb light and convert it into heat. We measured temperature increases of approximately 5 °C within the myotube using a fluorescent thermometer. This temperature increase was sufficient to trigger muscle contraction, even in the absence of Ca^2 signaling, which is the traditional pathway for muscle activation as mentioned in the previous section. This indicates that the heat released by NSs directly affects the interaction between actin and myosin filaments, the molecular machinery responsible for muscle contraction. It was demonstrated that all the myotubes within the field of view contracted in a synchronous manner upon the NIR radiation.

The study also investigated the long-term effects of repeated NIR light exposure during the differentiation process from pre-differentiated state to matured myotubes. We found that chronic remote thermal stimulation increased the production of mRNA for heat shock proteins, the molecules involved in cellular stress response and protection, and sirtuin 1, a protein that promotes the creation of new mitochondria within cells. These findings suggest that NIR-mediated muscle stimulation may have additional benefits beyond remotely stimulating muscle cells using light, potentially including muscle tissue engineering, regenerative medicine, and bionics.

Precision in Heating at the Subcellular Scale

Instead of thermally stimulating cells in wide volume, we have also developed a new method for manipulating cellular activity with high precision using light-induced subcellular heating [7]. We fabricated a nanoheater-thermometer (nanoHT), a polymeric nanoparticle with a diameter of about 150 nm that contains both fluorescent temperature probes and photothermal dyes in its inside. When illuminated with a 808 nm NIR laser light, nanoHT generates a subcellular-sized heat spot within a living cell. This unique characteristic allows for highly localized heating with simultaneous monitoring of temperature changes using fluorescence temperature imaging under the confocal fluorescence microscope.

We explored the potential of nanoHT for manipulating various cellular functions. We first demonstrated that nanoHT can efficiently induce cell death in HeLa cells. Within seconds of NIR laser exposure, a localized temperature increase of approximately 11.4 °C from the base temperature (37 °C) triggered cell death specifically at the location of the heat spot. Next, we investigated the use of nanoHT for inducing muscle contraction in C2C12 myotubes as described in the previous section. By generating a heat spot with nanoHT, we were able to successfully stimulate muscle contractions, but only at the vicinity of the intracellular nanoHT.

In overall, the study highlights the potential of nanoHT for developing new strategies for targeted cell ablation, as well as for activating specific cellular processes at the subcellular scale with high spatial and temporal control. The ability to generate subcellular heat spots and to monitor temperature changes using nanoHT opens up exciting possibilities for various biological and medical applications, potentially including targeted cell ablation in cancer treatment and for stimulating specific cellular responses in regenerative medicine.

Conclusion and Perspectives

In this paper, I introduced the applications of two functional nanoparticles as optothermal agents. They are demonstrated to be used in the heat induced muscle contraction. Then, the potential uses in tissue engineering and biomedical applications were proposed. However, the limitations of the current study and the potential future directions for research should also be explored in more detail. For instance, it would be necessary to investigate the biocompatibility of these nanoparticles and its potential side effects for a long time over weeks and months. Exploring the use of NSs and nanoHT for manipulating other cellular processes beyond cell death and muscle contraction could be a promising area for future research.

In terms of advances in optical heating technology, we have recently achieved controlling spatial and temporal gradients of the temperature. Using the aforementioned local heating methods, we can directly apply either a wide field heating to the inside of cells, or do so at the subcellular spatial resolution with a single heat spot. However, controllable multiple heat spots were not previously possible. Also, the re-cooling procedure was solely dependent on the heat diffusion to the surrounding media. Our novel method enables the creation of more than two diffraction limited heat sources, and change their locations in an arbitrary manner. Furthermore, one can control the rate of re-cooling. I envisage that our new method will pave the way for our deeper understanding of the thermal sensitivities in many biological systems.

Acknowledgments

The author gratefully acknowledges the contributions of A. Marino, G. Ciofani, Ferdinandus, K. Oyama, S. Arai, S. Ishii and N. Fukuda.

References

[1] K. Oyama, S. Ishii and M. Suzuki, "Opto-Thermal Technologies for

Microscopic Analysis of Cellular Temperature-sensing Systems," Biophys. Rev., 14, pp. 41-54 (2022).

[2] M. Suzuki, C. Liu, K. Oyama and T. Yamazawa, "Trans-Scale Thermal Signaling in Biological Systems," J. Biochem., 174, pp. 217-225 (2023).

[3] K. Oyama, A. Mizuno, S. A. Shintani, H. Itoh, T. Serizawa, N. Fukuda, M. Suzuki and S. Ishiwata, "Microscopic Heat Pulses Induce Contraction of Cardiomyocytes without Calcium Transients," Biochem. Biophys. Res. Commun., 417, pp. 607-612 (2012).

[4] S. Ishii, K. Oyama, F. Kobirumaki-Shimozawa, T. Nakanishi, N. Nakahara, M. Suzuki, S. Ishiwata, N. Fukuda, "Myosin and Tropomyosin–Troponin Complementarily Regulate Thermal Activation of Muscles," J. Gen. Physiol., 155, pp. e202313414 (2023).

[5] S. Ishii, K. Oyama, T. Arai, H. Itoh, S. A. Shintani, M. Suzuki, F. Kobirumaki-Shimozawa, T. Terui, N. Fukuda and S. Ishiwata, "Microscopic Heat Pulses Activate Cardiac Thin Filaments," J. Gen Physiol., 151, pp. 860-869 (2019).

[6] A. Marino, S. Arai, Y. Hou, A. Degl'Innocenti, V. Cappello, B. Mazzolai, Y.-T. Chang, V. Mattoli, M. Suzuki and G. Ciofani, "Gold Nanoshell-Mediated Remote Myotube Activation," ACS Nano, 11, pp. 2494-2508 (2017).

[7] Ferdinandus, M. Suzuki, C. Q. Vu, Y. Harada, S. R. Sarker, S. Ishiwata, T. Kitaguchi and S. Arai, "Modulation of Local Cellular Activities using a Photothermal Dye-Based Subcellular-Sized Heat Spot," ACS Nano, 16, pp. 9004-9018 (2022).

An energy-efficient microwave magnetic field generator for NV center quantum magnetometers based on an array of four injection-locked VCOs

Hadi Lotfi[1], Qing Yang[1], Tarek Elrifai[1], Michal Kern[1], and Jens Anders[1,2*]

[1]Institute of Smart Sensors, University of Stuttgart, D-70569 Stuttgart, Germany
[2]Institute for Microelectronics Stuttgart (IMS CHIPS), D-70569 Stuttgart, Germany

*Email: jens.anders@iis.uni-stuttgart.de

ABSTRACT

This paper proposes the use of chip-integrated LC voltage-controlled oscillators (VCOs) as the B_1 field source in NV diamond-based magnetometers. VCOs provide a compact, energy-efficient solution by directly converting DC power into the microwave B_1 field required for NV center control, bypassing conventional RF amplification stages. Additionally, the VCO's intrinsic frequency agility, enabled by the varactor inside the LC tank, which simultaneously changes the resonant frequency of the LC resonator and the oscillation frequency, ensures a near-constant B_1 field across the tuning range. Current state-of-the-art chip-integrated diamond magnetometers, while showing promising results, typically make use of classical transmitter architectures, in general, providing a lower energy efficiency compared to the presented VCO-based approach. With its very good energy efficiency, the proposed VCO-based approach offers a path to scalable, low-power NV-based magnetometers suitable for portable and embedded applications. To provide a sufficiently large active area for typical diamond samples, the presented VCO-based B_1 generator uses an array of four injection-locked VCOs. An on-chip phase-locked loop (PLL) precisely defines the frequency of the B_1 field from an external reference. The design is validated using electrical measurements that reflect the key metrics of the target quantum magnetometry application.

Keywords: NV center magnetometer, B_1 source, VCO, injection-locked VCO array

INTRODUCTION

Magnetometers based on nitrogen-vacancy (NV) centers in diamond have emerged as powerful tools for detecting and measuring magnetic fields, driven by applications spanning medicine, material characterization, and fundamental physics research [1], [2]. Their unique ability to achieve high sensitivity at room temperature and in a compact form factor renders them advantageous over

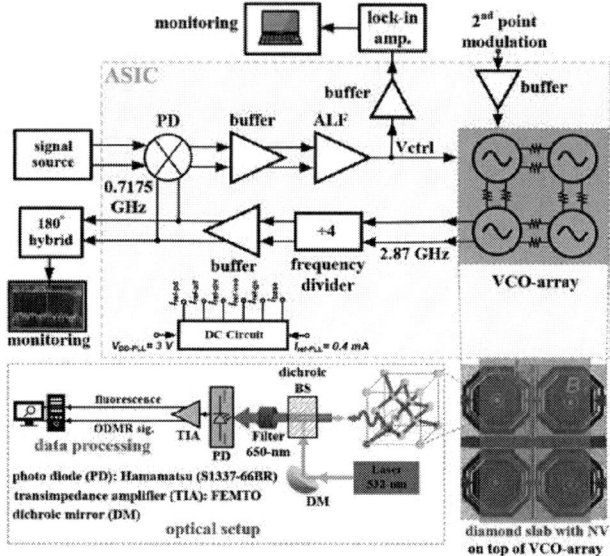

Fig. 1. Top: Architecture of the presented B_1 generator chip. Bottom: Illustration of the use of the B_1 generator chip in an ODMR setup.

classical magnetic field sensors and other quantum magnetometers, such as superconducting quantum interference devices (SQUIDs) and atomic vapor magnetometers [3], [4], where the latter are often constrained by size, cooling or heating requirements, and/or operational complexity. However, developing energy-efficient, portable NV-based magnetometers necessitates innovation in all aspects of NV center miniaturization, including photonic and microelectronic integration [5]. Here, one important aspect is the generation of the microwave B_1 field that is required to control the electron spin of the NV center.

In conventional NV magnetometers, the B_1 field is typically generated using a microwave resonator or a simple wire, which is driven by off-the-shelf power amplifiers supplied by a microwave generator. Therefore, the conventional approach limits the magnetometer's miniaturization and energy efficiency capabilities. To mitigate this, in this paper, we propose a com-

979-8-3315-0417-5/25 $31.00 © 2025 IEEE

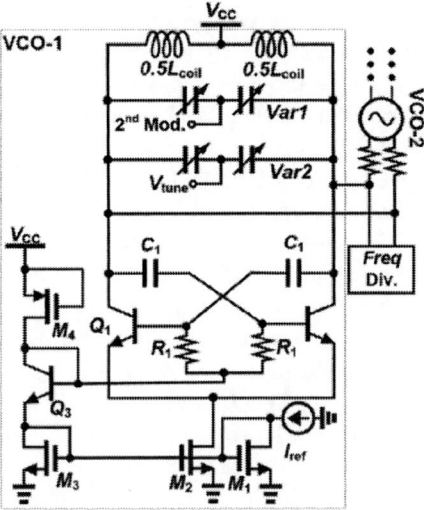

Fig. 2. Schematic of the individual LC tank VCOs.

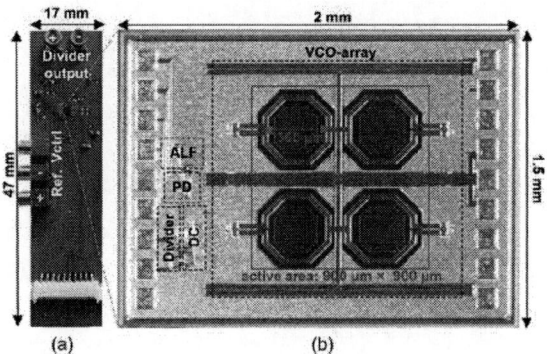

Fig. 3. Photograph of the PCB-based probe head used for electrical characterization with an enlarged view of the presented, chip-integrated B_1 source.

pact, energy-efficient solution by employing an array of injection-locked voltage-controlled oscillators (VCOs) as a source for the microwave B_1 field in future miniaturized and scalable NV magnetometers. An LC tank VCO directly converts DC power to a variable-frequency microwave current running through its tank inductor through the VCO's nonlinear operation principle. The current through the tank inductor then produces the B_1 field. This eliminates the need for a power-hungry power amplifier. Importantly, frequency modulation, which is required for lock-in detection, can be directly introduced in the VCO-based approach by modulating the VCO tuning voltage accordingly. Overall, the proposed VCO-based approach reduces both energy consumption and system complexity, presenting a promising route toward scalable, low-power NV magnetometers.

CHIP ARCHITECTURE

The architecture of the presented VCO-based B_1 source for optically-detected magnetic resonance (ODMR) is shown as part of Fig. 1. The chip uses an array of four injection-locked VCOs, depicted in Fig. 2, both to increase the active volume of the B_1 source and to lower the resulting frequency noise by a factor of $N_{ch} = 4$ (in power) compared to a single VCO [6]. The schematic of an individual VCO is shown in Fig. 2. Injection locking is achieved using resistive coupling elements between the individual VCO outputs in a circular fashion, see Figs. 1 and 2. The active area of the presented B_1 source is approximately

$900 \times 900 \ \mu m^2$. The VCO array is embedded into a PLL to derive the frequency of the B_1 field precisely from an external reference around 700 MHz. The PLL allows for frequency modulating the B_1 field for lock-in detection in continuous-wave (cw) ODMR experiments by modulating the PLL reference at rates within the PLL bandwidth. Additionally, the PLL features two-point modulation outside the PLL bandwidth for more advanced B_1 profiles, such as fast frequency chirps and pulsed excitation waveforms. Fig. 1 also shows the remaining external components required to perform an ODMR experiment with the presented B_1 source. The setup is similar to the one presented in [7].

MEASUREMENTS

Fig. 3 shows a micrograph of the VCO-array chip fabricated in a 130-nm SiGe BiCMOS technology, with a die area of $1.5 \times 2 \ mm^2$. The chip consumes an electrical power of 290 mW at a carrier frequency of 3.1 GHz. For the following electrical characterization, the chip was directly wirebonded onto a PCB-based probe head made from FR4. The electrical characterization focused on the key performance metrics for the target ODMR application, including the frequency tuning range of the B_1 field, as well as its phase and frequency noise characteristics. The latter two are important to ensure precise control of the NV center's electron spin. To measure the frequency tuning range, we monitored the divided-down VCO output with a signal source analyzer (FSWP8, Rohde&Schwarz). The measured tuning range extends from 2.86 GHz to 3.38 GHz, corresponding to a large relative tuning range of 16.7% around a center frequency of 3.12 GHz, sufficient for wideband NV-based magnetometry. The measured tuning range is shown in

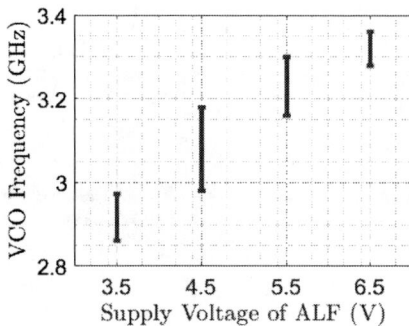

Fig. 4. Measured PLL lock-in range for different ALF supply voltages.

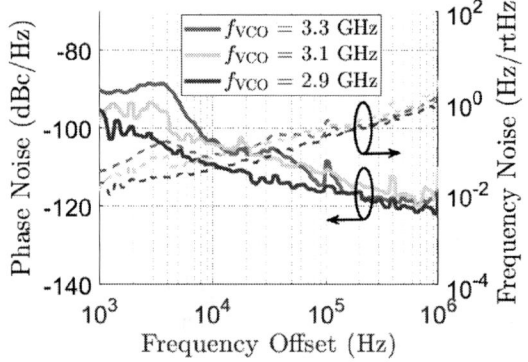

Fig. 5. Measured phase and frequency noise vs. offset frequency from different carrier frequencies.

Fig. 4. Here, it should be noted that the supply voltage of the analog loop filter (ALF) inside the PLL is adjusted to maximize the tuning range by covering the full range of the varactor of the VCO array with sufficient loop gain, see Fig. 1. The phase and frequency noise of the presented B_1 source were measured using a signal source analyzer (FSWP8, Rohde&Schwartz). The corresponding results are shown in Fig. 5. The measured closed-loop phase noise levels at offsets of 100 kHz and 1 MHz from a 3.1 GHz carrier are -111 dBc/Hz and -119 dBc/Hz, respectively. The measured frequency noise levels at the same offset frequencies from a 3.1 GHz carrier are $0.3\,\mathrm{Hz}/\sqrt{\mathrm{Hz}}$ and $2\,\mathrm{Hz}/\sqrt{\mathrm{Hz}}$, respectively.

CONCLUSION AND OUTLOOK

In this paper, we have presented a compact and energy-efficient approach for generating the B_1 field in NV center-based magnetometers using a chip-integrated, injection-locked array of VCOs. Unlike conventional techniques that require multiple conversion steps and external power-hungry components, the VCO-based solution directly converts DC energy to the necessary B_1 field through the inherent nonlinear operation of electronic oscillators. This direct DC-to-B_1 conversion significantly enhances energy efficiency, making this approach well-suited for portable and low-power applications. Furthermore, the LC tank circuit embedded within each VCO ensures a near-constant B_1 field across its tuning range, as the LC tank inside the VCO is inherently tuned to the current oscillation frequency. Electrical measurements of the most relevant performance metrics for the target application demonstrate the validity of the proposed approach. The B_1 field generation using chip-integrated, injection-locked VCO arrays holds considerable potential to advance the design scalability and energy efficiency of NV center magnetometers, particularly in applications requiring minimized form factors and optimized energy usage.

ACKNOWLEDGMENT

The authors gratefully acknowledge funding by the BMBF and the DFG under grant numbers 03ZU1110CB, 03ZU1110DC, 03ZU1110FE, 03ZU1110GA, AN 984/24-1, INST 121384/270-1, AN 984/27-1, and GRK2642.

REFERENCES

[1] A. Kuwahata *et al.*, "Magnetometer with nitrogen-vacancy center in a bulk diamond for detecting magnetic nanoparticles in biomedical applications," *Scientific Reports*, vol. 10, no. 1, p. 2483, 2020.

[2] J. Zhang *et al.*, "Blueprint for nv center ensemble based magnetometer: precise diamond sensor material characterization," 2024.

[3] M. Schmelz and R. Stolz, *Superconducting Quantum Interference Device (SQUID) Magnetometers*. Cham: Springer International Publishing, 2017, pp. 279–311.

[4] A. Fabricant, I. Novikova, and G. Bison, "How to build a magnetometer with thermal atomic vapor: a tutorial," *New Journal of Physics*, vol. 25, no. 2, p. 025001, Feb. 2023.

[5] F. M. Stürner *et al.*, "Integrated and portable magnetometer based on nitrogen-vacancy ensembles in diamond," *Advanced Quantum Technologies*, vol. 4, no. 4, p. 2000111, 2021.

[6] A. Chu *et al.*, "An 8-channel 13ghz esr-on-a-chip injection-locked vco-array achieving 200m-concentration sensitivity," in *2018 IEEE International Solid-State Circuits Conference - (ISSCC)*, 2018, pp. 354–356.

[7] H. Lotfi *et al.*, "A four-channel bicmos transmitter for a quantum magnetometer based on nitrogen-vacancy centers in diamond," *IEEE Journal of Solid-State Circuits*, vol. 59, no. 5, pp. 1421–1432, 2024.

Theoretical Design of Silicon-Based Nanostructures for Spin Qubits

Yang Liu[1], Shan Guan[1], and Jun-Wei Luo[1,2,*]

[1]State Key Laboratory of Superlattices and Microstructures, Institute of Semiconductors, Chinese Academy of Sciences, Beijing 100083, China
[2]Center of Materials Science and Optoelectronics Engineering, University of Chinese Academy of Sciences, Beijing 100049, China
[*]email: jwluo@semi.ac.cn

ABSTRACT

Silicon spin quantum bits (qubits) are promising for large-scale integration in fault-tolerant quantum computers. However, key spin qubit properties are closely linked to the atom-level characteristics of Si-based nanostructures. This paper discusses our progress in using the atomistic empirical pseudopotential method (EPM) to study and tailor Si-based nanostructures with excellent properties suitable for spin qubits, particularly involving valleys of electron qubits and spin-orbit coupling (SOC) of hole qubits.

Keywords: Silicon spin qubit, Valley and SOC

INTRODUCTION

Quantum computing represents a transformative leap in computing technology, capable of solving problems beyond the limits of classical computing. Silicon spin qubits, fabricated using standard complementary metal-oxidesemiconductor (CMOS) techniques, offer a promising approach for realizing quantum computing, with the potential for scalable integration similar to conventional chips [1]. Silicon spin qubits depend on the confinement of electrons or holes within quantum dots (QDs) created on Si-based substrates, such as Si metal-oxide-semiconductor (MOS) structures, Si/SiGe quantum wells (QWs), and Ge/SiGe QWs, among others [2]. These material platforms hosting electron or hole qubits include a variety of one- and two-dimensional nanostructures, whose atomic-level properties directly affect the performance of qubits. Therefore, atomic-level theoretical studies of these Si-based low-dimensional nanostructures provide valuable insights for interpreting experiments and exploring the physical principles of these quantum devices, while also aiding in the design and optimization of these nanostructures to achieve high-quality qubits [2].

In this work, we present the detailed implementation of the atomistic EPM, a powerful tool for investigating the physical properties of Si-based nanostructures related to spin qubits. Our method accurately captures atomic-level details in QWs, such as interfaces and disorder, and their influence on the valley splitting and SOC, which are critical factors for electron and hole qubits, respectively. Our results contribute to the advancement of Si spin qubits toward the development of practical, scalable quantum computing systems.

METHOD

First, we construct a supercell containing the nanostructure, and allow the atoms within the supercell to relax in order to minimize strain energy using the valence force field method. The electronic band structures of Si-based nanostructures are then calculated by directly diagonalizing the Hamiltonian, $-\frac{1}{2}\nabla^2 + V(\boldsymbol{r})$, where $V(\boldsymbol{r})$ incorporates both a local component in reciprocal space and a nonlocal spin-orbit interaction term. Diagonalization is carried out in a plane-wave basis using the folded spectrum method, with periodic boundary conditions applied to the supercell. To minimize "LDA errors" in bulk crystal properties, the pseudopotentials are fitted to accurately reproduce band gaps, effective masses, offsets, and spin-orbit splittings.

VALLEY SPLITTING

The electronic ground state energy levels of Si-based QDs are not solely determined by spin but are also influenced by valley degrees of freedom, causing leakage and decoherence [2]. A sharp interface between the well and barrier layers can lift valley degeneracy, resulting in valley splitting. Typically, valley splitting must be large enough to prevent quantum information leakage. However, the valley splitting E_v is often quite small, reaching only up to 0.2 meV in Si/SiGe QWs [1], and can be further diminished by interfacial disorder, such as atomic steps and alloy disorder. Our EPM method accurately captures the effect of atomic-scale disorder on valley splitting. Figure 1 shows the valley splitting magnitude (E_{VS}) in two types of interface step

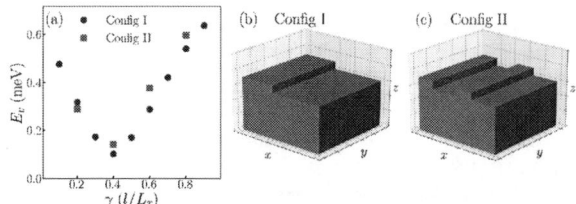

Fig. 1. (a) E_v for two step configurations (I and II) with varying interface step distributions γ, calculated through EPM. (b-c) Schematic diagrams of the structures of the step configurations I and II.

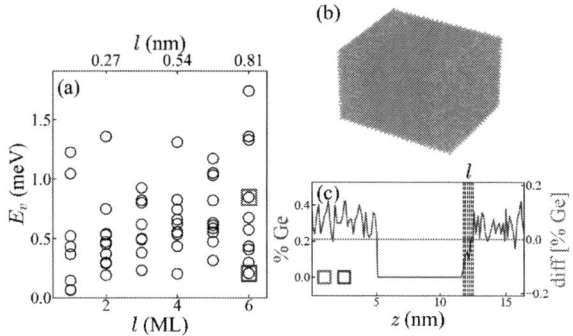

Fig. 2. Atomistic calculations of E_v in Si/SiGe QWs against the variety of interface broadening length (l). (b) The point cloud image for the Si/SiGe QWs punctuated in (a). (c) The in-plane averaged Ge concentration for the two specific Si/SiGe QWs.

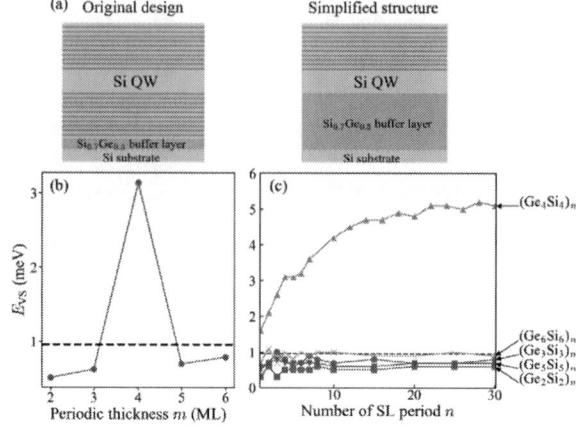

Fig. 3. (a) Schematic sketch of Si QW structure in the original design (left) and simplified Si QW structure (right). (b) Computationally predicted valley splitting of $Si_{40}QW/(Ge_mSi_m)_4$ SL barrier hybrid systems as varying period thickness m from 2 to 6. (c) Computationally predicted valley splitting of $Si_{40}QW/(Ge_mSi_m)_n$ SL barrier hybrid systems. Adapted from Ref. [4].

configurations, Config I and Config II, under different distributions of interface steps. Here, a one-monolayer-tall ($a/4$) interface atomic step edge is positioned along the y-direction at $x = \gamma L_x$ (where a is the Si lattice constant and L_x is the supercell size in the x-direction). It can be observed that, compared to an ideal QW, the presence of steps significantly reduces the valley splitting, reaching its lowest value when the areas of the two regions separated by the step are approximately equal. Figure 2 illustrates the significant fluctuations in E_v caused by alloy disorder in the SiGe barrier, where intrinsic Ge concentration fluctuations in the vertical direction and atomic configuration variations in the lateral direction are present. Surprisingly, we demonstrated that a substantial spread of E_v can arise solely from the in-plane random distribution of Si and Ge atoms within the SiGe barrier, representing the lower bound of the wide spectrum of E_v observed in various Si/SiGe QWs.

Maximizing valley splitting is critical for silicon electron spin qubits, which can be achieved through material design. By combining atomistic EPM with genetic algorithm optimization, we identified an optimal Ge/Si layering sequence, $Ge_4Si_6Ge_2Si_6Ge_4Si_4Ge_4Si_4$, which enhances valley splitting by an order of magnitude compared to alloy barriers (Fig. 3a) [3]. Notably, since all optimal configurations begin with a Ge_4 sublayer, the complex sequence can be simplified to a $(Ge_4Si_4)_n$ superlattice barrier, with valley splitting reduced from 8.7 to 5.2 meV [4]. As shown in Fig. 3c, even with the minimum superlattice period ($n = 1$), a substantial valley splitting of 1.6 meV is achieved.

SPIN-ORBIT COUPLING

Spin-orbit coupling links spin and orbital degrees of freedom, enabling all-electric, rapid, and scalable spin control via electric dipole spin resonance (EDSR), a crucial factor behind recent advancements in Ge hole-spin qubits. However, this raises a puzzle, as centrosymmetric Ge lacks Dresselhaus SOC—a critical element in the original hole-based EDSR proposal. Using the EPM method, we uncover finite k-linear spin splitting in Ge QWs (Fig. 4). This suggests the existence of long-neglected k-linear SOC at the valence band maximum of the Ge two-dimensional hole gas, which originates from interface-induced heavy-hole-light-hole (HH-LH) mixing and direct dipole coupling to an external electric field. This recently uncovered finite k-linear SOC offers fast hole-spin control via EDSR with Rabi frequencies in excellent agreement with experimental results over a wide range of driving fields.

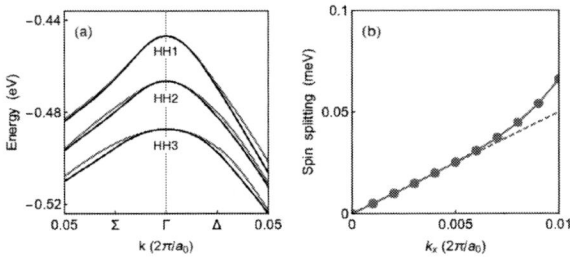

Fig. 4. (a) Band structure and (b) spin splitting of the valence band maximum in the [001]-oriented $Ge_{120}Si_{60}$ QW.

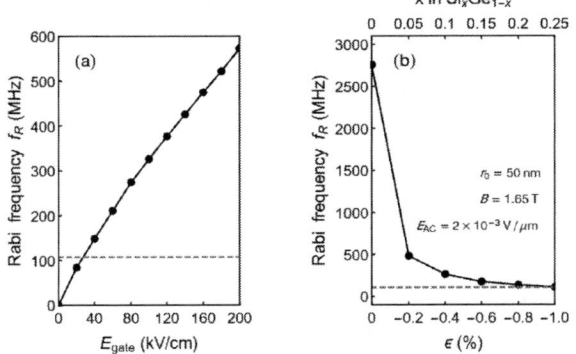

Fig. 5. (a) The predicted Rabi frequency as a function of gate electric field. (b) The corresponding Rabi frequency as a function of biaxial strain in the Ge layer.

EPM also enables us to propose optimization strategies for further enhancing the hole-spin Rabi frequency. The strength of the SOC is electrically tunable through biased gates, and as shown in Fig. 5a, the Rabi frequency increases linearly with the gate electric field. The Rabi frequency boosts to 600 MHz at a 200 kV/cm gate electric field. Moreover, the linear SOC is related to the HH-LH mixing at the interface. Reducing strain enhances this coupling by decreasing the HH-LH bandgap, while also increasing the effective mass, thereby leading to an increase in the Rabi frequency, as shown in Fig. 5b. In addition, narrow odd-ML QWs, lacking spatial inversion symmetry, exhibit a larger Rabi frequency (Fig. 6a). As shown in Fig. 6b, the strong intrinsic HH-LH mixing in [110] oriented QWls allows for a 16-fold increase in SOC, leading to control speeds up to the GHz range. These strategies further accelerate spin manipulation in Ge planar QDs, offering more efficient control options.

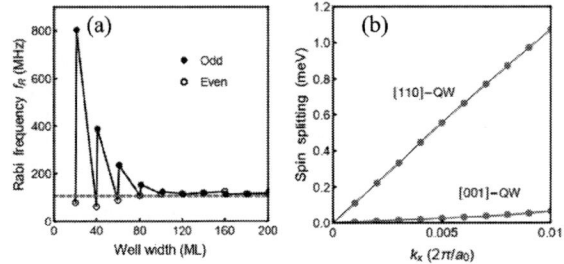

Fig. 6. (a) The Rabi frequency as a function of the QW width. (b) The calculated spin splitting of the [110]- and [001]-oriented QW.

CONCLUSION

In summary, atomistic EPM, as an advanced ab initio material modeling approach, offers a quantitative description of valley splitting and SOC, which not only enhances our understanding of the microscopic physics in Si-based low-dimensional nanostructures but also supports the design of new material platforms to host high-quality qubits. This method provides an ideal foundation for establishing a "Quantum-CAD" framework for the optimization and realization of spin qubits, akin to the well-established TCAD strategy in microelectronics.

ACKNOWLEDGMENT

The work was supported by the National Science Fund for Distinguished Young Scholars under Grant No. 11925407, the Basic Science Center Program of the National Natural Science Foundation of China (NSFC) under Grant No. 61888102, and the Key Research Program of Frontier Sciences, CAS under Grant No. ZDBS-LY-JSC019 and CAS Project for Young Scientists in Basic Research under Grant No. YSBR-026.

REFERENCES

[1] G. Burkard, T. D. Ladd, A. Pan, J. M. Nichol, and J. R. Petta, "Semiconductor spin qubits," *Reviews of Modern Physics*, vol. 95, no. 2, p. 025003, 2023.

[2] Y. Liu, S. Guan, J.-W. Luo, and S.-S. Li, "Progress of gate-defined semiconductor spin qubit: Host materials and device geometries," *Advanced Functional Materials*, vol. 34, no. 19, p. 2304725, 2024.

[3] L. Zhang, J.-W. Luo, A. Saraiva, B. Koiller, and A. Zunger, "Genetic design of enhanced valley splitting towards a spin qubit in silicon," *Nature communications*, vol. 4, no. 1, p. 2396, 2013.

[4] G. Wang, Z.-G. Song, J.-W. Luo, and S.-S. Li, "Origin of giant valley splitting in silicon quantum wells induced by superlattice barriers," *Physical Review B*, vol. 105, no. 16, p. 165308, 2022.

Integration of 2D Ultrafast Flash Memory: From Device to Chip

Zhenyuan Cao[1], Chunsen Liu[1,*]

[1]State Key Laboratory of Integrated Chip and System, School of Microelectronics, Frontier Institute of Chip and System, Fudan University, Shanghai 200433, China

Abstract

The introduction of 2D materials has brought a major breakthrough in the performance of flash memory. 2D flash memory can simultaneously achieve ultrafast programming speed and non-volatile memory, bridging the gap between DRAM and silicon-based flash memory. This paper discusses the progress of 2D flash memory in terms of performance breakthrough, integration and channel length scaling. We also give an outlook on future chip-level integration and fabricate a NAND flash circuit for preliminary verification.

Keywords: Flash Memory, 2D Materials, Integration

Introduction

The advent of the big data era has led to an increased demand for high-performance and high-density memory. With its simple mechanism, 10-year retention and 3D stacking capability [1], flash memory has become the mainstream non-volatile memory. However, due to the high energy barrier of tunneling, the programming speed of traditional silicon-based flash memory is still in the range of $10 \sim 100$ μs [2], which limits its application in the field of high energy efficiency. 2D materials have unique electrical properties and rich energy band structures [3], giving them the potential for high performance. In addition, with the atomic thickness, 2D materials have the advantage of overcoming the short channel effect [4], which is beneficial for realizing high-density memory. With these benefits, 2D flash memories with ultrafast programming speed are proposed [5], [6], overcoming one of the main drawbacks of flash memory. In this article, we review the progress of 2D flash memory in terms of performance breakthrough, large-scale integration and channel length miniaturization. We also fabricate a NAND circuit based on 2D flash memory to verify the ultrafast programming characteristic at the circuit level. Finally, we give an outlook on the chip-level integration of 2D ultrafast flash memory.

Performance Breakthrough

Using atomically thin 2D materials as channels (Fig. 1a), two triangular barriers are formed between the contact metal and the floating gate or charge trapping layer (Fig. 1b), increasing the efficiency of charge tunneling. Taking advantage of this, 2D flash memory achieves a programming speed of 20 ns while maintaining a 10-year retention characteristic [5], [6]. This bridges the gap between DRAM and silicon-based flash memory (Fig. 2). With the ultrafast programming speed, 2D flash memory suffers less damage during programming, which is beneficial for higher endurance. By using $Al_2O_3/HfO_2/h$-BN as the charge trap memory stack, flash memory with 8×10^6 endurance has been proposed, which is 10^2 times higher than silicon-based flash memory [7]. Similar performance has also been achieved through phase-engineered edge contacts [8]. These works show that 2D flash memory can simultaneously achieve ultrafast speed, non-volatile memory and high robustness, demonstrating its potential in next-generation high-performance memory.

Integration and Scaling

The performance of flash memory has made breakthroughs through the introduction of 2D materials. Another important issue is how to achieve large-scale preparation of 2D flash memory while maintaining high performance and high yield. The quality of the materials in the different layers of flash memory has a great impact on the performance of flash memory [9]. The CVD process for large-scale MoS_2 has become increasingly mature [10], [11] making it a suitable choice for the channel material. H-BN has been used as the tunnelling layer material in previous ultrafast flash works. However, the large-scale growth of h-BN remains challenging. An alternative high-quality, large-area tunneling layer material should be introduced. What's more, since the growth temperature of CVD 2D materials is relatively high [10], [11] an ultra-clean, damage-free, large-scale 2D material transfer technology is needed to meet the thermal budget requirement. Some works on the integration of 2D flash memory have been reported [12], but the programming speed in these works is still at the level of milliseconds. Recently, a high-yield, low-thermal-budget, large-scale fabrication process for ultrafast flash memory has been reported (Fig. 3a) [13]. CVD MoS_2 and 3 nm Pt were used as the channel and floating gate materials, respectively. HfO_2/Al_2O_3 was used as the tunneling and blocking layer materials. To improve the cleanliness and flatness of the channel/tunnelling layer stack, soaking in NMP and annealing in N_2 were employed. The integration scale reaches 1 Kb, and the device yield reaches 98.4% (Fig. 3b). This work demonstrates the integrability of 2D ultrafast flash memory.

Miniaturization of the device feature size is necessary to achieve high-density storage. Affected by the short channel effect, the performance of the device will degrade when the channel size is reduced to a certain extent. 2D materials have the advantage of overcoming the short channel effect [4], giving them potential for channel length scaling down. Using a shadow evaporation method, a flash memory with 8 nm channel has been proposed (Fig. 4) [13], exceeding the limit of silicon-based flash memory (~15 nm). Programming speed of 30 ns, >10^5 endurance and over 4-bit memory states were achieved, demonstrating that the excellent performance of 2D flash memory can be extended to the sub-10 nm scale.

NAND Circuit Integration

Towards the practical application and high-density integration of 2D ultrafast flash memory, memory circuits with some kinds of interconnection structure should be formed. NAND is one of the mainstream structures of flash memory and occupies most of the memory market share. Compared to NOR structure, memory circuits with NAND structure can achieve higher parallelism and smaller footprint. Here, we fabricated a NAND circuit based on 2D flash memory to verify the ultrafast programming speed at the circuit level (Fig. 5). CVD MoS$_2$ and HfO$_2$/Pt/HfO$_2$ are used as the channel and memory stack, respectively. The cells sharing the same WL form a page, and the cells sharing the same BL and SL form a NAND string (Fig. 5a). In this work, a NAND string consists of two flash memory cells and two selective transistors at the edge. The NAND string is turned on only when all memory cells and selective transistors are turned on. Two selective transistors are used to select a particular NAND string during operation. Their conduction state is controlled by SSL and GSL, respectively.

When the programming operation is performed, the WL of the selected cell is applied with a programming voltage (V_{pro}), and the BL and SL of the selected string are grounded (Fig. 5b). A pass voltage ($V_{pass-pro}$) is applied to the other unselected WL, GSL and selected SSL to turn on the selected NAND string. Therefore, there is sufficient voltage difference between the gate and channel of the selected cell to achieve FN tunneling. When the reading operation is performed, a reading voltage (V_{read}) is applied to the BL of the selected string and a smaller pass voltage ($V_{pass-read}$) is applied to the unselected WL, GSL and selected SSL. Because of the application of the pass voltage, crosstalk may occur in other unselected cells (Cell$_{crosstalk}$) that share the same BL and SL with the selected cell (Cell$_{program}$). It is necessary to select an appropriate pass voltage to make a trade-off between conductance and crosstalk. To verify the ultrafast programming characteristic of the NAND flash circuit, we applied a programming pulse of 23 V, 100 ns to the selected WL, a pass pulse of 10 V, 100 ns to the unselected WL, GSL and selected SSL. The memory state of the selected cell is programmed from state "1" to state "0", and the unselected cell is still in state "1" while crosstalk occurs (Fig. 5c). The results prove that the ultrafast programming speed can be maintained at the circuit level, while more appropriate programming algorithms should be explored to reduce crosstalk.

Outlook for Chip-Level Integration

2D flash memory has made great progress in recent years, achieving high programming speed, high endurance and long retention simultaneously. The high performance of 2D flash memory has been further extended to the 1 Kb scale and sub-10 nm channel length, demonstrating its potential for energy-efficient and high-density storage. Despite these advances, chip-level integration of 2D ultrafast memory remains challenging. Much effort is needed to improve device performance, optimize large-scale integration process and co-design at system level.

In terms of performance improvement, the programming voltage of 2D flash memory is still high, which will increase the complexity of peripheral logic circuits and the instability of operation. Appropriate design of the memory stack and GCR can be useful to reduce the operating voltage of 2D flash memory.

In a flash memory chip, all cells operate according to the same scheme. And one failed cell can cause the entire row, column or array to fail. Therefore, chip-level integration places higher demands on the uniformity and yield of memory cells. In the current 2D flash memory integration process, 2D materials are mainly transferred manually through a transfer platform, which can easily cause the damage and wrinkle of material and the instability of process. The more automated, ultra-clean, damage-free and wafer-scale transfer technology should be explored and adopted.

In terms of system-level co-design, new operation scheme suitable for 2D ultrafast flash memory should be explored to realize the trade-off between performance, parallelism and crosstalk. In addition, as the scale of integration increases, the parasitic effects in the array will become more apparent, causing performance degradation and non-uniformity. It is important to construct the model of the entire 2D flash memory circuit to study the effects of different operating scheme and capacitance distribution simultaneously.

Acknowledgments

References

[1] W. Jung et al., "13.3 A 280-Layer 1Tb 4b/cell 3D-NAND Flash Memory with a 28.5Gb/mm2 Areal Density and a 3.2GB/s High-Speed IO Rate," in 2024 IEEE International Solid-State Circuits Conference (ISSCC), pp. 236-237 (2024).

[2] T. Kouchi et al., "13.5 A 128Gb 1b/Cell 96-Word-Line-Layer 3D Flash Memory to Improve Random Read Latency with tPROG=75μs and tR=4μs," in 2020 IEEE International Solid-State Circuits Conference - (ISSCC), pp. 226-228 (2020).

[3] S. Pinilla, J. Coelho, K. Li, J. Liu, and V. Nicolosi, "Two-dimensional material inks," Nature Reviews Materials, vol. 7, no. 9, pp. 717-735 (2022).

[4] J. Jiang, L. Xu, C. Qiu, and L.-M. Peng, "Ballistic two-dimensional InSe transistors," Nature, vol. 616, no. 7957, pp. 470-475 (2023).

[5] L. Liu et al., "Ultrafast non-volatile flash memory based on van der Waals heterostructures," Nature Nanotechnology, vol. 16, no. 8, pp. 874-881 (2021).

[6] L. Wu et al., "Atomically sharp interface enabled ultrahigh-speed non-volatile memory devices," Nature Nanotechnology, vol. 16, no. 8, pp. 882-887 (2021).

[7] X. Huang, C. Liu, Z. Tang, S. Zeng, S. Wang, and P. Zhou, "An ultrafast bipolar flash memory for self-activated in-memory computing," Nature Nanotechnology, vol. 18, no. 5, pp. 486-492 (2023).

[8] J. Yu et al., "Simultaneously ultrafast and robust two-dimensional flash memory devices based on phase-engineered edge contacts," Nature Communications, vol. 14, no. 1, p. 5662 (2023).

[9] G. M. Marega et al., "How to Achieve Large-Area Ultra-Fast Operation of MoS2 Monolayer Flash Memories?," IEEE Nanotechnology Magazine, vol. 17, no. 5, pp. 39-43 (2023).

[10] J. Kwon et al., "200-mm-wafer-scale integration of polycrystalline molybdenum disulfide transistors," Nature Electronics, vol. 7, no. 5, pp. 356-364 (2024).

[11] Y. Xia et al., "12-inch growth of uniform MoS2 monolayer for integrated circuit manufacture," Nature Materials, vol. 22, no. 11, pp. 1324-1331 (2023).

[12] G. Migliato Marega et al., "A large-scale integrated vector–matrix multiplication processor based on monolayer molybdenum disulfide memories," Nature Electronics, vol. 6, no. 12, pp. 991-998 (2023).

[13] Y. Jiang et al., "A scalable integration process for ultrafast two-dimensional flash memory," Nature Electronics, vol. 7, no. 10, pp. 868-875 (2024).

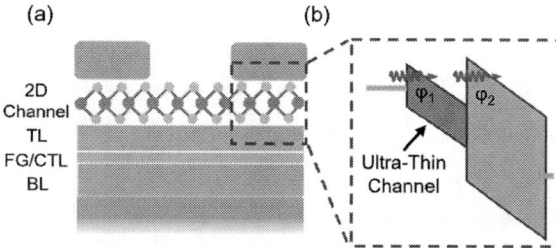

Fig. 1: (a) The schematic diagram of 2D flash memory. (b) Energy band structure of 2D flash memory. Two triangular barriers are formed due to the ultra-thin 2D channel. TL: tunneling layer. FG: floating gate. CTL: charge trapping layer. BL: blocking layer.

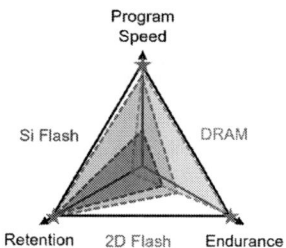

Fig. 2: Performance comparison of DRAM, silicon-based Flash, and 2D Flash.

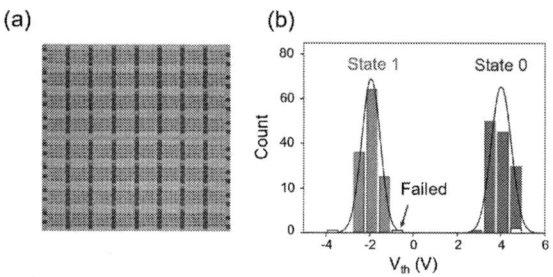

Fig. 3: (a) Optical image and (b) statistical transfer curves of 1 Kb 2D flash memory array. 98.4% yield has been achieved. Reproduce with permission from [13].

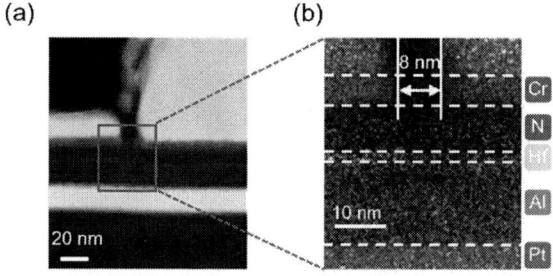

Fig. 4: (a) TEM and (b) EDS mapping of 2D flash memory with sub-10 nm channel length. Reproduce with permission from [13].

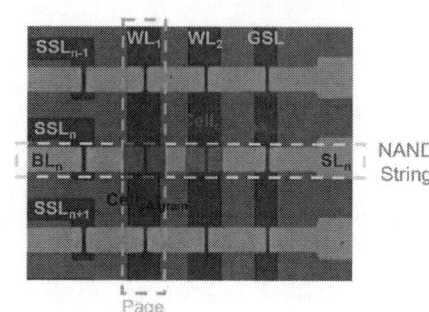

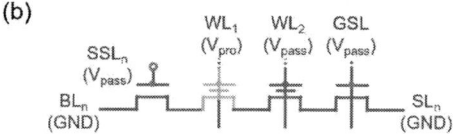

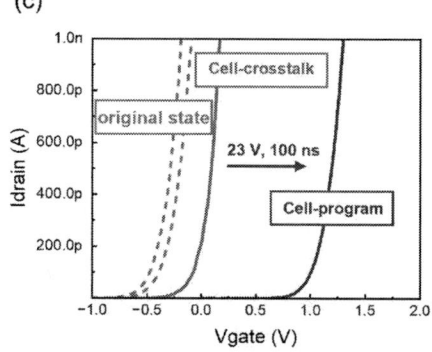

Fig. 5: (a) Optical image of the 2D NAND Flash circuit. (b) Diagram of the programming scheme. (c) Transfer characteristic curve of $Cell_{program}$ and $Cell_{crosstalk}$ before and after programming.

979-8-3315-0417-5/25 $31.00 © 2025 IEEE

Robust OS-FeFETs with Crystallized Anatase-TiO$_2$ Channel-Hafnia Ferroelectric Layer Stack for Integration of 3D Memory Applications

J.F. Kang*[1,2], X.J. Song[1,2], C.X. Yu[1,2], D.J. Sun[1,2], J.J. Zhang[1,2], S.Z. Li[1,2], and X.Y. Liu[1,2]

[1]School of Integrated Circuits, Peking University, 100871, Beijing, China; *kangjf@pku.edu.cn
[2]Beijing Advanced Innovation Center for Integrated Circuits, Beijing 100871, China

Abstract

In this study, the planar and vertical OS-FeFETs were fabricated based on a stack of TiO$_2$-channel/HfLaO ferroelectric layers without thermal budget mismatch. The fabricated FeFETs demonstrate exceptional performance and robust stability, including: 1) Memory properties with an on/off ratio exceeding 10^6, a memory window of over 1V, and endurance surpassing 10^8 cycles; 2) High-temperature operational stability up to 475K and resistance to ambient exposure for over a year. The excellent memory performances and thermal-stability achieved in OS-FeFETs could be owed to material- and process-compatibility between the crystalized TiO$_2$ channel/HfLaO ferroelectric layers that results in a thermally-stable stack with negligible interfacial interdiffusion.

(Keywords: Oxide Semiconductor (OS), Ferroelectric Field Effect Transistor (FeFET), Anatase TiO$_2$, HfLaO, Thermal stability, High-temperature operation capacity)

Introduction

Rapid advances in artificial intelligence (AI) have enabled computing systems to process massive data sets more quickly and efficiently but suffered from a 'memory wall' bottleneck between processing and memory units [1]. To overcome the bottleneck, the possible solutions include: 1) to achieve much denser and more fine-grained connectivity between memory and processing units by using three-dimensional (3D) integration approaches; 2) to explore the novel architectures and devices such as computing in-memory with emerged memory devices [1-3]. Recent studies show that the oxide semiconductor channel–based FeFETs (OS-FeFETs) with hafnia-based ferroelectric gates have emerged as one of the promising candidates for the solutions, which have exhibited the fantastic features such as high speed and low-power operation capability, three-dimensional (3D) integration potential, and CMOS compatibility [4-6]. However, the instability of the typical amorphous oxide semiconductors such as IGZO along with the high density of defects and the limits of low temperature budget for fabrication process and operation conditions, poses significant challenges [7,8]. Facing the instability issues of OS-FeFETs, we proposed and demonstrated a thermally-stable stack of crystalized TiO$_2$-channel/HfLaO ferroelectric layers for the fabrication and integration of OS-FeFETs. The excellent FeFET performances with high stability against the fabrication &

operation temperatures and the ambient exposure time were achieved in both the planar and vertical OS-FeFETs [9-12]. In this study, we will review the characteristics of the fabricated planar and vertical OS-FeFETs, especially focusing on the performance stability against the operation temperatures and ambient exposure for future 3D integration application scenarios of OS-FeFETs based on the proposed stack.

Fabrications of OS-FeFETs

This study fabricated planar and vertical OS-FeFETs with back gates, using a TiO$_2$ channel and HfLaO ferroelectric stack, as illustrated in Fig. 1. The stack of TiO$_2$/HfLaO were fabricated via continuous ALD-based deposition for the stacking FE–OS channel layers followed by a one-step rapid thermal annealing (RTA) process [10].

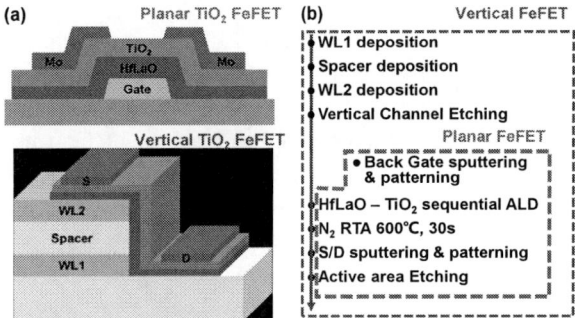

Fig.1 (a) Schematic structures of the planar and vertical OS-FeFETs; (b) Corresponding process flows of the planar and vertical OS-FeFETs.

Results and Discussion

In order to quantify the characteristics of the planar and vertical OS-FeFETs, the performances of the fabricated devices, including the properties of OS-FeFETs and their ambient stability & thermal robustness against high operating-temperatures, were measured and evaluated. Furthermore, the microstructure features of the TiO$_2$/HfLaO stack integrated in the planar and vertical OS-FeFETs were observed by HRTEM, trying to explore their impacts on the performances of OS-FeFETs.

A. Performances of FeFETs

At first, the typical device characteristics of the fabricated planar and vertical OS-FeFETs, including the ferroelectricity, I-V curves, and reliability behaviors were measured. Fig.2 indicates that the sufficient ferroelectric polarization

979-8-3315-0417-5/25 $31.00 © 2025 IEEE

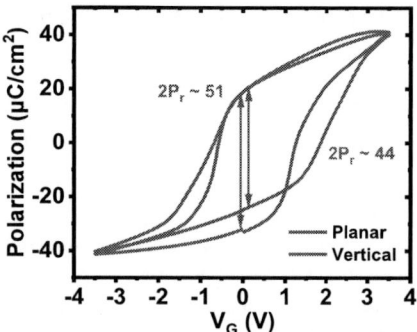

Fig.2 Measured PV curves of HfLaO ferroelectric layer in the planar and vertical OS-FeFETs

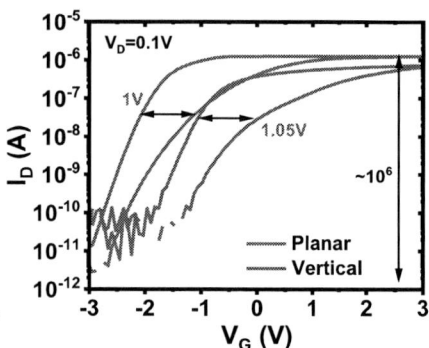

Fig.3 Measured I_d-V_g curves in the fabricated planar and vertical FeFETs.

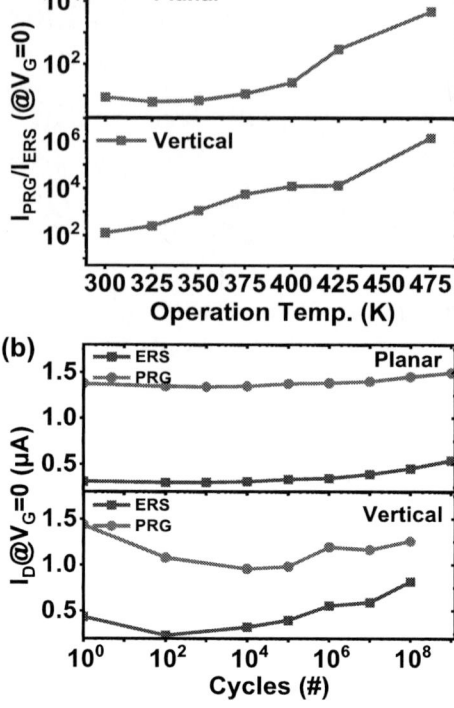

Fig.4 Measured high-temperature operation stability of planar and vertical OS-FeFETs, including (a) I_{PRG}/I_{ERS} ratio from 300K to 475K, and (b) endurance at 475K

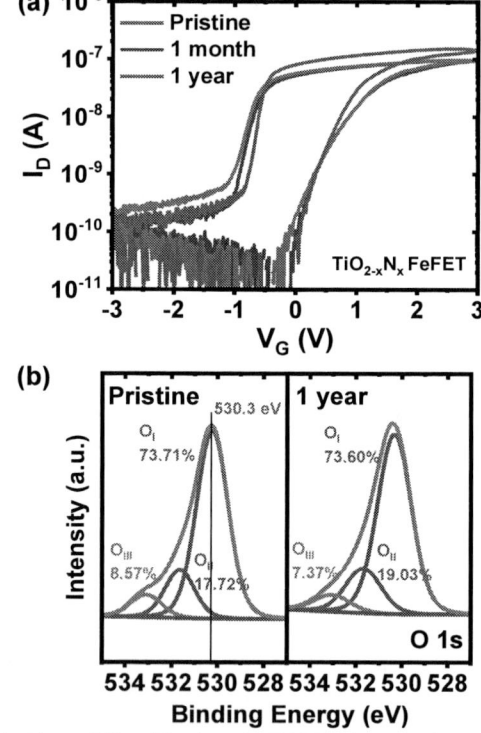

Fig.5 Ambient stability of the planar OS-FeFETs (a) Measured Id-Vg curves of the device across 1 year under air exposure conditions (b) XPS spectra (O 1s peak) of theTiO$_2$ channel layer (O$_I$ – lattice oxygen, O$_{II}$- oxygen vacancy, O$_{III}$- -OH)

properties with $2P_r>44$ $\mu C/cm^2$ can be realized in the planar and vertical ferroelectric layers. As shown in Fig.3, the excellent memory and switching performances such as sufficient memory window (> 1V) and large on/off ratio more than 10^6 with over 10^8 cycling endurance can be achieved in the fabricated OS-FeFETs with planar and vertical structured TiO2/HfLaO stack. Those results indicate that the demonstrated TiO$_2$/HfLaO stack can be integrated into both planar and vertical structured OS-FeFETs without the performance degradation. Such a feature is beneficial for the high density of integration with both 3D NAND and M3D

B. Thermal Stability of High Temperature Operations

In 3D systems, stacked multi-layers often result in poor heat dissipation, creating high-temperature operational environments. In this case, the high-temperature stability of OS-FeFETs is a crucial issue for 3D integrated memory and computing systems. The temperature-dependent characteristics of the fabricated planar and vertical OS-FeFETs from 300K to 475K were measured. As shown in Fig.4, the excellent device performances including I_{PRG}/I_{ERS} ratio and endurance over 10^8 are maintained under high temperatures up to 475K, indicating the robust high-temperature operation stability of the fabricated planar and vertical OS-FeFETs based on the TiO$_2$/HfLaO stack.

C. Ambient Stability

For the typical OS-FeFETs with a typical amorphous-oxide-semiconductor channel such as InGaZnO, the ambient stability is a serious concern issue [7, 8]. To evaluate ambient stability of the fabricated devices, they were subjected to aging tests under ambient air conditions. As shown in Fig.5, the FeFET exhibits relatively stable transfer characteristics after one year of exposure to room air, with the chemical stability of the channel confirmed by XPS analysis [10].

D. Microstructure Characteristics of TiO₂/HLO Stack

The observed memory switching and reliability characteristics in the planar and vertical structured OS-FeFETs are strongly correlated with the features of the TiO₂/HfLaO stack fabricated. In order to explore the origins of the excellent thermal stability of OS-FeFETs, the microstructure-features of the TiO₂/HfLaO stack fabricated @600°C process were investigated by HRTEM observations, as shown in Fig.6. From TEM image and FFT pattern of TiO₂/HfLaO stack, the crystallized anatase TiO₂ and monoclinic HfLaO layers were fabricated in the stack. From HADDF and HR-EDS images, stable interface between TiO₂ and HfLaO layers without obvious interdiffusion was formed. These results indicate that a thermally stable stack of crystallized anatase TiO₂ and monoclinic HfLaO layers is formed, even after undergoing a high-temperature process up to 600°C, due to the material and process compatibility between the anatase TiO₂ and monoclinic HfLaO layers. In this thermally stable stack, the interdiffusion of oxygen vacancies between the crystallized TiO₂ and HfLaO layers is

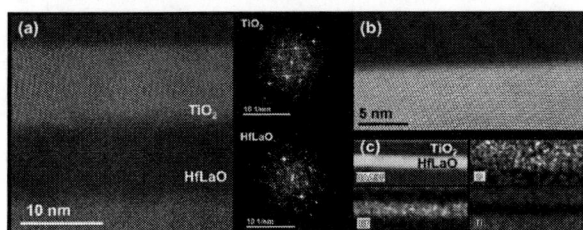

Fig.6 (a) HRTEM image and FFT pattern of TiO₂/HfLaO stack with crystallized anatase TiO₂ and HfLaO layers; (b) HADDF image of the TiO₂/HfLaO interface; (c) HR-EDS image of the TiO₂/HfLaO

suppressed, which enables the stable ferroelectricity of HfLaO and the performance of OS-FeFETs [13].

Conclusion

In this study, a thermally-stable stack of the crystallized TiO₂ channel and HfLaO ferroelectric layers is presented due to their process- and material-compatibility. This TiO₂/HfLaO stack enables the fabrication of planar and vertically structured OS-FeFETs with strong memory performance and robust stability under high-temperature operation and ambient exposure conditions. These demonstrated characteristics show that the thermally-stable TiO₂/HfLaO stack is promising for the monolithic 3D and 3D vertical NAND integration of OS-FeFETs for future memory/computing in-memory system applications.

Acknowledgments

This work is supported by the National Natural Science Foundation of China (Grant No. 92064001). The authors acknowledge the National Micro/Nano Fabrication Laboratory of Peking University for assistance in device fabrication.

References

[1] A. Sebastian, M.L. Gallo, R. Khaddam-Aljameh, and E. Eleftheriou, "Memory devices and applications for in-memory computing", Nature Nanotechnology, vol.15, p529–544, (2020).

[2] A. I. Khan, A. Keshavarzi, and S. Datta, "The future of ferroelectric field-effect transistor technology", Nature Electronics, vol 3, p. 588–597, (2020).

[3] I-J. Kim and J-S. Lee, "Ferroelectric Transistors for Memory and Neuromorphic Device Applications", Advanced Materials, vol.35, 2206864, (2023).

[4] G. Kim, H. Shin, T. Eom, M. Jung, T. Kim, S. Lee, M. Kim, Y. Jeong, J.-S. Kim, K.-J. Nam, B. J. Kuh, and S. Jeon, "Design Guidelines of Thermally Stable Hafnia Ferroelectrics for the Fabrication of 3D Memory Devices," International Electron Devices Meeting (IEDM), pp. 5.4.1-5.4.4, (2022).

[5] Z. Liang, K. Tang, J. Dong, Q. Li, Y. Zhou, R. Zhu, Y. Wu, D. Han, and R. Huang, "A Novel High-Endurance FeFET Memory Device Based on ZrO2 Anti-ferroelectric and IGZO Channel," International Electron Devices Meeting (IEDM), pp. 17.3.1-17.3.4, (2021).

[6] C.-K. Chen, Z. Fang, S. Hooda, M. Lal, U. Chand, Z. Xu, J. Pan, S.-H. Tsai, E. Zamburg, and A. V.-Y. Thean, "First Demonstration of Ultra-Low Dit Top-Gated Ferroelectric Oxide-Semiconductor Memtransistor With Record Performance by Channel Defect Self-Compensation Effect for BEOL-Compatible Non-volatile Logic Switch," International Electron Devices Meeting (IEDM), pp. 6.1.1-6.1.4, (2022).

[7] H.-W. Lee and W.-J. Cho, "Effects of Vacuum Rapid Thermal Annealing on the Electrical Characteristics of Amorphous Indium Gallium Zinc Oxide Thin Films," AIP Advances, vol. 8, no. 1, p. 015007, (2018).

[8] W. Kim, J. Kim, D. Ko, J.-H. Cha, G. Park, Y. Ahn, J.-Y. Lee, M. Sung, H. Choi, S. W. Ryu, S. Kim, M. Na, and S. Cha, "Demonstration of Crystalline IGZO Transistor with High Thermal Stability for Memory Applications" Symposium on VLSI Technology and Circuits (VLSI Technology and Circuits), pp. 1-2, (2023).

[9] X. Song, S. Li, D. Sun, X. Liu, J. Kang, "High Performance HfLaO-based TiO₂-Channel FE V-NAND with High Consistency and Low Operation Voltage", Silicon Nanoelectronics Workshop (SNW), p.85, (2024).

[10] X. Song, D. Sun, C. Yu, S. Li, Z. Zhou, X. Liu, J. Kang, "Optimized MFS Stack with N-Doped TiO₂ Channel and La-Doped HfO₂ Ferroelectric Layer for Highly Stable FeFETs", IEEE Electron Device Letters, 45, p. 2213 (2024).

[11] X. Song, D. Sun, C. Yu, H. Zhu, S. Li, X. Liu, J. Kang, "High-Temperature Performance and Reliability of MFS structured NTO Channel FeFETs for Extreme Environments Applications", submitted to IRPS2025.

[12] X. Song, D. Sun, X. Liu, J. Kang, "Thermal Stability of TiO₂ Channel FE-VNAND: From Fabrication to High-Temperature Operation", submitted to EDTM2025

[13] C. Yu, H. Ma, M. Li, F. Liu, X. Ding, Y. Zhao, H. Li, X. Song, F. Liu, W. Yang, J. Xu, J. Zhang, X. Hao, L. Liu, P. Huang, P. Gao, and J. Kang, "Insights into the origin of robust ferroelectricity in HfO₂-based thin films from the order-disorder transition driven by vacancies", Physical Review Applied, vol.22, No.024028 (2024).

979-8-3315-0417-5/25 $31.00 © 2025 IEEE

2D Novel Antiferroelectric Materials for Neuromorphic Computing

Dongliang Yang[1,2], Linfeng Sun[1,2]*

[1]Centre for Quantum Physics Key Laboratory of Advanced Optoelectronic Quantum Architecture and Measurement (MOE) School of Physics Beijing Institute of Technology Beijing 100081, China,
[2]Beijing Key Lab of Nanophotonics & Ultrafine Optoelectronic Systems, School of Physics, Beijing Institute of Technology, Beijing, 100081, China

Abstract

Neuromorphic computing emulates biological neural networks for efficient, low-power information processing. In this study, we use the novel 2D antiferroelectric device ($CuBiP_2Se_6$, CBPS) for neuromorphic computing that exhibits tunable synaptic behavior under electrical and optical stimuli, successfully mimicking Pavlovian conditioning. Through innovatively integrating multi-modal information, this research has opened up a new path for constructing more flexible and efficient brain-like computing architectures.

Keywords: Neuromorphic computing, Antiferroelectric, and $CuBiP_2Se_6$

Introduction

The rapid development of artificial intelligence and the demand for efficient, low-power computation drive the research in neuromorphic computing[1-4]. Traditional CMOS-based systems, limited by the von Neumann architecture (i.e., the separation of memory and processing units), are less efficient in emulating complex brain-like functions. In contrast, neuromorphic computing integrates memory and processing units, similar to the brain's functionality, significantly reducing energy consumption and enhancing parallel processing efficiency[5-8]. A core challenge in this field is developing materials capable of emulating synaptic plasticity, allowing for dynamic adjustment of connection strength between neurons in response to external stimuli[9, 10].

Two-dimensional materials, with their unique physical phenomena and atomically thin structures, have become key candidates for neuromorphic devices[11, 12]. Compared to conventional antiferroelectric materials, 2D antiferroelectric materials exhibit a more gradual polarization response and higher energy storage density under applied electric fields. Notably, when the external field is removed, their polarization decays slowly rather than vanishing immediately, making antiferroelectric ideal for emulating synaptic plasticity in neuromorphic computing[13-15]. $CuBiP_2Se_6$ (CBPS) [16-18], as an emerging van der Waals antiferroelectric material, also demonstrates excellent photoresponse across a wide spectral range[19], showcasing the strong potential for optoelectronic-neuromorphic integration. In this study, CBPS-based devices were fabricated to verify their antiferroelectric properties and assess both electrical and optical responses. Results show the capability of CBPS devices to simulate Pavlovian conditioning, highlighting the significant potential of antiferroelectric CBPS devices for multimodal integration (e.g., optoelectronic co-stimulation) in brain-inspired computing applications.

Results

In this work, we fabricated an Au/CBPS/Au neuromorphic device using micro-nanofabrication techniques, with its schematic structure shown in **Fig. 1a**. The device thickness, measured by AFM, is approximately 8 nm. We obtained the P-E hysteresis loop of CBPS at room temperature under a range of applied electric fields. As shown in **Fig. 1b**, the results indicate that an external electric field induces spontaneous polarization in CBPS (field-induced ferroelectric state), which returns to zero once the field is removed, confirming its antiferroelectric behavior.

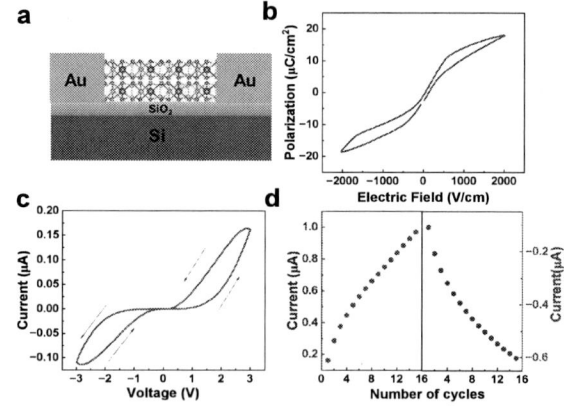

Fig. 1 (a) Schematic illustration of CBPS-based devices simulating biological synapses. (b) P–E hysteresis loops measured on the CBPS antiferroelectric capacitor. (c) I–V curve of the CBPS device. (d) Current statistics at ±3 V over multiple I–V cycles in the CBPS device.

We also measured the I-V characteristics of the device (**Fig. 1c**), where the current increases nonlinearly with the applied voltage, attributed primarily to internal domain switching under the electric field, which alters the polarization state of the material and, consequently, the current. Additionally, during the reverse voltage sweep, the current does not retrace the forward path, showing a hysteresis effect. Once the electric field is removed, the current returns to its original state, forming a closed loop with zero remanent polarization. The same behavior was observed under negative voltage sweeps. Furthermore, multiple I-V scans were performed, and the current at ±3 V was recorded over cycles. As shown in **Fig. 1d**, the current increases linearly with the number of scans.

To investigate the potential of the CBPS device as a synapse, we applied electrical and optical pulses to observe its response. In **Fig. 2a**, a series of 10 pulses with an amplitude of 1 V and a read voltage of 0.1 V were applied to the device. The current increased with the number of pulses, reaching a maximum of 18 nA. This effect is likely due to the internal domains in the antiferroelectric material being gradually "trained" or reoriented by repeated electric field pulses, causing a cumulative change in the overall polarization state. Although antiferroelectric materials tend to return to their original polarization upon removal of the field, multiple pulses may partially fix or locally shift some domains, resulting in a progressively higher current response with subsequent pulses. Moreover, CBPS has been shown to exhibit excellent optoelectronic properties and can operate across a broad visible wavelength range, indicating its potential as an artificial photosensitive synaptic material for simulating neural signal processing and transmission in visual systems. Accordingly, we applied optical pulses to the CBPS device (**Fig. 2b**), which demonstrated an impressive photoresponse. Notably, the device did not immediately return to its initial state after optical stimulation, likely due to surface and defect states capturing photogenerated charge carriers and delaying the recovery of material resistance.

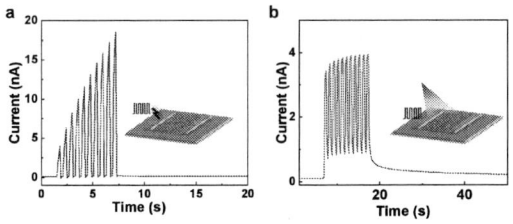

Fig. 2 (a) Stimulation response of the device to voltage pulses. (b) Stimulation response of the device to light pulses.

The experimental paradigm of Pavlov's dog provides a specific model system for the functional verification of neuromorphic devices[20]. Analogies can be drawn between neuromorphic devices and the learning and memory processes of dogs. By constructing similar conditioned reflex experimental scenarios and using neuromorphic devices to simulate the brain functions of dogs, it is possible to examine whether the devices can correctly implement behaviors such as learning, memory, and conditioned reflexes, thereby verifying the effectiveness and performance of the devices. Consequently, the CBPS devices successfully realized the classic conditioned reflex experiment, i.e., Pavlov's dog experiment under the modulation of photoelectric pulses, as shown in **Fig. 3**. The electrical pulse (1 V) represents food (unconditioned stimulus) and elicits an unconditioned response, while the light pulse (650 nm) serves as the bell (conditioned stimulus) and generates a conditioned response. The set threshold current is 4.5 nA. Under the stimulation of 10 light pulses (the bell), the excitatory postsynaptic current (EPSC) only increases to 4 nA, which is lower than the threshold current, indicating that the bell stimulus does not trigger the salivary secretion response. In contrast, when 10 electrical pulses (food) are applied to the device, the EPSC is much higher than the threshold, suggesting that dogs secrete saliva instinctively when they see food. Meanwhile, during the training process, light and electrical pulses are applied to establish the associative reflex between the bell and food, triggering the conditioned reflex. After training, when the light pulse is used as the conditioned stimulus signal, the EPSC exceeds the threshold, indicating that dogs secrete saliva in the conditioned response to the bell. When the light pulse is used as the conditioned stimulus signal after being trained by electrical pulses, the obtained EPSC exceeds the threshold, indicating that the conditioned reflex has been successfully established within the CBPS-based neuromorphic device.

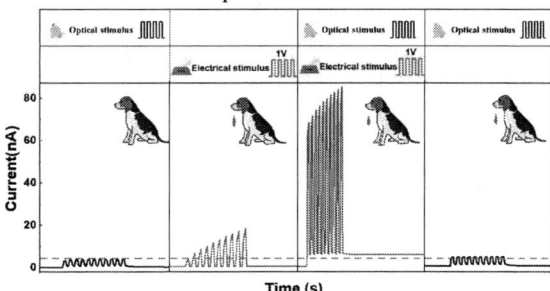

Fig. 3 Simulation of light adaptation in the human visual system and Pavlov's dog experiment.

Conclusion

In this work, CBPS can not only simulate synaptic-like plasticity through reversible polarization switching under electrical stimulation but also extend its functions by responding to optical stimuli across a wide spectrum. These capabilities enable CBPS devices to mimic the key properties of biological synapses. The adjustable synaptic behaviors exhibited under electrical and optical stimuli successfully simulate the classic neural response model—Pavlov's conditioned reflex. This paves a new way for constructing more flexible and efficient brain-like computing architectures with antiferroelectric materials, thus holding the promise of achieving further breakthroughs in the fields of multimodal data processing and intelligent computing.

Acknowledgments

This work was supported by National Key R&D Plan (2022YFA1405600) and Key Research Program of Beijing Natural Science Foundation (Grant No. Z210006).

References

[1] D. Liu, H. Yu, and Y. Chai, "Low-power computing with neuromorphic engineering," Advanced Intelligent Systems, vol. 3, no. 2, pp. 2000150, 2021.

[2] S. Pazos, X. Xu, T. Guo, K. Zhu, H. N. Alshareef, and M. Lanza, "Solution-processed memristors: performance and reliability," Nature Reviews Materials, pp. 1-16, 2024.

[3] K. Zhu, S. Pazos, F. Aguirre, Y. Shen, Y. Yuan, W. Zheng, O. Alharbi, M. A. Villena, B. Fang, and X. Li, "Hybrid 2D–CMOS microchips for memristive applications," Nature, vol. 618, no. 7963, pp. 57-62, 2023.

[4] X. Chen, D. Yang, G. Hwang, Y. Dong, B. Cui, D. Wang, H. Chen, N. Lin, W. Zhang, and H. Li, "Oscillatory Neural Network-Based Ising Machine Using 2D Memristors," ACS nano, vol. 18, no. 16, pp. 10758-10767, 2024.

[5] C. López, "Artificial intelligence and advanced materials," Advanced Materials, vol. 35, no. 23, pp. 2208683, 2023.

[6] Z. Zhang, D. Yang, H. Li, C. Li, Z. Wang, L. Sun, and H. Yang, "2D materials and van der Waals heterojunctions for neuromorphic computing," Neuromorphic Computing and Engineering, vol. 2, no. 3, pp. 032004, 2022.

[7] L. Ce, Y. Dong-Liang, and S. Lin-Feng, "Research progress of neuromorphic devices based on two-dimensional layered materials," ACTA PHYSICA SINICA, vol. 71, no. 21, 2022.

[8] D. Yang, H. Yang, X. Guo, H. Zhang, C. Jiao, W. Xiao, P. Guo, Q. Wang, and D. He, "Robust polyethylenimine electrolyte for high performance and thermally stable atomic switch memristors," Advanced Functional Materials, vol. 30, no. 50, pp. 2004514, 2020.

[9] H. Liu, Y. Qin, H. Y. Chen, J. Wu, J. Ma, Z. Du, N. Wang, J. Zou, S. Lin, and X. Zhang, "Artificial neuronal devices based on emerging materials: neuronal dynamics and applications,"

Advanced Materials, vol. 35, no. 37, pp. 2205047, 2023.

[10] L. Sun, Z. Wang, J. Jiang, Y. Kim, B. Joo, S. Zheng, S. Lee, W. J. Yu, B.-S. Kong, and H. Yang, "In-sensor reservoir computing for language learning via two-dimensional memristors," Science advances, vol. 7, no. 20, pp. eabg1455, 2021.

[11] Z. Huang, Y. Li, Y. Zhang, J. Chen, J. He, and J. Jiang, "2D multifunctional devices: from material preparation to device fabrication and neuromorphic applications," International Journal of Extreme Manufacturing, vol. 6, no. 3, pp. 032003, 2024.

[12] Y. Sun, R. Zhang, C. Teng, J. Tan, Z. Zhang, S. Li, J. Wang, S. Zhao, W. Chen, and B. Liu, "Internal ion transport in ionic 2D CuInP2S6 enabling multi-state neuromorphic computing with low operation current," Materials Today, vol. 66, pp. 9-16, 2023.

[13] F. Xue, Y. Ma, H. Wang, L. Luo, Y. Xu, T. D. Anthopoulos, M. Lanza, B. Yu, and X. Zhang, "Two-dimensional ferroelectricity and antiferroelectricity for next-generation computing paradigms," Matter, vol. 5, no. 7, pp. 1999-2014, 2022.

[14] J. R. Reimers, S. A. Tawfik, and M. J. Ford, "van der Waals forces control ferroelectric–antiferroelectric ordering in CuInP 2 S 6 and CuBiP 2 Se 6 laminar materials," Chemical Science, vol. 9, no. 39, pp. 7620-7627, 2018.

[15] T. Pan, J. Zhang, Z. N. Guan, Y. Yan, J. Ma, X. Li, S. Guo, J. Wang, and Y. Wang, "Enhanced Energy Density and Efficiency in Lead-Free Sodium Niobate-Based Relaxor Antiferroelectric Ceramics for Electrostatic Energy Storage Application," Advanced Electronic Materials, vol. 8, no. 12, pp. 2200793, 2022.

[16] X. Jiang, J. Tan, D. Liu, Y. Feng, K.-Q. Chen, E. A. Kazakova, A. S. Vasenko, and E. V. Chulkov, "Ferroelectric Polarization and Single-Atom Catalyst Synergistically Promoting CO2 Photoreduction of CuBiP2Se6," The Journal of Physical Chemistry Letters, vol. 15, no. 13, pp. 3611-3618, 2024.

[17] M. A. Gave, D. Bilc, S. Mahanti, J. D. Breshears, and M. G. Kanatzidis, "On the lamellar compounds CuBiP2Se6, AgBiP2Se6 and AgBiP2S6. Antiferroelectric phase transitions due to cooperative Cu+ and Bi3+ ion motion," Inorganic chemistry, vol. 44, no. 15, pp. 5293-5303, 2005.

[18] A. Dziaugys, J. Banys, J. Macutkevic, Y. Vysochanskii, I. Pritz, and M. Gurzan, "Phase transitions in CuBiP2Se6 crystals," Phase Transitions, vol. 84, no. 2, pp. 147-156, 2011.

[19] W. He, L. Kong, P. Yu, and G. Yang, "Record-High Work-Function p-Type CuBiP2Se6 Atomic Layers for High-Photoresponse van der Waals Vertical Heterostructure Phototransistor," Advanced Materials, vol. 35, no. 14, pp. 2209995, 2023.

[20] Z. Shang, L. Liu, G. Wang, H. Xu, Y. Cui, J. Deng, Z. Lou, Y. Yan, J. Deng, and S.-T. Han, "Ferroelectric Polarization Enhanced Optoelectronic Synaptic Response of a CuInP2S6 Transistor Structure," ACS nano, 2024.

Dual Driven Approaches for General Purposed Brain Inspired Computing

Luping Shi[1,2,3*], Yuqing Cong[1,4], and Wei Zhang[5,6]

[1]Center for Brain-Inspired Computing Research (CBICIR), China,

[2]Optical Memory National Engineering Research Center, & Department of Precision Instrument, Tsinghua University, Beijing 100084, China,

[3]IDG/McGovern Institute for Brain Research, Tsinghua University, Beijing 100084, China,

[4]Department of Computer Science and Technology, Tsinghua University, China,

[5]China Nanhu Academy of Electronics and Information Technology, Jiaxing, Zhejiang, China,

[6]China Electronics Technology HIK Group Co., Ltd, Hangzhou, Zhejiang, China

[*]Phone: +86 13910693300, Email: lpshi@tsinghua.edu.cn

Abstract

Modern Von Neumann-based computers are increasingly inadequate to meet current computational demands. Brain-inspired computing (BIC) offers a promising alternative, aiming to emulate the efficiency, adaptability, and learning capabilities of the human brain. Despite significant progress, a unified technological solution for BIC has yet to emerge. This talk reviews recent BIC advancements, covering theories, chips, software, and systems. Key challenges and potential solutions for robust BIC systems and the development of general purposed BIC are also examined.

Keywords: Brain-Inspired Computing (BIC), Neuromorphic computing

Introduction

Over the past half-century, the primary driven force behind von Neumann-based computer development has been scaling, following Moore's law. However, we are now approaching the physical limits of this approach. The separation of the processing and memory in this architecture creates a "memory wall," which severely limits computational efficiency and energy performance, especially for data-intensive tasks[1]. Inspired by the human brain, brain-inspired computing (BIC) chips have emerged, generally following two approaches: neuroscience-oriented and computer-science-oriented. The neuroscience-oriented approach seeks to closely mimic key features of the cerebral cortex of human brains, such as memory-computing integration and spatiotemporal dynamics, to enable more efficient and adaptable systems. Meanwhile, the computer-science-oriented approach relies on explicit algorithms executed on computers. Although both approaches have been effective in addressing specific sub-problems in data-rich, specialized domains, they struggle with complex, dynamic problems involving uncertainty or incomplete information.

In order to further advance the capabilities of BIC, an increasing trend is the incorporation of brain-inspired models and algorithms into prevailing neural networks, fostering a more explicit dialog between the two approaches[1-3]. This synthesis aims to harness synergistic interactions, alleviating the limitations of current BIC architectures by combining complementary advancements in brain and computer sciences. Together, these approaches offer a promising path toward breakthroughs that leverage the strengths of both fields, as depicted in Fig. 1.

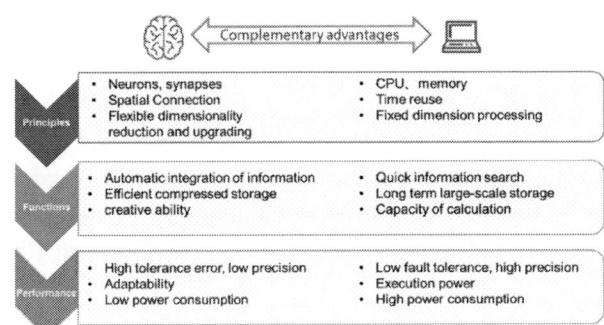

Figure 1. Comparison of the key features between Brain and Computer

Theory and experiments

The increasing success of BIC chips in practical applications suggests a future in which general-purpose BIC may serve as next-generation computing infrastructure. The dual-driven BIC paradigm shows promise in the development of GPBIC systems. To advance in this direction, it is imperative to establish a theoretical foundation that defines the scientific scope and provides guidance, such as computational completeness and formalized computing models. To establish generalized scientific principles, researchers need to not only explore the underlying intrinsic causes that contribute to the advantages of BIC but also examine the essential features of both the brain and computers.

Turing completeness lays the theoretical foundation for computers, and the hierarchical system structure decouples modern computers' software and hardware to ensure their independent progress as well as compatibility. However, the lack of guidance in computational completeness and system structure leads to the tight coupling development of dedicated software and hardware in BIC, hindering the establishment of GPBIC. Recently, neuromorphic completeness was proposed as a more adaptive and broader definition of completeness for BIC[2]. Under the guidance of this theory, a BIC system hierarchy is developed including Turing completeness software abstraction model and a Neuromorphic completeness hardware. The roles and optimization objectives of different levels of BIC hardware and software involved are also defined more clearly. Compatibility between different hardware and/or software designs could be improved, which is an urgent need for GPBIC.

Recently, we presented a foundational perspective on general-purpose brain-inspired computing (GPBIC)[4], proposing a promising approach that integrates insights from both neuroscience and computer science to achieve key attributes such as flexibility, adaptivity, and superior performance. This approach introduces hybrid mechanisms across computation, execution, and model dimensions to address fundamental challenges in representing, storing, computing, and transmitting heterogeneous information in BIC, as shown in Figure 2.

Computing capability differ between Turing-Machine (TM)-based modern computers and the brain. While computers emphasize temporal richness, the brain uses massive neurons with spatial parallelism and their connections to process complex information, demonstrating significant spatial richness. The "temporal richness" of BIC machine arises from neuron dynamics, which allow for multiple states and flexible state transitions within each unit, while the "spatial richness" is due to interconnections between neurons that enable information exchange among processing units. Integrating temporal and spatial richness can enhance the effectiveness of BIC machines. The elasticity is influenced by the spatial-temporal richness and coding, which enables the BIC machine to adapt to different computation scenarios with varying performance. Hence, a critical challenge of the dual-driven BIC is to balance spatial and temporal richness.

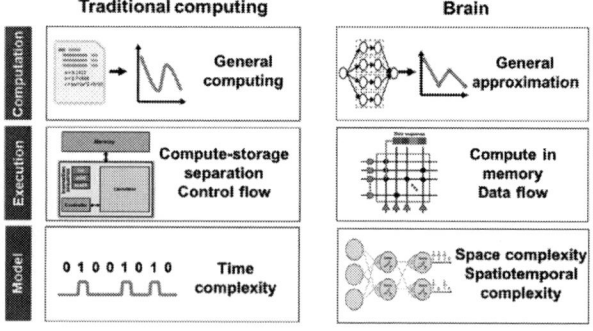

Figure 2. Dual-Driven Cross-Paradigm Hybrid BIC

A relationship between "general computation" and "general approximation" was established by neuromorphic completeness theory to enlarge the design space between cost and precision[2]. It can improve the versatility of the dual-driven BIC machine. Furthermore, the dual-driven BIC machine can be designed to follow precise logic defined by the program, which enables powerful and reliable computations, as well as the development of complex programs based on precise logic. Moreover, it should support approximation in two potential parts, namely the algorithm and language parts.

The approximation of the algorithm involves the use of algorithm approximations that may not fully conform to its exact logic. The approximation of the language part involves the use of approximate language, which may result in less precise encoding data for BIC machines. Specifically, the approximate input can be encoded to approximate the implementation of the algorithm.

In modern computer, control-flow and data-flow architectures are two commonly adopted architectures. The former relies on a sequential execution of instructions to determine the processing unit's function, while the latter usually takes an event-driven approach to achieve low latency and low power consumption. Most BIC machines currently use the data-flow architecture. Combining data-flow with control-flow has the potential to increase controllability and interactivity while retaining the performance advantages of data-flow architecture. To introduce control-flow, BIC system can establish synchronous groups or barriers to improve controllability in the original data-flow architecture, which typically adopts an asynchronous event-driven pattern.

In recent years, integrating BIC hardware with traditional general-purpose processors such as CPUs or FPGAs has become a popular method to increase flexibility. Such heterogenous structure offers several benefits, including high programmability and reuse of current computers. As BIC hardware continues to evolve and the number of applications utilizing BIC patterns grows, it is possible that BIC hardware may replace the CPU as a brain-inspired central processor (BCU).

As applications become more generalized, domain-agnostic but pattern-inclined intermediate programming languages and compilers can be developed for BIC pattern programs. A brain-inspired OS can run directly on the BCU through the scheduling ability of the hardware. A general-purpose BIC-only computer system can be enabled by developing a general approximation compiler, making a self-operating system that can adaptively manage the entire computer a possibility.

979-8-3315-0417-5/25 $31.00 © 2025 IEEE

With the continuous advancement of system software, the dual-driven BIC offers a systematic perspective on high-level concepts such as "general-purpose". The system software decouples the applications/algorithms and hardware for independent development, which is the foundation for a thriving ecology. Decoupling and co-evolving algorithm-software-hardware will make the dual-driven BIC a feasible and pragmatic solution, potentially becoming the next-generation general-purpose computing infrastructure in society.

A key milestone for GPBIC is the development of a versatile BIC chip capable of flexibly supporting a wide spectrum of parallel applications while maintaining high performance.

In this talk, we review recent advancements in the development of BIC toward GPBIC, including theories, chips, and software, and discuss how combining neuroscience and computer science offers an optimal pathway for advancing BIC, particularly for GPBIC. We also outline a roadmap for GPBIC development, addressing challenges and proposing solutions for integrating these two approaches in BIC's evolution.

The implementation of in/near-memory computing and in-sensor computing architectures using the post-CMOS technologies, such as memristors and non-volatile memories, has proven to significantly improve power efficiency and area utilization in some domain-specific hardware. Nevertheless, there is great potential in establishing a novel heterogeneous memory hierarchy that leverages the strengths of emerging technologies, promoting the development of GPBIC systems with high computing and power efficiency. This will also be addressed.

Conclusion

This talk reviewed recent advancements in the development of BIC toward GPBIC, highlighting the dual-driven approach that integrates neuroscience and computer science principles. In addition, key theories, chips, software and systems were explored, outlining challenges and potential solutions in the development of BIC.

Acknowledgements

This work was supported by National Nature Science Foundation of China (nos. 62088102) and the STI 2030—Major Projects 2021ZD0200300.

References

[1] Jing Pei, et al, "Towards artificial general intelligence with hybrid Tianjic chip architecture", Nature, 572(7767): 106-111. (2019)

[2] Youhui Zhang, et al, "A system hierarchy for brain-inspired computing", Nature, 586(7829): 378-384 (2020).

[3] Songchen Ma, et al, "Neuromorphic computing chip with spatiotemporal elasticity for multi-intelligent-tasking robots. Science Robotics", 7(67), eabk2948(2022).

[4] Weihao Zhang, et al, "The development of general-purpose brain-inspired computing", Nature Electronics, https://doi.org/10.1038/s41928-024-01277-y (2024).

Precise transmission matrix measurement of a multimode fiber and its applications

Yuwen Xiong[1,2,3], Zihao Ma[1,2,3], Yi Xu[1,2,3*], Yuwen Qin[1,2,3*]

[1]Institute of Advanced Photonics Technology, School of Information Engineering, Guangdong University of Technology, Guangzhou, 510006, China,

[2]Key Laboratory of Photonic Technology for Integrated Sensing and Communication, Ministry of Education of China, Guangdong University of Technology, Guangzhou, 510006, China,

[3]Guangdong Provincial Key Laboratory of Information Photonics Technology, Guangdong University of Technology, Guangzhou, 510006, China

Email: yixu@gdut.edu.cn, qinyw@gdut.edu.cn

Abstract

In this paper, we present our recent results on measurement of the transmission matrix for a multimode fiber (MMF) with potential applications in the multi-dimensional light field information transmission with near-unity fidelity and vectorial holography with arbitrary distributions of optical intensity and polarization over the MMF. Deterministic and deep learning types of methods are shown to be quite efficient to unscramble multiple scattering taking place when light passes through the MMF.

Keywords: Multimode fiber, transmission matrix, deep learning

Introduction

When light propagates through a scattering medium, both the amplitude and phase information encoded in the wavefront of light are very fragile, resulting in the seemly chaotic speckle intensity output from the scattering medium [1-3]. For example, although light propagation in an MMF is a deterministic process, the multiple scattering nature of the MMF inevitably imposes a great challenge to precisely control either the light field information to be delivered or the output morphology of light field. The transmission matrix (TM) method and the deep learning method are two popular methods for achieving these goals. Both of them are powerful tools to unscramble and even manipulate the multiple optical scattering [2,4], facilitating versatile functionalities in optical focusing [5-7], imaging [8-17], 3D holography [18-21] and fiber laser [22]. In this paper, we present our recent works of multi-dimensional light field information transmission with near-unity fidelity and vectorial holography with arbitrary distributions of optical intensity and polarization over an MMF by means of the precisely calibrated transmission matrix and a physically enhanced neural network based on the transmission matrix.

Experimental methods

As a typical multiple-input-multiple-output (MIMO) optical transmission system, the transmission of light in an MMF can be modeled mathematically with a linear relationship as follows:

$$E_{out} = T \cdot E_{in} \qquad (1)$$

where T is the TM (an $M \times N$ matrix), E_{in} (an $N \times 1$ vector) and E_{out} (an $M \times 1$ vector) are the input and output complex amplitudes of light field, respectively. Each element of T describes the mapping relationship between the spatial channels from each input port to the output port. As long as the TM is accurately measured, the information encoded in the input wavefront can be precisely retrieved by using $E_{in} = T^{-1} \cdot E_{out}$, where T^{-1} is the inversion or pseudo-inversion of the measured TM for the MMF, i.e. inverse transmission matrix method (ITM).

When the TM is accurately measured, one can also precisely manipulate the output light field of the MMF, such as holographic projection with arbitrary distributions of optical intensity [23] and polarization [24]:

$$E_{in} = [T^\dagger T + \sigma I]^{-1} T^\dagger E_{target} \qquad (2)$$

where $\dagger$ stands for conjugate transpose, σ represents the noise factor of regularization, I represents the identity matrix and E_{target} is a desired output light field. The reason for using a regularization term is because the matrix TT^{-1} in $E_{out} = TT^{-1}E_{target}$ is not an identity matrix if M>N even though the TM is accurately measured.

The optical setup for measuring the TM of an MMF is shown in Fig. 1, where a collimated CW laser is modulated by a digital micromirror device. A 4f system is used to filter the first-order diffracted light. The input light field is modulated in amplitude and phase simultaneously, which is then coupled into the proximal end of the MMF using an objective lens. The other objective lens collects the scattered light field at the distal end of the MMF, where the speckle is recorded by a complementary metal-oxide-semiconductor camera. The linear polarizer in front of the camera is used to analyze the polarization state of the captured speckle. The four-step phase shift method is used to calibrate the TM.

When the TM is combined with a deep neural network (DNN) as an empirical model, a physically enhanced neural network named deep empirical neural network (DENN) can be achieved. In DENN, the phase-encoded information can be evaluated by $\varphi = F_m(I)$, where F_m is the mapping function

979-8-3315-0417-5/25 $31.00 © 2025 IEEE

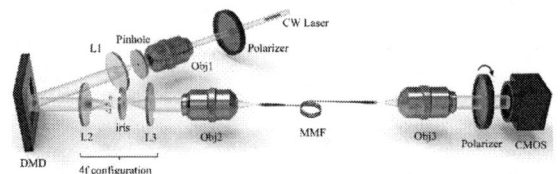

Fig.1. Optical setup for vectorial holography through an MMF. Obj denotes the microscopic objective lens. L, lens. DMD, digital micromirror device. The figure is modified based on the one in Ref. [24].

determined by the weights and biases of the DNN. The DENN can be updated under the physical guidance of a single intensity input I utilizing:

$$F_m^* = \arg\min\| |Te^{i\varphi\pi}|^2 - I \|^2 \quad (3)$$

where the empirical TM links the normalized phase-encoded information $\varphi = F_m(I)$ of multiple inputs with the speckle intensity I of multiple outputs. The schematics of the whole system is shown in Fig.2.

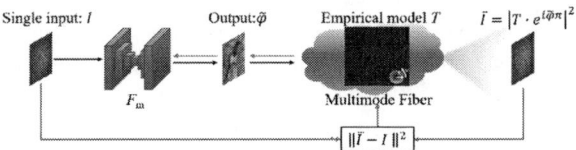

Fig.2. Schematics of the deep empirical neural network for the MMF system.

When the converged condition of Eq. (3) is met, the mapping function F_m^* can be used to retrieve the phase-encoded information for the input speckle I:

$$\varphi = F_m^*(I) \quad (4)$$

Experimental results

We experimentally demonstrate the transmission of optical information with phase encoded wavefronts over an MMF using the ITM method [25,26], where the results are shown in Fig. 3. Here, 256-level grayscale images of natural scene images are used as the encoded wavefronts. As can be seen in Fig.3, the averaged Pearson correlation coefficient (PCC) and structural similarity index measure (SSIM) are up to 0.99 and 0.98 in experiment, respectively, which demonstrates the ability of unscrambling multiple scattering and deliver multilevel digital information through MMF.

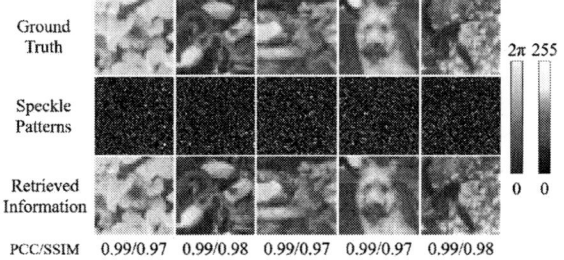

PCC/SSIM   0.99/0.97   0.99/0.98   0.99/0.97   0.99/0.97   0.99/0.98

Fig.3. Retrieval of 256-level gray-scale natural scene images with near-unity fidelity by the ITM method. Top row shows the ground truths, second row shows the corresponding speckle patterns when the central wavelength of

the laser is 850 nm, bottom row shows the retrieved information by ITM.

We also experimentally demonstrate the generation of light field with arbitrary intensity and polarization at the distal end of the MMF[24], as shown in Fig. 4. Parts of these patterns are adopted from the Fashion-MNIST dataset. The experimental results shown in Fig. 4 indicate that the proposed method can achieve high-fidelity vectorial holographic projection over the MMF. Under the analysis of a polarizer, the generated light field with targeted distributions of intensity and polarization can be validated.

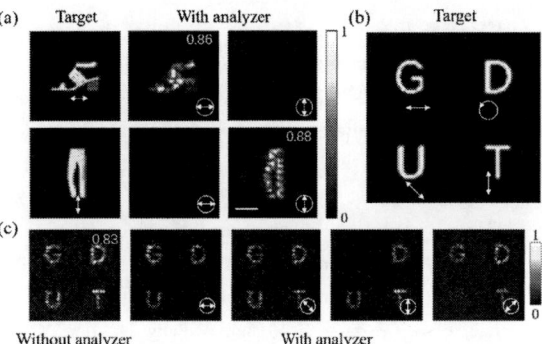

Fig.4. Experimental demonstration and evaluation of vectorial holography over an MMF. (a) Holographical display of the fashion patterns with specified polarization states. (b)The targeted patterns "GDUT" with four distinct polarization states. (c) Performance of vectorial holography using the proposed method. The white arrows indicate the corresponding polarization state of the polarizer. The corresponding PCCs are also provided. The figure is modified based on the one in Ref. [24].

Furthermore, by embedding TM into the DENN, optical phase retrieval over an MMF can be achieved in an untrained manner using DNN. The experimental results of the proposed DENN based on the empirical TM are shown in Fig. 5, where the maximum experimentally achieved PCC/SSIM are up to 0.97/0.84 for the natural scene images and accuracy are up to 99.2% for binary information, respectively.

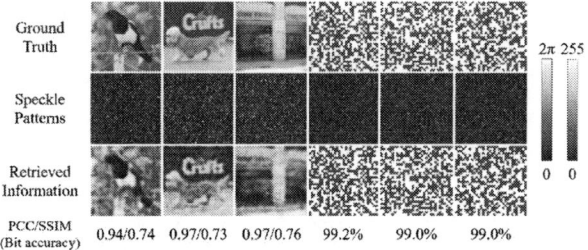

PCC/SSIM (Bit accuracy)   0.94/0.74   0.97/0.73   0.97/0.76   99.2%   99.0%   99.0%

Fig.5. Experiment retrieval results of phase-encoded natural scene images after passing by DENN. Top row shows the ground truths, the second row shows the corresponding speckle patterns when the central wavelength of the laser is 532 nm, bottom row shows retrieved information by DENN.

Conclusion

In summary, we experimentally demonstrate the multi-dimensional light field information transmission with near-unity fidelity and vectorial holography over an MMF by utilizing the ITM. The combination of TM and DNN can achieve the physically enhanced neural network under the guidance of empirical physics. The demonstrated method can also be generalized to other strongly scattering media.

Acknowledgments

This work was supported by National Natural Science Foundation of China under grant nos. 62222505 (Y.X.) and 62335005(Y.X.) and Guangdong Introducing Innovative, Entrepreneurial Teams of "The Pearl River Talent Recruitment Program" under grant no. 2019ZT08X340 (Y.Q.).

References

[1] H. Cao, A. P. Mosk, and S. Rotter, "Shaping the propagation of light in complex media," Nat. Phys. 18, 994 (2022).

[2] S. Yoon, M. Kim, M. Jang, Y. Choi, W. Choi, S. Kang, and W. Choi, "Deep optical imaging within complex scattering media," Nat. Rev. Phys. 2, 141 (2020).

[3] A. P. Mosk, A. Lagendijk, G. Lerosey, and M. Fink, "Controlling waves in space and time for imaging and focusing in complex media," Nat. Photonics 6, 283 (2012).

[4] Z. Yu, H. Li, T. Zhong, J.-H. Park, S. Cheng, C. M. Woo, Q. Zhao, J. Yao, Y. Zhou, X. Huang, W. Pang, H. Yoon, Y. Shen, H. Liu, Y. Zheng, Y. Park, L. V. Wang, and P. Lai, "Wavefront shaping: a versatile tool to conquer multiple scattering in multidisciplinary fields," The Innov. 3, 100292 (2022).

[5] I. M. Vellekoop, A. Lagendijk, and A. P. Mosk, "Exploiting disorder for perfect focusing," Nat. Photonics 4, 320 (2010).

[6] P. Lai, L. Wang, J. W. Tay, and L. V. Wang, "Photoacoustically guided wavefront shaping for enhanced optical focusing in scattering media," Nat. Photonics 9, 126 (2015).

[7] G. Huang, D. Wu, J. Luo, L. Lu, F. Li, Y. Shen, and Z. Li, "Generalizing the Gerchberg–Saxton algorithm for retrieving complex optical transmission matrices," Photonics Res. 9, 34 (2021).

[8] S. Popoff, G. Lerosey, M. Fink, A. C. Boccara, and S. Gigan, "Image transmission through an opaque material," Nat. Commun. 1, 81 (2010).

[9] Y. Choi, C. Yoon, M. Kim, T. D. Yang, C. Fang-Yen, R. R. Dasari, K. J. Lee, and W. Choi, "Scanner-free and wide-field endoscopic imaging by using a single multimode optical fiber," Phys. Rev. Lett. 109, 203901 (2012).

[10] D. Loterie, S. Farahi, I. Papadopoulos, A. Goy, D. Psaltis, and C. Moser, "Digital confocal microscopy through a multimode fiber," Opt. Express 23, 23845 (2015).

[11] K. Lee and Y. Park, "Exploiting the speckle-correlation scattering matrix for a compact reference-free holographic image sensor," Nat. Commun. 7, 13359 (2016).

[12] Z. Cai, J. Chen, G. Pedrini, W. Osten, X. Liu, and X. Peng, "Lensless light-field imaging through diffuser encoding," Light. Sci. Appl. 9, 143 (2020).

[13] Z. Wen, Z. Dong, C. Pang, C. F. Kaminski, Q. Deng, J. Xu, L. Wang, S. Liu, J. Tang, W. Chen, X. Liu, and Q. Yang, "Single multimode fiber for in vivo light-field encoded nano-imaging," arXiv:2207.03096 (2022).

[14] Z. Dong, Z. Wen, C. Pang, L. Wang, L. Wu, X. Liu, and Q. Yang, "A modulated sparse random matrix for high-resolution and high-speed 3D compressive imaging through a multimode fiber," Sci. Bull. 67, 1224 (2022).

[15] W. Li, B. Wang, T. Wu, F. Xu, and X. Shao, "Lensless imaging through thin scattering layers under broadband illumination," Photonics Res. 10, 2471 (2022).

[16] F. Wang, C. Wang, M. Chen, W. Gong, Y. Zhang, S. Han, and G. Situ, "Farfield super-resolution ghost imaging with a deep neural network constraint," Light. Sci. Appl. 11, 1 (2022).

[17] J. Oh, K. Lee, and Y. Park, "Single-shot reference-free holographic imaging using a liquid crystal geometric phase diffuser," Laser Photonics Rev. 16, 2100559 (2022).

[18] A. K. Singh, D. N. Naik, G. Pedrini, M. Takeda, and W. Osten, "Exploiting scattering media for exploring 3D objects," Light. Sci. Appl. 6, e16219 (2017).

[19] H. Yu, K. Lee, J. Park, and Y. Park, "Ultrahigh-definition dynamic 3D holographic display by active control of volume speckle fields," Nat. Photonics 11, 186 (2017).

[20] J. Park, K. Lee, and Y. Park, "Ultrathin wide-angle large-area digital 3D holographic display using a non-periodic photon sieve," Nat. Commun. 10, 1304 (2019).

[21] S. Li, C. Saunders, D. J. Lum, J. Murray-Bruce, V. K. Goyal, T. Čižmár, and D. B. Phillips, "Compressively sampling the optical transmission matrix of a multimode fibre," Light. Sci. Appl. 10, 88 (2021).

[22] X. Wei, J. C. Jing, Y. Shen, and L. V. Wang, "Harnessing a multi-dimensional fiber laser using genetic wavefront shaping," Light. Sci. Appl. 9, 149 (2020).

[23] D. Loterie, S. Farahi, I. Papadopoulos, A. Goy, D. Psaltis, and C. Moser, "Digital confocal microscopy through a multimode fiber," Opt. Express 23, 23845-23858 (2015)

[24] H. Liu, J. Ye, P. Xu, L. Wu, Y. Xu, and Y. Qin, "Vectorial holography over a multimode fiber," Opt. Lett. 49, 1798-1801 (2024).

[25] J. Ye, T. Pan, K. Zheng, Z. Luo, Y. Xu, S. Fu, Y. Wang, Y. Qin, "Light field information transmission through scattering media with high fidelity," Chin. Opt. Lett. 21, 121101 (2023).

[26] Z. Ma, H. Liu, J. Ye, Y. Xu, Y. Qin. "Computational imaging over a kilometer-scale multimode fiber based on transmission matrix (*invited*),". *Infrared and Laser Engineering*, 53, 20240348 (2024).

979-8-3315-0417-5/25 $31.00 © 2025 IEEE

Performance Evaluation of 6T-SRAM in Sub-3 nm Complementary FET

Anirban Kar[1,†], Mahdi Benkhelifa[1,†], Yogesh Singh Chauhan[2], and Hussam Amrouch[1,*]

[1]Technical University of Munich; TUM School of Computation, Information and Technology;
Chair of AI Processor Design; Munich Institute of Robotics and Machine Intelligence, Munich, Germany,

[2]Department of Electrical Engineering, Indian Institute of Technology Kanpur, India,

[†]Equal contribution; [*]Email: amrouch@tum.de

Abstract—**This work investigates the integration of comple-mentary FET (CFET) transistors within static random-access memory (SRAM) to deliver aggressive bitcell area scaling and substantial performance gains for deeply scaled CMOS nodes beyond 3nm. By vertically stacking the pFET atop the nFET, CFETs achieve a profound reduction in the cell area, presenting a viable pathway to meet extreme density demands. Utilizing the industry-standard BSIM-CMG model, we carefully calibrate the CFET device's electrical characteristics against measure-ments from fabricated devices. Our calibrated model is then applied to a 6T-SRAM cell and critical peripheral circuits, including pre-charge, sense amplifier, and latch configurations. Comprehensive SPICE simulations enable a detailed assessment of SRAM performance, quantifying noise margins, access delays, and power dissipation across both read and write cycles. Our analysis further dissects the delay and power contributions along the signal path, underscoring CFET's transformative potential in advancing SRAM scalability and efficiency in leading-edge technology nodes.**

Index Terms—**CFET, SRAM, BSIM-CMG, Reliability**

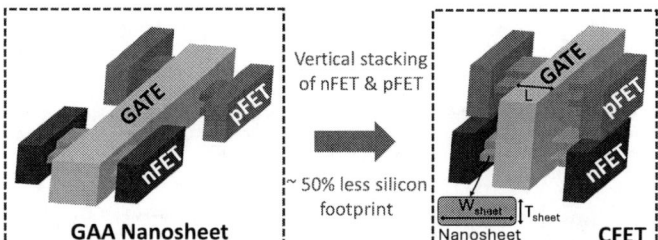

Fig. 1: Structure of the CFET. Vertical stacking of GAA nanosheet transistors, with pFET stacked above nFET, minimizes silicon footprint, enabling area-efficient standard cell layouts. This configuration reduces interconnect and routing complexity, allowing for the use of fewer metal tracks in standard cell designs.

I. INTRODUCTION

For more than five decades, the continuous scaling of semi-conductor devices has been a cornerstone of progress in the integrated circuit (IC) industry, enabling advancements in per-formance and functionality in modern electronic systems [1]. However, as technology nodes advance into the nanometer regime, conventional planar metal-oxide-semiconductor field-effect transistors (MOSFETs) encounter significant challenges in electrostatic control due to the progressively reduced chan-nel dimensions. This scaling limitation has driven the industry to shift from conventional planar technology and adopt 3-D architectures like FinFETs [2]. FinFETs, featuring a three-dimensional structure that provides gate control on three sides of the channel, improve leakage and scaling issues seen in planar MOSFETs, enabling scaling to sub-10nm nodes. However, beyond 5nm, FinFETs face challenges in device performance and VLSI layout-driven scaling [3]. A key issue is fin depopulation, where maintaining a compact cell height in scaled FinFET designs requires fewer fins per transistor while increasing fin height [4]. At 3nm, only structures with one fin per transistor are feasible, causing a significant performance limitation [5].

To enable scaling beyond FinFETs, Samsung introduced the nanosheet FET (NSFET) for the 3nm node, a major advancement in gate-all-around (GAA) transistor technology. The GAA structure, with the gate fully surrounding the channel, enhances electrostatic control beyond FinFET's three-sided contact, improving device efficiency. NSFETs feature ultra-thin, vertically stacked sheet-like channels that increase current-carrying capacity and scalability, while eliminating the lateral spacing required by FinFETs, thus optimizing chip area usage. Nonetheless, challenges in further scaling persist.

Recent analysis reveals that cell-level scaling is limited not just by transistor size but by the lateral separation between n-type and p-type devices, which is constrained by lithog-raphy in laterally placed transistors [6]. To address this, the complementary FET (CFET) architecture has been proposed, allowing advanced scaling with fewer metal tracks in standard cell layouts [7], [8]. As shown in Fig. 1, by vertically stacking pFETs over nFETs, CFET reduces the silicon footprint in standard cells by nearly 50% [9]. In high-density 6T-SRAM cells, cell height reduction has lagged logic scaling due to pin accessibility and routing congestion issues [10], [11]. The introduction of buried power rail (BPR) mitigates these issues, reducing front-side congestion and cell height [12]–[14]. When CFET is combined with BPR, leveraging vertical contacts and routing, it enables over 40% cell area reduction [15]–[18]. In this study, we have calibrated CFET device characteristics using the industry-standard BSIM-CMG model, as discussed in Section II. With this calibrated model, we have conducted SPICE simulations of a 6T-SRAM cell, incorporating the pre-charge circuit, sense amplifier, and an S-R latch – all imple-mented with CFET technology (Section III). The performance metrics evaluated in this study include delay, power dissipation during read and write operations, and noise margins of the SRAM cell. The analysis is done for high-density and high-performance design constraints. We have presented detailed

979-8-3315-0417-5/25 $31.00 © 2025 IEEE

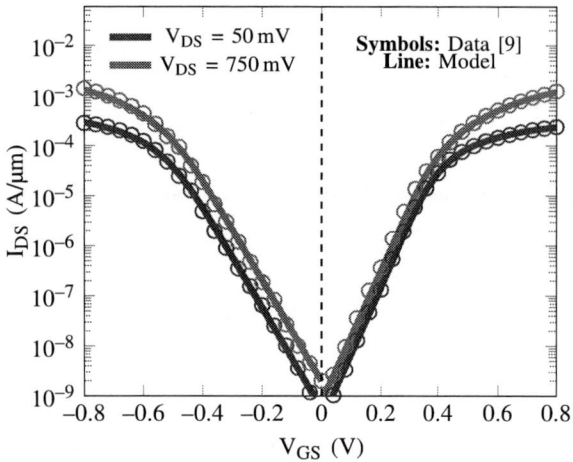

Fig. 2: Calibrated transfer characteristics of the CFET. The model parameters of the BSIM-CMG is extracted to accurately capture the I-V characteristics of both nFET and pFET at $V_{DS} = 50\,\text{mV}$ & $750\,\text{mV}$.

assessments of delay and power across different stages of the SRAM circuit—beginning with the individual SRAM cell and extending to associated sensing components, such as the sense amplifier and latch circuits. Finally, we conclude in Section IV.

II. CFET MODEL CALIBRATION

We employ the industry-standard BSIM-CMG compact model [19] for GAA transistor technology to simulate the nanosheet transistors that form the core of our CFET cell. The model is calibrated to the I-V characteristics of CFET devices fabricated by TSMC [9]. Our modeling approach begins by determining key process parameters, like equivalent oxide thickness (EOT), channel doping concentration (NSUB), channel length (L), nanosheet width (W_{sheet}), and nanosheet thickness (T_{sheet}). The CFET devices used in this work feature a channel length of 15 nm, W_{sheet} of 20 nm, and T_{sheet} of 6 nm. Once these structural parameters are set, we focus on extracting parameters that strongly influence the sub-threshold and linear behavior of the transistors. To achieve accurate sub-threshold behavior, we refine BSIM-CMG parameters such as PHIG (work function), CIT (interface traps), and CDSC (source/drain coupling capacitance).

To account for mobility degradation across operating regimes, we calibrate mobility-related parameters, including U0 (low-field mobility), UA and EU (for phonon and surface roughness scattering), UD (Coulomb scattering), and ETA-MOB (effective field parameter), ensuring accurate calibration in moderate inversion conditions at low drain biases. In the strong inversion regime, we extract source/drain series resistances to characterize linear operation accurately. At a high drain bias ($V_{DS} = 0.75\,\text{V}$), sub-threshold swing (SS) and drain-induced barrier lowering (DIBL) are modeled by tuning the CDSCD parameter to capture the drain bias sensitivity affecting SS and adjusting ETA0 to accurately represent the DIBL response. Furthermore, velocity saturation effects are incorporated by optimizing the VSAT parameter. The

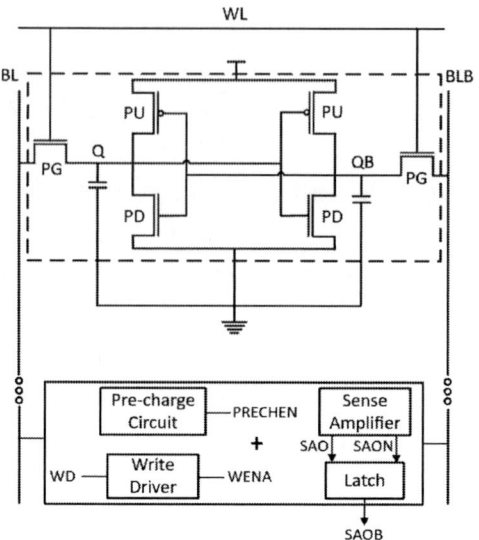

Fig. 3: Schematic of the 6T-SRAM cell along with the associated peripheral circuits. PU, PD, and PG refer to the pull-up, pull-down, and pass-gate (access) transistors, respectively. The bitline (BL) and bitline bar (BLB) are connected to the pre-charge circuit, write driver, and sense amplifier. The wordline (WL) is used to activate the PG transistors to connect the BL and BLB with the cross-coupled inverters of the SRAM.

model incorporates channel length modulation by adjusting the PCLM parameter at high drain bias. The accuracy of our model with the measured characteristics is shown in Fig. 2.

III. CFET-BASED SRAM EVALUATION

The calibrated CFET model is used to simulate the 6T-SRAM SPICE framework that was developed in [20]. The framework involves 32 SRAM cells in a single column connected to a set of peripheral circuits to perform the read and write operations. The peripherals include a pre-charge (PC) circuit to charge the BL/BLB capacitance for the read operation, a sense amplifier (SA) and latch (LA) to detect the fluctuation in BL/BLB voltage during a read operation, and a write driver to overwrite the cell during the write operation [20]. An overview of the SRAM cell is presented in Fig. 3. The framework evaluates SRAM performance in terms of noise margins, delay, and power dissipation for the hold, read, and write operations. The high-density cell with a "111" configuration prioritizes reduced silicon footprint over performance. The "111" refers to the ratio of the number of sheets in the pull-up (PU) to pull-down (PD) to pass-gate (PG) transistors. Scaling of sheets offers a trade-off between area and performance. Additional configurations, "121" and "132", are also analyzed to represent high-performance cells with stronger PD transistors.

A. Signal Noise Margin

SRAM cell resilience to noise and voltage variations is quantified by the Signal Noise Margin (SNM), which we estimate using the butterfly curve method [21]. In this work, we analyze the impact of supply voltage on the SNM of high-density cells, as shown in Fig. 4. For Hold SNM (HSNM), we observe a monotonic decrease as V_{DD} is reduced, with a

979-8-3315-0417-5/25 $31.00 © 2025 IEEE

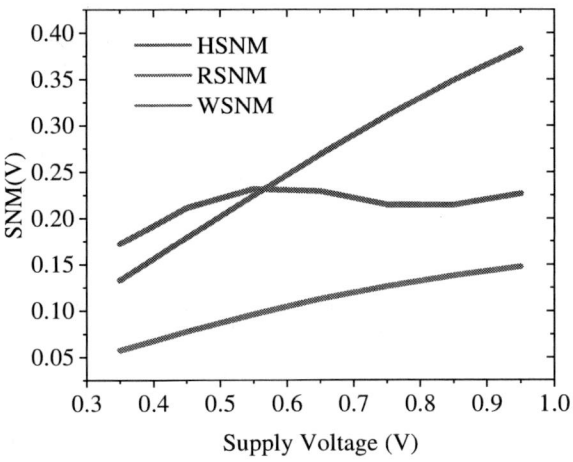

Fig. 4: Impact of variation of the supply voltage (V_{DD}) on the various noise margins of the SRAM cell.

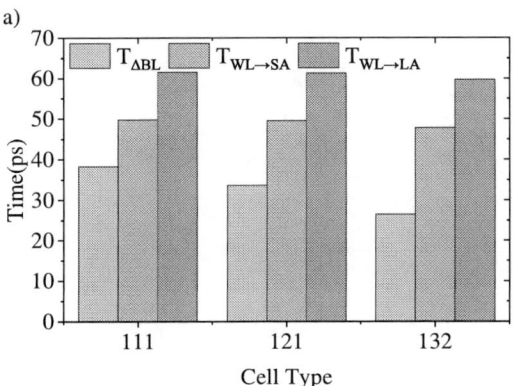

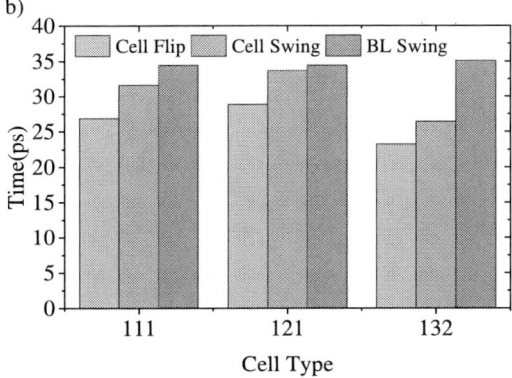

Fig. 5: Delays of the SRAM cell for different design topologies during (a) Read, and (b) Write.

value of 310 mV at the nominal operating voltage of 0.75 V. The CFET-based SRAM exhibits an excellent Read SNM (RSNM) of 126 mV at nominal voltage, indicating reliable non-destructive reads. The Write SNM (WSNM) shows minimal variation with supply voltage, which may impact performance at higher voltages. However, the WSNM of 214 mV at nominal voltage is sufficient for safe write operations.

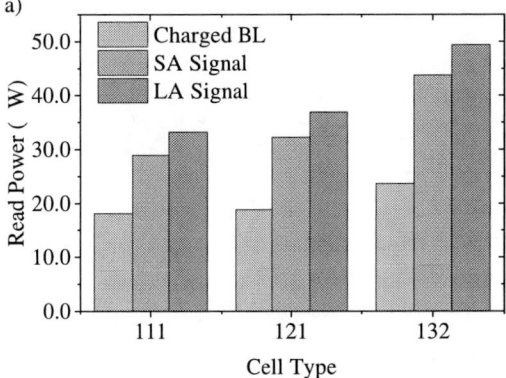

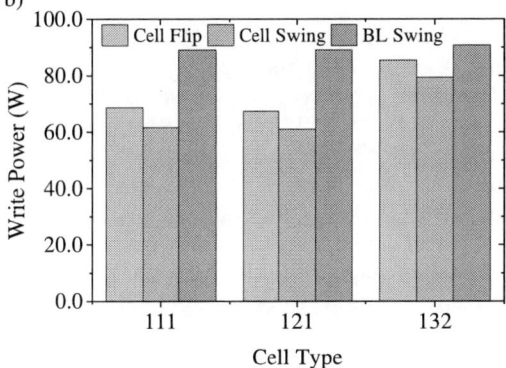

Fig. 6: Power dissipation of the SRAM cell for different design topologies during (a) Read, and (b) Write.

B. SRAM Delay & Power Dissipation

Typically, SRAM read access delay exceeds write access delay, making it a primary factor in defining the overall performance of an SRAM array [22]. The read delay timing is measured starting from the activation of the WL, with the SA enable signal triggered simultaneously with the WL signal. The delay $T_{\Delta BL}$ represents the time required for the SRAM cell to establish a 100 mV difference between BL and BLB during a read operation [23]. The high-density "111" cell exhibits a $T_{\Delta BL}$ of 38.31 ps, while the high-performance "121" and "132" cells demonstrate delay improvements of 12 % and 30.7 %, respectively, compared to the "111" cell, as shown in Fig. 5. The delay from WL activation to the SA output (denoted as $T_{WL \to SA}$) is measured from the moment the WL signal is activated until the SA output reaches 10 % of V_{DD}. Similarly, the delay from WL activation to the latch output ($T_{WL \to LA}$) is the time taken for the latch output to reach 90 % of its target voltage. For both of these delays, the performance is dominated by the SA and LA, leading to minimal differences across the cell types. We report average delays of 49.07 ps for $T_{WL \to SA}$ and 60.86 ps for $T_{WL \to LA}$ across the three cell types. Although the high-performance cells show improvements in cell-level delays, the overhead of the peripheral circuits results in nearly identical overall delay.

For write delays, we categorize the delays into cell flip, cell swing, and bitline swing delays. A cell is considered to have flipped when the output Q equals QB (Fig. 3). The cell swing delay is defined as the time it takes for Q to reach 90 % of its target voltage. Similar to read delays, the time measurement begins with the activation of the WL. However, the bitline swing delay has a different starting point, as the write driver (WD) is enabled prior to the WL activation to prepare the bitlines. As a result, the bitline swing delay is measured from the activation of the write enable (WENA) signal to the point when BL/BLB reaches 90 % of its target voltage. Fig. 6 presents the write delay results for the three cell types. Notably, the "121" cell exhibits the slowest performance for both cell flip and cell swing, being approximately 7 % slower than the high-density "111" cell. The high-performance "132" cell demonstrates a 13% reduction in cell flip delay and a 16% reduction in cell swing delay compared to the "111" cell. The bitline swing delay is consistent across the cell types, averaging 34.66 ps. While the "132" cell achieves faster operation, this comes at the cost of increased power consumption. As shown in Fig. 6, the "132" cell exhibits a 30 % increase in power consumption compared to the 28.94 µW of the "111" cell during the read operation. Additionally, the write power consumption for the cell flip operation in the "132" cell is 25 % higher than that of the other two cell types. Finally, advanced nodes like CFET might suffer from elevated temperatures and thermal management becomes essential [24].

IV. CONCLUSION

This study examines the performance of a 6T-SRAM cell designed using CFETs, with transistor characteristics modeled using the industry-standard BSIM-CMG model. The model is employed for SPICE simulations of the SRAM cell and its peripheral circuits. Key performance metrics, including SNMs, read/write delays, and power consumption, are evaluated to assess the performance of CFET-based SRAM cells at advanced technology nodes beyond 3 nm.

ACKNOWLEDGMENT

This work was supported by the German Research Foundation under grant AM 534/5-1 (NN-Thunder, 506419033).

REFERENCES

[1] M. T. Bohr, "Logic Technology Scaling to Continue Moore's Law," in *2018 IEEE 2nd Electron Devices Technology and Manufacturing Conference (EDTM)*, 2018, pp. 1–3. DOI: 10.1109/EDTM.2018.8421433.

[2] C. Auth, C. Allen, A. Blattner, *et al.*, "A 22nm high performance and low-power CMOS technology featuring fully-depleted tri-gate transistors, self-aligned contacts and high density MIM capacitors," in *2012 Symposium on VLSI Technology (VLSIT)*, 2012, pp. 131–132. DOI: 10.1109/VLSIT.2012.6242496.

[3] V. Moroz, J. Huang, and M. Choi, "FinFET/nanowire design for 5nm/3nm technology nodes: Channel cladding and introducing a "bottleneck" shape to remove performance bottleneck," in *2017 IEEE Electron Devices Technology and Manufacturing Conference (EDTM)*, 2017, pp. 67–69. DOI: 10.1109/EDTM.2017.7947509.

[4] M. G. Bardon, P. Schuddinck, P. Raghavan, *et al.*, "Dimensioning for power and performance under 10nm: The limits of FinFETs scaling," in *2015 International Conference on IC Design & Technology (ICICDT)*, 2015, pp. 1–4. DOI: 10.1109/ICICDT.2015.7165883.

[5] *IEEE International Roadmap for Devices and Systems*, 2021.

[6] P. Weckx, J. Ryckaert, E. D. Litta, *et al.*, "Novel forksheet device architecture as ultimate logic scaling device towards 2nm," in *2019 IEEE International Electron Devices Meeting (IEDM)*, 2019, pp. 36.5.1–36.5.4. DOI: 10.1109/IEDM19573.2019.8993635.

[7] E. Park and T. Song, "Complementary FET (CFET) Standard Cell Design for Low Parasitics and Its Impact on VLSI Prediction at 3-nm Process," *IEEE Transactions on Very Large Scale Integration (VLSI) Systems*, vol. 31, no. 2, pp. 177–187, 2023. DOI: 10.1109/TVLSI.2022.3220339.

[8] S. M. Y. Sherazi, J. K. Chae, P. Debacker, *et al.*, "CFET standard-cell design down to 3Track height for node 3nm and below," in *Design-Process-Technology Co-optimization for Manufacturability XIII*, J. P. Cain, Ed., International Society for Optics and Photonics, vol. 10962, SPIE, 2019, p. 1096206. DOI: 10.1117/12.2514571. [Online]. Available: https://doi.org/10.1117/12.2514571.

[9] S. Liao, L. Yang, T. Chiu, *et al.*, "Complementary Field-Effect Transistor (CFET) Demonstration at 48nm Gate Pitch for Future Logic Technology Scaling," in *2023 International Electron Devices Meeting (IEDM)*, 2023, pp. 1–4. DOI: 10.1109/IEDM45741.2023.10413672.

[10] S. Yu and T.-H. Kim, "Semiconductor Memory Technologies: State-of-the-Art and Future Trends," *Computer*, vol. 57, no. 4, pp. 150–154, 2024. DOI: 10.1109/MC.2024.3363269.

[11] V. Moroz, X.-W. Lin, P. Asenov, *et al.*, "DTCO Launches Moore's Law Over the Feature Scaling Wall," in *2020 IEEE International Electron Devices Meeting (IEDM)*, 2020, pp. 41.1.1–41.1.4. DOI: 10.1109/IEDM13553.2020.9372010.

[12] A. Gupta, O. V. Pedreira, G. Arutchelvan, *et al.*, "Buried Power Rail Integration With FinFETs for Ultimate CMOS Scaling," *IEEE Transactions on Electron Devices*, vol. 67, no. 12, pp. 5349–5354, 2020. DOI: 10.1109/TED.2020.3033510.

[13] A. Gupta, S. Kundu, L. Teugels, *et al.*, "High-Aspect-Ratio Ruthenium Lines for Buried Power Rail," in *2018 IEEE International Interconnect Technology Conference (IITC)*, 2018, pp. 4–6. DOI: 10.1109/IITC.2018.8430415.

[14] A. Gupta, H. Mertens, Z. Tao, *et al.*, "Buried Power Rail Integration with Si FinFETs for CMOS Scaling beyond the 5 nm Node," in *2020 IEEE Symposium on VLSI Technology*, 2020, pp. 1–2. DOI: 10.1109/VLSITechnology18217.2020.9265113.

[15] C. Liu and S. K. Lim, "Ultra-high density 3D SRAM cell designs for monolithic 3D integration," in *2012 IEEE International Interconnect Technology Conference*, 2012, pp. 1–3. DOI: 10.1109/IITC.2012.6251581.

[16] M. K. Gupta, P. Weckx, P. Schuddinck, *et al.*, "The Complementary FET (CFET) 6T-SRAM," *IEEE Transactions on Electron Devices*, vol. 68, no. 12, pp. 6106–6111, 2021. DOI: 10.1109/TED.2021.3121349.

[17] J. Ryckaert, A. Gupta, A. Jourdain, *et al.*, "Extending the roadmap beyond 3nm through system scaling boosters: A case study on Buried Power Rail and Backside Power Delivery," in *2019 Electron Devices Technology and Manufacturing Conference (EDTM)*, 2019, pp. 50–52. DOI: 10.1109/EDTM.2019.8731234.

[18] N. A. Lanzillo, R. Sengupta, and D. J. Dechene, "Comprehensive BEOL Performance Assessment: Interconnects Optimized for Signal Routing and Power Delivery in Advanced CMOS Technology Nodes (Invited)," in *2020 IEEE International Interconnect Technology Conference (IITC)*, 2020, pp. 34–36. DOI: 10.1109/IITC47697.2020.9515657.

[19] J. P. Duarte, S. Khandelwal, A. Medury, *et al.*, "BSIM-CMG: Standard FinFET compact model for advanced circuit design," in *Eur. Solid-State Circuits Conf.*, vol. 2015-October, 2015, pp. 196–201. DOI: 10.1109/ESSCIRC.2015.7313862.

[20] S. S. Parihar, V. M. van Santen, S. Thomann, G. Pahwa, Y. S. Chauhan, and H. Amrouch, "Cryogenic CMOS for Quantum Processing: 5-nm FinFET-Based SRAM Arrays at 10 K," *IEEE Transactions on Circuits and Systems I: Regular Papers*, vol. 70, no. 8, pp. 3089–3102, 2023. DOI: 10.1109/TCSI.2023.3278351.

[21] E. Seevinck, F. List, and J. Lohstroh, "Static-noise margin analysis of mos sram cells," *IEEE Journal of Solid-State Circuits*, vol. 22, no. 5, pp. 748–754, 1987. DOI: 10.1109/JSSC.1987.1052809.

[22] M. K. Gupta, P. Weckx, P. Schuddinck, *et al.*, "A Comprehensive Study of Nanosheet and Forksheet SRAM for Beyond N5 Node," *IEEE Transactions on Electron Devices*, vol. 68, no. 8, pp. 3819–3825, 2021. DOI: 10.1109/TED.2021.3088392.

[23] K. C. Chun, P. Jain, T.-H. Kim, and C. H. Kim, "A 667 MHz Logic-Compatible Embedded DRAM Featuring an Asymmetric 2T Gain Cell for High Speed On-Die Caches," *IEEE Journal of Solid-State Circuits*, vol. 47, no. 2, pp. 547–559, 2012. DOI: 10.1109/JSSC.2011.2168729.

[24] J. Henkel, T. Ebi, H. Amrouch, and H. Khdr, "Thermal management for dependable on-chip systems," in *2013 18th Asia and South Pacific Design Automation Conference (ASP-DAC)*, 2013, pp. 113–118. DOI: 10.1109/ASPDAC.2013.6509582.

Experiment Investigation on La₂O₃ Dipole-last Cap-less V_FB Tuning Technology Based on Nitrogen Atmosphere

Yu Wang[1,2,3], Jiaxin Yao[1,2]*, Yanzhao Wei[1,2,3], Qingzhu Zhang[1,2], and Huaxiang Yin[1,2,3]*

[1]Integrated Circuit Advanced Process R&D Center, Institute of Microelectronics of the Chinese Academy of Sciences, Beijing 100029, China

[2]Key Laboratory of Fabrication Technologies for Integrated Circuits, Chinese Academy of Sciences, Beijing 100029, China

[3]School of Integrated Circuits, University of Chinese Academy of Sciences, Beijing 100049, China

*Corresponding Author's Email: yaojiaxin@ime.ac.cn; yinhuaxiang@ime.ac.cn

Abstract

The effect of post-deposition anneal (PDA) temperature on flat-band voltage (V_{FB}) modulation using the La dipole-last cap-less technique is investigated. The results show that increasing the PDA temperature enhances V_{FB} modulation, reduces the equivalent oxide thickness (EOT), and achieves a 40% reduction in interface trap density (D_{it}). An La-assisted oxygen migration model, supported by energy-dispersive spectroscopy (EDS) and trap/detrap electron density (N_{ot}) results, is proposed, offering a promising method for multiple threshold voltage (multi-V_t) formation. Keywords: dipole-last cap-less, positive V_{FB} modulation, La₂O₃.

Introduction

To meet the low-power and high-performance demands of gate-all-around field effect transistor (GAA-FET) devices, multiple voltage (multi-V_t) technique is required.[1] Dipole engineering has become widely studied as a volume-less effective work function (EWF) tuning approach in contrast to traditional metal thickness EWF tuning.[2] La₂O₃ as a dipole-forming layer is particularly suitable for negative V_t modulation.[3] The primary methods for forming a La-dipole include the dipole-first and dipole-last processes.[4] For the dipole-first process, the dipole-forming layer was deposited on the SiO₂ interface layer (IL), followed by deposition of the high-κ (HK) layer.[5], [6] For the dipole-last process, the dipole-forming layer was deposited on the HK layer with a cap layer, then diffused into the HK/IL interface through drive-in annealing.[7] Zhu et al.[8] reported a typical dipole-last approach, where a cap layer was deposited on the HfO₂ high-κ layer, followed by a 1000°C annealing process. Finally, the α-Si and TiN cap layers were removed using wet chemistry, achieving a V_{FB} negative shift of approximately 250 mV. However, the high thermal budget and complex process flow present challenges for future device integration.

In this paper, we investigate the effect of post-deposition anneal (PDA) temperature on flat-band voltage (V_{FB}) in TiN/La₂O₃/HfO₂/IL/Si gate stacks for La dipole-last cap-less technique. This technique achieves an anomalous positive V_{FB} modulation, significantly reduces the effective oxide thickness (EOT), and effectively suppresses interface trap densities (D_{it}). A rational model is proposed to explain these experimental results.

Experiment and results

Fig. 1 illustrates the key process and the high-κ metal gate (HKMG) structure. An 8-inch n-type Si (100) wafer serves as the Si substrate. After removing the native oxide using a buffered oxide etchant (BOE), a 1-nm SiO₂ interfacial layer (IL) was grown via a chemical oxidation process. Next, a 3-nm HfO₂ high-κ layer was deposited by atomic layer deposition (ALD). For Split 1 and Split 2, a 9-cycle La₂O₃ dipole-forming layer was deposited via ALD, followed by PDA at 450°C and 650°C in an N₂ atmosphere, respectively. For the reference sample, only HfO₂ was used as the high-κ dielectric layer. Finally, a metal gate and a 100-nm tungsten (W) gap-filling layer were deposited, with electrode patterning and a metallization process at 400°C to improve interface quality. The capacitance-voltage (C-V) was measured using a Keysight B1500A semiconductor parameter analyzer in air at room temperature.

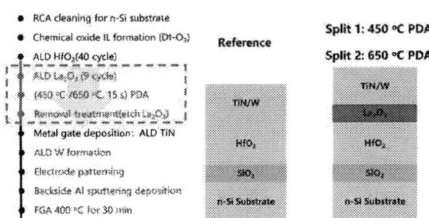

Fig. 1: Key flow and the gate stacks structure of samples.

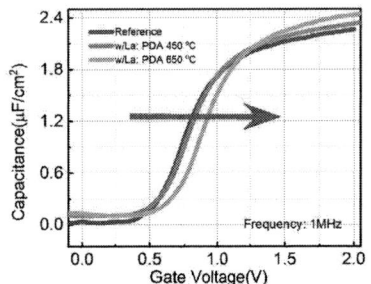

Fig. 2: C-V curves of different PDA temperature with La₂O₃ dipole-forming layer samples.

The C-V curves of the samples are shown in Fig. 2.

Compared to the reference, both split 1 and split 2 with the PDA process exhibit positive V_{FB} modulation, with modulation increasing at higher PDA temperature—an effect distinct from traditional La dipole-last technique—the V_{FB} values extracted from the C-V curves are presented in Fig. 3. The ΔV_{FB} results for split 1 and split 2, compared to the reference, show positive shift of 40 mV and 150 mV, respectively. Additionally, Fig. 3 displays the ΔEOT values for split 1 and split 2, measured at -0.3 Å and -0.9 Å, respectively, indicating that higher PDA temperature effectively reduce EOT. This reduction is particularly valuable for enhancing the gate control capability of the device.

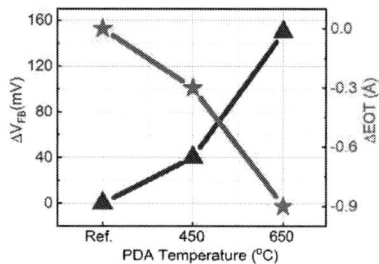

Fig. 3: The variations in V_{FB} and EOT among the samples illustrate the effect of different processing conditions on device performance.

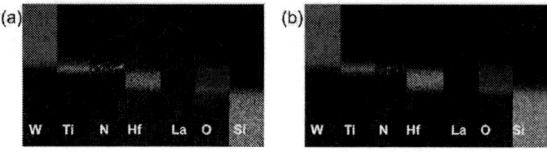

Fig. 4: The EDS mapping results for W, Ti, N, Si, La, Hf, and O elements in the gate stacks of Split 1 and Split 2

To investigate the mechanism behind the positive modulation achieved by the La dipole-last technique, energy-dispersive spectroscopy (EDS) was employed. Fig. 3 presents elemental mapping of Si, La, Hf, O, Ti, N, and W for split 2 and split 3. The EDS mapping shows negligible La distribution within the gate stacks, indicating minimal diffusion of La atom into the high-κ (HK) layer during the PDA process and the absence of residual La on the HK surface after the removal treatment. However, at high temperature, O atoms tend to diffuse. Thus, the increased formation of oxygen vacancies due to oxygen diffusion during PDA is the most likely cause of the V_{FB} positive shift.[9]

Further analysis involved evaluating the interface trap density (D_{it}) and trap/detrap electron density (N_{ot}) using C-V testing. Fig. 4(a) shows a slight increase in N_{ot} values, calculated from V_{FB} hysteresis, with rising PDA temperature, likely attributed to increased oxygen vacancy in the HK layer. Additionally, multi-frequency C-V measurements (1 kHz to 1 MHz) were used to extract D_{it} values, shown in Fig. 4(b).

The sample utilizing the La dipole-last cap-less technique exhibited a 40% reduction in D_{it} compared to the reference, demonstrating significant interface quality improvement.

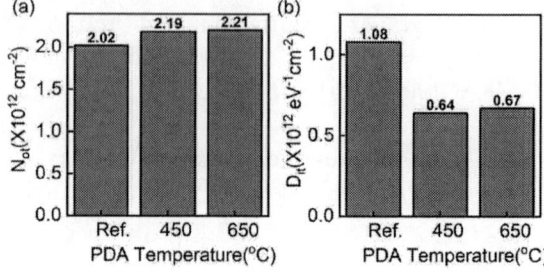

Fig. 5: The variation in the N_{ot} and D_{it} values among the samples highlights the impact of the La dipole-last cap-less technique and PDA conditions on trap characteristics.

Based on these findings, we propose an La-assisted oxygen migration model, as illustrated in Fig. 5. During the PDA process, oxygen atoms from the HK layer diffuse toward the La_2O_3 dipole-forming layer, creating oxygen vacancies within the HK. Subsequently, oxygen atoms from the IL/Si interface migrate to fill these vacancies in the HK. However, the rate of oxygen vacancy formation in the HK exceeds the rate of reduction, resulting in increased oxygen vacancies, which cause the observed V_{FB} positive modulation, IL thinning, EOT reduction, and interface quality improvement.

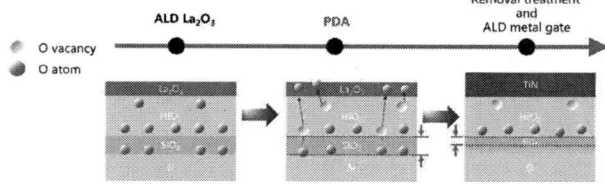

Fig. 6: Schematic of La dipole-last cap-less technology in TiN/La_2O_3/HfO_2/SiO_2/Si gate stacks.

Conclusion

In this paper, the impact of PDA temperature on V_{FB} modulation in TiN/La_2O_3/HfO_2/IL/Si gate stacks using the La dipole-last cap-less technique is investigated. A positive V_{FB} modulation effect is achieved with this technique, and the modulation strength significantly increases with higher PDA temperature. Additionally, EOT and D_{it} characteristics are comprehensively improved. Based on EDS mapping and N_{ot} results, an La-assisted oxygen migration model is proposed to explain the V_{FB} modulation mechanism. These findings demonstrate a promising approach for multi-Vt formation and interface quality improvement.

Acknowledgments

This work is supported by National Natural Science Foundation of China (No. 62304247), and Strategic Priority Research Program of Chinese Academy of Sciences (No.

XDA0330302). The authors would like to thank all staff at the Integrated Circuit Advanced Process Center of IMECAS, for their support in the fabrication on 8-inch CMOS pilot line.

References

[1] J. Yao *et al.*, "Record 7(N)+7(P) Multiple V_T s Demonstration on GAA Si Nanosheet n/pFETs using WFM-Less Direct Interfacial La/Al-Dipole Technique," in *2022 International Electron Devices Meeting (IEDM)*, San Francisco, CA, USA: IEEE, Dec. 2022, p. 34.2.1-34.2.4. doi: 10.1109/IEDM45625.2022.10019361.

[2] J. Zhang *et al.*, "High-k metal gate fundamental learning and multi-Vt options for stacked nanosheet gate-all-around transistor," in *2017 IEEE International Electron Devices Meeting (IEDM)*, San Francisco, CA, USA: IEEE, Dec. 2017, p. 22.1.1-22.1.4. doi: 10.1109/IEDM.2017.8268438.

[3] A. Fet, V. Häublein, A. J. Bauer, H. Ryssel, and L. Frey, "Lanthanum implantation for threshold voltage control in metal/high-k devices," *Microelectronic Engineering*, vol. 86, no. 7–9, pp. 1782–1785, Jul. 2009, doi: 10.1016/j.mee.2009.03.042.

[4] H. Arimura *et al.*, "Dipole-First Gate Stack as a Scalable and Thermal Budget Flexible Multi-Vt Solution for Nanosheet/CFET Devices," in *2021 IEEE International Electron Devices Meeting (IEDM)*, San Francisco, CA, USA: IEEE, Dec. 2021, p. 13.5.1-13.5.4. doi: 10.1109/IEDM19574.2021.9720527.

[5] R. Xu *et al.*, "Experimental Investigation of Ultrathin Al_2O_3 Ex-Situ Interfacial Doping Strategy on Laminated HKMG Stacks via ALD," *IEEE Trans. Electron Devices*, vol. 69, no. 4, pp. 1964–1971, Apr. 2022, doi: 10.1109/TED.2022.3152976.

[6] Y. Wei *et al.*, "Sub-5-Å La_2O_3 In Situ Dipole Technique for Large V_{FB} Modulation With EOT Reduction and Improved Interface for HKMG Technology," *IEEE Trans. Electron Devices*, vol. 71, no. 1, pp. 746–751, Jan. 2024, doi: 10.1109/TED.2023.3335900.

[7] Ruqiang Bao *et al.*, "Extendable and Manufacturable Volume-less Multi-Vt Solution for 7nm Technology Node and Beyond," *2018 IEEE International Electron Devices Meeting (IEDM)*, 2018.

[8] M. Zhu *et al.*, "Study of Lanthanum Diffusion in HfO_2 -Based High-k Gate Stack," *ECS Trans.*, vol. 85, no. 8, pp. 131–136, Apr. 2018, doi: 10.1149/08508.0131ecst.

[9] M. Togo *et al.*, "Novel N/PFET Vt control by TiN plasma nitridation for aggressive gate scaling," in *2016 IEEE Symposium on VLSI Technology*, Honolulu, HI, USA: IEEE, Jun. 2016, pp. 1–2. doi: 10.1109/VLSIT.2016.7573436.

979-8-3315-0417-5/25 $31.00 © 2025 IEEE

A Digital Twin for Advanced Manufacturing of Materials

Dipayan Sanpui[1,2], Anirban Chandra[1,2,*], Sukriti Manna[1,2] Henry Chan[1,2], and Subramanian K.R.S. Sankaranarayanan[1,2]

[1] Center for Nanoscale Materials, Argonne National Laboratory, Lemont, Illinois 60439, USA.

[2] Department of Mechanical and Industrial Engineering, University of Illinois, Chicago, Illinois 60607, USA.

*Currently at Shell International Exploration and Production Inc., Boston, Massachusetts, 02210, United States

Abstract

There is a clear need in advanced manufacturing to develop Digital Twins that combine AI/ML with physics-based multiscale models to understand structure-property-processing relationships in materials. Additive manufacturing (AM) is one such transformative technique that allows creation of components with complex geometries layer-by-layer that are prohibitively difficult to achieve with traditional manufacturing techniques. Existing commercial and open-source software allow users to carry out materials simulations but lack predictive power to model manufacturing processes with complex thermal history and do not allow users the flexibility to design microstructures and hence functionality tailored to suit their specific needs. We aim to overcome this critical barrier by developing a high-performance user-friendly Digital Twin that will enable end users to exhaustively explore, identify and design time-dependent AM protocols that achieve tailor-made microstructures. A subset of the processing conditions that lead to the most promising microstructures and functionality will then be experimentally manufactured and validated. Our Digital Twin employs a kinetic Monte Carlo (KMC) based model of the AM process to simulate microstructural evolution for a diverse set of experimentally relevant processing conditions and uses AI/ML to explore the relationship between microstructural features and processing conditions. We also present probabilistic machine learning methodologies, namely Gaussian Process Regression (GPR) and Probabilistic Bayesian Neural Networks (BNNs) algorithms for precise dimensional control and uncertainty quantification in additive manufacturing.

Keywords: Digital Twin, Artificial Intelligence, Additive Manufacturing, Microstructure, Defects

Introduction

Unlike conventional manufacturing, AM is a complex history-dependent process; a workpiece is built up from feedstock material, such as powder or wire, and a targeted-energy deposition source, such as an electron beam or laser [1]. The final microstructure and hence the mechanical properties of AM samples are a strong function of processing parameters. Setting AM process parameters is mostly done by trial-and-error which is time consuming, costly, highly subjective, and machine- and material-specific. Process control of the microstructure remains elusive. AM metal parts typically suffer from inconsistent quality issues which arise from difficult-to-measure features, such as internal residual stresses and inhomogeneous and anisotropic microstructure [2]. Machine-learning approaches are being used to relate multiple processing conditions to thin-film microstructures and create high-dimensional structure zone diagrams. Combined with other advanced techniques such as active-learning and high-throughput experiments, such approaches can automate and accelerate the discovery of targeted thin-film materials. Despite these advances, a key limitation of current process-structure mappings is that they do not yet consider processing conditions that vary during the AM process which precludes autonomous control. Such mappings are potentially missing out on families of AM protocols that could efficiently and accurately lead to the design of thin-film microstructures, as well as achieve complex, hierarchical microstructures with features across multiple length scales.

Our Digital Twin will allows users to deploy AI algorithms that learn both quasi-static and time-dependent protocols for understanding structure-processing relationships in AM processes and thus achieve precise control over complex, hierarchical microstructures with features across multiple length scales.

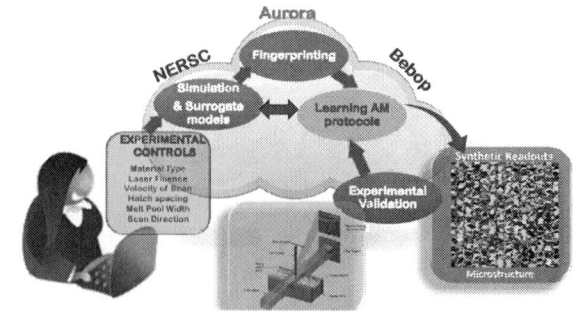

Fig. 1: Our Vision for a Digital Twin for Additive Manufacturing Processes

In the Digital Twin shown in Fig. 1, we envision that users

would be allowed two modes of operation (i) quasi-static (ii) time-dependent manufacturing protocols to perform both forward and inverse design tasks. The quasi-static mode would allow user to map multiple processing conditions to the final microstructures and create high-dimensional structure zone diagrams –this mapping may not be unique and poses challenges which would be addressed via autoencoders, attention networks and reinforcement learning with decision trees. Learning the time-dependent processing protocols is grand challenge in manufacturing science and we aim to address this by coupling Kinetic Monte Carlo, surrogate ML models, newer microstructure fingerprints and reinforcement learning to identify and design time-dependent deposition protocols that achieve tailor-made microstructures.

Forward Mesoscale Modeling of the AM Manufacturing Protocols: The Digital Twin relies on accurately modeling spatiotemporal additive manufacturing (AM) experiments in a virtual environment. Efficient and precise models are crucial for this purpose. To overcome the experimental challenges of exploring the impact of various parameters on microstructure, we propose creating a digital twin by simulating the AM process using a Kinetic Monte Carlo (KMC) based Potts model. The KMC Potts model describes grain evolution within the melt pool. By refining the computational model with experimental input, we can replicate the experimental AM processes successfully. A detailed description of this methodology can be found in our recent work [4] and a schematic of the workflow within our Digital Twin us shown in Fig. 2. This method has been tremendously successful in replicating experimental AM processes.

To initiate a digital twin for the AM process, we use a Kinetic Monte Carlo (KMC) simulation based on the Potts model. This on-lattice approach models local grain evolution as the molten pool moves across fixed lattice sites, with grain growth driven by minimizing grain boundary energy. Only lattice sites within the molten zone are active in the simulation, while those outside remain inactive. This approach effectively replicates experimental AM processes, with essential hyperparameters calibrated from Rodgers et al. [3]. We then analyze KMC-generated microstructure images with automated featurization algorithms to develop metrics that distinguish different microstructures. Hyperparameters that are necessary to define and calibrate this computational model will be refined based on experimental input. Melt pool morphologies visualized at APS during in-situ experiments under different process conditions (laser power, scanning speed, and number of laser passes) will be used to calibrate the molten boundary zone for a variety of materials (such as Ti64, aluminum at steel). Calibration of the molten boundary zone will be done using melt pool morphologies obtained from in-situ experiments, and featurization algorithms will be employed to analyze the resulting microstructures.

Surrogate ML Models to Overcome Spatiotemporal Challenges in AM Simulations & Experiments: Three-dimensional (3D) KMC simulations of AM processes demand substantial computational resources due to their layer-by-layer modeling of component evolution. The computational cost can take hours of wall time, even for 2D simulations that provide microstructural morphologies in the build direction. Obtaining reliable insights requires averaging over multiple independent KMC simulations for the same processing condition due to their stochastic nature. To overcome these spatiotemporal challenges, we are developing uncertainty-aware surrogate ML models. We will train variational autoencoders (VAEs) on a comprehensive dataset of 2D microstructural images, employing data augmentation strategies to increase training samples. The VAE-based ML framework will take input processing conditions, generate corresponding latent space encodings, and decode them to create the microstructure, capturing the stochasticity of KMC simulations. Complex architectures such as disentangled VAEs, conditional VAEs, GANs, and Diffusion models will also be explored. To ensure successful surrogate model training, our team will implement an active learning workflow, concurrently training, evaluating, and gathering data to fill blind spots. This approach maximizes the efficient use of HPC resources and minimizes the time between data acquisition and identification of the optimal next simulation.

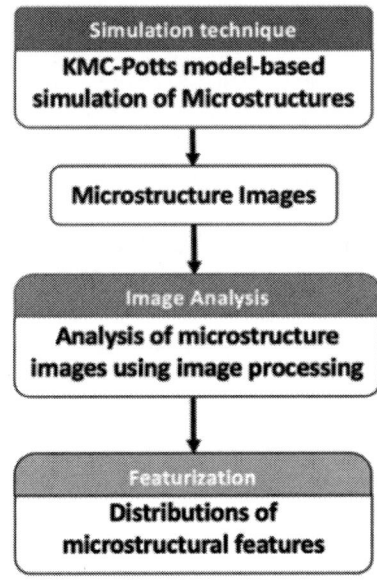

Fig. 2. Schematic representation of the overall computational workflow

Fingerprinting microstructures and feature extraction: In the digital twin, a robust computational model is crucial for simulating AM processes and enabling rapid design and testing of various components. However, to fully utilize computational methods, automated and efficient characterization or fingerprinting algorithms are necessary. Understanding key features in resulting structures is vital for assessing the impact of different manufacturing processing conditions. Automated featurization is essential for inverse design problems involving microstructures, enabling the definition of a scoring function and high-throughput exploration of the processing space [5].

Our workflow in Fig. 2 builds upon automated image processing techniques using Python libraries (OpenCV, NumPy, SciPy, Scikit-image, and Pandas) to analyze simulated images. Custom algorithms and off-the-shelf toolkits like the Watershed algorithm are used to extract descriptors and separate individual grains. The featurization workflow has been successfully applied to AM microstructural images and benchmarked against experimental results. Kernel density estimation (KDE) is employed for visualization and data analysis, approximating the histogram with a continuous function [6]. Importantly, our methodology is agnostic to the source of microstructural images, allowing extension to experimentally obtained images from synchrotron in-situ experiments, including porosity and cracking observations in AM microstructures.

Conclusion

We present a Digital Twins framework for featurizing microstructural images to identify key descriptors and relate them to input processing conditions. Our methodology is adaptable to any source of microstructural images. While our analysis focuses on images generated synthetically via KMC simulations, it can be easily extended to experimentally obtained data. Using high-throughput KMC simulations, we explore various AM processing parameters, including laser power, spot size, scan rate, and scan patterns, and analyze the resulting microstructures with an automated featurization workflow. We find that at least four descriptors—aspect ratio, perimeter-area ratio, grain orientation, and equivalent diameter—are essential for distinguishing microstructures generated under different conditions. Additionally, due to the stochastic nature of KMC simulations, averaging descriptors across independent runs is crucial for meaningful qualitative and quantitative insights. Our analysis demonstrates that for effective optimization or inverse design of microstructural images, at least four descriptors—orientation, aspect ratio,

perimeter-area ratio, and equivalent diameter—are essential. This study lays a foundational framework for developing customized metrics tailored to inverse design challenges, offering a promising direction for future research. Furthermore, our application of a Probabilistic Bayesian Neural Network (BNN) to an additive manufacturing dataset, which quantifies both epistemic and aleatoric uncertainties, enhances the robustness of manufacturing design. Together, these insights pave the way for more reliable, data-driven approaches to material and process design in advanced manufacturing.

Acknowledgments

This work performed at the Center for Nanoscale Materials, a U.S. Department of Energy Office of Science User Facility, was supported by the U.S. DOE, Office of Basic Energy Sciences, under Contract No. DE-AC02-06CH11357. This material is based on work supported by the DOE, Office of Science, BES Data, Artificial Intelligence, and Machine Learning at DOE Scientific User Facilities program (ML-Exchange and Digital Twins). All authors thank Suvo Banik and Partha Sarathi Dutta for their useful comments and discussion during the research activity. This research used resources of the National Energy Research Scientific Computing Center (NERSC), a US Department of Energy Office of Science User Facility located at Lawrence Berkeley National Laboratory, operated under Contract No. DE-AC02-05CH11231.

References

[1] C.Y. Yap, C.K. Chua, Z.L. Dong, Z.H. Liu, D.Q. Zhang, L.E. Loh, S. L. Sing. Review of selective laser melting: Materials and applications. Appl. Phys. Rev., 2 (4), 041101 (2015).

[2] F.R. Medina, R.B. Wicker. Metal fabrication by additive manufacturing using laser and electron beam melting technologies. J. Mater. Sci. Technol., 28 (1), 1-14 (2012).

[3] T. M. Rodgers, J. D. Madison, and V. Tikare. Simulation of metal additive manufacturing microstructures using kinetic Monte Carlo. Computational Materials Science, 135, 78-89 (2017).

[4] D. Sanpui, A. Chandra, S. Manna, P. S. Dutta, M. K.Y. Chan, H. Chan, S.K.R.S. Sankaranarayanan, Understanding structure-processing relationships in metal additive manufacturing via featurization of microstructural images, Computational Materials Science, 231, 112566 (2024).

[5] H. Chan, M. Cherukara, T.D. Loeffler, B. Narayanan, S.K. Sankaranarayanan. Machine learning enabled autonomous microstructural characterization in 3D samples. npj Comput. Mater., 6 (1), 1-9 (2020).

[6] Y.C. Chen. A tutorial on kernel density estimation and recent advances Biostatistics & Epidemiology, 1 (1), 161-187 (2017).

Strategies for Reliable Emerging Memories and Their Applications

Shinhyun Choi[1]

[1]The school of Electrical Engineering, Korea Advanced Institute of Science and Technology (KAIST),
Republic of Korea

Abstract

This talk will cover how to achieve reliable emerging memory devices and real-time demonstration of ReRAM-based hardware platform.

Keywords: ReRAM, PRAM, Emerging Memory

Introduction

Artificial intelligence (AI) will enable machines to think and solve complex tasks like human beings. In recent years, artificial neural networks have improved recognition and classification accuracy. However, state-of-the-art deep learning algorithms require large network models with multiple layers, which pose significant challenges for complementary metal-oxide-semiconductor (CMOS) implementation due to limitations in conjoining computation, memory, and communication requirements in large networks. As an alternative hardware platform, emerging memories have been proposed for weight storage and fast parallel neural computing with low power consumption. The parallelism property of the crossbar arrays for matrix-vector multiplication enables significant acceleration of core neural computations. In this talk, Prof. Choi will present a systematic study on the fundamental understanding of emerging memristive devices. He will talk about the approach how to achieve highly reliable artificial neurons and synapses for neuromorphic computing which can be a key step paving the way towards post von Neumann computing. In addition, he will also introduce the application of developed crossbar network, which suggests potential applications of emerging memory/computing device-based network to effective data processing for solving real-world problems. He will also talk about his recent work on phase change memory that shows low power consumption with cheap fabrication process. Finally, he will discuss the projections and future directions.

Acknowledgments

This work was supported by National R&D Program through the National Research Foundation of Korea (NRF) funded by Ministry of Science and ICT (2022M3I7A2078273, 2022M3F3A2A01072851, 2020R1C1C1007464, and 2021M3F3A2A01037858) and Nanomedical Devices Development Project of NNFC.

References

[1] S. Choi, S. H. Tan, Z. Li, Y. Kim, C. Choi, P. Y. Chen, H. Yeon, S. Yu and J. Kim, "SiGe epitaxial memory for neuromorphic computing with reproducible high performance based on engineered dislocations", Nature Materials, 17(4), pp. 335-340

[2] S. Choi, S. Park, S. Seo and S. Choi, "Reliable multilevel memristive neuromorphic devices based on amorphous matrix via quasi-1D filament confinement and buffer layer", Science Advances, 8(3), eabj7866 (2022)

[3] S. O. Park, H. Jeong, J. Park, J. Bae and S. Choi, "Experimental demonstration of highly reliable dynamic memristor for artificial neuron and neuromorphic computing", Nature Communications, 13(1), 2888 (2022)

[4] S. O. Park, S. Hong, S. J. Sung, D. Kim, S. Seo, H. Jeong, T. Park, W. J. Cho, J. Kim and S. Choi, "Phase-change memory via a phase-changeable self-confined nano-filament", Nature, 628(8007), pp. 293-298 (2024)

[5] J. Bae, C. Kwon, S. O. Park, H. Jeong, T. Park, T. Jang, Y. Cho, S. Kim and S. Choi, "Tunable ion energy barrier modulation through aliovalent halide doping for reliable and dynamic memristive neuromorphic systems", Science Advances, 10(23), eadm7221 (2024)

979-8-3315-0417-5/25 $31.00 © 2025 IEEE

A Hybrid Design Method of Lamb Wave Mode Filter Based on Machine Learning and COM Model

Lihang Liao[1], Chen Ma[1], Zhiyu Wang[1], Xiangyu Zou[3], Zhiyuan Wang[1], Xi He[1],
Feixuan Huang[1], Qinghua Ren[1], Fengyuan Yang[1], Yiming Ma[1], Jianlin Chen[1], and Nan Wang[1,2]

[1]School of Microelectronics, Shanghai University, Shanghai, China
[2]Collaborative Innovation Center of Intelligent Sensing Chip Technology,
Shanghai University, Shanghai, China
[3]Univista, Shanghai, China

ABSTRACT

This paper reports a acoustic filter design process using a hybrid approach based on machine learning and COM model. Machine learning is deployed to accurately predict the filter structure that meets performance requirements, while the COM model is used to greatly simplify the process of simulating impedance characteristics. The model is validated by a specific Lamb Wave Resonator (LWR) and a filter for the N1 frequency band, demonstrating great potential for fast filter design for 5G/6G applications.

Keywords:Machine-learning, neural networks, COM, Lamb Wave Resonator

INTRODUCTION

The field of Lamb wave resonator (LWR) has attracted a great deal of research attention, due to its capability of lithographic frequency tunability, which could potentially lead to multi-band filters on a single chip. However, the development of commercial products remains limited, primarily due to the inherently low coupling coefficient. Although recent studies have made advancements in enhancing the coupling coefficient [1]–[4], the realization of commercial devices remains challenging, largely due to the following two critical challenges encountered in the filter design process. First, it is not straight forward to derive the performance parameters of each resonator from the filter's specifications. Second, the LWR lack equivalent modeling determined by the device's structural parameters, which leads to the inability to perform simulations involving multiple LWR.

First, the performance of a filter can be regarded as a functional relationship between input and output, and the design parameters are those that determine this functional relationship. This process is called regression analysis. Machine learning has mature applications in this field, showing great potential in eliminating the dependence on designers' experience during the design

process [10]–[13]. Also, synthesis method can obtain the performance parameters of each resonator (resonant frequency, static capacitance C_0) based on the system function predicted by machine learning [5]–[9].

Secondly, because the LWR possess the same stopband dispersion characteristics as the coupling-of-modes (COM) model [14]–[16], the COM model can be used to describe the LWR. With the help of the COM model, obtaining the impedance characteristics of LWR does not require frequency domain simulation of finite element method (FEM), greatly reducing the difficulty of acquiring impedance characteristics.

As such, this paper introduces a new design procedure which utilizes machine learning to predict the filter function, while using synthesis method to convert the filter function into the performance parameters of the resonators. In addition, in response to the particularities of LWR, a LWR COM model method capable of rapidly extracting parameters is proposed.

MACHINE LEARNING IN FILTER DESIGN

A. Machine Learning Architecture

In this paper, the filter system function will be predicted through the S-parameter sequence. Since the input data and output data in the dataset correspond to each other, the back-propagation algorithm can be chosen to complete this prediction.

B. Dataset and normalization

In order for the dataset to be applicable to different frequency bands, the filter system functions are all low-pass filter system functions under normalized frequency conditions. The frequency transformation formula between band-pass filters and low-pass filters is as follows:

$$\Omega = \alpha\left(\frac{\omega}{\omega_0} - \frac{\omega_0}{\omega}\right); \alpha = \frac{\omega_0}{\omega_2 - \omega_1} \tag{1}$$

where α is the inverse of the relative bandwidth, ω is the unnormalized frequency variable and ω_0 is the center frequency of the band-pass. ω_1 and ω_2 are the pass-band edges.

The output data in the dataset is set as an array composed of all the zeros of the filter. A sufficient number of different zero distributions are generated through random arrays. The S-parameters of the filter will serve as the input data in the dataset. The generated zeros are used to calculate the S-parameter sequence of the filter through the Chebyshev function. To reduce the size of the dataset, it is necessary to sample the S-parameters. The S_{11} parameter selects the maximum value within the passband as the representative of return loss (RL). At the same time, the S_{21} parameter selects the maximum value within each sampling interval as the representative of the data in that interval. In summary, the sampled S-parameter sequences serve as the input data, and the zero-point distribution serves as the output data.

THE COM MODEL OF THE LWR

C. COM model

The COM model is a branch of wave propagation theory in periodic media. This theory proposes that the complexity of the problem can be simplified by considering only the harmonics that are strongly coupled in the periodic medium [15]. In a medium with a periodic grating, considering the excitation of voltage V and current I, the wave function can be represented as:

$$
\begin{cases}
\frac{\mathrm{d}R(x)}{\mathrm{d}x} = -i\delta R(x) + i\kappa S(x) + i\alpha V \\
\frac{\mathrm{d}S(x)}{\mathrm{d}x} = -i\kappa R(x) + i\delta S(x) - i\alpha V \\
\frac{\mathrm{d}I(x)}{\mathrm{d}x} = -2i\alpha R(x) - 2i\alpha S(x) - i\omega C V \\
\delta = \frac{\omega}{\nu} - \frac{\pi}{p} - i\gamma
\end{cases}
\tag{2}
$$

Here, $R(x)$ is the amplitude of the wave propagating in the forward direction, $S(x)$ is the amplitude of the wave resulting from reflection, κ is the reflection coefficient, α is the transmission coefficient, C is the static capacitance per unit length of the grating in one period, ν is the wave velocity, p is half the length of the grating period, and γ is the loss of the material.

From the equation, the admittance of the resonator can be derived as:

$$
Y = i\omega C_0 \left[1 - \frac{4|\alpha|^2}{\omega C} \frac{1}{\delta + \kappa} \right]
\tag{3}
$$

D. Parameter extraction

Taking the resonator structure in Fig. 1 as an example [3], The length of the electrode finger $L = 150\mu m$. In Eq.3, the parameters to be extracted are: κ, α, C, and ν, in which C can be obtained through Stationary simulations in COMSOL software.

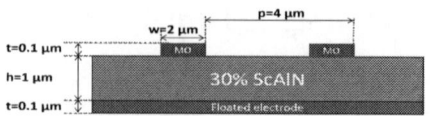

Fig. 1. Cross-section illustration of the structure a Lamb wave resonator

Due to the stopband effect, two characteristic modes as shown in the Fig. 2 will appear at the upper and lower edges of the stopband. This phenomenon occurs both when the electrode fingers are short-circuited and open-circuited. Here, the characteristic frequencies corresponding to the modes in the short-circuit condition are denoted as f_{sc-} and f_{sc+}, respectively. The characteristic frequencies corresponding to the modes in the open-circuit condition are denoted as f_{oc-} and f_{oc+}, respectively. The "+" represents the higher characteristic frequency.

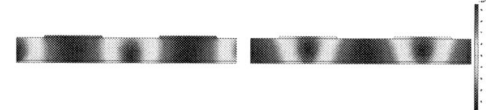

Fig. 2. Displacement diagram of a typical LWR

The parameters of the COM model and the four characteristic frequencies have the following relationship:

$$
\begin{cases}
\nu = p(f_{sc+} + f_{sc-}) \\
|\kappa| \lambda = 2\pi \frac{f_{sc+} - f_{sc-}}{f_{sc+} + f_{sc-}} \\
|\alpha| = \sqrt{\omega C \frac{\lambda^2 \pi}{W} \left(\frac{f_{oc+} + f_{oc-}}{f_{sc+} + f_{sc-}} - 1 \right)}
\end{cases}
\tag{4}
$$

The admittance characteristics obtained from the COM model and the results from FEM are compared as shown in the Fig3.

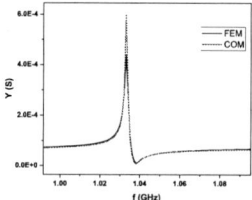

Fig. 3. Displacement diagram.

DESIGN EXAMPLE

In order to verify the proposed design process, a sixth-order ladder filter for the N1 frequency band based on the structure shown in Fig.1 is designed. Table I shows the input data.

TABLE I: Input data and predicted results

Ω	1-1.1	1.1-1.2	1.2-1.3	1.3-1.4
S_{21}(dB)	-5	-23	-34	-40
Ω	1.4-1.5	1.5-2	2-2.5	2.5-3
S_{21}(dB)	-54	-86	-63	-60
Ω	3-3.5	3.5-4	4-4.5	4.5-5
S_{21}(dB)	-54	-54	-51	-50
RL	10			

Each LWR only changes the parameter p and the number of electrode fingers (n), with all other parameters being kept constant, i.e., $h = 0.85\mu m$, $t = 0.9\mu m$, $L = 100\mu m$ and $W = p/2\mu m$. Table II shows the resonator structural parameters, and Fig. 4 shows the S-parameter plot of the filter.

TABLE II: Resonator structural parameters

$p(\mu m)$	1.45	1.61	1.44	1.6	1.44	1.58
n	88	337	38	314	38	154

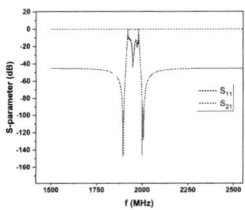

Fig. 4. S-parameter of the filter.

CONCLUSION

This paper introduces machine learning and the COM model into the design process, reducing the complexity and cost of Lamb wave filter design. This method has great potential for designing Lamb wave filters with different frequency bands and various resonator structures for 5G/6G applications.

REFERENCES

[1] C.Sun, B. Soon, Y. Zhu, N. Wang, S. Loke, X.Mu, J. Tao, A. Gu, "Methods for improving electromechanical coupling coefficient in two dimensional electric field excited AlN Lamb wave resonators," Appl. Phys. Lett. 22 June 2015; 106 (25): 253502.

[2] Y. Zhu, N. Wang, C. Sun, S. Merugu, N. Singh and Y. Gu, "A High Coupling Coefficient 2.3-GHz AlN Resonator for High Band LTE Filtering Application," in IEEE Electron Device Letters, vol. 37, no. 10, pp. 1344-1346, Oct. 2016, doi: 10.1109/LED.2016.2602852.

[3] Y. Zhu, N. Wang, G. Chua, C. Sun, N. Singh and Y. Gu, "ScAlN-Based LCAT Mode Resonators Above 2 GHz With High FOM and Reduced Fabrication Complexity," in IEEE Electron Device Letters, vol. 38, no. 10, pp. 1481-1484, Oct. 2017, doi: 10.1109/LED.2017.2747089.

[4] N. Wang, Y. Zhu, G. Chua, B. Chen, S. Merugu, N. Singh and Y. Gu, "Over 10% of k_{eff}^2 Demonstrated by 2-GHz Spurious Mode-Free $Sc_{0.12}Al_{0.88}N$ Laterally Coupled Alternating Thickness Mode Resonators," in IEEE Electron Device Letters, vol. 40, no. 6, pp. 957-960, June 2019, doi: 10.1109/LED.2019.2910836.

[5] A. Giménez, J. Verdú and P. De Paco Sánchez, "General Synthesis Methodology for the Design of Acoustic Wave Ladder Filters and Duplexers," in IEEE Access, vol. 6, pp. 47969-47979, 2018, doi: 10.1109/ACCESS.2018.2865808.

[6] S. -Y. Tseng and R. -B. Wu, "Synthesis of Chebyshev/Elliptic Filters Using Minimum Acoustic Wave Resonators," in IEEE Access, vol. 7, pp. 103456-103462, 2019, doi: 10.1109/ACCESS.2019.2930904.

[7] S. Cano, L. Acosta, C. Caballero, J. Verdú and P. de Paco, "Synthesis of Wideband Filters Based on Acoustic Wave Transversal and Ladder Topologies," 2024 IEEE MTT-S International Conference on Microwave Acoustics & Mechanics (IC-MAM), Chengdu, China, 2024, pp. 101-104, doi: 10.1109/IC-MAM60575.2024.10539025.

[8] E. Guerrero, L. Acosta, J. Verdú and P. de Paco, "Direct Synthesis of Acoustic Wave Multiplexers Built on Fully Canonical Multiport Functions," in IEEE Transactions on Microwave Theory and Techniques, vol. 71, no. 4, pp. 1391-1401, April 2023, doi: 10.1109/TMTT.2022.3222426.

[9] E. Guerrero, J. Verdú and P. de Paco, "Synthesis of Extracted Pole Filters With Transmission Zeros in Both Stopbands and Nonresonant Nodes of the Same Nature," in IEEE Microwave and Wireless Components Letters, vol. 31, no. 1, pp. 17-20, Jan. 2021, doi: 10.1109/LMWC.2020.3035848.

[10] R. Guo, R. Xu, Z. Wang, F. Sui and L. Lin, "Accelerating Mems Design Process Through Machine Learning from Pixelated Binary Images," 2021 IEEE 34th International Conference on Micro Electro Mechanical Systems (MEMS), Gainesville, FL, USA, 2021, pp. 153-156, doi: 10.1109/MEMS51782.2021.9375315.

[11] R. Guo, F. Sui, W. Yue, Z. Wang, S. Pala, k. Li, R. Xu and L. Lin, "Deep learning for non-parameterized MEMS structural design," Microsyst Nanoeng 8,91(2022). https://doi.org/10.1038/s41378-022-00432-9

[12] J. -l. An, T. -t. Liu and Y. Gao, "FBAR Filter Structural Parameters Optimizing With Deep-Learning Approach," 2020 15th Symposium on Piezoelectrcity, Acoustic Waves and Device Applications (SPAWDA), Zhengzhou, Henan Province, China, 2021, pp. 428-432, doi: 10.1109/SPAWDA51471.2021.9445561.

[13] M. Almalkawi and J. Caron, "Domain Decomposition Based Artificial Neural Networks (ANNs) Modeling of Acoustic Wave Resonators and Filters," 2021 IEEE 21st Annual Wireless and Microwave Technology Conference (WAMICON), Sand Key, FL, USA, 2021, pp. 1-4, doi: 10.1109/WAMICON47156.2021.9443601.

[14] V. Yantchev and I. Katardjiev, "Propagation characteristics of the fundamental symmetric Lamb wave in thin aluminum nitride membranes with infinite gratings," J. Appl. Phys. 15 October 2005; 98 (8): 084910.

[15] V. Plessky and J. Koskela, "Coupling-of-modes Analysis of SAW Devices," International Journal of High Speed Electronics and Systems 10 (2000): 867-947.

[16] V. Yantchev, L. Arapan and I. Katardjiev, "Coupled Mode approach to the analysis of thin film S0 lamb wave resonators," 2009 IEEE International Frequency Control Symposium Joint with the 22nd European Frequency and Time forum, Besancon, France, 2009, pp. 79-84, doi: 10.1109/FREQ.2009.5168146.

Engineered Substrates for 3D RF Front Ends

Luis ANDIA[1]

[1]Soitec Microelectronics, Singapore

Abstract

3D IC integration is playing an increasing role in ensuring modern semiconductor systems processing, sensing, memory and connectivity requirements. This paper discusses the drivers for 3D monolithic integration adoption for RF Front Ends (RFFE). It provides an overview of its adoption and guidelines for its future evolution from the perspective of the engineered substrates on top of which they are built.

Keywords: 3D integration, SOI substrates, wafer bonding, mobile communications, RFIC.

Introduction

Over the recent years, many techniques and technologies have been successfully introduced to stack patterned substrates containing dies of functional ICs. For the purposes of this paper we look at two identified families: 3D System-in-Package (3D-SiP) and 3D monolithic IC (3D-IC). The first one is often considered as an intermediate step between 2D and 3D – 2.5D – as individual 2D dies are assembled by packaging teams on an interposer used to interconnect them. 3D-IC, on the other hand, is considered as a monolithic integration and it's manufactured at semiconductor foundries using standard high volume manufacturing processing steps.

RFFE 2D, 2.5D and 3D integration

Due to their reduced volume footprint and very advanced processing, sensing, memory and connectivity requirements, smartphones have played a key role in the adoption and evolution of semiconductor integration density, starting from 2D, going to 2.5D and more recently 3D.

A. 2D RFFE integration

The cellular connectivity evolution from 3G to 4G introduced new frequency bands and the need for new RF circuitry to receive and transmit them. RF-SOI is considered as one essential technology in 4G, and later 5G, systems RFFEs as it has enabled the integration of low insertion loss, high isolation and high linearity switches routing the RF signal through the several paths in between antennas and receivers and transmitters. Thanks to this performance and other additional benefits, RF-SOI is today used to integrate RF control, Low Noise Amplifiers (LNA) and switches and even Power Amplifiers (PA) [1] in advanced RFFEs ICs. The continuous RF-SOI performance enhancement plays a significant role in the increasing 2D integration density.

B. 2.5D RFFE integration

With the adoption of 5G in the early 2020s, the need for higher RFFE integration increased drastically as new frequency bands were introduced. A modern 5G smartphone uses more than 8 antennas and contains 5 to 7 below-6GHz modules. 7 to 9 if we also consider mmWave spectrum.

To accommodate all this new content, in parallel to RF-SOI continuous improvements, packaging actors introduced a Double Sided Molded (DSM) package where dies and other RF components are assembled on top and bottom of a laminate interposer [2]. Today Land and Ball Grid Arrays versions of DSM packaging are available and widely used in smartphone's and IoT's RFFEs.

C. 3D RFFE Integration

With the introduction of new functionalities, such as AI, new form factors, such as two-folded and three-folded, and potentially new frequency bands and antennas, smartphones, but also wearables and other IoT devices, are pushing 2 and 2.5D integration density to its limits. 3D has emerged as a commercial offer for RFFEs [3]. Today available on homogeneous, RF-SOI on RF-SOI, technologies, it has the potential to be extended to heterogeneous, X on RF-SOI, technologies.

3D RFFE Engineered Substrates

A well known challenge for any electronic system is heat dissipation and this is not different for RFFEs. In wafer stacks heat generated thermal expansion leads to layers warpage and mechanical stress.

In 3D-SiP or 2.5D, the use of different materials with different Coefficients of Thermal Expansion (CTE) increase the reliability challenges compared to homogeneous 3D-IC where only RF-SOI silicon technologies – with the same CTE – are used. Heat dissipation remains a challenge in 3D-IC given that its layers still experience different thermal expansion magnitudes as temperatures vary among them. As suggested in [4] this should be carefully assessed from early stages of the stack design and manufacturing. Careful interconnects routing and appropriate placing of functional blocks help ensuring high 3D-IC reliability.

A. Current 3D-IC RF-SOI RFFE

Commercial homogeneous RF-SOI on RF-SOI substrate-to-substrate stacking is done using a hybrid bonding process. The top substrate handle wafer is typically removed using grinding, cleaning and polishing steps. As

---

979-8-3315-0417-5/25 $31.00 © 2025 IEEE   1122

this wafer is meant to be grinded out, it is made of a cost/performance-optimized material different from that of the bottom substrate which uses a High Resistivity (HR) wafer as handle. The bottom substrate HR wafer with Buried Oxide (BOX) and Trap-Rich (TR) layers ensures RF IC's vertical transistor isolation (no transistor latch-up) and high linearity. The top substrate BOX layer is also required to ensure vertical transistor isolation for the RF IC manufactured in the top substrate. At the same time, this BOX acts as a good quality Etch Stop Layer (ESL) for the process steps intended to remove the handle wafer.

Beyond SOI's substrate BOX, other options of ESL are adopted in different 3D-IC solutions. One that has significant adoption because of his compatibility with CMOS processes – among other reasons – is the use of SiGe as ESL. For RFFE 3D-IC this option lacks of interest as the RF-SOI BOX is required for transistor isolation as previously mentioned.

After stacking and grinding the total thickness of the 3D-IC remains similar to that of any 2D processed RF-SOI substrate. This is a considerable advantage over a 3D-SiP which total thickness is roughly twice that as shown in Figure 1. The area reduction added to the roughly unchanged vertical dimension makes the 3D-IC RF-SOI solution the most compact available today and have placed it in a very competitive position for RFFE modules used in volume constrained applications such as smartphones and wearables antenna tuners that benefit of being placed as close as possible to the antennas they tune [5]. They also show a lot of potential for RFFE modules used in volume and weight constrained high reliability industrial, automotive and other IoT devices as well as XR/VR and spatial computing headsets.

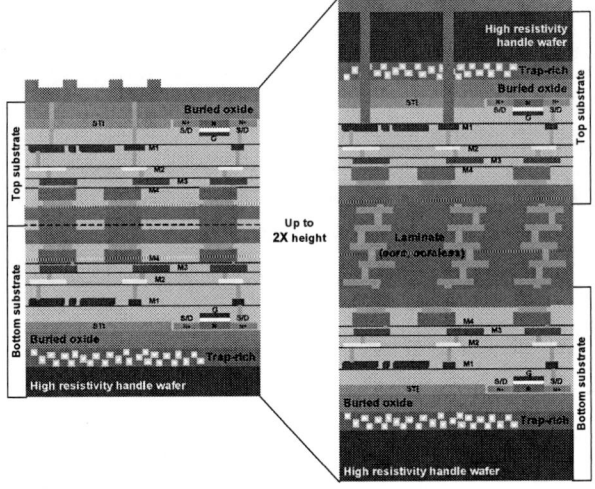

Fig. 1: 3D-IC (left) vs. 3D-SiP (right) height comparison

B. Future 3D-IC RFFE research and development directions

While top handle wafer removal by grinding is a well known and mastered process, it is recognized as a long and dirty one increasing manufacturing cycle times and

generating debris that must be properly disposed. An alternative solution is proposed in [4] and depicted in Figure 2 and consists on the use of a variant of the Smart Cut™ technique.

The use of low temperature to split the handle wafer from the BOX and active silicon in the top RF-SOI substrate allows removing it after hybrid bonding with reduced manufacturing cycle time and debris generation. Furthermore the handle wafer could be reconditioned and reused generating economies of scale. The then stacked active silicon layer could be used to manufacture at low temperatures (typically ≤ 600ºC) a new functional IC in a process often named as 3D Sequential Integration (3DSI) [6].

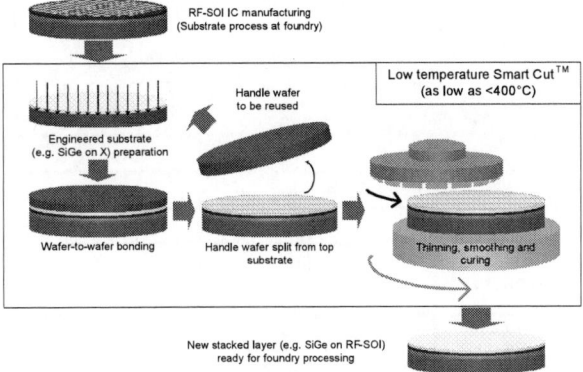

Fig. 2: 3D Sequential Integration by Low Temperature Smart Cut™

Beyond homogenous – RF-SOI on RF-SOI – integration, the use of low temperature (LT) Smart Cut™ helps enable several options for heterogeneous wafer stacks. A couple of examples are of particular interest for future systems RFFEs and several efforts for monolithic integration had been reported.

A first one consists of stacking SiGe BiCMOS on RF-SOI or FD-SOI. This stack enables the design and manufacturing of ultra-wideband RFFEs from HF to hundreds of GHz. LNAs designed and implemented in advanced RF SiGe technologies associated with RF-SOI switches provide high gain/low noise with low current consumption [7].

A second one consists of stacking filters on RF-SOI or FD-SOI. This stack enables an improved frequency management with reduced interconnect losses.

In both cases, stacking SiGe BiCMOS or filters on RF-SOI or FD-SOI enables the RF IC integration with high density digital CMOS electronics at the wafer scale to enable RFFE embedded digital intelligence. This is proved to be a very attractive advantage for 5G-A RFFE products and it will be one for 6G [8].

Figure 3 and table 1 provide a comparison between two different 3DSI process flows; one using a SOI substrate BOX as ESL and a second using low temperature Smart Cut™. From them we can observe that the second flow offers the potential of reduced manufacturing cycle time and optimized total cost of ownership.

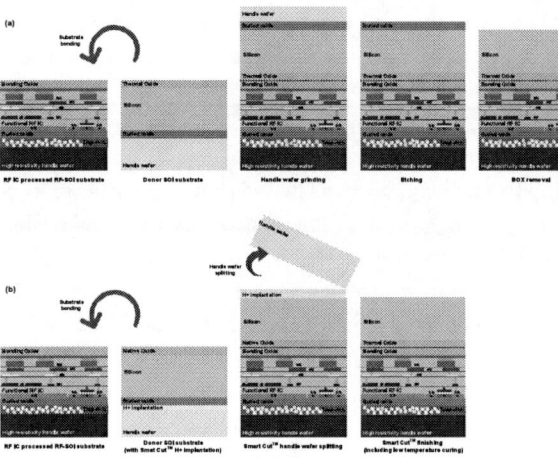

Fig. 2: (a) SOI plus grinding & etching, (b) LT Smart Cut™ 3DSI flows

Table 1: (a) SOI plus grinding & etching, (b) LT Smart Cut™ 3DSI flows main process steps comparison

	SOI BOX ESL	LT Smart Cut™
Donor substrate	RF-SOI substrate	RF-SOI substrate
	Oxide deposition	H+ implant
Bonding	Bonding	Bonding
	Post bonding anneal	
Thinning	Grinding	Splitting
	Handle wafer etch	
	Oxide wet etch	
	Thickness metrology	
Post thinning	RTA	Low temp curing
	APC Si wet	APC Si wet

A further evolution of the Smart Cut™ technique will allow independent processing - before bonding - of the substrates to be stacked. The two patterned substrates will be bond together before proceeding to split the top substrate handle wafer. The advantage of this new evolution is allowing standard temperature (typically up to ~1000ºC) CMOS processing in both substrates [9].

Conclusion

The RFFE 2D and 2.5D integration density enabled by the latest RF-SOI technology platforms is showing limitations for 5G-A and 6G systems. The innovations of thin layer transfer technologies are becoming more and more important in the 3D-IC integration future. We discussed and compared – from the engineered substrates point of view – the main technologies options advantages and inconvenient and pointed to future development directions.

Acknowledgments

The author gratefully acknowledges the fruitful discussions and contributions of Hankel Chang from Soitec Taiwan.

References

[1] D. Parat et al., "A Linear High-Power Reconfigurable SOI-CMOS Front-End Module for WI-FI 6/6E Applications," IEEE Radio Frequency Integrated Circuits Symposium (RFIC) (2022).

[2] M. Tsai et al., "Innovative Packaging Solutions of 3D Double Side Molding with System in Package for IoT and 5G Application," IEEE Electronic Components and Technology Conference (ECTC) (2019).

[3] R. Verma, "3D-IC for RF Front End Applications", semi Semicon® Taiwan 2024.

[4] H. Chang, "Thin Layer Transfer Technologies in Heterogeneous Integration," International VLSI Symposium on Technology, Systems and Applications (VLSI-TSA/VLSI-DAT) (2023)

[5] A. Kuchikulla, "How to Implement Aperture Tuning: Best Practices for 4G/5G Smartphones", Qorvo® e-guide 2020.

[6] J. Lugo-Alvarez et al., "First Radio-Frequency Circuits Fabricated in Top-Tier of a Full 3D Sequential Integration Process at mmW for 5G Applications," IEEE Symposium on VLSI Technology and Circuits (2024).

[7] V. Jain et al., "415/610GHz fT/fMAX SiGe HBTs Integrated in a 45nm PDSOI BiCMOS process," International Electron Devices Meeting (IEDM) (2022).

[8] C. -X. Wang et al., "On the Road to 6G: Visions, Requirements, Key Technologies, and Testbeds," IEEE Communications Surveys & Tutorials, vol. 25, no. 2, pp. 905-974, 2023.

[9] H. Chang et al., "Novel Ultra-Thin/Transistor Layer Transfer (TLT) Substrates based on Smart Cut™ technology and IR Laser Release for W2W and D2W Heterogeneous Integration Applications", unpublished.

Calculation Optimization of Double-Free-Layer Magnetic Tunnel Junction

Zifeng Wang[1,2] and Lang Zeng[1,2]

[1] Fert Beijing Institute, MIIT Key Laboratory of Spintronics, School of Integrated Circuit
Science and Engineering, Beihang University, Beijing 100191, China.
[2] National Key Laboratory of Spintronics, Hangzhou International Innovation Institute,
Beihang University, Hangzhou 311115, China.
Email: zenglang@buaa.edu.cn

ABSTRACT

Double-free-layer magnetic tunnel junctions (DMTJs) demonstrate significant advantages in enhancing thermal stability and reducing critical current. However, they also present challenges in modeling quantum transport within the device because of the quantum well, while few existing researches handle the problem correctly. To tackle these issues, we employ the non-equilibrium Green's function (NEGF) method, complemented by a specialized algorithm that balances both accuracy and efficiency.

Keywords: DMTJ, MRAM and NEGF

INTRODUCTION

Magnetic tunnel junctions (MTJs) are promising devices widely used in magnetic random access memory (MRAM) for their efficient spin transfer torque (STT) switching [1]-[4] . To achieve improved electrical performance including reduced switching current and enhanced thermal stability, various innovative MTJ structures have been proposed. Among them, DMTJ exhibits remarkable properties, which have been validated experimentally [5] . Due to the second free layer, the interface interaction between the oxide and the ferromagnetic layers is enhanced, leading to a stronger thermal stability while maintaining relatively short switching time [5] .

While the drift-diffusion (DD) model is commonly used for nano-electronic transport due to its low computational cost , it struggles to capture quantum effects as device dimensions scale down [1], [2], [6] . The non-equilibrium Green's function (NEGF) method, based on the Schrödinger equation, is better suited for quantum transport modeling [3], [7], [8].

In DMTJs, a quantum well forms between the two barriers and the first free layer, which may give rise to resonant states and complicate the energy-resolved integral in NEGF calculations. While existing modeling methods are sufficient for calculating features of single-free-layer magnetic tunnel junctions (SMTJs), computational methods for DMTJs remain limited and underdeveloped. In this paper, we propose a numerical solution within the NEGF framework that employs a dedicated sampling strategy to achieve a balance between computational efficiency and precision.

I. NEGF METHOD FOR SPIN QUANTUM TRANSPORT

Fig. 1 shows the device schematic of SMTJ and relevant energy band parameters. To accurately describe quantum transport, the NEGF method is employed to model the structure. The magnetic moment of the free layer rotates on a sphere and can therefore be represented in spherical coordinates as $m = (sin(\theta)cos(\phi), sin(\theta)sin(\phi), cos(\theta))$. To handle non-collinear states, the potential matrix needs to be transformed using a rotation matrix, which is defined as follows:

$$R = \begin{bmatrix} cos\frac{\theta}{2}e^{-i\frac{\phi}{2}} & -sin\frac{\theta}{2}e^{-i\frac{\phi}{2}} \\ sin\frac{\theta}{2}e^{i\frac{\phi}{2}} & cos\frac{\theta}{2}e^{i\frac{\phi}{2}} \end{bmatrix} \quad (1)$$

The Hamiltonian matrix is thus defined as

$$H_{i,j} = RH_0^{i,j}R^{\dagger} \quad (2)$$

where each $H_{i,j}$ is a 2×2 block matrix in spin space.

The retarded Green's function G is calculated at each longitudinal energy E_y (along the transport direction) and transverse energy $E_\perp$ (parallel to the cross-section), while the correlation function G^n for a specific E_y is calculated by an integral over $E_\perp$.

$$G(E_y, E_\perp) = [E - H(E_y, E_\perp) - \Sigma(E_y)]^{-1} \quad (3)$$

$$G^n(E_y) = \int D_{2D}G\Sigma^{in}G^{\dagger}dE_\perp \quad (4)$$

Once G^n is obtained, the current I and spin current $I_{s,k}(k = x, y, z)$ are calculated using the commutation relations in quantum mechanics and by integrating over E_y, as follows.

$$I_i = iqTr(\frac{H_{i,i+1}^{tot}G_{i+1,i}^n - G_{i,i+1}^n H_{i+1,i}^{tot}}{h}) \quad (5)$$

$$I_{s,k,i} = iqTr(\frac{H_{i,i+1}^{tot}G_{i+1,i}^n - G_{i,i+1}^n H_{i+1,i}^{tot}}{h}\sigma_k) \quad (6)$$

With the derived current, we examine the resistance's voltage dependence at various angles, shown in **Fig. 2**, where simulation results align closely with experimental data, validating the accuracy of our model [4]. Since the

979-8-3315-0417-5/25 $31.00 © 2025 IEEE

spin torque acting on each free layer results from the spin current loss while passing through the layer, it can be derived without additional computational cost:

$$T = \frac{\mu_B}{qM_s a}(I_{s,start} - I_{s,end}) \qquad (7)$$

The comparison of the voltage dependence of STT between simulation and experiment is also depicted in **Fig. 3**. The damping-like component aligns closely with experimental data, while the field-like component deviates, displaying an asymmetry in the simulation results compared to the symmetric experimental data.

II. MIXED SAMPLING REGARDING DMTJ

The device schematic for the DMTJ is illustrated in **Fig. 4**, which has similar dimensions and conduction band characteristics to the SMTJ. The DMTJ's double-barrier structure forms a quantum well in the first free layer, leading to significant oscillations in the voltage-dependence curves of current and STT when using the same linear energy lattice integration method as for the SMTJ, as shown in **Fig. 5(a)**. These oscillations are reduced when using a denser energy lattice with 10^5 points, as seen in **Fig. 5(b)**, although this approach significantly increases computational cost, with the computing time per data point rising from 0.34 s to 442.24 s.

The oscillations in **Fig. 5** do not reflect physical phenomena but are instead caused by integration errors in the energy-resolved charge/spin current. These errors stem from sharp peaks in the integrand due to resonant states, underscoring the need for an efficient sampling method for integration. We initially tested Gaussian Sampling (GS) and Latin Hypercube Sampling (LHS), shown in **Fig. 6(a)-(b)**. While both methods generally smooth the curves, some randomness remains. To analyze this further, we present an example in **Fig. 7**. The blue curve represents a densely sampled linear lattice with 2000 points, and the orange curve shows the sampled data from various methods. Notably, GS in **Fig. 7(a)** tends to concentrate points near previous peaks rather than distributing them evenly, while LHS provides a broader distribution of points in **Fig. 7(b)**. However, in energy regions away from the peaks, both methods sample fewer points, potentially impacting integration accuracy. LHS, in particular, does not adequately sample near the resonant peak. To address this, we introduce a fixed sampling approach that ensures sufficient sampling in underrepresented areas. As shown in **Fig. 7(a)**, the maximum integral error for each interval is calculated as $S_{error,max} = \Delta x \Delta y / 2$, with a threshold set at 1% of the overall integral. If $S_{error,max}$ exceeds this threshold, additional sampling is introduced in that interval. This sampling strategy has nothing to do with randomness, thus named as fixed sampling. As GS exhibits better performance in peak

detection, we combine the Gaussian random sampling and the fixed sampling method into a mixed sampling method, to integrate the curve and derive the voltage dependence of current and STT as illustrated by the red curves in **Fig. 6(a)-(b)**. **Fig. 7(c)** demonstrates improved sampling across the region. This method significantly reduces randomness, as shown in **Fig. 8**, with minimal variation across 1000 simulations using the same parameters. **Table. I** indicates that mixed sampling is computationally more efficient compared to GS and LHS , and achieves a substantially lower relative error than both methods when benchmarked against a dense 10^5 linear lattice sampling.

III. MAGNETODYNAMICS

To simulate magnetic dynamics, a coupled model is illustrated in **Fig. 9**. Quantum transport is solved using the NEGF method, and the resulting STT is used to update the magnetic state based on the Landau-Lifshitz-Gilbert (LLG) equation, which then feeds back into the STT term. The switching dynamics of SMTJ and DMTJ are shown in **Fig. 10** and **Fig. 11** respectively. The overall switching time of the DMTJ is approximately 45% shorter than that of the SMTJ, while also requiring a 33% to 50% lower current through the device. Besides, $m_{z,1}$ reverses much faster than $m_{z,2}$. This could be explained by the spatial distribution of spin current and STT within the DMTJ in **Fig. 12**, where both quantities are larger in the first free layer than in the second.

CONCLUSION

In this project, we propose a coupled model to describe the current, STT, and magnetodynamics of the DMTJ. An efficient sampling method is applied to handle the integration of energy-resolved charge and spin currents. This method achieves a sufficiently low relative error, benchmarked against integration over a dense linear lattice, and offers substantial computational cost savings compared to Gaussian and Latin hypercube sampling. Our model accurately computes the magnetic reversal dynamics, demonstrating that the critical current required for DMTJ is lower than that for SMTJ, with a faster reversal speed for DMTJ as well, offering a promising foundation for future device-technology co-optimization .

REFERENCES

[1] S. Fiorentini, et al. *Scientific Reports*, vol. 12, no. 1, p. 20958, 2022. [2] S. Fiorentini, et al. *ECS Transactions*, vol. 111, no. 1, p. 181, 2023. [3] P. Flauger, et al. *Physical Review B*, vol. 105, no. 13, p. 134407, 2022. [4] H. Kubota, et al. *Nature Physics*, vol. 4, no. 1, pp. 37–41, 2008. [5] B. Jinnai, et al. *IEEE International Electron Devices Meeting (IEDM). IEEE*, 2020, pp. 24–6. [6] S. Fiorentini, et al. in *ESSDERC 2022-IEEE 52nd European Solid-State Device Research Conference (ESSDERC). IEEE*, 2022, pp. 348–351. [7] S. Datta, New York: Cambridge University Press, 2005. [8] D. Deepanjan, Ph.D. dissertation, Purdue University, 2012.

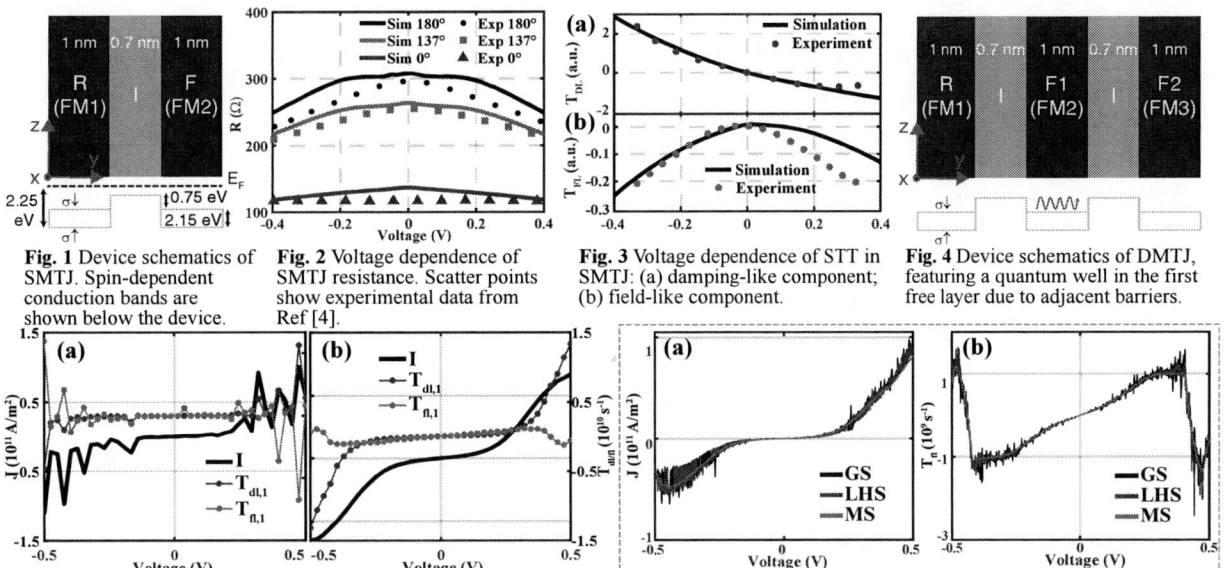

Fig. 1 Device schematics of SMTJ. Spin-dependent conduction bands are shown below the device.

Fig. 2 Voltage dependence of SMTJ resistance. Scatter points show experimental data from Ref [4].

Fig. 3 Voltage dependence of STT in SMTJ: (a) damping-like component; (b) field-like component.

Fig. 4 Device schematics of DMTJ, featuring a quantum well in the first free layer due to adjacent barriers.

Fig. 5 The voltage dependence of current and the two components of STT obtained by integrating over a linear energy lattice with (a) 100 energy points and (b) 100,000 energy points.

Fig. 6 (a) The voltage dependence of current at $\theta_1=\pi$, $\theta_2=\pi$; (b) The voltage dependence of field-like component of STT in the first free layer at $\theta_1=\pi/2$, $\theta_2=\pi$. (GS: Gaussian sampling; LHS: Latin hypercube sampling; MS: mixed sampling proposed by us.

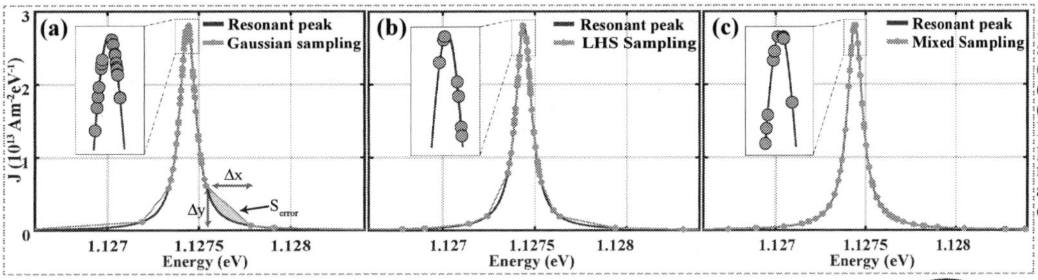

Fig. 7 Schematics of sampling methods: (a)Gaussian sampling; (b)LHS sampling; (c)Mixed sampling. In mixed sampling, points are distributed more uniformly,giving a more accurate image of the resonant peak.

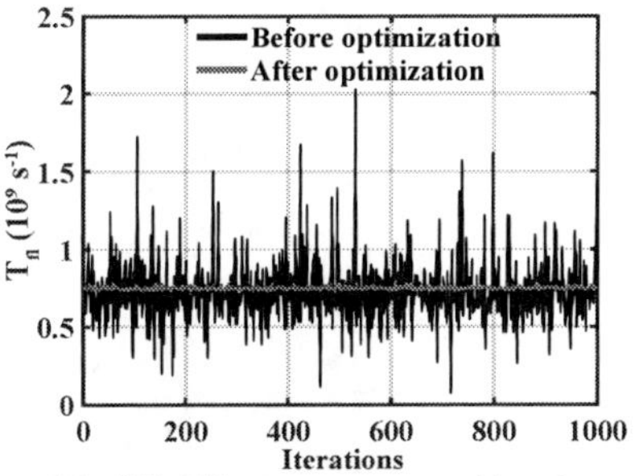

Fig. 8 Stability improvement achieved through the introduction of the fixed sampling mechanism.

(a)

	Time (s)	Relative Error
MS	1.853	0.40 %
GS	2.684	20.15 %
LHS	2.956	13.07 %

(b)

	Time (s)	Relative Error
MS	11.735	0.22 %
GS	36.082	4.68 %
LHS	41.784	2.90 %

Tab. I The comparison of calculation efficiency and accuracy among the three different methods for (a) current and (b) STT.

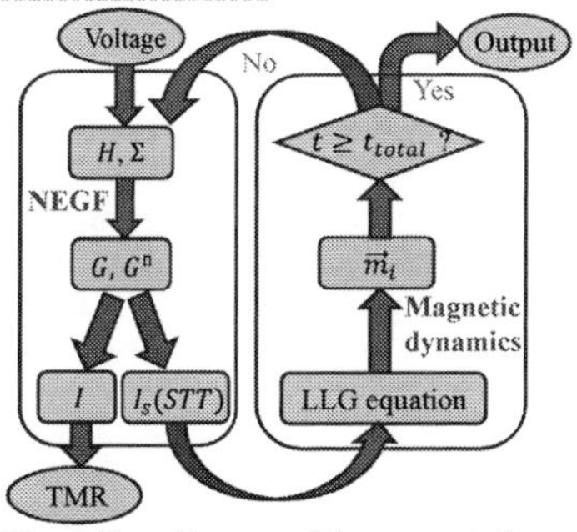

Fig. 9 Flow diagram of the computation algorithm, comprising two coupled parts: transport based on NEGF and magnetic dynamics based on the LLG equation.

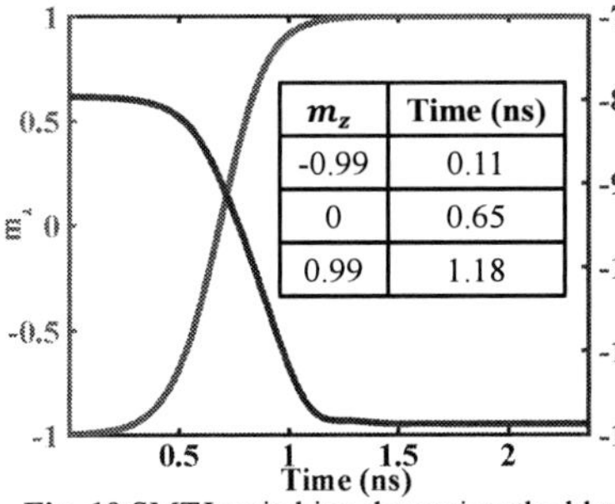

m_z	Time (ns)
-0.99	0.11
0	0.65
0.99	1.18

Fig. 10 SMTJ switching dynamics: the blue curve indicates the magnetic moment reversal, while the red curve represents corresponding current.

m_z	FL 1 (ns)	FL 2 (ns)
-0.99	0.04	0.20
0	0.23	0.37
0.99	0.61	0.65

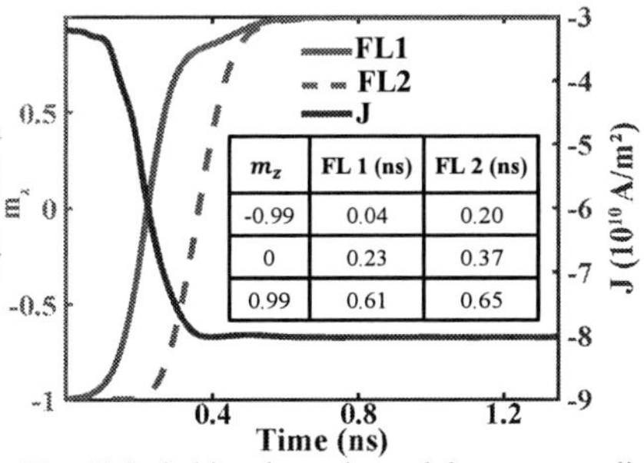

Fig. 11 Switching dynamics and the corresponding current change of DMTJ. The switching is noticeably faster than that of SMTJ.

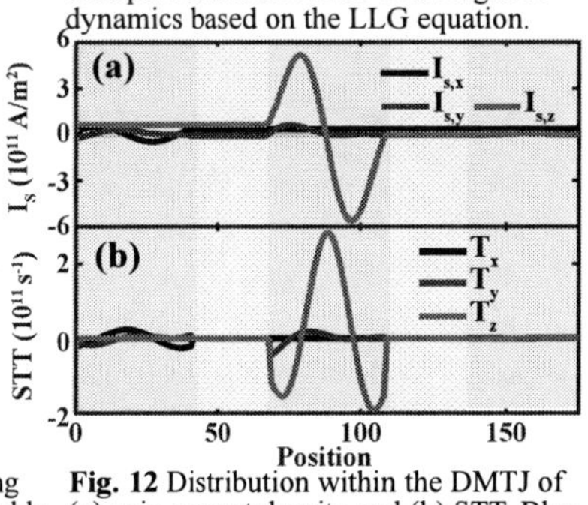

Fig. 12 Distribution within the DMTJ of (a) spin current density and (b) STT. Blue backgrounds represent FM areas, while yellow backgrounds indicate barriers.

Single Photon Devices Using Layered Materials

Mayank Chhaperwal, Nithin Abraham, and Kausik Majumdar
Department of Electrical Communication Engineering
Indian Institute of Science, Bangalore 560 012, India
e-mail: kausikm@iisc.ac.in; phone: +91-80-2293-2742

Abstract— **The vertical heterojunctions of two-dimensional layered materials allow us an unprecedented opportunity to manipulate electrons at the nanoscale, empowered by their atomically sharp interfaces. Here we show demonstrations of two important device applications, namely a single photon emitter and a single photon detector using such van der Waals heterojunctions. These devices could play important roles for interesting quantum technology applications.**

Keywords—2D materials, van der Waals heterojunction, single photon emitter, single photon detector.

I. INTRODUCTION

Single photon emitters (SPEs) [1] and single photon detectors (SPDs) [2] are crucial for the advancement of quantum technologies, including quantum computing, quantum communication, and quantum metrology. Transition metal dichalcogenides (TMDCs) have emerged as promising materials for optoelectronic devices due to their two-dimensional nature, which offers several advantages. These include ease of integration with photonic and plasmonic cavities, low outcoupling loss, and gate-induced spectral tunability [1,3]. Specifically for single photon emitters, despite the potential of TMDC-based SPEs, two significant challenges need to be addressed (see Fig. 1). First, the overall brightness of these emitters is generally low, often falling short of the requirements for practical quantum applications [4]. This low brightness is partly due to the early saturation of emission intensity, which is limited by Auger annihilation processes. Auger annihilation prevents the SPE from achieving its lifetime-limited emission rate, as it causes excitons to annihilate before they can be trapped by the single photon emitting defect state. Second, maintaining high

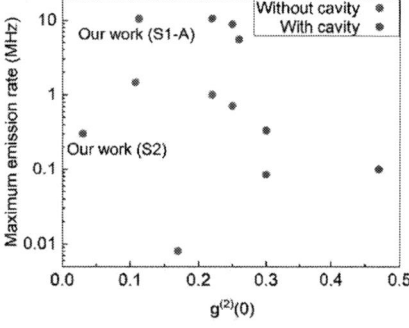

Fig. 1: Benchmarking the performance of our device against the reports from the literature. Adapted from [5].

single photon purity while increasing the emission rate is a significant hurdle. The diffraction-limited laser spot size is much larger than the nanopillar diameter, leading to considerable background emission from the surrounding

regions. This background emission is not bottlenecked by a single level and does not saturate with increasing power, thereby degrading the single photon purity at higher excitation powers. On the other hand, for single photon detectors, there has not been any significant work on detecting single photons in the 1550 nm wavelength range using 2D materials. Here we describe our recent works on both SPE and room temperature SPD at 1550 nm using van der Waals heterojunctions.

II. SINGLE PHOTON EMITTER

A. Design and Fabrication

To overcome these challenges, our work introduces a novel design methodology using high-aspect-ratio nanopillar arrays on a metal film (Figure 2a). The nanopillars are fabricated using a negative photoresist, which provides a softer and smoother structure compared to traditional materials like SiO_2. This approach minimizes the risk of damaging the TMDC monolayer during the transfer process, a common issue with high-aspect-ratio nanopillars made of

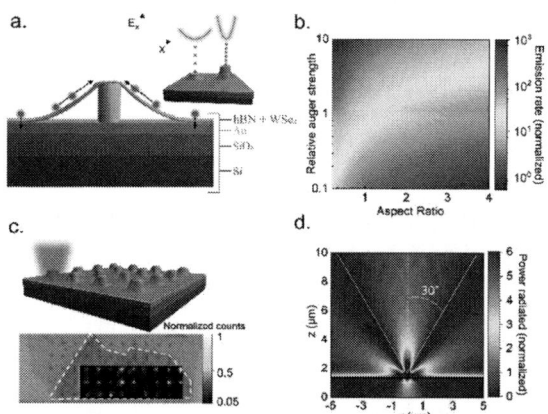

Fig 2: (a) Cross section of the device structure showing different layers and the exciton quenching mechanism. (b) Results from our model show the dependence of SPE emission rate on the aspect ratio of the pillar as well as the Auger strength. (c Top) Schematic of the device showing an array of nanopillars hosting SPEs, one of which is excited at a time by the laser spot. (c Bottom) PL map superimposed on the optical device of the device showing the functioning of gold as an excellent background suppressor. (d) Results from the FDTD simulations show the factor of total intensity collected by our objective with NA 0.5 (shown by white dashed lines). Adapted from [5].

hard materials [6]. The nanopillars have an aspect ratio

(height:diameter) of approximately 3, significantly higher than previous reports. The design is inspired by our model that predicts the brightness of the SPE as a function of Auger strength and pillar aspect ratio (Figure 2b). The results suggest that a higher aspect ratio helps overcome the limitation imposed by Auger-induced exciton annihilation through stronger exciton funnelling toward the pillar site.

The SPE array design includes several critical features:

(a) High-Aspect-Ratio Nanopillars: These pillars introduce nonuniform strain in the TMDC monolayer, reducing the local band gap and increasing the funnelling of excitons toward the pillar site. These excitons are captured by a defect state leading to single photon emission.

(b) Encapsulation with hBN: A few-layer hBN flake is transferred on top of the monolayer/nanopillar stack to smooth out inhomogeneous potential fluctuations and screen inter-exciton interactions. This reduces the Auger coefficient significantly, thereby suppressing the interaction among locally trapped excitons and enhancing the overall quantum efficiency of the SPE.

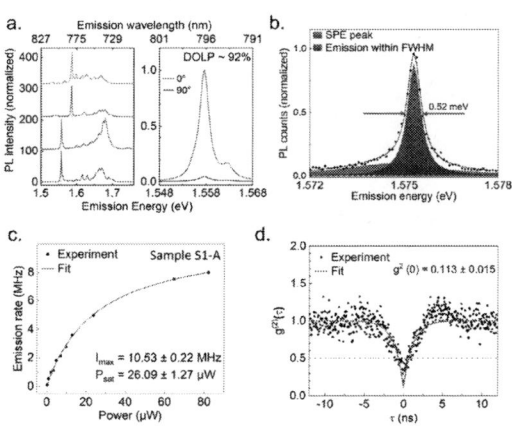

Fig 3: (a) PL spectra from four different pillars (with WSe$_2$ on top) showing sharp SPE emission (dashed vertical lines). (b) SPE peak fitted to remove the contribution from the background and neighbouring peaks. (c) Power-dependent SPE intensity fitted with a saturation equation to obtain the maximum collected count rate of 10.53 ± 0.22 MHz. (d) Second-order correlation was measured at an emission rate of 1 MHz to characterize the purity of the SPE. After deconvolution with the IRF of the setup, the g$^{(2)}$(0) value obtained is 0.113 ± 0.015. Adapted from [5].

(c) Gold-Coated Substrate: The nanopillars are fabricated on a gold-coated substrate, which acts as a back-reflector to improve photon out-coupling. The gold substrate also quenches emission from regions away from the nanopillars, enhancing single photon purity by selectively collecting emission only from the nanopillar region. This is supported by photoluminescence (PL) scans showing that the emission from the flake away from the nanopillars is almost completely quenched, while excitons funnel to the nanopillar locations, providing high brightness at the pillar site (Figure 2c).

B. Characterization

Figure 3a shows PL spectra from multiple pillar sites that have a similar sharp emission line in the 770-800 nm band, which acts as an SPE. The sharpest peak measured has a full-width-at-half-maximum (fwhm) of 520 µeV (Figure 3b). The fabricated SPEs exhibit several notable characteristics:

(a) High Emission Rate: The SPEs achieve a collected rate of over 10 MHz (Figure 3c) in the 770-800 nm band, which is compatible with quantum memory and repeater networks (Rb-87-D1/D2 lines) and satellite quantum communication. The integrated emission intensity is calculated by fitting the SPE peak with Voigt functions and removing contributions from side peaks and background emission (Figure 3b). The collection efficiency of the emitted photons is also a critical aspect. FDTD simulations predict that 33.93% of the total photons emitted by the dipole above the substrate surface are collected by the objective with a numerical aperture (NA) of 0.5. This gives an estimated emission rate of nearly 31 MHz above the surface of the substrate, highlighting the efficiency of the design in enhancing photon out-coupling (Figure 3d).

(b) High Single Photon Purity: The emitters show excellent single photon purity, as evidenced by second-order correlation measurements which give a g$^{(2)}$(0) value of 0.113 ± 0.015 when measured at a high emission rate of 1 MHz (Figure 3d). This proves that the use of a gold-coated substrate and high-aspect-ratio nanopillars helps in mitigating background emission and maintaining purity even at high emission rates.

(c) Linear Polarization: The emission peaks exhibit a high degree of linear polarization (about 92%) (Figure 3a, right panel), indicating the intrinsic linearly polarized nature of the SPEs. This polarization remains independent of the direction of the polarization of the excitation, further confirming the intrinsic properties of the SPE.

III. SINGLE PHOTON DETECTOR

A. Design and Fabrication

To achieve single photon detector operation at 1550 nm wavelength, we segregate the photon absorption from the electron manipulation region. The photon absorber uses a layer of black phosphorus (BP) of about 25 nm thickness, which is placed on a few-layer thick MoS$_2$, forming type-II heterojunction. The low bandgap of BP allows photon absorption at 1550 nm, and due to built-in field, the photoelectrons are transferred to the MoS$_2$ layer in a fast time scale, before their recombination in BP. The design of the device allows this electron to be trapped in the MoS$_2$ layer near the source end of a WSe$_2$ p-type transistor (see Figure 4). The trapped electron creates a local gating for the transistor, generating a fluctuation in the hole injection efficiency, eventually creating detectable signal when an electron is trapped. The physical size of the trapping region plays an important role in the single photon detection device.

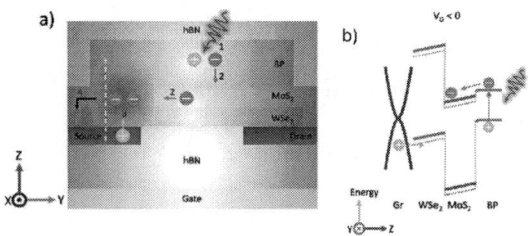

Fig 4: (a) A schematic view of the cross section of the single photon detector. (b) A schematic band diagram of the device. Adapted from [7].

B. Characterization

We use a home-built setup to characterize the SPD where

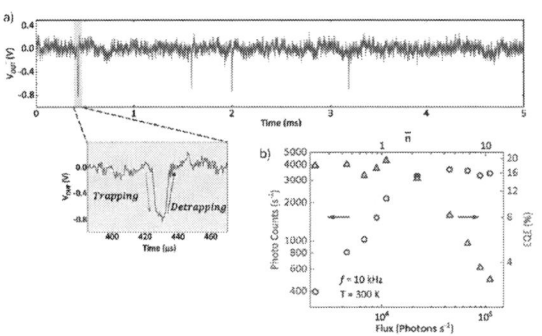

Fig 5: (a) Response of the device when electrons are trapped into the potential well. (b) The measured external quantum efficiency of the single photon detector plotted as a function of the photon flux. Adapted from [7].

incoming light from a laser source is attenuated heavily to generate a stream of single photons (the probability of occurrence of two or more photons is made extremely low). Figure 5 describes the output characteristics of the SPD when single electrons are trapped in the potential well (either by thermal generation or by photogeneration). At 300 K, the overall measured detection efficiency reaches about 20%, while the best dark count rate achieved is less than 1 kHz – making the device competitive against commercially available room temperature single photon detectors.

IV. CONCLUSION

The development of efficient single photon emitter and single photon detector using layered materials is expected to play a significant role in future quantum technological applications.

ACKNOWLEDGMENT

The authors acknowledge the support of K. Watanabe, T. Taniguchi, and S. A. Tongay. K. M. acknowledges the support of a Core Research Grant from the Science and Engineering Research Board (SERB) under the Department of Science and Technology (DST), a grant from the Indian Space Research Organization (ISRO), a grant from MHRD under STARS, a seed grant under Quantum Research Park (QuRP) from Karnataka Innovation and Technology Society (KITS), K-Tech, Government of Karnataka, a grant from I-HUB QTF, IISER Pune, and a grant from MHRD, MeitY and DST Nano Mission through NNetRA.

REFERENCES

[1] Luo, Y.; Shepard, G. D.; Ardelean, J. V.; Rhodes, D. A.; Kim, B.; Barmak, K.; Hone, J. C.; Strauf, S. Deterministic coupling of site-controlled quantum emitters in monolayer WSe₂ to plasmonic nanocavities. Nat. Nanotechnol. 2018, 13 (12), 1137.

[2] Marsili, F. et al, Detecting single infrared photons with 93% system efficiency. Nat. Photonics 2013, 7, 210.

[3] Guo, S.; Germanis, S.; Taniguchi, T.; Watanabe, K.; Withers, F.; Luxmoore, I. J. Electrically Driven Site-Controlled Single Photon Source. ACS Photonics 2023, 10 (8), 2549.

[4] Hoang, T. B.; Akselrod, G. M.; Mikkelsen, M. H. Ultrafast Room-Temperature Single Photon Emission from Quantum Dots Coupled to Plasmonic Nanocavities. Nano Lett. 2016, 16 (1), 270.

[5] Chhaperwal, M.; Tongale; H. M.; Hays, P.; Watanabe, K.; Taniguchi, T.; Tongay, S. A.; Majumdar, K. Simultaneously Enhancing Brightness and Purity of WSe 2 Single Photon Emitter Using High-Aspect-Ratio Nanopillar Array on Metal. Nano Lett. 2024, 24 (40), 12461.

[6] Palacios-Berraquero, C.; Kara, D. M.; Montblanch, A. R.-P.; Barbone, M.; Latawiec, P.; Yoon, D.; Ott, A. K.; Loncar, M.; Ferrari, A. C.; Atature, M. Large-scale quantum-emitter arrays in atomically thin semiconductors. Nat. Commun. 2017, 8 (1), 15093.

[7] Abraham, N.; Watanabe, K.; Taniguchi, T.; Majumdar, K. Room Temperature Single Photon Detection at 1550 nm using van der Waals Heterojunction. Advanced Functional Materials, 2024, 34, 2406510.

Impact of Dielectrics on Hysteresis and Bias Stress Stability in Oxide Semiconductor and 2D-Material Field-Effect Transistors

Alwin Daus[1], Sumaiya Wahid[2], Quỳnh Thị Phùng[3], Eric Pop[2]

[1]Institute of Semiconductor Eng., University of Stuttgart, Germany, [2]Department of Electrical Eng., Stanford University, USA, [3]Department of Microsystems Eng., University of Freiburg, Germany

Abstract

We provide an overview of mechanisms for hysteresis and bias stress instability in field-effect transistors (FETs). We discuss challenges and peculiarities of FETs based on emerging materials such as amorphous oxide and two-dimensional semiconductors. Therein we focus on the role of gate dielectrics and their interfaces. The understanding of such reliability aspects is important to improve practical prospects of these emerging material FETs for application in various fields.

Keywords: FET, bias stress, hysteresis, dielectric

Introduction

Semiconductors beyond conventional silicon are under investigation as channel materials in field-effect transistors (FETs) for a variety of applications such as future logic technology nodes, memory, back-end-of-line integration (BEOL) and flexible electronics [1-4]. Here, we discuss reliability aspects of FETs with amorphous oxide semiconductor and two-dimensional (2D) semiconductor channels. Specifically, we focus on the role of the gate dielectric on hysteresis and bias stress stability. While the conventional Si-to-SiO$_2$/high-k/metal gate stack has been optimized for many years, we still find ourselves in a situation, where charge trapping and/or mobile ions significantly affect the reliability of FETs with these alternative semiconductor channels [5]. This can be attributed to the non-ideal interfaces of semiconductor to dielectric (instead of the native SiO$_2$ on Si), difficulties in dielectric deposition on such channel materials, non-optimized dielectric deposition processes, or limitations in deposition temperature e.g., for flexible electronics, which can lead to large dielectric defect concentrations [5-7].

Thus, it is imperative to study performance and reliability parameters of FETs with emerging channel materials to understand the quality of the dielectrics, interfaces, and interaction of dielectric deposition process with the channel material. In the following, we will first introduce mechanisms for hysteresis and bias stress instability in FETs with oxide semiconductor and 2D material channels. Then we share our findings on how different dielectric materials, deposition processes, and device configurations impact such parameters.

Mechanisms for Hysteresis and Bias Stress Instability

The most common reason for hysteresis and bias stress instability in FETs is the trapping of channel carriers in the dielectric [5, 8]. For n-channel devices, this leads to the clockwise hysteresis sketched in Fig. 1a and a positive threshold voltage V_T shift, Fig. 1b, when positive bias stress (PBS) is applied. Minimizing dielectric defects that can actively trap and detrap channel carriers is thus necessary. It is also possible to obtain opposite behaviors with counterclockwise hysteresis and negative V_T shifts upon PBS (Fig. 1). The reason for this can be charge trapping of carriers from the gate electrode into the dielectric or the movement of mobile ions through the dielectric [8, 9]. However, it should be noted that reality is more complex: Mobile ions e.g., in electrolytic gate insulators may cause counterclockwise hysteresis but they do not always

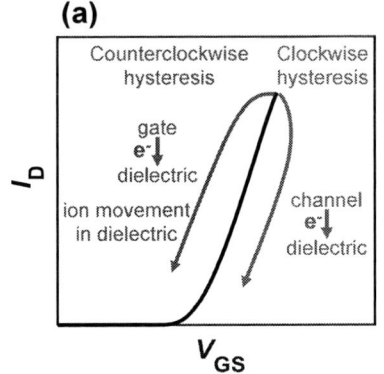
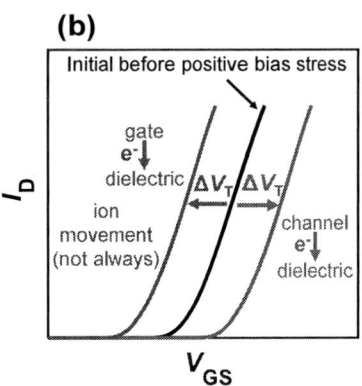

Fig. 1: Schematic illustrations of hysteresis and threshold voltage shift ΔV_T upon PBS when charge trapping or mobile ion movement occur in n-channel FETs. (a) Hysteresis of drain current I_D when the gate-source voltage V_{GS} is swept. Blue and red indicates possible mechanisms for counterclockwise and clockwise hysteresis, respectively. (b) I_D versus V_{GS} for PBS with positive and negative ΔV_T in red and blue color along with possible mechanisms.

979-8-3315-0417-5/25 $31.00 © 2025 IEEE

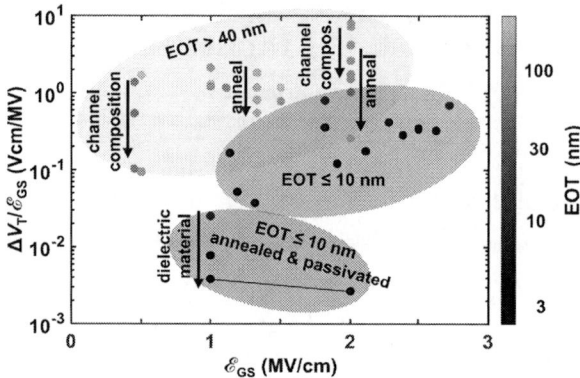

Fig. 2: Comparison of PBS stability (stress time 1000 s) for several FETs with oxide semiconductor channels. The color scale clearly shows that a lower EOT reduces the electric field-normalized threshold voltage shift ($\Delta V_T / \mathscr{E}_{GS}$). In addition, other factors are indicated that have been studied like the role of channel material composition, annealing and choice of dielectric material. Note that all devices here showed positive ΔV_T, consistent with channel-to-dielectric charge trapping. This figure was adapted from [11].

lead to negative V_T shifts upon PBS [10]. Furthermore, in a real device, multiple charge trapping and/or ion movement mechanisms can simultaneously exist, partially compensating each other, and exhibiting different time constants [5, 8, 10].

Oxide Semiconductor FETs

Depending on the dielectric/semiconductor stack, all above-mentioned phenomena can be observed in FETs with amorphous oxide semiconductor channels. Semiconductor-to-dielectric trapping is most commonly reported [11], but we have also observed gate-to-dielectric charge trapping in our and others' works [8, 11-13]. The latter seems to be more prevalent in low-temperature deposited dielectrics or generally more defective dielectrics [8, 11-13]. There have also been reports on the movement of ionic species like oxygen vacan-

cies or hydrogen (H) [9,14]. One particular aspect of amorphous oxide semiconductors is that hydrogen also acts as donor in the semiconductor, and thus H-migration can change the doping level and thus electrical parameters [14].

Recently, we have more closely investigated the bias stress stability in amorphous indium tin oxide (ITO) transistors, which are being considered for BEOL applications [2, 11]. In back-gated devices, we first elucidated the benefit of low equivalent oxide thickness (EOT) for improved PBS (Fig. 2) [11]. In addition, we studied top-gated and dual-gated ITO FETs under different annealing conditions and upon bias stressing, and were able to obtain extraordinarily good bias stress stability with V_T shifts ~3 mV/(MV/cm) at room temperature, and high drain current ~2 mA/μm at drain-source voltage of 1 V in short channel devices [15-17].

2D Material FETs

Transistors based on 2D transition metal dichalcogenide (TMD) channel materials, such as molybdenum disulfide (MoS_2) or tungsten diselenide (WSe_2), have fundamentally different interfacial properties compared to conventional (bulk) semiconductors. The 2D material exhibits covalent bonds in-plane and van der Waals gaps out-of-plane [3,4]. On one hand, this allows for integration on various substrates due to the weaker interaction with interfaces. On the other hand, it results in challenges when interfacing with common oxide dielectrics [5]. We have developed a fabrication approach for high-performance flexible 2D transistors where part of the device (2D material and contacts) are prepatterned on a rigid substrate and then the materials are directly transferred by coating with thin polyimide layers and subsequent release [10, 18]. This requires the fabrication of a top-gate dielectric onto the 2D material, which is enabled by using a thin evaporated Al seed layer followed by thermal atomic layer deposition

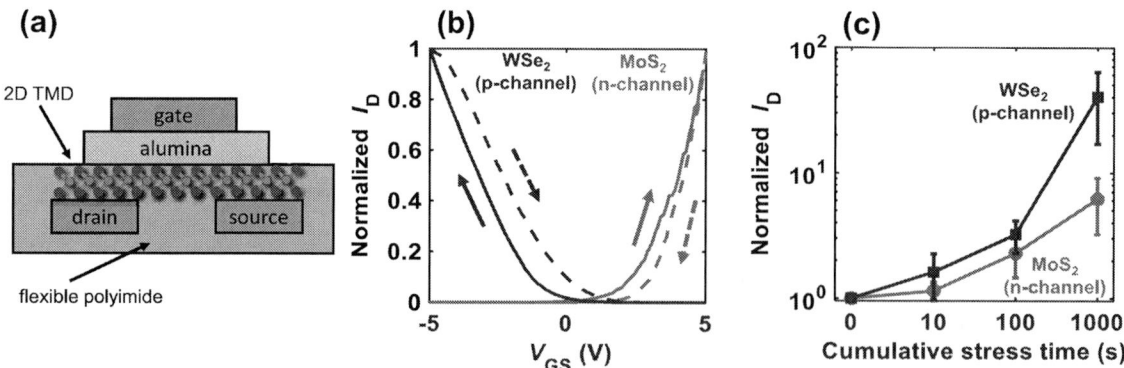

Fig. 3: Flexible 2D TMD FETs fabricated with substrate-embedded contacts [10, 18]. (a) Schematic device cross-section. (b) Normalized I_D at absolute drain-source voltage = 1 V for a WSe_2 p-channel FET and an MoS_2 n-channel FET. Both devices consist of multilayer TMD material grown by metal-organic chemical vapor deposition. The hysteresis indicates different charge trapping mechanisms dominant in each of the devices. The WSe_2 FET has increased I_D after reaching its "on-state" at V_{GS} = -5 V, whereas the MoS_2 FET has decreased I_D after reaching its "on-state" at V_{GS} = +5 V. (c) Normalized I_D after bias stress at each respective "on-state" ($V_{GS} = \pm 5$ V) at absolute drain-source voltage 0.1 V. More data and discussion in [10].

(ALD) of Al_2O_3 (see Fig. 3a for schematic device cross-section). Interestingly, we found that the hysteresis behaviors are different for MoS_2 and WSe_2 channel materials, while the bias stress behavior is in the same direction (Fig. 3b-c). In comparison, the bias stress instability is more severe for p-type WSe_2 transistors compared to n-type MoS_2 transistors. These differences could be related to the energy levels and band alignments within the respective gate stacks [10]. In fact, the dielectric defect bands and their alignments with 2D semiconductor conduction and valence bands have recently been considered as relevant design parameters to find appropriate combinations of gate stacks for bias stress stable FETs [19].

Conclusion

We have introduced different hysteresis and bias stress mechanisms and outlined their relevance to amorphous oxide and 2D semiconductor FETs. Then we discussed how the gate dielectric can impact device stability and which findings we have made for different device configurations.

Acknowledgments

A.D. and Q.T.P. gratefully acknowledge support by the German Research Foundation through the Emmy Noether Programme (506140715). S.W. and E.P. acknowledge partial support from PRISM, a JUMP 2.0 Center sponsored by the SRC and DARPA.

References

[1] M. C. Lemme, A. Daus, "Low-temperature MoS_2 growth on CMOS wafers", Nat. Nanotechnol., 18, pp. 446-447 (2023).

[2] H. Inoue, T. Hirose, T. Mizuguchi, Y. Komura, T. Saito, M. Ito, "Heterogeneous Oxide Semiconductor FETs Comprising Planar FET and Vertical Channel FETs Monolithically Stacked on Si CMOS, Enabling 1-Mbit 3D DRAM", IEEE International Memory Workshop (IMW) 2024.

[3] C.-C. Chiang, V. Ostwal, P. Wu, ,C.-S. Pang, F. Zhang, Z. Chen, J. Appenzeller, "Memory applications from 2D materials", Appl. Phys. Rev., 8, p. 021306 (2021).

[4] D. Akinwande, N. Petrone, J. Hone, "Two-dimensional flexible nanoelectronics", Nat. Commun. 5, p. 5678 (2014).

[5] Y. Y. Illarionov, T. Knobloch, M. Jech, M. Lanza, D. Akinwande, M. I. Vexler, T. Mueller, M. C. Lemme, G. Fiori, F. Schwierz, T. Grasser, "Insulators for 2D nanoelectronics: the gap to bridge", Nat. Commun., 11, p. 3385 (2020).

[6] H. G. Kim, H.-B.-R. Lee, "Atomic Layer Deposition on 2D Materials", Chem. Mater., 29, pp. 3809–3826 (2017).

[7] S. Kim, S.-H. Lee, I. H. Jo, J. Seo, Y.-E. Yoo, J. H. Kim, "Influence of growth temperature on dielectric strength of Al_2O_3 thin films prepared via atomic layer deposition at low temperature", Sci. Rep., 12, p. 5124 (2022).

[8] A. Daus, C. Vogt, N. Münzenrieder, L. Petti, S. Knobelspies, G. Cantarella, M. Luisier, G. A. Salvatore, G. Tröster, "Positive charge trapping phenomenon in n-channel thin-film transistors with amorphous alumina gate insulators", J. Appl. Phys., 120, p. 244501 (2016).

[9] P. Balakrishna Pillai, M. M. De Souza, "Nanoionics-based three-terminal synaptic device using zinc oxide", ACS Appl. Mater. Interfaces, 9, pp. 1609-1618 (2017).

[10] Q. T. Phùng, L. Völkel, A. Piacentini, A. Esteki, A. Grundmann, H. Kalisch, M. Heuken, A. Vescan, D. Neumaier, M. C. Lemme, A. Daus, "Flexible p-Type WSe_2 Transistors with Alumina Top-Gate Dielectric", ACS Appl. Mater. Interfaces, 16, pp. 60541–60547 (2024).

[11] A. Daus, L. Hoang, C. Gilardi, S. Wahid, J. Kwon, S. Qin, J. S. Ko, M. Islam, A. Kumar, K. M. Neilson, K.C. Saraswat, S. Mitra, H.-S. P. Wong, E: Pop, "Effect of Back-Gate Dielectric on Indium Tin Oxide (ITO) Transistor Performance and Stability", IEEE Trans. Electron Devices, 70, pp. 5685-5689 (2023).

[12] A. Zeumault, V. Subramanian, "Mobility Enhancement in Solution-Processed Transparent Conductive Oxide TFTs due to Electron Donation from Traps in High-k Gate Dielectrics", Adv. Funct. Mater., 26, pp. 955-963 (2016).

[13] S. Bolat, G. Torres Sevilla, A. Mancinelli, E. Gilshtein, J. Sastre, A. Cabas Vidani, D. Bachmann, I. Shorubalko, D. Briand, A. N. Tiwari, Y. E. Romanyuk, "Synaptic transistors with aluminum oxide dielectrics enabling full audio frequency range signal processing", Sci. Rep. 10, p. 16664 (2020).

[14] J. Li, Y. Zhang, J. Wang, H. Yang, X. Zhou, M. Chan, X. Wang, L. Lu, S. Zhang, "Near-ideal top-gate controllability of InGaZnO thin-film transistors by suppressing interface defects with an ultrathin atomic layer deposited gate insulator", ACS Appl. Mater. Interfaces, 15, pp. 8666-8675 (2023).

[15] S. Wahid, A. Daus, J. Kwon, S. Qin, J.-S. Ko, H.-S. P. Wong, E. Pop, "Effect of Top-Gate Dielectric Deposition on the Performance of Indium Tin Oxide Transistors", IEEE Electron Device Lett., 44, pp. 951-954 (2023).

[16] S. Wahid, L. Hoang, A. Daus, E. Pop, "Up to 100-fold Improvement of Threshold Voltage Stability in ITO Transistors", 81st Device Research Conference (DRC) (2023).

[17] S. Wahid, A. Daus, A. Kumar, H.-S. P. Wong, E. Pop, "First Demonstration of Dual-Gated Indium Tin Oxide Transistors with Record Drive Current ~2.3 mA/μm at L ≈ 60 nm and V_{DS} = 1 V", IEEE International Electron Devices Meeting (IEDM) (2022).

[18] A. Daus, S. Vaziri, V. Chen, Ç. Köroğlu, R. W. Grady, C. S. Bailey, H. R. Lee, K. Schauble, K. Brenner, E. Pop, "High-performance flexible nanoscale transistors based on transition metal dichalcogenides", Nat. Electron., 4, pp. 495-501 (2021).

[19] T. Knobloch, B. Uzlu, Y. Y. Illarionov, Z. Wang, M. Otto, L. Filipovic, M. Waltl, D. Neumaier, M. C. Lemme, T. Grasser, "Improving stability in two-dimensional transistors with amorphous gate oxides by Fermi-level tuning", Nat. Electron., 5, pp. 356-366 (2022).

Heterogeneous 3D CFET with Hybrid Channel Configuration

Sanghyeon Kim[1*], Seongkwang Kim[1], HyeongRak Lim[1], Jaeyong Jeong[1], Youngkeun Park[1], Jaejoong Jeong[1], Joonpyo Kim[1], Bongho Kim[1], Daemyeong Geum[2], Younghyun Kim[3], Byung Jin Cho[1]

*E-mail: shkim.ee@kaist.ac.kr

[1]KAIST, Korea, [2]Inha University, Korea, [3]Hanyang University, Korea

Abstract

Complementary field-effect transistors (CFETs) have been seriously studied for next-generation device architectures to improve PPA (power, performance, and area). However, many challenges remain, including process integration, structure optimization, implementation schemes (monolithic/sequential), etc. At the transistor level, unbalanced transport between n- and p-FET would be one of the most critical issues because CFETs inherently require the same width both for n- and p-FETs. Furthermore, new parameters such as spacing length between top and bottom FETs have emerged. Here, we discuss the opportunity for heterogeneous channel design to mitigate these issues.

Keywords: CFET, Si/Ge, Wafer bonding

Introduction

CFETs have been regarded as the next device architecture for logic technology [1]. PPA benefits would be the strong driver for this technology, thereby, various industry players have already reported scaled CFETs [2]-[5]. Nevertheless, there are still many issues one must consider for the device design of CFETs, as shown in Fig. 1. Even though the integration scheme has not been decided yet, both sequential and monolithic CFETs need layer stacking, providing high crystal quality and process applicability. Furthermore, due to its inherent structure, one of the most critical issues would be the unbalanced transport between n- and p-FETs. The transport characteristics of nFET on Si(100) will be much better than that of pFET on Si(100) due to the band anisotropy [6]. Also, the interlayer dielectric (ILD) between top and bottom FETs has become a new important device parameter in CFETs, but it has not been systematically investigated yet. When one discusses 3D structured devices, heat dissipation is always the issue to guarantee stable device performance and reliability. Therefore, those should be carefully investigated.

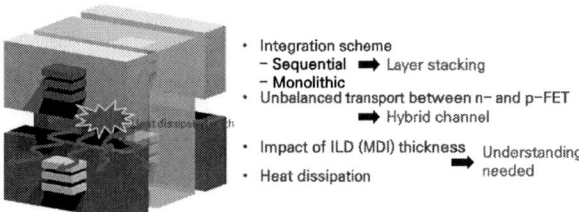

Fig. 1 Design considerations in CFET design

In this paper, we report and discuss our recent demonstration of CFETs and their component technology and investigation of the impact of the ILD thickness on electrical and thermal characteristics.

Results and Discussions

A. Layer stacking

As described above, layer stacking is needed to fabricate CFETs, regardless of their integration scheme whether sequential or monolithic. Even if epitaxial growth-based stacked structures have been popularly used in industrial CFETs, wafer bonding-based layer stacking will offer much more flexibility in process integration, materials choice, etc. For this, we have developed layer stacking based on heteroepitaxy and wafer bonding, as shown in Fig. 2 [7]-[10]. Starting from the Si substrate, Ge virtual substrate, SiGe etch stop and Ge channel layer subsequently grown. Then, plasma-assisted low-temperature wafer bonding was carried out with an oxide bonding medium. Finally, Si mother substrate, Ge, SiGe were selectively etched, resulting in Ge on insulator structure, which corresponds to the top structure of CFETs. It can be formed on the bottom devices for CFETs. This method offers flexible channel orientation capability using a Si substrate with different surface orientations and excellent thickness and bottom interface controllability. Ge films transferred using this method showed high crystallinity [7]-[10].

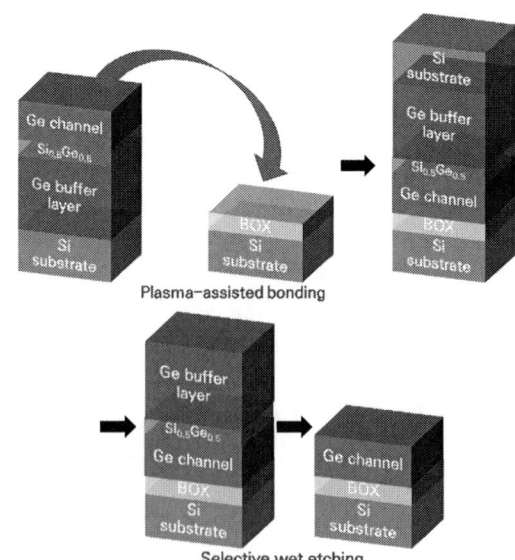

Fig. 2 Layer stacking process based on heteroepitaxy and wafer bonding

B. Potential of Ge nanosheet channel

With the Ge film transferred using the method above, we explored Ge MOSFETs. Specifically, to enhance the transport characteristics, we formed Ge(110)-OI using Si(110) mother substrate [7],[9]. From its interesting band structure, Ge(110) MOSFETs with channel direction <110> provided a significant mobility improvement (Fig. 3(c)) with decent I-V characteristics (Figs. 3(a), (3(b)), which will potentially mitigate the issue of unbalanced transport in CFETs.

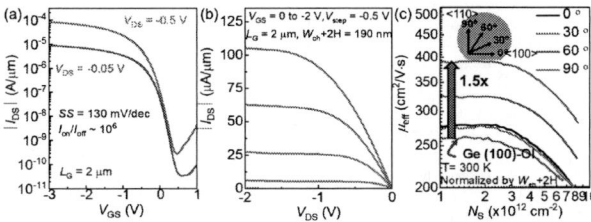

Fig. 3 The (a) transfer and (b) output curve of Ge(110) MOSFETs and (c) the channel orientation dependence of the effective mobility.

C. Impact of ILD thickness

Since CFETs have an ILD between the top and bottom FETs, so ILD thickness plays a role in electrical characteristics. The gate electrode of the bottom FETs is like a bottom gate for the top FETs in sequential CFETs, thereby, the body factor quickly increases with ILD thickness scaling, as shown in Fig. 4. The impact of this resulted in the improved transient response of CFET inverter by ILD thickness scaling, as shown in Fig. 5. This must be taken into account when we design CFETs for CMOS circuits.

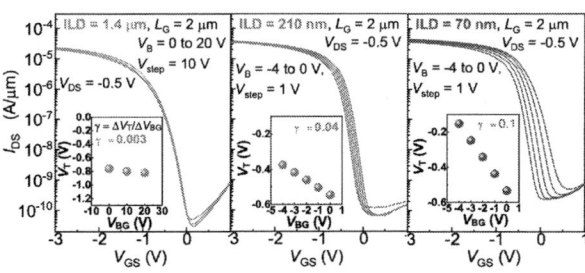

Fig. 4 The transfer curves of top Ge pFET with different gate bias to the gate of bottom Si nFET for different ILD thicknesses.

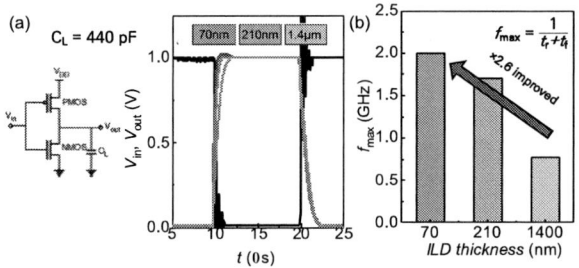

Fig. 5 (a) Transient voltage swing of the CFET inverters with different ILD thicknesses and the corresponding fmax.

D. Heat dissipation

Self-heating is a phenomenon during transistor operation, which cannot be avoidable. Therefore, heat dissipation properties are significantly important to guarantee the stable operation and reliability of the transistor and circuits, especially in 3D device architecture such as CFET. Moreover, the self-heating behaviors of CFETs have not been investigated much yet. We have investigated the self-heating behaviors of our Ge/Si heterogeneous CFETs using thermos-reflectance microscopy. Fig. 6 shows the measured temperature map of the top Ge pFETs. As predicted, the device temperature becomes hotter as the input power increases. Furthermore, it can be noticed that the ILD thickness affects the device temperature quite significantly. In turn, it provides the insight that ILD thickness scaling will be the key to promoting the heat dissipation of the top devices. Quantitatively, thermal resistance was reduced by 70% when the ILD thickness was scaled from 1400 nm to 70 nm.

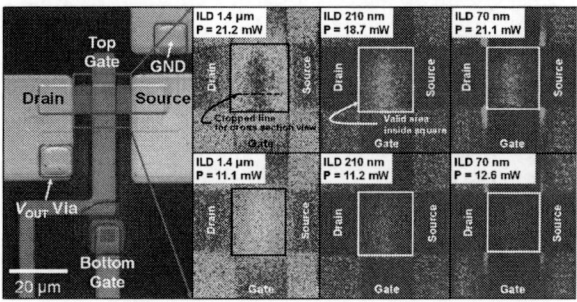

Fig. 6 Measured temperature map of the top Ge pFETs with different input power (top vs bottom) and ILD thicknesses (same raw)

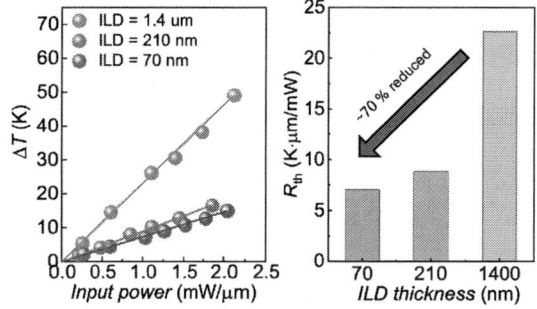

Fig. 7 Temperature increase as a function of the input power and estimated thermal resistance with different ILD thicknesses.

D. CFET demonstration and benchmarks

Leveraging layer stacking and low-temperature FET integration technologies, we have demonstrated CFETs with hybrid channels of Ge pFETs and Si nFETs. Fig 8 shows the I-V curves of CFETs and the voltage transfer curves of the inverter. The transfer curves show decent switching characteristics both for n- and p-FETs with reasonable subthreshold slope considering the EOT of the gate stack. Steep voltage transfer characteristics were also observed.

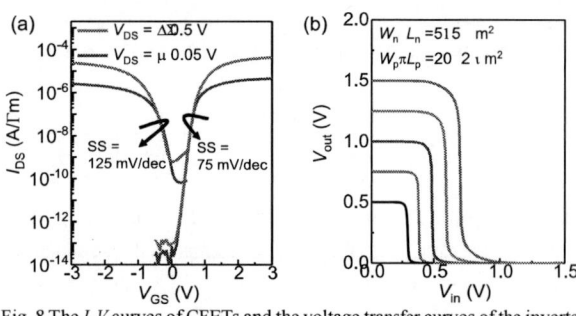

Fig. 8 The *I-V* curves of CFETs and the voltage transfer curves of the inverter with Ge pFET/Si nFET channels.

D. Future perspectives

There are still various technological issues to further explore CFETs. For example, the number of sheets should be carefully chosen considering the parasitic capacitance, resistance, and process complexity [11]. Also, multi-threshold voltage capability would be one of the important values in CMOS, thereby, threshold voltage controllability should be investigated considering the limited physical space between the sheets [12]. Furthermore, other mobility boosters such as strain, orientations, quantum confinement, etc. should be seriously considered in the future [13]. At the same time, not only CFET architecture itself but also the power delivery perspective needs to be considered considering the difficulty and increased complexity of accessing the transistors from the BEOL in CFETs [14].

Conclusion

In this paper, we discussed our recent progress on CFET including layer stacking technology based on wafer bonding, and non-Si, non-(100) Ge channel integration, which will potentially resolve the issues of drivability mismatch between nFET and pFET. With these channel configurations, we successfully demonstrated 3D CFETs with Ge/Si hybrid channels. Furthermore, we investigated the impact of ILD on the electrical properties and thermal properties of sequential CFETs.

Acknowledgments

This work was partly supported by the NRF of Korea (No. 2022M3F3A2A01065057, 2023R1A2C2002777, RS-2024-00407767, RS-2023-00224408) and Samsung Electronics.

References

[1] P. Schuddinck *et al.*, "PPAC of sheet-based CFET configurations for 4 track design with 16nm metal pitch," *IEEE Symposium on VLSI Technology and Circuits*, p. 365 (2022)

[2] V. VegaGonzalez *et al.*, "Integration of a Stacked Contact MOL for Monolithic CFET," *IEEE Symposium on VLSI Technology and Circuits*, p. T6-4 (2023)

[3] M. Rodosavljevic *et al.*, "Demonstration of a Stacked CMOS Inverter at 60nm Gate Pitch with Power Via and Direct Backside Device Contacts," *IEEE International Electron Devices Meeting (IEDM)*, 29-2 (2023)

[4] S. Sandy Liao *et al.*, "Complementary Field-Effect Transistor (CFET) Demonstration at 48nm Gate Pitch for Future Logic Technology Scaling," *IEEE International Electron Devices Meeting (IEDM)*, 29-6 (2023)

[5] J. -H. Park *et al.*, "First demonstration of 3-dimensional stacked FET with top/bottom source-drain isolation and stacked n/p metal gate ," *IEEE International Electron Devices Meeting (IEDM)*, 29-4 (2023)

[6] J. Cai, "CMOS Device Technology for the Next Decade," *IEEE Symposium on VLSI Technology and Circuits*, short course 1 (2021)

[7] S. -K. Kim *et al.*, "Heterogeneous 3D Sequential CFET with Ge (110) Nanosheet p-FET on Si (100) bulk n-FET by Direct Wafer Bonding," *IEEE International Electron Devices Meeting (IEDM)*, p. 471 (2022)

[8] S. -K. Kim *et al.*, "Role of Inter-Layer Dielectric on the Electrical and Heat Dissipation Characteristics in the Heterogeneous 3D Sequential CFETs with Ge p-FETs on Si n-FETs," *IEEE International Electron Devices Meeting (IEDM)*, 13-5 (2023)

[9] S. -K. Kim *et al.*, "Heterogeneous 3-D Sequential CFETs With Ge (110) Nanosheet p-FETs on Si (100) Bulk n-FETs," *IEEE Transactions on Electron Devices* 71, p. 393 (2024)

[10] S. -K. Kim *et al.*, "Ge(110) GAA Nanosheet / Si(100) Tri-gate Nanosheet Monolithic CFETs Featuring Record-high Hole Mobility," *IEEE Symposium on VLSI Technology and Circuits*, T5-3 (2024)

[11] W. -L. Sung *et al.*, "Characteristics of Stacked Gate-All-Around Si Nanosheet MOSFETs With Metal Sidewall Source/Drain and Their Impacts on CMOS Circuit Properties," *IEEE Transactions on Electron Devices* 68, p. 3124 (2021)

[12] N. Yoshida, "Advanced Logic Transistor Process Technology towards 1-nm Node," *IEEE Symposium on VLSI Technology and Circuits*, short course 1 (2023)

[13] C. - T. Chen *et al.*, "Hole Mobility Boosters of (110)-Oriented Extremely Thin Body SiGe-on- Insulator (SGOI) pMOSFETs," *IEEE Transactions on Electron Devices* 70, p. 3963 (2023)

[14] J. Jeong, C. Lee *et al.*, "Vertically Integrated Active Power Delivery Network (PDN) for Heterogenous 3D (H3D) Stacked Systems: 3D On-chip Integration of GaN Power Devices on PDN with Direct Heat Spreading Layer Bonding," *IEEE International Electron Devices Meeting (IEDM)*, 31-2 (2024)

Benchmarking of the BSIM-BULK for Cryo-CMOS Design

Wajid Manzoor, Nisha Manzoor, Yawar Hayat Zarkob, Debashish Nandi, Aloke K. Dutta
and Yogesh Singh Chauhan
Department of Electrical Engineering, Indian Institute of Technology, Kanpur, UP, India
Email: {wajid, aloke, and chauhan} @iitk.ac.in

ABSTRACT

This article presents a benchmarking study of compact models at cryogenic temperatures, employing BSIM-BULK as the core model. The BSIM-BULK model is validated with characterized data across multiple temperatures, showing a strong correlation with the experimental terminal characteristics. The robustness and qualitative accuracy of the model are assessed through a series of benchmark tests. The Gummel symmetry test is utilized for symmetry analysis, confirming the symmetry of the drain current and the continuity of its derivatives, down to cryogenic temperatures. The nonlinear behavior of the model is evaluated through harmonic balance tests, with slopes at each temperature aligning well with expected physical trends. Furthermore, we have applied tree-top and slope ratio tests to evaluate the smooth transition of the model from weak to strong inversion, demonstrating consistent and smooth behavior in all regions of MOSFET operation at all temperatures.

Keywords: BSIM, Benchmark test, Cryogenic Electronics, Cryo-CMOS, Compact Modeling

INTRODUCTION

As CMOS-based circuits have been identified as possible candidates for the reading, writing, and manipulation of qubits, the circuitry developed using MOSFET devices is estimated to work at cryogenic temperatures [1]. Therefore, the compact models need to predict the accurate behavior of circuits at these temperatures before fabrication. Compact models have to undergo benchmark tests before being approved as industry-standard models [2]. These tests check the symmetry, non-linearity, harmonic distortions, stabilization of bias points, etc, at different voltages. Since these tests determine the overall behavior of a model, therefore there is a need for temperature scalability of the benchmark tests of a compact model as well. In this work, we have focused on the temperature dependence of different benchmark tests that are often used to check the symmetry, non-linearity, differentiability, and continuity of a compact model.

We present the Benchmark testing of a cryo-enabled model that we have formulated using BSIM-BULK as the base model. The detailed analysis of the model is provided in our prior work [3]. Since the standard BSIM-BULK is invalid at cryogenic temperatures, it cannot stand with the changes observed in intrinsic carrier concentration, flatband voltage, mobility, mobility degradation, threshold voltage, and subthreshold swing saturation, etc [4]. We have considered all these changes in the modified model [3], and it is validated with the experimental data capturing all the temperature-dependent effects.

The symmetry of the compact model is assessed using the Gummel symmetry test, confirming that it displays the expected symmetry and its derivatives are consistent [2]. Non-linear behavior is analyzed through harmonic balance testing, and the transition from weak to strong inversion is examined via tree-top and slope ratio tests [2]. The robustness of our cryo-enabled model at different temperatures is evaluated through a series of these rigorous tests after validating it with experimental data, as discussed next.

CHARACTERIZATION AND MODELING

We characterized a low-threshold voltage (V_{TH}) device with width = 1 μm and length = 320 nm using the Lakeshore CRX-VF system and Keysight B1500A semiconductor parameter analyzer over a temperature range of 10 K to 300 K. The resulting data is modeled using a modified BSIM framework, discussed in [3]. The characterized data and its corresponding modeling are shown in Figs. 1 (a) and (b), demonstrating an excellent match between them. The drain-to-source current (I_{DS}) behavior can be described in two distinct regions based on the zero-temperature coefficient (ZTC) point. Below the ZTC point, I_{DS} decreases with decreasing temperature and reaches saturation below approximately 30 K, resulting in a sub-threshold saturation effect. Above the ZTC point, however, I_{DS} increases as temperature decreases, primarily due to the rise in carrier mobility. The parameter set extracted in ICCAP is used to perform different benchmark tests, as discussed in the next section.

BENCHMARK TESTS

A. Gummel Symmetry Test

Since BULK-MOSFET is a symmetric device, therefore exchanging source and drain terminals at a particular bias condition should give the rotational symmetry of the I_{DS} with respect to the origin [5]. A model must assure symmetry of higher order derivatives of current and terminal charges around $V_{DS} = 0$. Signals $+V_x$ and $-V_x$ are applied at the source and drain side, respectively. The applied signal makes I_{DS} an odd function. Therefore, even derivatives are zero around $V_x = 0$, and odd derivatives are non-zero continuous

979-8-3315-0417-5/25 $31.00 © 2025 IEEE

functions at $V_x = 0$ [6]. Our model shows I_{DS} and its even derivatives are odd functions around $V_x = 0$ at all the temperatures, as shown in Fig. 1(c) and Fig 2(a), and its odd derivatives are continuous down to 10 K, as shown in Figs 2 (b) and (c). The value of I_{DS} at a particular temperature depends on the values of V_x and V_{GS}. Therefore, we observe a slight variation in the value of I_{DS} in the figure. Since the rate of change of I_{DS} is different at different temperatures therefore, their derivatives with respect to V_x show increased slope as the temperature decreases. Our model shows the accurate results for even and odd derivatives, and the I_{DS} passes through the origin at zero V_x. Hence, the modified BSIM-BULK model passes the Gummel symmetry test accurately.

B. Harmonic Balance Test

Non-linearity and distortion analysis, important to radio frequency (RF) IC design, are significantly aided by Harmonic Balance (HB) simulations, which estimate the harmonic power levels generated by typically non-linear CMOS components. In common-gate circuit simulations, such as RF transmit/receive switches and passive mixers that transition through $V_{ds} = 0$ V, singularities in the terminal characteristics can lead to unphysical harmonic prediction [2], [7]. This becomes especially critical in cryo-CMOS applications, like deep space missions, where these common-gate RF circuits are frequently employed. Accurately predicting HB trends is vital for efficient cryo-CMOS IC design. An accurate compact model predicts the slopes of HB simulations correctly, i.e., the nth-order harmonic should exhibit a slope proportional to $n \times$ slope of the fundamental component. Achieving this relies on ensuring the physical correctness of derivatives and thorough continuity around $V_{ds} = 0$ V, in the expressions used to develop the cryogenic compact model. Fig. 3 illustrates the common-gate HB simulation results for the cryo-enhanced BSIM-BULK model under study [3], which accurately predicts the HB slopes up to 5th order across a wide range of temperatures, thus establishing the effectiveness of the developed model.

C. Tree-Top and Slope Ratio Tests

Tree-top and slope ratio tests are used in compact models to validate their physical behavior as the device operation shifts from weak to strong inversion. The tree-top test is given by the ratio of g_m and I_{DS} as shown in Fig 4(a). The main difference observed in the characteristics with respect to temperature is seen in the subthreshold region where the maximum value increases as the temperature decreases. The drain current varies exponentially in the subthreshold region, therefore the derivative of I_{DS} is given by [8]:

$$\frac{g_m}{I_{DS}} = \frac{1}{I_{DS}} \frac{dI_{DS}}{dV_{GS}} = \frac{1}{\phi_t} \frac{d\psi_s}{dV_{GS}} \quad (1)$$

Therefore, the theoretical maximum value of the tree-top test is $1/\phi_T$ where ϕ_T is the thermal voltage. Around the threshold voltage, the ratio saturates to the value of one at all temperatures, demonstrating the smooth transition from weak to strong inversion.

The slope ratio tests the transition of the model from weak to strong inversion using the following equation [2]:

$$S_R = \frac{(I_{DS2} + I_{DS1})(V_{DS1} - V_{DS2})}{(I_{DS2} - I_{DS1})(V_{DS1} + V_{DS2})} \quad (2)$$

where I_{DS1} and I_{DS2} are drain-to-source currents at V_{DS1} and V_{DS2} respectively. The difference between the V_{DS1} and V_{DS2} is 10 mV and the results obtained from the S_R at different temperatures are shown in Figs. 4(b) and (c). An increase in the value of S_R is expected in the weak inversion due to the exponential dependence of I_{DS1} on temperature and V_{DS} in weak inversion. These plots also saturate at the value of one at all temperatures, thus validating the S_R test. The temperature-invariant slope ratio test is not practical at cryogenic temperatures because the thermal voltage ϕ_T becomes very small. As a result, the difference between the drain voltages used in the test becomes extremely small, making it difficult to obtain meaningful measurements.

CONCLUSION

This work benchmarks the cryo-enhanced BSIM-BULK compact model using MOSFET device electrical characteristics, with a focus on its performance across a wide temperature range, including cryogenic conditions. The Gummel symmetry test confirms continuity and symmetry of terminal characteristics and its derivatives around $V_{ds} = 0$ V across varying temperatures. For RF applications, the harmonic balance test verifies consistent harmonic slopes, ensuring correct nonlinear behavior prediction even at cryogenic temperatures. Additionally, tree-top and slope ratio tests demonstrate a smooth transition from weak to strong inversion, which has paramount importance in cryogenic temperatures.

REFERENCES

[1] F. Jazaeri et al., "A review on quantum computing: From qubits to front-end electronics and cryogenic MOSFET physics," in IEEE MIXDES-26th, pp. 15-25, 2019.

[2] G. Gildenblat, "Compact Modeling: Principles, Techniques and Applications", Springer Netherlands, 2010.

[3] W. Manzoor et al., "Extending Standard BSIM-BULK Model to Cryogenic Temperatures," in IEEE TED, 71 (8), pp. 4510-6, 2024.

[4] Y. H. Zarkob et al., "BSIM-BULK 107.2.0 MOSFET Compact Model Technical Manual," UC Berkeley, Tech. Rep., 2024.

[5] H. Agarwal et al., "BSIM-Bulk MOSFET Model for IC Design-Digital, Analog, RF and High-voltage", Woodhead Publishing, pp. 245-54, 2023.

[6] P. Bendix et al., "RF distortion analysis with compact MOSFET models", in IEEE CICC, pp. 9-12, 2004.

[7] D. Nandi et al., "Validation of Dynamically Depleted Symmetric BSIM-SOI Compact model for RF SOI T/R Switch Applications", in IEEE EDTM, pp. 1-3, 2024.

[8] G. Ghibaudo et al., "A method for MOSFET parameter extraction at very low temperature," in SSE, 32 (3), pp. 221–3, 1989.

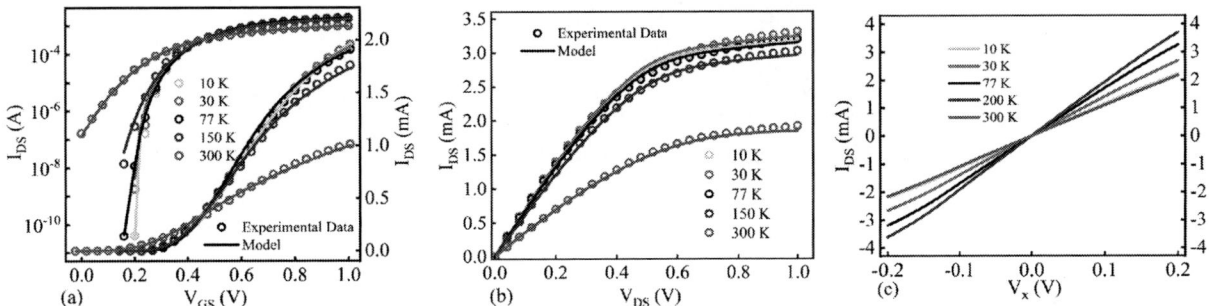

Fig. 1. (a) and (b) Experimental validation of I_{DS}-V_{GS} and I_{DS}-V_{DS} characteristics, respectively, of a bulk MOSFET at different temperatures, using our modified BSIM-BULK model. (c) Variation of the I_{DS} with respect to V_x demonstrating the symmetry at different temperatures.

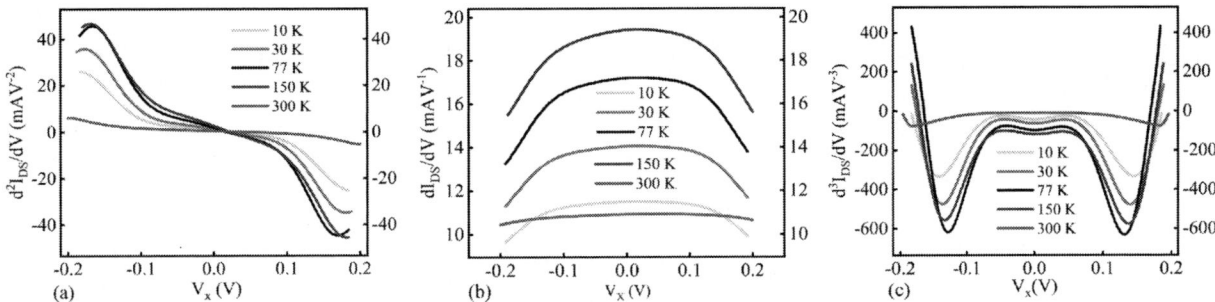

Fig. 2. (a) The variation of the second derivative of I_{DS} with respect to V_x at different temperatures behaves as an odd function, as expected from the Gummel symmetry test. (b) and (c) The variation of the odd derivatives of I_{DS} with respect to V_x demonstrates continuity at $V_x = 0$ across different temperatures, which confirms the validity of the Gummel symmetry test.

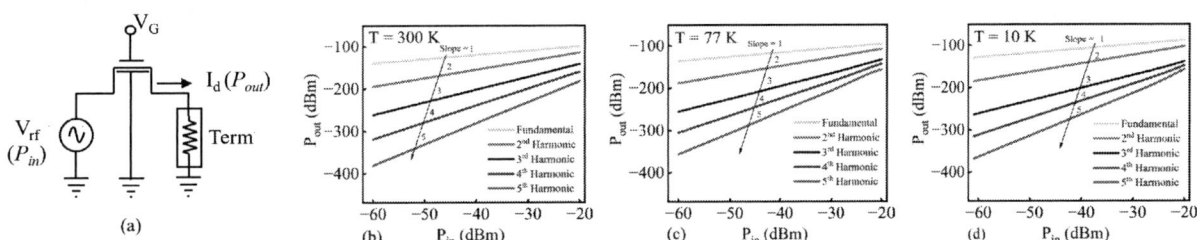

Fig. 3. (a) Simplified harmonic balance test setup where the bulk MOSFET is connected in a common-gate configuration. RF input (V_{rf}) is provided at the drain terminal while output is measured at the source. HB simulation result shows correct slopes up to 5th order where $f_0 = 1$ GHz, for (b) T = 300 K, (c) T = 77 K and (d) T = 10 K.

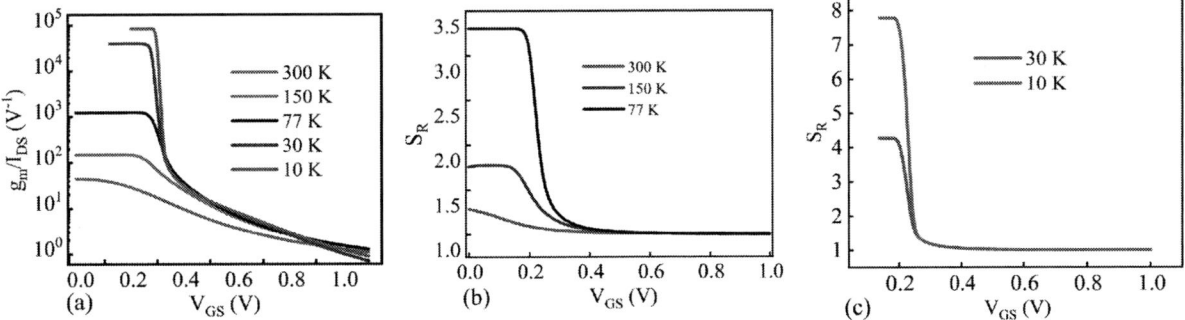

Fig. 4. (a) Validation of the tree-top test at varying temperatures reveals a smooth transition from weak to strong inversion. (b) and (c) Slope ratio test also demonstrates a smooth transition from weak to strong inversion at different temperatures.

979-8-3315-0417-5/25 $31.00 © 2025 IEEE

Oxidation of TiN Interface and Improvement of AlN Intercalation of ZrO$_2$ Capacitor in DRAM

Songming Miao[a], Xinyi Tang[a], Yuanbiao Li, Guangwei Xu, Di Lu and Shibing Long*

University of Science and Technology of China, Hefei 230026, China.

*Email: shibinglong@ustc.edu.cn, [a]These authors contributed equally to this work.

Abstract

In the DRAM capacitor field, due to the electrode TiN oxygen-gettering nature, an additional oxide layer at the interface of the capacitor was inevitably made. To prevent the formation of oxide layer, we adopted AlN as oxygen diffusion blocking layer (ODBL) and optimized the thickness of this interface layer. Moreover, AlN acted as ODBL could take into account both oxidation resistance and electrical properties, which positively contributed to the improvement of interface quality in the DRAM capacitor.

Keywords: DRAM Capacitors, Interface Oxidation, AlN

Introduction

The oxidation of electrode TiN in contact with oxide was a major challenge in engineering of TiN/high-k dielectric (HK) interfaces for DRAM capacitors [1]. Most TiN in the TiN/HK interface were oxidized to TiO$_2$ or TiON during HK film deposition and post-deposition annealing (PDA), which had low dielectric constant and could influence the growth of HK films [2, 3]. Moreover, with the development of Moore's law, the proportion of the interface layer in the MIM capacitors increased with the decrease of the dielectric film thickness [1, 4]. Thus, reducing the interfacial layer formation became more and more important. So far, various methods, such as changing bottom electrode to Ru, inserting Al$_2$O$_3$ as interface layer, had been proposed for improving the interface quality [3, 5]. But these methods might lower the total effective dielectric constant due to reducing the crystallization of HK films and low dielectric constant of the interface layer. In this study, we investigated the interfacial oxidation phenomenon and considered both inhibiting of the oxidation layer at the interface and improving the performance of the capacitor by using AlN as ODBL.

Device Structure and Fabrication

We fabricated metal-insulator-metal (MIM) bilayer capacitors using atomic layer deposition (ALD) process (**Fig. 1(a) and 1(b)**). To investigate the effect of oxygen interface layer, we carefully control the N$_2$ atmosphere with little oxygen for bottom electrode to amplifying the influence of oxygen diffusion. The silicon wafers covered with TiN film were went through a rapid thermal annealing process with the annealing temperature of 500 °C, 600 °C and 700 °C for 10 min. TEM image was shown in **Fig. 1(c)** to show more details about the interface between TiN and ZrO$_2$. All films were not annealed after depositing ZrO$_2$ and top electrode. The shape of the capacitor was a circle with a radius of 50 μm.

Results and Discussion

A. Effect of the Oxidation Process on Capacitors

To determine the influence of oxygen diffusion, the TiN films were characterized by KPFM, shown in **Fig. 2(a)**. The results showed that the work function of TiN without annealing was consistent with the annealed films and also the same as TiO$_2$ [6]. This phenomenon indicated that the TiN electrode was easily oxidized to TiO$_2$. For the surface morphology, the surface roughness (R_a) increased with the annealing temperature, shown in **Fig. 2(b)**, **2(c)** and **Fig. 3**. This trend was related to the formation of TiO$_2$ at the interface after annealing. In detail, **Fig. 4** showed the crystallization of the TiN films after different annealing temperature by GIXRD. For the TiN as deposited without annealing, the characteristic peaks at 36.7° corresponded to the (111) plane of the cubic (c) phase. For the films after annealing, each sample exhibited strong diffraction peak at 27.4°, corresponding to the (110) plane of the rutile (r) phase of TiO$_2$ [7]. And the oxidation degree increased with the annealing temperature.

Fig. 5(a) showed the relative dielectric constant-voltage (k-V) curve at various annealing temperatures of TiN electrode. The annealing process had a significant effect on the relative dielectric constant which seriously decreased (~27 as deposited and only ~8 after annealing at 700 °C). Meanwhile, electrical properties of capacitors were obtained by both Capacitance-Voltage (CV) and Quasi-Static Capacitance Voltage (QSCV) measurements. In **Fig. 5(b)**, the test results of CV were much lower than QSCV, showing that capacitors had a lot of defects probably in the TiO$_2$ interface layer formed by the oxidation of TiN.

B. Inhibiting the Oxidation Process by Intercalation

To solve the problem of interfacial oxidization, we proposed an AlN intercalation method, which AlN was inserted between the bottom electrode TiN and the dielectric film ZrO$_2$, shown in **Fig. 6**. For the surface morphology, the roughness of 3 nm AlN interface layer keep the same after the 600 °C annealing process, but become worse when the annealing temperature was up to 700 °C, shown in **Fig. 3**. The dielectric constant via AlN thickness (k-d) at various annealing temperature of the new MIM capacitor structure was shown in **Fig. 7**. With 3 nm AlN interface layer, the

979-8-3315-0417-5/25 $31.00 © 2025 IEEE

dielectric constant increased slightly after 600 °C annealing process, but became worse after 700 °C annealing process.

The current density for ZrO_2 MIM capacitors at various annealing temperature without (**Fig. 8(a)**) and with (**Fig. 8(b)**) AlN interface layer were shown. Without AlN films, the current density decreased after annealing in the low field region, but it was independent of the annealing temperature in the high field region. However, the current density with AlN films was different in the high negative bias region. The higher the annealing temperature, the lower the current density displayed. This trend mean that the leakage mechanism might change at the TiN/AlN/ZrO_2/TiN capacitors after annealing. With AlN films as ODBL, the capacitors could better meet the leakage target of DRAM. Therefore, the subsequent annealing process could have a larger process window.

C. XPS Analysis of the Films Using AlN as ODBL

Furthermore, it is important to determine the correlation between oxidization process and influence by analyzing XPS peaks at the interface of the TiN/ZrO_2 layer. **Fig. 9(a)** showed the full spectrum of XPS without AlN layer as deposited, showing the elemental composition of the film. And **Fig. 9(b)** showed the Ti 2p XPS spectra at the bottom interface of the MIM capacitors as deposited and 600 °C annealing without AlN films. Among the various chemical states of Ti, the binding energies of TiN, TiON and TiO_2 were located at 454.8 eV, 456.52 eV and 458.98 eV [3]. Even though the interface was not annealed, unamplifying the oxidization process, the XPS peak of TiON supported that the bottom electrode could be oxidized during the ALD process of ZrO_2 or just exposed to the air. And after 600 °C annealing, the XPS peak of Ti 2p supported that the interface of TiN was oxidized to TiO_2. **Fig. 9(c)** and **Fig. 9(d)** showed the Al 2p and Ti 2p XPS peaks at the bottom interface (TiN/AlN/ZrO_2) of the MIM capacitors as deposited and 600 °C annealing process with 3 nm AlN films. The same XPS peaks indicated that both the AlN film and bottom electrode TiN were not oxidized during annealing process, and the AlN interface layer could protect the TiN film well. Meanwhile, the severe dielectric constant reduction in **Fig. 5(a)** was due to the interface oxidization TiO_2 layer, in which the oxygen vacancies widely existed after annealing.

In order to better understand the role of AlN interface layer, we illustrated the schematics of the possible oxidization mechanism (**Fig. 10(a-d)**) from the combined results of electrical measurements, GIXRD patterns and XPS depth profile intensity. **Fig. 10(a)** and **Fig. 10(b)** described the method of TiN oxidization during low pressure oxygen atmosphere annealing without AlN interface layer. During this process, the precursor ozone or oxygen in the air could easily contact the bottom electrode TiN and then oxidized TiN to TiO_2. There would be more vacancies in the interfacial

layer, resulting in severe dielectric constant reduction. Meanwhile, oxidation could be inhibited by AlN ODBL and the oxygen vacancies were not being created. As shown in **Fig. 10(c)** and **Fig. 10(d)**, oxidation reaction will not occur in the case of inserting AlN film as interface layer. Thus, the degree and influence of interfacial oxidation were controlled simultaneously by using AlN layer in the DRAM capacitors.

Conclusion

In summary, we investigated the oxidation in two types of bilayer capacitors (TiN/ZrO_2/TiN and TiN/AlN/ZrO_2/TiN) by changing the annealing temperature for bottom electrode. There would be severe dielectric constant reduction and many defects after oxidation. Very precisely, the AlN interface layer acting as ODBL protected TiN from oxidation and inhibited the above phenomenon. More importantly, the dielectric constant of the entire capacitor remained unchanged and the leakage current was reduced. In the fabrication of DRAM, the conditions were not as extreme as our work, so the AlN layer could be thinner to meet the thickness requirements of the capacitor. We believe that this study could contribute to a better understanding of the oxidization in DRAM capacitors, which also could offer a way to control the oxidizing process for using TiN as electrode.

Acknowledgments

This work is supported by the NSFC under Grant Nos. U20A20207. This work was partially carried out at the Center for Micro and Nanoscale Research and Fabrication of University of Science and Technology of China.

References

[1] Silva J P B, Roadmap on ferroelectric hafnia-and zirconia-based materials and devices. APL Materials, 11(8) (2023).

[2] Jeon W, Recent advances in the understanding of high-*k* dielectric materials deposited by atomic layer deposition for dynamic random-access memory capacitor applications. Journal of Materials Research, 35(7): 775-794 (2020).

[3] Lee J H, Thompson D, et al. Improved interface quality of atomic-layer-deposited ZrO_2 metal-insulator-metal capacitors with Ru bottom electrodes. Thin Solid Films, 701: 137950 (2020).

[4] Lim H J, Study of Metal–Dielectric Interface for Improving Electrical Properties and Reliability of DRAM Capacitor. Advanced Materials Technologies, 8(20): 2200412 (2023).

[5] Chen H, Significant improvement of ferroelectricity and reliability in $Hf_{0.5}Zr_{0.5}O_2$ films by inserting an ultrathin Al_2O_3 buffer layer. Applied Surface Science, 542: 148737 (2021).

[6] Mansfeldova V, Work function of TiO_2 (anatase, rutile, and brookite) single crystals: effects of the environment. The Journal of Physical Chemistry C, 125(3): 1902-1912 (2021).

[7] Tai L, Towards Low-thermal-budget Processing in Ferroelectric $Hf_{0.5}Zr_{0.5}O_2$ Thin Films by Ozone Interface Oxidation. IEEE Electron Device Letters (2023).

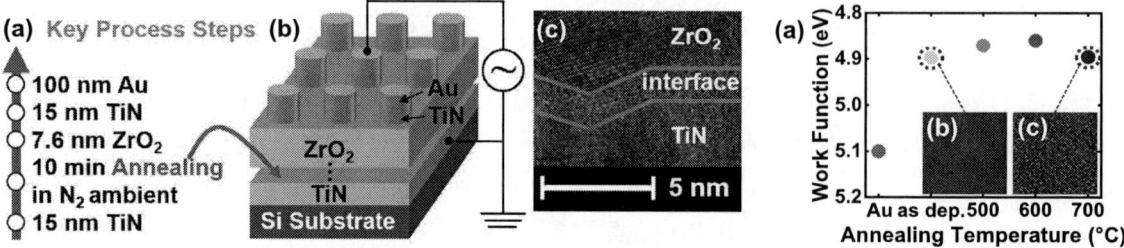

Fig. 1. Process integration steps (a) and schematic illustration of the device structure (b) for the TiN/ZrO₂/TiN capacitor. (c) TEM image of the interface layer between TiN and ZrO₂.

Fig. 2. Surface potential (a) obtained by KPFM for TiN films with different annealing temperature and AFM morphology images (b), (c).

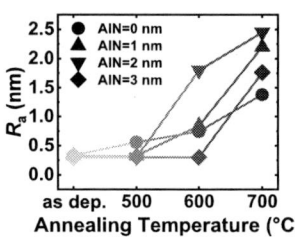

Fig. 3. Surface roughness vs annealing temperature with different AlN thickness.

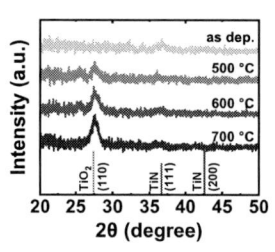

Fig. 4. GIXRD spectra of bottom electrode TiN with different annealing temperature.

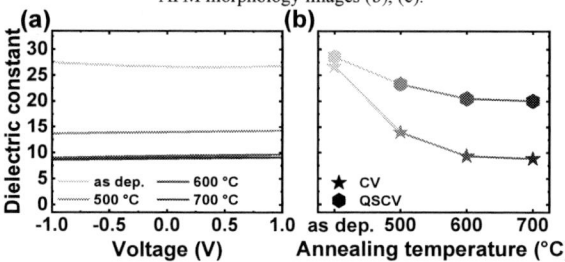

Fig. 5. (a) Dielectric constant with bias voltage for ZrO₂ films at various annealing temperature. (b) Dielectric constant at various annealing temperature by CV and QSCV test.

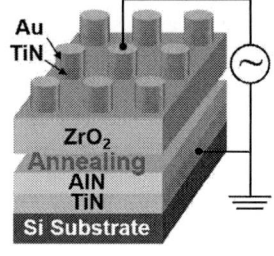

Fig. 6. Device structure and annealing process for the TiN/AlN/ZrO₂/TiN capacitor.

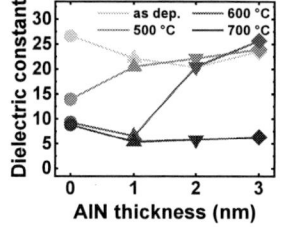

Fig. 7. Dielectric constant vs AlN thickness at various annealing temperature.

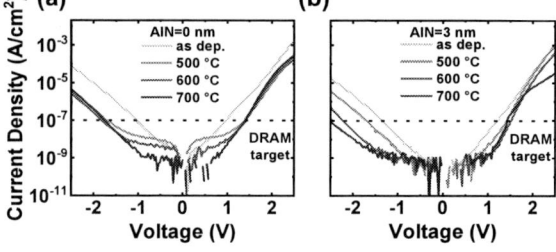

Fig. 8. The current density for ZrO₂ MIM capacitor at various annealing temperature with (a) and without (b) AlN films.

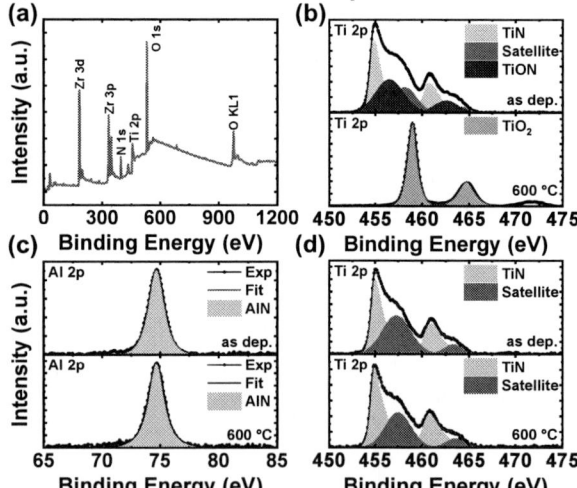

Fig. 9. (a) Full spectrum of XPS for TiN/ZrO₂ without annealing. (b) Ti 2p peaks for TiN/ZrO₂ as dep. and 600 °C annealing process. Al 2p (c) and Ti 2p (d) peaks for TiN/AlN/ZrO₂ as dep. and 600 °C annealing process.

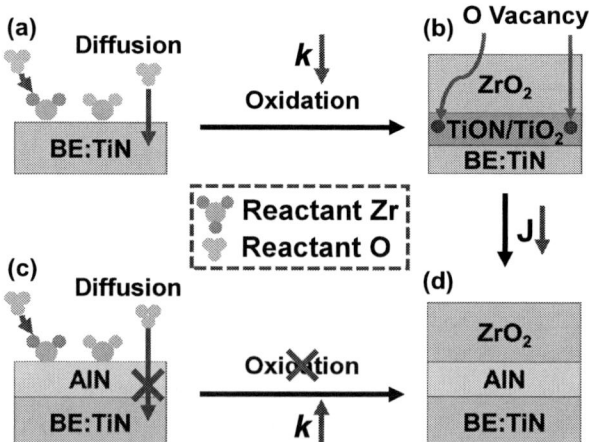

Fig. 10. Mechanism of the proposed interfacial oxidation controlling methods in the bilayer capacitors: (a) TiN film and oxidation process, (b) oxidation without ODBL, (c) TiN/AlN film and oxidation process, (d) inhibiting the oxidation process by AlN film.

979-8-3315-0417-5/25 $31.00 © 2025 IEEE

Heterogeneously Integrated Intelligent System for Learning at the Edge

Wenju Huo[1], Peng Chen[1,2], Peng Lin[1,2,*], Gang Pan[1,2,3]

[1] College of Computer Science and Technology, Zhejiang University, China, [2] State Key Laboratory of Brain Machine Intelligence, China, [3]MOE Frontier Science Center for Brain Science and Brain-Machine Integration, Zhejiang University, Hangzhou, China. *penglin@zju.edu.cn

Abstract

Edge learning deploys training directly on IoT devices to enable real-time, on-device adaptation. The edge learning reduces data movement and processing overhead, but puts high demand on computing capability of edge devices. In this paper, we discuss a unified heterogeneous integration architecture, building on our recent advancements in 3D memristors, ECRAM arrays, and 3D hetero-integration techniques. With 3D memristor arrays for convolutional preprocessing and ECRAM arrays for efficient in-situ learning, the system achieves high-throughput, low-power processing suitable for real-time applications.

Keywords: Edge learning, 3D stacking, Hetero-integration

Introduction

Incorporating sensors with computing units provides efficient, real-time and secure processing capability in the decentralized and intelligent system **(Fig. 1a)**. In traditional smart edge devices, data acquisition and processing are often separated **(Fig. 1b)**, requiring data shuttling between sensors, processors and memories. The inefficient data processing mechanism consume too much power, thus setting a limit on the system capacity, and making learning too costly to be implemented in edge devices.

Near/in-sensor computing with emerging smart devices offers a new solution by integrating sensors, memories and computing functionalities within a unified circuit [1][2], which provides unparallel processing efficiency and capabilities **(Fig. 1c)**. These new systems are well-suited for applications that demand rapid data processing and low-energy operation, such as real-time environmental monitoring, autonomous navigation systems, and intelligent sensing networks.

To advance the capabilities of edge computing system, we propose a unified sensing-learning architecture that combines high-performance sensors, 3D-integrated preprocessors, and in-memory learning units based on electrochemical random-accessed memory (ECRAM). This architecture leverages 3D stacking with heterogeneous integration to connect layered functional units vertically, building direct signal processing pathways between different modules, thus could significantly reduce the data communication overhead in conventional systems. The unique stacking capability is enabled by the mild device fabrication processes of memristors and ECRAMs suitable for 3D stacking, and advanced assembling technologies to integrate high quality sensors with the computing units. The flexible design of 3D array topologies matching those in neural networks could achieve natural and physical computing beyond simple regression and classifications. By enabling edge learning with high-throughput sensory-computing integration, this system could address the common limitations of traditional edge computing hardware.

The 3D integrated edge learning system

The proposed 3D edge learning system is illustrated in **Fig. 1d**, comprising three main components: sensory units, data preprocessors based on 3D memristor arrays, and learning chips based on ECRAM array, stacked on top of the other. The sensory units capture environmental data and encode them into electrical signals. The encoded data is forward to the memristor preprocessor with a 3D structure that can be tailored for unique functions such as convolution, and enabling dimension reduction and feature extraction of the raw inputs. Finally, the ECRAM array is used as output layer of the system with excellent programming capability, thus enabling efficient learning in edge devices. The integration of these components implements a compact, efficient system capable of high-throughput, low-power edge learning for real-time applications. The following sections discuss key technologies involved in the proposed 3D hetero-integrated systems.

3D hetero-integration technology

One prominent challenge of such system is to heterogeneously integrate different components together. While some devices such as memristors and ECRAM are stacking friendly, high-quality sensors and conventional complementary metal oxide semiconductor (CMOS) components normally demand extreme processes such as high temperature, making their monolithic integration difficult. Advanced fabrication methods such as transfer-printing could be used to prepare high sensitivity sensor material on a donor substrate and transfer to the 3D chip. In an earlier work, we built a heterogeneously integrated AI chip composed of single crystal InGaP light emitting diode (LED), GaAs photodetectors (PD) and AgCu-aSi based memristor arrays [3] **(Fig. 2)**. The InGaP and GaAs films were epitaxially grown at high temperatures before being transferred to the target substrate using epitaxial lift-off (ELO) process. The hetero-integrated chip achieved image sensing and processing within same chip and enabled efficient light communications between different vertically aligned chips,

979-8-3315-0417-5/25 $31.00 © 2025 IEEE   1143

offering a Lego-like reconfigurable functionality.

High-throughput preprocessing with 3D memristor

Preprocessing is a necessary step in complex tasks which effectively extract useful features from raw input. Applying preprocessing at the edge is important to avoid communication overhead to transfer large volume raw and often redundant data captured by sensory units. A standard preprocessing for vision task is filtering, or convolution, which can be used to extract features such as edges of the objects in images. A convolution is achieved through kernel operation scanning over the input images, which is repeated matrix computations between kernel weights and input pixels. The kernel operation can be implemented using traditional 2D memristor arrays (**Fig. 3**). In this setup, memristors at each cross-point represent kernel weights, and kernel output was directly obtained through analog multiply-accumulate operations based on Ohm's Law and Kirchhoff's Law [4]. While 2D arrays provide a feasible means for convolutional operations, it cannot achieve parallel kernel processing for different sensory pixels due to mismatch between topology of convolutions and 2D arrays. Therefore, connection between an image sensor and 2D arrays require additional buffers and processors, and are not suitable for the proposed 3D edge system.

To address this issue, we introduced a novel array design paradigm by tailoring 3D array topologies to match the network topologies of the image convolutions [5-8], allowing parallel processing of the sensory input. **Fig. 4** illustrates the schematic and micrograph of the 3D memristive circuits composed of eight layers of monolithic integrated HfO_2 memristors. This unique 3D design enables parallel convolution operations beyond the benefits of high integration density in traditional 3D architecture. Based on this design, preprocessing modules such as Prewitt filters for edge detection were implemented for video processing (**Fig. 5**), which may be used in the proposed 3D integrated system.

Learning with ECRAM arrays

In our proposed 3D edge learning system, ECRAM arrays serve as a key component to enable in-situ learning capabilities. Unlike conventional memristors, which need dozens of write cycles to achieve acceptable programming accuracy, ECRAM devices can accurately adjust conductance using single voltage pulse without the need of verify reads (**Fig. 6**). This is attributed to a quasi-linear relationship between the conduction and oxygen concentration of the channels, and a stable oxygen ion exchange rate at the channel/electrolyte interface, contributing to the linear and symmetric conductance modulation of ECRAM devices. To investigate the learning capability of ECRAM, we fabricated a 10×10 ECRAM array to demonstrate array-level analog weight updates [9].

Uniform conductance modulation of ECRAM devices was observed across the array, achieving open-loop pattern programming which was rarely achieved in other emerging devices. The ECRAM array was used to train a two-layer perceptron for classification tasks, which achieved fast convergence and 99.7% recognition accuracy (**Fig. 7**). The demonstration showed that addition of ECRAM arrays to the 3D architecture is a promising approach to enhancing efficient learning at the edge.

Conclusion

3D heterogeneous integration of sensing and in-memory learning units opens up the possibilities for high-throughput edge learning in resource-limited environment. Despite challenges such as communication fidelity between sensing and learning units, the scalability of 3D memristor and ECRAM integration, our approach points towards a feasible system-level solution.

Acknowledgments

This work was supported in part by National Key R&D Plan of China (2022YFB4500100), Natural Science Foundation of China (62404198), Major Program of Natural Science Foundation of Zhejiang Province in China (LDQ23F040001).

References

[1]. Zhou F and Chai Y, Near-sensor and in-sensor computing, Nature Electronics. 2020;3(7):664-71.

[2]. Chen P, Xiong X, Zhang B, Ye Y, Pan G and Lin P, Neuromorphic auditory classification based on a single dynamical electrochemical memristor, Neuromorphic Computing and Engineering. 2024, 4(1); 014012.

[3]. Choi C, Kim H, Lin P, Kim J et al., Reconfigurable heterogeneous integration using stackable chips with embedded artificial intelligence, Nature Electronics. 2022; 5:386-93

[4]. Yeon H, Lin P, Choi C, Tan SH, Park Y, Lee D et al., Alloying conducting channels for reliable neuromorphic computing, Nature Nanotechnology. 2020;15(7):574-9.

[5]. Lin P, Li C, Wang Z, Li Y, Jiang H, Song W et al., Three-dimensional memristor circuits as complex neural networks, Nature Electronics. 2020;3(4):225-32.

[6]. Lin P and Xia Q, Tutorial: Fabrication and three-dimensional integration of nanoscale memristive devices and arrays, Journal of Applied Physics. 2018;124(15):152001.

[7]. Lin P, Pi S and Xia Q, 3D integration of planar crossbar memristive devices with CMOS substrate, Nanotechnology. 2014; 25:405202.

[8]. Lin P, and Xia Q, Three-dimensional hybrid circuits: the future of neuromorphic computing hardware, Nano Express 2021; 2(3): 031003.

[9]. Chen P, Liu F, Lin P, Li P, Xiao Y, Zhang B, Pan G, Open-loop analog programmable electrochemical memory array, Nature Communications. 2023;14(1):6184.

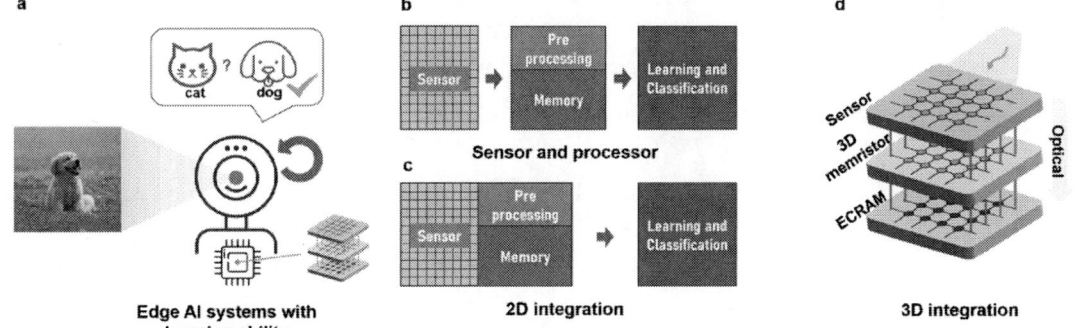

Figure 1. Evolution of intelligent system for learning at the edge. (a) Intelligent edge systems with sensing, computing, and learning capabilities. (b) Traditional edge computing systems consist of three separate parts: sensing, processing with memory and classification with learning ability, requiring data shuttling between sensors, processors and memories. (c) The near/in-sensor computing system combines sensing, memory and processing in a 2D integrated architecture, which provides unparallel processing efficiency and capabilities [1]. (d) The proposed 3D integrated edge learning system combines high-performance sensors, 3D-integrated preprocessors, and in-memory learning units based on electrochemical random-accessed memory (ECRAM).

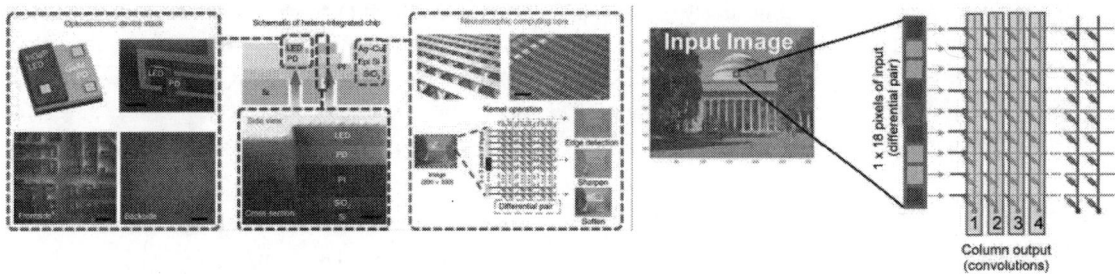

Figure 2. 3D hetero-integration technology of sensor computing systems for edge computing. A system with stackable and replaceable chips that can directly classify information from a light-based image source was created. Three different types of operation (edge detection, sharpen and soften) were performed in memristor crossbar arrays [3].

Figure 3. Memristor array implementation of convolutional kernels. Four convolutional kernels were programmed into four columns of the 32 × 32 array for parallel kernel operation [4].

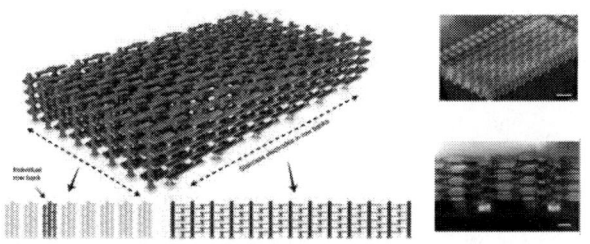

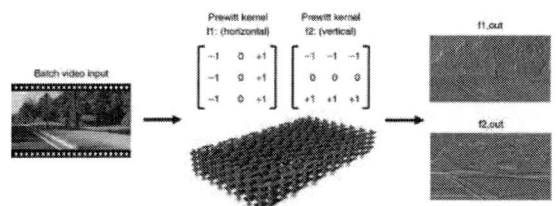

Figure 4. Extending the memristors into 3D provides substantial improvement to the speed and energy efficiency while running complex neural network models. An eight-layer prototype 3D array was built with purposely designed connections that are able to perform parallel convolutional kernel operations in the CNN [5].

Figure 5. Parallel video processing using the 3D circuits. Two sets of Prewitt kernels (f1 and f2) were programmed into the 3D circuits to perform parallel edge detection [5].

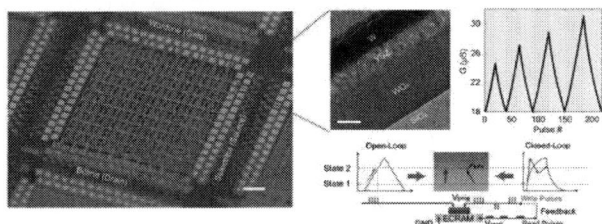

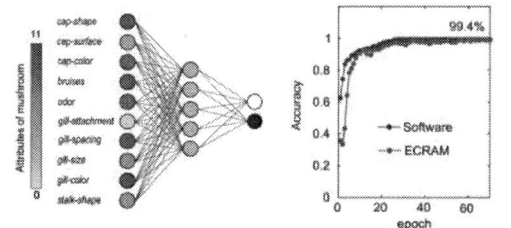

Figure 6. Open-loop analog programmable electrochemical memory array (ECRAM) provides a promising toolset for in-situ training tasks. Training neural networks with open-loop analog programming in ECRAM array can significantly reduce the design complexity of the system, which is highly advantageous over existing write-and-verify methods [9].

Figure 7. ECRAM can contribute to software-comparable training accuracy for neural networks. After the in-situ training, a bi-layer neural network reached a classification accuracy of 99.4% on the whole database [9].

979-8-3315-0417-5/25 $31.00 © 2025 IEEE

Novel GaN Integrated Photonics For Advanced Optical Communication and Imaging

Muhammad Hunain Memon, Huabin Yu, Haochen Zhang, and Haiding Sun*

*Email: Haiding@ust

iGaN Laboratory, School of Microelectronics, University of Science and Technology of China, Hefei, People's Republic of China.

Abstract

This work presents two significant advances in monolithic III-nitride optoelectronics for optical communication. First, optimized triangular micro-diodes are demonstrated to achieve high modulation bandwidth and responsivity, enabling efficient bidirectional data transmission. Second, a novel three-terminal diode architecture is introduced, offering tunable light emission and reconfigurable logic functionality. These innovations collectively enhance emission efficiency, detection speed, and multifunctionality, significantly advancing the development of integrated on-chip optoelectronic systems for next-generation optical communication technologies.

Keywords: III-nitride, Integrated optoelectronic and Optical communication

Introduction

Optical communication systems, which leverage light for transmitting information, are becoming increasingly important due to their high-speed data transfer capabilities and secure communication potential. These systems rely on critical optoelectronic devices such as light-emitting diodes (LEDs) and photodetectors (PDs), which serve as transmitters and receivers, respectively. While traditional optical communication has focused on visible and near-infrared wavelengths, advances in the development of new optoelectronic devices, especially in the ultraviolet spectrum, are opening new opportunities for faster, more secure, and higher-capacity communication systems [1-3]. However, several challenges persist in achieving high performance in both emission and detection, including limitations in modulation bandwidth, emission efficiency, and detection speed [4]. As the demand for more sophisticated communication systems grows, overcoming these challenges remains a key focus in the development of advanced optoelectronic devices [5].

This work introduces two key advancements in monolithic III-nitride optoelectronics for optical communication systems. First, we present optimized triangular micro-diodes (T-μ-diodes) that function as both light emitters and detectors on a single platform [6]. As emitter, the T-μ-diodes show an increase in output power and achieve a modulation bandwidth of 566 MHz. As detectors, T-μ-diodes exhibit a responsivity of 200 mA/W and a response time of 3.7 ns.

Second, we introduce a novel three-terminal diode architecture that integrates a traditional p-n diode with a third terminal (Tt) composed of metal/Al_2O_3 layers [7]. This design enables tunable light emission and reconfigurable logic functionality, boosting the modulation bandwidth to 264 MHz—compared to the 160 MHz of conventional two-terminal LEDs. Additionally, the three-terminal diode (TTD) allows the creation of reconfigurable optoelectronic logic gates, offering greater versatility in integrated systems. These device optimizations and multifunctional architecture enable simpler, more compact designs capable of dynamically switching between emitting and detecting functions, unlocking new possibilities for advanced, energy-efficient optoelectronic systems for advanced optical communication.

Results and Discussion

The AlGaN-based DUV MQW diode wafers were grown on 2-inch c-plane sapphire substrates using metal-organic chemical vapor deposition (MOCVD). The first design involves a triangular-shaped T-μ-diode, as shown in Fig. 1(a), with a chip size measuring 2826 μm². A scanning electron microscope (SEM) image of the fabricated device is presented in Fig. 1(b). The schematic representation and SEM image of the TTD are displayed in Figs. 1(c) and 1(d). To optimize current spreading and enhance light emission and detection performance, a finger-shaped p-electrode

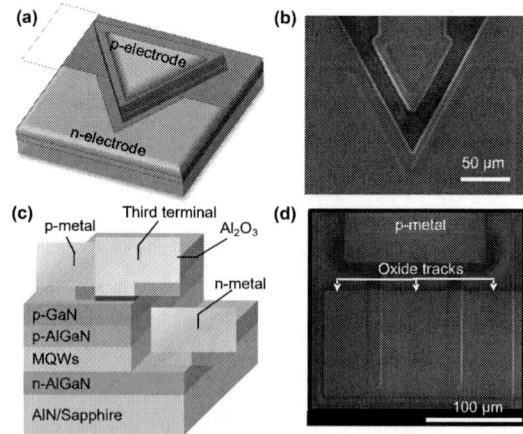

Fig. 1: (a) The 3D schematic illustration of triangular micro-diode (T-μ-diode). (b) SEM images of the fabricated T-μ-diode. (c) Schematic structure of three-terminal diode (TTD). (d) SEM captured image of the fabricated TTD.

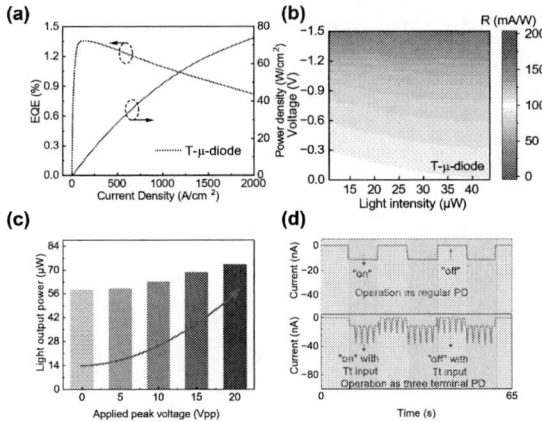

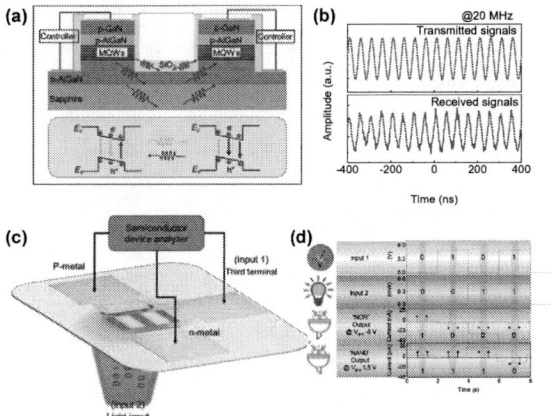

Fig. 2: (a) Power density, and EQE of T-µ-diode. (b) Color contour plot of the responsivity of T-µ-diode under 265 nm irradiation with different light intensity and various biases. (c) LOP of the TTD measured under AC sine signal of 10 MHz with different V_{pp} values of 0, 5, 10, 15, and 20 V applied to the Tt at a constant current of 2 mA. Time-dependent photoresponse of TTD.

Fig. 4: (a) The on-chip communication system using T-µ-diodes as the transmitter and receiver. (b) Transmitted and recivded signal via T-µ-diodes. (c) The testing system of TTD for optoelectronic logic gates. (d) Device operation mechanism for different arrangements on Vp-n with respect to light intensity represents the NOR and NAND logic working specifications.

configuration was adopted. Furthermore, metal/Al₂O₃ tracks, designed in a similar finger-like pattern, were incorporated between the p-metal finger. The light output power density and external quantum efficiency (EQE) are presented in Fig. 2(c). With increasing current density, the T-µ-diode exhibits lower series resistance, indicating improved thermal management under high injection currents. At an injection current density of 2 kA/cm², the T-µ-diode demonstrates exceptional performance. The device responsivity (R) was

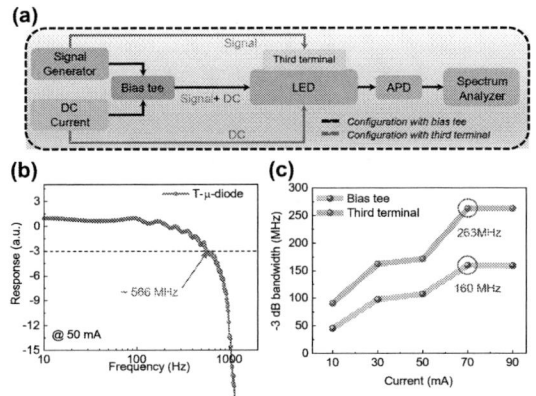

Fig. 3: (a) Optical communication bandwidth testing system. (b) Obtained bandwidth using T-µ diode at an applied current of 50 mA. (c) The -3 db bandwidth plot with bias tee and third terminal configuration at different applied current conditions.

evaluated using a 265 nm DUV light source, achieving a significantly high R of over 200 mA/W, as shown in Fig. 2(b). The TTD displayed tunable behavior when s an AC sine signal of 10 MHz with varying peak-to-peak voltages (Vpp) applied to the Tt, while keeping a constant injection current of 2 mA at the p- and n-metal, as depicted in Fig. 2(c). Additionally, the TTD was tested as a photodiode, and its

photoresponse was analyzed under DUV illumination (~260 nm) with and without a DC bias applied to the Tt. Without bias, it operates as a standard PD, whereas with a DC bias, the Tt enables tunable photo response, as shown in Fig. 2d. The optical communication system shown in the Fig. 3(a) with different electrical configuration the regular configuration with bias-tee and with Tt. Using the regular configuration, the T-µ-diode obtain the -3 db optical bandwidth of 566 MHz. Using two different configuration the obtained – 3 db optical bandwidth of TTD on different applied current conditions can be seen in Fig. 3(c). both of the diode shows significant obtained bandwidth. The testing setup for T-µ-diodes in on-chip communication is illustrated in Fig. 4(a). A 20 MHz signal was used for the evaluation, and the transmitted and received signals are shown in Fig. 4(b). Furthermore, the TTD was employed to configure optoelectronic logic gates. By altering the bias conditions, the configuration of gates, such as NOR and NAND, can be dynamically switched. The constant voltage applied to the third terminal, in conjunction with the optical signal, facilitates the creation of logic gates, as depicted in Fig. 4(d).

Conclusion

In summary, we have demonstrated two significant advances in III-nitride optoelectronic devices designed for optical communication systems. First, we present triangular micro-diodes that can both emit and detect light on a single platform. These diodes exhibit substantial performance improvements, including higher light output, a modulation bandwidth of 566 MHz, enhanced responsivity of 200 mA/W, and a faster response time of 3.7 ns, making them ideal for high-speed communication applications. Second, we introduce a novel three-terminal diode configuration that integrates a conventional p-n diode with an additional terminal made of metal/Al_2O_3 layers. This innovative design enables tunable light emission, reconfigurable logic functionality, and a modulation bandwidth increase to 264 MHz, along with the ability to implement optoelectronic logic gates. These optimizations and integrations result in efficient, compact, and multifunctional optoelectronic systems capable of dynamically switching between emitting and detecting, marking a significant advancement for next-generation optical communication technologies.

Acknowledgments

We would like to thank the Information Science Center of the University of Science and Technology of China for the hardware/software services and the China Postdoctoral Foundation for financial support.

References

[1] A. Vavoulas, H. G. Sandalidis, N. D. Chatzidiamantis, Z. Xu, and G. K. Karagiannidis, "A survey on ultraviolet C-band (UV-C) communications," *IEEE Communications Surveys & Tutorials,* vol. 21, no. 3, pp. 2111-2133, 2019.

[2] H. Yu *et al.*, "Miniaturized AlGaN-Based Deep-Ultraviolet Light-Emitting and Detecting Diode with Superior Light-Responsive Characteristics," *Advanced Optical Materials,* p. 2400499, 2024.

[3] S. Xiao *et al.*, "In-depth investigation of deep ultraviolet MicroLED geometry for enhanced performance," *IEEE Electron Device Letters,* 2023.

[4] Z. Xu and B. M. Sadler, "Ultraviolet communications: potential and state-of-the-art," *IEEE Communications Magazine,* vol. 46, no. 5, pp. 67-73, 2008.

[5] A. Ren *et al.*, "Emerging light-emitting diodes for next-generation data communications," *Nature Electronics,* vol. 4, no. 8, pp. 559-572, 2021.

[6] H. Yu *et al.*, "Dual-Functional Triangular-Shape Micro-Size Light-Emitting and Detecting Diode for On-Chip Optical Communication in the Deep Ultraviolet Band," *Laser & Photonics Reviews,* p. 2300789, 2024.

[7] M. H. Memon *et al.*, "A three-terminal light emitting and detecting diode," *Nature Electronics,* pp. 1-9, 2024.

Leveraging Mature Chip Manufacturing Techniques for Innovative Technology Development

Yunlong Li[1,2,3], Kai Xu[1,2], Dianyu Qi[1,2], Yishu Zhang[1,2], Ran Tao[1,2], Yongyu Wu[2], Dawei Gao[1,2]

[1]College of Integrated Circuits, Zhejiang University, Hangzhou, China

[2]Zhejiang ICsprout Semiconductor Co., Ltd, Hangzhou, China

[3]State Key Laboratory of Silicon and Advanced Semiconductor Materials, Zhejiang University,

Hangzhou, China

Abstract

This paper provides an in-depth look at the innovative advancements in chip manufacturing techniques, specifically focusing on the research and development efforts at Zhejiang ICsprout Semiconductor Co., Ltd's CMOS pilot line. The exploratory development encompasses high-reliability CMOS processes, next-generation power devices, novel embedded flash memory devices, and post-CMOS heterogeneous integration. By rethinking traditional CMOS fabrication processes, we showcase how these proven techniques can propel substantial technological progress and underscore the capabilities of a mature CMOS pilot line in improving reliability, performance, and functionality in emerging semiconductor technologies.

Keywords: CMOS Pilot Line, Embedded Flash Memory, Power Devices

Introduction

In recent decades, the semiconductor industry has rapidly progressed, with a focus on reducing critical dimensions to meet growing demands. However, as Moore's Law loses momentum, there is a shift in semiconductor research and development towards maximizing the power-performance-area-time (PPAT) matrix. The hardware and infrastructure for chip manufacturing are becoming more mature, allowing for a longer functional lifespan. It is imperative to leverage existing investments in new chip technology R&D to maximize the benefits of these advancements.

Specializing in high-reliability logic chips, cutting-edge memory devices, next-generation power devices, and innovative heterogeneous integration post-processing technologies, Zhejiang ICsprout Semiconductor Co., Ltd (ICS) operates a 12-inch CMOS pilot line to bring these advancements from lab to market.

In this paper, we explore the vast potential of enhancing the R&D capabilities of mature CMOS technology through a collaborative effort between Zhejiang University and ICS.

Main investigations

Over the past three years, ICS has made significant breakthroughs in CMOS device performance and various application fields by utilizing mature CMOS manufacturing techniques. These achievements serve as a strategic blueprint for future innovations that will be built upon the foundation of the CMOS pilot line.

A. CMOS baseline optimization

In order to create innovative chip technologies using established chip manufacturing methods, it is crucial to optimize current technologies to improve performance in such areas as process stability, consistency, and yield. Through the process optimization of mature 55-nm CMOS manufacturing techniques [Fig. 1], significant advancements have been made in enhancing the performance of logic devices. This includes a notable increase of 5.4% in NMOS saturation currents and 7.0% in PMOS saturation currents, while effectively managing low leakage currents [Fig. 2]. These optimizations not only strengthen the reliability and power efficiency of logic devices, but also establish a strong foundation for future application-driven advancements.

B. Advanced power device development

Utilizing the logic process platform, an advanced 180nm BCD (Bipolar-CMOS-DMOS) multi-device structure integration process has been successfully established, resulting in a full BCD manufacturing flow. Through simulation development and tape-out of structures like Shallow Trench Isolation (STI), Field Plate (FP), and contact structures, significant progress has been made. The enhancement of FP technology is particularly crucial, as high-voltage field plates enhance breakdown voltage through the refinement of the active region morphology and optimization of electric field distribution in the drift region. Building on these technical advancements, the platform has introduced a variety of devices for the 180-nm BCD platform. Key components encompass a range of MOS devices such as nLDMOS, pLDMOS, nEDMOS, and pEDMOS, in addition to CMOS devices, Zener diodes, Schottky diodes, and high-poly resistors. Furthermore, the platform features capacitive components like MOSCAPs and varactors, along with UTM and BJT devices, all supporting menu-based device selection. Fine-tuning of ion implantation concentration and energy, in conjunction with TCAD simulation, has unveiled the crucial impact of the overlap length between the gate and FP on

979-8-3315-0417-5/25 $31.00 © 2025 IEEE

device performance. Utilizing these insights, the platform has implemented a specialized split table to facilitate parameter optimization and tape-out validation for LDMOS devices. As a result, the platform has successfully broadened the range of device voltage options within the same device types, offering multiple nLDMOS and pLDMOS alternatives (Fig. 3).

On the 180nm BCD platform, the low-voltage (1.8 V) MOS, medium-voltage (5 V) MOS, and high-voltage (up to 40 V) LDMOS devices have achieved outstanding industry-leading performance. Notably, the 40 V core device exhibits an impressive maximum breakdown voltage (BV) of 62.5 V, accompanied by an exceptionally low R_{sp} of just 25 m$\Omega \cdot$mm2 (Fig. 4).

C. Embedded flash memory development

In order to enhance the performance of the MCU, it is imperative to incorporate novel embedded flash technologies. At the ICS pilot line, we have successfully developed two cutting-edge embedded flash memory devices that are poised to revolutionize the industry.

The first one is a novel split-gate "L-shaped" floating gate memory with a bit cell size of approximately 25.7 F² (Fig. 5). This device incorporates an efficient source-side hot-carrier injection (SSI) mechanism for programming (10 μs), a tip-to-tip Fowler-Nordheim (FN) tunneling mechanism for erasure (2 ms), and a low-voltage read (1.5 V, ≤20 ns). This floating-gate memory with a novel structure has demonstrated the capability to achieve 100K cycling stability and data retention of 10 years at 125°C.

The second one is a novel dual-bit Charge Trapping Flash (CTF) cell structure (Fig. 6), named PXFlash, which features a self-aligned middle select-gate and interwoven source/drain serpentine wiring design and achieves a minimum cell size of 0.03 μm² at the 55-nm technology node, reducing the size by 62% compared to existing eFlash cells (40eCT) [1]. A two-step Dual-Channel Hot Electron (DCHE) programming method and a Channel Hot Hole (CHH) erase mechanism are proposed, which effectively reduce the effects of localized Trapped Charge Mismatch Effect (TCME) through staged programming control and efficient erase processes. This significantly enhances the endurance and data retention capabilities of the memory cell. Experimental data show that under typical program/erase conditions, PXFlash cells can withstand up to 100K P/E cycles, while maintaining a data retention memory window of over 2.1V with only a 5% degradation after baking at 175°C for 12 hours. Fully verified in an 8Mb array at the 55-nm process node, PXFlash demonstrates exceptional program/erase characteristics, faster operational speeds, and excellent reliability. This technology meets the requirements for both high-capacity standalone and embedded storage applications, marking a significant breakthrough for future storage technologies in mobile, automotive electronics, and IoT devices.

D. Heterogeneous integration development

Integrated within the CMOS interconnect layers, additional functionalities can be achieved through heterogeneous integration. While most advancements are currently being developed in a laboratory setting, the transition to the pilot line is foreseeable. An example of this integration is the self-rectifying memristor (SRM), designed to offer a high rectification ratio and dynamic linearity for in-memory computing applications. Current study builds upon these findings by introducing a novel Pt/HfO$_2$/WO$_3$-x/TiN SRM structure that achieves a remarkable rectification ratio above 10^6 and showcases excellent dynamic linearity, making it a standout in the field of memristor research [Fig. 7]. The highlights of this research include the systematic investigation of the impact of the WO$_3$-x resistive layer thickness on the device's conductive behavior, revealing the synergistic effect of abundant traps in the WO$_3$-x layer and the insulating property of HfO$_2$ in suppressing negative current while promoting positive current. The proposed SRMs not only achieve a high rectification ratio but also demonstrate a remarkable linearity of 0.9973 in artificial synapses, which is crucial for deep learning applications. The study also simulates the scalability of the crossbar array based on these SRMs, indicating the potential for an array scale of over 21 Gbit. These results underscore the great potential of these SRMs for ultra-large-scale integration of neuromorphic hardware, providing a guide for future ultra-high-energy efficiency hardware with minimal circuit overhead.

Another representative application is the integration of metal oxide based TFT with CMOS (Fig. 8), which has the potential of integrating switches and memory arrays in interconnect [3].

Conclusion

The significant impact of mature CMOS manufacturing techniques on driving technological advancements is evident. ICS is at the forefront of leveraging established CMOS processes to explore cutting-edge memory devices, next-generation power devices, and advanced heterogeneous integration technologies. The initial findings highlight the importance of mature fabrication in fostering innovation and lay a solid foundation for future semiconductor breakthroughs.

Acknowledgments

The authors gratefully acknowledge the contributions of the engineering and operation teams at Zhejiang ICsprout Semiconductor Co., Ltd and the colleagues at College of Integrated Circuits, Zhejiang University.

References

[1] Z. Li et al., IEDM 2024 (accepted)

[2] G. Zhang, Z. Wang. APL. 2024. 125, 133501

[3] K. Chen et al., EDTM (submitted)

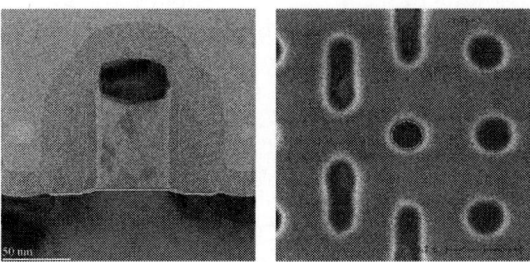

Fig. 1 Post optimization gate (left) and contact (right)

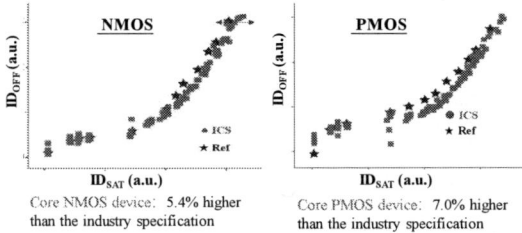

Core NMOS device: 5.4% higher than the industry specification

Core PMOS device: 7.0% higher than the industry specification

Fig. 1 The transfer characteristics curve of the core devices is improved after process optimization.

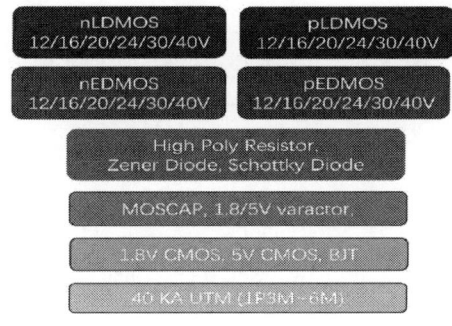

Fig. 3 ICS 180BCD 40V special process platform

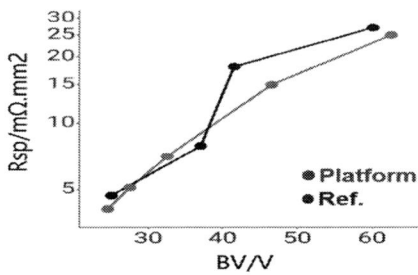

Fig. 4 Performance of the nLDMOS device: benchmarked against the industry's highest standards

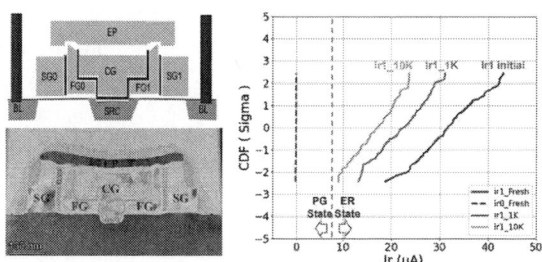

Fig. 5 A high-performance split-gate Flash Cell

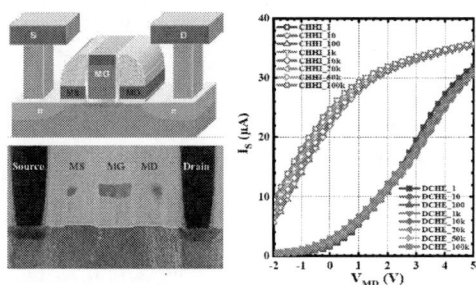

Fig. 6 A Novel Dual-Bit Charge Trapping Flash Cell

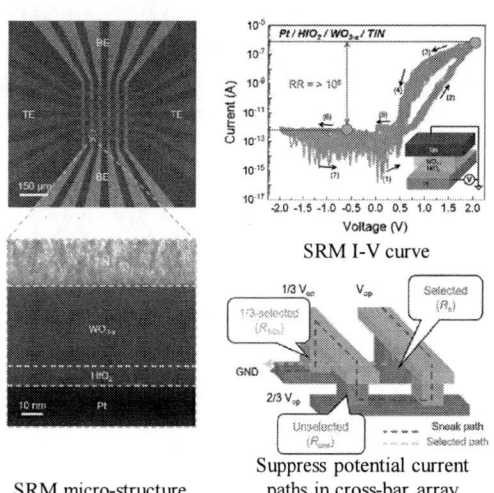

SRM I-V curve

SRM micro-structure

Suppress potential current paths in cross-bar array

Fig. 7 The layout, morphology and I-V characteristic curve of the SRM device in development

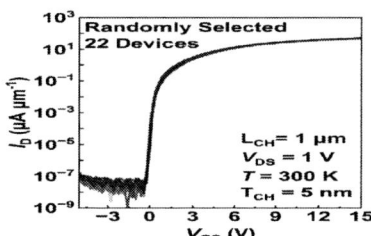

Fig. 8 Transfer curves of 22 IGZTO TFTs with 5 nm T_{CH} [3]

979-8-3315-0417-5/25 $31.00 © 2025 IEEE

Ultra-low Temperature Solution Processed Organic Thin-film Transistor for Flexible Integration

Zikang Mei, and Xiaojun Guo[*]

Department of Electronic Engineering, Shanghai Jiao Tong University, Shanghai 200240, China

[*]E-mail: x.guo@sjtu.edu.cn

Abstract

Abstract—A steep subthreshold organic thin-film transistor (OTFT) is processed at ultra-low temperature (≤80 °C) to add electronic functions on all kinds of surfaces of everyday objects or things and human body surfaces. "Flexible integration" based on the steep subthreshold OTFT is demonstrated, including direct integration on a temperature-sensitive function layer and implementation of flexible hybrid sensing system.

Keywords: Thin-Film Transistor, flexible hybrid integration, sensor, low temperature processes

Introduction

For the envisioned ubiquitous computing era, there is a demand to add electronic functions to various surfaces of everyday objects and human skin or body surfaces to non-invasively monitor conditions, predict health, better understand the environment, and receive interactive inputs [1]. To achieve this, the electronic functions need to be manufactured in a more flexible manner, and be thin and flexible in many cases [2]. To implement such 'flexible' integration, the electronic devices should be processed at low temperatures with reduced complexity.

The thin-film transistor (TFT) is a key component in realizing various electronic functions (**Fig. 1(a)**) of large area coverage for displays and imagers [3-4]. However, due to high processing temperature and high cost in infrastructure and maintenance, the inorganic TFT technologies used in display industry meet challenges for constructing those ubiquitous electronics functions. Moreover, display-oriented TFTs are not well-suited for low-voltage and low-power applications.

Compared to the inorganic counterparts, organic thin-film transistors (OTFTs) offer the advantages of very low processing temperature (below 120 ℃) with facile processes for possibility of direct integration on ubiquitous substrate surfaces (**Fig. 1(b)**) [5]. It is worth exploring the possibility for further reduction of the processing temperature, while achieving steep subthreshold swing (SS), for low power operation.

However, to form high-quality channels of low trap density of states (DOS), organic semiconductors are normally dissolved in solvent of relatively high boil point to slow down the crystallization rate of OSC solution [6-7]. As

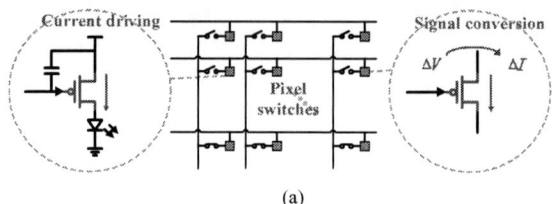

(a)

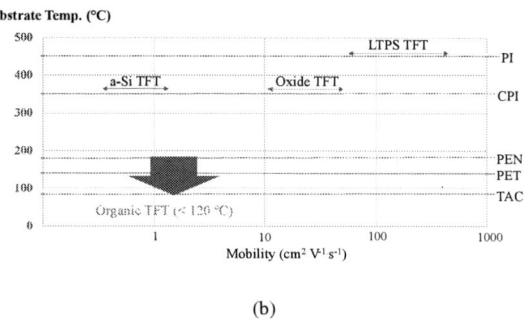

(b)

Fig. 1. (a) Illustration of using the TFT for different functions. (b) Comparison of the maximum temperature at the substrate required for fabricating different TFTs.

a result, high temperature post-deposition annealing is needed. Moreover, formation of the organic gate dielectric layer also requires high temperature to achieve good insulation, especially when high-k dielectrics are used for steep SS [8]. Therefore, it is challenging to fabricate OTFTs of steep SS with very low processing temperature.

This paper will discuss the recent progress on processing low power OTFTs at ultra-low temperature, as well as examples of "flexible integration", including direct integration on a temperature-sensitive function layer and implementation of flexible hybrid sensing system.

Ultra-low Temperature Processes

When a field effect transistor is used for the conversion of a sensed signal to a current output change, a steep SS is required for low power operation and high sensitivity [9]. A device technology is developed for fabricating OTFTs of steep SS and excellent operational stability with ultra-low temperature (≤80 °C) solution processes [10]. As shown in **Fig. 2(a)**, a bi-layer organic gate insulator (OGI) structure is designed to achieve good electrical insulation and provide a smooth and low-polar surface at low annealing temperature. A blended solution of small molecule OSC and polymer binder in alkyl solvent of suitable boil point is deposited on

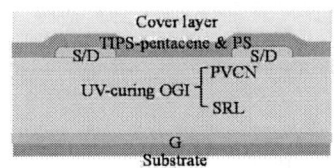

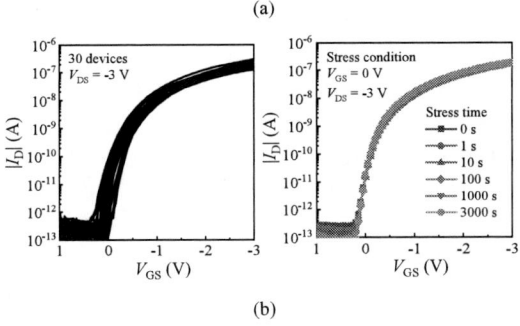

(a)

(b)

Fig. 2. (a) The device structure of the ultra-low temperature solution processed OTFT. (b) The measured transfer characteristics of the steep subthreshold swing OTFTs showing good uniformity and bias stress reliability.

top to form uniform crystalline channels and low trap DOS channel interface without the need for thermal annealing [11]. As a result, steep SS OTFTs are fabricated using small gate dielectric capacitance (low-k thick polymer dielectric layer) [12]. The devices also exhibit low OFF-state leakage current and excellent uniformity and operational stability (**Fig. 2(b)**). Attributed to ultra-low processing temperature, it is further demonstrated that the OTFTs can be directly fabricated on an

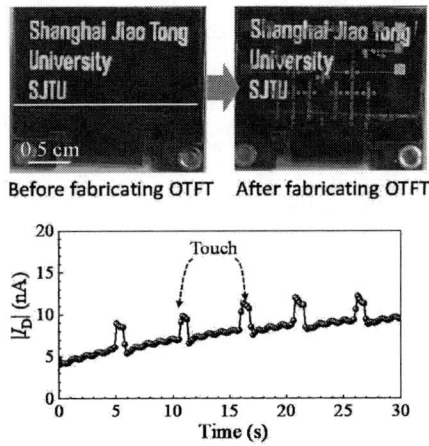

Fig. 3. Direct fabrication of the OTFTs on an OLED display for on-display touch sensing without affecting the display performance.

OLED display for on-display touch sensing without affecting the display performance (**Fig. 3**)

Flexible Hybrid Integration

To realize flexible sensor systems based the OTFT, one route is fully based on the OTFT-only processes for low cost

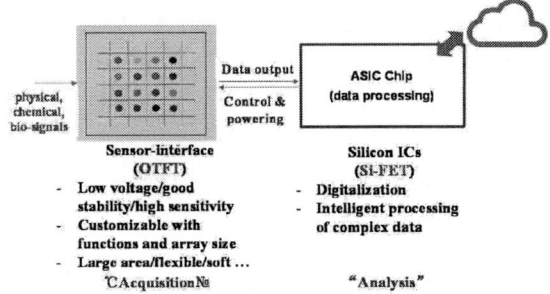

Fig. 4. Illustration of the flexible hybrid integration (FHI) architecture composed of the OTFT part and the silicon IC part.

and short design-to-product time [13]. However, it is difficult for OTFTs to achieve circuits of high complexity and performance due to low carrier mobility and large feature size. Therefore, the achievable performance and functionality of the reported OTFT-only "stand-along" systems are very limited.

A flexible hybrid integration (FHI) architecture is more realistic by combining the advantages of both the OTFT and the silicon technology as shown in **Fig. 4** [13]. In the architecture, the OTFT part is able to provide customizability in both sensor function integration and form factor, especially for flexible or conformable large area coverage on various surfaces. Through device and circuit co-design, the OTFT circuitry would be able to transduce different types of sensor signals sensor signals to a standard electrical output, The sensed signal can be further amplified locally with the OTFT circuitry. By using the OTFT as pixel switches, active-matrix sensor arrays can also be constructed to obtain spatial signal mapping. The general-purpose silicon ASIC performs signal acquisition and conditioning, data processing and transmission, and power management. The entire system is

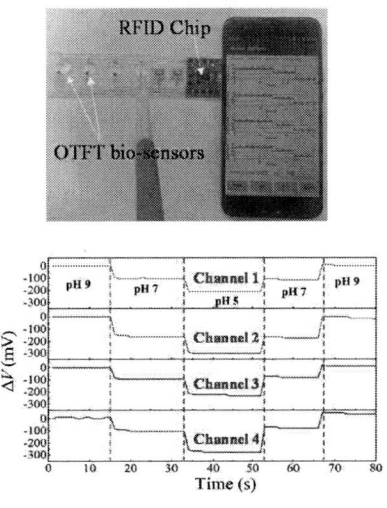

Fig. 5. Demonstration of a mobile ion-sensing platform by combining the OTFT bio-sensor technology and a 13.56 MHz RFID IC chip.

realized by integrating the OTFT with the function layer onto a common plastic substrate, and bonding the ASIC die onto the substrate.

A wireless powered mobile bio-sensing platform is demonstrated by integrating the OTFT based ion sensors with a 13.56 MHz silicon RFID chip on plastic substrate is demonstrated for with (**Fig. 5**) [14]. With limited energy, the sensed signals at the four test points are able to be collected through the OTFT circuitry and wireless connectivity to a smartphone. The readout results agree well with the pH value variations of the test buffer solutions.

Conclusion

This paper reviews the recent progress on processing low power OTFTs at ultra-low temperature, and examples of "flexible integration", including direct integration onto a temperature-sensitive functional layer and the implementation of flexible hybrid sensing systems. This OTFT device and integration technology could pave the way for adding electronic functions to all types of surfaces on everyday objects, things, and even human bodies.

Acknowledgments

The authors gratefully acknowledge funding support through the National Key R&D Program of China (Grant No. 2019YFA0706100).

References

[1] M. Sang, K. Kim, J. Shin, and K. J. Yu, "Ultra-thin flexible encapsulating materials for soft bio-integrated electronics," *Adv. Sci.,* vol. 9, no. 30, p. 2202980, 2022.

[2] X. Guo, Y. Xu, S. Ogier, T. N. Ng, M. Caironi, A. Perinot, L. Li, J. Zhao, W. Tang, and R. A. Sporea, "Current status and opportunities of organic thin-film transistor technologies," *IEEE Trans. Electron Devices,* vol. 64, no. 5, pp. 1906-1921, 2017.

[3] M.-G. Kim, M. G. Kanatzidis, A. Facchetti, and T. J. Marks, "Low-temperature fabrication of high-performance metal oxide thin-film electronics via combustion processing," *Nat. Mater.,* vol. 10, no. 5, pp. 382-388, 2011.

[4] X. Guo, L. Han, Y. Huang, and W. Tang, "Development of organic tft technology for active - matrix display backplane," in *SID Symp. Digest Tech. Papers,* 2021, pp. 9-12.

[5] H. Park, S. Yoo, M. H. Yi, Y. H. Kim, and S. Jung, "Flexible and stable organic field-effect transistors using low-temperature solution-processed polyimide gate dielectrics," *Org. Electron.,* vol. 68, pp. 70-75,

2019.

[6] W. Tang, L. Feng, P. Yu, J. Zhao, and X. Guo, "Highly efficient all-solution-processed low-voltage organic transistor with a micrometer-thick low-k polymer gate dielectric layer," *Adv. Electron. Mater.,* vol. 2, no. 5, p. 1500454, 2016.

[7] Y. Yuan, G. Giri, A. L. Ayzner, A. P. Zoombelt, S. C. Mannsfeld, J. Chen, D. Nordlund, M. F. Toney, J. Huang, and Z. Bao, "Ultra-high mobility transparent organic thin film transistors grown by an off-centre spin-coating method," *Nat. Commun.,* vol. 5, no. 1, p. 3005, 2014.

[8] Y. Huang, W. Tang, S. Chen, L. Han, X. Hou, and X. Guo, "Scalable processing of low voltage organic field effect transistors with a facile soft-contact coating approach," *IEEE Electron Device Lett.,* vol. 40, no. 12, pp. 1945-1948, 2019.

[9] W. Tang, Y. Fu, Y. Huang, Y. Li, Y. Song, X. Xi, Y. Yu, Y. Su, F. Yan, and X. Guo, "Solution processed low power organic field-effect transistor bio-chemical sensor of high transconductance efficiency," *npj Flexible Electron.,* vol. 6, no. 1, p. 18, 2022.

[10] X. Yin, W. Tang, Y. Huang, R. Shi, Y. Su, and X. Guo, "Ultra-low temperature solution processed steep subthreshold organic thin-film transistor for on-display integrated sensing," *IEEE Electron Device Lett.,* 2023.

[11] W. Tang, J. Li, J. Zhao, W. Zhang, F. Yan, and X. Guo, "High-performance solution-processed low-voltage polymer thin-film transistors with low-$ k $/high-$ k $ bilayer gate dielectric," *IEEE Electron Device Lett.,* vol. 36, no. 9, pp. 950-952, 2015.

[12] L. Feng, W. Tang, X. Xu, Q. Cui, and X. Guo, "Ultralow-voltage solution-processed organic transistors with small gate dielectric capacitance," *IEEE Electron Device Lett.,* vol. 34, no. 1, pp. 129-131, 2012.

[13] Y. Huang, W. Tang, L. Feng, S. Chen, J. Zhao, Z. Liu, L. Han, B. Ouyang, and X. Guo, "Printable low power organic transistor technology for customizable hybrid integration towards internet of everything," *IEEE J. Electron Devices Soc.,* vol. 8, pp. 1219-1226, 2020.

[14] B. Ouyang, Y. Song, W. Cai, Y. Tang, Y. Si, X. Yin, S. Chen, W. Tang, H. Zhou, and B. Huang, "Rf powered flexible printed ion-sensitive organic field effect transistor chip with design-to-manufacturing automation for mobile bio-sensing," in *2021 IEEE International Electron Devices Meeting (IEDM),* 2021, pp. 16.3. 1-16.3. 4.

A compact 256-channel CMOS brain surface recording and stimulation array with soft electrodes

Lachlan Fraser[1], Muhammad Saif Ul Islam[1], Peijun Qin[1], Nigel Lovell[1], and David Tsai[1,2]

[1]Graduate School of Biomedical Engineering, UNSW Sydney, [2]School of Electrical Engineering & Telecommunications, UNSW Sydney

ABSTRACT

The ideal neural interface should offer at-scale, high resolution interactions with the brain while minimizing biological invasiveness. Towards this goal, we are developing a 256-channel neural interface for the brain. The soft electrodes permit conformal placement over the cortical surface with minimal force. The custom integrated circuit with low-noise and low-power architecture enables neural recordings with high signal-to-noise ratio while offering stimulation capabilities. Our miniature system will enable new experimental possibilities for neuroscientists.

Keywords: CMOS, bioelectronics, brain machine interface

INTRODUCTION

Brain machine interfaces with high channel count are desirable for observing and manipulating neuronal activities at scale and with high spatiotemporal resolution. A range of neural probes, ranging from several hundred [1], [2] to one thousand channels [4], [4], [5], have been reported. Some of these devices are now commercially available. However, these silicon-based penetrating probes are very invasive. Advances in soft materials have allowed the development of mechanically flexible and biocompatible 256-electrode arrays for action potential recordings from the brain surface [6]. However, these passive (electrode only) devices require considerable external, tethered electronics. The resulting biological invasiveness limits such devices to one-off, acute studies, while their bulkiness restricts experimental subjects' mobility, hence limiting the addressable scope of scientific enquiry. To overcome these shortcomings, we are developing soft, electrophysiology arrays with foundry-grade CMOS (complementary metal-oxide semiconductor) microelectronics. Our devices significantly reduce invasiveness, improve usability for chronic experiments. Here, we report on our 256-channel prototype brain surface recording and stimulation devices, with purpose-built CMOS integrated circuits and custom-made soft

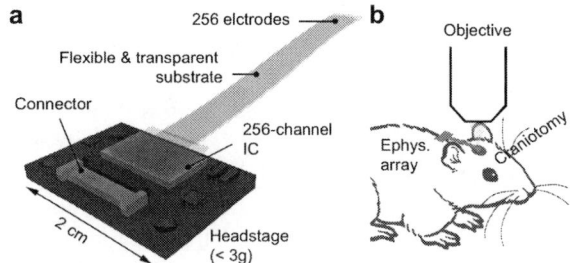

Fig. 1. A miniature, 256-channel neural interface for recording and stimulation on the brain surface. (a) A small, light-weight neural interface with custom CMOS integrated circuits and soft, flexible electrode array. (b) The device's compact device footprint is well suited for small-animal studies.

electrode arrays. These components are assembled using low-temperature flip-chip processes we refined in-house.

METHODS & RESULTS

Our 256-channel soft, brain-surface neural interface contains two major hardware components: (1) custom electronics for neural recording and stimulation, implemented as CMOS integrated circuits and supported by a miniature printed circuit board, and (2) a custom-fabricated soft electrode array tailored for brain surface operations. Fig. 1a illustrates the device architecture. The compact, sub-2-cm footprint and < 3 g weigh enables neural interfacing at significantly reduced invasiveness comparing to previous devices [5]. The soft electrode also mechanically decouples the electronics from the brain tissue, further minimizing invasiveness. These characteristics are particularly beneficial for small-animal studies (Fig. 1b).

A. CMOS electronics for neural recording and stimulation

Our system contains a custom CMOS integrated circuit for interacting with the brain. The architecture is summarized in Fig. 2. There are 256 analog frontends

979-8-3315-0417-5/25 $31.00 © 2025 IEEE

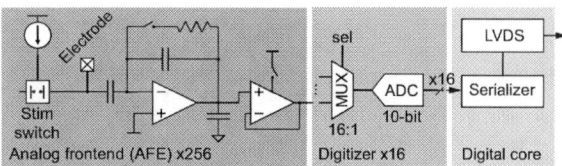

Fig. 2. Architecture of the CMOS integrated circuit for 256-channel neural recording and stimulation.

(AFEs), each capable of neural recording and electrical stimulation to elicit neural responses. Each AFE's input is electrically connected to one electrode of the 256-channel electrode array (see Section B) using flip-chip techniques (see Section C). To faithfully capture microvolt-amplitude neural signals, each AFE contains a low-noise amplifier with 46 dB gain. The amplifier's feedback loop and output are configured for bandpass filtering to extract signals of interest while suppressing dc-offset arising from the electrode, and to remove high-frequency thermal noise. To reduce heat dissipation, each amplifier consumes only 4.8 μW, achieved by operating the input transistors in weak inversion. Every AFE cell also contains stimulation capability to deliver constant-current biphasic pulses or exponentially decaying current pulses. The stimulation parameters (current amplitude, pulse duration, inter-pulse interval, and repetition rate) are user configurable.

There are 16 analog-to-digital converters in the chip, Sixteen AFEs are multiplexed to one 10-bit ADC for reduced chip footprint and lowering chip power budget. The ADCs are of the successive approximation (SAR) architecture for high precision at low power consumption. The second opamp within each AFE is used to charge the SAR ADC's sampling capacitors to achieve bandwidth needed for 20 kHz sampling of each AFE during 16:1 multiplexing.

To transfer the recorded neural data to an external PC-controlled acquisition system, previous systems need large amount of wiring, hindering subject mobility and restricting the investigation to only tethered scientific experiments. The data serializer and 8b10b line-coding in our CMOS chip address this problem by combining all ADC outputs, the data reading clock, and the synchronization patterns into one serial bit stream. Next, a low-voltage differential signal (LVDS) circuit transmits the data to an external FPGA-based acquisition board in TIA/EIA-644-compliant format, to minimize EMI/C concerns and to facilitates reliable data transmission and recovery over long wires.

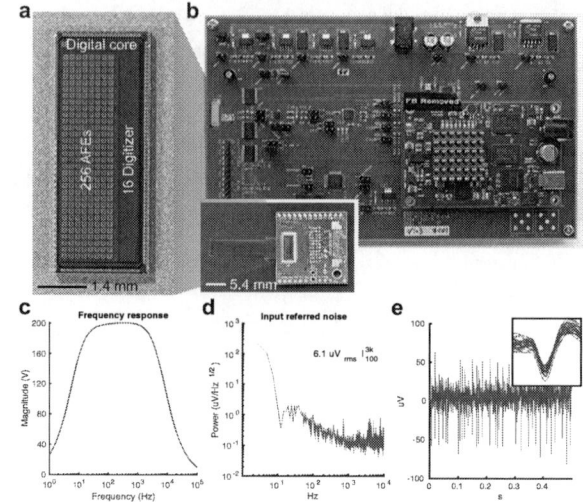

Fig. 3. Custom-built CMOS chip, miniature headstage and FPGA-based acquisition board. (a) Die photo. (b) Headstage and FPGA acquisition boards. (c) Bandwidth and (d) input referred noise of the recording circuit. (e) Example neural recordings. Inset: zoom-in view of spikes.

All operations with our IC are realized via 48-bit commands serially transmitted to the chip. The chip's digital core is responsible for interpreting and executing the commands. Fig. 3a is a photo of the IC die, fabricated in TSMC 180-nm, high-voltage process. The IC is wire-bonded to a small head-stage circuit board (Fig. 3b, bottom-right). Power and data lines (as described above) between the head-stage and the Artix-7 (AMD, USA) FPGA-based acquisition board (Fig. 3b top-left) is realized via a thin FFC (flexible flat cable) with matching connectors. Fig. 3c shows the measured AFE frequency response spanning 10 to 10k Hz, as required for action potential recordings. With an input referred noise of 6.1 μVrms over 100 to 3k Hz bandwidth (Fig. 3d), the system can record with high signal-to-noise ratio. Fig. 3e shows example neural recordings by our system. The neural spikes are easily distinguishable from background noise (inset).

B. Soft 256-channel electrode array

Soft electrodes fabricated using polymer substrate and thin-film metal deposition permit neural interfacing with less invasiveness comparing to conventional silicon penetrating shank electrodes. A photo of the 256-channel electrode array with 40 μm recording electrodes is shown in Fig. 4a. Similar to our commercial foundry-fabricated CMOS chips, the wafer-scale fabrication process [7]

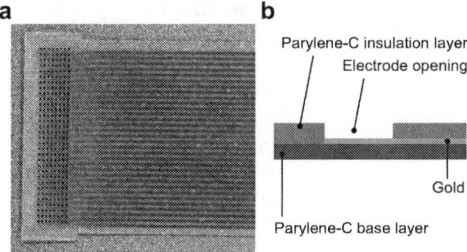

Fig. 4. 256-channel Parylene C soft electrode array for brain surface recording and stimulation. (a) Photo of electrodes. (b) Cross-section of electrode.

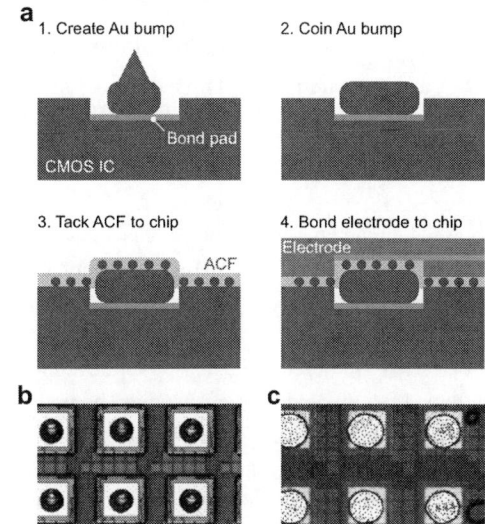

Fig. 5. Connecting the soft electrode array to the IC chip using low-temperature anisotropic conductive film. (a) Process steps. Photo of (b) bumped die and (c) coined bumps with tacked ACF.

enables volume production of these soft electrode arrays. Fig. 4b shows the device cross section. The process begins with depositing a 10-μm parylene C base layer on a SiO_2 carrier wafer. Next, electrodes, conductive tracks and contact pads are patterned on the base layer using photolithography and metal deposition. A 10-μm parylene C insulating layer is then uniformly deposited over the entire wafer, followed by etching to expose the electrodes and contact pads, and cutting-out the individual arrays.

C. Flip-chip packaging

While the softness of polymer-based electrode arrays is desirable for biological applications, packaging such arrays at scale (e.g. > 100 channels) presents unique challenges. The polymer's limited thermal precludes most well-established high-density microelectronic packaging techniques. Our strategy uses a combination of stud bumping and anisotropic conductive films with several in-house modifications to reduce process temperature and to optimise yield. We begin by bumping the CMOS chip's electrode pads using a thermosonic wire bonder to create a gold stud of 34 μm height. Next the studs are coined (planarized) into a disk by applying 125 N force uniformly across the bumped chip using a wafer bonder (EVG 520). A layer of anisotropic conductive film (ACF) with uniformly patterned 3-μm conductive particles (Dexerials, Japan) is tacked onto the chip surface using a flip-chip bonder (FinePlacer Lambda) at 80 degree C for several seconds. Finally, after alignment of chip pads against the 256-channel electrode array's pads using the flip-chip bonder, the two parts are pressed together, then heat-cured by applying 200 N force at 140 degree C for 280 seconds.

ACKNOWLEDGMENT

This research is support in part by NHMRC Ideas Grant APP1188414. Microfabrication supports were provided by the Australian National Fabrication Facility (ANFF) at UNSW, Australian National University, and University of Sydney. The electrode arrays described in this report were provided by the Polymer Implantable Electrode Foundry at the University of Southern California and funded under BRAIN Award U24NS113647.

REFERENCES

[1] J Jung et al., Fully integrated silicon probes for high-density recording of neural activity, 232–236 551, Nature (2017)

[2] A Paulk et al., arge-scale neural recordings with single neuron resolution using Neuropixels probes in human cortex, 252–263, Nature Neuroscience (2022)

[3] J Scholvin et al., Close-packed silicon microelectrodes for scalable spatially oversampled neural recording, 120–130, 63, IEEE Transactions on Biomedical Engineering (2015)

[4] G Rios et al., Nanofabricated neural probes for dense 3-D recordings of brain activity, 6857–6862, 16, Nano Letters (2016)

[5] J Shobe et al., Brain activity mapping at multiple scales with silicon microprobes containing 1,024 electrodes, 2043–2052, 114, Journal of Neurophysiology (2015)

[6] D Khodagholy et al., NeuroGrid: recording action potentials from the surface of the brain, 310–315, 18, Nature Neuroscience (2015)

[7] K Scholten et al., A shared resource for building polymer-based microelectrode arrays as neural interfaces, 2023 11th International IEEE/EMBS Conference on Neural Engineering (NER)

Hafnia-based XP-FeRAM: A Novel High-speed and Low-power Cross-point Ferroelectric Memory for Data-intensive Applications

Qianqian Huang[*], Shengjie Cao, Zhiyuan Fu, and Ru Huang

School of Integrated Circuits, Beijing Advanced Innovation Center for Integrated Circuits, Peking University, Beijing 100871, China ([*]E-mail: hqq@pku.edu.cn)

Abstract

Device and circuit co-optimization of a novel hafnia-based cross-point FeRAM (XP-FeRAM) are comprehensively investigated for high-density, high-speed and low-power memory applications. Planar and 3D vertical-stacked XP-FeRAM designs are demonstrated with application-specific device optimization, and the outstanding comprehensive performance are achieved with high $2P_r$, excellent disturbance immunity under low-voltage operation and fast switching at scaled size, as well as good device reliability. A modified V/2 operation scheme with in-situ write-back is further presented, leading to faster operation and lower power consumption than traditional 1T1C FeRAM. Combined with its excellent scalability, XP-FeRAM shows strong potential as a competitive candidate of future non-volatile memory technology for data-intensive demands.

(Keywords: HZO-based ferroelectric memory, cross-point, non-volatile random access)

Introduction

With the fast development of artificial intelligence and data-centric computing technology, date-intensive demands grow explosively, leading to a surge in research on high-density, low-power, and high-speed memory technologies [1-4]. Among various emerging non-volatile memories, hafnia-based ferroelectric (FE) memories have attracted tremendous attention in recent years due to the properties of fast polarization switching, low write energy, good scalability, and high CMOS-compatibility. Specifically, hafnia-based ferroelectric random-access memory (FeRAM) is regarded as a promising candidate for non-volatile RAM with DRAM-like 1T1C (1 transistor, 1 FE capacitor) or 1TnC (1 transistor, multiple FE capacitors) structure [5-6]. Compared with 1T1C and 1TnC structures, hafnia-based selector-less cross-point FeRAM (XP-FeRAM) has the potential for higher density, faster access and lower power consumption without the need of plateline and its charging process. Benefiting from the excellent scalability of hafnia-based materials, XP-FeRAM shows 3D-stackable potential and can achieve $4F^2/N$ cell size. However, due to the multi-grain and multi-domain nature of hafnia-based FE material with relatively gradual polarization switching, XP-FeRAM might be susceptible to writing disturbance during programming and suffer from array-level multi-disturb issue. On the other hand, the destructive readout of FE capacitors in FeRAM always requires write-back process for data restore, while the conventional access

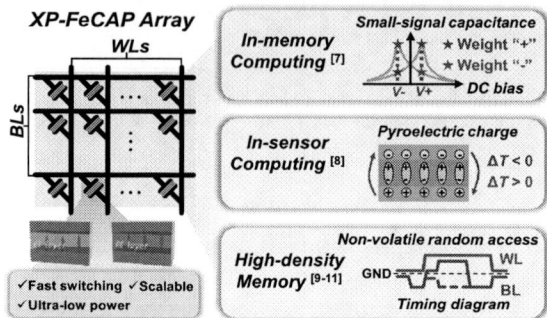

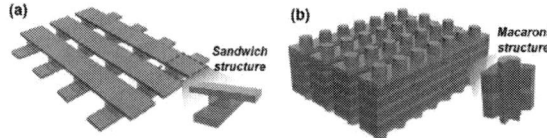

Fig. 1. FE capacitor cross-point array structure and applications.

Fig. 2. Two types of XP-FeRAM design for different applications. **(a)** PXP-FeRAM for embedded memory; **(b)** 3D VXP-FeRAM for standalone memory.

schemes for cross-point memory array are not applicable to XP-FeRAM.

Despite the various other applications of hafnia-based FE capacitor cross-point array in the field of in-memory computing [7] and in-memory sensor [8], in this work, we will focus on the high-density memory applications of FE capacitor cross-point array (**Fig. 1**). Through comprehensive design and optimization from both device and array perspective, 3D-stackable $Hf_{0.5}Zr_{0.5}O_2$ (HZO)-based XP-FeRAM with a modified V/2 operation scheme was demonstrated with excellent disturbance immunity, high performance and low power, showing its significant potential for data-intensive demands [9-11].

Device Structure of HZO-based XP-FeRAM

Fig. 2 shows two kinds of HZO-based XP-FeRAM structures for different memory applications [11]. For XP-FeRAM with planar FE capacitor (PXP-FeRAM), it is suitable for embedded memory applications with compact sandwich-like memory cell structure and cost-effective production method. As one of possible fabrication process, the FE film and top electrode (TE) layers are deposited sequentially on the etched bottom electrode (BE) layer. For standalone memory applications, 3D vertical XP-FeRAM (3D VXP-FeRAM) with macaroni-like memory cells

979-8-3315-0417-5/25 $31.00 © 2025 IEEE

presents a more dense approach by stacking multiple layers vertically. The FE film and electrodes can be deposited continuously with sacrifice process in this design.

Device Optimization for XP-FeRAMs

Device performance of HZO-based FE capacitors, such as switching speed, remnant polarization (P_r), disturbance immunity and endurance behavior, may directly influence the operation speed, memory window and array-level reliability of XP-FeRAM. Considering the different applications, the impacts of FE layer deposition process sequence should also be investigated for device optimization [11].

The FE capacitor in embedded PXP-FeRAM, which adopts post etching deposition technique for FE layer, may introduce extra interfacial layers (IL) between FE and electrodes after BE etching and annealing. Therefore, electrode material selection and interfacial layer engineering are important to control the formation of grain and IL. In our experiments, W as TE and Ti/Pt as contact layer are adopted for high stress [12] and IL optimization. Compared with the conventional TiN TE device, the optimized device has larger grain size (>30nm), tighter E_c distribution and much steeper FE switching, indicating improved uniformity in domain directions and reduced IL voltage drop [9-10]. Considering different process conditions, the annealing condition may have the significant impacts on the switching speed of FE capacitors due to the IL [11]. Optimized devices show the outstanding comprehensive performance with 1.5ns switching, $2P_r$ of $47\mu C/cm^2$ and the record best disturbance immunity among FE-HZO with an ultra-low α_{Dmin} of 0.14 [11] (α_{Dmin} is defined and extracted by "pulse-disturb" analysis method for quantitative evaluation of disturbance behavior [9-10]). Moreover, good reliability of extrapolated 10^{14} endurance cycles and 10 years data retention, as well as excellent device variability, can be also obtained.

As for standalone memory applications, the HZO-based FE capacitor are fabricated through continuous deposition method [11]. Compared with post etching deposition, this method mitigates the formation of IL, thus the switching speed is much faster and can be more stable under varied annealing temperatures. With area scaling, faster switching speed of sub-ns can be highly expected in sub-μm^2 FE capacitors for practical high-density XP-FeRAMs. Moreover, the device can achieve much better disturbance immunity and higher $2P_r$ with the increased annealing temperature under low-voltage operation voltage. The optimized device can achieve less disturbed polarization even under V/2 operation.

Operation Scheme Design and Array-level Evaluation

Cross-point arrays typically employ half-select scheme to reduce the write disturb of adjacent cells during array-level programming. Conventional methods [13] include V/3 and V/2 programming schemes as shown in **Fig. 3**. For V/3 scheme, a pair of V/3 and 2V/3 voltages is biased on

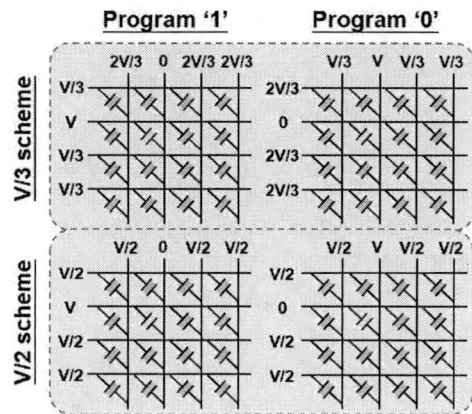

Fig. 3. Conventional V/3 and V/2 programming scheme.

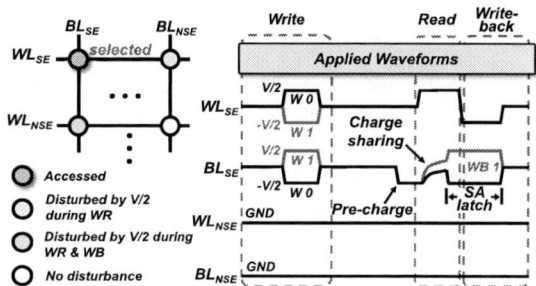

Fig. 4. A modified V/2 operation scheme with in-situ write-back for XP-FeRAM.

unselected cells to minimize the data disturbance, while the requirement for multiple voltage levels leads to increased periphery complexity and power consumption. In contrast, V/2 scheme is less complex from array-level perspective (only V and V/2 needed) but will introduce larger disturbance in large-scale arrays, and additional design optimization may be necessary to ensure data reliability. Focusing on the improvement of disturbance immunity, we have proposed a novel "disturb-recovery" programming scheme, which can effectively solve array-level multi-disturb issues and achieve stable memory window [9-10].

To further realize random-access operation with low-disturbance and data restore in XP-FeRAM, a new modified V/2 operation scheme is proposed with complete write, read and in-situ write-back process (**Fig. 4**) [11]. Based on conventional V/2 programming scheme, a pair of $\pm V/2$ bias is utilized as V_{DD} and V_{SS} in this design and unselected wordlines and bitlines keep grounded. During the readout process, selected bitline (BL_{SE}) is firstly pre-charged to $-V/2$ and then selected wordline (WL_{SE}) is pulled to V/2 to develop bitline voltage, which is then amplified to V_{DD} or V_{SS} according to the stored data. After reading out, WL_{SE} decreases to $-V/2$ and sense amplifier latches to write back the destructed datum. Benefiting from the in-situ write-back, XP-FeRAM with proposed modified V/2 operation scheme can achieve faster random-access speed compared with V/3 scheme or 1T1C FeRAM from array perspective.

979-8-3315-0417-5/25 $31.00 © 2025 IEEE   1159

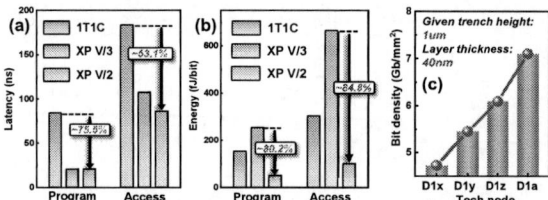

Fig. 5. Estimated **(a)** latency, **(b)** energy of three FeRAM designs at the same simulation setup, and **(c)** bit density of 3D VXP-FeRAM for high-density memory applications.

By device and circuit co-optimization, the performance of XP-FeRAM can be significantly improved for high-density memory. Simulation results [11] show that XP-FeRAMs can effectively improve the overall access latency and energy consumption incorporating in-situ write-back and simplified array-level operation (**Fig. 5ab**), making it ideal for high-speed and low-power memory applications. Referring to typical structure parameters of advanced DRAM nodes, the estimated bit density (**Fig. 5c**) shows the great potential of XP-FeRAM to meet high data-intensive demands.

Summary

XP-FeRAM has great potential to meet data-intensive demands as a novel non-volatile cross-point memory technology. Combining the intrinsic properties of HZO materials and application-specific device process optimizations, the optimized devices have achieved the best comprehensive performance. Further investigation of modified V/2 operation scheme promotes XP-FeRAMs as a promising candidate for high-density, high-speed and low-power memory applications.

Acknowledgments

This work was supported by NSFC (62374009, 61927901), Beijing SAMT Project (SAMT-BD-KT-22030101), 111 Project (B18001).

References

[1] T. Tran et al., "Making Memory Magic and the Economics Beyond Moore's Law," in 2023 International Electron Devices Meeting (IEDM), San Francisco, CA, USA: IEEE, Dec. 2023, pp. 1–4.

[2] S. Hong et al., "Extremely high performance, high density 20nm self-selecting cross-point memory for Compute Express Link," in 2022 International Electron Devices Meeting (IEDM), San Francisco, CA, USA: IEEE, Dec. 2022, p. 18.6.1-18.6.4.

[3] J. Yi et al., "The chalcogenide-based memory technology continues: beyond 20nm 4-deck 256Gb cross-point memory," in 2023 IEEE Symposium on VLSI Technology and Circuits (VLSI Technology and Circuits), Kyoto, Japan: IEEE, Jun. 2023, pp. 1–2.

[4] M. Kim et al., "First Demonstration of Fully Integrated 16 nm Half-Pitch Selector Only Memory (SOM) for Emerging CXL Memory," in 2024 IEEE Symposium on VLSI Technology and Circuits (VLSI Technology and Circuits), Honolulu, HI, USA: IEEE, Jun. 2024, pp. 1–2.

[5] N. Ramaswamy et al., "NVDRAM: A 32Gb Dual Layer 3D Stacked Non-volatile Ferroelectric Memory with Near-DRAM Performance for Demanding AI Workloads," in 2023 International Electron Devices Meeting (IEDM), San Francisco, CA, USA: IEEE, Dec. 2023, pp. 1–4.

[6] N. Haratipour et al., "Hafnia-Based FeRAM: A Path Toward Ultra-High Density for Next-Generation High-Speed Embedded Memory," in 2022 International Electron Devices Meeting (IEDM), San Francisco, CA, USA: IEEE, Dec. 2022, p. 6.7.1-6.7.4.

[7] W. Xu et al., "A Novel Small-Signal Ferroelectric Capacitance-Based Content Addressable Memory for Area- and Energy-Efficient Lifelong Learning," IEEE Electron Device Lett., vol. 45, no. 1, pp. 24–27, Jan. 2024.

[8] Z. Fu et al., "Novel Energy-efficient Hafnia-based Ferroelectric Processing-in-Sensor with in-situ Motion Detection and Four-quarter Multiplication," in 2022 International Electron Devices Meeting (IEDM), San Francisco, CA, USA: IEEE, Dec. 2022, p. 24.5.1-24.5.4.

[9] Z. Fu et al., "First Demonstration of Hafnia-based Selector-Free FeRAM with High Disturb Immunity through Design Technology Co-Optimization," in 2023 International Electron Devices Meeting (IEDM), San Francisco, CA, USA: IEEE, Dec. 2023, pp. 1–4.

[10] Z. Fu et al., "Hafnia-Based High-Disturbance-Immune and Selector-Free Cross-Point FeRAM," IEEE Trans. Electron Devices, vol. 71, no. 5, pp. 3358–3364, May 2024.

[11] S. Cao et al., "Comprehensive Performance Re-assessment of Hafnia-based Cross-point FeRAM with Ultra-fast and Low-power Operation from Device/Array Perspective." in 2024 International Electron Devices Meeting (IEDM), San Francisco, CA, USA: IEEE, Dec. 2024, pp. 37.3.1–37.34.

[12] S. Oh, H. Kim, A. Kashir, and H. Hwang, "Effect of dead layers on the ferroelectric property of ultrathin HfZrOx film," Applied Physics Letters, vol. 117, no. 25, p. 252906, Dec. 2020.

[13] H. Li et al., "3-D Resistive Memory Arrays: From Intrinsic Switching Behaviors to Optimization Guidelines," IEEE Trans. Electron Devices, vol. 62, no. 10, pp. 3160–3167, Oct. 2015.

Decoupling Polarization and Trap Charges by Direct V_{mid} Measurement for Insights into Dynamic Mechanisms of MFMIS-FeFET

Wenpu Luo[#1], Runteng Zhu[#1], Hanyong Shao[1], Yuejia Zhou[1], Ru Huang[1,2*], Kechao Tang[1,2*]

[1]School of Integrated Circuits, Peking University, Beijing 100871, China, [2]Beijing Advanced Innovation Center for Integrated Circuits, Beijing 100871, China
(*Email: tkch@pku.edu.cn, ruhuang@pku.edu.cn)

Abstract

In this work, the polarization switching (P_{FE}) and charge trapping (Q_{trap}) were decoupled by direct middle gate (MG) voltage (V_{mid}) measurements on MFMIS-FeFET. The non-ideal effects of this method were thoroughly analyzed and optimized. According to the decoupled data, the channel side injection (Ct-Inj) and gate-side injection (Gt-Inj) were both observed, consistent with the previous reports. New phenomena regarding the frequency dependent Q_{trap}, P_{FE} speed and V_{mid} amplitude were also revealed. We proposed that P_{FE} has a stronger gate voltage dependence than Ch-Inj, explaining all the findings. Based on these findings, guidelines for optimizing the writing frequency to enhance FeFET endurance were proposed.

Keywords: MFMIS-FeFET, FeFET dynamics decoupling.

Introduction

HfO$_2$-based ferroelectric field effect transistors (FeFETs) are strong candidates for future embedded non-volatile memory, 3D NAND storage and computing in memory [1], [2], [3]. In recent years, FeFETs with a gate stack of metal-ferroelectric(FE)-metal-insulator-semiconductor (MFMIS) seem promising. This device has advantages including large tunable memory window, good endurance, low write voltage and mitigated variation (Fig. 1(a))[4], [5]. The mechanism of charges of traps injected from both the channel and gate sides in MFMIS structures were observed previously using MW analysis and modeling [4]. However, deep insights into their dynamic behaviors alongside polarization switching (P_{FE}) remain elusive due to the inherent challenge of coupling between PFE and charge trapping (Q_{trap}). This coupling hinders the extraction of P_{FE} and Q_{trap} dynamics, complicating the understanding of their detailed charge behaviors during device switching (Fig. 1(b)).

In this work, we decoupled P_{FE} and Q_{trap} by direct middle gate (MG) voltage (V_{mid}) measurement. The parasitic effects during measurements were analyzed and optimized thoroughly. According to the decoupled data, the channel side injections (Ch-Inj)and gate side injections (Gt-Inj) were both observed, consistent to previous works. The decoupled data of various conditions newly reveals frequency dependent Q_{trap}, P_{FE} speed and V_{mid} amplitude. They were explained by the different gate voltage (V_g) dependence of P_{FE} and Ch-Inj. Based on these, guidelines on the writing frequency of FeFET

toward high endurance were provided.

Measurement and Decoupling

A. Direct V_{mid} Measurement Method

We designed and fabricated an MFMIS FeFET featuring a middle gate (MG) contact structure (Fig. 1). Fig. 2 illustrates our measurement and decoupling method, with the key being monitoring the $V_{mid}(t)$ by an oscilloscope, which enables decoupling by charge balance equation. However, this probing would disturb the operation of FeFET, wherein the MG remains floating. This disturbance is mainly caused by parasitic capacitance (Fig .2(a)), consisting of the input capacitance of the oscilloscope ($C_{in,OSC}$), the capacitance of the coaxial cable (C_{cable}) and other components. Among these, the C_{cable} can be optimized by minimizing its length (Fig .2(b)). We evaluated the degree of the disturbance by comparing the drain current (I_D) waveforms with/without probing the MG. With the shortest cable used, the measurement results demonstrated an acceptable disturbance. To verify that this disturbance is caused by C_{par}, a simulation of the MFM in series with a TFT structure, with and without C_{par}, was also conducted and produced I_D waveforms consistent with the measurements (Fig. 2(c)). In the future, the C_{par} issue could be totally addressed by an integrated on-chip amplifier, with a fF level input capacitance (Fig. 2(d)).

B. Decoupling Method

The decoupling procedures were illustrated on Fig. 4. The measured dynamic $V_{mid}(t)$ data allows real-time calculation and updating of device states (Fig. 4(a)), enabling the extraction of Q_{trap} through a charge balance equation. The MFM component and the channel charge of the thin-film transistor (TFT) are computed using the nucleation-limited switching (NLS) model and a TFT surface potential model respectively (Fig. 4(b)) [6]. The Q_{trap} calculated in this manner represents the net charge of the MG gate, as well as the defect charges within the IL layer and at the channel interface. Especially, the MG contact structure shown in Fig. 1(a) allows for in-situ device measurements for the models' calibration, enhancing the accuracy by minimizing the device-to-device variation (Fig 4(c)). For the MFM, Fig. 4(d) shows the calibrated ΔP_{FE} under various pulse conditions. For the TFT model, we additionally introduced interface

979-8-3315-0417-5/25 $31.00 © 2025 IEEE

states, achieving well-fitted I_d-V_g and C-V characteristics simultaneously (Fig. 4(e)).

Results and Discussion

A. Decoupled Results and Phenomena

Fig. 5 shows a set of decoupled data, and two basic features are observed. ①ΔQ_{trap} is roughly 90% of ΔP_{FE} during switching. ②Around the peak of V_g, Q_{trap} decreases as V_{mid} increases. For more details, results of different V_g waveforms were further compared, revealing two new phenomena. We found that: ③Q_{trap} always precedes P_{FE}, and the P_{FE} lag time, which is extracted by a fixed charge of 20pC (Fig. 6(a)), increases as the frequency decreases (Fig. 6(b)). ④V_{mid} amplitude increases as V_g frequency decreases (Fig. 6(c)). Fig. 7 supplies a summary of these phenomena.

B. Understanding of the Decoupled Results

Phenomenon ① and ② are consistent with previous reports, corresponding to the Ch-Inj [7], [8] and Gt-Inj [4] respectively. The Ch-Inj and P_{FE} are initiated as soon as V_g is applied(Fig. 8(a)). Around the peak of V_g, Gt-Inj becomes dominant over Ch-Inj, (Fig. 8(b)).

The phenomena ③④ offer a deeper understanding of the gate voltage dependence of polarization switching and charge trapping speed. We infer that P_{FE} is likely to exhibit a stronger V_g dependence than Ch-Inj, as illustrated in Fig. 9(a). On the one hand, the $\Delta E_{FE}/\Delta E_{IL}$ under different V_g is large due to the area ratio. On the other hand, according to the NLS model, the time constant of the P_{FE} exponentially depends on higher powers of $1/E_{FE}$ [6]. This dependence appears to be more pronounced than the exponential dependence of the defect capture time constant on E_{IL} [8].

As the frequency decreases, the reduced stressed V_g during the same initial duration slows down the P_{FE} more significantly than Ch-Inj. This results in a larger P_{FE} lag time, i.e. the phenomenon ③ (Fig. 9(b)). For the phenomenon ④, the stress duration of high V_g is longer under lower frequency conditions, during which P_{FE} and Gt-Inj dominate over Ch-Inj. This leads to a greater increase in P_{FE} plus Gt-Inj compared to Ch-Inj, resulting in a larger P_{FE} - Q_{trap}. Based on the charge balance equation in Fig. 4(a), V_{mid} and P_{FE} - Q_{trap} exhibit a positive correlation. Therefore, V_{mid} increases under lower frequency conditions (Fig. 9(c)).

C. Guidance on the Operation Optimization

The frequency dependence of V_{mid} discussed above plays a crucial role in optimizing endurance. A higher V_{mid} tends to degrade the IL layer more severely, whereas a larger $V_{FE} = V_g$ - V_{mid} can lead to a faster breakdown of MFM. The former mechanism is usually observed, but the latter can also occur in large area ratio MFMIS under large write voltage.

We measured the endurance of our device under triangular waves writing conditions of 10V/7V amplitudes and 800ns/8µs periods (Fig. 10). As for ±10V writing, the 800ns condition showed a faster breakdown before 1E4 cycles, indicating a smaller V_{mid}. As for ±7V writing, 8µs condition shows a much more sever memory window degration, indicating a larger V_{mid}. In summary, low-frequency writing is preferred to minimize FE breakdown, while high-frequency writing is preferred to mitigate IL degradation.

Conclusion

In this paper, we proposed a method to decouple the Q_{trap} and P_{FE} on MFMIS-FeFET by direct V_{mid} measurement. The non-ideal effects of this method were thoroughly analyzed and optimized to an acceptable level. Our decoupled data shows features of channel-injection and gate-injection consistent to previous work as a result. New phenomena of the frequency dependent P_{FE} lag time to Q_{trap} and V_{mid} amplitude were also observed, indicating that PFE has a stronger V_g dependence. Based on these findings, the guidelines of operation optimization of FeFETs towards high endurance were provided.

Acknowledgments

This work was supported by National Key R&D Program of China (2022YFB4400300), NSFC (62274003, 61927901 and 92164203) and 111 Project (B18001). *(Wenpu Luo and Runteng Zhu contributed equally to this work)*

References

[1] S. Mueller, "Ferroelectric HfO2 and Its Impact on the Memory Landscape," in *2018 IEEE International Memory Workshop (IMW)*, May 2018, pp. 1–4. doi: 10.1109/IMW.2018.8388831.

[2] S. Yoon *et al.*, "QLC Programmable 3D Ferroelectric NAND Flash Memory by Memory Window Expansion using Cell Stack Engineering," in *2023 IEEE Symposium on VLSI Technology and Circuits (VLSI Technology and Circuits)*, Jun. 2023, pp. 1–2. doi: 10.23919/VLSITechnologyandCir57934.2023.10185294.

[3] T. Soliman *et al.*, "First demonstration of in-memory computing crossbar using multi-level Cell FeFET," *Nat Commun*, vol. 14, no. 1, p. 6348, Oct. 2023, doi: 10.1038/s41467-023-42110-y.

[4] X. Wang *et al.*, "Deep insights into the Interplay of Polarization Switching, Charge Trapping, and Soft Breakdown in Metal-Ferroelectric-Metal-Insulator-Semiconductor Structure: Experiment and Modeling," in *2022 International Electron Devices Meeting (IEDM)*, Dec. 2022, p. 13.3.1-13.3.4. doi: 10.1109/IEDM45625.2022.10019390.

[5] S. Lee *et al.*, "Effect of Floating Gate Insertion on the Analog States of Ferroelectric Field-Effect Transistors," *IEEE Transactions on Electron Devices*, vol. 70, no. 1, pp. 349–353, Jan. 2023, doi: 10.1109/TED.2022.3223640.

[6] W. Zhang, J. Wang, C. Sun, Z. Wu, X. Gong, and X. Fong, "Modeling of Ferroelectric Thin Film Transistors with Amorphous Oxide Semiconductor Channel," in *2024 8th IEEE Electron Devices Technology & Manufacturing Conference (EDTM)*, Mar. 2024, pp. 1–3. doi: 10.1109/EDTM58488.2024.10511948.

[7] M. Passlack *et al.*, "Direct Quantitative Extraction of Internal Variables from Measured PUND Characteristics Providing New Key Insights into Physics and Performance of Silicon and Oxide Channel Ferroelectric FETs," in *2022 International Electron Devices Meeting (IEDM)*, Dec. 2022, p. 32.4.1-32.4.4. doi: 10.1109/IEDM45625.2022.10019459.

[8] Z. Wu *et al.*, "Characterizing and Modelling of the BTI Reliability in IGZO-TFT using Light-assisted I-V Spectroscopy," in *2022 International Electron Devices Meeting (IEDM)*, Dec. 2022, p. 30.1.1-30.1.4. doi: 10.1109/IEDM45625.2022.10019454.

979-8-3315-0417-5/25 $31.00 © 2025 IEEE

Device structure and decoupling method

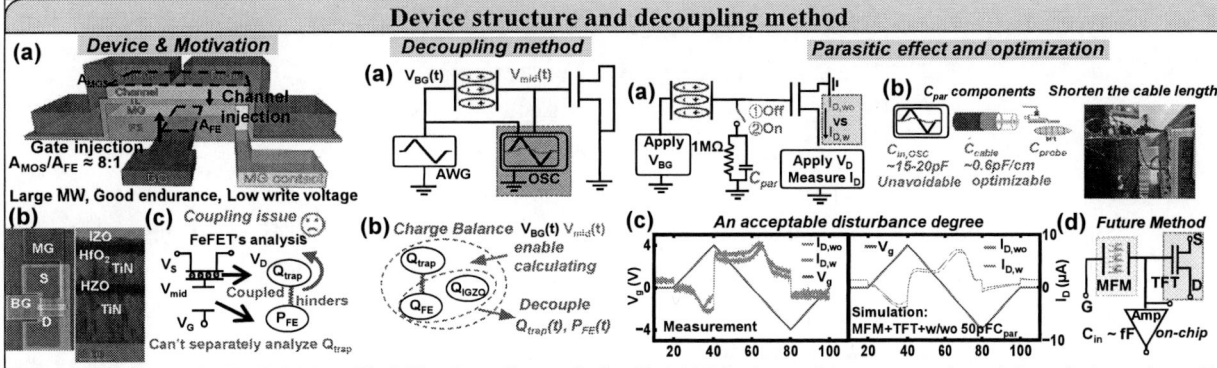

Fig.1 (a) The structure of MFMIS FeFET. Adjustable area and Gt-Inj result in its advantages. (b) SEM/TEM results. (c) Analysis challenge: the coupling of P_{FE} and Q_{trap}.

Fig.2 The decoupling method: (a) Direct $V_{mid}(t)$ measurement. (b) V_{mid} enables the decoupling by charge balance equation.

Fig.3 (a) Disturbance of measurement exists and the method to evaluate. (b) Components of C_{par}, where C_{cable} is optimizable by shortening the length. (c) Measurement and simulation shows an acceptable disturbance. (d) In the future, on-chip amplifier can decrease the C_{par} to fF.

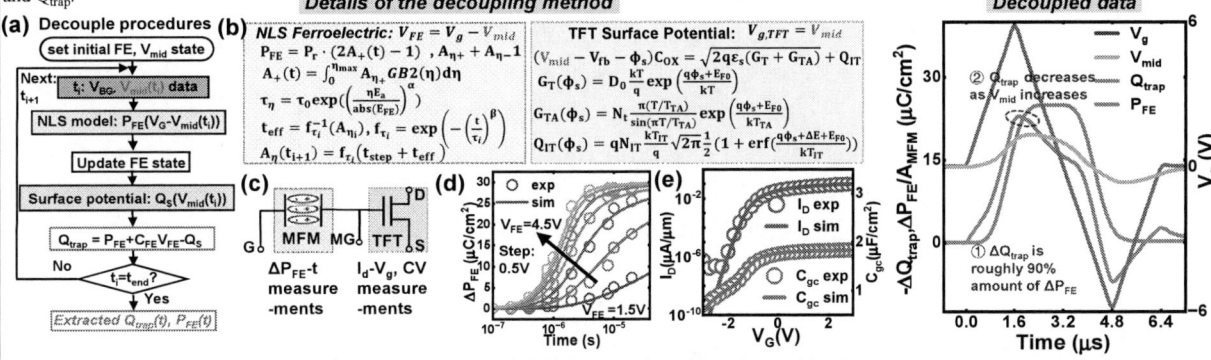

Fig.4. (a) Procedures to decouple Q_{trap} and P_{FE} based on $V_{mid}(t)$ (b) Brief explanation of the NLS model and surface potential TFT model. (c) Measurements for models' calibration. (d) FeCAP and (e) TFT calibration based on the measured data with NLS and TFT model. An excellent fitting is demonstrated.

Fig.5. Two basic features of decoupled data showing: (i) ΔQ_{trap} and is roughly 90% amount of ΔP_{FE}. (ii) A stage when Q_{trap} decreases as V_{mid} increases exists.

Decoupled results and discussions

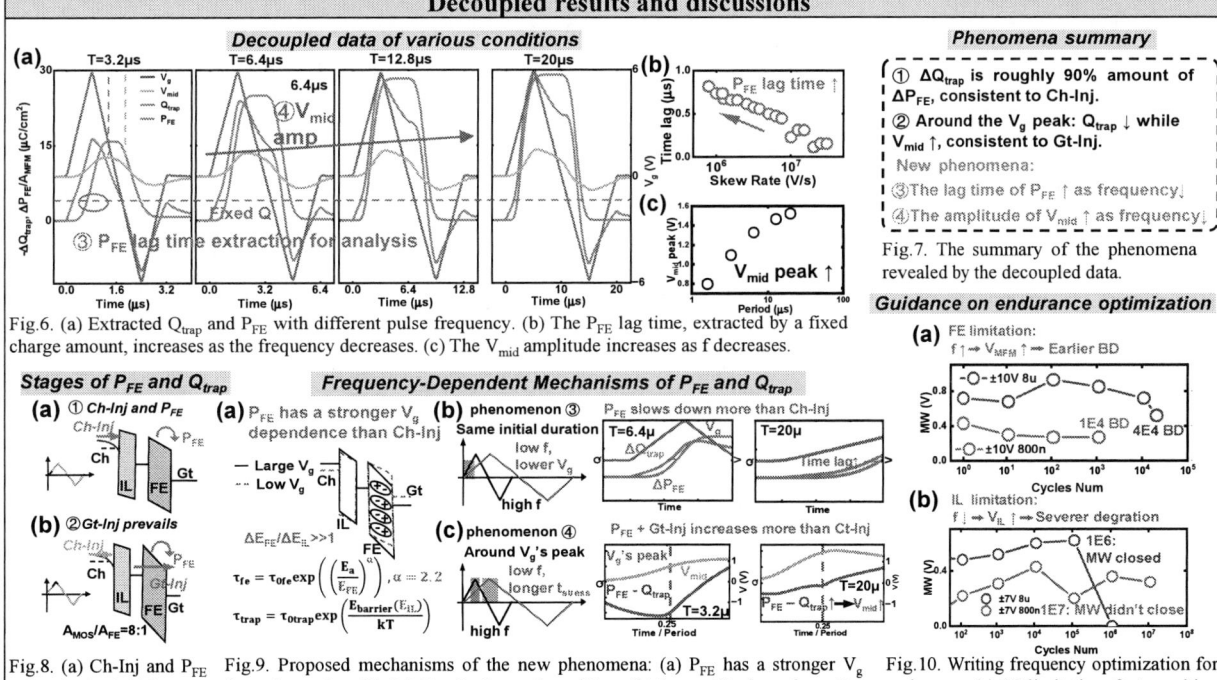

Fig.6. (a) Extracted Q_{trap} and P_{FE} with different pulse frequency. (b) The P_{FE} lag time, extracted by a fixed charge amount, increases as the frequency decreases. (c) The V_{mid} amplitude increases as f decreases.

Fig.7. The summary of the phenomena revealed by the decoupled data.

Fig.8. (a) Ch-Inj and P_{FE} exists in the initial stage. (b) Around V_g's peak, Gt-Inj and P_{FE} prevail.

Fig.9. Proposed mechanisms of the new phenomena: (a) P_{FE} has a stronger V_g dependence than Ch-Inj. For the lower f condition: (b) Lower V_g slows down P_{FE} more than Ch-Inj. (c) Longer t_{stress} leads to a more increase of P_{FE} plus Gt-Inj than Ch-Inj

Fig.10. Writing frequency optimization for endurance: (a) FE limitation: faster writing BDs earlier. (b) IL limitation: slower writing degrades IL more severely.

979-8-3315-0417-5/25 $31.00 © 2025 IEEE   1163

Multi-functional Flexible Intelligent Glove for Gesture Recognition and Combustible Detection

Jiachuang Wang[1,2,3], Fangyu Zhao[1,2], Wenyuan Liu[1,2], Nan Qin[1,2],
and Tiger H. Tao[1,2,3,4,5,6,7,8]

[1]State Key Laboratory of Transducer Technology, Shanghai Institute of Microsystem and
Information Technology, Chinese Academy of Sciences, Shanghai, CHINA
[2]School of Graduate Study, University of Chinese Academy of Sciences, Beijing, CHINA
[3]2020 X-Lab, Shanghai Institute of Microsystem and Information Technology,
Chinese Academy of Sciences, Shanghai, CHINA
[4]Center of Materials Science and Optoelectronics Engineering,
University of Chinese Academy of Sciences, Beijing, CHINA
[5]Center for Excellence in Brain Science and Intelligence Technology,
Chinese Academy of Sciences, Shanghai, CHINA
[6]Neuroxess Co., Ltd. (Jiangxi), Nanchang, Jiangxi, CHINA
[7]Guangdong Institute of Intelligence Science and Technology,
Hengqin, Zhuhai, Guangdong, CHINA and
[8]Tianqiao and Chrissy Chen Institute for Translational Research, Shanghai, CHINA

Abstract

We report a multi-functional flexible intelligent glove that integrates artificial tactile and olfactory sensor array. It is able to assist wearers and humanoid robots in tasks such as gesture recognition and combustion odor detection. Our proposed convolutional neural network algorithm can fuse multi-dimensional sensing information, distinguish typical gestures, and achieve an accuracy of 97.2% in recognizing four types of combustibles. It provides opportunities for rapid decision-making in the rescue fields of gas analysis and target recognition.
Keywords: Multi-dimensional fusion, intelligent glove, sensor array

Introduction

Multi-dimensional information fusion is an important solution for perceiving complex environments and one of the main research directions for humanoid robots [1-3]. Touch and smell are the most direct information feedback in the process of object interaction, which can be used for system integration of humanoid robots. Micro-electro-mechanical system (MEMS) technology can efficiently improve sensor performance by creating fine microstructures, converting physical signals in nature into quantifiable electrical signals for subsequent processing and analysis [4].
With the development of sensor performance and the

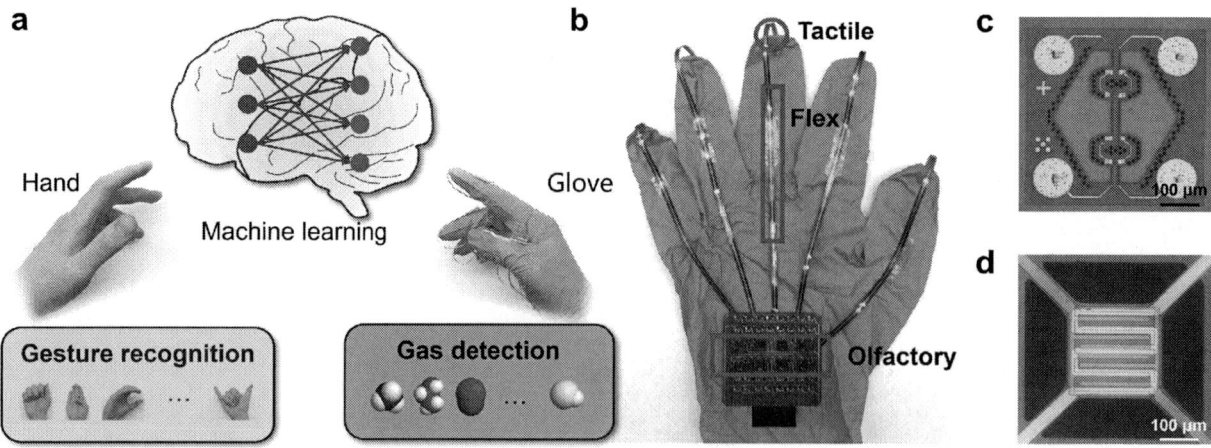

Figure 1: (a) Schematic diagram of biomimetic glove for gesture recognition and gas detection. (b) Optical diagram of flexible intelligent glove. (c) Optical microscope image of MEMS pressure sensor. (d) Optical microscope image of MEMS gas sensor.

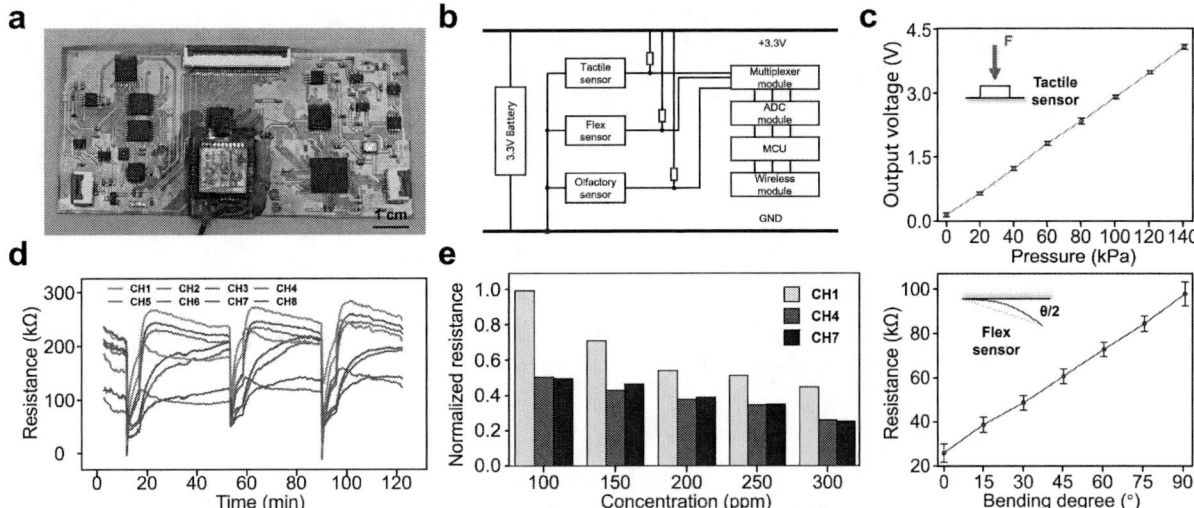

Figure 2: (a) A flexible data acquisition board used for collecting multi-channel sensor information. (b) A brief flowchart for achieving high-speed data acquisition and wireless transmission. (c) The characterization of tactile sensor and flex sensor for demonstrating the sensitivity. (d) Cycle experiment of gas sensor array on 80 ppm ammonia gas. (e) Normalized resistance of three different channel gas sensors as carbon monoxide concentration changes.

popularization of artificial intelligence, the combination of integrated sensor platforms and intelligent algorithms has received continuous attention from researchers [5-7]. In emergency rescue, obstructed vision and burning objects often make it difficult to detect trapped individuals. Fusion of touch and smell for gesture decision-making and recognition of combustion odors is a feasible solution that can quickly handle dangerous situations to reduce property damage and personnel injuries [8,9].

Design and Fabrication

Machine learning originates from the exploration and study of the biological principles and mechanisms of neural networks. We designed a bionic glove with the capabilities of recognizing the gestures of human hand and the odors of combustibles (Figure 1a). Specifically, we used Altium Designer software to design the overall flexible printed circuit (FPC) for attachment onto the experimental glove (Figure 1b). Firstly, the MEMS pressure sensor (Figure 1c) is integrated at the fingertip to provide feedback on the contact with the object and the grasping force. Secondly, the flex sensor is arranged at the finger to measure the hand posture when grasping the object. The artificial touch constructed through this layout is crucial for recognizing the shape and texture of objects. Then the MEMS gas sensor array (Figure 1d) is integrated in the palm area for recording the gas response. Moreover, necessary components are soldered in the surrounding area to form a voltage divider circuit and support the overall flexible printed circuit structure.

Experimental Results

The entire glove involves the collection and processing of data from three types of sensors, including pressure, flex and gas. Therefore, we prepared a data acquisition board based

on flexible printed circuit according to the requirement of sampling rate and power consumption (Figure 2a), which is connected to the front-end sensor through a 40 pin FPC ribbon cable. Specifically, since all sensors integrated into the flexible glove are passive, we use a 3.3 V portable lithium battery for power supply (Figure 2b). Further use a micro controller unit (MCU) to control the multiplexer modules and analog-to-digital converter (ADC) modules to achieve multi-channel sensor data acquisition, and then transmit the data to the terminal through a wireless module, achieving portability and wireless performance.

High sensitivity pressure sensors are able to provide accurate sampling data for determining the state of contact objects. As shown in Figure 2c, we set the pressure generator within the range of 0~140 kPa and measured its sensitivity to 28.37 mV kPa^{-1} through a signal amplifier. A stable and repeatable flex sensor is beneficial for accurately determining posture. We measured the resistance output multiple times within the bending range of 0~90 °. The above sensing information can accurately reflect various states during the contact process in typical object interaction sequences.

Ammonia is an important gas in the exhaled breath of human, so rapid detection of ammonia has the potential to identify trapped individuals in emergency rescue. We conducted a cycle experiment on 80 ppm ammonia gas (Figure 2d), and the gas sensor array showed great stability and repeatability. Carbon monoxide often appears in environments with insufficient combustion. It is colorless and odorless, but harmful to human health. We tested the response of different channels within the range of carbon monoxide concentration of 100~300 ppm (Figure 2e), which is an important basis for distinguishing the degree of object combustion.

Finally, we designed a convolutional neural network

979-8-3315-0417-5/25 $31.00 © 2025 IEEE   1165

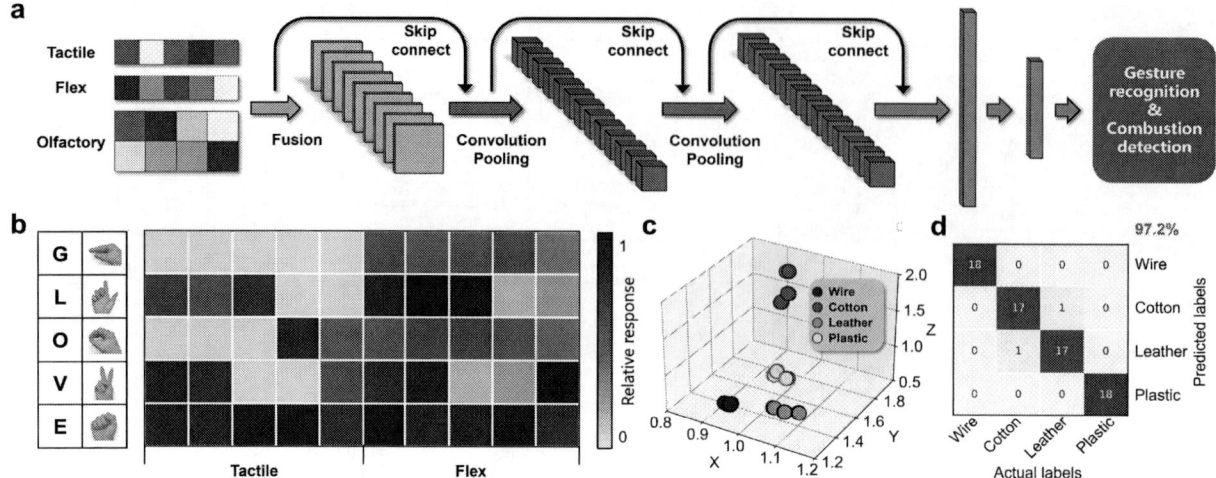

Figure 3: (a) Flowchart of convolutional neural network algorithm for gesture recognition and gas detection. (b) The relative resistance map output by the pressure sensor and the bending sensor when making different gestures. (c) The visualization result of the feature space obtained by fusing and reconstructing the tactile and olfactory information of typical combustibles. (d) Identification confusion matrix diagram of typical combustibles.

algorithm model that integrates tactile, flex, and olfactory sensor information for gesture recognition and combustion detection (Figure 3a). As shown in Figure 3b, our flexible glove can be used to recognize English letter gestures through algorithms and translate the word "GLOVE". We visualized the feature space of the fused tactile and olfactory information of typical combustibles (Figure 3c), including wire, leather, cotton, and plastic. With the fusion strategy, we achieved a high recognition accuracy of 97.2% (Figure 3d). It is expected to assist emergency operations such as fire rescue and on-site decision-making.

Conclusion

In summary, we report a biomimetic glove that integrates artificial touch and smell. Based on a convolutional neural network algorithm model, this multi-functional intelligent glove is capable of recognizing hand posture and combustibles odor without visual information input. The simulation experimental results validate the feasibility and superiority of our flexible glove in emergency rescue environments through high-sensitivity sensors and multidimensional information fusion.

Acknowledgments

This work was supported by National Science and Technology Major Project from the Minister of Science and Technology of China (grant no. 2018AAA0103100), National Natural Science Foundation of China (grant no. 62236005), National Science Fund for Excellent Young Scholars (grant no. 61822406), Key Research Program of Frontier Sciences, CAS (grant no. ZDBS-LY-JSC024) and Fund of Youth Innovation Promotion Association CAS (grant no. 2022234). These authors contributed equally: Jiachuang Wang, Fangyu Zhao and Wenyuan Liu.

Contact

*Tiger H. Tao, tel: +86-21-62511070,
tiger@mail.sim.ac.cn
*Nan Qin, qinnan@mail.sim.ac.cn

References

[1] L. Chen, et al. Spike timing–based coding in neuromimetic tactile system enables dynamic object classification, Science, 384, 660-665 (2024).

[2] Y. Liu, et al. Intelligent wearable olfactory interface for latency-free mixed reality and fast olfactory enhancement, Nature Communications, 15, 4474 (2024).

[3] M. Liu, et al. Robotic Manipulation under Harsh Conditions Using Self-Healing Silk-Based Iontronics, Advanced Science, 9, 2102596 (2022).

[4] C. Wang, et al. Biomimetic olfactory chips based on large-scale monolithically integrated nanotube sensor arrays, Nature Electronics, 7, 157-167 (2024).

[5] A. Padmanabha, et al. A multimodal sensing ring for quantification of scratch intensity, Communications Medicine, 3, 115 (2023).

[6] S. Zhang, et al. Body-Integrated, Enzyme-Triggered Degradable, Silk-Based Mechanical Sensors for Customized Health/Fitness Monitoring and In Situ Treatment, Advanced Science, 1903802 (2020).

[7] J. Broek, et al. Highly selective detection of methanol over ethanol by a handheld gas sensor, Nature Communications, 10, 4220 (2022).

[8] K. Kang, et al. Bionic artificial skin with a fully implantable wireless tactile sensory system for wound healing and restoring skin tactile function, Nature Communications, 15, 10 (2024).

[9] H. Kang, et al. Multiarray nanopattern electronic nose (e-nose) by high-resolution top-down nanolithography. Advance Functional Materials, 30, 2002486 (2020).

Piezoelectric Micromachined Ultrasonic Transducers for Advanced Sensing Applications

Bowen Sheng, Lei Zhao, Jinghan Gan, Yufeng Gao, Jiao Xia, Junhao Wang, Chenyuan Zhang and Yipeng Lu

Peking University, China

Abstract

This paper discusses the design, optimization, and applications of piezoelectric micromachined ultrasonic transducers (PMUTs) for ultrasonic sensing systems. PMUTs offer advantages such as miniaturization, large bandwidth, and high-yield process of transducer array. Key advancements, including non-uniform membrane design and multi-frequency operation, are highlighted for their role in enhancing PMUT performance and versatility. Applications in structural health monitoring, underwater communication, and imaging are also presented.

(Keywords: PMUT, ultrasonic sensing, structural health monitoring)

Introduction

Ultrasonic transducers are essential in applications such as non-destructive testing, structural health monitoring, and material characterization, etc. Traditional bulk piezoelectric ceramic transducers face challenges, including low fill-factor, complex electrical routing, and acoustic impedance mismatch. These issues can be addressed by micromachined ultrasonic transducers (MUTs) based on micro-electro-mechanical systems (MEMS) technology, offering benefits like miniaturization and better integration with CMOS circuits. MUTs are typically divided into capacitive (CMUTs) and piezoelectric (PMUTs) types. While CMUTs offer high bandwidth, their reliance on high bias voltages and nanoscale precision during fabrication introduces complexity. In contrast, PMUTs, which use piezoelectric thin films such as PZT and AlN, overcome these challenges. Advances in thin-film deposition have further improved PMUT performance, making them promising candidates for next-generation ultrasonic applications [1]. This paper focuses on optimizing PMUT designs to enhance their performance in advanced sensing applications, such as structural health monitoring and underwater communication.

PMUT Design and Optimization

PMUTs are highly valued for their compact size, low power consumption, and high sensitivity, making them ideal for various sensing applications. Optimizing PMUT design is essential to improve their performance further. Conventional PMUTs often utilize uniform membranes, which can limit their vibration efficiency and sensitivity. Advancing PMUT designs through non-uniform membrane optimization and multi-frequency operation is crucial for expanding functionality and enhancing overall performance.

A. Non-Uniform Membrane Modeling

Accurate modeling techniques capable of handling non-uniform membrane structures are critical for improving PMUT performance. A novel approach using the variational principle [2] enables accurate prediction of vibration modes for non-uniform PMUT membranes, as shown in Fig. 1(a). By employing a piecewise cubic polynomial fitting technique, the eigenmodes of the membrane were predicted with high accuracy, achieving less than 0.1% error compared to theoretical models for uniform membranes, as shown in Fig. 1(b) and 1(c). The approach also includes patterning the piezoelectric layer, which optimizes the vibration shape and enhances the effective acoustic transmission. This design refinement increases both the efficiency of energy conversion and the overall performance of the PMUTs.

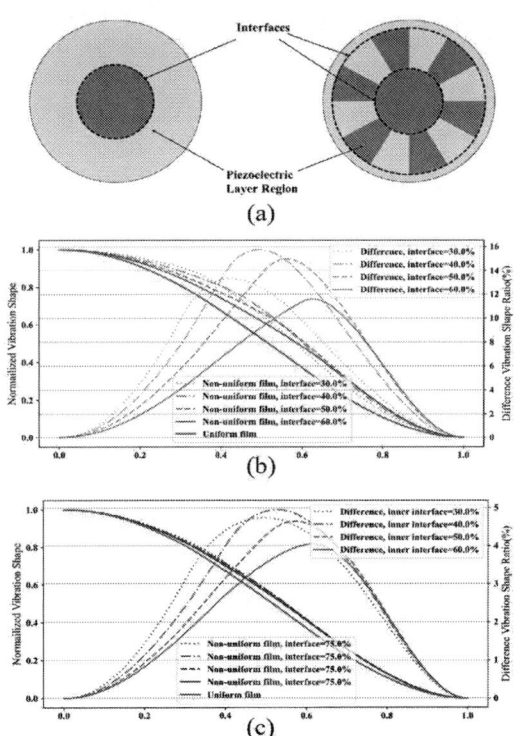

Fig. 1: Non-uniform membrane modeling [2]: (a) schematic diagram for patterned piezoelectric layer PMUT with annular and rib-structure design; vibration shape of (b) annular and (c) rib-structure design.

B. Harmonic Tuning for Performance Enhancement

Traditional PMUTs typically operate at a single frequency,

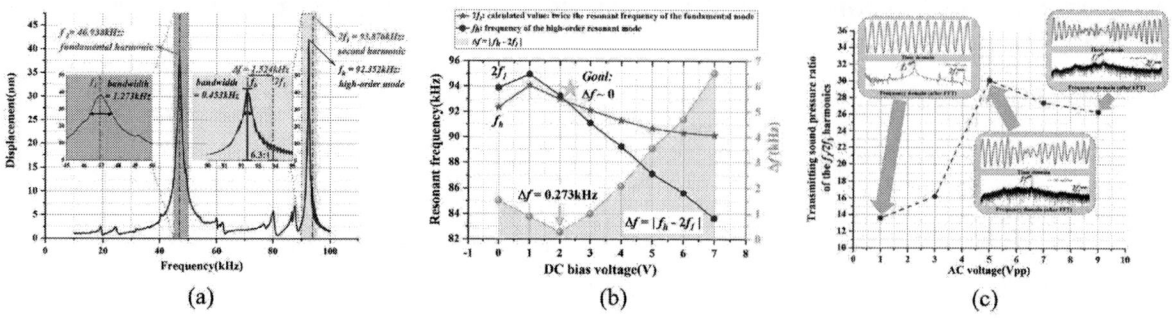

(a) (b) (c)

Fig. 2: Harmonic tuning for performance enhancement [3]: (a) frequency response of the PMUT with a high-order resonant frequency approaching twice that of the fundamental resonant frequency, facilitating accurate sensing with enhanced second harmonic; (b) DC bias voltage introduced energy modification of the fundamental and the second harmonic; (c) AC voltage introduced acoustic harmonic energy modification.

which limits the amount of sensing information that can be extracted. For applications that require higher accuracy, such as gas sensing, there is a need to expand frequency range without increasing design complexity. Multi-frequency operation, which uses multiple frequency channels within a single PMUT device, offers a promising solution [3]. The concept of harmonic energy modification was introduced using a PMUT with enhanced second harmonic to achieve this, as shown in Fig. 2(a). The energy distribution between the fundamental and second harmonic frequencies can be controlled by applying both DC and AC bias voltages, as shown in Fig. 2(b) and 2(c). The second harmonic is intentionally tuned to twice the fundamental frequency, maximizing the harmonic output. This allows the second harmonic to be used as an additional operational frequency, offering more detailed sensing capabilities without increasing die area or design complexity. By optimizing the resonant frequencies and maximizing harmonic energy, this method enhances the PMUT's ability to provide more accurate and reliable data, particularly in scenarios like gas sensing where multi-frequency precision is critical. This design enables the next generation of PMUT-based sensors to perform with higher accuracy and greater functionality, meeting the growing demand for sophisticated sensing technologies.

Applications

PMUTs are increasingly being used in various fields because of their compact size, high sensitivity, and versatility. This paper lists a few examples as below. In structural health monitoring, PMUTs can be used for real-time and non-destructive inspection. For underwater applications, PMUTs can be used for communication [4] and imaging by transmitting and receiving ultrasonic signals effectively in aquatic environments.

A. Structural Health Monitoring

PMUTs have been used in ultrasonic scanning in energy storage for lithium-ion battery health monitoring [5]. By analyzing ultrasonic phenomena during battery charging and discharging cycles, we can assess the state of charge (SOC) and physical condition of the battery, particularly at low temperatures [6]. In our experiments, changes in ultrasonic amplitude and time of flight (ToF) during cycles were notably influenced by ultrasound frequency and temperature. Lower temperatures weakened transmission signals, reflecting a decrease in state of health (SOH) from 84.5% to 71.7%. We achieved precise SOC and SOH estimations by tracking these changes and applying specialized algorithms. This approach presents a promising alternative to traditional battery monitoring, providing more accurate assessments, especially in challenging low-temperature environments.

PMUTs were proven to be effective in detecting surface icing on composite insulators when embedded in a silicone rubber layer [7]. Pulse-echo tests were conducted on a silicone oil-air interface with a 10 mm silicone oil layer. To improve testing accuracy, pulse-echo signals from various ice thicknesses were first adjusted by subtracting signals from thick ice layers (>50 mm). Absolute values and envelope processing were then applied to isolate the ice echo signals for different thicknesses. A thickness of 0.26 mm was identified as near the detection limit for the 3.7 MHz PMUTs, with thinner layers leading to decreased accuracy. This approach was demonstrated to reliably measure ice thickness and type, proving to be more effective than traditional detection techniques in complex outdoor environments.

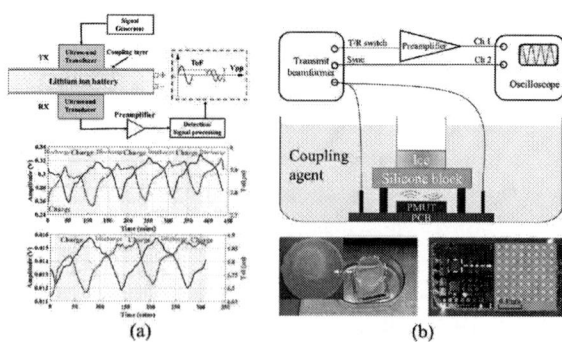

(a) (b)

Fig. 3: Structural Health Monitoring: (a) lithium-ion battery health monitoring [6] and (b) ice detection [7].

B. Underwater Communication

A novel ultrasound communication method was recently demonstrated by modulating the amplitude and phase of transmitted waves via DC bias voltage and pulse polarity [8]. When a DC bias is applied, it alters the membrane stress within PMUTs [9], thereby modulating the amplitude of the emitted ultrasound. Additionally, pulse signals with varied polarities can shift the phase of the acoustic waves, enabling effective phase modulation. This approach eliminates the need for high-frequency carriers and complex circuitry, lowering power consumption and enhancing interference resistance. In an underwater test with an 8×8 PMUT array operating at 1.06 MHz and -6 dB bandwidth of 71.4%. By employing varying levels of DC bias and short pulses of differing polarities, as shown in Fig. 4(a) both the amplitude and phase of the transmission signal were modulated. This method demonstrated promising results for low-power and interference-resilient ultrasound communication.

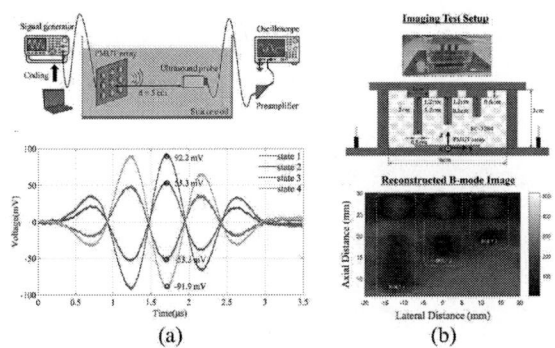

Fig. 4: Advanced sensing applications of PMUTs in (a) underwater ultrasound communication [8], and (b) ultrasonic imaging [11].

C. Imaging

Two critical factors for ultrasonic transducers are acoustic pressure output and bandwidth, as they determine imaging depth and axial resolution [10]. Acoustic pressure output is tied to the fill-factor of the transducer array. A high fill-factor ensures an effective radiating area per unit surface, leading to a concentrated acoustic pressure distribution and high sound pressure level (SPL) output. Optimizing fill-factor can be achieved by altering the diaphragm shape from circular to rectangular, which allows for fill factors exceeding 70% [11]. The high fill-factor and rectangular membrane structure yield a high SPL output of 1400 dB/V and a board fractional bandwidth of 72%. The PMUT array's performance was validated through imaging of bolt phantoms, with the final reconstructed B-mode image presented in Fig. 4(b).

Conclusion

This paper has explored key advancements in PMUT design, particularly the optimization of non-uniform membranes and the integration of multi-frequency operation, both of which are critical for enhancing performance in complex sensing systems. PMUTs have been demonstrated as a promising solution for advanced sensing applications, such as structural health monitoring, underwater communication, and imaging.

Acknowledgments

This research was supported by the National Key Research and Development Program under Grant 2023YFE0209600.

References

[1] K. Roy, J. E.-Y. Lee and C. Lee "Thin-film PMUTs: a review of over 40 years of research", *Microsystems & Nanoengineering*, 9(1), p. 95, 2023.

[2] J. Gan, Z. You, C. Yang, Y. Lu, "Modeling of PMUTs with Non-Uniform Membrane Based on The Variational Principle," *2024 IEEE UFFC-JS*, Taipei, September 22-26, 2024.

[3] Y. Gao, A. Bao, C. Yang, L. Zhao and Y. Lu, "Multi-frequency Harmonic Energy Modification of Piezoelectric Micromachined Ultrasonic Transducer (PMUT) for Advanced Sensing Applications," *2024 IEEE UFFC-JS*, Taipei, September 22-26, 2024.

[4] F. Pop, B. Herrera and M. Rinaldi, "Lithium Niobate Piezoelectric Micromachined Ultrasonic Transducers for high data-rate intrabody communication", *Nature Communications*, 13(1), p. 1782, 2022.

[5] Z. Deng, X. Lin, Z. Huang, J. Meng, Y. Zhong, G. Ma, Y. Zhou, Y. Shen, H. Ding and Y. Huang, "Recent Progress on Advanced Imaging Techniques for Lithium-Ion Batteries", *Advanced Energy Materials*, 11(2), p. 2000806, 2021.

[6] J. Xia, T. Xie, Y. Guo, J. Wang, Y. Lu, "Battery Status Monitoring Based on Advanced Ultrasonic Technology," *2024 IEEE UFFC-JS, Taipei*, September 22-26, 2024.

[7] T. Xie, J. Wang, C. Yang, J. Xia, W. Wang and Y. Lu, "Detection of Surface Icing Based on Piezoelectric Micromachined Ultrasonic Transducers (PMUTs)," *2024 IEEE UFFC-JS*, Taipei, September 22-26, 2024.

[8] Y. Guo, C. Zhang, C. Yang, J. Xia, W. Wang and Y. Lu, "Underwater Ultrasound Communication Based on PMUTs With DC Bias Voltage," *2024 IEEE UFFC-JS*, Taipei, September 22-26, 2024.

[9] Y. Gao, L. Zhao, C. Yang and Y Lu, "Characterization and Optimization of PZT-Based PMUTs With Wide Range Frequency Tuning", *Journal of Microelectromechanical Systems*, 33(4), pp. 427-437, 2024.

[10] L. Zhao, C. Yang, X. Zhang, Z. You and Y. Lu, "Design, Fabrication, and Characterization of High-Performance PMUT Arrays Based on Potassium Sodium Niobate," *Journal of Microelectromechanical Systems*, 33(4), pp. 438-445, 2024.

[11] L. Zhao, C. Yang, Y. Guo, Y. Lu, "Rectangular PMUT Phased Array with High Fill-Factor and High Sound Pressure Level," *2024 IEEE UFFC-JS*, Taipei, September 22-26, 2024.

Contact

Yipeng Lu, yplu@pku.edu

Epitaxial Growth of Stacking Faults-free Hexagonal Bilayer MoS$_2$

Cheol-Joo Kim[1,2]

[1] Center for Epitaxial van der Waals Quantum Solids, Institute for Basic Science (IBS), Pohang 37673, Republic of Korea,

[2] Department of Chemical Engineering, Pohang University of Science and Technology (POSTECH), Pohang 37673, Republic of Korea

Abstract

Deterministic growth of a single-phase material with a low defect density is essential for achieving phase-sensitive properties uniformly and on demand in a scalable manner. While layered materials exhibit various polytypes with significantly different crystal symmetries and properties, the small difference in free energies between polytypes makes phase control challenging [1]. Here, we report the deterministic growth of bilayer MoS$_2$ films with a hexagonal single phase over a large area using metalorganic chemical vapor deposition. In layer-by-layer growth, forming single Mo atom seeds on top of the first-grown MoS$_2$ layer prior to the second-layer growth ensures deterministic stacking of the second layer. Density functional theory suggests that these atomic seeds serve as nucleation sites, guiding the crystal growth of the hexagonal phase and preventing the formation of the rhombohedral phase.

Keywords: Layer-by-layer Epitaxy, MoS$_2$, MOCVD, Bilayer

References

[1] Q. Wang *et al.*, "Layer-by-layer epitaxy of multi-layer MoS$_2$ wafers" National Science Review, 9 (2022).

This page intentionally left blank.

979-8-3315-0417-5/25 $31.00 © 2025 IEEE

Exploring the Potential of Hafnium Oxide-Based Ferroelectric Memories for Next-Generation Storage Class Memories

Sourav De, IEEE Senior Member

College of Semiconductor Research, National Tsing Hua University, email: sourav.de@mx.nthu.edu.tw

Abstract: Hafnium oxide (HfO_2)-based ferroelectric memories are emerging as appealing candidates for storage class memory (SCM) applications due to their scalability, low energy consumption, and compatibility with complementary-metal-oxide-semiconductor (CMOS) technology. Their strong ferroelectric properties at nanoscale thicknesses make them suitable for high-density memory, effectively bridging the gap between volatile and non-volatile storage. Key factors influencing their performance include the selection of dopants, the quality of the interfaces, defect management, and overall reliability, all of which impact endurance, retention, and the device's stability.

Keywords: hafnium oxide, ferroelectric memory, storage-class memory, FeFET.

I. Introduction: Modern computing demands memory solutions that combine high performance with low power consumption to address the "memory wall" in von Neumann architectures. While technologies like SRAM, DRAM, and Flash are crucial, data-centric applications require faster and more efficient emerging non-volatile memory (eNVM). HfO_2-based ferroelectric memory is a promising option that offers rapid write speeds, low operating voltages, and high endurance.

Storage Class Memory (SCM) merges the advantages of DRAM and NAND Flash, providing high speed and non-volatility. Technologies such as Phase-Change Memory (PCM) and magneto-resistive RAM (MRAM) enhance endurance, speed, and power efficiency, making SCM vital for applications in AI and real-time data processing. Ideal SCM features include:
- Latency of 50 nanoseconds to 10 microseconds
- Minimum bandwidth of over 100 MB/s
- High endurance of over 100 million write cycles
- Low hard error rate of 1 in 10,000 bits per terabyte
- Lifespan of around 2 million hours
- Energy consumption of 100 mW during operation and under 1 mW on standby
- Cost under $5 per GB

Ferroelectric materials are promising for SCM due to their fast switching and high endurance. While lead zirconate titanate (PZT) has compatibility issues, HfO_2-based ferroelectric materials maintain strong properties while being more compatible with CMOS technology, making them a focus for future development in SCM applications.

II. A Tour d'horizon to Ferroelectricity:

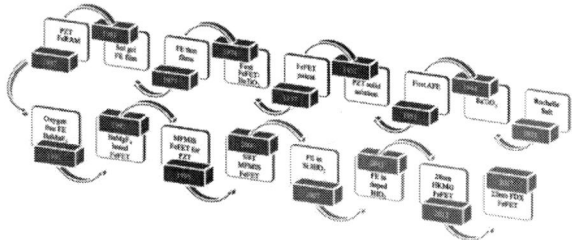

Fig. 1: History of ferroelectricity

In La Rochelle, France, the history of salt began with Jehan Seignette, who took over his father's pharmacy in 1592. His brother, Elie, created "sel polychreste" in 1665. Rochelle salt gained prominence in the 19th century when David Brewster discovered its pyroelectricity in 1824, and the Curie brothers termed it piezoelectric in 1880. Professor Peter Debye suggested that certain molecules have a permanent electric dipole, leading to infinite dielectric constants at specific temperatures. Erwin Schrödinger introduced "ferroelectric" in 1912, arguing that all solids could show ferroelectricity at low temperatures.

Joseph Valasek highlighted Rochelle salt's piezoelectric properties in 1920, noting its electric hysteresis and polarization. Paul Scherrer and Georg Busch acknowledged research linking ferroelectricity in potassium salts to mobile hydrogen atoms. The first artificial ferroelectric material, barium titanate ($BaTiO_3$), was discovered between 1942 and 1944, demonstrating ferroelectricity in simple oxides without hydrogen. In 2011, Tim Böscke found ferroelectricity in HfO_2, with films doped with less than 4% SiO_2 exhibiting ferroelectric properties after annealing, primarily in a meta-stable polar orthorhombic phase. [1-7]

Types of Ferroelectric Memories and Recent Progress:

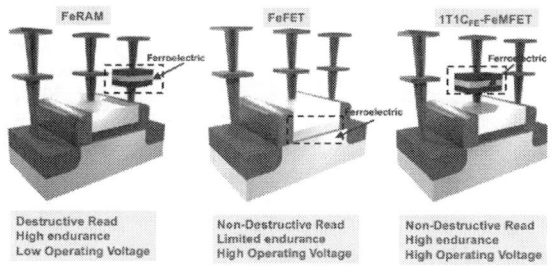

Fig. 2: Different types of ferroelectric memories

Among the various types, Ferroelectric Field-Effect Transistors (FeFETs), Ferroelectric Random-Access Memory (FeRAM), and 1T1C FeFET configurations are gaining significant attention due to their unique advantages and potential for integration into next-generation storage class memory systems.

FeRAM (Ferroelectric Random-Access Memory): In FeRAM, data is written by switching the polarization of the ferroelectric material while read operations detect the polarization state. FeRAM is known for its fast read/write speeds, low power consumption, and endurance compared to traditional non-volatile memories like Flash, though it is typically more expensive to manufacture.

In this series of studies on FeRAM, the advanced engineering of hafnium zirconium oxide (HZO) thin films and FeRAM architectures are demonstrated to improve retention, endurance, and switching characteristics for FeRAM and FeFET applications, with quantitative milestones showing the technology's scalability. A $Hf_{1-x}Zr_xO_2$ nanolaminate structure with thin tetragonal phase layers achieved over 10 years of retention at 85 °C and an endurance surpassing 10^{10} cycles at operating voltages as low as 1.0 V in 5.5 nm-thick devices. A ferroelectric-antiferroelectric layered design also enabled a coercive field reduction and an ultrafast switching time of 780 ps at 2 V, verified using a high-speed pulsed measurement system and nucleation-limited switching models. Comprehensive benchmarking of 1T1C FeRAM arrays fabricated with the *HfO₂* process showed fast switching, lower read/write latencies, and energy consumption, and detailed analysis of plate line driver and bit line capacitances highlighted key performance improvements. For film scaling studies, 8-nm *HfO₂* samples exhibited a higher tolerance to dielectric breakdown under cycling conditions, outperforming 10-nm samples with capacitance areas as small as 0.20 μm². Furthermore, a novel capacitor-under-bitline structure achieved a large remanent polarization ($2P^r > 40$ μC/cm²), projected endurance beyond 10^{11} cycles, and robust data retention over ten years at 85 °C, demonstrating stable CMOS-compatible integration with an operation voltage of 2.5 V and read/write speeds below 10 ns. These advances mark significant steps toward ultra-low-voltage, high-endurance ferroelectric memories, positioning them for future system-on-chip applications. [8-11]

Ferroelectric Capacitive Memory: Recent advancements in hafnium zirconate-based ferroelectric capacitors (FeCAPs) have led to energy-efficient, non-volatile memory elements that allow non-destructive read operations. The authors achieved a record capacitive memory window (CMW) of ~8.7 at 0 V in HZO-based metal-ferroelectric-metal (MFM) capacitors through interfacial engineering that introduced interface asymmetry, along with non-destructive reading capability ensures device reliability, with read endurance surpassing 10^{11} cycles. FeCAPs are a key for CIM applications, particularly in machine learning, as they enable efficient operations without additional selectors. While challenges like limited capacitance ratios may cause minor errors in computations, co-design approaches are being explored to address these issues. Enhancements in ferroelectric properties, such as a $2P^r$ of 66.5 μC/cm², were achieved by integrating novel interfacial layers, ensuring compatibility with CMOS technology. These developments highlight the transformative potential of hafnium zirconate-based FeCAPs in scalable and efficient memory technologies [12-16].

Silicon FeFET (Ferroelectric Field-Effect Transistor): Ferroelectric field-effect transistors (FeFETs), which use ferroelectric materials as gate dielectrics, are promising for eNVM and neuromorphic computing. They offer high-speed switching, low power consumption, and scalability, making them suitable for advanced computing. The polarization of the ferroelectric layer influences the transistor's threshold voltage, enabling persistent charge states for data storage. Integrating FeFETs with CMOS processes improves their use in sub-20nm technology. Recent developments show that FeFETs can maintain high-temperature retention and support distinct memory states for IoT devices. Additionally, FeFET-based crossbar architectures reduce variations in drain current, achieving about 97% accuracy on MNIST datasets—crucial for neuromorphic computing.

Silicon-based FeFETs, particularly Si-FeFinFETs, have garnered attention in the context of neuromorphic computing. These devices have exhibited 3-bit-per-cell functionality and have demonstrated high endurance, with all-ferroelectric neural networks achieving an impressive accuracy of 97.91%. Binary neural networks optimized with Fe-FinFETs have sustained 95.2% accuracy across an extensive temperature range, from −40°C to 125°C, effectively addressing the challenges associated with temperature stability. Implementing tri-gate FeFETs with 40nm gate lengths has revealed strong ferroelectric switching capabilities, rendering them suitable for online training tasks. The scalability, multilevel storage capacity, and temperature

resilience of Si-FeFETs position them as versatile candidates for next-generation computing architectures [17-24].

Metal-Oxide Channel FeFET: Recent advancements in metal oxide semiconductor channel-based FeFETs strengthen their potential for low-power, high-performance memory applications. The use of Hf₀.₅Zr₀.₅O₂ with back-end-of-line (BEOL) compatibility has led to monolithic 3D (M3D) architectures, achieving a remnant polarization density of 40 μC/cm², a 0.45V memory window in 20nm transistors, write speeds of about 100ns, and endurance over 10^8 cycles. M3D FeFETs offer three times the energy efficiency of static random-access memory (SRAM) in compute-in-memory (CIM) tasks, making them suitable for compact energy-efficient systems.

Additionally, IL-free BEOL FeFETs with amorphous indium tungsten oxide (IWO) channels feature a 1.6V memory window, read-after-write latency under 300ns, and endurance over 10^{11} cycles. Innovations like anti-ferroelectric zirconium dioxide (ZrO₂) with IGZO channels reduce coercive fields, enhance endurance (up to 10^9 cycles), and support ultra-low operating voltages of 2V, with retention exceeding 10 years. Collectively, these developments highlight the transformative potential of FeFETs in memory and computing, integrating seamlessly with standard CMOS processes and innovative architectures like M3D, positioning them as a disruptive technology for energy-efficient, high-density memory and neuromorphic computing [25,26,27].

1T1C FeFET (One-Transistor-One-Capacitor FeFET): The 1T1C FeFET architecture combines a single transistor with a ferroelectric capacitor, creating a compact and efficient solution for memory applications. This design enhances data retention and reliability by using the ferroelectric capacitor to maintain polarization while the transistor handles read and write operations. It effectively balances speed and scalability, making it ideal for high-density, low-power non-volatile memory systems.

Advancements in ferroelectric memory architectures, such as 2TnC, 1T1C, and 2T2C configurations, have improved compute-in-memory (CiM) systems for edge AI and embedded applications. The 2TnC design uses two transistors and multiple ferroelectric capacitors to enhance scalability and signal integrity, allowing simultaneous read and write operations and reducing the need for 3D integration.

The 2T2C memory cell supports dual functionality, operating as either ferroelectric RAM (FeRAM) or a memristive ferroelectric tunnel junction (FTJ), enabling independent programming of its devices while facilitating in-memory processing essential for edge computing.

Further advancements in the 1T1C architecture, specifically through the C²FeRAM approach, have improved endurance and energy efficiency, achieving four times better efficiency and a 200-fold increase in inference speed compared to traditional FeRAM. Additionally, integrating ferroelectric capacitors with HK/MG MOSFETs in FeMFET configurations enhances compatibility with standard CMOS processes at ultra-low write voltages, marking significant progress towards scalable and energy-efficient ferroelectric CiM systems for future intelligent devices [28,29].

Where do ferroelectric memories find a place in the memory hierarchy?: Ferroelectric memories exhibit exceptional promise for integration across various applications, particularly in storage-class memory (SCM), where they effectively bridge the performance gap between DRAM and NAND flash. While offering access times as low as 10–20 ns, DRAM is limited by its volatility. With higher latency (50–100 μs) and limited endurance (10^3–10^6 cycles), NAND flash struggles to meet the requirements of high-write

979-8-3315-0417-5/25 $31.00 © 2025 IEEE   1174

workloads. Ferroelectric memories, leveraging materials like doped HfO_2, achieve sub-10 ns access times, endurance over 10^{10}, and retention capabilities spanning multiple years, making them viable for SCM applications.

As nonvolatile DRAM (nvDRAM), these

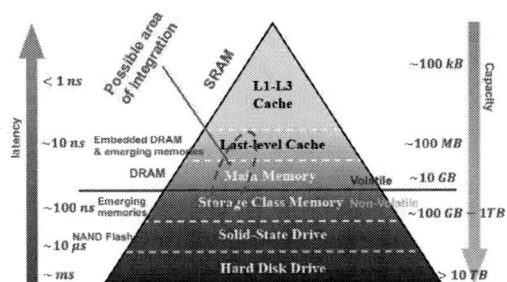

Fig. 3: Overview of semiconductor memory and possible areas of integration of ferroelectric memory

memories combine DRAM-equivalent speeds with nonvolatility, which is critical for servers, HPC systems, and real-time applications that demand high-speed data persistence during power interruptions. Furthermore, ferroelectric nonvolatile SRAM (nvSRAM) offers the speed (<5 ns) and energy efficiency of SRAM while enabling data retention, making it a compelling candidate for IoT and edge computing systems. With scalable 3D integration and densities rivaling NAND flash (terabit scale), ferroelectric memories can mitigate challenges like device scaling and thermal budgets, offering a pathway for next-generation memory solutions to replace or complement existing architectures. This technological leap positions ferroelectric memories as a cornerstone in evolving semiconductor memory hierarchies [30-35].

Acknowledgment: This work was funded by Taiwan's National Science and Technology Council's grant with number 114-2222-E-007-001.

References:

[1] Maurice Soenen, La Pharmacie à La Rochelle. Les Seignettes et le sel polychreste. Thèse de doctorat de l'Université de Bordeaux, 1910

[2] Jacques et Pierre Curie, Compt. rend. 91, 294-295, 383-387 (1880). Développement par pression de l'électricité polaire dans des crystaux hémiédriques à faces inclinées.

[3] P. Debye, Physik. Zeitschr. XIII, 97-100 (1912). Einige Resultate einer kinetischen Theorie der Isolatoren (Vorläufige Mitteilung).

[4] Erwin Schrödinger, Aus sen Sitzungsberichten der Kaiserl. Akademie der Wissenschaften in Wien. Mathem.-Naturw. Klasse; Bd. CXXI, Abt. Iia, November 1912. Studien üder Kinetik der Dielektrika, dam Schmelzpunkt, Pyro-und Piezoelekrizität.

[5] J. Valasek, Phys. Rev. 15, 537 (1920); 17, 475-481 (1920). Piezoelectric and allied phenomena in Rochelle salt.

[6] J. Valasek, Phys. Rev. 19, 478-491 (1922); Piezo-electric activity of Rochelle salt under various conditions.

[7] T. S. Böscke, et.al, Ferroelectricity in hafnium oxide thin films. Appl. Phys. Letter

[8] M. Adnaan, et.al, "Design Considerations for Sub-1-V 1T1C FeRAM Memory Circuits," in IEEE Journal on Exploratory Solid-State Computational Devices and Circuits

[9] J. Okuno et al., "Reliability Study of 1T1C FeRAM Arrays With Hf0.5Zr0.5O$_2$ Thickness Scaling," in IEEE JEDS.

[10] J. Okuno et al., "1T1C FeRAM Memory Array Based on Ferroelectric HZO With Capacitor Under Bitline," in IEEE JEDS.

[11] Hojung Jang et.al, "Demonstration of 1 V Reliable FeRAM Operation: V_c Engineering Using Quasi-Chirality of Hf$_{1-}$

$_x$Zr$_x$O$_2$ in a Nanolaminate Structure"ACS Applied Materials & Interfaces

[12] C.-Y. Cho et al., "Exploring BEOL-Compatible Ferroelectricity in Ultra-Thin Hafnium Zirconium Oxide: Thermal Budget, FTJ Characteristics, and Device Reliability," in IEEE JEDS

[13] C. -Y. Chiu, et al., "Trade-off Between Thermal Budget and Thickness Scaling: A Bottleneck on Quest for BEOL Compatible Ultra-Thin Ferroelectric Films Sub-5nm," IEEE EDTM.

[14] M. I. Popovici et al., "High performance La-doped HZO based ferroelectric capacitors by interfacial engineering," IEEE IEDM.

[15] S. Mukherjee et al., "Capacitive Memory Window With Non-Destructive Read in Ferroelectric Capacitors," in IEEE EDL.

[16] S. Mukherjee et al., "Pulse-Based Capacitive Memory Window with High Non-Destructive Read Endurance in Fully BEOL Compatible Ferroelectric Capacitors," in IEEE IEDM.

[17] F. Müller et al., "Multilevel Operation of Ferroelectric FET Memory Arrays Considering Current Percolation Paths Impacting Switching Behavior," in IEEE EDL.

[18] Masud Rana Sk, et al., "Ferroelectric Content-Addressable Memory Cells with IGZO Channel: Impact of Retention Degradation on the Multibit Operation" ACS Applied Electronic Materials

[19] S. De et al., "28 nm HKMG-Based Current Limited FeFET Crossbar-Array for Inference Application," in IEEE TED

[20] M. R. Sk et al., "1F-1T Array: Current Limiting Transistor Cascoded FeFET Memory Array for Variation Tolerant Vector-Matrix Multiplication Operation," in IEEE TNano

[21] S. De et al., "Robust Binary Neural Network Operation From 233 K to 398 K via Gate Stack and Bias Optimization of Ferroelectric FinFET Synapses," in IEEE EDL

[22] S. De et al., "Ultra-Low Power Robust 3bit/cell Hf0.5Zr0.5O2 Ferroelectric FinFET with High Endurance for Advanced Computing-In-Memory Technology," 2021 Symposium on VLSI Technology.

[23] S. De et al., "Demonstration of Multiply-Accumulate Operation With 28 nm FeFET Crossbar Array," in IEEE EDL.

[24] M. Jerry et al., "Ferroelectric FET analog synapse for acceleration of deep neural network training," 2017 IEDM.

[25] J. Lee et al., "Optimization of Backside of Silicon-Compatible High Voltage Superlattice Capacitor for 12V-to-6V On-Chip Voltage Conversion," 2024 DRC

[26] K. A. Aabrar et al., "BEOL-Compatible Superlattice FEFET Analog Synapse With Improved Linearity and Symmetry of Weight Update," in IEEE TED.

[27] Y. Luo et al., "Monolithic 3D Compute-in-Memory Accelerator with BEOL Transistor based Reconfigurable Interconnect," IEEE IEDM.

[28] X. Ma et al., "A 2-Transistor-2-Capacitor Ferroelectric Edge Compute-in-Memory Scheme With Disturb-Free Inference and High Endurance," in IEEE EDL.

[29] Y. Xiao et al., "Quasi-Nondestructive Read Out of Ferroelectric Capacitor Polarization by Exploiting a 2TnC Cell to Relax the Endurance Requirement," in IEEE EDL.

[30] L. Fernandes et al., "Material Choices for Tunnel Dielectric Layer and Gate Blocking Layer for Ferroelectric NAND Applications," in IEEE EDL.

[31] D. Das et al., "Experimental demonstration and modeling of a ferroelectric gate stack with a tunnel dielectric insert for NAND applications," IEEE IEDM

[32] I. O'Connor et al., "FVLLMONTI: The 3D Neural Network Compute Cube (N2C2) Concept for Efficient Transformer Architectures Towards Speech-to-Speech Translation," 2024 DATE.

[33] C. Marchand et al., "FeFET based Logic-in-Memory design methodologies, tools and open challenges," IFIP/IEEE VLSI-SoC

[34] N. Ramaswamy et al., "NVDRAM: A 32Gb Dual Layer 3D Stacked Non-volatile Ferroelectric Memory with Near-DRAM Performance for Demanding AI Workloads," 2023 IEDM.

[35] Y. Shuto et al., "HZO-based Nonvolatile SRAM Array with 100% Bit Recall Yield and Sufficient Retention Time at 85°C," IEEE Symposium on VLSI Technology and Circuits.

Simulation of Germanium-Tin-based n^+/i-Well Dot Single-Photon Avalanche Diode for Fiber-Optic Telecommunication Networks

Harshvardhan Kumar[1], Advaita Sinha[1], and P. Susthitha Menon[2]

[1]Department of Electronics and Communication Engineering, The LNM Institute of Information Technology, Jaipur, Rajasthan, 302031, India,
[2]Institute of Microengineering and Nanoelectronics (IMEN), Universiti Kebangsaan Malaysia (UKM), 43600 UKM Bangi, Selangor, Malaysia

Abstract

We propose novel $Ge_{1-x}Sn_x$-based CMOS-integrated dot single-photon avalanche diodes (SPADs), which comprises a diminutive core n^+/i-well hemispherical cathode/i-well configuration encircled by an anode ring. The multiplication layer comprises the fully depleted i-well region. The proposed design facilitates extensive light-sensitive region, yielding responsivity of ~389 AW^{-1} at 1.55 μm. Thus, the proposed structure may avoid the requirement for a multi-dot configuration, which consists of multiple cathode/i-well dots sharing a common anode, thereby lowering the fabrication cost.

Keywords: dot-SPAD, CMOS, high performance, $Ge_{1-x}Sn_x$ alloy

Introduction

Single photon avalanche diodes (SPADs) developed via complementary metal-oxide-semiconductor (CMOS) processes are increasingly utilized in various applications, including fiber-optic telecommunications[1], quantum photonics[2], biomedical imaging[3], and light detection and ranging (LiDAR)[4]. This is due to the fact that CMOS SPADs are less expensive to fabricate and can be seamlessly integrated with other CMOS circuits, resulting in complete and compact systems, unlike SPADs produced in custom technologies[3].

In Geiger-mode operation, the SPAD is typically biased above the breakdown voltage, necessitating the use of guard rings (GRs) to avert premature edge breakdown (PEB)[3], [5]. Nonetheless, GRs have the potential to expand the overall area of a SPAD, which may lead to a decrease in its fill factor[5]. The GR structure plays a crucial role in influencing the noise performance of the SPAD[3]. A recent report by Poushi et al. illustrates a dot avalanche-based APD that achieves a nearly perfect fill factor for the visible range[5]. This device design paves the way towards developing high-performance detectors that do not require the inclusion of GRs. Nonetheless, for applications necessitating a more extensive light-sensitive area, a multi-dot or array structure becomes essential to enhance the active area[5]. This, in turn, escalates both the complexity of the fabrication process and the associated costs. Furthermore, there have been limited reports on Si and Ge-based SPADs for short-wave infrared (SWIR) bands due to their large

bandgap and low absorption coefficient in these bands[6].

In recent years, the successful growth and tunable electrical and optical characteristics of $Ge_{1-x}Sn_x$ alloy render it a promising material for the short-wave infrared (SWIR) and mid-infrared (MIR) bands[7]. Recently, Chen et al. reported on a resonant-cavity-enhanced GeSn-based SPAD[8]. This device structure demonstrated encouraging outcomes pertaining to high efficiency at room temperature. Nevertheless, this device structure showed higher dark current and a significant dark count rate (DCR), which will result in degraded noise performance and lower responsivity. As a result, a novel device design is necessary for addressing the challenges of limited fill-factor, narrowed light-sensitive area, and higher value of DCR.

In this work, we propose a novel design of a $Ge_{1-x}Sn_x$-based n+/i-well dot SPAD featuring an active area exceeding 10 μm, aimed at optical fiber-based applications, where photons can be efficiently concentrated within a few-μm spot. Furthermore, the proposed structure is designed to potentially address the aforementioned issues and is capable of attaining high performance regarding (i) high responsivity, (ii) low dark current, and (iii) low DCR values at 1.55 μm.

Device Structure and Simulation

Fig. 1 illustrates the two-dimensional (2-D) schematic of the proposed single-dot SPAD (SD-SPAD) on the Si platform utilizing Ge virtual substrate (VS), and the designed device is simulated by using COMSOL Multiphysics. The device structure incorporates Ge VS to overcome the lattice mismatch between GeSn and Si layers. Therefore, owing to the reduced defects at the heterointerfaces, the device demonstrates a low dark current, leading to reduced DCR and noise values. A semi-hemispherical, highly doped n^+–Si region constitutes the cathode of this SPAD, with a doping concentration of 10^{20} cm^{-3}, embedded within a half-sphere intrinsic-well Si region acts as the multiplication layer (ML). The anode consists of a p^+–Si region with a doping concentration of 3×10^{19} cm^{-3}. In contrast, the $Ge_{1-x}Sn_x$ (x = 4%) absorber or active layer is lightly p-doped with a doping concentration of 10^{16} cm^{-3}. The $Ge_{1-x}Sn_x$ materials' parameters utilized in the simulation were taken and calculated from[9], [10]. To improve the accuracy of the simulation, the thermionic emission model

979-8-3315-0417-5/25 $31.00 © 2025 IEEE

has been selected for the heterojunction. Furthermore, the simulation incorporates trap-assisted recombination for domains and user-defined in the i-Si ML. The impact ionization coefficients for both electrons (α_n) and holes (α_p) can be determined using the following expressions[11];

$$\alpha_n = 2.4 \times 10^7 e^{(-1.6 \times 10^6/E)}; \tag{1}$$
$$\alpha_n = 1.8 \times 10^7 e^{(-3.2 \times 10^6/E)} \tag{2}$$

where, E denotes the electric field (in V/cm), obtained using COMSOL Multiphysics.

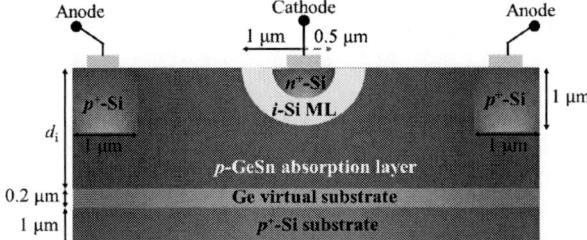

Fig. 1: 2-D schematic of the proposed $Ge_{1-x}Sn_x$–SPAD on the Si platform.

Fig. 2 illustrates the simulated 2-D spatial electric field (EF) distribution of the SPAD at a reverse voltage of 20.5 V, which is 5.5 V above the breakdown voltage. The EF meets the two design criteria: (1) The EF is strong and uniform within the i-Si ML; (2) a moderate electric field exists in the GeSn active layer to facilitate the drift of photo-carriers into the ML and initiate an avalanche event.

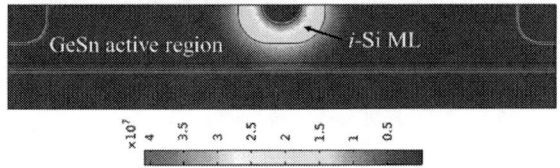

Fig. 2: Simulated electric field distribution in the proposed $Ge_{1-x}Sn_x$–

SPAD at reverse bias voltage of 20.5 V.

Results and Discussion
A. Electro-optical Characteristics of the SD-SPAD
Fig. 3 illustrates the simulated electro-optical characteristics and and optical responsivity of the n^+/i-well single-dot GeSn SPADs, which have diameters of 10 μm. First, the dark and optical current characteristics (I-V) with respect to the reverse voltage for the GeSn SD-SPAD working at room temperature are simulated and presented in Fig. 3(a). The dark current density of GeSn SPADs, particularly when biased above the punch-through voltage, escalates as a result of the amplified intrinsic carrier densities in GeSn caused by bandgap shrinkage. This finding aligns with the reported GeSn/Si APDs[12]. The I-V curves demonstrate the avalanche effects of the SD-SPADs, showing a breakdown voltage of 15 V. The punch-through voltage, defined as the voltage at which the electric field effectively penetrates into the GeSn absorption layer, leading to efficient collection of photo-carriers, is estimated to be approximately 10 V. It is noteworthy that the proposed device exhibits ~140% lower

breakdown voltage and ~130% lower punch-through voltage in comparison to the other GeSn SPAD[8]. The dark and optical current of the proposed device are ~2.7 pA and ~ 0.2 μA, respectively, at 90% of the breakdown voltage. The dark current value is significantly lower than the results of reported GeSn SPAD[8]. The ratio of photo-to-dark current (I_{opt}/I_{dark}) is 7.5×10^4 at $\lambda = 1.55$ μm with an optical power of 100 nW. The remarkably high ratio of I_{opt}/I_{dark} demonstrates that the device exhibits exceptional sensitivity to light. Next, the photodetection characterization of the SD-SPAD has been done at $\lambda = 1.55$ μm. Based on the optical current characteristics, the responsivity $(R = I_{ph}/P_{opt})$ as a function of reverse bias voltage is obtained, where dark current has been subtracted.

Next, Fig. 3(b) illustrates the responsivity in relation to the reverse bias voltage when the optical power is fixed at 100 nW. The SD-SPAD demonstrates a responsivity of 1.2 AW^{-1} when operating at low bias voltages (V = -1 V). It has been noted that the responsivity remains stable across a broad voltage range, subsequently increasing with a reverse bias voltage exceeding ~9 V. At the operating voltage of 15 V (breakdown voltage), the responsivity is 3.15 AW^{-1}, which increases to ~389 AW^{-1} when the bias voltage is increased to 20.5 V. Nonetheless, the highest possible responsivity is influenced by the optical power and decreases at higher optical power levels due to the saturation effect occurring in

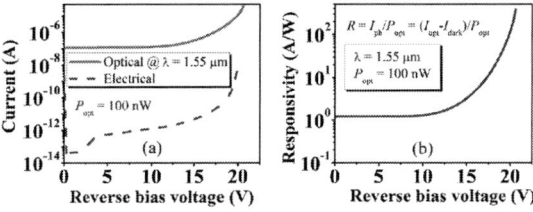

Fig. 3: (a) Simulated current-voltage characteristics of SD-SPAD under dark and illuminated conditions and (b) optical responsivity as a function of reverse bias voltage.

the multiplication process under such conditions. It is important to highlight that a diode can operate at lower voltages, working in an unamplified mode in conditions where the optical power is sufficiently high for detection without the need for amplification.

B. Dark Count Rate (DCR) of the SD-SPAD
In the dark condition, thermally generated carriers in SPAD can initiate avalanche events, a phenomenon referred to as dark counts. The DCR can be determined by[8]:

$$DCR = S \int (P_e + P_h - P_e P_h) G dz \tag{3}$$

Where, S represents the active area of the SPAD, while G denotes the net generation rate of the carriers. Conversely, P_e and P_h represent the probabilities that electrons and holes will be triggered, respectively. The DCR of SPAD is primarily influenced by three key generation mechanisms: thermal generation and recombination (SRH), trap-assisted

979-8-3315-0417-5/25 $31.00 © 2025 IEEE   1177

tunneling (TAT), and band-to-band tunneling (BTBT). This work considered all these generation mechanisms to determine the DCR. Theoretically, as outlined by the established Schockley-Read-Hall theory, the rate of thermal generation of carriers, taking into account the TAT effect and the influence of doping concentration, is defined by[13]. Furthermore, the generation rate of carriers for indirect bandgap semiconductors is described by Hurkx's model[14].

Subsequently, the key metric for SPAD, namely DCR, is calculated with respect to excess bias voltage above breakdown voltage at room temperature, considering various threading dislocations density (TDD) values. As illustrated in Fig. 4, it is evident that the DCR increases with an increase in excess bias voltage. One effective method to reduce the DCR is by cooling SPADs to minimize the generation of carriers due to thermal effects. Nonetheless, in practical applications, opting for SPAD without an additional cooling system is favored, and enhancing DCR through the engineering of GeSn material quality concerning TDD is the more effective approach. With the exception of highly defective GeSn layers (TDD$>10^{12}$ cm^{-2})[8], the majority of documented TDD values for GeSn on Si substrates range from 10^6 to 10^9 cm^{-2} [8]. Consequently, the TDD-dependent DCRs of Ge$_{0.96}$Sn$_{0.04}$ SD-SPAD under various excess voltages are calculated (Fig. 4). The simulation indicates that as the TDD decreases from 10^{10} to 10^6 cm^{-2}, there is a reduction in the DCR by five orders of magnitude at an excess voltage of 5 V. Furthermore, it is evident that this works presents a five orders lower magnitude in comparison to the previously reported GeSn SPAD for the same TDD[8]. This comparison highlights the effectiveness of the proposed SD-SPAD and paves the way for future design considerations for low DCR GeSn SPAD in the telecommunication band.

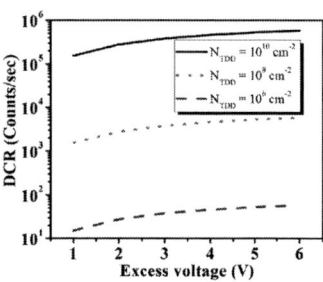

Fig. 4 Calculated DCR of GeSn SD-SPAD as a function of excess voltage for various TDD.

Conclusion

In summary, the electro-optical analysis of the proposed CMOS-integrated GeSn SD-SPAD is presented. Furthermore, the influence of TDD on DCR is examined, revealing that an increase in TDD corresponds to a reduction in DCR values. Furthermore, due to the extensive light-sensitive area and higher absorption coefficient of the GeSn alloy, the proposed device demonstrated an exceptionally high responsivity of ~389 AW^{-1} at 1.55 um with V = 20.5 V. Furthermore, the device demonstrated better results pertaining to dark current and DCR values when compared to previously reported GeSn SPAD. This analysis demonstrates the effectiveness and potential of the developed device for telecommunication applications.

References

[1] M. J. Deen and P. K. Basu, *Silicon Photonics: Fundamentals and Devices*. United Kingdom: John Wiley, 2012. [Online]. Available: http://onlinelibrary.wiley.com/doi/10.1002/9781119945161.fmatter/pdf

[2] F. Ceccarelli, G. Acconcia, A. Gulinatti, M. Ghioni, I. Rech, and R. Osellame, "Recent Advances and Future Perspectives of Single-Photon Avalanche Diodes for Quantum Photonics Applications," *Adv. Quantum Technol.*, vol. 4, no. 2, 2021, doi: 10.1002/qute.202000102.

[3] W. Jiang, R. Scott, and M. Jamal Deen, "Improved Noise Performance of CMOS Poly Gate Single-Photon Avalanche Diodes," *IEEE Photonics J.*, vol. 14, no. 1, 2022, doi: 10.1109/JPHOT.2021.3128055.

[4] J. F. Haase, S. Grollius, S. Grosse, A. Buchner, and M. Ligges, "A 32x24 pixel SPAD detector system for LiDAR and quantum imaging," in *Photonic Instrumentation Engineering VIII*, SPIE, 2021. doi: https://doi.org/10.1117/12.2578775.

[5] S. S. Kohneh Poushi, B. Goll, K. Schneider-Hornstein, M. Hofbauer, and H. Zimmermann, "Area and Bandwidth Enhancement of an n+/p-Well Dot Avalanche Photodiode in 0.35 μm CMOS Technology," *Sensors*, vol. 23, no. 7, 2023, doi: 10.3390/s23073403.

[6] E. Van Sieleghem *et al.*, "A Near-Infrared Enhanced Silicon Single-Photon Avalanche Diode with a Spherically Uniform Electric Field Peak," *IEEE Electron Device Lett.*, vol. 42, no. 6, pp. 879–882, 2021, doi: 10.1109/LED.2021.3070691.

[7] Z. Kong *et al.*, "Growth and Strain Modulation of GeSn Alloys for Photonic and Electronic Applications," *Nanomaterials*, vol. 12, no. 6, pp. 1–15, 2022, doi: 10.3390/nano12060981.

[8] Q. Chen, S. Wu, L. Zhang, W. Fan, and C. S. Tan, "Simulation of High-Efficiency Resonant-Cavity-Enhanced GeSn Single-Photon Avalanche Photodiodes for Sensing and Optical Quantum Applications," *IEEE Sens. J.*, vol. 21, no. 13, pp. 14789–14798, 2021, doi: 10.1109/JSEN.2021.3074407.

[9] H. Kumar and R. Basu, "Design of Mid-Infrared Ge1-xSnx Homojunction p-i-n Photodiodes on Si Substrate," *IEEE Sens. J.*, vol. 22, no. 8, pp. 7743–7751, 2022, doi: 10.1109/JSEN.2022.3159833.

[10] H. Kumar and M. Oehme, "Design of Mid-Infrared Ge$_{1-x}$Sn$_x$/Ge Heterojunction Photodetectors on GeSnOI Platform with a Bandwidth Exceeding 100 GHz," *IEEE J. Sel. Top. Quantum Electron.*, vol. 31, no. 1, pp. 1–8, 2024, doi: 10.1109/JSTQE.2024.3396608.

[11] C. Multiphysics, C. Software, and L. Agreement, "Radiation Effects in a PIN Diode." [Online]. Available: https://www.comsol.com/model/radiation-effects-in-a-pin-diode-74891

[12] A. Mosleh *et al.*, "Investigation on the Formation and Propagation of Defects in GeSn Thin Films," *ECS Meet. Abstr.*, vol. MA2014-02, no. 35, pp. 1845–1845, 2014, doi: 10.1149/ma2014-02/35/1845.

[13] W. J. Kindt and H. W. Van Zeijl, "Modelling and Fabrication of Geiger mode Avalanche Photodiodes," *IEEE Trans. Nucl. Sci.*, vol. 45, no. 3 PART 1, pp. 715–719, 1998, doi: 10.1109/23.682621.

[14] G. A. M. Hurkx, D. B. M. Klaassen, and M. P. G. Knuvers, "A New Recombination Model for Device Simulation Including Tunneling," *IEEE Trans. Electron Devices*, vol. 39, no. 2, pp. 331–338, 1992, doi: 10.1109/16.121690.

979-8-3315-0417-5/25 $31.00 © 2025 IEEE

High-Frequency Capacitance Measurement Techniques and Their Applications in Memory Technology Development

Jiang Qian[1], Dan Lv[2], Lu Wang[2], Ruiqi Ma[1], Shuai Kong[2], Shanting Zhang[2], Dongdong Li[2], and Liang Zhao[1*]

[1]College of Information Science and Electronic Engineering, Zhejiang University, Hangzhou 310000, China
[2]Zhangjiang Laboratory, Shanghai 201210, China
*Email: lzhao2020@zju.edu.cn

ABSTRACT

In the era of artificial intelligence and big data, the demand for high-speed and reliable memory technologies is growing rapidly. For semiconductor memories such as DRAM and FeRAM, metal-insulator-metal and metal-insulator-semiconductor capacitors play crucial roles, and their high-frequency characteristics (e.g. from 3MHz to 3GHz) significantly impact the device performance. However, conventional measurement techniques such as the bridge method struggle to provide accurate capacitance values at such frequencies, posing a challenge for memory technology development. In this work, we introduce our recent studies on two high-frequency capacitance measurement techniques: the network analysis method and the transient signal method. The initial results suggest both methods can accurately extract DUT capacitance up to 1GHz, offering valuable insights for memory device innovation.

Keywords: High-frequency capacitance measurement, network analysis method, transient signal method

I. INTRODUCTION

The rapid growth of artificial intelligence (AI) and big data has driven the demand for high performance and reliable memory devices. Among various semiconductor memories, dynamic random-access memory (DRAM) and ferroelectric memories have attracted wide interests for AI applications. DRAM offers high capacity, fast access speed and relatively low cost, while ferroelectric memories (e.g. FeRAM and FeFET) are non-volatile and offer high speed, good endurance and low power consumption. These important memory devices share a common core component: a capacitor based on metal-insulator-metal (MIM) or metal-insulator-semiconductor (MIS) structures.

Since advanced memories typically operate at high frequencies (from a few MHz to GHz), the characteristics of these capacitors at such frequencies are critical for the device performance, demanding accurate capacitance measurements. However, conventional techniques such as the bridge method [1] struggle with accuracy at high frequencies due to the parasitic effects and the measurement equipment's limited bandwidth.

In this work, we present our recent studies on two high-frequency capacitance measurement techniques:

the **network analysis method**, which uses S-parameters for impedance extraction; and the **transient signal method**, which works by injecting predefined waveforms into the device-under-test (DUT) and analyzing the output response to calculate device parameters. With appropriate setup, both methods can provide accurate high-frequency capacitance measurements, offering valuable insights for high-speed device analysis.

II. THE NETWORK ANALYSIS METHOD

The network analysis method treats the DUT as a circuit impedance network. By measuring the S-parameters of each port of the DUT and applying circuit analysis, the electrical parameters of the DUT, such as resistance, capacitance and inductance, can be deduced based on corresponding circuit models.

A. Basic Principles

For the network analysis method, the most crucial element is the scattering parameters (S-parameters). The S-parameter matrix reflects the relationship between the incident and reflected electromagnetic waves at each port of the DUT. For a two-port network, the definition of S-parameters is shown in Figure 1, and their mathematical equations are as follows:

$$\begin{bmatrix} b_1 \\ b_2 \end{bmatrix} = \begin{bmatrix} S_{11} & S_{12} \\ S_{21} & S_{22} \end{bmatrix} \begin{bmatrix} a_1 \\ a_2 \end{bmatrix} \tag{1}$$

$$S_{11} = \left. \frac{b_1}{a_1} \right|_{a_2=0}, \quad S_{12} = \left. \frac{b_1}{a_2} \right|_{a_1=0}, \\ S_{21} = \left. \frac{b_2}{a_1} \right|_{a_2=0}, \quad S_{22} = \left. \frac{b_2}{a_2} \right|_{a_1=0} \tag{2}$$

Here, a_1 and a_2 represent the sinusoidal signals (including both amplitude and phase information) entering into port 1 and port 2, while b_1 and b_2 represent the sinusoidal signals leaving port 1 and port 2, respectively.

Although the network analysis method is generally applicable, it also has some limitations: (1) the Vector Network Analyzer (VNA) required for S-parameter measurements are usually quite expensive; (2) the measurement of S-parameters relies on carefully-designed shield structures (e.g. the ground-signal-ground or GSG structures) to guarantee the accuracy of results. It cannot be directly applied on arbitrary DUT.

979-8-3315-0417-5/25 $31.00 © 2025 IEEE

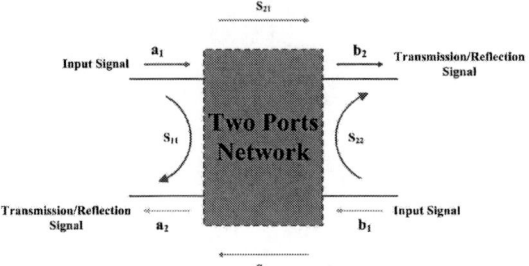

Fig. 1. Illustration of S-parameter definitions.

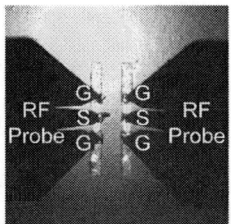

Fig. 2. Typical probing setup for GSG pads.

B. Extraction of Capacitance from S-Parameters

In order to accurately extract the impedance from S-parameters, shield structures (e.g. GSG structures and guard rings) for the DUT are typically required, along with impedance-matched RF probes and cables. Some guidelines on the design of GSG structures are provided in Ref [2]. A typical schematic of the GSG probing setup is shown in Figure 2. Also, it is worth mentioning that before actual data collection, the VNA must be calibrated to eliminate system errors. The specific calibration methods can be found in [3], [4].

For a two-port DUT with measured S_{11} and S_{21}, the impedance Z between the two ports can be derived as:

$$S_{11} = \frac{Z - Z_0}{Z + Z_0} \Longrightarrow Z = \frac{Z_0(1 + S_{11})}{1 - S_{11}} \quad (3)$$

$$S_{21} = \frac{2Z}{Z_0 + 2Z} \Longrightarrow Z = \frac{2Z_0(1 - S_{21})}{S_{21}} \quad (4)$$

Here, Z_0 is the characteristic impedance of 50Ω. The capacitance can be calculated based on the imaginary part of the impedance with the equation $Z = 1/j\omega C$, ignoring the inductive component of DUT. It is important to note that the raw data of S-parameters must be de-embedded to eliminate the parasitic effects introduced by the test fixtures, ensuring more accurate measurement results. Detailed instructions on how to perform de-embedding have been described in [5] [6]. Figure 3 demonstrates some examples of capacitance measurement results of MIM capacitors based on VNA, indicating measurement capabilities up to 300MHz.

III. THE TRANSIENT SIGNAL METHOD

The transient signal method is achieved by applying periodic excitation signals of varying frequencies across

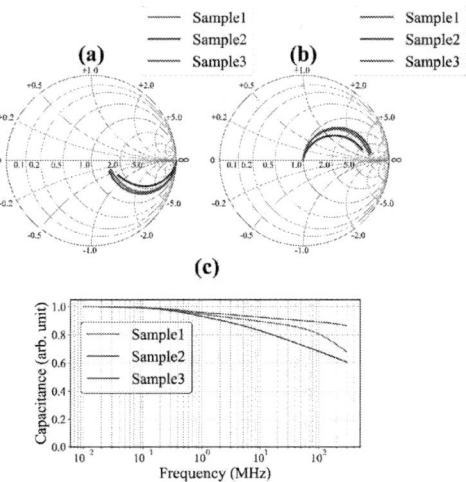

Fig. 3. Measurement results of the network analysis method: (a) S11 Smith chart, (b) S21 Smith chart and (c) corresponding Capacitance-Frequency curves of three tested samples.

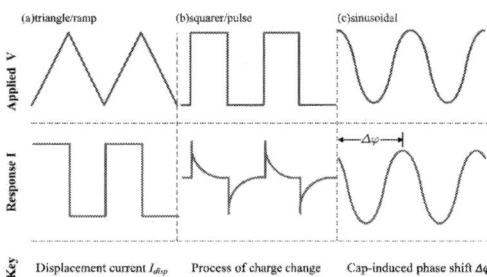

Fig. 4. Three types of waveforms for transient signal method [7].

the DUT, and collecting the transient waveforms at both the input and output ports. By comparing the amplitudes and phases of the waveforms and performing circuit analysis, the impedance information of the DUT can be obtained to further extract its capacitance.

A. Basic Principles

Figure 4 demonstrates some typical excitation signals used by the transient signal method [7]. Using 2-port DUT as an example (Figure 5), the measurement is achieved by applying high-frequency excitation signals to the input port of DUT using an arbitrary waveform generator (AWG), while the response waveforms are collected from the DUT using an oscilloscope (OSC). It is important to collect waveforms from both ports of the DUT synchronously, so that the impedance can be extracted using appropriate circuit models.

B. Extraction of Capacitance Values

Since the response waveforms are collected from the closest possible locations to the DUT, the parasitic effects on the measured impedance are suppressed. Using the triangle wave as an example, the expression for the response current at the output port is as follows:

$$I_{response} = I_{disp} = C \frac{dV}{dt} \quad (5)$$

979-8-3315-0417-5/25 $31.00 © 2025 IEEE   1181

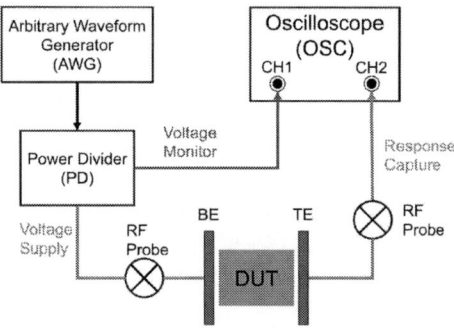

Fig. 5. Schematic diagram of the electrical connections of the transient signal method for a 2-port DUT.

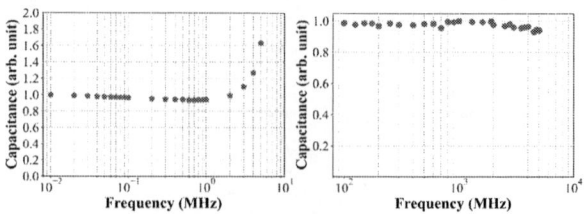

Fig. 6. Device capacitance measurement results by (a) CVU of a parameter analyzer and (b) the transient signal method.

Here I_{disp} corresponds to the displacement current, which characterizes the nature of the response signal. This current can be measured by the oscilloscope (OSC) through a fixed-value termination resistor. Subsequently, the DUT's capacitance can be extracted by dividing I_{disp} with the slope of input voltage (dV/dt). As for the sine wave, the following extraction equations apply:

$$Z = \frac{V_{A2} R_{\text{ref}}}{\sqrt{V_{A1}^2 - 2V_{A1}V_{A2}\cos\Delta\varphi + V_{A2}^2}} \quad (6)$$

$$\alpha = \Delta\varphi - \tan^{-1}\left[\frac{-V_{A2}\sin\Delta\varphi}{V_{A1} - V_{A2}\cos\Delta\varphi}\right] \quad (7)$$

$$R_{ESR} = Z\cos\alpha \quad (8)$$

$$C = \frac{-1}{2\pi f Z \sin\alpha} \quad (9)$$

Where V_{A1} and V_{A2} represent the total voltage drops across the test circuit and the DUT, respectively, $\Delta\varphi$ is the phase difference, α is the impedance angle of the DUT, R_{ESR} is the equivalent series resistance of the DUT, and C is the capacitance value to be determined. By selecting appropriate test voltage and internal resistance, the capacitance value of the device can be easily obtained based on the above equations.

Figure 6 presents an example of capacitance test results using the transient signal method (with sine wave), and compares it with the results from the Capacitance-Voltage Unit (CVU) of a semiconductor parameter analyzer. The results suggest that the CVU can only measure the capacitance accurately at low frequencies, and loses its precision above 1 MHz due to parasitic effects. In contrast, the transient signal method excels at higher frequencies up to 1GHz. Compared to the network analysis method, the transient signal method is less vulnerable to noise and are less expensive to implement, making it a promising alternative for high-frequency capacitance measurements when VNA and GSG structures are not available.

IV. CONCLUSION

This work presented our recent studies on two high-frequency capacitance measurement techniques: the network analysis method and the transient signal method. It is demonstrated that both methods can accurately extract DUT capacitances up to 1GHz, making them useful in the technology development of advanced memories.

ACKNOWLEDGMENT

This work was supported by the National Science and Technology Major Project of the Ministry of Science and Technology of China (Grant No.2024YFE0201100) and National Natural Science Foundation of China (Grant No. 62474162).

REFERENCES

[1] P. Holmberg, "Automatic balancing of linear ac bridge circuits for capacitive sensor elements," *IEEE Transactions on Instrumentation and Measurement*, vol. 44, no. 3, pp. 803–805, 1995.

[2] Q. Wang, L. Liu, Y. Feng, W. Zhou, A. Guo, H. Ye, and Y. Guo, "The gsg-pad layout optimization for test-key of mm-wave devices," in *2019 China Semiconductor Technology International Conference (CSTIC)*, 2019.

[3] A. Rumiantsev and N. Ridler, "Vna calibration," *IEEE Microwave Magazine*, vol. 9, no. 3, pp. 86–99, 2008.

[4] S. Fregonese, M. Deng, M. De Matos, C. Yadav, S. Joly, B. Plano, C. Raya, B. Ardouin, and T. Zimmer, "Comparison of on-wafer trl calibration to iss solt calibration with open-short de-embedding up to 500 ghz," *IEEE Transactions on Terahertz Science and Technology*, vol. 9, no. 1, pp. 89–97, 2019.

[5] Y. Wu, Y. Hao, J. Liu, C. Zhao, Y. Xu, W. Yin, and K. Kang, "An improved ultrawideband open-short de-embedding method applied up to 220 ghz," *IEEE Transactions on Components, Packaging and Manufacturing Technology*, vol. 8, no. 2, pp. 269–276, 2018.

[6] C. Yadav, S. Fregonese, M. Deng, M. Cabbia, M. De Matos, and T. Zimmer, "On the variation in short-open de-embedded s-parameter measurement of sige hbt upto 500 ghz," in *2019 12th German Microwave Conference (GeMiC)*, 2019, pp. 264–267.

[7] Y. Qu, Y. Shen, M. Su, J. Lu, and Y. Zhao, "Ghz c-v characterization methodology and its application for understanding polarization behaviors in high-k dielectric films," in *2022 IEEE International Reliability Physics Symposium (IRPS)*, 2022, pp. 3A.3.1–3A.3.6.

979-8-3315-0417-5/25 $31.00 © 2025 IEEE

Characterizing building blocks for optoelectronic devices based on two-dimensional materials by resonant Raman spectroscopy

Ping-Heng Tan[1]

[1]State Key Laboratory of Superlattices and Microstructures, Institute of Semiconductors, Chinese Academy of Sciences, Beijing 100083, China

Abstract

As is known, TMD, hBN and graphene have become essential building blocks in heterostructure-based devices. The identification of layer numbers and interlayer coupling of building blocks and the good interfacial coupling between different building blocks is essential for the device performance and property exploration. In this talk, I will discuss how to characterize building blocks for optoelectronic devices based on two-dimensional materials by resonant Raman spectroscopy, including layer number, interlayer coupling, interfacial coupling and electron-phonon coupling.

Keywords: building block, two-dimensional materials, resonant Raman spectroscopy

Introduction

Two-dimensional materials (2DMs), such as graphene and transition metal dichalcogenides (TMDs), have been under intensive investigation. The advent of 2DMs has provided unprecedented opportunities for novel heterostructures in the form of van der Waals stacks to form Van der Waals heterostructures (vdWHs). vdWHs with atomically clean interfaces provide an ideal platform for fundamental studies and novel device demonstration. With the development of integration techniques, the structure of vdWHs becomes complicated. vdWH-based optoelectronic devices composed of a semiconducting channel, metallic contacts, or an insulating dielectric/encapsulation, has already become a basic paradigm in related electronic and optoelectronic device engineering. Transition metal dichalcogenides (TMDs), graphene (Gr) and hexagonal boron nitride (hBN), which usually correspond to the three roles, in turn, have naturally become essential building blocks in vdWHs-based devices. These constituents in vdWHs are coupled together by van der Waals interaction, which induces the interfacial coupling for emerging superior properties. Furthermore, issues in interfacial coupling induced by stacking sequence, constituent thickness, and twist angle between the two adjacent constituents also significantly influence the physical properties of vdWHs and further affect the related device performance. Therefore, the understanding of basic properties of two-dimensional materials the interfacial coupling between constituents in vdWHs is critical to achieve demanded performance of novel vdWHs-based devices.

A variety of optical and electrical techniques have been used to characterize the interlayer coupling in layered systems. Among them, low-frequency (LF) Raman spectroscopy can directly detect the interlayer vibration modes, i.e., shear (S) modes andlayer-breathing (LB) modes in layered materials. Using the frequencies of the measured S or LB modes, the relevant force constants can be obtained based on the linear chain model (LCM), and then the interlayer coupling strength can be estimated quantitatively. In vdWHs, additional interlayer vibration modes have been observed due to the interfacial coupling between the constituents. The interfacial coupling in Gr/TMD and hBN/TMD binary vdWHs has been revealed by probing the corresponding interlayer vibration modes. The understanding and manipulation of the interlayer phonon behavior in 2DMS and the corresponding vdWHs also contribute to the optimization of electrical and thermal transport properties in the related devices. In this talk, I will discuss how to characterize layer number, interlayer coupling, interfacial coupling and electron-phonon coupling of 2DMs and the corresponding vdWHs by resonant Raman spectroscopy.

Results and Discussions

The Raman spectrum of graphite and MLG consists of two fundamentally different sets of peaks. Those, such as D, G and 2D, are present also in 1LG, which result from in-plane vibrations, and others, such as the shear (C) modes and layer-breathing (LB) modes, due to the relative motions of the planes themselves, are either perpendicular or parallel to their normal. Detecting of the C and LB modes allows one to directly probe the interlayer and interface interactions of MLGs and related vdWHs.

In tMLG, the C and LB modes are directly observed by resonant Raman spectroscopy. There is a significant enhancement of the C and LB modes due to resonance with new optically allowed electronic transitions, determined by the relative orientation of the layers. It indicates that the interlayer coupling at the interface between the twisted Bernal layers is only about 20% of the coupling in the Bernal-stacked layers. However, we find that twisting has a small effect on LB modes, quite different from the case of the C modes. This implies that the periodicity mismatch between two twisted layers mostly affects shear interactions. Because of the significant interfacial layer-breathing couplings

979-8-3315-0417-5/25 $31.00 © 2025 IEEE

between MoS2 and graphene flakes, a series of layer-breathing modes with frequencies dependent on their layer numbers are observed in the vdWHs. The interfacial layer-breathing force constant between MoS2 and graphene is comparable with the layer-breathing force constant of multilayer MoS2 and graphene.

To reveal the interfacial coupling between TMD and hBN, we fabricated the van der Waals heterostructures of WS2 and hBN. The layer number of WS2 and hBN can be identified by its interlayer mode and AFM. All the shear modes in heterostructures are located at the same position of the shear mode in the corresponding individual WS2 flakes, indicating that the interfacial shear coupling between WS2 and hBN is very weak, in this case, the shear modes in the heterostructures are mainly confined within the WS2 constituent. However, new LB modes were observed in heterostructures. Using the linear chain model proposed above, the interlayer layer-breathing coupling in hBN and interfacial layer-breathing coupling between hBN and WS2 can be fitted by the new LB modes. The fitted interlayer and interfacial coupling strengths are comparable to the interlayer coupling in WS2, which indicates that the LB modes are collective vibrations of all the stacking layers in van der Waals heterostructures.

We also found that, not all the layer-breathing modes in the heterostructures are observed, but those close to the Raman-active layer-breathing modes in trilayer WS2 exhibit strong intensity, no matter how much number of layers of the hBN flake is. We guess that there should be some peculiar electron-phonon coupling mechanism. Because of strong interfacial coupling, layer-breathing vibration can extend over tens to hundreds of layer thickness, exhibiting bulk-like 3D phonon feature. The band gap of WS2 is in the visible range, while that of hBN is in the DUV range. The band gap alignment of WS2/hBN heterostructure is type I. That is to say, the electronic states of WS2 constituent are confined within the WS2 flake itself, exhibiting 2D feature. The similar Raman excitation profile of LB modes between WS2/hBN heterostructure and individual WS2 flake suggest that, the two-dimensional electronic states are strongly coupled to the three-dimensional layer-breathing phonons.

To understand the peak intensity of each layer-breathing mode in heterostructures, we check the atomic displacement of different layer-breathing modes, we find that when the atomic displacement of standalone WS2 flakes of a Raman-active layer-breathing mode is similar to that of the WS2 constituent for a layer-breathing mode in vdWHs, the intensity of the layer-breathing mode will become stronger. Therefore, EPC strength in heterostructures can be understood by the weighting factor between interlayer displacements of layer-breathing phonons in heterostructures from all the layer-breathing modes (φ_j) in the standalone WS2 constituent, then we can estimate the Raman intensity

of the layer-breathing phonons by this kind of wavefunction projection. Indeed, the calculated results are in line with the experimental ones. Thus, the intensity of the layer-breathing modes in heterostructures induced by the cross-dimensional electron-phonon coupling can be reproduced by the wavefunction projection.

Now, we focus on the interfacial coupling between TMD and graphene by taking heterostructure of MoS2 and graphene as an example. We find that the annealing is necessary to form high quality heterostructure. In the low-wavenumber region, we find that the layer breathing mode of 2LM/2LG heterostructures are significantly different from the individual 2LM flake, indicating good contact and strong interfacial coupling. The Raman mapping shows that the position of layer breathing modes in the 2D vdWHs is significantly changed relative to the intrinsic bilayer MoS2 flakes.

We also transfer 2 layer MoS2 on n layer graphenes to form a set of heterostructures. Again, the results indicate that the shear modes in the heterostructures are mainly confined within the MoS2 constituent. However, the peak position of layer-breathing modes in heterostructures is very sensitive to the number of layers of graphene flakes within the heterostructure. We summary the data here. We consider the linear chain model for this system. Only one parameter of interfacial coupling strength can be used to well fit all the experimental data. The interfacial coupling strength is comparable to those of MoS2 and graphene flakes. Thus, like TMD/hBN heterostructures, the layer-breathing vibration modes are extended within the whole TMD/graphene heterostructures, not only in the constituents.

Having an insight into the observed LB modes, only LB modes with frequency close to Raman-active LB modes in four-layer MoS2 are observed in heterostructures. This is also like the result in hBN/WS2 heterostructures. The LB intensity calculated by the proposed wavefunction projection well reproduced the experimental data. Therefore, the intensity of LB modes in heterostructures can be quantitatively understand by the wavefunction projection.

Although graphene, hBN and TMD are essential building blocks in heterostructure-based devices, we highlight that the substrate, electrode or encapsulated layers can be involved into phonon vibrations of the studied two-dimensional materials and heterostructures, which can also modulate the electron-phonon coupling strength, and thus to the electrical and thermal transport properties in the related devices. More importantly, enhanced layer-breathing modes by electron-phonon coupling of TMD can help to probe the interlayer coupling in two-dimensional materials with Raman inactive layer-breathing modes.

Conclusion

First, Techniques of low-wavenumber Raman spectroscopy

are greatly improved, making the probe of low-energy phonons possible by a common single-stage Raman system; Second, The S modes in AB-stacked MLG and interlayer modes in TMDs and related vdWHs can be well understood by the linear chain model. Third, the shear modes are localized in the constituents while the layer-breathing modes are collective vibrations from all the layers in the heterostructures. Finally, The LB mode in TMD-based heterostructures can be enhanced by cross-dimensional electron-phonon coupling between LB phonons of heterostructures and the C exciton in TMDs. Raman intensity of the LB mode in heterostructures can be well reproduced by the wavefunction projection method.

Acknowledgments

We acknowledge the support from the National Key Research and Development Program of China (Grant No. 2023YFA1407000), the Strategic Priority Research Program of CAS (Grant No. XDB0460000), National Natural Science Foundation of China (Grant Nos. 12322401, 12127807 and 12393832).

References

[1] P. H. Tan, W. P. Han, W. J. Zhao, et al., "The shear mode of multilayer graphene", Nature Materials, 11, pp. 294-300(2012).

[2] Jiang-Bin Wu, Xin Zhang, Mari Ijas, Wen-Peng Han, Xiao-Fen Qiao, Xiao-Li Li, De-Sheng Jiang, Andrea C. Ferrari, and Ping-Heng Tan, "Resonant Raman spectroscopy of twisted multilayer graphene", Nature Communications, 5, pp. 5309, (2014).

[3] Jiang-Bin Wu, Zhi-Xin Hu, Xin Zhang, Wen-Peng Han, Yan Lu, Wei Shi, Xiao-Fen Qiao, Mari Ijas, Silvia Milana, Wei Ji, Andrea C. Ferrari, and Ping-Heng Tan, "Interface Coupling in Twisted Multilayer Graphene by Resonant Raman Spectroscopy of Layer Breathing Modes", ACS Nano, 9, pp. 7440-7449, (2015).

[4] Hai Li, Jiang-Bin Wu, Feirong Ran, Miao-Ling Lin, Xue-Lu Liu, Yanyuan Zhao, Xin Lu, Qihua Xiong, Jun Zhang, Wei Huang, Hua Zhang and Ping-Heng Tan, "Interfacial Interactions in van der Waals Heterostructures of MoS2 and Graphene", ACS Nano, 11, pp.11714-11723, (2017).

[5] Miao-Ling Lin, Yu Zhou, Jiang-Bin Wu, Xin Cong, Xue-Lu Liu, Jun Zhang, Hai Li, Wang Yao, and Ping-Heng Tan, "Cross-dimensional electron-phonon coupling in van der Waals heterostructures", Nature Communications, 10, pp.2419, (2019).

[6] Heng Wu, Miao-Ling Lin, Yu-Chen Leng, Xue Chen, Yan Zhou, Jun Zhang, and Ping-Heng Tan, "Probing the interfacial coupling in ternary van der Waals heterostructures", npj 2D Materials and Applications, 6, pp.87, (2022).

[7] Xin Zhang and Ping-Heng Tan, "Raman spectroscopy: Nanostructures", Encyclopedia of Condensed Matter Physics (Second Edition), 04, pp.160-172, (2024).

Nanorods-based Memristors
: Advancing Bio-inspired System and Neuromorphic Computing

Ji Eun Kim[1], Suk Yeop Chun[1], Keunho Soh[2], Jung Ho Yoon[2*]

[1] Electronic Materials Research Center, Korea Institute of Science and Technology (KIST), Seoul 02792 Republic of Korea

[2] School of Advanced Materials Science and Engineering, Sungkyunkwan University, Suwon 16419, Republic of Korea

Abstract

Memristors are promising next-generation devices due to their energy efficiency and high data processing capabilities. Recent studies have extensively reported using memristors to mimic the human sensory system and function as versatile simulation platforms for artificial intelligence applications. This review explores the development of high-performance memristors incorporating oxide nanorods and their applications in mimicking the human sensory system and advanced computing.

Keywords: Memristor, Nanorods, Bio-inspired system, Neuromorphic computing

Introduction

Real-time sensing and processing of information with conventional CMOS-based semiconductors face limitations due to data bottlenecks. Significant efforts are being focused on emulating biological neural systems with instantaneous data recognition and parallel processing capabilities.[1] The redox-based memristor, a next-generation semiconductor device, features a simple two-terminal structure. Its resistance changes through the formation and rupture of conductive paths in response to external stimuli and can be retained temporarily. This enables data processing and storage within a single device, making it a key focus for mimicking biological neural networks.[2] The switching characteristics of the memristor are determined by the ion dynamics—such as activation energy, mobility, and redox rate—that govern the formation of conductive paths.[3]

This review discusses the precise control of ion dynamics achieved through the integration of nanorods (NR) and explores the application of NR-based memristors as artificial biological neural networks.

The effect of Nanorods structure oxide

In conventional thin-film (TF)-based devices, the precise control of conductive path formation is challenging, and low ion mobility contributes to slower switching speeds and higher operating voltages. Consequently, TF-based memristors face fundamental limitations regarding reliability and energy efficiency.

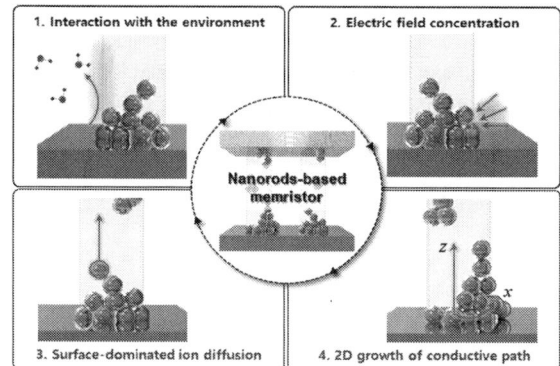

Fig. 1 Influence of nanorod integration on ion dynamics

The adoption of NR oxide layers in memristors presents a promising solution to the inherent limitations of TF, as shown in Fig. 1. Firstly, ion migration in NR structures is primarily governed along surfaces with relatively lower diffusion barriers compared to TF. Additionally, the structural characteristics of NRs concentrate the electric field at the edges, enhancing ion formation and surface migration along the NR. Confining the ion migration region to the surface enables precise control over the formation location. Secondly, a moisture layer formed on the surface of the porous NR facilitates stable redox reactions. This layer promotes rapid compensatory reactions to oxidation, stabilizing the redox process. Finally, interactions with the external environment enable surface chemistry modulation through surface treatments or promote reactions between external oxygen and oxygen vacancies in the oxide layer. These effects of nanorods enable more precise control over ion dynamics, supporting improved emulation of the human sensory system.

Neuromorphic computing

Biological systems sense, analyze and store external information in a parallel fashion while efficiently handling large-scale analog data through pre-processing mechanisms.

Sensory receptors convert external stimuli into electrical signals for the nervous system, adapting or maladapting based on stimulus duration to filter irrelevant inputs while maintaining sensitivity to critical stimuli.(Fig. 2a) Neurons generate spike signals as output only when the input

979-8-3315-0417-5/25 $31.00 © 2025 IEEE

transmitted from the synapses surpasses a certain threshold. The neuron achieves efficient data processing through spike-based signal transmission, significantly reducing energy usage.(Fig. 2b) Volatile memristors with threshold switching can be employed to emulate the characteristics of receptors and neurons.(Fig. 2c)The synapse transmits signals between neurons, playing an essential role in brain communication and information processing. Through synaptic plasticity, the synapse adjusts strength based on experience, supporting learning and memory storage.(Fig. 2d) To mimic the function of a synapse that stores responses to external stimuli, a memristor exhibits non-volatile properties to save its state even after the applied voltage is removed.(Fig. 2e)

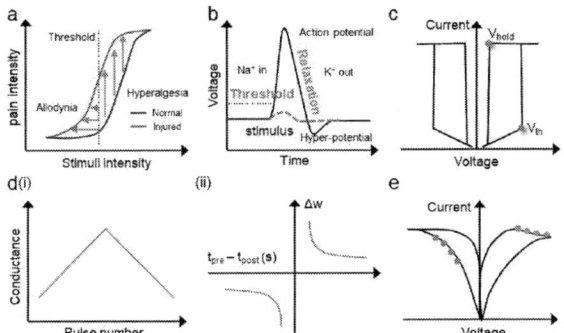

Fig. 2 Function of (a) sensory receptor and (b) neuron (c) Volatile (threshold) switching behavior (d) Function of the synapse (e) Non-volatile switching behavior

Artificial sensory receptor

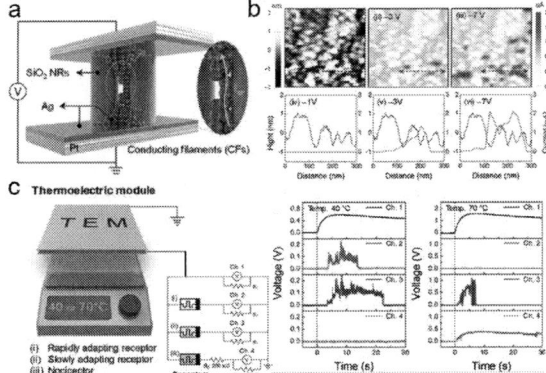

Fig. 3 (a) Schematic of NR-based memristor (b) Overlapped topographic and current images and line profiles (c) Artificial thermoreceptor using NR-based memristors[4] Copyright 2022, Wiley-VCH.

The artificial receptor using the Pt/Ag/SiO$_2$ NRs/Ag/Pt device was reported, as shown in Fig. 3a.[4] Ag-based volatile memristor exhibits threshold switching and relaxation characteristics. NRs were introduced to mimic sensory receptors effectively. Fig. 3b shows that the conducting paths align with the locally hollow regions, presumably at the edges of the SiO$_2$ NRs, in conductive atomic force microscopy analysis. The NRs induce surface diffusion, allowing for fast switching speeds and precise

control over the formation and growth of the conductive path. Furthermore, the absolute ion supply was adjusted through the thickness control of the metal electrode to fine-tune the size and strength of the conductive paths. As the rupture conditions of the conductive path vary with its size and strength, both adaptive and maladaptive characteristics can be achieved. (Fig. 3c)

Artificial neuron

Neuron spike can be implemented using the Leaky Integrate-and-Fire (LIF) model. R$_1$ and a capacitor represent the membrane, while the ion channels are represented by an Ag-based volatile memristor connected in parallel with R$_1$ and the capacitor, as shown in Fig. 4a. Previously reported studies demonstrated that in Ag and NR-based volatile memristors, the formation and migration of metal ions predominantly occur on the surface.[5] Thus, applying effective surface treatments through the large specific surface area of the NR led to control of the surface chemistry (Fig. 4b). Fig. 4c demonstrates that UVO treatment effectively induces the hydrophilic NR surface. The formed moisture layer facilitates redox reactions and lowers the surface diffusion energy barrier. The UVO-treated memristor can achieve high uniformity, stable LIF behavior, high spiking frequency, and significantly reduced energy consumption (Fig. 4d).

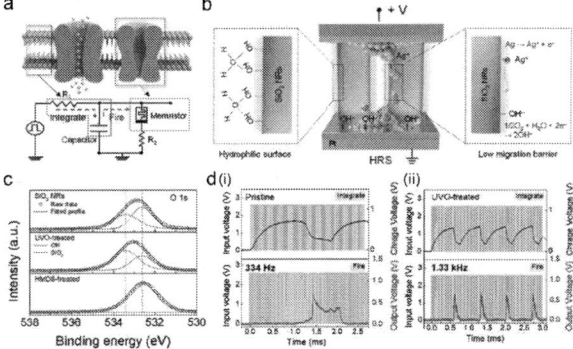

Fig. 4 (a) Schematic of a neuron and equivalent electrical LIF circuit (b) Surface chemistry upon treatment with hydrophilic (UVO) and hydrophobic (HMDS) (c) X-ray photoelectron spectra of the O 1s core level upon surface treatment (d) Neuron spike for pristine and hydrophilic surface[5] Copyright 2023 American Chemical Society

Artificial synapse

Memristors with a Pt/HfO$_2$ NRs/TiN structure exhibit excellent synaptic characteristics.[6] The high electric field at NR edges promotes oxygen vacancy formation and lowers the activation energy for ion migration along the surface. Additionally, oxygen gas generated at the interface can readily escape into the atmosphere through the porous NR structure. During conductive path rupture, external oxygen assists in recombining oxygen with vacancies, supporting stable and repeatable switching without

979-8-3315-0417-5/25 $31.00 © 2025 IEEE   1187

limitations on ion supply (Fig. 5a). This precise control of the conductive path enables excellent linearity in conductance modulation, corresponding to synaptic weight adjustment (Fig. 5b). Such linearity in conductance modulation allows for precise synaptic weight tuning, accurate inference of input signal intensity, and compatibility with existing algorithms and learning rules.(Fig. 5c) Additionally, Fig. 5d shows spike-timing-dependent plasticity (STDP), one of the most well-known Hebbian learning rules, implemented with the NR-based memristor.

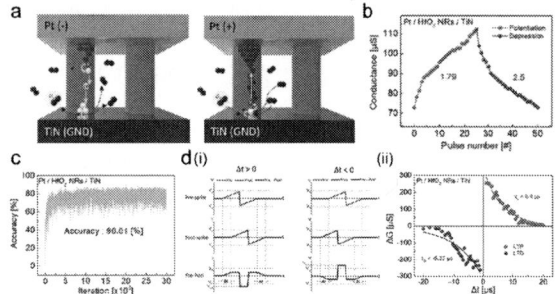

Fig. 5 (a) Illustration of conductive path growth mechanism of NR-based memristor (b) Conductance modulation for NR memristor (c) Pattern-recognition accuracy for NR memristors (d) STDP of the NR memristor[6] Copyright 2022 American Chemical Society.

In-sensor sensory system

Currently, most bio-inspired memristors necessitate additional hardware or circuits to detect and process external stimuli. To reduce complexity, there is a growing demand for in-sensor computing capable of converting continuous analog signals corresponding to external stimuli into electrical spikes without the assistance of additional parts.

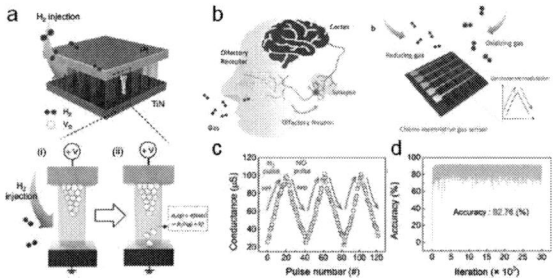

Fig. 6 (a) Schematic of the sensor response (b) Schematic of the mimicking of the biological olfactory cognitive process (c) The conductance modulations based on the target gas (d) Pattern-recognition accuracy[7] Copyright 2023, Wiley-VCH.

Recently, olfactory systems based on in-sensor computing have been reported using only a Pt/TiO$_2$/TiN memristor.[7] In this study, excellent sensing performance and hysteretic characteristics were realized through new dynamics distinct from conventional gas sensors, enabled through the generation of additional oxygen vacancies in the ruptured filament region (Fig. 6a). The NR-based memristor exhibits excellent gas response characteristics

due to its large specific surface area. These advantages of NR-based memristor make it highly effective in mimicking the human olfactory system, as shown in Fig. 6b. Fig. 6c and d demonstrate the effective emulation of synaptic weight changes induced by input gas stimuli and the high accuracy of gas pattern recognition.

Conclusion

This review summarizes the functionalities and key characteristics of nanorod-structured memristors for their application in artificial sensory systems and neuromorphic computing. Nanorods enable precise ion dynamics control, facilitating efficient emulation of receptor adaptation, neuronal spiking, and synaptic weight modulation. Additionally, mimicking human olfactory characteristics without external circuitry demonstrates the potential for efficient artificial sensory systems. Nanorods offer a practical pathway for high-density, reliable memristor-based sensory and computing architectures.

Acknowledgments

This research was supported by the National R&D Program through the National Research Foundation of Korea (NRF) and the Korea Basic Science Institute (National Research Facilities and Equipment Center), funded by the Ministry of Science and ICT (RS2024-00406418, RS-2024-00403917, and NRF-2022R1C1C1004176).

Reference

[1] X. Duan et al., "Memristor-Based Neuromorphic Chips," Advanced Materials, vol. 36, no. 14, pp. 2310704, 2024.

[2] J. Y. Kwon et al., "Artificial sensory system based on memristive devices," Exploration, vol. 4, no. 1, pp. 20220162, 2024.

[3] Y. Yang et al., "Electrochemical dynamics of nanoscale metallic inclusions in dielectrics," Nature Communications, vol. 5, no. 1, pp. 4232, 2014/06/23, 2014.

[4] Y. G. Song et al., "Artificial Adaptive and Maladaptive Sensory Receptors Based on a Surface-Dominated Diffusive Memristor," Advanced Science, vol. 9, no. 4, pp. 2103484, 2022.

[5] Y. G. Song et al., "Highly Reliable Threshold Switching Characteristics of Surface-Modulated Diffusive Memristors Immune to Atmospheric Changes," ACS Applied Materials & Interfaces, vol. 15, no. 4, pp. 5495-5503, 2023/02/01, 2023.

[6] J. U. Kwon et al., "Surface-Dominated HfO2 Nanorod-Based Memristor Exhibiting Highly Linear and Symmetrical Conductance Modulation for High-Precision Neuromorphic Computing," ACS Applied Materials & Interfaces, vol. 14, no. 39, pp. 44550-44560, 2022/10/05, 2022.

[7] S. Y. Chun et al., "An Artificial Olfactory System Based on a Chemi-Memristive Device," Advanced Materials, vol. 35, no. 35, pp. 2302219, 2023.

Novel Three-dimensional DRAM Cell Architectures with IGZO-channel and Key Technologies toward sub-10nm and Beyond

Wonsok Lee, Daewon Ha, Y. Lee, S. Yoo, M.H. Cho, K. Yoo, S.M. Lee,
S. Lee, M. Terai, T.H Lee, J.H. Bae, K.J. Moon, C. Sung, M. Hong, D.G. Cho,
K. Lee, S.W. Park, K. Park, B.J. Kuh, P. Yun, S. Hyun, S.J. Ahn, and J. Song

Semiconductor R&D Center, Samsung Electronics Co. Ltd., Gyeonggi-do, Korea, Email: vero.lee@samsung.com

Abstract

As DRAM device dimension approaches 10nm scale and below in buried cell array transistors (BCAT), it is indispensable to adopt innovative cell structures and/or new materials for the next generation DRAM architecture. In this paper, three promising candidates of IGZO-based DRAM cell architecture, including a vertical channel transistor (VCT), a vertical stacked cell array transistor (VS-CAT), and capacitor-less two transistors (2T0C) are introduced. In addition, key technologies to be considered for DRAM application are discussed.

Keywords: DRAM, innovative cell structure, VCT, VS-CAT, 2T0C, IGZO

Introduction

From PCs to smartphones, IoT, and AI applications such as ChatGPT, the number of connected devices is growing rapidly and digital data generation is exploding. Accordingly, high-performance computing capability and large-scale storage capacity are required for both edge devices and data centers, providing greater opportunities for semiconductor industry [1]. The "miniaturization" of DRAM has been based on advanced DRAM cell technologies BCAT, storage capacitors with high aspect ratio, and processes like EUV as shown in **fig. 1** [2]. However, as the scaling of the $6F^2$ Si-based DRAM cell reaches its physical limitations in terms of dimensions, resulting in degradation of the cell's electrical properties, it is indispensable to develop an innovative DRAM cell architecture using alternative channel materials. Recently, "deposition-able" IGZO-based transistors have received a lot of attention due to their extremely low off-state leakage current (I_{OFF}), resulting from a large bandgap, and relatively high electron mobility even in amorphous phase, enabled by the overlap of metal s-orbitals [3-5]. Unlike in the three-terminal Si devices, it also has robustness against floating body effect (FBE) due to negligible electron-hole pair (EHP) generation and charge-up in the channel. In this paper, we discuss IGZO based innovative DRAM architectures and recent advances in key technologies.

Candidates for DRAM Cell Architecture

Promising architectures were selected in consideration of three main criteria: manufacturability, manufacturing cost, and technological longevity [6]. The proposed structures are suitable for utilizing the advantages of IGZO as they can be stacked monolithically and a cell-over-peripheral (COP) architecture can be applied without wafer bonding.

A. Vertical Channel Transistor (VCT)

In VCT architecture, a bit-line (BL) and a storage node contact can be placed out of plane, as shown in **fig. 2(a)**, this unique structural benefit makes its miniaturization easier than conventional BCAT structure. Accordingly, BL pitch can be reduced from 3F to 2F, enabling a $4F^2$ architecture. Short channel effects (SCE) can be suppressed by adjusting gate length which does not directly affect the unit cell area. In addition, IGZO VCT devices are robust against disturbances from adjacent cells, such as row hammer interference, because the structure does not share an active region with other cells. There can be various configurations for the VCT structure (**fig. 3**), and each has its own advantages and disadvantages due to the nature of the structure as discussed in [4]. Even though a double-gated VCT shows the highest drive current and a negligible passing gate effects (PGE), an asymmetric single-gated VCT can be an appropriate selection considering scalability with manageable interference as shown in **fig. 3** [4]. When choosing the optimal structure, both electrical performance and complexity of process integration should be considered. One of the examples of COP DRAM based on single-gated VCT is shown in **fig. 4**.

B. Vertically Stacked Cell Array Transistor (VS-CAT)

VS-CAT structure is a structure in which both cell transistor and storage capacitor are placed on the same plane as shown in **fig. 2 (b)** and a plurality of layers are stacked upward on top of core/periphery transistors to reduce the unit area of DRAM cell as shown in **fig. 5**. Several structural options are available including single-, double-, and gate-all-around configurations. In order to maintain scalability across generations, it is important to reduce the unit volume of each layer by reducing the stack height as well as minimizing the horizontal dimensions. By using an IGZO channel instead of Si-channel, the stack height can be reduced, since IGZO allows for a single-gate stack as in **fig. 5 (b)**. In addition, lateral dimension can also be scaled down by reducing the area for a storage capacitor due to the extremely low I_{OFF}. It is well known that Si based VS-CAT is vulnerable to FBE [8-9], because of off-state EHP generation. However, with the

979-8-3315-0417-5/25 $31.00 © 2025 IEEE

application of IGZO-channel, FBE can be effectively suppressed owing to the large bandgap and negligible EHP generations [6].

C. Capacitor-less Two Transistors (2T0C)

A 2T0C cell consists of two transistors without a structural storage capacitor as shown in **fig. 2(c)**. Instead, a storage node capacitance consists of a parasitic capacitance and a gate capacitance of a read transistor as shown in circuit diagram in **fig. 6 (b)**. The structure itself, also known as the "gain-cell", has been introduced long time ago, but initially received not much attention due to the poor retention characteristics of Si-based transistors. However, in recent years, with the emergence of IGZO channels boosting extremely low I_{OFF}, 2T0C has gained great attention, and a variety of new architectures have been introduced [10, 11]. However, 2T0C DRAM requires complex contacts and routings to form an array and enable its operation, which increases unit cell area and cost. A vertical channel-all-around (CAA) structure can be one of the potential candidates to solve the scalability issue in conventional planar 2T0C configuration [11]. The design of a cell array and its operation scheme needs to be carefully considered. Challenges include interference from adjacent or connected cells, capacitive coupling of the storage node with write and read operations. To address these issues, various schemes have been introduced, including hybrid gain cells and modifications to cell operations [12,13].

Key Technologies

For the application of a new channel material and novel structures, it is required to develop key process technologies. In this section, recent understandings and advances in key technologies will be discussed.

A. Deposition and patterning of IGZO

For the formation of IGZO layer in the proposed structures, either conventional physical vapor deposition (PVD) or atomic layer deposition (ALD) can be applied. Especially, for the highly scaled devices, ALD is preferred due to its superior step coverage and capability of adjusting cation and anion concentrations, which determines the electrical properties and reliability of IGZO devices [14]. **Fig. 7** shows the comparison of transfer characteristics between PVD and ALD IGZO. Besides, it is necessary to optimize dry etching techniques to achieve anisotropic etching profile, fast etching speed, and selectivity for patterning [15].

B. Interface engineering : Oxide and IGZO channel

To investigate the electrostatic and transport characteristics in the IGZO channel, electron concentration was analyzed as a function of the gate voltage (V_{GS}) using TCAD simulation. As shown in **fig. 8(a)** a higher concentration of electrons are observed near the interface between the bottom dielectric and IGZO channel than the front interface at $V_{GS}=V_T$. It can be explained by the energy band diagram across the gate stack. Ionic impurities and traps at the bottom interface are likely to have more significant impact on device performance, particularly in sub-V_T operation regime. Our experimental results, as shown in **fig. 8 (b)**, indicates that V_T variability can be greatly improved by proper passivation of bottom interface traps, which concurrently improves I_{ON} and sub-V_T slope (SS) characteristics (not shown).

C. Channel and contact resistance

Controlling the concentration of carriers in IGZO is very important engineering technique because ion implantation, as widely adopted in Si devices, is generally not applicable in IGZO devices. Instead, as illustrated in **fig.9**, defects and impurities such as oxygen vacancies (V_O) and various hydrogen species are of great importance for the determination of in-channel turn-on characteristics, extension resistance in source and drain regions, and contact resistance between IGZO and contact metals. Due to the nature of Schottky characteristics, carrier transmission at the interface is greatly affected by V_{GS} and V_{DS} as shown in **fig. 10** [16]. And an asymmetry between source and drain interface with metal contacts makes it more complicated to define the contact resistance. Accordingly, a simple yet practical approach to analyze contact resistance is required as proposed in [17]. Our experimental results reveal that the mismatch on a performance can be severe as shown in **fig. 11 (a)** and it can be solved by the optimization of contact metal and interface engineering (**fig. 11 (b)**).

Summary

It is expected that scaling of DRAM below 10nm for high memory capacity and bandwidth will be sustained with the combination of deposition-able IGZO and novel structures which can utilize the merit of the material.

References

[1] K. Kim, IEDM 2021, pp. 1.1.1-1.1.8

[2] D. Ha and H-S. Kim, VLSI Symp., 2022, pp. 417-418

[3] A. Belmonte, et. al., IEDM 2020, pp. 28.2.1-28.2.4

[4] D. Ha et. al., IEDM 2023, pp. 6.3.1-6.3.4

[5] K. Ide et al. physica status solidi (a) 2019, pp. 1800372

[6] D. Ha et. al., IMW 2024, pp. T1-2

[7] J. Han, et. al., VLSI Symp., 2023, TFS1-1

[8] F. Morishita et. al., VLSI Symp., 1995, pp.141-142

[9] Y. Cho et. al., IEEE Elec. Dev. Lett., 2018, pp.1860-1863

[10] W. Lu, et. Al., IEDM 2022, PP. T26-4

[11] X. Dual, et. al., IEEE Trans. Electron Dev., pp. 2196-2202.

[12] S. Liu, et. al., VLSI Symp., 2024, pp. TFS1-4

[13] M. Shi, et. al., IEDM 2023, pp. T14-2

[14] D.G. Kim, et. al., ACS Appl. Mater. Int. 2023, 15, 26, 31652-31663

[15] J.W. Hong, et. al., Appl. Surf. Sci., 2024, vol. 671, 30, 160692

[16] W. Chakraborty, et. al., VLSI Symp., 2020, pp. TH2-1

[17] S.W. Yoo, et. al., VLSI Symp., 2024, pp. T16-5

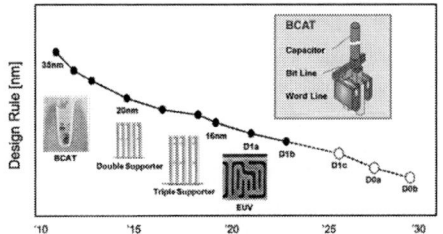

Fig. 1. Scaling trend of BCAT based DRAM cell design rule and applied key technologies.

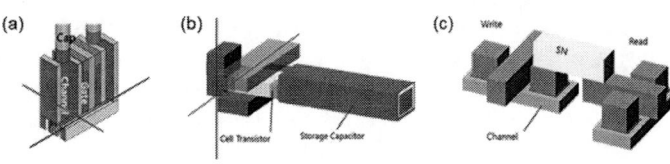

Fig. 2. Candidates for sub-10nm DRAM cell array transistors. (a) VCT (b) VS-CAT (c) 2T0C DRAM. For all the structures, oxide semiconductors such as IGZO can be applied to utilize the properties of extremely low I_{OFF}.

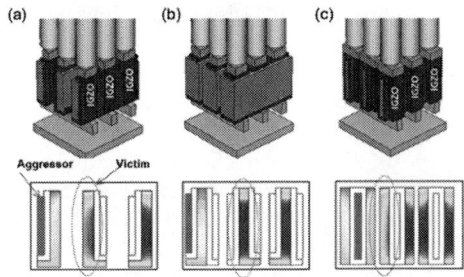

Fig. 3. IGZO-based structures for (a) single-, (b) double-, and (c) inner-gated VCTs and disturbed potential profiles inside IGZO channel. Bright blue indicates higher interferences from adjacent cells.

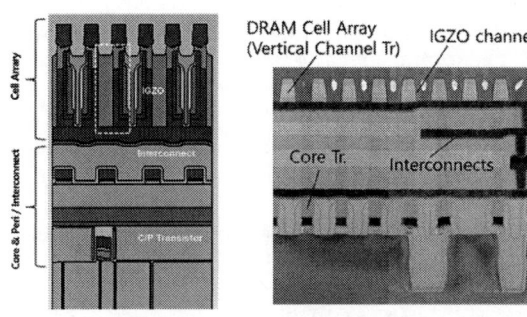

Fig. 4. An example of cell-over-peripheral (COP) DRAM architecture based on 4F² single-gated VCT. IGZO enables the monolithic stacking of DRAM cell array transistors on top of core/peripheral (C/P) transistors and local interconnects without the need for wafer bonding.

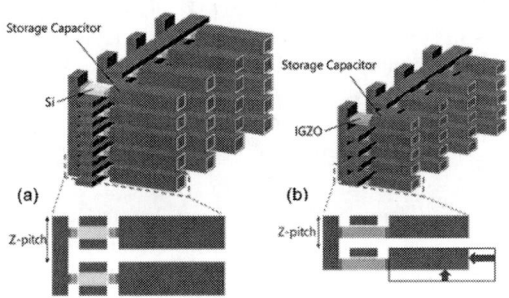

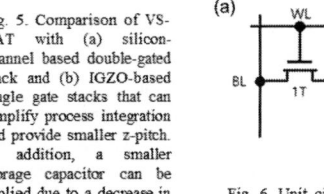

Fig. 5. Comparison of VS-CAT with (a) silicon-channel based double-gated stack and (b) IGZO-based single gate stacks that can simplify process integration and provide smaller z-pitch. In addition, a smaller storage capacitor can be applied due to a decrease in leakage current in IGZO devices.

Fig. 6. Unit circuit diagram for (a) conventional 1T1C DRAM and (b) capacitor-less 2T (2T0C) DRAM. A data sensing scheme also needs to be determined in consideration of write and read operations.

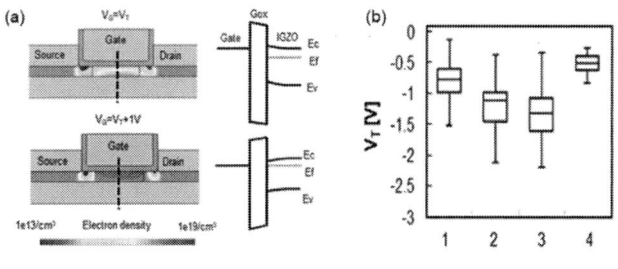

Fig. 7. Measured I_{DS}-V_{GS} transfer curves and trans-conductance for ALD- and PVD-IGZO deposited single-gated VCTs for V_{DS} = 1.0V at 85°C. ALD shows a steeper sub-V_T slope (SS) and lower I_{OFF}.

Fig. 8. Simulated (a) electron density profiles in IGZO and corresponding energy band diagram as a function of V_{GS}. (b) impact of pre-deposition cleaning processes (denoted by process 1,2,3 and 4) of IGZO channel on V_T.

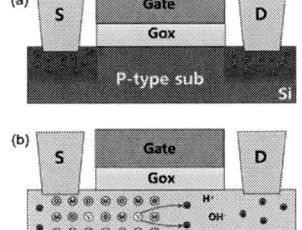

Fig. 9. Illustration of (a) Si-based and (b) IGZO-based NMOS transistors. Unlike (a), control of carrier is more complicated in (b) and needs understanding on the role of defects and impurities.

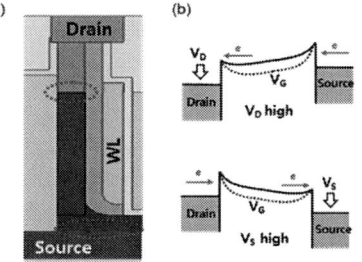

Fig. 10. Schematic of (a) A vertical structure for asymmetric single-gated VCT and (b) Energy band diagram for the cases of V_D=high and V_S=high.

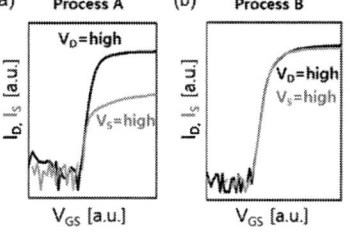

Fig. 11. Comparison of transfer curves for the cases of contact Process A and B. (a) A severe mismatch observed between I_D and I_S and (b) with the optimization of contact engineering, the mismatch can be removed.

979-8-3315-0417-5/25 $31.00 © 2025 IEEE

Evaluation of Insulator Candidates for Nanoelectronics Based on 2D Materials

Mina Bahrami*, Theresia Knobloch*, Pedram Khakbaz*, Mohammad Rasool Davoudi*,
Alexander Karl*, Seyed Mehdi Sattari-Esfahlan *, Dominic Waldhoer*, and Tibor Grasser*

* Institute for Microelectronics, Technische Universität Wien, Gußhausstraße 27–29, 1040 Vienna, Austria

Email: bahrami | knobloch | grasser@iue.tuwien.ac.at

ABSTRACT

Despite considerable advancements in electronic devices based on 2D materials, their performance suffers from the lack of suitable gate insulators. Here, we focus on two promising insulators: Strontium titanate ($SrTiO_3$), a high-κ perovskite used with MoS_2 to create high-performance n-type field-effect transistors (FETs) and bismuth oxyselenide (Bi_2O_2Se), whose native oxide (Bi_2SeO_5) serves as a gate dielectric. We evaluate the main performance criteria and suitability of these insulators as gate dielectrics based on multi-scale simulations (DFT and TCAD).

INTRODUCTION

Scaling down electronic devices to channel lengths smaller than $5\,\mathrm{nm}$ necessitates an aggressive reduction in channel thickness to mitigate short-channel effects. However, in traditional three-dimensional (3D) semiconductors, reducing the thickness below $3\,\mathrm{nm}$ leads to a substantial degradation in carrier mobility [1]. In this context, two-dimensional (2D) semiconductors offer a promising solution to these challenges. Since their thickness can be reduced down to a monolayer, 2D materials can be used as atomically thin channels that provide excellent electrostatic control [2,3]. Furthermore, many 2D materials exhibit high theoretical mobilities [4] and can be flexibly stacked to form van der Waals heterostructures [5]. However, a major obstacle to their use in ultra-scaled nanoelectronics for commercial applications is the lack of suitable gate insulators [6]–[8]. In the following, we aim to provide criteria for identifying suitable gate insulators for 2D field-effect transistors (FETS). These criteria will be used to analyze the performance of two promising insulators: (i) bismuth perselenate (Bi_2SeO_5), which is the native oxide of the two-dimensional semiconductor bismuth oxyselenide (Bi_2O_2Se), and (ii) strontium titanate ($SrTiO_3$). The performance of 2D FETs inherently depends on the material properties of the chosen gate stack, most importantly on their respective dielectric constants, band gaps, band offsets, tunnel masses and the quality of the interface formed between the insulator and the 2D semiconducting channel. We will use a multi-scale modeling approach for our performance analysis, first evaluating the material properties based on density functional theory (DFT) calculations. Next, the material properties will serve as input to clarify the performance of different insulators at the device level using technology computer-aided design (TCAD) simulations in combination with experimental data from prototype devices.

INSULATORS

Strontium titanate ($SrTiO_3$, STO) crystallizes in a cubic perovskite structure at room temperature (see Fig. 1-top. Its structure consists of TiO_6 octahedra, with Sr atoms occupying the interstitial sites. This arrangement results in high bulk permittivity that exceeds 300 at room temperature [9]. The other candidate, bismuth oxoselenate(Bi_2SeO_5, BSO), also has a high dielectric constant of about 20 even in ultra-thin layers. It is formed by oxidizing the 2D semiconductor bismuth oxyselenide (Bi_2O_2Se) which offers high electron mobilities of up to 812 $cm^2V^{-1}s^{-1}$ at room temperature [10]. Also, the layered structure of Bi_2SeO_5 is similar to that of its corresponding 2D semiconductor, as illustrated in Fig. 1-bottom. The main difference is that it contains additional oxygen atoms between the layers bonded to selenium (Se) atoms [11]. This layered structure results in a high-quality zipper interface formed between BSO and Bi_2O_2Se [12].

For these two gate insulators, we performed DFT simulations with CP2K [13]. We used the non-local hybrid functional PBE0-TC-LRC [14] and a mixing factor $\alpha = 0.25$ for the Hartree-Fock(HF) exchange. The permittivities were evaluated by the change in polarization after applying a finite electric field using the Berry phase formalism [15]. We calculated the most relevant material properties of bulk $SrTiO_3$ and Bi_2SeO_5 for their application in 2D FETs through DFT simulations. The results we obtained are summarized in Tab. I. Our calculated results are in good agreement with previously reported experimental data [9,16].

INSULATOR IMPACT ON DEVICE PERFORMANCE

Critical performance aspects of 2D FETs are defined by the insulators that form the gate stack. First of all, the gate dielectric needs to provide sufficient capacitive gate control, hence, the layer needs to be thin while also effectively blocking leakage currents. Furthermore, the density of interface traps needs to be as small as possible, so as not to degrade the subthreshold slope. Moreover, the densities of charge traps within the insulator should be minimized, as well as the densities of mobile ions or charged defect complexes, as this contributes to the hysteresis. In the following, we evaluate the performance potential of STO and BSO as gate insulators based on a combination of TCAD simulations and electrical characterization of prototype devices. The first prototype device is a back-gated FET featuring a molybdenum disulfide (MoS_2) channel and $SrTiO_3$ as a gate insulator, which was epitaxially grown on a sacrificial layer and transferred using a PDMS stamp (see Fig. 2-top) [17]. The second prototype is a planar top-gated FET consisting of a Bi_2O_2Se channel grown by chemical vapor deposition on top of a strontium titanate ($SrTiO_3$) substrate. Part of the Bi_2O_2Se was oxidized to form the gate oxide Bi_2SeO_5 in a UV-assisted process at about 120 °C (see Fig. 2-bottom) [18]. Device performance was modeled using the open-source simulation framework Comphy [19]. In Fig. 3-left, the modeled $I_D(V_G)$ characteristics of $MoS_2/SrTiO_3$ FETs are shown for a device with $W = 6$ μm and $L = 2.5$ μm. In the simulation, we considered interface defects with a density of $N = 6 \times 10^{12} \, cm^{-2}$, yielding good agreement with experimental data. Similarly, the $I_D(V_G)$ characteristics for the Bi_2O_2Se/Bi_2SeO_5 device are presented for a device with $W = 7.9$ μm and $L = 3.2$ μm in Fig. 3-right with a higher interface defect density of $N = 10^{13} \, cm^{-2}$.

Gate Leakage Currents

For modeling gate leakage currents Comphy uses a semi-classical Tsu-Esaki model [20] which we used to extrapolate the best-case performance of the leakage current through a scaled gate stack, corresponding to an equivalent oxide thickness (EOT) of 1 nm [21]. The leakage currents are modeled for gate stack consisting of Au as a gate metal, an insulator, and MoS_2 or Bi_2O_2Se as a channel material, respectively, as illustrated by the band diagram shown in Fig. 4. Both material combinations $MoS_2/SrTiO_3$ and Bi_2O_2Se/Bi_2SeO_5 stay below the low power limit of $1.5 \times 10^{-2} \, Acm^{-2}$, see Fig. 5.

Hysteresis in the Transfer Characteristics

Fig. 6-right exhibits the normalized hysteresis width for two different devices at different temperatures (298K and 363K for BSO, and 300K and 350K for STO). We observe clockwise (CW) hysteresis in Bi_2O_2Se/Bi_2SeO_5 devices and counter-clockwise (CCW) hysteresis in $MoS_2/SrTiO_3$ device. The CW hysteresis in Bi_2O_2Se/Bi_2SeO_5 is caused by the charge transfer from the Bi_2O_2Se conduction band to the electron trapping band in the oxide. By increasing the gate bias, the Fermi level rises, leading to the capture of electrons and resulting in a positive shift of the threshold voltage (V_{th}). Both electron traps (acceptor-like) and hole traps (donor-like) contribute to the clockwise (CW) hysteresis through charge exchange with the channel. As the gate voltage (V_G) increases during the up-sweep, the Fermi level rises, resulting in the traps becoming populated with electrons or emitting holes. This process causes a positive shift in the threshold voltage (V_{th}). On the other hand, in the $MoS_2/SrTiO_3$ device, we observe counter-clockwise (CCW) hysteresis. The mechanisms responsible for CCW hysteresis are currently under debate. One potential explanation is charge trapping at the gate interface or the diffusion of positively charged ions within the dielectric layer. These mobile charges can originate either from external contaminants in the insulator or from intrinsic charged defects. Fig. 6-left

illustrates the observed hysteresis in both technologies measured at the same sweep frequency. The red curves represent the transfer characteristics during the up-sweep of the gate voltage, while the blue curves correspond to the down- sweep. It is evident that the $SrTiO_3$ device exhibits counter-clockwise hysteresis, whereas the Bi_2SeO_5 device displays clockwise hysteresis.

CONCLUSIONS

We have investigated the potential of two promising insulators—strontium titanate (STO) and β-bismuth oxoselenide (BSO)—for their use in two-dimensional field-effect transistors (2D FETs). By employing a comprehensive multi-scale modeling approach, we gained valuable insights into their material properties and the device performance of 2D FETs based on these insulators. Both insulators were compared in terms of material characteristics and device performance. Gate leakage currents for both materials were found to be below the low-power limit of $1.5 \times 10^{-2} \, A/cm^2$ for EOT=1 nm, indicating effective insulation properties at scaled equivalent oxide thicknesses. A notable difference between the two devices was observed in the hysteresis behavior of their transfer characteristics. Overall, STO and BSO demonstrate promising properties as gate insulators for 2D FETs, with high dielectric constants and compatibility with 2D semiconductors that enable effective gate control and low leakage currents.

ACKNOWLEDGMENTS

This work was supported by the European Research Council under grant agreement no. 101055379. Support from Huawei Europe under Research Agreement TC20220601024 is kindly acknowledged. The support from the Vienna Scientific Cluster for providing resources on the Austrian high-performance clusters VSC4 and VSC5 is gratefully acknowledged.

REFERENCES

[1] K. Uchida *et al.*, IEDM (2002).
[2] Wanrong *et al.*, Nanoscale (2019).
[3] R. Chau, IEEE International Electron Devices Meeting (IEDM) (2019).
[4] A. V. Kretinin *et al.*, Nano letters (2014).
[5] Y. Liu *et al.*, Nature (2019).
[6] Y. Y. Illarionov *et al.*, Nature communications (2020).
[7] M. R. Osanloo *et al.*, Nature Communications (2021).
[8] S. Yang *et al.*, Advanced Materials (2022).
[9] R. Neville *et al.*, Journal of Applied Physics (1972).
[10] C. Zhang *et al.*, Nature materials (2023).
[11] P. Khakbaz *et al.*, (2024).
[12] Q. Wei *et al.*, ACS Nano (2019).
[13] T. D. Kühne *et al.*, The Journal of Chemical Physics (2020).
[14] M. Guidon *et al.*, Journal of chemical theory and computation (2009).
[15] R. Resta *et al.*, (2007).
[16] K. Van Benthem *et al.*, Journal of applied physics (2001).
[17] m. A. J. Yang, Nature Elec. (2022).
[18] Y. Zhang *et al.*, Nature Electronics (2022).
[19] W. et al., Dominic, Microelectronics Reliability (2023).
[20] R. Tsu *et al.*, Applied Physics Letters (1973).
[21] T. Knobloch *et al.*, Nature Electronics (2021).
[22] L. Cao et al., physica status solidi (a) (2000).
[23] D. Gryaznov *et al.*, The Journal of Physical Chemistry C (2013).

Table I: Material properties of SrTiO$_3$ and β-Bi$_2$SeO$_5$ calculated by DFT.

Property	SrTiO$_3$	Lit.	[Ref.]	β-Bi$_2$SeO$_5$
Lattice Constant [Å]	3.9	3.9	[22]	8.1, 8.1, 15
Indirect Bandgap E_G [eV]	3.30	3.25	[16]	3.50
Optical Permittivity ϵ_∞ [1]	4.31	4.64	[23]	10.36
Static Permittivity ϵ_0 [1]	214	300	[9]	35.30

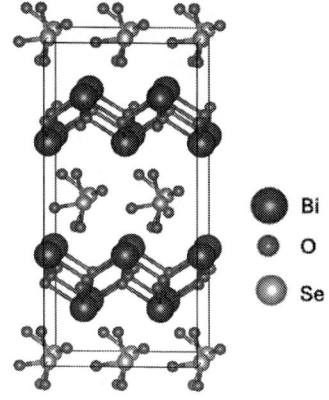

Figure 1: Crystal structure of the cubic SrTiO$_3$ perovskite (top), and of β-Bi$_2$SeO$_5$ (bottom).

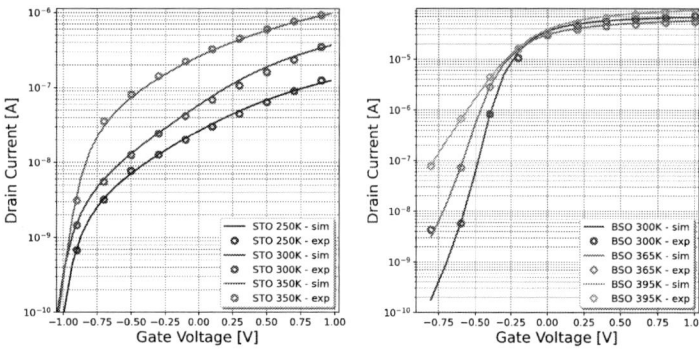

Figure 2: Schematic device geometry of the MoS$_2$/SrTiO$_3$ [17] (top) and of the Bi$_2$O$_2$Se/Bi$_2$SeO$_5$ devices [18] (bottom).

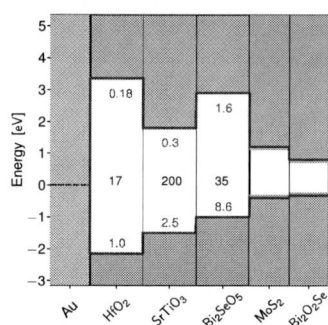

Figure 3: Transfer characteristics at 250 K, 300 K, and 350 K for an STO device (left) and for a Bi$_2$O$_2$Se/Bi$_2$SeO$_5$ device at 300 K, 365 K and 395 K (right).

Figure 4: Band alignments, effective masses, and dielectric constants of potential gate insulators.

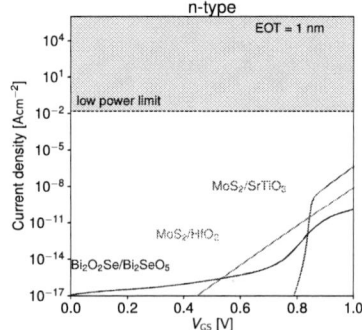

Figure 5: Calculated gate leakage currents for scaled insulators with an EOT of 1 nm.

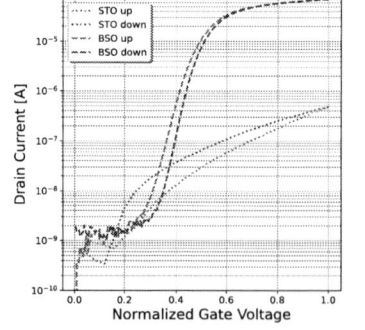
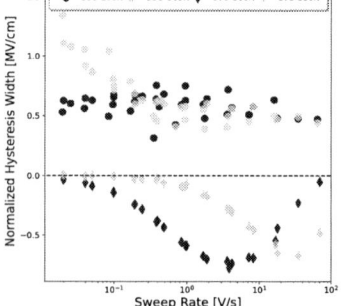

Figure 6: CW hysteresis in Bi$_2$O$_2$Se/Bi$_2$SeO$_5$ and CCW hysteresis in STO over normalized gate bias (left). Hysteresis widths normalized by EOT at two temperatures for each device type (right).

Physics of Operation of GaN Power Devices: Modeling Device and Circuit Effects using MIT Virtual Source GaNFET (MVSG) Model

Ujwal Radhakrishna, Daiyao Xu, Kaiman Chan, Tim Merkin, Ryan Fang, Johan Alant, Lan Wei

Texas Instruments Inc, USA, University of Waterloo, CA

Abstract

We present a framework to capture essential physics of operation of devices built on Gallium Nitride technology for High-Voltage (HV) and power applications. Devices include Unidirectional (UDS) and Bi-directional Switches (BDS) that span voltages between 60-650V. The modeling framework is based on well-known industry standard MVSG Verilog-A compact model platform. The model is proven to distill essential device physics to predict device-circuit interactions in power conversion applications. In this work, we show the model comparison against DC-IVs, CVs, switching measurements of UDS and BDS along with a demonstration of slew-rate prediction; a key device-circuit interaction effect. Secondary effects such as self-heating, charge-trapping, P-GaN gate-dynamics are also highlighted. **Keywords:** GaN-HEMT, Power Electronics, Compact Modeling, MVSG, UDS, BDS.

Introduction

GaN-technology based power electronics is making significant inroads in automotive, industrial motor drives, solar inverters, EV-chargers, and server applications. At the component level, GaN UDS and BDS-FETs are the engines driving this technology. A thorough understanding of the physics of operation of these devices in applications is required to design such power-converter systems optimized for efficiency and reliability. The MVSG framework is a proven industry-standard physics-based compact modeling framework that enables GaN device-circuit co-design [1]. This work is a summary of MVSG-model's capability to describe device-effects in both MBS and BDS relevant to HV applications, such as: Device currents and charges, Multi-temperature scaling, static and transient device-heating, Schottky-gate dynamics, trapping and dynamic-R_{DSON}.

Modeling MVSG channel-currents and charges

Prior works have detailed derivations of MVSG FET-currents and charges [1]-[3]. Only a snapshot of the core-equations is provided in Table I. The model is charge-based and is symmetric with respect to source and drain terminals (two sources in the case of common-drain BDS). Self-consistent currents and charge formulations reduce the number of parameters required to fit both static and transient characteristics simultaneously.

A. MVSG currents

The charge-based current formulation is an extension of the well-known Virtual Source (VS) concept of charge-transport

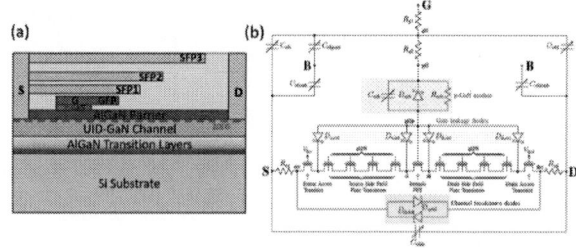

Fig. 1: (a) Typical cross-section of a power GaN-UDS that shows multiple field plates (FPs) that are either Gate-connected (GFP) or Source-connected (SFP). The gate-construction is typically a Schottky P-GaN junction to achieve enhancement-mode (E-mode) operation with low gate-leakage. (b) MVSG sub-circuit that represents different physical phenomena of UDS-FET such as: Core and FP transistors, gate-leakage diodes (de-biased through contacts or FPs), dynamic P-GaN gate behavior modeled using diode-R-C approach, channel-breakdown diodes, and parasitic fringing and cross-coupled capacitors.

in scaled Si-FETs. In power GaN-HEMTs, the model extends to velocity-saturated or mobility-dominant regime and resembles EKV model. The log-exp function format ensures smooth weak-to-strong accumulation transition with infinite derivates - a key requirement for compact models used in circuit simulators. The linear-to-saturation transition is governed via F_{vsat} in Table I. The model includes velocity- and mobility-reduction with gate-charge. The core-current model is used for each transistor of the sub-circuit shown in Fig. 1b that represents gated-region, FPs and access regions of Fig. 1a. Typical quality of current-fits of MVSG model against measurements are shown in Fig. 2(a)-(f) highlighting the model's ability to capture currents, transconductance and output-conductance across all bias-regimes.

B. MVSG charges

Table I. also shows charge formulation calculated from the same terminal charges used for current computation. Using Ward-Dutton charge partitioning, the channel charge is apportioned to source and drain nodes of each transistor of Fig. 1b. Further, cross-coupled charge due to fringing field and body charge terms associated with each FP are implemented in the same log-exp format. FP-transistors deplete at their threshold voltages in off-state and yield non-linear transitions in the device-CVs as shown in Fig. 2(g)-(h).

Multi-temperature modeling

Thermal effects are critical for power conversion applications as they have direct impact on efficiency and lifetime reliability. GaN-HEMTs in power-conversion applications can operate over a wide range of temperatures (typically from

979-8-3315-0417-5/25 $31.00 © 2025 IEEE

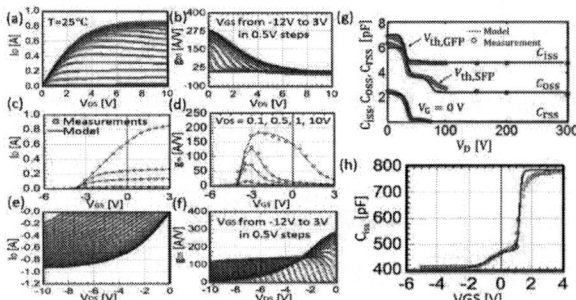

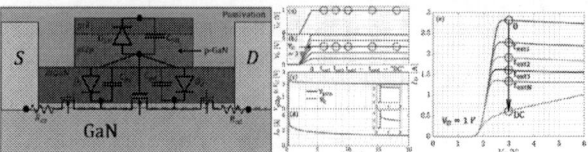

Fig. 2: Demonstration of MVSG-model's accuracy in (a) first-quadrant output and (b) its derivative characteristics, (c) transfer curves and (d) their derivatives. Third-quadrant device-model is critical for power-FETs and is also self-consistently captured as shown in (e) output curves and (f) its derivatives. (g)-(h) Accuracy to predict non-linear capacitance behavior showing FP-transitions in C_{iss}, C_{rss} and C_{oss} are shown.

-40°C to 150°C). This necessitates modification of device-transport as a function of junction temperatures (T_j) [4]. To accomplish this, the model captures temperature coefficients (*Tempcos*) of key metrics such as mobility, velocity, subthreshold slope, contact- and sheet-resistance, threshold voltages. The parameters are extracted using IV-fits at three T_j and verified against fourth, as shown in Fig. 3a. In addition, thermal circuit of Fig. 3b is used to capture static self-heating (via R_{th}) and transient-heating (via C_{th}) calibrated against pulsed-IV data of the figure. Heat dissipation through package/board is captured using Θ_J connected to thermal node dt and heating from adjacent FETs is captured using P_{ext}-source, both defined at the netlist level.

Floating P-GaN Schottky-gate model

E-mode operation requires a P-GaN Schottky gate that results in a gate architecture with two back-to-back diodes shown in

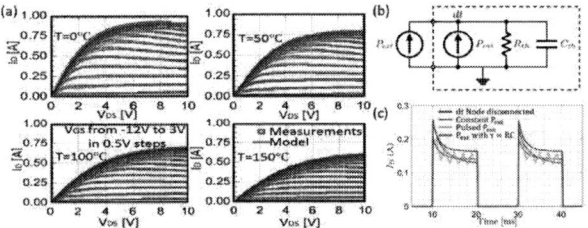

Fig. 4a [5]. The diode-drops reduce effective gate-overdrive

Fig. 3: Temperature-related effects are modeled through three aspects in MVSG framework: device Tempcos that can be extracted from IV-measurements at multiple junction-temperatures (T_j), device self-heating done using a thermal RC-sub-circuit, and impacts of ambient temperature change and external heating. (a) Tempcos are extracted from output curves at three- T_j and the model is able to match IVs at the fourth Tj. (b) Thermal RC-sub-circuit implementation to mimic both static and dynamic Tj-rise during switching operation. (c) Transient current simulations show the impact of self-heating as well as external-heating with a constant and a time-varying P_{ext} source. Impacts of ambient temperature change and external heating can be included via P_{ext} term in (b). This captures system-level

thermal impedance (Θ_J) governed by package/board-level heatsinks.

Fig. 4: Typical E-mode Power UDS-FETs have P-GaN Schottky gate which results in a back-to-back diode configuration shown in the cross-section schematic. The metal/P-GaN junction (C_{SH} and D_{SH}) with the gate-diodes (D_s/D_d) create a floating P-GaN node that has transient lag in its response to external gate-signal. This creates an enhanced channel turn-on at short switching timescales but a degraded gate-overdrive (and hence I_{DSAT}) at longer times. (a)-(b) Represent classic Single-Pulse-Test (SPT) and the device response in terms of (c) internal gate-voltages and (d) currents. (e) The transient IDVG response is amplified vs. steady state, yielding higher I_D near hard-switching transitions of GaN-based power converters. MVSG model's ability to capture this dynamic behavior is crucial to estimate power converter's heating, efficiency-penalty (from overlap losses) and enhanced hot-carrier injection related reliability issues.

and hence channel current in DC-operation. The diode-drops reduce effective gate-overdrive and hence channel current in DC-operation. Further, a finite time is needed to deplete or charge this 'floating' P-GaN region, causing dynamic V_T-shifts along with a boost in current in transient switching conditions (due to capacitance divider in Fig. 4a). Single-pulse-test (SPT) simulations of Fig. 4(b)-(e) illustrate this.

Emulating charge-trapping effect

Trapping effects in power GaN HEMTs cause R_{DSON} to increase from DC to switching conditions, increasing static losses. The phenomenon is also referred as drain/gate-lag and is modeled using sub-circuit of Fig. 5.

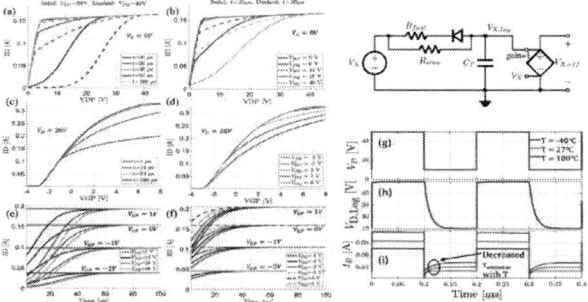

Fig. 5: Charge-trapping induced dynamic R_{DSON} is captured in MVSG framework using a diode-R-C sub-circuit shown. There are two time-constants associated with traps: capture times ($R_{fast}C_T$) and emission times ($R_{slow}C_T$). Two copies of the sub-circuit are implemented to mimic drain-lag (with $V_X=-V_D$) and gate-lag effects (with $V_X=V_G$), and their output voltages ($V_{X,eff}$) are fed into drain- and gate-voltage input-terms for charge/current calculations. (a)-(b) Pulsed output-IVs show impact of increased V_{DQ} (from 28V to 40V) and reduced pulse-times on degradation of device R_{DSON} and I_{DSAT} due to drain-lag. (c)-(d) Similar impact from V_{GQ} and pulse-times is seen in transfer-IVs due to gate-lag. Transient response from drain/gate-lag circuits on I_D post device-turn-on is evident from (e)-(f). The model also includes Tempcos of capture and emission times, as demonstrated using

979-8-3315-0417-5/25 $31.00 © 2025 IEEE

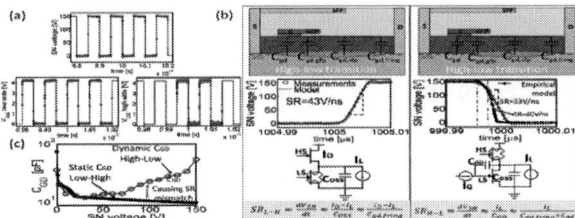

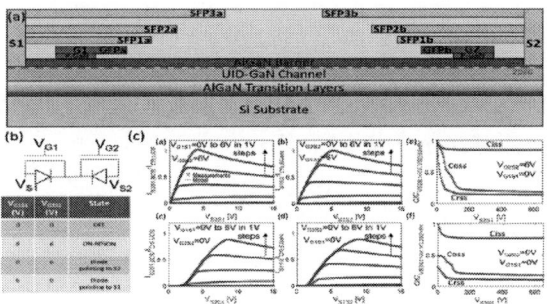

multi-Temperature transient I_D-response in (g)-(h).

Fig. 6: Commercial GaN hard-switched half-bridge is benchmarked with circuit simulations that use MVSG model. (a) Simulated switching waveforms vs. measurements demonstrate convergence robustness of the model. (b) The physics of device-operation in both switching transitions are illustrated. (i) High-to-low transition is soft-switching where the device is off (channel fully depleted). The switch-node slew-rate (SR) is governed by off-state C_{oss} determined by fringing-capacitance ($C_{gd,fring}$). (ii) Low-to-high transition is hard-switching and the device is partially turned on during slew-time. Device cross-section illustrates the effect of turning on core-transistor while the FET still blocks high V_{DS}. C_{oss} governing the SR is now higher due to the added core-transistor capacitance (C_{gd}) as shown in (c). Empirical models that use static off-state CV values (at the same V_{DS}) underestimate C_{oss} and hence predict faster SR while MVSG's distributed sub-circuit charge model predicts correct dynamic-C_{oss} and hence SR.

Quiescent-voltages (V_{DQ} and V_{GQ}) are fed to this network and its output is used to reduce Q_{is} and Q_{id} (of Table I) to change transient I_D [6]. The approach enables MVSG model to mimic pulsed output and transfer plots of Fig. 5(a)-(i) that exhibit the well-known 'knee-walkout' effect as well as a changed dynamic output-/trans-conductance in the UDS-FET.

Prediction of Switching Slew-Rates

Commercial GaN-based Half-Bridge converter simulation using MVSG model in Fig. 6a demonstrates its convergence capability in transient switching along with accuracy of estimating Switch (SW)-node SRs. Both hard- and soft-SW SRs are accurately estimated as shown in Fig. 6b, thanks to the distributed charge-model in MVSG [7]. The potential re-distribution between core- and FP-transistors in hard-switching effectively increases C_{GD} (Miller capacitance) compared to static CV (Fig. 6c), yielding lower SRs. Empirical models predict faster SRs, underestimating the overlap-losses, associated heating and reliability issues.

Extending MVSG framework to GaN-BDS

While GaN power-FETs typically refer to UDS, a symmetric common-drain BDS shown in Fig. 7a is gaining traction for AC-DC power conversion due to its RSP benefits vs. two back-to-back UDS. The MVSG-framework of Fig. 1b can be extended to describe the operational modes of this device. The states of high-side and low-side V_{GS} determine if the device is ON, OFF or in diode-modes (Fig. 7b). Typical BDS DC-IV and CV fits using MVSG model are shown in Fig. 7c. Symmetric forward- and reverse-mode behavior, FET-IV vs.

Fig. 7: Cross-section schematic of a symmetric BDS with common-drain configuration, typically used as AC-switches in AC-converters. MVSG-UDS framework can be easily extended to this device. (b) the device has four operating modes depending on the states of both gate-overdrive voltages: V_{G1S1} and V_{G2S2}. The model accurately captures typical output-IV curves in all four states. (a)-(b) Regular FET-output curves are observed when $V_{S2}>V_{S1}$ with $V_{G2S2}=6V$, or when $V_{S1}>V_{S2}$ with $V_{G1S1}=6V$ (high-side gate ON): so-called FET-mode. (c)-(d) IV-curves with diode-drop emerge when $V_{S2}>V_{S1}$ with $V_{G2S2}=0V$, or when $V_{S1}>V_{S2}$ with $V_{G1S1}=0V$ (high-side gate OFF): so-called diode-mode. The curves are symmetric between forward-mode ($V_{S2S1}>0V$) and reverse-modes ($V_{S1S2}>0V$) and the model matches well with measurements. CV benchmarking in both FET-mode in (e) and diode-mode in (f) shows excellent match of the three CV-metrics including non-linear voltage dependence arising from FPs and fringing fields.

diode-IV (with cut-in voltage in linear region), FET-CV vs. diode-CV (with reduced capacitance) are captured accurately.

$$I_{DS} = W \frac{Q_{is} + Q_{id}}{2} v, \quad v = \left[\frac{FF 2\phi_T \mu_0}{L_G} + (1 - FF) v_{x0} \right] F_{VSAT}$$

$$F_{VSAT} = \frac{Q_{VSAT}}{(1 + Q_{VSAT}^\beta)^{\frac{1}{\beta}}}, \quad Q_{VSAT} = \frac{(Q_{is} - Q_{id})}{C_{INV} V_{DSAT}}, \quad V_{DSAT} = \frac{v_{sat} L_G}{\mu_{eff}}$$

$$Q_{is} = 2 C_{INV} n \phi_T \log \left(1 + \exp(\frac{V_{GD} - V_{SX} - V_T}{2 n \phi_T}) \right)$$

$$Q_{id} = 2 C_{INV} n \phi_T \log \left(1 + \exp(\frac{V_{GS} - V_{DX} - V_T}{2 n \phi_T}) \right)$$

$$Q_s = \frac{2 W L_G}{(Q_{is}^2 - Q_{id}^2)^2} \left[-Q_{id}^2 \frac{Q_{is}^3 - Q_{id}^3}{3} + \frac{Q_{is}^5 - Q_{id}^5}{5} \right], \quad C_{gs} = -\partial Q_s / \partial V_G$$

$$Q_d = \frac{2 W L_G}{(Q_{is}^2 - Q_{id}^2)^2} \left[Q_{is}^2 \frac{Q_{is}^3 - Q_{id}^3}{3} - \frac{Q_{is}^5 - Q_{id}^5}{5} \right], \quad C_{gd} = -\partial Q_d / \partial V_G$$

Table. I

Conclusion

This work summarizes the present capability of MVSG modeling framework to aid GaN power circuit design. A full suite of physical effects that govern GaN-device operation in realistic power electronic systems is captured by the model. Device-level and circuit-level accuracy is demonstrated by benchmarking model results against measurements.

References

[1]   U. Radhakrishna et.al., PSS-C, vol. 11, no. 3, pp. 848-852, Mar. 2014.

[2]   U. Radhakrishna et.al., IEEE-TED, vol. 66, no. 1, pp. 95-105, Jan. 2019.

[3]   R. Fang et.al., IEEE-BCICTS 2022, Phoenix, AZ, 2022, pp. 1–4.

[4]   U. Radhakrishna et.al., IEDM 2015, pp.9.6.1-9.6.4, DOI: 10.1109.

[5]   R. Fang et.al., ISPSD 2023, pp. 123-126, DOI: 10.1109/ISPSD57135.2023.

[6]   L. Wei et. al., BCICTS invited paper 2024.

[7]   U. Radhakrishna et.al, IEDM, vol.66, no.1, pp.95-105, ISSN:1557-9646.

979-8-3315-0417-5/25 $31.00 © 2025 IEEE

Magnetic Resonance based Soft Electronic Implant for Wireless Electrotherapy and Thermal Ablation

Sicheng Xing, Wubin Bai

Department of Applied Physical Sciences, University of North Carolina at Chapel Hill, Chapel Hill, NC 27599, USA.

Email: wbai@unc.edu

Abstract

This study explores the potential of magnetic resonance wireless power transfer (MR-WPT) for non-invasive medical applications, addressing limitations of traditional power sources in bioelectronic implants. By leveraging resonant coupling, MR-WPT achieves efficient energy transfer across biological barriers with minimal attenuation. A flexible printed circuit board (fPCB)-based MR-WPT network is developed to deliver wireless, battery-free power to implants, demonstrating resilience to geometric changes, such as separation, offset, and tilt. Experiments highlight its capability to maintain high power transfer efficiency under varied conditions and validate its functionality in targeted thermal ablation and customized electrical stimulation. The results suggest that MR-WPT could revolutionize bioelectronic implant design, offering enhanced safety, reliability, and therapeutic efficacy.

keywords are: Magnetic Resonance Wireless Power Transfer (MR-WPT); Bioelectronic implants; battery-free electrotherapy

Introduction

Non-invasive therapies like thermal ablation for oncology[1], [2], [3], [4], [5], [6] and electrotherapy for neurological and cardiovascular conditions[7], [8], [9], [10] require precise energy delivery. Traditional power sources - tethered connections and batteries - have significant limitations: tethered devices restrict mobility and risk infection, while batteries require periodic replacement and risk leakage[11], [12].

Recent wireless power transfer (WPT) alternatives include non-resonant coupling[13], [14] and ultrasonic power transfer (UPT)[15]. However, non-resonant coupling suffers from low efficiency and limited range across varying tissue layers, while ultrasonic transfer faces tissue impedance mismatches.

Magnetic resonance wireless power transfer (MR-WPT) addresses these limitations through resonant coupling, achieving high energy transfer efficiency over moderate distances[16], [17], [18], [19]. By operating at specific resonant frequencies, it transfers power across biological barriers with minimal attenuation, reducing overheating risk and enabling continuous power supply.

This paper investigates MR-WPT's feasibility for thermal ablation and electrotherapy through benchtop and in-situ experiments, demonstrating how it can integrate with electronic implants while maintaining functionality in complex internal environments. Our work presents MR-WPT as a promising solution for enhancing the efficacy and safety of future electrotherapeutics.

Results and Discussion

Here, we present a flexible printed circuit board (fPCB) based MR-WPT network to achieve the delivery of electrical power to implanted devices to achieve wireless and battery free operations, as shown in Fig 1A. The system can be easily integrated into existing electrical designs of electrical implants and offers unique advantages as its low modulus allows adaptations in different internal environments and prevent mechanical mismatches between the system and tissue.

As shown in Fig 1B, our finite element analysis study showed effective coupling in magnetic fields allowing efficient power transmission. We also numerically studied the lumped circuit behavior of the system, as shown in Fig 1C. With the load resistance matches that of the internal resistance of the supply; by performing T-network transformation of mutual inductance, we can compute power delivery as a function of coupling factor as well as the quality factor of the network[13], [20]. This equation has shown that when k and Q are relatively small (kQ < 1) due to the small size of the coil, increasing Q and increasing k will lead to greater power delivered to the receiving coil at resonant frequency. For larger kQ values, deviation from the resonance frequency, known as frequency split, was observed.

$$P_L = V_{S_{RMS}}^2 \left(\frac{1}{\frac{1}{kQ} + kQ} \right)^2 \cdot \frac{1}{R}$$

For bioelectronic implants to function at optimal performance, surgeons often need to exercise judgement to individualize the placement to account for difference in patient anatomy as well as underlying pathogenesis. Additionally, the implants are often subjected to movement and deformation due to patient movement, as well as movement of the internal organs like the heart or the digestive tract as part of their normal function. These internal environmental factors require wireless power systems to remain effective despite sub-optimal and constantly changing system geometry[21].

Here, we experimentally studied the effect of changing system geometry on wireless power transmission. In order to quantify the systemic geometric changes encountered by the implants, we divided the relative displacement between coils into three categories: distance (separation between coils), offset (side-to-side positional difference between coils), and tilt (angular rotational displacement between coils), as shown in Figure 2A. Figure 2B and 2C showed the power transfer profile for different separation distances at 14MHz (resonance frequency) and 20MHz. Here, we demonstrated that, at resonance frequency, power transfer remains effective despite increasing separation distances, showing the versatility of MR-WPT in mid-range distances. We also observed that, at very small distances, we saw a significant decrease in resonance frequency power transmission but having a significant increase in power transmission at non-resonant frequencies, which is also demonstrated in Figure 2D. Figure 2E and 2F showed that despite small changes in offset and tilt angles, the power transfer efficiency remains unaffected, with maximum transfer efficiency at approximately 70% and highest power at 0.7W. We noticed that the power transfer profile is not linear, but instead reaching a plateau at higher powers, which can likely be attributed to distortions caused by intrinsic characteristics of power amplifiers used in this experiment.

To validate the efficacy of the system to be integrated as part of electrical implants, here, we present two devices that leverage the unique characteristics of MR-WPT, each representing unique challenges in wireless, battery-free implant designs. In Fig 3A, we presented an electrical implant that, leveraging the high transmission efficiency and total transmission power, enables the targeted thermal ablation at tissues. As shown in Fig 3B, no significant heat dissipation in the coil was observed while the heating pad reaching an effective thermal ablation temperature at greater than 50°C, allowing targeted therapy for local malignancy. By employing custom wave modulation, we were also able to create a wireless electrical stimulator that was able to customize stimulation waveform, as shown in Fig 3C. The electrical stimulator rectifies and filters the transmitted electrical signal (Fig 3D) and delivers it as waveform pulses to tissue (Fig 3E) for electrical stimulation to achieve effective electrical therapy, demonstrating its potential in neural modulation and cardiac rhythm management.

Conclusion

In conclusion, the proposed MR-WPT system demonstrates significant potential as a reliable, efficient, and safe method for wireless energy transfer in biomedical applications, overcoming the dependency on tethered connections or batteries. The system's adaptability to varying geometries and patient movement, and high efficiency at resonant frequencies establish its viability for integration into implantable medical devices. Furthermore, the applications in thermal ablation and electrical stimulation highlight its versatility and potential impact in diverse therapeutic contexts, such as oncology and neurology. Overall, MR-WPT offers a promising solution for enhancing patient safety and improving the efficacy of bioelectronic implants. Future work could focus on scaling this technology for broader clinical applications, further optimizing its design for specific anatomical and pathological conditions and exploring its long-term in-vivo performance.

Materials and Methods

fPCB Manufacturing: The fPCB is manufactured by ISO 9001 compliant commercial vendors. All circuit components were soldered using tin-lead solder paste by hot-air blowing.

Electrical Characteristic Studies: The inductance of the coil is experimentally measured using a benchtop LCR meter (Matrix MCR-5010). The resonant characteristics of the power transfer network are experimentally measured using a handheld vector network analyzer (NanoVNA).

Benchtop Characterizations: A 50Ω resistive load is connected to the receiving network to simulate the load in benchtop characterizations. The transmission network is connected to a radio frequency power source with a 50Ω characteristic source impedance that consists of a signal generator (Siglent SDG 2042X) connected to an RF power amplifier (EIN 503L). Two high-impedance oscilloscope probes are connected to the transmission network and the resistive load respectively to measure the time-domain waveform using a digital oscilloscope (Siglent SDS 1104X-E). Infrared images were taken using FLIR camera to study power transfer associated heating.

In-situ electrical stimulation studies: In-situ electrical stimulation experiments were performed using meat products as tissue phantoms, using commercial cardiac electrodes as recording electrodes and recorded using PowerLab DAQ.

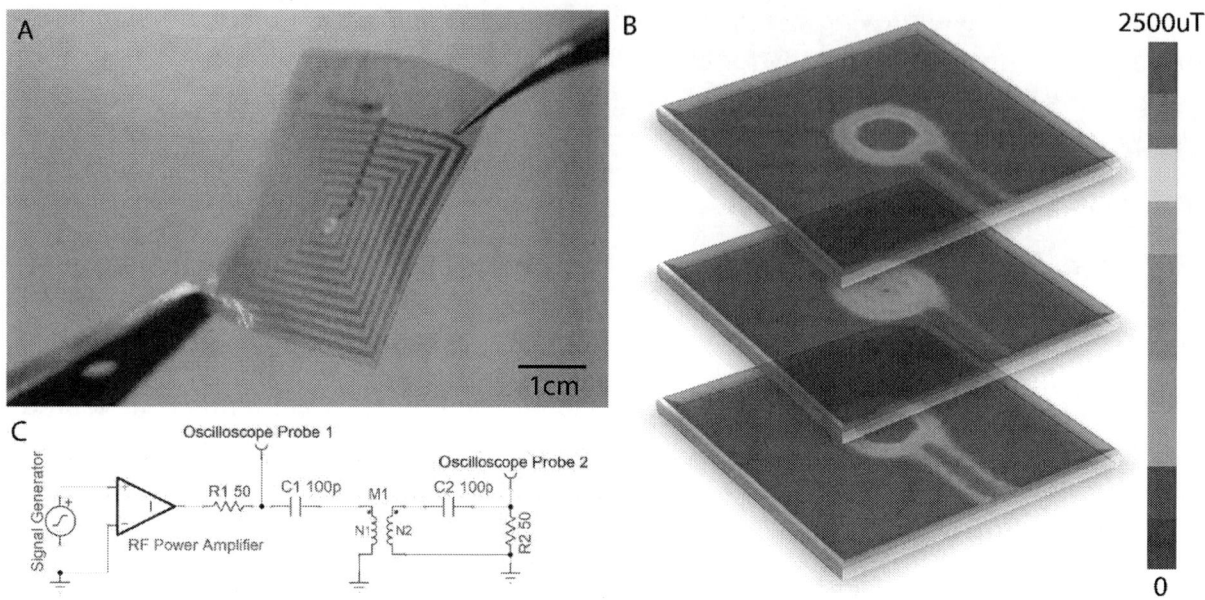

Figure 1. Design and Analysis of the MR-WPT System (A) Optical Image of a Flexible printed circuit board (fPCB) based MR-WPT network (B) Finite element analysis showing magnetic field coupling (C) Equivalent circuit diagram of the proposed system with signal generator, power amplifier, and the power transfer network.

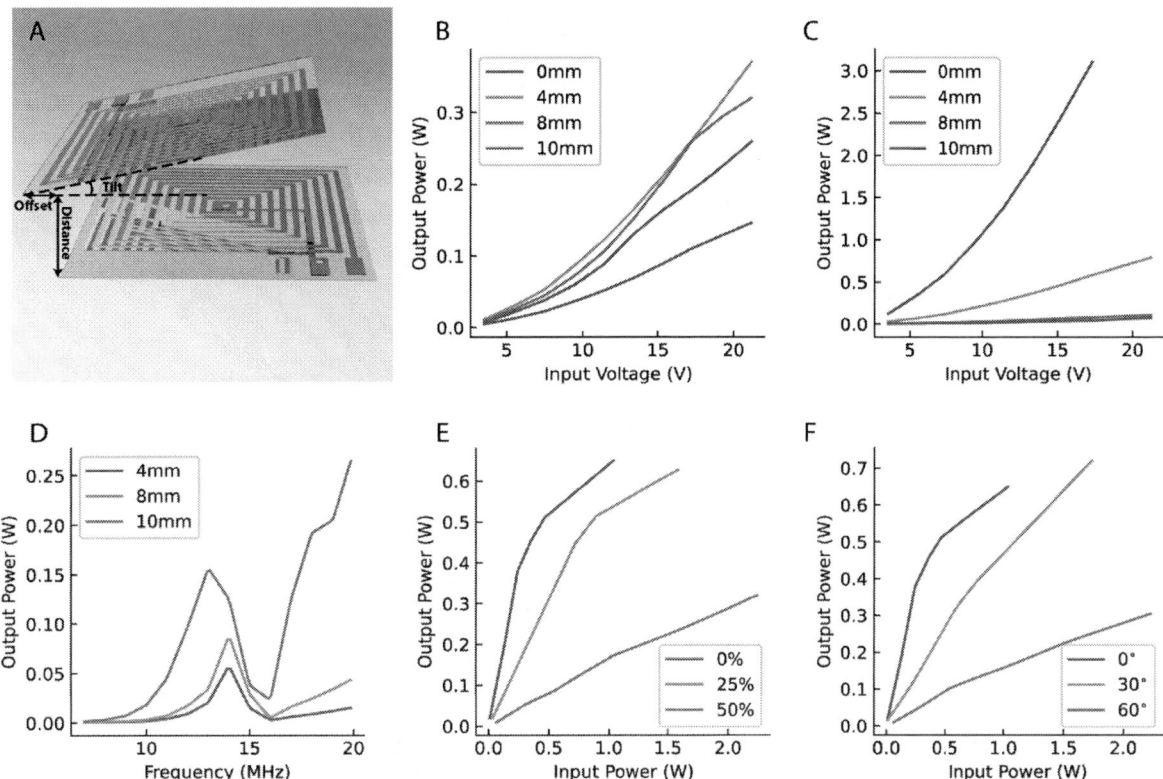

Figure 2. Effect of Geometric Changes on Power Transfer (A) Categories of relative displacement between coils: distance, offset, and tilt (B) Power transfer profile at 14MHz (resonance frequency) with varying separation distances (C) Power transfer profile at 20MHz with varying separation distances (D) Frequency response at varying separation distances (E) Power transfer efficiency with offset changes (F) Power transfer efficiency with tilt angle changes

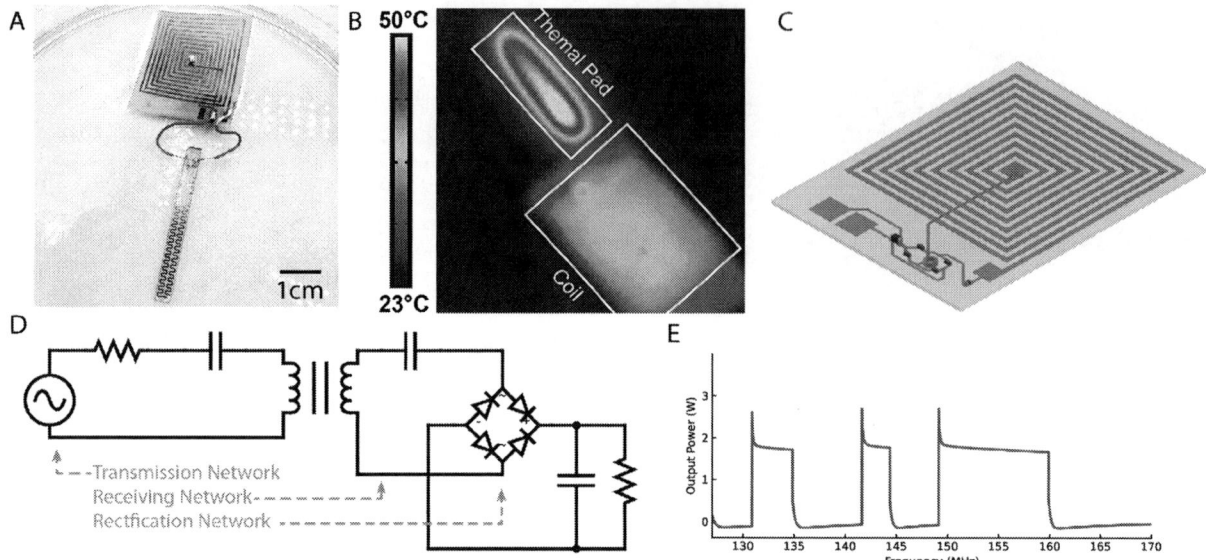

Figure 3. Applications of MR-WPT in Medical Implants (A) Optical Image of a battery-free, wireless thermal ablation implant (B) Infrared image of the thermal ablation implant. (C) Design of a wireless electrical stimulator (D) Circuit diagram of the wireless electrical stimulator (E) In-situ tissue recordings of electrical pulses delivered by the wireless electrical stimulator.

[1] B. Al-Sakere and others, 'Tumor ablation with irreversible electroporation', *PLoS One*, vol. 2, 2007, doi: 10.1371/journal.pone.0001135.

[2] C. Brace, 'Thermal tumor ablation in clinical use', *IEEE Pulse*, vol. 2, pp. 28–38, 2011, doi: 10.1109/MPUL.2011.942603.

[3] R. Habash and others, 'Thermal therapy, Part III: ablation techniques', *Crit Rev Biomed Eng*, vol. 35, no. 1–2, pp. 37–121, 2007, doi: 10.1615/CRITREVBIOMEDENG.V35.I1-2.20.

[4] Z. Zhao and F. Wu, 'Minimally-invasive thermal ablation of early-stage breast cancer: a systematic review', *European Journal of Surgical Oncology*, vol. 36, no. 12, pp. 1149–1155, 2010, doi: 10.1016/j.ejso.2010.09.012.

[5] M. Prantner and N. Parspour, 'Eddy current heating of implanted devices for tumor ablation', *IEEE Access*, vol. 11, pp. 52088–52100, 2023, doi: 10.1109/ACCESS.2023.3280052.

[6] A. Gillams and others, 'Thermal ablation of colorectal liver metastases: a position paper', *Eur Radiol*, vol. 25, pp. 3438–3454, 2015, doi: 10.1007/s00330-015-3779-z.

[7] D. Kucharzyk, 'Implantable electrical stimulation in spinal fusion', *Spine (Phila Pa 1976)*, vol. 24, no. 5, pp. 465–468, 1999, doi: 10.1097/00007632-199903010-00012.

[8] D. Rushton, 'Electrical stimulation in the treatment of pain', *Disabil Rehabil*, vol. 24, pp. 407–415, 2002, doi: 10.1080/09638280110108832.

[9] A. Salter and others, 'First clinical experience with BION implants for electrical stimulation', *Neuromodulation: Technology at the Neural Interface*, vol. 7, 2004, doi: 10.1111/j.1525-1403.2004.04005.x.

[10] P. Wang and others, 'Vascular electrical stimulation reduces atherosclerotic plaque formation', *Small*, p. e2300584, 2023, doi: 10.1002/smll.202300584.

[11] K. Poudel and M. Pant, 'Wireless Power Transfer for Medical Implants', in *USNC-URSI Radio Science Meeting*, 2019, pp. 57–58. doi: 10.1109/USNC-URSI.2019.8861793.

[12] K. Agarwal, R. Jegadeesan, Y. Guo, and N. Thakor, 'Wireless Power Transfer Strategies for Implantable Bioelectronics', *IEEE Rev Biomed Eng*, vol. 10, pp. 136–161, 2017, doi: 10.1109/RBME.2017.2683520.

[13] T. Imura, *Wireless power transfer: Using magnetic and electric resonance coupling techniques*. Springer Singapore, 2020. doi: 10.1007/978-981-15-4580-1.

[14] I. Mayordomo, T. Drager, P. Spies, J. Bernhard, and A. Pflaum, 'An overview of technical challenges and advances of inductive wireless power transmission', 2013, *Institute of Electrical and Electronics Engineers Inc.* doi: 10.1109/JPROC.2013.2243691.

[15] Z. Chen, Y. Xu, and J. Y. et al., 'Ultrasound Wireless Power Transfer for Cardiac Pacemaker Applications', *IEEE Trans Biomed Eng*, vol. 67, no. 3, pp. 1058–1066, 2020, doi: 10.1109/TBME.2019.2944135.

[16] R.-F. Xue, K.-W. Cheng, and M. Je, 'High-Efficiency Wireless Power Transfer for Biomedical Implants by Optimal Resonant Load Transformation', *IEEE Transactions on Circuits and Systems I: Regular Papers*, vol. 60, pp. 867–874, 2013, doi: 10.1109/TCSI.2012.2209297.

[17] X. Li, C. Tsui, and W. Ki, 'A 13.56 MHz Wireless Power Transfer System With Reconfigurable Resonant Regulating Rectifier and Wireless Power Control for Implantable Medical Devices', *IEEE J Solid-State Circuits*, vol. 50, pp. 978–989, 2015, doi: 10.1109/JSSC.2014.2387832.

[18] Y. Zeng, D. Qiu, and X. M. et al., 'Optimized Design of Coils for Wireless Power Transfer in Implanted Medical Devices', *IEEE J Electromagn RF Microw Med Biol*, vol. 2, pp. 277–285, 2018, doi: 10.1109/JERM.2018.2863955.

[19] A. S. Hossain, P. Mohseni, and H. Lavasani, 'A Resonant Capacitive Wireless Power and Data Transfer Link with 52% PTE and 6.5 Mbps Data Rate for Biomedical Implants', in *2022 IEEE Biomedical Circuits and Systems Conference (BioCAS)*, 2022, pp. 292–296. doi: 10.1109/BioCAS54905.2022.9948595.

[20] M. Kiani and M. Ghovanloo, 'The circuit theory behind coupled-mode magnetic resonance-based wireless power transmission', *IEEE Transactions on Circuits and Systems I: Regular Papers*, vol. 59, no. 9, pp. 2065–2074, 2012, doi: 10.1109/TCSI.2011.2180446.

[21] L. Zhang *et al.*, 'Skin-inspired, sensory robots for electronic implants', *Nat Commun*, vol. 15, no. 1, Dec. 2024, doi: 10.1038/s41467-024-48903-z.

Characterization of self-heating using the AC conductance method

(Invited Paper)

A.J. Scholten[*], *Fellow, IEEE*, R.M.T. Pijper[*], and T.V. Dinh[†], *Member, IEEE*

[*]NXP Semiconductors, Eindhoven, The Netherlands; [†]NXP Semiconductors, Austin, USA

Abstract

The AC conductance method to determine the thermal resistance R_{th} is reviewed and demonstrated for several device types. A comparison is made with pulsed measurements, which are typically found to be too slow for reliable R_{th} determination of modern devices.

Keywords: self-heating, thermal resistance, compact models

Introduction

Due to ever increasing power densities, the topic of self-heating, traditionally restricted to LDMOS [1], SOI [2], and bipolar devices [3, 4], is becoming an increasing concern in advanced CMOS technologies such as FinFET [5, 6], FDSOI [7, 8], and GAA [9]. In this paper, we review the AC conductance method, with the help of which one can determine the thermal resistance R_{th} of a device, as well as its dynamic thermal behavior. The assumptions underlying the method are listed, and three examples (FinFET, FDSOI, and NPN) are given. Even for the thermally slowest of these three devices, 200-ns pulsed measurements are shown to be too slow to determine R_{th} reliably.

The AC conductance method

The AC conductance method relies on the different thermal response at 'low' and 'high' frequencies [10]. When an AC excitation is applied to a device, its temperature can follow that AC excitation only at very low frequencies. In case of a MOSFET, where the drain current decreases with temperature (at sufficiently high V_{GS}), this leads to a reduction of the output conductance, which can even become negative under certain conditions. At very high frequencies, on the other hand, the instantaneous device temperature can no longer follow the AC excitation, and the output conductance is only affected by the DC heating associated to the device operating point. This difference in output conductance at low and high frequencies can be exploited to derive the thermal resistance. To achieve this, the following assumptions are made:

- The device heating is characterized by a single, 'lumped', value of ΔT. Needless to say, this simplifying assumption violates physical reality. However, ΔT can be interpreted as a suitably weighted average over the device [11].
- ΔT is given by $\Delta T = R_{\text{th}} \cdot P_{\text{diss}}$, where P_{diss} is the dissipated power and R_{th}, the thermal resistance of the device, is assumed to be independent of device and ambient temperature.
- For the power dissipation, the (usually very small) contributions of gate/base and substrate currents are neglected.
- In the absence of self-heating, the device conductances are assumed to be independent of frequency. In particular, this means that trapping effects must be negligible [12].

Now, without further mathematical approximations, we can extract the thermal resistance using the real part of the output conductance, i.e. $\text{Re}(Y_{\text{DD}})$ for a MOSFET and $\text{Re}(Y_{\text{CC}})$ for a bipolar transistor [3, 6, 13, 14]:

$$R_{\text{th,g}_{\text{out}}} = \frac{g_{\text{out,DC}} - g_{\text{out,HF}}}{S_{\text{I}} \cdot (I_{\text{out}} + V \cdot g_{\text{out,HF}})} \, . \qquad (1)$$

Here, $g_{\text{out,DC}}$ and $g_{\text{out,HF}}$ are the values of g_{out}, extracted at 'low' and 'high' frequencies, respectively, where 'low' and 'high' mean far below and far above the thermal cut-off frequency, respectively. Moreover, I_{out} and V are the DC drain current and drain-source voltage (MOSFET case), or the DC collector current and collector-emitter voltage (bipolar case). Finally, S_{I} denotes the derivative of I_{out} w.r.t. ambient temperature T_{amb}. Note that in some papers [15, 16], the quantity $g_{\text{out,HF}}$ in the denominator of Eq. (1) has been replaced erroneously by $g_{\text{out,DC}}$.

Alternatively, one can use the frequency dependence of the transconductance, i.e. $\text{Re}(Y_{\text{DG}})$ for a MOSFET and $\text{Re}(Y_{\text{CB}})$ for a bipolar transistor, to extract the thermal resistance [6, 17]:

$$R_{\text{th,g}_{\text{m}}} = \frac{g_{\text{m,DC}} - g_{\text{m,HF}}}{S_{\text{I}} \cdot V \cdot g_{\text{m,HF}}} \, , \qquad (2)$$

where, $g_{\text{m,DC}}$ and $g_{\text{m,HF}}$ are the values of g_{m}, extracted at 'low' and 'high' frequencies, respectively, and S_{I} has the same meaning as before.

As a side remark, we mention that not only g_{out} and g_{m}, but also low-frequency capacitances are strongly affected by the self-heating effect [3, 4, 6]. Actually, this effect can also be used to extract thermal resistances/capacitances [6]. That is, however, beyond the scope of the present paper.

Like the AC conductance method, most compact models also describe the self-heating effect using a single, 'lumped', value of ΔT. As a consequence, the AC conductance method is ideal for compact model parameter extraction purposes. On top of that, we can use compact model simulations (in particular MEXTRAM [18] and PSP [19]) to verify that Eqs. (1) and (2) are correct. This is done by inserting a known parameter value for R_{th}, setting its temperature dependencies to zero, and carrying out the necessary simulations (Y-parameter vs frequency, and DC currents vs ambient temperature). Next, we evaluate Eqs. (1) and (2) and obtain *exactly* the same values that were entered into the model. This confirms that the AC conductance method has *no mathematical approximations* on top of the assumptions listed.

Experiments

We present examples of self-heating characterization using the AC conductance technique for an n-channel FinFET, an n-channel FDSOI transistor, and an NPN bipolar transistor from a BiCMOS technology [20].

The measured DC output characteristics of these transistors are shown in Fig. 1. Note that for the MOS devices (FinFET and FDSOI), the shape of the output characteristics looks very regular; no direct evidence for self-heating is seen. As we will see, this does not mean that the amount of self-heating is low; instead, the effect of the self-heating on the output characteristics is rather modest. For the NPN transistor (right frame in Fig. 1), the situation is rather different: we observe a distinct curling-up of the I_{C}-V_{CE} curve, which is a direct consequence of the self-heating.

Low-frequency RF measurements were carried out in the frequency ranges 9 kHz - 8 GHz and 100 MHz - 50 GHz

979-8-3315-0417-5/25 $31.00 © 2025 IEEE

range, with the help of an Agilent E5071C ENA and a Keysight N5227A PNA network analyzer, respectively. The measurements were de-embedded down to the device level using open, short, and load dummies. The results, for FinFET, FDSOI, and NPN, are shown in Fig. 2, and have been measured at the DC bias conditions marked with a cross in Fig. 1. In all three cases we plot the output conductance g_{out} and the transconductance g_{m} as a function of frequency.

The results for FinFET and FDSOI (left and middle frames in Fig. 2) look very similar: we observe a low-frequency plateau, followed by an *increase* in both g_{out} and g_{m} in the range from 10 MHz to 1 GHz, followed by a high-frequency plateau around 10 GHz. This behavior is typical for self-heating. At even higher frequencies, g_{out} shows a sharp increase and g_{m} shows a sharp decrease, due to effects of the gate resistance. The result for the NPN transistor (right frames in Fig. 2) is different from the MOS results in two ways. First, both g_{out} and g_{m} show a *decrease* with frequency. This is due to the fact that the temperature dependence of the collector current (S_I) is positive, as opposed to the negative temperature sensitivity for the MOS devices (at the DC bias condition measured). Second, the transition from low- to high frequency plateaus takes place at much lower frequency than for the MOS devices measured. This is attributed to the larger device volume, and thus larger thermal capacitance, of the NPN device, compared to the MOS devices measured.

Next, we extract the thermal resistance of the three devices with the help of Eqs. (1) and (2). The low- and high frequency points used for this extraction, are indicated as the orange markers in Fig. 2. The value of S_I is determined by measuring the DC current for different chuck temperatures and taking the derivative $\mathrm{d}I/\mathrm{d}T_{\mathrm{amb}}$.

Theoretically, the R_{th} values extracted from Eqs. (1) and (2) should be the same. In practice, however, we observe differences up to approximately 20%; see the FinFET example in Fig. 3. These differences are due to (i) measurement uncertainties, and *(ii)* parasitic-resistance effects, that start to affect the measurements at the high-frequency thermal plateau (see Fig. 2). From the two measured values $R_{\mathrm{th,g_{out}}}$ and $R_{\mathrm{th,g_m}}$ we determine the average

$$R_{\mathrm{th}} = \frac{R_{\mathrm{th,g_{out}}} + R_{\mathrm{th,g_m}}}{2} \qquad (3)$$

and an estimated uncertainty

$$\sigma_{R_{\mathrm{th}}} = \frac{|R_{\mathrm{th,g_{out}}} - R_{\mathrm{th,g_m}}|}{2} + 0.05 \cdot R_{\mathrm{th}} , \qquad (4)$$

where the last term represents the uncertainty in the determination of S_I.

Using the above procedure, we determine the R_{th} and ΔT from the measurements of Fig. 2; see Table I. Interestingly, for both the MOSFET devices, the extracted values for ΔT are quite substantial, although the effect of self-heating is not directly evident from their DC IV curves (Fig. 1). As explained above, this is due to the low values of S_I. For the NPN measurement, we also see significant heating of almost 100 °C. In this case, however, the self-heating is also very evident in the DC IV curve, because the NPN collector current is much more temperature dependent than the MOSFET drain current. Finally note that the extractions in Table I are not meant for, and can not be used to, compare the thermal behavior of the different technologies; neither

device areas, nor dissipated powers are equal for the three experiments considered.

Table I: Extracted values of R_{th} and ΔT.

device	bias	$R_{\mathrm{th}}(10^3$ K/W)	$\Delta T(°C)$
FinFET	$V_{\mathrm{GS}} = V_{\mathrm{DS}} = 1$ V	18.3 ± 1.5	146 ± 11
FDSOI	$V_{\mathrm{GS}} = V_{\mathrm{DS}} = 0.8$ V	11.1 ± 1.2	87 ± 9
NPN	$V_{\mathrm{BE}} = 0.85$ V; $V_{\mathrm{CE}} = 1.5$ V	2.3 ± 2.6	93 ± 10

More results are shown in Fig. 4, where the extracted FinFET R_{th} is plotted as a function of the drain-source voltage V_{DS}. As expected, no significant V_{DS}-dependence is observed.

In Fig. 5, the extracted NPN R_{th} is plotted vs the collector-emitter voltage V_{CE}, for different values of V_{BE}. For most data points, we observe a bias-independent R_{th}, similar to the FinFET case (Fig. 4). For the two bias points with the highest P_{diss}, however, the extracted R_{th} is starting to deviate. This is the subject of further investigations.

Comparison with pulsed measurements

Of the three examples in Fig. 2, the NPN bipolar has the slowest thermal response by far. Therefore, for this device, we have also attempted to extract R_{th} using pulsed measurements. This was done using a Keithley 4200 SCS parameter analyzer with 4225-PMU pulsed measurement units. In Fig. 6, the measured I_{C}-V_{CE} curves are shown for different pulse widths. For long pulses, the pulsed data converge to the DC measurement, as expected. For the three shortest pulses (500, 300, and 200 ns), we clearly see the curling-up of the I_{C}-V_{CE} curve to become less, due to less self-heating. Nevertheless, we do not observe any convergence, which means that, even at 200 ns pulse width, the device is not in the isothermal regime yet. Unfortunately, shorter pulses did not yield useful results, because they were dominated by charging currents and unwanted reflections.

Neglecting this observation, and — for the moment — pretending that the 200 ns pulse width *is* short enough, we can proceed and extract the thermal resistance by determining crossing points of the 200-ns data with the DC data. The results, for a range of temperatures, are depicted in Fig. 7. The extracted values for R_{th} are approximately a factor of 2 lower than the (true) ones extracted using the AC conductance method (see Table I). This once more confirms that 200 ns pulses are not short enough for this purpose. Needless to say, this is even more true for the MOS devices in Fig. 2, which exhibit a much faster thermal response than the NPN. For FinFETs, this was demonstrated previously in Ref. [6].

Conclusion

In conclusion, we have reviewed the AC conductance method and its underlying assumptions. We have shown examples of its application to FinFET, FDSOI, and NPN bipolar transistors. Furthermore, we have demonstrated that the dynamic thermal behavior of modern devices is typically too fast to assess self-heating with standard pulsed measurement solutions.

Acknowledgment The authors gratefully acknowledge Oswald Moonen and Oliver Dieball (NXP) for their help in the measurements. Also, the authors are grateful to Lisa Tondelli and Luca Selmi (Università di Modena e Reggio Emilia, Italy) for stimulating discussions.

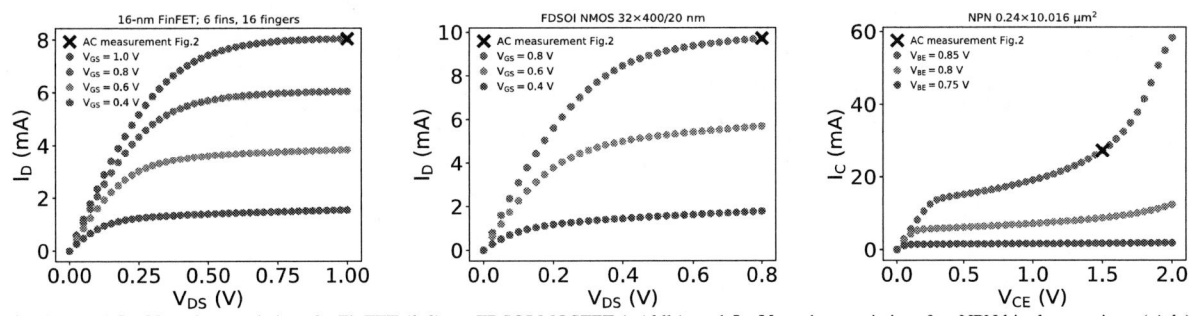

Fig. 1: Measured I_D-V_{DS} characteristics of a FinFET (*left*), an FDSOI MOSFET (*middle*), and I_C-V_{CE} characteristics of an NPN bipolar transistor (*right*), all at $T_{amb} = 25$ °C. Black crosses indicate DC bias points used in the AC measurements of Fig. 2.

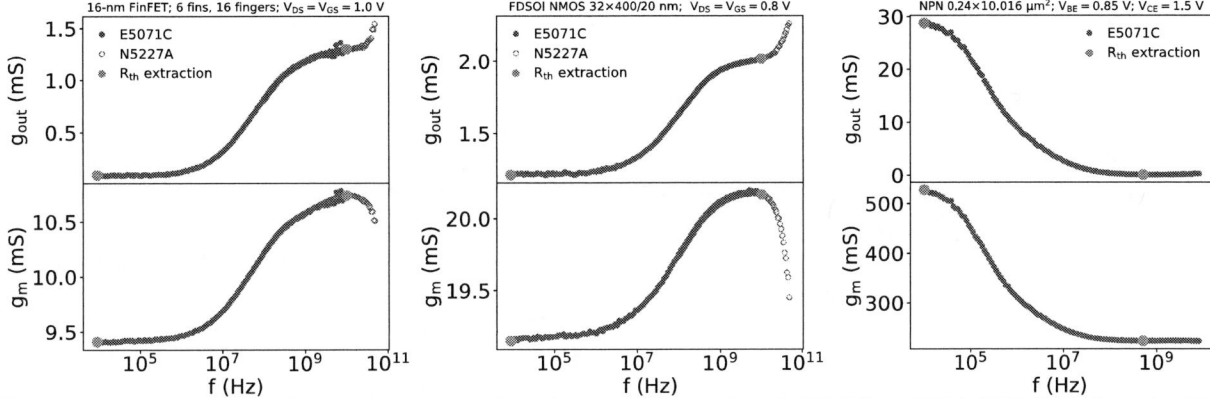

Fig. 2: Output conductance (*top*) and transconductance (*bottom*) as a function of frequency, for a FinFET (*left*), an FDSOI MOSFET (*middle*), and an NPN bipolar transistor (*right*), all at $T_{amb} = 25$ °C. Filled blue markers represent measurements with the Agilent E5071C ENA network analyzer (9 kHz - 8 GHz), and open blue markers represent measurements with the Keysight N5227A PNA network analyzer (100 MHz - 50 GHz). The orange markers are the low- and high-frequency data points chosen for R_{th} extraction.

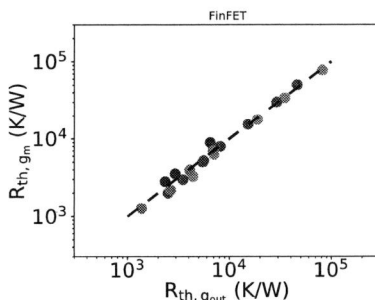

Fig. 3: FinFET thermal resistances ($T_{amb} = 25$ °C) extracted from g_m, plotted against the corresponding values extracted from g_{out}, for various NMOS and PMOS device flavors (blue shades and red shades, respectively). The dashed line represents the expected $R_{th,g_m} = R_{th,g_{out}}$ behavior.

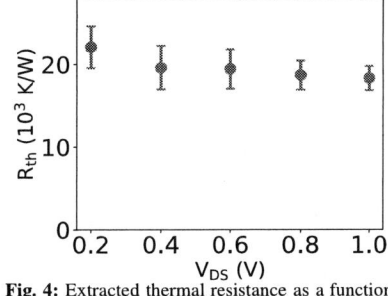

Fig. 4: Extracted thermal resistance as a function of V_{DS}, for an $L = 16$ nm FinFET biased at $V_{GS} = 1$ V at $T_{amb} = 25$ °C.

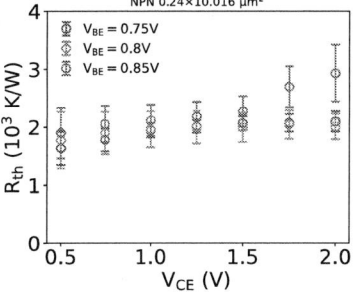

Fig. 5: Extracted thermal resistance as a function of V_{CE}, for three different values of V_{BE}, as indicated by different colors. The device is a 0.24×10.016 μm^2 NPN from a BiCMOS process [20].

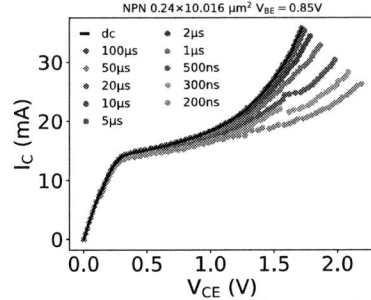

Fig. 6: DC measurements (solid line) and pulsed measurements (markers) of collector current as a function of V_{CE}, at $V_{BE} = 0.85$ V and $T_{amb} = 25$ °C. Different pulse widths are represented by different colors, as indicated in the figure. The device is the same as in Fig. 5.

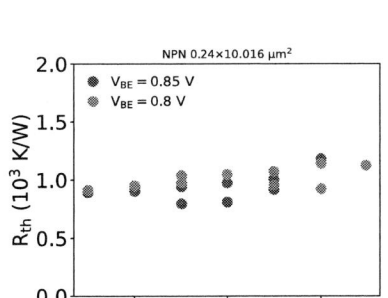

Fig. 7: R_{th} as a function of ambient temperature, extracted from pulsed measurements (see Fig. 6). Two V_{BE} values were used, as indicated in the figure. Note that the extracted values are roughly 2 times smaller than the ones in Fig. 5, due to insufficiently short pulse width.

References

[1] C. Anghel *et al.*, EDL **25** (2004).
[2] W. Redman-White *et al.*, El. Letters **29** (1993).
[3] N. Rinaldi, T-ED **48** (2001).
[4] S.F. Shams *et al.*, BCTM 2002.
[5] S. Kolluri *et al.*, IEDM 2007.
[6] A.J. Scholten *et al.*, IEDM 2009.
[7] S. Makovejev *et al.*, EuroSOI-ULIS 2015.
[8] K. Triantopoulos *et al.*, T-ED **66** (2019).
[9] R. Wang *et al.*, EDL **30** (2009).
[10] A. Caviglia *et al.*, IEEE Int. SOI Conf. 1992.
[11] L. Tondelli *et al.*, T-ED **71** (2024).
[12] L. Tondelli *et al.*, ESSERC 2024.
[13] S. Makovejev *et al.*, T-ED **60** (2013).
[14] B. Gonzalez *et al.*, T-ED **69** (2022).
[15] U.S. Kumar *et al.*, T-ED **64** (2017).
[16] C. Mukherjee *et al.*, T-ED **70** (2023).
[17] Z. Chen *et al.*, T-ED **65** (2018).
[18] www.eng.auburn.edu/~niuguof/mextram
[19] www.cea.fr/cea-tech/leti/pspsupport
[20] P.H.C. Magnée *et al.*, BCICTS 2024.

979-8-3315-0417-5/25 $31.00 © 2025 IEEE

Reliability in Heterogeneous Integration: A Theoretical View (Invited)

Zhiping Xu

Department of Engineering Mechanics, Tsinghua University, Beijing 100084, China

ABSTRACT

Heterogeneous integration is challenged by its multi-scale, and multi-physics complexities that extend beyond traditional integrated circuit design. Tackling these issues demands efficient modeling techniques and design workflows that are capable of bridging field variables across different scales, from individual components and their interfaces to packaging. This paper presents theoretical insights on the reliability issues in heterogeneous integration, which can be addressed by a framework combining modeling and experiments.

Keywords Multiscale Modeling, Multiphysics Coupling, Heterogeneous Integration, Advanced Packaging, Reliability, Interfaces

INTRODUCTION

Moore's Law and Dennard scaling face limitations due to leakage currents and thermal issues. Heterogeneous integration (HI) offers a flexible framework to rethink the spatial arrangement and interconnection of diverse components, extending the benefits of post-scaling while mitigating efficiency and performance barriers [1]. HI enables efficient scaling and IP reuse through modular design and split capabilities, which allow companies to build adaptable, high-performance systems across product lines, support rapid design iterations, and optimize manufacturing, fostering both innovation and economic sustainability.

Implementing HI faces challenges of their structural and material heterogeneity, and multi-physics interactions. Integrating components of different sizes, shapes, and materials within a single package complicates electromagnetic analysis and imposes thermal stress, increasing risks of warpage and failure. The close proximity of components in HI exacerbates thermal hotspots and stress-induced deformation, while varied material interfaces impede thermal dissipation and create weak points, often leading to mechanical failure through delamination or cracking.

Multi-physics modeling has been proposed to address these challenges spanning across multiple spatiotemporal scales. Effective methodologies like submodeling and reduced-order models capture critical interactions with minimal computational demand, enabling efficient design optimization for reliability and performance. However, difficulties persist in accurately mapping field variables from the component to package level and in managing complex microstructures and interfaces. Scalable, adaptable models are essential, though industry adoption remains slow due to limited experimental validation and the need for practical workflow refinement. Different HI architectures require flexible approaches, and while system technology co-optimization (STCO) offers a promising framework, full integration of circuit design with multi-physics modeling is still evolving.

MULTISCALE FEATURES IN HETEROGENEOUS INTEGRATION

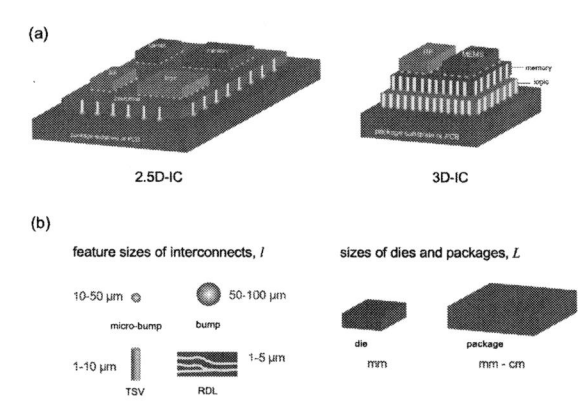

Fig. 1. **Heterogeneous Integration (HI).** (a) Representative 2.5D and 3D-ICs, showing the heterostructural and heterogeneous nature of integrated components. (b) Sizes of key components, highlighting dimensional mismatch that challenges the multiscale problem solvers.

HI integrates dies in representative 2.5D and 3D-IC setups, with their characteristic structural components and material interfaces shown in Fig. 1. The sizes of dies and the package usually range from milimeters to centimeters (L), while feature scales in interconnects such as the diameters of bumps and vias as well as their pitches and depths are usually micrometers (l). The

979-8-3315-0417-5/25 $31.00 © 2025 IEEE

size mismatch (L/l) is significant. To estimate warpage and stress, thermal distribution in a typical HI setup, the degrees of freedom in a direct, full-scale simulation could reach 10^9, challenging the state-of-the-art limits of computational power. Fortunately, observation of the displacement, temperature and stress distributions of typical multi-die and multi-layer designs suggests that one could separate the scales into a coarse one (e.g., for displacements) and a fine one (e.g., for electromagnetic, temperature, and stress fields near interconnects and interfaces) (Fig. 2). A high-fidelity mapping between these two scales offers a practical solution for analysis.

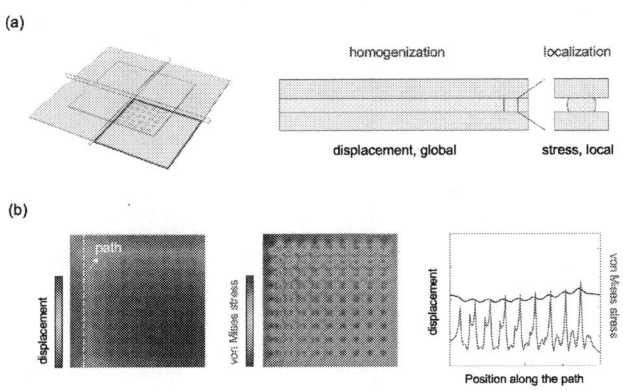

Fig. 2. **Multiscale and Multiphysics Modeling.** (a) The homogenization-localization mapping from slowly-varying displacement fields to rapidly-varying stress fields. (b) Displacement and stress distributions a bilayer with explicit bumps.

Stress transmission and thermal dissipation could be locally perturbed by the interfaces, which carries essential material information beyond geometrical boundaries of simulation domains. Stress coupling at the interface is usually transmitted by the cohesion and shear responses of the interface, or the adhesive layers, where the normal (σ) and shear (τ) stresses across the interface can be described through a trapezoidal force-displacement relation in the cohesive zone model (CZM) [2]. Meanwhile, the thermal resistance of an interface introduces thermal barriers. For weak interfaces, the thermal conductance κ_c scales almost linearly with the interfacial energy density γ as $\kappa_c \approx 100\gamma$ MW/JK, rather than depending on the thermal conductivity of the materials themselves [3]. Consequently, the effective thickness of the thermal interface, or namely the Kapitza length of a nanometer-thick interface ($l_k = \kappa/\kappa_c$) could reach one micrometer.

In HI, structural components with distinct geometries and material properties must conform to compatibility requirements, placing constraints on structural deformation and stress distribution. Multi-die design often uses dummy dies to mitigate warpage, a common issue arising from mismatched thermal expansion coefficients and structural asymmetry. In stacked multilayers, the out-of-plane displacement of dies are further constrained. However, while effective in reducing out-of-plane deformation, these constraints can inadvertently increase in-plane stress, especially at material interfaces where differences in stiffness and thermal expansion are most pronounced. Elevated stress at these critical regions may activate mechanisms of failure, including cracking and delamination, which undermine yield and elevate reliability concerns. This underscores the need for optimized design and material selection to balance warpage control with stress management, minimizing the risk of interfacial degradation and enhancing long-term performance.

COMPLEXITIES FROM MULTIPHYSICS COUPLING

From the perspective of STCO, a holistic approach to the multiphysics fields is essential, encompassing electromagnetic, thermal, and stress fields. The primary challenge lies in accurately capturing the coupling between these fields (Fig. 3), which spans over multiple timescales and encompasses long-term effects such as electromigration, creep, and fatigue . Properly modeling these interactions requires high-fidelity constitutive models and accurate material parameters to account for time-dependent and non-linear behaviors. Yet, the presence of size-dependent effects, such as surface and interface effects, single-crystalline anisotropy and microstructural complexities, intensifies the challenge. Imperfections introduced in the fabrication and packaging processes should also be aware.

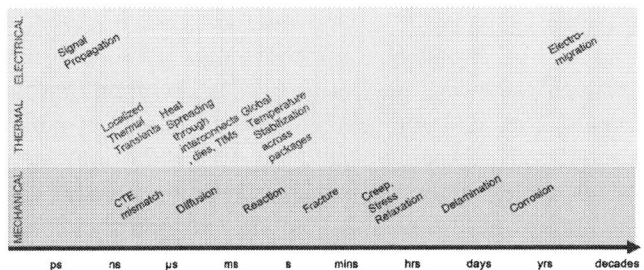

Fig. 3. **Time scales of Multiphysics Processes.**

It is important to note that warpage and stress concentration in the packaging process conditions or reliability tests (e.g., thermal cycling) may not be the same as those under power-on conditions. The thermal and mechanical issues of reliability induced by Joule heating are especially pronounced in dense, high-power designs, where local heating exacerbates stresses at critical interfaces, potentially accelerating failure mechanisms like electromigration and stress-induced failure. Reliable modeling for operational conditions thus requires dynamically coupling thermal, electrical, and mechanical fields to account for these power-driven shifts in thermal and stress profiles, underscoring the need for enhanced predictive models that incorporate in situ power-on effects.

IMPLEMENTATION OF THE WORKFLOW

Incorporating multiphysics models into electronic design automation (EDA) in a cross-stage manner aligns with the principles of hardware integration. This modular approach allows the use of simulation tools for individual physical processes. To facilitate HI modeling, there is a growing emphasis on adopting standardized data exchange protocols to enhance interoperability in multivendor environments [4]. Open-source initiatives such as OpenROAD [5] emerge as collaborative platforms that aim to streamline these processes across the design ecosystem, allowing for reproducibility and accessibility in the workflow.

The HI community also recognizes the importance of digital libraries for the geometrical models of structural components and physical properties of materials and their interfaces. Comprehensive databases enable reusability, facilitate rapid design iterations, and can incorporate data from both simulation and experimental results, enhancing design accuracy and efficiency. However, material and interface properties are often inadequately characterized, and processing conditions are typically only coarsely defined, necessitating rigorous uncertainty quantification to validate models that involve multiple factors with varying levels of complexity. More importantly, strong physical constraints within the theoretical framework pose challenges to incorporating the uncertainties and compensating for deviations in model assumptions. Experimental studies from the individual components and their interfaces to the dies and packages are important, which can help to constrain the uncertainty propagation across the scales.

Integrating empirical and experimental knowledges into mathematical models will thus significantly improve our ability to connect microscale material and interface behaviors to system-level performance predictions, thereby addressing critical reliability challenges in HI.

The constructed databases and libraries support AI-driven modeling, allowing model-based signoff instead of rule-based design, and even digital twins to incorporate diverse material behaviors and geometric variations as seen in actual designs and fabrication processes. The shift to model- and system-driven design is motivated by the need for enhanced yield, reliability, and performance across the lifecycle of products, to address the 'heterogeneity curse'. Collaborative efforts between academia and industry are crucial to make advances, in *an open models, private data* ecosystem.

Finally, the integrated platform embeds physical information into functional blueprints, with their geometries and material properties encoded in the process design kit (PDK) linking design to modeling. This prompts a vital question: how can the exploitation-exploration approach redefine chip design in the physical space of materials and the mathematical space of arrangement and connectivity?

ACKNOWLEDGMENT

Z.X. thanks help from Yingjie Zhao and Zhaoheng Zhang, and the financial support from National Natural Science Foundation of China (12425201, 52090032).

REFERENCES

[1] M. Ahmad, J. DeLaCruz, and A. Ramamurthy, "Heterogeneous integration of chiplets: Cost and yield tradeoff analysis," in *2022 23rd International Conference on Thermal, Mechanical and Multi-Physics Simulation and Experiments in Microelectronics and Microsystems (EuroSimE)*. IEEE, 2022, pp. 1–9.

[2] S. Wang, M. Zhang, J. Feng, and Z. Xu, "Mechanical performance of the polymer-inorganic interfaces: Molecular mechanisms and modes of failure," *Mechanics of Materials*, vol. 175, p. 104479, 2022.

[3] Y. Wang, Z. Qin, M. J. Buehler, and Z. Xu, "Intercalated water layers promote thermal dissipation at bio-nano interfaces," *Nature Communications*, vol. 7, no. 1, p. 12854, 2016.

[4] A. Mastroianni, B. Kerr, J. Nasrullah, K. Cameron, H. J. Wong, D. Ratchkov, and J. Reynick, "Proposed standardization of heterogenous integrated chiplet models," in *2021 IEEE International 3D Systems Integration Conference (3DIC)*. IEEE, 2021, pp. 1–8.

[5] V. A. Chhabria, W. Jiang, A. B. Kahng, R. Liang, H. Ren, S. S. Sapatnekar, and B.-Y. Wu, "OpenROAD and CircuitOps: Infrastructure for ML EDA research and education," in *2024 IEEE 42nd VLSI Test Symposium (VTS)*. IEEE, 2024, pp. 1–4.

979-8-3315-0417-5/25 $31.00 © 2025 IEEE

Fundamentals and Future Challenges of SiC Power Devices

T. Kimoto[1], R. Ishikawa[1], K. Tachiki[1], X. Chi[1], K. Mikami[1], and M. Kaneko[1]

[1]Graduate School of Engineering, Kyoto University, Japan

Abstract

Silicon carbide (SiC) exhibits unique anisotropy in carrier transport such as higher electron mobility (> 1,200 cm²/Vs) and higher critical electric field strength (2.5–5.0 MV/cm) along <0001> than those perpendicular to <0001>. Although SiC power MOSFETs still suffer from low channel mobility, recent innovations in both gate-oxide formation and device designs enable significant performance improvement. An oxidation-minimizing process is promising for mobility enhancement (> 100 cm²/Vs) of SiC MOSFETs.

Keywords: SiC, power device, MOSFET

Introduction

Commercialization of large-diameter (150–200 mm) SiC (4H polytype) wafers and progress in device technology have accelerated market penetration of 600–3,300 V SiC power devices (power MOSFETs and Schottky barrier diodes (SBDs)) [1]. The major application fields include EVs, fast chargers, photovoltaic power conditioners, robotics, and railcars. Despite rapid growth of SiC market, however, fundamental problems of SiC MOSFETs, such as a high density of interface defects, have remained unsolved for more than 20 years, leaving the performance of SiC power MOSFETs far from the ideal potential [2,3]. In addition, many of important physical properties of SiC are still unknown, making accurate device simulation difficult. In this paper, fundamental studies on the intrinsic material properties and MOS interface of SiC are described. Impacts of the obtained results on SiC power devices are discussed together with future challenges.

Anisotropic carrier-transport properties of SiC

A. High critical electric field strength along <0001>

Breakdown voltage of power devices is dominated by carrier multiplication under high electric field (avalanche breakdown), and this phenomenon is determined by impact ionization coefficients. The authors' group extracted impact ionization coefficients of electron (α_n) and hole (α_p) in SiC including their anisotropy and temperature dependence [4,5]. Fig. 1 shows the electron and hole impact ionization coefficients along <0001> (c-axis) and <11$\bar{2}$0> (a-axis) plotted against the inverse of the electric field strength. Note that the electron impact ionization coefficient along <0001> is smaller than the others by one or two orders of magnitude. In SiC, several minigaps ranging from 0.5 to 1.4 eV are present inside the conduction band only along the M–L direction in the first Brillouin zone, which corresponds to electron transport along <0001>. These minigaps work as a severe obstacle for electron acceleration in SiC only along <0001>, which is the main reason why α_n along this direction

is unusually small [6]. As a result, the critical (breakdown) electric field strength is significantly higher along <0001> (2.5 MV/cm at 1×10^{16} cm^{-3} and 5.0 MV/cm at 1×10^{18} cm^{-3}), which is beneficial for fabrication of vertical power devices on SiC (0001) wafers.

B. High electron mobility along <0001>

Although numerous studies on bulk mobility in SiC have been reported, they exclusively investigated "in-plane" mobility on SiC (0001) samples [7,8]. For vertical power devices on SiC (0001), however, the on-resistance, another important figure of power devices, is determined by the electron mobility along <0001> (not "in-plane" mobility). The authors fabricated Hall bar structures on custom-made SiC (11$\bar{2}$0) epitaxial substrates and conducted Hall effect measurements in a wide temperature range (140–600 K), the result of which is plotted in Fig. 2 [9]. It is notable that the electron mobility along <0001> is consistently higher than that perpendicular to <0001>, and a mobility of 1,210 cm²/Vs acquired from a lightly-doped epilayer at room temperature is the highest ever reported for SiC.

Fig. 3 shows the trade-off relationship between the specific on-resistance and breakdown voltage for SiC unipolar devices ("SiC unipolar limit") calculated by using the updated critical electric field and mobility along <0001>. In this figure, the trade-off for Si unipolar devices and data of actual SiC devices reported in literature are also plotted. The updated SiC unipolar limit demonstrates a great promise [10], being almost identical to the GaN unipolar limit calculated by using the latest material properties [11].

Evolution and challenges of SiC MOSFETs

A. Present status of SiC power MOSFETs

Fig. 4 shows the major territories of Si, SiC, and GaN power transistors plotted against the blocking voltage. Regarding 600–1,200 V devices, which are used for volume markets, Si super-junction MOSFETs, Si IGBTs, SiC power MOSFETs, and GaN HEMTs (on Si) will compete. The main features of SiC power MOSFETs include low loss and fast switching as compared with Si devices and superior high-power (high-current) capability as compared with GaN devices.

1 kV-class SiC power MOSFETs exhibit much higher on-resistance than the SiC unipolar limit, as shown in Fig. 3, due to the poor channel mobility (typically 10–20 cm²/Vs in actual devices). In other words, the on-resistance of SiC power MOSFETs is limited by the MOS-channel resistance. Therefore, mobility improvement in SiC MOSFETs is of importance to reduce the on-resistance and cost. Mobility improvement will also enable more flexibility in device designs, leading to improved reliability and switching

performance [2].

B. Defects near SiC MOS interface

Despite extensive studies on SiC MOS interface defects, the defect origins are still unidentified. In the meantime, the authors' group discovered that carbon vacancy defects in SiC can be mostly eliminated (density: from 1×10^{13} cm^{-3} to below 3×10^{10} cm^{-3}) by thermal oxidation. This carbon vacancy elimination is interpreted by a phenomenon that excess carbon atoms are emitted at the oxidation front and the emitted carbon atoms diffuse into the bulk region, leading to filling of carbon vacancy defects [12]. The authors also found that the carbon atom density in a SiO$_2$ layer formed by oxidation of SiC remarkably increased from $< 10^{18}$ cm^{-3} to 10^{20} cm^{-3} simply by subsequent annealing in pure Ar at 1300°C [13], indicating that a high density of carbon atoms exist near the SiO$_2$/SiC interface after oxidation. A recent first-principles calculation study revealed that several carbon-related defects may be easily generated not only at the interface but also inside the SiC bulk region [14].

Fig. 5 shows the DLTS (deep level transient spectroscopy) spectra acquired from Ni Schottky structures formed on n-type SiC epilayers with and without oxidation at 1300°C followed by HF etching [15]. After oxidation, large and continuous DLTS signals are observed in the low-temperature side from the oxidized samples, indicating that multiple shallow defect levels near the conduction band edge are created by oxidation of SiC. Since these DLTS spectra were acquired from SiC Schottky structures without SiO$_2$, the defects responsible for the observed DLTS signals are located near the surface of SiC (inside SiC, not at the SiO$_2$/SiC interface). The total density of the observed defects is roughly estimated to be in a mid 10^{12} cm^{-2} range, which is the same order as the sheet electron density in an inversion layer.

C. Mobility enhancement of SiC MOSFETs

As described in the last section, thermal oxidation of SiC generates a high density of carbon-related defects in the sub-surface region of SiC. Base on these results, an oxidation-minimizing process for gate-oxide formation was proposed, as shown in Fig. 6 [16]. This process consists of three steps, namely 1) high-temperature H$_2$ treatment under slightly Si-rich condition, 2) oxide deposition, and 3) annealing in NO. In this process, the quality of a deposited oxide is improved by the subsequent annealing in NO at 1200–1300°C, and a high oxide breakdown field over 11 MV/cm is achieved. The interface trap density (D_{it}) extracted by a high-low method is as low as 6×10^{10} cm^{-2}eV^{-1} on SiC (0001) and approximately 3×10^{10} cm^{-2}eV^{-1} on SiC (11$\bar{2}$0) and (1$\bar{1}$00) [16,17].

Fig. 7 plots the channel mobility of SiC n-channel MOSFETs versus the acceptor density of the p-body, where MOSFETs were fabricated with the proposed process on (0001), (11$\bar{2}$0), and (1$\bar{1}$00) [17]. The mobility data for (0001) MOSFETs fabricated with the conventional process (oxidation followed by NO annealing [18]) are also shown. It is notable that high mobilities over 130 cm^2/Vs were obtained for (11$\bar{2}$0) and (1$\bar{1}$00) MOSFETs with a heavily-doped p-body (acceptor density: 10^{17} cm^{-3}). More recently, two-fold improvement of channel mobility on the trench sidewalls has been reported by using an oxidation-minimizing process [19].

Another approach for mobility enhancement is adoption of thin fin channels (FinFETs) [20]. In the fin channel with a sufficiently thin width, the effective normal field near the MOS interface is remarkably reduced owing to the gate bias on both sides of the fin sidewalls. As a result, the wavefunction of electrons is expanded toward the bulk region from the MOS interface, leading to reduced Coulomb scattering by the interface charges (fixed charge and trapped electrons). By using a 55-nm-thick fin channel, a high mobility over 250 cm^2/Vs has been achieved. Combination of the fin channel and super-junction structure may be the ideal structure of SiC power MOSFETs in a future.

Conclusion

Anisotropic carrier-transport properties of SiC make this material extremely suitable for vertical high-voltage power devices. Defect generation inside SiC via thermal oxidation and a few methods to solve this problem were presented.

Acknowledgments

This work was supported by KAKENHI Grant (# 21H05003) from the Japan Society for the Promotion of Science and the ALCA-Next Program (# 24-241042748) from the Japan Science and Technology Agency.

References

[1] T. Kimoto, "High-voltage SiC power devices for improved energy efficiency," *Proc. Japan Academy, Ser. B*, vol. 98, pp. 161-189, 2022.

[2] T. Kimoto and H. Watanabe, "Defect engineering in SiC technology for high-voltage power devices," *Appl. Phys. Express*, vol. 13, 120101, 2020.

[3] G. Liu, B.R. Tuttle, and S. Dhar, "Silicon carbide: A unique platform for metal-oxide-semiconductor physics," *Appl. Phys. Rev.*, vol. 2, 021307, 2015.

[4] T. Kimoto et al., "Carrier lifetime and breakdown phenomena in SiC power device material," *J. Phys. D: Appl. Phys.*, vol. 51, 363001, 2018.

[5] D. Stefanakis et al., "Experimental determination of impact ionization coefficients along <11$\bar{2}$0> in 4H-SiC," *IEEE Trans. on Electron Devices*, vol. 67, pp. 3740-3744, 2020.

[6] H. Tanaka, T. Kimoto, and N. Mori, "Theoretical study on high-field carrier transport and impact ionization coefficients in 4H-SiC," *Mater. Sci. in Semicond. Process.*, vol. 173, 108126, 2024.

[7] J. Pernot et al., "Electrical transport in n-type 4H silicon carbide," *J. Appl. Phys.* vol. 90, pp. 1869–1878, 2001.

[8] H. Tanaka, S. Asada, T. Kimoto, and J. Suda, "Theoretical analysis of Hall factor and hole mobility in p-type 4H-SiC considering anisotropic valence band structure," *J. Appl. Phys.*, vol. 123, 245704, 2018.

[9] R. Ishikawa et al., "Electron mobility along <0001> and <1$\bar{1}$00> directions in 4H-SiC over a wide range of donor concentration and temperature," *Appl. Phys. Express*, vol. 14, 061005, 2021.

[10] T. Kimoto, "Updated trade-off relationship between specific on-resistance and breakdown voltage in 4H-SiC{0001} unipolar devices," *Jpn. J. Appl. Phys.*, vol. 58, 018002, 2019.

[11] T. Maeda et al., "Impact ionization coefficients and critical electric field in GaN," *J. Appl. Phys.*, vol. 129, 185702, 2021.

[12] T. Hiyoshi and T. Kimoto, "Reduction of deep levels and improvement of carrier lifetime in n-type 4H-SiC by thermal oxidation," *Appl. Phys. Express*, vol. 2, 041101, 2009.

[13] T. Kobayashi and T. Kimoto, "Carbon ejection from a SiO$_2$/SiC(0001) interface by annealing in high-purity Ar," *Appl. Phys. Lett.*, vol. 111, 062101, 2017.

[14] T. Kobayashi and Y. Matsushita, "Structure and energetics of carbon defects in SiC(0001)/SiO$_2$ systems at realistic temperatures: Defects in SiC, SiO$_2$, and at their interface," *J. Appl. Phys.*, vol. 126, 145302, 2019.

[15] H. Fujii, M. Kaneko, and T. Kimoto, "Generation of deep levels near the 4H-SiC surface by thermal oxidation," *Appl. Phys. Express*, vol. 17, 041004, 2024.

[16] K. Tachiki, M. Kaneko, and T. Kimoto, "Mobility improvement of 4H-SiC (0001) MOSFETs by a three-step process of H_2 etching, SiO_2 deposition, and interface nitridation," *Appl. Phys. Express*, vol. 14, 031001, 2021.

[17] K. Tachiki et al., "Mobility enhancement in heavily doped 4H-SiC (0001), (11$\bar{2}$0), and (1$\bar{1}$00) MOSFETs via an oxidation-minimizing process," *Appl. Phys. Express*, vol.15, 071001, 2022.

[18] G. Y. Chung et al., "Improved inversion channel mobility for 4H-SiC MOSFETs following high temperature anneals in nitric oxide," *IEEE Electron Device Lett.*, vol. 22, pp. 176-178, 2001.

[19] H. Tomita et al., "4H-SiC vertical trench power MOSFET fabricated by oxidation-minimizing process," *Ext. Abstr. Int. Conf. on Silicon Carbide and Related Materials 2024* (Raleigh, USA), 8B, 2024.

[20] T. Kato et al., "Enhanced performance of 50 nm ultra-narrow-body silicon carbide MOSFETs based on FinFET effect," *Proc. IEEE 32nd Int. Symp. on Power Semicond. Devices & ICs* (Vienna, Austria), pp. 62-65, 2020.

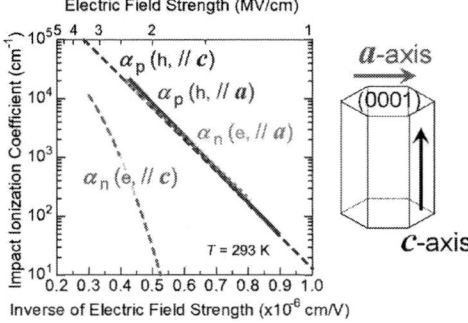

Fig. 1: Impact ionization coefficients of electron (α_n) and hole (α_p) along <0001> (*c*-axis) and <11$\bar{2}$0> (*a*-axis).

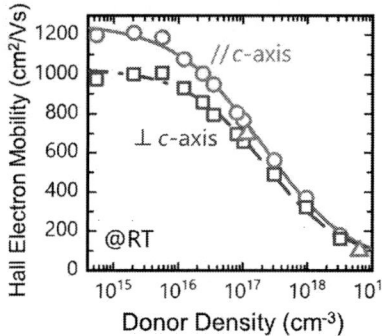

Fig. 2: Electron mobility versus donor density at room temperature. The mobility data along and perpendicular to <0001> are shown [9].

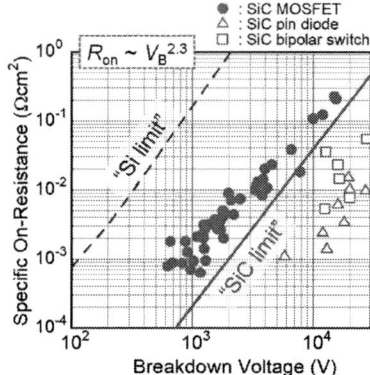

Fig. 3: Trade-off relationship between the specific on-resistance and breakdown voltage for SiC unipolar devices ("SiC unipolar limit") calculated by using the updated physical properties. Actual device data reported in literature are also plotted.

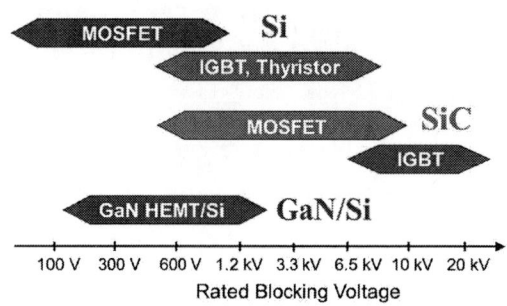

Fig. 4: Major territories of Si, SiC, and GaN power transistors plotted against the blocking voltage of devices.

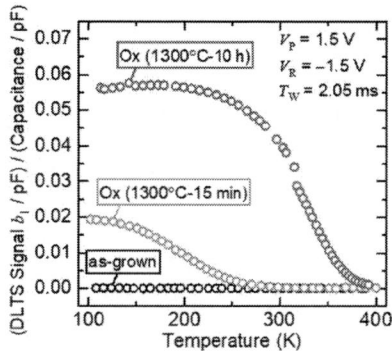

Fig. 5: DLTS spectra acquired from Ni Schottky structures formed on n-type SiC epilayers with and without oxidation at 1300°C followed by HF etching [15].

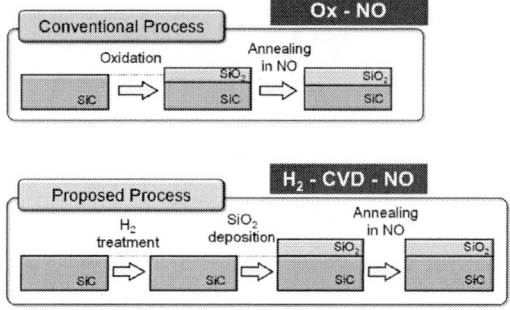

Fig. 6: Conventional and proposed (oxidation-minimizing) processes for gate-oxide formation.

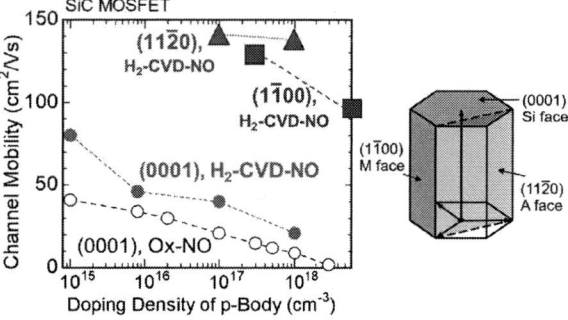

Fig. 7: Channel mobility of SiC n-channel MOSFETs versus acceptor density of the p-body, where MOSFETs were fabricated with the proposed process on (0001), (11$\bar{2}$0), and (1$\bar{1}$00) [17].

How semiconductor industry new challenges will foster Engineered Substrates?

C. Figuet[1]

[1]SOITEC, domaine des fontaines, Bernin, France,

Abstract

In this article we present some key ways in which Engineered Substrates will impact the future of the semiconductor industry as well as our daily lives. It already happens to automotive revolutions with wide bandgap materials and vehicle electrification, as well as FDSOI for edge computing and will happen to the new emerging technology breakthrough for advanced CMOS transistors as well as new materials for mobility. Soitec's proprietary Smart Cut™ technology is a versatile platform for combining all types of materials to improve device performance and meet new challenges in the semiconductor industry.

Keywords: Engineered Substrate, SmartSiC, FDSOI

Introduction

Semiconductors are at the heart of the major technological revolutions of our time. They are usually made of a single material which allows growing them as single crystal but also limits the versatility of different material combination. Semiconductor Engineered substrates are specially designed materials that can be used to create more efficient, powerful, and versatile semiconductor devices and they are becoming increasingly important in the field of electronics and technology. Their development and implementation could significantly shape our future across various industries, from computing (AI) to energy efficiency, mobility like smart phones, smart watches and beyond. The new requirements and advance performance these technologies need, foster new materials or the combination of materials that were not able to be assembled before. Engineered Substrates made using Smart Cut™ technology can combine the best properties of different materials to make an unprecedented high performance device.

Engineered Substrates technology:

Layer transfer & Smart Cut™

Engineered Substrates consist of multiple materials that are stacked together to combine the very best of each material property like silicon-on-insulator (SOI) substrate which has a thin silicon layer on top of a buried oxide (BOX) supported by a bulk silicon base substrate.

Smart Cut™ technology as described in [1] for SOI wafers allows transferring, thanks to hydrogen ion implantation and wafer bonding, a very thin silicon layer onto a bulk silicon wafer with an intermediate oxide layer.

Depending on the final application, the thickness of each layer can be tailored to meet the different device requirements from several nanometers (FDSOI) up to several microns for the top silicon layer (power-SOI). The buried oxide layer can be adjusted as well from few tens of nanometers (FDSOI) up to several microns for photonics applications (photonics SOI).

Engineered substrates for automotive revolutions

The automotive industry is making at the same time two major revolutions; the first one is switching from thermal to electrical propulsion to reduce carbon emission and the second one is adding driving assistance from basic assistance up to full autonomous vehicle. Both revolutions need new Engineered Substrates to support the main challenges these changes will bring to people's life.

Wide band gap materials such as gallium nitride (GaN) and silicon carbide (SiC) are crucial for allowing higher operating temperatures and reducing the switching losses in power electronics for a broad range of power systems, such as Smartphone/tablet chargers, power supplies for servers, etc… and, very important, electric vehicles, thus increasing their range. The key devices in this domain are inverters which are needed in electrical vehicles for different functions. They play a crucial role by converting direct current (DC) electricity from the vehicle's battery into alternative current (AC) needed for the electric motor and also reversely from the power network to the vehicle battery for charging.

979-8-3315-0417-5/25 $31.00 © 2025 IEEE   1216

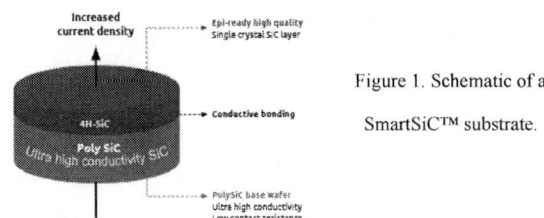

Figure 1. Schematic of a SmartSiC™ substrate.

Smart Cut™ technology brings new performance for inverters by combining the benefit of high conductivity poly-SiC and high crystal quality mono-crystalline Silicon Carbide namely SmartSiC substrates (figure 1). These substrates consist of the conductive bonding of a thin mono-crystalline 4H-SiC layer on top of polycrystalline SiC substrate. This substrate has lower resistivity than bulk mono-SiC improving nominal current per square millimeter leading to better Electrical Vehicle range (figure 2) [2].

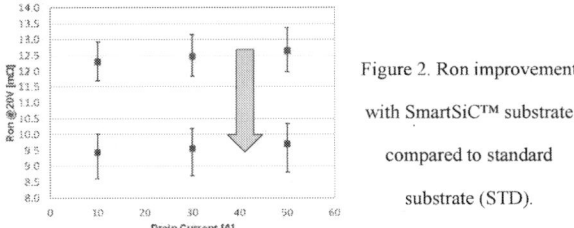

Figure 2. Ron improvement with SmartSiC™ substrate compared to standard substrate (STD).

The autonomous vehicle revolution is accompanied by a significant need for sensors for Advanced Driver Assistance Systems (ADAS) to replace human senses. Besides sensors, information must be processed as close as possible to it, not only for high-performance and energy efficiency but also for reliability and security, as the criticality of automotive applications leaves zero margins for error: this is edge computing or edge AI. It enables real-time decision-making, essential for vehicle safety and performance, by leveraging vision processors, microcontroller units (MCUs), radars, and zonal controller. These two revolutions, electrification and autonomous vehicles, advancing in parallel require both performance and low energy consumption. Soitec's FD-SOI technology is well-positioned to power AI in Advanced Driver Assistance Systems (ADAS) and autonomous driving applications, as demonstrated by its integration into several high-performance automotive platforms; from processors like the Mobileye EyeQ4 and STMicroelectronics' Stellar MCUs to Lattice's Nexus FPGA platform and Bosch Radar. The Mobileye EyeQ4 SoC, for example, leverages FD-SOI

to achieve significant computational efficiency, all while maintaining a low power consumption of approximately 3 watts—an essential factor in automotive applications where managing heat and power is crucial. [3].

Engineered substrates to support advanced CMOS roadmap

Whereas for edge AI power consumption is critical, the performance is the main driver of advanced CMOS roadmap. In this way, Engineered Substrates are becoming a key enabler in the HPC roadmap supporting complementary field effect transistors (CFET) developments. This new transistor integration scheme is becoming a viable evolution of nanosheet CMOS transistors [4] [5] [6]. CFET are made of semiconductor nanosheet based CMOS transistors for which nMOS and pMOS transistors are stacked on top of each other (figure 3a). In between, a thin layer of dielectric material (middle dielectric insulator: MDI) ensure the insulation between n and pMOS transistors.

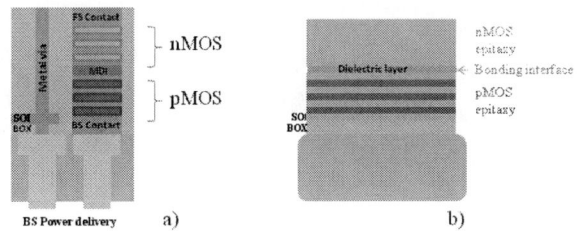

Figure 3. a) Schematic of a CFET transistor stack, and b) possible integration scheme on SOI for different material or substrate orientation.

Stacking the materials together using wafer bonding and layer transfer allows a new degree of freedom to leverage the benefit of different materials in order to have the best pMOS and the best nMOS transistor whatever the material and/or crystal orientation needed for each. For example, one can imagine using (100) strained silicon for nMOS transistor and (110) germanium for pMOS transistors [7]. This integration scheme (figure 3b) can highly benefit from the layer transfer techniques and SOI. This advantage can be also used for a new paradigm which is ensuring the power delivery from the backside to simplify the frontside layout. Smart Cut™ technology can be also attractive to transfer one of the two epitaxial stacks either from bulk substrate or from SOI substrate to the other epitaxial stack as proposed in figure 3b. In this case, the dielectric layer ensures the bonding between the two stacks and can be made as thin as a few nanometers.

From silicon to oxide based Engineered Substrates: The mobility evolution

Other examples of Engineered Substrates which involve even non semiconducting materials are already shaping the future of our daily life. With the increasing demands for 5G telecommunications services, RF device manufacturers require more and more ultra wide band filters. For this purpose acoustic wave devices have been in commercial use for over 60 years. Most used applications are typically surface acoustic wave (SAW) and bulk acoustic wave (BAW) devices, based on piezoelectric material. The transfer of thin mono-crystal piezoelectric layers on specific base substrate is an innovative, efficient and industrial Engineered Substrate solution (figure 4). Indeed, they could answer the main SAW device requirement: obtain a low Temperature Coefficient of Frequency (TCF) to ensure the stability of the bandwidth over a large range of operating temperatures [8]. The low thermal coefficient of thermal expansion of silicon as compared to bulk piezoelectric materials allows reducing the frequency shift related to thermal expansion which is a key feature for 5G communications and beyond because of the high density of frequency bands used.

Figure 4. Piezo-electric on insulator substrate for low TCF SAW filters.

From SOI to Anything-on-anything

At the heart of every device, materials play a crucial role and can highly improve their performance. Layer transfer in general and Smart Cut™ in particular are versatile technologies that can apply to a very broad range of materials to get the best of different worlds. This technology opens new possibilities overcoming the limitations of lattice matched epitaxy. For example, Indium Phosphide layers for SWIR detectors or laser can be transferred on bare silicon substrate or processed silicon substrate without any crystal defect unlike epitaxy which leads to a high density of dislocations. InPOSi substrates have been demonstrated from 100mm bare InP substrate with the same quality of further Multiple

Quantum Wells (MQW) epitaxy as compared to the same epitaxy on bulk InP [9].

All combination of materials that can be imagined is now within easy reach. This opens the era of anything on anything to solve any integration limitation. Moreover, the 40 years of wafer bonding history is also crucial to be able to bond very thin oxide layers down to few nanometers or even without any intermediate layer.

Conclusion

Semiconductor Engineered substrates are a foundational technology with the potential to drive innovations across a wide range of industries. Their ability to enhance efficiency, reduce power consumption, and enable new computing paradigms will be crucial in shaping the future of technology. By enabling smaller, faster, and more specialized electronic components, Engineered Substrates are poised to support advancements that will impact nearly every aspect of modern life, from consumer electronics to healthcare, transportation, and beyond.

References

[1] https://www.soitec.com/en/products/smart-cut/

[2] G. Picun, L. Zumbo, E. Guiot, et al. "Engineered Substrates with ultra-low resistivity polycrystalline SiC Base", Bodo's power Journal, June 2024, p28 (2024).

[3] https://www.mobileye.com/technology/eyeq-chip/

[4] S. Liao, L. Yang, T.K. Chiu, et al. "Complementary Field-Effect Transistor (CFET) Demonstration at 48nm Gate Pitch for Future Logic Technology Scaling", IEDM Tech. Dig., pp. 1-4 (2023).

[5] J. Park, W. Kim, S. Park, et al. "First demonstration of 3-dimensional stacked FET with top/bottom source-drain isolation and stacked n/p metal gate", IEDM Tech. Dig., pp. 1-4 (2023).

[6] M. Radosavljević, C.-Y. Huang, R. Galatage,.et al. "Demonstration of a Stacked CMOS Inverter at 60nm Gate Pitch with Power Via and Direct Backside Device Contacts", IEDM Tech. Dig., pp. 1-4 (2023).

[7] S. K. Kim et al., "Ge(110) GAA Nanosheet / Si(100) Tri-gate Nanosheet Monolithic CFETs Featuring Record-High Hole Mobility," 2024 IEEE Symposium on VLSI Technology and Circuits, Honolulu, USA, pp. 1-2 (2024).

[8] C. Deguet, M. Pijolat, N. Blanc, et al., "LiNbO3 Single Crystal Layer Transfer Techniques for High Performances RF Filters ", 2010 ECS Trans. 33 225 (2010).

[9] B. Ghyselen, F.X. Darras, O. Mourey, et al., "Large Diameter Epi-Ready InP on Si (InPOSi) Substrates" Digests of 2023 International Conference on Compound Semiconductor CSMANTECH (2023).

Advanced electronic devices empowering opto-sensors for imaging and perception

Wen Pan†, Lai Wang†*, Member, IEEE, Jinpu Tang, Zhibiao Hao, Changzheng Sun, Member, IEEE, Bing Xiong, Jian Wang, Yanjun Han, Hongtao Li, Lin Gan, and Yi Luo

†These authors contributed equally to the work. *Correspondence to Lai Wang (e-mail: wanglai@tsinghua.edu.cn).

Department of Electronic Engineering, Tsinghua University, Beijing National Research Center for Information Science and Technology (BNRist), Beijing 100084, China

Abstract

In addition to covering applications of 2D opto-sensors with processing capabilities, we summarize possible architectures of 2D opto-sensors based on advanced electronic devices. Several innovative works using these architectures have been reviewed in this article, in which the functions of opto-sensors go beyond those of traditional concepts. These works may revolutionize sensors in terms of imaging and perception.
Keywords: sensor array, electronic devices, optoelectronics

Introduction

The recent progress in 2D opto-sensors has been enabled by innovations in device physics and the implementation of machine learning algorithms. Advanced electronic devices provide a hardware foundation for the design of possible architectures. As increasingly more integration technologies and computing resources become available, opto-sensors are becoming more highly compact, multifunctional, and intelligent under the development of novel schemes. Based on the above, 2D opto-sensors will have stronger processing capabilities while sensing, enriching their performance and expanding applications. As shown in Fig. 1, applications include pre-process (image enhancement, edge detection, etc.), perception (object recognition, motion detect, target tracking, etc.), imaging (information compression, fast imaging, etc.), machine (robot vision, driverless navigation, edge devices, etc.).

Here, we summarize possible architectures of 2D opto-sensors based on advanced electronic devices, including modulating-type and programming-type opto-sensors. Several innovative works have been reviewed here (encoded photodetector array for compressed sensing and computational imaging, PD-RRAM array for in-sensor computing, etc.) using these architectures in our previous works[1]. The applications of different architectures are analyzed according to their characteristics

Architectures of 2D Opto-sensors

So far, many architectures have been proposed for constructing 2D opto-sensors that could process image information on chip, including near- and in-sensor computing[2], etc. Thanks to improvements in device physics, many architectures based on processing in the physical analog domain have gradually been proposed. Advanced algorithms also help to enrich the way of processing, no longer only used for low-level processing. Moreover, the physical domain computing method based on Kirchhoff's laws also promotes the integration of algorithms and hardware, accelerating developments of computing in opto-sensors. One of the most important works is to integrate memory cells near or within sensors to enable interpretation and efficient processing of amounts of sensing data.

A. Modulating-type opto-sensors based on adjustable optoelectronic devices and arrays.

With the development of optoelectronics over years, many studies have reported various solutions for opto-sensors. Among them, the modulating-type opto-sensors based on adjustable optoelectronic devices and arrays are quite appropriate for computing in opto-sensors, especially devices with linear modulations[3]. The modulation information needs to be stored in the external memory, which is beneficial for achieving more compact opto-sensors (shown in Fig. 2a). However, the external memory and opto-sensor are separated from each other, and the interconnections of data transmission pose bottlenecks to the overall processing speed. To address this issue, it is necessary to make the designed applications more in line with hardware implementation to achieve the best optimized performance. According to the characteristics of modulating-type opto-sensors, processing methods such as edge, convolution, and column-parallel computing are easier to implement.

B. Programming-type opto-sensor based-on programmable devices with detection and storage functions

In recent years, the programming-type opto-sensors based on some advanced electronic devices perfectly solve the bottlenecks of the processing speed. Some devices that have both storage and detection capabilities provide new solutions for inner memory of sensors (shown in Fig. 2b), revolutionizing 2D opto-sensors and far surpassing traditional CMOS sensors in terms of the integration density, fill factor and scalability. Due to the inner memory[4], data transmission will be significantly reduced, greatly improving processing speed and efficiency. And with the improvements of the computing power, the functions of 2D opto-sensors are gradually enriched, enabling more high-level processing methods (such as object recognition, intelligent perception, motion detect, machine learning, neuromorphic computing, etc.).

Opto-sensors for Imaging and Perception

A. Encoded photodetector array for compressed sensing and computational imaging.

An encoded photodetector array that precisely utilizes the advantages of modulating-type opto-sensors, is previously proposed for compressed sensing and computational imaging[5]. The opto-sensor acts as an encoder that converts high-dimensional physical quantities into the array outputs. This process can be regarded as a mapping from one vector space to the high-dimensional vector spaces in the measurement and the decoding process deciphers the encoded high-dimensional information by using prior knowledge. This method can be engineered to capture different kinds of information (spatial, spectral, polarization, temporal information, etc.) and allows for the multiple functionality multiplexing. Based on the characteristics of modulating-type opto-sensors, column-parallel computing helps to implement compressed sensing in the encoded photodetector array, improving the speed of imaging. The photo-responsivies of each row can be electrically modulated and the measurement process would be finished by refreshing. The result of image reconstructions is highly consistent with the simulation, and the frame rate could reach 7.14 kfps under an 8 MHz clock. The proposed approach for compressed sensing and computational imaging shows a promising future of constructing opto-sensors that deliver high-speed imaging capability of 10000 fps or higher, which might enable advanced medical, visual, and communication techniques. Though facing numerous challenges, it remains a worthy pursuit for future investigation.

B. PD-RRAM array utilized in pattern recognition for in-sensor computing.

As well, we proposed an optoelectronic in-sensor computing array that integrates photodiodes (PDs) with resistive random access memories (RRAMs). It's a typical work based on programming-type opto-sensors. This approach embodies "front-end computation", shifting processing to the sensor source and eliminating reliance on traditional data transfers in von Neumann architectures. Its inner memory enables processing of short-distance sensing data, which can utilize actual physical signals for in-sensor computing, effectively reducing issues such as signal attenuation, latency, and energy consumption. This paradigm is ideal for the Internet of Things (IoT) and edge computing, where low power and real-time processing are critical. And in recent experiments, the PD-RRAM array utilized in analogue multiply-accumulate computation (MAC) operations has been verified at the pixel level, which provides a foundation for object recognition and other processing. It has demonstrated its capability to perform precise image processing (10-level states for the accuracy of 95%~100%), self-powered inference (photo-responsivities ~ 0.2 A/W), high speed recognition (response time ~ 30 ns @Si-PD with pixel-level parallel computing). And the miniaturization of RRAMs also brings great potential of improving integration degree, significantly increasing the storage capacity of opto-sensors. Despite its promise, PD-RRAM arrays still face significant challenges in scaling from experimental setups to practical deployment. Thermal management and interconnect complexity further limit its scalability and adaptability. Addressing these challenges requires coordinated efforts in hardware design (devices, integration, interconnects and fabrication, etc.), algorithmic co-development (lightweight neural networks, pruned and quantized models, dynamic weight adjustment capabilities, etc.), and system integration. In the long term, in-sensor computing will go beyond technical optimization to redefine its role in computational ecosystems. These innovations will transform PD-RRAM arrays from a promising experimental technology into a foundation of next-generation intelligent hardware.

Conclusion

Hybrid architectures of 2D opto-sensors will not only reshape imaging and computing processes but also redefine how systems interact with and adapt to the complexities of an increasingly intelligent and connected world. Future highly compact, multifunctional, reconfigurable, and intelligent 2D opto-sensors will find applications in medical imaging, environmental monitoring, and many other areas of our daily lives, especially in the mobile domain and the IoT.

Acknowledgments

All the authors gratefully acknowledge the National Key Research and Development Program (2021YFA0716400), the National Natural Science Foundation of China (62225405, 62350002, 61991443, 62031011, 62127814, 61927811), and the Collaborative Innovation Centre of Solid-State Lighting and Energy-Saving Electronics.

References

[1] W. Pan, J. Zheng, L. Wang, and Y. Luo, "A Future Perspective on In-Sensor Computing," Engineering, vol. 14, pp. 19-21 (2022).

[2] F. Zhou and Y. Chai, "Near-sensor and in-sensor computing," Nature Electronics, vol. 3, pp. 664-671 (2020).

[3] L. Mennel, J. Symonowicz, S. Wachter, D. K. Polyushkin, A. J. Molina-Mendoza, and T. Mueller, "Ultrafast machine vision with 2D material neural network image sensors," Nature, vol. 579, pp. 62-66 (2020).

[4] T. Komuro, S. Kagami, and M. Ishikawa, "A Dynamically Reconfigurable SIMD Processor for a Vision Chip," IEEE Journal of Solid-State Circuits, vol. 39, pp. 265-268 (2004).

[5] W. Pan, Z. Yang, Z. Hao, C. Sun, B. Xiong, J. Wang, et al., "An Encoded Photodetector Array for High-frame-rate Compressed Sensing and Computational Imaging," IEEE Sensors Journal (2024). http://dx.doi.org/10.1109/JSEN.2024.3506937 (In press)

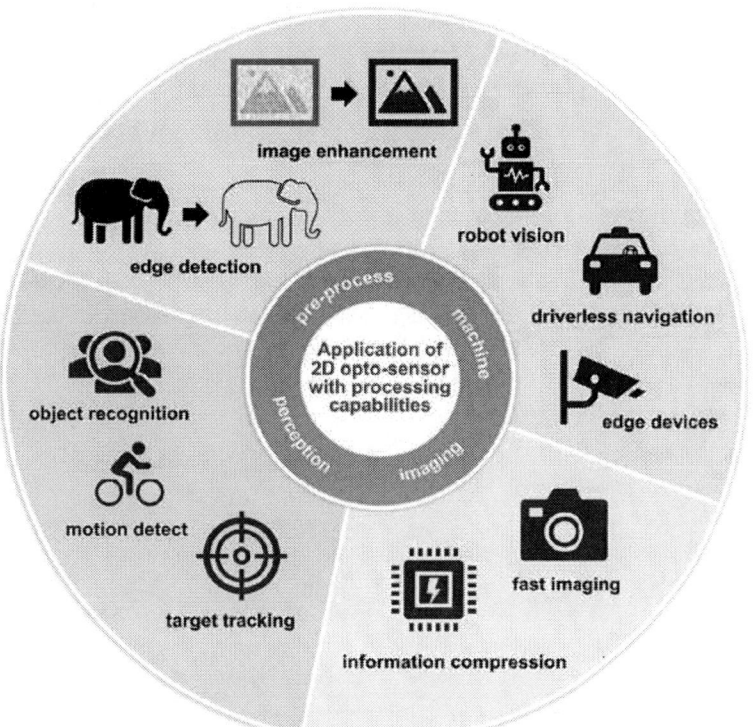

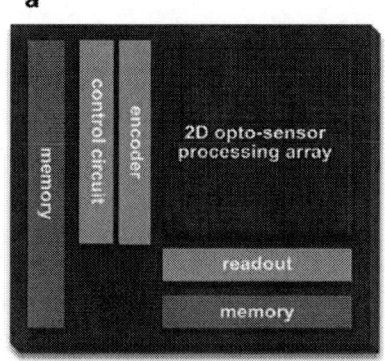

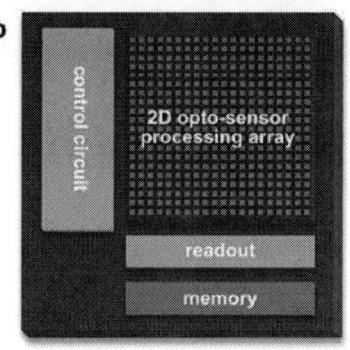

Fig. 1: Applications of 2D opto-sensors with processing capabilities. It particularly includes pre-process (image enhancement, edge detection, etc.), perception (object recognition, motion detect, target tracking, etc.), imaging (information compression, fast imaging, etc.), machine (robot vision, driverless navigation, edge devices, etc.).

Fig. 2: Possible architectures of 2D opto-sensors with processing capabilities. a) modulating-type sensor. External memory is used for modulating. b) programming-type sensor. Inner memory is used for programming.

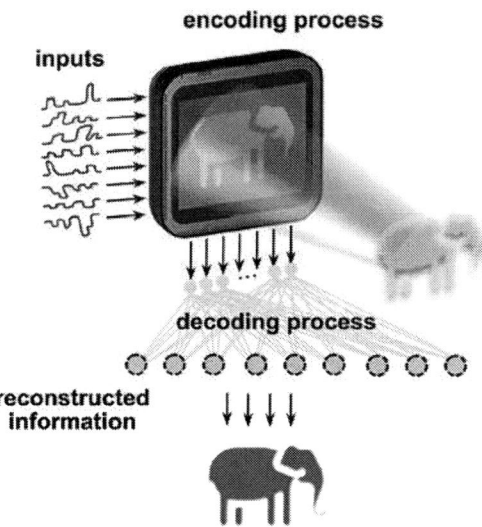

Fig. 3: Schematic of the encoded photodetector array for compressed sensing and computational imaging.

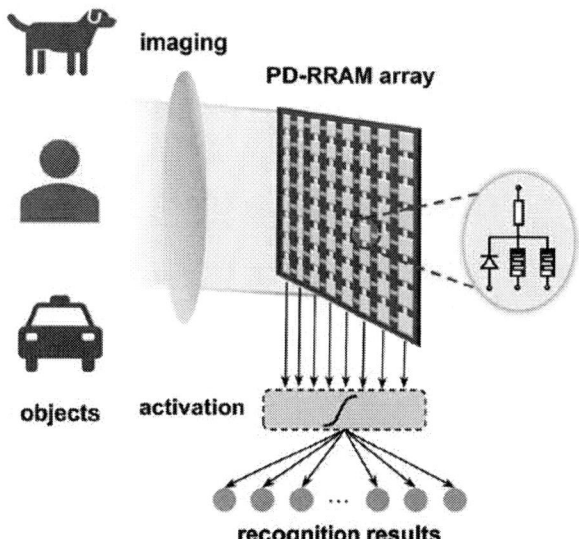

Fig. 4: Schematic of the PD-RRAM array utilized in pattern recognition for in-sensor computing.

979-8-3315-0417-5/25 $31.00 © 2025 IEEE

Neuromorphic Multisensory Numerosity Perception Enhanced by a Tactile Glove

Hongwei Tan[1,2], Syed Ashraf [1], Sebastian Hannula[1], Zhong-Peng Lv[1], Bo Peng[1,3], Sebastiaan van Dijken[1]

[1]Department of Applied Physics, Aalto University, 02150 Espoo, Finland
[2]Max Planck Institute for Polymer Research, 55128 Mainz, Germany
[3]Department of Materials Science, Advanced Coatings Research Center of Ministry of Education of China, Fudan University, 200433, Shanghai, China

Abstract

Neuromorphic numerosity perception utilizes bioinspired sensory systems to process and interpret numerical information, mimicking how the human brain perceives quantity. Here, we present a tactile glove with gesture-learning capability to enhance neuromorphic numerosity perception, generated by vision or auditory inputs. Equipped with 10 flexible pressure sensors strategically positioned on finger joints, the glove accurately recognizes hand gestures representing numbers. By integrating these tactile signals with visual and auditory cues through multisensory integration, the system significantly improves numerosity perception.

Keywords: tactile glove, multisensory integration, neuromorphic numerosity perception

Introduction

Numerosity perception, the intuitive ability to estimate and process numerical quantities, is a fundamental cognitive function of the human brain [1-2]. Mimicking this ability through bioinspired sensory systems holds potential for applications in robotics, human-computer interaction, and assistive technologies. Recent advancements in neuromorphic systems with multisensory integration [3-10] have demonstrated the potential to emulate the brain's capability to integrate information across modalities such as vision, touch, and sound [11-13]. These multisensory systems enable applications like accurate object recognition, navigation, and decision-making.

In this study, we present a neuromorphic tactile glove designed to enhance numerosity perception through bioinspired multisensory integration. The glove is equipped with 10 distributed pressure sensors placed on the finger joints to detect bending movements, enabling gesture recognition and learning. With the aid of an artificial neural network, the system achieves efficient recognition of number expressed through American Sign Language. Our results demonstrate the systems' capability to improve numerosity perception accuracy to 96.3%, significantly outperforming unisensory accuracies: 36.4% for vision, 83.2% for touch, and 68.9% for audio. This highlights the benefits of multisensory integration in achieving superior performance.

The proposed tactile glove system has promising applications in self-learning robotics, industrial manufacturing, social robotics, and intuitive human-machine interactions.

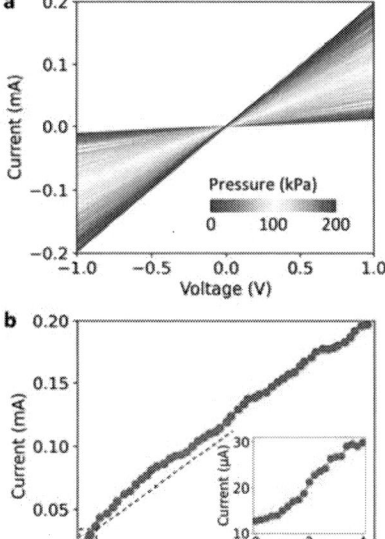

Fig. 1: Pressure sensor performance characterization. a, Current-voltage (I-V) characteristics of the pressure sensor measured under applied pressures ranging from 0 to 200 kPa. b, Current output values of the pressure sensor at a fixed voltage of 1 V, derived from the data in (a).

Results

Pressure sensor

A two dimensional (2D) MXene-based freestanding film, optimized based our previous work [14-16], was used as the pressure-sensitive medium. This MXene film was placed on a cross-finger-structured Au/Ti electrode deposited on flexible PET substrates. The electrodes were fabricated via magnetron sputtering using a shadow mask. The working area of the pressure sensor is 5 mm × 5 mm. The pressure sensor operates effectively within a range of 0 to 200 kPa (Figure 1), encompassing the working range of biological mechanoreceptors. Moreover, the sensor demonstrates high sensitivity to low pressures (0-4 kPa, inset of Figure 1b), highlighting its wide working range and suitability for tactile applications.

Tactile glove for gesture recognition

Ten pressure sensors were positioned on the joint areas of a glove to enable gesture detection. The normalized response of these sensors is shown in Figure 2 for American Sign Language numbers 1 to 10.

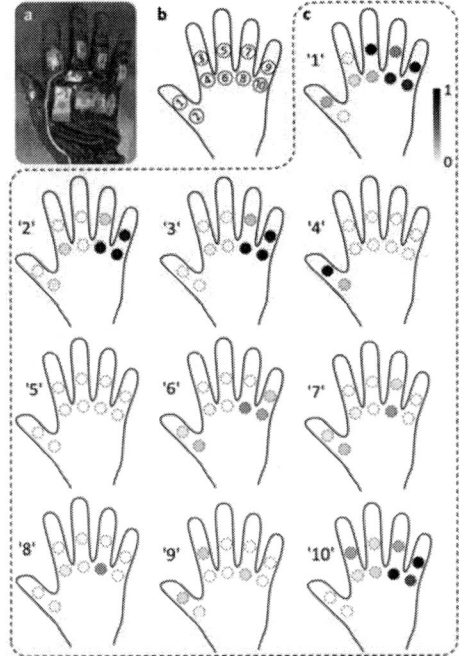

Fig. 2: Tactile glove with MXene-based pressure sensors positioned on joint areas. a, Photograph of the tactile glove. b, Schematic illustration of sensor positions distributed across the tactile glove. c, Tactile response patterns corresponding to gestures representing American Sign Language numbers 1 to 10.

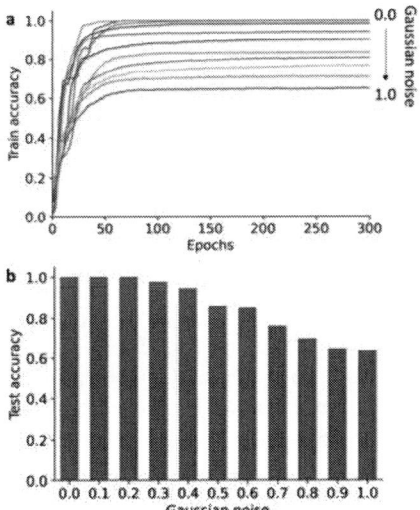

Fig. 3: Training and testing performance of the system for recognizing American Sign Language numbers 1 to 10. a, Training accuracy across 300 epochs with Gaussian noise levels ranging from 0% to 100% added to tactile datasets. b, Test accuracies of the ANN trained on tactile datasets with varying levels of Gaussian noise.

A simple artificial neural network (ANN) was employed to classify the tactile patterns corresponding to these numbers. The ANN comprises 10 inputs neurons (representing the 10 tactile sensors), 15 hidden neurons, and 10 output neurons (corresponding to the 10 numbers). A dataset of 2000 samples was used for training, augmented with Gaussian noise to enhance system robustness. After 300 training epochs, the system achieved 100% accuracy with <20% Gaussian noise and ~64% accuracy even with 100% Gaussian noise. These results demonstrate the robust gesture recognition capability of the tactile glove.

Tactile-enhanced multisensory numerosity perception

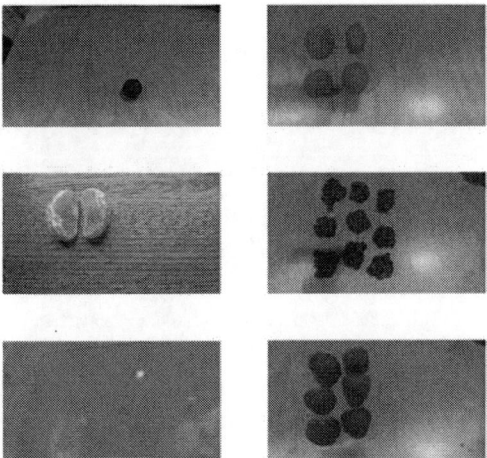

Fig. 4: Representative visual image datasets used for testing visual numerosity perception.

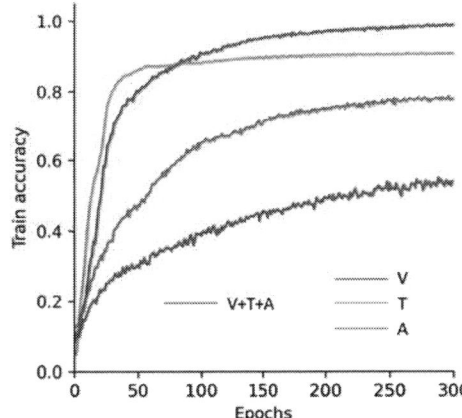

Fig. 5: Training accuracies for numerosity perception using unisensory inputs - visual (V), tactile (T), auditory (A) - and multisensory integration (V+T+A).

To improve numerosity perception under unisensory visual or auditory inputs, we mimicked biological multisensory integration principles [11-13] by incorporating tactile information. For example, unisensory vision-based numerosity perception yielded a low accuracy of 36.4%, tested through object quantity recognition using real image datasets (Figure 4). Similarly, unisensory audio-based numerosity perception achieved an accuracy of 68.9%.

To improve performance, we integrated tactile, visual, and auditory data into a combined database of 2000 samples per quantity (1 to 10). An ANN with identical hidden and output layers was trained using this multimodal dataset. After 300 training epochs, the system achieved an accuracy of 99%, significantly outperforming unisensory results: 53% for vision, 91% for touch, and 78% for audio (Figure 5). The test accuracy was also markedly higher at 96.3%, compared to 36.4% (vision), 83.2% (touch), and 68.9% (audio) for unisensory perception (Figure 6).

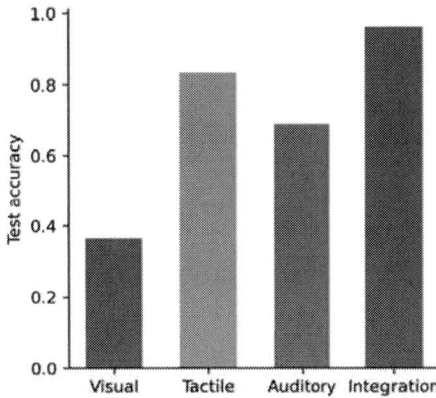

Fig. 6: Test accuracies for numerosity perception with unisensory inputs and multisensory integration.

Conclusion

We developed a neuromorphic tactile glove with MXene-based pressure sensors for accurate recognition of American Sign Language numbers. By integrating tactile, visual, and auditory inputs, we emulated cortical multisensory integration to achieve superior neuromorphic multisensory numerosity perceptions. This tactile-enhanced system significantly exceeds unisensory performance in recognition accuracy and demonstrates potential for applications in cross-modal self-learning, industrial manufacturing, social robotics, and intuitive human-machine interaction. Additionally, the tactile glove offers promising opportunities for advancing neuromorphic vision [17] and other sensory technologies.

Acknowledgments

We gratefully acknowledge E.I. Kauppinen for providing infrastructure support for the electrical measurements. The project made use of the OtaNano—Micronova Nanofabrication Center and the OtaNano—Nanomicroscopy Center, supported by Aalto University. This work was supported by the Research Council of Finland (Grant no. 316973 H. T.).

References

[1] W. S. Jevgns, "The Power of Numerical Discrimination", Nature, 3, pp. 281-282 (1871).

[2] S. J. Cheyette, and S. T. Piantadosi, "A unified account of numerosity perception", Nat. Hum. Behav., 4, pp. 1265–1272 (2020).

[3] N. Fazeli, M. Oller, J. Wu, Z. Wu, J. B. Tenenbaum, and A. Rodriguez, "See, feel, act: Hierarchical learning for complex manipulation skills with multisensory fusion", Sci. Robot., 4, pp. eaav3123 (2019).

[4] C. Wan, P. Cai, X. Guo, M. Wang, N. Matsuhisa, L. Yang, Z. Lv, Y. Luo, X. J. Loh, and X. Chen, "An artificial sensory neuron with visual-haptic fusion", Nat. Commun., 11, pp. 4602 (2020).

[5] M. Wang, Z. Yan, T. Wang, P. Cai, S. Gao, Y. Zeng, C. Wan, H. Wang, L. Pan, J. Yu, S. Pan, K. He, J. Lu, and X. Chen, "Gesture recognition using a bioinspired learning architecture that integrates visual data with somatosensory data from stretchable sensors", Nat. Electron., 3, pp. 563–570 (2020).

[6] H. Tan, Y. Zhou, Q. Tao, J. Rosen, and S. van Dijken, "Bioinspired multisensory neural network with crossmodal integration and recognition", Nat. Commun., 12, pp. 1120 (2021).

[7] J. Yu, Y. Wang, S. Qin, G. Gao, C. Xu, Z. L. Wang, and Q. Sun, "Bioinspired interactive neuromorphic devices", Mater. Today, 60, pp. 158-182 (2022).

[8] F. Yu, Y. Wu, S. Ma, M. Xu, H. Li, H. Qu, C. Song, T. Wang, R. Zhao, and L. Shi, "Brain-inspired multimodal hybrid neural network for robot place recognition", Sci. Robot., 8, pp. eabm6996 (2023).

[9] X. Liu, C. Sun, X. Ye, X. Zhu, C. Hu, H. Tan, S. He, M. Shao, and R.-W. Li, "Neuromorphic Nanoionics for Human–Machine Interaction: From Materials to Applications", Adv. Mater., 36, pp. 2311472 (2024).

[10] Q. Mao, Z. Liao, J. Yuan, and R. Zhu, "Multimodal tactile sensing fused with vision for dexterous robotic housekeeping", Nat. Commun., 15, pp. 6871 (2024).

[11] B. E. Stein, and M. A. Meredith, "Multisensory Integration", Science, 261, pp. 928-929 (1993).

[12] J. J. McDonald, W. A. Teder-Sälejärvi, and L. M. Ward, "Multisensory Integration and Crossmodal Attention Effects in the Human Brain", Science, 292, pp. 1791 (2001).

[13] Z. Okray, P. F. Jacob, C. Stern, K. Desmond, N.Otto, C. B. Talbot, P. Vargas-Gutierrez, and S. Waddell, "Multisensory learning binds neurons into a cross-modal memory engram", Nature, 617, pp. 777-784 (2023).

[14] H. Tan, Q. Tao, I. Pande, S. Majumdar, F. Liu, Y. Zhou, P. O.Å. Persson, J. Rosen, and S. van Dijken, "Tactile sensory coding and learning with bio-inspired optoelectronic spiking afferent nerves", Nat. Commun., 11, pp. 1369 (2020).

[15] H. Tan, and S. van Dijken, "Dynamic machine vision with retinomorphic photomemristor-reservoir computing", Nat. Commun., 14, pp. 2169 (2023).

[16] X. Hong, Z. Xu, Z.-P. Lv, Z. Lin, M. Ahmadi, L. Cui, V. Liljeström, V. Dudko, J. Sheng, X. Cui, A. P Tsapenko, J. Breu, Z.Sun, Q. Zhang, E. Kauppinen, B. Peng, and Olli Ikkala. "High‐permittivity Solvents Increase MXene Stability and Stacking Order Enabling Ultraefficient Terahertz Shielding", Adv. Sci. 11, pp. 2305099 (2024).

[17] H. Tan, and S. van Dijken, "A universal neuromorphic vision processing system", Nat. Electron., 7, pp. 946-947 (224).

2D MoS$_2$ Thin-Film Transistors for Large-Area, Flexible Electronics

Jong-Hyun Ahn[1]

1School of Electrical and Electronic Engineering, Yonsei University, Republic of Korea

Abstract

2D MoS$_2$, a promising TMD semiconductor, offers superior mechanical flexibility over traditional materials like amorphous-Si, poly-Si, and IGZO. This makes it ideal for large-area, flexible electronics. We present a fabrication process for MoS$_2$-based backplane TFTs, synthesized via modified MOCVD and fabricated using conventional photolithography and etching. This approach enables the development of flexible devices like wearable OLEDs, micro-LED displays, and X-ray detectors.

Keywords: MoS2, TFT and Flexible electronics

Introduction

Thin-Film Transistors (TFTs) are key components in large-area electronic systems, including displays, sensors and digital X-ray detectors. Conventional backplane circuitry often relies on semiconductors like amorphous-Si, poly-Si, and IGZO, which are selected for their high carrier mobility, substantial on/off ratio, and robust stability.

To advance the development of mechanically flexible devices, two-dimensional (2D) transition metal dichalcogenide (TMD) semiconductors, have emerged as promising alternatives [1]. This work presents the fabrication process of representative 2D TMD, MoS$_2$-based backplane TFTs, specifically designed to power large-area displays and digital X-ray detectors. High-quality MoS$_2$ is synthesized via a modified metal-organic chemical vapor deposition process, followed by the fabrication of TFT arrays using conventional photolithography and etching techniques. The resulting MoS$_2$-based TFTs exhibit remarkable electrical performance, including good carrier mobility, steep subthreshold swings, and high on/off current ratios. Moreover, the exceptional mechanical flexibility of MoS$_2$ enables the realization of various flexible and wearable electronic devices. The potential applications of MoS$_2$-based TFTs are vast, ranging from wearable OLED displays to flexible tactile sensors and digital X-ray detectors.

Result and Discussion

A high-quality bilayer MoS$_2$ film was synthesized on a 4-inch SiO$_2$/Si wafer using metal-organic chemical vapor deposition (MOCVD) (2,3). This technique allowed for precise control over the gas precursors. The MoS$_2$ film was then transferred onto a 6 mm-thick polyethylene terephthalate (PET) substrate coated with a 30 nm Al$_2$O$_3$ layer deposited by atomic layer deposition (ALD).

The resulting MoS$_2$ transistor array featured a unique structure with an Al$_2$O$_3$ encapsulation, which improved metal contact and carrier mobility through n-doping effects. The high uniformity of the MOCVD-grown MoS$_2$ film enabled the formation of a highly uniform TFT array, essential for stable display operation.

The MoS$_2$ transistors exhibited excellent performance with a mobility of approximately 18 cm²/Vs at a drain voltage of 1 V and an on/off ratio exceeding 10⁷. The threshold voltage (V_{th}) was measured to be 5 V, ensuring efficient device operation.

To demonstrate the potential of large-area electronics, we fabricated wearable OLED displays and flexible digital X-ray detectors using MoS$_2$ TFT arrays on plastic substrates.

High-performance MoS$_2$ backplane arrays were used to create large-area AMOLEDs on ultra-thin PET substrates (fig 1). The OLED pixels, operating via bottom emission, exhibited consistent and uniform brightness across the entire display. A 18x18 array of 324 pixels was integrated to demonstrate full-color capabilities. The device's flexibility and low stiffness allowed for seamless integration onto curved surfaces, such as human skin, without compromising performance (4).

We also developed a flexible X-ray detector comprising a MoS$_2$ TFT array and a graphene/MoS$_2$ photodetector (PD) array, featuring 3,600 pixels across a 3 cm x 3 cm area (5). Traditional X-ray detectors often rely on rigid glass substrates, limiting their flexibility. In contrast, our MoS$_2$-based backplane offers excellent mechanical flexibility, enabling the detector to conform to non-planar objects.

The flexible X-ray detector consists of a large-area active-matrix 2D material-based backplane and a scintillator film. The backplane utilizes a 1T-1R configuration of MoS$_2$ TFTs and graphene/MoS$_2$ PDs to generate current upon exposure to visible light emitted from the scintillator. A 300 μm-thick Gd$_2$O$_2$S scintillator film converts X-ray radiation to visible light. Bilayer MoS$_2$ channels were employed in both TFTs and PDs to optimize electrical and optical performance. The PD array, composed of MoS$_2$ channels and optically transparent graphene interdigitated electrodes, efficiently absorbs photons, ensuring high-contrast images even under low X-ray conditions.

To demonstrate the reduction of projection-based image distortion, we prepared a metal cylinder with a '+' shaped hole and simulated industrial pipeline inspection. X-ray radiation was directed at a detector placed behind the tilted cylinder.

A flat detector captured a significantly distorted image, exhibiting tilting and shrinkage in the horizontal direction. However, by using a curved detector that matched the cylinder's curvature, the captured image showed minimal distortion, accurately representing the object's shape. This highlights the detector's ability to mitigate projection-based distortion and accurately capture images of irregular surfaces.

Conclusion

We successfully synthesized high-quality MoS_2 thin-film transistors (TFTs) using a modified metal-organic chemical vapor deposition (MOCVD) technique. These MoS_2 TFTs exhibited excellent electrical performance, including high mobility and on/off ratios. The exceptional mechanical properties of MoS_2 enabled the development of flexible electronic devices. Moreover, we demonstrated the potential of MoS_2-based TFTs by fabricating wearable OLED displays and flexible X-ray detectors. These devices showcased improved performance and flexibility compared to traditional technologies, highlighting the promising future of 2D materials in electronics.

Acknowledgments

This work is supported by the National Research Foundation of Korea (RS-2024-00435661).

References

[1] A. Katiyar, et al., "2D materials in flexible electronics: recent advances and future prospectives", Chem. Rev. 124, 328 (2023).

[2] K. Kang et al., "High mobility three atom thick semiconducting films with wafer-scale homogeneity", *Nature* 520, 656 (2015)

[3] S Hwangbo, L Hu, A. T. Hoang, J. Y. Choi, J. H. Ahn, "Wafer-scale monolithic integration of full-colour micro-LED display using MoS_2 transistor", Nature Nanotechnology 17, 500 (2022)

[4] M Choi et al., "Full-color active-matrix organic light-emitting diode display on human skin based on a large-area MoS_2 backplane", Science advances 6, eabb5898 (2020)

[5] B. Kim et al., "A flexible active-matrix X-ray detector with a backplane based on two-dimensional materials", Nature electronics, in press

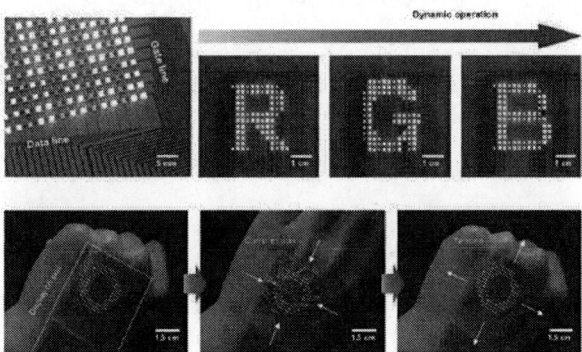

Fig. 1: A red, green, and blue OLED display integrated with an array of 324 MoS$_2$ TFTs fabricated on a 4cm x 4cm plastic substrate (top), and a display image showing stable operation when attached to the skin (bottom).

Spintronic foundation cells for scalable unconventional computing

Zhihua Xiao[1,2], Qiming Shao[1,2#]

[1]The Hong Kong University of Science and Technology, Hong Kong, China;
[2]AI Chip Center for Emerging Smart Systems, Hong Kong, China
#Email: eeqshao@ust.hk

Abstract

With the ending of Moore's law and the thriving of unconventional computing, there is a growing demand for the development of emerging device technologies. However, it is anticipated that completely replacing complementary metal-oxide-semiconductor (CMOS) technology will not be feasible in the foreseeable future. Thus, in this paper, we showed how one of the emerging technologies, spintronic devices can be integrated with CMOS technologies as the foundation cells for scalable unconventional computing.

Keywords: Spintronics, CMOS, Neuromorphic Computing and Probabilistic Computing

Introduction

The concept of computing has been completely reshaped by the recent unprecedented development of intelligence systems, leading to changes in both the form of computing and the demand for resources. Although the Von-Neumann architecture and CMOS technology successfully empowered the information era in the 20th century, they now face significant challenges, such as the memory wall and physical limitations. Consequently, there is a growing interest in unconventional computing methods to support the development of modern intelligent systems[1].

Unconventional computing methods, including neuromorphic computing, probabilistic computing, quantum computing, etc., necessitate the support of emerging device technologies for efficient execution. Spintronic devices offer a novel dimension of functionality by utilizing the spin degree of freedom in addition to traditional charge-based electronics, providing rich mechanisms and dynamics. Over the past two decades, switching mechanisms have evolved from spin-transfer torque (STT) to spin-orbit torque (SOT) and voltage-controlled magnetic anisotropy (VCMA). These devices can be fabricated using a variety of new materials, such as antiferromagnetic, topological, and two-dimensional (2D) materials[2]. Researchers are actively exploring the opportunities presented by these new physics and devices to enhance performance in unconventional computing[3].

Though seems promising, spintronic devices still require integration with CMOS technology for scalable applications. Scalable CMOS logic circuit designs rely on the integration of foundation cells such as the seven basic logic gates. On a larger scale, multiple digital circuits are integrated with each other and with analog circuits via intellectual property (IP) cores and analog-to-digital (ADC) or digital-to-analog (DAC) converters. These design principles remain relevant in the era of unconventional computing. New devices and physics must build fundamental cells that can be seamlessly integrated with CMOS technology or other cells for scalability[4].

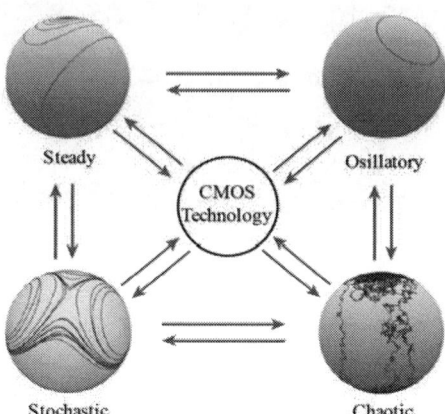

Fig. 1. The spintronic fundamental cells are based on four dynamics of spintronic devices that can interact with each other and with CMOS technologies.

Spintronic fundamental cells can be built based on the four dynamic properties (**Fig. 1**): steady, oscillatory, stochastic, and chaotic. Based on different dynamics, the fundamental cells are used for applications such as data storage, radio frequency coupling, true random number generation, reservoir computing, probabilistic computing, and more. This paper showcased some of the spintronic fundamental cell designs and demonstrated their applications.

Spintronic Foundation Cells

Spintronic foundation cells serve as a bridge between spintronic devices and CMOS systems, necessitating cross-layer co-design involving materials, devices, circuits, and systems. This section presents examples of co-designs of spintronic foundation cells.

A. Basic Computing Cells

The design of separated memory and computing units in conventional Von Neumann architecture has increasingly become a significant bottleneck in computational efficiency. To address this issue, a novel computing paradigm known as the Compute-In-Memory (CIM) framework has been introduced. The core structure of CIM is the crossbar array that can perform multiply and accumulation (MAC) within the array, thereby minimizing data transfer requirements. Like other memristor devices, spintronic devices can serve as emerging devices for CIM (**Fig. 2**). The resistance state of a magnetic tunnel junction (MTJ) is determined by the spin polarization direction of the free layer and the pinned layer. This configuration results in a distinctive characteristic known as spatially random yet temporally fixed variation. Essentially, MTJs exhibit extremely low variation from cycle to cycle, and with appropriate sensing and adjustment techniques, any fixed device-to-device variation can be

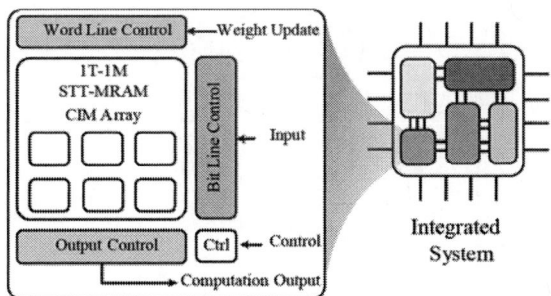

Fig. 2. Schematic of a basic computing cell that can integrated with other cells via digital interconnections.

effectively managed[5], [6]. To streamline the design process, calibration circuits are integrated within the fundamental cells, thereby shielding designers who may be unfamiliar with the properties of MTJs from these complexities. Consequently, designers can utilize these fundamental cells as they would conventional digital cells, while simultaneously benefiting from the enhanced performance capabilities offered by emerging device technologies.

B. Cryogenic Computing Cells

Quantum computing has emerged as the next generation of computing methods. Quantum systems are typically cooled to cryogenic temperatures to maintain functionality, and the control of quantum bits (Q-bits) relies on cryogenic CMOS technologies. Recent advancements in reinforcement learning have surpassed conventional Q-bit control algorithms, making cryogenic CIM systems a promising solution for low-latency, high-efficiency, and accurate Q-bit control[7]. Among memristor candidates, magnetic topologic insulator (MTI) outperforms other devices in terms of energy, speed, reliability, and endurance. The Hall resistance R_H has the unique ability to implement positive, negative, and zero weights, which would otherwise require differential devices. While traditional ferromagnets exhibit R_H in the range of several ohms, MTIs can achieve increased R_H with decreasing temperature and be quantized at 25.6 kΩ (quantum anomalous Hall effect) at ultra-low temperatures, making them ideal for cryogenic CIM[8].

The Hall bar device is a four-terminal conductor, posing challenges for direct application in MAC operations due to the sneak paths. To address this issue, we employed a transverse-read Hall-current-based MAC to construct the fundamental cell for cryogenic CIM (**Fig. 3a**). In **Fig. 3b**, the MAC operation is performed by summing the Hall current from each device, with the measured summation equating to the sum of individual devices. Validated by device measurement and circuit simulation, MTI-based spintronic fundamental cells (**Fig. 3c**) deliver a throughput of 724 TOPS/W at 2K temperature.

C. Reservoir Computing Cells

Modern AI models are based on the combination of nonlinear functions to model arbitrary functions and distributions. Reservoir computing cells can provide highly nonlinear dynamics with much less hardware cost. A spin-torque oscillator (STO) serves as a physical reservoir device by tuning its oscillation using the spin-transfer torque effect (**Fig. 4a**) in the free layer of an MTJ[9]. The magnetization in the free layer is described by the Landau-Lifshitz-Gilbert (LLG) equation incorporating the spin-transfer damping-like torque in Eq. (1):

$$\frac{\partial m}{\partial t} = -\gamma m \times H_{eff} + \alpha m \times \frac{\partial m}{\partial t} + \gamma a_j m \times (m \times e_x) \quad (1)$$

Where γ and α are gyromagnetic ratios and damping factors. H_{eff} is an effective magnetic field, including the

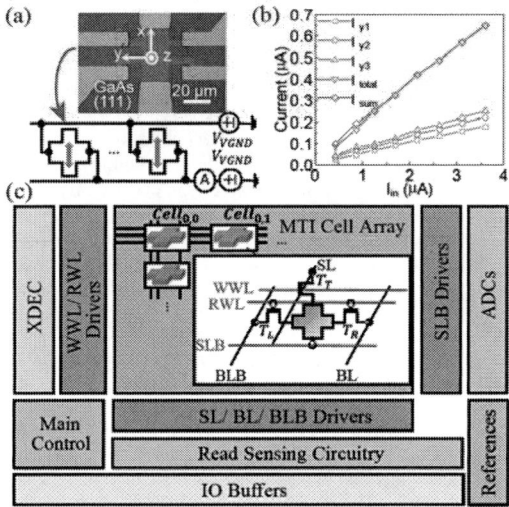

Fig. 3. (a) Optical image of the MTI Hall bar device and the parallel connection for Hall current summation. **(b)** Experimental measurement of the Hall current MAC operation. **(c)** Schematic of the Hall current MAC cryogenic computing cell.

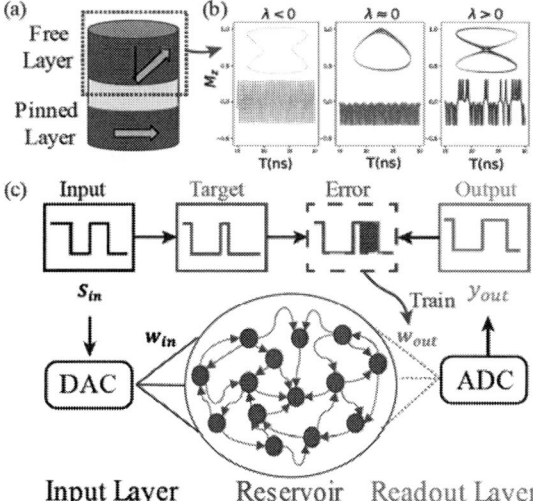

Fig. 4. (a) Schematic of reservoir computing framework, the physical reservoir is the free layer in an MTJ. **(b)** The diverse spin oscillation behavior in the free layer is driven by different AC/DC components. **(c)** Schematic of the reservoir computing cell.

979-8-3315-0417-5/25 $31.00 © 2025 IEEE

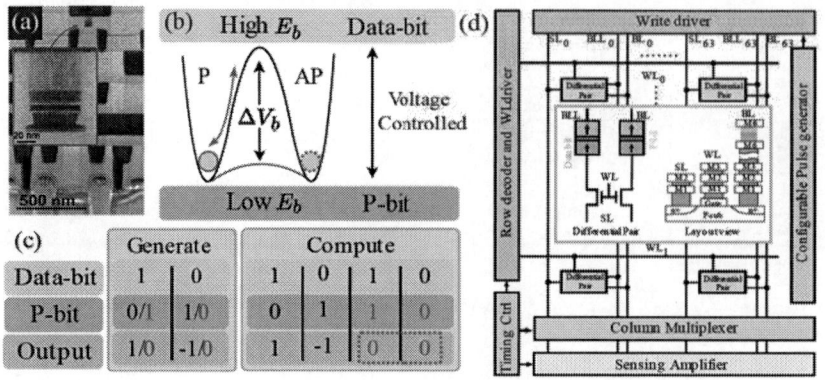

Fig. 5. (a) TEM image of the CMOS integrated VCMA-MTJ. **(b)** The VCMA effect allows direct control of the energy barrier of the device to switch it between data-bit and P-bit. **(c)** The control scheme of the P-bit. The P-bit first generates random states and then sampled to determine switching or un-switching states. The switched cells are set to zero. **(d)** The schematic of in-memory probabilistic computing cell.

external field and demagnetization field. a_j is the STT magnetic field induced by charge current, including direct current (DC) and alternating current (AC) components. A continuous precession of magnetization in the free layer will occur when the amplitude of spin-transfer damping-like torque induced by the DC charge current is balanced with the Gilbert-damping torque. Applying additional AC components in the input stream, three degrees of freedom (two from magnetization) allow chaotic behavior in the magnetization dynamic space (**Fig. 4b**). Therefore, by tuning the AC and DC configurations, the input can be encoded to the reservoir computing cells for the information process (**Fig. 4c**).

D. Probabilistic Computing Cells

Probabilistic computing provides a complementary approach to conventional deterministic computing for addressing computationally challenging tasks. The fundamental cell for probabilistic computing is the probabilistic bit (P-bit). Developing an efficient P-bit enhances the performance and efficiency of probabilistic computing. Unlike conventional CMOS devices, which require complex circuits to harness intrinsic stochasticity, emerging technologies can achieve this with fewer devices.

However, existing P-bits based on emerging technologies are mainly placed outside the computing units. The ex-situ P-bits generally follow the generate-sample-transfer-compute scheme[10]. This brings a limitation to the existing probabilistic computing since although the emerging devices can generate random states efficiently, the sampling circuits and transfer process are either energy-consuming or slow. Simplifying the process of probabilistic computing can further improve efficiency.

This issue can be addressed by spintronic devices since they exhibit stochastic behavior when the energy barrier is low due to thermal noise while remaining stable when the energy barrier is high. Leveraging this insight, we designed and fabricated an in-memory P-bit using voltage-controlled magnetic anisotropy (VCMA) magnetic tunnel junctions

(MTJs) as shown in **Fig. 5a**. The energy barrier of the VCMA-MTJ can be modulated by external voltage, allowing the device to transit between stable and stochastic states (**Fig. 5b**). The cell is compact, as device behavior is controlled by voltage pulses, and is fully compatible with existing differential cells for CIM. As in **Fig. 5c**, a probabilistic differential cell is set to zero if the P-bit switches, with switching probability controlled by pulse duration and amplitude. This fundamental cell (**Fig. 5d**) achieves a true random number generation throughput of 29.8 bits/ns without additional area cost, reducing total area by 1.83 times compared to state-of-the-art CMOS systems[11].

Conclusion

The success of very large-scale integrated circuits is based on CMOS fundamental cells. By adhering to this design principle, spintronic foundation cells reduce the design complexity of scalable unconventional computing. Emerging device technologies, beyond spintronics, must also establish their fundamental cells for scalability designs to integrate with existing CMOS designs and other devices. Co-design, where hardware and software are developed in tandem, plays a crucial role in this process.

Acknowledgments: This work was supported by ACCESS – AI Chip Center for Emerging Smart Systems, sponsored by InnoHK funding, Hong Kong SAR and National Key R&D Program of China (Grants No.2021YFA1401500)

References: [1] C. D. Schuman *et al.*, *Nat. Comput. Sci. 2022 21*, vol. 2, no. 1, pp. 10–19, Jan. 2022, doi: 10.1038/s43588-021-00184-y. [2] Q. Shao *et al.*, Dec. 2021, Available: https://arxiv.org/abs/2112.02879v2. [3] Z. Wang *et al.*, *Nat. Rev. Mater.*, 2020, doi: 10.1038/s41578-019-0159-3. [4] Q. Shao, *et al.*, *Nat. Rev. Electr. Eng. 2024 111*, vol. 1, no. 11, pp. 694–695, Oct. 2024, doi: 10.1038/s44287-024-00106-w. [5] Z. Xiao *et al.*, in *Technical Digest - International Electron Devices Meeting, IEDM*, 2022. doi: 10.1109/IEDM45625.2022.10019482. [6] Z. Xiao *et al.*, *Sci. Adv.*, vol. 10, no. 38, p. eadp3710, Sep. 2024, doi: 10.1126/SCIADV.ADP3710/SUPPL_FILE/SCIADV.ADP3710_SM.PDF. [7] Y. Liu *et al.*, 2022. doi: 10.48550/arXiv.2209.09443. [8] K. Qian *et al.*, in *Technical Digest - International Electron Devices Meeting, IEDM*, Dec. 2024. [9] X. Wu *et al.*, *Phys. Rev. Appl.*, vol. 20, no. 2, p. 024069, Aug. 2023,doi:10.1103/PHYSREVAPPLIED.20.024069/FIGURES/14/MEDIU M. [10] N. S. Singh *et al.*, in *Technical Digest - International Electron Devices Meeting, IEDM*, 2023. doi: 10.1109/IEDM45741.2023.10413686. [11] Z. Xiao *et al.*, in *Technical Digest - International Electron Devices Meeting, IEDM*, Dec. 2024.

An Efficient Pipeline Programming Scheme
Based on 40nm PCM Compute-in-Memory Chip for CNNs

Xile Wang[1#], Longhao Yan[1#], Xi Li[5], Yaoyu Tao[1,3], Zhitang Song[5], and Yuchao Yang[1,2,3,4*]

[1]Beijing Advanced Innovation Center for Integrated Circuits, School of Integrate Circuits, Peking University, Beijing, China. [2]Center for Brain Inspired Intelligence, Chinese Institute for Brain Research (CIBR), Beijing, China. [3]Center for Brain Inspired Chips, Institute for Artificial Intelligence, Peking University, Beijing, China. [4]Guangdong Provincial Key Laboratory of In-Memory Computing Chips, School of Electronic and Computer Engineering, Peking University, Shenzhen, China. [5]State Key Laboratory of Functional Materials for Informatics, Shanghai Institute of Microsystem and Information Technology, Chinese Academy of Sciences.
[#]These authors contributed equally. *Corresponding Author's Email: yuchaoyang@pku.edu.cn

Abstract

Phase change memory has emerged as a promising candidate for in-memory computing. The conventional write-verify programming scheme for multilevel cells necessitates numerous iterative programming cycles and is susceptible to conductance drift effects. This paper presents the adaptive pulse width programming and pipeline programming scheme to reduce programming cycles by 71.2% and mitigate the conductance drift. Furthermore, we validate the proposed method through image classification tasks, achieving an accuracy of 91.7% on a PCM-based compute-in-memory chip.

Keywords: multilevel cells, phase change memory, compute-in-memory

Introduction

Emerging non-volatile devices have been widely researched in various computing applications [1-3], particularly to enable in-memory computing, which aims to overcome the limitations of the traditional von Neumann architecture. Phase-change memory (PCM), characterized by its high density and non-volatile properties, is considered a key storage medium for in-memory computing. Moreover, PCM supports multi-level cells (MLCs), which further increases memory density. Despite these advantages, MLC programming still encounters challenges such as slow programming speed and significant conductance drift [4].

Here, we fabricate the PCM-based compute-in-memory chip using the 40 nm CMOS process and design Adaptive Pulse Width Programming Scheme (APWPS) which refines the conventional Write-Verify (W&V) scheme by adjusting pulse widths dynamically. Additionally, we develop a Pipeline Programming Scheme (PPS) which mitigates conductance drift, achieving a more compact and stable conductance distribution. Ultimately, this method significantly reduces programming cycles by 71.2% and achieves a near-software-level accuracy of 91.7% in CNN-based image classification tasks.

Adaptive Pulse Width Programming Scheme

We present APWPS for the 40nm PCM compute-in-memory chip. Figure 4 illustrates the microphotograph and the chip architecture. The chip integrates a 144Kb PCM 1T1R array alongside peripheral circuits, including programming circuits, drivers, and successive approximation analog-to-digital converter (SAR-ADC). The programming circuit, as depicted in Figure 5, consists of a decoder, ten sets of current mirror and transmission gates. The decoder's inputs, CK and WORD<3:0>, control the programming current magnitude by switching the transmission gates. The RESET programming pulse is a single current rectangular pulse designed to reset the PCM cell to a high-resistance state. In contrast, the SET programming pulse is a step pulse composed of four current pulses with decreasing amplitudes, intending to switch the cell from a high-resistance state to a low-resistance state.

During the programming process, a RESET pulse is first applied to the cell to reset it to a high-resistance state, followed by SET pulses with varying pulse widths to program the cell into different conductance states. We test the proposed programming approach on 1024 cells, as shown in Figure 6. The results demonstrate that the change of PCM conductance can be quantified as a function of the SET programming pulse width, exhibiting good continuity. Compared to programming schemes that adjust pulse amplitude, adjusting the pulse width yields a monotonic and smoother programming curve, significantly enhancing the convergence speed of the W&V programming algorithm. Furthermore, Figure 7 shows the success rate of programming cells to MLC with a single pulse. The results indicate that the programming success rate is nearly 100% under the Single-Level Cell (SLC) storage scheme. However, as the number of conductance states increases, the success rate gradually decreases.

To address this issue, we design an adaptive pulse width write-verify programming scheme, as detailed in Figure 8. Initially, the target conductance of PCM is input into the programming model to determine the initial SET programming pulse width, followed by an initial programming pulse, consisting of a RESET pulse and a SET pulse. Subsequently, the PCM conductance state is verified. If the conductance of the cell falls within the target range [$G_{target-}$, $G_{target+}$], the programming is

979-8-3315-0417-5/25 $31.00 © 2025 IEEE

successful. Otherwise, the error between the measured conductance and the target conductance, $\triangle$G, is input into the programming model to update the SET programming pulse width, and the cell is reprogrammed. This process is repeated until the programming is successful or the maximum number of programming attempts is reached. Compared to traditional W&V programming schemes with a fixed step size, APWPS rapidly converges to the optimal programming conditions, significantly reducing the number of iterative programming attempts. Besides accelerating the programming speed, it also effectively improves the programming success rate, as demonstrated in Figures 9 and 10.

Pipeline Programming Scheme

In this work, we propose a pipeline programming scheme for PCM compute-in-memory chip, aimed at mitigating the PCM conductance drift effects. Figure 11 depicts the cumulative distribution of eight representative conductance states programmed by the W&V scheme. The results indicate that while the conductance distributions are well-separated immediately after programming, they overlap after a 12-hour period due to conductance drift effects. Studies have shown that increasing the delay time between programming and verification operations in the W&V process can effectively suppress the impact of conductance drift effects [5]. We research the impact of different W&V delay time on MLC programming outcomes, as shown in Figure 12. The results demonstrate that a delay of T_{delay}=6s significantly suppresses conductance reduction caused by PCM conductance drift and broadens conductance distribution due to relaxation effects, compared to a delay of T_{delay}=6ms. Consequently, we propose PPS that effectively inhibits PCM conductance drift effects while maintaining programming speed. Figure 13 illustrates the detailed step of the PPS. Once the cell is programmed, its address is added to the rear of the programming queue. When the queue is full, the address at the front of the queue is removed and the conductance state of the corresponding device is verified. If the programming fails, i.e. G is not within [$G_{target-}$, $G_{target+}$], the device is reprogrammed, and its address is re-added to the rear of the queue. Otherwise, a new device from the PCM array is selected for programming, and its address is added to the rear of the programming queue. Notably, in our programming scheme, the duration of single programming and verification is fixed to ensure that the delay time from programming to verification is the same for each device. We apply this method to a 144Kb PCM compute-in-memory chip for 4-bit programming test, with results shown in Figure 14. As the length of the pipeline programming queue increases, the standard deviation of cell conductance distribution and the shift caused by drift gradually decrease. Additionally, we compare the programming speed of schemes with and without the use of PPS and APWPS. The results indicate that using PPS and APWPS significantly enhances the bulk programming speed for the chip.

To further illustrate the impact of the PPS on the inference accuracy of Convolutional Neural Networks (CNNs), we implement the Conv2d_15 layer of ResNet-20 on the chip, while the remaining layers are implemented in software, and test its performance on the CIFAR-10 dataset. Figure 15 shows the mapping scheme of the CNN layer to the chip, where each weight in the network is differentially represented by two PCM cells in the array. Figure 16 presents our board-level test system and experimental results, demonstrating that as the length of the pipeline programming queue increases, the inference accuracy improves from 89.45% (LP=1.5K) to 91.7% (LP=120K).

Conclusion

In this paper, we propose an efficient programming pipeline scheme (APWPS+PPS) based on 40nm PCM compute-in-memory chip for CNNs. This programming scheme significantly enhances the programming efficiency and reliability of MLC PCM, achieving a 71.2% reduction in programming cycles and an 8% increase in programming success rates compared with conventional W&V programming scheme. The inference accuracy on the CIFAR-10 dataset reaches 91.7%, approaching the software result.

Acknowledgments

This work is supported by the National Key R&D Program of China (2023YFB4502200), National Natural Science Foundation of China (61925401, 92064004, 61927901, 8206100486, 92164302), Beijing Natural Science Foundation (L234026) and the 111 Project (B18001).

References

[1] K. Liu et al., "Tuning the ferroelectricity of $Hf_{0.5}Zr_{0.5}O_2$ with alloy electrodes," Science China Information Sciences, vol. 67, no. 182402 (2024).

[2] R. Yuan et al., "Efficient 16 Boolean logic and arithmetic based on bipolar oxide memristors", Science China Information Sciences, vol. 63, no. 202401 (2020).

[3] L. Yan et al., "Uncertainty Quantification Based on Multilevel Conductance and Stochasticity of Heater Size Dependent C-doped $Ge_2Sb_2Te_5$ PCM Chip", IEDM, 28.2. 1-28.2. 4 (2021).

[4] Q. Wu et al., "Hybrid Program Algorithm Enables Significant Reduction in Write Latency and Power Consumption for Multilevel Phase Change Memory", TED, pp. 4145 - 4149 (2023).

[5] S. R. Nandakumar et al., "Precision of synaptic weights programmed in phase change memory devices for deep learning inference", IEDM, 29.4. 1-29.4. 4 (2020).

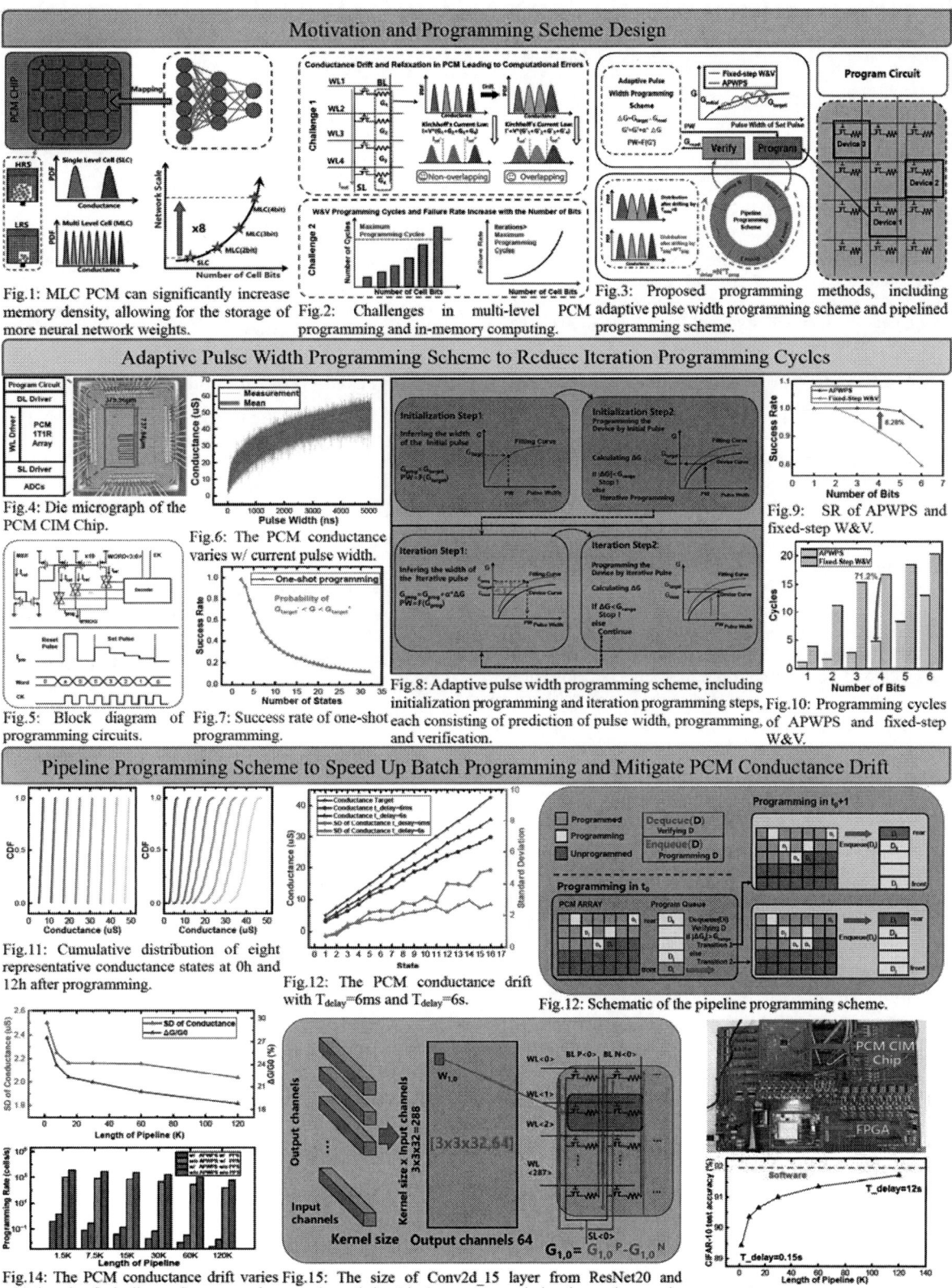

Motivation and Programming Scheme Design

Fig.1: MLC PCM can significantly increase memory density, allowing for the storage of more neural network weights.

Fig.2: Challenges in multi-level PCM programming and in-memory computing.

Fig.3: Proposed programming methods, including adaptive pulse width programming scheme and pipelined programming scheme.

Adaptive Pulse Width Programming Scheme to Reduce Iteration Programming Cycles

Fig.4: Die micrograph of the PCM CIM Chip.

Fig.6: The PCM conductance varies w/ current pulse width.

Fig.5: Block diagram of programming circuits.

Fig.7: Success rate of one-shot programming.

Fig.8: Adaptive pulse width programming scheme, including initialization programming and iteration programming steps, each consisting of prediction of pulse width, programming, and verification.

Fig.9: SR of APWPS and fixed-step W&V.

Fig.10: Programming cycles of APWPS and fixed-step W&V.

Pipeline Programming Scheme to Speed Up Batch Programming and Mitigate PCM Conductance Drift

Fig.11: Cumulative distribution of eight representative conductance states at 0h and 12h after programming.

Fig.12: The PCM conductance drift with T_{delay}=6ms and T_{delay}=6s.

Fig.12: Schematic of the pipeline programming scheme.

Fig.14: The PCM conductance drift varies with length of pipeline and programming rate by different scheme.

Fig.15: The size of Conv2d_15 layer from ResNet20 and mapping relationship between convolutional neural networks and PCM compute-in-memory chip.

Fig.16: Testing system and inference accuracy of CIFAR-10.

979-8-3315-0417-5/25 $31.00 © 2025 IEEE

Investigation on the Effect of Self-clocking in MEMS Gyroscope

Xuewen Liu[1], Zhiyuan Wang[1], Hongsheng Li[1]

[1]School of Instrument Science and Engineering, Southeast University, Nanjing 210096, China
e-mail:230208842@seu.edu.cn;220223314@seu.edu.cn;hsli@seu.edu.cn

Abstract—This paper investigates the mechanism of self-clocking in digital systems of MEMS gyroscopes, aiming to achieve stable phase-frequency characteristics by synchronizing the sampling rate with the input signal frequency. Experimental validations are performed to demonstrate the effectiveness of self-clocking in improving the temperature stability of gyroscopes. The results show that self-clocking indeed contributes to minimizing the bias drift with temperature variations, thereby improving the overall performance and reliability. Moreover, a comprehensive analysis of various self-clocking application scenarios is presented, including the principles of the self-clocking and the specific functions of each clock. These findings provide valuable insights for implementing self-clocking in digital MEMS gyroscope systems.

Keywords—MEMS Gyroscope; Digital system; Self-clocking; Temperature stability

I. INTRODUCTION

Microelectromechanical systems (MEMS) gyroscopes are extensively utilized in various domains such as automotive electronics, industrial robotics, and aerospace navigation due to their compact size, easy integration, and low power consumption [1]. Rapid advances and integration capabilities of Application Specific Integrated Circuits (ASICs) have made ASIC-based MEMS gyroscope digital control systems the dominant approach in this field [2]. However, the sensitivity of gyroscope outputs to ambient temperature poses a challenge for their further application. One of the primary errors affecting the bias temperature stability is the quadrature error caused by the stiffness coupling between two modes [3]. Phase-sensitive demodulation is typically employed to suppress the quadrature error signal due to its inherent 90° phase difference from the Coriolis signal. However, the signal processing stage often introduces additional phase delay [4]. The delay error varies with temperature and causes the quadrature error to leak into the Coriolis signal during demodulation [5], which impacts the overall performance of the gyroscope system. To eliminate the quadrature error, orthogonal electrostatic forces 90° out of phase with the Coriolis force are typically applied to force feedback electrodes [6]. This technique, known as orthogonal force feedback control, necessitates precise phase requirements for the applied orthogonal feedback forces. Another method involves the design of a dedicated orthogonal stiffness correction electrode within the mechanical structure of the gyroscope to compensate for the orthogonal error by introducing electrostatic stiffness [7]. However, this method requires a dedicated electrode for quadrature error correction,

which is not applicable to all gyroscope structures. In MEMS gyroscope systems based on Sigma-Delta modulators, the system clock and ADC sampling clock commonly derive from a self-clocking scheme where the gyroscope's drive resonance frequency drives all the clocks [8]. This methodology ensures long-term stability in temperature-varying environments.

II. THEORY

A. Self Clocking Principle

In the open-loop control mode, the zero bias output of a MEMS gyroscope can be simplified as follows

$$\Omega_0 = \Omega_i cos(\varphi_e) + \Omega_q sin(\varphi_e) \tag{1}$$

where φ_e is the demodulation phase error, the Ω_q is expressed as an quadrature error, Ω_i represent the in-phase channel errors. In general, the phase delay of the signal processing loop before demodulation changes with the temperature T, due to the drift of the resonance frequency f_d with T. The architecture of the gyroscope system in this paper adopts a mostly-digital implementation, where the signal processing is implemented in the digital domain. The signal processing flow is illustrated in Fig.1.

The phase delay of the signal processing before demodulation, without distinguishing between discrete and continuous domain, can be expressed as.

$$\varphi_e = \varphi_{cv}(w_d) + \varphi_z(w_d) - \varphi_d \tag{2}$$

the terms $\varphi_{cv}(w)$ and $\varphi_z(w)$ represent phase-frequency characteristics of the CV conversion and digital signal processing stages respectively. φ_d is the demodulation phase can be adjusted in digital system.

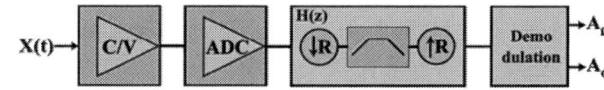

Fig. 1. Gyroscope signal processing block diagram

Fig.2 shows the simulation of the variation of phase delay in the signal processing module with driving frequency. The simulation results indicate that as the resonant frequency of the gyroscope changes due to environmental variations, the phase delay in the signal processing also varies accordingly. Within the range of gyroscope frequency variation, there exists a linear relationship between the phase delay and the input signal frequency and the phase delay decreases as the input frequency increases.

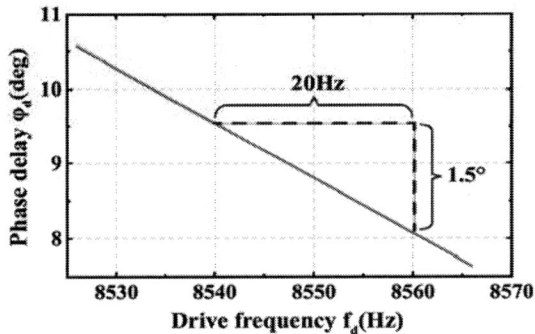

Fig. 2. Simulation Results of Phase Delay in Signal Processing

According to the definition of the z-transform, the analog frequency f corresponds to the digital angular frequency w as follows

$$w = 2\pi f(\frac{1}{f_s}) = 2\pi \cdot f/f_s \qquad (3)$$

where w represents the digital angular frequency, f_s is the data sampling rate, and f denotes the input signal frequency. When the sampling frequency of the signal processing stage is an integer multiple of the gyroscope signal frequency, the phase-frequency characteristic $\varphi_z(w)$ of the digital signal processing stage is

$$\varphi_z(w) = \varphi_z(2\pi \cdot \frac{f_d}{N f_d}) = \varphi_z(2\pi/N) \qquad (4)$$

Due to the fixed filter coefficients in the digital system, the phase delay of the digital system processing remains unchanged in the case of sampling self-clocking. The variation in phase error will significantly decrease with temperature changes, thereby reducing the leakage of orthogonal errors into the zero-rate signal and improving the bias stability.

B. Different Self-clocking Cases Analysis

When dealing with the self-clocking scheme, the difference between sampling clock multiplication and system clock multiplication has not been discussed. To provide a better understanding of the principles and working mechanisms of self-clocking, a detailed analysis of the various self-clocking scenarios is conducted in this paper.

According to the mainstream types of ADCs and sampling methods in MEMS gyroscope control systems, the clock multiplication scenario is illustrated in Fig.3. For the triggered-sampling mode at low sampling rates, the sampling rate $f2$ of the ADC is significantly lower than the digital system control clock frequency $f1$. In this case, the sampling clock Clk2 determines the sampling rate of the digital signal processing stage $h(z)$, which needs to be multiplied to an integer multiple of the input signal frequency. Similarly, for continuous sampling mode, the sampling clock Clk2 is derived from the ADC control clock Clk3. So it is sufficient to multiply the ADC clock Clk3 to achieve the desired sampling clock frequency.

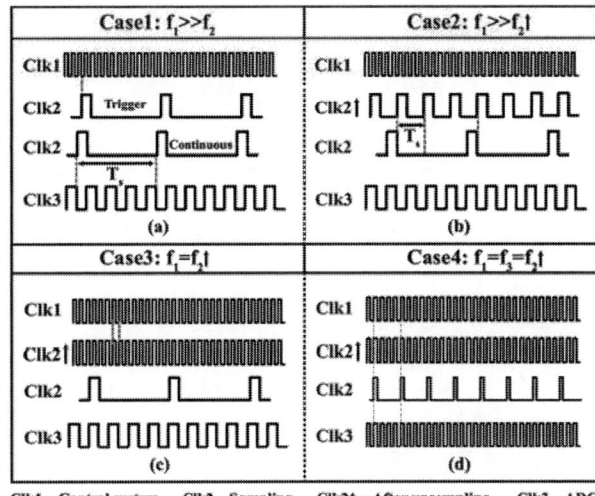

Fig. 3. Various Self-clocking Scenarios

As shown in case 2, in the case of low-speed sampling followed by upsampling within the system, if the upsampling clock frequency $f2 \uparrow$ is lower than the system clock frequency $f1$, the data update rate is still independent of the system clock Clk1. The difference here from case 1 is that the internal upsampling module also require self-clocking. For case 3, the upsampling clock $f2 \uparrow$ is close to the system clock, the sampling rate of $H(z)$ is still determined by Clk2. However, in this case, considering the clock synchronization between the system control clock Clk1 and the data update rate clock Clk2↑, Clk1 and Clk2↑ must share the same clock, which means that the system control clock and the upsampling clock are equal and both locked to an integer multiple of the drive resonance frequency.

Case 4 illustrates the sampling method of the sigma-delta ADC in this paper. Typically, the clock frequency of the Sigma-Delta modulator is comparable to the system control clock frequency. Therefore, according to the aforementioned principles, considering self-clocking to eliminate phase drift and achieve clock synchronization, both the system control clock Clk1 and the ADC control clock Clk3 all need to be locked to the fixed radio of the drive resonance frequency.

III. EXPERIMENTS

The generation of the multiplied clock is implemented by the circuit shown in Fig.4 (a), the drive detection signal $X(t)$ is first converted into a square wave signal at the drive frequency using a low-pass filter and comparator, then the square signal is processed through a PLL and a frequency division loop within the digital system to generate the system control clock. and the frequency of the drive mode and the system is collected by frequency meter for verifying the effectiveness of the scheme.

Fig.5 shows the digital signal obtained by the Signaltap in FPGA of the zero-rate displacement signal converted by the ADC and digital system. This illustrates the effect of the

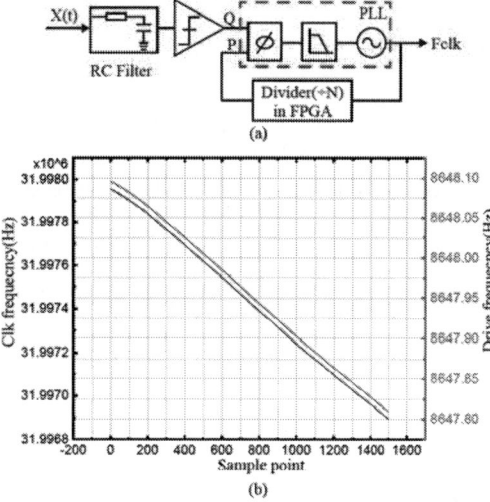

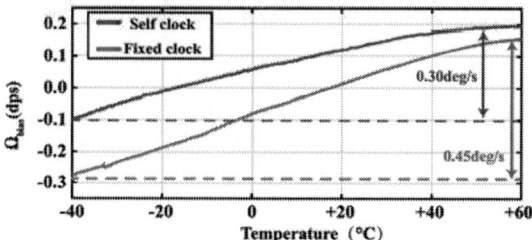

in this work.

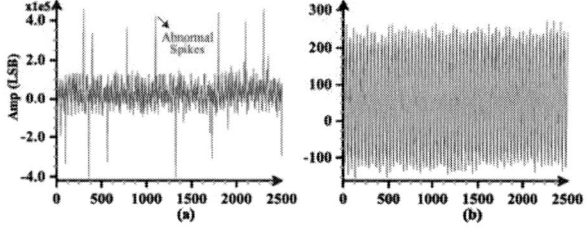

Fig. 4. (a) Self-clocking scheme and (b) drive frequency and system clock frequency test results

Fig. 6. Comparison of the bias output with and without selfclock

CONCLUSION

This paper explores the principle of self-clocking in digital systems of MEMS gyroscopes, which achieves stable phase-frequency characteristics by synchronizing the sampling rate with the input signal frequency. A zero bias test was conducted over the temperature range from $-40°$ to $+60°$ to demonstrate the effectiveness of self-clocking in improving the temperature stability of gyroscopes. Furthermore, the paper presents a comprehensive analysis of various self-clocking application scenarios, including the choice of clock in self-clocking scheme and the specific functions of each clock. And the influence of clock synchronization on signal processing verifies the importance of high-speed sampling clock and system clock synchronization. This research provides valuable knowledge and understanding of self-clocking mechanisms, offering contributions to the practical implementation of self-clocking in MEMS gyroscope digital systems.

synchronization of the system clock and ADC clock on digital signal conversion. As discussed in Fig.3 (d), When the two clocks are not synchronized, which leads to timing confusion in data processing and resulting in abnormal data. As shown in Fig.5(a), the converted digital quantity deviates from the actual value and exhibits abnormal noise values, resulting in a serious decrease in the signal-to-noise ratio. In Fig.5 (b), the conversion of the zero-rate displacement signal is normal when the two clocks are synchronized.

REFERENCES

[1] N. Yazdi, F. Ayazi, K. Najafi. "Micromachined inertial sensors," Proceedings of the IEEE, 1998, 86(8): 1640-1659.

[2] Y. Wang, X. Liu, W. Zhang, et al. "A Mode-Matched DRG Realized by Monolithic Integrated Interface Circuit Design," IEEE Sensors Journal, 2022, 22(19): 18658-18668.

[3] J. Cui, Q. Zhao. "Bias thermal stability improvement of MEMS gyroscope with quadrature motion correction and temperature self-sensing compensation," Micro & Nano Letters, 2020, 15(4): 234-238.

[4] LG. Pagani , L. Guerinoni, L. Falorni, et al. "Direct phase measurement and compensation to enhance MEMS gyroscopes ZRO stability,". Journal of Microelectromechanical Systems, 2021, 30(5): 703-711.

[5] H. Wu, X. Zheng, X. Wang, et al. "Effects of both the drive-and sense-mode circuit phase delay on MEMS gyroscope performance and real-time suppression of the residual fluctuation phase error," Journal of Micromechanics and Microengineering, 2021, 31(5): 055006.

[6] J. Seeger, A. J. Rastegar, and M. T. Tormey, "Method and apparatus for electronic cancellation of quadrature error," U.S. Patent 0 180 908 A1, Aug. 9, 2007

[7] H. Li, H. Cao, Y. Ni. "Electrostatic stiffness correction for quadrature error in decoupled dual-mass MEMS gyroscope," Journal of Micro/Nanolithography, MEMS, and MOEMS, 2014, 13(3): 033003-033003.

[8] A. Elsayed, A. Elshennawy, A. Elmallah, et al. "A self-clocked ASIC interface for MEMS gyroscope with $1 m°/s/\sqrt{Hz}$ noise floor," 2011 IEEE Custom Integrated Circuits Conference (CICC). IEEE, 2011: 1-4.

Fig. 5. Output of the detected displacement signal under (a) asynchronous clock and (b) synchronous clock conditions.

In order to verify the effectiveness of the self-clock in improving stability, the bias outputs of the gyroscope over $-40°$ to $+60°$ temperature range with selfclock and fixed clock are collected by the LabVIEW. As shown in the Fig.6. The temperature drift of zero bias is measured to be 0.3 deg/s (dps) and 0.45 deg/s respectively, which is improved by 33% without any compensation. Based on the simulation results of the phase delay variation with resonant frequency in Fig.2, when the resonant frequency changes by 17Hz within the temperature range from -40 to $+60$, Equivalent quadrature coupling error is 0.26 dps when the equivalent quadrature error is 12 dps. Although the bias temperature drift still exists, One possible reason could be attributed to the lower precision of the selfclock. As a result, there is a need for further improvement

High Density and High Reliability (H²DR) RRAM for Advanced Memory Technology

Zongwei Wang[1], Zimeng Wu[1], Lin Bao[2], Qishen Wang[1], Yuhang Yang[1], Shengyu Bao[1], Jingwei Sun[1], Cuimei Wang[1] & Yimao Cai[1], Ru Huang[1]

[1] School of Integrated Circuits, Beijing Advanced Innovation Center for Integrated Circuits, Peking University, Beijing 100871, P. R. China
[2] State Key Laboratory of Information Photonics and Optical Communications, Beijing University of Posts and Telecommunications, Beijing 100876, P. R. China

Abstract

The rapid advancement of emerging fields such as artificial intelligence has significantly increased the demand for high-performance embedded memory. Resistive random-access memory (RRAM) has received substantial attention due to its simple structure and exceptional integration density. This paper addresses the critical requirements of high-density and high-reliability (H²DR) RRAM, providing an in-depth analysis of key technologies for both 2D and 3D RRAM. Involving oxygen vacancy modulation, innovative memory cell designs, and advanced programming methodologies. In combination, these advancements establish a technical foundation for the large-scale integration of RRAM in future applications.

Keywords: RRAM, High Density, High Reliability

Introduction

The development of emerging fields such as artificial intelligence, autonomous driving, and the Internet of Things (IoT) has necessitated heightened demands on the memory density of embedded memories [1]. In recent times, CMOS-based logic processes have achieved remarkable progress, with technology nodes scaling down to 5nm. Conversely, the miniaturization trend of embedded flash memory technology has reached a plateau at the 55nm process node, resulting in a significant mismatch in integration with advanced logic circuits [2]. To address the requirement for high-density and high-reliability embedded memories in advanced system on chip (SoC), numerous novel memory technologies have emerged, such as RRAM, PCRAM and FeRAM [3]. Among these, resistive random-access memory (RRAM) has gained extensive attention due to its simple structure, high integration density, fast operational speed, and exceptional compatibility with in-memory computing (IMC) paradigms [4]. In this paper, to meet the high-density & high-reliability (H²DR) demands of RRAM, we proposed comprehensive optimization spanning the device-cell-array hierarchy. In device level, RRAM faces the challenge of oxygen vacancies migration, we introduce a physics model to predict the migration and SPS technology to constrain the migration. In cell level, against the mismatch between transistor drive current and RRAM switching current, we put forward dynamic-gate technology and self-selective cell to realize high density memory. In array level, we proposed novel algorithm and thermal-effect model for high programming

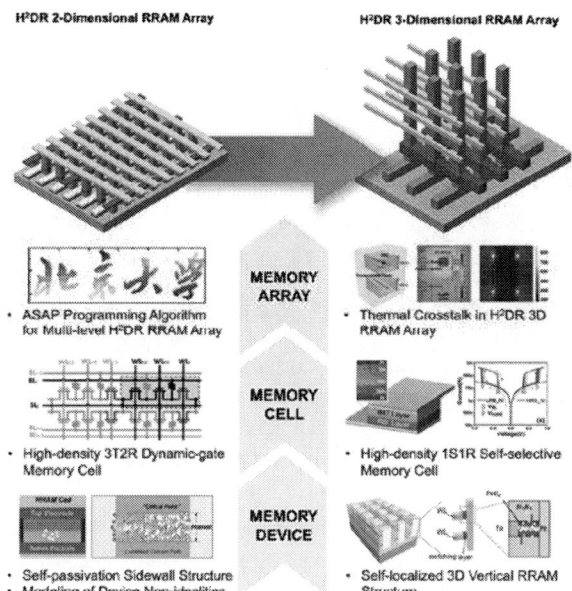

Fig. 1: H²DR RRAM technology: from 2D RRAM to 3D RRAM.

efficiency and high reliability.

Two-Dimensional (2D) H²DR RRAM Technology

A. Device Optimization for 2D H²DR RRAM

The distribution and migration of oxygen vacancies play a vital role in determining the reliability of RRAM devices. Modeling and regulating the microscopic migration behavior of these vacancies are crucial steps towards enhancing the performance of RRAM. Herein, we introduce a physical model designed to predict the migration behavior of oxygen vacancies in RRAM, as illustrated in Fig. 2(a) [5]. This model establishes a correlation between the microscopic migration of oxygen vacancies and the electrical performance of the device, thereby offering valuable insights for regulating the migration of oxygen vacancies. Based on this physical model, we propose the SPS self-oxidizing sidewall process, which effectively constrains the migration of oxygen vacancies and sidewall defects. The device structure and corresponding Transmission Electron Microscope (TEM) images are shown in Fig. 2(b) & (c). The SPS technology uses the self-passivation sidewall formed during the etching process to hinder the exchange or migration of oxygen vacancies, without the need for high-temperature annealing. This ensures the reliability of the fabricated devices. Experimental results demonstrate that the cumulative distribution of all dies

979-8-3315-0417-5/25 $31.00 © 2025 IEEE

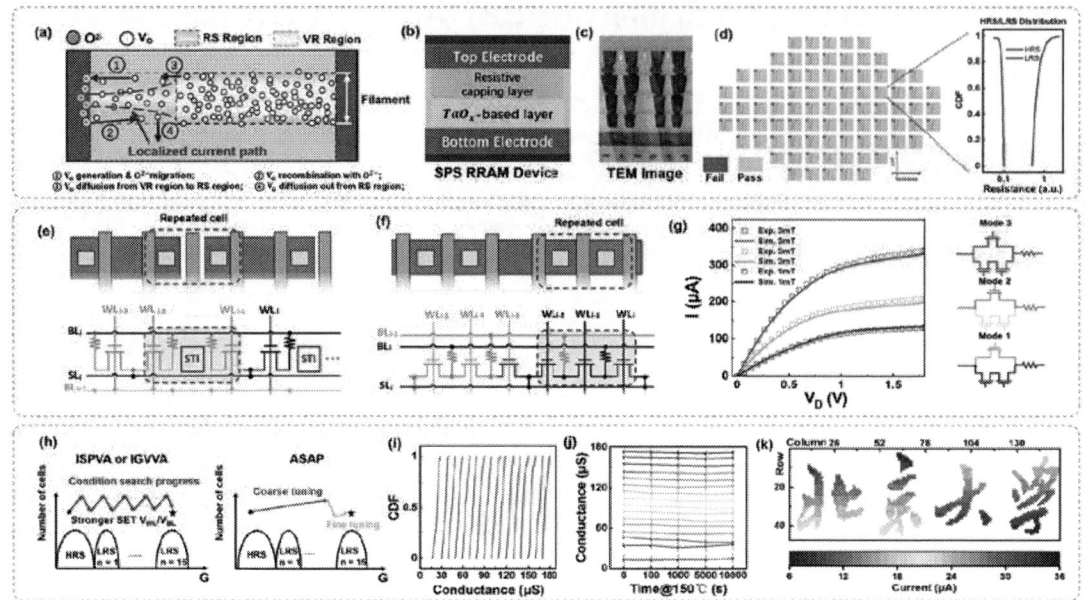

Fig. 2: (a) Physical model of RRAM. (b), (c) The structure of SPS RRAM device and corresponding TEM image. (d) The devices have excellent uniformity on a 12-inch wafer. (e), (f) The conventional 2T2R cell and novel STI-less dynamic-gate cell. (g) Multiple drive modes of dynamic-gate cell. (h) ASAP programming strategy. (i), (j), (k) The RRAM array can realize 4-bit information storage after adopting ASAP algorithm.

on a 12-inch wafer retains excellent uniformity after 50 operation cycles, validating the effectiveness of our proposed approach (Fig. 2(d)). After applying thermal stress conditions of 150 °C and 10^4 seconds, SPS cells show high MLC reliability (Fig. 2(j)).

B. Dynamic-gate Cell for 2D H²DR RRAM

The co-optimization of transistor and RRAM devices is essential for enhancing both cell density and reliability. As process nodes continue to scale down, the conventional 2T2R cell faces challenges due to the limited drive current of the transistor (Fig. 2(e)). This significantly inhibits the potential increase in memory density. To address this issue, we propose a 3T2R STI-less dynamic-gate (DG) cell, as depicted in the Fig. 2(f) [6]. In this design, the STI dummy gate is replaced by a transistor, enabling the RRAM device to be driven by multiple transistors. This configuration results in a cell density increase of approximately 24% compared to the conventional 2T2R structure. Furthermore, by dynamically adjusting the drive modes of the multiple gates, multiple drive currents can be realized. This provides high flexibility in specific current compliance, allowing it to effectively adapt to various application scenarios (Fig. 2(g)).

C. Fast Programming Strategy for 2D H²DR RRAM Array

The memory density of RRAM can be significantly enhanced through the development of multi-level cell (MLC) technology. However, achieving efficient and reliable multi-level programming within arrays poses a major challenge for MLC technology. To tackle this bottleneck, we introduce the Adaptive Step Adjustment Programming (ASAP) programming algorithm, as illustrated in the Fig. 2(h) [7]. In contrast to traditional incremental step pulse verification

algorithm (ISPVA) and incremental gate voltage verification algorithm (IGVVA) methods, ASAP allows for dynamic initial operation conditions and step sizes, leading to great improvements in the programming efficiency of MLCs within arrays. The effectiveness of the ASAP algorithm is demonstrated in Fig. 2(i), (j) & (k), where 7.3k cells were selected from the array to realize 16-level conductance levels.

Three-Dimensional (3D) H²DR RRAM Technology

A. Encapsulated Device for 3D H²DR RRAM

In conventional 3D vertical RRAM (VRRAM) cells, oxygen vacancy migration between devices is a main cause of endurance degradation [8]. However, achieving physical isolation between vertical devices is a complex process with excessive process cost. To address this issue, we proposed encapsulated 3D VRRAM devices [9]. These devices utilize a thermal oxidation process to grow self-limiting TaO_x resistive layers on the sidewall Ta electrodes, effectively achieving physical isolation between adjacent devices (Fig. 3(a)). Electrical test results demonstrate that the device on/off ratio remains above 10 even after 10^7 cycles, indicating excellent endurance characteristics (Fig. 3(b) & (c)).

B. Self-selective Cell for 3D H²DR RRAM

The incorporation of transistors in 3D VRRAM as selectors presents significant integration challenges and process costs. To overcome these obstacles, we propose self-selecting RRAM cells that integrate threshold switching (TS) layer and resistive switching (RS) layer stacks [10]. The device structure, depicted in Fig. 3(d), features VO_2 as the TS layer and HfO_2 as the resistive layer. The hybrid switching mechanism can be attributed to the coexistence of nonvolatile resistive transitions dominated by oxygen vacancies and

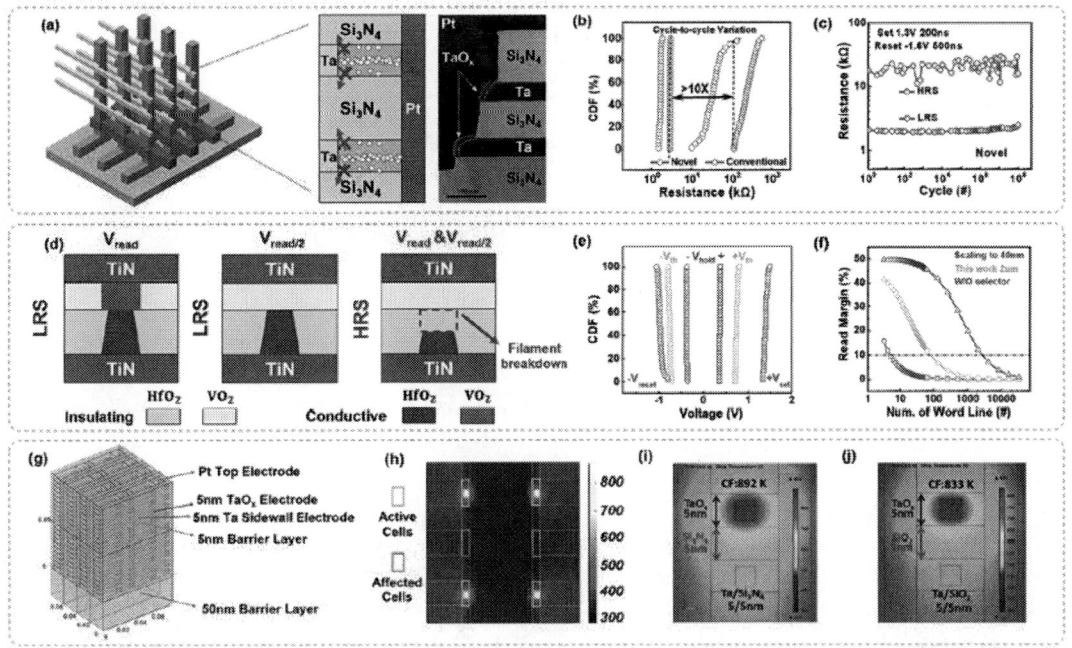

Fig. 3: (a) The structure of encapsulated 3D VRRAM devices and corresponding TEM image. (b), (c) The device on/off ratio remains above 10 even after 10^7 cycles, indicating excellent endurance characteristics. (d) The schematic of hybrid switching mechanism in self-selective RRAM cell. (e) The cell maintains exceptional parameter consistency after 50 cycles. (f) Calculated read margin as a function of array size based on measured data. (g) The model of 3D VRRAM array for temperature simulations. (h) Calculated temperature profile when programing 3D VRRAM array. (i) & (j) Temperature distribution of the array with Si_3N_4 and SiO_2 interlayer isolation materials.

insulator-metal-transition triggered by an electric field. Measurements reveal that the devices maintain exceptional parameter consistency after 50 cycles (Fig. 3(e)). Furthermore, array simulation results show that the self-selective cell array size can be enhanced to $10^3 \times 10^3$ when scaled to 40 nm node. (Fig. 3(f)).

C. Thermal-effect Modeling for 3D H²DR RRAM Array

The temperature field crucially influences the distribution and migration of oxygen vacancies. In high-density 3D VRRAM arrays, the limited heat dissipation capability of the array can lead to heat accumulation and severe self-heating effects, ultimately compromising device reliability. To assess the self-heating effect in these arrays, we developed a model of 3D VRRAM array and conducted array-level temperature simulations using a finite element analysis method (Fig. 3(g)) [5, 11]. The results illustrate the temperature profile during array operation (Fig. 3(h)). Additionally, Fig. 3(i) & (j) depicts the temperature distribution of the array with various interlayer isolation materials. These results helped us to propose thermal-aware technology optimization methods. In terms of device structure, we proposed SPS technology, in device material, our results suggests that good vacancy/ion-block capability and high thermal conductivity materials are preferred.

Conclusion

This paper addresses the growing demand for density and reliability in advanced memory and key technologies in both planar and three-dimensional RRAM, laying a technical foundation for the subsequent large-scale production and application of RRAM.

Acknowledgments

This work was supported by National Natural Science Foundation of China under Grant 62322401, 62025401, 62341407, and 61927901, and in part by Beijing Nova Program under Grant 20220484113, "111" Project under grant B18001.

References

[1] Q. Xia et al. "Memristive crossbar arrays for brain-inspired computing," Nature Electron., vol. 18, pp. 309–323, Mar. 2019.

[2] Y. Chen, "ReRAM: History, status, and future". IEEE TED, 2020, vol. 67, pp. 1420-1433, Jan. 2020.

[3] M. A. Zidan, et al. "The future of electronics based on memristive systems", Nature Electron., vol. 1, pp. 22–29, Jan. 2018

[4] D. Ielmini, et al. "In-memory computing with resistive switching devices", Nature Electron., vol. 1, pp. 333–343, Jun. 2018.

[5] Y. Cai, et al. "Technology-array-algorithm co-optimization of RRAM for storage and neuromorphic computing: Device non-idealities and thermal cross-talk". IEEE IEDM, 2020: 13.4.1-13.4.4.

[6] Q. Wang, et al. "A Logic-Process Compatible RRAM with 15.43 Mb/mm 2 Density and 10years @ 150°C retention using STI-less Dynamic-Gate and Self-Passivation Sidewall", IEEE IEDM, 2023: 1-4.

[7] J. Sun, et al. "ASAP: An Efficient and Reliable Programming Algorithm for Multi-level RRAM Cell", IEEE IRPS, 2024: 1-4

[8] Y. Chen, et al. "Improvement of data retention in HfO_2/Hf 1T1R RRAM cell under low operating current", IEEE IEDM, 2013: 10.1. 1-10.1. 4.

[9] M. Yu, et al. "Novel vertical 3D structure of TaO_x-based RRAM with self-localized switching region by sidewall electrode oxidation", Scientific reports, vol. 6, pp. 21020, Feb. 2016.

[10] Z. Wang, et al. "Self-selective resistive device with hybrid switching mode for passive crossbar memory application", IEEE EDL, vol. 41, pp. 1009-1012, Jul. 2020

[11] M. Yu, et al. "Encapsulation layer design and scalability in encapsulated vertical 3D RRAM", Nanotechnology, vol. 27, pp. 205202, Apr. 2016

4H-SiC semiconductor for EV and beyond

Hongchao LIU

Anhui YOFC Advanced Semiconductor Co., Ltd

Abstract

4H-SiC (SiC) is an emerging semiconductor material. SiC power devices provide advantages of high efficiency, high voltage, high temperature, high thermal dissipation, and relatively high frequency, which make SiC a "chosen" material for electrical power conversion, a critical material to achieve the goal of Net-Zero Carbon Emission. As a matter of fact, industry is pouring huge fund to build up SiC semiconductor eco-system, the market is also growing extremely fast at CAGR of more than 60% in the past several years. However, there are still some challenges for SiC to penetrate more applications, including the growth rate and defects within SiC material, gate oxide process of SiC MOSFET, reliability and ruggedness of SiC devices, materials system enabling SiC devices for high temperature applications. With innovation of process and device technology, cost down and yield up of SiC industry chain, SiC is poised to play a more and more important role in the energy conversion of electrical vehicle (EV), green energy, track transportation, smart grid, and the emerging AI computing and data storage.

Keywords: 4H-SiC MOSFET, growth rate, crystal defect, reliability, gate oxide, mobility, instability

Introduction

Efficient renewable electricity generation, conversion, and delivery are crucial for achieving the goal of Net-Zero Carbon Emission, which targets global temperature rise to below 2 °C by 2050. Silicon Carbide of polytype 4H (4H-SiC) is emerging as one of the most critical semiconductor materials for power efficient solutions, owing to its exceptional physical and electronic properties. The advantages of 4H-SiC power devices include working at higher temperatures and higher voltages with lower conducting and switching losses, minimized form factor and weight. With maturing of technology and cost down of the supply chain, especially the contribution of substrates from Chinese suppliers, SiC power devices are quickly introduced into several strategic market sectors, e.g. energy conversion in automotive and renewable energies, smart grids, transportation, etc. The electric vehicle (EV) market is now moving toward SiC inverters after the pioneering work of Toyota and Tesla. Now SiC power devices is pervading in the EV as Xiaomi's ubiquitous adoption in power converting applications. SiC device market is in the size of $4-5B, expected to reach between $11 billion and $14 billion in 2030 at an estimated CAGR of 26% [1]. SiC power devices covers voltage rating from 600V to 3.3 kV with potential up to tens of kV. They are poised to enable new applications, such as the replacement of the traditional 60 Hz oil-cooled bulky transformer with small size Solid-State Transformer (SST). Showcases of high-power applications illustrated that SiC-based power conversion system is rapidly approaching a nearly 100 % efficiency [2]. The EVs' accelerating adoption and the increasingly vital role of SiC have SiC makers pouring around $50B for capacity expansion, scaling to 8" fab, and technology development. However, for further expanding the applications and replacing traditional power devices, several challenges must be further addressed. These includes improving scalability of substrate manufacturing in terms of materials growth rate and defects, mobility and stability of MOSFET channel, device ruggedness and reliability, enabling SiC in high temperature applications, as well as cost-performance in the manufacturing chain.

Material and defects

The challenges in scalability of SiC crystal are growth rate and defects. Now SiC industry is transferring to 8 inch, which makes the diameter of substrate is less important. The most critical issue comes from the low growth rate. Currently, almost all SiC substrates are obtained through physical vapor transport (PVT) method. The typical growth rate of PVT is about 0.2–0.5 mm/h, only 1/100 of that of Si ingots. So, ten times faster method must be developed to make the SiC semiconductor affordable and sustainable. The ongoing efforts are being made to refine the PVT growth technology and attempting new growth technologies by Top Seed Solution growth (TSSG) and high temperature chemical vapor deposition (HT-CVD) methods. Using TSSG, the SiC boule can be obtained with growth rate up to mm/h, but the inclusion of solvent remains challenging for semiconductor uses. HT-CVD was early studied in Linkoping with temperature around 2200°C, the latest progress is using much higher temperature in the range of 2500-2600°C, as called gas method in Japan. Vertical-type reactors capable of growing SiC bulk crystals up to 200 mm in diameter have been developed, high growth rate exceeding 3 mm/h has been achieved. The dislocation densities were reported to decline significantly in the direction of growth under proper growth conditions [3].

Crystal defects like basal plane dislocations in SiC is another key issue. As it is known, silicon can be produced perfect without dislocation, however, the best quality of SiC is with dislocation densities in the order of 10^3 cm^{-2} up to date.

979-8-3315-0417-5/25 $31.00 © 2025 IEEE

These defects have adverse effects on wafer device making and device performance, as well as final yield. Furthermore, their densities and defect types vary dramatically between wafers, which makes the process control complicated.

Crystal defects can also be introduced during device manufacturing processes, mainly from epi, ion implant, and activation. Epi defects can be killer defects, which have a high probability to kill the devices, lower final product yield, and even impact reliability. Ion implantation of 4H-SiC is one of the crucial steps in fabricating SiC devices., but this process step results in the generation of electrically active defects. After ion implanted, the anneal process is needed for activating the dopants at temperatures of 1600–1700°C. High temperature process can make Si evaporated, which leads to point defects and surface roughness in the device area. In additional, the high temperature and point defects can trigger stacking faults, which may seriously impact the device performance and reliability [4].

Mobility and instability

Mobility and instability are correlated to the critical gate oxide of SiC MOSFET. Mobility hereafter refers to the surface inversion mobility of MOSFET channel, while the instability refers the shift and drift of MOSFET's V_{th}.

The bulk mobility of the SiC is around 800-900cm^2/VS, which is comparable to that of Si. However, the surface inversion mobility of channel is quite different, 10s cm^2/Vs or less of SiC MOSFET to 500 cm^2/Vs of Si. The low channel mobility of SiC MOSFET is attributed to off-angle cutting of SiC wafer and carbon residual during oxidation of SiC. A post oxide anneal is indispensable for SiC MOSFET gate oxide process, normally nitration by $NO/NO_2/N_2$, which could enhance the mobility up to middle of 20 cm^2/Vs. For a 1200V device, the channel resistance takes up to 40% of total resistance of SiC MOSFET. Therefore, increasing the channel mobility is quite important to the device performance, lower the on specific on-state resistance, which could help to improve of $R_{DS(on)}$, reduce conduction losses. Higher mobility could also be beneficial for using thicker gate oxide, which can improve reliability of the device. Many studies have managed to increase the channel mobility of SiC MOSFET. Using the (1-100) or (11-20) like in the trench SiC MOSFET could increase the mobility to higher than 20 cm^2/VS. Introducing phosphorous or alkaline metals as alternative post oxidation anneal can lead to much higher mobility in the range of 50-100 cm^2/Vs. However, these approaches could make V_{th} worse. Using deposited oxide instead of in-situ oxidation seems to be much practical and a good compromise of enhancement of mobility and stability of V_{th}, even in this case, a post anneal is indispensable [5].

Instability of V_{th} is highly correlated to charges in gate oxide process. Due to various trapping mechanism of interface or near interface, or even in bulk of gate oxide, V_{th} can be shifted with temperature, voltage applied, or drifted after long time stress. This kind of change could lead to malfunction of the devices in applications, which can be the reason of system failure. The stability of V_{th} can be improved through device design and process tune. Recently, an oxidation-minimizing gate oxidation process was proposed with a low subthreshold swing, small threshold voltage shift, and high mobility, especially at low temperatures, about 200 cm^2/Vs at 100 K [6].

Robustness, Ruggedness, and Reliability

In additional to the reliability issues encountered in Si devices, SiC power device has its special reliability problems. As mentioned before, contrary to Si, SiC materials is have a lot of defects, which may affect device reliability, the so-called bipolar degrading is just one kind of it [7]. Bipolar degrading has been observed not only from basal plane dislocations (BPDs) in epitaxial layers but also from BPDs in SiC substrates due to hole injection through epitaxial layers. Under working condition of SiC MOSFET, holes are injected into SiC, causing stacking faults to expand and degrade the device performance [8]. Ion (H^+ or Al^{3+}) implantation was reported to be effective to suppress the expansion of stacking faults in SiC epi layer, which help to alleviate bipolar degrading [9,10].

Besides the bipolar degrading, gate oxide reliability, threshold voltage instability has been intensively studied. Gate oxide in SiC MOSFET is exposed to high voltage stress, which is much more vulnerable than that in Si devices, especially in the trench SiC MOSFET. A positive V_{th} shift attributed to electron trapping leads to an increase in on-resistance and a reduction in power efficiency during application. A negative V_{th} shift due to hole trapping causes enhanced leakage current during the off-state of SiC MOSFET, which has a negative impact on long-term reliability [8]. At high stress fields, holes may be generated via impact ionization in the oxide, and be subsequently trapped in the oxide, yielding to a negative V_{th}' shift. In the meantime, SiC MOSFET normally are used in power module, many devices are configured in parallel, so V_{th} needs to be super stable and consistent in value [11].

As SiC devices move to higher density with reduced pitch size, the ruggedness and robustness of the SiC MOSFET becomes important. The energy density of SiC power devices is much higher than that of Si counterparts, SiC devices has some robustness and ruggedness weakness, namely, the short circuit withstand time (Tsc). Currently, Tsc is around 2-3μs in market available device, there is a need to trade-off between Ron performance and device's robustness when designing devices and process [12].

Enabling high temperature applications

The SiC power device and power IC have the potential to

work under extreme high temperatures and other harsh environment. Si-based devices have the physical limitations to work up to 200°C due to leakage and reliability issue, although some Si-SOI based devices offers slight higher temperature. Since its beginning, SiC has been recognized as a semiconductor capable of working at extreme high temperature [13]. SiC can resist higher temperature, and it's a great thermal conductor. Some recent researches have demonstrated SiC MOSFET functionable at 540°C. But there are thermal management concerns with increasing power density, which impacts the future development of process and packaging technologies. Currently, top metal of device, encapsulated epoxy, and soldering materials system was inherited from Si-based devices, there is a lot of work to be done to make the SiC devices workable at high temperature. Integrating the temperature sensor in chip will also help to monitor current sharing and temperature of each die in large power module [14,15].

Conclusion remarks

Nowadays, SiC is not just limited to EV and new energy. With the accelerating of Artificial Intelligence, computing is becoming energy-thirsty which makes the advanced power system and energy conversion efficiency much more important. SiC power devices are poised to play a key role in those emerging applications. In the meantime, cost-down of SiC eco-system through device and process innovations, yield-up will pave the way of 4H-SiC power device affordable in many cost-sensitive applications. Definitely, SiC will shape the future of electrical power conversion.

References

[1] https://www.mckinsey.com/~/media/mckinsey/ industries/semiconductors/our insights/new silicon carbide prospects emerge as market adapts to ev expansion/

[2] Z. Chen, and A. Q. Huang, "Extreme high efficiency enabled by silicon carbide (SiC) power devices", Materials Science in Semiconductor Processing, 172, 108052 (2024).

[3] H. Tsuchida, and T. Kanda, "Advances in fast 4H–SiC crystal growth and defect reduction by high-temperature gas-source method", Materials Science in Semiconductor Processing, 176, 108315 (2024).

[4] P. Kumar, M.I.M. Martins, M.E. Bathen, T. Prokscha, and U. Grossner, "Al-implantation induced damage in 4H-SiC", Materials Science in Semiconductor Processing, 174, 108241 (2024).

[5] K. Mikami, K. Tachiki, M. Kaneko, and T. Kimoto, "Insight into Mobility Improvement by the Oxidation-Minimizing Process in SiC MOSFETs", IEEE Transactions on Electron Devices,71, pp. 931-934, (2024).

[6] K. Tachiki, M. Kaneko, and T. Kimoto, "Mobility improvement of 4H-SiC (0001) MOSFETs by a three-step process of H_2 etching, SiO_2 deposition, and interface nitridation", Appl. Phys. Express, 14, 031001 (2021).

[7] A. Agarwal, H. Fatima, S. Haney, and S.-H. Ryu, "A new degradation mechanism in high-voltage SiC power MOSFETs", IEEE Electron Device Lett., 28, pp. 587–589 (2007).

[8] M. Kato, O. Watanabe, S. Harada, and H. Sakane, "Effects of proton implantation into 4H-SiC substrate: Stacking faults in epilayer on the substrate", Materials Science in Semiconductor Processing, 175, 108264(2024).

[9] T. Kimoto, and J. A. Cooper, "Fundamentals of Silicon Carbide Technology: Growth, Characterization, Devices and Applications", Singapore, Wiley, pp. 197–200 (2014).

[10] M. Kato, O. Watanabe, T. Mii, H. Sakane, and S. Harada, "Suppression of stacking-fault expansion in 4H-SiC PiN diodes using proton implantation to solve bipolar degradation", Scientific Reports, 12, 18790 (2022).

[11] L. Shi, J. Qian, M. Jin, M. Bhattacharya, H. Yu, A. Shimbori, M. H. White, and A. K. Agarwal, "Investigation on gate oxide reliability under gate bias screening for commercial SiC planar and trench MOSFETs", Materials Science in Semiconductor Processing, 174, 108194 (2024).

[12] R. Yu, S. Jahdi, O. Alatise, J. Ortiz-Gonzalez, S. P. Munagala, and N. Simpson, "Measurements and Review of Failure Mechanisms and Reliability Constraints of 4H-SiC Power MOSFETs Under Short Circuit Events", IEEE Transactions on Device and Materials Reliability, 23, pp. 544-563 (2023).

[13] J. R. O'connor, and J. Smiltens, "Silicon Carbide, A high temperature semiconductor, Proceedings of the conference on Silicon Carbide", Pergamon Press, Oxford (1960).

[14] S. Isukapati, H. Zhang, T. Liu, U. Gupta, E. Ashik, A. Morgan, S. Jang, B. Lee, W. Sung, A. Fayed, and A. Agarwal "Design and experimental demonstration of high-voltage lateral nMOSFETs and high-temperature CMOS ICs", Materials Science in Semiconductor Processing, 169, 107921 (2024).

[15] F. La Via, D. Alquier, F. Giannazzo, and T. Kimoto, P. Neudeck, H. Ou, A. Roncaglia, S. E. Saddow, and S. Tudisco, "Emerging SiC applications beyond power electronic devices", Micromachines, 14, 1200 (2023).

The Atomic Layer Etching Technique with Low Damage for p-GaN/AlGaN/GaN Structure

Xinyi Tang[1, #], Honghao Lu[1, #], Chun Fu[1], Chuying Tang[1],

Fangzhou Du[1], Qing Wang[1,2,3, b)], Hongyu Yu[1,2,3, a)]

[1]School of Microelectronics, Southern University of Science and Technology, Shenzhen 518055, Guangdong, China

[2]Engineering Research Center of Integrated Circuits for Next-Generation Communications, Ministry of Education, Southern University of Science and Technology, Shenzhen, 518055, China

[3]The Key Laboratory of the Third Generation Semiconductor, Southern University of Science and Technology, Shenzhen, 518055, China

[#]Xinyi Tang and Honghao Lu contribute equally to this work

a) E-mail: yuhy@sustech.edu.cn

b) E-mail: wangq7@sustech.edu.cn

Abstract

In this work, an atomic layer etching (ALE) technique is proposed for p-GaN/AlGaN/GaN. The surface roughness was significantly reduced from 0.5 nm to 0.357 nm, after an ALE process with 15 s O_2 modification step and 32 s BCl_3 removal step. Additionally, the oxides product in O_2 modification step can be effectively etched or removed. These findings indicate that O_2/BCl_3-based ALE technique is a promising approach for fabricating a variety of high-performance GaN devices that contain the p-GaN layer, such as p-GaN Gate HEMTs and GaN p-FETs.

Keywords: Atomic layer etching, p-GaN and low damage

Introduction

Gallium nitride-based high electron mobility transistors (GaN-based HEMTs) show great potential for power and radio frequency applications owing to their superior material properties [1]. Currently, p-GaN gate HEMT is a common E-mode device scheme, and GaN p-channel field-effect transistor (GaN p-FET) is the key to develop GaN CMOS applications [2-3]. Therefore, the etching process of p-GaN layer plays a critical role in the fabrication of GaN devices, directly affecting the device performance.

The conventional p-GaN etching process is conducted by inductively coupled plasma etching (ICP). However, it is challenging to control the etching rate and etch the p-GaN layer completely without causing damage to the surface [4-5], leading to degraded device performance, such as high leakage current and poor reliability. Digital etching is another method of etching p-GaN. However, this method requires frequent use of solution soaking, resulting in a complicated complex and time-consuming process [6-7].

Atomic layer etching (ALE) is an atomic-level film etching technique, which can significantly reduce the plasma damage of continuous dry etching and improve the surface morphology [8]. As shown in Fig. 1 (a)-(c), the ALE process usually consists of two steps, which are the modification step and removal step. Recently, O_2/BCl_3-based ALE process has been applied to AlGaN/GaN and InAlN/GaN structures [9–10], demonstrating the advantages of ALE process, such as precise etching and the achievement of low-damage surfaces. However, few studies have focused on the ALE technique for p-GaN/AlGaN/GaN structure [11], and the time and effects of O_2 in modification step remain unclear.

In this work, an O_2/BCl_3-based ALE etching technique for p-GaN/AlGaN/GaN structure was studied. The samples were systematically investigated through atomic force microscopy (AFM), a root-mean-square (RMS) roughness of 0.357 nm was obtained, which was much lower than that of the as-grown p-GaN surface (0.50 nm), and the p-GaN etching depth could be precisely controlled by ALE process. The X-ray photoelectron spectroscopy (XPS) results indicate that the oxides generated in the O_2 modification can be effectively etched or removed by buffered oxide etch (BOE) treatment.

Experiments

The structure of the p-GaN epitaxial wafer is shown in Fig. 1 (d). It consisted of a 90 nm p-GaN layer doped with Mg at a concentration of 5×10^{19} cm^{-3}, a 20 nm $Al_{0.25}Ga_{0.75}N$ barrier layer, a 0.8 nm AlN insert layer, a 1.1 μm GaN buffer layer, and a 3.5 μm AlGaN buffer layer with a graded Al composition, grown on the Si substrate.

The process flow is illustrated in Fig. 1(e). After wafer cleaning, a SiO_2 layer was deposited by plasma enhanced chemical vapor deposition (PECVD) to serve as the hard mask for the ALE process. The SiO_2 mask was selectively removed by inductively coupled plasma-reactive ion etching (ICP-RIE) after photolithography. The exposed p-GaN was then etched in two steps: the first step was oxidation modification by O_2 plasma, using 50 sccm O_2 flow, 40 W RF power and 200 W ICP power. The time for O_2 to reach the maximum depth of self-limiting modification was determined by experiment. The second step was BCl_3

979-8-3315-0417-5/25 $31.00 © 2025 IEEE

plasma dry etching, using 50 sccm BCl₃ flow, 40 W RF power and 200 W ICP power. The BCl₃ plasma removal time was fixed at 32 s based on our previous experience.

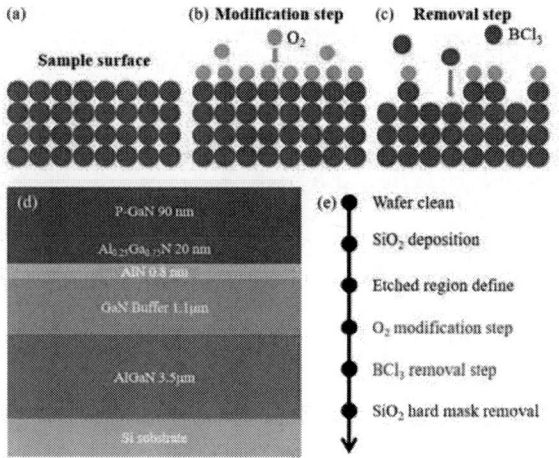

Fig. 1: (a) As-grown sample, (b) modification step, and (c) removal step of ALE theoretical process; (d) p-GaN epitaxial wafer structure diagram and (e) ALE process flow.

AFM was used to obtain the surface RMS roughness and EPC of etched samples. XPS was performed to investigate the influence of O_2 modification on the surface.

Results and discussion

The average etching depth and the etching rate of the samples with different O_2 modification times (5, 15, 30 and 45 s) and etching cycles (9,12 and 15 cycles) are listed in Table 1.

Table 1: The process parameters, etching depth, EPC and RMS roughness of p-GaN ALE process.

Etching Cycles	O_2 modification time (s)	Etching depth (nm)	EPC (nm/cycle)	RMS (nm)
9	5	19.54	2.171	0.553
9	15	23.83	2.648	0.412
9	30	23.66	2.629	0.406
9	45	23.85	2.65	0.382
12	5	26.99	2.249	0.568
12	15	31.85	2.654	0.357
12	30	31.94	2.662	0.454
12	45	31.77	2.647	0.449
15	5	32.93	2.195	0.601
15	15	40.06	2.671	0.418
15	30	39.48	2.632	0.433
15	45	39.15	2.61	0.377

As shown in Fig. 2, the etching cycles have little impact on the etching rate of p-GaN. With the increasing O_2 modification time, the etching rate initially rose and then stabilized. The etching rate was approximately 2.2 nm/cycle for an O_2 modification time of 5 s, and increased to 2.6 nm/cycle when the O_2 modification time was extended to 15

s. However, when the modification time was further extended to 30 s and 45 s, the etching rate remained relatively stable, indicating that the etching depth could be precisely controlled. This is because when the modification time is long enough, the depth of the oxidized p-GaN exceeds the etching depth in each cycle, leading to a constant etching depth per cycle.

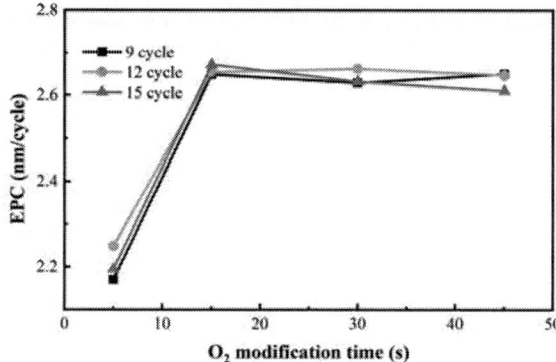

Fig. 2: EPC of p-GaN for ALE approach with different etching cycles and various O_2 modification time.

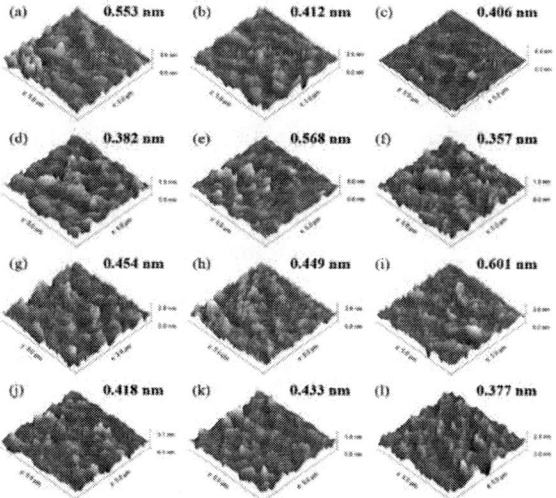

Fig. 3: AFM images of p-GaN surface of $5 \times 5\ \mu m^2$ area after 9 cycles for (a) 5 s, (b) 15 s, (c) 30 s, (d) 45 s O_2 modification; 12 cycles for (e) 5 s, (f) 15 s, (g) 30 s, (h) 45 s O_2 modification; 15 cycles for (i) 5 s, (j) 15 s, (k) 30 s, (l) 45 s O_2 modification.

Fig. 3 shows the 3D surface morphology of p-GaN samples with different O_2 modification time after 9, 12 and 15 etching cycles measured by AFM. The surface roughness was 0.55-0.6 nm after ALE process with 5 s O_2 modification, which was slightly higher than that of the non-etched sample (0.50 nm). However, when the modification time exceeded 15 s, the surface roughness improved significantly, reaching 0.357 nm. Typically, continuous etching with long etching time leads to high roughness. However, when the O_2 modification time is sufficient, the surface roughness with ALE process remains lower than that of the non-etched sample even if the etching cycle is further increased. These results suggest that the ALE cycle process effectively reduces etching damage, leading to a significant improvement in surface smoothness.

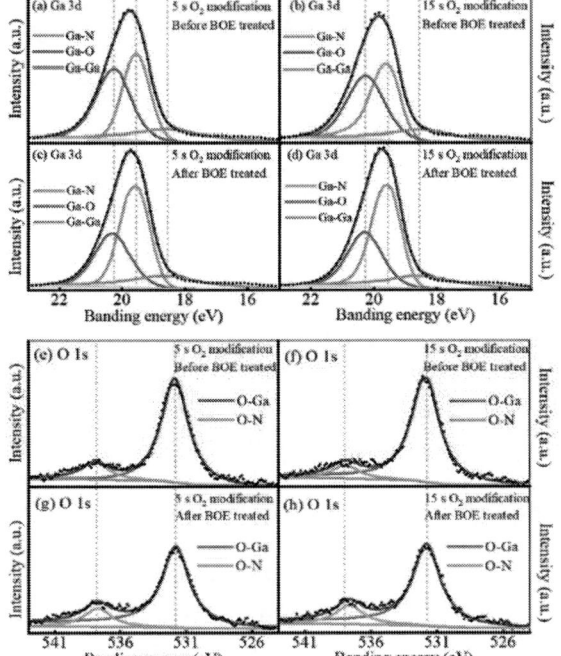

Fig. 4: XPS results of Ga 3d for (a) 5 s and (b) 15s O_2 modification before BOE treatment, (c) 5 s and (d) 15 s O_2 modification after BOE treatment; O 1s for (e) 5 s (f) 15 s O_2 modification before BOE treatment, (g) 5 s and (h) 15 s O_2 modification after BOE treatment.

To investigate the surface chemical composition of the ALE samples, the peak position of the Ga and O elements were analyzed by XPS. As shown in Figs. 4(a), (b), (e) and (f), when the O_2 modification time was 5 s or 15 s, the content of Ga-O bond was significantly high. In figs. 4(c), (d), (g) and (h), after BOE treatment, the Ga-O bonds of both samples were notably reduced and became nearly identical, indicating that the BOE treatment effectively removed the oxide layer formed during etching process. Additionally, the peak position of Ga-O bond did not shift significantly in these samples, suggesting that no new oxides that difficult to be removed by BOE were formed on the samples surface during ALE process with different O_2 modification time.

Conclusion

In this work, the ALE process of p-GaN with varying O_2 modification time was investigated. The results show that the etching rate increased with longer O_2 modification times, which maintained at approximately 2.6 nm/cycle for 15 s or longer O_2 modification time. When the O_2 modification time is sufficient, the etching rate remains constant even as the etching cycle increases, indicating that ALE process can precisely control the etching depth of p-GaN. Additionally, the ALE process significantly reduced surface roughness to 0.357 nm compared to the non-etched sample (0.50 nm). The XPS analysis confirmed that the oxides produced in O_2 modification step could be etched by the following BCl_3

removal step or removed by BOE treatment. These findings suggest that the O_2/BCl_3 ALE process is an effective technique for precisely etching p-GaN while improving surface quality, making it a promising approach in fabrication of high-quality p-GaN Gate HEMTs, GaN p-FETs, and other GaN devices which include p-GaN layer.

Acknowledgments

This research was funded by National Natural Science Foundation of China (Grant No:62274082), Research on mechanism of Source/Drain ohmic contact and the related GaN p-FET(Grant No: 2023A1515030034), Research on high-reliable GaN power device and the related industrial power system (Grant No: HZQB-KCZYZ-2021052), Study on the reliability of GaN power devices (Grant No: JCYJ20220818100605012), Research on novelty low-resistance Source/Drain ohmic contact for GaN p-FET (Grant No: CY20220530115411025), "5G Frontier" project Micro-Nano processing platform (Grant No: K2023390010), High level of special funds (G03034K004).

References

[1] Zeng, Fanming, et al. "A Comprehensive Review of Recent Progress on GaN High Electron Mobility Transistors: Devices, Fabrication and Reliability." Electronics 7.12(2018).

[2] Roccaforte, et al. "An Overview of Normally-Off GaN-Based High Electron Mobility Transistors." Materials 12.10(2019):1599.

[3] Zheng, Zheyang, et al. "Gallium nitride-based complementary logic integrated circuits." Nature Electronics.

[4] Andrzej Taube, et al. "Selective etching of p-GaN over Al 0.25 Ga 0.75 N in Cl 2 /Ar/O 2 ICP plasma for fabrication of normally-off GaN HEMTs." Materials Science in Semiconductor Processing 122.

[5] Toprak, Ahmet, et al. "Selectively dry etched of p-GaN/InAlN heterostructures using BCl 3 -based plasma for normally-off HEMT technology." Materials Research Express (2021).

[6] Jiang, Yang, et al. "A Novel Oxygen-Based Digital Etching Technique for p-GaN/AlGaN Structures without Etch-Stop Layers." Chinese Physics Letters 37.6(2020):068503.

[7] Wu, Jingyi, et al. "Oxygen-plasma-based digital etching for GaN/AlGaN high electron mobility transistors." IEEE (2019).

[8] Kanarik, Keren J., et al. "Overview of atomic layer etching in the semiconductor industry." Journal of Vacuum Science & Technology A Vacuum Surfaces & Films 33.2(2015):20802-20801.

[9] Burnham, Shawn D., et al. "Gate-recessed normally-off GaN-on- Si HEMT using a new O2-BCl3 digital etching technique." Physica Status Solidi 7.7-8(2011):2010-2012.

[10] Du, Fangzhou, et al. "Atomic layer etching technique for InAlN/GaN heterostructure with AlN etch-stop layer." Materials Science in Semiconductor Processing 143(2022):106544-.

[11] Gao, Xiaoxiao, et al. "Research on p-GaN/AlGaN high sidewall verticality and low damage etching applied to enhanced GaN HEMTs devices." 2023 24th International Conference on Electronic Packaging Technology (ICEPT) Shihezi City, China, 2023, pp. 1-5.

Device Considerations for GaN Power Switching Transistors from Application and Reliability Perspectives (Invited)

Zhikai Tang[1], Maik Peter Kaufmann[2], Sandeep Bahl[1], Chang Soo Suh[3], Jungwoo Joh[3], Dong Seup Lee[3], Tim Merkin[3], Jeffrey Morroni[3]

[1]Texas Instruments Inc., Santa Clara, CA 95054, USA
[2]Texas Instruments Inc., 85356 Freising, Germany
[3]Texas Instruments Inc., Dallas, TX 75243, USA
zhikaitang@ti.com

Abstract

In this work, detailed analysis and discussion on the design of state-of-the-art enhancement and depletion-mode (E/D-mode) GaN power switching transistors considering application and reliability are presented. Key challenges and issues during the development phase are elaborated with a summary of potential solutions at device and circuit system levels providing useful technical reference for a designer using GaN device for targeting power electronics application with high energy efficiency and robustness.

Keywords: Power device, GaN, figure of merit (FOM), reliability, high frequency, power switching, enhancement mode, depletion mode

Introduction

Power semiconductor devices based on the wide bandgap material GaN have been successfully commercialized in high-volume production for its very first killer application of fast charging power supply in consumer electronics with many more on the way towards higher energy efficiency and enhanced switching performance [1], [2]. For instance, by integrating the GaN technology, a 650 V intelligent power module (IPM) with more than 99% inverter efficiency for motor drive applications in appliances and HVAC systems has been recently introduced [3]. Since the first experimental demonstration of AlGaN/GaN HEMT in 1993 [4], the device technology has gone through multiple generations of advancement on the roadmap to significantly outperform the state-of-the-art Si power MOSFET counterparts in terms of power device figure of merit (FOM) such as specific on-resistance (Ron, sp) and Ron·Qg. In addition to the low Ron enabled by the high-density high-mobility 2D electron gas (2DEG) induced by strong polarization effects [5], the initial key technical challenges such as dynamic Ron (dRon) and current collapse due to charge trapping within the device during high-voltage switching operation have been well dealt with by design innovation and steady improvement in epitaxy growth and fab processes [6], [7]. In order to design a successful device product that would meet end users' increasingly stringent mission profile requirement, one needs to consider not only device performance at transistor level, but also switching behavior under real application scenarios as well as robustness and reliability under different types of electrical, thermal, and mechanical stresses [8].

For power converters, normally off or E-mode power switch is highly desirable and necessary in most use cases owing to fail-safe operation and easier gate drive design. Currently there are mainly two technical schemes to realize E-mode operation, namely, the single-chip E-mode GaN HEMT with p-type GaN gate (Fig. 1) and the D-mode GaN co-packaged with Si MOSFET or controller (Fig. 2) [9], [10]. Although these different solutions provide flexibility for the power device option in a circuit/system design, they also present challenges on how to make such a choice. Therefore, in this paper, we focus on discussion of commercially available GaN power switching transistors from various critical design angles in order to provide some potentially useful reference for a designer to make an informed decision when it comes to choosing the appropriate GaN solution for a specific application.

Device Design and Reliability

Due to the 2DEG channel naturally formed at the AlGaN/GaN heterojunction interface, the lateral GaN HEMT is inherently D-mode. Significant efforts have been made during the last two decades in gate engineering to enable E-mode operation with the p-GaN gate chosen as the first for industry adoption owing to technology maturity and ease of fab process control. Meanwhile, D-mode GaN MIS-HEMT integrated with low-voltage Si FET or controller in a package has also been implemented in high-voltage power converters as an alternative. For D-mode device, it is technically easier to obtain a better Ron, sp since the 2DEG density (Ns) can be higher without concern of lowering threshold voltage (Vt). However, the higher Ns translates into higher electric field in the channel in high-voltage off state, which makes it more challenging to effectively prevent premature breakdown and dRon degradation. Moreover, cost advantage resulting from lower Ron, sp becomes smaller when the D-mode FET is configured in series with Si FET. Nevertheless, both E- and D-mode GaN transistors face

979-8-3315-0417-5/25 $31.00 © 2025 IEEE

the same challenges of minimizing both static and dynamic Rsp, on and enhancing breakdown voltage by optimized device design including device architecture, epitaxy and fab processes.

Focused research works on surface passivation and field plate engineering have demonstrated GaN HEMT power devices with excellent dRon and breakdown voltage performance [7], [11]. With cell pitch reduction, optimized surface passivation and epi buffer growth (Fig. 4), low and stable dRon were achieved at both device and circuit system levels, as shown in Fig. 3 and Fig. 5 [12], [13]. For D-mode, there is no concern on gate overdrive voltage stress as the device is on with 0 V gate bias. However, for the E-mode FET with p-GaN gate, the reverse biased metal/p-GaN Schottky junction when $V_G > 0$ in the on state would require special reliability attention in terms of gate overshoot and TDDB. Fig. 6 illustrates the simulated gate overshoot/ringing and wearout rate during device turn-on of discrete and co-package solutions under 480 V 500 kHz hard switching with Vgs_on = 6 V. The gate wearout rate noticeably increases with larger gate ringing due to larger gate loop inductance as calculated using the method in [14] and reliability model from [15], indicating the advantage of co-package integration over its discrete counterpart.

User consideration in applications

A single-chip E-mode p-GaN gate HEMT is usually favored by a circuit designer owing to straightforward slew rate control and low-voltage (e.g., 6 V) gate driver. However, it normally has lower saturation current density (Id, sat) and thus more limited Id, sat/Ioff ratio than the D-mode. Meanwhile, the electron trapping induced adverse effects of dRon would normally be more prominent due to lower Ns in the channel. As a consequence, D-mode device with either cascode or direct drive configuration can be preferred in certain applications where e.g., higher Id, sat and lower Ron, sp are critical. However, co-package solution of D-mode GaN with Si FET and control does have its own design tradeoff that needs to be taken into consideration. In a cascode configuration, the D-mode GaN FET with a fixed gate potential at ground is connected in series with a 20-30 V silicon FET, whose gate is switched to turn the integrated device on and off. Special care on slew rate control to prevent Si FET from avalanche breakdown and relatively higher Coss are the main challenges for this approach. With direct drive, the D-mode gate is directly controlled by the gate drive with excellent slew rate control capability, where a negative gate drive voltage is generated by a power converter IC. A 20-30 V Si FET is employed as a safety switch in series. It should also be noted that the short-circuit capability may decrease with higher Id, sat and thus faster FET turn off by the protection circuit is necessary to accommodate smaller short-circuit withstand time. A comparison of key device FOM's taking into account Ron, Id, sat, Qoss, Qrr, and Qg for multiple 650 V class GaN power products is summarized in Table 1 illustrating influence of device architecture.

Conclusion

In this paper, we have reviewed and discussed the design of commercial GaN power switching devices from various angles at component and circuit/system levels with reliability considerations aiming at providing useful technical reference for proper GaN device choice for a specific targeting power switching application.

References

[1] S. Pendharkar, "GaN and SiC enable increased energy efficiency in power supplies", TI Technical Journal, SSZY033, pp.1-6 (2018).

[2] K. J. Chen et al., "GaN-on-Si power technology: Devices and applications," IEEE Trans. Electron Devices, pp. 779–795 (2017).

[3] C. Munoz, "Achieving household energy efficiency and cost savings with GaN-based motor system designs", TI Technical Journal, SSZTD41, pp.1-4 (2024).

[4] M. A. Khan et al., "High electron mobility transistor based on a GaN-Al$_x$Ga$_{1-x}$N heterojunction", Appl. Phys. Lett. 63, 1214–1215 (1993).

[5] O. Ambacher, et al., "Two-dimensional electron gases induced by spontaneous and piezoelectric polarization charges in N- and Ga-face AlGaN/GaN heterostructures", J. Appl. Phys. 85, 3222–3233 (1999).

[6] D. Marcon et al., "200mm GaN-on-Si epitaxy and e-mode device technology," IEDM Tech. Dig., pp. 414–417 (2015).

[7] T. Hashizume et al., "Surface passivation of GaN and GaN/AlGaN heterostructures by dielectric films and its application to insulated-gate heterostructure transistors", J. Vac. Sci. Technol. B, 1828–1838 (2003).

[8] J. P. Kozak et al., "Stability, reliability, and robustness of GaN power devices: A review," IEEE Trans. Power Electron., pp. 8442–8471 (2023).

[9] P. L. Brohlin et al., "Direct-drive configuration for GaN devices," TI Technical Journal, SLPY008A, pp.1-6 (2018).

[10] Y. Uemoto et al., "Gate injection transistor (GIT)—A normally-off AlGaN/GaN power transistor using conductivity modulation," IEEE Trans. Electron Devices, pp. 3393–3399 (2007).

[11] W. Saito et al., "Field-plate structure dependence of current collapse phenomena in high-voltage GaN-HEMTs," in IEEE Electron Device Lett., pp. 659-661 (2010).

[12] J. Joh et al., "Key challenges in process development for future high voltage GaN roadmap", Proc. CS MANTECH (2024).

[13] S. Bahl et al., "Application reliability validation of GaN power devices", IEDM Tech. Dig., pp. 544-547 (2016).

[14] S. Bahl et al., "Mission profile approach for the calculation of GaN FET reliability in power supply applications, IEEE IRPS 2024, pp. 2C.3-2-2C.3-7.

[15] J. He et al., "Frequency- and Temperature-Dependent Gate Reliability of Schottky-Type p-GaN Gate HEMTs", IEEE Trans. Electron Devices, pp. 3453-3458 (2019).

979-8-3315-0417-5/25 $31.00 © 2025 IEEE

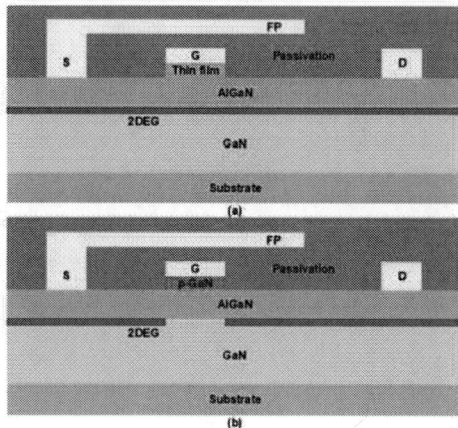

Fig 1. (a) D-mode GaN device with MIS gate and (b) E-mode GaN device with p-GaN gate.

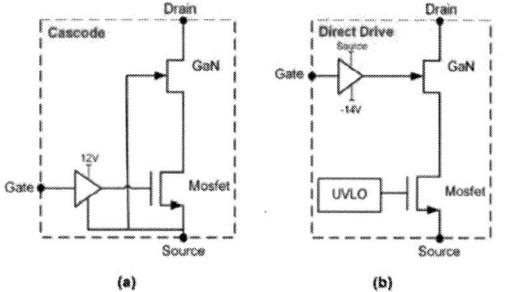

Fig 2. Approach of converting D-mode GaN device into E-mode in a package: (a) cascode and (b) direct drive.

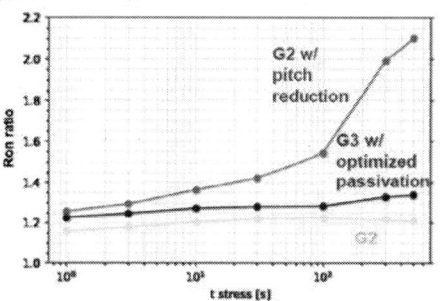

Fig. 3. Ron ratio vs stress time with device cell pitch reduction and optimized surface passivation of D-mode GaN FET.

Table 1. Comparison of key FOM's of E-mode and D-mode GaN power transistors for power switching applications.

Part	LMG3622	Vendor A	LMG3522R050	Vendor B
E-mode technology	p-type gate	p-type gate	D-mode GaN direct drive	D-mode GaN cascode
Voltage rating (V)	650	650	650	650
Ron (mΩ)	120	78	50	72
No of transistors in power path	1	1	2	2
Drive Voltage (V)	N/A	0-6	-14-0	0-10
Ron·Id, sat (Ω·A)	2.7	2.7	5.3	8.6
Qoss·Ron (nC·Ω)	3.4	2.9	5.0	5.6
Qrr·Ron (nC·Ω)	0	0	0	0
Qg·Ron (nC·Ω)	N/A	0.3	N/A	0.6

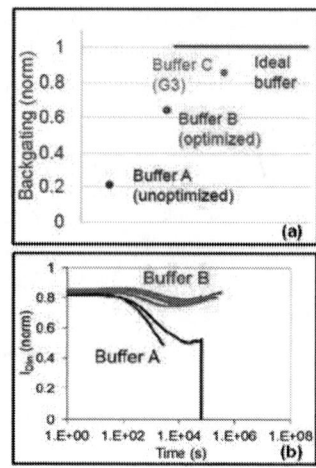

Fig 4. (a) Backgating characteristics for 3 epi buffer designs and (b) current collapse performance comparison between Buffer A and B under 600 V 150°C hard switching stress indicating influence of backgating.

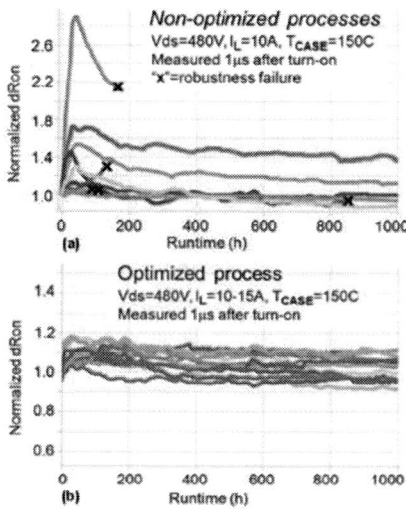

Fig. 5. in-situ dRon under accelerated hard switching stress with inductive load at Vds = 480 V and 150 °C of D-mode GaN FET.

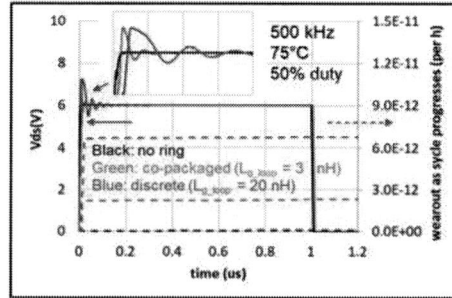

Fig. 6. Simulated E-mode p-GaN gate overshoot/ringing and wearout rate during device turn-on of discrete vs co-package solution under 480 V 500 kHz hard switching with Vgs_on = 6 V.

HfO₂-Based Ferroelectric Field-Effect Transistors for Next-Generation Storage and In-Memory Computing Applications

Genquan Han[1,2,*], Chengji Jin[1,2,*], Jiajia Chen[1,2], Xiao Yu[1,2], Jiuren Zhou[1,2], Siying Zheng[1,2], Haoji Qian[1,2], Ran Cheng[3], Bing Chen[1,2] and Yan Liu[1,2]

[1] Hangzhou Institute of Technology, Xidian University, Hangzhou 311231, China,

[2] Faculty of Integrated Circuits, Xidian University, Xi'an 710071, China,

[3] College of Integrated Circuits, Zhejiang University, Hangzhou 311200, China.

*Email: gqhan@xidian.edu.cn; cjjin@ieee.org

Abstract

HfO₂-based Ferroelectric field-effect transistor (FeFET) is a promising technology for advanced memory and computing systems, offering CMOS compatibility, low-power consumption, and scalability. This paper highlights our recent findings in FeFET reliability, including endurance enhancement at cryogenic temperatures, polarization-induced temperature instability (PTI) and disturb issue in a NAND array. Possible solutions are proposed to mitigate these reliability issues. In addition, FeFET demonstrates significant potential in multi-level content addressable memory (CAM), hybrid-precision synapses for neuromorphic computing, and logic-in-memory (LiM).

Keywords: FeFET, reliability and emerging computing systems

Introduction

HfO₂-based Ferroelectric field-effect transistor (FeFET) has emerged as a promising technology for next-generation memory and computing architectures [1, 2]. Its unique combination of fast switching, low-power consumption, and complementary metal-oxide-semiconductor (CMOS) compatibility makes them as an ideal candidate for a variety of applications, including high-density memory, in-memory computing, and neuromorphic systems. However, challenges in device reliability and the need for innovative architectures require continued research and optimization.

This paper summarizes finding from our recent research works, focusing on two critical aspects: reliability challenges of HfO₂-based FeFET and its applications in advanced computing systems. We demonstrate how FeFET is tackling high-temperature stress, disturb, and other reliability issues while unlocking new possibilities in neuromorphic computing, multi-level cell (MLC) memories, and logic-in-memory (LiM) systems.

Reliability Challenges and Improvements of FeFET

A. Enhanced Performance at Cryogenic Temperatures

FDSOI FeFET show significantly improved endurance and stability at 5 K compared to 300 K. Oxygen vacancy migration, a major cause of polarization fatigue, is suppressed at cryogenic temperatures, resulting in robust polarization states. Endurance tests reveal negligible fatigue up to 10^7 cycles at 5 K, with polarization retention showing little degradation (Fig. 1) [1]. This makes cryogenic FeFET ideal for aerospace, quantum computing, and other low-temperature applications. Meanwhile, FeFET exhibits better multi-level cell (MLC) performance under cryogenic conditions, providing highly stable retention and wider memory window (MW), as shown in Fig. 2. These improvements highlight the potential for cryogenic memory systems with enhanced reliability.

B. Polarization-Induced Temperature Instability

FeFETs face reliability challenges when exposed to high-temperature stress (HTS). Since the existence of spontaneous polarization in ferroelectric, an internal bias emerges even without applied gate voltage, which results in performance degradation that can be accelerated under HTS. The polarization-induced temperature instability (PTI) of ferroelectric HfO₂ layers directly impacts the performance including MW, endurance, subthreshold swing (SS), and on-state current (I_{on}), as shown in Fig. 3(a)-(g) [4]. Positive PTI (PPTI) is identified as a more dominant than negative PTI (NPTI). Prolonged HTS accelerates the degradation, significantly limiting MW and endurance. The degradation mechanisms are studied using quasi-static capacitance-voltage (QSCV) methods, which indicates the transition from stable polarization (P_{Sta}) to reversible polarization (P_{Rev}) is responsible for the performance degradation by PTI (Fig. 3(h)). The transition from P_{Sta} to P_{Rev} by the trap site generation during HTS is confirmed to be the origin of performance degradation by PTI (Fig. 3(i)).

C. Disturb in NAND-Type FeFET Arrays

In FeFET-based memory arrays, program disturb and read disturb present significant challenges to the reliability of MLC operations. These issues arise primarily due to unintended polarization switching in non-selected cells during array operations. Disturb-free operations of MLC FeFET in a NAND array is experimentally investigated [5]. Optimized write schemes for various states of MLC FeFET are determined and Fig. 4(a) plots the read I_d-V_g curves for 4 V_{th} states. Both program and read disturbs are systematically characterized in a 2×2 FeFET-based NAND array (Fig. 4(b)). Margins of program inhibition voltage (V_{inhib}) and pass voltage (V_{pass}) are determined from the measurement results of the worst case. V_{inhib} higher than 2 V should be satisfied to prevent program disturb (Fig. 4(c)). Meanwhile, V_{inhib} should

---

979-8-3315-0417-5/25 $31.00 © 2025 IEEE

be lower than a certain level, or V_{th} shift by drain erase would happen. As for the V_{pass} margin, V_{pass} no higher than 2 V is required for read disturb-free operations in this study (Fig. 4(d)), while a sufficiently large V_{pass} should be applied to achieve low channel resistance of pass transistors.

Applications of FeFET in Emerging Computing Systems

A. Multi-Level Content Addressable Memory (CAM)

A novel multi-bit CAM cell based on two series connected FeFETs is proposed [6]. The CAM cells can be integrated in a form of 3D vertical FeFET NAND, which enables ultra-high density and energy efficiency at low cost. In our design, the capability of multi-bit data search in a CAM cell originates from MLC operations of FeFET. Fig. 5(a) illustrates operation mechanisms of MLC FeFET. Four polarization states (i.e., 'S0', 'S1', 'S2', and 'S3'), and thus threshold voltages (V_{TH}) can be tuned by different program pulse amplitudes or widths. Fig. 5(b) shows the proposed CAM cell with two FeFETs connected in series, where T0 and T1 are upper and lower FeFETs with threshold voltages V_{TH0} and V_{TH1}, respectively. By utilizing MLC operations of FeFET, multi-bit CAM functions can be realized, if we assume the input voltages of SL and $\overline{SL}$ (V_{SL} and $V_{\overline{SL}}$) in a cell are complementary. Fig. 5(c) plots the input V_{SL} waveform and measured I_{ML} in search operations of a multi-bit CAM cell. Parallel search of a query with 4-bits data in a whole NAND block is experimentally demonstrated by a 4×4 FeFET NAND array. Figs. 5(d) and (e) are the schematic and the microscope image of a fabricated 4×4 FeFET NAND array. Fig. 5(f) shows the 2D mapping of measured I_{ML} with various input queries and stored entries. The CAM design proposed in this work shows great potential for future data-centric computing systems.

B. Logic-in-Memory with 1FeFET-1RRAM Architectures

A 1FeFET-1RRAM (1FeFET1R) design is proposed and experimentally demonstrated to realize logic gates by utilizing the non-volatile storage function of the FeFET for logic operations and the switchable resistance of the RRAM for the logic reconfiguration [7]. Wafer-level 1FeFET1R logic units as well as the 2-level and 3-level cascaded logic gates are fabricated, with the complete Boolean logic functions demonstrated through electrical tests. Fig. 6 shows an example of XNOR and XOR logic realized by the 3-level cascaded circuit. The speed and stability of the 1FeFET1R circuits are also investigated. The nonvolatile 1FeFET1R-based logic circuits exhibit smaller area consumption, higher computing efficiency, simpler and CMOS compatible process, and field reconfigurability for possible logic-in-memory (LiM) applications.

C. Hybrid-Precision Synapses for Neuromorphic Computing

We characterized the effect of the back gate (BG) on the memory properties under TG read mode, and the corresponding synaptic performance of FDSOI FeFET [8]. BG of device acts as the "V_T shifter" and "amplification/reduction" of post-synaptic current. Based on the dynamic range (> 150×) tuning of post-synaptic current by V_{BG} as shown in Fig. 7, we proposed a hybrid-precision synapse cell with multiple FDSOI FeFET, which is expected to generate multi-level (> 1000) weights and realize ultra-high precision synapse. Hybrid-precision synapses are well-suited for neuromorphic hardware, enabling energy-efficient training and inference in edge devices.

Conclusion

On the reliability front, addressing PTI and read disturb issues is critical for ensuring stable, long-term operation. Meanwhile, innovative architectures such as multi-bit CAMs and hybrid 1FeFET1R designs pave the way for in-memory and logic-in-memory applications with unprecedented density, efficiency, and scalability. Continued efforts in material engineering, device integration, and circuit optimization will unlock the full potential of FeFETs in modern computing paradigms.

Acknowledgments

This work is supported in part by the National Key R&D Program of China (No. 2023YFB4402303); in part by the National Natural Science Foundation of China (Grant No. 62025402, 62204228, 62204229 and 91964202).

References

[1] H. Mulaosmanovic, E. T. Breyer, S. Dünkel, S. Beyer, T. Mikolajick, and S. Slesazeck, "Ferroelectric field-effect transistors based on HfO₂: a review," Nanotechnology, vol. 32, no. 50, p. 502002, Dec. 2021, doi: 10.1088/1361-6528/ac189f.

[2] U. Schroeder, M. H. Park, T. Mikolajick, and C. S. Hwang, "The fundamentals and applications of ferroelectric HfO₂," Nat Rev Mater, vol. 7, no. 8, pp. 653–669, Mar. 2022, doi: 10.1038/s41578-022-00431-2.

[3] M. Zhang, H. Qian, J. Xu, M. Ma, R. Shen, G. Lin, J. Gu, Y. Liu, C. Jin, J. Chen, and G. Han, "Enhanced Endurance and Stability of FDSOI Ferroelectric FETs at Cryogenic Temperatures for Advanced Memory Applications," IEEE Trans. Electron Devices, vol. 71, no. 11, pp. 6680–6685, Nov. 2024, doi: 10.1109/TED.2024.3456763.

[4] C. Jin, J. Xu, J. Gu, J. Chen, H. Zhang, H. Liu, H. Qian, B. Chen, R. Cheng, Y. Liu, X. Yu, and G. Han, "Polarization-Induced Temperature Instability of HfO₂-Based Ferroelectric FET," IEEE Electron Device Lett., vol. 45, no. 1, pp. 32–35, Jan. 2024, doi: 10.1109/LED.2023.3332649.

[5] C. Jin, J. Xu, J. Gu, J. Zhao, X. Jia, J. Chen, H. Liu, M. Zhang, Y. Peng, B. Chen, R. Cheng, Y. Liu, X. Yu, and G. Han, "Disturb-Free Operations of Multilevel Cell Ferroelectric FETs for NAND Applications," IEEE Trans. Electron Devices, vol. 70, no. 4, pp. 1653–1658, Apr. 2023, doi: 10.1109/TED.2023.3242922.

[6] C. Jin, J. Xu, J. Zhao, J. Gu, J. Chen, H. Liu, H. Qian, M. Zhang, B. Chen, R. Cheng, Y. Liu, X. Yu, and G. Han, "A Multi-Bit CAM Design With Ultra-High Density and Energy Efficiency Based on FeFET NAND," IEEE Electron Device Lett., vol. 44, no. 7, pp. 1104–1107, Jul. 2023, doi: 10.1109/LED.2023.3277845.

[7] Z. Ding, X. Li, C. Jin, X. Yu, B. Chen, R. Cheng, and G. Han, "Experimental Demonstration of Non-Volatile Boolean Logic With Field Configurable 1FeFET-1RRAM Technology," IEEE Electron Device Lett., vol. 45, no. 6, pp. 1084–1087, Jun. 2024, doi: 10.1109/LED.2024.3390403.

[8] M. Zhang, L. Chou, H. Zhang, H. Qian, C. Jin, B. Chen, R. Cheng, Y. Liu, G. Bernard, S. Loubriat, B.-Y. Nguyen, X. Yu, J. Chen, and G. Han, "Hybrid-Precision Synapse Based-on FDSOI FeFETs With Back-Gate Bias Modulation," IEEE Trans. Electron Devices, vol. 71, no. 6, pp. 3994–3997, Jun. 2024, doi: 10.1109/TED.2024.3388414.

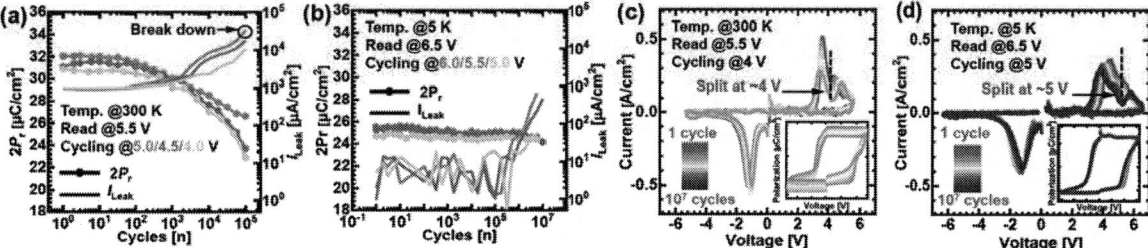

Fig.1 Comparison of endurance characteristics at 5 K and 300 K with various applied voltage amplitude. The $2P_r$ and leakage current density at 3 V during different voltage cycling, extracted by PUND measurements at (a) 300 K and (b) 5 K, respectively. The corresponding evolution of I–V curves at (c) 300 K (10^5 cycles) and at (d) 5 K (10^7 cycles), respectively. The insets are the corresponding evolution of P–V curves [3]. Copyright 2024, IEEE.

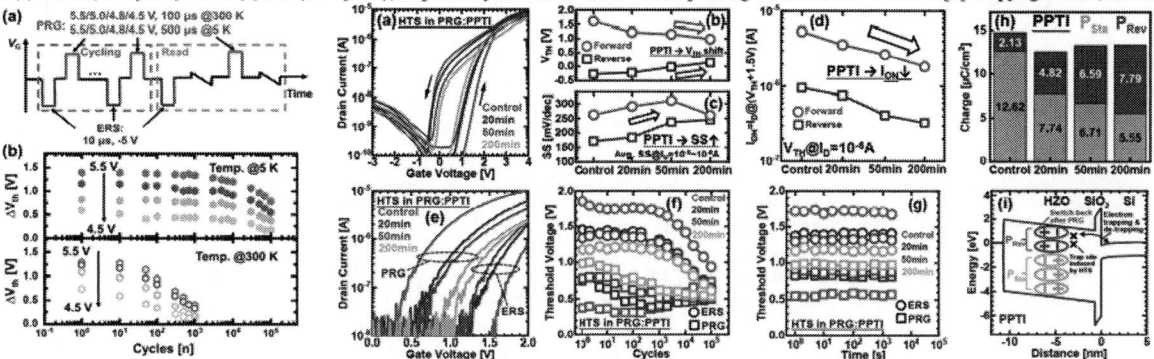

Fig.2 Comparison of endurance for FDSOI FeFET at 5 K and 300 K. (a) The pulse sequence of measurements. (b) Endurance of the four-level PRG states at 5 K and 300 K [3]. Copyright 2024, IEEE.

Fig. 3. (a) Dual DC I_D-V_G of FeFET cells before and after HTS. (b) Significant V_{TH} shift is observed for both forward and reverse sweeps after HTS. (c) SS and (d) I_{ON} also show degradation after HTS. (e) Fast I_D-V_G (10μs) after write (+4/-3V, 100μs) of FeFET cells before and after HTS. (f) Endurance characteristic of FeFET cells after HTS. (g) Retention is not affected by PPTI even after 200min HTS. (h) The evolution of P_{Sta} and P_{Rev} for PPTI. (i) Illustration of PPTI mechanisms [4]. Copyright 2023, IEEE.

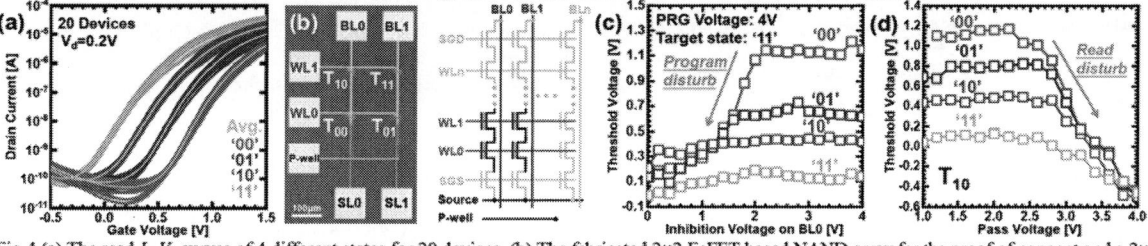

Fig.4 (a) The read I_d-V_g curves of 4 different states for 20 devices. (b) The fabricated 2×2 FeFET-based NAND array for the proof of concept and a 2×2 FeFET test structure in a typical NAND array. (c) Measured V_{th} after 4 V program operations as a function of V_{inhib} on BL0. 4 V_{th} states are well separated at high V_{inhib}, indicating successful MLC program operations at the array level. (d) Readout V_{th} for T_{10} [5]. Copyright 2023, IEEE.

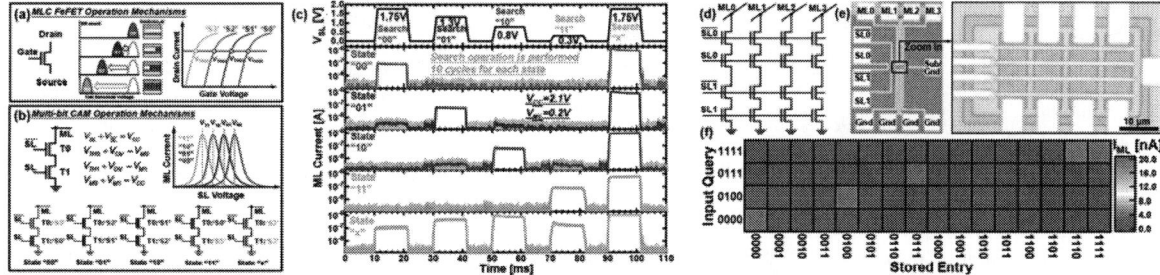

Fig.5 (a) Operation mechanisms of MLC FeFET. (b) Operation mechanisms of the proposed multi-bit CAM cell based on two FeFETs connected in series. (c) The input V_{SL} waveform and measured I_{ML} in search operations of a multi-bit CAM cell. (d) The schematic and (e) the microscope image of a 4×4 FeFET NAND array. (f) The 2D mapping of measured I_{ML} with various input queries and stored entries [6]. Copyright 2023, IEEE.

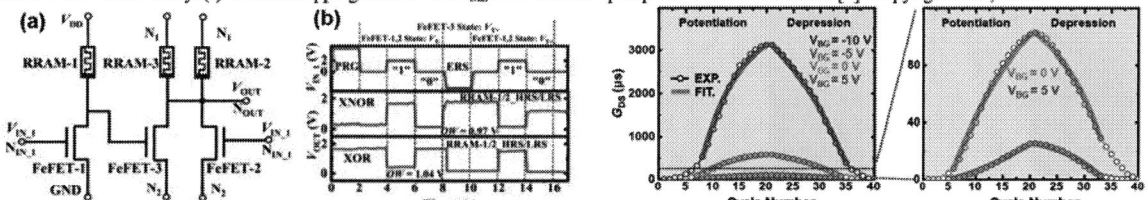

Fig.6 (a) Circuit schematics of a 3-level cascaded circuit. (b) The timing diagram of V_{IN} and V_{OUT} for the (d) XNOR and XOR logic realized by the 3-level cascaded circuit [7]. Copyright 2024, IEEE.

Fig.7 (a) Potentiation and depression curves with various V_{BG} (5.0/-5/-10V) under variable amplitude pulses scheme [8]. Copyright 2024, IEEE.

979-8-3315-0417-5/25 $31.00 © 2025 IEEE

Silicon Doping in Amorphous Gallium Oxide Films by Plasma-Enhanced Atomic Layer Deposition for Dielectric and Optoelectronic Applications

Yongjie He
School of Microelectronics
Fudan University
Shanghai, China
yjhe24@m.fudan.edu.cn

Jingxuan Wei
School of Microelectronics
Fudan University
Shanghai, China
weijx9695@126.com

Jining Yang
School of Microelectronics
Fudan University
Shanghai, China
jnyang22@m.fudan.edu.cn

Rognxu Bai
School of Microelectronics
Fudan University
Shanghai, China
rongxu_bai@fudan.edu.cn

Hao Zhu
School of Microelectronics
Fudan University
Shanghai, China
hao_zhu@fudan.edu.cn

Qingqing Sun
School of Microelectronics
Fudan University
Shanghai, China
qqsun@fudan.edu.cn

Abstract—**Gallium oxide is a representative of the fourth-generation semiconductor. The silicon doping of gallium oxide plays a crucial role in optimizing performance, with PEALD as a viable deposition technique due to the advantages of high coverage and precise composition control. In this study, various compositions of amorphous Si-doped gallium oxide thin films were first deposited via PEALD with systematic characterizations. Subsequently, the electrical and optical characterizations were performed incorporating subcycle ratio and Si concentration analysis, paving the way for dielectric and optoelectronic applications of the amorphous Si-GaO$_x$ thin film.**

Keywords—silicon doping, PEALD, amorphous gallium oxide

I. INTRODUCTION

Gallium oxide (GaO$_x$) is a material representative of the fourth-generation semiconductor applied in power electronics [1], dielectrics [2], and optoelectronics [3]. To expand these applications, doping GaO$_x$ is a critical approach for property modification. Various elements including silicon [4][5], tin [6], indium [7], and zinc [8] have been doped into GaO$_x$, leading to characteristic variations. In particular, Si-doped GaO$_x$ (Si-GaO$_x$) emerges not only as a semiconductor candidate but as an applicable insulator, indicating potential applications in dielectrics and optoelectronics [4][5]. Among the synthesis and doping techniques of GaO$_x$ thin films [9][10][11], atomic layer deposition (ALD) has garnered significant attention due to the advantages of high coverage and low deposition temperature [4][11][12]. Moreover, Si doping concentration in GaO$_x$ can be controlled relatively precisely by the subcycle ratio.

In this study, amorphous Si-GaO$_x$ thin films with varying concentrations were first deposited by plasma-enhanced ALD (PEALD) with characterizations to evaluate dielectric and optoelectronic applications. The compositional, structural, and surface characterizations were conducted to assess the quality of the as-deposited thin film and interpret the influences on parameters by the PEALD processes and Si concentrations. Based on the deposited thin films, the electrical characterizations reveal that the Si-GaO$_x$ thin film with a subcycle ratio 1/9 has the highest breakdown voltage and a typical dielectric constant suitable for dielectric applications. Moreover, the optical characterizations of the Si-GaO$_x$ thin films of varying Si concentrations indicate their potential applicability for ultraviolet detectors.

II. EXPERIMENTS

The PEALD process characterized by various subcycles and supercycles is illustrated in Fig. 1. A standard PEALD subcycle comprises four stages including a precursor pulse (50 ms), Ar purge, oxidant dosing (15 s), and Ar purge. For Si doping, the depositions of GaO$_x$ and SiO$_x$ were governed by M subcycles and N subcycles in a supercycle, respectively, leading to an increase in Si concentration by the ratio N/M. The overall thickness of the Si-GaO$_x$ thin film was dependent on the number of supercycles. Herein, triethylgallium (TEGa) and diisopropylaminosilane (DIPAS) were selected as precursors of gallium and silicon with O$_2$ plasma as oxidant at a deposition temperature of 200°C. Moreover, all PEALD processes were performed on the p-type Si <100> substrate with a resistivity of 1~10 Ω·cm.

TABLE I presents the sample ID associated with various PEALD processes by the increase in Si concentration from Sample 1 to 5. Notably, Sample 1 serves as a control, featuring a GaO$_x$ thin film without doping. For comparison, the product of the sum of subcycles (M+N) and the number of supercycles was standardized to be 200.

III. RESULTS AND DISCUSSIONS

A. The Surface, Compositional and Structural Analysis

Fig. 2 presents the atomic force microscope (AFM) images with average roughness analysis, with Fig. 2(a)~2(e) corresponding to Sample 1~5. As shown in Fig. 2(f), all as-deposited thin films of varying ALD processes demonstrate an average roughness of <0.6 nm with a fluctuation of <0.1 nm, indicating high smoothness in the thin films. The X-ray reflectivity (XRR) characterization was conducted to analyze the density, thickness, and roughness, as shown in Fig. 3. Compared with the β-Ga$_2$O$_3$ crystal density (~6.44 g/cm^3) [6], amorphous Si-GaO$_x$ has a lower density (~5.3 g/cm^3) attributable to a looser atomic structure than that in crystals.

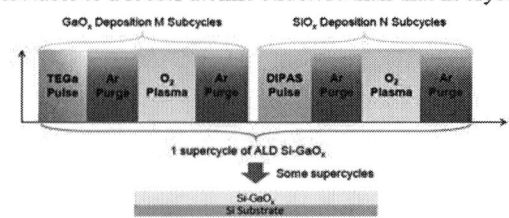

Fig. 1. The ALD process of Si-GaO$_x$ thin films.

TABLE I.   THE SAMPLE ID OF VARIOUS PEALD PROCESSES

Sample ID	PEALD processes		
	Ga₂O₃ cycles (M)	SiO₂ cycles (N)	Supercycles
1	200	0	1
2	19	1	10
3	9	1	20
4	4	1	40
5	1	1	100

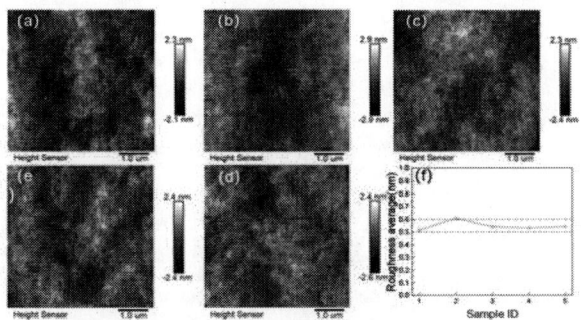

Fig. 2. The AFM Image and the roughness of the surfaces of all samples, (a)~(e) corresponding to Sample 1~5, and (f) shows the roughness averages, where all scanning regions are 5×5µm squares.

As SiO_x has a lower density of ~2.64 g/cm³, the thin films exhibit a reduced density by the increase of Si concentration (Fig. 3(b)). The XRR fitting results exhibit a roughness <0.62 nm (Fig. 3(c)), consistent with Fig. 2(f). Moreover, the thicknesses by XRR and spectroscopy ellipsometer (SE) fitting are shown in Fig. 3(d), revealing a thickness range of 10.3 nm~16.3 nm. Although discrepancies in thickness arise due to different models, an identical trend is observed of an initial decrease followed by an increase in Si concentration.

Fig. 4 shows the X-ray photoelectron spectroscopy (XPS) characterizations for composition characterization. The extraction of XPS data for Ga 2p exhibits increasingly pronounced peaks for Ga 2p³ and Ga2p¹ from Sample 5 to 1, directing an increase in Ga concentration (Fig. 4(a)). Since XPS characterizes the optoelectrons from the surface to a depth of ~10 nm, the XPS Si 2p data were influenced by the Si substrate. To avoid the disturbance, the XPS peak of the Si substrate was truncated, isolating the peak of the Si-O bond in Si 2p data (Fig. 4(b)), where the Si-O peak becomes more pronounced with the increase in the subcycle N and Si concentration. From the O 1s data in Fig. 4(c), the sharpness of all samples is consistent while the O 1s peak shifts from the Ga-O peak (~529.8 eV) to the Si-O peak (~532 eV) from Sample 1 to 5, reflecting the variation of the chemical states of oxygen atoms and an increase in Si concentration. By Fig. 4(a) and 4(b), the Si/Ga atomic ratios were calculated in Fig. 4(d). Although the trend in Si/Ga atomic ratio corresponds with the trend in N/M, the ratio is not strictly proportional to N/M, arising from a different deposition mechanism of Si-GaO_x compared with that of GaO_x or SiO_x.

Fig. 5(a) shows the grazing incidence X-ray diffraction (GIXRD) characterization to confirm the non-crystalline structure, as evidenced by the presence of the only peak corresponding to the Si <100> substrate. Moreover, Fig. 5(b) shows the secondary ion mass spectroscopy (SIMS) of Sample 3, dividing distinct interfaces among the surface, the as-deposited thin film, the native oxide, and the substrate with each isotope exhibiting a unique profile in the thin film. In summary, these characterizations demonstrate the influence on GaO_x thin film properties of Si concentration and PEALD

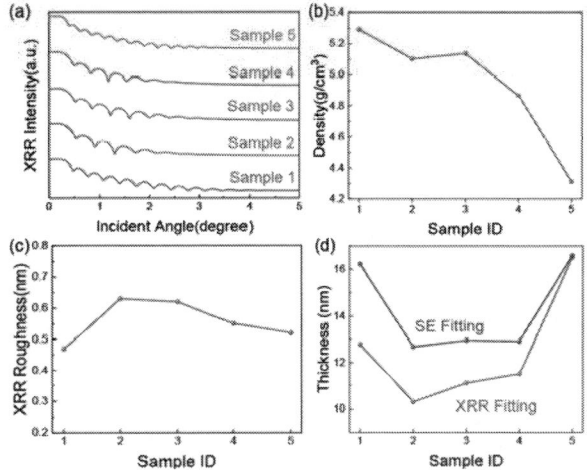

Fig. 3. The (a) XRR characterizations of the as-deposited samples with the(b) density, (c) roughness and (d) thickness analysis with SE fitting thickness.

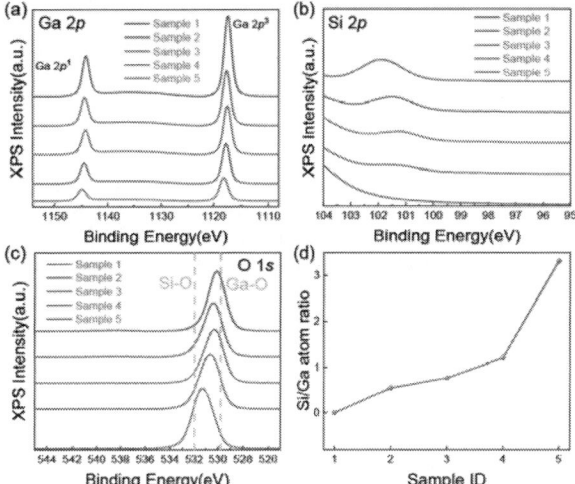

Fig. 4. The XPS characterizations of (a) Ga 2p, (b) Si 2p and (c) O 1s with (d) the Si/Ga atom ratio analysis.

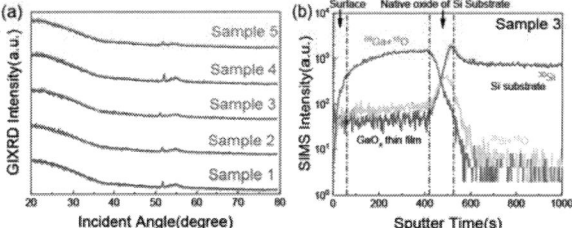

Fig. 5. The (a) GIXRD characterization of all samples and (b) SIMS characterizations of Sample 3 of ³⁰Si, ⁶⁹Ga+¹⁸O cluster and ²⁸Si +¹⁸O cluster.

processes and the high quality of the thin films, facilitating advancements in optoelectronic and dielectric applications.

B. The Optical and Electrical Analysis for Dielectric and Optoelectronic Applications

Fig. 6 presents the optical characterizations for optoelectronic applications. Fig. 6(a) shows the refractive index n across a wavelength range of 300~800 nm. In ultraviolet and visible light spectra, the n varies from 1.4 to 2.4 with an initial decrease followed by an increase in stability by wavelength in total. Moreover, the n shows a decrease

979-8-3315-0417-5/25 $31.00 © 2025 IEEE   1253

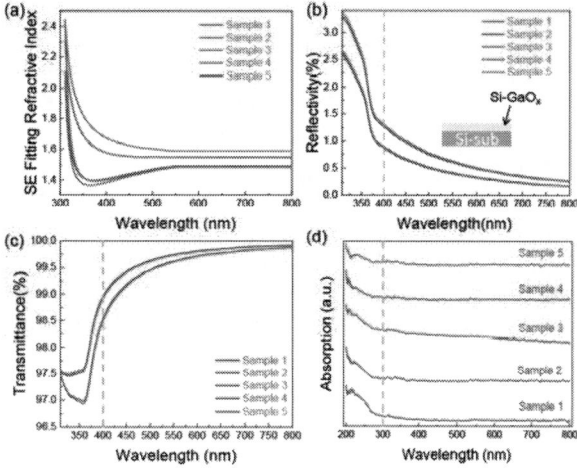

Fig. 6. The optical characterizations of (a) refractive index, (b) reflection and (c) transmittance with (d) an absorption spectrum.

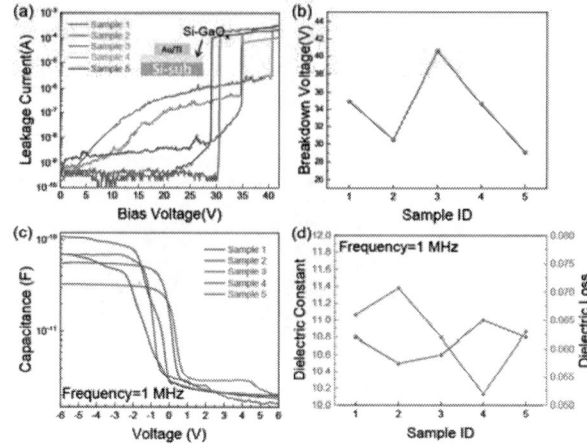

Fig. 7. The electric characterizations of the as-deposited thin films. (a) The breakdown characterizations with (b) the breakdown voltage. (c) The C-V characterizations with (d) the dielectric constant and dielectric loss

and then an increase with the increase in Si concentration. Fig. 6(b) and 6(c) show the reflection and transmittance characterizations, respectively. As transparent thin films, Si-GaO_x thin films exhibit low reflection <3.3% and high transmittance ~100%, indicating superior transparency. The transmittance increases by wavelength with a contrast trend of reflection. Fig. 6(d) shows the ultraviolet absorption spectrum across the wavelength range of 200~800 nm, revealing significant absorption at the 200~300 nm range uniformly for all samples. The pronounced absorption wavelength range in the ultraviolet region suggests potential applications for ultraviolet detectors [3].

Fig. 7 presents the breakdown and capacitance-voltage (C-V) characterizations that were conducted for dielectric applications. A square Au/Ti electrode of 100×100 μm was fabricated by lithography and DC sputtering for electrical characterizations (Fig. 7(a)). Due to the Si substrate conductivity, Fig. 7(a) shows the leakage current of the thin films under high bias voltages. Samples 1, 3, and 4 exhibit soft breakdown while other samples demonstrate hard breakdown. The extracted breakdown voltage is shown in Fig. 7(b), with Sample 3 of subcycle ratio 1/9 achieving the highest breakdown voltage >40 V and directing a superior dielectric application. Moreover, the C-V characterizations at 1 MHz are shown in Fig. 7(c) with the Si substrate functioning as a semiconductor, resulting in a metal-oxide-semiconductor C-V curve. Sample 3 shows a relatively high capacitance (~70 pF) with a typical dielectric constant (~10.6) and a dielectric loss of ~0.062, as shown in Fig.7(d). In summary, Sample 3 with the subcycle ratio of 1/9 exhibits a comprehensively superior performance for dielectric applications.

IV. CONCLUSION

In this study, high-quality amorphous Si-GaO_x thin films with varying concentrations were first deposited by PEALD with systematic characterizations. Comprehensive analysis showcases the performance of the as-deposited thin film and the influence of the parameters by PEALD processes and Si concentrations, revealing certain trends and regularities of ALD Si-GaO_x thin films. The electrical characterizations indicate the Si-GaO_x with a subcycle ratio of 1/9 exhibiting the highest breakdown voltage and a relatively increased

dielectric constant for dielectric applications. Moreover, by optical characterizations, the Si-GaO_x thin films of various Si concentrations were analyzed, highlighting a potential application in ultraviolet detection.

REFERENCES

[1] M. Higashiwaki, A. Kuramata, H. Murakami, and Y. Kumagai, "State-of-the-art technologies of gallium oxide power devices," J. Phys. D: Appl. Phys., vol. 50, no. 33, p. 333002, Jul. 2017.

[2] Yuan, L., Li, S., Song, G., Sun X. and Zhang X. "Solution-processed amorphous gallium oxide gate dielectric for low-voltage operation oxide thin film transistors." J Mater Sci: Mater Electron 32, 8347–8353 (2021).

[3] X. Chen, F. Ren, S. Gu, and J. Ye, "Review of gallium-oxide-based solar-blind ultraviolet photodetectors," Photon. Res., PRJ, vol. 7, no. 4, pp. 381–415, Apr. 2019.

[4] X.-Y. Zhang et al., "Growth and characterization of Si-doped Ga2O3 thin films by remote plasma atomic layer deposition: Toward UVC-LED application," Surface and Coatings Technology, vol. 435, p. 128252, Apr. 2022.

[5] Liu J., Gao S., Li W., Dai J., Suo Z. and Suo Z. "First-Principles Calculations of Electronic Structure and Optical Properties of Si-Doped and Vacancy β-Ga2O3." Crystal Research and Technology, 2022, 57(1): 2100126.

[6] Shen Y. et al. "Atomic-level Sn doping effect in Ga2O3 films using plasma-enhanced atomic layer deposition". Nanomaterials, 2022, 12(23): 4256.

[7] J. Sheng, E. J. Park, B. Shong, and J.-S. Park, "Atomic Layer Deposition of an Indium Gallium Oxide Thin Film for Thin-Film Transistor Applications," ACS Appl. Mater. Interfaces, vol. 9, no. 28, pp. 23934–23940, Jul. 2017.

[8] M.-I. Chen, A. K. Singh, J.-L. Chiang, R.-H. Horng, and D.-S. Wuu, "Zinc Gallium Oxide—A Review from Synthesis to Applications," Nanomaterials, vol. 10, no. 11, Art. no. 11, Nov. 2020.

[9] Terasako T, Ichinotani H, Yagi M. "Growth of β-gallium oxide films and nanostructures by atmospheric-pressure CVD using gallium and water as source materials," physica status solidi (c), 2015, 12(7): 985-988.

[10] C. V. Ramana et al., "Chemical bonding, optical constants, and electrical resistivity of sputter-deposited gallium oxide thin films," Journal of Applied Physics, vol. 115, no. 4, p. 043508, Jan. 2014.

[11] X. Liu et al., "Growth characteristics and properties of Ga 2 O 3 films fabricated by atomic layer deposition technique," Journal of Materials Chemistry C, vol. 10, no. 43, pp. 16247–16264, 2022.

[12] Mahmoodinezhad A, Janowitz C, Naumann F, et al. "Low-temperature growth of gallium oxide thin films by plasma-enhanced atomic layer deposition," Journal of Vacuum Science & Technology A, 2020, 38(2).

On the scalability of nanosheet oxide semiconductor transistors

Masaharu Kobayashi[1,2], Kaito Hikake[2], Xingyu Huang[2], Sunghun Kim[2], Kota Sakai[2], Zhuo Li[2], Tomoko Mizutani[2], Takuya Saraya[2], Toshiro Hiramoto[2], Takanori Takahashi[3], Mutsunori Uenuma[3], Yukiharu Uraoka[3]

[1]d.lab, School of Engineering, The University of Tokyo, [2]Institute of Industrial Science, The University of Tokyo, [3]Nara Institute of Science and Technology

Abstract

We have investigated the scaling potential of nanosheet oxide semiconductor FETs (NS OS FETs) for monolithic 3D integration in terms of ALD material engineering, high-field transport, and statistical variability. The highlights are: (1) systematic comparison among InGaO, InZnO and InGaZnO grown by ALD, (2) demonstration of unsaturated carrier velocity behavior in sub-100nm gate length, (3) comparable variability of NS OS FETs against Si CMOS. This work provides insights for designing scaled NS OS FETs.
Keywords: Nanosheet, Oxide semiconductor, monolithic 3D integration, ALD

Introduction

Growing demands for AI technologies require high performance computing and high memory bandwidth with high energy-efficiency. Monolithic 3D integration (M3D) of memory arrays on a computing unit is a promising solution to minimize energy cost for data transfer because of its proximity [1]. For M3D, BEOL-compatible transistor technology is needed. Oxide semiconductor (OS) is a promising channel material because of its low temperature formation, its high mobility, low leakage and high thermal stability [2-3]. Atomic layer deposition (ALD) method is a key to enable ultrathin OS layer growth and conformal deposition on 3D structure [4]. Although short channel OS FETs have been demonstrated with nanosheet (NS) OS [5-10], high-field transport and statistical variability are not fully explored yet. These are important for device modeling and performance prediction, and manufacturability.

In this paper, we discuss the scaling potential of NS OS FETs fabricated with ALD, regarding high-field transport, and statistical variability.

Experimental method

We developed a device fabrication process of long and short channel NS OS FETs. Both devices have TiN bottom gate, ALD HfO_2 gate insulator, and OS channel layer. OS layer was deposited by thermal ALD process at 250°C using alkyl-based precursor and O_3. EB lithography and lift-off process were used for Ni S/D electrodes formation of the short channel OS FETs. We fabricated two types of TEGs for short channel OS FETs. One TEG has various gate length (L_g) of OS FETs from 60nm to 1μm for L_g dependence study. **Fig. 1** shows the cross-sectional TEM images of the fabricated short L_g OS FETs. The other TEG has >1k OS FETs with the same L_g of 60nm for statistical variability. For reference, we also have a chip of foundry's 65nm bulk Si FETs [11].

ALD material engineering

First, we systematically investigated ALD-grown InGaO (IGO), InZnO (IZO) and InGaZnO (IGZO) FETs as a continuous work from our previous report [4]. Thickness and composition ratio are varied. μ_{eff}, V_{th}, and bias stress V_{th} shift (ΔV_{th}) were characterized and compared. There is a universal trade-off among them for IGO, IZO and IGZO FETs as shown in **Fig. 2**: As In% decreases, V_{th} increases, μ_{eff} decreases and ΔV_{th} increases. This is due to the reduction of oxygen vacancy, termination of conduction path by InOx network, and increase of excess oxygen induced by oxygen-rich chemical reaction in ALD. The difference in oxygen dissociation energy for Ga-O and Zn-O makes the difference in the composition dependence. IGZO shows higher μ_{eff} at the same V_{th} and higher V_{th} at the same μ_{eff}, which indicates IGZO provides a good balance among them.

High-field transport in NS OS FETs

We fabricated both long- and short-channel IGO FETs with varied Ga% and 8nm-thick IGO NS. Transistors behave well even in sub-100nm L_g. **Fig. 3** shows I_d-V_g curves and I_d-V_d curves for long L_g and short L_g IGO FETs. Subthreshold slope is maintained because of the nanosheet structure, while Vth lowering and DIBL are observed. In the I_d-V_d curves, both long L_g and short L_g FETs show the quadratic increase of I_d as a function of V_g. This indicates that NS OS FETs do not show velocity saturation. V_{th} roll-off curves were obtained, which are nearly the same for Ga%. Transconductance (g_m) is used as a metric for high-field transport. We carefully correct g_m by external resistance and extract intrinsic g_m (g_m') [12] in **Fig. 4**. While bulk Si FETs show velocity saturation behavior even after parasitic correction, NS OS FETs show unsaturated velocity behavior. In fact, I_d-V_d curves shows parabolic increment even for short channel. Thus, OS FET performance can get closer to Si FET by further L_g scaling and reducing parasitic resistance.

Statistical variability of NS OS FETs

We fabricated the TEG of 1k (1024) NS IGO FETs with designed L_g of 60nm, channel width (W_{ch}) of 140nm, and channel thickness of 7.5nm in our university lab. We measured I_d-V_g curves of the FETs as shown in **Fig. 5**. Cumulative distributions of V_{th}, DIBL and I_{on} are plotted in

Fig.6. Tight V_{th} distributions of IGO FETs with $\sigma = 20mV$ and $21mV$ for $V_{th}(lin)$ and $V_{th}(sat)$ are obtained, which is comparable to those of Si FETs. This is mainly because IGO FETs have donor concentration $\sim 10^{17}cm^{-3}$ with thin body structure, while Si FETs have channel doping concetntration $> 10^{18}cm^{-3}$. The small average DIBL value of 18.7mV/V and tight distribution with $\sigma = 3.5mV/V$ are obtained in IGO FETs. This is simply because IGO FETs have thin boy structure and drain field is not penetrated into channel region. I_{on} normalized by the mean value of IGO FETs has $\sigma = 4.8\%$, which is 1.6% smaller than Si FETs. This is a consequence of low V_{th} variability.

Summary

We systematically compared ALD-grown IGO, IZO and IGZO FETs. IGZO has a good balance among μ_{eff}, V_{th} and ΔV_{th} by suppressing excess oxygen in oxygen-rich ALD process. We fabricated sub-100nm L_g NS OS FETs and demonstrated good short channel control and unsaturated carrier velocity behavior, which is encouraging in that OS FETs performance can get closer to Si FETs performance by scaling with parasitic resistance reduction. We also obtained statistical variability data of NS OS FETs, which is comparable against Si CMOS and promising for manufacturability of OS FETs.

Acknowledgments

This work was supported by JST CREST (23830112), JST ASPIRE (23836464), JSPS KAKENHI (21H04549, 24H00309), JST SHW (23924565), TSMC Advanced Semiconductor Research Project.

References

[1] W. Gomes, "Beyond Exascale: A Paradigm shift for AI and HPC", IEDM Tech. Dig., pp. 1-4 (2023).

[2] K. Nomura, H. Ohta, A. Takagi, T. Kamiya, M. Hirano, and H. Hosono, "Room-temperature fabrication of transparent flexible thin-film transistors using amorphous oxide semiconductors", nature, 432, pp. 488-492 (2004).

[3] H. Kunitake, K. Ohshima, K. Tsuda, N. Matsumoto, H. Sawai, Y. Yanagisawa, S. Saga, R. Arasawa, T. Seki, R. Tokumaru, T. Atsumi, K. Kato, and S. Yamazaki, "High thermal tolerance of 25-nm c-axis aligned crystalline In-Ga-Zn oxide FET", IEDM Tech. Dig., pp. 312-315 (2018).

[4] K. Hikake, Z. Li, J. Hao, C. Pandy, T. Saraya, T. Hiramoto, T. Takahashi, M. Uenuma, Y. Uraoka, and M. Kobayashi, "A Nanosheet Oxide Semiconductor FET Using ALD InGaOx Channel for 3-D Integrated Devices", IEEE Transactions on Electron Devices, 71, 4 pp. 2373-2379 (2024).

[5] W. Chakraborty, B. Grisafe, H. Ye, I. Lightcap, K. Ni, and S. Datta, "BEOL Compatible Dual-Gate Ultra Thin-Body W-Doped Indium-Oxide Transistor with Ion = $370\mu A/\mu m$, SS = 73mV/dec and I_{on}/I_{off} ratio $> 4x10^9$", VLSI Symposium, TH2-1 (2020).

[6] S. Subhechha, N. Rassoul, A. Belmonte, R. Delhougne, K. Banerjee, G. L. Donadio, H. Dekkers, M. J. van Setten, H. Puliyalil, M. Mao, S. Kundu, M. Pak, L. Teugels, D. Tsvetanova, L. Klijs, H. Hody, A. Chasin, J. Hejlen, L. Goux, and G. S. Kar, "First demonstration of sub-12 nm L_g gate last IGZO-TFTs with oxygen tunnel architecture for front gate devices", VLSI Symposium, T10-5 (2021).

[7] S. Samanta , K. Han , C. Sun, C. Wang, A. Kumar, A. V. -Y. Thean, and Xiao Gong, "Amorphous InGaZnO Thin-Film Transistors With Sub-10-nm Channel Thickness and Ultrascaled Channel Length", IEEE Transactions on Electron Devices, 68, 3, pp. 1050-1056 (2021).

[8] S. Hooda, C. -K. Chen, M. Lal, S. -H. Tsai, E. Zamburg, A. V. -Y. Thean, "Overcoming Negative nFET VTH by Defect-Compensated Low-Thermal Budget ITO IGZO Hetero-Oxide Channel to Achieve Record Mobility and Enhancement-mode Operation", VLSI Symposium, T17-1 (2023).

[9] L. Xu, K. Chen, Z. Li, J. Guo, L. Wang, Y. Zhao, S. Huang, Z. Zhou, C. Dou, G. Yang, L. Wang, L. Li, and M. Liu, "Reliability-Aware Ultra-Scaled IDG-InGaZnO-FET Compact Model to Enable Cross-layer Co-design for Highly Efficient Analog Computing in 2T0C-DRAM", IEDM, 24-5 (2023).

[10] C. Niu, Z. Lin, Z. Zhang, P. Tan, M. Si, Z. Shang, Y. Zhang, H. Wang, and P. D. Ye, "Record-Low Metal to Semiconductor Contact Resistance in Atomic-Layer-Deposited In_2O_3 TFTs Reaching the Quantum Limit", IEDM, 37-2 (2023).

[11] T. Mizutani, Y. Yamamoto, H. Makiyama, T. Tsunomura, T. Iwamatsu, H. Oda, N. Sugii, and T. Hiramoto, "Reduced Drain Current Variability in Fully Depleted Silicon-on-Thin-BOX (SOTB) MOSFETs", Silicon Nanoelectronics Workshop, pp. 71-72 (2012).

[12] Y. Taur and T. Ning, "Fundamentals of Modern VLSI Devices – 2nd edition", Cambridge University Press (2019).

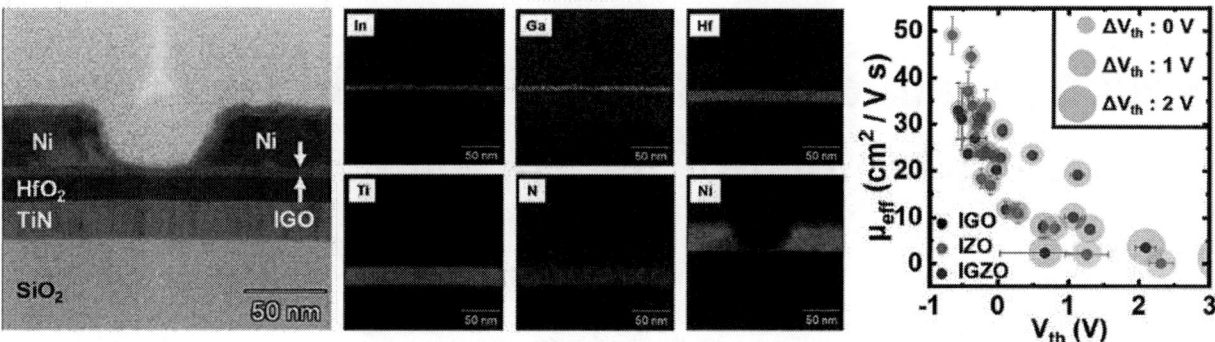

Fig. 1 Cross-sectional TEM image and EDX elemental mapping images of the fabricated short L_g OS FETs. The IGO layer was uniformly formed. No significant interdiffusion was observed among the layers of the device.

Fig. 2 Relationship among mobility, V_{th} and V_{th} shift by positive bias stress for IGO, IZO, and IGZO FETs.

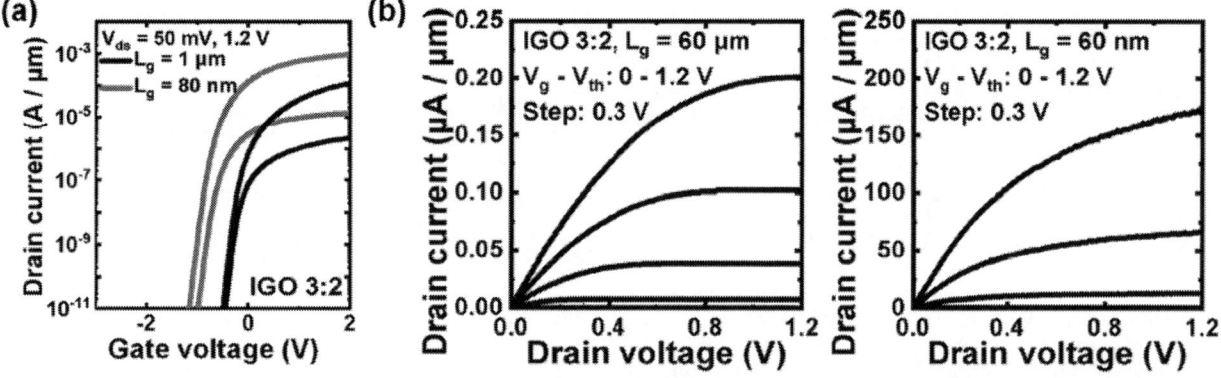

Fig. 3 (a) Measured I_d-V_g curves of long and short L_g. V_{th} is lowered and DIBL occurs, but subthreshold slope is well maintained in sub-100nm L_g. and (b) Measured I_d-V_d curves of the fabricated long and short L_g IGO NS FETs, both of which shows quadratic increase.

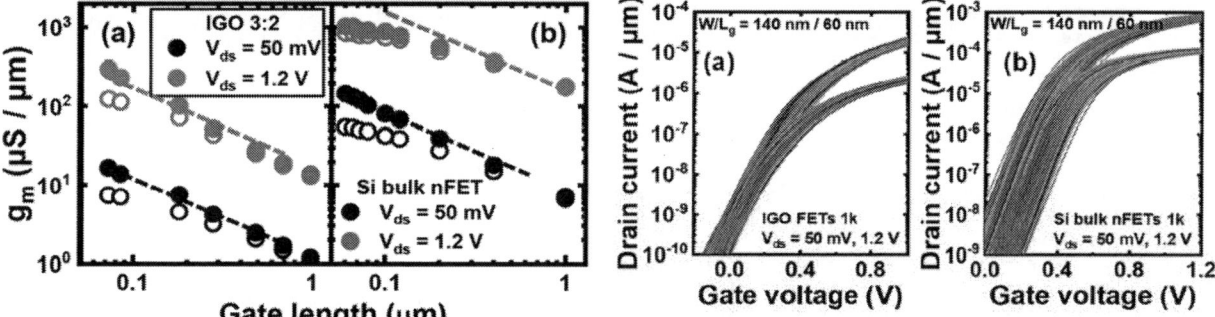

Fig. 4 Extracted transconductance for linear and saturation region for (a) IGO FETs and (b) 60nm bulk Si FETs with (close symbol) and without (open symbol) parasitic resistance correction.

Fig. 5 Measured 1k I_d-V_g curves for (left) IGO NS FETs and for 65nm bulk Si MOSFETs at L_g=60nm. V_{th} variability is well suppressed in IGO NS FETs.

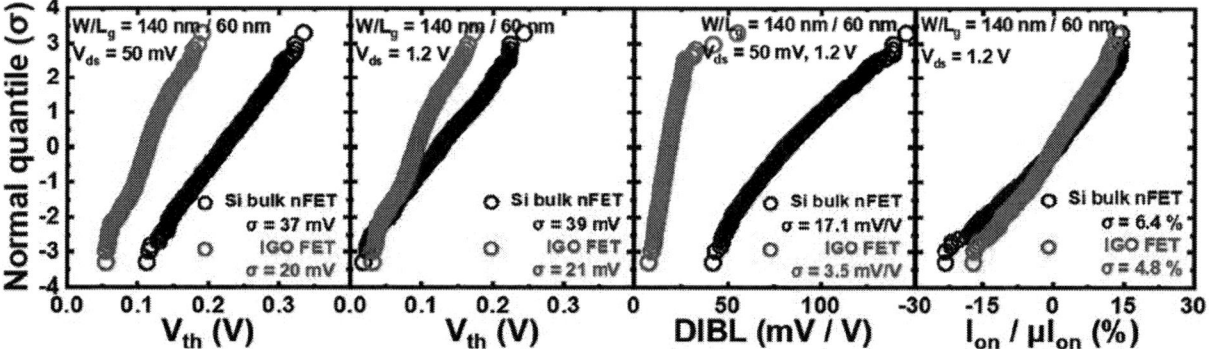

Fig. 6 Cumulative distribution plots for V_{th} (lin), V_{th} (sat), DIBL and normalized Ion for IGO NS FETs and 65nm bulk-Si FETs. V_{th} variability is well suppressed and DIBL is small in IGO NS FETs due to the low donor concentration and thin body structure.

979-8-3315-0417-5/25 $31.00 © 2025 IEEE

Reconfigurable magnonic devices for spin-wave manipulation on the nanoscale

Huajun Qin

School of Physics and Technology
Wuhan University
Wuhan, China
qinhuajun@whu.edu.cn

Abstract—**Magnonics, which harnesses the unique properties of spin waves, offers a promising prospect for low-power, wave-based parallel computing. Active control of propagating spin waves is essential for practical realization of magnonic technology. Here we introduce nanoscale magnonic Fabry-Pérot (FP) resonators, waveguides, and magnonic crystals for low-loss spin-wave guiding and manipulation. These devices are made of an array of ferromagnetic metal (FM) nanostripes patterned on a low-damping yttrium iron garnet (YIG) film, where chiral dipolar coupling between the YIG film and the FM stripe produces asymmetrical spin-wave dispersion, which plays a central role in the spin-wave control. In an array of magnonic FP resonators where two edges of each FM stripe act as two sharp magnetic interfaces, propagating spin waves coherently circulate between the two magnetic interfaces of the resonator and interfere destructively when the phase accumulation is an odd number of π, forming transmission gaps with the depth down to the background level. Thus-formed gaps, called as FP gaps, can be reconfigured by switching the magnetization states and efficiently tuned over a wide frequency range via the variation of the stripe width, strip number, and film thickness. When the period between FP resonators meets the Bragg condition, magnonic crystals (MCs) with deep gaps and low-loss minibands are produced. The MC gaps can also be reconfigured by switching the magnetization states. Moreover, these FP and MC gaps are strongly hybridized when approaching each other, resulting in the gap anti-crossing. In the magnonic waveguide, the downshift of spin-wave dispersion due to the small effective magnetic field in the YIG/FM region enables the guiding of spin-wave propagation along the nanostripe. We report on low-loss spin-wave channeling in straight and curved nanoscopic waveguides, and further demonstrate reversible control of the spin-wave propagation by changing the frequency and magnetic bias field in a hybrid waveguiding structure. Based on these results, spin-wave filter, multiplexer, and beam splitter using low-loss YIG and FM hybrid structures have been proposed.**

Index Terms—**spin waves, magnonic Fabry-Pérot resonators, magnonic crystals, wave-based computing**

I. INTRODUCTION

Magnonics using spin waves - collective spin excitations in magnetically ordered materials - as information carriers provides a potential platform for low-power, wave-based logic devices and parallel computing [1], [2]. Unlike CMOS technology based on the charge transport, magnonic devices, which transfer spin angular momentum via spin waves, can be operated without producing Joule-heating. Moreover, spin waves have micrometer-nanometer wavelengths, facilitating device miniaturization and enabling on-chip integration. In addition, large spin-wave nonlinearity and nonreciprocity hold promise for non-Hermitian and chiral measurements [3]–[5]. The key challenges in realizing viable magnonic technology are active control of multifunctional magnonic devices and efficient communication between multiple magnonic units in an integrated magnonic circuit. Various magnonic devices including multiplexers [6], transistors [7], valves [8], half-adders [9], majority gates [10], and logic gates [11] have been fabricated using ferromagnetic metals (FM) or insulating yttrium iron garnet (YIG) films. Compared to these two materials, the FM metal exhibits high magnetic damping, limiting the spin-wave propagation distance, but can be easily nanopatterned, enabling high-density integration and compatibility with CMOS technology. The YIG film holds the record for low Gilbert damping, but is difficult for nanopatterning because its damping parameter is easily deteriorated by defect formation during patterning or milling. Straightforward nanofabrication of an FM structure on a continuous low-damping YIG film can circumvent these hurdles and combine the merits of both materials. In this paper, we report on the experimental design of nanoscale magnonic FP resonators, MCs, and spin-wave filters, multiplexers and beam splitters using YIG/FM hybrid structures, and demonstrate active control of spin-wave propagation in these magnonic devices [12], [13].

II. RESULTS

A. Magnonic Fabry-Pérot resonators

Figure 1a shows the schematic of an array of magnonic FP resonators, consisting of periodic FM (CoFeB or Py) stripes patterned on a low-damping YIG film of nanometer thickness. A thin TaO_x nonmagnetic spacer is inserted between the YIG and FM layer to suppress the interfacial exchange interaction. The spin waves, excited by a microwave antenna patterned on the YIG film, propagate across the YIG/FM bilayer region, and are detected by the other antenna behind the stripe region or are imaged by Super-Nyquist-Sampling magneto-optical Kerr microscopy (SNS-MOKE). In the hybrid structure, nanometer-thick YIG films were grown on (111)-oriented $Gd_3Ga_5O_{12}$ substrates using pulsed laser deposition technique and the 30-nm-thick FM films were grown by magnetron sputtering. The Gilbert damping constant (α) and saturation magnetization (M_s) of as-grown YIG film is $\alpha = 0.0005$ and $M_s = 175$ kA/m, extracted from ferromagnetic

979-8-3315-0417-5/25 $31.00 © 2025 IEEE

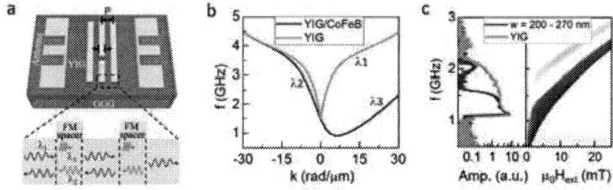

Fig. 1. (a) Schematic of magnonic Fabry-Pérot resonators, consisting of an array of FM nanostripes patterned on a continuous YIG film with low-damping. (b) Spin-wave dispersion curves for an uncovered YIG film (λ_1) and a 100 nm YIG/30 nm CoFeB bilayer with a 5 nm spacer (λ_2 and λ_3). (c) Spin-wave transmission spectra recorded at 6 mT and contour plot of the S_{12} amplitude as a function of magnetic bias field for an array of resonators. The CoFeB stripe widths increase from 200 to 270 nm in 10 nm steps.

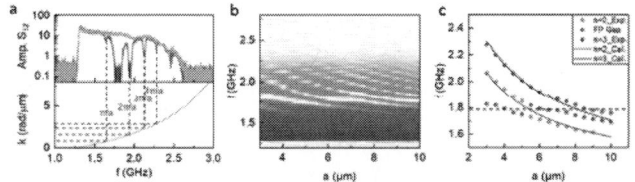

Fig. 2. (a) Spin-wave transmission spectra in an array of Fabry-Pérot resonators (blue) and in the bare YIG film (orange). The Bragg conditions, marked by the dashed lines, match well to spin-wave dispersion for the bare YIG film (orange line). The stripe period is 4 μm and the stripe width is 250 nm. (b) Contour plot of spin-wave spectra measured in an array of Fabry-Pérot resonators with the period increasing from 3 μm to 10 μm in steps of 0.5 μm. (c) Measured (circles) and calculated (solid lines) gap center as a function of period. The dashed line marks the center of the FP gap. The external magnetic field of 10 mT is applied parallel to the stripe in the measurements.

resonance spectra measured by broadband spin-wave spectroscopy. During the measurements, an external magnetic field was applied parallel to the FM stripe, establishing the Damon-Eshbach geometry.

In the hybrid structure shown in Fig. 1a, chiral dipolar coupling between the YIG and FM layers produces an asymmetric spin-wave dispersion relation [see Fig. 1b]. Two edges of each FM stripe in the bilayer region act as two sharp magnetic interfaces where the wavelength of the spin wave converts upon reflection or transmission [see Fig. 1a]. Following the dispersion, the spin wave with the wavelength of λ_1 in the bare YIG converts to the λ_3 wave after entering the bilayer region. At the other edge of the bilayer, the λ_3 wave is reflected to λ_2 wave or is reverted back to the λ_1 wave across the bilayer region. Because of low-damping YIG/FM bilayer, the λ_3 and λ_2 waves circulate and accumulate the phase $\Delta\phi$. When $\Delta\phi = (k_2 + k_3)w = (2n + 1)\pi$, where $k_2 = 2\pi/\lambda_2$, $k_3 = 2\pi/\lambda_3$ and w is the width of each stripe, destructive interference between transmitted waves after the bilayer region produces transmission gaps at discrete frequencies, while constructive interference occurs between reflected waves before the bilayer region. As shown in Fig. 1c, an array of FP resonators made of eight CoFeB stripes with varying w create the tens of MHz wide gap with the depth down to the measurement level. Outside the gap, no extra loss of spin-wave transmission was observed, compared

to the reference signal measured in the bare YIG film. The gap size is largely tuned by changing the width, thickness, and number of the FM stripe. For instance, the resonator with an array of CoFeB stripes of varying widths can create the gap with the size up to hundreds of MHz, orders of magnitude larger than the MC gap in the YIG film. Further, via switching magnetization between parallel and antiparallel states, which changes the bilayer dispersion, reconfigurable control of spin-wave transport in the FP resonator has been achieved [12].

B. Low-loss magnonic crystals

When the Bragg condition of $k = n\pi/P$ is satisfied in an array of FP resonators with the period P, the resonant reflection of spin waves propagating across the YIG/FM stripe region leads to minimum transmission and the formation of MC gaps. Figure 2a shows spin-wave transmission in the MC crystal made of ten Py stripes with $w = 250$ nm and $P = 4$ μm. The external magnetic field of 10 mT is applied parallel to the stripe. We observed the deep FP gap with the size of about 400 MHz and four MC gaps with the size of tens of MHz. All gap positions match well to the Bragg condition. Remarkably, the depth of the $n = 1$ gap is much smaller than the $n = 2 - 4$ gaps. With increasing the P up to 10 μm shown in Fig. 2b, more MC gaps are formed and their positions change, while the position of the FP gap remains unchanged due to the same stripe width. When the MC gap approaches to the FP gap, strong gap hybridization leads to the anti-crossing behavior [see Fig. 2c], which forbids the formation of the MC gaps with frequencies below the FP gap.

C. Nanoscale magnonic waveguides

We now discuss hybrid YIG/FM waveguiding structures for low-loss spin-wave guiding [13]. Figure 3a shows the schematic of the device, which consists of the CoFeB nanostripe intersected with a micrometer bar at the right angle. In the structure, the demagnetization field of the CoFeB nanostripe reduces the effective magnetic field in the YIG/CoFeB bilayer region, resulting in the downshift of the asymmetric spin-wave dispersion [see Fig. 3b], enabling the transverse spin-wave confinement and low-loss propagation along the nanochannel for frequencies below the FMR frequency of the YIG film. Because the demagnetization field produced by the micrometer bar is smaller than that by the nanostripe, a difference in the spin-wave frequency downshift allows for control of waveguide spin-wave propagation [see Fig. 3b]. We show the waveguide spin waves can be redirected efficiently by stray-field-induced bends, interfere in the Y-shaped waveguide, and reversible control of the spin-wave propagation by changing the spin-wave frequency, magnetic field strength, and field direction, as shown in Fig. 3c.

III. CONCLUSION

We experimentally demonstrate nanoscale magnonic Fabry-Pérot resonators and hybrid waveguiding structures for low-loss spin-wave guiding and manipulation. The straightforward device nanofabrication i.e., direct patterning FM structures

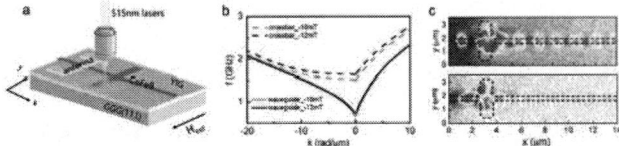

Fig. 3. (a) Schematic of of a waveguiding structure, consisting of a straight FM nanostripe intersected with a micrometer-size bar at the right angle, patterned on a continuous YIG film. The width of the stripe is 240 nm. The bar is 1 μm wide and 2 μm long. An external magnetic field is perpendicular to the stripe. (b) Spin-wave dispersion relations for the hybrid bar (dashed lines) and the stripe (solid lines) regions for 10 and 12 mT. (c) The SNS-MOKE image of spin-wave propagation in the 250-nm-wide YIG/CoFeB hybrid waveguide with the bar under 10-mT and 12-mT external fields. The dashed lines indicate the CoFeB structure on the YIG film.

on top of a continuous YIG film, highlights a new approach toward the implementation of reconfigurable spin-wave filter, multiplexer, and beam splitter, which are building blocks for compact magnonic integrated circuits, used for spin-wave computing.

REFERENCES

[1] A. Barman, G. Gubbiotti, S. Ladak, A. O. Adeyeye, M. Krawczyk, J. Gräfe, C. Adelmann, S. Cotofana, A. Naeemi, V. I. Vasyuchka *et al.*, "The 2021 magnonics roadmap," *Journal of Physics: Condensed Matter*, vol. 33, no. 41, p. 413001, 2021.

[2] A. Chumak, P. Kabos, M. Wu, C. Abert, C. Adelmann, A. Adeyeye, J. Åkerman, F. Aliev, A. Anane, A. Awad *et al.*, "Advances in magnetics roadmap on spin-wave computing," *IEEE Transactions on Magnetics*, vol. 58, no. 6, pp. 1–72, 2022.

[3] V. V. Kruglyak, "Chiral magnonic resonators: Rediscovering the basic magnetic chirality in magnonics," *Appl. Phys. Lett.*, vol. 119, no. 20, p. 200502, Nov. 2021.

[4] T. Yu, Z. Luo, and G. E. W. Bauer, "Chirality as generalized spin-orbit interaction in spintronics," *Physics Reports*, vol. 1009, pp. 1–115, 2023.

[5] S. Zheng, Z. Wang, Y. Wang, F. Sun, Q. He, P. Yan, and H. Y. Yuan, "Tutorial: Nonlinear magnonics," *J. Appl. Phys.*, vol. 134, no. 15, p. 151101, Oct. 2023.

[6] K. Vogt, F. Fradin, J. Pearson, T. Sebastian, S. Bader, B. Hillebrands, A. Hoffmann, and H. Schultheiss, "Realization of a spin-wave multiplexer," *Nature Communications*, vol. 5, no. 1, Apr. 2014.

[7] A. V. Chumak, A. A. Serga, and B. Hillebrands, "Magnon transistor for all-magnon data processing," *Nature Communications*, vol. 5, no. 1, p. 4700, 2014. [Online]. Available: https://doi.org/10.1038/ncomms5700

[8] H. Wu, L. Huang, C. Fang, B. S. Yang, C. H. Wan, G. Q. Yu, J. F. Feng, H. X. Wei, and X. F. Han, "Magnon valve effect between two magnetic insulators," *Phys. Rev. Lett.*, vol. 120, p. 097205, Mar 2018. [Online]. Available: https://link.aps.org/doi/10.1103/PhysRevLett.120.097205

[9] Q. Wang, M. Kewenig, M. Schneider, R. Verba, F. Kohl, B. Heinz, M. Geilen, M. Mohseni, B. Lägel, F. Ciubotaru *et al.*, "A magnonic directional coupler for integrated magnonic half-adders," *Nat. Elect.*, vol. 3, no. 12, pp. 765–774, 2020.

[10] G. Talmelli, T. Devolder, N. Träger, J. Förster, S. Wintz, M. Weigand, H. Stoll, M. Heyns, G. Schütz, I. P. Radu *et al.*, "Reconfigurable submicrometer spin-wave majority gate with electrical transducers," *Science Advances*, vol. 6, no. 51, p. eabb4042, 2020.

[11] T. Schneider, A. A. Serga, B. Leven, B. Hillebrands, R. L. Stamps, and M. P. Kostylev, "Realization of spin-wave logic gates," *Appl. Phys. Lett.*, vol. 92, no. 2, p. 022505, Jan. 2008. [Online]. Available: https://doi.org/10.1063/1.2834714

[12] H. Qin, R. B. Holländer, L. Flajšman, F. Hermann, R. Dreyer, G. Woltersdorf, and S. van Dijken, "Nanoscale magnonic fabry-pérot resonator for low-loss spin-wave manipulation," *Nat. Commun.*, vol. 12, no. 1, pp. 1–10, 2021.

[13] H. Qin, R. B. Holländer, L. Flajsman, and S. van Dijken, "Low-loss nanoscopic spin-wave guiding in continuous yttrium iron garnet films," *Nano Letters*, vol. 22, no. 13, pp. 5294–5300, 2022.

The IHP OpenPDK Initiative:
the status and roadmap

Wladek Grabinski (presenter)[12], Mustafa Alchalabi[6], Sergei Andreev[2], Luca Benini[9]
Mike Brinson[5], Matthias Bucher[4], Anton Datsuk[2], Frank K. Gurkaynak[9] Norbert Herfurth[2],
Krzysztof Herman[2], Sebastien Martinie[3], Vinayak Pachkawade[2], Harald Pretl[8], Christoph Sandner[7],
Rene Scholz[2], Jan Taro Svejda[6], Frank Vater[2], Christian Wittke[2]

1 MOS-AK (EU), 2 IHP (D), 3 CEA Leti (F), 4 TU Creta (GR), 5 LondonMET (UK),
6 Uni. Duisburg-Essen (D), 7 Infineon Villach (A), 8 JKU Linz (A), 7 ETH Zurich (CH)

Abstract—**The IHP OpenPDK initiative aligns with the EU Chips Act, which provides a fully open and manufacturable SG13G2 BiCMOS technology platform for analog/RF, mixed-signal, and digital IC applications. This initiative aims to bridge the gap between academia, startups, and the semiconductor industry, promoting open collaboration and overcoming economic and technical barriers in semiconductor innovation.**

INTRODUCTION

In the field of semiconductor technology, compact modeling, and IC designs, the OpenPDK Initiative provides an international platform for discussing advanced technologies, fostering collaboration among industry and academic leaders in electronic design automation (EDA). We review selected R&D topics presented at a recent event by prominent academic researchers and industrial professionals who presented and discussed innovative approaches in CAD/EDA tools, techniques including compact/SPICE modeling, and IC design that address the demands of emerging semiconductor technology applications.

I. IHP OPENPDK

IHP's OpenPDK effort aims to create open source Process Design Kits (PDKs) to boost access and cooperation in chip design [1]-[6]. Using RF and terahertz technologies, the IHP initiative encourages the adoption of open source electronic design automation (EDA) tools and intends to provide free multiproject wafer (MPW as in Tab:I) runs during 2024 and 2025. It aims to lower entry hurdles for educational and non-for-profit endeavors, promote collaborative development with the open source community, and support ongoing PDK maintenance within Europe. This mission serves as a foundation for future collaboration and innovation in open source hardware design.

TABLE I: Tapeout dates, technologies, and area specifications

Date	22/11/24	01/03/25	09/05/25	18/07/25	15/09/25
Technology	SG13G2	SG13G2	SG13G2	SG13G2	SG13C
Area [mm²]	20	140	30	30	220

II. ADVANCED COMPACT/SPICE MODELS

A. Advance HBT Compact Model in IHP OpenPDK

The IHP OpenPDK is based on a SiGe:C BiCMOS 130 nm technology, which features high-performance heterojunction bipolar transistors (HBTs) offering 350/450 GHz f_T/f_{max}. Figure 1 shows the temperature dependences of the transit frequency f_T as a function of the HBT collector current, which is particularly useful for high end GHz analog/RF IC designs [5]. To achieve optimal speed, it is advisable for designers to bias the devices with a collector current of approximately 2.5mA per HBT emitter stripe.

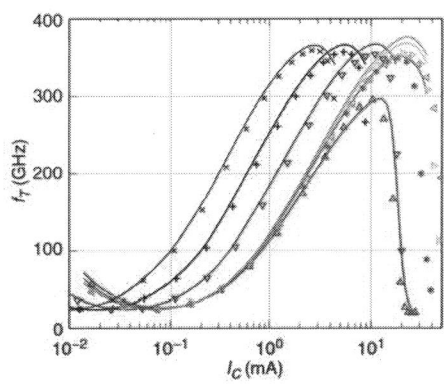

Fig. 1: The IHP SG13G2 HBT transit frequency $f_T = f(I_C)$

B. PSP Compact Model in IHP OpenPDK

The PSP (Physics-Based Scalable PSP) model is essential to accurately simulate CMOS devices in modern semiconductor design, addressing high-frequency performance and behavior at nanoscale dimensions critical for analog, RF and mixed-signal circuits [8]. Enhancements to the PSP model improve predictive accuracy and scalability, supporting design optimization for future semiconductor technologies. Key improvements include better accuracy for nanoscale CMOS devices (eg: Fig. 2), enhanced modeling for high-frequency and RF applications, and scalability across different technology nodes.

979-8-3315-0417-5/25 $31.00 © 2025 IEEE

These advances aim to improve predictability in mixed signal and analog circuit designs, especially under the constraints of small geometries.

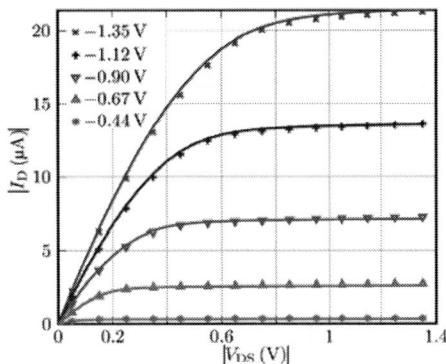

Fig. 2: LV p-MOSFET IdVd with W=0.6um and L=1.2um at 300K

C. Alternative Compact Models

The EKV3 model [7] is a charge-based MOS transistor model for OpenPDK frameworks. EKV3 supports various operating conditions in a unified g_m/I_d methodology. Its parameters are directly related to the mobile charge in the channel, allowing consistent modeling across all MOSFET regions without distinct formulations for each operating ultra low power modes.

Advancements in the UT SOI model [8] enhance accuracy in modeling ultra-thin SOI devices, especially regarding short-channel effects and scalability for advanced nodes. These improvements lead to more precise performance predictions for SOI transistors in analog/RF, digital in cryo domain, as in Fig. 3.

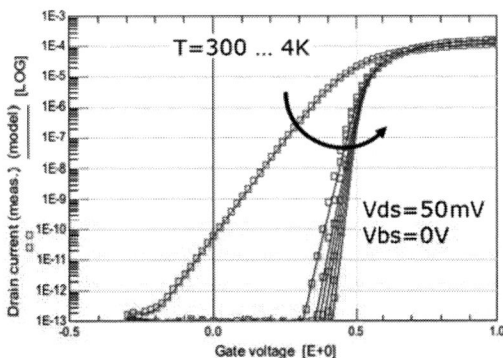

Fig. 3: L-UTSOI 102.7 Ids-Vgs at Vds=50 mV and Vds=0V, extracted model vs FDSOI experiment data down to 4K

III. SELECTED OPEN SOURCE TOOLS

A. Quite Universal Circuit Simulator with SPICE

QUCS-S (Quite Universal Circuit Simulator with SPICE) [9] supports the IHP 130nm BiCMOS open source PDK, enabling analog, digital, and mixed-signal simulations with various open source tools. Its database includes model symbols, test benches for MOS devices, and Monte Carlo analysis

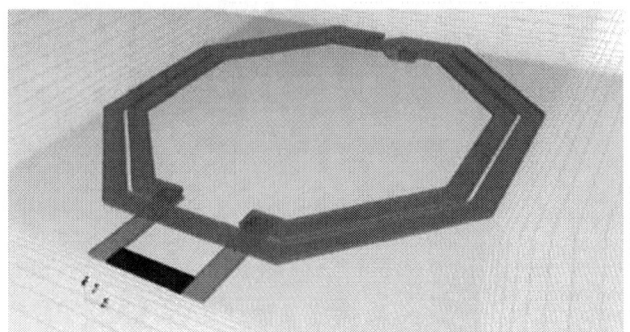

Fig. 4: RF spiral inductor in openEMS 3D environment

for DC and AC simulations. QUCS-S also supports digital logic simulations via XSPICE and RF/microwave simulations through QucsRF and openEMS [?]. This setup offers a complete open source ecosystem for analo/RF IC design, with future support for physical testing by IHP to validate simulation accuracy. This collaboration highlights the rapid advances in FOSS tools, meeting the semiconductor industry needs for flexible and accessible design tools.

B. openEMS Electromagnetic Simulator

openEMS [10] is an open-source electromagnetic simulator for RF component analysis in the IHP OpenPDK environment. It uses an EC-FDTD approach to efficiently model electromagnetic fields across a wide frequency range, includes structures, meshing, visualization, and parameter extraction targeting mmW RF IC applications. Future developments under the DI-DEMICO project aim to improve openEMS within the IHP OpenPDK ecosystem.

IV. VALUE AND OPPORTUNITIES IN OPEN SOURCE FOR CIRCUIT DESIGN

The transformative potential of open source approaches in hardware design has been discussed [11] to draw parallels to the successes in software domain. The urgent need for innovation in the semiconductor industry is emphasized as challenges such as workforce shortages and high entry barriers are faced. Four foundational pillars are identified: education, open source PDKs, open source IP, and automation in analog design. Open source methodologies can democratize IC design, increasing accessibility and reducing costs. Examples like SkyWater's 130nm PDK and IHP's 130nm BiCMOS platform show how open-access resources mitigate hardware development obstacles. The MOSAIC framework fosters collaboration between academia and industry. While tools like OpenROAD and Chipyard advance IC design, challenges in analog design automation persist. Promoting cooperative innovation between industry and academia is essential for a sustainable semiconductor technology trajectory.

A. Open source analog/RF design flow

The advancements and challenges of using open-source EDA tools for analog and RF have been explored [12]. The availability of open-source resources, such as CAD/EDA tools

979-8-3315-0417-5/25 $31.00 © 2025 IEEE   1262

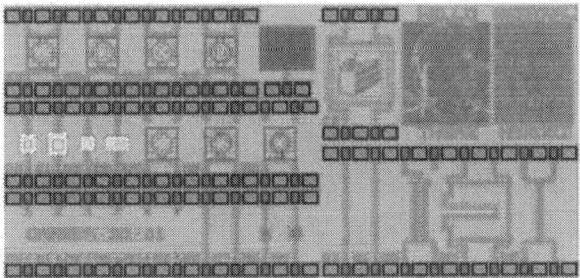

Fig. 5: mm-wave Layout gdsfactory autogenerated

and OpenPDKs, enables the design, simulation, and fabrication of analog/RF chips without proprietary software. Key tools like Xschem, Ngspice, kLayout, Magic, and OpenROAD support various aspects of mixed-signal and RF IC design. Recent projects, such as SAR ADCs and analog layouts with Python libraries, show these alternatives can meet professional needs. However, gaps remain, including improved parasitic extraction, fast and accurate analog simulation, and better tool interoperability. Ongoing development is essential, especially for complex mixed-signal designs. OpenPDKs also initiate educational programs to equip new engineers with vital skills, enhancing accessibility and fostering innovation in analog/RF IC design.

Fig. 6: Completely open SoC design

B. Open source digital design flow

The readiness of open source EDA for large-scale, commercial digital designs was reviewed. Initiatives like IHP OpenPDKs have democratized IC design through accessible fabrication. 'Basilisk', a complex SoC design with over a million gates, shows improvements in timing and area efficiency. However, challenges remain in back-end processes due to PDK access and third-party IP constraints. Key components like DRC and LVS often require commercial support, indicating inconsistent high-quality results in open source flows. The need for advanced PDKs, especially below 130nm, is emphasized. Although promising, further refinement is needed for open source digital design flows to meet the demands of high-complexity projects.

V. OUTLOOK

The IHP Open PDK initiative aims to promote collaboration in chip design by offering open source electronic design automation (EDA) and open process design kits (PDKs), particularly in RF and terahertz (THz) technologies. The roadmap emphasizes user feedback integration, tool compatibility enhancement, and fostering collaborations. Future goals include expanding support for RF designs, developing parameterized cells, and achieving industrial-grade quality assurance. Aligning with the EU Chips Act, the IHP Open PDK underscores the importance of open collaboration in overcoming economic and technical barriers to semiconductor innovation.

ACKNOWLEDGMENT

The authors would like to thank to all the open source community, SemiMod(D), ETH Zurich (CH), Uni Linz (A), CEA-Leti (F), TU Crete (GR), UDE/CENIDE Duisburg (D), PULP project team (CH) ChipFlow, Staf Verhaegen (PDKMaster), M. Koefferlein (kLayout), D. Warning, H. Vogt (ngspice), M. Brinson (Qucs-S), Flowspace, OCDCpro, DIDEMICO, and many more; All that is genroucelly supported by public-founded German projects: VE-HEP (16KIS1339K), IHP Open130-G2 (16ME0852), FMD-QNC (16ME0831), including (DE:Sign) Design tools for sovereign chip development with open source, started May'24.

REFERENCES

[1] IHP OpenPDK Networking Workshop FMD-QNC, (2023)
[2] W. Grabinski at al. *FOSS CAD/EDA tools supporting the open access PDK initiative* FOSDEM 2024; Brussels (2024)
[3] W. Grabinski et al., *FOSS CAD for the Compact Verilog-A Model Standardization in Open Access PDKs* 8th EDTM, Bangalore (2024)
[4] K. Herman (IHP, Germany) Reflections on the First European Open Source PDK by IHP – experiences after one year and future activities (invited talk) MIXDES, Gdańsk (2024)
[5] K. Herman et al., "On the Versatility of the IHP BiCMOS Open Source and Manufacturable PDK: A step towards the future where anybody can design and build a chip," in IEEE SSC Magazine, 16/2, (2024)
[6] R. Scholz, *The IHP OpenPDK Initiative*, MOS-AK/ESSERC Bruges (2024)
[7] M. Bucher, *A review of charge-based MOS Transistor modeling* , MOS-AK/ESSERC Bruges (2024)
[8] S. Martinie, *Overview and latest updates of PSP and L-UTSOI standard models*, MOS-AK/ESSERC Bruges (2024)
[9] M. Brinson, *QUCS-S —a central tool in the openPDK IC design flow*, MOS-AK/ESSERC Bruges (2024)
[10] Mu. Alchalabi and J.T. Svejda, *openEMS as a versatile tool in the framework of mm-wave openPDK-based RF chip design*, MOS-AK/ESSERC Bruges (2024)
[11] Ch. Sandner, *Value and Opportunities in Open-Source for Circuit Design*, MOS-AK/ESSERC Bruges (2024)
[12] H. Pretl, *Designing Analog/RF Chips Using Open PDKs and Open-Source Tools*, MOS-AK/ESSERC Bruges (2024)
[13] F.K. Gurkaynak, *Are open source digital design flows ready for mainstream?*, MOS-AK/ESSERC Bruges (2024)

Heterogeneous Integration of Compound Semiconductor Materials and Devices by Ion-Cutting Technique

Tiangui You[1,2], Tian Liang[1], Jiaxin Ding[1,2], Jialiang Sun[1], Shangyu Yang[1], Xin Ou[1,2]

[1] State Key Laboratory of Materials for Integrated Circuits, Shanghai Institute of Microsystem and Information Technology, Chinese Academy of Sciences, China, [2]Center of Materials Science and Optoelectronics Engineering, University of Chinese Academy of Sciences, China
E-mail: t.you@mail.sim.ac.cn, ouxin@mail.sim.ac.cn

Abstract

The heterogeneous integration of compound semiconductors with Si substrates can combine the cost advantage and maturity of Si technology with the superior performance of compound semiconductors, enabling a new class of high-performance integrated circuits with multiple functionalities for the post-Moore era. This work provides a brief description of the ion-cutting technique, which basically consists of ion implantation, wafer bonding, annealing to achieve exfoliation, and polishing. In addition to the industrial production of silicon-on-insulator (SOI) wafers with diameters up to 300 mm, several non-silicon semiconductor thin films, such as III-V compound semiconductors and wide bandgap semiconductors, have been exfoliated and transferred to foreign substrates using the ion-cutting technique.

Keywords: Heterogeneous Integration, Compound Semiconductor, Ion-Cutting technique

Introduction

Along with the developments in microelectronics and optoelectronics towards miniaturization and integration, microsystem chips have become more diversified and sophisticated, which has placed great demands on heterogeneous integration technology. Heterogeneous integration, in turn, will open up new pathways for the development of microelectronics technology in the post-Moore era. Based on the current feature sizes, developing heterogeneous materials and integrating various functional devices will enable functional diversification on a single chip, especially for the single-chip integration of optoelectronics, micro-energy, analog, RF, passive components, and MEMS devices. The heterogeneous integration of devices and systems relies on heterogeneous integration materials. Conventionally, the heterogeneous integration of different semiconductors with Si has been achieved through hetero-epitaxy growth techniques, such as Molecular Beam Epitaxy (MBE), Metalorganic Chemical Vapor Deposition (MOCVD), and Metalorganic Vapor Phase Epitaxy (MOVPE), which date back over 40 years.[1] However, challenges such as lattice mismatch, thermal expansion coefficient mismatch, crystal structure mismatch, and anti-phase domains remain to be overcome in order to hetero-epitaxially grow high-quality single-crystalline non-silicon thin films on Si substrates. Moreover, it is virtually impossible to fabricate "on-insulator" structures using hetero-epitaxy growth techniques, in which a buried oxide layer (e.g., SiO_2) exists between the non-silicon thin films and the Si substrate. "On-insulator" structures, often referred to as "XOI"[2], inherit the merits of silicon-on-insulator (SOI) technology, such as low parasitic capacitance and power, low leakage current, high speed, reduced short-channel effects, latch-up effects, and radiation tolerance. These advantages are beneficial in enhancing device performance.

On the other hand, the direct wafer bonding technique has been developed as a promising solution for combining dissimilar materials, regardless of their structures (monocrystalline, polycrystalline, amorphous), crystallographic orientations, or lattice parameters. Following the direct wafer bonding process, thinning the donor wafer to the desired thin-film thickness is usually required for subsequent device fabrication. The most straightforward approaches include mechanical grinding/polishing, reactive ion etching, and wet etching. The obvious drawback of these approaches is the loss of the complete donor wafer, which is not cost-effective for expensive wafers, such as high-purity SiC and free-standing GaN. Another issue is the deterioration of thin-film uniformity during the thinning processes. Alternatively, the ion-cutting technique, commercially known as Smart-Cut™,[3] has been proven to be a more elegant method for releasing a thin layer from the donor wafer. First proposed in 1995, this technique is now used in the industrial production of SOI wafers with diameters up to 300 mm. In recent years, the ion-cutting technique has been extensively studied and used to exfoliate non-silicon thin films, which can then be transferred to foreign substrates to form heterogeneous integration materials. In this work, we intend to provide a comparatively recent progress on the underlying physic mechanisms of the ion-implantation induced thin film exfoliation, and the fabrication and engineering of the heterogeneous integration materials and devices for the non-silicon semiconductors.

Procedure of Ion-Cutting Technique

The process flow of the ion-cutting technique is schematically illustrated in Fig. 1, and it basically consists of ion implantation, wafer bonding, annealing to achieve

979-8-3315-0417-5/25 $31.00 © 2025 IEEE

exfoliation, and polishing. Firstly, light ion implantation (hydrogen and/or helium implantation) is performed on the donor wafer to create a damage layer beneath the implanted surface. The depth of the damage layer depends on the ion implantation energy, which can be calculated using SRIM software[4]. During the ion implantation process, the donor wafers are typically slightly tilted from the normal to minimize the ion channeling effect. The implanted donor wafer is then bonded to a handle wafer, and the bonding pair is thermally annealed. A high bonding energy is required for the wafer bonding, so that the bonding pair can withstand the thermal strain induced by the thermal expansion coefficient mismatch of the donor wafer during the subsequent annealing process. During annealing, the interaction of the implanted ions with the implantation-induced damage results in the formation of extended internal surfaces, in the form of pressurized micro-cracks along the damage layer. After the annealing process, a thin film is exfoliated from the donor wafer and transferred to the handle wafer, thus creating a heterogeneous integration material. In general, a residual damage layer remains on both the transferred thin film and the remaining donor wafer, which can be removed through chemical mechanical polishing (CMP). After the CMP process, the remaining donor wafer can be recycled, which helps reduce material costs. A further annealing process is usually required for the as-fabricated heterogeneous integration material to recover from the implantation-induced damage in the transferred thin film. Moreover, this annealing process also increases the bonding strength between the transferred thin film and the handle wafer, preventing the thin film from peeling off during subsequent device fabrication.

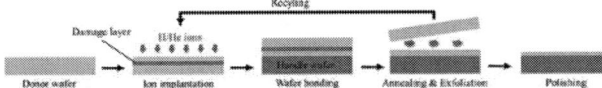

Fig. 1 Process flow of the ion-cutting technique.

Results

In addition to SOI, several non-silicon semiconductor thin films, such as InP, Ga_2O_3, SiC, GaN, $LiNbO_3$ (LN), and $LiTaO_3$ (LT), have been exfoliated and transferred to foreign substrates using the ion-cutting technique to form heterogeneous integration materials like InPOI[5,6], Ga_2O_3-OI[7-9], SiCOI[10,11], GaNOI[12-14], LNOI[15], LTOI[16] as shown in Fig. 2. In comparison with silicon, H and/or He implantation-induced blistering/exfoliation in non-silicon semiconductors is more sensitive to implantation parameters, such as implantation temperature, fluence, flux, and post-implantation annealing temperature and time. This may be due to the larger diffusivity of H ions in non-silicon semiconductors and the stronger bond strength of these materials. A convenient method to optimize the implantation-induced layer exfoliation process is to inspect the surface blistering in the semiconductors after post-implantation

annealing, without bonding the implanted wafer to a handle wafer.

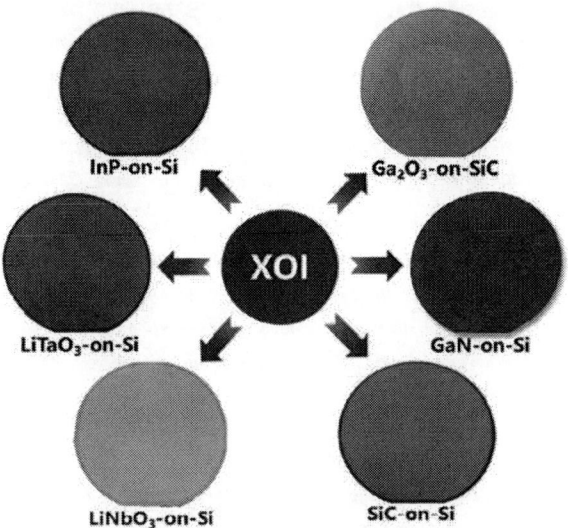

Fig. 2 XOI wafers fabricated by ion-cutting technique

The fabricated XOI wafers can be further used as a template for the epitaxy growth of compound semiconductor thin films. With the regrowth of III-V on the InP-on-Si heterogeneous substrate, integrated lasers on Si substrates can be achieved.[17,18] Using this approach, we achieved state-of-the-art performance with an electrically pumped, continuous-wave (CW) 1.55-μm Si-based laser, featuring a room-temperature threshold current density of 0.65 kA/cm² and an output power exceeding 155 mW per facet without facet coating in CW mode. CW lasing at 120°C and pulsed lasing at over 130°C were achieved. This generic approach is also applied to other material systems, providing better performance and more functionalities for photonics and microelectronics.

Conclusion

In this work, we have presented an overview of the fabrication of heterogeneous integration materials by exfoliating and transferring single-crystalline non-silicon semiconductor thin films onto foreign substrates using the ion-cutting technique. Essentially, the ion-cutting technique consists of wafer bonding and thin-film exfoliation induced by light ion implantation. The implantation-induced layer exfoliation process can be optimized by inspecting the surface blistering in the semiconductors after post-implantation annealing, without bonding the implanted wafer to a handle wafer. Thanks to the availability of high-quality, free-standing wafers and advanced wafer bonding technology, wafer-scale heterogeneous integration materials, such as InPOI, Ga_2O_3-OI SiCOI, GaNOI, LNOI and LTOI, have been achieved using the ion-cutting technique. It has been demonstrated that these heterogeneous materials can be used as templates for further layer growth and device fabrication.

Acknowledgments

This work was supported by the National Key Research and Development of China (No. 2021YFB3601900), National Natural Science Foundation of China (No. 62293521, 62174167), Strategic Priority Research Program of Chinese Academy of Sciences (No. XDB0670202), Shanghai Rising-Star Program (No. 22QA1410700), The Shanghai Science and Technology Innovation Action Plan (No. 24520714800).

References

[1] S. Irvine and P. Capper, "Metalorganic Vapor Phase Epitaxy (MOVPE): Growth, Materials Properties, and Applications."

[2] H. Ko et al., "Ultrathin compound semiconductor on insulator layers for high-performance nanoscale transistors," Nature, vol. 468, pp. 286-289, 2010

[3] M. Bruel et al., in 1995 IEEE International SOI Conference Proceedings, 1995, pp. 178-179.

[4] J. F. Ziegler, M. D. Ziegler, and J. P. Biersack, "SRIM – The stopping and range of ions in matter (2010)," Nuclear Instrum. Methods Phys. Res. Sect. B: Beam Interact. Mater. Atoms, vol. 268, pp. 1818-1823, 2010.

[5] J. Lin et al., "Wafer-scale heterogeneous integration InP on trenched Si with a bubble-free interface," APL Materials, vol. 8, pp. 051110, 2020.

[6] J. Lin et al., "Efficient ion-slicing of InP thin film for Si-based hetero-integration," Nanotechnology, vol. 29, pp. 504002, 2018.

[7] W. Xu et al., "Current transport mechanism of lateral Schottky barrier diodes on β-Ga$_2$O$_3$/SiC structure with atomic level interface," Appl. Phys. Lett., vol. 124, pp. 112102, 2024.

[8] W. Xu et al., "First demonstration of wafer-scale heterogeneous integration of Ga$_2$O$_3$ MOSFETs on SiC and Si substrates by ion-cutting process," 2019 IEEE International Electron Devices Meeting (IEDM), 2019, pp. 12.5.1-12.5.4.

[9] Z. Shen et al., "Wafer-scale single-crystalline β-Ga$_2$O$_3$ thin film on SiC substrate by ion-cutting technique with hydrophilic wafer bonding at elevated temperatures," Science China Materials, vol. 66, pp. 756, 2023.

[10] A. Yi et al., "Wafer-scale 4H-silicon carbide-on-insulator (4H–SiCOI) platform for nonlinear integrated optical devices," Optical Materials, vol. 107, pp. 109990, 2020.

[11] A. Yi, C. Wang, L. Zhou, Y. Zhu, S. Zhang, T. You, J. Zhang, and X. Ou, "Silicon carbide for integrated photonics," Appl. Phys. Rev., vol. 9, pp. 031302, 2022.

[12] H. Shi et al., "Elucidating the formation mechanisms of the parasitic channel with buffer-free GaN/Si hetero-bonding structures," Appl. Phys. Lett., vol. 124, pp. 192103, 2024.

[13] H. Shi et al., "Defect evolution in GaN thin film heterogeneously integrated with CMOS-compatible Si(100) substrate by ion-cutting technology," Science China Information Sciences, vol. 66, pp. 219403, 2023.

[14] H. Shi et al., "Realization of wafer-scale single-crystalline GaN film on CMOS-compatible Si (100) substrate by ion-cutting technique," Semicond. Sci. Technol., vol. 35, pp. 125004, 2020.

[15] Y. Chen et al., "Wafer-scale fabrication of silicon film on lithium niobate on insulator (LNOI)," Crystals, vol. 12, pp. 1477, 2022.

[16] Y. Yan, K. Huang, H. Zhou, et al., "Wafer-scale fabrication of 42° rotated Y-cut LiTaO$_3$-on-insulator (LTOI) substrate for a SAW resonator," ACS Appl. Electron. Mater., vol. 1, , pp. 1660-1666, 2019.

[17] J. Sun et al., "High-power, electrically-driven continuous-wave 1.55-µm Si-based multi-quantum well lasers with a wide operating temperature range grown on wafer-scale InP-on-Si (100) heterogeneous substrate," Light: Sci. Appl., vol. 13, pp. 71, 2024.

[18] J. Lin et al., "2.1 µm multi-quantum well laser epitaxially grown on on-axis (001) InP/SiO$_2$/Si substrate fabricated by ion-slicing," Opt. Express, vol. 32, pp. 19655, 2024.

Impact of Charge Trapping at Defects on the Robustness of Electronic Circuits

Michael Waltl[1,2], Bernhard Stampfer[2], and Roberto Orio[2]

[1]Christian Doppler Laboratory for Single-Defect Spectroscopy at the
[2]Institute for Microelectronics, TU Wien, Vienna, Austria

ABSTRACT

Charge trapping at oxide defects is a significant reliability issue in MOS transistors and becomes even more prominent in scaled technology nodes. As devices shrink, the impact of charge capture and emission events of oxide and interface defects on performance becomes more severe. Our work demonstrates how charge trapping affects variability in scaled Silicon and 2D technologies, highlighting the necessity of low trap density to achieve high yield and ensure robust electronic circuit designs.

INTRODUCTION

The proper functionality and robustness of complex integrated electronic circuits are based on the well-defined interaction of various electronic devices, with MOSFETs serving as key components in modern electronic applications. Significant efforts to enhance the performance of integrated MOS transistors have led to devices with outstanding electrical characteristics and geometry, e.g., FinFETs. As an innovative solution for next-generation chips, the so-called complementary FETs and 3D integration (low thermal budget oxides) are currently under research. Despite continuous advancements in fabrication equipment and processes, critical parameters such as the threshold voltage, sub-threshold slope, and on-resistance remain heavily influenced by atomic-level defects in the transistor structure.

CHARGE TRAPPING IN TRANSISTORS

To unravel the chemical nature of these defects and understand their charge trapping behavior, numerous theoretical studies using DFT methods have been conducted [1] , in conjunction with the development of innovative measurement techniques [2], [3] as well as physical models [4] to explain the defect-induced impact on the device behavior. The methodology has finally been used to successfully explain charge trapping in various Si technologies [5]–[7], but also in devices employing wide band-gap materials such as SiC [8] and

GaN [9], [10] and more exotic transistors based on 2D materials [11], [12].

For Si technology, the initial focus has been placed on large-area devices, where the collective electrical response of many defects was analyzed, see Figure 1. However, with the development of scaled transistors,

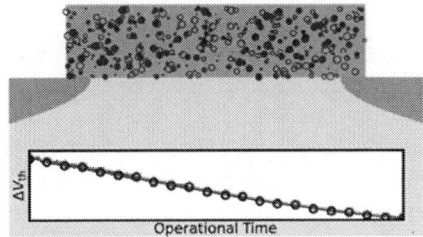

Fig. 1. Large-area devices contain numerous defects, where charge trapping at these defects can be observed as a continuous drift in the drain-source current.

individual defects have emerged as the main contributors to the reliability challenges in integrated MOS devices, as shown in Figure 2. Numerous studies used scaled

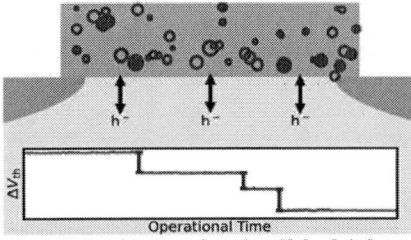

Fig. 2. In scaled transistors, only a handful of defects are present. However, a change in the charge state of a defect appears as a discrete step in the drain-source current. These steps can be utilized to investigate the physical mechanisms of charge trapping in the underlying device technology.

devices and focused on analyzing individual defects, e.g., by evaluating random telegraph noise (RTN) signals [13]–[15]. An advancement in RTN measurements is provided by the so-called time-dependent defect spectroscopy (TDDS) technique [6], which enables exploring the full charge trapping kinetics of defects. The application of TDDS was complemented by the development of a sophisticated defect-probing instrument capable of detecting threshold voltage changes smaller than one milli-

979-8-3315-0417-5/25 $31.00 © 2025 IEEE

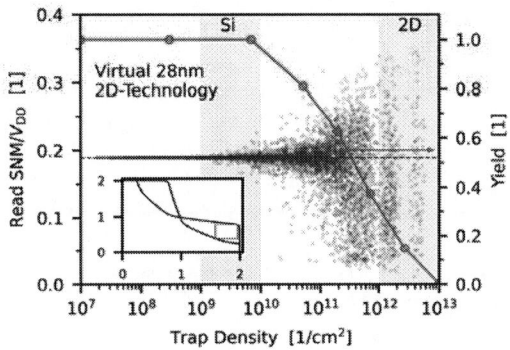

Fig. 3. At a trap density above 10^{10}cm^{-2} a large variability in the read static noise margin can be observed. This is a consequence of the distribution of the impact of the defects on the device characteristics.

volt [16]. Although TDDS provides detailed insights into the charge-trapping kinetics of defects, the measurement process is often time-intensive [17]. Therefore, advanced methods are crucial for efficiently extracting trap-specific parameters from TDDS data, which are found by the effective single-defect decomposition method [18].

The importance of analyzing device reliability at the single defect level has been shown in various studies [19], which have clearly demonstrated that as miniaturization progresses, the impact of individual defects on transistor behavior increases considerably. At the same time, variability between components also increases. Both of these factors must be carefully considered in circuit design.

IMPACT OF CHARGE TRAPPING ON CIRCUITS

Charge trapping at defects also affects the robustness of electronic circuits and must be considered properly during circuit design. To simulate circuits, compact models that describe the behavior of the components are typically used. Such models typically rely on mean values for the characteristic electrical parameters, for example, the threshold voltage, of a transistor [20], [21]. However, in several simulation approaches, also device-to-device variations are considered by adding large guard bands to the respective mean values [22].

In the example shown in this work, we implemented the two-state defect model proposed in [4] into the OpenSource circuit simulator ngspice and evaluated the impact of charge trapping on static noise margins of an NMOS only SRAM cell considering a virtual 28 nm 2D technology (circuit not explicitly shown). It should be noted that the model has been carefully calibrated to statistical data extracted from various 2D technologies [12], [23]. One can observe a dramatic increase for the read static noise margin from Figure 3. In our analysis, we

compare conventional Si technology and 2D technology, and the data reveal that for a trap density exceeding 10^{10}cm^{-2} the noise margin increases considerably, and also the yield drops significantly.

This is an example that highlights the importance of accurately describing charge trapping and its impact on circuits. It should be noted that in digital applications, variability effects can be mitigated to some extent by using a larger overdrive bias, which, however, counteracts the scaling of V_{DD}. In contrast, defects in analog circuits have a serious detrimental effect on the signal-to-noise ratio (SNR), making an accurate description of these effects indispensable in order to create optimized chip designs.

CONCLUSION

In our work, we address the fact that charge trapping at oxide and interface defects poses a critical reliability challenge for MOS transistors, particularly in scaled technology nodes. As device sizes continue to shrink and geometries become more complex, the effects of charge trapping on device behavior and variability become increasingly significant. Our study highlights the impact of charge trapping in both scaled Si and 2D technologies, emphasizing that minimizing trap density is essential to achieve high yield and ensuring the robustness of electronic circuits. Addressing this issue during the design process is crucial for the continued advancement and reliability of semiconductor technologies.

ACKNOWLEDGMENT

The financial support by the Austrian Federal Ministry of Labour and Economy, the National Foundation for Research, Technology and Development, and the Christian Doppler Research Association is gratefully acknowledged.

REFERENCES

[1] J. Cottom, G. Gruber, G. Pobegen, T. Aichinger, and A. L. Shluger, "Recombination defects at the 4H-SiC/SiO2 interface investigated with electrically detected magnetic resonance and ab initio calculations," *JAP*, vol. 124, no. 4, p. 045302, 2018.

[2] B. Kaczer, T. Grasser, P. J. Roussel, J. Franco, R. Degraeve, L.-A. Ragnarsson, E. Simoen, G. Groeseneken, and H. Reisinger, "Origin of NBTI Variability in Deeply Scaled pFETs," in *IRPS*, May 2010, pp. 26–32.

[3] H. Reisinger, O. Blank, W. Heinrigs, W. Gustin, and C. Schlunder, "A comparison of very fast to very slow components in degradation and recovery due to nbti and bulk hole trapping to existing physical models," *IEEE Transactions on Device and Materials Reliability*, vol. 7, no. 1, pp. 119–129, 2007.

[4] G. Rzepa, J. Franco, B. O'Sullivan, A. Subirats, M. Simicic, G. Hellings, P. Weckx, M. Jech, T. Knobloch, M. Waltl, P. J. Roussel, D. Linten, B. Kaczer, and T. Grasser, "Comphy – A Compact-Physics Framework for Unified Modeling of BTI," *MR*, vol. 85, pp. 49–65, 2018, (invited).

[5] J. Michl, A. Grill, D. Waldhoer, W. Goes, B. Kaczer, D. Linten, B. Parvais, B. Govoreanu, I. Radu, T. Grasser, and M. Waltl, "Efficient modeling of charge trapping at cryogenic temperatures—part ii: Experimental," *IEEE Transactions on Electron Devices*, vol. 68, no. 12, 2021.

[6] T. Grasser, H. Reisinger, P.-J. Wagner, F. Schanovsky, W. Goes, and B. Kaczer, "The Time Dependent Defect Spectroscopy (TDDS) for the Characterization of the Bias Temperature Instability," in *IRPS*, May 2010, pp. 16–25.

[7] T. Grasser, M. Waltl, Y. Wimmer, W. Goes, R. Kosik, G. Rzepa, H. Reisinger, G. Pobegen, A. El-Sayed, A. Shluger, and B. Kaczer, "Gate-Sided Hydrogen Release as the Origin of "Permanent" NBTI Degradation: From Single Defects to Lifetimes," in *IEDM*, 2015.

[8] C. Schleich, D. Waldhoer, K. Waschneck, M. W. Feil, H. Reisinger, T. Grasser, and M. Waltl, "Physical modeling of charge trapping in 4h-sic dmosfet technologies," *IEEE Transactions on Electron Devices*, vol. 68, no. 8, pp. 4016–4021, 2021.

[9] A. Grill, B. Stampfer, M. Waltl, K.-S. Im, J.-H. Lee, C. Ostermaier, H. Ceric, and T. Grasser, "Characterization and Modeling of Single Defects in GaN/AlGaN Fin-MIS-HEMTs," in *IRPS*, 2017.

[10] R. Stradiotto, G. Pobegen, C. Ostermaier, M. Waltl, A. Grill, and T. Grasser, "Characterization of interface defects with distributed activation energies in gan-based mis-hemts," *IEEE Transactions on Electron Devices*, vol. 64, no. 3, pp. 1045–1052, 2017.

[11] M. Waltl, T. Knobloch, K. Tselios, L. Filipovic, B. Stampfer, Y. Hernandez, D. Waldhör, Y. Illarionov, B. Kaczer, and T. Grasser, "Perspective of 2d integrated electronic circuits: Scientific pipe dream or disruptive technology?" *Advanced Materials*, vol. 34, no. 48, p. 2201082, 2022.

[12] B. Stampfer, F. Zhang, Y. Y. Illarionov, T. Knobloch, P. Wu, M. Waltl, A. Grill, J. Appenzeller, and T. Grasser, "Characterization of Single Defects in Ultra-Scaled MoS$_2$ Field-Effect Transistors," *ACS Nano*, vol. 12, pp. 5368–5375, 2018.

[13] M. J. Uren, D. J. Day, and M. J. Kirton, "1/f and Random Telegraph Noise in Silicon Metal-Oxide-Semiconductor Field-Effect Transistors," *APL*, vol. 47, no. 11, 1985.

[14] D. Veksler, G. Bersuker, L. Vandelli, A. Padovani, L. Larcher, A. Muraviev, B. Chakrabarti, E. Vogel, D. C. Gilmer, and P. D. Kirsch, "Random telegraph noise (RTN) in scaled RRAM devices," in *IRPS*, 2013, pp. MY.10.1–MY.10.4.

[15] H. Miki, N. Tega, M. Yamaoka, D. J. Frank, A. Bansal, M. Kobayashi, K. Cheng, C. P. D'Emic, Z. Ren, S. Wu, J. Yau, Y. Zhu, M. A. Guillorn, D.-G. Park, W. Haensch, E. Leobandung, and K. Torii, "Statistical measurement of random telegraph noise and its impact in scaled-down high-k/metal-gate MOSFETs," in *IEDM*, 2012, pp. 19.1.1–19.1.4.

[16] M. Waltl, "Ultra-low noise defect probing instrument for defect spectroscopy of mos transistors," *IEEE Transactions on Device and Materials Reliability*, vol. 20, no. 2, pp. 242–250, 2020.

[17] M. Waltl, B. Stampfer, G. Rzepa, B. Kaczer, and T. Grasser, "Separation of electron and hole trapping components of pbti in sion nmos transistors," *MR*, vol. 114, p. 113746, 2020.

[18] D. Waldhoer, C. Schleich, J. Michl, B. Stampfer, K. Tselios, E. G. Ioannidis, H. Enichlmair, M. Waltl, and T. Grasser, "Toward automated defect extraction from bias temperature instability measurements," *IEEE Transactions on Electron Devices*, vol. 68, no. 8, pp. 4057–4063, 2021.

[19] M. Waltl, D. Waldhoer, K. Tselios, B. Stampfer, C. Schleich, G. Rzepa, H. Enichlmair, E. G. Ioannidis, R. Minixhofer, and T. Grasser, "Impact of single-defects on the variability of cmos inverter circuits," *Microelectronics Reliability*, vol. 126, p. 114275, 2021.

[20] G. Gildenblat, X. Li, W. Wu, H. Wang, A. Jha, R. Van Langevelde, G. D. Smit, A. J. Scholten, and D. B. Klaassen, "Psp: An advanced surface-potential-based mosfet model for circuit simulation," *IEEE Transactions on Electron Devices*, vol. 53, no. 9, 2006.

[21] J. R. Brews, "A charge-sheet model of the mosfet," *Solid-State Electronics*, vol. 21, no. 2, pp. 345–355, 1978.

[22] H. Li, Z. Jiang, P. Huang, Y. Wu, H.-Y. Chen, B. Gao, X. Liu, J. Kang, and H.-S. Wong, "Variation-aware, reliability-emphasized design and optimization of rram using spice model," in *2015 Design, Automation & Test in Europe Conference & Exhibition (DATE)*. IEEE, 2015, pp. 1425–1430.

[23] T. Knobloch, J. Michl, D. Waldhör, Y. Illarionov, B. Stampfer, A. Grill, R. Zhou, P. Wu, M. Waltl, J. Appenzeller, and T. Grasser, "Analysis of Single Electron Traps in Nano-scaled MoS$_2$ FETs at Cryogenic Temperatures," in *Device Research Conference (DRC)*, 2020, pp. 52–53.

Solution-Processed Reduced-Dimensional Cesium Lead Halide Perovskites

Gaukhar Nigmetova,[1] Zhuldyz Yelzhanova,[1] Hryhorii Parkhomenko,[2] Meruyert Tilegen,[3] Askhat N. Jumabekov,[2] Tri T. Pham,[3] Annie Ng[1,*]

[1]Department of Electrical and Computer Engineering, Nazarbayev University, Astana 010000, Kazakhstan

[2]Department of Physics, Nazarbayev University, Astana 010000, Kazakhstan

[3]Department of Biology, Nazarbayev University, Astana 010000, Kazakhstan

Phone: +7 775 1930118; E-mail: annie.ng@nu.edu.kz

Abstract

This work introduces a solution-processed approach for growing Cs_2PbX_4 nanostructures on $CsPbI_2Br$ thin films, achieving effective control over interface morphology. The surface of $CsPbI_2Br$ thin films is modified through CsCl post-treatment using various organic solvents. Careful optimization of post-treatment parameters, including solvent selection and processing conditions, ensures uniform nanostructure growth and improved interfacial properties. This strategy significantly reduces defects and enhances charge transport, leading to efficient and stable Cs-based perovskite solar cells (PSCs).

Keywords: inorganic halide perovskites, low-dimensional perovskites, interface engineering

Introduction

Hybrid halide perovskite materials have emerged as promising light absorbers among third-generation solar cells. However, long-term stability remains a significant challenge to the commercialization of perovskite solar cells (PSCs). Among various halide perovskites, all-inorganic Cs-based perovskites have attracted increasing attention due to their superior thermal stability compared to conventional organic cation-based PSCs [1]. Moreover, these materials are particularly suitable for integration into tandem devices with narrow band-gap sub-cells. This study focuses on $CsPbI_2Br$ perovskite, which has a bandgap of 1.9 eV. A key strategy of this work is to stabilize $CsPbI_2Br$ thin film by using low-dimensional Cs_2PbX_4 for surface modification, in contrast to the commonly employed organic cation treatments in PSCs [2,3]. By forming low-dimensional Cs-based perovskite nanostructures on $CsPbI_2Br$ thin films, we demonstrate an effective surface modification technique that significantly improves the performance of PSCs.

The morphology optimization of these low-dimensional perovskite nanostructures through solution processing is critical for achieving these improvements. Unlike other studies that focus on organic cation components, such as phenylethylammonium iodide (PEAI) [4] or butylammonium iodide (BAI) [5], this work elucidates the less-explored area of solution processing conditions for growing all-inorganic nanostructures. Understanding these conditions is essential for future advancements in large-area manufacturing of Cs-based perovskite devices, ensuring scalability and practical application. The insights gained from this work will contribute to the development of efficient solution processing methods and interface engineering techniques, providing valuable guidance for the future development of PSCs.

Experimental

The perovskite precursor solution was prepared by dissolving 0.9 M CsI, 0.45 M PbI_2, and 0.45 M $PbBr_2$ in a 4:1 DMF:DMSO solvent mixture, which was stirred overnight in an N_2 glove box. $CsPbI_2Br$ films were then deposited on FTO-coated glass substrates using spin-coating set at 1000 rpm for 10 s, followed by 5000 rpm for 30 s. The films were sequentially annealed at 40°C for 2 min and 250°C for 10 min. For the 2D perovskite layer formation, a CsCl solution (0.45 M) in different alcohols was spin-coated onto the $CsPbI_2Br$ films and then heated at 100°C with controlled external flow to vary the evaporation rate. Device fabrication was carried out in an N_2-filled glovebox, following an n-i-p architecture (FTO/ETL/Perovskite/HTL/Au). The resulting device configuration is represented schematically in Figure 1. The FTO-coated glass substrates were cleaned using a series of solvents, followed by UV-ozone treatment. The TiO_2 ETL was obtained by spin-coating the precursor solution onto the substrate and sintering the samples at 500°C for 2 hours.

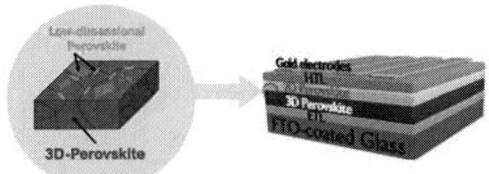

Figure 1. The schematic diagram of the $CsPbI_2Br$ PSC.

Then, the ZnO ink was spin-coated and annealed at 110°C for 10 min. After the CsPbI$_2$Br perovskite film was obtained and treated with CsCl, the spiro-MeOTAD solution was spin-coated as a hole-transport layer (HTL). Finally, gold electrodes were thermally evaporated onto the HTL, achieving a thickness of 70 nm.

Results and Discussion

The top-view images of the pristine CsPbI$_2$Br thin film and the samples after CsCl post-treatment using different solvents are shown in Figure 2. It is found that the growth of Cs$_2$PbX$_4$ nanostructures on CsPbI$_2$Br is strongly affected by the post-treatment conditions. The control sample without post-treatment (Figure 2a) displays a smooth CsPbI$_2$Br perovskite surface with no evident nanostructures. The sample post-treated with CsCl dissolved in methanol (Figure 2b) exhibits relatively small and sparsely distributed nanostructures on the surface. In contrast, treatment with CsCl in isopropanol (Figure 2c) results in the formation of uniform and moderately sized nanostructures across the surface of CsPbI$_2$Br films. When the samples are treated with CsCl in butanol (Figure 2d) the largest and most extended nanostructures are observed. The size and distribution of these low-dimensional nanostructures vary significantly depending on the solvent used for CsCl deposition. These observations suggest that the choice of solvent during CsCl post-treatment plays a critical role in controlling the morphology of the resulting nanostructures. The solvents used in this work have different boiling points, resulting in

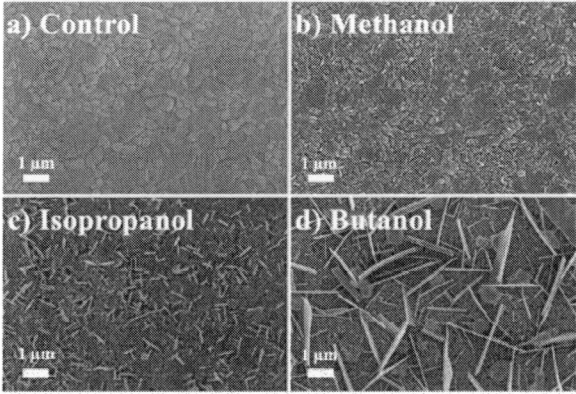

Figure 2. SEM images of the top-view of CsPbI$_2$Br perovskite film a) before and after CsCl post-treatment using different solvents such as b) methanol, c) isopropanol and d) butanol, forming the Cs$_2$PbX$_4$ nanostructures on the surface of CsPbI$_2$Br thin films.

varied solvent evaporation rates that influence the crystallization dynamics of Cs$_2$PbX$_4$ nanostructures, ultimately determining the size and uniformity of the nanostructures formed on the CsPbI$_2$Br films. Therefore, the solvent effects of CsCl post-treatment should be considered as an important processing parameter for optimizing the morphology and interfacial properties of CsPbI$_2$Br films

during device fabrication.

Atomic force microscopy (AFM) was performed to examine the surface roughness of the control sample and the samples post-treated with CsCl under fast and slow solvent evaporation conditions. The corresponding AFM images are presented in Figure 3. The control sample (Figure 3a) exhibits a smooth CsPbI$_2$Br surface with a root-mean-square (RMS) roughness of 0.23 μm. Post-treatment with CsCl under fast solvent evaporation conditions reduces the RMS roughness to 0.19 μm, consistent with the formation of smaller Cs$_2$PbX$_4$ nanostructures observed in SEM images. In contrast, slow solvent evaporation increases the RMS roughness to 0.29 μm, corresponding to the larger and more densely distributed low-dimensional nanostructures as-grown on the CsPbI$_2$Br surface.

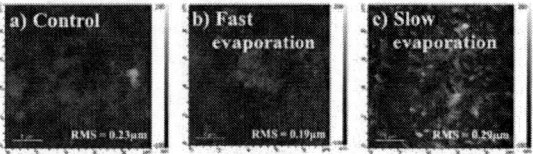

Figure 3. Surface topology and corresponding RMS surface roughness values of CsPbI$_2$Br perovskite films: a) control sample without post-treatment, b) fast solvent evaporation, and c) slow solvent evaporation

The PSCs with pristine and post-treated CsPbI$_2$Br thin films were fabricated in a glovebox to evaluate the effects of interfacial engineering on device performance. The representative current density-voltage (*J-V*) curves are shown in Figure 4. The device with pristine CsPbI$_2$Br film (control sample) exhibits a V_{oc} of 1.11 V, J_{sc} of 15.7 mA/cm², FF of 0.61, and PCE of 10.54%. After CsCl post-treatment under the controlled solvent evaporation condition, the device performance improves substantially, with a V_{oc} of 1.20 V, J_{sc} of 16.3 mA/cm², FF of 0.65, and PCE of 12.8%. These results indicate the substantial enhancement in device performance achieved by interfacial engineering. The improvement in photovoltaic performance can be attributed to the optimized formation of Cs$_2$PbX$_4$ nanostructures at the interface, which enhances charge transport and reduces recombination losses, improving the overall device performance. However, under slow evaporation conditions, the formation of large and extensively distributed Cs$_2$PbX$_4$ nanostructures results in less favorable interfacial properties for PSCs, hindering device performance. These preliminary findings highlight the critical role of controlling solvent evaporation dynamics during the post-treatment to achieve optimal interfacial engineering. For future advancements in solution-processed interfacial engineering for PSCs, precise modulation of interfacial nanostructures will be essential to achieve high-performance devices with good reproducibility in manufacturing.

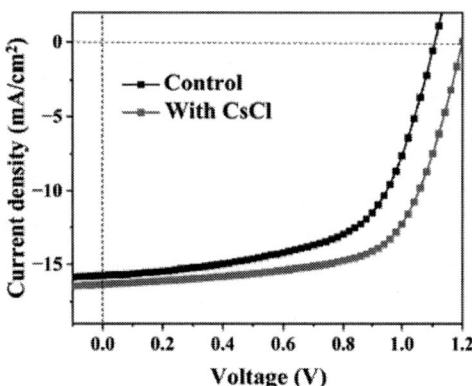

Figure 4. The representative *J-V* curves of the control device and the device with CsCl post-treatment.

Conclusion

The formation of Cs_2PbX_4 nanostructures on the surface of $CsPbI_2Br$ thin films is achieved through CsCl post-treatment. The use of various alcohols as solvents for CsCl results in different extents of low-dimensional Cs_2PbX_4 growth on $CsPbI_2Br$. By carefully controlling the solvent evaporation conditions, the nanostructures of Cs_2PbX_4 can be tuned, leading to a substantial enhancement in PSC device performance. This interfacial engineering approach offers a promising strategy for future solution-processed manufacturing and performance optimization.

Acknowledgments

A.N. acknowledges the financial support from the Science Committee of the Ministry of Science and Higher Education of the Republic of Kazakhstan (Scientific Research Grant № AP19576154, AP14869983, and AP14869871). This research is also funded by Nazarbayev University under faculty-development competitive research grants program for 2024-2026 Grant № 201223FD8801 (A.N.) and Collaborative Research Program for 2024-2026 Grant № 211123CRP1613 (A.N.).

References

[1] R. J. Sutton, G. E. Eperon, L. Miranda, E. S. Parrott, B. A. Kamino, J. B. Patel, M. T. Hörantner, M. B. Johnston, A. A. Haghighirad, D. T. Moore, and H. J. Snaith, "Bandgap-Tunable Cesium Lead Halide Perovskites with High Thermal Stability for Efficient Solar Cells", Adv. Energy Mater., 6, 1502458 (2016).

[2] B. P. Kore, W. Zhang, B. W. Hoogendoorn, M. Safdari, and J. M. Gardner, "Moisture tolerant solar cells by encapsulating 3D perovskite with long-chain alkylammonium cation-based 2D perovskite", Commun. Mater., 2, 100 (2021).

[3] Q. Jiang, Y. Zhao, X. Zhang, X. Yang, Y. Chen, Z. Chu, Q. Ye, X. Li, Z. Yin, and J. You, "Surface passivation of perovskite film for efficient solar cells", Nat. Photonics, 13, pp. 460–466 (2019).

[4] T. Liu, J. Zhang, M. Qin, X. Wu, F. Li, X. Lu, Z. Zhu, and A. K.-Y. Jen, "Modifying Surface Termination of CsPbI3 Grain Boundaries by 2D Perovskite Layer for Efficient and Stable Photovoltaics", Adv. Funct. Mater., 31, 2009515 (2021).

[5] M. Tai, Y. Zhou, X. Yin, J. Han, Q. Zhang, Y. Zhou, and H. Lin, "In situ formation of a 2D/3D heterostructure for efficient and stable CsPbI2Br solar cells", J. Mater. Chem. A, 7, 22675–22682 (2019).

Knowledge discovery from microscopic image data using an explainable AI "extended free energy model"

Masato Kotsugi[1]

[1]Tokyo University of Science, Faculty of Advanced Engineering, Japan

Abstract

This study places a spotlight on the role of explainable AI, embodied in the extended Landau free-energy model [1-5], as a transformative tool for understanding and designing magnetic systems. By integrating physics-based features with data-driven methodologies, this model facilitates transparent, interpretable, and actionable insights into magnetization reversal processes, addressing longstanding challenges in mesoscale physics.

Magnetization reversal in nanomagnets is a cornerstone phenomenon for advancing spintronic devices, demanding novel analytical approaches to decode its complexity. Leveraging the capabilities of the extended Landau free-energy model alongside topological data analysis (TDA), this research bridges the gap between microstructures and macroscopic properties. Persistent homology (PH), ridge regression (RR), and principal component analysis (PCA) are employed to quantify magnetic domain complexities, identify energy contributions, and visualize magnetization dynamics. This integrated framework not only reveals the causal relationships driving energy barriers and domain transitions but also offers a pathway for energy-efficient and highly optimized magnetic device design.

Introduction

The extended Landau free-energy model, equipped with explainable AI capabilities, represents a paradigm shift in materials analysis, enabling researchers to uncover hidden causal relationships in magnetization reversal. This study integrates this advanced model with TDA to address critical limitations in traditional analyses, which often rely on qualitative interpretations and overlook the stochastic behaviors inherent in magnetic systems. By combining PH to quantify domain complexities and PCA to map structural trends, the research establishes a robust framework for elucidating the intricate interplay between microscopic domain structures and macroscopic energy landscapes.

Methods

Micromagnetic Simulations and PH Analysis: Micromagnetic simulations based on the Landau-Lifshitz-Gilbert (LLG) equation provided datasets of permalloy nanodots under varying magnetic fields and defect positions.

PH quantified the topological features of magnetic domains, capturing essential structural information for subsequent analysis. RR linked these features to specific energy terms, while PCA reduced dimensionality to visualize and interpret energy landscapes.

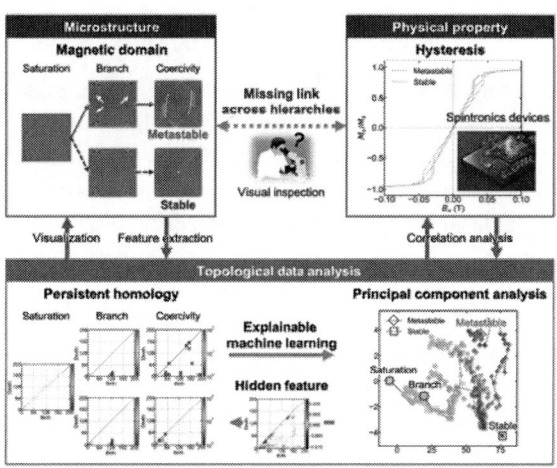

Explainable AI and Causal Analysis: The extended Landau model introduced a physics-informed, data-driven approach to construct interpretable energy landscapes. By employing gradient analysis, the model identified critical energy barriers and decomposed contributions from exchange and demagnetization energies. The Hadamard product was used to map energy contributions back to real-space magnetic domain structures, enabling direct visualization of causality and facilitating inverse design.

Results and Discussion:

Magnetic Domain Structures and Reversal Dynamics: Persistent homology revealed key features of magnetic domain evolution, distinguishing stable and metastable states. Simulations captured critical magnetic states, including C-states, Landau patterns, and vortex structures, while PCA mapped their transitions onto an interpretable energy landscape.

Energy Contributions and Insights through Explainable AI: The extended Landau model demonstrated its utility as an explainable AI tool by isolating and quantifying energy contributions. Demagnetization energy was identified as the

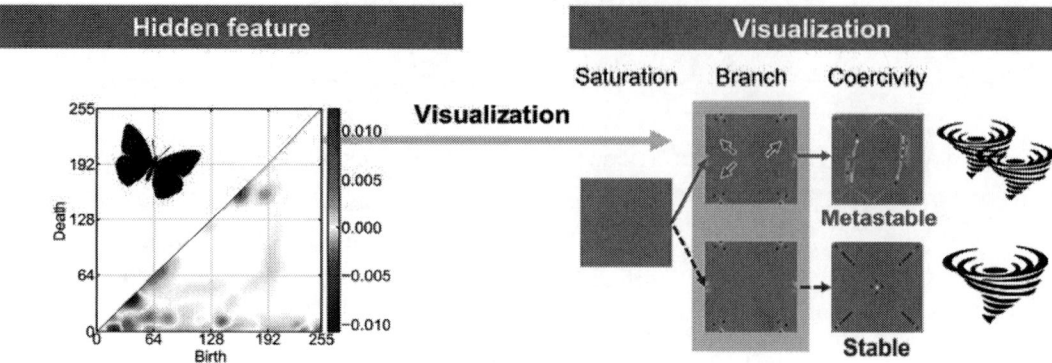

Determining the stability of the magnetic domains. Difficult to recognize by human eye.

Detecting the causes of stable/metastable processes in the early stage.

dominant factor in pinning phenomena, while exchange energy played a stabilizing role. Gradient analysis highlighted the energy barriers critical for domain transitions, with the Hadamard product offering a spatially resolved view of these interactions. This transparency enabled direct connections between microscopic features and macroscopic energy dynamics.

TDA and Super-Hierarchical Analysis: TDA provided a comprehensive visualization of the magnetization reversal process, with PC1 correlating strongly with magnetization and PC2 uncovering subtle variations in exchange energy. These insights emphasized the extended Landau model's capacity to extract actionable knowledge, demonstrating how structural and energetic factors converge to influence magnetization dynamics.

Implications for Magnetic Device Design: By bridging the gap between microstructural complexity and macroscopic performance, the research demonstrated how explainable AI frameworks could guide the optimization of spintronic devices. Defect positioning, informed by visualized energy landscapes, was shown to significantly affect energy barriers, providing actionable insights for reducing energy consumption in magnetic storage technologies.

Conclusions

The extended Landau free-energy model, enhanced by explainable AI capabilities, offers a transformative framework for causal and super-hierarchical analysis of magnetization reversal. By integrating TDA, PH, and PCA, the research delivers unprecedented transparency and interpretability, enabling the extraction of critical knowledge for materials optimization. These advancements highlight the synergy of physics-based and data-driven methodologies, paving the way for energy-efficient and innovative magnetic devices.

	Device A	Device B	Device C
Visualization	400 nm		◆ Concentration of energy barrier ■ Defect
Demagnetization	1.2×10^{-10}	1.1×10^{-10}	8.2×10^{-11} Min
Exchange	6.5×10^{-12}	1.6×10^{-12} Min	1.8×10^{-12}
Total Energy	1.3×10^{-10}	1.2×10^{-10}	0.8×10^{-10} Min

Acknowledgments

We extend our heartfelt gratitude to Y. Hiraoka, I. Obayashi, K. Akagi, and H. Fukuyama for their insightful discussions and contributions to this study. This research was supported by the Japan Society for the Promotion of Science (JSPS) under KAKENHI Grant Numbers [21H04656, 19K22117, 22K14590]. Computational resources and persistent homology analysis were facilitated by the open-source HomCloud platform, for which we acknowledge its development team. Special thanks are extended to the members of Kotsugi lab of Tokyo University of Science for their institutional support and their constructive feedback throughout the project.

Reference

[1] S. Kunii, A. Foggiatto, C. Mitsumata, and M. Kotsugi*, Science and Technology of Advanced Materials: Methods, 2 (2022) 445

[2] S. Kunii, K. Masuzawa, A. Fogiatto, C. Mitsumata, and M. Kotsugi*, Scientific Reports, 12, (2022) 29892

[3] A. Foggiatto, S. Kunii, C. Mitsumata, and M. Kotsugi, Communications Physics, 5 (2022) 277.

[4] M. Taniwaki, .F L. Alexandre, C. Mitsumata, T. Yamazaki, I. Obayashi, Y. Hiraoka, Y. Igarashi, Y. Mizutori, S. A. Hossein, T. Ohkubo, M. Kotsugi*, 2023 IEEE International Magnetic Conference - Short Papers (INTERMAG Short Papers) 2023, pp. 1-2

[5] R. Nagaoka, K. Masuzawa, M. Taniwaki, A. L. Foggiatto, T. Yamazaki, I. Obayashi, Y. Hiraoka, C. Mitsumata, M. Kotsugi*, IEEE Transaction on Magnetics, 60, 9, pp. 1-5, 4000305 (2024)

Ferroelectric 3D NAND Storage
(Invited)

Prasanna Venkatesan[1], Asif Khan[1,2]

[1]School of Electrical and Computer Engineering, Georgia Tech, Atlanta, GA, USA.
[2]School of Materials Science and Engineering Engineering, Georgia Tech, Atlanta, GA, USA.

ABSTRACT

In this work, we report a multi-pronged optimization of band engineered ferroelectric field effect transistors with dielectric inserts for NAND solutions for in-storage compute applications: (1) design space exploration for optimizing memory window (MW) and (2) retention alongside (3) a novel disturb mitigation scheme to reduce pass disturb. The optimized device is then leveraged to demonstrate a high density in-storage compute solution for protein identification using open modification search.

Keywords: Ferroelectrics, Disturb, Retention, In-memory compute, 3D-NAND

INTRODUCTION

Recent advances in artificial intelligence and its broad range of applications have pushed the need for compute involving large datasets. The training of such large AI models and inferencing using them is limited by the latency and energy consumption associated with moving data between the memory and compute hardware. To this end, 3D NAND with its high memory capacity and possibility for parallel operation offers a compelling in-storage compute solution [1]–[4].

While conventional 3D-NAND has scaled to 20-30 GB/mm^2 in data density over the last decade, reliability challenges linked with z-scaling and high write voltages hinder further scaling [1]–[4]. Ferroelectric field effect transistors have emerged as a solution for further z-scaling. However, in order for the ferroelectrics to be a drop-in solution to conventional charge trap layer, FE-FETs are required to meet the specifications of vertical NAND process flow: (1) MW $\geq$ 7.5 V and (2) thickness $\leq$ 20 nm, along with robust retention and resilience against disturb.

Over the last two years, dielectric inserts have been shown to increase the MW of standard FEFETs. This MW enhancement could be achived by laminating the dielectric layer in the middle of the ferroelectric gate stack (Tunnel dielectric layer, TDL) or placing it next to the gate to act as a gate blocking layer (GBL). In this

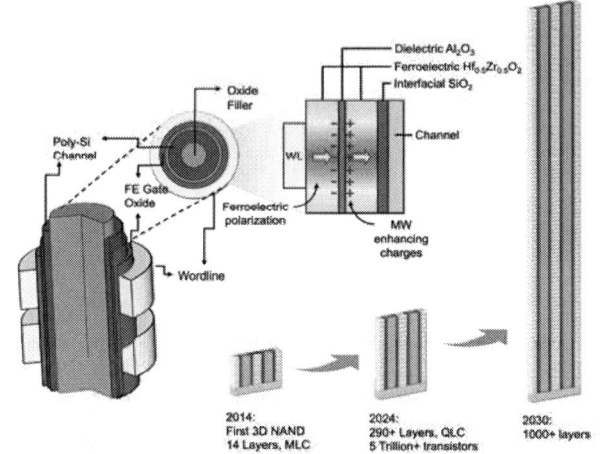

Fig. 1. Ferroelectrics have emerged as an alternative for the charge trap flash layer in 3D-NAND to enable continue z-scaling.

work, we explore different dielectrics and geometries to optimize for large MW enhancement, study retention of FEFETs with TDL and GBL gate stacks, and characterize disturb in the FEFET with the best retention. In the last section, we benchmark these FE-NAND devices against the incumbent solutions for open modification search for protein identification [5].

MW OPTIMIZATION

FE-MOSCAPs with different dielectric inserts and geometries are fabricated using the process flow outlined in [6], [7] to study the role of different dielectrics and their position in the ferroelectric gate stack. The MW is extracted from the C-V curves as shown in Fig. 2(a) [8]. It is identified that Al_2O_3 as a TDL and SiO_2 as a GBL exhibit the largest MWs. A hybrid gate stack with both TDL and GBL exhibits further MW enhancement as high as 11 V. The origin of these enhanced MWs have been described in detail in [3]

Based on the insights from the FE-MOSCAPs, FE-FETs with reference (19 nm HZO), TDL (8/3(Al_2O_3)/8) and GBL (14/4(SiO_2)) gate stacks are fabricated following the process flow shown in [9], [10] for the retention and disturb characterization.

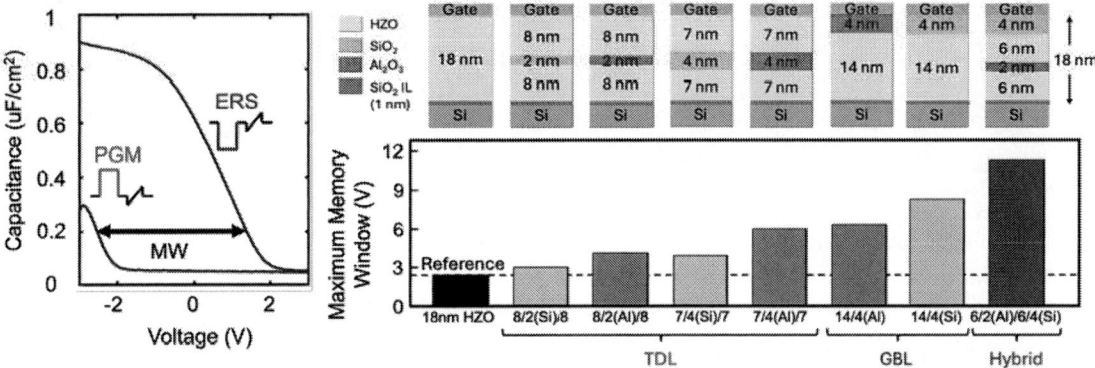

Fig. 2. (a) The MW of the ferroelectric gate stacks are optimized on FE-MOSCAPs by extracting the MW from the C-V curves. (b) Al_2O_3 acts as the best TDL while SiO_2 is better as a GBL. The hybrid gate stack with a Al_2O_3 TDL and SiO_2 GBL is shown to achieve a large MW as high as 11 V.

RETENTION

The retention characteristics are measured at RT using the pulse scheme shown in Fig. 3(a). The evolution of the threshold voltages over times for the TDL FEFETs show less than 1% retention loss akin to that of the reference FEFET. However, the GBL FEFETs show significant retention loss resulting from detrapping of the MW enhancing charges trapped at the FE-GBL interface through the GBL [11].

DISTURB AND ITS MITIGATION

In order to characterize the influence of the pass voltage on the FEFETs, pass disturb was measured on the FEFET with 8/3(Al)/8 gate stack as shown in Fig. 4. It is observed that there is significant shift in V_T (28%) under 50 μs pass disturb pulses of $V_{pass} = V_T + 2$ V. This positive shift in V_T is hypothesized to be due to

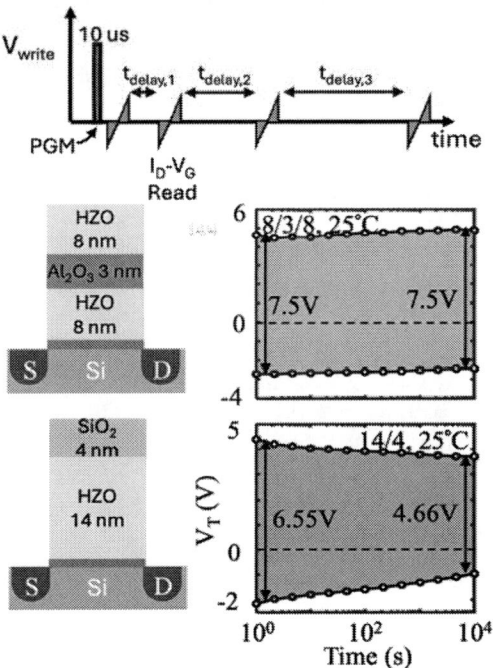

Fig. 3. (a) Pulse scheme for retention characterization. (b) Retention at RT in the TDL and GBL FEFETs shows that TDL FEFETs demonstrate robust retention while GBL FEFETs show a 29% retention loss after 1e4 s.

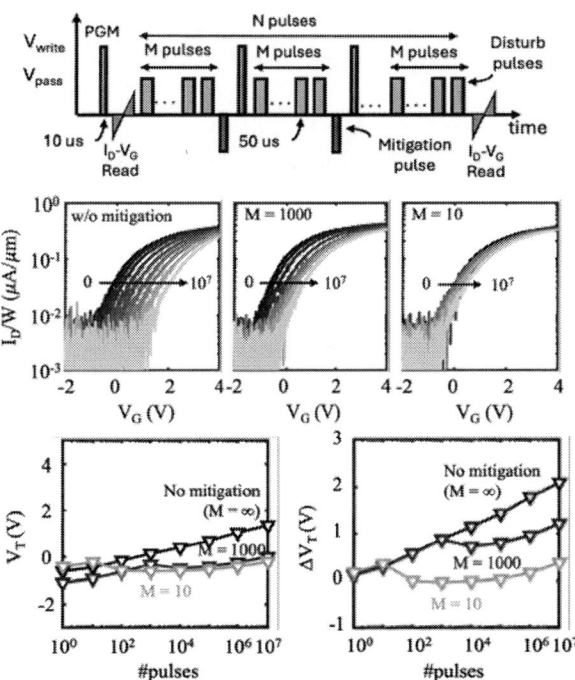

Fig. 4. (a) Disturb mitigation scheme for reducing pass disturb to acceptable levels. (b) Evolution of transfer characteristics of the PGM state in the 8/3(Al)/8 FEFET after increasing number of pass disturb pulses ($V_{pass} = V_T + 2V$) with no mitigation and mitigation pulses applied every 1000 and 10 cycles. (c&d) V_T and ΔV_T shows that applying the mitigation pulse reduces pass disturb from 28% down to 4%.

979-8-3315-0417-5/25 $31.00 © 2025 IEEE

Architecture	GPU [13]	DRAM [14]	MLC ReRAM [15]	3D NAND [16]	FeNAND (this work)
Compute type	-	Near-memory	Near-memory	In-memory	In-memory
Algorithm	HOMS-TC	HyperOMS-PIM-DRAM	HyperOMS-PIM-ReRAM	HyperOMS-3DNAND	HyperOMS
Technology Node	RTX 4090 (5nm)	22nm DRAM (DDR4) 28nm node-compute	130nm RRAM with 3M cells	NAND: 14nm ASIC: 7nm FinFET	FeNAND: 14nm ASIC:7nm FinFET
Speed	1x (23min)	2.43x	1.71	423x	737x
Energy Efficiency	1x(454kJ)	101x	516x	7230x	22146x
Capacity Limit(per module)	24GB	128GB	3M cells	16TB	>100TB
Data density	-	-	-	20-30 GB/mm2	>100 GB/mm2

TABLE I: As a benchmarking standard, we assume a repository with one billion reference HVs and 15k query HVs. The QLC FE-NAND devices retain the advantages arising from the parallelism and read energy efficiency of CTF NAND while achieving significantly higher data densities.

electron trapping during the positive pass disturb pulses. Based on this, we proposed a mitigation scheme where a periodic refresh is applied every M pass disturb pulses to detrap the electrons and reduce V_T shift. It is observed that for $M = 1000$ and 10, the pass disturb decreases to 16% and 4%, respectively, thereby, highlighting the efficacy of the mitigation scheme [12].

IN-MEMORY COMPUTE

The speed and energy consumption for in-storage protein identification using OMS is estimated and benchmarked against the other techniques in Table 1. It is to be noted that FE-NAND is more energy efficient and faster than conventional 3D NAND while offering increased data density and hence using lesser number of tiles [13]–[16].

CONCLUSION

Ferroelectric NAND devices can be optimized to enable ultra-high density in-storage compute in scales previously unprecedently, potentially allowing for petabyte scale memories. Such parallelization and in-memory operation coupled with the low write energies and latencies compared to CTF NAND devices, make FE-NAND devices as a suitable candidate for large dataset processing.

ACKNOWLEDGMENT

This work was supported by Samsung Electronics and SUPREME, one of the seven SRC-DARPA JUMP 2.0 centers. Fab was done at the IEN, supported by the NSF-NNCI program (ECCS-1542174).

REFERENCES

[1] J. Han, S. Kang, K. Kim, J. Jang and J. Song, "Fundamental Issues in VNAND Integration Toward More Than 1K Layers," International Electron Devices Meeting (IEDM), San Francisco, CA, USA, (2023).

[2] S. Lim et al., "Comprehensive Design Guidelines of Gate Stack for QLC and Highly Reliable Ferroelectric VNAND," International Electron Devices Meeting (IEDM), San Francisco, CA, USA, (2023).

[3] D. Das et al., "Experimental demonstration and modeling of a ferroelectric gate stack with a tunnel dielectric insert for NAND applications," International Electron Devices Meeting (IEDM), San Francisco, CA, USA (2023).

[4] G. Kim et al., "In-depth Analysis of the Hafnia Ferroelectrics as a Key Enabler for Low Voltage & QLC 3D VNAND Beyond 1K Layers: Experimental Demonstration and Modeling," 2024 IEEE Symposium on VLSI Technology and Circuits (VLSI Technology and Circuits), Honolulu, HI, USA (2024).

[5] J. Kang, W. Xu, W. Bittremieux, and T. Rosing. Massively Parallel Open Modification Spectral Library Searching with Hyperdimensional Computing. In Proceedings of the International Conference on Parallel Architectures and Compilation Techniques (2022).

[6] L. Fernandes et al., "Material Choices for Tunnel Dielectric Layer and Gate Blocking Layer for Ferroelectric NAND Applications," in IEEE Electron Device Letters (2024).

[7] L. Fernandes et al., "Optimizing Memory Window for Ferroelectric Nand Applications: An Experimental Study on Dielectric Material Selection and Layer Positioning," in IEEE Transactions on Electron Devices (2024).

[8] D. Das et al., "Ferroelectric Gate Stack Engineering with Tunnel Dielectric Insert for Achieving High Memory Window in FEFETs for NAND Applications," 2024 8th IEEE Electron Devices Technology & Manufacturing Conference (EDTM), Bangalore, India (2024).

[9] N. Tasneem et al., "Trap Capture and Emission Dynamics in Ferroelectric Field-Effect Transistors and their Impact on Device Operation and Reliability," 2021 IEEE International Electron Devices Meeting (IEDM), San Francisco, CA, USA (2021).

[10] C. Park et al., "Plasma-Enhanced Atomic Layer Deposition-Based Ferroelectric Field-Effect Transistors," in IEEE Journal of the Electron Devices Society (2024).

[11] P. Venkatesan et al., "Demonstration of Robust Retention in Band engineered FEFETs for NAND Storage Applications using Tunnel Dielectric Layer," in IEEE Electron Device Letters (2025).

[12] P. Venkatesan et al., "Disturb and its Mitigation in Ferroelectric Field-Effect Transistors With Large Memory Window for NAND Flash Applications," in IEEE Electron Device Letters (2024).

[13] J. Kang, W. Xu, W. Bittremieux, N. Moshiri, T. Rosing, Accelerating open modification spectral library searching on tensor core in high-dimensional space, Bioinformatics (2023).

[14] J. Kang, W. Xu, W. Bittremieux, N. Moshiri and T. Rosing, "DRAM-Based Acceleration of Open Modification Search in Hyperdimensional Space," in IEEE Transactions on Computer-Aided Design of Integrated Circuits and Systems (2024).

[15] K. Fan, W. Chen, S. Pinge, H.S.P. Wong, and T. Rosing. Efficient Open Modification Spectral Library Searching in High-Dimensional Space with Multi-Level-Cell Memory. In Proceedings of the 61st ACM/IEEE Design Automation Conference (2024).

[16] P. Hsu, W. Xu, T. Rosing, and S. Yu. An In-Storage Processing Architecture with 3D NAND Heterogeneous Integration for Spectra Open Modification Search. In Proceedings of the International Symposium on Memory Systems (2023).

2D Materials for Neuromorphic Computing Devices

Max C. Lemme[1,2], Lukas Völkel[1], Sofia Cruces[1], Jimin Lee[1], Yuan Fa[1,2]

[1] Chair of Electronic Devices, RWTH Aachen University, Otto-Blumenthal-Str. 25, 52074 Aachen, Germany,

[2] AMO GmbH, Advanced Microelectronic Center Aachen, Otto-Blumenthal-Str. 25, 52074 Aachen, Germany

Abstract

Two-dimensional materials show promise for resistive switching devices as building blocks for neuromorphic computing. We explore MoS_2 in different device configurations utilizing ion transport along van der Waals gaps, revealing reduced variability and high yields. We demonstrate volatile switching in vertical bilayer stacks of SiO_x and MoS_2. We also show lateral MoS_2 devices with ultra-low threshold voltages. Temperature-dependent current-voltage data on h-BN memristors elucidate conduction mechanisms for a deeper understanding of the resistive switching mechanism.

Keywords: Memristor, Neuromorphic Computing, 2D Materials

Introduction

2D materials could make a difference in various fields of future microelectronics, either as a continuation of "device scaling" (i.e. Moore's Law), or by adding functionality with sensing [1], [2], photonics [3], [4], [5], [6], or future electronics enabled by quantum or neuromorphic computing [7]. Resistive switching devices are considered game-changers for the latter, and several material classes are investigated for such "memristors". They can be switched between two or more resistive states in a volatile or non-volatile manner [8]. They may be used in computing-in-memory architectures [9], cross-bar arrays [10], or as electronic synapses [11].

Semiconducting molybdenum disulfide (MoS_2) has been investigated intensively as a resistive switching material. The physical origin of the switching has been attributed to different mechanisms: bias-induced migration of sulfur vacancies and grain boundaries, lattice distortion, reversible modulation of MoS_2 phases, and ion migration. Additionally, hexagonal boron nitride (h-BN) is gaining popularity for resistive switching due to its wide bandgap high thermal conductivity, stability and mechanical flexibility. The resistive switching origin is mainly attributed to metal filament formation under voltage stress.

Here, we will first discuss MoS_2 as the resistive switching material in vertical and lateral device configurations. Second, we will present a study on the current conduction mechanism in memristors with h-BN as the active material. The devices were electrically measured in their low and high resistance states in a wide range of temperatures.

Results and Discussion

Cross-point devices, where two metal electrodes are separated by one or several MoS_2 layers oriented perpendicular to the electrodes, are probably the best-studied 2D memristors. Here, we present a wafer-scale fabrication process based on metal-organic chemical vapor deposited (MOCVD) MoS_2, with silver (Ag) and palladium (Pd) as the top and bottom electrodes, respectively (Fig. 1a). Direct current (DC) current-voltage (I-V) measurements (Fig. 1b) reveal that these memristors exhibit both volatile and nonvolatile resistive switching, controlled by the current compliance (I_{cc}). Notably, the devices switch between an intermediate and a low resistive state when I_{cc} is set to 1 mA, likely due to the formation of multiple filaments. Our MOCVD MoS_2 memristors not only show bifunctional switching behavior but also exhibit excellent endurance and retention properties in the nonvolatile mode. These characteristics position them as promising candidates for advanced applications, such as artificial neurons and neuromorphic computing.

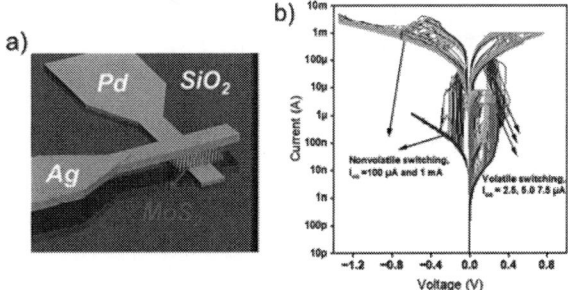

Fig. 1: a) Schematic of a vertical structure of Ag/MoS2/Pd memristor with laterally aligned MoS_2. b) I-V curves of memristor, showing the memristor can achieve volatile and nonvolatile switching by controlling I_{cc}.

In addition to lateral devices, vertical MoS_2 films can be used as a resistive switching material. Vertically aligned MoS_2 films can be grown under certain conditions by thermal conversion of thin molybdenum films. Here, the van der Waals gaps between the vertical layers can facilitate electrically controlled ion movement [12], leading to non-volatile resistive switching. We demonstrate volatile RS in devices combining a SiO_x layer with vertically aligned MoS_2 (VAMoS_2). The devices exhibit repeatable threshold switching with low on-threshold ($V_{t,on}$) and hold voltages (V_{hold}) below 1 V, switching times below 400 ns, and endurance of 8000 cycles [13].

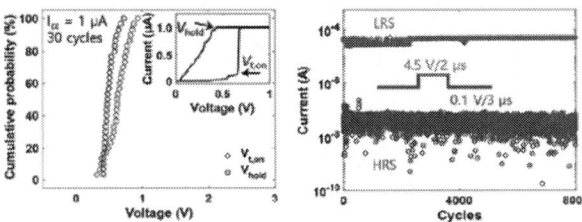

Fig. 2: a) Cumulative probability distribution of switching voltages ($V_{t,on}$ and V_{hold}) derived from 30 DC I-V sweeps for an Ag/SiO_x/VAMoS_2/Au device with vertically aligned MoS_2. Inset: An exemplary DC I-V characteristics of the SiO_x/VAMoS_2 device at 1 μA current compliance. b) Pulsed endurance test of devices. Inset: Programming pulse followed by a read pulse.

Fig. 2a shows a cumulative probability distribution of the switching voltages ($V_{t,on}$, V_{hold}) for an Ag/SiO_x/VAMoS_2/Au device measured over 30 resistive switching cycles. The inset shows exemplary DC I-V characteristics, demonstrating volatile resistive switching behavior with a positive voltage applied to the top Ag electrode. Fig. 2b shows the endurance plot under pulsed voltage stress with 8000 cycles. The programming pulse was 4.5 V for 2 μs followed by a read pulse of 0.1 V for 3 μs (inset).

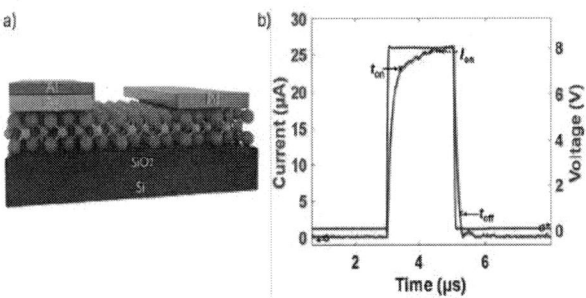

Fig. 3: a) Schematic of a MoS_2 memristor with Ag/Al and Pd top contacts and laterally aligned MoS_2. b) Transient current response under pulsed voltage stress of a Al/Ag/MoS_2/Pd lateral memristor.

A similar effect of ion transport in van der Waals gaps can be observed in lateral two-terminal devices based on MOCVD MoS_2. We will show experiments on silver (Ag) ion migration in lateral MoS_2 memristors that do not require higher "forming" voltages for their initial resistive switching event. These palladium/MoS_2/Ag devices show repeatable volatile switching with ultra-low threshold voltages of 0.3 V and 100-600 ns switching [14]. Fig. 3a shows a schematic drawing of a lateral MoS_2 memristor with Ag/Al and Pd top contacts. Fig. 3b shows the transient current response of a device under pulsed voltage stress. The device exhibits

threshold switching with a switching time (t_{on}) in the ns regime. The switching time (t_{on}) is defined as the time needed to reach 90% of the ON current (I_{on}), whereas the recovery time (t_{off}) indicates the time required to reach 10% of the OFF current (I_{off}) when reverting to the HRS (compare Fig. 3b).

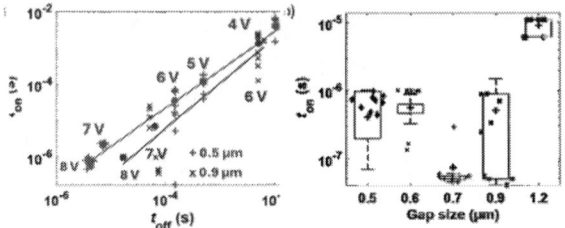

Fig. 4: a) Double-logarithmic plot of the switching time vs. the recovery time for two different gap sizes and several voltage pulse heights.

Furthermore, we analyzed the timing parameters depending on the gap size. For example, Fig. 4a shows the dependency of the switching time on the recovery time for two different gap sizes (0.5 μm and 0.9 μm) and several voltage pulse amplitudes. In the double-logarithmic plot, both parameters reveal a linear dependency. Fig. 4b shows the dependency of the switching time on the gap size for gap sizes between 0.5 μm and 1.2 μm. While the switching time for gap sizes between 0.5 μm and 0.9 μm is in the same range, the switching time is considerably larger for a gap size of 1.2 μm.

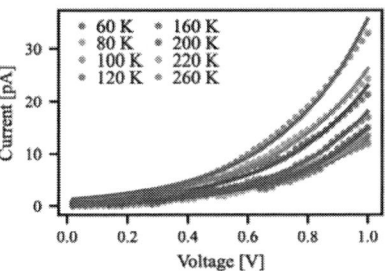

Fig. 5: Temperature-dependent current-voltage measurements of a h-BN based memristor with fits of the hopping conduction mechanism (solid lines).

Understanding the resistive switching mechanisms of emerging 2D-materials-based memristors is crucial for optimizing their device performance. This is mainly done using transmission electron microscopy (TEM) [15] or conductive atomic force microscopy (C-AFM) [16]. Both methods can provide detailed information about local structural or electrical changes during RS with high spatial resolution, but they are also mostly destructive. Here, we thoroughly study the current conduction mechanisms of Ni/h-BN/Ni threshold memristors in low and high resistance states (LRS and HRS) through temperature-dependent current voltage measurements [17]. Fig. 5 shows an exemplary plot of temperature-dependent current-voltage measurements in the HRS state. The solid lines are fits of the hopping conduction mechanism to the measurement data.

979-8-3315-0417-5/25 $31.00 © 2025 IEEE

Conclusion

We highlight the potential of 2D materials, specifically MoS_2 and h-BN, for resistive switching applications in neuromorphic computing. Vertical and lateral MoS_2 devices demonstrate promising performance with low variability, high endurance, and ultra-low threshold voltages, driven by mechanisms such as ion migration and van der Waals gap effects. Furthermore, temperature-dependent analyses of h-BN memristors provide deeper insights into conduction mechanisms. These findings advance the understanding and optimization of 2D-material-based memristors for future computing technologies.

Acknowledgments

We acknowledge funding from the European Union's Horizon 2020 research and innovation programme under grant agreements 952792 (2D-EPL), 881603 (Graphene Flagship Core 3), and 101194458 (ENERGIZE), and the German Ministry of Education and Research (BMBF) under grant agreements 16ES1121 (NobleNEMS), 03ZU1106AA/ 03ZU2106AE / 03ZU2106AA (NeuroSys) and 16ME0399 / 16ME0400 (NEUROTEC II).

References

[1] M. C. Lemme et al., "Nanoelectromechanical Sensors Based on Suspended 2D Materials," *Research*, vol. 2020, p. 8748602, Jul. 2020, doi: 10.34133/2020/8748602.

[2] S. Lukas et al., "High-Yield Large-Scale Suspended Graphene Membranes over Closed Cavities for Sensor Applications," *ACS Nano*, vol. 18, no. 37, pp. 25614–25624, Sep. 2024, doi: 10.1021/acsnano.4c06827.

[3] D. Schall et al., "50 GBit/s Photodetectors Based on Wafer-Scale Graphene for Integrated Silicon Photonic Communication Systems," *ACS Photonics*, vol. 1, no. 9, pp. 781–784, Sep. 2014, doi: 10.1021/ph5001605.

[4] M. Romagnoli et al., "Graphene-based integrated photonics for next-generation datacom and telecom," *Nat. Rev. Mater.*, vol. 3, no. 10, Art. no. 10, Oct. 2018, doi: 10.1038/s41578-018-0040-9.

[5] S. Parhizkar et al., "Two-Dimensional Platinum Diselenide Waveguide-Integrated Infrared Photodetectors," *ACS Photonics*, vol. 9, no. 3, pp. 859–867, Mar. 2022, doi: 10.1021/acsphotonics.1c01517.

[6] N. Negm et al., "Graphene Thermal Infrared Emitters Integrated into Silicon Photonic Waveguides," *ACS Photonics*, vol. 11, no. 8, pp. 2961–2969, Aug. 2024, doi: 10.1021/acsphotonics.3c01892.

[7] M. C. Lemme, D. Akinwande, C. Huyghebaert, and C. Stampfer, "2D materials for future heterogeneous electronics," *Nat. Commun.*, vol. 13, no. 1, Art. no. 1, Mar. 2022, doi: 10.1038/s41467-022-29001-4.

[8] D. B. Strukov, G. S. Snider, D. R. Stewart, and R. S. Williams, "The missing memristor found," *Nature*, vol. 453, no. 7191, Art. no. 7191, May 2008, doi: 10.1038/nature06932.

[9] D. Ielmini and H.-S. P. Wong, "In-memory computing with resistive switching devices," *Nat. Electron.*, vol. 1, no. 6, Art. no. 6, Jun. 2018, doi: 10.1038/s41928-018-0092-2.

[10] S. Chen et al., "Wafer-scale integration of two-dimensional materials in high-density memristive crossbar arrays for artificial neural networks," *Nat. Electron.*, vol. 3, no. 10, Art. no. 10, Oct. 2020, doi: 10.1038/s41928-020-00473-w.

[11] G. Zhou et al., "Volatile and Nonvolatile Memristive Devices for Neuromorphic Computing," *Adv. Electron. Mater.*, vol. 8, no. 7, p. 2101127, 2022, doi: 10.1002/aelm.202101127.

[12] M. Belete et al., "Nonvolatile Resistive Switching in Nanocrystalline Molybdenum Disulfide with Ion-Based Plasticity," *Adv. Electron. Mater.*, vol. 6, no. 3, p. 1900892, 2020, doi: 10.1002/aelm.201900892.

[13] Lee, Jimin et al., "Threshold Resistive Switching in SiOx/Vertically Aligned MoS2 Devices based on Silver (Ag) Ion Migration," in *2024 Device Research Conference (DRC)*, College Park, Maryland, USA, 2024.

[14] S. Cruces et al., "Volatile MoS2 Memristors with Lateral Silver Ion Migration for Artificial Neuron Applications," *Small Science*, 2024, doi: 10.1002/smsc.202400523.

[15] B. Yuan et al., "150 nm × 200 nm Cross-Point Hexagonal Boron Nitride-Based Memristors," *Adv. Electron. Mater.*, vol. 6, no. 12, p. 1900115, 2020, doi: https://doi.org/10.1002/aelm.201900115.

[16] A. Ranjan et al., "Conductive Atomic Force Microscope Study of Bipolar and Threshold Resistive Switching in 2D Hexagonal Boron Nitride Films," *Sci. Rep.*, vol. 8, no. 1, Art. no. 1, Feb. 2018, doi: 10.1038/s41598-018-21138-x.

[17] L. Völkel et al., "Resistive Switching and Current Conduction Mechanisms in Hexagonal Boron Nitride Threshold Memristors with Nickel Electrodes," *Adv. Funct. Mater.*, vol. 34, no. 15, p. 2300428, Apr. 2024, doi: 10.1002/adfm.202300428.

979-8-3315-0417-5/25 $31.00 © 2025 IEEE

Insulators for Devices Based on 2D Materials

Tibor Grasser, Dominic Waldhoer, and Theresia Knobloch

Institute for Microelectronics, TU Wien, Austria

ABSTRACT

Despite the breathtaking progress already achieved for 2D electronic devices, they are still far from realizing their predicted performance potential. Serious challenges which have been discussed at length in literature are the low mobility often observed in polycrystalline layers of the 2D semiconductors, as well as high contact resistances. An issue which has not yet received a similar amount of attention is the lack of scalable insulators, which would go along with 2D materials as nicely as SiO_2 goes with silicon. As a result, there is no commercially competitive 2D transistor technology available today. Here we will address the current state-of-the-art on insulators for 2D nanoelectronics and summarize the main problems together with potential solutions.

INTRODUCTION

A significant amount of attention has been dedicated to the choice and fabrication of the best 2D semiconductor material to possibly replace silicon for end-of-the-roadmap technologies [1]. On the other hand, even though all electronic devices require insulating layers, the selection of suitable insulators for 2D nanoelectronics has not been at the center of attention, although it represents a formidable challenge. The problem of finding a suitable insulator for 2D materials is therefore of key importance, in particular because scaling of 2D semiconductor devices towards sub-10nm channel lengths is only possible with high-quality gate insulators scalable down to sub-1 nm equivalent oxide thicknesses (EOT). In order to achieve competitive device performance, these insulators need to meet stringent requirements regarding *(i)* low gate leakage currents [2], *(ii)* low density of interface and border traps [3] and *(iii)* their ability to withstand large electric fields [4], all while *(iv)* offering high thermal conductivity and *(v)* maintaining high charge carrier mobilities in the adjacent 2D semiconductors.

INSULATOR OPTIONS

The insulators typically used for 2D electronic devices are amorphous 3D oxides known from Si technologies (SiO_2, HfO_2, Al_2O_3), while native 2D oxides (MoO_3, WO_3 and Bi_2SeO_5), layered 2D crystals (hBN, mica) and 3D crystals like fluorides (CaF_2, SrF_2, or MgF_2) or perovskites ($SrTiO_3$ or $BaTiO_3$) are receiving increasing attention.

3D oxides

While it is relatively straight-forward to transfer a 2D semiconductor on top of an existing 3D insulator, like SiO_2 or HfO_2, and thereby create back-gated devices, the deposition of 3D insulators on top of 2D channel materials for the creation of top gates has been found to be a real challenge. Irrespective of the deposition method used, the resulting insulating material often forms poor quality interfaces with 2D semiconductors. The resulting border trap densities can be very large and severely perturb stable device operation. While the earliest 2D transistors employed SiO_2 as a back-gate insulator, the problems resulting from the poor interface were soon identified and alternatives such as crystalline hBN were proposed [5]. Still, 3D oxides currently appear to be the most compatible with conventional fab processing (e.g. ALD deposition) and industrial papers (Intel, TSMC, imec) follow this route. Astonishingly good results have already been obtained [6]–[8] by creatively addressing the problem of nucleating an ALD layer on top of the chemically inert, hydrophobic surface of 2D TMDs with a seeding promoter or via a plasma activation.

Native oxides

One reason for the success of conventional Si technology is the excellent interface of Si with its native insulator SiO_2. In fact, there are many semiconductors with better properties than Si, e.g. larger mobilities (Ge, GaAs) or larger bandgaps (GaN, SiC). However, with the exception of SiC, none of them possess a sufficiently stable native insulator and the deposition of other insulating materials has turned into a veritable show-stopper. It therefore appears tempting to transfer this idea to 2D materials. Indeed, a few 2D semiconductors which can be oxidized into a material with a wider bandgap have already been identified, e.g. Bi_2O_2Se to Bi_2SeO_5 [9], HfS_2 [10] or $HfSe_2$ [11] to HfO_2, as well as $ZrSe_2$ to ZrO_2 [11] or MoS_2 to MoO_3 and WS_2 and WSe_2 to WO_3. With the exception of Bi_2O_2Se, these native oxides often appear non-stoichiometric and amorphous, possibly due to the lack of well-adjusted oxidation methods and thus are not yet of good quality. Also, many of these native oxides possess inherently narrow bandgaps, which might lead to unacceptably large leakage currents. The currently most advanced material system seems to be Bi_2O_2Se/Bi_2SeO_5, which can even be grown vertically for the creation of finFETs [12]. In general, the class of bismuth oxycalcogenides, Bi_2O_2X/Bi_2XO_5, with X being not just Se but also S [13] and Te [14], appears to be promising. In addition, elements from other groups have been used, e.g. $X =$Si [15]. However, none of these systems has yet reached the maturity of the Bi_2O_2Se/Bi_2SeO_5 system and it has not always been possible to create both semiconductor and insulator. Also, the bandgaps of the insulators are typically small, e.g. around 3 eV, technically qualifying them as wide-bandgap semiconductors rather than insulators. Still, since the permittivity is large (> 20) and thus the layer can be made very thick, and the conduction band offsets larger than 1 eV [16], the experimentally observed leakage currents have been found to be acceptable. Another highly relevant feature of bismuth oxycalcogenides is that in addition to MBE, they can be grown by CVD, thereby rendering them fab-compatible.

Another native insulator option that has recently emerged is the oxidation of 2D GaS [17] into Ga_2O_3, the native oxide of gallium [18]. Ga_2O_3 is also a wide-bandgap semiconductor ($E_g \approx 5$ eV) that is becoming increasingly important for power devices [19].

Layered materials

As the best known candidate, the layered 2D insulator hBN forms excellent van der Waals interfaces with 2D semiconductors, but has only mediocre dielectric properties resulting in excessive leakage currents for sub-1nm EOT [2]. While the scaling potential of other layered materials like mica is currently unclear, first devices have been demonstrated [20]. Recently, numerous other layered materials which exhibit higher dielectric constants and reduced leakage currents compared to hBN have been theoretically identified, including LaOCl, LaOBr, $Sr_2I_2F_2$, $Sn_4P_4O_{12}F_4$, and $Sb_2P_2O_8$ [21], [22] and partially experimentally validated [23]. Contrary to native insulators, these layered materials will have to be either transferred or directly grown on 2D channel materials, which poses an enormous experimental challenge. On the other hand, assuming that one day high-quality layer transfer should become possible, particularly by avoiding contamination through polymer residues [24], [25], the enormous flexibility and versatility previously envisaged [26] could be realized.

Perovskites

Perovskites form a large group of materials which might potentially be used as insulators for 2D material based transistors. Many of these perovskites can have gigantic permittivities > 200. In particular, strontium titanate (STO) has been demonstrated to form a stable high quality insulator [27]. Many other perovskites have been suggested, e.g. $LaNb_2O_7$, $CaLaNb_2TiO_{10}$, $Ca_2Ta_2TiO_{10}$ and $Ca_2Nb_3O_{10}$ [28] as well as $BaTiO_3$ or $PbZrTiO_3$ (even though these options are ferroelectric). However, so far only limited information is available regarding the performance of these more complex perovskite compounds in 2D devices.

Fluorides

Finally, promising insulators for 2D electronics are 3D ionic crystals like CaF_2 which form well-defined interfaces to 2D channel materials [3], see Fig. 1. In contrast to hBN, fluorides have good dielectric properties and thus exhibit low gate leakage currents [2], [4]. The biggest challenge for fluorides is currently the growth of good quality top gates on top of 2D semiconductors. Rather than using MBE, recent attempts used thermal evaporation to create polycrystalline films of 20 different fluorides [29]. Alternatively, magnetron sputtering has been used for the fabrication of CaF_2-based top gate stacks [30]. Whether these films are of sufficient quality for high-performance logic remains to be seen [31].

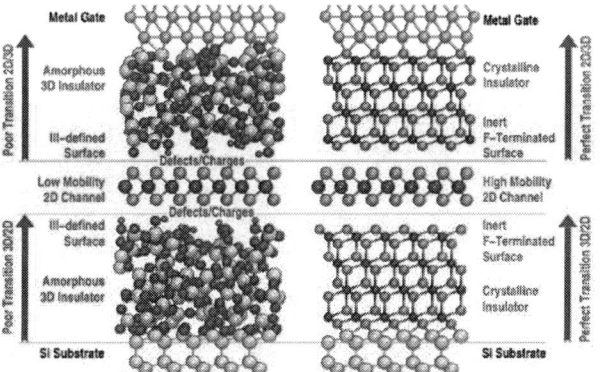

Fig. 1. **Left**: Amorphous insulators have ill-defined surfaces which contain a lot of dangling bonds and trapped charges which are difficult to control. **Right**: Fluorides are crystalline and offer an inert F-terminated surface which forms quasi-vdW interfaces to 2D materials.

MATERIAL MATCHING

A particular aspect that requires consideration for the choice of a good insulator is the matching of the defect bands in the insulator to the conduction (nMOS) or valence (pMOS) bands in the semiconductor. This is because as soon as the Fermi-level crosses a defect level in the insulator during operation, a charge trapping/detrapping event can be triggered, which can result for instance in hysteresis or bias temperature instability (BTI) [32]–[34]. As a general rule, the farther away these defect bands are from the band edges, the better the device performance will be in terms of hysteresis and

979-8-3315-0417-5/25 $31.00 © 2025 IEEE

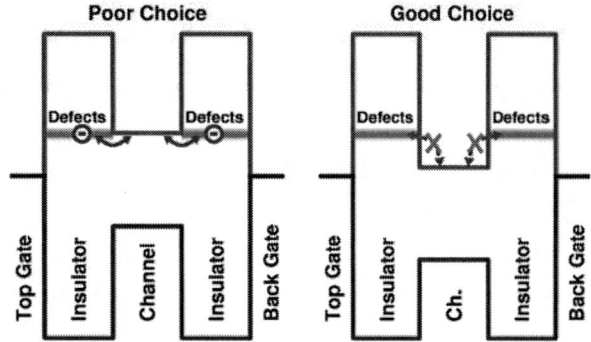

Fig. 2. **Left**: Schematic view of a poorly matched material system with the conduction band of the semiconductor closely aligned with the defect band in the insulator, which can lead to enormous charge trapping. **Right**: By selecting a different 2D semiconductor with a lower conduction band edge, charge trapping can be significantly suppressed even though the defect band in the insulator remains the same.

reliability [35], [36], see Fig. 2. The most important conclusion here is that it is not sufficient to optimize semiconductor and insulator independently, they have to work together as a "team" to result in the best overall device performance.

BENCHMARKING

For benchmarking different insulators, we primarily focus on detailed hysteresis measurements, as those provide information about the distribution of defects inside the material system [37], as well as the stability of the material and defect distribution over time. For an accurate assessment it is imperative to vary the parameters of the measurements, including the sweep rate, sweep voltage range, and temperature [3]. Different sweep rates are required to properly sample the broad distribution of defect time constants [34], [37], while different sweep voltage ranges can be used to probe different energetic regions of the insulator. Since all defect time constants are strongly temperature dependent, using different temperatures broadens the range of accessible defects considerably. In addition to charge trapping, various other mechanisms can contribute to hysteresis, most notably the presence of mobile ions (impurities or mobile defects, such as oxygen vacancies in STO [38]). Since the response of mobile ions can have a significantly different temperature dependence as charge trapping, experiments at different temperatures can be used to partially separate these effects.

In addition, detailed modeling of the data can reveal crucial defect parameters. Here we use our open-source tool Comphy [39] for efficient parameter extraction (a release dedicated to 2D materials is currently under preparation). Given these extracted parameters, like trap levels, relaxation energies, or diffusion barriers, a comparison to DFT simulations can then be used for defect identification [40]–[42].

CONCLUSIONS

Identification of the best matching semiconductor/insulator pair is a major challenge and will in the end determine the quality and stability of transistors made from 2D materials. Various options are currently being pursued and their properties need to be carefully evaluated.

REFERENCES

[1] S. Das *et al.*, Nature Electronics 4, 786–799 (2021).
[2] T. Knobloch *et al.*, Nature Electronics 4, 98 (2021).
[3] Y. Illarionov *et al.*, Nature Communications 11, (2020).
[4] C. Wen *et al.*, Advanced Materials 2002525 (2020).
[5] C. R. Dean *et al.*, Nature Nanotechnology 5, 722 (2010).
[6] C. Dorow *et al.*, in IEDM (2022), pp. 1–4.
[7] T.-E. Lee *et al.*, in IEDM (2022), pp. 1–4.
[8] A. Provias *et al.*, in IEDM (IEEE, 2023), pp. 1–4.
[9] T. Li *et al.*, Nature Electronics 3, 473 (2020).
[10] N. Peimyoo *et al.*, Science Advances 5, (2019).
[11] M. J. Mleczko *et al.*, Science Advances 3, (2017).
[12] C. Tan *et al.*, Nature 616, 66 (2023).
[13] X. Yang *et al.*, ACS Applied Materials & Interfaces 14, 7175 (2022).
[14] X. Zou *et al.*, Advanced Functional Materials 33, (2023).
[15] J. Chen *et al.*, Nature Communications 14, (2023).
[16] J. Robertson, Rep.Prog.Phys. 69, 327 (2006).
[17] A. AlMutairi *et al.*, Advanced Materials Interfaces (2024).
[18] K. Yi *et al.*, Nature Electronics (2024).
[19] A. J. Green *et al.*, APL Materials 10, (2022).
[20] X. Zou *et al.*, Nanotechnology 31, (2020).
[21] M. R. Osanloo, M. L. Van de Put, A. Saadat, and W. G. Vandenberghe, Nature Communications 12, (2021).
[22] Y. Li *et al.*, Nature Communications 15, (2024).
[23] A. Söll *et al.*, ACS Nano 18, 10397 (2024).
[24] R. Tilmann *et al.*, ACS Nano 17, 10617 (2023).
[25] W. Wang *et al.*, Nature Electronics 6, 981 (2023).
[26] K. S. Novoselov, A. Mishchenko, A. Carvalho, and A. H. Castro Neto, Science 353, (2016).
[27] J.-K. Huang *et al.*, Nature 605, 262 (2022).
[28] M. Osada and T. Sasaki, Advanced Materials 24, 210–228 (2011).
[29] K. Meng *et al.*, Nature Nanotechnology 19, 932 (2024).
[30] Y. Gao *et al.*, Applied Physics Letters 124, (2024).
[31] T. Grasser, M. Waltl, and T. Knobloch, Nature Nanotechnology 19, 880 (2024).
[32] T. Aichinger, M. Nelhiebel, and T. Grasser, in IRPS (2009), pp. 2–7.
[33] T. Grasser *et al.*, T-ED 58, 3652 (2011).
[34] T. Grasser, MR 52, 39 (2012).
[35] J. Franco *et al.*, in IEDM (IEEE, 2013).
[36] T. Knobloch *et al.*, Nature Electronics 5, 356 (2022).
[37] T. Knobloch *et al.*, IEEE Journal of the Electron Devices Society 6, 972 (2018).
[38] T. Knobloch *et al.*, in IEDM (IEEE, 2023), p. 1–4.
[39] D. Waldhoer *et al.*, Microelectronics Reliability 146, 115004 (2023).
[40] F. Schanovsky, W. Goes, and T. Grasser, JVST B 29, 01A2011 (2011).
[41] W. Goes *et al.*, MR 82, 1 (2018).
[42] C. Wilhelmer *et al.*, MR 139, 114801 (2022).

Heterointegrated Ga$_2$O$_3$-on-SiC RF MOSFETs

Xinxin Yu[1,2], Wenhui Xu[3], Rui Shen[2], Bing Qiao[2], Zhonghui Li[2], Xin Ou[3], Jiandong Ye[1]

[1]School of Electronic Science and Engineering, Nanjing University, China, [2] CETC Key Laboratory of Carbon-based Electronics, Nanjing Electronic Devices Institute, China, 3 Shanghai Institute of Microsystem and Information Technology, CAS, China.

Abstract

β-Ga$_2$O$_3$ radio-frequency (RF) MOSFETs with high frequency performances were fabricated on the highly thermal conductive 4H-SiC substrate through an ion-cutting process. The MOSFET yields a high current density of 661 mA/mm, a high record-high current cut-off frequency (f_T) of 47 GHz and maximum oscillation frequency (f_{max}) of 51 GHz. Furthermore, an output power density of 296 mW/mm and a high power gain of 11 dB at 2 GHz in continue wave (CW) mode.

Keywords: Ga$_2$O$_3$ on SiC, RF MOSFET, Smart ion-cut

Introduction

β-Ga$_2$O$_3$ is a promising candidate for high frequency and high power RF applications owing to its large breakdown electric field, high saturation electron velocity (v_{sat}) of 2×10^7 cm/s and excellent Johnson's figure of Merit (JFOM) [1-6]. However, the output current density, frequency performance, and output powers are far from expected due to its low carrier density, low channel mobility, severe short channels effects, as well as poor thermal conductivity [7-14].

To counter this hurdle, we demonstrated high performance RF MOSFETs based on the shallowly implanted Ga$_2$O$_3$ nanomembranes, which are hetero-integrated on SiC substrate through the smart ion-cutting technology. Owing to the strongly confined 2DEG-like channel in Ga$_2$O$_3$ nanomembrane hetero-integrated on SiC, the resultant device shows a high current density (I_{Dsat}) of 661 mA/mm and a high g_m of 57 mS/mm. The recorded high f_T of 47 GHz and f_{max} of 51 GHz were achieved with the deeply L_G scaling down to 0.1 μm. Furthermore, the device with L_G=0.1 μm operated in a CW mode delivers an output power density of 296 mW/mm and a high power gain of 11dB at 2 GHz thanks to the enhanced thermal conducting capability.

Device Structure and Fabrication

Ga$_2$O$_3$ MOSFETs were fabricated on an unintentional doped (UID) (-201) 50 nm-thick β-Ga$_2$O$_3$ nanomembrane which is hetero-integrated on the thermally conductive and electrically semi-insulating 4H-SiC substrate through the well-established smart ion-cutting process, as schematically shown in Fig.1 (a). The Ga$_2$O$_3$ surface was shallowly implanted by Si ions with a low energy of 15 keV and a target dose of 3×10^{13} cm^{-2}. The activation was performed by rapid thermal annealing (RTA) at 950°C in N$_2$ for 2 minutes to achieve high conductive electron channel formed near Ga$_2$O$_3$ surface. The shallowly implanted channel exhibits a sheet resistance (R_{sh}) of 9.3 kΩ/sq measured by the transmission line method (TLM) in Fig.1 (d).

Fig. 1 (c) shows the scanning electron microscopy (SEM) image of a representative MOSFET device structure. The source and drain electrodes were defined by photo lithography, and a Ti/Au (20/200 nm) metal stack was deposited by electron beam evaporation (EBE). After the RTA at 470 °C in N$_2$ ambient for 3 minutes, a low ohmic contact resistance (R_c) of 2.0 Ω·mm was achieved with its corresponding specific contact resistance (ρ_c) of 4.3×10^{-6} Ω·cm^2, which was extracted from the TLM result in Fig.1 (d). The Ga$_2$O$_3$ nanomembrane outside the active regions were etched away by inductively coupled plasma (ICP) to realize device electric isolations. A 15-nm-thick Al$_2$O$_3$ gate dielectric was deposited by the atomic layer deposition at 320 °C. Subsequently, the Ni/Au (20/500 nm) T-shaped gate electrodes were fabricated. The source-to-drain spacing (L_{SD}) and the gate width (W_G) were fixed at 2 and 200 μm, respectively. Gates were slightly closer to the source electrode with varied gate length L_G of 0.1, 0.16, 0.35 and 0.5 μm.

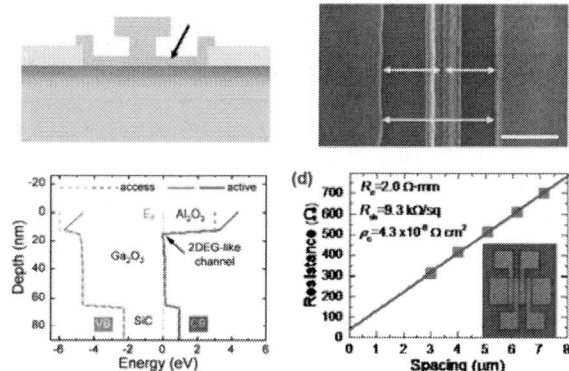

Fig. 1 (a) Schematic structure, (b) energy band diagram and (c) SEM image of the β-Ga2O3/SiC RF MOSFET with a shallowly implanted channel. (d) TLM plot of the implanted layer after the RTA process.

Results and Discussion

Fig.2 (a) shows current output characteristic curves of the Ga$_2$O$_3$-on-SiC RF MOSFET (L_G =0.1 μm) and the controlled device on bulk Ga$_2$O$_3$ substrate. For the Ga$_2$O$_3$-on-SiC RF MOSFET, a high current density (I_{Dsat}) of 661 mA/mm was achieved with a low on-resistance (R_{on}) of 24 Ω·mm at V_{GS}=8 V and V_{DS}=20 V. The device pinched off well at V_{GS}=-8 V and no obvious short channel effect was observed. As

---

979-8-3315-0417-5/25 $31.00 © 2025 IEEE

comparing with the controlled device on bulk Ga_2O_3 substrate, the self-heating effect was well suppressed in such Ga_2O_3-on-SiC architecture, which is extremely important for high power RF transistors. The transfer curve measured at V_{DS}=15 V is shown in Fig. 2 (b), exhibiting a threshold voltage of -7.5 V and a peak g_m of 57 mS/mm, which are benefited from the excellent gate control capability and low gate leakage. With L_G increasing, the threshold voltage was slightly positively shifted. As indicated in Fig.2 (c), the off-state I_{DS} and I_{GS} at V_{GS}=-12 V were extremely small and fell below the detection limit of the employed instrument before reaching the breakdown point at V_{DS} of about 90 V.

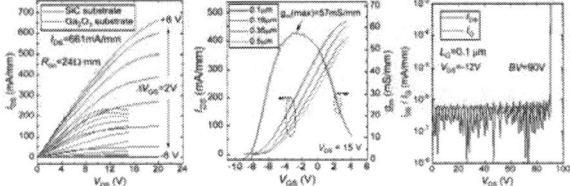

Fig. 2 (a) DC output, (b) transfer and (c) breakdown characteristics of the Ga_2O_3-on-SiC RF MOSFET. The DC-output characteristics of the controlled device on bulk Ga_2O_3 substrate is shown in (a) for comparison.

Small-signal RF gain performances of the highly L_G-scaled MOSFETs were measured at V_{DS} of 15 V by a Keysight N5227B vector network analyzer. S-parameters for open and short structures fabricated on the same substrate were measured to de-embed parasitic pad capacitances and inductances. Fig. 3 (a) shows the current gain (h_{21}) and unilateral power gain (U) of the MOSFET with a L_G of 0.1 μm. A record high f_T of 47 GHz and f_{max} of 51 GHz was achieved at V_{GS}=-5 V. Fig. 3 (b) shows the RF performances at various V_{GS}. As V_{GS} was swept from -7 V to 0 V, both f_T and f_{max} were in the range of 25 to 51 GHz, implying its operation capability with a wide frequency band.

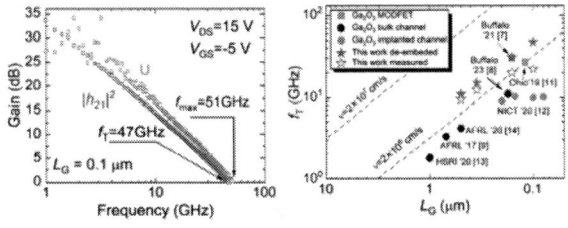

Fig. 3 (a) Small-signal RF gain and (b) the frequency performance of the β-Ga_2O_3 RF MOSFET (L_G=0.1 μm) at various V_{GS}. (c) and (d) Benchmarks of β-Ga_2O_3 RF MOSFETs in the f_T-L_G and f_{max}-L_G plots, respectively.

Fig. 4 (a) shows large-signal power output characteristics of the MOSFET with L_G of 0.1 μm, which was conducted using an on-wafer load-pull setup at a frequency of 2 GHz, operated in a continuous wave (CW) mode with V_{DS}= 15 V and V_{GS}=0 V. A saturated output power (P_{out}) of 296 mW/mm, an associated power gain of 11 dB and a maximum power-added efficiency (PAE) of 25.7% was achieved. It is noteworthy that the power gain remained consistently high and exhibited no

apparent decline with increasing input power (P_{in}).

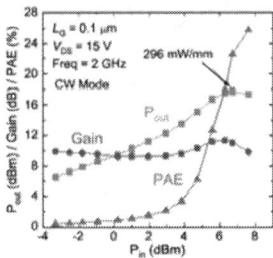

Fig. 4 Large-signal performance of the RF Ga_2O_3 MOSFET with a L_G of 0.1 μm at 2 GHz measured in CW mode.

Conclusion

This work showcases the high RF performance MOSFETs built upon shallowly implanted Ga_2O_3 nanomembranes on 4H-SiC through the smart ion-cutting technology. The resultant Ga_2O_3-on-SiC RF MOSFET featuring a gate length of 0.1 μm has a high output current density of 661 mA/mm, resulting in a record high f_T of 47 GHz and f_{max} of 51 GHz. Operated at 2 GHz in CW mode, the device delivers notable metrics including a high output power density of 296 mW/mm, a power gain of 11 dB and maximum PAE of 25.7 %. This strategy holds promise for enabling the hetero-integration of Ga_2O_3 nanomembrane transistors across diverse platforms, thereby paving the way for compelling high-power and high-frequency applications.

Acknowledgments

The authors gratefully acknowledge xxxx.

References

[1] Green A J, Speck J, Xing G, et al. β-Gallium oxide power electronics [J]. APL Materials, 2022, 10(2): 029201.

[2] Yadava N, Chauhan R K. Review—Recent Advances in Designing Gallium Oxide MOSFET for RF Application [J]. ECS Journal of Solid State Science and Technology, 2020, 9(6): 065010.

[3] Zhang M, Liu Z, Yang L, et al. β-Ga_2O_3-Based Power Devices: A Concise Review [J]. Crystals, 2022, 12(3): 406.

[4] Singh R, Lenka T R, Velpula R T, et al. A novel β-Ga_2O_3 HEMT with f_T of 166 GHz and X-band P_{OUT} of 2.91 W/mm [J]. International Journal of Numerical Modelling: Electronic Networks, Devices and Fields, 2020, 34(1): 1-11.

[5] Singh R, Lenka T R, Panda D K, et al. The dawn of Ga_2O_3 HEMTs for high power electronics - A review [J]. Materials Science in Semiconductor Processing, 2020, 119: 105216.

[6] Donato N, Rouger N, Pernot J, et al. Diamond power devices: state of the art, modelling, figures of merit and future perspective [J]. Journal of Physics D: Applied Physics, 2020, 53(9): 093001.

[7] Vaidya A, Saha C N, Singisetti U. Enhancement Mode β-(Al$_x$Ga$_{1-x}$)$_2$O$_3$/Ga$_2$O$_3$ Heterostructure FET (HFET) With High Transconductance and Cutoff Frequency [J]. IEEE Electron Device Letters, 2021, 42(10): 1444-1447.

[8] Green A J, Chabak K D, Baldini M, et al. β-Ga_2O_3 MOSFETs for

Radio Frequency Operation [J]. IEEE Electron Device Letters, 2017, 38(6): 790-793.

[9] Singh M, Casbon M A, Uren M J, et al. Pulsed Large Signal RF Performance of Field-Plated Ga$_2$O$_3$ MOSFETs [J]. IEEE Electron Device Letters, 2018, 39(10): 1572-1575.

[10] Xia Z, Xue H, Joishi C, et al. β-Ga$_2$O$_3$ Delta-Doped Field-Effect Transistors With Current Gain Cutoff Frequency of 27 GHz [J]. IEEE Electron Device Letters, 2019, 40(7): 1052-1055.

[11] Kamimura T, Nakata Y, Higashiwaki M. Delay-time analysis in radio-frequency β-Ga$_2$O$_3$ field effect transistors [J]. Applied Physics Letters, 2020, 117(25): 253501.

[12] Lv Y, Liu H, Wang Y, et al. Oxygen annealing impact on β-Ga$_2$O$_3$ MOSFETs: Improved pinch-off characteristic and output power density [J]. Applied Physics Letters, 2020, 117(13): 133503.

[13] Moser N, Liddy K, Islam A, et al. Toward high voltage radio frequency devices in β-Ga$_2$O$_3$ [J]. Applied Physics Letters, 2020, 117(24): 242101.

[14] Yu X, Gong H, Zhou J, et al. RF performance enhancement in sub-μm scaled β-Ga2O3 tri-gate FinFETs [J]. Applied Physics Letters, 2022, 121(7): 072102.

Benefits of Using High-Resistivity Substrates for RF ICs

Xunyu Li[1], Cesar Roda Neve[2], Ionut Radu[2] and Albert Wang[1]

[1]Dept. of ECE, University of California, Riverside, USA; [2]Soitec, France

Abstract

Compared to low-resistivity wafers commonly used in CMOS, high-resistivity substrates offer many advantages for radio-frequency integrated circuits. This comprehensive study reveals various benefits of using high-resistivity substrates for radio-frequency integrated circuits, validated in both passive devices and circuits implemented in a foundry 22nm fully-depleted silicon-on-insulator technology.
Keywords: High-resistivity, HR, substrate, SOI and RF IC

Introduction

Substrates play a key role in IC technologies. Mainstream CMOS technologies, particularly at advanced nodes, typically use low-resistivity (LR; e.g., 1-25Ω-cm) substrates to achieve high performance of logic ICs while avoiding some problems, e.g., latch-up effect inherent to CMOS. On the other hand, LR wafers can be disadvantageous for analog, mixed-signal (AMX) and RF ICs, e.g., noise coupling through an LR substrate. Particularly, low substrate resistivity can seriously affect integrated passive devices, e.g., on-chip inductors that are critical to RF ICs [1]. Specifically, quality factor (Q-factor) of inductors made in LR substrates is rather low due to signal energy loss caused by capacitive coupling into the LR substrate. Reduction in Q-factor is more significant for stacked spiral inductors using metal layers closer to Si substrate due to stronger capacitive coupling. Substrate engineering can improve inductor performance, e.g., local materials removal [2], but leading to higher process complexity and costs. High-resistivity (HR) substrates offer a solution for making high-Q inductors [1]. This paper reports a comparison study of passive devices and their RF ICs, in both LR and HR substrates, revealing major benefits of using HR substrates and offering guidelines for balancing foundry CMOS processes, IC performance and costs.

Substrate Impacts on RF Passives

This study evaluates comprehensively substrate impacts on typical passive devices for RF ICs in both LR and HR substrates. The RF passives studied include coplanar waveguide lines (CPW), spiral inductors and transformers. For practical designs, a foundry 22nm fully-depleted SOI (FDSOI) CMOS technology was used as a process platform [3]. Limited by foundry supports, HFSS full-wave electromagnetic simulation was used to evaluate the RF passive devices designed in 22nm FDSOI across wide substrate resistivity range (1-5000Ω-cm) and frequency range (to 60GHz). Substrate impacts on RF ICs were studied by device-circuit co-simulation. The commercial 22nm FDSOI process used offers many metal layers, shown in Table 1, which can be used to design RF passive devices. Typically, thicker top metal layers are used to design transmission lines and stacked spiral inductors, while thinner low metals are avoided due to high resistance and metal-substrate coupling effect. For mass production, it is important to balance process complexity, fabrication costs and IC performance. For example, though using thicker, top metals improves high-Q of inductors, it dramatically increases process complexity (e.g., requiring CMP for surface uniformity), fabrication cycle and IC production costs.

Table 1 Metal stacks in a foundry 22nm FDSOI process [3]

Metal Layer Codes	Metal Thickness (µm)	Materials
LB	2.80	Al
OI	3.30	Cu
QA/QB	3.00	Cu
JA/JB	1.13	Cu
IA/IB	0.81	Cu
BA/BB	0.17	Cu
Cx (1-6)	0.087	Cu
M2	0.07	Cu
M1	0.07	Cu

A. Substrate Impacts on Coplanar Waveguide Line

Fig. 1 inset shows an exemplar CPW structure designed in 22nm FDSOI in substrates of various resistivities. The CPW structure uses the topmost, thick metal (LB) with signal line width of 2µm and signal-ground spacing of 2µm. Fig. 1 depicts S21-frequency characteristic for two substrate splits: LR = 7Ω-cm and HR = 1000Ω-cm. It is readily observed that CPW in HR substrate significantly reduces signal loss (insertion loss, IL) along the CPW line, e.g., from -2.5dB to -1.4dB at 40GHz and from -4.2dB to -2.5dB at 60GHz.

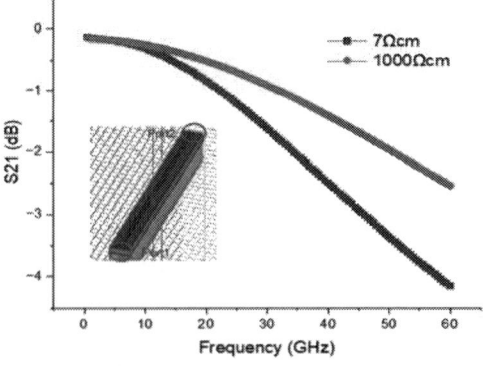

Fig. 1: CPW line (inset) made in LR and HR substrates and extracted signal loss..

B. Substrate Impacts on Spiral Inductors

Spiral inductors of various design splits in LR and HR substrates are evaluated. Fig. 2 depicts an exemplar single-metal (OI layer) symmetric spiral inductor of 425pH (inner diameter 54.4μm, coil width 3.5μm, spacing 4μm). Inductor performance is evaluated against a wide substrate resistivity range, from LR=7Ω-cm (nominal for 22nm FDSOI) to HR up to 5000Ω-cm. It shows the extracted Q-factor at 30GHz against varying substrate resistivity, which clearly increases substantially as substrate resistivity increases. However, Q-factor improvement tends to saturate when resistivity reaches to around 300Ω-cm, offering a guideline in selecting HR substrates considering higher process costs for HR wafers. In RF ICs, stacked-spiral inductors are typically used. For comparison study, two-layer inductors of 1nH (3 spiral turns, inner diameter ~72.5μm, coil width ~6μm) are designed using combined LB+OI and OI+JA metal stacks, both in LR (7Ω-cm) and HR (500Ω-cm) substrates. The extracted Q-factors are given in Fig. 3 (LB+OI) and Fig. 4 (OI+JA). Significant Q-improvement is readily observed, i.e., peak-Q from 16.4 to 21.5 (~31%) for LB+OI inductor and 16.2 to 22.5 (~39%) for OI+JA inductor. More importantly, it is found that, using HR substrate, inductor using lower (OI+JA) and thinner (JA=1.13μm) metals can deliver same Q-factor as its counterpart using higher (LB+OI) and thicker (LB=2.8μm) metals. This offers important guidelines for choosing suitable HR substrate to balance process complexity, IC performance and product costs, i.e., in HR substrates, it is possible to design stacked-spiral inductors using lower, thinner metals to deliver similar/same/better Q-factor. This means that the thicker, upper metals (e.g., LB) may be replaced by a combination of lower, thinner metals for comparable performance in HR substrates, which leads to dramatically reduction in process complexity (e.g., no CMP), cycle time (e.g., fewer metal layers) and costs.

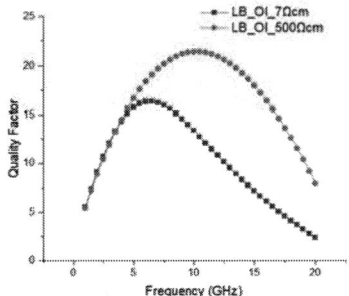

Fig. 3: Q-factor for the 1nH LB-OI inductor.

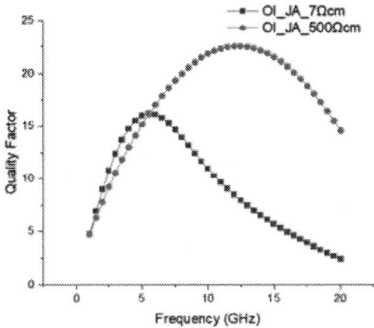

Fig. 4: Q-factor for the 1nH OI+JA inductor.

Substrate Impacts on RF ICs

Impacts of HR substrates are further studied for RF ICs including low-noise amplifier (LNA) and RF switch circuits containing multiple inductors.

A. Impacts on 12GHz LNA

Fig. 5 shows schematic of a 12GHz LNA circuit featuring a cascode topology. This LNA uses multiple inductors for impedance matching at input (L_{gate} and L_{deg}) and output (L_D) ports. Noise figure (NF) is one of the most important specification in LNA circuits, which is directly affected by inductors [2]. Specifically, higher Q-factor can reduce NF of a common-source LNA [4, 5]. In this design, L_{gate} of 1nH

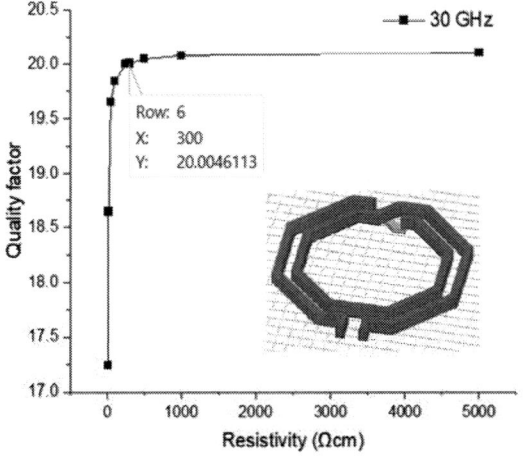

Fig. 2: Q-factor versus substrate resistivities at 30 GHz for a single-layer spiral inductor.

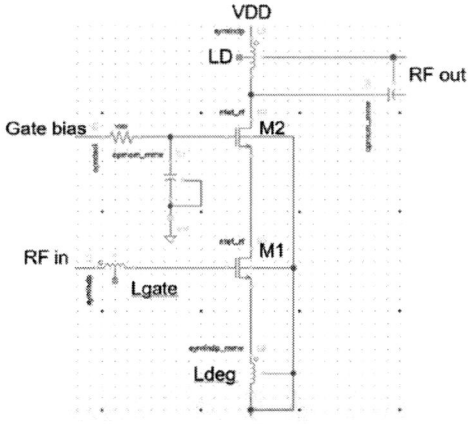

Fig. 5: Schematic for LNA.

used for input impedance matching is much larger than the source-degeneration inductor L_{deg} (50pH) and output drain matching inductor L_D. Hence, HR impact on NF of LNA mainly come from L_{gate}. Co-simulation is conducted for LNA where inductor is simulated by HFSS and the HFSS data file is directly imported into LNA circuit for circuit level simulation using Cadence Virtuoso simulator. HFSS simulation shows that Q-factor of L_{gate} in HR substrate (1000Ω-cm) improves by ~30% over that in LR substrate (7Ω-cm). Fig. 6 depicts NF for LNA from 10GHz to 15GHz, in both LR and HR substrates. It is readily observed that, using HR substrate, NF of LNA reduces significantly, from ~2dB in LR to ~1.3dB in HR at the center frequency of 12GHz, i.e., about 35% reduction in noise figure.

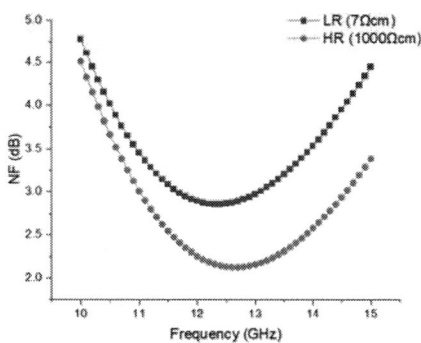

Fig. 6: Noise figure of LNA in LR and HR substrates.

A. Impacts on 38GHz SPDT

Fig. 7 shows a series-shunt single-pole-double-through (SPDT) RF switch widely used in RF SoC. SPDT can switch between ON and OFF by controlling the series and shunt MOSFETs. This SPDT circuit includes inductors for impedance matching at input and output. The inductors used are stacked-spiral inductors made in LB+OI stack (150pH, Q=20.5 in LR and Q=23.3 in HR). Due to impacts of substrate resistivity on inductors, SPDT will be affected by substrates. This work compares SPDT performance in both LR (7Ω-cm) and HR (1000Ω-cm) substrates. The key SPDT circuit parameter affected directly by substrate resistivity is

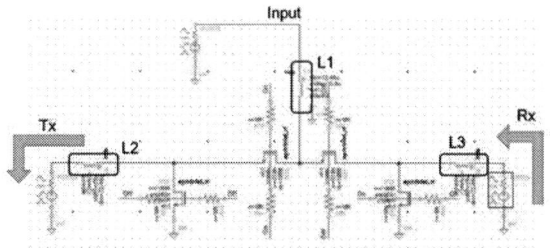

Fig. 7: A series-shunt SPDT RF switch.

insertion loss (IL) in the ON mode reflecting the RF signal integrity, which is evaluated by analyzing S-parameters. Fig. 8 depicts the extracted IL for SPDT in both LR and HR substrates across 1GHz to 60GHz, which clearly shows the significant reduction in signal loss from LR split to HR split, i.e., 0.1dB at 40GHz and 0.62dB at 60GHz.

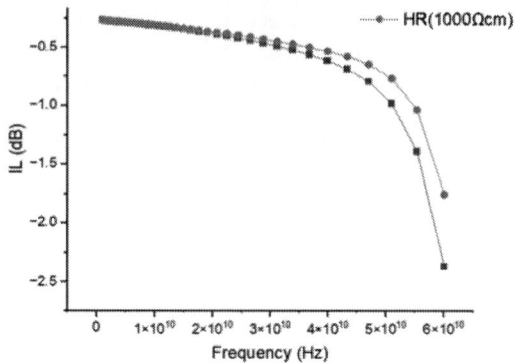

Fig. 8: Insertion loss of SPDT switch in LR and HR substrates.

Conclusion

A comprehensive comparison study reveals that HR substrates can significantly improve RF passive devices, which in turn, contributes to performance improvement of RF ICs. It shows that using suitable HR substrates, lower, thinner metals can be combined together to deliver similar or better RF passive devices. This study suggests that HR substrates can be used to balance process complexity, fabrication cycle time, RF IC performance and product costs.

Acknowledgments

The authors gratefully acknowledge Soitec for supports.

References

[1] K. B. Ali, C. R. Neve, A. Gharsallah and J. -P. Raskin, "RF Performance of SOI CMOS Technology on Commercial 200-mm Enhanced Signal Integrity High Resistivity SOI Substrate" , *IEEE Trans. Electron Devices, vol. 61, no. 3*, pp. 722-728, March 2014.

[2] A. Bhaskar, J. Philippe, V. Avramovic, F. Braud, J-F. Robillard, C. Durand, D. Gloria, C. Gaquiere and E. Dubois, "Substrate Engineering of Inductors on SOI for Improvement of Q-Factor and Application in LNA", *IEEE J. the Electron Devices Society, vol. 8*, pp. 959-969, 2020.

[3] 22nm fully-depleted SOI process by GlobalFoundries.

[4] D. K. Shaeffer and T. H. Lee, "A 1.5-V, 1.5-GHz CMOS Low Noise Amplifier" *IEEE J. Solid-State Circuits, vol. 32, no. 5*, pp. 745–759, May 1997.

[5] L. Belostotski and J. W. Haslett, "Noise Figure Optimization of Inductively Degenerated CMOS LNAs with Integrated Gate Inductors", *IEEE Trans. Circuits and Systems I: Regular Papers, vol. 53, no. 7*, pp. 1409-1422, July 2006.

979-8-3315-0417-5/25 $31.00 © 2025 IEEE

2D Materials for Future Physical Computing

Feng Miao
Nanjing University, China

Abstract

In this talk, I will show how 2D materials open up unprecedented opportunities for harnessing new physics to advance physical computing. I will present our findings on Wigner crystals and ferroelectricity in graphene moiré systems, and show how adjusting the interface potential barrier in 2D heterostructures can lead to the development of future computing devices. Our initial explorations on new physical computing schemes and our vision of future physical computing will also be discussed.

Keywords: 2D materials, future computing, neuromorphic computing

Introduction

The continuous enhancement of computational power is crucial for driving societal progress. Currently, this improvement heavily relies on the integration of transistors. As this integration level nears its limit, marking the end of Moore's Law, the growth in hardware computational power has slowed and been struggling to meet the exponential data processing needs of the AI era. This presents a significant challenge. To overcome it, we need to explore entirely new computing approaches to process information. Unlike traditional digital computing, which relies on abstract symbolic representation and operates at the CMOS circuit level, physical computing processes information at the device level by leveraging material-specific physical processes, thus offering ongoing improvements in computational power. Two-dimensional (2D) materials, with their atomic-layer thickness, enable precise control of physical properties using external fields, creating a superior platform for future physical computing. In this talk, I will show how 2D materials open up unprecedented opportunities for harnessing new physics to advance physical computing. I will begin by presenting our findings on Wigner crystals and ferroelectricity in graphene moiré systems and discuss how these properties can be applied to build basic solid-state quantum simulators [1], moiré synaptic transistors [2], and noise-resistant neuromorphic devices [3]. I will also show how adjusting the interface potential barrier in heterostructures composed of 2D materials can lead to the development of reconfigurable retinomorphic sensors [4-6], visual motion perceptrons [7], and in-sensor dynamic

computing [8]. Finally, I will present our initial explorations on new physical computing schemes [9-10] and share our vision of future physical computing.

References

[1] Q. Li, B. Cheng, M. Chen, B. Xie, Y. Xie, P. Wang, F. Chen, Z. Liu, K. Watanabe, T. Taniguchi, S. -J Liang, D. Wang, C. Wang, Q. -H Wang, J. Liu & F. Miao, "Tunable quantum criticalities in an isospin extended Hubbard model simulator", Nature 609, pp. 479, 2022.

[2] P. Wang, M. Chen, Y. Xie, C. Pan, K. Watanabe, T. Taniguchi, B. Cheng, S.-J Liang and F. Miao, "Moiré Synaptic Transistor for Homogeneous-Architecture Reservoir Computing", Chinese Physics Letters 40, pp. 117201, 2023.

[3] M. Chen, Y. Xie, B. Cheng, Z. Yang, X. Li, F. Chen, Q. Li, J. Xie, K. Watanabe, T. Taniguchi, W. He, M. Wu, S.-J Liang, and F. Miao, "Selective and quasi-continuous switching of ferroelectric Chern insulator devices for neuromorphic computing", Nature Nanotechnology 19, 962, 2024.

[4] C. Wang, S. -J Liang, S. Wang, P. Wang, Z. Li, Z. Wang, A. Gao, C. Pan, C. Liu, J. Liu, H. Yang, X. Liu, W. Song, C. Wang, B. Cheng, X. Wang, K. Chen, Z. Wang, K. Watanabe, T. Taniguchi, J. Yang, F. Miao, "Gate-tunable van der Waals heterostructure for reconfigurable neural network vision sensor", Science Advances 6, eaba6173, 2020.

[5] L. Pi, P. Wang, S. -J Liang, P. Luo, H. Wang, D. Li, Z. Li, P. Chen, X. Zhou, F. Miao, T. Zhai, "Broadband convolutional processing using band-alignment-tunable heterostructures", Nature Electronics 5, 248, 2022.

[6] S. Wang, C. Wang, P. Wang, C. Wang, Z. Li, C. Pan, Y. Dai, A. Gao, C. Liu, J. Liu, H. Yang, X. Liu, B. Cheng, K. Chen, Z. Wang, K. Watanabe, T. Taniguchi, S. -J Liang, F. Miao, "Networking retinomorphic sensor with memristive crossbar for brain-inspired visual perception", National Science Review 8, nwaa172, 2021.

[7] X. Pan, J. Shi, P. Wang, S. Wang, C. Pan, W. Yu, B. Cheng, S.-J Liang and F. Miao, "Parallel perception of visual motion using light-tunable memory matrix", Science Advances 9, eadi4083, 2023.

[8] Y. Yang, C. Pan, Y. Li, X.-J. Yangdong, P. Wang, Z. Li, S. Wang, W. Yu, G. Liu, B. Cheng, Z. Di, S-J Liang, and F. Miao, "In-sensor Dynamic Computing for Intelligent Machine Vision", Nature Electronics 7, 225, 2024.

[9] C. Wang, S.-J Liang, C.-Y. Wang, Z. Yang, Y. Ge, C. Chen, X. Shen, W. Wei, Y. Zhao, Z. Zhang, B. Cheng, C. Zhang, F. Miao, "Scalable massively parallel computing using continuous-time data representation in nanoscale crossbar array", Nature Nanotechnology 16, 1079, 2021.

[10] C. Wang, G. Ruan, Z. Yang, X.-J. Yangdong, Y. Li, L. Wu, Y. Ge, Y. Zhao, C. Pan, W. Wei, L. Wang, B. Cheng, Z. Zhang, C. Zhang, S.-J Liang and F. Miao, "Parallel in-memory wireless computing", Nature Electronics 6, 381, 2023.

CMOS BEOL-Compatible Three-Dimensional Heterogeneous Integration with Emerging Devices for Advanced Information Processing Systems

Heyi Huang[1,2], Feixiong Wang[1,2], Shuang Liu[1,2], Yanqin Li[1,2], Huaxiang Yin[1,2*], Xiaolei Wang[1,2],
Jun Luo[1,2*]

[1]Affiliation, Country, [2]Affiliation, Country[1]State Key Laboratory of Fabrication Technologies for Integrated Circuits, Institute of Microelectronics of the Chinese Academy of Sciences. [2]School of Integrated Circuits, University of Chinese Academy of Sciences.

Abstract

Hardware system developments by ICs based on traditional Moore's Law are approaching their physical limits, posing significant challenges to advanced information processing efficiency in the era of artificial intelligence. Three-dimensional heterogeneous stacking technology on Si-based CMOS circuits offers flexible vertical integration benefits of different processing modules. This paper focuses on a series of key technologies of CMOS back-end-of-line (BEOL)-compatible three-dimensional heterogeneous integration (3DHI) approaches by different emerging materials, devices or circuits with low thermal budgets and good compatibility with traditional Si CMOS processes.

Keywords: Three-dimensional heterogeneous integration (3DHI), Low thermal budget, CMOS back-end-of-line (BEOL).

Introduction

As the scaling down of the device approaches the physical limits, Moore's Law is gradually slowing down towards the end of the decade. The process technology based on three-dimensional (3D) stacking represents a significant innovation in the "More Moore" era[1]. Three-dimensional heterogeneous integration (3DHI) on top of silicon-based CMOS logic circuits enables high-density device integration, greatly enhancing the processing efficiency of the chip. Simultaneously, it can also be used to expand the application range of the chip. For instance, sensors, memory, and computing can all be integrated into one chip through three-dimensional heterogeneous integration (**Fig.1**). This further reduces processing latency and achieves high-bandwidth information processing, providing an efficient hardware foundation for intelligent task processing[2].

There are several ways to achieve three-dimensional integration, including die-to-die integration, wafer-to-wafer interconnection, and monolithic three-dimensional integration. The first two techniques have been significantly improved in recent years, benefiting from the high density and enhanced reliability of the through-silicon via (TSV) process technology, which increases the inter-layer communication bandwidth and reduces the distance[3]. Currently, these technologies are mainly applied in commercial CMOS image sensors, where sensors, flash memories, and Dynamic Random Access Memories (DRAMs) are stacked in three dimensions. The third monolithic integration technique is to vertically stack devices with different functions on the same wafer. In this method, the circuit routing and functional configuration are relatively flexible, and the density and functionality can be further improved; however, its main challenges lie in that the thermal budget of the device fabrication process on the upper layer of the silicon substrate is limited to 450°C[4]. Additionally, mechanical and stress issues also need to be addressed. Two solutions have emerged to solve these problems. One is to adopt a silicon device process with a low thermal budget and improve device performance through post-treatment annealing. The other is to utilize emerging devices, such as oxides (IGZO, ZnO, HfO_x), carbon nanotubes, and two-dimensional materials[5]. These materials have a low thermal budget and multiple configurable functions, providing important solutions for monolithic 3D integration[6].

In this paper, we summarize the heterogeneous 3D integration process and its application directions based on the previous work. Specifically, discussions are conducted on the emerging device process technology compatible with the back-end of CMOS.

BEOL-compatible 3DHI by Emerging Devices

A. **Low thermal budget Si-based transistor** is the first choice for 3D integration for its CMOS integrability, high mobilities and excellent robustness. However, fabricating top-tier devices requires a low thermal budget to prevent degradation in the performance and reliability of underlying transistors and intermediate interconnects. A thermal budget of 500°C for 2 hours is typically considered safe for CMOS integration. To enable low-temperature Si transistor fabrication, advanced dopant activation techniques have been developed to minimize the impact on bottom devices [7]. However, these techniques often introduce process complexity and reduce yield, complicating the realization of multilayer monolithic 3D (M3D) integration. We investigated sub-500 ℃ Schottky Barrier transistor with Nickle Silicide source/drain region, emphasizing the simplicity of process integration, shown in **Fig.2** [8]. Dopant segregation was achieved using long-pulse green laser annealing, effectively modulating the Schottky barrier with a minimal thermal budget of 500°C for 30 seconds and nanosecond laser pulses. High-performance characteristics, including excellent subthreshold properties, were demonstrated. Experimental CMOS inverters and NAND/NOR logic gates highlight this approach's promise for 3D integration.

B. Low-dimensional nanomaterial devices such as CNTs offer great advantages in terms of thermal budget and multifunctional applications[9, 10]. The fabrication process of CNFET can be divided into nine modules, as depicted in **Fig. 3a**, and none of the process temperatures exceed 300 ℃. The insert picture presents a SEM image of the CNT film in the channel region, which reveals minimal polymer residue and extended lengths The device's characteristic curve is shown in **Fig. 3b**, with an on-state current of more than 10μA. In addition, the device presents different current magnitudes when stimulated with different wavelengths of light, and the optical response of the device increases with decreasing light wavelength. This exhibits promising sensing characteristics. **Fig. 3c** illustrates the output characteristic curve of the CNFET, highlighting its response to varying gate voltages. Based on the sensing function of the device, it also has excellent multi-bit memory function. **Fig. 3d** and **3e** present TEM and EDS characterizations of the CNFET, confirming its exemplary fabrication. And it lays a solid foundation for subsequent in-memory computation. Furthermore, the CNFET can realize the logic functions of "AND" and "OR". To conclude, CNFETs integrate sensing, memory, computing, and logic functions, positioning them as a highly promising candidate for future 3D systems.

C. Oxide semiconductor materials and devices represent a promising solution compatible with CMOS processes, particularly due to their low thermal budget, typically not exceeding 350°C. These materials are widely utilized in the fabrication of thin-film transistors (TFTs) or resistive random-access memory (RRAM) devices, enabling either logic or memory functionalities. For instance, ZnO-based RRAM can be fabricated at room temperature to form a memristor with a metal-insulator-metal (MIM) structure[11]. During electrical operation, the migration of oxygen vacancies is induced, resulting in the formation of conductive filaments (**Fig.4a**), which exhibit reconfigurable analog resistive switching behavior. The memristor can be integrated in the back end of the line (BEOL) above CMOS logic circuits (**Fig.4b-d**), which with excellent multi-level uniformity (**Fig.4e-f**). This architecture supports in-memory computing capabilities, enabling convolutional operations to be efficiently performed.

Conclusion

In conclusion, we primarily discuss the key challenges and major technological solutions in CMOS BEOL-compatible three-dimensional heterogeneous integration techniques.

Among these, thermal budget remains one of the primary issues in stacking. We have introduced a series of emerging material devices compatible with Si CMOS process, including low-thermal-budget silicon-based devices, carbon nanotube materials, and oxide materials. The fabrication technologies, device performance optimization methods, and functional applications in advanced information process systems are delivered, which are expected to not only offer flexible and low-cost fabrication methods but also exhibit richer functionalities compared to traditional silicon-based CMOS circuits for next-generation information processing technologies in the post-Moore era.

Acknowledgments

This work was supported by the State Key Laboratory of Fabrication Technologies for Integrated Circuits, Institute of Microelectronics of the Chinese Academy of Sciences.

References

[1] Radosavljevic, 3D-stacked CMOS takes Moore's law to new heights. IEEE Spectrum https://spectrum.ieee.org/3d-cmos (2022).

[2] D. Jayachandran et al., "Three-dimensional integration of two-dimensional field-effect transistors," Nature, 625, pp. 276-281 (2024).

[3] Ghosh, S., Zheng, Y., Zhang, Z. et al. Monolithic and heterogeneous three-dimensional integration of two-dimensional materials with high-density vias. Nat Electron 7, pp.892–903 (2024).

[4] A. Thean et al., "Low-Thermal-Budget BEOL-Compatible Beyond-Silicon Transistor Technologies for Future Monolithic-3D Compute and Memory Applications," 2022 International Electron Devices Meeting (IEDM), pp. 12.2.1-12.2.4 (2022).

[5] F. M. Lee et al., "3D Monolithically Integrated Device of Si CMOS Logic, IGZO DRAM-like, and 2D MoS2 Phototransistor for Smart Image Sensing," 2023 International Electron Devices Meeting (IEDM), pp. 1-4 (2023)

[6] Shuye Zhang, et al., "Challenges and recent prospectives of 3D heterogeneous integration, e-Prime - Advances in Electrical Engineering, Electronics and Energy, 2, pp.100052 (2022)

[7] D. Bosch, A. Viey, et.al., "Breakthrough Processes for Si CMOS Devices with BEOL Compatibility for 3D Sequential Integrated more than Moore Analog Applications," in 2024 IEEE Symposium on VLSI Technology and Circuits (VLSI), Jun. 2024, pp. 1–2 (2024).

[8] Feixiong Wang, et.al., "Laser-Induced Low Temperature Dopant Segregation Schottky Barrier MOSFET for Monolithic-3D," in 2025 IEEE Electron Devices Technology and Manufacturing (EDTM), pp. 1-3 (2025).

[9] M. M. Shulaker et al., "Carbon nanotube computer," Nature,.501, pp. 526-30 (2013).

[10] L.-M. Peng, Z. Zhang, and C. Qiu, "Carbon nanotube digital electronics," Nature Electronics, 2, pp. 499-505 (2019).

[11] Huang, H., Liang, X., Wang, Y. et al. Fully integrated multi-mode optoelectronic memristor array for diversified in-sensor computing. Nat. Nanotechnol. 20, pp. 93–103 (2025).

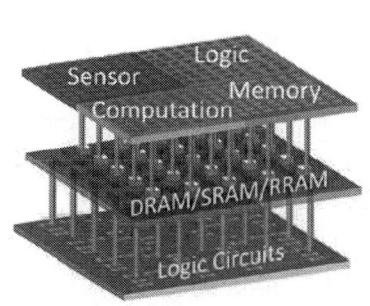

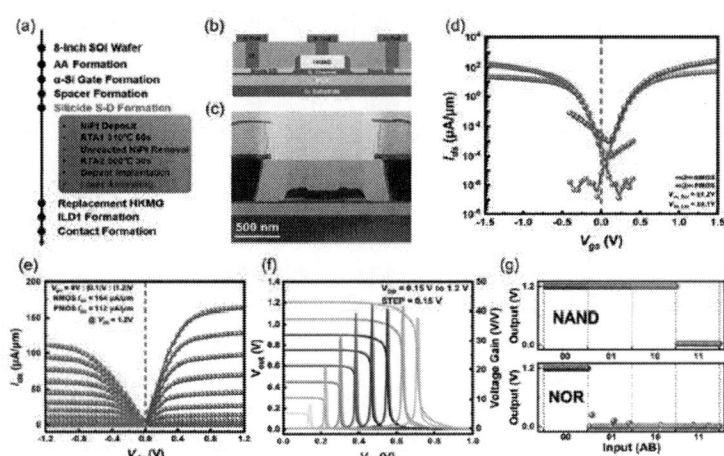

Fig.1 Schematic diagram of a three-dimensional heterogeneous integration architecture, which allows for the flexible stacking of various functional modules above the silicon CMOS logic layer, featuring high-density interconnect communication in the vertical direction between layers.

Fig.2 (a) Fabrication process flow of simulated Top-Tier Schottky Barrier MOSFET. (b) Schematic diagram of Top-Tier Schottky Barrier MOSFET. (c) Cross-Section TEM of Schottky Barrier MOSFET. (d) and (e) Measured Ids-Vds of fabricated MOSFET. (f) Voltage transfer characteristics (VTC) and Voltage Gain of a fabricated inverter. (g) Electrical measurement of NAND and NOR logic gates.

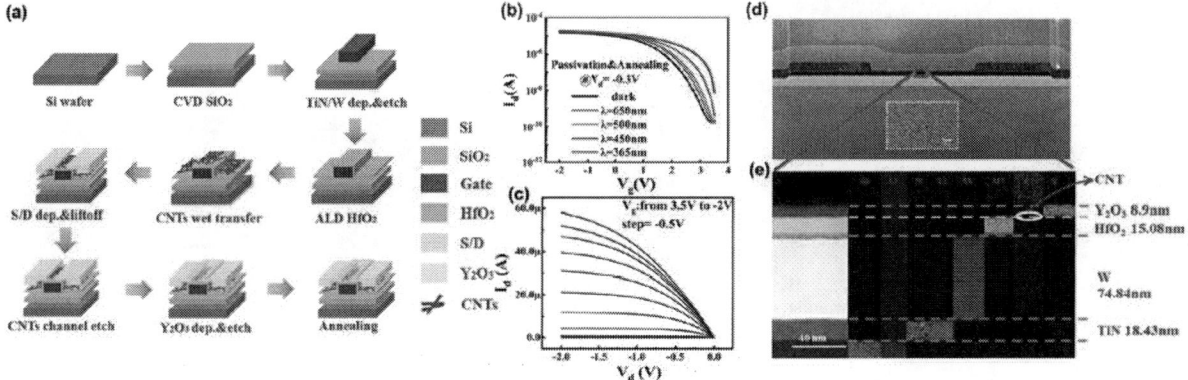

Fig 3. Structure and electrical and optoelectronic characteristics of CNFETs. (a) Fabrication process flow of CNFFET. (b) Transfer characteristic curves of the CNFET under different wavelengths of light stimulation. (c)The output characteristic curve of the CNFET in the dark. (d) TEM image of the device and (e) EDS elemental analysis elucidates the elemental composition across the device's channel.

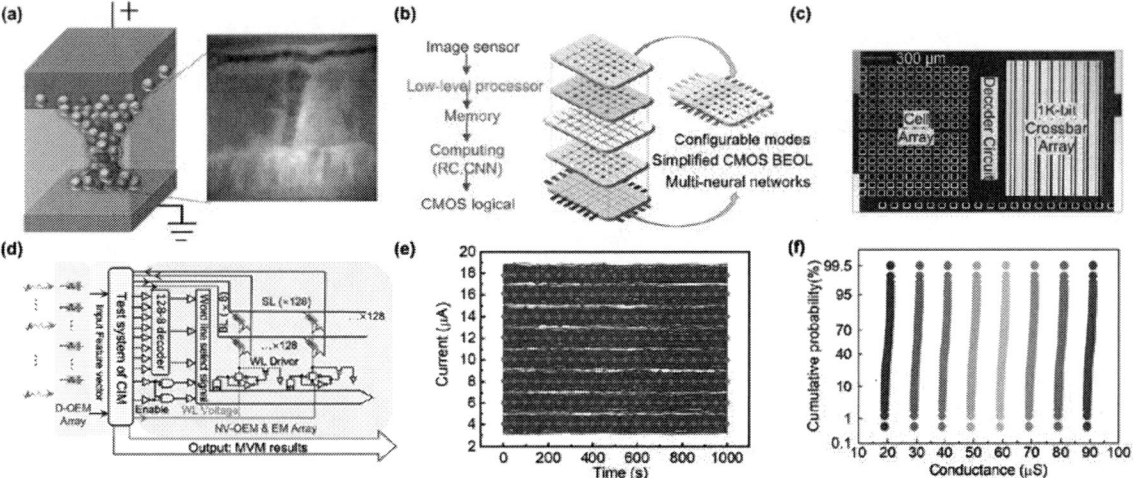

Fig 4. Characteristics and Functions of Oxide Memristors. (a) Structure of the memristor and morphology of the conductive filaments. (b) Memristor array integrated on a silicon-based logic control circuit. (c) An optical micrograph of the 128 × 8 crossbar array. (d) Circuit architecture diagram of the array. (e) Retention characteristics of the memristor array. (f) Probability distribution of different resistance states.

979-8-3315-0417-5/25 $31.00 © 2025 IEEE

Phase Change Memory: From Technological Challenges to Materials Science

Ruobing Wang, Yichen Song, Xilin Zhou*, Zhitang Song*

[1]State Key Laboratory of Materials for Integrated Circuits, Shanghai Institute of Microsystem and Information Technology, Chinese Academy of Sciences, Shanghai 200050, China

*Email: xilinzhou@mail.sim.ac.cn; ztsong@mail.sim.ac.cn

Abstract

Nonvolatile phase-change memory (PCM), which utilizes rapid and reversible electrothermal-induced phase transitions, is emerging as a key candidate for storage class memory to close the access-time gap between conventional memory and storage in computer systems. In this talk an alternative view on the phase transition mechanism of phase change memory will be proposed, starting from the octahedral structure motifs. Based on this mechanism, high performance phase change materials are developed for storage-class memory and embedded data storage applications.

Keywords: phase change memory, storage-class memory, embedded applications

Introduction

The main part of phase change memory (PCM) is a phase change material based on chalcogenide compounds, which uses the resistance difference between the amorphous and crystalline states of the phase change material to achieve information storage [1]. In 1968, Ovshinsky reported the phenomenon of transition between high and low resistance of chalcogenide compounds under electric field excitation [2]. The amorphous state of chalcogenide materials shows high resistance, corresponding to logic "0". When a long low-voltage pulse is applied to it, the material will be affected by the heat generated by the electric pulse and change from the amorphous state to the low-resistance crystalline state (corresponding to logic "1"). When a high enough energy electric pulse is applied to the crystalline material, the material will be melt-quenched to the high-resistance amorphous state [3].

Based on this principle, Gordon Moore developed the first PCM storage array in 1970 [4]. However, operating the storage array required a voltage of 25 V and a current of 200 mA, which was an unimaginable power consumption for the memory industry. It was not until the 21st century that semiconductor technology began to develop rapidly, and Intel took the lead in using the 180 nm CMOS process to prepare a 4 Mb PCM with a 1T1R (one transistor one memory) cell structure. It realized small-size and low-power PCM which brought the PCM technology returned to the research hotspot again [5]. In 2008, the process entered 90 nm, Samsung adopted a 1D1R (one diode one memory) cell structure to achieve 512 Mb PCM [6]. At the end of 2009, Numonyx released 1 Gb PCM at 45 nm node [7]. Intel and Micron jointly released 3D-Xpoint memory in 2015, which revolutionary used ovonic threshold switch (OTS) material as the gate to form a 1S1R (one selector one memory) structure [8]. It claims that the write speed and endurance are 1000 times better than that of NAND Flash. In 2018, ST Microelectronics released the embedded PCM products based on 28 nm FD-SOI process for the automotive and IoT applications which is followed by the PCM based on 18nm FD-SOI in 2024 [9]. In 2023, SK Hynix demonstrated a second-generation 3D-Xpoint PCM chip with a capacity of 256 Gb using a 20 nm node and four-layer stack [10]. As shown in Fig. 1, the roadmap of PCM in the past fifty years is outlined.

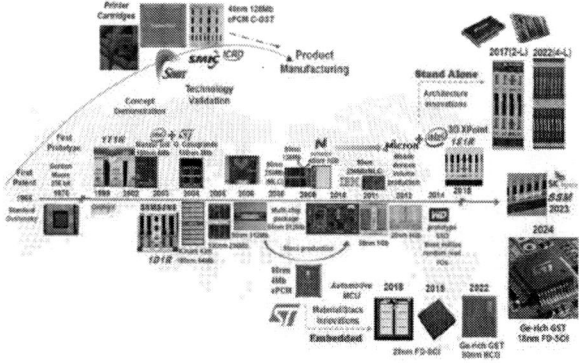

Fig. 1 The roadmap of phase change memory in the past fifty years. (Compiled from publicly available data)

Challenges for phase change memory technology

PCM is expected to replace conventional embedded memory due to its good scalability and compatibility with CMOS processes. The key challenge is to realize high performance PCM after the thermal treatment under high temperature during soldering reflow. Figure 2 shows the specifications of relevant embedded applications for memory products, based on data from [11]. So far, ST Microelectronics is the first player to announce the mass production of embedded phase change memory (ePCM) for automotive micro-controller units (MCUs). Among all resistive memories proposed to replace embedded Flash memory, ePCM is the only one that has been proven to meet the most stringent requirements of the automotive industry. The 100-year data retention temperature (T_{10}) of ePCM is up to 150 °C. The ePCM is able to survive after reflow soldering (curve peak temperature 260°C), with not only fast programming, bit alterability, and low energy cost, but also sufficient reliability.

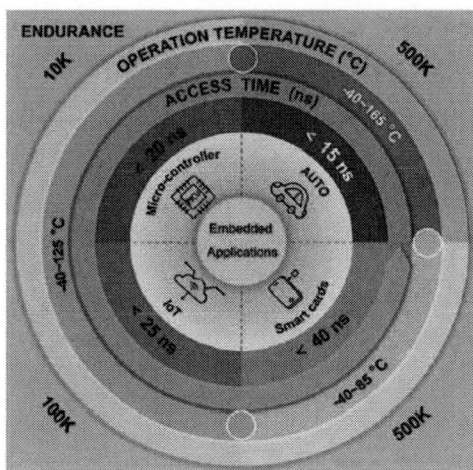

Fig. 2 The specifications of relevant embedded applications for phase change memory.

The device structure determines the performance of phase change memory. The main purpose of improving the cell structure is to reduce the power consumption of phase change memory and increase the storage density. Figure 3 shows the schematic diagram of the 1T1R PCM cells with L-type bottom heater. One of the technological challenges in fabricating such memory array is trimming the size of the L-type bottom heater, usually expected to below 10 nm.

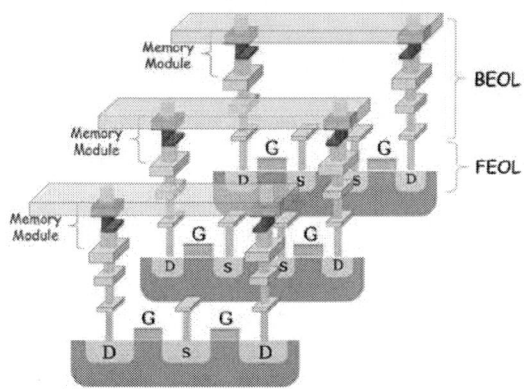

Fig. 3 Schematic diagram of the 1T1R PCM cells with L-type bottom heater.

Key to achieving high performance PCM is the ability to secure consistent and reliable switching electrical properties for commercial using under ultrahigh temperature. And this is associated to the functional property of phase change materials. It is because that voids and phase segregations are always observed after cyclic reversible switching under programming electrical current.

The "gene" of phase change materials

Mature phase change material $Ge_2Sb_2Te_5$ (GST) with good performance is derived from the compatibility between GeTe (GT) and Sb_2Te_3 (ST) materials. The crystalline phases of GST are divided into two: metastable and stable phase [8]. The first transition from the amorphous to metastable crystalline phase occurs around 150 °C, which is insufficient

to meet the thermal stability in embedded data storage application. Moreover, the metastable crystalline phase causes severe problems related to phase stability, such as phase separation and element segregation during repeated switching operations. So far, doping methods have been employed to solve the problems related in conjunction with the crystalline structure and phase transition process and to explore characteristics of PCM alloys. Previous works on developing phase change material by doping have made great improvements on the switching performances of PCM. The PCM cell based on Sc-Sb-Te material achieved a 700 ps writing by introducing geometrically matched Sc-Te motif [12]. Ti-Sb-Te PCM cell presented one order faster switching [13]. But the large mismatch between Ti-Te and Sb-Te result in the phase segregation in Ti-Sb-Te [14]. Similar results have been found in Ta-Sb-Te film where Ta-Te lamellae occurs at the grain boundaries due to the large lattice distortion [15]. The formation of telluride lamellae reduces the grains size of Sb-Te to improve the thermal stability, leading to the unsatisfied device reliability and performance after annealing process. Figure 4 shows the lattice mismatch between Sb-Te and the dopants octahedral motifs.

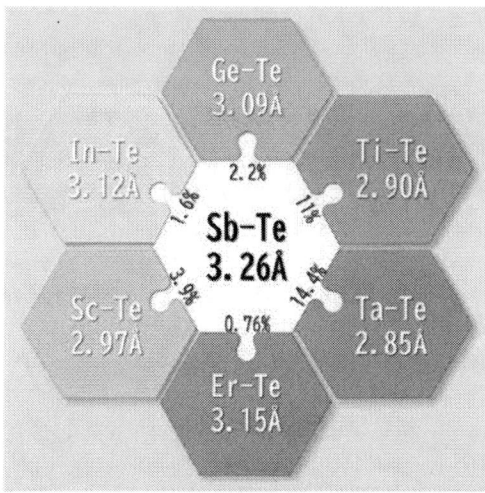

Fig. 4 The phase change material design rules based on octahedral motifs.

Based on this view of the matching octahedral motifs, a homogeneous phase change material In-Ge-Sb-Te material is developed by constructing three matched octahedrons In-Te, Ge-Te, and Sb-Te [16]. As shown in Fig. 5, compared to traditional GST material, The In-Ge-Sb-Te PCM cell achieved an overall improvement in performance, presenting 180 °C 10-year data retention temperature, 6 ns operation speed, one order of magnitude longer life time and 75% reduced power consumption [17]. The performance of In-Ge-Sb-Te PCM cell remains after annealing at 400 °C for 30 min. The stable and good performances after high temperature treatment, providing a viable solution for PCM featuring high thermal stability in embedded data storage applications.

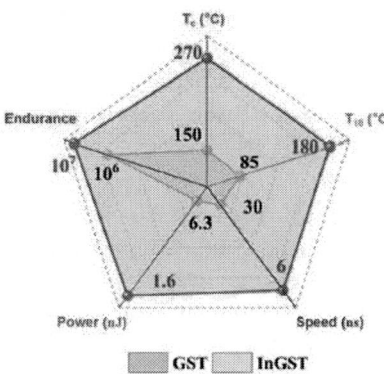

Fig. 5 The performance comparison between the Ge-Sb-Te and In-Ge-Sb-Te phase change materials.

Conclusion

In this talk, the main technological challenges of PCM towards the further developments in SCM and embedded data storage applications are discussed. PCM is suitable for high-performance embedded systems due to the discovery of phase change material with high thermal stability. Research on the design rules of phase change materials has made a breakthrough. However, PCM technology still faces challenges in terms of reliable switching, high-speed read, write circuit design, high-density process integration, and energy-efficient in-memory computing chips, which require more endeavor in material engineering and device fabrication in the futher.

Acknowledgments

This work is supported by the National Key Research and Development Program of China (2023YFB4404500), National Natural Science Foundation of China (62174168, 62404233, 92164302), Strategic Priority Research Program of the Chinese Academy of Sciences (XDB0670000), Science and Technology Council of Shanghai (23XD1404700, 23JC1400900), Autonomous deployment project of State Key Laboratory of Materials for Integrated Circuits (SKLJC-Z2024-A01), China Postdoctoral Science Foundation (2023TQ0363, 2024M753370), and Postdoctoral Fellowship Program of CPSF (GZC20232839).

References

[1] M. Wuttig and N. Yamada, 'Phase-change materials for rewriteable data storage', Nat. Mater., vol. 6, no. 11, Art. no. 11, Nov. 2007, doi: 10.1038/nmat2009.

[2] S. R. Ovshinsky, 'Reversible Electrical Switching Phenomena in Disordered Structures', Phys. Rev. Lett., vol. 21, no. 20, pp. 1450–1453, Nov. 1968, doi: 10.1103/PhysRevLett.21.1450.

[3] A. V. Kolobov, P. Fons, A. I. Frenkel, A. L. Ankudinov, J. Tominaga, and T. Uruga, 'Understanding the phase-change mechanism of rewritable optical media', Nat. Mater., vol. 3, no. 10, Art. no. 10, Oct. 2004, doi: 10.1038/nmat1215.

[4] R. G. Neale and J. A. Aseltine, 'The application of amorphous materials to computer memories', IEEE Trans. Electron Devices, vol. 20, no. 2, pp. 195–205, Feb. 1973, doi: 10.1109/T-ED.1973.17628.

[5] W. Y. Cho et al., 'A 0.18 /spl mu/m 3.0 V 64 Mb non-volatile phase-transition random-access memory (PRAM)', in 2004 IEEE International Solid-State Circuits Conference (IEEE Cat. No.04CH37519), Feb. 2004, pp. 40-512 Vol.1. doi: 10.1109/ISSCC.2004.1332583.

[6] K.-J. Lee et al., 'A 90 nm 1.8 V 512 Mb Diode-Switch PRAM With 266 MB/s Read Throughput', IEEE J. Solid-State Circuits, vol. 43, no. 1, pp. 150–162, Jan. 2008, doi: 10.1109/JSSC.2007.908001.

[7] G. Servalli, 'A 45nm generation Phase Change Memory technology', in 2009 IEEE International Electron Devices Meeting (IEDM), Dec. 2009, pp. 1–4. doi: 10.1109/IEDM.2009.5424409.

[8] D. Kau et al., 'A stackable cross point Phase Change Memory', in 2009 IEEE International Electron Devices Meeting (IEDM), Dec. 2009, pp. 1–4. doi: 10.1109/IEDM.2009.5424263.

[9] F. Arnaud et al., 'Truly Innovative 28nm FDSOI Technology for Automotive Micro-Controller Applications embedding 16MB Phase Change Memory', in 2018 IEEE International Electron Devices Meeting (IEDM), Dec. 2018, p. 18.4.1-18.4.4. doi: 10.1109/IEDM.2018.8614595.

[10] J. Yi et al., 'The chalcogenide-based memory technology continues: beyond 20nm 4-deck 256Gb cross-point memory', in 2023 IEEE Symposium on VLSI Technology and Circuits (VLSI Technology and Circuits), Jun. 2023, pp. 1–2. doi: 10.23919/VLSITechnologyandCir57934.2023.10185210.

[11] F. Ottogalli et al., 'Phase-change memory technology for embedded applications', in Proceedings of the 30th European Solid-State Circuits Conference (IEEE Cat. No.04EX850), Sep. 2004, pp. 293–296. doi: 10.1109/ESSDER.2004.1356547.

[12] F. Rao et al., 'Reducing the stochasticity of crystal nucleation to enable subnanosecond memory writing', Science, vol. 358, no. 6369, pp. 1423–1427, Dec. 2017, doi: 10.1126/science.aao3212.

[13] M. Zhu et al., 'One order of magnitude faster phase change at reduced power in Ti-Sb-Te', Nat. Commun., vol. 5, no. 1, p. 4086, Sep. 2014, doi: 10.1038/ncomms5086.

[14] F. Rao et al., 'Direct observation of titanium-centered octahedra in titanium–antimony–tellurium phase-change material', Nat. Commun., vol. 6, no. 1, p. 10040, Dec. 2015, doi: 10.1038/ncomms10040.

[15] Y. Xue et al., 'Phase change memory based on Ta–Sb–Te alloy – Towards a universal memory', Mater. Today Phys., vol. 15, p. 100266, Dec. 2020, doi: 10.1016/j.mtphys.2020.100266.

[16] Z. Song, R. Wang, Y. Xue, and S. Song, 'The "gene" of reversible phase transformation of phase change materials: Octahedral motif', Nano Res., vol. 15, no. 2, pp. 765–772, Feb. 2022, doi: 10.1007/s12274-021-3570-1.

[17] R. Wang et al., 'Phase-change memory based on matched Ge-Te, Sb-Te, and In-Te octahedrons: Improved electrical performances and robust thermal stability', InfoMat, vol. 3, no. 9, pp. 1008–1015, Sep. 2021.

Advancing Emerging Device and Architecture Innovations in the AI Era

Ming Liu

State Key Laboratory of Integrated Chips and Systems, Frontier Institute of Chip and System, Fudan University, Shanghai 200433, China.

Email: liuming@fudan.edu.cn

Abstract

The rapid advancement of artificial intelligence (AI) has brought new opportunities for our society, yet the ever-increasing demand for computational power presents significant challenges in the design of integrated circuit and system. This paper provides an overview of the current status of devices and architectures, highlighting the emerging innovations in technology, advanced memory devices and the new computing paradigms in the AI era.

Keywords: AI, chiplet integration, 3D stack, emerging memory, computing-in-memory

Introduction

The information age, featuring the advent of smartphone, Internet, private computer and cloud, has significantly transformed our life in the past few years. At the same time, the scale of the semiconductor industry, as a medium and carrier of the information, has also been expanding. Recently, the global semiconductor industry surpassed $600 billion, with projections indicating it could exceed one trillion dollars by 2030, driven in large part by the continued advancement of AI technology [1].

The modern deep-learning-based AI algorithm began to thrive around 2010, with the introduction of famous AlexNet, which efficiently addressed challenges such as ImageNet. After that, the scaling law became an important rule in the development of AI models. From AlexNet with only 10M parameters to the recently launched DeepSeek v3 with ~100GB parameters, the scale of AI models has grown exponentially, expanding their applications across diverse fields. However, the growth in model parameters inevitably brings the increase in computational demand, pushing traditional hardware beyond its limits. This led to the development of AI-specific hardware such as Graphics Processing Unit (GPU), Neural-network Processing Unit (NPU) and AI accelerators, etc., designed to meet the performance needs of models with GBs parameter.

There are three key specifications directly affecting AI chips' performance: computing power, memory capacity and interconnect bandwidth. State-of-the-arts AI chips keep advancing the three specifications. Take the most advanced Nvidia B200 GPU product on the market at present as an example. Its efficiency reached 4500 TFLOPS (BF16), on-chip storage capacity reached 192 GB, and its memory bandwidth reached 8000 GBPS, improved by 7.2×, 2.4× and 5.1×, respectively, than the A100 series chips released two years ago. However, the ongoing scaling of the above three elements faces significant challenges: (1) AI algorithms require computing power (TOPS/TFLOPS) that is 300× faster than Moore's Law, which is approaching its physical limit at 2 nm technology. (2) The rapid growth of Large Language Model (LLM) has driven memory requirements from 100MB to 100GB within just 3 years, while the scaling down of embedded SRAM has stalled after 5nm. (3) The scaling of DRAM memory access and GPUs' interconnect bandwidth has not kept pace, resulting in one of the most significant bottlenecks in the entire computing system.

In response to the above three major challenges, this paper focus on the iteration of hardware driven by the huge demand brought by AI. We summarize the state-of-the-art research in technology, memory, and new computing architectures, and looked ahead to their development prospects.

Key Factors Affecting the Performance of the Chip

A. Technology for Computing Logic Scaling

Over the past few decades, the shrinking of chip and integrate circuit strictly follows Moore's Law, which means its performance doubles every 18 to 24 months. During this period, technologies based on planar processes, such as strained Si and high-K Gate, as well as those based on three-dimensional processes, including FinFET, GAA, and CFET technologies, have been successively proposed, driving the CMOS technology scaling from 90nm to 2nm. The shift from FinFET to GAA transistors enables higher device density and better electrostatic control, boosting the gates count in computing logic. However, these advancements are encountering physical limitations as they approach the 2nm node. Thus, researchers are exploring new methods to manufacture integrated circuits at even finer scales.

At the device level, Si/SiGe stack is proposed to deal with the challenges of sustaining high drive current and reducing leakage when scaling below 5nm for silicon. The superlattice stack improves carrier mobility through strain engineering, enhancing electrostatic control and reducing short-channel effects. Experimental results shows that 2nm/18A logic technology delivers a ~10x performance improvement within same area compared to 28nm logic technology [1].

At the chip packaging level, chiplet technology is regarded as the most effective way to break the area limitations of a single chip. Due to the limitation of the reticle size, under the current circumstance, the maximum size of a

979-8-3315-0417-5/25 $31.00 © 2025 IEEE

single chip is restricted to less than 858mm², seriously restricting its functional expansion. Fortunately, the chiplet technology provides a "Lego-like" scaling solution by allowing multiple dies to be packaged onto a single substrate, effectively increasing the maximum chip area several times.

Interposer plays an important role in 2.5D integration, including both Silicon Interposer and RDL/FanOut Interposer. Silicon Interposer is used in 2.5D integration to connect multiple chiplet dies to a common substrate by through-silicon vias (TSVs) to make a larger integrated chip. Redistribution Layer (RDL) interposers provide wider line widths and larger via diameter due to the photolithography limitations of organic materials. Silicon bridges embedded in RDL interposer offers locally fine-pitch interconnect with lower cost compared to silicon interposers.

B. Advanced Memory Devices & Technology

As AI models continue to scale, the demands on memory are becoming increasingly stringent. On one hand, the memory needs to have the highest feasible capacity to store gigabyte-scale parameters. On the other hand, it should also be designed with low latency and low power consumption to reduce the overheads during the training and inference.

Generally, different types of memory are employed in different scenarios to maximize their functional advantages. SRAM is typically utilized to store the local weights within a single layer or to serve as a KV buffer to feed compute blocks. DRAM, with high capacity, is employed to store all the parameters of an entire model, which requires about 500 GB of capacity. However, SRAM has its limitations. It is unable to process nodes below 5nm, which substantially restricts its density. Moreover, the storage paradigm of DRAM entails frequent refreshing, leading to a significant power consumption overhead that cannot be ignored. Consequently, both the academic and industrial community have made numerous attempts to improve or replace the existing memory to further enhance its performance.

For SRAM, complementary FET (CFET) technology is one of the most significant innovations in recent years. It is regarded as the potential device architecture for logic technology nodes beyond 1nm. CFET reduces SRAM bit-cell area by stacking NMOS/PMOS devices. Compared to nanosheet technology, it can reduce nearly half of the area in the memory cell, thus meeting the high - density requirement.

Embedded NVM devices featuring high density and low power consumption, such as RRAM, PCM, MRAM, and FeRAM, are being adopted as alternatives to SRAM. For example, RRAM employs a 1T1R (1 transistor and 1 resistor) cell and stores data through the resistance value of the device. The RRAM cells are fabricated by a fully CMOS - compatible technology and exhibit a density approximately twice that of SRAM. TSMC has demonstrated that RRAM memory has achieved a capacity of 32 Mb with a density of 20.1 Mb/mm², showing the remarkable advantage of RRAM memory in terms of storage density [2].

In practical applications, RRAM/SRAM hybrid accelerators are explored to overcome the limitations of single devices, such as the write endurance of RRAM and the low density of SRAM. RRAM and SRAM are used for fixed - weight storage and dynamic key - value (KV) storage in transformers, respectively. This hybrid storage mode reduces the storage overhead of AI accelerators and promotes the edge - side deployment of CNN/Transformer algorithms.

The scaling down of DRAM is also a major concern for researcher. Although the size of DRAM cell is smaller than that of SRAM, its technology stays at 10nm technology for a long time. It is believed that the horizontal scaling is limited by patterning and cell transistor/cap. To achieve a higher capacitor within small area, the structure of cell capacitor structure has evolved from cylindrical to pillar designs, and high-k dielectric materials have been integrated to increase capacitance. 3D stacking technology is another new type of DRAM manufacturing process. By stacking several layers of transistors in the vertical direction, the storage density of DRAM has been significantly increased. With the development of high - performance, 3D - stackable access transistors and lateral capacitors, vertically stacked 3D-DRAM is promising as next-generation DRAM with higher density and lower cost.

The emergence of new storage materials is also an important trend in the optimization of DRAM. Hafnium-based ferroelectric materials and IGZO are representative examples. Ferroelectric materials can be polarized in changing electric field, allowing information to be written/read by applying an external electric field, achieving low-power-consumption and high-density NVM. However, the existence of high coercive electric field may degrade the power and read/write speed. To overcome this problem, excessive Hf(Zr) metal elements are introduced to provide lattice internal stress to stable r-phase, obtaining low coercive field and high reliable ferroelectric materials [3]. The latest research shows that Hf-based FeRAM chip can achieve 10^{12} endurance cycles with 7ns write/5ns read latency, enabling direct deployment in nonvolatile DRAM architectures [4].

To further improve the density of FeRAM, CMOS-compatible ferroelectric capacitors are stacked with poly-Si vertical access transistors. Micron has proposed its world's first dual-layer nonvolatile ferroelectric memory at IEDM 2023, achieving 32 Gb capacity with near-DRAM speed and endurance [5]. Such performance demonstrates its potential to be used in the next-generation hardware of AI.

IGZO is another material that is used to save the area of DRAM cells. IMECAS has shown its capacitor-less 2T0C DRAM cell at IEDM 2021, where IGZO provides extremely low I_{off} and fully compatible with BEOL fabrication. A $4F^2$ bit-cell can be obtained by stacking two Channel-All-Around

(CAA) FETs. A high current density of 32.8 μA/μm was obtained in the CAA-IGZO FET with a 50 nm CD [6].

C. CIM: Breaking the Bandwidth Bottleneck

The training and inference of LLM model rely heavily on a vast amount of data calculations. However, under the traditional von Neumann architecture, data inevitably has to be transferred between the memory and the computing unit. Statistics show that during the Llama2 deployment, the majority of latency and power consumption occurs in the data movement, resulting so-called "memory wall".

In response to this issue, computing-in-memory (CIM) emerges as a groundbreaking solution. It shatters the traditional architecture that separates storage and computing, integrating computation directly within memory to alleviate the bandwidth problem. The integration granularity varies from the package level to the device level. During this process, the differences among different models in the mapping process cannot be ignored. For example, transformer-based LLM focus on the storage density since the model only computes 1 input vector per weight load. in contrast, convolution-based CNNs are sensitive to compute density as they compute multiple input vectors per weight load. Thus, different AI algorithms require a heterogeneous architecture to address varying storage-to-computation ratios, from storage-intensive to computation-intensive.

3D stacking technology offers great convenience for the CIM architecture. It was originally introduced in HBM technology for 2.5D integration, aiming to meet the growth demands of weight scale and access bandwidth. The DRAM die used for storing weights and the logic die used for performing calculations are integrated together to provide a storage-intensive (GB+) memory with computing capability. The most advanced 16 - layer DRAM proposed by SK Hynix can provide more than 48 GB capacity and a bandwidth of 1.6TB/s [7]. Moreover, with the future Hybrid Bonding technology, the communication bandwidth can be further increased to 10TB/s, which will greatly reduce the time and power consumption required for LLM training and inference.

The CIM architecture can be applied to different types of memory, including SRAM, DRAM, RRAM, and FLASH. Among them, SRAM is widely used due to its mature process. For example, the 6T-SRAM cell is applied for fast memory access, where the classical 6T-cell is used for storage and two additional transistors perform binary analog or digital computing. The fully-digital domain CIM macros are compatible with advanced technology to further improve the energy efficiency of the processor.

Emerging NVM devices, such as RRAM, can always be used in the CIM macro. Due to its CMOS capability, RRAM-based macros are frequently used in AI accelerator. RRAM not only stores weight but also perform MAC through Ohm's Law, thus achieving high compute density. The latest research shows that RRAM CIM technology has evolved from array fabrication to full SoCs. The state-of-the-art RRAM-based AI accelerator has achieved an up-to-100× improvement in energy efficiency for CIM macros through cross-layer co-design of 2T1R cell, sparse-mapping computation and ADC circuits [8].

Conclusion

This paper summarizes the advancements in devices and architectures that have been proposed to meet the increasing demands for computing power, memory capacity and interconnect bandwidth in the AI era. Compute logic scaling is achieved by further shrinking devices after FinFET, or integrating more chiplets through advanced packaging. Both embedded memory and DRAM scaling faces significant challenges, requiring innovative device or technology to address power, speed, and capacity limitations. CIM technologies, including DRAM-logic 3D stacking systems and NVM based macros, offer promising solutions to solve bandwidth bottlenecks and improve computing efficiency.

Acknowledgments

This work was supported by the National Natural Science Foundation of China under Grant No. 62488101.

References

[1] Y. -J. Mii, "Semiconductor Industry Outlook and New Technology Frontiers", IEDM, pp.1-6 (2024).

[2] Y. -C. Huang et al., "15.7 A 32Mb RRAM in a 12nm FinFet Technology with a 0.0249 μm² Bit-Cell, a 3.2 GB/S Read Throughput, a 10 K Cycle Write Endurance and a 10-Year Retention at 105°C", ISSCC, pp.288-290 (2024).

[3] Y. wang et al., "A stable rhombohedral phase in ferroelectric Hf(Zr)$_{1-x}$O$_2$ capacitor with ultralow coercive field", Science, 381, pp. 558-563 (2023).

[4] J. Yang et al., "A 9 Mb HZO-Based Embedded FeRAM with 10^{12}-Cycle Endurance and 5/7ns Read/Write using ECC-Assisted Data Refresh and Offset-Canceled Sense Amplifier", ISSCC, pp.1-3 (2023).

[5] N. Ramaswamy et al., "NVDRAM: A 32Gb Dual Layer 3D Stacked Non-volatile Ferroelectric Memory with Near-DRAM Performance for Demanding AI Workloads", IEDM, pp.1-4 (2023).

[6] C. Chen et al., "Inter-Layer Dielectric Engineering for Monolithic Stacking 4F2-2 T0C DRAM with Channel-All-Around (CAA) IGZO FET to Achieve Good Reliability (>10^4 s Bias Stress, >10^{12} Cycles Endurance)", IEDM, pp.26.5.1-26.5.4 (2022).

[7] https://www.skhynix.com/

[8] W. Ye et al., "A 28-nm RRAM Computing-in-Memory Macro Using Weighted Hybrid 2T1R Cell Array and Reference Subtracting Sense Amplifier for AI Edge Inference", IEEE Journal of Solid-State Circuits, 58, pp. 2839-2850 (2023).

AUTHOR INDEX

Abraham, Nithin 1128
Agarwal, Harshit 793, 799, 802
Agarwal, Tarun 925
Aharonovich, Igor 998
Ahmad, Naef 426
Ahmed, Faisal 986
Ahn, Jong-Hyun 1225
Ahn, P.-H. 291
Ahn, S. J. 1189
Ahn, Yonghwan 652
Ai, Hao 1003
Akshay, K 540
Alant, Johan 1195
Alchalabi, Mustafa 1261
Amara, Selma 288
Amaram, Ashutosh Krishna 925
Amrouch, Hussam 1107
An, Jiayi 786
An, Xia 453, 471, 500
Anders, Jens 1086
Andia, Luis 1122
Ando, Koji 897
Andreev, Sergei 1261
Ang, Kah-Wee 257
Anjali, A. 1057
Anwar, Muhammad Abid 285, 491, 528, 868
Ao, Chengkang 369, 438
Ao, Mingrui 1059, 1062, 1065
Arabhavi, A. M. 7
Arakawa, Takaki 558
Ashraf, Syed 1222
Bae, J. H. 1189
Baek, Rock-Hyun 652, 691
Bagga, Navjeet 706, 796
Bahl, Sandeep 1246
Bahrami, Mina 1192
Bai, Jing 357
Bai, Mingkai 276
Bai, Rognxu 1252
Bai, Wubin 1198
Bao, Lin 841, 1237
Bao, Shengyu 485, 1237
Bao, Wenzhong 1059, 1062, 1065
Bao, X. 977
Bao, Yunjiao 336, 546
Barbot, Justine 399
Ben, Jianwei 360
Benini, Luca 1261

Benkhelifa, Mahdi 1107
Bernert, Kerstin 399
Bestelink, Eva 242
Bhat, Aditya K. 1057
Bhattacharjee, N. 28
Bi, Ran 500, 744
Bian, Zheng 552
Biswas, Anmol 965
Bodepudi, Srikrishna Chanakya 285, 491, 528, 868
Bolognesi, C. R. 7
Braun, Dennis 497
Brinson, Mike 1261
Bu, Saiyu 130
Bu, Weihai 474, 601
Bucher, Matthias 1261
Burgt, Yoeri Van De 759
Butler, Keith T. 895
Cai, Bozhou 503
Cai, Hao 366, 661
Cai, Hecheng 303
Cai, Jian 980
Cai, Jing 688
Cai, Kaiming 429
Cai, Puyang 459
Cai, Weiwei 986
Cai, Yimao 485, 841, 1021, 1024, 1237
Cao, Bingyang 175
Cao, Shengjie 934, 956, 1158
Cao, Xianmao 157
Cao, Yingnan 139
Cao, Zhenyuan 1092
Chai, Junshuai 34, 40, 103, 267, 276
Chai, Y. J. 214
Chai, Yang 172, 239, 333, 339, 396, 721, 783, 903
Chakarov, Ivan 181
Chan, Chak Lam Jonathan 531
Chan, Henry 1114
Chan, Kaiman 1195
Chan, Mansun 309, 414, 640, 989
Chan, Paddy K. L. 139
Chand, Rakesh 121
Chandra, Anirban 1114
Chang, Chin-Yu 121
Chang, Fu-Shen 771
Chang, Hao 236, 513
Chang, Hsiang-Hung 943
Chang, Jonathan 151
Chang, Yii-Tay 771

Chang, Yu-Ming 257
Charan, Vanjari Sai 1057
Charbon, Edoardo 838
Chatterjee, Payel 718
Chau, Yu Foon 482
Chauhan, Amit Kumar Singh 330
Chauhan, Yogesh Singh 462, 1080, 1107, 1137
Chen, Aobei 52
Chen, Bing 390, 1249
Chen, Chih 943
Chen, Chun-Zhang 82
Chen, Danyang 378, 962
Chen, Gang 865
Chen, H.-M. 900
Chen, Haibao 459
Chen, Hai-Bao 628
Chen, Han 130
Chen, Hao 411
Chen, Haojie 1059, 1062, 1065
Chen, Hongda 372
Chen, Hongzhen 372
Chen, Jen-Hao 55
Chen, Jiajia 1249
Chen, Jianjun 423
Chen, Jianlin 1119
Chen, Jie Wei 172
Chen, Jiewei 721
Chen, Jiezhi 85
Chen, Jingyang 679
Chen, Jun 937
Chen, Junting 300, 622
Chen, Kai 321
Chen, Kuan-Ting 121
Chen, Kun 817
Chen, Liang 956
Chen, Long 82
Chen, Nanbo 679
Chen, Peng 859, 1143
Chen, Qi 184
Chen, Qiang 780
Chen, Qingming 37
Chen, Rongsheng 22
Chen, Ruiqi 1003
Chen, Shuai 883
Chen, Shuaiyu 573
Chen, Sihao 823
Chen, Siyu 133, 205
Chen, Sujie 616
Chen, T.-Y. 151
Chen, Tiwei 64
Chen, Wanjun 124, 865
Chen, Weiliang 233

Chen, Xiaojin 912
Chen, Xin 411
Chen, Xin-Ru 10
Chen, Xuanqi 405
Chen, Y.-C. 880
Chen, Yance 285, 491
Chen, Yang 774
Chen, Yifan 471
Chen, Yitong 88, 260
Chen, Yiyang 1003
Chen, Yuefeng 441, 595
Chen, Yung-Hsiang 121
Chen, Zeqi 34, 103
Chen, Zequan 1057
Chen, Zerui 1045
Chen, Zheng 1027
Chen, Zhenyu 1030
Chen, Zhongming 109
Chen, Zhuoya 1024
Chen, Zimin 519
Chen, ... 522
Cheng, Che-Chi 771
Cheng, Fang 555
Cheng, Guoyao 309
Cheng, Jiangong 411
Cheng, Kai Wen 151
Cheng, Ran 390, 1249
Cheng, Shengliang 519
Cheng, Yu Jian 903
Cheng, Zengguang 408
Cheng, Zhe 494
Chhaperwal, Mayank 1128
Chi, X. 1213
Chi, Yaqing 423
Chih, Yu-Der 151
Chiu, C.-Y. 886
Chiu, Wei-Lan 943
Cho, Byung Jin 1134
Cho, D. G. 1189
Cho, Kyeongrae 652
Cho, M. H. 1189
Choi, Shinhyun 1117
Chong, Chen 76
Chowbhury, S. 977
Chu, Wen-Ting 151
Chu, Yanbang 601, 747
Chun, Suk Yeop 1186
Chung, Jerry 151
Chunxia, Li 226
Ciabattini, F. 7
Cong, Yuqing 1101
Cruces, Sofia 1279

Cruces, Sofía 497
Cui, Boyao 459, 634, 1048
Cui, Jiawei 236, 513
Cui, Jie 393
Cui, Peng 604
Cui, Tianning 962
Cui, Xiaoqi 986
Cui, Yan 429
Cui, Yanxia 522
Cui, Zhi-Li 694
Cunman, Liang 217
Cüppers, Felix 832
Dai, Jie-Ni 820
Dai, Saifei 267
Dai, Shun-Qi 643
Dai, Xianqi 64
Dai, Xinyue 208
Dai, Yue 285
Dai, Zhongbin 384
Dan, Yaping 534
Das, Saptarshi 97
Das, Susobhan 986
Das, Tanay 426
Dasgupta, S. 706, 796
Datsuk, Anton 1261
Daus, Alwin 497, 1131
Davoudi, Mohammad Rasool 1192
Dayte, I. 977
De, Sourav 1173
Deen, M. Jamal 4
Deng, Chenkai 297
Deng, L. 479
Deng, Meng 163
Deng, Minyue 956
Deshmukh, Shreyas 564, 965
Deshpande, Veeresh 564
Devi, Reshma 895
Dijken, Sebastiaan Van 1222
Ding, Chenchen 543
Ding, Guanglong 348
Ding, Jiaxin 1042, 1264
Ding, Mingzheng 546
Ding, Rongzheng 619, 631
Ding, Shenlei 387
Ding, Xiaolei 285, 491
Ding, Yajing 103
Ding, Yi 847
Ding, Yian 390
Ding, Yichong 871
Dinh, T. V. 1207
Divyanshu, ... 288
Dixit, Ankit 796

Dong, Daoyi 727
Dong, Junchen 229
Dong, Xiangqi 1059, 1062, 1065
Dong, Yulong 378, 962
Đonko, Dženana 670
Dou, Xiaoyu 85
Drescher, M. 28
Du, Fangzhou 58, 67, 1243
Du, Fengyu 928
Du, Hanghai 912
Du, Haoran 661
Du, Peiran 282
Du, Peiyuan 390
Du, Pengbo 1027
Du, Yiwei 70
Du, Yu 561
Du, Zhiyuan 139
Duan, Jiahui 267
Dutta, Aloke K. 462, 1137
Ebrahimi, M. 7
Elrifai, Tarek 1086
Enz, Christian 838
Eom, Seungjoon 691
Ervin, Joseph 181
Esseni, David 829
Estrada, Cristine Jin 989
Fa, Yuan 1279
Fan, Shi Quan 342
Fan, Xuemeng 73, 127
Fan, Yijia 70
Fan, Yutong 312
Fan, Yuyan 378
Fan, Zhengfang 534
Fan, Zhiyong 465, 531
Fang, Chi 595
Fang, Cize 906
Fang, Haotian 196
Fang, Ryan 1195
Fang, Shengli 342
Fang, Tong 732
Fang, Wencheng 715
Fang, Wenzhang 285, 491
Fang, Wuqing 196
Fang, Yilong 549
Fariborzi, Hossein 288
Fatin, Mohammad Ajmain 756
Feng, Dahu 561, 579
Feng, Ning 835, 1036
Feng, Peng 679
Feng, Qian 345
Feng, Yulin 1003
Feng, ... 16

Figuet, C.	1216	Guo, D. Y.	664
Florica, Camelia	288	Guo, Dengyao	928
Fraser, Lachlan	1155	Guo, Fuqiang	573, 786
Fu, Chun	1243	Guo, Gaofu	64
Fu, Jianghao	193	Guo, Guoping	441, 595
Fu, Jun	393	Guo, Haowen	919
Fu, Lan	727	Guo, Hengxu	441, 595
Fu, Sulei	765, 862	Guo, J. K.	664
Fu, Xingyu	573	Guo, Jian-Bin	10
Fu, Yunyi	46, 1024	Guo, Jianmiao	333
Fu, Yu-Yang	682	Guo, Jingkai	928
Fu, Zhiyuan	805, 934, 1045, 1158	Guo, Junwei	459
Fujiwara, M.	558	Guo, Min	613
Fujiwara, Masazumi	115	Guo, Qijun	381
Galderisi, G.	28	Guo, Qinan	124
Gan, Jinghan	1167	Guo, Qinhua	1018
Gan, Lin	1219	Guo, Rui	747
Gan, Xuetao	477	Guo, Taoming	637, 953
Gan, Zhouchao	555, 947	Guo, X.	479
Ganguly, Udayan	564, 965	Guo, Xiaojun	616, 1152
Gao, Bin	70, 133, 205, 506, 729, 847	Guo, Xiaokun	435
Gao, Chao	762, 811, 937	Guo, Xinrui	100, 270, 468
Gao, Dawei	19, 127, 1149	Guo, Yeye	983, 1012
Gao, Gc	52	Guo, Zhanfeng	735
Gao, Guoyun	139	Guo, Zihan	1021
Gao, Jiajun	765, 862	Gupta, Manish	892
Gao, Jianfeng	429, 753	Gurkaynak, Frank K.	1261
Gao, Ju	369, 438	Ha, Daewon	223, 1189
Gao, Wenze	582	Hajare, Hemant	564, 965
Gao, Xinguo	573	Hamzeloui, S.	7
Gao, Yi	841	Han, Chuan Yu	342
Gao, Yufeng	1167	Han, Cong	992
Geng, Di	874	Han, Dedong	229
Geng, Li	342	Han, Genquan	390, 906, 1249
Georgiev, Vihar	796	Han, Hongyan	474
Geum, Dae-Myeong	1039	Han, Hung-Chi	838
Geum, Daemyeong	1134	Han, Jiachen	411
Goel, Sanket	187	Han, Kaizhen	405
Gong, Heng Yue	625	Han, Pengfei	303
Gong, Xiao	405, 850	Han, Runhao	34, 103
Gopalakrishnan, Sai Gautam	895	Han, Songjia	697
Goshima, D.	712	Han, Su-Ting	303, 348
Goshima, Daiki	598	Han, Yanjun	1219
Gou, Saifei	1059, 1062, 1065	Hannula, Sebastian	1222
Grabinski, Wladek	1261	Hao, Hongwei	522
Grasser, Tibor	1192, 1282	Hao, Yue	312, 345, 372, 390, 444, 447, 585, 667, 871, 906, 912
Grenouillet, Laurent	516	
Grothe, M.	28	Hao, Zhibiao	1219
Gu, Chen	874	Hara, A.	712
Gu, Denshun	768	Hara, Akito	598
Gu, Jiangmin	387	Hasan, Tawfique	986
Guan, Shan	1089	Hashimoto, S.	315

Havel, V. 28
He, Gufeng 1033
He, Jiaheng 494
He, Jiayin 369, 438
He, Jin 528
He, Ming 100, 270, 468, 1071
He, Minghao 82
He, Nan 883
He, Shikun 381, 429
He, Xi 1119
He, Xiaohan 372, 447
He, Xinliu 1059, 1062, 1065
He, Y. 28
He, Yang 853
He, Yaoyu 85
He, Yongjie 1252
He, Youshui 868
He, Yuhui 184, 339, 396
He, Zhaolong 321
He, Zong Rui 903
Herfurth, Norbert 1261
Herman, Krzysztof 1261
Hikake, Kaito 1255
Hiramoto, Toshiro 1255
Ho, Yenshih 121
Hoentschel, J. 28
Hoffmann-Eifert, Susanne 832
Hong, M. 1189
Hong, Peiyan 417
Hong, S.-M. 291
Horiike, Ryota 31
Hossain, Mainul 756
Hou, Bin 444, 582, 585
Hou, Jiahao 306
Hou, Yilin 85
Hsiang, Kuo-Yu 771
Hsieh, E Ray 151
Hsieh, Y.-S. 151
Hsiung, M.-H. 880
Hsu, J.-F. 977
Hu, C. 61
Hu, Chenming 55
Hu, Chun 52
Hu, Guohua 655
Hu, Haodong 906
Hu, Hengyi 220
Hu, Huan 285, 491
Hu, K.-K. 977
Hu, Qianlan 1077
Hu, Runyu 408
Hu, Ruofei 133, 205
Hu, Tao 34, 103, 276

Hu, Tong 709
Hu, V. P.-H. 886
Hu, Vita Pi-Ho 892
Hu, Weida 959
Hu, Yan 1059, 1062, 1065
Hu, Yilin 1006
Hu, Youfan 567
Hu, Yuqing 312
Hu, Zheyuan 64
Hu, Ziyang 679
Hua, Mengyuan 300, 622
Hua, Qilin 109
Huang, Feixuan 1119
Huang, Haoxin 112
Huang, Heyi 1293
Huang, Jing-Kai 94, 257
Huang, Peng 1003
Huang, Qi 88, 199, 260
Huang, Qianqian 279, 474, 805, 934, 956, 1045, 1158
Huang, Ru 91, 100, 254, 270, 273, 279, 309, 453,
......459, 468, 471, 474, 485, 500, 525, 601, 628, 634, 676, 744,
747, 805, 823, 835, 841, 934, 956, 1006, 1021, 1024, 1036,
1045, 1048, 1051, 1071, 1158, 1161, 1237
Huang, S.-T. 900
Huang, Sen 573, 604, 786
Huang, Shan 16
Huang, Tengyan 43
Huang, Tzu-Yun 118
Huang, Wei 22, 814
Huang, Xiao-Di 703
Huang, Xingyu 1255
Huang, Yi 613
Huang, Yilong 133, 205
Hui, Fei 1009
Huo, Wenju 1143
Huo, Yikang 106
Hwang, Jason C. C. 94
Hyun, S. 1189
Illarionov, Yu. Yu. 214
Im, Changhyeok 166
Ishikawa, R. 1213
Islam, Muhammad Saif Ul 1155
Ito, Y. 712
Iwata, J. 315
Jain, Khushi 796
James, Jerry Joseph 658
Jang, G.-T. 291
Jang, M.-S. 291
Jansen, S. 28
Jena, Swadhin Kumar 540
Jeong, Jaejoong 1134
Jeong, Jaeyong 1039, 1134

Jhang, J.-H.	977	Kaneko, M.	1213
Ji, Botao	260	Kang, Bryan	688
Ji, Haotian	387	Kang, J. F.	1095
Ji, Ning	835	Kang, Jiachen	1033
Ji, Zhenghui	381	Kang, Jin	601
Ji, Zhigang	91, 459, 628, 634, 673, 1006, 1048	Kang, Jinfeng	354, 1003
Ji, Zhongchen	786	Kang, Junzhe	19
Jia, Mao	444	Kang, Kaixiang	163
Jia, Rundong	474	Kang, Song	937
Jia, Xiaole	906	Kang, Yuye	405
Jia, Xinpei J	103	Kang, Zhuodong	729
Jia, Xinpei	34, 267, 276	Kar, Anirban	1107
Jia, Yueyang	750, 856	Karasawa, Hajime	31
Jia, Yuping	360	Karl, Alexander	1192
Jian, Ming	685	Kasperovich, A.	977
Jiang, Biyi	49	Kaufmann, Maik Peter	1246
Jiang, Bowei	811	Ke, Mengnan	390
Jiang, Chen	233, 637, 953	Ke, Xiaoyu	40, 267
Jiang, Chunsheng	106	Kern, Michal	1086
Jiang, Hao	139	Keskin, Batuhan	838
Jiang, Haodong	916, 919	Khakbaz, Pedram	1192
Jiang, Huaxing	519	Khan, Asif	1276
Jiang, Jianbo	324	Khan, Imtiyaz Ahmad	330
Jiang, Jianhua	432	Ki, Dong-Kcun	139, 257
Jiang, Jianpeng	688	Kim, Bong Ho	1039
Jiang, Jingru	676	Kim, Bongho	1134
Jiang, Jinxia	607	Kim, Cheol-Joo	1170
Jiang, Ke	360, 732	Kim, Choul-Young	1039
Jiang, Mingrui	139	Kim, I. K.	291
Jiang, Pinfeng	549	Kim, J.	977
Jiang, Qimeng	208, 573, 786	Kim, Ji Eun	1186
Jiang, Renjie	537	Kim, Jongmin	1039
Jiang, Songyi	753	Kim, Joon Pyo	1039
Jiang, Wei	4	Kim, Joonpyo	1134
Jiang, Wenfeng	874	Kim, Kihyun	1039
Jiang, Xixi	172	Kim, Minchan	652
Jiang, Yang	58, 67	Kim, Mingyu	223
Jiang, Yi	321	Kim, Sanghyeon	1039, 1134
Jiang, Yuelin	133, 205	Kim, Seong Kwang	1039
Jiang, Yunzhou	387	Kim, Seongkwang	1134
Jiang, Zhou	169	Kim, Sunghun	1255
Jin, Chengji	390, 1249	Kim, Sungju	166
Jin, Minhyun	588	Kim, Younghyun	1134
Jin, Xin	762, 811	Kimoto, T.	1213
Jin, Yufeng	576	Kinjo, K.	558
Joh, Jungwoo	1246	Knobloch, Theresia	1192, 1282
Jumabekov, Askhat N.	1270	Kobayashi, Masaharu	1255
Jung, S.-W.	291	Kondo, T.	315
Kadam, Abhishek	564, 965	Kong, Shuai	1180
Kado, M.	315	Kopperberg, Nils	456
Kamada, F.	558	Kotsugi, Masato	1273
Kämpfe, Thomas	399	Krivic, Senka	670

Kuang, Renfei 37
Kuang, Zhipeng 729
Kuball, Martin 1057
Kuh, B. J. 1189
Kumar, A. 28
Kumar, Harshvardhan 1176
Kumar, Naveen 796
Kumar, Sandeep 706, 796
Kumar, Shivam 465
Kunal, ... 330
Kurihara, A. 712
Kurihara, Akito 598
Kwok, Hoi Sing 160
Laber, Andreas 670
Lai, Yuru 519
Lam, Sang 646
Lan, Gongpeng 196
Lao, Yunhong 236
Lashkare, Sandip 426, 570, 965
Lee, Chan Jik 1039
Lee, Dong Seup 1246
Lee, Jia-Yang 771
Lee, Jimin 497, 1279
Lee, Jiwon 588
Lee, Jongwon 1039
Lee, Jooseok 1039
Lee, Jooyoung 25
Lee, Junjong 652
Lee, K. 1189
Lee, K.-W. 291
Lee, Kyunghwan 223
Lee, Min-Hung 771
Lee, S. M. 1189
Lee, S. 1189
Lee, Sanguk 652, 691
Lee, Seunghwan 652
Lee, T. H 1189
Lee, Won-Chul 1039
Lee, Wonsok 1189
Lee, Y. 1189
Lee, Zhi-Qiang 121
Lehninger, David 399
Lei, Yuan 643
Lemme, Max C. 497, 1279
Leng, Yan-Bing 303
Leroux, Nathan 909
Li, Ang 607
Li, Bingbai 510
Li, Bochang 906
Li, Can 139
Li, Chenxi 519
Li, Chenyang 257

Li, Chunyang 688
Li, Dabing 360, 732, 774
Li, Dan 360
Li, Dapeng 52
Li, Dingwei 13, 88, 260
Li, Dongdong 522, 1030, 1180
Li, Dongya 390
Li, Fanfan 13, 260
Li, Guohui 522
Li, Haixia 453, 471, 744
Li, Haiyang 208
Li, Hang 348
Li, Hao 525, 835
Li, Haolin 877
Li, Hongsheng 1234
Li, Hongtao 1219
Li, Hongyi 561
Li, J. 479
Li, Ji 510
Li, Jia Cheng 625
Li, Jian-Cong. 682, 703
Li, Jiancong 709
Li, Jiaqi 163
Li, Jic 768
Li, Jilin 160
Li, Jinghua 1074
Li, Jinxian 408
Li, Jiye 136
Li, Juan 282
Li, Junfeng 777
Li, Junjie 753
Li, Junkang 321
Li, K. 28
Li, Lain-Jong 139, 257
Li, Lianlian 537
Li, Ling 874
Li, Lingqi 257
Li, Mengdi 871
Li, Mengjie 555
Li, Ming 22, 453, 471, 500, 525, 601, 676, 744, 747
Li, Minghua 741
Li, Ming-Huang 118, 121
Li, Muchan 46
Li, Mujun 82
Li, Peihong 859
Li, Pengtao 73
Li, Qingxiu 381
Li, Qinkun 537
Li, S. Z. 1095
Li, S. 479
Li, Sean 94, 257
Li, Sheng 402

Li, Shiming	372, 447, 585
Li, Shiyu	983, 1012
Li, Shuang	700
Li, Siying	616
Li, Taige	1015
Li, Tiaoyang	844
Li, Tiefu	649
Li, Xi	715, 1231
Li, Xiangdong	912
Li, Xianming	937
Li, Xiao	43
Li, Xiaopeng	85
Li, Xin	342
Li, Xiuyan	378, 962
Li, Xuefei	417
Li, Xueqing	233
Li, Xufan	874
Li, Xun	768
Li, Xunyu	1288
Li, Yan	294, 375
Li, Yang	193
Li, Yanqin	1293
Li, Yanru	429
Li, Yi	220, 682, 694, 703, 709
Li, Yida	202
Li, Yongliang	294, 375
Li, Yu	1036
Li, Yuanbiao	1068, 1140
Li, Yuanzhe	169
Li, Yuchun	634
Li, Yun	94, 971
Li, Yunlong	321, 1149
Li, Zhe	727
Li, Zhengyue	853
Li, Zhenyuan	912
Li, Zhiyuan	248
Li, Zhonghui	1285
Li, Zhucheng	64
Li, Zhuo	1255
Li, Zi	883
Li, Zichun	16
Li, Zongwen	285, 491, 528, 868
Liang, Bin	423
Liang, Caijing	950
Liang, Chao	510
Liang, Cunman	169
Liang, Jiaqiao	245
Liang, Jing	688
Liang, Lei	318
Liang, Li-Wei	826
Liang, Renrong	423
Liang, Shijie	37

Liang, Tian	1264
Liang, Xiaoci	239, 488, 613
Liang, Xifa	37
Liang, Zhengyu	324
Liang, Zhongqi	576
Liang, Zhongxin	254
Liao, Chun-Yu	771
Liao, Hongxu	453, 744
Liao, Lihang	1119
Liao, Meiyong	145
Liao, Min	40, 267
Liao, S. Sandy	977
Liapis, Andreas C.	986
Likhitkar, Praful	570
Lim, Hyeongrak	1134
Lim, Jeong-Taek	1039
Lin, Cheng-Chien	121
Lin, Ching-Ju	151
Lin, Haobo	459
Lin, Huai-En	943
Lin, Huamao	741
Lin, J.-P.	977
Lin, Peng	859, 1143
Lin, T.-Y.	886
Lin, Xi	94
Lin, Yibo	601
Lin, Yu-Cheng	151
Lin, Yudeng	729
Lin, Yu-Hsien	151
Lin, Yuxin	1045
Lin, Zhaojun	604
Lin, Zhiyu	154, 178, 378, 1030
Lin, Ziyuan	333
Ling, Haifeng	814
Lipsanen, Harri	986
Liu, Bo	661
Liu, Bowen	637
Liu, Can	306
Liu, Chao	604
Liu, Chen	741
Liu, Cheng-Hong	771
Liu, Chenyu	906
Liu, Chuan	239, 488, 613, 697
Liu, Chunsen	1092
Liu, Dingyao	357
Liu, Enlong	381
Liu, Fangze	453
Liu, Fanyu	375
Liu, Fei	46
Liu, Fucai	318
Liu, Guolei	13, 260
Liu, Guozhu	79

Liu, Haoyan	294
Liu, Hongchao	1240
Liu, Hongxia	76
Liu, Hsien-Yang	516
Liu, Huan	390
Liu, Huanan	847
Liu, Jiacheng	889
Liu, Jian	257, 679
Liu, Jinbiao	546
Liu, Jingquan	378, 962
Liu, Jinhao	459, 634, 1048
Liu, Jun	193
Liu, Junling	100, 270, 468, 1071
Liu, Kai	130
Liu, Kaimeng	133, 205
Liu, Lifeng	229, 1003
Liu, Liwei	263
Liu, Liyuan	679
Liu, Meijie	79
Liu, Ming	1299
Liu, Mingrui	360
Liu, Mingxu	688
Liu, Peisen	765, 862
Liu, Pcisong	1009
Liu, Peng	741
Liu, Pengyu	655
Liu, Qi	937
Liu, Qianqian	753, 777
Liu, Qihan	139
Liu, Qihui	411
Liu, Qing	318
Liu, Shangming	1015
Liu, Shanjing	282
Liu, Shuang	546, 1293
Liu, Shuiren	1009
Liu, Shuo	100, 270, 468, 1071
Liu, Si Rui	903
Liu, Sihang	236
Liu, Siyang	402
Liu, Siyuan	327
Liu, Weihua	342
Liu, Wen	387
Liu, Wenjun	983, 1012
Liu, Wenyuan	1054, 1164
Liu, X. Y.	1095
Liu, Xiaohua	832
Liu, Xiaolin	762, 811
Liu, Xiaosen	236
Liu, Xiaoyan	354, 420, 877, 931, 1003
Liu, Xin	1009
Liu, Xinghui	780
Liu, Xinyu	786

Liu, Xiwen	414, 1018
Liu, Xuewen	1234
Liu, Yan	390, 700, 906, 1249
Liu, Yang	1089
Liu, Yanming	148, 685
Liu, Yanyan	393
Liu, Yi	251, 992
Liu, Yibo	16
Liu, Yichen	411
Liu, Yichun	142
Liu, Yinchi	983, 1012
Liu, Yizhan	420
Liu, Yu	747
Liu, Yuan	610
Liu, Yumeng	534
Liu, Yuzhuo	856
Liu, Zhaojun	16
Liu, Zhengwu	543
Liu, Zhihong	312, 912
Liu, Ziheng	369, 438
Liu, Zuheng	790
Lizzit, Daniel	829
Lone, Aijaz H.	288
Long, Shibing	1068, 1140
Long, Yanxi	685
Long, Yinfeng	130
Long, Zhenghao	465, 531
Lotfi, Hadi	1086
Lou, Qiang	724
Lovell, Nigel	1155
Low, Kain Lu	387
Lu, Albert	482
Lu, Di	1068, 1140
Lu, Donglin	610
Lu, Hao	585
Lu, Haoran	601, 676, 747
Lu, Honghao	1243
Lu, Hongliang	634
Lu, Lei	43, 136, 160
Lu, Xing	519
Lu, Xun	983, 1012
Lu, Yipeng	1167
Lu, Yupeng	336
Lu, Yuyao	729
Luo, Chao	441, 595
Luo, Deng	423
Luo, Huaizhi	375
Luo, Jie	336
Luo, Jin	805, 1045
Luo, Jun	267, 429, 753, 777, 1293
Luo, Jun-Wei	1089
Luo, Peiwen	591

Luo, Wenpu	1161
Luo, Xiao	318
Luo, Xin	411
Luo, Yi	1219
Luo, Yiyang	488
Luo, Zhao-Feng	771
Lv, Bingchen	774
Lv, Dan	1030, 1180
Lv, Shunpeng	360
Lv, Y. Z.	214
Lv, Yuanjie	604
Lv, Yudong	619, 631
Lv, Zhong-Peng	1222
Lv, Ziyu	303
Lyu, Haiyuan	327
Ma, Awang	729
Ma, Chen	1119
Ma, Chengxiang	133, 205
Ma, Hanbin	637
Ma, Ruiqi	1180
Ma, Xiaohua	372, 444, 447, 585, 871
Ma, Yanfeng	402
Ma, Yiming	1119
Ma, Yuan Xiao	625
Ma, Yuan	528, 868
Ma, Yunfei	124
Ma, Zichao	989
Ma, Zihao	1104
Ma, Zongmin	193
Machhiwar, Yogendra	799, 802
Maeki, M.	558
Mahapatra, Souvik	718
Maheshwari, Navin	570
Mähne, Hannes	399
Mai, Zhancheng	363, 780
Majumdar, Kausik	1128
Malakoutian, M.	977
Malik, Muhammad	285, 491, 868
Manea, Paul-Philipp	909
Manhas, Sanjeev Kumar	330
Manna, Sukriti	1114
Manzoor, Nisha	462, 1137
Manzoor, Wajid	462, 1137
Mao, Jiayi	199
Mao, Peiyu	912
Mao, Ruibin	139
Martinie, Sebastien	1261
Matsui, Chihiro	897, 940
Mayer, Joachim	497
Mei, Zikang	1152
Mei, Ziqi	510
Memon, Muhammad Hunain	1146

Meng, Lingxian	1009
Menon, P. Susthitha	1176
Menzel, Stephan	456
Merkin, Tim	1195, 1246
Metze, C.	28
Mevic, Amina	670
Mi, Changxin	494
Miao, Feng	1291
Miao, Jialei	552
Miao, Songming	1068, 1140
Miao, Xiang-Shui	184, 682, 694, 703
Miao, Xiangshui	220, 248, 339, 396, 549, 555, 709,
	947
Miao, Yechen	351
Mikami, K.	1213
Mikolajick, T.	28
Misawa, Naoko	897, 940
Miyamura, Yoshiji	503
Miyano, S.	315
Mizutani, Tomoko	1255
Mohapatra, Nihar Ranjan	965
Moon, K. J.	1189
Morroni, Jeffrey	1246
Mou, Xing	729
Nagahashi, Tomoya	31
Nandi, Debashish	462, 1137
Nazir, Mohammad Sajid	1080
Neve, Cesar Roda	1288
Ng, Annie	1270
Ni, Siyuan	324
Ni, Zhao	217
Nielinger, Dennis	832
Nigmatulin, Fedor	986
Nigmetova, Gaukhar	1270
Ning, Jing	667
Nishizawa, Shin-Ichi	503
Nomura, M.	977
Nugraha, Ferris Prima	889
Ogier, S.	479
Oh, Saeroonter	1
Oh, T.	291
Ootera, Y.	315
Orio, Roberto	1267
Oshimi, K.	558
Oshiyama, Atsushi	31
Ostinelli, O.	7
Ota, Takashi	897
Ou, Xin	853, 916, 919, 1042, 1264, 1285
Ouyang, Bangsen	783
Pachkawade, Vinayak	1261
Padhee, Chiranjibi	540
Padiyal, Adil	940

Padovani, Andrea	516
Pahwa, Girish	802
Pai, C.-S.	900
Pampori, Ahtisham	55, 1080
Pan, Feng	765, 862
Pan, Gang	859, 1143
Pan, Wen	1219
Pan, Zhoujie	148
Pan, Zijian	528
Park, Jaehyun	223
Park, K.	1189
Park, Minsik	1039
Park, S. W.	1189
Park, Sungil	223
Park, Youngkeun	1134
Parkhomenko, Hryhorii	1270
Patil, Deven H.	706, 796
Patil, Shubham	564, 965
Pei, Huai-Zhi	703
Pei, Mengjiao	971
Pei, Yanli	488, 519
Peller, Markus	399
Peng, Baokang	309, 1051
Peng, Bin	591
Peng, Bo	1222
Peng, Hao	136
Peng, Hongjie	369, 438
Peng, Pei	46
Peng, Wanyue	601, 676, 747
Peng, Xiao	411
Pey, Kin Leong	516
Pham, Tri T.	1270
Phùng, Qu?nh Th?	1131
Pijper, R. M. T.	1207
Plakhotnik, Taras	1000
Poddar, Swapnadeep	465
Pomeroy, James	1057
Pop, Eric	1131
Pradeep, Yelehanka Ramachandramurthy	121
Pretl, Harald	1261
Puglisi, Francesco Maria	516
Pun, Kong-Pang	655
Qi, Dianyu	1149
Qi, Yihong	811
Qi, Yixin	169
Qian, Haoji	1249
Qian, He	70, 133, 205, 729, 847
Qian, Jiang	1180
Qian, Xuanyu	4
Qian, You	741
Qian, Yuchen	871
Qiao, Bing	1285

Qin, Haiming	251
Qin, Huajun	1258
Qin, Lingjie	871
Qin, Nan	1054, 1164
Qin, Peijun	1155
Qin, Yuwen	1104
Qiu, Guoxiu	381
Qiu, Jiajun	835
Qiu, Liwen	724
Qiu, Shirong	169
Qiu, Xiao	531
Qu, Hanbin	1027
Qu, Junle	263
Qu, Yuwei	85
Qu, Zhenyu	916, 919
Quan, Dechang	387
Quinsat, M.	315
Radhakrishna, Ujwal	1195
Radu, Ionut	1288
Raghavan, Nagarajan	516, 658
Rahimi, Daniel N.	288
Rai, S.	28
Raja, Danish	799, 802
Raju, Harsh	330
Ramya, K	187
Ran, Ke	497
Rathore, Sunil	706, 796
Ren, Huihui	88, 260
Ren, Liming	46
Ren, Pengpeng	91, 459, 673, 1006, 1048
Ren, Qinghua	1119
Ren, Sheng-Guang	682, 694
Ren, Tian-Ling	735, 826
Ren, Xuanhui	193
Ren, Yuan	543
Ren, Zhipeng	1027
Ren, Zhongyang	46
Rheem, Nahyun	1039
Robert, Isaac Emanuel	399
Rossi, Chiara	829
Ruma, S R	892
Ruttloff, K.	28
Saha, Arpan	756
Saito, Wataru	503
Saitoh, M.	315
Sakai, Kota	1255
Sakhuja, Jayatika	965
Salahuddin, S.	61
Salahuddin, Sayeef	55
Sandner, Christoph	1261
Sang, Pengpeng	85
Sankaranarayanan, Subramanian K. R. S.	1114

Sanpui, Dipayan	1114
Sara, M.	558
Saraya, Takuya	1255
Sattari-Esfahlan, Seyed Mehdi	1192
Satyam, Parlapalli Venkata	540
Scholten, A. J.	1207
Scholz, Rene	1261
Seidel, A.-S.	28
Seidel, Konrad	399
Seong, Minkyoung	1039
Sessi, V.	28
Setti, Gianluca	288
Shafi, Abde Mayeen	986
Shahriar, Md.	756
Shakir, Mohd.	796
Shan, Linbo	485, 1021, 1024
Shang, Jieya	1059, 1062, 1065
Shang, Zongwei	471, 525
Shao, Feng	688
Shao, Hanyong	1161
Shao, He	814
Shao, Qiming	889, 1228
Shao, Rui	405
Shao, Xianzhou	267
Shayoub, Mir Mohammad	1080
Shen, Bo	236, 513, 573
Shen, Bowen	429, 847
Shen, Guozhen	109
Shen, Jiabin	408
Shen, Minliang	856
Shen, Rui	1285
Sheng, Bowen	1167
Sheng, Chuming	1065
Sheng, Haoyu	595
Shi, Chunzhou	582
Shi, Luping	1101
Shi, Mingcheng	70, 847
Shi, Mingmin	471, 500
Shi, R.	479
Shi, Ruipeng	429
Shi, Runxiao	160
Shi, Yuanyuan	196
Shi, Yunfei	777
Shi, Zhiming	360
Shibata, Koki	897
Shih, C.-C.	977
Shim, Joonsup	1039
Shimomura, N.	315
Shin, Hyungcheol	25, 166, 223
Shiraishi, Kenji	31
Shuyan, Zhu	226
Si, Mengwei	154, 178, 378, 962, 1030
Si, Yuying	616
Simon, M.	28
Sin, Stanislav	1
Singh, Ajay Kumar	965
Singh, Harshita	793
Singhal, Anant	793
Sinha, Advaita	1176
Slesazeck, S.	28
Smith, Matthew D.	1057
Soh, Keunho	1186
Soman, R.	977
Somappa, Laxmeesha	426
Sonawane, Jay	564
Song, Changming	980
Song, Cheng	765, 862
Song, J.	1189
Song, Lekai	655
Song, Peixuan	282
Song, Q. W.	664
Song, Qingwen	928
Song, X. J.	1095
Song, Xujin	354
Song, Yichen	1296
Song, Yixian	19
Song, Yufei	1059, 1062, 1065
Song, Zhitang	715, 1231, 1296
Sporea, Radu A.	242
Stampfer, Bernhard	1267
Strachan, John Paul	909
Su, C.-J.	886
Su, Chang	956
Su, Dongyue	239
Su, Huan-Hsiang	151
Su, Jiaqi	94
Su, P.	886
Su, Pin	820
Su, Qiyi	780
Su, Xiangwei	950
Su, Yahui	282
Su, Yanbo	70, 847
Su, Yongquan	411
Su, Zhijuan	534
Su, Zijia	450
Su, Zi-Jia	826
Suh, Chang Soo	1246
Suh, Yoon-Je	1039
Sui, Nianzi	163
Sui, Zhiyuan	79
Sul, Woo-Suk	1039
Sun, Changzheng	1219
Sun, D. J.	1095
Sun, Dijiang	354

Sun, Dongdong	196	Tang, Jinjin	622
Sun, Haiding	450, 1146	Tang, Jinpu	1219
Sun, Hanhan	423	Tang, Jinyao	139
Sun, Haolun	447	Tang, Kechao	254, 273, 1161
Sun, Huarui	853	Tang, Meng	288
Sun, Jiacheng	601, 676, 747	Tang, Nan	1003
Sun, Jiahao	750	Tang, Ning	573
Sun, Jialiang	1042, 1264	Tang, W.	479
Sun, Jia-Yi	694	Tang, Wei	248, 1033
Sun, Jingwei	1237	Tang, X. Y.	664
Sun, Ke	351	Tang, Xiaoyan	928
Sun, L. J.	664	Tang, Xinyi	58, 67, 1068, 1140, 1243
Sun, Lejia	928	Tang, Xiyuan	1024
Sun, Lifei	181	Tang, Y.-T.	880, 900
Sun, Linfeng	1098	Tang, Yanbo	619, 631
Sun, Liuyang	324	Tang, Yingjie	88, 260
Sun, Longyu	294	Tang, Younian	239
Sun, Maojun	724	Tang, Zhidong	70
Sun, Mingyang	753, 777	Tang, Zhikai	1246
Sun, Qicheng	1059, 1062, 1065	Tao, Lu-Qi	826
Sun, Qingqing	172, 1252	Tao, Quanyao	610
Sun, Ruize	124, 865	Tao, Ran	1149
Sun, Weifeng	402	Tao, Tiger H.	324, 351, 1054, 1164
Sun, Wen	847	Tao, Yaoyu	1231
Sun, Xiaojuan	360, 732, 774	Teng, Qiao	19
Sun, Xiaoqing	34, 40, 103, 267, 276	Teng, Zilin	432
Sun, Yi	351	Terai, M.	1189
Sun, Yiyuan	850	Tewari, Mohit	925
Sun, Yuhua	64	Thakor, Karansingh	718
Sun, Zhengjie	1062	Thiem, Steffen	399
Sun, Zhengzong	1065	Tian, Feng	528, 868
Sun, Zhibo	531	Tian, Fengbin	267
Sun, Zhipei	477, 986	Tian, He	148, 685, 735
Sünbül, Ayse	399	Tian, Jiaojiao	46
Sung, C.	1189	Tian, Jing	703
Suzuki, Madoka	1083	Tian, Qiaoling	142
Svejda, Jan Taro	1261	Tian, Ruijuan	477
Szedmak, Sandor	670	Tian, Xinyu	357
Tachiki, K.	1213	Tian, Yuchen	1059, 1062, 1065
Takahashi, Takanori	1255	Tilegen, Meruyert	1270
Takeuchi, Ken	897, 940	Toda, Masaya	211
Tan, Chaoliang	112	Tokeshi, M.	558
Tan, Hongwei	1222	Tokuhira, H.	315
Tan, Lin	980	Tong, Anyu	1077
Tan, Ping-Heng	1183	Tong, Hao	184
Tan, Tiang Teck	516	Tong, Yi	251, 883, 992
Tan, Xiaohong	576	Torraca, Paolo La	516
Tan, Xiaojun	1062	Tripathi, Karunesh Kumar	799, 802
Tang, Chuying	1243	Trommer, J.	28
Tang, Huansong	49	Tsai, C.-Y.	880
Tang, Huawei	619, 631	Tsai, David	1155
Tang, Jianshi	70, 133, 205, 729, 847	Tsai, Y.-T.	880, 900

Tseng, Jen-Chou 151
Tu, Quanyi 414
Tung, C. T. 61
Tung, Chien-Ting 55
Uddin, Md Gius 986
Ueda, Y. 315
Uenuma, Mutsunori 1255
Umetsu, N. 315
Uraoka, Yukiharu 1255
Vater, Frank 1261
Vaziri, S. 977
Veliadis, Victor 974
Venkatesan, Prasanna 1276
Vincent, Benjamin 181
Völkel, Lukas 497, 1279
Wahid, Sumaiya 1131
Waldhoer, Dominic 1192, 1282
Waltl, Michael 1267
Wan, Changjin 971
Wan, Tianqing 333
Wan, Yi 257
Wan, Yuxi 208, 865
Wan, Yu-Xi 700
Wan, Ziqi 715
Wang, Albert 1288
Wang, Bingxiang 732
Wang, Chang-Hao 10
Wang, Chen 154, 178, 369, 438, 817
Wang, Chengcai 622
Wang, Chenguang 263
Wang, Chien-Fan 151
Wang, Chunxiu 245
Wang, Cong 333
Wang, Cuimei 1237
Wang, Di 390
Wang, Fang 351
Wang, Feixiong 546, 1293
Wang, Gang 519, 679
Wang, Guilei 688
Wang, Han 139
Wang, Hongyue 369
Wang, Huan 220
Wang, Hui 193, 282
Wang, Jiachuang 1054, 1164
Wang, Jiahao 1059
Wang, Jialiang 239, 783
Wang, Jian 1219
Wang, Jianhuan 500
Wang, Jingli 172
Wang, Jinyan 369, 438, 513
Wang, Junhao 1167
Wang, Kai 363, 762, 780, 811, 937

Wang, Kaifeng 474
Wang, Kuan 184
Wang, Lai 1219
Wang, Le 814
Wang, Lei 883, 986
Wang, Letian 549
Wang, Lihao 411
Wang, Lin 130
Wang, Lingfei 874
Wang, Long 345
Wang, Lu 1180
Wang, Lun 184
Wang, Maojun 236, 513
Wang, Mengye 239, 488
Wang, Ming 429
Wang, Mingyan 604
Wang, Nan 411, 1119
Wang, Peiran 297
Wang, Peng 336, 537, 1015
Wang, Qian 980
Wang, Qing 58, 67, 82, 297, 471, 744, 1243
Wang, Qingpeng 181
Wang, Qishen 485, 1024, 1237
Wang, Rui 765, 862
Wang, Runsheng 309, 459, 525, 601, 628, 634, 676, 747, 753, 823, 835, 1006, 1036, 1048, 1051
Wang, Ruobing 715, 1296
Wang, Ruqi 37
Wang, Sen 1059
Wang, Shuang 233
Wang, Shuying 91, 628, 673
Wang, Shuyu 661
Wang, Tao 940
Wang, Tong 912
Wang, Weibiao 765, 862
Wang, Weisheng 387
Wang, Wen 661
Wang, Wenwu 34, 40, 103, 267, 276, 753, 777
Wang, Wenxiao 721
Wang, Wenyu 94
Wang, Xiangsheng 688
Wang, Xiaochen 528
Wang, Xiaohui 82
Wang, Xiaolei 34, 40, 85, 103, 267, 276, 777, 1293
Wang, Xiaoming 865
Wang, Xiaoping 208, 865
Wang, Xiayu 510
Wang, Xile 1231
Wang, Xin 369, 438
Wang, Xinghua 741
Wang, Xingsheng 549, 555, 947
Wang, Xinhe 79

Wang, Xinhua	786
Wang, Xinpeng	251, 883, 992
Wang, Xinwei	43
Wang, Xuemei	318
Wang, Xuepei	459, 634, 1048
Wang, Xueying	324
Wang, Yan	88, 260, 865
Wang, Yang	808, 959
Wang, Yao	345
Wang, Yaoyi	983, 1012
Wang, Yasai	339, 396
Wang, Ye Liang	625
Wang, Yeliang	916, 919
Wang, Yi	549
Wang, Yibo	906
Wang, Yikang	306
Wang, Yimeng	747
Wang, Yiru	814
Wang, Yizhuo	534
Wang, Yu	922, 1111
Wang, Yunda	1018
Wang, Yuxuan	405
Wang, Yuyan	847
Wang, Zcbin	163
Wang, Zexi	76
Wang, Zhen	49, 73
Wang, Zhenxin	983, 1012
Wang, Zhiyao	546
Wang, Zhiyu	1077, 1119
Wang, Zhiyuan	1119, 1234
Wang, Zhongju	727
Wang, Zhongqiang	142
Wang, Zhongrui	58, 537
Wang, Zhuming	817
Wang, Zifeng	1125
Wang, Ziheng	378
Wang, Zijian	73, 127
Wang, Ziyang	58, 67
Wang, Ziyi	94
Wang, Zongwei	485, 841, 1021, 1024, 1237
Wanghe, Jianhui	1033
Waser, Rainer	832
Wei, H.-K.	977
Wei, Hu	912
Wei, Jianyong	534
Wei, Jin	236, 300, 438, 513
Wei, Jinghe	79
Wei, Jingxuan	1252
Wei, Ke	786
Wei, Lan	1195
Wei, Shaojun	688
Wei, Tiantian	688

Wei, Xiaoling	324
Wei, Xiaoyu	774
Wei, Yanzhao	922, 1111
Wei, Yanzhuo	522
Wei, Yidan	79
Wei, Yingqiang	79
Wen, Bo	139
Wen, Kangyao	58
Wen, Liaoyong	528
Wen, Xinyu	184
Wen, Yichen	459, 628, 634, 1006, 1048
Weng, Xiaoyu	263
Wiefels, Stefan	456, 832
Wijvliet, M.	28
Wittke, Christian	1261
Wong, Hiu Yung	482
Wong, Kwok-Ho	640
Wong, Man Hoi	16, 160
Wong, Ngai	543
Woon, W.-Y.	977
Wu, C.-H.	886
Wu, Chunlei	983, 1012
Wu, Daixuan	148, 685
Wu, Dong	70, 133, 205
Wu, F.	886
Wu, Haidi	667
Wu, Heng	453, 500, 601, 676, 747
Wu, Honglin	823
Wu, Hongzhao	950
Wu, Huaqiang	70, 133, 205, 729, 847
Wu, Jack	601
Wu, Jianting	613
Wu, Jixuan	85
Wu, Maokun	459, 628, 634, 1006, 1048
Wu, Mei	372, 447, 585
Wu, Nanjian	679
Wu, Qian	613
Wu, Renjie	124
Wu, Shenghao	685
Wu, Tian-Li	516
Wu, Xiaopeng	306
Wu, Xing	525
Wu, Xuankun	494
Wu, Y.	977
Wu, Ya Fei	903
Wu, Yanqing	844, 1077
Wu, Yishan	459, 634, 1048
Wu, Yiting	157
Wu, Yiyang	906
Wu, Yongbo	980
Wu, Yongyu	1149
Wu, Yuchen	375

Wu, Zhenyu	411
Wu, Zhida	1027
Wu, Zimeng	1237
Xi, Bin	239
Xi, Peng	263
Xi, Qi	79
Xia, Chenhao	46
Xia, Jiao	1167
Xia, Wenbo	438
Xia, Yang Hui	625
Xia, Yun	865
Xiang, Jinjuan	40
Xiang, Ke	768
Xiao, Boyuan	765, 862
Xiao, Fan	576
Xiao, Kai	738
Xiao, Qian	444
Xiao, Zhihua	1228
Xie, Haorong	937
Xie, Jiawei	405
Xie, Lanyi	453
Xie, Li	715
Xie, Maosong	750, 856
Xie, Qin	106
Xie, Rongbo	510
Xie, Shujie	494
Xie, Yinfei	853
Xie, Yunfei	285, 528, 868
Xie, Zhi-Fei	826
Xing, Shengpeng	73
Xing, Sicheng	1198
Xing, Weichuan	912
Xiong, Bing	1219
Xiong, Weiwei	339, 396
Xiong, Yifeng	947
Xiong, Yuwen	1104
Xiu, Huixin	239
Xu, Daiyao	1195
Xu, Dengqin	229
Xu, Gaobo	336
Xu, Guangwei	1068, 1140
Xu, Haiyang	142
Xu, Hang	10
Xu, Hao	34, 40, 103, 267, 276
Xu, Jianbin	435
Xu, Jiaqiang	109
Xu, Jinghan	931
Xu, Jun	441
Xu, Juyan	79
Xu, Kai	19, 1149
Xu, Lei	1071
Xu, Ming	220

Xu, Mingkun	561
Xu, Qiufeng	765, 862
Xu, Shaodi	934
Xu, Shuo	429
Xu, Weikai	279
Xu, Wenhui	853, 916, 919, 1285
Xu, Wenyang	768
Xu, Xiaoyan	453, 471, 500
Xu, Yang	285, 491, 528, 868
Xu, Yi	1104
Xu, Ying	850
Xu, Yitong	780
Xu, Yu	768
Xu, Yunsong	387
Xu, Zhengde	190
Xu, Zheqi	980
Xu, Zhiping	1210
Xu, Zihan	1065
Xu, Ziqiao	601, 747
Xue, Hongxia	139
Xue, Renhao	414
Xue, Yi-Bai	694, 703
Xue, Yibai	709
Xue, Yiheng	432
Xue, Yongkang	91, 673
Xue, Zhipeng	962
Xue, Zhiqiang	444
Yamane, Daisuke	995
Yan, Bei-Ping	643
Yan, Jianmin	333, 339
Yan, Longhao	1231
Yan, Weikang	937
Yan, Yong	450
Yang, Benjamin	847
Yang, Bowen	429, 585
Yang, Changhui	679
Yang, Chunzhen	239
Yang, Dandan	381
Yang, Dongliang	1098
Yang, Fan	318, 555, 947
Yang, Fengyuan	1119
Yang, Guanhua	874
Yang, Guoshen	576
Yang, Han	513
Yang, Haozhang	1003
Yang, Heng	351
Yang, Hong	753, 777
Yang, Huan	43, 136
Yang, Huazhong	233
Yang, Hui Xia	625
Yang, Huiran	324
Yang, Jia	34, 103

Yang, Jining	983, 1012, 1252
Yang, Junjie	236, 438, 513
Yang, Ling	444, 447, 582, 585
Yang, Mei	22
Yang, Meiyin	429
Yang, Ni	139, 257
Yang, Qing	1086
Yang, Qingyuan	561
Yang, Rui	248, 750, 790, 856
Yang, Shangyu	1264
Yang, Sheng	91, 673
Yang, Shengjie	43
Yang, Shuai	753
Yang, Siyao	506
Yang, Tao	777
Yang, Wenlong	381
Yang, Xiaolei	429
Yang, Xiaoman	628
Yang, Xuelin	236, 513, 573
Yang, Yafen	10
Yang, Yanyu	336
Yang, Yi	826
Yang, Yifan	549
Yang, Yingjic	193
Yang, Yintang	306
Yang, Yuchao	1231
Yang, Yuhang	485, 1009, 1237
Yang, Zongyin	986
Yao, Jiaping	248
Yao, Jiaxin	922, 1111
Yao, Peng	729
Yao, Rui Ray	646
Yao, Zikang	962
Ye, Changqing	525
Ye, Chenglin	420
Ye, Jiandong	1285
Ye, Ran	402
Ye, Sheng	459, 634, 1006, 1048
Ye, Tianchun	34, 103, 276
Ye, Zhuangzhuang	429
Yelzhanova, Zhuldyz	1270
Yeon, Deuk Ho	223
Yeung, Fion Sze Yan	160
Yi, Tingchen	229
Yin, Huaxiang	327, 336, 537, 546, 777, 922, 1111, 1293
Yin, Youyi	513
Yip, Pak San	139
Yoo, Jinil	25
Yoo, K.	1189
Yoo, S.	1189
Yoon, Hoon Hahn	986

Yoon, Jung Ho	1186
Yoshikawa, M.	315
You, Rui	510
You, Tiangui	853, 916, 919, 1042, 1264
Yu, Bin	285, 491, 528, 868
Yu, C. X.	1095
Yu, Chen-Yu	516
Yu, Chunxiao	906
Yu, Guofang	393, 423
Yu, Guohao	607
Yu, Hao	351
Yu, Haozhe	82
Yu, Hongyu	58, 67, 82, 297, 1243
Yu, Huabin	1146
Yu, Jiexun	980
Yu, Jingjing	236, 513
Yu, Peiyue	429
Yu, Qian	582
Yu, Shaofeng	619, 631
Yu, Xiao	390, 1249
Yu, Xinxin	1285
Yu, Ying-Jie	682
Yu, Yingjie	709
Yu, Yiwcn	983, 1012
Yu, Yong	688
Yu, Yue	934
Yu, Zongguang	79
Yuan, Jiahui	369
Yuan, Jian	847
Yuan, Jiuyang	503
Yuan, Mengqiang	312
Yuan, Miaojia	1006
Yuan, Shuai	790
Yuan, Wenhan	124
Yuan, Zhengnan	531
Yue, Han	607
Yue, Yuanyuan	774
Yun, P.	1189
Zang, Hang	360
Zarkob, Yawar Hayat	1137
Zeng, Chunhong	64
Zeng, Fei	765, 862
Zeng, Lang	1125
Zeng, Liang	251
Zeng, Min	1077
Zeng, Ruli	983, 1012
Zeng, Shicheng	1065
Zeng, Wei	348
Zeng, Xinlong	950
Zeng, Zhongming	607
Zeun, A.	28
Zha, Jiajia	112

Zha, Xian-Hu	700
Zhai, Yongbiao	303
Zhan, Rui	432
Zhan, Xuepeng	85
Zhang, A. Ping	1015
Zhang, Bailin	679
Zhang, Baoshun	64, 607
Zhang, Baotong	744
Zhang, Beining	248
Zhang, Bin	369
Zhang, Bo	124
Zhang, Bosen	724
Zhang, Bowen	871
Zhang, Boyang	727
Zhang, Changcheng	1027
Zhang, Chenyang	91, 673
Zhang, Chenyuan	1167
Zhang, Chi	510
Zhang, Conghui	1009
Zhang, Dao Hua	700
Zhang, David Wei	10, 634, 817
Zhang, Guanjun	591
Zhang, Guobin	127
Zhang, Guohua	303
Zhang, Haisu	1024
Zhang, Hang	537
Zhang, Hao	883, 983, 992, 1012
Zhang, Haochen	1146
Zhang, Heng	552
Zhang, Heng-Feng	694
Zhang, Hongjie	738
Zhang, Hongrui	390
Zhang, J. J.	1095
Zhang, Ji	257
Zhang, Jiajun	1018
Zhang, Jianjun	453, 500
Zhang, Jiayi	294
Zhang, Jieyin	453
Zhang, Jin	774
Zhang, Jincheng	312, 345, 667, 912
Zhang, Jingyang	16
Zhang, Jinshu	1059, 1062, 1065
Zhang, Li	1027
Zhang, Lian	494
Zhang, Lijie	601
Zhang, Lining	309, 823, 835, 989, 1036, 1051
Zhang, Long	582
Zhang, Meng	582, 585
Zhang, Miaocheng	992
Zhang, Min	245, 968
Zhang, Mingchen	871
Zhang, Panpan	157, 432
Zhang, Pengcheng	790
Zhang, Pu-Yi	682
Zhang, Qianqian	528, 868
Zhang, Qingxin	741
Zhang, Qingzhu	537, 546, 777, 922, 1111
Zhang, Qirui	318
Zhang, Qiuqi	616
Zhang, Renlong	263
Zhang, Rong-Jun	700
Zhang, Rui	321
Zhang, Shanting	1180
Zhang, Shan-Ting	522
Zhang, Shengdong	43, 136, 160
Zhang, Shiming	357
Zhang, Shiyi	378
Zhang, Shiyu	130
Zhang, Shuai	765, 862
Zhang, Shukui	172
Zhang, Tianjiao	552
Zhang, Ting	1006
Zhang, Wei	1101
Zhang, Weihang	312, 912
Zhang, Weizhe	679
Zhang, Wenjing	697
Zhang, Wenxu	591
Zhang, Xi	294
Zhang, Xiaodong	64
Zhang, Xiaohan	300
Zhang, Xin	601
Zhang, Xing	229, 1021
Zhang, Xue	190
Zhang, Xuning	983, 1012
Zhang, Y. M.	664
Zhang, Yachao	312, 345
Zhang, Yadong	546
Zhang, Yan	510
Zhang, Yewei	91, 673
Zhang, Yibei	70
Zhang, Yijian	750, 856
Zhang, Ying	741
Zhang, Yishu	73, 127, 1149
Zhang, Yu	477, 694
Zhang, Yuanke	441, 595
Zhang, Yuhan	136, 160
Zhang, Yuhang	43
Zhang, Yuming	928
Zhang, Yun	494
Zhang, Yu-Qi	303
Zhang, Zeyu	279
Zhang, Zhaohao	327, 546
Zhang, Zhejia	1059, 1062, 1065
Zhang, Zhekang	37

Zhang, Zhi-Xiang	285, 491, 528, 868	Zhou, Hang	576, 724
Zhang, Zichong	549	Zhou, Haobin	637, 953
Zhang, Zihan	148	Zhou, Haoxian	263
Zhang, Zijing	375	Zhou, Heng	604
Zhao, Chao	688	Zhou, Jiajun	543
Zhao, Dengrui	64	Zhou, Jiang	217
Zhao, Fangyu	1054, 1164	Zhou, Jin	912
Zhao, Haolin	363, 780	Zhou, Jingjie	1062
Zhao, Haoran	500	Zhou, Jingyi	49
Zhao, Jiaming	417	Zhou, Jiuren	1249
Zhao, Jianwen	163	Zhou, Kaiyuan	381
Zhao, Jinxiu	154, 178	Zhou, Leidang	519
Zhao, Jiyu	348	Zhou, Menglong	688
Zhao, Junlei	300	Zhou, Min	916, 919
Zhao, Lei	429, 1167	Zhou, Peng	408, 959
Zhao, Liang	1180	Zhou, Wenli	646
Zhao, Lu	393	Zhou, Wenyong	543
Zhao, Ni	169	Zhou, Xijun	453, 471
Zhao, Rong	561, 579	Zhou, Xilin	715, 1296
Zhao, Shujing	342	Zhou, Xinchen	765, 862
Zhao, Tiancheng	916, 919	Zhou, Xinlong	1012
Zhao, Wei	79	Zhou, Y.	664
Zhao, Wenjie	1077	Zhou, Ya Dong	625
Zhao, Xiaoguang	510	Zhou, Yaoqiang	435
Zhao, Xiaoning	142	Zhou, Ye	303, 348
Zhao, Yapeng	844	Zhou, Yiming	729
Zhao, Yuanyuan	40	Zhou, Yingxi	474
Zhao, Yuda	552, 868, 950	Zhou, Yu	928
Zhao, Yue	393, 874	Zhou, Yue	339, 721
Zhao, Yusi	619, 631	Zhou, Yuejia	254, 273, 1161
Zheng, Dezhi	52	Zhou, Yuxi	871
Zheng, Doudou	193	Zhou, Zheng	229, 420, 877, 931, 1003
Zheng, Fangyuan	257	Zhou, Zhitao	324
Zheng, Gerui	405	Zhou,	762, 811
Zheng, Hao	934	Zhu, Bowen	13, 88, 199, 260
Zheng, Huajian	613	Zhu, Chaoyi	339
Zheng, Jun	906	Zhu, Fangchen	808
Zheng, Qianze	133, 205	Zhu, Handong	983, 1012
Zheng, Siying	1249	Zhu, Hao	1252
Zheng, Xuefeng	1027	Zhu, Jiayan	956
Zheng, Yaojie	953	Zhu, Jiefei	968
Zheng, Zijie	405, 850	Zhu, Jiejie	871
Zhong, Kun	327	Zhu, Mingde	549
Zhong, Yujia	181	Zhu, Qihang	220
Zhou, Bin	432	Zhu, Quanhai	366
Zhou, Changjian	968	Zhu, Runteng	1161
Zhou, Dayu	251	Zhu, Shirui	303
Zhou, Enze	510	Zhu, Yanda	94
Zhou, Falong	601	Zhu, Yao	741
Zhou, Feichi	202	Zhu, Yaochen	411
Zhou, Guangdong	768	Zhu, Yu	100, 270, 468
Zhou, Guopei	324	Zhu, Yuxuan	1059, 1062, 1065

Zhu, Yuzhe 1077
Zhu, Zhifeng 190
Zhu, Ziyi 324
Zhuang, Kai 363, 780
Zier, M. 28
Zou, Dujuan 324
Zou, Xiangyu 1119
Zou, Xinbo 916, 919
Zou, Xu 582
Zou, Yating 838
Zou, Zhili 64
Zuo, Chengjie 384, 450
Zuo, Wen-Bin 694
Zuo, Wenbin 709

IEEE
445 Hoes Lane
Piscataway, NJ 08854-4141

ISBN 979-8-3315-0417-5

9 798331 504175